COMPUTING HANDBOOK

THIRD EDITION

Computer Science and Software Engineering

COMPUTING HANDBOOK

THIRD EDITION

Computer Science and Software Engineering

EDITED BY

Teofilo Gonzalez
University of California
Santa Barbara, California, USA

Jorge Díaz-Herrera
Keuka College
Keuka Park, New York, USA

EDITOR-IN-CHIEF

Allen Tucker
Bowdoin College
Brunswick, Maine, USA

CRC Press
Taylor & Francis Group
Boca Raton London New York

CRC Press is an imprint of the
Taylor & Francis Group, an **informa** business
A CHAPMAN & HALL BOOK

CRC Press
Taylor & Francis Group
6000 Broken Sound Parkway NW, Suite 300
Boca Raton, FL 33487-2742

Printed on acid-free paper
Version Date: 20140117

International Standard Book Number-13: 978-1-4398-9852-9 (Hardback)

Library of Congress Cataloging-in-Publication Data

Computer science and engineering handbook
Computing handbook : computer science and software engineering / editors, Allen Tucker, Teofilo Gonzalez, Jorge Diaz-Herrera. -- Third edition.
volumes cm
Originally published as: The computer science and engineering handbook. c1992, and The computer science handbook, c2004, both edited by Allen B. Tucker.
Includes bibliographical references and index.
ISBN 978-1-4398-9852-9 (hardback)
1. Computer science--Handbooks, manuals, etc. I. Tucker, Allen B. II. Gonzalez, Teofilo F. III. Diaz-Herrera, Jorge L., 1950- IV. Title.

QA76.C57315 2014
004--dc23

2013049505

Visit the Taylor & Francis Web site at
http://www.taylorandfrancis.com

and the CRC Press Web site at
http://www.crcpress.com

Contents

PART III Architecture and Organization

PART IV Computational Science and Graphics

PART V Intelligent Systems

PART VI Networking and Communication

PART VII Operating Systems

PART X Software Quality and Measurement

PART XI Software Development Management: Processes and Paradigms

PART XII Software Modeling, Analysis, and Design

Preface to the *Computing Handbook Set*

The purpose of the *Computing Handbook Set* is to provide a single, comprehensive reference for specialists in computer science, information systems, information technology, software engineering, and other fields who wish to broaden or deepen their understanding in a particular subfield of the computing discipline. Our goal is to provide up-to-date information on a wide range of topics in a form that is accessible to students, faculty, and professionals.

The discipline of computing has developed rapidly since CRC Press published the second edition of the *Computer Science Handbook* in 2004 (Tucker, 2004). Indeed, it has developed so much that this third edition requires repartitioning and expanding the topic coverage into a two-volume set.

The need for two volumes recognizes not only the dramatic growth of computing as a discipline but also the relatively new delineation of computing as a family of five separate disciplines, as described by their professional societies—The Association for Computing Machinery (ACM), The IEEE Computer Society (IEEE-CS), and The Association for Information Systems (AIS) (Shackelford et al., 2005).

These separate disciplines are known today as computer engineering, computer science, information systems, information technology, and software engineering. These names more or less fully encompass the variety of undergraduate and graduate degree programs that have evolved around the world, with the exception of countries where the term *informatics* is used for a subset of these disciplines. The document "Computing curricula 2005: The overview report" describes computing this way (Shackelford et al., 2005, p. 9):

> In a general way, we can define computing to mean any goal-oriented activity requiring, benefiting from, or creating computers. Thus, computing includes designing and building hardware and software systems for a wide range of purposes; processing, structuring, and managing various kinds of information; doing scientific studies using computers; making computer systems behave intelligently; creating and using communications and entertainment media; finding and gathering information relevant to any particular purpose, and so on.

To add much flesh to the bones of this very broad definition, this handbook set describes in some depth what goes on in research laboratories, educational institutions, and public and private organizations to advance the effective development and utilization of computers and computing in today's world. The two volumes in this set cover four of the five disciplines in the following way:*

1. Volume I: Computer Science and Software Engineering
2. Volume II: Information Systems and Information Technology

* The fifth discipline, computer engineering, is the subject of a separate handbook published by CRC Press in 2008.

This set is not designed to be an easy read, as would be gained by browsing a collection of encyclopedia entries on computing and its various subtopics. On the contrary, it provides deep insights into the subject matter through research-level survey articles. Readers who will benefit most from these articles may be undergraduate or graduate students in computing or a related discipline, researchers in one area of computing aiming to expand their knowledge of another area, or other professionals interested in understanding the principles and practices that drive computing education, research, and development in the twenty-first century.

This set is designed as a professional reference that serves the interests of readers who wish to explore the subject matter by moving directly to a particular part and chapter of the appropriate volume. The chapters are organized with minimal interdependence, so that they can be read in any order. To facilitate rapid inquiry, each volume also contains a table of contents and a subject index, thus providing access to specific topics at various levels of detail.

The Preface to Volume I provides a more detailed overview of the organization and content of this volume. A similar overview of the coverage of information systems and information technology appears in the Preface to Volume II.

Preface to Volume I: Computer Science and Software Engineering

This volume is organized to mirror the modern taxonomy of computer science and software engineering as described by the ACM and IEEE-CS (Joint Task Force, 2012):

Algorithms and Complexity
Architecture and Organization
Computational Science
Discrete Structures
Graphics and Visual Computing
*Human–Computer Interaction
*Security and Information Assurance
*Information Management
Intelligent Systems
Networking and Communication
Operating Systems
Platform-Based Development
Parallel and Distributed Computing
Programming Languages
Software Development Fundamentals
Software Engineering
Systems Fundamentals
Social and Professional Issues

To avoid redundancy, the three starred (*) topics in this list (Human–Computer Interaction, Security and Information Assurance, and Information Management) are covered in Volume II, while the remaining topics are covered in this volume.

The first eight parts of this volume cover the discipline of computer science (edited by Teofilo Gonzalez), while the last four parts cover software engineering (edited by Jorge Díaz-Herrera). Allen Tucker is the editor in chief of the two volumes.

Computer Science

Implicit in the aforementioned taxonomy (and, indeed, in all current thinking about the nature of computer science) is the idea of a dynamic discipline with an increasingly broad reach. For example, the *computational science* area noted earlier emphasizes the necessity of interaction among computer scientists and natural scientists in the many research and educational areas where their work now overlaps.

A more elaborate description of this idea is expressed in the following definition of computer science, taken from the 2005 Joint Task Force Report (Shackelford, 2005, p. 13):

> Computer science spans a wide range, from its theoretical and algorithmic foundations to cutting-edge developments in robotics, computer vision, intelligent systems, bioinformatics, and other exciting areas. We can think of the work of computer scientists as falling into three categories.
>
> 1. They design and implement software. Computer scientists take on challenging programming jobs. They also supervise other programmers, keeping them aware of new approaches.
> 2. They devise new ways to use computers. Progress in the CS areas of networking, database, and human-computer-interface enabled the development of the World Wide Web. Now CS researchers are working with scientists from other fields to make robots become practical and intelligent aides, to use databases to create new knowledge, and to use computers to help decipher the secrets of our DNA.
> 3. They develop effective ways to solve computing problems. For example, computer scientists develop the best possible ways to store information in databases, send data over networks, and display complex images. Their theoretical background allows them to determine the best performance possible, and their study of algorithms helps them to develop new approaches that provide better performance.

The computer science chapters are organized into eight parts: (I) Overview of Computer Science; (II) Algorithms and Complexity; (III) Architecture and Organization; (IV) Computational Science and Graphics; (V) Intelligent Systems; (VI) Networking and Communication; (VII) Operating Systems; and (VIII) Programming Languages. A brief summary of what you can expect to find in each part is provided in the following.

Overview of Computer Science

Part I consists of two very important chapters. In Chapter 1, Peter Denning discusses the 50-year evolution of computer science as a discipline distinct from the fields from which it emerged: mathematics, science, and engineering. In Chapter 2, Valerie Barr introduces the idea of "computational thinking," a relatively new term that is helping to shape modern thinking about the expanding presence of computer science, especially in the K-12 and undergraduate college curricula.

Algorithms and Complexity

This part discusses the algorithmic and computational complexity issues that underlie all of computer science. Chapter 3 covers important ways of organizing data inside computers to facilitate their retrieval. It discusses the time and space trade-offs for implementing the most important algorithms used in almost every application. The different design paradigms behind these algorithms are discussed in Chapter 4. The chapter covers greedy methods, divide-and-conquer, and dynamic programming, as well as other techniques that have been applied successfully in the design of efficient algorithms. Graphs and networks have been used to model a wide range of computer science problems as well as problems in different disciplines. Chapter 5 discusses graph and network techniques that have been devised to efficiently solve these problems. In some applications, computational problems can be simplified by formulating them as problems in two-dimensional space. Computational geometry problems include those that have intrinsic geometric properties. Chapter 6 discusses the techniques that have been applied successfully in this domain.

Chapter 7 discusses computational complexity classes. This is a grouping of problems according to the time and space required to solve them. The major classes are called P and NP, and they help characterize most of the computational challenges that arise in practice. A fundamental problem in computer science is to prove or disprove whether P = NP. In fact, it is so important that a so-called "Millennium Prize" consisting of $1,000,000 has been offered by the Clay Institute to anyone who solves this problem. At a more abstract level, Chapter 8 reviews formal methods and computability. This chapter explores the limits of what is computable within the most basic computing models that are equivalent to current computer architectures. One of the most complex areas of research is cryptography. Chapter 9 provides a comprehensive overview of this very important area. Since millions of private transactions take place on the Internet every second, we need ways to keep them secure and guarantee their integrity. Solutions in this area require sophisticated mathematical analysis techniques. Chapter 10 addresses the area of algebraic algorithms. The basic applications in this area are in science and engineering.

As the input size of most computationally complex problems continues to grow (sometimes referred to as the phenomenon of *big data*), the study of randomized algorithms has grown considerably in recent years. Chapter 11 discusses techniques to design such algorithms by presenting algorithms to solve some specific problems. Another way of dealing with computational complexity is to aim for approximations rather than optimal solutions. Chapter 12 discusses approximation algorithms and metaheuristics. The different methodologies used in these two areas are surveyed and applied to problems from different disciplines. Algorithms that take advantage of the combinatorial structure of the feasible solution space are discussed in Chapter 13. Another important algorithmic challenge is how to visualize data and their structure. Chapter 14 covers the area of graph drawing, which tries to design algorithms that display information in a clear way according to given objective functions. In this era of big data, we also need to discover ways of storing, retrieving, and transmitting it efficiently. Chapter 15 covers strategies for data compression and finding patterns in data quickly. Chapter 16 studies computing in a landscape where the computations are distributed among many computers. In this scenario, we need to address the issue of fault-tolerant computing; as the number of independent devices grows, the likely failure of at least one of the components increases substantially.

Architecture and Organization

Part III takes us closer to computing machines themselves by examining their architecture and organization. Starting at the most basic organizational level, Chapter 17 covers digital logic, how it is implemented, and how logical components are designed to carry out basic computations. Chapter 18 covers the organization of memory and its implementation, including the different levels of memory, access speeds, and implementation costs. Chapter 19 extends this discussion to the treatment of secondary storage. A main issue here is how to store files in disks, solid state devices, and flash memories to ensure reliability. Chapter 20 covers computer arithmetic, including the efficient implementation of the basic arithmetic and logical operations that are fundamental to all computing activities. Modern input and output devices are discussed in Chapter 21. As the way we interact with computers has changed over the years, input and output design remains at the forefront of research and development.

Chapter 22 discusses performance enhancements for instruction processing, such as the use of overlap and low-level concurrency. Focusing on the theme of design for efficiency, Chapter 23 surveys the different ways of organizing parallel computers. It begins with the standard types introduced several decades ago and concludes with the highly successful GP-GPUs as well as cloud computing. To circumvent the memory access and sequential performance walls, modern architectures include multiple processors on a single chip. Chapter 24 discusses the hardware and software challenges for the design and use of multicore architectures. Chapter 25 explains how DNA can be used to compute some logical functions and perform biological tests. Currently, performing these tests takes some time because it requires several computational facilities and involves human intervention.

Computational Science and Graphics

Some aspects of computational science have been part of computing since its beginning. Computational electromagnetics (Chapter 26) basically studies the computational solution of Maxwell equations. Analytical solutions to Maxwell equations are usually unavailable, and the few closed-form results that exist have very restrictive solutions. The use of computer-based methods for the prediction of fluid flows has seen tremendous growth in the past half century. Fluid dynamics (Chapter 27) has been one of the earliest, and most active, fields for the application of numerical techniques. Computational astrophysics (Chapter 28) covers astrophysics, a computationally intensive discipline. Physics and astronomy make heavy use of online access and data retrieval. Databases, data analysis, and theoretical modeling go hand in hand in this area. The gravitational n-body problem and its modeling form the basis for cosmology. Chapter 29 discusses the rich set of computational applications in chemistry. Computational biology/bioinformatics is an area that has grown tremendously over the last decade. Chapter 30 discusses the interplay between the genetic analysis provided by computational algorithms and the underlying biological processes. Chapter 31 on terrain modeling examines the computational problems and new insights arising from the analysis of GIS data.

Computer graphics provides the underlying tools for scientific visualization and animation of a wide variety of real and imaginary phenomena. Chapter 32 discusses the basic geometric primitives manipulated by typical graphics systems. These primitives allow for the efficient representation and manipulation of visual objects in a computer. Chapter 33 covers computer animation, which has become widely used in the production of movies and videos.

Intelligent Systems

Mathematical logic and declarative programming provide theoretical foundations for artificial intelligence applications, which are called intelligent systems. Chapter 34 explores the use of logic to allow a system to work in spite of contradictions, dilemmas, and conflicts. Chapter 35 discusses qualitative reasoning, which supports reasoning with very little information. This is the first approach that one uses to understand a problem before developing a more formal or quantitative strategy. Machine learning, discussed in Chapter 36, is concerned with identifying and exploiting patterns in data to simulate human learning. Applications of machine learning to real-world problems are widely used on the Internet, for example, in marketing. Chapter 37 introduces explanation-based learning, which adds inferential domain theory to expand the reach of statistical machine learning. This theory adds expressiveness to statistical methods such as Bayesian or Markov models.

Chapter 38 describes searching, which is a basic operation in most computing areas and is particularly important in intelligent systems where the search domain is often complex. This chapter explores both blind exhaustive search methods using informed heuristics and optimal heuristics. Chapter 39 covers planning (the process of creating an organized set of actions) and scheduling (the process of assigning a set of actions to resources over time). These are basic operations arising in all sorts of environments and have been studied over the years in several disciplines. Chapter 40 explores natural language processing, or developing systems to understand human languages. In 2011, IBM's Watson program convincingly won the popular Jeopardy game over the most highly skilled human competitors. This was a major accomplishment for natural language processing. Spoken language understanding, covered in Chapter 41, is an emerging field between the areas of speech processing and natural language processing. This area leverages technologies from machine learning to create systems that would facilitate communication with users. Chapter 42 covers neural networks, which have been used to solve a wide range of problems in modeling intelligent behavior. Chapter 43 deals with cognitive modeling and surveys the methods and models used to understand human cognition. The semantic basis and properties of graphic models (Bayesian Networks) are explored in Chapter 44. Applications to reasoning and planning are discussed in this chapter.

Networking and Communication

In the last 20 years, the use of computer networks for communication has become an integral part of our daily life. An incredible volume of information is transferred constantly worldwide through computer networks, and these must operate efficiently and reliably so that modern business and commerce can function effectively. Chapter 45 examines the communications software needed to interconnect computers, servers, and all sort of devices across networks. The unique aspects of wireless networks are discussed in Chapter 61 in Part VII. Chapter 46 discusses the routing algorithms that manage information flow across networks, including the trade-offs that exist among different alternatives. Chapter 47 discusses access control, which is fundamental for protecting information against improper disclosure or modification.

Compression of images and videos needed for the efficient storage and transmission of these types of data is covered in Chapter 48. Chapter 49 introduces underwater sensor networks, whose challenges and techniques are very different from those in other types of networks. Chapter 50 surveys ideas for transforming the World Wide Web to facilitate interoperability and integration of multiauthored, multithematic, and multiperspective information services seamlessly. Chapter 51 deals with search engines, which have become indispensable in almost all aspects of our daily life. It discusses the fundamental issues involved in designing and constructing search engines that interact with enormous amounts of information and digest it for users.

Operating Systems

An operating system is the fundamental software that controls all aspects of a computer, its applications, and its user interface. Chapter 52 discusses processes, the basic software activities that an operating system manages, and the mechanisms needed for communication and synchronization. Chapter 53 covers thread management, where threads are basic sequences of code that comprise a program running in a system. Breaking programs and systems into threads is particularly useful in multicore machines, the modern organization of computers. Virtual memory, discussed in Chapter 54, explains the way in which computer memory is organized for efficient access in computers. This chapter also discusses how these techniques apply to other levels of storage and systems. The organization of file systems and secondary storage is discussed in Chapter 55. The main concern is the use of secondary storage space for its use as long storage media.

Chapter 56 covers performance modeling of computer systems. Different computer architectures have different performance limits. The chapter discusses different ways to analyze computer system performance. Chapter 57 deals with a similar topic but for interconnected distributed computer systems such as clusters, grids, and clouds. An important issue here is maintaining quality of service during contention management. The architecture and special applications of real-time systems are discussed in Chapter 58. In these systems, response must occur within specific deadlines; it is a challenge for the operating system to schedule processes so that these deadlines are met. The design of distributed operating systems faces two main challenges: scheduling, covered in Chapter 59, and file systems, covered in Chapter 60. Mobile devices pose new challenges for operating system design, and these are addressed in Chapter 61. The proliferation of smart phones and tablets has brought this area to the forefront of research and development. Chapter 62 provides an introduction to service-oriented operating systems and concludes this part.

Programming Languages

Programming language research has a rich history, and it has remained vibrant with the development of modern architectures, networks, and applications. Chapter 63 covers imperative programming languages, the oldest and probably the most prolific style of programming. These types of languages are in widespread use and are familiar to software developers in all areas of computing. The object-oriented programming paradigm discussed in Chapter 64 has become popular as it facilitates modeling real-world software systems using object decomposition. Another form of programming, logic programming discussed in Chapter 65, has major applications in artificial intelligence and natural language processing research. Recent developments have been especially active in the areas of multiparadigm programming languages and scripting languages described in Chapter 66 and Chapter 67, respectively.

Chapter 68 deals with compilers and interpreters, which provide the interface between programs written in higher-level programming languages and an equivalent encoding that a computer can understand and carry out. Many aspects of this process have been automated, and the techniques and limitations of these processes are also discussed. Every programming language has syntax and semantics. Syntax refers to the valid strings of characters that define valid programs in the language, and semantics refers to the meaning of such strings. Chapter 69 covers formal semantics for programming languages. Chapter 70 describes type systems, which are used in modern programming languages to guarantee that programs are free of type-related errors. This chapter discusses the formal underpinnings of type systems as well as their implementations. Chapter 71 covers formal methods, a set of notations and tools that provide a mathematical basis for proving the correctness of a program with regard to its formal specification.

Software Engineering

As an academic field of study, software engineering was introduced in the early 1970s as a specialization of computer science and engineering, first as master-level programs. In the 1990s, the field proliferated in North America, Europe, and Australia, first in the form of BS programs and, more recently, programs at the PhD level; that is, software engineering programs separate from computer science programs (Lethbridge et al., 2006). ***What is the difference and why do we care?***

Software is of critical importance in today's world and of increasingly serious public significance in our daily lives for our safety and security. Although the voicing of concerns about the critical consequences of not considering software development as a more rigorous profession dates back to the late 1960s (Naur, 1969), today software engineering *is still not a fully formed, mature professional discipline.* This seems more of an elusive goal than a concrete reality or even a future possibility. Some wonder if trying to retrofit software engineering with the traditional *engineering* paradigm has hindered its acceptance as a professional discipline. Particularly lacking is a widespread recognition of software developers as professionals in the full sense of the word.*

Over the past decades, we have identified scientific and engineering body of knowledge needed to solve technical software problems that meet user requirements (functional and otherwise) and deliver software within budget, on time, as well as within its economic, legal, and social considerations. The study of engineering requires a strong background in mathematics and the physical sciences. Although this is not different for software engineers, the kind of mathematics and the specific science may be different from more traditional engineering disciplines. However, a legitimate research question remains on whether there could be a scientific basis for understanding the complexity of software such that it can be *engineered* to have predictable quality and behavior.

Ian Sommerville (1999) introduced the idea that software engineering is perhaps something else: "In essence, the type of systems which we are interested in are *socio-technical software-intensive systems....* Systems, therefore, always include computer hardware, software which may be specially designed or bought-in as off-the-shelf packages, policies and procedures, and people who may be end-users and producers/consumers of information used by the system." Furthermore, Sommerville and his colleagues claim that "there is no technical solution to software complexity" (Sommerville, 2012).

The fact that *computing* has become a distinct branch of knowledge, that is, a discipline in its own right different from science and engineering—an idea first formally introduced by Denning (1998) and further developed by Denning and Freeman (Denning, 2009)—fits this notion as has been demonstrated by the creation of a unifying and enclosing academic entity that embraces the science, engineering, and technology of computing: the computing college, "standing alone, headed by its own dean, with the

* It is a fact of life that many of the software industry professionals are graduates from computer science and related fields. This is of no surprise since in most advertisements for such positions, the educational requirements often listed are degrees in computer science. However, once hired, employers give them the title of software engineer.

same stature as traditional colleges" [op. cit.]. In this way, computer science is a different kind of science and software engineering a different kind of engineering.

The chapters in the last four parts consolidate our current understanding of the discipline of software engineering and its effect on the practice of software development and the education of software professionals. There are four main parts. (1) Discipline of Software Engineering; (2) Software Quality and Measurement; (3) Software Development Management: Processes and Paradigms; and (4) Software Modeling, Analysis, and Design (notice that programming topics were amply covered in Part VIII, programming languages).

Discipline of Software Engineering

This part includes five chapters. It starts with Chapter 72, which provides an overview of software engineering. The chapter delineates the historical development of the discipline and provides an analysis of accepted definitions and an outline of the elements of the field. It looks at generally accepted definitions of engineering and shows the relation of certain elements to software development. It also points out important differences and demonstrates through a detailed analysis how prominent features that cut across all engineering disciplines can reasonably be mapped to software development. In this regard, the two most fundamental aspects are problem solving and engineering design. Mastery of *problem solving* involves a combination of proper judgment, experience, common sense, and know-how that must be used to reduce a real-world problem to such a form that *science* can be applied to find its solution. *Engineering design*, on the other hand, is the process of devising a system, component, or process to meet desired needs by finding technical solutions to specific practical problems, while taking into account economic, legal, and ecological considerations. Among the fundamental elements of this process are the establishment of objectives and criteria, synthesis, analysis, construction, testing, and evaluation, all of which have a counterpart in software engineering.

Chapters 73 and 74 deal with *professionalism and certification* issues and the *code of ethics and professional conduct*, respectively. Chapter 73 addresses software engineering as a profession from the point of view of the body of knowledge, accreditation of academic programs, and issues on licensure and certification of software engineering professionals. It provides a good account of the political realities of making software development a professional discipline and concludes that "it seems clear that the future of software engineering professionalism will be driven by industry and government," although the uptake of credentials has been driven by practitioners' interest (e.g., IEEE-CS CSDP).

Concomitant with the development of software engineering as a professional discipline is the introduction of a code of ethics, and in late 1999 the Association for Computing Machinery and the IEEE Computer Society approved the Software Engineering Code of Ethics and Professional Practice. In Chapter 74, Gotterbarn, a principal architect of the code, provides a rich historical account of its development and a thorough description of its purpose and principles, together with simple cases as examples and practical guidance for public accountability and educational purposes.

Chapters 75 and 76 address *software IT business and economics* and *open source and governance*. Chapter 75 provides an overview of the business of software from an IT and economics point of view, with a focus on the success rate of projects in terms of schedule predictability and quality, while emphasizing available solutions and best practices to practically implement this framework. It concludes with a look at business trends, naturally influenced and determined by external factors such as demand for value, fashion, individualism, ever-changing expectations, demand for ubiquitous services, global competition, economic and ecologic behaviors, and the need for security and stability, which in turn influence trends for software and IT business.

Chapter 76 on open source and governance is an authoritative manifesto in the subject, providing a comprehensive set of underlying principles regarding copyright law and license terms, including the peculiarities of COTS (commercial off-the-shelf). It also contains numerous examples of best practices covering a myriad of situations in which companies "desiring to supplement and expand the capabilities of their existing software" can find themselves.

Software Quality and Measurement

This part highlights an important aspect that makes software engineering a distinct computing discipline and that lends it much scientific credibility. The first two chapters focus on *evidence-informed software engineering* and *empirical software engineering*. Chapter 77 introduces the evidence-based paradigm, which originally emerged in medicine, together with a *systematic literature review* of its application in software engineering. The idea, first advocated by Kitchenham et al. in 2004, suggests that with some adaptation to fit the nature of software engineering studies, this could be a valuable way of consolidating our empirical knowledge about what works, when, and where. Chapter 78 provides an overview of how the ideas and practices associated with empirical software engineering have evolved. It represents an invaluable aid for identifying suitable forms of study for a particular need and to provide enough background for software engineers to appreciate the strengths and limitations of empirical knowledge and practice. This is particularly relevant if software engineering is to become a professional discipline; in this way, its practices, methods, and tools would need to be informed by evidence from well-conducted empirical studies.

Chapter 79 looks at quality in general and provides a comprehensive survey and analysis of quality under the umbrella of software process improvement. It differentiates software quality attributes from the point of view of the end user to IT operations, through marketing and projects and technical concerns. It also describes five perspective approaches to quality, that is, transcendental, product, user, manufacturing, and value based, with corresponding mappings to ISO quality standards. The chapter then juxtaposes process and quality and cost and quality. It provides an overview of best practices covering the various approaches to software process maturity such as CMM and SPICE together with practical implementation implications. It also identifies issues associated with small organizations and Agile development, as well as measurement and the use of statistical process control.

Chapter 80 focuses on *software metrics and measurements* and hence follows naturally here. This is an important component of software engineering as a professional discipline. The chapter demonstrates the value of software measurement and the role of international standards. These standards embody principles and best practices as defined and codified by experts in the field. Several approaches to implement measurement are discussed, including CMMI measurement and analysis (M&A), measurement and analysis infrastructure diagnostic (MAID), and goal–question–metric (GQM) among others. The chapter also covers predictive models and indicators, developing benchmarks and heuristics and improving data quality, concluding that "software measurement is required to quantify quality and performance and to provide an empirical and objective foundation for decision making—a foundation that is a necessary element if software engineering is truly to be a disciplined field of engineering."

Software Development Management: Processes and Paradigms

The chapters in this part address the development of software from a higher-level managerial, rather than code development point of view. The first three chapters deal with management and organizational issues. Chapter 81 argues that there is no one best way to develop software and manage projects for all kinds of applications but that there are some basic principles that can be applied to a variety of projects within their contexts. It takes us through a series of high-level process concepts, innovation and design, and architecture strategies and how they are used in different situations. It concludes with a brief analysis of the results of a survey on global differences in practices and performance metrics in different regions of the world.

Chapter 82 covers issues related to *project personnel and organization* such as how to put together a team and how to decide on the types of personnel that you will need in your team based on the different types of roles and responsibilities. The chapter also explains the different ways you can organize your personnel along with the advantages and disadvantages of each approach, as well as underlying principles and best practices.

Chapter 83 discusses *project and process control*. It makes a distinction between the project manager's control activities in executing a project, that is, *project control*, and the control of the underlying software development processes, or *process control*. Controlling a project and controlling the underlying processes are very different activities and, consequently, they require very different techniques to make them effective. The chapter describes the underlying mathematical and engineering foundations of two well-known control methods that can be applied to controlling software development projects and to their underlying software development processes. The first is closely related to traditional engineering methods for controlling manufacturing and communications systems using feedback. The second is known as statistical process control (SPC). The chapter concludes with a brief discussion of the extensive literature describing the application of SPC methods to software projects and the successes and failures that they have reported.

The last three chapters in this part address more recent development frameworks. Chapter 84 provides a thorough introduction to *Agile* methods, an umbrella term popularized with the publication of the Agile manifesto, reflecting an intention to differentiate Agile approaches from a trend in software development methodologies that had gained dominance in previous decades. It describes the set of 12 principles underlying the Agile approach as provided by the Agile manifesto. Specific implementations of the approach are described as best practices, including extreme programming, scrum, etc., together with an analysis of their specific incorporation, or not, of the 12 principles. The chapter concludes with a discussion of plan-driven approaches' compatibility, such as CMMIs, with Agile methods and thorough analyses of important research directions.

Chapter 85 presents a process for developing software based on service-oriented architectures. The fundamental building blocks of the approach are services, commonly viewed as self-contained software application modules exposed through interfaces over a distributed environment. The chapter introduces related elements such as component-based development, distributed components, and web services. The underlying principles of the approach are described together with the service infrastructure needed. In terms of best practices, the chapter outlines reference models and reference architectures and standards as well as the enterprise service bus idea and compares the specific implementations of the web services SOAP and REST.

Chapter 86 on *software product lines* (SPL) provides an introduction to the topic and a thorough analysis of its research challenges. SPL is one of the most active research areas of software engineering, and despite remarkable contributions, which are briefly highlighted, important technical problems remain. The chapter reports quantitative information, based on a secondary study and literature review, on these remaining problems, which slow down widespread adoption of SPL. The results provided are supported by a survey among longtime and active SPL researchers and practitioners. The top two reported research problems were variability management and evolution.

Software Modeling, Analysis, and Design

The last part of this section focuses on specific software development activities starting with requirements elicitation and specification, followed by model checking, design strategies, and software architecture and finally dealing with more overarching topics such human–computer interfaces and software assurance.

Chapter 87 covers *requirements elicitation*, a subprocess of requirements engineering concerned with gathering information about a computer-based system to be built. This is generally the first step in the software development life cycle regardless of the process model, whether it be waterfall, spiral, or even Agile. The chapter considers the different scenarios, reviews all available techniques with their pros and cons, and provides sound advice on what could go wrong during requirements elicitation.

Chapter 88 focuses on *specifications* for software systems and the *software requirements specification* document (SRS) document capturing them. Software requirement specifications are discussed at different stages from a very informal description to inform users, to descriptions in unambiguous terms, written in a rather formal (i.e., precise and unambiguous) manner. The chapter provides an overview of best practices for producing an SRS possessing all the desirable quality attributes and characteristics. It also provides a survey of the various specification languages and lists outstanding research issues such as completeness, nonfunctional requirements, cost, program construction and proof, and correctness-preserving program transformations.

Chapter 89 provides an overview of techniques and tools for analyzing the correctness of software systems using model checking. Two classes of techniques are covered, namely, counterexample-guided abstraction refinement (CEGAR) using predicate abstraction, and bounded model checking; other key model checking techniques that are geared toward software verification analysis are also covered. The chapter concludes with a brief survey of tools.

The next two chapters are about software design. Chapter 90 on *design strategies* leads us directly from requirements as "one assumption underlying any design is that the design will satisfy some set of requirements." A set of underlying principles is discussed, including the notion of views, architectural patterns and tactics, decomposition and refinements, and code generation and testing. Design best practices are presented together with an analysis of important research concerns such as automatic/semiautomatic design, the accuracy of models, and the use and appropriate place for tools.

Chapter 91 on *software architecture* focuses exclusively on the main product of design: an architectural design of the software system to solve a specified problem. Various architectural frameworks are contrasted from their specific viewpoints, as well as a survey of design methods. A metaprocess is described, starting with concern analysis and domain analysis and concluding with architecture design, evaluation, and realization. Research topics in software architectures presented include modeling quality and issues related to model-driven design.

Chapter 92 on *human–computer interfaces* in software engineering uses speech applications (both input and synthesis) from requirements to design and implementation to illustrate principles and best practices and describe difficult research issues that lie ahead on the implication of human–computer interaction on software development. The latter discusses inherent current limits in speech applications design.

Chapter 93 deals with *software assurance*. It discusses the impact of the risks associated with software vulnerabilities and introduces basic definitions of software assurance. It also presents modern principles of software assurance and identifies a number of relevant process models, frameworks, and best practices. The chapter concludes with a research framework to support and identify gaps for future research and includes a description of the knowledge areas for a Master of Software Assurance as mapped to maturity levels for building assured systems.

References

Denning, P., Computing the profession: An invitation for computer scientists to cross the chasm. *Educom Review*, 33(6), 1998:26–30;46–59.

Denning, P. and P. Freeman. Computing's paradigm. *Communications of the ACM*, December 2009.

Joint Task Force on Computing Curricula, *Computer Science Curricula 2013*, Association for Computing Machinery (ACM) and the IEEE-Computer Society (IEEE-CS), December 2013. See http://www.acm.org/education/curricula-recommendations.

Lethbridge, T.C., R.J. LeBlanc Jr., A.E. Kelley Sobel, T.B. Hilburn, and J.L. Díaz-Herrera. SE2004: Curriculum recommendations for undergraduate software engineering programs. *IEEE Software*, Nov/Dec 2006, 19–25.

Lunt, B.M., J.J. Ekstrom, S. Gorka, G. Hislop, R. Kamali, E. Lawson, R. LeBlanc, J. Miller, and H. Reichgelt. *IT 2008: Curriculum Guidelines for Undergraduate Degree Programs in Information Technology*. ACM and IEEE Computer Society, 2008.

Naur, P. and B. Randell (eds.). *Software Engineering*, Report on a Conference Sponsored by the NATO Science Committee, October 1968.

Shackelford, R., J. Cross, G. Davies, J. Impagliazzo, R. Kamali, R. LeBlanc, B. Lunt, A. McGettrick, R. Sloan, and H. Topi. *Computing Curricula 2005: The Overview Report*, A cooperative project of the ACM, AIS, and IEEE Computer Society, 30 September, 2005. See http://www.acm.org/education/curricula-recommendations.

Sommerville, I. Systems engineering for software engineers. *Annals of Software Engineering*, 6(1–4), April 1999: 111–129.

Sommerville, I., D. Cliff, R. Calinescu, J. Keen, T. Kelly, M. Kwiatkowska, J. Mcdermid, and R. Paige. Large-scale complex IT systems. *Communications of the ACM*, 55(7), July 2012: 71–77.

Topi, H., J. Valacich, R. Wright, K. Kaiser, J. Nunamaker, J. Sipior, and G. Jan de Vreede. IS 2010: Curriculum guidelines for undergraduate degree programs in information systems. *Communications of the Association for Information Systems*, 26(18). http://aisel.aisnet.org/cais/vol26/iss1/18.

Tucker, A. (ed.). *Computer Science Handbook*, 2nd edn., Chapman & Hall/CRC Press, in cooperation with the Association for Computing Machinery (ACM), Boca Raton, FL, 2004.

Acknowledgments

This volume would not have been possible without the tireless and dedicated work of all the authors and reviewers. We appreciate the willingness of both established experts and influential younger researchers to join this effort and contribute to this work. Even though we have been working for decades in the computing field, our work on this volume has enriched substantially our knowledge and understanding of computing.

We would also like to thank the editorial staff at Chapman & Hall/CRC Press, particularly the persistent and gentle leadership of Randi Cohen, computer science acquisitions editor, whose role has extended beyond any reasonable call of duty. Bob Stern, as the original computer science editor who helped create this handbook in its original and second editions, also deserves recognition for his continued support over the years.

Teo acknowledges the support of the University of California, Santa Barbara (UCSB). Without his sabbatical leave, this project would not have been possible. He also wants to thank his wife, Dorothy, and children, Jeanmarie, Alexis, Julia, Teofilo, and Paolo, for their love, encouragement, understanding, patience, and support throughout the project and his tenure at UCSB.

Jorge acknowledges the support of Dr. Özlem Albayrak during the entire project. He is grateful to his team at Keuka College, particularly Lori Haines and Sandra Wilmott, for their moral support and encouragement as he took the helm of the College concomitant with the start of this project.

Allen acknowledges Bowdoin College for providing him the time and support needed to develop the first, second, and third editions of this handbook set over the last 15 years. He also thanks the many colleagues who have contributed to earlier editions of this handbook, without which the current edition would not have been possible. Finally, he thanks his wife, Meg, and children, Jenny and Brian, for their constant love and support over a lifetime of teaching and learning.

We hope that this volume will be actively and frequently used both by academics who want to explore the current status and new opportunities within specific research areas and by advanced practitioners who want to understand how they can benefit from the latest thinking within the disciplines of computer science and software engineering. We also warmly welcome your feedback.

Teofilo Gonzalez
University of California, Santa Barbara

Jorge Díaz-Herrera
Keuka College

Allen B. Tucker
Bowdoin College

Editors

Allen B. Tucker is the Anne T. and Robert M. Bass Professor Emeritus in the Department of Computer Science at Bowdoin College, Brunswick, Maine, where he served on the faculty from 1988 to 2008. Prior to that, he held similar positions at Colgate and Georgetown Universities. Overall, he served for 18 years as a department chair and two years as an associate dean of the faculty. While at Colgate, he held the John D. and Catherine T. MacArthur Chair in Computer Science.

Professor Tucker earned a BA in mathematics from Wesleyan University in 1963 and an MS and PhD in computer science from Northwestern University in 1970. He is the author or coauthor of several books and articles in the areas of programming languages, natural language processing, and software engineering. He has given many talks, panel discussions, and workshop presentations in these areas and has served as a reviewer for various journals, NSF programs, and curriculum projects. He has also served as a consultant to colleges, universities, and other institutions in the areas of computer science curriculum, software development, programming languages, and natural language processing applications. Since retiring from his full-time academic position, Professor Tucker continues to write, teach, and develop open source software for nonprofit organizations.

A fellow of the ACM, Professor Tucker coauthored the 1986 Liberal Arts Model Curriculum in Computer Science and cochaired the ACM/IEEE-CS Joint Curriculum Task Force that developed Computing Curricula 1991. For these and related efforts, he received the ACM's 1991 Outstanding Contribution Award, shared the IEEE's 1991 Meritorious Service Award, and received the ACM SIGCSE's 2001 Award for Outstanding Contributions to Computer Science Education. In 2001, he was a Fulbright Lecturer at the Ternopil Academy of National Economy in Ukraine, and in 2005, he was an Erskine Lecturer at the University of Canterbury in New Zealand. Professor Tucker has been a member of the ACM, the NSF CISE Advisory Committee, the IEEE Computer Society, Computer Professionals for Social Responsibility, the Liberal Arts Computer Science (LACS) Consortium, and the Humanitarian Free and Open Source Software (HFOSS) Project.

Dr. Jorge L. Díaz-Herrera became the 19th president of Keuka College on July 1, 2011. He was professor and founding dean of the B. Thomas Golisano College of Computing and Information Sciences at Rochester Institute of Technology in Rochester, New York, since July 2002. Prior to this appointment, he was professor of computer science and department head at SPSU in Atlanta and Yamacraw project coordinator with Georgia Tech. He became the chair of the first Software Engineering Department in the United States at Monmouth University in New Jersey. He has had other academic appointments with Carnegie Mellon's Software Engineering Institute, George Mason University in Virginia, and at SUNY Binghamton in New York.

Dr. Díaz-Herrera has conducted extensive consulting services with a number of firms and government agencies, including New York Stock Exchange (SIAC), MITRE Corp., the Institute for Defense Analysis, General Electric, Singer-Link, TRW, EG&G, and IBM among others. He has participated in technical meetings and provided professional expertise to international organizations, including the

European Software Institute, Australian Defense Science and Technology Office, Kyoto Computing Gaikum, Kuwait University, Cairo University, Instituto Politecnico Santo Domingo (INTEC), Bilkent University in Turkey, and Malaysia University of Technology among others. He has chaired several national and international conferences and has been a technical reviewer for the National Science Foundation, the American Society for Engineering Education, and several conferences and journals. He has more than 90 publications.

Dr. Díaz-Herrera is a senior member of the IEEE, and was a member of IEEE-CS Distinguished Visitor Program for nine consecutive years, a leading writer of the Software Engineering Professional Examination, and coeditor of the *Software Engineering* volume of the Computing Curricula project. He was an active member of the CRA-Deans Group of the Computer Research Association in Washington, DC and serves, and has served, on various technical advisory committees and governing boards, including the Technical Advisory Group, SEI, Carnegie Mellon University, Pittsburgh, Pennsylvania; Advisory Committee for CISE, National Science Foundation, Washington, DC; Advisory Committee for GPRA/PA, National Science Foundation, Washington, DC; Board of Trustees, Strong National Museum of Play, Rochester, New York; Board of Trustees, Gallaudet University, Washington, DC; NY State Universal Broadband Council, Action Team on Digital Literacy, Albany, New York; and Research Board, Instituto Technológico de las Américas (ITLA), Dominican Republic.

Dr. Díaz-Herrera completed his undergraduate education in Venezuela and holds both a master's degree and a PhD in computing studies from Lancaster University in the United Kingdom. He completed two graduate certificates in management leadership in education from Harvard University's Graduate School of Education. He recently received a doctor honoris causa from his alma mater, Universidad Centro Occidental Lisandaro Alvarado, in Barquisimeto, Venezuela.

Professor Teofilo F. Gonzalez was one of the first ten students who received a computer science undergraduate degree in Mexico (ITESM 1972). He received his PhD from the University of Minnesota in 1975. He has been member of the faculty at OU, Penn State, and UT Dallas and has spent sabbatical leave at the ITESM (Monterrey, Mexico) and U. Utrecht (the Netherlands). Since 1984, he has served as professor of computer science at the University of California, Santa Barbara.

Professor Gonzalez's main research contributions include the development of exact and approximation algorithms as well as establishing complexity results for problems in several research areas, including message dissemination (multicasting), scheduling, VLSI placement and routing, computational geometry, graph problems, clustering, etc.

Professor Gonzalez edited the *Handbook on Approximation Algorithms and Metaheuristics* (Chapman & Hall/CRC, 2007). This handbook was the first one in the area and included contributions from top researchers worldwide. His work has been published in top journals in computer science, operations research and computer engineering, as well as in research books and conference proceedings. Professor Gonzalez has served as editor and guest editor of several publications. Currently, he serves as associate editor of *IEEE Transactions on Computers and ACM Computing Surveys*. He has been program committee chair and general chair for many conferences. He has received the Outstanding Computer Science Professor of the Year Award three times. He also received the Outstanding Performance as PC Chair for PDCS several times. Professor Gonzalez is a IASTED fellow. Since 2002, he has participated in the ABET accreditation process as a CAC program evaluator and commissioner.

Contributors

Jemal Abawajy
School of Information Technology
Deakin University
Melbourne, Victoria, Australia

Özlem Albayrak
School of Applied Technology and Management
Bilkent University
Ankara, Turkey

Eric W. Allender
Department of Computer Science
Rutgers University
New Brunswick, New Jersey

Dan Alistarh
Computer Science and Artificial Intelligence Lab
Massachusetts Institute of Technology
Cambridge, Massachusetts

Hanan Alnizami
School of Computing
Clemson University
Clemson, South Carolina

Ignacio Alvarez
School of Computing
Clemson University
Clemson, South Carolina

Thomas E. Anderson
University of Washington
Seattle, Washington

Marvin Andujar
School of Computing
Clemson University
Clemson, South Carolina

Valerie Barr
Department of Computer Science
Union College
Schenectady, New York

Roman Barták
Faculty of Mathematics and Physics
Charles University in Prague
Prague, Czech Republic

Sanjoy Baruah
Department of Computer Science
The University of North Carolina, Chapel Hill
Chapel Hill, North Carolina

Len Bass
National ICT Australia Ltd
Sydney, New South Wales, Australia

Anne Benoit
Laboratoire d'Informatique de Paris
Ecole Normale Supérieure de Lyon
Lyon, France

Daniel M. Berry
David R. Cheriton School of Computer Science
University of Waterloo
Waterloo, Ontario, Canada

Brian N. Bershad
Google
Seattle, Washington

Christopher M. Bishop
Microsoft Research
Cambridge, United Kingdom

Daniel Rubio Bonilla
High Performance Computing Centre
Stuttgart, Germany

J. Phillip Bowen
Department of Pharmaceutical Sciences
College of Pharmacy
Mercer University
Atlanta, Georgia

Jonathan P. Bowen
Museophile Limited
London, United Kingdom

Pearl Brereton
School of Computing and Mathematics
Keele University
Staffordshire, United Kingdom

James Brock
Cognitive Electronics
Boston, Massachusetts

Kim Bruce
Department of Computer Science
Pomona College
Claremont, California

David Budgen
School of Engineering and Computing Sciences
Durham University
Durham, United Kingdom

Hieu Bui
Department of Computer Science
Duke University
Durham, North Carolina

Rajkumar Buyya
Cloud Computing and Distributed Systems Laboratory
Department of Computing and Information Systems
The University of Melbourne
Melbourne, Victoria, Australia

Derek Buzasi
Department of Physics and Astronomy
Florida Gulf Coast University
Fort Myers, Florida

David A. Caughey
Sibley School of Mechanical and Aerospace Engineering
Cornell University
Ithaca, New York

Harish Chandran
Google Corporation
Mountain View, California

Vijay Chandru
National Institute of Advanced Studies
Bangalore, India

Gustavo Chávez
Strategic Initiative in Extreme Computing
King Abdullah University of Science and Technology
Thuwal, Saudi Arabia

Baozhi Chen
Department of Electrical and Computer Engineering
Rutgers University
New Brunswick, New Jersey

Eric Chown
Bowdoin College
Brunswick, Maine

Jacques Cohen
Brandeis University
Waltham, Massachusetts

James L. Cox
Brooklyn College
and
Graduate Center
The City University of New York
Brooklyn, New York

Maxime Crochemore
King's College London
London, United Kingdom

and

Université Paris-Est Marne-la-Vallée
Paris, France

Michael A. Cusumano
Engineering Systems Division
Sloan School of Management
Massachusetts Institute of Technology
Cambridge, Massachusetts

Shelby S. Darnell
School of Computing
Clemson University
Clemson, South Carolina

Sabrina De Capitani di Vimercati
Dipartimento di Informatica
Università degli Studi di Milano
Milan, Italy

Gerald DeJong
Department of Computer Science
University of Illinois at Urbana Champaign
Champaign, Illinois

Onur Demirörs
Informatics Institute
Middle East Technical University
Ankara, Turkey

Peter J. Denning
Naval Postgraduate School in Monterey
Cebrowski Institute for Innovation and Information Superiority
Monterey, California

Jorge L. Díaz-Herrera
Office of the President
Keuka College
Keuka Park, New York

Thomas W. Doeppner
Department of Computer Science
Brown University
Providence, Rhode Island

Alastair Donaldson
Department of Computing
Imperial College London
London, United Kingdom

Joseph Dumas
Department of Computer Science and Engineering
University of Tennessee, Chattanooga
Chattanooga, Tennessee

Christof Ebert
Vector Consulting Services GmbH
Stuttgart, Germany

Raimund K. Ege
Northern Illinois University
DeKalb, Illinois

Josh Ekandem
School of Computing
Clemson University
Clemson, South Carolina

Ioannis Z. Emiris
Department of Informatics and Telecommunications
University of Athens
Athens, Greece

John Favaro
Consulenza Informatica
Pisa, Italy

Michael J. Flynn
Department of Electrical Engineering
Stanford University
Stanford, California

Kenneth Forbus
Northwestern University
Evanston, Illinois

Peter A. Freeman
College of Computing
Georgia Institute of Technology
Atlanta, Georgia

Brian M. Gaff
McDermott Will and Emery, LLP
Boston, Massachusetts

Sudhanshu Garg
Department of Computer Science
Duke University
Durham, North Carolina

Apostolos Gerasoulis
Rutgers University
New Brunswick, New Jersey

Juan E. Gilbert
School of Computing
Clemson University
Clemson, South Carolina

Teofilo F. Gonzalez
Department of Computer Science
University of California, Santa Barbara
Santa Barbara, California

Nikhil Gopalkrishnan
Department of Computer Science
Duke University
Durham, North Carolina

Don Gotterbarn
Department of Computing
East Tennessee State University
Johnson City, Tennessee

Rachid Guerraoui
Distributed Programming Lab
École Polytechnique Fédérale de Lausanne
Lausanne, Switzerland

Dilek Hakkani-Tür
Conversational Systems Research Center
Microsoft Research
Mountain View, California

Jon Hakkila
College of Charleston
Charleston, South Carolina

Michael Hanus
Institute of Informatics
Christian-Albrechts-University of Kiel
Kiel, Germany

Herman Haverkort
Technische Universiteit Eindhoven
Eindhoven, the Netherlands

Frederick J. Heldrich
Department of Chemistry and Biochemistry
College of Charleston
Charleston, South Carolina

Michael G. Hinchey
Lero
University of Limerick
Limerick, Ireland

Ken Hinckley
Microsoft Research
Redmond, Washington

Pascal Hitzler
Department of Computer Science and Engineering
Wright State University
Dayton, Ohio

Nitin Indurkhya
School of Computer Science and Engineering
The University of New South Wales
Sydney, New South Wales, Australia

France Jackson
School of Computing
Clemson University
Clemson, South Carolina

Robert J.K. Jacob
Tufts University
Medford, Massachusetts

Melva James
School of Computing
Clemson University
Clemson, South Carolina

Krzysztof Janowicz
Department of Geography
University of California, Santa Barbara
Santa Barbara, California

Tao Jiang
University of California, Riverside
Riverside, California

Michael J. Jipping
Department of Computer Science
Hope College
Holland, Michigan

Michael I. Jordan
University of California, Berkeley
Berkeley, California

Konstantinos G. Kakoulis
Department of Industrial Design Engineering
Technological Educational Institute of West Macedonia
Kozani, Greece

Jonathan Katz
Department of Computer Science
University of Maryland
College Park, Maryland

Samir Khuller
University of Maryland
College Park, Maryland

Yoongu Kim
Department of Electrical and Computer Engineering
Carnegie Mellon University
Pittsburgh, Pennsylvania

Barbara Kitchenham
School of Computing and Mathematics
Keele University
Staffordshire, United Kingdom

Danny Kopec
Brooklyn College
and
Graduate Center
The City University of New York
Brooklyn, New York

Marc van Kreveld
Department of Information and Computing Sciences
Utrecht University
Utrecht, the Netherlands

Brian R. Landry
Saul Ewing, LLP
Boston, Massachusetts

Trishan de Lanerolle
Trinity College
Hartford, Connecticut

Edward D. Lazowska
University of Washington
Seattle, Washington

Thierry Lecroq
University of Rouen
Mont-Saint-Aignan, France

Miriam Leeser
Department of Electrical and Computer Engineering
Northeastern University
Boston, Massachusetts

Henry M. Levy
University of Washington
Seattle, Washington

Ming Li
University of Waterloo
Waterloo, Ontario, Canada

Kenneth C. Louden
Department of Computer Science
San Jose State University
San Jose, California

Michael C. Loui
Department of Electrical and Computer Engineering
University of Illinois at Urbana Champaign
Champaign, Illinois

Stephen Lucci
The City College of New York
The City University of New York
Brooklyn, New York

Naja Mack
School of Computing
Clemson University
Clemson, South Carolina

Ronald Mak
Department of Computer Science
San Jose State University
San Jose, California

Loris Marchal
Laboratoire d'Informatique de Paris
Ecole Normale Supérieure de Lyon
Lyon, France

Stephen Marsland
School of Engineering and Advanced Technology
Massey University
Palmerston North, New Zealand

Tom McBride
University of Technology, Sydney
Sydney, Australia

James McDonald
Department of Computer Science and Software Engineering
Monmouth University
West Long Branch, New Jersey

Andrew McGettrick
Department of Computer and Information Sciences
University of Strathclyde
Glasgow, Scotland

Marshall Kirk McKusick
Marshall Kirk McKusick Consultancy
Berkeley, California

Paul McMahon
PEM Systems
Binghamton, New York

Nancy R. Mead
Software Engineering Institute
Carnegie Mellon University
Pittsburgh, Pennsylvania

Clyde R. Metz
Department of Chemistry and Biochemistry
College of Charleston
Charleston, South Carolina

Michael Mitzenmacher
School of Engineering and Applied Sciences
Harvard University
Cambridge, Massachusetts

Reem Mokhtar
Department of Computer Science
Duke University
Durham, North Carolina

Dekita Moon
School of Computing
Clemson University
Clemson, South Carolina

Ralph Morelli
Trinity College
Hartford, Connecticut

Onur Mutlu
Department of Electrical and Computer Engineering
and
Department of Computer Science
Carnegie Mellon University
Pittsburgh, Pennsylvania

Kazumi Nakamatsu
School of Human Science and Environment
University of Hyogo
Himeji, Japan

Ö. Ufuk Nalbantoğlu
Occult Information Laboratory
Department of Electrical Engineering
University of Nebraska-Lincoln
Lincoln, Nebraska

Robert E. Noonan
College of William and Mary
Williamsburg, Virginia

Bariş Özkan
Informatics Institute
Middle East Technical University
Ankara, Turkey

Victor Y. Pan
Department of Mathematics and Computer Science
Lehman College
The City University of New York
New York, New York

Robert M. Panoff
Shodor and the National Computational Science Institute
Durham, North Carolina

Judea Pearl
Cognitive Systems Laboratory
Department of Computer Science
University of California, Los Angeles
Los Angeles, California

Melvin Pérez-Cedano
Construx Software
Seattle, Washington

Radia Perlman
Intel Laboratories
Pittsburgh, Pennsylvania

Gregory J. Ploussios
Edwards Wildman Palmer, LLP
Boston, Massachusetts

Dario Pompili
Department of Electrical and Computer Engineering
Rutgers University
New Brunswick, New Jersey

Raphael Poss
Institute for Informatics
University of Amsterdam
Amsterdam, the Netherlands

Balaji Raghavachari
The University of Texas, Dallas
Richardson, Texas

M.R. Rao
Indian School of Business
Hyderabad, India

Bala Ravikumar
University of Rhode Island
Kingston, Rhode Island

Kenneth W. Regan
Department of Computer Science and Engineering
University at Buffalo
State University of New York
Buffalo, New York

John Reif
Duke University
Durham, North Carolina

and

King Abdulaziz University
Jeddah, Saudi Arabia

Edward M. Reingold
Department of Computer Science
Illinois Institute of Technology
Chicago, Illinois

Yves Robert
Laboratoire d'Informatique de Paris
Ecole Normale Supérieure de Lyon
Lyon, France

and

University of Tennessee
Knoxville, Tennessee

Alyn Rockwood
Geometric Modeling and Scientific Visualization Center
King Abdullah University of Science and Technology
Thuwal, Saudi Arabia

Kevin W. Rudd
Department of Electrical and Computer Engineering
United States Naval Academy
Annapolis, Maryland

Mohsen Amini Salehi
Cloud Computing and Distributed Systems Laboratory
Department of Computing and Information Systems
The University of Melbourne
Melbourne, Victoria, Australia

Pierangela Samarati
Dipartimento di Informatica
Università degli Studi di Milano
Milan, Italy

Ravi Sandhu
Institute for Cyber Security
The University of Texas, San Antonio
San Antonio, Texas

K. Sayood
Occult Information Laboratory
Department of Electrical Engineering
University of Nebraska-Lincoln
Lincoln, Nebraska

David A. Schmidt
Department of Computing and Information Sciences
Kansas State University
Manhattan, Kansas

Lutz Schubert
Institute for Organisation and Management of Information Systems
University Ulm
Ulm, Germany

Stephen B. Seidman
Texas State University
San Marcos, Texas

Shawn C. Sendlinger
Department of Chemistry
North Carolina Central University
Durham, North Carolina

J.S. Shang
Wright State University
Dayton, Ohio

Dan Shoemaker
Center for Cyber Security and Intelligence Studies
University of Detroit Mercy
Detroit, Michigan

and

International Cyber Security Education Coalition
Ann Arbor, Michigan

Tianqi Song
Department of Computer Science
Duke University
Durham, North Carolina

Lee Staff
Geometric Modeling and Scientific Visualization Center
King Abdullah University of Science and Technology
Thuwal, Saudi Arabia

William Stallings
Independent Consultant

Earl E. Swartzlander, Jr.
Department of Electrical and Computer Engineering
The University of Texas, Austin
Austin, Texas

Bedir Tekinerdogan
Department of Computer Engineering
Bilkent University
Ankara, Turkey

Robert J. Thacker
St. Mary's University
Halifax, Nova Scotia, Canada

Daniel Thalmann
Institute for Media Innovation
Nanyang Technological University
Singapore, Singapore

and

École Polytechnique Fédérale de Lausanne
Lausanne, Switzerland

Nadia Magnenat Thalmann
Institute for Media Innovation
Nanyang Technological University
Singapore, Singapore

and

MIRALab
University of Geneva
Geneva, Switzerland

Alexander Thomasian
Thomasian & Associates
Pleasantville, New York

Steven M. Thompson
BioInfo 4U
Valdosta, Georgia

Ioannis G. Tollis
Department of Computer Science
University of Crete and ICS-FORTH
Crete, Greece

Laura Toma
Bowdoin College
Brunswick, Maine

Neal E. Tonks
Department of Chemistry and Biochemistry
College of Charleston
Charleston, South Carolina

Elias P. Tsigaridas
Laboratoire d' Informatique de Paris
Université Pierre and Marie Curie
and
Institut National de Recherche en Informatique et en Automatique Paris-Rocquencourt
Paris, France

Gokhan Tur
Conversational Systems Research Center
Microsoft Research
Mountain View, California

Bora Uçar
Laboratoire d'Informatique de Paris
Ecole Normale Supérieure de Lyon
Lyon, France

Eli Upfal
Department of Computer Science
Brown University
Providence, Rhode Island

Frédéric Vivien
Laboratoire d'Informatique de Paris
Ecole Normale Supérieure de Lyon
Lyon, France

Andy Wang
Southern Illinois University Carbondale
Carbondale, Illinois

Ye-Yi Wang
Online Services Division
Microsoft
Bellevue, Washington

Colin Ware
University of New Hampshire
Durham, New Hampshire

Stephanie Weirich
University of Pennsylvania
Philadelphia, Pennsylvania

Mark Allen Weiss
School of Computing and Information Sciences
Florida International University
Miami, Florida

Stefan Wesner
Institute for Organisation and Management of Information Systems
University Ulm
Ulm, Germany

Daniel Wigdor
University of Toronto
Toronto, Ontario, Canada

Craig E. Wills
Worcester Polytechnic Institute
Worcester, Massachusetts

Jacob O. Wobbrock
University of Washington
Seattle, Washington

Carol Woody
Software Engineering Institute
Carnegie Mellon University
Pittsburgh, Pennsylvania

Tao Yang
University of California, Santa Barbara
Santa Barbara, California

Guangzhi Zheng
Southern Polytechnic State University
Marietta, Georgia

David Zubrow
Software Engineering Institute
Carnegie Mellon University
Pittsburgh, Pennsylvania

I

Overview of Computer Science

1

Structure and Organization of Computing*

Peter J. Denning
Naval Postgraduate School in Monterey

Computing is integral to science—not just as a tool for analyzing data but also as an agent of thought and discovery.

It has not always been this way. Computing is a relatively young discipline. It started as an academic field of study in the 1930s with a cluster of remarkable papers by Kurt Gödel, Alonzo Church, Emil Post, and Alan Turing. The papers laid the mathematical foundations that would answer the question, "what is computation?" and discussed schemes for its implementation. These men saw the importance of automatic computation and sought its precise mathematical foundation. The various schemes they each proposed for implementing computation were quickly found to be equivalent, as a computation in any one could be realized in any other. It is all the more remarkable that their models all led to the same conclusion that certain functions of practical interest—such as whether a computational algorithm (a method of evaluating a function) will ever come to completion instead of being stuck in an infinite loop—cannot be answered computationally.

In the time that these men wrote, the terms "computation" and "computers" were already in common use but with different connotations from today. Computation was taken to be the mechanical steps followed to evaluate mathematical functions. Computers were people who did computations. In recognition of the social changes they were ushering in, the designers of the first digital computer projects

* An earlier version of this chapter, without the section on technology view of the field, was published in *American Scientist 98* (September–October 2010), pp. 198–202. It was reprinted in *Best Writings on Mathematics 2011* (M. Pitici, ed.), Princeton University Press (2011). Copyright is held by the author.

all named their systems with acronyms ending in "-AC," meaning automatic computer or something similar—resulting in names such as ENIAC, UNIVAC, and EDSAC.

At the start of World War II, the militaries of the United States and the United Kingdom became interested in applying computation to the calculation of ballistic and navigation tables and the cracking of ciphers. They commissioned projects to design and build electronic digital computers. Only one of the projects completed before the war was over. That was the top-secret project at Bletchley Park in England, which cracked the German Enigma cipher using methods designed by Alan Turing.

Many people involved in those projects went on to start computer companies in the early 1950s. The universities began offering programs of study in the new field in the late 1950s. The field and the industry have grown steadily into a modern behemoth whose Internet data centers are said to consume almost 3% of the world's electricity.

During its youth, computing was an enigma to the established fields of science and engineering. At first, it looked like only the technology applications of math, electrical engineering, or science, depending on the observer. However, over the years, computing seemed to provide an unending stream of new insights, and it defied many early predictions by resisting absorption back into the fields of its roots. By 1980, computing had mastered algorithms, data structures, numerical methods, programming languages, operating systems, networks, databases, graphics, artificial intelligence, and software engineering. Its great technology achievements—the chip, the personal computer, and the Internet—brought it into many lives. These advances stimulated more new subfields, including network science, web science, mobile computing, enterprise computing, cooperative work, cyberspace protection, user-interface design, and information visualization. The resulting commercial applications have spawned new research challenges in social networks, endlessly evolving computation, music, video, digital photography, vision, massive multiplayer online games, user-generated content, and much more.

The name of the field changed several times to keep up with the flux. In the 1940s, it was called *automatic computation*, and in the 1950s, *information processing*. In the 1960s, as it moved into academia, it acquired the name *computer science* in the United States and *informatics* in Europe. By the 1980s, the computing field comprised a complex of related fields including computer science, informatics, computational science, computer engineering, software engineering, information systems, and information technology. By 1990, the term *computing* became the standard for referring to this core group.

1.1 Computing Paradigm

Traditional scientists frequently questioned the name *computer science*. They could easily see an engineering paradigm (design and implementation of systems) and a mathematics paradigm (proofs of theorems) but they could not see much of a science paradigm (experimental verification of hypotheses). Moreover, they understood science as a way of dealing with the natural world, and computers looked suspiciously artificial.

The word "paradigm" for our purposes means a belief system and its associated practices, defining how a field sees the world and approaches the solutions of problems. This is the sense that Thomas Kuhn used in his famous book, *The Structure of Scientific Revolutions* (1962). Paradigms can contain subparadigms: thus, engineering divides into electrical, mechanical, chemical, civil, etc., and science divides into physical, life, and social sciences, which further divide into separate fields of science. Table 1.1 outlines the three paradigms that combined to make the early computing field.

The founders of the field came from all three paradigms. Some thought computing was a branch of applied mathematics, some a branch of electrical engineering, and some a branch of computational-oriented science. During its first four decades, the field focused primarily on engineering: The challenges of building reliable computers, networks, and complex software were daunting and occupied almost everyone's attention. By the 1980s, these challenges largely had been met and computing was

TABLE 1.1 Subparadigms Embedded in Computing

	Math	Science	Engineering
1. Initiation	Characterize objects of study (definition).	Observe a possible recurrence or pattern of phenomena (hypothesis).	Create statements about desired system actions and responses (requirements).
2. Conceptualization	Hypothesize possible relationships among objects (theorem).	Construct a model that explains the observation and enables predictions (model).	Create formal statements of system functions and interactions (specifications).
3. Realization	Deduce which relationships are true (proof).	Perform experiments and collect data (validate).	Design and implement prototypes (design).
4. Evaluation	Interpret results.	Interpret results.	Test the prototypes.
5. Action	Act on results (apply).	Act on results (predict).	Act on results (build).

spreading rapidly into all fields, with the help of networks, supercomputers, and personal computers. During the 1980s, computers had become powerful enough that science visionaries could see how to use them to tackle the hardest, "grand challenge" problems in science and engineering. The resulting "computational science" movement involved scientists from all countries and culminated in the US Congress's adopting the High Performance Computing and Communications (HPCC) act of 1991 to support research on a host of large computational problems.

Today, there is agreement that computing *exemplifies* science and engineering and that neither science nor engineering *characterizes* computing. Then what does? What is computing's paradigm?

The leaders of the field struggled with the paradigm question ever since the beginning. Along the way, there were three waves of attempts to unify views. Newell et al. (1967) led the first one. They argued that computing was unique among all the sciences in its study of information processes. Simon (1996), a Nobel laureate in Economics, went so far as to call computing a science of the artificial. Amarel (1971) endorsed this basic idea and added an emphasis on interactions with other fields. A catchphrase of this wave was that "computing is the study of phenomena surrounding computers."

The second wave focused on programming, the art of designing algorithms that produced information processes. In the early 1970s, computing pioneers Edsger Dijkstra and Donald Knuth took strong stands favoring algorithm analysis as the unifying theme. A catchphrase of this wave was "computer science equals programming." In recent times, this view has foundered because the field has expanded well beyond programming, whereas public understanding of a programmer has narrowed to just those who write code.

The third wave came as a result of the NSF-funded Computer Science and Engineering Research Study (COSERS), led by Bruce Arden in the late 1970s. Its catchphrase was "computing is the automation of information processes." Although its final report successfully exposed the science of computing and explained many esoteric aspects to the layperson, its central view did not catch on.

An important aspect of all three definitions was the positioning of the computer as the object of attention. The computational science movement of the 1980s began to step away from that notion, adopting the view that computing is not only a tool for science but also a new method of thought and discovery in science. The process of dissociating from the computer as the focal center came to completion in the late 1990s when leaders of the field of biology—epitomized by Nobel laureate David Baltimore (2001) and echoing cognitive scientist Douglas Hofstadter (1985)—said that biology had become an information science and DNA translation is a natural information process. Many computer scientists have joined biologists in research to understand the nature of DNA information processes and to discover what algorithms might govern them.

Take a moment to savor this distinction that biology makes. First, some information processes are natural. Second, we do not know whether all natural information processes are produced by algorithms.

The second statement challenges the traditional view that algorithms (and programming) are at the heart of computing. Information processes may be more fundamental than algorithms.

Scientists in other fields have come to similar conclusions. They include physicists working with quantum computation and quantum cryptography, chemists working with materials, economists working with economic systems, cognitive scientists working with brain processes, and social scientists working with networks. All have said that they discovered information processes in their disciplines' deep structures. Stephen Wolfram (2002), a physicist and creator of the software program *Mathematica*, went further, arguing that information processes underlie every natural process in the universe.

All this leads us to the modern catchphrase: "Computing is the study of information processes, natural and artificial." The computer is a tool in these studies but is not the object of study. Dijkstra once said: "Computing is no more about computers than astronomy is about telescopes."

The term *computational thinking* has become popular to refer to the mode of thought that accompanies design and discovery done with computation (Wing 2006). This term was originally called *algorithmic thinking* in the Newell et al. (1960) and was widely used in the 1980s as part of the rationale for computational science. To think computationally is to interpret a problem as an information process and then seek to discover an algorithmic solution. It is a very powerful paradigm that has led to several Nobel Prizes.

All this suggests that computing has developed a paradigm all its own (Denning and Freeman 2009). Computing is no longer just about algorithms, data structures, numerical methods, programming languages, operating systems, networks, databases, graphics, artificial intelligence, and software engineering, as it was prior to 1990. It now also includes exciting new subjects including Internet, web science, mobile computing, cyberspace protection, user-interface design, and information visualization. The resulting commercial applications have spawned new research challenges in social networking, endlessly evolving computation, music, video, digital photography, vision, massive multiplayer online games, user-generated content, and much more.

The computing paradigm places a strong emphasis on the scientific (experimental) method to understand computations. Heuristic algorithms, distributed data, fused data, digital forensics, distributed networks, social networks, and automated robotic systems, to name a few, are often too complex for mathematical analysis but yield to the scientific method. These scientific approaches reveal that *discovery* is as important as *construction* or *design*. Discovery and design are closely linked: the behavior of many large designed systems (such as the web) is discovered by observation; we design simulations to imitate discovered information processes. Moreover, computing has developed search tools that are helping make scientific discoveries in many fields.

The central focus of the computing paradigm can be summarized as information processes—natural or constructed processes that transform information. They can be discrete or continuous.

Table 1.2 summarizes the computing paradigm with this focus. While it contains echoes of engineering, science, and mathematics, it is distinctively different because of its central focus on information processes (Denning and Freeman 2009). It allows engineering and science to be present together without having to choose.

There is an interesting distinction between computational expressions and the normal language of engineering, science, and mathematics. Engineers, scientists, and mathematicians endeavor to position themselves as outside observers of the objects or systems they build or study. Outside observers are purely representational. Thus, traditional blueprints, scientific models, and mathematical models are not executable. (However, when combined with computational systems, they give automatic fabricators, simulators of models, and mathematical software libraries.) Computational expressions are not constrained to be outside the systems they represent. The possibility of self-reference makes for very powerful computational schemes based on recursive designs and executions and also for very powerful limitations on computing, such as the noncomputability of halting problems. Self-reference is common in natural information processes; the cell, for example, contains its own blueprint.

TABLE 1.2 The Computing Paradigm

	Computing
1. Initiation	Determine if the system to be built (or observed) can be represented by information processes, either finite (terminating) or infinite (continuing interactive).
2. Conceptualization	Design (or discover) a computational model (e.g., an algorithm or a set of computational agents) that generates the system's behaviors.
3. Realization	Implement designed processes in a medium capable of executing its instructions. Design simulations and models of discovered processes. Observe behaviors of information processes.
4. Evaluation	Test the implementation for logical correctness, consistency with hypotheses, performance constraints, and meeting original goals. Evolve the realization as needed.
5. Action	Put the results to action in the world. Monitor for continued evaluation.

1.2 Two Views of Computing

Part of a scientific paradigm is a description of the knowledge of the field, often referred to as the "body of knowledge." Within the computing paradigm, two descriptions of the computing body of knowledge have grown up. They might be called a technology interpretation and a principles interpretation.

Before 1990, most computing scientists would have given a technological interpretation, describing the field in terms of its component technologies. After 1990, the increasingly important science aspect began to emphasize the fundamental principles that empower and constrain the technologies.

In reality, these two interpretations are complementary. They both see the same body of knowledge, but in different ways. The technological view reflects the way the field has evolved around categories of technology; many of these categories reflect technical specialties and career paths. The science view reflects a deeper look at timeless principles and an experimental outlook on modeling and validation in computing.

These two views are discussed in Sections 1.3 and 1.4.

1.3 View 1: Technologies of Computing

Over the years, the ACM and Institute of Electrical and Electronics Engineers Computer Society (IEEECS) collaborated on a computing body of knowledge and curriculum recommendations for computer science departments. The milestones of this process give a nice picture of the technological development of the field.

1.3.1 First Milestone: Curriculum 68

In the mid-1960s, the ACM (with help from people in IEEECS) undertook the task to define curriculum recommendations for schools that wished to offer degrees in the new field of computer science (ACM 1968). Their report said that the field consisted of three main parts:

Information structures and processes
Information processing systems
Methodologies

The methodologies included design approaches for software and applications. The core material was mostly the mathematical underpinnings for the parts listed earlier:

Algorithms
Programming
Data structures
Discrete math
Logic circuits
Sequential machines
Parsing
Numerical methods

Many computer science departments adopted these recommendations.

1.3.2 Second Milestone: Computing as a Discipline

The ACM and IEEECS formally joined forces in 1987 to defend computing curricula from a bastardized view that "CS = programming." Around Donald Knuth and Edsger Dijkstra (1970) started making strong and eloquent cases for formal methods of software design, analysis, and construction. They said "we are all programmers" trying to employ powerful intellectual tools to tame complexity and enable correct and dependable software. Although computer scientists understood a programmer as highly skilled expert at these things, the public view of programmers was narrowing to low-level coders, who occasionally caused trouble by hacking into other people's systems.

The committee laid out a model of the computing field that emphasized its breadth, showing that it is much richer than simply programming (Denning et al. 1989). Table 1.3 depicts the 9 × 3 matrix model of the computing field offered by the committee. Theory, abstraction, and design were used in the report for the mathematics, science, and engineering paradigms, respectively. The report gave details about what ideas and technologies fit into each of the 27 boxes in the matrix. It became the basis for a major ACM/IEEE curriculum revision in 1991.

Although this effort had a strong internal influence on the curriculum, it had little external influence on the perception that "CS=programming." In fact, that perception was alive and well in the early 2000s when enrollments declined by over 50%.

1.3.3 Third Milestone: Information Technology Profession

In 1998, the ACM launched an "IT profession" initiative, based on a widely held perception that the field had evolved from a discipline to a profession (Denning 1998, Denning 2001, Holmes 2000). The initiative responded to three trends: the growing interest in the industry for professional standards (especially in safety-critical systems), organized professional bodies representing various specialties,

TABLE 1.3 Matrix Model of Computing Discipline, 1989

Topic Area	Theory	Abstraction	Design
1. Algorithms and data structures			
2. Programming languages			
3. Architecture			
4. Operating systems and networks			
5. Software engineering			
6. Databases and information retrieval			
7. Artificial intelligence and robotics			
8. Graphics			
9. Human–computer interaction			

TABLE 1.4 The Profession of Information Technology

IT-Core Disciplines	IT-Intensive Disciplines	IT-Supportive Occupations
Artificial intelligence	Aerospace engineering	Computer technician
Computer science	Bioinformatics	Help desk technician
Computer engineering	Cognitive science	Network engineer
Computational science	Cryptography	Professional IT trainer
Database engineering	Digital library science	Security specialist
Graphics	E-commerce	System administrator
Human–computer interaction	Economics	Web services designer
Network engineering	Genetic engineering	Web identity designer
Operating systems	Information science	Database administrator
Performance engineering	Information systems	
Robotics	Public policy and privacy	
Scientific computing	Quantum computing	
Software architecture	Instructional design	
Software engineering	Knowledge engineering	
System security	Management information systems	
	Material science	
	Multimedia design	
	Telecommunications	

and a university movement to establish degree programs in information technology. The ACM leadership concluded that the computing field met the basic criteria for a profession and that it was time for ACM to configure itself accordingly.

Table 1.4 is an inventory ACM made of the organized groups in the field. They saw IT professionals as a much larger and more diverse group than computer scientists and engineers, with at least 42 organized affinity groups in three categories. The first category comprises the major technical areas of IT and spans the intellectual core of the field. The second category comprises other well-established fields that are intensive users of IT; they draw heavily on IT and often make novel contributions to computing. The third category comprises areas of skill and practice necessary to keep and support the IT infrastructures that everyone uses. Allen Tucker and Peter Wegner (1996) also noted the dramatic growth and professionalization of the field and its growing influence on many other fields.

Unfortunately, the talk about "profession" led to a new round of terminological confusion. A profession is a social structure that includes many disciplines, but it is not a discipline in its own right. IT is not a field of research; the core disciplines (left column) and partner disciplines (middle column) attend to the research. To what does the term "computing field" refer in this context?

A decade later, it was clear that this interpretation of the field did not match what had actually evolved (Denning and Freeman 2009). The popular label IT did not reconcile the three parts of the computing field under a single umbrella unique to computing. IT now connotes technological infrastructure and its financial and commercial applications, but not the core technical aspects of computing.

1.3.4 Fourth Milestone: Computing Curriculum 2001

The ACM and IEEECS Education Boards were more cautious than ACM leadership in embracing an IT profession when they undertook a curriculum review and revision in 1999. They focused on the core specialties (first column in Table 1.2) and identified the computing discipline with these six academic specialties:

EE—Electrical engineering
CE—Computer engineering
CS—Computer science

SWE—Software engineering
IS—Information systems
IT—Information technology

It was understood that students interested in hardware would enroll in an EE or CE program; students interested in software in a CE, CS, or SWE program; and students interested in organizational and enterprise aspects would enroll in IS or IT programs. Here, the term "IT" is far from what the IT profession initiative envisioned—it refers simply to a set of degree programs that focus on organizational applications of computing technology.

The CC2001 committee organized the body of knowledge into 14 main categories, as follows:

Algorithms and complexity
Architecture and organization
Computational science
Discrete structures
Graphics and visual computing
Human–computer interaction
Information management
Intelligent systems
Net-centric computing
Operating systems
Programming fundamentals
Programming languages
Social and professional issues
Software engineering

There were a total of 130 subcategories. The body of knowledge had 50% more categories than a decade before!

1.3.5 Fifth Milestone: Computing Curriculum 2013

The ACM and IEEECS again collaborated on a ten-year review of the computing curriculum. They learned that the field had grown from 14 to 18 knowledge areas since the 2001 review:

Algorithms and complexity
Architecture and organization
Computational science
Discrete structure
Graphics and visual computing
Human–computer interaction
Information assurance and security
Information management
Intelligent systems
Networking and communications
Operating systems
Platform-based development
Parallel and distributed computing
Programming languages
Software development fundamentals
Software engineering
Systems fundamentals
Social and professional issues

The committee was concerned about the pressure to increase the size of the computer science core. They calculated that the 2001 curriculum recommended 280 core hours and an update in 2008 increased that to 290. The core hours to cover the list earlier would be 305. They divided the core into two parts. Tier 1, the "must have" knowledge, and Tier 2, the "good to have" knowledge. They recommended that individual departments choose at least 80% of the Tier 2 courses, for a total of 276 hours, leaving plenty of time for electives in a student's specialization area.

When ACM issued Curriculum 68, most of us believed that every computer scientist should know the entire core. Today, that is very difficult, even for seasoned computer scientists, since the field has grown so much since 1968.

1.4 View 2: Great Principles of Computing

The idea of organizing the computing body of knowledge around the field's fundamental principles is not new. Many of the field's pioneers were deeply concerned about why computing seemed like a new field, not a subset of other fields like mathematics, engineering, or science. They spent considerable effort to explain what they were doing in terms of the fundamental principles they worked with. Prominent examples are Turing's paper (1937); the essays of Newell et al. (1967); Simon's book (1996); and Arden's COSERS report (1971, 1983). In subsets of the field, thinkers ferreted out the fundamental principles. Examples are Coffman and Denning (1973) on operating systems, Kleinrock (1975) on queueing systems, Hillis (1999) on the nature of computing machines, and Harel (2003) on algorithms and limits of computing.

This viewpoint, however, stayed in the background. I think the reason was simply that for many years, we were concerned with the engineering problems of constructing computers and networks that worked reliably. Most computer scientists were occupied solving engineering problems. The ones most interested in fundamental principles were the ones interested in theory. By 1990, we had succeeded beyond our wildest dreams with the engineering. However, our descriptions of the field looked like combinations of engineering and mathematics. Many outsiders wondered what the word "science" was doing in our title.

When the computational science movement began in the 1980s, many computer scientists felt like they were being excluded. Computational scientists, for their part, did not realize that computer scientists were interested in science. A growing number of us became interested in articulating the science side of computing. It was not easy, because many scientists agreed with Herb Simon (1996), that we are at best a science of the artificial, but not a real science. Real sciences, in their opinions, dealt with naturally occurring processes.

But by 1990, prominent scientists were claiming to have discovered natural information processes, such as in biology, quantum physics, economics, and chemistry. This gave new momentum to our efforts to articulate a science-oriented view of computing (Denning 2005, Denning 2007).

Inspired by the great principles work of James Trefil and Robert Hazen (1996) for science, my colleagues and I have developed the Great Principles of Computing framework to accomplish this goal (Denning 2003, Denning and Martell 2004). Computing principles fall into seven categories: computation, communication, coordination, recollection, automation, evaluation, and design (Table 1.5).

Each category is a perspective on computing: a window into the computing knowledge space. The categories are not mutually exclusive. For example, the Internet can be seen as a communication system, a coordination system, or a storage system. We have found that most computing technologies use principles from all seven categories. Each category has its own weight in the mixture, but they are all there.

In addition to the principles, which are relatively static, we need to take account of the dynamics of interactions between computing and other fields. Scientific phenomena can affect each other in one of two ways: implementation and influence. A combination of existing things implements a phenomenon by generating its behaviors. Thus, digital hardware physically implements computation, artificial intelligence implements aspects of human thought, a compiler implements a high-level language with machine code, hydrogen and oxygen implement water, and complex combinations of amino acids implement life.

TABLE 1.5 Great Principles of Computing

Category	Focus	Examples
Computation	What can and cannot be computed	Classifying complexity of problems in terms of the number of computational steps to achieve a solution. Is P=NP? Quantum computation.
Communication	Reliably moving information between locations	Information measured as entropy. Compression of files, error-correcting codes, cryptography.
Coordination	Achieving unity of operation from many autonomous computing agents	Protocols that eliminate conditions that cause indeterminate results. Choice uncertainty: cannot choose between two near simultaneous signals within a deadline. Protocols that lead the parties to common beliefs about each other's system.
Recollection	Representing, storing, and retrieving information from media	All storage systems are hierarchical, but no storage system can offer equal access time to all objects. Locality principle: all computations favor subsets of their data objects in any time interval. Because of locality, no storage system can offer equal access time to all objects.
Automation	Discovering algorithms for information processes	Most heuristic algorithms can be formulated as searches over enormous data spaces. Human memory and inference are statistical phenomena described by Bayes Rule. Many human cognitive processes can be modeled as information processes.
Evaluation	Predicting performance of complex systems	Most computational systems can be modeled as networks of servers whose fast solutions yield close approximations of real throughput and response time.
Design	Structuring software systems for reliability and dependability	Complex systems can be decomposed into interacting modules and virtual machines following the principles of information hiding and least privilege. Modules can be stratified by layers corresponding to time scales of events that manipulate objects.

Influence occurs when two phenomena interact with each other. Atoms arise from the interactions among the forces generated by protons, neutrons, and electrons. Galaxies interact via gravitational waves. Humans interact with speech, touch, and computers. Interactions exist across domains as well as within domains. For example, computation influences physical action (electronic controls), life processes (DNA translation), and social processes (games with outputs). Table 1.6 illustrates interactions between computing and each of the physical, life, and social sciences as well as within computing itself. There can be no question about the pervasiveness of computing in all fields of science.

1.5 Relation between the Views

The technology and the principles views discussed earlier are two different interpretations of the same knowledge space. They are alternatives for expressing the computing body of knowledge.

The same principle may appear in several technologies, and a particular technology likely relies on several principles. The set of active principles (those used in at least one technology) evolves much more slowly than the technologies.

TABLE 1.6 Examples of Computing Interacting with Other Domains

	Physical	Social	Life	Computing
Implemented by	Mechanical, optical, electronic, quantum, and chemical computing	Wizard of Oz, mechanical robots, human cognition, games with inputs and outputs	Genomic, neural, immunological, DNA transcription, evolutionary computing	Compilers, OS, emulation, reflection, abstractions, procedures, architectures, languages
Implements	Modeling, simulation, databases, data systems, quantum cryptography	Artificial intelligence, cognitive modeling, autonomic systems	Artificial life, biomimetics, systems biology	
Influenced by	Sensors, scanners, computer vision, optical character recognition, localization	Learning, programming, user modeling, authorization, speech understanding	Eye, gesture, expression, and movement tracking, biosensors	Networking, security, parallel computing, distributed systems, grids
Influences	Locomotion, fabrication, manipulation, open-loop control	Screens, printers, graphics, speech generation, network science	Bioeffectors, haptics, sensory immersion	
Bidirectional influence	Robots, closed-loop control	Human–computer interaction, games	Brain–computer interfaces	

While the two styles of framework are different, they are strongly connected. To see the connection, imagine a 2D matrix. The rows name technologies, and the columns name categories of principles. The interior of the matrix is the knowledge space of the field.

Imagine someone who wants to enumerate all the principles involved with a technology. If the matrix is already filled in, the answer is simply to read the principles from the row of the matrix. Otherwise, fill it in by analyzing the technology for principles in each of the seven categories. In the figure later, we see that the security topic draws principles from all seven categories.

	Computation	Communication	Coordination	Recollection	Automation	Evaluation	Design
Security	O(.) of encryption functions	Secrecy authentication covert channels	Key distr protocol zero knowl proof	Confinement partitioning for MLS reference monitor	Instrusion detection biometric id	Protocol perform under various loads	End-to-end layered functions virtual machines

Within the principles framework, someone can enumerate all the technologies that employ a particular principle. In the example later, we see that the coordination category contributes principles to all the technologies listed.

	Computation	Communication	Coordination	Recollection	Automation	Evaluation	Design
Architecture			Hardware handshake				
Internet			TCP and IP protocols				
Security			Key distr protocol zero knowl proof				
Virtual memory			Page fault interrupt				
Database			Locking protocol				
Programming language			Semaphores monitors				

1.6 What Are Information Processes?

There is a potential difficulty with defining computation in terms of information. Information seems to have no settled definition. Claude Shannon the father of information theory, in 1948 defined information as the expected number of yes–no questions one must ask to decide which message was sent by a source. This definition describes the inherent information of a source before any code is applied; all codes for the course contain the same information. Shannon purposely skirted the issue of the meaning of bit patterns, which seems to be important to defining information. In sifting through many published definitions, Paolo Rocchi (2010) concluded that definitions of information necessarily involve an objective component, signs and their referents or, in other words, symbols and what they stand for, and a subjective component, meanings. How can we base a scientific definition of information on something with such an essential subjective component?

Biologists have a similar problem with "life." Life scientist Robert Hazen (2007) notes that biologists have no precise definition of life, but they do have a list of seven criteria for when an entity is living. The observable affects of life, such as chemistry, energy, and reproduction, are sufficient to ground the science of biology. In the same way, we can ground a science of information on the observable affects (signs and referents) without a precise definition of meaning.

A representation is a pattern of symbols that stands for something. The association between a representation and what it stands for can be recorded as a link in a table or database or as a memory in people's brains. There are two important aspects of representations: *syntax* and *stuff*. Syntax is the rules for constructing patterns; it allows us to distinguish patterns that stand for something from patterns that do not. Stuff is measurable physical states of the world that hold representations, usually in media or signals. Put these two together and we can build machines that can detect when a valid pattern is present.

A representation that stands for a method of evaluating a function is called an algorithm. A representation that stands for values is called data. When implemented by a machine, an algorithm controls the transformation of an input data representation to an output data representation. The distinction between the algorithm and the data representations is pretty weak; the executable code output by a compiler looks like data to the compiler and algorithm to the person running the code.

Even this simple notion of representation has deep consequences. For example, as Gregory Chaitin (2006) has shown, there is no algorithm for finding the shortest possible representation of something.

Some scientists leave open the question of whether an observed information process is actually controlled by an algorithm. DNA translation can be called an information process; if someone discovers a controlling algorithm, it could be also called a computation.

Some mathematicians define computation separate from implementation. They do this by treating computations as logical orderings of strings in abstract languages and are able to determine the logical limits of computation. However, to answer questions about running time of observable computations, they have to introduce costs representing the time or energy of storing, retrieving, or converting representations. Many real-world problems require exponential-time computations as a consequence of these implementable representations. I still prefer to deal with implementable representations because they are the basis of a scientific approach to computation.

These notions of representation are sufficient to give us the definitions we need for computing. An information process is a sequence of representations. (In the physical world, it is a continuously evolving, changing representation.) A computation is an information process in which the transitions from one element of the sequence to the next are controlled by a representation. (In the continuous world, we would say that each infinitesimal time and space step is controlled by a representation.)

1.7 Where Computing Stands

Computing as a field has come to exemplify good science as well as engineering. The science is essential to the advancement of the field because many systems are so complex that experimental methods are the only way to make discoveries and understand limits. Computing is now seen as a broad field that studies information processes, natural and artificial.

This definition is broad enough to accommodate three issues that have nagged computing scientists for many years: Continuous information processes (such as signals in communication systems or analog computers), interactive processes (such as ongoing web services), and natural processes (such as DNA translation) all seemed like computation but did not fit the traditional algorithmic definitions.

The great principles framework reveals a rich set of rules on which all computation is based. These principles interact with the domains of the physical, life, and social sciences, as well as with computing technology itself.

Computing is not a subset of other sciences. None of those domains is fundamentally concerned with the nature of information processes and their transformations. Yet this knowledge is now essential in all the other domains of science. Computer scientist Paul Rosenbloom (2012) of the University of Southern California argued that computing is a new great domain of science. He is on to something.

References

Amarel, S. Computer science: A conceptual framework for curriculum planning. *ACM Communications* 14(6): 391–401 (June 1971).

Arden, B. W. (ed.) The computer science and engineering research study. *ACM Communications* 19(12): 670–673 (December 1971).

Arden, B. W. (ed.) *What Can Be Automated: Computer Science and Engineering Research Study* (*COSERS*). MIT Press, Cambridge, MA (1983).

Atchison, W. F. et al. (eds.) ACM Curriculum 68. *ACM Communications* 11: 151–197 (March 1968).

Baltimore, D. How biology became an information science. In *The Invisible Future* (P. Denning, ed.), McGraw-Hill, New York (2001), pp. 43–56.

Chaitin, G. *Meta Math! The Quest for Omega*. Vintage, New York (2006).

Coffman, E. G. and P. Denning. *Operating Systems Theory*. Prentice-Hall, Englewood Cliffs, NJ (1973).

Denning, P. Computing the profession. *Educom Review* 33: 26–30, 46–59 (1998). http://net.educause.edu/ir/library/html/erm/erm98/erm9862.html (accessed on October 15, 2013).

Denning, P. Who are we? *ACM Communications* 44(2): 15–19 (February 2001).

Denning, P. Great principles of computing. *ACM Communications* 46(11): 15–20 (November 2003).
Denning, P. Is computer science science? *ACM Communications* 48(4): 27–31 (April 2005).
Denning, P. Computing is a natural science. *ACM Communications* 50(7): 15–18 (July 2007).
Denning, P., D. E. Comer, D. Gries, M. C. Mulder, A. Tucker, A. J. Turner, and P. R. Young. Computing as a discipline. *ACM Communications* 32(1): 9–23 (January 1989).
Denning, P. and P. Freeman. Computing's paradigm. *ACM Communications* 52(12): 28–30 (December 2009).
Denning, P. and C. Martell. Great principles of computing web site (2004). http://greatprinciples.org.
Denning, P. and P. Rosenbloom. The fourth great domain of science. *ACM Communications* 52(9): 27–29 (September 2009).
Forsythe, G., B. A. Galler, J. Hartmanis, A. J. Perlis, and J. F. Traub. Computer science and mathematics. *ACM SIGCSE Bulletin* 2 (1970). http://doi.acm.org/10.1145/873661.873662 (accessed on October 15, 2013).
Harel, D. *Computers, Ltd.* Oxford University Press, Oxford, U.K. (2003).
Hazen, R. *Genesis: The Scientific Quest for Life's Origins.* Joseph Henry Press, Washington, DC (2007).
Hillis, D. *The Pattern on the Stone.* Basic Books, New York (1999).
Hofstadter, D. *Metamagical Themas: Questing for the Essence of Mind and Pattern.* Basic Books, New York (1985). See his essay on "The Genetic Code: Arbitrary?"
Holmes, N. Fashioning a foundation for the computing profession. *IEEE Computer* 33: 97–98 (July 2000).
Kleinrock, L. *Queueing Systems.* John Wiley & Sons, New York (1975).
Newell, A., A. J. Perlis, and H. A. Simon. Computer science. *Science* 157(3795): 1373–1374 (September 1967).
Rocchi, P. *Logic of Analog and Digital Machines.* Nova Publishers, New York (2010).
Rosenbloom, P. S. A new framework for computer science and engineering. *IEEE Computer* 31–36: 23–28 (November 2004).
Rosenbloom, P. S. *On Computing: The Fourth Great Scientific Domain.* MIT Press (2012).
Shannon, C. and W. Weaver. *The Mathematical Theory of Communication.* University of Illinois Press, Champaign, IL (1948). First edition 1949. Shannon's original paper is available on the Web: http://cm.bell-labs.com/cm/ms/what/shannonday/paper.html (accessed on October 15, 2013).
Simon, H. *The Sciences of the Artificial.* MIT Press, Cambridge, MA (1st edn. 1969, 3rd edn. 1996).
Trefil, J. and R. Hazen. *Science: An Integrated Approach.* John Wiley & Sons, Chichester, U.K. (1st edn. 1996, 6th edn. 2009).
Tucker, A. and P. Wegner. Computer science and engineering: The discipline and its impact. Chapter 1, in *Handbook of Computer Science and Engineering*, A. Tucker (ed.), pp. 1–15. CRC Press, Boca Raton, FL (1996).
Turing, A. M. On computable numbers with an application to the Entscheidungsproblem, *Proceedings of the London Mathematical Society*, Series 2, Vol. 42, pp. 230–265, London, U.K. (1937).
Wing, J. Computational thinking. *ACM Communications* 49(3): 33–35 (March 2006).
Wolfram, S. *A New Kind of Science.* Wolfram Media, Champaign, IL (2002).

2

Computational Thinking

Valerie Barr
Union College

2.1 Introduction

The early twenty-first century is a time of two trends in computing. One trend is that computing technology has become less expensive and more ubiquitous. The second trend is "big data," both the existence of extremely large amounts of data and a push to be able to process large amounts of data. For example, in 2012 the Big Data R&D Initiative committed six US federal agencies and departments to "improve our ability to extract knowledge and insights" from large quantities of digital data [1].

Concurrent with these trends, the term "computational thinking" (CT) has come into use. This frequently refers to a set of computational concepts and skills that should be widely taught. These concepts and skills are generally seen as important and necessary additions to the mathematical, linguistic, and logical thinking skills already taught [2]. There have been several approaches to this in the past, including teaching students basic computer literacy (usually identified as being able to use the computer as a tool), teaching programming in specific languages, or teaching programming through specific application areas [2]. Many computer scientists, however, generally feel that these approaches do not lead to a clear understanding of the shifts in thinking and approach that are necessary and facilitated when computing technology is incorporated into problem solving. Yet with the increased use of the term "computational thinking," there has arisen considerable debate about exactly what it means, who should be expected to do it, and in what contexts. The confusion about what CT means is even apparent in nontechnical discourse. An example of this is found in a March 2012 article in the *New York Times* [3] in which the author conflates "understanding of computational processes" and "the general concepts programming languages employ."

In 2010, the National Research Council's Committee for the Workshops on Computational Thinking identified an ongoing need in the field to "build consensus on the scope, nature, and structure of computational thinking." To some extent, the lack of consensus stems from disagreement about the level of granularity used in the definition. This has led to a set of questions that are at the heart of discussions about CT. Does CT refer to the ways in which we change our thoughts about a problem and its possible solution when we know we can incorporate computing? Is CT about managing layers of abstraction through use of computation, or should it explicitly involve detailed computing concepts such as state machines, formal correctness, recursion, and optimization [2]? Is there a set of analytical skills, unique to computing disciplines, that *all* people should develop? Is there a distinction between doing CT and effectively applying computation to solve problems? Is CT something other than simply saying "let's apply computing to solve a problem in X (e.g., biology, economics, literary analysis, etc.)"? Finally, who is best equipped or best suited to teach CT to those outside of computer science? Should it be computer scientists, or should it be disciplinary specialists?

2.2 Definitions of Computational Thinking

The term "computational thinking" was first introduced by Seymour Papert [4] in a 1996 paper about math education. It was next brought into use by Jeanette Wing in 2006 [5] and subsequently clarified by her in later work (e.g., [6,7]). Since 2006, there have been various attempts to clarify and redefine the term, as well as efforts to articulate the ways in which CT should shape K-12 and college level education and pedagogy. There is not yet consensus about what is meant by the term, especially across educational levels, though a wide range of activities have evolved that are enriching educational practices in computer science and related fields.

In his paper, Papert [4] laid out several approaches to solving a geometric problem, all of which could be implemented using software packages or programming approaches: a probabilistic approach, a standard Euclidean solution, and a heuristic approach that was used to create a supporting visualization. This led Papert to identify two ways in which computers are typically used to contribute to problem solving. The first, evident in the probabilistic approach, is to use the computer as a high-speed data analysis tool, able to evaluate many values and generate an answer. The second, in the Euclidean approach, is to use the computer to provide an alternative to building geometric representations by hand. In both the cases, the computer quickly and effectively generates a solution, but does so using representations that have no clear connection to the original problem and do not bring clarity to the solution. Papert argued that a computationally thoughtful approach would allow one to address his fundamentally geometric problem with a combination of different kinds of computer approaches, making it possible to use the computer to solve problems in ways that "forge ideas," that allow one to carry out heuristic approaches, and that actually help people better analyze and explain the problem, the solution, and the connection between them.

At her most basic, Wing defines CT as a problem-solving and system design approach that "draws on concepts fundamental to computing" [6] and as the "thought processes involved in formulating problems and their solutions so that the solutions are represented in a form that can be effectively carried out by an information-processing agent" [7]. Key to CT, in Wing's view, are the layers of symbolic abstractions of data, processes, and equipment that are used when problems are encoded so that they can be solved using a computer. She argues that the underlying computing device forces a strictly computational problem-solving approach, yet she labels as computational various concepts and skills that predate the advent of computing and are used across numerous disciplines. By contrast, Papert argues that the use of computing devices, in combination with other fields of knowledge and *their* problem-solving approaches, enables rich, creative problem-solving methods that are new and not merely the computerization of an algorithm that already exists in another discipline.

The fundamental differences between Papert's and Wing's definitions lie in the relationship between computing and the problem domain, and in the genesis of the problem-solving strategy. Wing's is a unidirectional view, basically asking the question "What would I have to do to get a computer to implement an existing solution to this problem?" [6]. In this view, the computer has a set of capabilities. Based on these capabilities, problems are expressed algorithmically, as a set of specific instructions that the computer can execute. Papert's approach is bidirectional, using computing in concert with other problem-solving approaches to evolve new solution methods. He interprets "computation" broadly, not just numerically, and sees the computer as an agent that can provide us with useful capabilities and encourage new ways of viewing a problem. It is this rich bidirectional perspective that allows us to address problems that lie at the intersection of disciplines, and is reflected in the development of new fields, such as bioinformatics, computational linguistics, and lexomics [8].

It is when Wing makes a distinction between shallow and deep CT [6] that she begins to approach Papert's view. While still rooted in a largely unidirectional view (choice of abstractions, modeling larger order-of-magnitude problems, hierarchical decomposition), she also considers that initial problem-solving steps can prompt new questions and new ways of viewing data. However, it is not clear that computing is necessarily the best means for teaching concepts such as abstraction, for example. In fact, abstraction is used quite differently in computing than in mathematics. In computing, abstraction is used as a way to express details so that they are manipulable by the computer, whereas in mathematics, underlying detail can be completely ignored or forgotten once the abstraction is created.

Much of the discussion about CT following Wing's 2006 article has followed a unidirectional approach, focusing on whether there is a set of experiences, knowledge, concepts, and skills that are necessary in order for people to be able to apply algorithmic solutions to problems in non-computer science fields. Is CT a definable set of concepts that can be taught? Are these concepts taught, by default, to computer science students, or is CT relevant only in the context of non-computer science disciplines? Should it be taught explicitly as a set of concepts and skills, or only implicitly through the combination of disciplinary knowledge and computational methods? These questions have arisen repeatedly in discussions of pedagogy and curriculum, particularly in K-12 where decisions ultimately depend on the age of the students as well as the context in which the teaching is being carried out. At the college level, we have to ask whether the goal is to expect a single person to have sufficient depth in both computing and another discipline, or whether the goal is to prepare people to bring their respective knowledge and skills to a conversation that can examine how computing and another discipline influence and serve each other. Overall, the discussion of CT has largely ignored the question of what level of exposure to computer science concepts and skills is necessary in order for people to achieve the bidirectional problem-solving insights that Papert presumed can take place.

Several writers have critiqued Wing's original definition of CT or tried to pose definitions that will lead more directly to curriculum and classroom practices. In many cases, a fundamental question raised is whether her definition is at all different than "algorithmic thinking" or "logical thinking." The critiques are summarized and expanded on by David Hemmendinger [9] who cautions against easy distortions of CT. Echoing Papert, he argues that computational metaphors may inspire new problem solutions (and the solving of new problems) in non-computer science disciplines. Hemmendinger points out that "thinking with computation" requires use of algorithms and a consideration of resources that would not be necessary if one were not planning on a computer implementation. Yet he cautions that a push for "computational thinking" is easily permuted into a call to make everyone think like computer scientists, when the goal should be to promote instances in which applications of computing facilitate explicative problem solutions.

At the 2012 ACM Turing Centenary [10], Turing Award recipient Dana Scott posed the question "Is computational thinking or algorithmic thinking something that one knows when he sees it, or can it be explained?" Interestingly, none of the responses delved into any distinction between computational and algorithmic thinking. Donald Knuth commented that algorithmic thinking has been present in works of past centuries, including thirteenth century Sanskrit writings about permutations in music. He discussed the importance of being able to fluently jump levels of abstraction, "seeing things in the small and the large simultaneously, and being able to go effortlessly to many levels in between." Christos Papadimitriou followed Knuth by saying that algorithmic thinking meant "being comfortable with detail, understanding that it is detail, being able to push it away and work without it, and then come back to it." Les Valiant added that we had to recognize the latent computational content of scientific objects and processes. This is an example of the implicit view of CT, not arguing that we should impose on a problem a computer science decision about the appropriate algorithmic solution, but rather suggesting that we ask what data and algorithmic content is already present. This is particularly the case for biological applications in which nature has already provided considerable data encoding.

Papert introduced CT in an education-oriented context, based on the idea that "every time you learn something, you learn two things, the other being the model of learning you just used" [4]. The computer science community has understandably embraced CT, in part because we want people to learn *about* computing by learning via computing. Wing has also discussed CT in a context of core concepts and tools that should be taught to children in K-12, particularly taking advantage of children's willingness to explore new technology devices. Subsequent to Wing's call for "computational thinking for everyone," many efforts were undertaken to incorporate CT into the K-12 educational landscape and into the college experience. In the following sections, we discuss a number of these efforts.

Despite the lack of consensus within the community about what CT is, there does seem to be consensus that students at all levels will benefit from exposure to and increased practice with computing-related concepts and skills. Depending on the degree and type of exposure, students will better understand the role that computing can play in the formulation and execution of problem solutions. This will ultimately help them engage in collaborative problem solving across a range of disciplines, either as a disciplinary specialist with an understanding of computation or as a computer scientist who can collaborate with disciplinary specialists. Certainly, there is no denying that the increasing use of computational methods in many disciplines, and the development of new interdisciplinary fields that include computing, requires that we educate students in a manner that prepares them to do that work.

This still leaves open, however, the question of whether there truly are core aspects of computing that everyone needs to understand. The National Science Foundation (NSF), in their 2009 CPATH solicitation (CISE Pathways to Revitalized Undergraduate Computing Education) [11], loosely defined CT as "competencies in computing concepts, methods, technologies, and tools" [12]. On their FAQ site, they said "NSF has described a broad CT framework, but does not attempt to define the computing constructs. You are invited to join in the definition and articulation of the computing constructs that have permeated and transformed our modern world and to use these to develop methods, curricula, and pedagogies to assure that all students are able to think computationally." The CPATH program focused primarily on college and university interventions, but also considered K-12 interventions in situations where they were tied to transition into an undergraduate program that involved CT. By contrast, the subsequent NSF CE21 solicitation (Computing Education for the twenty-first century) gives no definition of CT at all, simply saying that "Efforts can focus on computational thinking as taught in computing courses or infused across the curriculum" and will "increase ... knowledge of computational thinking concepts and skills" [13].

2.3 College and University

There are numerous examples of the infusion of computing and computer science content into non-computer science courses and non-computer science disciplines. While many do not explicitly mention CT, these efforts generally have a goal of exposing students to interdisciplinary applications of computing. We present a few examples, both individual courses at different curricular levels and efforts at larger programmatic or institutional change.

2.3.1 Individual Courses

2.3.1.1 Computing for Poets

Professor Mark LeBlanc of Wheaton College (Norton, MA) developed the course *Computing for Poets* [14], which focuses on the use of computers to analyze written text. It involves a number of "computational thinking skills," including problem decomposition, algorithmic thinking, and experimental design. A key focus is on developing programs that can address new questions about texts, rather than relying solely on existing text analysis tools. This course has been developed in the context of the Lexomics project [15] in which various computational and statistical methods are combined with existing text analysis approaches, leading to methods for answering new questions about ancient texts. The course is connected to two English courses, one on Anglo-Saxon Literature and another on J.R.R. Tolkien. This connection of courses gives students firsthand experience devising problem-solving methods that combine traditional and computational approaches, allowing them to ask new questions, and then ask whether bringing the power of computing to literary analysis leads to the formulation of new answers.

2.3.1.2 Computational Microeconomics

Professor Vincent Conitzer of Duke University developed the course *Computational Microeconomics* [16], which was intended to introduce undergraduates to work at the intersection of economics and computer science. The course is usually taken by students from computer science, economics, math, and electrical and computer engineering. It has minimal prerequisites and starts with an introduction to linear and integer programming, along with the MathProg language and the GNU Linear Programming Kit. The remainder of the course focuses on economics topics such as expressive marketplaces, game theory, Bayesian games, and auction design. The students work on both written and programming assignments and do a small (possibly team) project that involves programming. This course provides students with firsthand experience in working at the intersection of two disciplines.

2.3.1.3 Advanced Algorithmic Techniques for GPUs

In the context of a course on Advanced Algorithmic Techniques for GPUs (graphical processing units), two researchers from the visual computing technology company NVIDIA, Wen-mei Hwu and David Kirk, present a lengthy list of areas in which GPU computing is being used. They include a range of computing, business, scientific, and engineering applications such as financial analysis, digital audio and video processing, scientific simulation, computer vision, and biomedical informatics. The inference is that people from a range of backgrounds and disciplines are increasingly in a position to try to use GPUs, or to coordinate with computer scientists who will use GPUs. This is underscored when they put forward a definition of CT skills: "The ability to translate/formulate domain problems into computational models that can be solved efficiently by available computing resources" [17]. The goal of their course is to make commonly used many-core GPU techniques accessible to scientists and engineers, claiming that "computational thinking is not as hard as you may think it is." They further define CT skills as

- Understanding the relationship between the domain problem and the computational models
- Understanding the strengths and limitations of the computing devices
- Defining problems and models to enable efficient computational solutions

2.3.2 Institutional Change

2.3.2.1 Bowdoin College

Bowdoin College is beginning to take steps toward institution-wide change, with initial discussion of the importance of some form of CT across the curriculum. Bowdoin's president, Dr. Barry Mills, addressed this in his 2012 baccalaureate address to graduates [18]. Speaking about the role of information and the ability to interpret information, Mills addressed CT as a "mode of inquiry" in the curriculum. While citing Wing's 2006 definition, Mills put forth that it is possible to "engage computational thinking and analysis" across disciplines that span the sciences, social sciences, and humanities. Rather than arguing for the direct application of computing, he argues for the integration of disciplines with "a rapidly changing landscape of data, information, and knowledge," seeing this as an important step for the continued development of liberal arts education.

The perspective put forward by Dr. Mills is an indication of the extent to which leaders of educational institutions are beginning to grapple with some definition of CT as an educational priority.

2.3.2.2 DePaul University: Computational Thinking across the Curriculum

Faculty at DePaul University, led by Professors Amber Settle and Ljubomir Perkovic, undertook a project [19] designed to incorporate CT into the DePaul curriculum at all levels, using their Liberal Studies (general education) courses as the vehicle to accomplish this. The DePaul project leaders recognized that Wing's original definition of CT was problematic, in part because she gave no examples from outside the sciences and quantitative social sciences. They began their project by defining CT as "the intellectual and reasoning skills that a professional needs to master in order to apply computational techniques or computer applications to the problems and projects in their field." This definition hews closely to Wing's approach, though they argue in their project documents that hers is not a new way of thinking, pointing to a number of examples that significantly predate her article. However, they make a distinction between the implicit use of CT within various fields and the explicit recognition of the use of computationally oriented reasoning skills. The underlying principle of the DePaul project was that it could only be successful if faculty across disciplines agreed on a useful framework for implementing CT in their courses.

In order to convince faculty from other fields to participate, the project leaders developed a set of examples from a range of disciplines, based on categories of CT that they defined using Peter Denning's "Great Principles of Computing" [20]. Denning, preceding Wing's article by several years, put forward a set of principles with the goal of showing people in other disciplines "how to map the principles into their own fields." The DePaul project offered definitions of each principle (computation, communication, coordination, recollection, automation, evaluation, and design) and gave examples of how CT has been used in a number of disciplines. Ultimately, the project included 18 faculty at DePaul who modified 19 Liberal Studies courses. The project leaders also collaborated with six teachers at the University of Chicago Lab Schools in order to also explore middle school and high school course modifications [21].

2.3.2.3 Union College: Creating a Campus-Wide Computation Initiative

Union College [22], with NSF funding, carried out a five-year project to create a campus-wide computation initiative. The goal of the project was to provide an appropriate foundation in computing for non-computer science students who are or should be using computing in the context of their discipline. There were two key elements:

1. Provide engaging introductory computer science courses, open to all students on campus. The department offers six theme-based courses: big data, robotics, game development, artificial intelligence, media computation, and programming for engineers.
2. Support infusion of a computing component into non-computer science courses.

To date, 24 faculty from 15 disciplines, working with summer research students, have developed computational components for over two dozen courses. These efforts were also extended to three other undergraduate institutions, with faculty–student teams at Mount Holyoke College (Music), Denison College (Geology), and Bard College (Computer Science and Literature). The nature of the computational modules is quite varied, with no common set of key concepts or skills across all of the courses. The key result, however, is that a majority of students taking the infused courses have reported that the computational component helped them understand the disciplinary material. In addition, in line with Papert's definition of CT, faculty and students alike have been able to move forward with new research efforts and new class exercises because incorporating a computational element allows them to ask new questions and combine data in new ways. Very few of the infused courses, however, have an explicit programming component, or any explicit discussion of concepts such as abstraction, problem decomposition, etc., beyond what would normally take place in the context of that discipline.

2.4 K-12

Considerable effort has been made to address content and pedagogy issues regarding CT in K-12. Despite the lack of consensus about exactly what CT is, numerous attempts have been made to bring component elements into the K-12 classroom. Each K-12 effort reflects an inherent definition of CT, with a resulting view of what should be introduced and how it should be incorporated into the classroom setting (as separate content, or integrated with existing curricular content). Concepts often considered part of CT that have been introduced across the array of K-12 activities include iterative refinement, problem decomposition, connections between social systems and physical systems, modeling of abstraction, continuous versus discrete data, Boolean operations, causal relationships, searching for patterns, data mining, and complexity. These concepts are introduced either through general subject areas or through domain-specific content, including scientific models (simple models of physical phenomena for younger children; more complex models for older students), visualization and large data sets, computational media, game development, genetics, and environmental simulations. Throughout the K-12 work, an important issue is the role of age and grade level in determining what students can do, and the transition from prebuilt examples and models to independent building or programming. Of course, competing pedagogies lead to different approaches based on how accessible the pedagogy makes different concepts and skills.

K-12 projects that are focused on teaching computer science can effectively sidestep the issue of how to define CT. Such courses (e.g., see Section 2.4.4) will cover concepts frequently considered part of CT because those are part and parcel of computer science. The Computer Science Teachers Association (CSTA) addresses grade level and implementation issues through a comprehensive set of curricular standards for K-12 computer science education [23]. These standards, divided into three levels, cover the introduction of fundamental concepts of computing at all levels, as well as additional secondary-level computer science courses. The curriculum is built around five complementary strands, one of which is CT (the others are collaboration; computing practice; computers and communication devices; and community, global, and ethical impacts).

The focus of the CT strand is to promote problem solving, system design, knowledge creation, and improved understanding of the power and limitations of computing. The CSTA uses the definition developed during a joint project with the International Society for Technology in Education (ISTE) (Section 2.4.1). The interdisciplinary view that CT is important for combining computer science with all disciplines can actually facilitate teaching it in lower grades since it does not necessitate specific detailed knowledge of computing on the part of teachers. For example, in grades K-3, the standard for CT is met if students can use certain technology resources to solve age-appropriate problems, use various tools to illustrate stories, and understand how to sort information. In grades 3–6, the standards are met by students understanding and using the basic steps of algorithmic problem solving, developing a simple understanding of an algorithm, and making a list of subproblems that contribute to the solution of a larger problem. These abilities continue to scale up for grades 6–9 and 9–12 so that eventually students can describe the software development process, understand sequence, selection, iteration, and recursion, compare techniques for analyzing massive data collections, etc.

2.4.1 CSTA and ISTE

The CSTA and the ISTE, with support from the NSF, undertook development of an operational (rather than formal) definition of CT for K-12 [24]. This effort, involving K-12 teachers, college faculty, and educational administrators, focused on a set of questions relevant to the K-12 setting (from [24]):

- What would CT look like in the classroom?
- What are the skills that students would demonstrate?
- What would a teacher need in order to put CT into practice?
- What are teachers already doing that could be modified and extended?

The consensus reached was that CT, in the K-12 setting, involves solving problems such that the solution can be implemented with a computer. Students will use concepts such as abstraction, recursion, and iteration, applying these across many subjects. This led to the identification of key concepts and capabilities, with examples of how these could be embedded in activities across the K-12 curriculum. This effort addresses an issue raised by Lu and Fletcher [25]. While agreeing with Wing's basic definition of CT, they argue for the gradual introduction of CT concepts through consistent terminology. Lu and Fletcher propose a *computational thinking language* of vocabulary and symbols that can be used from grades 3–12 to refer to computation, abstraction, information, execution, etc. Many disciplines involve problem solving and promote logical and algorithmic thinking, so active engagement of these practices across the disciplines will reinforce students' developing capabilities (such as designing solutions to problems, implementing designs, testing and debugging, etc.). In this sense, linguistic analysis of sentences in a Language Arts class may not seem inherently computational, but it will reinforce patterns of thinking that can be applied to problem solving in other fields. The full set of concepts and capabilities identified in the CSTA–ISTE effort includes data collection, data analysis, data representation, problem decomposition, abstraction, algorithms and procedures, automation, parallelization, and simulation. Examples are provided from computer science, math, science, social studies, and language arts. Finally, a set of resources provide concrete examples for classroom teachers [26,27].

2.4.2 UK Computing: A Curriculum for Schools

In 2012, the British Computing at School Working Group published the report *Computer Science: A curriculum for schools* [28]. The purpose of the report was to address the question of what constitutes computer science as a school subject. The authors present the following definition of CT, very much based on Wing: "a mode of thought that goes well beyond software and hardware, and that provides a framework within which to reason about systems and problems." As might be expected in a document focused on computer science learning requirements, they see CT in the unidirectional way, as "the process of *recognising* aspects of computation in the world that surrounds us, and *applying* tools and techniques from computing to understand and reason about … systems and processes." They further argue that students are empowered by the combination of CT, computing principles, and a computational approach to problem solving.

Like the CSTA–ISTE project, the Computing at School group presents a detailed breakdown of what students should know about computing and computer science, and what they should be able to do at each academic level (roughly equivalent to grades K-2, 2–5, 5–8, and 8–10). The set of concepts is roughly equivalent to the CSTA–ISTE set, focusing on algorithms, programs, data, computers, and communication. Whereas the CSTA–ISTE project identifies ways in which the key concepts and capabilities could be addressed in non-computing disciplines, the Computing at School group does not address how the material should be taught. They do, however, provide a very clear set of "attainment targets" for each level student, demonstrating increased mastery as students progress.

2.4.3 Computational Science and Computational Science Thinking

The slogan of the Shodor Foundation [29] is "Transforming Learning Through Computational Thinking." Their focus is on computational science, combining quantitative reasoning, analogical thinking, algorithmic thinking, and multiscale modeling. They want students to develop the math and computing knowledge and skills necessary to generate and understand computational results across a number of disciplines. (Part IV of this volume describes advances in computational science, including computational biology, bioinformatics, astrophysics, chemistry, scientific visualization, and other areas.) The Shodor Foundation provides resources for both faculty and students, including free online workshop materials that focus on model building, simulation, and model-based reasoning in various math and science disciplines (such as astronomy and astrophysics, biomedical science, computational chemistry, environmental science, forensic science, and physics). Without an overt emphasis on CT, Shodor also provides instructional materials for various levels of K-12 that support student facility with underlying math and technology (databases and spreadsheets, graphics and visualization, and scientific computing, for example). Given the nature of their materials, much can be used either inside the classroom in a formal setting or outside the classroom in an informal education setting.

For college and university faculty, Shodor provides discipline-specific workshops that focus on the use of computation (e.g., computational biology, computational chemistry) as well as workshops that focus more generally on the use of parallel and cluster computing in computational science. A related effort is found in the *INSTANCES* project, Incorporating Computational Scientific Thinking Advances into Education and Science Courses [30]. The focus of this project is to introduce K-12 teachers, through a series of modules, to computing in the context of scientific problems, viewing computational science as the solution of scientific and engineering problems using the "combination of techniques, tools, and knowledge from multiple disciplines," combined with computer simulations. Like the Shodor Foundation (part of the INSTANCES project), they argue that the problems that are central to computational science form an important core for science, math, and computer science education. They define computational science thinking (CST) as "using simulation and data processing to augment the scientific method's search for scientific truth for the realities hidden within data and revealed by abstractions." Each module details the learning goal and objectives, activities, products that will be generated by students, and discussion of the embedded CST. The modules cover areas such as exponential spontaneous decay, exponential biological growth, bug population dynamics, and projectile drag.

2.4.4 Exploring Computer Science, CS Principles

A number of efforts are aimed at increasing high school–level interest in computer science, particularly in ways that will encourage students to continue with computer science study in college. These efforts address CT to the extent that a student just learning about computer science would be encouraged to consider applications of computing to other disciplines.

Exploring Computer Science (ECS) [31] is a one-year college-preparatory course that evolved out of efforts to involve in computer science a larger number of underrepresented students in the Los Angeles Unified School District. The course is focused on computer science concepts, with the following key instructional units:

- Human Computer Interaction
- Problem Solving
- Web Design
- Introduction to Programming
- Data Analysis
- Robotics

The course is designed to address "real world, socially relevant, interdisciplinary, and creative applications of computing" [31]. In that context, the developers identify three CT practices: algorithm development, problem solving, and programming (specifically related to real-life problems faced by students).

The goal of the CS Principles project [32] is the development of a new Advanced Placement Computer Science course. A set of six CT practices are listed as part of the Principles project, including analyzing problems and artifacts, communicating, and collaborating. The three practices that most directly involve computing are

1. Connecting computing, specifically identifying its impacts, describing connections between people and computing, and explaining connections between computing concepts
2. Developing computational artifacts, using appropriate techniques and appropriate algorithmic and information management principles
3. Abstracting, which involves representation, identification of and use of abstractions in computation or modeling

These are reasonable, and not unexpected, practices to incorporate into a computer science course, and will be taught to students who are already committed to studying at least introductory computer science.

2.5 Changes to Computer Science

As discussed earlier, a bidirectional view of CT has led to the development of new disciplines, and it has also led to changes in computer science as a field. Computer science has become a driving force of innovative practices and new knowledge. The combination of computing power, existing and new computing techniques, and disciplinary knowledge leads to the identification of new problems and the development of new solution methods. An early example (1988) of this is the Human Genome Project, when computer science was faced with a "big data" problem. This led to efforts to build parallel sequence analysis algorithms, better visualization tools, mechanisms to distribute databases, new pattern recognition algorithms, and gene modeling algorithms. Today, we may think of this work within the disciplines of genomics or computational biology, but its origins were in efforts to determine how best to combine computing with biological knowledge and methods to solve a new problem. Similarly, data mining and information retrieval methods were advanced in the context of the genome project, and today are employed to solve problems relevant to a host of disciplines.

Beyond the above connections, the relationship between biology and computer science is extremely bidirectional. Initially, there were two separate efforts, with computational methods to solve biological problems and the study of biological systems to derive new optimization methods for computing. Increasingly, however, computing and biology have supported each other in new ways, based on a view of biological processes as algorithms and information processing systems that solve real-world problems. In an extensive survey [33], Navlakha and Bar-Joseph focus on coordination, networks, tracking, and vision as examples of areas in which an information processing view has led to better understanding of biological properties *and* led to the improvement of computational systems. They report on research on problems that can be addressed both computationally and biologically. A model constructed for the biological problem can then be used to improve the solution to the computational problem. For example, they cite research in visual processing computational neuroscience that is leading to new models of human visual processing, which in turn is leading to better algorithms for computer vision. The bidirectional connections between biology and computer science demonstrate that commonalities between computational and biological processes allow solutions to problems in one area to apply to problems in the other.

Terrain analysis (see Chapter 46) is another domain in which new computer science methods had to be developed if computing was to be used effectively as part of new problem-solving approaches. Terrain analysis involves applying computational models to very large global positioning system (GPS) data sets (replacing surveyor data) in order to analyze terrain features, particularly important for environmental modeling and land use management. The use of digital, remotely acquired, data for terrain analysis allows for the analysis of very remote, relatively unexplored areas for which actual on-site surveying would be prohibitive. The use of digital data for terrain analysis, however, has necessitated evolution of interpolation methods, visualization of the relevant GPS data, animation, and other areas of computer graphics. In general, the sizable number of tools available ([34] gives examples) is a clear indication of the extent to which the analysis of geographic information system data has become a critical part of certain fields, and computer science has developed the techniques necessary to meet the analysis needs.

There are also humanities disciplines and humanities research that present opportunities for computing to support research and for the demands of research to push developments in computing. One such area, underscored by a NSF and National Endowment for the Humanities (NEH) grant program, is documenting endangered languages [35]. The goal is to "exploit advances in information technology to build computational infrastructure for endangered language research." The solicitation is explicit that funding can support fieldwork efforts related to the recording, documenting, and archiving of languages, but also will support preparation of "lexicons, grammars, text samples, and databases," with an interest in data management and archiving. This is clear recognition that the necessary computing infrastructure to address the problem of endangered languages may not yet exist. Therefore, the aggregate of projects in this area have to address both the humanities side and the computing side of the problem, and that work on both sides will be informed by understanding and knowledge of the other.

2.6 Summary

The increasing ubiquity of computing and increased focus on "big data" make it possible for almost every discipline to consider new questions or address longstanding problems. In some instances, the methods for solving those problems are already clear, requiring straightforward applications of computing in order to implement existing algorithms. In other situations both disciplinary and computing knowledge, the bidirectional form of CT, must be utilized to develop new solutions. This will continue to lead to changes in existing fields, and to the formulation of new disciplines. There is not yet agreement on a single set of CT concepts and skills that all students should be taught. There is clearly increasing interest, however, in broadening student exposure to elements of computing, preparing those in computer science and those in other fields to better contribute to interdisciplinary applications.

Acknowledgments

The author thanks Chris Stephenson (Executive Director, CSTA), Amber Settle (Associate Professor, DePaul University), and David Hemmendinger (Professor Emeritus, Union College) for their thoughtful feedback. Allen Tucker (editor and Professor Emeritus, Bowdoin College) provided a number of useful examples of ways in which computer science has been pushed by other fields.

References

1. White House Office of Science and Technology Policy. OSTP Website: Big Data Press Releast. http://www.whitehouse.gov/sites/default/files/microsites/ostp/big_data_press_release_final_2.pdf, 2012 (accessed on October 9, 2013).
2. Committee for the Workshops on Computational Thinking. Report of a Workshop on the Scope and Nature of Computational Thinking. Technical report, National Research Council, 2010.
3. R. Stross. Computer Science for the Rest of Us. *New York Times*, April 2012.
4. S. Papert. An exploration in the space of mathematics educations. *International Journal of Computers for Mathematical Learning*, 1(1):95–123, 1996.
5. J. Wing. Computational thinking. *Communications of the ACM*, 49(3):33–35, 2006.
6. J. Wing. Computational thinking and thinking about computing. *Philosophical Transactions of the Royal Society A*, 366:3717–3725, 2008.
7. J. Wing. Computational thinking: what and why? http://www.cs.cmu.edu/~CompThink/papers/TheLinkWing.pdf (accessed on October 9, 2013).
8. B. Dyer and M. LeBlanc. Wheaton College Genomics group. http://wheatoncollege.edu/genomics/lexomics/ (accessed on October 9, 2013).
9. D. Hemmendinger. A plea for modesty. *ACM Inroads*, 1(2):4–7, 2010.
10. Association of Computing Machinery. ACM A.M. Turing Centenary Celebration Webcast. http://amturing.acm.org/acm_tcc_webcasts.cfm (accessed on October 9, 2013).
11. National Science Foundation Directorate for Computer & Information Science & Engineering. CISE pathways to revitalized undergraduate computing education. http://www.nsf.gov/cise/funding/cpath_faq.jsp, 2009.
12. National Science Foundation Directorate for Computer & Information Science & Engineering. CISE pathways to revitalized undergraduate computing education. http://www.nsf.gov/pubs/2009/nsf09528/nsf09528.htm, 2009.

13. National Science Foundation Directorate for Computer & Information Science & Engineering. Computing Education for the 21st Century (CE21). http://www.nsf.gov/pubs/2012/nsf12527/nsf12527.htm, 2012.
14. M. LeBlanc, Computing for Poets. http://cs.wheatoncollege.edu/~mleblanc/131/ (accessed on October 9, 2013).
15. M. Drout and M. Leblanc. Wheaton College Lexomics group. http://wheatoncollege.edu/lexomics/ (accessed on October 9, 2013)
16. V. Conitzer. An undergraduate course in the intersection of computer science and economics. In *Proceedings of the Third AAAI Symposium on Educational Advances in Artificial Intelligence (EAAI-12)*, Toronto, Ontario, Canada, 2012.
17. Wen mei Hwu and David Kirk. Berkeley winter school advanced algorithmic techniques for gpus. http://iccs.lbl.gov/assets/docs/2011-01-24/lecture1_computational_thinking_Berkeley_2011.pdf, 2011 (accessed on October 9, 2013).
18. B. Mills. Barry Mills: Baccalaureate 2012 Address. http://www.bowdoindailysun.com/2012/05/barry-mills-baccalaureate-2012-address/ (accessed on October 9, 2013).
19. L. Perkovi and A. Settle. Computational Thinking Across the Curriculum: A Conceptual Framework. Technical report 10-001, College of Computing and Digital Media, January 2010.
20. P. Denning. Great principles of computing. *Communications of the ACM*, 46(11):15–20, 2003.
21. A. Settle, B. Franke, R. Hansen, F. Spaltro, C. Jurisson, C. Rennert-May, and B. Wildeman. Infusing computational thinking into the middle- and high-school curriculum. In *Proceedings of ITiCSE 2012: The 17th Annual Conference on Innovation and Technology in Computer Science Education*, Haifa, Israel, July 2012.
22. V. Barr. *Disciplinary thinking; computational doing*. 2012 (under review).
23. Computer Science Teachers Association, CSTA K-12 Computer Science Standards. http://www.csta.acm.org/Curriculum/sub/K12Standards.html (accessed on October 9, 2013).
24. V. Barr and C. Stephenson. Bringing computational thinking to K-12: What is involved and what is the role of the computer science education community? *ACM Inroads*, 2(1):48–54, 2011.
25. J. Lu and G. Fletcher. Thinking about computational thinking. In *Proceedings of SIGCSE 2009: The 40th ACM Technical Symposium on Computer Science Education*, Chattanooga, Tennessee, 2009.
26. Computer Science Teachers Association. Computational thinking resources. http://csta.acm.org/Curriculum/sub/CompThinking.html (accessed on October 9, 2013).
27. International Society for Technology in Education. Computational thinking toolkit. http://www.iste.org/learn/computational-thinking/ct-toolkit (accessed on October 9, 2013).
28. Computing at School Working Group. Computer science: A curriculum for schools. http://www.computingatschool.org.uk/data/uploads/ComputingCurric.pdf, 2012.
29. Shodor Foundation. http://www.shodor.org (accessed on October 9, 2013).
30. R. Landau et al., Incorporating computational scientific thinking advances into education and science courses. http://www.physics.orst.edu/~rubin/INSTANCES (accessed on October 9, 2013).
31. D. Bernier et al., Exploring computer science. http://www.exploringcs.org (accessed on October 9, 2013).
32. O. Astrachan, A. Briggs, L. Diaz., Exploring computer science. http://www.exploringcs.org (accessed on October 9, 2013).
33. S. Navlakha and Z. Bar-Joseph. Algorithms in nature: the convergence of systems biology and computational thinking. *Molecular Systems Biology*, 7(546), November 2011.
34. D. Wright. Davey Jones Locker Treasure Chest of Web Links, http://MarineCoastalGIS.net (accessed on October 9, 2013).
35. Behavioral & Economic Sciences Division of Behavioral National Science Foundation Directorate for Computer & Information Science & Engineering, Directorate for Social and Office of Polar Programs Cognitive Sciences. Documenting Endangered Languages: data, infrastructure and computational methods. http://www.nsf.gov/pubs/2011/nsf11554/nsf11554.htm, 2012.

II

Algorithms and Complexity

3

Data Structures

Mark Allen Weiss
Florida International University

A *data structure* is a method for organizing large amounts of data so that it can be accessed efficiently. Any piece of software will almost certainly make use of fundamental data structures, and proper choices of data structures can have a significant impact on the speed of an application, while poor choices can make the application impractical for even modest-sized inputs. Consequently, the study of data structures is one of the classic topics in computer science. Most data structures can be viewed as *containers* that store collections of items.

3.1 Types of Containers

All basic containers support retrieving, inserting, and removing an item, and the basic containers differ in which items can be efficiently accessed. Containers also often provide a mechanism for viewing every item; in languages such as C++, C#, and Java, this is implemented with an *iterator* and a language *for loop construct*.

A *sequence container* stores items in a linear order, and an item is inserted at a specified position. Examples of sequence containers include *fixed-sized arrays*, *variable-length arrays*, and *linked lists*. Typically, retrieval is based on finding an item given a position; testing whether some item present in the list is likely to devolve into a sequential search and be inefficient.

A *set* container stores items without linear order and thus can provide for more efficient retrieval. Some sets will guarantee that items can be viewed in sorted order; others provide no such guarantees. Examples of sets include *search trees*, *skip lists*, and *hash tables*. In some languages, sets can have duplicates.

A *map* container allows searching for key/value pairs by key. Although the interface is different from the set container, the underlying data structures used for efficient implementation are the same.

A *priority queue* container provides access only to the minimum item. An efficient implementation of a priority queue is the *binary heap*.

Some containers are designed specifically for searching of text. Examples include *suffix trees* and *radix search trees* [1–3].

3.2 Organization of the Chapter

The remainder of this chapter focuses primarily on the fundamental container types, namely, the sequence containers, ordered and unordered sets, and the priority queue. We also discuss one data structure that is not a container: the *union/find data structure*. After examining the fundamental data structures, we discuss the recurring themes and interesting research challenges seen in the design of data structures.

3.2.1 Sequence Containers

A *sequence container* stores items in a linear order, and an item is inserted at a specified position. Typically, retrieval is based on finding an item given a position; testing whether some item present in the list is likely to devolve into a sequential search and be inefficient. If items can be accessed, inserted, and removed from only one end, the container is a *stack*. If items can be inserted at one end (typically the back) and removed only from the other end (the front), then the container is a *queue*. If all operations are allowed at both ends, the container is a *double-ended queue* [4].

3.2.1.1 Fixed-Size Arrays

A fixed-sized array is usually a language primitive array, as in C++ or Java. Items are accessed by the array-indexing operator, and the fixed-size array is often used as a building block for resizable arrays. In many cases, particularly when insertions and deletions are rare, fixed-sized arrays are the fastest sequence container.

3.2.1.2 Resizable Arrays

Resizable arrays allow efficient access to an item by specifying its position (an array index) and also efficient adding and removal of an item, provided that the position is at or very close to the high index. Inserting new items near the low index is inefficient because it requires pushing all the subsequent items one index higher in the array; similar logic applies for removing items near the front. Resizable arrays are often implemented by using a primitive array and maintaining a current size; the primitive array is replaced with a larger array if capacity is reached. If the array size is increased by a multiplicative factor (e.g., it is doubled) then the amortized overhead incurred by the expansion is constant time. The C++ `vector` and Java `ArrayList` are examples of resizable arrays.

3.3 Linked Lists

In a linked list, items are not maintained in (logically) contiguous order; this potentially avoids the problem of having to move large numbers of items during an insertion or deletion. In a *singly linked list*, a *node* stores an item along with a link to the next node in the list. Additionally, the size of the list and a link to the first item are stored. This gives $O(1)$ access time for adding, removing, or accessing the first item. If a link to the last item is also stored, we obtain $O(1)$ access time for adding or accessing the last item. If each node contains both the next and previous item in the list, we obtain a *doubly linked list*. Although the doubly linked list uses more space than the singly linked list, it supports the removal of the last item in $O(1)$ time and allows bidirectional traversal of the list.

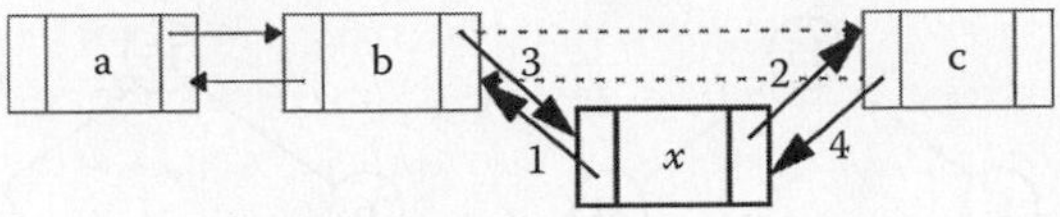

FIGURE 3.1 Insertion into a doubly linked list.

	Resizable Array	Singly-Linked List	Doubly-Linked List
addFront	$O(N)$	$O(1)$	$O(1)$
addBack	$O(1)$	$O(1)$	$O(1)$
addAtIndex	$O(N)$	$O(N)$	$O(N)$
addAfterIterator	$O(N)$	$O(1)$	$O(1)$
addBeforeIterator	$O(N)$	$O(N)$	$O(1)$
removeFront	$O(N)$	$O(1)$	$O(1)$
removeBack	$O(1)$	$O(N)$	$O(1)$
removeAtIndex	$O(N)$	$O(N)$	$O(N)$
removeIterator	$O(N)$	$O(1)$*	$O(1)$
getByIndex	$O(1)$	$O(N)$	$O(N)$
advanceIterator	$O(1)$	$O(1)$	$O(1)$
retreatIterator	$O(1)$	$O(N)$	$O(1)$
contains	$O(N)$	$O(N)$	$O(N)$
remove	$O(N)$	$O(N)$	$O(N)$

FIGURE 3.2 Running times for operations on sequential containers. *represents constant time if we are presented with the node prior to the one actually being removed; $O(N)$ otherwise.

Figure 3.1 shows how a new node is spliced into a doubly linked list (the reverse process also shows the removal).

Accessing of an item by specifying its index position is inefficient if the index is not near one of the list ends because the item can only be reached by sequentially following links.

Items can be inserted into the middle of a linked list with only a constant amount of data movement by splicing in a new node. However, an insertion based on providing an index is inefficient; rather these kinds of insertions are usually done by providing a link to the node itself using some form of an iterator mechanism.

Figure 3.2 summarizes the cost of the basic operations on the sequential containers.

3.4 Ordered Sets and Maps (Search Trees)

An *ordered set* stores a collection of unique items and allows for their efficient insertion, removal, and access. Iteration over the set is expected to view the items in their sorted order. An implementation using a sorted array allows for access of an item by using *binary search*. In a binary search, we compare the target item x with the item in the middle of the array. This either locates x or reduces to a search in either the lower or upper half of the array that can be resolved recursively. Since the recursion reaches a base case after in approximately $\log N$ steps,* the cost of an access is in $O(\log N)$ time; however, adding or removing items will take $O(N)$ time, because all items larger than the item being manipulated will need to be shifted one position in the array. Using a sorted linked list appears problematic because binary search requires accessing the middle item, and that operation is not efficiently supported by linked lists. Most set implementation use a different linked structure based on trees; typically, these are binary trees.

* Throughout, $\log N$ means $\log_2 N$.

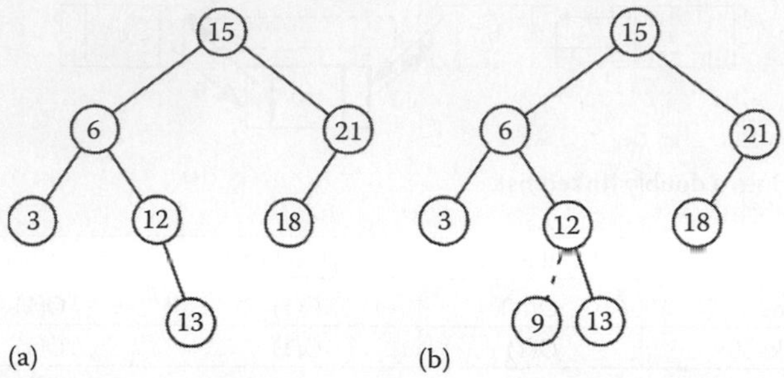

FIGURE 3.3 Binary search tree (a) before and (b) after insertion of 9.

3.4.1 Binary Search Trees

In a *binary search tree*, each node stores an item and also contains a link to a left subtree and a right subtree. All items in the left subtree are smaller than the value in the node; all items in the right subtree are larger than the item in the node. The size of the tree and a link to the root node are also stored.

Figure 3.3a shows a binary search tree (the root is the node with item 15). Searching in a binary search tree starts at the root and branches either left or right depending on whether the item in a node is smaller or larger than the target. Either the target is found or we eventually reach the bottom. Thus, the search for 9 in the binary search tree in Figure 3.3a reports that 9 is not present, after visiting nodes containing 15, 6, and 12 in that order. Insertion of item x into a binary search tree proceeds in the same manner, by searching for x. If the item is found, the insertion will fail because duplicates are not allowed. Otherwise, the new item can be inserted at the exact point at which the search terminated, as shown in the insertion of 9 in Figure 3.3b. Removal of an item is slightly more complicated. If the item to remove is at the bottom of the tree (i.e., it is a *leaf*, having no children), it can be removed from the tree. If the item to remove has only one child (e.g., node 12 in Figure 3.3a), then it can be bypassed, as would be the case in a singly linked list. If the item to remove has two children, we leave the node in place and replace its value with the smallest value in the right subtree. This preserves the binary search tree order but creates a duplicate. The removal can be completed by removing the smallest value from the right subtree recursively.

The cost of any of insertion, removal, or access is dependent on the depth of the tree. If the tree is created by random insertions, the depth is $O(\log N)$ on average. However, for nonrandom insertion sequences, such as 1, 2, 3, 4, …, N that typically arise, the depth is $O(N)$, and thus, binary search trees are not suitable for general-purpose library implementations. Instead, slightly more advanced techniques are required in order to assure $O(\log N)$ behavior, which can be either a worst-case bound (balanced binary search trees), an expected-time bound (skip lists), or an amortized bound (splay trees).

3.4.2 Balanced Binary Search Trees

A *balanced binary search tree* is a binary search tree with a *balance condition* that guarantees that the depth of the tree is $O(\log N)$. With this balance condition in force, searching of a balanced binary search tree is identical to searching of a binary search tree but has an $O(\log N)$ worst-case guarantee. Insertion and removal in a balanced binary search tree are also identical to operations on a binary search tree, except that all nodes on the path from the accessed node back to the root must have their balance maintained and possibly restored. Examples of balanced binary search trees include *AVL trees* and *red–black trees*.

3.4.2.1 AVL Trees

The first balanced search tree was the *AVL tree* [5], and it is also one of the simplest to describe and implement. In an AVL tree, each node stores balance information (for instance its height), and the balancing property is that for every node, its left and right subtrees must have heights that differ from each other by at most 1.

The balancing property implies that the height of an AVL tree is at most approximately $\log_{\varphi} N$, where $\varphi = 1.618$ is the golden ratio, though in practice it is much closer to the optimal $\log N$.

Insertion of a new value x proceeds in the usual fashion as a binary search tree. This step will be $O(\log N)$ worst case, since the depth of the tree is now guaranteed. However, it potentially changes the height information of all nodes on the path from x to the root, and it may also violate the balance requirement. Thus, we follow the path from x back to the root, updating height information. If at any point a balance violation is detected, we repair the node by using a local *rotation* of tree nodes. Let α be the first node on the return path to the root whose balance is violated. There are four possible cases:

1. x was inserted into the left subtree of the left child of α.
2. x was inserted into the right subtree of the right child of α.
3. x was inserted into the right subtree of the left child of α.
4. x was inserted into the left subtree of the right child of α.

Note that cases 1 and 2 are mirror-image symmetries, as are cases 3 and 4. Hence, we focus on cases 1 and 3 only.

Case 1 is repaired with a single rotation as shown in Figure 3.4

Here, k_2's left subtree is two levels deeper than k_2's right subtree. By changing two links (k_2's left child and k_1's right child), the subtree becomes balanced. Once this subtree is reattached to the parent, the heights for the parent and all other nodes on the path to the root are all guaranteed to be unchanged from prior to the insertion of x.

Case 3 is repaired with a double rotation as shown in Figure 3.5

In this case, x is inserted into either subtree B or C, and one of those trees is two levels deeper than D (possibly both subtrees are deeper in the special case that all of A, B, C, and D are empty). B and C are

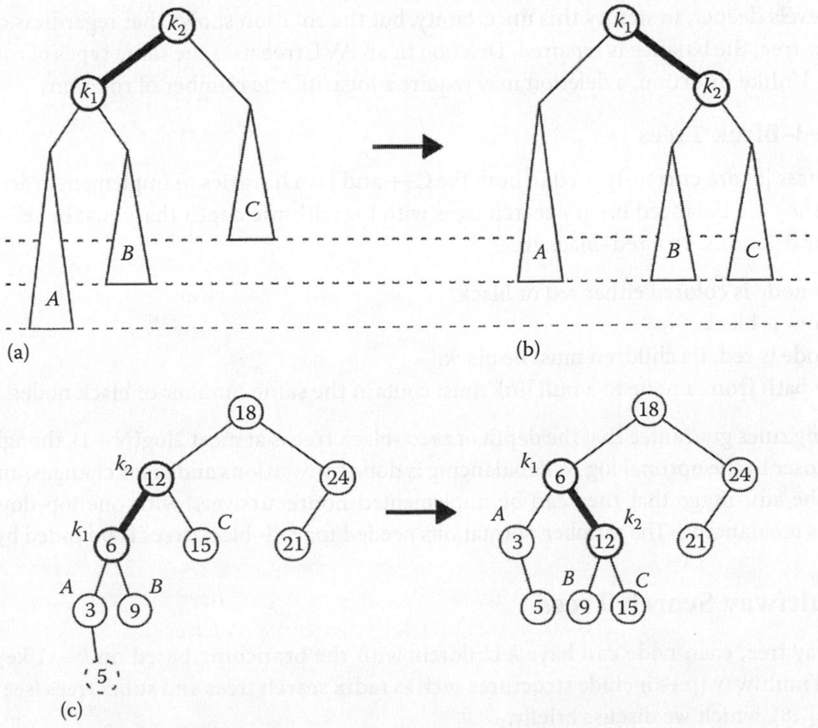

FIGURE 3.4 Single rotation (case 1) to fix an AVL tree. (a) Generic case before rotation, (b) generic case after rotation, and (c) rotation applied after insertion of 5.

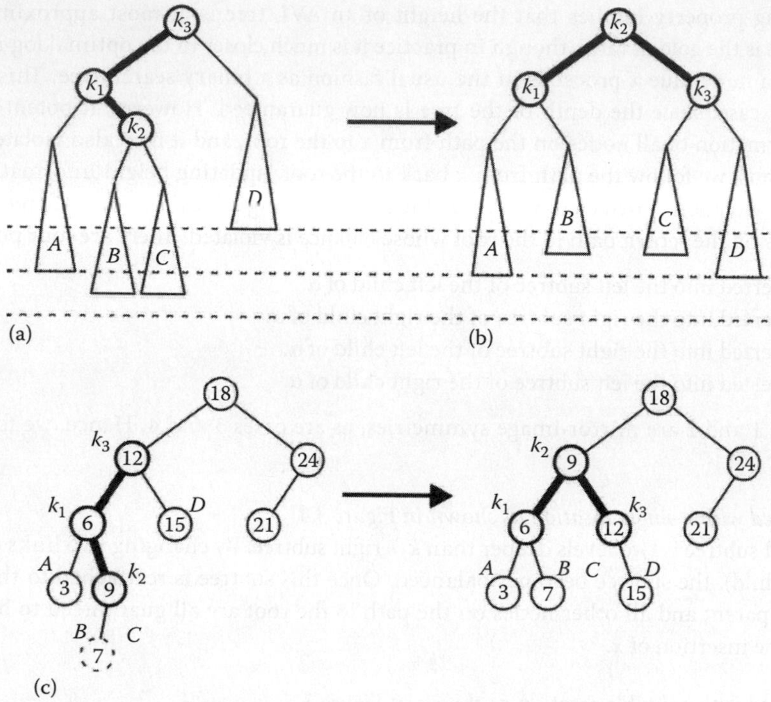

FIGURE 3.5 Double rotation (case 3) to fix an AVL tree. (a) Generic case before rotation, (b) generic case after rotation, and (c) rotation applied after insertion of 7.

drawn 1½ levels deeper, to signify this uncertainty, but the rotation shows that regardless of which tree is the deeper tree, the balance is repaired. Deletion in an AVL tree uses the same types of rotations, with more cases. Unlike insertion, a deletion may require a logarithmic number of rotations.

3.4.2.2 Red–Black Trees

Red–black trees [6] are currently used in both the C++ and Java libraries to implement ordered sets. Like AVL trees, they are balanced binary search trees with logarithmic depth that must be rebalanced after insertions or deletions. In a red–black tree,

1. Every node is colored either red or black.
2. The root is black.
3. If a node is red, its children must be black.
4. Every path from a node to a null link must contain the same number of black nodes.

The balancing rules guarantee that the depth of a red–black tree is at most $2\log(N+1)$, though in practice it is much closer to the optimal $\log N$. Rebalancing is done by rotations and color changes, and red–black trees have the advantage that they can be implemented nonrecursively, with one top-down pass that incorporates rebalancing. The number of rotations needed for red–black trees is bounded by a constant.

3.4.3 Multiway Search Trees

In a multiway tree, each node can have k children, with the branching based on $k - 1$ keys per node. Examples of multiway trees include structures such as radix search trees and suffix trees (see Chapter 15) and *B*-trees [7,8], which we discuss briefly.

In a *B-tree* of order m, each non-leaf node has between $\lceil m/2 \rceil$ and m children (except that the root can have between 2 and m children if it is not a leaf) and all leaves are at the same level. Special cases of $m = 3$

and $m = 4$ are 2–3 trees and 2–3–4 trees. Interestingly, 2–3–4 trees are equivalent to red–black trees: a four-node is replaced with a black node and two red children; a three-node is replaced with a black node and one red child, and a two-node is replaced with a black node. If insertion of a new node would create an $(m + 1)$th child for the node's potential parent, that parent is split into two nodes containing $\lceil m/2 \rceil$ and $\lfloor m/2 \rfloor$ children. As this adds a new child to the grandparent, the splitting process can potentially repeat all the way up until the root, in the worst case resulting in the creation of a new root with only two children. Deletion works in a similar manner by combining nodes. *B*-trees are widely used in the implementation of databases because large values of m will yield very shallow trees and minimize disk accesses.

3.4.3.1 Skip Lists

A *skip list* [9] implements an ordered set using a randomization strategy. As we described earlier in the section, binary search on a singly linked list does not work because it is difficult to get to the middle of the list. In a skip list, each node potentially stores many links. A level k node has k links, each storing a link to the next level i node for $1 \leq i \leq k$. As shown in the top of Figure 3.6, nodes 3, 9, 15, and 21 are level 1 nodes; nodes 6 and 18 are level 2 nodes; node 12 is a level 3 node; and there are currently no level 4 nodes. A header node is also provided, large enough to store the largest level.

To search the skip list, we use the highest levels to advance quickly, dropping down to lower levels when it is apparent that we have advanced too far. If levels are randomly assigned, and the probability that a node's level k is 2^{-k}, then on average there will be O(log N) levels, and we will spend only a constant amount of time on any level, yielding O(log N) expected performance for a search (here, the expectation is based on the randomness of the creating a level, not the particular input; see Chapter 11 for a discussion of randomized algorithms). Insertion is straightforward; as shown in the bottom of Figure 3.6, we create a new node with a randomly chosen level (level k is selected with probability 2^{-k}) and then it is spliced in at the points where search switched levels.

The primary advantage of the skip list is that it can be efficiently implemented in a multithreaded environment [10] and it is used in the standard Java library.

3.4.3.2 Splay Trees

A *splay tree* [11] is a binary search tree that stores no balance information yet provides a guarantee of O(log N) amortized cost per basic operation of insertion, removal, or access. In a splay tree, after a node x is accessed, it is repeatedly rotated toward the root until it becomes the root, a process that is called *splaying*. If the node is a child of the root, this is accomplished with a single rotation, which we call a *zig rotation* and which terminates the splaying step. Otherwise, let the node's parent be p, and its grandparent be g. Then we have four cases that are similar to an AVL tree:

1. x is the left child of p and p is the left child of g.
2. x is the right child of p and p is the right child of g.
3. x is the right child of p and p is the left child of g.
4. x is the left child of p and p is the right child of g.

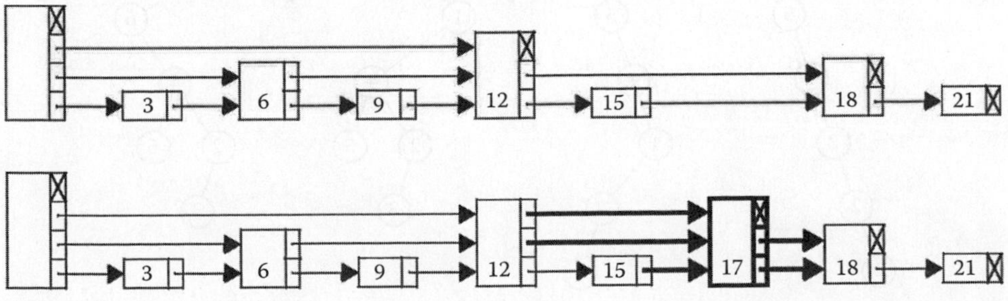

FIGURE 3.6 Insertion into a skip list.

In cases 1 and 2, which are symmetric, we rotate p with g and then x with p. This is a *zig–zig rotation*. In cases 3 and 4, which are also symmetric, we rotate x with p and then x with g (this is a basic double rotation, known as a *zig–zag rotation* for splay trees). If this makes x the root, we are done; otherwise, we continue the splaying step.

Figure 3.7 shows the three types of rotations that can be involved in splaying, and Figure 3.8 shows how node 1 is splayed to the root. Splaying has the effect of roughly halving the length of long access paths.

A search in a splay tree is performed by a usual binary search tree search followed by a splay of the last accessed node. An insertion is performed by a usual binary search tree insertion, and then the inserted node is splayed to the root. A deletion is performed by first performing a search. This brings

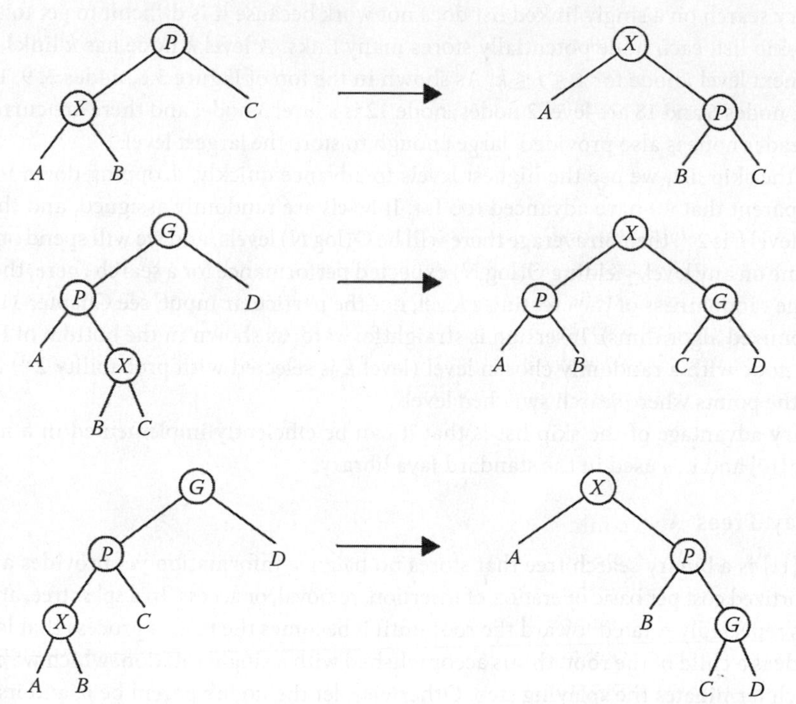

FIGURE 3.7 Rotations in a splay tree: zig (top, a basic single rotation), zig–zag (middle, a basic double rotation), and zig-zig (bottom, unique to the splay tree).

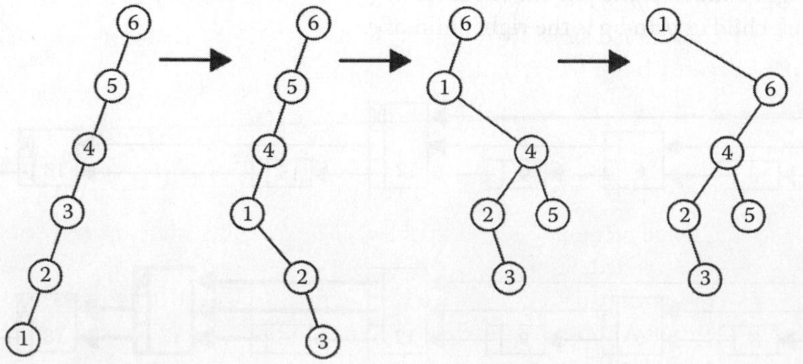

FIGURE 3.8 Splaying node 1 to the root (two zig–zigs and one zig).

the item to delete to the root, at which point it can be removed, leaving just a left subtree and a right subtree. If either is empty, the deletion is easy to complete. Otherwise, the largest item in the left subtree can be accessed, bringing it to the root by a splay, and the right subtree can be attached to the new root of the left subtree.

The proof that splay trees take $O(\log N)$ amortized cost per operation is one of the classic results in amortized analysis.

3.5 Unordered Sets and Maps (Hashing)

An *unordered set* stores a collection of unique items and allows for their efficient insertion, removal, and access. Unlike an ordered set, iteration over the set is not expected to view the items in their sorted order, and thus, the items do not require the ability to be comparison ordered. Because the operations required for an unordered set are a subset of the ordered set operations, it is reasonable to expect that it can be implemented more efficiently.

3.5.1 Separate Chaining

The most common implementation of an unordered set is a *separate chaining hash table* [12]. In a separate chaining hash table, we maintain a collection of M buckets and provide a *hash function* that maps any item into a bucket index. If the hash function is reasonably chosen, then each bucket will have only N/M items on average. If N is $O(M)$, then each bucket will have constant size on average, and can be searched sequentially, and then updated as needed, all in constant average time. The items in the bucket will typically be maintained in a singly linked list.

3.5.2 Linear Probing

An alternative to separate chaining is *linear probing* [13]; we maintain a larger table (typically of size roughly $2N$) but only allow one item per bucket, thus avoiding the need to maintain linked lists. Let p be the position computed by the hash function for item x. When inserting, if position p is already occupied, we sequentially probe positions $p + 1$, $p + 2$, etc., until a vacant position is found (wrapping around to the start of the bucket array if needed). The average number of probes is determined by the load factor $\lambda = N/M$ and is approximately $(1 + 1/(1 - \lambda))/2$ for a successful search and $(1 + 1/(1 - \lambda)^2)/2$ for an insertion (or unsuccessful search) [14]. This is higher than the naïve estimate of $1/(1 - \lambda)$ that one would obtain if each probe was an independent random event, and translates to an average of 2.5 rather than 2.0 probes for insertion of an item into a half-full hash table, but significantly worse performance at higher load factors. This phenomenon is known as *primary clustering*: items that probe to different array slots (e.g., slots 6, 7, and 8) nonetheless attempt to resolve collisions in the same manner, thus creating long chains of occupied slots. Primary clustering was long thought to make linear probing a poor choice for hash tables, but it has recently reemerged as a popular choice due to its excellent locality of reference.

3.5.3 Quadratic Probing

Quadratic probing avoids the primary clustering of linear probing by using positions $p + 1^2$, $p + 2^2$, etc., again wrapping around as needed. Since $p + i^2 = (p + (i - 1)^2) + (2i - 1)$, the ith position can be computed from the $(i - 1)$th position with two additions and a bit shift and, possibly, a subtraction if a wraparound calculation is needed [15,16]. Although quadratic probing has not been fully analyzed, its observed performance is similar to what would be expected without primary clustering, it is easy to implement, and it has locality properties that are similar to linear probing.

3.5.4 Alternative Hashing Schemes

Perfect hash tables [17] provide guarantees of constant access time, assuming that the items are known in advance. Using a random hash function, a primary hash table is constructed with N buckets; however, for each bucket i, if the bucket contains b_i items, the items will be stored in a secondary hash table of size b_i^2, using a randomly generated hash function (each secondary hash table will use its own random hash function). This two-level scheme is shown in Figure 3.9.

If the total size of all the secondary hash tables exceeds $4N$, then we have picked an unlucky (primary) random hash function, and we simply pick a different hash function until the total size is no larger than $4N$. This will require only a constant number of attempts on average. In each secondary hash table, since there are b_i items and b_i^2 buckets, a collision occurs with probability less than ½. Hence, only a constant number of attempts are needed to choose each secondary hash function until we obtain a collision-free secondary hash table. Once all this is done, at total cost of $O(N)$ to build all the tables, an access can be done in two table lookups. The perfect hash table can be extended to support insertions and deletions [18].

Cuckoo hash tables [19] provide $O(1)$ worst-case access using a relatively simple algorithm. In cuckoo hashing, we maintain two tables, each more than half empty, and we have two independent hash functions that can assign each item to one position in each table. Cuckoo hashing maintains the invariant that an item is always stored in one of its two hash locations. Consequently, the cost of a search is guaranteed to be $O(1)$, and the remove operation is likewise very simple. There are several alternative algorithms to insert an item x, all use the same basic concepts. If either of x's valid locations is empty, then we are done. Otherwise, place x in one of the occupied locations, displacing some other element a. Then place a in a's other valid location; if that other valid location was empty we are done; otherwise, we have displaced another element b, and we continue the same process until either we are able to place the most recently displaced item or we detect a cycle (or reach a threshold of displacement that indicates a high probability of a cycle). In this case, we choose a different random hash function and rebuild the table. If the tables are less than half empty, it is unlikely that a new random hash function is needed; however, the hash functions must be carefully chosen.

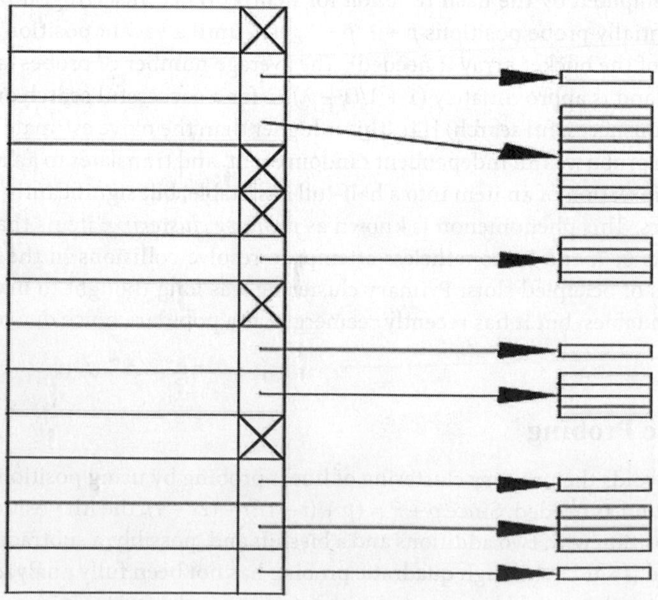

FIGURE 3.9 Perfect hash table.

3.5.5 Hash Functions

All implementations of hash tables presume the existence of a hash function that returns a suitable array index, which in modern languages will range from 0 to $M - 1$. Any hash function should satisfy basic properties:

1. If $x = y$, then $hash(x) = hash(y)$.
2. If $x \neq y$, then the collision probability that $hash(x) = hash(y)$ should be approximately $1/M$.

The hash function should be relatively easy to compute, since if it is a good function, the cost of computing the hash function will be a significant cost of any hash table operation.

Normally, the input values can be considered to be a long sequence of bits, so hash functions can be modeled as a mathematical operation on large integers. In such a case, $hash(x) = x \bmod M$ generates a hash code with the desired properties if the input values have a significantly larger range than 0 to $M - 1$ and if they are randomly distributed in that range. However, this hash function can be an unsuitable choice if, for example, M is a power of 2 and the low-order bits of all inputs are likely to be similar.

When random functions are needed, an alternative is to use a *universal hash function*, which provides the needed hash properties for all pairs x and y, where the collision probability for any pair (x, y) is taken over all choices of random functions [17]. One set of universal hash functions can be obtained by choosing a prime $p \geq M$ and then any random pair (a,b) such that $1 \leq a < p$ and $0 \leq b < p$, yielding

$$hash_{(a,b)}(x) = ((ax + b) \bmod p) \bmod M$$

An alternative scheme that is more efficient in practice is the multiply shift in [20].

An important use case for hash tables is storing strings. There have been many different hash functions proposed. The simplest that provides very good performance is the algorithm shown in Figure 3.10, implemented in C++. This function returns an unsigned integral type that can be brought into the proper range with a mod operation by the hash table implementation. Many choices of (a, h_0) have been used including (31, 0) used in both the Java library and in [21]; other implementations replace addition with a bitwise-exclusive OR, some implementations perform the addition (or bitwise-exclusive OR) before the multiply, and implementations on 64-bit machines are likely to use large prime numbers for a.

Hash functions that are used in hash tables are generally reversible in the sense that it is possible to produce several inputs that map to the same output. *Cryptographic hash functions* are used to authenticate documents and store passwords; for these hash functions, it is infeasible to modify x without changing $hash(x)$, it is infeasible to produce an input y such that $hash(y)$ is some specified value, it is infeasible to recover x from $hash(x)$ even if a large number of x's bits are known, and it is infeasible to produce a pair of inputs that compute the same output. Because cryptographic hash functions are expensive to compute, their use is primarily in computer security.

Chapter 11, which discusses randomized algorithms, has other material related to hashing (e.g., tabulation hashing, hashing with choices, and the analysis of cuckoo hashing).

```
size_t hash ( const string & key )
{
    size_t hashVal = h0;

    for ( char ch : key )
    hashVal = a * hashVal + ch;

    return hashVal;
}
```

FIGURE 3.10 Simple C++ hash function.

3.6 Priority Queues

A *priority queue* stores a collection of items and allows for their efficient insertion but allows removal (deleteMin) and access (findMin) of the minimum item only. Alternatively, sometimes priority queues are defined to allow removal and access of the maximum item only. Priority queues are used in diverse applications such as job scheduling, discrete-event simulation, sorting, and graph algorithms. As an example, suppose we want to merge k sorted lists containing a total of N items into one sorted list of N items. We can place the first item of each list in a priority queue and use deleteMin to select the smallest. We then insert the next item from the appropriate list in the priority queue, and repeat. After N insert and N deleteMin operations in a priority queue that never contains more than k items, we have successfully merged all sorted lists into one. This algorithm takes $O(N \log k)$ if priority queue operations are logarithmic.

In fact, priority queues can be implemented with a balanced binary search tree at logarithmic worst-case cost per operation. An alternative implementation that is likely to be faster and more space efficient is the binary heap.

3.6.1 Binary Heaps

A *binary heap* [22] stores items in a *complete binary tree*. In a complete binary tree, all levels are full except possibly the last level, which is filled in left to right with no missing nodes. A complete binary tree has at most $\lceil \log(N + 1) \rceil$ nodes on the path from the root to a bottom leaf and can be easily represented without using the normal links associated with a binary tree. Specifically, if we store the nodes in level order in an array whose index begins with 1, then the root is stored at index 1, and a node that is stored at index i will have its left child, right child, and parent stored at indices $2i$, $2i + 1$, and $\lfloor i/2 \rfloor$, respectively.

The items in the binary heap will also satisfy *heap order*: the value stored in any node is never smaller than the value stored in its parent. Figure 3.11 shows a binary heap and its representation in an array.

In a binary heap, the minimum item must be at the root. We can insert a new item x by creating an empty new node at the next available position and then placing x in the sorted position along the path from the newly created node to the root. Algorithmically, this involves moving the vacant node's parent into the vacant node until x can be placed into the vacant node without violating heap order and is easily seen to require at most $\log N$ comparisons. In this step, the vacant node moves up toward the root until x is finally placed in it. Figure 3.12 shows the insertion of 13 into a binary heap.

Figure 3.13 shows that removing the minimum item follows a similar process. The minimum item, which is at the root, is removed from the heap, thus vacating the root. The last node in the binary heap is removed, thus leaving one value, x, homeless. x is placed in the sorted position along the path from the root through smaller children. Algorithmically, this involves moving the vacant node's smaller child into the vacant node until x can be placed into the vacant node without violating heap order and requires at most $2\log N$ comparisons.

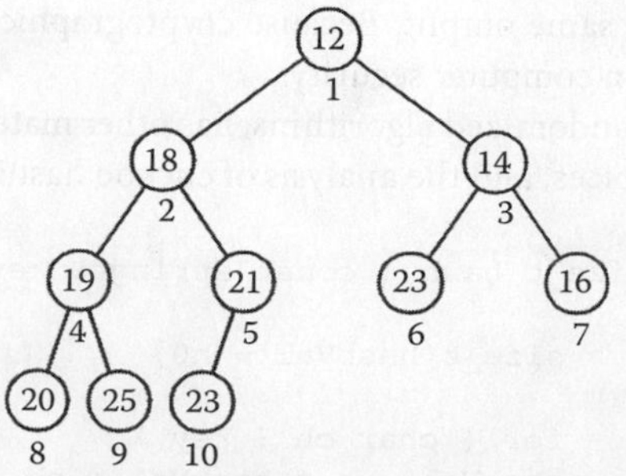

	12	18	14	19	21	23	16	20	25	23			
0	1	2	3	4	5	6	7	8	9	10	11	12	13

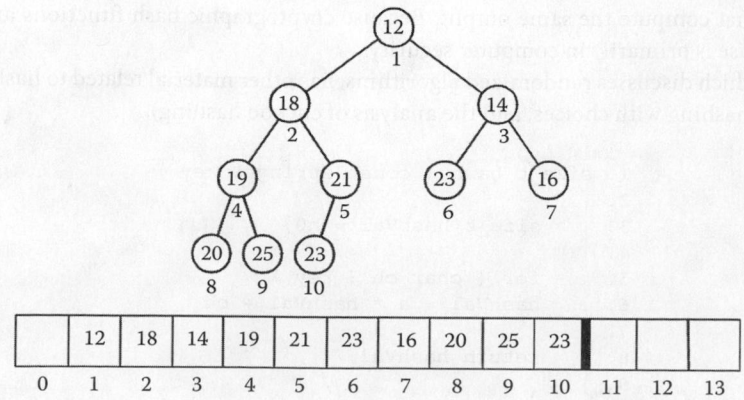

FIGURE 3.11 Binary heap and an array representation.

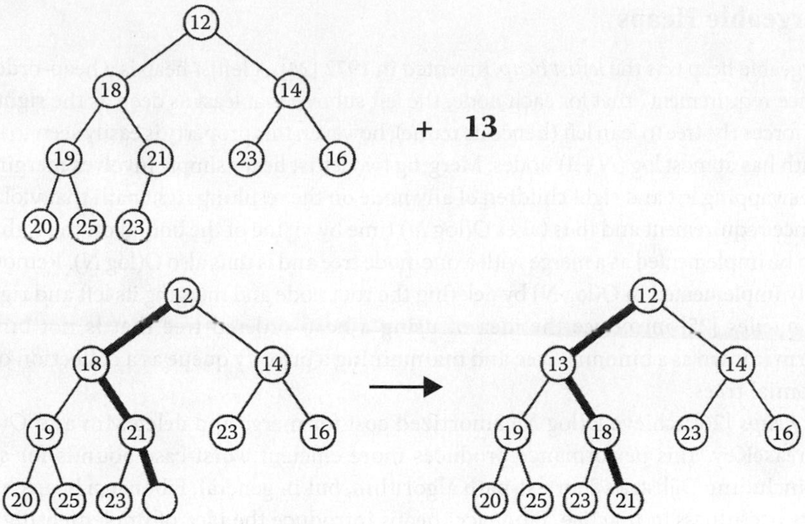

FIGURE 3.12 Initial binary heap; inserting 13 shows creation of new vacant node and the path to the root; 13 is inserted onto this path.

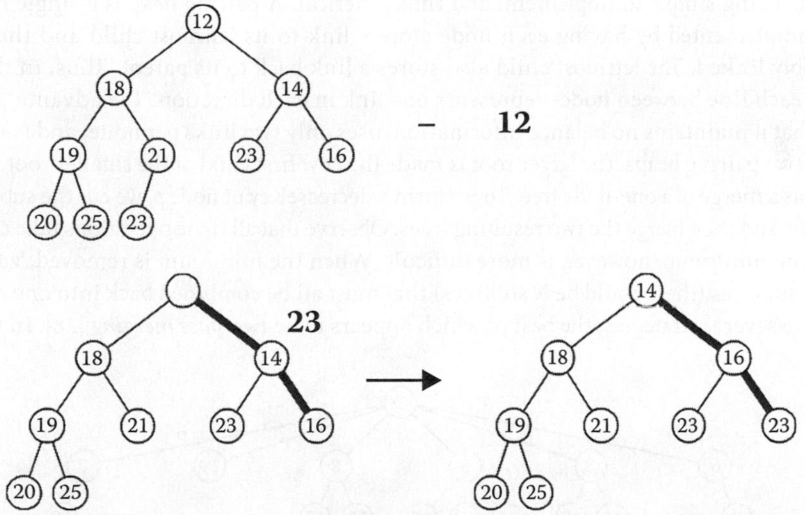

FIGURE 3.13 Initial binary heap; removing the minimum (12) vacates the root and displaces last item (23) path of minimum children is shown; 23 is placed by following path of minimum children.

A binary heap can also be constructed in linear time by placing the items in the binary tree without heap order and then repairing the heap order bottom-up [23].

If the position of an item is known, the item's value can be either increased or decreased by generalizing the deletion or insertion algorithm; decreasing a value (decreaseKey) involves moving it up the tree toward the root, whereas increasing a value involves moving it down the tree through a path of minimum children; this operation takes $O(\log N)$ time. In some algorithms that use priority queues, such as computation of shortest paths, implementing the decreaseKey operation in $o(\log N)$ yields a faster running time. Thus, alternative priority queue schemes have been proposed, and most use linked heap-ordered tree structures that can be easily merged. Several examples include leftist heaps, Fibonacci heaps, and pairing heaps.

3.6.2 Mergeable Heaps

The first mergeable heap was the *leftist heap*, invented in 1972 [24]. A leftist heap is a heap-ordered tree with the "imbalance requirement" that for each node, the left subtree is at least as deep as the right subtree. The requirement forces the tree to lean left (hence its name); however, this property is easily seen to imply that the rightmost path has at most log $(N + 1)$ nodes. Merging two leftist heaps simply involves merging right paths together and swapping left and right children of any node on the resulting right path that violates the leftist heap imbalance requirement and thus takes $O(\log N)$ time by virtue of the bound on the right path length. Insertion can be implemented as a merge with a one-node tree and is thus also $O(\log N)$. Removing the minimum is easily implemented in $O(\log N)$ by deleting the root node and merging its left and right subtrees.

Binomial queues [25] introduce the idea of using a heap-ordered tree that is not binary but of a restricted form known as a binomial tree and maintaining a priority queue as a collection of these heap-ordered binomial trees.

Fibonacci heaps [26] achieve $O(\log N)$ amortized cost for merge and deleteMin and $O(1)$ amortized cost for decreaseKey. This performance produces more efficient worst-case bounds for several graph algorithms, including Dijkstra's shortest-path algorithm, but in general, Fibonacci heaps have overhead that limits its usefulness in practice. Fibonacci heaps introduce the idea of implementing decreaseKey by cutting a tree into two and also introduce the idea that there can be as many as $\Theta(N)$ trees while maintaining amortized efficiency.

Pairing heaps [27] were designed in the hope of matching Fibonacci heaps theoretical performance bounds while being simple to implement, and thus practical. A pairing heap is a single heap-ordered tree that is implemented by having each node store a link to its leftmost child and then having all siblings doubly linked. The leftmost child also stores a link back to its parent. Thus, in the bottom of Figure 3.14, each line between nodes represents one link in each direction. The advantage of the pairing heap is that it maintains no balance information, uses only two links per node, and is only one tree.

To merge two pairing heaps, the larger root is made the new first child of the smaller root. We can treat an insertion as a merge of a one-node tree. To perform a decreaseKey of node p, we cut the subtree rooted at p from the tree and then merge the two resulting trees. Observe that all these operations take constant time.

Deleting the minimum, however, is more difficult. When the minimum is removed, what is left is a collection of subtrees (there could be N subtrees) that must all be combined back into one heap-ordered tree. There are several strategies, the best of which appears to be *two-pass merging* [28]. In the first pass,

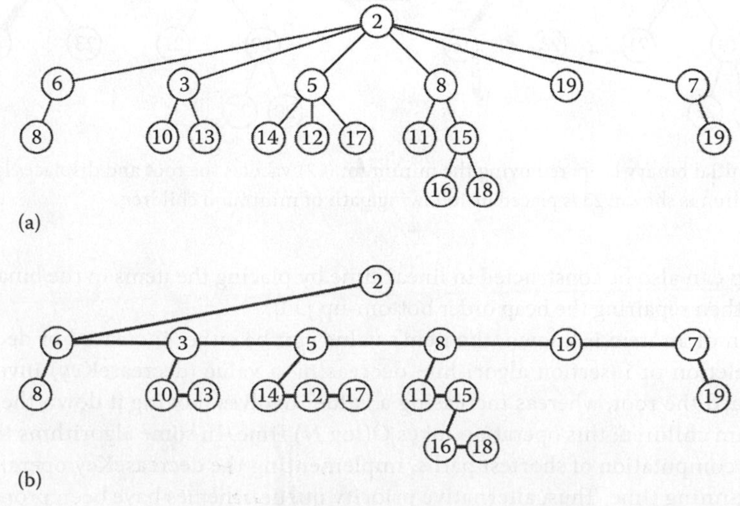

FIGURE 3.14 (a) Pairing heap as a tree, and (b) pairing heap in left child, right sibling representation, using bidirectional links everywhere.

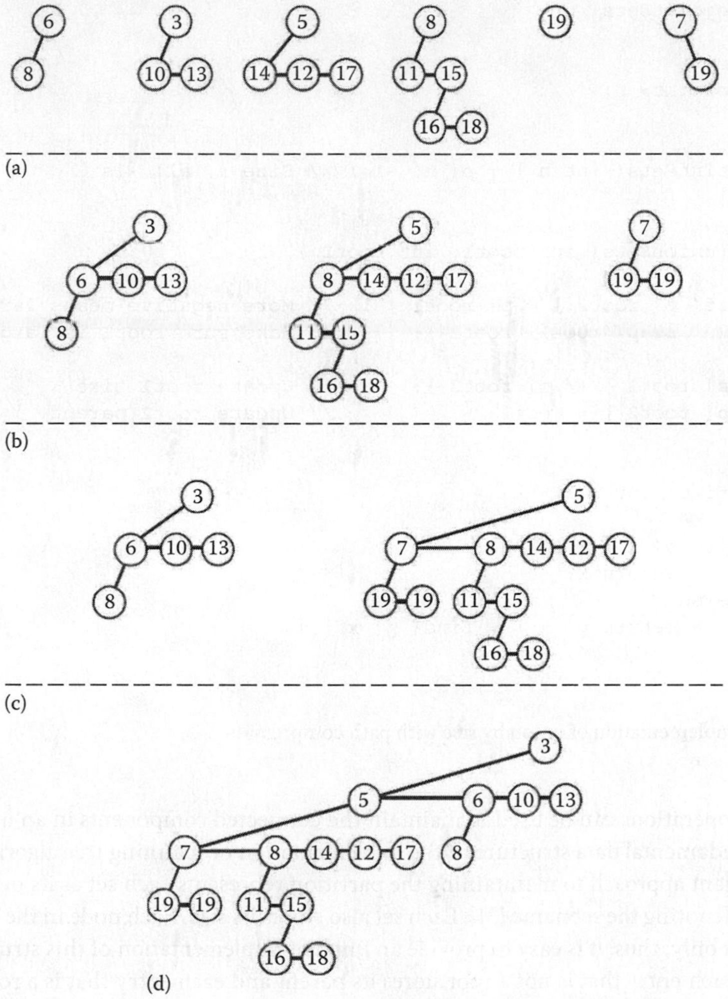

FIGURE 3.15 (a) Trees resulting from removal of root in Figure 3.14, (b) after first pass halves the number of trees, (c) first merge of the second pass, and (d) second merge of the second pass.

we merge each pair of heaps (see Figure 3.15b); then we work backward, merging the last two trees, then that result with the next tree, and so on (Figure 3.15c and d) until we obtain only one tree.

Although the performance of the pairing heap is excellent, the analysis of the pairing heap is still unresolved. It has been shown that if the amortized cost of merge and deleteMin are $O(\log N)$, decreaseKey is not $O(1)$. For a wide range of merging strategies, decrease Key's amortized cost will be $\Omega(\log \log N)$ and for two-pass merging, the cost has recently been bounded by $O\left(2^{2\sqrt{\log \log N}}\right)$ [29,30].

3.7 Union/Find Data Structure

The *union/find data structure* is a classic solution to the problem of maintaining a partition of {1, 2, 3, ..., *N*} under the following operations:

- *initialize*: Creates *N* single element sets.
- *union*(s_1, s_2): Combines two sets s_1 and s_2, destroying the original (thus maintaining the partition).
- *find*(x): Returns the set containing x (at the time of the query); most importantly, *find*(x) and *find*(y) return the same result if and only if x and y are in the same set.

```
class DisjointSets
{
  private:
    vector<int> p;

  public:
    DisjointSets( int n ) : p( n, -1 ) // Size n, all -1s
      { }

    void unionSets( int root1, int root2 )
    {
        if( p[ root2 ] < p[ root1 ] )  // More negative means larger root
            swap( root1, root2 );      // Make sure root1 is larger (or same)

        p[ root1 ] += p[ root2 ];      // Update root1 size
        p[ root2 ] = root1;            // Update root2 parent
    }

    int find( int x )
    {
        if( p[ x ] < 0 )
            return x;
        else
            return p[ x ] = find( p[ x ] );
    }
};
```

FIGURE 3.16 Implementation of union by size with path compression.

The union/find operations can be used to maintain the connected components in an undirected graph and thus is a fundamental data structure in the implementation of spanning tree algorithms.

The most elegant approach to maintaining the partition represents each set as its own tree, with the root of the tree denoting the set name [31]. Each set also stores its size. Each node in the tree maintains a link to its parent only; thus, it is easy to provide an implicit implementation of this structure in a single array in which each entry that is not a root stores its parent and each entry that is a root stores its size information as a negative number.

To perform a union given two tree roots, we make the smaller tree a child of the larger tree and update the larger tree's size. This is clearly a constant-time operation. To perform a find, we simply follow parent links until we reach a root. The cost per operation is the number of nodes on the path to the root, which will be $O(\log N)$ if the union is done by size (because a node gets one level deeper only when its tree is combined into a larger tree, the size of the node's tree doubles every time it gets one level deeper). A simple optimization on the find known as *path compression* changes the parent of every node on the find search path to be the eventual root. With this optimization and union using the size rule (or a similar height rule), the amortized cost of a find operation is nearly constant and bounded by $O(\alpha(N))$, where $\alpha(N)$ is the exceptionally slow-growing inverse Ackermann function [32–34]. Figure 3.16 shows a complete C++ implementation (minus error checking) of the union/find data structure.

3.8 Recurring Themes and Ongoing Research

The study of data structures involves many recurring issues, most of which are still being actively researched, many of which lead into each other. We briefly discuss some of these, pointing especially to examples that we have already seen.

3.8.1 Static, Dynamic, and Persistent Structures

Static data structures restrict operations to access on the data and do not support updates. Examples in this chapter include the binary search and perfect hash table. *Dynamic data structures* support updates and are thus more complex; most of the data structures we have seen are of this type. Advanced data structures have been designed to maintain connected components and shortest paths in graphs even as the underlying structure of the graph can change. *Persistent data structures* support updates and allow operations on prior versions of the data structure. We have not examined any persistent data structures.

3.8.2 Implicit versus Explicit Structures

An *implicit data structure* stores the data using little extra memory, in such a way that the structure of the items can be implied without the need for extra storage for links between the items. Examples of implicit data structures include the binary heap and the union/find data structure. An explicit data structure is more typical and uses extra memory to link items together structurally.

3.8.3 Space versus Time Considerations

A recurring theme in data structures and algorithm design is the ability to trade space for time. The hash table is an example of a data structure in which the ability to use more space can save time; similarly, a doubly linked list will use more space than a singly linked list but has the advantage that some operations can be made faster because of the additional links.

3.8.4 Theory versus Practice

Although most data structures have theoretical asymptotic performance guarantees, usually expressed in Big-Oh notation, it is sometimes the case that the constants hidden in the Big-Oh notation can make it difficult to compare data structures with identical Big-Oh running times. Additionally, some bounds such as worst-case bounds can be overly pessimistic, leading to situations in which a theoretically inferior data structure performs better in practice than its counterpart with better run time guarantees. Examples include hash tables that are very difficult to compare based on the $O(1)$ run time per operation and also priority queues, in which the Fibonacci heap despite having the best time bound is generally outperformed by the pairing heap and binary heap (e.g., see [35]).

3.8.5 Modern Architecture Considerations

Because of many advances in computer hardware including multilevel memories and CPU design, experimental results for data structures can be very machine dependent, often requiring the designer to have several competitive alternatives to benchmark. For example, linear probing performs much better relative to other hash table algorithm on modern machines than on older computers. Multicore CPUs will require a reexamination of fundamental data structures that for the most part have been designed for single threaded efficiency (e.g., see [36]).

3.8.6 Data Structures for Specialized Types

Containers that work on types such as integers and strings have for many years been an active area of research. In particular, there has been much work designing priority queues for integers and containers for strings [37].

Key Terms

AVL tree: The first balanced binary search tree. Has a balance requirement that each node's left and right subtrees can have heights that differ by at most 1.

Balanced binary search tree: A binary search tree that adds a balance requirement to every node that guarantees the tree has $O(\log N)$ depth. Typically, this means that insertions and deletions must provide algorithms to maintain the balance requirement.

Binary heap: An implementation of a priority queue that uses only a simple array and supports insertion and deleteMin in logarithmic time.

Binary search: An algorithm to search a sorted array in $O(\log N)$ time by searching in the middle and recursing.

Binary search tree: An implementation of an ordered sequence collection that generally uses two links per node; without balancing, performance can be linear in the worst case but logarithmic on average.

B-tree: A multiway search tree often used for database systems.

Complete binary tree: A binary tree in which all levels are full, except possibly for the last level that is left-filled with no missing nodes. Used in the binary heap.

Container: Stores a collection of items.

Cuckoo hash table: A hash table implementation that uses two tables and two hash functions and supports constant worst-case access.

Double-ended queue: A sequence container in which operations are restricted to the front and back.

Fibonacci heap: An implementation of a priority queue that provides $O(1)$ amortized cost for decreaseKey.

Hash function: Maps items to array indices, with the property that if $x = y$, then $hash(x) = hash(y)$ and if $x \neq y$, then the collision probability that $hash(x) = hash(y)$ should be approximately $1/M$.

Hash table: An implementation of an unordered sequence collection that typically provides constant time search and update on average.

Heap order: In a priority queue that allows access to the minimum item, heap order means that every node's value is at least as large as its parent's value.

Leftist heap: The first efficient mergeable priority queue.

Linear probing: A hash table implementation that uses only a simple table and tries array slots sequentially starting from the hash value position until an empty slot is found.

Linked list: A sequence container in which the items are linked; can be singly linked, storing a link to the next item, or doubly linked, storing links to both the previous and next items.

Map: Stores key value pairs.

Pairing heap: An implementation of a priority queue that provides $o(\log N)$ amortized cost for decreaseKey, uses two links per node, does not require balance information, and performs very well in practice.

Path compression: In the union/find data structure, the process of changing the parent of every node on a find path to the root.

Perfect hash table: A hash table scheme that supports constant time access by using hash tables to resolve hash table collisions.

Primary clustering: A phenomena in linear probing in which keys with different hash values attempt to resolve to similar alternate locations, potentially resulting in poor performance.

Priority queue: A container in which only the minimum can be accessed and removed.

Quadratic probing: A hash table implementation that uses only a simple table and tries slots sequentially starting from the hash value position plus i^2 (starting with $i = 0$), until an empty slot is found.

Queue: A sequence container in which insertions are restricted to the back and access and removal is restricted to the front.

Red–black tree: A balanced search tree currently used in both the C++ and Java library.

Rotation: A process by which parent/child relations among a few nodes are changed, while retaining binary search tree order. Examples include single and double rotations for AVL trees and a zig–zig rotation for splay trees.

Search tree: An implementation of an ordered sequence collection that generally uses either binary trees or multiway trees.

Separate chaining hash table: A hash table scheme in which collisions are resolved by singly linked lists.

Sequence container: Stores items in a linear order, with items inserted at specified positions.

Skip list: An ordered container that uses linked lists with multiple forward pointers per node. A concurrent version is implemented as part of the Java library.

Splay tree: A binary search tree that maintains no balance information but that has $O(\log N)$ amortized cost per operation.

Splaying: The process in splay trees by which a node is rotated toward the root using zig, zig–zig, or zig–zag rotations.

Stack: A sequence container in which operations are restricted to one end.

Union/find data structure: Maintains a partition of $\{1, 2, \ldots, N\}$ under a sequence of union and find operations at only slightly more than constant cost per find.

Universal hash function: A collection of hash functions such that for any specific $x \neq y$, only $O(1/M)$ of the hash functions in the collection yield $hash(x) = hash(y)$.

Further Information

Material in this chapter is based on the presentations in Weiss [15,16]. Other popular textbooks on data structures and algorithms include Cormen et al. [38], Sahni [39], and Sedgewick and Wayne [40].

References

1. U. Manber and G. Myers, Suffix arrays: A new method for on-line string searches, *SIAM Journal on Computing*, 22 (1993), 935–948.
2. E. M. McCreight, A space-economical suffix tree construction algorithm, *Journal of the ACM*, 23 (1976), 262–272.
3. D. R. Morrison, PATRICIA—Practical algorithm to retrieve information coded in alphanumeric, *Journal of the ACM*, 15 (1968), 514–534.
4. D. E. Knuth, *The Art of Computer Programming: Volume 1: Fundamental Algorithms*, 3rd edn., Addison-Wesley, Reading, MA (1997).
5. G. M. Adelson-Velskii and E. M. Landis, An algorithm for the organization of information, *Soviet Mathematics Doklady*, 3 (1962), 1259–1263.
6. L. J. Guibas and R. Sedgewick, A dichromatic framework for balanced trees, *Proceedings of the Nineteenth Annual IEEE Symposium on Foundations of Computer Science*, Ann Arbor, MI, pp. 8–21 (1978).
7. R. Bayer and E. M. McCreight, Organization and maintenance of large ordered indices, *Acta Informatica*, 1 (1972), 173–189.
8. D. Comer, The ubiquitous B-tree, *Computing Surveys*, 11 (1979), 121–137.
9. W. Pugh, Skip lists: A probabilistic alternative to balanced trees, *Communications of the ACM*, 33 (1990), 668–676.
10. K. Fraser, Practical lock-freedom, PhD thesis, University of Cambridge, Cambridge, U.K. (2004).
11. D. D. Sleator and R. E. Tarjan, Self-adjusting binary search trees, *Journal of the ACM*, 32 (1985), 652–686.
12. I. Dumey, Indexing for rapid random-access memory, *Computers and Automation*, 5 (1956), 6–9.
13. W. W. Peterson, Addressing for random access storage, *IBM Journal of Research and Development*, 1 (1957), 130–146.
14. D. E. Knuth, *The Art of Computer Programming: Volume 3: Sorting and Searching*, 2nd edn., Addison-Wesley, Reading, MA (1997).
15. M. A. Weiss, *Data Structures and Algorithm Analysis in C++*, 4th edn., Pearson, Boston, MA (2013).
16. M. A. Weiss, *Data Structures and Problem Solving Using Java*, 4th edn., Pearson, Boston, MA (2010).

17. J. L. Carter and M. N. Wegman, Universal classes of hash functions, *Journal of Computer and System Sciences*, 18 (1979), 143–154.
18. M. Dietzfelbinger, A. R. Karlin, K. Melhorn, F. Meyer auf def Heide, H. Rohnert, and R. E. Tarjan, Dynamic perfect hashing: Upper and lower bounds, *SIAM Journal on Computing*, 23 (1994), 738–761.
19. A. Pagh and F. F. Rodler, Cuckoo hashing, *Journal of Algorithms*, 51 (2004), 122–144.
20. M. Dietzfelbinger, T. Hagerup, J. Katajainen, and M. Penttonen, A reliable randomized algorithm for the closest-pair problem, *Journal of Algorithms*, 25 (1997), 19–51.
21. B. Kernighan and D. M. Ritchie, *The C Programming Language*, 2nd edn., Prentice-Hall, Englewood Cliffs, NJ (1988).
22. J. W. J. Williams, Algorithm 232: Heapsort, *Communications of the ACM*, 7 (1964), 347–348.
23. R. W. Floyd, Algorithm 245: Treesort 3, *Communications of the ACM*, 7 (1964), 701.
24. C. A. Crane, Linear lists and priority queues as balanced binary trees, Technical report STAN-CS-72–259, Computer Science Department, Stanford University, Palo Alto, CA (1972).
25. J. Vuillemin, A data structure for manipulating priority queues, *Communications of the ACM*, 21 (1978), 309–314.
26. M. L. Fredman and R. E. Tarjan, Fibonacci heaps and their uses in improved network optimization algorithms, *Journal of the ACM*, 34 (1987), 596–615.
27. M. L. Fredman, R. Sedgewick, D. D. Sleator, and R. E. Tarjan, The pairing heap: A new form of self-adjusting heap, *Algorithmica*, 1 (1986), 111–129.
28. J. T. Stasko and J. S. Vitter, Pairing heaps: Experiments and analysis, *Communications of the ACM*, 32 (1987), 234–249.
29. M. L. Fredman, On the efficiency of pairing heaps and related data structures, *Journal of the ACM*, 46 (1999), 473–501.
30. S. Pettie, Towards a final analysis of pairing heaps, *Proceedings of the 46th Annual IEEE Symposium on Foundations of Computer Science*, Pittsburgh, PA (2005), pp. 174–183.
31. B. A. Galler and M. J. Fischer, An improved equivalence algorithm, *Communications of the ACM*, 7 (1964), 301–303.
32. R. E. Tarjan, Efficiency of a good but not linear set union algorithm, *Journal of the ACM*, 22 (1975), 215–225.
33. R. E. Tarjan, A class of algorithms which require nonlinear time to maintain disjoint sets, *Journal of Computer and System Sciences*, 18 (1979), 110–127.
34. R. Seidel and M. Sharir, Top-down analysis of path compression, *SIAM Journal of Computing*, 34 (2005), 515–525.
35. B. M. E. Moret and H. D. Shapiro, An empirical analysis of algorithms for constructing a minimum spanning tree, *Proceedings of the Second Workshop on Algorithms and Data Structures*, Ottawa, Ontario, Canada (1991), pp. 400–411.
36. M. Herlihy, N. Shavit, and M. Tzafrir, Hopscotch hashing, *Proceedings of the Twenty Second International Symposium on Distributed Computing*, Boston, MA (2008), pp. 350–364.
37. D. Gusfield, *Algorithms on Strings, Trees and Sequences: Computer Science and Computational Biology*, Cambridge University Press, Cambridge, U.K. (1997).
38. T. Cormen, C. Leiserson, R. Rivest, and C. Stein, *Introduction to Algorithms*, 3rd edn., MIT Press, Cambridge, MA (2009).
39. S. Sahni, *Data Structures, Algorithms, and Applications in Java*, 2nd edn., Silicon Press, Summit, NJ (2004).
40. R. Sedgewick and K. Wayne, *Algorithms*, 4th edn, Addison-Wesley Professional, Boston, MA (2011).

4

Basic Techniques for Design and Analysis of Algorithms

Edward M. Reingold
Illinois Institute of Technology

We outline the basic methods of algorithm design and analysis that have found application in the manipulation of discrete objects such as lists, arrays, sets, and graphs, and geometric objects such as points, lines, and polygons. We begin by discussing recurrence relations and their use in the analysis of algorithms. Then we discuss some specific examples in algorithm analysis, sorting, and priority queues. In the final three sections, we explore three important techniques of algorithm design—divide-and-conquer, dynamic programming, and greedy heuristics.

4.1 Analyzing Algorithms

It is convenient to classify algorithms based on the relative amount of time they require: how fast does the time required grow as the size of the problem increases? For example, in the case of arrays, the "size of the problem" is ordinarily the number of elements in the array. If the size of the problem is measured by a variable n, we can express the time required as a function of n, $T(n)$. When this function $T(n)$ grows rapidly, the algorithm becomes unusable for large n; conversely, when $T(n)$ grows slowly, the algorithm remains useful even when n becomes large.

We say an algorithm is $\Theta(n^2)$ if the time it takes quadruples when n doubles; an algorithm is $\Theta(n)$ if the time it takes doubles when n doubles; an algorithm is $\Theta(\log n)$ if the time it takes increases by a constant, independent of n, when n doubles; an algorithm is $\Theta(1)$ if its time does not increase at all when n increases. In general, an algorithm is $\Theta(T(n))$ if the time it requires on problems of size n grows proportionally to $T(n)$ as n increases. Table 4.1 summarizes the common growth rates encountered in the analysis of algorithms.

The analysis of an algorithm is often accomplished by finding and solving a recurrence relation that describes the time required by the algorithm. The most commonly occurring families of recurrences in the analysis of algorithms are linear recurrences and divide-and-conquer recurrences. In the following

TABLE 4.1 Common Growth Rates of Times of Algorithms

Rate of Growth	Comment	Examples
$\Theta(1)$	Time required is constant, independent of problem size	Expected time for hash searching
$\Theta(\log \log n)$	Very slow growth of time required	Expected time of interpolation search
$\Theta(\log n)$	Logarithmic growth of time required—doubling the problem size increases the time by only a constant amount	Computing x^n; binary search of an array
$\Theta(n)$	Time grows linearly with problem size—doubling the problem size doubles the time required	Adding/subtracting n-digit numbers; linear search of an n-element array
$\Theta(n \log n)$	Time grows worse than linearly, but not much worse—doubling the problem size more than doubles the time required	Merge sort; heapsort; lower bound on comparison-based sorting
$\Theta(n^2)$	Time grows quadratically—doubling the problem size quadruples the time required	Simple-minded sorting algorithms
$\Theta(n^3)$	Time grows cubically—doubling the problem size results in an eight-fold increase in the time required	Ordinary matrix multiplication
$\Theta(c^n)$	Time grows exponentially—increasing the problem size by 1 results in a c-fold increase in the time required; doubling the problem size *squares* the time required	Traveling salesman problem

subsection, we describe the "method of operators" for solving linear recurrences; in the next subsection, we describe how to transform divide-and-conquer recurrences into linear recurrences by substitution to obtain an asymptotic solution.

4.1.1 Linear Recurrences

A *linear recurrence with constant coefficients* has the form

$$c_0 a_n + c_1 a_{n-1} + c_2 a_{n-2} + \cdots + c_k a_{n-k} = f(n), \tag{4.1}$$

for some constant k, where each c_i is constant. To solve such a recurrence for a broad class of functions f (i.e., to express a_n in closed form as a function of n) by the *method of operators*, we consider two basic operators on sequences: S, which shifts the sequence left,

$$\mathcal{S}\langle a_0, a_1, a_2, \ldots\rangle = \langle a_1, a_2, a_3, \ldots\rangle,$$

and C, which, for any constant C, multiplies each term of the sequence by C:

$$C\langle a_0, a_1, a_2, \ldots\rangle = \langle Ca_0, Ca_1, Ca_2, \ldots\rangle.$$

Then, given operators A and B, we define the sum and product

$$(A+B)\langle a_0, a_1, a_2, \ldots\rangle = A\langle a_0, a_1, a_2, \ldots\rangle + B\langle a_0, a_1, a_2, \ldots\rangle,$$

$$(AB)\langle a_0, a_1, a_2, \ldots\rangle = A(B\langle a_0, a_1, a_2, \ldots\rangle).$$

Thus, for example,

$$(\mathcal{S}^2 - 4)\langle a_0, a_1, a_2, \ldots\rangle = \langle a_2 - 4a_0, a_3 - 4a_1, a_4 - 4a_2, \ldots\rangle,$$

which we write more briefly as

$$(S^2 - 4)\langle a_i \rangle = \langle a_{i+2} - 4a_i \rangle.$$

With the operator notation, we can rewrite Equation 4.1 as

$$P(S)\langle a_i \rangle = \langle f(i) \rangle,$$

where

$$P(S) = c_0 S^k + c_1 S^{k-1} + c_2 S^{k-2} + \cdots + c_k$$

is a polynomial in S.

Given a sequence $\langle a_i \rangle$, we say that the operator $P(S)$ *annihilates* $\langle a_i \rangle$ if $P(S)\langle a_i \rangle = \langle 0 \rangle$. For example, $S^2 - 4$ annihilates any sequence of the form $\langle u2^i + v(-2)^i \rangle$, with constants u and v. In general,

The operator $S^{k+1} - c$ annihilates $\langle c^i \times$ a polynomial in i of degree $k \rangle$.

The *product* of two annihilators annihilates the *sum* of the sequences annihilated by each of the operators—that is, if A annihilates $\langle a_i \rangle$ and B annihilates $\langle b_i \rangle$, then AB annihilates $\langle a_i + b_i \rangle$. Thus, determining the annihilator of a sequence is tantamount to determining the sequence; moreover, it is straightforward to determine the annihilator from a recurrence relation.

For example, consider the Fibonacci recurrence

$$F_0 = 0,$$

$$F_1 = 1,$$

$$F_{i+2} = F_{i+1} + F_i.$$

The last line of this definition can be rewritten as $F_{i+2} - F_{i+1} - F_i = 0$, which tells us that $\langle F_i \rangle$ is annihilated by the operator

$$S^2 - S - 1 = (S - \phi)\left(\frac{S+1}{\phi}\right),$$

where $\phi = (1+\sqrt{5})/2$. Thus we conclude that

$$F_i = u\phi^i + v(-\phi)^{-i},$$

for some constants u and v. We can now use the initial conditions $F_0 = 0$ and $F_1 = 1$ to determine u and v. These initial conditions mean that

$$u\phi^0 + v(-\phi)^{-0} = 0$$

$$u\phi^1 + v(-\phi)^{-1} = 1$$

and these linear equations have the solution

$$u = v = \frac{1}{\sqrt{5}},$$

and hence

$$F_i = \frac{\phi^i}{\sqrt{5}} + \frac{(-\phi)^{-i}}{\sqrt{5}}.$$

In the case of the similar recurrence,

$$G_0 = 0,$$

$$G_1 = 1,$$

$$G_{i+2} = G_{i+1} + G_i + i,$$

the last equation tells us that

$$(S^2 - S - 1)\langle G_i \rangle = \langle i \rangle,$$

so the annihilator for $\langle G_i \rangle$ is $(S^2 - S - 1)(S - 1)^2$ since $(S - 1)^2$ annihilates $\langle i \rangle$ (a polynomial of degree 1 in i) and hence the solution is

$$G_i = u\phi^i + v(-\phi)^{-i} + (\text{a polynomial of degree 1 in } i),$$

that is,

$$G_i = u\phi^i + v(-\phi)^{-i} + wi + z.$$

Again, we use the initial conditions to determine the constants u, v, w, and x.

In general, then, to solve the recurrence (4.1), we factor the annihilator

$$P(S) = c_0 S^k + c_1 S^{k-1} + c_2 S^{k-2} + \cdots + c_k,$$

multiply it by the annihilator for $\langle f(i) \rangle$, write down the form of the solution from this product (which is the annihilator for the sequence $\langle a_i \rangle$), and use the initial conditions for the recurrence to determine the coefficients in the solution.

4.1.2 Divide-and-Conquer Recurrences

The divide-and-conquer paradigm of algorithm construction that we discuss in Section 4.3 leads naturally to divide-and-conquer recurrences of the type

$$T(n) = g(n) + uT(n/v),$$

for constants u and v, $v > 1$, and sufficient initial values to define the sequence $\langle T(0), T(1), T(2), \ldots \rangle$. The growth rates of $T(n)$ for various values of u and v are given in Table 4.2. The growth rates in this table are derived by transforming the divide-and-conquer recurrence into a linear recurrence for a subsequence of $\langle T(0), T(1), T(2), \ldots \rangle$.

To illustrate this method, we derive the penultimate line in Table 4.2. We want to solve

$$T(n) = n^2 + v^2 T(n/v),$$

TABLE 4.2 Rate of Growth of the Solution to the Recurrence $T(n) = g(n) + uT(n/v)$, the Divide-and-Conquer Recurrence Relations

$g(n)$	u, v	Growth Rate of $T(n)$
$\Theta(1)$	$u = 1$	$\Theta(\log n)$
	$u \neq 1$	$\Theta(n^{\log_v u})$
$\Theta(\log n)$	$u = 1$	$\Theta[(\log n)^2]$
	$u \neq 1$	$\Theta(n^{\log_v u})$
$\Theta(n)$	$u < v$	$\Theta(n)$
	$u = v$	$\Theta(n \log n)$
	$u > v$	$\Theta(n^{\log_v u})$
$\Theta(n^2)$	$u < v^2$	$\Theta(n^2)$
	$u = v^2$	$\Theta(n^2 \log n)$
	$u > v^2$	$\Theta(n^{\log_v u})$

Note: u and v are positive constants, independent of n, and $v > 1$.

so we want to find a subsequence of $\langle T(0), T(1), T(2), \ldots\rangle$ that will be easy to handle. Let $n_k = v^k$; then,

$$T(n_k) = n_k^2 + v^2 T\left(\frac{n_k}{v}\right),$$

or

$$T(v^k) = v^{2k} + v^2 T(v^{k-1}).$$

Defining $t_k = T(v^k)$,

$$t_k = v^{2k} + v^2 t_{k-1}.$$

The annihilator for t_k is then $(S - v^2)^2$ and thus

$$t_k = v^{2k}(ak + b),$$

for constants a and b. Expressing this in terms of $T(n)$,

$$T(n) \approx t_{\log_v n} = v^{2\log_v n}(a \log_v n + b) = an^2 \log_v n + bn^2,$$

or

$$T(n) = \Theta(n^2 \log n).$$

4.2 Some Examples of the Analysis of Algorithms

In this section, we introduce the basic ideas of analyzing algorithms by looking at some data structure problems that occur commonly in practice, problems relating to maintaining a collection of n objects and retrieving objects based on their relative size. For example, how can we determine the smallest of the elements? Or, more generally, how can we determine the kth largest of the elements? What is the running time of such algorithms in the worst case? Or, on the average, if all $n!$ permutations of the input are equally likely? What if the set of items is dynamic—that is, the set changes through insertions and deletions—how efficiently can we keep track of, say, the largest element?

4.2.1 Sorting

The most demanding request that we can make of an array of n values `x[1]`, `x[2]`, ..., `x[n]` is that they be kept in perfect order so that `x[1]` $\leq$ `x[2]` $\leq \ldots \leq$ `x[n]`. The simplest way to put the values in order is to mimic what we might do by hand: take item after item and insert each one into the proper place among those items already inserted:

```
void insert (float x[], int i, float a) {
  // Insert a into x[1] ... x[i]
  // x[1] ... x[i-1] are sorted; x[i] is unoccupied
  if (i == 1 || x[i-1] <= a)
    x[i] = a;
  else {
    x[i] = x[i-1];
    insert(x, i-1, a);
  }
}

void insertionSort (int n, float x[]) {
  // Sort x[1] ... x[n]
  if (n > 1) {
    insertionSort(n-1, x);
    insert(x, n, x[n]);
  }
}
```

To determine the time required in the worst case to sort n elements with `insertionSort`, we let t_n be the time to sort n elements and derive and solve a recurrence relation for t_n. We have,

$$t_n \begin{cases} \Theta(1) & \text{if } n = 1, \\ t_{n-1} + s_{n-1} + \Theta(1) & \text{otherwise,} \end{cases}$$

where s_m is the time required to insert an element in place among m elements using `insert`. The value of s_m is also given by a recurrence relation:

$$s_m \begin{cases} \Theta(1) & \text{if } m = 1, \\ s_{m-1} + \Theta(1) & \text{otherwise.} \end{cases}$$

The annihilator for $\langle s_i \rangle$ is $(S - 1)^2$, so $s_m = \Theta(m)$. Thus the annihilator for $\langle t_i \rangle$ is $(S - 1)^3$, so $t_n = \Theta(n^2)$. The analysis of the average behavior is nearly identical; only the constants hidden in the Θ-notation change.

We can design better sorting methods using the divide-and-conquer idea of the next section. These algorithms avoid $\Theta(n^2)$ worst-case behavior, working in time $\Theta(n\log n)$. We can also achieve time $\Theta(n\log n)$ by using a clever way of viewing the array of elements to be sorted as a tree: consider `x[1]` is the root of the tree and, in general, `x[2*i]` is the root of the left subtree of `x[i]` and `x[2*i+1]` is the root of the right subtree of `x[i]`. If we further insist that parents be greater than or equal to children, we have a *heap*; Figure 4.1 shows a small example.

A heap can be used for sorting by observing that the largest element is at the root, that is, `x[1]`; thus to put the largest element in place, we swap `x[1]` and `x[n]`. To continue, we must restore the heap property which may now be violated at the root. Such restoration is accomplished by swapping `x[1]` with its larger child, if that child is larger than `x[1]`, and the continuing to swap it downward until either it reaches the bottom or a spot where it is greater or equal to its children. Since the tree-cum-array has height $\Theta(\log n)$, this restoration process takes time $\Theta(\log n)$. Now, with the heap in `x[1]` to `x[n-1]` and

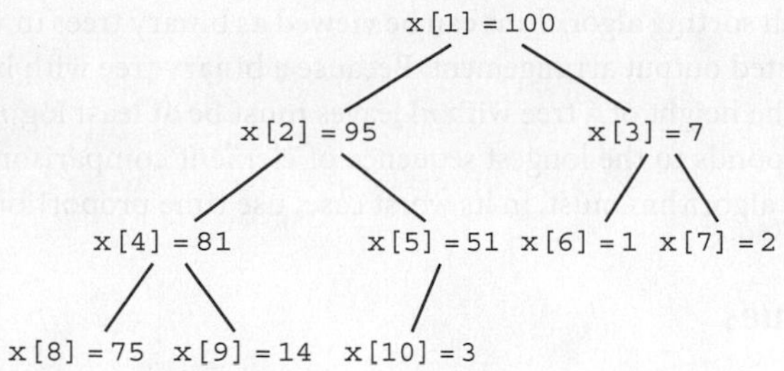

FIGURE 4.1 A heap—that is, an array, interpreted as a binary tree.

x[n] the largest value in the array, we can put the second largest element in place by swapping x[1] and x[n–1]; then we restore the heap property in x[1] to x[n–2] by propagating x[1] downward—this takes time $\Theta(\log(n-1))$. Continuing in this fashion, we find we can sort the entire array in time

$$\Theta(\log n + \log(n-1) + \cdots + \log 1) = \Theta(n \log n).$$

The initial creation of the heap from an unordered array is done by applying the restoration process successively to x[n/2], x[n/2–1], …, x[1] which takes time $\Theta(n)$.

Hence we have the following $\Theta(n\log n)$ sorting algorithm:

```
void heapify (int n, float x[], int i) {
  // Repair heap property below x[i] in x[1] … x[n]
  int largest = i; // largest of x[i], x[2*i], x[2*i+1]
  if (2*i <= n && x[2*i] > x[i])
    largest = 2*i;
  if (2*i+1 <= n && x[2*i+1] > x[largest])
    largest = 2*i+1;
  if (largest != i) {
    // swap x[i] with larger child and repair heap below
    float t = x[largest]; x[largest] = x[i]; x[i] = t;
    heapify(n, x, largest);
  }
}

void makeheap (int n, float x[]) {
  // Make x[1] ... x[n] into a heap
  for (int i=n/2; i>0; i--)
    heapify(n, x, i);
}

void heapsort (int n, float x[]) {
  // Sort x[1] ... x[n]
  float t;
  makeheap(n, x);
  for (int i=n; i>1; i--) {
    // put x[1] in place and repair heap
    t = x[1]; x[1] = x[i]; x[i] = t;
    heapify(i-1, x, 1);
  }
}
```

Can we find sorting algorithms that take time less than $\Theta(n \log n)$? The answer is no if we are restricted to sorting algorithms that derive their information from comparisons between the values of elements.

The flow of control in such sorting algorithms can be viewed as binary trees in which there are $n!$ leaves, one for every possible sorted output arrangement. Because a binary tree with height h can have at most 2^h leaves, it follows that the height of a tree with $n!$ leaves must be at least $\log_2 n! = \Theta(n \log n)$. Since the height of this tree corresponds to the longest sequence of element comparisons possible in the flow of control, any such sorting algorithm must, in its worst case, use time proportional to $n \log n$.

4.2.2 Priority Queues

Aside from its application to sorting, the heap is an interesting data structure in its own right. In particular, heaps provide a simple way to implement a *priority queue*—a priority queue is an abstract data structure that keeps track of a dynamically changing set of values allowing the operations

`create`: Create an empty priority queue.
`insert`: Insert a new element into a priority queue.
`decrease`: Decrease an element in a priority queue.
`minimum`: Report the minimum element in a priority queue.
`deleteMinimum`: Delete the minimum element in a priority queue.
`delete`: Delete an element in a priority queue.
`merge`: Merge two priority queues.

A heap can implement a priority queue by altering the heap property to insist that parents are less than or equal to their children, so that the smallest value in the heap is at the root, that is, in the first array position. Creation of an empty heap requires just the allocation of an array, an $\Theta(1)$ operation; we assume that once created, the array containing the heap can be extended arbitrarily at the right end. Inserting a new element means putting that element in the $(n + 1)$st location and "bubbling it up" by swapping it with its parent until it reaches either the root or a parent with a smaller value. Since a heap has logarithmic height, insertion to a heap of n elements thus requires worst-case time $O(\log n)$. Decreasing a value in a heap requires only a similar $O(\log n)$ "bubbling up." The minimum element of such a heap is always at the root, so reporting it takes $\Theta(1)$ time. Deleting the minimum is done by swapping the first and last array positions, bubbling the new root value downward until it reaches its proper location, and truncating the array to eliminate the last position. Delete is handled by decreasing the value so that it is the least in the heap and then applying the `deleteMinimum` operation; this takes a total of $O(\log n)$ time.

The merge operation, unfortunately, is not so economically accomplished—there is little choice but to create a new heap out of the two heaps in a manner similar to the `makeheap` function in heap sort. If there are a total of n elements in the two heaps to be merged, this recreation will require time $O(n)$.

There are better data structures than a heap for implementing priority queues, however. In particular, the *Fibonacci heap* provides an implementation of priority queues in which the delete and `deleteMinimum` operations take $O(\log n)$ time and the remaining operations take $\Theta(1)$ time, *provided we consider the times required for a sequence of priority queue operations, rather than individual times.* That is, we must consider the cost of the individual operations *amortized over the sequence of operations*: Given a sequence of n priority queue operations, we will compute the total time $T(n)$ for all n operations. In doing this computation, however, we do not simply add the costs of the individual operations; rather, we subdivide the cost of each operation into two parts, the *immediate cost* of doing the operation and the *long-term savings* that result from doing the operation—the long-term savings represent costs *not* incurred by later operations as a result of the present operation. The immediate cost minus the long-term savings give the amortized cost of the operation.

It is easy to calculate the immediate cost (time required) of an operation, but how can we measure the long-term savings that result? We imagine that the data structure has associated with it a bank account; at any given moment, the bank account must have a nonnegative balance. When we do an operation that will save future effort, we are making a deposit to the savings account and when, later on, we derive the benefits of that earlier operation, we are making a withdrawal from the savings account.

Let $\mathcal{B}(i)$ denote the balance in the account after the ith operation, $\mathcal{B}(0) = 0$. We define the amortized cost of the ith operation to be

$$\text{amortized cost of } i\text{th operation} = (\text{immediate cost of } i\text{th operation}) + (\text{change in bank account})$$
$$= (\text{immediate cost of } i\text{th operation}) + (\mathcal{B}(i) - \mathcal{B}(i-1)).$$

Since the bank account $\mathcal{B}$ can go up or down as a result of the ith operation, the amortized cost may be less than or more than the immediate cost. By summing the previous equation, we get

$$\sum_{i=1}^{n} (\text{amortized cost of } i\text{th operation}) = \sum_{i=1}^{n} (\text{immediate cost of } i\text{th operation}) + (\mathcal{B}(n) - \mathcal{B}(0))$$
$$= (\text{total cost of all } n \text{ operations}) + \mathcal{B}(n)$$
$$\geq \text{total cost of all } n \text{ operations}$$
$$= T(n),$$

because $\mathcal{B}(i)$ is nonnegative. Thus defined, the sum of the amortized costs of the operations gives us an upper bound on the total time $T(n)$ for all n operations.

It is important to note that the function $\mathcal{B}(i)$ is not part of the data structure, but is just our way to measure how much time is used by the sequence of operations. As such, we can choose *any rules* for $\mathcal{B}$, provided $\mathcal{B}(0) = 0$ and $\mathcal{B}(i) \geq 0$ for $i \geq 1$. Then the sum of the amortized costs defined by

$$\text{amortized cost of } i\text{th operation} = (\text{immediate cost of } i\text{th operation}) + (\mathcal{B}(i) - \mathcal{B}(i-1))$$

bounds the overall cost of the operation of the data structure.

Now, we apply this method to priority queues. A *Fibonacci heap* is a list of heap-ordered trees (not necessarily binary); since the trees are heap ordered, the minimum element must be one of the roots and we keep track of which root is the overall minimum. Some of the tree nodes are *marked*. We define

$$\mathcal{B}(i) = (\text{number of trees after the } i\text{th operation})$$
$$+2 \times (\text{number of marked nodes after the } i\text{th operation}).$$

The clever rules by which nodes are marked and unmarked, and the intricate algorithms that manipulate the set of trees, are too complex to present here in their complete form, so we just briefly describe the simpler operations and show the calculation of their amortized costs:

`create`: To create an empty Fibonacci heap, we create an empty list of heap-ordered trees. The immediate cost is $\Theta(1)$; since the numbers of trees and marked nodes are zero before and after this operation, $\mathcal{B}(i) - \mathcal{B}(i-1)$ is zero, and the amortized time is $\Theta(1)$.

`insert`: To insert a new element into a Fibonacci heap, we add a new one-element tree to the list of trees constituting the heap and update the record of what root is the overall minimum. The immediate cost is $\Theta(1)$. $\mathcal{B}(i) - \mathcal{B}(i-1)$ is also 1 since the number of trees has increased by 1, while the number of marked nodes is unchanged. The amortized time is thus $\Theta(1)$.

`decrease`: Decreasing an element in a Fibonacci heap is done by cutting the link to its parent, if any, adding the item as a root in the list of trees, and decreasing its value. Furthermore, the marked parent of a cut element is itself cut, propagating upward in the tree. Cut nodes become unmarked, and the unmarked parent of a cut element becomes marked. The immediate cost of this operation is $\Theta(c)$, where c is the number of cut nodes. If there were t trees and m marked elements before this operation, the value of $\mathcal{B}$ before the operation was $t + 2m$. After the operation, the value of $\mathcal{B}$ is $(t + c) + 2(m - c + 2)$ so $\mathcal{B}(i) - \mathcal{B}(i - 1) = 4 - c$. The amortized time is thus $\Theta(c) + 4 - c = \Theta(1)$ *by changing the definition of* $\mathcal{B}$ *by a multiplicative constant large enough to dominate the constant hidden in* $\Theta(c)$.

`minimum`: Reporting the minimum element in a Fibonacci heap takes time $\Theta(1)$ and does not change the numbers of trees and marked nodes, the amortized time is thus $\Theta(1)$.

`deleteMinimum`: Deleting the minimum element in a Fibonacci heap is done by deleting that tree root, making its children roots in the list of trees. Then, the list of tree roots is "consolidated" in a complicated $O(\log n)$ operation that we do not describe. The result takes amortized time $O(\log n)$.

`delete`: Deleting an element in a Fibonacci heap is done by decreasing its value to $-\infty$ and then doing a `deleteMinimum`. The amortized cost is the sum of the amortized cost of the two operations, $O(\log n)$.

`merge`: Merging two Fibonacci heaps is done by concatenating their lists of trees and updating the record of which root is the minimum. The amortized time is thus $\Theta(1)$.

Notice that the amortized cost of each operation is $\Theta(1)$ except `deleteMinimum` and `delete`, both of which are $O(\log n)$.

4.3 Divide-and-Conquer Algorithms

One approach to the design of algorithms is to decompose a problem into subproblems that resemble the original problem, but on a reduced scale. Suppose, for example, that we want to compute x^n. We reason that the value we want can be computed from $x^{\lfloor n/2 \rfloor}$ because

$$x^n = \begin{cases} 1 & \text{if } n = 0, \\ \left(x^{\lfloor n/2 \rfloor}\right)^2 & \text{if } n \text{ is even,} \\ x \times \left(x^{\lfloor n/2 \rfloor}\right)^2 & \text{if } n \text{ is odd.} \end{cases}$$

This recursive definition can be translated directly into

```
1   int power (float x, int n) {
2     // Compute the n-th power of x
3     if (n == 0)
4       return 1;
5     else {
6       int t = power(x, floor(n/2));
7       if ((n % 2) == 0)
8         return t*t;
9       else
10        return x*t*t;
11    }
12  }
```

To analyze the time required by this algorithm, we notice that the time will be proportional to the number of multiplication operations performed in lines 8 and 10, so the divide-and-conquer recurrence

$$T(n) = 2 + T\left(\lfloor n/2 \rfloor\right),$$

with $T(0) = 0$, describes the rate of growth of the time required by this algorithm. By considering the subsequence $n_k = 2^k$, we find, using the methods of the previous section, that $T(n) = \Theta(\log n)$. Thus above algorithm is considerably more efficient than the more obvious

```
int power (int k, int n) {
 // Compute the n-th power of k
  int product = 1;
  for (int i = 1; i <= n; i++)
    // at this point power is k*k*k*...*k (i times)
    product = product * k;
  return product;
}
```

which requires time $\Theta(n)$.

An extremely well-known instance of the divide-and-conquer algorithm is *binary search* of an ordered array of n elements for a given element—we "probe" the middle element of the array, continuing in either the lower or upper segment of the array, depending on the outcome of the probe:

```
int binarySearch (int x, int w[], int low, int high) {
 // Search for x among sorted array w[low..high]. The integer returned
 // is either the location of x in w, or the location where x belongs.
 if (low > high) // Not found
    return low;
 else {
    int middle := (low+high)/2;
    if (w[middle] < x)
      return binarySearch(x, w, middle+1, high);
    else if (w[middle] == x)
      return middle;
    else
      return binarySearch(x, w, low, middle-1);
  }
 }
```

The analysis of binary search in an array of n elements is based on counting the number of probes used in the search, since all remaining work is proportional to the number of probes. But, the number of probes needed is described by the divide-and-conquer recurrence

$$T(n) = 1 + T\left(\frac{n}{2}\right),$$

with $T(0) = 0$, $T(1) = 1$. We find from Table 4.2 (the top line) that $T(n) = \Theta(\log n)$. Hence binary search is much more efficient than a simple linear scan of the array.

To multiply two very large integers x and y, assume that x has exactly $l \geq 2$ digits and y has at most l digits. Let $x_0, x_1, x_2, \ldots, x_{l-1}$ be the digits of x and $y_0, y_1, \ldots, y_{l-1}$ be the digits of y (some of the significant digits at the end of y may be zeros, if y is shorter than x), so that

$$x = x_0 + 10x_1 + 10^2 x_2 + \cdots + 10^{l-1} x_{l-1},$$

and

$$y = y_0 + 10y_1 + 10^2 y_2 + \cdots + 10^{l-1} y_{l-1}.$$

We apply the divide-and-conquer idea to multiplication by chopping x into two pieces, the leftmost n digits and the remaining digits:

$$x = x_{\text{left}} + 10^n x_{\text{right}},$$

where $n = l/2$. Similarly, chop y into two corresponding pieces:

$$y = y_{\text{left}} + 10^n y_{\text{right}},$$

because y has at most the number of digits that x does, y_{right} might be 0. The product $x \times y$ can be now written

$$\begin{aligned} x \times y &= (x_{\text{left}} + 10^n x_{\text{right}}) \times (y_{\text{left}} + 10^n y_{\text{right}}), \\ &= x_{\text{left}} \times y_{\text{left}} + 10^n (x_{\text{right}} \times y_{\text{left}} + x_{\text{left}} \times y_{\text{right}}) + 10^{2n} x_{\text{right}} \times y_{\text{right}}. \end{aligned}$$

If $T(n)$ is the time to multiply two n-digit numbers with this method, then

$$T(n) = kn + 4T\left(\frac{n}{2}\right);$$

the kn part is the time to chop up x and y and to do the needed additions and shifts; each of these tasks involves n-digit numbers and hence $\Theta(n)$ time. The $4T(n/2)$ part is the time to form the four needed subproducts, each of which is a product of about $n/2$ digits.

The line for $g(n) = \Theta(n)$, $u = 4 > v = 2$ in Table 4.2 tells us that $T(n) = \Theta(n^{\log_2 4}) = \Theta(n^2)$, so the divide-and-conquer algorithm is no more efficient than the elementary-school method of multiplication. However, we can be more economical in our formation of subproducts:

$$\begin{aligned} x \times y &= (x_{\text{left}} + 10^n x_{\text{right}}) \times (y_{\text{left}} + 10^n y_{\text{right}}), \\ &= B + 10^n C + 10^{2n} A, \end{aligned}$$

where

$$\begin{aligned} A &= x_{\text{right}} \times y_{\text{right}}, \\ B &= x_{\text{left}} \times y_{\text{left}}, \\ C &= (x_{\text{left}} + x_{\text{right}}) \times (y_{\text{left}} + y_{\text{right}}) - A - B. \end{aligned}$$

The recurrence for the time required changes to

$$T(n) = kn + 3T\left(\frac{n}{2}\right).$$

The kn part is the time to do the two additions that form $x \times y$ from A, B, and C and the two additions and the two subtractions in the formula for C; each of these six additions/subtractions involves

n-digit numbers. The $3T(n/2)$ part is the time to (recursively) form the three needed products, each of which is a product of about $n/2$ digits. The line for $g(n) = \Theta(n)$, $u = 3 > v = 2$ in Table 4.2 now tells us that

$$T(n) = \Theta(n^{\log_2 3}).$$

Now,

$$\log_2 3 = \frac{\log_{10} 3}{\log_{10} 2} \approx 1.5849625\cdots,$$

which means that this divide-and-conquer multiplication technique will be faster than the straightforward $\Theta(n^2)$ method for large numbers of digits.

Sorting a sequence of n values efficiently can be done using the divide-and-conquer idea. Split the n values arbitrarily into two piles of $n/2$ values each, sort each of the piles separately, and then merge the two piles into a single sorted pile. This sorting technique, pictured in Figure 4.2, is called *merge sort*. Let $T(n)$ be the time required by merge sort for sorting n values. The time needed to do the merging is proportional to the number of elements being merged, so that

$$T(n) = cn + 2T(n/2),$$

because we must sort the two halves (time $T(n/2)$ each) and then merge (time proportional to n). We see by Table 4.2 that the growth rate of $T(n)$ is $\Theta(n \log n)$, since $u = v = 2$ and $g(n) = \Theta(n)$.

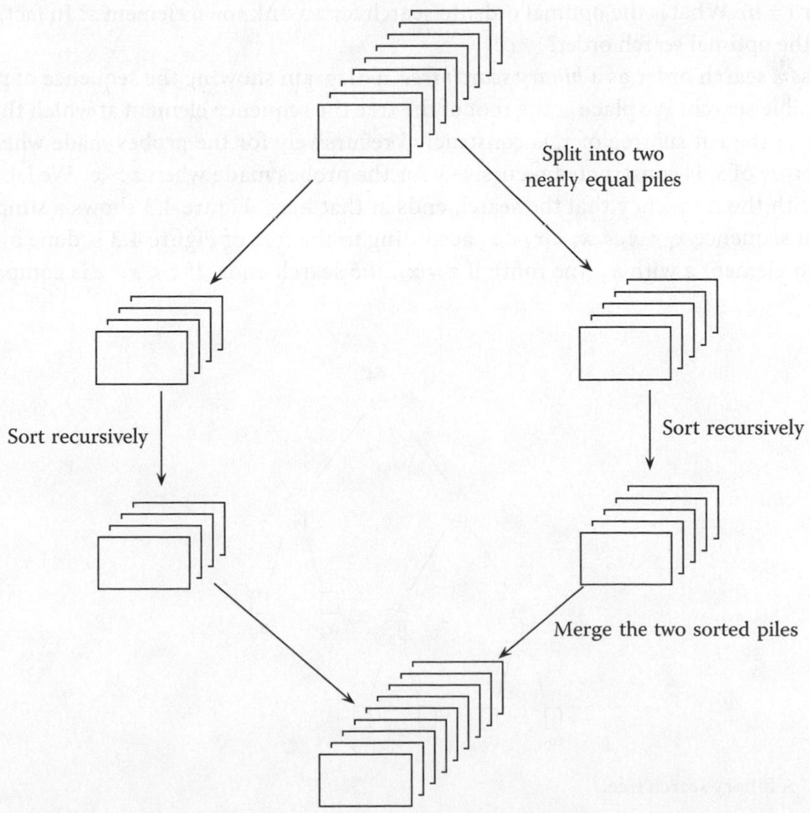

FIGURE 4.2 Schematic description of merge sort.

4.4 Dynamic Programming

In the design of algorithms to solve optimization problems, we need to make the optimal (lowest cost, highest value, shortest distance, and so on) choice from among a large number of alternative solutions; *dynamic programming* is an organized way to find an optimal solution by systematically exploring all possibilities without unnecessary repetition. Often, dynamic programming leads to efficient, polynomial-time algorithms for problems that appear to require searching through exponentially many possibilities.

Like the divide-and-conquer method, dynamic programming is based on the observation that many optimization problems can be solved by solving similar subproblems and composing the solutions of those subproblems into a solution for the original problem. In addition, the problem is viewed as a sequence of decisions, each decision leading to different subproblems; if a wrong decision is made, a suboptimal solution results, so all possible decisions need to be accounted for.

As an example of dynamic programming, consider the problem of constructing an optimal search pattern for probing an ordered sequence of elements. The problem is similar to searching an array—in the previous section, we described binary search in which an interval in an array is repeatedly bisected until the search ends. Now, however, suppose we know the frequencies with which the search will seek various elements (both in the sequence and missing from it). For example, if we know that the last few elements in the sequence are frequently sought—binary search does not make use of this information—it might be more efficient to begin the search at the right end of the array, not in the middle. Specifically, we are given an ordered sequence $x_1 < x_2 < \cdots < x_n$ and associated frequencies of access $\beta_1, \beta_2, \ldots, \beta_n$, respectively; furthermore, we are given $\alpha_0, \alpha_1, \ldots, \alpha_n$ where α_i is the frequency with which the search will fail because the object sought, z, was missing from the sequence, $x_i < z < x_{i+1}$ (with the obvious meaning when $i = 0$ or $i = n$). What is the optimal order to search for an unknown element z? In fact, how should we describe the optimal search order?

We express a search order as a *binary search tree*, a diagram showing the sequence of probes made in every possible search. We place at the root of the tree the sequence element at which the first probe is made, say x_i; the left subtree of x_i is constructed recursively for the probes made when $z < x_i$ and the right subtree of x_i is constructed recursively for the probes made when $z > x_i$. We label each item in the tree with the frequency that the search ends at that item. Figure 4.3 shows a simple example. The search of sequence $x_1 < x_2 < x_3 < x_4 < x_5$ according to the tree of Figure 4.3 is done by comparing the unknown element z with x_4 (the root); if $z = x_4$, the search ends. If $z < x_2$, z is compared with x_2

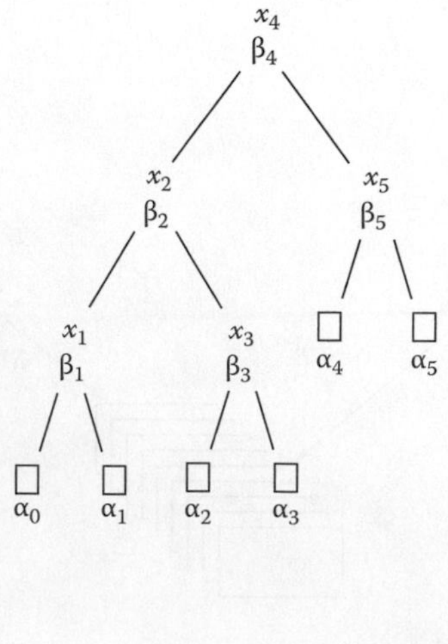

FIGURE 4.3 A binary search tree.

(the root of the left subtree); if $z = x_2$, the search ends. Otherwise, if $z < x_2$, z is compared with x_1 (the root of the left subtree of x_2); if $z = x_1$, the search ends. Otherwise, if $z < x_1$, the search ends unsuccessfully at the leaf labeled α_0. Other results of comparisons lead along other paths in the tree from the root downward. By its nature, a binary search tree is *lexicographic* in that for all nodes in the tree, the elements in the left subtree of the node are smaller and the elements in the right subtree of the node are larger than the node.

Because we are to find an optimal search pattern (tree), we want the cost of searching to be minimized. The cost of searching is measured by the *weighted path length* of the tree:

$$\sum_{i=1}^{n} \beta_i \times [1 + \text{level}(\beta_i)] + \sum_{i=0}^{n} \alpha_i \times \text{level}(\alpha_i),$$

defined formally as

$$W\left(T = \underset{T_l \;\; T_r}{\bigwedge}\right) = W(T_l) + W(T_r) + \sum \alpha_i + \sum \beta_i,$$

where the summations $\sum \alpha_i$ and $\sum \beta_i$ are over all α_i and β_i in T. Since there are exponentially many possible binary trees, finding the one with minimum weighted path length could, if done naïvely, take exponentially long.

The key observation we make is that a *principle of optimality* holds for the cost of binary search trees: subtrees of an optimal search tree must themselves be optimal. This observation means, for example, if the tree shown in Figure 4.3 is optimal, then its left subtree must be the optimal tree for the problem of searching the sequence $x_1 < x_2 < x_3$ with frequencies $\beta_1, \beta_2, \beta_3$ and $\alpha_0, \alpha_1, \alpha_2, \alpha_3$. (If a subtree in Figure 4.3 were *not* optimal, we could replace it with a better one, reducing the weighted path length of the entire tree because of the recursive definition of weighted path length.) In general terms, the principle of optimality states that subsolutions of an optimal solution must themselves be optimal.

The optimality principle, together with the recursive definition of weighted path length, means that we can express the construction of an optimal tree recursively. Let $C_{i,j}$, $0 \le i \le j \le n$, be the cost of an optimal tree over $x_{i+1} < x_{i+2} < \cdots < x_j$ with the associated frequencies $\beta_{i+1}, \beta_{i+2}, \ldots, \beta_j$ and $\alpha_i, \alpha_{i+1}, \ldots, \alpha_j$. Then,

$$C_{i,i} = 0,$$

$$C_{i,j} = \min_{i<k\le j}(C_{i,k-1} + C_{k,j}) + W_{i,j},$$

where

$$W_{i,i} = \alpha_i,$$

$$W_{i,j} = W_{i,j-1} + \beta_j + \alpha_j.$$

These two recurrence relations can be implemented directly as recursive functions to compute $C_{0,n}$, the cost of the optimal tree, leading to the following two functions:

```
1   int W (int i, int j) {
2     if (i == j)
3       return alpha[j];
4     else
5       return W(i,j-1) + beta[j] + alpha[j];
6   }
7
8   int C (int i, int j) {
9     if (i == j)
10      return 0;
11    else {
12      int minCost = MAXINT;
13      int cost;
14      for (int k = i+1; k <= j; k++) {
15        cost = C(i,k-1) + C(k,j) + W(i,j);
16        if (cost < minCost)
17          minCost = cost;
18      }
19      return minCost;
20    }
21  }
```

These two functions correctly compute the cost of an optimal tree; the tree itself can be obtained by storing the values of `k` when `cost < minCost` in line 16.

However, the above functions are unnecessarily time-consuming (requiring exponential time) because the same subproblems are solved repeatedly. For example, each call `W(i, j)` uses time Θ(`j – i`) and such calls are made repeatedly for the same values of `i` and `j`. We can make the process more efficient by caching the values of `W(i, j)` in an array as they are computed and using the cached values when possible:

```
int W[n][n];
for (int i = 0; i < n; i++)
  for (int j = 0; j < n; j++)
    W[i][j] = MAXINT;

int W (int i, int j) {
  if (W[i][j] = MAXINT)
    if (i == j)
      W[i][j] = alpha[j];
    else
      W[i][j] = W(i,j-1) + beta[j] + alpha[j];
  return W[i][j];
}
```

In the same way, we should cache the values of `C(i, j)` in an array as they are computed:

```
int C[n][n];
for (int i = 0; i < n; i++)
  for (int j = 0; j < n; j++)
    C[i][j] = MAXINT;

int C (int i, int j) {
```

```
    if (C[i][j] == MAXINT)
      if (i == j)
          C[i][j] = 0;
      else {
        int minCost = MAXINT;
        int cost;
        for (int k = i+1; k <= j; k++) {
          cost = C(i,k-1) + C(k,j) + W(i,j);
          if (cost < minCost)
            minCost = cost;
        }
        C[i][j] = minCost;
      }
    return C[i][j];
  }
```

The idea of caching the solutions to subproblems is crucial to making the algorithm efficient. In this case, the resulting computation requires time $\Theta(n^3)$; this is surprisingly efficient, considering that an optimal tree is being found from among exponentially many possible trees.

By studying the pattern in which the arrays `C` and `W` are filled in, we see that the main diagonal `C[i][i]` is filled in first, then the first upper super-diagonal `C[i][i + 1]`, then the second upper super-diagonal `C[i][i + 2]`, and so on until the upper right corner of the array is reached. Rewriting the code to do this directly, and adding an array `R[][]` to keep track of the roots of subtrees, we obtain

```
int W[n][n];
int R[n][n];
int C[n][n];

// Fill in main diagonal
for (int i = 0; i < n; i++) {
  W[i][i] = alpha[i];
  R[i][i] = 0;
  C[i][i] = 0;
}

int minCost, cost;
for (int d = 1; d < n; d++)
  // Fill in d-th upper super-diagonal
  for (i = 0; i < n-d; i++) {
   W[i][i+d] = W[i][i+d-1] + beta[i+d] + alpha[i+d];
   R[i][i+d] = i+1;
   C[i][i+d] = C[i][i] + C[i+1][i+d] + W[i][i+d];;
   for (int k = i+2; k <= i+d; k++) {
     cost = C[i][k-1] + C[k][i+d] + W[i][i+d];
     if (cost < C[i][i+d]) {
      R[i][i+d] = k;
      C[i][i+d] = cost;
     }
   }
  }
```

which more clearly shows the $\Theta(n^3)$ behavior.

As a second example of dynamic programming, consider the *traveling salesman problem* in which a salesman must visit n cities, returning to his starting point, and is required to minimize the cost of the trip. The cost of going from city i to city j is $C_{i,j}$. To use dynamic programming, we must specify an optimal tour in a recursive framework, with subproblems resembling the overall problem. Thus we define

$$T(i; j_1, j_2, \ldots, j_k) = \begin{cases} \text{cost of an optimal tour from city } i \text{ to city} \\ \text{1 that goes through each of the cities } j_1, \\ j_2, \ldots, j_k \text{ exactly once, in any order, and} \\ \text{through no other cities.} \end{cases}$$

The principle of optimality tells us that

$$T(i; j_1, j_2, \ldots, j_k) = \min_{1 \le m \le k} \{C_{i,j_m} + T(j_m; j_1, j_2, \ldots, j_{m-1}, j_{m+1}, \ldots, j_k)\},$$

where, by definition,

$$T(i; j) = C_{i,j} + C_{j,1}.$$

We can write a function T that directly implements the above recursive definition, but as in the optimal search tree problem, many subproblems would be solved repeatedly, leading to an algorithm requiring time $\Theta(n!)$. By caching the values $T(i; j_1, j_2, \ldots, j_k)$, we reduce the time required to $\Theta(n^2 2^n)$, still exponential, but considerably less than without caching.

4.5 Greedy Heuristics

Optimization problems always have an objective function to be minimized or maximized, but it is not often clear what steps to take to reach the optimum value. For example, in the optimum binary search tree problem of the previous section, we used dynamic programming to examine systematically all possible trees; but perhaps there is a simple rule that leads directly to the best tree—say by choosing the largest β_i to be the root and then continuing recursively. Such an approach would be less time-consuming than the $\Theta(n^3)$ algorithm we gave, but it does not necessarily give an optimum tree (if we follow the rule of choosing the largest β_i to be the root, we get trees that are no better, on the average, than a randomly chosen trees). The problem with such an approach is that it makes decisions that are *locally optimum*, though perhaps not *globally optimum*. But, such a "greedy" sequence of locally optimum choices does lead to a globally optimum solution in some circumstances.

Suppose, for example, $\beta_i = 0$ for $1 \le i \le n$, and we remove the lexicographic requirement of the tree; the resulting problem is the determination of an optimal prefix code for $n + 1$ letters with frequencies $\alpha_0, \alpha_1, \ldots, \alpha_n$. Because we have removed the lexicographic restriction, the dynamic programming solution of the previous section no longer works, but the following simple greedy strategy yields an optimum tree: Repeatedly combine the two lowest-frequency items as the left and right subtrees of a newly created item whose frequency is the sum of the two frequencies combined. Here is an example of this construction; we start with five leaves with weights

First, combine leaves $\alpha_0 = 25$ and $\alpha_5 = 21$ into a subtree of frequency $25 + 21 = 45$:

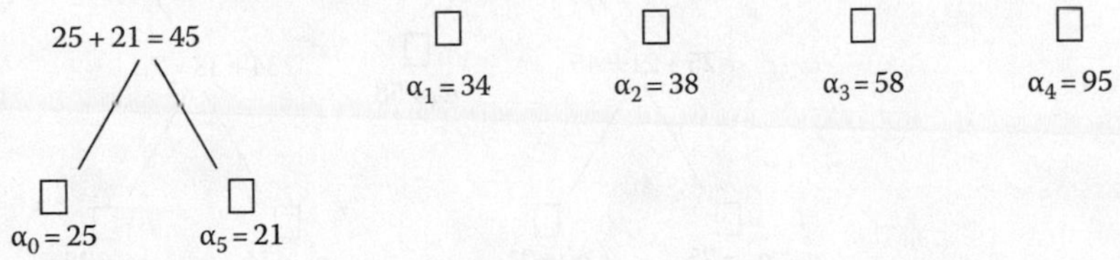

Then combine leaves $\alpha_1 = 34$ and $\alpha_2 = 38$ into a subtree of frequency $34 + 38 = 72$:

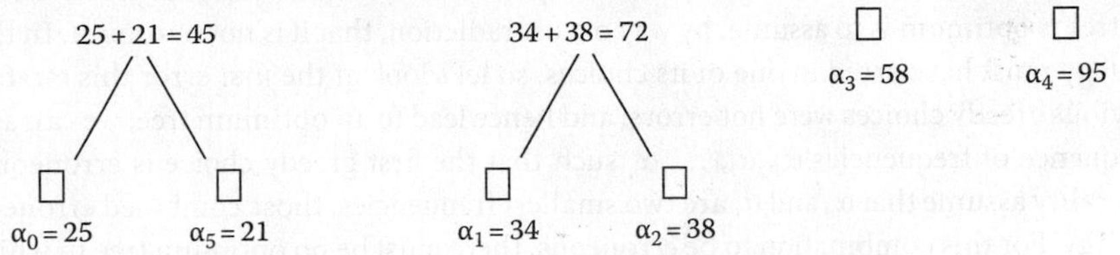

Next, combine the subtree of frequency $\alpha_0 + \alpha_5 = 45$ with $\alpha_3 = 58$:

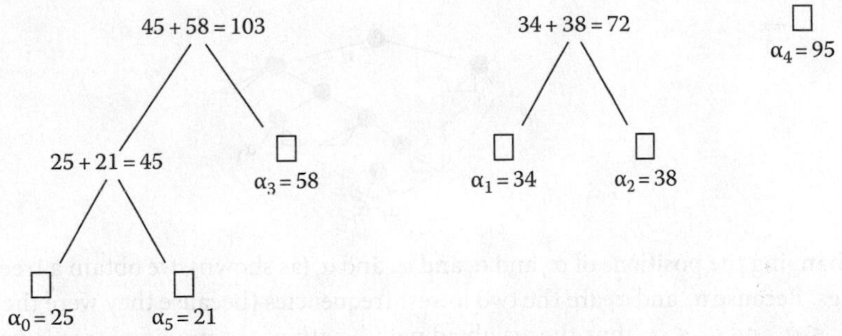

Then, combine the subtree of frequency $\alpha_1 + \alpha_4 = 72$ with $\alpha_4 = 95$:

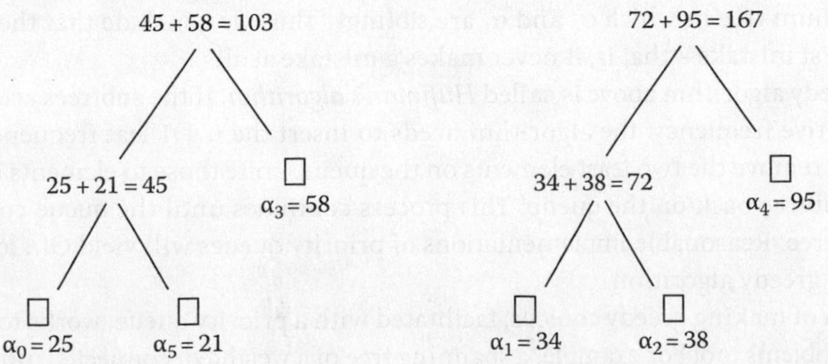

Finally, combine the only two remaining subtrees:

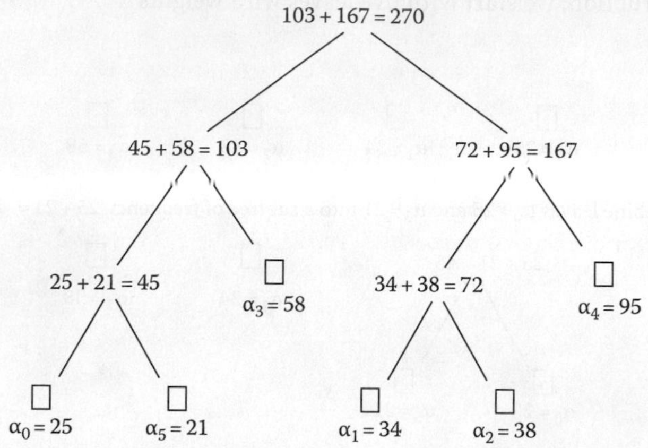

How do we know that the above-outlined process leads to an optimum tree? The key to proving that the tree is optimum is to assume, by way of contradiction, that it is not optimum. In this case, the greedy strategy must have erred in one of its choices, so let's look at the *first* error this strategy made. Since all previous greedy choices were not errors, and hence lead to an optimum tree, we can assume that we have a sequence of frequencies $\alpha_0, \alpha_1, \ldots, \alpha_n$ such that the first greedy choice is erroneous—without loss of generality assume that α_0 and α_1 are two smallest frequencies, those combined erroneously by the greedy strategy. For this combination to be erroneous, there must be no optimum tree in which these two leaves are siblings, so consider an optimum tree, the locations of α_0 and α_1, and the location of the two deepest leaves in the tree, α_i and α_j:

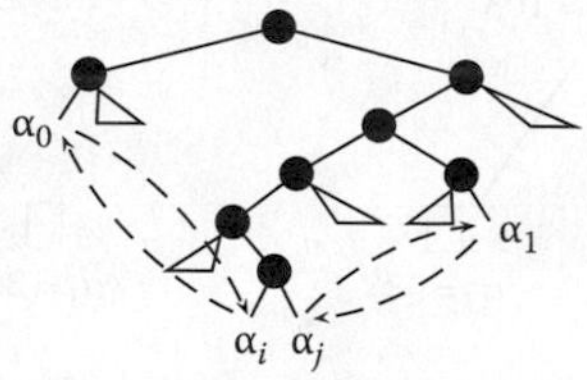

By interchanging the positions of α_0 and α_i and α_1 and α_j (as shown), we obtain a tree in which α_0 and α_1 are siblings. Because α_0 and α_1 are the two lowest frequencies (because they were the greedy algorithm's choice) $\alpha_0 \le \alpha_i$ and $\alpha_1 \le \alpha_j$, thus the weighted path length of the modified tree is no larger than before the modification since level$(\alpha_0) \ge$ level(α_i), level$(\alpha_1) \ge$ level(α_j) and hence

$$\text{level}(\alpha_i)\times\alpha_0 + \text{level}(\alpha_j)\times\alpha_1 \le \text{level}(\alpha_0)\times\alpha_0 + \text{level}(\alpha_1)\times\alpha_1.$$

In other words, the first so-called mistake of the greedy algorithm was in fact not a mistake since there is an optimum tree in which α_0 and α_1 are siblings. Thus we conclude that the greedy algorithm never makes a first mistake—that is, it never makes a mistake at all!

The greedy algorithm above is called *Huffman's algorithm*. If the subtrees are kept on a priority queue by cumulative frequency, the algorithm needs to insert the $n + 1$ leaf frequencies onto the queue, and repeatedly remove the two least elements on the queue, unite those to elements into a single subtree, and put that subtree back on the queue. This process continues until the queue contains a single item, the optimum tree. Reasonable implementations of priority queues will yield $O(n \log n)$ implementations of Huffman's greedy algorithm.

The idea of making greedy choices, facilitated with a priority queue, works to find optimum solutions to other problems too. For example, a spanning tree of a weighted, connected, undirected graph $G = (V, E)$ is a subset of $|V| - 1$ edges from E connecting all the vertices in G; a spanning tree is minimum if the sum of the weights of its edges is as small as possible. *Prim's algorithm* uses a sequence of greedy choices to determine a minimum spanning tree: Start with an arbitrary vertex $v \in V$ as the spanning-tree-to-be. Then, repeatedly add the cheapest edge connecting the spanning-tree-to-be to a vertex not yet in it. If the vertices not yet in the tree are stored in a priority queue implemented by a Fibonacci heap, the total time required by Prim's algorithm will be $O(|E| + |V|\log |V|)$. But why does the sequence of greedy choices lead to a minimum spanning tree?

Suppose Prim's algorithm does *not* result in a minimum spanning tree. As we did with Huffman's algorithm, we ask what the state of affairs must be when Prim's algorithm makes its first mistake; we will see that the assumption of a first mistake leads to a contradiction, proving the correctness of Prim's algorithm. Let the edges added to the spanning tree be, in the order added, $e_1, e_2, e_3, \ldots$, and let e_i be the first mistake. In other words, there is a minimum spanning tree $T_{\min}$ containing $e_1, e_2, \ldots, e_{i-1}$, but no minimum spanning tree contains $e_1, e_2, \ldots, e_i$. Imagine what happens if we add the edge e_i to $T_{\min}$: since $T_{\min}$ is a spanning tree, the addition of e_i causes a cycle containing e_i. Let $e_{\max}$ be the highest-cost edge on that cycle. Because Prim's algorithm makes a greedy choice—that is, chooses the lowest-cost available edge—the cost of $e_{\max}$ is at least that of e_i, so the cost of the spanning $T_{\min} - \{e_{\max}\} \cup \{e_i\}$ is at most that of $T_{\min}$; in other words, $T_{\min} - \{e_{\max}\} \cup \{e_i\}$ is also a minimum spanning tree, contradicting our assumption that the choice of e_i is the first mistake. Therefore, the spanning tree constructed by Prim's algorithm must be a minimum spanning tree.

We can apply the greedy heuristic to many optimization problems, and even if the results are not optimal, they are often quite good. For example, in the n-city traveling salesman problem, we can get near-optimal tours in time $O(n^2)$ when the intercity costs are symmetric ($C_{i,j} = C_{j,i}$ for all i and j) and satisfy the triangle inequality($C_{i,j} \le C_{i,k} + C_{k,j}$ for all i, j, and k). The *closest insertion algorithm* starts with a "tour" consisting of a single, arbitrarily chosen city, and successively inserts the remaining cities to the tour, making a greedy choice about which city to insert next and where to insert it: the city chosen for insertion is the city not on the tour but closest to a city on the tour; the chosen city is inserted adjacent to the city on the tour to which it is closest.

Given an $n \times n$ symmetric distance matrix C that satisfies the triangle inequality, let I_n be th\e tour of length $|I_n|$ produced by the closest insertion heuristic and let O_n be an optimal tour of length $|O_n|$. Then,

$$\frac{|I_n|}{|O_n|} < 2.$$

This bound is proved by an incremental form of the optimality proofs for greedy heuristics we have seen above: we ask not where the first error is, but by how much we are in error at each greedy insertion to the tour—we establish a correspondence between edges of the optimal tour and cities inserted on the closest insertion tour. We show that at each insertion of a new city to the closest insertion tour, the cost of that insertion is at most twice the cost of corresponding edge of the optimal tour.

To establish the correspondence, imagine the closest insertion algorithm keeping track not only of the current tour, but also of a spiderlike configuration including the edges of the current tour (the body of the spider) and pieces of the optimal tour (the legs of the spider). We show the current tour in solid lines and the pieces of optimal tour as dotted lines:

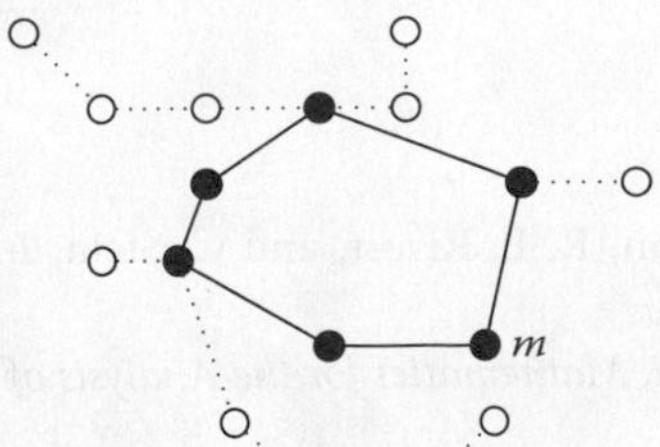

Initially, the spider consists of the arbitrarily chosen city with which the closest insertion tour begins and the legs of the spider consist of all the edges of the optimal tour *except* for one edge eliminated arbitrarily. As each city is inserted into the closest insertion tour, the algorithm will delete from the spider-like configuration one of the dotted edges from the optimal tour. When city k is inserted between cities l and m, the edge deleted is the one attaching spider to the leg containing the city inserted (from city x to city y), shown here in bold:

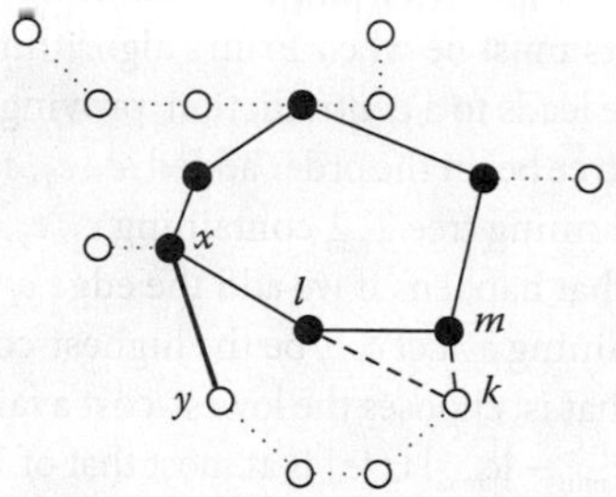

Now,

$$C_{k,m} \le C_{x,y},$$

because of the greedy choice to add city k to the tour and not city y. By the triangle inequality,

$$C_{l,k} \le C_{l,m} + C_{m,k},$$

and by symmetry we can combine these two inequalities to get

$$C_{l,k} \le C_{l,m} + C_{x,y}.$$

Adding this last inequality to the first one,

$$C_{l,k} + C_{k,m} \le C_{l,m} + 2C_{x,y},$$

that is,

$$C_{l,k} + C_{k,m} - C_{l,m} \le 2C_{x,y}.$$

Thus, adding city k between cities l and m adds no more to I_n than $2C_{x,y}$. Summing these incremental amounts over the cost of the entire algorithm tells us

$$I_n \le 2O_n,$$

as we claimed.

Further Reading

1. Cormen, T. H., C. E. Leiserson, R. L. Rivest, and C. Stein, *Introduction to Algorithms*, 3rd edn., McGraw-Hill, New York, 2009.
2. Greene, D. H. and D. E. Knuth, *Mathematics for the Analysis of Algorithms*, 3rd rev. ed., Birkhäuser, Boston, MA, 2007.
3. Kleinberg, J. and É. Tardos, *Algorithm Design*, Addison-Wesley, Reading, MA, 2005.

4. Knuth, D. E., *The Art of Computer Programming, Volume 1: Fundamental Algorithms,* 3rd edn., Addison-Wesley, Reading, MA, 1997.
5. Knuth, D. E., *The Art of Computer Programming, Volume 2: Seminumerical Algorithms,* 3rd edn., Addison-Wesley, Reading, MA, 1997.
6. Knuth, D. E., *The Art of Computer Programming, Volume 3: Sorting and Searching,* 2nd edn., Addison-Wesley, Reading, MA, 1998.
7. Knuth, D. E., *The Art of Computer Programming, Volume 4A: Combinatorial Algorithms, Part 1,* Addison-Wesley, Reading, MA, 2011.
8. Lueker, G. S., Some techniques for solving recurrences, *Computing Surveys* 12 (1980), 419–436.
9. Reingold, E. M., J. Nievergelt, and N. Deo, *Combinatorial Algorithms: Theory and Practice*, Prentice-Hall, Englewood Cliffs, NJ, 1977.

5 Graph and Network Algorithms

Samir Khuller
University of Maryland

Balaji Raghavachari
The University of Texas, Dallas

5.1 Introduction

Graphs are useful in modeling many problems from different scientific disciplines because they capture the basic concept of objects (vertices) and relationships between objects (edges). Indeed, many optimization problems can be formulated in graph-theoretic terms. Hence, algorithms on graphs have been widely studied. In this chapter, a few fundamental graph algorithms are described. For a more detailed treatment of graph algorithms, the reader is referred to textbooks on graph algorithms (Cormen et al. 2001, Even 1979, Gibbons 1985, Tarjan 1983).

An undirected *graph* $G = (V, E)$ is defined as a set V of *vertices* and a set E of *edges*. An edge $e = (u, v)$ is an unordered pair of vertices. A *directed graph* is defined similarly, except that its edges are ordered pairs of vertices; that is, for a directed graph, $E \subseteq V \times V$. The terms *nodes* and vertices are used interchangeably.

In this chapter, it is assumed that the graph has neither self-loops, edges of the form (v, v), nor multiple edges connecting two given vertices. A graph is a *sparse graph* if $|E| \ll |V|^2$.

Bipartite graphs form a subclass of graphs and are defined as follows. A graph $G = (V, E)$ is bipartite if the vertex set V can be partitioned into two sets X and Y such that $E \subseteq X \times Y$. In other words, each edge of G connects a vertex in X with a vertex in Y. Such a graph is denoted by $G = (X, Y, E)$. Because bipartite graphs occur commonly in practice, algorithms are often specially designed for them.

A vertex w is *adjacent* to another vertex v if $(v, w) \in E$. An edge (v, w) is said to be *incident* on vertices v and w. The *neighbors* of a vertex v are all vertices $w \in V$ such that $(v, w) \in E$. The number of edges incident to a vertex v is called the *degree* of vertex v. For a directed graph, if (v, w) is an edge, then we say that the edge goes from v to w. The *out-degree* of a vertex v is the number of edges from v to other vertices. The *in-degree* of v is the number of edges from other vertices to v.

A *path* $p = [v_0, v_1, \ldots, v_k]$ from v_0 to v_k is a sequence of vertices such that (v_i, v_{i+1}) is an edge in the graph for $0 \leq i < k$. Any edge may be used only once in a path. A *cycle* is a path whose end vertices are the same, that is, $v_0 = v_k$. A path is *simple* if all its internal vertices are distinct. A cycle is *simple* if every node has exactly two edges incident to it in the cycle. A *walk* $w = [v_0, v_1, \ldots, v_k]$ from v_0 to v_k is a sequence of vertices such that (v_i, v_{i+1}) is an edge in the graph for $0 \leq i < k$, in which edges and vertices may be repeated. A walk is *closed* if $v_0 = v_k$. A graph is *connected* if there is a path between every pair of vertices. A directed graph is *strongly connected* if there is a path between every pair of vertices in each direction. An acyclic, undirected graph is a *forest*, and a *tree* is a connected forest. A directed graph without cycles is known as a *directed acyclic graph* (DAG). Consider a binary relation C between the vertices of an undirected graph G such that for any two vertices u and v, uCv if and only if there is a path in G between u and v. It can be shown that C is an equivalence relation, partitioning the vertices of G into equivalence classes, known as the connected components of G.

There are two convenient ways of representing graphs on computers. We first discuss the *adjacency list* representation. Each vertex has a linked list: there is one entry in the list for each of its adjacent vertices. The graph is thus represented as an array of linked lists, one list for each vertex. This representation uses $O(|V| + |E|)$ storage, which is good for sparse graphs. Such a storage scheme allows one to scan all vertices adjacent to a given vertex in time proportional to its degree. The second representation, the *adjacency matrix*, is as follows. In this scheme, an $n \times n$ array is used to represent the graph. The $[i, j]$ entry of this array is 1 if the graph has an edge between vertices i and j, and 0 otherwise. This representation permits one to test if there is an edge between any pair of vertices in constant time. Both these representation schemes can be used in a natural way to represent directed graphs. For all algorithms in this chapter, it is assumed that the given graph is represented by an adjacency list.

Section 5.2 discusses various types of tree traversal algorithms. Sections 5.3 and 5.4 discuss depth-first search (DFS) and breadth-first search (BFS) techniques. Section 5.5 discusses the single-source shortest path problem. Section 5.6 discusses minimum spanning trees (MSTs). Section 5.7 discusses the bipartite matching problem and the single commodity maximum-flow problem. Section 5.8 discusses some traversal problems in graphs, and the Further Information section concludes with some pointers to current research on graph algorithms.

5.2 Tree Traversals

A tree is *rooted* if one of its vertices is designated as the root vertex and all edges of the tree are oriented (directed) to point away from the root. In a rooted tree, there is a directed path from the root to any vertex in the tree. For any directed edge (u, v) in a rooted tree, u is v's *parent* and v is u's *child*. The *descendants* of a vertex w are all vertices in the tree (including w) that are reachable by directed paths starting at w. The *ancestors* of a vertex w are those vertices for which w is a descendant. Vertices that have no children are called *leaves*. A *binary tree* is a special case of a rooted tree in which each node has at most two children, namely, the left child and the right child. The trees rooted at the two children of a node are called the *left subtree* and *right subtree*.

In this section, we study techniques for processing the vertices of a given binary tree in various orders. We assume that each vertex of the binary tree is represented by a record that contains fields to hold attributes of that vertex and two special fields *left* and *right* that point to its left and right subtree, respectively.

The three major tree traversal techniques are *preorder*, *inorder*, and *postorder*. These techniques are used as procedures in many tree algorithms where the vertices of the tree have to be processed in a specific order. In a preorder traversal, the root of any subtree has to be processed *before* any of its descendants. In a postorder traversal, the root of any subtree has to be processed *after* all of its descendants. In an inorder traversal, the root of a subtree is processed after all vertices in its left subtree have been processed but before any of the vertices in its right subtree are processed. Preorder and postorder traversals generalize to arbitrary rooted trees. In the example to follow, we show how postorder can be used to count the number of descendants of each node and store the value in that node. The algorithm runs in linear time in the size of the tree:

Postorder Algorithm. *PostOrder* (*T*):

```
if T ≠ nil then
   lc ← PostOrder (left[T]).
   rc ← PostOrder (right[T]).
   desc[T] ← lc + rc + 1.
   return desc[T].
else
   return 0.
end-if
end-proc
```

5.3 Depth-First Search

DFS is a fundamental graph searching technique (Hopcroft and Tarjan 1973, Tarjan 1972). Similar graph searching techniques were given earlier by Tremaux (see Fraenkel 1970, Lucas 1882). The structure of DFS enables efficient algorithms for many other graph problems such as biconnectivity, triconnectivity, and planarity (Even 1979).

The algorithm first initializes all vertices of the graph as being unvisited. Processing of the graph starts from an arbitrary vertex, known as the root vertex. Each vertex is processed when it is first discovered (also referred to as *visiting* a vertex). It is first marked as visited, and its adjacency list is then scanned for unvisited vertices. Each time an unvisited vertex is discovered, it is processed recursively by DFS. After a node's entire adjacency list has been explored, that invocation of the DFS procedure returns. This procedure eventually visits all vertices that are in the same connected component of the root vertex. Once DFS terminates, if there are still any unvisited vertices left in the graph, one of them is chosen as the root and the same procedure is repeated.

The set of edges such that each one led to the discovery of a new vertex form a maximal forest of the graph, known as the *DFS forest*; a *maximal forest* of a graph G is an acyclic subgraph of G such that the addition of any other edge of G to the subgraph introduces a cycle. The algorithm keeps track of this forest using parent pointers. In each connected component, only the root vertex has a *nil* parent in the DFS tree.

5.3.1 Depth-First Search Algorithm

DFS is illustrated using an algorithm that labels vertices with numbers 1, 2, ... in such a way that vertices in the same component receive the same label. This labeling scheme is a useful preprocessing step in many problems. Each time the algorithm processes a new component, it numbers its vertices with a new label.

Depth-First Search Algorithm. *DFS-Connected Component* (*G*):

```
c ← 0.
for all vertices v in G do
    visited[v] ← false.
    finished[v] ← false.
    parent[v] ← nil.
end-for
for all vertices v in G do
    if not visited[v] then
        c ← c + 1.
        DFS (v, c).
    end-if
end-for
end-proc
```

DFS (*v*, *c*):

```
visited[v] ← true.
component[v] ← c.
for all vertices w in adj[v] do
    if not visited[w] then
        parent[w] ← v.
        DFS (w, c).
    end-if
end-for
finished[v] ← true.
end-proc
```

5.3.2 Sample Execution

Figure 5.1 shows a graph having two connected components. DFS was started at vertex *a*, and the DFS forest is shown on the right. DFS visits the vertices *b*, *d*, *c*, *e*, and *f*, in that order. DFS then

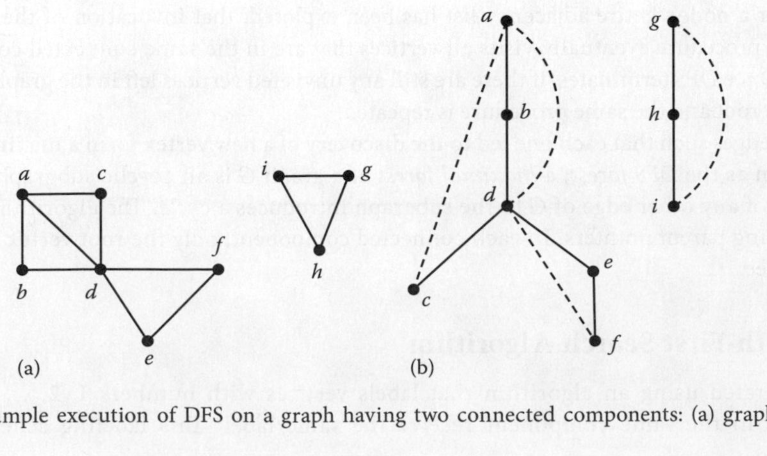

FIGURE 5.1 Sample execution of DFS on a graph having two connected components: (a) graph and (b) DFS forest.

continues with vertices *g*, *h*, and *i*. In each case, the recursive call returns when the vertex has no more unvisited neighbors. Edges (*d*, *a*), (*c*, *a*), (*f*, *d*), and (*i*, *g*) are called *back edges* (these do not belong to the DFS forest).

5.3.3 Analysis

A vertex *v* is processed as soon as it is encountered, and therefore at the start of DFS (*v*), *visited*[*v*] is *false*. Since *visited*[*v*] is set to true as soon as DFS starts execution, each vertex is visited exactly once. DFS processes each edge of the graph exactly twice, once from each of its incident vertices. Since the algorithm spends constant time processing each edge of *G*, it runs in $O(|V| + |E|)$ time.

Remark 5.1

In the following discussion, there is no loss of generality in assuming that the input graph is connected. For a rooted DFS tree, vertices *u* and *v* are said to be *related*, if either *u* is an ancestor of *v*, or vice versa.

DFS is useful due to the special way in which the edges of the graph may be classified with respect to a DFS tree. Notice that the DFS tree is not unique, and which edges are added to the tree depends on the order in which edges are explored while executing DFS. Edges of the DFS tree are known as *tree* edges. All other edges of the graph are known as *back* edges, and it can be shown that for any edge (*u*, *v*), *u* and *v* must be related. The graph does not have any *cross* edges, edges that connect two vertices that are unrelated. This property is utilized by a DFS-based algorithm that classifies the edges of a graph into *biconnected* components, maximal subgraphs that cannot be disconnected by the removal of any single vertex (Even 1979).

5.3.4 Directed Depth-First Search

The DFS algorithm extends naturally to directed graphs. Each vertex stores an adjacency list of its outgoing edges. During the processing of a vertex, first mark it as visited, and then scan its adjacency list for unvisited neighbors. Each time an unvisited vertex is discovered, it is processed recursively. Apart from tree edges and back edges (from vertices to their ancestors in the tree), directed graphs may also have *forward* edges (from vertices to their descendants) and *cross* edges (between unrelated vertices). There may be a cross edge (*u*, *v*) in the graph only if *u* is visited after the procedure call DFS (*v*) has completed execution.

5.3.5 Sample Execution

A sample execution of the directed DFS algorithm is shown in Figure 5.2. DFS was started at vertex *a*, and the DFS forest is shown on the right. DFS visits vertices *b*, *d*, *f*, and *c* in that order. DFS then returns and continues with *e* and then *g*. From *g*, vertices *h* and *i* are visited in that order. Observe that (*d*, *a*) and (*i*, *g*) are back edges. Edges (*c*, *d*), (*e*, *d*), and (*e*, *f*) are cross edges. There is a single forward edge (*g*, *i*).

5.3.6 Applications of Depth-First Search

Directed DFS can be used to design a linear-time algorithm that classifies the edges of a given directed graph into *strongly connected* components: maximal subgraphs that have directed paths connecting any pair of vertices in them, in each direction. The algorithm itself involves running DFS twice, once on the original graph and then a second time on G^R, which is the graph obtained by reversing the

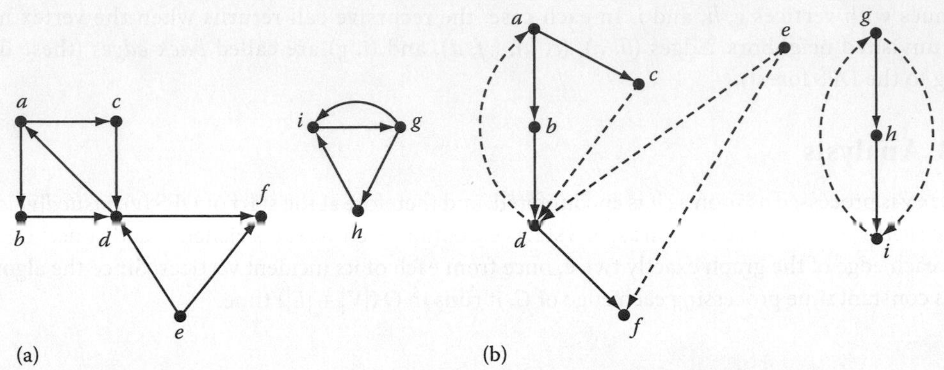

FIGURE 5.2 Sample execution of DFS on a directed graph: (a) graph and (b) DFS forest.

direction of all edges in *G*. During the second DFS, we are able to obtain all of the strongly connected components. The proof of this algorithm is somewhat subtle, and the reader is referred to Cormen et al. (2001) for details.

Checking if a graph has a cycle can be done in linear time using DFS. A graph has a cycle if and only if there exists a back edge relative to any of its DFS trees. A directed graph that does not have any cycles is known as a DAG. DAGs are useful in modeling precedence constraints in scheduling problems, where nodes denote jobs/tasks, and a directed edge from u to v denotes the constraint that job u must be completed before job v can begin execution. Many problems on DAGs can be solved efficiently using dynamic programming.

A useful concept in DAGs is that of a *topological order*: a linear ordering of the vertices that is consistent with the partial order defined by the edges of the DAG. In other words, the vertices can be labeled with distinct integers in the range $[1 \ldots |V|]$ such that if there is a directed edge from a vertex labeled i to a vertex labeled j, then $i < j$. The vertices of a given DAG can be ordered topologically in linear time by a suitable modification of the DFS algorithm. We keep a counter whose initial value is $|V|$. As each vertex is marked finished, we assign the counter value as its topological number and decrement the counter. Observe that there will be no back edges and that for all edges (u, v), v will be marked finished before u. Thus, the topological number of v will be higher than that of u. Topological sort has applications in diverse areas such as project management, scheduling, and circuit evaluation.

5.4 Breadth-First Search

BFS is another natural way of searching a graph. The search starts at a root vertex r. Vertices are added to a queue as they are discovered and processed in first-in–first-out (FIFO) order.

Initially, all vertices are marked as unvisited, and the queue consists of only the root vertex. The algorithm repeatedly removes the vertex at the front of the queue and scans its neighbors in the graph. Any neighbor not visited is added to the end of the queue. This process is repeated until the queue is empty. All vertices in the same connected component as the root are scanned and the algorithm outputs a spanning tree of this component. This tree, known as a *breadth-first tree*, is made up of the edges that led to the discovery of new vertices. The algorithm labels each vertex v by $d[v]$, the distance (length of a shortest path) of v from the root vertex, and stores the BFS tree in the array p, using parent pointers. Vertices can be partitioned into levels based on their distance from the root. Observe that edges not in the BFS tree always go either between vertices in the same level or between vertices in adjacent levels. This property is often useful.

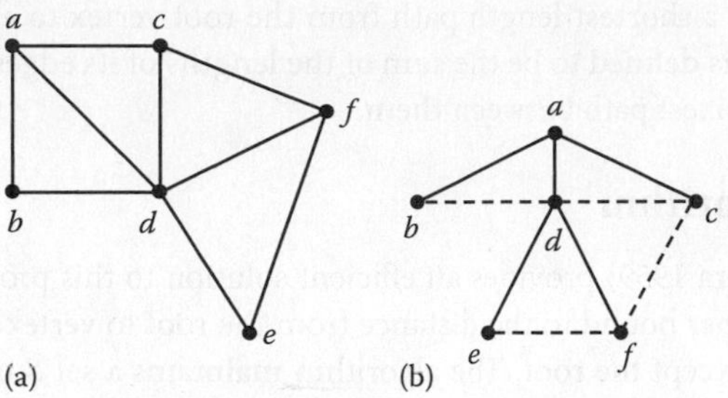

FIGURE 5.3 Sample execution of BFS on a graph: (a) graph and (b) BFS tree.

Breadth-First Search Algorithm. *BFS-Distance* (*G*, *r*):

```
MakeEmptyQueue (Q).
for all vertices v in G do
    visited[v] ← false.
    d[v] ← ∞.
    p[v] ← nil.
end-for
visited[r] ← true.
d[r] ← 0.
Enqueue (Q, r).
while not Empty (Q) do
    v ← Dequeue (Q).
    for all vertices w in adj[v] do
        if not visited[w] then
            visited[w] ← true.
            p[w] ← v.
            d[w] ← d[v] + 1.
            Enqueue (Q, w).
        end-if
    end-for
end-while
end-proc
```

5.4.1 Sample Execution

Figure 5.3 shows a connected graph on which BFS was run with vertex *a* as the root. When *a* is processed, vertices *b*, *d*, and *c* are added to the queue. When *b* is processed, nothing is done since all its neighbors have been visited. When *d* is processed, *e* and *f* are added to the queue. Finally *c*, *e*, and *f* are processed.

5.4.2 Analysis

There is no loss of generality in assuming that the graph *G* is connected, since the algorithm can be repeated in each connected component, similar to the DFS algorithm. The algorithm processes each vertex exactly once and each edge exactly twice. It spends a constant amount of time in processing each edge. Hence, the algorithm runs in $O(|V| + |E|)$ time.

5.5 Single-Source Shortest Paths

A natural problem that often arises in practice is to compute the shortest paths from a specified node to all other nodes in an undirected graph. BFS solves this problem if all edges in the graph have the same length. Consider the more general case when each edge is given an arbitrary, nonnegative length, and one needs to calculate a shortest-length path from the root vertex to all other nodes of the graph, where the length of a path is defined to be the sum of the lengths of its edges. The distance between two nodes is the length of a shortest path between them.

5.5.1 Dijkstra's Algorithm

Dijkstra's algorithm (Dijkstra 1959) provides an efficient solution to this problem. For each vertex v, the algorithm maintains an upper bound to the distance from the root to vertex v in $d[v]$; initially $d[v]$ is set to infinity for all vertices except the root. The algorithm maintains a set S of vertices with the property that for each vertex $v \in S$, $d[v]$ is the length of a shortest path from the root to v. For each vertex u in $V - S$, the algorithm maintains $d[u]$, the shortest known distance from the root to u that goes entirely within S, except for the last edge. It selects a vertex u in $V - S$ of minimum $d[u]$, adds it to S, and updates the distance estimates to the other vertices in $V - S$. In this update step, it checks to see if there is a shorter path to any vertex in $V - S$ from the root that goes through u. Only the distance estimates of vertices that are adjacent to u are updated in this step. Because the primary operation is the selection of a vertex with minimum distance estimate, a priority queue is used to maintain the d-values of vertices. The priority queue should be able to handle a DecreaseKey operation to update the d-value in each iteration. The next algorithm implements Dijkstra's algorithm.

Dijkstra's Algorithm. *Dijkstra-Shortest Paths* (*G*, *r*):

```
 1 for all vertices v in G do
 2     visited[v] ← false.
 3     d[v] ← ∞.
 4     p[v] ← nil.
 5 end-for
 6 d[r] ← 0.
 7 BuildPQ(H, d).
 8 while not Empty(H) do
 9     u ← DeleteMin(H).
10     visited[u] ← true.
11     for all vertices v in adj[u] do
12         Relax (u, v).
13     end-for
14 end-while
end-proc
```

```
Relax (u, v)
1 if not visited[v] and d[v] > d[u] + w(u, v) then
2     d[v] ← d[u] + w(u, v).
3     p[v] ← u.
4     DecreaseKey (H, v, d[v]).
5 end-if
end-proc
```

5.5.1.1 Sample Execution

Figure 5.4 shows a sample execution of the algorithm. The column titled Iter specifies the number of iterations that the algorithm has executed through the *while* loop in step 8. In iteration 0, the initial values of the distance estimates are ∞. In each subsequent line of the table, the column marked u shows the vertex that was chosen in step 9 of the algorithm and the change to the distance estimates at the end of that iteration of the *while* loop. In the first iteration, vertex r was chosen, after that a was chosen because

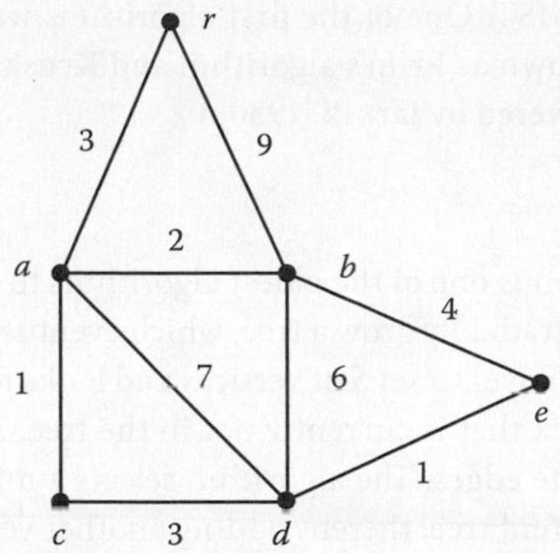

Iter	u	$d[a]$	$d[b]$	$d[c]$	$d[d]$	$d[e]$
0	—	∞	∞	∞	∞	∞
1	r	3	9	∞	∞	∞
2	a	3	5	4	10	∞
3	c	3	5	4	7	∞
4	b	3	5	4	7	9
5	d	3	5	4	7	8
6	e	3	5	4	7	8

FIGURE 5.4 Dijkstra's shortest path algorithm.

it had the minimum distance label among the unvisited vertices, and so on. The distance labels of the unvisited neighbors of the visited vertex are updated in each iteration.

5.5.1.2 Analysis

The running time of the algorithm depends on the data structure that is used to implement the priority queue H. The algorithm performs $|V|$ DeleteMin operations and, at most, $|E|$ DecreaseKey operations. If a binary heap is used to update the records of any given vertex, each of these operations runs in $O(\log |V|)$ time. There is no loss of generality in assuming that the graph is connected. Hence, the algorithm runs in $O(|E| \log |V|)$. If a Fibonacci heap is used to implement the priority queue, the running time of the algorithm is $O(|E| + |V| \log |V|)$. Although the Fibonacci heap gives the best asymptotic running time, the binary heap implementation is likely to give better running times for most practical instances.

5.5.2 Bellman–Ford Algorithm

The shortest path algorithm described earlier directly generalizes to directed graphs, but it does not work correctly if the graph has edges of negative length. For graphs that have edges of negative length, but no cycles of negative length, there is a different algorithm due to Bellman (1958) and Ford and Fulkerson (1962) that solves the single-source shortest paths problem in $O(|V||E|)$ time.

The key to understanding this algorithm is the RELAX operation applied to an edge. In a single scan of the edges, we execute the RELAX operation on each edge. We then repeat the step $|V| - 1$ times. No special data structures are required to implement this algorithm, and the proof relies on the fact that a shortest path is simple and contains at most $|V| - 1$ edges (see Cormen et al. 2001 for a proof).

This problem also finds applications in finding a feasible solution to a system of linear equations, where each equation specifies a bound on the difference of two variables. Each constraint is modeled by an edge in a suitably defined directed graph. Such systems of equations arise in real-time applications.

5.6 Minimum Spanning Trees

The following fundamental problem arises in network design. A set of sites needs to be connected by a network. This problem has a natural formulation in graph-theoretic terms. Each site is represented by a vertex. Edges between vertices represent a potential link connecting the corresponding nodes. Each edge is given a nonnegative cost corresponding to the cost of constructing that link. A tree is a minimal network that connects a set of nodes. The cost of a tree is the sum of the costs of its edges. A minimum-cost tree connecting the nodes of a given graph is called a minimum-cost spanning tree, or simply an **MST**. The problem of computing an MST arises in many areas and as a subproblem in combinatorial and geometric problems. MSTs can be computed efficiently using algorithms that are greedy in nature, and there are several different algorithms for finding an MST. One of the first algorithms was due to Boruvka (1926). The two algorithms that are popularly known as Prim's algorithm and Kruskal's algorithm are described here. (Prim's algorithm was first discovered by Jarnik [1930].)

5.6.1 Prim's Algorithm

Prim's (1957) algorithm for finding an MST of a given graph is one of the oldest algorithms to solve the problem. The basic idea is to start from a single vertex and gradually grow a tree, which eventually spans the entire graph. At each step, the algorithm has a tree that covers a set S of vertices and looks for a *good* edge that may be used to extend the tree to include a vertex that is currently not in the tree. All edges that go from a vertex in S to a vertex in $V - S$ are candidate edges. The algorithm selects a minimum-cost edge from these candidate edges and adds it to the current tree, thereby adding another vertex to S.

As in the case of Dijkstra's algorithm, each vertex $u \in V - S$ can attach itself to only one vertex in the tree (so that cycles are not generated in the solution). Because the algorithm always chooses a minimum-cost edge, it needs to maintain a minimum-cost edge that connects u to some vertex in S as the candidate edge for including u in the tree. A priority queue of vertices is used to select a vertex in $V - S$ that is incident to a minimum-cost candidate edge.

Prim's Algorithm. *Prim-MST* (G, r):

```
for all vertices v in G do
    visited[v] ← false.
    d[v] ← ∞.
    p[v] ← nil.
end-for
d[r] ← 0.
BuildPQ (H, d).
while not Empty (H) do
    u ← DeleteMin (H).
    visited[u] ← true.
    for all vertices v in adj[u] do
        if not visited[v] and d[v] > w(u, v) then
            d[v] ← w(u, v).
            p[v] ← u.
            DecreaseKey (H, v, d[v]).
        end-if
    end-for
end-while
end-proc
```

5.6.1.1 Analysis

First observe the similarity between Prim's and Dijkstra's algorithms. Both algorithms start building the tree from a single vertex and grow it by adding one vertex at a time. The only difference is the rule for deciding when the current label is updated for vertices outside the tree. Both algorithms have the same structure and therefore have similar running times. Prim's algorithm runs in $O(|E| \log |V|)$ time if the priority queue is implemented using binary heaps, and it runs in $O(|E| + |V| \log |V|)$ if the priority queue is implemented using Fibonacci heaps.

5.6.2 Kruskal's Algorithm

Kruskal's (1956) algorithm for finding an MST of a given graph is another classical algorithm for the problem and is also greedy in nature. Unlike Prim's algorithm, which grows a single tree, Kruskal's algorithm grows a forest. First, the edges of the graph are sorted in nondecreasing order of their costs. The algorithm starts with the empty spanning forest (no edges). The edges of the graph are scanned in sorted order, and if the addition of the current edge does not generate a cycle in the current forest, it is added to the forest. The main test at each step is: does the current edge connect two vertices in the same connected component? Eventually, the algorithm adds $|V| - 1$ edges to make a spanning tree in the graph.

The main data structure needed to implement the algorithm is for the maintenance of connected components, to ensure that the algorithm does not add an edge between two nodes in the same connected component. An abstract version of this problem is known as the union–find problem for a collection of disjoint sets. Efficient algorithms are known for this problem, where an arbitrary sequence of UNION and FIND operations can be implemented to run in almost linear time (Cormen et al. 2001, Tarjan 1983).

Kruskal's Algorithm. *Kruskal-MST* (*G*):

```
1 T ← ϕ.
2 for all vertices v in G do
3    Makeset(v).
4 Sort the edges of G by nondecreasing order of costs.
5 for all edges e = (u, v) in G in sorted order do
6    if Find (u) ≠ Find (v) then
7       T ← T ∪ (u, v).
8       Union (u, v).
9 end-proc
```

5.6.2.1 Analysis

The running time of the algorithm is dominated by step 4 of the algorithm in which the edges of the graph are sorted by nondecreasing order of their costs. This takes $O(|E| \log |E|)$ [which is also $O(|E| \log |V|)$] time using an efficient sorting algorithm such as heapsort. Kruskal's algorithm runs faster in the following special cases: if the edges are presorted, if the edge costs are within a small range, or if the number of different edge costs is bounded by a constant. In all of these cases, the edges can be sorted in linear time, and the algorithm runs in near-linear time, $O(|E|\ \alpha\ (|E|, |V|))$, where $\alpha(m, n)$ is the inverse Ackermann function (Tarjan 1983).

Remark 5.2

The MST problem can be generalized to directed graphs. The equivalent of trees in directed graphs are called *arborescences* or *branchings*; and because edges have directions, they are rooted spanning trees. An incoming branching has the property that every vertex has a unique path to the root. An outgoing branching has the property that there is a unique path from the root to each vertex in the graph. The input is a directed graph with arbitrary costs on the edges and a root vertex *r*. The output is a minimum-cost branching rooted at *r*. The algorithms discussed in this section for finding MSTs do not directly extend to the problem of finding optimal branchings. There are efficient algorithms that run in $O(|E| + |V| \log |V|)$ time using Fibonacci heaps for finding minimum-cost branchings (Gabow et al. 1986, Gibbons 1985). These algorithms are based on techniques for weighted matroid intersection (Lawler 1976). Almost linear-time deterministic algorithms for the MST problem in undirected graphs are also known (Fredman and Tarjan 1987).

5.7 Matchings and Network Flows

Networks are important both for electronic communication and for transporting goods. The problem of efficiently moving entities (such as bits, people, or products) from one place to another in an underlying network is modeled by the *network flow* problem. The problem plays a central role in the fields of operations research and computer science, and much emphasis has been placed on the design of efficient algorithms for solving it. Many of the basic algorithms studied earlier in this chapter play an important role in developing various implementations for network flow algorithms.

First, the *matching* problem, which is a special case of the flow problem, is introduced. Then, the *assignment problem*, which is a generalization of the matching problem to the weighted case, is studied. Finally, the network flow problem is introduced and algorithms for solving it are outlined.

The maximum matching problem is studied here in detail only for bipartite graphs. Although this restricts the class of graphs, the same principles are used to design polynomial time algorithms for graphs that are not necessarily bipartite. The algorithms for general graphs are complex due to the presence of structures called *blossoms*, and the reader is referred to Papadimitriou and Steiglitz (1982, Chapter 10) or Tarjan (1983, Chapter 9) for a detailed treatment of how blossoms are handled. Edmonds (see Even 1979) gave the first algorithm to solve the matching problem in polynomial time. Micali and Vazirani (1980) obtained an $O\left(\sqrt{|V|}|E|\right)$ algorithm for nonbipartite matching by extending the algorithm by Hopcroft and Karp (1973) for the bipartite case.

5.7.1 Matching Problem Definitions

Given a graph $G = (V, E)$, a matching M is a subset of the edges such that no two edges in M share a common vertex. In other words, the problem is that of finding a set of independent edges that have no incident vertices in common. The cardinality of M is usually referred to as its *size*.

The following terms are defined with respect to a matching M. The edges in M are called *matched edges* and edges not in M are called *free edges*. Likewise, a vertex is a *matched vertex* if it is incident to a matched edge. A *free vertex* is one that is not matched. The *mate* of a matched vertex v is its neighbor w that is at the other end of the matched edge incident to v. A matching is called *perfect* if all vertices of the graph are matched in it. The objective of the maximum matching problem is to maximize $|M|$ the size of the matching. If the edges of the graph have weights, then the *weight* of a matching is defined to be the sum of the weights of the edges in the matching. A path $p = [v_1, v_2, \ldots, v_k]$ is called an *alternating path* if the edges (v_{2j-1}, v_{2j}), $j = 1, 2, \ldots$, are free and the edges (v_{2j}, v_{2j+1}), $j = 1, 2, \ldots$, are matched. An *augmenting path* $p = [v_1, v_2, \ldots, v_k]$ is an alternating path in which both v_1 and v_k are free vertices.

Observe that an augmenting path is defined with respect to a specific matching. The symmetric difference of a matching M and an augmenting path P, $M \oplus P$, is defined to be $(M - P) \cup (P - M)$. The operation can be generalized to the case when P is any subset of the edges.

5.7.2 Applications of Matching

Matchings are the underlying basis for many optimization problems. Problems of assigning workers to jobs can be naturally modeled as a bipartite matching problem. Other applications include assigning a collection of jobs with precedence constraints to two processors, such that the total execution time is minimized (Lawler 1976). Other applications arise in chemistry, in determining structure of chemical bonds, matching moving objects based on a sequence of photographs, and localization of objects in space after obtaining information from multiple sensors (Ahuja et al. 1993).

5.7.3 Matchings and Augmenting Paths

The following theorem gives necessary and sufficient conditions for the existence of a perfect matching in a bipartite graph.

Theorem 5.1 (Hall's Theorem)

A bipartite graph $G = (X, Y, E)$ with $|X| = |Y|$ has a perfect matching if and only if $\forall S \subseteq X$, $|N(S)| \geq |S|$, where $N(S) \subseteq Y$ is the set of vertices that are neighbors of some vertex in S.

Although Theorem 5.1 captures exactly the conditions under which a given bipartite graph has a perfect matching, it does not lead directly to an algorithm for finding maximum matchings. The following lemma shows how an augmenting path with respect to a given matching can be used to increase the size of a matching. An efficient algorithm that uses augmenting paths to construct a maximum matching incrementally is described later.

Lemma 5.1

Let P be the edges on an augmenting path $p = [v_1, \ldots, v_k]$ with respect to a matching M. Then, $M' = M \oplus P$ is a matching of cardinality $|M| + 1$.

Proof Since P is an augmenting path, both v_1 and v_k are free vertices in M. The number of free edges in P is one more than the number of matched edges. The symmetric difference operator replaces the matched edges of M in P by the free edges in P. Hence, the size of the resulting matching, $|M'|$, is one more than $|M|$.

The following theorem provides a necessary and sufficient condition for a given matching M to be a maximum matching.

Theorem 5.2

A matching M in a graph G is a maximum matching if and only if there is no augmenting path in G with respect to M.

Proof If there is an augmenting path with respect to M, then M cannot be a maximum matching, since by Lemma 5.1 there is a matching whose size is larger than that of M. To prove the converse, we show that if there is no augmenting path with respect to M, then M is a maximum matching. Suppose that there is a matching M' such that $|M'| > |M|$. Consider the set of edges $M \oplus M'$. These edges form a subgraph in G. Each vertex in this subgraph has degree at most two, since each node has at most one edge from each matching incident to it. Hence, each connected component of this subgraph is either a path or a simple cycle. For each cycle, the number of edges of M is the same as the number of edges of M'. Since $|M'| > |M|$, one of the paths must have more edges from M' than from M. This path is an augmenting path in G with respect to the matching M, contradicting the assumption that there were no augmenting paths with respect to M.

5.7.4 Bipartite Matching Algorithm

5.7.4.1 High-Level Description

The algorithm starts with the empty matching $M = \emptyset$ and augments the matching in phases. In each phase, an augmenting path with respect to the current matching M is found, and it is used to increase the size of the matching. An augmenting path, if one exists, can be found in $O(|E|)$ time, using a procedure similar to BFS described in Section 5.4.

The search for an augmenting path proceeds from the free vertices. At each step when a vertex in X is processed, all its unvisited neighbors are also searched. When a matched vertex in Y is considered, only its matched neighbor is searched. This search proceeds along a subgraph referred to as the *Hungarian tree*.

Initially, all free vertices in X are placed in a queue that holds vertices that are yet to be processed. The vertices are removed one by one from the queue and processed as follows. In turn, when vertex v is removed from the queue, the edges incident to it are scanned. If it has a neighbor in the vertex set Y that is free, then the search for an augmenting path is successful; procedure AUGMENT is called to update the matching, and the algorithm proceeds to its next phase. Otherwise, add the mates of all of the matched neighbors of v to the queue if they have never been added to the queue, and continue the search for an augmenting path. If the algorithm empties the queue without finding an augmenting path, its current matching is a maximum matching and it terminates.

The main data structure that the algorithm uses consists of the arrays *mate* and *free*. The array *mate* is used to represent the current matching. For a matched vertex $v \in G$, *mate*[v] denotes the matched neighbor of vertex v. For $v \in X$, *free*[v] is a vertex in Y that is adjacent to v and is free. If no such vertex exists, then *free*[v] = 0.

Bipartite Matching Algorithm. *Bipartite Matching* ($G = (X, Y, E)$):

```
for all vertices v in G do
    mate[v] ← 0.
end-for
found ← false.
while not found do
    Initialize.
    MakeEmptyQueue (Q).
    for all vertices x ∈ X do
        if mate[x] = 0 then
            Enqueue (Q, x).
            label[x] ← 0.
        endif
    end-for
    done ← false.
    while not done and not Empty (Q) do
        x ← Dequeue (Q).
        if free[x] ≠ 0 then
            Augment (x).
            done ← true.
        else
            for all edges (x, x') ∈ A do
                if label[x'] = 0 then
                    label[x'] ← x.
                    Enqueue (Q, x').
                end-if
            end-for
        end-if
        if Empty (Q) then
            found ← true.
        end-if
    end-while
end-while
end-proc
```

```
Initialize:
for all vertices x ∈ X do
    free[x] ← 0.
end-for
A ← Ø.
for all edges (x, y) ∈ E do
    if mate[y] = 0 then free[x] ← y
    else if mate[y] ≠ x then A ← A ∪ (x, mate[y]).
    end-if
end-for
end-proc

Augment(x):
if label[x] = 0 then
    mate[x] ← free[x].
    mate[free[x]] ← x
else
    free[label[x]] ← mate[x]
    mate[x] ← free[x]
    mate[free[x]] ← x
    Augment (label[x])
end-if
end-proc
```

5.7.4.2 Sample Execution

Figure 5.5 shows a sample execution of the matching algorithm. We start with a partial matching and show the structure of the resulting Hungarian tree. An augmenting path from vertex b to vertex u is found by the algorithm.

5.7.4.3 Analysis

If there are augmenting paths with respect to the current matching, the algorithm will find at least onc of them. Hence, when the algorithm terminates, the graph has no augmenting paths with respect to the current matching and the current matching is optimal. Each iteration of the main *while* loop of the algorithm runs in $O(|E|)$ time. The construction of the auxiliary graph A and computation of the array *free* also take $O(|E|)$ time. In each iteration, the size of the matching increases by one and thus there are, at most, $\min(|X|, |Y|)$ iterations of the *while* loop. Therefore, the algorithm solves the matching problem for bipartite graphs in time $O(\min(|X|, |Y|)|E|)$. Hopcroft and Karp (1973) showed how to improve the running time by finding a maximal set of shortest disjoint augmenting paths in a single phase in $O(|E|)$ time. They also proved that the algorithm runs in only $O\left(\sqrt{|V|}\right)$ phases.

5.7.5 Assignment Problem

We now introduce the assignment problem, which is that of finding a maximum-weight matching in a given bipartite graph in which edges are given nonnegative weights. There is no loss of generality in assuming that the graph is complete, since zero-weight edges may be added between pairs of vertices that are nonadjacent in the original graph without affecting the weight of a maximum-weight matching. The minimum-weight perfect matching can be reduced to the maximum-weight matching problem as follows: choose a constant M that is larger than the weight of any edge. Assign each edge a new weight of $w'(e) = M - w(e)$. Observe that maximum-weight matchings with the new weight function are minimum-weight perfect

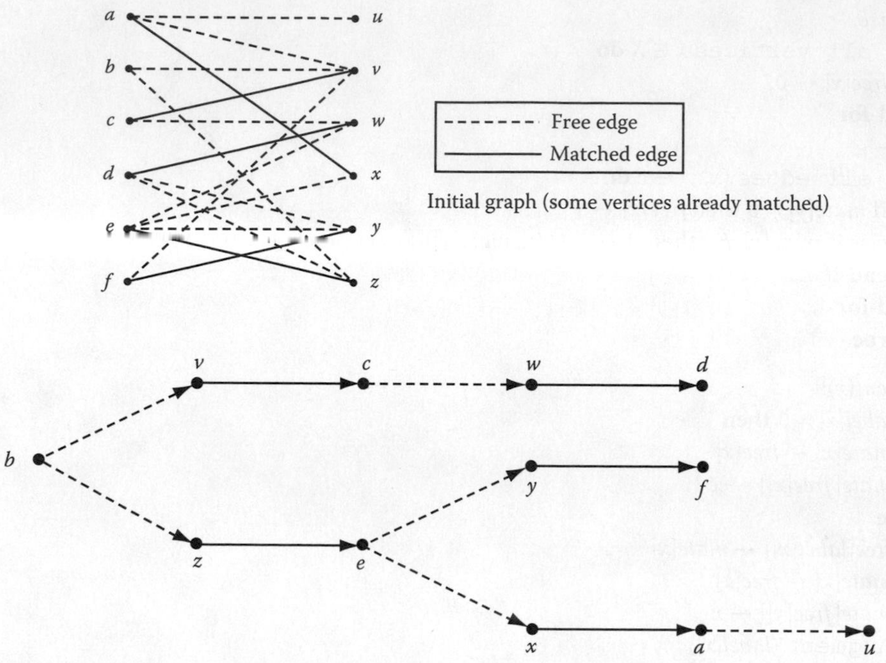

FIGURE 5.5 Sample execution of matching algorithm.

matchings with the original weights. We restrict our attention to the study of the maximum-weight matching problem for bipartite graphs. Similar techniques have been used to solve the maximum-weight matching problem in arbitrary graphs (see Lawler 1976, Papadimitriou and Steiglitz 1982).

The input is a complete bipartite graph $G = (X, Y, X \times Y)$ and each edge e has a nonnegative weight of $w(e)$. The following algorithm, known as the Hungarian method, was first given by Kuhn (1955). The method can be viewed as a primal–dual algorithm in the linear programming framework (Papadimitriou and Steiglitz 1982). No knowledge of linear programming is assumed here.

A *feasible vertex-labeling* ℓ is defined to be a mapping from the set of vertices in G to the real numbers such that for each edge (x_i, y_j) the following condition holds:

$$\ell(x_i) + \ell(y_j) \geq w(x_i, y_j)$$

The following can be verified to be a feasible vertex labeling. For each vertex $y_j \in Y$, set $\ell(y_j)$ to be 0; and for each vertex $x_i \in X$, set $\ell(x_i)$ to be the maximum weight of an edge incident to x_i:

$$\ell(y_j) = 0,$$

$$\ell(x_i) = \max_j w(x_i, y_j)$$

The *equality subgraph*, G_ℓ, is defined to be the subgraph of G, which includes all vertices of G but only those edges (x_i, y_j) that have weights such that

$$\ell(x_i) + \ell(y_j) = w(x_i, y_j)$$

The connection between equality subgraphs and maximum-weighted matchings is established by the following theorem.

Theorem 5.3

If the equality subgraph, G_ℓ, has a perfect matching, M^*, then M^* is a maximum-weight matching in G.

Proof Let M^* be a perfect matching in G_ℓ. By definition,

$$w(M^*) = \sum_{e \in M^*} w(e) = \sum_{v \in X \cup Y} \ell(v)$$

Let M be any perfect matching in G. Then,

$$w(M) = \sum_{e \in M} w(e) \leq \sum_{v \in X \cup Y} \ell(v) = w(M^*)$$

Hence, M^* is a maximum-weight perfect matching.

5.7.5.1 High-Level Description

Theorem 5.3 is the basis of the algorithm for finding a maximum-weight matching in a complete bipartite graph. The algorithm starts with a feasible labeling and then computes the equality subgraph and a maximum cardinality matching in this subgraph. If the matching found is perfect, by Theorem 5.3, the matching must be a maximum-weight matching and the algorithm returns it as its output. Otherwise, more edges need to be added to the equality subgraph by *revising* the vertex labels. The revision keeps edges from the current matching in the equality subgraph. After more edges are added to the equality subgraph, the algorithm grows the Hungarian trees further. Either the size of the matching increases because an augmenting path is found or a new vertex is added to the Hungarian tree. In the former case, the current phase terminates and the algorithm starts a new phase, because the matching size has increased. In the latter case, new nodes are added to the Hungarian tree. In n phases, the tree includes all of the nodes, and therefore there are at most n phases before the size of the matching increases.

It is now described in more detail how the labels are updated and which edges are added to the equality subgraph G_ℓ. Suppose M is a maximum matching in G_ℓ found by the algorithm. Hungarian trees are grown from all the free vertices in X. Vertices of X (including the free vertices) that are encountered in the search are added to a set S, and vertices of Y that are encountered in the search are added to a set T. Let $\bar{S} = X - S$ and $\bar{T} = Y - T$. Figure 5.6 illustrates the structure of the sets S and T. Matched edges are shown in bold; the other edges are the edges in G_ℓ. Observe that there are no edges in the equality subgraph from S to $\bar{T}$, although there may be edges from T to $\bar{S}$. Let us choose δ to be the smallest value such that some edge of $G - G_\ell$ enters the equality subgraph. The algorithm now revises the labels as follows. Decrease all of the labels of vertices in S by δ and increase the labels of the vertices in T by δ. This ensures that edges in the matching continue to stay in the equality subgraph. Edges in G (not in G_ℓ) that go from vertices in S to vertices in $\bar{T}$ are candidate edges to enter the equality subgraph, since one label is decreasing and the other is unchanged. Suppose this edge goes from $x \in S$ to $y \in \bar{T}$. If y is free, then an augmenting path has been found. On the other hand, if y is matched, the Hungarian tree is grown by moving y to T and its matched neighbor to S, and the process of revising labels continues.

5.7.6 B-Matching Problem

The b-matching problem is a generalization of the matching problem. In its simplest form, given an integer $b \geq 1$, the problem is to find a subgraph H of a given graph G such that the degree of each vertex is

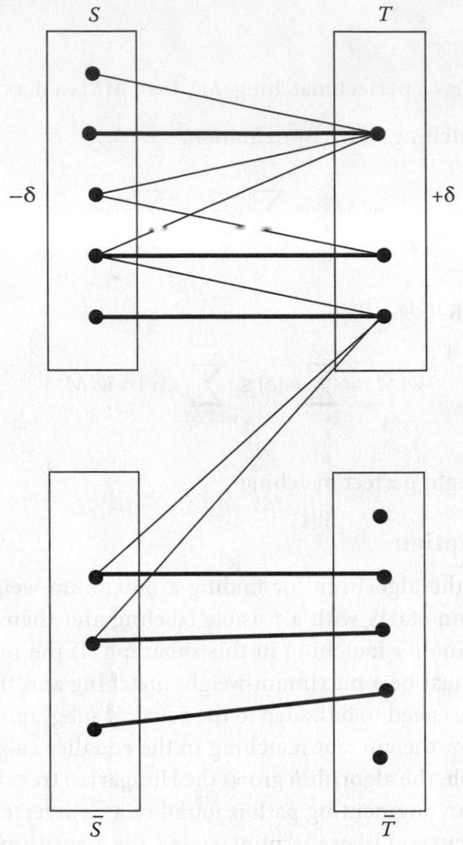

FIGURE 5.6 Sets S and T as maintained by the algorithm. Only edges in G_ℓ are shown.

exactly equal to b in H (such a subgraph is called a *b-regular subgraph*). The problem can also be formulated as an optimization problem by seeking a subgraph H with most edges, with the degree of each vertex to be at most b in H. Several generalizations are possible, including different degree bounds at each vertex, degrees of some vertices unspecified, and edges with weights. All variations of the b-matching problem can be solved using the techniques for solving the matching problem.

In this section, we show how the problem can be solved for the unweighted b-matching problem in which each vertex v is given a degree bound of $b[v]$, and the objective is to find a subgraph H in which the degree of each vertex v is exactly equal to $b[v]$. From the given graph G, construct a new graph G_b as follows. For each vertex $v \in G$, introduce $b[v]$ vertices in G_b, namely, $v_1, v_2, \ldots, v_{b[v]}$. For each edge $e = (u, v)$ in G, add two new vertices e_u and e_v to G_b, along with the edge (e_u, e_v). In addition, add edges between v_i and e_v, for $1 \le i \le b[v]$ (and between u_j and e_u, for $1 \le j \le b[u]$). We now show that there is a natural one-to-one correspondence between b-matchings in G and perfect matchings in G_b.

Given a b-matching H in G, we show how to construct a perfect matching in G_b. For each edge $(u, v) \in H$, match e_u to the next available u_j and e_v to the next available v_i. Since u is incident to exactly $b[u]$ edges in H, there are exactly enough nodes $u_1, u_2 \ldots u_{b[v]}$ in the previous step. For all edges $e = (u, v) \in G - H$, we match e_u and e_v. It can be verified that this yields a perfect matching in G_b.

We now show how to construct a b-matching in G, given a perfect matching in G_b. Let M be a perfect matching in G_b. For each edge $e = (u, v) \in G$, if $(e_u, e_b) \in M$, then do not include the edge e in the b-matching. Otherwise, e_u is matched to some u_j and e_v is matched to some v_i in M. In this case, we include e in our b-matching. Since there are exactly $b[u]$ vertices $u_1, u_2, \ldots u_{b[u]}$, each such vertex introduces an edge into the b-matching, and therefore the degree of u is exactly $b[u]$. Therefore, we get a b-matching in G.

5.7.7 Network Flows

A number of polynomial time flow algorithms have been developed over the past two decades. The reader is referred to Ahuja et al. (1993) for a detailed account of the historical development of the various flow methods. Cormen et al. (2001) review the preflow push method in detail; and to complement their coverage, an implementation of the blocking flow technique of Malhotra et al. (1978) is discussed here.

5.7.8 Network Flow Problem Definitions

First, the network flow problem and its basic terminology are defined.

Flow network: A flow network $G = (V, E)$ is a directed graph, with two specially marked nodes, namely, the source s and the sink t. There is a *capacity* function $c: E \mapsto R^+$ that maps edges to positive real numbers.

Max-flow problem: A flow function $f: E \mapsto R$ maps edges to real numbers. For an edge $e = (u, v)$, $f(e)$ refers to the flow on edge e, which is also called the net flow from vertex u to vertex v. This notation is extended to sets of vertices as follows: If X and Y are sets of vertices, then $f(X, Y)$ is defined to be $\sum_{x \in X} \sum_{y \in Y} f(x, y)$. A flow function is required to satisfy the following constraints:

- *Capacity constraint*. For all edges e, $f(e) \le c(e)$.
- *Skew symmetry constraint*. For an edge $e = (u, v)$, $f(u, v) = -f(v, u)$.
- *Flow conservation*. For all vertices $u \in V - \{s, t\}$, $\sum_{v \in V} f(u, v) = 0$.

The capacity constraint says that the total flow on an edge does not exceed its capacity. The skew symmetry condition says that the flow on an edge is the negative of the flow in the reverse direction. The flow conservation constraint says that the total net flow out of any vertex other than the source and sink is zero.

The *value* of the flow is defined as

$$|f| = \sum_{v \in V} f(s, v)$$

In other words, it is the net flow out of the source. In the *maximum-flow problem*, the objective is to find a flow function that satisfies the three constraints and also maximizes the total flow value $|f|$.

Remark 5.3

This formulation of the network flow problem is powerful enough to capture generalizations where there are many sources and sinks (single commodity flow) and where both vertices and edges have capacity constraints.

First, the notion of cuts is defined, and the max-flow min-cut theorem is introduced. Then, residual networks, layered networks, and the concept of blocking flows are introduced. Finally, an efficient algorithm for finding a blocking flow is described.

An *s–t cut* of the graph is a partitioning of the vertex set V into two sets S and $T = V - S$ such that $s \in S$ and $t \in T$. If f is a flow, then the net flow across the cut is defined as $f(S, T)$. The capacity of the cut is similarly defined as $c(S, T) = \sum_{x \in X} \sum_{y \in Y} c(x, y)$. The net flow across a cut may include negative net flows between vertices, but the capacity of the cut includes only nonnegative values, that is, only the capacities of edges from S to T.

Using the flow conservation principle, it can be shown that the net flow across an s–t cut is exactly the flow value $|f|$. By the capacity constraint, the flow across the cut cannot exceed the capacity of the cut. Thus, the value of the maximum flow is no greater than the capacity of a minimum s–t cut. The well-known *max-flow min-cut theorem* (Elias et al. 1956, Ford and Fulkerson 1962) proves that the two numbers are actually equal. In other words, if f^* is a maximum flow, then there is some cut $(X, \bar{X})$ such that $|f^*| = c(X, \bar{X})$. The reader is referred to Cormen et al. (2001) and Tarjan (1983) for further details.

The *residual capacity* of a flow f is defined to be a function on vertex pairs given by $c'(v, w) = c(v, w) - f(v, w)$. The residual capacity of an edge (v, w), $c'(v, w)$, is the number of additional units of flow that can be pushed from v to w without violating the capacity constraints. An edge e is *saturated* if $c(e) = f(e)$, that is, if its residual capacity, $c'(e)$, is zero. The residual graph $G_R(f)$ for a flow f is the graph with vertex set V, source and sink s and t, respectively, and those edges (v, w) for which $c'(v, w) > 0$.

An augmenting path for f is a path P from s to t in $G_R(f)$. The residual capacity of P, denoted by $c'(P)$, is the minimum value of $c'(v, w)$ over all edges (v, w) in the path P. The flow can be increased by $c'(P)$, by increasing the flow on each edge of P by this amount. Whenever $f(v, w)$ is changed, $f(w, v)$ is also correspondingly changed to maintain skew symmetry.

Most flow algorithms are based on the concept of augmenting paths pioneered by Ford and Fulkerson (1956). They start with an initial zero flow and augment the flow in stages. In each stage, a residual graph $G_R(f)$ with respect to the current flow function f is constructed and an augmenting path in $G_R(f)$ is found to increase the value of the flow. Flow is increased along this path until an edge in this path is saturated. The algorithms iteratively keep increasing the flow until there are no more augmenting paths in $G_R(f)$ and return the final flow f as their output.

The following lemma is fundamental in understanding the basic strategy behind these algorithms.

Lemma 5.2

Let f be any flow and f^* a maximum flow in G, and let $G_R(f)$ be the residual graph for f. The value of a maximum flow in $G_R(f)$ is $|f^*| - |f|$.

Proof Let f' be any flow in $G_R(f)$. Define $f + f'$ to be the flow defined by the flow function $f(v, w) + f'(v, w)$ for each edge (v, w). Observe that $f + f'$ is a feasible flow in G of value $|f| + |f'|$. Since f^* is the maximum flow possible in G, $|f'| \leq |f^*| - |f|$. Similarly define $f^* - f$ to be a flow in $G_R(f)$ defined by $f^*(v, w) - f(v, w)$ in each edge (v, w), and this is a feasible flow in $G_R(f)$ of value $|f^*| - |f|$, and it is a maximum flow in $G_R(f)$.

Blocking flow: A flow f is a *blocking flow* if every path in G from s to t contains a saturated edge.

It is important to note that a blocking flow is not necessarily a maximum flow. There may be augmenting paths that increase the flow on some edges and decrease the flow on other edges (by increasing the flow in the reverse direction).

Layered networks: Let $G_R(f)$ be the residual graph with respect to a flow f. The level of a vertex v is the length of a shortest path (using the least number of edges) from s to v in $G_R(f)$. The level graph L for f is the subgraph of $G_R(f)$ containing vertices reachable from s and only the edges (v, w) such that $\text{dist}(s, w) = 1 + \text{dist}(s, v)$. L contains all shortest-length augmenting paths and can be constructed in $O(|E|)$ time.

The maximum-flow algorithm proposed by Dinitz (1970) starts with the zero flow and iteratively increases the flow by augmenting it with a blocking flow in $G_R(f)$ until t is not reachable from s in $G_R(f)$.

At each step, the current flow is replaced by the sum of the current flow and the blocking flow. Since in each iteration the shortest distance from s to t in the residual graph increases, and the shortest path from s to t is at most $|V| - 1$, this gives an upper bound on the number of iterations of the algorithm.

An algorithm to find a blocking flow that runs in $O|V|^2$ time is described here, and this yields an $O(|V|^3)$ max-flow algorithm. There are a number of $O(|V|^2)$ blocking flow algorithms available (Karzanov 1974, Malhotra et al. 1978, Tarjan 1983), some of which are described in detail in Tarjan (1983).

5.7.9 Blocking Flows

Dinitz's algorithm to find a blocking flow runs in $O(|V||E|)$ time (Dinitz 1970). The main step is to find paths from the source to the sink and saturate them by pushing as much flow as possible on these paths. Every time the flow is increased by pushing more flow along an augmenting path, one of the edges on this path becomes saturated. It takes $O(|V|)$ time to compute the amount of flow that can be pushed on the path. Since there are $|E|$ edges, this yields an upper bound of $O(|V||E|)$ steps on the running time of the algorithm.

Malhotra–Kumar–Maheshwari Blocking Flow Algorithm. The algorithm has a current flow function f and its corresponding residual graph $G_R(f)$. Define for each node $v \in G_R(f)$, a quantity $tp[v]$ that specifies its maximum throughput, that is, either the sum of the capacities of the incoming arcs or the sum of the capacities of the outgoing arcs, whichever is smaller. $tp[v]$ represents the maximum flow that could pass through v in any feasible blocking flow in the residual graph. Vertices for which the throughput is zero are deleted from $G_R(f)$.

The algorithm selects a vertex u for which its throughput is a minimum among all vertices with nonzero throughput. It then greedily pushes a flow of $tp[u]$ from u toward t, level by level in the layered residual graph. This can be done by creating a queue, which initially contains u and which is assigned the task of pushing $tp[u]$ out of it. In each step, the vertex v at the front of the queue is removed and the arcs going out of v are scanned one at a time, and as much flow as possible is pushed out of them until v's allocated flow has been pushed out. For each arc (v, w) that the algorithm pushed flow through, it updates the residual capacity of the arc (v, w) and places w on a queue (if it is not already there) and increments the net incoming flow into w. Also, $tp[v]$ is reduced by the amount of flow that was sent through it now. The flow finally reaches t, and the algorithm never comes across a vertex that has incoming flow that exceeds its outgoing capacity since u was chosen as a vertex with the smallest throughput. The preceding idea is again repeated to pull a flow of $tp[u]$ from the source s to u. Combining the two steps yields a flow of $tp[u]$ from s to t in the residual network that goes through u. The flow f is augmented by this amount. Vertex u is deleted from the residual graph, along with any other vertices that have zero throughput.

This procedure is repeated until all vertices are deleted from the residual graph. The algorithm has a blocking flow at this stage since at least one vertex is saturated in every path from s to t. In the algorithm, whenever an edge is saturated, it may be deleted from the residual graph. Since the algorithm uses a greedy strategy to send flows, at most $O(|E|)$ time is spent when an edge is saturated. When finding flow paths to push $tp[u]$, there are at most n times, one each per vertex, when the algorithm pushes a flow that does not saturate the corresponding edge. After this step, u is deleted from the residual graph. Hence, in $O(|E| + |V|^2) = O(|V|^2)$ steps, the algorithm to compute blocking flows terminates.

Goldberg and Tarjan (1988) proposed a preflow push method that runs in $O(|V||E|\log|V|^2/|E|)$ time without explicitly finding a blocking flow at each step.

5.7.10 Applications of Network Flow

There are numerous applications of the maximum-flow algorithm in scheduling problems of various kinds. See Ahuja et al. (1993) for further details.

5.8 Tour and Traversal Problems

There are many applications for finding certain kinds of paths and tours in graphs. We briefly discuss some of the basic problems.

The *traveling salesman problem* (*TSP*) is that of finding a shortest tour that visits all of the vertices in a given graph with weights on the edges. It has received considerable attention in the literature (Lawler et al. 1985). The problem is known to be computationally intractable (NP-hard). Several heuristics are known to solve practical instances. Considerable progress has also been made for finding optimal solutions for graphs with a few thousand vertices.

One of the first graph-theoretic problems to be studied, the *Euler tour problem*, asks for the existence of a closed walk in a given connected graph that traverses each edge exactly once. Euler proved that such a closed walk exists if and only if each vertex has even degree (Gibbons 1985). Such a graph is known as an *Eulerian graph*. Given an Eulerian graph, an Euler tour in it can be computed using DFS in linear time. Given an edge-weighted graph, the *Chinese postman problem* is that of finding a shortest closed walk that traverses each edge at least once. Although the problem sounds very similar to the TSP problem, it can be solved optimally in polynomial time by reducing it to the matching problem (Ahuja et al. 1993).

Key Terms

Assignment problem: That of finding a perfect matching of maximum (or minimum) total weight.
Augmenting path: An alternating path that can be used to augment (increase) the size of a matching.
Biconnected graph: A graph that cannot be disconnected by the removal of any single vertex.
Bipartite graph: A graph in which the vertex set can be partitioned into two sets X and Y, such that each edge connects a node in X with a node in Y.
Blocking flow: A flow function in which any directed path from s to t contains a saturated edge.
Branching: A spanning tree in a rooted graph, such that the root has a path to each vertex.
Chinese postman problem: Asks for a minimum length tour that traverses each edge at least once.
Connected: A graph in which there is a path between each pair of vertices.
Cycle: A path in which the start and end vertices of the path are identical.
Degree: The number of edges incident to a vertex in a graph.
DFS forest: A rooted forest formed by DFS.
Directed acyclic graph: A directed graph with no cycles.
Eulerian graph: A graph that has an Euler tour.
Euler tour problem: Asks for a traversal of the edges that visits each edge exactly once.
Forest: An acyclic graph.
Leaves: Vertices of degree one in a tree.
Matching: A subset of edges that do not share a common vertex.
Minimum spanning tree: A spanning tree of minimum total weight.
Network flow: An assignment of flow values to the edges of a graph that satisfies flow conservation, skew symmetry, and capacity constraints.
Path: An ordered list of edges such that any two consecutive edges are incident to a common vertex.
Perfect matching: A matching in which every node is matched by an edge to another node.
Sparse graph: A graph in which $|E| \ll |V|^2$.
s–t cut: A partitioning of the vertex set into S and T such that $s \in S$ and $t \in T$.
Strongly connected: A directed graph in which there is a directed path in each direction between each pair of vertices.
Topological order: A linear ordering of the edges of a DAG such that every edge in the graph goes from left to right.

Traveling salesman problem: Asks for a minimum length tour of a graph that visits all of the vertices exactly once.

Tree: An acyclic graph with $|V| - 1$ edges.

Walk: An ordered sequence of edges (in which edges could repeat) such that any two consecutive edges are incident to a common vertex.

Acknowledgments

Samir Khuller's research is supported by National Science Foundation (NSF) Awards CCR-9820965 and CCR-0113192. Balaji Raghavachari's research is supported by the National Science Foundation under grant CCR-9820902.

Further Information

The area of graph algorithms continues to be a very active field of research. There are several journals and conferences that discuss advances in the field. Here, we name a partial list of some of the important meetings: *ACM Symposium on Theory of Computing, IEEE Conference on Foundations of Computer Science, ACM–SIAM Symposium on Discrete Algorithms, the International Colloquium on Automata, Languages and Programming, and the European Symposium on Algorithms.* There are many other regional algorithms/theory conferences that carry research papers on graph algorithms. The journals that carry articles on current research in graph algorithms are the *Journal of the ACM, SIAM Journal on Computing, SIAM Journal on Discrete Mathematics, Journal of Algorithms, Algorithmica, Journal of Computer and System Sciences, Information and Computation, Information Processing Letters,* and *Theoretical Computer Science.*

To find more details about some of the graph algorithms described in this chapter, we refer the reader to the books by Cormen et al. (2001), Even (1979), and Tarjan (1983). For network flows and matching, a more detailed survey regarding various approaches can be found in Tarjan (1983). Papadimitriou and Steiglitz (1982) discuss the solution of many combinatorial optimization problems using a primal–dual framework.

Current research on graph algorithms focuses on approximation algorithms (Hochbaum 1996), dynamic algorithms, and in the area of graph layout and drawing (DiBattista et al. 1994).

References

Ahuja, R.K., Magnanti, T., and Orlin, J. 1993. *Network Flows*. Prentice Hall, Upper Saddle River, NJ.

Bellman, R. 1958. On a routing problem. *Q. Appl. Math.*, 16(1): 87–90.

Boruvka, O. 1926. O jistem problemu minimalnim. *Praca Moravske Prirodovedecke Spolecnosti*, 3: 37–58 (in Czech).

Cormen, T.H., Leiserson, C.E., Rivest, R.L., and Stein, C. 2001. *Introduction to Algorithms,* 2nd edn., The MIT Press, Cambridge, MA.

DiBattista, G., Eades, P., Tamassia, R., and Tollis, I. 1994. Annotated bibliography on graph drawing algorithms. *Comput. Geom. Theory Appl.*, 4: 235–282.

Dijkstra, E.W. 1959. A note on two problems in connexion with graphs. *Numerische Mathematik*, 1: 269–271.

Dinitz, E.A. 1970. Algorithm for solution of a problem of maximum flow in a network with power estimation. *Soviet Math. Dokl.*, 11: 1277–1280.

Elias, P., Feinstein, A., and Shannon, C.E. 1956. Note on maximum flow through a network. *IRE Trans. Inf. Theory*, IT-2: 117–119.

Even, S. 1979. *Graph Algorithms*. Computer Science Press, Potomac, MD.

Ford, L.R. Jr. and Fulkerson, D.R. 1956. Maximal flow through a network. *Can. J. Math.*, 8: 399–404.

Ford, L.R. Jr. and Fulkerson, D.R. 1962. *Flows in Networks*. Princeton University Press, Princeton, NJ.

Fraenkel, A.S. 1970. Economic traversal of labyrinths. *Math. Mag.*, 43: 125–130.

Fredman, M. and Tarjan, R.E. 1985. Fibonacci heaps and their uses in improved network optimization algorithms. *J. ACM*, 34(3): 596–615.

Gabow, H.N., Galil, Z., Spencer, T., and Tarjan, R.E. 1986. Efficient algorithms for finding minimum spanning trees in undirected and directed graphs. *Combinatorica*, 6(2): 109–122.

Gibbons, A.M. 1985. *Algorithmic Graph Theory*. Cambridge University Press, New York.

Goldberg, A.V. and Tarjan, R.E. 1988. A new approach to the maximum-flow problem. *J. ACM*, 35: 921–940.

Hochbaum, D.S. Ed. 1996. *Approximation Algorithms for NP-Hard Problems*. PWS Publishing, Boston, MA.

Hopcroft, J.E. and Karp, R.M. 1973. An $n^{2.5}$ algorithm for maximum matching in bipartite graphs. *SIAM J. Comput.*, 2(4): 225–231.

Hopcroft, J.E. and Tarjan, R.E. 1973. Efficient algorithms for graph manipulation. *Commun. ACM*, 16: 372–378.

Jarnik, V. 1930. O jistem problemu minimalnim. *Praca Moravske Prirodovedecke Spolecnosti*, 6: 57–63 (in Czech).

Karzanov, A.V. 1974. Determining the maximal flow in a network by the method of preflows. *Soviet Math. Dokl.*, 15: 434–437.

Kruskal, J.B. 1956. On the shortest spanning subtree of a graph and the traveling salesman problem. *Proc. Am. Math. Soc.*, 7: 48–50.

Kuhn, H.W. 1955. The Hungarian method for the assignment problem. *Nav. Res. Logist. Q.*, 2: 83–98.

Lawler, E.L. 1976. *Combinatorial Optimization: Networks and Matroids*. Holt, Rinehart and Winston, New York.

Lawler, E.L., Lenstra, J.K., Rinnooy Kan, A.H.G., and Shmoys, D.B. 1985. *The Traveling Salesman Problem: A Guided Tour of Combinatorial Optimization*. Wiley, New York.

Lucas, E. 1882. *Recreations Mathematiques*. Gauthier-Villars et fils, Paris, France.

Malhotra, V.M., Kumar, M.P., and Maheshwari, S.N. 1978. An $O(|V|^3)$ algorithm for finding maximum flows in networks. *Inf. Process. Lett.*, 7: 277–278.

Micali, S. and Vazirani, V.V. 1980. An $O\left(\sqrt{|V|}\,|E|\right)$ algorithm for finding maximum matching in general graphs. In *Proceedings of the 21st Annual Symposium on Foundations of Computer Science*, New York, pp. 17–25.

Papadimitriou, C.H. and Steiglitz, K. 1982. *Combinatorial Optimization: Algorithms and Complexity*. Prentice Hall, Upper Saddle River, NJ.

Prim, R.C. 1957. Shortest connection networks and some generalizations. *Bell Syst. Tech. J.*, 36: 1389–1401.

Tarjan, R.E. 1972. Depth first search and linear graph algorithms. *SIAM J. Comput.*, 1: 146–160.

Tarjan, R.E. 1983. *Data Structures and Network Algorithms*. SIAM, Philadelphia.

6

Computational Geometry

Marc van Kreveld
Utrecht University

6.1 Introduction

Computational geometry is the branch of algorithms research that deals with problems that are geometric in nature. Its goal is to develop algorithms that are always correct and efficient. Theoretically, efficiency refers to the scaling behavior of an algorithm when the input size gets larger and larger. Furthermore, we are interested in worst-case analysis, where an upper bound on the scaling behavior is to be determined irrespective of the specific input values. The basic problem of sorting a set of values can be seen as a one-dimensional problem. Although computational geometry includes one-dimensional problems, its emphasis is on two-, three-, and higher-dimensional problems. In most cases, the dimension is constant.

Typical geometric objects involved in geometric computations are sets of points, line segments, lines or hyperplanes, sets of circles or spheres, sets of triangles or simplices, or simple polygons or polyhedra. Relationships between these objects include distances, intersections, and angles.

Geometric problems that require computation show up in many different application areas. In robot motion planning, a path must be determined that avoids walls and other obstacles. In automated cartography, text labels must be placed on a map next to the corresponding feature without overlapping other labels or other important map information. In data analysis, clusters of objects from a collection must be formed using similarity measures that involve distance in some space. To form a complex between two proteins, a good local geometric fit is required. The list of examples of such questions where geometric computations on data are necessary is endless.

6.1.1 Basic Algorithmic Problems

To get a quick feel of which problems belong to computational geometry, we list a number of these problems in this section; see also Figure 6.1.

Convex hull: Given a set P of n points in the plane or d-dimensional space, compute a representation of the boundary of the minimum convex set that contains these points.

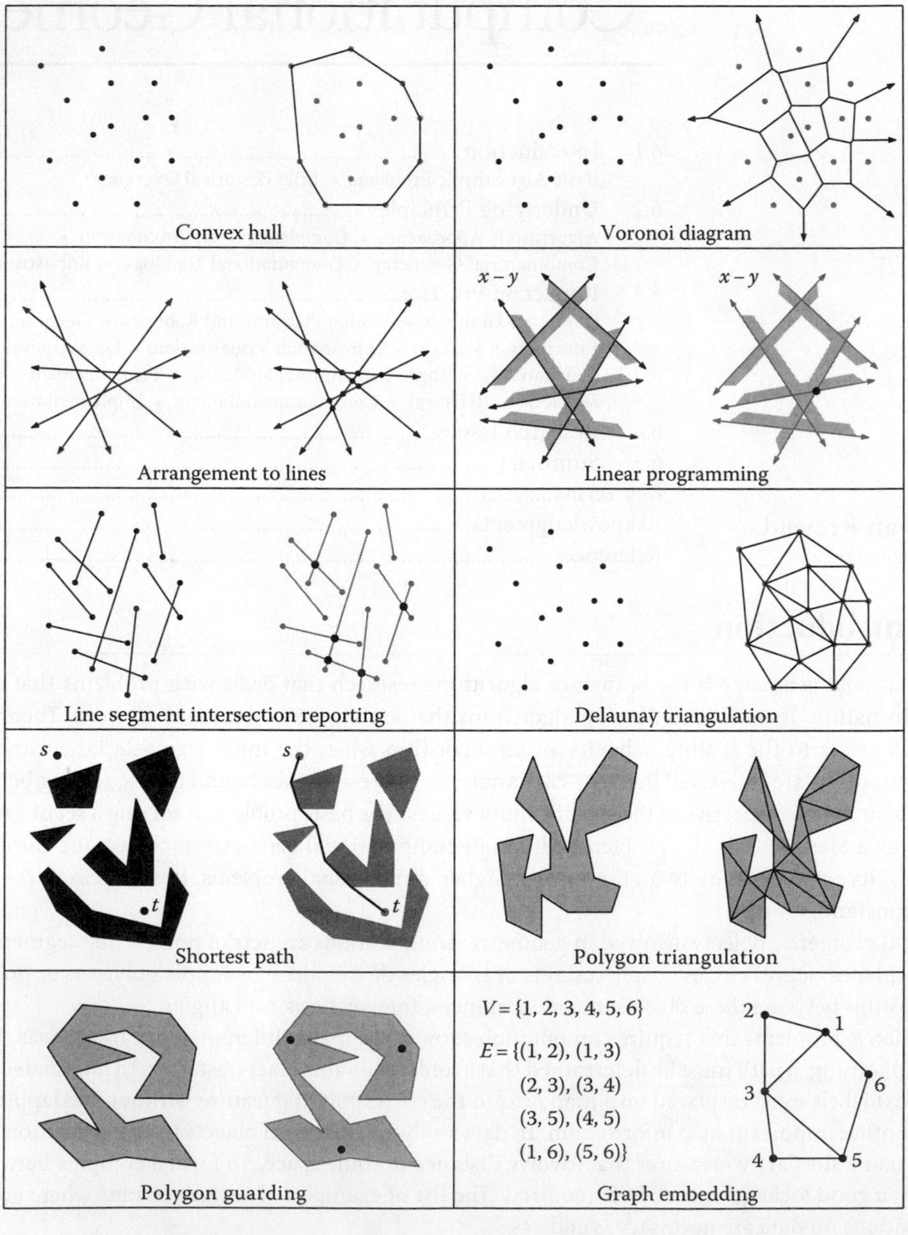

FIGURE 6.1 Ten basic problems in computational geometry. To the left the input is shown in black, and to the right the computed output is shown in black.

This problem can be solved in $O(n \log n)$ time when $d = 2$ [6,38,56,86] or $d = 3$ [35,38,86], and in $O(n^{\lfloor d/2 \rfloor})$ time when $d \geq 4$ [24].

Voronoi diagram: Given a set P of n points in the plane or d-dimensional space, compute a subdivision into cells so that any single cell represents exactly that part of the space where a single point is closest, among the points in P.

This problem can be solved in $O(n \log n)$ time when $d = 2$ [38,50,95], and in $O(n^{\lceil d/2 \rceil})$ time when $d \geq 3$ [22,35,94].

Arrangement of hyperplanes: Given a set H of n hyperplanes in d-dimensional space (lines in the plane, planes in three-dimensional space, etc.), compute a representation of the subdivision they induce.

This problem can be solved in $O(n^d)$ time [42,43].

Linear programming: Given a set H of n half-spaces in d-dimensional space and a linear function in the d coordinates, determine if the common intersection of the half-spaces of H is nonempty, and if so, determine a point that minimizes the linear function or report that arbitrarily low values can be realized.

This problem can be solved in $O(n)$ time [34,76,94] when d is a constant.

Intersection reporting: Given a set of n objects (line segments, rectangles, circles) in the plane, report every intersection point or every pair of intersecting objects.

This problem can be solved in $O(n \log n + k)$ time, where k is the number of pairs of intersecting objects [25,35,79].

Delaunay triangulation: Given a set P of n points in the plane or d-dimensional space, compute a maximal subdivision into triangles (simplices) whose vertices come from P such that no two triangles (simplices) intersect in their interior and the circumcircle of each triangle (circumsphere of each simplex) does not have any points of P in its interior.

This problem can be solved in $O(n \log n)$ time when $d = 2$ [38,50,95], and in $O(n^{\lceil d/2 \rceil})$ time when $d \geq 3$ [22,35,94].

Shortest path: Given a set P of disjoint polygons in the plane with n vertices in total and two points s and t, determine a shortest path between s and t that does not intersect the interior of any polygon of P or report that none exists.

This problem can be solved in $O(n \log n)$ time [63]. The three-dimensional version is NP-hard [19].

Polygon triangulation: Given a simple polygon P with n vertices, partition its interior into triangles using edges that connect pairs of vertices, lie inside the polygon, and do not mutually intersect.

This problem can be solved in $O(n)$ time [23], although a much simpler algorithm exists that is only slightly less efficient [93]. The three-dimensional problem is tetrahedralizing a simple polyhedron. Not all simple polyhedra can be tetrahedralized without using extra vertices, and as many as $\Omega(n^2)$ of these may be necessary for some polyhedra [21].

Polygon guarding: Given a simple polygon P with n vertices, determine a minimum-size set S of points such that any point in the interior of P can be connected to at least one point of P with a line segment that does not intersect the exterior of P.

No polynomial-time algorithm for this problem is known. The *Art Gallery Theorem* states that $\lfloor n/3 \rfloor$ point guards are always sufficient and sometimes necessary [32,82].

Graph embedding: Given a graph $G = (V, E)$, compute an embedding in the plane by assigning coordinates to each vertex, such that no two vertices have the same location, no edge has a vertex in its interior, and no pair of edges intersect in their interiors, or decide that this is not possible.

This problem can be solved in $O(n)$ time, where n is the size of the graph [12,39]. The vertices can be placed on an integer grid of size $2n - 4$ by $n - 2$.

The problems listed in the preceding text are all of the single computation type. Within computational geometry, geometric data structures have also been developed. These structures are either used to represent a subdivision of the plane or higher-dimensional space or to allow efficient query answering. A common data structure of the former type is the doubly-connected edge list [38,78]. It records incidences of edges with vertices, faces, and other edges in a pointer-based structure, and allows easy traversal of the boundary of a face or the edges incident to some vertex. In higher dimensions we can store facets of all dimensionalities and the incidences among them [17,40].

We list a few basic data structuring problems for efficient query answering. The most important properties of the structures are the amount of storage they require and the time needed to answer a query.

Planar point location: Given a planar subdivision S represented by n vertices and edges, where each face has a label, store S in a data structure so that the following type of query can be answered efficiently: for a query point p, report the label of the face that contains p.

This problem can be solved with a data structure that uses $O(n)$ storage and answers queries in $O(\log n)$ time [37,41,66,79,91].

Orthogonal range searching: Given a set P of n points in the plane, store them in a data structure so that the following type of query can be answered efficiently: for an axis-aligned query rectangle, report all points of P that lie inside it.

This problem can be solved with a kd-tree, which requires $O(n)$ storage and can answer queries in $O(\sqrt{n}+k)$ time, where k is the number of points reported [13]. The range tree requires $O(n \log n)$ storage and answers queries in $O(\log n + k)$ time [72,109].

The d-dimensional version of the problem is solved by a kd-tree with $O(n)$ storage and query time $O(n^{1-\frac{1}{d}}+k)$, and by a range tree with $O(n \log^{d-1} n)$ storage and $O(\log^{d-1} n + k)$ query time.

Simplex range searching: Given a set P of n points in the plane or d-dimensional space, store them in a data structure so that the following type of query can be answered efficiently: for a query triangle or simplex, report all points of P that lie inside it.

This problem can be solved with $O(n)$ storage and query time $O(\sqrt{n}+k)$ [73]. Alternatively, it can be solved with $O(n^d)$ storage and query time $O(\log^{d+1} n + k)$ [73].

Algorithms for many of the problems listed in the preceding text and their running time analyses can be found in textbooks in the area of computational geometry [38,83,87]. Many more results and references can be found in the *Handbook of Discrete and Computational Geometry* [54].

6.1.2 Brief Historical Overview

There are examples of geometric algorithms developed before the 1970s, but it was in the 1970s when most of the basic geometric problems were solved in an algorithmically efficient manner. This is true for many of the problems just listed, although some, like the shortest path problem, defied an efficient solution until the 1990s. In 1985 the first textbook on computational geometry was published, and that same year the annual *ACM Symposium on Computational Geometry* was held for the first time. It was realized early that implementing geometric algorithms is highly nontrivial, since robust computation tends to be much harder than for graph algorithms, for example, or for one-dimensional problems. Another early realization was that efficiency analyses of geometric algorithms often depend on combinatorial results in geometry. For this reason, combinatorial geometry has always been considered an integral part of computational geometry.

While the range of geometric problems under consideration continues to expand to this day, much effort has also been directed to putting the theory to use in the application areas. Researchers from computational geometry have contributed to areas like computer graphics, motion planning, pattern recognition, and geographic information systems (GISs) for a long time now. Involvement in three-dimensional reconstruction and computational biology are of a more recent date, where the recent trend toward computational topology has played an important role.

6.2 Underlying Principles

Computational geometry revolves around approaches to develop geometric algorithms, analysis of these algorithms in computational models, optimization of geometric problems, and applications and software development. In this section, we treat the first three aspects; applications and software development follow in the next section.

6.2.1 Algorithmic Approaches

Several algorithmic approaches used in computational geometry are also standard approaches in other algorithmic areas. For example, incremental algorithms, divide-and-conquer algorithms, brute-force algorithms, greedy algorithms, randomized algorithms, and dynamic programming are all used to solve geometric problems.

6.2.1.1 Plane-Sweep Algorithms

One very important and typically geometric algorithmic approach is plane sweep. It is an incremental approach, but one with a specific structure that we explain next.

A plane-sweep algorithm takes a set of objects in the plane as its input. The idea is to use an imaginary line—the sweep line—that moves over all objects in the plane and computes whatever is necessary when it reaches special situations. For example, the sweep line could be a horizontal line that starts above all objects and moves downward until it is below all objects. The sweep line stops only at certain positions (y-coordinates) where something of interest happens. These positions are called *events*. At the same time, the algorithm must maintain certain essential information to be able to generate the desired output. This information changes with the position of the sweep line and is called the *status* (Figure 6.2).

In the example of computing the intersecting pairs among a set of line segments in the plane, the events occur when the sweep line starts to intersect a line segment (at the y-coordinate of its upper endpoint), when the sweep line stops intersecting a line segment (at the y-coordinate of its lower endpoint), and when two line segments intersect (at the y-coordinate of the intersection point). In a plane-sweep algorithm the sweep line moves only in one direction, so the events will occur in the order of decreasing y-coordinate. Note that the events coming from line segment endpoints can be obtained easily from the input, but the events coming from intersection points are not known yet. In fact, the whole purpose of the algorithm is to determine these intersection points (and report the intersecting pairs).

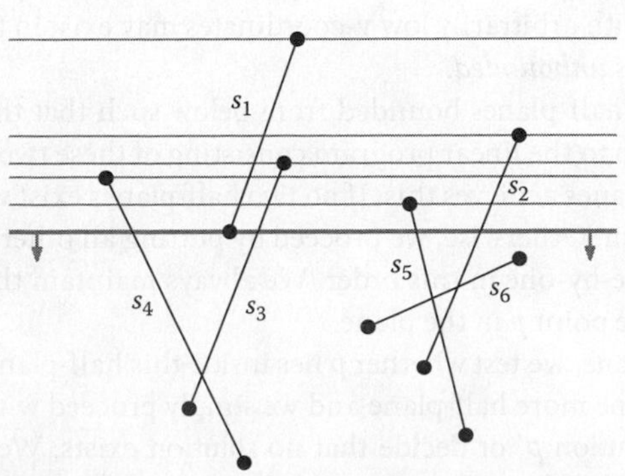

FIGURE 6.2 Plane-sweep algorithm. Positions where the sweep line has stopped are shown in gray. If the current position is the gray, bold line, then the status is $\langle s_4, s_1, s_3, s_5, s_2 \rangle$, and s_1 will be deleted at this event. Also, s_4 and s_3 become horizontally adjacent; their intersection is detected and inserted in the event structure.

The status is the situation at the current position of the sweep line. For our example, it is the set of line segments that intersect the sweep line, ordered from left to right. Observe that the status changes exactly at the events that we just listed.

The events and the status are each stored in a suitable data structure. In our example we use balanced binary search trees. The event structure stores the (known) events by y-coordinate in the tree. The status structure stores the intersected line segments by x-coordinate (valid for the current position of the sweep line) in the tree. Initially, the status structure is empty because the sweep line starts above all line segments. The event structure is initialized with all endpoints of the line segments.

The core of a plane-sweep algorithm is the handling of the events. This is where the algorithm actually does its work. The imaginary sweep line jumps from event to event and handles each one by updating the status, detecting new events that were not known yet, and generating output.

We must detect intersection point events before the sweep line reaches them. As soon as the algorithm finds out that two line segments will intersect later, the event can be inserted in the event structure. The basic idea that makes the algorithm work is the following: before two line segments can intersect, they must become horizontally adjacent first. Two line segments can become horizontally adjacent only at other, higher events (recall that the sweep line goes from top to bottom). So during the handling of any event, we also check if there are new horizontal adjacencies caused by that event, if the involved line segments intersect below the sweep line, and if so, insert the intersection point in the event structure.

The scheme described in the preceding text solves the line segment intersection reporting problem by a sequence of binary search tree operations and geometric tests. Any event is handled by a small, constant number of binary search tree operations. If the input consists of n line segments and there are k intersecting pairs that will be reported, then the algorithm runs in $O((n + k) \log n)$ time, because there are $O(n + k)$ events [14]. Unless there are many intersecting pairs, this compares favorably with the brute-force approach of testing each pair for intersection (taking $O(n^2)$ time).

Plane-sweep algorithms have been developed for computing Voronoi diagrams, Delaunay triangulations, the area of the union of a set of axis-parallel rectangles, as a step in simple polygon triangulation, and many other problems. The three-dimensional version, space sweep with a sweep plane, also exists.

6.2.1.2 Randomized Incremental Construction

Another well-known technique is *randomized incremental construction*. We illustrate this technique by two-dimensional linear programming: Given a set of n half-planes in the plane, determine a point in the common intersection of these half-planes with the lowest y-coordinate, if such a point exists. There are two reasons why such a point would not exist. Firstly, the common intersection of the half-planes may be empty. Secondly, points with arbitrarily low y-coordinates may exist in the common intersection: we say that the linear program is *unbounded*.

To initialize, we find two half-planes bounded from below such that the intersection point of their bounding lines is the solution to the linear program consisting of these two half-planes only. A straightforward scan over the half-planes achieves this. If no two half-planes exist with this property, there is no solution to the linear program. Otherwise, we proceed by putting all other $n - 2$ half-planes in random order and handling them one-by-one in this order. We always maintain the solution to the half-planes handled so far, which is some point p in the plane.

To handle the next half-plane, we test whether p lies inside this half-plane. If so, then p is the solution to the linear program with one more half-plane and we simply proceed with the next half-plane. If not, then we must find a new solution p' or decide that no solution exists. We can argue that if a solution exists, then it must lie on the line bounding the half-plane that we are handling. Hence, we can search for p' on this line, which is a one-dimensional problem that can be solved by considering all previously handled half-planes once more.

The algorithm outlined earlier takes $O(n)$ expected time in the worst case. "Worst case" refers to the fact that the time holds for any set of half-planes. "Expected" refers to the fact that the time bound depends on the random choices made during the algorithm execution. If these choices are unlucky, then the algorithm may take quadratic time, but this is very unlikely.

The argument to prove the running time bound is as follows. We observe that if the next half-plane contains p, then this half-plane is handled in $O(1)$ time, but if it does not, then this half-plane is handled in time proportional to the number of half-planes that have been handled before. Suppose we are considering the ith half-plane. We ask ourselves what the probability is that it does not contain point p. This probability is the same as the following: after adding the ith half-plane we may have a solution p', what is the probability that p' is no longer the solution when we remove a random half-plane from the $i-2$ half-planes (the two initial half-planes were not put in random order so we cannot remove them). This probability is at most $2/(i-2)$, which can be shown by studying a few geometric situations. If the probability of the expensive case, requiring us to do $O(i)$ work, is at most $2/(i-2)$, then the expected time for handling the ith half-plane is $O(1)$. In short, every one of the $n-2$ half-planes that is handled requires $O(1)$ expected time, so $O(n)$ time is expected to be used overall. This argument is called *backward analysis* [94].

Other geometric problems that can be solved efficiently with randomized incremental construction include computing the smallest enclosing disk, Voronoi diagram, Delaunay triangulation, convex hull, constructing a planar point location structure, deciding on red–blue point set separability, and many more [54,79,107].

6.2.1.3 Using Data Structures

Many computational-geometry problems can be solved by using data structures and querying them. As an example, we study the problem of computing the Hausdorff distance between a set R of n red points and a set B of n blue points in the plane.

The *Hausdorff distance* $H(R, B)$ between sets R and B is defined as follows:

$$H(R,B) = \max\{\max_{r\in R}(\min_{b\in B} d(r,b)), \max_{b\in B}(\min_{r\in R} d(b,r))\}.$$

Intuitively, it is the maximum among all closest point distances from one set to the other or vice versa. The Hausdorff distance is used as a similarity measure between two point sets or simple polygons.

An algorithm to compute the Hausdorff distance between R and B is easy to describe, using the results given before. We first build the Voronoi diagram of R and preprocess it for efficient planar point location. Then we query with each point of B to find the closest point of R, and keep the maximum of these distances. Then we do the same but with the roles of R and B reversed. The overall maximum is the Hausdorff distance.

It takes $O(n \log n)$ time to build the Voronoi diagram of a set of n points. The Voronoi diagram consists of $O(n)$ line segments, and preprocessing it for planar point location takes $O(n \log n)$ time as well. We query with each point of the other set, taking n times $O(\log n)$ time. So this is $O(n \log n)$ total query time as well. Hence, we can compute the Hausdorff distance of two sets of n points in the plane in $O(n \log n)$ time.

6.2.1.4 Dynamic Programming

Dynamic programming is an important algorithmic technique for optimization problems. It can often be used if the problem has the *optimal substructure property:* an optimal solution of any subproblem is composed of optimal solutions to smaller subproblems. We discuss the problem of triangulating a simple polygon with minimum total edge length [67].

Let P be a simple polygon with vertices $v_0, \ldots, v_{n-1}$ listed clockwise. We make a table T of size $n \times n$ with an entry for each ordered pair (v_i, v_j) of vertices. This entry will store (i) whether $\overline{v_i v_j}$ is interior to P and does not intersect any of its edges. If the answer is yes we mark the pair VALID and we also store

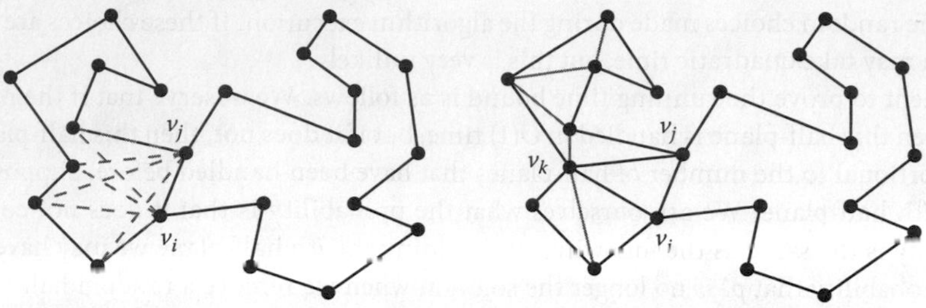

FIGURE 6.3 To compute L_{ij}, there are four valid choices for v_k. The optimal choice of v_k and the optimal triangulation to the left of $\overline{v_i v_j}$ are shown to the right.

(ii) how many vertices N_{ij}, the subpolygon of P bounded by $\overline{v_i v_j}$ and to the left of $\overrightarrow{v_i v_j}$, has, and (iii) the minimum total edge length L_{ij} required to triangulate this subpolygon, excluding $\overline{v_i v_j}$.

For the pairs (v_i, v_j) where $j = i + 1 \pmod n$ we also mark them VALID with $N_{ij} = 2$ and $L_{ij} = 0$. For the pairs (v_j, v_i) where $j = i + 1 \pmod n$ we also mark them VALID with $N_{ji} = n$ and initially we set $L_{ji} = +\infty$. At the end of the algorithm, all such L_{ji} will contain the total edge length of the optimal triangulation.

The first and second pieces of information can be precomputed easily in $O(n^3)$ time overall, for all $O(n^2)$ pairs. The third piece is initialized to $+\infty$. Next we sort the ordered pairs by the value of (ii) in increasing order, ties broken arbitrarily. We treat the ones with $N_{ij} > 2$ only.

We iterate over all pairs of edges in the sorted order. Suppose we are treating a pair (v_i, v_j), see Figure 6.3.

$$L_{ij} = \min_{1 \le k \le n}\{L_{ik} + L_{kj} + \| \overline{v_i v_k} \| + \| \overline{v_k v_j} \| \mid (v_i, v_k) \text{ and } (v_k, v_j) \text{ are valid and } N_{ik}, N_{kj} < N_{ij}\}.$$

The recurrence above defines the optimal substructure that polygon triangulation with minimum total edge length has. If (v_i, v_j) is valid, then there must be a triangle in the triangulation to the left of $\overrightarrow{v_i v_j}$. The recurrence optimizes over all choices of such triangles and selects the one that yields a minimum total edge length. If this triangle is $\Delta v_i v_j v_k$, then the optimal triangulation of the polygon left of $\overline{v_i v_j}$ always uses optimal triangulations of the polygons left of $\overrightarrow{v_i v_k}$ and $\overrightarrow{v_k v_j}$, and these are polygons with fewer vertices.

The algorithm essentially fills in the L-values of the table T in the order of increasing N_{ij}. To determine L_{ij} we simply read for each k the information in the entries for pairs (v_i, v_k) and (v_k, v_j) in table T. The optimal triangulation length can be found in any L_{ji} where $j = i + 1 \pmod n$.

Filling one entry takes $O(n)$ time, and hence the whole algorithm takes $O(n^3)$ time.

Other applications of dynamic programming in geometric algorithms include optimization or approximation for label placement, packing and covering, simplification, and matching.

6.2.1.5 Other Algorithmic Techniques in Computational Geometry

There are several other techniques that can be employed to develop algorithms for geometric problems. These include random sampling, prune-and-search, divide-and-conquer, and parametric search. Other techniques of importance are geometric transformations to relate problems to other problems. These include geometric duality, inversion, and Plücker coordinates. For geometric data structures, the technique of fractional cascading can help to make queries more efficient [26,27], while the topics of dynamization [30,31,84] and kinetic data structures [60,101] are also important research areas.

6.2.2 Complexity

6.2.2.1 Machine Model

The standard model of computation assumed in computational geometry is the *real RAM*. In the real RAM model, a real number can be stored in $O(1)$ storage space, and any basic operation or analytic function can be applied to real numbers in $O(1)$ time. Obviously, we can evaluate a polynomial for any real value in $O(1)$ time only if the polynomial has constant degree.

Since algorithms that handle large amounts of data are typically slowed down by disk accesses, the I/O complexity model may sometimes be more appropriate [3,106].

6.2.2.2 Hardness and Reductions

Hardness of a computational problem refers to the (im)possibility of solving that problem efficiently or solving it at all. Hardness results may be lower bounds on the time complexity to solve a problem. These can be proved directly in a model of computation or obtained using a reduction from another problem for which a lower bound is already known. It is essential to specify the model of computation in which a lower bound holds. Many models do not allow rounding of reals to the nearest integer, for example, and algorithms that make use of rounding can be more efficient than what a lower bound claims is possible (in a model without rounding).

Lower bounds on the time complexity of computational problems can be proved using *algebraic decision trees* [88,89]. Branches of such a tree represent comparisons that an algorithm might make based on the input. The longest path in a decision tree corresponds to the worst-case running time of an algorithm. Therefore, a lower bound on the depth of a decision tree for any algorithm that solves a problem is a lower bound for that problem in the decision tree model. Algebraic decision trees give an $\Omega(n \log n)$-time lower bound for the very basic one-dimensional problem of deciding whether a set of real numbers has two elements that are the same.

A two-dimensional problem that has an $\Omega(n^2)$-time lower bound is the following [48]: Given a set of n points in the plane, are there three of them that lie on a line? In higher dimensions, the problem of deciding whether a set of n points contains a subset of $d + 1$ points that lie on a common hyperplane has a lower bound of $\Omega(n^d)$ time.

Another type of hardness is referred to as *numerical nonrobustness*. Let P be a set of n points in the plane. To solve the facility location problem of computing a point that minimizes the sum of distances to the points of P, we must minimize an expression that is the sum of n terms with square roots. This is not possible in the usual model of computation, and hence we can only approximate the solution [10,16]. If we were interested in minimizing the sum of *squared* distances, then it would have been easy to compute the optimal location: its coordinates are the mean x and mean y values of the input point set P.

A number of geometric optimization problems are *NP-hard*. Some well-known examples are Euclidean traveling salesperson in the plane [53], Euclidean minimum Steiner tree in the plane [53], minimum-weight triangulation of a planar point set [80], and shortest path in three-dimensional space amid tetrahedra [19]. There are many more examples. Proofs of NP-hardness may use reductions from any of the known NP-hard problems. It turns out that PLANAR 3-SAT is one of the more useful problems to reduce from. PLANAR 3-SAT is a special case of the satisfiability problem where a formula has three literals per clause, and the bipartite graph that has a node for every variable and for every clause, and an edge between them if the variable occurs in the clause, is planar. See Figure 6.4.

Reductions from other problems establish that a geometric problem is at least as hard. For example, the sorting problem can be reduced to planar convex hull computation, and hence any algorithm must take $\Omega(n \log n)$ time to compute a planar convex hull in the algebraic decision tree model [110]. The 3SUM-problem is the problem of deciding, for a given set of n integers, whether it contains three elements that sum up to zero. No algorithm is known for 3SUM that is more efficient than quadratic time. It turns out that many geometric problems can be used to solve 3SUM—in other words, a reduction from 3SUM

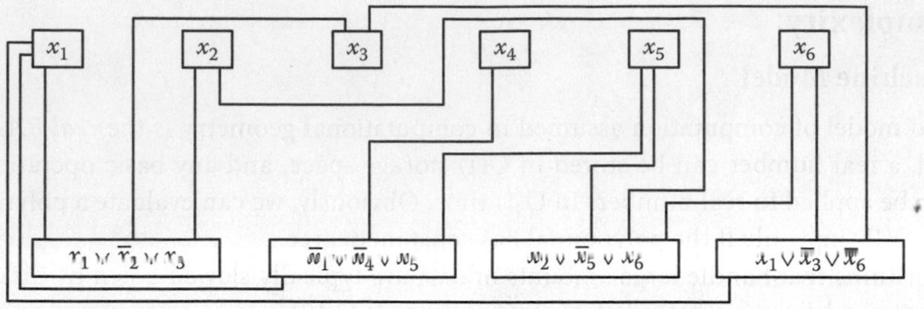

FIGURE 6.4 Planar layout of the 3-SAT expression $(x_1 \vee \bar{x}_2 \vee x_3) \wedge (\bar{x}_1 \vee \bar{x}_4 \vee x_5) \wedge (x_2 \vee \bar{x}_5 \vee x_6) \wedge (x_1 \vee \bar{x}_3 \vee \bar{x}_6)$.

to such geometric problems exists—and therefore these geometric problems are unlikely to be solvable more efficiently than in quadratic time [52]. Examples of such problems include deciding whether a planar point set contains three points that are collinear, deciding whether a line segment can be moved with translations and rotations to a given position without colliding with obstacles, and deciding whether a set of triangles in three-dimensional space occludes another triangle for a given viewpoint [52].

6.2.2.3 Realistic Input Models

It is possible that the union of a set of n triangles in the plane has descriptive complexity $\Theta(n^2)$; see Figure 6.5. Therefore, any algorithm that computes the union of n triangles must take at least quadratic time because it may have to report a shape of quadratic complexity. Now consider the problem of deciding whether a set of n triangles in the plane cover the unit square $[0:1] \times [0:1]$ completely. This problem has constant output size (yes or no), but still all known algorithms take quadratic time or more. The problem is as difficult as 3SUM [52].

In practice, however, unions of n triangles hardly ever have quadratic complexity. It is known that if there is some constant $\alpha > 0$, and all triangles have their three angles at least α, then the union complexity cannot be quadratic but always is $O(n \log \log n)$ [75]. Such triangles are called *fat*. The problem of deciding whether n fat triangles cover the unit square can be solved much faster than quadratic time. We have made an assumption on the input that (possibly) corresponds to real-world situations and stated a worst-case running time that is valid if the assumption holds. Such an assumption is called a *realistic input model.*

Other realistic input models may refer to relative sizes of the objects in the input. We could assume, for example, that a constant $c > 0$ exists so that the largest diameter of an input object is no more than c times the smallest diameter of an input object. Yet another realistic input model is the *bounded spread* of a point set: the distance between the closest two points is at least $f(n)$ times the diameter of the point set, where $f(n)$ is some function of n. With a packing argument, one can show that any d-dimensional point set has spread $\Omega(n^{1/d})$, so we cannot assume $f(n)$ to be constant.

Realistic input models are important to explain the complexity of structures or running times of problems in practice, since it is often different from the worst-case situation. For example, it is known that a three-dimensional point set may have a Delaunay tetrahedrilization of size $\Theta(n^2)$, although

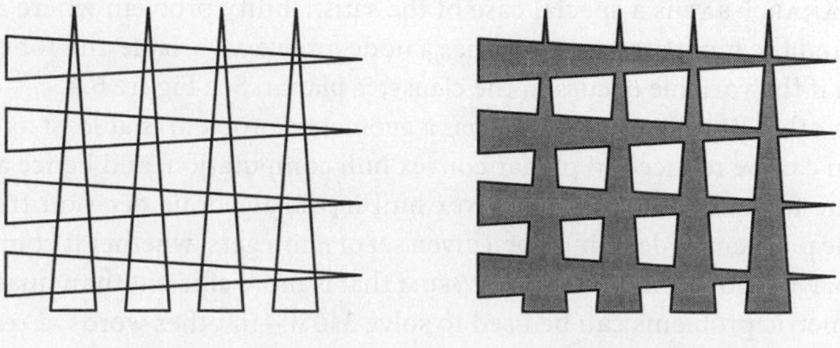

FIGURE 6.5 The union of n triangles can have $\Omega(n^2)$ vertices and edges in its boundary.

three-dimensional point sets that occur in practice often have a Delaunay tetrahedrilization of linear size or slightly more. It has been shown that for three-dimensional point sets with spread $O(n^{1/3})$, the Delaunay tetrahedrilization has size $O(n)$ in the worst case [46,47].

6.2.3 Approximation

Geometric algorithms where optimization of a measure appears inefficient may be tackled using approximation algorithms. This is the case for NP-hard problems, but approximation algorithms have also been developed to improve on quadratic running times, or even to reduce $O(n \log n)$ time bounds to linear.

An algorithm for a minimization problem A is a *c-approximation algorithm* for some $c > 1$ if the algorithm outputs a solution for any instance I whose value is at most c times larger than the minimum possible for instance I. For a maximization problem a *c-approximation algorithm* always gives a solution that has value at least the maximum divided by c. Approximation algorithms occur in all algorithms areas, but for geometric problems the proofs of approximation typically use geometric arguments involving distances or areas (packing arguments).

A geometric setting often makes a problem easier to approximate. For example, consider traveling salesperson on a graph and on a set of points in the plane. Both versions are NP-hard. Traveling salesperson on a graph cannot be approximated within a constant factor in polynomial time (unless $P = NP$), but for Euclidean traveling salesperson a simple $O(n \log n)$-time 2-approximation algorithm exists (polynomial-time approximation schemes exist as well [7,77]).

Similarly, when a graph is induced by a geometric configuration, approximation is often simpler than on general graphs. The maximum independent set problem cannot be approximated within a constant factor, while for a graph that is the intersection graph of a set of disks in the plane, a simple 5-approximation algorithm exists. The algorithm selects the smallest disk D and puts it into the independent set, then removes all disks that intersect D and D itself from consideration, and then iterates until no disks remain. The reason why this gives a 5-approximation is the following. The smallest disk D can intersect many disks of the set, but it can intersect no more than five disks that mutually do not intersect each other because these are at least as large as D, and the *kissing number* of disks implies this fact. So we may have chosen one disk in the independent set while five were possible, but it cannot be worse. The kissing number argument is an example of a packing argument.

Approximation can also be used to deal with numerical nonrobustness. For example, the facility location problem of computing a point that minimizes the distance to the input points can be approximated well: for any $\varepsilon > 0$, one can compute a point in $O(n)$ time whose sum of distances to the input points is at most $1 + \varepsilon$ times as large as the minimum possible [16]. The idea of this solution is to replace the Euclidean distance by a different distance function that does not have algebraic hardness, and in which all distances are within a factor $1 + \varepsilon$ of the Euclidean distance (we use a polygonal convex distance function).

When approximate solutions are allowed to certain queries, data structuring problems may have much more efficient solutions. Two examples are approximate range searching and approximate nearest neighbor searching. In approximate range searching, points that lie close to the boundary of the range may or may not be reported, where "close" is related to the size of the range [8]. An approximate nearest neighbor searching data structure may return a point that is a factor $1 + \varepsilon$ further than the real nearest neighbor [9].

6.2.4 Combinatorial Geometry

Combinatorial bounds relating to geometric situations can help to prove efficiency bounds on geometric algorithms. For this reason, many combinatorial geometric results have been published in the area of computational geometry. We discuss some of the main results.

The number of faces of all dimensions in an arrangement of n hyperplanes in d-dimensional space is $\Theta(n^d)$ in the worst case. The total complexity of all faces incident to a single hyperplane in an arrangement is $O(n^{d-1})$, which is the most important result upon which the optimal algorithm to construct an arrangement relies [43].

The convex hull of a set of n points in d-dimensional space obviously has at most n vertices, but the total number of faces of all dimensions is $O(n^{\lfloor d/2 \rfloor})$. This result is known as the *Upper Bound Theorem*.

The maximum complexity of a single cell in an arrangement of n line segments in the plane is $\Theta(n\alpha(n))$ in the worst case, where $\alpha(n)$ is the extremely slowly growing functional inverse of the Ackermann function [59]. The upper bound proof makes use of *Davenport–Schinzel sequences*: an (n, s)-Davenport–Schinzel sequence is a sequence of symbols from an alphabet of size n such that for any two symbols a, b, there is no subsequence of the form …, a, …, b, …, a, …, b, … (with symbols a and b alternating; in total $s + 2$ symbols a or b), and no two equal symbols are adjacent. The concept is purely combinatorial, but one can show that the total complexity of a face in an arrangement of line segments is bounded by a function linear in the maximum length of an $(n, 3)$-Davenport–Schinzel sequence [97]. Since such a sequence has length $O(n\alpha(n))$, the upper bound on the complexity of a face follows. The matching lower bound is obtained by a clever construction [108].

The maximum number of ways to divide a set of n points in the plane by a line into a set of k points and a set of $n - k$ points is $\Omega(n \log k)$ and $O(nk^{1/3})$ in the worst case; the true bound is unknown (a slightly better lower bound, $\Omega(n2^{c\sqrt{\log k}})$, is known [102]). This is known as the *k-set problem*. The maximum number of ways to separate at least one and at most k points, the $(\leq k)$-sets, is $\Theta(nk)$ in the worst case.

For a set of n points in the plane, each moving with constant speed along some line, the maximum number of times the Voronoi diagram (or Delaunay triangulation) changes structurally is $\Omega(n^2)$ and $O(n^3 2^{\alpha(n)})$ [4].

6.2.5 Computational Topology

In topology, metric aspects of geometry like distances, angles, areas, and volumes are not relevant. Computational topology deals with algorithms for problems relating to incidence structures and other objects that are invariant under continuous deformations. It is often the combination of geometry and topology that leads to new, practically relevant approaches.

As an example of such an approach, we consider *α-complexes*. Given a set of points in the plane, we can imagine growing a disk centered on each point. At first the disks will not intersect, but when they do we can imagine a subdivision inside the union of the disks that is the Voronoi diagram of the points. For any radius of the disks we have such a subdivision, revealing more and more of the Voronoi diagram (Figure 6.6).

We can define a structure dual to the Voronoi diagram inside the union of disks for any given radius $1/\alpha$. This structure is the α-complex and consists of vertices, edges, and triangles. The vertices are simply the points of the set. Two vertices are connected by an edge in the α-complex if the union of disks

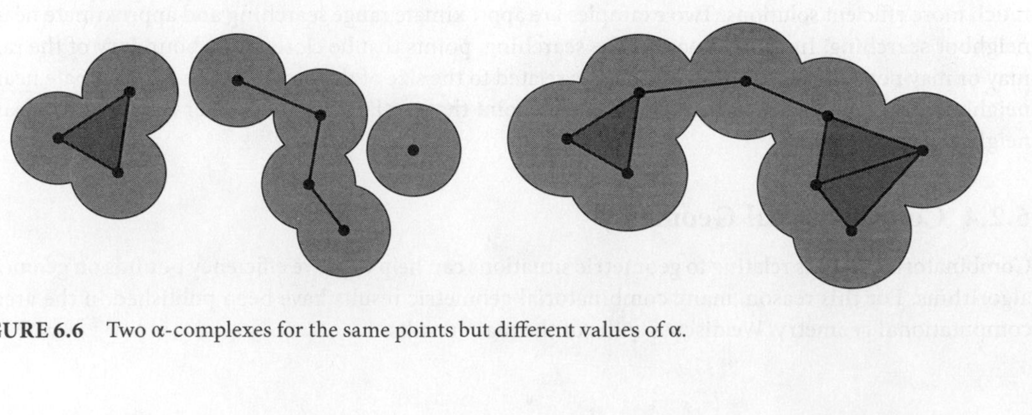

FIGURE 6.6 Two α-complexes for the same points but different values of α.

contains a part of their bisector in the Voronoi diagram. Three vertices form a triangle in the α-complex if their points define a Voronoi vertex in the union of the disks.

When the radius of the disks grows from 0 to ∞, the α-complex changes from the Delaunay triangulation to a set of isolated points. At certain critical radii, new edges and new triangles appear in the complex. The changes to the α-complex are topological, but they are induced by a geometric process.

The boundary of the α-complex—known as α-shape—is a simple and good shape descriptor for a set of points that could have been samples from that shape. The radius of the disks corresponds to the level of detail at which we regard the shape. A high level of detail can be realized only if the sampling density is sufficiently high.

6.2.6 Robustness

Nonrobustness is caused by using fixed-precision numbers and computations when an algorithm designed for the real RAM is implemented. It can cause an algorithm to produce a slightly incorrect value, but it can also cause inconsistent results and program crashes. Geometric algorithms appear to be more susceptible to nonrobustness than other algorithmic areas [112].

For example, when computing the Delaunay triangulation of a planar point set P, the most important test is deciding whether the circle through three points p, q, r contains any other points of the set P, because p, q, r define a Delaunay triangle if and only if this circle does not contain any other points. If P contains four points p, q, r, s that are nearly cocircular and the computation is not exact, then the test for emptiness of the circle may return true for the triples (p, q, r), (q, r, s), (r, s, p), and (s, p, q). This means that the quadrilateral p, q, r, s is covered by two pairs of triangles that overlap, and the computed Delaunay triangulation will not be a proper triangulation. It will not even be a planar graph. Other algorithms that rely on the planarity of the Delaunay triangulation may crash due to this robustness problem.

One approach to obtain robustness is using fixed precision [55,57,61]. Points and other objects must be represented using grid points with a fixed precision in some special way. For example, when *snap-rounding* a set of line segments [62,85], their endpoints and intersections must all lie on the grid. Each line segment becomes a polygonal line whose vertices lie on the grid. A caveat is that a geometric object is moved or deformed.

Another approach is exact geometric computation [51,70,111]. Here a number may be represented implicitly by constructs that use several numbers and operations like $\sqrt{\ }$. Any comparison that influences a branching decision in the algorithm must be exact. Since branching typically depends on comparisons using an inequality, we may achieve this by incrementally computing bits of the values on the left-hand side and right-hand side of the inequality until we can evaluate the comparison correctly.

6.3 Impact on Practice

Geometric computation is needed in many areas of science, and the range of applications is large. Fortunately, there are structures and concepts of general use for which geometric algorithms have been developed. However, due to their generality, such structures are hardly ever directly useful in a practical situation. To transfer general, theoretical results to applications, domain knowledge of the application is needed to understand when abstraction is allowed and when it would make the result lose its relevance.

We overview a number of research areas where computational geometry has made important contributions and highlight these. We also discuss the development of a general purpose, geometric algorithms library called CGAL.

6.3.1 Computer Graphics

When three-dimensional scenes that are represented in models must be displayed on a two-dimensional screen, one must compute what is visible from a given viewpoint. In the end, every pixel gets a color based on what object is visible "beyond" that pixel and how it is illuminated. Scenes are often composed of many triangles that represent the boundaries of the objects in the scene. One of the main tasks in computer graphics is *rendering*, which is the process of getting the objects in a three-dimensional scene onto a two-dimensional image display for a given viewpoint. The operations needed are done partly in the three-dimensional object space and partly in the two-dimensional image space. Computational geometry is involved in those tasks that occur in object space.

Since graphics processors have become a lot more powerful and allow massive parallelization, there has been a trend in computer graphics to do more and more in image space. The z-buffer method is clearly the most popular method for deciding which parts of a scene are visible, especially in dynamic or interactive situations.

On the other hand, object-space computations allow a more realistic visualization via ray tracing [99], and object-space computations may still be important to reduce the number of objects to be rendered, using simplification algorithms [33]. Furthermore, morphing between two shapes can be done better in object space [36].

6.3.2 Motion Planning and Robotics

In motion planning, a robot has a number of degrees of freedom depending on its motions. For example, a robot may move around and rotate in the plane, giving it three degrees of freedom. A robot arm with four angular joints and one telescoping piece has five degrees of freedom. The different states of such a robot or robot arm can be described by a number of real values equal to the number of degrees of freedom. In an environment with other objects—referred to as obstacles—some states may not be possible because the robot would intersect an obstacle. We can capture this situation in the *configuration space*, whose dimensionality is the same as the number of degrees of freedom. A point in the configuration space is a potential state of the robot, which may be possible or impossible. All possible states form a set of points called the *free configuration space*.

Bringing a robot from one position or state to another comes down to determining a path between two points that lies completely in the free configuration space. To decide if this is possible, we can compute the cell of the configuration space that contains the starting configuration, and test whether the end configuration lies inside that cell as well.

In the case of two-dimensional translational motion planning for a polygonal robot amid polygonal obstacles, we need to solve the computational-geometry problem of computing a single cell in an arrangement of line segments [59,96]. If we want to compute the full free configuration space, we need to solve the well-known problems of computing the *Minkowski sum* of two polygons and computing the union of a set of polygonal regions [65,96].

6.3.3 Geographic Information Systems

Many computations that are done in a GIS are geometric [103]. One example is the map overlay operation, which is possibly the most important operation in a GIS. Map overlay takes two or more different thematic map layers covering the same region and performs a "spatial join." For example, if the one map layer is current landuse and the other layer soil type (sand, clay, …), then the overlay is a subdivision where map space is partitioned into regions with some landuse on some soil type. The standard map layer representation in GIS is similar to the doubly-connected edge list, and the standard algorithm for map overlay is based on the plane-sweep paradigm as explained in Section 6.2.1.1.

Spatial interpolation is needed when a type of data is acquired using point measurements, and we are also interested in the values at places in between where no measurements were taken. Think of data like depth to groundwater obtained from bore holes, temperature obtained from thermometers, and ground pollution obtained from soil samples. Spatial interpolation between point-based measurements can be based on a triangulation of the points, where we can use linear interpolation over each triangle. Alternatively, one can use *natural neighbor interpolation* based on Voronoi diagrams [100].

6.3.4 Information Visualization

Information visualization is about how graphs, maps, charts, and diagrams can communicate information effectively to humans. Graph visualization occurs in workflow diagrams and VLSI design, for example. The area of graph drawing is concerned with the drawings of all types of graphs in theory and practice. Quality criteria like few edge–edge intersections, no very long edges, and large enough angles between edges incident to the same vertex determine how good a drawing of a graph is. To draw a graph, we must choose a position and shape for each vertex and edge, and methods from computational geometry are often applied.

Specialized maps like metro maps and flow maps require a highly schematized representation of cartographic information that satisfies certain conventions associated with the type of map [18,81]. The computation of these schematized representations while keeping some degree of spatial correctness requires geometric algorithms that often have a multiobjective flavor. The same is true for map annotation, also called label placement [2]. Here, text must be placed with features on a map, diagram, or graph without creating confusion as to which text belongs to which feature and without overlaps of the different texts.

Other types of information visualization involve hierarchical or clustered representations. The final geometry of the visualization should highlight the hierarchy or clustering, so again geometric computations are needed.

6.3.5 Data Analysis and Statistics

In data analysis, the objective is to find dependencies or structure in large sets of data. One of the main types of structure in data are clusters. For example, for a set of epicenters of earthquakes, one can wonder if there are subsets of epicenters that are close together, closer than would occur by a random process. Similarly, one could be interested in clusters of motion paths of hurricanes in the Atlantic. Since clustering in general involves distance in some space, it requires geometric computations, and since data sets can be large, efficient geometric algorithms are needed. Given a distance measure, there are several well-known clustering algorithms, like single-link clustering, complete-link clustering, and k-means clustering [64].

Outliers are data elements that are significantly different from the rest of the data. Outliers may be of special interest because of this, but they may also be due to erroneous measurements. In any case, it is important to be able to detect outliers, which again requires a notion of distance.

Instead of detecting and eliminating outliers, one can also devise robust statistics. The best-known example is the median of a set of numbers. A few erratic values in the set do not influence the value of the median much, so it is a robust statistic for the concept of center. For multivariate statistics, robust methods can often be seen as geometric tasks for which efficient algorithms need to be developed [90].

6.3.6 Three-Dimensional Modeling

Several data acquisition methods yield three-dimensional point clouds. These could come from an object scanned by a three-dimensional scanner or from an urban environment scanned by LiDAR methods. The construction of a three-dimensional model from such a point cloud uses geometric algorithms of various sorts. For free-form shapes, approaches may use the medial axis and prove reconstruction correctness by relating the sampling density to the curvature [5]. Often, techniques from computational topology are applied, like persistence [20].

In urban reconstruction, one can use the fact that shape information is known. In urban environments many points are approximately coplanar because they were sampled from the same facade or roof.

This makes detection of planes that lie close to many points, one of the basic steps in reconstruction [92]. Once these planes are found they must be connected where appropriate, and remaining gaps should be filled to obtain a good three-dimensional model [11,104].

6.3.7 Finite Element Methods

The *finite element method* is used to approximately solve partial differential equations, which is useful for computing and visualizing air flow around an airplane wing, stress of an object when forces are applied, heat distribution on a printed circuit board, and similar situations. One of the main ingredients for finite element methods is a subdivision of the space of interest into elements of a simple shape (triangles or convex quadrilaterals in the plane). This subdivision should respect the environment (the airplane wing for air flow, the conductive tracks on a printed circuit board, etc.), meaning that no element intersects any boundary of the environment. This subdivision is called a *mesh*, and the process is called *mesh generation* [15,98]. The quality of a mesh for the finite element method depends on the shape of its elements. Very small or large angles should be avoided. Hence, mesh generation is concerned with producing meshes with such geometric quality criteria.

6.3.8 Retrieval

Both database search and information retrieval have a geometric version of the problem domain. To search in multimedia databases of shapes, the query object may be a shape as well, and a full or partial similarity matching of the query shape and all stored shapes determines which shapes are returned and in what order [69]. The similarity measure used may be the Hausdorff distance, Fréchet distance, or Earth movers distance, for instance [105]. In geographic information retrieval the spatial component of information, often distance, is used to decide which documents or web pages are most relevant to a query [68].

6.3.9 Computational Biology

Within the area of computational biology, molecular simulation is concerned with the modeling of three-dimensional molecules and proteins, and their interactions [44]. Molecules and proteins are made up of atoms, and a standard geometric model is to take a ball for each atom and the union of these balls for the molecule or protein. The radii of the balls are often chosen to be the van der Waals radii. The shape of proteins determines for a large part its function. In particular, accessibility for solvents and docking mechanisms of proteins depend on their shape.

6.3.10 Implementation

Providing correct implementations of geometric algorithms has proved to be a challenge in the past, and it still is. This is due to two main issues. Firstly, implementations should also handle all degenerate cases, while algorithm descriptions often ignore these, and secondly, robustness of the computations should always be considered to avoid that rounding errors dramatically influence the outcome of an implementation.

In the mid-1990s, a considerable effort has been made to develop a computational-geometry algorithms library called CGAL, which allows robust number types and includes many basic operations that are necessary in algorithms (compute the intersection point of two line segments, construct the circle through three given points, etc.). In a layer on top of these basic operations, many data structures and algorithms are implemented in the library. These include Voronoi diagrams, Delaunay triangulations, smallest enclosing circle, the doubly-connected edge list, arrangement construction, α-shapes, and much more. The library is made accessible by extensive documentation, including manuals, tutorials, and books [49].

6.4 Research Issues

Although computational geometry deals with constant-dimensional computational problems, the curse of dimension is still an issue. Most problems are considerably harder in three-dimensional space than in the plane, and higher-dimensional problems are even harder and their solutions less well understood. It should be mentioned that they are less common in applications, and therefore the research focus has been on two- and three-dimensional problems.

A major open problem is developing a d-dimensional linear programming method that runs in time $O(f(d) \cdot n)$, where n is the number of constraints and f is a polynomial function in d. A solution exists where f is subexponential but superpolynomial [74].

There are many planar and three-dimensional problems whose algorithmic solutions are not known to be optimal. In computational geometry, there is an ongoing quest to improve existing running time bounds of algorithms, while providing better lower bounds is of equal interest but apparently much harder.

One can expect that realistic input models for three-dimensional data can lead to more results that are provable and useful; it is an important area for future research.

In combinatorial geometry, two very important open problems are finding tight bounds on the number of k-sets in the plane and on the number of changes in the Voronoi diagram or Delaunay triangulation of moving points. There are several other interesting open problems where better combinatorial upper bounds lead to more efficient algorithms.

A research direction with a lot of potential is computational topology [45]. In the last years, the interest in the area has been growing steadily, also because it provides new insight into many different application areas.

Computational geometry has not been involved much in the computations necessary in various processes or simulations. Some exceptions are forest growth simulation [1], bird flocking behavior [28,29], and docking mechanisms in computational biology [44]. An expansion of the research in this direction is challenging and exciting.

There are many challenges in the technology transfer of the ideas and methods developed in computational geometry into practice. This goes much further than making practitioners aware of existing ideas and methods in computational geometry. Similarly, there are challenges in the right formalizations of practical problems into versions that can be addressed by computational geometry but are still relevant to practice.

A general concern from practice is data imprecision. It is important to know to what extent data imprecision influences further processing, and hence a computational model of data imprecision is needed. Such a model can be statistical or combinatorial [71]. Both offer opportunities for important research to which computational geometry can contribute.

6.5 Summary

The area of computational geometry is a challenging and useful research area that includes both fundamental and applied questions. It plays a role whenever geometric data is present and methods are needed to deal with such data. Since geometric data appears in many application areas, and new functionality is needed in these applications, a need remains for new, advanced methods to supply this functionality. At the same time, much research within computational geometry is motivated by curiosity from within, and a search for the best algorithmic solution for any geometric problem.

The overview given in this chapter is just a small sample of all known results and research directions. The *Handbook of Discrete and Computational Geometry* [54] provides a much more extensive list of concepts and results, while the textbook by de Berg et al. [38] provides a good introduction in the area. The annual *Symposium on Computational Geometry* is the main event where the latest trends and results are published. Several dozens of other conferences—algorithmic and applied—regularly publish papers on computational geometry as well.

Key Terms

α-complex: Structure with vertices, edges, and simplicial facets of all dimensions whose adjacencies are defined by a disk of radius 1/α.

α-shape: Structure with vertices and edges (for the planar case) defined for a point set, where the points are the vertices and two points are connected by an edge if and only if there is a disk of radius 1/α through both points, and with no points of the set inside it.

Algebraic decision tree: Tree representing the flow of a program or algorithm where nodes represent conditional statements or tests, which determine which branch under the node is taken next. These tests are algebraic equations.

Backward analysis of a randomized incremental algorithm: Argument used to prove the expected cost of an incremental step by considering the situation after that step and thinking backwards (What could have led to this situation?).

Configuration space: Space in a robot motion planning problem that has one coordinate axis for every degree of freedom of the robot.

Convex hull: Smallest convex set that contains a given set of points or other objects.

Davenport–Schinzel sequence: Sequence of symbols over an alphabet where no two adjacent symbols may be equal, and any two different symbols can occur in alternation in a limited manner.

Delaunay triangulation of a set of points: Maximal planar subdivision of the plane where the points are the vertices and all bounded faces are triangles, such that the circumcircle of the three vertices of any triangle does not have any of the input points inside.

Doubly-connected edge list: Structure to represent planar subdivisions that allows navigation to incident features (e.g., from an edge to the incident faces and vertices).

Dynamic programming: Algorithmic design paradigm that exploits the fact that some optimization problems have solutions that are composed of optimal solutions to smaller problems. These optimal solutions to smaller problems are often stored in tables.

Graph embedding: Layout of a graph where the nodes and arcs are represented as geometric objects; nodes are often points specified by their coordinates.

Greedy algorithm: Algorithmic design paradigm for optimization problems where locally optimal choices are made repeatedly.

Hausdorff distance of two subsets of the plane: Maximum distance from any point in one subset to the nearest point in the other subset.

Minkowski sum: Binary operator that takes two geometric objects and returns the object composed of the union of all vectors that are the sum of a vector to a point in the one object and a vector to a point in the other object.

NP-hardness: Informally, feature of a computational problem describing that any algorithm that solves it requires more than polynomially many operations in the worst case. However, there is no proof known that algorithms with polynomially many operations for NP-hard problems do not exist.

Numerical nonrobustness: Issue appearing in the design of an implementation due to the fact that computers cannot store real numbers exactly.

Planar 3-SAT: Satisfiability expression with three literals per clause that allows a planar layout of the graph where literals and clauses are nodes and occurrences of a literal in a clause are arcs.

Plane sweep: Algorithmic design paradigm where a computational problem is solved by imagining the sweep of a line over the plane.

Randomized incremental construction: Algorithmic design paradigm where a problem is solved incrementally using a randomized order.

Real RAM: Model of computation that allows the storage of any real number in a constant amount of memory and that has random access memory.

Realistic input model: Assumption or set of assumptions that rules out pathological input situations, in order to prove better complexity bounds.

Simple polygon: Bounded planar shape whose boundary consists of a cyclic sequence of line segments where adjacent line segments meet in a common endpoint, and no other line segments have any point in common.

Simple polyhedron: Bounded solid shape in three-dimensional space with linear boundaries that is a generalization of a simple polygon.

Simplex: Polytope in d-dimensional space that is the convex hull of $d + 1$ linearly independent points. In the plane, it is a triangle and in three-dimensional space it is a tetrahedron.

Spread of a point set: Ratio of the largest and smallest point-to-point distance.

Sweep line: Imaginary line that sweeps over the plane, for instance a horizontal line that goes from top to bottom, while computing a solution to a problem.

Upper Bound Theorem: Combinatorial result that states that the maximum number of facets of all dimensions of a d-dimensional polytope bounded by n hyperplanes is $O(n^{\lfloor d/2 \rfloor})$.

Acknowledgments

The author thanks Maarten Löffler and Teofilo Gonzalez for comments that helped to improve the presentation of this chapter.

References

1. P. K. Agarwal, T. Mølhave, H. Yu, and J. S. Clark. Exploiting temporal coherence in forest dynamics simulation. In: *Proceedings of the 27th ACM Symposium on Computational Geometry*, Paris, France, pp. 77–86, 2011.
2. P. K. Agarwal, M. J. van Kreveld, and S. Suri. Label placement by maximum independent set in rectangles. *Comput. Geom.*, 11(3–4):209–218, 1998.
3. A. Aggarwal and J. S. Vitter. The input/output complexity of sorting and related problems. *Commun. ACM*, 31(9):1116–1127, 1988.
4. G. Albers, L. J. Guibas, J. S. B. Mitchell, and T. Roos. Voronoi diagrams of moving points. *Int. J. Comput. Geom. Appl.*, 8(3):365–380, 1998.
5. N. Amenta, S. Choi, and R. K. Kolluri. The power crust, unions of balls, and the medial axis transform. *Comput. Geom.*, 19(2–3):127–153, 2001.
6. A. M. Andrew. Another efficient algorithm for convex hulls in two dimensions. *Inform. Process. Lett.*, 9(5):216–219, 1979.
7. S. Arora. Polynomial time approximation schemes for Euclidean traveling salesman and other geometric problems. *J. ACM*, 45(5):753–782, 1998.
8. S. Arya and D. M. Mount. Approximate range searching. *Comput. Geom.*, 17(3–4):135–152, 2000.
9. S. Arya, D. M. Mount, N. S. Netanyahu, R. Silverman, and A. Y. Wu. An optimal algorithm for approximate nearest neighbor searching fixed dimensions. *J. ACM*, 45(6):891–923, 1998.
10. C. L. Bajaj. The algebraic degree of geometric optimization problems. *Discrete Comput. Geom.*, 3:177–191, 1988.
11. G. Barequet and M. Sharir. Filling gaps in the boundary of a polyhedron. *Comput. Aided Geom. Design*, 12(2):207–229, 1995.
12. G. Di Battista, P. Eades, R. Tamassia, and I. G. Tollis. *Graph Drawing: Algorithms for the Visualization of Graphs*. Prentice Hall, Upper Saddle River, NJ, 1999.
13. J. L. Bentley. Multidimensional binary search trees used for associative searching. *Commun. ACM*, 18(9):509–517, 1975.
14. J. L. Bentley and T. A. Ottmann. Algorithms for reporting and counting geometric intersections. *IEEE Trans. Comput.*, C-28:643–647, 1979.
15. M. Bern, D. Eppstein, and J. Gilbert. Provably good mesh generation. *J. Comput. Syst. Sci.*, 48:384–409, 1994.

16. P. Bose, A. Maheshwari, and P. Morin. Fast approximations for sums of distances, clustering and the Fermat–Weber problem. *Comput. Geom.*, 24(3):135–146, 2003.
17. E. Brisson. Representing geometric structures in *d* dimensions: Topology and order. *Discrete Comput. Geom.*, 9:387–426, 1993.
18. K. Buchin, B. Speckmann, and K. Verbeek. Flow map layout via spiral trees. *IEEE Trans. Vis. Comput. Graph.*, 17(12):2536–2544, 2011.
19. J. Canny. *The Complexity of Robot Motion Planning*. MIT Press, Cambridge, MA, 1987.
20. F. Chazal and S. Oudot. Towards persistence-based reconstruction in Euclidean spaces. In *Proceedings of the 24th ACM Symposium on Computational Geometry*, College Park, MD, pp. 232–241, 2008.
21. B. Chazelle. Convex partitions of polyhedra: A lower bound and worst-case optimal algorithm. *SIAM J. Comput.*, 13:488–507, 1984.
22. B. Chazelle. An optimal convex hull algorithm and new results on cuttings. In *Proceedings of the 32nd Annual IEEE Symposium on Foundations of Computer Science*, San Juan, pp. 29–38, 1991.
23. B. Chazelle. Triangulating a simple polygon in linear time. *Discrete Comput. Geom.*, 6:485–524, 1991.
24. B. Chazelle. An optimal convex hull algorithm in any fixed dimension. *Discrete Comput. Geom.*, 10:377–409, 1993.
25. B. Chazelle and H. Edelsbrunner. An optimal algorithm for intersecting line segments in the plane. *J. ACM*, 39:1–54, 1992.
26. B. Chazelle and L. J. Guibas. Fractional cascading: I. A data structuring technique. *Algorithmica*, 1:133–162, 1986.
27. B. Chazelle and L. J. Guibas. Fractional cascading: II. Applications. *Algorithmica*, 1:163–191, 1986.
28. B. Chazelle. Natural algorithms. In *Proceedings of the 20th Annual ACM-SIAM Symposium on Discrete Algorithms, SODA 2009*, New York, pp. 422–431, 2009.
29. B. Chazelle. A geometric approach to collective motion. In *Proceedings of the 26th ACM Symposium on Computational Geometry*, Snowbird, UT, pp. 117–126, 2010.
30. Y.-J. Chiang, F. P. Preparata, and R. Tamassia. A unified approach to dynamic point location, ray shooting, and shortest paths in planar maps. *SIAM J. Comput.*, 25:207–233, 1996.
31. Y.-J. Chiang and R. Tamassia. Dynamic algorithms in computational geometry. *Proc. IEEE*, 80(9):1412–1434, 1992.
32. V. Chvátal. A combinatorial theorem in plane geometry. *J. Combin. Theory Ser. B*, 18:39–41, 1975.
33. P. Cignoni, C. Montani, and R. Scopigno. A comparison of mesh simplification algorithms. *Comput. Graph.*, 22(1):37–54, 1998.
34. K. L. Clarkson. Las vegas algorithms for linear and integer programming when the dimension is small. *J. ACM*, 42:488–499, 1995.
35. K. L. Clarkson and P. W. Shor. Applications of random sampling in computational geometry, II. *Discrete Comput. Geom.*, 4:387–421, 1989.
36. D. Cohen-Or, A. Solomovici, and D. Levin. Three-dimensional distance field metamorphosis. *ACM Trans. Graph.*, 17(2):116–141, 1998.
37. R. Cole. Searching and storing similar lists. *J. Algorithm*, 7:202–220, 1986.
38. M. de Berg, O. Cheong, M. van Kreveld, and M. Overmars. *Computational Geometry: Algorithms and Aplications*, 3rd edn. Springer, Berlin, Germany, 2008.
39. H. de Fraysseix, J. Pach, and R. Pollack. How to draw a planar graph on a grid. *Combinatorica*, 10(1):41–51, 1990.
40. D. P. Dobkin and M. J. Laszlo. Primitives for the manipulation of three-dimensional subdivisions. *Algorithmica*, 4(1):3–32, 1989.
41. H. Edelsbrunner, L. J. Guibas, and J. Stolfi. Optimal point location in a monotone subdivision. *SIAM J. Comput.*, 15:317–340, 1986.

42. H. Edelsbrunner, J. O'Rourke, and R. Seidel. Constructing arrangements of lines and hyperplanes with applications. *SIAM J. Comput.*, 15:341–363, 1986.
43. H. Edelsbrunner, R. Seidel, and M. Sharir. On the zone theorem for hyperplane arrangements. *SIAM J. Comput.*, 22(2):418–429, 1993.
44. H. Edelsbrunner. Biological applications of computational topology. In: J. E. Goodman and J. O'Rourke, eds., *Handbook of Discrete and Computational Geometry*, 2nd edn., pp. 1395–1412. Chapman & Hall/CRC, Boca Raton, FL, 2004.
45. H. Edelsbrunner and J. L. Harer. *Computational Topology—An Introduction*. American Mathematical Society, Providence, RI, 2010.
46. J. Erickson. Nice point sets can have nasty Delaunay triangulations. *Discrete Comput. Geom.*, 30(1):109–132, 2003.
47. J. Erickson. Dense point sets have sparse Delaunay triangulations or "... but not too nasty". *Discrete Comput. Geom.*, 33(1):83–115, 2005.
48. J. Erickson and R. Seidel. Better lower bounds on detecting affine and spherical degeneracies. *Discrete Comput. Geom.*, 13:41–57, 1995.
49. E. Fogel, D. Halperin, and R. Wein. *CGAL Arrangements and Their Applications—A Step-by-Step Guide, Geometry and Computing*, Vol. 7. Springer, Berlin, Germany, 2012.
50. S. J. Fortune. A sweepline algorithm for Voronoi diagrams. *Algorithmica*, 2:153–174, 1987.
51. S. Fortune and C. J. Van Wyk. Efficient exact arithmetic for computational geometry. In *Proceedings of the 9th Annual ACM Symposium on Computational Geometry*, San Diego, CA, pp. 163–172, 1993.
52. A. Gajentaan and M. H. Overmars. On a class of $o(n^2)$ problems in computational geometry. *Comput. Geom.*, 5:165–185, 1995.
53. M. R. Garey and D. S. Johnson. *Computers and Intractability*. Freeman, San Francisco, CA, 1979.
54. J. E. Goodman and J. O'Rourke, eds. *Handbook of Discrete and Computational Geometry*, 2nd edn. Chapman & Hall/CRC, Boca Raton, FL 2004.
55. M. T. Goodrich, L. J. Guibas, J. Hershberger, and P. J. Tanenbaum. Snap rounding line segments efficiently in two and three dimensions. In *Proceedings of the 13th Annual ACM Symposium on Computational Geometry*, Nice, France, pp. 284–293, 1997.
56. R. L. Graham. An efficient algorithm for determining the convex hull of a finite planar set. *Inform. Process. Lett.*, 1:132–133, 1972.
57. D. H. Greene and F. F. Yao. Finite-resolution computational geometry. In *Proceedings of the 27th Annual Symposium on Foundations of Computer Science*, Toronto, Ontario, Canada, pp. 143–152, 1986.
58. L. J. Guibas, D. E. Knuth, and M. Sharir. Randomized incremental construction of Delaunay and Voronoi diagrams. *Algorithmica*, 7:381–413, 1992.
59. L. J. Guibas, M. Sharir, and S. Sifrony. On the general motion planning problem with two degrees of freedom. *Discrete Comput. Geom.*, 4:491–521, 1989.
60. L. J. Guibas. Modelling motion. In: J. E. Goodman and J. O'Rourke, eds., *Handbook of Discrete and Computational Geometry*, 2nd edn., pp. 1117–1134. Chapman & Hall/CRC, Boca Raton, FL, 2004.
61. L. J. Guibas, D. Salesin, and J. Stolfi. Epsilon geometry: Building robust algorithms from imprecise computations. In: *Symposium on Computational Geometry*, pp. 208–217. ACM Press, Saarbrücken, Germany, 1989.
62. J. Hershberger. Stable snap rounding. In: F. Hurtado and M. J. van Kreveld, eds., *Symposium on Computational Geometry*, Paris, France, pp. 197–206. ACM Press, New York, 2011.
63. J. Hershberger and S. Suri. An optimal algorithm for Euclidean shortest paths in the plane. *SIAM J. Comput.*, 28(6):2215–2256, 1999.
64. A. K. Jain, M. N. Murty, and P. J. Flynn. Data clustering: A review. *ACM Comput. Surv.*, 31(3):264–323, 1999.
65. K. Kedem, R. Livne, J. Pach, and M. Sharir. On the union of Jordan regions and collision-free translational motion amidst polygonal obstacles. *Discrete Comput. Geom.*, 1:59–71, 1986.
66. D. G. Kirkpatrick. Optimal search in planar subdivisions. *SIAM J. Comput.*, 12:28–35, 1983.

67. G. T. Klincsek. Optimal triangulations of polygonal domains. *Discrete Math.*, 9:121–123, 1980.
68. R. R. Larson and P. Frontiera. Spatial ranking methods for geographic information retrieval (gir) in digital libraries. In: Research and Advanced Technology for Digital Libraries, *8th European Conference, ECDL 2004, Lecture Notes in Computer Science*, Vol. 3232. Springer, Berlin, Germany, pp. 45–56. 2004.
69. M. S. Lew, N. Sebe, C. Djeraba, and R. Jain. Content-based multimedia information retrieval: State of the art and challenges. *TOMCCAP*, 2(1):1–19, 2006.
70. C. Li, S. Pion, and C. K. Yap. Recent progress in exact geometric computation. *J. Log. Algebr. Program.*, 64(1):85–111, 2005.
71. M. Löffler. Data Imprecision in Computational Geometry. PhD thesis, Department of Information and Computing Sciences, Utrecht University, Utrecht, The Netherlands, 2009.
72. G. S. Lueker. A data structure for orthogonal range queries. In *Proceedings of the 19th Annual IEEE Symposium on Foundations of Computer Science*, Ann Arbor, MI, pp. 28–34, 1978.
73. J. Matoušek. Range searching with efficient hierarchical cuttings. *Discrete Comput. Geom.*, 10(2):157–182, 1993.
74. J. Matoušek, M. Sharir, and E. Welzl. A subexponential bound for linear programming. *Algorithmica*, 16:498–516, 1996.
75. J. Matoušek, J. Pach, M. Sharir, S. Sifrony, and E. Welzl. Fat triangles determine linearly many holes. *SIAM J. Comput.*, 23:154–169, 1994.
76. N. Megiddo. Linear programming in linear time when the dimension is fixed. *J. ACM*, 31:114–127, 1984.
77. J. S. B. Mitchell. Guillotine subdivisions approximate polygonal subdivisions: A simple polynomial-time approximation scheme for geometric TSP, k-MST, and related problems. *SIAM J. Comput.*, 28(4):1298–1309, 1999.
78. D. E. Muller and F. P. Preparata. Finding the intersection of two convex polyhedra. *Theor. Comput. Sci.*, 7:217–236, 1978.
79. K. Mulmuley. A fast planar partition algorithm, I. *J. Symb. Comput.*, 10:253–280, 1990.
80. W. Mulzer and G. Rote. Minimum-weight triangulation is NP-hard. *J. ACM*, 55(2), 2008.
81. M. Nöllenburg and A. Wolff. Drawing and labeling high-quality metro maps by mixed-integer programming. *IEEE Trans. Vis. Comput. Graph.*, 17(5):626–641, 2011.
82. J. O'Rourke. *Art Gallery Theorems and Algorithms*. Oxford University Press, New York, 1987.
83. J. O'Rourke. *Computational Geometry in C*, 2nd edn. Cambridge University Press, Cambridge, NY, 1998.
84. M. H. Overmars. *The Design of Dynamic Data Structures, Lecture Notes in Computer Science*, Vol. 156. Springer-Verlag, Heidelberg, Germany, 1983.
85. E. Packer. Iterated snap rounding with bounded drift. *Comput. Geom.*, 40(3):231–251, 2008.
86. F. P. Preparata and S. J. Hong. Convex hulls of finite sets of points in two and three dimensions. *Commun. ACM*, 20:87–93, 1977.
87. F. P. Preparata and M. I. Shamos. *Computational Geometry: An Introduction*. Springer-Verlag, New York, 1985.
88. M. O. Rabin. Proving simultaneous positivity of linear forms. *J. Comput. Syst. Sci.*, 6(6):639–650, 1972.
89. E. M. Reingold. On the optimality of some set algorithms. *J. ACM*, 19(4):649–659, 1972.
90. P. J. Rousseeuw and A. Struyf. Computation of robust statistics. In: J. E. Goodman and J. O'Rourke, eds., *Handbook of Discrete and Computational Geometry*, 2nd edn., pp. 1279–1292. Chapman & Hall/CRC, Boca Raton, FL, 2004.
91. N. Sarnak and R. E. Tarjan. Planar point location using persistent search trees. *Commun. ACM*, 29:669–679, 1986.
92. R. Schnabel, R. Wahl, and R. Klein. Efficient RANSAC for point-cloud shape detection. *Comput. Graph. Forum*, 26(2):214–226, 2007.

93. R. Seidel. A simple and fast incremental randomized algorithm for computing trapezoidal decompositions and for triangulating polygons. *Comput. Geom. Theory Appl.*, 1:51–64, 1991.
94. R. Seidel. Small-dimensional linear programming and convex hulls made easy. *Discrete Comput. Geom.*, 6:423–434, 1991.
95. M. I. Shamos and D. Hoey. Closest-point problems. In: *Proceedings of the 16th Annual IEEE Symposium on Foundations of Computer Science*, Berkeley, CA, pp. 151–162, 1975.
96. M. Sharir. Algorithmic motion planning. In: Goodman and O'Rourke [54], chapter 47.
97. M. Sharir and P. K. Agarwal. *Davenport–Schinzel Sequences and Their Geometric Applications.* Cambridge University Press, Cambridge, NY, 1995.
98. J. R. Shewchuck. Delaunay refincment algorithms for for triangular mesh generation. *Comput. Geom. Theory Appl.*, 22:21–74, 2002.
99. P. Shirley and R. K. Morley. *Realistic Ray Tracing*, 2nd edn. A.K. Peters, Natick, MA, 2003.
100. R. Sibson. A brief description of natural neighbour interpolation. In: Vic Barnet, ed., *Interpreting Multivariate Data*, pp. 21–36. John Wiley & Sons, Chichester, U.K., 1981.
101. B. Speckmann. Kinetic data structures. In: M.-Y. Kao, ed., *Encyclopedia of Algorithms*, pp. 417–419. Springer, New York, 2008.
102. G. Tóth. Point sets with many *k*-sets. *Discrete Comput. Geom.*, 26(2):187–194, 2001.
103. M. van Kreveld. Geographic information systems. In: J. E. Goodman and J. O'Rourke, eds., *Handbook of Discrete and Computational Geometry*, 2nd edn., pp. 1293–1314. Chapman & Hall/CRC, Boca Raton, FL, 2004.
104. M. J. van Kreveld, T. van Lankveld, and R. C. Veltkamp. On the shape of a set of points and lines in the plane. *Comput. Graph. Forum*, 30(5):1553–1562, 2011.
105. R. C. Veltkamp. Shape matching: Similarity measures and algorithms. In: *Shape Modeling International*, Los Alamitos, CA, pp. 188–197. IEEE Press, Genova, Italy, 2001.
106. J. S. Vitter. External memory algorithms and data structures. *ACM Comput. Surv.*, 33(2):209–271, 2001.
107. E. Welzl. Smallest enclosing disks (balls and ellipsoids). In: H. Maurer, ed., *New Results and New Trends in Computer Science, Lecture Notes in Computer Science*, Vol. 555. Springer-Verlag, New York, pp. 359–370, 1991.
108. A. Wiernik and M. Sharir. Planar realizations of nonlinear Davenport–Schnitzel sequences by segments. *Discrete Comput. Geom.*, 3:15–47, 1988.
109. D. E. Willard. Predicate-oriented Database Search Algorithms. PhD. thesis, Aiken Comput. Lab., Harvard Univ., Cambridge, MA, 1978. Report TR-20-78.
110. A. C. Yao. A lower bound to finding convex hulls. *J. ACM*, 28:780–787, 1981.
111. C. Yap. Towards exact geometric computation. *Comput. Geom. Theory Appl.*, 7(1):3–23, 1997.
112. C. K. Yap. Robust geometric computation. In: J. E. Goodman and J. O'Rourke, eds., *Handbook of Discrete and Computational Geometry*, 2nd edn. Chapman & Hall/CRC, Boca Raton, FL, 2004.

7 Complexity Theory

Eric W. Allender*
Rutgers University

Michael C. Loui†
University of Illinois at Urbana Champaign

Kenneth W. Regan‡
State University of New York

7.1 Introduction

Computational complexity is the study of the difficulty of solving computational problems, in terms of the required computational resources, such as time and space (memory). Whereas the analysis of algorithms focuses on the time or space requirements of an *individual* algorithm for a *specific* problem (such as sorting), complexity theory focuses on the **complexity class** of problems solvable in the same amount

* Supported by the National Science Foundation under Grants CCF-0832787 and CCF-1064785.

† Supported by the National Science Foundation under Grants IIS-0832843, CNS-0851957, and DUE-1044207.

‡ Supported by the National Science Foundation under Grant CCR-9821040.

of time or space. Most common computational problems fall into a small number of complexity classes. Two important complexity classes are P, the set of problems that can be solved in polynomial time, and NP, the set of problems whose solutions can be verified in polynomial time.

By quantifying the resources required to solve a problem, complexity theory has profoundly affected our thinking about computation. Computability theory establishes the existence of undecidable problems, which cannot be solved in principle regardless of the amount of time invested. However, computability theory fails to find meaningful distinctions among decidable problems. In contrast, complexity theory establishes the existence of decidable problems that, although solvable in principle, cannot be solved in practice, because the time and space required would be larger than the age and size of the known universe (Stockmeyer and Chandra, 1979). Thus, complexity theory characterizes the computationally feasible problems.

The quest for the boundaries of the set of feasible problems has led to the most important unsolved question in all of computer science: Is P different from NP? Hundreds of fundamental problems, including many ubiquitous optimization problems of operations research, are **NP-complete**; they are the hardest problems in NP. If someone could find a polynomial-time algorithm for any one NP-complete problem, then there would be polynomial-time algorithms for all of them. Despite the concerted efforts of many scientists over several decades, no polynomial-time algorithm has been found for any NP-complete problem. Although we do not yet know whether P is different from NP, showing that a problem is NP-complete provides strong evidence that the problem is computationally infeasible and justifies the use of heuristics for solving the problem.

In this chapter, we define P, NP, and related complexity classes. We illustrate the use of **diagonalization** and **padding** techniques to prove relationships between classes. Next, we define NP-completeness, and we show how to prove that a problem is NP-complete. Finally, we define complexity classes for probabilistic and interactive computations.

Throughout this chapter, all numeric functions take integer arguments and produce integer values. All logarithms are taken to base 2. In particular, log *n* means $\lceil \log_2 n \rceil$.

7.2 Models of Computation

To develop a theory of the difficulty of computational problems, we need to specify precisely what a problem is, what an algorithm is, and what a measure of difficulty is. For simplicity, complexity theorists have chosen to represent problems as languages (i.e., as sets of strings of symbols), model algorithms by off-line multitape **Turing machines**, and measure computational difficulty by the time and space required by a Turing machine. To justify these choices, some theorems of complexity theory show how to translate statements about, say, the time complexity of language recognition by Turing machines into statements about computational problems on more realistic models of computation. These theorems imply that the principles of complexity theory are not artifacts of Turing machines, but instead are intrinsic properties of computation.

This section defines different kinds of Turing machines. The *deterministic* Turing machine models actual computers. The *nondeterministic* Turing machine is not a realistic model, but it helps classify the complexity of important computational problems. The *alternating* Turing machine models a form of parallel computation, and it helps elucidate the relationship between time and space.

7.2.1 Computational Problems and Languages

Computer scientists have invented many elegant formalisms for representing data and control structures. Fundamentally, all representations are patterns of symbols. Therefore, we represent an instance of a computational problem as a sequence of symbols.

Let Σ be a finite set, called the *alphabet*. A *word* over Σ is a finite sequence of symbols from Σ. Sometimes a word is called a *string*. Let Σ^* denote the set of all words over Σ. For example, if $\Sigma = \{\mathbf{0}, \mathbf{1}\}$, then

$$\Sigma^* = \{\lambda, \mathbf{0}, \mathbf{1}, \mathbf{00}, \mathbf{01}, \mathbf{10}, \mathbf{11}, \mathbf{000}, \ldots\}$$

is the set of all binary words, including the empty word λ. The *length* of a word w, denoted by $|w|$, is the number of symbols in w. A *language* over Σ is a subset of Σ^*.

A *decision problem* is a computational problem whose answer is simply yes or no. For example: Is the input graph connected? or Is the input a sorted list of integers? A decision problem can be expressed as a membership problem for a language A: for an input x, does x belong to A? For a language A that represents connected graphs, the input word x might represent an input graph G, and $x \in A$ if and only if G is connected.

For every decision problem, the representation should allow for easy parsing, to determine whether a word represents a legitimate instance of the problem. Furthermore, the representation should be concise. In particular, it would be unfair to encode the answer to the problem into the representation of an instance of the problem; for example, for the problem of deciding whether an input graph is connected, the representation should not have an extra bit that tells whether the graph is connected. A set of integers $S = \{x_1,\ldots, x_m\}$ is represented by listing the binary representation of each x_i, with the representations of consecutive integers in S separated by a nonbinary symbol. A graph is naturally represented by giving either its adjacency matrix or a set of adjacency lists, where the list for each vertex v specifies the vertices adjacent to v.

Whereas the solution to a decision problem is yes or no, the solution to an optimization problem is more complicated; for example, determine the shortest path from vertex u to vertex v in an input graph G. Nevertheless, for every optimization (minimization) problem, with objective function g, there is a corresponding decision problem that asks whether there exists a feasible solution z such that $g(z) \le k$, where k is a given target value. Clearly, if there is an algorithm that solves an optimization problem, then that algorithm can be used to solve the corresponding decision problem. Conversely, if algorithm solves the decision problem, then with a binary search on the range of values of g, we can determine the optimal value. Moreover, using a decision problem as a subroutine often enables us to construct an optimal solution; for example, if we are trying to find a shortest path, we can use a decision problem that determines if a shortest path starting from a given vertex uses a given edge. Therefore, there is little loss of generality in considering only decision problems, represented as language membership problems.

7.2.2 Turing Machines

This section and the next three give precise, formal definitions of Turing machines and their variants. These sections are intended for reference. For the rest of this chapter, the reader need not understand these definitions in detail, but may generally substitute "program" or "computer" for each reference to "Turing machine."

A k-worktape **Turing machine** M consists of the following:

- A finite set of states Q, with special states q_0 (initial state), q_A (accept state), and q_R (reject state).
- A finite alphabet Σ, and a special blank symbol $\square \notin \Sigma$.
- The $k + 1$ linear tapes, each divided into cells. Tape 0 is the *input tape*, and tapes 1, …, k are the *worktapes*. Each tape is infinite to the left and to the right. Each cell holds a single symbol from $\Sigma \cup \{\square\}$. By convention, the input tape is read only. Each tape has an access head, and at every instant, each access head scans one cell. See Figure 7.1.
- A finite transition table δ, which comprises tuples of the form

$$(q, s_0, s_1, \ldots, s_k, q', s_1', \ldots, s_k', d_0, d_1, \ldots, d_k)$$

where $q, q' \in Q$, each $s_i, s_i' \in \Sigma \cup \{\square\}$, and each $d_i \in \{-1, 0, +1\}$.

A tuple specifies a step of M: if the current state is q, and s_0, s_1, …, s_k are the symbols in the cells scanned by the access heads, then M replaces s_i by s_i' for $i = 1, \ldots, k$ simultaneously, changes state to q', and

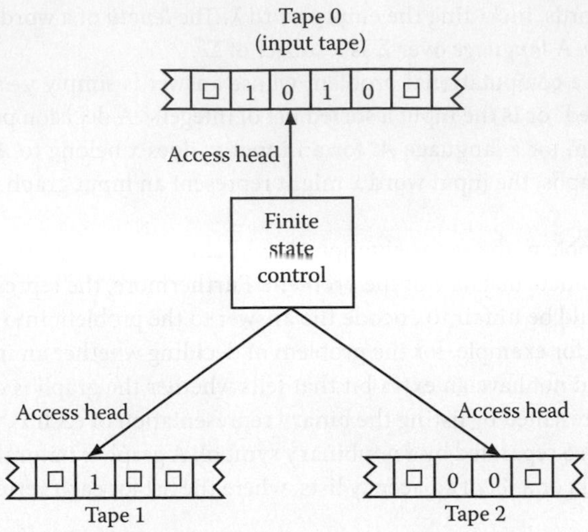

FIGURE 7.1 A two-tape Turing machine.

moves the head on tape i one cell to the left ($d_i = -1$) or right ($d_i = +1$) or not at all ($d_i = 0$) for $i = 0, \ldots, k$. Note that M cannot write on tape 0, that is, M may write only on the worktapes, not on the input tape.

- No tuple contains q_A or q_R as its first component. Thus, once M enters state q_A or state q_R, it stops.
- Initially, M is in state q_0, an input word in Σ^* is inscribed on contiguous cells of the input tape, the access head on the input tape is on the leftmost symbol of the input word, and all other cells of all tapes contain the blank symbol $\square$.

The Turing machine M that we have defined is *nondeterministic:* δ may have several tuples with the same combination of state q and symbols $s_0, s_1, \ldots, s_k$ as the first $k + 2$ components, so that M may have several possible next steps. A machine M is *deterministic* if for every combination of state q and symbols $s_0, s_1, \ldots, s_k$, at most one tuple in δ contains the combination as its first $k + 2$ components. A deterministic machine always has at most one possible next step.

A *configuration* of a Turing machine M specifies the current state, the contents of all tapes, and the positions of all access heads.

A *computation path* is a sequence of configurations $C_0, C_1, \ldots, C_t, \ldots$, where C_0 is the initial configuration of M, and each C_{j+1} follows from C_j in one step by applying the changes specified by a tuple in δ. If no tuple is applicable to C_t, then C_t is *terminal*, and the computation path is *halting*. If M has no infinite computation paths, then M *always halts*.

A halting computation path is *accepting* if the state in the last configuration C_t is q_A; otherwise it is *rejecting*. By adding tuples to the program if needed, we can ensure that every rejecting computation ends in state q_R. This leaves the question of computation paths that do not halt. In complexity theory we rule this out by considering only machines whose computation paths *always halt*. M *accepts* an input word x if there exists an accepting computation path that starts from the initial configuration in which x is on the input tape. For nondeterministic M, it does not matter if some other computation paths end at q_R. If M is deterministic, then there is at most one halting computation path, hence at most one accepting path.

The *language accepted by M*, written $L(M)$, is the set of words accepted by M. If $A = L(M)$, and M always halts, then M *decides A*.

In addition to deciding languages, deterministic Turing machines can compute functions. Designate tape 1 to be the *output tape*. If M halts on input word x, then the nonblank word on tape 1 in the final configuration is the output of M. A function f is *computable* if there exists a deterministic Turing machine M that always halts such that for each input word x, the output of M is the value of $f(x)$.

Almost all results in complexity theory are insensitive to minor variations in the underlying computational models. For example, we could have chosen Turing machines whose tapes are restricted to be only one-way infinite or whose alphabet is restricted to $\{\mathbf{0}, \mathbf{1}\}$. It is straightforward to simulate a Turing machine as defined by one of these restricted Turing machines, one step at a time: each step of the original machine can be simulated by $O(1)$ steps of the restricted machine.

7.2.3 Universal Turing Machines

Chapter 8 states that there exists a *universal Turing machine* U, which takes as input a string $\langle M, x\rangle$ that encodes a Turing machine M and a word x, and simulates the operation of M on x, and U accepts $\langle M, x\rangle$ if and only if M accepts x. For our purposes here, we also need that U simulates M *efficiently*: A theorem of Hennie and Stearns (1966) implies that the machine U can be constructed to have only two worktapes, such that U can simulate any t steps of M in only $O(t \log t)$ steps of its own, using only $O(1)$ times the worktape cells used by M. The constants implicit in these big-O bounds may depend on M.

We can think of U with a fixed M as a machine U_M and define $L(U_M) = \{x : U \text{ accepts } \langle M, x\rangle\}$. Then $L(U_M) = L(M)$. If M always halts, then U_M always halts, and if M is deterministic, then U_M is deterministic.

7.2.4 Alternating Turing Machines

By definition, a nondeterministic Turing machine M accepts its input word x if there exists an accepting computation path, starting from the initial configuration with x on the input tape. Let us call a *configuration* C accepting if there is a computation path of M that starts in C and ends in a configuration whose state is q_A. Equivalently, a configuration C is accepting if either the state in C is q_A or there exists an accepting configuration C' reachable from C by one step of M. Then, M accepts x if the initial configuration with input word x is accepting.

The *alternating Turing machine* generalizes this notion of acceptance. In an alternating Turing machine M, each state is labeled either existential or universal. (Do not confuse the universal state in an alternating Turing machine with the universal Turing machine.) A nonterminal configuration C is existential (respectively, universal) if the state in C is labeled existential (universal). A terminal configuration is accepting if its state is q_A. A nonterminal existential configuration C is accepting if there *exists* an accepting configuration C' reachable from C by one step of M. A nonterminal universal configuration C is accepting if for *every* configuration C' reachable from C by one step of M, the configuration C' is accepting. Finally, M accepts x if the initial configuration with input word x is an accepting configuration.

A nondeterministic Turing machine is thus a special case of an alternating Turing machine in which every state is existential.

The computation of an alternating Turing machine M alternates between existential states and universal states. Intuitively, from an existential configuration, M guesses a step that leads toward acceptance; from a universal configuration, M checks whether each possible next step leads toward acceptance—in a sense, M checks all possible choices in parallel. An alternating computation captures the essence of a two-player game: player 1 has a winning strategy if there exists a move for player 1 such that for every move by player 2, there exists a subsequent move by player 1, etc., such that player 1 eventually wins.

7.2.5 Oracle Turing Machines

Complexity theoreticians have found it very useful to have a mechanism to describe when it is helpful to have a subroutine for some problem A, when trying to solve another problem. The language A is called an **oracle**. Conceptually, an algorithm queries the oracle whether a word w is in A, and it receives the correct answer in one step.

An *oracle Turing machine* is a Turing machine M with a special *oracle tape* and special states QUERY, YES, and NO. The computation of the oracle Turing machine M^A, with oracle language A, is the same as that of an ordinary Turing machine, except that when M enters the QUERY state with a word w on the oracle tape, in one step, M enters either the YES state if $w \in A$ or the NO state if $w \notin A$. Furthermore, during this step, the oracle tape is erased, so that the time for setting up each query is accounted separately.

7.3 Resources and Complexity Classes

In this section, we define the measures of difficulty of solving computational problems. We introduce complexity classes, which enable us to classify problems according to the difficulty of their solution.

7.3.1 Time and Space

We measure the difficulty of a computational problem by the running time and the space (memory) requirements of an algorithm that solves the problem.

We express the complexity of a problem, in terms of the growth of the required time or space, as a function of the length n of the input word that encodes a problem instance. We consider the worst-case complexity, that is, for each n, the maximum time or space required among all inputs of length n.

Let M be a Turing machine that always halts. The *time* taken by M on input word x, denoted by $\text{Time}_M(x)$, is defined as follows:

- If M accepts x, then $\text{Time}_M(x)$ is the number of steps in the shortest accepting computation path for x.
- If M rejects x, then $\text{Time}_M(x)$ is the number of steps in the longest computation path for x.

For a deterministic machine M, for every input x, there is at most one halting computation path, and its length is $\text{Time}_M(x)$. For a nondeterministic machine M, if $x \in L(M)$, then M can guess the correct steps to take toward an accepting configuration, and $\text{Time}_M(x)$ measures the length of the path on which M always makes the best guess.

The *space* used by a Turing machine M on input x, denoted $\text{Space}_M(x)$, is defined as follows. The space used by a halting computation path is the number of worktape cells visited by the worktape heads of M during the computation path. Because the space occupied by the input word is not counted, a machine can use a sublinear ($o(n)$) amount of space.

- If M accepts x, then $\text{Space}_M(x)$ is the minimum space used among all accepting computation paths for x.
- If M rejects x, then $\text{Space}_M(x)$ is the maximum space used among all computation paths for x.

The **time complexity** of a machine M is the function

$$t(n) = \max\{\text{Time}_M(x) : |x| = n\}$$

We assume that M reads all of its input word and the blank symbol after the right end of the input word, so $t(n) \geq n + 1$. The **space complexity** of M is the function

$$s(n) = \max\{\text{Space}_M(x) : |x| = n\}$$

Because few interesting languages can be decided by machines of sublogarithmic space complexity, we henceforth assume that $s(n) \geq \log n$.

A function $f(x)$ is *computable in polynomial time* if there exists a deterministic Turing machine M of polynomial-time complexity such that for each input word x, the output of M is $f(x)$.

7.3.2 Complexity Classes

Having defined the time complexity and space complexity of individual Turing machines, we now define classes of languages with particular complexity bounds. These definitions will lead to definitions of P and NP.

Let $t(n)$ and $s(n)$ be numeric functions. Define the following classes of languages:

- DTIME[$t(n)$] is the class of languages decided by deterministic Turing machines of time complexity $O(t(n))$.
- NTIME[$t(n)$] is the class of languages decided by nondeterministic Turing machines of time complexity $O(t(n))$.
- DSPACE[$s(n)$] is the class of languages decided by deterministic Turing machines of space complexity $O(s(n))$.
- NSPACE[$s(n)$] is the class of languages decided by nondeterministic Turing machines of space complexity $O(s(n))$.

We sometimes abbreviate DTIME[$t(n)$] to DTIME[t] (and so on) when t is understood to be a function, and when no reference is made to the input length n.

The following are the **canonical complexity classes:**

- L = DSPACE[$\log n$] (deterministic log space)
- NL = NSPACE[$\log n$] (nondeterministic log space)
- P = DTIME[$n^{O(1)}$] = $\bigcup_{k\geq 1}$ DTIME[n^k] (polynomial time)
- NP = NTIME[$n^{O(1)}$] = $\bigcup_{k\geq 1}$ NTIME[n^k] (nondeterministic polynomial time)
- PSPACE = DSPACE[$n^{O(1)}$] = $\bigcup_{k\geq 1}$ DSPACE[n^k] (polynomial space)
- E = DTIME[$2^{O(n)}$] = $\bigcup_{k\geq 1}$ DTIME[k^n]
- NE = NTIME[$2^{O(n)}$] = $\bigcup_{k\geq 1}$ NTIME[k^n]
- EXP = DTIME[$2^{n^{O(1)}}$] = $\bigcup_{k\geq 1}$ DTIME[2^{n^k}](deterministic exponential time)
- NEXP = NTIME[$2^{n^{O(1)}}$] = $\bigcup_{k\geq 1}$ NTIME[2^{n^k}] (nondeterministic exponential time)

The class PSPACE is defined in terms of the DSPACE complexity measure. By Savitch's Theorem (see Theorem 7.2), the NSPACE measure with polynomial bounds also yields PSPACE.

The class P contains many familiar problems that can be solved efficiently, such as (decision problem versions of) finding shortest paths in networks, parsing for context-free languages, sorting, matrix multiplication, and linear programming. Consequently, P has become accepted as representing the set of computationally feasible problems. Although one could legitimately argue that a problem whose best algorithm has time complexity $\Theta(n^{99})$ is really infeasible, in practice, the time complexities of the vast majority of natural polynomial-time problems have low degrees: they have algorithms that run in $O(n^4)$ time or less. Moreover, P is a robust class: though defined by Turing machines, P remains the same when defined by other models of sequential computation. For example, random-access machines (RAMs) (a more realistic model of computation defined in Chapter 8) can be used to define P, because Turing machines and RAMs can simulate each other with polynomial-time overhead.

The class NP can also be defined by means other than nondeterministic Turing machines. NP equals the class of problems whose solutions can be *verified* quickly, by deterministic machines in polynomial time. Equivalently, NP comprises those languages whose membership proofs can be checked quickly.

For example, one language in NP is the set of satisfiable Boolean formulas, called SAT. A Boolean formula ϕ is satisfiable if there exists a way of assigning true or false to each variable such that under this truth assignment, the value of ϕ is true. For example, the formula $x \wedge (\bar{x} \vee y)$ is satisfiable, but $x \wedge \bar{y} \wedge (\bar{x} \vee y)$ is not satisfiable. A nondeterministic Turing machine M, after checking the syntax of ϕ and counting the number n of variables, can nondeterministically write down an n-bit 0-1 string a on its tape, and then deterministically (and easily) evaluate ϕ for the truth assignment denoted by a.

The computation path corresponding to each individual a accepts if and only if $\phi(a) = \texttt{true}$, and so M itself accepts ϕ if and only if ϕ is satisfiable; that is, $L(M) = \texttt{SAT}$. Again, this checking of given assignments differs significantly from trying to *find* an accepting assignment.

Another language in NP is the set of undirected graphs with a *Hamiltonian circuit*, that is, a path of edges that visits each vertex exactly once and returns to the starting point. If a solution exists and is given, its correctness can be verified quickly. Finding such a circuit, however, or proving one does not exist, appears to be computationally difficult.

The characterization of NP as the set of problems with easily verified solutions is formalized as follows: $A \in \texttt{NP}$ if and only if there exist a language $A' \in \texttt{P}$ and a polynomial p such that for every x, $x \in A$ if and only if there exists a y such that $|y| \leq p(|x|)$ and $(x, y) \in A'$. Here, whenever x belongs to A, y is interpreted as a positive solution to the problem represented by x, or equivalently, as a proof that x belongs to A. The difference between P and NP is that between solving and checking, or between finding a proof of a mathematical theorem and testing whether a candidate proof is correct. In essence, NP represents all sets of theorems with proofs that are short (i.e., of polynomial length) and checkable quickly (i.e., in polynomial time), while P represents those statements that can proved or refuted quickly from scratch.

Further motivation for studying L, NL, and PSPACE comes from their relationships to P and NP. Namely, L and NL are the largest space-bounded classes known to be contained in P, and PSPACE is the smallest space-bounded class known to contain NP. (It is worth mentioning here that NP does not stand for "nonpolynomial time"; the class P is a subclass of NP.) Similarly, EXP is of interest primarily because it is the smallest deterministic time class known to contain NP. The closely related class E is not known to contain NP.

7.4 Relationships between Complexity Classes

The P vs. NP question asks about the relationship between these complexity classes: Is P a proper subset of NP, or does P = NP? Much of complexity theory focuses on the relationships between complexity classes, because these relationships have implications for the difficulty of solving computational problems. In this section, we summarize important known relationships. We demonstrate two techniques for proving relationships between classes: diagonalization and padding.

7.4.1 Constructibility

The most basic theorem that one should expect from complexity theory would say, "If you have more resources, you can do more." Unfortunately, if we are not careful with our definitions, then this claim is false:

Theorem 7.1 (Gap Theorem)

There is a computable, strictly increasing time bound $t(n)$ such that $\texttt{DTIME}[t(n)] = \texttt{DTIME}[2^{2^{t(n)}}]$ (Borodin, 1972).

That is, there is an empty gap between time $t(n)$ and time doubly exponentially greater than $t(n)$, in the sense that anything that can be computed in the larger time bound can already be computed in the smaller time bound. That is, even with much more time, you cannot compute more. This gap can be made much larger than doubly exponential; for any computable r, there is a computable time bound t such that $\texttt{DTIME}[t(n)] = \texttt{DTIME}[r(t(n))]$. Exactly analogous statements hold for the NTIME, DSPACE, and NSPACE measures.

Fortunately, the gap phenomenon cannot happen for time bounds t that anyone would ever be interested in. Indeed, the proof of the Gap Theorem proceeds by showing that one can define a time bound t such that no machine has a running time that is between $t(n)$ and $2^{2^{t(n)}}$. This theorem indicates the need for formulating only those time bounds that actually describe the complexity of some machine.

A function $t(n)$ is **time-constructible** if there exists a deterministic Turing machine that halts after exactly $t(n)$ steps for every input of length n. A function $s(n)$ is **space-constructible** if there exists a deterministic Turing machine that uses exactly $s(n)$ worktape cells for every input of length n. (Most authors consider only functions $t(n) \geq n + 1$ to be time-constructible, and many limit attention to $s(n) \geq \log n$ for space bounds. There do exist sublogarithmic space-constructible functions, but we prefer to avoid the tricky theory of $o(\log n)$ space bounds.)

For example, $t(n) = n + 1$ is time-constructible. Furthermore, if $t_1(n)$ and $t_2(n)$ are time-constructible, then so are the functions $t_1 + t_2$, t_1t_2, $t_1^{t_2}$, and c^{t_1} for every integer $c > 1$. Consequently, if $p(n)$ is a polynomial, then $p(n) = \Theta(t(n))$ for some time-constructible polynomial function $t(n)$. Similarly, $s(n) = \log n$ is space-constructible, and if $s_1(n)$ and $s_2(n)$ are space-constructible, then so are the functions $s_1 + s_2$, s_1s_2, $s_1^{s_2}$, and c^{s_1} for every integer $c > 1$. Many common functions are space-constructible: for example, $n \log n$, n^3, 2^n, $n!$

Constructibility helps eliminate an arbitrary choice in the definition of the basic time and space classes. For general time functions t, the classes DTIME[t] and NTIME[t] may vary depending on whether machines are required to halt within t steps on all computation paths or just on those paths that accept. If t is time-constructible and s is space-constructible, however, then DTIME[t], NTIME[t], DSPACE[s], and NSPACE[s] can be defined without loss of generality in terms of Turing machines that always halt.

As a general rule, any function $t(n) \geq n + 1$ and any function $s(n) \geq \log n$ that one is interested in as a time or space bound is time- or space-constructible, respectively. As we have seen, little of interest can be proved without restricting attention to constructible functions. This restriction still leaves a rich class of resource bounds.

7.4.2 Basic Relationships

Clearly, for all time functions $t(n)$ and space functions $s(n)$, DTIME[$t(n)$] $\subseteq$ NTIME[$t(n)$] and DSPACE[$s(n)$] $\subseteq$ NSPACE[$s(n)$], because a deterministic machine constitutes a special case of a nondeterministic machine. Furthermore, DTIME[$t(n)$] $\subseteq$ DSPACE[$t(n)$] and NTIME[$t(n)$] $\subseteq$ NSPACE[$t(n)$], because at each step, a k-tape Turing machine can write on at most $k = O(1)$ previously unwritten cells. The next theorem presents additional important relationships between classes.

Theorem 7.2

Let $t(n)$ be a time-constructible function, and let $s(n)$ be a space-constructible function, $s(n) \geq \log n$.

(a) NTIME[$t(n)$] $\subseteq$ DTIME[$2^{O(t(n))}$].
(b) NSPACE[$s(n)$] $\subseteq$ DTIME[$2^{O(s(n))}$].
(c) NTIME[$t(n)$] $\subseteq$ DSPACE[$t(n)$].
(d) **(Savitch's Theorem)** NSPACE[$s(n)$] $\subseteq$ DSPACE[$s(n)^2$] (Savitch, 1970).

As a consequence of the first part of this theorem, NP $\subseteq$ EXP. No better general upper bound on deterministic time is known for languages in NP. See Figure 7.2 for other known inclusion relationships between canonical complexity classes.

Although we do not know whether allowing nondeterminism strictly increases the class of languages decided in polynomial time, Savitch's Theorem says that for space classes, nondeterminism does not help by more than a polynomial amount.

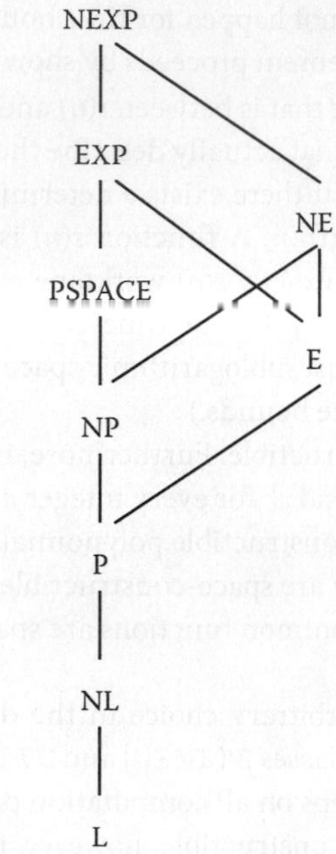

FIGURE 7.2 Inclusion relationships among the canonical complexity classes.

7.4.3 Complementation

For a language A over an alphabet Σ, define $\bar{A}$ to be the complement of A in the set of words over Σ: that is, $\bar{A} = \Sigma^* - A$. For a class of languages $\mathcal{C}$, define co-$\mathcal{C} = \{\bar{A} : A \in \mathcal{C}\}$. If $\mathcal{C}$ = co-$\mathcal{C}$, then $\mathcal{C}$ is **closed under complementation.**

In particular, co-NP is the class of languages that are complements of languages in NP. For the language SAT of satisfiable Boolean formulas, $\overline{\text{SAT}}$ is essentially the set of unsatisfiable formulas, whose value is false for every truth assignment, together with the syntactically incorrect formulas. A closely related language in co-NP is the set of Boolean tautologies, namely, those formulas whose value is true for every truth assignment. The question of whether NP equals co-NP comes down to whether every tautology has a short (i.e., polynomial-sized) proof. The only obvious general way to prove a tautology ϕ in m variables is to verify all 2^m rows of the truth table for ϕ, taking exponential time. Most complexity theorists believe that there is no general way to reduce this time to polynomial, hence that NP $\neq$ co-NP.

Questions about complementation bear directly on the P vs. NP question. It is easy to show that P is closed under complementation (see Theorem 7.3). Consequently, if NP $\neq$ co-NP, then P $\neq$ NP.

Theorem 7.3 (Complementation Theorems)

Let t be a time-constructible function, and let s be a space-constructible function, with $s(n) \geq \log n$ for all n. Then,

1. DTIME[t] *is closed under complementation.*
2. DSPACE[s] *is closed under complementation.*
3. NSPACE[s] *is closed under complementation* (Immerman, 1988; Szelepcsényi, 1988).

The Complementation Theorems are used to prove the Hierarchy Theorems in the next section.

7.4.4 Hierarchy Theorems and Diagonalization

A hierarchy theorem is a theorem that says "If you have more resources, you can compute more." As we saw in Section 7.4.1, this theorem is possible only if we restrict attention to constructible time and space bounds. Next, we state hierarchy theorems for deterministic and nondeterministic time and space classes. In the following, ⊂ denotes *strict* inclusion between complexity classes.

Theorem 7.4 (Hierarchy Theorems)

Let t_1 and t_2 be time-constructible functions, and let s_1 and s_2 be space-constructible functions, with $s_1(n), s_2(n) \geq \log n$ for all n.

(a) If $t_1(n)\log t_1(n) = o(t_2(n))$, then $\mathtt{DTIME}[t_1] \subset \mathtt{DTIME}[t_2]$.
(b) If $t_1(n + 1) = o(t_2(n))$, then $\mathtt{NTIME}[t_1] \subset \mathtt{NTIME}[t_2]$ (Seiferas et al., 1978).
(c) If $s_1(n) = o(s_2(n))$, then $\mathtt{DSPACE}[s_1] \subset \mathtt{DSPACE}[s_2]$.
(d) If $s_1(n) = o(s_2(n))$, then $\mathtt{NSPACE}[s_1] \subset \mathtt{NSPACE}[s_2]$.

As a corollary of the Hierarchy Theorem for DTIME,

$$\mathtt{P} \subseteq \mathtt{DTIME}\,[n^{\log n}] \subset \mathtt{DTIME}\,[2^n] \subseteq \mathtt{E}$$

hence we have the strict inclusion P ⊂ E. Although we do not know whether P ⊂ NP, there exists a problem in E that cannot be solved in polynomial time. Other consequences of the Hierarchy Theorems are NE ⊂ NEXP and NL ⊂ PSPACE.

In the Hierarchy Theorem for DTIME, the hypothesis on t_1 and t_2 is $t_1(n)\log t_1(n) = o(t_2(n))$, instead of $t_1(n) = o(t_2(n))$, for technical reasons related to the simulation of machines with multiple worktapes by a single universal Turing machine with a fixed number of worktapes. Other computational models, such as RAMs, enjoy tighter time hierarchy theorems.

All proofs of the Hierarchy Theorems use the technique of **diagonalization**. For example, the proof for DTIME constructs a Turing machine M of time complexity t_2 that considers all machines $M_1, M_2, \ldots$ whose time complexity is t_1; for each i, the proof finds a word x_i that is accepted by M if and only if $x_i \notin L(M_i)$, the language decided by M_i. Consequently, $L(M)$, the language decided by M, differs from each $L(M_i)$, hence $L(M) \notin \mathtt{DTIME}[t_1]$. The diagonalization technique resembles the classic method used to prove that the real numbers are uncountable, by constructing a number whose jth digit differs from the jth digit of the jth number on the list. To illustrate the diagonalization technique, we outline the proof of the Hierarchy Theorem for DSPACE. In this section, $\langle i, x\rangle$ stands for the string 0^i1x, and *zeroes*(y) stands for the number of 0's that a given string y starts with. Note that *zeroes*$(\langle i, x\rangle) = i$.

Proof. (of the DSPACE Hierarchy Theorem)

We construct a deterministic Turing machine M that decides a language A such that $A \in \mathtt{DSPACE}[s_2] - \mathtt{DSPACE}[s_1]$.

Let U be a deterministic universal Turing machine, as described in Section 7.2.3. On input x of length n, machine M performs the following:

1. Lay out $s_2(n)$ cells on a worktape.
2. Let i = *zeroes*(x).
3. Simulate the universal machine U on input $\langle i, x\rangle$. Accept x if U tries to use more than s_2 worktape cells. (We omit some technical details, such as the way in which the constructibility of s_2 is used to ensure that this process halts.)
4. If U accepts $\langle i, x\rangle$, then reject; if U rejects $\langle i, x\rangle$, then accept.

Clearly, M always halts and uses space $O(s_2(n))$. Let $A = L(M)$.

Suppose $A \in \texttt{DSPACE}[s_1(n)]$. Then there is some Turing machine M_j accepting A using space at most $s_1(n)$. Since the space used by U is $O(1)$ times the space used by M_j, there is a constant k depending only on j (in fact, we can take $k = |j|$), such that U, on inputs z of the form $z = \langle j, x\rangle$, uses at most $ks_1(|x|)$ space.

Since $s_1(n) = o(s_2(n))$, there is an n_0 such that $ks_1(n) \leq s_2(n)$ for all $n \geq n_0$. Let x be a string of length greater than n_0 such that the first $j + 1$ symbols of x are 0^j1. Note that the universal Turing machine U, on input $\langle j, x\rangle$, simulates M_j on input x and uses space at most $ks_1(n) \leq s_2(n)$. Thus, when we consider the machine M defining A, we see that on input x the simulation does not stop in step 3 but continues on to step 4, and thus $x \in A$ if and only if U rejects $\langle j, x\rangle$. Consequently, M_j does not accept A, contrary to our assumption. Thus $A \notin \texttt{DSPACE}[s_1(n)]$. □

Although the diagonalization technique successfully separates some pairs of complexity classes, diagonalization does not seem strong enough to separate P from NP (see Theorem 7.11).

7.4.5 Padding Arguments

A useful technique for establishing relationships between complexity classes is the **padding argument.** Let A be a language over alphabet Σ, and let # be a symbol not in Σ. Let f be a numeric function. The ***f*-padded version of** L is the language

$$A' = \{x\#^{f(n)} : x \in A \text{ and } n = |x|\}$$

That is, each word of A' is a word in A concatenated with $f(n)$ consecutive # symbols. The padded version A' has the same information content as A, but because each word is longer, the computational complexity of A' is smaller.

The proof of the next theorem illustrates the use of a padding argument.

Theorem 7.5

If P = NP, then E = NE (Book, 1974).

Proof. Since E ⊆ NE, we prove that NE ⊆ E.

Let $A \in \texttt{NE}$ be decided by a nondeterministic Turing machine M in at most $t(n) = k^n$ time for some constant integer k. Let A' be the $t(n)$-padded version of A. From M, we construct a nondeterministic Turing machine M' that decides A' in linear time: M' checks that its input has the correct format, using the time-constructibility of t; then M' runs M on the prefix of the input preceding the first # symbol. Thus, $A' \in \texttt{NP}$.

If P = NP, then there is a deterministic Turing machine D' that decides A' in at most $p'(n)$ time for some polynomial p'. From D', we construct a deterministic Turing machine D that decides A, as follows. On input x of length n, since $t(n)$ is time-constructible, machine D constructs $x\#^{t(n)}$, whose length is $n + t(n)$, in $O(t(n))$ time. Then D runs D' on this input word. The time complexity of D is at most $O(t(n)) + p'(n + t(n)) = 2^{O(n)}$. Therefore, NE ⊆ E. □

A similar argument shows that the E = NE question is equivalent to the question of whether NP − P contains a subset of 1^*, that is, a language over a single-letter alphabet.

7.5 Reducibility and Completeness

In this section, we discuss relationships between problems: informally, if one problem reduces to another problem, then, in a sense, the second problem is harder than the first. The hardest problems in NP are the NP-complete problems. We define NP-completeness precisely, and we show how to prove that a problem is

NP-complete. The theory of NP-completeness, together with the many known NP-complete problems, is perhaps the best justification for interest in the classes P and NP. All of the other canonical complexity classes listed earlier have natural and important problems that are complete for them; we give some of these as well.

7.5.1 Resource-Bounded Reducibilities

In mathematics, as in everyday life, a typical way to solve a new problem is to reduce it to a previously solved problem. Frequently, an instance of the new problem is expressed completely in terms of an instance of the prior problem, and the solution is then interpreted in the terms of the new problem. For example, the maximum matching problem for bipartite graphs reduces to the network flow problem (see Chapter 5). This kind of reduction is called **many-one reducibility** and is defined after the next paragraph.

A different way to solve the new problem is to use a subroutine that solves the prior problem. For example, we can solve an optimization problem whose solution is feasible and maximizes the value of an objective function g by repeatedly calling a subroutine that solves the corresponding decision problem of whether there exists a feasible solution x whose value $g(x)$ satisfies $g(x) \geq k$. This kind of reduction is called **Turing reducibility** and is also defined after the next paragraph.

Let A_1 and A_2 be languages. A_1 is many-one reducible to A_2, written $A_1 \leq_m A_2$, if there exists a computable function f such that for all x, $x \in A_1$ if and only if $f(x) \in A_2$. The function f is called the **transformation function**. A_1 is Turing reducible to A_2, written $A_1 \leq_T A_2$, if A_1 can be decided by a deterministic oracle Turing machine M using A_2 as its oracle, that is, $A_1 = L(M^{A_2})$. (Computable functions and oracle Turing machines are defined in Section 7.2.) The oracle for A_2 models a hypothetical efficient subroutine for A_2.

If f or M above consumes too much time or space, the reductions they compute are not helpful. To study complexity classes defined by bounds on time and space resources, it is natural to consider resource-bounded reducibilities. Let A_1 and A_2 be languages.

- A_1 is **Karp reducible** to A_2, written $A_1 \leq_m^p A_2$, if A_1 is many-one reducible to A_2 via a transformation function that is computable deterministically in polynomial time.
- A_1 is **log-space reducible** to A_2, written $A_1 \leq_m^{\log} A_2$, if A_1 is many-one reducible to A_2 via a transformation function that is computable deterministically in $O(\log n)$ space.
- A_1 is **Cook reducible** to A_2, written $A_1 \leq_T^p A_2$, if A_1 is Turing reducible to A_2 via a deterministic oracle Turing machine of polynomial-time complexity.

The term "polynomial-time reducibility" usually refers to Karp reducibility. If $A_1 \leq_m^p A_2$ and $A_2 \leq_m^p A_1$, then A_1 and A_2 are **equivalent** under Karp reducibility. Equivalence under Cook reducibility is defined similarly.

Karp and Cook reductions are useful for finding relationships between languages of high complexity, but they are not useful at all for distinguishing between problems in P, because all problems in P are equivalent under Karp (and hence Cook) reductions. (Here and later we ignore the special cases $A_1 = \emptyset$ and $A_1 = \Sigma^*$, and consider them to reduce to any language.) Since there are interesting distinctions to be made among problems in P, it is useful to have a more restrictive notion of reducibility.

Log-space reducibility (Jones, 1975) is just such a notion. Although it is not known whether log-space reducibility is different from Karp reducibility (just as it is not known whether L is a proper subset of P), this is widely believed to be true. (However, in practice, most problems that can be shown to be NP-complete under Karp reductions remain complete under log-space reductions.)

Theorem 7.6

Log-space reducibility implies Karp reducibility, which implies Cook reducibility:

1. If $A_1 \leq_m^{\log} A_2$, then $A_1 \leq_m^p A_2$.
2. If $A_1 \leq_m^p A_2$, then $A_1 \leq_T^p A_2$.

Theorem 7.7

Log-space reducibility, Karp reducibility, and Cook reducibility are transitive:

1. If $A_1 \leq_m^{\log} A_2$ and $A_2 \leq_m^{\log} A_3$, then $A_1 \leq_m^{\log} A_3$.
2. If $A_1 \leq_m^p A_2$ and $A_2 \leq_m^p A_3$, then $A_1 \leq_m^p A_3$.
3. If $A_1 \leq_T^p A_2$ and $A_2 \leq_T^p A_3$, then $A_1 \leq_T^p A_3$.

The key property of Cook and Karp reductions is that they preserve polynomial-time feasibility. Suppose $A_1 \leq_m^p A_2$ via a transformation f. If M_2 decides A_2, and M_f computes f, then to decide whether an input word x is in A_1, we may use M_f to compute $f(x)$, and then run M_2 on input $f(x)$. If the time complexities of M_2 and M_f are bounded by polynomials t_2 and t_f, respectively, then on each input x of length $n = |x|$, the time taken by this method of deciding A_1 is at most $t_f(n) + t_2(t_f(n))$, which is also a polynomial in n. In summary, if A_2 is feasible, and there is an efficient reduction from A_1 to A_2, then A_1 is feasible. Although this is a simple observation, this fact is important enough to state as a theorem (Theorem 7.8). First, though, we need the concept of "closure."

A class of languages $\mathcal{C}$ is **closed under a reducibility** $\leq_r$ if for all languages A_1 and A_2, whenever $A_1 \leq_r A_2$ and $A_2 \in \mathcal{C}$, necessarily $A_1 \in \mathcal{C}$.

Theorem 7.8

1. P is closed under log-space reducibility, Karp reducibility, and Cook reducibility.
2. NP is closed under log-space reducibility and Karp reducibility.
3. L and NL are closed under log-space reducibility.

We shall see the importance of closure under a reducibility in conjunction with the concept of completeness, which we define in the next section.

7.5.2 Complete Languages

Let $\mathcal{C}$ be a class of languages that represent computational problems. A language A_0 is $\mathcal{C}$-**hard** under a reducibility $\leq_r$ if for all A in $\mathcal{C}$, $A \leq_r A_0$. A language A_0 is $\mathcal{C}$-**complete** under $\leq_r$ if A_0 is $\mathcal{C}$-hard, and $A_0 \in \mathcal{C}$. Informally, if A_0 is $\mathcal{C}$-hard, then A_0 represents a problem that is at least as difficult to solve as any problem in $\mathcal{C}$. If A_0 is $\mathcal{C}$-complete, then in a sense, A_0 is one of the most difficult problems in $\mathcal{C}$.

There is another way to view completeness. Completeness provides us with tight lower bounds on the complexity of problems. If a language A is complete for complexity class $\mathcal{C}$, then we have a lower bound on its complexity. Namely, A is as hard as the most difficult problem in $\mathcal{C}$, assuming that the complexity of the reduction itself is small enough not to matter. The lower bound is tight because A is in $\mathcal{C}$; that is, the upper bound matches the lower bound.

In the case $\mathcal{C}$ = NP, the reducibility $\leq_r$ is usually taken to be Karp reducibility unless otherwise stated. Thus we say

- A language A_0 is **NP-hard** if A_0 is NP-hard under Karp reducibility.
- A_0 is **NP-complete** if A_0 is NP-complete under Karp reducibility.

Note, however, that some sources use the term "NP-hard" to refer to Cook reducibility.

Many important languages are now known to be NP-complete. Before we get to them, let us discuss some implications of the statement "A_0 is NP-complete," and also some things this statement does *not* mean.

The first implication is that if there exists a deterministic Turing machine that decides A_0 in polynomial time—that is, if $A_0 \in$ P—then because P is closed under Karp reducibility (Theorem 7.8) it would follow that NP ⊆ P, hence P = NP. In essence, the question of whether P is the same as NP comes down to the question of whether any particular NP-complete language is in P. Put another way, *all* of the NP-complete languages stand or fall together: if one is in P, then all are in P; if one is not, then all are not. Another implication, which follows by a similar closure argument applied to co-NP, is that if $A_0 \in$ co-NP then NP = co-NP. It is also believed unlikely that NP = co-NP, as was noted in Section 7.4.3 in connection with whether all tautologies have short proofs.

A common misconception is that the above property of NP-complete languages is actually their definition, namely: if $A \in$ NP, and $A \in$ P implies P = NP, then A is NP-complete. This "definition" is wrong if P ≠ NP. A theorem due to Ladner (1975) shows that P ≠ NP if and only if there exists a language A' in NP − P such that A' is not NP-complete. Thus, if P ≠ NP, then A' is a counterexample to the "definition."

Another common misconception arises from a misunderstanding of the statement "If A_0 is NP-complete, then A_0 is one of the most difficult problems in NP." This statement is true on one level: if there is any problem at all in NP that is not in P, then the NP-complete language A_0 is one such problem. However, note that there are NP-complete problems in NTIME[n]—and these problems are, in some sense, much *simpler* than many problems in NTIME[$n^{10^{500}}$].

7.5.3 Cook–Levin Theorem

Interest in NP-complete problems started with a theorem of Cook (1971), proved independently by Levin (1973). Recall that SAT is the language of Boolean formulas $\phi(z_1,\ldots, z_r)$ such that there exists a truth assignment to the variables $z_1,\ldots, z_r$ that makes ϕ true.

Theorem 7.9 (Cook–Levin Theorem)

SAT *is* NP-*complete.*

Proof. We know already that SAT is in NP, so to prove that SAT is NP-complete, we need to take an arbitrary given language A in NP and show that $A \leq_m^p$ SAT. Take N to be a nondeterministic Turing machine that decides A in polynomial time. Then the relation $R(x, y)$ = "y is a computation path of N that leads it to accept x" is decidable in deterministic polynomial time depending only on $n = |x|$. We can assume that the length m of possible y's encoded as binary strings depends only on n and not on a particular x.

It is straightforward to show that there is a polynomial p and for each n a Boolean circuit C_n^R with $p(n)$ wires, with $n + m$ input wires labeled $x_1,\ldots, x_n, y_1,\ldots, y_m$ and one output wire w_0, such that $C_n^R(x,y)$ outputs 1 if and only if $R(x, y)$ holds. (We describe circuits in more detail in the following text, and state a theorem for this principle as Part 1 of Theorem 7.15) Importantly, C_n^R itself can be designed in time polynomial in n, and – by the universality of NAND – may be composed entirely of binary NAND gates. Label the wires by variables $x_1,\ldots, x_n, y_1,\ldots, y_m, w_0, w_1,\ldots, w_{p(n)-n-m-1}$. These become the variables of our Boolean formulas. For each NAND gate g with input wires u and v, and for each output wire w of g, write down the subformula

$$\phi_{g,w} = (u \vee w) \wedge (v \vee w) \wedge (\bar{u} \vee \bar{v} \vee \bar{w}).$$

This subformula is satisfied by precisely those assignments to u, v, w that give $w = u$ NAND v. The conjunction ϕ_0 of $\phi_{g,w}$ over the polynomially many gates g and their output wires w thus is satisfied only by assignments that set every gate's output correctly given its inputs. Thus for any binary strings x and y of lengths n, m respectively, the formula $\phi_1 = \phi_0 \wedge w_0$ is satisfiable by a setting of the wire variables $w_0, w_1,\ldots, w_{p(n)-n-m-1}$ if and only if $C_n^R(x,y) = 1$—that is, if and only if $R(x, y)$ holds.

Now given any fixed x and taking $n = |x|$, the Karp reduction computes ϕ_1 as mentioned earlier and finally outputs the Boolean formula ϕ obtained by substituting the bit-values of x into ϕ_1. This ϕ has variables $y_1,\ldots, y_m, w_0, w_1,\ldots, w_{p(n)-n-m-1}$, and the computation of ϕ from x runs in deterministic polynomial time. Then $x \in A$ if and only if N accepts x, if and only if there exists y such that $R(x, y)$ holds, if and only if there exists an assignment to the variables $w_0, w_1,\ldots, w_{p(n)-n-m-1}$ *and* $y_1,\ldots, y_m$ that satisfies ϕ, if and only if $\phi \in$ SAT. This shows $A \leq_m^p$ SAT. □

We have actually proved that SAT remains NP-complete even when the given instances ϕ are *restricted* to Boolean formulas that are a conjunction of *clauses*, where each clause consists of (here, at most three) disjuncted literals. Such formulas are said to be in *conjunctive normal form*. Theorem 7.9 is also commonly known as **Cook's Theorem.**

7.5.4 Proving NP-Completeness

After one language has been proved complete for a class, others can be proved complete by constructing transformations. For NP, if A_0 is NP-complete, then to prove that another language A_1 is NP-complete, it suffices to prove that $A_1 \in$ NP and to construct a polynomial-time transformation that establishes $A_0 \leq_m^p A_1$. Since A_0 is NP-complete, for every language A in NP, $A \leq_m^p A_0$, hence, by transitivity (Theorem 7.7), $A \leq_m^p A_1$.

Beginning with Cook (1971) and Karp (1972), hundreds of computational problems in many fields of science and engineering have been proved to be NP-complete, almost always by reduction from a problem that was previously known to be NP-complete. The following NP-complete decision problems are frequently used in these reductions—the language corresponding to each problem is the set of instances whose answers are yes.

- 3-Satisfiability (3SAT)
  Instance: A Boolean expression ϕ in conjunctive normal form with three literals per clause [e.g., $(w \vee x \vee \bar{y}) \wedge (\bar{x} \vee y \vee z)$]
  Question: Is ϕ satisfiable?
- Vertex Cover
  Instance: A graph G and an integer k
  Question: Does G have a set W of k vertices such that every edge in G is incident on a vertex of W?
- Clique
  Instance: A graph G and an integer k
  Question: Does G have a set U of k vertices such that every two vertices in U are adjacent in G?
- Independent Set
  Instance: A graph G and an integer k
  Question: Does G have a set U of k vertices such that no two vertices in U are adjacent in G?
- Hamiltonian Circuit
  Instance: A graph G
  Question: Does G have a circuit that includes every vertex exactly once?
- Three-Dimensional Matching
  Instance: Sets W, X, Y with $|W| = |X| = |Y| = q$ and a subset $S \subseteq W \times X \times Y$
  Question: Is there a subset $S' \subseteq S$ of size q such that no two triples in S' agree in any coordinate?
- Partition
  Instance: A set S of positive integers
  Question: Is there a subset $S' \subseteq S$ such that the sum of the elements of S' equals the sum of the elements of $S - S'$?

Note that our ϕ in the above proof of the Cook–Levin Theorem already meets a form of the definition of 3SAT relaxed to allow "at most 3 literals per clause." Padding ϕ with some extra variables to bring up the number in each clause to exactly three, while preserving whether the formula is satisfiable or not, is not difficult and establishes the NP-completeness of 3SAT. Below we use this in showing the NP-completeness of INDEPENDENT SET.

Some NP-completeness proofs require only "minor adjustments" even between different-looking problems. Consider CLIQUE, INDEPENDENT SET, and VERTEX COVER. A graph G has a clique of size k if and only if its complementary graph G' has an independent set of size k. It follows that the function f defined by $f(G, k) = (G', k)$ is a Karp reduction from INDEPENDENT SET to CLIQUE. To forge a link to the VERTEX COVER problem, note that all vertices *not* in a given vertex cover form an independent set, and vice versa. Thus a graph G on n vertices has a vertex cover of size at most k if and only if G has an independent set of size at least $n - k$. Hence the function $g(G, k) = (G, n - k)$ is a Karp reduction from INDEPENDENT SET to VERTEX COVER.

Here is another example of an NP-completeness proof, for the following decision problem:

- TRAVELING SALESMAN PROBLEM (TSP)
  Instance: A set of m "cities" $C_1, \ldots, C_m$, with an integer distance $d(i, j)$ between every pair of cities C_i and C_j, and an integer D.
  Question: Is there a tour of the cities whose total length is at most D, that is, a permutation $c_1, \ldots, c_m$ of $\{1, \ldots, m\}$, such that

$$d(c_1, c_2) + \cdots + d(c_{m-1}, c_m) + d(c_m, c_1) \le D?$$

First, it is easy to see that TSP is in NP: a nondeterministic Turing machine simply guesses a tour and checks that the total length is at most D.

Next, we construct a reduction from Hamiltonian Circuit to TSP. (The reduction goes from the known NP-complete problem, Hamiltonian Circuit, to the new problem, TSP, not vice versa.)

From a graph G on m vertices $v_1, \ldots, v_m$, define the distance function d as follows:

$$d(i, j) = \begin{cases} 1 & \text{if } (v_i, v_j) \text{ is an edge in } G \\ m+1 & \text{otherwise} \end{cases}$$

Set $D = m$. Clearly, d and D can be computed in polynomial time from G. Each vertex of G corresponds to a city in the constructed instance of TSP.

If G has a Hamiltonian circuit, then the length of the tour that corresponds to this circuit is exactly m. Conversely, if there is a tour whose length is at most m, then each step of the tour must have distance 1, not $m + 1$. Thus, each step corresponds to an edge of G, and the corresponding sequence of vertices in G is a Hamiltonian circuit.

7.5.5 NP-Completeness by Combinatorial Transformation

The following example shows how the combinatorial mechanism of one problem (here, 3SAT) can be *transformed* by a reduction into the seemingly much different mechanism of another problem.

Theorem 7.10

INDEPENDENT SET is NP-complete. Hence also CLIQUE and VERTEX COVER are NP-complete.

Proof. We have remarked already that the languages of these three problems belong to NP and shown already that INDEPENDENT SET $\le_m^p$ CLIQUE and INDEPENDENT SET $\le_m^p$ VERTEX COVER. It suffices to show that 3SAT $\le_m^p$ INDEPENDENT SET.

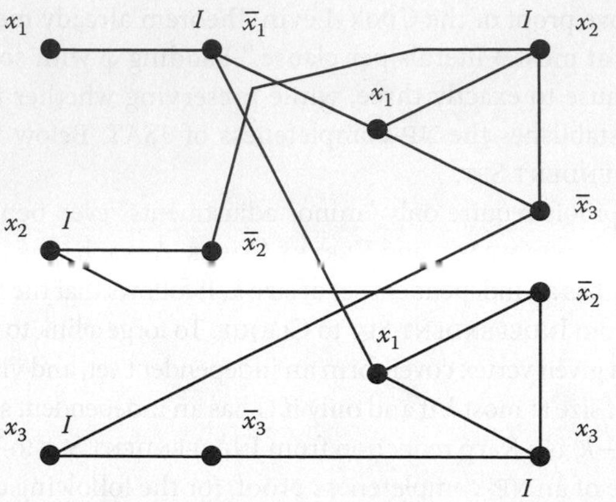

FIGURE 7.3 Construction in the proof of NP-completeness of Independent Set for the formula $(x_1 \vee x_2 \vee \bar{x}_3) \wedge (x_1 \vee \bar{x}_2 \vee x_3)$. The independent set of size 5 corresponding to the satisfying assignment x_1 = `false`, x_2 = `true`, and x_3 = `true` is shown by nodes marked *I*.

Construction. Let the Boolean formula ϕ be a given instance of 3SAT with variables $x_1,\ldots, x_n$ and clauses $C_1,\ldots, C_m$. The graph G_ϕ we build consists of a "ladder" on $2n$ vertices labeled $x_1, \bar{x}_1,\ldots, x_n, \bar{x}_n$, with edges $(x_i, \bar{x}_i)$ for $1 \leq i \leq n$ forming the "rungs," and m "clause components." Here the component for each clause C_j has one vertex for each literal x_i or $\bar{x}_i$ in the clause, and all pairs of vertices within each clause component are joined by an edge. Finally, each clause-component node with a label x_i is connected by a "crossing edge" to the node with the opposite label $\bar{x}_i$ in the ith "rung," and similarly each occurrence of $\bar{x}_i$ in a clause is joined to the rung node x_i. This finishes the construction of G_ϕ. See Figure 7.3.

Also set $k = n + m$. Then the reduction function f is defined for all arguments ϕ by $f(\phi) = (G_\phi, k)$.

Complexity. It is not hard to see that f is computable in polynomial time given (a straightforward encoding of) ϕ.

Correctness. To complete the proof, we need to argue that ϕ is satisfiable if and only if G_ϕ has an independent set of size $n + m$. To see this, first note that any independent set I of that size must contain exactly one of the two nodes from each "rung," and exactly one node from each clause component—because the edges in the rungs and the clause component prevent any more nodes from being added. And if I selects a node labeled x_i in a clause component, then I must also select x_i in the ith rung. If I selects $\bar{x}_j$ in a clause component, then I must also select $\bar{x}_j$ in the rung. In this manner I induces a truth assignment in which x_i = `true` and x_j = `false`, and so on for all variables. This assignment satisfies ϕ, because the node selected from each clause component tells how the corresponding clause is satisfied by the assignment. Going the other way, if ϕ has a satisfying assignment, then that assignment yields an independent set I of size $n + m$ in like manner. □

7.5.6 Complete Problems for Other Classes

Besides NP, the following canonical complexity classes have natural complete problems. The three problems now listed are complete for their respective classes under log-space reducibility.

- NL: Graph Accessibility Problem
  Instance: A directed graph G with nodes $1,\ldots, N$
  Question: Does G have a directed path from node 1 to node N?
- P: Circuit Value Problem
  Instance: A Boolean circuit (see Section 7.9) with output node u, and an assignment I of $\{\mathbf{0}, \mathbf{1}\}$ to each input node
  Question: Is $\mathbf{1}$ the value of u under I?

- PSPACE: Quantified Boolean Formulas
  Instance: A Boolean expression with all variables quantified with either ∀ or ∃ [e.g., $\forall x \forall y \exists z(x \wedge (\bar{y} \vee z))$]
  Question: Is the expression `true`?

These problems can be used to prove that other problems are NL-complete, P-complete, and PSPACE-complete, respectively.

Stockmeyer and Meyer (1973) defined a natural decision problem that they proved to be complete for NE. If this problem were in P, then by closure under Karp reducibility (Theorem 7.8), we would have NE ⊆ P, a contradiction of the hierarchy theorems (Theorem 7.4). Therefore, this decision problem is infeasible: it has no polynomial-time algorithm. In contrast, decision problems in NEXP − P that have been constructed by diagonalization are artificial problems that nobody would want to solve anyway. Although diagonalization produces unnatural problems by itself, the combination of diagonalization and completeness shows that *natural* problems are intractable.

The next section points out some limitations of current diagonalization techniques.

7.6 Relativization of the P vs. NP Problem

Let A be a language. Define P^A (respectively, NP^A) to be the class of languages accepted in polynomial time by deterministic (nondeterministic) oracle Turing machines with oracle A.

Proofs that use the diagonalization technique on Turing machines without oracles generally carry over to oracle Turing machines. Thus, for instance, the proof of the DTIME hierarchy theorem also shows that, for *any* oracle A, $\mathrm{DTIME}^A[n^2]$ is properly contained in $\mathrm{DTIME}^A[n^3]$. This can be seen as a *strength* of the diagonalization technique, since it allows an argument to "relativize" to computation carried out relative to an oracle. In fact, there are examples of lower bounds (for deterministic, "unrelativized" circuit models) that make crucial use of the fact that the time hierarchies relativize in this sense.

But it can also be seen as a weakness of the diagonalization technique. The following important theorem demonstrates why.

Theorem 7.11

There exist languages A and B such that $\mathrm{P}^A = \mathrm{NP}^A$, and $\mathrm{P}^B \neq \mathrm{NP}^B$ (Baker et al., 1975).

This shows that resolving the P vs. NP question requires techniques that do not relativize, that is, that do not apply to oracle Turing machines too. Thus, diagonalization as we currently know it is unlikely to succeed in separating P from NP, because the diagonalization arguments we know (and in fact *most* of the arguments we know) relativize. Important nonrelativizing proof techniques were discovered in the early 1990s, in connection with interactive proof systems (Section 7.12.1).

7.7 The Polynomial Hierarchy

Let C be a class of languages. Define

- $\mathrm{NP}^C = \bigcup_{A \in C} \mathrm{NP}^A$
- $\Sigma_0^P = \Pi_0^P = \mathrm{P}$

and for $k \geq 0$, define

- $\Sigma_{k+1}^P = \mathrm{NP}^{\Sigma_k^P}$
- $\Pi_{k+1}^P = \text{co-}\Sigma_{k+1}^P$

Observe that $\Sigma_1^P = \mathrm{NP}^{\mathrm{P}} = \mathrm{NP}$, because each of polynomially many queries to an oracle language in P can be answered directly by a (nondeterministic) Turing machine in polynomial time. Consequently, $\Pi_1^P = \text{co-NP}$. For each k, $\Sigma_k^P \cup \Pi_k^P \subseteq \Sigma_{k+1}^P \cap \Pi_{k+1}^P$, but this inclusion is not known to be strict.

The classes Σ_k^P and Π_k^P constitute the **polynomial hierarchy**. Define

$$\mathrm{PH} = \bigcup_{k \geq 0} \Sigma_k^P$$

It is straightforward to prove that PH ⊆ PSPACE, but it is not known whether the inclusion is strict. In fact, if PH = PSPACE, then the polynomial hierarchy collapses to some level, that is, $\mathrm{PH} = \Sigma_m^P$ for some m. In the next section, we define the polynomial hierarchy in two other ways, one of which is in terms of alternating Turing machines.

7.8 Alternating Complexity Classes

In this section, we define time and space complexity classes for alternating Turing machines, and we show how these classes are related to the classes introduced already. The possible computations of an alternating Turing machine M on an input word x can be represented by a tree T_x in which the root is the initial configuration, and the children of a nonterminal node C are the configurations reachable from C by one step of M. For a word x in $L(M)$, define an **accepting subtree** S of T_x to be a subtree of T_x with the following properties:

- S is finite.
- The root of S is the initial configuration with input word x.
- If S has an existential configuration C, then S has exactly one child of C in T_x; if S has a universal configuration C, then S has all children of C in T_x.
- Every leaf is a configuration whose state is the accepting state q_A.

Observe that each node in S is an accepting configuration.

We consider only alternating Turing machines that always halt. For $x \in L(M)$, define the time taken by M to be the height of the shortest accepting tree for x, and the space to be the maximum number of worktape cells visited along any path in an accepting tree that minimizes this number. For $x \notin L(M)$, define the time to be the height of T_x, and the space to be the maximum number of worktape cells visited along any path in T_x.

Let $t(n)$ be a time-constructible function, and let $s(n)$ be a space-constructible function. Define the following complexity classes:

- ATIME$[t(n)]$ is the class of languages decided by alternating Turing machines of time complexity $O(t(n))$.
- ASPACE$[s(n)]$ is the class of languages decided by alternating Turing machines of space complexity $O(s(n))$.

Because a nondeterministic Turing machine is a special case of an alternating Turing machine, for every $t(n)$ and $s(n)$, NTIME$[t]$ ⊆ ATIME$[t]$ and NSPACE$[s]$ ⊆ ASPACE$[s]$. The next theorem states further relationships between computational resources used by alternating Turing machines and resources used by deterministic and nondeterministic Turing machines.

Theorem 7.12 (Alternation Theorems)

Let $t(n)$ be a time-constructible function, and let $s(n)$ be a space-constructible function, $s(n) \geq \log n$ (Chandra et al., 1981).

(a) NSPACE[$s(n)$] ⊆ ATIME[$s(n)^2$].
(b) ATIME[$t(n)$] ⊆ DSPACE[$t(n)$].
(c) ASPACE[$s(n)$] ⊆ DTIME[$2^{O(s(n))}$].
(d) DTIME[$t(n)$] ⊆ ASPACE[$\log t(n)$].

In other words, space on deterministic and nondeterministic Turing machines is polynomially related to time on alternating Turing machines. Space on alternating Turing machines is exponentially related to time on deterministic Turing machines. The following corollary is immediate.

Theorem 7.13

(a) ASPACE[$O(\log n)$] = P.
(b) ATIME[$n^{O(1)}$] = PSPACE.
(c) ASPACE[$n^{O(1)}$] = EXP.

In Section 7.7, we defined the classes of the polynomial hierarchy in terms of oracles, but we can also define them in terms of alternating Turing machines with restrictions on the number of alternations between existential and universal states. Define a *k-alternating Turing machine* to be a machine such that on every computation path, the number of changes from an existential state to a universal state, or from a universal state to an existential state, is at most $k - 1$. Thus, a nondeterministic Turing machine, which stays in existential states, is a 1-alternating Turing machine.

Theorem 7.14

For any language A, the following are equivalent (Stockmeyer, 1976; Wrathall, 1976):

1. $A \in \Sigma_k^P$.
2. A is decided in polynomial time by a k-alternating Turing machine that starts in an existential state.
3. There exists a language B in P and a polynomial p such that for all x, $x \in A$ if and only if

$$(\exists y_1 : |y_1| \leq p(|x|))(\forall y_2 : |y_2| \leq p(|x|)) \cdots (Qy_k : |y_k| \leq p(|x|))[(x, y_1, \ldots, y_k) \in B]$$

where the quantifier Q is $\exists$ if k is odd, $\forall$ if k is even.

Alternating Turing machines are closely related to Boolean circuits, which are defined in Section 7.9.

7.9 Circuit Complexity

The hardware of electronic digital computers is based on digital logic gates, connected into combinational circuits (see Chapter 17). Here, we specify a model of computation that formalizes the combinational circuit.

A *Boolean circuit* on n input variables $x_1, \ldots, x_n$ is a directed acyclic graph with exactly n input nodes of indegree 0 labeled $x_1, \ldots, x_n$ and other nodes of indegree 1 or 2, called *gates*, labeled with the Boolean operators in $\{\wedge, \vee, \neg\}$. One node is designated as the output of the circuit. See Figure 7.4. Without loss of generality, we assume that there are no extraneous nodes; there is a directed path from each node to the output node. The indegree of a gate is also called its *fan-in*.

An *input assignment* is a function I that maps each variable x_i to either 0 or 1. The value of each gate g under I is obtained by applying the Boolean operation that labels g to the values of the immediate predecessors of g. The function computed by the circuit is the value of the output node for each input assignment.

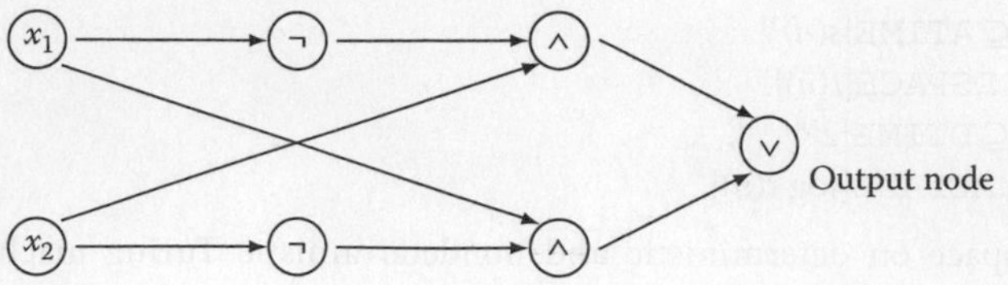

FIGURE 7.4 A Boolean circuit.

A Boolean circuit computes a finite function: a function of only n binary input variables. To decide membership in a language, we need a circuit for each input length n.

A *circuit family* is an infinite set of circuits $C = \{c_1, c_2, \ldots\}$ in which each c_n is a Boolean circuit on n inputs. C *decides* a language $A \subseteq \{\mathbf{0}, \mathbf{1}\}^*$ if for every n and every assignment $a_1, \ldots, a_n$ of $\{\mathbf{0}, \mathbf{1}\}$ to the n inputs, the value of the output node of c_n is $\mathbf{1}$ if and only if the word $a_1 \cdots a_n \in A$. The *size complexity* of C is the function $z(n)$ that specifies the number of nodes in each c_n. The *depth complexity* of C is the function $d(n)$ that specifies the length of the longest directed path in c_n. Clearly, since the fan-in of each gate is at most 2, $d(n) \geq \log z(n) \geq \log n$. The class of languages decided by polynomial-size circuits is denoted by P/poly.

With a different circuit for each input length, a circuit family could solve an undecidable problem such as the halting problem (see Chapter 8). For each input length, a table of all answers for machine descriptions of that length could be encoded into the circuit. Thus, we need to restrict our circuit families. The most natural restriction is that all circuits in a family should have a concise, uniform description, to disallow a different answer table for each input length. Several uniformity conditions have been studied; the following is the easiest to present:

A circuit family $\{c_1, c_2, \ldots\}$ of size complexity $z(n)$ is *log-space uniform* if there exists a deterministic Turing machine M such that on each input of length n, machine M produces a description of c_n, using space $O(\log z(n))$.

Now we define complexity classes for uniform circuit families and relate these classes to previously defined classes. Define the following complexity classes:

- SIZE$[z(n)]$ is the class of languages decided by log-space uniform circuit families of size complexity $O(z(n))$.
- DEPTH$[d(n)]$ is the class of languages decided by log-space uniform circuit families of depth complexity $O(d(n))$.

In our notation, SIZE$[n^{O(1)}]$ equals P, which is a proper subclass of P/poly.

Theorem 7.15

1. If $t(n)$ is a time-constructible function, then DTIME$[t(n)] \subseteq$ SIZE$[t(n) \log t(n)]$ (Pippenger and Fischer, 1979).
2. SIZE$[z(n)] \subseteq$ DTIME$[z(n)^{O(1)}]$.
3. If $s(n)$ is a space-constructible function and $s(n) \geq \log n$, then NSPACE$[s(n)] \subseteq$ DEPTH$[s(n)^2]$ (Borodin, 1977).
4. If $d(n) \geq \log n$, then DEPTH$[d(n)] \subseteq$ DSPACE$[d(n)]$ (Borodin, 1977).

The next theorem shows that size and depth on Boolean circuits are closely related to space and time on alternating Turing machines, provided that we permit sublinear running times for alternating Turing machines, as follows. We augment alternating Turing machines with a random-access input capability.

To access the cell at position j on the input tape, M writes the binary representation of j on a special tape, in $\log j$ steps, and enters a special reading state to obtain the symbol in cell j.

Theorem 7.16

(Ruzzo, 1979). Let $t(n) \geq \log n$ and $s(n) \geq \log n$ be such that the mapping $n \mapsto (t(n), s(n))$ (in binary) is computable in time $s(n)$.

1. Every language decided by an alternating Turing machine of simultaneous space complexity $s(n)$ and time complexity $t(n)$ can be decided by a log-space uniform circuit family of simultaneous size complexity $2^{O(s(n))}$ and depth complexity $O(t(n))$.
2. If $d(n) \geq (\log z(n))^2$, then every language decided by a log-space uniform circuit family of simultaneous size complexity $z(n)$ and depth complexity $d(n)$ can be decided by an alternating Turing machine of simultaneous space complexity $O(\log z(n))$ and time complexity $O(d(n))$.

The correspondence between time on alternating Turing machines and depth on log-space uniform circuits breaks down for time and depth less than $\log^2 n$, primarily because L is not known to be contained in ATIME($o(\log^2 n)$). This has motivated the adoption of more restrictive notions of uniformity (see, e.g., Ruzzo, 1981; Vollmer, 1999). When researchers mention small uniform circuit classes, they usually are referring to these more restrictive notions of uniformity. (For depths at least $\log n$, the complexity classes obtained via these different notions of uniformity coincide.)

In a sense, the Boolean circuit family is a model of parallel computation, because all gates compute independently, in parallel. For each $k \geq 0$, NC^k denotes the class of languages decided by uniform bounded-fan-in circuits of polynomial size and depth $O((\log n)^k)$, and AC^k is defined analogously for unbounded fan-in circuits. In particular, AC^k is the same as the class of languages decided by a parallel machine model called the CRCW PRAM with polynomially many processors in parallel time $O((\log n)^k)$ (Stockmeyer and Vishkin, 1984).

7.10 Probabilistic Complexity Classes

Since the 1970s, with the development of randomized algorithms for computational problems (see Chapter 11), complexity theorists have placed randomized algorithms on a firm intellectual foundation. In this section, we outline some basic concepts in this area.

A **probabilistic Turing machine** M can be formalized as a nondeterministic Turing machine with exactly two choices at each step. During a computation, M chooses each possible next step with independent probability 1/2. Intuitively, at each step, M flips a fair coin to decide what to do next. The probability of a computation path of t steps is $1/2^t$. The probability that M accepts an input string x, denoted by $p_M(x)$, is the sum of the probabilities of the accepting computation paths.

Throughout this section, we consider only machines whose time complexity $t(n)$ is time-constructible. Without loss of generality, we may assume that every computation path of such a machine halts in exactly t steps.

Let A be a language. A probabilistic Turing machine M decides A with

		for all $x \in A$	for all $x \notin A$
Unbounded two-sided error	if	$p_M(x) > 1/2$	$p_M(x) \leq 1/2$
Bounded two-sided error	if	$p_M(x) > 1/2 + \varepsilon$	$p_M(x) < 1/2 - \varepsilon$
		for some positive constant ε	
One-sided error	if	$p_M(x) > 1/2$	$p_M(x) = 0$.

Many practical and important probabilistic algorithms make one-sided errors. For example, in the primality testing algorithm of Solovay and Strassen (1977), when the input x is a prime number, the algorithm *always* says "prime"; when x is composite, the algorithm *usually* says "composite," but may occasionally say "prime." Using the definitions above, this means that the Solovay–Strassen algorithm is a one-sided error algorithm for the set A of composite numbers. It also is a bounded two-sided error algorithm for $\bar{A}$, the set of prime numbers.

These three kinds of errors suggest three complexity classes:

- PP is the class of languages decided by probabilistic Turing machines of polynomial-time complexity with unbounded two-sided error.
- BPP is the class of languages decided by probabilistic Turing machines of polynomial-time complexity with bounded two-sided error.
- RP is the class of languages decided by probabilistic Turing machines of polynomial-time complexity with one-sided error.

In the literature, RP has also sometimes been called R.

A probabilistic Turing machine M is a PP-**machine** (respectively, a BPP-**machine**, an RP-**machine**) if M has polynomial-time complexity, and M decides with two-sided error (bounded two-sided error, one-sided error).

Through repeated Bernoulli trials, we can make the error probabilities of BPP-machines and RP-machines arbitrarily small, as stated in the following theorem. (Among other things, this theorem implies that RP ⊆ BPP.)

Theorem 7.17

If $A \in$ BPP, then for every polynomial $q(n)$, there exists a BPP-machine M such that $p_M(x) > 1 - 1/2^{q(n)}$ for every $x \in A$, and $p_M(x) < 1/2^{q(n)}$ for every $x \notin A$.

If $L \in$ RP, then for every polynomial $q(n)$, there exists an RP-machine M such that $p_M(x) > 1 - 1/2^{q(n)}$ for every x in L.

It is important to note just how minuscule the probability of error is (provided that the coin flips are truly random). If the probability of error is less than $1/2^{5000}$, then it is less likely that the algorithm produces an incorrect answer than that the computer will be struck by a meteor. An algorithm whose probability of error is $1/2^{5000}$ is essentially as good as an algorithm that makes no errors. For this reason, many computer scientists consider BPP to be the class of practically feasible computational problems.

Next, we define a class of problems that have probabilistic algorithms that make no errors. Define

- ZPP = RP ∩ co-RP.

The letter Z in ZPP is for zero probability of error, as we now demonstrate. Suppose $A \in$ ZPP. Here is an algorithm that checks membership in A. Let M be an RP-machine that decides A, and let M' be an RP-machine that decides $\bar{A}$. For an input string x, alternately run M and M' on x, repeatedly, until a computation path of one machine accepts x. If M accepts x, then accept x; if M' accepts x, then reject x. This algorithm works correctly because when an RP-machine accepts its input, it does not make a mistake. This algorithm might not terminate, but with very high probability, the algorithm terminates after a few iterations.

The next theorem expresses some known relationships between probabilistic complexity classes and other complexity classes, such as classes in the polynomial hierarchy. See Section 7.7 and Figure 7.5.

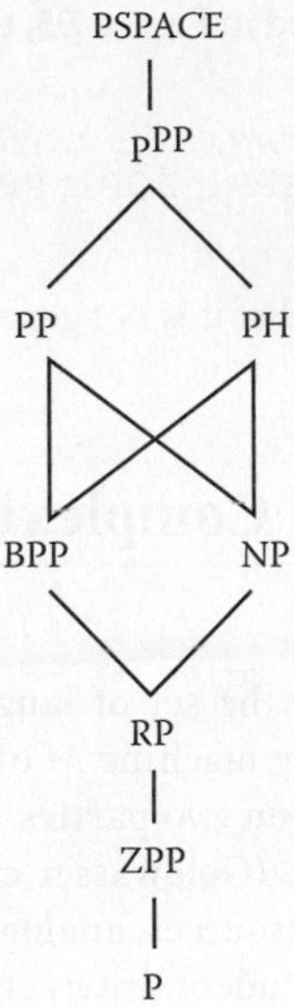

FIGURE 7.5 Probabilistic complexity classes.

Theorem 7.18

(a) $\mathrm{P} \subseteq \mathrm{ZPP} \subseteq \mathrm{RP} \subseteq \mathrm{BPP} \subseteq \mathrm{PP} \subseteq \mathrm{PSPACE}$ (Gill, 1977).
(b) $\mathrm{RP} \subseteq \mathrm{NP} \subseteq \mathrm{PP}$ (Gill, 1977).
(c) $\mathrm{BPP} \subseteq \Sigma_2^P \cap \Pi_2^P$ (Lautemann, 1983; Sipser, 1983).
(d) $\mathrm{BPP} \subset \mathrm{P/poly}$.
(e) $\mathrm{PH} \subseteq \mathrm{P}^{\mathrm{PP}}$ (Toda, 1991).

An important recent research area called **de-randomization** studies whether randomized algorithms can be converted to deterministic ones of the same or comparable efficiency. For example, if there is a language in E that requires Boolean circuits of size $2^{\Omega(n)}$ to decide it, then BPP = P (Impagliazzo and Wigderson, 1997).

7.11 Quantum Computation

A probabilistic computer enables one to sample efficiently over exponentially many possible computation paths. However, if BPP = P then the added capability is weak. A *quantum computer*, however, can harness parallelism and randomness in ways that appear theoretically to be much more powerful, ways that are able to solve problems believed not to lie in BPP. A key difference is that by quantum *interference*, opening up new computation paths may cause the probabilities of some other paths to *vanish*, whereas for a BPP-machine those probabilities would remain positive.

- BQP is the class of languages decided by quantum machines of polynomial-time complexity with bounded two-sided error.

A prime motivation for studying BQP is that it includes language versions of the integer factoring and discrete logarithm problems, which are not known to belong to BPP. The public-key cryptosystems in common use today rely on the presumed intractability of these problems and are theoretically breakable by eavesdroppers armed with quantum computers. This is one of several reasons why the interplay of quantum mechanics and computational complexity is important to cryptographers.

In terms of the complexity classes depicted in Figure 7.5, the only inclusions that are known involving BQP are

$$\mathrm{BPP} \subseteq \mathrm{BQP} \subseteq \mathrm{PP}.$$

In particular, it is important to emphasize that it is *not* generally believed that quantum computers can solve NP-complete problems quickly.

7.12 Interactive Models and Complexity Classes

7.12.1 Interactive Proofs

In Section 7.3.2, we characterized NP as the set of languages whose membership proofs can be checked quickly, by a deterministic Turing machine M of polynomial-time complexity. A different notion of proof involves interaction between two parties, a prover P and a verifier V, who exchange messages. In an **interactive proof system** (Goldwasser et al., 1989), the prover is an all-powerful machine, with unlimited computational resources, analogous to a teacher. The verifier is a computationally limited machine, analogous to a student. Interactive proof systems are also called "Arthur–Merlin games": the wizard Merlin corresponds to P and the impatient Arthur corresponds to V (Babai and Moran, 1988).

Formally, an **interactive proof system** comprises the following:

- A read-only input tape on which an input string x is written.
- A *verifier* V, which is a probabilistic Turing machine augmented with the capability to send and receive messages. The running time of V is bounded by a polynomial in $|x|$.
- A *prover* P, which receives messages from V and sends messages to V.
- A tape on which V writes messages to send to P, and a tape on which P writes messages to send to V. The length of every message is bounded by a polynomial in $|x|$.

A computation of an interactive proof system (P, V) proceeds in rounds, as follows. For $j = 1, 2,\ldots$, in round j, V performs some steps, writes a message m_j, and temporarily stops. Then P reads m_j and responds with a message m'_j, which V reads in round $j + 1$. An interactive proof system (P, V) **accepts** an input string x if the probability of acceptance by V satisfies $p_V(x) > 1/2$.

In an interactive proof system, a prover can convince the verifier about the truth of a statement without exhibiting an entire proof, as the following example illustrates.

Consider the graph nonisomorphism problem: the input consists of two graphs G and H, and the decision is yes if and only if G is not isomorphic to H. Although there is a short proof that two graphs *are* isomorphic (namely: the proof consists of the isomorphism mapping G onto H), nobody has found a general way of proving that two graphs are *not* isomorphic that is significantly shorter than listing all $n!$ permutations and showing that each fails to be an isomorphism. (That is, the graph nonisomorphism problem is in co-NP but is not known to be in NP.) In contrast, the verifier V in an interactive proof system is able to take statistical evidence into account, and determine "beyond all reasonable doubt" that two graphs are nonisomorphic, using the following protocol.

In each round, V randomly chooses either G or H with equal probability; if V chooses G, then V computes a random permutation G' of G, presents G' to P, and asks P whether G' came from G or from H (and similarly H' if V chooses H). If G and H are nonisomorphic, then P always answers correctly. If G and H are isomorphic, then after k consecutive rounds, the probability that P always answers correctly is $1/2^k$. (To see this, it is important to understand that the prover P does not see the coins that V flips in making its random choices; P sees only the graphs G' and H' that V sends as messages.) V accepts the interaction with P as "proof" that G and H are nonisomorphic if P is able to pick the correct graph for 100 consecutive rounds. Note that V has ample grounds to accept this as a convincing demonstration: if the graphs are indeed isomorphic, the prover *P* would have to have an incredible streak of luck to fool *V*.

It is important to comment that de-randomization techniques applied to these proof systems have shown that under plausible hardness assumptions, proofs of nonisomorphism of subexponential length (or even polynomial length) do exist (Klivans and van Melkebeek, 2002). Thus many complexity theoreticians now conjecture that the graph isomorphism problem lies in NP ∩ co-NP.

The complexity class IP comprises the languages *A* for which there exists a verifier *V* and a positive ε such that

- There exists a prover $\hat{P}$ such that for all x in A, the interactive proof system $(\hat{P}, V)$ accepts x with probability greater than $1/2 + \varepsilon$
- For every prover P and every $x \notin A$, the interactive proof system (P, V) rejects x with probability greater than $1/2 + \varepsilon$

By substituting random choices for existential choices in the proof that $\mathtt{ATIME}(t) \subseteq \mathtt{DSPACE}(t)$ (Theorem 7.12), it is straightforward to show that IP ⊆ PSPACE. It was originally believed likely that IP was a small subclass of PSPACE. Evidence supporting this belief was the construction of an oracle language B for which $\text{co-NP}^B - \mathtt{IP}^B \neq \emptyset$ (Fortnow and Sipser, 1988), so that $\mathtt{IP}^B$ is strictly included in $\mathtt{PSPACE}^B$. Using a proof technique that does not relativize, however, Shamir (1992) proved that in fact, IP and PSPACE are the same class.

Theorem 7.19

IP = PSPACE (Shamir, 1992).

If NP is a proper subset of PSPACE, as is widely believed, then Theorem 7.19 says that interactive proof systems can decide a larger class of languages than NP.

7.12.2 Probabilistically Checkable Proofs

In an interactive proof system, the verifier does not need a complete conventional proof to become convinced about the membership of a word in a language, but uses random choices to query parts of a proof that the prover may know. This interpretation inspired another notion of "proof": a proof consists of a (potentially) large amount of information that the verifier need only inspect in a few places in order to become convinced. The following definition makes this idea more precise.

A language *A* has a **probabilistically checkable proof** if there exists an oracle BPP-machine *M* such that

- For all $x \in A$, there exists an oracle language B_x such that M^{B_x} accepts x with probability 1.
- For all $x \notin A$, and for every language B, machine M^B accepts x with probability strictly less than 1/2.

Intuitively, the oracle language B_x represents a proof of membership of x in A. Notice that B_x can be finite since the length of each possible query during a computation of M^{B_x} on x is bounded by the running time of M. The oracle language takes the role of the prover in an interactive proof system—but in contrast to an interactive proof system, the prover cannot change strategy adaptively in response to the questions that the verifier poses. This change results in a potentially stronger system, since a machine M that has bounded error probability relative to all languages B might not have bounded error probability relative to some adaptive prover. Although this change to the proof system framework may seem modest, it leads to a characterization of a class that seems to be much larger than PSPACE.

Theorem 7.20

A has a probabilistically checkable proof if and only if $A \in$ NEXP (Babai et al., 1991).

Although the notion of probabilistically checkable proofs seems to lead us away from feasible complexity classes, by considering natural restrictions on how the proof is accessed, we can obtain important insights into familiar complexity classes.

Let PCP[$r(n)$, $q(n)$] denote the class of languages with probabilistically checkable proofs in which the probabilistic oracle Turing machine M makes $r(n)$ random binary choices and queries its oracle $q(n)$ times. (For this definition, we assume that M has either one or two choices for each step.) It follows from the definitions that BPP = PCP[$n^{O(1)}$, 0], and NP = PCP[0, $n^{O(1)}$].

Theorem 7.21 (The PCP Theorem)

NP = PCP[$O(\log n)$, $O(1)$] (Arora et al., 1998).

Theorem 7.21 asserts that for every language A in NP, a proof that $x \in A$ can be encoded so that the verifier can be convinced of the correctness of the proof (or detect an incorrect proof) by using only $O(\log n)$ random choices and inspecting only a *constant* number of bits of the proof.

7.13 Kolmogorov Complexity

Until now, we have considered only dynamic complexity measures, namely, the time and space used by Turing machines. Kolmogorov complexity is a static complexity measure that captures the difficulty of describing a string. For example, the string consisting of three million zeroes can be described with fewer than three million symbols (as in this sentence). In contrast, for a string consisting of three million randomly generated bits, with high probability there is no shorter description than the string itself.

Our goal will be to equate the complexity of a string with the length of its shortest "description"—but this raises the question of what class of "descriptions" should be considered. Given any Turing machine M, one can define a measure $C_M(x)$ to be the length of the shortest input d (the "description") such that $M(d) = x$. At first, this might not seem to be a very promising approach, because using different machines M_1 and M_2 will yield measures C_{M_1} and C_{M_2} that differ wildly (or worse, are undefined for those x that are never produced as output). Fortunately, the next theorem shows that if one chooses a *universal* Turing machine U to define a measure $C_U(x)$ in this way, the resulting measure is roughly the same as that obtained by using any other universal machine.

Theorem 7.22 (Invariance Theorem)

There exists a universal Turing machine U such that for every universal Turing machine U', there is a constant c such that for all y, $C_U(y) \le C_{U'}(y) + c$.

Henceforth, let C be defined by the universal Turing machine of Theorem 7.22. For every integer n and every binary string y of length n, because y can be described by giving itself explicitly, $C(y) \le n + c'$ for a constant c'. Call y **incompressible** if $C(y) \ge n$. Since there are 2^n binary strings of length n and only $2^n - 1$ possible shorter descriptions, there exists an incompressible string for every length n.

Kolmogorov complexity gives a precise mathematical meaning to the intuitive notion of "randomness." If someone flips a coin 50 times and it comes up "heads" each time, then intuitively, the sequence of flips is not random—although from the standpoint of probability theory the all-heads sequence is precisely as likely as any other sequence. Probability theory does not provide the tools for calling one sequence "more random" than another; Kolmogorov complexity theory does.

Kolmogorov complexity provides a useful framework for presenting combinatorial arguments. For example, when one wants to prove that an object with some property P exists, then it is sufficient to show that any object that does *not* have property P has a short description; thus any incompressible (or "random") object must have property P. This sort of argument has been useful in proving lower bounds in complexity theory.

7.14 Research Issues and Summary

The core research questions in complexity theory are expressed in terms of separating complexity classes:

- Is `L` different from `NL`?
- Is `P` different from `RP` or `BPP`?
- Is `P` different from `NP`?
- Is `NP` different from `PSPACE`?

Motivated by these questions, much current research is devoted to efforts to understand the power of nondeterminism, randomization, and interaction. In these studies, researchers have gone well beyond the theory presented in this chapter:

- Beyond Turing machines and Boolean circuits, to restricted and specialized models in which nontrivial lower bounds on complexity can be proved
- Beyond deterministic reducibilities, to nondeterministic and probabilistic reducibilities, and refined versions of the reducibilities considered here
- Beyond worst-case complexity, to average case complexity

Recent research in complexity theory has had direct applications to other areas of computer science and mathematics. Probabilistically checkable proofs were used to show that obtaining approximate solutions to some optimization problems is as difficult as solving them exactly. Complexity theory has provided new tools for studying questions in finite model theory, a branch of mathematical logic. Fundamental questions in complexity theory are intimately linked to practical questions about the use of cryptography for computer security, such as the existence of one-way functions and the strength of public-key cryptosystems.

This last point illustrates the urgent practical need for progress in computational complexity theory. Many popular cryptographic systems in current use are based on unproven assumptions about the difficulty of computing certain functions (such as the factoring and discrete logarithm problems). All of these systems are thus based on wishful thinking and conjecture. Research is needed to resolve these open questions and replace conjecture with mathematical certainty.

Key Terms

Complexity class: A set of languages that are decided within a particular resource bound. For example, $\texttt{NTIME}(n^2 \log n)$ is the set of languages decided by nondeterministic Turing machines within $O(n^2 \log n)$ time.

Constructibility: A function $f(n)$ is time (respectively, space) constructible if there exists a deterministic Turing machine that halts after exactly $f(n)$ steps [after using exactly $f(n)$ worktape cells] for every input of length n.

Diagonalization: A technique for constructing a language A that differs from every $L(M_i)$ for a list of machines $M_1, M_2, \ldots$.

NP-complete: A language A_0 is NP-complete if $A_0 \in \mathrm{NP}$ and $A \leq_m^p A_0$ for every A in NP; that is, for every A in NP, there exists a function f computable in polynomial time such that for every x, $x \in A$ if and only if $f(x) \in A_0$.

Oracle: An oracle is a language A to which a machine presents queries of the form "Is w in A" and receives each correct answer in one step.

Padding: A technique for establishing relationships between complexity classes that uses padded versions of languages, in which each word is padded out with multiple occurrences of a new symbol—the word x is replaced by the word $x\#^{f(|x|)}$ for a numeric function f—in order to artificially lower the complexity of the language.

Reduction: A language A reduces to a language B if a machine that decides B can be used to decide A efficiently.

Time and space complexity: The time (respectively, space) complexity of a deterministic Turing machine M is the maximum number of steps taken (worktape cells visited) by M among all input words of length n.

Turing machine: A Turing machine M is a model of computation with a read-only input tape and multiple worktapes. At each step, M reads the tape cells on which its access heads are located, and depending on its current state and the symbols in those cells, M changes state, writes new symbols on the worktape cells, and moves each access head one cell left or right or not at all.

Acknowledgments

Donna Brown, Bevan Das, Raymond Greenlaw, Lane Hemaspaandra, John Jozwiak, Sung-il Pae, Leonard Pitt, Michael Roman, and Martin Tompa read earlier versions of this chapter and suggested numerous helpful improvements. Karen Walny checked the references.

Further Information

This chapter is a short version of three chapters written by the same authors for the *Algorithms and Theory of Computation Handbook* (Allender et al., 1999).

The formal theoretical study of computational complexity began with the paper of Hartmanis and Stearns (1965), who introduced the basic concepts and proved the first results. For historical perspectives on complexity theory, see Stearns (1990), Sipser (1992), Hartmanis (1994), Wigderson (2006), Allender (2009), and Fortnow (2009).

Contemporary textbooks on complexity theory are by Moore and Mertens (2011), Homer and Selman (2011), Arora and Barak (2009), Goldreich (2008), while venerable texts include Sipser (2005), Hopcroft et al. (2000), Hemaspaandra and Ogihara (2002), and Papadimitriou (1994). A good general reference is the *Handbook of Theoretical Computer Science* (van Leeuwen, 1990), volume A.

The text by Garey and Johnson (1979) explains NP-completeness thoroughly, with examples of NP-completeness proofs, and a collection of hundreds of NP-complete problems. Li and Vitányi (2008) provide a comprehensive scholarly treatment of Kolmogorov complexity, with many applications.

Five organizations and mainstay websites that promote research in computational complexity online sources are the ACM Special Interest Group on Algorithms and Computation Theory (SIGACT), the European Association for Theoretical Computer Science (EATCS), the annual IEEE Conference on Computational Complexity, the "Complexity Zoo," and the theory section of `StackExchange`:

http://sigact.org/
http://www.eatcs.org/
http://www.computationalcomplexity.org/

http://www.complexityzoo.com/
http://cstheory.stackexchange.com/

The *Electronic Colloquium on Computational Complexity* maintained at the University of Trier, Germany at http://eccc.hpi-web.de/ and http://arxiv.org/ run by Cornell University include downloadable current research papers in the field, often with updates and revisions. Most individual researchers have websites with their work, and some also write blogs where news and ideas can be found. Among the most prominent blogs are:

Computational Complexity, http://blog.computationalcomplexity.org/, by Lance Fortnow and William Gasarch.
Shtetl-Optimized, http://www.scottaaronson.com/blog/, by Scott Aaronson.
Gödel's Lost Letter and P=NP, http://rjlipton.wordpress.com/, by Richard Lipton and the third author of this chapter.

Research papers on complexity theory are presented at several annual conferences, including the ACM Symposium on Theory of Computing (STOC); the IEEE Symposium on Foundations of Computer Science (FOCS), the International Colloquium on Automata, Languages, and Programming (ICALP) sponsored by EATCS, and the Symposium on Theoretical Aspects of Computer Science (STACS). The annual Conference on Computational Complexity (formerly Structure in Complexity Theory), also sponsored by the IEEE, is entirely devoted to complexity theory. Journals in which complexity research often appears include: *Chicago Journal on Theoretical Computer Science, Computational Complexity, Information and Computation, Journal of the ACM, Journal of Computer and System Sciences, SIAM Journal on Computing, Theoretical Computer Science, Theory of Computing Systems* (formerly *Mathematical Systems Theory*), *ACM Transactions on Computation Theory*, and *Theory of Computing*. Each issue of *ACM SIGACT News* and *Bulletin of the EATCS* contains a column on complexity theory.

References

Allender, E. 2009. A status report on the P versus NP question. *Adv. Comput.* 77:118–147.

Allender, E., Loui, M.C., and Regan, K.W. 1999. Chapter 23: Complexity classes, Chapter 24: Reducibility and completeness, Chapter 25: Other complexity classes and measures. In: M.J. Atallah and M. Blanton, eds., *Algorithms and Theory of Computation Handbook*, 2nd edn., Vol. 1: *General Concepts and Techniques*. Chapman & Hall/CRC, Boca Raton, FL, 2009.

Arora, S. and Barak, B. 2009. *Complexity Theory: A Modern Approach*. Cambridge University Press, Cambridge, U.K.

Arora, S., Lund, C., Motwani, R., Sudan, M., and Szegedy, M. 1998. Proof verification and hardness of approximation problems. *J. Assoc. Comput. Mach.* 45(3):501–555.

Babai, L., Fortnow, L., and Lund, C. 1991. Nondeterministic exponential time has two-prover interactive protocols. *Comput. Complex.* 1:3–40.

Babai, L. and Moran, S. 1988. Arthur–Merlin games: A randomized proof system, and a hierarchy of complexity classes. *J. Comput. Syst. Sci.* 36(2):254–276.

Baker, T., Gill, J., and Solovay, R. 1975. Relativizations of the P=NP? question. *SIAM J. Comput.* 4(4):431–442.

Book, R.V. 1974. Comparing complexity classes. *J. Comput. Syst. Sci.* 9(2):213–229.

Borodin, A. 1972. Computational complexity and the existence of complexity gaps. *J. Assoc. Comput. Mach.* 19(1):158–174.

Borodin, A. 1977. On relating time and space to size and depth. *SIAM J. Comput.* 6(4):733–744.

Chandra, A.K., Kozen, D.C., and Stockmeyer, L.J. 1981. Alternation. *J. Assoc. Comput. Mach.* 28(1):114–133.

Cook, S.A. 1971. The complexity of theorem-proving procedures. In: *Proceedings of the third Annual ACM Symposium on Theory of Computing*, pp. 151–158. Shaker Heights, OH.

Fortnow, L. 2009. The history and status of the P vs. NP problem. *Commun. ACM*, 52(9):78–86.

Fortnow, L. and Sipser, M. 1988. Are there interactive protocols for co-NP languages? *Inform. Process. Lett.* 28(5):249–251.

Garey, M.R. and Johnson, D.S. 1979. *Computers and Intractability: A Guide to the Theory of NP-Completeness.* W.H. Freeman, San Francisco, CA.

Gill, J. 1977. Computational complexity of probabilistic Turing machines. *SIAM J. Comput.* 6(4):675–695.

Goldreich, O. 2008. *Computational Complexity: A Conceptual Perspective*. Cambridge University Press, Cambridge, U.K.

Goldwasser, S., Micali, S., and Rackoff, C. 1989. The knowledge complexity of interactive proof systems. *SIAM J. Comput.* 18(1):186–208.

Hartmanis, J. and Stearns, R.E. 1965. On the computational complexity of algorithms. *Trans. Am. Math. Soc.* 117:285–306.

Hartmanis, J. 1994. On computational complexity and the nature of computer science. *Commun. ACM* 37(10):37–43.

Hemaspaandra, L.A. and Ogihara, M. 2002. *The Complexity Theory Companion*. Springer-Verlag, Berlin, Germany.

Hennie, F. and Stearns, R.A. 1966. Two-way simulation of multitape Turing machines. *J. Assoc. Comput. Mach.* 13(4):533–546.

Homer, S. and Selman, A. 2011. *Computability and Complexity Theory,* 2nd edn. Springer-Verlag, Berlin, Germany.

Hopcroft, J., Motwani, R., and Ullman, J. 2000. *Introduction to Automata Theory, Languages, and Computation,* 2nd edn., Addison-Wesley, Reading, MA.

Immerman, N. 1988. Nondeterministic space is closed under complementation. *SIAM J. Comput.* 17(5):935–938.

Impagliazzo, R. and Wigderson, A. 1997. P = BPP if E requires exponential circuits: Derandomizing the XOR lemma. In: *Proceedings of the 29th Annual ACM Symposium on Theory of Computing*, pp. 220–229. ACM Press, El Paso, TX.

Jones, N.D. 1975. Space-bounded reducibility among combinatorial problems. *J. Comput. Syst. Sci.* 11(1):68–85; Corrigendum *J. Comput. Syst. Sci.* 15(2):241, 1977.

Karp, R.M. 1972. Reducibility among combinatorial problems. In: R.E. Miller and J.W. Thatcher, eds., *Complexity of Computer Computations*, pp. 85–103. Plenum Press, New York.

Klivans, A.R. and van Melkebeek, D. 2002. Graph nonisomorphism has subexponential size proofs unless the polynomial-time hierarchy collapses. *SIAM J. Comput.* 31(5):1501–1526.

Ladner, R.E. 1975. On the structure of polynomial-time reducibility. *J. Assoc. Comput. Mach.* 22(1):155–171.

Lautemann, C. 1983. BPP and the polynomial hierarchy. *Inform. Process. Lett.* 17(4):215–217.

Li, M. and Vitányi, P.M.B. 2008. *An Introduction to Kolmogorov Complexity and Its Applications*, 3rd edn. Springer-Verlag, Berlin, Germany.

Moore, C. and Mertens, S. 2011. *The Nature of Computation*. Oxford University Press, Oxford, U.K.

Papadimitriou, C.H. 1994. *Computational Complexity*. Addison-Wesley, Reading, MA.

Pippenger, N. and Fischer, M. 1979. Relations among complexity measures. *J. Assoc. Comput. Mach.* 26(2):361–381.

Ruzzo, W.L. 1981. On uniform circuit complexity. *J. Comput. Syst. Sci.* 22(3):365–383.

Savitch, W.J. 1970. Relationship between nondeterministic and deterministic tape complexities. *J. Comput. Syst. Sci.* 4(2):177–192.

Seiferas, J.I., Fischer, M.J., and Meyer, A.R. 1978. Separating nondeterministic time complexity classes. *J. Assoc. Comput. Mach.* 25(1):146–167.

Shamir, A. 1992. IP = PSPACE. *J. Assoc. Comput. Mach.* 39(4):869–877.

Sipser, M. 1983. Borel sets and circuit complexity. In: *Proceedings of the 15th Annual ACM Symposium on the Theory of Computing*, pp. 61–69. ACM Press, Boston, MA.

Sipser, M. 1992. The history and status of the P versus NP question. In: *Proceedings of the 24th Annual ACM Symposium on the Theory of Computing*, pp. 603–618. ACM Press, Victoria, British Columbia, Canada.

Sipser, M. 2005. *Introduction to the Theory of Computation*. Thomson, Boston, MA.

Solovay, R. and Strassen, V. 1977. A fast Monte-Carlo test for primality. *SIAM J. Comput.* 6(1):84–85.

Stearns, R.E. 1990. Juris Hartmanis: The beginnings of computational complexity. In: A.L. Selman, ed., *Complexity Theory Retrospective*, pp. 5–18. Springer-Verlag, Berlin, Germany.

Stockmeyer, L.J. 1976. The polynomial time hierarchy. *Theor. Comput. Sci.* 3(1):1–22.

Stockmeyer, L.J. and Chandra, A.K. 1979. Intrinsically difficult problems. *Sci. Am.* 240(5):140–159.

Stockmeyer, L.J. and Meyer, A.R. 1973. Word problems requiring exponential time: Preliminary report. In: *Proceedings of the fifth Annual ACM Symposium on the Theory of Computing*, pp. 1–9. ACM Press, Austin, TX.

Stockmeyer, L.J. and Vishkin, U. 1984. Simulation of parallel random access machines by circuits. *SIAM J. Comput.* 13(2):409–422.

Szelepcsényi, R. 1988. The method of forced enumeration for nondeterministic automata. *Acta Inform.* 26(3):279–284.

Toda, S. 1991. PP is as hard as the polynomial-time hierarchy. *SIAM J. Comput.* 20(5):865–877.

van Leeuwen, J. 1990. *Handbook of Theoretical Computer Science*, Volume A: Algorithms and Complexity. Elsevier Science, Amsterdam, the Netherlands.

Wigderson, A. 2006. P, NP, and mathematics—A computational complexity perspective. In: *Proceedings of the International Congress of Mathematicians*, pp. 665–712. Madrid, Spain, Vol. 1, EMS Publishing House, pp. 665–712.

Wrathall, C. 1976. Complete sets and the polynomial-time hierarchy. *Theor. Comput. Sci.* 3(1):23–33.

8

Formal Models and Computability

Tao Jiang
University of California, Riverside

Ming Li
University of Waterloo

Bala Ravikumar
University of Rhode Island

8.1 Introduction

The concept of *algorithms* is perhaps almost as old as human civilization. The famous Euclid algorithm is more than 2000 years old. Angle trisection, solving Diophantine equations, and finding polynomial roots in terms of radicals of coefficients are some well-known examples of algorithmic questions. However, until the 1930s, the notion of algorithms was used informally (or rigorously but in a limited context). It was a major triumph of logicians and mathematicians of the last century to offer a rigorous definition of this fundamental concept. The revolution that resulted in this triumph was a collective achievement of many mathematicians, notably Church, Gödel, Kleene, Post, and Turing. Of particular interest is a machine model proposed by Turing (1936), which has come to be known as the *Turing machine* (*TM*).

This particular achievement had numerous significant consequences. It led to the concept of a general-purpose computer or universal computation, a revolutionary idea originally anticipated by Babbage in the 1800s. It is widely acknowledged that the development of a universal TM was prophetic of the modern all-purpose digital computer and played a key role in the thinking of pioneers in the development of modern computers such as von Neumann (Davis 1980). From a mathematical point of view, however, a more interesting consequence was that it was now possible to show the *nonexistence* of algorithms, hitherto impossible due to their elusive nature. In addition, many apparently different definitions of an algorithm proposed by different researchers in different continents turned out to be equivalent (in a precise technical sense, explained later). This equivalence led to the widely held hypothesis known as the *Church–Turing thesis* that mechanical solvability is the same as solvability on a TM.

Formal languages are closely related to algorithms. They were introduced as a way to convey mathematical proofs without errors. Although the concept of a formal language dates back at least to the time of Leibniz, a systematic study of them did not begin until the beginning of the last century. It became a vigorous field of study when Chomsky formulated simple grammatical rules to describe the syntax of a language (Chomsky 1956). *Grammars* and *formal languages* entered into computability theory when Chomsky and others found ways to use them to classify algorithms.

The main theme of this chapter is about formal models, which include TMs (and their variants) as well as grammars. In fact, the two concepts are intimately related. Formal computational models are aimed at providing a framework for computational problem solving, much as electromagnetic theory provides a framework for problems in electrical engineering. Thus, formal models guide the way to build computers and the way to program them. At the same time, new models are motivated by advances in the technology of computing machines. In this chapter, we will discuss only the most basic computational models and use these models to classify problems into some fundamental classes. In doing so, we hope to provide the reader with a conceptual basis with which to read other chapters in this handbook.

8.2 Computability and a Universal Algorithm

Turing's notion of mechanical computation was based on identifying the basic steps of such computations. He reasoned that an operation such as multiplication is not primitive because it can be divided into more basic steps such as digit-by-digit multiplication, shifting, and adding. Addition itself can be expressed in terms of more basic steps such as add the lowest digits, compute, carry, and move to the next digit. Turing thus reasoned that the most basic features of mechanical computation are the abilities to read and write on a storage medium (which he chose to be a linear tape divided into cells or squares) and to make some simple logical decisions. He also restricted each tape cell to hold only one among a finite number of symbols (which we call the *tape alphabet*).* The decision step enables the computer to control the sequence of actions. To make things simple, Turing restricted the next action to be performed on a cell neighboring the one on which the current action occurred. He also introduced an instruction that told the computer to stop. In summary, Turing proposed a model to characterize mechanical computation as being carried out as a sequence of instructions of the following form: write a symbol (such as 0 or 1) on the tape cell, move to the next cell, observe the symbol currently scanned and choose the next step accordingly, or stop.

These operations define a language we call the GOTO language.† Its instructions are

PRINT i (i is a tape symbol)
GO RIGHT
GO LEFT
GO TO STEP j IF i IS SCANNED
STOP

A *program* in this language is a sequence of instructions (written one per line) numbered $1 - k$. To run a program written in this language, we should provide the *input*. We will assume that the input is a string of symbols from a finite input alphabet (which is a subset of the tape alphabet), which is stored on the tape before the computation begins. How much memory should we allow the computer to use? Although we do not want to place any bounds on it, allowing an infinite tape is not realistic. This problem is circumvented by allowing *expandable memory*. In the beginning, the tape containing the input defines its boundary. When the machine moves beyond the current boundary, a new memory cell will

* This bold step of using a discrete model was perhaps the harbinger of the digital revolution that was soon to follow.

† Turing's original formulation is closer to our presentation in Section 8.5. But the GOTO language presents an equivalent model.

```
1   PRINT 0
2   GO LEFT
3   GO TO STEP 2 IF 1 IS SCANNED
4   PRINT 1
5   GO RIGHT
6   GO TO STEP 5 IF 1 IS SCANNED
7   PRINT 1
8   GO RIGHT
9   GO TO STEP 1 IF 1 IS SCANNED
10  STOP
```

FIGURE 8.1 The doubling program in the GOTO language.

be attached with a special symbol *B* (blank) written on it. Finally, we define the result of computation as the contents of the tape when the computer reaches the STOP instruction.

We will present an example program written in the GOTO language. This program accomplishes the simple task of doubling the number of 1s (Figure 8.1). More precisely, on the input containing *k* 1s, the program produces 2*k* 1s. Informally, the program achieves its goal as follows. When it reads a 1, it changes the 1 to 0, moves left looking for a new cell, writes a 1 in the cell, returns to the starting cell and rewrites as 1, and repeats this step for each 1. Note the way the GOTO instructions are used for repetition. This feature is the most important aspect of programming and can be found in all of the imperative style programming languages.

The simplicity of the GOTO language is rather deceptive. There is strong reason to believe that it is powerful enough that any mechanical computation can be expressed by a suitable program in the GOTO language. Note also that the programs written in the GOTO language may not always halt, that is, on certain inputs, the program may never reach the STOP instruction. In this case, we say that the output is undefined.

We can now give a precise definition of what an algorithm is. An algorithm is any program written in the GOTO language with the additional property that it halts on all inputs. Such programs will be called *halting programs*. Throughout this chapter, we will be interested mainly in computational problems of a special kind called *decision problems* that have a yes/no answer. We will modify our language slightly when dealing with decision problems. We will augment our instruction set to include ACCEPT and REJECT (and omit STOP). When the ACCEPT (REJECT) instruction is reached, the machine will output yes or 1 (no or 0) and halt.

8.2.1 Some Computational Problems

We will temporarily shift our focus from the tool for problem solving (the computer) to the problems themselves. Throughout this chapter, a computational problem refers to an input/output relationship. For example, consider the problem of squaring an integer input. This problem assigns to each integer (such as 22) its square (in this case 484). In technical terms, this input/output relationship defines a function. Therefore, solving a computational problem is the same as computing the function defined by the problem. When we say that an algorithm (or a program) solves a problem, what we mean is that, for all inputs, the program halts and produces the correct output. We will allow inputs of arbitrary size and place no restrictions. A reader with primary interest in software applications is apt to question the validity (or even the meaningfulness) of allowing inputs of arbitrary size because it makes the set of all *possible* inputs infinite, and thus unrealistic, in real-world programming. But there are no really good alternatives. Any finite bound is artificial and is likely to become obsolete as the technology and our requirements change. Also, in practice, we do not know how to take advantage of restrictions on the size of the inputs. (See the discussion about nonuniform models in Section 8.5.) Problems (functions) that can be solved by an algorithm (or a halting GOTO program) are called *computable*.

As already remarked, we are interested mainly in decision problems. A decision problem is said to be decidable if there is a halting GOTO program that solves it correctly on all inputs. An important class of problems called *partially decidable decision problems* can be defined by relaxing our requirement a little bit; a decision problem is partially decidable if there is a GOTO program that halts and outputs 1 on all inputs for which the output should be 1 and either halts and outputs 0 or loops forever on the other inputs.

This means that the program may never give a wrong answer but is not required to halt on negative inputs (i.e., inputs with 0 as output).

We now list some problems that are fundamental either because of their inherent importance or because of their historical roles in the development of computation theory:

Problem 1 (halting problem). The input to this problem is a program *P* in the GOTO language and a binary string *x*. The expected output is 1 (or yes) if the program *P* halts when run on the input *x*, 0 (or no) otherwise.

Problem 2 (universal computation problem). A related problem takes as input a program *P* and an input *x* and produces as output what (if any) *P* would produce on input *x*. (Note that this is a decision problem if *P* is restricted to a yes/no program.)

Problem 3 (string compression). For a string *x*, we want to find the shortest program in the GOTO language that when started with the empty tape (i.e., tape containing one *B* symbol) halts and prints *x*. Here shortest means the total number of symbols in the program is as small as possible.

Problem 4 (tiling). A tile* is a square card of unit size (i.e., 1×1) divided into four quarters by two diagonals, each quarter colored with some color (selected from a finite set of colors). The tiles have fixed orientation and cannot be rotated. Given some finite set *T* of such tiles as input, the program is to determine if finite rectangular areas of all sizes (i.e., $k \times m$ for all positive integers *k* and *m*) can be tiled using only the given tiles such that the colors on any two touching edges are the same. It is assumed that an unlimited number of basic tiles of each type are available. Figure 8.2b shows how the base set of tiles given in Figure 8.2a can be used to tile a 5×5 square area.

Problem 5 (linear programming). Given a system of linear inequalities (called constraints), such as $3x - 4y \leq 13$ with integer coefficients, the goal is to find if the system has a solution satisfying all of the constraints.

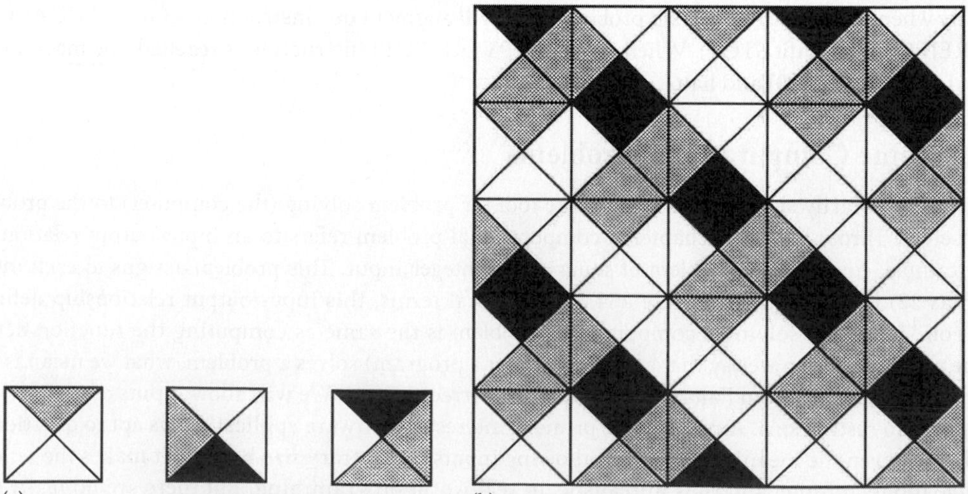

(a) (b)

FIGURE 8.2 An example of tiling. (a) Basic tiles and (b) 5×5 square composed of basic tiles.

* More precisely, a Wang tile, after Hao Wang, who wrote the first research paper on it.

Some remarks must be made about the preceding problems. The problems in our list include non-numerical problems and *meta-problems*, which are problems about other problems. The first two problems are motivated by a quest for reliable program design. An algorithm for problem 1 (if it exists) can be used to test if a program contains an infinite loop. Problem 2 is motivated by an attempt to design a *universal algorithm*, which can simulate any other. This problem was first attempted by Babbage, whose analytical engine had many ingredients of a modern electronic computer (although it was based on mechanical devices). Problem 3 is an important problem in information theory and arises in the following setting. Physical theories are aimed at creating simple laws to explain large volumes of experimental data. A famous example is Kepler's laws, which explained Tycho Brahe's huge and meticulous observational data. Problem 3 asks if this compression process can be automated. When we allow the inference rules to be sufficiently strong, this problem becomes *undecidable*. We will not discuss this problem further in this section but will refer the reader to some related formal systems discussed in Li and Vitányi (1993). The tiling problem is not merely an interesting puzzle. It is an art form of great interest to architects and painters. Tiling has recently found applications in crystallography. Linear programming is a problem of central importance in economics, game theory, and operations research.

In the remainder of the section, we will present some basic algorithm design techniques and sketch how these techniques can be used to solve some of the problems listed (or their special cases). The main purpose of this discussion is to present techniques for showing the decidability (or partial decidability) of these problems. The reader can learn more advanced techniques of algorithm design in some later sections of this chapter as well as in many other chapters of this volume.

8.2.1.1 Table Lookup

The basic idea is to create a table for a function *f*, which needs to be computed by tabulating in one column an input *x* and the corresponding $f(x)$ in a second column. Then the table itself can be used as an algorithm. This method cannot be used directly because the set of all inputs is infinite. Therefore, it is not very useful, although it can be made to work in conjunction with the technique described subsequently.

8.2.1.2 Bounding the Search Domain

The difficulty of establishing the decidability of a problem is usually caused by the fact that the object we are searching for may have no known upper limit. Thus, if we can place such an upper bound (based on the structure of the problem), then we can reduce the search to a finite domain. Then table lookup can be used to complete the search (although there may be better methods in practice). For example, consider the following special case of the tiling problem: Let k be a fixed integer, say 1000. Given a set of tiles, we want to determine whether all rectangular rooms of shape $k \times n$ can be tiled for all n. (Note the difference between this special case and the general problem. The general one allows k and n both to have unbounded value. But here we allow only n to be unbounded.) It can be shown (see Section 8.5 for details) that there are two bounds n_0 and n_1 (they depend on k) such that if there is at least one tile of size $k \times t$ that can be tiled for some $n_0 \leq t \leq n_1$, then every tile of size $k \times n$ can be tiled. If no $k \times t$ tile can be tiled for any t between n_0 and n_1, then obviously the answer is no. Thus, we have reduced an infinite search domain to a finite one.

As another example, consider the linear programming problem. The set of possible solutions to this problem is infinite, and thus a table search cannot be used. But it is possible to reduce the search domain to a finite set using the geometric properties of the set of solutions of the linear programming problem. The fact that the set of solutions is convex makes the search especially easy.

8.2.1.3 Use of Subroutines

This is more of a program design tool than a tool for algorithm design. A central concept of programming is repetitive (or iterative) computation. We already observed how GOTO statements can be used to perform a sequence of steps repetitively. The idea of a subroutine is another central concept of programming. The idea is to make use of a program *P* itself as a single step in another program Q. Building programs from simpler programs is a natural way to deal with the complexity of programming tasks. We will illustrate the idea with a simple example. Consider the problem of multiplying two positive integers *i* and *j*. The input to the problem will be the form 11 … 1011 … 1 (*i* 1s followed by a 0, followed by *j* 1s) and the output will be $i \times j$ 1s (with possibly some 0s on either end). We will use the notation 1^i01^j to denote the starting configuration of the tape. This just means that the tape contains *i* 1s followed by a 0 followed by *j* 1s.

The basic idea behind a GOTO program for this problem is simple; add *j* 1s on the right end of tape exactly $i - 1$ times and then erase the original sequence of *i* 1s on the left. A little thought reveals that the subroutine we need here is to duplicate a string of 1s so that if we start with $x\,02^k\,1^j$ a call to the subroutine will produce $x\,02^{k+j}\,1^j$. Here *x* is just any sequence of symbols. Note the role played by the symbol 2. As new 1s are created on the right, the old 1s change to 2s. This will ensure that there are exactly *j* 1s on the right end of the tape all of the time. This duplication subroutine is very similar to the doubling program, and the reader should have very little difficulty writing this program. Finally, the multiplication program can be done using the copy subroutine $(i - 1)$ times.

8.2.2 A Universal Algorithm

We will now present in some detail a (partial) solution to problem 2 by arguing that there is a program *U* written in the GOTO language, which takes as input a program *P* (also written using the GOTO language) and an input *x* and produces as output *P* (*x*), the output of *P* on input *x*. For convenience, we will assume that all programs written in the GOTO language use a fixed alphabet containing just 0, 1, and *B*. Because we have assumed this for all programs in the GOTO language, we should first address the issue of how an input to program *U* will look. We cannot directly place a program *P* on the tape because the alphabet used to write the program *P* uses letters G, O, T, O, etc. This minor problem can be easily circumvented by coding. The idea is to represent each instruction using only 0 and 1. One such coding scheme is shown in Table 8.1.

To encode an entire program, we simply write down in order (without the line numbers) the code for each instruction as given in the table. For example, here is the code for the doubling program shown in Figure 8.1:

0001001011110110001101001111110110001101001110111100

Note that the encoded string contains all of the information about the program so that the encoding is completely reversible. From now on, if *P* is a program in the GOTO language, then code(*P*) will denote its binary code as just described. When there is no confusion, we will identify *P* and code(*P*).

TABLE 8.1 Coding the GOTO Instructions

Instruction	Code
PRINT *i*	0001^{i+1}
GO LEFT	001
GO RIGHT	010
GO TO *j* IF *i* IS SCANNED	0111^j01^{i+1}
STOP	100

Before proceeding further, the reader may want to test his/her understanding of the encoding/decoding process by decoding the following string: 010011101100.

The basic idea behind the construction of a universal algorithm is simple, although the details involved in actually constructing one are enormous. We will present the central ideas and leave out the actual construction. Such a construction was carried out in complete detail by Turing himself and was simplified by others.* U has as its input code(P) followed by the string x. U simulates the computational steps of P on input x. It divides the input tape into three segments, one containing the program P, the second one essentially containing the contents of the tape of P as it changes with successive moves, and the third one containing the line number in program P of the instruction being currently simulated (similar to a *program counter* in an actual computer).

We now describe a *cycle* of computation by U, which is similar to a central processing unit (CPU) cycle in a real computer. A single instruction of P is implemented by U in one cycle. First, U should know which location on the tape that P is currently reading. A simple artifact can handle this as follows: U uses in its tape alphabet two special symbols $0'$ and $1'$. U stores the tape of P in the tape segment alluded to in the previous paragraph exactly as it would appear when the program P is run on the input x with one minor modification. The symbol currently being read by program P is stored as the *primed version* ($0'$ is the primed version of 0). For example, suppose after completing 12 instructions P is reading the fourth symbol (from left) on its tape containing 01001001. Then, the tape region of U after 12 cycles looks like $0100'1001$. At the beginning of a new cycle, U uses a subroutine to move to the region of the tape that contains the ith instruction of program P where i is the value of the program counter. It then decodes the ith instruction. Based on what type it is, U proceeds as follows: If it is a PRINT i instruction, then U scans the tape until the unique primed symbol in the *tape region* is reached and rewrites it as instructed. If it is a GO LEFT or GO RIGHT symbol, U locates the primed symbol, unprimes it, and primes its left or right neighbor, as instructed. In both cases, U returns to the program counter and increments it. If the instruction is GO TO i IF j IS SCANNED, U reads the primed symbol, and if it is j', U changes the program counter to i. This completes a cycle. Note that the three regions may grow and contract while U executes the cycles of computation just described. This may result in one of them running into another. U must then shift one of them to the left or right and make room as needed.

It is not too difficult to see that all of the steps described can be done using the instructions of the GOTO language. The main point to remember is that these actions will have to be coded as a single program, which has nothing whatsoever to do with program P. In fact, the program U is totally independent of P. If we replace P with some other program Q, it should simulate Q as well. The preceding argument shows that problem 2 is partially decidable. But it does not show that this problem is decidable. Why? It is because U may not halt on all inputs; specifically, consider an input consisting of a program P and a string x such that P does not halt on x. Then, U will also keep executing cycle after cycle the moves of P and will never halt. In fact, in Section 8.3, we will show that problem 2 is not decidable.

8.3 Undecidability

Recall the definition of an undecidable problem. In this section, we will establish the undecidability of problem 2, Section 8.2. The simplest way to establish the existence of undecidable problems is as follows: There are more problems than there are programs, the former set being uncountable, whereas the latter is countably infinite.† But this argument is purely existential and does not identify any specific problem as undecidable. In what follows, we will show that problem 2 introduced in Section 8.2 is one such problem.

---

* A particularly simple exposition can be found in Robinson (1991).

† The reader who does not know what countable and uncountable infinities are can safely ignore this statement; the rest of the section does not depend on it.

8.3.1 Diagonalization and Self-Reference

Undecidability is inextricably tied to the concept of self-reference, and so we begin by looking at this rather perplexing and sometimes paradoxical concept. The idea of self-reference seems to be many centuries old and may have originated with a barber in ancient Greece who had a sign board that read: "I shave all those who do not shave themselves." When the statement is applied to the barber himself, we get a self-contradictory statement. Does he shave himself? If the answer is yes, then he is one of those who shaves himself, and so the barber should not shave him. The contrary answer no is equally untenable. So neither yes nor no seems to be the correct answer to the question; this is the essence of the paradox. The barber's paradox has made entry into modern mathematics in various forms. We will present some of them in the next few paragraphs.*

The first version, called Berry's paradox, concerns English descriptions of natural numbers. For example, the number 7 can be described by many different phrases: seven, six plus one, the fourth smallest prime, etc. We are interested in the *shortest* of such descriptions, namely, the one with the fewest letters in it. Clearly, there are (infinitely) many positive integers whose shortest descriptions exceed 100 letters. (A simple counting argument can be used to show this. The set of positive integers is infinite, but the set of positive integers with English descriptions in fewer than or equal to 100 letters is finite.) Let D denote the set of positive integers that do not have English descriptions with fewer than 100 letters. Thus, D is not empty. It is a well-known fact in set theory that any nonempty subset of positive integers has a smallest integer. Let x be the smallest integer in D. Does x have an English description with fewer than or equal to 100 letters? By the definition of the set D and x, we have the following: x is "the smallest positive integer that cannot be described in English in fewer than 100 letters." This is clearly absurd because part of the last sentence in quotes is a description of x and it contains fewer than 100 letters in it. A similar paradox was found by the British mathematician Bertrand Russell when he considered the set of all sets that do not include themselves as elements, that is, $S = \{x \mid x \notin x\}$. The question "Is $S \in S$?" leads to a similar paradox.

As a last example, we will consider a charming self-referential paradox due to mathematician William Zwicker. Consider the collection of all two-person games (such as chess, tic-tac-toe) in which players make alternate moves until one of them loses. Call such a game *normal* if it has to end in a finite number of moves, no matter what strategies the two players use. For example, tic-tac-toe must end in at most nine moves and so it is normal. Chess is also normal because the 50-move rule ensures that the game cannot go forever. Now here is *hypergame*. In the first move of the hypergame, the first player calls out a normal game, and then the two players go on to play the game, with the second player making the first move. The question is: "Is hypergame normal?" Suppose it is normal. Imagine two players playing hypergame. The first player can call out hypergame (since it is a normal game). This makes the second player call out the name of a normal game; hypergame can be called out again and they can keep saying hypergame without end, and this contradicts the definition of a normal game. On the other hand, suppose it is not a normal game. But now in the first move, player 1 cannot call out hypergame and would call a normal game instead, and so the infinite move sequence just given is not possible and so hypergame is normal after all!

In the rest of the section, we will show how these paradoxes can be modified to give nonparadoxical but surprising conclusions about the decidability of certain problems. Recall the encoding we presented in Section 8.2 that encodes any program written in the GOTO language as a binary string. Clearly this encoding is reversible in the sense that if we start with a program and encode it, it is possible to decode it back to the program. However, not every binary string corresponds to a program because there are many strings that cannot be decoded in a meaningful way, for example, 11010011000110. For the purposes

* The most enchanting discussions of self-reference are due to the great puzzlist and mathematician R. Smullyan who brings out the breadth and depth of this concept in such delightful books as *What is the name of this book?* published by Prentice–Hall in 1978 and *Satan, Cantor, and Infinity* published by Alfred A. Knopf in 1992. We heartily recommend them to anyone who wants to be amused, entertained, and, more importantly, educated on the intricacies of mathematical logic and computability.

of this section, however, it would be convenient if we can treat *every* binary string as a program. Thus, we will simply stipulate that any undecodable string be decoded to the program containing the single statement.

1. REJECT

In the following discussion, we will identify a string x with a GOTO program to which it decodes. Now define a function f_D as follows: $f_D(x) = 1$ if x, decoded into a GOTO program, does not halt when started with x itself as the input. Note the self-reference in this definition. Although the definition of f_D seems artificial, its importance will become clear in the next section when we use it to show the undecidability of problem 2. First, we will prove that f_D is not computable. Actually, we will prove a stronger statement, namely, that f_D is not even partially decidable. [Recall that a function is partially decidable if there is a GOTO program (not necessarily halting) that computes it. An important distinction between computable and semicomputable functions is that a GOTO program for the latter need not halt on inputs with output = 0.]

Theorem 8.1

Function f_D is not partially decidable

The proof is by contradiction. Suppose a GOTO program P' computes the function f_D. We will modify P' into another program P in the GOTO language such that P computes the same function as P' but has the additional property that it will never terminate its computation by ending up in a REJECT statement.* Thus, P is a program with the property that it computes f_D and halts on an input y if and only if $f_D(y) = 1$. We will complete the proof by showing that there is at least one input in which the program produces a wrong output, that is, there is an x such that $f_D(x) \neq P(x)$.

Let x be the encoding of program P. Now consider the following question: Does P halt when given x as input? Suppose the answer is yes. Then, by the way we constructed P, here $P(x) = 1$. On the other hand, the definition of f_D implies that $f_D(x) = 0$. (This is the punch line in this proof. We urge the reader to take a few moments and read the definition of f_D a few times and make sure that he or she is convinced about this fact!) Similarly, if we start with the assumption that $P(x) = 0$, we are led to the conclusion that $f_D(x) = 1$. *In both cases,* $f_D(x) \neq P(x)$ and thus P is not the correct program for f_D. Therefore, P' is not the correct program for f_D either because P and P' compute the same function. This contradicts the hypothesis that such a program exists, and the proof is complete.

Note the crucial difference between the paradoxes we presented earlier and the proof of this theorem. Here, we do not have a paradox because our conclusion is of the form $f_D(x) = 0$ if and only if $P(x) = 1$ and not $f_D(x) = 1$ if and only if $f_D(x) = 0$. But in some sense, the function f_D was motivated by Russell's paradox. We can similarly create another function f_Z (based on Zwicker's paradox of hypergame). Let f be any function that maps binary strings to $\{0, 1\}$. We will describe a method to generate successive functions f_1, f_2, etc., as follows: Suppose $f(x) = 0$ for all x. Then we cannot create any more functions, and the sequence stops with f. On the other hand, if $f(x) = 1$ for some x, then choose one such x and decode it as a GOTO program. This defines another function; call it f_1 and repeat the same process with f_1 in the place of f. We call f a *normal function* if no matter how x is selected at each step the process terminates after a finite number of steps. A simple example of a nonnormal function is as follows: Suppose $P(Q) = 1$ for some program P and input Q and at the same time $Q(P) = 1$ (note that we are using a program and its code interchangeably), then it is easy to see that the functions defined by both P and Q are not normal. Finally, define $f_Z(X) = 1$ if X is a normal program, 0 if it is not. We leave it as an instructive exercise to the reader

* The modification needed to produce P from P' is straightforward. If P' did not have any REJECT statements at all, then no modification would be needed. If it had, then we would have to replace each one by a looping statement, which keeps repeating the same instruction forever.

to show that f_Z is not semicomputable. A perceptive reader will note the connection between Berry's paradox and problem 3 in our list (string compression problem) just as f_Z is related to Zwicker's paradox. Such a reader should be able to show the undecidability of problem 3 by imitating Berry's paradox.

8.3.2 Reductions and More Undecidable Problems

Theory of computation deals not only with the behavior of individual problems but also with relations among them. A *reduction* is a simple way to relate two problems so that we can deduce the (un)decidability of one from the (un)decidability of the other. Reduction is similar to using a subroutine. Consider two problems A and B. We say that problem A can be reduced to problem B if there is an algorithm for B provided that A has one. To define the reduction (also called a *Turing reduction*) precisely, it is convenient to augment the instruction set of the GOTO programming language to include a new instruction CALL X, i, j where X is a (different) GOTO program and i and j are line numbers. In detail, the execution of such augmented programs is carried out as follows: When the computer reaches the instruction CALL X, i, j, the program will simply start executing the instructions of the program from line 1, treating whatever is on the tape currently as the input to the program X. When (if at all) X finishes the computation by reaching the ACCEPT statement, the execution of the original program continues at line number i and, if it finishes with REJECT, the original program continues from line number j.

We can now give a more precise definition of a reduction between two problems. Let A and B be two computational problems. We say that A is reducible to B if there is a halting program Y in the GOTO language for problem A in which calls can be made to a halting program X for problem B. The algorithm for problem A described in the preceding reduction does not assume the availability of program X and cannot use the details behind the design of this algorithm. The right way to think about a reduction is as follows: Algorithm Y, from time to time, needs to know the solutions to different instances of problem B. It can query an algorithm for problem B (as a black box) and use the answer to the query for making further decisions. An important point to be noted is that the program Y actually can be implemented even if program X was never built as long as someone can correctly answer some questions asked by program Y about the output of problem B for certain inputs. Programs with such calls are sometimes called *oracle programs*. Reduction is rather difficult to assimilate at the first attempt, and so we will try to explain it using a puzzle. How do you play two chess games, one each with Kasparov and Anand (two of the world's best players), and ensure that you get at least one point? (You earn one point for a win, 0 for a loss, and 1/2 for a draw.) Because you are a novice and are pitted against two Goliaths, you are allowed a concession. You can choose to play white or black on either board. The well-known answer is the following: Take white against one player, say Anand, and black against the other, namely, Kasparov. Watch the first move of Kasparov (as he plays white) and make the same move against Anand, get his reply and play it back to Kasparov, and keep playing back and forth like this. It takes only a moment's thought that you are guaranteed to win (exactly) 1 point. The point is that your game involves taking the position of one game, applying the algorithm of one player, getting the result, and applying it to the other board, and you do not even have to know the rules of chess to do this. This is exactly how algorithm Y is required to use algorithm X.

We will use reductions to show the undecidability as follows: Suppose A can be reduced to B as in the preceding definition. If there is an algorithm for problem B, it can be used to design a program for A by essentially imitating the execution of the augmented program for A (with calls to the oracle for B) as just described. But we will turn it into a negative argument as follows: If A is undecidable, then so is B. Thus, a reduction from a problem known to be undecidable to problem B will prove B's undecidability.

First we define a new problem, problem 2′, which is a special case of problem 2. Recall that in problem 2 the input is (the code of) a program P in GOTO language and a string x. The output required is $P(x)$. In problem 2′, the input is (only) the code of a program P and the output required is $P(P)$, that is, instead of requiring P to run on a given input, this problem requires that it be run on its own code. This is clearly a special case of problem 2. The reader may readily see the self-reference in problem 2′ and suspect that it may be undecidable; therefore, the more general problem 2 may be undecidable as well. We will establish these claims more rigorously as follows.

We first observe a general statement about the decidability of a function f (or problem) and its *complement*. The complement function is defined to take value 1 on all inputs for which the original function value is 0 and vice versa. The statement is that a function f is decidable if and only if the complement $\bar{f}$ is decidable. This can be easily proved as follows. Consider a program P that computes f. Change P into $\bar{P}$ by interchanging all of the ACCEPT and REJECT statements. It is easy to see that $\bar{P}$ actually computes $\bar{f}$. The converse also is easily seen to hold. It readily follows that the function defined by problem 2′ is undecidable because it is, in fact, the complement of f_D.

Finally, we will show that problem 2 is uncomputable. The idea is to use a reduction from problem 2′ to problem 2. (Note the direction of reduction. This always confuses a beginner.) Suppose there is an algorithm for problem 2. Let X be the GOTO language program that implements this algorithm. X takes as input code(P) (for any program P) followed by x, produces the result P (x), and halts. We want to design a program Y that takes as input code(P) and produce the output P (P) using calls to program X. It is clear what needs to be done. We just create the input in proper form code(P) followed by code (P) and call X. This requires first duplicating the input, but this is a simple programming task similar to the one we demonstrated in our first program in Section 8.2. Then a call to X completes the task. This shows that problem 2′ reduces to problem 2, and thus the latter is undecidable as well.

By a more elaborate reduction (from f_D), it can be shown that tiling is not partially decidable. We will not do it here and refer the interested reader to Harel (1992). But we would like to point out how the undecidability result can be used to infer a result about tiling. This deduction is of interest because the result is an important one and is hard to derive directly. We need the following definition before we can state the result. A different way to pose the tiling problem is whether a given set of tiles can tile *an entire plane* in such a way that all of the adjacent tiles have the same color on the meeting quarter. (Note that this question is different from the way we originally posed it: Can a given set of tiles tile any *finite* rectangular region? Interestingly, the two problems are identical in the sense that the answer to one version is yes if and only if it is yes for the other version.) Call a tiling of the plane periodic if one can identify a $k \times k$ square such that the entire tiling is made by repeating this $k \times k$ square tile. Otherwise, call it *aperiodic*. Consider the following question: Is there a (finite) set of unit tiles that can tile the plane but only aperiodically? The answer is yes and it can be shown from the total undecidability of the tiling problem. Suppose the answer is no. Then, for any given set of tiles, the entire plane can be tiled if and only if the plane can be tiled periodically. But a periodic tiling can be found, if one exists, by trying to tile a $k \times k$ region for successively increasing values of k. This process will eventually succeed (in a finite number of steps) if the tiling exists. This will make the tiling problem partially decidable, which contradicts the total undecidability of the problem. This means that the assumption that the entire plane can be tiled if and only if some $k \times k$ region can be tiled is wrong. Thus, there exists a (finite) set of tiles that can tile the entire plane but only aperiodically.

8.4 Formal Languages and Grammars

The universe of strings is probably the most general medium for the representation of information. This section is concerned with sets of strings called *languages* and certain systems generating these languages such as *grammars*. Every programming language including Pascal, C, or Fortran can be precisely described by a grammar. Moreover, the grammar allows us to write a computer program (called the *lexical analyzer* in a compiler) to determine if a piece of code is syntactically correct in the programming language. Would not it be nice to also have such a grammar for English and a corresponding computer program that can tell us what English sentences are grammatically correct?* The focus of this brief

* Actually, English and the other natural languages have grammars; but these grammars are not precise enough to tell apart the correct and incorrect sentences with 100% accuracy. The main problem is that *there is no universal agreement* on what are grammatically correct English sentences.

exposition is the formalism and mathematical properties of various languages and grammars. Many of the concepts have applications in domains including natural language and computer language processing and string matching. We begin with some standard definitions about languages.

Definition 8.1

An *alphabet* is a finite nonempty set of *symbols*, which are assumed to be *indivisible*.

For example, the alphabet for English consists of 26 uppercase letters $A, B, \ldots, Z$ and 26 lowercase letters $a, b, \ldots, z$. We usually use the symbol Σ to denote an alphabet.

Definition 8.2

A *string* over an alphabet Σ is a finite sequence of symbols of Σ.

The number of symbols in a string x is called its *length*, denoted $|x|$. It is convenient to introduce an empty string, denoted ε, which contains no symbols at all. The length of ε is 0.

Definition 8.3

Let $x = a_1 a_2 \ldots a_n$ and $y = b_1 b_2 \ldots b_m$ be two strings. The *concatenation* of x and y, denoted xy, is the string $a_1 a_2 \ldots a_n b_1 b_2 \ldots b_m$.

Thus, for any string x, $\varepsilon x = x\varepsilon = x$. For any string x and integer $n \geq 0$, we use x^n to denote the string formed by sequentially concatenating n copies of x.

Definition 8.4

The set of all strings over an alphabet Σ is denoted Σ^* and the set of all nonempty strings over Σ is denoted Σ^+. The empty set of strings is denoted $\varnothing$.

Definition 8.5

For any alphabet Σ, a *language* over Σ is a set of strings over Σ. The members of a language are also called the *words* of the language.

Example 8.1

The sets $L_1 = \{01, 11, 0110\}$ and $L_2 = \{0^n 1^n \mid n \geq 0\}$ are two languages over the binary alphabet $\{0, 1\}$. The string 01 is in both languages, whereas 11 is in L_1 but not in L_2.

Because languages are just sets, standard set operations such as union, intersection, and complementation apply to languages. It is useful to introduce two more operations for languages: *concatenation* and *Kleene closure*.

Definition 8.6

Let L_1 and L_2 be two languages over Σ. The concatenation of L_1 and L_2, denoted $L_1 L_2$, is the language $\{xy \mid x \in L_1, y \in L_2\}$.

Definition 8.7

Let L be a language over Σ. Define $L^0 = \{\varepsilon\}$ and $L^i = LL^{i-1}$ for $i \geq 1$. The Kleene closure of L, denoted L^*, is the language

$$L^* = \bigcup_{i \geq 0} L^i$$

and the *positive closure* of L, denoted L^+, is the language

$$L^+ = \bigcup_{i \geq 1} L^i$$

In other words, the Kleene closure of language L consists of all strings that can be formed by concatenating some words from L. For example, if $L = \{0, 01\}$, then $LL = \{00, 001, 010, 0101\}$ and L^* includes all binary strings in which every 1 is preceded by a 0. L^+ is the same as L^* except it excludes ε in this case. Note that, for any language L, L^* always contains ε and L^+ contains ε if and only if L does. Also note that Σ^* is in fact the Kleene closure of the alphabet Σ when viewed as a language of words of length 1, and Σ^+ is just the positive closure of Σ.

8.4.1 Representation of Languages

In general, a language over an alphabet Σ is a subset of Σ^*. How can we describe a language rigorously so that we know if a given string belongs to the language or not? As shown in the preceding paragraphs, a finite language such as L_1 in Example 8.1 can be explicitly defined by enumerating its elements, and a simple infinite language such as L_2 in the same example can be described using a rule characterizing all members of L_2. It is possible to define some more systematic methods to represent a wide class of languages. In the following, we will introduce three such methods: regular expressions, pattern systems, and grammars. The languages that can be described by this kind of system are often referred to as *formal languages.*

Definition 8.8

Let Σ be an alphabet. The *regular expressions* over Σ and the languages they represent are defined inductively as follows:

1. The symbol $\varnothing$ is a regular expression, denoting the empty set.
2. The symbol ε is a regular expression, denoting the set $\{\varepsilon\}$.
3. For each $a \in \Sigma$, a is a regular expression, denoting the set $\{a\}$.
4. If r and s are regular expressions denoting the languages R and S, then $(r + s)$, (rs), and (r^*) are regular expressions that denote the sets $R \cup S$, RS, and R^*, respectively.

For example, $((0(0 + 1)^*) + ((0 + 1)^*0))$ is a regular expression over $\{0, 1\}$, and it represents the language consisting of all binary strings that begin or end with a 0. Because the set operations union and concatenation are both associative, many parentheses can be omitted from regular expressions if we assume that the Kleene closure has higher precedence than concatenation and concatenation has higher precedence than union. For example, the preceding regular expression can be abbreviated as $0(0 + 1)^* + (0 + 1)^*0$. We will also abbreviate the expression rr^* as r^+. Let us look at a few more examples of regular expressions and the languages they represent.

Example 8.2

The expression $0(0 + 1)^*1$ represents the set of all strings that begin with a 0 and end with a 1.

Example 8.3

The expression $0 + 1 + 0(0 + 1)^*0 + 1(0 + 1)^*1$ represents the set of all nonempty binary strings that begin and end with the same bit.

Example 8.4

The expressions 0^*, 0^*10^*, and $0^*10^*10^*$ represent the languages consisting of strings that contain no 1, exactly one 1, and exactly two 1s, respectively.

Example 8.5

The expressions $(0 + 1)^*1(0 + 1)^*1(0 + 1)^*$, $(0 + 1)^*10^*1(0 + 1)^*$, $0^*10^*1(0 + 1)^*$, and $(0 + 1)^*10^*10^*$ all represent the same set of strings that contain at least two 1s.

For any regular expression r, the language represented by r is denoted as $L(r)$. Two regular expressions representing the same language are called *equivalent*. It is possible to introduce some identities to algebraically manipulate regular expressions to construct equivalent expressions, by tailoring the set identities for the operations union, concatenation, and Kleene closure to regular expressions. For more details, see Salomaa (1966). For example, it is easy to prove that the expressions $r(s + t)$ and $rs + rt$ are equivalent and $(r^*)^*$ is equivalent to r^*.

Example 8.6

Let us construct a regular expression for the set of all strings that contain no consecutive 0s. A string in this set may begin and end with a sequence of 1s. Because there are no consecutive 0s, every 0 that is not the last symbol of the string must be followed by at least a 1. This gives us the expression $1^*(01^+)^*1^*(\varepsilon + 0)$. It is not hard to see that the second 1^* is redundant, and thus the expression can in fact be simplified to $1^*(01^+)^*(\varepsilon + 0)$.

Regular expressions were first introduced in Kleene (1956) for studying the properties of neural nets. The preceding examples illustrate that regular expressions often give very clear and concise representations of languages. Unfortunately, not every language can be represented by regular expressions. For example, it will become clear that there is no regular expression for the language $\{0^n 1^n \mid n \geq 1\}$. The languages represented by regular expressions are called the *regular languages*. Later, we will see that regular languages are exactly the class of languages generated by the so-called *right-linear grammars.* This connection allows one to prove some interesting mathematical properties about regular languages as well as to design an efficient algorithm to determine whether a given string belongs to the language represented by a given *regular expression*.

Another way of representing languages is to use *pattern systems* (Angluin 1980, Jiang et al. 1995).

Definition 8.9

A *pattern system* is a triple (Σ, V, p), where Σ is the alphabet, V is the set of *variables* with $\Sigma \cap V = \varnothing$, and p is a string over $\Sigma \cup V$ called the *pattern*.

An example pattern system is $(\{0, 1\}, \{v_1, v_2\}, v_1 v_1 0 v_2)$.

Definition 8.10

The language generated by a pattern system (Σ, V, p) consists of all strings over Σ that can be obtained from p by replacing each variable in p with a string over Σ.

For example, the language generated by $(\{0, 1\}, \{v_1, v_2\}, v_1 v_1 0 v_2)$ contains words 0, 00, 01, 000, 001, 010, 011, 110, etc., but does not contain strings, 1, 10, 11, 100, 101, etc. The pattern system $(\{0, 1\}, \{v_1\}, v_1 v_1)$ generates the set of all strings, which is the concatenation of two equal substrings, that is, the set $\{xx \mid x \in \{0, 1\}^*\}$. The languages generated by pattern systems are called the *pattern languages*.

Regular languages and pattern languages are really different. One can prove that the pattern language $\{xx \mid x \in \{0, 1\}^*\}$ is not a regular language and the set represented by the regular expression 0^*1^* is not a pattern language. Although it is easy to write an algorithm to decide if a string is in the language generated by a given pattern system, such an algorithm most likely would have to be very inefficient (Angluin 1980).

Perhaps the most useful and general system for representing languages is based on grammars, which are extensions of the pattern systems.

Definition 8.11

A grammar is a quadruple (Σ, N, S, P), where

1. Σ is a finite nonempty set called the alphabet. The elements of Σ are called the *terminals*
2. N is a finite nonempty set disjoint from Σ. The elements of N are called the *nonterminals* or *variables*
3. $S \in N$ is a distinguished nonterminal called the *start symbol*
4. P is a finite set of *productions* (or *rules*) of the form

$$\alpha \rightarrow \beta$$

where $\alpha \in (\Sigma \cup N)^* N(\Sigma \cup N)^*$ and $\beta \in (\Sigma \cup N)^*$, that is, α is a string of terminals and nonterminals containing at least one nonterminal and α is a string of terminals and nonterminals.

Example 8.7

Let $G_1 = (\{0, 1\}, \{S, T, O, I\}, S, P)$, where P contains the following productions:

$$S \to OT$$
$$S \to OI$$
$$T \to SI$$
$$O \to 0$$
$$I \to 1$$

As we shall see, the grammar G_1 can be used to describe the set $\{0^n 1^n \mid n \geq 1\}$.

Example 8.8

Let $G_2 = (\{0, 1, 2\}, \{S, A\}, S, P)$, where P contains the following productions:

$$S \to 0S\,A2$$
$$S \to \varepsilon$$
$$2A \to A2$$
$$0A \to 01$$
$$1A \to 11$$

This grammar G_2 can be used to describe the set $\{0^n 1^n 2^n \mid n \geq 0\}$.

Example 8.9

To construct a grammar G_3 to describe English sentences, the alphabet Σ contains all words in English.

N would contain nonterminals, which correspond to the structural components in an English sentence, for example, ⟨sentence⟩, ⟨subject⟩, ⟨predicate⟩, ⟨noun⟩, ⟨verb⟩, and ⟨article⟩. The start symbol would be ⟨sentence⟩. Some typical productions are

⟨sentence⟩ → ⟨subject⟩⟨predicate⟩
⟨subject⟩ → ⟨noun⟩
⟨predicate⟩ → ⟨verb⟩⟨article⟩⟨noun⟩
⟨noun⟩ → mary
⟨noun⟩ → algorithm
⟨verb⟩ → wrote
⟨article⟩ → an

The rule ⟨sentence⟩ → ⟨subject⟩⟨predicate⟩ follows from the fact that a sentence consists of a subject phrase and a predicate phrase. The rules ⟨noun⟩ → mary and ⟨noun⟩ → algorithm mean that both mary and algorithms are possible nouns.

To explain how a grammar represents a language, we need the following concepts.

Definition 8.12

Let (Σ, N, S, P) be a grammar. A *sentential form* of G is any string of terminals and nonterminals, that is, a string over $\Sigma \cup N$.

Definition 8.13

Let (Σ, N, S, P) be a grammar and γ_1 and γ_2 two sentential forms of G. We say that γ_1 *directly derives* γ_2, denoted $\gamma_1 \Rightarrow \gamma_2$, if $\gamma_1 = \sigma\alpha\tau$, $\gamma_2 = \sigma\beta\tau$, and $\alpha \to \beta$ is a production in P.

For example, the sentential form $00S11$ directly derives the sentential form $00OT11$ in grammar G_1, and $A2A2$ directly derives $AA22$ in grammar G_2.

Definition 8.14

Let γ_1 and γ_2 be two sentential forms of a grammar G. We say that γ_1 *derives* γ_2, denoted $\gamma_1 \overset{*}{\Rightarrow} \gamma_2$, if there exists a sequence of (zero or more) sentential forms $\sigma_1, \ldots, \sigma_n$ such that

$$\gamma_1 \Rightarrow \sigma_1 \Rightarrow \cdots \Rightarrow \sigma_n \Rightarrow \gamma_2$$

The sequence $\gamma_1 \Rightarrow \sigma_1 \Rightarrow \cdots \Rightarrow \sigma_n \Rightarrow \gamma_2$ is called a derivation from γ_1 to γ_2. For example, in grammar G_1, $S \overset{*}{\Rightarrow} 0011$ because

$$S \Rightarrow \underline{O}T \Rightarrow 0\underline{T} \Rightarrow 0S\underline{I} \Rightarrow 0\underline{S}1 \Rightarrow 0\underline{O}I1 \Rightarrow 00\underline{I}1 \Rightarrow 0011$$

and in grammar G_2, $S \overset{*}{\Rightarrow} 001122$ because

$$S \Rightarrow 0\underline{S}A2 \Rightarrow 00\underline{S}A2A2 \Rightarrow 0\underline{0A}2A2 \Rightarrow 001\underline{2A}2 \Rightarrow 00\underline{11A}22 \Rightarrow 001122$$

Here the left-hand side of the relevant production in each derivation step is underlined for clarity.

Definition 8.15

Let (Σ, N, S, P) be a grammar. The language generated by G, denoted $L(G)$, is defined as

$$L(G) = \{x \mid x \in \Sigma^*, S \overset{*}{\Rightarrow} x\}$$

The words in $L(G)$ are also called the *sentences* of $L(G)$.

Clearly, $L(G_1)$ contains all strings of the form $0^n 1^n$, $n \geq 1$, and $L(G_2)$ contains all strings of the form $0^n 1^n 2^n$, $n \geq 0$. Although only a partial definition of G_3 is given, we know that $L(G_3)$ contains sentences such as "mary wrote an algorithm" and "algorithm wrote an algorithm" but does not contain sentences such as "an wrote algorithm."

The introduction of formal grammars dates back to the 1940s (Post 1943), although the study of rigorous description of languages by grammars did not begin until the 1950s (Chomsky 1956). In the Section 8.4.2, we consider various restrictions on the form of productions in a grammar and see how these restrictions can affect the power of a grammar in representing languages. In particular, we will know that regular languages and pattern languages can all be generated by grammars under different restrictions.

8.4.2 Hierarchy of Grammars

Grammars can be divided into four classes by gradually increasing the restrictions on the form of the productions. Such a classification is due to Chomsky (1956, 1963) and is called the *Chomsky hierarchy*.

Definition 8.16

Let $G = (\Sigma, N, S, P)$ be a grammar.

1. G is also called a *type-0 grammar* or an *unrestricted grammar.*
2. G is *type-*1 or *context sensitive* if each production $\alpha \to \beta$ in P either has the form $S \to \varepsilon$ or satisfies $|\alpha| \le |\beta|$.
3. G is *type-*2 or *context-free* if each production $\alpha \to \beta$ in P satisfies $|\alpha| = 1$, that is, α is a nonterminal.
4. G is *type-*3 or right linear or regular if each production has one of the following three forms:

$$A \to aB, \quad A \to a, \quad A \to \varepsilon$$

where A and B are nonterminals and a is a terminal.

The language generated by a type-i grammar is called a type-i language, $i = 0, 1, 2, 3$. A type-1 language is also called a *context-sensitive language* and a type-2 language is also called a *context-free language.* It turns out that every type-3 language is in fact a regular language, that is, it is represented by some regular expression and vice versa. See the Section 8.4.3 for the proof of the equivalence of type-3 (right-linear) grammars and regular expressions.

The grammars G_1 and G_3 given in the last subsection are context-free and the grammar G_2 is context sensitive. Now we give some examples of unrestricted and right-linear grammars.

Example 8.10

Let $G_4 = (\{0, 1\}, \{S, A, O, I, T\}, S, P)$, where P contains

$$\begin{array}{ll} S \to AT & \\ A \to 0AO & A \to 1AI \\ O0 \to 0O & O1 \to 1O \\ I0 \to 0I & I1 \to 1I \\ OT \to 0T & IT \to 1T \\ A \to \varepsilon & T \to \varepsilon \end{array}$$

Then G_4 generates the set $\{xx \mid x \in \{0, 1\}^*\}$. For example, we can derive the word 0101 from S as follows:

$$S \Rightarrow \underline{A}T \Rightarrow 0\underline{A}OT \Rightarrow 01\underline{A}IOT \Rightarrow 01\underline{IO}T \Rightarrow 01\underline{I0}T \Rightarrow 010\underline{IT} \Rightarrow 0101\underline{T} \Rightarrow 0101$$

Example 8.11

We give a right-linear grammar G_5 to generate the language represented by the regular expression in Example 8.3, that is, the set of all nonempty binary strings beginning and ending with the same bit. Let $G_5 = (\{0, 1\}, \{S, O, I\}, S, P)$, where P contains

$$\begin{array}{ll} S \to 0O & S \to 1I \\ S \to 0 & S \to 1 \\ O \to 0O & O \to 1O \\ I \to 0I & I \to 1I \\ O \to 0 & I \to 1 \end{array}$$

The following theorem is due to Chomsky (1956, 1963).

Theorem 8.2

For each $i = 0, 1, 2$, the class of type-i languages properly contains the class of type-$(i + 1)$ languages.

For example, one can prove by using a technique called *pumping* that the set $\{0^n 1^n \mid n \geq 1\}$ is context-free but not regular, and the sets $\{0^n 1^n 2^n \mid n \geq 0\}$ and $\{xx \mid x \in \{0, 1\}^*\}$ are context sensitive but not context-free (Hopcroft and Ullman 1979). It is, however, a bit involved to construct a language that is of type-0 but not context sensitive. See, for example, Hopcroft and Ullman (1979) for such a language.

The four classes of languages in the Chomsky hierarchy also have been completely characterized in terms of TMs and their restricted versions. We have already defined a TM in Section 8.2. Many restricted versions of it will be defined in the next section. It is known that type-0 languages are exactly those recognized by TMs, context-sensitive languages are those recognized by TMs running in linear space, context-free languages are those recognized by TMs whose work tapes operate as pushdown stacks [called *pushdown automata* (PDA)], and regular languages are those recognized by TMs without any work tapes (called *finite-state machine* or *finite automata*) (Hopcroft and Ullman 1979).

Remark 8.1

Recall our definition of a TM and the function it computes from Section 8.2. In the preceding paragraph, we refer to *a language recognized* by a TM. These are two seemingly different ideas, but they are essentially the same. The reason is that the function f, which maps the set of strings over a finite alphabet to $\{0, 1\}$, corresponds in a natural way to the language L_f over Σ defined as: $L_f = \{x \mid f(x) = 1\}$. Instead of saying that a TM computes the function f, we say equivalently that it recognizes L_f.

Because $\{xx \mid x \in \{0, 1\}^*\}$ is a pattern language, the preceding discussion implies that the class of pattern languages is not contained in the class of context-free languages. The next theorem shows that the class of pattern languages is contained in the class of context-sensitive languages.

Theorem 8.3

Every pattern language is context sensitive.

The theorem follows from the fact that every pattern language is recognized by a TM in linear space (Angluin 1980) and linear space-bounded TMs recognize exactly context-sensitive languages. To show the basic idea involved, let us construct a context-sensitive grammar for the pattern language $\{xx \mid x \in \{0, 1\}^*\}$. The grammar G_4 given in Example 8.10 for this language is almost context sensitive. We just have to get rid of the two ε-productions: $A \to \varepsilon$ and $T \to \varepsilon$. A careful modification of G_4 results in the following grammar: $G_6 = (\{0, 1\}, \{S, A_0, A_1, O, I, T_0, T_1\}, S, P)$, where P contains

$$
\begin{array}{ll}
S \to \varepsilon & \\
S \to A_0 T_0 & S \to A_1 T_1 \\
A_0 \to 0 A_0 O & A_0 \to 1 A_0 I \\
A_1 \to 0 A_1 O & A_1 \to 1 A_1 I \\
A_0 \to 0 & A_1 \to 1 \\
O0 \to 0O & O1 \to 1O \\
I0 \to 0I & I1 \to 1I \\
OT_0 \to 0T_0 & IT_0 \to 1T_0 \\
OT_1 \to 0T_1 & IT_1 \to 1T_1 \\
T_0 \to 0 & T_1 \to 1,
\end{array}
$$

which is context sensitive and generates $\{xx \mid x \in \{0, 1\}^*\}$. For example, we can derive 011011 as

$$\Rightarrow \underline{A_1}T_1 \Rightarrow 0\underline{A_1}OT_1 \Rightarrow 01\underline{A_1}IOT_1$$

$$\Rightarrow 011I\underline{OT_1} \Rightarrow 011\underline{I0}T_1 \Rightarrow 0110\underline{IT_1} \Rightarrow 01101\underline{T_1} \Rightarrow 011011$$

For a class of languages, we are often interested in the so-called *closure properties* of the class.

Definition 8.17

A class of languages (e.g., regular languages) is said to be *closed* under a particular operation (e.g., union, intersection, complementation, concatenation, Kleene closure) if each application of the operation on language(s) of the class results in a language of the class.

These properties are often useful in constructing new languages from existing languages as well as proving many theoretical properties of languages and grammars. The closure properties of the four types of languages in the Chomsky hierarchy are now summarized (Gurari 1989, Harrison 1978, Hopcroft and Ullman 1979).

Theorem 8.4

1. The class of type-0 languages is closed under union, intersection, concatenation, and Kleene closure but not under complementation.
2. The class of context-free languages is closed under union, concatenation, and Kleene closure but not under intersection or complementation.
3. The classes of context-sensitive and regular languages are closed under all five of the operations.

For example, let $L_1 = \{0^m 1^n 2^p \mid m = n \text{ or } n = p\}$, $L_2 = \{0^m 1^n 2^p \mid m = n\}$, and $L_3 = \{0^m 1^n 2^p \mid n = p\}$. It is easy to see that all three are context-free languages. (In fact, $L_1 = L_2 \cup L_3$.) However, intersecting L_2 with L_3 gives the set $\{0^m 1^n 2^p \mid m = n = p\}$, which is not context-free.

We will look at context-free grammars more closely in the Section 8.4.3 and introduce the concept of *parsing* and ambiguity.

8.4.3 Context-Free Grammars and Parsing

From a practical point of view, for each grammar $G = (\Sigma, N, S, P)$ representing some language, the following two problems are important:

1. (Membership) Given a string over Σ, does it belong to $L(G)$?
2. (Parsing) Given a string in $L(G)$, how can it be derived from S?

The importance of the membership problem is quite obvious: given an English sentence or computer program, we wish to know if it is grammatically correct or has the right format. Parsing is important because a derivation usually allows us to interpret the meaning of the string. For example, in the case of a Pascal program, a derivation of the program in Pascal grammar tells the compiler how the program should be executed. The following theorem illustrates the decidability of the membership problem for the four classes of grammars in the Chomsky hierarchy. The proofs can be found in Chomsky (1963), Harrison (1978), and Hopcroft and Ullman (1979).

Theorem 8.5

The membership problem for type-0 grammars is undecidable in general and is decidable for any context-sensitive grammar (and thus for any context-free or right-linear grammars).

Because context-free grammars play a very important role in describing computer programming languages, we discuss the membership and parsing problems for context-free grammars in more detail. First, let us look at another example of context-free grammar. For convenience, let us abbreviate a set of productions with the same left-hand side nonterminal

$$A \to \alpha_1, \ldots, A \to \alpha_n$$

as

$$A \to \alpha_1 \mid \ldots \mid \alpha_n$$

Example 8.12

We construct a context-free grammar for the set of all valid Pascal real values. In general, a real constant in Pascal has one of the following forms:

$$m.n, \quad meq, \quad m.neq,$$

where m and q are signed or unsigned integers and n is an unsigned integer. Let $\Sigma = \{0, 1, 2, 3, 4, 5, 6, 7, 8, 9, \mathbf{e}, +, -, .\}$, $N = \{S, M, N, D\}$, and the set P of the productions contain

$$S \to M.N \mid MeM \mid M.NeM$$
$$M \to N \mid +N \mid -N$$
$$N \to DN \mid D$$
$$D \to 0 \mid 1 \mid 2 \mid 3 \mid 4 \mid 5 \mid 7 \mid 8 \mid 9$$

Then the grammar generates all valid Pascal real values (including some absurd ones like 001.200e000). The value 12.3**e** − 4 can be derived as

$$S \Rightarrow \underline{M}.NeM \Rightarrow \underline{N}.NeM \Rightarrow \underline{D}N.NeM \Rightarrow 1\underline{N}.NeM \Rightarrow 1\underline{D}.NeM$$

$$\Rightarrow 12.\underline{N}eM \Rightarrow 12.\underline{D}eM \Rightarrow 12.3\mathbf{e}\underline{M} \Rightarrow 12.3\mathbf{e}{-}\underline{N} \Rightarrow 12.3\mathbf{e}{-}\underline{D} \Rightarrow 12.3\mathbf{e}{-}4$$

Perhaps the most natural representation of derivations for a context-free grammar is *a derivation tree* or *a parse tree*. Each *internal node* of such a tree corresponds to a nonterminal and each *leaf* corresponds to a terminal. If A is an internal node with children $B_1, \ldots, B_n$ ordered from left to right, then $A \to B_1 \ldots B_n$ must be a production. The concatenation of all leaves from left to right yields the string being derived. For example, the derivation tree corresponding to the preceding derivation of 12.3e−4 is given in Figure 8.3. Such a tree also makes possible the extraction of the parts 12, 3, and −4, which are useful in the storage of the real value in a computer memory.

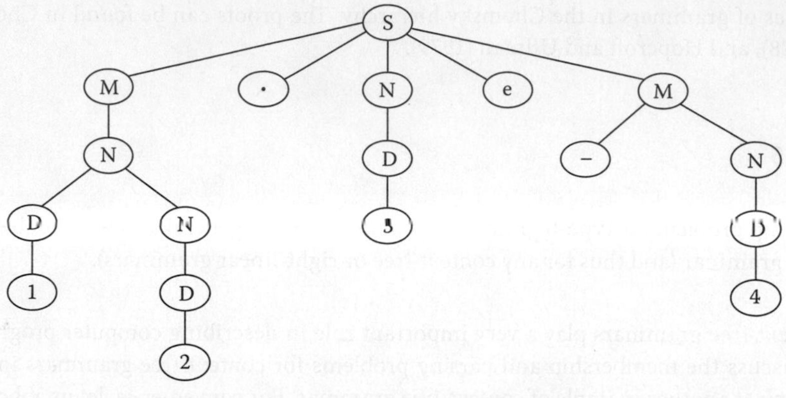

FIGURE 8.3 The derivation tree for 12.3**e** – 4.

Definition 8.18

A context-free grammar G is *ambiguous* if there is a string $x \in L(G)$, which has two distinct derivation trees. Otherwise G is *unambiguous.*

Unambiguity is a very desirable property to have as it allows a unique interpretation of each sentence in the language. It is not hard to see that the preceding grammar for Pascal real values and the grammar G_1 defined in Example 8.7 are all unambiguous. The following example shows an ambiguous grammar.

Example 8.13

Consider a grammar G_7 for all valid arithmetic expressions that are composed of unsigned positive integers and symbols +, *, (,). For convenience, let us use the symbol n to denote any unsigned positive integer.

This grammar has the productions

$$S \rightarrow T + S \mid S + T \mid T$$

$$T \rightarrow F * T \mid T * F \mid F$$

$$F \rightarrow n \mid (S)$$

Two possible different derivation trees for the expression 1 + 2 * 3 + 4 are shown in Figure 8.4. Thus, G_7 is ambiguous. The left tree means that the first addition should be done before the second addition and the right tree says the opposite.

Although in the preceding example different derivations/interpretations of any expression always result in the same value because the operations addition and multiplication are associative, there are situations where the difference in the derivation can affect the final outcome. Actually, the grammar G_7 can be made unambiguous by removing some (redundant) productions, for example, $S \rightarrow T + S$ and $T \rightarrow F * T$.

This corresponds to the convention that a sequence of consecutive additions (or multiplications) is always evaluated from left to right and will not change the language generated by G_7. It is worth noting that there are context-free languages that cannot be generated by any

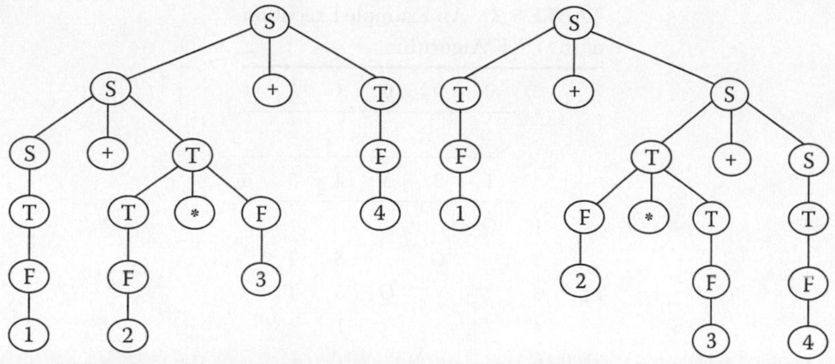

FIGURE 8.4 Different derivation trees for the expression 1 + 2 * 3 + 4.

unambiguous context-free grammar (Hopcroft and Ullman 1979). Such languages are said to be *inherently ambiguous*. An example of inherently ambiguous languages is the set

$$\{0^m1^m2^n3^n \mid m, n > 0\} \cup \{0^m1^n2^m3^n \mid m, n > 0\}$$

We end this section by presenting an efficient algorithm for the membership problem for context-free grammars. The algorithm is due to Cocke, Younger, and Kasami (Hopcroft and Ullman 1979) and is often called the CYK algorithm. Let $G = (\Sigma, N, S, P)$ be a context-free grammar. For simplicity, let us assume that G does not generate the empty string ε and that G is in the so-called *Chomsky normal form* (Chomsky 1963), that is, every production of G is either in the form $A \to BC$ where B and C are nonterminals, or in the form $A \to a$ where a is a terminal. An example of such a grammar is G_1 given in Example 8.7. This is not a restrictive assumption because there is a simple algorithm that can convert every context-free grammar that does not generate ε into one in the Chomsky normal form.

Suppose that $x = a_1 \dots a_n$ is a string of n terminals. The basic idea of the CYK algorithm, which decides if $x \in L(G)$, is *dynamic programming*. For each pair i, j, where $1 \le i \le j \le n$, define a set $X_{i,j} \subseteq N$ as

$$X_{i,j} = \{A \mid A \overset{*}{\Rightarrow} a_i \dots a_j\}$$

Thus, $x \in L(G)$ if and only if $S \in X_{1,n}$. The sets $X_{i,j}$ can be computed inductively in the ascending order of $j - i$. It is easy to figure out $X_{i,i}$ for each i because $X_{i,i} = \{A \mid A \to a_i \in P\}$. Suppose that we have computed all $X_{i,j}$ where $j - i < d$ for some $d > 0$. To compute a set $X_{i,j}$, where $j - i = d$, we just have to find all of the nonterminals A such that there exist some nonterminals B and C satisfying $A \to BC \in P$ and for some k, $i \le k < j$, $B \in X_{i,k}$, and $C \in X_{k+1,j}$. A rigorous description of the algorithm in a Pascal-style pseudocode is given as follows.

Algorithm CYK($x = a_1 \cdots a_n$):

```
for i ← 1 to n do
    X_{i,i} ← {A | A → a_i ∈ P}
for d ← 1 to n − 1 do
    for i ← 1 to n − d do
        X_{i,i+d} ← Ø
        for t ← 0 to d − 1 do
            X_{i,i+d} ← X_{i,i+d} ∪ {A | A → BC ∈ P for some B ∈ X_{i,i+t} and C ∈ X_{i+t+1,i+d}}
```

Table 8.2 shows the sets $X_{i,j}$ for the grammar G_1 and the string $x = 000111$. It just so happens that every $X_{i,j}$ is either empty or a singleton. The computation proceeds from the main diagonal toward the upper-right corner.

TABLE 8.2 An Example Execution of the CYK Algorithm

		0	0	0	1	1	1
				$j \rightarrow$			
		1	2	3	4	5	6
	1	O					$\mathcal{S}$
	2		O		S	T	
$i\downarrow$	3			O	S	T	
	4				I		
	5					I	
	6						I

8.5 Computational Models

In this section, we will present many restricted versions of TMs and address the question of what kinds of problems they can solve. Such a classification is a central goal of computation theory. We have already classified problems broadly into (totally) decidable, partially decidable, and totally undecidable. Because the decidable problems are the ones of most practical interest, we can consider further classification of decidable problems by placing two types of restrictions on a TM. The first one is to restrict its structure. This way we obtain many machines of which a finite automaton and a pushdown automaton are the most important. The other way to restrict a TM is to bound the amount of resources it uses, such as the number of time steps or the number of tape cells it can use. The resulting machines form the basis for *complexity theory.*

8.5.1 Finite Automata

The finite automaton (in its deterministic version) was first introduced by McCulloch and Pitts (1943) as a logical model for the behavior of neural systems. Rabin and Scott (1959) introduced the nondeterministic version of the finite automaton and showed the equivalence of the nondeterministic and deterministic versions. Chomsky and Miller (1958) proved that the set of languages that can be recognized by a finite automaton is precisely the regular languages introduced in Section 8.4. Kleene (1956) showed that the languages accepted by finite automata are characterized by regular expressions as defined in Section 8.4.

In addition to their original role in the study of neural nets, finite automata have enjoyed great success in many fields such as sequential circuit analysis in circuit design (Kohavi 1978), asynchronous circuits (Brzozowski and Seger 1994), lexical analysis in text processing (Lesk 1975), and compiler design. They also led to the design of more efficient algorithms. One excellent example is the development of linear-time string-matching algorithms, as described in Knuth et al. (1977). Other applications of finite automata can be found in computational biology (Searls 1993), natural language processing, and distributed computing.

A finite automaton, as in Figure 8.5, consists of an input tape that contains a (finite) sequence of input symbols such as *aabab* ⋯, as shown in the figure, and a finite-state control. The tape is read by the one-way *read-only* input head from left to right, one symbol at a time. Each time the input head reads an input symbol, the finite control changes its state according to the symbol and the current state of the machine. When the input head reaches the right end of the input tape, if the machine is in a final state, we say that the input is accepted; if the machine is not in a final state, we say that the input is rejected. The following is the formal definition.

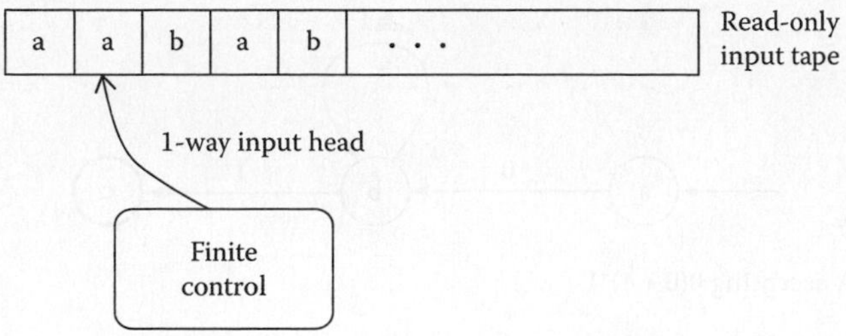

FIGURE 8.5 A finite automaton.

Definition 8.19

A *nondeterministic finite automaton* (NFA) is a quintuple $(Q, \Sigma, \delta, q_0, F)$, where

- Q is a finite set of *states*
- Σ is a finite set of *input symbols*
- δ, the *state transition function*, is a mapping from $Q \times \Sigma$ to subsets of Q
- $q_0 \in Q$ is the *initial state* of the NFA
- $F \subseteq Q$ is the set of *final states*

If δ maps $Q \times \Sigma$ to singleton subsets of Q, then we call such a machine a *deterministic finite automaton* (DFA).

When an automaton, M, is nondeterministic, then from the current state and input symbol, it may go to one of several different states. One may imagine that the device goes to all such states in parallel. The DFA is just a special case of the NFA; it always follows a single deterministic path. The device M *accepts* an input string x if, starting with q_0 and the read head at the first symbol of x, one of these parallel paths reaches an accepting state when the read head reaches the end of x. Otherwise, we say M *rejects* x. A language, L, is accepted by M if M accepts all of the strings in L and nothing else, and we write $L = L(M)$. We will also allow the machine to make *ε-transitions*, that is, changing state without advancing the read head. This allows transition functions such as $\delta(s, \varepsilon) = \{s^1\}$. It is easy to show that such a generalization does not add more power.

Remark 8.2

The concept of a nondeterministic automaton is rather confusing for a beginner. But there is a simple way to relate it to a concept that must be familiar to all of the readers. It is that of a solitaire game. Imagine a game like *Klondike*. The game starts with a certain arrangement of cards (the input) and there is a well-defined final position that results in success; there are also dead ends where a further move is not possible; you lose if you reach any of them. At each step, the precise rules of the game dictate how a new arrangement of cards can be reached from the current one. But the most important point is that there are many possible moves at each step. (Otherwise, the game would be no fun!) Now consider the following question: What starting positions are *winnable*? These are the starting positions for which *there is a winning move sequence*; of course, in a typical play, a player may not achieve it. But that is beside the point in the definition of what starting positions are winnable. The connection between such games and a nondeterministic automaton should be clear. The multiple choices at each step are what make it *nondeterministic*. Our definition of winnable positions is similar to the concept of acceptance of a string by a nondeterministic automaton. Thus, an NFA may be viewed as a formal model to define solitaire games.

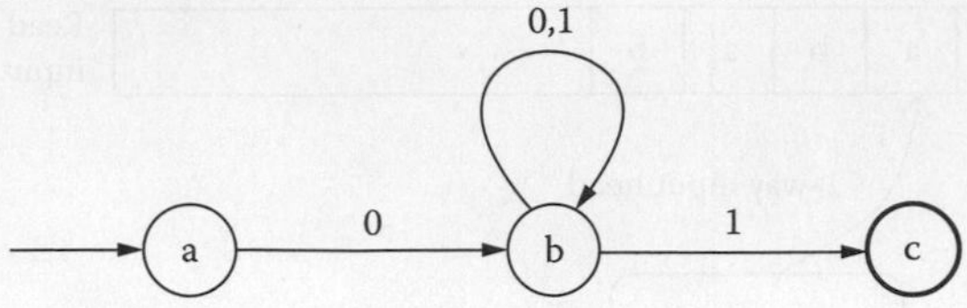

FIGURE 8.6 An NFA accepting 0(0 + 1)*1.

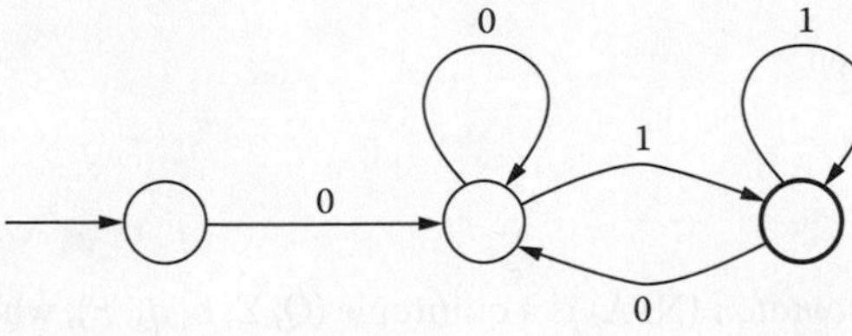

FIGURE 8.7 A DFA accepting 0(0 + 1)*1.

Example 8.14

We design a DFA to accept the language represented by the regular expression 0(0 + 1)*1 as in Example 8.2, that is, the set of all strings in {0, 1} that begin with a 0 and end with a 1. It is usually convenient to draw our solution as in Figure 8.6. As a convention, each circle represents a state; the state a, pointed at by the initial arrow, is the initial state. The darker circle represents the final states (state c). The transition function δ is represented by the labeled edges. For example, δ(a, 0) = {b}. When a transition is missing, for example, on input 1 from a and on inputs 0 and 1 from c, it is assumed that all of these lead to an implicit nonaccepting trap state, which has transitions to itself on all inputs.

The machine in Figure 8.6 is nondeterministic because from b on input 1 the machine has two choices: stay at b or go to c.

Figure 8.7 gives an equivalent DFA, accepting the same language.

Example 8.15

The DFA in Figure 8.8 accepts the set of all strings in {0, 1}* with an even number of 1s. The corresponding regular expression is (0*10*1)*0*.

Example 8.16

As a final example, consider the special case of the tiling problem that we discussed in Section 8.2. This version of the problem is as follows: Let k be a fixed positive integer. Given a set of unit tiles, we want to know if they can tile any $k \times n$ area for all n. We show how to deal with the case $k = 1$ and leave it as an exercise to generalize our method for larger values of k. Number the quarters

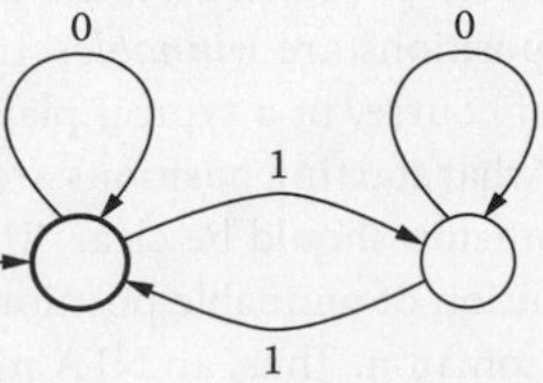

FIGURE 8.8 A DFA accepting (0*10*1)*0*.

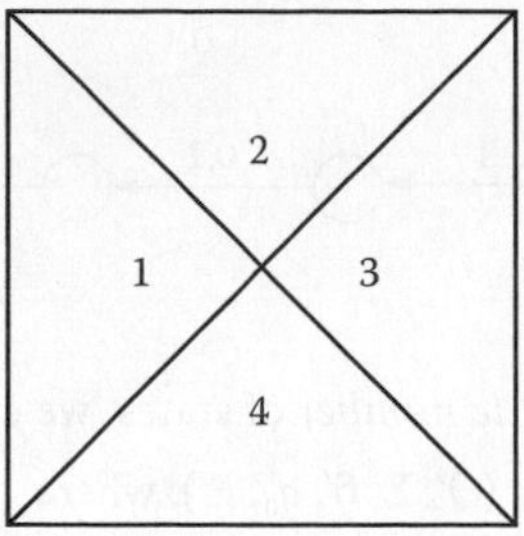

FIGURE 8.9 Numbering the quarters of a tile.

of each tile as in Figure 8.9. The given set of tiles will tile the area if we can find a sequence of the given tiles $T_1, T_2, \ldots, T_m$ such that (1) the third quarter of T_1 has the same color as the first quarter of T_2 and the third quarter of T_2 has the same color as the first quarter of T_3, etc., and (2) the third quarter of T_m has the same color as T_1. These conditions can be easily understood as follows. The first condition states that the tiles T_1, T_2, etc., can be placed adjacent to each other along a row in that order. The second condition implies that the whole sequence $T_1\ T_2 \cdots T_m$ can be replicated any number of times. And a little thought reveals that this is all we need to answer yes on the input. But if we cannot find such a sequence, then the answer must be no. Also note that in the sequence no tile needs to be repeated and so the value of m is bounded by the number of tiles in the input. Thus, we have reduced the problem to searching a finite number of possibilities and we are done.

How is the preceding discussion related to finite automata? To see the connection, define an alphabet consisting of the unit tiles and define a language $L = \{T_1\ T_2 \ldots T_m \mid T_1\ T_2 \ldots T_m \text{ is a valid tiling}, m \geq 0\}$. We will now construct an NFA for the language L. It consists of states corresponding to *distinct* colors contained in the tiles plus two states, one of them the start state and another state called the dead state. The NFA makes transitions as follows: From the start state, there is an ε-transition to each color state, and all states except the dead state are accepting states. When in the state corresponding to color i, suppose it receives input tile T. If the first quarter of this tile has color i, then it moves to the color of the third quarter of T; otherwise, it enters the dead state. The basic idea is to remember the only relevant piece of information after processing some input. In this case, it is the third quarter color of the last tile seen. Having constructed this NFA, the question we are asking is if the language accepted by this NFA is infinite. There is a simple algorithm for this problem (Hopcroft and Ullman 1979).

The next three theorems show a satisfying result that all the following language classes are identical:

- The class of languages accepted by DFAs
- The class of languages accepted by NFAs
- The class of languages generated by regular expressions, as in Definition 8.8
- The class of languages generated by the right-linear, or type-3, grammars, as in Definition 8.16

Recall that this class of languages is called the *regular languages* (see Section 8.4).

Theorem 8.6

For each NFA, there is an equivalent DFA.

Proof An NFA might look more powerful because it can carry out its computation in parallel with its nondeterministic branches.

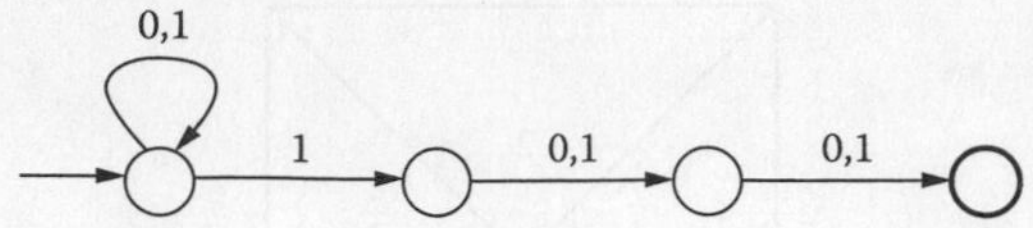

FIGURE 8.10 An NFA accepting L_3.

But because we are working with a *finite number* of states, we can simulate an NFA $M = (Q, \Sigma, \delta, q_0, F)$ by a DFA $M' = (Q', \Sigma, \delta', q_0', F')$, where

- $Q' = \{[S] : S \subseteq Q\}$
- $q_0' = [\{q_0\}]$
- $\delta'([S], a) = [S'] = [\cup_{q_l \in S} \delta(q_l, a)]$
- F' is the set of all subsets of Q containing a state in F

It can now be verified that $L(M) = L(M')$.

Example 8.17

Example 8.1 contains an NFA and an equivalent DFA accepting the same language. In fact, the proof provides an effective procedure for converting an NFA to a DFA. Although each NFA can be converted to an equivalent DFA, the resulting DFA might be exponentially large in terms of the number of states, as we can see from the previous procedure. This turns out to be the best thing one can do in the worst case. Consider the following language: $L_k = \{x: x \in \{0, 1\}^*$ and the kth letter from the right of x is a 1$\}$. An NFA of $k + 1$ states (for $k = 3$) accepting L_k is given in Figure 8.10. A counting argument shows that any DFA accepting L_k must have at least 2^k states.

Theorem 8.7

L is generated by a right-linear grammar if it is accepted by an NFA.

Proof Let L be accepted by a right-linear grammar $G = (\Sigma, N, S, P)$. We design an NFA $M = (Q, \Sigma, \delta, q_0, F)$ where $Q = N \cup \{f\}$, $q_0 = S$, $F = \{f\}$. To define the δ function, we have $C \in \delta(A, b)$ if $A \rightarrow bC$. For rules $A \rightarrow b$, $\delta(A, b) = \{f\}$. Obviously, $L(M) = L(G)$.

Conversely, if L is accepted by an NFA $M = (Q, \Sigma, \delta, q_0, F)$, we define an equivalent right-linear grammar $G = (\Sigma, N, S, P)$, where $N = Q$, $S = q_0$, $q_i \rightarrow aq_j \in N$ if $q_j \in \delta(q_i, a)$, and $q_j \rightarrow \varepsilon$ if $q_j \in F$. Again it is easily seen that $L(M) = L(G)$.

Theorem 8.8

L is generated by a regular expression if it is accepted by an NFA.

Proof (Idea) **Part 1.** We inductively convert a regular expression to an NFA that accepts the language generated by the regular expression as follows:

- Regular expression ε converts to $(\{q\}, \Sigma, \varnothing, q, \{q\})$.
- Regular expression $\varnothing$ converts to $(\{q\}, \Sigma, \varnothing, q, \varnothing)$.
- Regular expression a, for each $a \in \Sigma$ converts to $(\{q, f\}, \Sigma, \delta(q, a) = \{f\}, q, \{f\})$.
- If α and β are regular expressions, converting to NFAs M_α and M_β, respectively, then the regular expression $\alpha \cup \beta$ converts to an NFA M, which connects M_α and M_β in parallel: M has an

initial state q_0 and all of the states and transitions of M_α and M_β; by ε-transitions, M goes from q_0 to the initial states of M_α and M_β.

- If α and β are regular expressions, converting to NFAs M_α and M_β, respectively, then the regular expression α ∪ β converts to NFA M, which connects M_α and M_β sequentially: M has all of the states and transitions of M_α and M_β, with M_α's initial state as M's initial state, ε-transition from the final states of M_α to the initial state of M_β, and M_β's final states as M's final states.
- If α is a regular expression, converting to NFA M_α, then connecting all of the final states of M_α to its initial state with ε-transitions gives α^+. Union of this with the NFA for ε gives the NFA for α^*.

Part 2. We now show how to convert an NFA to an equivalent regular expression. The idea used here is based on Brzozowski and McCluskey (1963); see also Brzozowski and Seger (1994) and Wood (1987).

Given an NFA M, expand it to M' by adding two extra states i, the initial state of M', and t, the only final state of M', with ε-transitions from i to the initial state of M and from all final states of M to t. Clearly, $L(M) = L(M')$. In M', remove states other than i and t one by one as follows. To remove state p, for each triple of states q, p, q' as shown in Figure 8.11a, add the transition as shown in Figure 8.11b. If p does not have a transition leading back to itself, then β = ε. After we have considered all such triples, delete state p and transitions related to p. Finally, we obtain Figure 8.12 and $L(\alpha) = L(M)$.

Apparently, DFAs cannot serve as our model for a modern computer. Many extremely simple languages cannot be accepted by DFAs. For example, $L = \{xx : x \in \{0, 1\}^*\}$ cannot be accepted by a DFA. One can prove this by counting or using the so-called pumping lemmas; one can also prove this by arguing that x contains more information than a *finite*-state machine can *remember*. We refer the interested readers to textbooks such as Hopcroft and Ullmann (1979), Gurari (1989), Wood (1987), and Floyd and Beigel (1994) for traditional approaches and to Li and Vitányi (1993) for a nontraditional approach. One can try to generalize the DFA to allow the input head to be *two way* but still read only. But such machines are not more powerful; they can be simulated by normal DFAs. The next step is apparently to add *storage* space such that our machines can *write* information in.

8.5.2 Turing Machines

In this section, we will provide an alternative definition of a TM to make it compatible with our definitions of a DFA, PDA, etc. This also makes it easier to define a nondeterministic TM. But this formulation (at least the deterministic version) is essentially the same as the one presented in Section 8.2.

A TM, as in Figure 8.13, consists of a *finite control*, an infinite *tape* divided into cells, and a read/write *head* on the tape. We refer to the two directions on the tape as *left* and *right*. The finite control can be in any one of a finite set Q of states, and each tape cell can contain a 0, a 1, or a *blank* B. Time is

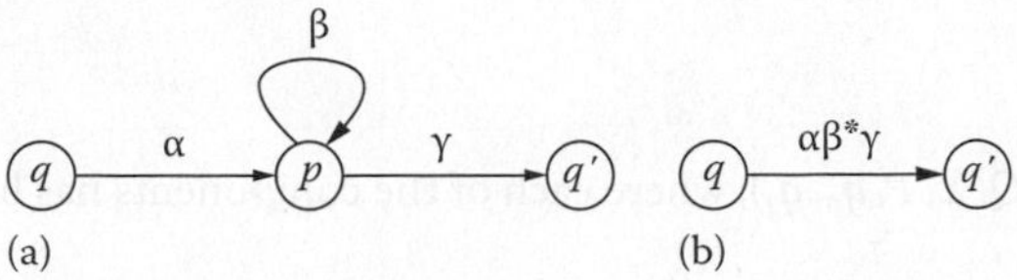

FIGURE 8.11 Converting an NFA to a regular expression.

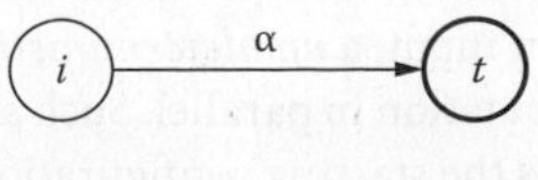

FIGURE 8.12 The reduced NFA.

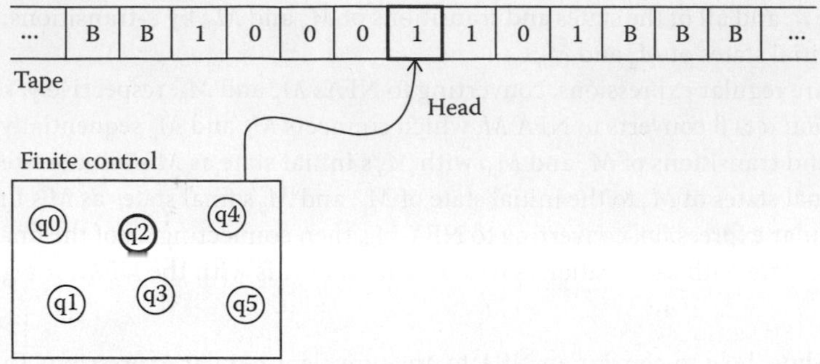

FIGURE 8.13 A Turing machine.

discrete and the time instants are ordered 0, 1, 2, … with 0 the time at which the machine starts its computation. At any time, the head is positioned over a particular cell, which it is said to *scan*. At time 0, the head is situated on a distinguished cell on the tape called the *start cell*, and the finite control is in the initial state q_0. At time 0, all cells contain *B*s, except a contiguous finite sequence of cells, extending from the start cell to the right, which contain 0s and 1s. This binary sequence is called the *input*.

The device can perform the following basic operations:

1. It can write an element from the tape alphabet $\Sigma = \{0, 1, B\}$ in the cell it scans.
2. It can shift the head one cell left or right.

Also, the device executes these operations at the rate of one operation per time unit (a *step*). At the conclusion of each step, the finite control takes on a state in Q. The device operates according to a finite set P of *rules*.

The rules have format (p, s, a, q) with the meaning that if the device is in state p and s is the symbol under scan, then write a if $a \in \{0, 1, B\}$ or move the head according to a if $a \in \{L, R\}$ and the finite control changes to state q. At some point, if the device gets into a special *final* state q_f, the device stops and accepts the input.

If every pair of distinct quadruples differs in the first two elements, then the device is *deterministic*. Otherwise, the device is *nondeterministic*. Not every possible combination of the first two elements has to be in the set; in this way, we permit the device to perform *no* operation. In this case, we say the device *halts*. In this case, if the machine is not in a final state, we say that the machine *rejects* the input.

Definition 8.20

A TM is a quintuple $M = (Q, \Sigma, P, q_0, q_f)$, where each of the components has been described previously.

Given an input, a deterministic TM carries out a uniquely determined succession of operations, which may or may not terminate in a finite number of steps. If it terminates, then the nonblank symbols left on the tape are the output. Given an input, a *nondeterministic TM* behaves much like an NFA. One may imagine that it carries out its computation in parallel. Such a computation may be viewed as a (possibly infinite) tree. The root of the tree is the starting configuration of the machine. The children of each node are all possible configurations one step away from this node. If any of the branches terminates in the final state q_f, we say the machine accepts the input. The reader may want to test understanding this new formulation of a TM by redoing the doubling program on a TM with states and transitions (rather than a GOTO program).

A TM *accepts* a language L if $L = \{w : M \text{ accepts } w\}$. Furthermore, if M halts on all inputs, then we say that L is *Turing decidable* or *recursive*. The connection between a recursive language and a decidable problem (function) should be clear. It is that function f is decidable if and only if L_f is recursive. (Readers who may have forgotten the connection between function f and the associated language L_f should review Remark 8.1.)

Theorem 8.9

All of the following generalizations of TMs can be simulated by a one-tape deterministic TM defined in Definition 8.20.

- Larger tape alphabet Σ
- More work tapes
- More access points, read/write heads, on each tape
- 2D or more dimensional tapes
- Nondeterminism

Although these generalizations do not make a TM compute more, they do make a TM more efficient and easier to program. Many more variants of TMs are studied and used in the literature. Of all simulations in Theorem 8.9, the last one needs some comments. A nondeterministic computation branches like a tree. When simulating such a computation for n steps, the obvious thing for a deterministic TM to do is to try all possibilities; thus, this requires up to c^n steps, where c is the maximum number of nondeterministic choices at each step.

Example 8.18

A DFA is an extremely simple TM. It just reads the input symbols from left to right. TMs naturally accept more languages than DFAs can. For example, a TM can accept

$L = \{xx : x \in \{0, 1\}^*\}$ as follows:

- Find the middle point first: It is trivial by using two heads; with one head, one can mark one symbol at the left and then mark another on the right and go back and forth to eventually find the middle point.
- Match the two parts: With two heads, this is again trivial; with one head, one can again use the marking method matching a pair of symbols each round; if the two parts match, accept the input by entering q_f.

There are types of storage media other than a tape:

- A *pushdown store* is a semi-infinite work tape with one head such that each time the head moves to the left, it erases the symbol scanned previously; this is a last-in first-out storage.
- A *queue* is a semi-infinite work tape with two heads that move only to the right, the leading head is write only and the trailing head is read only; this is a first-in first-out storage.
- A *counter* is a pushdown store with a single-letter alphabet (except its one end, which holds a special marker symbol). Thus, a counter can store a nonnegative integer and can perform three operations.

A queue machine can simulate a normal TM, but the other two types of machines are not powerful enough to simulate a TM.

Example 8.19

When the TM tape is replaced by a pushdown store, the machine is called a *pushdown automaton.* Pushdown automata have been thoroughly studied because they accept the class of context-free languages defined in Section 8.4. More precisely, it can be shown that if L is a context-free language, then it is accepted by a PDA, and if L is accepted by a PDA, then there is a CFG generating L. Various types of PDAs have fundamental applications in compiler design.

The PDA is more restricted than a TM. For example, $L = \{xx : x \in \{0, 1\}^*\}$ cannot be accepted by a PDA, but it can be accepted by a TM as in Example 8.18. But a PDA is more powerful than a DFA. For example, a PDA can accept the language $L' = \{0^k 1^k : k \geq 0\}$ easily. It can read the 0s and push them into the pushdown store; then, after it finishes the 0s, each time the PDA reads a 1, it removes a 0 from the pushdown store; at the end, it accepts if the pushdown store is empty (the number of 0s matches that of 1s). But a DFA cannot accept L', because after it has read all of the 0s, it cannot remember k when k has higher information content than the DFA's finite control.

Two pushdown stores can be used to simulate a tape easily. For comparisons of powers of pushdown stores, queues, counters, and tapes, see van Emde Boas (1990) and Li and Vitányi (1993).

The idea of the universal algorithm was introduced in Section 8.2. Formally, a *universal TM, U,* takes an encoding of a pair of parameters (M, x) as input and simulates M on input x. U accepts (M, x) if M accepts x. The universal TMs have many applications. For example, the definition of the Kolmogorov complexity (Li and Vitányi 1993) fundamentally relies on them.

Example 8.20

Let $L_u = \{\langle M, w\rangle : M \text{ accepts } w\}$. Then, L_u can be accepted by a TM, but it is not Turing decidable. The proof is omitted.

If a language is Turing acceptable but not Turing decidable, we call such a language *recursively enumerable* (r.e.). Thus, L_u is r.e. but not recursive. It is easily seen that if both a language and its complement are r.e., then both of them are recursive. Thus, $\bar{L}_u$ is not r.e.

8.5.2.1 Time and Space Complexity

With TMs, we can now formally define what we mean by *time and space complexities.* Such a formal investigation by Hartmanis and Stearns (1965) marked the beginning of the field of *computational complexity.* We refer the readers to Hartmanis' Turing Award lecture (Hartmanis 1994) for an interesting account of the history and the future of this field.

To define the space complexity properly (in the sublinear case), we need to slightly modify the TM of Figure 8.13. We will replace the tape containing the input by a read-only input tape and give the TM some extra work tapes.

Definition 8.21

Let M be a TM. If, for each n, for each input of length n, and for each sequence of choices of moves when M is nondeterministic M makes at most $T(n)$ moves, we say that M is of *time complexity* $T(n)$; similarly, if M uses at most $S(n)$ tape cells of the work tape, we say that M is of space complexity $S(n)$.

Theorem 8.10

Any TM using $s(n)$ space can be simulated by a TM, with just one work tape, using $s(n)$ space. If a language is accepted by *a* k-tape TM running in time $t(n)$ [space $s(n)$], then it also can be accepted by another k-tape TM running in time $ct(n)$ [space $cs(n)$], for any constant $c > 0$.

To avoid writing the constant c everywhere, we use the standard big-O notation: we say $f(n)$ is $O(g(n))$ if there is a constant c such that $f(n) \leq cg(n)$ for all but finitely many n. The preceding theorem is called the linear speedup theorem; it can be proved easily by using a larger tape alphabet to encode several cells into one and hence compress several steps into one. It leads to the following definitions.

Definition 8.22

DTIME[$t(n)$] is the set of languages accepted by multitape deterministic TMs in time $O(t(n))$.
NTIME[$t(n)$] is the set of languages accepted by multitape nondeterministic TMs in time $O(t(n))$.
DSPACE[$s(n)$] is the set of languages accepted by multitape deterministic TMs in space $O(s(n))$.
NSPACE[$s(n)$] is the set of languages accepted by multitape nondeterministic TMs in space $O(s(n))$.
P is the complexity class $U_{c \in N}$ DTIME[n^c].
NP is the complexity class $U_{c \in N}$ NTIME[n^c].
PSPACE is the complexity class $U_{c \in N}$ DSPACE[n^c].

Example 8.21

We mentioned in Example 8.18 that $L = \{xx : x \in \{0, 1\}^*\}$ can be accepted by a TM. The procedure we have presented in Example 8.18 for a one-head one-tape TM takes $O(n^2)$ time because the single head must go back and forth marking and matching. With two heads, or two tapes, L can be easily accepted in $O(n)$ time.

It should be clear that any language that can be accepted by a DFA, an NFA, or a PDA can be accepted by a TM in $O(n)$ time. The type-1 grammar in Definition 8.16 can be accepted by a TM in $O(n)$ space. Languages in P, that is, languages acceptable by TMs in *polynomial* time, are considered as *feasibly* computable. It is important to point out that all generalizations of the TM, except the nondeterministic version, can all be simulated by the basic one-tape deterministic TM with at most polynomial slowdown. The class NP represents the class of languages accepted in polynomial time by a nondeterministic TM. The nondeterministic version of PSPACE turns out to be identical to PSPACE (Savitch 1970). The following relationships are true:

$$P \subseteq NP \subseteq PSPACE$$

Whether or not either of the inclusions is proper is one of the most fundamental open questions in computer science and mathematics. Research in computational complexity theory centers around these questions. To solve these problems, one can identify the hardest problems in NP or PSPACE. These topics will be discussed in Chapter 8. We refer the interested reader to Gurari (1989), Hopcroft and Ullman (1979), Wood (1987), and Floyd and Beigel (1994).

8.5.2.2 Other Computing Models

Over the years, many alternative computing models have been proposed. With reasonable complexity measures, they can all be simulated by TMs with at most a polynomial slowdown. The reference van Emde Boas (1990) provides a nice survey of various computing models other than TMs. Because of limited space, we will discuss a few such alternatives very briefly and refer our readers to van Emde Boas (1990) for details and references.

Random Access Machines. The *random access machine* (RAM) (Cook and Reckhow 1973) consists of a finite control where a program is stored, with several arithmetic registers and an infinite collection of memory registers $R[1], R[2], \ldots$. All registers have an unbounded word length. The basic instructions for the program are LOAD, ADD, MULT, STORE, GOTO, ACCEPT, REJECT, etc. Indirect addressing is also used. Apparently, compared to TMs, this is a closer but more complicated approximation of modern computers. There are two standard ways for measuring time complexity of the model:

- The *unit-cost RAM*: in this case, each instruction takes one unit of time, no matter how big the operands are. This measure is convenient for analyzing some algorithms such as sorting. But it is unrealistic or even meaningless for analyzing some other algorithms, such as integer multiplication.
- The *log-cost RAM*: each instruction is charged for the sum of the lengths of all data manipulated implicitly or explicitly by the instruction. This is a more realistic model but sometimes less convenient to use.

Log-cost RAMs and TMs can simulate each other with polynomial overheads. The unit-cost RAM might be exponentially (but unrealistically) faster when, for example, it uses its power of multiplying two large numbers in one step.

Pointer Machines. The pointer machines were introduced by Kolmogorov and Uspenskii (1958) (also known as the Kolmogorov–Uspenskii machine) and by Schönhage in 1980 (also known as the storage modification machine, see Schönhage [1980]). We informally describe the pointer machine here. A pointer machine is similar to a RAM but differs in its memory structure. A pointer machine operates on a storage structure called a Δ structure, where Δ is a finite alphabet of size greater than one. A Δ-structure S is a finite directed graph (the Kolmogorov–Uspenskii version is an undirected graph) in which each node has $k = |\Delta|$ outgoing edges, which are labeled by the k symbols in Δ. S has a distinguished node called the *center*, which acts as a starting point for addressing, with words over Δ, other nodes in the structure. The pointer machine has various instructions to redirect the pointers or edges and thus modify the storage structure. It should be clear that turning machines and pointer machines can simulate each other with at most polynomial delay if we use the log-cost model as with the RAMs. There are many interesting studies on the efficiency of the preceding simulations. We refer the reader to van Emde Boas (1990) for more pointers on the pointer machines.

Circuits and Nonuniform Models. A *Boolean circuit* is a finite, labeled, directed acyclic graph. Input nodes are nodes without ancestors; they are labeled with input variables $x_1, \ldots, x_n$. The internal nodes are labeled with functions from a finite set of Boolean operations, for example, {and, or, not} or $\{\oplus\}$. The number of ancestors of an internal node is precisely the number of arguments of the Boolean function that the node is labeled with. A node without successors is an output node. The circuit is naturally evaluated from input to output: at each node, the function labeling the node is evaluated using the results of its ancestors as arguments. Two cost measures for the circuit model are the following:

- *Depth:* the length of a longest path from an input node to an output node
- *Size:* the number of nodes in the circuit

These measures are applied to a family of circuits $\{C_n : n \geq 1\}$ for a particular problem, where C_n solves the problem of size n. If C_n can be computed from n (in polynomial time), then this is a *uniform measure.* Such circuit families are equivalent to TMs. If C_n cannot be computed from n, then such measures are *nonuniform* measures, and such classes of circuits are more powerful than TMs because they simply can compute any function by encoding the solutions of all inputs for each n. See van Emde Boas (1990) for more details and pointers to the literature.

Key Terms

Algorithm: A finite sequence of instructions that is supposed to solve a particular problem.

Ambiguous context-free grammar: For some string of terminals, the grammar has two distinct derivation trees.

Chomsky normal form: Every rule of the context-free grammar has the form $A \to BC$ or $A \to a$, where A, B, and C are nonterminals and a is a terminal.

Computable or decidable function/problem: A function/problem that can be solved by an algorithm (or equivalently, a TM).

Context-free grammar: A grammar whose rules have the form $A \to B$, where A is a nonterminal and B is a string of nonterminals and terminals.

Context-free language: A language that can be described by some context-free grammar.

Context-sensitive grammar: A grammar whose rules have the form $\alpha \to \beta$, where α and β are strings of nonterminals and terminals and $|\alpha| \leq |\beta|$.

Context-sensitive language: A language that can be described by some context-sensitive grammar.

Derivation or parsing: An illustration of how a string of terminals is obtained from the start symbol by successively applying the rules of the grammar.

Finite automaton or finite-state machine: A restricted TM where the head is read only and shifts only from left to right.

(Formal) grammar: A description of some language typically consisting of a set of terminals, a set of nonterminals with a distinguished one called the start symbol, and a set of rules (or productions) of the form $\alpha \to \beta$, depicting what string α of terminals and nonterminals can be rewritten as another string β of terminals and nonterminals.

(Formal) language: A set of strings over some fixed alphabet.

Halting problem: The problem of deciding if a given program (or TM) halts on a given input.

Nondeterministic TM: A TM that can make any one of a prescribed set of moves on a given state and symbol read on the tape.

Partially decidable decision problem: There exists a program that always halts and outputs 1 for every input expecting a positive answer and either halts and outputs 0 or loops forever for every input expecting a negative answer.

Program: A sequence of instructions that is not required to terminate on every input.

Pushdown automaton: A restricted TM where the tape acts as a pushdown store (or a stack).

Reduction: A computable transformation of one problem into another.

Regular expression: A description of some language using operators union, concatenation, and Kleene closure.

Regular language: A language that can be described by some right-linear/regular grammar (or equivalently by some regular expression).

Right-linear or regular grammar: A grammar whose rules have the form $A \to aB$ or $A \to a$, where A, B are nonterminals and a is either a terminal or the null string.

Time/space complexity: A function describing the maximum time/space required by the machine on any input of length n.

Turing machine: A simplest formal model of computation consisting of a finite-state control and a semi-infinite sequential tape with a read/write head. Depending on the current state and symbol read on the tape, the machine can change its state and move the head to the left or right.

Uncomputable or undecidable function/problem: A function/problem that cannot be solved by any algorithm (or equivalently, any TM).

Universal algorithm: An algorithm that is capable of simulating any other algorithms if properly encoded.

Acknowledgment

We would like to thank John Tromp and the reviewers for reading the initial drafts and helping us to improve the presentation.

Further Information

The fundamentals of the theory of computation, automata theory, and formal languages can be found in many text books including Floyd and Beigel (1994), Gurari (1989), Harel (1992), Harrison (1978), Hopcroft and Ullman (1979), and Wood (1987). The central focus of research in this area is to understand the relationships between the different resource complexity classes. This work is motivated in part by some major open questions about the relationships between resources (such as time and space) and the role of control mechanisms (nondeterminism/randomness). At the same time, new computational models are being introduced and studied. One such recent model that has led to the resolution of a number of interesting problems is the interactive proof systems. They exploit the power of randomness and interaction. Among their applications are new ways to encrypt information as well as some unexpected results about the difficulty of solving some difficult problems even approximately. Another new model is the quantum computational model that incorporates quantum-mechanical effects into the basic move of a TM. There are also attempts to use molecular or cell-level interactions as the basic operations of a computer. Yet another research direction motivated in part by the advances in hardware technology is the study of neural networks, which model (albeit in a simplistic manner) the brain structure of mammals. The following chapters of this volume will present state-of-the-art information about many of these developments. The following annual conferences present the leading research work in computation theory: Association of Computer Machinery (ACM) *Annual Symposium on Theory of Computing; Institute of Electrical and Electronics Engineers (IEEE) Symposium on the Foundations of Computer Science; IEEE Conference on Structure in Complexity Theory; International Colloquium on Automata, Languages and Programming; Symposium on Theoretical Aspects of Computer Science; Mathematical Foundations of Computer Science; and Fundamentals of Computation Theory.* There are many related conferences such as *Computational Learning Theory* and *ACM Symposium on Principles of Distributed Computing*, where specialized computational models are studied for a specific application area. Concrete algorithms is another closely related area in which the focus is to develop algorithms for specific problems. A number of annual conferences are devoted to this field. We conclude with a list of major journals whose primary focus is in theory of computation: *The Journal of the Association of Computer Machinery, SIAM Journal on Computing, Journal of Computer and System Sciences, Information and Computation, Mathematical Systems Theory, Theoretical Computer Science, Computational Complexity, Journal of Complexity, Information Processing Letters, International Journal of Foundations of Computer Science*, and *ACTA Informatica*.

References

Angluin, D. 1980. Finding patterns common to a set of strings. *J. Comput. Syst. Sci.* 21: 46–62.

Brzozowski, J. and McCluskey, E. Jr. 1963. Signal flow graph techniques for sequential circuit state diagram. *IEEE Trans. Electron. Comput.* EC-12(2): 67–76.

Brzozowski, J. A. and Seger, C.-J. H. 1994. *Asynchronous Circuits*. Springer–Verlag, New York.

Chomsky, N. 1956. Three models for the description of language. *IRE Trans. Inf. Theory* 2(2): 113–124.

Chomsky, N. 1963. Formal properties of grammars. In *Handbook of Mathematical Psychology*, Vol. 2, pp. 323–418. John Wiley & Sons, New York.

Chomsky, N. and Miller, G. 1958. Finite-state languages. *Inf. Control* 1: 91–112.

Cook, S. and Reckhow, R. 1973. Time bounded random access machines. *J. Comput. Syst. Sci.* 7: 354–375.

Davis, M. 1980. What is computation? In *Mathematics Today-Twelve Informal Essays*, L. Steen (ed.), pp. 241–259. Vintage Books, New York.

Floyd, R. W. and Beigel, R. 1994. *The Language of Machines: An Introduction to Computability and Formal Languages*. Computer Science Press, New York.

Gurari, E. 1989. *An Introduction to the Theory of Computation*. Computer Science Press, Rockville, MD.

Harel, D. 1992. *Algorithmics: The Spirit of Computing*. Addison–Wesley, Reading, MA.

Harrison, M. 1978. *Introduction to Formal Language Theory*. Addison–Wesley, Reading, MA.

Hartmanis, J. 1994. On computational complexity and the nature of computer science. *Commun. ACM* 37(10): 37–43.

Hartmanis, J. and Stearns, R. 1965. On the computational complexity of algorithms. *Trans. Am. Math. Soc.* 117: 285–306.

Hopcroft, J. and Ullman, J. 1979. *Introduction to Automata Theory, Languages and Computation*. Addison–Wesley, Reading, MA.

Jiang, T., Salomaa, A., Salomaa, K., and Yu, S. 1995. Decision problems for patterns. *J. Comput. Syst. Sci.* 50(1): 53–63.

Kleene, S. 1956. Representation of events in nerve nets and finite automata. In *Automata Studies*, C. E Shannon and J. McCarthy (eds.), pp. 3–41. Princeton University Press, Princeton, NJ.

Knuth, D., Morris, J., and Pratt, V. 1977. Fast pattern matching in strings. *SIAM J. Comput.* 6: 323–350.

Kohavi, Z. 1978. *Switching and Finite Automata Theory*. McGraw–Hill, New York.

Kolmogorov, A. and Uspenskii, V. 1958. On the definition of an algorithm. *Usp. Mat. Nauk.* 13: 3–28.

Lesk, M. 1975. LEX–a lexical analyzer generator. Technical Report 39. Bell Laboratories, Murray Hill, NJ.

Li, M. and Vitányi, P. 1993. *An Introduction to Kolmogorov Complexity and Its Applications*. Springer–Verlag, Berlin, Germany.

McCulloch, W. and Pitts, W. 1943. A logical calculus of ideas immanent in nervous activity. *Bull. Math. Biophys.* 5: 115–133.

Post, E. 1943. Formal reductions of the general combinatorial decision problems. *Am. J. Math.* 65: 197–215.

Rabin, M. and Scott, D. 1959. Finite automata and their decision problems. *IBM J. Res. Dev.* 3: 114–125.

Robinson, R. 1991. Minsky's small universal Turing machine. *Int. J. Math.* 2(5): 551–562.

Salomaa, A. 1966. Two complete axiom systems for the algebra of regular events. *J. ACM* 13(1): 158–169.

Savitch, J. 1970. Relationships between nondeterministic and deterministic tape complexities. *J. Comput. Syst. Sci.* 4(2): 177–192.

Schönhage, A. 1980. Storage modification machines. *SIAM J. Comput.* 9: 490–508.

Searls, D. 1993. The computational linguistics of biological sequences. In *Artificial Intelligence and Molecular Biology*, L. Hunter (ed.), pp. 47–120. MIT Press, Cambridge, MA.

Turing, A. 1936. On computable numbers with an application to the Entscheidungsproblem. *Proc. London Math. Soc., Ser. 2* 42: 230–265.

van Emde Boas, P. 1990. Machine models and simulations. In *Handbook of Theoretical Computer Science*, J. van Leeuwen (ed.), pp. 1–66. Elsevier, Amsterdam, the Netherlands.

Wood, D. 1987. *Theory of Computation*. Harper and Row, New York.

9 Cryptography

Jonathan Katz
University of Maryland

9.1 Introduction

Cryptography is a vast subject, addressing problems as diverse as e-cash, remote authentication, fault-tolerant distributed computing, and more. We cannot hope to give a comprehensive account of the field here. Instead, we will narrow our focus to those aspects of cryptography most relevant to the problem of *secure communication*. Broadly speaking, secure communication encompasses two complementary goals: the **secrecy** and **integrity** of communicated data. These terms can be illustrated using the simple example of a user A attempting to transmit a message m to a user B over a public channel. In the simplest sense, techniques for data secrecy ensure that an eavesdropping adversary (i.e., an adversary who sees all communication occurring on the channel) cannot learn any information about the underlying message m. Viewed in this way, such techniques protect against a *passive* adversary who listens to—but does not otherwise interfere with—the parties' communication. Techniques for data integrity, on the other hand, protect against an *active* adversary who may arbitrarily modify the information sent over the channel or may inject new messages of his own. Security in this setting requires that any such modifications or insertions performed by the adversary will be detected by the receiving party.

In the cases of both secrecy and integrity, two different assumptions regarding the initial setup of the communicating parties can be considered. In the **private-key setting** (also known as the "shared-key," "secret-key," or "symmetric-key" setting), which was the setting used exclusively for cryptography until the mid-1970s, parties A and B are assumed to have shared some secret information—a **key**—in advance. This key, which is completely unknown to the adversary, is then used to secure their future communication. (We do not comment further on how such a key might be generated and shared; for

TABLE 9.1 Overview of the Topics Covered in This Survey

	Private-Key Setting	Public-Key Setting
Secrecy	Private-key encryption (Section 9.2)	Public-key encryption (Section 9.4)
Integrity	Message authentication codes (Section 9.3)	Digital signature schemes (Section 9.5)

our purposes, it is simply an assumption of the model.) Techniques for secrecy in this setting are called **private-key encryption** schemes, and those for data integrity are termed **message authentication codes (MACs)**.

In the **public-key setting**, one (or both) of the parties generate a pair of keys: a *public key* that is widely disseminated, and an associated *private key* which is kept secret. The party generating these keys may now use them as a receiver to ensure message secrecy using a **public-key encryption** scheme, or as a sender to provide data integrity using a **digital signature scheme**. Table 9.1 gives an overview of these different cryptographic tools.

In addition to showcasing the above primitives, and explaining how they should be used and how they can be constructed, the treatment here will also introduce a bit of the methodology of modern (i.e., post-1980s) cryptography. This includes an emphasis on formal *definitions* of security that pin down exactly what goals a scheme is intended to achieve; precisely stated *assumptions* (if needed) regarding the hardness of certain mathematical problems; and rigorous *proofs* of security that a cryptographic construction meets some definition of security given a particular assumption. This approach to designing and analyzing cryptosystems is much preferred to the heuristic, "ad-hoc" approach used in the past.

We warn the reader in advance that it is not the intention of this survey to cover the precise details of schemes used in practice today, nor will the survey be comprehensive. Rather, the aim of the survey is to provide the reader with an appreciation for the problem of secure communication along with an explanation of the core techniques used to address it. The reader seeking further details is advised to consult the references listed at the end of this chapter.

9.2 Private-Key Setting and Private-Key Encryption

We begin by discussing the private-key setting, where two parties share a random, secret key k that will be used to secure their future communication. Let us jump right in by defining the syntax of private-key encryption. Formally, a private-key encryption scheme consists of a pair of algorithms (`Enc`, `Dec`). The encryption algorithm `Enc` takes as input a key k and a message m (sometimes also called the **plaintext**), and outputs an encrypted version of the message called the **ciphertext** that will be denoted by c. The decryption algorithm `Dec` takes as input a key and a ciphertext, and outputs a message. The basic correctness requirement is that for every key k and message m (in some allowed set of messages) we have $m = \mathtt{Dec}_k(\mathtt{Enc}_k(m))$. Since it will come up later, we mention here that `Enc` may be a *randomized* algorithm, so that multiple ciphertexts can potentially be output when encrypting a given message with some key, in which case the preceding correctness condition is required to hold with probability 1. Without loss of generality, we may assume `Dec` is deterministic.

A private-key encryption scheme is used in the following way. Let k be the key shared by two parties A and B. For A to send a message m to B, she first **encrypts** the message by computing $c \leftarrow \mathtt{Enc}_k(m)$ (we use "$\leftarrow$" to explicitly indicate that `Enc` may be randomized); the ciphertext c is then transmitted over the public channel to B. Upon receiving this ciphertext, B **decrypts** the ciphertext to recover the message by computing $m := \mathtt{Dec}_k(c)$.

A private-key encryption scheme can also be used by a single party A to encrypt data (say, on a hard drive) that will be accessed by A herself at a later point in time. Here, A encrypts the data using a key k that she stores securely somewhere else. When A later wants to read the data, she can recover it by decrypting using k. Here, the "channel" is the hard drive itself and, rather than being separated in *space*, encryption and decryption are now separated in *time*. Everything we say in what follows will apply to either usage of private-key encryption. (Message authentication codes, discussed in Section 9.3, can also be used for either of these canonical applications.)

The above discussion says nothing about the security provided by the encryption scheme. We consider this aspect in the following sections.

9.2.1 Perfect Secrecy

The goal of a private-key encryption scheme is to ensure the secrecy of m from an eavesdropping adversary $\mathcal{A}$ who views c, but does not know k. How should secrecy be defined formally?

A natural first attempt would be to say that an encryption scheme is secure if $\mathcal{A}$ cannot recover the message m in its entirety (assuming, say, m is chosen uniformly). A little thought shows that such a definition is inadequate: What if the distribution of m is not uniform? And surely we would not consider secure a scheme that always leaks the first bit of the message (without revealing anything about the rest of m)!

A better definition, introduced by Shannon [38] and termed *perfect secrecy*, is that the ciphertext c should leak no information whatsoever about m, regardless of its distribution. This is formalized by requiring that the *a posteriori* distribution of m (after $\mathcal{A}$ observes the ciphertext) should be equivalent to the *a priori* distribution of m (reflecting the adversary's prior knowledge about the distribution of the message). Namely, a scheme (Enc, Dec) is perfectly secret if for any distribution $\mathcal{M}$ over the space of possible messages, any message m, and any ciphertext c, it holds that

$$\Pr[M = m \mid C = c] = \Pr[M = m],$$

where M (respectively, C) denotes the random variable taking the value of the actual message of the sender (respectively, actual ciphertext transmitted).

Before giving an example of a scheme satisfying the above definition, we stress that the adversary (implicit in the above) is assumed to know the full details of the encryption scheme being used by the honest parties. (This is called *Kerckhoffs's principle* [29].) It is a mistake to require the details of a cryptosystem scheme to be hidden in order for it to be secure, and modern schemes are designed to be secure even when the full details of all algorithms are publicly known. The only thing unknown to the adversary is the key itself. This highlights the necessity of choosing the key at random, and the importance of keeping it completely secret.

Perfect secrecy can be achieved by the **one-time pad** encryption scheme, which works as follows. Let ℓ be the length of the message to be transmitted, where the message is viewed as a binary string. (Note that all potential messages are assumed to have the same length.) The parties share in advance a uniformly distributed, ℓ-bit key $k \in \{0, 1\}^\ell$. To encrypt message m the sender computes $c := m \oplus k$, where $\oplus$ represents bit-wise exclusive-or. Decryption is performed by setting $m := c \oplus k$. Clearly, decryption always recovers the original message.

To see that this scheme is perfectly secret, fix any initial distribution over messages and let K be the random variable denoting the key. For any message m and observed ciphertext c, we have

$$\Pr[M = m \mid C = c] = \frac{\Pr[C = c \mid M = m]\Pr[M = m]}{\Pr[C = c]} = \frac{\Pr[C = c \mid M = m]\Pr[M = m]}{\sum_{m'} \Pr[C = c \mid M = m'] \cdot \Pr[M = m']}, \quad (9.1)$$

where the summation is over all possible messages m'. Moreover, for any c, m' (including $m' = m$) we have

$$\Pr[C = c \mid M = m'] = \Pr[K = c \oplus m'] = 2^{-\ell},$$

since k is a uniform ℓ-bit string. Substituting into (9.1), we have $\Pr[M = m \mid C = c] = \Pr[M = m]$ as desired.

Although the one-time pad is perfectly secret, it is of limited value in practice. For one, *the length of the shared key is equal to the length of the message.* Thus, the scheme becomes impractical when long messages are to be sent. Second, it is easy to see that the scheme provides secrecy *only when a given key is used to encrypt a single message* (hence the name "one-time pad"). This will not do in typical scenarios where A and B wish to share a single key that they can use to send multiple messages. Unfortunately, it can be shown that both these limitations are inherent for schemes achieving perfect secrecy.

9.2.2 Computational Secrecy

At the end of the previous section, we observed some fundamental limitations of perfect secrecy. To obtain reasonable solutions, we thus need to (slightly) relax our definition of secrecy. This is not too bad, however, since perfect secrecy may be considered to be unnecessarily strong: it requires *absolutely no information* about m to be leaked, even to an *all-powerful* eavesdropper. Arguably, it would be sufficient to leak a *tiny amount of information* and to restrict attention to eavesdroppers having some *bounded amount of computing power.* To take some concrete numbers, we may be satisfied with an encryption scheme that leaks a most 2^{-60} bits of information (on average) to any eavesdropper that invests at most 100 years of computational effort (on a standard desktop PC, for instance). Definitions of this latter sort are termed **computational**, to distinguish them from notions (like perfect secrecy) that are **information-theoretic** in nature. As initiated in the work of [11,36,33,35,44,9,22] and others, computational security is now the default way security is defined for cryptographic primitives.

A drawback of computational notions of security is that, given the current state of our knowledge, proofs that a given scheme satisfies any such definition must necessarily rely on (unproven) assumptions regarding the computational hardness of certain problems. Our confidence in the security of a particular scheme can be no better than our confidence in the underlying assumption(s) it is based on.

We illustrate the above by introducing a basic definition of computational secrecy. Fix some $t, \varepsilon \geq 0$. Private-key encryption scheme ($\mathtt{Enc}$, $\mathtt{Dec}$) is (t, ε)-*indistinguishable* if for every eavesdropper $\mathcal{A}$ running in time at most t, and for all (equal-length) messages m_0, m_1, we have

$$\left|\Pr\left[\mathcal{A}(\mathtt{Enc}_k(m_b)) = b\right] - \frac{1}{2}\right| \leq \varepsilon.$$

The probability above is taken over uniform choice of the key k and the bit b, as well as any randomness used by $\mathtt{Enc}$. In words: we choose a random key k, encrypt one of m_0 or m_1 (each with equal probability), and give the resulting ciphertext to $\mathcal{A}$; the scheme is indistinguishable if any $\mathcal{A}$ running in time t cannot determine which message was encrypted with probability ε-better than a random guess.

Perfect secrecy can be shown to be equivalent to $(\infty, 0)$-indistinguishability, and so the above is a natural relaxation of perfect secrecy. Taking t bounded and ε strictly positive exactly corresponds to our intuitive idea of relaxing perfect secrecy by placing a bound t on the running time of $\mathcal{A}$, and being content with possibly allowing a tiny amount ε of "information leakage."

9.2.3 Security against Chosen-Plaintext Attacks

Rather than give a construction satisfying the above definition, we immediately introduce an even stronger definition that is better suited for practical applications. Thus far, our security definitions have been restricted to consideration of an adversary who eavesdrops on a *single* ciphertext and, as we have mentioned, this restriction is essential in the context of perfect secrecy. As noted earlier, however, the honest parties would prefer to encrypt multiple messages using the same shared key; we would like to guarantee secrecy in this setting as well. Moreover, it may be the case that the adversary already knows some of the messages being encrypted; in fact, the adversary might even be able to influence some of the messages the parties send. This latter scenario, where the adversary can cause the parties to encrypt plaintexts of the adversary's choice, is termed a **chosen-plaintext attack**. The one-time pad is trivially insecure against a chosen-plaintext attack: given a ciphertext $c = k \oplus m$ for a known message m, the adversary can easily recover the key k.

To capture the above attack scenarios, we modify the previous definition by additionally giving $\mathcal{A}$ access to an *encryption oracle* $\mathtt{Enc}_k(\cdot)$. This oracle allows the adversary to obtain the encryption of any message(s) of its choice using the key k shared by the parties. This oracle is meant to model the real-world capabilities of an adversary who can control what messages get encrypted by the parties; of course, if the adversary has only partial control over what messages get encrypted, then this only makes the adversary weaker. We say that an encryption scheme (Enc, Dec) is (t, ε)*-indistinguishable against a chosen-plaintext attack* (or (t, ε)*-CPA secure*) if for every adversary $\mathcal{A}$ running in time at most t

$$\left|\Pr\left[\mathcal{A}^{\mathtt{Enc}_k(\cdot)}(\mathtt{Enc}_k(m_b)) = b\right] - \frac{1}{2}\right| \leq \varepsilon,$$

where the probability space is as before.

It is not entirely obvious that CPA-security implies security when multiple messages are encrypted, but this can be shown to be the case [27].

Any easy, but important, observation is that any CPA-secure encryption scheme must have randomized encryption. This is true even if only security for multiple encrypted messages is desired: if the encryption scheme is deterministic, then given any two ciphertexts the adversary can tell whether or not they are encryptions of the same message, an undesirable leakage of information.

9.2.4 Block Ciphers

As a step toward constructing a CPA-secure private-key encryption scheme, we first introduce an important primitive for this application. A *keyed function* $F : \{0, 1\}^n \times \{0, 1\}^\ell \to \{0, 1\}^\ell$ is an efficiently computable function that maps two inputs to a single output; we treat the first input to F as a key that will be chosen at random and then fixed, and define $F_k(x) = F(k, x)$. (We have assumed for simplicity that the input and output of F_k have the same length ℓ, but this is not essential.) We call n the *key length* and ℓ the *block length*. Informally, F is pseudorandom if the function F_k, for a randomly chosen k, is "indistinguishable" from a completely random function f with the same domain and range. That is, consider an adversary $\mathcal{A}$ who can send inputs to and receive outputs from a box that contains either F_k (for a random $k \in \{0, 1\}^n$) or f (for a random function f); the keyed function F is pseudorandom if $\mathcal{A}$ cannot tell which is the case with probability significantly better than random guessing. Formally, F is a (t, ε)-**pseudorandom function** [20] if for any adversary $\mathcal{A}$ running in time t

$$\left|\Pr[\mathcal{A}^{F_k(\cdot)} = 1] - \Pr[\mathcal{A}^{f(\cdot)} = 1]\right| \leq \varepsilon,$$

where in the first case k is chosen uniformly from $\{0, 1\}^n$ and in the second case f is a completely random function.

If, for all k, the function F_k is a permutation (i.e., bijection) over $\{0, 1\}^\ell$ and moreover the inverse F_k^{-1} can be computed efficiently (given the key k), then F is called a keyed *permutation*. A pseudorandom function that is also a keyed permutation is called a pseudorandom permutation or a **block cipher**. (Technically, in this case F_k should be indistinguishable from a random permutation, but for large enough ℓ this is equivalent to being indistinguishable from a random function.)

A long sequence of theoretical results culminating in [20,24,31] shows that pseudorandom functions and block ciphers can be constructed from rather minimal cryptographic assumptions, and thus in particular from the cryptographic assumptions we will introduce in Section 9.4. Such constructions are rather inefficient. In practice, dedicated and highly efficient block ciphers are used instead; although the security of these block ciphers cannot be cleanly reduced to a concise mathematical assumption, several design principles used in their construction can be given theoretical justification. More importantly, such block ciphers have been subjected to intense scrutiny by the cryptographic community for several years, and thus it is not unreasonable to view the assumption that these block ciphers are secure as being on par with other assumptions used in cryptography.

The most popular block cipher today is the Advanced Encryption Standard (AES) [10], which was standardized by NIST in 2001 after a multi-year, public competition. AES supports 128-, 192-, or 256-bit keys, and has a 128-bit block length. It superseded the Data Encryption Standard (DES) [13], which had been standardized by the US government in 1977. DES is still in wide use, but is considered insecure due to its relatively short key length (56 bits) and block length (64 bits). Further details and additional block ciphers are discussed in [30].

9.2.5 A CPA-Secure Scheme

Given any pseudorandom function $F : \{0, 1\}^n \times \{0, 1\}^\ell \rightarrow \{0, 1\}^\ell$ with ℓ sufficiently long, it is possible to construct a private-key encryption scheme (`Enc`, `Dec`) that is CPA-secure [21]. To encrypt a message $m \in \{0, 1\}^\ell$ using a key k, first choose a random string $r \in \{0, 1\}^\ell$; then output the ciphertext $\langle r, F_k(r) \oplus m \rangle$. Note that encryption here is randomized, and there are many possible ciphertexts associated with a given key and message. Decryption of a ciphertext $c = \langle c_1, c_2 \rangle$ using key k is performed by computing $m := c_2 \oplus F_k(c_1)$. It can be verified that correctness holds.

Intuitively, the sender and receiver are using $F_k(r)$ as a "one-time pad" to encrypt m. Although $F_k(r)$ is not random, it *is* pseudorandom and one can show that this is enough for security to hold. In a bit more detail: say the sender encrypts multiple messages $m_1, m_2, \ldots, m_q$ using random strings $r_1, r_2, \ldots, r_q$, respectively. The facts that F is a pseudorandom function and k is unknown to the adversary imply that $F_k(r_1), F_k(r_2), \ldots, F_k(r_q)$ are indistinguishable from independent, uniform strings of length ℓ unless $r_i = r_j$ for some $i \neq j$ (in which case, $F_k(r_i)$ and $F_k(r_j)$ are, of course, equal). Assuming this does not occur, then, usage of the encryption scheme is equivalent to using the one-time pad encryption scheme with q *independent* keys, and is thus secure. The probability that there exist distinct i, j with $r_i = r_j$ can be bounded by $q^2/2^\ell$, which is small for typical values of q, ℓ. A full proof can be found in [27, Chapter 3].

The scheme described above can be applied to messages of arbitrary length by encrypting in a block-by-block fashion. This results in a ciphertext whose length is twice that of the original plaintext. More efficient **modes of encryption** [14,27] are used in practice to encrypt long messages. As an example, counter mode (CTR-mode) encryption of a message $m = \langle m_1, \ldots, m_t \rangle$ (with $m_i \in \{0, 1\}^\ell$) using a key k is done by choosing a random r as above, and then computing the ciphertext

$$\langle r, F_k(r+1) \oplus m_1, F_k(r+2) \oplus m_2, \ldots, F_k(r+t) \oplus m_t \rangle.$$

(A proof of security for this scheme follows along similar lines as above.) The ciphertext is now only a single block longer than the plaintext.

9.2.6 Stronger Security Notions

Even the notion of CPA security considers only a passive adversary who eavesdrops on the public channel, but not an active adversary who interferes with the communication between the parties. (Although we will treat active attacks in the next section, there our concern will primarily be integrity rather than secrecy.) Encryption schemes providing security against an active adversary are available; see Section 9.3.2 and Section 9.3.7 [27] for further discussion.

9.3 Message Authentication Codes

The preceding section discussed how to achieve *secrecy*; we now discuss techniques for ensuring *integrity* (sometimes also called *authenticity*). Here, the problem is as follows: parties A and B share a key k in advance, and then communicate over a channel that is under the complete control of an adversary. When B receives a message, he wants to ensure that this message indeed originated from A and was not, for example, injected by the adversary, or generated by modifying the real message sent by A.

Secrecy and integrity are incomparable goals, and it is possible to achieve either one without the other. In particular, the one-time pad—which achieves perfect secrecy—provides no integrity whatsoever since *any* ciphertext c of the appropriate length decrypts to some valid message. Even worse, if c represents the encryption of some (possibly unknown) message m, then flipping the first bit of c has the predictable effect of flipping the first bit of the resulting decrypted message. This illustrates that *integrity is not implied by secrecy, and if integrity is required then specific techniques to achieve it must be used.*

In the private-key setting, the right tool in this context is a MAC. We first define the syntax. A MAC consists of two algorithms ($\mathtt{Mac}$, $\mathtt{Vrfy}$). The *tag-generation algorithm* $\mathtt{Mac}$ takes as input a key k and a message m, and outputs a tag $\mathtt{tag}$; although it is possible for this algorithm to be randomized, there is not much loss of generality in assuming that it is deterministic, and so we write this as $\mathtt{tag} := \mathtt{Mac}_k(m)$. The *verification algorithm* $\mathtt{Vrfy}$ takes as input a key, a message, and a tag; it outputs a single bit b with the intention that $b = 1$ denotes "validity" and $b = 0$ indicates "invalidity." (If $\mathtt{Vrfy}_k(m, \mathtt{tag}) = 1$, then we say that $\mathtt{tag}$ is a *valid tag* for message m with respect to key k.) An honestly generated tag on a message m should be accepted as valid, and so correctness requires that for any key k and message m, we have $\mathtt{Vrfy}_k(m, \mathtt{Mac}_k(m)) = 1$.

Defining security for MACs is relatively simple, and there is only one widely accepted definition [3,23]. At a high level, the goal is to prevent an adversary from generating a valid tag on a message that was never previously authenticated by one of the honest parties. This should hold even if the adversary observes valid tags on several other messages of its choice. Formally, consider an adversary $\mathcal{A}$ who is given access to an oracle $\mathtt{Mac}_k(\cdot)$; the adversary can submit messages of its choice to this oracle and obtain the corresponding valid tags. $\mathcal{A}$ succeeds if it can then output $(m, \mathtt{tag})$ such that (1) m was not one of the messages $m_1, \ldots$ that $\mathcal{A}$ had previously submitted to its oracle, and (2) $\mathtt{tag}$ is a valid tag for m; that is, $\mathtt{Vrfy}_k(m, \mathtt{tag}) = 1$. A MAC is (t, ε)-*secure* if for all $\mathcal{A}$ running in time at most t, the probability with which $\mathcal{A}$ succeeds in this experiment (where the probability is taken over random choice of k) is at most ε. This security notion is also called *existential unforgeability under adaptive chosen-message attack.*

One attack not addressed in the above discussion is a *replay attack*, whereby an adversary resends an honestly generated message m along with its valid tag. The receiver, in general, has no way of knowing whether the legitimate sender has simply sent m again, or whether this second instance of m was injected by the adversary. MACs, as defined, are stateless and so are unable to prevent such an attack. This is by choice: since repeated messages may be legitimate in some contexts, any handling of replay attacks must be taken care of at a higher level.

We now show a simple construction of a secure MAC based on any pseudorandom function/block cipher $F : \{0, 1\}^n \times \{0, 1\}^\ell \to \{0, 1\}^\ell$ with ℓ sufficiently long [21]. To generate a tag on a message $m \in \{0, 1\}^\ell$ using key k, simply compute $\mathtt{tag} := F_k(m)$. Verification is done in the obvious way: $\mathtt{Vrfy}_k(m, \mathtt{tag})$ outputs 1 if and only if $\mathtt{tag} \stackrel{?}{=} F_k(m)$. We sketch the proof of security for this construction. Let $m_1, \ldots$ denote those messages for which adversary $\mathcal{A}$ has requested a tag. Since F is a pseudorandom function, $\mathtt{Mac}_k(m) = F_k(m)$ "looks random" for any $m \notin \{m_1, \ldots\}$, the probability with which $\mathcal{A}$ can correctly predict the value of $F_k(m)$, then, is roughly $2^{-\ell}$. For ℓ large enough, the probability of a successful forgery is very small.

9.3.1 MACs for Long Messages

The construction in the previous section assumed that messages to be authenticated had length ℓ, the block length of the underlying pseudorandom function. Practical block ciphers have relatively short block length (e.g., $\ell \approx 128$ bits), which means that only very short messages can be authenticated by this construction. In this section, we explore two approaches for doing better.

The first approach is a specific construction called the *cipher-block chaining MAC* (CBC-MAC), which can be based on any pseudorandom function/block cipher F as before. Here, we assume that the length of the messages to be authenticated is some *fixed* multiple of the block length ℓ. The tag on a message $M = \langle m_1, m_2, \ldots, m_L \rangle$ (with $m_i \in \{0, 1\}^\ell$) using key k is computed as follows:

$\mathtt{tag}_0 = 0^\ell$
For $i = 1$ to L:
  $\mathtt{tag}_i = F_k(m_i \oplus \mathtt{tag}_{i-1})$
Output $\mathtt{tag}_L$

Verification of a tag $\mathtt{tag}$ on a message $M = \langle m_1, \ldots, m_L \rangle$ is done by recomputing $\mathtt{tag}_L$ as above and outputting 1 if and only if $\mathtt{tag} = \mathtt{tag}_L$.

CBC-MAC is known to be secure if F is a pseudorandom function [3]. This is true only as long as fixed-length messages (i.e., when the number of message blocks L is fixed) are authenticated, and there are several known attacks on the basic CBC-MAC presented above when this is not the case. Subsequent work has shown how to extend basic CBC-MAC to allow authentication of arbitrary-length messages [8,25,34].

A second approach to authenticating arbitrary-length messages is generic, in that it gives a way to modify *any* MAC for short messages so as to handle longer messages. As we will present it here, however, this approach requires an additional cryptographic primitive called a **collision-resistant hash function**. Although hash functions play an important role in cryptography, our discussion will be brief and informal since they are used sparingly in the remainder of this survey.

A **hash function** H is a function that *compresses* an arbitrary-length input to a short, fixed-length string. Hash functions are widely used in many areas of computer science, but cryptographic hash functions have some special requirements that are typically not needed for other applications. The most important such requirement, and the only one we will discuss, is *collision resistance*. Informally, H is collision resistant if it is infeasible for an efficient adversary to find a *collision* in H, where a collision is a pair of distinct inputs x, x' with $H(x) = H(x')$. If H is collision resistant, then the hash of a long message serves as a "secure digest" of that message, in the following sense: for any value y (whether produced by an adversary or not), an adversary can come up with at most one x such that $H(x) = y$. The output length of a hash function fixes an upper bound on the computational difficulty of finding a collision: if H has output length ℓ, then a collision can always be found in $\mathcal{O}(2^{\ell/2})$ steps (see [27]).

As in the case of block ciphers, collision-resistant hash functions can be constructed from number-theoretic assumptions but such constructions are inefficient. (Interestingly, the precise assumptions needed to construct collision-resistant hash functions appear to be stronger than what is necessary to construct block ciphers.) Several dedicated, efficient constructions of hash functions are known; the most popular ones are currently given by the SHA family of hash functions [15], which have output lengths ranging from 160 to 512 bits. As of the time of this writing, however, NIST is running a public competition to choose a replacement (see http://www.nist.gov/hash-competition).

Returning to our discussion of MACs, let $H: \{0, 1\}^* \to \{0, 1\}^\ell$ be a collision-resistant hash function, and let ($\mathtt{Mac}$, $\mathtt{Vrfy}$) be a secure MAC for messages of length ℓ. Then we can construct a message authentication code ($\mathtt{Mac}'$, $\mathtt{Vrfy}'$) for arbitrary-length messages as follows. To authenticate m, first hash it to an ℓ-bit digest, and then authenticate the digest using the original MAC, that is, $\mathtt{Mac}'_k(m) \overset{\text{def}}{=} \mathtt{Mac}_k(H(m))$. Verification is done in the natural way, with $\mathtt{Vrfy}'_k(m, \mathtt{tag}) \overset{\text{def}}{=} \mathtt{Vrfy}_k(H(m), \mathtt{tag})$. It is not difficult to show that this construction is secure. The standardized HMAC message authentication code [2,17] can be viewed as following the above paradigm.

9.3.2 Joint Secrecy and Integrity

When communicating over a public channel, it is usually the case that both secrecy and integrity are required. Schemes achieving both these properties are called **authenticated encryption schemes** [4,28]. The natural way to achieve both these properties is to combine a private-key encryption scheme with a MAC; there are a number of subtleties in doing so, and the reader is referred elsewhere for a more in-depth treatment [4,27]. More efficient constructions of authenticated encryption schemes are also known.

9.4 Public-Key Setting and Public-Key Encryption

The private-key setting we have been considering until now requires the honest parties to share a secret key in advance in order to secure their communication. Historically, the private-key setting was the only one considered in cryptography. In the mid-1970s [11,33,35,36], however, the field of cryptography was revolutionized by the development of *public-key cryptography* which can enable secure communication between parties who share no secret information in advance and carry out all their communication over a public channel. The only requirement is that there is a way for one party to reliably send a copy of their public key to the other.

In the setting of public-key cryptography, any party A generates a pair of keys (pk, sk) on its own; the private key sk is held privately by A, while the public key pk must be obtained by any other party B who wishes to communicate with A. The first party can either send a copy of its public key directly to B, if an authenticated (but not necessarily private!) channel is available between A and B; alternately, A can place a copy of its public key in a public directory (or on her webpage) and B can then obtain pk when needed. Regardless of which method is used, when we analyze security of public-key cryptosystems we simply assume that parties are able to obtain authentic copies of each others' public keys. (In practice, certification authorities and a public-key infrastructure are used to reliably distribute public keys; a discussion is outside the scope of this chapter.) We assume the adversary also knows all parties' public keys; this makes sense since parties make no effort to keep their public keys secret.

Public-key cryptography has a number of advantages relative to private-key cryptography. As we have already discussed, the use of public-key cryptography can potentially simplify key distribution since a private channel between the communicating users is not needed (as it is in the private-key setting). Public-key cryptography can also simplify key management in large systems. For example, consider a company with N employees where each employee should be able to communicate securely with any other employee. Using a private-key solution would require each user to share a (unique) key with every other user, requiring each employee to store $N - 1$ secret keys. In contrast, with a public-key solution, each user would need to know their own private key and the $N - 1$ public keys of the other employees; these public keys, however, could be stored in a public directory or other nonprivate storage. Finally, public-key cryptography is more suitable for "open systems" such as the Internet where the parties who need to communicate securely may have no prior trust relationship, as is the case when a (new) customer wants to encrypt their credit card prior to sending it to an online merchant.

The primary disadvantage of public-key cryptography is that it is less efficient than private-key cryptography. An exact comparison depends on many factors, but as a rough estimate, when encrypting "short" messages (say, less than 10 kB) public-key cryptosystems are 500–1000 times slower than comparable private-key cryptosystems and use roughly a factor of 10 more bandwidth. The power consumption needed to use public-key cryptography can also be an issue when cryptography is used in low-power devices (sensors, RFID tags, etc.). Thus, when private-key cryptography is an option, it is preferable to use it.

Let us formalize the syntax of public-key encryption. A public-key encryption scheme is composed of three algorithms (`Gen`, `Enc`, `Dec`). The *key-generation algorithm* `Gen` is a randomized algorithm that outputs a pair of keys (pk, sk) as described earlier. The *encryption algorithm* `Enc` takes as input a public key pk and a message m, and outputs a ciphertext c. The *decryption algorithm* Dec takes as input a private key sk and a ciphertext c, and outputs a message m. (We highlight here that the decryption algorithm and the encryption algorithm use *different* keys.) Correctness requires that for all (pk, sk) output by Gen, and any message m, we have $\text{Dec}_{sk}(\text{Enc}_{pk}(m)) = m$.

Our definition of security is patterned on the earlier definition given for private-key encryption schemes; in fact, the only difference is that the adversary is given the public key. That is, public-key encryption scheme (Gen, Enc, Dec) is (t, ε)*-indistinguishable* if for every (probabilistic) eavesdropper $\mathcal{A}$ running in time at most t, and for all (equal-length) messages m_0, m_1, we have

$$\left|\Pr\left[\mathcal{A}(pk, \text{Enc}_{pk}(m_b)) = b\right] - \frac{1}{2}\right| \le \varepsilon,$$

where the probability is taken over random generation of (pk, sk) by Gen, uniform choice of b, and any randomness used by the encryption algorithm itself.

Interestingly, in the public-key setting, the above definition implies security against chosen-plaintext attacks (and hence also security when multiple messages are encrypted) and for this reason we will also use the above as our definition of CPA-security; this is in contrast to the private-key setting where security against chosen-plaintext attacks is stronger than indistinguishability. A bit of reflection shows why: in the public-key setting access to an encryption oracle is superfluous, since an adversary who is given the public key can encrypt any messages it likes by itself. A consequence is that, in order to meet even a minimal notion of security in the public-key setting, encryption must be randomized.

In contrast to the private-key setting, public-key cryptography seems to inherently rely on number-theoretic techniques. The two most commonly used techniques are explored in the following two subsections. These sections assume some basic mathematical background on the part of the reader; the necessary background is covered in [27].

9.4.1 RSA Encryption

We discuss the general case of **RSA cryptography**, followed by its application to the particular case of public-key encryption.

Rivest et al. [36] introduced the concept of RSA-based cryptography in 1978. Security here is ultimately based on (though not equivalent to) the assumption that factoring large numbers is hard, even though multiplying large numbers is easy. This, in turn, gives rise to problems that are easy if the factorization of some modulus N is known, but that are believed to be hard when the factors of N are unknown. This asymmetry can be exploited to construct public-key cryptosystems.

Specifically, let $N = pq$ be the product of two large primes p and q. (For concreteness, one may take p and q to be 1000-bit integers.) Let $\mathbb{Z}_N^*$ denote the set of integers between 1 and $N - 1$ that are invertible modulo N; that is, $x \in \mathbb{Z}_N^*$ if there exists an integer x^{-1} such that $x \cdot x^{-1} = 1 \bmod N$. It is known that $\mathbb{Z}_N^*$ consists precisely of those integers between 1 and $N - 1$ that are *relatively prime* to (i.e., have no factor in common with) N. Using this, one can show that

$$|\mathbb{Z}_N^*| = (p-1)\cdot(q-1).$$

Define $\varphi(N) \stackrel{\text{def}}{=} |\mathbb{Z}_N^*|$. Basic group theory implies that for any integers e, d with $ed = 1 \bmod \varphi(N)$ and any $x \in \mathbb{Z}_N^*$, it holds that

$$(x^e)^d = x \bmod N; \tag{9.2}$$

in particular, this means that the function $f_{N,e} : \mathbb{Z}_N^* \to \mathbb{Z}_N^*$ defined by

$$f_{N,e}(x) = x^e \bmod N$$

is a bijection. Anyone given N, e, and x can easily compute $f_{N,e}(x)$ using standard algorithms for efficient modular exponentiation. The *RSA problem* is the problem of inverting this function: namely, given N, e, y with $y \in \mathbb{Z}_N^*$, to find a value $x \in \mathbb{Z}_N^*$ such that $x^e = y \bmod N$.

If the factors of N are known, then it is easy to compute $\varphi(N)$ and hence for any e relatively prime to $\varphi(N)$ the value $d \stackrel{\text{def}}{=} e^{-1} \bmod \varphi(N)$ can be efficiently computed. This value d can then be used to invert $f_{N,e}$ using Equation 9.2, and so in this case the RSA problem is easily solved. On the other hand, there is no known efficient algorithm for inverting $f_{N,e}$ given only N and e (and, in particular, without the factorization of N). The *RSA assumption* formalizes the apparent computational difficulty of solving the RSA problem. Let RSAGen be a randomized algorithm that outputs (N, e, d) where N is a product of two random, large primes and $ed = 1 \bmod \varphi(N)$. (We do not discuss how such an algorithm can be constructed; suffice it to say that efficient algorithms with the required behavior are known.) We say *the RSA problem is* (t, ε)*-hard for* RSAGen if, for all algorithms $\mathcal{A}$ running in time at most t,

$$\Pr[\mathcal{A}(N, e, y) = x \text{ s.t. } x^e = y \bmod N] \le \varepsilon,$$

where the probability is taken over the randomness of RSAGen as well as uniform choice of $y \in \mathbb{Z}_N^*$. It is clear that if the RSA problem is hard for RSAGen, then factoring moduli N output by RSAGen must be hard; the converse is not known, but hardness of the RSA problem for moduli generated appropriately (i.e., the RSA assumption) is widely believed to hold.

The above discussion should naturally motivate our first candidate public-key encryption scheme that we call "textbook RSA encryption." For this scheme, the key-generation algorithm runs RSAGen to obtain (N, e, d); the public key consists of (N, e), while the private key contains (N, d). To encrypt a message $m \in \mathbb{Z}_N^*$ using the public key (N, e), the sender computes $c := m^e \bmod N$. Given a ciphertext c thus computed, the message can be recovered by using the corresponding private key to compute

$$c^d \bmod N = (m^e)^d = m \bmod N.$$

It follows directly from the RSA assumption that if m is chosen uniformly from $\mathbb{Z}_N^*$, then an adversary who observes the ciphertext c cannot compute m in its entirety. But this, alone, is not a very satisfying guarantee! For one thing, a real-life message is unlikely to be random; in particular, it may very well correspond to some structured text. Furthermore, even in the case that m is random, the textbook RSA scheme provides no assurance that an adversary cannot deduce some partial information about m (and in fact, recovering some partial information about m is known to be possible). Finally, this scheme does not have randomized encryption; textbook RSA encryption thus cannot possibly be CPA-secure.

Randomized variants of textbook RSA, however, are used extensively in practice. For simplicity, we describe here a variant that roughly corresponds to the RSA PKCS #1 v1.5 standard [37]. (This standard has since been superseded by a later version which should be used instead.) Here, the message is randomly padded during encryption. That is, if N is a 2000-bit modulus, then encryption of an 800-bit message m can be done by choosing a random 200-bit string r and then computing

$$c = (r \,\|\, m)^e \bmod N;$$

decryption is done in the natural way. It is conjectured that this scheme is CPA-secure, though no proof based on the RSA assumption is known.

9.4.2 El Gamal Encryption

A second class of problems that can be used for public-key cryptography is related to the presumed hardness of the *discrete logarithm problem* in certain groups. This idea was introduced, in somewhat different form, by Diffie and Hellman in 1976 [11]. Our treatment will be abstract, though after introducing the basic ideas we will discuss some concrete instantiations.

Let $\mathbb{G}$ be a finite, cyclic group of large prime order q, and let $g \in \mathbb{G}$ be a generator of $\mathbb{G}$. The fact that g is a generator implies that for any element $h \in \mathbb{G}$ there exists an $x \in \{0,\ldots, q-1\} = \mathbb{Z}_q$ such that $g^x = h$. By analogy with logarithms over the real numbers, in this case we write $x = \log_g h$. The *discrete logarithm problem in* $\mathbb{G}$ is to compute $\log_g h$ given h. For many groups (see below), this problem appears to be computationally infeasible. We may formalize this, the *discrete logarithm assumption in* $\mathbb{G}$, as follows: the discrete logarithm problem in $\mathbb{G}$ is (t, ε)-hard if for all algorithms $\mathcal{A}$ running in time at most t it holds that

$$\Pr[\mathcal{A}(h) = x \text{ s.t. } g^x = h] \leq \varepsilon,$$

where the probability is taken over uniform choice of $h \in \mathbb{G}$.

For cryptographic applications, a stronger assumption is often needed. The **decisional Diffie–Hellman** problem is to distinguish tuples of the form (g^x, g^y, g^{xy}) (where x, y are chosen uniformly from $\mathbb{Z}_q$) from tuples of the form (g^x, g^y, g^z) (where x, y, z are chosen uniformly from $\mathbb{Z}_q$). We say the *decisional Diffie–Hellman problem is* (t, ε)*-hard in* $\mathbb{G}$ if for all algorithms $\mathcal{A}$ running in time at most t we have

$$\left|\Pr[\mathcal{A}(g^x, g^y, g^{xy}) = 1] - \Pr[\mathcal{A}(g^x, g^y, g^z) = 1]\right| \leq \varepsilon,$$

where the probability space in each case is as described earlier. One way to solve the decisional Diffie–Hellman problem given a candidate tuple (g_1, g_2, g_3) is to compute $x = \log_g g_1$ and then check whether $g_3 \stackrel{?}{=} g_2^x$. We thus see that the decisional Diffie–Hellman problem in some group $\mathbb{G}$ is no harder than the discrete logarithm problem in the same group. The converse is not, in general, true as there are candidate groups in which the discrete logarithm problem is hard but the decisional Diffie–Hellman problem is not. Nevertheless, for the groups discussed below, the stronger decisional Diffie–Hellman problem is widely believed to hold.

A classical example of a class of groups for which the above assumptions are believed to hold is given by large prime-order subgroups of $\mathbb{Z}_p^*$ for p prime. As one concrete example, let $p = 2q + 1$, where both p and q are prime. Then the set of *quadratic residues* modulo p (namely, elements of $\mathbb{Z}_p^*$ that can be written as squares of other elements) constitutes a subgroup $\mathbb{G} \subset \mathbb{Z}_p^*$ of order q. Another important example is given by the group of points over certain elliptic curves [43].

In their initial paper [11], Diffie and Hellman suggested a key-exchange protocol based on (what would later come to be called) the decisional Diffie–Hellman problem. This protocol was later adapted [12] to give the *El Gamal encryption scheme* that we describe now. Fix a group $\mathbb{G}$ of prime order q, and generator $g \in \mathbb{G}$ as above. The key-generation algorithm simply chooses a random $x \in \mathbb{Z}_q$ which will be the private key; the public key is $h = g^x$. To encrypt a message $m \in \mathbb{G}$ with respect to the public key h, the sender chooses a random $y \in \mathbb{Z}_q$ and outputs the ciphertext $\langle g^y, h^y \cdot m\rangle$. Decryption of a ciphertext $\langle c_1, c_2\rangle$ using the private key x is done by computing c_2/c_1^x. This recovers the correct result since

$$\frac{c_2}{c_1^x} = \frac{h^y \cdot m}{(g^y)^x} = \frac{(g^x)^y \cdot m}{g^{yx}} = \frac{g^{xy} \cdot m}{g^{yx}} = m.$$

It can be shown that the El Gamal encryption scheme is CPA-secure if the decisional Diffie–Hellman problem is hard in the underlying group $\mathbb{G}$.

9.4.3 Hybrid Encryption

In contrast to private-key encryption, which can be based on highly efficient block ciphers, current public-key encryption schemes involve arithmetic operations with "very large" integers. (Appropriate choice of parameters depends on numerous factors; the reader can think of operations on ≈1000-bit numbers though that need not always be the case.) As a consequence, naive public-key encryption is orders of magnitude slower than private-key encryption.

When encrypting very long messages, however, *hybrid encryption* can be used to obtain the functionality of public-key encryption with the (asymptotic) efficiency of private-key encryption. Let Enc denote the encryption algorithm for some public-key scheme, and let Enc′ denote the encryption algorithm for a private-key scheme using keys of length n. Hybrid encryption works as follows: to encrypt a (long) message m, the sender first chooses a random secret key $k \in \{0, 1\}^n$, encrypts k using Enc and the public key pk, and then encrypts m using Enc′ and the key k. That is, the entire ciphertext is $\langle \mathtt{Enc}_{pk}(k), \mathtt{Enc}'_k(m) \rangle$. Decryption is done in the natural way, by recovering k from the first component of the ciphertext and then using k to decrypt the second part of the ciphertext. Note that the public-key scheme Enc is used only to encrypt a short key, while the bulk of the work is done using the more efficient private-key scheme.

One can show that hybrid encryption is CPA-secure if (1) the underlying public-key encryption scheme is CPA-secure and (2) the underlying private-key encryption scheme is indistinguishable. (CPA-security of the private-key encryption scheme is not necessary; intuitively, this is because the key k is generated freshly at random each time encryption is done, so any particular key is used only once.)

9.4.4 Stronger Security Notions

Our treatment of public-key encryption has only focused on security against passive attacks. As in the case of private-key encryption, however, in many scenarios security against an active adversary is also required. In fact, active attacks are arguably an even greater concern in the public-key setting where a recipient uses the same public key pk to communicate with multiple senders, some of whom may be malicious. (We refer to [27] for examples of some potential attacks.) To defend against this possibility, public-key encryption schemes satisfying stronger notions of security should be used; see [27] for details.

9.5 Digital Signature Schemes

As public-key encryption is to private-key encryption, so are digital signature schemes to MACs. That is, signature schemes allow a *signer* who has established a public key to "sign" messages in a way which is verifiable to anyone who knows the signer's public key. Furthermore (by analogy with MACs), no adversary can forge a valid signature on any message that was not explicitly authenticated (i.e., signed) by the legitimate signer.

Formally, a signature scheme is comprised of three algorithms (Gen, Sign, Vrfy). The *key-generation algorithm* Gen is a probabilistic algorithm that outputs a pair of public and private keys (pk, sk) exactly as in the case of public-key encryption. The *signing algorithm* Sign takes as input a private key sk and a message m, and outputs a *signature* σ; we denote this by $\sigma := \mathtt{Sign}_{sk}(m)$. (One may assume this algorithm is deterministic without loss of generality.) Finally, the *verification algorithm* Vrfy takes as input a public key, a message, and a signature and outputs 0 or 1. If $\mathtt{Vrfy}_{pk}(m, \sigma) = 1$, then we say that σ is a *valid signature* on m (with respect to pk). Analogous to the case of MACs, the correctness requirement is that for every (pk, sk) output by Gen, and for any message m, an honestly generated signature $\mathtt{Sign}_{sk}(m)$ is always a valid signature on m.

A signature scheme can be used in the following way. User A locally generates (pk, sk) and widely publicizes her public key pk. She can then authenticate any desired message m by computing $\sigma := \mathtt{Sign}_{sk}(m)$. Any other user B who has obtained a legitimate copy of A's public key can then verify that this message was indeed certified by A by checking that $\texttt{Vrfy}_{pk}(m, \sigma) \stackrel{?}{=} 1$. Note that in this case, in contrast to the case of public-key encryption, the owner of the public key acts as the sender rather than as the receiver.

Security for digital signature schemes is defined in a manner completely analogous to that of MACs. As there, we consider an adversary $\mathcal{A}$ who is given access to an oracle $\texttt{Sign}_{sk}(\cdot)$; the adversary can submit messages of its choice to this oracle and obtain the corresponding valid signatures. In addition, here the adversary is also given the public key pk. Adversary $\mathcal{A}$ succeeds if it can then output (m, σ) such that (1) m was not one of the messages $m_1, \ldots$ that $\mathcal{A}$ had previously submitted to its oracle, and (2) σ is a valid signature on m; that is, $\texttt{Vrfy}_{pk}(m, \sigma) = 1$. A signature scheme is (t, ε)-*secure* if for all $\mathcal{A}$ running in time at most t, the probability with which $\mathcal{A}$ succeeds in this experiment (where the probability is taken over generation of (pk, sk)) is at most ε. This security notion is also called *existential unforgeability under adaptive chosen-message attack* [23].

Signature schemes offer several advantages relative to MACs. A basic advantage is that, once a signer's public key is widely distributed, that signer's signatures can be verified by *anyone*. (Contrast this to MACs, where a tag can only be verified by the party who shares the key with the sender.) As a result, signature schemes are suitable when "one-to-many" communication is required, with a canonical example being a software company releasing signed updates of its software. The very fact that signatures are publicly verifiable also implies their usefulness for *nonrepudiation*. That is, once a signer has issued a signature on a message, a recipient can prove that fact to someone else (a judge, say) by simply showing them the signature. (This makes digital signatures a potential replacement for handwritten signatures on legally binding documents.) On the other hand, for authenticating "short" messages, a signature scheme may be orders of magnitude less efficient than a MAC.

In the following sections, we briefly discuss two signature schemes in wide use today. Other schemes, some with better theoretical properties that those discussed here, can be found in [26]. Note that it suffices to construct a signature scheme for "short" messages; collision-resistant hashing can then be used, as in Section 9.3.1, to extend any such scheme for signing arbitrary-length messages.

9.5.1 RSA Signatures

We can use the RSA assumption introduced in Section 9.4.1 to construct signature schemes. We begin with a simple scheme that is insecure, and then discuss how to adapt it so as to make it resistant to attack.

At an intuitive level, signing a message must involve some operation that is "easy" for the signer who knows the private key corresponding to its own public key, but "difficult" for anyone who does *not* know the signer's private key. At the same time, the signature thus computed must be easy to verify for anyone knowing only the signer's public key. The RSA problem discussed earlier provides exactly this. Key generation is exactly the same as for RSA public-key encryption; that is, the algorithm runs $\texttt{RSAGen}$ to obtain (N, e, d), and outputs (N, e) and the public key, and (N, d) as the private key. To sign a message $m \in \mathbb{Z}_N^*$ using its private key, the signer computes $\sigma := m^d \bmod N$. Verification of a purported signature σ on a message m with respect to a public key (N, e) is done by checking whether $\sigma^e \stackrel{?}{=} m \bmod N$. Correctness holds for an honestly generated signature since

$$\sigma^e = (m^d)^e = m \bmod N.$$

Hardness of the RSA problem implies that it is infeasible for an adversary who knows only the signer's public key to generate a valid signature on a random message $m \in \mathbb{Z}_N^*$. But this is insufficient for security! In fact, the "textbook RSA" signature scheme just described is not secure. Here are two attacks:

1. An adversary knowing only the public key can easily construct a forgery by choosing an arbitrary $\sigma \in \mathbb{Z}_N^*$ and outputting this as the signature on the message $m = \sigma^e \bmod N$. Although m computed in this way may not be "meaningful," this attack is already enough to violate the

security definition. (In fact, the definition of what constitutes a "meaningful" message is context-dependent and so such an attack may be problematic in practice. Moreover, by repeatedly choosing random σ and following the same steps, the adversary may end up hitting upon a meaningful message.)

2. Perhaps even worse, an adversary can forge a valid signature on an arbitrary message $\hat{m}$ if it can get the signer to sign a single (different) message of the adversary's choice. This attack begins by first having the adversary compute (m, σ) as above. Assume the adversary can then obtain a signature σ' (from the legitimate signer) on the message $m' = (\hat{m}/m) \bmod N$. The adversary then computes the signature $\hat{\sigma} = \sigma \cdot \sigma' \bmod N$ on $\hat{m}$. Note that $\hat{\sigma}$ is a valid signature on $\hat{m}$ since

$$\hat{\sigma}^e = (\sigma \cdot \sigma')^e = \sigma^e \cdot (\sigma')^e = m \cdot m' = \hat{m} \bmod N.$$

The above shows that textbook RSA signatures are not secure. Fortunately, a simple modification prevents the above attacks: *hash* the message before signing it. That is, the signature on a message $m \in \{0, 1\}^*$ is now

$$\sigma = H(m)^d \bmod N,$$

where H is a hash function mapping arbitrary-length inputs to $\mathbb{Z}_N^*$. Verification is done in the natural way. It can be shown [6] that this scheme is secure if the RSA problem is hard and if, in addition, H "acts like a random function" in a sense we do not define here; see [5,27] for further discussion.

9.5.2 The Digital Signature Standard

The Digital Signature Algorithm (DSA), included as part of the Digital Signature Standard (DSS) [1,16], is another widely used signature scheme. Its security is related (though not known to be equivalent) to the hardness of computing discrete logarithms in a specific group $\mathbb{G}$. Specifically, $\mathbb{G}$ is taken to be a subgroup of $\mathbb{Z}_p^*$ of order q where both p and q are large primes. Let g be a generator of $\mathbb{G}$, and let H be a cryptographic hash function that, for simplicity, we view as mapping arbitrary-length inputs to $\mathbb{Z}_q$. (As will be clear, for security to hold H must at least be collision resistant.) Key generation works by choosing a random $x \in \mathbb{Z}_q$ and setting $y = g^x \bmod p$; the public key is y and the private key is x. To sign a message $m \in \{0, 1\}^*$, the signer generates a random $k \in \mathbb{Z}_q$ and computes

$$r = (g^k \bmod p) \bmod q$$
$$s = (H(m) + xr) \cdot k^{-1} \bmod q;$$

the signature is (r, s). Verification of signature (r, s) on message m with respect to public key y is done by checking that $r, s \in \mathbb{Z}_q^*$ and

$$r \stackrel{?}{=} (g^{H(m)s^{-1}} \cdot y^{rs^{-1}} \bmod p) \bmod q.$$

No proof of security for DSA is known; we refer the reader to a survey article by Vaudenay [42] for further discussion. See [26] and references therein for other signature schemes whose security can be proved equivalent to the discrete logarithm or Diffie–Hellman problems.

Key Terms

Block cipher: An efficient instantiation of a pseudorandom function that has the additional property of being efficiently invertible.

Chosen-plaintext attack: An attack on an encryption scheme in which an adversary causes the sender to encrypt messages of the adversary's choice.

Ciphertext: The result of encrypting a message.

Collision-resistant hash function: A hash function for which it is infeasible to find two different inputs mapping to the same output.

Computational security: Provides security against computationally bounded adversaries, assuming the computational hardness of some problem.

Data integrity: Ensuring that modifications to a communicated message are detected.

Data secrecy: Hiding the contents of a communicated message.

Decrypt: To recover the original message from the ciphertext.

Digital signature scheme: A method for providing data integrity in the public-key setting.

Encrypt: To apply an encryption scheme to a plaintext message.

Hash function: A cryptographic function mapping arbitrary-length inputs to fixed-length outputs.

Information-theoretic security: Provides security even against computationally unbounded adversaries.

Key: The secret shared by parties in the private-key setting. Also used for the public and private values generated by a party in the public-key setting.

Message authentication code: A method for providing data integrity in the private-key setting.

Mode of encryption: A method for using a block cipher to encrypt arbitrary-length messages.

One-time pad: A private-key encryption scheme achieving perfect secrecy.

Plaintext: The communicated data, or message.

Private-key encryption: Technique for ensuring data secrecy in the private-key setting.

Private-key setting: Setting in which communicating parties share a secret in advance of their communication.

Pseudorandom function: A keyed function which is indistinguishable from a truly random function.

Public-key encryption: Technique for ensuring data secrecy in the public-key setting.

Public-key setting: Setting in which parties generate public/private keys and widely disseminate their public key.

RSA cryptography: Cryptography based on the hardness of computing eth roots modulo a number N that is the product of two large primes.

Further Information

The textbook by this author and Yehuda Lindell [27] covers the material of this survey, and more, in extensive detail. The online notes of Bellare and Rogaway [7] are also quite useful. The books by Smart [39] and Stinson [41] give slightly different perspectives on the material. Readers looking for a more advanced treatment may consult the texts by Goldreich [18,19]. More applied aspects of cryptography can be found in the book by Stallings [40] or the comprehensive (though now somewhat outdated) *Handbook of Applied Cryptography* [32].

The International Association for Cryptologic Research (IACR) sponsors the *Journal of Cryptology* along with a number of conferences, with Crypto and Eurocrypt being the best known. Proceedings of these conferences (dating, in some cases, to the early 1980s) are published as part of Springer-Verlag's *Lecture Notes in Computer Science*. Research in theoretical cryptography often appears at the ACM Symposium on Theory of Computing, the IEEE Symposium on Foundations of Computer Science, and elsewhere; more applied aspects of cryptography are frequently published in security conferences including the ACM Conference on Computer and Communications Security. Articles on cryptography also appear in the *Journal of Computer and System Sciences*, the *Journal of the ACM*, and the *SIAM Journal on Computing*.

References

1. ANSI X9.30. 1997. Public key cryptography for the financial services industry, part 1: The digital signature algorithm (DSA). American National Standards Institute. American Bankers Association.
2. Bellare, M., Canetti, R., and Krawczyk, H. 1996. Keying hash functions for message authentication. *Advances in Cryptology—Crypto '96*, Lecture Notes in Computer Science, vol. 1109, N. Koblitz, Ed., Springer-Verlag, Berlin, pp. 1–15.
3. Bellare, M., Kilian, J., and Rogaway, P. 2000. The security of the cipher block chaining message authentication code. *Journal of Computer and System Sciences* 61(3): 362–399.
4. Bellare, M. and Namprempre, C. 2000. Authenticated encryption: Relations among notions and analysis of the generic composition paradigm. *Advances in Cryptology—Asiacrypt 2000*, Lecture Notes in Computer Science, vol. 1976, T. Okamoto, Ed., Springer-Verlag, Berlin, pp. 531–545.
5. Bellare, M. and Rogaway, P. 1993. Random oracles are practical: A paradigm for designing efficient protocols. *First ACM Conference on Computer and Communications Security*, ACM, New York, NY, pp. 62–73.
6. Bellare, M. and Rogaway, P. 1996. The exact security of digital signatures: How to sign with RSA and Rabin. *Advances in Cryptology—Eurocrypt '96*, Lecture Notes in Computer Science, vol. 1070, U. Maurer, Ed., Springer-Verlag, Berlin, pp. 399–416.
7. Bellare, M. and Rogaway, P. 2005. Introduction to modern cryptography (course notes). Available at http://www.cs.ucsd.edu/users/mihir/cse207/classnotes.html (accessed October 03, 2013).
8. Black, J. and Rogaway, P. 2005. CBC MACs for arbitrary-length messages: The three-key constructions. *Journal of Cryptology* 18(2): 111–131.
9. Blum, M. and Micali, S. 1984. How to generate cryptographically strong sequences of pseudorandom bits. *SIAM Journal of Computing* 13(4): 850–864.
10. Daemen, J. and Rijmen, V. 2002. *The Design of Rijndael: AES—the Advanced Encryption Standard.* Springer-Verlag, Berlin.
11. Diffie, W. and Hellman, M. 1976. New directions in cryptography. *IEEE Transactions on Information Theory* 22(6): 644–654.
12. El Gamal, T. 1985. A public-key cryptosystem and a signature scheme based on discrete logarithms. *IEEE Transactions on Information Theory* 31(4): 469–472.
13. *Federal Information Processing Standards* publication #46-3. 1999. Data encryption standard (DES). U.S. Department of Commerce/National Institute of Standards and Technology.
14. *Federal Information Processing Standards* publication #81. 1980. DES modes of operation. U.S. Department of Commerce/National Bureau of Standards.
15. *Federal Information Processing Standards* publication #180-2. 2002. Secure hash standard. U.S. Department of Commerce/National Institute of Standards and Technology.
16. *Federal Information Processing Standards* publication #186-2. 2000. Digital signature standard (DSS). U.S. Department of Commerce/National Institute of Standards and Technology.
17. *Federal Information Processing Standards* publication #198. 2002. The Keyed-Hash Message Authentication Code (HMAC). U.S. Department of Commerce/National Institute of Standards and Technology.
18. Goldreich, O. 2001. *Foundations of Cryptography, vol. 1: Basic Tools.* Cambridge University Press, New York, NY.
19. Goldreich, O. 2004. *Foundations of Cryptography, vol. 2: Basic Applications.* Cambridge University Press, New York, NY.
20. Goldreich, O., Goldwasser, S., and Micali, S. 1986. How to construct random functions. *Journal of the ACM* 33(4): 792–807.
21. Goldreich, O., Goldwasser, S., and Micali, S. 1985. On the cryptographic applications of random functions. *Advances in Cryptology—Crypto '84*, Lecture Notes in Computer Science, vol. 196, G.R. Blakley and D. Chaum, Eds., Springer-Verlag, Berlin, pp. 276–288.
22. Goldwasser, S. and Micali, S. 1984. Probabilistic encryption. *Journal of Computer and System Sciences* 28(2): 270–299.

23. Goldwasser, S., Micali, S., and Rivest, R. 1988. A digital signature scheme secure against adaptive chosen-message attacks. *SIAM Journal of Computing* 17(2): 281–308.
24. Håstad, J., Impagliazzo, R., Levin, L., and Luby, M. 1999. A pseudorandom generator from any one-way function. *SIAM Journal of Computing* 28(4): 1364–1396.
25. Iwata, T. and Kurosawa, K. 2003. OMAC: One-Key CBC MAC. *Fast Software Encryption—FSE 2003*, Lecture Notes in Computer Science, vol. 2887, T. Johansson, Ed., Springer-Verlag, Berlin, pp. 129–153.
26. Katz, J. 2010. *Digital Signatures*. Springer, Berlin.
27. Katz, J. and Lindell, Y. 2007. *Introduction to Modern Cryptography*. Chapman and Hall/CRC Press.
28. Katz, J. and Yung, M. 2000. Unforgeable encryption and chosen ciphertext secure modes of operation. *Fast Software Encryption—FSE 2000*, Lecture Notes in Computer Science, vol. 1978, B. Schneier, Ed., Springer-Verlag, Berlin, pp. 284–299.
29. Kerckhoffs, A. 1883. La cryptographie militaire. *Journal des Sciences Militaires*, 9th Series, pp. 161–191.
30. Knudsen, L.R. and Robshaw, M.J.B. 2011. *The Block Cipher Companion*. Springer, Berlin.
31. Luby, M. and Rackoff, C. 1988. How to construct pseudorandom permutations from pseudorandom functions. *SIAM Journal of Computing* 17(2): 412–426.
32. Menezes, A.J., van Oorschot, P.C., and Vanstone, S.A. 2001. *Handbook of Applied Cryptography*. CRC Press.
33. Merkle, R. and Hellman, M. 1978. Hiding information and signatures in trapdoor knapsacks. *IEEE Transactions on Information Theory* 24: 525–530.
34. Petrank, E. and Rackoff, C. 2000. CBC MAC for real-time data sources. *Journal of Cryptology* 13(3): 315–338.
35. Rabin, M.O. 1979. Digitalized signatures and public key functions as intractable as factoring. MIT/LCS/TR-212, MIT Laboratory for Computer Science.
36. Rivest, R., Shamir, A., and Adleman, L.M. 1978. A method for obtaining digital signatures and public-key cryptosystems. *Communications of the ACM* 21(2): 120–126.
37. RSA Laboratories. 1993. RSA laboratories public-key cryptography standard #1, version 1.5. Available from at http://www.emc.com/emc-plus/rsa-labs/standards-initiatives/pkcs-rsa-cryptography-standard.htm
38. Shannon, C.E. 1949. Communication theory of secrecy systems. *Bell System Technical Journal* 28(4): 656–715.
39. Smart, N. 2004. *Cryptography: An Introduction*. McGraw-Hill, Berkshire, UK.
40. Stallings, W. 2010. *Cryptography and Network Security: Principles and Practice, 5th edition*. Prentice Hall, Englewood Cliffs, NJ.
41. Stinson, D.R. 2005. *Cryptography: Theory and Practice, 3rd edition*. Chapman & Hall/CRC Press.
42. Vaudenay, S. 2003. The security of DSA and ECDSA. *Public-Key Cryptography—PKC 2003*, Lecture Notes in Computer Science, vol. 2567, Y. Desmedt, Ed., Springer-Verlag, Berlin, pp. 309–323.
43. Washington, L. 2008. *Elliptic Curves: Number Theory and Cryptography*. Chapman and Hall/CRC Press.
44. Yao, A.C. 1982. Theory and application of trapdoor functions. *Proceedings of the 23rd Annual Symposium on Foundations of Computer Science*, IEEE, pp. 80–91.

10

Algebraic Algorithms*

Ioannis Z. Emiris
University of Athens

Victor Y. Pan
The City University of New York

Elias P. Tsigaridas
Université Pierre and Marie Curie and Institut National de Recherche en Informatique et en Automatique Paris-Rocquencourt

10.1 Introduction

Algebraic algorithms deal with numbers, vectors, matrices, polynomials, formal power series, exponential and differential polynomials, rational functions, algebraic sets, curves, and surfaces. In this vast area, manipulation with matrices and polynomials is fundamental for modern computations in sciences and engineering. The list of the respective computational problems includes the solution of a polynomial equation and linear and polynomial systems of equations, univariate and multivariate polynomial evaluation, interpolation, factorization and decompositions, rational interpolation, computing matrix factorization and decompositions (which in turn include various triangular and orthogonal factorizations such as LU, PLU, QR, QRP, QLP, CS, LR, Cholesky factorizations, and eigenvalue and singular value decompositions), computation of the matrix characteristic and minimal polynomials, determinants, Smith and Frobenius normal forms, ranks, and (generalized) inverses, univariate and

* This material is based on work supported in part by the European Union through Marie-Curie Initial Training Network "SAGA" (ShApes, Geometry, Algebra), with FP7-PEOPLE contract PITN-GA-2008-214584 (Ioannis Z. Emiris), by NSF Grant CCF-1116736 and PSC CUNY Awards 63153–0041 and 64512–0042 (Victor Y. Pan), by the Danish Agency for Science, Technology and Innovation (postdoctoral grant), Danish NRF and NSF of China (grant 61061130540), CFEM, and the Danish Strategic Research Council (Elias P. Tsigaridas). Sections 10.3.5, 10.5, and "Further information" have been written jointly by all authors, Section 10.4 has been contributed essentially by the first author, the other sections by the second author.

multivariate polynomial resultants, Newton's polytopes, greatest common divisors (GCDS), and least common multiples as well as manipulation with truncated series and algebraic sets.

Such problems can be solved by using the error-free symbolic computations with infinite precision. Computer algebra systems such as Maple and Mathematica compute the solutions based on various nontrivial computational techniques such as modular computations, the Euclidean algorithm and continuous fraction approximation, Hensel's and Newton's lifting, Chinese remainder algorithm, elimination and resultant methods, and Gröbner bases computation. The price to achieve perfect accuracy is the substantial memory space and computer time required to support the computations.

The alternative numerical methods rely on operations with binary or decimal numbers truncated or rounded to a fixed precision. Operating with the IEEE standard floating point numbers represented with single or double precision enables much faster computations using much smaller memory but requires theoretical and/or experimental study of the impact of rounding errors on the output. The study involves forward and backward error analysis, linear and nonlinear operators, and advanced techniques from approximation and perturbation theories. Solution of some problems involves more costly computations with extended precision. The resulting algorithms support high performance libraries and packages of subroutines such as those in Matlab, NAG SMP, LAPACK, ScaLAPACK, ARPACK, PARPACK, MPSolve, and EigenSolve.

In this chapter, we cover both approaches, whose combination frequently increases their power and enables more effective computations. We focus on the algebraic algorithms in the large, popular, and highly important fields of matrix computations and root-finding for univariate polynomials and systems of multivariate polynomials. We cover part of these huge subjects and include basic bibliography for further study. To meet space limitation we cite books, surveys, and comprehensive articles with pointers to further references, rather than including all the original technical papers. Our expositions in Sections 10.2 and 10.3 follow the line of the first surveys in this area in [163,168,173–175].

We state the complexity bounds under the RAM (random access machine) model of computation [1,96]. In most cases we assume the *arithmetic* model, that is we assign a unit cost to addition, subtraction, multiplication, and division of real numbers, as well as to reading or writing them into a memory location. This model is realistic for computations with a fixed (e.g., the IEEE standard single or double) precision, which fits the size of a computer word, and then the arithmetic model turns into the *word* model [96]. In other cases we allow working with extended precision and assume the *Boolean* or *bit* model, assigning a unit cost to every Boolean or bitwise operation. This accounts for both arithmetic operations and the length (precision) of the operands. We denote the bounds for this complexity by $\tilde{O}_B(\cdot)$. We explicitly specify whether we use the arithmetic, word, or Boolean model unless this is clear from the context.

We write *ops* for "arithmetic operations," "log" for "$\log_2$" unless specified otherwise, and $\tilde{O}_B(\cdot)$ to show that we are ignoring logarithmic factors.

10.2 Matrix Computations

Matrix computations are the most popular and a highly important area of scientific and engineering computing. Most frequently they are performed numerically, with values represented using the IEEE standard single or double precision. In the chapter of this size we must omit or just barely touch on many important subjects of this field. The reader can find further material and bibliography in the surveys [163,168] and the books [6,8,21,56,62,64,103,110,178,222,227,239] and for more specific subject areas in [6,103,222,227,237,239] on eigendecomposition and SVD, [8,56,62,103,110,222,227] on other numerical matrix factorizations, [23,130] on the over- and underdetermined linear systems, their least-squares solution, and various other numerical computations with singular matrices, [106] on randomized matrix computations, [114,178] on structured matrix computations, [21,103,172,210] on parallel matrix algorithms, and [43,47,66,67,96,97,115,116,169,172,183,202,226,238] on "Error-free Rational Matrix Computations," including computations over finite fields, rings, and semirings that produce solutions to linear systems of equations, matrix inverses, ranks, determinants, characteristic and minimal polynomials, and Smith and Frobenius normal forms.

10.2.1 Dense, Sparse, and Structured Matrices: Their Storage and Multiplication by Vectors

An $m \times n$ matrix $A = [a_{i,j}, i = 1,\ldots, m; j = 1,\ldots, n]$ is also denoted $[a_{i,j}]_{i,j=1}^{m,n}$ and $[\mathbf{A}_1 \mid \ldots \mid \mathbf{A}_m]$; it is a two-dimensional array with the (i, j)th entry $[A]_{i,j} = a_{i,j}$ and the jth column $\mathbf{A}_j$. A^T is the transpose of A. Matrix A is a column vector if $n = 1$ and a row vector if $m = 1$. Vector $\mathbf{v} = [v_i]_{i=1}^n$ is an nth dimensional column vector. The straightforward algorithm computes the product $A\mathbf{v}$ by performing $(2n - 1)m$ ops; this is optimal for general (dense unstructured) $m \times n$ matrices, represented with their entries, but numerous applications involve structured matrices represented with much fewer than mn scalar values. A matrix is singular if its product by some vectors vanish; they form its *null space*.

An $m \times n$ matrix is *sparse* if it is filled mostly with zeros, having only $\phi = o(mn)$ nonzero entries. An important class is the matrices associated with graphs that have *families of small separators* [102,134]. This includes *banded* matrices $[b_{i,j}]_{i,j}$ with small *bandwidth* $2w + 1$ such that $b_{i,j} = 0$ unless $|i - j| \le w$. A sparse matrix can be stored economically by using appropriate data structures and can be multiplied by a vector fast, in $2\phi - m$ ops. Sparse matrices arise in many important applications, for example, to solving ordinary and partial differential equations (ODEs and PDEs) and graph computations.

Dense structured $n \times n$ matrices are usually defined by $O(n)$ parameters, and one can apply FFT to multiply such matrix by a vector by using $O(n \log n)$ or $O(n \log^2 n)$ ops [178]. Such matrices are omnipresent in applications in signal and image processing, coding, ODEs, PDEs, particle simulation, and Markov chains. Most popular among them are the *Toeplitz matrices* $T = [t_{i,j}]_{i,j=1}^{m,n}$ and the *Hankel matrices* $H = [h_{i,j}]_{i,j=1}^{m,n}$ where $t_{i,j} = t_{i+1,j+1}$ and $h_{i,j} = h_{i+1,j-1}$ for all i and j in the range of their definition. Each such matrix is defined by $m + n - 1$ entries of its first row and first or last column. Products $T\mathbf{v}$ and $H\mathbf{v}$ can be equivalently written as polynomial products or vector convolutions; their FFT-based computation takes $O((m + n) \log (m + n))$ ops per product [1,21,178]. Many other fundamental computations with Toeplitz and other structured matrices can be linked to polynomial computations enabling acceleration in both areas of computing [17–21,24,81,85,150–152,168,172,178,187,207,208]. Similar properties hold for Vandermonde matrices $V = [v_i^j]_{i,j=0}^{m-1,n-1}$ and Cauchy matrices $C = \left[\frac{1}{s_i - t_j}\right]_{i,j=1}^{m,n}$ where s_i and t_j denote $m + n$ distinct scalars.

One can extend the structures of *Hankel, Bézout, Sylvester, Frobenius (companion), Vandermonde,* and *Cauchy* matrices to more general classes of matrices by associating linear displacement operators. (See [21,178] for the details and the bibliography.) The important classes of *semiseparable, quasiseparable,* and other *rank structured* $m \times n$ matrices generalize banded matrices and their inverses; they are expressed by $O(m + n)$ parameters and can be multiplied by vectors by performing $O(m + n)$ ops [68, 231].

10.2.2 Matrix Multiplication, Factorization, and Randomization

The straightforward algorithm computes the $m \times p$ product AB of $m \times n$ by $n \times p$ matrices by using $2mnp - mp$ ops, which is $2n^3 - n^2$ if $m = n = p$. This upper bound is not sharp. Strassen decreased it to $O(n^{2.81})$ ops in 1969. His result was first improved in [162] and 10 times afterward, most recently by Coppersmith and Winograd in [48], Stothers in [224], and Vasilevska Williams in [233], who use Cn^ω ops for $\omega < 2.376$, $\omega < 2.374$, and $\omega < 2.3727$, respectively. Due to the huge overhead constants C, however, we have that $Cn^\omega < 2n^3$ only for enormous values n. The well-recognized group-theoretic techniques [44] enable a distinct description of the known matrix multiplication algorithms, but so far have only supported the same upper bounds on the complexity as the preceding works. References [224,233] extend the algorithms given in Reference [48], which in turn combines arithmetic progression technique with the previous advanced techniques. Each technique, however, contributes to a dramatic increase of the overhead constant that makes the resulting algorithms practically noncompetitive.

The only exception is the *trilinear aggregating* technique of [161] (cf. [163]), which alone supports the exponent 2.7753 [128] and together with the any precision approximation techniques of [163] was an indispensable ingredient of all algorithms that have beaten Strassen's exponent 2.81 of 1969. The triple product property, which is the basis of [44], may very well have a natural link to trilinear aggregating, although the descriptions available for the two approaches are distinct. For matrices of realistic sizes the numerical algorithms in [118], relying on trilinear aggregating, use about as many ops as the algorithms of Strassen 1969 and Winograd 1971 but need substantially less memory space and are more stable numerically.

The exponent ω of matrix multiplication is fundamental for the theory of computing because $O(n^{\omega})$ or $O(n^{\omega} \log n)$ bounds the complexity of many important matrix computations such as the computation of det A, the *determinant* of an $n \times n$ matrix A; its *inverse* A^{-1} (where det $A \neq 0$); its *characteristic polynomial* $c_A(x) = \det(xI - A)$ and *minimal polynomial* $m_A(x)$, for a scalar variable x; the Smith and Frobenius normal forms; the *rank*, rank A; a submatrix of A having the maximal rank, the solution vector $\mathbf{x} = A^{-1}\, \mathbf{v}$ to a nonsingular *linear system of equations* $A\, \mathbf{x} = \mathbf{v}$, and various *orthogonal* and *triangular factorizations* of the matrix A, as well as various *computations with singular matrices* and seemingly unrelated combinatorial and graph computations, for example, pattern recognition or computing all pair shortest distances in a graph [21, p. 222] or its transitive closure [1]. Consequently, all these operations use $O(n^{\omega})$ ops where theoretically $\omega < 2.3727$ [1, Chapter 6], [21, Chapter 2]. In practice, however, the solution of all these problems takes order of n^3 ops, because of the huge overhead constant C of all known algorithms that multiply $n \times n$ matrices in Cn^{ω} ops for $\omega < 2.775$, the overhead of the reduction to a matrix multiplication problem, the memory space requirements, and numerical stability problems [103].

Moreover, the straightforward algorithm for matrix multiplication remains the users' choice because it is highly effective on parallel and pipeline architectures [103,210]; on many computers it supersedes even the so-called "superfast" algorithms, which multiply a pair of $n \times n$ structured matrices in nearby linear arithmetic time, namely, by using $O(n \log n)$ or $O(n \log^2 n)$ ops, where both input and output matrices are represented with their short generator matrices having $O(n)$ entries [178].

Numerous important practical problems have been reduced to matrix multiplication because it is so effective. This has also motivated the development of block matrix algorithms (called *level-three BLAS*, which is the acronym for Basic Linear Algebra Subprograms).

Devising asymptotically fast matrix multipliers, however, had independent technical interest. For example, trilinear aggregating was a nontrivial decomposition of the three-dimensional tensor associated with matrix multiplication, and [161] was the first of now numerous examples where nontrivial tensor decompositions enable dramatic acceleration of important matrix computations [124,137,160].

The two basic techniques below extend matrix multiplication. Hereafter O denotes matrices filled with zeros; I is the square identity matrices, with ones on the diagonal and zeros elsewhere.

Suppose we seek the *Krylov sequence* or *Krylov matrix* $[B^i \mathbf{v}]_{i=0}^{k-1}$ for an $n \times n$ matrix B and an n-dimensional vector $\mathbf{v}$ [103,104,238]; in block Krylov computations the vector $\mathbf{v}$ is replaced by a matrix. The straightforward algorithm uses $(2n-1)n(k-1)$ ops, that is about $2n^3$ for $k = n$. An alternative algorithm first computes the matrix powers

$$B^2, B^4, B^8, \ldots, B^{2^s}, \qquad s = \lceil \log k \rceil - 1$$

and then the products of $n \times n$ matrices B^{2^i} by $n \times 2^i$ matrices, for $i = 0, 1, \ldots, s$:

$$\begin{array}{ll} B & \mathbf{v}, \\ B^2 & [\mathbf{v}, B\mathbf{v}] = [B^2\mathbf{v}, B^3\mathbf{v}], \\ B^4 & [\mathbf{v}, B\mathbf{v}, B^2\mathbf{v}, B^3\mathbf{v}] = [B^4\mathbf{v}, B^5\mathbf{v}, B^6\mathbf{v}, B^7\mathbf{v}] \\ \vdots & \end{array}$$

The last step completes the evaluation of the Krylov sequence in $2s + 1$ matrix multiplications, by using $O(n^{\omega} \log k)$ ops overall.

Special techniques for parallel computation of Krylov sequences for sparse and/or structured matrices A can be found in [170]. According to these techniques, Krylov sequence is recovered from the solution to the associated linear system $(I - A)\,\mathbf{x} = \mathbf{v}$, which is solved fast in the case of a special matrix A.

Another basic idea of matrix algorithms is to represent the input matrix A as a block matrix and to operate with its blocks rather than with its entries. For example, one can compute $\det A$ and A^{-1} by first factorizing A as a 2×2 block matrix,

$$A = \begin{bmatrix} I & O \\ A_{1,0}A_{0,0}^{-1} & I \end{bmatrix} \begin{bmatrix} A_{0,0} & O \\ O & S \end{bmatrix} \begin{bmatrix} I & A_{0,0}^{-1}A_{0,1} \\ O & I \end{bmatrix} \tag{10.1}$$

where $S = A_{1,1} - A_{1,0}A_{0,0}^{-1}A_{0,1}$. The 2×2 block triangular factors are readily invertible, $\det A = (\det A_{0,0})\det S$ and $(BCD)^{-1} = D^{-1}C^{-1}B^{-1}$, and so the cited tasks for the input A are reduced to the same tasks for the half-size matrices $A_{0,0}$ and S. It remains to factorize them recursively. The northwestern blocks (such as $A_{0,0}$), called leading principal submatrices, must be nonsingular throughout the recursive process, but this property holds for the highly important class of *symmetric positive definite* matrices $A = C^TC$, $\det C \neq 0$, and can be also achieved by means of symmetrization, pivoting, or randomization [1, Chapter 6], [21, Chapter 2], [178, Sections 5.5 and 5.6]). Recursive application of (10.1) should produce the LDU factorization $A = LDU$ where the matrices L and U^T are lower triangular and D diagonal. Having this factorization computed, we can readily solve linear systems $A\mathbf{x}_i = \mathbf{b}_i$ for various vectors $\mathbf{b}_i$, by using about $2n^2$ ops for each i, rather than $\frac{2}{3}n^3 + O(n^2)$ in Gaussian elimination.

Factorizations (including PLU, QR, QRP, QLP, CS, LR, Cholesky factorizations, and eigenvalue and singular value decompositions) are the most basic tool of matrix computations (see, e.g., [222]), recently made even more powerful with *randomization* (see [106,186,188,193–196,198], and the bibliography therein). It is well known that random matrices tend to be nonsingular and well conditioned (see, e.g., [217]), that is they lie far from singular matrices and therefore [103,110,222] are not sensitive to rounding errors and are suitable for numerical computations. The solution $\mathbf{x} = A^{-1}\mathbf{b}$ of a nonsingular linear system $A\mathbf{x} = \mathbf{b}$ of n equations can be obtained with a precision p_{out} in $O_{\tilde{B}}(n^3p + n^2p_{\text{out}})$ Boolean time for a fixed low precision p provided the matrix A is well conditioned; that accelerates Gaussian elimination by an order of magnitude for large $n + p_{\text{out}}$. Recent randomization techniques in [106,186,188,193–196,198] extend this property to much larger class of linear systems and enhance the power of various other matrix computations with singular or ill-conditioned matrices, for example, their approximation by low-rank matrices, computing a basis for the null space of a singular matrix, and approximating such bases for nearly singular matrices. Similar results have been proved for rectangular and Toeplitz matrices.

We refer the reader to [99,106,217] on impressive progress achieved in many other areas of matrix computations by means of randomization techniques.

10.2.3 Solution of Linear Systems of Equations

The solution of a linear system of n equations, $A\mathbf{x} = \mathbf{b}$, is the most frequent operation in scientific and engineering computations and is highly important theoretically. Gaussian elimination solves such a system by applying $(2/3)n^3 + O(n^2)$ ops.

Both Gaussian elimination and *(Block) Cyclic Reduction* use $O(nw^2)$ ops for banded linear systems with bandwidth $O(w)$. One can solve rank structured linear systems in $O(n)$ ops [68,231]; generalized nested dissection uses $O(n^{1.5})$ flops for the inputs associated with small separator families [134,169,202].

Likewise, we can dramatically accelerate Gaussian elimination for dense structured input matrices represented with their short generators, defined by the associated *displacement operators*.

This includes Toeplitz, Hankel, Vandermonde, and Cauchy matrices as well as matrices with similar structures. The MBA divide-and-conquer "superfast" algorithm (due to Morf 1974/1980 and Bitmead and Anderson 1980) solves nonsingular structured linear systems of n equations in $O(n \log^2 n)$ ops by applying the recursive 2×2 block factorization (10.1) and preserving matrix structure [21,178,191,205]. In the presence of rounding errors, however, Gaussian elimination, the MBA and Cyclic Reduction algorithms easily fail unless one applies pivoting, that is interchanges the equations (and sometimes unknowns) to avoid divisions by absolutely small numbers. A by-product is the factorization $A = PLU$ or $A = PLUP'$, for lower triangular matrices L and U^T and permutation matrices P and P'.

Pivoting, however, takes its toll. It "usually degrades the performance" [103, p. 119] by interrupting the string of arithmetic computations with the foreign operations of comparisons, is not friendly to block matrix algorithms and updating input matrices, hinders parallel processing and pipelining, and tends to destroy structure and sparseness, except for the inputs that have Cauchy-like and Vandermonde-like structure. The latter exceptional classes have been extended to the inputs with structures of Toeplitz/Hankel type by means of *displacement transformation* [167,178]. The users welcome this numerical stabilization, even though it slows down the MBA algorithm by a factor of $n/\log^2 n$, that is from "superfast" to "fast," which is still by a factor of n faster than the solution for general unstructured inputs, which takes order n^3 ops.

Can we avoid pivoting in numerical algorithms with rounding for general, sparse and structured linear systems to achieve both numerical stability and superfast performance? Yes, for the important classes where the input matrices $A = (a_{ij})_{i,j}$ are diagonally dominant, that is $|a_{ii}| > \sum_{i \neq j} |a_{ij}|$ or $|a_{ii}| > \sum_{j \neq i} |a_{ij}|$ for all i, or symmetric positive definite, that is $A = C^T C$ for a nonsingular matrix C. To these input classes Gaussian elimination, cyclic reduction, and the MBA algorithm can be safely applied with rounding and with no pivoting. For some other classes of sparse and positive definite linear systems, pivoting has been modified into nested dissection, Markowitz heuristic rule, and other techniques that preserve sparseness during the elimination yielding faster solution without causing numerical problems [62,101,134,169,202]. Can we extend these benefits to other input matrix classes?

Every nonsingular linear system $A\mathbf{x} = \mathbf{b}$ is equivalent to the symmetric positive definite ones $A^T A\mathbf{x} = A^T\mathbf{b}$ and $A A^T\mathbf{y} = \mathbf{b}$ where $\mathbf{x} = A\mathbf{y}$, but great caution is recommended in such symmetrizations because the condition number $\kappa(A) = \|A\|_2 \|A^{-1}\|_2 \geq 1$ is squared in the transition to the matrices $A^T A$ and AA^T, which means growing propagation and magnification of rounding errors.

There are two superior directions. The algorithms of [195,196,199] avoid pivoting for general and structured linear systems by applying randomization. These techniques are recent, proposed in [186, Section 12.2], but their effectiveness has formal and experimental support.

A popular classical alternative to Gaussian elimination is the iterative solution, for example, by means of the conjugate gradient and GMRES algorithms [10,103,104,232]. They compute sufficiently long Krylov sequences (defined in the previous section) and then approximate the solution with linear combinations $\sum_i c_i A^i \mathbf{b}$ or $\sum_i c_i (A^T A)^i A^T \mathbf{b}$ for proper coefficients c_i. The cost of computing the product of the matrix A or $A^T A$ by a vector is dominant, but it is small for structured and sparse matrices A. One can even call a matrix sparse or structured if and only if it can be multiplied by a vector fast.

Fast convergence to the solution is critical. It is not generally guaranteed but proved for some important classes of input matrices. The major challenges are the extension of these classes and the design of powerful methods for special input classes, notably *multilevel methods* (based on the *algebraic multigrid*) [140,149,201] and tensor decompositions [124,160], highly effective for many linear systems arising in discretization of ODEs, PDEs, and integral equations.

Preconditioning of the input matrices at a low computational cost accelerates convergence of iterations for many important classes of sparse and structured linear systems [10,104], and more recently, based on randomized preconditioning, for quite general as well as structured linear systems [186,188,193–196,198].

One can iteratively approximate the inverse or pseudoinverse of a matrix [103, Section 5.5.4] by means of Newton's iteration $X_{i+1} = 2X_i - X_iMX_i$, $i = 0, 1, \ldots$. We have $I - MX_{i+1} = (I - MX_i)^2 = (I - MX_0)^{2^{i+1}}$; therefore, the residual norm $\|I - MX_i\|$ is squared in every iteration step, $\|I - MX_i\| \le \|I - MX_0\|^{2^i}$ for $i = 1, 2, \ldots$, and so convergence is very fast unless $\|I - MX_0\| \ge 1$ or is near 1. The cost of two matrix multiplications is dominant per an iteration step; this makes the computation fast on multiprocessors as well as in the case of structured matrices M and X_i. See more on Newton's iteration, including the study of its initialization, convergence, and preserving displacement matrix structure, in [178, Chapters 4 and 6], [182,185,189,200,203,204].

10.2.4 Symbolic Matrix Computations

Rational matrix computations for a rational or integer input (such as the solution of a linear system and computing the determinant of a matrix) can be performed with no errors. To decrease the computational cost, one should control the growth of the precision of computing. Some special techniques achieve this in rational Gaussian elimination [7,97]. As a more fundamental tool one can reduce the computations modulo a sufficiently large integer m to obtain the rational or integer output values $z = p/q$ (e.g., the solution vector for a linear system) modulo m. Then we can recover z from two integers m and $z \bmod m$ by applying the continued fraction approximation algorithm, in other contexts called Euclidean algorithm [96,236]. Instead we can readily obtain $z = z \bmod m$ if $z \bmod m < r$ or $z = -m + z \bmod m$ if $z \bmod m < r$ otherwise, provided we know that the integer z lies in the range $[-r, r]$ and if $m > 2r$.

Computing the determinant of an integer matrix, we can choose the modulus m based on Hadamard's bound. A nonsingular linear system $A\mathbf{x} = \mathbf{v}$ can become singular after the reduction modulo a prime p but only with a low probability for a random choice of a prime p in a fixed sufficiently large interval as well as for a reasonably large power of two and a random integer matrix [205].

One can choose $m = m_1m_2 \cdots m_k$ for pairwise relatively prime integers $m_1, m_2, \ldots, m_k$ (we call them *coprimes*), then compute z modulo all these coprimes, and finally recover z by applying the Chinese remainder algorithm [1,96]. The error-free computations modulo m_i require the precision of $\log m_i$ bits; the cost of computing the values $z \bmod m_i$ for $i = 1, \ldots, k$ dominates the cost of the subsequent recovery of the value $z \bmod m$.

Alternatively one can apply *p-adic (Newton–Hensel) lifting* [96]. For solving linear systems of equations and matrix inversion they can be viewed as the symbolic counterparts to iterative refinement and Newton's iteration of the previous section, both well known in numerical linear algebra [183].

Newton's lifting begins with a prime p, a larger integer k, an integer matrix M, and its inverse $Q = M^{-1} \bmod p$, such that $I - QM \bmod p = 0$. Then, one writes $X_0 = Q$, recursively computes the matrices $X_j = 2X_{j-1} - X_{j-1}MX_{j-1} \bmod (p^{2^j})$, notes that $I - X_jM = 0 \bmod (p^{2^j})$ for $j = 1, 2, \ldots, k$, and finally recovers the inverse matrix M^{-1} from $X_k = M^{-1} \bmod p^{2^k}$.

Hensel's lifting begins with the same input complemented with an integer vector $\mathbf{b}$. Then one writes $\mathbf{r}^{(0)} = \mathbf{b}$, recursively computes the vectors

$$\mathbf{u}^{(i)} = Q\mathbf{r}^{(i)} \bmod p, \quad \mathbf{r}^{(i+1)} = (\mathbf{r}^{(i)} - M\mathbf{u}^{(i)})/p, \quad i = 0, 1, \ldots, k-1$$

and $\mathbf{x}^{(k)} = \sum_{i=0}^{k-1} \mathbf{u}^{(i)} p^i$ such that $M\mathbf{x}^{(k)} = \mathbf{b} \bmod (p^k)$, and finally recovers the solution $\mathbf{x}$ to the linear system $M\mathbf{x} = \mathbf{b}$ from the vector $\mathbf{x}^{(k)} = \mathbf{x} \bmod (p^k)$.

Newton's and Hensel's lifting are particularly powerful where the input matrices M and M^{-1} are sparse and/or structured, for example, Toeplitz, Hankel, Vandermonde, Cauchy. Hensel's lifting enables the solution in nearly optimal time under both Boolean and word models [183]. We can choose p being a power of two and use computations in the binary mode. Reference [69] discusses lifting for sparse linear systems.

10.2.5 Computing the Sign and the Value of a Determinant

The value or just the sign of det A, the determinant of a square matrix A, is required in some fundamental geometric and algebraic/geometric computations such as the computation of convex hulls, Voronoi diagrams, algebraic curves and surfaces, multivariate and univariate resultants, and Newton's polytopes. Faster numerical methods are preferred as long as the correctness of the output can be certified. In the customary *arithmetic filtering* approach, one applies fast numerical methods as long as they work and, in the rare cases when they fail, shifts to the slower symbolic methods. For fast numerical computation of det A one can employ factorizations $A = PLUP'$ (see Section 10.2.2) or $A = QR$ [45,103], precondition the matrix A [186], and then certify the output sign [206].

If A is a rational or integer matrix, then the Chinese remainder algorithm of the previous subsection is highly effective, particularly using heuristics for working modulo m for m much smaller than Hadamard's bound on $|\det A|$ [26].

Alternatively [70,165,166], one can solve linear systems $A\mathbf{y}(i) = \mathbf{b}(i)$ for random vectors $\mathbf{b}(i)$ and then apply Hensel's lifting to recover det A as a least common denominator of the rational components of all $\mathbf{y}(i)$.

Storjohann in [223] advanced randomized Newton's lifting to yield det A more directly in the optimal asymptotic Boolean time $\tilde{O}_B(n^{\omega+1})$ for $\omega < 2.3727$. Wiedemann in 1986, Coppersmith in 1994, and a number of their successors compute det A by extending the Lanczos and block Lanczos classical algorithms. This is particularly effective for sparse or structured matrices A and in further extension to multivariate determinants and resultants (cf. [85,86,117,180]).

10.3 Polynomial Root-Finding and Factorization

10.3.1 Computational Complexity Issues

Approximate solution of an nth degree polynomial equation,

$$p(x) = \sum_{i=0}^{n} p_i x^i = p_n \prod_{j=1}^{n} (x - z_j) = 0, \quad p_n \neq 0 \tag{10.2}$$

that is the approximation of the roots $z_1,\ldots, z_n$ for given coefficients $p_0,\ldots, p_n$, is a classical problem that has greatly influenced the development of mathematics and computational mathematics throughout four millennia, since the Sumerian times [173,174]. The problem remains highly important for the theory and practice of the present day algebraic and algebraic/geometric computation, and new root-finding algorithms appear every year [141–144].

To approximate even a single root of a monic polynomial $p(x)$ within error bound 2^{-b} we must process at least $(n + 1)nb/2$ bits of the input coefficients $p_0,\ldots, p_{n-1}$. Indeed perturb the x-free coefficient of the polynomial $(x - 6/7)^n$ by 2^{-bn}. Then the root $x = 6/7$ jumps by 2^{-b}, and similarly if we perturb the coefficients p_i by $2^{(i-n)b}$ for $i = 1,\ldots, n - 1$. Thus to ensure the output precision of b bits, we need an input precision of at least $(n - i)b$ bits for each coefficient p_i, $i = 0, 1,\ldots, n - 1$. We need at least $\lceil (n + 1)nb/4 \rceil$ bitwise operations to process these bits, each operation having at most two input bits.

It can be surprising, but we can approximate all n roots within 2^{-b} by using bn^2 Boolean (bit) operations up to a polylogarithmic factor for b of order $n \log n$ or higher, that is we can approximate all roots about as fast as we write down the input. We achieve this by applying the *divide-and-conquer algorithms* in [171,173,179] (see [123,157,218] on the related works). The algorithms first compute a sufficiently wide root-free annulus A on the complex plane, whose exterior and interior contain comparable numbers of the roots, that is the same numbers up to a fixed constant factor. Then the two factors of $p(x)$ are numerically computed, that is $F(x)$, having all its roots in the interior of the annulus, and $G(x) = p(x)/F(x)$, having no roots there. Then the polynomials $F(x)$ and $G(x)$ are recursively factorized until factorization of $p(x)$ into the product of linear factors is computed numerically. From this factorization, approximations to all roots of $p(x)$ are obtained. For approximation of a single root see the competitive algorithms of [177].

It is interesting that, up to polylog factors, both lower and upper bounds on the Boolean time decrease to bn [179] if we only seek the factorization of $p(x)$, that is, if instead of the roots z_j, we compute scalars a_j and b_j such that

$$\| p(x) - \prod_{j=1}^{n} (a_j x - c_j) \| < 2^{-h} \| p(x) \| \tag{10.3}$$

for the polynomial norm $\left\| \sum_i q_i x^i \right\| = \sum_i |q_i|$.

The *isolation of the zeros* of a polynomial $p(x)$ of (10.2) having integer coefficients and simple zeros is the computation of n disjoint discs, each containing exactly one root of $p(x)$. This can be a bottleneck stage of root approximation because one can contract such discs by performing a few subdivisions and then apply numerical iterations (such as Newton's) that would very rapidly approximate the isolated zeros within a required tolerance. Reference [184] yields even faster refinement by extending the techniques of [171,173,179].

Based on the classical gap theorem (recently advanced in [84]), Schönhage in [218, Section 20] the isolation problem has been reduced to computing factorization (10.3) for $b = \lceil (2n + 1)(l + 1 + \log(n + 1)) \rceil$ where l is the maximal coefficient length, that is the minimum integer such that $|\Re(p_j)| < 2^l$ and $|\Im(p_j)| < 2^l$ for $j = 0, 1, \ldots, n$. Combining the cited algorithms of [171,173,179] with this reduction yields

Theorem 10.1

Let polynomial $p(x)$ of (10.2) have n distinct simple zeros and integer coefficients in the range $[-2^\tau, 2^\tau]$. Then one can isolate the n zeros of $p(x)$ from each other at the Boolean cost $\tilde{O}_B(n^2\tau)$.

The algorithms of [171,173,179] incorporate the techniques of [157,218], but advance them and support substantially smaller upper bounds on the computational complexity. In particular these algorithms decrease by a factor of n the estimates of [218, Theorems 2.1, 19.2 and 20.1] on the Boolean complexity of polynomial factorization, root approximation, and root isolation.

10.3.2 Root-Finding via Functional Iterations

About the same record complexity estimates for root-finding would be also supported by some *functional iteration* algorithms if one assumes their convergence rate defined by ample empirical evidence, although never proved formally. The users accept such an evidence instead of the proof and prefer the latter algorithms because they are easy to program and have been carefully implemented; like the algorithms of [171,173,177,179] they allow tuning the precision of computing to the precision required for every output root, which is higher for clustered and multiple roots than for single isolated roots.

For approximating a single root z, the current practical champions are modifications of *Newton's iteration*, $z(i + 1) = z(i) - a(i)p(z(i))/p'(z(i))$, $a(i)$ being the step-size parameter [136], *Laguerre's method* [94,107], and the *Jenkins–Traub algorithm* [112]. One can deflate the input polynomial via its numerical division by $x - z$ to extend these algorithms to approximating a small number of other roots. If one deflates many roots, the coefficients of the remaining factor can grow large as, for example, in the divisor of the polynomial $p(x) = x^{1000} + 1$ that has degree 498 and shares with $p(x)$ all its roots having positive real parts.

For the approximation of all roots, a good option is the Weierstrass–Durand–Kerner's (hereafter *WDK*) algorithm, defined by the recurrence

$$z_j(l+1) = z_j(l) - \frac{p(z_j(l))}{p_n \prod_{i \neq j} (z_j(l) - z_i(l))}, \quad j = 1, \ldots, n, \quad l = 0, 1, \ldots \tag{10.4}$$

It has excellent empirical global convergence. Reference [208] links it to polynomial factorization and adjusts it to approximating a single root in $O(n)$ ops per step.

A customary choice of n initial approximations $z_j(0)$ to the n roots of the polynomial $p(x)$ (see [16] for a heuristic alternative) is given by $z_j(0) = r\,t\exp\left(2\pi\sqrt{-1}/n\right)$, $j = 1, \ldots, n$. Here, $t > 1$ is a fixed scalar and r is an upper bound on the root radius, such that all roots z_j lie in the disc $\{x : |x| = r\}$ on the complex plane. This holds, for example, for

$$r = 2\max_{i<n} \left|p_i/p_n\right|^{\frac{1}{n-i}} \tag{10.5}$$

For a fixed l and for all j the computation in (10.4) uses $O(n^2)$ ops. We can use just $O(n \log^2 n)$ ops if we apply fast multipoint polynomial evaluation algorithms based on fast FFT-based polynomial division [1,21,25,178,190], but then we would face numerical stability problems.

As with Newton's, Laguerre's, Jenkins–Traub's algorithms and the inverse power iteration in [17,207], one can employ this variant of the WDK to approximate many or all roots of $p(x)$ without deflation. Toward this goal, one can concurrently apply the algorithm at sufficiently many distinct initial points $z_j(0) = r\,t\exp\left(2\pi\sqrt{-1}/N\right)$, $j = 1, \ldots, N \geq n$ (on a large circle for large t) or according to [16]. The work can be distributed among processors that do not need to interact with each other until they compute the roots.

See [141–144,173] and references therein on this and other effective functional iteration algorithms. Reference [16] covers MPSolve, the most effective current root-finding subroutines, based on Ehrlich–Aberth's algorithm.

10.3.3 Matrix Methods for Polynomial Root-Finding

By cautiously avoiding numerical problems [103, Section 7.4.6], one can approximate the roots of $p(x)$ as the eigenvalues of the associated (generalized) companion matrices, that is matrices having characteristic polynomial $p(x)$. Then one can employ numerically stable methods and the excellent software available for matrix computations, such as the QR celebrated algorithm. For example, Matlab's subroutine *roots* applies it to the companion matrix of a polynomial. Malek and Vaillancourt (1995), and Fortune [93] and in his root-finding package EigenSolve, apply it to other generalized companion matrices and update them when the approximations to the roots are improved.

The algorithms of [15,17–19,181,197,207,230] exploit the structure of (generalized) companion matrices, for example, where they are diagonal plus rank-one (hereafter *DPR1*) matrices, to accelerate the eigenvalue computations. The papers [17,207] apply and extend the inverse power method [103, Section 7.6.1]; they exploit matrix structure, simplify the customary use of Rayleigh quotients for updating approximate eigenvalues, and apply special preprocessing techniques. For both companion and DPR1 inputs the resulting algorithms use linear space and linear arithmetic time per iteration step, enable dramatic parallel acceleration, and deflate the input in $O(n)$ ops; for DPR1 matrices repeated deflation can produce all n roots with no numerical problems.

The algorithms of [15,18,19,230] employ the QR algorithm but decrease the arithmetic time per iteration step from quadratic to linear by exploiting the rank matrix structure of companion matrices.

Substantial further refinement of these techniques is required to make them competitive with MPSolve. See [243] on recent progress.

The papers [181,197] advance Cardinal's polynomial root-finders of 1996, based on repeated squaring. Each squaring is reduced to performing a small number of FFTs and thus uses order $n \log n$ ops. One can weigh potential advantage of convergence to nonlinear factors of $p(x)$, representing multiple roots or root clusters, at the price of increasing the time per step by a factor of $\log n$ versus the inverse power method, advanced for root-finding in [17,207].

10.3.4 Extension to Approximate Polynomial GCDs

Reference [176] combines polynomial root-finders with algorithms for bipartite matching to compute approximate univariate polynomial GCD of two polynomials, that is, the GCD of the maximum degree for two polynomials of the same or smaller degrees lying in the ϵ-neighborhood of the input polynomials for a fixed positive ϵ. Approximate GCDs are required in computer vision, algebraic geometry, computer modeling, and control. For a single example, GCD defines the intersection of two algebraic curves defined by the two input polynomials, and approximate GCD does this under input perturbations of small norms. See [14] in the bibliography on approximate GCDs, but see [167,178,195] on the structured matrix algorithms involved.

10.3.5 Univariate Real Root Isolation and Approximation

In some algebraic and geometric computations, the input polynomial $p(x)$ has real coefficients, and only its real roots must be approximated. One of the fastest real root-finders in the current practice is still MPSolve, which uses almost the same running time for real roots as for all complex roots. This can be quite vexing, because very frequently the real roots make up only a small fraction of all roots [77]. Recently, however, the challenge was taken in the papers [197,207], whose numerical iterations are directed to converge to real and nearly real roots. This promises acceleration by a factor of d/r where the input polynomial has d roots, of which r roots are real or nearly real. In the rest of this section we cover an alternative direction, that is real root-finding by means of isolation of the real roots of a polynomial.

We write $p(x) = a_d x^d + \cdots + a_1 x + a_0$, assume integral coefficients with the maximum bit size $\tau = 1 + \max_{i \leq d}\{\lg|a_i|\}$, and seek isolation of real roots, that is seek real line intervals with rational endpoints, each containing exactly one real root. We may seek also the root's multiplicity. We assume rational algorithms, that is, error-free algorithms that operate with rational numbers.

If all roots of $p(x)$ are simple, then the minimal distance between them, the *separation bound*, is at most $b = d^{-(d+2)/2}(d+1)^{(1-d)/2}2^{\tau(1-d)}$, or roughly $2^{-\tilde{O}(d\tau)}$ (e.g., [147]), and we isolate real roots as soon as we approximate them within less than $b/2$. Effective solution algorithms rely on continued fractions (see the following text), having highly competitive implementation in SYNAPS [109,153] and its descendant REALROOT, a package of MATHEMAGIX, on the Descartes' rule of signs, and the Sturm or Sturm–Habicht sequences.

Theorem 10.2

The rational algorithms discussed in the sequel isolate all r real roots of $p(x)$ in $\tilde{O}_B(d^4\tau^2)$ bitwise ops. Under certain probability distributions for the coefficients, they are expected to use $\tilde{O}_B(d^3\tau)$ or $\tilde{O}_B(rd^2\tau)$.

The bounds exceed those of Theorem 10.1, has changed this, by closing the gap. Moreover rational solvers are heavily in use [245], have long and respected history, and are of independent technical interest. Most popular are the subdivision algorithms, such as STURM, DESCARTES, and BERNSTEIN.

By mimicking binary search, they repeatedly subdivide an initial interval that contains all real roots until every tested interval contains at most one real root. They differ in the way of counting the real roots in an interval.

The algorithm STURM (due to Sturm 1835) is the closest to binary search; it produces isolating intervals and root multiplicities at the cost $\tilde{O}_B(d^4\tau^2)$ [63,83]; see [77] on the decrease of the expected cost to $\tilde{O}_B(rd^2\tau)$.

The complexity of both algorithms DESCARTES and BERNSTEIN is $\tilde{O}_B(d^4\tau^2)$ [72,83]. Both rely on Descartes' rule of sign, but the BERNSTEIN algorithm also employs the Bernstein basis polynomial representation. See [2,234] on the theory and history of DESCARTES, [46,71,146,215,216] on its modern versions, and [83,156] and the references therein on the BERNSTEIN algorithm.

The continued fraction algorithm, CF, computes the continued fraction expansions of the real roots of the polynomial. The first formulation of the algorithm is due to Vincent. By Vincent's theorem repeated transforms $x \mapsto c + \frac{1}{x}$ eventually yield a polynomial with zero or one sign variation and thus (by Descartes' rule) with zero or resp. one real root in $(0, \infty)$. In the latter case the inverse transformation computes an isolating interval. Moreover, the c's in the transform correspond to the partial quotients of the continued fraction expansion of the real root. Variants differ in the way they compute the partial quotients.

Recent algorithms control the growth of coefficient bit-size and decrease the bit-complexity from exponential (of Vincent) to $\tilde{O}_B(d^3\tau)$ expected and $\tilde{O}_B(d^4\tau^2)$ worst-case bit complexity. See [145,219,228,229] and the references therein on these results, history, and variants of CP algorithms.

10.4 Systems of Nonlinear Equations

Given a system $\{p_1(x_1, \ldots, x_n), \ldots, p_r(x_1, \ldots, x_n)\}$ of nonlinear polynomials with rational coefficients, the n-tuple of complex numbers $(a_1, \ldots, a_n)$ is a solution of the system if $p_i(a_1, \ldots, a_n) = 0$, $1 \le i \le r$. Each $p_i(x_1, \ldots, x_n)$ is said to be an element of $\mathbf{Q}[x_1, \ldots, x_n]$, the ring of polynomials in $x_1, \ldots, x_n$ over the field of rational numbers. In this section, we explore the problem of solving a well-constrained system of nonlinear equations, namely when $r = n$, which is the typical case in applications. We also indicate how an initial phase of exact algebraic computation leads to certain numerical methods that can approximate all solutions; the interaction of symbolic and numeric computation is currently an active domain of research, for example [22,82,125]. We provide an overview and cite references to different symbolic techniques used for solving systems of algebraic (polynomial) equations. In particular, we describe methods involving *resultant* and *Gröbner basis* computations.

Resultants, as explained in the following text, formally express the solvability of algebraic systems with $r = n + 1$; solving a well-constrained system reduces to a resultant computation as illustrated in the sequel. The *Sylvester resultant method* is the technique most frequently utilized for determining a common root of two polynomial equations in one variable. However, using the Sylvester method successively to solve a system of multivariate polynomials proves to be inefficient.

It is more efficient to eliminate n variables together from $n + 1$ polynomials, thus, leading to the notion of the *multivariate resultant*. The three most commonly used multivariate resultant matrix formulations are those named after *Sylvester or Macaulay* [36,38,135], those named after *Bézout or Dixon* [33,60,121], or the *hybrid formulation* [57,113,122]. Extending the Sylvester–Macaulay type, we shall emphasize also *sparse resultant* formulations [37,98,225]. For a unified treatment, see [81].

The theory of Gröbner bases provides powerful tools for performing computations in multivariate polynomial rings. Formulating the problem of solving systems of polynomial equations in terms of polynomial ideals, we will see that a Gröbner basis can be computed from the input polynomial set, thus, allowing for a form of back substitution in order to compute the common roots.

Although not discussed, it should be noted that the *characteristic set algorithm* can be utilized for solving polynomial systems. Although introduced for studying algebraic differential equations [213], the method was converted to ordinary polynomial rings when developing an effective method for automatic theorem proving [241]. Given a polynomial system P, the characteristic set algorithm computes a new system in triangular form, such that the set of common roots of P is equivalent to the set of roots of the triangular system [120]. *Triangular systems* have k_1 polynomials in a specific variable, k_2 polynomials in this and one more variable, k_3 polynomials in these two and one more variable, and so on, for a total number of $k_1 + \cdots + k_n$ polynomials.

10.4.1 Resultant of Univariate Systems

The question of whether two polynomials $f(x), g(x) \in \mathbf{Q}[x]$,

$$f(x) = f_n x^n + f_{n-1}x^{n-1} + \cdots + f_1 x + f_0$$

$$g(x) = g_m x^m + g_{m-1}x^{m-1} + \cdots + g_1 x + g_0$$

have a common root leads to a condition that has to be satisfied by the coefficients of f, g. Using a derivation of this condition due to Euler, the *Sylvester matrix* of f, g (which is of dimension $m + n$) can be formulated. The vanishing of the determinant of the Sylvester matrix, known as the *Sylvester resultant,* is a necessary and sufficient condition for f, g to have common roots in the algebraic closure of the coefficient ring.

As a running example, let us consider the following bivariate system [131]:

$$f = x^2 + xy + 2x \qquad + y - 1 = 0$$

$$g = x^2 \qquad + 3x - y^2 + 2y - 1 = 0$$

Without loss of generality, the roots of the Sylvester resultant of f and g treated as polynomials in y, whose coefficients are polynomials in x, are the x-coordinates of the common roots of f, g. More specifically, the Sylvester resultant with respect to y is given by the following determinant:

$$\det \begin{bmatrix} x+1 & x^2+2x-1 & 0 \\ 0 & x+1 & x^2+2x-1 \\ -1 & 2 & x^2+3x-1 \end{bmatrix} = -x^3 - 2x^2 + 3x$$

An alternative matrix of order $\max\{m, n\}$, named after Bézout, yields the same determinant.

The roots of the Sylvester determinant are $\{-3, 0, 1\}$. For each x value, one can substitute the x value back into the original polynomials yielding the solutions $(-3, 1)$, $(0, 1)$, $(1, -1)$. More practically, one can use the Sylvester matrix to reduce system solving to the computation of eigenvalues and eigenvectors as explained in Section 10.4.3.

The Sylvester formulations have led to a *subresultant theory,* which produced an efficient algorithm for computing the GCD of univariate polynomials and their resultant, while controlling intermediate expression swell [133,212]. Subresultant theory has been generalized to several variables, for example [32,53].

10.4.2 Resultants of Multivariate Systems

The solvability of a set of nonlinear multivariate polynomials is determined by the vanishing of a generalization of the resultant of two univariate polynomials. We examine two generalizations: the classical and the sparse resultants. Both generalize the determinant of $n + 1$ *linear* polynomials in n variables.

The *classical resultant* of a system of $n + 1$ polynomials with symbolic coefficients in n variables vanishes exactly when there exists a common solution in the *projective* space over the algebraic closure of the coefficient ring [50]. The *sparse (or toric) resultant* characterizes solvability of the same overconstrained system over a smaller space, which coincides with affine space under certain genericity conditions [51,98,225]. The main algorithmic question is to construct a matrix whose determinant is the resultant or a nontrivial multiple of it.

Cayley, and later Dixon, generalized Bézout's method to a set

$$\{p_1(x_1, \ldots, x_n), \ldots, p_{n+1}(x_1, \ldots, x_n)\}$$

of $n + 1$ polynomials in n variables. The vanishing of the determinant of the Bézout–Dixon matrix is a necessary and sufficient condition for the polynomials to have a nontrivial projective common root and also a necessary condition for the existence of an affine common root [33,60,81,121]. A nontrivial resultant multiple, known as the *projection operator,* can be extracted via a method discussed in [41, Theorem 3.3.4]. This article, along with [73], explains the correlation between residue theory and the Bézout–Dixon matrix; the former leads to an alternative approach for studying and approximating all common solutions.

Macaulay [135] constructed a matrix whose determinant is a multiple of the classical resultant; he stated his approach for a well-constrained system of n homogeneous polynomials in n variables. The Macaulay matrix simultaneously generalizes the Sylvester matrix and the coefficient matrix of a system of linear equations. Like the Dixon formulation, the Macaulay determinant is a multiple of the resultant. Macaulay, however, proved that a certain minor of his matrix divides the matrix determinant to yield the exact resultant in the case of generic coefficients. To address arbitrary coefficients, Canny [36] proposed a general method that perturbs any polynomial system and extracts a nontrivial projection operator from Macaulay's construction.

By exploiting the structure of polynomial systems by means of sparse elimination theory, a matrix formula for computing the sparse resultant of $n + 1$ polynomials in n variables was given in [37] and consequently improved in [40,76]. Like the Macaulay and Dixon matrices, the determinant of the sparse resultant matrix, also known as Newton matrix, only yields a projection operation. However, in certain cases of bivariate and multihomogeneous systems, determinantal formulae for the sparse resultant have been derived [57,80,122]. To address degeneracy issues, Canny's perturbation has been extended in the sparse context [54]. D'Andrea [52] extended Macaulay's rational formula for the resultant to the sparse setting, thus defining the sparse resultant as the quotient of two determinants; see [79] for a simplified algorithm in certain cases.

Here, sparsity means that only certain monomials in each of the $n + 1$ polynomials have nonzero coefficients. Sparsity is measured in geometric terms, namely, by the **Newton polytope** of the polynomial, which is the convex hull of the exponent vectors corresponding to nonzero coefficients. The **mixed volume** of the Newton polytopes of n polynomials in n variables is defined as an integer-valued function that bounds the number of toric common roots of these polynomials [13]. This remarkable bound is the cornerstone of sparse elimination theory. The mixed volume bound is significantly smaller than the classical Bézout bound for polynomials with small Newton polytopes but they coincide for polynomials whose Newton polytope is the unit simplex multiplied by the polynomial's total degree. Since these bounds also determine the degree of the sparse and classical resultants, respectively, the latter has larger degree for sparse polynomials. Last, but not least, the classical resultant can identically vanish over sparse systems, whereas the sparse resultant can still yield the desired information about their common roots [51].

10.4.3 Polynomial System Solving by Using Resultants

Suppose we are asked to find the common roots of a set of n polynomials in n variables $\{p_1(x_1, \ldots, x_n), \ldots, p_n(x_1, \ldots, x_n)\}$. By augmenting this set by a generic linear polynomial [36,51], we construct the *u-resultant* of a given system of polynomials. The u-resultant is named after the indeterminates u, traditionally used to represent the generic coefficients of the additional linear polynomial. The u-resultant factors into linear factors over the complex numbers, providing the common roots of the given polynomials equations. The method relies on the properties of the multivariate resultant, and hence, can be constructed using either Macaulay's, Dixon's, or sparse formulations. An alternative approach is to *hide* a variable in the coefficient field [74,81,138].

Consider the previous example augmented by a generic linear form:

$$p_1 = x^2 + xy + 2x \quad + y - 1 = 0$$

$$p_2 = x^2 \quad + 3x - y^2 + 2y - 1 = 0$$

$$p_l = \quad ux \quad + vy + w = 0$$

As described in [38], the following (transposed) Macaulay matrix M corresponds to the u-resultant of the above system of polynomials:

$$M = \begin{bmatrix} 1 & 0 & 0 & 1 & 0 & 0 & 0 & 0 & 0 & 0 \\ 1 & 1 & 0 & 0 & 1 & 0 & u & 0 & 0 & 0 \\ 2 & 0 & 1 & 3 & 0 & 1 & 0 & u & 0 & 0 \\ 0 & 1 & 0 & -1 & 0 & 0 & v & 0 & 0 & 0 \\ 1 & 2 & 1 & 2 & 3 & 0 & w & v & u & 0 \\ -1 & 0 & 2 & -1 & 0 & 3 & 0 & w & 0 & u \\ 0 & 0 & 0 & 0 & -1 & 0 & 0 & 0 & 0 & 0 \\ 0 & 1 & 0 & 0 & 2 & -1 & 0 & 0 & v & 0 \\ 0 & -1 & 1 & 0 & -1 & 2 & 0 & 0 & w & v \\ 0 & 0 & -1 & 0 & 0 & -1 & 0 & 0 & 0 & w \end{bmatrix}$$

It should be noted that

$$\det(M) = (u - v + w)(-3u + v + w)(v + w)(u - v)$$

corresponds to the affine solutions (1, −1), (−3, 1), (0, 1), whereas one solution at infinity corresponds to the last factor.

Resultant matrices can also reduce polynomial system solving to a regular or generalized eigenproblem (cf. "Matrix Eigenvalues and Singular Values Problems"), thus, transforming the nonlinear question to a problem in linear algebra. This is a classical technique that enables us to numerically approximate all solutions [4,39,41,74,81]. For demonstration, consider the previous system and its resultant matrix M. The matrix rows are indexed by the following row vector of monomials in the eliminated variables:

$$\mathbf{v} = \left[x^3, x^2y, x^2, xy^2, xy, x, y^3, y^2, y, 1\right]$$

Vector $\mathbf{v}M$ expresses the polynomials indexing the columns of M, which are multiples of the three input polynomials by various monomials. Let us specialize variables u and v to random values. Then, M contains a single variable w and is denoted $M(w)$. Solving the linear system $\mathbf{v}M(w) = \mathbf{0}$ in vector $\mathbf{v}$ and in scalar w is a generalized eigenproblem, since $M(w)$ can be represented as $M_0 + wM_1$, where M_0 and M_1 have numeric entries. If, moreover, M_1 is invertible, we arrive at the following eigenproblem:

$$\mathbf{v}(M_0 + wM_1) = \mathbf{0} \Leftrightarrow \mathbf{v}\left(-M_1^{-1}M_0 - wI\right) = \mathbf{0} \Leftrightarrow \mathbf{v}\left(-M_1^{-1}M_0\right) = w\mathbf{v}$$

For every solution (a, b) of the original system, there is a vector $\mathbf{v}$ among the computed eigenvectors, which we evaluate at $x = a$, $y = b$ and from which the solution can be recovered by division [74]. As for the eigenvalues, they correspond to the values of w at the solutions; see [75] on numerical issues and an implementation.

An alternative method for approximating or isolating all real roots of the system is to use the so-called rational univariate representation (RUR) of algebraic numbers [35,214]. This allows us to express each root coordinate as the value of a univariate polynomial, evaluated over an algebraic number, which is specified as a solution of a single polynomial equation. All polynomials involved in this approach are derived from the resultant.

The resultant matrices are sparse and have quasi Toeplitz/Hankel structure (also called multilevel Toeplitz/Hankel structure), which enables their fast multiplication by vectors. By combining the latter property with various advanced nontrivial methods of multivariate polynomial root-finding, substantial acceleration of the construction and computation of the resultant matrices and approximation of the system's solutions was achieved in [24,85,86,150–152].

A comparison of the resultant formulations can be found, for example, in [81,120,138]. The multivariate resultant formulations have been used for diverse applications such as *algebraic and geometric reasoning* [41,59,138], including separation bounds for the isolated roots of arbitrary polynomial systems [84], *robot kinematics* [55,138,211], and *nonlinear computational geometry, computer-aided geometric design and, in particular, implicitization* [32,42,78,87,111].

10.4.4 Gröbner Bases

Solving systems of nonlinear equations can be formulated in terms of polynomial ideals [50,105,127]. The *ideal* generated by a system of polynomials $p_1, \ldots, p_r$ over $\mathbf{Q}[x_1, \ldots, x_n]$ is the set of all linear combinations

$$(p_1, \ldots, p_r) = \{h_1 p_1 + \cdots + h_r p_r \mid h_1, \ldots, h_r \in \mathbf{Q}[x_1, \ldots, x_n]\}$$

The algebraic variety of $p_1, \ldots, p_r \in \mathbf{Q}[x_1, \ldots, x_n]$ is the set of their common roots,

$$V(p_1, \ldots, p_r) = \{(a_1, \ldots, a_n) \in \mathbf{C}^n \mid p_1(a_1, \ldots, a_n) = \ldots = p_r(a_1, \ldots, a_n) = 0\}$$

A version of the *Hilbert Nullstellensatz* states that

$$V(p_1, \ldots, p_r) = \text{the empty set } \varnothing \Leftrightarrow 1 \in (p_1, \ldots, p_r) \text{ over } \mathbf{Q}[x_1, \ldots, x_n]$$

which relates the solvability of polynomial systems to the ideal membership problem.

A term $t = x_1^{e_1} x_2^{e_2} \ldots x_n^{e_n}$ of a polynomial is a product of powers with $\deg(t) = e_1 + \cdots + e_n$. In order to add needed structure to the polynomial ring we will require that the terms in a polynomial be ordered in an admissible fashion [50,97]. Two of the most common admissible orderings are the **lexicographic order** ($<_l$), where terms are ordered as in a dictionary, and the **degree order** ($<_d$), where terms are first compared by their degrees with equal degree terms compared lexicographically. A variation to the lexicographic order is the *reverse lexicographic order*, where the lexicographic order is reversed.

Much like a polynomial remainder process, the process of polynomial reduction involves subtracting a multiple of one polynomial from another to obtain a smaller degree result [50,105,127]. A polynomial g is said to be reducible with respect to a set $P = \{p_1,\ldots, p_r\}$ of polynomials if it can be reduced by one or more polynomials in P. When g is no longer reducible by the polynomials in P, we say that g is *reduced* or is *a normal form* with respect to P.

For an arbitrary set of basis polynomials, it is possible that different reduction sequences applied to a given polynomial g could reduce to different normal forms. A basis $G \subseteq \mathbf{Q}[x_1,\ldots, x_n]$ is a *Gröbner basis* if and only if every polynomial in $\mathbf{Q}[x_1,\ldots, x_n]$ has a unique normal form with respect to G. Buchberger [27–29] showed that every basis for an ideal $(p_1,\ldots, p_r)$ in $\mathbf{Q}[x_1,\ldots, x_n]$ can be converted into a Gröbner basis $\{p_1^*,\ldots,p_s^*\} = GB(p_1,\ldots,p_r)$, concomitantly designing an algorithm that transforms an arbitrary ideal basis into a Gröbner basis. Another characteristic of Gröbner bases is that by using the abovementioned reduction process we have

$$g \in (p_1\ldots,p_r) \Leftrightarrow g \bmod \left(p_1^*,\ldots,p_s^*\right) = 0$$

Further, by using the Nullstellensatz it can be shown that $p_1\ldots, p_r$ viewed as a system of algebraic equations is solvable if and only if $1 \notin GB(p_1,\ldots, p_r)$.

Depending on which admissible term ordering is used in the Gröbner bases construction, an ideal can have different Gröbner bases. However, an ideal cannot have different (reduced) Gröbner bases for the same term ordering. Any system of polynomial equations can be solved using a lexicographic Gröbner basis for the ideal generated by the given polynomials. It has been observed, however, that Gröbner bases, more specifically lexicographic Gröbner bases, are hard to compute [139]. In the case of zero-dimensional ideals, those whose varieties have only isolated points, a change of basis algorithm was outlined in [90], which can be utilized for solving: one computes a Gröbner basis for the ideal generated by a system of polynomials under a degree ordering. The so-called *change of basis algorithm* can then be applied to the degree ordered Gröbner basis to obtain a Gröbner basis under a lexicographic ordering. Significant progress has been achieved in the algorithmic realm by Faugère [88,89].

Another way to finding all common real roots is by means of RUR; see the previous section. All polynomials involved in this approach can be derived from the Gröbner basis. A rather recent development concerns the generalization of Gröbner bases to *border bases*, which contain all information required for system solving but can be computed faster and seem to be numerically more stable [127,154,155,221].

Turning to Lazard's example in form of a polynomial basis,

$$\begin{array}{rclllll} p_1 & = & x^2 & +xy & +2x & & +y & -1 \\ p_2 & = & x^2 & & +3x & -y^2 & +2y & -1 \end{array}$$

one obtains (under lexicographical ordering with $x <_l y$) a Gröbner basis in which the variables are triangulated such that the finitely many solutions can be computed via back substitution:

$$\begin{array}{rcllll} p_1^* & = & x^2 & +3x & +2y & -2 \\ p_2^* & = & xy & -x & -y & +1 \\ p_3^* & = & & & y^2 & -1 \end{array}$$

The final univariate polynomial has minimal degree, whereas the polynomials used in the back substitution have total degree no larger than the number of roots. As an example, x^2y^2 is reduced with respect to the previously computed Gröbner basis $\{p_1^*, p_2^*, p_3^*\} = GB(p_1, p_2)$ along two distinct reduction paths, both yielding $-3x - 2y + 2$ as the normal form.

There is a strong connection between lexicographic Gröbner bases and the previously mentioned resultant techniques. For some types of input polynomials, the computation of a reduced system via resultants might be much faster than the computation of a lexicographic Gröbner basis.

Gröbner bases can be used for many polynomial ideal theoretic operations [29,49]. Other applications include computer-aided geometric design [111], polynomial interpolation [129], coding and cryptography [92], and robotics [91].

10.5 Research Issues and Summary

Algebraic algorithms deal with numbers, vectors, matrices, polynomials, formal power series, exponential and differential polynomials, rational functions, algebraic sets, curves, and surfaces. In this vast area, manipulations with matrices and polynomials, in particular the solution of a polynomial equation and linear and polynomial systems of equations, are most fundamental in modern computations in sciences, engineering, and signal and image processing. We reviewed the state of the art for the solution of these three tasks and gave pointers to the extensive bibliography.

Among numerous interesting and important research directions of the topics in Sections 10.2 and 10.3, we wish to cite computations with structured matrices, including their applications to polynomial root-finding, currently of growing interest, and new techniques for randomized preprocessing for matrix computations, evaluation of resultants and polynomial root-finding.

Section 10.4 of this chapter has briefly reviewed polynomial system solving based on resultant matrices as well as Gröbner bases. Both approaches are currently active. This includes practical applications to small and medium-size systems. Efficient implementations that handle the nongeneric cases, including multiple roots and nonisolated solutions, are probably the most crucial issues today in relation to resultants. The latter are also studied in relation to a more general object, namely the discriminant of a well-constrained system, which characterizes the existence of multiple roots. Another interesting current direction is algorithmic improvement by exploiting the structure of the polynomial systems, including sparsity, or the structure of the encountered matrices, for both resultants and Gröbner bases.

Key Terms

Characteristic polynomial: Shift an input matrix A by subtracting the identity matrix xI scaled by variable x. The determinant of the resulting matrix is the characteristic polynomial of the matrix A. Its roots coincide with the eigenvalues of the shifted matrix $A - xI$.

Condition number of a matrix is a scalar κ which grows large as the matrix approaches a singular matrix; then numeric inversion becomes an ill-conditioned problem. κ `OUTPUT ERROR NORM` ≈ `INPUT ERROR NORM`.

Degree order: An order on the terms in a multivariate polynomial; for two variables x and y with $x < y$ the ascending chain of terms is $1 < x < y < x^2 < xy < y^2 \ldots$.

Determinant: A polynomial in the entries of a square matrix whose value is invariant in adding to a row (resp. column) any linear combination of other rows (resp. columns). $\det(AB) = \det A \cdot \det B$ for a pair of square matrices A and B, $\det B = -\det A$ if the matrix B is obtained by interchanging a pair of adjacent rows or columns of a matrix A, $\det A \neq 0$ if and only if a matrix A is invertible. Determinant of a block diagonal or block triangular matrix is the product of the diagonal blocks, and so $\det A = (\det A_{0,0}) \det S$ under (10.1). One can compute a determinant by using these properties and matrix factorizations, for example, recursive factorization (10.1).

Gröbner basis: Given a term ordering, the Gröbner basis of a polynomial ideal is a generating set of this ideal, such that the (multivariate) division of any polynomial by the basis has a unique remainder.

Lexicographic order: An order on the terms in a multivariate polynomial; for two variables x and y with $x < y$ the ascending chain of terms is $1 < x < x^2 < \cdots < y < xy < x^2y \cdots < y^2 < xy^2 \ldots$.

Matrix eigenvector: A column vector **v** such that $A\mathbf{v} = \lambda\mathbf{v}$, for a square matrix A and the associated eigenvalue λ. A generalized eigenvector **v** satisfies the equation $A\mathbf{v} = \lambda B\mathbf{v}$ for two square matrices A and B and the associated eigenvalue λ. Both definitions extend to row vectors that premultiply the associated matrices.

Mixed volume: An integer-valued function of n convex polytopes in n-dimensional Euclidean space. Under proper scaling, this function bounds the number of toric complex roots of a well-constrained polynomial system, where the convex polytopes are defined to be the Newton polytopes of the given polynomials.

Newton polytope: The convex hull of the exponent vectors corresponding to terms with nonzero coefficients in a given multivariate polynomial.

Ops: Arithmetic operations, that is, additions, subtractions, multiplications, or divisions; as in **flops,** that is, floating point operations.

Resultant: A polynomial in the coefficients of a system of n polynomials with $n + 1$ variables, whose vanishing is the minimal necessary and sufficient condition for the existence of a solution of the system.

Separation bound: The minimum distance between two (complex) roots of a univariate polynomial.

Singularity: a square matrix is singular if its product with some nonzero matrix is the zero matrix. Singular matrices do not have inverses.

Sparse matrix: a matrix whose zero entries are much more numerous than its nonzero entries.

Structured matrix: A matrix whose every entry can be derived by a formula depending on a smaller number of parameters, typically on $O(m + n)$ parameters for an $m \times n$ matrix, as opposed to its mn entries. For instance, an $m \times n$ Cauchy matrix has $\frac{1}{s_i - t_j}$ as the entry in row i and column j and is defined by $m + n$ parameters s_i and t_j, $i = 1,\ldots, m; j = 1,\ldots, n$. Typically a structured matrix can be multiplied by a vector in nearly linear arithmetic time.

Further Information

The books and special issues of journals [1,5,21,25,31,58,82,97,178,221,244] provide a broader introduction to the general subject and further bibliography.

There are well-known libraries and packages of subroutines for the most popular numerical matrix computations, in particular, [61] for solving linear systems of equations, [95,220], ARPACK, and PARPACK for approximating matrix eigenvalues, and [3] for both of the two latter computational problems. Comprehensive treatment of numerical matrix computations and extensive bibliography can be found in [103,222], and there are many more specialized books on them [6,8,62,100,104,110,209,227,239] as well as many survey articles [108,159,168] and thousands of research articles. Further applications to the graph and combinatorial computations related to linear algebra are cited in "Some Computations Related to Matrix Multiplication" and [169].

On parallel matrix computations, see [101,103,115,116,192] assuming general input matrices, [101,108,169,202] assuming sparse inputs, [62] assuming banded inputs, and [21,172,178] assuming dense structured inputs. On Symbolic-Numeric algorithms, see the books [21,178,235], surveys [168,173,175], special issues [22,82,125,126], and the bibliography therein. For the general area of exact computation and the theory behind algebraic algorithms and computer algebra, see [9,30,50,51,58,96, 97,147,148,240,242,244].

There is a lot of generic software packages for exact computation, SYNAPS [253], a C++ open source library devoted to symbolic and numeric computations with polynomials, algebraic numbers, and polynomial systems, which has been evolving into the REALROOT package of the open source computer algebra system MATHEMAGIX; NTL a high-performance C++ library providing data structures and algorithms for vectors, matrices, and polynomials over the integers and finite fields, and EXACUS [11], a C++ library for curves and surfaces that provides exact methods for solving polynomial equations. A highly efficient tool is FGb for Gröbner basis, and RS for the rational univariate representation, and real solutions of systems of polynomial equations and inequalities. Finally, Lin Box [65] is a C++ library that provides exact high-performance implementations of linear algebra algorithms.

This chapter does not cover the area of polynomial factorization. We refer the interested reader to [96,132,158], and the bibliography therein.

The *SIAM Journal on Matrix Analysis and Applications* and *Linear Algebra and Its Applications* are specialized on Matrix Computations, *Mathematics of Computation* and *Numerische Mathematik* are leading among numerous other good journals on numerical computing.

The *Journal of Symbolic Computation* and the *Foundations of Computational Mathematics* specialize on topics in Computer Algebra, which are also covered in the *Journal of Computational Complexity*, the *Journal of Pure and Applied Algebra* and, less regularly, in the *Journal of Complexity*. *Mathematics for Computer Science* and *Applicable Algebra in Engineering, Communication and Computing* are currently dedicated to the subject of the chapter as well. *Theoretical Computer Science* has become more open to algebraic–numerical and algebraic–geometric subjects [22,34,82,125].

The annual *International Symposium on Symbolic and Algebraic Computation (ISSAC)* is the main conference in computer algebra; these topics are also presented at the bi-annual Conference *MEGA* and the newly founded SIAM conference on Applications of Algebraic Geometry. They also appear in the annual *ACM Conference on Computational Geometry*, as well as at various Computer Science conferences, including SODA, FOCS, and STOC.

Among many conferences on numerical computing, most comprehensive ones are organized under the auspices of SIAM and ICIAM. The International Workshop on Symbolic-Numeric Algorithms can be traced back to 1997 (SNAP in INRIA, Sophia Antipolis) and a special session in IMACS/ACA98 Conference in Prague, Czech Republic, in 1998 [175]. It restarted in Xi'an, China, 2005; Timishiora, Romania, 2006 (supported by IEEE), and London, Ontario, Canada, 2007 (supported by ACM). The topics of Symbolic-Numerical Computation are also represented at the conferences on the *Foundations of Computational Mathematics (FoCM)* (meets every 3 years) and quite often at ISSAC.

References

1. Aho, A., Hopcroft, J., Ullman, J., *The Design and Analysis of Algorithms*. Addison-Wesley, Reading, MA, 1974.
2. Alesina, A., Galuzzi, M., A new proof of Vincent's theorem. *L'Enseignement Mathématique*, 44, 219–256, 1998.
3. Anderson, E., Bai, Z., Bischof, C., Blackford, S., Demmel, J., Dongarra, J., Du Croz, J., Greenbaum, A., Hammarling, S., McKenney, A., Sorensen, D., *LAPACK Users' Guide*, 3rd edn., SIAM, Philadelphia, PA, 1999.
4. Auzinger, W., Stetter, H.J., An elimination algorithm for the computation of all zeros of a system of multivariate polynomial equations. In: *Proceedings of the International Conference on Numerical Mathematics, International Series of Numerical Mathematics*, 86, 12–30. Birkhäuser, Basel, Switzerland, 1988.
5. Bach, E., Shallit, J., *Algorithmic Number Theory, Volume 1: Efficient Algorithms*. The MIT Press, Cambridge, MA, 1996.
6. Bai, Z., Demmel, J., Dongarra, J., Ruhe, A., van der Vorst, H., eds., *Templates for the Solution of Algebraic Eigenvalue Problems: A Practical Guide*. SIAM, Philadelphia, PA, 2000.
7. Bareiss, E.H., Sylvester's identity and multistep integers preserving Gaussian elimination. *Math. Comput.*, 22, 565–578, 1968.
8. Barrett, R., Berry, M.W., Chan, T.F., Demmel, J., Donato, J., Dongarra, J., Eijkhout, V., Pozo, R., Romine, C., Van Der Vorst, H., *Templates for the Solution of Linear Systems: Building Blocks for Iterative Methods*. SIAM, Philadelphia, PA, 1993.
9. Basu, S., Pollack, R., and Roy, M.-F., *Algorithms in Real Algebraic Geometry, Algorithms and Computation in Mathematics*, Vol. 10, Springer, Berlin, Germany, 2003.

10. Benzi, M., Preconditioning techniques for large linear systems: A survey. *J. Comput. Phys.*, 182, 418–477, 2002.
11. Berberich, E., Eigenwillig, A., Hemmer, M., Hert, S., Kettner, L., Mehlhorn, K., Reichel, J., Schmitt, S., Schömer, E., Wolpert, N., EXACUS: Efficient and exact algorithms for curves and surfaces. In: *ESA*, *LNCS*, Vol. 1669, 155–166, Springer, Heidelberg, Germany, 2005.
12. Berlekamp, E.R., Factoring polynomials over large finite fields. *Math. Comput.*, 24, 713–735, 1970.
13. Bernshtein, D.N., The number of roots of a system of equations. *Funct. Anal. Appl.*, 9(2), 183–185, 1975.
14. Bini, D.A., Boito, P., A fast algorithm for approximate polynomial GCD based on structured matrix computations. *Operator Theory: Advances and Applications*, Vol. 199, 155–173, Birkhäuser, Basel, Switzerland, 2010.
15. Bini, D.A., Boito, P., Eidelman, Y., Gemignani, L., Gohberg, I., A fast implicit QR algorithm for companion matrices. *Linear Algebra Appl.*, 432, 2006–2031, 2010.
16. Bini, D.A., Fiorentino, G., Design, analysis, and implementation of a multiprecision polynomial rootfinder. *Numer. Algorithms*, 23, 127–173, 2000.
17. Bini, D.A., Gemignani, L., Pan, V.Y., Inverse power and Durand/Kerner iteration for univariate polynomial root-finding. *Comput. Math. (with Appl.)*, 47(2/3), 447–459, 2004.
18. Bini, D.A., Gemignani, L., Pan, V.Y., Fast and stable QR eigenvalue algorithms for generalized companion matrices and secular equation. *Numer. Math.*, 3, 373–408, 2005.
19. Bini, D.A., Gemignani, L., Pan, V.Y., Improved initialization of the accelerated and robust QR-like polynomial root-finding. *Electron. Trans. Numer. Anal.*, 17, 195–205, 2004.
20. Bini, D., Pan, V.Y., Polynomial division and its computational complexity. *J. Complexity*, 2, 179–203, 1986.
21. Bini, D., Pan, V.Y., *Polynomial and Matrix Computations, Volume 1, Fundamental Algorithms*. Birkhäuser, Boston, MA, 1994.
22. Bini, D.A., Pan, V.Y., Verschelde, J., eds., Special Issue on Symbolic–Numerical Algorithms. *Theor. Comput. Sci.*, 409, 2, 255–268, 2008.
23. Björck, Å., *Numerical Methods for Least Squares Problems*. SIAM, Philadelphia, PA, 1996.
24. Bondyfalat, D., Mourrain, B., Pan, V.Y., Computation of a specified root of a polynomial system of equations using eigenvectors. *Linear Algebra Appl.*, 319, 193–209, 2000. *Proceedings of ISSAC*, 252–259, ACM Press, New York, 1998.
25. Borodin, A., Munro, I., *Computational Complexity of Algebraic and Numeric Problems*. American Elsevier, New York, 1975.
26. Brönnimann, H., Emiris, I.Z., Pan, V.Y., Pion, S., Sign determination in residue number systems. *Theor. Comput, Sci.*, 210(1), 173–197, 1999.
27. Buchberger, B., *Ein Algorithmus zum Auffinden der Basiselemente des Restklassenringes nach einem nulldimensionalen Polynomideal*. Dissertation, University of Innsbruck, Austria, 1965.
28. Buchberger, B., A theoretical basis for the reduction of polynomials to canonical form. *ACM SIGSAM Bull.*, 10(3), 19–29, 1976.
29. Buchberger, B., Gröbner bases: An algorithmic method in polynomial ideal theory. In: Bose, N.K., ed., *Recent Trends in Multidimensional Systems Theory*, pp. 184–232. D. Reidel, Dordrecht, The Netherlands, 1985.
30. Buchberger, B., Collins, G.E., Loos, R., Albrecht, R., eds. *Computer Algebra: Symbolic Algebraic Computation*. 2nd edn., Springer, Wien, Austria, 1983.
31. Bürgisser, P., Clausen, M., Shokrollahi, M.A., *Algebraic Complexity Theory*. Springer, Berlin, 1997.
32. Busé, L., D'Andrea, C., Inversion of parameterized hypersurfaces by means of subresultants. In: *Proceedings of ISSAC*, pp. 65–71, ACM Press, New York, 2004.
33. Busé, L., Elkadi, M., Mourrain, B., Residual resultant of complete intersection. *J. Pure Appl. Algebra*, 164, 35–57, 2001.
34. Busé, L., Elkadi, M., Mourrain, B., eds. Special Issue on Algebraic–Geometric Computations. *Theor. Comp. Sci.*, 392 (1–3), 1–178, 2008.

35. Canny, J., Some Algebraic and Geometric Computations in PSPACE. In: *Proceedings of the ACM Symposium on the Theory of Computing*, New York, 460–467, 1988.
36. Canny, J., Generalized characteristic polynomials. *J. Symb. Comput.*, 9(3), 241–250, 1990.
37. Canny, J., Emiris, I.Z., A subdivision-based algorithm for the sparse resultant. *J. ACM*, 47(3), 417–451, 2000.
38. Canny, J., Kaltofen, E., Lakshman, Y., Solving systems of non-linear polynomial equations faster. In: *Proceedings of the ISSAC*, pp. 121–128, ACM, New York, 1989.
39. Canny, J., Manocha, D., Efficient techniques for multipolynomial resultant algorithms. In: *Proc. ISSAC*, pp. 85–95, ACM Press, New York, 1991.
40. Canny, J., Pedersen, P., An algorithm for the Newton resultant. Technical Report 1394, Computer Science Department, Cornell University, 1993.
41. Cardinal, J.-P., Mourrain, B., Algebraic approach of residues and applications. In: *The Math. of Numerical Analysis, Lects. in Applied Math.*, Vol. 32, 189–210, AMS, Providence, RI, 1996.
42. Chen, F., Cox, D.A., Liu, Y., The mu-basis and implicitization of a rational parametric surface. *J. Symb. Comput.*, 39(6), 689–706, 2005.
43. Chen, Z., Storjohann, A., A BLAS based C library for exact linear algebra on integer matrices. In: *Proceedings of ISSAC*, pp. 92–99, ACM Press, New York, 2005.
44. Cohn, H., Kleinberg, R., Szegedy, B., Umans, C., Group-theoretic algorithms for matrix multiplication. In: *Proceedings of the IEEE FOCS*, pp. 379–388, Pittsburg, PA, 2005.
45. Clarkson, K.L., Safe and effective determinant evaluation. *Proceedings of IEEE FOCS*, pp. 387–395, IEEE Computer Society Press, Pittsburg, PA, 1992.
46. Collins, G.E., Akritas, A., Polynomial real root isolation using Descartes' rule of signs. In: *SYMSAC '76*, 272–275, ACM Press, New York, 1976.
47. Coppersmith, D., Solving homogeneous linear equations over GF(2) via block Wiedemann algorithm. *Math. Comput.*, 62(205), 333–350, 1994.
48. Coppersmith, D., Winograd, S., Matrix multiplication via arithmetic progressions. *J. Symb. Comput.*, 9(3), 251–280, 1990.
49. Cox, D.A., Gröbner bases: A sampler of recent developments. In: *Proc. ISSAC*, Waterloo, Ontario, Canada, pp. 387–388, ACM, 2007.
50. Cox, D., Little, J., O'Shea, D. *Ideals, Varieties, and Algorithms*, 2nd edn. Undergraduate Texts in Mathematics, Springer, New York, 1997.
51. Cox, D., Little, J., O'Shea, D. *Using Algebraic Geometry*, 2nd edn. Graduate Texts in Mathematics, 185, Springer, New York, 2005.
52. D'Andrea, C., Macaulay-style formulas for the sparse resultant. *Trans. AMS*, 354, 2595–2629, 2002.
53. D'Andrea, C., Krick, T., Szanto, A., Multivariate subresultants in roots. *J. Algebra*, 302(1), 16–36, 2006.
54. D'Andrea, C., Emiris, I.Z., Computing sparse projection operators. In: *Symbolic Computation: Solving Equations in Algebra, Geometry, and Engineering*, pp. 121–139, AMS, Providence, RI, 2001.
55. Daney, D., Emiris, I.Z., Robust parallel robot calibration with partial information. In: *Proceedings of the IEEE International Conference on Robotics & Automation*, Seoul, South Korea, 3262–3267, 2001.
56. Demmel, J.J.W., *Applied Numerical Linear Algebra*. SIAM, Philadelphia, PA, 1997.
57. Dickenstein, A., Emiris, I.Z., Multihomogeneous resultant formulae by means of complexes. *J. Symb. Comput.*, 36, 317–342, 2003.
58. Dickenstein, A., Emiris, I.Z., eds., Solving Polynomial Equations: Foundations, Algorithms and Applications. In: *Algorithms and Computation in Mathematics*, Vol. 14, Springer, Berlin, Germany, 2005.
59. Dickenstein, A., Sturmfels, B., Elimination theory in codimension 2. *J. Symb. Comput.*, 34(2):119–135, 2002.
60. Dixon, A.L., The elimination of three quantics in two independent variables. *Proc. Lond. Math. Soc.*, 6, 468–478, 1908.
61. Dongarra, J., Bunch, J., Moler, C., Stewart, P., *LINPACK Users' Guide*. SIAM, Philadelphia, PA, 1978.

62. Dongarra, J.J., Duff, I.S., Sorensen, D.C., Van Der Vorst, H.A., *Numerical Linear Algebra for High-Performance Computers*. SIAM, Philadelphia, PA, 1998.
63. Du, Z., Sharma, V., Yap, C.K., Amortized bound for root isolation via Sturm sequences. [235], pp. 113–129.
64. Duff, I.S., Erisman, A.M., Reid, J.K., *Direct Methods for Sparse Matrices*. Clarendon Press, Oxford, England, 1986.
65. Dumas, J.-G., Gautier, T., Giesbrecht, M., Giorgi, P., Hovinen, B., Kaltofen, E., Saunders, B.D., Turner, W.J., Villard, G., LinBox. A generic library for exact linear algebra. In: Cohen, A.M., Gao, X.-S., Takayama, N., eds., *Proceedings of ICMS 200*, pp. 40–50, Beijing, China, 2002.
66. Dumas, J-G., Gautier, T., Pernet, C., Finite field linear algebra subroutines. In: *Proceedings of ISSAC*, pp. 63–74, ACM Press, New York, 2002.
67. Dumas, J-G., Giorgi, P., Pernet, C., Finite field linear algebra package. In: *Proceedings of ISSAC*, pp. 118–126, ACM Press, New York, 2004.
68. Eidelman, Y., Gohberg, I., *On a New Class of Structured Matrices, Integral Equations & Operator Theory*, Vol. 34, 293–324, Birkhäuser, Basel, 1999.
69. Eberly, W., Giesbrecht, M., Giorgi, P., Storjohann, A., Villard, G., Faster inversion and other black box matrix computations using efficient block projections. In: *Proceedings of ISSAC*, pp. 143–150, ACM Press, New York, 2007.
70. Eberly, W., Giesbrecht, M., Villard, G., On computing the determinant and Smith form of an integer matrix. *Proceedings of IEEE FOCS*, pp. 675–685, IEEE Computer Society Press, Los Alamitos, CA, 2000.
71. Eigenwillig, A., Kettner, L., Krandick, W., Mehlhorn, K., Schmitt, S., Wolpert, N., A descartes algorithm for polynomials with bit-stream coefficients. In: *CASC'2005*, *LNCS*, Vol. 3718, 38–149. Springer, Heidelberg, Germany, 2005.
72. Eigenwillig, A., Sharma, V., Yap, C.K., Almost tight recursion tree bounds for the Descartes method. *Proceedings of ISSAC*, pp. 71–78, ACM, New York, 2006.
73. Elkadi, M., Mourrain, B., Algorithms for residues and Lojasiewicz exponents. *J. Pure Appl. Algebra*, 153, 27–44, 2000.
74. Emiris, I.Z., On the complexity of sparse elimination. *J. Complexity*, 12, 134–166, 1996.
75. Emiris, I.Z., Matrix methods for solving algebraic systems, In: *Symbolic Algebraic Methods and Verification Methods*, Springer, Wien, Austria, pp. 69–78, 2001. Also `arxiv.org/abs/1201.5810`, 2011.
76. Emiris, I.Z., Canny, J.F., Efficient incremental algorithms for the sparse resultant and the mixed volume, *J. Symb. Comput.*, 20(2), 117–149, 1995.
77. Emiris, I.Z., Galligo, A., Tsigaridas, E.P., Random polynomials and expected complexity of bisection methods for real solving, *Proceedings of ISSAC*, pp. 235–242, ACM Press, New York, 2010.
78. Emiris, I.Z., Kalinka, T., Konaxis, C., Luu-Ba, T., Implicitization of curves and surfaces using predicted support, in [126].
79. Emiris, I.Z., Konaxis, C., Single-lifting Macaulay-type formulae of generalized unmixed sparse resultants, *J. Symb. Comput.*, 46(8), 919–942, 2011.
80. Emiris, I.Z., Mantzaflaris, A., Multihomogeneous resultant matrices for systems with scaled support. In *Proceedings of ISSAC*, pp. 143–150, ACM, New York, 2009.
81. Emiris, I.Z., Mourrain, B., Matrices in elimination theory. *J. Symb. Comput.*, 28, 3–44, 1999.
82. Emiris, I.Z., Mourrain, B., Pan, V.Y., eds. Special Issue on Algebraic and numerical algorithms, *Theor. Comput. Sci.*, 315, 307–672, 2004.
83. Emiris, I.Z., Mourrain, B., Tsigaridas, E.P., Real algebraic numbers: Complexity analysis and experimentation. In: *Reliable Implementations of Real Number Algorithms: Theory and Practice*, LNCS, Springer, Berlin, Germany, 2007.
84. Emiris, I.Z., Mourrain, B., Tsigaridas, E.P., The DMM bound: Multivariate (aggregate) separation bound. In: *Proceedings of ISSAC*, pp. 242–250, ACM Press, New York, 2010.
85. Emiris, I.Z., Pan, V.Y., Symbolic and numeric methods for exploiting structure in constructing resultant matrices. *J. Symb. Comput.*, 33, 393–413, 2002.

86. Emiris I.Z., Pan, V.Y., Improved algorithms for computing determinants and resultants. *J. Complexity*, 21 (1), 43–71, 2005. Also E. W. Mayr, V. G. Ganzha, E. V. Vorozhtzov, eds. *Proc. CASC'03*, 81–94, Technische Univ. Muenchen, Germany, 2003.
87. Emiris, I.Z., Tzoumas, G.M., Exact and efficient evaluation of the InCircle predicate for parametric ellipses and smooth convex objects, *Comput. Aided Design*, 40(6), 691–700, 2008.
88. Faugère, J.-C., A new efficient algorithm for computing Gröbner bases (F4). *J. Pure Appl. Algebra*, 139, 61–88, 1999
89. Faugère, J.-C., A new efficient algorithm for computing Gröbner bases without reduction to zero (F5). In: *Proceedings of ISSAC*, pp. 75–83, ACM, New York, 2002.
90. Faugère, J.-C., Gianni, P., Lazard, D., Mora, T., Efficient computation of zero-dimensional Gröbner bases by change of ordering. *J. Symb. Comput.*, 16(4), 329–344, 1993.
91. Faugère, J-C., Lazard, D., The combinatorial classes of parallel manipulators. *Mech. Mach. Theory*, 30, 765–776, 1995.
92. Faugère, J.-C., Levy-dit-Vehel, F., Perret, L., Cryptanalysis of MinRank. In: *Proceedings of CRYPTO*, 280–296, Springer, Heidelberg, Germany, 2008.
93. Fortune, S., An iterated eigenvalue algorithm for approximating roots of univariate polynomials. *J. Symb. Comput.*, 33(5), 627–646, 2002.
94. Foster, L.V., Generalizations of Laguerre's method: Higher order methods. *SIAM J. Numer. Anal.*, 18, 1004–1018, 1981.
95. Brian, S., James, B., Jack, D., Burton, G., Ikebe, Y., Klema, V., and Cleve, M., *Matrix Eigensystem Routines: EISPACK Guide Extension*, Lecture Notes in Computer Science, Vol. 6. Springer Verlag, New York, 1972.
96. von zur Gathen, J., Gerhard, J., *Modern Computer Algebra*. 2nd edn. Cambridge U. Press, Cambridge, 2003.
97. Geddes, K.O., Czapor, S.R., Labahn, G., *Algorithms for Computer Algebra*. Kluwer Academic, Boston, MA, 1992.
98. Gelfand, I.M., Kapranov, M.M., Zelevinsky, A.V., *Discriminants, Resultants and Multidimensional Determinants*. Birkhäuser, Boston, 1994.
99. Gilbert, A., Indyk, P., Sparse recovery using sparse matrices. *Proc. IEEE*, 98(6), 937–947, 2010.
100. George, A., Liu, J.W.-H., *Computer Solution of Large Sparse Positive Definite Linear Systems*. Prentice Hall, Englewood Cliffs, NJ, 1981.
101. Gilbert, J.R., Schreiber, R., Highly parallel sparse Cholesky factorization. *SIAM J. Sci. Comput.*, 13, 1151–1172, 1992.
102. Gilbert, J.R., Tarjan, R.E., The analysis of a nested dissection algorithm. *Numer. Math.*, 50, 377–404, 1987.
103. Golub, G.H., Van Loan, C.F., *Matrix Computations*, 3rd edn., Johns Hopkins University Press, Baltimore, MD, 1996.
104. Greenbaum, A., *Iterative Methods for Solving Linear Systems*. SIAM Publications, Philadelphia, PA, 1997.
105. Greuel, G.-M., Pfister, G., *A Singular Introduction to Commutative Algebra* (with contributions by O. Bachmann, C. Lossen, H. Schönemann). Springer, Berlin, Germany, 2002.
106. Halko, N., Martinsson, P.G., Tropp, J.A., Finding structure with randomness: Probabilistic algorithms for constructing approximate matrix decompositions, *SIAM Rev.*, 53(2), 217–288, 2011.
107. Hansen, E., Patrick, M., Rusnak, J., Some modifications of Laguerre's method, *BIT*, 17, 409–417, 1977.
108. Heath, M.T., Ng, E., Peyton, B.W., Parallel algorithms for sparse linear systems, *SIAM Rev.*, 33, 420–460, 1991.
109. Hemmer, M., Tsigaridas, E.P., Zafeirakopoulos, Z., Emiris, I.Z., Karavelas, M., Mourrain, B., Experimental evaluation and cross-benchmarking of univariate real solvers, In: *Proceedings of SNC'09*, Kyoto, Japan, 2009.
110. Higham, N.J., *Accuracy and Stability of Numerical Algorithms*. 2nd edn. SIAM, Philadelphia, 2002.
111. Hoffmann, C.M., Sendra, J.R., Winkler, F., Special Issue on parametric algebraic curves and applications, *J. Symb. Comput.*, 23, 1997.

112. Jenkins, M.A., Traub, J.F., A three-stage variable-shift iteration for polynomial zeros and its relation to generalized Rayleigh iteration, *Numer. Math.*, 14, 252–263, 1970.
113. Jouanolou, J.-P., Formes d'Inertie et Résultant : Un Formulaire. *Adv. Math.*, 126:119–250, 1997. Also TR 499/P-288, IRMA, Strasbourg, 1992.
114. Kailath, T., Sayed, A., eds. *SIAM Volume on Fast Reliable Algorithms for Matrices with Structure*, SIAM Publications, Philadelphia, 1999.
115. Kaltofen, E., Pan, V.Y., Processor efficient parallel solution of linear systems over an abstract field. In: *Proceedings of SPAA '91*, pp. 180–191, ACM, New York, 1991.
116. Kaltofen, E., Pan, V.Y., Processor-efficient parallel solution of linear systems II: The positive characteristic and singular cases. In: *Proceedings of FOCS '92*, pp. 714–723, IEEE Computer Society, Los Alamitos, CA, 1992.
117. Kaltofen, E., Villard, G., Computing the sign or the value of the determinant of an integer matrix, a complexity survey. *J. Comput. Appl. Math.*, 162(1), 133–146, 2004.
118. Kaporin, I., The aggregation and cancellation techniques as a practical tool for faster matrix multiplication. [82], pp. 469–510.
119. Kapur, D., Geometry theorem proving using Hilbert's Nullstellensatz. *J. Symb. Comput.*, 2, 399–408, 1986.
120. Kapur, D., Lakshman, Y.N., Elimination methods an introduction. In: Donald, B., Kapur, D., Mundy, J., eds., *Symbolic and Numerical Computation for Artificial Intelligence*. Academic Press, Orlando, FL, 1992.
121. Kapur, D., Saxena, T., Comparison of various multivariate resultant formulations. In: *Proceedings of ISSAC*, pp. 187–195, ACM, New York, 1995.
122. Khetan, A., The resultant of an unmixed bivariate system, *J. Symb. Comput.*, 36, 425–442, 2003.
123. Kirrinnis, P., Polynomial factorization and partial fraction decomposition by simultaneous Newton's iteration, *J. Complexity*, 14, 378–444, 1998.
124. Kolda, T. G., Bader, B. W., Tensor decompositions and applications, *SIAM Rev.*, 51(3), 455–500, 2009.
125. Kotsireas, I., Mourrain, B., Pan, V. Y., eds., Special Issue on algebraic and numerical algorithms. *Theor. Comput. Sci.*, 412(16), 1443–1543, 2011.
126. Kotsireas, I., Mourrain, B., Pan, V. Y., Zhi, L., eds., Special Issue on symbolic-numerical algorithms. *Theor. Comput. Sci.*, 479, 1–186, 2013.
127. Kreuzer, M., Robbiano, L., *Computational Commutative Algebra 1*. Springer Verlag, Heidelberg, Germany, 2000.
128. Laderman, J., Pan, V.Y., Sha, H.X., On practical algorithms for accelerated matrix multiplication. *Linear Algebra Appl.*, 162–164, 557–588, 1992.
129. Lakshman, Y.N., Saunders, B.D., Sparse polynomial interpolation in non-standard bases. *SIAM J. Comput.*, 24(2), 387–397, 1995.
130. Lawson, C.L., Hanson, R.J., *Solving Least Squares Problems*. Prentice-Hall, New Jersey, 1974, and (with a survey of recent developments) SIAM, 1995.
131. Lazard, D., Resolution des systemes d'equation algebriques. *Theor. Comput. Sci.*, 15, 77–110, 1981 (In French).
132. Lenstra, A.K., Lenstra, H.W., Lovász, L., Factoring polynomials with rational coefficients. *Math. Ann.*, 261, 515–534, 1982.
133. Lickteig, T., Roy, M.-F., Sylvester–Habicht sequences and fast Cauchy index computation, *J. Symb. Comput.*, 31(3), 315–341, 2001.
134. Lipton, R.J., Rose, D., Tarjan, R.E., Generalized nested dissection. *SIAM J. Numer. Anal.*, 16(2), 346–358, 1979.
135. Macaulay, F.S., Algebraic theory of modular systems. Cambridge Tracts 19, University Press, Cambridge, U.K., 1916.
136. Madsen, K., A root-finding algorithm based on Newton's method. *BIT*, 13, 71–75, 1973.
137. Mahoney, M. W., Maggioni, M., Drineas, P., Tensor-CUR decompositions for tensor-based data, *SIAM J. Matrix Anal. Appl.*, 30(2), 957–987, 2008.

138. Manocha, D., *Algebraic and Numeric Techniques for Modeling and Robotics*. Ph.D. thesis, CSD, DEECS, UC, Berkeley, CA, 1992.
139. Mayr, E.W., Meyer, A.R., The complexity of the finite containment problem for petri nets. *J. ACM*, 28(3), 561–576, 1981.
140. McCormick, S., ed., *Multigrid Methods*. SIAM, Philadelphia, PA, 1987.
141. McNamee, J.M., A 2000 updated supplementary bibliography on roots of polynomials, *J. Comput. Appl. Math.*, 142, 433–434, 2002.
142. McNamee, J.M., *Numerical Methods for Roots of Polynomials (Part 1)*, Elsevier, Amsterdam, The Netherlands, 2007.
143. McNamee, J.M., Pan, V.Y., Efficient polynomial root-refiners: A survey and new record estimates. *Comput. Math. (Appl.)*, 63, 239–254, 2012.
144. McNamee, J.M., Pan, V.Y., *Numerical Methods for Roots of Polynomials, Part 2*, 780+XIX pages, submitted to Elsevier publishers, 2013.
145. Mehlhorn, K., Ray, S., Faster algorithms for computing Hong's bound on absolute positiveness, *J. Symb. Comput.*, 45(6), 677–683, 2010.
146. Mehlhorn, K., Sagraloff, M., A deterministic algorithm for isolating real roots of a real polynomial, *J. Symb. Comput.*, 46(1), 70–90, 2011.
147. Mignotte, M., *Mathematics for Computer Algebra*. Springer-Verlag, New York, 1992.
148. Mignotte, M., Stefanescu, D., *Polynomials: An Algorithmic Approach*. Springer, Singapore, 1999.
149. Miranker, W.L., Pan, V.Y., Methods of Aggregations. *Linear Algebra Appl.*, 29, 231–257, 1980.
150. Mourrain, B., Pan, V.Y., Asymptotic acceleration of solving polynomial systems. *Proceedings of STOC'98*, pp. 488–496, ACM Press, New York, 1998.
151. Mourrain, B., Pan, V.Y., Multivariate polynomials, duality and structured matrices. *J. Complexity*, 16(1), 110–180, 2000.
152. Mourrain, B., Pan, V.Y., Ruatta, O., Accelerated solution of multivariate polynomial systems of equations, *SIAM J. Comput.*, 32(2), 435–454, 2003.
153. Mourrain, B., Pavone, J.-P., Trébuchet, P., Tsigaridas, E.P., SYNAPS: A library for symbolic–numeric computing. In: *Proceedings of the 8th International Symposium on Effective Methods in Algebraic Geometry (MEGA)*, Italy, 2005 (software presentation).
154. Mourrain, B., Trébuchet, P., Solving projective complete intersection faster. *J. Symb. Comput.*, 33(5), 679–699, 2002.
155. Mourrain, B., Trébuchet, P., Stable normal forms for polynomial system solving, *Theor. Comput. Sci.*, 409(2), 229–240, 2008.
156. Mourrain, B., Vrahatis, M., Yakoubsohn, J.C., On the complexity of isolating real roots and computing with certainty the topological degree, *J. Complexity*, 18(2), 2002.
157. Neff, C.A., Reif, J.H., An $O(nl+\varepsilon)$ algorithm for the complex root problem. In: *Proceedings of IEEE FOCS*, pp. 540–547, IEEE Computer Society Press, Los Alamitos, CA, 1994.
158. Nguyen, P.Q., Valle, B., eds., *The LLL Algorithm, Survey and Applications. Series: Information Security and Cryptography*, XIV, 496 pp., Springer, Dordrecht, The Netherlands, 2010, ISBN 978-3-642-02294-4.
159. Ortega, J.M., Voight, R.G., Solution of partial differential equations on vector and parallel computers. *SIAM Rev.*, 27(2), 149–240, 1985.
160. Oseledets, I.V., Tyrtyshnikov, E.E., Breaking the curse of dimensionality, or how to use SVD in many dimensions. *SISC*, 31(5), 3744–3759, 2009.
161. Pan, V.Y., On schemes for the evaluation of products and inverses of matrices, *Uspekhi Mat. Nauk*, 27, 5(167), 249–250, 1972 (in Russian).
162. Pan, V.Y., Strassen's algorithm is not optimal. Trilinear technique of aggregating. *Proceedings of IEEE FOCS*, 166–176, IEEE, Ann Arbor, MI, 1978.
163. Pan, V.Y., How can we speed up matrix multiplication? *SIAM Rev.*, 26(3), 393–415, 1984.

164. Pan, V.Y., *How to Multiply Matrices Faster, Lecture Notes in Computer Science*, Vol. 179, Springer Verlag, Berlin, Germany, 1984.
165. Pan, V.Y., Complexity of parallel matrix computations. *Theor. Comput. Sci.*, 54, 65–85, 1987.
166. Pan, V.Y., Computing the determinant and the characteristic polynomials of a matrix via solving linear systems of equations. *Inform. Process. Lett.*, 28, 71–75, 1988.
167. Pan, V.Y., On computations with dense structured matrices. *Math. Comput.*, 55(191), 179–190, 1990. Also *Proceedings of ISSAC*, pp. 34–42, ACM, New York, 1989.
168. Pan, V.Y., Complexity of computations with matrices and polynomials. *SIAM Rev.*, 34(2), 225–262, 1992.
169. Pan, V.Y., Parallel solution of sparse linear and path systems. In: Reif, J.H., ed., *Synthesis of Parallel Algorithms,* chapter 14, 621–678. Morgan Kaufmann, San Mateo, CA, 1993.
170. Pan, V.Y., Parallel computation of a Krylov matrix for a sparse and structured input. *Math. Comput. Model.*, 21(11), 97–99, 1995.
171. Pan, V.Y., Optimal and nearly optimal algorithms for approximating polynomial zeros, *Comput. Math. (with Applications),* 31(12), 97–138, 1996. Also STOC'95, 741–750, ACM, Press, New York, 1995.
172. Pan, V.Y., Parallel computation of polynomial GCD and some related parallel computations over abstract fields, *Theor. Comput. Sci.*, 162(2), 173–223, 1996.
173. Pan, V.Y., Solving a polynomial equation: Some history and recent progress, *SIAM Rev.*, 39(2), 187–220, 1997.
174. Pan, V.Y., Solving polynomials with computers. *Am. Sci.*, 86, 62–69, January–February 1998.
175. Pan, V.Y., Some recent algebraic/numerical algorithms. *Electronic Proceedings of IMACS/ACA*, Prague, Czech Republic, 1998. www-troja.fjfi.cvut.cz/aca98/sessions/approximate
176. Pan, V.Y., Numerical computation of a polynomial GCD and extensions. *Information and Computation*, 167(2), 71–85, 2001. Also *Proceedings of SODA '98*, 68–77, ACM Press, New York and SIAM Publications, Philadelphia, PA, 1998.
177. Pan, V.Y., On approximating complex polynomial zeros: Modified quadtree (Weyl's) construction and improved Newton's iteration. *J. Complexity*, 16(1), 213–264, 2000.
178. Pan, V.Y., *Structured Matrices and Polynomials: Unified Superfast Algorithms.* Birkhäuser/Springer, Boston/New York, 2001.
179. Pan, V.Y., Univariate polynomials: Nearly optimal algorithms for factorization and rootfinding. *J. Symb. Comput.*, 33(5), 701–733, 2002.
180. Pan, V.Y., On theoretical and practical acceleration of randomized computation of the determinant of an integer matrix. *Zapiski Nauchnykh Seminarov POMI (in English)*, 316, 163–187, St. Petersburg, Russia, 2004. Also available at http://comet.lehman.cuny.edu/vpan/
181. Pan, V.Y., Amended DSeSC power method for polynomial root-finding. *Comput. Math. (with Applications)*, 49(9–10), 1515–1524, 2005.
182. Pan, V.Y., Newton's iteration for matrix inversion, advances and extensions. In: Olshevsky V., Tyrtyshnikov, E., eds., *Matrix Methods: Theory, Algorithms and Applications* pp. 364–381, World Scientific, Hackensack, NJ, 2010.
183. Pan, V.Y., Nearly optimal solution of rational linear systems of equations with symbolic lifting and numerical initialization. *Comput. Math. Appl.*, 62, 1685–1706, 2011.
184. Pan, V.Y., Root-refining for a polynomial equation. *Proceedings of CASC 2012, Maribor, Slovenia, Lecture Notes in Computer Science*, Springer, Berlin, Germany, 2012.
185. Pan, V.Y., Branham, S., Rosholt, R., Zheng, A., Newton's iteration for structured matrices and linear systems of equations. [114], Chapter 7, 189–210.
186. Pan, V.Y., Grady, D., Murphy, B., Qian, G., Rosholt, R.E., Schur aggregation for linear systems and determinants, pp. 255–268 in [22].
187. Pan, V.Y., Ivolgin, D., Murphy, B., Rosholt, R.E., Tang, Y., Wang, X., Root-finding with eigen-solving. [235], pp. 185–210.

188. Pan, V.Y., Ivolgin, D., Murphy, B., Rosholt, R.E., Tang, Y., Yan, X., Additive preconditioning for matrix computations. *Linear Algebra Appl.*, 432, 1070–1089, 2010.
189. Pan, V.Y., Kunin, M., Rosholt, R.E., Kodal, H., Homotopic residual correction processes. *Math. Comput.*, 75, 345–368, 2006.
190. Pan, V.Y., Landowne, E., Sadikou, A., Univariate polynomial division with a remainder by means of evaluation and interpolation. *Inform. Process. Lett.*, 44, 149–153, 1992.
191. Pan, V.Y., Murphy, B., Rosholt, R.E., Unified nearly optimal algorithms for structured integer matrices. *Operator Theory: Advances and Applications*, 199, 359–375, Birkhaüser, Basel, Switzerland, 2010.
192. Pan, V.Y., Preparata, F.P., Work-preserving speed-up of parallel matrix computations, *SIAM J. Comput.*, 24(4), 811–821, 1995.
193. Pan, V.Y., Qian, G., Randomized preprocessing of homogeneous linear systems of equations, *Linear Algebra Appl.*, 432, 3272–3318, 2010.
194. Pan, V.Y., Qian, G., Solving linear systems of equations with randomization, augmentation and aggregation, *Linear Algebra Appl.*, 437, 2851–2876, 2012.
195. Pan, V.Y., Qian, G., Zheng, A., Randomized preconditioning of the MBA algorithm. *Proceedings of ISSAC*, San Jose, CA, pp. 281–288, ACM, 2011.
196. Pan, V.Y., Qian, G., Zheng, A., Randomized preprocessing versus pivoting. *Linear Algebra and Its Applications*, 438, 4, 1883–1889, 2013.
197. Pan, V.Y., Qian, G., Zheng, A., Real and complex polynomial root-finding via eigen-solving and randomization. *Proceedings of CASC 2012, Maribor, Slovenia, LNCS*, 2012.
198. Pan, V.Y., Qian, G., Zheng, A., Randomized matrix computations, Tech. Report TR 2012005, *Ph.D. Program in Computer Science, Graduate Center, the City University of New York*, 2012. Available at http://www.cs.gc.cuny.edu/tr/techreport.php?id=432.
199. Pan, V.Y., Qian, G., Zheng, A., Chen, Z., Matrix computations and polynomial root-finding with preprocessing, *Linear Algebra Appl.*, 434, 854–879, 2011.
200. Pan, V.Y., Rami, Y., Wang, X., Structured matrices and Newton's iteration: unified approach, *Linear Algebra Appl.*, 343/344, 233–265, 2002.
201. Pan, V.Y., Reif, J.H., Compact multigrid, *SIAM J. Sci. Stat. Comput.*, 13(1), 119–127, 1992.
202. Pan, V.Y., Reif, J.H., Fast and efficient parallel solution of sparse linear systems, *SIAM J. Comput.*, 22(6), 1227–1250, 1993.
203. Pan, V.Y., Schreiber, R., An improved Newton iteration for the generalized inverse of a matrix, with applications, *SIAM J. Sci. Stat. Comput.*, 12(5), 1109–1131, 1991.
204. Pan, V.Y., Wang, X., Inversion of displacement operators, *SIAM J. Matrix Anal. Appl.*, 24(3), 660–677, 2003.
205. Pan, V.Y., Wang, X., Degeneration of integer matrices modulo an integer. *Linear Algebra Appl.*, 429, 2113–2130, 2008.
206. Pan, V.Y., Yu, Y., Certification of numerical computation of the sign of the determinant of a matrix. *Algorithmica*, 30, 708–724, 2001. Also *Proceedings of SODA '99*, pp. 715–724, ACM New York/SIAM, Philadelphia, PA, 1999.
207. Pan, V.Y., Zheng, A., New progress in real and complex polynomial root-finding. *Comput. Math. (with Applications)* 61, 1305–1334. Also *Proc. ISSAC*, 219–226, ACM Press, New York, 2010.
208. Pan, V.Y., Zheng, A., Root-finding by expansion with independent constraints, *Comput. Math. (with Applications)*, 62, 3164–3182, 2011.
209. Parlett, B., *Symmetric Eigenvalue Problem*. Prentice Hall, Englewood Cliffs, NJ, 1980.
210. Quinn, M.J., *Parallel Computing: Theory and Practice*. McGraw-Hill, New York, 1994.
211. Raghavan M., Roth, B., Solving polynomial systems for the kinematics analysis and synthesis of mechanisms and robot manipulators, *Trans. ASME, Special Issue*, 117, 71–79, 1995.

212. Reischert, D., Asymptotically fast computation of subresultants. In: *Proceedings of ISSAC*, 233–240, ACM, New York, 1997.
213. Ritt, J.F., *Differential Algebra*. AMS, New York, 1950.
214. Rouillier, F., Solving zero-dimensional systems through the rational univariate representation, *AAECC J.*, 9, 433–461, 1999.
215. Rouillier, F., Zimmermann, P., Efficient isolation of polynomial's real roots, *J. Comput. Appl. Math.*, 162(1):33–50, 2004.
216. Sagraloff, M., When Newton meets Descartes: A simple and fast algorithm to isolate the real roots of a polynomial. *CoRR*, abs/1109.6279, 2011.
217. Sankar, A., Spielman, D., Teng, S.-H., Smoothed analysis of the condition numbers and growth factors of matrices, *SIAM J. Matrix Anal.*, 28(2), 446–476, 2006.
218. Schönhage, A., *The Fundamental Theorem of Algebra in Terms of Computational Complexity*. Math. Dept., Univ. Tübingen, Germany, 1982.
219. Sharma, V., Complexity of real root isolation using continued fractions, *Theor. Comput. Sci.*, 409(2):292–310, 2008.
220. Smith, B.T. et al., *Matrix Eigensystem Routines: EISPACK Guide,* 2nd edn. Springer, New York, 1970.
221. Stetter, H., *Numerical Polynomial Algebra*. SIAM, Philadelphia, PA, 2004.
222. Stewart, G.W., *Matrix Algorithms, Vol I: Basic Decompositions. Vol II: Eigensystems*. SIAM, Philadelphia, PA, 1998.
223. Storjohann, A., The shifted number system for fast linear algebra on integer matrices, *J. Complexity*, 21(4), 609–650, 2005.
224. Stothers, A.J., On the Complexity of Matrix Multiplication. Ph.D. thesis, University of Edinburgh, Edinburgh, U.K., 2010.
225. Sturmfels, B., Sparse elimination theory. Eisenbud, D., Robbiano, L., eds., In: *Proceedings of the computational algebraic geometry and commutative algebra*, Cortona, Italy, 1991.
226. Tarjan, R.E., A unified approach to path problems. *J. ACM*, 28(3), 577–593 and 594–614, 1981.
227. Trefethen, L.N., Bau III, D., *Numerical Linear Algebra*. SIAM, Philadelphia, PA, 1997.
228. Tsigaridas, E.P., Improved complexity bounds for real root isolation using Continued Fractions. In: S. Ratschan, ed., *Proceedings of the Fourth International Conference on Mathematical Aspects of Computer Information Sciences (MACIS)*, 226–237, Beijing, China, 2011.
229. Tsigaridas, E.P., Emiris, I.Z., On the complexity of real root isolation using continued fractions. *Theor. Comput. Sci.*, 392, 158–173, 2008.
230. Van Barel, M., Vandebril, R., Van Dooren, P., Frederix, K., Implicit double shift QR-algorithm for companion matrices. *Numer. Math.* 116(2), 177–212, 2010.
231. Vandebril, R., Van Barel, M., Mastronardi, N., *Matrix Computations and Semiseparable Matrices: Linear Systems* (Volume 1). The Johns Hopkins University Press, Baltimore, MD, 2007.
232. van der Vorst, H.A., *Iterative Krylov Methods for Large Linear Systems*. Cambridge University Press, Cambridge, U.K., 2003.
233. Vassilevska Williams, V., Multiplying matrices faster than Coppersmith–Winograd. *Proceedings of STOC 2012*, New York, 887–898, 2012.
234. Vincent, A.J.H., Sur la résolution des équations numériques. *J. Math. Pure Appl.*, 1, 341–372, 1836.
235. Wang, D., Zhi, L. eds. *Symbolic-Numeric Computation*. Birkhäuser, Basel/Boston, 2007.
236. Wang, X., Pan, V.Y., Acceleration of euclidean algorithm and rational number reconstruction. *SIAM J. Comput.*, 32(2), 548–556, 2003.
237. Watkins, D.S., *The Matrix Eigenvalue Problem: GR and Krylov Subspace Methods*. SIAM, Philadelphia, PA, 2007.
238. Wiedemann, D., Solving sparse linear equations over finite fields. *IEEE Trans. Inf. Theory* IT–32, 54–62, 1986.

239. Wilkinson, J.H., *The Algebraic Eigenvalue Problem*. Clarendon Press, Oxford, U.K., 1965.
240. Winkler, F., *Polynomial Algorithms in Computer Algebra*. Springer, Wien, Austria, 1996.
241. Wu, W., Basis principles of mechanical theorem proving in elementary geometries. *J. Syst. Sci. Math Sci.*, 4(3), 207–235, 1984.
242. Yap, C.K., *Fundamental Problems of Algorithmic Algebra*. Oxford University Press, New York, 2000.
243. Zhlobich, P., Differential qd algorithm with shifts for rank-structured matrices. *SIAM J. Matrix Anal. Appl.*, 33(4), 1153–1171, 2012.
244. Zippel, R., *Effective Polynomial Computations*. Kluwer Academic, Boston, MA, 1993.
245. Pan, V.Y., Tsigaridas E. P., On the Boolean complexity of the real root refinement. *Proceedings of the ISSAC*, 299–306, ACM Press, 2013.

11 Some Practical Randomized Algorithms and Data Structures

Michael Mitzenmacher
Harvard University

Eli Upfal
Brown University

From an engineering standpoint, "randomness" sounds like a bad idea; when we build things, we want them to be predictable. It is therefore somewhat surprising how important randomness has become in the design of algorithms and data structures. A randomized algorithm or data structure is allowed to make random choices during its execution.*

The reason to allow randomness in the design is that it provides a larger trade-off space for qualities we may desire. Typically, when we talk about analyzing algorithms and data structures, we are concerned with the two basic measures we want to optimize: worst-case time and space. Allowing randomness allows us to consider other measures, including the following:

- *Expected time and space*: We might allow worst-case time or space to be large if they are not large very frequently.
- *Correctness*: We might allow our algorithm or data structure to occasionally give a wrong answer or otherwise fail, again if it does not happen frequently, in return for gains in average or worst-case time or space. While it may be counterintuitive to treat correctness as just another performance metric, this idea has proven quite powerful in both theory and practice.

* A closely related but different line of work addresses the probabilistic analysis of algorithms (and data structures), where one assumes the *input* is random according to some distribution, rather than allowing the *algorithm* to make random choices. In the setting of probabilistic analysis, there can still be bad but rare *inputs* to the algorithm; in studying randomized algorithms, we are concerned with bad but rare *random choices* made by the algorithm.

Finally, randomized algorithms and data structures often provide another less tangible benefit: simplicity. Simplicity can be useful not just for theoretical analysis; if we think of programmer time as a measurable and important resource, a simple randomized algorithm may be more useful than a more complex deterministic algorithm, even if it does not gain by other more standard metrics.

The field of randomized algorithms and data structure is so large, it would be impossible to consider summarizing the entire area in a short survey. There are entire books devoted the subject, notably the general texts by Motwani and Raghavan [47] and Mitzenmacher and Upfal [43], and there are several texts that cover specialized subbranches of the area as well, such as [21,46,60,66].

In what follows, we provide some basic examples of how randomness is used in algorithms and data structures. We have chosen to emphasize approaches that have proven useful in practice. We assume that the reader has had an appropriate undergraduate-level training in algorithms, complexity, and probability theory.

Outline: Hashing is one of the most fundamental uses of randomness in computer science. In Section 11.1, we discuss several variations of hash tables. Tabulation hashing gives an efficient implementation of a hash function using a small amount of space with desirable theoretical and practical properties. We then explore the power of building hash tables with multiple hash functions for items to choose from and in particular examine cuckoo hashing, which allows items to move among its choices even after they were initially inserted to the table. Section 11.2 applies hashing to create a sketch of the data set; we consider two sketches, the Bloom filter and the count-min sketch. A further application of hashing is presented in Section 11.3, namely, the use of hash functions to create concise *fingerprints* of items. We demonstrate two applications of fingerprints in text analysis: the Karp–Rabin string matching algorithm and the shingling algorithm for identifying similar documents. Randomization is also an essential component of modern public-key cryptosystems. Section 11.4 demonstrates how randomness is utilized in the RSA cryptosystem and in a related algorithm for choosing large random prime numbers. We also present two practical applications of randomization that do not have (yet) a complete theoretical analysis: the naive Bayesian classifier (Section 11.5) and randomized greedy heuristics (Section 11.6).

11.1 Hashing

A hash function h maps a domain to a range so that the result "looks random." Of course, any particular function is fully deterministic. In analyzing hashing, we assume that the hash function h is chosen at random from a family of hash functions $\mathcal{H}$. To simplify the analysis we often assume that the hash function is fully random, that is, chosen at random from the family of all functions, although in practice fully random hash functions are not possible. The number of random bits needed to represent a fully random hash function mapping $\{0, 1\}^n$ to $\{0, 1\}^m$ would be $2^n \times m$ bits, which is far too large for any reasonable n. In practice we use hash functions chosen from smaller families that have concise representation and are easy to compute. Constructing such families of functions that look sufficiently random is a major theoretical and practical research area, as we discuss in Section 11.1.1.

A common use for hashing is to build a hash table; the hash of an item—which is more commonly called a *key* in this context—provides a location, often called a bin or bucket, for where to find the key. If we have n keys and a hash table of n buckets, on average, each bucket will have a load of 1 key; however, even with a fully random hash function, the bucket with the most keys (or maximum load) will have $(1 + o(1))(\log n/\log \log n)$ keys with high probability [30].

We use this result as a baseline. To start, we consider one method for coming up with good hash functions using fewer random bits. It yields hash tables with a maximum load similar in size to fully random hash functions. Then, we show some useful ways to make better hash tables, with much smaller maximum load per bucket, by using multiple hash functions.

11.1.1 Tabulation Hashing

Tabulation hashing is a fairly old scheme based on lookup tables currently enjoying a renaissance based on new results regarding its performance. Variations (such as Zobrist hashing [67], which was used to hash positions in board games to bit strings) were known in the late 1960s; it was also considered by Carter and Wegman [10]. Our description is based primarily on recent work by Pătraşcu and Thorup [51,52].

We assume the keys are a fixed number of bits b. We view b as a vector of t characters, each of c bits, so that $b = ct$. Suppose that we want our output to consist of a bits; that is, we want a hash function $h: \{0, 1\}^b \to \{0, 1\}^a$. Our hash function is derived from a lookup table with $2^c \times t$ entries; each entry is initialized with an a-bit random value. Hence, the space required is $at2^c$ bits in total and our analysis assumes that the bits of the table are chosen uniformly at random from all 2^{at2^c} possibilities.

Let our lookup table T be such that $T[i, j]$ provides the a bits when the ith character takes on the value j, where $i \in [1, t]$ and $j \in [0, 2^c - 1]$. Let our key x be given by $x = x_1 \ldots x_t$, where the x_k are the t characters constituting x. Then

$$h(x) = T[1, x_1] \oplus T[2, x_2] \ldots \oplus T[t, x_t],$$

where $\oplus$ is the bitwise exclusive-or operation. Naturally, a key advantage of tabulation hashing is that it is very fast, if one can choose the table parameters to fit into memory. For example, suppose one wants to hash 128 bit keys (such as IPv6 addresses) into 32 bits. Using $c = 8$ and $t = 16$, the table requires 2^{17} bits, small enough to fit into even fairly small caches, and only a small number of simple operations.

Figure 11.1 that follows shows how 32-bit keys can be hashed into 12-bit values using just 384 random bits.

Tabulation hashing seems weak under some standard measures of approximate independence of hash values. For example, it is three-wise independent but not four-wise independent. That is, for any three keys x, y, and z, we have that $h(x)$, $h(y)$, and $h(z)$ are all mutually independent, but if we have four keys w, x, y, and z, the values $h(w)$, $h(x)$, $h(y)$, and $h(z)$ need not be mutually independent. For example, if w, x, y, and z are all the same except for the last two characters, so that for a prefix p we have $w = pc_1c_3$, $x = pc_2c_4$, $y = pc_1c_4$, and $w = pc_2c_3$, then $h(w) \oplus h(x) \oplus h(y) \oplus h(z) = 0$. More specifically the strings

00000000000000000000000000000000, 00000000000000000000000011111111
00000000000000001111111100000000, 00000000000000001111111111111111

arc not linearly independent; the hash of each is the exclusive-or of the hashes of the other three.

	Char 1	Char 2	Char 3	Char 4
00000000	***101001010101***	000011010101	111100111100	010011110001
00000001	110100101010	101010101010	101101010101	***101010000111***
00000010	001001011010	***101010100000***	000010110101	100101000111
00000011	010100001100	001010100100	101010010111	101010101100
00000100	010100101000	101010100111	000011001011	010100110101
00000101	010101010000	000010010101	111010100001	101001010001
00000110	110011001110	110010101110	000110101010	101011110101
00000111	101010100001	111000011110	***000011010000***	101010100010

.....

Hash(00000000000000100000011100000001)
= 101001010101 xor 101010100000 xor 000011010000 xor 101010000111

FIGURE 11.1 The first several rows of a table for tabulation hashing. In the example, a 32-bit string is broken into 4 characters of 8 bits each. By looking at each character in the appropriate column—shown by the bold an italicized entries—we find the 12-bit strings that are XOR'ed together to form the final hash value.

But recent work shows tabulation hashing is much stronger than the earlier result suggests. For example, one standard question is how well a hash function distributes n keys into a hash table with m buckets. In this respect, tabulation hashing does quite well: when $m = O(n)$, the most loaded bucket in the table has $O(\log n/\log\log n)$ keys with high probability, matching the same bounds as perfect random hashing. This follows from the following theorem (taken from Theorem 1 in [51]), which shows that tabulation hashing yields useful Chernoff-like bounds:

Theorem 11.1

Consider hashing n keys into $m \geq n^{1-1/(2c)}$ buckets using tabulation hashing, where c is a constant and $a = \log m$. Let X denote the number of keys that hash into a specific bucket (e.g., bucket 1). Let $\mu = E[X] = n/m$. The following probability bounds hold for any constant γ:

$$\forall\delta \leq 1: \mathbf{Pr}\left[\,|\,X - \mu\,|\,\right] \geq \delta\mu] \leq 2e^{-\Omega(\delta^2\mu)} + m^{-\gamma};$$

$$\forall\delta = \Omega(1): \mathbf{Pr}[X \geq (1+\delta)\mu] \leq (1+\delta)^{-\Omega((1+\delta)\mu)} + m^{-\gamma}.$$

The result follows by taking $m = O(n)$ and using a union bound.

Other known methods for generating hash values have not been shown to have (and often do not have) this useful property. The simplicity and small space usage of tabulation hashing, combined with performance guarantees such as this one, make it a strong contender for many practical applications. Indeed, tabulation hashing is stronger than standard pairwise-independent hashing, where any pair of items appear to be hashed independently even though larger collections of items may have dependencies, but very competitive in terms of computation. The tabulation hashing paper [51] as well as other works [62,63] provide useful insights regarding implementing both tabulation hashing and other universal hashing schemes, including comparisons.

11.1.2 Hashing with Choices

One way to break the $O(\log n/\log\log n)$ maximum load for hashing when n keys are hashed into n buckets is to give each key being hashed more than one location to be placed—that is, give the keys a choice. The main result behind multiple-choice hashing was presented in a seminal work by Azar et al. [4], who showed the following: suppose that n keys are hashed sequentially into n buckets by hashing each key d times to obtain d choices of a bucket for each key and placing each key in the choice with the smallest current number of keys (or *load*). As we have noted, when $d = 1$, which is standard hashing, then the maximum load grows like $(1 + o(1))(\log n/\log\log n)$ with high probability [30]; the result of [4] is that when $d \geq 2$, the maximum load grows like $\log\log n/\log d + O(1)$ with high probability, which even for two choices gives a maximum load of 5 in most practical scenarios. Notice that the gap in going from one to two choices is a different type of behavior; going from two to three choices just changes a constant factor. The upshot is that by giving keys just a small amount of choice in where they are placed, the maximum load can be greatly reduced; the cost is that now d locations have to be checked when trying to look up the key, which is usually a small price to pay in systems where the d locations can be looked up in parallel.

The intuition behind the proof is very easy (although the probabilistic formalism is somewhat more complicated). Suppose that we know that the fraction of buckets with load at least i after hashing all n keys is given by p_i. Then, at any time in the process, the probability that a key hashes to d locations that all have load at least i is at most $(p_i)^d$. Hence, the expected number of keys that cause a bucket to have load at least $i + 1$ is at most $n(p_i)^d$, so the fraction p_{i+1} of buckets with load at least $i + 1$ should satisfy $p_{i+1} \leq (p_i)^d$.

TABLE 11.1 Results for the Basic *d*-Choice Scheme (S) vs. Vöcking's Scheme (V)

	$d = 2$		$d = 3$		$d = 4$	
i	S	V	S	V	S	V
1	7.6×10^{-1}	7.7×10^{-1}	8.2×10^{-1}	8.4×10^{-1}	8.6×10^{-1}	8.8×10^{-1}
2	2.3×10^{-1}	2.2×10^{-1}	1.8×10^{-1}	1.6×10^{-1}	1.4×10^{-1}	1.2×10^{-1}
3	8.9×10^{-3}	4.4×10^{-3}	5.1×10^{-4}	1.1×10^{-5}	2.3×10^{-5}	7.8×10^{-11}
4	6.0×10^{-6}	5.2×10^{-8}	3.9×10^{-12}	4.5×10^{-31}	4.0×10^{-21}	2.3×10^{-141}

Note: Numbers represent the fraction of buckets of each load under fluid limit model, when n keys are placed in n buckets. These numbers are very accurate even for moderate values of n (in the thousands).

Let us assume that this equation holds for all i. Now clearly $p_2 \leq 1/2$, since at most 1/2 the buckets can have load at least 2, and by induction

$$p_k \leq 2^{-d^{k-2}}.$$

In particular, when $k > \log\log n/\log d + 2$, we have $p_k < 1/n$, so the expected number of buckets of this load is less than 1. The proof of [4] formalizes this argument more carefully, proving the maximum load in a bucket is $\log\log n/\log d + O(1)$ with high probability.

One interesting aspect of multiple-choice hashing in practice is that very precise estimates of the load distribution on the buckets can be obtained using fluid limit models, which give a simple family of differential equations that accurately describe the behavior of the hash table. Details appear in [41,45].

A variant later introduced by Vöcking [64] both gives slightly improved performance and is particularly amenable to parallelization. The hash table is split into d equal-sized subtables, which we think of as being ordered from left to right; when inserting a key, one bucket is chosen uniformly and independently from each subtable as a possible location; the key is placed in the least loaded bucket, breaking ties to the left. (The scheme is often called the *d*-left scheme.) This combination of splitting and tiebreaking reduces the maximum load to $\log\log n/(d\phi_d) + O(1)$, where ϕ_d is the asymptotic growth rate of the dth order Fibonacci numbers [64]. Again, very precise numerical estimates can be found using fluid limit models [45]. Table 11.1, for example, shows the fraction of buckets with loads from 1 to 4 for the basic *d*-choice scheme and for Vöcking's variation for $d = 2$, 3, and 4 in the asymptotic limit with n balls and n buckets. As can be seen, the loads with Vöcking's scheme fall even more quickly than the standard scheme. For example, under Vöcking's scheme with three choices, and practical sizes for n, we would never expect to see any bucket with load 4 or more; for the standard scheme, with three choices, a bucket with load 4 would be a rare but not outstandingly unusual event.

In practice, the size of the most loaded bucket is generally highly concentrated around a specific value. As noted for example in [7], this effectively means that *d*-left hash tables can provide an "almost perfect" hash table in many settings. That is, the hash table is highly loaded with little empty space, insertions and lookups are guaranteed to take constant time, and the probability of failure (such as some bucket becoming overloaded) is very small. Such hash tables can then be used to bootstrap further data structures. They are particularly important for fast lookups, a common need in network processing tasks.

Several other variations of this process have been considered, such as when items are both inserted and deleted, when a "stash" is used to handle items that overflow buckets, and when placements are done in parallel; some of these variations are covered in various surveys [34,42].

11.1.3 Cuckoo Hashing: Adding Moves

Cuckoo hashing [50] extends multiple-choice hashing schemes by allowing keys to move among their choices in order to avoid collisions. In the original description, a key can be placed in one of two possible

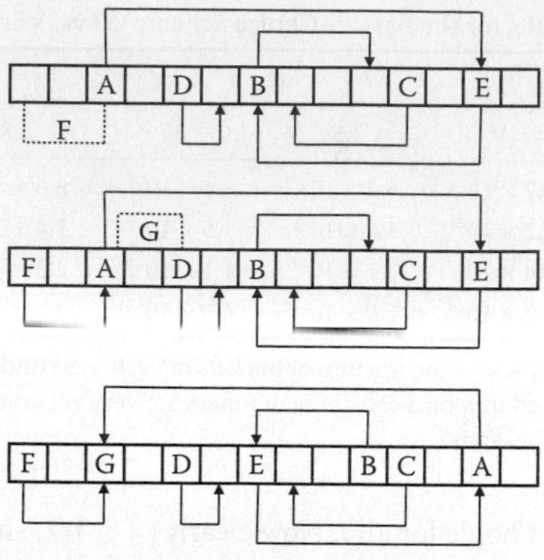

FIGURE 11.2 A cuckoo hash table example. When key F is inserted, it is placed in an empty bucket. When key G is inserted, it causes a sequence of moves in order to find a place.

buckets; each bucket can hold a single key. If both of the buckets already hold keys, we *move the key* in one of those buckets to the other of its two choices. If this other bucket is itself full, this move may in turn require the move of this other element, and so on until an empty spot for the current key is found. The name cuckoo hashing comes from the cuckoo bird in nature, which kicks other birds out of their nest, much like the hashing scheme recursively kicks keys out of their location as needed.

Figure 11.2 shows how one view of how a cuckoo hash table with two choices per key can be represented. An arrow represents the other choice for each key, so key A is placed in one bucket, but its other choice initially contains the key E. When key F is inserted, it is placed into an empty bucket. However, when key G is inserted, it takes the place of A, which then displaces E, which then displaces B to an empty bucket. It is unfortunately possible that a key cannot be placed, because the process of moving keys leads to a cycle; this is a cuckoo hash table failure.

In the case of two bucket choices per key, we can naturally view the process in the settings of random graphs. Let us assume that the two bucket choices for an element are always chosen to be distinct. Then we can think of the m buckets as vertices of a graph, and each of the n keys as being a random edge, with its two bucket choices being the endpoints of the edge. The question then becomes whether the edges of the graph can be *oriented* so each points to one of its two endpoints, in such a way that no endpoint is pointed to by two edges. Such an orientation would correspond to a successful placement of keys into buckets. Of course, beyond asking whether there *exists* a valid orientation, it is important to determine whether the cuckoo hashing algorithm outlined earlier can find it.

Hence, for two choices, we can consider the basic setting of random graphs, where there are very natural properties we can use.

Successfully placing an element corresponds to finding an augmenting path in the underlying graph where buckets are nodes and elements correspond to edges between nodes. When there are n keys to be placed in $2(1 + \varepsilon)n$ buckets, that is, when the load of the table is less than 1/2, all such augmenting paths are $O(\log n)$ in length with high probability. We therefore can declare that a failure occurs if a key cannot be placed after $c \log n$ steps for an appropriately chosen constant c.

Theorem 11.2

Consider a cuckoo hash table with $2(1 + \varepsilon)n$ buckets and two choices per key for n keys. Then with probability $1 - O(1/n)$, when keys are placed sequentially, all keys are successfully placed, with each key taking at most $O(\log n)$ steps to be placed. The expected time to place each key is $O(1)$.

Cuckoo hashing can naturally be generalized to situations with more than two choices per bucket and more than one key per bucket [17,26]. The advantage of these generalizations is that much higher loads can be obtained. If everything can be placed successfully, then lookups take time $O(1)$ as long as the number of choices d is constant. For $d > 2$, there is a correspondence to random hypergraphs, and when a bucket can hold $k > 1$ keys, we seek orientations that allow at most k edges assigned to each vertex. While these settings are substantially more technically challenging, a great deal has been learned about such schemes in the last few years. For example, when $d = 3$ choices are used, a cuckoo hash table can handle (asymptotically) a load of over 91%; with $d = 4$ choices, a load of over 97% can be handled [16,27,28]. Also, the use of a small "stash" to hold keys that would not otherwise be able to be placed can dramatically reduce the probability of failure in practical settings [35].

Cuckoo hashing is a fairly new hashing method, but several applications of the idea have already been suggested, including for storage schemes using flash memory [13], hash tables for graphical processing units and associated graphics applications [3], scalable distributed directories [25], and in hardware caches [58].

11.2 Sketches

A large number of data structures have arisen that make use of hashing to create a *sketch* of a data set. A sketch can be seen as a useful small summary of a data set that allows specific operations on the data. Here we look at some simple sketches that provide a basic introduction to the general methodology.

11.2.1 Bloom Filters

Bloom filters are named after Burton Bloom, who first described them in a paper in 1970 [5]. A Bloom filter is a randomized data structure that succinctly represents a set and supports membership queries. For a set S, if $x \in S$, the Bloom filter will always return 1. If $x \notin S$, the Bloom filter will return 0 with probability close to 1.

We review the necessary background, based on the presentation of the survey [8], which covers more details and some basic applications of Bloom filters. A Bloom filter representing a set $S = \{x_1, x_2, \ldots, x_n\}$ of n keys from a universe U consists of an array B of m bits, initially all set to 0. The filter uses k hash functions $h_1, \ldots, h_k$ with range $\{0, \ldots, m - 1\}$. Here (and throughout in our applications unless otherwise stated), we assume that our hash functions are fully random, so that each hash value is uniform on the range and independent of other hash values. Technically, this is a strong assumption, but it proves both useful and generally accurate in practice. (See [44] for possible reasons for this.)

For each key $x \in S$, the bits $B[h_i(x)]$ are set to 1 for $1 \le i \le k$ in the array B. (A bit may be set to 1 multiple times.) To check if an key y is in S, we check whether all array bits $B[h_i(y)]$ are set to 1 for $1 \le i \le k$. If not, then clearly y is not a member of S. If all $h_i(y)$ are set to 1, we assume that y is in S. This may be incorrect; an element $y \notin S$ may hash to bits in the array that are set to 1 by elements of S. Hence, a Bloom filter may yield a *false-positive* error. Figure 11.3 shows further how Bloom filters function and how false-positives can occur.

The probability of a false-positive can be estimated in a straightforward fashion, under our assumption that the hash functions are fully random. Suppose that after all n keys of S are hashed into the Bloom filter, a fraction ρ of the bits is 0. Then by independence of our hash values, the probability that an element y gives a false-positive is just $(1 - \rho)^k$. What can we say about ρ? The expected value of ρ is just the probability that any specific bit remains 0; a bit has to be missed by kn hashes for that to happen, and that occurs with probability

$$p' = \left(1 - \frac{1}{m}\right)^{kn} \approx e^{-kn/m}.$$

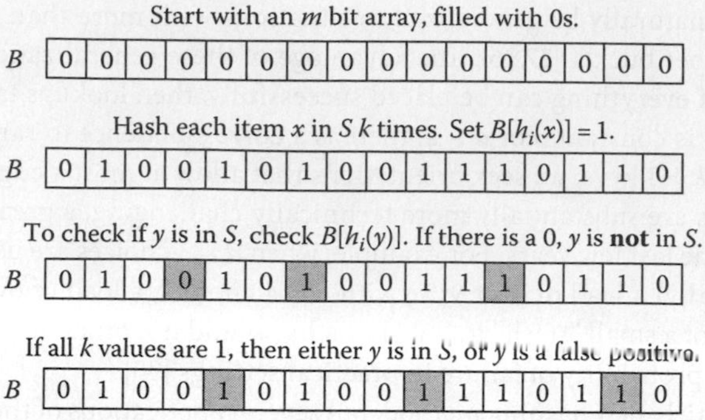

FIGURE 11.3 The basics of Bloom filters.

Now ρ can be shown (using various concentration methods) to be closely concentrated around its expectation p'. Hence, a clean formula that very closely approximates the false-positive probability f is

$$f \approx (1 - e^{-kn/m})^k.$$

What is the best choice for the number of hash functions? Using calculus, it is easy to show that f is minimized when $k = \ln 2 \cdot (m/n)$, in which case

$$f \approx (1 - e^{-kn/m})^k = \left(\frac{1}{2}\right)^k \approx (0.6185)^{m/n}.$$

Of course, in practice, k must be an integer.

The utility of Bloom filters stems from the fact that the false-positive probability falls exponentially in m/n, the number of bits per key. When m/n is 8, the false-positive probability is just over 2%; when m/n is 10, it falls below 1%. Hence, when elements are large, Bloom filters allow one to save significant space at the cost of some false-positives. For many applications, that trade-off is appealing. In particular, Bloom filters have been suggested for a number of network processing tasks, such as longest prefix matching, or matching of signature strings within packets [14,15,61]. In the case of packet signatures, we may read the packet byte by byte and in parallel consider strings of various lengths ending at that byte. The signatures might live in a slower memory hash table, one for each signature length, or other lookup structure. By using a smaller Bloom filter in faster memory to record the set of signature strings in each table, the number of lookups can be reduced; in the common case, no slower lookups will be required, and false-positives will only minimally affect performance if they are kept sufficiently small. This use of a Bloom filter as a method to filter out requests to a slower or otherwise more costly lookup structure when we expect most lookups to be unsuccessful provides a common application of Bloom filters in real systems. Another notable early use of Bloom filters in networking is to provide a sketch of the contents of a web cache, to allow cooperative caching [23,56]. Instead of passing around lists of URLs, machines can pass around more space-efficient Bloom filters. The effects of false-positives, which can make a machine look to a local neighbor for a document that is not actually stored before then going to the web for it, can be made suitably small to still allow for improved performance.

11.2.2 Count-Min Sketches

Bloom filters only provide information about set membership, but the same basic framework—hash keys into multiple buckets in a way that preserves information about the elements despite collisions—has now been used for many other data structures. As one example, we present here the *count-min sketch* [12] of Cormode and Muthukrishnan.* The count-min sketch is designed for settings where elements have associated counts. For example, in a network, a source–destination pair might have an associated number of bytes sent between them. We might want to track these counts, approximately, with an eye toward finding the *heavy hitters*, which are the flows sending the largest number of bytes. Tracking byte counts for all flows within a router would be too expensive, so we seek a small data structure that allows for accurate detection. Such structures also provide a building block for many other types of queries and applications, in databases and in various data stream settings. For example, they can be used to effectively estimate inner products of vectors [12].

We use the following framework. The input is a stream of *updates* in the form of order pairs (i_t, c_t), starting from $t = 1$, where each *key* i_t is a member of a universe $U = \{1,\ldots, n\}$, and each *count* c_t is an integer. The state of the system at time T is given by a vector $\vec{a}(T) = (a_1(T),\ldots, a_n(T))$, where $a_j(T)$ is the total count associated with key j at time T, or

$$a_j(T) = \sum_{t \le T : i_t = j} c_t.$$

We will assume in what follows the typical case where we have the guarantee that counts are nonnegative, so $a_j(T) \ge 0$ for every j and T.

It is slightly more convenient for our analysis to split our structure into a 2D array Count with k rows and w columns, with each array entry being a counter. (There is essentially no difference in theory or practice if we just use a single array, but the argument is slightly harder.) As we did with Bloom filters, here we use k independent hash functions $h_1,\ldots, h_k$: $\{1,\ldots, n\} \to \{1,\ldots, w\}$; there is one for each row. (Actually, it is enough to assume that the hash functions are universal—also called pairwise independent—as shown in [12]; our arguments also hold with this assumption.) Every entry of the array is initialized to 0.

When a pair (i, c) arrives, we add c to the counters Count$[1, h_1(i)],\ldots,$ Count$[k, h_k(i)]$. As long as all counts are nonnegative, then each entry Count$[j, h_j(i)]$ will be a *lower bound* on $a_i(T)$. Intuitively, our hope is that, for at least one of these counters, no or at least very few other keys land in the same array counter. In that case, the minimum of the values Count$[j, h_j(i)]$ will actually be very close to the true count $a_i(T)$. That is, we use of $\hat{a}_i = \min_{j \in \{1,\ldots,d\}}$ Count$[j, h_j(i)]$ as an estimate of $a_i(T)$.

We now show that $\hat{a}_i$ is a good bound for $a_i(T)$, as long as $a_i(T)$ is reasonably large and we size our array of counters appropriately. For $j \in \{1,\ldots, k\}$, let

$$X_{i,j} = \sum_{i' \ne i : h_j(i') = h_j(i)} a_{i'}(T).$$

Since the hash functions are fully random, $\mathbf{E}[X_{i,j}] \le \|\vec{a}\|/w$, where $\|\vec{a}\| = \sum_{x \in [1,n]} a_x(T)$ (the L_1 norm of $\vec{a}$, again assuming all entries are nonnegative). Markov's inequality then implies that for any threshold value $\theta > 0$, we have $\mathbf{Pr}(X_{i,j} \ge \theta) \le \|\vec{a}\|/(w\theta)$. By definition, our estimate satisfies $\hat{a}_i = a_i + \min_{j \in 1,\ldots,d} X_{i,j}$, and by independence of the h_j's, we can conclude that

$$\mathbf{Pr}(\hat{a}_i \ge a_i + \theta) \le \left(\frac{\|\vec{a}\|}{w\theta}\right)^k.$$

* Estan and Varghese introduced the same data structure at essentially the same time and used it for applications in network measurement [22]. They refer to the data structure as a *parallel multistage filter*.

In particular, if we fix some parameters ε, δ > 0 and set $w = \lceil e/\varepsilon \rceil$, $d = \lceil \ln(1/\delta) \rceil$, and $\theta = \varepsilon \| \vec{a} \|$, then we obtain

$$\Pr\left(\hat{a}_i \geq a_i + \varepsilon \| \vec{a} \|\right) \leq \left(\frac{1}{e}\right)^{\ln(1/\delta)} = \delta.$$

In essence, we have that $\hat{a}_i$ is likely to be a fairly good estimate of a_i—within ε times the total count over all keys. As long as a_i is not too small for example, it is a constant fraction of the total count—this can be a useful estimate.

One practical improvement, which unfortunately has resisted much theoretical analysis, uses a technique called *conservative updating* [22]. Here, we need to assume that all increment values c_t are nonnegative. The basic idea is to never increment a counter more than is strictly necessary in order to guarantee that $\hat{a}_i \geq a_i(T)$. Formally, to process an update (i, c), we let $c' = \min_{j \in \{1,\ldots,k\}} \mathsf{Count}[j, h_j(i)]$ and then set $\mathsf{Count}[j, h_j(i)] = \max(c + c', \mathsf{Count}[j, h_j(i)])$ for $j = 1, \ldots, k$. Notice that a counter is not updated if it is already larger than the previous estimated count associated with the key, which is c' (the minimum over all the counters associated with i before the update), plus the update value c. It is easy to see that at every time step we still have $\hat{a}_i \geq a_i(T)$ using this conservative updating, but now the counters are not incremented as much. Intuitively, this technique further reduces the effects of keys colliding in our array of counters. In particular, keys with small count are much less likely to have any effect on estimates for large keys. Experiments show that the improvement can be substantial [22].

11.3 String Matching and Document Similarity

Hash functions can be used to create succinct *fingerprints* of keys, so that with high probability most keys have unique fingerprints. We present two applications of this tool in the context of text analysis.

11.3.1 Karp–Rabin String Matching Algorithm

Given a long string of n characters $X[1, n] = x_1,\ldots, x_n$ and a shorter string of m characters $Y[1, m] = y_1,\ldots, y_m$, we want to find all the occurrences of $Y[1, m]$ in $X[1, n]$, that is, all indices $i_1,\ldots, i_k$, such that the m character substring $X[i_j, i_j + m - 1]$ is identical to $Y[1, m]$. The problem is easily solved in $O(mn)$ time. The randomized Karp–Rabin algorithm [33,29] always gives the correct answer and terminates in expected $O(n + km)$ time, where k is the number of occurrences of the string $Y[1, m]$ in $X[1, n]$. (In the worst case, the algorithm still takes only $\Omega(mn)$ time.)

Without lost of generality, we can assume that the characters in the strings are from the integer domain$\{0,\ldots, b - 1\}$. Alternatively, we can view an m character string as a number written in radix b. The Karp–Rabin algorithm is based on two ideas:

1. The m characters strings can be hashed into a small domain of integers to speed up the comparisons of two strings.
2. The hash value of the string $X[i + 1, i + m]$ can be computed with a constant number of operations from the hash value of the string $X[i, i + m - 1]$.

The second idea is key; if we simply hashed each m character substring of $X[1, m]$ in $O(m)$ time per hash, there would be no gain. An example of a hash function that satisfies the earlier requirements is the radix b representation of a string of m characters modulo a prime integer p:

$$h\left(X[i, i+m-1]\right) = \left(\sum_{j=0}^{m-1} x_{i+j} b^{m-j-1}\right) \bmod p.$$

To see how to go from one hash value to the next, assume that we have computed $h(X[i, i + m - 1])$. Then

$$h(X[i+1, i+m]) = \left(b\left(h(X[i, i+m-1]) - x_i b^{m-1}\right) + x_{i+m}\right) \bmod p,$$

which requires only $O(1)$ operations modulo p for each new value.

The algorithm begins by computing

$$h(Y[1, m]) = \left(\sum_{j=1}^{m} y_j b^{m-j}\right) \bmod p$$

in $O(m)$ operations, and then for $i = 1, \ldots, m - n + 1$ it computes and tests if

$$h(X[i, i+m-1]) = h(Y[1, m]),$$

which requires a total of $O(n)$ operations overall.

If $h(X[i, i + m - 1]) \neq h(Y[1, m])$, then the two strings are not equal. If $h(X[i, i + m - 1]) = h(Y[1, m])$, the algorithm compares the m characters in the two string to verify the identity. Thus, the algorithm always gives the correct answer.

The expected run-time of the algorithm depends on the expected number of times in which $h(X[i, i + m - 1]) = h(Y[1, m])$, since in those cases the algorithm needs to individually check each of the m characters in the two strings. In particular, we are interested in minimizing the number of times in which the two hash values are equal but the two strings are not identical. In that case for some integer $r \neq 0$.

$$\sum_{j=0}^{m-1} x_{i+j} b^{m-j-1} - \sum_{j=1}^{m} y_j b^{m-j} = rp \neq 0. \quad (11.1)$$

Algorithm 11.1 Karp–Rabin Multiple Pattern Search

```
input: String X [1, n]; ℓ; patterns Y_1 [1, m], ..., Y_ℓ [1, m], a random prime p and hash function h as
   defined in the text;
for j = 0, ... p − 1 do
   TABLE[j] = φ;
end for
Add 1 to the bucket TABLE[h(X [1, m])];
for i = 1, ..., n − m + 1 do
   h(X [i + 1, i + m]) = (b(h(X [i, i + m − 1]) − x_i b^{m−1}) + x_{i+m}) mod p
   Add i + 1 to the bucket TABLE[h(X[i + 1, i + m])];
end for
for j = 1, ... ℓ; do
   for k ∈ TABLE[h(Y_j [1, m])] do
      if X [k, k + m − 1] = Y_j [1, m] then
         OUTPUT: "Pattern j appears in X [k, k + m − 1]"
      end if
   end for
end for
```

For a given p, there are many sequences that satisfy (1). However, if we choose p at random from a sufficiently large collection of prime numbers, then the probability of satisfying (1) is small. Specifically, an integer n has no more than $\log n$ distinct primes factors, while there are asymptotically $n/\log n$ primes in the range $[1, n]$ (see, e.g., [48,59] [Theorem 5.14] or other references to the Prime Number Theorem). Since $\sum_{j=0}^{m-1} x_{i+j} b^{m-j-1} \leq b^m$, it is divisible by no more than $m \log b$ prime numbers. Thus, if we choose p uniformly at random from all the primes in the range $[0, nm^2 \log b \log(nm^2 \log b)]$, the expected number of false equalities in the running of the algorithm is bounded by

$$(n-m+1)(m \log b)\frac{\log(nm^2 \log b \log(nm^2 \log b))}{nm^2 \log b \log(nm^2 \log b)} = O\left(\frac{1}{m}\right).$$

In this case, the expected extra work from testing false identities is $O(1)$, and the expected run-time (except for verifying correct matches) is $O(n + m)$ steps.

The Karp–Rabin method is particularly useful to search for multiple patterns simultaneously (see Algorithm 11.1). Assume that we have ℓ equal-length patterns $Y_1[1, m], \ldots, Y_\ell[1, m]$, and we want to find any occurrence of any of these patterns in a long string $X[1, n]$. Such searches are useful for plagiarism detection and DNA motif discovery [57]. (If the ℓ patterns do not have the same length, we can use prefixes of the longer patterns.) Comparing the hash value of each of the strings $X[i, i + m - 1]$, $1 \leq i \leq n - m + 1$ to the hash value of each of the ℓ patterns in the natural manner could require $\Omega(\ell n)$ time. However, suppose that we instead store the hash values of the strings in a hash table with an appropriate pointer back to the corresponding string location. For simplicity, we assume that the hash table has size p; however, similar results can be proven for a table of $O(n)$ entries. We test for each of the ℓ patterns if its hash value coincides with any stored pointer in the table (an entry of the table can store a set of indices). The total work in populating the hash table is $O(n)$, and computing the hash values for the ℓ patterns takes $O(\ell m)$ time. If a substring matches the pattern hash, we need to verify the string matching by comparing the m characters. If we choose a prime p at random from an interval $[0, nm^2\ell \log b \log(nm^2\ell \log b)]$, the probability of a false identity is $O(1/m)$, and the expected run-time (except for verifying correct matches) is $O(n + \ell m)$. (To reduce the hash table size from modulo p to modulo n, we must keep both the string location and the corresponding original hash value modulo p; we then only do the $O(m)$ work to check the pattern if the hash values modulo p match as well.)

11.3.2 Detecting Similar Documents

A major computational problem in indexing web pages for search engines is eliminating near-duplicate documents. A near-duplicate document is not an exact duplicate; exact duplicates are actually fairly easy to detect. For example, one could hash each document to a fingerprint of 128 bits and keep a sorted list of the fingerprints. To check for an exact duplicate, one could just search the fingerprint list; with a fingerprint of 128 bits, with high probability, two distinct documents should never obtain the same fingerprint hash value. In contrast, near-duplicate documents often differ in just a few locations, such as the date of last update. But even small differences will yield distinct fingerprints, so we cannot simply hash the pages as we can to check for exact duplicates.

It is estimated that about 40% of the pages on the web are duplicates or near-duplicates of other pages [38]. Eliminating near-duplicate documents increases the efficiency of the search engine and in general provides better results to the users. The *shingling technique* [9] provides an efficient and practical solution to this problem. Similar techniques are also used in other settings, such as plagiarism detectors [11].

Given a document, the process usually begins by putting it in a canonical form, by removing all punctuation, capitalization, stop words, and other extraneous information, so that the document is reduced to a sequence of terms (words). Then, given a document of n terms $X = X[1, n]$ and an integer

$k > 0$, a *k-shingle* in X is any subsequence of k consecutive terms $X[j, j + k - 1]$ in X. Let S_X be the collections of all k-shingles in X:

$$S_X = \{X[i, i+k-1], 1 \le i \le n-k+1\}.$$

We measure the similarity of two documents X and Y by the *Jaccard coefficient* [31] of their respective k-shingle sets, S_X and S_Y:

$$Jaccard(S_X, S_Y) = \frac{|S_X \cap S_Y|}{|S_X \cup S_Y|}.$$

The Jaccard coefficient has properties that make it a useful measure of similarity. In particular, the Jaccard coefficient is 1 for identical sets and 0 for disjoint sets. If $Jaccard(S_X, S_Y) \ge r$ for some predefined threshold r (say 0.9), we consider the two documents similar and remove one of them. The challenge is to implement this computation efficiently so that it can scale to the size of web search engines. This is achieved through two randomization steps.

We first map each shingle, using a hash function $h()$, into a relatively large space of numbers (say 64 bits numbers). Let $h(s)$ be the hash value of shingle s and denote by H_X the collection of hash values of the shingles in X:

$$H_X = \{h(s) \mid s \text{ is a shingle in } X\}.$$

This mapping allows us to efficiently compare two shingles with relatively small probability of a false-positive. However, typical documents have a large number of shingles, so computing the Jaccard coefficient of H_X and H_Y for every pair of documents X and Y is still too expensive.

To speed up the computation of the Jaccard coefficient, we use the following randomized estimate. Assume that the hash function maps the shingles into the range $\{0, \ldots, b - 1\}$ (where again, in practice $b = 2^{64}$ might be a suitable choice). Let $\Sigma(b)$ be the set of $b!$ permutations on $0, \ldots, b - 1$. We choose a random permutation $\sigma \in \Sigma(b)$ and permute the two sets H_X and H_Y. The following theorem shows that the probability that that the minimum values in the two permuted sets are equal is an unbiased estimate for the Jaccard coefficient of X and Y.

Theorem 11.3

Consider two documents X and Y and their collections of hashed shingles values H_X and H_Y. Choose $\sigma \in \Sigma(b)$ uniformly at random, let

$$m_X(\sigma) = \min\{\sigma(h(s)) \mid h(s) \in H_X\},$$

and similarly define m_Y. Then,

$$Jaccard(S_X, S_Y) = \mathbf{Pr}(m_X(\sigma) = m_Y(\sigma)).$$

Proof: Let $u = |H_X \cup H_Y|$ and $t = |H_X \cap H_Y|$. There are $u!$ permutations of $H_X \cup H_Y$ (we can ignore the values that do not appear in $H_X \cup H_Y$). Out of the $u!$ permutations, there are $t(u - 1)!$ permutations in which the minimum elements in the two sets are the same, since in this case there are t possible first elements one can choose first to be the minimum. Thus,

$$\mathbf{Pr}(m_X = m_Y) = \frac{t(u-1)!}{u!} = \frac{t}{u} = \frac{|H_X \cap H_Y|}{|H_X \cup H_Y|} = Jaccard(S_X, S_Y).$$

To estimate the probability that $\mathbf{Pr}(m_X = m_Y)$, we choose s permutations and count the number of permutations that give $m_x = m_Y$. Thus, our estimate of $J(X, Y) = Jaccard(H_X, H_Y)$ is

$$\tilde{J}(X,Y) = \frac{1}{s}\sum_{i=1}^{s} 1_{m_X(\sigma_i) = m_Y(\sigma_i)}.$$

Applying a Chernoff type bound [43] (Exercise 1.15), we have

$$\mathbf{Pr}\left(|\tilde{J}(X,Y) - J(X,Y)| \geq \varepsilon J(X,Y)\right) \leq 2e^{-2s(J-\varepsilon)^2}.$$

Thus, for any $\varepsilon > 0$, the failure probability decreases exponentially with s. Choosing s to be, for example, 100 in practice generally performs quite well.

The major advantage of the shingling algorithm arises when comparing all pairs of a set of n documents. We first choose a set of s random permutations. We then process each document and store its list of s minimal values under the s permutations. Comparing two documents requires comparisons of s pairs of integers. Thus, most of the computation is done in processing the individual documents and therefore is linear in n. The work to compare all documents can still be quadratic in n, which is efficient for small n. Further methods, such as using hashing to place items into buckets according to their s shingle values, can reduce the time further by avoiding the need to explicitly compare the s values for every pair.

11.4 RSA and Primality Testing

Randomization is an essential component of any modern cryptography system. As an example, we will discuss the RSA encryption system and the related randomized primality testing algorithm. While these cryptographic schemes rely on arguments from number theory, we focus here the how randomness is used within the algorithms and therefore quote without proof the relevant number theory results as needed.

11.4.1 Public Key Encryption System

To receive a secure message in a *public key encryption system* [18], Bob runs a protocol that generates two keys: a public key *Pub_Key* and a private key *Prv_Key*. Bob publishes the public key and keeps the private key secret. Associated with the public key is an encryption function *Enc*(), and associated with the private key is a decryption function *Dec*(). To send a secure message x to Bob, Alice computes *Enc*(x) and sends it to Bob. The scheme is an efficient and secure encryption system if (1) *Dec*(*Enc*(x)) = x, that is, decoding an encoded message yields the original message; (2) the functions *Enc*() and *Dec*() can be computed efficiently (in time polynomial in the message's length); but (3) there is no efficient algorithm that computes x from *Enc*(x) (decrypts the encryption function) without the knowledge of the private key *Prv_Key*. In practice, the algorithm that Bob uses to generate his private and public keys is itself public; the security of the encryption scheme is not based on obscuring how the encryption algorithm works. What keeps the private key secret is that it should be generated by a randomized algorithm, from a large domain of possible private keys. Indeed, the number of possible private keys must be superpolynomial in the size of the message. Otherwise, an adversary could try every possible private key to obtain a polynomial time algorithm. The random choice of private key becomes the essential obstacle that prevents an adversary from trivially replicating Bob's encryption process.

11.4.2 RSA

In the RSA system [55], the protocol for generating the public and private keys begins by generating two large random prime numbers p and q and an integer e, such that the greatest common divisor of $(p - 1)(q - 1)$ and e is 1. (This is commonly written as $gcd((p - 1)(q - 1), e) = 1$. One RSA variant is to let $e = 3$, which means that p and q cannot be integers of the form $3x + 1$ for an integer x. Let $n = pq$ and $d = e^{-1} \bmod (p - 1)(q - 1)$. It is a basic fact from number theory that e has a multiplicative inverse modulo $(p - 1)(q - 1)$ if we choose e so that $gcd((p - 1)(q - 1), e) = 1$. The public key that Bob publishes is (n, e), and

$$Enc(x) = x^e \bmod n.$$

The private key that Bob keeps secret is (p, q, d), and

$$Dec(y) = y^d \bmod n.$$

We first show that the decryption function correctly decodes the encrypted message. To decode Alice's message, Bob computes

$$Dec(Enc(x)) = x^{de} \bmod n.$$

Since d and e are multiplicative inverses modulo $(p - 1)(q - 1)$, we have that $de = 1 + k(p - 1)(q - 1)$ for some $k \neq 0$. Hence, for this k,

$$Dec(Enc(x)) = x^{de} \bmod n = x^{1+k(p-1)(q-1)} \bmod n. \tag{11.2}$$

We would like to show that the last expression is equal to x. This follows from a well-known number-theoretic fact, Fermat's little theorem.

Theorem 11.4 (Fermat's little theorem)

If p is prime and $a \neq 0 \bmod p$, then

$$a^{p-1} = 1 \bmod p.$$

Applying Fermat's little theorem to (2), we have $x^{1+k(p-1)(q-1)} = x \bmod p$. This holds trivially when $x = 0 \bmod p$ and by Fermat's little theorem otherwise. Similarly, $x^{1+k(p-1)(q-1)} = x \bmod q$. It follows again from the number theory that if $x = a \bmod p$ and $x = a \bmod q$ for primes p and q that $x = a \bmod pq$; thus,

$$x^{1+k(p-1)(q-1)} = x \bmod pq = x \bmod n,$$

proving that the RSA protocol correctly decodes the encoded message. Furthermore, both the encoding and decoding are efficiently computable functions using repeated squaring for modular exponentiations. (To compute a number like $x^{16} \bmod p$, you repeatedly square to compute $x^2 \bmod p$, $x^4 \bmod p$, $x^8 \bmod p$, and $x^{16} \bmod p$; one can compute $x^n \bmod p$ in $O(\log n)$ multiplications modulo p.)

How secure is the RSA cryptosystem? First, notice that if an adversary can factor n, she can compute p and q and obtain Bob's private key. The RSA system depends on the assumption that it is computationally hard to factor a product of two large primes. It is widely believed that factoring a sufficiently large number is computationally infeasible, and so this line of attack is ineffective, barring a breakthrough in factorization algorithms. Typically, in practice it is recommended that each prime is at least 768 bit [1], but longer keys can and should be used for greater security. If you want your RSA message to be unbreakable for the next 20 or 30 years, and you expect improvements in factoring algorithms over that time, you may want to use bigger primes.

As we have stated, in order for the adversary not to know what primes have been chosen for the private key, they should be chosen randomly; we discuss how to do this in the following. Even in choosing large random primes, it is important that the protocol avoids "easy" cases. For example, if one of the primes it small, the number is easier to factor, and that is why both primes in the key should be large. Similarly, if the two primes are close together, the factoring problem becomes easier. Finally, one must be sure to really choose the prime randomly; recently, an attack on RSA broke a large number of private keys, making use of the fact that several implementations did not choose their primes in a sufficiently random manner, allowing private key values n to be factored [36].

When the set of possible messages from Alice to Bob is small, an alternative line of attack for an adversary is to keep a table of the encrypted version of all possible messages. To defend against this weakness in the protocol, current implementations randomize the original message by adding a number of (easy to discard) extra random characters. This randomization phase—generally called "salting" [32]—significantly increases the domain of all possible messages, thus making this line of attack less feasible.

11.4.3 Primality Testing

The RSA and similar encryption systems require the generation of random large primes. By the prime number (e.g., Theorem 5.14 of [59]), the probability that a random integer in the range $[1, N]$ is prime is asymptotically $1/\ln N$. Thus, if we choose $\ln N$ random integers in the range $[1, N]$, we are more likely than not to have at least one prime number among these random numbers. (If we are particularly clever, and avoid numbers that are, for example, obviously divisible by 2, 3, and 5, we can find a prime number faster.) The challenge is to identify a prime number, that is, to efficiently test for primality of large integers. We present here an efficient randomized algorithm for this problem. Interestingly, the question of whether there is an efficient deterministic algorithm for primality testing has only relatively recently been affirmatively solved [2]. Even though deterministic algorithms are now known, this randomized algorithm is particularly quick and effective.

11.4.4 Miller–Rabin Primality Test

Algorithm 11.2 Miller–Rabin Primality Test

input: An odd integer $n > 1$.
Choose $w \in \{2, \ldots, n-1\}$
$s \leftarrow 0$; $r \leftarrow n - 1$;
while s is even **do**
    $s = s + 1$
    $r \leftarrow r/2$
end while
Compute $w^r, w^{2r}, \ldots, w^{2^s r} = w^{n-1}$
for $i = 1, \ldots, s$ **do**
    if $w^{2^i t} = 1 \bmod n$ **and** $w^{2^{i-1} t} \neq \pm 1 \bmod n$ **then**
        Return **composite**
    end if
end for
Return **prime**

The Miller–Rabin algorithm [53] is a randomized test with one-sided error. If the input is a prime number, the test always correctly confirms that the input is prime; if the input is a composite number, then with small probability the test can make a mistake and identify the input as prime. The error probability can be made arbitrarily small by repeating the test sufficiently number of times.

Given an input n, the test tries to randomly find a *witness* w demonstrating that the input n is composite. An integer $w \in \{2, \ldots, n-1\}$ is a witness for this primality test if it satisfies at least one of the following conditions:

1. $w^{n-1} \neq 1 \bmod n$
2. $w^{2^i r} \neq \pm 1 \bmod n$, and $w^{2^{i-1} r} = 1 \bmod n$, for some integers r and $0 < i \leq s$, such that $r2^s = n - 1$

Condition (1) follows from Fermat's little theorem; if $w^{n-1} \neq 1 \bmod n$, Fermat's little theorem says that n cannot be prime, so w provides a witness for the compositeness of n. Condition (2) gives an integer $x = w^{2^{i-1} r}$ such that $x^2 = 1 \bmod n$ and $x \neq \pm 1 \bmod n$. Thus, $x^2 - 1 = (x - 1)(x + 1)$ is divisible by n, but both $x - 1$ and $x + 1$ are each not themselves divisible by n. This is impossible when n is prime, so n must be composite.

The efficiency of the earlier conditions for screening out non-prime numbers is due to the following fact (see [59] for a detailed discussion and in particular Theorem 10.3):

Theorem 11.5

If n is not a prime number and w is chosen uniformly at random in the range $\{2, \ldots, n-1\}$, then with probability at least 3/4, w is a witness according to the conditions earlier.

Hence, with probability 3/4, each time we run the test, we will catch that n is composite. Repeating the test with k independently chosen random integers in the range $\{2, \ldots, n-1\}$, the error probability of the Miller–Rabin test is at most $(1/4)^k$.

11.5 Naive Bayesian Classifier

A naive Bayesian classifier [39] is a supervised learning algorithm that classifies objects by estimating conditional probabilities using the Bayesian rule in a simplified ("naive") probabilistic model. In spite of its oversimplified assumptions, this method is very effective in many practical applications such as subject classification of text documents and junk e-mail filtering (see [40] and the references there). Strictly speaking, the algorithm is deterministic, and it makes no random steps. We include it here because it is one of the most practical and often-used algorithms based on a fundamental probabilistic concept.

Assume that we are given a collection of n training examples

$$\{(D_i(x_{i,1}, \ldots, x_{m,i}), c_i) \mid i = 1, \ldots, n\},$$

where $D_i = D_i(x_{1,i}, \ldots, x_{m,i})$ is an object with features $x_{1,i}, \ldots, x_{m,i}$ and c_i is the classification of the object.

Let C be the set of possible classes and X_i the set of values of the ith feature. For example, $D_i(x_{1,i}, \ldots, x_{m,i})$ can be a text document with Boolean features $x_{1,i}, \ldots, x_{m,i}$, where $x_{i,j} = 1$ if keyword j appears in document i, otherwise $x_{i,j} = 0$, and C can be a set of possible subjects or a {"spam," "no-span"} classification.

The classification paradigm assumes that the training set is a sample from an unknown distribution in which the classification of an object is a (possibly probabilistic) function of the m features, that is, the classification of a new object $D(x_1,\dots x_m)$ is fully characterized by the conditional probabilities

$$\mathbf{Pr}(c \in C \mid (x_1,\dots,x_m)) = \frac{\mathbf{Pr}((x_1,\dots,x_m) \mid c)\mathbf{Pr}(c)}{\mathbf{Pr}((x_1,\dots,x_m))}.$$

The main difficulty in applying this model is that we need to learn a large collection of conditional probabilities, corresponding to all possible combination of values of the m features. Even if each feature has just two values, we need to estimate 2^m conditional probabilities per class, which requires $\Omega(|C|2^m)$ samples. The training process is faster and requires significantly fewer examples if we assume a "naive" model in which the m features are independent. In that case,

$$\mathbf{Pr}((x_1,\dots,x_m) \mid c) = \prod_{j=1}^{m} \mathbf{Pr}(x_i \mid c), \quad \text{and} \quad \mathbf{Pr}((x_1,\dots,x_m)) = \prod_{j=1}^{m} \mathbf{Pr}(x_i).$$

With a constant number of possible values per feature, we only need to learn $O(m(|C| + 1) + |C|)$ probabilities. The training process is simple: $\mathbf{Pr}(x_i)$ is the fraction of objects where the ith feature equals x_i, $\mathbf{Pr}(x_i|c)$ is the fraction of objects classified c where the ith feature equals x_i, and $\mathbf{Pr}(c)$ is the fraction of objects classified c in the training set. (In practice, one often adds 1/2 to the numerator in each of the fractions to guarantee that no probability equals 0.)

Once we train the classifier, the classification of a new document $D(x_1,\dots x_m)$ is computed using the Bayesian rule:

$$\mathbf{Pr}(c \mid (x_1,\dots,x_m)) = \frac{\mathbf{Pr}(c)\prod_{j=1}^{m} \mathbf{Pr}(x_i \mid c)}{\prod_{j=1}^{m} \mathbf{Pr}(x_i)}.$$

Depending on the application, we can use the classification that maximizes the conditional probability (maximum likelihood classifier) or use the distribution over all possible classifications of D defined by the conditional probabilities.

The naive Bayesian classifier is efficient and simple to implement due to the "naive" assumption of independence. This assumption may lead to misleading outcomes when the classification depends on combinations of features. As a simple example, consider a collection of items characterized by two Boolean features X and Y. If $X = Y$, the item is in class A, and otherwise it is in class B. Assume further that for each value of X and Y the training set has an equal number of items in each class. All the conditional probabilities computed by the classifiers equal 0.5, and therefore the classifier is not better than a coin flip in this example. In practice, such phenomena are rare and the naive Bayesian classifier is often very effective. Verifying and explaining the practical success of this method is an active research area [19,20,49,65].

11.6 Randomized Greedy Heuristics

As a final example of how to use randomness, we offer another heuristic that remains open to theoretical analysis but proves useful in practice. Several greedy algorithms can be placed in the following natural framework. We have a collection of n elements that need to be placed (for some notion of placed); their placement will determine how good the solution is. Our greedy algorithm first orders the n elements into some intuitively good ordering, and then elements are greedily placed one at a time according to some process.

A standard example is the first-fit decreasing algorithm for bin packing. We have n items $a_1, a_2, \ldots, a_n$ with sizes in the range [0, 1], and several bins all of size 1, which we call $b_1, b_2, \ldots$. We order the items in decreasing order in terms of their size and then place the items one at a time, placing each item in the lowest-numbered bin in which it can fit; that is, we place the item in bin b_k where k is the smallest number where the item can be placed given the placement of the preceding items. The goal is to minimize the number of nonempty bins, that is, the number of bins actually used. Intuitively, placing items in decreasing order is a good approach, as larger items are less easily placed, and first fit is a natural process for placing items. (Best fit, or placing each item in the bin with the least amount of free space where it can fit, is also possible.) As a more complex example of the same type, in the strip-packing problem, our items are rectangular boxes, and the goal is to fit the boxes into a larger rectangular box of fixed width W and minimum height. Again, a natural approach is to sort the items in decreasing order of area and to place items according to the bottom-left heuristic: place each box as far down vertically and then as much to the left as it can be placed. In some cases, these greedy heuristics can be shown to guarantee an approximation ratio; even in the worst case, they are guaranteed to be within some factor of the optimal solution.

The downside of using a greedy heuristic in this way is that it provides a single solution, but that is all. If time is available, one might hope to continue computing to find an improved solution. Obviously, one could adopt an entirely different approach, developing an algorithm based on techniques such as simulating annealing or genetic algorithms. But here we are focused on simplicity: can we better use the greedy algorithm we have in hand?

If the placement function is coded so that it takes the list of sorted elements as input, we can obtain additional possible solutions by simply providing the elements in a different order. But what order should we choose? Given that this chapter is based on randomized algorithms, one might be tempted to suggest we try orderings of elements chosen uniformly at random. But that seems like a poor idea; our heuristic started with an intuitively good ordering, and throwing that intuition away by using a completely random ordering is unlikely to be helpful.

Instead, we want a random ordering that is near our good ordering. There are multiple approaches that can be used to obtain such orderings. One natural approach is to repeatedly choose an element uniformly at random from the top k remaining elements for some parameter k. If the elements were sorted, this gives an ordering that is "close to" sorted, keeping the intuition intact but offering the flexibility of many more orderings [24,54]. An arguably more theoretically appealing approach is to repeatedly invoke the following process: take a coin that comes up heads with probability p, where p is a chosen parameter. Flip it, and if it comes up heads, take the next element in the sorted order for your permutation; otherwise, go on to the next element and continue [37]. (One can cycle back to the beginning if needed.) At each step, this is equivalent to choosing the ith element in the remaining ordered elements with probability proportional to $(1 - p)^i$. (The process can be programmed more efficiently in various ways.)

One advantage of this approach is that all orderings are possible to achieve. Indeed, more importantly, the probability an ordering is chosen is proportional to $(1 - p)^D$ where D is the Kendall tau distance, or the number of transpositions performed in a bubble sort, between the chosen ordering and the true sorted ordering. The original greedy ordering is the most likely, but it's exponentially unlikely to be chosen. Instead, the outcome will almost surely be a sequence that is "mostly ordered."

There are multiple variations on this theme. Many greedy algorithms change the order at each step. For example, the standard greedy algorithm for set cover at each step chooses to add to the cover the set that covers the most uncovered remaining elements. Randomized greedy algorithms can work here as well, although one must recompute the ordering at each step. Also, one need not keep the original greedy ordering as the baseline. A natural approach is to take the best ordering found thus far and use that as the ordering to randomize from. Finally, more can be done by adding local search as well.

A randomized greedy approach can be used to develop a variety of initial solutions, from which local search can be applied to look for locally optimal solutions. This general type of metaheuristic algorithm is known as a greedy randomized adaptive search procedure, or GRASP [24,54].

Improvements from randomized greedy schemes are highly problem and instance dependent. Results for several problems are given in [37], where the approach yields improvements of several percent after several hundred orderings and further improvements up through thousands of orderings. While gains for some problems over the basic greedy algorithm might be minimal, a key point in favor of randomized greedy approaches is the simplicity with which it can be implemented on top of the greedy algorithm. One loses very little in programming time for potentially interesting gains.

There is essentially nothing known that is theoretically provable about randomized greedy algorithms of this form. A better understanding of deterministic greedy algorithms of this form, which have been dubbed priority algorithms, was initiated by [6], but little is known about randomized variants.

Of course, the basic concept of doing multiple repeated randomized attempts in order to improve an algorithm has appeared in other settings. For example, one other potentially useful practical application arises in the randomized rounding of linear programs. In many cases, one can turn an optimization problem into an integer linear program, relax the integer constraint to solve the corresponding linear program, and then round the results of the linear program to obtain an integer program with the hope of obtaining a solution that is close to the optimal integer solution. In cases where an integer solution is not obtained, randomized rounding can be used: for example, a variable x with value 0.3 in the linear program solution can be rounded to 1 with probability 0.3 and 0 with probability 0.7, and similarly all variables can be independently randomly rounded. Repeated randomized rounding offers the opportunity to obtain multiple solutions, all governed by the same intuition of being "near" the solution from the linear program.

Acknowledgment

Mitzenmacher's work was supported by NSF grants CNS-1228598, IIS-0964473, and CCF-0915922. Upfal's work was supported by NSF grants IIS-0905553 and IIS-1247581.

References

1. RSA Labs. How large a key should be used in the RSA cryptosystem? http://www.rsa.com/rsalabs/node.asp?id=2218 (accessed on October 7, 2013).
2. M. Agrawal, N. Kayal, and N. Saxena. PRIMES is in P. *Annals of Mathematics*, 160(2):781–793, 2004.
3. D.A. Alcantara, A. Sharf, F. Abbasinejad, S. Sengupta, M. Mitzenmacher, J.D. Owens, and N. Amenta. Real-time parallel hashing on the GPU. *ACM Transactions on Graphics (TOG)*, 28(5):154, 2009.
4. Y. Azar, A.Z. Broder, A.R. Karlin, and E. Upfal. Balanced allocations. *SIAM Journal on Computing*, 29:180, 1999.
5. B.H. Bloom. Space/time trade-offs in hash coding with allowable errors. *Communications of the ACM*, 13(7):422–426, 1970.
6. A. Borodin, M.N. Nielsen, and C. Rackoff. (Incremental) priority algorithms. *Algorithmica*, 37(4):295–326, 2003.
7. A. Broder and M. Mitzenmacher. Using multiple hash functions to improve IP lookups. In *Proceedings of INFOCOM 2001*, vol. 3, pp. 1454–1463, IEEE, Shanghai, China, 2001.
8. A. Broder and M. Mitzenmacher. Network applications of Bloom filters: A survey. *Internet Mathematics*, 1(4):485–509, 2004.
9. A.Z. Broder, S.C. Glassman, M.S. Manasse, and G. Zweig. Syntactic clustering of the web. In *Proceedings of the World Wide Web Conference*, pp. 391–404, Santa Clara, CA, 1997.

10. J.L. Carter and M.N. Wegman. Universal classes of hash functions. *Journal of Computer and System Sciences*, 18(2):143–154, 1979.
11. C. Chen, J. Yeh, and H. Ke. Plagiarism detection using ROUGE and WordNet. *Journal of Computing*, 2(3), March 2010, ISSN 2151-9617. https://sites.google.com/site/journalofcomputing/ 2010.
12. G. Cormode and S. Muthukrishnan. An improved data stream summary: The count-min sketch and its applications. *Journal of Algorithms*, 55(1):58–75, 2005.
13. B. Debnath, S. Sengupta, and J. Li. Chunkstash: Speeding up inline storage deduplication using flash memory. In *Proceedings of the 2010 USENIX Annual Technical Conference*, p. 16, USENIX Association, Boston, MA, 2010.
14. S. Dharmapurikar, P. Krishnamurthy, T. Sproull, and J. Lockwood. Deep packet inspection using parallel Bloom filters. In *Proceedings of the 11th Symposium on High Performance Interconnects*, pp. 44–51, IEEE, Stanford, CA, 2003.
15. S. Dharmapurikar, P. Krishnamurthy, and D.E. Taylor. Longest prefix matching using Bloom filters. *IEEE/ACM Transactions on Networking*, 14(2):397–409, 2006.
16. M. Dietzfelbinger, A. Goerdt, M. Mitzenmacher, A. Montanari, R. Pagh, and M. Rink. Tight thresholds for cuckoo hashing via XORSAT. *Proceedings of the 37th International Colloquium on Automata, Languages and Programming*, 213–225, Springer-VerlagBerlin, Heidelberg, Germany, 2010.
17. M. Dietzfelbinger and C. Weidling. Balanced allocation and dictionaries with tightly packed constant size bins. *Theoretical Computer Science*, 380(1):47–68, 2007.
18. W. Diffie and M.E. Hellman. New directions in cryptography. *IEEE Transactions on Information Theory*, IT-22:644–654, 1976.
19. P. Domingos and M. Pazzani. On the optimality of the simple Bayesian classifier under zero-one loss. *Machine Learning*, 29(2–3):103, November 1997.
20. P. Domingos and M. Pazzani. Beyond independence: Conditions for the optimality of the simple Bayesian classifier. In L. Saitta, ed., *International Conference on Machine Learning*, pp. 105–112, Bari, Italy, Morgan Kaufmann, 1996.
21. D. Dubhashi and A. Panconesi. *Concentration of Measure for the Analysis of Randomized Algorithms*, vol. 19, Cambridge University Press, U.K., 2009.
22. C. Estan and G. Varghese. New directions in traffic measurement and accounting: Focusing on the elephants, ignoring the mice. *ACM Transactions on Computer Systems (TOCS)*, 21(3):270–313, 2003.
23. L. Fan, P. Cao, J. Almeida, and A.Z. Broder. Summary cache: A scalable wide-area web cache sharing protocol. *ACM SIGCOMM Computer Communication Review*, 28(4):254–265, 1998.
24. T.A. Feo and M.G.C. Resende. Greedy randomized adaptive search procedures. *Journal of Global Optimization*, 6(2):109–133, 1995.
25. M. Ferdman, P. Lotfi-Kamran, K. Balet, and B. Falsafi. Cuckoo directory: A scalable directory for many- core systems. In *Proceedings of the 17th Annual International Symposium on High Performance Computer Architecture (HPCA)*, pp. 169–180, IEEE, San Antonio, TX, 2011.
26. D. Fotakis, R. Pagh, P. Sanders, and P. Spirakis. Space efficient hash tables with worst case constant access time. *Theory of Computing Systems*, 38(2):229–248, 2005.
27. N. Fountoulakis and K. Panagiotou. Orientability of random hypergraphs and the power of multiple choices. *Proceedings of the 37th International Colloquium on Automata, Languages and Programming*, 348–359, Springer-VerlagBerlin, Heidelberg, Germany, 2010.
28. A. Frieze and P. Melsted. Maximum matchings in random bipartite graphs and the space utilization of cuckoo hash tables. *Random Structures and Algorithms*. 41(3), 334–364, John Wiley & Sons, Inc. New York, 2012.
29. G.H. Gonnet and R.A. Baeza-Yates. An analysis of the Karp–Rabin string matching algorithm. *Information Processing Letters*, 34:271–274, 1990.

30. G.H. Gonnet. Expected length of the longest probe sequence in hash code searching. *Journal of the ACM (JACM)*, 28(2):289–304, 1981.
31. P. Jaccard. Etude comparative de la distribution florale dans une portion des alpes et du jura. *Bulletin de la Socit Vaudoise des Sciences Naturelles*, 37(9):547–549, 1901.
32. J. Jonsson and B. Kaliski. Public-key cryptography standards (PKCS)# 1: RSA cryptography specifications version 2.1. Technical report, RFC 3447, February, 2003. http://tools.ietf.org/html/rfc3447
33. R. Karp and M. Rabin. Efficient randomized pattern-matching algorithms. *IBM Journal of Research and Development*, 31:249–260, 1987.
34. A. Kirsch, M. Mitzenmacher, and G. Varghese. Hash-based techniques for high-speed packet processing. In G. Cormode and M. Thottan, eds., *Algorithms for Next Generation Networks*, pp. 181–218, Springer, London, U.K., 2010.
35. A. Kirsch, M. Mitzenmacher, and U. Wieder. More robust hashing: Cuckoo hashing with a stash. In *Proceedings of the European Symposium on Algorithms*, pp. 611–622, Universität Karlsruhe, Germany, 2008.
36. A.K. Lenstra, J. Hughes, M. Augier, J. Bos, T. Kleinjung, and C. Wachter. Ron was wrong, Whit is right. *IACR Cryptology*, 2012:64, 2012.
37. N. Lesh and M. Mitzenmacher. Bubblesearch: A simple heuristic for improving priority-based greedy algorithms. *Information Processing Letters*, 97(4):161–169, 2006.
38. C. Manning, P. Raghavan, and H. Schutze. *Introduction to Information Retrieval*, Cambridge University Press, U.K., 2008.
39. M.E. Maron. Automatic indexing: An experimental inquiry. *Journal of the ACM*, 8(3):404–417, July 1961.
40. T. Mitchell. *Machine Learning*, McGraw Hill, New York, 1997.
41. M. Mitzenmacher. Studying balanced allocations with differential equations. *Combinatorics, Probability and Computing*, 8(05):473–482, 1999.
42. M. Mitzenmacher, A. Richa, and R.K. Sitaraman. The power of two random choices: A survey of the techniques and results. In P. Pardalos, S. Rajasekaran, and J. Rolim, eds., *Handbook of Randomized Computing*, vol. 1, pp. 255–305, Kluwer Press, Boston, MA, July 2001.
43. M. Mitzenmacher and E. Upfal. *Probability and Computing: Randomized Algorithms and Probabilistic Analysis*, Cambridge University Press, U.K., 2005.
44. M. Mitzenmacher and S. Vadhan. Why simple hash functions work: Exploiting the entropy in a data stream. In *Proceedings of the Nineteenth Annual ACM-SIAM Symposium on Discrete Algorithms*, pp. 746–755, San Francisco, CA, Society for Industrial and Applied Mathematics, 2008.
45. M. Mitzenmacher and B. Vöcking. The asymptotics of selecting the shortest of two, improved. In *Proceedings of the 37th Allerton Conference on Communication, Control, and Computing*, Monticello, IL, 1999.
46. C. Moore and S. Mertens. *The Nature of Computation*, Oxford University Press, Oxford, U.K., 2011.
47. R. Motwani and P. Raghavan. *Randomized Algorithms*, Cambridge University Press, U.K., 1995.
48. D.J. Newman. Simple analytic proof of the prime number theorem. *The American Mathematical Monthly*, 87(9):693–696, 1980.
49. A. Ng and M. Jordan. On discriminative vs. generative classifiers: A comparison of logistic regression and naive Bayes. In T. G. Dietterich, S. Becker, and Z. Ghahramani, eds., *Advances in Neural Information Processing System*, pp. 841–848, MIT Press, San Mateo, CA, 2001.
50. R. Pagh and F.F. Rodler. Cuckoo hashing. *Journal of Algorithms*, 51(2):122–144, 2004.
51. M. Pătraşcu and M. Thorup. The power of simple tabulation hashing. http://arxiv.org/abs/1101.5200 (accessed on October 7, 2013).
52. M. Pătraşcu and M. Thorup. The power of simple tabulation hashing. In *Proceedings of the 43rd ACM Symposium on Theory of Computing (STOC)*, pp. 1–10, 2011.
53. M. Rabin. Probabilistic algorithm for testing primality. *Journal of Number Theory*, 12(1):128–138, 1980.

54. M. Resende and C. Ribeiro. Greedy randomized adaptive search procedures. *Handbook of Metaheuristics*, pp. 219–249, Springer, 2003.
55. R. Rivest, A. Shamir, and L. Adleman. A method for obtaining digital signatures and public-key cryptosystems. *Communications of the ACM*, 21:120–126, 1978.
56. A. Rousskov and D. Wessels. Cache digests. *Computer Networks and ISDN Systems*, 30(22–23):2155–2168, 1998.
57. L. Salmela, J. Tarhio, and J. Kytöjoki. Multipattern string matching with q-grams. *Journal of Experimental Algorithmics*, 11, 1–19, February 2007.
58. D. Sanchez and C. Kozyrakis. The Z-Cache: Decoupling ways and associativity. In *43rd Annual IEEE/ACM International Symposium on Microarchitecture*, pp. 187–198, IEEE, Los Alamitos, CA, 2010.
59. V. Shoup. *A Computational Introduction to Number Theory and Algebra*. Cambridge University Press, Cambridge, U.K., 2009.
60. A. Sinclair. *Algorithms for Random Generation and Counting: A Markov Chain Approach*, vol. 7, Birkhauser, Boston, MA, 1993.
61. H. Song, S. Dharmapurikar, J. Turner, and J. Lockwood. Fast hash table lookup using extended Bloom filter: An aid to network processing. *ACM SIGCOMM Computer Communication Review*, 35(4):181–192, 2005.
62. M. Thorup. Even strongly universal hashing is pretty fast. In *Proceedings of the 11th ACM-SIAM Symposium on Discrete Algorithms*, pp. 496–497, San Francisco, CA, 2000.
63. M. Thorup and Y. Zhang. Tabulation based 4-universal hashing with applications to second moment estimation. In *Proceedings of the 15th ACM-SIAM Symposium on Discrete Algorithms*, pp. 615–624, New Orleans, LA, 2004.
64. B. Vöcking. How asymmetry helps load balancing. *Journal of the ACM*, 50(4):568–589, 2003.
65. G. Webb, J. Boughton, and Z. Wang. Not so naive Bayes: Aggregating one-dependence estimators. *Machine Learning*, 58(1):5–24, 2005.
66. D.P. Williamson and D.B. Shmoys. *The Design of Approximation Algorithms*. Cambridge University Press, Cambridge, U.K., 2011.
67. A.L. Zobrist. A new hashing method with application for game playing. *ICCA Journal*, 13(2):69–73, 1970.

12

Approximation Algorithms and Metaheuristics

Teofilo F. Gonzalez
University of California, Santa Barbara

12.1 Underlying Principles

12.1.1 Introduction

The concept of approximation has been around for centuries, and it is used in all quantitative-based research areas. Examples include approximation of numbers (e.g., *pi*, *e*, $\sqrt{2}$, etc.) as well as mathematical functions, physical laws, geometrical objects, etc. Modeling of physical phenomena is by approximating its behavior. Problems in many areas are very difficult to solve analytically, and sometimes they are impossible to solve with the current knowledge, skills, and tools. Therefore, one needs to resort to approximations. In this chapter, we concentrate on "relatively" new research areas for approximations, which are called *approximation algorithms* and *metaheuristics*.

Even before the 1960s, researchers in applied mathematics and graph theory had established upper and lower bounds for certain properties of graphs. For example, bounds had been established for the chromatic number, achromatic number, chromatic index, maximum clique, maximum independent set (MIS), etc. Some of these results were implementable and could be transformed into efficient algorithms that may be seen as the precursors of approximation algorithms. By the 1960s, it was understood that there were problems that computationally could be solved efficiently, whereas for other problems all the known algorithms required exponential time with respect to the input size. Heuristics were being developed to find quick solutions to problems that appeared to be computationally difficult to solve. Researchers were experimenting with heuristics, branch-and-bound procedures, and iterative improvement frameworks to evaluate their performance when solving actual problem instances. There were many claims being made, not all of which could be substantiated, about the performance of the procedures being developed to generate optimal and suboptimal solutions to combinatorial optimization problems.

Approximation algorithms, as we know them now, were formally introduced in the 1960s by Graham [1] and their most important feature is that they guarantee near-optimal solutions to *all* instances of the optimization problem being solved. These optimization problems could not be solved to optimality by the computing machines and algorithmic techniques available at that time. In the past 50 years there have been exponential improvements in the computer devices available; computing machines are orders of magnitude faster, can store significantly larger amount of data, and their cost and power consumption has decreased significantly. Also, in the past 50 years there has been tremendous growth of research in algorithmic techniques. These developments seem to imply that problems are nowadays more likely to be solved optimally than before, and that the need for approximation algorithms should have vanished, or at least decreased significantly by now. But this is not the case! There are two main reasons why solving a large class of problems to optimality is currently impossible and will remain in this state for the near future. First the size of the problem instances we are trying to solve has grown excessively over the past 50 years. New applications require the solution of problems with massive amounts of data, for example, DNA sequencing, VLSI design automation, weather prediction, Internet searching, etc. Second, most optimization problems that arise in practice are computationally intractable. This means that there are no known efficient algorithms for their solution and it has been conjectured that no such algorithms exist.

In the early 1970s, computer scientists developed the theory of NP-completeness (see Chapter 7) and showed that an extremely large number of decision problems fall into this class of computationally intractable problems. This theory tells us that the problems in this class are either all solvable via algorithms that take a polynomial number of steps for their solutions or none of them can. It is conjectured that all of the problems in this class do not have polynomial time algorithms for their solution. In practical terms this implies that even for relatively small problem instances, all algorithms for their solution take in the worst case exponential time with respect to the instance (or input) size. The fastest algorithms to solve NP-Complete problems take time $O(2^{n/2})$, $O(2^n)$, $O(n!)$, etc., where n is the number of inputs (see Sahni [3] for the definition of the Big-Oh notation). For most NP-Complete problems their best known algorithms take $O(2^n)$ time. Increasing the input size by one, for an algorithm that takes 2^n time, results in the doubling of the time needed to solve the problem instance. Even when n is 50 or 100, it would take, in the worst case, years for the fastest computer available today to solve the problem instance with this type of algorithms! For n in the millions, the time required to solve problems would be prohibitively large for most instances of the problem, not just the problem instances for which algorithms take the longest time to solve. This will remain unchanged even if new technologies were developed that would allow us to construct considerably faster computing machines. On the other hand, suppose that all NP-Complete problems can be solved in polynomial time. There is no guarantee that these polynomial time algorithms are low order polynomial time algorithms with small constants. In other words, these algorithms might not be practical. As we said before, a huge set of optimization problem that arise in practice have the property that their corresponding decision problems are NP-Complete problems. These optimization problems are called *NP-Hard* problems because computationally they are at least as hard to solve as the NP-Complete problems. But problem instances need to be solved every day when flights have to scheduled, messages have to be routed through the Internet, computer code needs to be optimized, deadlocks in an operating system have to detected, investment portfolios have to optimized, energy consumption needs to be minimized, etc. In our everyday world we need to be able to cope with computational intractability.

We cope with NP-Completeness through restriction, randomization, and approximations. For example, one of the most basic NP-Complete problem is called *3-SAT* (satisfiability of a Boolean expression in which all clauses have three literals, see Chapter 7). Formal verification for program correctness as well as other properties of computer programs can be reduced to 3-SAT problem instances. Most of these instances can be solved quickly. However this does not necessarily contradict the theory of NP-Completeness because these restricted instances of 3-SAT might fall into a class of instances that can be solved in polynomial time. So one way to cope with intractability is by characterizing the instances that arise in practice (restriction) and then developing efficient algorithms to solve such instances.

Another way to cope with intractability is via randomization (see Chapter 11). The idea is to design algorithms that make choices by flipping digital coins. For some problems such algorithms have been shown to generate near-optimal solutions with high probability. The third approach to cope with intractability is via approximation algorithms and metaheuristics, the main topic of this chapter.

One justification for approximations is based on computer intractability, but one can go even further and argue that one needs to apply approximation techniques even when the problems being solved are computationally tractable. The reason for this is that the input size may be very large or simply because we need to solve thousands of instances each fraction of a second. Also, in some applications the input has noise, so there is no need to solve a problem instance to optimality as the instance does not represent exactly the actual instance we want to solve. There are problems that are harder to solve than NP-Complete problems, so approximations for these problems are also required. In fact the situation is even worse, as there is a class of decision problems that are undecidable, that is, there is no algorithm to solve these problems. Chapter 7 discusses P, NP, NP-Completeness, and other complexity classes in more detail.

Our previous discussion gives the impression that approximation algorithms provide a solution to all of our problems, but that is not really the case. As it was established in the 1970s, for some problems one can generate near optimal solutions quickly, while for other problems generating provably good suboptimal solutions is computationally as difficult as generating optimal ones! In other words, generating provably good suboptimal solutions to some problems is as hard as generating optimal solutions to these problems (see [2]). To deal with the inapproximable problems there were a few techniques introduced in the 1980s and 1990s. These methodologies have been referred to as metaheuristics and include simulated annealing (SA), ant colony optimization (ACO), evolutionary computation (EC), tabu search (TS), memetic algorithms (MA), etc. There has been a tremendous amount of research in metaheuristics during the past two decades. Other previously established methodologies like local search, backtracking, and branch-and-bound have also been explored in the context of approximations. These techniques have been evaluated experimentally and have demonstrated their usefulness for solving large problem instances that arise in real world applications.

During the last two decades approximation algorithms have attracted considerable attention. This was a result of a stronger inapproximability methodology that could be applied to a wider range of problems and the development of new approximation algorithms for problems arising in established as well as in emerging application areas. Today approximation algorithms enjoy a stature comparable to that of algorithms in general and the area of metaheuristics has established itself as a fundamental area. The new stature is a byproduct of a natural expansion of research into more practical areas where solutions to real world problems are expected, as well as by the higher level of sophistication required to design and analyze these new procedures. The goal of approximation algorithms and metaheuristics is to provide the best possible solutions with the resources available and to guarantee that such solutions satisfy certain important properties. The research area of approximation algorithms and metaheuristics is huge, and we only have a few pages to talk about it.

12.1.2 Criteria to Judge Approximations

One can use many different criteria to judge approximation algorithms and metaheuristics. To name a few, we could use the quality of solutions generated, and the time and space complexity needed to generate it. One may measure the criteria in different ways, for example, we could use the worst case, average case, median case, etc. The evaluation could be analytic or experimental. Additional criteria include: characterization of data sets where the algorithm performs very well or very poorly; comparison with other algorithms using benchmarks or data sets arising in practice; tightness of bounds (bounds for quality of solution, time and space complexity); the value of the constants associated with the time complexity bound including the ones for the lower order terms; and so on. For some researchers the most important aspect of an approximation algorithm is that it is complex to analyze, but for others it is more important that the algorithm be complex and involve the use of sophisticated data structures.

For researchers working on problems directly applicable to the "real world," experimental evaluation or evaluation on benchmarks is a more important criterion. Clearly, there is a wide variety of criteria one can use to evaluate approximation algorithms.

For any given optimization problem P, let $A_1, A_2, \ldots$ be the set of current algorithms that generate a feasible solution for each instance of problem P. Suppose that we select a set of performance criteria C and measure each one in a way we feel it is most important. How can we decide which algorithm is best for problem P with respect to C? For example the criteria could be time and space complexity and in both cases we measure the worst case (asymptotic). We may visualize every algorithm as a point (or region) in multidimensional space. Now the basic approach used to compare feasible solutions for multiobjective function problems can also be used in this case to label some of the algorithms as *current Pareto optimal* with respect to C. Algorithm A is said to be *dominated* by algorithm B with respect to C, if for each criteria $c \in C$ algorithm B is "not worse" than A, and for at least one criteria $c \in C$ algorithm B is "better" than A. An algorithm A is said to be a *current Pareto optimal* algorithm with respect to C if none of the current algorithms dominates it. Later on in this chapter we apply this metric to specific algorithms.

For some problems it is very hard to judge the quality of the solution generated. For example, color quantization, which essentially means to approximate colors, can only be judged by viewing the resulting images and that is very subjective. Problems arising in practice (e.g., in the areas of bioinformatics, VLSI CAD, etc.) have multiobjective optimization functions, and therefore it is not simple to judge the quality of the solutions generated.

The main criterion used to compare approximation algorithms has been the quality of the solution generated. An important secondary criterion is the worst case time complexity. We are mainly interested in the worst case scenario for the solutions generated by our algorithms. There are many justification for concentrating on the worst case behavior. The book entitled "Anitfragile," by N. N. Taleb (Random House, 2012) mentions an interesting one relating to underestimation of events. He identified this issue as "... the Lucretius problem, after the Latin poetic philosopher who wrote that the fool believes that the tallest mountain in the world will be equal to the tallest one he has observed."

12.1.3 Approximation Algorithms

The basic approximation methodologies are restriction (solution space and algorithmic), relaxation, rounding, partial brute force, randomization, and combinations of these methodologies. In what follows we discuss these techniques as well as formal notation, inapproximability and approximation classes.

12.1.3.1 Restriction (Solution Space)

The idea is to generate a solution to a given problem P by providing an optimal or suboptimal solution to a subproblem of P. A subproblem of a problem P means restricting the solution space for P by disallowing a subset of the feasible solutions. The idea is to restrict the solution space so that the resulting problem has some structure which can be exploited by an efficient algorithm that solves the problem optimally or suboptimally. For this approach to be effective, the subproblem must have the property that, for every problem instance, its optimal or suboptimal solution has an objective function value that is "close" to the optimal one for the original problem instance. The most common approach is to solve just one subproblem, but there are algorithms where more than one type of subproblems are solved and then the best of the solutions computed is the solution generated. In what follows we apply this technique to a couple of classical problems.

12.1.3.1.1 Application to Traveling Salesperson and Steiner Tree Problems

The traveling salesperson, Steiner tree, and spanning tree problems have a large number of applications. The *Steiner tree* problem consists of a graph $G = (V, E, W)$, where V is the set of vertices, T is a subset of V $(T \subseteq V)$, E is the set of edges, and W is a function that assigns a positive value to each edge in such a way that for every pair of vertices a and b the weight of the edge $\{a, b\}$ is at most the weight of any path from a to b. The problem is to find a tree that includes all the vertices in T plus some other vertices in the graph such that the sum of the weight of the edges in the tree is least possible. When $T = V$ the problem is called the *minimum weight (or cost) spanning tree* problem. The *traveling salesperson problem*, called the traveling salesman problem before the 1970s, is given an edge weighted complete graph, find a minimum weight tour that starts and ends at vertex one and visits every vertex *exactly* once. The weight of a tour is the sum of the weight of the edges in the tour.

Even before the 1960s, there were several well known polynomial time algorithms to construct minimum weight spanning trees for edge weighted graphs [3]. These simple greedy algorithms have low order polynomial time complexity bounds. It was known at that time that the same type of procedures would not generate an optimal tour for the traveling salesperson problem and would not construct optimal Steiner trees. However, in 1968 Moore [4] showed that for any set of points P in metric space $L_M \le L_T \le 2L_S$, where L_M, L_T, and L_S are the weights of a minimum weight spanning tree, a minimum weight tour (solution) for the traveling salesperson problem (TSP) and minimum weight Steiner tree for P, respectively.* Since every spanning tree for P is a Steiner tree for P and $L_M \le L_T \le 2L_S$ we know that a minimum weight spanning tree provides a solution to the Steiner tree problem whose total weight is at most twice the weight of an optimal Steiner tree.

In other words, any algorithm that generates a minimum weight spanning tree for the set of vertices T is a 2-approximation algorithm for the Steiner tree problem. This is called the *approximation ratio* or *approximation factor* for the algorithm. This criterion for measuring the quality of the solutions generated by an algorithm remains the most important one in use today. The approximation technique used in this case is restriction of the solution space. We are restricting the solution space by just allowing Steiner trees without any Steiner (additional) points. In fact we are constructing the best tree without any Steiner points, which is simply a minimum weight spanning tree for T.

The time complexity for our approximation method is the time complexity for solving the minimum weight spanning tree problem [3]. The bounds $L_M \le L_T \le 2L_S$ are established by first defining a simple transformation from any minimum weight Steiner tree into a TSP tour in such a way that $L_T \le 2L_S$ [4]. Then by observing that the deletion of an edge in an optimum tour for the TSP problem results in a spanning tree, it follows that $L_M \le L_T$. The *Steiner ratio* is defined as L_S/L_M. The above arguments show that the Steiner ratio is at least 1/2. Gilbert and Pollak [4] conjectured that the Steiner ratio in the Euclidean plane equals $\sqrt{3}/2$ (the .86603 … conjecture). The proof of this conjecture and improved approximation algorithms for the Steiner tree problem are discussed in [5].

The above approach can also be used to establish a 2-approximation algorithm for the TSP problem for metric graphs as follows. First, construct a minimum weight spanning tree. Then the graph (actually multigraph) with all the vertices in G and two copies of each edge in the minimum weight spanning tree has always an Euler circuit. An *Euler circuit* is a path that starts and ends at the same vertex and visits all the edges in the graph exactly once. This is a relaxed tour that visits every city at least once. Since graph G is metric, this relaxed tour can be easily transformed into one that visits every city at most once without increasing the tour length. From the above analysis we can show that the cost of the tour constructed is at most twice the cost of an optimal tour. The algorithm takes polynomial time and results in a 2-approximation algorithm for the TSP problem. This approximation algorithm for the TSP is also referred to as the *double spanning tree algorithm*. The approximation technique behind this algorithm is also based on restriction as initially we are finding a relaxed tour the uses the least number of different edges, though each of those edges is traversed twice.

The best approximation algorithm known for the TSP defined over metric graphs is the one by Christofides [6]. The approximation ratio for this algorithm is 3/2, which is smaller than the approximation ratio of 2 for the above algorithm initially reported in [4,7]. However, looking at the bigger picture that includes the time complexity of the approximation algorithm, Christofides algorithm is not of the

* Note that for the MWST and the TSP problem the only "edges" allowed are those between any pair of points in P, but for the Steiner tree problem any point in the metric space may be a vertex in the Steiner tree.

same order as the ones given in [4,7]. Therefore, neither of these two approximation algorithms dominates the other as one has a smaller time complexity bound, whereas the other (Christofides algorithm) has a smaller worst case approximation ratio. Both of these algorithms are said to be Pareto optimal with respect to the worst case approximation ratio and worst case time complexity.

A closely related approach to restriction is *transformation-restriction*. The idea is to transform the problem instance to a restricted instance of the same problem. The difference is that the restricted problem instance is not a sub-problem of original problem instance as in the case of restriction, but it is a "simpler" problem of the same type. This approach has been applied to the problems of routing multiterminal nets and embedding hyperedges in a cycle.

12.1.3.2 Restriction (Algorithmic)

Algorithmic restriction (or restriction via a procedure) is a technique similar to restricting the solution space. The restriction is limiting the solutions to those generated by an algorithm or a class of algorithms. Even though this restricts the solution space, in many situations it is difficult to characterize the restricted solution space. This is why we call this technique *algorithmic restriction*. Greedy methods fall into this category. Greedy algorithms generate a solution by making a sequence of irrevocable decisions. Each of these decisions is a best possible choice at that point, for example, select an edge of least weight, select the vertex of highest degree, or select the task with largest processing time requirement.

12.1.3.2.1 Application to Scheduling on Identical Machines

Let us now apply this methodology to a couple of scheduling problems. The first problem is scheduling a set of tasks on a set of identical machines (or processors) to minimize the makespan. The problem is given a set of n tasks denoted by $T_1, T_2, \ldots, T_n$ with processing time requirements $t_1, t_2, \ldots, t_n$ to be processed by a set of m identical machines. A partial order C is defined over the set of tasks to enforce a set of precedence constraints, or task dependencies. The partial order specifies that a machine cannot commence the processing of a task until all of its predecessors have been completed. Each task T_i has to be processed for t_i units of time by one of the machines. A (nonpreemptive) schedule is an assignment of tasks to time intervals on the machines in such a way that (1) each task is processed continuously for t_i units of time by one of the machines; (2) each machine processes at most one task at a time; and (3) the precedence constraints are satisfied. The *makespan* of a schedule is the latest time at which a task is being processed. The problem is to construct a minimum makespan schedule for any given set of partially ordered tasks to be processed by a set of identical machines.

Example 12.1

The number of tasks, n, is 8 and the number of machines, m, is 3. The processing time requirements for the tasks and the precedence constraints are given in Figure 12.1 where a directed graph is used to represent the task dependencies. Vertices represent tasks and the directed edges represent task dependencies. The integers next to the vertices represent the task processing requirements. Figure 12.2 depicts two schedules for this problem instance.

The *list scheduling* procedure constructs a schedule given an ordering of the tasks specified by a priority list L, which may be any permutation of the set of tasks. The list scheduling procedure finds the earliest time t when a machine is idle and an unassigned task ready to be processed is *available* (i.e., all its predecessors have completed). It then assigns the leftmost available task in the list L to an idle machine at time t and this step is repeated until all the tasks are scheduled.

For example, tasks may be ordered in decreasing (actually nonincreasing) order of their processing times. In this case, it is a greedy method with respect to the processing time requirements. The list could

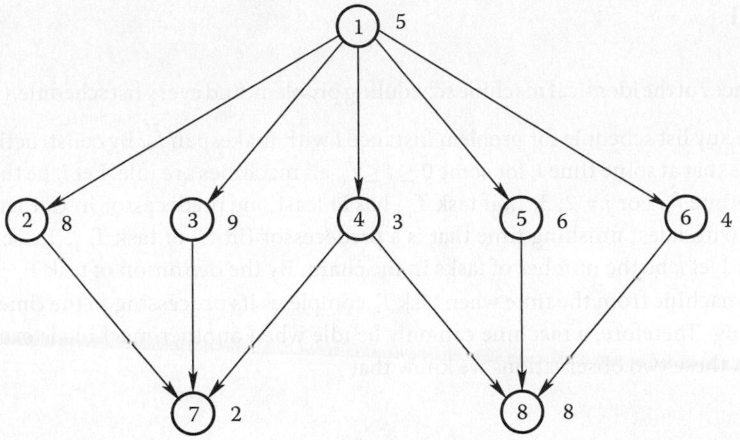

FIGURE 12.1 Precedence constraints and processing time requirements for Example 12.1. (From Gonzalez, T.F., *Handbook on Approximation Algorithms and Metaheuristics*, Chapman & Hall / CRC, Boca Raton, FL, 2007.)

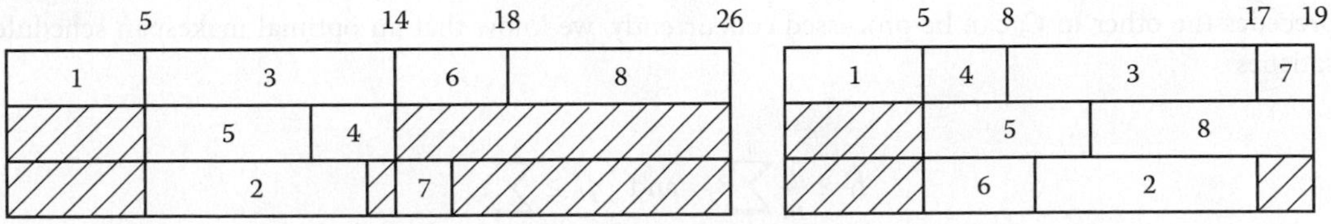

(a) (b)

FIGURE 12.2 (a) List schedules. Schedule (b) is an optimal schedule for Example 12.1. (From Gonzalez, T.F., Handbook on Approximation Algorithms and Metaheuristics, Chapman & Hall/CRC, Boca Raton, FL, 2007.)

be any ordering of the tasks. So this is algorithmic restriction. In this case, one can characterize the solution space. The solution space is restricted to schedules without forced "idle time," that is, each feasible schedule is such that for every task *i* there is no idle time from the time at which all the predecessors of task *i* (in C) are completed to the time when the processing of task *i* begins. This is clearly a restriction of the solution space. Is the value of the optimal solution of the original problem equal to the one of the restricted problem? Figure 12.2 suggests that an optimal schedule can be generated by just finding a clever method to break ties, that is, by finding an optimal permutation of the tasks. Unfortunately one cannot prove that this is always the case because there are problem instances for which all optimal schedules for the original problem are not list schedules, that is, they all have forced idle time at some point! That is, all optimal schedules have idle time at some point even though at that point there is a task that is available (all its predecessors have been completed). Our technique is to generate one schedule without any forced idle time.

The main result in [1] is showing that for every problem instance, *I*, the schedule generated by this policy has a makespan that is bounded above by (2 – 1/*m*) times the optimal makespan for the instance. This is expressed by

$$\frac{\hat{f}_I}{f_I^*} \leq 2 - \frac{1}{m}$$

where $\hat{f}_I$ is the makespan of any possible list schedule for problem instance *I*, and f_I^* is the makespan of an optimal schedule for *I*.

This property is established in the following theorem.

Theorem 12.1

For every instance I of the identical machine scheduling problem, and every list schedule, $(\hat{f}_I/f_I^*) \le 2-(1/m)$

Proof: Let S be any list schedule for problem instance I with makespan $\hat{f}_I$. By construction of list schedules it cannot be that at some time t, for some $0 \le t \le \hat{f}_I$, all machines are idle. Let i_1 be the index of a task that finishes at time $\hat{f}_I$. For $j = 2, 3,\ldots$, if task $T_{i_{j-1}}$ has at least one predecessor in C, then define i_j as the index of a task with latest finishing time that is a predecessor (in C) of task $T_{i_{j-1}}$. These tasks identified form a *chain* and let k be the number of tasks in the chain. By the definition of task T_{i_j}, it cannot be that there is an idle machine from the time when task T_{i_j} completes its processing to the time when task $T_{i_{j-1}}$ begins processing. Therefore, a machine can only be idle when another machine is executing a task in the chain. From these two observations we know that

$$m\hat{f}_I \le (m-1)\sum_{j=1}^{k} t_{i_j} + \sum_{j=1}^{n} t_j$$

Since no machine can process more than one task at a time, and since no two tasks, one of which precedes the other in C, can be processed concurrently, we know that an optimal makespan schedule satisfies

$$f_I^* \ge \frac{1}{m}\sum_{j=1}^{n} t_j \quad \text{and} \quad f_I^* \ge \sum_{j=1}^{k} t_{i_j}$$

Substituting in the above inequality, we know that $\hat{f}_I/f_I^* \le 2-(1/m)$.

The second contribution in [1] was showing that the approximation ratio (2 – 1/*m*) is the best possible for list schedules. This implies that the analysis of the approximation ratio for this algorithm cannot be improved. This was established by constructing problem instances (for all m and $n = 2m + 1$) and lists for which the schedule generated by the procedure has a makespan equal to $2 - 1/m$ times the optimal makespan for the instance.

The third important aspect of the results in [1] was showing that list scheduling has anomalies. To explain this we need to define some terms. The makespan of the list schedule for instance I using list L is denoted by $f_L(I)$. Suppose that instance I' is instance I, except that I' has an additional machine. Intuitively $f_L(I') \le f_L(I)$ because with one additional machine all tasks should be finished earlier or, in the worst case, at the same time as $f_L(I)$. But this is not always the case for list schedules! There are problem instances and lists for which $f_L(I') > f_L(I)$. This is called an *anomaly*, because it is counterintuitive. Our expectation would be reasonable if list scheduling would generate minimum makespan schedules for all problem instances. But we have a procedure that for the most part generates suboptimal solutions. Guaranteeing that $f_L(I') \le f_L(I)$ for all instances is not always possible in this type of environment. List scheduling suffers from other anomalies. For example, relaxing the precedence constraints, or decreasing the execution time of the tasks. In both of these cases one would expect schedules with smaller or the same makespan. But, this is not always the case. The main reason for discussing anomalies is that even today numerous papers are being published and systems are being deployed where "common sense" based procedures are being developed without any analytic justification and/or thorough experimental validation. Anomalies show that since we live for the most part in a "suboptimal" world, the effect of our decisions is not always the intended one; there are unintended consequences when one does not fully analyze the effects of approximation algorithms. An anomaly tends to arise in situations where its negative effect has the most impact.

It is interesting to note that the worst case behavior of list schedules arises when the task with largest processing requirement is being processed while the rest of the machines are idle. Can a significantly better approximation bound be established when ties are broken in favor of a task with largest processing time (*LPT ordered list schedules* or simply *LPT schedules*). The answer is no.

The problem with the LPT rule is that it only considers the processing requirements of the tasks ready to process but ignores the processing requirements of the tasks that follow it. We define the *weight of a directed path* as the sum of the processing time requirements of the tasks in the path. Any directed path that starts at task *t* with maximum total weight among all paths that start at task *t* is called a *critical path* (CP) *for task t*. The *CP* schedule is defined as a list schedule where the decision of which task to process next is a task whose CP "total time" is largest among the ready to be processed tasks. A CP schedule is optimal for many problem instances, however, Graham constructed problem instances for which the makespan of the CP schedule is 2 – 1/*m* times the length of an optimal schedule.

12.1.3.2.2 Application to Scheduling Independent Tasks

A restricted version of the above scheduling problem that has received considerable attention is when the tasks are independent, that is, there are no precedence constraints. Graham's [8] elegant analysis for LPT scheduling has become a classic. In fact quite a few subsequent exact and approximation scheduling algorithms used his analysis technique.

Let I be any problem instance with n independent tasks and m identical machines. We use $\hat{f}_I$ as the makespan for an LPT schedule for I and f_I^* as the one for an optimal schedule. In the next theorem we establish the approximation ratio for LPT schedules.

Theorem 12.2

For every scheduling problem instance I with n independent tasks and m identical machines, every LPT schedule satisfies $\hat{f}_I/f_I^* \leq (4/3) - (1/3m)$.

Proof: It is clear that LPT schedules are optimal for $m = 1$. Assume that $m \geq 2$. The proof is by contradiction. Suppose the above bound does not hold. Let I be a problem instance with the least number of tasks for which $\hat{f}_I/f_I^* > (4/3) - (1/3m)$. Let n be the number of tasks in I, m be the number of machines and assume that $t_1 \geq t_2 \geq \ldots \geq t_n$. Let k be the smallest index of a task that finishes at time $\hat{f}_I$. It cannot be that $k < n$, as otherwise the problem instance $T_1, T_2, \ldots, T_k$ is also a counter-example and it has fewer tasks than instance I, but by assumption problem instance I is a counter example with the least number of tasks. Therefore, k must be equal to n.

By the definition of LPT schedules, we know that there cannot be idle time before task T_n begins execution. Therefore,

$$\sum_{i=1}^{n} t_i + (m-1)t_n \geq m\hat{f}_I$$

This is equivalent to

$$\hat{f}_I \leq \frac{1}{m}\sum_{i=1}^{n} t_i + \left(1 - \frac{1}{m}\right)t_n$$

Since each machine cannot process more than one task at a time, we know that $f_I^* \geq \sum_{i=1}^{n} t_i / m$. Combining these two bounds, we have

$$\frac{\hat{f}_I}{f_I^*} \leq 1 + \left(1 - \frac{1}{m}\right)\frac{t_n}{f_I^*}$$

Since I is a counter example for the theorem, this bound must be greater than (4/3) – (1/3m). Simplifying we know that $f_I^* < 3t_n$. Since t_n is the task with smallest processing time requirement it must be that in an optimal schedule for instance I none of the machines can process three or more tasks. Therefore, the number of tasks n is at most $2m$. But one can show that LPT is an optimal schedule when $n \leq 2m$ and there is an optima schedule where every machine has at most two tasks. A contradiction.

For all m there are problem instances for which the ratio given by Theorem 2 is tight. One can prove that LPT schedules do not have the anomalies reported for list schedules.

Other approximation algorithms with improved performance were subsequently developed. Coffman et al. [9] introduced the multifit, MF, approach. A k attempt MF approach is denoted by MF_k. The MF_k procedure performs k binary search steps to find the smallest capacity c such that all the tasks can be packed into a set of m bins when packing using first fit with the tasks sorted in nondecreasing order of their processing times. The tasks assigned to bin i correspond to machine i and c is the makespan of the schedule. The approximation ratio has been shown to be $1.22 + 2^-k$ and the time complexity of the algorithm is $O(n \log n + kn \log m)$. Subsequent improvements to $1.2 + 2^-k$ [10] and $(72/61) + (1/2^k)$ [11] were possible within the same time complexity bound. However, the latter algorithm has a very large constant associated with the Big "oh" time complexity bound.

12.1.3.3 Relaxation and Related Methodologies

This approach is in a sense the opposite of "restricting the solution space." In this case one augments the feasible solution space by including previously infeasible solutions. In this case one needs to solve a superproblem of P. An algorithm for P solves the superproblem (optimally or suboptimally) and then transforms such solution to one that is feasible for P. Approximation algorithms based on the linear programming (LP) approach fall under this category.

Khachiyan [12] developed a polynomial time algorithm for LP problems in the late 1970s. This result had a tremendous impact on research as it started a new wave of approximation algorithms. Two subsequent research accomplishments were at least as significant as Khachiyan's [12] result. The first one was a faster polynomial time algorithm for solving LP problems developed by Karmakar [13]. The other major accomplishment was the work of Grötschel et al. [14,15]. They showed that it is possible to solve a LP problem with an exponential number of constraints (with respect to the number of variables) in time which is polynomial in the number of variables and the number of bits used to describe the input, given a *separation oracle* plus a bounding ball and a lower bound on the volume of the feasible solution space. Given a solution, the separation oracle determines in polynomial time whether or not the solution is feasible, and if it is not it finds a constraint that is violated. This technique has been shown to solve more general problems. Important developments have taken place during the past 20 years. Books [15,16] as well as Chapter 13 are excellent references for LP theory, algorithms, and applications. Among the first to apply this approach for the set and vertex cover problems include Lovász [17], Chvátal [18], Hochbaum [19], and Bar-Yehuda and Even [20] for the set and vertex cover problems and Gonzalez [21] for VLSI CAD problems.

In what follows we discuss several direct and indirect applications of relaxation to the vertex cover problem. These approaches include LP relaxation followed by deterministic or randomized rounding, α-vector, local ratio, and primal-dual.

12.1.3.3.1 Linear Programming and Rounding

Let us now consider the minimum weight vertex cover, which is a fundamental problem in the study of approximation algorithms. As input you are given a vertex weighted undirected graph G with the set of vertices $V = \{v_1, v_2, \ldots, v_n\}$, edges $E = \{e_1, e_2, \ldots, e_m\}$, and a positive real number (weight) w_i assigned to each vertex v_i. The problem is to find a minimum weight vertex cover, that is, a subset of vertices $C \subset V$ such that every edge is incident to at least one vertex in C. The weight of the vertex cover C is the sum of the weight of the vertices in C.

It is well known that the minimum weight vertex cover problem is an NP-hard problem. One can show that greedy methods based on selecting the vertex with minimum weight, maximum weight, minimum weight divided by the number of edges incident to it, or combinations of these approaches do not result in constant ratio approximation algorithms, that is, a *c*-approximation algorithm for some fixed constant *c*. Other approaches to solve the problem can also be shown to fail to provide a constant ratio approximation algorithm for the weighted vertex cover.

Another way to view the minimum weight vertex cover is by defining a 0/1 variable x_i for each vertex v_i in the graph. The 0/1 vector X defines a subset of vertices C as follows. Vertex v_i is in C if and only if $x_i = 1$. The set of vertices C defined by X is a vertex cover if and only if for every edge $\{i, j\}$ in the graph $x_i + x_j \geq 1$. Using the earlier notation, the vertex cover problem is formulated as a 0/1 integer linear programming (ILP) as follows:

$$\text{minimize} \quad \sum_{i \in V} w_i x_i \tag{12.1}$$

$$\text{subject to} \quad x_i + x_j \geq 1 \quad \forall \{i, j\} \in E \tag{12.2}$$

$$x_i \in \{0, 1\} \quad \forall i \in V \tag{12.3}$$

The 0/1 ILP is also an NP-hard problem.

Let us apply the relaxation technique. In this case, one relaxes the integer constraints for the x_i values. That is, we replace Constraint 12.3 ($x_i = \{0, 1\}$) by $0 \leq x_i \leq 1$ (or simply $x_i \geq 0$, which in this case is equivalent). This means that we are augmenting the solution space by adding solutions that are not feasible for the original problem. This approach will at least provide us with what appears to be a good lower bound for the value of an optimal solution of the original problem, since every feasible solution to the original problem is a feasible solution to the relaxed problem (but the converse is not true). This relaxed problem is an instance of the LP problem which can be solved in polynomial time. Let X^* be an optimal solution to the LP problem. Clearly, X^* might not be a vertex cover as the x_i^* values can be fractional. The previous interpretation for the X^* values has been lost because it does not make sense to talk about a fractional part of a vertex being part of a vertex cover. To circumvent this situation, we need to use the X^* vector to construct a 0/1 vector $\hat{X}$ that represents a vertex cover. In order for a vector $\hat{X}$ to represent a vertex cover it needs to satisfy Inequality 12.2 (i.e., $\hat{x}_i + \hat{x}_j \geq 1$), for every edge $e_k = \{i, j\} \in E$. Clearly, the inequalities hold for X^*. This means that for each edge $e_k = \{i, j\} \in E$ at least one of x_i^* or x_j^* has value greater than or equal to 1/2. The vector $\hat{X}$ defined from X^* as $\hat{x}_i = 1$ if $x_i^* \geq 1/2$ (rounding up) and $\hat{x}_i = 0$ if $x_i^* < 1/2$ (rounding down) represents a vertex cover. Furthermore, because of the rounding up the objective function value for the vertex cover $\hat{X}$ is at most $2\sum w_i x_i^*$. Since $\sum w_i x_i^*$ value is a lower bound for an optimal solution to the weighted vertex cover problem, we know that this procedure generates a vertex cover whose weight is at most twice the weight of an optimal cover, that is, it is a 2-approximation algorithm. This rounding process is called (deterministic) *LP-rounding*, and it has been applied to a large class of problems.

12.1.3.3.2 Randomized Rounding

Another way to round is via randomization, which means in this case that we flip a biased coin (with respect to the value of x_i^* [vertex cover problem] and perhaps other factors) to decide the value for $\hat{x}_i$. The probability of $\hat{X}$ is a vertex cover and its expected weight can be computed analytically. By repeating this randomization process several times, one can show that a cover with weight at most twice the optimal one will be generated with very high probability. Since we have deterministic rounding for the vertex cover problem, it is clear that randomized rounding is not justified. However, for other problems it is justified. There is more complex randomized rounding for semidefinite programming (SDP).

Raghavan and Thompson [22] were the first to apply randomized rounding to solutions of relaxed LP problems to find near-optimal solutions. Since then this field has grown tremendously.

These rounding methods have the disadvantage that a LP problem needs to be solved. Experimental evaluations over several decades have shown that the simplex method solves quickly (in polynomial time) the LP problem. But the worst case time complexity is exponential with respect to the problem size. The ellipsoid algorithm and other more recent ones solve LP problems in polynomial time. Even though these algorithms have polynomial time complexity, there is a term that depends on the number of bits needed to represent the input. Much progress has been made in speeding-up these procedures. Additional information can be found in Chapter 13. Promising new LP packages include *CoinCLP* and *GUROBI* to solve large LP problem instances faster than other packages available for free to academicians.

12.1.3.3.3 α-Vector

All the remaining methods in this subsection are "independent" of LP, in the sense that the LP problem does not need to be solved. However, these methods are based on LP formulations to some extent. The first approach is called the *α-vector* method and we now apply it to the weighted vertex cover problem. For every vertex $i \in V$ we define $\delta(i)$ as the set of edges incident to vertex i. Let $\alpha = (\alpha_1, \alpha_2, \ldots, \alpha_m)$ be any vector of m nonnegative real values, where $m = |E|$ is the number of edges in the graph. For all j multiply the jth edge inequality for the earlier LP formulation by α_j

$$\sum_{\{i,j\} \in e_k} (\alpha_k x_i + \alpha_k x_j) \geq \alpha_k \quad \forall e_k \in E \tag{12.4}$$

The total sum of these inequalities can be expressed as

$$\sum_{i \in V} \sum_{e_k \in \delta(i)} \alpha_k x_i \geq \sum_{e_k \in E} \alpha_k \tag{12.5}$$

Define $\beta_i = \sum_{e_k \in \delta(i)} \alpha_k$ for every vertex $i \in V$. In other words, let β_i be the sum of the α values of all the edges incident to vertex i. Substituting in the earlier inequality we know that

$$\sum_{i \in V} \beta_i x_i \geq \sum_{e_k \in E} \alpha_k \tag{12.6}$$

Suppose that the α vector is such that $w_i \geq \beta_i$ for all i. Then it follows that

$$\sum_{i \in V} w_i x_i \geq \sum_{i \in V} \beta_i x_i \geq \sum_{e_k \in E} \alpha_k \tag{12.7}$$

In other words, any vector α such that the resulting vector β computed from it satisfies $w_i \geq \beta_i$ provides us with the lower bound $\sum_{e_k \in E} \alpha_k$ for every vector X that represent a vertex cover. In other words if we assign a nonnegative weight to each edge in such a way that the sum of the weight of the edges incident to each vertex i is at most w_i, then the sum of the weights of the edges is a lower bound for an optimal solution.

This is a powerful lower bound. To get maximum strength we need to find a vector α such that $\sum_{e_k \in E} \alpha_k$ is maximum. But finding this vector is as hard as solving the LP problem described before. We will explain why this is the case later on. What if we find a maximal vector α, that is, a vector that cannot possibly be increased in any of its components? This is a simple task. It is just matter of starting with an α vector with all entries being zero and then increasing one of its components until it is no longer possible to do so. We keep on doing this until there are no edges whose α value can be increased. In this maximal solution, we know that for each edge in the graph at least one of its endpoints has the property that $\beta_i = w_i$, as otherwise the maximality of α is contradicted. Define the vector $\hat{X}$ from the α vector as follows: $x_i = 1$ if $\beta_i = w_i$, and $x_i = 0$, otherwise. Clearly, $\hat{X}$ represents a vertex cover because for every edge in the graph we know that for at least one of its vertices has $\beta_i = w_i$. What is the weight of the vertex cover represented by $\hat{X}$? We know that $\sum w_i \hat{x}_i = \sum \beta_i \hat{x}_i \leq 2 \sum \alpha_k$ because each α_k can contribute its value to at most two β_is. Therefore we have a simple 2-approximation algorithm for the weighted vertex cover problem. Furthermore, the procedure to construct the vertex cover takes linear time with respect to the number of vertices and edges in the graph. The above algorithm can be proved to be a 2-approximation algorithm without using the ILP formulation. That is, the same result can be established by using simple combinatorial arguments [23].

12.1.3.3.4 Local Ratio

The *local ratio* approach was developed by Bar-Yehuda and Even [24]. Let us now apply it to the weighted vertex problem.

Initially, each vertex is assigned a cost which is simply its weight and it is referred to as the *remaining cost*. At each step the algorithm makes a "down payment" on a pair of vertices. This has the effect of decreasing the remaining cost of each of the two vertices. Label the edges in the graph $\{e_1, e_2, \ldots, e_m\}$. The algorithm considers one edge at a time using this ordering. When the kth edge $e_k = \{i, j\}$ is considered, define γ_k as the minimum of the remaining cost of vertex i and vertex j. The edge makes a down payment of γ_k to each of its two endpoints and each of the two vertices has its remaining cost decreased by γ_k. The procedure stops when we have considered all the edges. All the vertices whose remaining cost is zero have been paid for completely and they are yours to keep. The remaining ones have not been paid for and there are "no refunds" (not even if you talk to the store manager). The vertices that have been paid for completely form a vertex cover. The weight of all the vertices in the cover generated by the procedure is at most twice $\sum_{e_k \in E} \gamma_k$, which is simply the sum of the down payments made. What is the weight of an optimal vertex cover? One can show it is at least $\sum_{e_k \in E} \gamma_k$. The reason is simple. Consider the first step when we introduce γ_1 for edge e_1. Let I_0 be the initial problem instance and I_1 be the resulting instance after deleting edge e_1 and reducing the cost of the two endpoints of edge e_1 by γ_1. One can easily establish that $f^*(I_0) \geq f^*(I_1) + \gamma_1$ and inductively that $f^*(I_0) \geq \sum_{e_k \in E} \gamma_k$ [25]. The algorithm is a 2-approximation algorithm for the weighted vertex cover. The approach is called *local ratio* because at each step one adds $2\gamma_k$ to the value of the solution generated and one accounts for γ_k value of an optimal solution. This local ratio approach has been successfully applied to solve optimally and suboptimally quite a few problems. The nice feature is that it is very simple to understand and does not require any LP knowledge or background.

12.1.3.3.5 Primal-Dual

The primal-dual approach is similar to the previous ones, but it uses the foundations of LP theory. Let us now apply it to the weight vertex cover problem using our previous formulation. The LP relaxation problem is

$$\text{minimize} \sum_{i \in V} w_i x_i \tag{12.8}$$

$$\text{subject to} \quad x_i + x_j \geq 1 \quad \forall e_k = \{i, j\} \in E \tag{12.9}$$

$$x_i \geq 0 \quad \forall i \in V \tag{12.10}$$

The LP problem is called the *primal* problem. The corresponding dual problem is

$$\text{maximize} \quad \sum_{e_k \in E} y_k \tag{12.11}$$

$$\text{subject to} \quad \sum_{e_k \in \delta(i)} y_k \le w_i \quad \forall i \in V \tag{12.12}$$

$$y_k \ge 0 \quad \forall e_k \in E \tag{12.13}$$

As you can see the Y vector is simply the α vector defined before, and the dual is to find a Y vector with maximum $\sum_{i \in V} y_i$. Linear programming theory [15,16] states that any feasible solution X to the primal problem and any feasible solution Y to the dual problem are such that

$$\sum_{e_k \in E} y_k \le \sum_{i \in V} w_i x_i$$

This is called *weak duality. Strong duality* states that

$$\sum_{e_k \in E} y_i^* = \sum_{i \in V} w_i x_i^*$$

where

X^* is an optimal solution to the primal problem
Y^* is an optimal solution to the dual problem

Note that the dual variables are multiplied by weights which are the right hand side of the constraints in the primal problem. In this case all of them are one.

The primal-dual approach is based on the weak duality property. The idea is to first construct a feasible solution to the dual problem. This solution will give us a lower bound for the value of an optimal vertex cover. Then we use this solution to construct a solution to the primal problem. The idea is that the difference of the objective function value between the primal and dual solutions we constructed is "small." In this case we construct a maximal vector Y (as we did with the α vector before). Then we note that since the Y vector is maximal, then for at least one of the endpoints (say i) of every edge must satisfy Inequality 12.12 tight, that is, $\sum_{e_k \in \delta(i)} y_k = w_i$. Now define vector X with $x_i = 1$ if Inequality 12.12 is tight in the dual solution. Clearly, X represent a feasible solution to the primal problem and its objective function value is at most $2\sum_k y_k$. It then follows by weak duality that an optimal weighted vertex cover has value at least $\sum_k y_k$ and we have a 2-approximation algorithm for the weighted vertex cover. It is simple to see that the algorithm takes linear time (with respect to the number of vertices and edges in the graph) to solve the problem.

There are other ways to construct a solution to the dual problem. For example, by increasing all the y values until one inequality becomes tight. Then we fix a set of y values and increase all the remaining ones until another inequality becomes tight. We repeat this process until all the values of the y are fixed [26]. The only issue with this procedure is that it takes $O(m^2)$ time, rather than just $O(m)$ as the previous one.

It is interesting to note that these relaxation techniques do not work for most problems. This is the case even when these problems are very similar in nature to the ones for which the techniques can be applied successfully. Consider the *MIS* problem. As input you are given an undirected graph $G(V, E)$. The problem is to find a largest set of vertices such that no two vertices are adjacent. The problem can be formulated as an ILP problem as the vertex cover problem, except that it is a maximization problem and the Constraint 12.2 is $x_i + x_j \le 0$. Applying the same arguments as in the case of the vertex cover problem does not result in a constant ratio approximation! The interesting fact is that if there is a way to make this work and produce a constant ratio approximation in polynomial time, then all NP-Complete problems can be solved in polynomial time.

Linear programming has also been used as a tool to compute the approximation ratio of some algorithms. This type of research may eventually be called the *automatic analysis of approximation algorithms*.

Algorithms based on the primal-dual approach are for the most part faster, since they normally take (low) polynomial time with respect to the number of "objects" in the input. However, the LP-rounding approach can be applied to a much larger class of problems and it is more robust since the technique is more likely to be applicable after changing the objective function and/or constraints for a problem.

There are many other very interesting results that have been published in the last 15 years. Goemans and Williamson [27] developed improved approximation algorithms for the maxcut and satisfiability problems using *SDP*. This seminal work opened a new venue for the design of approximation algorithms. Goemans and Williamson [28] also developed powerful techniques for designing approximation algorithms based on the primal-dual approach. The dual-fitting and factor revealing approach is used in [29].

Approximations algorithms that are based on restriction and relaxation exist. These algorithms first restricts the solution space and then relaxes it resulting in a solution space that is different from the original one. Gonzalez and Gonzalez [30] have applied this approach successfully to the minimum edge length corridor problem.

12.1.3.4 Formal Notation

Let P be an optimization problem and let A be an algorithm that generates a feasible solution for every instance I of problem P. We use $\hat{f}_A(I)$ to denote the objective function value of the solution generated by algorithm A for instance I. We drop A and use $\hat{f}(I)$ when it is clear which algorithm is being used. Let $f^*(I)$ be the objective function value of an optimal solution for instance I. Note that normally we do not know the value of $f^*(I)$ exactly, but we use lower bounds which should be as tight as possible.

Sahni [31] defines as an ϵ-approximation algorithm for problem P an algorithm that generates a feasible solution for every problem instance I of P such that

$$\left|\frac{\hat{f}(I) - f^*(I)}{f^*(I)}\right| \le \varepsilon$$

It is assumed that $f^*(I) > 0$. For a minimization problem $\varepsilon > 0$ and for a maximization problem $0 < \varepsilon < 1$. In both cases ε represents the percentage of error. The algorithm is called an ε-approximation algorithm and the solution is said to be an ϵ-approximate solution. Graham's list scheduling algorithm [1] is a $1 - 1/n$ approximation algorithm, and Sahni and Gonzalez [2] algorithm, which we discuss later on, for the k-maxcut problem is a $1/k$-approximation algorithm. Note that this notation is different from the one discussed in previous sections. The difference is 1 unit, that is, the ϵ in this notation corresponds to $1 + \varepsilon$ in the other. Note that, as in the research literature, we use both notations interchangeably. However, readers can easily disambiguate from the context.

Johnson [32] used a slightly different but equivalent notation. He uses the approximation ratio ρ to mean that for every problem instance I of P, the algorithm satisfies $(\hat{f}(I)/f^*(I)) \le \rho$ for minimization problems, and $(f^*(I)/\hat{f}(I)) \le \rho$ for maximization problems. The one for minimization problems is the same as the one we have been using [1]. The value for ρ is always greater than one, and the closer to one, the better the solution generated by the algorithm. One refers to ρ as the *approximation ratio* and the algorithm is a ρ-approximation algorithm. The list scheduling algorithm in the previous section is a $2 - (1/m)$-approximation algorithm and the algorithm given in Section 12.1.3 for the k-maxcut problem is a $k/(k-1)$-approximation algorithm. Some researchers use $1/\rho$ as the approximation ratio for maximization problems. Using this notation, the algorithm in Section 12.1.3 for the k-maxcut problem is a $1-(1/k)$-approximation algorithm.

All the above forms are in use today. The most popular ones are ρ for minimization and 1/ρ for maximization. These are referred to as approximation ratios or approximation factors. We refer to all of these algorithms as ε-approximation algorithms. The point to remember is that one needs to be aware of the differences and be alert when reading the research literature.

In general it is preferred that ε and ρ are constants independent of the problem size. But, they can be dependent on the size of the problem instance *I*. For example, it may be ln *n*, or n^ϵ for some problems, where *n* is some parameter of the problem that depends on *I*, for example, the number of nodes in the input graph, and ϵ depends on the algorithm being used to generate the solutions.

Normally one prefers an algorithm with a smaller approximation ratio. However it is not always the case that an algorithm with smaller approximation ratio always generates solutions closer to optimal than one with a larger approximation ratio. The main reason is that the notation is for the worst case ratio and the worst case does not always occur. But there are other subtle reasons too. For example, the lower bound for the optimal solution value used in the analysis of two different algorithms may be different. Let *P* be the shortest path minimization problem, given an edge weighted (positive integers) graph find a shortest path between vertices *s* and *t*. Let *A* be an algorithm with approximation ratio 2. In this case we use *d* as the lower bound for $f^*(I)$, where *d* is some parameter of the problem instance. Algorithm *B* is a 1.5-approximation algorithm, but $f^*(I)$ used to establish it is the exact optimal solution value. Suppose that for problem instance *I* the value of *d* is 5 and $f^*(I) = 8$. Algorithm *A* will generate a path with weight at most 10, whereas algorithm *B* will generate one with weight at most $1.5 \times 8 = 12$. So the solution generated by Algorithm *B* may be worse than the one generated by *A* even when both algorithms generate the worst possible solution for the instance.

One could argue that the average "error" makes more sense than worst case. The problem is to define and establish bounds for average "error." There are many other pitfalls when using worst case ratios. It is important to keep in mind all of this when making comparisons between algorithms. In practice one may run several different approximation algorithms concurrently and output the best of the solutions. This has the disadvantage that the running time of this compound algorithm will be the one for the slowest algorithm. The analysis quickly becomes very difficult, but the solutions generated are in many cases significantly better than the one generated by one strategy.

More elaborate approximation algorithms have been developed that generate a solution for any fixed constant ϵ. Formally, a *Polynomial Time Approximation Scheme (PTAS)* for problem *P* is an algorithm *A* that given any fixed constant $\epsilon > 0$ it constructs a solution to problem *P* such that $|(\hat{f}(I) - f^*(I))/f^*(I)| \le \epsilon$ in polynomial time with respect to the length of the instance *I*. Note that the time complexity may be exponential with respect to 1/ϵ. For example, the time complexity could be $O(n^{(1/\epsilon)})$ or $O(n + 4^{O(1/\epsilon)})$. Equivalent PTAS is also defined using different notation, for example based on $(\hat{f}(I)/f^*(I)) \le 1 + \epsilon$ for minimization problems. We will see examples of PTAS later on in this chapter.

Clearly, with respect to approximation ratio a PTAS is far better than an ϵ-approximation algorithms for some fixed ϵ. But the main drawback of all known PTAS is that they are not practical because the time complexity is exponential on 1/ϵ or the constants associated with time complexity bound are huge. There may be PTAS that is practical for "natural" occurring problems, but none has been found yet. On the positive side, a PTAS establishes that a problem can be approximated for all fixed constants.

A PTAS is said to be a *Fully Polynomial Time Approximation Scheme (FPTAS)* if its time complexity is polynomial with respect to *n* (the problem size) and 1/ϵ. FPTAS are for the most part practical algorithms. We discuss one such FPTAS later on in this section.

There are a few problems for which a large approximation ratio applies only to problem instances where the value of the optimal solution is small. For these problems we use the following ratio. Informally, ρ_A^∞ is the smallest constant such that there exists a constant $K < \infty$ for which

$$\hat{f}(I) \le \rho_A^\infty f^*(I) + K$$

The *asymptotic approximation ratio* is the multiplicative constant and it hides the additive constant K. This is most useful when K is small. The asymptotic notation is mainly used for the bin packing problem and its variants.

Approximation schemes based on asymptotic approximation algorithms have been developed. The first APTAS (asymptotic PTAS) was developed by Fernandez de la Vega and Lueker [33] for the bin packing problem. The first AFPTAS (asymptotic FPTAS) for the same problem was developed by Karmakar and Karp [34]. Randomized PTAS and FPTAS have also been developed [5].

12.1.3.5 Inapproximability and Approximation Classes

Sahni and Gonzalez [2] were the first to show that there are NP-hard optimization problems for which the existence of a constant ratio polynomial time approximation algorithm implies the existence of a polynomial time algorithm to generate an optimal solution. In other words, for these problems the complexity of generating a constant ratio approximation and an optimal solution are computationally equivalent problems. For these problems the approximation problem is NP-hard or simply inapproximable (under the assumption that $P \neq NP$). Later on, this notion was extended to mean that there is no polynomial time algorithm with approximation ratio r for a problem under some complexity theoretic hypothesis. The approximation ratio r is called the *inapproximability ratio* [5].

Consider the k-min-cluster problem. Given an edge-weighted undirected graph the k-min-cluster problem is to partition the set of vertices into k sets so as to minimize the sum of the weight of the edges with endpoints in the same group. The k-maxcut problem is defined as the k-min-cluster problem, except that the objective is to maximize the sum of the weight of the edges with endpoints in different sets.

Even though the k-min-cluster and the k-maxcut problems have exactly the same set of feasible and optimal solutions, there is a linear time algorithm for the k-maxcut problem that generates k-partitions with weight at least $(k-1)/k$ times the weight of an optimal k-cut [2], whereas approximating the k-min-cluster problem is a computationally intractable problem [2]. The former problem has the property that a near optimal solution may be obtained as long as partial decisions are made optimally, whereas for the k-min-cluster an optimal partial decision may turn out to force a terrible overall solution (one local optimal decision may turn out to have devastating effects to the quality of the final solution). In other words, one strike and you are out.

Another interesting problem whose approximation problem is NP-hard is the TSP problem [2]. This is not exactly the same version of the TSP problem discussed earlier which we said has several constant ratio polynomial time approximation algorithms. The difference is that the graph may have any set of arbitrary weights on the edges, in other words, the graph is not restricted to be metric. On the other hands, the approximation algorithms given in [4,7] can be adapted easily to provide a constant ratio approximation to the version of the TSP problem for arbitrarily weighted graphs when a tour is defined as visiting each vertex in the graph *at least* once. Since Moore's approximation algorithms for the metric Steiner tree and metric TSP are based on the same idea, one would expect that the Steiner tree problem defined over arbitrarily weighted graphs is NP-hard to approximate. However, this is not the case. Moore's algorithm [4] can be modified to be a 2-approximation algorithm for this more general Steiner tree problem.

As pointed out in [35], Levner and Gens [36] added a couple of problems to the list of problems that are NP-hard to approximate. Garey and Johnson [37] show that the max clique problem has the property that if for some constant r there is a polynomial time r-approximation algorithm, then there is a polynomial time r'-approximation for any constant r'. Since then researchers had tried many different algorithms for the clique problem without success, it was conjectured that none existed, under the assumption that $P \neq NP$.

Garey and Johnson [38] showed that a class of NP-hard optimization problems that satisfy certain properties and whose corresponding decision problems are strongly NP-Complete have the property that if any one of them has a FPTAS, then $P = NP$. The properties are that the objective function value of every feasible solution is a positive integer, and the problem is *strongly* NP-hard. This means that the problem is NP-hard even when the magnitude of the maximum number in the input is bounded by a polynomial on the input length. For example, the TSP problem is strongly NP-hard, whereas the knapsack problem is not, under the assumption that $P \neq NP$.

Informally, the class of *NP optimization problems, NPO*, is the set of all optimization problems Π which can be "recognized" in polynomial time. In the late 1980s Papadimitriou and Yannakakis [39] defined *MAXSNP* as a subclass of NPO. These problems can be approximated within a constant factor and have a nice logical characterization. They showed that if max3sat, vertex cover, maxcut, and some other problems in the class could be approximated in polynomial time with an arbitrary precision, then all MAXSNP problems have the same property. This fact was established by using *approximation preserving* reductions. In the 1990s Arora et al. [40], using complex arguments, showed that max3sat is hard to approximate within a factor of $1 + \varepsilon$ for some $\varepsilon > 0$ unless $P = NP$. Thus, all problems in MAXSNP do not admit a PTAS unless $P = NP$. This work led to major developments in the area of approximation algorithms, including inapproximability results for other problems, approximation preserving reductions, discovery of new inapproximability classes, and construction of approximation algorithms achieving optimal or near optimal ratios.

Feige et al. [41] showed that the clique problem could not be approximated to within some constant value. Applying the previous results in [38] it showed that the clique problem is inapproximable to within any constant. Feige [42] showed that the set cover is inapproximable within $\ln n$. Other inapproximable results appear in [43,44].

Ausiello et al. [45] introduced and studied the *differential ratio* error measurement. Informally, an algorithm is said to be a δ differential ratio approximation algorithm if for every instance I of P

$$\frac{\omega(I) - \hat{f}(I)}{\omega(I) - f^*(I)} \leq \delta$$

where $\omega(I)$ is the value of a worst solution for instance I. Differential ratio has some interesting properties for the complexity of the approximation problems. For some problems $\omega(I)$ may be unbounded in which case one has to limit its value. For example, when scheduling independent tasks on identical processors the worst schedule would postpone indefinitely the processing of a task. But for this problem one could limit $\omega(I)$ by restricting the feasible schedules to those without idle time on all machines at the same time.

Differential ratio destroys the artificial dissymmetry between "equivalent" minimization and maximization problems (for example, the k-max-cut and the k-min cluster discussed earlier) when it comes to approximation. This ratio uses the difference between the worst possible solution minus the solution generated by the algorithm, divided by the difference between the worst solution minus the best solution. Cornuejols et al. [46] also discussed a variation of differential ratio approximations since for the problem they studied they wanted the ratio to satisfy the following property: "A modification of the data that adds a constant to the objective function value should also leave the error measure unchanged." That is, the "error" by the approximation algorithm should be the same as before. Differential ratio along with other similar notions are discussed in [45]. Ausiello et al. [45] also introduced *reductions that preserve approximability.* Since then there have been several new types of approximation preserving reductions. The main advantage of these reductions is that they enable us to define large classes of optimization problems that behave in the same way with respect to approximation. An *NPO* problem Π is said to be in *APX*, if it has a constant approximation ratio polynomial time algorithm. The class *PTAS* consists of all *NPO* problems which have PTAS. The class *FPTAS* is defined similarly. Other classes, *Poly-APX*, *Log-APX*, and *Exp-APX*, have also been defined.

As said before there are many different criteria to compare algorithms. What if we use both the approximation ratio and time complexity? For example, the simple linear time approximation algorithm for the k-maxcut problem in [2] and the complex one given in [27]. Both of these algorithms are current Pareto optimal as one is faster, but the other has worst case ration that is smaller.

The best algorithm to use also depends on the instance being solved. It makes a difference whether we are dealing with an instance of the TSP with optimal tour cost equal to one billion dollars and one with optimal cost equal to just a few pennies. Though, it also depends on the number of such instances being solved.

12.1.3.6 Partial Brute Force

The idea here is to solve by enumeration (or other techniques) part of the problem and then the remaining decisions can be made via a simple (perhaps greedy) procedure. The approach is to solve optimally the "most important part of the problem" and then solve suboptimally the remaining portion of the problem.

As mentioned before, Graham [8] studied the problem of scheduling tasks on identical machines, but restricted to the case where all the tasks are independent, that is, the set of precedence constraints is empty. We described and analyzed before an LPT procedure that guarantees solutions that are within a fixed ϵ value. We now apply the partial brute force approach to obtain a PTAS for this problem.

Graham [8], following a suggestion by Kleitman and Knuth, considered list schedules where the first portion of the list L consists of k tasks with the largest processing times arranged by their starting times in an optimal schedule for these k tasks (only). Then the list L has the remaining $n - k$ tasks in any order. The approximation ratio for this list schedule using list L is $1+(1-1/m)/\left(1+\lceil k/m \rceil\right)$. An optimal schedule for the largest k tasks can be constructed in $O(km^k)$ time by a straight forward branch and bound algorithm. In other words, this algorithm has approximation ratio $1 + \epsilon$ and time complexity $O(n + m^{(m-1-\epsilon m)/\epsilon})$. For any fixed constants m and ϵ the algorithm constructs in polynomial (linear) time with respect to n a schedule with makespan at most $1 + \epsilon$ times the optimal makespan. Note that for a fixed constant m the time complexity is polynomial with respect to n, but it is not polynomial with respect to $1/\epsilon$. This was the first algorithm of its kind and later on it was referred to as a *Polynomial Time Approximation Scheme (PTAS)*.

Clearly, with respect to approximation ratio, a PTAS is better than an ϵ-approximation algorithm for some fixed value ϵ. But their main drawback is that PTAS is not useful in practice because the time complexity is exponential on $1/\epsilon$ or the constants associated with the time complexity bound are large. This does not preclude the existence of a practical PTAS for "natural" occurring problem. However, a PTAS establishes that a problem can be approximated for all fixed constants.

12.1.3.7 Rounding

In the rounding technique one modifies the parameters of the problem instance to obtain another instance that can be easily solved optimally or suboptimally. We solve such instance and use it as a suboptimal solution to the original instance. Before applying this technique to obtain a FPTAS to the knapsack problem, additional notation needs to be defined.

The knapsack problem consists of n items with item i having a profit $p(i)$ and a weight $w(i)$, where each $p(i)$ and $w(i)$ is an integer. One is also given the weight capacity of the knapsack. The problem is to find a subset of items with total weight at most C and maximum total profit.

Via dynamic programming one can show that the knapsack problem can be solved in time $O\left(n * min\left\{2^n, \sum p(i), C\right\}\right)$. Clearly, when $min\left\{\sum p(i), C\right\} \le poly(n)\right\}$, where $poly(n)$ is a polynomial on n, the problem can be solved in polynomial time with respect to n, the problem size. The problem is computationally intractable when the capacity of the knapsack and the sum of the profits are large provided that there is no optimal solution with a "small" number of items.

When given an instance I, the approach is to construct an instance I' which can be solved in polynomial time with respect to n, the number of items. Every feasible solution to I' is also a feasible one to I and the percentage difference between the optimal solutions to these two problems is bounded by $1 + \epsilon$. The rounding that we use leaves the weights of the items untouched, but the profits are rounded proportionally to $1/\epsilon$. For further details see Chapter 10 in [5].

The first FPTAS was developed by Ibarra and Kim [47] for the knapsack problem. Sahni [48] developed three different techniques based on rounding, interval partitioning, and separation to construct FPTAS for sequencing and scheduling problems. Horowitz and Sahni [49] developed FPTAS for scheduling on machines with different processing speeds. Gens and Levner [35] discuss a simple $O(n^3/\epsilon)$ FPTAS for the knapsack problem developed by Babat [50,51]. Lawler [52] developed techniques to speed up FPTAS for the knapsack and related problems. Additional results are discussed in [5].

12.1.3.8 Additional Classical Results

In the early 1970s, Garey et al. [53] as well as Johnson [54,55] developed the first set of polynomial time approximation algorithms for the bin packing problem. In the bin packing problem we are given a set of n items $a_1, a_2, \ldots, a_n$, with size $0 < s(a_i) < 1$, and an unlimited number of bins with capacity 1. The goal is to find the minimum number of bins in such a way that the sum of the size of the items assigned to each bin is at most one. Research on the bin packing problem and its variants has attracted very talented investigators who have generated more than 650 papers, most of which deal with approximations. This work has been driven by numerous applications in engineering and information sciences. Interestingly several algorithmic methodologies were applied for the first time to the bin packing problem and then to problems in other areas.

12.1.4 Local Search and Metaheuristics

Local search techniques have a long history; they range from simple constructive and iterative improvement algorithms to rather complex methods that require significant fine-tuning and large search neighborhoods. Local search is perhaps one of the most natural ways to attempt to find an optimal (or suboptimal) solution to an optimization problem. The idea of "local" search is simple: Start from a solution and improve it by making local changes until no further progress is possible. Local search normally terminates at a local optimal. To escape from a local optimal solution several extensions have been developed. One very natural one is to expand the size of the neighborhood and make it very large. But finding the best neighbor to move to is most of the time an NP-hard problem. Therefore, an approximation solution is needed at this step. Stochastic local search algorithms use randomized movements in the neighborhood.

The study of local search sparked the study of modern heuristics, which have evolved and are now called *metaheuristics*. The term metaheuristics was coined by Glover [56] in 1986 and in general means "to find beyond in an upper level." The main idea is to automate mechanisms to escape local optima. As discussed in Section 12.1.1, metaheuristics methodologies include TS, SA, ACO, EC, ILC (Iterated Local Search), MA, etc. One of the motivations for the study of metaheuristics is that it was recognized early on that constant ratio polynomial time approximation algorithms are not likely to exist for a large class of practical problems [2]. Metaheuristics do not guarantee that optimal or near optimal solutions can be found quickly for all problem instances. However, these programs do find near optimal solutions for many problem instances that arise in practice. These procedures are general in the sense that they apply to a wide range of problems. Metaheuristics include mechanisms to avoid being trapped in local optimal like local search. These are the most appealing aspects of metaheuristics. At some level metaheuristics explore the solution space (diversification) and then use the accumulated search experience to guide the search (intensification). They also include short term strategies tied to randomness, and medium and long term strategies based on the use of memory. Different metaheuristics make use of these search strategies at different levels. The idea is to explore the solution space and move to better neighborhoods with better local optimal solutions.

Metaheuristics fall into two categories: single point search and population-based strategies. TS, SA, and local search methods fall into the single point search strategy. The idea is that at different points they behave like local search, that is, move to better solutions in the local neighborhood. However, sometimes they jump out of the local neighborhood and continue from that point. These jumps are guided from past experience in the search. Methods like MA, genetic algorithms (GAs), and ACO are population-based. This means that at any given point a set of solutions are in existence. Then at each step the set evolves into another set and the process continues. The evolution is to local and nonlocal neighborhoods.

Metaheuristics can also be categorized as nature versus nonnature inspired. EC and ACO are nature inspired methodologies. Whereas TS and MA are not nature inspired methods. Some methodologies use memory to guide the search process and other methods are memoryless. Different levels of randomization are used in metaheuristics. In what follows we briefly discuss these methodologies.

12.1.4.1 Local Search

Local search is perhaps one of the most natural ways to attempt to find an optimal or suboptimal solution to an optimization problem. The idea of local search is simple: Start from a solution and improve it by making local changes until no further progress is possible. Most of the time it is simple to find an initial feasible solution, but there are problems where finding an initial feasible solution is a computationally intractable problem. In this case the problem is inapproximable in polynomial time unless $P \neq NP$. When the neighborhood to search for the next solution is very large, finding the best neighbor to move to is many times an NP-hard problem. Therefore, an approximation solution is needed at this step. Reactive search advocates the use of simple symbolic machine learning to automate the parameter tuning process and make it an integral (and fully documented) part of the algorithm. Parameters are normally tuned through a feedback loop that many times depend on the user. Reactive search attempts to mechanize this process. Stochastic local search algorithms have been introduced that make stochastic choices.

Lin and Kernighan [57] developed local search heuristics that established experimentally that instances of the TSP with up to 110 cities can be solved to optimality with 95% confidence in $O(n^2)$ time. This local search based procedure was applied to a set of randomly selected feasible solutions. The process was to perform $k = 2$ pairs of link (edge) interchanges that improved the length of the tour. On the other hand, Papadimitriou and Steiglitz [58] showed that for the TSP no local optimum of an efficiently searchable neighborhood can be within a constant factor of the optimal value unless $P = NP$. Since then there has been quite a bit of research activity in this area.

12.1.4.2 Tabu Search

The term *tabu search* (TS) was coined by Glover [56]. TS is based on *adaptive memory* and *responsive exploration*. The former allows for the effective and efficient search of the solution space. The latter is used to guide the search process by imposing restraints and inducements based on the information collected. Intensification and diversification are controlled by the information collected, rather than by a random process.

12.1.4.3 Simulated Annealing

In the early 1980s Kirkpatrick et al. [59] and independently Černý [60] introduced SA as a randomized local search algorithm to solve combinatorial optimization problems. *SA* is a local search algorithm, which means that it starts with an initial solution and then searches through the solution space by iteratively generating a new solution that is "near" to it. Sometimes the moves are to a worse solution to escape local optimal solutions. The probability of such moves decreases with time. This method is based on statistical mechanics (Metropolis algorithm). It was heavily inspired by an analogy between the physical annealing process of solids and the problem of solving large combinatorial optimization problems.

12.1.4.4 Evolutionary Computation

Evolutionary computation (EC) is a metaphor for building, applying, and studying algorithms based on Darwinian principles of natural selection. Algorithms that are based on evolutionary principles are called *Evolutionary Algorithms* (EAs). They are inspired by nature's capability to evolve living beings well adapted to their environment. There have been a variety of slightly different EAs proposed over the years. Three different strands of EAs were developed independently of each other over time. These are *Evolutionary Programming* introduced by Fogel [61] and Fogel et al. [62], *Evolutionary Strategies* proposed by Rechenberg [63] and *GAs* initiated by Holland [64]. GAs are mainly applied to solve discrete problems. *Genetic Programming* and *Scatter Search* are more recent members of the EA family. EAs can be understood from a unified point of view with respect to their main components and the way they explore the search space. The two main factors are recombination (crossover) which combine two or more individuals (solutions) and mutation (modification) of an individual (solution).

12.1.4.5 Ant Colony Optimization

This metaheuristic is inspired by the behavior of real ants, it was proposed by Dorigo and colleagues [65] in the early 1990s as a method for solving hard combinatorial optimization problems. The basic idea is to simulate the use of pheromone for finding shortest paths like the ones ants use to locate food. But also use random walks, like ants, to find other sources of food. These algorithms may be considered to be part of *swarm intelligence*, the research field that studies algorithms inspired by the observation of the behavior of *swarms*. Swarm intelligence algorithms are made up of simple individuals that cooperate through self-organization.

12.1.4.6 Memetic Algorithms

Moscato [66] introduced MA in the late 1980s to denote a family of metaheuristics which can be characterized as the hybridization of different algorithmic approaches for a given problem. It is a population-based approach in which a set of cooperating and competing agents are engaged in periods of individual improvement of the solutions while they sporadically interact. An important component is *problem and instance-dependent knowledge* which is used to speed-up the search process.

12.2 Impact on Practice

Approximation algorithms and metaheuristics have been developed to solve a wide variety of problems. A good portion of these results have only theoretical value due to the fact that the time complexity function for these algorithms is a high order polynomial and/or have huge constants associated with their time complexity bounds. However, these results are important because they establish what is possible now, and it may be that in the near future these algorithms will be transformed into practical ones. The rest of the approximation algorithms do not suffer from this pitfall, but some were designed for problems with limited applicability. However, the remaining approximation algorithms have real world applications. On the other hand, there is a huge number of important application areas, including new emerging ones, where approximation algorithms and metaheuristics have barely penetrated and we believe there is an enormous potential for approximation algorithms and metaheuristics in these areas. Our goal is to expose the reader to the areas of the approximation algorithms and metaheuristics as well as to the different methodologies used to design these algorithms. These areas are relatively young so it is fair to repeat the lyrics of an old popular song "The best is yet to come, you ain't seen nothing yet."

In the last couple of decades we have seen approximation algorithms being applied to traditional combinatorial optimization problems as well as problems arising in other areas of research activity. These areas include: VLSI design automation, networks (wired, sensor, and wireless), bioinformatics, game theory, computational geometry, graph problems, etc.

Most real world problems are solved via heuristics, without any significant analysis. However, more problems are being solved via approximation algorithms and metaheuristics as industry is being populated by computer science graduates. For example, content delivery networks, like the one built by *Akami*, have been using approximation algorithms to decide where data must reside for in order to facilitate its retrieval by users. Similar algorithmic problems arise when designing the software behind search engines (see Chapter 51). Manufacturing computing environments give rise to a large number of sequencing and scheduling problems. There is a rich set of approximation algorithms for this types of sequencing and scheduling problems [67].

Approximation algorithms and metaheuristics for graph, Steiner tree, TSP, clustering, logic, data mining, CAD placement, routing, and synchronization problems and their variations are extensively used in a variety of industrial settings.

12.3 Research Issues

There are many important research issues that need to be resolved. The most important one is to develop new methodologies to design approximation algorithms and metaheuristics. These new methodologies should be applicable to a wide range of problems and be amenable for analysis. Tools and techniques for the analysis of sophisticated approximation algorithms need to be developed.

A methodology for the experimental evaluation of approximation algorithms and metaheuristics needs to be developed. This area is in its infancy. Automatic tools for testing and verifying approximation claims are for the most part nonexistent. The research approaches seem to be divided into theory and experiments. There is a need to combine them into just one that deals with both aspects: theoretical analysis and experimental evaluation.

We also need to find the point where approximability ends and inapproximability begins for most problems. We need insights into how to find these points quickly. These break points are only known for a few problems.

The efficient use of energy when computing is a relatively new area of research where algorithmic strategies can be designed and used to significantly decrease energy consumption for computing.

12.4 Summary

In this chapter we have discussed the different methodologies to design approximation algorithms and metaheuristics. Approximation algorithms based on restriction, relaxation, partial enumeration, rounding, and local search were explored. We applied these techniques to well-known problems. We discussed the basic metaheuristics. These include: TS, SA, ACO, EC, and MA. We discussed the history of approximation algorithms and metaheuristics.

Acknowledgments

Preliminary versions of portions of this chapter and the two figures appeared in Chapters 1 through 3 of the *Handbook on Approximation Algorithms and Metaheuristics* [5].

Defining Terms

ε-approximation algorithm: For problem P it is an algorithm that generates a feasible solution for every problem instance I of P such that

$$\left| \frac{\hat{f}(I) - f^*(I)}{f^*(I)} \right| \le \varepsilon$$

It is assumed that $f^*(I) > 0$. There are variations of this notation.

Approximation Methodologies: Restriction, relaxation, partial enumeration, rounding, and local search.

APX: A problem with a polynomial time ε-approximation algorithm.

FPTAS: A PTAS whose time complexity is polynomial with respect to the length of the instance I and $1/\varepsilon$.

MAXSNP: Problems that can be approximated within a constant factor and have a nice logical characterization.

P, NP, and NP-Completeness: See Chapter 7.

PTAS: For problem P it is an algorithm A that given any fixed constant $\varepsilon > 0$ it constructs a solution to problem P such that $\left| \frac{\hat{f}(I) - f^*(I)}{f^*(I)} \right| \le \varepsilon$ in polynomial time with respect to the length of the instance I.

Further Information

Additional information may be found in the *Handbook on Approximation Algorithms and Metaheuristics* [5], as well as in the textbooks by Vazirani [26], Hochbaum (*Approximation Algorithms for NP-Hard problems* by D. Hochbaum (ed.), PWS Publishing, 1997.), and Sahni [3]). Most journals and conference proceedings publish papers in the area of approximation algorithms and metaheuristics. Below you will find the main ones. *Journal of the ACM, ACM Transactions on Algorithms, SIAM Journal on Computing, INFORMS Journal on Computing, European Journal of Operational Research, International Journal of Metaheuristics, International Journal of Applied Metaheuristic Computing*, and conferences and workshops include *ACM-SIAM Symposium on Discrete Algorithms, Symposium on the Theory of Computing, IEEE Symposium on Foundations of Computer Science, International Workshop on Approximation Algorithms for Combinatorial Optimization Problems, International Workshop on Randomization and Computation and Metaheuristics International Conference.*

References

1. Graham, R.L., Bounds for certain multiprocessing anomalies, *Bell Syst. Tech. J*, 45, 1563, 1966.
2. Sahni, S. and Gonzalez, T., P-complete approximation problems, *J. Assoc. Comput. Mach.*, 23, 555, 1976.
3. Sahni, S., *Data Structures, Algorithms, and Applications in C++*, 2nd edn., Silicon Press, Summit, NJ, 2005.
4. Gilbert, E.N. and Pollak, H.O., Steiner minimal trees, *SIAM J. Appl. Math.*, 16(1), 1, 1968.
5. Gonzalez, T.F., *Handbook on Approximation Algorithms and Metaheuristics*, Chapman & Hall/CRC, Boca Raton, FL, 2007.
6. Christofides, N., Worst-case analysis of a new heuristic for the traveling salesman problem. Technical report 338, Grad School of Industrial Administration, CMU, Pittsburgh, PA, 1976.
7. Rosenkrantz, R., Stearns, R., and Lewis, L., An analysis of several heuristics for the traveling salesman problem. *SIAM J. Comput.*, 6(3), 563, 1977.
8. Graham, R.L., Bounds on multiprocessing timing anomalies, *SIAM J. Appl. Math.*, 17, 263, 1969.
9. Coffman Jr., E.G., Garey, M.R., and Johnson, D.S., An application of bin-packing to multiprocessor scheduling, *SIAM J. Comput.*, 7, 1, 1978.
10. Friesen, D.K., Tighter bounds for the multifit processor scheduling algorithm, *SIAM J. Comput.*, 13, 170, 1984.
11. Friesen, D.K. and Langston, M.A., Bounds for multifit scheduling on uniform processors, *SIAM J. Comput.*, 12, 60, 1983.
12. Khachiyan, L.G., A polynomial algorithms for the linear programming problem, *Dokl. Akad. Nauk SSSR*, 244(5), 1093, 1979 (in Russian).
13. Karmakar, N., A new polynomial-time algorithm for linear programming, *Combinatorica*, 4, 373, 1984.
14. Grötschel, M., Lovász, L., and Schrijver, A., The ellipsoid method and its consequences in combinatorial optimization, *Combinatorica*, 1, 169, 1981.
15. Schrijver, A., *Theory of Linear and Integer Programming*, Wiley-Interscience Series in Discrete Mathematics and Optimization, John Wiley, Chichester, U.K., 2000.
16. Vanderbei, R.J., *Linear Programming Foundations and Extensions*, Series: International Series in Operations Research & Management Science, 3rd edn., Vol. 37, Springer, New York, 2008.
17. Lovász, L., On the ratio of optimal integral and fractional covers, *Discrete Math.*, 13, 383, 1975.
18. Chvátal, V., A greedy heuristic for the set-covering problem, *Math. Oper. Res.*, 4(3), 233, 1979.
19. Hochbaum, D.S., Approximation algorithms for set covering and vertex covering problems, *SIAM J. Comput.*, 11, 555, 1982.

20. Bar-Yehuda, R. and Even, S., A linear time approximation algorithm for the weighted vertex cover problem, *J. Algorithm*, 2, 198, 1981.
21. Gonzalez, T.F., An approximation algorithm for the multi-via assignment problem, *IEEE Trans. Comput. Aid. Des. Integr. Circ. Syst.*, CAD, 3(4), 257–264, 1984.
22. Raghavan, R. and Thompson, C., Randomized rounding: A technique for provably good algorithms and algorithmic proof, *Combinatorica*, 7, 365, 1987.
23. Gonzalez, T.F., A simple LP-free approximation algorithm for the minimum web pages vertex cover problem, *Inform. Proc. Lett.*, 54(3), 129, 1995.
24. Bar-Yehuda, R. and Even, S., A local-ratio theorem for approximating the weighted set cover problem, *Ann. Disc. Math.*, 25, 27, 1985.
25. Bar-Yehuda, R. and Bendel, K., Local ratio: A unified framework for approximation algorithms, *ACM Comput. Surv.*, 36(4), 422, 2004.
26. Vazirani, V.V., *Approximation Algorithms*, Springer-Verlag, Berlin, Germany, 2001.
27. Goemans, M.X. and Williamson, D.P., Improved approximation algorithms for maximum cut and satisfiability problems using semi-definite programming, *J. Assoc. Comput. Mach.*, 42(6), 1115, 1995.
28. Goemans, M.X. and Williamson, D.P., A general approximation technique for constrained forest problems, *SIAM J. Comput.*, 24(2), 296, 1995.
29. Jain, K., Mahdian, M., Markakis, E., Saberi, A., and Vazirani, V.V., Approximation algorithms for facility location via dual fitting with factor-revealing LP, *J. Assoc. Comput. Mach.*, 50, 795, 2003.
30. Gonzalez, A. and Gonzalez, T.F., Approximating corridors and tours via restriction and relaxation techniques, *ACM Trans. Algorithm*, 6(3), 2010.
31. Sahni, S., Approximate algorithms for the 0/1 knapsack problem, *J. Assoc. Comput. Mach.*, 22(1), 115, 1975.
32. Johnson, D.S., Approximation algorithms for combinatorial problems, *J. Comput. Syst. Sci.*, 9, 256, 1974
33. Fernandez de la Vega, W. and Lueker, G.S., Bin Packing can be solved within $1 + \varepsilon$ in linear time, *Combinatorica*, 1, 349, 1981.
34. Karmakar, N. and Karp, R.M., An efficient approximation scheme for the one-dimensional bin packing problem, in: *Proceedings of Foundations of Computer Science*, New York, 1982, p. 312.
35. Gens, G.V. and Levner, E., Complexity of approximation algorithms for combinatorial problems: A survey, *SIGACT News*, 52, 1980.
36. Levner, E. and Gens, G.V., Discrete Optimization Problems and Efficient Approximation Algorithms, Central Economic and Meth. Inst., Moscow, Russia, 1978 (in Russian).
37. Garey, M.R. and Johnson, D.S., The complexity of near-optimal graph coloring, *SIAM J. Comput.*, 4, 397, 1975.
38. Garey, M.R. and Johnson, D.S., Strong NP-completeness results: Motivations, examples, and implications, *J. Assoc. Comput. Mach.*, 25, 499, 1978.
39. Papadimitriou, C.H. and Yannakakis, M., Optimization, approximation and complexity classes, *J. Comput. Syst. Sci.*, 43, 425, 1991.
40. Arora, S., Lund, C., Motwani, R., Sudan, M., and Szegedy, M., Proof verification and hardness of approximation problems, in: *Proceedings of Foundations of Computer Science*, Pittsburg, PA, 1992.
41. Feige, U., Goldwasser, S., Lovasz, L., Safra, S., and Szegedy, M., Interactive proofs and the hardness of approximating cliques, *J. Assoc. Comput. Mach.*, 43, 1996.
42. Feige, U., A threshold of ln *n* for approximating set cover, *J. Assoc. Comput. Mach.*, 45(4), 634, 1998. (Prelim. version in STOC'96.)
43. Engebretsen, L. and Holmerin, J., Towards optimal lower bounds for clique and chromatic number, *Theor. Comput. Sci.*, 299, 2003.
44. Hastad, J., Some optimal inapproximability results, *J. Assoc. Comput. Mach.*, 48, 2001. (Prelim. version in STOC'97.)

45. Ausiello, G., D'Atri, A., and Protasi, M., On the structure of combinatorial problems and structure preserving reductions, in: *Proceedings of ICALP'77*, LNCS, Turku, Finland, Vol. 52, Springer-Verlag, 1977, p. 45.
46. Cornuejols, G., Fisher, M.L., and Nemhauser, G.L., Location of bank accounts to optimize float: An analytic study of exact and approximate algorithms, *Manage. Sci.*, 23(8), 789, 1977.
47. Ibarra, O. and Kim, C., Fast approximation algorithms for the knapsack and sum of subset problems, *J. Assoc. Comput. Mach.*, 22(4), 463, 1975.
48. Sahni, S., Algorithms for scheduling independent tasks, *J. Assoc. Comput. Mach.*, 23(1), 116, 1976.
49. Horowitz, E. and Sahni, S., Exact and approximate algorithms for scheduling nonidentical processors, *J. Assoc. Comput. Mach.*, 23(2), 317, 1976.
50. Babat, L.G., Approximate computation of linear functions on vertices of the unit *N*-dimensional cube, in: Fridman, A.A., ed., *Studies in Discrete Optimization*, Nauka, Moscow, Russia, 1976 (in Russian).
51. Babat, L.G., A fixed-charge problem, *Izv. Akad. Nauk SSR, Techn, Kibernet.*, 3, 25, 1978 (in Russian).
52. Lawler, E., Fast approximation algorithms for knapsack problems, *Math. Oper. Res.*, 4, 339, 1979.
53. Garey, M.R., Graham, R.L., and Ullman, J.D., Worst-case analysis of memory allocation algorithms, in: *Proceeding of Symposium on Theory of Computing*, ACM, New York, 1972, 143.
54. Johnson, D.S., *Near-Optimal Bin Packing Algorithms*, PhD thesis, Massachusetts Institute of Technology, Department of Mathematics, Cambridge, MA, 1973.
55. Johnson, D.S., Fast algorithms for bin packing, *J. Comput. Syst. Sci.*, 8, 272, 1974.
56. Glover, F., Future paths for integer programming and links to artificial intelligence, *Comput. Oper. Res.*, 13, 533, 1986.
57. Lin, S., and Kernighan, B.W., An effective heuristic algorithm for the traveling salesman problem, *Oper. Res.*, 21(2), 498, 1973.
58. Papadimitriou, C.H. and Steiglitz, K., On the complexity of local search for the traveling salesman problem, *SIAM J. Comput.*, 6, 76, 1977.
59. Kirkpatrick, S., Gelatt Jr., C.D., and Vecchi, M.P., Optimization by simulated annealing, *Science*, 220, 671, 1983.
60. Černý, V., Thermodynamical approach to the traveling salesman problem: An efficient simulation algorithm, *J. Optimiz. Theory Appl.*, 45, 41, 1985.
61. Fogel, L.J., Toward inductive inference automata, in *Proceedings of the International Federation for Information Processing Congress*, Munich, Germany, 1962, p. 395.
62. Fogel, L.J., Owens, A.J., and Walsh, M.J., *Artificial Intelligence through Simulated Evolution*. Wiley, New York, 1966.
63. Rechenberg, I., *Evolutionsstrategie: Optimierung technischer Systeme nach Prinzipien der biologischen Evolution*, Frommann-Holzboog, Stuttgart, Germany, 1973.
64. Holland, J.H., *Adaption in Natural and Artificial Systems*, The University of Michigan Press, Ann Harbor, MI, 1975.
65. Dorigo, M., Maniezzo, V., and Colorni, A., Positive feedback as a search strategy, Technical Report 91-016, Dipartimento di Elettronica, Politecnico di Milano, Milan, Italy, 1991.
66. Moscato, P., On genetic crossover operators for relative order preservation, C3P Report 778, California Institute of Technology, Pasadena, CA, 1989.
67. Leung, J. Y-T., ed., *Handbook of Scheduling: Algorithms, Models, and Performance Analysis*, Chapman & Hall/CRC, Boca Raton, FL, 2004.

13 Combinatorial Optimization

Vijay Chandru
National Institute of Advanced Studies

M.R. Rao
Indian School of Business

13.1 Introduction

Bin packing, routing, scheduling, layout, and network design are generic examples of combinatorial optimization problems that often arise in computer engineering and decision support. Unfortunately, almost all interesting generic classes of combinatorial optimization problems are $\mathcal{NP}$-hard. The scale at which these problems arise in applications and the explosive exponential complexity of the search spaces preclude the use of simplistic enumeration and search techniques. Despite the worst-case intractability of combinatorial optimization, in practice we are able to solve many large problems and often with off-the-shelf software. Effective software for combinatorial optimization is usually problem specific and based on sophisticated algorithms that combine approximation methods with search schemes that exploit mathematical (not just syntactic) structure in the problem at hand.

TABLE 13.1 Paradigms in Combinatorial Optimization

Paradigm	Representation	Methodology
Integer programming	Linear constraints Linear objective integer variables	Linear programming and extensions
Search	State space discrete control	Dynamic programming A*
Local improvement	Neighborhoods fitness functions	Hill climbing Simulated annealing Tabu search Genetic algorithms
Constraint logic programming	Horn rules	Resolution Constraint solvers

Multidisciplinary interests in combinatorial optimization have led to several fairly distinct paradigms in the development of this subject. Each paradigm may be thought of as a particular combination of a *representation scheme* and a *methodology* (see Table 13.1). The most established of these, the *integer programming* paradigm, uses implicit algebraic forms (linear constraints) to represent combinatorial optimization and **linear programming** and its extensions as the workhorses in the design of the solution algorithms. It is this paradigm that forms the central theme of this chapter.

Other well-known paradigms in combinatorial optimization are *search*, *local improvement*, and *constraint logic programming* (CLP). Search uses state-space representations and partial enumeration techniques such as $\mathcal{A}^*$ and dynamic programming. Local improvement requires only a representation of neighborhood in the solution space, and methodologies vary from simple hill climbing to the more sophisticated techniques of simulated annealing, tabu search, and genetic algorithms. CLP uses the syntax of Horn rules to represent combinatorial optimization problems and uses resolution to orchestrate the solution of these problems with the use of domain-specific constraint solvers. Whereas integer programming was developed and nurtured by the mathematical programming community, these other paradigms have been popularized by the artificial intelligence community.

An abstract formulation of combinatorial optimization is

$$\text{(CO)}\quad \min\{f(I) : I \in \mathcal{I}\}$$

where $\mathcal{I}$ is a collection of subsets of a finite ground set $E = \{e_1, e_2, \ldots, e_n\}$ and f is a criterion (objective) function that maps 2^E (the power set of E) to the reals. A *mixed integer linear program* (MILP) is of the form

$$\text{(MILP)}\quad \min_{x \in \Re^n}\{\mathbf{cx} : A\mathbf{x} \geq \mathbf{b}, \mathbf{x}_j \text{ integer } \forall\, j \in J\}$$

which seeks to minimize a linear function of the decision vector $\mathbf{x}$ subject to linear inequality constraints and the requirement that a subset of the decision variables is integer valued. This model captures many variants. If $J = \{1, 2, \ldots, n\}$, we say that the integer program is *pure*, and *mixed* otherwise. Linear equations and bounds on the variables can be easily accommodated in the inequality constraints. Notice that by adding in inequalities of the form $0 \leq \mathbf{x}_j \leq 1$ for a $j \in J$ we have forced $\mathbf{x}_j$ to take value 0 or 1. It is such Boolean variables that help capture combinatorial optimization problems as special cases of MILP.

Pure integer programming with variables that take arbitrary integer values is a class which has strong connections to number theory and particularly the geometry of numbers and Presburgher arithmetic. Although this is a fascinating subject with important applications in cryptography, in the interests of brevity we shall largely restrict our attention to MILP where the integer variables are Boolean.

The fact that MILPs subsume combinatorial optimization problems follows from two simple observations. The first is that a collection $\mathcal{I}$ of subsets of a finite ground set E can always be represented by a corresponding collection of incidence vectors, which are {0, 1}-vectors in $\Re^E$. Further, arbitrary nonlinear functions can be represented via piecewise linear approximations by using linear constraints and mixed variables (continuous and Boolean).

The next section contains a primer on linear inequalities, polyhedra, and linear programming. These are the tools we will need to analyze and solve integer programs. Section 13.4 is a testimony to the earlier cryptic comments on how integer programs model combinatorial optimization problems. In addition to working a number of examples of such integer programming formulations, we shall also review a formal representation theory of (Boolean) MILPs.

With any mixed integer program, we associate a *linear programming relaxation* obtained by simply ignoring the integrality restrictions on the variables. The point being, of course, that we have polyno mial-time (and practical) algorithms for solving linear programs. Thus, the linear programming relaxation of (MILP) is given by

$$\text{(LP)} \quad \min_{x \in \Re^n} \{\mathbf{cx} : A\mathbf{x} \geq \mathbf{b}\}$$

The thesis underlying the integer linear programming approach to combinatorial optimization is that this linear programming relaxation retains enough of the structure of the combinatorial optimization problem to be a useful weak representation. In Section 13.5, we shall take a closer look at this thesis in that we shall encounter special structures for which this relaxation is *tight*. For general integer programs, there are several alternative schemes for generating linear programming relaxations with varying qualities of approximation. A general principle is that we often need to disaggregate integer formulations to obtain higher quality linear programming relaxations. To solve such huge linear programs, we need specialized techniques of large-scale linear programming. These aspects will be the content of Section 13.3.

The reader should note that the focus in this chapter is on solving hard combinatorial optimization problems. We catalog the special structures in integer programs that lead to tight linear programming relaxations (Section 13.5) and hence to polynomial-time algorithms. These include structures such as network flows, matching, and matroid optimization problems. Many hard problems actually have pieces of these nice structures embedded in them. Practitioners of combinatorial optimization have always used insights from special structures to devise strategies for hard problems.

The computational art of integer programming rests on useful interplays between search methodologies and linear programming relaxations. The paradigms of branch and bound (B-and-B) and branch and cut (B-and-C) are the two enormously effective partial enumeration schemes that have evolved at this interface. These will be discussed in Section 13.6. It may be noted that all general purpose integer programming software available today uses one or both of these paradigms.

The inherent complexity of integer linear programming has led to a long-standing research program in approximation methods for these problems. Linear programming relaxation and Lagrangian relaxation are two general approximation schemes that have been the real workhorses of computational practice. Primal–dual strategies and semidefinite relaxations are two recent entrants that appear to be very promising. Section 13.7 reviews these developments in the approximation of combinatorial optimization problems.

We conclude the chapter with brief comments on future prospects in combinatorial optimization from the algebraic modeling perspective. The original version of this chapter was written in the late 1990s. In revising the chapter and resurveying the field, we have had an opportunity to describe some significant advances that have happened in resolving some open conjectures as well as in new applications and improved computational techniques. However, to a large extent, none of the foundations have been shaken by these advances and we are confident that the original framework we laid out is still a good approach for appreciating the beauty and the nuances of this approach to combinatorial optimaization.

13.2 Primer on Linear Programming

Polyhedral combinatorics is the study of embeddings of combinatorial structures in Euclidean space and their algebraic representations. We will make extensive use of some standard terminology from polyhedral theory. Definitions of terms not given in the brief review below can be found in Nemhauser and Wolsey (1988).

A (convex) *polyhedron* in $\Re^n$ can be algebraically defined in two ways. The first and more straightforward definition is the *implicit* representation of a polyhedron in $\Re^n$ as the solution set to a finite system of linear inequalities in n variables. A single linear inequality $\mathbf{ax} \le a_0$; $\mathbf{a} \ne \mathbf{0}$ defines a *half-space* of $\Re^n$. Therefore, geometrically a polyhedron is the intersection set of a finite number of half-spaces.

A *polytope* is a bounded polyhedron. Every polytope is the convex closure of a finite set of points. Given a set of points whose convex combinations generate a polytope, we have an explicit or *parametric* algebraic representation of it. A *polyhedral cone* is the solution set of a system of homogeneous linear inequalities. Every (polyhedral) cone is the conical or positive closure of a finite set of vectors. These generators of the cone provide a parametric representation of the cone. And finally, a polyhedron can be alternatively defined as the Minkowski sum of a polytope and a cone. Moving from one representation of any of these polyhedral objects to another defines the essence of the computational burden of polyhedral combinatorics. This is particularly true if we are interested in *minimal* representations.

A set of points $\mathbf{x}^1, \ldots, \mathbf{x}^m$ is *affinely independent* if the unique solution of $\sum_{i=1}^m \lambda_i \mathbf{x}^i = 0, \sum_{i=1}^m \lambda_i = 0$ is $\lambda_i = 0$ for $i = 1, \ldots, m$. Note that the maximum number of affinely independent points in $\Re^n$ is $n + 1$. A polyhedron P is of *dimension* k, $\dim P = k$, if the maximum number of affinely independent points in P is $k + 1$. A polyhedron $P \subseteq \Re^n$ of dimension n is called *full dimensional*. An inequality $\mathbf{ax} \le a_0$ is called *valid* for a polyhedron P if it is satisfied by all $\mathbf{x}$ in P. It is called *supporting* if in addition there is an $\tilde{\mathbf{x}}$ in P that satisfies $\mathbf{a}\tilde{\mathbf{x}} = a_0$. A *face* of the polyhedron is the set of all $\mathbf{x}$ in P that also satisfies a valid inequality as an equality. In general, many valid inequalities might represent the same face. Faces other than P itself are called *proper*. A *facet* of P is a maximal nonempty and proper face. A facet is then a face of P with a dimension of $\dim P - 1$. A face of dimension zero, that is, a point v in P that is a face by itself, is called an *extreme point* of P. The extreme points are the elements of P that cannot be expressed as a strict convex combination of two distinct points in P. For a full-dimensional polyhedron, the valid inequality representing a facet is unique up to multiplication by a positive scalar, and facet-inducing inequalities give a minimal implicit representation of the polyhedron. Extreme points, on the other hand, give rise to minimal parametric representations of polytopes.

The two fundamental problems of linear programming (which are polynomially equivalent) follow:

1. *Solvability*: This is the problem of checking if a system of linear constraints on real (rational) variables is solvable or not. Geometrically, we have to check if a polyhedron, defined by such constraints, is nonempty.
2. *Optimization*: This is the problem (LP) of optimizing a linear objective function over a polyhedron described by a system of linear constraints.

Building on polarity in cones and polyhedra, duality in linear programming is a fundamental concept which is related to both the complexity of linear programming and to the design of algorithms for solvability and optimization. We will encounter the solvability version of duality (called *Farkas' Lemma*) while discussing the Fourier elimination technique subsequently. Here, we will state the main duality results for optimization. If we take the *primal* linear program to be

$$(P)\ \min_{x \in \Re^n} \{\mathbf{cx} : A\mathbf{x} \ge \mathbf{b}\}$$

there is an associated *dual* linear program

$$(D)\ \max_{y \in \Re^m}\{\mathbf{b}^T\mathbf{y} : A^T\mathbf{y} = \mathbf{c}^T, \mathbf{y} \geq \mathbf{0}\}$$

and the two problems satisfy the following:

1. For any $\hat{\mathbf{x}}$ and $\hat{\mathbf{y}}$ feasible in *(P)* and *(D)* (i.e., they satisfy the respective constraints), we have $\mathbf{c}\hat{\mathbf{x}} \geq \mathbf{b}^T\hat{\mathbf{y}}$ (*weak duality*). Consequently, *(P)* has a finite optimal solution if and only if *(D)* does.
2. The pair $\mathbf{x}^*$ and $\mathbf{y}^*$ are optimal solutions for *(P)* and *(D)*, respectively, if and only if $\mathbf{x}^*$ and $\mathbf{y}^*$ are feasible in *(P)* and *(D)* (i.e., they satisfy the respective constraints) and $\mathbf{cx}^* = \mathbf{b}^T\mathbf{y}^*$ (*strong duality*).
3. The pair $\mathbf{x}^*$ and $\mathbf{y}^*$ are optimal solutions for *(P)* and *(D)*, respectively, if and only if $\mathbf{x}^*$ and $\mathbf{y}^*$ are feasible in *(P)* and *(D)* (i.e., they satisfy the respective constraints) and $(A\mathbf{x}^* - \mathbf{b})^T\mathbf{y}^* = 0$ (*complementary slackness*).

The strong duality condition gives us a good stopping criterion for optimization algorithms. The complementary slackness condition, on the other hand, gives us a constructive tool for moving from dual to primal solutions and vice versa. The weak duality condition gives us a technique for obtaining lower bounds for minimization problems and upper bounds for maximization problems.

Note that the properties just given have been stated for linear programs in a particular form. The reader should be able to check that if, for example, the primal is of the form

$$(P')\ \min_{\mathbf{x} \in \Re^n}\{\mathbf{cx} : A\mathbf{x} = \mathbf{b}, \mathbf{x} \geq \mathbf{0}\}$$

then the corresponding dual will have the form

$$(D')\ \max_{\mathbf{y} \in \Re^m}\{\mathbf{b}^T\mathbf{y} : A^T\mathbf{y} \leq \mathbf{c}^T\}$$

The tricks needed for seeing this are that any equation can be written as two inequalities, an unrestricted variable can be substituted by the difference of two nonnegatively constrained variables, and an inequality can be treated as an equality by adding a nonnegatively constrained variable to the lesser side. Using these tricks, the reader could also check that duality in linear programming is involutory (i.e., the dual of the dual is the primal).

13.2.1 Algorithms for Linear Programming

We will now take a quick tour of some algorithms for linear programming. We start with the classical technique of Fourier (1824) which is interesting because of its really simple syntactic specification. It leads to simple proofs of the duality principle of linear programming (solvability) that has been alluded to. We will then review the simplex method of linear programming, a method that has been finely honed over almost five decades. We will spend some time with the ellipsoid method and, in particular, with the polynomial equivalence of solvability (optimization) and separation problems, for this aspect of the ellipsoid method has had a major impact on the identification of many tractable classes of combinatorial optimization problems. We conclude the primer with a description of Karmarkar's (1984) breakthrough, which was an important landmark in the brief history of linear programming. A noteworthy role of interior point methods has been to make practical the theoretical demonstrations of tractability of various aspects of linear programming, including solvability and optimization, that were provided via the ellipsoid method.

13.2.2 Fourier's Scheme for Linear Inequalities

Constraint systems of linear *inequalities* of the form $A\mathbf{x} \le \mathbf{b}$, where A is an $m \times n$ matrix of real numbers, are widely used in mathematical models. Testing the solvability of such a system is equivalent to linear programming.

Suppose we wish to eliminate the first variable $\mathbf{x}_1$ from the system $A\mathbf{x} \le \mathbf{b}$. Let us denote

$$I^+ = \{i : A_{i1} > 0\} \quad I^- = \{i : A_{i1} < 0\} \quad I^0 = \{i : A_{i1} = 0\}$$

Our goal is to create an equivalent system of linear inequalities $\tilde{A}\tilde{\mathbf{x}} \le \tilde{\mathbf{b}}$ defined on the variables $\tilde{\mathbf{x}} = (\mathbf{x}_2, \mathbf{x}_3, \ldots, \mathbf{x}_n)$:

- If I^+ is empty then we can simply delete all the inequalities with indices in I^- since they can be trivially satisfied by choosing a large enough value for $\mathbf{x}_1$. Similarly, if I^- is empty we can discard all inequalities in I^+.
- For each $k \in I^+, l \in I^-$ we add $-A_{l1}$ times the inequality $A_k\mathbf{x} \le \mathbf{b}_k$ to A_{k1} times $A_l\mathbf{x} \le \mathbf{b}_l$. In these new inequalities, the coefficient of $\mathbf{x}_1$ is wiped out, that is, $\mathbf{x}_1$ is eliminated. Add these new inequalities to those already in I^0.
- The inequalities $\{\tilde{A}_i\tilde{\mathbf{x}} \le \tilde{\mathbf{b}}_i\}$ for all $i \in I^0$ represent the equivalent system on the variables $\tilde{\mathbf{x}} = (\mathbf{x}_2, \mathbf{x}_3, \ldots, \mathbf{x}_n)$.

Repeat this construction with $\tilde{A}\tilde{\mathbf{x}} \le \tilde{\mathbf{b}}$ to eliminate $\mathbf{x}_2$ and so on until all variables are eliminated. If the resulting $\tilde{\mathbf{b}}$ (after eliminating $\mathbf{x}_n$) is nonnegative, we declare the original (and intermediate) inequality systems as being consistent. Otherwise,* $\tilde{\mathbf{b}} \ngeq 0$ and we declare the system inconsistent.

As an illustration of the power of elimination as a tool for theorem proving, we show now that Farkas Lemma is a simple consequence of the correctness of Fourier elimination. The lemma gives a direct proof that solvability of linear inequalities is in $\mathcal{NP} \bigcap co\mathcal{NP}$.

Farkas Lemma 13.1 (Duality in Linear Programming: Solvability)

Exactly one of the alternatives

$$\text{I.} \quad \exists\, \mathbf{x} \in \Re^n : A\mathbf{x} \le \mathbf{b}$$

$$\text{II.} \quad \exists\, \mathbf{y} \in \Re^m_+ : \mathbf{y}^t A = \mathbf{0}, \mathbf{y}^t\mathbf{b} < 0$$

is true for any given real matrices A, $\mathbf{b}$.

Proof 13.1

Let us analyze the case when Fourier elimination provides a proof of the inconsistency of a given linear inequality system $A\mathbf{x} \le \mathbf{b}$. The method clearly converts the given system into $RA\mathbf{x} \le R\mathbf{b}$ where RA is zero and $R\mathbf{b}$ has at least one negative component. Therefore, there is some row of R, say, $\mathbf{r}$, such that $\mathbf{r}A = \mathbf{0}$ and $\mathbf{rb} < 0$. Thus ¬I implies II. It is easy to see that I and II cannot both be true for fixed A, $\mathbf{b}$.

* Note that the final $\tilde{\mathbf{b}}$ may not be defined if all of the inequalities are deleted by the monotone sign condition of the first step of the construction described. In such a situation, we declare the system $A\mathbf{x} \le \mathbf{b}$ *strongly consistent* since it is consistent for any choice of $\mathbf{b}$ in $\Re^m$. To avoid making repeated references to this exceptional situation, let us simply assume that it does not occur. The reader is urged to verify that this assumption is indeed benign.

In general, the Fourier elimination method is quite inefficient. Let k be any positive integer and n the number of variables be $2^k + k + 2$. If the input inequalities have left-hand sides of the form $\pm\mathbf{x}_r \pm\mathbf{x}_s \pm\mathbf{x}_t$ for all possible $1 \le r < s < t \le n$, it is easy to prove by induction that after k variables are eliminated, by Fourier's method, we would have at least $2^{n/2}$ inequalities. The method is therefore exponential in the worst case, and the explosion in the number of inequalities has been noted, in practice as well, on a wide variety of problems. We will discuss the central idea of minimal generators of the projection cone that results in a much improved elimination method.

First, let us identify the set of variables to be eliminated. Let the input system be of the form

$$P = \{(\mathbf{x}, \mathbf{u}) \in \Re^{n_1+n_2} \mid A\mathbf{x} + B\mathbf{u} \le \mathbf{b}\}$$

where $\mathbf{u}$ is the set to be eliminated. The projection of P onto $\mathbf{x}$ or equivalently the effect of eliminating the $\mathbf{u}$ variables is

$$P_{\mathbf{x}} = \{\mathbf{x} \in \Re^{n_1} \mid \exists \mathbf{u} \in \Re^{n_2} \text{ such that } A\mathbf{x} + B\mathbf{u} \le \mathbf{b}\}$$

Now W, the *projection cone* of P, is given by

$$W = \{\mathbf{w} \in \Re^{m} \mid \mathbf{w}B = \mathbf{0}, \mathbf{w} \ge \mathbf{0}\}$$

A simple application of Farkas Lemma yields a description of $P_{\mathbf{x}}$ in terms of W.

Projection Lemma 13.2

Let G be any set of generators (e.g., the set of extreme rays) of the cone W. Then $P_{\mathbf{x}} = \{\mathbf{x} \in \Re^{n_1} \mid (\mathbf{g}A)x \le \mathbf{g}b \forall \mathbf{g} \in G\}$.

The lemma, sometimes attributed to Černikov (1961), reduces the computation of $P_{\mathbf{x}}$ to enumerating the extreme rays of the cone W or equivalently the extreme points of the polytope $W \cap \left\{\mathbf{w} \in \Re^m \mid \sum_{i=1}^{m} \mathbf{w}_i = 1\right\}$.

13.2.3 Simplex Method

Consider a polyhedron $\mathcal{K} = \{\mathbf{x} \in \Re^n : A\mathbf{x} = \mathbf{b}, \mathbf{x} \ge \mathbf{0}\}$. Now $\mathcal{K}$ cannot contain an infinite (in both directions) line since it is lying within the nonnegative orthant of $\Re^n$. Such a polyhedron is called a *pointed* polyhedron. Given a pointed polyhedron $\mathcal{K}$, we observe the following:

- If $\mathcal{K} \neq \emptyset$, then $\mathcal{K}$ has at least one extreme point.
- If $\min\{\mathbf{cx} : A\mathbf{x} = \mathbf{b}, \mathbf{x} \ge \mathbf{0}\}$ has an optimal solution, then it has an optimal extreme point solution.

These observations together are sometimes called the *fundamental theorem* of linear programming since they suggest simple finite tests for both solvability and optimization. To generate all extreme points of $\mathcal{K}$, in order to find an optimal solution, is an impractical idea. However, we may try to run a partial search of the space of extreme points for an optimal solution. A simple local improvement search strategy of moving from extreme point to adjacent extreme point until we get to a local optimum is nothing but the simplex method of linear programming. The local optimum also turns out to be a global optimum because of the convexity of the polyhedron $\mathcal{K}$ and the linearity of the objective function $\mathbf{cx}$.

The simplex method walks along edge paths on the combinatorial graph structure defined by the boundary of convex polyhedra. Since these graphs are quite dense (Balinski's theorem states that the graph of d-dimensional polyhedron must be d-connected [Ziegler, 1995]) and possibly large (the Lower Bound Theorem states that the number of vertices can be exponential in the dimension [Ziegler, 1995]), it is indeed somewhat of a miracle that it manages to get to an optimal extreme point as quickly as it does. Empirical and probabilistic analyses indicate that the number of iterations of the simplex method is just slightly more than linear in the dimension of the primal polyhedron. However, there is no known variant of the simplex method with a worst-case polynomial guarantee on the number of iterations. Even a polynomial bound on the diameter of polyhedral graphs is not known.

Procedure 13.1 Primal Simplex ($\mathcal{K}$, c):

0. **Initialize:**

    $\mathbf{x}_0$:= an extreme point of $\mathcal{K}$

    $k := 0$

1. **Iterative step:**

    do

    If for all edge directions $\mathcal{D}_k$ at $\mathbf{x}_k$, the objective function is nondecreasing, i.e.,

$$\mathbf{cd} \geq 0 \quad \forall\, \mathbf{d} \in \mathcal{D}_k$$

    then exit and return optimal $\mathbf{x}_k$.

    Else pick some $\mathbf{d}_k$ in $\mathcal{D}_k$ such that $\mathbf{cd}_k < 0$.

    If $\mathbf{d}_k \geq 0$ **then** declare the linear program unbounded in objective value and exit.

    Else $\mathbf{x}_{k+1} := \mathbf{x}_k + \theta_k * \mathbf{d}_k$, where

$$\theta_k = \max\{\theta : \mathbf{x}_k + \theta * \mathbf{d}_k \geq 0\}$$

    $k := k + 1$

    od

2. **End**

Remark 13.1

In the initialization step, we assumed that an extreme point $\mathbf{x}_0$ of the polyhedron $\mathcal{K}$ is available. This also assumes that the solvability of the constraints defining $\mathcal{K}$ has been established. These assumptions are reasonable since we can formulate the solvability problem as an optimization problem, with a self-evident extreme point, whose optimal solution either establishes unsolvability of $A\mathbf{x} = \mathbf{b}, \mathbf{x} \geq \mathbf{0}$ or provides an extreme point of $\mathcal{K}$. Such an optimization problem is usually called a phase I model. The point being, of course, that the simplex method, as just described, can be invoked on the *phase I model* and, if successful, can be invoked once again to carry out the intended minimization of $\mathbf{cx}$. There are several different formulations of the phase I model that have been advocated. Here is one:

$$\min\{v_0 : A\mathbf{x} + \mathbf{b}v_0 = \mathbf{b}, \mathbf{x} \geq \mathbf{0}, v_0 \geq 0\}$$

The solution $(\mathbf{x}, v_0)^T = (0,\ldots, 0, 1)$ is a self-evident extreme point and $v_0 = 0$ at an optimal solution of this model is a necessary and sufficient condition for the solvability of $A\mathbf{x} = \mathbf{b}, \mathbf{x} \geq \mathbf{0}$.

Remark 13.2

The scheme for generating improving edge directions uses an algebraic representation of the extreme points as certain bases, called *feasible bases*, of the vector space generated by the columns of the matrix A. It is possible to have linear programs for which an extreme point is geometrically overdetermined (degenerate), that is, there are more than d facets of $\mathcal{K}$ that contain the extreme point, where d is the dimension of $\mathcal{K}$. In such a situation, there would be several feasible bases corresponding to the same extreme point. When this happens, the linear program is said to be *primal degenerate.*

Remark 13.3

There are two sources of nondeterminism in the primal simplex procedure. The first involves the choice of edge direction $\mathbf{d}_k$ made in step 1. At a typical iteration, there may be many edge directions that are improving in the sense that $\mathbf{cd}_k < 0$. Dantzig's rule, the maximum improvement rule, and steepest descent rule are some of the many rules that have been used to make the choice of edge direction in the simplex method. There is, unfortunately, no clearly dominant rule and successful codes exploit the empirical and analytic insights that have been gained over the years to resolve the edge selection nondeterminism in simplex methods. The second source of nondeterminism arises from degeneracy. When there are multiple feasible bases corresponding to an extreme point, the simplex method has to pivot from basis to adjacent basis by picking an entering basic variable (a pseudoEdge direction) and by dropping one of the old ones. A wrong choice of the leaving variables may lead to cycling in the sequence of feasible bases generated at this extreme point. Cycling is a serious problem when linear programs are highly degenerate as in the case of linear relaxations of many combinatorial optimization problems. The lexicographic rule (perturbation rule) for the choice of leaving variables in the simplex method is a provably finite method (i.e., all cycles are broken). A clever method proposed by Bland (cf. Schrijver, 1986) preorders the rows and columns of the matrix A. In the case of nondeterminism in either entering or leaving variable choices, Bland's rule just picks the lowest index candidate. All cycles are avoided by this rule also.

The simplex method has been the veritable workhorse of linear programming for four decades now. However, as already noted, we do not know of a simplex method that has worst-case bounds that are polynomial. In fact, Klee and Minty exploited the sensitivity of the original simplex method of Dantzig, to projective scaling of the data, and constructed exponential examples for it. Recently, Spielman and Tang (2001) introduced the concept of smoothed analysis and smoothed complexity of algorithms, which is a hybrid of worst-case and average-case analysis of algorithms. Essentially, this involves the study of performance of algorithms under small random Gaussian perturbations of the coefficients of the constraint matrix. The authors show that a variant of the simplex algorithm, known as the *shadow vertex simplex algorithm* (Gass and Saaty, 1955), has polynomial smoothed complexity.

The ellipsoid method of Shor (1970), was devised to overcome poor scaling in convex programming problems and, therefore, turned out to be the natural choice of an algorithm to first establish polynomial-time solvability of linear programming. Later, Karmarkar (1984) took care of both projection and scaling simultaneously and arrived at a superior algorithm.

13.2.4 Ellipsoid Algorithm

The ellipsoid algorithm of Shor (1970) gained prominence in the late 1970s when Hačijan (1979) (pronounced Khachiyan) showed that this convex programming method specializes to a polynomial-time algorithm for linear programming problems. This theoretical breakthrough naturally led to intense study of this method and its properties. The survey paper by Bland et al. (1981) and the monograph by Akgül (1984) attest to this fact. The direct theoretical consequences for combinatorial optimization problems were independently documented by Padberg and Rao (1981), Karp and Papadimitriou (1982), and Grötschel et al. (1988). The ability of this method to implicitly handle linear programs with an exponential list of constraints and maintain polynomial-time convergence is a characteristic that is the key to its applications in combinatorial optimization. For an elegant treatment of the many deep theoretical consequences of the ellipsoid algorithm, the reader is directed to the monograph by Lovász (1986) and the book by Grötschel et al. (1988).

Computational experience with the ellipsoid algorithm, however, showed a disappointing gap between the theoretical promise and practical efficiency of this method in the solution of linear programming problems. Dense matrix computations as well as the slow average-case convergence properties are the reasons most often cited for this behavior of the ellipsoid algorithm. On the positive side though, it has been noted (cf. Ecker and Kupferschmid, 1983) that the ellipsoid method is competitive with the best known algorithms for (nonlinear) convex programming problems.

Let us consider the problem of testing if a polyhedron $\mathcal{Q} \in \Re^d$, defined by linear inequalities, is nonempty. For technical reasons, let us assume that $\mathcal{Q}$ is rational, that is, all extreme points and rays of $\mathcal{Q}$ are rational vectors or, equivalently, that all inequalities in some description of $\mathcal{Q}$ involve only rational coefficients. The ellipsoid method does not require the linear inequalities describing $\mathcal{Q}$ to be explicitly specified. It suffices to have an oracle representation of $\mathcal{Q}$. Several different types of oracles can be used in conjunction with the ellipsoid method (Padberg and Rao, 1981; Karp and Papadimitriou, 1982; Grötschel et al., 1988). We will use the *strong separation oracle:*

Oracle: **Strong Separation**($\mathcal{Q}$, **y**)

Given a vector $\mathbf{y} \in \Re^d$, decide whether $\mathbf{y} \in \mathcal{Q}$, and if not find a hyperplane that separates $\mathbf{y}$ from $\mathcal{Q}$; more precisely, find a vector $\mathbf{c} \in \Re^d$ such that $\mathbf{c}^T\mathbf{y} < \min\{\mathbf{c}^T\mathbf{x} \mid \mathbf{x} \in \mathcal{Q}\}$.

The ellipsoid algorithm initially chooses an ellipsoid large enough to contain a part of the polyhedron $\mathcal{Q}$ if it is nonempty. This is easily accomplished because we know that if $\mathcal{Q}$ is nonempty then it has a rational solution whose (binary encoding) length is bounded by a polynomial function of the length of the largest coefficient in the linear program and the dimension of the space.

The center of the ellipsoid is a feasible point if the separation oracle tells us so. In this case, the algorithm terminates with the coordinates of the center as a solution. Otherwise, the separation oracle outputs an inequality that separates the center point of the ellipsoid from the polyhedron $\mathcal{Q}$. We translate the hyperplane defined by this inequality to the center point. The hyperplane slices the ellipsoid into two halves, one of which can be discarded. The algorithm now creates a new ellipsoid that is the minimum volume ellipsoid containing the remaining half of the old one. The algorithm questions if the new center is feasible and so on. The key is that the new ellipsoid has substantially smaller volume than the previous one. When the volume of the current ellipsoid shrinks to a sufficiently small value, we are able to conclude that $\mathcal{Q}$ is empty. This fact is used to show the polynomial-time convergence of the algorithm.

The crux of the complexity analysis of the algorithm is on the a priori determination of the iteration bound. This in turn depends on three factors. The volume of the initial ellipsoid E_0, the rate of volume shrinkage $(vol(E_{k+1})/vol(E_k) < e^{-\frac{1}{(2d)}})$, and the volume threshold at which we can safely conclude that $\mathcal{Q}$ must be empty. The assumption of $\mathcal{Q}$ being a rational polyhedron is used to argue that $\mathcal{Q}$ can be modified into a full-dimensional polytope without affecting the decision question: "Is $\mathcal{Q}$ non-empty ?" After careful accounting for all of these technical details and some others (e.g., compensating for the roundoff errors caused by the square root computation in the algorithm), it is possible to establish the following fundamental result.

Theorem 13.1

There exists a polynomial $g(d, \phi)$ such that the ellipsoid method runs in time bounded by $T\, g(d, \phi)$, where ϕ is an upper bound on the size of linear inequalities in some description of $\mathcal{Q}$ and T is the maximum time required by the oracle Strong Separation($\mathcal{Q}$, y) on inputs y of size at most $g(d, \phi)$.

The size of a linear inequality is just the length of the encoding of all of the coefficients needed to describe the inequality. A direct implication of the theorem is that solvability of linear inequalities can be checked in polynomial time if strong separation can be solved in polynomial time. This implies that the standard linear programming solvability question has a polynomial-time algorithm (since separation can be effected by simply checking all of the constraints). Happily, this approach provides polynomial-time algorithms for much more than just the standard case of linear programming solvability. The theorem can be extended to show that the optimization of a linear objective function over $\mathcal{Q}$ also reduces to a polynomial number of calls to the strong separation oracle on $\mathcal{Q}$. A converse to this theorem also holds, namely, separation can be solved by a polynomial number of calls to a solvability/optimization oracle (Grötschel et al., 1982). Thus, optimization and separation are polynomially equivalent. This provides a very powerful technique for identifying tractable classes of optimization problems. Semidefinite programming (SDP) and submodular function minimization are two important classes of optimization problems that can be solved in polynomial time using this property of the ellipsoid method.

13.2.5 Semidefinite Programming

The following optimization problem defined on symmetric $(n \times n)$ real matrices

$$\text{(SDP)} \quad \min_{X \in \Re^{n \times n}} \left\{ \sum_{ij} C \cdot X : A \cdot X = B, X \succeq 0 \right\}$$

is called a *semidefinite program*. Note that $X \succeq 0$ denotes the requirement that X is a positive semidefinite matrix, and $F \cdot G$ for $n \times n$ matrices F and G denotes the product matrix $(F_{ij} * G_{ij})$. From the definition of positive semidefinite matrices, $X \succeq 0$ is equivalent to

$$\mathbf{q}^T X \mathbf{q} \geq 0 \quad \text{for every } \mathbf{q} \in \Re^n$$

Thus, SDP is really a linear program on $O(n^2)$ variables with an (uncountably) infinite number of linear inequality constraints. Fortunately, the strong separation oracle is easily realized for these constraints. For a given symmetric X, we use Cholesky factorization to identify the minimum eigenvalue $\lambda_{\min}$. If $\lambda_{\min}$ is nonnegative then $X \succeq 0$ and if, on the other hand, $\lambda_{\min}$ is negative we have a separating inequality

$$\gamma_{\min}^T X \gamma_{\min} \geq 0$$

where $\boldsymbol{\gamma}_{\min}$ is the eigenvector corresponding to $\lambda_{\min}$. Since the Cholesky factorization can be computed by an $O(n^3)$ algorithm, we have a polynomial-time separation oracle and an efficient algorithm for SDP via the ellipsoid method. Alizadeh (1995) has shown that interior point methods can also be adapted to solving SDP to within an additive error ε in time polynomial in the size of the input and $\log 1/\epsilon$.

This result has been used to construct efficient approximation algorithms for maximum stable sets and cuts of graphs, Shannon capacity of graphs, and minimum colorings of graphs. It has been used to define hierarchies of relaxations for integer linear programs that strictly improve on known exponential-size linear programming relaxations. We shall encounter the use of SDP in the approximation of a maximum weight cut of a given vertex-weighted graph in Section 13.7.

13.2.6 Minimizing Submodular Set Functions

The minimization of submodular set functions is another important class of optimization problems for which ellipsoidal and projective scaling algorithms provide polynomial-time solution methods.

Definition 13.1

Let N be a finite set. A real valued set function f defined on the subsets of N is submodular if $f(X \cup Y) + f(X \cap Y) \leq f(X) + f(Y)$ for $X, Y \subseteq N$.

Example 13.1

Let $G = (V, E)$ be an undirected graph with V as the node set and E as the edge set. Let $c_{ij} \geq 0$ be the weight or capacity associated with edge $(ij) \in E$. For $S \subseteq V$, define the cut function $c(S) = \sum_{i \in S, j \in V \setminus S} c_{ij}$. The cut function defined on the subsets of V is submodular since $c(X) + c(Y) - c(X \cup Y) - c(X \cap Y) = \sum_{i \in X \setminus Y, j \in Y \setminus X} 2c_{ij} \geq 0$.

The optimization problem of interest is

$$\min\{f(X) : X \subseteq N\}$$

The following remarkable construction that connects submodular function minimization with convex function minimization is due to Lovász (see Grötschel et al., 1988).

Definition 13.2

The Lovász extension $\hat{f}(.)$ of a submodular function $f(.)$ satisfies

- $\hat{f} : [0, 1]^N \to \Re$.
- $\hat{f}(\mathbf{x}) = \sum_{I \in \mathcal{I}} \lambda_I f(\mathbf{x}_I)$ where $\mathbf{x} = \sum_{I \in \mathcal{I}} \lambda_I \mathbf{x}_I$, $\mathbf{x} \in [0, 1]^N$, $\mathbf{x}_I$ is the incidence vector of I for each $I \in \mathcal{I}$, $\lambda_I > 0$ for each I in $\mathcal{I}$, and $\mathcal{I} = \{I_1, I_2, \ldots, I_k\}$ with $\varnothing \neq I_1 \subset I_2 \subset \cdots \subset I_k \subseteq N$. Note that the representation $\mathbf{x} = \sum_{I \in \mathcal{I}} \lambda_I \mathbf{x}_I$ is unique given that the $\lambda_I > 0$ and that the sets in $\mathcal{I}$ are nested.

It is easy to check that $\hat{f}(.)$ is a convex function. Lovász also showed that the minimization of the submodular function $f(.)$ is a special case of convex programming by proving

$$\min\{f(X) : X \subseteq N\} = \min\{\hat{f}(\mathbf{x}) : \mathbf{x} \in [0, 1]^N\}$$

Further, if $\mathbf{x}^*$ is an optimal solution to the convex program and

$$\mathbf{x}^* = \sum_{I \in \mathcal{I}} \lambda_I \mathbf{x}_I$$

then for each $\lambda_I > 0$, it can be shown that $I \in \mathcal{I}$ minimizes f. The ellipsoid method can be used to solve this convex program (and hence submodular minimization) using a polynomial number of calls to an oracle for f [this oracle returns the value of $f(X)$ when input X].

13.2.7 Interior Point Methods

The announcement of the polynomial solvability of linear programming followed by the probabilistic analyses of the simplex method in the early 1980s left researchers in linear programming with a dilemma. We had one method that was good in a theoretical sense but poor in practice and another that was good in practice (and on average) but poor in a theoretical worst-case sense. This left the door wide open for a method that was good in both senses. Narendra Karmarkar closed this gap with a breathtaking new projective scaling algorithm. In retrospect, the new algorithm has been identified with a class of nonlinear programming methods known as logarithmic barrier methods. Implementations of a primal–dual variant of the logarithmic barrier method have proven to be the best approach at present. It is this variant that we describe.

It is well known that moving through the interior of the feasible region of a linear program using the negative of the gradient of the objective function, as the movement direction, runs into trouble because of getting *jammed* into corners (in high dimensions, corners make up most of the interior of a polyhedron). This jamming can be overcome if the negative gradient is balanced with a *centering* direction. The centering direction in Karmarkar's algorithm is based on the *analytic center* $\mathbf{y}_c$ of a full-dimensional polyhedron $\mathcal{D} = \{\mathbf{y} : A^T\mathbf{y} \le c\}$ which is the unique optimal solution to

$$\max\left\{\sum_{j=1}^{n} \ell n(\mathbf{z}_j) : A^T\mathbf{y} + \mathbf{z} = \mathbf{c}\right\}$$

Recall the primal and dual forms of a linear program may be taken as

$$(P) \quad \min\{\mathbf{cx} : A\mathbf{x} = \mathbf{b}, \mathbf{x} \ge \mathbf{0}\}$$

$$(D) \quad \max\{\mathbf{b}^T\mathbf{y} : A^T\mathbf{y} \le \mathbf{c}\}$$

The logarithmic barrier formulation of the dual (D) is

$$(D_\mu) \quad \max\left\{\mathbf{b}^T\mathbf{y} + \mu\sum_{j=1}^{n} \ell n(\mathbf{z}_j) : A^T\mathbf{y} + \mathbf{z} = \mathbf{c}\right\}$$

Notice that (D_μ) is equivalent to (D) as $\mu \to 0^+$. The optimality (Karush–Kuhn–Tucker) conditions for (D_μ) are given by

$$D_\mathbf{x} D_\mathbf{z} \mathbf{e} = \mu\mathbf{e}$$

$$A\mathbf{x} = \mathbf{b} \tag{13.1}$$

$$A^T\mathbf{y} + \mathbf{z} = \mathbf{c}$$

where $D_\mathbf{x}$ and $D_\mathbf{z}$ denote $n \times n$ diagonal matrices whose diagonals are $\mathbf{x}$ and $\mathbf{z}$, respectively. Notice that if we set μ to 0, the above conditions are precisely the primal–dual optimality conditions: complementary slackness, primal and dual feasibility of a pair of optimal (*P*) and (*D*) solutions. The problem has been reduced to solving the equations in **x**, **y**, **z**. The classical technique for solving equations is Newton's method, which prescribes the directions,

$$\Delta \mathbf{y} = -(AD_\mathbf{x}D_\mathbf{z}^{-1}A^T)^{-1}AD_\mathbf{z}^{-1}(\mu\mathbf{e} - D_\mathbf{x}D_\mathbf{z}\mathbf{e})\Delta\mathbf{z} = -A^T\Delta\mathbf{y}\,\Delta\mathbf{x}$$
$$= D_\mathbf{z}^{-1}(\mu\mathbf{e} - D_\mathbf{x}D_\mathbf{z}\mathbf{e}) - D_\mathbf{x}D_\mathbf{z}^{-1}\Delta\mathbf{z} \qquad (13.2)$$

The strategy is to take one Newton step, reduce μ, and iterate until the optimization is complete. The criterion for stopping can be determined by checking for feasibility ($\mathbf{x}, \mathbf{z} \geq \mathbf{0}$) and if the duality gap ($\mathbf{x}^t\mathbf{z}$) is close enough to 0. We are now ready to describe the algorithm.

Procedure 13.2 Primal–Dual Interior:

0. **Initialize:**
   $\mathbf{x}_0 > 0, \mathbf{y}_0 \in \Re^m, \mathbf{z}_0 > 0, \mu_0 > 0, \epsilon > 0, \rho > 0$
   $k := 0$
1. **Iterative step:**
   do
   Stop if $A\mathbf{x}_k = \mathbf{b}, A^T\mathbf{y}_k + \mathbf{z}_k = \mathbf{c}$ **and** $\mathbf{x}_k^T\mathbf{z}_k \leq \epsilon$.
   $\mathbf{x}_{k+1} \leftarrow \mathbf{x}_k + \alpha_k^P \Delta\mathbf{x}_k$
   $\mathbf{y}_{k+1} \leftarrow \mathbf{y}_k + \alpha_k^D \Delta\mathbf{y}_k$
   $\mathbf{z}_{k+1} \leftarrow \mathbf{z}_k + \alpha_k^D \Delta\mathbf{z}_k$
   /* $\Delta\mathbf{x}_k, \Delta\mathbf{y}_k, \Delta\mathbf{z}_k$ are the Newton directions from (1) */
   $\mu_{k+1} \leftarrow \rho\mu_k$
   $k := k + 1$
   od
2. **End**

Remark 13.4

The step sizes α_k^P and α_k^D are chosen to keep $\mathbf{x}_{k+1}$ and $\mathbf{z}_{k+1}$ strictly positive. The ability in the primal–dual scheme to choose separate step sizes for the primal and dual variables is a major advantage that this method has over the pure primal or dual methods. Empirically, this advantage translates to a significant reduction in the number of iterations.

Remark 13.5

The stopping condition essentially checks for primal and dual feasibility and near complementary slackness. Exact complementary slackness is not possible with interior solutions. It is possible to maintain primal and dual feasibility through the algorithm, but this would require a phase I construction via artificial variables. Empirically, this feasible variant has not been found to be worthwhile. In any case, when the algorithm terminates with an interior solution, a post-processing step is usually invoked to obtain optimal extreme point solutions for the primal and dual. This is usually called the *purification* of solutions and is based on a clever scheme described by Megiddo (1991).

Remark 13.6

Instead of using Newton steps to drive the solutions to satisfy the optimality conditions of (D_μ), Mehrotra (1992) suggested a predictor–corrector approach based on power series approximations. This approach has the added advantage of providing a rational scheme for reducing the value of μ. It is the predictor-corrector-based primal–dual interior method that is considered the current winner in interior point methods. The OB1 code of Lustig et al. (1994) is based on this scheme.

Remark 13.7

CPLEX, a general purpose linear (and integer) programming solver, contains implementations of interior point methods. Computational studies in the late 1990s of parallel implementations of simplex and interior point methods on the SGI power challenge (SGI R8000) platform indicated that on all but a few small linear programs in the NETLIB linear programming benchmark problem set, interior point methods dominated the simplex method in run times. New advances in handling Cholesky factorizations in parallel were apparently the reason for this exceptional performance of interior point methods. For the simplex method, CPLEX incorporated efficient methods of solving triangular linear systems and faster updating of reduced costs for identifying improving edge directions. For the interior point method, the same code included improvements in computing Cholesky factorizations and better use of level-two cache available in contemporary computing architectures. Using CPLEX 6.5 and CPLEX 5.0, Bixby et al. (2001) carried out extensive computational testing comparing the two codes with respect to the performance of the Primal simplex, Dual simplex, and Interior Point methods as well as a comparison of the performance of these three methods. While CPLEX 6.5 considerably outperformed CPLEX 5.0 for all the three methods, the comparison among the three methods was inconclusive. However, as stated by Bixby et al. (2001), the computational testing was biased against interior point method because of the inferior floating point performance of the machine used and the nonimplementation of the parallel features on shared memory machines. In a recent review of interior point methods, Jacek Gondzio (2012) compared CPLEX 11.0.1 with matrix-free interior point methods for addressing LP relaxations of quadratic assignment problems and some challenging linear optimization problems in quantum physics. Gondzio concludes that the iterative methods of linear algebra are worth considering in situations where direct matrix factorizations are impractical.

Remark 13.8

Karmarkar (1990) has proposed an interior-point approach for integer programming problems. The main idea is to reformulate an integer program as the minimization of a quadratic energy function over linear constraints on continuous variables. Interior-point methods are applied to this formulation to find local optima.

13.3 Large-Scale Linear Programming in Combinatorial Optimization

Linear programming problems with thousands of rows and columns are routinely solved either by variants of the simplex method or by interior point methods. However, for several linear programs that arise in combinatorial optimization, the number of columns (or rows in the dual) are too numerous to be enumerated explicitly. The columns, however, often have a structure which is exploited to generate the columns as and when required in the simplex method. Such an approach, which is referred to as *column generation*, is illustrated next on the *cutting stock problem* (Gilmore and Gomory, 1963), which is also known as the *bin packing problem* in the computer science literature.

13.3.1 Cutting Stock Problem

Rolls of sheet metal of standard length L are used to cut required lengths l_i, $i = 1, 2,\ldots, m$. The jth cutting pattern should be such that a_{ij}, the number of sheets of length l_i cut from one roll of standard length L, must satisfy $\sum_{i=1}^{m} a_{ij} l_i \le L$. Suppose n_i, $i = 1, 2,\ldots, m$ sheets of length l_i are required. The problem is to find cutting patterns so as to minimize the number of rolls of standard length L that are used to meet the requirements. A linear programming formulation of the problem is as follows.

Let $\mathbf{x}_j$, $j = 1, 2,\ldots, n$, denote the number of times the jth cutting pattern is used. In general, $\mathbf{x}_j$, $j = 1, 2,\ldots, n$ should be an integer but in the next formulation the variables are permitted to be fractional.

$$\text{(P1)} \quad \text{Min} \sum_{j=1}^{n} \mathbf{x}_j$$

$$\text{Subject to} \sum_{j=1}^{n} a_{ij}\mathbf{x}_j \ge n_i \quad i = 1, 2,\ldots, m$$

$$\mathbf{x}_j \ge 0 \quad j = 1, 2,\ldots, n$$

$$\text{where} \sum_{i=1}^{m} l_i a_{ij} \le L \quad j = 1, 2,\ldots, n$$

The formulation can easily be extended to allow for the possibility of p standard lengths L_k, $k = 1, 2,\ldots, p$, from which the n_i units of length l_i, $i = 1, 2,\ldots, m$, are to be cut.

The cutting stock problem can also be viewed as a bin packing problem. Several bins, each of standard capacity L, are to be packed with n_i units of item i, each of which uses up capacity of l_i in a bin. The problem is to minimize the number of bins used.

13.3.2 Column Generation

In general, the number of columns in (P1) is too large to enumerate all of the columns explicitly. The simplex method, however, does not require all of the columns to be explicitly written down. Given a basic feasible solution and the corresponding simplex multipliers $\mathbf{w}_i$, $i = 1, 2,\ldots, m$, the column to enter the basis is determined by applying dynamic programming to solve the following knapsack problem:

$$\text{(P2)} \quad z = \text{Max} \sum_{i=1}^{m} \mathbf{w}_i a_i$$

$$\text{Subject to} \sum_{i=1}^{m} l_i a_i \le L$$

$$a_i \ge 0 \text{ and integer,} \quad \text{for } i = 1, 2,\ldots, m$$

Let a_i^*, $i = 1, 2,\ldots, m$, denote an optimal solution to (P2). If $z > 1$, the kth column to enter the basis has coefficients $a_{ik} = a_i^*$, $i = 1, 2,\ldots, m$.

Using the identified columns, a new improved (in terms of the objective function value) basis is obtained, and the column generation procedure is repeated. A major iteration is one in which (P2) is solved to identify, if there is one, a column to enter the basis. Between two major iterations, several minor iterations may be performed to optimize the linear program using only the available (generated) columns.

If $z \leq 1$, the current basic feasible solution is optimal to (P1). From a computational point of view, alternative strategies are possible. For instance, instead of solving (P2) to optimality, a column to enter the basis can be indentified as soon as a feasible solution to (P2) with an objective function value greater than 1 has been found. Such an approach would reduce the time required to solve (P2) but may increase the number of iterations required to solve (P1).

A column once generated may be retained, even if it comes out of the basis at a subsequent iteration, so as to avoid generating the same column again later on. However, at a particular iteration some columns, which appear unattractive in terms of their reduced costs, may be discarded in order to avoid having to store a large number of columns. Such columns can always be generated again subsequently, if necessary. The rationale for this approach is that such unattractive columns will rarely be required subsequently.

The dual of (P1) has a large number of rows. Hence, column generation may be viewed as row generation in the dual. In other words, in the dual we start with only a few constraints explicitly written down. Given an optimal solution **w** to the current dual problem (i.e., with only a few constraints which have been explicitly written down) find a constraint that is violated by **w** or conclude that no such constraint exists. The problem to be solved for identifying a violated constraint, if any, is exactly the separation problem that we encountered in the section on algorithms for linear programming.

13.3.3 Decomposition and Compact Representations

Large-scale linear programs sometimes have a block diagonal structure with a few additional constraints linking the different blocks. The linking constraints are referred to as the master constraints and the various blocks of constraints are referred to as subproblem constraints. Using the representation theorem of polyhedra (see, for instance, Nemhauser and Wolsey, 1988), the decomposition approach of Dantzig and Wolfe (1961) is to convert the original problem to an equivalent linear program with a small number of constraints but with a large number of columns or variables. In the cutting stock problem described in the preceding section, the columns are generated, as and when required, by solving a knapsack problem via dynamic programming. In the Dantzig–Wolfe decomposition scheme, the columns are generated, as and when required, by solving appropriate linear programs on the subproblem constraints.

It is interesting to note that the reverse of decomposition is also possible. In other words, suppose we start with a statement of a problem and an associated linear programming formulation with a large number of columns (or rows in the dual). If the column generation (or row generation in the dual) can be accomplished by solving a linear program, then a *compact* formulation of the original problem can be obtained. Here compact refers to the number of rows and columns being bounded by a polynomial function of the input length of the original problem. This result due to Martin (1991) enables one to solve the problem in the polynomial time by solving the compact formulation using interior point methods.

13.4 Integer Linear Programs

Integer linear programming problems (ILPs) are linear programs in which all of the variables are restricted to be integers. If only some but not all variables are restricted to be integers, the problem is referred to as a mixed integer program. Many combinatorial problems can be formulated as integer linear programs in which all of the variables are restricted to be 0 or 1. We will first discuss several examples of combinatorial optimization problems and their formulation as integer programs. Then we will review a general representation theory for integer programs that gives a formal measure of the expressiveness of this algebraic approach. We conclude this section with a representation theorem due to Benders (1962), which has been very useful in solving certain large-scale combinatorial optimization problems in practice.

13.4.1 Example Formulations

13.4.1.1 Covering and Packing Problems

A wide variety of location and scheduling problems can be formulated as set covering or set packing or set partitioning problems. The three different types of *covering and packing* problems can be succinctly stated as follows: Given (1) a finite set of elements $\mathcal{M} = \{1, 2,\ldots, m\}$, and (2) a family F of subsets of $\mathcal{M}$ with each member F_j, $j = 1, 2,\ldots, n$ having a profit (or cost) $\mathbf{c}_j$ associated with it, find a collection, S, of the members of F that maximizes the profit (or minimizes the cost) while ensuring that every element of $\mathcal{M}$ is in one of the following:

(P3): at most one member of S (set packing problem)
(P4): at least one member of S (set covering problem)
(P5): exactly one member of S (set partitioning problem)

The three problems (P3), (P4), and (P5) can be formulated as ILPs as follows:
Let A denote the $m \times n$ matrix where

$$A_{ij} = \begin{cases} 1 & \text{if element } i \in F_j \\ 0 & \text{otherwise} \end{cases}$$

The decision variables are $\mathbf{x}_j$, $j = 1, 2,\ldots, n$ where

$$\mathbf{x}_{ij} = \begin{cases} 1 & \text{if } F_j \text{ is chosen} \\ 0 & \text{otherwise} \end{cases}$$

The set packing problem is

$$\text{(P3)} \quad \text{Max } \mathbf{cx}$$

$$\text{Subject to} \quad A\mathbf{x} \le \mathbf{e}_m$$

$$\mathbf{x}_j = 0 \text{ or } 1, \quad j = 1, 2, \ldots, n$$

where $\mathbf{e}_m$ is an m-dimensional column vector of ones.

The set covering problem (P4) is (P3) with less than or equal to constraints replaced by greater than or equal to constraints and the objective is to minimize rather than maximize. The set partitioning problem (P5) is (P3) with the constraints written as equalities. The set partitioning problem can be converted to a set packing problem or set covering problem (see Padberg, 1995) using standard transformations. If the right-hand side vector $\mathbf{e}_m$ is replaced by a nonnegative integer vector $\mathbf{b}$, (P3) is referred to as the generalized set packing problem.

The airline crew scheduling problem is a classic example of the set partitioning or the set covering problem. Each element of $\mathcal{M}$ corresponds to a flight segment. Each subset F_j corresponds to an acceptable set of flight segments of a crew. The problem is to cover, at minimum cost, each flight segment exactly once. This is a set partitioning problem. If *dead heading* of crew is permitted, we have the set covering problem.

13.4.1.2 Packing and Covering Problems in a Graph

Suppose A is the node-edge incidence matrix of a graph. Now, (P3) is a weighted matching problem. If in addition, the right-hand side vector $\mathbf{e}_m$ is replaced by a nonnegative integer vector $\mathbf{b}$, (P3) is referred to as a weighted $\mathbf{b}$-matching problem. In this case, each variable $\mathbf{x}_j$ which is restricted to be an integer may have a positive upper bound of u_j. Problem (P4) is now referred to as the weighted edge covering problem. Note that by substituting for $\mathbf{x}_j = 1 - \mathbf{y}_j$, where $\mathbf{y}_j = 0$ or 1, the weighted edge covering problem is transformed to a weighted $\mathbf{b}$-matching problem in which the variables are restricted to be 0 or 1.

Suppose A is the edge-node incidence matrix of a graph. Now, (P3) is referred to as the weighted vertex packing problem and (P4) is referred to as the weighted vertex covering problem. The *set packing* problem can be transformed to a weighted vertex packing problem in a graph G as follows:

> G contains a node for each $\mathbf{x}_j$ and an edge between nodes j and k exists if and only if the columns $A_{\cdot j}$ and $A_{\cdot k}$ are not orthogonal. G is called the *intersection graph* of A. The set packing problem is equivalent to the weighted vertex packing problem on G. Given G, the complement graph $\bar{G}$ has the same node set as G and there is an edge between nodes j and k in $\bar{G}$ if and only if there is no such corresponding edge in G. A clique in a graph is a subset, k, of nodes of G such that the subgraph induced by k is complete. Clearly, the weighted vertex packing problem in G is equivalent to finding a maximum weighted clique in $\bar{G}$.

13.4.1.3 Plant Location Problems

Given a set of customer locations $N = \{1, 2,\ldots, n\}$ and a set of potential sites for plants $M = \{1, 2,\ldots, m\}$, the plant location problem is to identify the sites where the plants are to be located so that the customers are served at a minimum cost. There is a fixed cost $\mathbf{f}_i$ of locating the plant at site i and the cost of serving customer j from site i is $\mathbf{c}_{ij}$. The decision variables are: $\mathbf{y}_i$ is set to 1 if a plant is located at site i and to 0 otherwise; $\mathbf{x}_{ij}$ is set to 1 if site i serves customer j and to 0 otherwise.

A formulation of the problem is

$$\text{(P6)}\quad \text{Min} \sum_{i=1}^{m}\sum_{j=1}^{n} \mathbf{c}_{ij}\mathbf{x}_{ij} + \sum_{i=1}^{m} \mathbf{f}_i\mathbf{y}_i$$

$$\text{subject to} \quad \sum_{i=1}^{m} \mathbf{x}_{ij} = 1 \qquad j = 1, 2,\ldots, n$$

$$\mathbf{x}_{ij} - \mathbf{y}_i \le 0 \qquad i = 1, 2,\ldots, m;\quad j = 1, 2,\ldots, n$$

$$\mathbf{y}_i = 0 \text{ or } 1 \quad i = 1, 2,\ldots, m$$

$$\mathbf{x}_{ij} = 0 \text{ or } 1 \quad i = 1, 2,\ldots, m;\quad j = 1, 2,\ldots, n$$

Note that the constraints $\mathbf{x}_{ij} - \mathbf{y}_i \le 0$ are required to ensure that customer j may be served from site i only if a plant is located at site i. Note that the constraints $\mathbf{y}_i = 0$ or 1 force an optimal solution in which $\mathbf{x}_{ij} = 0$ or 1. Consequently, the $\mathbf{x}_{ij} = 0$ or 1 constraints may be replaced by nonnegativity constraints $\mathbf{x}_{ij} \ge 0$.

The linear programming relaxation associated with (P6) is obtained by replacing constraints $\mathbf{y}_i = 0$ or 1 and $\mathbf{x}_{ij} = 0$ or 1 by nonnegativity contraints on $\mathbf{x}_{ij}$ and $\mathbf{y}_i$. The upper bound constraints on $\mathbf{y}_i$ are not required provided $\mathbf{f}_i \ge 0$, $i = 1, 2,\ldots, m$. The upper bound constraints on $\mathbf{x}_{ij}$ are not required in view of constraints $\sum_{i=1}^{m} \mathbf{x}_{ij} = 1$.

Remark 13.9

It is frequently possible to formulate the same combinatorial problem as two or more different ILPs. Suppose we have two ILP formulations (F1) and (F2) of the given combinatorial problem with both (F1) and (F2) being minimizing problems. Formulation (F1) is said to be stronger than (F2) if (LP1), the linear programming relaxation of (F1), always has an optimal objective function value which is greater than or equal to the optimal objective function value of (LP2), which is the linear programming relaxation of (F2).

It is possible to reduce the number of constraints in (P6) by replacing the constraints $\mathbf{x}_{ij} - \mathbf{y}_i \leq 0$ by an aggregate:

$$\sum_{j=1}^{n} \mathbf{x}_{ij} - n\mathbf{y}_i \leq 0 \quad i = 1, 2, \ldots, m$$

However, the disaggregated (P6) is a stronger formulation than the formulation obtained by aggregating the constraints as previously. By using standard transformations, (P6) can also be converted into a set packing problem.

13.4.1.4 Satisfiability and Inference Problems

In propositional logic, a truth assignment is an assignment of true or false to each atomic proposition $\mathbf{x}_1$, $\mathbf{x}_2, \ldots \mathbf{x}_n$. A literal is an atomic proposition $\mathbf{x}_j$ or its negation $\neg\mathbf{x}_j$. For propositions in conjunctive normal form, a clause is a disjunction of literals and the proposition is a conjunction of clauses. A clause is obviously satisfied by a given truth assignment if at least one of its literals is true. The satisfiability problem consists of determining whether there exists a truth assignment to atomic propositions such that a set S of clauses is satisfied.

Let T_i denote the set of atomic propositions such that if any one of them is assigned true, the clause $i \in S$ is satisfied. Similarly, let F_i denote the set of atomic propositions such that if any one of them is assigned false, the clause $i \in S$ is satisfied.

The decision variables are

$$\mathbf{x}_j = \begin{cases} 1 & \text{if atomic proposition } j \text{ is assigned true} \\ 0 & \text{if atomic proposition } j \text{ is assigned false} \end{cases}$$

The satisfiability problem is to find a feasible solution to

$$\text{(P7)} \quad \sum_{j \in T_i} \mathbf{x}_j - \sum_{j \in F_i} \mathbf{x}_j \geq 1 - |F_i| \quad i \in S$$

$$\mathbf{x}_j = 0 \text{ or } 1 \quad \text{for } j = 1, 2, \ldots, n$$

By substituting $\mathbf{x}_j = 1 - \mathbf{y}_j$, where $\mathbf{y}_j = 0$ or 1, for $j \in F_i$, (P7) is equivalent to the set covering problem

$$\text{(P8)} \quad \text{Min} \sum_{j=1}^{n} (\mathbf{x}_j + \mathbf{y}_j) \tag{13.3}$$

$$\text{subject to} \quad \sum_{j \in T_i} \mathbf{x}_j + \sum_{j \in F_i} \mathbf{y}_j \geq 1 \quad i \in S \tag{13.4}$$

$$\mathbf{x}_j + \mathbf{y}_j \geq 1 \quad j = 1, 2, \ldots, n \tag{13.5}$$

$$\mathbf{x}_j, \mathbf{y}_j = 0 \text{ or } 1 \quad j = 1, 2, \ldots, n \tag{13.6}$$

Clearly, (P7) is feasible if and only if (P8) has an optimal objective function value equal to n.

Given a set S of clauses and an additional clause $k \notin S$, the logical inference problem is to find out whether every truth assignment that satisfies all of the clauses in S also satisfies the clause k. The logical inference problem is

$$\text{(P9)} \quad \text{Min} \quad \sum_{j \in T_k} \mathbf{x}_j - \sum_{j \in F_k} \mathbf{x}_j$$

$$\text{subject to} \quad \sum_{j \in T_i} \mathbf{x}_j - \sum_{j \in F_i} \mathbf{x}_j \geq 1 - |F_i| \quad i \in S$$

$$\mathbf{x}_j = 0 \text{ or } 1 \quad j = 1, 2, \ldots, n$$

The clause k is implied by the set of clauses S, if and only if (P9) has an optimal objective function value greater than $-|F_k|$. It is also straightforward to express the MAX-SAT problem (i.e., find a truth assignment that maximizes the number of satisfied clauses in a given set S) as an integer linear program.

13.4.1.5 Multiprocessor Scheduling

Given n jobs and m processors, the problem is to allocate each job to one and only one of the processors so as to minimize the make span time, that is, minimize the completion time of all of the jobs. The processors may not be identical and, hence, job j if allocated to processor i requires p_{ij} units of time. The multiprocessor scheduling problem is

$$\text{(P10)} \quad \text{Min} \quad T$$

$$\text{subject to} \quad \sum_{i=1}^{m} \mathbf{x}_{ij} = 1 \quad j = 1, 2, \ldots, n$$

$$\sum_{j=1}^{n} \mathbf{p}_{ij} \mathbf{x}_{ij} - T \leq 0 \quad i = 1, 2, \ldots, m$$

$$\mathbf{x}_{ij} = 0 \text{ or } 1$$

Note that if all $\mathbf{p}_{ij}$ are integers, the optimal solution will be such that T is an integer.

Remark 13.10

Protein Folding is a challenging optimization problem in biology which has seen a great deal of research by computational scientists and structural biologists. The computational complexity has been established even for very restricted lattice models to be NP-Hard and even hard to approximate within a guaranteed factor. Chandru et al. (2003) survey the hardness results for lattice embeddings of the folding problem and also review the results on approximation methods with a guarantee of $\frac{1}{3}$ of optimal due to Newman for folding in the plane and $\frac{3}{8}$ for folding in 3D which is due to hart and Istrail for the so-called *Hydrophobic-Hydrophylic (HP) model* of energy minimization. For the same HP model, Chandru et al. (2004) show that the lattice version of the protein folding problem can be formulated as an Integer Programming problem and show that B-and-B methods can improve on the approximations produced by the approximation methods mentioned above.

13.4.2 Jeroslow's Representability Theorem

Jeroslow (1989), building on joint work with Lowe (see Jeroslow and Lowe [1984]), characterized subsets of n-space that can be represented as the feasible region of a mixed integer (Boolean) program. They proved that a set is the feasible region of some mixed integer/linear programming problem (MILP) if and only if it is the union of finitely many polyhedra having the same recession cone (defined subsequently). Although this result is not widely known, it might well be regarded as the fundamental theorem of mixed integer modeling.

The basic idea of Jeroslow's results is that any set that can be represented in a mixed integer model can be represented in a disjunctive programming problem (i.e., a problem with either/or constraints). A *recession direction* for a set S in n-space is a vector $\mathbf{x}$ such that $s + \alpha\mathbf{x} \in S$ for all $s \in S$ and all $\alpha \geq 0$. The set of recession directions is denoted $rec(S)$. Consider the general mixed integer constraint set

$$\mathbf{f}(\mathbf{x}, \mathbf{y}, \boldsymbol{\lambda}) \leq \mathbf{b}$$

$$\mathbf{x} \in \mathfrak{R}^n, \quad \mathbf{y} \in \mathfrak{R}^p \tag{13.7}$$

$$\boldsymbol{\lambda} = (\lambda_1, \ldots, \lambda_k), \quad \text{with} \quad \lambda_j \in \{0, 1\} \quad \text{for } j = 1, \ldots, k$$

Here $\mathbf{f}$ is a vector-valued function, so that $\mathbf{f}(\mathbf{x}, \mathbf{y}, \boldsymbol{\lambda}) \leq \mathbf{b}$ represents a set of constraints. We say that a set $S \subset \mathfrak{R}^n$ is *represented* by Equation 13.7 if,

$$\mathbf{x} \in S \quad \text{if and only if } (\mathbf{x}, \mathbf{y}, \boldsymbol{\lambda}) \text{ satisfies Equation 13.7 for some } y, \lambda.$$

If $\mathbf{f}$ is a linear transformation, so that Equation 13.7 is a MILP constraint set, we will say that S is *MILP representable*. The main result can now be stated.

Theorem 13.2

A set in n-space is *MILP* representable if and only if it is the union of finitely many polyhedra having the same set of recession directions (Jeroslow and Lowe, 1984; Jeroslow, 1989).

13.4.3 Benders's Representation

Any MILP can be reformulated so that there is only one continuous variable. This reformulation, due to Benders (1962), will in general have an exponential number of constraints. Analogous to column generation, discussed earlier, these rows (constraints) can be generated as and when required.

Consider the (MILP)

$$\max\{\mathbf{cx} + \mathbf{dy} : A\mathbf{x} + G\mathbf{y} \leq \mathbf{b}, \mathbf{x} \geq \mathbf{0}, \mathbf{y} \geq \mathbf{0} \text{ and integer}\}$$

Suppose the integer variables **y** are fixed at some values. Then, the associated linear program is

$$\text{(LP)} \quad \max\{\mathbf{cx} : \mathbf{x} \in \mathcal{P} = \{\mathbf{x} : A\mathbf{x} \le \mathbf{b} - G\mathbf{y}, \mathbf{x} \ge \mathbf{0}\}\}$$

and its dual is

$$\text{(DLP)} \quad \min\{\mathbf{w}(\mathbf{b} - G\mathbf{y}) : \mathbf{w} \in \mathcal{Q} = \{\mathbf{w} : \mathbf{w}A \ge \mathbf{c}, \mathbf{w} \ge \mathbf{0}\}\}$$

Let $\{\mathbf{w}^k\}$, $k = 1, 2, \ldots, K$ be the extreme points of $\mathcal{Q}$ and $\{\mathbf{u}^j\}$, $j = 1, 2, \ldots, J$ be the extreme rays of the recession cone of $\mathcal{Q}$, $\mathcal{C}_{\mathcal{Q}} = \{\mathbf{u} : \mathbf{u}A \ge \mathbf{0}, \mathbf{u} \ge \mathbf{0}\}$. Note that if $\mathcal{Q}$ is nonempty, the $\{\mathbf{u}^j\}$ are all of the extreme rays of $\mathcal{Q}$.

From linear programming duality, we know that if $\mathcal{Q}$ is empty and $\mathbf{u}^j(\mathbf{b} - G\mathbf{y}) \ge 0$, $j = 1, 2, \ldots, J$ for some $\mathbf{y} \ge \mathbf{0}$ and integer then (LP) and consequently (MILP) have an unbounded solution. If $\mathcal{Q}$ is nonempty and $\mathbf{u}^j(\mathbf{b} - G\mathbf{y}) \ge 0$, $j = 1, 2, \ldots, J$ for some $\mathbf{y} \ge \mathbf{0}$ and integer then (LP) has a finite optimum given by

$$\min_k \{\mathbf{w}^k(\mathbf{b} - G\mathbf{y})\}$$

Hence, an equivalent formulation of (MILP) is

$$
\begin{aligned}
&\text{Max } \alpha \\
&\alpha \le \mathbf{dy} + \mathbf{w}^k(\mathbf{b} - G\mathbf{y}), \quad k = 1, 2, \ldots, K \\
&\mathbf{u}^j(\mathbf{b} - G\mathbf{y}) \ge 0, \quad j = 1, 2, \ldots, J \\
&y \ge \mathbf{0} \text{ and integer} \\
&\alpha \quad \text{unrestricted}
\end{aligned}
$$

which has only one continuous variable α as promised.

13.5 Polyhedral Combinatorics

One of the main purposes of writing down an algebraic formulation of a combinatorial optimization problem as an integer program is to then examine the linear programming relaxation and understand how well it represents the discrete integer program. There are somewhat special but rich classes of such formulations for which the linear programming relaxation is sharp or tight. These correspond to linear programs that have integer valued extreme points. Such polyhedra are called *integral polyhedra*.

13.5.1 Special Structures and Integral Polyhedra

A natural question of interest is whether the LP associated with an ILP has only integral extreme points. For instance, the linear programs associated with matching and edge covering polytopes in a bipartite graph have only integral vertices. Clearly, in such a situation, the ILP can be solved as LP. A polyhedron or a polytope is referred to as being integral if it is either empty or has only integral vertices.

Definition 13.3

A 0, ±1 matrix is totally unimodular if the determinant of every square submatrix is 0 or ±1.

Theorem 13.3

Let

$$A = \begin{pmatrix} A_1 \\ A_2 \\ A_3 \end{pmatrix}$$

be a 0, ±1 matrix and

$$\mathbf{b} = \begin{pmatrix} \mathbf{b}_1 \\ \mathbf{b}_2 \\ \mathbf{b}_3 \end{pmatrix}$$

be a vector of appropriate dimensions (Hoffman and Kruskal, 1956). Then A is totally unimodular if and only if the polyhedron

$$P(A, \mathbf{b}) = \{\mathbf{x} : A_1\mathbf{x} \le \mathbf{b}_1; A_2\mathbf{x} \ge \mathbf{b}_2; A_3\mathbf{x} = \mathbf{b}_3; \mathbf{x} \ge 0\}$$

is integral for all integral vectors **b**.

The constraint matrix associated with a network flow problem (see, for instance, Ahuja et al., 1993) is totally unimodular. Note that for a given integral **b**, $P(A, \mathbf{b})$ may be integral even if A is not totally unimodular.

Definition 13.4

A polyhedron defined by a system of linear constraints is totally dual integral (TDI) if for each objective function with integral coefficient the dual linear program has an integral optimal solution whenever an optimal solution exists.

Theorem 13.4

(Edmonds and Giles, 1977). If $P(A) = \{\mathbf{x} : A\mathbf{x} \le \mathbf{b}\}$ is *TDI* and **b** is integral, then $P(A)$ is integral.

Hoffman and Kruskal (1956) have, in fact, shown that the polyhedron $P(A, \mathbf{b})$ defined in Theorem 13.3 is TDI. This follows from Theorem 13.3 and the fact that A is totally unimodular if and only if A^T is totally unimodular.

Balanced matrices, first introduced by Berge (1972), have important implications for packing and covering problems (see also Berge and Las Vergnas, 1970).

Definition 13.5

A 0, 1 matrix is balanced if it does not contain a square submatrix of odd order with two 1s per row and column.

Theorem 13.5

Let A be a balanced 0, 1 matrix (Berge, 1972; Fulkerson et al., 1974). Then the set packing, set covering, and set partitioning polytopes associated with A are integral, that is, the polytopes

$$P(A) = \{\mathbf{x} : \mathbf{x} \geq 0;\ A\mathbf{x} \leq 1\}$$

$$Q(A) = \{\mathbf{x} : 0 \leq \mathbf{x} \leq 1;\ A\mathbf{x} \geq 1\}$$

$$R(A) = \{\mathbf{x} : \mathbf{x} \geq 0;\ A\mathbf{x} = 1\}$$

are integral.

Let

$$A = \begin{pmatrix} A_1 \\ A_2 \\ A_3 \end{pmatrix}$$

be a balanced 0, 1 matrix. Fulkerson et al. (1974) have shown that the polytope $P(A) = \{\mathbf{x} : A_1\mathbf{x} \leq \mathbf{1};\ A_2\mathbf{x} \geq \mathbf{1};\ A_3\mathbf{x} = \mathbf{1};\ \mathbf{x} \geq \mathbf{0}\}$ is TDI and by the theorem of Edmonds and Giles (1977) it follows that $P(A)$ is integral.

Truemper (1992) has extended the definition of balanced matrices to include 0, ±1 matrices.

Definition 13.6

A 0, ±1 matrix is balanced if for every square submatrix with exactly two nonzero entries in each row and each column, the sum of the entries is a multiple of 4.

Theorem 13.6

Suppose A is a balanced 0, ±1 matrix (Conforti and Cornuejols, 1992b). Let $\mathbf{n}(A)$ denote the column vector whose ith component is the number of $-1s$ in the ith row of A. Then the polytopes

$$P(A) = \{\mathbf{x}: A\mathbf{x} \leq \mathbf{1} - \mathbf{n}(A); \mathbf{0} \leq \mathbf{x} \leq \mathbf{1}\}$$

$$Q(A) = \{\mathbf{x}: A\mathbf{x} \geq \mathbf{1} - \mathbf{n}(A); \mathbf{0} \leq \mathbf{x} \leq \mathbf{1}\}$$

$$R(A) = \{\mathbf{x}: A\mathbf{x} = \mathbf{1} - \mathbf{n}(A); \mathbf{0} \leq \mathbf{x} \leq \mathbf{1}\}$$

are integral.

Note that a 0, ±1 matrix A is balanced if and only if A^T is balanced. Moreover, A is balanced (totally unimodular) if and only if every submatrix of A is balanced (totally unimodular). Thus, if A is balanced (totally unimodular), it follows that Theorem 13.6 (Theorem 13.3) holds for every submatrix of A.

Totally unimodular matrices constitute a subclass of balanced matrices, that is, a totally unimodular 0, ±1 matrix is always balanced. This follows from a theorem of Camion (1965), which states that a

$$A = \begin{bmatrix} 1 & 1 & 0 & 0 \\ 1 & 1 & 1 & 1 \\ 1 & 0 & 1 & 0 \\ 1 & 0 & 0 & 1 \end{bmatrix} \quad A = \begin{bmatrix} 1 & 1 & 0 \\ 0 & 1 & 1 \\ 1 & 0 & 1 \\ 1 & 1 & 1 \end{bmatrix}$$

FIGURE 13.1 A balanced matrix and a perfect matrix. (From Chandru, V. and Rao, M. R., *ACM Comput. Survey*, 28, 55, 1996.)

0, ±1 is totally unimodular if and only if for every square submatrix with an even number of nonzero entries in each row and in each column, the sum of the entries equals a multiple of 4. The 4 × 4 matrix in Figure 13.1 illustrates the fact that a balanced matrix is not necessarily totally unimodular. Balanced 0, ±1 matrices have implications for solving the satisfiability problem. If the given set of clauses defines a balanced 0, ±1 matrix, then as shown by Conforti and Cornuejols (1992b), the satisfiability problem is trivial to solve and the associated MAXSAT problem is solvable in polynomial time by linear programming. A survey of balanced matrices is in Conforti et al. (1994a,b).

Definition 13.7

A 0, 1 matrix A is perfect if the set packing polytope $P(A) = \{\mathbf{x} : A\mathbf{x} \leq \mathbf{1}; \mathbf{x} \geq \mathbf{0}\}$ is integral.

The chromatic number of a graph is the minimum number of colors required to color the vertices of the graph so that no two vertices with the same color have an edge incident between them. A graph G is perfect if for every node induced subgraph H, the chromatic number of H equals the number of nodes in the maximum clique of H.

Recall that the complement graph of G is the graph $\bar{G}$ with the same node set as G but the edge set comprises only of all the edges not in G. The perfect graph conjecture due to Berge (1961) is that G is perfect if and only if $\bar{G}$ is perfect. This conjecture was proved by Lovász (1972). The strong perfect graph conjecture, also due to Berge (1961), is that G is perfect if and only if it contains no induced sub graph which is an odd hole of length of 5 or more, or an odd antihole which is the complement of an odd hole of length 5 or more. A proof of this conjecture was announced by Chudnovsky, Robertson, Seymour and Thomas in 2002 and published in 2006 (Chudnovsky et al., 2006).

The connections between the integrality of the set packing polytope and the notion of a perfect graph, as defined by Berge (1961, 1970), are given in Fulkerson (1970), Lovasz (1972), Padberg (1974), and Chvátal (1975).

Theorem 13.7

Let A be 0, 1 matrix whose columns correspond to the nodes of a graph G and whose rows are the incidence vectors of the maximal cliques of G (Fulkerson, 1970; Lovasz, 1972; Chvátal, 1975). The graph G is perfect if and only if A is perfect.

Let G_A denote the intersection graph associated with a given 0, 1 matrix A (see Section 13.4). Clearly, a row of A is the incidence vector of a clique in G_A. In order for A to be perfect, every maximal clique of G_A must be represented as a row of A because inequalities defined by maximal cliques are facet defining. Thus, by Theorem 13.7, it follows that a 0, 1 matrix A is perfect if and only if the undominated (a row of

A is dominated if its support is contained in the support of another row of A) rows of A form the clique-node incidence matrix of a perfect graph.

Balanced matrices with 0, 1 entries constitute a subclass of 0, 1 perfect matrices, that is, if a 0, 1 matrix A is balanced, then A is perfect. The 4×3 matrix in Figure 13.1 is an example of a matrix that is perfect but not balanced.

Definition 13.8

A 0, 1 matrix A is ideal if the set covering polytope

$$Q(A) = \{\mathbf{x} : A\mathbf{x} \geq \mathbf{1}; \mathbf{0} \leq \mathbf{x} \leq \mathbf{1}\}$$

is integral.

Properties of ideal matrices are described by Lehman (1979), Padberg (1993), and Cornuejols and Novick (1994). The notion of a 0, 1 perfect (ideal) matrix has a natural extension to a 0, ±1 perfect (ideal) matrix. Some results pertaining to 0, ±1 ideal matrices are contained in Hooker (1992), whereas some results pertaining to 0, ±1 perfect matrices are given in Conforti et al. (1993).

An interesting combinatorial problem is to check whether a given 0, ±1 matrix is totally unimodular, balanced, or perfect. Seymour's (1980) characterization of totally unimodular matrices provides a polynomial-time algorithm to test whether a given 0, 1 matrix is totally unimodular. Conforti et al. (1999) give a polynomial-time algorithm to check whether a 0, 1 matrix is balanced. This has been extended by Conforti et al. (1994a,b) to check in polynomial time whether a 0, ±1 matrix is balanced. An open problem is that of checking in polynomial time whether a 0, 1 matrix is perfect. For linear matrices (a matrix is linear if it does not contain a 2×2 submatrix of all ones), this problem has been solved by Fonlupt and Zemirline (1981) and Conforti and Rao (1993).

13.5.2 Matroids

Matroids and submodular functions have been studied extensively, especially from the point of view of combinatorial optimization (see, for instance, Nemhauser and Wolsey, 1988). Matroids have nice properties that lead to efficient algorithms for the associated optimization problems. One of the interesting examples of a matroid is the problem of finding a maximum or minimum weight spanning tree in a graph. Two different but equivalent definitions of a matroid are given first. A greedy algorithm to solve a linear optimization problem over a matroid is presented. The matroid intersection problem is then discussed briefly.

Definition 13.9

Let $N = \{1, 2, \cdot, n\}$ be a finite set and let $\mathcal{F}$ be a set of subsets of N. Then $I = (N, \mathcal{F})$ is an independence system if $S_1 \in \mathcal{F}$ implies that $S_2 \in \mathcal{F}$ for all $S_2 \subseteq S_1$. Elements of $\mathcal{F}$ are called *independent sets*. A set $S \in \mathcal{F}$ is a maximal independent set if $S \cup \{j\} \notin \mathcal{F}$ for all $j \in N \backslash S$. A maximal independent set T is a maximum if $|T| \geq |S|$ for all $S \in \mathcal{F}$.

The rank $r(Y)$ of a subset $Y \subseteq N$ is the cardinality of the maximum independent subset $X \subseteq Y$. Note that $r(\phi) = 0$, $r(X) \leq |X|$ for $X \subseteq N$ and the rank function is nondecreasing, that is, $r(X) \leq r(Y)$ for $X \subseteq Y \subseteq N$.

A *matroid* $M = (N, \mathcal{F})$ is an independence system in which every maximal independent set is a maximum.

Example 13.2

Let $G = (V, E)$ be an undirected connected graph with V as the node set and E as the edge set.

1. Let $I = (E, \mathcal{F})$ where $F \in \mathcal{F}$ if $F \subseteq E$ is such that at most one edge in F is incident to each node of V, that is, $F \in \mathcal{F}$ if F is a matching in G. Then $I = (E, \mathcal{F})$ is an independence system but not a matroid.
2. Let $M = (E, \mathcal{F})$ where $F \in \mathcal{F}$ if $F \subseteq E$ is such that $G_F = (V, F)$ is a forest, that is, G_F contains no cycles. Then $M = (E, \mathcal{F})$ is a matroid and maximal independent sets of M are spanning trees.

An alternative but equivalent definition of matroids is in terms of submodular functions.

Definition 13.10

A nondecreasing integer valued submodular function r defined on the subsets of N is called a *matroid rank function* if $r(\phi) = 0$ and $r(\{j\}) \leq 1$ for $j \in N$. The pair (N, r) is called a matroid.

A nondecreasing, integer-valued, submodular function f, defined on the subsets of N, is called a *polymatroid function* if $f(\phi) = 0$. The pair (N, r) is called a *polymatroid*.

13.5.2.1 Matroid Optimization

To decide whether an optimization problem over a matroid is polynomially solvable or not, we need to first address the issue of representation of a matroid. If the matroid is given either by listing the independent sets or by its rank function, many of the associated linear optimization problems are trivial to solve. However, matroids associated with graphs are completely described by the graph and the condition for independence. For instance, the matroid in which the maximal independent sets are spanning forests, the graph $G = (V, E)$ and the independence condition of no cycles describes the matroid.

Most of the algorithms for matroid optimization problems require a test to determine whether a specified subset is independent. We assume the existence of an oracle or subroutine to do this checking in running time, which is a polynomial function of $|N| = n$.

13.5.2.1.1 Maximum Weight Independent Set

Given a matroid $M = (N, \mathcal{F})$ and weights w_j for $j \in N$, the problem of finding a maximum weight independent set is $\max_{F \in \mathcal{F}} \left\{ \sum_{j \in F} w_j \right\}$. The greedy algorithm to solve this problem is as follows:

Procedure 13.3 Greedy:

0. **Initialize:** Order the elements of N so that $w_i \geq w_{i+1}$, $i = 1, 2, \ldots, n-1$. Let $T = \phi$, $i = 1$.
1. **If** $w_i \leq 0$ or $i > n$, **stop** T is optimal, i.e., $x_j = 1$ for $j \in T$ and $x_j = 0$ for $j \notin T$. If $w_i > 0$ and $T \cup \{i\} \in \mathcal{F}$, add element i to T.
2. **Increment** i by 1 and return to step 1.

Edmonds (1970, 1971) derived a complete description of the *matroid polytope*, the convex hull of the characteristic vectors of independent sets of a matroid. While this description has a large (exponential) number of constraints, it permits the treatment of linear optimization problems on independent sets of matroids as linear programs. Cunningham (1984) describes a polynomial algorithm to solve the separation problem for the matroid polytope. The matroid polytope and the associated greedy algorithm have been extended to polymatroids (Edmonds, 1970; McDiarmid, 1975).

The separation problem for a polymatroid is equivalent to the problem of minimizing a submodular function defined over the subsets of N (see Nemhauser and Wolsey, 1988). A class of submodular functions that have some additional properties can be minimized in polynomial time by solving a maximum flow problem (Rhys, 1970; Picard and Ratliff, 1975). The general submodular function can be minimized in polynomial time by the ellipsoid algorithm (Grötschel et al., 1988).

The uncapacitated plant location problem formulated in Section 13.4 can be reduced to maximizing a submodular function. Hence, it follows that maximizing a submodular function is $\mathcal{NP}$-hard.

13.5.2.2 Matroid Intersection

A matroid intersection problem involves finding an independent set contained in two or more matroids defined on the same set of elements.

Let $G = (V_1, V_2, E)$ be a bipartite graph. Let $M_i = (E, \mathcal{F}_i)$, $i = 1, 2$, where $F \in \mathcal{F}_i$ if $F \subseteq E$ is such that no more than one edge of F is incident to each node in V_i. The set of matchings in G constitutes the intersection of the two matroids M_i, $i = 1, 2$. The problem of finding a maximum weight independent set in the intersection of two matroids can be solved in polynomial time (Lawler, 1975; Edmonds, 1970, 1979; Frank, 1981). The two (poly) matroid intersection polytope has been studied by Edmonds (1979).

The problem of testing whether a graph contains a Hamiltonian path is $\mathcal{NP}$-complete. Since this problem can be reduced to the problem of finding a maximum cordinality independent set in the intersection of three matroids, it follows that the matroid intersection problem involving three or more matroids is $\mathcal{NP}$-hard.

13.5.3 Valid Inequalities, Facets, and Cutting Plane Methods

Earlier in this section, we were concerned with conditions under which the packing and covering polytopes are integral. But, in general, these polytopes are not integral, and additional inequalities are required to have a complete linear description of the convex hull of integer solutions. The existence of finitely many such linear inequalities is guaranteed by Weyl's (1935) Theorem.

Consider the feasible region of an ILP given by

$$P_I = \{\mathbf{x} : A\mathbf{x} \le \mathbf{b};\ \mathbf{x} \ge \mathbf{0} \text{ and integer}\} \tag{13.8}$$

Recall that an inequality $\mathbf{fx} \le f_0$ is referred to as a valid inequality for P_I if $\mathbf{fx}^* \le f_0$ for all $\mathbf{x}^* \in P_I$. A valid linear inequality for $P_I(A, \mathbf{b})$ is said to be facet defining if it intersects $P_I(A, \mathbf{b})$ in a face of dimension one less than the dimension of $P_I(A, \mathbf{b})$. In the example shown in Figure 13.2, the inequality $\mathbf{x}_2 + \mathbf{x}_3 \le 1$ is a facet defining inequality of the integer hull.

Let $\mathbf{u} \ge 0$ be a row vector of appropriate size. Clearly $\mathbf{u}A\mathbf{x} \le \mathbf{ub}$ holds for every $\mathbf{x}$ in P_I. Let $(\mathbf{u}A)_j$ denote the jth component of the row vector $\mathbf{u}A$ and $\lfloor(\mathbf{u}A)_j\rfloor$ denote the largest integer less than or equal to $(\mathbf{u}A)_j$. Now, since $\mathbf{x} \in P_I$ is a vector of nonnegative integers, it follows that $\sum_j \lfloor(\mathbf{u}A)_j\rfloor \mathbf{x}_j \le \lfloor \mathbf{ub} \rfloor$ is a valid inequality for P_I. This scheme can be used to generate many valid inequalities by using different $\mathbf{u} \ge \mathbf{0}$. Any set of generated valid inequalities may be added to the constraints in Equation 13.7 and the process of generating them may be repeated with the enhanced set of inequalities. This iterative procedure of generating valid inequalities is called *Gomory–Chvátal (GC) rounding*. It is remarkable that this simple

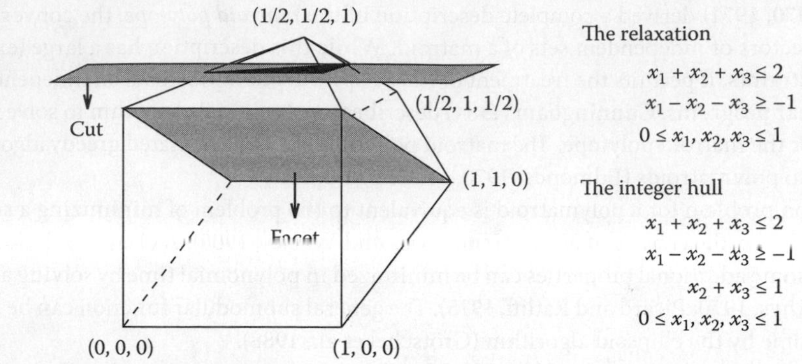

FIGURE 13.2 Relaxation, cuts, and facets. (From Chandru, V. and Rao, M. R., *ACM Comput. Survey*, 28, 55, 1996.)

scheme is complete, that is, every valid inequality of P_I can be generated by finite application of GC rounding (Chvátal, 1973; Schrijver, 1986).

The number of inequalities needed to describe the convex hull of P_I is usually exponential in the size of A. But to solve an optimization problem on P_I, one is only interested in obtaining a partial description of P_I that facilitates the identification of an integer solution and prove its optimality. This is the underlying basis of any cutting plane approach to combinatorial problems.

13.5.3.1 Cutting Plane Method

Consider the optimization problem

$$\max\{\mathbf{cx} : \mathbf{x} \in P_I = \{\mathbf{x} : A\mathbf{x} \le \mathbf{b}; \mathbf{x} \ge \mathbf{0} \text{ and integer}\}\}$$

The generic *cutting plane* method as applied to this formulation is given as follows.

Procedure 13.4 Cutting Plane:

1. Initialize $A' \leftarrow A$ and $\mathbf{b}' \leftarrow \mathbf{b}$.
2. Find an optimal solution $\bar{\mathbf{x}}$ to the linear program

$$\max\{\mathbf{cx} : A'\mathbf{x} \le \mathbf{b}'; \mathbf{x} \ge \mathbf{0}\}$$

   If $\bar{\mathbf{x}} \in P_I$, stop and return $\bar{\mathbf{x}}$.

3. Generate a valid inequality $\mathbf{fx} \le f_0$ for P_I such that $\mathbf{f}\bar{\mathbf{x}} > f_0$ (the inequality "cuts" $\bar{\mathbf{x}}$).
4. Add the inequality to the constraint system, update

$$A' \leftarrow \begin{pmatrix} A' \\ \mathbf{f} \end{pmatrix}, \qquad \mathbf{b}' \leftarrow \begin{pmatrix} \mathbf{b}' \\ f_0 \end{pmatrix}$$

   Go to step 2.

In step 3 of the cutting plane method, we require a suitable application of the GC rounding scheme (or some alternative method of identifying a cutting plane). Notice that while the GC rounding scheme will generate valid inequalities, the identification of one that cuts off the current solution to the linear programming relaxation is all that is needed. Gomory (1958) provided just such a specialization of the rounding scheme that generates a cutting plane. Although this met the theoretical challenge of designing a sound and complete cutting plane method for integer linear programming, it turned out to be a weak method in practice. Successful cutting plane methods, in use today, use considerable additional insights into the structure of facet-defining cutting planes. Using facet cuts makes a huge difference in the speed of convergence of these methods. Also, the idea of combining cutting plane methods with search methods has been found to have a lot of merit. These B-and-C methods will be discussed in the next section.

13.5.3.2 b-Matching Problem

Consider the **b**-matching problem:

$$\max\{\mathbf{cx} : A\mathbf{x} \le \mathbf{b}, \mathbf{x} \ge \mathbf{0} \text{ and integer}\} \tag{13.9}$$

where A is the node–edge incidence matrix of an undirected graph and **b** is a vector of positive integers. Let G be the undirected graph whose node-edge incidence matrix is given by A and let $W \subseteq V$ be any subset of nodes of G (i.e., subset of rows of A) such that

$$\mathbf{b}(W) = \sum_{i \in W} \mathbf{b}_i$$

is odd. Then the inequality

$$\mathbf{x}(W) = \sum_{e \in E(W)} \mathbf{x}_e \le \frac{1}{2}(\mathbf{b}(W) - 1) \tag{13.10}$$

is a valid inequality for integer solutions to Equation 13.8 where $E(W) \subseteq E$ is the set of edges of G having both ends in W. Edmonds (1965) has shown that the inequalities Equations 13.8 and 13.9 define the integral **b**-matching polytope. Note that the number of inequalities Equation 13.9 is exponential in the number of nodes of G. An instance of the successful application of the idea of using only a partial description of P_I is in the blossom algorithm for the matching problem, due to Edmonds (1965).

As we saw, an implication of the ellipsoid method for linear programming is that the linear program over P_I can be solved in polynomial time if and only if the associated separation problem (also referred to as the constraint identification problem, see Section 13.2) can be solved in polynomial time, see Grötschel et al. (1982), Karp and Papadimitriou (1982), and Padberg and Rao (1981). The separation problem for the **b**-matching problem with or without upper bounds was shown by Padberg and Rao (1982), to be solvable in polynomial time. The procedure involves a minor modification of the algorithm of Gomory and Hu (1961) for multiterminal networks. However, no polynomial (in the number of nodes of the graph) linear programming formulation of this separation problem is known. A related unresolved issue is whether there exists a polynomial size (compact) formulation for the **b**-matching problem. Yannakakis (1988) has shown that, under a symmetry assumption, such a formulation is impossible.

However, Kaibel et al. (2012) have shown that the symmetry assumption can make a big difference. Considering matchings of size of $\lfloor log(n) \rfloor$, where n is the number of vertices in the graph, they show that there is a polynomial size nonsymmetric extended formulation while there is no compact symmetric extended formulation. Fiorini et al. (2012) have shown that the max-cut, traveling salesman and stable set problems cannot have a polynomial size extended formulation. But it is still an open question whether the perfect matching problem has a compact representation. Conforti et al. (2012) provide a survey on extended formulations.

13.5.3.3 Other Combinatorial Problems

Besides the matching problem, several other combinatorial problems and their associated polytopes have been well studied and some families of facet defining inequalities have been identified. For instance, the set packing, graph partitioning, plant location, max-cut, traveling salesman, and Steiner tree problems have been extensively studied from a polyhedral point of view (see, for instance, Nemhauser and Wolsey, 1988).

These combinatorial problems belong to the class of $\mathcal{NP}$-complete problems. In terms of a worst-case analysis, no polynomial-time algorithms are known for these problems. Nevertheless, using a cutting plane approach with B-and-B or B-and-C (see Section 13.6), large instances of these problems have been successfully solved, see Crowder et al. (1983), for general 0–1 problems, Barahona et al. (1989) for the max-cut problem, Padberg and Rinaldi (1991) for the traveling salesman problem (TSP), and Chopra et al. (1992) for the Steiner tree problem.

13.6 Partial Enumeration Methods

In many instances, to find an optimal solution to ILP, the structure of the problem is exploited together with some sort of partial enumeration. In this section, we review the B-and-B and B-and-C methods for solving an ILP.

13.6.1 Branch and Bound

The B-and-B method is a systematic scheme for implicitly enumerating the finitely many feasible solutions to an ILP. Although, theoretically the size of the enumeration tree is exponential in the problem parameters, in most cases, the method eliminates a large number of feasible solutions. The key features of B-and-B method are

1. *Selection/removal* of one or more problems from a candidate list of problems
2. *Relaxation* of the selected problem so as to obtain a lower bound (on a minimization problem) on the optimal objective function value for the selected problem
3. *Fathoming*, if possible, of the selected problem
4. *Branching* strategy is needed if the selected problem is not fathomed. Branching creates subproblems, which are added to the candidate list of problems

The four steps are repeated until the candidate list is empty. The B-and-B method sequentially examines problems that are added and removed from a candidate list of problems.

Initialization. Initially, the candidate list contains only the original ILP, which is denoted as

$$(P) \quad \min\{\mathbf{cx} : A\mathbf{x} \le \mathbf{b}, \mathbf{x} \ge \mathbf{0} \text{ and integer}\}$$

Let $F(P)$ denote the feasible region of (P) and $z(P)$ denote the optimal objective function value of (P). For any $\bar{\mathbf{x}}$ in $F(P)$, let $z_P(\bar{\mathbf{x}}) = \mathbf{c}\bar{\mathbf{x}}$.

Frequently, heuristic procedures are first applied to get a good feasible solution to (P). The best solution known for (P) is referred to as the current incumbent solution. The corresponding objective function value is denoted as z_I. In most instances, the initial heuristic solution is neither optimal nor at least immediately certified to be optimal. Thus, further analysis is required to ensure that an optimal solution to (P) is obtained. If no feasible solution to (P) is known, z_I is set to ∞.

Selection/removal. In each iterative step of B-and-B, a problem is selected and removed from the candidate list for further analysis. The selected problem is henceforth referred to as the candidate problem (CP). The algorithm terminates if there is no problem to select from the candidate list. Initially, there is no issue of selection since the candidate list contains only the problem (P). However, as the algorithm proceeds, there would be many problems on the candidate list and a selection rule is required.

Appropriate selection rules, also referred to as branching strategies, are discussed later. Conceptually, several problems may be simultaneously selected and removed from the candidate list. However, most sequential implementations of B-and-B select only one problem from the candidate list and this is assumed henceforth. Parallel aspects of B-and-B on 0–1 integer linear programs are discussed in Cannon and Hoffman (1990) and for the case of TSP in Applegate et al. (1994).

The computational time required for the B-and-B algorithm depends crucially on the order in which the problems in the candidate list are examined. A number of clever heuristic rules may be employed in devising such strategies. Two general purpose selection strategies that are commonly used are as follows:

1. Choose the problem that was added last to the candidate list. This last-in–first-out rule (LIFO) is also called *depth first search* (DFS) since the selected candidate problem increases the depth of the active enumeration tree.
2. Choose the problem on the candidate list that has the least lower bound. Ties may be broken by choosing the problem that was added last to the candidate list. This rule would require that a lower bound be obtained for each of the problems on the candidate list. In other words, when a problem is added to the candidate list, an associated lower bound should also be stored. This may be accomplished by using ad hoc rules or by solving a relaxation of each problem before it is added to the candidate list.

Rule 1 is known to empirically dominate rule 2 when storage requirements for candidate list and computation time to solve (*P*) are taken into account. However, some analysis indicates that rule 2 can be shown to be superior if minimizing the number of candidate problems to be solved is the criterion (see Parker and Rardin, 1988).

Relaxation. In order to analyze the selected candidate problem (*CP*), a **relaxation** (CP_R) of (*CP*) is solved to obtain a lower bound $z(CP_R) \le z(CP)$. (CP_R) is a relaxation of (*CP*) if

1. $F(CP) \subseteq F(CP_R)$
2. For $\bar{\mathbf{x}} \in F(CP)$, $z_{CP_R}(\bar{\mathbf{x}}) \le z_{CP}(\bar{\mathbf{x}})$
3. For $\bar{\mathbf{x}}, \mathbf{x} \in F(CP)$, $z_{CP_R}(\bar{\mathbf{x}}) \le z_{CP_R}(\mathbf{x})$ implies that $z_{CP}(\bar{\mathbf{x}}) \le z_{CP}(\hat{\mathbf{x}})$

Relaxations are needed because the candidate problems are typically hard to solve. The relaxations used most often are either linear programming or Lagrangian relaxations of (*CP*), see Section 13.7 for details. Sometimes, instead of solving a relaxation of (*CP*), a lower bound is obtained by using some ad hoc rules such as penalty functions.

Fathoming. A candidate problem is fathomed if

(FC1) analysis of (CP_R) reveals that (*CP*) is infeasible. For instance, if $F(CP_R) = \phi$, then $F(CP) = \phi$.

(FC2) analysis of (CP_R) reveals that (*CP*) has no feasible solution better than the current incumbent solution. For instance, if $z(CP_R) \ge z_I$, then $z(CP) \ge z(CP_R) \ge z_I$.

(FC3) analysis of (CP_R) reveals an optimal solution of (*CP*). For instance, if the optimal solution, $\mathbf{x}_R$, to (CP_R) is feasible in (*CP*), then ($\mathbf{x}_R$) is an optimal solution to (*CP*) and $z(CP) = \mathbf{c}\mathbf{x}_R$.

(FC4) analysis of (CP_R) reveals that (*CP*) is dominated by some other problem, say, CP^*, in the candidate list. For instance, if it can shown that $z(CP^*) \le z(CP)$, then there is no need to analyze (*CP*) further.

If a candidate problem (*CP*) is fathomed using any of the preceding criteria, then further examination of (*CP*) or its descendants (subproblems) obtained by separation is not required. If (FC3) holds, and $z(CP) < z_I$, the incumbent is updated as $\mathbf{x}_R$ and z_I is updated as $z(CP)$.

Separation/branching. If the candidate problem (*CP*) is not fathomed, then *CP* is separated into several problems, say, (CP_1), (CP_2),…, (CP_q), where $\bigcup_{t=1}^{q} F(CP_t) = F(CP)$ and, typically,

$$F(CP_i) \cap F(CP_j) = \phi \ \forall i \ne j$$

For instance, a separation of (CP) into (CP_i), $i = 1, 2,\ldots, q$, is obtained by fixing a single variable, say, $\mathbf{x}_j$, to one of the q possible values of $\mathbf{x}_j$ in an optimal solution to (CP). The choice of the variable to fix depends on the separation strategy, which is also part of the branching strategy. After separation, the subproblems are added to the candidate list. Each subproblem (CP_t) is a restriction of (CP) since $F(CP_t) \subseteq F(CP)$. Consequently, $z(CP) \leq z(CP_t)$ and $z(CP) = \min_t z(CP_t)$.

The various steps in the B-and-B algorithm are outlined as follows.

Procedure 13.5 B-and-B:

0. **Initialize:** Given the problem (P), the incumbent value z_I is obtained by applying some heuristic (if a feasible solution to (P) is not available, set $z_I = +\infty$). Initialize the candidate list $C \leftarrow \{(P)\}$.
1. **Optimality:** If $C = \emptyset$ and $z_I = +\infty$, then (P) is infeasible, stop. Stop also if $C = \emptyset$ and $z_I < +\infty$, the incumbent is an optimal solution to (P).
2. **Selection:** Using some candidate selection rule, select and remove a problem $(CP) \in C$.
3. **Bound:** Obtain a lower bound for (CP) by either solving a relaxation (CP_R) of (CP) or by applying some ad-hoc rules. If (CP_R) is infeasible, return to Step 1. Else, let x_R be an optimal solution of (CP_R).
4. **Fathom:** If $z(CP_R) \geq z_I$, return to step 1. Else if x_R is feasible in (CP) and $z(CP) < z_I$, set $z_I \leftarrow z(CP)$, update the incumbent as x_R and return to step 1. Finally, if x_R is feasible in (CP) but $z(CP) \geq z_I$, return to step 1.
5. **Separation:** Using some separation or branching rule, separate (CP) into (CP_i), $i = 1, 2,\ldots, q$ and set $C \leftarrow C \cup \{CP_1), (CP_2),\ldots, (CP_q)\}$ and return to step 1.
6. **End Procedure.**

Although the B-and-B method is easy to understand, the implementation of this scheme for a particular ILP is a nontrivial task requiring the following:

1. A relaxation strategy with efficient procedures for solving these relaxations
2. Efficient data structures for handling the rather complicated bookkeeping of the candidate list
3. Clever strategies for selecting promising candidate problems
4. Separation or branching strategies that could effectively prune the enumeration tree

A key problem is that of devising a relaxation strategy, that is, to find *good relaxations*, which are significantly easier to solve than the original problems and tend to give sharp lower bounds. Since these two are conflicting, one has to find a reasonable tradeoff.

13.6.2 Branch and Cut

In the past few years, the B-and-C method has become popular for solving NP-complete combinatorial optimization problems. As the name suggests, the B-and-C method incorporates the features of both the B-and-B method just presented and the cutting plane method presented previously. The main difference between the B-and-C method and the general B-and-B scheme is in the bound step (step 3).

A distinguishing feature of the B-and-C method is that the relaxation (CP_R) of the candidate problem (CP) is a linear programming problem, and, instead of merely solving (CP_R), an attempt is made to solve (CP) by using cutting planes to tighten the relaxation. If (CP_R) contains inequalities that are valid for (CP) but not for the given ILP, then the GC rounding procedure may generate inequalities that are valid for (CP) but not for the ILP. In the B-and-C method, the inequalities that are generated are always valid for the ILP and hence can be used globally in the enumeration tree.

Another feature of the B-and-C method is that often heuristic methods are used to convert some of the fractional solutions, encountered during the cutting plane phase, into feasible solutions of the (*CP*) or more generally of the given ILP. Such feasible solutions naturally provide upper bounds for the ILP. Some of these upper bounds may be better than the previously identified best upper bound and, if so, the current incumbent is updated accordingly.

We thus obtain the B-and-C method by replacing the bound step (step 3) of the B-and-B method by steps 3(a) and 3(b) and also by replacing the fathom step (step 4) by steps 4(a) and 4(b) given subsequently.

3(a) **Bound:** Let (CP_R) be the LP relaxation of (*CP*). Attempt to solve (*CP*) by a cutting plane method which generates valid inequalities for (*P*). Update the constraint System of (*P*) and the incumbent as appropriate.

Let $F\mathbf{x} \leq \mathbf{f}$ denote all of the valid inequalities generated during this phase. Update the constraint system of (*P*) to include all of the generated inequalities, that is, set $A^T \leftarrow (A^T, F^T)$ and $\mathbf{b}^T \leftarrow (\mathbf{b}^T, \mathbf{f}^T)$. The constraints for all of the problems in the candidate list are also to be updated.

During the cutting plane phase, apply heuristic methods to convert some of the identified fractional solutions into feasible solutions to (*P*). If a feasible solution, $\bar{\mathbf{x}}$, to (*P*), is obtained such that $\mathbf{c}\bar{\mathbf{x}} < z_I$, update the incumbent to $\bar{\mathbf{x}}$ and z_I to $\mathbf{c}\bar{\mathbf{x}}$. Hence, the remaining changes to B-and-B are as follows:

3(b) **If** (*CP*) is solved go to step 4(a). **Else**, let $\hat{\mathbf{x}}$ be the solution obtained when the cutting plane phase is terminated (we are unable to identify a valid inequality of (*P*) that is violated by $\hat{\mathbf{x}}$). Go to step 4(b).

4(b) **Fathom by Optimality:** Let $\mathbf{x}^*$ be an optimal solution to (*CP*). If $z(CP) < z_I$, set $\mathbf{x}_I \leftarrow z(CP)$ and update the incumbent as $\mathbf{x}^*$. Return to step 1.

4(b) **Fathom by Bound:** If $\mathbf{c}\hat{\mathbf{x}} \geq z_I$, return to Step 1.
Else go to step 5.

The incorporation of a cutting plane phase into the B-and-B scheme involves several technicalities which require careful design and implementation of the B-and-C algorithm. Details of the state of the art in cutting plane algorithms including the B-and-C algorithm are reviewed in Jünger et al. (1995).

13.7 Approximation in Combinatorial Optimization

The inherent complexity of integer linear programming has led to a long-standing research program in approximation methods for these problems. Linear programming relaxation and Lagrangian relaxation are two general approximation schemes that have been the real workhorses of computational practice. Semidefinite relaxation is a recent entrant that appears to be very promising. In this section, we present a brief review of these developments in the approximation of combinatorial optimization problems.

In the past few years, there has been significant progress in our understanding of performance guarantees for approximation of $\mathcal{NP}$-hard combinatorial optimization problems. A ρ-*approximate* algorithm for an optimization problem is an approximation algorithm that delivers a feasible solution with objective value within a factor of ρ of optimal (think of minimization problems and $\rho \geq 1$). For some combinatorial optimization problems, it is possible to *efficiently* find solutions that are arbitrarily close to optimal even though finding the true optimal is hard. If this were true of most of the problems of interest, we would be in good shape. However, the recent results of Arora et al. (1992) indicate exactly the opposite conclusion.

A polynomial-time approximation scheme (PTAS) for an optimization problem is a family of algorithms, A_ρ, such that for each $\rho > 1$, A_ρ is a polynomial-time ρ-approximate algorithm. Despite concentrated effort spanning about two decades, the situation in the early 1990s was that for many combinatorial optimization problems, we had no PTAS and no evidence to suggest the nonexistence of such schemes either. This led Papadimitriou and Yannakakis (1991) to define a new complexity class (using reductions that preserve approximate solutions) called MAXSNP, and they identified several complete languages in this class. The work of Arora et al. (1992) completed this agenda by showing that, assuming $\mathcal{P} \neq \mathcal{NP}$, there is no PTAS for a MAXSNP-complete problem.

An implication of these theoretical developments is that for most combinatorial optimization problems, we have to be quite satisfied with performance guarantee factors ρ that are of some small fixed value. (There are problems, like the general TSP for which there are no ρ-approximate algorithms for any finite value of ρ, assuming of course that $\mathcal{P} \neq \mathcal{NP}$.) Thus, one avenue of research is to go problem by problem and knock ρ down to its smallest possible value. A different approach would be to look for other notions of good approximations based on probabilistic guarantees or empirical validation. Let us see how the polyhedral combinatorics perspective helps in each of these directions.

13.7.1 LP Relaxation and Randomized Rounding

Consider the well-known problem of finding the *smallest weight vertex cover* in a graph. We are given a graph $G(V, E)$ and a nonnegative weight $\mathbf{w}(v)$ for each vertex $v \in V$. We want to find the smallest total weight subset of vertices S such that each edge of G has at least one end in S. (This problem is known to be MAXSNP-hard.) An integer programming formulation of this problem is given by

$$\min\left\{\sum_{v\in V}\mathbf{w}(v)\mathbf{x}(v) : \mathbf{x}(u)+\mathbf{x}(v)\geq 1,\quad \forall (u,v)\in E,\quad \mathbf{x}(v)\in\{0,1\}\ \forall v\in V\right\}$$

To obtain the linear programming relaxation, we substitute the $\mathbf{x}(v) \in \{0, 1\}$ constraint with $\mathbf{x}(v) \geq 0$ for each $v \in V$. Let $\mathbf{x}^*$ denote an optimal solution to this relaxation. Now let us round the fractional parts of $\mathbf{x}^*$ in the usual way, that is, values of 0.5 and up are rounded to 1 and smaller values down to 0. Let $\hat{\mathbf{x}}$ be the 0–1 solution obtained. First note that $\hat{\mathbf{x}}(v) \leq 2\mathbf{x}^*(v)$ for each $v \in V$. Also, for each $(u, v) \in E$, since $\mathbf{x}^*(u) + \mathbf{x}^*(v) \geq 1$, at least one of $\hat{\mathbf{x}}(u)$ and $\hat{\mathbf{x}}(v)$ must be set to 1. Hence $\hat{\mathbf{x}}$ is the incidence vector of a vertex cover of G whose total weight is within twice the total weight of the linear programming relaxation (which is a lower bound on the weight of the optimal vertex cover). Thus, we have a 2-approximate algorithm for this problem, which solves a linear programming relaxation and uses rounding to obtain a feasible solution.

The deterministic rounding of the fractional solution worked quite well for the vertex cover problem. One gets a lot more power from this approach by adding in randomization to the rounding step. Raghavan and Thompson (1987) proposed the following obvious randomized rounding scheme. Given a 0–1 integer program, solve its linear programming relaxation to obtain an optimal $\mathbf{x}^*$. Treat the $\mathbf{x}_j^* \in [0,1]$ as probabilities, that is, let probability $\{\mathbf{x}_j = 1\} = \mathbf{x}_j^*$, to randomly round the fractional solution to a 0–1 solution. Using Chernoff bounds on the tails of the binomial distribution, Raghavan and Thompson (1987) were able to show, for specific problems, that with high probability, this scheme produces integer solutions which are close to optimal. In certain problems, this rounding method may not always produce a feasible solution. In such cases, the expected values have to be computed as conditioned on feasible solutions produced by rounding. More complex (nonlinear) randomized rounding schemes have been recently studied and have been found to be extremely effective. We will see an example of nonlinear rounding in the context of semidefinite relaxations of the max-cut problem in the following.

13.7.2 Primal–Dual Approximation

The linear programming relaxation of the vertex cover problem, as we saw previously, is given by

$$(P_{VC})\quad \min\left\{\sum_{v\in V}\mathbf{w}(v)\mathbf{x}(v) : \mathbf{x}(u)+\mathbf{x}(v)\geq 1,\quad \forall (u,v)\in E,\quad \mathbf{x}(v)\geq 0\ \forall v\in V\right\}$$

and its dual is

$$(D_{VC})\quad \max\left\{\sum_{(u,v)\in E} \mathbf{y}(u,v):\sum_{(u,v)\in E} \mathbf{y}(u,v)\le w(v),\quad \forall v\in V,\quad \mathbf{y}(u,v)\ge 0\ \forall (u,v)\in E\right\}$$

The primal–dual approximation approach would first obtain an optimal solution $\mathbf{y}^*$ to the dual problem (D_{VC}). Let $\hat{V}\subseteq V$ denote the set of vertices for which the dual constraints are tight, that is,

$$\hat{V}=\left\{v\in V:\sum_{(u,v)\in E}\mathbf{y}^*(u,v)=\mathbf{w}(v)\right\}$$

The approximate vertex cover is taken to be $\hat{V}$. It follows from complementary slackness that $\hat{V}$ is a vertex cover. Using the fact that each edge (u, v) is in the star of at most two vertices (u and v), it also follows that $\hat{V}$ is a 2-approximate solution to the minimum weight vertex cover problem.

In general, the primal–dual approximation strategy is to use a dual solution to the linear programming relaxation, along with complementary slackness conditions as a heuristic to generate an integer (primal) feasible solution, which for many problems turns out to be a good approximation of the optimal solution to the original integer program.

Remark 13.11

A recent survey of primal-dual approximation algorithms and some related interesting results are presented in Williamson (2000).

13.7.3 Semidefinite Relaxation and Rounding

The idea of using SDP to solve combinatorial optimization problems appears to have originated in the work of Lovász (1979) on the Shannon capacity of graphs. Grötschel et al. (1988) later used the same technique to compute a maximum stable set of vertices in perfect graphs via the ellipsoid method. Recently, Lovász and Schrijver (1991) resurrected the technique to present a fascinating theory of semidefinite relaxations for general 0–1 integer linear programs. We will not present the full-blown theory here but instead will present an excellent application of this methodology to the problem of finding the maximum weight cut of a graph. This application of semidefinite relaxation for approximating MAXCUT is due to Goemans and Williamson (1994).

We begin with a quadratic Boolean formulation of MAXCUT

$$\max\left\{\frac{1}{2}\sum_{(u,v)\in E}\mathbf{w}(u,v)(1-\mathbf{x}(u)\mathbf{x}(v)):\mathbf{x}(v)\in\{-1,1\}\ \forall\, v\in V\right\}$$

where $G(V, E)$ is the graph and $\mathbf{w}(u, v)$ is the nonnegative weight on edge (u, v). Any $\{-1, 1\}$ vector of $\mathbf{x}$ values provides a bipartition of the vertex set of G. The expression $(1 - \mathbf{x}(u)\mathbf{x}(v))$ evaluates to 0 if u and v are on the same side of the bipartition and to 2 otherwise. Thus, the optimization problem does indeed represent exactly the MAXCUT problem.

Next we reformulate the problem in the following way:

- We square the number of variables by substituting each $\mathbf{x}(v)$ with $\chi(v)$ an n-vector of variables (where n is the number of vertices of the graph).

- The quadratic term $\mathbf{x}(u)\mathbf{x}(v)$ is replaced by $\chi(u) \cdot \chi(v)$, which is the inner product of the vectors.
- Instead of the {−1, 1} restriction on the $\mathbf{x}(v)$, we use the Euclidean normalization $\|\chi(v)\| = 1$ on the $\chi(v)$.

Thus, we now have a problem

$$\max\left\{\frac{1}{2}\sum_{(u,v)\in E} w(u,v)(1-\chi(u)\ \chi(v)) : \|\chi(v)\| = 1\ \forall\ v \in V\right\}$$

which is a relaxation of the MAXCUT problem (note that if we force only the first component of the $\chi(v)$ to have nonzero value, we would just have the old formulation as a special case).

The final step is in noting that this reformulation is nothing but a semidefinite program. To see this we introduce $n \times n$ Gram matrix Y of the unit vectors $\chi(v)$. So $Y = X^T X$ where $X = (\chi(v) : v \in V)$. Thus, the relaxation of MAXCUT can now be stated as a semidefinite program,

$$\max\left\{\frac{1}{2}\sum_{(u,v)\in E} w(u,v)(1-Y_{(u,v)}) : Y \succeq 0,\quad Y_{(u,v)} = 1\ \forall\ v \in V\right\}$$

Recall from Section 13.2 that we are able to solve such semidefinite programs to an additive error ε in time polynomial in the input length and log 1/ε by using either the ellipsoid method or interior point methods.

Let χ^* denote the near optimal solution to the SDP relaxation of MAXCUT (convince yourself that χ^* can be reconstructed from an optimal Y^* solution). Now we encounter the final trick of Goemans and Williamson. The approximate maximum weight cut is extracted from χ^* by randomized rounding. We simply pick a random hyperplane H passing through the origin. All of the $v \in V$ lying to one side of H get assigned to one side of the cut and the rest to the other. Goemans and Williamson observed the following inequality.

Lemma 13.3

For χ_1 and χ_2, two random n-vectors of unit norm, let $\mathbf{x}(1)$ and $\mathbf{x}(2)$ be ±1 values with opposing signs if H separates the two vectors and with same signs otherwise. Then $\tilde{E}(1 - \chi_1 \cdot \chi_2) \leq 1.1393 \cdot \tilde{E}(1 - \mathbf{x}(1)\mathbf{x}(2))$ where $\tilde{E}$ denotes the expected value.

By linearity of expectation, the lemma implies that the expected value of the cut produced by the rounding is at least 0.878 times the expected value of the semidefinite program. Using standard conditional probability techniques for derandomizing, Goemans and Williamson show that a deterministic polynomial-time approximation algorithm with the same margin of approximation can be realized. Hence, we have a cut with value at least 0.878 of the maximum cut value.

Remark 13.12

For semidefinite relaxations of mixed integer programs in which the integer variables are restricted to be 0 or 1, Iyengar and Cezik (2002) develop methods for generating Gomory–Chavatal and disjunctive cutting planes that extends the work of Balas et al. (1993). Ye (2000) shows that strengthened semidefinite relaxations and mixed rounding methods achieve superior performance guarantee for some discrete optimization problems. A recent survey of SDP and applications is in Wolkowicz et al. (2000).

13.7.4 Neighborhood Search

A combinatorial optimization problem may be written succinctly as

$$\min\{f(x): x \in X\}$$

The traditional neighborhood method starts at a feasible point x_0 (in X), and iteratively proceeds to a neighborhood point that is better in terms of the objective function $f(.)$ until a specified termination condition is attained. While the concept of neighborhood $N(x)$ of a point x is well defined in calculus, the specification of $N(x)$ is itself a matter of consideration in combinatorial optimization. For instance, for the travelling salesman problem, the so-called *k-opt heuristic* (see Lin and Kernighan, 1973) is a neighborhood search method which for a given tour considers "neighborhood tours" in which k variables (edges) in the given tour are replaced by k other variables such that a tour is maintained. This search technique has proved to be effective though it is quite complicated to implement when k is larger than 3.

A neighborhood search method leads to a local optimum in terms of the neighborhood chosen. Of course, the chosen neighborhood may be large enough to ensure a global optimum, but such a procedure is typically not practical in terms of searching the neighborhood for a better solution. Recently Orlin (2000) has presented very large scale neighborhood search algorithms in which the neighborhood is searched using network flow or dynamic programming methods. Another method advocated by Orlin (2000) is to define the neighborhood in such a manner that the search process becomes a polynomially solvable special case of a hard combinatorial problem.

To avoid getting trapped at a local optimum solution, different strategies such as Tabu Search (see for instance Glover and Laguna, 1997), simulated annealing (see, for instance, Aarts and Korst, 1989), genetic algorithms (see, for instance, Whitley, 1993), and neural networks have been developed. Essentially, these methods allow for the possibility of sometimes moving to an inferior solution in terms of the objective function or even to an infeasible solution. While there is no guarantee of obtaining a global optimal solution, computational experience in solving several difficult combinatorial optimization problems has been very encouraging. However, a drawback of these methods is that performance guarantees are not typically available.

13.7.5 Lagrangian Relaxation

We end our discussion of approximation methods for combinatorial optimization with the description of Lagrangian relaxation. This approach has been widely used for about two decades now in many practical applications. Lagrangian relaxation, like linear programming relaxation, provides bounds on the combinatorial optimization problem being relaxed (i.e., lower bounds for minimization problems).

Lagrangian relaxation has been so successful because of a couple of distinctive features. As was noted earlier, in many hard combinatorial optimization problems, we usually have embedded some nice tractable subproblems which have efficient algorithms. Lagrangian relaxation gives us a framework to *jerry-rig* an approximation scheme that uses these efficient algorithms for the subproblems as subroutines. A second observation is that it has been empirically observed that well-chosen Lagrangian relaxation strategies usually provide very tight bounds on the optimal objective value of integer programs. This is often used to great advantage within partial enumeration schemes to get very effective pruning tests for the search trees.

Practitioners also have found considerable success with designing heuristics for combinatorial optimization by starting with solutions from Lagrangian relaxations and constructing good feasible solutions via so-called *dual ascent* strategies. This may be thought of as the analogue of rounding strategies for linear programming relaxations (but with no performance guarantees, other than empirical ones).

Consider a representation of our combinatorial optimization problem in the form

$$(P) \quad z = \min\{\mathbf{cx} : A\mathbf{x} \geq \mathbf{b}, \mathbf{x} \in X \subseteq \mathfrak{R}^n\}$$

Implicit in this representation is the assumption that the explicit constraints ($A\mathbf{x} \geq \mathbf{b}$) are *small* in number. For convenience, let us also assume that X can be replaced by a finite list $\{\mathbf{x}^1, \mathbf{x}^2, \ldots, \mathbf{x}^T\}$.

The following definitions are with respect to (*P*):

- Lagrangian. $L(\mathbf{u}, \mathbf{x}) = \mathbf{u}(A\mathbf{x} - \mathbf{b}) + \mathbf{cx}$ where $\mathbf{u}$ are the Lagrange multipliers.
- Lagrangian-dual function. $\mathcal{L}(\mathbf{u}) = \min_{\mathbf{x} \in X}\{L(\mathbf{u}, \mathbf{x})\}$.
- Lagrangian-dual problem. $(D)\ d = \max_{\mathbf{u} \geq 0}\{\mathcal{L}(\mathbf{u})\}$.

It is easily shown that (*D*) satisfies a weak duality relationship with respect to (*P*), that is, $z \geq d$. The discreteness of X also implies that $\mathcal{L}(\mathbf{u})$ is a piecewise linear and concave function (see Shapiro, 1979). In practice, the constraints X are chosen such that the evaluation of the Lagrangian dual function $\mathcal{L}(\mathbf{u})$ is easily made (i.e., the *Lagrangian subproblem* $\min_{\mathbf{x}} \in X\{L(\mathbf{u}, \mathbf{x})\}$ is easily solved for a fixed value of $\mathbf{u}$).

Example 13.3 Traveling salesman problem (TSP).

For an undirected graph *G*, with costs on each edge, the TSP is to find a minimum cost set *H* of edges of *G* such that it forms a Hamiltonian cycle of the graph. *H* is a Hamiltonian cycle of *G* if it is a simple cycle that spans all the vertices of *G*. Alternatively, *H* must satisfy: (1) exactly two edges of *H* are adjacent to each node, and (2) *H* forms a connected, spanning subgraph of *G*.

Held and Karp (1970) used these observations to formulate a Lagrangian relaxation approach for TSP that relaxes the degree constraints (1). Notice that the resulting subproblems are minimum spanning tree problems which can be easily solved.

The most commonly used general method of finding the optimal multipliers in Lagrangian relaxation is subgradient optimization (cf. Held et al., 1974). Subgradient optimization is the non-differentiable counterpart of steepest descent methods. Given a dual vector $\mathbf{u}^k$, the iterative rule for creating a sequence of solutions is given by:

$$\mathbf{u}^{k+1} = \mathbf{u}^k + t_k \gamma(\mathbf{u}^k)$$

where t_k is an appropriately chosen step size, and $\gamma(\mathbf{u}^k)$ is a subgradient of the dual function $\mathcal{L}$ at $\mathbf{u}^k$. Such a subgradient is easily generated by

$$\gamma(\mathbf{u}^k) = A\mathbf{x}^k - \mathbf{b}$$

where $\mathbf{x}^k$ is a maximizer of $\min_{\mathbf{x} \in X}\{L(\mathbf{u}^k, \mathbf{x})\}$.

Subgradient optimization has proven effective in practice for a variety of problems. It is possible to choose the step sizes $\{t_k\}$ to guarantee convergence to the optimal solution. Unfortunately, the method is not finite, in that the optimal solution is attained only in the limit. Further, it is not a pure descent method. In practice, the method is heuristically terminated and the best solution in the generated sequence is recorded. In the context of nondifferentiable optimization, the ellipsoid algorithm was devised by Shor (1970) to overcome precisely some of these difficulties with the subgradient method.

The ellipsoid algorithm may be viewed as a scaled subgradient method in much the same way as variable metric methods may be viewed as scaled steepest descent methods (cf. Akgul, 1984). And if we use the ellipsoid method to solve the Lagrangian dual problem, we obtain the following as a consequence of the polynomial-time equivalence of optimization and separation.

Theorem 13.8

The Lagrangian dual problem is polynomial-time solvable if and only if the Lagrangian subproblem is. Consequently, the Lagrangian dual problem is $\mathcal{NP}$-hard if and only if the Lagrangian subproblem is.

The theorem suggests that, in practice, if we set up the Lagrangian relaxation so that the subproblem is tractable, then the search for optimal Lagrangian multipliers is also tractable.

13.8 Prospects in Integer Programming

The TSP has been the subject of considerable research in terms of developing specialized software incorporating several theoretical results. The largest problem, in terms of the number of cities, that has been solved is a 85,900 cities problem that arises in the process of making a computer chip, see Cook (2012).

The current emphasis in software design for integer programming is in the development of shells (for example, CPLEX 6.5 [1999], MINTO [Savelsbergh et al., 1994], and OSL [1991]) wherein a general purpose solver like B-and-C is the driving engine. Problem-specific codes for generation of cuts and facets can be easily interfaced with the engine. Recent computational results (Bixby et al., 2001) suggests that it is now possible to solve relatively large size integer programming problems using general purpose codes. Bixby and his team have gone on to create a new optimization engine called GUROBI (GUROBI, 2013) which appears to be the new benchmark. We believe that this trend will eventually lead to the creation of general purpose problem-solving languages for combinatorial optimization akin to AMPL (Fourer et al., 1993) for linear and nonlinear programming.

A promising line of research is the development of an empirical science of algorithms for combinatorial optimization (Hooker, 1993). Computational testing has always been an important aspect of research on the efficiency of algorithms for integer programming. However, the standards of test designs and empirical analysis have not been uniformly applied. We believe that there will be important strides in this aspect of integer programming and more generally of algorithms. Hooker argues that it may be useful to stop looking at algorithmics as purely a deductive science and start looking for advances through repeated application of "hypothesize and test" paradigms, that is, through empirical science. Hooker and Vinay (1995) develop a science of selection rules for the Davis–Putnam–Loveland scheme of theorem proving in propositional logic by applying the empirical approach.

The integration of logic-based methodologies and mathematical programming approaches is evidenced in the recent emergence of CLP systems (Borning, 1994; Saraswat and Van Hentenryck, 1995) and logico-mathematical programming (Jeroslow, 1989; Chandru and Hooker, 1991). In CLP, we see a structure of Prolog-like programming language in which some of the predicates are constraint predicates whose truth values are determined by the solvability of constraints in a wide range of algebraic and combinatorial settings. The solution scheme is simply a clever orchestration of constraint solvers in these various domains and the role of conductor is played by resolution. The clean semantics of logic programming is preserved in CLP. A bonus is that the output language is symbolic and expressive. An orthogonal approach to CLP is to use constraint methods to solve inference problems in logic. Imbeddings of logics in mixed integer programming sets were proposed by Williams (1987) and Jeroslow (1989). Efficient algorithms have been developed for inference algorithms in many types and fragments of logic, ranging from Boolean to predicate to belief logics (Chandru and Hooker, 1999).

A persistent theme in the integer programming approach to combinatorial optimization, as we have seen, is that the representation (formulation) of the problem deeply affects the efficacy of the solution methodology. A proper choice of formulation can therefore make the difference between a successful solution of an optimization problem and the more common perception that the problem is insoluble and one must be satisfied with the best that heuristics can provide. Formulation of integer programs has been treated more as an art form than a science by the mathematical programming community. (See Jeroslow (1989) for a refreshingly different perspective on representation theories for mixed integer programming.) We believe that progress in representation theory can have an important influence on the future of integer programming as a broad-based problem solving methodology in combinatorial optimization.

Key Terms

Column generation: A scheme for solving linear programs with a huge number of columns.

Cutting plane: A valid inequality for an integer polyhedron that separates the polyhedron from a given point outside it.

Extreme point: A corner point of a polyhedron.

Fathoming: Pruning a search tree.

Integer polyhedron: A polyhedron, all of whose extreme points are integer valued.

Linear program: Optimization of a linear function subject to linear equality and inequality constraints.

Mixed integer linear program: A linear program with the added constraint that some of the decision variables are integer valued.

Packing and covering: Given a finite collection of subsets of a finite ground set, to find an optimal subcollection that is pairwise disjoint (packing) or whose union covers the ground set (covering).

Polyhedron: The set of solutions to a finite system of linear inequalities on real-valued variables. Equivalently, the intersection of a finite number of linear half-spaces in $\Re^n$.

Relaxation: An enlargement of the feasible region of an optimization problem. Typically, the relaxation is considerably easier to solve than the original optimization problem.

ρ-Approximation: An approximation method that delivers a feasible solution with an objective value within a factor ρ of the optimal value of a combinatorial optimization problem.

References

Aarts, E. H. L. and Korst, J. H. 1989. *Simulated Annealing and Boltzmann Machines: A Stochastic Approach to Combinatorial Optimization and Neural Computing*. Wiley, New York.

Ahuja, R. K., Magnati, T. L., and Orlin, J. B. 1993. *Network Flows: Theory, Algorithms and Applications*. Prentice-Hall, Englewood Cliffs, NJ.

Akgul, M. 1984. *Topics in Relaxation and Ellipsoidal Methods, Research Notes in Mathematics*. Pitman, Boston, MA.

Alizadeh, F. 1995. Interior point methods in semidefinite programming with applications to combinatorial optimization. *SIAM J. Opt.* 5(1):13–51.

Applegate, D., Bixby, R. E., Chvátal, V., and Cook, W. 1994. Finding cuts in large TSP's. *Technical Report*. AT&T Bell Laboratories, Murray Hill, NJ, Aug.

Arora, S., Lund, C., Motwani, R., Sudan, M., and Szegedy, M. 1992. Proof verification and hardness of approximation problems. In *Proceedings 33rd IEEE Symposium on Foundations of Computer Sci.*, Pittsburgh, PA, pp. 14–23.

Balas, E., Ceria, S. and Cornuejols,G. 1993. A lift and project cutting plane algorithm for mixed 0-1 programs. *Math. Program.* 58:295–324.

Barahona, F., Jünger, M., and Reinelt, G. 1989. Experiments in quadratic 0 – 1 programming. *Math. Program.* 44:127–137.

Benders, J. F. 1962. Partitioning procedures for solving mixed-variables programming problems. *Numer. Math.* 4:238–252.

Berge, C. 1961. Farbung von Graphen deren samtliche bzw. deren ungerade Kreise starr sind (Zusammenfassung). Wissenschaftliche Zeitschrift, Martin Luther Universitat Halle-Wittenberg, Mathematisch-Naturwiseenschaftliche Reihe. pp. 114–115.

Berge, C. 1970. Sur certains hypergraphes generalisant les graphes bipartites. In *Combinatorial Theory and its Applications I*. P. Erdos, A. Renyi, and V. Sos, eds., Colloq Math. Soc. Janos Bolyai, 4, pp. 119–133. North Holland, Amsterdam, the Netherlands.

Berge, C. 1972. Balanced matrices. *Math. Program.* 2:19–31.

Berge, C. and Las Vergnas, M. 1970. Sur un theoreme du type Konig pour hypergraphes. In *Int. Conf. Combinatorial Math.*, pp. 31–40. Ann. New York Acad. Sci. 175.

Bixby, R. E., Fenelon, M., Gu, Z., Rothberg, E., and Wunderling, R. 2001. M.I.P: Theory and practice-Closing the gap; Paper presented at Padberg-Festschrift, Berlin, Germany.

Bland, R., Goldfarb, D., and Todd, M. J. 1981. The ellipsoid method: A survey. *Oper. Res.* 29:1039–1091.

Borning, A., ed. 1994. *Principles and Practice of Constraint Programming*, LNCS. Vol. 874, Springer-Verlag, Berlin, Germany.

Camion, P. 1965. Characterization of totally unimodular matrices. *Proc. Am. Math. Soc.* 16:1068–1073.

Cannon, T. L. and Hoffman, K. L. 1990. Large-scale zero-one linear programming on distributed workstations. *Ann. Oper. Res.* 22:181–217.

Černikov, R. N. 1961. The solution of linear programming problems by elimination of unknowns. *Dokl. Akad. Nauk* 139:1314–1317 (translation in 1961. *Sov. Mathemat. Dokl.* 2:1099–1103).

Chandru,V., Dattasharma,A., Kumar V.S.A. 2003. The algorithmics of folding proteins on lattices. *Discrete Appl. Math.* 127(1):145–161.

Chandru, V. and Hooker, J. N. 1991. Extended Horn sets in propositional logic. *JACM* 38:205–221.

Chandru, V. and Hooker, J. N. 1999. *Optimization Methods for Logical Inference.* Wiley Interscience, New York.

Chandru, V. and Rao, M. R. Combinatorial optimization: An integer programming perspective. *ACM Comput. Survey* 28:55–58.

Chandru, V. Rao, M. R. and Swaminathan, G. 2004. Folding proteins on lattices: an integer programming approach. In *The Sharpest Cut: Festschrift fr Manfred Padberg*, pp. 185–196. Springer Verlag, Berlin, Germany.

Chopra, S., Gorres, E. R., and Rao, M. R. 1992. Solving Steiner tree problems by branch and cut. *ORSA J. Comput.* 3:149–156.

Chudnovsky, M., Robertson, N., Seymour, P., and Thomas, R. 2006. The strong perfect graph theorem. *Ann. Math* 164(1):51–229.

Chvátal, V. 1973. Edmonds polytopes and a hierarchy of combinatorial problems. *Discrete Math.* 4:305–337.

Chvátal, V. 1975. On certain polytopes associated with graphs. *J. Comb. Theory B* 18:138–154.

Conforti, M. and Cornuejols, G. 1992a. *Balanced 0, ±1 Matrices, Bicoloring and Total Dual Integrality. Preprint*. Carnegie Mellon University.

Conforti, M. and Cornuejols, G. 1992b. A class of logical inference problems solvable by linear programming. *FOCS* 33:670–675.

Conforti, M., Cornuejols, G., and De Francesco, C. 1993. *Perfect 0, ±1 Matrices. Preprint*. Carnegie Mellon University, Pittsburgh, PA.

Conforti, M., Cornuejols, G., Kapoor, A., and Vuskovic, K. 1994a. *Balanced 0, ±1 Matrices. Pts. I–II.* Preprints. Carnegie Mellon University, Pittsburgh, PA.

Conforti, M., Cornuejols, G., Kapoor, A. Vuskovic, K., and Rao, M. R. 1994b. Balanced matrices. In *Mathematical Programming, State of the Art 1994.* J. R. Birge and K. G. Murty, eds., University of Michigan, Ann Arbor, MI.

Conforti, M., Cornuejols, G., and Rao, M. R. 1999. Decomposition of balanced 0, 1 matrices. *J. Comb. Theory B* 77:292–406.

Conforti, M., Corneujols, G. and Zambelli, G., 2012. Extended formulations in combinatorial optimization. Earlier version appeared in 4OR. *Q. J. Oper. Res.* 8:1–48.

Conforti, M. and Rao, M. R. 1993. Testing balancedness and perfection of linear matrices. *Math. Program.* 61:1–18.

Cook, W., Lovász, L., and Seymour, P., eds. 1995. *Combinatorial Optimization: Papers from the DIMACS Special Year.* Series in Discrete Mathematics and Theoretical Computer Science, Vol. 20. AMS.

Cook, W. J. 2012. *In Pursuit of the Traveling Salesman.* Princeton University Press, Princeton, NJ.

Cornuejols, G. and Novick, B. 1994. Ideal 0, 1 matrices. *J. Comb. Theory* 60:145–157.

CPLEX 6.5 1999. Using the CPLEX Callable Library and CPLEX Mixed Integer Library, Ilog Inc.

Crowder, H., Johnson, E. L., and Padberg, M. W. 1983. Solving large scale 0–1 linear programming problems. *Oper. Res.* 31:803–832.

Cunningham, W. H. 1984. Testing membership in matroid polyhedra. *J. Comb. Theory* 36B:161–188.

Dantzig, G. B. and Wolfe, P. 1961. The decomposition algorithm for linear programming. *Econometrica* 29:767–778.

Ecker, J. G. and Kupferschmid, M. 1983. An ellipsoid algorithm for nonlinear programming. *Math. Program.* 27:1–19.

Edmonds, J. 1965. Maximum matching and a polyhedron with 0–1 vertices. *J. Res. Nat. Bur. Stand.* 69B:125–130.

Edmonds, J. 1970. Submodular functions, matroids and certain polyhedra. In *Combinatorial Structures and Their Applications*, R. Guy, ed., pp. 69–87. Gordon Breach, New York.

Edmonds, J. 1971. Matroids and the greedy algorithm. *Math. Program.* 127–136.

Edmonds, J. 1979. Matroid intersection. *Ann. Discrete Math.* 4:39–49.

Edmonds, J. and Giles, R. 1977. A min-max relation for submodular functions on graphs. *Ann. Discrete Math.* 1:185–204.

Fiorini, S., Massar, S., Pokutta, S., Tiwary, H. R., and de Wolf, R. 2012. Linear vs. semidefinite extended formulations: Exponential separation and strong lower bounds, In *Proc. 44th symp. Theory Comput.*, Karloff, H. J., and Pitassi, T. eds., pp. 95–106. ACM.

Fonlupt, J. and Zemirline, A. 1981. A polynomial recognition algorithm for $K_4\backslash e$-free perfect graphs. *Research Report*, University of Grenoble, Grenoble, France.

Fourer, R., Gay, D. M., and Kernighian, B. W. 1993. *AMPL: A Modeling Language for Mathematical Programming.* Scientific Press, San Francisco, CA.

Fourier, L. B. J. 1824. In: Analyse des travaux de l'Academie Royale des Sciences, pendant l'annee 1824, Partie mathematique. *Histoire de l'Academie Royale des Sciences de l'Institut de France 7* (1827) xlvii-lv (partial english translation Kohler, D. A. 1973. *Translation of a Report by Fourier on his Work on Linear Inequalities. Opsearch* 10:38–42).

Frank, A. 1981. A weighted matroid intersection theorem. *J. Algorithms* 2:328–336.

Fulkerson, D. R. 1970. The perfect graph conjecture and the pluperfect graph theorem. In *Proc. 2nd Chapel Hill Conf. Combinatorial Math. Appl.* R. C. Bose et al., eds., pp. 171–175.

Fulkerson, D. R., Hoffman, A., and Oppenheim, R. 1974. On balanced matrices. *Math. Program. Stud.* 1:120–132.

Gass, S. and Saaty, T. 1955. The computational algorithm for the parametric objective function. *Nav. Res. Log. Q.* 2:39–45.

Gilmore, P. and Gomory, R. E. 1963. A linear programming approach to the cutting stock problem. Pt. I. *Oper. Res.* 9:849–854; Pt. II. *Oper. Res.* 11:863–887.

Glover, F. and Laguna, M. 1997. *Tabu Search*. Kluwer Academic Publishers.

Goemans, M. X. and Williamson, D. P. 1994. .878 approximation algorithms MAX CUT and MAX 2SAT. In *Proc. ACM STOC*. pp. 422–431, Montreal, Quebec, Canada.

Gomory, R. E. 1958. Outline of an algorithm for integer solutions to linear programs. *Bull. Am. Math. Soc.* 64:275–278.

Gomory, R. E. and Hu, T. C. 1961. Multi-terminal network flows. *SIAM J. Appl. Math.* 9:551–556.

Gondzio, J. 2012. Interior point methods 25 years later. *Eur. J. Oper. Res.* 218:587–601.

Grötschel, M., Lovasz, L., and Schrijver, A. 1982. The ellipsoid method and its consequences in combinatorial optimization. *Combinatorica* 1:169–197.

Grötschel, M., Lovász, L., and Schrijver, A. 1988. *Geometric Algorithms and Combinatorial Optimization*. Springer–Verlag, Berlin, Germany.

GUROBI 2013. http://www.gurobi.com/documentation/5.1/reference-manual/.

Hacijan, L. G. 1979. A polynomial algorithm in linear programming. *Sov. Math. Dokl.* 20:191–194.

Held, M. and Karp, R. M. 1970. The travelling-salesman problem and minimum spanning trees. *Oper. Res.* 18:1138–1162, Pt. II. 1971. *Math. Program.* 1:6–25.

Held, M., Wolfe, P., and Crowder, H. P. 1974. Validation of subgradient optimization. *Math. Program.* 6:62–88.

Hoffman, A. J. and Kruskal, J. K. 1956. Integral boundary points of convex polyhedra. In *Linear Inequalities and Related Systems*. H. W. Kuhn and A. W. Tucker, eds., pp. 223–246. Princeton University Press, Princeton, NJ.

Hooker, J. N. 1992. *Resolution and the Integrality of Satisfiability Polytopes*. Preprint, GSIA, Carnegie Mellon University, Pittsburgh, PA.

Hooker, J. N. 1993. Towards and empirical science of algorithms. *Oper. Res.* 42:201–212.

Hooker, J. N. and Vinay, V. 1995. Branching rules for satisfiability. *Autom. Reason.* 15:359–383.

Iyengar, G. and Cezik, M.T. 2002. Cutting planes for mixed 0-1 semidefinite programs. *Proc. VIII IPCO Conf.*

Jeroslow, R. E. and Lowe, J. K. 1984. Modeling with integer variables. *Math. Program. Stud.* 22:167–184.

Jeroslow, R. G. 1989. *Logic-Based Decision Support: Mixed Integer Model Formulation*. Ann. Discrete Mathematics, Vol. 40, North–Holland.

Jünger, M., Reinelt, G., and Thienel, S. 1995. Practical problem solving with cutting plane algorithms. In *Combinatorial Optimization: Papers from the DIMACS Special Year*. Series in Discrete Mathematics and Theoritical Computer Science, Vol. 20, pp. 111–152. AMS.

Kaibel, V., Pashkovich, K., and Theis, D.O. 2012. Symmetry matters for sizes of extended formulations. SIAM Journal on Discrete Mathematics, 26(3), 1361–1382.

Karmarkar, N. K. 1984. A new polynomial-time algorithm for linear programming. *Combinatorica* 4:373–395.

Karmarkar, N. K. 1990. An interior-point approach to NP-complete problems—Part I. *Contemporary Mathematics*, Vol. 114, pp. 297–308.

Karp, R. M. and Papadimitriou, C. H. 1982. On linear characterizations of combinatorial optimization problems. *SIAM J. Comput.* 11:620–632.

Lawler, E. L. 1975. Matroid intersection algorithms. *Math. Program.* 9:31–56.

Lehman, A. 1979. On the width-length inequality, mimeographic notes (1965). *Math. Program.* 17:403–417.

Lin, S. and Kernighan, B. W. 1973. An effective heuristic algorithm for the travelling salesman problem. *Oper. Res.* 21:498–516.

Lovasz, L. 1972. Normal hypergraphs and the perfect graph conjecture. *Discrete Math.* 2:253–267.

Lovasz, L. 1979. On the Shannon capacity of a graph. *IEEE Trans. Inf. Theory* 25:1–7.

Lovasz, L. 1986. *An Algorithmic Theory of Numbers, Graphs and Convexity*. SIAM Press, Philadelphia, PA.

Lovasz, L. and Schrijver, A. 1991. Cones of matrices and setfunctions. *SIAM J. Optim.* 1:166–190.

Lustig, I. J., Marsten, R. E., and Shanno, D. F. 1994. Interior point methods for linear programming: Computational state of the art. *ORSA J. Comput.* 6(1):1–14.

Martin, R. K. 1991. Using separation algorithms to generate mixed integer model reformulations. *Oper. Res. Lett.* 10:119–128.

McDiarmid, C. J. H. 1975. Rado's theorem for polymatroids. *Proc. Cambridge Philos. Soc.* 78:263–281.

Megiddo, N. 1991. On finding primal- and dual-optimal bases. *ORSA J. Comput.* 3:63–65.

Mehrotra, S. 1992. On the implementation of a primal-dual interior point method. *SIAM J. Optim.* 2(4):575–601.

Nemhauser, G. L. and Wolsey, L. A. 1988. *Integer and Combinatorial Optimization*. Wiley, New York.

Orlin, J. B. 2000. Very Large-scale neighborhood search techniques. *Featured Lecture at the International Symposium on Mathematical Programming*. Atlanta, GA.

Optimization Subroutine Library (OSL). 1991. Optimization Subroutine Library Guide and Reference. *IBM Systems Journal*, Vol. 31.

Padberg, M. W. 1974. Perfect zero-one matrices. *Math. Program.* 6:180–196.

Padberg, M. W. 1993. Lehman's forbidden minor characterization of ideal 0, 1 matrices. *Discrete Math.* 111:409–420.

Padberg, M. W. 1995. *Linear Optimization and Extensions*. Springer–Verlag, Berlin, Germany.

Padberg, M. W. and Rao, M. R. 1981. *The Russian Method for Linear Inequalities. Part III, Bounded Integer Programming*. Preprint, New York University, New York.

Padberg, M. W. and Rao, M. R. 1982. Odd minimum cut-sets and b-matching. *Math. Oper. Res.* 7:67–80.

Padberg, M. W. and Rinaldi, G. 1991. A branch and cut algorithm for the resolution of large scale symmetric travelling salesman problems. *SIAM Rev.* 33:60–100.

Papadimitriou, C. H. and Yannakakis, M. 1991. Optimization, approximation, and complexity classes. *J. Comput. Syst. Sci.* 43:425–440.

Parker, G. and Rardin, R. L. 1988. *Discrete Optimization*. Wiley, New York.

Picard, J. C. and Ratliff, H. D. 1975. Minimum cuts and related problems. *Networks* 5:357–370.

Raghavan, P. and Thompson, C. D. 1987. Randomized rounding: A technique for provably good algorithms and algorithmic proofs. *Combinatorica* 7:365–374.

Rhys, J. M. W. 1970. A selection problem of shared fixed costs and network flows. *Manage. Sci.* 17:200–207.

Saraswat, V. and Van Hentenryck, P., eds. 1995. *Principles and Practice of Constraint Programming*. MIT Press, Cambridge, MA.

Savelsbergh, M. W. P., Sigosmondi, G. S., and Nemhauser, G. L. 1994. MINTO, a mixed integer optimizer. *Oper. Res. Lett.* 15:47–58.

Schrijver, A. 1986. *Theory of Linear and Integer Programming*. Wiley, Chichester, U.K.

Seymour, P. 1980. Decompositions of regular matroids. *J. Comb. Theory B* 28:305–359.

Shapiro, J. F. 1979. A survey of lagrangian techniques for discrete optimization. *Ann. Discrete Math.* 5:113–138.

Shor, N. Z. 1970. Convergence rate of the gradient descent method with dilation of the space. *Cybernetics* 6.

Spielman, D. A., and Tang, S. H. 2001. Smoothed analysis of algorithms: Why the simplex method usually takes polynomial time. *Proc. 33rd Annual ACM Symp. Theory Comput.*, pp. 296–305.

Truemper, K. 1992. Alpha-balanced graphs and matrices and GF(3)-representability of matroids. *J. Comb. Theory B* 55:302–335.

Weyl, H. 1935. Elemetere Theorie der konvexen polyerer. *Comm. Math. Helv.* Vol. pp. 3–18 (English translation 1950. *Ann. Math. Stud.* 24, Princeton).

Whitley, D. 1993. *Foundations of Genetic Algorithms 2*. Morgan Kaufmann, San Mateo, CA.

Williams, H. P. 1987. Linear and integer programming applied to the propositional calculus. *Int. J. Syst. Res. Inf. Sci.* 2:81–100.

Williamson, D. P. 2000. The primal-dual method for approximation algorithms. *Proc. Int. Symp. Math. Program*. Atlanta, GA.

Wolkowicz, W., Saigal, R., and Vanderberghe, L., eds. 2000. *Handbook of Semidefinite Programming*. Kluwer Academic Publisher, Boston, MA.

Yannakakis, M. 1988. Expressing combinatorial optimization problems by linear programs. In *Proc. ACM Symp. Theory Comput*. pp. 223–228, Chicago, IL.

Ye, Y. 2000. Semidefinite programming for discrete optimization: Approximation and Computation. *Proc. Int. Symp. Math. Program*. Atlanta, GA.

Ziegler, M. 1995. *Convex Polytopes*. Springer–Verlag, Berlin, Germany.

14

Graph Drawing

Ioannis G. Tollis
University of Crete and ICS-FORTH

Konstantinos G. Kakoulis
Technological Educational Institute of West Macedonia

14.1 Introduction

Graph drawing addresses the problem of visualizing information by constructing geometric representations of conceptual structures that are modeled by graphs, networks, or hypergraphs. Graphs may be used to represent any information that can be modeled as objects and connections between those objects. The area of graph drawing has grown significantly in the recent years motivated mostly by applications in information visualization and in key computer technologies, including software engineering (call graphs, class hierarchies), visual interfaces, database systems (entity-relationship diagrams), digital libraries and World Wide Web authoring and browsing (hypermedia documents), VLSI (symbolic layout), electronic systems (block diagrams, circuit schematics), project management (PERT diagrams, organization charts), telecommunications (ring covers of networks), business, and social sciences (social network analysis).

14.2 Underlying Principles

A graph is drawn in two or three dimensions. Usually, the nodes are represented by circles or boxes, and the edges are represented by simple open curves between the nodes. A graph $G = (V, E)$ consists of a set V of nodes and a set E of edges, that is, unordered pairs of nodes. Given a graph G, the number of nodes is denoted by n and the number of edges is denoted by m.

The objective of a graph drawing algorithm is to take a graph and produce a nice drawing of it. In order to define the requirements for a nice drawing, Di Battista et al. [28] distinguish three different concepts: (i) drawing conventions, (ii) aesthetics, and (iii) constraints. In the following, we give a brief overview of them.

14.2.1 Drawing Conventions

Various drawing standards have been proposed for drawing graphs, depending on the application. Drawing conventions provide the basic rules that the drawing must satisfy to be admissible. These rules deal with representation and placement of nodes and edges. Examples of drawing conventions are the following:

Straight-line drawing: Each edge is represented as a straight-line segment.

Polyline drawing: Each edge is represented as a polygonal chain.

Orthogonal drawing: Each edge is drawn as a sequence of alternating horizontal and vertical line segments.

Grid drawing: Nodes, edges, and edge bends have integer coordinates.

Planar drawing: No two edges cross each other.

Upward drawing: Edges are drawn as curves that are monotonically nondecreasing in the vertical direction. A drawing is strictly upward if each edge is drawn as a curve strictly increasing in the vertical direction.

14.2.2 Aesthetics

One of the fundamental characteristics of a graph drawing is its readability, that is, the capability of conveying clearly the information associated with the graph. Readability issues are expressed by means of aesthetics, which can be formulated as optimization goals for the graph drawing algorithms. Some aesthetic criteria are general, others depend on the drawing convention adopted and on the particular class of graphs considered.

Further information on aesthetics can be found in [21,98,99,119,122]. It should be noted that aesthetics are subjective and may need to be tailored to suit personal preferences, applications, traditions, and culture. Next, we present some of the most important aesthetics:

Area: The area of a drawing is the area of the smallest rectangle that encloses the drawing and should be minimized.

Aspect ratio: The ratio of the longest to the shortest side of the smallest rectangle enclosing the drawing should be balanced.

Bends: The number of edge bends especially for orthogonal drawings should be minimized.

Edge crossings: The number of crossings in a drawing should be minimized.

Edge length: The total length of all edges, the maximum length of a single edge, or the variance of the lengths of the edges should be minimized.

Convexity: Faces of planar drawings should be drawn as convex polygons.

Symmetry: The symmetries of a graph must be displayed in the drawing.

Angular resolution: The angle between any two consecutive edges of a node should be maximized.

It is not generally possible to fulfill all aesthetics simultaneously. For example, in Figure 14.1, two embeddings of K_4 are displayed, the one on the left minimizes the number of edge crossings, whereas

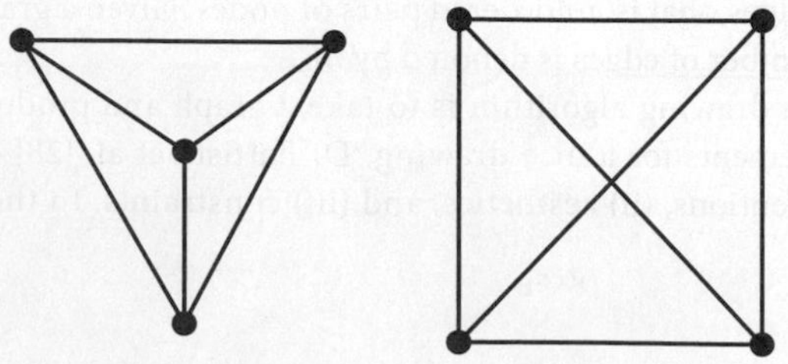

FIGURE 14.1 Drawings of a K_4 graph.

the one on the right maximizes the display of symmetries. Additionally, it is hard to achieve optimum results since most of the problems described earlier are computationally hard. For example, the following problems related to aesthetics are NP-complete or NP-hard: drawing a graph with the minimum number of crossings [55], embedding a graph in a grid of prescribed size [47,75], determining the minimum number of bends for orthogonal drawings [56], and drawing a graph symmetrically [81,82].

14.2.3 Constraints

While drawing conventions and aesthetics apply to an entire drawing, drawing constraints only apply to specific parts. They usually emphasize semantic aspects of a graph (e.g., by placing related nodes close to each other) and are given as additional input to the layout algorithm. Some examples of drawing constraints are the following:

Center: Place a given node near the center of the drawing.

External: Place a given node at the boundary of the drawing.

Cluster: Place a given subset of nodes close to each other.

Shape: Draw a subgraph with a given shape.

Sequence: Place a given subset of nodes along a straight line.

Order: Place a given node to the left, right, top, or bottom of another given node.

Position: Place a given node at a given position.

Direction: Draw a given subset of edges monotonically in a prescribed direction.

14.3 Graph Drawing Algorithms

Graph drawing algorithms may be classified according to the style of drawing they generate (e.g., hierarchical, orthogonal, force-directed, and circular), the methodology on which they are based (e.g., planarization, and divide, and conquer), and the class of graphs they can be applied to (e.g., planar, directed acyclic, and trees).

Many algorithmic techniques have been devised in order to draw graphs. Due to space restrictions, the main concepts of the most important techniques are presented here. A detailed description of graph drawing algorithms can be found in the literature, see for example [28,71].

14.3.1 Graph Drawing Layout Styles

There are algorithms that produce drawings that attempt to highlight hierarchy, symmetry, grouping structure, or orthogonality. Next, we present the most interesting techniques with respect to the flavor of the resulting layout of a graph drawing.

14.3.1.1 Hierarchical Layout

The hierarchical layout is ideal for displaying hierarchies. Directed graphs that contain an inherent hierarchy are drawn using hierarchical layout algorithms. This layout style is used in a variety of applications, for example, in PERT diagrams, organization charts, scheduling and logistic diagrams, and class hierarchies in software engineering. Figure 14.2 shows an example of a hierarchical drawing.

The basic method for hierarchical graph drawings is presented in [113], upon which all other methods are based. It accepts, as input, a directed graph and produces, as output, a layered drawing, that is, a representation where the nodes are placed on horizontal layers, and the edges are drawn as polygonal chains connecting their end nodes. In hierarchical drawings preferably most edges flow

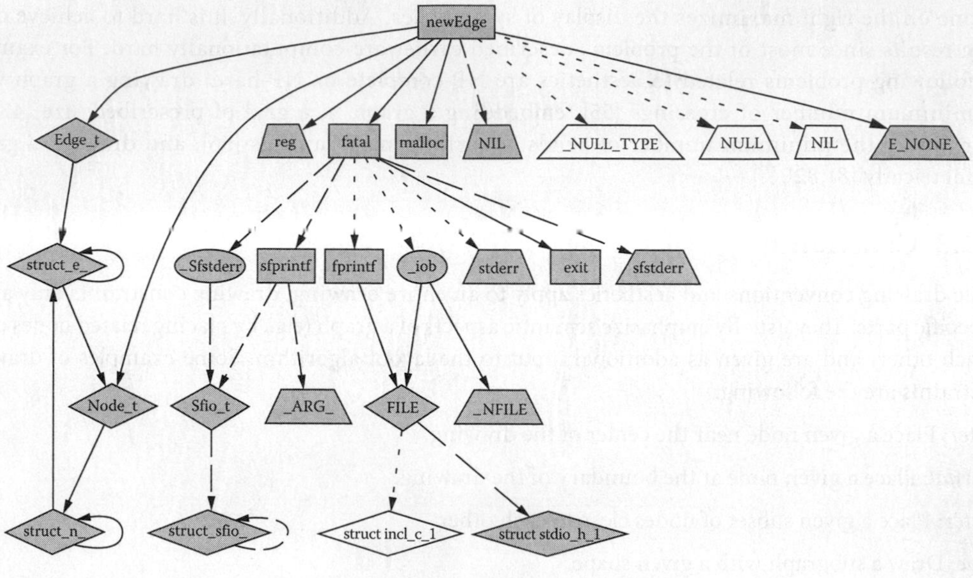

FIGURE 14.2 A hierarchical drawing that displays the reachability analysis of the C function newEdge. (Courtesy of R. Chen, E. Koutsofios, and S. North, AT&T Research Labs.)

in the same direction, for example, from top to bottom. The methodology based on this hierarchical layout algorithm consists of the following steps:

Cycle removal: If the input graph contains directed cycles, then the cycles are removed by temporarily reversing the direction of some edges.

Layer assignment: The nodes are assigned to layers such that edges point from lower to higher layers. Dummy nodes are inserted along the edges that cross layers, so that each edge in the resulting digraph connects nodes in consecutive layers.

Crossing reduction: The nodes are ordered on each layer to reduce the number of edge crossings.

Horizontal coordinate assignment: The nodes (original and dummy) are assigned coordinates such that we get only few bend points and short edges.

Reversing the minimum number of edges in order to *remove cycles* is an NP-hard problem. We can perform a Depth-First-Search (DFS) of the graph and reverse all back edges to remove cycles. This simple approach removes all cycles; however, in the worst case this may reverse as many as $m - n - 1$ edges. In [40], a greedy approach guarantees the reversal of at most $m/2 - n/6$ edges.

In most applications, the nodes need to be *assigned to layers*. The longest path layering algorithm is the most commonly used [38]. All sources are assigned to the lowest level L_0; each remaining node v is assigned to level L_p where p is the length of the longest path from any source to v. This algorithm runs in linear time and produces the minimum number of layers. However, it tends to produce drawings that are wide towards the bottom. It is NP-hard [38] to minimize both the number of layers and the number of nodes on each layer. For more work on layer assignment see [18,54,104].

The number of *edge crossings* depends on the ordering of the nodes within each layer. This problem is NP-hard even if we restrict it to bipartite (two-layered) graphs [55]. It remains NP-hard even if there are only two layers and the permutation of the nodes on one of them is fixed [39]. Several heuristics for the crossing minimization problem have been proposed (e.g., [62,88,110]).

The *horizontal coordinate assignment* reduces the number of bends by positioning the nodes on each layer without changing their order established in the crossing reduction step. This is an optimization problem and the goal is to minimize the total amount by which an edge deviates from being a straight line [9,15,54].

14.3.1.2 Force-Directed Layout

Force-directed layout produces graph drawings that are aesthetically pleasing, exhibit (*i*) symmetries and (*ii*) clustered structure of the underlying graphs. It is a popular choice for drawing general graphs, because it produces good drawings (well-balanced distribution of nodes and few edge crossings) and is relatively easy to understand and implement. However, force-directed drawings are suboptimal and often unpredictable; in addition, there is no performance guarantee and the running time of this technique is typically high. It has applications in many areas of visualization of complex structures, such as bioinformatics, enterprise networking, knowledge representation, system management, WWW visualization, and Mesh visualization. Figure 14.3 shows an example of a force-directed drawing.

In general, a force-directed layout algorithm simulates a system of forces defined on an input graph, and outputs a locally minimum energy configuration. This approach consists of two basic steps:

- A model consisting of physical objects (representing the elements of the graph) and interactions between these objects
- An algorithm for finding a locally minimum energy configuration

In a seminal paper, Eades [35] presented the first force-directed graph drawing algorithm. A graph is modeled as a physical system where the nodes are rings and the edges are springs that connect pairs of rings. The rings are placed in some initial layout and the spring forces move the system to a local minimum of energy. Attractive forces are calculated using a logarithmic function in place of Hooke's law for spring forces and are applied only between pairs of adjacent nodes. Repulsive forces act between every pair of nodes in order to avoid node overlaps. A node will move to its new location based on the sum of those attractive and repulsive forces applied to the node. Most force-directed algorithms adopt the physical model of [35] and propose variants of the definition of the forces [23,50,52,69,112]. For example, in [112], the concept of magnetic springs is introduced to control the orientation of the edges in the drawing. Edges are modeled with three types of springs: non-magnetic springs, unidirectional magnetic springs, and bidirectional magnetic springs. Various magnetic fields are considered: three standard magnetic fields, namely, parallel, polar, and concentric, and two compound magnetic fields, namely, orthogonal and polar-concentric. This model is suited for drawing directed graphs, trees, and graphs with more than one type of edges. In [52], attractive forces take into account the optimal distance between nodes in the drawing which is defined as a function of the number of nodes in the graph and the size of the drawing window. In [69], the forces are chosen so that the Euclidean distance between any two nodes in the drawing is proportional to the graph-theoretic distance between the corresponding nodes in the graph. In [23], the drawing of a graph is formulated as an optimization problem and the solution is found

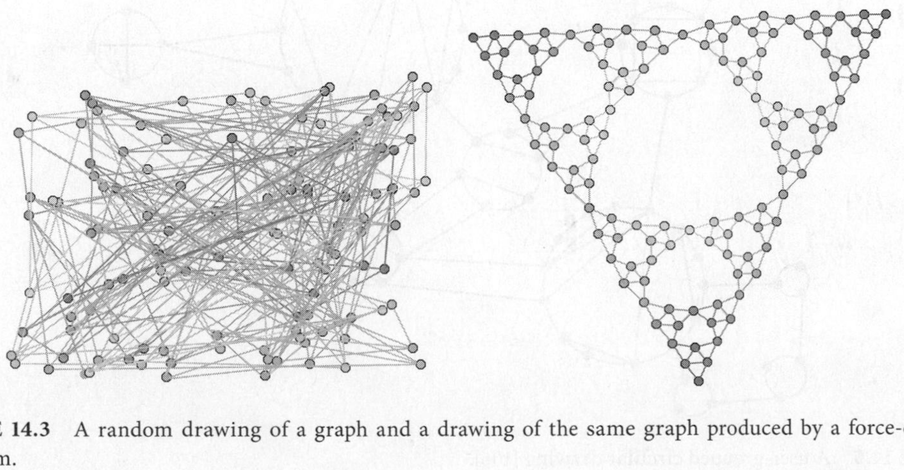

FIGURE 14.3 A random drawing of a graph and a drawing of the same graph produced by a force-directed algorithm.

using simulated annealing. The goal is to minimize a cost function that is a linear combination of energy functions, each defined to optimize a certain aesthetic criterion, such as uniform distribution of nodes, uniform distribution of edge lengths, number of edge-crossings, and distance between nodes and edges.

14.3.1.3 Circular Layout

Circular layout produces graph drawings that emphasize group structures that are either inherent in a graph's topology or user-defined. Circular layout has many applications, for example, in social networking (organization charts, citation networks, and collaboration networks), security (case information diagrams, fraud networks, and felony event flow diagrams), network management (telecommunication and computer networks, LAN diagrams, and traffic visualization), WWW visualization (visitor movement graphs and internet traffic), and eCommerce. A circular graph drawing (see Figures 14.4 and 14.5 for examples) is a visualization of a graph with the following characteristics:

- The graph is partitioned into groups.
- The nodes of each group are placed onto the circumference of an embedding circle.
- Each edge is drawn as a straight line.

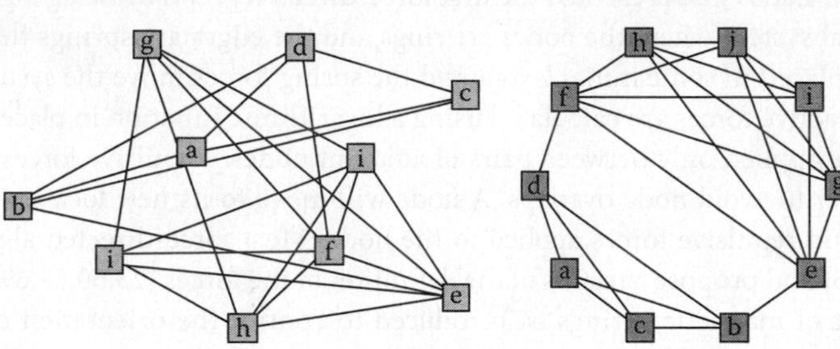

FIGURE 14.4 A random drawing of a graph and a drawing of the same graph produced by a circular layout algorithm [107].

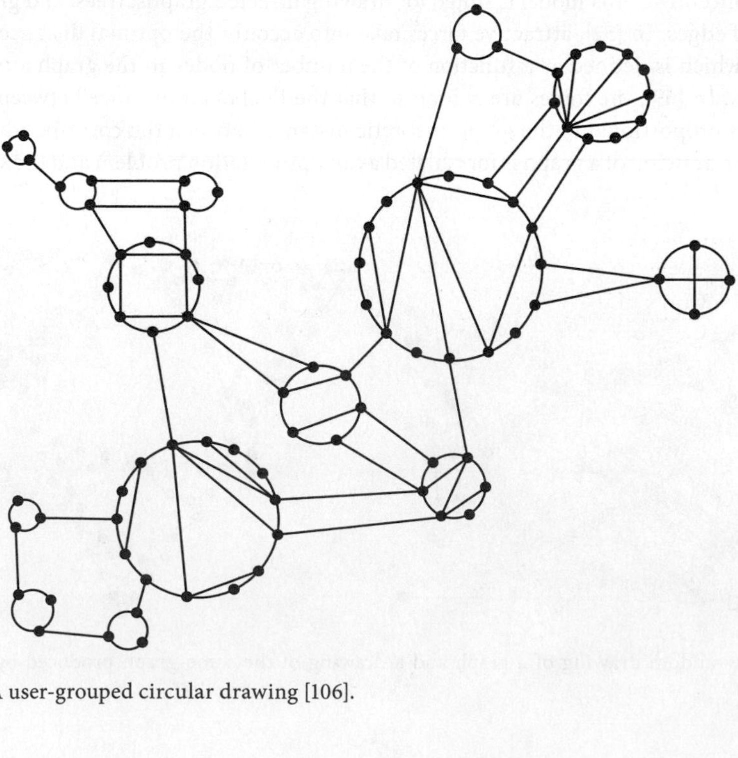

FIGURE 14.5 A user-grouped circular drawing [106].

Node partitions are created by either analyzing the connectivity structure of the underlying graph, or are user-defined. The partitioning of the graph into clusters can show structural information such as biconnectivity, or the clusters can highlight semantic qualities of the graph. An example of an application in which user-grouped circular drawing would be useful is a computer network management system. It would be very helpful to allow the user to group the computers by IP addresses, IP subnet masks, or other criteria.

It is essential that the number of edge crossings within each cluster remain low in order to reduce the visual complexity of the resulting drawings. However, the problem of finding a minimum crossing embedding such that all the nodes are placed onto the circumference of a circle and all edges are represented with straight lines is NP-complete [84].

A technique for producing circular drawings of biconnected graphs on a single embedding circle is presented in [105]. In order to produce circular drawings with few crossings, the algorithm tends to place edges toward the outside of the embedding circle. Also, nodes are placed near their neighbors. In fact, this algorithm tries to maximize the number of edges appearing toward the periphery of the embedding circle. It selectively removes some edges and then builds a DFS based node ordering of the resulting graph. This technique visits the nodes in a wave-like fashion, looking for pairs of edges (edges incident to two nodes which share at least one neighbor) which are then removed. Sometimes, triangulation edges are added to aid this process. It is the selective edge removal that causes many edges to be placed toward the periphery of the embedding circle. Subsequent to the edge removal, a DFS is performed on the reduced graph. The longest path of the resulting DFS tree is placed on the embedding circle and the remaining nodes are nicely merged into this ordering. The worst-case time requirement of this technique is $O(m)$ (recall that m is the number of edges). An important property of this technique is the guarantee that if a zero-crossing drawing exists for a given biconnected graph then this technique will produce it. An alternative approach where selected edges are drawn outside the embedding circle is described in [53].

The earlier presented algorithm can be extended in order to produce circular drawings of nonbiconnected graphs on a single embedding circle. Given a nonbiconnected graph, it can be decomposed into its biconnected components. The algorithm layouts the resulting block-cutpoint tree on a circle and then it layouts each biconnected component.

A framework for producing circular drawings of nonbiconnected graphs on multiple circles is presented in [107]. It decomposes a nonbiconnected graph into biconnected components in $O(m)$ time. Next, it produces a layout of the block-cutpoint tree using a radial layout technique [28,36] (see also Section 14.3.1.5). Finally, each biconnected component of the graph is drawn on a single embedding circle with the circular layout style.

A circular graph drawing technique is presented in [106] where the node grouping is user-defined. It draws each group of nodes efficiently and effectively, and visualizes the superstructure well. This technique produces drawings in which the user-defined groupings are highly visible, each group is laid out with a low number of edge crossings, and the number of crossings between intragroup and intergroup edges is low. First, it produces a layout of the superstructure graph of the input graph by using a basic force-directed technique (see Section 14.3.1.2 for details). Each group of nodes of the input graph is a node in the superstructure and for each edge of the input graph, which is incident to nodes in two different node groups, there is an edge between nodes representing the respective groups in the superstructure graph. Next, each group of nodes is drawn on a single embedding circle with the circular layout style. Finally, in order to reduce the number of intragroup and intergroup edge crossings, a specialized force-directed algorithm is applied, in which the nodes are restricted to appear on circular tracks.

Several linear time algorithms for the visualization of survivable telecommunication networks were introduced in [120]. Given the ring covers of a network, these algorithms create circular drawings such that the survivability of the network is very visible.

An interesting application of circular graph drawings was presented in [76]. The tool InFlow is used to visualize human networks and produces diagrams and statistical summaries to pinpoint the strengths and weaknesses within an organization. The usually unvisualized characteristics of self-organization, emergent structures, knowledge exchange, and network dynamics can be seen in the drawings of InFlow. Resource bottlenecks, unexpected work flows, and gaps within the organization are clearly visible in circular drawings.

A circular drawing technique and tool for network management was presented in [70]. In [32], an advanced version of this technique was introduced. It partitions nodes into groups based on a predefined method, such as biconnectivity of the graph or it is user-defined. Each group of nodes is placed on radiating circles based on their logical interconnection.

Much work (e.g., [4,53,72,114]) has been done on circular layouts where all nodes are placed on a single embedding circle.

14.3.1.4 Orthogonal Layout

An orthogonal drawing is a polyline drawing that maps each edge into a chain of horizontal and vertical segments. Orthogonal drawings are widely used for visualizing diagrams, such as Entity-Relationship and Data-Flow diagrams in database conceptual design, circuit schematics in VLSI design, and UML diagrams in software engineering. Figure 14.6 shows an example of an orthogonal drawing.

A popular and effective technique for computing an orthogonal drawing of a graph is the so-called topology-shape-metrics approach [28]. This technique is well suited for medium-sized sparse graphs. It produces compact drawings with no overlaps, few crossings, and few bends. The topology-shape-metrics approach consists of three phases:

Planarization: Determines the topology of the drawing which is described by a planar embedding. If the input graph is not planar, a set of dummy nodes is added to replace edge crossings. This transforms the input not planar graph to an augmented planar graph.

Orthogonalization: An orthogonal representation H of the graph is computed within the planar embedding found in the planarization step.

Compaction: The final drawing that preserves the orthogonal representation of the second phase is computed by assigning integer coordinates to nodes and bends. Finally, the inserted dummy nodes are removed.

The distinct phases of the topology-shape-metrics approach have been extensively studied in the literature. The first phase is explained in detail in the planarization graph drawing technique, which is presented in the following section.

In the second phase, an orthogonal representation is computed from the planar representation of the previous phase. The goal is to keep the number of bends as small as possible. In [115], a very elegant

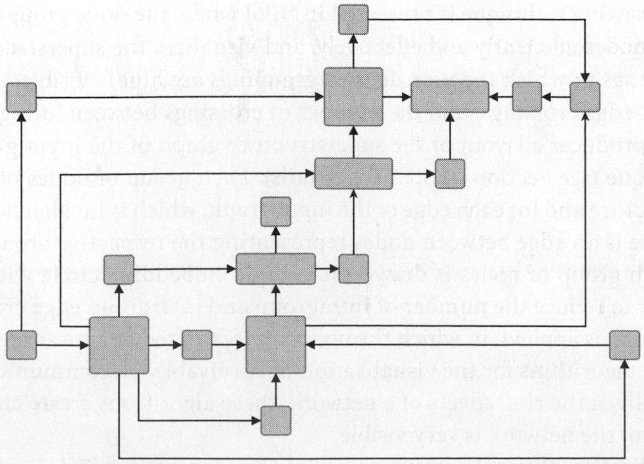

FIGURE 14.6 An orthogonal drawing.

network-flow method for constructing an orthogonal representation of an embedded planar graph with nodes having at most four incident edges is presented. It guarantees the minimum number of bends and runs in $O(n^2 \log n)$ time. In [118], a heuristic guarantees a constant number of bends on each edge and runs in linear time.

During the compaction phase, most approaches try to minimize the area and the total edge length of the drawing. However, compacting an orthogonal representation in such a way that it has either minimum area or minimum total edge length is an NP-hard problem [97]. The compaction problem has been studied extensively especially in the context of circuit design [78]. Compaction heuristics in the context of graph drawing can be found in [13,73,115].

A pairing technique for producing orthogonal drawings with maximum degree 4 is presented in [93]. It is the first orthogonal drawing algorithm that produces drawings with less than n^2 area. The drawing is constructed incrementally. The nodes of the graph are grouped into pairs. Then, node pairs are inserted into the drawing according to an st-numbering of the graph. Nodes of the same pair are placed on the same row or column. This technique runs in linear time, requires provably $0.76n^2$ area, and produces drawings with a linear number of bends and at most two bends per edge. A similar approach has been taken in [6].

In [108], a postprocessing technique is introduced that refines and significantly improves aesthetic qualities of a given orthogonal graph drawing such as area, bends, crossings, and total edge length.

For more work on orthogonal drawings see [8,27,29,48,92,118].

14.3.1.5 Tree Layout

Trees are a special case of planar graphs that are commonly used in many applications. Tree layouts have applications in algorithm animation, circuit design, visualization of class hierarchies, flowcharts, project management diagrams, and syntax trees.

The most popular method for drawing a rooted tree is presented in [100]. It draws binary trees in the hierarchical layout style, nodes are placed on horizontal layers, and it is based on the divide and conquer principle. The idea is to recursively draw the left and right subtrees independently in a bottom-up manner, then shift the two drawings in the x-direction as close to each other as possible, and center the parent of the two subtrees between their roots. This method runs in linear time, requires $O(n^2)$ area, and produces drawings with a high degree of symmetry (identical subtrees are drawn identically). In addition, the width of the drawing is relatively small. This algorithm can be easily extended to general rooted trees. Figure 14.7a shows an example of a tree drawing produced by this technique.

Hierarchical layout algorithms can be used to draw trees, since trees are acyclic graphs with edges pointing away from the root. However, the resulting drawings are wide and the parent nodes are not necessarily centered with respect to their subtrees [28].

Another way of drawing trees is the so-called radial layout [28,36]. In the radial layout, the root of the tree is placed at the center and all descendant nodes are placed on concentric circles according to their depth in the tree. To avoid crossings between subtrees, a sector called annulus wedge is assigned to every subtree. In principle, these sectors are proportional in size to the number of nodes in the subtree.

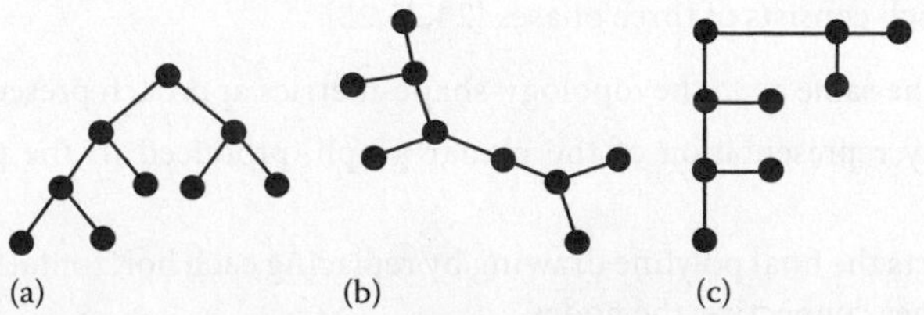

FIGURE 14.7 Three representations of a tree: (a) layered (hierarchical) drawing, (b) radial drawing, and (c) HV drawing.

Algorithms for constructing hierarchical drawings can be adapted to draw trees radially. Figure 14.7b shows an example of the radial layout.

Figure 14.7c represents the drawing of a tree in the HV-drawing style. The HV method (HV stands for horizontal–vertical) can be used on both binary and general trees. A divide and conquer strategy is used to recursively construct an HV drawing of a tree. The root of the tree is placed on the top-left corner, and the drawings of its left and right subtrees are placed one next to the other (horizontal drawing) or one below the other (vertical drawing). Each subtree is assigned its own area so that no overlapping occurs.

14.3.2 Graph Drawing Techniques

The following techniques are frequently used to implement graph drawing algorithms [28]:

Planarization: The planarization approach is motivated by the availability of many efficient graph drawing algorithms for planar graphs. If the graph is nonplanar, it is transformed into a planar graph by introducing dummy nodes at edge crossings. Then, graph drawing algorithms for planar graphs can be used to draw the transformed nonplanar graph.

In [61], a linear time algorithm tests if a graph is planar. A planar embedding of a planar graph can be found in linear time [16,85]. Finding the minimum number of crossings [55] or a maximum planar subgraph [80] are NP-hard problems. Hence, heuristics are used to solve the planarization problem. A natural approach to this problem is to first determine a large planar subgraph and then insert the remaining edges, while trying to keep a minimal number of crossings. An annotated bibliography on planarization techniques is presented in [79].

The orthogonal drawing algorithm (topology-shape-metrics approach), presented in the previous section, is based on the planarization technique.

Divide and conquer: Graphs are broken down into smaller components for which smaller layouts are computed recursively. The smaller layouts are assembled to form the overall layout [28]. This technique is often used when drawing trees.

Flow methods: Bend minimization in orthogonal drawings can efficiently be solved by reduction to a network flow problem, at least if the topology of the embedding is fixed [115]. The same techniques can be used to maximize angles between edges [28].

Visibility approach: The visibility approach is a general purpose technique for producing polyline drawings of graphs and it is based on the visibility representation of planar graphs. Linear time algorithms for producing visibility representations of planar graphs were presented in [103,117].

In a visibility representation of a planar graph, nodes are represented as horizontal segments and edges as vertical segments such that each edge segment has its endpoints on the segments associated with its incident nodes and does not cross any other node segment. A visibility representation can be considered to be a skeleton or a sketch of the final drawing. In [25], a technique is presented, which constructs visibility representations so that some prespecified edges are vertically aligned. These visibility representations, called constrained visibility representations, can be used as starting points for obtaining drawings of graphs with interesting properties. An example of a visibility representation of a planar graph is given in Figure 14.8.

The visibility approach consists of three phases [24,25,28]:

Planarization: This is the same as in the topology-shape-metrics approach presented earlier.

Visibility: The visibility representation of the planar graph, produced in the planarization phase, is computed.

Replacement: Constructs the final polyline drawing by replacing each horizontal line by its node and the vertical lines by polylines connecting the nodes.

In the replacement phase, there are several strategies in order to construct planar polyline drawings from a visibility representation [28]. In [118], a linear time algorithm is presented for constructing

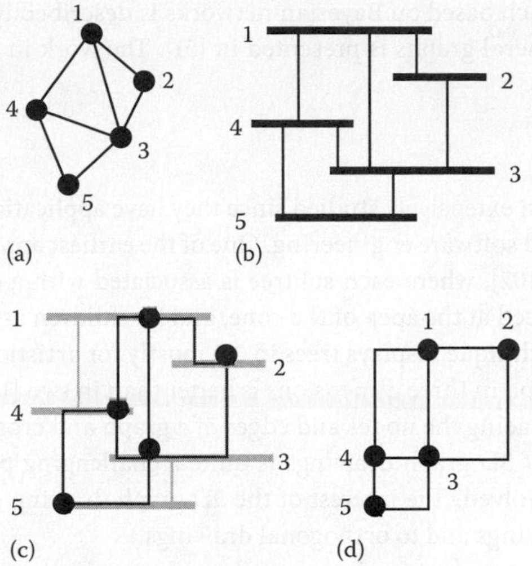

FIGURE 14.8 (a) A planar graph, (b) its visibility representation, (c) a preliminary orthogonal drawing derived from a visibility representation, and (d) the final orthogonal drawing after modifications.

orthogonal drawings which is based on the visibility approach. In the replacement phase of this algorithm, the visibility representation of the graph is transformed into an orthogonal grid embedding. A number of local transformations are applied to the orthogonal representation of the grid embedding in order to reduce the number of bends. Finally, an orthogonal grid embedding with at most $2.4n + 2$ bends is computed from the orthogonal representation. Figure 14.8 shows the steps of this algorithm.

14.3.2.1 Interactive (Dynamic) Drawings

Many applications require the layout of dynamic graphs, representing changing relational information over time. Examples of applications include software engineering visualizations, changing Internet usage, and changes in social network structures. Dynamic graph drawing involves laying out graphs which evolve over time by continuously adding or deleting nodes and edges. The challenge in dynamic graph drawing is to preserve the mental map of the drawing each time we perform an addition or deletion. The notion of the mental map of a drawing has been introduced in [86]. By preserving the user's mental map of a drawing, we generally mean that a user that has studied a drawing will not have to reinvest significant effort to visualize or understand the resulting drawing after some changes.

Traditional (static) layout techniques may not be suited for drawing dynamic graphs. When static layout algorithms are employed, insertion or deletion of even one node or edge can significantly change the layout and thus alter the user's mental map of the drawing.

Many algorithms address the problem of offline dynamic graph drawing, where the entire sequence of graphs to be drawn is known in advance. This gives the layout algorithm information about future changes in the graph, which allows it to optimize the layouts generated across the entire sequence. For example, the algorithm can reserve place in order to accommodate a node that appears later in the sequence. In [30], a meta-graph built, using information from the entire graph sequence, is used in order to maintain the mental map. In [77], an abstracted version of the graph is used. The nodes are topologically sorted into a tree structure in order to expose interesting features. An offline force-directed algorithm is used in [44] in order to create two-dimensional (2D) and three-dimensional (3D) animations of evolving graphs.

The more interesting problem of online dynamic graph drawing where the graph sequence is not known in advance has been studied in depth. Sophisticated incremental algorithms for orthogonal graph drawings are presented in [7,12,94,96]. An efficient incremental algorithm for online hierarchical graph drawings is presented in [90]. An algorithm for visualizing dynamic social networks is discussed in [87]. An approach based on Bayesian networks is described in [10]. An algorithm for producing online layouts of general graphs is presented in [51]. The work in [19] presents data structures for dynamic graph drawings.

14.3.2.2 3D Drawings

3D graph drawings have been extensively studied since they have applications in information visualization, VLSI circuit design, and software engineering. One of the earliest approaches in 3D graph drawing is the cone tree algorithm [102], where each subtree is associated with a cone, such that the vertex at the root of the subtree is placed at the apex of the cone, and its children are circularly arranged around the base of the cone. This technique displays trees in 3D mostly for artistic purposes. Empirical studies suggest that displaying a graph in three dimensions is better than in two [101,124]. The extra dimension gives greater flexibility for placing the nodes and edges of a graph and crossings can be always avoided. However, producing efficient 3D graph drawings is quite a challenging problem due to the occlusion and navigation problems involved. The interest of the 3D graph drawing community has been mainly devoted to straight-line drawings and to orthogonal drawings.

Straight-line drawings map nodes to points and edges to straight-line segments. In [20], it is shown that every graph admits a straight-line, crossing-free 3D drawing on an integer grid of $O(n^3)$ volume, and proved that this is asymptotically optimal. It distributes the nodes of the graph along a "momentum curve," so that there are not crossings among the edges. In [45], it is shown that all outerplanar graphs can be drawn in a restricted integer 3D grid of $O(n)$ volume consisting of three parallel lines at distance 1 from each other. In [34], it is shown that all planar graphs can be drawn with an upper bound of $O(n^{\frac{3}{2}})$ volume. A technique for drawing 3D hierarchical graph drawings that combines 2D drawings with a lifting transformation is presented in [57]. First, a 2D non-upward representation of a directed acyclic graph is created and then the nodes are lifted along a third dimension. The lifting height of the nodes reflects the hierarchy. Force-directed techniques have been extended to create 3D graph drawing (e.g., [14,22,91]). 3D visibility representations of graphs have been studied in [3,58,109].

Orthogonal drawings map nodes to points and edges to polygonal chains composed of segments that are parallel to the axes. In [5], a linear time algorithm is presented that draws a graph in $O(n^2)$ volume with at most 14 bends per edge. The drawing is obtained by placing all the nodes on a certain horizontal plane and by assigning a further horizontal plane to every edge. In [41], a $O(n^{\frac{3}{2}})$ running time algorithm is proposed that is based on augmenting the graph to an Eulerian graph. The algorithm produces drawings that have $O(n^{\frac{3}{2}})$ volume and at most 16 bends per edge. The algorithm in [42] requires $O(n^{\frac{3}{2}})$ time and volume and introduces at most seven bends per edge. In the same paper, a second algorithm is presented whose complexity is linear, while its volume is $27n^3$, and at most three bends per edge are introduced. In [95], a linear time algorithm is presented that draws a graph of maximum degree 6 in $O(n^3)$ volume with at most three bends per edge. It is incremental and can be extended to draw graphs with nodes of arbitrary degree. The construction starts from a first pair of adjacent nodes, and then it adds one node at a time with its incident edges. In [125], an algorithm for maximum degree 5 graphs that requires $O(n^3)$ volume and at most two bends per edge is presented.

In [11,60,123], a different approach has been taken to construct 3D drawings. More specifically, this framework divides a graph into a set of subgraphs, and then draws the subgraphs in different parallel planes, using well-known 2D drawing algorithms, but with interdependent layouts.

14.3.2.3 Labeling

An important aspect of graph drawing is the automatic placement of text or symbol labels corresponding to nodes and edges of a drawing. Labels are textual descriptions which convey information or clarify

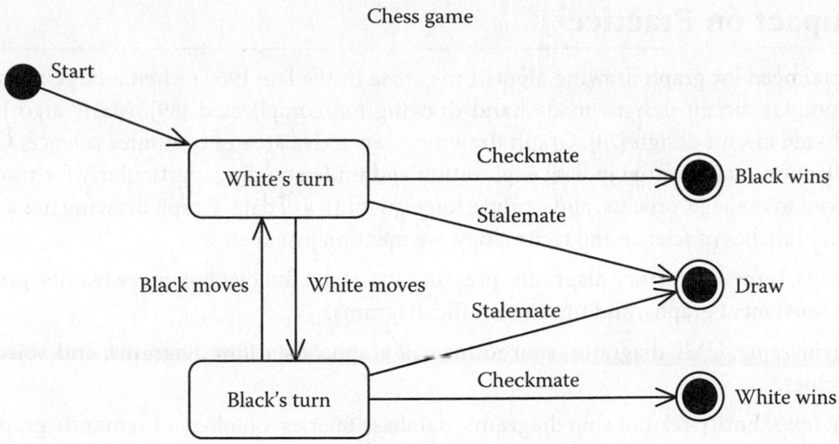

FIGURE 14.9 A UML state diagram that makes use of labels for states and transitions as well as the title of the diagram [33].

the meaning of the elements of a drawing. The problem of positioning labels corresponding to graphical objects of drawings is called automatic label placement or simply graph labeling. The need for displaying node or edge labels is essential in many applications, for example in online Geographic Information Systems, Internet-based map search, UML diagrams, and engineering diagrams.

Labeling of edges, nodes, and graphs themselves aims to communicate the attributes of these graph objects in the most convenient way. This requires that labels be positioned in the most appropriate places and follows some basic rules [66]. It is worth noting that the problems of labeling nodes [83] and edges [65] are NP-hard. Because automatic labeling is a very difficult problem, we rely on heuristics to provide practical solutions for real world problems.

Most of the research on labeling addresses the problem of assigning labels to a set of points or nodes [17,46,59,126]. The problem of assigning labels to a set of lines or edges has been studied in [33,64,121]. For the general labeling problem (i.e., the labeling of nodes, edges, and graphs), most labeling algorithms are based on local and exhaustive search algorithms [1,31,43,49].

In [66], the general labeling problem is transformed into a matching problem. The general framework of this techniques is flexible and can be adjusted for particular labeling requirements. In Figure 14.9, the labeling assignment is produced by this technique. The basic steps for this labeling technique are the following:

1. A set of discrete potential label solutions for each object is carefully selected.
2. This set of labels is reduced by removing heavily overlapping labels. The remaining labels are assigned to groups, such that if two labels overlap, then they belong to the same group.
3. Finally, labels are assigned by solving a variant of the matching problem, where at most one label position from each group is part of the solution.

In many practical applications, each graphical feature may have more than one label. The need for assigning multiple labels is necessary not only when objects are large or long, but also when it is necessary to display different attributes of an object. This problem has been addressed in [49,67].

Since we rely on heuristics to solve the labeling problem, there are cases where the best methods available do not always produce an overlap free label assignment. In order to resolve label overlaps, one may modify minimally the drawing. An algorithm that modifies an orthogonal drawing of a graph, while preserving the orthogonal representation of the drawing, in order to resolve label overlaps is presented in [68]. It remains an open problem, for the other layout styles, how one can modify a drawing to resolve label overlaps.

14.4 Impact on Practice

The industrial need for graph drawing algorithms arose in the late 1960s when a large number of elements in complex circuit designs made hand-drawing too complicated [89]. Many algorithms were developed to aid circuit design [78]. Graph drawing is an active area in computer science. Graphs and their visualizations are essential in data exploration and understanding, particularly for those applications that need to manage, process, and analyze huge quantities of data. Graph drawing has applications in almost all branches of science and technology, we mention just a few:

Web: Site maps, browsing history diagrams, presentation and refinement of query results, product catalogs, visitor movement graphs, and Internet traffic diagrams.

Software engineering: UML diagrams, subroutine call graphs, data-flow diagrams, and object-oriented class hierarchies.

Database systems: Entity-relationship diagrams, database queries, ontology of semantic graphs, dependency graphs, and database schema diagrams.

Real-time systems: Petri nets and state-transition diagrams.

Networking: Telecommunication and computer network management, network visualization, network analysis, and LAN diagrams.

Engineering: Circuit schematics, CAD, and finite state diagrams.

Social Sciences: Social networks, citation networks, co-authorship networks, and collaboration networks.

Business: Business process diagrams, organization charts, event-driven process chains, and scheduling charts.

Biomedical technology: Gene regulatory networks, metabolic networks, protein–protein interaction networks, evolutionary trees, phylogenetic trees, molecular maps, genetic maps, biochemical pathways, and neuronal networks.

Security: Criminal networks, fraud networks, felony event flow diagrams, case information diagrams, attack trees, network monitoring, Internet traffic control, traffic visualization, and traffic analysis.

14.5 Research Issues and Summary

Current research in graph drawing addresses both theoretical and applied issues. On the theoretical side, researchers have investigated issues ranging from combinatorial properties of graphs to algorithms that produce drawings with provably good upper bounds on the area, number of bends, etc., to lower bounds. On the systems side, there are several graph drawing tools and commercial software products (e.g., GraphViz from AT&T, GDToolkit, AGD, Graphlet, Tom Sawyer Software, yWorks, and ILOG). For a compilation of results in graph drawing in excellent tabular exposition, see [116].

Peter Eades [37] points out that the scale problem currently drives much of Computer Science. Data sets are growing at a faster rate than the human ability to understand them. A challenge for the graph drawing community is to explore more efficient ways to draw large graphs. Three approaches have been proposed to solve the scale problem. (*i*) Dynamic drawings: View only part of the drawing, (*ii*) Cluster drawings: View an abstraction of the drawing, and (*iii*) 3D drawings: Spread the drawing over a third dimension. The first two approaches have the drawback that present part of the data in order to reduce visual complexity. Graph drawing algorithms for 3D layouts have mostly failed to untangle the complexity of large graphs [37]. Promising approaches for 3D drawings, which need further investigation, have been presented in [11,60,123]. In [74], a new approach in drawing 3D orthogonal drawings is presented. It combines dominance and row/column reuse and simplifies the visual confirmation of the existence of an edge and/or path between any two nodes. This is a promising technique for producing simple drawings of large graphs that needs further investigation.

Many open graph drawing problems originating in applied bioinformatics and network biology are presented in [2]. Traditional graph drawing techniques do not adhere to the special drawing conventions of the life science community. Thus, there is a need for specialized layout and visualization algorithms to present, explore, evaluate, and compare biological network data.

Key Terms

Acyclic digraph: A digraph without directed cycles.

Biconnected component: A maximal biconnected subgraph.

Biconnected graph: A graph with the property that for every three distinct nodes *u*, *v*, and *w* there is a path from *u* to *w* not containing *v*.

Binary tree: A rooted tree where each node has at most two children.

Connected component: A maximal connected subgraph.

Connected graph: An undirected graph that has a path between every pair of nodes.

Digraph: A graph where the edges have a direction associated with them (also called *directed* graph).

Embedded graph: A planar graph with a prespecified topological embedding (i.e., set of faces), which must be preserved in the drawing.

Face: A planar drawing of a graph divides the plane into regions called *faces*. The unbounded region is called the **external face**.

Orthogonal representation: A representation of an orthogonal drawing in terms of bends along each edge and angles around each node.

Outerplanar graph: A planar graph which admits an embedding such that all the nodes are on the same face.

Rooted tree: A directed tree with a distinguished node, the *root*, such that each node lies on a directed path to the root.

Sink node: A node of a digraph without outgoing edges (also called *target* node).

Source node: A node of a digraph without incoming edges.

St-digraph: An acyclic digraph with exactly one source and one sink, which are joined by an edge (also called *bipolar digraph*).

Tree: A connected graph without cycles.

Further Information

Three books have been published in the area of graph drawing [28,71,111], one book on Graph Drawing Software [63], and one book on Planar Graph Drawing [89]. A comprehensive bibliography on graph drawing algorithms [26] cites more than 300 papers written before 1994. An International Symposium on Graph Drawing is being held annually and the proceedings are published by Springer-Verlag. *The Journal of Graph Algorithms and Applications* (*JGAA*) is an online journal which publishes graph drawing-related work. In addition, journals in the area of Computational Geometry (Computational Geometry: Theory and Applications, Algorithmica, etc.) are of particular interest, since graph drawing has strong relationships with computational geometry. Online graph drawing resources can be found in: (i) http://graphdrawing.org (a collection of resources mostly related to the annual International Symposium on Graph Drawing), (ii) http://gdea.informatik.uni-koeln.de (an electronic repository and archive for research materials on the topic of graph drawing), and (iii) http://www.ics.uci.edu/eppstein/gina/gdraw.html (the section of the Geometry in Action pages devoted to graph drawing and maintained by David Eppstein).

References

1. J. Ahn and H. Freeman. A program for automatic name placement. *Cartographica*, 21(2 and 3):101–109, 1984.
2. M. Albrecht, A. Kerren, K. Klein, O. Kohlbacher, P. Mutzel, W. Paul, F. Schreiber, and M. Wybrow. On open problems in biological network visualization. In D. Eppstein and E. Gansner, eds., *Graph Drawing (Proc. GD 2009)*, vol. 5849 of *Lecture Notes in Computer Science*, pp. 256–267. Springer-Verlag, 2010.
3. H. Alt, M. Godau, and S. Whitesides. Universal 3-dimensional visibility representations for graphs. In F. Brandenburg, ed., *Graph Drawing (Proc. GD 1995)*, vol. 1027 of *Lecture Notes in Computer Science*, pp. 8–19. Springer-Verlag, 1996.
4. M. Baur and U. Brandes. Crossing reduction in circular layouts. In *Proceedings of the 30th International Workshop on Graph-Theoretic Concepts in Computer-Science (WG 2004)*, pp. 332–343, Springer-Verlag, 2004.
5. T. Biedl. Heuristics for 3D-orthogonal graph drawings. In *Proceedings of the Fourth Twente Workshop on Graphs and Combinatorial Optimization*, pp. 41–44, Universiteit Twente, The Netherlands, 1995.
6. T. Biedl and G. Kant. A better heuristic for orthogonal graph drawings. *Computational Geometry*, 9(3):159–180, 1998.
7. T. Biedl and M. Kaufmann. Area-efficient static and incremental graph drawings. In *Proceedings of the Fifth Annual European Symposium on Algorithms*, ESA '97, pp. 37–52. Springer-Verlag, 1997.
8. T. Biedl, B. Madden, and I. G. Tollis. The three-phase method: A unified approach to orthogonal graph drawing. In G. Di Battista, ed., *Graph Drawing (Proc. GD 1997)*, vol. 1353 of *Lecture Notes in Computer Science*, pp. 391–402. Springer-Verlag, 1997.
9. U. Brandes and B. Kopf. Fast and simple horizontal coordinate assignment. In S. Leipert, M. Junger, and P. Mutzel, eds., *Graph Drawing (Proc. GD 2001)*, vol. 2265 of *Lecture Notes in Computer Science*, pp. 31–44. Springer-Verlag, 2002.
10. U. Brandes and D. Wagner. A bayesian paradigm for dynamic graph layout. In G. Di Battista, ed., *Graph Drawing (Proc. GD 1996)*, vol. 1353 of *Lecture Notes in Computer Science*, pp. 236–247. Springer-Verlag, 1997.
11. U. Brandes, T. Dwyer, and F. Schreiber. Visual understanding of metabolic pathways across organisms using layout in two and a half dimensions. *Journal of Integrative Bioinformatics*, 2, 2004.
12. S. Bridgeman, J. Fanto, A. Garg, R. Tamassia, and L. Vismara. Interactive Giotto: An algorithm for interactive orthogonal graph drawing. In G. Di Battista, ed., *Graph Drawing (Proc. GD 1997)*, vol. 1353 of *Lecture Notes in Computer Science*, pp. 303–308. Springer-Verlag, 1997.
13. S. Bridgeman, G. Di Battista, W. Didimo, G. Liotta, R. Tamassia, and L. Vismara. Turn-regularity and optimal area drawings of orthogonal representations. *CGTA*, 16:53–93, 2000.
14. I. Bruß and A. Frick. Fast interactive 3-D graph visualization. In F. Brandenburg, ed., *Graph Drawing (Proc. GD 1995)*, vol. 1027 of *Lecture Notes in Computer Science*, pp. 99–110. Springer-Verlag, 1996.
15. C. Buchheim, M. Junger, and S. Leipert. A fast layout algorithm for k-level graphs. In J. Marks, ed., *Graph Drawing (Proc. GD 2000)*, vol. 1984 of *Lecture Notes in Computer Science*, pp. 31–44. Springer-Verlag, 2001.
16. N. Chiba, T. Nishizeki, S. Abe, and T. Ozawa. A linear algorithm for embedding planar graphs using PQ-trees. *Journal of Computer and System Sciences*, 30(1):54–76, 1985.
17. J. Christensen, J. Marks, and S. Shieber. An empirical study of algorithms for Point Feature Label Placement. *ACM Transactions on Graphics*, 14(3):203–232, 1995.
18. E. G. Coffman and R. L. Graham. Optimal scheduling for two processor systems. *Acta Informatica*, 1:200–213, 1972.
19. R. Cohen, G. Di Battista, R. Tamassia, and I. G. Tollis. Dynamic graph drawings: Trees, series-parallel digraphs, and planar ST-digraphs. *SIAM Journal on Computing*, 24(5):970–1001, 1995.

20. R. Cohen, P. Eades, T. Lin, and F. Ruskey. Three-dimensional graph drawing. *Algorithmica*, 17:199–208, 1997.
21. M. Coleman and P. Stott. Aesthetics-based graph layout for human consumption. *Software: Practice and Experience*, 26(12):1415–1438, 1996.
22. I. Cruz and J. Twarog. 3D graph drawing with simulated annealing. In F. Brandenburg, ed., *Graph Drawing (Proc. GD 1995)*, vol. 1027 of *Lecture Notes in Computer Science*, pp. 162–165. Springer-Verlag, 1996.
23. R. Davidson and D. Harel. Drawing graphs nicely using simulated annealing. *ACM Transactions on Graphics*, 15(4):301–331, 1996.
24. G. Di Battista and R. Tamassia. Algorithms for plane representations of acyclic digraphs. *Theoretical Computer Science*, 61:175–198, 1988.
25. G. Di Battista, R. Tamassia, and I. G. Tollis. Constrained visibility representations of graphs. *Information Processing Letters*, 41:1–7, 1992.
26. G. Di Battista, P. Eades, R. Tamassia, and I. G. Tollis. Algorithms for drawing graphs: An annotated bibliography. *Computational Geometry: Theory and Applications*, 4:235–282, 1994.
27. G. Di Battista, A. Garg, G. Liotta, R. Tamassia, E. Tassinari, and F. Vargiu. An experimental comparison of four graph drawing algorithms. *Computational Geometry*, 7(56):303–325, 1997.
28. G. Di Battista, P. Eades, R. Tamassia, and I. G. Tollis. *Graph Drawing: Algorithms for the Visualization of Graphs*. Prentice Hall, Upper Saddle River, NJ, 1998.
29. G. Di Battista, W. Didimo, M. Patrignani, and M. Pizzonia. Orthogonal and Quasi-upward drawings with vertices of prescribed size. In J. Kratochviyl, ed., *Graph Drawing (Proc. GD 1999)*, vol. 1731 of *Lecture Notes in Computer Science*, pp. 297–310. Springer-Verlag, 1999.
30. S. Diehl and C. Görg. Graphs, they are changing-dynamic graph drawing for a sequence of graphs. In M. Goodrich and S. Kobourov, eds., *Graph Drawing (Proc. GD 2002)*, vol. 2528 of *Lecture Notes in Computer Science*, pp. 23–31. Springer-Verlag, 2002.
31. J. S. Doerschler and H. Freeman. A rule based system for dense map name placement. *Communications of ACM*, 35(1):68–79, 1992.
32. U. Doğrusöz, B. Madden, and P. Madden. Circular layout in the graph layout toolkit. In S. North, ed., *Graph Drawing (Proc. GD 1996)*, vol. 1190 of *Lecture Notes in Computer Science*, pp. 92–100. Springer-Verlag, 1997.
33. U. Doğrusöz, K. G. Kakoulis, B. Madden, and I. G. Tollis. On labeling in graph visualization. *Special Issue on Graph Theory and Applications, Information Sciences Journal*, 177(12):2459–2472, 2007.
34. V. Dujmovic and D. Wood. Three-dimensional grid drawings with sub-quadratic vol.. In J. Pach, ed., *Towards a Theory of Geometric Graphs*, vol. 342 of *Contemporary Mathematics*, pp. 55–66. American Mathematical Society, Providence, RI, 2004.
35. P. Eades. A heuristic for graph drawing. *Congressus Numerantium*, 42:149–160, 1984.
36. P. Eades. Drawing free trees. *Bulletin of the Institute for Combinatorics and Its Applications*, 5:10–36, 1992.
37. P. Eades. Invited talk on the future of graph drawing. In *Symposium on Graph Drawing*, GD 2010, Konstanz, Germany, 2010.
38. P. Eades and K. Sugiyama. How to draw a directed graph. *Journal of Information Processing*, 13:424–437, 1990.
39. P. Eades and N. C. Wormald. Edge crossings in drawings of bipartite graphs. *Algorithmica*, 11(4):379–403, 1994.
40. P. Eades, X. Lin, and W. Smyth. A fast and effective heuristic for the feedback arc set problem. *Information Processing Letters*, 47:319–323, 1993.
41. P. Eades, C. Stirk, and S. Whitesides. The techniques of Kolmogorov and Bardzin for three dimensional orthogonal graph drawings. *Informations Processing Letters*, 60:97–103, 1996.
42. P. Eades, A. Symvonis, and S. Whitesides. Two algorithms for three dimensional orthogonal graph drawing. In S. North, ed., *Graph Drawing (Proc. GD 1996)*, vol. 1190 of *Lecture Notes in Computer Science*, pp. 139–154. Springer-Verlag, 1997.

43. S. Edmondson, J. Christensen, J. Marks, and S. Shieber. A general cartographic labeling algorithm. *Cartographica*, 33(4):321–342, 1997.
44. C. Erten, P. Harding, S. Kobourov, K. Wampler, and G. Yee. Graphael: Graph animations with evolving layouts. In G. Liotta, ed., *Graph Drawing (Proc. GD 2003)*, vol. 2912 of *Lecture Notes in Computer Science*, pp. 98–110. Springer-Verlag, 2004.
45. S. Felsner, G. Liotta, and S. Wismath. Straight-line drawings on restricted integer grids in two and three dimensions. *Journal of Graph Algorithms and Applications*, 7(4):363–398, 2003.
46. M. Formann and F. Wagner. A packing problem with applications to lettering of maps. In *Proceedings of the Seventh Annual ACM Symposium on Computational Geometry*, pp. 281–288, New York, 1991.
47. M. Formann and F. Wagner. The VLSI layout problem in various embedding models. In *Proceedings of the 16th International Workshop on Graph-Theoretic Concepts in Computer Science*, vol. 484 of *Lecture Notes in Computer Science*, pp. 130–139, Springer-Verlag, 1991.
48. U. Fößmeier and M. Kaufmann. Drawing high degree graphs with low bend numbers. In F. Brandenburg, ed., *Graph Drawing (Proc. GD 1995)*, vol. 1027 of *Lecture Notes in Computer Science*, pp. 254–266. Springer-Verlag, 1996.
49. H. Freeman and J. Ahn. On the problem of placing names in a geographical map. *International Journal of Pattern Recognition and Artificial Intelligence*, 1(1):121–140, 1987.
50. A. Frick, A. Ludwig, and H. Mehldau. A fast adaptive layout algorithm for undirected graphs. In R. Tamassia and I. G. Tollis, eds., *Graph Drawing (Proc. GD 1994)*, vol. 894 of *Lecture Notes in Computer Science*, pp. 135–146. Springer-Verlag, 1995.
51. Y. Frishman and A. Tal. Online dynamic graph drawing. *Visualization and Computer Graphics, IEEE Transactions on*, 14(4):727–740, 2008.
52. T. Fruchterman and E. Reingold. Graph drawing by force-directed placement. *Software: Practice and Experience*, 21(11):1129–1164, 1991.
53. E. Gansner and Y. Koren. Improved Circular Layouts. In M. Kaufmann and D. Wagner, eds., *Graph Drawing (Proc. GD 2006)*, vol. 4372 of *Lecture Notes in Computer Science*, pp. 386–398. Springer-Verlag, 2007.
54. E. Gansner, E. Koutsofios, S. North, and K. P. Vo. A technique for drawing directed graphs. *IEEE Transactions on Software Engineering*, SE-19(3):214–230, 1993.
55. M. R. Garey and D. S. Johnson. Crossing number is NP-complete. *SIAM Journal on Algebraic Discrete Methods*, 4(3):312–316, 1983.
56. A. Garg and R. Tamassia. On the computational complexity of upward and rectilinear planarity testing. In R. Tamassia and I. G. Tollis, eds., *Graph Drawing (Proc. GD 1994)*, vol. 894 of *Lecture Notes in Computer Science*, pp. 286–297. Springer-Verlag, 1995.
57. A. Garg and R. Tamassia. GIOTTO3D: A system for visualizing hierarchical structures in 3D. In S. North, ed., *Graph Drawing (Proc. GD 1996)*, vol. 1190 of *Lecture Notes in Computer Science*, pp. 193–200. Springer-Verlag, 1997.
58. A. Garg, R. Tamassia, and P. Vocca. Drawing with colors. In J. Diaz and M. Serna, eds., *Algorithms—ESA '96*, vol. 1136 of *Lecture Notes in Computer Science*, pp. 12–26. Springer-Verlag, 1996.
59. S. A. Hirsch. An algorithm for automatic name placement around point data. *The American Cartographer*, 9(1):5–17, 1982.
60. S. Hong. Multiplane: A new framework for drawing graphs in three dimensions. In P. Healy and N. S. Nikolov, eds., *Graph Drawing (Proc. GD 2005)*, vol. 3843 of *Lecture Notes in Computer Science*, pp. 514–515. Springer-Verlag, 2005.
61. J. Hopcroft and R. E. Tarjan. Efficient planarity testing. *Journal of the ACM*, 21(4):549–568, 1974.
62. M. Junger and P. Mutzel. 2-layer straightline crossing minimization: Performance of exact and heuristic algorithms. *Journal of Graph Algorithms and Applications*, 1(1):1–25, 1997.
63. M. Junger and P. Mutzel, eds. *Graph Drawing Software*. Springer-Verlag, 2004.

64. K. G. Kakoulis and I. G. Tollis. An algorithm for labeling edges of hierarchical drawings. In G. Di Battista, ed., *Graph Drawing (Proc. GD 1997)*, vol. 1353 of *Lecture Notes in Computer Science*, pp. 169–180. Springer-Verlag, 1998.
65. K. G. Kakoulis and I. G. Tollis. On the complexity of the edge label placement problem. *Computational Geometry*, 18(1):1–17, 2001.
66. K. G. Kakoulis and I. G. Tollis. A unified approach to automatic label placement. *International Journal of Computational Geometry and Applications*, 13(1):23–60, 2003.
67. K. G. Kakoulis and I. G. Tollis. Algorithms for the multiple label placement problem. *Computational Geometry*, 35(3):143–161, 2006.
68. K. G. Kakoulis and I. G. Tollis. Placing edge labels by modifying an orthogonal graph drawing. In U. Brandes and S. Cornelsen, eds., *Graph Drawing (Proc. GD 2010)*, vol. 6502 of *Lecture Notes in Computer Science*, pp. 395–396. Springer-Verlag, 2011.
69. T. Kamada and S. Kawai. An algorithm for drawing general undirected graphs. *Informations Processing Letters*, 31(1):7–15, 1989.
70. G. Kar, B. Madden, and R. Gilbert. Heuristic layout algorithms for network management presentation services. *Network IEEE*, 2(6):29–36, 1988.
71. M. Kaufmann and D. Wagner, eds. *Drawing Graphs Methods and Models*. Springer-Verlag, 2001.
72. M. Kaufmann and R. Wiese. Maintaining the mental map for circular drawings. In M. Goodrich and S. Kobourov, eds., *Graph Drawing (Proc. GD 2002)*, vol. 2528 of *Lecture Notes in Computer Science*, pp. 12–22. Springer-Verlag, 2002.
73. G. Klau and P. Mutzel. Optimal compaction of orthogonal grid drawings. In G. Cornuejols, R. E. Burkard, and G. J. Woeginger, eds., *Integer Programming and Combinatorial Optimization (IPCO '99)*, vol. 1610 of *Lecture Notes in Computer Science*, pp. 304–319. Springer-Verlag, 1999.
74. E. Kornaropoulos and I. G. Tollis. Overloaded orthogonal drawings. In M. van Kreveld and B. Speckmann, eds., *Graph Drawing (Proc. GD 2011)*, vol. 7034 of *Lecture Notes in Computer Science*, pp. 242–253. Springer-Verlag, 2012.
75. M. R. Kramer and J. van Leeuwen. The complexity of wire-routing and finding minimum area layouts for arbitrary VLSI circuits. In F. P. Preparata, ed., *Adv. Comput. Res.*, vol. 2, pp. 129–146. JAI Press, Greenwich, CT, 1985.
76. V. Krebs. Visualizing human networks. In *Release 1.0: Esther Dyson's Monthly Report*, February 12, Edventure Holdings Inc., New York, 1996.
77. G. Kumar and M. Garland. Visual exploration of complex time-varying graphs. *Visualization and Computer Graphics, IEEE Transactions on*, 12(5):805–812, 2006.
78. T. Lengauer. *Combinatorial Algorithms for Integrated Circuuit Layout*. Wiley-Teubner, New York, 1990.
79. A. Liebers. Planarizing graphs-a survey and annotated bibliography. *Journal of Graph Algorithms and Applications*, 5(1):1–74, 2001.
80. P. Liu and R. Geldmacher. On the deletion of nonplanar edges of a graph. In *Proceedings of the 10th Southeastern Conference on Combinatorics, Graph Theory, and Computing*, pp. 727–738. Boca Raton, FL, 1977.
81. A. Lubiw. Some NP-complete problems similar to graph isomorphism. *SIAM Journal on Computing*, 10(1):11–21, 1981.
82. J. Manning. Computational complexity of geometric symmetry detection in graphs. In N. Sherwani, E. de Doncker, and J. Kapenga, eds., *Computing in the 90's*, vol. 507 of *Lecture Notes in Computer Science*, pp. 1–7. Springer-Verlag, 1991.
83. J. Marks and S. Shieber. The computational complexity of cartographic label placement. Technical Report 05-91, Harvard University, Cambridge, MA, 1991.
84. S. Masuda, T. Kashiwabara, K. Nakajima, and T. Fujisawa. On the NP-completeness of a computer network layout problem. In *Proceedings of the IEEE 1987 International Symposium on Circuits and Systems*, Philadelphia, PA, pp. 292–295, 1987.

85. K. Mehlhorn and P. Mutzel. On the embedding phase of the Hopcroft and Tarjan planarity testing algorithm. *Algorithmica*, 16:233–242, 1996.
86. K. Misue, P. Eades, W. Lai, and K. Sugiyama. Layout adjustment and the mental map. *Journal of Visual Languages and Computing*, 6:183–210, 1995.
87. J. Moody, D. McFarland, and S. Bender-deMoll. Dynamic network visualization. *American Journal of Sociology*, 110(4):1206–1241, 2005.
88. P. Mutzel. An alternative method to crossing minimization on hierarchical graphs. *SIAM Journal of Optimization*, 11(4):1065–1080, 2001.
89. T. Nishizeki and Md. S. Rahman. *Planar Graph Drawing*. World Scientific, Singapore, 2004.
90. S. North and G. Woodhull. Online hierarchical graph drawing. In P. Mutzel, M. Junger, and S. Leipert, eds., *Graph Drawing (Proc. GD 2001)*, vol. 2265 of *Lecture Notes in Computer Science*, pp. 77–81. Springer-Verlag, 2002.
91. D. I. Ostry. Some three-dimensional graph drawing algorithms. In *M.Sc. thesis*. Department of Computer Science and Software Engineering, University of Newcastle, Newcastle, New South Wales, Australia, 1996.
92. A. Papakostas and I. G. Tollis. Orthogonal drawing of high degree graphs with small area and few bends. In F. Dehne, A. Rau-Chaplin, J. Sack, and R. Tamassia, eds., *Algorithms and Data Structures*, vol. 1272 of *Lecture Notes in Computer Science*, pp. 354–367. Springer-Verlag, 1997.
93. A. Papakostas and I. G. Tollis. Algorithms for area-efficient orthogonal drawings. *Computational Geometry*, 9(12):83–110, 1998.
94. A. Papakostas and I. G. Tollis. Interactive orthogonal graph drawing. *Computers, IEEE Transactions on*, 47(11):1297–1309, 1998.
95. A. Papakostas and I. G. Tollis. Algorithms for incremental orthogonal graph drawing in three dimensions. *Journal of Graph Algorithms and Applications*, 3(4):81–115, 1999.
96. A. Papakostas, J. Six, and I. G. Tollis. Experimental and theoretical results in interactive orthogonal graph drawing. In S. North, ed., *Graph Drawing (Proc. GD 1996)*, vol. 1190 of *Lecture Notes in Computer Science*, pp. 371–386. Springer-Verlag, 1997.
97. M. Patrignani. On the complexity of orthogonal compaction. *CGTA*, 19:47–67, 2001.
98. H. Purchase. Effective information visualisation: A study of graph drawing aesthetics and algorithms. *Interacting with Computers*, 13(2):147–162, 2000.
99. H. Purchase, C. Pilcher, and B. Plimmer. Graph drawing aesthetics-created by users, not algorithms. *Visualization and Computer Graphics, IEEE Transactions on*, 18(1):81–92, 2012.
100. E. Reingold and J. Tilford. Tidier drawing of trees. *IEEE Transactions on Software Engineering*, SE-7(2):223–228, 1981.
101. K. Risden, M. Czerwinski, T. Munzner, and D. Cook. An initial examination of ease of use for 2D and 3D information visualizations of web content. *International Journal of Human-Computer Studies*, 53(5):695–714, 2000.
102. G. G. Robertson, J. D. Mackinlay, and S. K. Card. Cone trees: Animated 3D visualizations of hierarchical information. In *Proceedings of the SIGCHI Conference on Human Factors in Computing Systems: Reaching through Technology*, CHI '91, pp. 189–194, ACM, New York, 1991.
103. P. Rosenstiehl and R. Tarjan. Rectilinear planar layouts and bipolar orientations of planar graphs. *Discrete and Computational Geometry*, 1:343–353, 1986.
104. G. Sander. Graph layout for applications in compiler construction. *Theoretical Computer Science*, 217(2):175–214, 1999.
105. J. Six and I. G. Tollis. Circular drawings of biconnected graphs. In M. Goodrich and C. McGeoch, eds., *Algorithm Engineering and Experimentation (Proc. ALENEX 1999)*, vol. 1619 of *Lecture Notes in Computer Science*, pp. 57–73. Springer-Verlag, 1999.
106. J. Six and I. G. Tollis. A framework for user-grouped circular drawings. In G. Liotta, ed., *Graph Drawing (Proc. GD 2003)*, vol. 2912 of *Lecture Notes in Computer Science*, pp. 135–146. Springer-Verlag, 2004.

107. J. Six and I. G. Tollis. A framework and algorithms for circular drawings of graphs. *Journal of Discrete Algorithms*, 4(1):25–50, 2006.
108. J. Six, K. G. Kakoulis, and I. G. Tollis. Techniques for the refinement of orthogonal graph drawings. *Journal of Graph Algorithms and Applications*, 4(3):75–103, 2000.
109. J. Stola. 3D visibility representations of complete graphs. In G. Liotta, ed., *Graph Drawing (Proc. GD 2003)*, vol. 2912 of *Lecture Notes in Computer Science*, pp. 226–237. Springer-Verlag, 2004.
110. M. Suderman and S. Whitesides. Experiments with the fixed-parameter approach for two-layer planarization. *Journal of Graph Algorithms and Applications*, 9(1):149–163, 2005.
111. K. Sugyiama. *Graph Drawing and Applications for Software and Knowledge Engineers.* World Scientific, Singapore, 2002.
112. K. Sugiyama and K. Misue. Graph drawing by the magnetic spring model. *Journal of Visual Languages and Computing*, 6(3):217–231, 1995.
113. K. Sugiyama, S. Tagawa, and M. Toda. Methods for visual understanding of hierarchical systems. *IEEE Transactions on Systems, Man, and Cybernetics*, SMC-11(2):109–125, 1981.
114. A. Symeonidis and I. G. Tollis. Visualization of biological information with circular drawings. In J. Barreiro et al., ed., *Proceedings of the Biological and Medical Data Analysis (ISBMDA04)*, vol. 3337 of *Lecture Notes in Computer Science*, pp. 468–478. Springer-Verlag, 2004.
115. R. Tamassia. On embedding a graph in the grid with the minimum number of bends. *SIAM Journal on Computing*, 16(3):421–444, 1987.
116. R. Tamassia and G. Liotta. Graph drawing. In J. E. Goodman and J. O'Rourke, eds., *Handbook of Discrete and Computational Geometry*, 2nd edn., chapter 52. CRC Press LLC, Boca Raton, FL, 2004.
117. R. Tamassia and I. G. Tollis. A unified approach to visibility representations of planar graphs. *Discrete and Computational Geometry*, 1:321–341, 1986.
118. R. Tamassia and I. G. Tollis. Planar grid embedding in linear time. *IEEE Transactions on Circuits and Systems*, CAS-36(9):1230–1234, 1989.
119. R. Tamassia, G. Di Battista, and C. Batini. Automatic graph drawing and readability of diagrams. *IEEE Transactions on Systems, Man, and Cybernetics*, SMC-18(1):61–79, 1988.
120. I. G. Tollis and C. Xia. Drawing telecommunication networks. In R. Tamassia and I. G. Tollis, eds., *Graph Drawing (Proc. GD 1994)*, vol. 894 of *Lecture Notes in Computer Science*, pp. 206–217. Springer-Verlag, 1995.
121. J. W. van Roessel. An algorithm for locating candidate labeling boxes within a polygon. *The American Cartographer*, 16(3):201–209, 1989.
122. L. Vismara, G. Di Battista, A. Garg, G. Liotta, R. Tamassia, and F. Vargiu. Experimental studies on graph drawing algorithms. *Software: Practice and Experience*, 30(11):1235–1284, 2000.
123. C. Ware. Designing with a 2 1/2D Attitude. *Information Design Journal*, 10(3):171–182, 2001.
124. C. Ware and G. Franck. Evaluating stereo and motion cues for visualizing information nets in three dimensions. *ACM Transactions on Graphics*, 15(2):121–140, 1996.
125. D. Wood. An algorithm for three-dimensional orthogonal graph drawing. In S. Whitesides, ed., *Graph Drawing (Proc. GD 1997)*, vol. 1547 of *Lecture Notes in Computer Science*, pp. 332–346. Springer-Verlag, 1998.
126. S. Zoraster. The solution of large 0-1 integer programming problems encountered in automated cartography. *Operation Research*, 38(5):752–759, 1990.

15

Pattern Matching and Text Compression Algorithms

Maxime Crochemore
King's College London
and
Université Paris-Est Marne-la-Vallée

Thierry Lecroq
University of Rouen

15.1 Processing Texts Efficiently

This present chapter describes a few standard algorithms used for processing texts. They apply, for example, to the manipulation of texts (text editors), to the storage of textual data (text compression), and to data retrieval systems. The algorithms of this chapter are interesting in different respects. First, they are basic components used in the implementation of practical software. Second, they introduce programming methods that serve as paradigms in other fields of computer science (system or software design). Third, they play an important role by providing challenging problems in theoretical computer science.

Although data are stored in various ways, text remains the main form of exchanging information. This is particularly evident in literature or linguistics where data are composed of huge corpora and dictionaries.

This applies as well to computer science where a large amount of data are stored in linear files. And this is also the case in molecular biology where biological molecules can often be approximated as sequences of nucleotides or amino acids. Moreover, the quantity of available data in these fields tends to double every 18 months. This is the reason why algorithms should be efficient even if the speed of computers increases at a steady pace.

Pattern matching is the problem of locating a specific pattern inside raw data. The pattern is usually a collection of strings described in some formal language. Two kinds of textual patterns are presented: single strings and approximated strings. Two algorithms for matching patterns in images that are extensions of string-matching algorithms are also presented.

In several applications, texts need to be structured before being searched. Even if no further information is known about their syntactic structure, it is possible and indeed extremely efficient to build a data structure that supports searches. From among several existing data structures equivalent to represent indexes, suffix trees and suffix arrays are presented, along with their construction algorithms.

The comparison of strings is implicit in the approximate pattern searching problem. Since it is sometimes required to compare just two strings (files or molecular sequences), a basic method based on longest common subsequences is discussed.

Finally, the chapter includes two classical text compression algorithms. Variants of these algorithms are implemented in practical compression software, in which they are often combined together or with other elementary methods. An example of mixing different methods is presented.

The efficiency of algorithms is assessed by their asymptotic running times, and sometimes by the amount of memory space they require at run time as well.

15.2 String-Matching Algorithms

String matching is the problem of finding one or, more generally, all the **occurrences** of a pattern in a text. The pattern and the text are both strings built over a finite alphabet (a finite set of symbols). Each algorithm of this section outputs all occurrences of the pattern in the text. The pattern is denoted by $x = x[0 \ldots m-1]$; its length is equal to m. The text is denoted by $y = y[0 \ldots n-1]$; its length is equal to n. The alphabet is denoted by Σ and its size is equal to σ.

String-matching algorithms of the present section work as follows: they first align the left ends of the pattern and the text, then compare the aligned symbols of the text and the pattern—this specific task is called an attempt or a scan—and after a whole match of the pattern or after a mismatch they shift the pattern to the right. They repeat the same procedure again until the right end of the pattern goes beyond the right end of the text. This is called the scan and shift mechanism. Each attempt is associated with the position j in the text, when the pattern is aligned with $y[j \ldots j+m-1]$.

The brute force algorithm consists in checking, at all positions in the text between 0 and $n - m$, whether an occurrence of the pattern starts there or not. Then, after each attempt, it shifts the pattern exactly one position to the right. This is the simplest algorithm, which is described in Figure 15.1.

The time complexity of the brute force algorithm is $O(mn)$ in the worst case, but its behavior in practice is often linear on specific data.

```
BF(x, m, y, n)
▷ Searching
for j ← 0 to n − m
    do i ← 0
        while i < m and x[i] = y[i + j]
            do i ← i + 1
        if i ≥ m
            then OUTPUT(j)
```

FIGURE 15.1 Brute force string-matching algorithm. Symbol ▷ introduces comments.

15.2.1 Karp–Rabin Algorithm

Hashing provides a simple method for avoiding a quadratic number of symbol comparisons in most practical situations. Instead of checking at each position of the text whether the pattern occurs, it seems to be more efficient to check only if the portion of the text aligned with the pattern "looks like" the pattern. In order to check the resemblance between these portions a hashing function is used. To be helpful for the string-matching problem, the hashing function should have the following properties:

- Efficiently computable
- Highly discriminating for strings
- $hash(y[j+1 \ldots j+m])$ must be easily computable from $hash(y[j \ldots j+m-1])$

  $hash(y[j+1 \ldots j+m]) = \text{REHASH}(y[j], y[j+m], hash(y[j \ldots j+m-1]))$

For a word w of length k, its symbols can be considered as digits, and function $hash(w)$ is defined by

$$hash(w[0 \ldots k-1]) = (w[0] \times 2^{k-1} + w[1] \times 2^{k-2} + \cdots + w[k-1]) \bmod q$$

where q is a large number. Then, REHASH has a simple expression

$$\text{REHASH}(a,b,h) = ((h - a \times d) \times 2 + b) \bmod q$$

where $d = 2^{k-1}$ and q are the computer word-size (see Figure 15.2).

During the search for the pattern x, $hash(x)$ is compared with $hash(y[j-m+1 \ldots j])$ for $m-1 \leq j \leq n-1$. If an equality is found, it is still necessary to check the equality $x = y[j-m+1 \ldots j]$ symbol by symbol.

In the algorithms of Figures 15.2 and 15.3 all multiplications by 2 are implemented by shifts (operator $\ll$). Furthermore, the computation of the modulus function is avoided by using the implicit modular arithmetic given by the hardware that forgets carries in integer operations. Thus, q is chosen as the maximum value of an integer of the system.

The worst-case time complexity of the Karp–Rabin algorithm is quadratic (as it is for the brute force algorithm), but its expected running time is $O(m+n)$.

Example 15.1

Let x = **ing**. Then $hash(x) = 105 \times 2^2 + 110 \times 2 + 103 = 743$ (symbols are assimilated with their ASCII codes).

y	=	s	t	r	i	n	g		m	a	t	c	h	i	n	g
$hash$	=			806	797	776	743	678	585	443	746	719	766	709	736	743

15.2.2 Knuth–Morris–Pratt Algorithm

This section presents the first discovered linear-time string-matching algorithm. Its design follows a tight analysis of the brute force algorithm, and especially of the way this latter algorithm wastes the information gathered during the scan of the text.

```
REHASH(a, b, h)
1  return ((h − a × d) << 1) + b
```

FIGURE 15.2 Function REHASH.

```
KR(x, m, y, n)
  ▷ Preprocessing
  d ← 1
  for i ← 1 to m – 1
      do d ← d ≪ 1
  h_x ← 0
  h_y ← 0
  for i ← 0 to m – 1
      do h_x ← (h_x ≪ 1) + x[i]
         h_y ← (h_y ≪ 1) + y[i]
  ▷ Searching
  if h_x = h_y and x = y[0 … m – 1]
     then OUTPUT(0)
  j ← m
  while j < n
      do h_y ← REHASH(y[j – m], y[j], h_y)
         if h_x = h_y and x = y[j – m + 1 … j]
            then OUTPUT(j – m + 1)
         j ← j + 1
```

FIGURE 15.3 Karp–Rabin string-matching algorithm.

Let us look more closely at the brute force algorithm. It is possible to improve the length of shifts and simultaneously remember some portions of the text that match the pattern. This saves comparisons between characters of the text and of the pattern, and consequently increases the speed of the search.

Consider an attempt at position j, that is, when the pattern $x[0 \ldots m-1]$ is aligned with the segment $y[j \ldots j+m-1]$ of the text. Assume that the first mismatch (during a left to right scan) occurs between symbols $x[i]$ and $y[i+j]$ for $0 \leq i < m$. Then, $x[0 \ldots i-1] = y[j \ldots i+j-1] = u$ and $a = x[i] \neq y[i+j] = b$. When shifting, it is reasonable to expect that a **prefix** v of the pattern matches some **suffix** of the portion u of the text. Moreover, if one wants to avoid another immediate mismatch, the letter following the prefix v in the pattern must be different from a. (Indeed, it should be expected that v matches a suffix of ub, but elaborating along this idea goes beyond the scope of the chapter.) The longest such prefix v is called the **border** of u (it occurs at both ends of u). This introduces the notation: let $next[i]$ be the length of the longest (proper) border of $x[0 \ldots i-1]$ followed by a character c different from $x[i]$. Then, after a shift, the comparisons can resume between characters $x[next[i]]$ and $y[i+j]$ without missing any occurrence of x in y and having to backtrack on the text (see Figure 15.4).

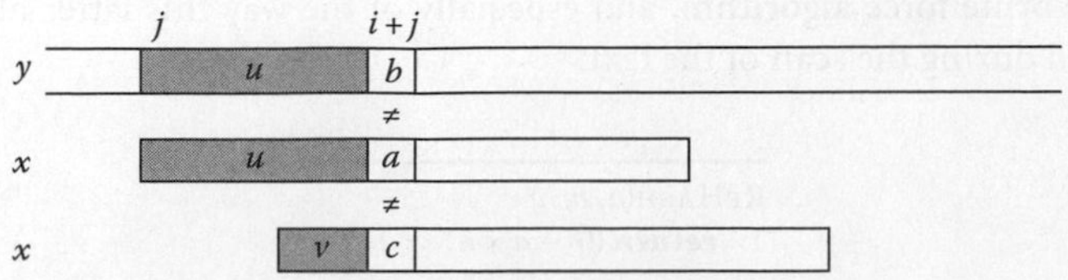

FIGURE 15.4 Shift in the Knuth–Morris–Pratt algorithm (v suffix of u).

Example 15.2

Here

$y =$	.	.	.	**a**	**b**	**a**	**b**	**a**	**a**	**b**	.				
$x =$				**a**	**b**	**a**	**b**	**a**	**b**	**a**					
$x =$									**a**	**b**	**a**	**b**	**a**	**b**	**a**

Compared symbols are underlined. Note that the empty string is the suitable border of **ababa**. Other borders of **ababa** are **aba** and **a**.

The Knuth–Morris–Pratt (KMP) algorithm is displayed in Figure 15.5. The table *next* it uses is precomputed in $O(m)$ time before the search phase, applying the same searching algorithm to the pattern itself, as if $y = x$ (see Figure 15.6). The worst-case running time of the algorithm is $O(m + n)$ and it requires $O(m)$ extra space. These quantities are independent of the size of the underlying alphabet.

```
KMP(x, m, y, n)
  ▷ Preprocessing
  next ← PREKMP(x, m)
  ▷ Searching
  i ← 0
  j ← 0
  while j < n
      do while i > – 1 and x[i] ≠ y[j]
             do i ← next[i]
         i ← i + 1
         j ← j + 1
         if i ≥ m
            then OUTPUT(j – i)
                 i ← next[i]
```

FIGURE 15.5 Knuth–Morris–Pratt string-matching algorithm.

```
PREKMP(x, m)
  i ← –1
  j ← 0
  next[0] ← –1
  while j < m
      do while i > –1 and x[i] ≠ x[j]
             do i ← next[i]
         i ← i + 1
         j ← j + 1
         if x[i] = x[j]
            then next[j] ← next[i]
            else next[j] ← i
  return next
```

FIGURE 15.6 Preprocessing phase of the Knuth–Morris–Pratt algorithm: computing *next*.

15.2.3 Boyer–Moore Algorithm

The Boyer–Moore (BM) algorithm is considered the most efficient string-matching algorithm in usual applications. A simplified version of it, or the entire algorithm, is often implemented in text editors for the search and substitute commands.

The algorithm scans the characters of the pattern from right to left beginning with the rightmost symbol. In case of a mismatch (or a complete match of the whole pattern), it uses two precomputed functions to shift the pattern to the right. These two shift functions are called the *bad-character shift* and the *good-suffix shift*. They are based on the following observations.

Assume that a mismatch occurs between the character $x[i] = a$ of the pattern and the character $y[i + j] = b$ of the text during an attempt at position j. Then, $x[i + 1 \ldots m - 1] = y[i + j + 1 \ldots j + m - 1] = u$ and $x[i] \neq y[i + j]$. The good-suffix shift consists in aligning the **segment** $y[i + j + 1 \ldots j + m - 1]$ with its rightmost occurrence in x that is preceded by a character different from $x[i]$ (see Figure 15.7). If there exists no such segment, the shift consists in aligning the longest suffix v of $y[i + j + 1 \ldots j + m - 1]$ with a matching prefix of x (see Figure 15.8).

Example 15.3

y =	.	.	.	a	b	b	a	a	b	b	a	b	b	a	.	.	.
x =	a	b	b	a	a	b	b	a	b	b	a						
x =				a	b	b	a	a	b	b	a	b	b	a			

The shift is driven by the suffix **abba** of x found in the text. After the shift, the segment **abba** in the middle of y matches a segment of x as in Figure 15.7. The same mismatch does not recur.

Example 15.4

y =	.	.	.	a	b	b	a	a	b	b	a	b	b	a	b	b	a	.	.
x =			b	b	a	b	b	a	b	b	a								
x =									b	b	a	b	b	a	b	b	a		

The segment **abba** found in y partially matches a prefix of x after the shift, as in Figure 15.8.

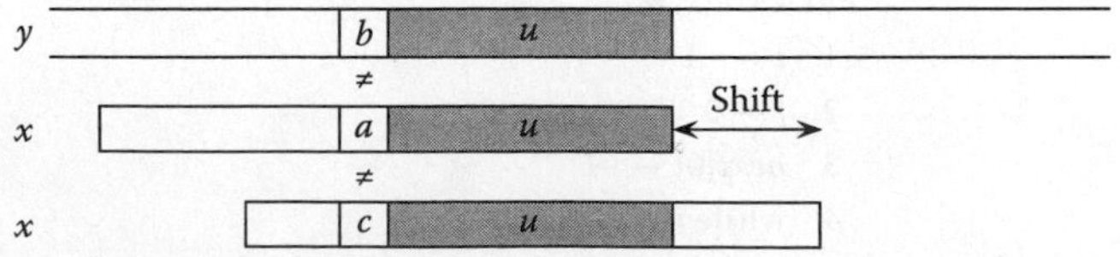

FIGURE 15.7 The good-suffix shift, when u reappears preceded by a character different from a.

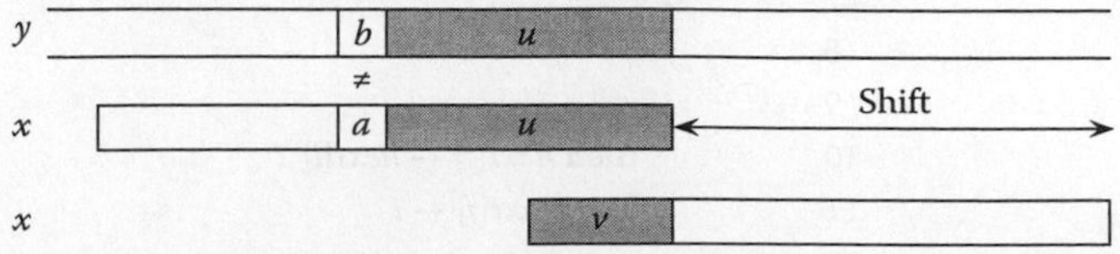

FIGURE 15.8 The good-suffix shift, when the situation of Figure 15.7 does not happen. Only a suffix of u reappears as a prefix of x.

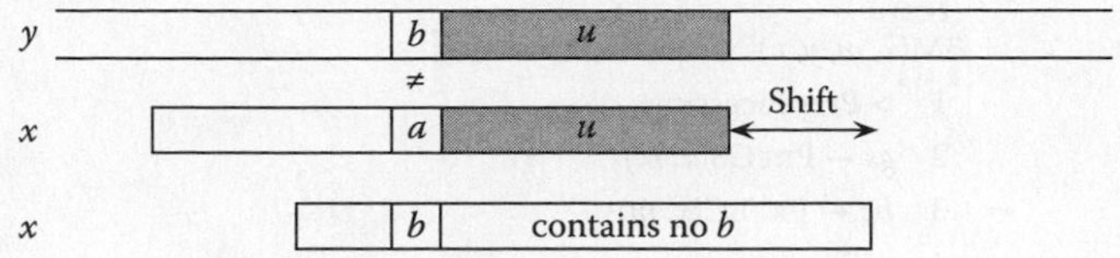

FIGURE 15.9 The bad-character shift, b appears in x.

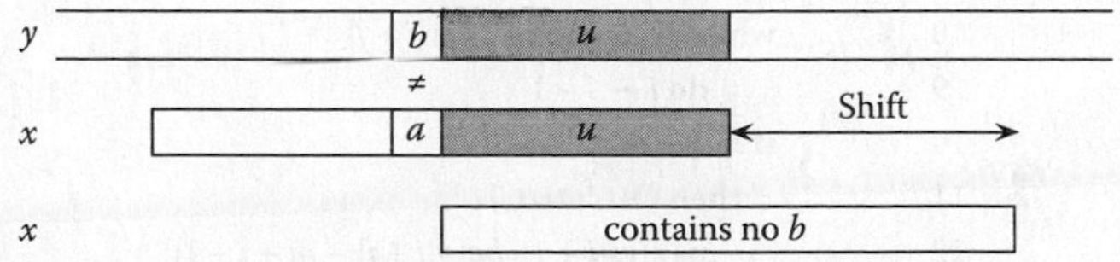

FIGURE 15.10 The bad-character shift, b does not appear in x (except possibly at $m - 1$).

The bad-character shift consists in aligning the text character $y[i + j]$ with its rightmost occurrence in $x[0 \ldots m - 2]$ (see Figure 15.9). If $y[i + j]$ does not appear in the pattern x, no occurrence of x in y can overlap the symbol $y[i + j]$, then the left end of the pattern is aligned with the character at position $i + j + 1$ (see Figure 15.10).

Example 15.5

$y =$	.	.	.	.	.	.	a	b	c	d	.	.	.	.
$x =$	c	d	a	h	g	f	e	b	c	d				
$x =$					c	d	a	h	g	f	e	b	c	d

The shift aligns the symbol **a** in x with the mismatch symbol **a** in the text y (Figure 15.9).

Example 15.6

$y =$	.	.	.	.	.	a	b	c	d	.	.	.	.	.	
$x =$	c	d	h	g	f	e	b	c	d						
$x =$							c	d	h	g	f	e	b	c	d

The shift positions the left end of x right after the symbol **a** of y (Figure 15.10).

The BM algorithm is shown in Figure 15.11. For shifting the pattern, it applies the maximum between the bad-character shift and the good-suffix shift. More formally, the two shift functions are defined as follows. The bad-character shift is stored in a table bc of size σ and the good-suffix shift is stored in a table gs of size $m + 1$. For $a \in \Sigma$

$$bc[a] = \begin{cases} \min\{i \mid 1 \le i < m \text{ and } x[m-1-i] = a\} & \text{if } a \text{ appears in } x, \\ m & \text{otherwise.} \end{cases}$$

```
BM(x, m, y, n)
▷ Preprocessing
gs ← PreGS(x, m)
bc ← PreBC(x, m)
▷ Preprocessing
j ← 0
while j ≤ n − m
      do i ← m − 1
         while i ≥ 0 and x[i] = y[i + j]
               do i ← i − 1
         if i < 0
            then Output(j)
         j ← max{gs[i + 1], bc[y[i + j] − m + i + 1]}
```

FIGURE 15.11 BM string-matching algorithm.

Let us define two conditions,

$$\begin{cases} cond_1(i,s): & \text{for each } k \text{ such that } i < k < m,\ s \geq k \text{ or } x[k-s] = x[k], \\ cond_2(i,s): & \text{if } s < i \text{ then } x[i-s] \neq x[i]. \end{cases}$$

Then, for $0 \leq i < m$,

$$gs[i+1] = \min\{s > 0 \mid cond_1(i,s) \text{ and } cond_2(i,s) \text{ hold}\}$$

and $gs[0]$ is defined as the length of the smallest period of x.

To compute the table gs, a table $suff$ is used. This table can be defined as follows: for $i = 0, 1, \ldots, m-1$,

$$suff[i] = \text{longest common suffix between } x[0\ldots i] \text{ and } x.$$

It is computed in linear time and space by the function Suffixes (see Figure 15.12).

```
Suffixes(x, m)
suff[m − 1] ← m
g ← m − 1
for i ← m − 2 downto 0
    do if i > g and suff[i + m − 1 − f] ≠ i − g
          then suff[i] ← min{suff[i + m − 1 − f], i − g}
          else if i < g
                  then g ← i
               f ← i
               while g ≥ 0 and x[g] = x[g + m − 1 − f]
                     do g ← g − 1
               suff[i] ← f − g
return suff
```

FIGURE 15.12 Computation of the table *suff*.

```
PreBC(x, m)
for a ← firstLetter to lastLetter
    do bc[a] ← m
for i ← 0 to m − 2
    do bc[x[i]] ← m − 1 − i
return bc
```

FIGURE 15.13 Computation of the bad-character shift.

```
PreGS(x, m)
gs ← Suffixes(x, m)
for i ← 0 to m − 1
    do gs[i] ← m
j ← 0
for i ← m − 1 downto −1
    do if i = −1 or suff[i] = i + 1
        then while j < m − 1 − i
            do if gs[j] = m
                then gs[j] ← m − 1 − i
            j ← j + 1
for i ← 0 to m − 2
    do gs[m − 1 − suff[i]] ← m − 1 − i
return gs
```

FIGURE 15.14 Computation of the good-suffix shift.

Tables *bc* and *gs* can be precomputed in time $O(m + \sigma)$ before the search phase and require an extra space in $O(m + \sigma)$ (see Figures 15.13 and 15.14). The worst-case running time of the algorithm is quadratic. However, on large alphabets (relative to the length of the pattern), the algorithm is extremely fast. Slight modifications of the strategy yield linear-time algorithms (see the References Section). When searching for a^m in $(a^{m-1}b)^{\lfloor n/m \rfloor}$ the algorithm makes only $O(n/m)$ comparisons, which is the absolute minimum for any string-matching algorithm in the model where the pattern only is preprocessed.

15.2.4 Quick Search Algorithm

The bad-character shift used in the BM algorithm is not very efficient for small alphabets, but when the alphabet is large compared with the length of the pattern, as it is often the case with the ASCII table and ordinary searches made under a text editor, it becomes very useful. Using it alone produces a very efficient algorithm (in practice) that is described now.

After an attempt where x is aligned with $y[j \ldots j + m - 1]$, the length of the shift is at least equal to one. Thus, the character $y[j + m]$ is necessarily involved in the next attempt, and thus can be used for the bad-character shift of the current attempt. In the present algorithm, the bad-character shift is slightly modified to take into account the observation as follows ($a \in \Sigma$):

$$bc[a] = 1 + \begin{cases} \min\{i \mid 0 \le i < m \text{ and } x[m-1-i] = a\} & \text{if } a \text{ appears in } x, \\ m & \text{otherwise.} \end{cases}$$

Indeed, the comparisons between text and pattern characters during each attempt can be done in any order. The algorithm of Figure 15.15 performs the comparisons from left to right. It was called Quick Search (QS) by its inventor and has a quadratic worst-case time complexity but good practical behavior.

Example 15.7

$y =$	s	t	r	i	n	g	-	m	a	t	c	h	i	n	g
$x =$	i	n	g												
$x =$				i	n	g									
$x =$								i	n	g					
$x =$												i	n	g	
$x =$													i	n	g

The QS algorithm makes only nine comparisons to find the two occurrences of **ing** inside the text of length 15.

15.2.5 Experimental Results

Figures 15.16 and 15.17 present the running times of three string-matching algorithms: the BM algorithm, the QS algorithm, and the Backward Oracle Matching (BOM) algorithm. The BOM algorithm can be viewed as a variation of the BM algorithm where factors (segments) rather than suffixes of the pattern are recognized. The BOM algorithm uses a data structure to store all the factors of the reversed pattern: a factor oracle but a suffix automaton or a **suffix tree** (see section 15.4) can also be used.

Tests have been performed on various types of texts. Figure 15.16 shows the results when the text is a DNA sequence on the four-letter alphabet of nucleotides **A**, **C**, **G**, **T**. In Figure 15.17, English text is considered.

For each pattern length, a large number of searches with random patterns were run. The average time according to the length is shown in the two figures. The running times of both preprocessing

```
QS(x, m, y, n)
▷ Preprocessing
for a ← firstLetter to lastLetter
        do bc[a] ← m + 1
for i ← 0 to m − 1
        do bc[x[i]] ← m − i
▷ Searching
j ← 0
while j ≤ n − m
        do i ← 0
            while i ≥ 0 and x[i] = y[i + j]
                do i ← i + 1
            if i ≥ m
                then OUTPUT(j)
            j ← bc[y[j + m]]
```

FIGURE 15.15 QS string-matching algorithm.

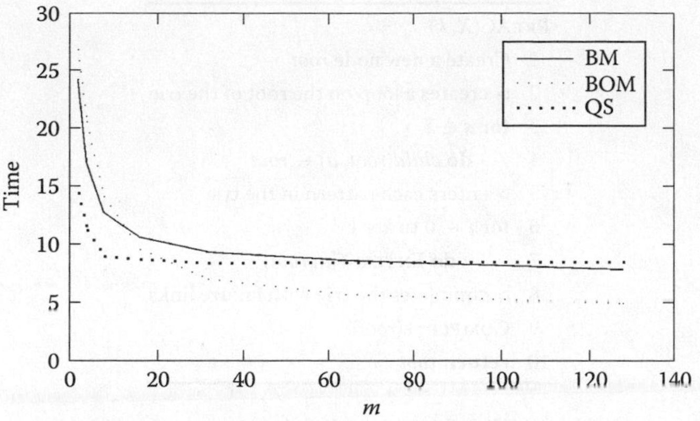

FIGURE 15.16 Running times for a DNA sequence.

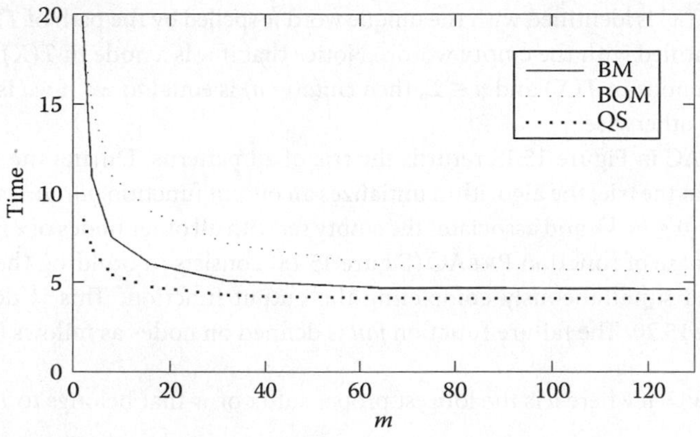

FIGURE 15.17 Running times for an English text.

and searching phases are added. The three algorithms are implemented in a homogeneous way in order to keep the comparison significant.

For the genome, as expected, the QS algorithm is the best for short patterns. But for long patterns, it is eventually less efficient than the BM algorithm. In this latter case, the BOM algorithm achieves the best results. For rather large alphabets, the three algorithms have similar behaviors; however, the QS is always faster.

15.2.6 Aho–Corasick Algorithm

The UNIX operating system provides standard text (or file) facilities. Among them is the series of **grep** commands that locate patterns in files. The algorithm underlying the **fgrep** command of UNIX is described in this section. It searches files for a finite set of strings, and can, for instance, output lines containing at least one of the strings.

If one is interested in searching for all occurrences of all patterns taken from a finite set of patterns, a first solution consists in repeating some string-matching algorithm for each pattern. If the set contains k patterns, this search runs in time $O(kn)$. The solution described in the present section and designed by Aho and Corasick runs in time $O(n \log \sigma)$. The algorithm is a direct extension of the KMP algorithm, and the running time is independent of the number of patterns.

Let $X = \{x_0, x_1, \ldots, x_{k-1}\}$ be the set of patterns, and let $|X| = |x_0| + |x_1| + \cdots + |x_{k-1}|$ be the total size of the set X. The Aho–Corasick algorithm first builds a **trie** $T(X)$, a digital tree recognizing the patterns of X.

```
PreAC(X, k)
Create a new node root
▷ creates a loop on the root of the trie
for a ∈ Σ
      do child(root, a) ← root
▷ enters each pattern in the trie
for i ← 0 to k − 1
      do Enter(X[i], root)
▷ completes the trie with failure links
Complete(root)
return root
```

FIGURE 15.18 Preprocessing phase of the Aho–Corasick algorithm.

The trie $T(X)$ is a tree in which edges are labeled by letters and in which branches spell the patterns of X. A node p in the trie $T(X)$ is identified with the unique word w spelled by the path of $T(X)$ from its root to p. The root itself is identified with the empty word ε. Notice that if w is a node in $T(X)$, then w is a prefix of some $x_i \in X$. If w is a node in $T(X)$ and $a \in \Sigma$, then *child*(w, a) is equal to wa if wa is a node in $T(X)$; it is equal to UNDEFINED otherwise.

The function PreAC in Figure 15.18 returns the trie of all patterns. During the second phase, where patterns are entered in the trie, the algorithm initializes an output function *out*. It associates the singleton $\{x_i\}$ with the nodes x_i $(0 \le i < k)$, and associates the empty set with all other nodes of $T(X)$ (see Figure 15.19).

Finally, the last phase of function PreAC (Figure 15.18) consists in building the failure link of each node of the trie, and simultaneously completing the output function. This is done by the function Complete in Figure 15.20. The failure function *fail* is defined on nodes as follows (w is a node):

$$fail(w) = u \text{ where } u \text{ is the longest proper suffix of } w \text{ that belongs to } T(X).$$

Computation of failure links is done during a breadth-first traversal of $T(X)$. Completion of the output function is done while computing the failure function *fail* using the following rule:

$$\text{if } fail(w) = u \text{ then } out(w) = out(w) \cup out(u).$$

```
Enter(x, root)
r ← root
i ← 0
▷ follows the existing edges
while i < |x| and child(r, x[i]) ≠ UNDEFINED and child(r, x[i]) ≠ root
      do r ← child(r, x[i])
         i ← i + 1
▷ creates new edges
while i < |x|
      do Create a new node s
         child(r, x[i]) ← s
         r ← s
         i ← i + 1
out(r) ← {x}
```

FIGURE 15.19 Construction of the trie.

```
COMPLETE(root)
  q ← empty queue
  ℓ ← list of the edges (root, a, p) for any character a ∈ Σ
          and any node p ≠ root
  while the list ℓ is not empty
      do (r, a, p) ← FIRST(ℓ)
         ℓ ← NEXT(ℓ)
         ENQUEUE(q, p)
         fail(p) ← root
  while the queue q is not empty
      do r ← DEQUEUE(q)
         ℓ ← list of the edges (root, a, p) for any character a ∈ Σ
                 and any node p
         while the list ℓ is not empty
             do (r, a, p) ← FIRST(ℓ)
                ℓ ← NEXT(ℓ)
                ENQUEUE(q, p)
                s ← fail(r)
                while child(s, a) = UNDEFINED
                    do s ← fail(s)
                fail(p) ← child(s, a)
                out(p) ← out(p) ∪ out(child(s, a))
```

FIGURE 15.20 Completion of the output function and construction of failure links.

Example 15.8

$X = \{$**search, ear, arch, chart**$\}$

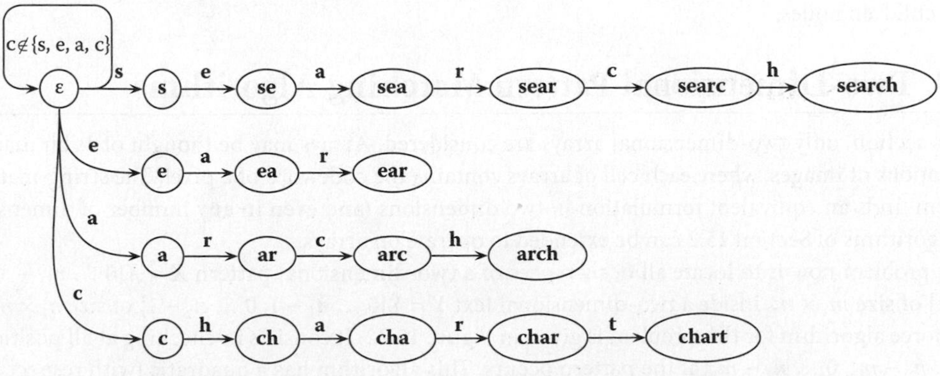

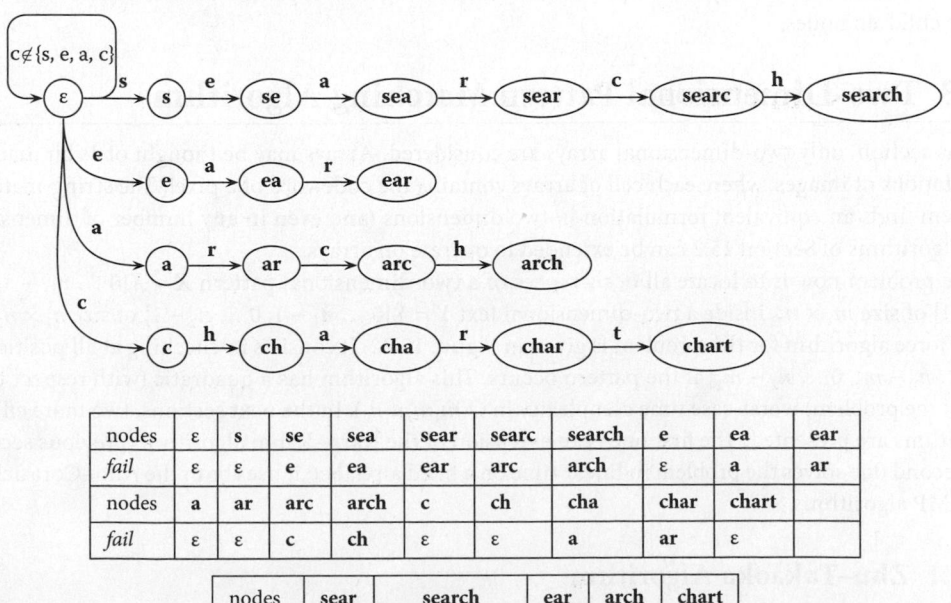

nodes	ε	s	se	sea	sear	searc	search	e	ea	ear
fail	ε	ε	e	ea	ear	arc	arch	ε	a	ar
nodes	a	ar	arc	arch	c	ch	cha	char	chart	
fail	ε	ε	c	ch	ε	ε	a	ar	ε	

nodes	sear	search	ear	arch	chart
out	ear	{search, arch}	ear	arch	chart

```
AC(X, k, y, n)
▷ Preprocessing
r ← PREAC(X, k)
▷ Searching
for j ← 0 to n − 1
      do while child(r, y[j]) = UNDEFINED
              do r ← fail(r)
         r ← child(r, y[j])
         if out(r) ≠ ∅
            then OUTPUT((out(r), j))
```

FIGURE 15.21 Complete Aho–Corasick algorithm.

To stop going back with failure links during the computation of the failure links, and also to overpass text characters for which no transition is defined from the root, a loop is added on the root of the trie for these symbols. This is done at the first phase of function PREAC.

After the preprocessing phase is completed, the searching phase consists in parsing all the characters of the text y with $T(X)$. This starts at the root of $T(X)$ and uses failure links whenever a character in y does not match any label of outgoing edges of the current node. Each time a node with a nonempty output is encountered, this means that the patterns of the output have been discovered in the text, ending at the current position. Then, the position is output.

An implementation of the Aho–Corasick algorithm from the previous discussion is shown in Figure 15.21. Note that the algorithm processes the text in an online way, so that the buffer on the text can be limited to only one symbol. Also, note that the instruction $r \leftarrow fail(r)$ in Figure 15.21 is the exact analogue of instruction $i \leftarrow next[i]$ in Figure 15.5. A unified view of both algorithms exists but is beyond the scope of the chapter.

The entire algorithm runs in time $O(|X| + n)$ if the *child* function is implemented to run in constant time. This is the case for any fixed alphabet. Otherwise a $\log \sigma$ multiplicative factor comes from access to the children nodes.

15.3 Two-Dimensional Pattern Matching Algorithms

In this section, only two-dimensional arrays are considered. Arrays may be thought of as bit map representations of images, where each cell of arrays contains the codeword of a pixel. The string-matching problem finds an equivalent formulation in two dimensions (and even in any number of dimensions), and algorithms of Section 15.2 can be extended to operate on arrays.

The problem now is to locate all occurrences of a two-dimensional pattern $X = X[0 \ldots m_1 - 1, 0 \ldots m_2 - 1]$ of size $m_1 \times m_2$ inside a two-dimensional text $Y = Y[0 \ldots n_1 - 1, 0 \ldots n_2 - 1]$ of size $n_1 \times n_2$. The brute force algorithm for this problem is given in Figure 15.22. It consists in checking at all positions of $Y[0 \ldots n_1 - m_1, 0 \ldots n_2 - m_2]$ if the pattern occurs. This algorithm has a quadratic (with respect to the size of the problem) worst-case time complexity in $O(m_1 m_2 n_1 n_2)$. In the next sections, two more efficient algorithms are presented. The first one is an extension of the Karp–Rabin algorithm (previous section). The second one solves the problem in linear time on a fixed alphabet; it uses both the Aho–Corasick and the KMP algorithms.

15.3.1 Zhu–Takaoka Algorithm

As for one-dimensional string matching, it is possible to check if the pattern occurs in the text only if the *aligned* portion of the text looks like the pattern. To do that, the idea is to use vertically the hash

```
BF2D(X, m1, m2, Y, n1, n2)
▷ Searching
for j1 ← 0 to n1 − m1
    do for j2 ← 0 to n2 − m2
        do i ← 0
            while i < m1
                    and x[i, 0 ... m2 − 1] = y[j1 + i, j2 ... j2 + m2 − 1]
                do i ← i + 1
            if i ≥ m1
                then OUTPUT(j1, j2)
```

FIGURE 15.22 Brute force two-dimensional pattern matching algorithm.

function method proposed by Karp and Rabin. To initialize the process, the two-dimensional arrays X and Y are translated into one-dimensional arrays of numbers x and y. The translation from X to x is done as follows ($0 \le i < m_2$):

$$x[i] = hash(X[0,i]X[1,i]\cdots X[m_1-1,i])$$

and the translation from Y to y is done by ($0 \le i < m_2$):

$$y[i] = hash(Y[0,i]Y[1,i]\cdots Y[m_1-1,i]).$$

The fingerprint y helps to find occurrences of X starting at row $j = 0$ in Y. It is then updated for each new row in the following way ($0 \le i < m_2$):

$$hash(Y[j+1,i]Y[j+2,i]\cdots Y[j+m_1,i])$$
$$= \text{REHASH}(Y[j,i],Y[j+m_1,i],hash(Y[j,i]Y[j+1,i]\cdots Y[j+m_1-1,i]))$$

(functions *hash* and REHASH are described in Section 15.2.1).

Example 15.9

$X =$

a	a	a
b	b	a
a	a	b

$Y =$

a	b	a	b	a	b	b
a	a	a	a	b	b	b
b	b	b	a	a	a	b
a	a	a	b	b	a	a
b	b	a	a	a	b	b
a	a	b	a	b	a	a

$x =$

681	681	680

$y =$

680	684	680	683	681	685	686

```
KMP-IN-LINE(X, m1, m2, Y, n1, n2, x, y, next, j1)
i2 ← 0
j2 ← 0
while j2 < n2
      do while i2 > −1 and x[i2] ≠ y[j2]
            do i2 ← next[i2]
         i2 ← i2 + 1
         j2 ← j2 + 1
         if i2 ≥ m2
            then DIRECT-COMPARE(X, m1, m2, Y, n1, n2, j1, j2 − 1)
                 i2 ← next[m2]
```

FIGURE 15.23 Search for x in y using KMP algorithm.

```
DIRECT-COMPARE(X, m1, m2, Y, row, column)
j1 ← row − m1 + 1
j2 ← column − m2 + 1
for i1 ← 0 to m1 − 1
      do for i2 ← 0 to m2 − 1
            do if X[i1, i2] ≠ Y[i1 + j1, i2 + j2]
                  then return
OUTPUT(j1, j2)
```

FIGURE 15.24 Naive check of an occurrence of x in y at position (*row, column*).

Next value of y is | 681 | 681 | 681 | 680 | 684 | 683 | 685 |. The occurrence of x at position 1 on y corresponds to an occurrence of X at position (1, 1) on Y.

Since the alphabet of x and y is large, searching for x in y must be done by a string-matching algorithm for which the running time is independent of the size of the alphabet: the KMP suits this application perfectly. Its adaptation is shown in Figure 15.23.

When an occurrence of x is found in y, then one still has to check if an occurrence of X starts in Y at the corresponding position. This is done naively by the procedure of Figure 15.24.

The Zhu–Takaoka algorithm as explained is displayed in Figure 15.25. The search for the pattern is performed row by row starting at row 0 and ending at row $n_1 - m_1$.

15.3.2 Bird–Baker Algorithm

The algorithm designed independently by Bird and Baker for the two-dimensional pattern matching problem combines the use of the Aho–Corasick algorithm and the KMP algorithm. The pattern X is divided into its m_1 rows $R_0 = X[0, 0 \ldots m_2 - 1]$ to $R_{m_1-1} = x[m_1 - 1, 0 \ldots m_2 - 1]$. The rows are preprocessed into a trie as in the Aho–Corasick algorithm described earlier.

```
ZT(X, m1, m2, Y, n1, n2)
 ▷ Preprocessing
 ▷ Computes x
 for i2 ← 0 to m2 − 1
     do x[i2] ← 0
        for i1 ← 0 to m1 − 1
            do x[i2] ← (x[i2] ≪ 1) + X[i1, i2]
 ▷ Computes the first value of y
 for j2 ← 0 to n2 − 1
     do y[j2] ← 0
        for j1 ← 0 to m1 − 1
            do y[j2] ← (y[j2] ≪ 1) + Y[j1, j2]
 d ← 1
 for i ← 1 to m1 − 1
     do d ← d ≪ 1
 next ← PREKMP(X', m2)
 ▷ Searching
 j1 ← m1 − 1
 while j1 < n1
     do KMP-IN-LINE(X, m1, m2, Y, n1, n2, x, y, next, j2)
        if j1 < n1 − 1
          then for j2 ← 0 to n2 − 1
                   do y[j2] ← REHASH(Y[j1 − m1 + 1, j2],
                                     Y[j1 + 1, j2], y[j2])
        j1 ← j1 + 1
```

FIGURE 15.25 Zhu–Takaoka two-dimensional pattern matching algorithm.

Example 15.10

Pattern X and the trie of its rows:

$X =$

b	a	a
a	b	b
b	a	a

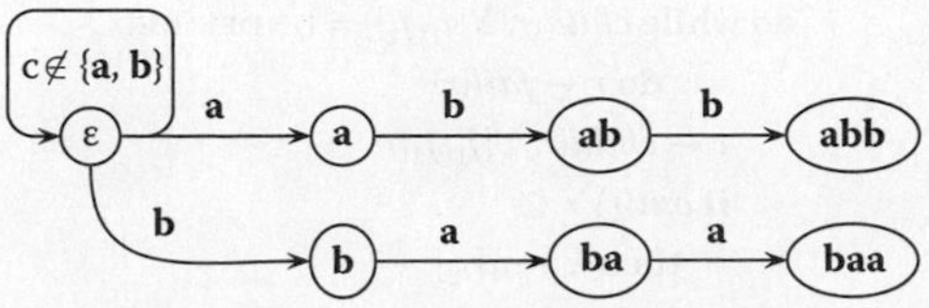

The search proceeds as follows. The text is read from the upper left corner to the bottom right corner, row by row. When reading the character $Y[j_1, j_2]$ the algorithm checks whether the portion $Y[j_1, j_2 - m_2 + 1 \ldots j_2] = R$ matches any of $R_0, \ldots, R_{m_1 - 1}$ using the Aho–Corasick machine. An additional one-dimensional array a of size n_1 is used as follows: $a[j_2] = k$ means that the $k - 1$ first rows $R_0, \ldots, R_{k-2}$ of the pattern match, respectively, the portions of the text: $Y[j_1 - k + 1, j_2 - m_2 + 1 \ldots j_2], \ldots, Y[j_1 - 1, j_2 - m_2 + 1 \ldots j_2]$. Then, if $R = R_{k-1}$, $a[j_2]$ is incremented to $k + 1$. If not, $a[j_2]$ is set to $s + 1$ where s is the maximum i such that

$$R_0 \cdots R_i = R_{k-s+1} \cdots R_{k-2} R.$$

The value s is computed using the KMP algorithm vertically (in columns). If there exists no such s, $a[j_2]$ is set to 0. Finally, if at some point $a[j_2] = m_1$ an occurrence of the pattern appears at position $(j_1 - m_1 + 1, j_2 - m_2 + 1)$ in the text.

The Bird-Baker algorithm is presented in Figures 15.26 and 15.27. It runs in time $O((n_1 n_2 + m_1 m_2) \log \sigma)$.

```
PRE-KMP-FOR-B(X, m1, m2)
i ← 0
next[0] ← −1
j ← −1
while i < m1
    do while j > −1 and X[i, 0 ... m2 − 1] ≠ X[j, 0 ... m2 − 1]
        do j ← next[j]
    i ← i + 1
    j ← j + 1
    if X[i, 0 ... m2 − 1] ≠ X[j, 0 ... m2 − 1]
        then next[i] ← next[j]
        else next[i] ← j
return next
```

FIGURE 15.26 Computes the function *next* for rows of X.

```
B(X, m1, m2, Y, n1, n2)
▷ Preprocessing
for i ← 0 to m2 − 1
    do a[i] ← 0
root ← PREAC(m1)
next ← PRE-KMP-FOR-B(X, m1, m2)
for j1 ← 0 to n1 − 1
    do r ← root
       for j2 ← 0 to n2 − 1
           do while child(r, Y[j1, j2]) = UNDEFINED
                  do r ← fail(r)
              r ← child(r, Y[j1, j2])
              if out(r) ≠ ∅
                 then k ← a[j2]
                      while k > 0 and X[k, 0 . . m2 − 1] = out(r)
                          do k ← next[k]
                      a[j2] ← k + 1
                      if k ≥ m1 − 1
                         then OUTPUT(j1 − m1 + 1, j2 − m2 + 1)
                 else a[j2] ← 0
```

FIGURE 15.27 Bird–Baker two-dimensional pattern matching algorithm.

15.4 Suffix Trees

The suffix tree $S(y)$ of a string y is a trie (as described earlier) containing all the suffixes of the string, and having the properties described subsequently. This data structure serves as an index on the string: it provides a direct access to all segments of the string, and gives the positions of all their occurrences in the string.

Once the suffix tree of a text y is built, searching for x in y remains to spell x along a branch of the tree. If this walk is successful, the positions of the pattern can be output. Otherwise, x does not occur in y.

Any kind of trie that represents the suffixes of a string can be used to search it. But the suffix tree has additional features which imply that its size is linear. The suffix tree of y is defined by the following properties:

- All branches of $S(y)$ are labeled by all suffixes of y.
- Edges of $S(y)$ are labeled by strings.
- Internal nodes of $S(y)$ have at least two children (when y is not empty).
- Edges outgoing an internal node are labeled by segments starting with different letters.
- The preceding segments are represented by their starting positions on y and their lengths.

Moreover, it is assumed that y ends with a symbol occurring nowhere else in it (the dollar sign is used in examples). This avoids marking nodes, and implies that $S(y)$ has exactly n leaves (number of nonempty suffixes). The other properties then imply that the total size of $S(y)$ is $O(n)$, which makes it possible to design a linear-time construction of the trie. The algorithm described in the present section has this time complexity provided the alphabet is fixed, or with an additional multiplicative factor $\log \sigma$ otherwise.

The algorithm inserts all nonempty suffixes of y in the data structure from the longest to the shortest suffix, as shown in Figure 15.28. Two definitions to explain how the algorithm works are introduced:

- $head_j$ is the longest prefix of $y[j \ldots n-1]$ which is also a prefix of $y[i \ldots n-1]$ for some $i < j$.
- $tail_j$ is the word such that $y[j \ldots n-1] = head_j\, tail_j$.

The strategy to insert the ith suffix in the tree is based on these definitions and described in Figure 15.29.

The second step of the insertion (Figure 15.29) is clearly performed in constant time. Thus, finding the node h is critical for the overall performance of the algorithm. A brute-force method to find it consists in spelling the current suffix $y[j \ldots n-1]$ from the root of the tree, giving an $O(|head_j|)$ time complexity for the insertion at step j, and an $O(n^2)$ running time to build $S(y)$. Adding short cut links leads to an overall $O(n)$ time complexity, although there is no guarantee that insertion at step j is realized in constant time.

```
SUFFIX-TREE(y, n)
T_{-1} ← one node tree
for j ← 0 to n − 1
    do T_j ← INSERT(T_{j−1}, y[j ... n − 1])
return T_{n−1}
```

FIGURE 15.28 Construction of a suffix tree for y.

```
INSERT(T_{j−1}, y[j ... n − 1])
1  locate the node h associated with head_j in T_{j−1}, possibly breaking an edge
2  add a new edge labeled tail_j from h to a new leaf representing y[j ... n − 1]
3  return the modified tree
```

FIGURE 15.29 Insertion of a new suffix in the tree.

Example 15.11

The different tries during the construction of the suffix tree of y = **CAGATAGAG**. Leaves are black and labeled by the position of the suffix they represent. Plain arrows are labeled by pairs: the pair (j, ℓ) stands for the segment $y[j \ldots j + \ell - 1]$. Dashed arrows represent the nontrivial suffix links.

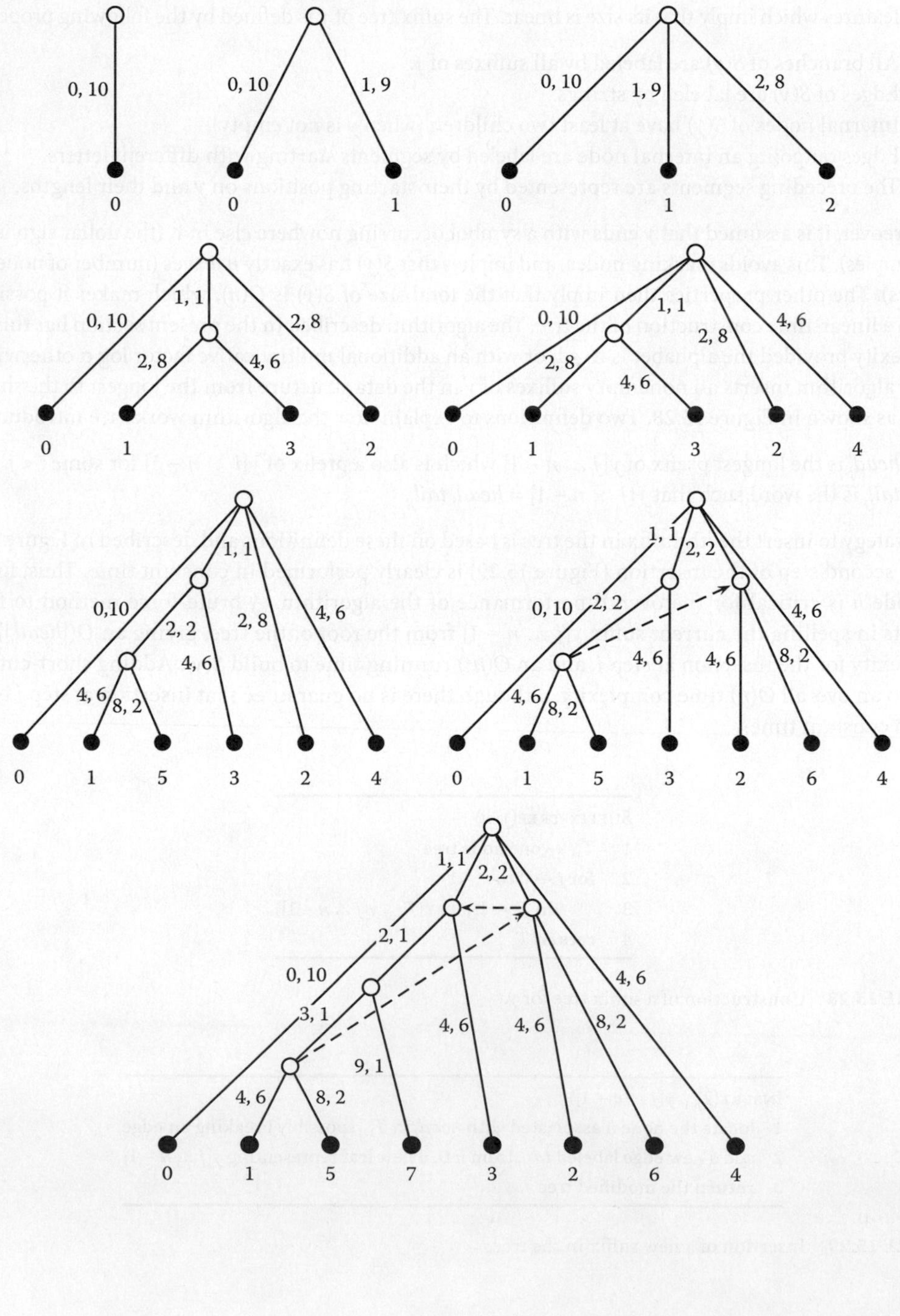

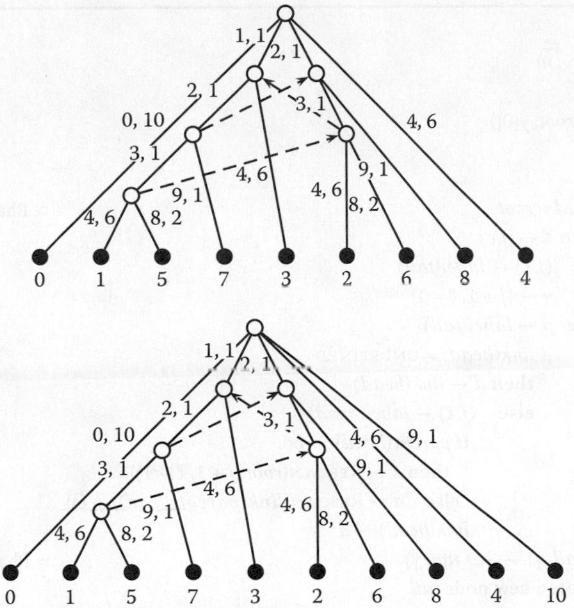

15.4.1 McCreight Algorithm

The key to get an efficient construction of the suffix tree $S(y)$ is to add links between nodes of the tree: they are called *suffix links*. Their definition relies on the relationship between $head_{j-1}$ and $head_j$: if $head_{j-1}$ is of the form az ($a \in \Sigma$, $z \in \Sigma^*$), then z is a prefix of $head_j$. In the suffix tree, the node associated with z is linked to the node associated with az. The suffix link creates a shortcut in the tree that helps with finding the next head efficiently. The insertion of the next suffix, namely, $head_j\ tail_j$, in the tree reduces to the insertion of $tail_j$ from the node associated with $head_j$.

The following property is an invariant of the construction: in T_j, only the node h associated with $head_j$ can fail to have a valid suffix link. This effectively happens when h has just been created at step j. The procedure to find the next head at step j is composed of two main phases:

A Rescanning: Assume that $head_{j-1} = az$ ($a \in \Sigma$, $z \in \Sigma^*$) and let d' be the associated node. If the suffix link on d' is defined, it leads to a node d from which the second step starts. Otherwise, the suffix link on d' is found by rescanning as follows. Let c' be the parent of d', and let (j, ℓ) be the label of edge (c', d'). For the ease of the description, assume that $az = av(y[j \ldots j + \ell - 1])$ (it may happen that $az = y[j \ldots j + \ell - 1]$). There is a suffix link defined on c' and going to some node c associated with v. The crucial observation here is that $y[j \ldots j + \ell - 1]$ is the prefix of the label of some branch starting at node c. Then, the algorithm rescans $y[j \ldots j + \ell - 1]$ in the tree: let e be the child of c along that branch, and let (k, m) be the label of edge (c, e). If $m < \ell$, then a recursive rescan of $q = y[j + m \ldots j + \ell - 1]$ starts from node e. If $m > \ell$, the edge (c, e) is broken to insert a new node d; labels are updated correspondingly. If $m = \ell$, d is simply set to e. If the suffix link of d' is currently undefined, it is set to d.

B Scanning: A downward search starts from d to find the node h associated with $head_j$. The search is dictated by the characters of $tail_{j-1}$ one at a time from left to right. If necessary, a new internal node is created at the end of the scanning.

After the two phases A and B are executed, the node associated with the new head is known, and the tail of the current suffix can be inserted in the tree.

To analyze the time complexity of the entire algorithm, one mainly has to evaluate the total time of all scannings, and the total time of all rescannings. Assume that the alphabet is fixed, so that branching from a node to one of its children can be implemented to take constant time. Thus, the time spent for

```
M(y, n)
root ← INIT(y, n)
head ← root
tail ← child(root, y[0])
n ← n − 1
while n > 0
      do if head = root                                                   ▷ Phase A (rescanning)
            then d ← root
                  (j, ℓ) ← label(tail)
                  γ ← (j + 1, ℓ − 1)
            else γ ← label(tail)
                  if link(head) ≠ UNDEFINED
                     then d ← link(head)
                     else (j, ℓ) ← label(head)
                           if parent(head) = root
                              then d ← RESCAN(root, j + 1, ℓ − 1))
                              else d ← RESCAN(link(parent(head)), j, ℓ))
                           link(head) ← d
            (head, γ) ← SCAN(d, γ)                                        ▷ Phase B (scanning)
            create a new node tail
            parent(tail) ← head
            label(tail) ← γ
            (j, ℓ) ← γ
            child(head, y[j]) ← tail
            n ← n − 1
return root
```

FIGURE 15.30 Suffix tree construction.

all scannings is linear because each letter of y is scanned only once. The same holds true for rescannings because each step downward (through node e) increases strictly the position of the segment of y considered there, and this position never decreases.

An implementation of McCreight's algorithm is shown in Figure 15.30. The next figures (Figures 15.31 through 15.34) give the procedures used by the algorithm, especially procedures RESCAN and SCAN.

The following notation will be used:

- *parent*(c) is the parent node of the node c.
- *label*(c) is the pair (i, l) if the edge from the parent node of c to c itself is associated with the factor $y[i \ldots i + l - 1]$.
- *child*(c, a) is the only node that can be reached from the node c with the character a.
- *link*(c) is the suffix node of the node c.

```
INIT(y, n)
create a new node root
create a new node c
parent(root) ← UNDEFINED
parent(c) ← root
child(root, y[0]) ← c
label(root) ← UNDEFINED
label(c) ← (0, n)
return root
```

FIGURE 15.31 Initialization procedure.

```
Rescan(c, j, ℓ)
(k, m) ← label(child(c, y[j]))
while ℓ > 0 and ℓ ≥ m
    do c ← child(c, y[j])
       ℓ ← ℓ − m
       j ← j + m
       (k, m) ← label(child(c, y[j]))
if ℓ > 0
    then return Break-edge(child(c, y[j]), ℓ)
    else return c
```

FIGURE 15.32 The crucial rescan operation.

```
Break-edge(c, k)
create a new node g
parent(g) ← parent(c)
(j, ℓ) ← label(c)
child(parent(c), y[j]) ← g
label(g) ← (j, k)
parent(c) ← g
label(c) ← (j + k, ℓ − k)
child(g, y[j + k]) ← c
link(g) ← UNDEFINED
return g
```

FIGURE 15.33 Breaking an edge.

```
Scan(d, γ)
(j, ℓ) ← γ
while child(d, y[j]) ≠ UNDEFINED
    do g ← child(d, y[j])
       k ← 1
       (s, lg) ← label(g)
       s ← s + 1
       ℓ ← ℓ − 1
       j ← j + 1
       while k < lg and y[j] = y[s]
           do j ← j + 1
              s ← s + 1
              k ← k + 1
              ℓ ← ℓ − 1
       if k < lg
           then return (Break-edge(g, k), (j, ℓ))
       d ← g
return (d, (j, ℓ))
```

FIGURE 15.34 The scan operation.

15.5 Suffix Arrays

Suffix trees are very powerful; however, they are very space consuming. In most cases, they can be replaced by suffix arrays that are more space economical. The technique involves binary search in the sorted list of the text suffixes. In this section, we show how to lexicographically sort the suffixes of a string y of length n. The resulting permutation, as an array, constitutes the suffix array of the string. It is usually enhanced with the LCP array described further in the section to get an efficient string searching algorithm.

The goal of the sorting is to compute a permutation p of the indices on y that satisfies the condition

$$y[p[0]\ldots n-1] < y[p[1]\ldots n-1] < \cdots < y[p[n-1]\ldots n-1]. \tag{15.1}$$

Note that the inequalities are strict since two suffixes occurring at distinct positions cannot be identical.

The implementation of a standard lexicographic sorting method (repeated radix sort) leads to an algorithm whose execution time is $O(n^2)$ since the sum of the lengths of the suffixes of y is quadratic.

Actually, the ordering is not entirely sufficient to get an efficient search. The precomputation and the utilization of common prefixes to the suffixes are extra elements that make the technique very efficient. The search for a string of length m in a string length n then takes $O(m + \log n)$ time.

To describe the sorting algorithm, for $u \in \Sigma^*$, we denote by

$$first_k(u) = \begin{cases} u & \text{if } |u| \le k, \\ u[0\ldots k-1] & \text{otherwise,} \end{cases}$$

the beginning of order k of the string u.

15.5.1 Kärkkäinen–Sanders Algorithm

In this section, the alphabet of y is assumed to be a bounded segment of integers, as it can be considered in most real applications. With this condition, not only letters can be sorted in linear time but suffixes of y as well.

The present algorithm for sorting the suffixes proceeds in four steps as follows, where k is an appropriate nonnegative integer.

Step 1 Positions i on the input string y are sorted for two-third of them, namely for $i = 3k$ or $i = 3k + 1$, according to $first_3(y[i \ldots n-1])$.
Let $t[i]$ be the rank of i in the sorted list.

Step 2 Suffixes of the 2/3-shorter word $z = t[0]t[3] \cdots t[3k] \cdots t[1]t[4] \cdots t[3k+1] \cdots$ are recursively sorted.

Let $s[i]$ be the rank of suffix at position i on y in the sorted list of them derived from the sorted list of suffixes of z.

Step 3 Suffixes $y[j \ldots n-1]$, for j of the form $3k + 2$, are sorted using the table s.

Step 4 The final step consists in merging the ordered lists obtained at the second and third steps.

A careful implementation of the algorithm leads to a linear running time. It relies on the following elements. The first step can be executed in linear time by using a mere radix sort. Since the order of suffixes $y[j+1 \ldots n-1]$ is already known from s, the third step can be done in linear time by just radix sorting pairs $(y[j], s[j+1])$. Comparing suffixes at positions i ($i = 3k$ or $i = 3k + 1$ for the first list) and j ($j = 3k + 2$ for the second list) remains to compare pairs of the form $(y[i], s[i+1])$ and $(y[j], s[j+1])$ if $i = 3k$ or pairs of the form $(y[i]y[i+1], s[i+2])$ and $(y[j]y[j+1], s[j+2])$ if $i = 3k + 1$. This is done in constant time and then the merge at the fourth step can thus be realized in linear time.

An example of a run of the algorithm is shown in Figure 15.35: (1) String y = **abaaabaaabb** and its two sets of positions P_{01} and P_2. (2) Step 1. Strings $first_3(y[i \ldots n-1])$ for $i \in P_{01}$ and their ranks: $t[i]$ is the

	i	0	1	2	3	4	5	6	7	8	9	10
(1)	$y[i]$	**a**	**b**	**a**	**a**	**a**	**b**	**a**	**a**	**a**	**b**	**b**

$P_{01} = \{0, 1, 3, 4, 6, 7, 9, 10\}$ and $P_2 = \{2, 5, 8\}$

		$i \bmod 3 = 0$				$i \bmod 3 = 1$			
(2)	i	0	3	6	9	1	4	7	10
	$first_3(y[i\,..\,n-1])$	**aba**	**aab**	**aaa**	**bb**	**baa**	**aba**	**aab**	**b**
	$t[i]$	2	1	0	5	4	2	1	3

$z = 2\,1\,0\,5\,4\,2\,1\,3$ and $L_{01} = (6, 3, 7, 0, 4, 10, 1, 9)$

(3)	$s[i]$	3	1	0	7	6	4	2	5

		$j \bmod 3 = 2$		
(4)	j	2	5	8
	$(y[j], s[j+1])$	(**a**, 1)	(**b**, 0)	(**a**, 7)

$L_2 = (2, 8, 5)$

		$i \bmod 3 = 0$ or 1							
	i	6	3	7	0	4	10	1	9
(5)	$(y[i]y[i+1], s[i+2])$			(**aa**, 7)		(**ab**, 0)	(**b**, −1)	(**ba**, 1)	
	$(y[i], s[i+1])$	(**a**, 2)	(**a**, 4)		(**a**, 6)				(**b**, 5)

		$j \bmod 3 = 2$		
	j	2	8	5
	$(y[j]y[j+1], s[j+2])$	(**aa**, 4)	(**ab**, 5)	(**ba**, 2)
	$(y[j], s[j+1])$	(**a**, 1)	(**a**, 7)	(**b**, 0)

(6)	i	0	1	2	3	4	5	6	7	8	9	10
	$p[i]$	2	6	3	7	0	4	8	10	1	5	9

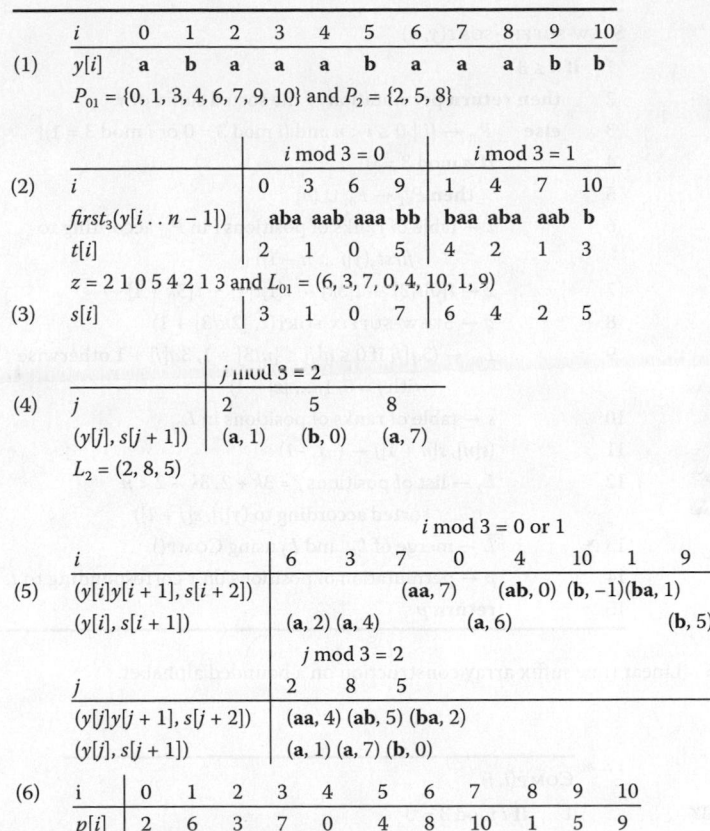

FIGURE 15.35 Building the suffix array of y = **abaaabaaabb.**

rank of i in the sorted list. (3) Step 2. Positions in P_{01} sorted according to their associated suffixes in z, resulting in L_{01} and the table of ranks s. (4) Step 3. Positions j in P_2 sorted according to pairs $(y[j], s[j+1])$ resulting in L_2. (5) Step 4. Pairs used for comparing positions when merging the sorted lists L_{01} and L_2. (6) Permutation p corresponding to the sorted suffixes of y.

The algorithm Skew-suffix-sort (Figure 15.36) describes the method presented above in a more precise way. To shorten the presentation of the algorithm, the definition of s (see Line 11) is extended to positions n and $n+1$ that are considered Lines 12 and 13 (call to Comp, Figure 15.37).

The recursivity of the algorithm (Line 8) yields the recurrence relation $T(n) = T(2n/3) + O(n)$, for $n > 3$, and $T(n) = O(1)$, for $n \leq 3$, because all other lines execute in constant time, that is, $O(n)$ time. The recurrence has solution $T(n) = O(n)$, Algorithm Skew-suffix-sort computes the suffix array of a string of length n in time $O(n)$.

15.5.2 Substring Search

Searching for a pattern x of length m in the string y using its suffix array can be done by a simple binary search (see Figure 15.38). It returns:

- $(-1, 0)$ if $x < y[p[0] \ldots n-1]$.
- $(n-1, n)$ if $x > y[p[n-1] \ldots n-1]$.
- i if x is a prefix of $x[p[i] \ldots n-1]$.
- $(i, i+1)$ if $y[p[i] \ldots n-1] < x < y[p[i+1] \ldots n-1]$.

```
Skew-suffix-sort(y, n)
if n ≤ 3
   then return permutation of the sorted suffix of y
   else    P01 ← {i | 0 ≤ i < n and (i mod 3 = 0 or i mod 3 = 1)}
           if n mod 3 = 0
              then P01 ← P01 ∪ {n}
           t ← table of ranks of positions i in P01 according to
                       first3(y[i … n − 1])
           z ← t[0]t[3] ⋯ t[3k] ⋯ t[1]t[4] ⋯ t[3k + 1] ⋯
           q ← Skew-suffix-sort(z, ⌊2n/3⌋ + 1)
           L01 ← (3q[j] if 0 ≤ q[j] ≤ ⌊n/3⌋ + 1, 3q[j] + 1 otherwise
                       with j = 0, 1,…, |z| − 1)
           s ← table of ranks of positions in L01
           (s[n], s[n + 1]) ← (−1, −1)
           L2 ← list of positions j = 3k + 2, 3k + 2 < n
                       sorted according to (y[j], s[j + 1])
           L ← merge of L01 and L2 using Comp()
           p ← permutation of positions on y corresponding to L
           return p
```

FIGURE 15.36 Linear time suffix array construction on a bounded alphabet.

```
Comp(i, j)
if i mod 3 = 0
   then if (y[i], s[i + 1]) < (y[j], s[j + 1])
           then return −1
           else return 1
   else if (y[i … i + 1], s[i + 2]) < (y[j … j + 1], s[j + 2])
           then return −1
           else return 1
```

FIGURE 15.37 Constant time comparison of $y[i \ldots n-1]$ and $y[j \ldots n-1]$ during the merge.

```
Binary-Search(y, n, p, x, m)
ℓ ← −1
r ← n
while ℓ + 1 < r
   do i ← ⌊(ℓ + r)/2⌋
      k ← |lcp(x, y[p[i] … n − 1])|
      if k = m
         then return i
      elseif x ≤ y[p[i] … n − 1]
         then r ← i
      else    ℓ ← i
return (ℓ, r)
```

FIGURE 15.38 Binary search of x is the suffix array of y.

The loop line 3 iterates at most log n times and each test line 8 can perform at most m character comparisons. Thus the time complexity of algorithm is $O(m \times \log n)$.

15.5.3 Longest Common Prefixes

In this section, the second element that constitutes a suffix array is described: the array *LCP* storing the maximal lengths of prefixes common to suffixes in the sorted list. We first introduce the following notation for two strings $u, v \in \Sigma^*$:

$$lcp(u,v) = \text{longest common prefix of } u \text{ and } v.$$

We start by the first half of the array *LCP* storing the maximal lengths of prefixes common to consecutive suffixes in the sorted list. It is defined as follows, for $1 \leq i \leq n - 1$:

$$LCP[i] = lcp(y[p[i-1]\ldots n-1], y[p[i]\ldots n-1]).$$

A direct computation using letter by letter comparisons leads to an execution time in $O(n^2)$ since the sum of lengths of the suffixes is quadratic.

We describe an algorithm that performs the computation in linear time. Indeed, the suffixes of y are not independent of each other. This dependence allows to reduce the computation time by the mean of a quite simple algorithm, based on the following property: let i, j, i' be positions on y for which $j = p[i]$ and $j - 1 = p[i']$, then $LCP[i'] - 1 \leq LCP[i]$.

Using this result, in order to compute $LCP[i]$, that is, the length of the longest common prefix between $y[p[i] \ldots n - 1]$ and $y[p[i - 1] \ldots n - 1]$ when $0 < i \leq n$, one can start the letter by letter comparison exactly at the position at which stops the previous computation, the one of $LCP[i']$. Knowing that, it is sufficient to proceed by considering the suffixes from the longest to the shortest, and not in the lexicographic order that seems more natural. This is what Algorithm Def-half-LCP (see Figure 15.39) realizes. It computes the values $LCP[i]$ for $0 \leq i \leq n$. Note that to determine the position i associated with position j, the algorithm uses the reverse of permutation p which is computed in the first step (Lines 1–2). This function is represented by the table called *Rank* since it indicates the rank of each suffix in the sorted list of suffixes of y.

```
DEF-HALF-LCP(y, n, p)
1   for i ← 0 to n − 1
2         do Rank[p[i]] ← i
3   ℓ ← 0
4   for j ← 0 to n − 1
5       do ℓ ← max{0, ℓ − 1}
6          i ← Rank[j]
7          if i ≠ 0
8             then j′ ← p[i − 1]
9                  while j + ℓ < n and j′ + ℓ < n
                          and y[j + ℓ] = y[j′ + ℓ]
10                      do ℓ ← ℓ + 1
11            else ℓ ← 0
12         LCP[i] ← ℓ
13  return LCP
```

FIGURE 15.39 Linear computation of the first LCP values.

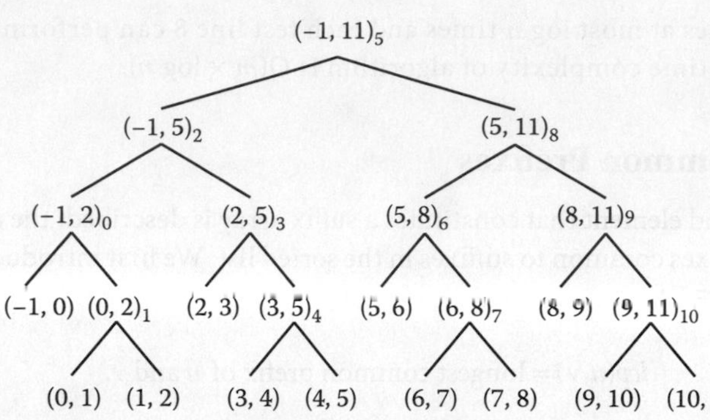

FIGURE 15.40 Pairs considered during the binary search in a string of length 11.

Applied to string y of length n and to the permutation p of it suffixes, Algorithm DEF-HALF-LCP computes the *LCP* array in time $O(n)$.

The number of pairs (ℓ, r) considered by algorithm BINARY-SEARCH(y, n, p, x, m) is bounded by $2n + 1$: there are $n + 1$ pairs of the form $(i, i + 1)$ that constitute the leaves of a binary tree; thus, there are at most n internal nodes. Figure 15.40 shows an example for a string of length 11. Small numbers close to pairs (ℓ, r) are $i = \lfloor(\ell + r)/2\rfloor$.

It is thus possible to extend array *LCP* as follows:

$$LCP[n+1+i] = lcp(p[\ell]\ldots n-1], y[p[r]\ldots n-1])$$
$$\text{where } i = \lfloor(\ell + r)/2\rfloor.$$

The length of the longest common prefix between $y[p[\ell] \ldots n - 1]$ and $y[p[r] \ldots n - 1]$ when $\ell < r - 1$ is equal to the minimum value among all $LCP[i]$ where $\ell < i \le r$. Then these values can be computed by a call to the algorithm DEF-LCP(y, 0, $n - 1$, *LCP*) (see Figure 15.41).

15.5.4 Substring Search with the Length of the Common Prefixes

Let ℓ, r, i be three integers, $0 \le \ell < i < r < n$. If $y[p[\ell] \ldots n - 1] < x < y[p[r] \ldots n - 1]$ let $lb = |lcp(x, y[p[\ell] \ldots n - 1])|$ and $le = |lcp(x, y[p[r] \ldots n - 1])|$ satisfying $lb \le le$. We then have:

$$|lcp(y[p[i]\ldots n-1], y[p[r]\ldots n-1])| < le \text{ implies } y[p[i]\ldots n-1] < x < y[p[r]\ldots n-1]$$

```
DEF-LCP(y, ℓ, r, LCP)
  ▷ ℓ < r
  if ℓ + 1 = r
     then return LCP[r]
     else i ← ⌊(ℓ + r)/2⌋
          LCP[n + 1 + i] ← min{DEF-LCP(y, ℓ, i, LCP),
                               DEF-LCP(y, i, r, LCP)}
          return LCP[n + 1 + i]
```

FIGURE 15.41 Linear computation of the last LCP values.

```
SEARCH(y, n, p, LCP, x, m)
(ℓ, lb) ← (−1, 0)
(r, le) ← (n, 0)
while ℓ + 1 < r
      do i ← ⌊(ℓ + r)/2⌋
         if ℓ + 1 = i
           then g ← LCP[i]
           else g ← LCP[n + 1 + ⌊(ℓ + i)/2⌋]
         if i + 1 = r
           then h ← LCP[r]
           else h ← LCP[n + 1 + ⌊(i + r)/2⌋]
         if lb ≤ h and h < le
           then (ℓ, lb) ← (i, h)
         elseif lb ≤ le and le < h
           then (r, le) ← (i, h)
         elseif le ≤ g and g < lb
           then (r, le) ← (i, g)
         elseif le ≤ lb and lb < g
           then (ℓ, lb) ← (i, g)
         else   ℓ ← max{lb, le}
                ℓ ← ℓ + |lcp(x[ℓ . . m − 1], y[p[i] + [ℓ . . n − 1])|
                if ℓ = m
                  then return i
                elseif x < y[p[i] . . n − 1]
                  then (r, le) ← (i, ℓ)
                else    (ℓ, lb) ← (i, ℓ)
return (ℓ, r)
```

FIGURE 15.42 Search for x in the suffix array of y using the length of the common prefixes.

and

$$|lcp(y[p[i]\ldots n-1], y[p[r]\ldots n-1])| > le \text{ implies } y[p[\ell]\ldots n-1] < x < y[p[i]\ldots n-1].$$

The search can then now be done in $O(m + \log n)$ with the algorithm SEARCH(y, n, p, LCP, x, m) (see Figure 15.42).

15.6 Alignment

Alignments are used to compare strings. They are widely used in computational molecular biology. They constitute a mean to visualize resemblance between strings. They are based on notions of distance or similarity. Their computation is usually done by dynamic programming. A typical example of this method is the computation of the longest common subsequence of two strings. The reduction of the memory space presented on it can be applied to similar problems. Three different kinds of alignment of two strings x and y are considered: global alignment (that consider the whole strings x and y), local alignment (that enable to find the segment of x that is closer to a segment of y), and the longest common subsequence of x and y.

An **alignment** of two strings x and y of length m and n, respectively, consists in aligning their symbols on vertical lines. Formally, an alignment of two strings $x, y \in \Sigma^*$ is a word w on the alphabet $(\Sigma \cup \{\varepsilon\}) \times (\Sigma \cup \{\varepsilon\}) \setminus (\{(\varepsilon, \varepsilon)\}$ (ε is the empty word) whose projection on the first component is x and whose projection of the second component is y.

Thus, an alignment $w = (\bar{x}_0, \bar{y}_0)(\bar{x}_1, \bar{y}_1) \cdots (\bar{x}_{p-1}, \bar{y}_{p-1})$ of length p is such that $x = \bar{x}_0\bar{x}_1 \cdots \bar{x}_{p-1}$ and $y = \bar{y}_0\bar{y}_1 \cdots \bar{y}_{p-1}$ with $\bar{x}_i \in \Sigma \cup \{\varepsilon\}$ and $\bar{y}_i \in \Sigma \cup \{\varepsilon\}$ for $0 \le i \le p - 1$. The alignment is represented as follows:

$\bar{x}_0$	$\bar{x}_1$	$\cdots$	$\bar{x}_{p-1}$
$\bar{y}_0$	$\bar{y}_1$	$\cdots$	$\bar{y}_{p-1}$

with the symbol—instead of the symbol ε.

Example 15.12

A	**C**	**G**	**–**	**–**	**A**
A	**T**	**G**	**C**	**T**	**A**

is an alignment of **ACGA** and **ATGCTA**.

15.6.1 Global Alignment

A global alignment of two strings x and y can be obtained by computing the distance between x and y. The notion of distance between two strings is widely used to compare files. The **diff** command of UNIX operating system implements an algorithm based on this notion, in which lines of the files are treated as symbols. The output of a comparison made by **diff** gives the minimum number of operations (substitute a symbol, insert a symbol, or delete a symbol) to transform one file into the other.

Let us define the edit distance between two strings x and y as follows: it is the minimum number of elementary edit operations that enable to transform x into y. The elementary edit operations are:

- The substitution of a character of x at a given position by a character of y.
- The deletion of a character of x at a given position.
- The insertion of a character of y in x at a given position.

A cost is associated to each elementary edit operation. For $a, b \in \Sigma$:

- *Sub*(a, b) denotes the cost of the substitution of the character a by the character b.
- *Del*(a) denotes the cost of the deletion of the character a.
- *Ins*(a) denotes the cost of the insertion of the character a.

This means that the costs of the edit operations are independent of the positions where the operations occur. The edit distance of two strings x and y can now be defined by

$$edit(x, y) = \min\{\text{cost of } \gamma \mid \gamma \in \Gamma_{x,y}\}$$

where $\Gamma_{x,y}$ is the set of all the sequences of edit operations that transform x into y, and the cost of an element $\gamma \in \Gamma_{x,y}$ is the sum of the costs of its elementary edit operations.

In order to compute $edit(x, y)$ for two strings x and y of length m and n, respectively, a two-dimensional table T of $m + 1$ rows and $n + 1$ columns is used such that

$$T[i, j] = edit(x[i], y[j])$$

for $i = 0, \ldots, m - 1$ and $j = 0, \ldots, n - 1$. It follows $edit(x, y) = T[m - 1, n - 1]$.

The values of the table T can be computed by the following recurrence formula:

$$T[-1,-1]=0,$$

$$T[i,-1]=T[i-1,-1]+Del(x[i]),$$

$$T[-1,j]=T[-1,j-1]+Ins(y[j]),$$

$$T[i,j]=\min\begin{cases}T[i-1,j-1]+Sub(x[i],y[j]),\\ T[i-1,j]+Del(x[i]),\\ T[i,j-1]+Ins(y[j]),\end{cases}$$

for $i = 0, 1,\ldots, m - 1$ and $j = 0, 1,\ldots, n - 1$.

The value at position (i, j) in the table T only depends on the values at the three neighbor positions $(i - 1, j - 1)$, $(i - 1, j)$, and $(i, j - 1)$.

The direct application of the recurrence formula mentioned earlier gives an exponential time algorithm to compute $T[m - 1, n - 1]$. However, the whole table T can be computed in quadratic time, technique known as "dynamic programming." This is a general technique that is used to solve the different kinds of alignments.

The computation of the table T proceeds in two steps. First, it initializes the first column and first row of T; this is done by a call to a generic function MARGIN which is a parameter of the algorithm and that depends on the kind of alignment that is considered. Second, it computes the remaining values of T; this is done by a call to a generic function FORMULA which is a parameter of the algorithm and that depends on the kind of alignment that is considered. Computing a global alignment of x and y can be done by a call to GENERIC-DP with the following parameters (x, m, y, n, MARGIN-GLOBAL, FORMULA-GLOBAL) (see Figures 15.43 through 15.45). The computation of all the values of the table T can thus be done in quadratic space and time: $O(m \times n)$.

```
GENERIC-DP(x, m, y, n, MARGIN, FORMULA)
MARGIN(T, x, m, y, n)
for j ← 0 to n − 1
    do for i ← 0 to m − 1
        do T[i, j] ← FORMULA(T, x, i, y, j)
return T
```

FIGURE 15.43 Computation of the table T by dynamic programming.

```
MARGIN-GLOBAL(T, x, m, y, n)
T[−1, −1] ← 0
for i ← 0 to m − 1
    do T[i, −1] ← T[i − 1, −1] + Del(x[i])
for j ← 0 to n − 1
    do T[−1, j] ← T[−1, j − 1] + Ins(y[j])
```

FIGURE 15.44 Margin initialization for the computation of a global alignment.

```
FORMULA-GLOBAL(T, x, i, y, j)
                 ⎧ T[i − 1, j − 1] + Sub(x[i], y[j])
1 return min     ⎨ T[i − 1, j] + Del(x[i])
                 ⎩ T[i, j − 1] + Ins(y[j])
```

FIGURE 15.45 Computation of $T[i, j]$ for a global alignment.

```
One-Alignment(T, x, i, y, j)
if i = −1 and j = −1
    then return (ε, ε)
    else  if i = −1
              then return One-alignment(T, x, −1, y, j − 1) · (ε, y[j])
          elseif j = −1
              then return One-alignment(T, x, i − 1, y, −1) · (x[i], ε)
          else    if T[i, j] = T[i − 1, j − 1] + Sub(x[i], y[j])
                      then return One-alignment(T, x, i − 1, y, j − 1).
                                  (x[i], y[j])
                  elseif T[i, j] = T[i − 1, j] + Del(x[i])
                      then return One-alignment(T, x, i − 1, y, j).
                                  (x[i], ε)
                  else    return One-alignment(T, x, i, y, j − 1).
                                  (ε, y[j])
```

FIGURE 15.46 Recovering an optimal alignment.

An optimal alignment (with minimal cost) can then be produced by a call to the function

$$\text{One-alignment}(T, x, m-1, y, n-1)$$

(see Figure 15.46). It consists in tracing back the computation of the values of the table T from position $[m-1, n-1]$ to position $[-1, -1]$. At each cell $[i, j]$, the algorithm determines among the three values $T[i-1, j-1] + Sub(x[i], y[j])$, $T[i-1, j] + Del(x[i])$, and $T[i, j-1] + Ins(y[j])$, which has been used to produce the value of $T[i, j]$. If $T[i-1, j-1] + Sub(x[i], y[j])$ has been used, it adds $(x[i], y[j])$ to the optimal alignment and proceeds recursively with the cell at $[i-1, j-1]$. If $T[i-1, j] + Del(x[i])$ has been used, it adds $(x[i], -)$ to the optimal alignment and proceeds recursively with cell at $[i-1, j]$. If $T[i, j-1] + Ins(y[j])$ has been used it adds $(-, y[j])$ to the optimal alignment and proceeds recursively with cell at $[i, j-1]$. Recovering all the optimal alignments can be done by a similar technique.

Example 15.13

T	*j*	−1	0	1	2	3	4	5
i		*y*[*j*]	**A**	**T**	**G**	**C**	**T**	**A**
−1	*x*[*i*]	0	1	2	3	4	5	6
0	**A**	1	0	1	2	3	4	5
1	**C**	2	1	1	2	2	3	4
2	**G**	3	2	2	1	2	3	4
3	**A**	4	3	3	2	2	3	3

The values of the above table have been obtained with the following unitary costs: $Sub(a, b) = 1$ if $a \neq b$ and $Sub(a, a) = 0$, $Del(a) = Ins(a) = 1$ for $a, b \in \Sigma$.

15.6.2 Local Alignment

A local alignment of two strings x and y consists in finding the segment of x that is closer to a segment of y. The notion of distance used to compute global alignments cannot be used in that case since the segments of x closer to segments of y would only be the empty segment or individual characters. This is why a notion of similarity is used based on a scoring scheme for edit operations.

A score (instead of a cost) is associated to each elementary edit operation. For $a, b \in \Sigma$:

- $Sub_S(a, b)$ denotes the score of substituting the character b for the character a.
- $Del_S(a)$ denotes the score of deleting the character a.
- $Ins_S(a)$ denotes the score of inserting the character a.

This means that the scores of the edit operations are independent of the positions where the operations occur. For two characters a and b, a positive value of $Sub_S(a, b)$ means that the two characters are close to each other, and a negative value of $Sub_S(a, b)$ means that the two characters are far apart.

The edit score of two strings x and y can now be defined by

$$sco(x, y) = \max\{\text{score of } \gamma \mid \gamma \in \Gamma_{x,y}\}$$

where $\Gamma_{x,y}$ is the set of all the sequences of edit operations that transform x into y and the score of an element $\sigma \in \Gamma_{x,y}$ is the sum of the scores of its elementary edit operations.

In order to compute $sco(x, y)$ for two strings x and y of length m and n, respectively, a two-dimensional table T of $m + 1$ rows and $n + 1$ columns is used such that

$$T[i, j] = sco(x[i], y[j])$$

for $i = 0,\ldots, m - 1$ and $j = 0,\ldots, n - 1$. Therefore $sco(x, y) = T[m - 1, n - 1]$.

The values of the table T can be computed by the following recurrence formula:

$$T[-1,-1] = 0,$$
$$T[i,-1] = 0,$$
$$T[-1, j] = 0,$$
$$T[i, j] = \max\begin{cases} T[i-1, j-1] + Sub_S(x[i], y[j]), \\ T[i-1, j] + Del_S(x[i]), \\ T[i, j-1] + Ins_S(y[j]), \\ 0, \end{cases}$$

for $i = 0, 1,\ldots, m - 1$ and $j = 0, 1,\ldots, n - 1$.

Computing the values of T for a local alignment of x and y can be done by a call to Generic-DP with the following parameters (x, m, y, n, Margin-local, Formula-local) in $O(mn)$ time and space complexity (see Figures 15.43, 15.47, and 15.48). Recovering a local alignment can be done in a way similar to what is done in the case of a global alignment (see Figure 15.46) but the trace back procedure must start at a position of a maximal value in T rather than at position $[m - 1, n - 1]$.

```
MARGIN-LOCAL(T, x, m, y, n)
T[−1, −1] ← 0
for i ← 0 to m − 1
    do T[i, −1] ← 0
for j ← 0 to n − 1
    do T[−1, j] ← 0
```

FIGURE 15.47 Margin initialization for computing a local alignment.

FORMULA-LOCAL(T, x, i, y, j)

return $\max \begin{cases} T[i-1, j-1] + Sub_S(x[i], y[j]) \\ T[i-1, j] + Del_S(x[i]) \\ T[i, j-1] + Ins_S(y[j]) \\ 0 \end{cases}$

FIGURE 15.48 Recurrence formula for computing a local alignment.

Example 15.14

Computation of an optimal local alignment of
x = **EAWACQGKL** and y = **ERDAWCQPGKWY** with scores:

$Sub_S(a, a) = 1$, $Sub_S(a, b) = -3$ and $Del_S(a) = Ins_S(a) = -1$ for $a, b \in \Sigma$, $a \neq b$.

T	*j*	−1	0	1	2	3	4	5	6	7	8	9	10	11
i		*y*[*j*]	**E**	**R**	**D**	**A**	**W**	**C**	**Q**	**P**	**G**	**K**	**W**	**Y**
−1	*x*[*i*]	0	0	0	0	0	0	0	0	0	0	0	0	0
0	**E**	0	1	0	0	0	0	0	0	0	0	0	0	0
1	**A**	0	0	0	0	1	0	0	0	0	0	0	0	0
2	**W**	0	0	0	0	0	2	1	0	0	0	0	1	0
3	**A**	0	0	0	0	1	1	0	0	0	0	0	0	0
4	**C**	0	0	0	0	0	0	2	1	0	0	0	0	0
5	**Q**	0	0	0	0	0	0	1	3	2	1	0	0	0
6	**G**	0	0	0	0	0	0	0	2	1	3	2	1	0
7	**K**	0	0	0	0	0	0	0	1	0	2	4	3	2
8	**L**	0	0	0	0	0	0	0	0	0	1	3	2	1

The corresponding optimal local alignment is

A	**W**	**A**	**C**	**Q**	—	**G**	**K**
A	**W**	—	**C**	**Q**	**P**	**G**	**K**

15.6.3 Longest Common Subsequence of Two Strings

A subsequence of a word x is obtained by deleting zero or more characters from x. More formally $w[0 \ldots i-1]$ is a subsequence of $x[0 \ldots m-1]$ if there exists an increasing sequence of integers $(k_j \mid j = 0, \ldots, i-1)$ such that for $0 \leq j \leq i-1$, $w[j] = x[k_j]$. A word is an lcs(x, y) if it is a **longest common subsequence**

```
FORMULA-LCS(T, x, i, y, j)
if x[i] = y[j]
    then return T[i − 1, j − 1] + 1
    else return max{T[i − 1, j], T[i, j − 1]}
```

FIGURE 15.49 Recurrence formula for computing an *lcs*.

of the two words *x* and *y*. Note that two strings can have several longest common subsequences. Their common length is denoted by llcs(*x*, *y*).

A brute-force method to compute an lcs(*x*, *y*) would consist in computing all the subsequences of *x*, checking if they are subsequences of *y*, and keeping the longest one. The word *x* of length *m* has 2^m subsequences, and so this method could take $O(2^m)$ time, which is impractical even for fairly small values of *m*.

However, llcs(*x*, *y*) can be computed with a two-dimensional table *T* by the following recurrence formula:

$$T[-1,-1]=0,$$

$$T[i,-1]=0,$$

$$T[-1,j]=0,$$

$$T[i,j]=\begin{cases}T[i-1,j-1]+1 & \textit{if } \mathrm{x[i]=y[j]},\\ \max(T[i-1,j],T[i,j-1]) & \textit{otherwise},\end{cases}$$

for $i = 0, 1,\ldots, m - 1$ and $j = 0, 1,\ldots, n - 1$. Then $T[i, j] = \mathrm{llcs}(x[0 \ldots i], y[0 \ldots j])$ and $\mathrm{llcs}(x, y) = T[m - 1, n - 1]$.

Computing $T[m - 1, n - 1]$ can be done by a call to GENERIC-DP with the following parameters (*x*, *m*, *y*, *n*, MARGIN-LOCAL, FORMULA-LCS) in *O*(*mn*) time and space complexity (see Figures 15.43, 15.47, and 15.49).

It is possible afterward to trace back a path from position $[m - 1, n - 1]$ in order to exhibit an lcs(*x*, *y*) in a similar way as for producing a global alignment (see Figure 15.46).

Example 15.15

The value $T[4, 8] = 4$ is llcs(*x*, *y*) for *x* = **AGCGA** and *y* = **CAGATAGAG**. String **AGGA** is an lcs of *x* and *y*.

T	*j*	−1	0	1	2	3	4	5	6	7	8
i		*y*[*j*]	**C**	**A**	**G**	**A**	**T**	**A**	**G**	**A**	**G**
−1	*x*[*i*]	0	0	0	0	0	0	0	0	0	0
0	**A**	0	0	1	1	1	1	1	1	1	1
1	**G**	0	0	1	2	2	2	2	2	2	2
2	**C**	0	1	1	2	2	2	2	2	2	2
3	**G**	0	1	1	2	2	2	2	3	3	3
4	**A**	0	1	2	2	3	3	3	3	4	4

15.6.4 Reducing the Space: Hirschberg Algorithm

If only the length of an lcs(x, y) is required, it is easy to see that only one row (or one column) of the table T needs to be stored during the computation. The space complexity becomes $O(\min(m, n))$ as can be checked on the algorithm of Figure 15.50. Indeed, the Hirschberg algorithm computes an lcs(x, y) in linear space and not only the value llcs(x, y). The computation uses the algorithm of Figure 15.50.

Let us define

$$T^*[i,n]=T^*[m,j]=0,\quad \text{for } 0\le i\le m \quad\text{and}\quad 0\le j\le n$$

$$T^*[m-i,n-j]=\text{llcs}((x[i\ldots m-1])^R,(y[j\ldots n-1])^R)$$

$$\text{for } 0\le i\le m-1 \quad\text{and}\quad 0\le j\le n-1$$

and

$$M(i)=\max_{0\le j<n}\{T[i,j]+T^*[m-i,n-j]\}.$$

where the word w^R is the reverse (or mirror image) of the word w. The following property is the key observation to compute an lcs(x, y) in linear space:

$$M(i)=T[m-1,n-1],\quad \text{for } 0\le i<m.$$

In the algorithm shown in Figure 15.51, the integer j is chosen as $n/2$. After $T[i, j-1]$ and $T^*[m-i, n-j]$ $(0 \le i < m)$ are computed, the algorithm finds an integer k such that $T[i, k] + T^*[m-i, n-k] = T[m-1, n-1]$. Then, recursively, it computes an lcs($x[0 \ldots k-1]$, $y[0 \ldots j-1]$) and an lcs($x[k \ldots m-1]$, $y[j \ldots n-1]$), and concatenates them to get an lcs(x, y).

The running time of the Hirschberg algorithm is still $O(mn)$, but the amount of space required for the computation becomes $O(\min(m, n))$ instead of being quadratic when computed by dynamic programming.

```
LLCS(x, m, y, n)
for i ← −1 to m − 1
    do C[i] ← 0
for j ← 0 to n − 1
    do last ← 0
       for i ← −1 to m − 1
           do if last > C[i]
                 then C[i] ← last
              elseif last < C[i]
                 then last ← C[i]
              elseif x[i] = y[j]
                 then C[i] ← C[i] + 1
                      last ← last + 1
return C
```

FIGURE 15.50 $O(m)$-space algorithm to compute llcs(x, y).

```
HIRSCHBERG(x, m, y, n)
if m = 0
   then return ε
   else  if m = 1
            then if x[0] ∈ y
                    then return x[0]
                    else  return ε
            else  j ← ⌊n/2⌋
                  C ← LLCS(x, m, y[0 ... j − 1], j)
                  C* ← LLCS(x^R, m, y[j ... n − 1]^R, n − j)
                  k ← m − 1
                  M ← C[m − 1] + C*[m − 1]
                  for j ← −1 to m − 2
                      do if C[j] + C*[j] > M
                            then M ← C[j] + C*[j]
                                 k ← j
                  return HIRSCHBERG(x[0 ... k − 1], k, y[0 . . j − 1], j).
                         HIRSCHBERG(x[k ... m − 1], m − k,
                         y[j ... n − 1], n − j)
```

FIGURE 15.51 $O(\min(m, n))$-space computation of lcs(x, y).

15.7 Approximate String Matching

Approximate string matching is the problem of finding all approximate occurrences of a pattern x of length m in a text y of length n. Approximate occurrences of x are segments of y that are close to x according to a specific distance: the distance between segments and x must be not greater than a given integer k. Two distances are considered in this section: the **Hamming distance** and the **Levenshtein distance**.

With the Hamming distance, the problem is also known as approximate string matching with k mismatches. With the Levenshtein distance (or edit distance), the problem is known as approximate string matching with k differences.

The Hamming distance between two words w_1 and w_2 of the same length is the number of positions with different characters. The Levenshtein distance between two words w_1 and w_2 (not necessarily of the same length) is the minimal number of differences between the two words. A difference is one of the following operations:

- A substitution: a character of w_1 corresponds to a different character in w_2.
- An insertion: a character of w_1 corresponds to no character in w_2.
- A deletion: a character of w_2 corresponds to no character in w_1.

The *Shift-Or algorithm* of the next section is a method that is both very fast in practice and very easy to implement. It solves the Hamming distance and the Levenshtein distance problems. The method for the exact string-matching problem is initially described and then it is presented how to handle the cases of k mismatches and of k differences. The method is flexible enough to be adapted to a wide range of similar approximate matching problems.

15.7.1 Shift-Or Algorithm

An algorithm to solve the exact string-matching problem is first presented that uses a technique different from those developed in Section 15.2, but which extends readily to the approximate string-matching problem.

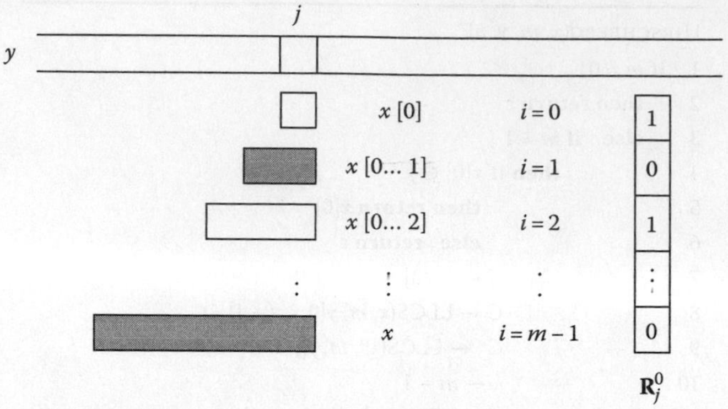

FIGURE 15.52 Meaning of vector $\mathbf{R}_j^0$.

Let $\mathbf{R}^0$ be a bit array of size m. Vector $\mathbf{R}_j^0$ is the value of the entire array $\mathbf{R}^0$ after text character $y[j]$ has been processed (see Figure 15.52). It contains information about all matches of prefixes of x that end at position j in the text. It is defined, for $0 \le i \le m-1$, by

$$\mathbf{R}_j^0[i] = \begin{cases} 0 & \text{if } x[0 \ldots i] = y[j-i \ldots j], \\ 1 & \text{otherwise.} \end{cases}$$

Therefore, $\mathbf{R}_j^0[m-1] = 0$ is equivalent to saying that an (exact) occurrence of the pattern x ends at position j in y.

The vector $\mathbf{R}_j^0$ can be computed after $\mathbf{R}_{j-1}^0$ by the following recurrence relation:

$$\mathbf{R}_j^0[i] = \begin{cases} 0 & \text{if } \mathbf{R}_{j-1}^0[i-1] = 0 \text{ and } x[i] = y[j], \\ 1 & \text{otherwise,} \end{cases}$$

and

$$\mathbf{R}_j^0[0] = \begin{cases} 0 & \text{if } x[0] = y[j], \\ 1 & \text{otherwise.} \end{cases}$$

The transition from $\mathbf{R}_{j-1}^0$ to $\mathbf{R}_j^0$ can be computed very fast as follows. For each $a \in \Sigma$, let S_a be a bit array of size m defined, for $0 \le i \le m-1$, by

$$S_a[i] = 0 \quad \text{iff} \quad x[i] = a.$$

The array S_a denotes the positions of the character a in the pattern x. All arrays S_a are preprocessed before the search starts. And the computation of $\mathbf{R}_j^0$ reduces to two operations, SHIFT and OR:

$$\mathbf{R}_j^0 = \text{SHIFT}(\mathbf{R}_{j-1}^0) \quad \text{OR} \quad S_{y[j]}.$$

Example 1.16

String x = **GATAA** occurs at position 2 in y = **CAGATAAGAGAA**.

S_A	S_C	S_G	S_T
1	1	0	1
0	1	1	1
1	1	1	0
0	1	1	1
0	1	1	1

	C	**A**	**G**	**A**	**T**	**A**	**A**	**G**	**A**	**G**	**A**	**A**
G	1	1	0	1	1	1	1	0	1	0	1	1
A	1	1	1	0	1	1	1	1	0	1	0	1
T	1	1	1	1	0	1	1	1	1	1	1	1
A	1	1	1	1	1	0	1	1	1	1	1	1
A	1	1	1	1	1	1	0	1	1	1	1	1

15.7.2 String Matching with *k* Mismatches

The Shift-Or algorithm easily adapts to support approximate string matching with k mismatches. To simplify the description, the case where at most one substitution is allowed is presented first.

Arrays $\mathbf{R}^0$ and S are used as before, and an additional bit array $\mathbf{R}^1$ of size m. Vector $\mathbf{R}^1_{j-1}$ indicates all matches with at most one substitution up to the text character $y[j-1]$. The recurrence on which the computation is based splits into two cases:

- There is an exact match on the first i characters of x up to $y[j-1]$ (i.e., $\mathbf{R}^0_{j-1}[i-1]=0$). Then, substituting $x[i]$ to $y[j]$ creates a match with one substitution (see Figure 15.53). Thus,

$$\mathbf{R}^1_j[i]=\mathbf{R}^0_{j-1}[i-1].$$

- There is a match with one substitution on the first i characters of x up to $y[j-1]$ and $x[i]=y[j]$. Then, there is a match with one substitution of the first $i+1$ characters of x up to $y[j]$ (see Figure 15.54). Thus,

$$\mathbf{R}^1_j[i]=\begin{cases}\mathbf{R}^1_{j-1}[i-1] & \text{if } x[i]=y[j],\\ 1 & \text{otherwise.}\end{cases}$$

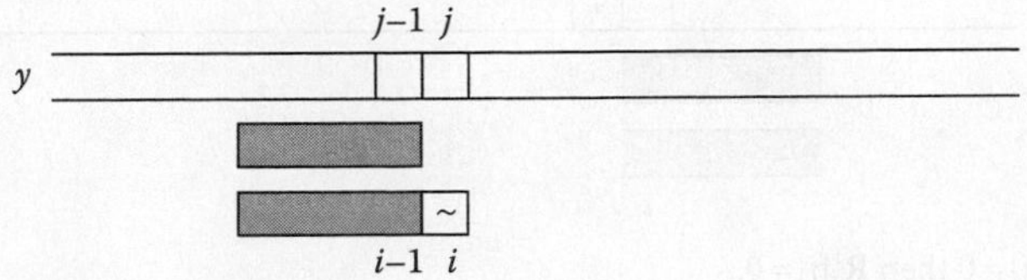

FIGURE 15.53 If $\mathbf{R}^0_{j-1}[i-1]=0$ then $\mathbf{R}^1_j[i]=0$.

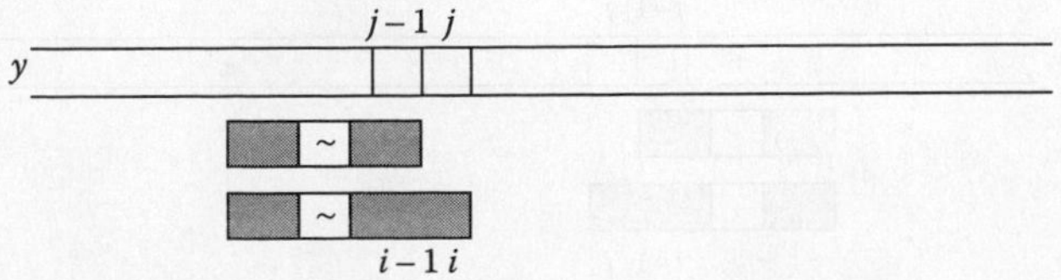

FIGURE 15.54 $\mathbf{R}^1_j[i]=\mathbf{R}^1_{j-1}[i-1]$ if $x[i]=y[j]$.

This implies that $\mathbf{R}_j^1$ can be updated from $\mathbf{R}_{j-1}^1$ by the relation:

$$\mathbf{R}_j^1 = (\text{SHIFT}(\mathbf{R}_{j-1}^1) \quad \text{OR} \quad S_{y[j]}) \quad \text{AND} \quad \text{SHIFT}(R_{j-1}^0).$$

Example 1.17

String x = **GATAA** occurs at positions 2 and 7 in y = **CAGATAAGAGAA** with no more than one mismatch.

	C	A	G	A	T	A	A	G	A	G	A	A
G	0	0	0	0	0	0	0	0	0	0	0	0
A	1	0	1	0	1	0	0	1	0	1	0	0
T	1	1	1	1	0	1	1	1	1	0	1	0
A	1	1	1	1	1	0	1	1	1	1	0	1
A	1	1	1	1	1	1	0	1	1	1	1	0

15.7.3 String Matching with *k* Differences

In this section, it is shown how to adapt the Shift-Or algorithm to the case of only one insertion, and then dually to the case of only one deletion. The method is based on the following elements.

One insertion is allowed: here, vector $\mathbf{R}_{j-1}^1$ indicates all matches with at most one insertion up to text character $y[j-1]$. Entry $\mathbf{R}_{j-1}^1[i-1] = 0$ if the first i characters of x ($x[0 \ldots i-1]$) match i symbols of the last $i+1$ text characters up to $y[j-1]$. Array $\mathbf{R}^0$ is maintained as before, and array $\mathbf{R}^1$ is maintained as follows. Two cases arise:

- There is an exact match on the first $i+1$ characters of x ($x[0 \ldots i]$) up to $y[j-1]$. Then inserting $y[j]$ creates a match with one insertion up to $y[j]$ (see Figure 15.55). Thus,

$$\mathbf{R}_j^1[i] = \mathbf{R}_{j-1}^0[i].$$

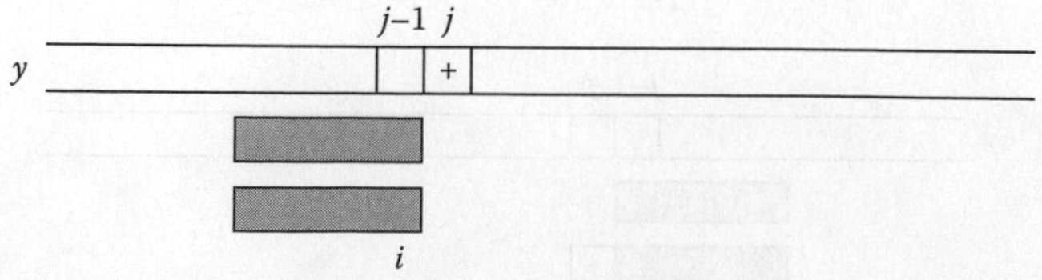

FIGURE 15.55 If $\mathbf{R}_{j-1}^0[i] = 0$ then $\mathbf{R}_j^1[i] = 0$.

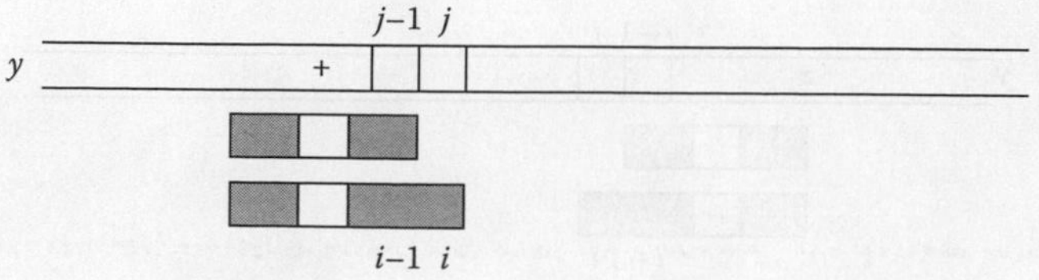

FIGURE 15.56 $\mathbf{R}_j^1[i] = \mathbf{R}_{j-1}^1[i-1]$ if $x[i] = y[j]$.

- There is a match with one insertion on the i first characters of x up to $y[j-1]$. Then if $x[i] = y[j]$, there is a match with one insertion on the first $i+1$ characters of x up to $y[j]$ (see Figure 15.56). Thus,

$$\mathbf{R}_j^1[i] = \begin{cases} \mathbf{R}_{j-1}^1[i-1] & \text{if } x[i] = y[j], \\ 1 & \text{otherwise.} \end{cases}$$

This shows that $\mathbf{R}_j^1$ can be updated from $\mathbf{R}_{j-1}^1$ with the formula

$$\mathbf{R}_j^1 = (\text{SHIFT}(\mathbf{R}_{j-1}^1) \quad \text{OR} \quad S_{y[j]}) \quad \text{AND} \quad \mathbf{R}_{j-1}^0.$$

Example 1.18

GATAAG is an occurrence of x = **GATAA** with exactly one insertion in y = **CAGATAAGAGAA**

	C	A	G	A	T	A	A	G	A	G	A	A
G	1	1	1	0	1	1	1	1	0	1	0	1
A	1	1	1	1	0	1	1	1	1	0	1	0
T	1	1	1	1	1	0	1	1	1	1	1	1
A	1	1	1	1	1	1	0	1	1	1	1	1
A	1	1	1	1	1	1	1	0	1	1	1	1

One deletion is allowed: assume here that $\mathbf{R}_{j-1}^1$ indicates all possible matches with at most one deletion up to $y[j-1]$. As in the previous solution, two cases arise:

- There is an exact match on the first $i+1$ characters of x ($x[0 \ldots i]$) up to $y[j]$ (i.e., $\mathbf{R}_j^0[i] = 0$). Then, deleting $x[i]$ creates a match with one deletion (see Figure 15.57). Thus,

$$\mathbf{R}_j^1[i] = \mathbf{R}_j^0[i].$$

- There is a match with one deletion on the first i characters of x up to $y[j-1]$ and $x[i] = y[j]$. Then, there is a match with one deletion on the first $i+1$ characters of x up to $y[j]$ (see Figure 15.58). Thus,

$$\mathbf{R}_j^1[i] = \begin{cases} \mathbf{R}_{j-1}^1[i-1] & \text{if } x[i] = y[j], \\ 1 & \text{otherwise.} \end{cases}$$

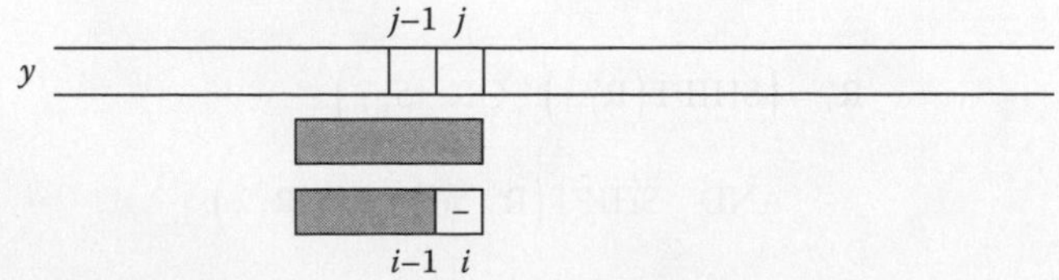

FIGURE 15.57 If $\mathbf{R}_j^0[i] = 0$ then $\mathbf{R}_j^1[i] = 0$.

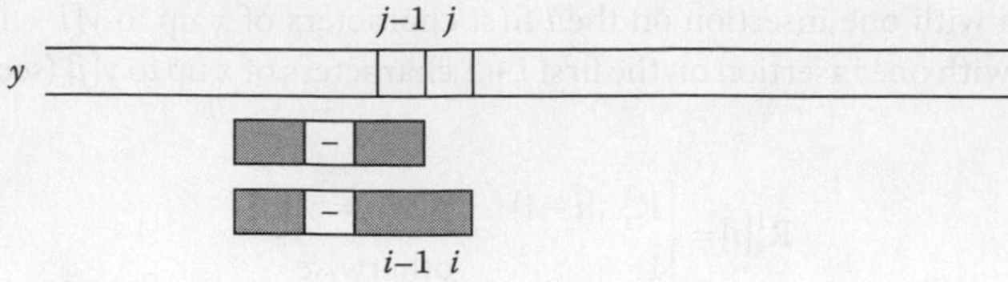

FIGURE 15.58 $\mathbf{R}^1_j[i] = \mathbf{R}^1_{j-1}[i-1]$ if $x[i] = y[j]$.

The discussion provides the following formula used to update $\mathbf{R}^1_j$ from $\mathbf{R}^1_{j-1}$:

$$\mathbf{R}^1_j = (\text{SHIFT}(\mathbf{R}^1_{j-1}) \ \text{ OR } \ S_{y[j]}) \ \text{ AND } \ \text{SHIFT}(\mathbf{R}^0_j).$$

Example 1.19

GATA and ATAA are two occurrences with one deletion of x = GATAA in y = CAGATAAGAGAA

	C	A	G	A	T	A	A	G	A	G	A	A
G	0	0	0	0	0	0	0	0	0	0	0	0
A	1	0	0	0	1	0	0	0	0	0	0	0
T	1	1	1	0	0	1	1	1	0	1	0	1
A	1	1	1	1	0	0	1	1	1	1	1	0
A	1	1	1	1	1	0	0	1	1	1	1	1

15.7.4 Wu–Manber Algorithm

In this section, a general solution for the approximate string-matching problem with at most k differences of the types: insertion, deletion, and substitution is presented. It is an extension of the problems presented above. The following algorithm maintains $k + 1$ bit arrays $\mathbf{R}^0, \mathbf{R}^1, \ldots, \mathbf{R}^k$ that are described now. The vector $\mathbf{R}^0$ is maintained similarly as in the exact matching case (Section 15.7.1). The other vectors are computed with the formula ($1 \le \ell \le k$)

$$\begin{aligned}\mathbf{R}^\ell_j = &\left(\text{SHIFT}\left(\mathbf{R}^\ell_{j-1}\right) \ \text{ OR } \ S_{y[j]}\right)\\ &\text{AND } \ \text{SHIFT}\left(\mathbf{R}^{\ell-1}_j\right)\\ &\text{AND } \ \text{SHIFT}\left(\mathbf{R}^{\ell-1}_{j-1}\right)\\ &\text{AND } \ \mathbf{R}^{\ell-1}_{j-1}\end{aligned}$$

which can be rewritten into

$$\begin{aligned}\mathbf{R}^\ell_j = &\left(\text{SHIFT}\left(\mathbf{R}^\ell_{j-1}\right) \ \text{ OR } \ S_{y[j]}\right)\\ &\text{AND } \ \text{SHIFT}\left(\mathbf{R}^{\ell-1}_j \ \text{ AND } \ \mathbf{R}^{\ell-1}_{j-1}\right)\\ &\text{AND } \ \mathbf{R}^{\ell-1}_{j-1}.\end{aligned}$$

```
WM(x, m, y, n, k)
for each character a ∈ Σ
    do S_a ← 1^m
for i ← 0 to m − 1
    do S_x[i][i] ← 0
R^0 ← 1^m
for ℓ ← 1 to k
    do R^ℓ ← SHIFT(R^(ℓ−1))
for j ← 0 to n − 1
    do T ← R^0
       R^0 ← SHIFT(R^0)  OR  S_y[j]
       for ℓ ← 1 to k
           do T′ ← R^ℓ
              R^ℓ ← (SHIFT(R^ℓ)  OR  S_y[j])  AND
                    (SHIFT(T  AND  R^(ℓ−1)))  AND  T
              T ← T′
       if R^k[m − 1] = 0
          then OUTPUT(j)
```

FIGURE 15.59 Wu–Manber approximate string-matching algorithm.

Example 15.20

x = **GATAA** and y = **CAGATAAGAGAA** and $k = 1$. The output 5, 6, 7, and 11 corresponds to the segments **GATA**, **GATAA**, **GATAAG**, and **GAGAA** which approximate the pattern **GATAA** with no more than one difference.

	C	A	G	A	T	A	A	G	A	G	A	A
G	0	0	0	0	0	0	0	0	0	0	0	0
A	1	0	0	0	0	0	0	0	0	0	0	0
T	1	1	1	0	0	0	1	1	0	0	0	0
A	1	1	1	1	0	0	0	1	1	1	0	0
A	1	1	1	1	1	0	0	0	1	1	1	0

The method, called the Wu–Manber algorithm, is implemented in Figure 15.59. It assumes that the length of the pattern is no more than the size of the memory word of the machine, which is often the case in applications.

The preprocessing phase of the algorithm takes $O(\sigma m + km)$ memory space, and runs in time $O(\sigma m + k)$. The time complexity of its searching phase is $O(kn)$.

15.8 Text Compression

This section is interested in algorithms that compress texts. Compression serves both to save storage space and to save transmission time. Assume that the uncompressed text is stored in a file. The aim of compression algorithms is to produce another file containing the compressed version of the same text. Methods in this section work with no loss of information, so that decompressing the compressed text restores exactly the original text.

Two main strategies to design the algorithms are applied. The first strategy is a statistical method that takes into account the frequencies of symbols to build a uniquely decipherable code optimal with respect to the compression. The code contains new codewords for the symbols occurring in the text. In this method, fixed-length blocks of bits are encoded by different codewords. *A contrario* the second strategy encodes variable-length segments of the text. To put it simply, the algorithm, while scanning the text, replaces some already read segments just by a pointer to their first occurrences.

Text compression software often use a mixture of several methods. An example of that is given in Section 15.8.3 which contains in particular two classical simple compression algorithms. They compress efficiently only a small variety of texts when used alone. But they become more powerful with the special preprocessing presented there.

15.8.1 Huffman Coding

The Huffman method is an optimal statistical coding. It transforms the original code used for characters of the text (ASCII code on 8 bits, for instance). Coding the text is just replacing each symbol (more exactly each occurrence of it) by its new codeword. The method works for any length of blocks (not only 8 bits), but the running time grows rapidly with the length. In the following, one assumes that symbols are originally encoded on 8 bits to simplify the description.

The Huffman algorithm uses the notion of **prefix code**. A prefix code is a set of words containing no word that is a prefix of another word of the set. The advantage of such a code is that decoding is immediate. Moreover, it can be proved that this type of code does not weaken the compression.

A prefix code on the binary alphabet {0, 1} can be represented by a trie (see Section 15.2.6) that is a binary tree. In the present method, codes are complete: they correspond to complete tries (internal nodes have exactly two children). The leaves are labeled by the original characters, edges are labeled by 0 or 1, and labels of branches are the words of the code. The condition on the code implies that codewords are identified with leaves only. The convention is adopted that, from an internal node, the edge to its left child is labeled by 0, and the edge to its right child is labeled by 1.

In the model where characters of the text are given new codewords, the Huffman algorithm builds a code that is optimal in the sense that the compression is the best possible (the length of the compressed text is minimum). The code depends on the text, and more precisely on the frequencies of each character in the uncompressed text. The more frequent characters are given short codewords, whereas the less frequent symbols have longer codewords.

15.8.1.1 Encoding

The coding algorithm is composed of three steps: count of character frequencies, construction of the prefix code, and encoding of the text.

The first step consists in counting the number of occurrences of each character in the original text (see Figure 15.60). A special end marker (denoted by END) is used, which (virtually) appears only once at the end of the text. It is possible to skip this first step if fixed statistics on the alphabet are used. In this case, the method is optimal according to the statistics, but not necessarily for the specific text.

```
COUNT(fin)
for each character a ∈ Σ
    do freq(a) ← 0
while not end of file fin and a is the next symbol
    do freq(a) ← freq(a) + 1
END ← 1
```

FIGURE 15.60 Counts the character frequencies.

The second step of the algorithm builds the tree of a prefix code using the character frequency *freq*(a) of each character a in the following way:

- Create a one-node tree t for each character a, setting *weight*(t) = *freq*(a) and *label*(t) = a.
- Repeat (1), extract the two least-weighted trees t_1 and t_2 (2) Create a new tree t_3 having left subtree t_1, right subtree t_2, and weight *weight*(t_3) = *weight*(t_1) + *weight*(t_2).
  until only one tree remains.

The tree is constructed by the algorithm BUILD-TREE in Figure 15.61. The implementation uses two linear lists. The first list contains the leaves of the future tree, each associated with a symbol. The list is sorted in the increasing order of the weight of the leaves (frequency of symbols). The second list contains the newly created trees. Extracting the two least weighted trees consists in extracting the two least weighted trees among the two first trees of the list of leaves and the two first trees of the list of created trees. Each new tree is inserted at the end of the list of the trees. The only tree remaining at the end of the procedure is the coding tree.

After the coding tree is built, it is possible to recover the codewords associated with characters by a simple depth-first search of the tree (see Figure 15.62); *codeword*(a) is then the binary code associated with the character a.

In the third step, the original text is encoded. Since the code depends on the original text, in order to be able to decode the compressed text, the coding tree and the original codewords of symbols must

```
BUILD-TREE()
for each character a ∈ Σ ∪ {END}
    do if freq(a) ≠ 0
          then create a new node t
               weight(t) ← freq(a)
               label(t) ← a
lleaves ← list of all the nodes in increasing order of weight
ltrees ← empty list
while LENGTH(lleaves) + LENGTH(ltrees) > 1
    do (ℓ, r) ← extract the two nodes of smallest weight (among the two nodes
                 at the beginning of lleaves and the two nodes at the beginning of ltrees)
       create a new node t
       weight(t) ← weight(ℓ) + weight(r)
       left(t) ← ℓ
       right(t) ← r
       insert t at the end of ltrees
return t
```

FIGURE 15.61 Builds the coding tree.

```
BUILD-CODE(t, length)
if t is not a leaf
   then temp[length] ← 0
        BUILD-CODE(left(t), length + 1)
        temp[length] ← 1
        BUILD-CODE(right(t), length + 1)
   else codeword(label(t)) ← temp[0 ... length − 1]
```

FIGURE 15.62 Builds the character codes from the coding tree.

```
CODE-TREE(fout, t)
if t is not a leaf
  then write a 0 in the file fout
       CODE-TREE(fout, left(t))
       CODE-TREE(fout, right(t))
  else write a 1 in the file fout
       write the original code of label(t) in the file fout
```

FIGURE 15.63 Memorizes the coding tree in the compressed file.

```
CODE-TEXT(fin, fout)
while not end of file fin and a is the next symbol
    do write codeword(a) in the file fout
write codeword(END) in the file fout
```

FIGURE 15.64 Encodes the characters in the compressed file.

```
CODING(fin, fout)
COUNT(fin)
t ← BUILD-TREE()
BUILD-CODE(t, 0)
CODE-TREE(fout, t)
CODE-TEXT(fin, fout)
```

FIGURE 15.65 Complete function for Huffman coding.

be stored with the compressed text. This information is placed in a header of the compressed file, to be read at decoding time just before the compressed text. The header is made via a depth-first traversal of the tree. Each time an internal node is encountered a 0 is produced. When a leaf is encountered, a 1 is produced followed by the original code of the corresponding character on 9 b (so that the end marker can be equal to 256 if all the characters appear in the original text). This part of the encoding algorithm is shown in Figure 15.63.

After the header of the compressed file is computed, the encoding of the original text is realized by the algorithm of Figure 15.64.

A complete implementation of the Huffman algorithm, composed of the three steps just described, is given in Figure 15.65.

Example 1.21

y = **CAGATAAGAGAA**. The length of $y = 12 \times 8 = 96$ bits (assuming an 8-bit code). The character frequencies are

A	C	G	T	END
7	1	3	1	1

The different steps during the construction of the coding tree are

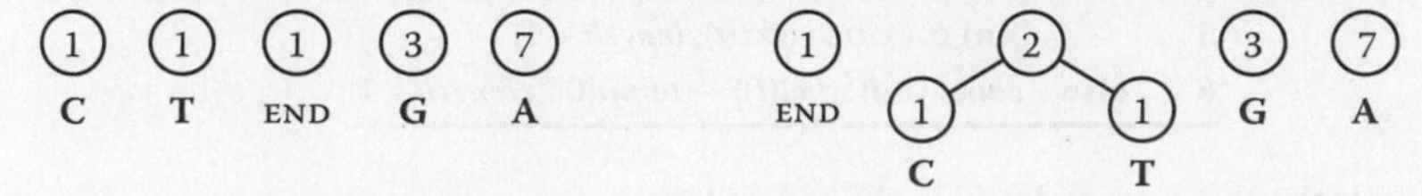

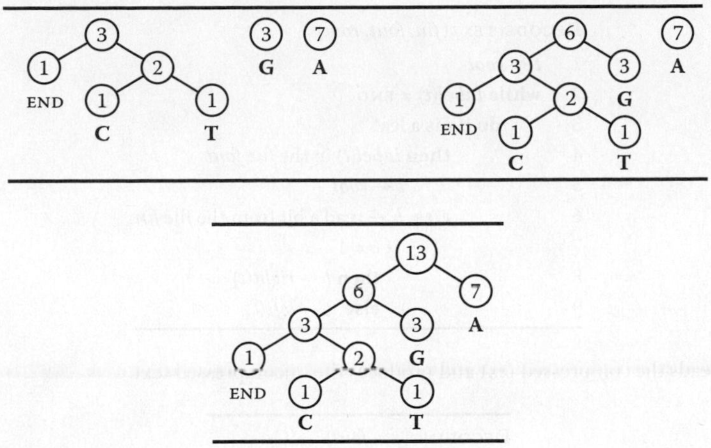

character codewords:

The encoded tree is **0001** binary (END, 9)**01**binary (C, 9)**1**binary(T, 9) **1**binary (G, 9)**1**binary (A, 9), which produces a header of length 54 bits,

0001 100000000 01 001000011 1 001010100 1 001000111 1 001000001

The encoded text

0010 1 01 1 0011 1 1 01 1 01 1 000

is of length 24 bits. The total length of the compressed file is 78 bits.

A	C	G	T	END
1	0010	01	0011	000

The construction of the tree takes $O(\sigma \log \sigma)$ time if the sorting of the list of the leaves is implemented efficiently. The rest of the encoding process runs in linear time in the sum of the sizes of the original and compressed texts.

15.8.1.2 Decoding

Decoding a file containing a text compressed by the Huffman algorithm is a mere programming exercise. First, the coding tree is rebuilt by the algorithm of Figure 15.66. Then, the uncompressed text is

```
REBUILD-TREE(fin, t)
b ← read a bit from the file fin
if b = 1                                   ▷ leaf
   then left(t) ← NIL
        right(t) ← NIL
        label(t) ← symbol corresponding to
                   the 9 next bits in the file fin
   else create a new node ℓ
        left(t) ← ℓ
        REBUILD-TREE(fin, ℓ)
        create a new node r
        right(t) ← r
        REBUILD-TREE(fin, r)
```

FIGURE 15.66 Rebuilds the tree read from the compressed file.

```
Decode-Text(fin, fout, root)
t ← root
while label(t) ≠ END
    do if t is a leaf
          then label(t) in the file fout
               t ← root
          else b ← read a bit from the file fin
               if b = 1
                  then t ← right(t)
                  else t ← left(t)
```

FIGURE 15.67 Reads the compressed text and produces the uncompressed text.

```
Decoding(fin, fout)
create a new node root
Rebuild-Tree(fin, root)
Decode-Text(fin, fout, root)
```

FIGURE 15.68 Complete function for Huffman decoding.

recovered by parsing the compressed text with the coding tree. The process begins at the root of the coding tree and follows a left edge when a 0 is read or a right edge when a 1 is read. When a leaf is encountered, the corresponding character (in fact, the original codeword of it) is produced and the parsing phase resumes at the root of the tree. The parsing ends when the codeword of the end marker is read. An implementation of the decoding of the text is presented in Figure 15.67.

The complete decoding program is given in Figure 15.68. It calls the preceding functions. The running time of the decoding program is linear in the sum of the sizes of the texts it manipulates.

15.8.2 Lempel–Ziv–Welsh (LZW) Compression

Ziv and Lempel designed a compression method using encoding segments. These segments are stored in a dictionary that is built during the compression process. When a segment of the dictionary is encountered later while scanning the original text, it is substituted by its index in the dictionary. In the model where portions of the text are replaced by pointers on previous occurrences, the Ziv–Lempel compression scheme can be proved to be asymptotically optimal (on large enough texts satisfying good conditions on the probability distribution of symbols).

The dictionary is the central point of the algorithm. It has the property of being prefix closed (every prefix of a word of the dictionary is in the dictionary), so that it can be implemented as a tree. Furthermore, a hashing technique makes its implementation efficient. The version described in this section is called the Lempel–Ziv–Welsh (LZW) method after several improvements introduced by Welsh. The algorithm is implemented by the **compress** command existing under the UNIX operating system.

15.8.2.1 Compression Method

The scheme of the compression method is first described. The dictionary is initialized with all the characters of the alphabet. The current situation is when one has just read a segment w in the text. Let a be the next symbol (just following w). Then the methods proceed as follows:

- If wa is not in the dictionary, write the index of w to the output file, and add wa to the dictionary. Then w is reset to a and process the next symbol (following a).
- If wa is in the dictionary, process the next symbol, with segment wa instead of w.

Initially, the segment w is set to the first symbol of the source text.

Example 1.22

Here y = **CAGTAAGAGAA**

C	**A**	**G**	**T**	**A**	**A**	**G**	**A**	**G**	**A**	**A**		w	out	added
	↑											**C**	67	**CA**, 257
		↑										**A**	65	**AG**, 258
			↑									**G**	71	**GT**, 259
				↑								**T**	84	**TA**, 260
					↑							**A**	65	**AA**, 261
						↑						**A**		
							↑					**AG**	258	**AGA**, 262
								↑				**A**		
									↑			**AG**		
										↑		**AGA**	262	**AGAA**, 262
											↑	**A**		
													65	
													256	

15.8.2.2 Decompression Method

The decompression method is symmetrical to the compression algorithm. The dictionary is recovered while the decompression process runs. It is basically done in this way:

- Read a code c in the compressed file.
- Write in the output file the segment w which has index c in the dictionary.
- Add to the dictionary the word wa where a is the first letter of the next segment.

In this scheme, a problem occurs if the next segment is the word which is being built. This arises only if the text contains a segment *azazax* for which *az* belongs to the dictionary but *aza* does not. During the compression process, the index of *az* is written into the compressed file, and *aza* is added to the dictionary. Next, *aza* is read and its index is written into the file. During the decompression process, the index of *aza* is read while the word *az* has not been completed yet: the segment *aza* is not already in the dictionary. However, since this is the unique case where the situation arises, the segment *aza* is recovered taking the last segment *az* added to the dictionary concatenated with its first letter *a*.

Example 15.23

Here the decoding is 67, 65, 71, 84, 65, 258, 262, 65, 256

read	written	added
67	**C**	
65	**A**	**CA**, 257
71	**G**	**AG**, 258
84	**T**	**GT**, 259
65	**A**	**TA**, 260
258	**AG**	**AA**, 261
262	**AGA**	**AGA**, 262
65	**A**	**AGAA**, 263
256		

```
COMPRESS(fin, fout)
count ← -1
for each character a ∈ Σ
    do count ← count + 1
        HASH-INSERT(D, (-1, a, count))
count ← count + 1
HASH-INSERT(D, (-1, END, count))
p ← -1
while not end of file fin
    do a ← next character of fin
        q ← HASH-SEARCH(D, (p, a))
        if q = NIL
            then write code(p) on 1 + log(count) bits in fout
                count ← count + 1
                HASH-INSERT(D, (p, a, count))
                p ← HASH-SEARCH(D, (-1, a))
            else p ← q
write code(p) on 1 + log(count) bits in fout
write code(HASH-SEARCH(D, (-1, END))) on 1 + log(count) bits in fout
```

FIGURE 15.69 LZW compression algorithm.

15.8.2.3 Implementation

For the compression algorithm shown in Figure 15.69, the dictionary is stored in a table D. The dictionary is implemented as a tree; each node z of the tree has the three following components:

- *parent*(z) is a link to the parent node of z.
- *label*(z) is a character.
- *code*(z) is the code associated with z.

The tree is stored in a table that is accessed with a hashing function. This provides fast access to the children of a node. The procedure HASH-INSERT((D, (p, a, c))) inserts a new node z in the dictionary D with *parent*(z) = p, *label*(z) = a, and *code*(z) = c. The function HASH-SEARCH((D, (p, a))) returns the node z such that *parent*(z) = p and *label*(z) = a.

For the decompression algorithm, no hashing technique is necessary. Having the index of the next segment, a bottom-up walk in the trie implementing the dictionary produces the mirror image of the segment. A stack is used to reverse it. Assume that the function *string*(c) performs this specific work for a code c. The bottom-up walk follows the parent links of the data structure. The function *first*(w) gives the first character of the word w. These features are part of the decompression algorithm displayed in Figure 15.70.

The Ziv–Lempel compression and decompression algorithms run both in linear time in the sizes of the files provided a good hashing technique is chosen. Indeed, it is very fast in practice. Its main advantage compared to Huffman coding is that it captures long repeated segments in the source file.

15.8.3 Mixing Several Methods

Simple compression methods, basis of the popular **bzip** software are described and then an example of a combination of several of them.

```
UNCOMPRESS(fin, fout)
count ← −1
for each character a ∈ Σ
    do count ← count + 1
        HASH-INSERT(D, (−1, a, count))
count ← count + 1
HASH-INSERT(D, (−1, END, count))
c ← first code on 1 + log(count) bits in fin
write string(c) in fout
a ← first(string(c))
while TRUE
    do d ← next code on 1 + log(count) bits in fin
        if d > count
            then count ← count + 1
                parent(count) ← c
                label(count) ← a
                write string(c)a in fout
                c ← d
            else a ← first(string(d))
                if a ≠ END
                    then count ← count + 1
                        parent(count) ← c
                        label(count) ← a
                        write string(d) in fout
                        c ← d
                    else break
```

FIGURE 15.70 LZW decompression algorithm.

15.8.3.1 Run Length Encoding

The aim of Run Length Encoding (RLE) is to efficiently encode repetitions occurring in the input data. Let us assume that it contains a good quantity of repetitions of the form $aa \ldots a$ for some character a ($a \in \Sigma$). A repetition of k consecutive occurrences of letter a is replaced by &ak, where the symbol & is a new character (& $\notin \Sigma$).

The string &ak that encodes a repetition of k consecutive occurrences of a is itself encoded on the binary alphabet {0, 1}. In practice, letters are often represented by their ASCII code. Therefore, the codeword of a letter belongs to $\{0, 1\}^k$ with $k = 7$ or 8. Generally, there is no problem in choosing or encoding the special character &. The integer k of the string &ak is also encoded on the binary alphabet, but it is not sufficient to translate it by its binary representation, because one would be unable to recover it at decoding time inside the stream of bits. A simple way to cope with this is to encode k by the string $0^{\ell}\text{bin}(k)$, where bin(k) is the binary representation of k, and ℓ is the length of it. This works well because the binary representation of k starts with a 1 so there is no ambiguity to recover ℓ by counting during the decoding phase. The size of the encoding of k is thus roughly $2 \log k$. More sophisticated integer representations are possible, but none is really suitable for the present situation. Simpler solution consists in encoding k on the same number of bits as other symbols, but this bounds values of ℓ and decreases the power of the method.

15.8.3.2 Move to Front

The Move To Front (MTF) method may be regarded as an extension of RLE or a simplification of Ziv–Lempel compression. It is efficient when the occurrences of letters in the input text are localized into relatively short segment of it. The technique is able to capture the proximity between occurrences of symbols and to turn it into a short encoded text.

Letters of the alphabet Σ of the input text are initially stored in a list that is managed dynamically. Letters are represented by their rank in the list, starting from 1, rank that is itself encoded as described above for RLE.

Letters of the input text are processed in an online manner. The clue of the method is that each letter is moved to the beginning of the list just after it is translated by the encoding of its rank.

The effect of MTF is to reduce the size of the encoding of a letter that reappears soon after its preceding occurrence.

15.8.3.3 Integrated Example

Most compression software combine several methods to be able to compress efficiently a large range of input data. An example of this strategy, implemented by the UNIX command **bzip**, is presented.

Let $y = y[0]y[1] \cdots y[n-1]$ be the input text. The k-th rotation (or conjugate) of y, $0 \le k \le n-1$, is the string $y_k = y[k]y[k+1] \cdots y[n-1]y[0]y[1] \cdots y[k-1]$.

The BW transformation is defined as $BW(y) = y[p_0]y[p_1] \cdots y[p_{n-1}]$, where $p_i + 1$ is such that y_{p_i+1} has rank i in the sorted list of all rotations of y.

It is remarkable that y can be recovered from both $BW(y)$ and a position on it, starting position of the inverse transformation (see Figure 15.71). This is possible due to the following property of the transformation. Assume that $i < j$ and $y[p_i] = y[p_j] = a$. Since $i < j$, the definition implies $y_{p_i+1} < y_{p_j+1}$. Since $y[p_i] = y[p_j]$, transferring the last letters of y_{p_i+1} and y_{p_j+1} to the beginning of these words does not change the inequality. This proves that the two occurrences of a in $BW(y)$ are in the same relative order as in the sorted list of letters of y. Figure 15.71 illustrates the inverse transformation. Top line is $BW(y)$ and bottom line the sorted list of letters of it. Top-down arrows correspond to succession of occurrences in y. Each bottom-up arrow links the same occurrence of a letter in y. Arrows starting from equal letters do not cross. The circular path is associated with rotations of the string y. If the starting point is known, the only occurrence of letter **b** here, following the path, produces the initial string y.

Transformation BW obviously does not compress the input text y. But $BW(y)$ is compressed more efficiently with simple methods. This is the strategy applied for the command **bzip**. It is a combination of the BW transformation followed by MTF encoding and RLE encoding. Arithmetic coding, a method providing compression ratios slightly better than Huffman coding, may also be used.

Table 15.1 contains a sample of experimental results showing the behavior of compression algorithms on different types of texts from the Calgary Corpus: `bib` (bibliography), `book1` (fiction book), `news` (USENET batch file), `pic` (black and white fax picture), `progc` (source code in C), and `trans` (transcript of terminal session).

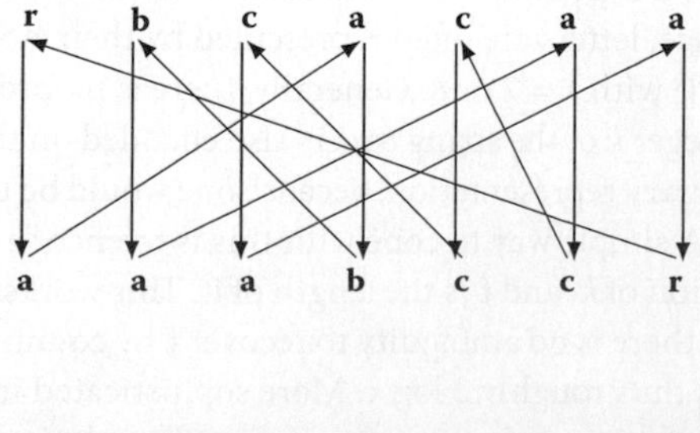

FIGURE 15.71 *BW*(**baccara**).

TABLE 15.1 Compression Results with Three Algorithms: Huffman Coding (**pack**), Ziv–Lempel Coding (**gzip-b**), and Burrows–Wheeler Coding (**bzip2-1**)

Bytes	111, 261	768, 771	377, 109	513, 216	39, 611	93, 695	
Texts	`bib`	`book1`	`news`	`pic`	`progc`	`trans`	Average
pack	5.24	4.56	5.23	1.66	5.26	5.58	4.99
gzip-b	2.51	3.25	3.06	0.82	2.68	1.61	2.69
bzip2-1	2.10	2.81	2.85	0.78	2.53	1.53	2.46

The compression algorithms reported in the table are: the Huffman coding algorithm implemented by **pack**, the Ziv–Lempel algorithm implemented by **gzip-b**, and the compression based on the *BW* transform implemented by **bzip2-1**. Figures give the number of bits used per character (letter). They show that **pack** is the less efficient method and that **bzip2-1** compresses a bit more than **gzip-b**. Additional compression results can be found at http://corpus.canterbury.ac.nz.

15.9 Research Issues and Summary

The algorithm for string searching by hashing was introduced by Harrison in 1971, and later fully analyzed by Karp and Rabin (1987).

The linear-time string-matching algorithm of Knuth, Morris, and Pratt is from 1976. It can be proved that, during the search, a character of the text is compared to a character of the pattern no more than $\log_\Phi(|x| + 1)$ (where Φ is the golden ratio $(1+\sqrt{5})/2$). Simon (1993) gives an algorithm similar to the previous one but with a delay bounded by the size of the alphabet (of the pattern x). Hancart (1993) proves that the delay of Simon's algorithm is, indeed, no more than $1 + \log_2|x|$. He also proves that this is optimal among algorithms searching the text through a window of size 1.

Galil (1981) gives a general criterion to transform searching algorithms of that type into real-time algorithms.

The BM algorithm was designed by Boyer and Moore (1977). The first proof of its linearity when restricted to the search for the first pattern occurrence is by Knuth et al. (1977). Cole (1994) proves that the maximum number of symbol comparisons is bounded by $3n$, and that this bound is tight.

Knuth et al. (1977) consider a variant of the BM algorithm in which all previous matches inside the current window are memorized. Each window configuration becomes the state of what is called the BM automaton. It is still unknown whether the maximum number of states of the automaton is polynomial or not.

Several variants of the BM algorithm avoid the quadratic behavior when searching for all occurrences of the pattern. Among the more efficient in terms of the number of symbol comparisons are: the algorithm of Apostolico and Giancarlo (1986), Turbo–BM algorithm by Crochemore *et al.* (1992) (the two algorithms are analyzed in Lecroq (1995)), and the algorithm of Colussi (1994).

The general bound on the expected time complexity of string matching is $O(|y| \log |x|/|x|)$. The probabilistic analysis of a simplified version of the BM algorithm, similar to the QS algorithm of Sunday (1990) described in the chapter, was studied by several authors.

String searching can be solved by a linear-time algorithm requiring only a constant amount of memory in addition to the pattern and the (window on the) text. This can be proved by different techniques presented in Crochemore and Rytter (2002).

Experimental results presented in this chapter have been realized using SMART (String Matching Algorithms Research Tool, http://www.dmi.unict.it/~faro/smart/) by Faro and Lecroq (2011). An exhaustive experimental study can be found in Faro and Lecroq (2010).

The Aho–Corasick algorithm is from Aho and Corasick (1975). It is implemented by the **fgrep** command under the UNIX operating system. Commentz-Walter (1979) has designed an extension of the BM algorithm to several patterns. It is fully described in Aho (1990).

On general alphabets, the two-dimensional pattern matching can be solved in linear time, whereas the running time of the Bird–Baker algorithm (Baker, 1978; Bird, 1977) has an additional $\log \sigma$ factor. It is still unknown whether the problem can be solved by an algorithm working simultaneously in linear time and using only a constant amount of memory space (see Crochemore and Rytter, 2002).

The suffix tree construction of Section 15.4 is by McCreight (1976). An online construction is given by Ukkonen (1995). Other data structures to represent indexes on text files are: direct acyclic word graph (Blumer et al., 1985), suffix automata (Crochemore, 1986), and suffix arrays (Manber and Myers, 1993). All these techniques are presented in Crochemore and Rytter (2002). The Zhu-Takakoka was published in Zhu and Takaoka (1989). The data structures implement full indexes with standard operations, whereas applications sometimes need only incomplete indexes. The design of compact indexes is still unsolved.

The suffix array of a string, with the associated search algorithm based on the knowledge of the common prefixes (Section 15.5), is from Manber and Myers (1993). The suffix array construction in Section 15.5.1 is from Kärkkäinen and Sanders (2003) (see also Kim et al., 2003; Ko and Aluru, 2003). The method used in Section 15.5.3 to compute the longest common prefixes of sorted suffixes is from Kasai et al. (2001).

First algorithms for aligning two sequences are by Needleman and Wunsch (1970) and Wagner and Fischer (1974). Idea and algorithm for local alignment is by Smith and Waterman (1981). Hirschberg (1975) presents the computation of the lcs in linear space. This is an important result because the algorithm is classically run on large sequences. Another implementation is given in Durbin et al. (1998). The quadratic time complexity of the algorithm to compute the Levenshtein distance is a bottleneck in practical string comparison for the same reason.

Approximate string searching is a lively domain of research. It includes, for instance, the notion of regular expressions to represent sets of strings. Algorithms based on regular expression are commonly found in books related to compiling techniques. The algorithms of Section 15.7 are by Baeza-Yates and Gonnet (1992) and Wu and Manber (1992).

The statistical compression algorithm of Huffman (1951) has a dynamic version where symbol counting is done at coding time. The current coding tree is used to encode the next character and then updated. At decoding time, a symmetrical process reconstructs the same tree, so the tree does not need to be stored with the compressed text, see Knuth (1985). The command **compact** of UNIX implements this version.

Several variants of the Ziv and Lempel algorithm exist. The reader can refer to Bell et al. (1990) for a discussion on them (see Welch [1984], for instance). Nelson (1992) presents practical implementations of various compression algorithms. The *BW* transform is from Burrows and Wheeler (1994).

Key Terms

Alignment: An alignment of two strings x and y is a word of the form $(\bar{x}_0, \bar{y}_0)(\bar{x}_1, \bar{y}_1) \cdots (\bar{x}_{p-1}, \bar{y}_{p-1})$ where each $(\bar{x}_i, \bar{y}_i) \in (\Sigma \cup \{\varepsilon\}) \times (\Sigma \cup \{\varepsilon\}) \backslash (\{(\varepsilon, \varepsilon)\}$ for $0 \le i \le p-1$ and both $x = \bar{x}_0\bar{x}_1 \cdots \bar{x}_{p-1}$ and $y = \bar{y}_0\bar{y}_1 \cdots \bar{y}_{p-1}$.

Border: A word $u \in \Sigma^*$ is a border of a word $w \in \Sigma^*$ if u is both a prefix and a suffix of w (there exist two words $v, z \in \Sigma^*$ such that $w = vu = uz$). The common length of v and z is a period of w.

Edit distance: The metric distance between two strings that counts the minimum number of insertions and deletions of symbols to transform one string into the other.

Hamming distance: The metric distance between two strings of same length that counts the number of mismatches.

Levenshtein distance: The metric distance between two strings that counts the minimum number of insertions, deletions, and substitutions of symbols to transform one string into the other.

Occurrence: An occurrence of a word $u \in \Sigma^*$, of length m, appears in a word $w \in \Sigma^*$, of length n, at position i if for $0 \le k \le m-1$, $u[k] = w[i+k]$.

Prefix: A word $u \in \Sigma^*$ is a prefix of a word $w \in \Sigma^*$ if $w = uz$ for some $z \in \Sigma^*$.

Prefix code: Set of words such that no word of the set is a prefix of another word contained in the set. A prefix code is represented by a coding tree.

Segment: A word $u \in \Sigma^*$ is a segment of a word $w \in \Sigma^*$ if u occurs in w (see occurrence), that is, $w = vuz$ for two words $v, z \in \Sigma^*$ (u is also referred to as a factor or a subword of w).

Subsequence: A word $u \in \Sigma^*$ is a subsequence of a word $w \in \Sigma^*$ if it is obtained from w by deleting zero or more symbols that need not be consecutive (u is sometimes referred to as a subword of w, with a possible confusion with the notion of segment).

Suffix: A word $u \in \Sigma^*$ is a suffix of a word $w \in \Sigma^*$ if $w = vu$ for some $v \in \Sigma^*$.

Suffix tree: Trie containing all the suffixes of a word.

Trie: Tree in which edges are labeled by letters or words.

Further Information

Problems and algorithms presented in the chapter are just a sample of questions related to pattern matching. They share the formal methods used to design solutions and efficient algorithms. A wider panorama of algorithms on texts may be found in a few books such as:

Adjeroh, D., Bell, T., and Mukherjee, A. 2008. *The Burrows-Wheeler Transform: Data Compression, Suffix Arrays, and Pattern Matching*. Springer Verlag, Heidelberg, Germany.

Apostolico, A. and Galil, Z., eds. 1997. *Pattern Matching Algorithms*. Oxford University Press, New York.

Bell, T. C., Cleary, J. G., and Witten, I. H. 1990. *Text Compression*. Prentice-Hall, Englewood Cliffs, NJ.

Crochemore, M., Hancart, C., and Lecroq, T. 2007. *Algorithms on Strings*. Cambridge University Press, Cambridge, U.K.

Crochemore, M. and Rytter, W. 2002. *Jewels of Stringology*. World Scientific, Singapore.

Gusfield D. 1997. *Algorithms on Strings, Trees and Sequences: Computer Science and Computational Biology*. Cambridge University Press, Cambridge, U.K.

Navarro, G. and Raffinot M. 2002. *Flexible Pattern Matching in Strings: Practical On-line Search Algorithms for Texts and Biological Sequences*. Cambridge University Press, Cambridge.

Nelson, M. 1992. *The Data Compression Book*. M&T Books, New York.

Salomon, D. 2000. *Data Compression: The Complete Reference*. Springer Verlag, New York.

Smyth, W. F. 2003. *Computing Patterns in Strings*. Addison-Wesley Longman, Harlow, U.K.

Stephen, G. A. 1994. *String Searching Algorithms*. World Scientific Press, Singapore.

Research papers in pattern matching are disseminated in a few journals, among which are: *Communications of the ACM, Journal of the ACM, Theoretical Computer Science, Algorithmica, Journal of Algorithms, SIAM Journal on Computing, Journal of Discrete Algorithms.*

Finally, three main annual conferences present the latest advances of this field of research: Combinatorial Pattern Matching, which started in 1990. Data Compression Conference, which is regularly held at Snowbird. The scope of SPIRE (String Processing and Information Retrieval) includes the domain of data retrieval.

General conferences in computer science often have sessions devoted to pattern matching algorithms.

Several books on the design and analysis of general algorithms contain chapters devoted to algorithms on texts. Here is a sample of these books:

Cormen, T. H., Leiserson, C. E., and Rivest, R. L. 1990. *Introduction to Algorithms*. MIT Press, Boston, MA.

Gonnet, G. H. and Baeza-Yates, R. A. 1991. *Handbook of Algorithms and Data Structures*. Addison-Wesley, Boston, MA.

Animations of selected algorithms may be found at:

- http://www-igm.univ-mlv.fr/~lecroq/string/ (Exact String Matching Algorithms),
- http://www-igm.univ-mlv.fr/~lecroq/seqcomp/ (Alignments).

References

Apostolico, A., Giancarlo, R. 1986. The Boyer-Moore-Galil string searching strategies revisited. *SIAM Journal on Computing* 15(1):98–105.

Aho, A. V. 1990. Algorithms for finding patterns in strings. In *Handbook of Theoretical Computer Science*, vol. A. *Algorithms and Complexity*, J. van Leeuwen, ed., pp. 255–300. Elsevier, Amsterdam, The Netherlands.

Aho, A. V. and Corasick, M. J. 1975. Efficient string matching: An aid to bibliographic search. *Comm. ACM* 18(6):333–340.

Baeza-Yates, R. A. and Gonnet, G. H. 1992. A new approach to text searching. *Comm. ACM* 35(10):74–82.

Baker, T. P. 1978. A technique for extending rapid exact-match string matching to arrays of more than one dimension. *SIAM J. Comput.* 7(4):533–541.

Bell, T. C., Cleary, J. G., and Witten, I. H. 1990. *Text Compression*. Prentice-Hall, Englewood Cliffs, NJ.

Bird, R. S. 1977. Two-dimensional pattern matching. *Inf. Process. Lett.* 6(5):168–170.

Blumer, A., Blumer, J., Ehrenfeucht, A., Haussler, D., Chen, M. T., and Seiferas, J. 1985. The smallest automaton recognizing the subwords of a text. *Theor. Comput. Sci.* 40:31–55.

Boyer, R. S. and Moore, J. S. 1977. A fast string searching algorithm. *Comm. ACM* 20(10):762–772.

Burrows, M. and Wheeler, D. 1994. A block sorting lossless data compression algorithm. Technical Report 124, Digital Equipment Corporation, Maynard, MA.

Cole, R. 1994. Tight bounds on the complexity of the Boyer-Moore pattern matching algorithm. *SIAM J. Comput.* 23(5):1075–1091.

Colussi, L. 1994. Fastest pattern matching in strings. *J. Algorithm.* 16(2):163–189.

Crochemore, M. 1986. Transducers and repetitions. *Theor. Comput. Sci.* 45(1):63–86.

Crochemore, M. and Rytter, W. 2002. *Jewels of Stringology*. World Scientific, Singapore.

Crochemore, M., Czumaj, A., Gasieniec, L., Jarominek, S., Lecroq, T., Plandowski, W., Rytter, W. 1994. Speeding up two string matching algorithms. *Algorithmica* 12(4/5):247–267.

Commentz-Walter, B. 1979. A String Matching Algorithm Fast on the Average. *Proceedings of the 6th Colloquium on Automata, Languages and Programming*, Graz, Austria. H. A. Maurer, ed. Lecture Notes in Computer Science 71, pp. 118–132. Springer-Verlag, Berlin, Germany.

Durbin, R., Eddy, S., Krogh, A., and Mitchison G. 1998. *Biological Sequence Analysis Probabilistic Models of Proteins and Nucleic Acids*. Cambridge University Press, Cambridge, U.K.

Faro, S. and Lecroq, T. 2010. The exact string matching problem: A comprehensive experimental evaluation. Report arXiv:1012.2547.

Faro, S. and Lecroq, T. 2011. SMART: a string matching algorithm research tool. http://www.dmi.unict.it/~faro/smart/ (accessed October 7, 2013).

Galil, Z. 1981. String matching in real time. *J. ACM* 28(1):134–149.

Hancart, C. 1993. On Simon's string searching algorithm. *Inf. Process. Lett.* 47(2):95–99.

Hirschberg, D. S. 1975. A linear space algorithm for computing maximal common subsequences. *Comm. ACM* 18(6):341–343.

Huffman, D.A. 1951. A method for the construction of minimum-redundancy codes. *Proceedings of the I.R.E*, pp. 1098–1102.

Kärkkäinen, J. and Sanders, P. 2003. Simple linear work suffix array construction. *Proceedings of the 30th International Colloquium on Automata, Languages and Programming* Eindhoven, The Netherlands, J. C. M. Baeten, J. K. Lenstra, J. Parrow, and G. J. Woeginger, eds. Lecture Notes in Computer Science 2719, pp. 943–955, Springer-Verlag, Berlin, Germany.

Karp, R. M. and Rabin, M. O. 1987. Efficient randomized pattern-matching algorithms. *IBM J. Res. Dev.* 31(2):249–260.

Kasai, T., Lee, G., Arimura, H., Arikawa, S., and Park, K. 2001. Linear-time longest-common-prefix computation in suffix arrays and its applications. *Proceedings of the 12th Annual Symposium on Combinatorial Pattern Matching*, Jerusalem, Israel, 2001. A. Amir and G. M. Landau, eds. Lecture Notes in Computer Science 2089, pp. 181–192, Springer-Verlag, Berlin, Germany.

Kim, D. K., Sim, J. S., Park, H., and Park, K. 2003. Linear-time construction of suffix arrays. *Proceedings of the 14th Annual Symposium on Combinatorial Pattern Matching*, Morelia, Michocán, Mexico, 2003. R. A. Baeza-Yates, E. Chávez, and M. Crochemore, eds. Lecture Notes in Computer Science 2676, pp. 186–199. Springer-Verlag, Berlin, Germany.

Knuth, D. E. 1985. Dynamic Huffman coding. *J. Algorithm.* 6(2):163–180.

Knuth, D. E., Morris, J. H., Jr, and Pratt, V. R. 1977. Fast pattern matching in strings. *SIAM J. Comput.* 6(1):323–350.

Ko, P. and Aluru, S. 2003. Space efficient linear time construction of suffix arrays. *Proceedings of the 14th Annual Symposium on Combinatorial Pattern Matching*, Morelia, Michocán, Mexico, 2003. R. A. Baeza-Yates, E. Chávez, and M. Crochemore, eds. Lecture Notes in Computer Science 2676, pp. 200–210. Springer-Verlag, Berlin, Germany.

Lecroq, T. 1995. Experimental results on string-matching algorithms. *Software—Pract. Exp.* 25(7):727–765.

Manber, U. and Myers, G. 1993. Suffix arrays: A new method for on-line string searches. *SIAM J. Comput.* 22(5):935–948.

McCreight, E. M. 1976. A space-economical suffix tree construction algorithm. *J. Algorithm.* 23(2):262–272.

Needleman, S. B. and Wunsch, C. D. 1970. A general method applicable to the search for similarities in the amino acid sequence of two proteins. *J. Mol. Biol.* 48:443–453.

Nelson, M. 1992. *The Data Compression Book*. M&T Books, New York.

Simon, I. 1993. String matching algorithms and automata. In *First American Workshop on String Processing*, Baeza-Yates and Ziviani, eds., pp. 151–157. Universidade Federal de Minas Gerais.

Smith, T. F. and Waterman, M. S. 1981. Identification of common molecular sequences. *J. Mol. Biol.* 147:195–197.

Sunday, D. M. 1990. A very fast substring search algorithm. *Comm. ACM* 33(8):132–142.

Ukkonen, E. 1995. On-line construction of suffix trees. *Algorithmica* 14(3):249–260.

Wagner, R. A. and Fischer, M. 1974. The string-to-string correction problem. *J. ACM* 21(1):168–173.

Welch, T. 1984. A technique for high-performance data compression. *IEEE Comput.* 17(6):8–19.

Wu, S. and Manber, U. 1992. Fast text searching allowing errors. *Comm. ACM* 35(10):83–91.

Zhu, R. F. and Takaoka, T. 1989. A technique for two-dimensional pattern matching. *Comm. ACM* 32(9):1110–1120.

16 Distributed Algorithms

Dan Alistarh
Massachusetts Institute of Technology

Rachid Guerraoui
École Polytechnique Fédérale de Lausanne

16.1 Introduction

One of the key trends in computing over the past two decades, both at the level of large-scale and small-scale systems, has been *distribution*. At one extreme, given the advent of new paradigms such as cloud computing, utility computing, and social networks, large-scale software systems are now routinely distributed over several machines or clusters of machines, which can in turn be spread over large geographic regions.

At the other extreme, that is, at the processor level, recent years have shown an industry shift towards *multicore* architectures. In particular, due to technological limitations, processor manufacturers were forced to stop increasing processor clock speeds and instead started providing more processing units or *cores* as part of a single processor. This change, known as the *multicore revolution*, implies that software applications must be designed with parallelism in mind, to exploit the extra computing power of multiple cores.

Given this broad shift toward distributed computing, understanding the power and limitations of computation in a distributed setting is one of the major challenges in modern Computer Science. This challenge has prompted researchers to look for common frameworks to specify distributed algorithms and compare their performance. For classical centralized algorithms such analytical tools have already been in place for decades. This guided the design of efficient algorithms for practical problems and provided an understanding of the inherent limitations of classical computing in terms of lower bounds and NP-completeness.

Research in distributed computing aims to provide the formal models and analytic tools for understanding problems arising when computation is performed on a set of devices, connected through a network or located on the same chip. In particular, distributed computing identifies fundamental problems that appear in practical distributed systems, states them formally, and provides algorithms and lower bounds for these problems.

One of the key differences from centralized computing is that there is no single, universal model of computation for distributed computing. This is because distributed systems vary significantly in terms of the communication primitives or timing guarantees that the system provides. Another important distinction is *fault-tolerance*: In centralized computing, it is generally assumed that the underlying (Turing) machine always functions properly, following its algorithm. On the other hand, distributed algorithms model the fact that single computing agents may be unreliable and are usually designed to deal with crashes of the individual units of computation or even with agents that act maliciously in order to influence the global computation. In some sense, the multiplicity of Turing machines calls for a new notion of computation that tolerates the unreliability of some of these machines.

Even though there may never be a single unified model for distributed computing, there does, however, exist a set of core problems that keep reoccurring in distributed systems, irrespective of their architecture. The focus of research is on the solvability and complexity of such fundamental problems under various timing and failure guarantees.

In this chapter, we give an overview of the core principles of distributed computing and present some of the field's fundamental results. We then survey some of the practical applications of the research and give an overview of the major challenges that the field will face in the near future. Since distributed computing is an extremely active research area, our presentation is not meant to be exhaustive. We just scratch the surface, in the hope that the reader will find it appealing to explore the field in more depth.

16.2 Underlying Principles

Over the past 30 years, distributed computing has emerged as a coherent research field. In general, the style of work follows a pattern: The field first abstracts problems and models relevant for practical systems, then develops algorithms to solve these problems, or proves impossibility results that characterize the feasibility and complexity of the problem at hand. In case the problem is proved to be unsolvable in a given setting, research focuses on ways to circumvent the impossibility by restricting the model or relaxing the problem requirements.

In the rest of this section, we present some of the fundamental notions underlying distributed computing and isolate the key questions that the field aims to answer. We illustrate the pattern of research outlined above by discussing some canonical distributed problems and outline some of the main results and research directions in the area.

16.2.1 Basic Notions

The fundamental concept in distributed computing is the *process*, that is, a sequential computing device, which executes a distributed algorithm.* In particular, a process is the functional equivalent of a sequential Turing machine; by extension, a distributed system can be seen as a collection of communicating Turing machines. In a distributed system, each process runs an instance of a distributed algorithm, with a limited amount of information about the state of the whole system.

Processes communicate with each other and gain information about the system by performing operations on *shared objects*, that is, abstractions that can be accessed by multiple processes. Examples of elementary shared objects are *registers*, that is, units of shared memory holding information, which can be read or written by multiple processes, and *communication channels*, which processes can use to send messages to each other. These basic communication primitives provided by hardware are also called *base objects*. Depending on the type of communication primitives available, distributed systems can be classified as either *shared-memory* or *message-passing* systems.†

* Some texts, for example [80], use the term *processor* instead of process.

† Under certain conditions, shared-memory and message-passing systems can be shown to be computationally equivalent [9].

Processes may be *faulty*, that is, may deviate from their algorithm. Such faulty behavior includes *crash faults*, where the process stops executing the algorithm after a certain point. Failures can also be *Byzantine*, in which case processes may behave arbitrarily; in particular, they might behave maliciously, attempting to manipulate the distributed computation. Research also considers failures of the communication mechanisms, such as message loss or memory corruption.

The timing behavior of processes and their faults are typically assumed to be under the control of an external abstraction called a *scheduler*, which is independent from the processes. In particular, the scheduler decides the order in which processes take steps and the timing of their faults. Since we focus on the correctness and worst-case behavior of algorithms, the scheduler is generally assumed to be *adversarial*. The most general scheduler is *asynchronous*, that is, does not assume any bound on the relative speeds of processes, and thus allows arbitrary interleavings of process steps.

Algorithms that guarantee termination in an asynchronous model in which all but one process may crash are called *wait-free*; such algorithms ensure progress without making any assumptions on the timing behavior of processes. Even though in typical real-world systems it is possible to make stronger timing assumptions, wait-free algorithms have the advantage of being general and portable; in particular, they are guaranteed to run correctly in systems with arbitrary timing guarantees.

16.2.2 Fundamental Questions

Roughly speaking, the key question in distributed computing is that of *computability*:

> *Given a set of processes communicating through shared objects, what abstractions can be implemented on top of these objects?*

In a broad sense, the objective is to raise the level of abstraction that a distributed system can provide, beyond the fundamental message-passing or shared-memory primitives generally available in hardware.* A distributed algorithm typically has at its disposal a set of *base objects* (basic primitives ensured by shared-memory or communication channels) and lower level *shared objects* (in turn, these may be composed of other base and shared objects, and so on). The problem is to correctly implement a higher-level abstraction on top of these objects, in spite of timing anomalies and process faults. This would then allow programmers to use such abstractions to build specific applications.

Once we know what abstractions can be implemented in a distributed system, a second fundamental question is that of *complexity*:

> *Given a distributed abstraction, what is the minimal cost of implementing it out of a set of lower-level shared objects?*

Several cost metrics may be of interest, such as the number of shared object accesses (the counterpart of sequential time complexity) or the number of shared objects that need to be allocated (the counterpart of sequential space complexity). Complexity is characterized by giving algorithms that minimize the cost and lower bounds that demonstrate inherent cost requirements. In general, distributed computing is interested in characterizing *worst-case* complexity, that is, the executions in which the cost is maximized; practical implementations sometimes focus on improving complexity in the average case or in certain common cases.

16.2.3 Fundamental Results

As we pointed out in the previous section, research in distributed computing follows a general pattern. In particular, it isolates models and problems of interest for distributed systems practitioners, abstracts them, and then studies the solvability and complexity of fundamental abstractions in general models. In case an abstraction cannot be implemented, research investigates ways to circumvent the impossibility.

---

* In the sequential world, the same role is played by data structures such as queues, stacks, sets, etc.

In the following, we illustrate this pattern by considering *state machine replication* and *consensus* as examples of such canonical problems. Along the way, we will describe some of the fundamental techniques and research directions underlying distributed computing.

16.2.3.1 Identifying the Practical Problem

A replicated state machine (RSM) [73,87] is the distributed equivalent of a highly available Turing machine, as it requires a set of processes to maintain a coherent state machine in a distributed, fault-tolerant fashion.

This problem is motivated by the observation that distributed software is often structured in terms of *clients* and *services*. A software service, such as a bank's web service, comprises one or more servers and exports operations that clients invoke by making requests. To ensure availability under high load or failures, multiple servers that fail independently have to be used, maintaining a consistent version of the service via a distributed protocol. These independent servers are called *replicas*, implementing the distributed service that is maintained as a RSM. (See Figure 16.1 for an illustration.)

16.2.3.2 Determining the Fundamental Abstraction

Several fundamental results, for example, [61,73], show that a RSM is computationally equivalent to the *consensus* problem. In this problem, a set of processes, each proposing a value, must *agree* on the same output from the values proposed. Formally, consensus is defined as follows.

Definition 16.1 (Consensus)

The consensus problem exports a propose operation, which takes as input an integer value. Each process p_i calls propose with its proposal value v_i as argument. The consensus object requires every nonfaulty process to eventually output a single output (the termination condition), such that all outputs are equal (the agreement condition). To avoid trivial solutions, the common output decided has to be the input of some process (the validity condition).

The intuition behind the transformation is given in Figure 16.2. The equivalence between RSM and consensus is important, since it suggests that, given consensus, any object representable as a state machine has a distributed implementation.

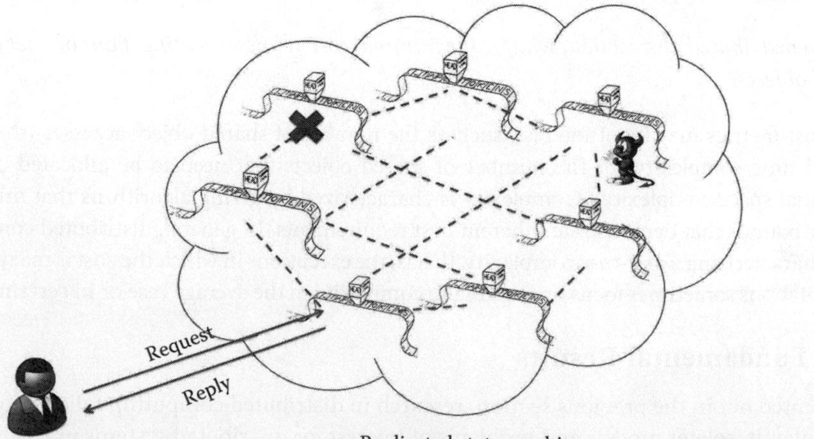

FIGURE 16.1 A distributed Turing machine implemented as a replicated state machine (RSM). The protocol must maintain the illusion of a centralized Turing machine, even though the implementation is distributed over multiple sequential machines, some of which may be subject to crash faults or malicious (Byzantine) faults.

```
Shared Objects:
  Cons, an array of consensus objects
Local Variables: /* at process i */
  Requests_i, a list of requests from clients
  stateMachine_i, a state machine
  index_i, a counter variable, initially 0

procedure ReplicatedStateMachine()
while Requests_i is not empty do
  request ← Requests_i.next()
  repeat
    cons ← Cons[index]
    decision ← cons.propose(request)
    stateMachine_i.perform(decision)
    index ← index + 1
  until request = decision
```

FIGURE 16.2 A basic RSM implementation using consensus objects. The replicas receive requests from clients, and use the consensus objects to agree on the order in which requests are committed.

16.2.3.3 Establishing a Fundamental Limitation

On the other hand, surprisingly, consensus cannot be implemented using elementary base objects such as read-write registers or message-passing communication channels in an asynchronous system in which a single process may crash. This result, proved by Fischer, Lynch, and Paterson [47], abbreviated as FLP, is probably the most celebrated result in distributed computing. By reduction, we obtain the same impossibility for implementations of a RSM and for many other abstractions, which can be used to implement consensus.

Theorem 16.1 (FLP [47])

Given an asynchronous message-passing system or an asynchronous shared-memory system with read-write registers, no consensus protocol is totally correct in spite of one crash fault.

The intuition behind the proof is the following. The argument proceeds by contradiction and assumes there exists an algorithm that solves consensus in the given setting. It then builds a valid execution of the algorithm in which no process can decide. The construction relies on two fundamental notions: one is the *bivalence* of an execution, that is, the property that an execution prefix can be continued by arranging process steps in two different ways, where each extension reaches a distinct decision. The second is the notion of *indistinguishability*: two system configurations are *indistinguishable* to a process if the values of all shared variables and the state of the process are the same in both configurations. Both these notions are key for proving lower bounds and impossibility results in distributed computing. The FLP argument starts by proving that any consensus algorithm has a bivalent initial configuration; it then shows that each bivalent configuration can be extended to a configuration that is still bivalent by scheduling an additional process step. Iterating this argument, we obtain an execution of infinite length in which each configuration is bivalent, therefore no process may decide in that execution.

16.2.3.4 Generalizing the Result

Besides the connection to the RSM problem, another strong argument for the centrality of consensus in distributed computing is Herlihy's *consensus hierarchy* [61]. Herlihy showed that every shared object can be assigned a *consensus number*, which is the maximum number of processes for which the object can solve wait-free consensus in an asynchronous system. The resulting hierarchy has a key separation property: No object with consensus number $x + 1$ can be implemented out of objects of consensus number at most x.

TABLE 16.1 Structure of Herlihy's Consensus Hierarchy [61]

Consensus Number	Shared Object
1	Read-write registers
2	Test-and-set, swap, fetch-and-add, queue, stack
⋮	⋮
∞	Augmented queue, compare-and-swap, sticky byte

For example, read-write registers have consensus number 1, since they can solve consensus for one process (trivially) but cannot solve consensus for two processes (by the FLP result). Herlihy also showed that shared objects such as *queues*, *stacks*, and *fetch-and-add* counters have consensus number 2. Shared-memory primitives such as *compare-and-swap* or *sticky bytes* have consensus number ∞, since they can solve consensus for any number of processes. (See Table 16.1). In turn, consensus is *universal*, that is, it may implement any other shared object wait-free.

This elegant result gives a way of characterizing the computational power of shared objects and suggests that distributed systems should provide universal primitives, which can be used to build wait-free distributed implementations of any sequential object.

16.2.3.5 Circumventing the Impossibility

Given the centrality of consensus in distributed computing, considerable research effort has been invested in understanding the limits of the FLP impossibility.

16.2.3.5.1 More Choices

One approach has been to relax the problem specification from consensus to *k-set agreement*, which allows k distinct decision values from the $n \geq k$ values proposed. This line of work led to a stronger impossibility: for any $k \geq 1$, k-set agreement is impossible in an asynchronous system in which k processes may fail by crashing.

This result was proved concurrently by three independent groups of researchers: Herlihy and Shavit [63], Borowsky and Gafni [19], and Saks and Zaharoglou [86], revealing a deep connection between distributed computing and a branch of mathematics called algebraic topology. This fundamental result has been extended to other models by reduction [6,50,51], while the topological tools have been developed further to characterize the solvability of other fundamental distributed tasks such as *renaming* [22,23,63].

16.2.3.5.2 Stronger Timing Assumptions

Another approach for circumventing the FLP impossibility has been to identify the additional amount of system information needed to solve consensus in a distributed system. This additional information can be provided in through a distinct shared object called a *failure detector* [27]. Whereas, in general, shared objects are used for interprocess communication, a failure detector acts as an oracle, providing extra information about process crashes or their relative speeds. These ideas have been developed into a coherent theory, which allows to identify the minimal amount of synchronization, that is, weakest failure detector, needed to implement abstractions such as consensus [26] or set agreement [52].

The FLP result holds in an *asynchronous* system, in which processes may be delayed for arbitrarily long periods of time. Real systems, however, exhibit bounded delays in most executions and unbounded delays only in the worst case. It is therefore natural to ask whether one can mitigate the impossibility of consensus by designing an algorithm that decides quickly when the system behaves well and maintains correctness (does not break agreement) in executions with failures or asynchrony. This idea was exploited by Dwork et al. [41], who showed that if the maximum delay between two process steps becomes bounded eventually, then consensus can be solved once this delay is respected. Further research, for example [12,39], extended this idea to other distributed problems and introduced frameworks for exploiting periods with bounded delays.

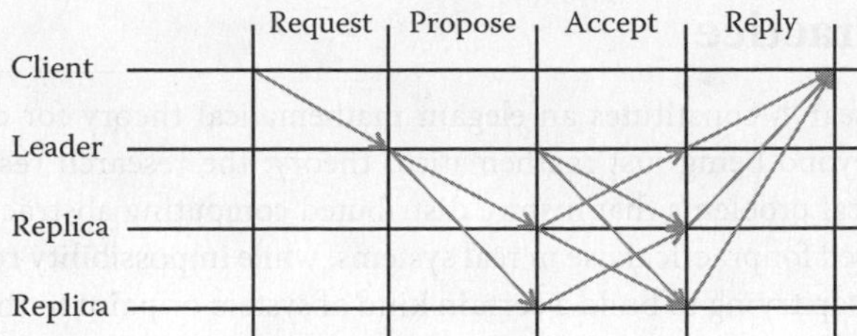

FIGURE 16.3 The message pattern in a failure-free execution of the Paxos protocol [74]. In the absence of failures, the protocol solves consensus among the replicas within two message delays.

16.2.3.5.3 Randomization

An important tool for circumventing the FLP impossibility is the use of *randomization*. The idea is to allow processes to flip coins during the execution and bias their decisions based on the results of the coin flips. The resulting probability space assigns weight to particular executions. One can then relax the problem specification to allow processes to terminate *with probability* 1—therefore, nonterminating executions are still possible in theory, but occur with probability 0. The other correctness properties must be preserved in every execution. Ben-Or [17] was the first to use this approach to show that randomized consensus can be solved in an fault-prone asynchronous system. Further research characterized the expected complexity of asynchronous consensus, for example [8,11], and applied randomization to other distributed problems, such as renaming [4,5,84] or concurrent data structures, for example [89].

16.2.3.6 Complexity of Consensus

The research described above aimed to isolate the minimal set of assumptions needed to solve consensus in a distributed system. On the other hand, another aspect of the quest for understanding the limitations of the FLP result has been understanding the *complexity* of consensus in a practical setting, for example [13,32,37,40,45].

A key approach in this direction has been to design *speculative* distributed algorithms, which decide quickly in well-behaved executions, for example, in which processes are relatively synchronous or there are no crashes and delay decision during periods with crashes or delays. The classic example is the *Paxos* protocol, for example [33,74,75], which solves consensus in a constant number of message delays if the system is well behaved (see Figure 16.3 for an illustration). Paxos does not contradict the FLP impossibility, since there exist worst-case executions in which it is prevented from terminating indefinitely. However, the protocol performs well in practice, since worst-case executions occur very rarely, and maintains correctness in all executions.

16.2.4 More Abstractions

To illustrate the pattern of work in distributed computing, we have considered consensus and state machine replication as examples of fundamental distributed abstractions. Of course, research in the field goes beyond these canonical problems and explores a wide variety of other research directions: for example, problems related to concurrent data structures, for example [7,42,46,68], distributed graph algorithms, for example [16,72,79], mutual exclusion, for example [14,18,91], distributed information dissemination, for example [30,36,69], clock synchronization, for example [44,78], distributed storage, for example [3,35], distributed task allocation, for example [20,71], distributed knowledge, for example [43,58], or issues related to decidability and checkability in a distributed setting, for example [48,49].

16.3 Impact on Practice

Distributed computing research constitutes an elegant mathematical theory for distributed systems. However, the field goes beyond being just mathematical theory: the research results can be applied back to the original practical problems that inspire distributed computing abstractions. In particular, algorithms can be engineered for practical use in real systems, while impossibility results can suggest to system designers when to stop trying to build a certain kind of system or point to the requirements that need to be relaxed to gain efficiency. In the following, we give a few examples of the impact of the field on practical aspects of computer systems.

16.3.1 Fault-Tolerant Computing

Chronologically, one of the first prominent practical applications of distributed computing results has been the *state machine replication* approach for large-scale distributed systems [73,87]. State machine replication is a general method of implementing a fault-tolerant distributed service by replicating servers and coordinating the interactions between replicas and clients. Lamport was the first to propose this approach in a seminal 1984 paper [73]; the approach was later elaborated by Schneider [87].

The Paxos family of protocols, for example [74,75], is at the core of many RSM implementations. The original protocol was proposed by Lamport [74] as a means to reach consensus among a set of servers in an asynchronous system. Although no deterministic consensus algorithm may solve consensus in an asynchronous system with faults [47], Paxos guarantees safety (no two processes decide distinct values) and may delay termination under asynchrony or failures, conditions that occur relatively rarely in practice [74,76].

Several variants of the original protocol were developed, tailored to different contexts such as disk storage [53] or fault-tolerant distributed databases [25]. Google uses the Paxos algorithm in the Chubby distributed lock service [25]; in turn, Chubby is used by BigTable, a high performance data storage system employed by several Google applications such as Gmail, Google Maps, or Youtube and available externally through the Google App Engine [29]. Microsoft also uses Paxos in the Autopilot cluster management service of the Bing search engine [67].

The Paxos protocol was generalized by Castro and Liskov in the context of malicious (Byzantine) failures among the replicas, to obtain a high-performance Byzantine fault-tolerant algorithm [24]. Their protocol has been later refined in terms of performance [57,70] and in terms of fault-tolerance [31]. Elements of Byzantine fault-tolerant protocols are used in the Bitcoin online currency system [55].

16.3.2 Concurrent Programming

The wide-spread introduction of multicore processors has brought problems related to concurrency in a distributed system at the forefront of software design. Even though the code executed by a single processor core may have few lines, the fact that many cores execute the program in parallel, with arbitrary interleavings of steps, implies that there is a variety of ways in which the program may behave, even given the same inputs. This makes concurrent programming an exceedingly difficult task. Further, classic interprocess synchronization primitives, such as locks or barriers, are increasingly believed not to be suitable for widespread use in large-scale multicore systems, for example [88].

Given the practical need for efficient concurrent code and the challenges posed by the use of classic synchronization primitives, significant research effort went into devising abstractions that make concurrent programming easier, while ensuring scalability of the software in a multicore environment.

16.3.2.1 Transactional Memory

One of the most promising solutions to this problem is *transactional memory* (TM). First introduced by Herlihy and Moss [62], the idea behind this abstraction is to enable processes to execute concurrently through the use of lightweight, in-memory *transactions*. A transaction is a sequence of operations that should be executed atomically, similarly to the code in the critical section of a conventional lock. Unlike locks, transactions may *abort*, in which case their operations are rolled back and invisible to any other processes. To gain performance, transactions only *appear* to have executed sequentially, while they might in fact execute concurrently.

TM is easy to use, since the programmer should simply convert the blocks of code that have to be executed atomically into transactions. The burden of executing the code correctly and efficiently is taken by the TM implementation, whose contention management mechanisms are hidden from the programmer. TM can be seen as an adaptation of the notion of database transaction [54] to handle concurrency in multicore systems.

Since its introduction in 1993 [62], significant research effort has gone into implementing efficient TM systems and formalizing its properties. Particular attention has been given to *software* TM [90], that is, implementations where contention management is performed in software, without additional architectural support, for example [34,38]; hardware TM implementations also exist, for example [65,82].

TM is currently migrating from being a research construct to adoption in production environments. In particular, Intel has recently announced that its new Haswell microprocessor architecture will feature Transactional Synchronization Extensions [66]. IBM's BlueGene/Q processor, used in the Sequoia supercomputer, will also support TM [65]. The Rock processor by Oracle also has hardware support for TM [82]. Practical software TM implementations include the Glasgow Haskell Compiler [2].

16.3.2.2 Concurrent Data Structures

A related research direction is the design of efficient *concurrent* versions of practical data structures, such as stacks, queues, sets, or hash tables. Research has focused on *lock-free* versions of these data structures, which generally employ read-modify-write hardware primitives such as *compare-and-swap* instead of locks. The key property of such implementations is that they guarantee *system-wide progress*; in particular, by contrast to lock-based techniques, the suspension of one or more processes does not prevent the whole system from making progress. Efficient lock-free implementations of many practical data structures, such as queues [81], lists [60,89], or counters [42], have been developed. Lock-free implementations of these data structures are available through several libraries [1,83].

16.4 Research Issues

We presented a few fundamental results in the area, which only scratch the surface of a deeper theory for distributed computation. As such, distributed computing will be more and more a field of intense research activity, stimulated by the proliferation of distributed systems in the context of large-scale Internet services and multicore architectures.

16.4.1 Unified Theory

Given the wide variety of models and abstractions for distributed computing, a major research challenge is isolating a core set of fundamental models, abstractions, and results for the field. Such results should express fundamental notions and limitations of distributed computing and could be extended to new settings introduced by new architectures or new programming paradigms. A growing body of evidence points to *wait-free computation* in asynchronous systems as being such a fundamental abstraction: Several central results proved for wait-free asynchronous computation can be extended to yield new lower bounds in other models. For example, it has been shown that the impossibility of asynchronous set agreement [19,63,86] can be extended through simulation to yield time complexity lower bounds in synchronous [50,51] and eventually synchronous models [6]. Identifying such fundamental notions and developing them into a unified theory of distributed computation remain a major open question.

16.4.2 Understanding Speculation

Another important research question is the structured understanding of *algorithmic speculation* in the context of distributed systems. As noted before, many distributed algorithms are based on the idea of taking advantage of common well-behaved executions, while covering worst-case scenarios via several intricate backup procedures. This is the case of several protocols of practical interest, such as *Paxos* [74] or PBFT [24]. However, due to their ad hoc structure, such algorithms are exceedingly difficult to design, implement, and analyze, for example, [25,57]. Thus, a key challenge is to develop a structured approach for building and analyzing such algorithms. Given a specific problem, such approaches should identify common scenarios in which the algorithm has to speculate and provide a framework for composing speculative algorithmic modules with the fail-safe backup protocols correctly and efficiently. Questions such as the inherent complexity overhead of speculation or formally specifying algorithmic composition are still not well understood and present interesting research opportunities.

Furthermore, significant research effort has been invested in characterizing the complexity of *wait-free* implementations of fundamental abstractions. Together with practical observations, this research suggests that wait-free deterministic implementations of many common objects may have prohibitive cost in the worst case and even in the average case. On the other hand, the per-thread progress ensured by wait-free implementations may not be necessary in all situations; implementations that guarantee *global* progress, such as *lock-free (nonblocking)* implementations, may suffice in many practical scenarios. However, the complexity requirements of such implementations are not known for many fundamental abstractions.

16.4.3 Consistency Conditions and Randomization

With the advent of large-scale distribution, research started exploring *relaxed consistency conditions* for shared object implementations. The goal of such an approach is to obtain a gain in performance by weakening some of the formal requirements for object implementations. A simple example of this approach is the use of *unordered* pool or hash map data structures in concurrent programs [88], instead of queue or stack data structures, which do maintain a precise order between elements, but have been shown to have high complexity overhead [5,46,68].

A related research direction is the use of *randomization* in shared object implementations to ensure good performance with high probability. The use of relaxed consistency conditions and randomization appears promising for some implementations; however, a major open question is to fully exploit the benefits of these approaches for practical use and to characterize their limitations in the context of distributed computation.

16.4.4 New Computing Paradigms

Finally, a major challenge that the field faces periodically is adapting to new trends in distributed computation, dictated by new computing paradigms and new architectures. Currently, such changes are brought about by developments such as the pervasiveness of mobile wireless devices, new distributed computing paradigms such as *big data* processing, the advent of social networks and crowdsourcing, or the introduction of new architectures and programming paradigms for mass-produced microprocessors.

16.5 Summary

In this chapter, we have given an overview of some basic principles and fundamental results that underlie the study of distributed algorithms. In particular, we considered consensus and state machine replication as examples of fundamental abstractions and gave an overview of the core results and research directions in distributed computing, together with some practical applications of results in the field. Given the continuing trend toward distribution, both at the macrolevel, through large-scale distributed services, and at the microlevel, through multicore processor architectures, we predict that distributed computing will be an area of intense research activity for years to come.

Key Terms

Adversary: The interleaving of process steps and the failures of the processes in a distributed system are assumed to be under the control of an entity called *adversary* or *scheduler*. It is assumed that the goal of the adversary is to maximize the cost of the algorithm (under a given cost metric), leading to worst case executions of the distributed algorithm.

Asynchronous system: A system is said to be *asynchronous* if there are no bounds on the speeds at which processes take steps. In particular, processes may take steps in arbitrary order and at arbitrary relative speeds.

Byzantine failure: A process failure is said to be *Byzantine* [77] if the process stops following its prescribed algorithm and instead may behave arbitrarily. In particular, the process may behave maliciously, attempting to influence the global computation.

Consensus: Consensus [77,85] is a fundamental problem in distributed computing. In consensus, each process p_i proposes a value v_i. Every nonfaulty process has to eventually return a value (the *termination* condition), which has to be the same for all nonfaulty processes (the *agreement* condition). Further, the value returned has to be one of the values proposed (the *validity* condition).

Crash failure: A process is said to *crash* if the process stops executing the algorithm after some point in the execution.

Eventual synchrony: A system is said to exhibit *eventual synchrony* [41] if, in any execution of an algorithm in the system, there exists a point in time after which the relative speeds of processes are bounded. Note that the time bound may be unknown to the processes.

Execution: An *execution* is a sequence of operations performed by a set of processes. To represent an execution, we usually assume discrete time, where at every time unit there is one active process. Thus, an execution can be represented as a string of process identifiers.

Failure: Distributed algorithms may run on hardware that exhibits faulty behavior; such behavior is modeled as process crashes or process Byzantine failures, but also failures of the communication mechanisms, such as message loss or duplication.

Failure detector: A *failure detector* [28] is an abstraction that encapsulates information about process failures in a distributed system.

Lock-free: An object implementation is *lock-free* if it always guarantees global progress; in particular, the failure of a process does not prevent the system from fulfilling client requests.

Process: The *process* is a sequential unit of computation; it can be seen as a Turing machine. Distributed computing considers the computational power of several connected processes (i.e., several connected Turing machines) under various model assumptions.

Renaming: Renaming [10] is a fundamental problem in distributed computing in which a group of processes need to pick unique names from a small namespace.

Scheduler: Synonymous to *adversary*.

Synchronous system: A system is *synchronous* if nonfaulty processes perform communication and computation in perfect lock-step synchrony.

Transactional memory: Transactional memory [62] is a concurrent programming abstraction, which allows the programmer to specify regions of sequential code that are guaranteed to be executed atomically (i.e., the operations appear to occur instantaneously).

Wait-free: An object implementation is *wait-free* [61] if it guarantees per-process progress; in particular, any nonfaulty process returns from any invocation on the object within a finite number of steps.

Further Information

Distributed computing research is presented at several dedicated conferences. The Association for Computing Machinery (ACM) sponsors the annual Symposium on Principles of Distributed Computing (PODC), and publishes its proceedings. The European Association for Theoretical Computer Science (EATCS) sponsors the annual International Symposium on Distributed Computing (DISC), whose proceedings are published by Springer-Verlag. These are the two major conferences of the field. Results of general interest in distributed computing are published in Computer Science theory conferences, such as the ACM Symposium on Theory of Computing (STOC) or the IEEE Symposium on the Foundations of Computer Science (FOCS), and in the ACM Symposium on Parallel Algorithms and Architectures (SPAA). Complete versions of results can be found in journals such as the *Distributed Computing* journal by Springer, the *Journal of the ACM*, the *SIAM Journal on Computing*, or the *Algorithmica* journal.

There are several excellent textbooks presenting several aspects of the field at an introductory level. One of the main references is the *Distributed Algorithms* book by Lynch [80]. Other textbooks focusing on various aspects of distributed systems are *Reliable and Secure Distributed Programming*, by Cachin et al. [21], and *Distributed Computing: Fundamentals, Simulations, and Advanced Topics* by Attiya and Welch [15]. A recent book by Herlihy and Shavit entitled *The Art of Multiprocessor Programming* [64] provides an up-to-date presentation of shared-memory programming. Harris et al. give an overview of transactional memory research in [59]. The theoretical foundations of transactional memory are explored in [56].

References

1. Noble: A library of non-blocking synchronization protocols. Available at: http://www.cse.chalmers.se/research/group/noble/.
2. The Glasgow Haskell Compiler. 2011. Available at: http://www.haskell.org/ghc/.
3. M. K. Aguilera, I. Keidar, D. Malkhi, and A. Shraer. Dynamic atomic storage without consensus. *Journal of the ACM*, 58(2):7, 2011.
4. D. Alistarh, J. Aspnes, K. Censor-Hillel, S. Gilbert, and M. Zadimoghaddam. Optimal-time adaptive strong renaming, with applications to counting. In: *Proceedings of the 30th Annual ACM Symposium on Principles of Distributed Computing (PODC)*, San Jose, CA, pp. 239–248, 2011.
5. D. Alistarh, J. Aspnes, S. Gilbert, and R. Guerraoui. The complexity of renaming. In: *Proceedings of the 52nd IEEE Symposium on Foundations of Computer Science (FOCS)*, Palm Springs, CA, USA, pp. 718–727, 2011.
6. D. Alistarh, S. Gilbert, R. Guerraoui, and C. Travers. Of choices, failures and asynchrony: The many faces of set agreement. *Algorithmica*, 62(1–2):595–629, 2012.
7. J. Aspnes, H. Attiya, and K. Censor. Polylogarithmic concurrent data structures from monotone circuits. *Journal of the ACM*, 59(1):2:1–2:24, 2012.
8. J. Aspnes and O. Waarts. Randomized consensus in expected $O(n \log^2 n)$ operations per processor. *SIAM Journal on Computing*, 25(5):1024–1044, 1996.
9. H. Attiya, A. Bar-Noy, and D. Dolev. Sharing memory robustly in message-passing systems. In: *Proceedings of the 9th Annual ACM Symposium on Principles of Distributed Computing (PODC)*, Quebec, Canada, pp. 363–375. ACM Press, New York, 1990.
10. H. Attiya, A. Bar-Noy, D. Dolev, D. Peleg, and R. Reischuk. Renaming in an asynchronous environment. *Journal of the ACM*, 37(3):524–548, 1990.
11. H. Attiya and K. Censor. Tight bounds for asynchronous randomized consensus. *Journal of the ACM*, 55(5):1–26, 2008.

12. H. Attiya, C. Dwork, N. Lynch, and L. Stockmeyer. Bounds on the time to reach agreement in the presence of timing uncertainty. *Journal of the ACM*, 41(1):122–152, 1994.
13. H. Attiya, R. Guerraoui, D. Hendler, and P. Kuznetsov. The complexity of obstruction-free implementations. *Journal of the ACM*, 56(4):1–33, 2009.
14. H. Attiya, D. Hendler, and P. Woelfel. Tight RMR lower bounds for mutual exclusion and other problems. In: *Proceedings of the 40th Annual ACM Symposium on Theory of Computing (STOC)*, Victoria, British Columbia, Canada, pp. 217–226. ACM, 2008.
15. H. Attiya and J. Welch. *Distributed Computing. Fundamentals, Simulations, and Advanced Topics*. McGraw-Hill, Hightstown, NJ, 1998.
16. L. Barenboim and M. Elkin. Distributed ($\Delta + 1$)-coloring in linear (in Δ) time. In: *Proceedings of the 41st Annual ACM Symposium on Theory of Computing (STOC)*, Bethesda, MD, pp. 111–120. ACM, 2009.
17. M. Ben-Or. Another advantage of free choice: Completely asynchronous agreement protocols. In: *Proceedings of the 2nd Annual ACM Symposium on Principles of Distributed Computing (PODC)*, Montreal, Quebec, Canada, pp. 27–30, 1983.
18. M. A. Bender and S. Gilbert. Mutual exclusion with $O(\log^2 \log n)$ amortized work. In: *Proceedings of the 52nd IEEE Symposium on Foundations of Computer Science (FOCS)*, Palm Springs, CA, pp. 728–737, 2011.
19. E. Borowsky and E. Gafni. Generalized FLP impossibility result for t-resilient asynchronous computations. In: *Proceedings of the 25th Annual ACM Symposium on Theory of Computing (STOC)*, San Diego, CA, pp. 91–100. ACM Press, New York, May 1993.
20. J. F. Buss, P. C. Kanellakis, P. Ragde, and A. A. Shvartsman. Parallel algorithms with processor failures and delays. *Journal of Algorithms*, 20(1):45–86, 1996.
21. C. Cachin, R. Guerraoui, and L. Rodrigues. *Introduction to Reliable and Secure Distributed Programming* (2nd edn.). Springer, Berlin, Germany, 2011.
22. A. Castañeda and S. Rajsbaum. New combinatorial topology bounds for renaming: The lower bound. *Distributed Computing*, 22(5–6):287–301, 2010.
23. A. Castañeda and S. Rajsbaum. New combinatorial topology bounds for renaming: The upper bound. *Journal of the ACM*, 59(1):3, 2012.
24. M. Castro and B. Liskov. Practical byzantine fault tolerance and proactive recovery. *ACM Transactions on Computer Systems (TOCS)*, 20(4):398–461, 2002.
25. T. D. Chandra, R. Griesemer, and J. Redstone. Paxos made live: An engineering perspective. In: *Proceedings of the 26th Annual Symposium on Principles of Distributed Computing (PODC)*, Portland, OR, pp. 398–407, ACM. New York, 2007.
26. T. D. Chandra, V. Hadzilacos, and S. Toueg. The weakest failure detector for solving consensus. *Journal of the ACM*, 43(4):685–722, 1996.
27. T. D. Chandra and S. Toueg. Unreliable failure detectors for asynchronous systems (preliminary version). In: *Proceedings of the 10th Annual ACM Symposium on Principles of Distributed Computing (PODC)*, Montreal, Quebec, Canada, pp. 325–340, August 1991.
28. T. D. Chandra and S. Toueg. Unreliable failure detectors for reliable distributed systems. *Journal of the ACM*, 43(2):225–267, 1996.
29. F. Chang, J. Dean, S. Ghemawat, W. C. Hsieh, D. A. Wallach, M. Burrows, T. Chandra, A. Fikes, and R. E. Gruber. Bigtable: A distributed storage system for structured data. *ACM Transactions on Computer System*, 26(2), 2008.
30. F. Chierichetti, S. Lattanzi, and A. Panconesi. Rumor spreading in social networks. *Theoretical Computer Science*, 412(24):2602–2610, 2011.
31. A. Clement, E. L. Wong, L. Alvisi, M. Dahlin, and M. Marchetti. Making byzantine fault tolerant systems tolerate byzantine faults. In: *Proceedings of the 6th USENIX Symposium on Networked Systems Design and Implementation (NSDI)*, Boston, MA, pp. 153–168, 2009.

32. F. Cristian. Synchronous and asynchronous. *Communications of the ACM*, 39(4):88–97, 1996.
33. R. De Prisco, B. Lampson, and N. A. Lynch. Revisiting the Paxos algorithm. *Theoretical Computer Science*, 243:35–91, 2000.
34. D. Dice, O. Shalev, and N. Shavit. Transactional Locking II. In: *Proceedings of the 20th International Symposium on Distributed Computing (DISC)*, Stockholm, Sweden, pp. 194–208, 2006.
35. D. Dobre, R. Guerraoui, M. Majuntke, N. Suri, and M. Vukolic. The complexity of robust atomic storage. In: *Proceedings of the 30th Annual ACM Symposium on Principles of Distributed Computing (PODC)*, San Jose, CA, pp. 59–68, 2011.
36. B. Doerr, M. Fouz, and T. Friedrich. Social networks spread rumors in sublogarithmic time. In: *Proceedings of the 43rd Annual ACM Symposium on Theory of Computing (STOC)*, San Jose, CA, pp. 21–30. ACM Press, New York, 2011.
37. D. Dolev, C. Dwork, and L. Stockmeyer. On the minimal synchronism needed for distributed consensus. *Journal of the ACM*, 34(1):77–97, 1987.
38. A. Dragojevic, R. Guerraoui, and M. Kapalka. Stretching transactional memory. In: *Proceedings of the ACM Conference on Programming Language Design and Implementation (PLDI)*, Dublin, Ireland, pp. 155–165, 2009.
39. P. Dutta and R. Guerraoui. The inherent price of indulgence. *Distributed Computing*, 18(1):85–98, 2005.
40. P. Dutta, R. Guerraoui, and I. Keidar. The overhead of consensus failure recovery. *Distributed Computing*, 19(5–6):373–386, 2007.
41. C. Dwork, N. A. Lynch, and L. Stockmeyer. Consensus in the presence of partial synchrony. *Journal of the ACM*, 35(2):288–323, 1988.
42. F. Ellen, Y. Lev, V. Luchangco, and M. Moir. SNZI: Scalable NonZero Indicators. In: *Proceedings of the 26th Annual ACM Symposium on Principles of Distributed Computing (PODC)*, Portland, OR, pages 13–22, ACM, New York, 2007.
43. R. Fagin, J. Y. Halpern, and M. Y. Vardi. What can machines know? On the properties of knowledge in distributed systems. *Journal of the ACM*, 39(2):328–376, 1992.
44. R. Fan and N. A. Lynch. Gradient clock synchronization. *Distributed Computing*, 18(4):255–266, 2006.
45. F. Fich, M. Herlihy, and N. Shavit. On the space complexity of randomized synchronization. *Journal of the ACM*, 45(5):843–862, 1998.
46. F. E. Fich, D. Hendler, and N. Shavit. Linear lower bounds on real-world implementations of concurrent objects. In: *Proceedings of the 46th IEEE Symposium on Foundations of Computer Science (FOCS)*, Pittsburg, PA, pp. 165–173, 2005.
47. M. J. Fischer, N. A. Lynch, and M. S. Paterson. Impossibility of distributed consensus with one faulty process. *Journal of the ACM*, 32(2):374–382, 1985.
48. P. Fraigniaud, A. Korman, and D. Peleg. Local distributed decision. In: Rafail Ostrovsky, editor, *IEEE 52nd Annual Symposium on Foundations of Computer Science*, Palm Springs, CA, pp. 708–717. IEEE, 2011.
49. P. Fraigniaud, S. Rajsbaum, and C. Travers. Locality and checkability in wait-free computing. In: *Proceedings of the 25th International Symposium on Distributed Computing (DISC)*, Rome, Italy, pp. 333–347, 2011.
50. E. Gafni. Round-by-round fault detectors (extended abstract): Unifying synchrony and asynchrony. In: *Proceedings of the 17th Symposium on Principles of Distributed Computing (PODC)*, Puerto Vallarta, Mexico, pp. 143–152, 1998.
51. E. Gafni, R. Guerraoui, and B. Pochon. The complexity of early deciding set agreement. *SIAM Journal on Computing*, 40(1):63–78, 2011.
52. E. Gafni and P. Kuznetsov. The weakest failure detector for solving k-set agreement. In: *Proceedings of the 28th ACM Symposium on the Principles of Distributed Computing (PODC)*, Calgary, AB, pp. 83–91, 2009.
53. E. Gafni and L. Lamport. Disk paxos. *Distributed Computing*, 16(1):1–20, 2003.

54. J. Gray. The transaction concept: Virtues and limitations (invited paper). In: *Proceedings of the 7th International Symposium on Very Large Data Bases (VLDB)*, Cannes, France, pp. 144–154, 1981.
55. R. Greenberg. Bitcoin: An innovative alternative digital currency. *Social Science Research Network Working Paper Series*, pp. 1–45, 2011.
56. R. Guerraoui and M. Kapalka. *Principles of Transactional Memory*. Synthesis Lectures on Distributed Computing Theory. Morgan & Claypool Publishers, 2010.
57. R. Guerraoui, N. Knezevic, V. Quéma, and M. Vukolic. The next 700 BFT protocols. In: *Proceedings of the 5th European Conference on Computer Systems (EuroSys)*, Paris, France, pp. 363–376, 2010.
58. J. Y. Halpern and Y. Moses. Knowledge and common knowledge in a distributed environment. *Journal of the ACM*, 37(3):549–587, 1990.
59. T. Harris, J. R. Larus, and R. Rajwar. *Transactional Memory, 2nd edn*. Synthesis Lectures on Computer Architecture. Morgan & Claypool Publishers, 2010.
60. T. L. Harris. A pragmatic implementation of non-blocking linked-lists. In: *Proceedings of the 15th International Conference on Distributed Computing (DISC)*, Lisbon, Portugal, pp. 300–314, Springer-Verlag, London, UK, 2001.
61. M. Herlihy. Wait-free synchronization. *ACM Transactions on Programming Languages and Systems*, 13(1):123–149, 1991.
62. M. Herlihy and J. E. B. Moss. Transactional memory: Architectural support for lock-free data structures. *SIGARCH Computer Architecture News*, 21(2):289–300, 1993.
63. M. Herlihy and N. Shavit. The topological structure of asynchronous computability. *Journal of the ACM*, 46(2):858–923, 1999.
64. M. Herlihy and N. Shavit. *The Art of Multiprocessor Programming*. Morgan Kaufmann, Amsterdam, 2008.
65. IBM. The IBM BlueGene System. 2011.
66. Intel. Transactional Synchronization in Haswell. 2012.
67. M. Isard. Autopilot: Automatic data center management. *Operating Systems Review*, 41(2):60–67, 2007.
68. P. Jayanti, K. Tan, and S. Toueg. Time and space lower bounds for nonblocking implementations. *SIAM Journal on Computing*, 30(2):438–456, 2000.
69. R. Karp, C. Schindelhauer, S. Shenker, and B. Vocking. Randomized rumor spreading. In: *Proceedings of the 41st IEEE Symposium on Foundations of Computer Science (FOCS)*, Los Angeles, CA, pp. 565–574, 2000.
70. R. Kotla, L. Alvisi, M. Dahlin, A. Clement, and E. L. Wong. Zyzzyva: Speculative byzantine fault tolerance. *ACM Transactions on Computer Systems*, 27(4), 2009.
71. D. R. Kowalski and A. A. Shvartsman. Writing-all deterministically and optimally using a nontrivial number of asynchronous processors. *ACM Transactions on Algorithms*, 4(3):1–22, 2008.
72. F. Kuhn. Weak graph colorings: Distributed algorithms and applications. In: *Proceedings of the 21st Annual Symposium on Parallelism in Algorithms and Architectures (SPAA)*, Calgary, Alberta, Canada, pp. 138–144. ACM, 2009.
73. L. Lamport. Using time instead of timeout for fault-tolerant distributed systems. *ACM Transactions on Programming Languages and Systems*, 6(2):254–280, 1984.
74. L. Lamport. The part-time parliament. *ACM Transactions on Computer Systems*, 16(2):133–169, 1998.
75. L. Lamport. Fast paxos. *Distributed Computing*, 19(2):79–103, 2006.
76. L. Lamport and M. Massa. Cheap paxos. In: *DSN*, Florence, Italy, pp. 307–314. IEEE Computer Society, 2004.
77. L. Lamport, R. Shostak, and M. Pease. The byzantine generals problem. *ACM Transactions on Programming Languages and Systems*, 4(3):382–401, 1982.
78. C. Lenzen, T. Locher, and R. Wattenhofer. Tight bounds for clock synchronization. *Journal of the ACM*, 57(2), 2010.
79. N. Linial. Locality in distributed graph algorithms. *SIAM Journal on Computing*, 21(1):193–201, 1992.

80. N. A. Lynch. *Distributed Algorithms*. Morgan Kaufmann, San Mateo, CA, 1996.
81. M. M. Michael and M. L. Scott. Simple, fast, and practical non-blocking and blocking concurrent queue algorithms. In: *Proceedings of the 15th Annual ACM Symposium on Principles of Distributed Computing (PODC)*, Philadelphia, PA, pp. 267–275, 1996.
82. Sun Microsystems. Rock: A SPARC CMT Processor. 2008.
83. Oracle. Package java.util.concurrent. Available at: http://docs.oracle.com/javase/6/docs/api/java/util/concurrent/.
84. A. Panconesi, M. Papatriantafilou, P. Tsigas, and P. M. B. Vitányi. Randomized naming using wait-free shared variables. *Distributed Computing*, 11(3):113–124, 1998.
85. M. Pease, R. Shostak, and L. Lamport. Reaching agreement in the presence of faults. *Journal of the ACM*, 27(2):228–234, 1980.
86. M. Saks and F. Zaharoglou. Wait-free *k*-set agreement is impossible: The topology of public knowledge. In: *Proceedings of the 25th ACM Symposium on Theory of Computing (STOC)*, San Diego, CA, pp. 101–110. ACM Press, New York, May 1993.
87. F. B. Schneider. Implementing fault-tolerant services using the state machine approach: A tutorial. *ACM Computing Surveys*, 22(4):299–319, 1990.
88. N. Shavit. Data structures in the multicore age. *Communications of the ACM*, 54(3):76–84, 2011.
89. N. Shavit and I. Lotan. Skiplist-based concurrent priority queues. In: *Proceedings of the 14th International Parallel and Distributed Processing Symposium (IPDPS)*, Cancun, Mexico, pp. 263–268, 2000.
90. N. Shavit and D. Touitou. Software transactional memory. *Distributed Computing*, 10(2):99–116, 1997.
91. J.-H. Yang and J. H. Anderson. A fast, scalable mutual exclusion algorithm. *Distributed Computing*, 9(1):51–60, 1995.

III

Architecture and Organization

17

Digital Logic

Miriam Leeser
Northeastern University

James Brock
Cognitive Electronics

17.1 Introduction

This chapter explores combinational and sequential Boolean logic design as well as technologies for implementing efficient, high speed digital circuits. Some of the most common devices used in computers and general logic circuits are described. Sections 17.2 through 17.4 introduce the fundamental concepts of logic circuits and in particular the rules and theorems upon which *combinational logic*, logic with no internal memory, is based. Section 17.5 describes in detail some frequently used combinational logic components and shows how they can be combined to build the arithmetic and logical unit (ALU) for a simple calculator. Section 17.6 introduces the subject of *sequential logic*; logic in which feedback and thus internal memory exist. Two of the most important elements of sequential logic design, the *data flip-flop* (DFF) and the *register*, are introduced. Memory elements are combined with the ALU to complete the design of a simple calculator. The final section of the chapter examines field-programmable gate arrays (FPGAs) which now provide fast, economical solutions for implementing large logic designs for solving diverse problems.

17.2 Overview of Logic

Logic has been a favorite academic subject, certainly since the Middle Ages and arguably since the days of the greatness of Athens. That use of *logic* connoted the pursuit of orderly methods for defining theorems and proving their consistency with certain accepted propositions. In the middle of the nineteenth century, George Boole put the whole subject on a sound mathematical basis and spread "logic" from the Philosophy Department into Engineering and Mathematics. (Boole's original writings have recently been reissued, Boole, 2009.) Specifically, what Boole did was to create an algebra of two-valued (*binary*) variables. Initially designated as *true* or *false*, these two values can represent any parameter that has two clearly defined states. Boolean algebras of more than two values have been explored, but the original binary variable of Boole dominates the design of circuitry for reasons that we will explore. This chapter presents some of the rules and methods of binary Boolean algebra and shows how it is used to design digital hardware to meet specific engineering applications.

A system that is well adapted to digital (discrete) representation is one that spends little time in a state of ambiguity. All digital systems spend some time in indeterminate states when switching between values. One very common definition of the two states is made for systems operating between 2.5 volts (V) and ground. It is shown in Figure 17.1. One state, usually called *one*, is defined as any voltage greater than 1.7 V. The other state, usually called *zero*, is defined as any voltage less than 0.8 V.

The gray area in the middle is *ambiguous*. When an input signal is between 0.8 and 1.7 V in a 2.5 V complementary metal oxide semiconductor (CMOS) digital circuit, you cannot predict the output value. Most of what you will read in this chapter assumes that input variables are clearly assigned to the state *one* or the state *zero*. In real designs, there are always moments when the inputs are ambiguous. A good design is one in which the system never makes decisions based on ambiguous data. Such requirements limit the speed of response of real systems; they must wait for the ambiguities to settle out.

17.3 Concept and Realization of Digital Gate

A *gate* is the basic building block of digital circuits. A gate is a circuit with one or more inputs and a single output. From a logical perspective in a binary system, any input or output can take on only the values *one* and *zero*. From an analog perspective, the gates make transitions through the ambiguous region with great rapidity and quickly resolve to an unambiguous state.

In Boolean algebra, a good place to begin is with three operations: NOT, AND, and OR. These have similar meaning to their meaning in English. Given two input variables, called *A* and *B*, and an output variable *X*, X = NOT *A* is true when *A* is false, and false when *A* is true. When both inputs are true, X = *A* AND *B* is true, and when either *A* or *B* is true (or both are true), X = *A* OR *B* is true. This is called an *inclusive or* function because it includes the case where *A* and *B* are both true. There is another Boolean operator, *exclusive or*, that is true when either *A* or *B*, but not both, is true. In fact there are 16 Boolean functions of two variables. The more useful functions are shown in truth table form in Table 17.1. These functions can be generalized to more than one variable, as is shown in Table 17.2.

Functions AND, OR, and NOT are sufficient to describe all Boolean logic functions. Why do we need all these other operators?

Logic gates are themselves an abstraction. The actual physical realization of logic gates is with transistors. Most digital designs are implemented in CMOS technology. In CMOS and most other transistor

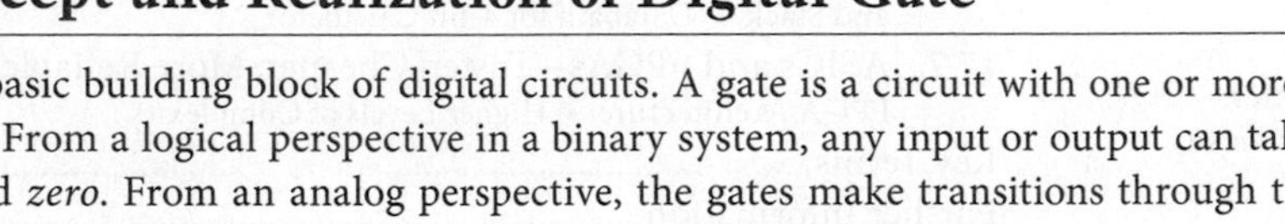

FIGURE 17.1 The states zero and one as defined in 2.5 V CMOS logic.

TABLE 17.1 Boolean Operators of Two Input Variables

Inputs A B	True	False	A	NOT(A)	AND	OR	XOR	NAND	NOR	XNOR
0 0	1	0	0	1	0	0	0	1	1	1
0 1	1	0	0	1	0	1	1	1	0	0
1 0	1	0	1	0	0	1	1	1	0	0
1 1	1	0	1	0	1	1	0	0	0	1

TABLE 17.2 Boolean Operators Extended to More Than Two Inputs

Operation	Input Variables	Operator Symbol	Output = 1 if
NOT	A	$\bar{A}$	$A = 0$
AND	$A, B, \ldots$	$A \cdot B \cdots$	All of the set $[A, B, \ldots]$ are 1.
OR	$A, B, \ldots$	$A + B + \cdots$	Any of the set $[A, B, \ldots]$ are 1.
NAND	$A, B, \ldots$	$\overline{(A \cdot B \cdots)}$	Any of the set $[A, B, \ldots]$ are 0.
NOR	$A, B, \ldots$	$\overline{(A + B + \cdots)}$	All of the set $[A, B, \ldots]$ are 0.
XOR	$A, B, \ldots$	$A \oplus B \oplus \cdots$	The set $[A, B, \ldots]$ contains an odd number of 1's.
XNOR	$A, B, \ldots$	$A \odot B \odot \cdots$	The set $[A, B, \ldots]$ contains an even number of 1's.

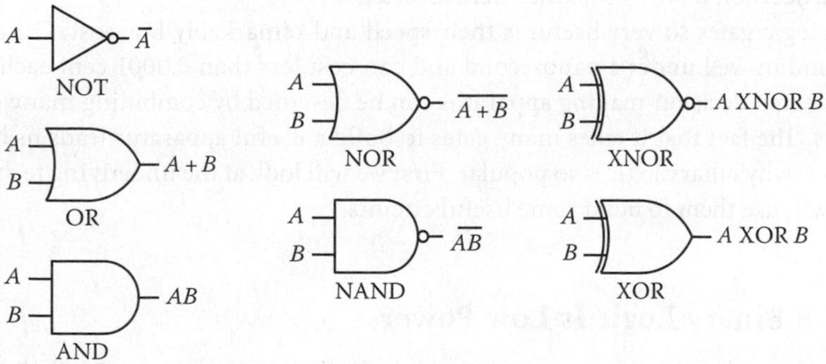

FIGURE 17.2 Commonly used graphical symbols for seven of the gates defined in Table 17.1.

technologies, logic gates are naturally inverting. In other words, it is very natural to build NOT, NAND, and NOR gates, even if it is more natural to think about positive logic: AND and OR. Neither XOR nor XNOR are natural building blocks of CMOS technology. They are included for completeness. As we shall see, AND and OR gates are implemented with NAND and NOR gates.

There are widely used graphical symbols for these same operations. These are presented in Figure 17.2. The symbol for NOT includes both a buffer (the triangle) and the actual inversion operation (the open circle). Often, the inversion operation alone is used, as seen in the outputs of NAND, NOR, and XNOR. In writing Boolean operations we use the symbols $\bar{A}$ for NOT A, $A + B$ for A OR B, and $A \cdot B$ for A AND B. $A + B$ is called the *sum* of A and B and $A \cdot B$ is called the *product*. The operator for AND is often omitted, and the operation is implied by adjacency, just like in multiplication. To illustrate the use of these symbols and operators, Figure 17.3 shows two constructs made from the gates of Figure 17.2. These two examples show how to build the expression $AB + CD$ and how to construct an XOR from the basic gates AND, OR, and NOT.

The first construct of Figure 17.3 would fit the logic of the sentence: "I will be content if my federal and state taxes are lowered (A and B, respectively), or if the money that I send is spent on reasonable things

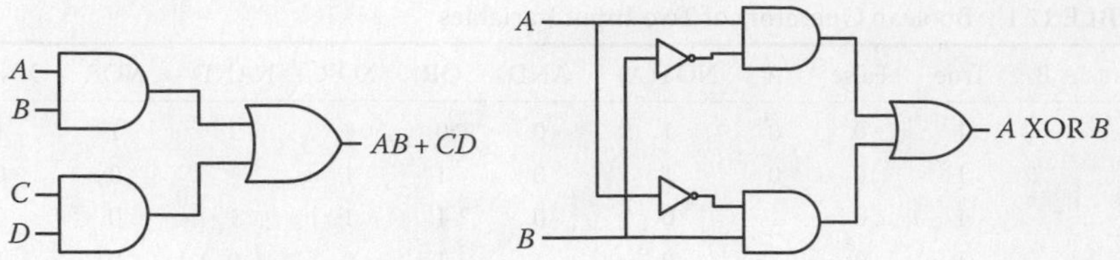

FIGURE 17.3 Two constructs built from the gates in column 1 of Figure 17.2. The first is a common construct in which if either of two paired propositions is TRUE, the output is TRUE. The second is XOR constructed from the more primitive gates, AND, OR, and NOT.

and spent effectively (*C* and *D*, respectively)." You would certainly expect the speaker to be content if either pair is TRUE and most definitely content if both are TRUE. The output on the right side of the construct is TRUE if either or both of the inputs to the OR is TRUE. The outputs of the AND gates are TRUE when both of their inputs are TRUE. In other words, both state and federal taxes must be reduced to make the top AND's output TRUE.

The right construct in Figure 17.3 gives an example of how one can build one of the basic logic gates, in this case the XOR gate, from several of the others. Let us consider the relationship of this construct to common speech. The sentence: "With the time remaining, we should eat dinner or go to a movie." The implication is that one cannot do both. The circuit on the right of Figure 17.3 would indicate an acceptable decision (TRUE if acceptable) if either movie or dinner were selected (*asserted* or made TRUE) but an unacceptable decision if both or neither were asserted.

What makes logic gates so very useful is their speed and remarkably low cost. On-chip logic gates today can respond in well under a nanosecond and can cost less than 0.0001 cent each. Furthermore, a rather sophisticated decision-making apparatus can be designed by combining many simple-minded binary decisions. The fact that it takes many gates to build a useful apparatus leads us back directly to one of the reasons why binary logic is so popular. First we will look at the underlying technology of logic gates. Then we will use them to build some useful circuits.

17.3.1 CMOS Binary Logic Is Low Power

A modern microcomputer chip contains on the order of billions of logic gates. If all of those gates were generating heat at all times, the chip would melt. Keeping them cool is one of the most critical issues in computer design. Good thermal designs were significant parts of the success of Cray, IBM, Intel, and Sun. One of the principal advantages of CMOS binary logic is that it can be made to expend much less energy to generate the same amount of calculation as other forms of circuitry.

Gates are classified as *active logic* or *saturated logic* depending on whether they control the current continuously or simply switch it on or off. In active logic, the gate has a considerable voltage across it and conducts current in all of its states. The result is that power is continually being dissipated. In saturated logic, the TRUE–FALSE dichotomy has the gate striving to be perfectly connected to the power bus when the output voltage is high and perfectly connected to the ground bus when the voltage is low. These are zero-dissipation ideals that are not achieved in real gates, but the closer one gets to the ideal, the better the gate. When you start with more than one million gates per chip, small reductions in power dissipation make the difference between usable and unusable chips.

Saturated logic is *saturated* because it is driven hard enough to ensure that it is in a minimum-dissipation state. Because it takes some effort to bring such logic out of saturation, it is a little slower than active logic. Active logic, on the other hand, is always dissipative. It is very fast, but it is always getting hot. Although it has often been the choice for the most active circuits in the fastest computers, active logic has never been a major player, and it owns a diminishing role in today's designs. This chapter focuses on today's dominant family of binary, saturated logic, which is CMOS.

17.3.2 CMOS Switching Model for NOT, NAND, and NOR

The metal–oxide–semiconductor (MOS) transistor is the oldest transistor in concept and still the best in one particular aspect: its control electrode—also called a *gate* but in a different meaning of that word from *logic gate*—is a purely capacitive load. Holding it at constant voltage takes no energy whatsoever. These MOS transistors, like most transistors, come in two types. One turns on with a positive voltage; the other turns off with a positive voltage. This pairing allows one to build *complementary* gates, which have the property that they dissipate no energy except when switching. Given the large number of logic gates and the criticality of power reduction, it is small wonder that the CMOS gate dominates today's digital technology.

Consider how we can construct a set of primitive gates in the CMOS family. The basic element is a pair of switches in series, the NOT gate. This basic building block is shown in Figure 17.4. The switching operation is shown in the two drawings to the right. If the input is low, the upper switch is closed and the lower one is open—complementary operation. This connects the output to the high side. Apart from voltage drops across the switch itself, the output voltage becomes the voltage of the high bus. If the input now goes high, both switches flip and the output is connected, through the resistance of the switch, to the ground bus, high-in, low-out, and vice versa. We have an inverter. Only while the switches are switching is there significant current flowing from one bus to the other. Thus, in the static state, these devices dissipate almost no power at all. Once one has the CMOS switch concept, it is easy to show how to build NAND and NOR gates with multiple inputs.

Let us look at the switching structure of a 3-input NAND and 3-input NOR, just to show how multiple-input gates are created. The basic inverter or NOT gate of Figure 17.4 is our paradigm; if the lower switch is closed, the upper one is open, and vice versa. To go from NOT to an *N*-input NAND, make the single lower switch in the NOT a series of *N* switches, so only one of these need be open to open the circuit. Then change the upper complementary switch in the NOT into *N* parallel switches. With these, only one switch need be closed to connect the circuit. Such an arrangement with $N = 3$ is shown on the left in Figure 17.5. On the left, if any input is low, the output is high. On the right is the construction for NOR. All three inputs must be low to drive the output high.

An interesting question at this point is: How many inputs can such a circuit support? The answer is called the *fan-in* of the circuit. The fan-in depends mostly on the resistance of each switch in the series string. In most cases, six or seven inputs would be considered a reasonable limit. The number of outputs a gate can support is the *fan-out*. Logic gates can be designed with a considerably higher fan-out than fan-in.

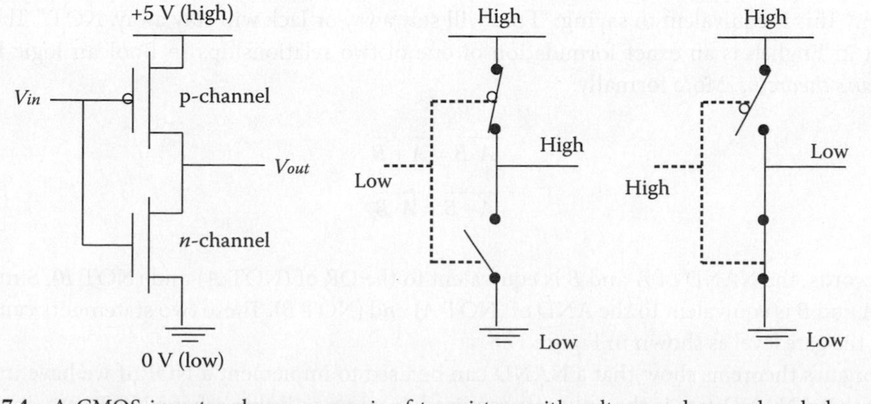

FIGURE 17.4 A CMOS inverter shown as a pair of transistors with voltage and ground and also as pairs of switches with logic levels. The open circle indicates logical negation (NOT).

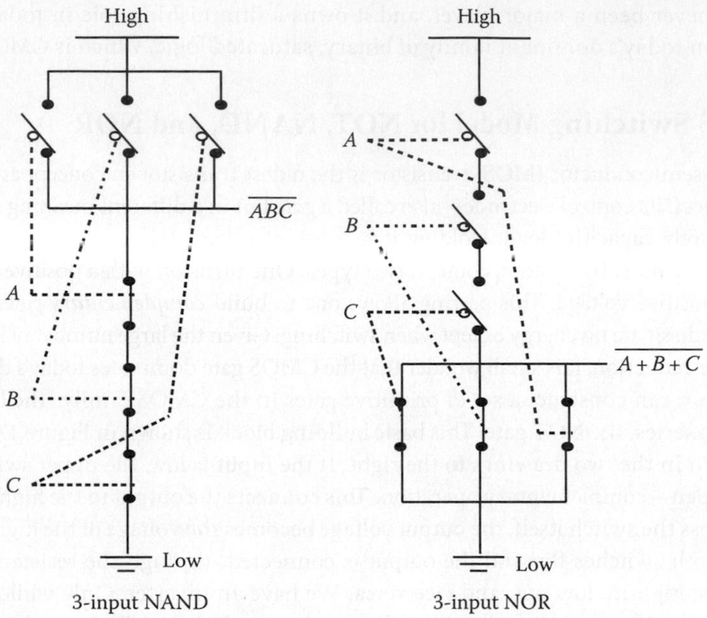

FIGURE 17.5 Three pairs of CMOS switches arranged on the left to execute the 3-input NAND function and on the right the 3-input NOR. The switches are shown with all the inputs high, putting the output in the low state.

17.3.3 Doing It All with NAND

We think of the basic logic operators as being NOT, AND, and OR, since these seem to be the most natural. When it comes to building logic gates out of CMOS transistor technology, as we have just seen, the "natural" logic gates are NOTs, NANDs, and NORs.

To build an AND or an OR gate, you take a NAND or NOR and add an inverter. The more primitive nature of NAND and NOR comes about because transistor switches are inherently inverting. Thus, a single-stage gate will be NAND or NOR; AND and OR gates require an extra stage. If this is the way one were to implement a design with a million gates, a million extra inverters would be required. Each extra stage requires extra area and introduces longer propagation delays. Simplifying logic to eliminate delay and unnecessary heat are two of the most important objectives of logic design. Instead of using an inverter after each NAND or NOR gate, most designs use the inverting gates directly. We will see how Boolean logic helps as to do this. Consider the declaration: "Fred and Jack will come over this afternoon." This is equivalent to saying: "Fred will stay away or Jack will stay away, NOT." This strange construct in English is an exact formulation of one of two relationships in Boolean logic known as *De Morgan's theorems*. More formally:

$$\overline{A \cdot B} = \overline{A} + \overline{B}$$

$$\overline{A + B} = \overline{A} \cdot \overline{B}$$

In other words, the NAND of *A* and *B* is equivalent to the OR of (NOT *A*) and (NOT *B*). Similarly, the NOR of *A* and *B* is equivalent to the AND of (NOT *A*) and (NOT *B*). These two statements can be represented at the gate level as shown in Figure 17.6.

De Morgan's theorems show that a NAND can be used to implement a NOR if we have inverters. It turns out that a NAND gate is the only gate required. Next we will show that a NOT gate (inverter) can be constructed from a NAND. Once we have shown that NORs and NOTs can be constructed out of

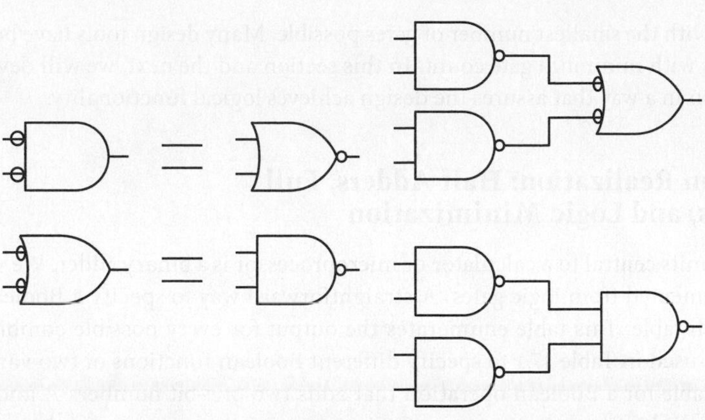

FIGURE 17.6 On the left, the two forms of De Morgan's theorem in logic gates. On the right, the two forms of the circuit on the left of Figure 17.3. In the upper form, we have replaced the lines between the ANDs and OR with two inverters in series. Then, we have used the lower form of De Morgan's theorem to replace the OR and its two inverters with a NAND. The resulting circuit is all-NAND and is simpler to implement than the construction from AND and OR in Figure 17.3.

NANDs, only NAND gates are required. An AND gate is a NAND followed by a NOT and an OR gate is a NOR followed by a NOT. Thus, all other logic gates can be implemented from NANDs. The same is true of NOR gates; all other logic gates can be implemented from NORs.

Take a NAND gate and connect both inputs to the same input A. The output is the function $(A \cdot A)$. Since A AND A is TRUE only if A is TRUE ($AA = A$), we have just constructed our inverter. If we actually wanted an inverter, we would not use a 2-input gate where a 1-input gate would do. But we could. This exercise shows that the minimal number of distinct logic gates required to implement all Boolean logic functions is one. In reality we use AND, OR, and NOT when using positive logic and NAND, NOR, and NOT when using negative logic or thinking about how logic gates are implemented with transistors.

17.4 Rules and Objectives in Combinational Design

Once the concept of a logic gate is established, the next natural question is: What useful devices can you build with them? We will look at a few basic building blocks and show how you can put them together to build a simple calculator.

The components of digital circuits can be divided into two classes. The first class of circuits has outputs that are simply some logical combination of their inputs. Such circuits are called *combinational*. Examples include the gates we have just looked at and those that we will examine in this section and in Section 17.5. The other class of circuits, constructed from combinational gates, but with the addition of internal feedback, have the property of *memory*. Thus, their output is a function not only of their inputs but also of their previous state(s). Since such circuits go through a sequence of states, they are called *sequential*. These will be discussed in Section 17.6.

The two principal objectives in digital design are functionality and minimum cost. Functionality requires not only that the circuit generates the correct outputs for any possible inputs but also that those outputs be available quickly enough to serve the application. Minimum cost must include both the design effort and the cost of production and operation. For very small production runs (<10,000), one wants to "program" off-the-shelf devices. For very large runs, costs focus mostly on manufacture and operation. The *operation* costs are dominated by cooling or battery drain, where these necessary peripherals add weight and complexity to the finished product. To fit in off-the-shelf devices, to reduce delays between input and output, and to reduce the gate count and thus the dissipation for a given functionality, designs must be realized with the smallest number of gates possible. Many design tools have been developed for achieving designs with minimum gate count. In this section and the next, we will develop the basis for such minimization in a way that assures the design achieves logical functionality.

17.4.1 Boolean Realization: Half Adders, Full Adders, and Logic Minimization

One of the basic units central to a calculator or microprocessor is a binary adder. We will consider how an adder is implemented from logic gates. A straightforward way to specify a Boolean logic function is by using a truth table. This table enumerates the output for every possible combination of inputs. Truth tables were used in Table 17.1 to specify different Boolean functions of two variables. Table 17.3 shows the truth table for a Boolean operation that adds two one-bit numbers *A* and *B* and produces two outputs: the sum bit *S* and the carry-out *C*. Since binary numbers can only have the values 1 or 0, adding two binary numbers each of value 1 will result in there being a carry-out. This operation is called a *half adder*.

To implement the half adder with logic gates, we need to write Boolean logic equations that are equivalent to the truth table. A separate Boolean logic equation is required for each output. The most straightforward way to write an equation from the truth table is to use sum of products (SOP) form to specify the outputs as a function of the inputs. An SOP expression is a set of "products" (ANDs) which are "summed" (ORed) together. Note that any Boolean formula can be expressed in SOP or POS (product of sums) form.

Let us consider output *S*. Every line in the truth table that has a 1 value for an output corresponds to a term that is ORed with other terms in SOP form. This term is formed by ANDing together all of the input variables. If the input variable is a 1 to make the output 1, the variable appears as is in the AND term. If the input is a zero to make the output 1, the variable appears negated in the AND term. Let us apply these rules to the half adder. The *S* output has two combinations of inputs that result in its output being 1, therefore its SOP form has two terms ORed together. The *C* output only has one AND or product term, since only one combination of inputs results in a 1 output. The entire truth table can be summarized as

$$S = \overline{A}\cdot B + A\cdot\overline{B}$$

$$C = A\cdot B$$

Note that we are implicitly using the fact that A and B are Boolean inputs. The equation for C can be read "*C* is 1 when *A* and *B* are both 1." We are assuming that *C* is zero in all other cases. From the Boolean logic equations, it is straightforward to implement *S* and *C* with logic gates, as shown in Figure 17.7. The logical function for *S* is that of an XOR gate, so we show *S* as an XOR gate in the figure.

The half adder is a building block in an *n*-bit binary adder. An *n*-bit binary adder adds *n*-bit numbers represented in base 2. Table 17.4 shows the representation of three-bit, unsigned binary numbers.

TABLE 17.3 Truth Table for a Half Adder

Inputs		Outputs	
A	*B*	*S*	*C*
0	0	0	0
0	1	1	0
1	0	1	0
1	1	0	1

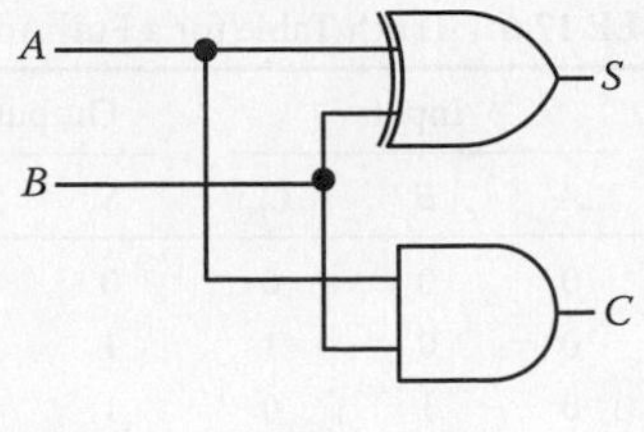

FIGURE 17.7 The gate level implementation of a half adder.

TABLE 17.4 Binary Representation of Decimal Numbers

Decimal	Binary
0	000
1	001
2	010
3	011
4	100
5	101
6	110
7	111

TABLE 17.5 Adding 3-Bit Binary Numbers

1	0		Carry Bits
0	1	0	2
0	1	1	+3
1	0	1	=5

The left-most bit is the most significant bit. It is in the 4's place. The middle bit represents the 2's place and the rightmost bit represents the 1's place. The largest representable number, 111_2, represents 4 + 2 + 1, or 7 in decimal.

Let us examine adding two binary numbers. Table 17.5 shows the operation 2 + 3 = 5 in binary. The top row is the carry-out from the addition in the previous bit location. Notice that there is a carry-out bit with value 1 from the second position to the third (leftmost) bit position.

The half adder described in the preceeding text has two inputs: A and B. This can be used for the rightmost bit where there is no carry-in bit. For other bit positions we use a *full adder* with inputs A, B, and C_{in} and outputs S and C_{out}. The truth table for the full adder is given in Table 17.6. Note that a full adder adds one bit position; it is NOT an n-bit adder.

To realize the full adder as a circuit we need to design it using logic gates. We do this in the same manner as with the half adder, by writing a logic equation for each of the outputs separately. For each 1 in the truth table on the output of the function there is an AND term in the SOP representation. Thus, there are four AND terms for the S equation, and four AND terms for the C_{out} equation. These equations are given as follows:

$$S = \overline{A} \cdot \overline{B} \cdot C_{in} + \overline{A} \cdot B \cdot \overline{C_{in}} + A \cdot \overline{B} \cdot \overline{C_{in}} + ABC_{in}$$

$$C_{out} = \overline{A} \cdot BC_{in} + A \cdot \overline{B} \cdot C_{in} + AB \cdot \overline{C_{in}} + ABC_{in}$$

TABLE 17.6 Truth Table for a Full Adder

	Inputs			Outputs	
	A	B	C_{in}	S	C_{out}
0	0	0	0	0	0
1	0	0	1	1	0
2	0	1	0	1	0
3	0	1	1	0	1
4	1	0	0	1	0
5	1	0	1	0	1
6	1	1	0	0	1
7	1	1	1	1	1

These equations for S and C_{out} are logically correct, but we would also like to use the minimum number of logic gates to implement these functions. The fewer gates used, the fewer gates that need to switch, and hence the smaller amount of power that is dissipated. Next, we will look at applying the rules of Boolean logic to minimize our logic equations.

17.4.2 Axioms and Theorems of Boolean Logic

Our goal is to use the minimum number of logic gates to implement a design. We use logic rules or axioms. These were first described by George Boole, hence the term Boolean algebra. Many of the axioms and theorems of Boolean algebra will seem familiar because they are similar to the rules you learned for algebra in high school. Let us be formal here and state the axioms:

1. Variables are binary: This means that every variable in the algebra can take on one of two values and these two values are not the same. Usually, we will choose to call the two values 1 and 0, but other binary pairs such as TRUE and FALSE, and HIGH and LOW are widely used and often more descriptive. Two binary operators, AND (·) and OR (+), and one unary operator, NOT, can transform variables into other variables. These operators were defined in Table 17.2.
2. Closure: The AND or OR of any two variables is also a binary variable.
3. Commutativity: $A \cdot B = B \cdot A$ and $A + B = B + A$.
4. Associativity: $(A \cdot B) \cdot C = A \cdot (B \cdot C)$ and $(A + B) + C = A + (B + C)$.
5. Identity elements: $A \cdot 1 = 1 \cdot A = A$ and $A + 0 = 0 + A = A$.
6. Distributivity: $A \cdot (B + C) = A \cdot B + A \cdot C$ and $A + (B \cdot C) = (A + B) \cdot (A + C)$. (The usual rules of algebraic hierarchy are used here where · is done before +.)
7. Complementary pairs: $A \cdot \bar{A} = 0$ and $A + \bar{A} = 1$.

These are the axioms of this algebra. They are used to prove further theorems. Each algebraic relationship in Boolean algebra has a *dual*. To get the dual of an axiom or a theorem, one simply interchanges AND and OR as well as 0 and 1. Because of this principle of *duality*, Boolean algebra axioms and theorems come in pairs. The principle of duality tells us that if a theorem is true, then its dual is also true.

In general, one may prove a Boolean theorem by exhaustion—that is, by listing all of the possible cases—although more abstract algebraic reasoning may be more efficient. Here is an example of a pair of theorems based on the axioms given earlier:

Theorem 17.1 (Idempotency)

$A \cdot A = A$ and $A + A = A$.

Proof 17.1

The definition of AND in Table 17.1 can be used with exhaustion to complete the proof for the first form.

$$A \text{ is } 1: 1 \cdot 1 = 1 = A$$

$$A \text{ is } 0: 0 \cdot 0 = 0 = A$$

The second form follows as the dual of the first.

Now let us consider reducing the expression from the previous section:

$$C_{out} = \overline{A} \cdot BC_{in} + A \cdot \overline{B} \cdot C_{in} + AB \cdot \overline{C_{in}} + ABC_{in}$$

First, we apply idempotency twice to triplicate the last term on the right and put the extra terms after the first and second terms by repeated application of axiom 3:

$$C_{out} = \overline{A} \cdot BC_{in} + ABC_{in} + A \cdot \overline{B} \cdot C_{in} + ABC_{in} + AB \cdot \overline{C_{in}} + ABC_{in}$$

Now we apply axioms 4, 3, and 6 to obtain:

$$C_{out} = (\overline{A} + A)BC_{in} + A(\overline{B} + B)C_{in} + AB(\overline{C} + C)$$

And finally, we apply axioms 7 and 5 to obtain:

$$C_{out} = AB + AC_{in} + BC_{in}$$

The reduced equation certainly looks simpler; let us consider the gate representation of the two equations. This is shown in Figure 17.8. From four 3-input ANDs to three 2-input ANDs and from a 4-input OR to a 3-input OR is a major saving in a basically simple circuit.

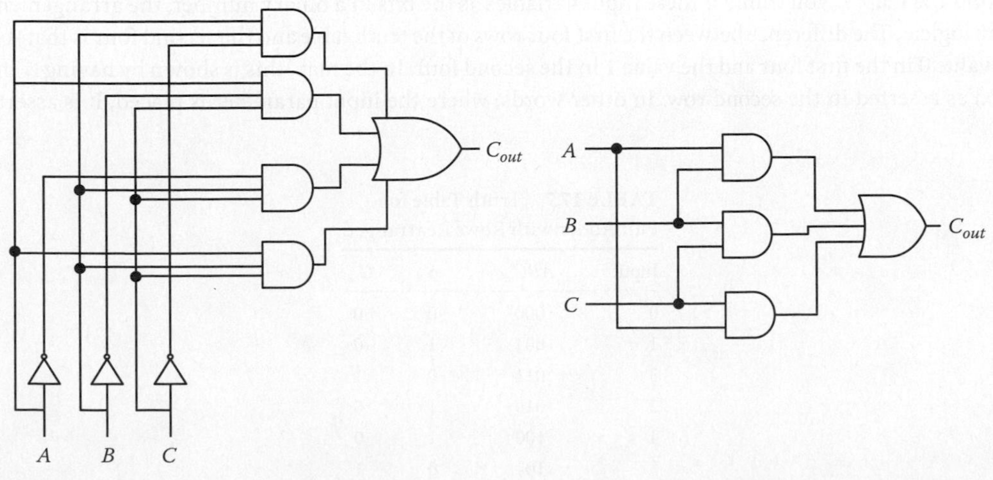

FIGURE 17.8 The direct and reduced circuits for computing the carry-out from the three inputs to the full adder.

The reduction is clear. The savings in a chip containing more than a million gates should build some enthusiasm for gate simplification. What is probably not so clear is how you could know that the key to all of this saving was knowing to make two extra copies of the fourth term in the direct expression. It turns out that there is a fairly direct way to see what you have to do, one that takes advantage of the eye's remarkable ability to see a pattern. This tool, the *Karnaugh map*, is the topic of the next section.

17.4.3 Design, Gate-Count Reduction, and SOP/POS Conversions

The *truth table* for the full adder was given in Table 17.6. All possible combinations of the three input bits appear in the second through fourth columns. Note that the first column is the *numerical value* if the three bits are interpreted as an unsigned binary number. So 000 is the value 0, 101 is the value 5, etc.

Let us rearrange the rows of the truth table, so that rather than being in increasing numerical order, the truth table values are listed in a way that each row differs from its neighbors by only one bit value. (Note that the fourth and fifth row (entries for 2 and 4) differ by more than one bit value.) It should become apparent soon why you would want to do this. The result will be Table 17.7.

Consider the last two lines in Table 17.7, corresponding to 6 and 7. Both have $C_{out} = 1$. On the input side, the pair is represented as $AB\bar{C}_{in} + ABC_{in} = AB(\bar{C}_{in} + C_{in})$.

The algebraic reduction operation shows up as adjacency in the table. In the same way, the 5,7 pair can be reduced. The two are adjacent and both C_{out} outputs are 1. It is less obvious in the truth table, but notice that 3,7 also forms just such a pair. In other words, all of the steps proposed in algebra are "visible" in this truth table. To make adjacency even clearer, we arrange the groups of four, one above the other, in a table called a *Karnaugh map* after its inventor, M. Karnaugh (1953). In this map, each possible combination of inputs is represented by a box. The contents of the box are the output for that combination of inputs. Adjacent boxes all have numerical values exactly one bit different from their neighbors on any side. It is customary to mark the asserted outputs (the 1's) but to leave the unasserted cells blank (for improved readability). The tables for S and C_{out} are shown in Figure 17.9. The two rows are just the first and second group of four from the truth table with the output values of the appropriate column. First, convince yourself that each and every cell differs from any of its neighbors (no diagonals) by precisely one bit. The neighbors of an outside cell include the opposite outside cell. That is, they wrap around. Thus, 2 and 0 or 4 and 6 are neighbors. The Karnaugh map (or K-map) simply shows the relationships of the outputs of conjugate pairs, which are sets of inputs that differ in exactly one bit location. The item that most people find difficult about K-maps is the meaning and arrangement of the input variables around the map. If you think of these input variables as the bits in a binary number, the arrangement is more logical. The difference between the first four rows of the truth table and the second four is that A has the value 0 in the first four and the value 1 in the second four. In the map, this is shown by having A indicated as asserted in the second row. In other words, where the input parameter is placed, it is asserted.

TABLE 17.7 Truth Table for Full Adder with Rows Rearranged

Input	ABC_{in}	S	C_{out}
0	000	0	0
1	001	1	0
3	011	0	1
2	010	1	0
4	100	1	0
5	101	0	1
7	111	1	1
6	110	0	1

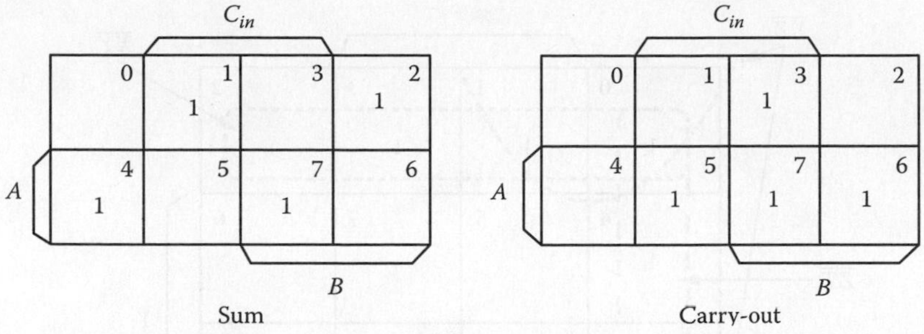

FIGURE 17.9 Karnaugh maps for sum and carry-out. The numbers in the cell corners give the bit patterns of ABC_{in}. The cells whose outputs are 1 are marked; those whose outputs are 0 are left blank.

Where it is not placed, it is unasserted. Accordingly, the middle two columns are those cells which have C_{in} asserted. The right two columns have B asserted. Column 3 has both B and C_{in} asserted.

Let us look at how the carry-out map implies gate reduction while sum's K-map shows that no reduction is possible. Since we are looking for conjugate pairs of asserted cells, we simply look for adjacent pairs of 1's. The carryout map has three such pairs; sum has none. We take pairs, pairs of pairs, or pairs of pairs of pairs—any rectangular grouping of 2^n cells with all 1's. With carryout, this gives us the groupings shown in Figure 17.10.

The three groupings in Figure 17.10 do the three things that we must always achieve:

1. The groups must cover all of the 1's (and none of the 0's).
2. Each group must include at least one cell not included in any other group.
3. Each group must be as large a rectangular box of 2^n cells as can be drawn.

Once we fulfill these three rules, we are assured of a minimal set, which is our goal. Although there is no ambiguity in the application of these rules in this example, there are other examples where more than one set of groups results in a correct, minimal set. K-maps can be used for functions of up to six input variables and are useful aids for humans to minimize logic functions. Computer-aided design programs use different techniques to accomplish the same goal for greater than six input variables.

Writing down the solution once you have done the groupings is done by reading the specification of the groups. The vertical pair in Figure 17.10 is BC_{in}. In other words, that pair of cells is uniquely defined as having B and C_{in} both 1. The other two groups are indicated in the figure. The sum of those three (where "+" is OR) is the very function we derived algebraically in the last section. Notice how you could know to twice replicate cell 7. It occurs in three different groups. It is important to keep in mind that the Karnaugh map simply represents the algebraic steps in a highly visual way. It is not intrinsically different from the algebra.

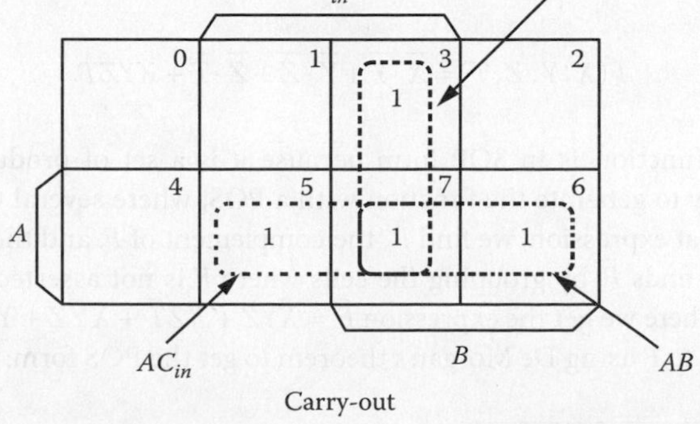

FIGURE 17.10 The groupings of conjugate pairs in carry-out.

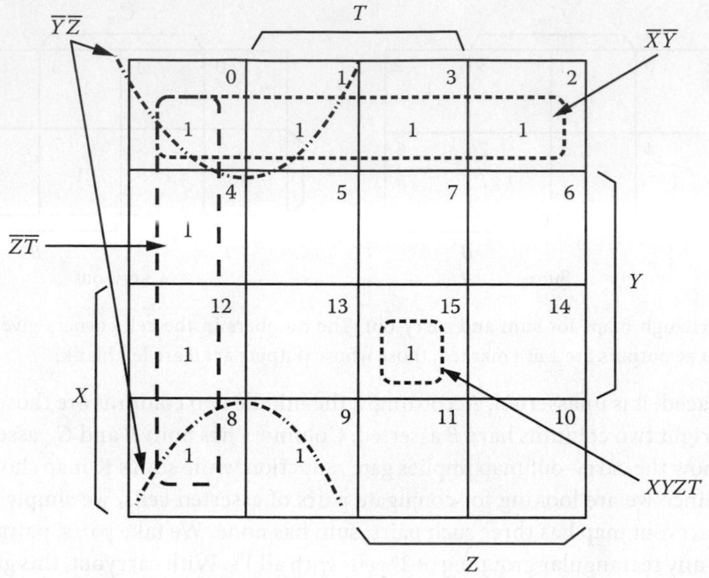

FIGURE 17.11 The K-map for $F(X, Y, Z, T) = \Sigma(0, 1, 2, 3, 4, 8, 9, 12, 15)$ with the minterm groupings shown.

We have used the word "cell" to refer to a single box in the K-map. The formal name for a cell whose value is 1 is the *minterm* of the function. Its counterpart, the *maxterm*, comprises all the cells which represent an output value of 0. Note that all cells are both possible minterms and possible maxterms.

Two more examples will complete our coverage of K-maps. One way to specify a function is to list the minterms in the form of a summation, for example, $C_{out} = \Sigma(2, 5, 6, 7)$. Consider the arbitrary 4-input function $F(X, Y, Z, T) = \Sigma(0, 1, 2, 3, 4, 8, 9, 12, 15)$. With four input variables, there are sixteen possible input states, and every minterm must contact four neighbors. That can be accomplished in a 4 × 4 array of cells as shown in Figure 17.11. Convince yourself that each cell is properly adjacent to its neighbors. For example, 11_{10} (1011) is adjacent to 15 (1111), 9 (1001), 10_{10} (1010), and 3 (0011) with each neighbor differing by one bit. Now consider the groupings. Minterm 15 has no neighbors whose value is 1. Hence it forms a group on its own, represented by the AND of all four inputs. The top row and first columns can each be grouped as a pair of pairs. It takes only two variables to specify such a group. For example, the top row includes all terms of the form 00*xx*, and the first column includes all the terms of the form *xx*00. This leaves us but one uncovered cell, 9. You might be tempted to group it with its neighbor, 8, but rule 3 demands that we make as large a covering as possible. We can make a group of four by including the neighbors 0 and 1 on top. Had we not done that, the bottom pair would be $X \cdot \overline{Y} \cdot \overline{Z}$, but by increasing the coverage, we get that down to $\overline{Y} \cdot \overline{Z}$, a 2-input AND versus a 3-input AND. The final expression is

$$F(X, Y, Z, T) = \overline{X} \cdot \overline{Y} + \overline{Y} \cdot \overline{Z} + \overline{Z} \cdot \overline{T} + XYZT$$

The aforementioned function is in SOP form because it is a set of products which are summed together. It is just as easy to generate the function with a POS, where several OR gates are joined by a single AND. To get to that expression, we find $\overline{F}$, the complement of *F*, and then convert to *F* using De Morgan's theorem. One finds $\overline{F}$ by grouping the cells where *F* is not asserted—the zero cells. This is shown in Figure 17.12, where we get the expression $\overline{F} = \overline{X}YZ + XZ\overline{T} + X\overline{Y}Z + Y\overline{Z}T$.

Let us convert from $\overline{F}$ to *F* using De Morgan's theorem to get the POS form:

$$F = \overline{(\overline{X}YZ + XZ\overline{T} + X\overline{Y}Z + Y\overline{Z}T)} = (X + \overline{Y} + \overline{Z})(\overline{X} + \overline{Z} + T)(\overline{X} + Y + \overline{Z})(\overline{Y} + Z + \overline{T})$$

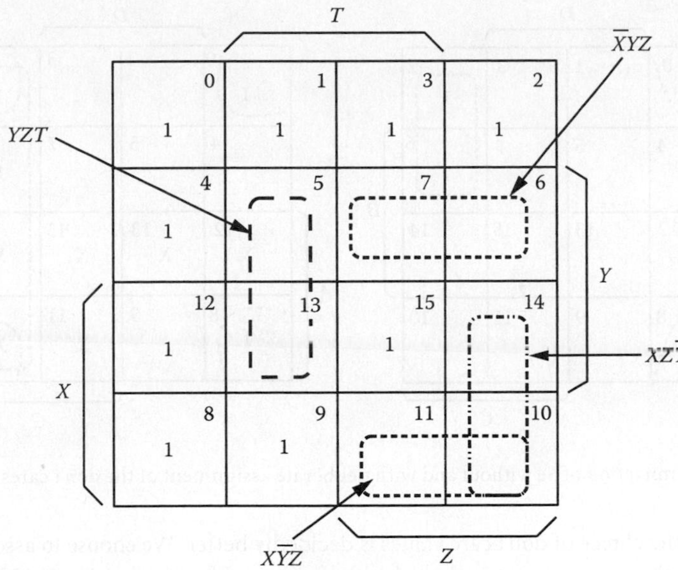

FIGURE 17.12 The K-map for the complement of *F* from Figure 17.11.

Why would one want to do this? Economy of gates. Sometimes the SOP form has fewer gates, sometimes the POS form does. In this example, the SOP form is somewhat more economical.

17.4.4 Minimizing with Don't Cares

Sometimes, we can guarantee that some combination of inputs will never occur. I don't care what the output value is for that particular combination of inputs because I know that the output can never occur. This is known as an "output" don't care. I can set these outputs to any value I want. The best way to do this is to set these outputs to values that will minimize the gate count of the entire circuit.

An example is the classic seven-segment numerical display that is common in watches, calculators, and other digital displays. The input to a seven-segment display is a number coded in binary-coded-decimal, a 4-bit representation with 16 possible input combinations, but only the 10 numbers 0, ..., 9 ever occur. The states 10, ..., 15 are called *don't cares*. One can assign them to achieve minimum gate count. Consider the entire number set that one can display using seven line segments. We will consider the one line segment indicated by the arrows in Figure 17.13. It is generally referred to as "segment *e*," and it is asserted only for the numbers 0, 2, 6, and 8.

Now we will minimize Se(*A*, *B*, *C*, *D*) with and without the use of the don't cares. We put an "X" wherever the don't cares may lie in the K-map and then treat each one as either 0 or 1 in such a way as to minimize the gate count. This is shown in Figure 17.14.

We are not doing something intrinsically different on the right and left. On the left, all of the don't cares are assigned to 0. In other words, if someone enters a 14 into this 0 : 9 decoder, it will not light up segment *e*. But since this is a don't care event, we examine the map to see if letting it light up on 14 will help.

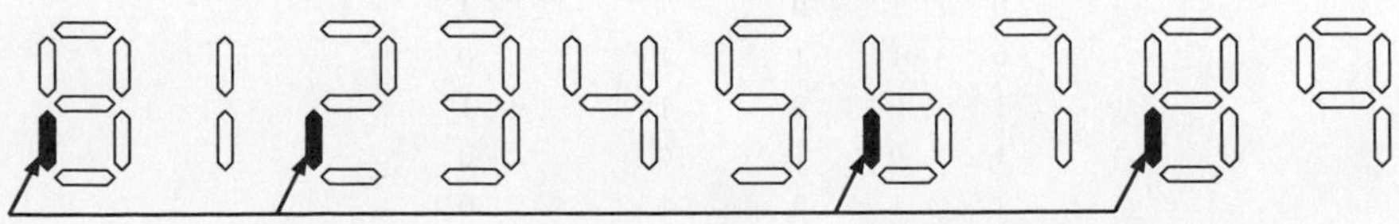

FIGURE 17.13 Segment *e* of the seven-segment display whose decoder we are going to minimize.

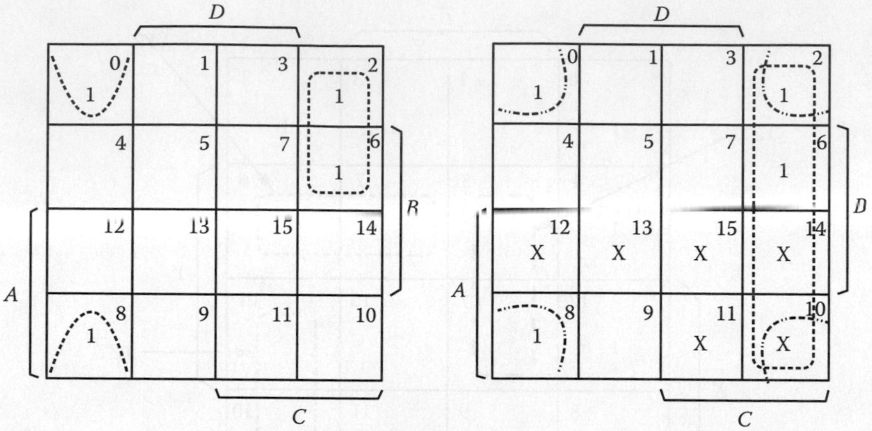

FIGURE 17.14 Minimization of Se without and with deliberate assignment of the don't cares.

The grouping with the choice of don't care values is decidedly better. We choose to assert *e* only for *don't cares* 10 and 14, but those assignments reduce the gates required from two 3-input ANDs to two 2-input ANDs. For this little circuit, that is a substantial reduction.

17.4.5 Adder/Subtractor

Let us return to the design of the full adder. A full (1-bit) adder can be implemented out of logic gates by implementing the equations for *C* and *S*. As we have seen, the simplified version for *C* is

$$C_{out} = AB + AC_{in} + BC_{in}$$

One cannot simplify *S* using K-maps. Instead, we will simplify *S* by inspection of the truth table for the full adder given in Table 17.6. Note that *S* is high when exactly one of the three inputs is high or when all the inputs are high. This is the same functionality as a 3-input XOR gate, as shown in Table 17.8. Thus, we can implement *S* with the following equation:

$$S = A \oplus B \oplus C_{in}$$

This completes our design of a full adder. Its implementation in logic gates is shown in Figure 17.15.

TABLE 17.8 Truth Table for 3-Input XOR

Inputs			Outputs	
A	B	C	$A \oplus B$	$A \oplus B \oplus C$
0	0	0	0	0
0	0	1	0	1
0	1	0	1	1
0	1	1	1	0
1	0	0	1	1
1	0	1	0	0
1	1	0	0	0
1	1	1	0	1

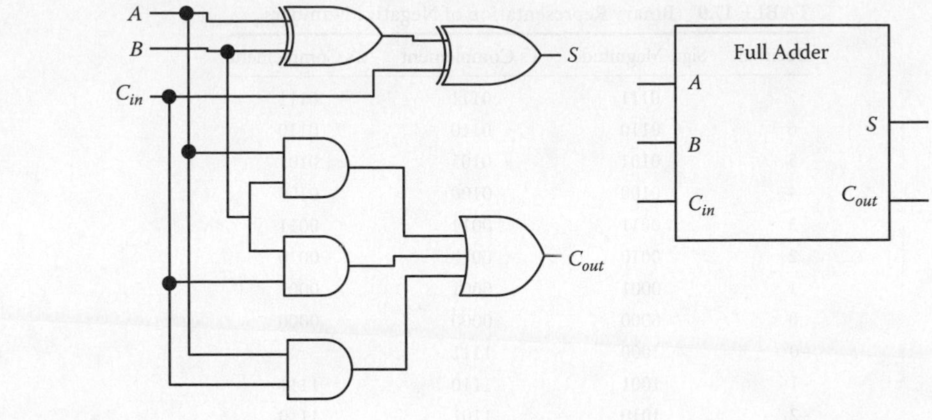

FIGURE 17.15 Implementation of full adder from logic gates on the left. Symbol of full adder on the right.

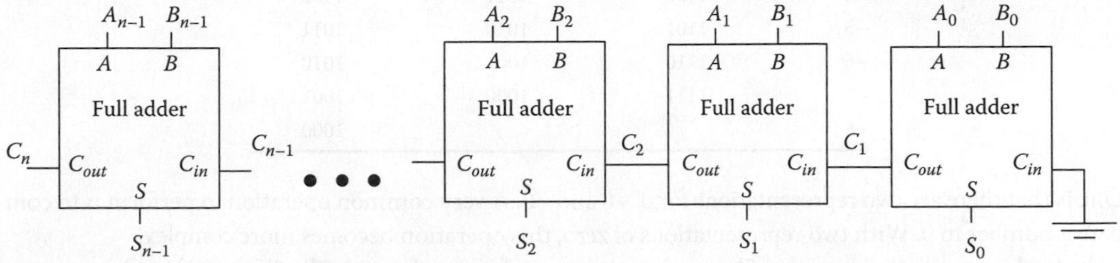

FIGURE 17.16 Implementation of an *n*-bit adder from full adders.

We would like to add *n*-bit binary numbers. To accomplish this we will connect *N* full adders as shown in Figure 17.16. This configuration connects the carry-out of one bit to the carry-in of the next bit and is called a *ripple-carry adder*. This design is small but slow because the carry must ripple through from the least significant bit to the most significant bit. For designs where speed is of the essence, such as modern high speed microprocessors, various techniques are used to speed up the calculation of the carry chain. Such techniques are beyond the scope of this chapter.

17.4.6 Representing Negative Binary Numbers

We would like to add negative numbers as well as positive numbers, and we would like to subtract numbers as well as add them. To do this we need a way of representing negative numbers in base 2. Two common methods are used: *sign-magnitude* and *2's complement*. A third method, *1's complement*, will also be described to aid in the explanation of 2's complement.

In Table 17.9, four bits are used to represent the values 7–0. The three different methods for representing negative numbers are shown. Note that for *all* three methods the positive numbers have the same representation. Also, the leftmost bit is always the sign bit. It is 0 for a positive number and 1 for a negative number. It is important to note that all these representations are different ways that a *human* interprets the bit patterns. The bit patterns are not what is changing; the interpretations are. Given a bit pattern, you cannot tell which system is being used unless someone tells you. This discussion can be extended to numbers represented with any number of bits.

The sign-magnitude method is the closest to the method we use for representing positive and negative numbers in decimal. The sign bit indicates whether it is positive or negative, and the remaining bits represent the magnitude or value of the number. So for example, to get the binary value of negative 3, you take the positive value 0011 and flip the sign bit to get 1011. While this is the most intuitive for humans to understand, it is not the easiest representation for computers to manipulate. This is true for several reasons.

TABLE 17.9 Binary Representation of Negative Numbers

Decimal	Sign-Magnitude	1's Complement	2's Complement
7	0111	0111	0111
6	0110	0110	0110
5	0101	0101	0101
4	0100	0100	0100
3	0011	0011	0011
2	0010	0010	0010
1	0001	0001	0001
0	0000	0000	0000
−0	1000	1111	
−1	1001	1110	1111
−2	1010	1101	1110
−3	1011	1100	1101
−4	1100	1011	1100
−5	1101	1010	1011
−6	1110	1001	1010
−7	1111	1000	1001
−8			1000

One is that there are two representations for 0, +0 and −0. A very common operation to perform is to compare a number to 0. With two representations of zero, this operation becomes more complex.

Instead, a representation called *2's complement* is more frequently used. The 2's complement representation has the feature that it has only one representation for zero. It has other advantages as well, including the fact that addition and subtraction are straightforward to implement; subtraction is the same as adding a negative number. If you add a number and its complement, the result is zero with no carry-out as one would expect. To form the negative value of a positive number in 2's complement, simply invert all the bits and add 1. Inverting the bits results in the 1's complement, as shown in Table 17.7. The 1's complement representation has many of the advantages of 2's complement, except that it still has two representations of zero. The 2's complement is formed by adding 1 to the 1's complement of a number. Number 3 in binary is 0011. Its one's complement is 1100 and its 2's complement is 1101. Given a negative number, how can I tell its value? Due to properties of 2's complement numbers, if I take the 2's complement of a negative number, I get its positive value. Given 1101, I invert the bits to get 0010 and then add 1 to get 0011, its positive value 3. Note that the 2's complement representation has one representation for zero, and it is the value represented by all zeros. Since I can represent 16 values with 4 bits, this leaves me with a nonsymmetric range. In other words, I can represent one more negative number 1000 than positive number. This number is −8. Its positive value cannot be represented in 4 bits. What happens if I take the 2's complement of −8 ? The 1's complement is 0111. When I add 1 to form the 2's complement, I get 1000. What is really happening is that the true value, +8, cannot be represented in the number of bits I have available. It overflows the range for the representation I am using.

Representing numbers in 2's complement makes it easy to do subtraction. To subtract two n-bit numbers $A - B$ you simply invert the bits of B and add 1 when you are summing $A + \bar{B}$.

We are now ready to expand our n-bit ripple-carry adder to an n-bit adder/subtractor. Just as we do addition one digit at a time, the adder circuit handles two input bits, A_i and B_i, plus a carry-in C_{in_i}. We can arrange as many of these circuits in parallel as we have bits. The ith circuit gets the ith bits of two operands plus the carry-out of the previous stage. It is straightforward to modify the full adder to be a one bit adder/subtractor. A one bit adder/subtractor performs the following tasks:

1. Choose B_i or the complement of B_i as the B input.
2. Form the sum of the three input bits, $S_i = A_i + B_i + C_{in_i}$.
3. Form the carry-out of the three bits, $C_{out_i} = f(A_i, B_i, C_{in_i})$.

TABLE 17.10 Choosing the *B* Input for an Adder/Subtractor

SB	B_i	Result
0	0	0
0	1	1
1	0	1
1	1	0

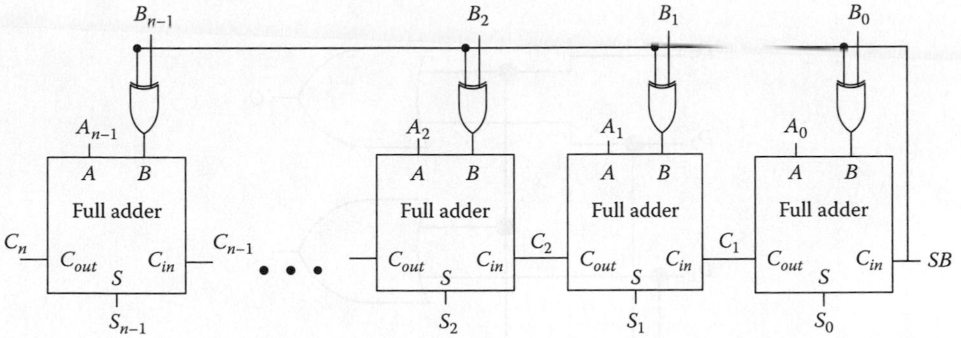

FIGURE 17.17 Connection of *n* full adders to form an *N*-bit ripple-carry adder/subtractor. At the rightmost adder, the subtract line (SB) is connected to C_{in_0}.

On a bit by bit basis, the complement of B_i is just $\overline{B_i}$. For an *n*-bit number, the two's complement is formed by taking the bit by bit complement and adding 1. I can add 1 to an *n*-bit subtraction by setting the carry-in bit, C_{in_0} to 1. In other words, I want $C_{in_0} = 1$ when subtract is true, and $C_{in_0} = 0$ when subtract is false. This is accomplished by connecting the control signal for subtracting (*SB*) to C_{in_0}. Similarly when $SB = 0$ I want to use B_i as the input to my full adder. When $SB = 1$, I want to use $\overline{B_i}$ as input. This is summarized in Table 17.10. By inspection, the desired *B* input bit to the *i*th full adder is the XOR of *SB* and B_i. If I put all the components together—*n* full adders, the carry-in of the LSB set to *SB*, and the XOR of *SB* and B_i to form the complement of *B*, I get an *n*-bit ripple-carry adder/subtractor as shown in Figure 17.17.

17.5 Frequently Used Digital Components

Many components such as full adders, half adders, 4-bit adders, 8-bit adders, etc. are used over and over again in digital logic design. These are usually stored in a design library to be used by designers. In some libraries these components are parameterized. For example, a generator for creating an *n*-bit adder may be stored. When the designer wants a 6-bit adder, he or she must instantiate the specific bit width for the component.

Many other, more complex components are stored in these libraries as well. This allows components to be designed efficiently once and reused many times. These include encoders, multiplexers, demultiplexers, and decoders. Such designs are described in more detail in the following text. Later, we will use them in the design of a calculator datapath.

17.5.1 Elementary Digital Devices: ENC, DEC, MUX, DEMUX

17.5.1.1 ENC

An ENCODER circuit has 2^n input lines and *n* output lines. The output is the number of the input line that is asserted. Such a circuit *encodes* the asserted line. The truth table of a 4-to-2 encoder is shown in Table 17.11. The inputs are D_0, D_1, D_2, and D_3, and the outputs are Q_0 and Q_1. Note that we assume at

TABLE 17.11 Truth Table for a 4-to-2 Encoder

D_0	D_1	D_2	D_3	Q_1	Q_0	V
0	0	0	0	0	0	0
1	0	0	0	0	0	1
0	1	0	0	0	1	1
0	0	1	0	1	0	1
0	0	0	1	1	1	1

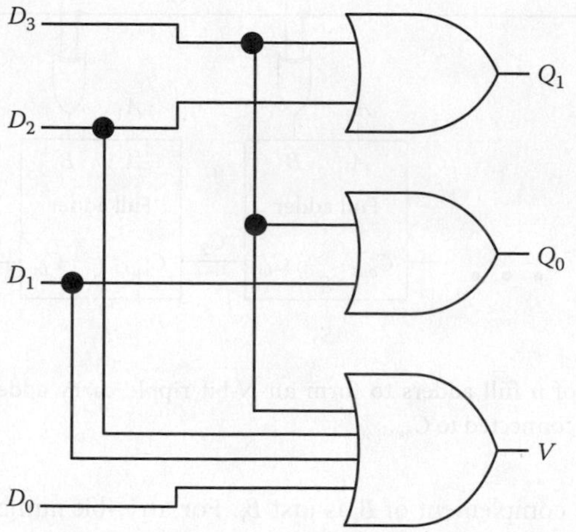

FIGURE 17.18 A four-to-two encoder with outputs Q_0 and Q_1 and valid signal.

most one input can be high at any given time. More complicated encoders, such as priority encoders, that allow more than one input to be high at a time, are described in the following.

An encoder, like all other components, is built out of basic logic gates. The equations for the Q outputs can be determined by inspection. Output Q_0 is 1 if D_1 is 1 or D_3 is 1. Output Q_1 is 1 if D_2 is 1 or D_3 is 1.

Note that there is no difference in the outputs of this circuit if *no* input is asserted or if input D_0 is asserted. To distinguish between these two cases, an output that indicates the Q outputs are valid, V is added. Output V is 1 if any of the inputs are 1 and 0 otherwise. V is considered a control output rather than a data output. A logic diagram of an encoder is shown in Figure 17.18.

17.5.1.2 DEC

A decoder performs the opposite function of an encoder. Exactly one of the outputs is true if the circuit is enabled. That output is the *decoded* value of the input, if I think of the input as a binary number. When the enable input (EN) is 0, the circuit is disabled, and all outputs are 0. The truth table of a 2-to-4 decoder is given in Table 17.12.

Note the use of X values for inputs in the first line of the truth table. Here, X stands for "don't care." In other words, I don't care what the value of the inputs A and B are. If $EN = 0$, the outputs will always have the value 0. I am using the don't care symbol, X, as a shorthand for the four combinations of input values for A and B. We have already used don't cares for minimizing circuits with K-maps mentioned earlier. In that case the don't cares were "output" don't cares. The don't cares in the truth table for the encoder are "input" don't cares. They are shorthand for several combinations of inputs.

TABLE 17.12 Truth Table for a 2-to-4 Decoder with Enable

EN	*A*	*B*	Q_0	Q_1	Q_2	Q_3
0	X	X	0	0	0	0
1	0	0	1	0	0	0
1	1	0	0	1	0	0
1	0	1	0	0	1	0
1	1	1	0	0	0	1

17.5.1.3 MUX

Many systems have multiple inputs, which are handled one at a time. *Call waiting* is an example. You are talking on one connection when a clicking noise signals that someone else is calling. You switch to the other, talk briefly, and then switch back. You can toggle between calls as often as you like. Since you are using one phone to talk on two different circuits, you need a *multiplexer* or MUX to choose between the two. There is also an inverse MUX gate called either a DEMUX or a DECODER. Again, a telephone example is the selection of an available line among, say, eight lines between two exchanges. That is, you have one line in and eight possible out, but only one output line is connected at any time. An algorithm based on which lines are currently free determines the choice. Let us design these two devices, beginning with a two-to-one MUX.

What we want in a 2-input MUX is a circuit with one output. The value of that output should be the same as the input that we select. We will call the two inputs *A* and *B* and the output *Q*. The select input *S* chooses which input to steer to the output.

Logically, I can think of *Q* being equal to *A* when $S = 0$ and *Q* being equal to *B* when $S = 1$. I can write this as the Boolean equation: $Q = (A \cdot \overline{S}) + (B \cdot S)$. You can use a truth table to convince yourself that this equation captures the behavior of a 2-input MUX. Note that we now have a new dichotomy of inputs. We call some of them *inputs* and the others *controls*. They are not inherently different, but from the human perspective, we would like to separate them. In logic circuit drawings, *inputs* come in from the left and outputs go out to the right. *Controls* are brought in from top or bottom. The select input for our multiplexer is a control input. Note that, even though I talk about S being a control, in the logic equation it is treated the same as an input. An enable is another kind of control input. A *valid* signal can be viewed as a control output. A realization of a 2-to-1 Multiplexer with enable is shown in Figure 17.19.

The 2-to-1 MUX circuit is quite useful and is found in many design libraries. Other similar circuits are the 4-to-1 MUX and the 8-to-1 MUX. In general, you can design *n*-to-1 MUX circuits, where *n* is a power of 2. The number of select bits needed for an *n*-to-1 MUX is $log_2(n)$.

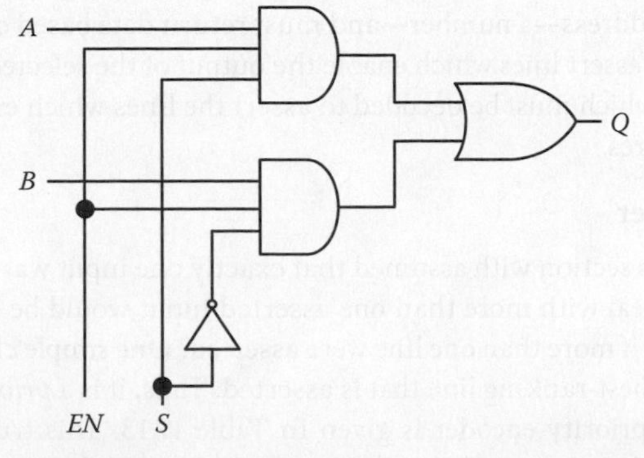

FIGURE 17.19 A two-to-one MUX with enable. If the enable is asserted, this circuit delivers at its output, *Q*, the value of *A* or the value of *B*, depending on the value of *S*. In this sense, the output is "connected" to one of the input lines. If the enable is not asserted, the output *Q* is low.

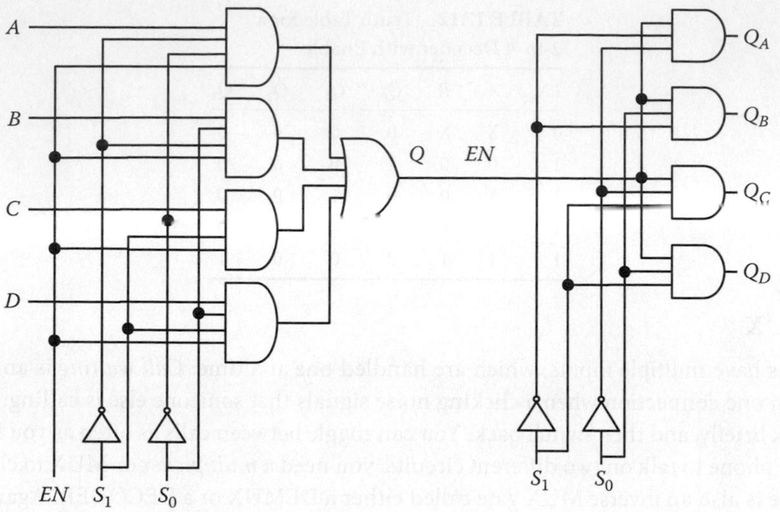

FIGURE 17.20 A four-to-one MUX feeding a one-to-four DEMUX. The value on MUX select lines S_1:S_0 determines the input connected to *Q*. *EN*, in turn, is connected to the output of choice by S_1:S_0 on the DEMUX.

Figure 17.20 shows a 4-to-1 multiplexer on the left and a 1-to-4 demultiplexer on the right. The MUX chooses one of four inputs using the two select lines: *S*1 and *S*0. If we view the values on the select lines as the binary numbers 0, …, 3, we understand the selection process as enabling the top AND when the input is 00 and then progressively lower ANDs as the numbers become 01, 10, and 11. Essentially, there is a decoder circuit within the four-to-one multiplexer.

17.5.1.4 DEMUX/DECODER

The inverse circuit to a multiplexer is a demultiplexer or DEMUX. A DEMUX has one line in, which it switches to one of the *n* possible output lines. A one-to-four DEMUX, used in conjunction with the four-to-one MUX, is shown on the right in Figure 17.20.

Note that the *Q* output line on the multiplexer is labeled as the *EN* input line on the DEMUX. If I treat this line as an *enable*, the DEMUX becomes a DECODER, in the sense that, when *EN* is asserted, one and only one of the four outputs is asserted, that being the output selected by the number on S1:S0. So a decoder and a DEMUX are the same circuit. You usually do not find a DEMUX in a design library. Rather, what you find is called a DECODER or sometimes a DEMUX/DECODER.

Decoding is an essential function in many places in computer design. For example, random-access memory (RAM) is fed an address—a number—and must return data based on that number. It does this by decoding the address to assert lines which enable the output of the selected data. Similarly, computer instructions are numbers which must be decoded to assert the lines which enable the specific hardware that each instruction requires.

17.5.1.5 Priority Encoder

The encoder we started this section with assumed that exactly one input was asserted at any given time. An encoder which could deal with more than one asserted input would be even more useful, but how would we define the output if more than one line were asserted? One simple choice is to have the encoder deliver the value of the highest-ranking line that is asserted. Thus, it is a *priority encoder.*

The truth table for the priority encoder is given in Table 17.13. This truth table has a lot of similarities to the simple encoder we started this section with. The valid output *V* tells us if any input is asserted. The output Q_0 is true if the only input asserted is D_1. The circuit differs in that more than one input may be asserted. In this case, the output encodes the value of the highest input that is asserted.

TABLE 17.13 Truth Table for a 4-to-2 Priority Encoder

D_0	D_1	D_2	D_3	Q_0	Q_1	V
0	0	0	0	0	0	0
1	0	0	0	0	0	1
X	1	0	0	1	0	1
X	X	1	0	0	1	1
X	X	X	1	1	1	1

So, for example, if D_0 and D_1 are both asserted, the output Q_0 is asserted. I don't care if the D_0 input is asserted or not, because the D_1 input has higher priority. Here once again, the don't cares are used as short hand to cover several different inputs. If I listed all possible combinations of inputs in the truth table, my truth table would have $2^4 = 16$ lines. Using don't cares makes the truth table more compact and readable.

17.5.2 Calculator Arithmetic and Logical Unit

Let us look at putting some of these components together to do useful work. The core of a calculator or microprocessor is its ALU. This is the part of the calculator that implements the arithmetic functions, such as addition, subtraction, and multiplication. In addition, logic functions are also implemented such as ANDing inputs and ORing inputs. A microprocessor may have several, sophisticated ALUs. We will examine the design of a very simple ALU for a 4-bit calculator. Our ALU will perform four different operations: AND, OR, addition, and subtraction on two 4-bit inputs, *A* and *B*. Two input control signals, I_1 and I_0, will be used to choose between the operations, as shown in Table 17.14. We will call the 4 bit result *R*.

A common way to implement such an ALU is to implement the various functions in parallel, then choose the requested result based on the setting of the control inputs. The 4-bit AND requires four AND gates, one for each bit position. Similarly, the 4-bit OR requires four OR gates. We will implement the adder and subtractor using a single adder/subtractor unit, since using two would waste area. The adder/subtractor unit in Figure 17.17 is perfect for our purposes here. Note, in Table 17.14, that I_0 is 0 for adding A and B and 1 for subtracting A and B, so we can use I_0 for the *SB* input to the adder/subtractor. These units will always operate on the *A* and *B* inputs; a multiplexer will select the correct output based on the values of I_1 and I_0. Four, 4-to-1 multiplexers are used, one for each output bit. 4-to-1 multiplexers are used because there are no 3-to-1 muxes. We will use the fourth input to pass the *A* input to the output *R*. The reason for doing this will become apparent when we use the ALU in a calculator datapath. To keep the diagram readable, we use the convention that signals with the same name are wired together. The resulting ALU implementation is shown in Figure 17.21. The symbol for this ALU is shown in Figure 17.22. We will use this symbol when we incorporate the ALU into a calculator datapath.

TABLE 17.14 ALU Instructions for Calculator

I_1	I_0	Result
0	0	*A* AND *B*
0	1	*A* OR *B*
1	0	$A + B$
1	1	$A - B$

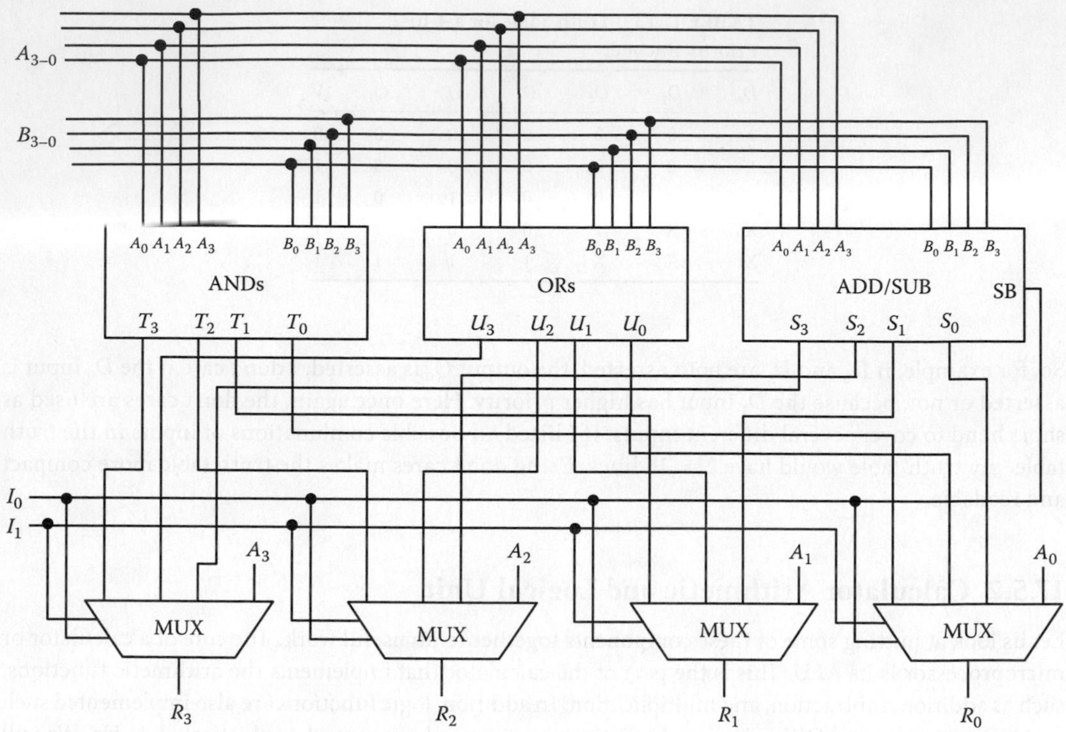

FIGURE 17.21 Implementation of an ALU from other components.

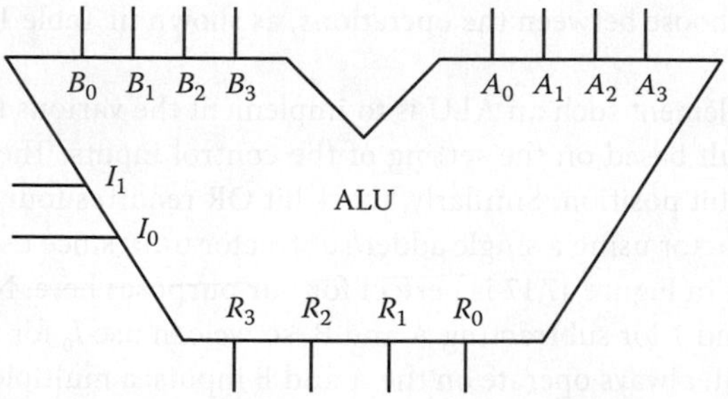

FIGURE 17.22 Symbol of an ALU component.

17.6 Sequential Circuits

17.6.1 Concept of a Sequential Device

So far, all the circuits we have discussed have been combinational. The current output can be determined by knowing the current inputs. Sequential circuits differ from combinational circuits because they have memory. For a circuit with memory, the current outputs depend on the current inputs *and* on the past history of the circuit. Memory elements dramatically change the way a circuit operates.

One of the oldest and most familiar sequential devices is a clock. In its modern implementation, a vibrating crystal of piezoelectric material has a mechanical resonance that is used to create an electrical signal with a very precise frequency. Because clock (CLK) signals oscillate between logic high and logic low states very quickly, and typically appear as a square wave. Figure 17.23 shows an example of a CLK signal. Except during transitions from one state to its successor, the clock is always in a discrete state.

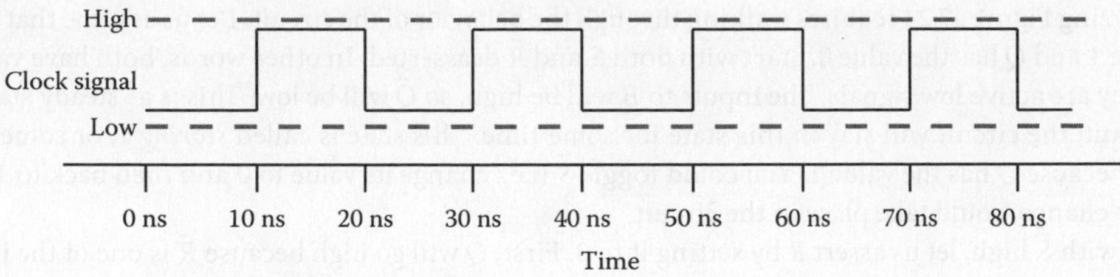

FIGURE 17.23 Example waveform of an oscillating clock signal with a period of 20 ns and frequency 50 MHz.

To be in a discrete state requires some form of memory. I can only know the current output of my clock if I know what its previous output was.

One of the most ubiquitous and essential memory elements in the digital world is the latch or flip-flop (FF). It snaps from one position to the other (storing a 1 or storing a 0) and retains memory of its current position. We shall see how to build such a latch out of logic gates.

Like clocks, computers and calculators are *finite state machines*. All of the states of a computer can be enumerated. Saying this does not in any way restrict what you can compute. The states of a finite state machine capture the history of the behavior of the circuit up to the current state.

By linking memory elements together, we can build predictable sequential machines that do important and interesting tasks. Only the electronic "latch" and the datapath of our simple calculator are included in this short chapter, but from the sequential elements presented here complex machines can be built. There are two kinds of sequential circuits, called *clocked* or synchronous circuits and *asynchronous* circuits. The clocked circuits are built from components such as the FF, which are synchronized to a common CLK signal. In asynchronous circuits, the "memory" is the intrinsic delay between input and output. To maintain an orderly sequence of events, they depend on knowing precisely how long it takes for a signal to get from input to output. Although that sounds difficult to manage in a very complex device, it turns out that keeping a common clock synchronized over a large and complex circuit is nontrivial as well. We will limit our discussion to clocked sequential circuits.

17.6.2 Data Flip-Flop and Register

17.6.2.1 *SR* Latch. Set, Reset, Hold, and Muddle

In all the circuits we have looked at so far, there was a clear distinction between inputs and outputs. Now we will erase this distinction by introducing positive feedback; we will *feed back* the outputs of a circuit to the inputs of the same circuit. In an electronic circuit, positive feedback can be used to force the circuit into a "stable state." Since saturated logic goes into such states quite normally, it is a very small step to generate an electronic latching circuit from a pair of NAND or NOR gates. The simplest such circuit is shown in Figure 17.24.

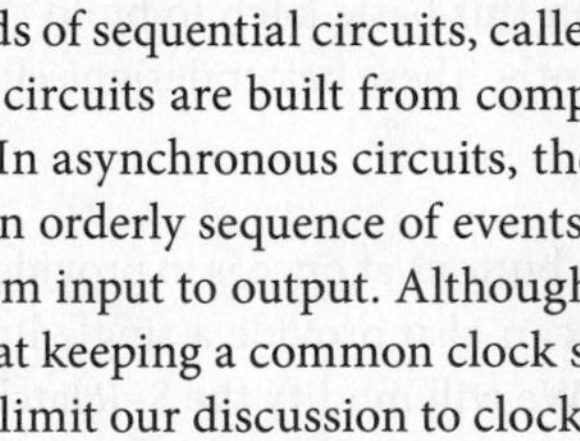

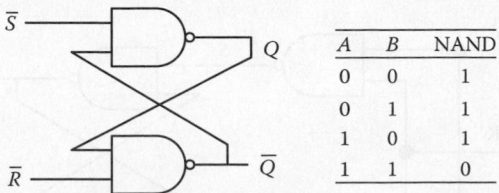

A	*B*	NAND
0	0	1
0	1	1
1	0	1
1	1	0

FIGURE 17.24 The basic set/reset (*SR*) latch is shown on the left. If $\overline{S}$ is asserted, *Q* is asserted (set). If $\overline{R}$ is asserted, *Q* is deasserted (reset). If neither $\overline{S}$ nor $\overline{R}$ is asserted (both high), the latch retains its current state. If both are asserted, the latch goes into a *muddle* state where *Q* is asserted and $\overline{Q}$ is deasserted (both high), but upon simultaneous release of the inputs, the next state is unpredictable. The truth table for an NAND gate is shown on the right.

Analyzing Figure 17.24 requires walking through the behavior of the circuit. Let us assume that Q has the value 1 and $\bar{Q}$ has the value 0. Start with both $\bar{S}$ and $\bar{R}$ deasserted. In other words, both have value 1, since they are active low signals. The inputs to B will be high, so $\bar{Q}$ will be low. This is a "steady state" of this circuit; the circuit will stay in this state for some time. This state is called *storing 1*, or sometimes just "1" because Q has the value 1. You could toggle $\bar{S}$ (i.e., change its value to 0 and then back to 1) and no other change would take place in the circuit.

Now, with $\bar{S}$ high, let us assert $\bar{R}$ by setting it to 0. First, $\bar{Q}$ will go high because $\bar{R}$ is one of the inputs to B, and the NAND of 0 with anything is 1. This makes both of the inputs to A high, so Q goes low. Now the upper input to B is low, so deasserting $\bar{R}$ (setting it to 1) will have no effect. Thus, asserting $\bar{R}$ has reset the latch. The latch is in the other steady state, "storing 0" or "0."

At this point, asserting $\bar{S}$ will set the latch, or put it back into the state "1." For this reason, the S input is the "set" input to the latch, and the R input is the "reset" input.

What happens if both $\bar{S}$ and $\bar{R}$ are asserted at the same time? The initial result is to have both Q and $\bar{Q}$ go high simultaneously. Now, deassert both inputs simultaneously. What happens? You cannot tell. It may go into either the set or the reset state. Occasionally, the circuit may even oscillate, although this behavior is rare. For this reason, it is usually understood that the designer is *not allowed* to assert both $\bar{S}$ and $\bar{R}$ at the same time. This means that, if the designer asserts both $\bar{S}$ and $\bar{R}$ at the same time, the future behavior of the circuit cannot be guaranteed, until it is set or reset again into a known state.

There is another problem with this circuit. To hold its value, both $\bar{S}$ and $\bar{R}$ must be continuously deasserted. Glitches and other noise in a circuit might cause the state to flip when it should not.

With a little extra logic, we can improve upon this basic latch to build circuits less likely to go into an unknown state, oscillate, or switch inadvertently. These better designs eliminate the muddle state.

17.6.2.2 Transparent *D*-Latch

A simple way to avoid having someone press two buttons at once is to provide them with a toggle switch. You can push it only one way at one time. We can also provide a single line to enable the latch. This enable control signal is usually called the *clock*. We will modify the *S–R* latch above. First, we will combine the *S* and *R* inputs into one input called the *data* or *D* input. When the *D* line is a one we will set the latch. When the *D* line is a zero we will reset the latch. Second, we will add a CLK signal to control when the latch updates. With the addition of two NANDs and an inverter, we can accomplish both purposes, as shown in Figure 17.25.

Note that we tie the data line, D, to the top NAND gate and the inverse or $\bar{D}$ to the bottom NAND gate. This assures that only one of the two NAND outputs can be low at one time. The CLK signal allows us to open the latch (let data through) or latch the data at will. This device is called a *transparent D-latch* and is found in many digital design libraries. This latch is called *transparent* because the current value of D appears at Q if the CLK signal is high. If CLK is low, then the latch retains the last value D had when CLK was high.

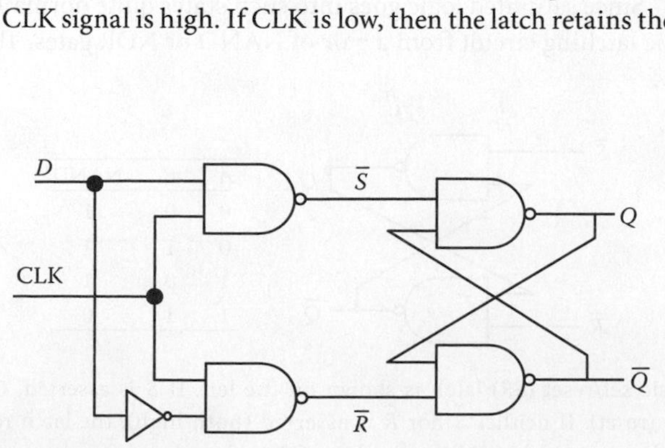

FIGURE 17.25 The transparent *D*-latch. The circuit is transparent when CLK is high (i.e., the current value *D* appears at *Q*) and latched when CLK is low (the value of *D* when the clock went low is held at *Q*).

Has this device solved all of the problems we described for the *SR*-latch? No. Consider what might happen if *D* changes from low to high just as the clock changes from high to low. For the brief period before the change has propagated through the *D*-inverter, both NANDs see both inputs high. Thus, at least briefly, both $\bar{R}$ and $\bar{S}$ are asserted. This is the very situation we wanted to avoid. This muddled situation would last only for the propagation time of the inverter, but then the CLK signal arrives and drives both $\bar{S}$ and $\bar{R}$ high. The latch might oscillate or flip either way. In any case, it will be unpredictable.

There is another problem with this circuit. It is indeed transparent during the high CLK signal. This means that *Q* will mirror *D* while CLK is high. If *D* changes rapidly, so will *Q*. Sometimes you may want transparency. However, frequently you want to be able to guarantee when the output will change and to only allow one transition on the output per clock cycle. In that case you do not want transparency; what you really want is a different circuit: an FF.

17.6.2.3 Master–Slave DFF to Eliminate Transparency

The problem with transparent gates is not a new one. A solution which first appeared in King Solomon's time (ninth century B.C.E.) will work here as well. The Solomonic gate was a series pair of two quite ordinary city gates. They were arranged so that both were never open at the same time. You entered the first and it was shut behind you. While you were stuck between the two gates, a well-armed, suspicious soldier asked your business. Only if you satisfied him was the second gate opened. The solution of putting out-of-phase transparent latches between input and output is certainly one obvious solution to generating a DFF. Such an arrangement of two *D*-latches is shown in Figure 17.26.

The latch on the left is called the *master*; that on the right is called the *slave*. This master–slave (MS) DFF solves the transparency problem but does nothing to ameliorate the timing problems. While timing problems are not entirely solvable in any FF, accommodating the number of delays in this circuit tends to make the MSFF a slow device and thus a less attractive solution. Why should it be slow? The issue is that to be sure that you do not put either of these devices into a metastable or oscillatory state, you must hold *D* constant for a relatively long setup time (before the clock edge) and continue it past the clock transition for a sufficient hold time. This accommodation limits the speed with which the whole system can switch.

Can we do better? Yes, not perfect, but better. The device of choice is the edge-triggered DFF. We will not go into the details of the implementation of an edge-triggered FF as it is considerably more complicated than the circuits considered so far. An edge-triggered FF is designed to pass the input datum to the output during a very short amount of time defined by a clock edge. Edge triggered FFs can either be *rising edge*-triggered or *falling edge*-triggered. A rising edge-triggered FF will only change its output on the rising edge of a clock. There is still a setup time and a hold time; a small amount of time right around the clock edge during which the input datum must be stable in order to avoid the FF becoming metastable. The advantage of the edge-triggered design is that there is only one brief moment when any critical changes take place. This improves synchronization and leads to faster circuits.

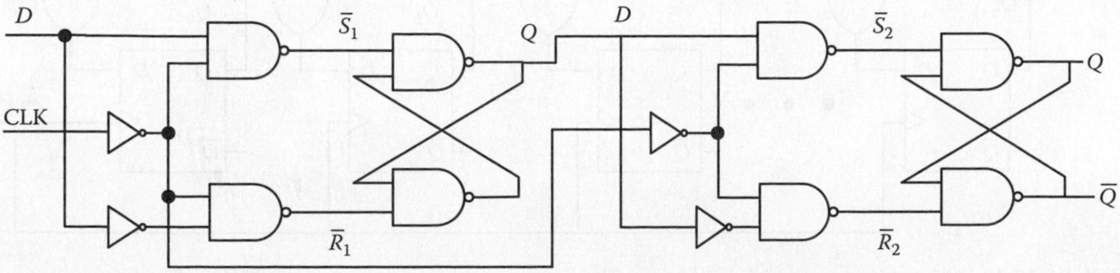

FIGURE 17.26 The master–slave data flip-flop constructed of two *D*-latches in series.

There are other types of FFs as well, but the DFF is widely used and useful for building more complex circuits. We will consider these as we build more complex circuits. Most designs of a positive-edge-triggered DFF include two additional **asynchronous** inputs, *preset* and *clear*. Asynchronous inputs cause the output to change independently of the clock input. The *D* input is a **synchronous** input; changes on the *D* input are only reflected at the output during a clock edge. An active signal on the preset input causes the output of the FF to be set to 1; an active signal on the clear input causes the output to be cleared or set to 0. These inputs are useful for putting FFs in a design into a known state.

17.6.3 From DFF to Data Register, Shift Register, and Stack

The simplest and most useful device we can build with DFFs is a data register. A data register stores data. The simplest data register stores new data on every clock edge. Its design is shown in Figure 17.27. It is useful to control when the data register stores new data. We can extend the data register by adding a control signal called *load*. Now our register will load new data only when *load* is high. Otherwise it will store its old data. The new register design is shown in Figure 17.28.

We can also build a *shift* register, which shifts data, one bit at a time, into a parallel register. Shift registers are useful for converting serial data to parallel data. For example, you may receive data one bit at a time from a serial connection to your computer but want to operate on the data in parallel.

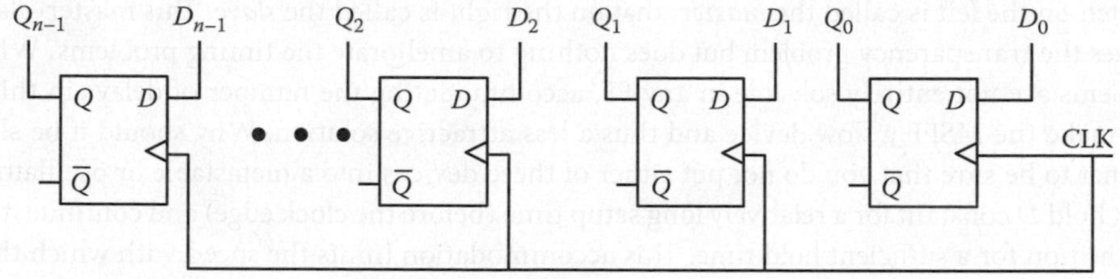

FIGURE 17.27 An *n*-bit data register built from *n* DFFs.

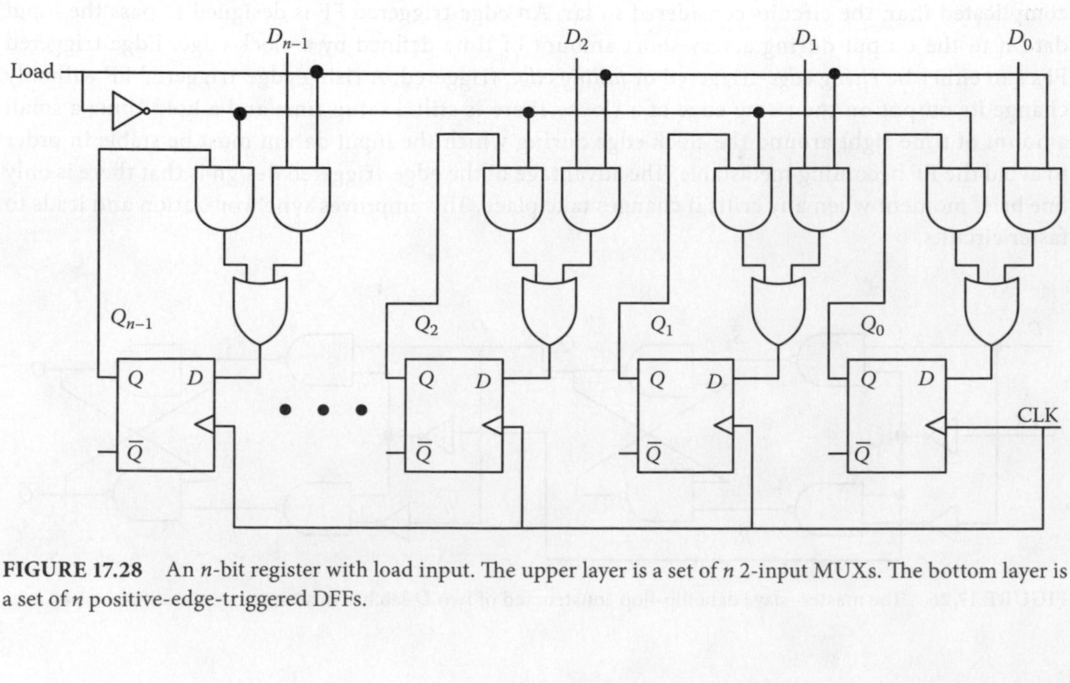

FIGURE 17.28 An *n*-bit register with load input. The upper layer is a set of *n* 2-input MUXs. The bottom layer is a set of *n* positive-edge-triggered DFFs.

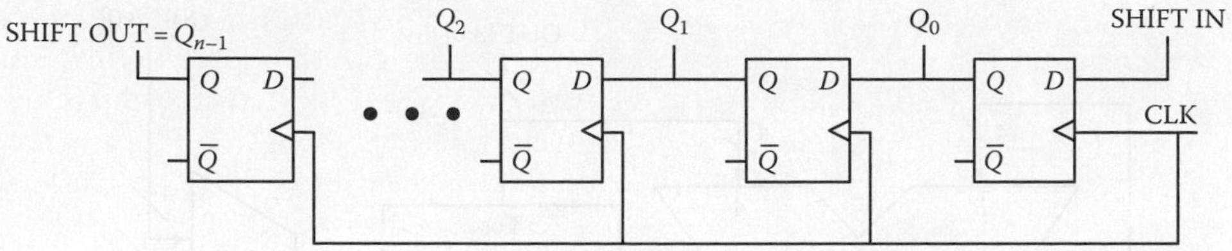

FIGURE 17.29 An n-bit shift register with serial input, parallel, and serial output. This register shifts one bit to the left at the positive edge of every clock cycle.

Shift registers can sometimes be loaded in parallel also, just like the data register. We will keep things simple; a serial-in parallel-out shift register is shown in Figure 17.29. This shift register also has a serial output.

17.6.3.1 Stack for Holding Data

Our calculator requires memory to store variables that need to be manipulated. We could implement a RAM with addressing, read and write capabilities. Instead, to keep things simple, we will implement a *stack* for storing variables and results. A stack is a group of memory locations with limited access to its contents. At any given time, only the top of the stack can be read. This makes its implementation simple because general addressing does not need to be supported. Our stack supports three operations: push, pop, and hold. When a value is pushed onto the stack, it becomes the value at the top of the stack. All values already in the stack are pushed down by one. If the stack was full *before* the push operation, then the oldest (first in) value on the stack is lost. In a pop operation, the top of the stack is deleted, and all other values move up in the stack by one position. Note that this operation differs from a software *pop* operation. In software, the popped value is stored in a register. In our hardware implementation there is no register storing the removed value, so this value is lost. A stack is sometimes called a Last In First Out because that is the order in which values are accessed. The hold operation ensures that the current contents of the stack are retained. It is important to explicitly support this so that the contents of the stack are not changed during operation. Push, pop, and hold all happen on a clock edge. By default, the stack holds its contents when there is no clock edge.

We will implement a stack to hold 4-bit variables. Our stack will contain four locations. One can implement this stack as four shift registers (one for each bit position) with each shift register containing four FFs. The total memory contents of our stack is held in 16 DFFs. In a real calculator implementation, the stack would be implemented with memory cells which use fewer transistors and consume less power than DFFs, but for our small design, DFFs will suffice.

17.6.4 Datapath for 4-Bit Calculator

Let us put the pieces together and implement the datapath of a 4-bit calculator. We will use the stack described in the preceeding text as our memory, and the ALU described in the previous section to implement logical operations. We need to add connections and a way to input data values. We also need an *instruction* to tell the calculator datapath what to do. We will use a 7-bit input that can be connected to toggle switches to provide the data and instructions. The datapath for the calculator is shown in Figure 17.30. The output is the top of stack (TOS). This is represented as the 4 bits `OUT[3:0]`. These could be hooked to a seven-segment display to display the results.

Our calculator will support the instructions: *push, pop, and, or, add*, and *sub*tract. Note that the last four are the operations that our ALU supports. These are operations that take two 4-bit input values and produce a 4-bit result. The operands will be found on the stack. The input values will be popped off the stack, and the output value will be pushed onto the top of the stack. For a push operation, the

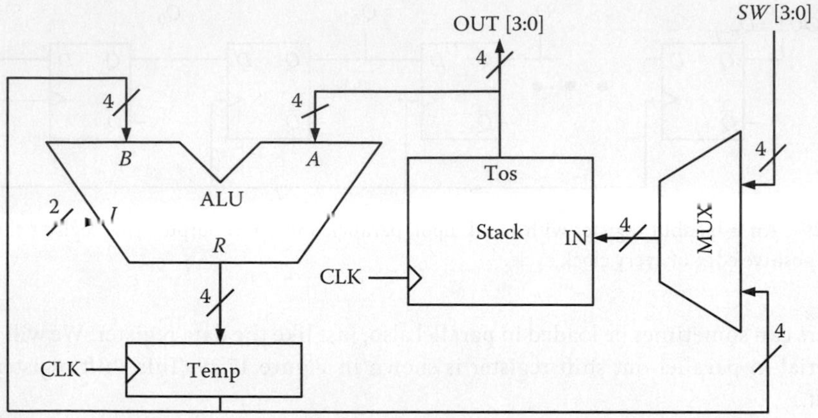

FIGURE 17.30 Calculator datapath. The temp register stores the ALU results every CLK edge.

value to be pushed onto the stack will be entered via the input switches. Suppose we want to calculate: 4 + (5 − 3). The operations required are:

```
PUSH 4
PUSH 3
PUSH 5
SUB
ADD
```

The contents of the stack for this sequence of operations are shown in Figure 17.31.

The format of the instructions and data on the input switches are shown in Table 17.15. Note that only the PUSH instruction uses external data.

What is missing from our calculator design? A controller to make sure that the right operations are performed on the datapath in the correct sequence for each instruction. While the design of such a controller

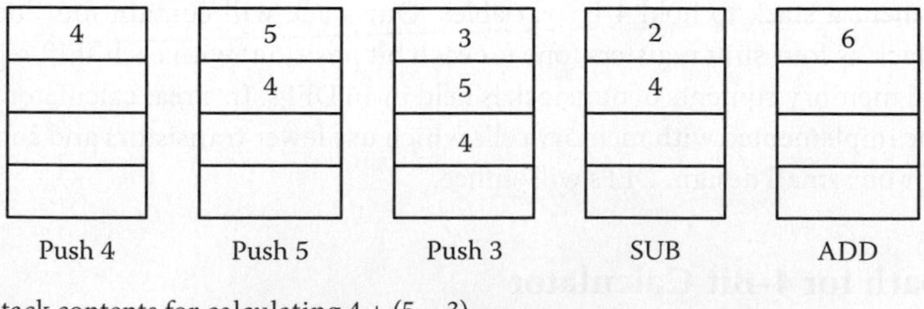

FIGURE 17.31 Stack contents for calculating 4 + (5 − 3).

TABLE 17.15 Switch Settings for Calculator Instructions

Instruction	$SW_6 \dots SW_4$	$SW_3 \dots SW_0$
AND	000	Unused
OR	001	Unused
ADD	010	Unused
SUB	011	Unused
PUSH	101	Data
POP	110	Unused

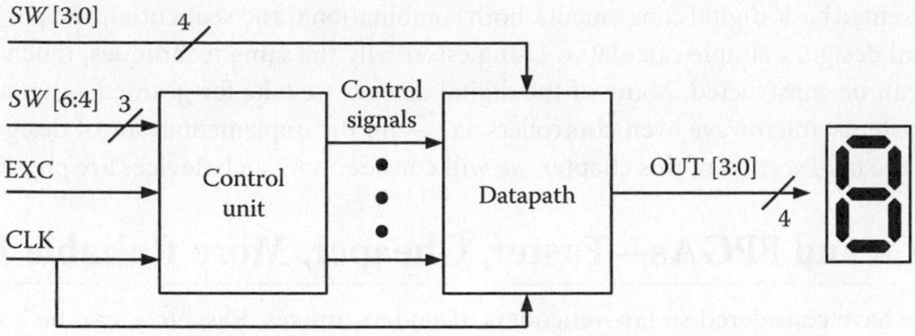

FIGURE 17.32 Connecting the controller, datapath, inputs, and outputs of the calculator.

is beyond the scope of this chapter, we will specify its behavior. For each instruction, the datapath goes through a sequence of operations or states. There are a finite number of such states to describe the behavior of the controller; hence, the controller is a finite state machine. Such a finite state machine can be implemented with FFs to hold the state and combinational logic to generate the next state and the outputs required. The inputs to the state machine are the current state and, for the calculator, the instruction which is entered via the switches. There is also a button the user needs to be able to press that indicates that the next instruction is ready to execute. The outputs are the control signals for the datapath. Our controller needs to output two bits for the ALU to tell it what to do, and two bits for the stack to generate the push, pop, or hold inputs, and 1 bit for the MUX to tell it which input to use. The temporary register requires no control signals—it will be updated every clock cycle. The entire calculator design is shown in Figure 17.32.

Let us look at how the instructions are executed on our calculator. There are three *classes* of instructions: *push, pop,* and ALU (*add, sub, and, or*). The *pop* instruction is the simplest. All that is needed is one state to pop the stack. In that state, the control inputs to the stack should be "pop." We don't care what the ALU or MUX control inputs are set to. Operation *push* is also a one state instruction. In that state the data on the switches is pushed onto the stack. So the control input for the MUX should be set to select the input switches and the control input for the stack should be set to "push."

The ALU instructions are more complicated; they require three control states. We assume that our calculator has the two operands on the stack before an ALU instruction is executed. In the first state the data value on the top of the stack is stored in the temp register and the stack is popped. How can we accomplish this in one state? Note that the ALU is a combinational circuit and the TOS is already the A input to the ALU. If we select the ALU operation to pass A through to the output of the ALU, then on the clock edge, we will simultaneously store the current TOS value in the temp register and pop the value off the stack. Now, at the start of the second state, the first operand is at input B of the ALU because the temp register output is connected to input B. The second operand is at input A of the ALU because the TOS is connected to input A. In this state we will execute the correct ALU instruction, store the result in the temp register, and pop the A operand off the stack. Finally, in the third state the contents of the temp register, that is the results of the ALU operation, are pushed onto the stack. At the end of an ALU operation, the two operands have been popped from the stack, and the result pushed on the stack as required. To summarize, the three states of an ALU operation are as follows:

```
state 1: temp <- operand A; pop
state 2: temp <- A op B   ; pop
state 3:                    push temp
```

We have discussed what happens when an instruction is executing. What happens between instructions? The stack should "hold" its contents. It does not matter what the ALU is doing, or what input the MUX passes through. The memory of this design is in the stack. As long as the stack holds its contents, the memory is maintained. The temporary register is updated every clock cycle, but the results are not saved, so this does not affect the correct operation of the calculator datapath. Note that the output is always active, and it always shows what is currently on the top of the stack.

We have presented basic digital components, both combinational and sequential, and put them together to build a useful design: a simple calculator. Using essentially the same techniques, much more complicated devices can be constructed. Many of the digital devices we take for granted—digital clocks, anti-lock braking systems, microwave oven controllers, etc.—are the implementations of designs using these techniques. In the final section of this chapter, we will consider how such devices are physically realized.

17.7 ASICs and FPGAs—Faster, Cheaper, More Reliable Logic

The circuits we have considered so far—encoders, decoders, muxes, FFs, etc.—can be bought in packages with several to a chip. For example, a 14-pin package containing four 2-input NANDs per package has been available for more than 30 years. These chips are called *small scale integrated circuits*. No one who has watched the astonishing decline in the cost of digital electronics brought about by very large scale integrated circuits (VLSI) would expect to find engineers generating circuits by hooking up vast arrays of such packages. In a world where powerful computer chips roll off the line with more than a billion transistors all properly connected and functioning at clock speeds in excess of 1 GHz, why would we be manually hooking up hundreds of these small packages with 16 transistors in a chip that takes 15–20 ns to get a signal through a single NAND? Today, the equivalent of many pages of random logic circuit diagrams can be implemented in a single chip. There are many different ways to specify such designs and many different ways to physically realize them. The dominant method of design entry is using a hardware description language (HDL). HDLs resemble software programming languages with added features specifically for describing hardware such as bitwidth, I/O ports, and controller state specifications. Design tools translate HDL descriptions of a circuit to the final implementation. One of the goals of an HDL design is to separate the design from the implementation technology. The same HDL description, in theory at least, can be mapped to different target technologies.

There are also many technologies for physically realizing digital logic designs. Application-specific integrated circuits (ASICs) can be implemented as VLSI circuits where millions of transistors are realized on a single chip resulting in very high speeds and very low power dissipation. Such designs are manufactured at a foundry and cannot be changed after they have been implemented. Since high performance VLSI designs are very expensive to manufacture, they are increasingly used only for very large volume designs and designs where low power is critical. For example, VLSI ASIC chips are found in mobile phones, tablets, and other handheld devices that meet these criteria.

Designers are increasingly turning to *programmable* and *reconfigurable* devices for realizing their designs. These devices are manufactured in large quantities using VLSI techniques with the latest technology. They are specialized to a particular design after the fabrication process, hence they are programmable. Devices that can be reprogrammed to fix errors or update functionality are also called *reconfigurable*. One of the most popular of these devices is the FPGA. Modern FPGAs can implement designs with the equivalent of millions of transistors on a single chip and operate with clock speeds of several hundred MegaHertz. For many designs, FPGAs are much more cost effective than ASICs.

For both FPGAs and ASICs, all of the steps that take the initial design to finished chip can be automated. The initial design can be described as a schematic, similar to the diagrams in this chapter, or using an HDL. In the case of FPGAs, design tools translate this specification into programming data that can be downloaded to the chip. As we shall see, FPGA chips are based on memory technology. Rather than downloading a data file to memory, you download a configuration file to an FPGA that changes the way the hardware functions. This *programming* of the chip is very rapid. One can make a change in a complex design and have a working realization in less than an hour. By comparison, ASIC fabrication can take several weeks. By tightening the design cycle, such rapid prototyping has dramatically reduced the cost of designing and producing complex circuits for specific applications.

The underlying technology of an FPGA, and what makes it programmable and reconfigurable, is memory. Writing HDL programs and programming the FPGA makes the design process sound more like software than hardware development. The major difference is that the underlying structures being programmed implement the hardware structures we have been discussing in this chapter. In this section, we introduce FPGA technology and explain how digital designs are mapped onto the underlying structures. There are several companies that design and manufacture FPGAs. We will use the architecture of the Xilinx FPGA as an example.

17.7.1 FPGA Architecture

Let us consider the architecture of Xilinx FPGAs. Our objective is not to learn how to program them—a task normally accomplished by software—but rather to show the relationship of these sophisticated chips to the logic we have already developed. The Xilinx chip is made up of three basic building blocks:

CLB the configurable logic blocks are where the computation of the user's circuit takes place.
IOB the input/output blocks connect I/O pins to the circuitry on the chip.
Interconnect interconnect is essential for wiring between CLBs and from IOBs to CLBs.

The Xilinx chip is organized with its configurable logic blocks (CLBs) in the middle, its I/O blocks (IOBs) on the periphery, and lots of different types of interconnect. Wiring is essential to support the versatility of the chips to implement different designs efficiently and to ensure that the resources on the FPGA can be utilized efficiently. The overview of a Xilinx chip is presented in Figure 17.33. Each CLB

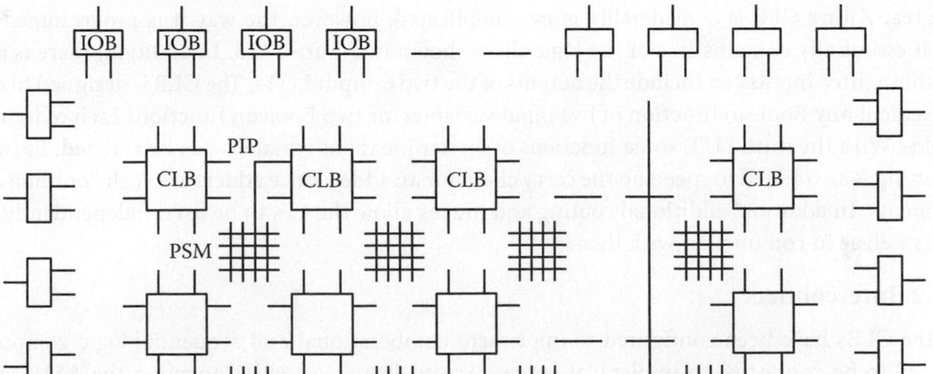

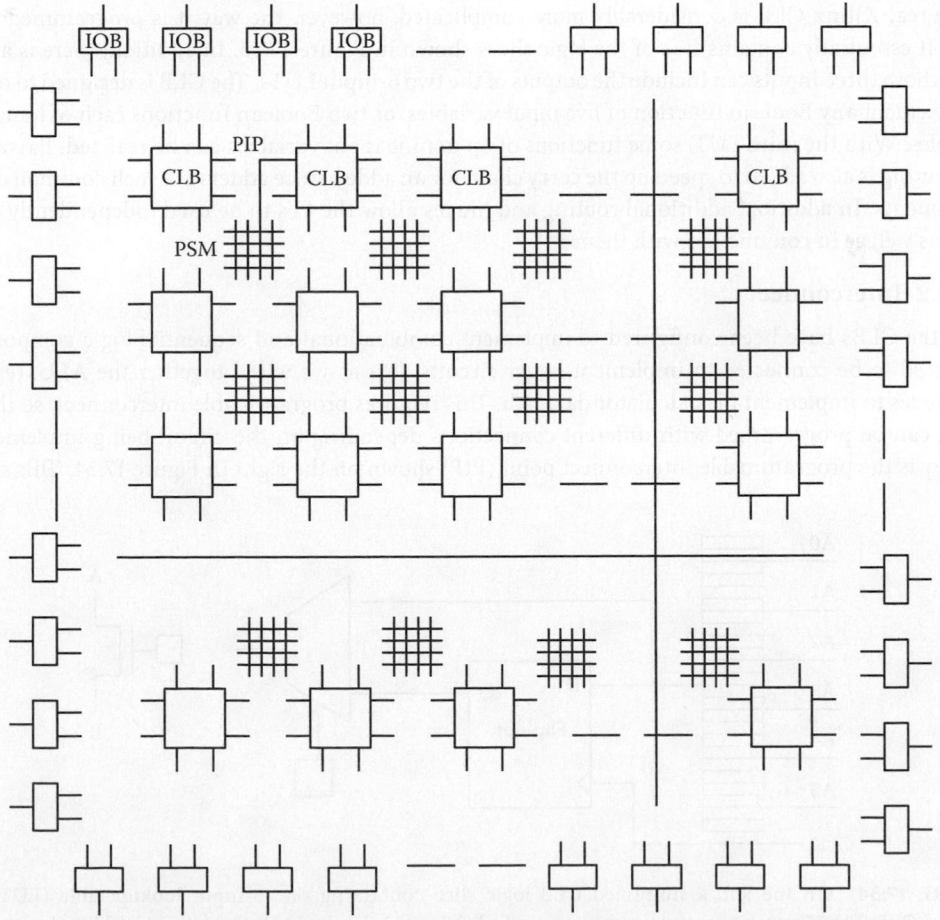

FIGURE 17.33 Overview of the Xilinx FPGA. I/O blocks (IOBs) are connected to pads on the chip which are connected to the chip-carrier pins. Several different types of interconnect are shown, including programmable interconnect points (PIPs), programmable switch matrices (PSMs), and long line interconnect.

is programmable and can implement combinational or sequential logic or both. Data enter or exit the chip through the IOBs. The interconnect can be programmed so that the desired connections are made. Distributed configuration memory (not shown in the figure) controls the functionality of the CLBs and IOBs, as well as the wiring connections. Implementation of the CLBs, interconnect, and IOBs is described in more detail in the following text.

17.7.1.1 Configurable Logic Block

How do you use memory to implement different Boolean logic functions? You download the truth table of the function to the memory. Changing the contents of the memory cells changes the functionality of the hardware. One set of memory cells for a one-bit output are referred to as a Look Up Table (LUT) because you "look up" the result. The basic Xilinx logic slice contains a 6-input LUT for realizing combinational logic. The result of this combinational function may or may not be stored in a DFF. The implementation of a logic slice, with 6-input LUT and optional FF on the LUT output, is shown on the left in Figure 17.34. Note that the multiplexer can be configured to output the combinational result of the LUT or the result of the LUT after it has been stored in the FF by setting the memory bit attached to the MUX's select line. The logic is configured by downloading 17 bits of memory: the 16 bits in the lookup table and the one select bit for the multiplexer. Using this simple structure, all of the designs described so far in this chapter can be implemented.

The real Xilinx CLB is considerably more complicated; however, the way it is programmed is the same. It essentially contains two of the logic slices shown in Figure 17.35. In addition, there is a third LUT whose three inputs can include the outputs of the two 6-input LUTs. The CLB is designed to be able to implement any Boolean function of five input variables, or two Boolean functions each of four input variables. With the third LUT, some functions of up to nine input variables can be realized. Extra logic and routing is also added to speed up the carry chain for an adder, since adders are such common digital components. In addition, additional routing and muxes allow the FFs to be used independently of the LUTs as well as in conjunction with them.

17.7.1.2 Interconnect

Once the CLBs have been configured to implement combinational and sequential logic components, they need to be connected to implement larger circuits, just as we wired together the ALU, register, and muxes to implement the calculator datapath. This requires programmable interconnect, so that an FPGA can be programmed with different connections depending on the circuit being implemented. The key is the programmable interconnect point (PIP) shown on the right in Figure 17.34. This simple

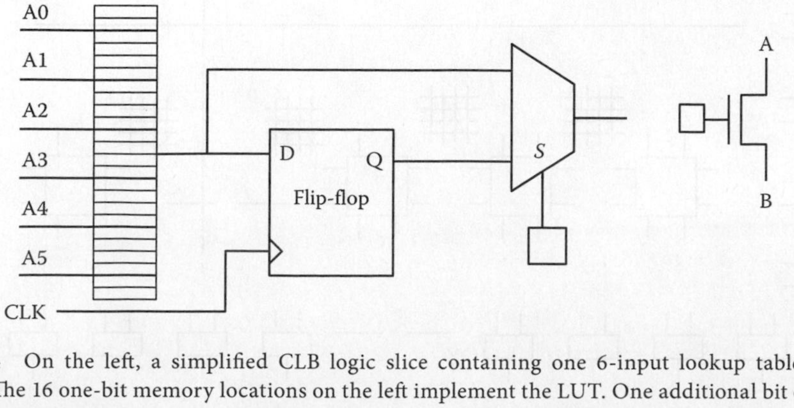

FIGURE 17.34 On the left, a simplified CLB logic slice containing one 6-input lookup table (LUT) and optional DFF. The 16 one-bit memory locations on the left implement the LUT. One additional bit of memory is used to configure the MUX so the output comes either directly from the LUT or from the DFF. On the right is a programmable interconnect point (PIP). LUTs, PIPs, and MUXes are three of the components that make FPGA hardware programmable.

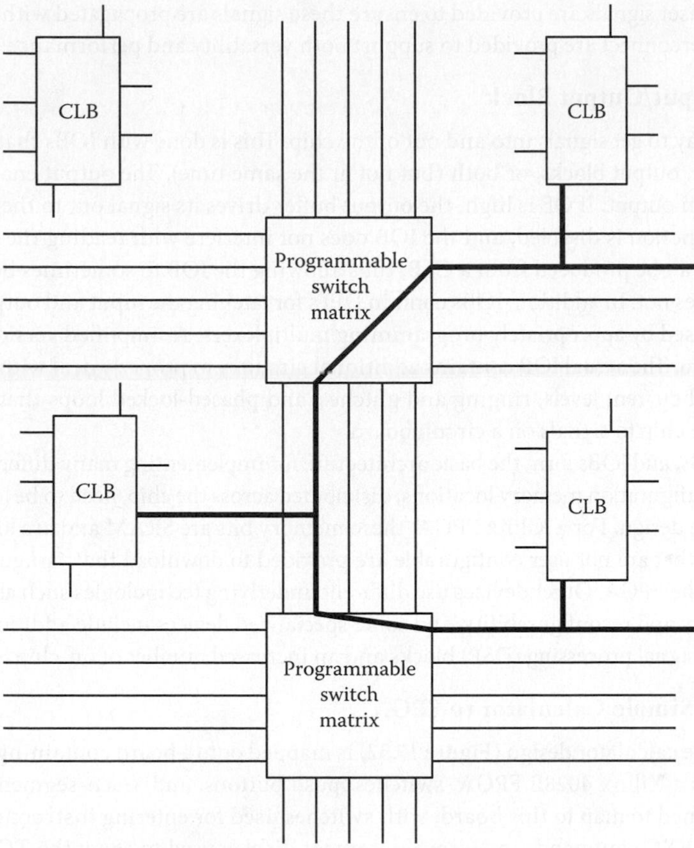

FIGURE 17.35 Programmable interconnect including two programmable switch matrices (PSMs) for connecting the output of one CLB to the input of two other CLBs.

device is a pass transistor with its gate connected to a memory bit. If that memory bit contains a one, the two ends of the transistor are logically connected; if the memory bit contains a zero, no connection is made. By appropriately loading these memory bits, different wiring connections can be realized. Note that there is considerably more delay across the PIP than across a simple metal wire on a chip. This is the tradeoff when using programmable interconnect.

Our FPGA architecture has CLBs arranged in a matrix over the surface of a chip, with routing channels for wiring between the CLBs. Programmable switch matrices (PSMs) are implemented at the intersection between a row and column of routing. These switch matrices support multiple connections, including signals on a row and a column, signals passing through on a row, and signals passing through on a column. Figure 17.35 shows a signal output from one CLB connecting to the inputs of two others. This signal passes through two PSMs and three PIPs, one for each CLB connection.

While programmable interconnect makes the FPGA versatile, each active device in the interconnection fabric slows the signal being routed. For this reason, early FPGA devices, where all the interconnect went through PIPs and PSMs, implemented designs that were considerably slower than their ASIC counterparts. More recent FPGA architectures have recognized the fact that high speed interconnect is essential to high performance designs. In addition to PIPs and PSMs, many other types of interconnect have been added. Many architectures have nearest neighbor connections, where wires connect from one CLB to its neighbors without going through a PIP. Lines that skip PSMs have been added. For example, double lines go through every other PSM in a row or a column. Long lines have been added to support signals that span the chip. Special channels for fast carry chains are available. Finally, global lines that transmit CLK and reset signals are provided to ensure these signals are propagated with little delay. All of these types of interconnect are provided to support both versatility and performance.

17.7.1.3 Xilinx Input/Output Block

Finally, we need a way to get signals into and out of the chip. This is done with IOBs that can be configured as input blocks, output blocks, or both (but not at the same time). The output enable (OE) signal enables the IOB as an output. If OE is high, the output buffer drives its signal out to the I/O pad. If OE is low, the output function is disabled, and the IOB does not interfere with reading the input from the pad. The OE signal can be produced from a CLB, thus allowing the IOB to sometimes be enabled as an output and sometimes not. In addition, IOBs contain DFFs for latching the input and output signals. The latches can be bypassed by appropriately programming multiplexers. A simplified version of the IOB is shown in Figure 17.36. The actual IOB contains additional circuitry to properly deal with such electrical issues as voltage and current levels, ringing and glitches, and phased-locked loops that are important when interfacing the chip to signals on a circuit board.

Interconnect, CLBs, and IOBs form the basic architecture for implementing many different designs in a single FPGA. The configuration memory locations, distributed across the chip, need to be loaded to implement the appropriate design. For a Xilinx FPGA, these memory bits are SRAM and are loaded on power up. Special I/O pins that are not user configurable are provided to download the configuration bits that define the design to the FPGA. Other devices use different underlying technologies such as SRAM to provide programmability and reconfigurability, and some specialized devices include additional specialized circuitry like digital signal processing (DSP) blocks and an increased number of on-chip SRAM blocks.

17.7.1.4 Mapping Simple Calculator to FPGA

Let us look at how the calculator design (Figure 17.32) is mapped onto a board containing an FPGA. The board used contains a Xilinx 4028E FPGA, switches, push buttons, and seven-segment displays. The calculator was designed to map to this board, with switches used for entering instructions and data, a push button for the EXC command, and a seven-segment display used to show the TOS. The logic of the calculator is mapped to CLBs. The controller is made up of boolean logic and DFFs to hold the state. The datapath is made up of the components developed in this chapter and mapped to LUTs and DFFs.

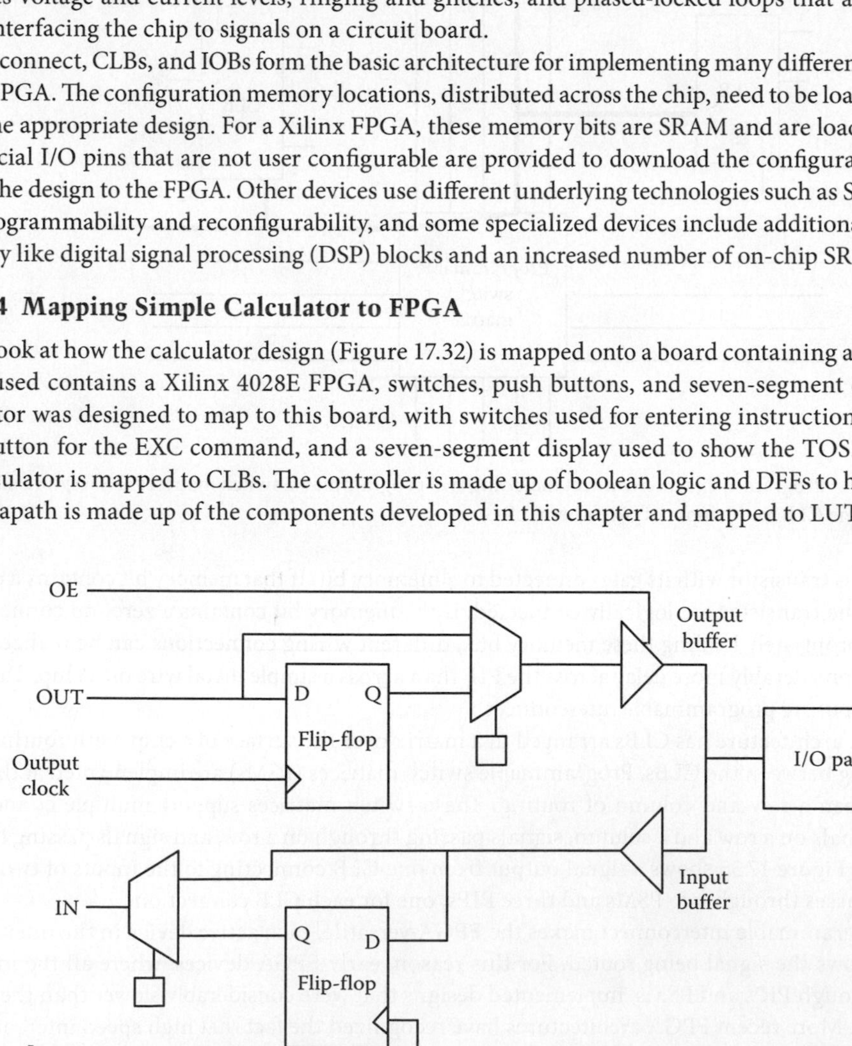

FIGURE 17.36 Simplified version of the IOB. The IOBs can be configured to input or output signals to the FPGA. When OE is high, the output buffer is enabled so the output signal is driven on the I/O pad. When OE is low, the IOB functions as an input block. Buffers handle electrical issues with signals from the I/O pad.

This calculator was developed as an undergraduate laboratory experiment. Students enter the design using a schematic capture tool, which involves drawing diagrams like the ones in this chapter. Synthesis tools translate the design to LUTs and FFs, breaking the logic up into 4-input chunks of logic, each of which is implemented with one truth table. Alternatively, the design can be described using a HDL to specify behavior. A different synthesis tool is involved, but the end result is a set of LUTs and DFFs that implement the design. Placement tools map these components to CLBs on the chip, and routing tools route the connections making use of the various kinds of interconnect available. The tools translate the logic design to a bitstream that is downloaded to the board from a PC through a download cable. The result is a functioning calculator on an FPGA board.

An advantage of this design flow is that designers can migrate their designs to the newest chip architecture without changing the specification. Only the tools need to change to target a faster or cheaper device.

17.7.2 Higher Levels of Complexity

Integrating functionality on a single chip allows for higher performance and smaller packages. As more and more transistors can be realized on a single chip, and functionality increases, it also has become increasingly clear that one particular structure for implementing a design does not suit all needs. While many digital designs can be implemented using FPGA structures, others are less well suited to this technology. For example, hardware multipliers are particularly inefficient when mapped to CLBs. Certain functions perform better on a microprocessor or a programmable DSP than in digital hardware. For this reason, FPGA manufacturers have begun integrating large functionality blocks on FPGAs. For example, both Xilinx and Altera, two of the major FPGA manufacturers, have introduced FPGAs with embedded multipliers, embedded RAM blocks, and embedded processors. Altera calls this approach "System on a Programmable Chip". Similarly, to support reconfigurability after manufacturing, ASIC designers are increasingly adding blocks of FPGA logic to their designs. It is clear that the future will bring more complex chips with more functionality, higher clock speeds, and more types of logic integrated on a single chip. Digital logic and reconfigurable hardware will be part of these designs for the foreseeable future.

Key Terms

Active logic: Digital logic which operates all of the time in the active, dissipative region of the electronic amplifiers from which it is constructed. Such logic is generally faster than saturated logic, but it dissipates much more energy. See *saturated logic*.

ASIC: Application-specific integrated circuit. Integrated circuits that are designed for a specific application. The term is used to describe VLSI circuits which can be configured before manufacturing to meet the specific needs of the application. High performance ASICs are low power and high speed but also expensive.

CLB: Configurable logic block in a Xilinx FPGA. This is where the computation occurs in an FPGA. CLBs implement truth tables and DFFs.

Clock: The input which provides the timing signal for a circuit. In general, the oscillator circuit which generates a synchronization signal.

CMOS: Complementary metal–oxide–semiconductor. The dominant family of binary, saturated logic. CMOS circuits are built from MOS transistors in complementary pairs. When one transistor is open, its complement is closed, ensuring that no current flows through the switch itself. Such a configuration has minimum power dissipation in any static state.

Combinational circuit: A logic circuit whose output is a function only of its inputs. Apart from propagation delays, the output always represents a logical combination of its present inputs. See sequential circuit.

Critical race: In a sequential circuit, a situation in which the "next state" is determined by which of two (or more) internal gates is first to reach saturated state. Such a race is dependent on minor variations in circuit parameters and on temperature, making the circuit unpredictable and possibly even unstable.

Decoder: A logic circuit with N inputs and 2^N outputs, one and only one of which is asserted to indicate the numerical value of the N input lines read as a binary number. A decoder and demultiplexer use the same internal circuitry.

Demultiplexer (DEMUX): A logic circuit with K control inputs which steers the data input to the one of the 2^K outputs selected by the control inputs.

D flip-flop (DFF): A fundamental sequential circuit whose output changes only upon a CLK signal and whose output represents the data on its input at the time of the last clock.

Don't care: In a truth table or Karnaugh map, a state which is irrelevant to the correct functioning of the circuit (e.g., because it never occurs in the intended application). Thus, the designer "doesn't care" whether that state is asserted, and he or she may choose the output that best minimizes the number of gates.

Edge-triggered FF: An FF which changes state on a clock transition from low to high or high to low rather than responding to the level of the CLK signal. Contrast to MS FF.

Encoder: A logic circuit with 2^N inputs and N outputs, the outputs indicating the binary number of the one input line that is asserted. See also *priority encoder.*

Flip-flop: Any of several related bistable circuits which form the memory elements in clocked, sequential circuits.

FPGA: Field-programmable gate array. VLSI chips with a large number of reconfigurable gates that can be "programmed" to function as complex logic circuits. The programming can be done on site and in some cases may be dynamically (in circuit) reprogrammable.

Glitch: A transient transition between logic states caused by different delays through parallel paths in a logic circuit. They are unintentional transitions, so they do not correctly represent the logic of the intended design.

HDL: Hardware Description Language. A language that resembles a programming language with added features for specifying hardware designs.

IOB: I/O block in a Xilinx FPGA. Block on the periphery of an FPGA that supports the input and/or output of a signal from the FPGA to its external environment.

Karnaugh map: A mapping of a truth table into a rectangular array of cells in which the nearest neighbors of any cell differ from that cell by exactly one binary input variable. K-maps are useful for minimizing Boolean logic functions.

Master–slave FF: An FF which changes state when the clock voltage reaches a threshold level. Contrast to edge-triggered FF.

Multiplexer (MUX): A circuit with N control inputs to select among one of 2^N data inputs and connect the appropriate data input to the single output line.

PIP: Programmable interconnect point on a Xilinx FPGA. A pass transistor with a memory bit connected to its gate terminal. If the memory is loaded with a "1" the two ends are connected; if loaded with a "0" the two ends are not connected. This is the basis of programmable interconnect.

Priority encoder: An encoder with the additional property that if several inputs are asserted simultaneously, the output number indicates the numerically highest input that is asserted. For example, if lines 1 and 3 were both asserted, the output value would be 3.

Saturated logic: Logic gates whose output is fully on or fully off. Saturated logic dissipates no power except while switching. The opposite of *saturated logic* is *active logic.*

Sequential circuit: A circuit which goes through a sequence of stable states, transitioning between such states at times determined by a CLK signal. The output of a sequential circuit depends both on its current inputs and its history, which is captured in the states. Contrast with *combinational* circuit.

Transparent latch: Essentially, an FF which continuously passes the input to the output (thus *transparent*) when the clock is high (low) but holds the last output during any interval when the clock is low (high). The circuit is said to have *latched* when it is holding its output constant regardless of the value of the input.

VLSI: Very large scale integrated circuit. A semiconductor device that integrates millions of transistors on a single chip. VLSI chips are typically very high speed and have very high power dissipation.

Further Information

This is a very quick pass through digital circuit design. What has been covered in this chapter provides a good overview of the principles as well as information to help the reader understand the chapter on computer architecture in this volume.

There are many textbooks devoted to the subject of digital logic design. Wakerly (2005) emphasizes basic principles and the underlying technologies. Other digital logic texts emphasize logic design tools (Katz, 2004) and computer design fundamentals (Mano and Kime, 2007).

There have also been many volumes published on design entry, tools for automating the digital design process, and mapping designs onto Field Programmable Logic (FPL). The Hardware Description Languages (HDLs) most widely used today are VHDL (Ashenden, 2008) and Verilog (Moorby and Thomas, 2008). The interested reader may also wish to pursue the topic of design with field programmable logic (Zeidman, 2002). The subjects of logic design, FPL and HDLs, are brought together in one volume (Salcic and Smailagic, 2000).

Ongoing research in this area is concerned with design entry, automation of the design process and new architectures, and technologies for implementing digital designs. The research in design entry is focused on raising the level of specification of digital logic designs. New HDLs based on high level languages such as Java and C are being developed. Another approach is design environments which incorporate sophisticated libraries of very complex, parameterized components such as digital filters, ALUs, and Ethernet controllers. The user can customize these blocks for their specific design. One such widely used environment is Simulink, a product of Mathworks.

Along with higher levels of design specification, researchers are investigating more sophisticated design automation tools. The goal is to have designers specify the functionality of their designs and to use synthesis tools to automatically translate that functionality to efficient hardware implementations.

References

Ashenden, P. J. 2008. *The Designer's Guide to VHDL*, 3rd edn. Elsevier Science, Burlington, MA.

Boole, G. 2009. *The Mathematical Analysis of Logic*. Reissue. Cambridge University Press, Cambridge, U.K.

Karnaugh, M. 1953. A map method for synthesis of combinational circuits. *Trans. AIEE Commun. Electron*. 72(1):593–599.

Katz, R. H. 2004. *Contemporary Logic Design*, 2nd edn. Prentice Hall, Upper Saddle River, NJ.

Mano, M. M. and Kime, C. R. 2007. *Logic and Computer Design Fundamentals*, 4th edn. Prentice Hall, Lebanon, IN.

Moorby, P. R. and Thomas, D. E. 2008. *The Verilog Hardware Description Language*, 5th edn. Springer-Verlag, New York.

Salcic, Z. and Smailagic, A. 2000. *Digital Systems Design and Prototyping Using Field Programmable Logic and Hardware Description Language*, 2nd edn. Springer-Verlag, Boston, MA.

Wakerly, J. F. 2005. *Digital Design: Principles and Practice*, 4th edn. Prentice Hall, Lebanon, IN.

Zeidman, R. 2002. *Designing with FPGAs and CPLDs*. CMP Books, Lawrence, KS.

18

Memory Systems

Yoongu Kim
Carnegie Mellon University

Onur Mutlu
Carnegie Mellon University

18.1 Introduction

As shown in Figure 18.1, a computing system consists of three fundamental units: (i) units of computation to perform operations on data (e.g., processors), (ii) units of storage (or memory) that store data to be operated on or archived, (iii) units of communication that communicate data between computation units and storage units. The storage/memory units are usually categorized into two: (1) *memory system*, which acts as a *working storage area*, storing the data that is currently being operated on by the running programs, and (2) the backup *storage system*, for example, the hard disk, which acts as a *backing store*, storing data for a longer term in a persistent manner. This chapter will focus on the "working storage area" of the processor, that is, the memory system.

The memory system is the repository of data from where data can be retrieved and updated by the processor (or processors). Throughout the operation of a computing system, the processor reads data from the memory system, performs computation on the data, and writes the modified data back into the memory system—continuously repeating this procedure until all the necessary computation has been performed on all the necessary data.

18.1.1 Basic Concepts and Metrics

The *capacity* of a memory system is the total amount of data that it can store. Every piece of data stored in the memory system is associated with a unique *address*. For example, the first piece of data has an

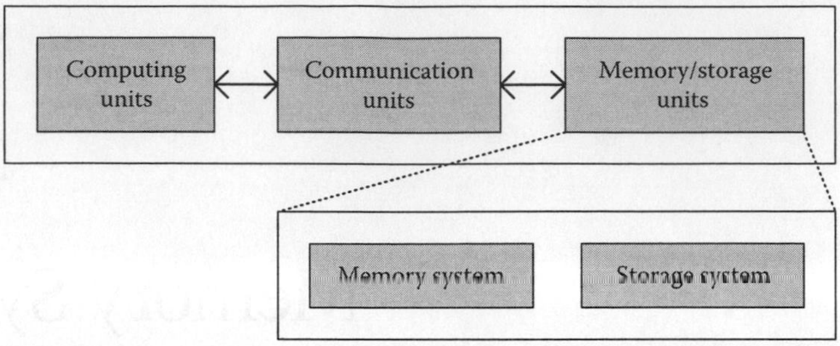

FIGURE 18.1 Computing system.

address of 0, whereas the last piece of data has an address of *capacity* − 1. The full range of possible addresses, spanning from 0 to *capacity* − 1, is referred to as the *address space* of the memory system. Therefore, in order to access a particular piece of data from the memory system, the processor must supply its address to the memory system.

The *performance* of a memory system is characterized by several important metrics: (i) latency, (ii) bandwidth, and (iii) parallelism. A high-performance memory system would have low latency, high bandwidth, and high parallelism. *Latency* is the amount of time it takes for the processor to access one piece of data from the memory system. *Bandwidth*, also known as *throughput*, is the rate at which the processor can access pieces of data from the memory system. At first blush, latency and bandwidth appear to be inverses of each other. For example, if it takes time T seconds to access one piece of data, then it would be tempting to assume that $1/T$ pieces of data can be accessed over the duration of 1 s. However, this is not always true. To fully understand the relationship between latency and bandwidth, we must also examine the third metric of a memory system's performance. *Parallelism* is the number of accesses to the memory system that can be served concurrently. If a memory system has a parallelism of 1, then all accesses are served one-at-a-time, and this is the only case in which bandwidth is the inverse of latency. But if a memory system has a parallelism of more than 1, then multiple accesses to different addresses can be served concurrently, thereby overlapping their latencies. For example, when the parallelism is equal to two, during the amount of time required to serve one access (T), another access is served in parallel. Therefore, while the latency of an individual access remains unchanged at T, the bandwidth of the memory system doubles to $2/T$. More generally, the relationship among latency, bandwidth, and parallelism of a memory system can be expressed as follows:

$$\text{Bandwith [accesses/time]} = \frac{\text{Parallelism [unitless]}}{\text{Latency [time/access]}}$$

While the bandwidth can be measured in the unit of accesses-per-time (preceding equation), another way of expressing bandwidth is in the unit of bytes-per-time—that is, the-amount-of-data-accessed-per-time (the following equation):

$$\text{Bandwidth [bytes/time]} = \frac{\text{Parallelism [unitless]}}{\text{Latency [time/access]}} \times \text{DataSize [bytes/access]}$$

An additional characteristic of a memory system is cost. The *cost* of a memory system is the capital expenditure required to implement it. Cost is closely related to the capacity and performance of the memory system: increasing the capacity and performance of a memory system *usually* also makes it more expensive.

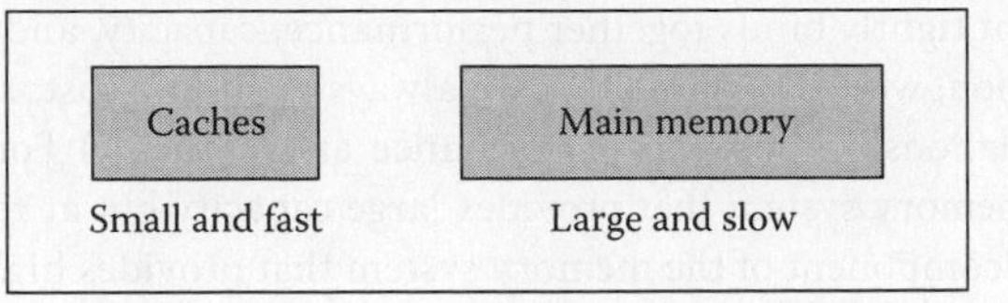

FIGURE 18.2 Memory system.

18.1.2 Two Components of the Memory System

When designing a memory system, computer architects strive to optimize for all of the three characteristics explained in the preceding text: large capacity, high performance, and low cost. However, a memory system that has both large capacity and high performance is also very expensive. For example, when one wants to increase the capacity of memory, it is almost always the case that the latency of memory also increases. Therefore, when designing a memory system within a reasonable cost budget, there exists a fundamental trade-off relationship between capacity and performance: it is possible to achieve either large capacity or high performance but not both at the same time in a cost-effective manner.

As a result of the trade-off between capacity and performance, a modern memory system typically consists of two components, as shown in Figure 18.2: (i) a *cache* [44] (pronounced "cash"), a component that is small but relatively fast-to-access and (ii) *main memory*, a component that is large but relatively slow-to-access. Between the two, main memory is the all-encompassing (i.e., "main") repository whose capacity is usually determined to minimize accesses to the storage system. If the memory system were to consist only of main memory, then all accesses to the memory system would be served by main memory, which is slow. That is why a memory system also contains a cache: Although a cache is much smaller than main memory, it is fast (exactly because it is smaller). The memory system utilizes the cache by taking a subset of the data from main memory and placing it into the cache—thereby enabling some of the accesses to the memory system to be served quickly by the cache. The data stored in the cache are replicas of the data that are already stored in main memory. Hence, the capacity of the memory system as a whole is defined by the capacity of only the main memory while ignoring the capacity of the cache. In other words, while the processor can always access the data it wants from main memory, the cache exists to *expedite* some of those accesses as long as they are to data that are replicated in the cache. In fact, most memory systems employ not just one but *multiple* caches, each of which provides a different trade-off between capacity and performance. For example, there can be two caches, one of which is 3 smaller and faster than the other. In this case, the data in the smaller cache is a subset of the data in the larger cache, similar to how the larger cache is a subset of the main memory.

Note that the structure and operation of the hardware components that make up the cache and main memory can be similar (in fact, they can be exactly the same). However, the structure and operation of cache and memory components are affected by (i) the function of the respective components and (ii) the technology in which they are implemented. The main function of caches is to store a small amount of data such that it can be accessed quickly. Traditionally, caches have been designed using the SRAM technology (so that they are fast), and main memory has been designed using the DRAM technology (so that it has large capacity). As a result, caches and main memory have evolved to be different in structure and operation, as we describe in Sections 18.4 and 18.5.

18.2 Memory Hierarchy

An ideal memory system would have the following three properties: high performance (i.e., low latency and high bandwidth), large capacity, and low cost. Realistically, however, technological and physical constraints limit the design of a memory system that achieves all three properties at the same time. Rather, one property can be improved only at the expense of another—that is, there exists a fundamental *trade-off* relationship that tightly binds together performance, capacity, and cost of a memory system. (To simplify the discussion, we will assume that we always want low cost, restricting ourselves to the "zero-sum" trade-off relationship between performance and capacity.) For example, main memory is a component of the memory system that provides large capacity but at relatively low performance, while a cache is another component of the memory system that provides high performance but at relatively small capacity. Therefore, caches and main memory lie at opposite ends of the trade-off spectrum between performance and capacity.

We have mentioned earlier that modern memory systems consist of both caches and main memory. The reasoning behind this is to achieve the best of both worlds (performance and capacity)—that is, a memory system that has the high performance of a cache and the large capacity of main memory. If the memory system consists only of main memory, then *every* access to the memory system would experience the high latency (i.e., low performance) of main memory. However, if caches are used in addition to main memory, then *some* of the accesses to the memory system would be served by the caches at low latency, while the remainder of the accesses are assured to find their data in main memory due to its large capacity—albeit at high latency. Therefore, as a net result of using both caches and main memory, the memory system's *effective latency* (i.e., average latency) becomes lower than that of main memory, while still retaining the large capacity of main memory.

Within a memory system, caches and main memory are said to be part of a *memory hierarchy*. More formally, a memory hierarchy refers to how multiple components (e.g., cache and main memory) with different performance/capacity properties are combined together to form the memory system. As shown in Figure 18.3, at the "top level" of the memory hierarchy lies the fastest (but the smallest) component, whereas at the "bottom level" of the memory hierarchy lies the slowest (but the largest) component. Going from top to bottom, the memory hierarchy typically consists of multiple levels of caches (e.g., L1, L2, L3 caches in Figure 18.3, standing for, respectively, level-1, level-2, level-3)—each of whose capacity is larger than the one above it—and a single level of main memory at the bottom whose capacity is the largest. The bottom-most cache (e.g., the L3 cache in Figure 18.3, which lies immediately above main memory) is also referred to as the *last-level cache*.

When the processor accesses the memory system, it typically does so in a sequential fashion by first searching the top component to see whether it contains the necessary data. If the data is found, then the access is said to have "hit" in the top component and the access is served there on-the-spot without having to search the lower components of the memory hierarchy. Otherwise, the access is said to have "missed" in the top component and it is passed down to the immediately lower component in the memory hierarchy, where the access may again experience either a hit or a miss. As an access goes lower into the hierarchy, the probability of a hit becomes greater due to the increasing capacity of the lower components, until it reaches the bottom level of the hierarchy where it is guaranteed to always hit (assuming the data is in main memory).

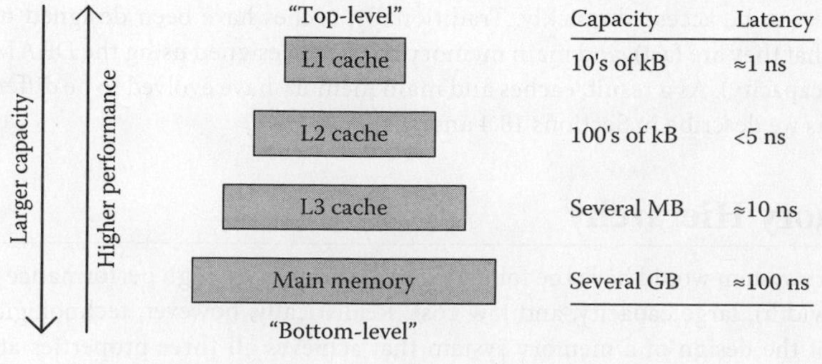

FIGURE 18.3 Example of a memory hierarchy with three levels of caches.

Assuming a two-level memory hierarchy with a single cache and main memory, the effective latency of the memory system can be expressed by the following equation. In the equation, $P_{\text{cache}}^{\text{hit}}$ denotes the probability that an access would hit in the cache, also known as the cache *hit-rate*:

$$\text{Latency}_{\text{effective}} = P_{\text{cache}}^{\text{hit}} \times \text{Latency}_{\text{cache}} + (1 - P_{\text{cache}}^{\text{hit}}) \times \text{Latency}_{\text{main\_memory}}$$

- $0 \le P_{\text{cache}}^{\text{hit}} \le 1$
- $\text{Latency}_{\text{cache}} \ll \text{Latency}_{\text{main\_memory}}$

Similarly, for a three-level memory hierarchy with two caches and main memory, the effective latency of the memory system can be expressed by the following equation.

$$\text{Latency}_{\text{effective}} = P_{\text{cache1}}^{\text{hit}} \times \text{Latency}_{\text{cache1}} + \left(1 - P_{\text{cache1}}^{\text{hit}}\right) \times \left\{P_{\text{cache2}}^{\text{hit}} \times \text{Latency}_{\text{cache2}} + \left(1 - P_{\text{cache2}}^{\text{hit}}\right) \times \text{Latency}_{\text{main\_memory}}\right\}$$

- $0 \le P_{\text{cache1}}^{\text{hit}}, P_{\text{cache2}}^{\text{hit}} \le 1$
- $\text{Latency}_{\text{cache1}} < \text{Latency}_{\text{cache2}} \ll \text{Latency}_{\text{main\_memory}}$

As both equations show, a high hit-rate in the cache implies a lower effective latency. In the best case, when all accesses hit in the cache $\left(P_{\text{cache}}^{\text{hit}} = 1\right)$, then the memory hierarchy has the lowest effective latency, equal to that of the cache. While having a high hit-rate is always desirable, the actual value of the hit-rate is determined primarily by (i) the cache's size and (ii) the processor's memory access behavior. First, compared to a small cache, a large cache is able to store more data and has a better chance that a given access will hit in the cache. Second, if the processor tends to access a small set of data over and over again, the cache can store those data so that subsequent accesses will hit in the cache. In this case, a small cache would be sufficient to achieve a high hit-rate.

Fortunately, many computer programs—that the processor executes—access the memory system in this manner. In other words, many computer programs exhibit *locality* in their memory access behavior. Locality exists in two forms: *temporal locality* and *spatial locality*. First, given a piece of data that has been accessed, temporal locality refers to the phenomenon (or memory access behavior) in which the same piece of data is likely to be accessed again in the near future. Second, given a piece of data that has been accessed, spatial locality refers to the phenomenon (or memory access behavior) in which neighboring pieces of data (i.e., data at nearby addresses) are likely to be accessed in the near future. Thanks to both temporal and spatial locality, the cache—and, more generally, the memory hierarchy—is able to reduce the effective latency of the memory system.

18.3 Managing the Memory Hierarchy

As mentioned earlier, every piece of data within a memory system is associated with a unique address. The set of all possible addresses for a memory system is called the address space and it ranges from 0 to *capacity* − 1, where *capacity* is the size of the memory system. Naturally, when software programmers compose computer programs, they rely on the memory system and assume that specific pieces of the program's data can be stored at specific addresses without any restriction, as long as the addresses are valid—that is, the addresses do not overflow the address space provided by the memory system. In practice, however, the memory system does *not* expose its address space directly to computer programs. This is due to two reasons.

First, when a computer program is being composed, the modern software programmer has no way of knowing the capacity of the memory system on which the program will run. For example, the program may run on a computer that has a very large memory system (e.g., 1 TB capacity) or a very small memory system (e.g., 1 MB capacity). While an address of 1 GB is perfectly valid for the first memory system, the same address is invalid for the second memory system, since it exceeds the maximum bound (1 MB) of the address space. As a result, if the address space of the memory system is directly exposed to the program, the software programmer can never be sure which addresses are valid and can be used to store data for the program she is composing.

Second, when a computer program is being composed, the software programmer has no way of knowing which other programs will run simultaneously with the program. For example, when the user runs the program, it may run on the same computer as many other different programs, all of which may happen to utilize the same address (e.g., address 0) to store a particular piece of their data. In this case, when one program modifies the data at that address, it overwrites another program's data that was stored at the same address, even though it should not be allowed to do so. As a result, if the address space of the memory system is directly exposed to the program, then multiple programs may overwrite and corrupt each other's data, leading to incorrect execution for all of the programs.

18.3.1 Virtual vs. Physical Address Spaces

As a solution to these two problems, a memory system does not directly expose its address space but instead provides the illusion of an extremely large address space that is separate for each individual program. While the illusion of a *large* address space solves the first problem of different capacities across different memory systems, the illusion of a *separate* large address space for each individual program solves the second problem of multiple programs modifying data at the same address. Since such large and separate address spaces are only an illusion provided to the programmer to make her life easier in composing programs, they are referred to as *virtual address spaces*. In contrast, the actual underlying address space of the memory system is called the *physical address space*. To use concrete numbers from today's systems (circa 2013), the physical address space of a typical 8 GB memory system ranges from 0 to 8 GB − 1, whereas the virtual address space for a program typically ranges from 0 to 256 TB − 1.*

A virtual address, just by itself, does not represent any real storage location unless it is actually backed up by a physical address. It is the job of the operating system—a program that manages all other programs as well as resources in the computing system—to substantiate a virtual address by mapping it to a physical address. For example, at the very beginning of when a program is executed, none of its virtual addresses are mapped to physical addresses. At this point, if the program attempts to store a piece of data to a particular virtual address, the operating system must first intervene on behalf of the program: the operating system maps the virtual address to a physical address that is *free*—that is, a physical address that is not yet mapped to another virtual address. Once this mapping has been established, it is memorized by the operating system and used later for "translating" any subsequent access to that virtual address to its corresponding physical address (where the data is stored).

18.3.2 Virtual Memory System

The entire mechanism, described so far, in which the operating system maps and translates between virtual and physical addresses is called *virtual memory*. There are three considerations when designing a virtual memory system as part of an operating system: (i) when to map a virtual address to a physical address, (ii) the mapping granularity, and (iii) what to do when physical addresses are exhausted.

First, most virtual memory systems map a virtual address when it is accessed for the very first time—that is, *on-demand*. In other words, if a virtual address is never accessed, it is never mapped

---

* For example, the current generation of mainstream x86-64 processors manufactured by Intel uses 48-bit virtual addresses—that is, a 256 TB virtual address space ($2^{48} = 256\ T$) [13].

to a physical address. Although the virtual address space is extremely large (e.g., 256 TB), in practice, only a small fraction of it is actually utilized by most programs. Therefore, mapping the entirety of virtual address space to the physical address space is wasteful, because the overwhelming majority of the virtual addresses will never be accessed. Not to mention the fact that the virtual address space is much larger than the physical address space such that it is not possible to map all virtual addresses to begin with.

Second, a virtual memory system must adopt a granularity at which it maps addresses from the virtual address space to the physical address space. For example, if the granularity is set equal to 1 byte, then the virtual memory system evenly divides the virtual/physical address into 1-byte virtual/physical "chunks," respectively. Then, the virtual memory system can arbitrarily map a 1-byte virtual chunk to any 1-byte physical chunk, as long as the physical chunk is free. However, such a fine division of the address spaces into large numbers of small chunks has a major disadvantage: it increases the complexity of the virtual memory system. As we recall, once a mapping between a pair of virtual/physical chunks is established, it must be memorized by the virtual memory system. Hence, large numbers of virtual/physical chunks imply a large number of possible mappings between the two, which increases the bookkeeping overhead of memorizing the mappings. To reduce such an overhead, most virtual memory systems coarsely divide the address spaces into fewer of chunks, where a chunk is called a *page* and whose typical size is 4 kB. As shown in Figure 18.4, a 4 kB chunk of the virtual address space is referred to as a *virtual page*, whereas a 4 kB chunk of the physical address is referred to as a *physical page* (alternatively, a *frame*). Every time a virtual page is mapped to a physical page, the operating system keeps track of the mapping by storing it in a data structure called the *page table*.

Third, as a program accesses a new virtual page for the very first time, the virtual memory system maps the virtual page to a free physical page. However, if this happens over and over, the physical address space may become exhausted—that is, none of the physical pages are free since all of them have been mapped to virtual pages. At this point, the virtual memory system must "create" a free physical page by reclaiming one of the mapped physical pages. The virtual memory system does so by *evicting* a physical page's data from main memory and "unmapping" the physical page from its virtual page. Once a free physical page is created in such a manner, the virtual memory system can map it to a new virtual page. More specifically, the virtual memory system takes the following three steps in order to reclaim a physical page and map it to a new virtual page. First, the virtual memory system selects the physical page that will be reclaimed, that is, the *victim*. The selection process of determining the victim is referred to as the *page replacement policy* [5]. While the simplest policy is randomly selecting any physical page, such a policy may significantly degrade the performance of the computing system. For example, if a

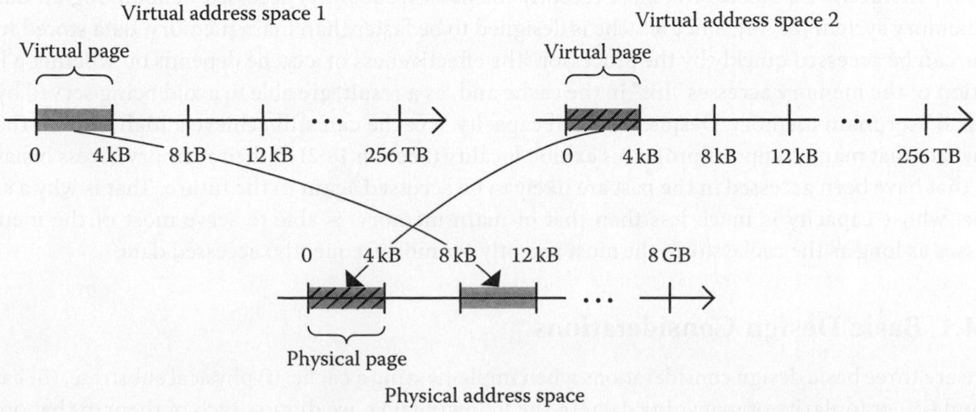

FIGURE 18.4 Virtual memory system.

very frequently accessed physical page is selected as the victim, then a future access to that physical page would be served by the hard disk. However, since a hard disk is extremely slow compared to main memory, the access would incur a very large latency. Instead, virtual memory systems employ more sophisticated page-replacement policies that try to select a physical page that is unlikely to be accessed in the near future, in order to minimize the performance degradation. Second, after a physical page has been selected as the victim, the virtual memory system decides whether the page's data should be migrated out of main memory and into the hard disk. If the page's data had been modified by the program while it was in main memory, then the page must be written back into the hard disk—otherwise, the modifications that were made to the page's data would be lost. On the other hand, if the page's data had *not* been modified, then the page can simply be evicted from main memory (without being written into the hard disk) since the program can always retrieve the page's original data from the hard disk. Third, the operating system updates the page table so that the virtual page (that had previously mapped to the victim) is now mapped to the hard disk instead of a physical page in main memory. Finally, now that a physical page had been reclaimed, the virtual memory system maps the free physical page to a new virtual page and updates the page table accordingly.

After the victim has been evicted from main memory, it would be best if the program does not access the victim's data ever again. This is because accessing the victim's data incurs the large latency of the hard disk where it is stored. However, if the victim is eventually accessed, then the virtual memory system brings the victim's data back from the hard disk and places it into a free physical page in main memory. Unfortunately, if main memory has no free physical pages remaining at this point, then another physical page must be chosen as a victim and be evicted from main memory. If this happens repeatedly, different physical pages are forced to *ping-pong* back and forth between main memory and hard disk. This phenomenon, referred to as *thrashing*, typically occurs when the capacity of the main memory is not large enough to accommodate all of the data that a program is actively accessing (i.e., its working set). When a computing system experiences thrashing, its performance degrades since it must constantly access the extremely slow hard disk instead of the faster main memory.

18.4 Caches

Generally, a cache is any structure that stores data that is likely to be accessed again (e.g., frequently accessed data or recently accessed data) in order to avoid the long latency operation required to access the data from a much slower structure. For example, web servers on the internet typically employ caches that store the most popular photographs or news articles so that they can be retrieved quickly and sent to the end user. In the context of the memory system, a cache refers to a small but fast component of the memory hierarchy that stores the most recently (or most frequently) accessed data among all data in the memory system [26,44]. Since a cache is designed to be faster than main memory, data stored in the cache can be accessed quickly by the processor. The effectiveness of a cache depends on whether a large fraction of the memory accesses "hit" in the cache and, as a result, are able to avoid being served by the much slower main memory. Despite its small capacity, a cache can still achieve a high hit-rate thanks to the fact that many computer programs exhibit locality (Section 18.2) in their memory access behavior: data that have been accessed in the past are likely to be accessed again in the future. That is why a small cache, whose capacity is much less than that of main memory, is able to serve most of the memory accesses as long as the cache stores the most recently (or most frequently) accessed data.

18.4.1 Basic Design Considerations

There are three basic design considerations when implementing a cache: (i) physical substrate, (ii) capacity, and (iii) granularity of managing data. In the following text, we discuss each of them in that order.

First, the physical substrate used to implement the cache must be able to deliver much lower latencies than that used to implement main memory. That is why *SRAM* (static random-access memory, pronounced "es-ram") has been—and continues to be—by far the most dominant physical substrate for caches. The primary advantage of SRAM is that it can operate at very high speeds that are on par with the processor itself. In addition, SRAM consists of the same type of semiconductor-based transistors that make up the processor. (This is not the case with DRAM, which is used to implement main memory, as we will see in Section 18.5.). So an SRAM-based cache can be placed—at low cost—side-by-side with the processor on the same semiconductor chip, allowing it to achieve even lower latencies. The smallest unit of SRAM is an *SRAM cell*, which typically consists of six transistors. The six transistors collectively store a single bit of data (0 or 1) in the form of electrical voltage ("low" or "high"). When the processor reads from an SRAM cell, the transistors feed the processor with the data value that corresponds to their voltage levels. On the other hand, when the processor writes into an SRAM cell, the voltage of the transistors is appropriately adjusted to reflect the updated data value that is being written.

Second, when determining the capacity of a cache, one must be careful to balance the need for high hit-rate, low cost, and low latency. While a large cache is more likely to provide a high hit-rate, it also has two shortcomings. First, a large cache has high cost due to the increased number of transistors that are required to implement it. Second, a large cache is likely to have a higher latency. This is because orchestrating the operation of many transistors is a complex task that introduces extra overhead delays (e.g., it may take a longer time to determine the location of an address). In practice, a cache is made large enough to achieve a sufficiently high hit-rate but not large enough to incur significantly high cost and latency.

Third, a cache is divided into many small pieces, called *cache blocks*, as shown in Figure 18.5. A cache block is the granularity in which the cache manages data. The typical size of a cache block is 64-bytes in modern memory systems. For example, a 64 kB cache consists of 1024 separate cache blocks. Each of these cache blocks can store data that corresponds to an arbitrary 64-byte "chunk" of the address space—that is, any given 64-byte cache block can contain data from address 0, or address 64, or address 128, address 192, and so on. Therefore, a cache block requires some sort of a "label" that conveys the address of the data stored in the cache block. For exactly this purpose, every cache block has its own *tag* where the address of the data (not the data itself) is stored. When the processor accesses the cache for a piece of data at a particular address, it searches the cache for the cache block whose tag matches the address. If such a cache block exists, then the processor accesses the data contained in the cache block—as explained earlier, this is called a cache hit. In addition to its address, a cache block's tag may also store other types of information about the cache block. For example, whether the cache block is empty, whether the cache block has been written to, or how recently the cache block has been accessed. These topics and more will soon be discussed in this section.

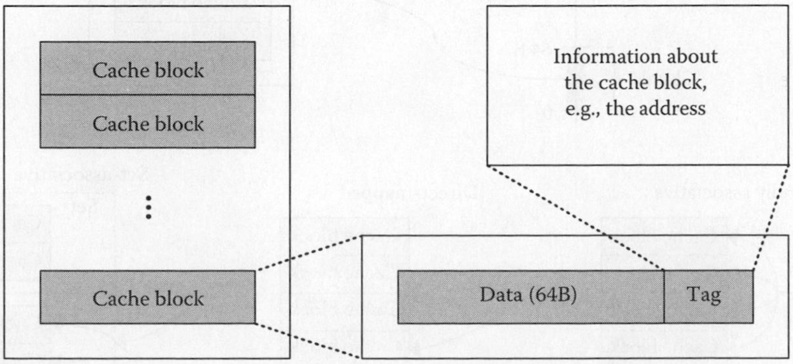

FIGURE 18.5 Cache.

18.4.2 Logical Organization

In addition to the basic considerations discussed in the preceding text, the computer architect must decide how a cache is logically organized—that is, how it maps 64-byte chunks of the address space onto its 64-byte cache blocks. Since the memory system's address space is much larger than the cache's capacity, there are many more chunks than there are cache blocks. As a result, the mapping between chunks and cache blocks is necessarily "many-to-few." However, for each individual chunk of the address space, the computer architect can design a cache that has the ability to map the chunk to (i) any of its cache blocks or (ii) only a specific cache block. These two different logical organizations represent two opposite extremes in the degree of freedom when mapping a chunk to a cache block. This freedom, in fact, presents a crucial trade-off between the complexity and utilization of the cache, as described in the following text.

On the one hand, a cache that provides the greatest freedom in mapping a chunk to a cache block is said to have a *fully associative* organization (Figure 18.6, lower-left). When a new chunk is brought in from main memory, a fully associative cache does not impose any restriction on where the chunk can be placed—that is, the chunk can be stored in *any* of the cache blocks. Therefore, as long as the cache has at least one empty cache block, the chunk is guaranteed to be stored in the cache without having to make room for it by evicting an already occupied (i.e., nonempty) cache block. In this regard, a fully associative cache is the best at efficiently utilizing all the cache blocks in the cache. However, the downside of a fully associative cache is that the processor must exhaustively search *all* cache blocks whenever it accesses the cache, since any one of the cache blocks may contain the data that the processor wants. Unfortunately, searching through all cache blocks not only takes a long time (leading to high access latency), but also wastes energy.

On the other hand, a cache that provides the least freedom in mapping a chunk to a cache block is said to have a *direct-mapped* organization (Figure 18.6, lower-middle). When a new chunk is brought in from main memory, a direct-mapped cache allows the chunk to be placed in only *a specific* cache block. For example, let us assume a 64 kB cache consisting of 1024 cache blocks (64 bytes). A simple implementation of a direct-mapped cache would map every 1024th chunk (64 bytes) of the address space to the same

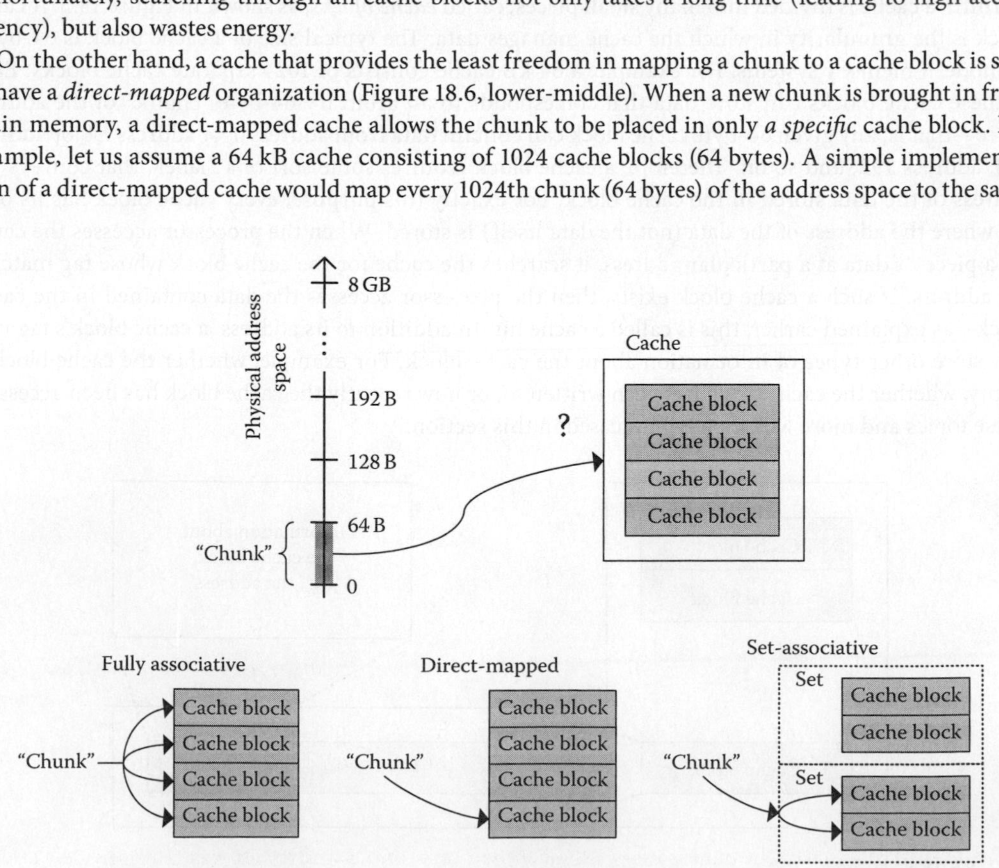

FIGURE 18.6 Logical organization.

cache block—for example, chunks at address 0, address 64K, address 128K, etc. would all map to the 0th cache block in the cache. But if the cache block is already occupied with a different chunk, then the old chunk must first be evicted before a new chunk can be stored in the cache block. This is referred to as a *conflict*—that is, when two different chunks (corresponding to two different addresses) contend with each other for the same cache block. In a direct-mapped cache, conflicts can occur at one cache block even when all other cache blocks are empty. In this regard, a direct-mapped cache is the worst at efficiently utilizing all the cache blocks in the cache. However, the upside of a direct-mapped cache is that the processor can simply search only *one* cache block to quickly determine whether the cache contains the data it wants. Hence, the access latency of a direct-mapped cache is low.

As a middle ground between the two organizations (fully associative vs. direct-mapped), there is a third alternative called the *set-associative* organization [12] (Figure 18.6, lower-right), which allows a chunk to map to one of multiple (but not all) cache blocks within a cache. If a cache has a total of N cache blocks, then a fully associative organization would map a chunk to any of the N cache blocks, while a direct-mapped organization would map a chunk to only one specific cache block. A set-associative organization, in contrast, is based on the concept of *sets*, which are small nonoverlapping groups of cache blocks. A set-associative cache is similar to a direct-mapped cache in that a chunk is mapped to only one specific set. However, a set-associative cache is also similar to a fully associative cache in that the chunk can map to any cache block that belongs to the specific set. For example, let us assume a set-associative cache in which each set consists of two cache blocks, which is called a *two-way* set-associative cache. Initially, such a cache maps a chunk to one specific set out of all $N/2$ sets. Then, within the set, the chunk can map to either of the two cache blocks that belong to the set. More generally, a W-way set-associative cache ($1 < W < N$) directly maps a chunk to one specific set, while fully associatively mapping a chunk to any of the W cache block within the set. For a set-associative cache, the value of W is fixed when the cache is designed and cannot be changed afterward. However, depending on the value of W, a set-associative cache can behave similarly to a fully associative cache (for large values of W) or a direct-mapped cache (for small values of W). In fact, an N-way set-associative cache degenerates into a fully associative cache, whereas a one-way set-associative cache degenerates into a direct-mapped cache.

18.4.3 Management Policies

If a cache has unlimited capacity, it would have the highest hit-rate since it can store all of the processor's data that is in memory. Realistically, however, a cache has only a limited capacity. Therefore, it needs to be selective about which data it should store—that is, the cache should store only the data that is the most likely to be accessed by the processor in the near future. In order to achieve that goal, a cache employs various management policies that enable it to make the best use of its small capacity: (i) allocation policy and (ii) replacement policy. In the following text, we discuss each of them, respectively.

First, the cache's *allocation policy* determines how its cache blocks become populated with data. Initially, before the execution of any program, all of the cache blocks are empty. As a program starts to execute, it accesses new chunks from the memory system that are not yet stored in the cache. Whenever this happens (i.e., a cache miss), the cache's allocation policy decides whether the new chunk should be stored in one of the empty cache blocks. Since many programs exhibit locality in their memory accesses, a chunk that is accessed (even for the first time) is likely to be accessed again in the future. That is why *always-allocate* is one of the most popular allocation policies: for every cache miss, the always-allocate policy populates an empty cache block with the new chunk. (On the other hand, a different allocation policy may be more discriminative and prevent certain chunks from being allocated in the cache.) However, when the cache has no empty cache blocks left, a new chunk cannot be stored in the cache unless the cache "creates" an empty cache block by reclaiming one of the occupied cache blocks. The cache does so by *evicting* the data stored in an occupied cache block and *replacing* it with the new chunk, as described next.

Second, when the cache does not have an empty cache block where it can store a new chunk, the cache's *replacement policy* selects one of the occupied cache blocks to evict, that is, the *victim* cache block. The replacement policy is invoked when a new chunk is brought into the cache. Depending on the cache's logical organization, the chunk may map to one or more cache blocks. However, if all such cache blocks are already occupied, the replacement policy must select one of the occupied cache blocks to evict from the cache. For a direct-mapped cache, the replacement policy is trivial: since a chunk can be mapped to only one specific cache block, then there is no choice but to evict that specific cache block if it is occupied. Therefore, a replacement policy applies only to set-associative or fully associative caches, where a chunk can potentially be mapped to multiple cache blocks—any one of which may become the victim if all of those cache blocks are occupied. Ideally, the replacement policy should select the cache block that is expected to be accessed the farthest away in the future, such that evicting the cache block has the least impact on the cache's hit-rate [3]. That is why one of the most common replacement policies is the *LRU* (*least-recently used*) policy, in which the cache block that has been the least recently accessed is selected as the victim. Due to the principle of locality, such a cache block is less likely to be accessed in the future. Under the LRU policy, the victim is the least recently accessed cache block among (i) all cache blocks within a set (for a set-associative cache) or (ii) among all cache blocks within the cache (for a fully associative cache). To implement the LRU policy, the cache must keep track of each block in terms of the last time when it was accessed. A similar replacement policy is the *LFU* (*least-frequently used*) policy, in which the cache block that has been the least frequently accessed is selected as the victim. We refer the reader to the following works for more details on cache replacement policies: Liptay [26] and Qureshi et al. [36,39].

18.4.4 Managing Multiple Caches

Until now, we have discussed the design issues and management policies for a single cache. However, as described previously, a memory hierarchy typically consists of more than just one cache. In the following, we discuss the policies that govern how multiple caches within the memory hierarchy interact with each other: (i) inclusion policy, (ii) write handling policy, and (iii) partitioning policy.

First, the memory hierarchy may consist of multiple levels of caches in addition to main memory. As the lowest level, main memory always contains the superset of all data stored in any of the caches. In other words, main memory is *inclusive* of the caches. However, the same relationship may not hold between one cache and another cache depending on the *inclusion policy* employed by the memory system [2]. There are three different inclusion policies: (i) inclusive, (ii) exclusive, and (iii) noninclusive. First, in the *inclusive* policy, a piece of data in one cache is guaranteed to be also found in all lower levels of caches. Second, in the *exclusive* policy, a piece of data in one cache is guaranteed *not* to be found in all lower levels of caches. Third, in the *noninclusive* policy, a piece of data in one cache may or may not be found in lower levels of caches. Among the three policies, the inclusive and exclusive policies are opposites of each other, while all other policies between the two are categorized as noninclusive. On the one hand, the advantage of the exclusive policy is that it does not waste cache capacity since it does not store multiple copies of the same data in all of the caches. On the other hand, the advantage of the inclusive policy is that it simplifies searching for data when there are multiple processors in the computing system. For example, if one processor wants to know whether another processor has the data it needs, it does not need to search all levels of caches of that other processor but instead search only the lowest-level cache. (This is related to the concept of *cache coherence*, which is not covered in this chapter.) Lastly, the advantage of the noninclusive policy is that it does not require the effort to maintain a strict inclusive/exclusive relationship between caches. For example, when a piece of data is inserted into one cache, inclusive or exclusive policies require that the same piece of data be inserted into or evicted from other levels of caches. In contrast, the noninclusive policy does not have this requirement.

Second, when the processor writes new data into a cache block, the data stored in the cache block is modified and becomes different from the data that was originally brought into the cache. While the cache contains the newest copy of the data, all lower levels of the memory hierarchy (i.e., caches and main memory) still contain an old copy of the data. In other words, when a write access hits in a cache, a discrepancy arises between the modified cache block and the lower levels of the memory hierarchy. The memory system resolves this discrepancy by employing a *write handling policy.* There are two types of write handling policies: (i) write-through and (ii) write-back. First, in a *write-through policy*, every write access that hits in the cache is propagated down to the lower levels of the memory hierarchy. In other words, when a cache at a particular level is modified, the same modification is made for all lower levels of caches and for main memory. The advantage of the write-through policy is that it prevents any data discrepancy from arising in the first place. But, its disadvantage is that every write access is propagated through the entire memory hierarchy (wasting energy and bandwidth), even when the write access hits in the cache. Second, in a *write back policy*, a write access that hits in the cache modifies the cache block at only that cache, without being propagated down to the rest of the memory hierarchy. In this case, however, the cache block contains the only modified copy of the data, which is different from the copies contained in lower levels of caches and main memory. To signify that the cache block contains modified data, a write-back cache must have a *dirty flag* in the tag of each cache block: when set to "1," the dirty flag denotes that the cache block contains modified data. Later on, when the cache block is evicted from the cache, it must be written into the immediately lower level in the memory hierarchy, where the dirty flag is again set to "1." Eventually, through a cascading series of evictions at multiple levels of caches, the modified data is propagated all the way down to main memory. The advantage of the write-back policy is that it can prevent write accesses from always being written into all levels of the memory hierarchy—thereby conserving energy and bandwidth. Its disadvantage is that it slightly complicates the cache design since it requires additional dirty flags and special handling when modified cache blocks are evicted.

Third, a cache at one particular level may be partitioned into two smaller caches, each of which is dedicated to two different types of data: (i) instruction and (ii) data. Instructions are a special type of data that tells the computer how to manipulate (e.g., add, subtract, move) other data. Having two separate caches (i.e., an instruction cache and a data cache) has two advantages. First, it prevents one type of data from monopolizing the cache. While the processor needs both types of data to execute a program, if the cache is filled with only one type of data, the processor may need to access the other type of data from lower levels of the memory hierarchy, thereby incurring a large latency. Second, it allows each of the caches to be places closer to the appropriate parts of the processor—lowering the latency to supply instructions and data to the processor. Typically, one part of the processor (i.e., the instruction fetch engine) accesses instructions, while another part of the processor (i.e., the data fetch engine) accesses noninstruction data. In this case, the two caches can each be colocated with the part of the processor that accesses its data—resulting in lower latencies and potentially higher operating frequency for the processor. For this reason, usually only the highest level of the cache in the memory hierarchy, which is *directly* accessed by the processor, is partitioned into an instruction cache and a data cache.

18.4.5 Specialized Caches for Virtual Memory

Specialized caches have been used to accelerate address translation in virtual memory systems (discussed in Section 18.3). The most commonly used such cache is referred to as a *TLB* (translation lookaside buffer). The role of a TLB is to cache the recently used virtual to physical address translations by the processor such that the translation is quick for the virtual addresses that hit in the TLB. Essentially, a TLB is a cache that caches the parts of the page table that have been used by the processor.

18.5 Main Memory

While caches are predominantly optimized for high speed, in contrast, the foremost purpose of main memory is to provide as large capacity as possible at low cost and at reasonably low latency. That is why the preferred physical substrate for implementing main memory is *DRAM* (dynamic random-access-memory, pronounced "dee-ram") [6]. The smallest unit of DRAM is a *DRAM cell*, consisting of one transistor (1T) and one capacitor (1C). A DRAM cell stores a single bit of data (0 or 1) in the form of electrical charge in its capacitor ("discharged" or "charged"). DRAM's primary advantage over SRAM lies in its small cell size: a DRAM cell requires fewer electrical components (1T and 1C) than an SRAM cell (6T). Therefore, many more DRAM cells can be placed in the same amount of area on a semiconductor chip, enabling DRAM-based main memory to achieve a much larger capacity for approximately the same amount of cost.

Typically, SRAM-based caches are integrated on the same semiconductor chip as the processor. In contrast, DRAM-based main memory is implemented using one or more dedicated DRAM chips that are separate from the processor chip. This is due to two reasons. First, a large capacity main memory requires such a large number of DRAM cells that they cannot all fit on the same chip as the processor. Second, the process technology needed to manufacture DRAM cells (with their capacitors) is not compatible at low cost with the process technology needed to manufacture processor chips. Therefore, placing DRAM and logic together would significantly increase cost in today's systems. Instead, main memory is implemented as one or more DRAM chips that are dedicated for the large number of DRAM cells. From the perspective of the processor, since DRAM-based main memory is not on the same chip as the processor, it is sometimes referred to as "off-chip" main memory.

A set of electrical wires, called the *memory bus*, connects the processor to the DRAM chips. Within a processor, there is a *memory controller* that communicates with the DRAM chips using the memory bus. In order to access a piece of data from the DRAM chips, the memory controller sends/receives the appropriate electrical signals to/from the DRAM chips through the memory bus. There are three types of signals: (i) address, (ii) command, and (iii) data. The address conveys the location of the data being accessed, the command conveys how the data is being accessed (e.g., read or write), and the data is the actual value of the data itself. Correspondingly, a memory bus is divided into three smaller sets of wires, each of which is dedicated to a specific type of signal: (i) address bus, (ii) command bus, and (i) data bus. Among these three, only the data bus is bidirectional since the memory controller can either send/receive data from the DRAM chips, whereas the address and command buses are unidirectional since only the memory controller sends the address and command to the DRAM chips.

18.5.1 DRAM Organization

As shown in Figure 18.7, a DRAM-based main memory system is logically organized as a hierarchy of (i) channels, (ii) ranks, and (iii) banks. *Banks* are the smallest memory structures that can be accessed in parallel with respect to each other. This is referred to as *bank-level parallelism* [34]. Next, a *rank* is a collection of DRAM chips (and their banks) that operate in lockstep. A DRAM rank typically consists of eight DRAM chips, each of which has eight banks. Since the chips operate in lockstep, the rank has only eight independent banks, each of which is the set of all *i*th bank across all chips. Banks in different ranks are fully decoupled with respect to their chip-level electrical operation and, consequently, offer better bank-level parallelism than banks in the same rank. Lastly, a *channel* is the collection of all banks that share the same memory bus (address, command, data buses). While banks from the same channel experience contention at the memory bus, banks from different channels can be accessed completely independently

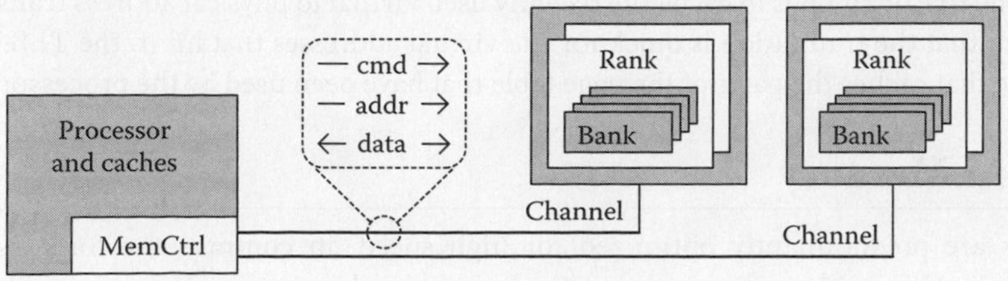

FIGURE 18.7 DRAM organization.

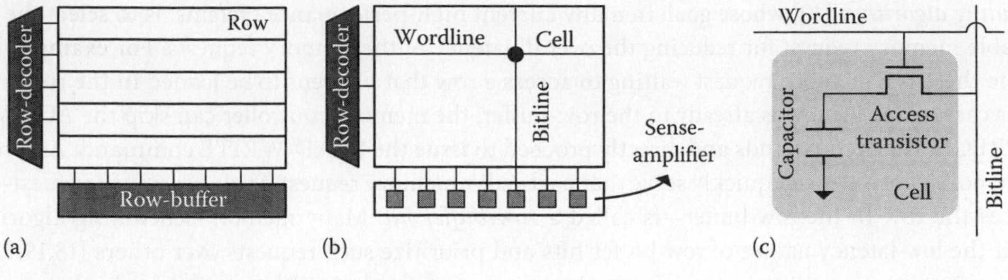

FIGURE 18.8 Bank organization. (a) High-level view, (b) low-level view, and (c) cell.

of each other. Although the DRAM system offers varying degrees of parallelism at different levels in its organization, two memory requests that access the same bank must be served one after another. To understand why, let us examine the logical organization of a DRAM bank as seen by the memory controller.

Figure 18.8 presents the logical organization of a DRAM bank. A DRAM bank is a two-dimensional array of capacitor-based DRAM cells. It is viewed as a collection of *rows*, each of which consists of multiple *columns*. Therefore, every cell is identified by a pair of addresses: a row address and a column address. Each bank contains a *row-buffer* which is an array of sense-amplifiers. The purpose of a *sense-amplifier* is to read from a cell by reliably detecting the very small amount of electrical charge stored in the cell. When writing to a cell, on the other hand, the sense-amplifier acts as an electrical driver and programs the cell by filling or depleting its stored charge. Spanning a bank in the columnwise direction are wires called the *bitlines*, each of which can connect a sense-amplifier to any of the cells in the same column. A wire called the *wordline* (one for each row) determines whether or not the corresponding row of cells is connected to the bitlines.

18.5.2 Bank Operation

To serve a memory request that accesses data at a particular row and column address, the memory controller issues three commands to a bank in the order as shown in the following list. Each command triggers a specific sequence of events within the bank to access the cell(s) associated with the address:

1. ACTIVATE (issued with a row address): Load the entire row into the row-buffer.
2. READ/WRITE (issued with a column address): From the row-buffer, access the data stored in a column. For a READ, transfer the data out of the row-buffer to the processor. For a WRITE, modify the data in the row-buffer according to the data received from the processor.
3. PRECHARGE: Clear the row-buffer.

Each DRAM command incurs a latency while it is processed by the DRAM chip. Undefined behavior may arise if a command is issued before the previous command is fully processed. To prevent such occurrences, the memory controller must obey a set of *timing constraints* while issuing commands to a DRAM chip [16]. These constraints define when a command becomes ready to be scheduled depending on all other commands issued before it to the same channel, rank, or bank. The exact values of the timing constraints are specified by the DRAM chip's datasheet and are different on a chip-by-chip basis. However, a rule of thumb is that the three DRAM commands (ACTIVATE, READ/WRITE, and PRECHARGE) each take about 15 ns [16]. For more information on the organization and operation of a DRAM bank, we refer the reader to Kim et al. [20].

18.5.3 Memory Request Scheduling

When there are multiple memory requests waiting to access DRAM, the memory controller must choose one of the memory requests to schedule next. The memory controller does so by employing a *memory scheduling algorithm* [40] whose goal, in many current high-performance systems, is to select the most favorable memory request for reducing the overall latency of the memory requests. For example, let us assume there is a memory request waiting to access a row that happens to be loaded in the row-buffer. In this case, since the row is already in the row-buffer, the memory controller can skip the ACTIVATE and PRECHARGE commands and directly proceed to issue the READ/WRITE command. As a result, the memory controller can quickly serve that particular memory request. Such a memory request—that accesses the row in the row-buffer—is called a *row-buffer hit*. Many memory scheduling algorithms exploit the low latency nature of row-buffer hits and prioritize such requests over others [18,19,33,34]. For more information on memory request scheduling, we refer the reader to recent works that explored the topic [18,19,33,34].

18.5.4 Refresh

A DRAM cell stores data as charge on a capacitor. Over time, this charge steadily leaks, causing the data to be lost. That is why DRAM is named "dynamic" RAM; its charge changes over time. In order to preserve data integrity, the charge in each DRAM cell must be periodically restored or *refreshed*. DRAM cells are refreshed at the granularity of a row by reading it out and writing it back in—which is equivalent to issuing an ACTIVATE and a PRECHARGE to the row in succession.

In modern DRAM chips, all DRAM rows must be refreshed once every 64 ms [16], which is called the *refresh interval*. The memory controller internally keeps track of time to ensure that it refreshes all DRAM rows before their refresh interval expires. When the memory controller decides to refresh the DRAM chips, it issues a REFRESH command. Upon receiving a REFRESH command, a DRAM chip internally refreshes a few of its rows by activating and precharging them. A DRAM chip refreshes only a few rows at a time since it has a very large number of rows and refreshing all of them would incur a very large latency. Since a DRAM chip cannot serve any memory requests while it is being refreshed, it is important that the refresh latency is kept short such that no memory request is delayed for too long. So instead of refreshing *all* rows at the end of each 64 ms interval, throughout a given 64 ms time interval, the memory controller issues many REFRESH commands, each of which triggers the DRAM chip to refresh only a subset of rows. However, the memory controller ensures that REFRESH commands are issued frequently enough such that all rows eventually do become refreshed before 64 ms has passed. For more information on DRAM refresh (and methods to reduce its effect on performance and energy), we refer the reader to Liu et al. [27].

18.6 Current and Future Research Issues

The memory system continues to be a major bottleneck in almost all computing systems (especially in terms of performance and energy/power consumption). It is becoming even more of a computing system bottleneck today and looking into the future: more and increasingly diverse processing cores and agents are sharing parts of the memory system; applications that run on the cores are becoming increasingly data and memory intensive; memory is consuming significant energy and power in modern systems; and, there is increasing difficulty scaling the well-established memory technologies, such as DRAM, to smaller technology nodes. As such, managing memory in significantly better ways at all levels of the transformation hierarchy, including both the software and hardware levels, is becoming even more important. Techniques or combinations of techniques that integrate the best ideas cooperatively at multiple levels together appear promising to solve the difficult performance, energy efficiency, correctness, security, and reliability problems we face in designing and managing memory systems today. In this section, we briefly discuss some of the major research problems (as we see them) related to caches and main memory and describe recent potential solution directions. Note that a comprehensive treatment of all research issues is out of the scope of this chapter, and neither is it our intent. For an illustrative treatment of some of the major research issues in memory systems, we refer the reader to Mutlu [32] (http://users.ece.cmu.edu/~omutlu/pub/onur-ismm-mspc-keynote-june-5-2011-short.pptx).

18.6.1 Caches

Efficient Utilization. To better utilize the limited capacity of a cache, many replacement policies have been proposed to improve upon the simple LRU policy. The LRU policy is not always the most beneficial across all memory-access patterns that have different amounts of locality. For example, in the LRU policy, the most-recently accessed data is always allocated in the cache even if the data has low-locality (i.e., unlikely to be ever accessed again). To make matters even worse, the low-locality data is unnecessarily retained in the cache for a long time. This is because the LRU policy always *inserts* data into the cache as the most-recently accessed and evicts the data only after it becomes the least-recently accessed. As a solution, researchers have been working to develop sophisticated replacement policies using a combination of three approaches. First, when a cache block is allocated in the cache, it should not always be inserted as the most-recently accessed. Instead, cache blocks with low-locality should be inserted as the least-recently accessed, so that they are quickly evicted from the cache to make room for other cache blocks that may have more locality (e.g., [15,35,41]). Second, when a cache block is evicted from the cache, it should not always be the least-recently accessed cache block. Ideally, "dead" cache blocks (which will never be accessed in the future) should be evicted—regardless of whether they are least-recently-accessed or not. Third, when choosing the cache block to evict, there should also be consideration for how costly it is to refetch the cache block from main memory. Between two cache blocks, all else being equal, the cache block that is likely to incur the shorter latency to refetch from main memory should be evicted [36].

Quality-of-service. In a multicore system, the processor consists of many cores, each of which can independently execute a separate program. In such a system, parts of the memory hierarchy may be shared by some or all of the cores. For example, the last-level cache in a processor is typically shared by all the cores. This is because the last-level cache has a large capacity and, hence, it is expensive to have multiple last-level caches separately for each core. In this case, however, it is important to ensure that the shared last-level cache is utilized by all the cores in a fair manner. Otherwise, a program running on one of the cores may fill up the last-level cache with only its data and evict the data needed by programs running on the other cores. To prevent such occurrences, researchers have proposed mechanisms to provide quality-of-service when a cache is shared by multiple cores. One mechanism is to partition the shared cache among the cores in such a way that each core has a dedicated partition where only the core's data is stored (e.g., [14,17,25,37]). Ideally, the size of a core's partition should be just large enough to hold the data that the core is actively accessing (the core's working set) and not any larger. Furthermore, as the size of a core's working set changes over time, the size of the core's partition should also dynamically expand or shrink in an appropriate manner.

Low power. Increasing the energy efficiency of computing systems is one of the key challenges faced by computer architects. Caches are one of the prime targets for energy optimization since they require large numbers of transistors, all of which dissipate power. For example, researchers have proposed to reduce a cache's power consumption by lowering the operating voltage of the cache's transistors [43]. However, this introduces new trade-offs in designing a cache that must be balanced: while a low voltage cache consumes less power, it may also be slower and more prone to errors. Researchers have been examining solutions to enable low-voltage caches that provide acceptable reliability at acceptable cost [43].

18.6.2 Main Memory

Challenges in DRAM scaling. Primarily due to its low cost-per-bit, DRAM has long been the most popular physical substrate for implementing main memory. In addition, DRAM's cost-per-bit has continuously decreased as DRAM process technology scaled to integrate more DRAM cells into the same area on a semiconductor chip. However, improving DRAM cell density by reducing the cell size, as has been done traditionally, is becoming more difficult due to increased manufacturing complexity/cost and reduced cell reliability. As a result, researchers are examining alternative ways of enhancing the performance and energy-efficiency of DRAM while still maintaining low cost. For example, there have been recent proposals to reduce DRAM access latency [24], to increase DRAM parallelism [20], and to reduce the number of DRAM refreshes [27].

3D-Stacked memory. A DRAM chip is connected to the processor chip by a memory bus. In today's systems, the memory bus is implemented using electrical wires on a motherboard. But, this has three disadvantages: high power, low bandwidth, and high cost. First, since a motherboard wire is long and thick, it takes a lot of power to transfer an electrical signal through it. Second, since it takes even more power to transfer many electrical signals in quick succession, there is a limit on the bandwidth that a motherboard wire can provide. Third, the processor chip and the DRAM chip each have a stub (called a *pin*) to which either end of a motherboard wire is connected. Unfortunately, pins are expensive. Therefore, a memory bus that consists of many motherboard wires (which requires just as many pins on the chips) increases the cost of the memory system. As a solution, researchers have proposed to stack one more DRAM chips directly on top of a processor chip [4,28] instead of placing them side by side on the motherboard. In such a 3D-stacked configuration, the processor chip can communicate directly with the DRAM chip(s), thereby eliminating the need for motherboard wires and pins.

Emerging technology. DRAM is by far the most dominant technology for implementing main memory. However, there has been research to develop alternative main memory technologies that may replace DRAM or be used in conjunction with DRAM. For example, *PCM* (phase-change memory, pronounced "pee-cee-em") is a technology in which data is stored as a resistance value. A PCM cell consists of a small crystal whose resistance value can be changed by applying heat to it at different rates. If heat is abruptly applied by a short burst of high current, then the crystal is transformed into the amorphous state, which has high resistance. On the other hand, if heat is steadily applied by a long burst of low current, then the crystal is transformed into the crystalline state, which has low resistance. PCM has important advantages that make it an attractive candidate for main memory [21,38]: nonvolatility (i.e., data is not lost when the power is turned off and there is no need to refresh PCM cells) and scalability (i.e., compared to DRAM, it may be easier to make smaller PCM cells in the future). However, PCM also has the disadvantages of large latency (especially for writes) and limited write-endurance (i.e., the PCM cell becomes unreliable beyond a certain number of writes), the solution to which is the topic of on-going research. To overcome the shortcomings of different technologies, several recent works have examined the use of multiple different technologies (e.g., PCM and DRAM together) as part of main memory, an approach called *hybrid memory* or *heterogeneous memory* [7,29,38,45]. The key challenge in hybrid memory systems is to devise effective algorithms that place data in the appropriate technology such that the system can exploit the advantages of each technology while hiding the disadvantages of each [29,45].

Quality-of-service. Similar to the last-level cache, main memory is also shared by all the cores in a processor. When the cores contend to access main memory, their accesses may interfere with each other and cause significant delays. In the worst case, a memory-intensive core may continuously access main memory in such a way that all the other cores are denied service from main memory [30]. This would detrimentally degrade the performance of not only those particular cores, but also of the entire computing system. To address this problem, researchers have proposed mechanisms that provide quality-of-service to each core when accessing shared main memory. For example, memory request scheduling algorithms can ensure that memory requests from all the cores are served in a fair manner [18,19,33,34]. Another approach is for the user to explicitly specify memory service requirements of a program to the memory controller so that the memory scheduling algorithm can subsequently guarantee that those requirements are met [42]. Other approaches to quality-of-service include mechanisms proposed to map the data of those applications that significantly interfere with each other to different memory channels [31] and mechanisms proposed to *throttle down* the request rate of the processors that cause significant interference to other processors [8]. Request scheduling mechanisms that prioritize bottleneck threads in parallel applications have also been proposed [10]. The QoS problem gets exacerbated when the processors that share the main memory are different, for example, when main memory is shared by a CPU consisting of multiple processors and a GPU, and recent research has started to examine solutions to this [1]. Finally, providing QoS and high performance in the presence of different types of memory requests from multiple processing cores, such as speculative *prefetch* requests that aim to fetch the data from memory before it is needed, is a challenging problem that recent research has started providing solutions for [9,11,22,23].

18.7 Summary

The memory system is a critical component of a computing system. It serves as the repository of data from where the processor (or processors) can access data. An ideal memory system would have both high performance and large capacity. However, there exists a fundamental trade-off relationship between the two: it is possible to achieve either high performance or large capacity but not both at the same time in a cost-effective manner. As a result of the trade-off, a memory system typically consists of two components: caches (which are small but relatively fast-to-access) and main memory (which is large but relatively slow-to-access). Multiple caches and a single main memory, all of which strike a different balance between performance and capacity, are combined to form a memory hierarchy. The goal of the memory hierarchy is to provide the high performance of a cache at the large capacity of main memory. The memory system is cooperatively managed by both the operating system and the hardware.

This chapter provided an introductory level description of memory systems employed in modern computing systems, focusing especially on how the memory hierarchy, consisting of caches and main memory, is organized and managed. The memory system continues to be an even more critical bottleneck going into the future, as described in Section 18.6. Many problems abound, yet the authors of this chapter remain confident that promising solutions, some of which are also described in Section 18.6, will also abound and hopefully prevail.

References

1. R. Ausavarungnirun, K. K.-W. Chang, L. Subramanian, G. H. Loh, and O. Mutlu. Staged memory scheduling: achieving high performance and scalability in heterogeneous systems. In: *International Symposium on Computer Architecture*, Portland, OR, 2012.
2. J.-L. Baer and W.-H. Wang. On the inclusion properties for multi-level cache hierarchies. In: *International Symposium on Computer architecture*, Honolulu, HI, 1988.
3. L. A. Belady. A study of replacement algorithms for a virtual-storage computer. *IBM Systems Journal*, 5(2), 78–101, June 1966.
4. B. Black, M. Annavaram, N. Brekelbaum, J. DeVale, L. Jiang, G. H. Loh, D. McCaule, P. Morrow, D. W. Nelson, D. Pantuso, P. Reed, J. Rupley, S. Shankar, J. Shen, and C. Webb. Die stacking (3D) microarchitecture. In: *International Symposium on Microarchitecture*, Orlando, FL, 2006.
5. F. J. Corbató. A paging experiment with the multics system. *In Honor of P. M. Morse*, MIT Press, Cambridge, U.K., 1969.
6. R. H. Dennard. Field-Effect Transistor Memory. US Patent Number 3387286, 1968.
7. G. Dhiman, R. Ayoub, and T. Rosing. PDRAM: A hybrid PRAM and DRAM main memory system. In: *Design Automation Conference*, San Francisco, CA, 2009.
8. E. Ebrahimi, C. J. Lee, O. Mutlu, and Y. N. Patt. Fairness via source throttling: A configurable and high-performance fairness substrate for multi-core memory systems. In: *Architectural Support for Programming Languages and Operating Systems*, Pittsburgh, PA, 2010.
9. E. Ebrahimi, C. J. Lee, O. Mutlu, and Y. N. Patt. Prefetch-aware shared resource management for multi-core systems. In: *International Symposium on Computer Architecture*, San Jose, CA, 2011.

10. E. Ebrahimi, R. Miftakhutdinov, C. Fallin, C. J. Lee, J. A. Joao, O. Mutlu, and Y. N. Patt. Parallel application memory scheduling. In: *International Symposium on Microarchitecture*, Porto Alegre, Brazil, 2011.
11. E. Ebrahimi, O. Mutlu, C. J. Lee, and Y. N. Patt. Coordinated control of multiple prefetchers in multicore systems. In: *International Symposium on Microarchitecture*, New York, 2009.
12. M. D. Hill and A. J. Smith. Evaluating associativity in CPU caches. *IEEE Transactions on Computers*, 38(12), 1612–1630, December 1989.
13. Intel. Intel 64 and IA-32 Architectures Software Developers Manual, August 2012.
14. R. Iyer. CQoS: A framework for enabling QoS in shared caches of CMP platforms. In: *International Conference on Supercomputing*, Saint Malo, France, 2004.
15. A. Jaleel, W. Hasenplaugh, M. Qureshi, J. Sebot, S. Steely Jr., and J. Emer. Adaptive insertion policies for managing shared caches. In: *International Conference on Parallel Architectures and Compilation Techniques*, Toronto, Ontario, Canada, 2008.
16. Joint Electron Devices Engineering Council (JEDEC). DDR3 SDRAM Standard (JESD79-3F), 2012.
17. S. Kim, D. Chandra, and Y. Solihin. Fair cache sharing and partitioning in a chip multiprocessor architecture. In: *International Conference on Parallel Architectures and Compilation Techniques*, Antibes Juan-les-Pins, France, 2004.
18. Y. Kim, D. Han, O. Mutlu, and M. Harchol-Balter. ATLAS: A scalable and high-performance scheduling algorithm for multiple memory controllers. In: *International Symposium on High Performance Computer Architecture*, Bangalore, India, 2010.
19. Y. Kim, M. Papamichael, O. Mutlu, and M. Harchol-Balter. Thread cluster memory scheduling: Exploiting differences in memory access behavior. In: *International Symposium on Microarchitecture*, Atlanta, GA, 2010.
20. Y. Kim, V. Seshadri, D. Lee, J. Liu, and O. Mutlu. A case for exploiting subarray-level parallelism (SALP) in DRAM. In: *International Symposium on Computer Architecture*, Portland, OR, 2012.
21. B. C. Lee, E. Ipek, O. Mutlu, and D. Burger. Architecting phase change memory as a scalable DRAM alternative. In: *International Symposium on Computer Architecture*, Austin, TX, 2009.
22. C. J. Lee, O. Mutlu, V. Narasiman, and Y. N. Patt. Prefetch-aware DRAM controllers. In: *International Symposium on Microarchitecture*, Lake Como, Italy, 2008.
23. C. J. Lee, V. Narasiman, O. Mutlu, and Y. N. Patt. Improving memory bank-level parallelism in the presence of prefetching. In: *International Symposium on Microarchitecture*, New York, 2009.
24. D. Lee, Y. Kim, V. Seshadri, J. Liu, L. Subramanian, and O. Mutlu. Tiered-latency DRAM: A low latency and low cost DRAM architecture. In: *International Symposium on High Performance Computer Architecture*, Shenzhen, China, 2013.
25. J. Lin, Q. Lu, X. Ding, Z. Zhang, X. Zhang, and P. Sadayappan. Gaining insights into multicore cache partitioning: Bridging the gap between simulation and real systems. In: *International Symposium on High Performance Computer Architecture*, Salt Lake City, UT, 2008.
26. J. S. Liptay. Structural aspects of the system/360 Model 85: II The cache. *IBM Systems Journal*, 7(1), 15, March 1968.
27. J. Liu, B. Jaiyen, R. Veras, and O. Mutlu. RAIDR: Retention-aware intelligent DRAM refresh. In: *International Symposium on Computer Architecture*, Portland, OR, 2012.
28. G. H. Loh. 3D-stacked memory architectures for multi-core processors. In: *International Symposium on Computer Architecture*, Beijing, China, 2008.
29. J. Meza, J. Chang, H. Yoon, O. Mutlu, and P. Ranganathan. Enabling efficient and scalable hybrid memories using fine-granularity DRAM cache management. *IEEE Computer Architecture Letters*, 11(2), 61–64, July 2012.
30. T. Moscibroda and O. Mutlu. Memory performance attacks: Denial of memory service in multi-core systems. In: *USENIX Security Symposium*, Boston, MA, 2007.
31. S. P. Muralidhara, L. Subramanian, O. Mutlu, M. Kandemir, and T. Moscibroda. Reducing memory interference in multicore systems via application-aware memory channel partitioning. In: *International Symposium on Microarchitecture*, Porto Alegre, Brazil, 2011.

32. O. Mutlu. Memory systems in the many-core era: Challenges, opportunities, and solution directions. In: *International Symposium on Memory Management*, San Jose, CA, 2011. http://users.ece.cmu.edu/~omutlu/pub/onur-ismm-mspc-keynote-june-5-2011-short.pptx.
33. O. Mutlu and T. Moscibroda. Stall-time fair memory access scheduling for chip multiprocessors. In: *International Symposium on Microarchitecture*, Chicago, IL, 2007.
34. O. Mutlu and T. Moscibroda. Parallelism-aware batch scheduling: Enhancing both performance and fairness of shared DRAM systems. In: *International Symposium on Computer Architecture*, Beijing, China, 2008.
35. M. K. Qureshi, A. Jaleel, Y. N. Patt, S. C. Steely, and J. Emer. Adaptive insertion policies for high performance caching. In: *International Symposium on Computer Architecture*, San Diego, CA, 2007.
36. M. K. Qureshi, D. N. Lynch, O. Mutlu, and Y. N. Patt. A case for MLP-aware cache replacement. In: *International Symposium on Computer Architecture*, Boston, MA, 2006.
37. M. K. Qureshi and Y. N. Patt. Utility-based cache partitioning: A low-overhead, high-performance, runtime mechanism to partition shared caches. In: *International Symposium on Microarchitecture*, Orlando, FL, 2006.
38. M. K. Qureshi, V. Srinivasan, and J. A. Rivers. Scalable high performance main memory system using phase-change memory technology. In: *International Symposium on Computer Architecture*, Austin, TX, 2009.
39. M. K. Qureshi, D. Thompson, and Y. N. Patt. The V-way cache: Demand based associativity via global replacement. In: *International Symposium on Computer Architecture*, Madison, WI, 2005.
40. S. Rixner, W. J. Dally, U. J. Kapasi, P. Mattson, and J. D. Owens. Memory access scheduling. In: *International Symposium on Computer Architecture*, Vancouver, British Columbia, Canada, 2000.
41. V. Seshadri, O. Mutlu, M. A. Kozuch, and T. C. Mowry. The evicted-address filter: A unified mechanism to address both cache pollution and thrashing. In: *International Conference on Parallel Architectures and Compilation Techniques*, Minneapolis, MN, 2012.
42. L. Subramanian, V. Seshadri, Y. Kim, B. Jaiyen, and O. Mutlu. MISE: Providing performance predictability and improving fairness in shared main memory systems. In: *International Symposium on High Performance Computer Architecture*, Shenzhen, China, 2013.
43. C. Wilkerson, H. Gao, A. R. Alameldeen, Z. Chishti, M. Khellah, and S.-L. Lu. Trading off cache capacity for reliability to enable low voltage operation. In: *International Symposium on Computer Architecture*, Beijing, China, 2008.
44. M. V. Wilkes. Slave memories and dynamic storage allocation. *IEEE Transactions on Electronic Computers*, EC-14(2), 270–271, 1965.
45. H. Yoon, J. Meza, R. Ausavarungnirun, R. A. Harding, and O. Mutlu. Row buffer locality aware caching policies for hybrid memories. In: *International Conference on Computer Design*, Montreal, Quebec, Canada, 2012.

19

Storage Systems*

Alexander Thomasian
Thomasian & Associates

19.1 Introduction

Desirable properties for a storage system (in no particular order) are: (1) Low cost per GigaByte (GB), (2) nonvolatility, (3) reliability, (4) low access latency, (5) high access bandwidth, (6) low power consumption, (7) high recording density, small footprint, (8) data compression and deduplication, (9) data security via encryption, (10) data retention or longevity, (11) endurance or high maximum number of cycles, and (12) *Compound Annual Growth Rate (CAGR)*.

* The writing of this chapter was started when the author was affiliated with Shenzhen Institutes of Advanced Technology—SIAT, which is part of Chinese Academy of Sciences, in Shenzhen, China.

No single storage technology has all these properties, but a combination of storage technologies can be used to meet data storage requirements. Low access time is achieved at the processor level via a memory hierarchy with fast, small, high cost per GB static RAM (SRAM) cache memories at the highest levels, followed by larger, slower, and lower cost dynamic RAM (DRAM) memories at the lower levels. Memory hierarchies are beneficial because of the locality of reference principle.* Temporal locality implies that code and associated variables accessed once will be referenced repeatedly, such as those inside a loop. Spatial locality implies that data in close proximity to data being accessed has a higher probability of being accessed, such as successive elements of a one-dimensional array. The hierarchical reuse model implied by the fractal structure of data reference is a powerful tool for understanding cache and database behaviors on many scales of size and time [63]. Prefetching is applicable to many levels of the memory hierarchy, for example, at the CPU cache level in [38] and disk level in [49].

The DRAM technology for main memories is volatile so that permanent data is held on magnetic hard disk drives (HDDs). CPU caches, DRAM (main memories), and disks are reviewed in [49]. Magnetic tape drives which were used for backup and archival storage are being challenged by less expensive, low-performance, and high capacity *Serial Advanced Technology Attachment (SATA)* HDDs,† as opposed to more expensive, high-performance, and smaller capacity *Small Computer Systems Interface (SCSI)* HDDs [49].‡ SCSI and SATA drives are compared in [2]. SCSI drives are used in enterprise storage and SATA drives are used in personal computers, and archival storage systems.

DRAM caches in disk array controllers (DACs) and main memories are the best way to improve I/O performance by eliminating disk accesses altogether as much as possible. Part of DAC's cache can be turned into *non volatile storage (NVS)* by backing it up by uninterruptible power supply (UPS). NVS holds dirty data blocks, which have not been written to disk, to avert data loss in case of a power outage. Even without UPS, lost updates in *transaction (txn)* processing systems are averted by write ahead logging (WAL), which writes updated data to NVS, before a txn is committed [81].

Flash memories are gaining popularity as replacements for disks in handheld devices, but also as disk caches.§ This configuration is not suited for a write intensive workload with small writes (see Section 19.11). Flash memories outperform *Serial Attached SCSI (SAS)* drives and allow a significant reduction in power consumption (see Section 19.7).

Storage systems mainly in the form of HDDs constitute a significant fraction of information technology (IT) expenditure. The early IBM *Random Access Method for Accounting and Control (RAMAC)* computer was equipped with a 5 megabyte (MB) capacity disk drive, which cost $10,000 [62].¶ There has been a dramatic drop in HDD cost per GB due to the exponential growth in disk recording densities. HDDs are highly reliable with a *Mean Time to Failure (MTTF)* exceeding a million hours (114 years) [91]. The technological impact of HDDs on storage systems is discussed in [32].

HDDs have two weaknesses due to their electromechanical nature: (i) high power consumption to rotate disks and (ii) poor performance as a result of high access time. The projection in [27] that DRAM will surpass disks in cost has not come true, but with the advent of *Storage Class Memories (SCM)* the quest for replacing HDDs with solid-state disks (SSDs) is becoming a reality. Ten SCM technologies are currently under consideration and are expected to be fast, inexpensive, and power efficient [11]. Phase change memory is one of the more promising SCM technologies [80]. Using SCM as a disk drive replacement yields a performance for random requests that is orders of magnitude faster than comparable disk-based systems and requires much less space and power in the data center.

* http://en.wikipedia.org/wiki/Locality_of_reference.
† http://en.wikipedia.org/wiki/Sata.
‡ http://en.wikipedia.org/wiki/SCSI.
§ http://en.wikipedia.org/wiki/Flash_memory.
¶ http://en.wikipedia.org/wiki/IBM_305_RAMAC.

An extrapolation of disk and SCM technology trends up to 2020 in [25] concludes that there will be a 100- to 1000-fold advantage for SCM in terms of the data-center space and power.

This is a major rewriting of the chapter titled "Secondary Storage Systems" in the previous edition of the handbook. It consists of the following sections. Trends in tape storage based on [23] are reported in Section 19.2. Section 19.3 discusses various aspects of HDDs. HDD characteristics are discussed in Section 19.3.1. Disk arm scheduling and data placement are discussed in Section 19.3.2. The log-structured file systems (LFS) paradigm is discussed in Section 19.3.3, followed by active disks in Section 19.3.4. RAID with erasure coding is discussed in Section 19.4. The starting point is RAID5, RAID6, and their variations in Section 19.4.1. RAID operation in normal/degraded/rebuild modes is discussed in Sections 19.4.2 through 19.4.4. In Sections 19.4.4.2 and 19.4.4.3, we discuss rebuild performance analysis and methods to improve RAID5 reliability for rebuild processing. RAID1 or mirrored disks are discussed in Section 19.5. RAID1 organizations are discussed in Section 19.5.1. Scheduling of requests in RAID1 is discussed in Section 19.3.2. RAID reliability is discussed in Section 19.6. Section 19.6.1 is a general introduction to reliability analysis. Section 19.6.2 lists expressions for mirrored disks reliabilities. Section 19.6.3 provides the approximate reliability analysis of disk arrays with emphasis on RAID1. Section 19.6.4 provides an analysis of RAID1 arrays with repair. Power conservation in RAID is discussed in Section 19.7. Three interesting RAID arrays are discussed in Section 19.8. Grid codes are discussed in Section 19.8.1. Hierarchical RAID (HRAID) is discussed in Section 19.8.2. Heterogeneous disk arrays (HDAs) are discussed in Section 19.8.3. Storage area networks (SANs) and network attached storage (NAS) are discussed in Section 19.9. Data deduplication is discussed in Section 19.10. SSDs with emphasis on flash memories are discussed in Section 19.11. Section 19.11.1 describes a new sparing method to increase the lifetime of flash memories, while Section 19.11.2 describes the partial MDS (PMDS) for dealing with sector as well as component failures. Conclusions are presented in Section Conclusions.

19.2 Magnetic Tapes for Low-Cost Archival Storage

Magnetic tapes preceding HDDs in 1950s as a data storage medium were considerably less costly than the removable disk drives at the time [62]. Magnetic tapes record data on Mylar (a polyester film) coated with magnetic material. Data is recorded and read from tapes using *Read/Write (R/W)* heads against which tapes are moved at a fixed speed. Early tapes which came in open reels were susceptible to being torn and there was a later conversion to cartridges. The problem of the wrong tape being loaded by the operator was solved by automated tape libraries with robotic arms. IBM's 3850 *Mass Storage System (MSS)* introduced in 1974 is an early system. Tape storage is discussed in [62], but there have been numerous publications on this topic since then.*

Early data processing applications held customer records on a tape "masterfile," which was sorted according to customer number. Access to find a customer's record based on customer number was not attempted, because scanning half of the file on the average to find a record (the whole file if the customer number was not there) would take time linearly proportional on the number of records in the file. Updates to the masterfile were in batches sorted by customer number. For example, in a monthly billing application, typical of an electric utility, the sorted file of customer numbers and meter readings is applied to the masterfile to compute the monthly bill.

Clever algorithms for sorting on tape discussed in Knuth's 3rd volume on Sorting and Searching are now mainly of historical interest. Hashing and indexing methods for direct disk access and algorithms for relational database operations such as joins and sorting, which utilize large main memory buffers for sequentially accessed disk data, are described in texts on data base management systems *(DBMSs)* [81].

* http://en.wikipedia.org/wiki/Magnetic_tape_data_storage.

Digital magnetic tape recording is quite complex [62].* Data compression, deduplication, and encryption were introduced in magnetic tapes before disks. IBM system storage tape encryption solutions are described in [36]. Encryption is also provided at disk-resident databases, which is beyond the scope of this discussion.

Linear Tape Open Generation LTX-4 has a lifetime of 30+ years and an error rate an order-of-magnitude lower than SAS drives. When a tape drive fails, the tape may be read on another drive [23], which was also the case with early disk drives with removable disks [62]. The inconvenience of accessing tape files is addressed in LTX-5, which provides indexing with a convenient interface for downloading files.

Tapes do not consume power unless they are being read or written, so that it is estimated in [23] that tape-based archival systems consume 290 times less power than disk-based archives. It is also noted that although 3 terabytes (TB) disks (four times 750 GB or five times 600 GB) are available at a reasonable price ($250) at this time, there is the issue of *total cost of ownership* (*TCO*). Power consumption is 5 w/h for 3.5″ and 2 w for 2.5″ disks, while disk reads/writes are in progress.

Long-term data storage is rapidly gaining in importance. According to [10], the 2700 petabytes (PB) of long-term storage in 2008 is expected to increase 10-fold by 2010. The requirements for such storage systems are quoted as follows:

(1) Standard interfaces for ingestion and retrieval of documents
(2) Flexible and secure access to documents
(3) Efficient search mechanisms
(4) Policies for multitier storage management to meet service-level agreements
(5) Scalability
(6) Ability to archive original file and metadata with assistance for its specification, generation, and discovery
(7) Dynamically varying system scale based on business needs
(8) Security controls, nonrepudiation, and auditing to meet compliance and data governance requirements
(9) Business continuity and protections, media migration provisions as technology changes

Redundant array of independent libraries (RAIL) is a tertiary storage system architecture that couples multiple small and inexpensive "building block" tape libraries [24]. It is shown that RAIL has performance and availability characteristics superior to conventional tertiary storage systems, for almost the same dollar/megabyte cost. The log-structured library array (LSLA) allows data compression and alleviates certain inefficiencies associated with RAID5 described in Section 19.4.

There have been recent proposals for long-term archival storage systems based on low-cost SATA disks, such as Pergamum [96], which is discussed in more detail in Section 19.7. Encrypting the data offers a higher level of security in case data is lost, stolen, or accessed by an unauthorized person. Solutions to problems associated with long-term archival storage system are addressed in [97].

19.3 Hard Disk Drives (HDDs)

The section first describes HDD characteristics in Section 19.3.1. We then proceed to disk arm scheduling in Section 19.3.2. The LFS is described in Section 19.3.3, followed by active disks in Section 19.3.4.

19.3.1 HDD Characteristics

Disks consist of one or more circular platters, which are attached to a spindle rotated by a motor at a constant angular velocity (CAV). The spindles are made of a light-weight aluminum alloy, which is

* http://www.quadibloc.com/comp/tapeint.htm.

coated with magnetic material. One or both sides of a platter may be accessed, based on the availability of R/W heads on the surfaces. Data are written on HDD tracks as 512 byte (B) sectors. Advanced format 4096 B sector disks provide a higher recording efficiency.*

Disks associated with IBM mainframes had the *count key data (CKD)* format [62].† The key is a unique field per record in ISAM and VSAM files, where the latter is a variation on ubiquitous B-trees [81]. The key search in early IBM disk drives was carried out by the disk controller [62]. IBM 3390 disks with extended CKD (ECKD) are not manufactured anymore, but are simulated by DAC's software on fixed block architecture (FBA) disks.‡ The advantage of the latter format is that the data is addressable as an array of sectors by the logical block number (LBN).

When the radius of the outermost track is twice that of the innermost track, disk zoning increases the disk capacity by as much as 50% by maintaining approximately the same bits per inch (BPI) at all tracks. The bookkeeping for the varying numbers of sectors per track is simplified by zoned bit recording (ZBR), which at the cost of reduced storage efficiency assigns the same number of sectors to multiple contiguous tracks. Adaptive zoning is a relatively new trend described in [56], so that the disk transfer rate from the same relative disk (track) position may vary from 6% faster to 14% slower.

In addition to BPI, disk recording density is determined by tracks per inch (TPIs), which may exceed 100,000. The BPI versus TPI trade-off is discussed in Section 19.3 in [49]. Increases in BPI are preferable to TPI, since higher TPIs result in an increased head settling time. With *Perpendicular Magnetic Recording (PMR)* technology, 630 gigabits per square inch and even 1 terabit are possible from leading HDD manufacturers, such as Seagate.§

Smaller form factor HDDs with 3.5 and 2.5 in. diameters have been adopted, since with increasing recording densities sufficient capacity is attained even with one surface of a platter. Higher capacity is attainable by recording on both surfaces of a platter or stacking multiple disk platters, which however require more power for disk rotation. Tracks on one or both surfaces of a platter, which are placed on top of each other form a cylinder. These are accessible simultaneously (without moving the arm) by multiple R/W heads attached to a single disk arm. Files are therefore usually written on successive disk tracks and cylinders. The head switching time is accounted for by skewing the starting points across tracks and cylinders, so that no latency is incurred in accessing large blocks of data [49]. Track-aligned extents utilize disk-specific knowledge to match access patterns to disks, for example, placing disk blocks to avoid track boundaries and using zero latency reads (ZLR) to minimize latency [89]. ZLR allows data transfers to start at any sector of a requested block, so that the latency to read a full track is just one half the sector time.

The movement of R/W heads across disk tracks is called seeking. Seek time is a significant component of disk access time, which depends on the placement of blocks being accessed. Seek times have improved significantly over the years, but the year to year improvement is quite small. The seek time depends on the number of disk tracks traversed by the arm. The seek time characteristic (STC) $t_{seek}(d)$, $1 \leq c - 1$, where c is the number of disk cylinders obtained by subjecting the disk to seeks with various distances and averaging the seek time over multiple trials. The STC has several regions for the HDD considered in [90]. For $d \leq 100$, the seek time is a small constant, followed by a discontinuity, after which it is approximately the square root of the seek distance. $t_{seek}(d)$ is linear for $d \geq (c - 1)/3$. Because short seeks incur less delay than head-switching time, the "serpentine" organizations shown in Figure 4 in [90] places successive blocks of a file on the same surface, before proceeding to the next one.

* http://en.wikipedia.org/wiki/Advanced_Format.
† http://en.wikipedia.org/wiki/Count_Key_Data.
‡ http://en.wikipedia.org/wiki/Fixed_Block_Architecture.
§ http://en.wikipedia.org/Magnetic_storage.

The seek distance distribution for nonzoned disks with uniform accesses to c disk cylinders is as follows:

$$P[d] = (1-p)\frac{2(c-d)}{c(c-1)}, 1 \le d \le c-1, \ \ P[0] = p,$$

where p is the probability that no seek is required with $p = 1/c$ for uniform accesses $p[d]$ linearly drops from its maximum value at $d = 1$ to its minimum at $d = c - 1$. $t_{\text{seek}}(d)$, $1 \le d \le c - 1$ is obtained by exercising the disk drive. The mean disk seek time is then $\bar{t}_{\text{seek}} = \sum_{d=1}^{c-1} t_{\text{seek}}(d) P[d]$.

In zoned disks with uniform accesses to disk blocks, $p[d]$ is computed by noting that the probability of accessing a cylinder is proportional to the number of blocks that it holds [119]. The $p[d]$ in this case exceeds the straight line for nonzoned disks for smaller seek distances, but there is a crossover point at higher seek distances.

The HDD rotational speed, expressed in rotations per minute (RPM), has increased over the years: 2400, 3600, 7200, 10,000, 15,000, and beyond. The disk rotation time is $T_{\text{rot}} = 60000/\text{RPM}$ in ms (milliseconds), determines the rotational delay, the time for a requested block to become accessible by the R/W heads after a seek is completed. For accesses to small data blocks $\bar{x}_{\text{lat}} \approx T_{\text{rot}}/2$. The mean transfer time of small blocks equals the number of sectors per block (16 for 8 KB blocks) divided by the number of sectors on the track times T_{rot}. To summarize the mean disk access time is $\bar{x}_{\text{disk}} = \bar{x}_{\text{seek}} + \bar{x}_{\text{lat}} + \bar{x}_{\text{xfer}}$. Data read from disk is initially transferred to an onboard disk cache, which is also used for prefetching when a sequential access pattern is detected.

The mean response time for disk accesses is the sum of mean waiting time (W) and mean disk service time: $R = \bar{x}_{\text{disk}} + W$. When the arrival process of disk requests is Poisson with rate λ and assuming that x_{disk} is exponentially distributed then: $W = \rho\bar{x}_{\text{disk}}/(1-\rho)$, where $\rho = \lambda\bar{x}_{\text{disk}}$ is the disk utilization [55]. R_{disk} increases rapidly with increasing λ as $\rho \to 1$. With the implicit assumption of FCFS scheduling and for a general distribution of x_{disk}, only its mean and second moment: $\overline{x^i}_{\text{disk}}$, $i = 1, 2$ are required to determine W, which is given as

$$W = \frac{\lambda \overline{x^2}_{\text{disk}}}{2(1-\rho)} = \frac{\rho\bar{x}(1+c^2_{\text{disk}})}{2(1-\rho)}. \quad (19.1)$$

The squared coefficient of variation of service time: $c^2_{\text{disk}} = \overline{x^2}_{\text{disk}}/(\bar{x}_{\text{disk}})^2 - 1$ was found to be less than one for random accesses to small disk blocks in [112]. Since $c_{\text{disk}} = 1$ for the exponential distribution, W is overestimate with this assumption.

Disk reads which affect application response time can be processed at a higher priority than writes. Assuming that the fraction of reads in the arrival stream is f_R, the mean waiting time for reads is given as [55]: $W = \rho\bar{x}_{\text{disk}}/(1-\rho_R)$, where $\rho_R = \lambda f_R \bar{x}_{\text{disk}}$ is the disk utilization due to read requests only. That is, read requests are not affected by the disk utilization due to write requests.

19.3.2 Disk Arm Scheduling

The FCFS disk scheduling policy in early HDDs was later superseded by more sophisticated disk scheduling methods. An easy method to reduce the mean disk service time $\bar{x}_{\text{disk}}$ is to keep track of the disk arm by the operating system (OS) as in the case of the *Shortest Seek Time First (SSTF)* and SCAN scheduling [20]. The scheduling was to be carried out by the OS, which kept track of the position of the disk arm. SCAN reduces the seek time of outstanding disk requests by processing them as the arm is moved from the outermost to innermost tracks. A combination of SSTF and SCAN is described in [26], which alleviates SSTF's susceptibility to starvation. It allows accesses in the opposite direction of SCAN, provided that their seek distance plus the product $r \times c$ is smaller than the seek distance in the forward direction.

The number of cylinders is c and r, $0 \leq r \leq 1$ is a coefficient, whose optimal value was determined to be $r = 0.2$ for the workload under consideration with the target HDD.

The *Shortest Access Time First (SATF)* method processes requests, which minimize the positioning time (sum of projected seek time and latency) first [50]. SATF is implemented at the HDD controller, which can better estimate the positioning time than the OS. Improvements to SATF utilizing lookahead to minimize disk access time and incorporation of priorities in SATF are described in [105]. Lookahead takes into account the time to process additional requests beyond the first request, but at a discounted cost. Priority scheduling with SATF discounts the access time for higher priority requests. In fact, Equation 19.1 is applicable to scheduling methods that do not take into account service time.

In addition to disk arm scheduling, the placement of disk files affects the seek time. The organ-pipe arrangement reduces the seek distance by placing most frequently accessed disk files at middle disk cylinders, that is, at an equal distance from innermost and outermost tracks, surrounded by less frequently accessed files [49]. The *Automatically Locality Improving Storage (ALIS)* method described in [44], in addition to optimizing placement, coalesces disk requests to exploit the efficiency of sequential accesses.

Multimedia systems with fixed rate requests and deadlines for guaranteed completion of requests require specialized disk scheduling methods. SCAN-EDF scheduling combines the efficiency of SCAN with the *Earliest Deadline First (EDF)* real-time scheduling policy [69]. The performance metric is the number of streams that can be supported, which increases with larger buffer space and for deadlines exceeding interrequest periods.

Two notable studies of disk-scheduling performance are [131] and [43], while [105] is a recent survey of this subject, with emphasis on author's research in this area.

19.3.3 Log-Structured File System (LFS)

LFS optimizes writing to disk in an environment where read accesses to disk are uncommon due to the high hit rate of a large cache [84]. LFS accumulates modified files in a buffer and writes them to disk in large chunks called segments to reduces disk arm positioning overhead. Similarly to database logging [81], segments are written onto consecutive disk locations.

The fact that files are not written in place allows file compression, since otherwise there is the possibility that a file will not compress below its previous size. Disk space held by modified files is designated as free space. A background garbage collection process merges half-empty segments into full segments and creates empty segments for future destaging. For example, two half empty segments will result in one full and one empty segment. Files are located by inodes written to the log.

The write cost in LFS is a steeply increasing function of the utilization of disk capacity (u), which is the fraction of live data in segments (see Figure 3 in [84]). The crossover point with respect to UNIX's *Fast File System (FFS)* is at $u = 0.5$. The following issues related to segment cleaning policies are investigated in [84]: (1) When should the segment cleaner be executed? (2) How many segments should be cleaned? (3) Which segments should be cleaned? (4) How should be live blocks grouped into segments?

19.3.4 Active Disks

There have been numerous proposals to offload database operations to specialized processors associated with disks. Early work on database machines is reviewed in [41]. The Teradata *Data Base Computer - DBC*/1012 with 10^{12} bytes or a TB storage capacity. DBC/1012 was designed for the efficient processing of relational algebra operators, see, for example, [81].*

HDDs are equipped with relatively powerful microprocessors, which have low utilizations, so that this processing power can be utilized in carrying out background tasks. Freeblock scheduling (FS) is based on opportunistic disk accesses, as the arm moves in processing regular requests [61]. This is

* http://en.wikipedia.org/wiki/DBC_1012.

advantageous since due to their mechanical nature, disk arms constrain the access bandwidth to disks with ever-increasing capacities. FS can be used in conjunction with LFS to prepare free segments, process nearest-neighbor queries, carry out data mining [81] and disk scrubbing [53], etc. [83].

Off-loading database activities work well for simpler operations, such as SQL SELECTs, but this is not the case for more complex SQL operations, such as relational joins and star-joins in data warehousing, which may involve multiple disks [81]. The SmartStor project at Almaden provided multiprocessor access to multiple disks, so that more complex database operations can be undertaken [42]. While a performance advantage is demonstrable, there is the additional cost of developing software with parallel processing capabilities at the storage level.

There has been recent activity at IBM to offload database applications from mainframes, such that most of their processing capacity remains available for *OnLine Txn Processing (OLTP)* applications with high-performance requirements [12]. There is the cost saving due to having a single copy of the database with a single set of disks. A 70%–97% reduction in CPU times and 9%–78% reduction in bytes transferred were observed in [79]. A key issue is that the "backend" software has the capability to process legacy file systems, such as VSAM (IBM's version of B+ trees). Little interference is expected at the disks, since OLTP applications will access most of the required data in main memory, but require logging on dedicated disks. The need by queries to read consistent data from relational tables requires innovative concurrency control solutions, see for example, [99].

19.4 Redundant Arrays of Independent Disks (RAID)

The five-level Berkeley RAID classification introduced in late 1980s was extended to seven levels with the addition of RAID0 and RAID6 in [15]. RAID0 has no redundancy, but utilizes striping for load balancing. Striping partitions the data associated with the file system into fixed size stripe-units or strips, which are placed round-robin across the disks. The data strips in a row constitute a stripe. Mirroring or replication, which is classified as RAID1, is described in Section 19.5. RAID5 and RAID6 arrays with erasure coding are described in Section 19.4.1. In Sections 19.4.2 through 19.4.4, we describe RAID5 operation in normal, degraded, and rebuild modes. The four RAID levels are shown in Figure 19.1.

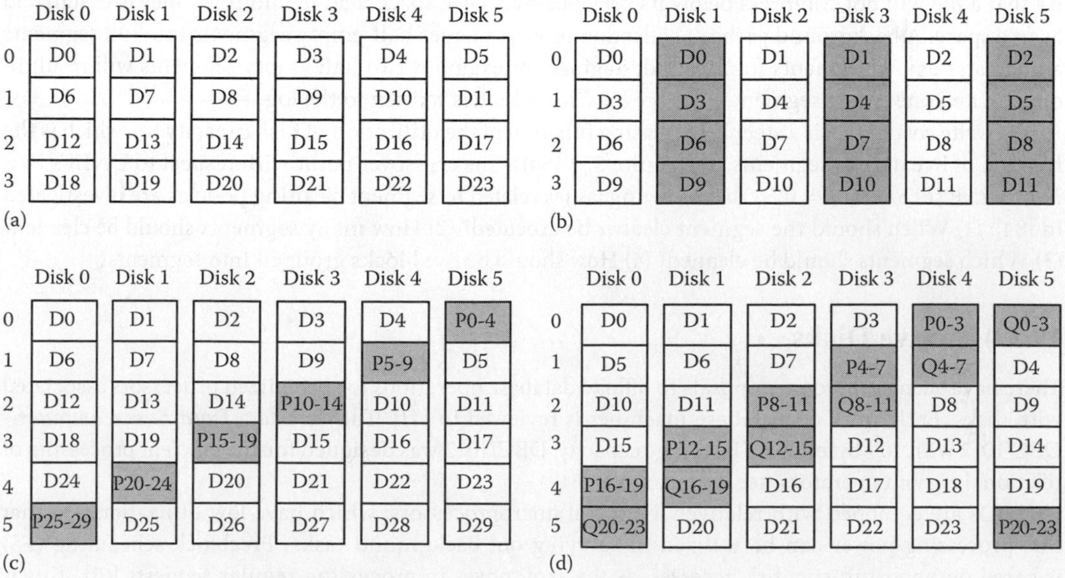

FIGURE 19.1 RAID0, RAID1, RAID5, and RAID6 layouts. (a) RAID0, (b) RAID1, (c) RAID5, and (d) RAID6 data layouts.

19.4.1 RAID5, RAID6, and Their Variations

RAID4 extends RAID0 with $N - 1$ disks, with an additional Nth disk dedicated to parity. Parity strips hold the *eXclusive-OR (XOR)* of the corresponding $N - 1$ data strips in the same stripe, for example, $P_N = \oplus_{i=1}^{N-1} D_i$. The strips of a failed disk (or their blocks) can be reconstructed by XORing the corresponding strips (or blocks) on surviving disks. For example, $D_1 = (\oplus_2^{N-1} D_i) \oplus P_N$. Two disadvantages of RAID4 are that the dedicated parity disk is not utilized at all by read requests and that it may constitute a bottleneck for write intensive workloads.

RAID5 disk array balances disk loads for processing updates by placing check strips in right to left diagonals, known as the left-symmetric organization [15], as shown in Figure 19.1. Diff-RAID, described in Section 19.11, utilizes an unbalanced allocation of parities to attain a higher reliability [6].

Parity striping is a variation on RAID5, which does not stripe the data at all, but dedicates $1/N$th of the capacity of all disks to parity. This placement sacrifices higher data transfer rates to the ability to access the strips of a file in parallel, thus increasing the throughput of the RAID5 array (see Section 19.4.2). The performance of RAID5 and parity striping is compared in [14].

In general, RAID$(4 + k)$, $k \geq 1$ arrays are extensions of RAID0 with k check strips, where k is also the number of failed disks that can be tolerated, hence the classification *k disk failure tolerant (kDFT)* [109]. Although, under the best circumstances, RAID1 arrays can tolerate up to one half of disk failures without data loss, as discussed in Section 19.5, a second disk failure may lead to data loss, hence RAID1 is a 1DFT.

Reed–Solomon (RS) codes are *maximum distance separable (MDS)* in that they incur the capacity equivalent of $k \geq 1$ redundant disks to tolerate as many disk failures [109]. RS is costly to implement, since it is based on Galois field finite arithmetic [75,76]. *EVENODD (EO)* is a 2DFT which uses parity coding with a significantly lower cost than RS for check code calculations [8], but is outperformed by the *Rotated Diagonal Parity (RDP)* invented 10 years later [18]. Figure 19.2 is an RDP array with n^2 data blocks and $n = 4$. The updating of almost all data blocks requires the updating of two blocks with the exception of one block in each row, that is, we need $2n^2 - 2n$ XORs to protect n^2 data blocks, or $2 - 2/n$ XORs per data block. This is smaller than $2 - 2/(n - 1)$ for EO [109]. A performance evaluation of storage erasure coding libraries is reported in [77].

X-code is another 2DFT MDS code [133], which dedicates two (horizontal) stripes to P and Q parities per segment, which is an $N \times N$ array of strips, so that N data stripes in $N - 2$ rows are protected. A shortcoming of X-code is that the number of disks (N) should be prime. The two parities cover strips placed in diagonals with positive and negative slopes, as shown in Figures 19.3 and 19.4.

RM2 is non-MDS but similarly to the X-code uses horizontally placed check codes. Each data strip is covered by two check strips, so that RM2 is 2DFT. Depending on the redundancy level $p = 1/M$, there are $M - 1$ data strips per parity stripe [74]. Based on constraints given in the paper, the smallest number of possible disks for $p = 1/3$ is $N = 7$, so that the redundancy level is higher than a RAID6 with seven disks, which is 2/7. In Figure 19.5 $P_0, \ldots, P_6$ are the parity strips for the seven parity groups. Data strip $D_{i,j}$ is protected by two parity strips P_i and P_j.

RAID operation in normal/degraded/rebuild modes are discussed in Sections 19.4.2 through 19.4.4.

Disk 0	Disk 1	Disk 2	Disk 3	Q Parity	P Parity
0	1	2	3	4	0
1	2	3	4	0	1
2	3	4	0	1	2
3	4	0	1	2	3

FIGURE 19.2 RDP array with $(p - 1)(p + 1)$ strips with $p = 5$.

	Disk Number						
Row	0	1	2	3	4	5	6
0	2	3	4	5	6	0	1
1	3	4	5	6	0	1	2
2	4	5	6	0	1	2	3
3	5	6	0	1	2	3	4
4	6	0	1	2	3	4	5
5	0	1	2	3	4	5	6
6							

FIGURE 19.3 P parities with positive slopes.

	Disk Number						
Row	0	1	2	3	4	5	6
0	5	6	0	1	2	3	4
1	4	5	6	0	1	2	3
2	3	4	5	6	0	1	2
3	2	3	4	5	6	0	1
4	1	2	3	4	5	6	0
5							
6	0	1	2	3	4	5	6

FIGURE 19.4 Q parities with negative slopes.

Disk0	Disk1	Disk2	Disk3	Disk4	Disk5	Disk6
P_0	P_1	P_2	P_3	P_4	P_5	P_6
$D_{2,3}$	$D_{3,4}$	$D_{4,5}$	$D_{5,6}$	$D_{0,6}$	$D_{0,1}$	$D_{1,2}$
$D_{1,4}$	$D_{2,5}$	$D_{3,6}$	$D_{4,0}$	$D_{5,1}$	$D_{6,2}$	$D_{0,3}$

FIGURE 19.5 RM2 disk layout for $N = 7$ disks.

19.4.2 RAID5 Operation in Normal Mode

The updating of each data block in RAID($k + 4$), $k \geq 1$ requires the reading and writing of k check blocks protecting a data block. For RAID5 with $k = 1$ given the updated data block d_{new}, the difference block is computed as the XOR of the new and old data block: $d_{\text{diff}} = d_{\text{old}} \oplus d_{\text{new}}$. The parity block is updated as $p_{\text{new}} = p_{\text{old}} \oplus d_{\text{diff}}$. If d_{old} and p_{old} are not cached, the updating of a single block in RAID5 requires the reading of the data and parity blocks and writing them. The four disk accesses are referred to as the small write penalty (SWP) in [15]. In the case of RS codes, d_{diff} is multiplied by an appropriate coefficient, before being XORed with the Q check blocks [75,76]. Disks are assumed to have an XOR capability and the XOR can be carried out as part of a read modify write (RMW) operation at the check disks consists of a read followed by an XOR and a disk rotation before the check blocks is overwritten. *Disk Architecture for Composite Operations (DACO)* is a proposal to place a write head followed by a read head to eliminate the disk rotation time [60]. The shortcomings of this approach are discussed in [106].

Optimum strip sizes, their layout, and efficient methods to update parities were areas of early RAID research, which are reviewed in [15]. Larger strip sizes facilitating single disk accesses to larger files are beneficial in a multiuser environment, but preclude parallel access in single user environments. Strip sizes should be large enough to accommodate most block sizes in order to preclude multiple disk accesses and unnecessary seeks, which would result in an unnecessary increase in disk utilization. The optimum strip size for database applications may be recommended by database companies to optimize the performance of their products.

A simple and pragmatic method for dealing with SWP is caching dirty blocks in NVS portion of DAC's cache [64]. Trace analysis shows that dirty blocks tend to be overwritten several times before they are destaged. Such blocks exhibit locality, for example, multiple blocks are destaged on the same track. This effect was taken into account in the performance analysis in [112]. Disk scheduling can be applied to reduce destage time. A detailed study of RAID5 cache behavior is reported in [122], which does not recommend the caching of parity blocks, since they do not contribute to the hit rate for read accesses.

The log structured array (LSA) is an extension of LFS to RAID5 according to [65]. In fact, LSA was first implemented in conjunction with StorageTek's Iceberg RAID6 disk array [15]. It has been implemented as the RAID5 portion of HP's AutoRAID disk array [127] (see Section 19.8.3). LSA eliminates SWP by utilizing full stripe writes, by first accumulating a sufficient volume of data in the form of modified files in DAC's cache. The check strips can be computed on the fly as data strips are written to consecutive disks. Similarly to LFS, the previous version of files is designated for garbage collection. This task is carried out periodically to coalesce n not fully utilizes stripes onto $m < n$ almost full stripes and $n-m$ empty stripes. Since files are not written in place, they may be compressed to attain a higher storage efficiency.

19.4.3 RAID5 Operation in Degraded Mode

RAID5 takes advantage of its redundancy to reconstruct disk blocks on demand. A data block on a failed disk or an unreadable block due to media failure requires the $N - 1$ corresponding blocks from surviving disks to be read and XORed. A read access to the failed disk is thus converted to an $(N - 1)$-way fork-join request, which is considered completed when the corresponding blocks from surviving disks are accessed and XORed. There is a doubling in disk loads due to for-join requests to reconstruct blocks on the failed disk, since in addition each disk processes its own read accesses. Read redirection accesses reconstructed blocks directly by simply keeping track of the progress made by the rebuild process, which systematically reconstructs the contents of a failed disk on a spare disk [68].

Clustered RAID (CRAID) or parity declustering is a data layout, which reduces the load increase in degraded mode [68]. This is accomplished by settling the number of disks in a parity group to G, which is smaller than the number of disks N. The load increase due to read requests is specified as the declustering ratio: $\alpha = (G - 1)/(N - 1)$. For $N = 10$ and $G = 4$, we have $\alpha = 1/3$ and 2.5 parity groups per stripe. Since as many strips are reconstructed per stripe, this will result in overloading of the spare disk. Adjusting the fraction of redirected reads to the spare disk is proposed as a solution to this problem in [68], since the increased load on surviving disks will result in reduced rate in rebuild reading (see Section 19.4.4). This may not be an adequate solution in all cases, since it may result in a backlog of rebuild units (RUs), which are the smallest block sizes for rebuild processing, in the rebuild buffer. Throttling of rebuild reads is then a more robust solution to this problem.

CRAID has a twofold effect on reducing the mean response time: (i) The mean response time at each disk lowers because of the smaller disk utilizations. (ii) We have a $(G - 1)$-way rather than an $(N - 1)$-way fork-join request, whose mean response time is proportional to H_{G-1}, rather than H_{N-1}, where $H_n = \sum_{i=1}^{n} 1/i$ is the Harmonic sum (strictly speaking this is true for exponential distributed disk response times). When the bandwidth of the rebuild buffer constitutes a bottleneck, CRAID, which requires G – 1-way, rather than an N – 1-way XOR processing with associated DRAM accesses, improves rebuild time, as shown in Figure 6 in [125].

Balanced incomplete block designs (BIBD) [72,39] and nearly random permutations (NRP) [67] are two efficient implementations of CRAID to balance disk update loads [109]. BIBD has the shortcoming that data layouts are not available for all values of N and α [109].

19.4.4 RAID5 Rebuild Processing

The discussion is organized in four parts. Section 19.4.4.1 is a general discussion of rebuild processing. Section 19.4.4.2 discusses rebuild performance. Section 19.4.4.3 discusses methods to improve RAID5 reliability by improving the chances that rebuild is completed successfully. Section 19.4.4.4 discusses methods to reduce rebuild cost.

19.4.4.1 General Discussion of Rebuild Processing

RAID5 rebuild is the systematic reconstruction of the contents of a failed disk on a spare disk. In the case of RAID5 with *N* disks successive RUs, for example, tracks, are read from the $N-1$ surviving disks, XORed to reconstruct the next missing RU, which is then written to disk.

In stripe-oriented rebuild (SOR), the strips in a stripe constitute RUs [39]. Since rebuild reading proceeds one stripe at a time, the synchronization at this level results in an unnecessary slowdown compared to disk-oriented rebuild (DOR), where contents of surviving disks are read in parallel into a rebuild buffer. They are then XORed to reconstruct the next RU to be written to disk. It follows from simulation results in [39] that DOR outperforms SOR.

In *distributed sparing* spare areas are distributed among the disks in the RAID array, so that the contents of a failed disk can be reconstructed on surviving disks [66,112]. It has the advantage over dedicated sparing in that the bandwidth of the hot spare is not wasted and conversely a higher access bandwidth is provided. Once a spare disk becomes available, the spare areas are copied onto this disk to free up space for further rebuilds, so that rebuild is a two-step process.

Parity sparing is another method, where two RAID5 arrays are merged into one, by utilizing one of two parities [66]. In the case of a single RAID5, *restriping* overwrites the parity strips, so that the disk array reverts to RAID0 [82]. In the case of RAID7, we have the following transitions:

$$\text{RAID7} \rightarrow \text{RAID6} \rightarrow \text{RAID5} \rightarrow \text{RAID0}.$$

Interestingly as check blocks are overwritten by restriping, there is an increase in the maximum disk I/Os per second (IOPS) with the progression of disk failures, since the effect of SWP is reduced as the redundancy level is lowered [120].

With the advent of multiterabyte disks, a substantial increase rebuild time is to be expected. Rebuild time can be reduced by carrying out rebuild processing at the level of identifiable objects, as discussed in Section 19.8.3.

19.4.4.2 Rebuild Performance Analysis

The vacationing server model (VSM) reduces the effect of rebuild reads on the response time of external requests, which are already higher in degraded operating mode. This is accomplished by starting rebuild reading when the disk becomes idle [111,112,118] and stopping it with the arrival of an external request. Provided disk requests arrive according to a Poisson process, each disk in a RAID5 array correspond to an M/G/1 queueing system, which has alternating busy and idle periods [55]. Disk accesses are processed in a busy period and an idle period starts when there are no requests pending. There is no forced idleness, although the latter has been shown to improve performance if a succeeding request in the same stream of requests is anticipated. The duration of an idle period is simply the interarrival time of external requests with rate λ, so that $\bar{t}_{idle} = 1/\lambda$. As shown in Figure 19.6, the first rebuild read requires a seek, while successive reads do not. It is assumed that the RU is a disk track.

To simplify the discussion, we assume all disk requests are reads with mean service time $\bar{x}_{disk}$, so that the disk utilization is $\rho = \lambda\bar{x}_{disk}$. The duration of the busy period $\bar{t}_{busy}$ can be obtained by noting that it is the fraction of time the system is busy: $\bar{t}_{busy}/(\bar{t}_{busy} + \bar{t}_{idle}) = \rho$, so that $\bar{t}_{busy} = \bar{x}_{disk}/(1-\rho)$ [55].

With rebuild processing, the duration of the busy period is modified since the service time of the first request starting the modified busy period is elongated by the residual lifetime of the rebuild read in progress: $\bar{z} = \bar{x}_{disk} + \bar{y}_r$, by the first request arriving during the final vacation during which a

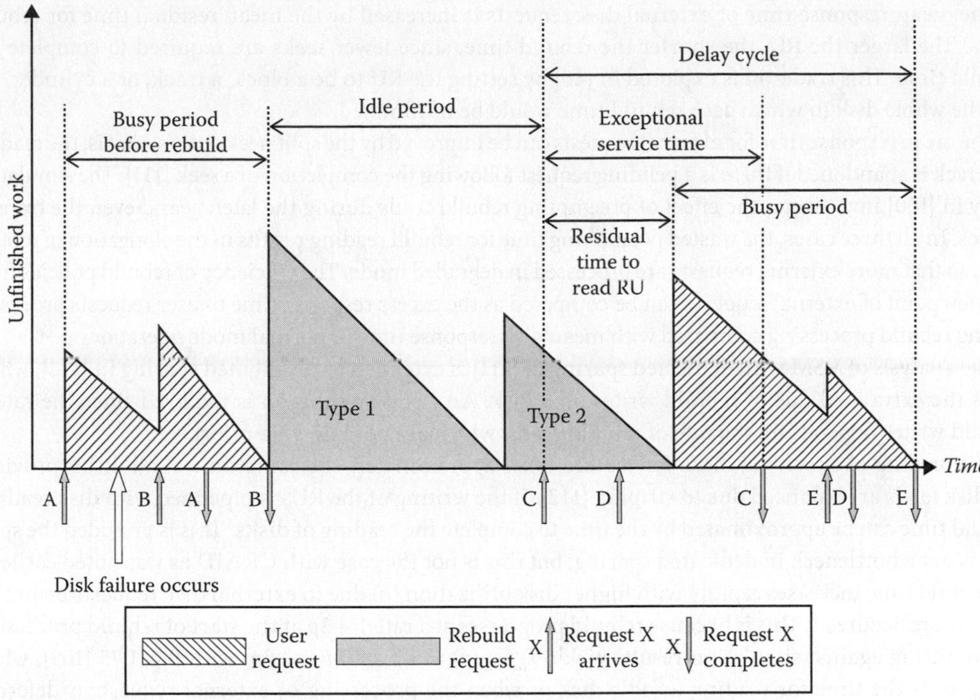

FIGURE 19.6 Key parameters associated with rebuild processing. (From Thomasian, A. et al., *Comput. J.*, 50(2), 217, March 2007.)

request arrives. To simplify the discussion, we assume that vacation times have a fixed duration $y = \nu$, so that mean remaining vacation time after an arrival in a vacation period is given as

$$\nu_r = T - \frac{\int_0^T \lambda e^{-\lambda t} dt}{\int_0^T e^{-\lambda t} dt} = \frac{\nu}{1 - e^{-\lambda T}} - \frac{1}{\lambda}. \quad (19.2)$$

The mean effective service time of the first request starting the next modified busy period, referred to as the delay cycle (with mean $\bar{d}$) is: $\bar{z} = \nu_r + \bar{x}$, so that [55]: $\bar{t}_{busy} = \bar{z}/(1 - \rho)$.

Similarly, ignoring the initial seek time the probability that a disk request arrives during a disk rotation is $p = 1 - e^{-\lambda T_{rot}} \approx \lambda T_{rot}$. One can estimate $\bar{z}$ more accurately by noting that it is the residual lifetime of the first arrival during rebuild reading that really matters [118]. The probability of n tracks being read in an idle period is: $p_n = p(1 - p)^{n-1}$, so that $\bar{n} = 1/p$. The rebuild reading time of a disk with N_{track} tracks is

$$T_{rebuild}(\rho) = [\bar{T}_{busy} + \bar{T}_{idle}] \frac{N_{track}}{\bar{n}},$$

where ρ denotes the initial disk utilization, before the failure of the disk, after which the load on surviving disks is doubled. The arrival rate of read requests at the start of rebuild processing is twice the rate in normal mode for read requests, but the arrival rate returns to normal mode as a result of read redirection. The variation in disk utilization in computing $T_{rebuild}(\rho)$ can be taken into account by computing rebuild processing time over k intervals [111]:

$$T_{rebuild}(\rho) = \sum_{i=1}^{k} T_{rebuild}(\rho_i), \quad \rho_i = \left[2 - \frac{2i-1}{2k}\right]\rho.$$

The mean response time of external disk requests is increased by the mean residual time for rebuild reads. The larger the RU, the shorter the rebuild time, since fewer seeks are required to complete the rebuild time. This trade-off is explored in [40] by setting the RU to be a block, a track, or a cylinder, but not the whole disk in which case rebuild time would be minimized.

The mean response time for external requests can be improved by the split-seek option, that is, the reading of a track is abandoned if there is a pending request following the completion of a seek [111]. The simulation study in [100] investigates the effect of preempting rebuild reads during the latency and even the transfer phases. In all three cases, the wasted positioning time for rebuild reading results in the elongation of rebuild time, so that more external requests are processed in degraded mode. The efficiency of rebuild policies from the viewpoint of external requests can be compared as the excess response time of user requests processed during rebuild processing compared with mean disk response time in normal mode operation.

The analysis of VSM with dedicated sparing in [111] is extended to distributed sparing in [112], which takes the extra load due to rebuild writes on a disk. An iterative solution is required since the rate of rebuild writes depends on the rate of rebuild reads, which are not known a priori.

The reading of surviving disks for rebuild takes approximately the same time at all disks, provided the disk loads are balanced due to striping [112]. If the writing of the RUs is pipelined with disk reading, rebuild time can be approximated by the time to complete the reading of disks. This is provided the spare disk is not a bottleneck in dedicated sparing, but this is not the case with CRAID as was noted earlier.

Rebuild time increases rapidly with higher disk utilization (ρ) due to external disk requests before the disk failure occurred. This is because rebuild progresses at a rate $1 - 2\rho$ at the start of rebuild processing. Curve-fitting against simulation results yields $T_{\text{rebuild}}(\rho) = T_{\text{rebuild}}(0)/(1 - \beta\rho)$ with $\beta = 1.75$ [109], where $T_{\text{rebuild}}(0)$ is the time for reading an idle disk or when the processing of external requests is deferred, which is not a practical solution.

The permanent customer model (PCM) for rebuild processing inserts a new rebuild read into the queue of pending disk requests as soon as another rebuild read is completed [67]. Given that disk requests are processed in FCFS order, the arrival of disk requests while a rebuild read is enqueued and being processed will result in more interruptions in rebuild reads than with VSM, which is only affected by arrivals while a rebuild read is being processed [109]. It is easy to see that VSM has less effect on external requests than PCM in that: (1) Rebuild reads are not processed ahead of external requests. (2) Rebuild reading with VSM takes less time than PCM, so that it interferes with fewer external requests.

The analysis in [112] is extended in [118] to zoned disks, with whole tracks constituting RUs. If a fixed size RU is selected, then partial tracks reads will result in multiple accesses to a track and additional latencies. Caching into disk's track buffer can be used to reduce the number of accesses to the same track. Lookahead reading into the track buffer while user requests are pending may, however, affect the response time of user requests.

19.4.4.3 Improved Reliability for Rebuild Processing

Rebuild may be unsuccessful due to: (1) A second disk failure before rebuild is completed. (2) Latent sector errors (LSEs) encountered, while rebuild is in progress. It was observed in [8] that 4% of unsuccessful rebuilds are due to LSEs and this was used as the motivation to develop EO. Using the data in [48], it can be shown that the probability of unsuccessful rebuilds due to LSEs is two orders of magnitude more likely than a second disk failure.

Intra-disk redundancy (IDR) is a novel method which applies redundancy at the level of disk sectors, so that LSEs are masked [22]. IDR utilized in several RAID products has been termed vertical coding, as opposed to horizontal coding at interdisk level in RAID5. The effectiveness of the single parity (SP), the RS, and interleaved parity (IP) codes are analyzed in [22]. It is shown that IP, which is less expensive to implement than RS, provides the same reliability level as RS. The reliability analysis in [22] shows that IDR+RAID5 yields the same reliability as RAID6. The small increase in disk block lengths due to IDR has a negligible effect on disk access time in RAID5, which provides a significant performance advantage with respect to RAID6 in that it updates a single check block instead of two [119].

Disk scrubbing is the periodic reading of disks to detect and correct LSEs using the redundancy provided by the RAID5 paradigm [53]. The advantage of disk scrubbing is that sector errors are detected and corrected early, before they hinder the completion of the rebuild process. Four disk scrubbing methods are proposed in [53]. The original *Disk Scanning Algorithm (DSA)* starts a scrubbing request when the disk is idle after a certain waiting time *wt* has expired. Adaptive DSA adjusts *wt* based on system activity, that is, *wt* is increased for heavier loads and decreased for lighter loads. Simplified DSA maintains two queues, one for user and one for scanning requests. After the user queue has been empty for a sufficiently long interval, a scanning request is issued. Rather than starting DSA right away, it can be issued periodically, for example, once a day. DSA with the *VERIFY* command transfers read data into internal disk buffers where the consistency of data is checked.

A straightforward analysis of disk scrubbing is reported in [48]. The probability of an error due to a write is P_w and writes constitute a fraction r_w of disk accesses with rate h. The probability of error in reading a sector without scrubbing is $P_e = r_w P_w$. The probability of sector failure for deterministic scrubbing and random (exponentially distributed) scrubbing period are $P_s = [1-(1-e^{-hT_s})/(hT_S)]p_e$ and $P_s = [hT_S/(1 + hT_S)]P_e$, respectively, where T_S denotes the mean scrub period. It follows that deterministic scrubbing is preferable, since random scrubbing has double the value for P_s. Furthermore $P_s \le P_e \le P_w$. The effectiveness of disk scrubbing and IDR for dealing with LSEs is investigated in [92] and [48], with the latter concluding that IDR is preferable to disk scrubbing.

When the memory bandwidth constitutes a bottleneck, rebuild time can be reduced via CRAID. This is because fewer RUs are XORed for rebuild and less memory bandwidth is required [125]. Taking into account hardware resource contention, rebuild time is the maximum of the delay due to disk and DRAM buffer transfer times [22].

19.4.4.4 Reducing the Cost of Rebuild Processing

The supplementary parity augmentation (SPA) method, in addition to the *P* parity which covers all disks, introduces the *S* parity which covers half of the disks [121]. This implies that two parities need to be updated for one half of the disks, but only one half of the disks are protected against two disk failures. There is a similarity to Pyramid codes proposed in [46]. Briefly, a disk array with three RS codes *P*, *Q*, and *R*, the *P* parity strips are replaced by two strips P_0 and P_1, so that given an even number of disks *N*, P_0 covers disks 1:*N*/2 and P_1 covers *N*/2 + 1: *N*. While SPA and Pyramid codes have the same level of redundancy, only one of two parities should be modified per update in the case of Pyramid codes. Both methods have the advantage of halving the volume of the data to be read for rebuild, which reduces the possibility of LSE being encountered during rebuild reading. Given that RAID5+IDR introduces the same level of reliability as RAID6, the overhead associated with SPA is hard to justify. A more detailed discussion of SPA appears in [107].

The *Local Reconstruction Codes (LRC)*, which was developed for *Windows Azure Storage (WAS)* similarly to Pyramid codes [46], partitions data fragments into groups to reduce rebuild processing cost [47]. As an example, we consider the (*k*, *l*, *r*) code with $k = 6$ data fragments, $l = 2$ local parities, and $r = 2$ global parities. LRC can tolerate arbitrary three failures by choosing the sets of coding coefficients:

$$q_{x,0} = \alpha_0 x_0 + \alpha_1 x_1 + \alpha_2 x_2, \quad q_{x,1} = \alpha_0^2 x_0 + \alpha_1^2 x_1 + \alpha_2^2 x_2, \quad q_{x,0} = x_0 + x_1 + x_2.$$

$$q_{y,0} = \beta_0 y_0 + \beta_1 y_1 + \beta_2 y_2, \quad q_{y,1} = \beta_0^2 y_0 + \beta_1^2 y_1 + \beta_2^2 y_2, \quad q_{y,0} = y_0 + y_1 + y_2.$$

The two local parities are $p_x = q_{x,2}$ and $p_y = q_{y,2}$. The two global parities are: $p_0 = q_{x,0} + q_{y,0}$ and $p_1 = q_{x,1} + q_{y,1}$. The choice of α's and β's from the Galois Field GF(2^4) is specified in the paper. The key properties of the LRC codes are: (i) single data fragments failure can be decoded from k/ℓ fragments; (ii) arbitrary failures up to $r + 1$ can be decoded. It is shown in the paper that for a set of typical parameters (6,2,2) outperforms the MTTF of (6,3) RS code by an order of magnitude.

Note that construction does not take into account the partial parities when computing the global parity, so that special cases need to be considered when decoding.

An important detail about WAS is that the data is initially written into three replicas, but periodically the replicas are sealed at the extent level (large chunks of data), which are not expected to be modified anymore, and converted to the LRC format for higher data efficiency. Unlike AutoRAID, this conversion is not reversible and the replicates are then deleted.

An improvement in the number of disk strips accessed for rebuilding a single disk block is possible by taking advantage of the fact that disk strips are protected by two parities. It is shown in [124] that in a distributed RDP storage system with two redundancy nodes, only 3/4 of the information needs to be transmitted to rebuild one node. In the case of X-code with p nodes, the "repair bandwidth" is at most $(3p^2 - 2p + 5)/4$ while the repair bandwidth for RDP and X-code is given as $3(p - 1)^2$. This is given as a lower bound for RDP disk arrays in [132], which adopts a symbol size equal to the strip size in a disk array context. RDP disk arrays similar to EO adopts small symbol sizes in the context of disk arrays, which makes their access pattern the same as RAID6.

Large symbol sizes equaling strip sizes are utilized by X-code [133], so that the method to reduce the cost of rebuilding a single disk is applicable to it [115]. Referring back to Figures 19.3 and 19.4 we assume that disk 0 has failed. Rather than reconstructing all blocks using P or Q parities, we reconstruct four blocks using P and the other three blocks using Q parities. In what follows we have specified block accesses to reconstruct strips (0,0), (1,0), (2,0), and (5,0) or $p(0)$ using P parities, followed by (3,0), (4,0), (6,0), or $q(0)$ using Q parities. With two disk failures, RDP with large symbol sizes requires a multistep recovery method [58], similarly to the X-code [115].

(0,0) using $p(2)$: (5,1), (4,2), (3.3), (2,4), $(1,5)_1$
(1,0) using $p(3)$: (5,3), (4,4), $(3,5)_2$, $(2,6)_3$, (0,1)
(2,0) using $p(4)$: (5,4), (4,5), $(3,6)_4$, (1,1), $(0,2)_5$
(5,0) using $p(0)$: $(4,1)_6$, (3,2), (2,3), $(1,4)_7$, (0,5)
(3,0) using $q(2)$: $(4,1)_{6'}$, (6,2), (0,4), $(1,5)_{1'}$, $(2,6)_{3'}$
(4,0) using $q(1)$: (6,1), (0,3), $(1,4)_{7'}$, (2,5), $(3,6)_{4'}$
(6,0) using $q(0)$: $(0,2)_{5'}$, (1,3), (2,4), $(3,5)_{2'}$, (4,6)

Repeated accesses to the same strip are specified by unprimed subscripts in a first occurrence and a primed subscript in a second occurrence. Rather than $N \times (N - 2) = 35$ accesses, 28 block accesses are required, provided seven strips are cached, after they are accessed, which amounts to a 20% decrease in the number of accessed strips.

This concept is extended to cloud storage in [54], where a rotated RS code is proposed such that strips are reused in the construction process, reducing its cost. We illustrate the savings in cost using Figure 7 in the paper, with $k = 6$ columns, $m = 3$ check strips, and $r = 4$ rows. All four rows are referred to as a stripe in this work. The coding for row 0 is then

$$c_{j,0} = \sum_{i=0}^{k-1} (2^j)^i d_{i,0}, \quad 0 \le j \le m-1.$$

Data strips from different rows are used in computing the check strips, according to the following equations, where $n = (kj/m)$ and $\%r$ stands for modulo r:

$$c_{j,b} = \sum_{i=0}^{n-1} (2^j)^i d_{(i,b+1)\%r} + \sum_{i=n}^{k} (2^j)^i d_{i,k}, \quad 0 \le k \le r-1, \quad 0 \le j \le m-1.$$

Four of the check strips required for reconstructing the four strips on disk 0 which has failed are as follows:

$$p_0 = \sum_{i=0}^{5} d_{i,0}, \quad q_0 = d_{1,0} + 2d_{1,1} + \sum_{i=2}^{5} 2^i d_{i,0} \quad p_2 = \sum_{i=0}^{5} 2^i d_{i,2}, \quad q_2 = d_{0,3} + 2d_{1,3} + \sum_{i=2}^{5} 2^i d_{i,2}.$$

It can be observed that only 16 symbols (strips) are required for recovery, versus 24 strips which would be otherwise required. For even r the cost of reconstructing a disk is $\frac{r}{2}\left(k + \left\lceil \frac{k}{m} \right\rceil\right)$ versus $k \times r$.

A disadvantage of this method is that $m \times r$ check strips need to be buffered when data is being written. It is assumed that rows constituting data strips are read one row at a time and applied to the check strips.

19.5 Mirrored Disks or RAID1

We start with a discussion of mirrored disk organizations in Section 19.5.1. This is followed by a discussion of efficient processing of disk requests without and with an NVS cache.

19.5.1 Mirrored Disk Organizations

Basic Mirroring (BM) with two disks is the simplest form of data replication, which was utilized in Tandem's (now HP) fault-tolerant computers, such as NonStop SQL, before its classification as RAID level 1 (RAID1) in [15]. For higher volumes of data, we have RAID1/0 with $N = 2M$ disks or M disk pairs, that is, two mirrored RAID0 arrays. Alternatively we have RAID0/1, a RAID0 array where each virtual disk is a mirrored pair, that is, striping is applied across mirrored pairs to balance disk loads. For RAID1/0 with $N = 8$, the four primary disks hold strips $(B_{1,1}, B_{1,2}, B_{1,3}, B_{1,4})$ and the secondary disks hold strips $(B'_{1,1}, B'_{1,2}, B'_{1,3}, B'_{1,4})$, where the indices denote the stripe (or row) and disk number, respectively. The failure of up to $M = N/2$ disks will not result in data loss in either case, as long as they do not constitute a pair, but the failure of only two disks constituting a pair will result in data loss, GRD, ID, and CD RAID1 arrays, described below, are 1DFTs, while LSI RAID1 is a 2DFT.

Group Rotate Declustering (GRD) is similar to RAID1/0, but rotates data on the secondary disks from row to row, so that the second row on the mirrors is: $(B'_{2,2}, B'_{2,3}, B'_{2,4}, B'_{2,1})$, and so on. The advantage of GRD over BM is that upon the failure of a single disk, its read load is evenly distributed over M disks at "the other side." Up to M disk failures will not lead to data loss as long as they are all on one side. The probability that a second disk failure leads to data loss is then $M/(2M - 1) > 0.5$, while it is only $1/(2M - 1)$ for BM.

Interleaved Declustering (ID) was introduced in Teradata's DBC/1012 computer. Disks are partitioned into equal-sized clusters, for example, for $N = 8$ and $c = 2$ clusters there are $K = N/c = 4$ disks per cluster. Disks are divided into primary and secondary areas, which replicate each other. We use capital letters for primary and small letters for secondary data blocks. Primary blocks B_i, $1 \leq i \leq 4$ have each $K - 1 = 3$ secondary blocks: $(b_{i,1}, b_{i,2}, b_{i,3})$ with $B_i = b_{i,1} \| b_{i,2} \| b_{i,3}$. The latter are placed in rotated manner at the other disks in the cluster, so that the read load of a failed disk is distributed evenly across the disks in the other cluster. The advantage of ID over BM is that the read load increase due to a single disk failure is $1/(K - 1) < 1/2$. Up to c disk failures can be tolerated by ID, as long as they are in different clusters.

Chained Declustering (CD) is an improvement over ID in terms of reliability, but unlike ID it requires careful routing of read request to balance disk loads. The space on D_i, $1 \leq i \leq N$ disks is partitioned into primary and secondary areas with equal sizes. Unprimed data blocks in the primary area of each disk are replicated on primed data blocks in the secondary area of the following disk, for example, disk D_i holds data blocks B_i, $B'_{(i-1) \bmod N}$, $1 \leq i \leq N$. Read requests can be accessed from the primary or secondary area, using static load balancing methods, which adjust the fraction of requests to balance disk loads as disk failures occur. Up to M disk failures can be tolerated by CD, as long as the two failed disks are not contiguous.

LSI RAID1. This is a hybrid disk array, which is described in a patent by LSI Logic Corp. [128] (LSI stands for Large Scale Integration). LSI RAID1 combined mirroring and parity, as shown in the following for $N = 8$ disks, four of which are data and four parity disks:

$$(D_1, P_{1,2}, D_2, P_{2,3}, D_3, P_{3,4}, D_4, P_{4,1})$$

Each *Parity disk (Pdisk)* is the XOR of the two neighboring *Data disks (Ddisks)*, for example, $P_{1,2} = D_1 \oplus D_2$. Ddisks can be reconstructed as follows, for example, $D_1 = P_{1,2} \oplus D_2 = D_4 \oplus P_{4,1}$. Three disks failures can be tolerated by LSI RAID1, unless they are three consecutive disk failures with a parity disk in the middle [113]. The failure of four Ddisks and also two Ddisk and two Pdisks can lead to data loss. Enumeration is used to extend the reliability analysis of mirrored disks to LSI RAID1 [113].

Survivable Storage using Parity in Redundant Array Layout (SSPiRAL) extends the LSI RAID paradigm to the case where Pdisks are computed over $m = 3$ Ddisks and each Ddisk is protected by $m = 3$ Pdisks, so that DoutD=PinD=3 [93]. For $N = 8$ the four Ddisks *A*, *B*, *C*, and *D* are protected by the four Pdisks holding $A \oplus B \oplus C$, $B \oplus C \oplus D$, $C \oplus D \oplus A$, and $D \oplus A \oplus B$. If Ddisks and Pdisks are placed in two rows, then each Pdisk is the XOR of the Ddisk above it and the two Ddisks that follow it, modulo $M = N/2$. It is easy to verify that up to three disk failures can be tolerated in all cases. If all Ddisks fail they can be reconstructed using Pdisks, which is not the case for LSI RAID. There is data loss if a Ddisk and the three Pdisks in which it participates fail. There are $\binom{8}{4} = 70$ configurations with four disk failures. Enumeration shows that for $N = 8$, data loss occurs in *1/5th* of four disk failure cases.

19.5.2 Routing of Disk Requests in RAID1

In addition to higher reliability, mirrored disks provide twice the access bandwidth of single disks for processing read requests, which tended to dominate write requests in many applications. This is less so with the availability of very large caches, which results in fewer read misses.

A reduction in seek distance for read requests can be achieved in BM by judicious routing. Routing policies can be classified as static and dynamic, where the latter take into account the states of the disks. Assuming that the R/W heads and disk blocks are positioned at random disk tracks, then selecting the R/W head closest to a requested block results in a mean seek distance $c/5$ for reads and $7c/25$ to complete both writes [7], versus $c/3$ for a single disk. There are many complications, such as the effect of two R/W heads being positioned on the same track after a write. Refinements to this performance analysis are discussed in [101]. The following static routing policies are considered in [101]:

Private Queue (PQ): Read requests are immediately routed to one of the disks according to some routing policy, while write requests are sent to both disks.

Shared Queue (SQ): Read requests are held in this queue and are routed to the first disk that becomes available. Dynamic routing can be pursued in this case.

Hybrid Queue (HQ): The routing of read requests from SQ to PQs can be deferred.

Routing of requests to PQ can be classified as static or dynamic [101]. Uniform and round-robin routing are two examples of static routing policies. Round-robin routing is simpler to implement than uniform routing. It improves the mean disk response time by making the arrival process more regular than exponential interarrival times with coefficient of variation $c_a = 1$, that is, Erlang-2 with $c_a = 1/\sqrt{2}$ [55,123].

The router in addition to checking whether a request is a read or a write can determine other request attributes, such as the addresses disk blocks organized as a single dimensional array. Read accesses with addresses below a certain value can be routed to one disk and others to another. Such affinity-based routing is beneficial for sequential accesses from a file, so that it will minimize the seek distance for such accesses. It is also possible to take advantage of data prefetching to the onboard buffer. This method works well if the disk loads are balanced.

A dynamic policy that takes into account the number of requests at each disk, such as *join the shortest queue (JSQ)*, may not be effective, since the queue length is not a good measure of pending processing times at the disk, because requests may be to highly variable file sizes. Performance can be improved by estimating the remaining service time at each disk, which is only feasible if requests are processed in FCFS order. In the case of SATF scheduling, the processing time changes with the arrival of each new request, although the number of requests under consideration may be limited to a finite number of requests.

Simulation studies have shown that the routing policy for PQ has a negligible effect on performance for a workload consisting of random disk accesses and that performance is mainly determined by the local scheduling policy. SQ provides more opportunities than PQ to improve performance, because more requests are available to carry out optimization, but then the PQ may be integrated with disk's controller, while SQ cannot. Mirrored disks with a single controller would be a solution to this problem. SATF scheduling with SQ provides better performance PQ, because SQ provides almost twice the requests for scheduling [101].

The performance of a mirrored disk system without an NVS cache can be improved by using a write anywhere policy on one disk to minimize disk arm movement and also susceptibility to data loss when an NVS cache is not available. Data is written in place later on the primary disk, which allows efficient sequential accesses, while a directory is required to keep track of blocks written anywhere. The distorted mirrors method described in [95] is one of several proposals for improving mirrored disk performance.

Similarly to RAID5, a large NVS caches can be used to improve the performance of mirrored disks. Prioritizing the processing of read requests yields a significant improvement in response time. Deferred writes can be processed efficiently in batches. The scheme proposed in [78] operates mirrored disks in two phases. While one disk is processing read requests, the other disk is processing writes in a batch mode using *Cyclical CSCAN (CSCAN)*, that is, SCAN in one direction only [105]. The performance of the above method can be improved as follows [110]: (1) Eliminating the forced idleness by processing write requests individually. This can be carried out opportunistically, as in FS. (2) Using SATF or exhaustive enumeration, which is only possible for sufficiently small batch sizes, instead of CSCAN, to find an optimal destaging sequence to minimize time. (3) Introducing a threshold for the number of read requests, which when exceeded defers the processing of write batches.

Rebuild processing in RAID1 with the BM organization, which is tantamount to copying the contents of the surviving disk on a spare, can be improved by carrying out the rebuild out-of-order, by opportunistic reading of unread tracks, which eliminates seek times [5]. Note that this method is not applicable to RAID5, because of excessive buffer space requirements. Results of an analytic and simulation study using the VSM an PCM methods described in Section 19.4.4.2 without and with opportunistic reading option are reported in [5].

19.6 RAID Reliability Analysis

This section is organized as follows: Section 19.6.1 is a general introduction to reliability analysis. Section 19.6.2 lists expressions for RAID reliabilities. Section 19.6.3 provides the asymptotic reliability analysis of disk arrays emphasizing RAID1. Section 19.6.4 provides an analysis of RAID5 arrays with repair.

19.6.1 General Discussion of Reliability Analysis

From a reliability viewpoint, systems can be classified into two extremes: series and parallel. In the former case, all components are required for the operation of the system, while in the latter case the system is not considered failed as long as one component survives.

Given n components with reliabilities R_i, $1 \le i \le n$, we have

$$R_{\text{series}} = \prod_{i=1}^{n} R_i, \quad R_{\text{parallel}} = 1 - \prod_{i=1}^{n} (1 - R_i).$$

Each component may be a complex system, whose reliability can be expressed as a function of its subcomponents. The reliability of a system may be specified as a logical expression specifying a minimal configuration for the operation of the system.

Component reliabilities are usually expressed as a function of time. Given the time to failure of the ith component is exponentially distributed with rate δ_i, that is, $R_i(t) = e^{-\delta t}$, then the reliability of a series system with n components can be expressed simply using the superposition property of Poisson processes as [123]

$$R_{\text{series}}(t) = \prod_{i=1}^{n} R_i(t) = \prod_{i=1}^{n} e^{-\delta_i t} = e^{-\Delta t}, \text{ where}: \Delta = \sum_{i=1}^{n} \delta_i.$$

Given that the $\text{MTTF}_{\text{single}} = 1/\delta$, then $\text{MTTF}_{\text{series}} = 1/(n\delta) = \text{MTTF}_{\text{single}}/n$. In the case of a parallel system, the time to failure is the maximum of the times to failure of its components [123]. Given two parallel components with reliabilities $R_i(t) = e^{-\delta_i t}$, $i = 1, 2$ we have

$$R_{par}(t) = 1 - \prod_{i=1}^{2} R_i(t) = e^{-\delta_1 t} + e^{-\delta_2 t} - e^{-(\delta_1 + \delta_2)t},$$

$$\text{MTTF}_2 = \int_{t=0}^{\infty} R_{par}(t)dt = \frac{1}{\delta_1} + \frac{1}{\delta_2} - \frac{1}{\delta_1 + \delta_2}.$$

For $\delta_1 = \delta_2$ $\text{MTTF}_2 = 1.5/\delta = H_2/\delta$, more generally using the substitution $u = 1 - e^{-\delta t}$ and noting that $\int_0^1 u^{i-1}du = 1/i$ and $du = \delta(1 - u)dt$:

$$\text{MTTF}_n = \int_0^{\infty} [1 - (1 - e^{-\delta t})^n]dt = \frac{1}{\delta} \int_0^1 \frac{1 - u^n}{1 - u} du = \frac{H_n}{\delta}.$$

It is easy to see that incremental increase in reliability decreases rapidly with n.

Most reliability analyses of disk arrays are concerned with disk reliabilities only, although it has been shown in [91] that a significant fraction of disk subsystem failures are due to components other than disks. A possible justification is the assumption that the rest of the systems are equally reliable and the interest is in comparing relative disk reliabilities. A comprehensive study of storage subsystem failure characteristics is reported in [51], which also spreads the blame for data loss to components other than disks.

The Weibull distribution has been determined to provide a good representation for the time to disk failure [27,91]. The exponential distribution is used instead, because of its mathematical tractability, in that it can be incorporated into continuous time Markov chain (CTMC) models [123]. A transient solution of the CTMC provides the time to data loss, which is the passage time from an initial fault-free state to a state where data loss has occurred. Rather than the distribution of time to data loss, most analyses report only the *Mean Time to Data Loss (MTTDL)* as a means of comparing the relative reliabilities of systems (see Section 19.6.4).

It is, however, the reliability of the system during its useful lifetime (t_u), rather than the time to failure that really matters. In a triple modular redundancy (TMR) system at least 2-out-of-3 components with reliability $r(t)$ are required for the correct operation of the system: $R_{TMR}(t) = r^3(t) + 3r^2(t)(1 - r(t))$. Given that $r(t) = e^{-\delta t}$ then $MTTF_{TMR} = 5/(6\delta)$, which is smaller than the MTTF of a single component: $MTTF_{single} = 1/\delta$. Setting $R_{TMR}(t) = R_{single}(t)$ yields the crossover point $t_c = ln(2)/\delta \approx 0.7/\delta$, preceding which $R_{TMR}(t) > R_{single}(t)$. The TMR system is preferable to a simplex system provided $t_u \leq t_c$.

The reliability analysis becomes complicated when dealing with nonexponential distributions, such as the Weibull. Estimating the MTTDL in RAID5 disk arrays using simulation may be quite costly, as explained below. Consider the simplified case, where data loss occurs when a second disk fails, while rebuild is in progress. This is a rare event since the disk MTTF is much larger than the *Mean Time to Repair (MTTR)* or rebuild time. Given the rarity of this event, very lengthy simulations are required to estimate the MTTDL (as reported in [121]). The importance sampling method for dealing with rare events in simulation should be adopted in this case to reduce the simulation cost [73].

19.6.2 Reliability Expressions for RAID Arrays

We use reliability analysis in [108] to compare the reliabilities of various RAID1 organizations. Let r denote the reliability of each disk. The maximum number of disk failures that can be tolerated for a RAID1 with $N = 2M$ disks is $I = M$ in all cases except ID, where it equals the number of clusters $I = c$. The general expression for RAID reliability is as follows:

$$R_{RAID}(N) = \sum_{i=0}^{I} A(N, i) r^{N-i}(1-r)^i, \tag{19.3}$$

where $A(N, i)$ is the number of cases out of $\binom{N}{i}$ that i disk failures do not lead to data loss $\left(A(N,0) = \binom{N}{0} = 1\right)$:

An interesting special case relevant to storage systems is RAID(4 + k), which does not sustain data loss as long as $N - k$ out of N disks survive:

$$A(N, i) = \binom{N}{i}, \quad 0 \leq i \leq k. \tag{19.4}$$

As far as RAID1 disk arrays are concerned, up to M disk failures can be tolerated by BM, as long as one disk in each pair survives:

$$A(N, i) = \binom{M}{i} 2^i, \quad 0 \leq i \leq M. \tag{19.5}$$

In the case of GRD up to M disks can fail, as long as they are all on one side:

$$A(N, i) = 2\binom{M}{i}, \quad 0 \leq i \leq M. \tag{19.6}$$

In the case of ID with c clusters and $K = N/c$ disks per cluster, only one disk failure per cluster can be tolerated so that $I = c$:

$$A(N, i) = \binom{c}{i} K^i, \quad 0 \leq i \leq c. \tag{19.7}$$

Data loss in CD is a result of two neighboring disk failures (modulo N). The expression for $A(N, i)$ for CD is derived in [108]:

$$A(N,i)=\binom{N-i-1}{i-1}+\binom{N-i}{i},\quad 1\le i\le M. \tag{19.8}$$

The behavior of the RAID1 system can be represented by a CTMC, where state S_i is indexed by the number of failed disks and $\mathcal{F}$ denotes the state with data loss:

$$S_0 \to S_1 \to S_2 \to \ldots \to S_I$$

$$S_i \to \mathcal{F},\ \text{for } 2\le i\le I.$$

Starting with the fault-free state: S_0, the mean number of visits to S_i is given as

$$V_i=\frac{A(N,i)}{\binom{N}{i}},\quad 1\le i\le I. \tag{19.9}$$

Assuming that $r = e^{-\delta t}$, the mean holding time at state S_i with i failed disks is $h_i = [(N - i)\delta]^{-1}$. The MTTDL is given as

$$\text{MTTDL}=\sum_{i=0}^{I}V_ih_i=\sum_{i=0}^{I}\frac{V_i}{(N-i)\delta}. \tag{19.10}$$

The MTTDL can be computed using integration by noting that $R(t) = 1 - F(t)$, where $F(t)$ is the distribution function [123]:

$$\text{MTTDL}=\int_0^\infty R_{\text{RAID}}(N)\mathrm{d}t=\int_0^\infty\sum_{i=0}^{I}A(N,i)r^{N-i}(1-r)^i\mathrm{d}t. \tag{19.11}$$

19.6.3 Approximate Reliability Analysis for RAID

A shortcut method to express the system reliability based on a power series expansion of disk unreliabilities: $\varepsilon = 1 - r$ is reported in [102]. For disks with exponential failure rate with MTTF = 10^6 h or 114 years after 3 years $r = e^{-3/114} = 0.975$, so that $\varepsilon = 0.025 \ll 1$. The reliability of RAID1 with two disks is: $R_{\text{RAID1}} = 1 - (1 - r)^2 = 1 - \varepsilon^2$. Generally, we need to retain the smallest power of ε, which represents the probability of data loss with a minimum number of disk failures:

$$R_{\text{RAID}}=1-\sum_{i>0}a_i\varepsilon^i\approx 1-a_J\varepsilon^J,$$

where j is the minimum i for which $a_i \geq 0$. This term can be interpreted as the probability of data loss with the minimum number of disks:

$$R_{\text{RAID5}} = r^N + N(1-r)r^{N-1} \approx 1 - \frac{N(N-1)}{2}\varepsilon^2 + \frac{N(N-1)(N-2)}{3}\varepsilon^3 - \cdots.$$

The approximate reliability equation for RAID(4 + k), $k \geq 1$ can be obtained by induction:

$$R_{\text{RAID}(4+k)} \approx 1 - \binom{N}{k+1}\varepsilon^{k+1} + (k+1)\binom{N}{k+2}\varepsilon^{k+2} - \cdots \tag{19.12}$$

The MTTDL for various RAID1 organizations, but also RAID5/6/7 for systems without repair are given in Table 19.1. It is interesting to note that LSI RAID1 has the highest MTTDL, followed by RAID7, although LSI does not tolerate all three disk failures. We also provide the first term in the series expansion for estimating reliabilities.

19.6.4 RAID5 Reliability Analysis with Repair

The reliability analysis of a storage system may include a repair process, that is, rebuild processing as discussed in detail in Section 19.4.4. In the case of repair on a spare disk, the assumption that an infinite number of spares is available may lead to misleading results. In the case of distributed sparing and restriping the number of rebuilds is constrained by the number of spare areas. RAID7 can tolerate three disk failures, but rather than operating in degraded mode, with an $(i + 1)$-fold reduction in its throughput in processing read requests it can be restriped into RAID6, RAID5, and RAID0 after each disk failure.

We consider a RAID5 array with N disks, where each disk has an exponentially distributed failure rate δ and an exponentially distributed repair rate μ. The assumption that the repair time is exponentially distributed is justifiable by the fact that as verified by simulation it yields a good approximation to MTTDL. The behavior of the system can be represented by a CTMC with three states: S_n, where $0 \leq n \leq 2$ is the number of failed disks., S_0 is the fault-free state, S_1 is the degraded state, and S_2 is the failed state. The transition $S_1 \rightarrow S_1$ has rate $N\delta$. The transition rate $S_1 \rightarrow S_0$ is μ. which is the inverse of mean rebuild time. Otherwise, we have the transition $S_1 \rightarrow S_2$ with rate $(N-1)\delta$, which leads to data loss. The probability of a successful rebuild is: $p = \mu/(N-1)\delta + \mu)$, so that the number of successful rebuilds

TABLE 19.1 Summary of MTTDLs and the First Term in Approximate Reliability Expression

RAID5	BM	CD	GRD	ID	RAID6	LSI	RAID7
$\frac{15}{56\delta}$	$\frac{163}{280\delta}$	$\frac{379}{840\delta}$	$\frac{3}{8\delta}$	$\frac{61}{168\delta}$	$\frac{73}{168\delta}$	$\frac{82}{105\delta}$	$\frac{533}{840\delta}$
$0.268\delta^{-1}$	$0.582\delta^{-1}$	$0.451\delta^{-1}$	$0.375\delta^{-1}$	$0.363\delta^{-1}$	$0.435\delta^{-1}$	$0.781\delta^{-1}$	$0.635\delta^{-1}$
$\frac{N(N-1)\varepsilon^2}{2}$	$\frac{N\varepsilon^2}{2}$	$N\varepsilon^2$	$\frac{N(N-1)\varepsilon^2}{4}$	$\frac{N(N-c)\varepsilon^2}{2c}$	$\frac{N(N-1)(N-2)\varepsilon^3}{6}$	$\frac{N\varepsilon^3}{2}$	$\frac{N(N-1)(N-2)(N-3)\varepsilon^4}{24}$

follows a geometric distribution: $P_n = (1-p)p^{n-1}$ with a mean $\bar{n} = (1-p)^{-1}$. Multiplied by the mean holding time in S_0 we obtain the MTTDL as

$$\text{MTTDL} \approx \frac{(N-1)\delta+\mu}{(N-1)\delta} \times \frac{1}{N\delta} = \frac{(N-1)\delta+\mu}{N\delta^2} \approx \frac{\text{MTTF}^2}{N(N-1)\text{MTTR}}. \quad (19.13)$$

This analysis is extended in [22] to include uncorrectable sector failures with an associated probability P_{uf}:

$$\text{MTTDL} = \frac{(2N-1)\delta+\mu}{N\delta[(N-1)\delta+\mu P_{uf}]} \quad (19.14)$$

Reliability expressions for RAID5 and RAID6 with IDR (intraDisk redundancy) are reported in [22], while the effect of disk scrubbing and IDR on RAID reliability is compared in [48].

19.7 Power Conservation in RAID

Power consumption in disks is mainly due to the 12 V motor for rotating disks, while less power is used by the 5 V electronics. The three states associated with a disk are: Active, Idle, and Standby. Reading and writing to disk is carried out in the Active state. The disk drive continues spinning in the idle state, but significant energy savings are only possible by spinning down the disk to a standby state. According to Table 1 in [21] 15K RPM IBM 36Z15 disks consume 13.5, 10.2, and 2.5 W in the three states, while the IBM 40GNX "mobile" disks consume only 3, 0.85, and 0.25 W in the three states.

Spinning down a disk to a standby state may be counterproductive if the disk does not remain in that state long enough, since spinning up the disks takes significant time and energy, for example, 11 s and 135 W. A timeout strategy is used by OS before spinning down a disk. Power consumption by a disk's motor is [38]

$$\text{Power} = D^2 \times \text{RPM}^{2.8} \times n,$$

where D is the disk diameter, and n is the number of the platters. Some disk drive models are available with a variable number of platters, with capacities proportional to the number of utilized platter surfaces.

Higher RPMs are required for higher performance by reducing latency for accesses to randomly placed disk blocks by OLTP applications. The power consumption of a SATA and SAS disk drives is compared in Figure 6.3 in [38], where the SAS drive consumes 25 W versus 12 for the SATA drive. The 7200 RPM SATA drives are used for archival storage, while 15,000 RPM SAS drives are used for high-performance storage systems.

Storage devices contribute up to 27% of the total energy consumed in data centers, so that it is important to reduce this power consumption. An early study is massive array of idle disks (MAID), where a small subset of "cache" disks store active files, while the remaining disks are on standby [17].

The Hybernator system takes advantage of Sony's multispeed disks, which can save more energy (29%) than previous solutions, while still providing a txn throughput for OLTP comparable to a RAID5 array with no energy management [134]. Variable speed disks are, however, not widely available.

Write off-loading redirects requests to spun-down disks to NVS [70]. This study shows a 28%–36% reduction in power consumption by spindown, which increases to 45%–60% with write off-loading.

It is argued in [96] that tape-based archival systems suffer from poor performance due to sequential nature of data access, while most disk systems are unsuited for long-term archival storage, because of their high energy demands. Pergamum is motivated by MAID, but in addition adds NVRAM at each node to store data signatures, metadata, etc. to allow deferred writes. Pergamum uses both IDR and inter-disk redundancy to guard against data loss and staggered rebuild to reduce peak energy usage, while rebuilding large redundancy stripes.

A recent study of practical power proportionality in storage systems addresses read and write availability, no performance degradation, consistency, and fault tolerance for general I/O workloads [98]. The power-aware layout used is a tradeoff between load balancing, rebuild parallelism, and power savings. A distributed virtual log is used to absorb updates to replicas. Recovery and migration techniques, and predictive gear scheduling are adopted. A large, real service (Hotmail) on a cluster shows power savings of 23%, but savings of 40%–50% are possible with more complex optimizations.

The issue of power consumption is currently addressed in the broader context of Green Computing, which is outside the scope of this discussion.

19.8 Miscellaneous RAID Arrays

The three topics discussed in this section are Grid codes in Section 19.8.1, *Hierarchical RAID (HRAID)* in Section 19.8.2, and *Heterogeneous Disk Arrays (HDAs)* in Section 19.8.3.

19.8.1 Grid Codes

Grid codes use coding in two dimensions to protect strips arranged in this manner to attain a higher reliability [59]. In two dimensions, there are two sets of erasure codes: Horizontal (H) and Vertical (V) codes or HVPC codes [59]. In the simplest case with the single parity code (SPC) over k_2 columns and k_1 rows we have

$$P_{i,*} = D_{i,0} \oplus D_{i,1} \oplus \ldots D_{i,k_2-1}, \quad i = 0, 1, \ldots, k_1 - 1;$$

$$P_{*,j} = D_{0,j} \oplus D_{1,j} \oplus \ldots D_{k_1-1,j}, \quad j = 0, 1, \ldots, k_2 - 1.$$

The diagonal parity $P_{*,*} = \oplus_{i=0}^{k_2-1} P_{i,*} = \oplus_{i=0}^{k_1-1} P_{*,i}$. Note that the parities constitute an extra row and column.

Given that H and V codes can protect against K and L strip failures, all $(K+1)(L+1) - 1$ strip failures can be tolerated. In other words, $t = (K+1)(L+1)$ is the lowest number of strip failures that will lead to data loss. For $K = L = 1$ data loss occurs if the four strips constitute a rectangle.

SPC and X-code [133] are two codes that can be utilized vertically, but X-codes have the limitation that the number of strips and hence the associated number of disks should be prime. Relevant horizontal codes are SPC, which is tantamount to RAID5, EO, and STAR (an extension of EO) [45], which can tolerate single, double, and triple disk failures, respectively. The two-dimensional GRID codes listed in [59] are summarized in Table 19.2. Since $k_1 = k_2$ needs to be a prime number for the X-code, we have modified the example in [59], so that instead of $k_1 = k_2 = 10$ we use $k_1 = k_2 = 11$ with the total number of strips: $d = k_1 \times k_2 = 121$. Then the redundancy ratio is the number of check strips divided by d.

The level of redundancy is quite significant, on the other hand higher levels of redundancy are to be expected.

19.8.2 Hierarchical RAID (HRAID)

It has been observed in [91] that the failure of components other than disks is a significant contributor to system failures and possibly data loss. Hierarchical RAID (HRAID) addresses whole storage node (SN) failures, as well as individual disk failures at each node [120]. The study has a similarity to [82], which is based on IBM's Collective Intelligent Bricks project [126]. There are N SNs and M disks per SN. The disks at

TABLE 19.2 Maximum Number of Failures (t) and the Redundancy Percent (r%) Associated with Miscellaneous Grid Codes

Codes	t	Redundancy Ratio (r%)
(SPC,SPC)	3	$(k_1 + k_2 - 1)/d = 19\%$
(SPC,EO)	5	$(2k_1 + k_2 - 2)/d = 28\%$
(SPC,STAR)	7	$(3k_1 + k_2 - 3)/d = 37\%$
(SPC, X-Code)	5	$(k_1 + 2k_2 - 2)/d = 28\%$
(EO,EO)	8	$(2k_1 + 2k_2 - 4)/d = 36\%$
(STAR,STAR)	15	$(3k_1 + 3k_2 - 9)/d = 51\%$
(STAR,X-Code)	11	$(3k_1 + 2k_2 - 6)/d = 44\%$

$D^1_{1,1}$	$D^1_{1,2}$	$P^1_{1,3}$	$Q^1_{1,4}$	$D^2_{1,1}$	$P^2_{1,2}$	$Q^2_{1,3}$	$D^2_{1,4}$	$P^3_{1,1}$	$Q^3_{1,2}$	$D^3_{1,3}$	$D^3_{1,4}$	$Q^4_{1,1}$	$D^4_{1,2}$	$D^4_{1,3}$	$P^4_{1,4}$
$D^1_{2,1}$	$P^1_{2,2}$	$Q^1_{2,3}$	$D^1_{2,4}$	$P^2_{2,1}$	$Q^2_{2,2}$	$D^2_{2,3}$	$D^2_{2,4}$	$Q^3_{2,1}$	$D^3_{2,2}$	$D^3_{2,3}$	$P^3_{2,4}$	$D^4_{2,1}$	$D^4_{2,2}$	$P^4_{2,3}$	$Q^4_{2,4}$
$P^1_{3,1}$	$Q^1_{3,2}$	$D^1_{3,3}$	$D^1_{3,4}$	$Q^2_{3,1}$	$D^2_{3,2}$	$D^2_{3,3}$	$P^2_{3,4}$	$D^3_{3,1}$	$D^3_{3,2}$	$P^3_{3,3}$	$Q^3_{3,4}$	$D^4_{3,1}$	$P^4_{3,2}$	$Q^4_{3,3}$	$D^4_{3,4}$
$Q^1_{4,1}$	$D^1_{4,2}$	$D^1_{4,3}$	$P^1_{4,4}$	$D^2_{4,1}$	$D^2_{4,2}$	$P^2_{4,3}$	$Q^2_{4,4}$	$D^3_{4,1}$	$P^3_{4,2}$	$Q^3_{4,3}$	$D^3_{4,4}$	$P^4_{4,1}$	$Q^4_{4,2}$	$D^4_{4,3}$	$D^4_{4,4}$

FIGURE 19.7 HRAID1/1 with $N = 4$ nodes and $M = 4$ disks per node.

a node are RAID level 0, 5, 6, or 7. The same RAID levels are considered at the internode level. An HRAID with redundancy level k and ℓ at the internode and internode level is specified as HRAIDk/ℓ. Figure 19.7 depicts the first four rows in HRAID1/1 with $N = 4$ SNs and $M = 4$ disks per SN, which utilizes a single parity P at the intra-node level and a single parity Q at the intra-node level ($k = \ell = 1$). Internode Q parities protect data strips at other nodes, while intranode P parities protect local data and Q parity strips at that SN.

To update data block $d^2_{4,1}$ in strip $D^2_{4,1}$, with $d^{2new}_{4,1}$, we have the following steps: (1) Read $d^{2old}_{4,1}$; (2) Compute the difference $d^{2diff}_{4,1} = d^{2new}_{4,1} \oplus d^{2old}_{4,1}$ and write $d^{2new}_{4,1}$; (3) Read $p^{2old}_{4,3}$ and update it: $p^{2new}_{4,3} = p^{2old}_{4,3} \oplus d^{2diff}_{4,1}$, write $p^{2new}_{4,3}$; (4) Send $d^{2diff}_{4,1}$ to Node 1, Read $q^{1old}_{4,1}$ and $p^{1old}_{4,4}$ and update them: $q^{1new}_{4,1} = q^{1old}_{4,1} \oplus d^{2diff}_{4,1}$, $p^{1new}_{4,4} = p^{1old}_{4,4} \oplus q^{1diff}_{4,1}$. Generally, HRAID$k/\ell$ requires $(k + 1)(\ell + 1)$ reads if the old data and check blocks are not cached and $(k + 1)(\ell + 1)$ writes to update a data block.

The minimum number of disk failures that can be tolerated by HRAIDk/ℓ is: $d_{min} = (k + 1)(\ell + 1) - 1$ and the maximum number $d_{max} = N\ell + (M - \ell)k$ or $d_{max} = N(k + \ell) - k\ell$ for $N = M$. For $N = M$ $k + \ell$ disk per node or $N(k + \ell)$ disks for the HRAID array are dedicated to check codes. Since at most d_{min} disk failures are tolerated, it follows that the coding in HRAID(k/ℓ) is not MDS.

The failure of the controller at an SN makes the disks controlled by it inaccessible. We consider three options for HRAID recovery. With Option I nodes operate independently and given that each node is an ℓDFT, data loss occurs if there are more than ℓ failed disks at the SN or the controller fails. With Option II internode recovery is allowed, so that the contents of failed disks, which are unrecoverable at the node level can be reconstructed via internode recovery. Rebuild processing is carried out via restriping both at disk and SN level. Option III does not carry out rebuild processing and instead carries out on demand multistep recovery.

19.8.3 Heterogeneous Disk Arrays (HDAs)

RAID level selection, most importantly RAID1 versus RAID5, was a difficult issue to resolve for early RAID users. RAID1 is suitable for high-performance OLTP applications, because it provides twice the access bandwidth for read requests. RAID5 provides the access parallelism required for high volumes of data accessed by data mining and decision support applications. AutoRAID developed by HP, which uses RAID1 as a cache to a RAID5 array at the lower level, was proposed as a solution to the RAID selection problem [127]. AutoRAID data is initially held with the space-inefficient RAID1 format, but when storage space is exhausted the system converts the less frequently used storage segments to RAID5/LSA format (see Section 19.4). The fact that RAID5/LSA introduces less load than RAID1 in dealing with large writes is used as the RAID level selection criterion in [116].

There was significant activity in the area of data allocation in storage systems at HP Labs, which led to the *Disk Array Designer (DAD)* [3]. IBM provisioning planners for storage allocation involve determining the number, size, and location of volumes with attention to access paths to meet performance requirements [10]. Disaster recovery is another aspect of the overall process. Introducing a new application into a data center is a complex task, which is however assisted by monitoring tools provided by IBM and EMC. These tools monitor SAN (storage area network) components to determine its configuration and performance. The volume planner takes new workloads space and performance requirements (I/O demand, R/W ratio, response time) and provides a recommendation. These information is used to determine disk utilizations. The selection process' configuration phase selects storage pools, which can potentially hold the new data. The load balancing phase uses a volume allocation algorithm, rather than using first-fit, best-fit, or worst-fit. The path planner determines paths to connect the volumes to the hosts. The zone planner restricts the connectivity to ensure security.

The Panasas *Parallel File System* (*PFS*) stripes files across *objects*, which are stored according to the RAID1 or RAID5 paradigm. Small files are stored as RAID1 and large files as RAID5, since they provide more parallelism for data access [125]. Note similarity to EMC's Centera file system for reliable storage of fixed content.* Four advantages associated with PFS are: (i) Scalable computing of parities for the files by clients. (ii) Capability for end-to-end data integrity checking. (iii) Parallel rebuild capability for files. (iv) Unrecoverable faults are restricted to individual files. The initial data placement is random. Passive balancing is used to place new data into empty nodes, while active balancing moves objects from node to node to eliminate hotspots.

In [117], we consider the allocation of *Virtual Arrays (VAs)* with different RAID levels in *Heterogeneous Disk Arrays (HDAs)*. HDAs are expected to providing cost savings by consolidating multiple disk arrays. The number of *Virtual Disks (VDs)* required to materialize a VA is determined by upper limits on VD bandwidth and space per physical disk drive in normal operating mode. The synthetic stream of allocation requests for VDs is used by a single-pass data allocation methods for HDA. The allocations are carried out in degraded mode, so that disk loads are not exceeded due to a single disk failure. Only 1DFT RAID1 and RAID5 arrays and single disk failures are considered in the simulation study, but we develop the analytic expressions for allocating other RAID levels. A VA allocation is successful if disks sharing VAs with a failed disk are not overloaded in degraded mode. Experimental results show that allocation methods minimizing the maximum disk bandwidth and capacity utilization or their variance across all disks yield the maximum number of allocated VAs. When the disk bandwidth is the bottleneck resource, the clustered RAID5 paradigm may be adopted to increase the number of allocated VAs (see Section 19.4.2).

19.9 Storage Area Network and Network Attached Storage

A storage area network (SAN) is defined as a set of disk drives connected by a network: a SCSI bus, an Ethernet switch, or a peer-to-peer network, where the disk drives are associated with cooperating servers. Requirements for SANs listed in [86] are summarized in the following:

1. *Space and access balance.* The requirement to store additional data is growing rapidly. Given that disk capacity is cheap, the storage capacity of the network can be easily expanded. There is a need to allocate data carefully to utilize disk capacities and balance their utilizations. This can be accomplished by striping and randomizing [87]. There is also the issue of data reorganization

* http://www.emc.com/products/family/emc-centera-family.htm.

to accommodate increased data volumes, larger capacity disks, etc. Given the characteristics of the disk system and the access rates of files, a nearly optimal assignment of files to disks is sought to optimize performance metrics, such as the mean disk response time [130]. Without such an optimized allocation, disks are susceptible to access skew.

2. *Availability.* High data availability can be attained by mirroring or erasure coding. The latter, which provides a much higher storage efficiency, has been adopted by the Zebra distributed file system based on the RAID5/LSA concept [37]. Data from various clients is combined into a single stream, which is written as one stripe, so that the parity strip is computed on the fly.
3. *Resource efficiency.* Higher availability implies wasted disk capacity in the form of check disks and replicated data. This overhead should be quantified against higher performance and availability.
4. *Access efficiency.* This is determined by the time and space requirements to access the data.
5. *Heterogeneity.* New disk drives, possibly with higher capacities, should be introduced with as little disruption as possible. It is not economically viable to replace all old disks with new disks, just for the sake of maintaining homogeneity. RAID5 data layout becomes more complicated with heterogeneous disks: variable stripe sizes and disk load imbalance for updating parity. Solutions to these and other problems are discussed in [19].
6. *Adaptivity.* This is a measure of how fast can the system adapt to increases in the data volume and the addition of more disk capacity. More specifically, a fraction of the existing data must be redistributed, and the efficiency of this process is called adaptivity. The faithfulness property is to balance disk capacity utilizations, so that given m data units and that the ith disk has a fraction d_i of the total capacity, then this disk's share of the data is $(d_i + \varepsilon)m$, where ε is a small value. The issue of access bandwidth should also be taken into consideration. A relevant study is [94], which poses the disk replacement problem (DRP), that is, minimizing the cost of migrating data following disk additions and removals.
7. *Locality.* This is a measure of the degree of communication required for accessing data.

Network attached storage (NAS) as opposed to direct attached storage (DAS) embeds the file system into storage, while a SAN does not [28]. Four typical NAS systems described in [86] are as follows:

- Storage appliances such as products from *Network Appliance (NetApp)* and SNAP.*
- *Network Attached Secure Devices (NASD)* project at CMU's *Parallel Data Laboratory (PDL)* [29].†
- The goal of the Petal project is to provide easily expandable storage with a block interface [57]. Incidentally, Petal adopts the CD mirroring data layout (see Section 19.5.1).
- The *Internet SCSI (iSCSI)* protocol, which implements the SCSI protocol on top of TCP/IP [88].

19.10 Data Deduplication

Data deduplication is an attempt to reduce wasted storage due to perceived high levels of redundancy in data storage systems. For example, the email server in a large company may hold thousands of copies of the same report sent to all of its employees as an attachment. Deduplication is known as intelligent compression, single-instance storage, data reduction, and capacity optimized storage, replaces the multiple copies to a pointer to a single copy. The need for deduplication could have been alleviated by sending a pointer to the message to start with. Data Domain, now part of EMC, introduced data deduplication software in 2004. Proprietary implementations of data deduplication for Data Domain and Symantec are reported in [135] and [34], respectively.

* http://www.netapp.com.
† http://www.pdl.cmu.edu/NASD/index.shtml.

Data deduplication is data compression at a coarse granularity, for example, files or blocks. Deduplication at a finer granularity may identify duplicates, which do not exist at a coarser level, as in the case of a lengthy files with minor modifications. File compression methods such as Lempel-Ziv [129], which remove short repeated strings result in a limited compression ratio, while deduplication tends to result in significant reduction in storage space requirements. There is an indirect reduction in power consumption, since fewer disks are required to store the data.

Cryptographic hash functions are applied to determine if two chunks of data are identical. There is a negligibly small possibility that the chunks are not identical due to the *pigeon hole principle*:* The *Secure Hashing Algorithm (SHA)*-1 returns a 20 byte hash value, and is more reliable than the earlier *Message Digest (MD)*-5, which returns 16 bytes. Data deduplication performance can be improved by first applying a weak hash function, followed by a strong hash function to attain a higher confidence. Some systems ascertain that two chunks are identical by a bit by bit comparison.

There have been several performance comparisons of deduplication implementations, with daily backup capacity (TB/day) as perhaps the most important single metric.†

19.11 Solid State Disks (SSDs)

Target specifications for an SCMs based on [25] are: Access time: 50–1000 nanoseconds (ns), Data rate: 100 MB/s, Endurance: 10^9–10^{12} cycles, Error rate per TB: 10^{-4}, mean time between failures (MTBF): 2 million hours, Data retention: 10 years, ON/standby power: 100/1 milliwatt (mW), Cost: $5 per GB, CAGR: 35%.

In what follows, we discuss the more popular NAND flash memories, as opposed to NOR flash, which are used to store code. The smallest unit of writing into a NAND-based flash storage is a 4 KB page. Before a page is written all its bits need to be set to one. This is done at the level of blocks, which are usually much larger than pages, Table II in [21] provides the characteristics of five typical flash memories, but more comprehensive data is provided in [33].

The two varieties of flash SSDs are single-level cell (SLC) and multi-level cell (MLC), where the former stores 1 bit per cell, while the latter stores multiple bits per cell, providing more storage capacity at a higher cost.

Design tradeoffs for flash memories are considered in [1]. A 4 GB Samsung NAND flash is used as an example in this paper (see Figure 1 and Table 1). There are two 2 GB dies packaged together. Data is organized as 256K blocks of thirty-two 4 KB pages. It takes 25 microseconds (μs) to read from a register or write to it. Block erase is a slow operation and takes 1.5 milliseconds (ms). It takes 100 μs to transfer a 4 KB page from an on-chip register. Pipelining the two operations leads to 40 MB/s. Writes take 100 μs per page for transfer time and 200 μs programming time. Given that there are two dies per chip these operations can be interleaved.

Writes cannot be performed in place in NAND flash and for higher efficiency should be performed in large chunks. Writes are carried out onto individual 4 KB pages from allocation pools, which may be an allocation plane or larger. Writing can only be carried out after the associated block is cleared, so that we maintain a free block list in each allocation pool.

Given that modified blocks in database applications are overwritten, it is advisable to hold such blocks in DRAM buffers or even HDDs, which act as caches for the flash memory. There are techniques to increase the lifespan of flash memories by enhancing the flash translation layer (FTL), which provides virtual to physical address translation. Since update in place is not possible, FTL implements updating out-of-place using an already erased page.

* http://en.wikipedia.org/wiki/Pigeonhole_principle.

† http://wikibon.org/wiki/v/Wikibon_Data_De-duplication_Performance_Tables.

A difficulty associated with FTL is the size of the SRAM cache, which stores the mapping table, for example, 32 MB cache is required for a 16 GB flash to map at the page level. Block-level mapping is costly due to wasted space and performance degradation due to garbage collection. A hybrid approach is used: data blocks are mapped at the block level, while the smaller number of update blocks are mapped at the page level. It is argued in [35] that this scheme suffers from poor garbage collection behavior, requires several workload specific tuning parameters, and does not exploit the temporal locality of most workloads. Four state-of-the-art FTLs are described in this paper and compared to *Demand-based FTL (DFTL)* proposed in this work.

Many existing approaches on flash-memory management are based on RAM-resident tables in which one single granularity size is used for both address translation and space management. As high-capacity flash memory is becoming more affordable than ever, the dilemma of how to manage the RAM space or how to improve the access performance is emerging for many vendors. A tree-based management scheme, which adopts multiple granularities in flash-memory management is proposed in [13], with the objective of not only reducing the run-time RAM footprint, but also managing the write workload, due to housekeeping. The proposed method was evaluated under realistic workloads, where significant advantages over existing approaches were observed, in terms of the RAM space, access performance, and flash-memory lifetime.

A *Fast Array of Wimpy Nodes (FAWN)* is an array of low power processors with local flash memories providing efficient parallel access to data [4]. FAWN provides an efficient implementation for Amazon's Dynamo key-storage system, which represents the key and the actual data in a key–value relationship. FAWN consists of frontend and backend nodes, for example, 80 versus 10,000 nodes Each frontend is responsible for a range of keys and directs a query to the appropriate backend.

HyStor integrates low-cost HDDs and the more expensive SSDs into a single layer block storage, so that no modifications are required to OS [16]. Blocks have a LBN on HDD and some of the blocks may be cached at the SDD, which is an order of magnitude smaller. The SDD has a remap and a writeback area, the latter area simply holds dirty blocks, while the former area is holds high-value, frequently accessed data blocks.

We are mainly interested in flash memories as a replacement for HDDs, which as SSDs are attractive from the viewpoint of speed, power consumption, and reliability. Flash memories are orders of magnitude more expensive than HDDs, and have a finite number of *Program/Erase (P/E)* cycles. Flash memories (32 GB, 64 GB, and 128 GB) have been introduced with R/W rates of 1400/2500 MB per second have been attained. While in recent years flash-based SSDs have grown enormously, both in capacity and popularity, their adoption in high-performance enterprise storage applications depends on exceeding disk performance, while closing the gap in cost per GB [33]. Although, the latter is becoming a reality, with increasing flash densities, flash reliability, and performance are declining. High-capacity and high-performance flash-based SSDs are not reliable enough to justify their cost in enterprise environments. The conclusions of this study are based on 45 flash chips from 6 manufacturers indicating that: "SSD manufacturers and users will face a tough choice in trading off between cost, performance, capacity and reliability."

We conclude the discussion of flash memories with two techniques to increase their reliability.

19.11.1 Differential RAID for Flash Memories

Diff-RAID is an unconventional RAID array whose design takes into account the fact that flash memories have very different failure characteristics from HDDs, that is, the *Bit Error Rate (BER)* of an SSD increases with more writes [6]. RAID arrays composed from SSDs are subject to correlated failures. By balancing writes evenly across the array, RAID schemes can wear out devices at similar rates, making the RAID susceptible to data loss. Diff-RAID reshuffles the parity distribution on each drive replacement to maintain an age differential when old devices are replaced by new ones. A simulator to evaluate Diff-RAID's reliability by using BERs from 12 flash chips showed that it is more reliable than RAID5 by orders of magnitude.

For MLC technology, the number of erasures is limited to 5,000–10,000 cycles per block. Diff-RAID differs significantly from RAID5 arrays on HDD in that it attains a higher reliability by distributing parity blocks *unevenly* [6]. Diff-RAID masks higher BERs on aging SSDs while: (1) Retaining the low overhead of RAID5; (2) extending the lifetime of commodity SSDs; (3) alleviating the need for expensive error correction hardware.

Load balancing can cause correlated failures, since with balanced writes all SSDs will be susceptible to failure when an SSD fails. Consider $n = 5$ SDDs with 80% of parity allocated on the first SDD and the rest of parities evenly distributed across remaining SDDs, so that we have the following configuration (80,5,5,5,5). After the first disk reaches its limit, it is replaced by the second SSD. Denoting the percentage of parity blocks on the ith and jth device by p_i and p_j, the aging rate of the two devices are

$$a_{i,j} = \frac{p_i(n-1)+(100-p_i)}{p_j(n-1)+(100-p_j)}$$

For example, for $n = 4$ and (70,10,10,10) the aging rate of the first device is twice that of the others. For $n = 5$ and with (80,5.5,5,5), after numerous replacements the ages of remaining devices at replacement time converge to (5750,4312.5,2875,1437.5). The implementation of device replacement can be accomplished via rebuild processing, while maintaining the same parity layout. The reduction in data loss probability, as given by Figures 6 and 8 in the paper, is significant.

19.11.2 Partial MDS Coding for SSDs

The *Partial MDS (PMDS)* method was developed in [9] to deal with sector failures in SSDs, since they are more prevalent than HDD sector failures. The sector is the data storage unit protected by a cyclic error correcting code (ECC) such as BCH.* Applying RS codes as in RAID6 is not appropriate for two reasons: (i) too expensive to dedicate an extra SSD to parity to handle sectors errors; (ii) RAID6 cannot handle more than two erased errors.

PMDS similarly to [47] relies on local and global parities, but is more complicated and does not require special consideration for parities being erased. Given an $m \times n$ array of sectors r erasures can be corrected in a horizontal row. As in the case of the EVENODD code $m = n - 1$ and that n should be prime, although this limitation can be easily circumvented [8]. If r columns are erased, an additional s global parities allow correcting s extra erasures. In the case $r = 1$ and $s = 2$ shown in Figure 19.8, necessary and sufficient conditions for finding codes and tables of codes are given in the paper. The r local parities can efficiently deal with failed sectors in a row and the extra s parities are invoked in rare occasions. Given six check sectors as many failed sectors can be corrected, say 4 sectors in a column and two other failed sectors.

In Figure 19.8, $a_{0,4}$, $a_{1,4}$, $a_{2,4}$, $a_{3,4}$ are the four local parities, and $a_{3,2}$ and $a_{3,3}$ placed somewhat arbitrarily are the global parities.† To simplify the notation, we superimpose a one-dimensional vector b_i, $0 \le i \le 19$,

$a_{0,0}$	$a_{0,1}$	$a_{0,2}$	$a_{0,3}$	$a_{0,4}$
$a_{1,0}$	$a_{1,1}$	$a_{1,2}$	$a_{1,3}$	$a_{1,4}$
$a_{2,0}$	$a_{2,1}$	$a_{2,2}$	$a_{2,3}$	$a_{2,4}$
$a_{3,0}$	$a_{3,1}$	$a_{3,2}$	$a_{3,3}$	$a_{3,4}$

FIGURE 19.8 Layout of the local and global parities for partial MDS with $r = 1$ and $s = 2$, The four local parities are $a_{i,4} = p_i$, $0 \le i \le 3$ and $a_{3,2} = g_1$ and $a_{3,3} = g_2$ are the global parities.

* http://en.wikipedia.org/wiki/BCH_code.

† The following discussion is partially based on Mario Blaum's presentation at Stanford University in July 2012.

which covers the blocks in Figure 19.8 one row at a time. The $m \times n$ codeword should satisfy the following six equations, with the first four for local parities and the last two for global parities:

$$S_j = b_{4j+0} \oplus b_{4j+1} \oplus b_{4j+2} \oplus b_{4j+3} = 0, \quad 0 \le j \le 3. \; S_4 = \sum_{i=0}^{19} \alpha^i b_i \;\; S_5 = \sum_{i=0}^{19} \alpha^{2i} b_i.$$

Given that disk two and two random blocks have failed, we designate them as $a_{0,2} = u$, $a_{1,2} = y$, $a_{2,2} = v$, $a_{3,2} = w$, and $a_{1,0} = x$, and $a_{1,3} = z$. The unknowns u, v, and w can be determined by using local parities: S_0, S_2, and S_3), while the following three equations can be used to determine x, y, and z:

$$x \oplus y \oplus z = S_1, \quad \alpha^5 x \oplus \alpha^7 y \alpha^8 z = S_4, \quad \alpha^{10} \oplus \alpha^{14} y \oplus \alpha^{16} z = S_5.$$

The set of linear equations has a unique solution provided the determinant of the matrix A defined in the following is not zero:

$$\mathcal{A} = \begin{pmatrix} 1 & 1 & 1 \\ \alpha^5 & \alpha^7 & \alpha^8 \\ \alpha^{10} & \alpha^{14} & \alpha^{16} \end{pmatrix}$$

$\mathcal{A}$ is a Vandermonde matrix since powers of α in its rows are different [75], so that $det(\mathcal{A}) \neq 0$ allowing x, y, and z to be determined.

As a second example, consider two sector errors: $x = a_{1,0}$ and $y = a_{3,1}$ and a failed SSD $u = a_{0,2}$, $z = a_{1,2}$, $v = a_{2,2}$, $w = a_{3,2}$. The values of u and v can be obtained using S_0 and S_2. We have the additional four equations:

$$x \oplus z = S_1, \quad y \oplus w = S_3, \quad \alpha^5 x + \alpha^7 z \oplus \alpha^{16} y \oplus \alpha^{17} w = S_4, \quad \alpha^{10} x + \alpha^{14} z + \alpha^{32} y + \alpha^{34} w = S_5$$

This system has a unique solution since the determinant is nonzero. A general condition is given by Theorem 4.1 for the necessary and sufficient conditions for the case of RAID5 (single global parity with $r = 1$); plus s global parities). In the special case of $s = 2$ Theorem 5,1 states that the code is PMDS depending on m, n, and the polynomial $f(x)$ defining the underlying field or ring. This polynomial $f(x)$ is not necessarily primitive,* but has a certain exponent, which has to exceed $m \times n$. For example for $f(x) = M_p(x) = 1 + x + x^2 + \cdots + x^{p-1}$, the polynomial has a prime exponent p, so that $m \times n$ should be smaller than p. Using Equation 16 in [9] a search is made for different parameters to obtain Table 19.1. The table is not extensive and does not consider all possible irreducible polynomials, but it gives a good idea.

19.12 Conclusions

In spite of the frequently used term "secondary storage," with main memory being the primary storage, secondary storage in the form of magnetic HDDs provides permanent storage for data. Unlike main memories, which are based on DRAM technology, HDDs are unaffected by occasional power outages and are made immune to rare disk failures using coding methods or replication. There are the additional levels of redundancy, such as replicated sites to protect against software errors, catastrophic failures (earthquakes, fires, floods, wars).

* http://en.wikipedia.org/wiki/Primitive_polynomial_(field_theory).

High-performance enterprise storage systems are used by applications such as OLTP, which incur high disk arm positioning time. Large-scale archival systems tend to have much higher capacities, but lower performance than enterprise drives. The recent indications are that while flash memories are not suitable replacements for HDDs, they may serve as cache storage with respect to HDDs, in what has been referred to as hybrid storage. According to [11] at least 10 SCMs are under consideration for this purpose.

We have discussed reliable storage systems based on erasure encoding, such as RAID5 and RAID6, and replication. In both cases, we discuss the issue of improving performance in normal, degraded, and rebuild mode of operations.

It is worthwhile to mention that it is easier to deal with erasures (broken disks) and unreadable sectors, than when the wrong data is read from disk or overwrites the wrong sectors. Higher level applications are separated from data being read/written to/from SDDs and HDDs, by many layers of software which complicates the issue of error detection. The narrow and well-defined interface to disk disallows the propagation writing of erroneous data, while faulty software may more easily destroy the contents of SSDs.

There is sufficient material in this chapter to serve as a full semester graduate course on storage systems, supplemented with appropriate reading material from the references.

Glossary

BIBD: Balanced Incomplete Block Design. A data layout method for clustered RAID, which has its origin in combinatorial mathematics and has been applied to design of experiments in statistics.

Clustered RAID (CRAID): A data layout to reduce the increase in disk load when operating in degraded mode, for example, one failed disk in RAID5.

Cryptographic Hash Functions: These are applied as part of data deduplication to determine if two chunks of data are identical. The *Secure Hashing Algorithm (SHA)*-1 returns a 20 byte hash value, and is more reliable than *Message Digest (MD)*-5, which returns 16 bytes. Data deduplication performance can be improved by first applying a weak hash function, followed by a stronger hash function to attain a higher confidence.

DAS: Direct Attached Storage has files directly accessible by a computer, rather than file servers attached to a network.

Dedicated/Distributed Sparing: A hot spare is provided in the case of dedicated sparing, which may be shared among several arrays. Sufficient spare capacity is provided at $N - 1$ out of N disks, to hold the data and parity blocks of a failed disk. Advantage with respect to dedicated sparing is that the bandwidth of the spare disk is not wasted.

Deduplication: Substituting duplicates with pointer to a single shared copy. Hash function are used to determine duplicates. Also known as intelligent compression, single instance storage, data reduction, and capacity optimized storage.

DRAM/SRAM: DRAM is more dense than SRAM and is used in main memories, but SRAM which is faster and more expensive per GB than DRAM is used in main memories.

ECKD: Extended count-key-data (CKD) is a format associated with IBM mainframes, starting with IBM S/360 series in 1960s. The IBM OS/390 or z/OS operates on ECKD disks such as 3990 and disk array controller (DAC) serves the role of simulating ECKD disks on disks with a fixed block architecture (FBA).

EVENODD (EO): Uses a horizontal and a diagonal parity to protect against two disk failure, hence more efficient code than RS for RAID6.

Freeblock Scheduling (FS): A proposal to utilize opportunistic disk accesses as the arm moves in processing regular requests to carry out other tasks. The goal of FS to maximize the efficiency of background processing.

Grid Codes: A combination of horizontal and vertical check codes protect against data strip failures in two dimensions.

HDA: Heterogeneous disk array allows the sharing of disk space among multiple virtual arrays with different levels.

HDD: Hard disk drive as opposed to floppy disk drive, which became popular in conjunction with personal computers.

HRAID*k*/ℓ: Hierarchical RAID uses two levels of erasure coding to tolerate k node failures and ℓ disk failures per node.

IDR: IntraDisk Redundancy a RAID paradigm at the level of disk blocks to deal with LSEs. IDR has been referred to as a vertical code in trade literature.

***k*DFT:** k Disk Failure tolerant. RAID1 disk arrays are 1DFT, LSI RAID is 2DFT, and RAID($4 + k$), $k \geq 1$ arrays are *k*DFT.

LFS: Log-structured file system is a scheme to improve write efficiency, by buffering modified files and writing them out sequentially in large chunks.

LSA: Log-structured arrays extend the LFS concept to RAID5/6. Full stripe writes are utilized in writing out batches of modified files. Data compression is possible because data is not written in place.

LSE: Latent sector error, which are unreadable and are the main cause of unsuccessful rebuilds.

MAID: Massive array of idle disks, where a small fraction of disks which hold active data remain online.

MB/GB/TB: Kilo/mega/giga/tera-byte, which are 10^3, 10^6, 10^9, 10^{12} bytes, respectively. These are followed by petabyte (10^{15}), exabyte (10^{18}), zettabyte (10^{21}), and yottabyte (10^{24}).

MDS: Maximum distance separable codes introduce the minimum level of redundancy for erasure coding, such that it is known which disk has failed. EO, RDP, and X-code in addition to Reed–Solomon codes are examples of MDS codes.

MTBF/MTTF/MTTDL: Mean time between failures/mean time to failure/mean time to data loss. MTBF implies that the failed component is repaired, but repair in the context of disk arrays implies the copying of a failed disk on a spare.

NAS: Network attached storage relies on a specialized computers, that is, network file system (NFS) for storing and serving files.

NVRAM/NVS: Non-volatile DRAM or storage. Nonvolatility is attained by backup batteries

Off/Near/Online Storage: Offline storage is a tape archive. Nearline storage is an automated tape library. Online storage is usually a SATA drive.

Parity Sparing: Combining two RAID5 arrays into one, by utilizing the parity strips of one of two arrays with the strips of the failed disk.

Parity Striping: A RAID5 data layout where data blocks are placed sequentially (without striping), but $1/N$ of the capacity of each disk is dedicated to parity blocks. In effect placing RAID5's parity strips on all disks horizontally, following $1 - 1/N$ of the space at each disk dedicated to data. Advantage that large chunks of the user data can be accessed efficiently without disruption.

Prefailure/Postfailure Replacement: The former can be carried out efficiently by utilizing Self-Monitoring Analysis and Reporting Technology (SMART),* so that the contents of a disk about to fail are copied, without incurring the usual rebuild process or postfailure replacement.

Primary/Secondary/Tertiary Storage: These refer to main memory, disk, and tape storage, respectively.

Partial MDS (PMDS) Codes: PMDS codes were developed for SDDs where hard sector errors are more common than in the case of HDDs. In an $m \times n$ array with $r = 1$ and $s = 2$, there is a single parity per row and $s = 2$ global parities. More general cases are considered in [9], while local reconstruction codes (LRCs) consider this special case.

Pyramid Codes: In RAID($4 + k$), $k > 1$ have the lowest order check strip, which is a parity, duplicated, such that each parity covers half of the disks in the array.

---

* http://en.wikipedia.org/wiki/S.M.A.R.T.

RAID: Redundant array of independent disks has seven levels: 0:6. RAID0 does not tolerate disk failures, RAID1 corresponds to mirrored disks, and RAID($4 + k$), $k \geq 1$ are kDFT arrays, so that RAID7 tolerates three disk failures.

RAID1: Mirrored disks have different data organizations. A more balanced load following disk failures is attainable by spreading the data of each disk over multiple disks, rather than one in basic mirroring, but this increases the chances of data loss when more disk failures occur. *Basic Mirroring (BM). Group Rotate Declustering (GRD), Interleaved Declustering (ID)*, and *Chained Declustering (CD)* are better known RAID1 organizations, while LSI RAID and SSPiRAL combine mirroring and parity coding.

ReConstruct Write (RCW): RCW reads the remaining strips in a stripe to compute and write check blocks.

Read-modify-write (RMW): RMW reads, modifies, and overwrites a disk block after a disk rotation. When the modification (an XOR) is done by the DAC, we have a single read followed by a single write.

Rebuild in RAID5: Postfailure rebuild is the systematic reconstruction of the contents of a failed disk in the form of rebuild units on a spare disk. See also prefailure replacement.

RDP: Rotated Diagonal Parity is a slight improvement on EO, developed at NetApp.

Reed-Solomon (RS) Code: An MDS code which is more expensive to implement than parity based codes such as EO and RDP, since it requires Galois field finite arithmetic.

Restriping: With successive disk failures in a RAID($4 + K$) array with $k > 1$, its check strips are overwritten with the strips of the failed disk.

RM2: A nonMDS 2DFT code with adjustable redundancy $p = 1/M$, so that one parity stripe per $M - 1$ data strips. Each data strip participates in two parity groups with $2M - 1$ strips and is hence a 2DFT but not MDS.

SAN: A Storage Area Network is a dedicated network that provides access to consolidated, block level data storage. Devices attached to SAN appear as if they are locally attached.

SAS: Serial Attached SCSI.

SATA: Serial Advanced Technology Attachment.

SCSI: Small Computer System Interface is a standard for connecting disks with computer systems.

Sparing Alternatives: Dedicated/distributed/parity-sparing/restriping are four sparing alternatives. Parity sparing combines two RAID5 arrays into one. These are discussed under separate headings.

SSD: Solid state disks are used to store persistent data and are much faster than HDDs.

Static/Dynamic Routing: The routing is in the context of mirrored disks.

Striping: Partitioning data into fixed sized strips, which are allocated round-robin across the disks in a disk array. The strips in a row constitute a stripe.

Small Write Penalty (SWP): Indicates the high cost of updating small data blocks in RAID($4 + k$), $k \geq 1$ disks arrays, where $k + 1$ data blocks, which have to be read (unless they were cached),

X-code: An MDS 2DFT code with two vertical parities placed horizontally as ($N - 1$)st and Nth stripes of $N \times N$ segments, where N is the number of disks.

The SNIA Dictionary is a glossary of storage networking, data, and information management terminology.*

Abbreviations in Paper

BER: Bit Error Rate. *BM:* Basic Mirroring. *CAGR:* Compound Annual Growth Rate. *CAV:* Constant Angular Velocity. *CD:* Chained Declustering. *CTMC:* Continuous Time Markov Chain. *DAC:* Disk Array Controller. *DAS:* Direct Attached Storage. *EDF:* Earliest Deadline First. *FTL:* Flash Translation Layer. *GRD:* Group Rotate Declustering. *HDA:* Heterogeneous Disk Array. *HDD:* Hard Disk Drive. *ID:* Interleaved Declustering. *IDR:* Intra-Disk Redundancy. *LBN:* Logical Block Number. *LFS:*

* http://www.snia.org/education/dictionary.

Log-structured File System. *LSA:* Log-Structured Array. *LSE:* Latent Sector Error. *MAID:* Massive Array of Idle Disks. *MDS:* Maximum Distance Separable. *MTTDL:* Mean Time to Data Loss. *MTTF:* Mean Time To Failure, *NAS:* Network Attached Storage. *PMDS:* Partial MDS. *RS*: Reed–Solomon (codes). *RU:* Rebuild Unit. *R/W:* Read/Write (head). *SAN:* Storage Area Network. *SAS:* Serial Attached SCSI. *SATA:* Serial Attached SCSI Attachment. *SCSI:* Small Computer System Interface. *SN:* Storage Node. *SSD:* Solid State Disk. *SPA:* Supplementary Parity Augmentation. *SWP:* Small write penalty. *TMR:* Triple Modular Redundancy. *VA:* Virtual Array. *VD:* Virtual Disk.

Abbreviations for Journal and Conference Titles

ACM: Association for Computing Machinery.* *ASPLOS:* Architectural Support for Programming Languages and Systems (ACM). *ATC:* Annual Technical Conference (USENIX). *CAN:* Computer Architecture News (ACM SIGARCH). *Special Interest Group (SIG). CACM:* Communications of the ACM. *DPDB:* Distributed and Parallel Databases. *EUROSYS:* European Conference on Computer Systems.† *FAST:* File and Storage Technologies Conference (USENIX). *HPCA:* High Performance Computer Architecture (IEEE International Symposium). *HPDC:* High Performance Distributed Computing (IEEE International Symposium). *ICDE:* International Conference on Data Engineering (IEEE). *IEEE:* Institute of Electrical and Electronics Engineers.‡ *IEEE Computer Society:*§ *ISCA:* International Symposium on Computer Architecture (ACM). *IBM J. R&D:* IBM Journal of Research and Development (since 2009 includes *IBM Systems Journal*). *ISCA:* International Conference on Computer Architecture (ACM). *JPDC:* Journal Parallel Distributed Computing (Elsevier). *NAS:* Networking, Architecture, and Storage (IEEE International Conference). *OSDI:* Operating System Design and Implementation (USENIX Symposium). *SIGMETRICS:* Conference on Measurement and Modeling of Computer Systems (ACM). *SIGMOD:* International Conference on Management of Data (ACM). *S. P. & E.:* Software Practice and Experience (Wiley). *SOSP:* Symposium on Operating Systems Principles (ACM). *TC:* Transactions on Computers (IEEE). *TDSC:* Transactions on Dependable and Secure Computing (IEEE). *TOCS:* Transactions on Computing Systems (ACM). *TOMCCAP:* Transactions on Multimedia Computing, Communications, and Applications (ACM). *TOS:* Transactions on Storage (ACM). *TPDS:* Transactions Parallel and Distributed Systems (IEEE). *VLDB:* Proceedings of Very Large Data Bases Conference.¶

Further Information

There have been many new recent developments in this field, since the publication of the second edition of the Handbook in 2004. Rather than making a minor modification to that chapter, it was completely rewritten. Sections dealing with performance analysis can be skipped by readers not interested in this topic. Conversely, readers are encouraged to read the chapter on *Performance Evaluation of Computer Systems* in the Handbook for a more detailed discussion of this topic.

Publications on storage systems and storage research in industry and universities are reviewed in [107] and [112]. ACM started publishing *Transactions on Storage (TOS)* in 2005, while articles on this topic appeared in ACM's *Trans. on Computer Systems (TOCS)*, before that year with some minor overlap. USENIX's conference on *File and Storage Technologies (FAST)* was started in 2002.

A good collection of papers on mass storage and parallel I/O is [52]. A new book with a good description of modern disks is [49]. but no space is dedicated to flash memories. A follow-up to [15] is [108], which emphasizes two-disk failure tolerant (2DFT) arrays and associated coding methods.

* http://www.acm.org.
† http://www.eurosys.org.
‡ http://www.ieee.org.
§ http://www.computer.org.
¶ http://www.vldb.org.

Storage Networking Industry Association (SNIA),* has numerous active technical projects in this area, including: Cloud storage, Hypervisor storage, Storage management, Object-based storage, IP storage Green storage, Solid state storage, Data integrity, and Long-term retention. Object-based storage devices (OSD) working group concerned with the creation of self-managed, heterogeneous, shared storage. This entails moving low-level storage functions to the device itself and providing the appropriate interface.

Web pages, if storage companies have white papers, reports, and presentations on their products. IBM publishes RedBooks online. Several storage-related articles are listed at http://www.redbooks.ibm.com/portals/storage.

The *Storage Performance Council (SPC)*† is an organization, which acts similarly to *Transaction Processing Council (TPC)*,‡, but develops storage performance and power efficiency benchmarks, while TPC deals with overall system processing database workloads.

This chapter touches upon the topic of Performance Evaluation of Computer Systems, covered elsewhere in this handbook. Also provided is a self-sufficient discussion of reliability analysis. Additional references on both topics are available from Appendix A: Bibliography in [122].

References

1. N. Agrawal, V. Prabhakaran, T. Wobber, J. D. Davis, M. S. Manasse, and R. Panigrahy. Design tradeoffs for SSD performance. *ATC'08*, Boston, MA, June 2008, pp. 57–70.
2. D. Anderson, J. Dykes, and E. Riedel. More than an interface - SCSI vs. ATA. *FAST'03*, San Francisco, CA, March 2003, pp. 245–257.
3. E. Anderson, S. Spence, R. Swaminathan, M. Kallahalla, and Q. Wang. Quickly finding near-optimal storage designs. *TOCS 234*: 337–374 (November 2005).
4. D. G. Anderson, J. Franklin, M. Kaminsky, A. Panishayee, L. Tan, and V. Vasudevan. FAWN: A fast array of Wimpy Nodes. *CACM 54*(7): 101–109 (July 2011).
5. E. Bachmat, and J. Schindler. Analysis of methods for scheduling low priority disk drive tasks. *SIGMETRICS'02*, Los Angeles, CA, June 2002, 55–65.
6. M. Balakrishnan, A. Kadav, V. Prabhakaran, and D. Malkhi. Differential RAID: Rethinking RAID for SSD reliability. *TOS 6*(2): Article 4 (July 2010).
7. D. Bitton, and J. Gray, Disk shadowing. *VLDB'88*, Los Angeles, CA, August 1988, pp. 331–338.
8. M. Blaum, J. Brady, J. Bruck, and J. Menon. EVENODD: An efficient scheme for tolerating double disk failures in RAID architectures. *TC* 44(2): 192–202 (February 1995).
9. M. Blaum, J. L. Hafner, and S. Hetzler. Partial-MDS codes and their application to RAID type of architectures. *IBM Research Report*, RJ 10498, 2012 (also *CoRR abs/1205.0997*: (2012)).
10. P. L. Bradshaw. Archive storage system design for long-term storage of massive amounts of data. *IBM J. R&D* 52(4/5): 449–464 (July 2008).
11. G. W. Burr, B. N. Kurdi, J. C. Scott, C. H. Lam, K. Gopalakrishnan, and R. S. Shenoy. Overview of candidate device technologies for storage-class memory. *IBM J. R&D* 52(4/5): 449–464 (July 2008).
12. D. D. Chambliss, P. Pandey, T. Thakur, A. Fleshler, T. Clark, J. A. Ruddy, K. D. Gougherty, M. Kalos, L. Merithew, J. G. Thompson, H. M. Yudenfriend. An Architecture for Storage-Hosted Application Extensions. *IBM J. R&D* 52(4/5): 427–437 (2008).
13. L. P. Chang, and T. W. Kuo. Efficient management for large-scale flash-memory storage systems with resource conservation. *TOS* 1(4): 381–418 (November 2005).
14. S. Chen, and D. F. Towsley. The design and evaluation of RAID 5 and parity striping disk array architectures. *JPDC* 17(1–2): 58–74 (January/February 1993).

* http://www.snia.org.
† http://www.spc.org.
‡ http://www.tpc.org.

15. P. M. Chen, E. K. Lee, G. A. Gibson, R. H. Katz, and D. A. Patterson. RAID: High-performance, reliable secondary storage. *ACM Comput. Surv.* 26(2): 145–185 (June 1994).
16. F. Chen, D. A. Koufaty, and X. Zhang. Hystor: Making the best use of solid state drives in high performance storage systems. *Proc. ACM/IEEE Int'l Conf. on Supercomputing (ICS)*, Tucson, AZ, May 2011, pp. 22–32.
17. D. Colarelli, and D. Grunwald. Massive arrays of idle disks for storage archives. *Proc. ACM/IEEE Int'l Conf. on Supercomputing (ICS)*, Baltimore, Maryland, USA, November 2002, pp. 1–11.
18. P. F. Corbett, R. English, A. Goel, T. Grcanac, S. Kleiman, J. Leong, and S. A. Sankar. Row-diagonal parity for double disk failure correction. *FAST'04*, San Francisco, CA, April 2004, pp. 1–14.
19. T. Cortes, and J. Labarta. Extending heterogeneity to RAID Level 5. *ATC'01*, Boston, MA, June 2001, pp. 119–132.
20. P. J. Denning. Effect of scheduling in File Memory Operations. *Proc. AFIPS Spring Joint Computer Conf.*, Atlantic City, NJ, April 1967, pp. 9–21.
21. Y. Deng. What is the future of disk drives, death or rebirth?. *ACM Comput. Surv.* 43(3): Article 23 (April 2011).
22. A. Dholakia, E. Eleftheriou, X.-Y. Hu, I. Iliadis, J. Menon, and K. K. Rao. A new intra-disk redundancy scheme for high-reliability RAID storage systems in the presence of unrecoverable errors. *ACM TOS* 4(1): Article 1 (May 2008).
23. E. Eleftheriou, R. Haas, J. Jelitto, M. A. Lantz, and H. Polizidis. Trends in storage technologies. *IEEE Data Eng. Bull.* 33(4): 4–13 (December 2010).
24. D. A. Ford, R. J. T. Morris, and A. E. Bell. Redundant arrays of independent libraries (RAIL): The StarFish tertiary storage system. *Parallel Comput.* 24(11): 45–64 (November 1998).
25. R. F. Frietas, and W. W. Wilcke. Storage class memory: The next generation of storage technology. *IBM J. R(4–5)*: 439–448 (2008).
26. R. Geist, and S. Daniel. A continuum of disk scheduling algorithms. *ACM TOCS* 5(1): 77–92 (February 1987).
27. G. A. Gibson. *Redundant Disk Arrays: Reliable Parallel Secondary Storage*, The MIT Press, 1992.
28. G. A. Gibson, and R. van Meter. Network attached storage architecture. *CACM* 43(11): 37–45 (November 2000).
29. G. A. Gibson, D. Nagle, K. Amiri, J. Butler, F. W. Chang, H. Gobioff, C. Hardin, E. Riedel, D. Rochberg, and J. Zelenka. A cost-effective, high-bandwidth storage architecture. *ASPLOS'98*, San Jose, CA, October 1998, pp. 92–103.
30. S. Gopisetty, E. Butler, S. Jacquet, M. R. Korupolu, T. K. Nayak, R. Routray. M. Seaman, A. Singh, C.-H. Tan, S. Uttumchandanni, and A. Verma. Automated planners for storage provisioning and disaster recovery. *IBM J. R(4/5)*: 353–366 (July 2008).
31. J. Gray, B. Horst, and M. Walker. Parity striping of disk arrays: Low-cost reliable storage with acceptable throughput. *VLDB'90*, Brisbane, Queensland, Australia, August 1990, pp. 148–161.
32. E. Grochowski, and R. D. Halem. Technological impact of magnetic hard disk drives on storage systems. *IBM Syst. J.* 42(2): 338–346 (November 2003).
33. L. M. Grupp, J. D. Davis, and S. Swanson. The bleak future of NAND flash memory. *FAST'12*, San Jose, CA, February 2012.
34. F. Guo, and P. Efstathopoulos. Building a high-performance deduplication system. *ATC'11*, Portland, OR, June 2011.
35. A. Gupta, Y. Kim, and B. Urgaonkar. DFTL: A flash translation layer employing demand-based selective caching of page-level address mapping. *ASPLOS'09*, Washington, DC, March 2009, pp. 229–240.
36. B. Haeuser, J. Narney, and A. Colvig. *IBM System Storage Tape Encryption Solutions*, IBM RedBooks, 2011. http://www.ibm.com/redbooks
37. J. H. Hartman, and J. K. Ousterhout. The Zebra Striped network file system. *TOCS* 13(3): 274–310 (August 1995).

38. J. L. Hennessy, and D. A. Patterson. *Computer Architecture: A Quantitative Approach, 4th edn.*, Elsevier, 2007 (5th edition, 2012).
39. M. Holland, G. A. Gibson, and D. P. Siewiorek. Architectures and algorithms for on-line failure recovery in redundant disk arrays. *DPDB* 2(3): 295–335 (July 1994).
40. R. Y. Hou, J. Menon, and Y. N. Patt. Balancing I/O response time and disk rebuild time in a RAID5 disk array. *Proc. 26th Hawaii Int'l Conf. System Science (HICSS), Vol. I* Honolulu, HI, January 1993, pp. 70–79.
41. D. K. Hsiao. *Advanced Database Machine Architectures*, Prentice-Hall, 1983.
42. W. W. Hsu, A. J. Smith, and H. C. Young. Projecting the performance of decision support workloads on systems with smart storage (SmartSTOR). *Proc. IEEE Int'l Conf. on Parallel and Distributed Systems ICPADS'00*, Iwate, Japan, July 2000, pp. 417–425.
43. W. W. Hsu, and A. J. Smith. The performance impact of I/O optimizations and disk improvements. *IBM J. R&D* 48(2): 255–289 (2004).
44. W. W. Hsu, A. J. Smith, and H. C. Young. The automatic improvement of locality in storage systems. *TOCS* 23(4): 424–473 (November 2005).
45. C. Huang, and L. Xu. An efficient coding scheme for correcting triple storage node failures. *FAST'05*, San Francisco, CA, December 2005, pp. 197–210.
46. C. Huang, M. Chen, and J. Li. Pyramid codes: Flexible schemes to trade space for access efficiency in reliable data storage systems. *Proc. 6th IEEE Int'l Symp. on Network Computing and Applications (NCA 2007)*, Cambridge, MA, July 2007, pp. 79–86.
47. C. Huang, H. Simitci, Y. Xu, A. Ogus, B. Calder, P. Goplana, J. Li, and S. Yekhanin. Erasure coding in Windows Azure storage. *ATC*, Boston, MA, June 2012.
48. I. Iliadis, R. Haas, X.-Y. Hu, and E. Eleftheriou. Disk scrubbing versus intradisk redundancy. *TOS* 7(2): Article 5 (July 2011).
49. B. L. Jacob S. W. Ng, and D. Y. Wang. *Memory Systems: Cache, DRAM, Disk*, Morgan-Kauffman/Elsevier, 2008.
50. D. M. Jacobson, and J. Wilkes. Disk scheduling algorithm based on rotational position. HP Labs Technical Report HPL-CSP-91-7rev1, March 1991.
51. W. Jiang, C. Hu, Y. Zhou, and A. Kanevsky. Are disks the dominant contributor for storage failures? Comprehensive study of storage subsystem failure characteristics. *TOS* 4(3): Article 7 (November 2008).
52. H. Jin, T. Cortes, and R. Buyya. *High Performance Mass Storage and Parallel I/O: Technologies and Applications*, Wiley, 2002.
53. H. H. Kari. Latent sector faults and relability of disk arrays. PhD Thesis, University of Helsinki, Finland, 1977.
54. O. Khan, J. Burns. J. S. Plank, W. Pierce, and C. Huang. Rethinking erasure codes for Cloud File Systems: Minimizing I/O for recovery and degraded reads. *FAST'12*, San Jose, CA, February 2012.
55. L. Kleinrock. *Queueing Systems: Vol. 1: Theory/ Vol. 2: Computer Applications*, Wiley-Interscience, 1975/6.
56. E. Krevat, J. Tucek, and G. R. Ganger. Disks are like snowflakes: No two are alike. *Proc. 13th Workshop on Hot Topics in Operating Systems (HotOS 2011)*, Napa Valley, CA, May 2011.
57. E. K. Lee, and C. A. Thekkath. Petal: Distributed virtual disks. *ASPLOS'96*, Cambridge, MA, October 1996, pp. 84–92.
58. C. Leuth. RAID-DP: Network Appliance Implementation of RAID Double Parity for Data Protection," *Technical Report No. 3298*, Network Appliance Inc., 2006. http://www.netapp.com/us/communities/tech-ontap/tot-sas-disk-storage-0911.html#author.
59. M. Li, J. Shu, and W. Zheng. Grid codes: Strip-based erasure codes with high fault tolerance for storage systems. *TOS* 4(4): Article 15 (January 2009).
60. M. Li, and J. Shu. DACO: A high-performance disk architecture designed specially for large-scale erasure-coded storage systems. *TC* 59(10): 1350–1362 (October 2010).
61. C. R. Lumb, J. Schindler, G. R. Ganger, D. Nagle, and E. Riedel. Towards higher disk head utilization: Extracting "Free" bandwidth from busy disk drives. *OSDI'00*, San Diego, CA, October 2010, pp. 87–102.

62. R. E. Matick. *Computer Storage Systems and Technology*, Wiley, 1977.
63. B. McNutt. *The Fractal Structure of Data Reference: Applications to the Memory Hierarchy*, Spinger, 2000.
64. J. Menon. Performance of RAID5 disk arrays with read and write caching. *DPDB* 2(3): 261–293 (1994).
65. J. Menon. Performance comparison of RAID-5 and log-structured arrays. *HPDC'95*, Washington, D.C., August 1995, pp. 167–178.
66. J. Menon, and D. Mattson. Comparison of sparing alternatives for disk arrays. *ISCA'92*, Gold Coast, Queensland, Australia, May 1992, pp. 318–329.
67. A. Merchant, and P. S. Yu. Analytic modeling of clustered RAID with mapping based on nearly random permutation. *TC* 45(3): 367–373 (March 1996).
68. R. Muntz, and J. C. S. Lui. Performance analysis of disk arrays under failure. *VLDB'90*, Brisbane, Queensland, Australia, August 1990, pp. 162–173.
69. A. L. Narasimha Reddy, J. C. Wyllie, and R. Wijayaratne. Disk scheduling in a multimedia I/O system. *TOMCCAP* 1(1): 37–59 (February 2005).
70. D. Narayanan, A. Donnelly, and A. I. T. Rowstron. Write off-loading: Practical power management in storage systems. *TOS* 4(3): 318–329 (November 2008).
71. D. Narayanan, E. Thereska, A. Donnelly, S. Elnikety, and A. I. T. Rowstron. Migrating server storage to SSDs: Analysis of tradeoffs. *EuroSys'09*, Nuremberg, Germany, April 2009, pp. 145–158
72. S. W. Ng, and R. L. Mattson. Uniform parity group distribution in disk arrays with multiple failures. *TC* 43(4): 501–506 (April 1994).
73. V. F. Nicola, M. Nakayama, P. Heidelberger, and A. Goyal. Fast simulation of highly dependable systems with general failure and repair processes. *TC* 42(12): 1440–1452 (December 1993).
74. C.-I. Park. Efficient placement of parity and data to tolerate two disk failures in disk array systems. *TPDS* 6(11): 1177–1184 (November 1995).
75. J. S. Plank. A tutorial on reed-Solomon coding for fault-tolerance in RAID-like systems. *S. P. &. E.* 27(9): 995–1012 (September 1997).
76. J. S. Plank, and Y. Ding. Note: Correction to the 1997 Tutorial on Reed-Solomon coding. *S. P. & E.* 35(2): 189–194 (February 2005).
77. J. S. Plank, J. Luo, C. D. Schuman, L. Xu, and Z. Wilcox-O'Hearn. A performance evaluation and examination of open-source erasure coding libraries for storage. *FAST'09*, San Francisco, CA, February 2009, pp. 253–265.
78. C. A. Polyzois, A. Bhide, and D. M. Dias. Disk mirroring with alternating deferred updates. *VLDB'93*, Dublin, Ireland, August 1993, pp. 604–617.
79. L. Qiao, V. Raman, I. Narang, P. Pandey, D. D. Chambliss, G. Fuh, J. A. Ruddy, Y.-L. Chen, K.-H. Yang, and F.-L. Ling. Integration of server, storage and database stack: Moving processing towards data. *ICDE'08*, Cancun, Mexico, April 2008, pp. 1200–1208.
80. M. K. Qureshi, S. Gurumurthi, and B. Rajendran. *Phase Change Memory: From Devices to Systems.* Morgan & Claypool Publishers, 2011.
81. R. Ramakrishnan, and J. Gehrke. *Database Management Systems, 3rd edn.*, McGraw-Hill, 2003.
82. K. K. Rao, J. L. Hafner, and R. A. Golding. Reliability for networked storage nodes. *TDSC* 8(3): 404–418 (May 2011).
83. E. Riedel, C. Faloutsos, G. R. Ganger, and D. F. Nagle. Active disks for large scale data processing. *IEEE Comput.* 34(6): 68–74 (June 2001).
84. M. Rosenblum, and J. K. Ousterhout. The design and implementation of a log-structured file system. *TOCS* 10(1): 26–52 (February 1992).
85. C. Ruemmler, and J. Wilkes. An introduction to disk drive modeling. *IEEE Comput.* 27(3): 17–28 (March 1994).
86. K. Salzwedel. Algorithmic approaches for storage networks. *Algorithms for Memory Hierarchies, Advanced Notes LNCS 2625*, 2003, pp. 251–272.
87. J. R. Santos, R. R. Muntz, and B. A. Ribeiro-Neto. Comparing random data allocation and data striping in multimedia servers. *SIGMETRICS'00*, Santa Clara, CA, June 2000, pp. 44–55.

88. P. Sarkar, K. Voruganti, K. Z. Meth, O. Biran, and J. Satran. Internet protocol storage area networks. *IBM Syst. J.* 42(2): 218–231 (2003)
89. J. Schindler, J. L. Griffin. C. R. Lumb, and H. R. Gregory. Track-aligned extents: Matching access patterns to disk drive characteristics. *FAST'02*, Monterey, CA, January 2002, pp. 259–274.
90. S. W. Schlosser, J. Schindler, S. Papadomanolakis, M. Shao, A. Ailamaki, C. Faloutsos, and G. R. Ganger. On multidimensional data and modern disks. *FAST'05*, San Francisco, CA, December 2005, pp. 225–238.
91. B. Schroeder, and G. A. Gibson. Understanding disk failure rates: What does an MTTF of 1,000,000 hours mean to you? *TOS* 3(3): Article 8 (October 2007).
92. B. Schroeder, S. Damouras, and P. Gill. Understanding latent sector errors and how to protect against them. *TOS* 6(3): Article 9 (September 2010).
93. T. J. F. Schwarz, D. D. E. Long, J. F. Paris, and A. Amer. Increased reliability with SSPIRAL data layouts. *Proc. 16th Int'l Symp. on Modeling, Analysis, and Simulation of Computer and Telecomm. Systems (MASCOTS'08)*, Baltimore, MD, September 2008, pp. 189–198.
94. B. Seo, and R. Zimmermann. Efficient disk replacement and data migration algorithms for large disk subsystems. *TOS* 1(3): 316–345 (August 2005).
95. J. A. Solworth, and C. U. Orji. Distorted mirrors. *Proc. Parallel and Distributed Information Systems (PDIS)*, Miami Beach, FL, December 1991, pp. 10–17.
96. M. W. Storer, K. M. Greenan, E. L. Miller, and K. Voruganti. Pergamum: Replacing tape with energy efficient, reliable, disk-based archival storage. *FAST'08*, San Jose, CA, February 2008, 1–16.
97. M. W. Storer, K. M. Greenan, E. L. Miller, and K. Voruganti. POTSHARDS—A secure, recoverable, long-term archival storage system. *TOS* 5(2): Article 5 (May 2009).
98. E. Thereska, A. Donnelly, and D. Narayanan. Sierra: Practical power-proportionality for data center storage. *EuroSys'11*, Salzburg, Austria, April 2011, pp. 169–182.
99. A. Thomasian. Concurrency control schemes to support the concurrent processing of update transactions and read-only queries. *IBM Research Report RC 12420,* Yorktown Heights, NY, December 1986.
100. A. Thomasian. Rebuild options in RAID5 disk arrays. *Proc. 7th IEEE Symp. on Parallel and Distributed Systems (SPDS'95)*, San Antonio, TX, October 1995, 511–518.
101. A. Thomasian. Mirrored disk routing and scheduling. *Cluster Comput.* 9(4): 475–484 (October 2006).
102. A. Thomasian. Shortcut method for reliability comparisons in RAID. *J. Syst. Softw. (JSS)* 79(11): 1599–1605 (November 2006).
103. A. Thomasian. Publications on storage and systems research. *CAN* 37(4): 1–26 (September 2009).
104. A. Thomasian. Storage research in industry and universities. *CAN* 38(2): 1–48 (May 2010).
105. A. Thomasian. Survey and analysis of disk scheduling methods. *CAN* 39(2): 8–25 (May 2011).
106. A. Thomasian. Comment on DACO: A High Performance Disk Architecture. *TC* 61(4): 588–590 (March 2012).
107. A. Thomasian. Rebuild processing in RAID5 with emphasis on supplementary parity augmentation method [121]. *CAN* 40(2): 18–27 (May 2012).
108. A. Thomasian, and M. Blaum. Mirrored disk organization reliability analysis. *IEEE TC* 55(12): 1640–1644 (December 2006).
109. A. Thomasian, and M. Blaum. Higher reliability redundant disk arrays: Organization, operation, and coding. *TOS* 5(3): Article 7 (November 2009).
110. A. Thomasian, and C. Liu. Performance comparison of mirrored disk scheduling methods with a shared non-volatile cache. *DPDB* 18(3): 253–281 (November 2005).
111. A. Thomasian, and J. Menon. Performance analysis of RAID5 disk arrays with a vacationing server model for rebuild mode operation. *ICDE'94*, Houston, TX, February 1994, pp. 111–119.
112. A. Thomasian, and J. Menon. RAID5 performance with distributed sparing. *TPDS* 8(6): 640–657 (June 1997).

113. A. Thomasian, and Y. Tang. Performance, reliability, and performability of hybrid mirrored disk arrays and comparison with traditional RAID1 arrays. *Cluster Comput.*, published online June 21, 2012.
114. A. Thomasian, and J. Xu. Reliability and performance of mirrored disk organizations. *Comput. J.* 51(6): 615–629 (November 2008).
115. A. Thomasian, and J. Xu. X-Code double parity array operation with two disk failures. *Inform. Process. Lett. (IPL)* 111(12): 568–574 (June 2011).
116. A. Thomasian, and J. Xu. RAID level selection for heterogeneous disk arrays. *Cluster Comput.* 14(2): 115–127 (June 2011).
117. A. Thomasian, and J. Xu. Data allocation in heterogeneous disk arrays (HDA). *NAS'11*, Dalian, China, July 2011, pp. 82–91.
118. A. Thomasian, G. Fu, and S. W. Ng. Analysis of rebuild processing in RAID5 disk arrays. *Comput. J.* 50(2): 217–231 (March 2007).
119. A. Thomasian, G. Fu, and C. Han. Performance of two-disk failure-tolerant disk arrays. *TC* 56(6): 799–814 (June 2007).
120. A. Thomasian, Y. Tang, and Y. Hu. Hierarchical RAID: Design, performance, reliability, and recovery. *JPDC*, published online July 2012.
121. L. Tian, Q. Cao, H. Jiang, D. Feng, C. Xie, and Q. Xin. Online availability upgrades for parity-based RAIDs through supplementary parity augmentations. *TOS* 6(4): Article 17 (May 2011).
122. K. Trieber, and J. Menon. Simulation study of cached RAID5 designs. *HPCA'95*, Raleigh, NC, January 1995, pp. 186–197.
123. K. S. Trivedi. *Probability and Statistics with Reliability, Queuing, and Computer Science Applications.* Wiley, 2001.
124. Z Wang, A. G. Dimakis, and J. Bruck. Rebuilding for array codes in distributed storage systems. *Workshop on Application of Communication Theory to Emerging Memory Technologies (ACTEMT)*, Orlando, FL, December 2010.
125. B Welch, M. Unangst, Z. Abbasi, G. A. Gibson, B. Mueller, J. Small, J. Zelenka, and B. Zhou. Scalable performance of the Panasas Parallel File System. *FAST'08*, San Jose, CA, February 2008, pp. 17–33.
126. W. W. Wilcke, R. B. Garner, C. Fleiner, R. F. Freitas, R. A. Golding, J. S. Glider, D. R. Kenchammana-Hosekote, J. L. Hafner, K. M. Mohiuddin, K. K. Rao, R. A. Becker- Szendy, T. M. Wong, O. A. Zaki, M. Hernandez, K. R. Fernandez, H. Huels, H. Lenk, K. Smolin, M. Ries, C. Goettert, T. Picunko, B. J. Rubin, H. Kahn, T. Loo. IBM intelligent brick project-petabytes and beyond. *IBM J. R&D.* 50(2–3): 181–197 (March–May 2006).
127. J. Wilkes, R. A. Golding, C. Staelin, and T. Sullivan. The HP AutoRAID hierarchical storage system. *TOCS* 14(1): 108–136 (February 1996).
128. A. Wilner. Multiple drive failure tolerant RAID system. US Patent 6,327,672, LSI Logic Corp., San Jose, CA, December 2001.
129. U. H. Witten, A. Moffat, and T. C. Bell. *Managing Gigabytes: Compressing and Indexing Documents and Images, 2nd edition*, Morgan-Kauffman/Elsevier, 1999.
130. J. L. Wolf. The placement optimization program: A practical solution to the disk file assignment problem. *SIGMETRICS'89*, Berkeley, CA, February 1989, pp. 1–10.
131. B. L. B. L. Worthington, G. R. Ganger, and Y. N. Patt. Scheduling algorithms for modern disk drives. *SIGMETRICS '94*, Nashville, TN, May 1004, pp. 241–252
132. L. Xiang, Y. Xu, J. C. S. Lui, and Q. Chang. Optimal recovery of single disk failure in RDP code storage systems. *SIGMETRICS'10*, New York, NY, June 2010, pp. 119–130.
133. L. Xu, and J. Bruck. X-Code: MDS array codes with optimal encoding. *IEEE Trans. Inform. Theory* 45(1): 272–276 (January 1999).
134. Q. Zhu, Z. Chen, L. Tan, Y. Zhou, K. Keeton, and J. Wilkes Hibernator: Helping disk arrays sleep through the winter. *SOSP'05*, Brighton, U.K, October 2005, pp. 177–190.
135. B. Zhu, K. Li. and H. Patterson. Avoiding the disk bottleneck in the data domain deduplication file system. *FAST'08*, San Jose, CA, February 2008, pp. 269–282.

20

High-Speed Computer Arithmetic

Earl E. Swartzlander, Jr.
The University of Texas, Austin

20.1 Introduction

The speeds of memory and the arithmetic units are the primary determinants of the speed of a computer. Whereas the speed of both units depends directly on the implementation technology, arithmetic unit speed also depends strongly on the logic design. Even for an integer adder, the speed can easily vary by an order of magnitude, while the complexity varies by less than 50%.

This chapter begins with a discussion of binary fixed point number systems in Section 20.2. Section 20.3 provides examples of fixed point implementations of the four basic arithmetic operations (i.e., add, subtract, multiply, and divide). Finally, Section 20.4 briefly describes algorithms for floating-point arithmetic.

Regarding notation, capital letters represent digital numbers (i.e., words), while subscripted lower case letters represent bits of the corresponding word. The subscripts range from $n - 1$ to 0 to indicate the bit position within the word (x_{n-1} is the most significant bit of X, x_0 is the least significant bit [lsb] of X, etc.). The logic designs presented in this chapter are based on positive logic with AND, OR, and invert operations. Depending on the technology used for implementation, different logical operations (such as NAND and NOR) or direct transistor realizations may be used, but the basic concepts do not change significantly.

20.2 Fixed-Point Number Systems

Most arithmetic is performed with fixed point numbers that have constant scaling (i.e., the position of the binary point is fixed). The numbers can be interpreted as fractions, integers, or mixed numbers, depending on the application. Pairs of fixed point numbers are used to create floating-point numbers, as discussed in Section 20.4.

TABLE 20.1 Example of 4 Bit Fractional Fixed-Point Numbers

Number	Two's Complement	Sign–Magnitude
+7/8	0111	0111
+5/8	0101	0101
+1/2	0100	0100
+3/8	0011	0011
+1/4	0010	0010
+1/8	0001	0001
+0	0000	0000
−0	N/A	1000
−1/8	1111	1001
−1/4	1110	1010
−3/8	1101	1011
−1/2	1100	1100
−5/8	1011	1101
−3/4	1010	1110
−7/8	1001	1111
−1	1000	N/A

At the present time, fixed point binary numbers are generally represented using the two's complement number system. This choice has prevailed over the sign–magnitude and other number systems, because the frequently performed operations of addition and subtraction are easily performed on two's complement numbers. Sign–magnitude numbers are more efficient for multiplication, but the lower frequency of multiplication and the development of efficient two's complement multiplication algorithms have resulted in the nearly universal selection of the two's complement number system for most applications. Most of the algorithms presented in this chapter assume the use of two's complement numbers.

Fixed point number systems represent numbers, for example, A, by n bits: a sign bit and $n - 1$ data bits. By convention, the most significant bit a_{n-1} is the sign bit, which is a 1 for negative numbers and a 0 for positive numbers. The $n - 1$ data bits are $a_{n-2}, a_{n-3}, \ldots, a_1, a_0$. In the following material, fixed point fractions will be described for the two's complement and sign–magnitude systems.

Table 20.1 compares 4 bit fractional fixed point numbers in the two number systems. Note that the sign–magnitude number system has two zeros (i.e., positive zero and negative zero) and that only two's complement is capable of representing −1. For positive numbers, both number systems use identical representations.

20.2.1 Two's Complement Number System

In the two's complement fractional number system, the value of a number is the sum of $n - 1$ positive binary fractional bits and a sign bit that has a weight of −1:

$$A = -a_{n-1} + \sum_{i=0}^{n-2} a_i 2^{i-n+1} \tag{20.1}$$

Two's complement numbers are negated by complementing all bits and adding a 1 to the lsb position, for example, using 4 bit numbers to form −3/8:

+3/8 = 0011
invert all bits = 1100
add 1 lsb 0001
1101 = −3/8

Check: invert all bits = 0010
Add 1 lsb 0001
0011 = 3/8

20.2.2 Sign–Magnitude Number System

Sign–magnitude numbers consist of a sign bit and $n - 1$ bits that express the magnitude of the number:

$$A = (1 - 2a_{n-1}) + \sum_{i=0}^{n-2} a_i 2^{i-n+1} \tag{20.2}$$

Sign–magnitude numbers are negated by complementing the sign bit. For example, to form −3/8,

+3/8 = 0011
invert sign bit = 1011 = −3/8

Check: invert sign bit = 0011 = 3/8

A significant difference between the two number systems is their behavior under truncation. Figure 20.1 shows the effect of truncating high-precision fixed point fractions X, to form three-bit fractions $T(X)$. Truncation of two's complement numbers never increases the value of the number (i.e., the truncated numbers have values that are unchanged or shift toward negative infinity), as can be seen from

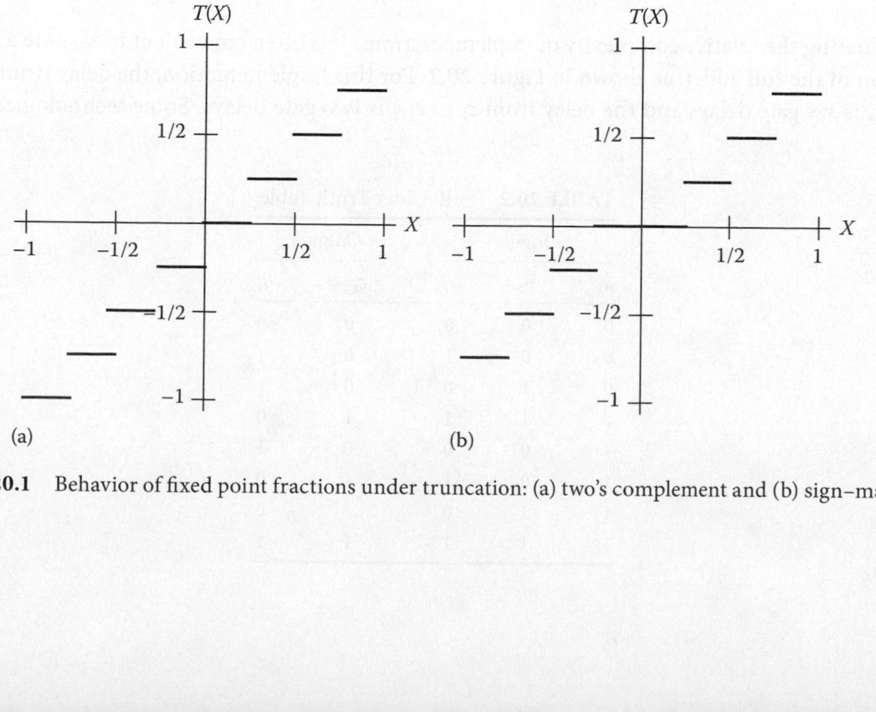

FIGURE 20.1 Behavior of fixed point fractions under truncation: (a) two's complement and (b) sign–magnitude.

Equation 20.1 since any truncated bits have positive weight. This bias can cause an accumulation of errors for computations that involve summing many truncated numbers (which may occur in scientific and signal processing applications). In the sign–magnitude number system, truncated numbers are unchanged or shifted toward zero, so that if approximately half of the numbers are positive and half are negative, the errors will tend to cancel.

20.3 Fixed-Point Arithmetic Algorithms

This section presents an assortment of typical fixed point algorithms for addition, subtraction, multiplication, and division.

20.3.1 Fixed-Point Addition

Addition is performed by summing the corresponding bits of the two n-bit numbers, including the sign bit. Subtraction is performed by summing the corresponding bits of the minuend and the two's complement of the subtrahend. Overflow is detected in a two's complement adder by comparing the carry signals into and out of the most significant adder stage (i.e., the stage that computes the sign bit). If the carries differ, an overflow has occurred and the result is invalid.

20.3.1.1 Full Adder

The full adder is the fundamental building block of most arithmetic circuits. Its operation is defined by the truth table shown in Table 20.2. The sum and carry outputs are described by the following equations:

$$s_k = c_{k+1} = a_k\overline{b_k}\,\overline{c_k} + \overline{a_k}b_k\overline{c_k} + \overline{a_k}\,\overline{b_k}c_k + a_kb_kc_k = a_k \oplus b_k \oplus c_k \tag{20.3}$$

$$c_{k+1} = \overline{a_k}b_kc_k + a_k\overline{b_k}c_k + a_kb_k\overline{c_k} + a_kb_kc_k = a_kb_k + a_kc_k + b_kc_k \tag{20.4}$$

where

a_k, b_k, and c_k are the inputs to the kth full adder stage
s_k and c_{k+1} are the sum and carry outputs, respectively
$\oplus$ is the exclusive-OR logic operation

In evaluating the relative complexity of implementations, it is often convenient to assume a nine gate realization of the full adder, as shown in Figure 20.2. For this implementation, the delay from either a_k or b_k to s_k is six gate delays and the delay from c_k to c_{k+1} is two gate delays. Some technologies, such as

TABLE 20.2 Full Adder Truth Table

Inputs			Outputs	
a_k	b_k	c_k	c_{k+1}	s_k
0	0	0	0	0
0	0	1	0	1
0	1	0	0	1
0	1	1	1	0
1	0	0	0	1
1	0	1	1	0
1	1	0	1	0
1	1	1	1	1

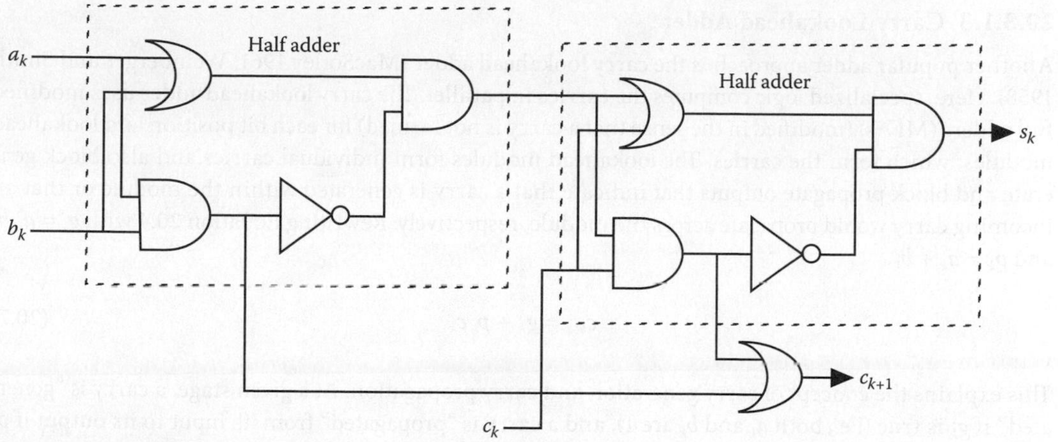

FIGURE 20.2 Nine gate full adder.

CMOS, form inverting gates (e.g., NAND and NOR gates) more efficiently than the noninverting gates that are assumed in this chapter. Circuits with equivalent speed and complexity can be constructed with inverting gates.

20.3.1.2 Ripple Carry Adder

A ripple carry adder for *n*-bit numbers can be implemented by concatenating *n* full adders, as shown in Figure 20.3. At the *k*th bit position, bits a_k and b_k of operands *A* and *B* and the carry signal from the preceding adder stage, c_k, are used to generate the *k*th bit of the sum, s_k, and the carry to the next stage, c_{k+1}. This is called a ripple carry adder, since the carry signals "ripple" from the lsb position to the most significant. If the ripple carry adder is implemented by concatenating *n* of the nine gate full adders, which were shown in Figure 20.2, an *n*-bit ripple carry adder requires $2n + 4$ gate delays to produce the most significant sum bit and $2n + 3$ gate delays to produce the carry output. A total of $9n$ logic gates are required to implement the *n*-bit ripple carry adder.

In comparing the delay and complexity of adders, the delay from data input to most significant sum output denoted by *DELAY* and the gate count denoted by *GATES* will be used. The *DELAY* and *GATES* values are subscripted by *RCA* to indicate that they are for a ripple carry adder. Although these simple metrics are suitable for first-order comparisons, more accurate comparisons require more exact modeling since the implementations may be effected with transistor networks (as opposed to gates), which will have different delay and complexity characteristics:

$$DELAY_{RCA} = 2n + 4 \tag{20.5}$$

$$GATES_{RCA} = 9n \tag{20.6}$$

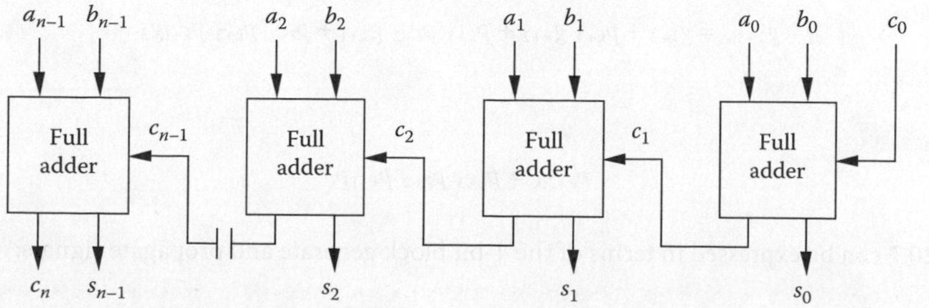

FIGURE 20.3 Ripple carry adder.

20.3.1.3 Carry Lookahead Adder

Another popular adder approach is the carry lookahead adder (MacSorley 1961, Weinberger and Smith 1958). Here, specialized logic computes the carries in parallel. The carry lookahead adder uses modified full adders (MFAs) (modified in the sense that a carry is not formed) for each bit position and lookahead modules, which form the carries. The lookahead modules form individual carries and also block generate and block propagate outputs that indicate that a carry is generated within the module or that an incoming carry would propagate across the module, respectively. Rewriting Equation 20.4 with $g_k = a_k\, b_k$ and $p_k = a_k + b_k$,

$$c_{k+1} = g_k + p_k\, c_k \tag{20.7}$$

This explains the concept of carry generation and carry propagation. At a given stage, a carry is "generated" if g_k is true (i.e., both a_k and b_k are 1), and a carry is "propagated" from its input to its output if p_k is true (i.e., either a_k or b_k is a 1). The nine gate full adder shown in Figure 20.2 has AND and OR gates that produce a_k and b_k with no additional complexity. Because the carry out is produced by the lookahead logic, the OR gate that produces c_{k+1} can be eliminated. The result is an eight gate MFA. Extending Equation 20.7 to a second stage,

$$\begin{aligned} c_{k+2} &= g_{k+1} + p_{k+1}c_{k+1} \\ &= g_{k+1} + p_{k+1}(g_k + p_k c_k) \\ &= g_{k+1} + p_{k+1}g_k + p_{k+1}p_k c_k \end{aligned} \tag{20.8}$$

Equation 20.8 results from evaluating Equation 20.7 for the $(k + 1)$th stage and substituting c_{k+1} from Equation 20.7. Carry c_{k+2} exits from stage $k + 1$ if: (1) a carry is generated there or (2) a carry is generated in stage k and propagates across stage $k + 1$ or (3) a carry enters stage k and propagates across both stages k and $k + 1$, etc. Extending to a third stage:

$$\begin{aligned} c_{k+3} &= g_{k+2} + p_{k+2}c_{k+2} \\ &= g_{k+2} + p_{k+2}(g_{k+1} + p_{k+1}g_k + p_{k+1}p_k c_k) \\ &= g_{k+2} + p_{k+2}g_{k+1} + p_{k+2}p_{k+1}g_k + p_{k+2}p_{k+1}p_k c_k \end{aligned} \tag{20.9}$$

Although it would be possible to continue this process indefinitely, each additional stage increases the size (i.e., the number of inputs) of the logic gates. Four inputs (as required to implement Equation 20.9) are frequently the maximum number of inputs per gate for current technologies. To continue the process, block generate and block propagate signals ($g_{k+3:k}$ and $p_{k+3:k}$), respectively, are defined over 4-bit blocks (stages k to $k + 3$):

$$g_{k+3:k} = g_{k+3} + p_{k+3}\, g_{k+2} + p_{k+3}\, p_{k+2}\, g_{k+1} + p_{k+3}\, p_{k+2}\, p_{k+1}g_k \tag{20.10}$$

and

$$p_{k+3:k} = p_{k+3}\, p_{k+2}\, p_{k+1}p_k \tag{20.11}$$

Equation 20.7 can be expressed in terms of the 4-bit block generate and propagate signals:

$$c_{k+4} = g_{k+3:k} + p_{k+3:k}c_k \tag{20.12}$$

Thus, the carry out from a 4-bit-wide block can be computed in only four gate delays (the first to compute p_i and g_i for $i = k$ through $k + 3$, the second to evaluate $p_{k+3:k}$, the second and third to evaluate $g_{k+3:k}$, and the third and fourth to evaluate c_{k+4} using Equation 20.12).

An n-bit carry lookahead adder implemented with r-bit blocks requires $\lceil (n - 1)/(r - 1) \rceil$ lookahead logic blocks. A 4 bit lookahead logic block is a direct implementation of Equations 20.7 through 20.11, requiring 14 logic gates. In general, an r-bit lookahead logic block requires 1/2 $(3r + r^2)$ logic gates. The Manchester carry chain (Kilburn et al. 1960) is an alternative switch-based technique for the implementation of a lookahead logic block.

Figure 20.4 shows the interconnection of 16 MFAs and five 4 bit lookahead logic blocks to realize a 16 bit carry lookahead adder. The sequence of events that occur during an add operation is as follows: (1) apply A, B, and carry in signals; (2) each adder computes P and G; (3) first-level lookahead logic computes the 4 bit propagate and generate signals; (4) second-level lookahead logic computes c_4, c_8, and c_{12}; (5) first-level lookahead logic computes the individual carries; and (6) each MFA computes the sum outputs. This process may be extended to larger adders by subdividing the large adder into 16 bit blocks and using additional levels of carry lookahead (e.g., a 64 bit adder requires three levels).

The delay of carry lookahead adders is evaluated by recognizing that an adder with a single level of carry lookahead (for 4 bit words if 4 bit blocks are used) has six gate delays and that each additional level of lookahead increases the maximum word size by a factor of four and adds four gate delays. More generally (Waser and Flynn 1982, pp. 83–88), the number of lookahead levels for an n-bit adder is $\lceil \log_r n \rceil$ where r is the maximum number of inputs per gate. Since an r-bit carry lookahead adder has six gate delays and there are four additional gate delays per carry lookahead level after the first,

$$DELAY_{CLA} = 2 + 4\lceil \log_r n \rceil \tag{20.13}$$

The complexity of an n-bit carry lookahead adder implemented with r-bit lookahead logic blocks is n MFAs (each of which requires eight gates) and $\lceil (n - 1)/(r - 1) \rceil$ lookahead logic blocks (each of which requires 1/2 $(3r + r^2)$ gates). In addition, two gates are used to calculate the carry out from the adder, c_n, from $p_{n-1:0}$ and $g_{n-1:0}$:

$$GATES_{CLA} = 8n + \frac{1}{2}(3r + r^2)\left\lceil \frac{n-1}{r-1} \right\rceil \tag{20.14}$$

For the common case of $r = 4$,

$$GATES_{CLA} \approx 12\tfrac{2}{3}n - 4\tfrac{2}{3} \tag{20.15}$$

The carry lookahead approach reduces the delay of adders from increasing in proportion to the word size (as is the case for ripple carry adders) to increasing in proportion to the logarithm of the word size. As with ripple carry adders, the carry lookahead adder complexity grows linearly with the word size (for $r = 4$, the complexity of a carry lookahead adder is about 40% greater than the complexity of a ripple carry adder). It is important to realize that most carry lookahead adders require gates with up to 4 inputs, while ripple carry adders use only inverters and 2 input gates.

20.3.1.4 Carry Skip Adder

The carry skip adder (Lehman and Burla 1961) divides the words to be added into blocks (like the carry lookahead adder). The basic structure of a 16 bit carry skip adder implemented with five 3 bit blocks and one 1-bit-wide block is shown in Figure 20.5. Within each block, a ripple carry adder is used to produce the sum bits and the carry (which is used as a block generate). In addition, an AND gate is used to form the block propagate signal. These signals are combined using Equation 20.12 to produce a fast carry signal.

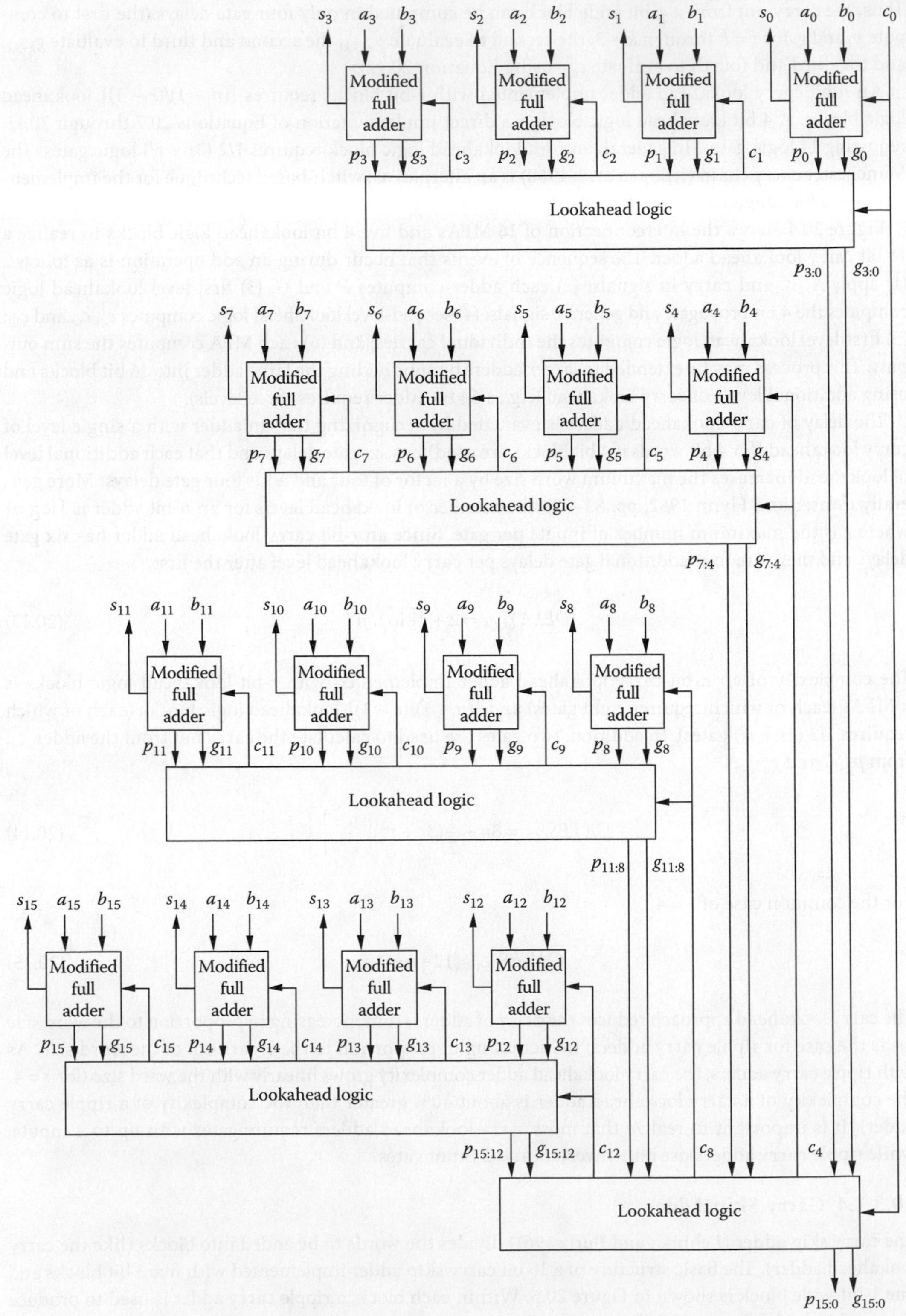

FIGURE 20.4 16 Bit carry lookahead adder.

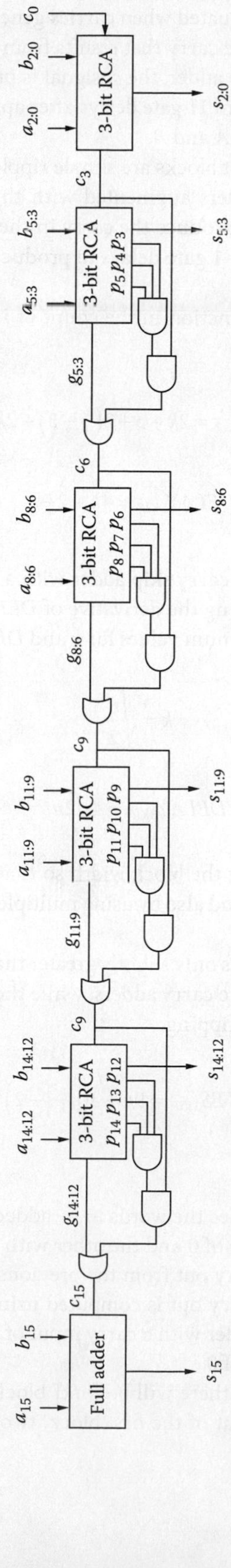

FIGURE 20.5 16 Bit carry skip adder.

For example, with $k = 4$, Equation 20.12 yields $c_8 = g_{7:4} + p_{7:4}\, c_4$. The carry out of the second ripple carry adder is a block generate signal if it is evaluated when carries generated by the data inputs (i.e., $a_{7:4}$ and $b_{7:4}$ in Figure 20.5) are valid, but before the carry that results from c_4. Normally, these two types of carries coincide in time, but in the carry skip adder, the c_4 signal is produced by a 4 bit ripple carry adder, so the carry output is a block generate from 11 gate delays after application of *A* and *B* until it becomes c_8 at 19 gate delays after the application of *A* and *B*.

In the carry skip adder, the first and last blocks are simple ripple carry adders, whereas the $\lceil n/k \rceil - 2$ intermediate blocks are ripple carry adders augmented with three gates. The delay of a carry skip adder is the sum of $2k + 3$ gate delays to produce the carry in the first block, two gate delays through each of the intermediate blocks, and $2k + 1$ gate delays to produce the most significant sum bit in the last block.

To simplify the analysis, the ceiling function in the count of intermediate blocks is ignored. If the block width is *k*,

$$DELAY_{SKIP} = 2k + 3 + 2\left(\tfrac{n}{k} - 2\right) + 2k + 1$$

$$DELAY_{SKIP} = 4k + 2\tfrac{n}{k} \tag{20.16}$$

where $DELAY_{SKIP}$ is the total delay of the carry skip adder with a single level of *k*-bit-wide blocks. The optimum block size is determined by taking the derivative of $DELAY_{SKIP}$ with respect to *k*, setting it to zero, and solving for *k*. The resulting optimum values for *k* and $DELAY_{SKIP}$ are

$$k = \sqrt{\frac{n}{2}} \tag{20.17}$$

$$DELAY_{SKIP} = 4\sqrt{2n} \tag{20.18}$$

Better results can be obtained by varying the block width so that the first and last blocks are smaller, while the intermediate blocks are larger, and also by using multiple levels of carry skip (Chen and Schlag 1990, Turrini 1989).

The complexity of the carry skip adder is only slightly greater than that of a ripple carry adder because the first and last blocks are standard ripple carry adders, while the intermediate blocks are ripple carry adders with three gates added for carry skipping.

$$GATES_{SKIP} = 9n + 3\left(\left\lceil \tfrac{n}{k} \right\rceil - 2\right) \tag{20.19}$$

20.3.1.5 Carry Select Adder

The carry select adder (Bedrij 1962) divides the words to be added into blocks and forms two sums for each block in parallel (one with a carry in of 0 and the other with a carry in of 1). As shown for a 16-bit carry select adder in Figure 20.6, the carry out from the previous block controls a multiplexer (MUX) that selects the appropriate sum. The carry out is computed using Equation 20.12, because the group propagate signal is the carry out of an adder with a carry input of 1 and the group generate signal is the carry out of an adder with a carry input of 0.

If a constant block width of *k* is used, there will be $\lceil n/k \rceil$ blocks, and the delay to generate the sum is $2k + 3$ gate delays to form the carry out of the first block, two gate delays for each of the $\lceil n/k \rceil - 2$

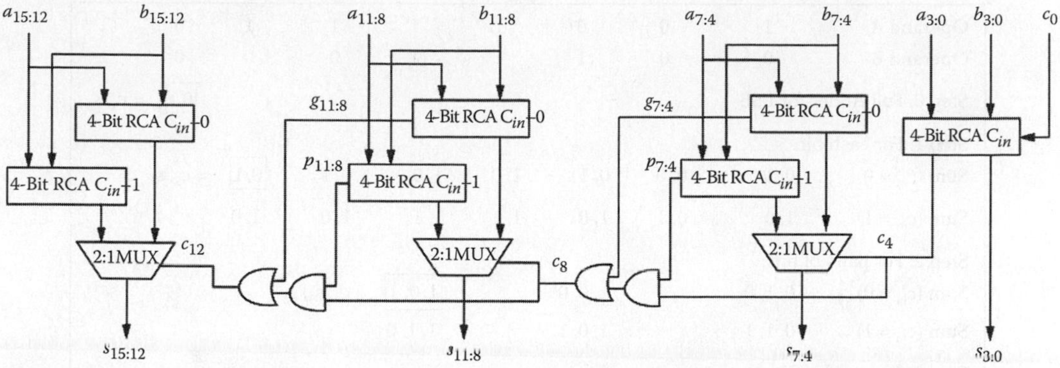

FIGURE 20.6 16 Bit carry select adder.

intermediate blocks, and three gate delays (for the MUX) in the final block. To simplify the analysis, the ceiling function in the count of intermediate blocks is ignored. The total delay is thus

$$DELAY_{C-SEL} = 8k + \frac{2n}{k} + 2 \tag{20.20}$$

where $DELAY_{C-SEL}$ is the total delay. The optimum block size is determined by taking the derivative of $DELAY_{C-SEL}$ with respect to k, setting it to zero, and solving for k. The result is

$$k = \sqrt{n} \tag{20.21}$$

$$DELAY_{C-SEL} = 2 + 4\sqrt{n} \tag{20.22}$$

As for the carry skip adder, better results can be obtained by varying the width of the blocks. In this case, the optimum is to make the two least significant blocks the same size and each successively more significant block 1 bit larger. In this configuration, the delay for each block's most significant sum bit will equal the delay to the MUX control signal (Goldberg 1990, p. A-38).

The complexity of the carry select adder is $2n - k$ ripple carry adder stages, the intermediate carry logic and $(\lceil n/k \rceil - 1)k$-bit-wide 2:1 MUXs:

$$GATES_{C-SEL} = 9(2n - k) + 2\left(\left\lceil \tfrac{n}{k} \right\rceil - 2\right) + 3(n - k) + \left\lceil \tfrac{n}{k} \right\rceil - 1$$

$$GATES_{C-SEL} = 21n - 12k + 3\left\lceil \tfrac{n}{k} \right\rceil - 5 \tag{20.23}$$

This is slightly more than twice the complexity of a ripple carry adder of the same size.

20.3.1.6 Conditional-Sum Adder

The conditional-sum adder (Sklansky 1960) can be viewed as a carry select adder where each block adder is constructed with smaller carry select adders that are constructed with still smaller carry select adders until the smallest carry select adders are 1 bit wide.

An example of the operation of an 8-bit conditional-sum adder is shown on Figure 20.7. In step 0, the lsb of the sum and c_1 is determined with a conventional full adder. In step 1, the conditional sums and carries for each remaining bit position are computed assuming a carry into the bit position of 0

Operand A	1	0	0	1	1	1	1	0	
Operand B	0	0	1	1	1	0	0	0	
Step 0: Full Adder for LSB								[0,0]	= c_1, s_0
Step 1: For each bit:									
Sum (c_{in} = 0)	0, 1	0, 0	0, 1	1, 0	1, 0	0, 1	[0,1]	= c_2, s_1	
Sum (c_{in} = 1)	1, 0	0, 1	1, 0	1, 1	1, 1	1, 0	1, 0		
Step 2: For pairs of bits:									
Sum (c_{in} = 0)	0, 1, 0		1, 0, 0		[1, 0, 1]	= c_4, s_3, s_2			
Sum (c_{in} = 1)	0, 1, 1		1, 0, 1		1, 1, 0				
Step 3: For quads:									
Sum (c_{in} = 0)	0, 1, 1, 0, 0								
Sum (c_{in} = 1)	[0, 1, 1, 0, 1] = c_8, s_7, s_6, s_5, s_4								
Thus: C_{out} = 0, Sum = 1, 1, 0, 1, 0, 1, 1, 0									

FIGURE 20.7 Conditional-sum addition example.

and assuming a carry into the bit position of 1. In step 2, pairs of conditional sums and carries from step 1 are combined to make groups of two adjacent sum bits and a carry out assuming a carry into the group of 0 and assuming a carry into the group of 1. This process of combining pairs of conditional-sum groups continues until there is only a single group.

As the groups are being computed, the correct least significant conditional group is selected by the carry from the previous stage.

Figure 20.8 shows an 8 bit conditional-sum adder. Along the top, there are seven modified half adders (MHAs), one at each bit position above the lsb, which has a full adder (FA). Each MHA produces two pairs of outputs, the sum, s_i^0, and carry, $c_{i+1}{}^0$, that are the sum and carry that would result if there is a carry of 0 into that bit (i.e., the sum and carry conditioned on a carry of 0 into the stage) and the sum, s_i^1, and carry, $c_{i+1}{}^1$, that are conditioned on a carry of 1 into the stage. Trees of MUXs are used to select the appropriate sum and carry outputs for the adder.

Assuming that the word size, n, is an integer power of 2, the delay of an n-bit conditional-sum adder, $DELAY_{COND}$, is given by

$$DELAY_{COND} = Delay\ C_1 + \log_2 n\ Delay\ MUX \tag{20.24}$$

The delay of the conditional-sum adder grows in proportion to the logarithm of the word size. This is attractive if fast (transistor level) MUXs are used. Although gate-level MUX designs have a delay of 3 from the select input to the output, transistor level designs can be significantly faster.

The complexity of the conditional-sum adder has two components, the MHAs that are proportional to the data word size and the MUXs that are roughly proportional to $n \log_2 n$ where n is the data word size.

20.3.2 Fixed-Point Subtraction

To produce an adder/subtracter, the adder is modified as shown in Figure 20.9 by including Exclusive-OR gates to complement operand B when performing subtraction. It forms either A + B or A − B. In the case of A + B, the mode selector is set to logic 0, which causes the Exclusive-OR gates to pass operand B through unchanged to the adder. The carry into the least significant adder stage is also set to ZERO, so standard addition occurs. Subtraction is implemented by setting the mode selector to logic ONE,

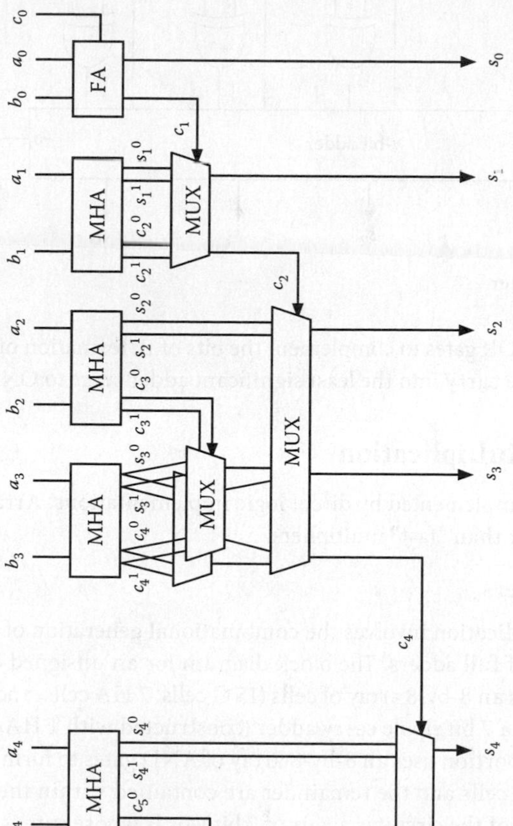

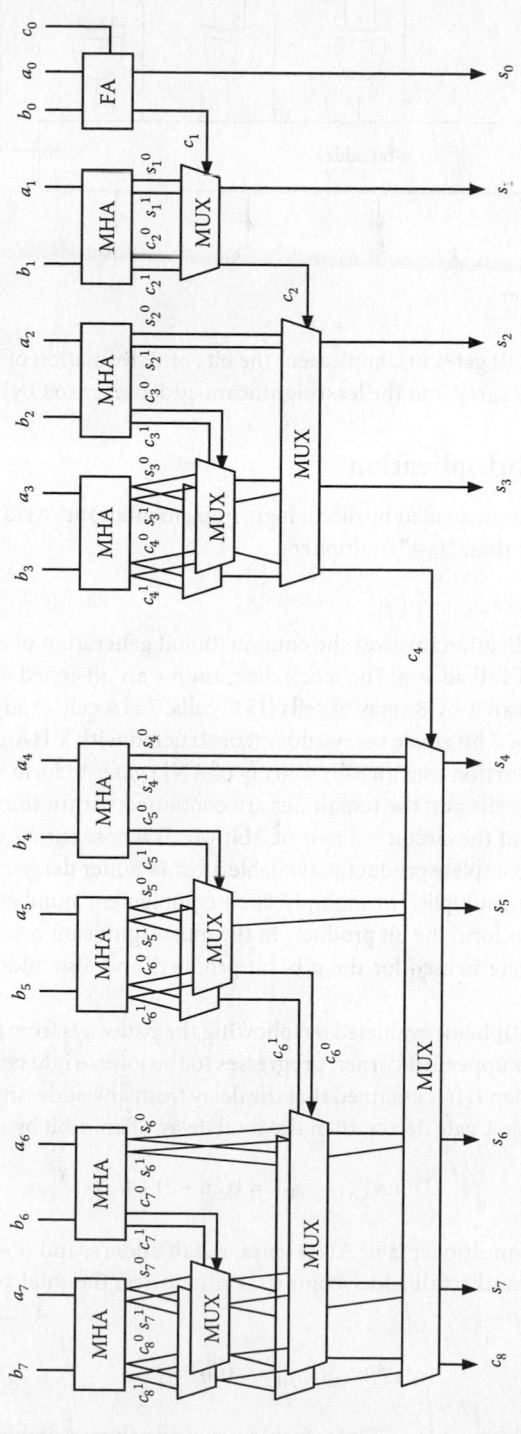

FIGURE 20.8 8-Bit conditional-sum adder.

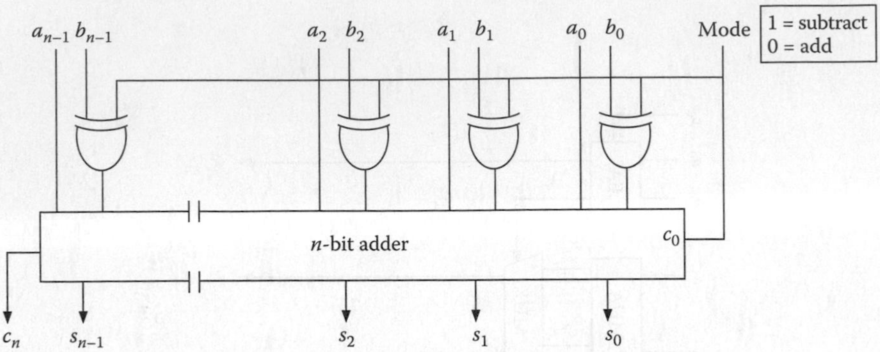

FIGURE 20.9 Adder/subtracter.

which causes the Exclusive-OR gates to complement the bits of B; formation of the two's complement of B is completed by setting the carry into the least significant adder stage to ONE.

20.3.3 Fixed-Point Multiplication

Multiplication is generally implemented by direct logic implementations. Array multipliers have a very regular layout but are slower than "fast" multipliers.

20.3.3.1 Array Multipliers

A direct approach to multiplication involves the combinational generation of all bit products and their summation with an array of full adders. The block diagram for an unsigned 8-by-8 array multiplier is shown in Figure 20.10. It has an 8-by-8 array of cells (15 G cells, 7 HA cells, and 42 FA cells) in the upper portion of the diagram and a 7 bit ripple carry adder (constructed with 1 HA cell and 6 FA cells) along the bottom row. The upper portion uses an 8-by-8 array of AND gates to form the bit products (some of the AND gates are in the G cells and the remainder are contained within the FA and HA blocks). The output of the upper portion of the circuit is a pair of 7 bit words whose sum is the most significant 8 bit of the product $p_8 - p_{15}$. The complete product is available after 14 adder delays.

Modification of the array multiplier to multiply two's complement numbers requires using NAND gates instead of AND gates to form the bit products in the most significant row (the b_7 row) and column (the a_7 column) (an AND gate is used for the a_7b_7 bit product) and also adding two correction terms (Parhami 2010, p. 223).

The delay of the array multiplier is evaluated by following the pathways from the inputs to the outputs. The longest path starts at the upper-left corner, progresses to the lower-right corner, and then across the bottom to the lower-left corner. If it is assumed that the delay from any adder input (for either half or full adders) to any adder output is k gate delays, then the total delay of an n-bit by n-bit array multiplier is

$$DELAY_{ARRAY\ MPY} = k(2n-2)+1 \tag{20.25}$$

The complexity of the array multiplier is n^2 AND gates, n half adders, and $n^2 - 2n$ full adders. If a half adder comprises four gates and a full adder comprises nine gates, the total complexity of an n-bit by n-bit array multiplier is

$$GATES_{ARRAY\ MPY} = 10n^2 - 14n \tag{20.26}$$

Array multipliers are easily laid out in a cellular fashion, making them suitable for very large scale integrated (VLSI) circuit implementation, where minimizing the design effort may be more important than maximizing the speed.

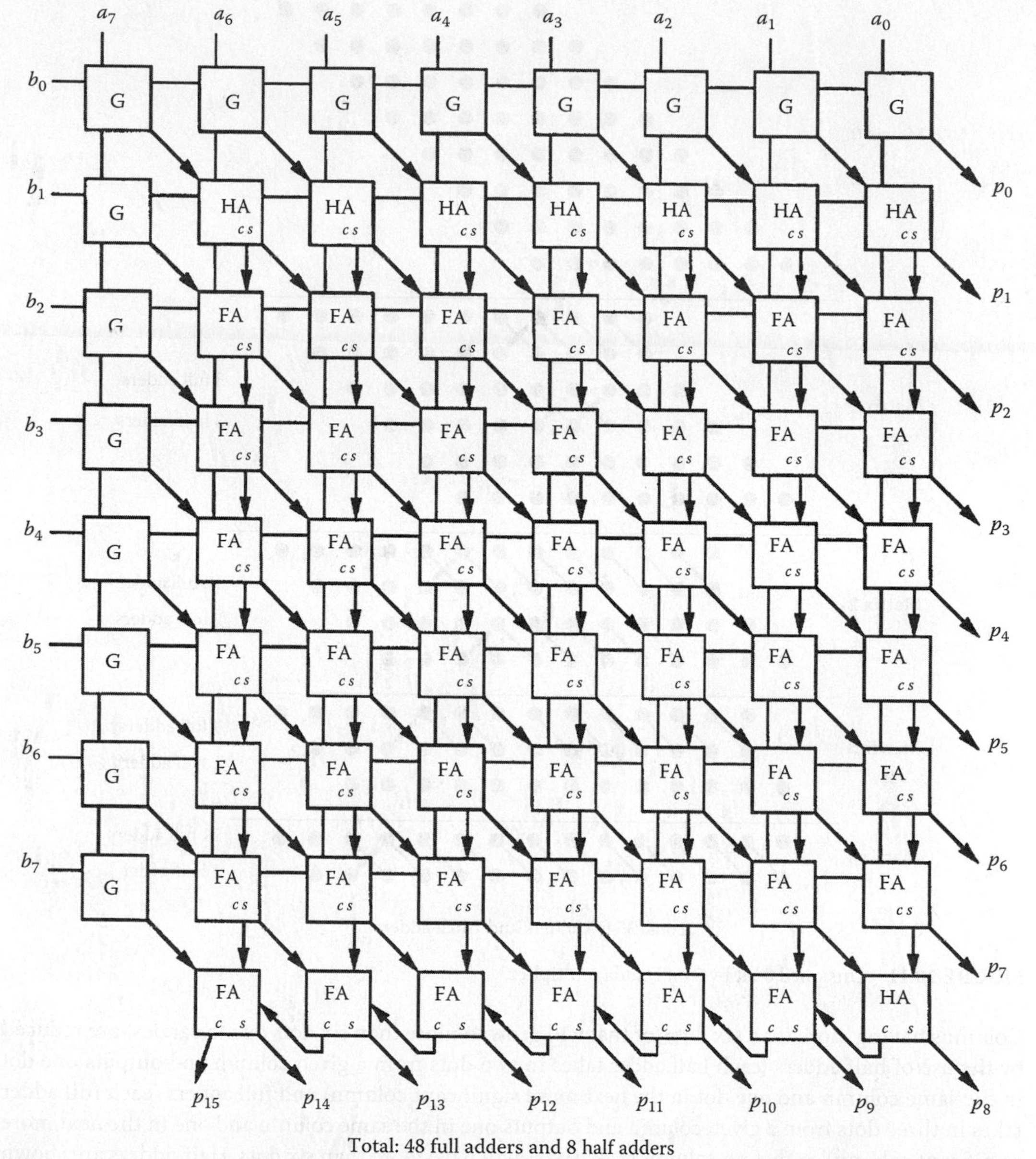

Total: 48 full adders and 8 half adders

FIGURE 20.10 Unsigned 8 bit by 8 bit array multiplier.

20.3.3.2 Wallace/Dadda Fast Multiplier

A method for fast multiplication was developed by Wallace (1964) and refined by Dadda (1965). With this method, a three-step process is used to multiply two numbers: (1) the bit products are formed, (2) the bit product matrix is reduced to a two-row matrix in which the sum of the rows equals the sum of the bit products, and (3) the two numbers are summed using a fast adder to produce the product. Although this may seem to be a complex process, it yields multipliers with delay proportional to the logarithm of the operand word size, which is faster than array multipliers that have delay that is proportional to the word size.

The second step in the fast multiplication process is shown for an unsigned 8-by-8 Dadda multiplier in Figure 20.11. An input 8-by-8 matrix of dots (each dot represents a bit product) is shown as Matrix 0.

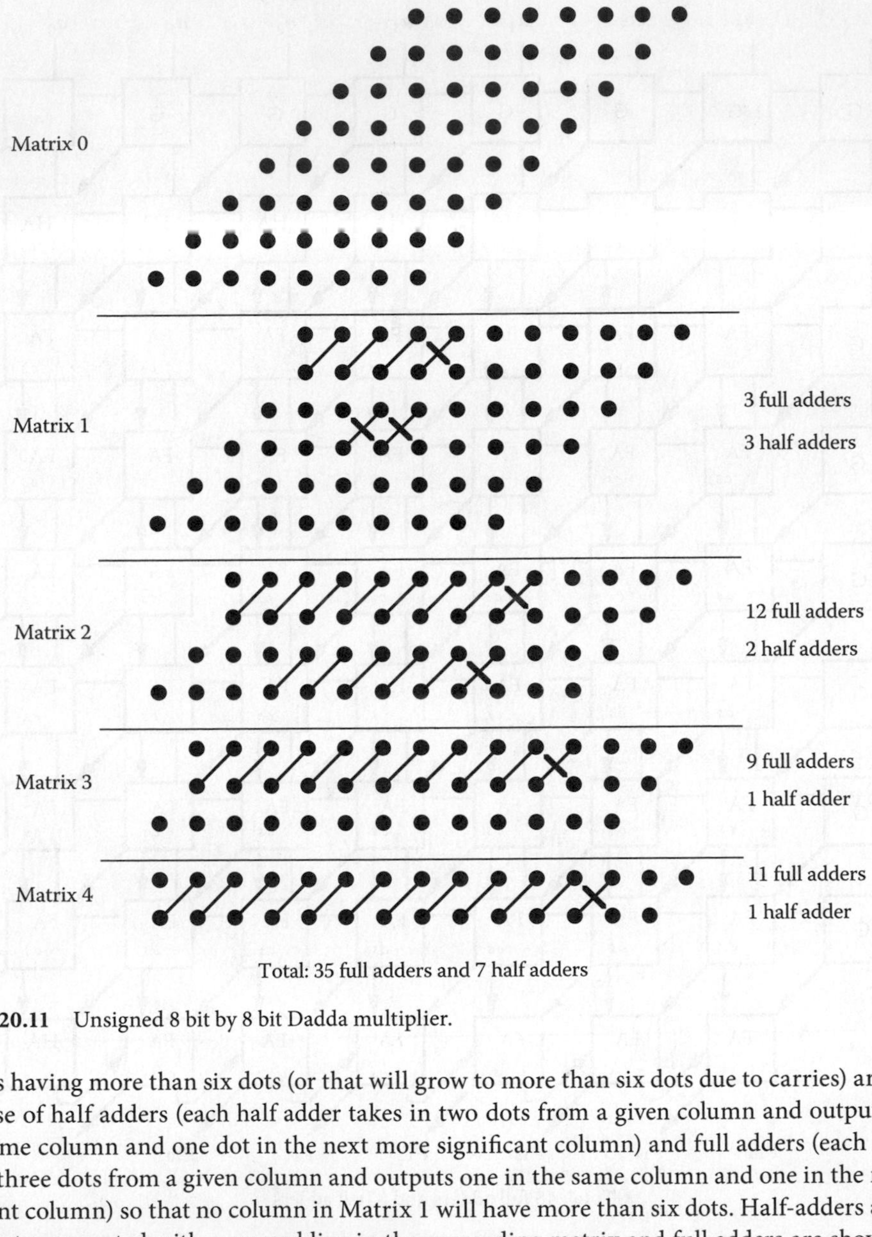

FIGURE 20.11 Unsigned 8 bit by 8 bit Dadda multiplier.

Columns having more than six dots (or that will grow to more than six dots due to carries) are reduced by the use of half adders (each half adder takes in two dots from a given column and outputs one dot in the same column and one dot in the next more significant column) and full adders (each full adder takes in three dots from a given column and outputs one in the same column and one in the next more significant column) so that no column in Matrix 1 will have more than six dots. Half-adders are shown by two dots connected with a crossed line in the succeeding matrix and full adders are shown by two dots connected with a line in the succeeding matrix. In each case, the rightmost dot of the pair connected by a line is in the column from which the inputs were taken for the adder. In the succeeding steps, reduction to Matrix 2 with no more than four dots per column, Matrix 3 with no more than three dots per column, and finally Matrix 4 with no more than two dots per column is performed. The height of the matrices is determined by working back from the final (two-row) matrix and limiting the height of each matrix to the largest integer that is no more than 1.5 times the height of its successor (i.e., 2, 3, 4, 6, 9, 13, 19, 28, 42, etc.). Each matrix is produced from its predecessor in one adder delay. Because the number of matrices is logarithmically related to the number of bits in the words to be multiplied, the delay of the matrix reduction process is proportional to log n. Because the adder that reduces the final two-row matrix can be implemented as a carry lookahead adder (which also has logarithmic delay), the total delay for this multiplier is proportional to the logarithm of the word size.

The delay of a Dadda multiplier is evaluated by following the pathways from the inputs to the outputs. The longest path starts at the center column of bit products (which require one gate delay to be formed) and progresses through the successive reduction matrices (which requires approximately $\log_{1.44}(n) - 2$ full adder delays) and finally through the $2n - 2$-bit carry propagate adder. If the delay from any adder input (for either half or full adders) to any adder output is k gate delays and if the carry propagate adder is realized with a carry lookahead adder implemented with 4-bit lookahead logic blocks (with delay given by Equation 20.13), the total delay (in gate delays) of an n-bit by n-bit Dadda multiplier is

$$DELAY_{DADDA\ MPY} = 1 + k(\log_{1.44}(n) - 2) + 2 + 4\lceil \log_r(2n-2) \rceil \quad (20.27)$$

The complexity of a Dadda multiplier is determined by evaluating the complexity of its parts. There are n^2 gates ($2n - 2$ are NAND gates, the rest are AND gates) to form the bit product matrix, $(n - 2)^2$ full adders, $n - 1$ half adders, and a $2n - 2$ bit carry propagate adder for the addition of the final two-row matrix. If the carry propagate adder is realized with a carry lookahead adder (implemented with 4 bit lookahead logic blocks) and if the complexity of a full adder is nine gates and the complexity of a half adder (either regular or special) is four gates, then the total complexity is

$$GATES_{DADDA\ MPY} = 10n^2 - 6\frac{2}{3}n - 26 \quad (20.28)$$

The Wallace multiplier is very similar to the Dadda multiplier, except that it does more reduction in the first stages of the reduction process, it uses more half adders, and it uses a slightly smaller carry propagating adder. A dot diagram for an unsigned 8 bit by 8 bit Wallace multiplier is shown on Figure 20.12. This reduction (which like the Dadda reduction requires four full adder delays) is followed by an 8-bit carry propagating adder. The total complexity of the Wallace multiplier is slightly greater than the total complexity of the Dadda multiplier. In most cases the Wallace and Dadda multipliers have the same delay.

20.3.4 Fixed-Point Division

There are two types of division algorithms in common use: digit recurrence and convergence methods. The digit recurrence approach computes the quotient on a digit-by-digit basis; hence, they have a delay proportional to the precision of the quotient. In contrast, the convergence methods compute an approximation that converges to the value of the quotient. For the common algorithms, the convergence is quadratic, meaning that the number of accurate bits approximately doubles on each iteration.

The digit recurrence methods that use a sequence of shift, add, or subtract and compare operations are relatively simple to implement. On the other hand, the convergence methods use multiplication on each cycle. This means higher hardware complexity, but if a fast multiplier is available, potentially higher speed.

20.3.4.1 Digit Recurrent Division

The digit recurrent algorithms (Robertson 1958) based on selecting digits of the quotient Q (where $Q = N/D$) to satisfy the following equation:

$$P_{k+1} = r P_k - q_{n-k-1} D \quad \text{for } k = 1, 2, \ldots, n-1 \quad (20.29)$$

where

P_k is the partial remainder after the selection of the kth quotient digit
$P_0 = N$ (subject to the constraint $|P_0| < |D|$)
r is the radix
q_{n-k-1} is the kth quotient digit to the right of the binary point
D is the divisor

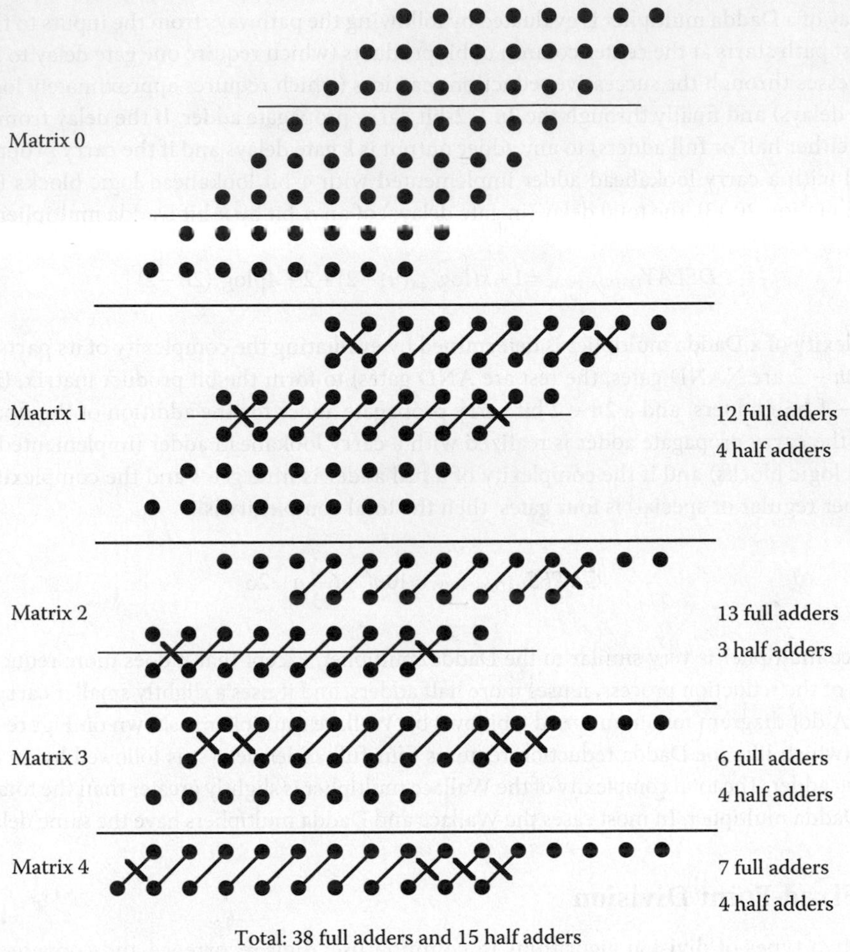

FIGURE 20.12 Unsigned 8 bit by 8 bit Wallace multiplier.

In this subsection, it is assumed that both N and D are positive; see Ercegovac and Lang (1994) for details on handling the general case.

20.3.4.2 Binary SRT Divider

The binary SRT division process (also known as radix-2 SRT division) selects the quotient from three candidate quotient digits {±1, 0}. The divisor is restricted to $0.5 \leq D < 1$. A flowchart of the basic binary SRT scheme is shown in Figure 20.13. Block 1 initializes the algorithm. In steps 3 and 5, $2P_k$ is used to select the quotient digit. In steps 4 and 6, $P_{k+1} = 2P_k - q\,D$. Step 8 tests whether all bits of the quotient have been formed and goes to step 2 if more need to be computed. Each pass through steps 2–8 forms one digit of the quotient. The result upon exiting from step 8 is a collection of n signed binary digits.

Step 9 converts the n digit signed digit number into an n-bit two's complement number by subtracting N, which has a 1 for each bit position where $q_i = -1$ and 0 elsewhere, from P, which has a 1 for each bit position where $q_i = 1$ and 0 elsewhere. For example,

$$Q = 0\,.1\;1\;-1\;0\;1 = 21/32$$

$$P = 0\,.1\;1\;0\;0\;1$$

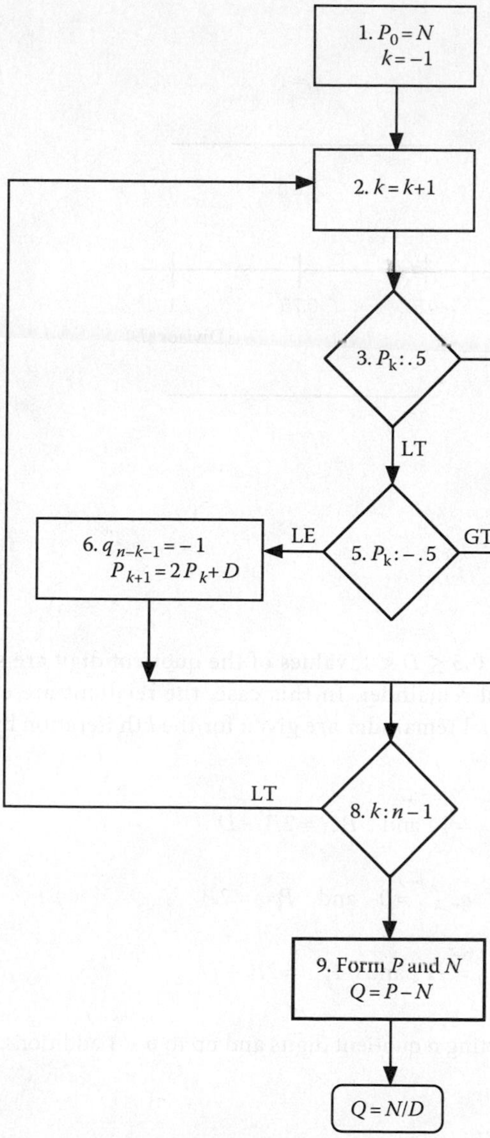

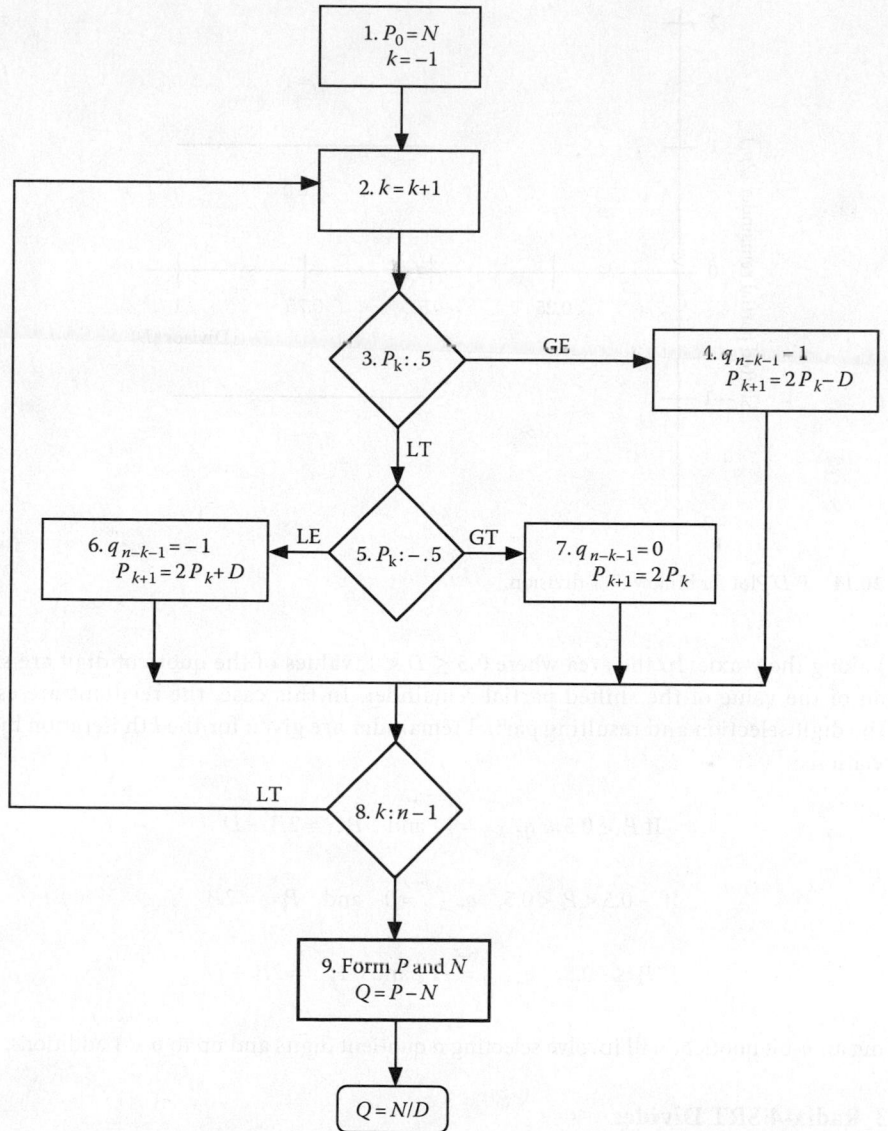

FIGURE 20.13 Flowchart of binary SRT division.

$$N = 0.0\,0\,1\,0\,0$$

$$Q = 0.1\,1\,0\,0\,1 - 0.0\,0\,1\,0\,0$$

$$Q = 0.1\,1\,0\,0\,1 + 1.1\,1\,1\,0\,0$$

$$Q = 0.1\,0\,1\,0\,1 = 21/32$$

The selection of the quotient digit can be visualized with a *P-D* plot such as the one shown in Figure 20.14. The plot shows the divisor along the x-axis and the shifted partial remainder (in this

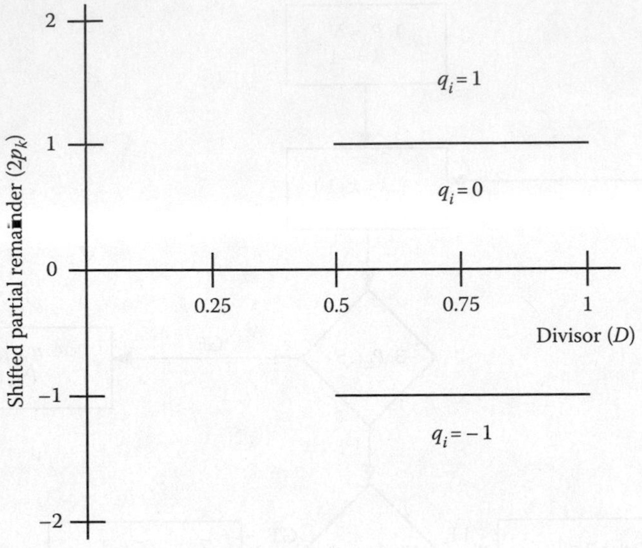

FIGURE 20.14 *P-D* Plot for binary SRT division.

case $2P_k$) along the y-axis. In the area where $0.5 \le D < 1$, values of the quotient digit are shown as a function of the value of the shifted partial remainder. In this case, the relations are especially simple. The digit selection and resulting partial remainder are given for the *k*th iteration by the following relations:

$$\text{If } P_k \ge 0.5, \quad q_{n-k-1} = 1 \quad \text{and} \quad P_{k+1} = 2P_k - D \tag{20.30}$$

$$\text{If } -0.5 < P_k < 0.5, \quad q_{n-k-1} = 0 \quad \text{and} \quad P_{k+1} = 2P_k \tag{20.31}$$

$$\text{If } P_k \le -0.5, \quad q_{n-k-1} = -1 \quad \text{and} \quad P_{k+1} = 2P_k + D \tag{20.32}$$

Computing an n-bit quotient will involve selecting n quotient digits and up to $n + 1$ additions.

20.3.4.3 Radix-4 SRT Divider

The higher radix SRT division process is similar to the binary SRT algorithm. Radix 4 is the most common higher radix SRT division algorithm with either a minimally redundant digit set of {±2, ±1, 0} or the maximally redundant digit set of {±3, ±2, ±1, 0}. The operation of the algorithm is similar to the binary SRT algorithm shown on Figure 20.13, except that in steps 3 and 5, $4P_k$ and D are used to determine the quotient digit. A *P-D* Plot is shown on Figure 20.15 for the maximum redundancy version of radix-4 SRT division. There are seven possible values for the quotient digit at each stage. The test for completion in step 8 becomes $k: n/2 - 1$. Also, the conversion to two's complement in step 9 is modified slightly since each quotient digit occupies two bits of the P and N numbers that are used to form the two's complement number.

20.3.4.4 Newton–Raphson Divider

The second category of division techniques uses a multiplication-based iteration to compute a quadratically convergent approximation to the quotient. In systems that include a fast multiplier, this process may be faster than the digit recurrent methods (Ferrari 1967). One popular approach is the Newton–Raphson

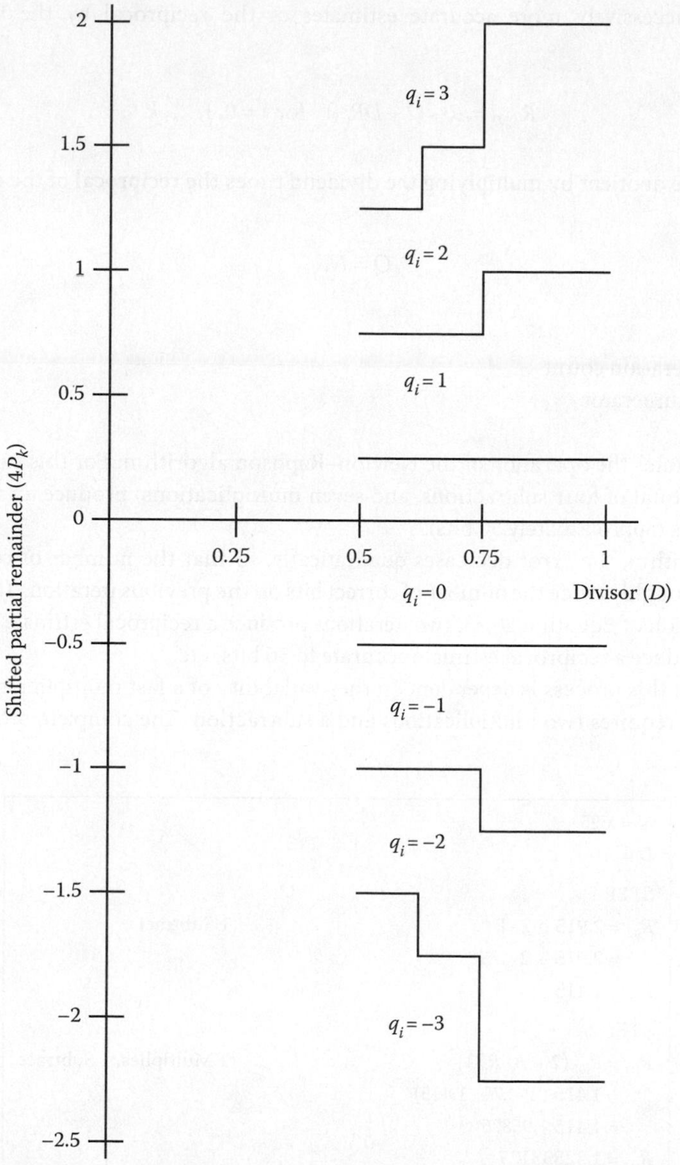

FIGURE 20.15 *P–D* Plot for radix-4 maximally redundant SRT division.

algorithm that computes an approximation to the reciprocal of the divisor that is then multiplied by the dividend to produce the quotient. The process to compute $Q = N/D$ consists of three steps:

1. Calculate a starting estimate of the reciprocal of the divisor, $R_{(0)}$. If the divisor, D, is normalized (i.e., $1/2 \leq D < 1$), then $R_{(0)} = 3 - 2D$ exactly computes $1/D$ at $D = .5$ and $D = 1$ and exhibits maximum error (of approximately 0.17) at $D = \sqrt{1/2}$. Adjusting $R_{(0)}$ downward to by half the maximum error gives

$$R_{(0)} = 2.915 - 2D \tag{20.33}$$

This produces an initial estimate, that is, within about 0.087 of the correct value for all points in the interval $1/2 \leq D < 1$.

2. Compute successively more accurate estimates of the reciprocal by the following iterative procedure:

$$R_{(i+1)} = R_{(i)}\ (2 - DR_{(i)}) \quad \text{for } i = 0, 1, \ldots, k \tag{20.34}$$

3. Compute the quotient by multiplying the dividend times the reciprocal of the divisor:

$$Q = NR_{(k)} \tag{20.35}$$

where
 k is the iteration count
 N is the numerator

Figure 20.16 illustrates the operation of the Newton–Raphson algorithm. For this example, three iterations (involving a total of four subtractions, and seven multiplications) produce an answer accurate to nine decimal digits (approximately 30 bits).

With this algorithm, the error decreases quadratically, so that the number of correct bits in each approximation is roughly twice the number of correct bits on the previous iteration. Thus, from the $3\frac{1}{2}$ bit initial approximation of Equation 20.33, two iterations produce a reciprocal estimate accurate to 14 bits, four iterations produce a reciprocal estimate accurate to 56 bits, etc.

The efficiency of this process is dependent on the availability of a fast multiplier, since each iteration of Equation 20.34 requires two multiplications and a subtraction. The complete process for the initial

$N = .625$
$D = .75$

STEP 1

$R_{(0)} = 2.915 - 2 \cdot B$ — 1 Subtract
$\quad = 2.915 - 2 \cdot .75$
$R_{(0)} = 1.415$

STEP 2

$R_{(1)} = R_{(0)}\ (2 - B \cdot R_{(0)})$ — 2 Multiplies, 1 Subtract
$\quad = 1.415\ (2 - .75 \cdot 1.415)$
$\quad = 1.415 \cdot .95875$
$R_{(1)} = 1.32833125$

$R_{(2)} = R_{(1)}\ (2 - B \cdot R_{(1)})$ — 2 Multiplies, 1 Subtract
$\quad = 1.32833125\ (2 - .75 \cdot 1.32833125)$
$\quad = 1.32833125 \cdot 1.00375156$
$R_{(2)} = 1.3333145677$

$R_{(3)} = R_{(2)}\ (2 - B \cdot R_{(2)})$ — 2 Multiplies, 1 Subtract
$\quad = 1.3333145677\ (2 - .75 \cdot 1.3333145677)$
$\quad = 1.3333145677 \cdot 1.00001407$
$R_{(3)} = 1.3333333331$

STEP 3

$Q = N \cdot R_{(3)}$ — 1 Multiply
$Q = .83333333319$

FIGURE 20.16 Example of Newton–Raphson division.

$N_0 = N = 0.625$	
$D_0 = D = 0.75$	
$F_1 = 1.3$	1 Table Look Up
$N_1 = N_0 \cdot F_1 = 0.625 \cdot 1.3 = 0.8125$	1 Multiply
$D_1 = D_0 \cdot F_1 = 0.75 \cdot 1.3 = 0.975$	1 Multiply
$F_2 = 2 - D_1 = 2 - 0.975 = 1.025$	1 Subtract
$N_2 = N_1 \cdot F_2 = 0.8125 \cdot 1.025 = 0.8328125$	1 Multiply
$D_2 = D_1 \cdot F_2 = 0.975 \cdot 1.025 = 0.999375$	1 Multiply
$F_3 = 2 - D_2 = 2 - 0.999375 = 1.000625$	1 Subtract
$N_3 = N_2 \cdot F_3 = 0.8328125 \cdot 1.000625 = 0.83333300781$	1 Multiply
$D_3 = D_2 \cdot F_3 = 0.999375 \cdot 1.000625 = 0.99999960937$	1 Multiply
$Q = N_3 = 0.83333300781$	

FIGURE 20.17 Example of Goldschmidt division.

estimate, three iterations, and the final quotient determination requires four subtraction operations and seven multiplication operations to produce a 16 bit quotient. This is faster than a conventional nonrestoring divider if multiplication is roughly as fast as addition, a condition that is satisfied for some systems that include a hardware multiplier.

20.3.4.5 Goldschmidt Divider

The Goldschmidt divider (Anderson et al. 1967) is another division algorithm that forms a quadratically convergent approximation to the quotient. It treats the division problem as a fraction whose numerator and denominator are multiplied by a succession of numbers so that the final denominator is approximately 1 and the final numerator is approximately equal to the quotient.

The steps to compute a quotient using the Goldschmidt method are as follows:

1. Scale the numerator, N_0, and the denominator, D_0, so that the value of the denominator is "close" to 1 (e.g., $\frac{3}{4} < D_0 < 1\frac{1}{4}$). Use a table to find a factor, F_1, that will make $D_0{*}F_1 \approx 1$. Set $i = 0$.
2. Multiply N_i and D_i by F_{i+1} forming N_{i+1} and D_{i+1}, respectively.
3. Increment i. Calculate $F_{i+1} = 2 - D_i$.

Repeat steps 2 and 3 until $D_n = 1$ within the desired precision.

This process is illustrated by the decimal example shown in Figure 20.17.

Like the Newton–Raphson algorithm, the value of N_k quadratically approaches the value of the quotient. If F_1 in step 1 results in $D_1 \approx 1$ within a maximum error, the number of iterations to achieve a specific accuracy of the quotient can be predetermined. In the example, D_1 is within 2.5% of 1, D_2 is within 0.06%, etc. Thus, each iteration (which requires 1 subtract and 2 multiplies) doubles the number of accurate bits to which N_k approximates the quotient. If the initial accuracy is known, the final multiplication that forms D_k is not needed.

20.4 Floating-Point Arithmetic

Recent advances in VLSI have increased the feasibility of hardware implementations of floating-point arithmetic units. The main advantage of floating-point arithmetic is that its wide dynamic range virtually eliminates overflow for most applications.

20.4.1 Floating-Point Number Systems

A floating-point number, A, consists of a significand (or mantissa), S_a, and an exponent, E_a. The value of a number, A, is given by the equation

$$A = S_a r^{Ea} \tag{20.36}$$

where r is the radix (or base) of the number system. The significand is generally normalized by requiring that the most significant digit be nonzero. Use of the binary radix (i.e., $r = 2$) gives maximum accuracy but may require more frequent normalization than higher radices.

The IEEE Standard 754 single precision (32-bit) floating-point format, which is widely implemented, has an 8 bit biased integer exponent that ranges between 0 and 255 (IEEE 2008). The exponent is expressed in excess 127 code so that its effective value is determined by subtracting 127 from the stored value. Thus, the range of effective values of the exponent is −127 to 128, corresponding to stored values of 0 to 255, respectively. A stored exponent value of 0 ($E_{\min}$) serves as a flag indicating that the value of the number is 0 if the significand is 0 and for denormalized numbers if the significand is nonzero. A stored exponent value of 255 ($E_{\max}$) serves as a flag indicating that the value of the number is infinity if the significand is 0 or "not a number" if the significand is nonzero. The significand is a 25 bit sign–magnitude mixed number (the binary point is to the right of the most significant bit). The leading bit of the significand is always a 1 (except for denormalized numbers). As a result, when numbers are stored, the leading bit can be omitted, giving an extra bit of precision. More details on floating-point formats and on the various considerations that arise in the implementation of floating-point arithmetic units are given in Gosling (1980), Goldberg (1990), Parhami (2010), Flynn and Oberman (2001), and Koren (2002).

20.4.1.1 Floating-Point Addition

A flow chart for the functions required for floating-point addition is shown in Figure 20.18. For this flowchart, the operands are assumed to be IEEE Standard 754 single precision numbers that have been "unpacked" and normalized with significand magnitudes in the range [1, 2) and exponents in the range [−126, 127].

On the flow chart, the operands are (E_a, S_a) and (E_b, S_b), the result is (E_s, S_s), and the radix is 2. In step 1, the operand exponents are compared; if they are unequal, the significand of the number with the smaller exponent is shifted right in step 3 or 4 by the difference in the exponent values to properly align the significands. For example, to add the decimal operands 0.867×10^5 and 0.512×10^4, the latter would be shifted right by one digit and 0.867 added to 0.0512 to give a sum of 0.9182×10^5. The addition of the significands is performed in step 5. Step 6 tests for overflow and step 7 corrects, if necessary, by shifting the significand one position to the right and incrementing the exponent. Step 9 tests for a zero significand. The loop of steps 10–11 scales unnormalized (but nonzero) significands upward to normalize the result. Step 12 tests for underflow.

It is important to recognize that the flowchart shows the basic functions that are required, but not necessarily the best way to perform those functions. For example, the loop in steps 10–11 is not a good way to normalize the significand of the sum. It is a serial loop that is executed a variable number of times depending on the data value. A much better way is to determine the number of leading zeros in the significand and then use a barrel shifter to do the shift in a single step in hardware.

Floating-point subtraction is implemented with a similar algorithm. Many refinements are possible to improve the speed of the addition and subtraction algorithms, but floating-point addition and subtraction will, in general, be much slower than fixed point addition as a result of the need for operand alignment and result normalization and rounding.

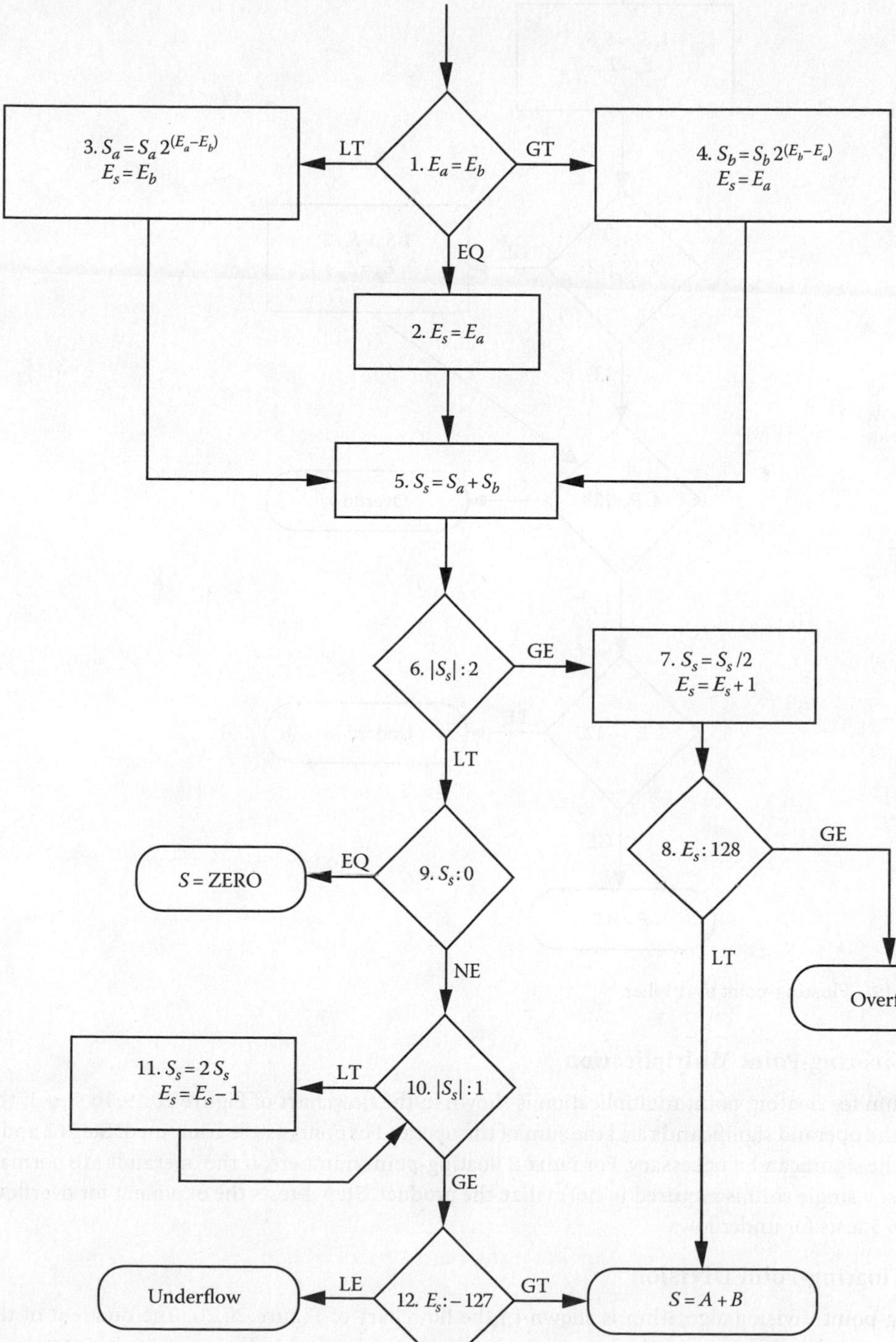

FIGURE 20.18 Floating-point adder.

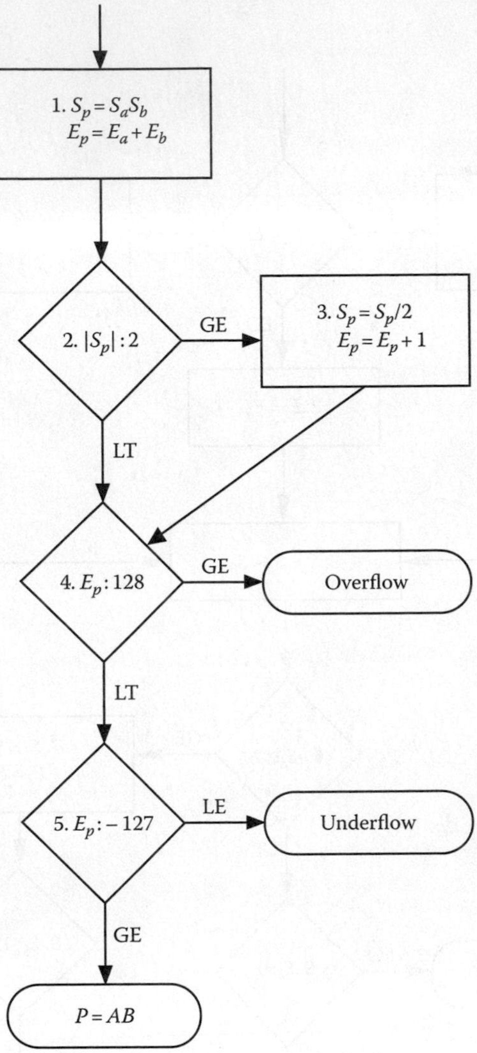

FIGURE 20.19 Floating-point multiplier.

20.4.1.2 Floating-Point Multiplication

The algorithm for floating-point multiplication is shown in the flowchart of Figure 20.19. In step 1, the product of the operand significands and the sum of the operand exponents are computed. Steps 2 and 3 normalize the significand if necessary. For radix-2 floating-point numbers, if the operands are normalized, at most a single shift is required to normalize the product. Step 4 tests the exponent for overflow. Finally, step 5 tests for underflow.

20.4.1.3 Floating-Point Division

The floating-point division algorithm is shown in the flowchart of Figure 20.20. The quotient of the significands and the difference of the exponents are computed in step 1. The quotient is normalized (if necessary) in steps 2 and 3 by shifting the quotient significand, while the quotient exponent is adjusted appropriately. For radix-2, if the operands are normalized, only a single shift is required to normalize the quotient. The computed exponent is tested for underflow in step 4. Finally, the fifth step tests for overflow.

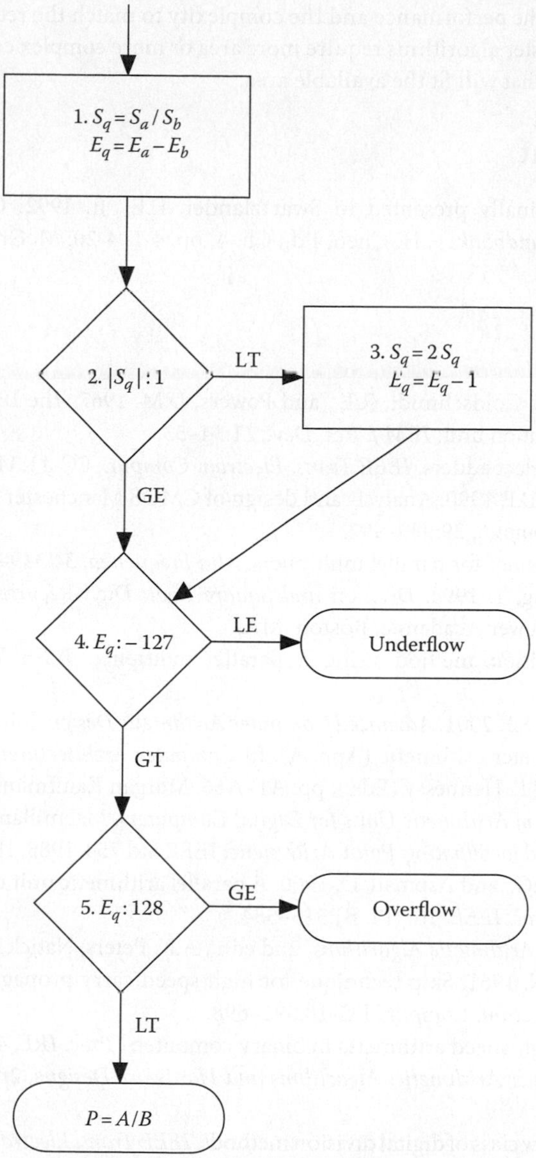

FIGURE 20.20 Floating-point divider.

20.4.1.4 Floating-Point Rounding

All floating-point algorithms may require rounding to produce a result in the correct format. A variety of alternative rounding schemes have been developed for specific applications. Round to nearest with ties to even, round to nearest with ties away from 0, round toward ∞, round toward −∞, and round toward 0 are all available in implementations of the IEEE floating-point standard (IEEE 2008).

20.5 Conclusions

This chapter has presented an overview of the two's complement number system; algorithms for the basic integer arithmetic operations of addition, subtraction, multiplication, and division; and a very brief discussion of floating-point operations. When implementing arithmetic units, there is often an opportunity to optimize the performance and the complexity to match the requirements of the specific application. In general, faster algorithms require more area or more complex control; it is often useful to use the fastest algorithm that will fit the available area.

Acknowledgment

Revision of chapter originally presented in Swartzlander, E.E., Jr. 1992. Computer arithmetic. In *Computer Engineering Handbook*. C.H. Chen, Ed., Ch. 4, pp. 4-1–4-20. McGraw-Hill, New York. With permission.

References

Anderson, S.F., Earle, J.G., Goldschmidt, R.E., and Powers, D.M. 1967. The IBM System/360 Model 91: Floating-point execution unit. *IBM J. Res. Dev.*, 11:34–53.

Bedrij, O. J. 1962. Carry select adders. *IEEE Trans. Electron. Comput.*, EC-11:340–346.

Chen, P.K. and Schlag, M.D.F. 1990. Analysis and design of CMOS Manchester adders with variable carry skip. *IEEE Trans. Comput.*, 39:983–992.

Dadda, L. 1965. Some schemes for parallel multipliers. *Alta Frequenza*, 34:349–356.

Ercegovac, M.D. and Lang, T. 1994. *Division and Square Root: Digit-Recurrence Algorithms and Their Implementations*. Kluwer Academic, Boston, MA.

Ferrari, D. 1967. A division method using a parallel multiplier. *IEEE Trans. Electron. Comput.*, EC-16:224–226.

Flynn, M.J. and Oberman, S.F. 2001. *Advanced Computer Arithmetic Design*. John Wiley & Sons, New York.

Goldberg, D. 1990. Computer arithmetic (App. A). In *Computer Architecture: A Quantitative Approach*, D.A. Patterson and J.L. Hennessy (Eds.), pp. A1–A66. Morgan Kauffmann, San Mateo, CA.

Gosling, J.B. 1980. *Design of Arithmetic Units for Digital Computers*. Macmillan, New York.

IEEE. 2008. *IEEE Standard for Floating-Point Arithmetic*. IEEE Std 754-1985, IEEE Press, New York.

Kilburn, T., Edwards, D.B.G., and Aspinall, D. 1960. A parallel arithmetic unit using a saturated transistor fast-carry circuit. *Proc. IEEE*, 107(Pt. B):573–584.

Koren, I. 2002. *Computer Arithmetic Algorithms*, 2nd edn., A.K. Peters, Natick, MA.

Lehman, M. and Burla, N. 1961. Skip technique for high speed carry propagation in binary arithmetic units. *IRE Trans. Electron. Comput.*, EC-10:691–698.

MacSorley, O.L. 1961. High-speed arithmetic in binary computers. *Proc. IRE*, 49:67–91.

Parhami, B. 2010. *Computer Arithmetic: Algorithms and Hardware Designs*, 2nd edn., Oxford University Press, New York.

Robertson, J.E. 1958. A new class of digital division methods. *IEEE Trans. Electron. Comput.*, EC-7:218–222.

Sklansky, J. 1960. Conditional-sum addition logic. *IEEE Trans. Electron. Comput.*, EC-9:226–231.

Turrini, S. 1989. Optimal group distribution in carry-skip adders. In *Proceedings of the 9th Symposium on Computer Arithmetic*. IEEE Computer Society Press, Los Alamitos, CA, pp. 96–103.

Wallace, C.S. 1964. A suggestion for a fast multiplier. *IEEE Trans. Electron. Comput.*, EC-13:14–17.

Waser, S. and Flynn, M.J. 1982. *Introduction to Arithmetic for Digital Systems Designers*. Holt, Rinehart and Winston, New York.

Weinberger, A. and Smith, J.L. 1958. A logic for high-speed addition. *Nat. Bur. Stand. Circular*, 591:3–12. National Bureau of Standards, Washington, DC.

21

Input/Output Devices and Interaction Techniques

Ken Hinckley
Microsoft Research

Robert J.K. Jacob
Tufts University

Colin Ware
University of New Hampshire

Jacob O. Wobbrock
University of Washington

Daniel Wigdor
University of Toronto

21.1 Introduction

The computing literature often draws a sharp distinction between input and output; computer scientists are used to regarding a screen as a passive output device and a mouse as a pure input device. However, nearly all examples of human–computer interaction (HCI) require *both* input and output to do anything useful. For example, what good would a mouse be without the corresponding feedback embodied by the cursor on the screen, as well as the sound and feel of the buttons when they are clicked? The distinction between output devices and input devices becomes even more blurred in the real world. A sheet of paper can be used to both record ideas (input) and display them (output). Clay reacts to the sculptor's fingers yet also provides feedback through the curvature and texture of its surface. Indeed, the complete and seamless integration of input and output is becoming a common research theme in advanced computer interfaces such as ubiquitous computing (Weiser, 1991) and tangible interaction (Ishii and Ullmer, 1997). And of course, with the modern commercial success of mobile touch screen devices such as the Apple iPhone and various Android units, people everywhere experience input and output as entirely inseparable and collocated concepts, their fingers acting directly on virtual widgets displayed on a responsive piece of glass. It is no accident that we should address input and output in the same chapter. Further, as the space required for a powerful computer has plummeted, new form factors have emerged. In mobile devices, the desire to provide large screens to maximize media consumption experiences has marginalized dedicated control surfaces, replacing physical keyboards and pointing devices with ad hoc screen-rendered controls manipulated through touch input. Although measurably inferior at nearly all input tasks, touch screens have nonetheless emerged as a dominant form of mobile computing.

Although inextricably linked, for the ease of discussion, we consider separately the issues of input and output. Input and output bridge the chasm between a computer's inner world of bits and the real world perceptible to the human senses. *Input* to computers consists of sensed information about the physical environment. Familiar examples include the mouse, which senses movement across a surface, and the keyboard, which detects a contact closure when the user presses a key. However, any sensed information about physical properties of people, places, or things can also serve as input to computer systems, for example the location of a user carrying a GPS-enabled mobile phone to inform recommendations for nearby restaurants. *Output* from computers can comprise any emission or modification to the physical environment, such as a display (including the cathode ray tube [CRT], flat-panel displays, or even light emitting diodes), speakers, or tactile and force feedback devices (sometimes referred to as *haptic displays*). There have even been olfactory displays such as the Smell Vision machine that debuted at the World's Fair in Stockholm in 1976. Finally, an *interaction technique* is a fusion of input and output, consisting of all hardware and software elements, that provides a way for the user to accomplish a low-level task. For example, in the traditional graphical user interface (GUI), users can scroll through a document by clicking or dragging the mouse (input) within a scrollbar displayed on the screen (output).

A fundamental task of HCI is to shuttle information between the brain of the user and the silicon world of the computer. Progress in this area attempts to increase the useful bandwidth across that interface by seeking faster, more natural, and more convenient means for users to transmit information to computers, as well as efficient, salient, and pleasant mechanisms to provide feedback to the user. On the user's side of the communication channel, interaction is constrained by the nature of human attention, cognition, and perceptual motor skills and abilities; on the computer side, it is constrained only by the technologies and methods that we can invent.

Research in input and output centers around the two ends of this channel: the devices and techniques computers can use for communicating with people and the perceptual abilities, processes, and organs people can use for communicating with computers.

It then attempts to find the common ground through which the two can be related by studying new modes of communication and expression that could be used for HCI and developing devices and techniques to use such modes. Basic research seeks theories and principles that inform us of the parameters of human cognitive and perceptual facilities, as well as models that can predict or interpret user performance in computing tasks (e.g., Wobbrock et al., 2008). Advances can be driven by the need for new modalities to support the unique requirements of specific application domains, by technological breakthroughs that HCI researchers apply to improving or extending the capabilities of interfaces, or by theoretical insights suggested by studies of human abilities and behaviors, or even problems uncovered during careful analyses of existing interfaces. These approaches complement one another, and all have their value and contributions to the field, but the best research seems to have elements of all of these.

21.2 Interaction Tasks, Techniques, and Devices

A designer looks at the *interaction tasks* necessary for a particular application (Foley et al., 1984). Interaction tasks are low-level primitive inputs required from the user, such as entering a text string or choosing a command. For each such task, the designer chooses an appropriate interaction technique. In selecting an interaction device and technique for each task in a human–computer interface, simply making an optimal choice for each task individually may lead to a poor overall design, with too many different or inconsistent types of devices or dialogues. Therefore, it is often desirable to compromise on the individual choices to reach a better overall design. Design is fundamentally about managing trade-offs in complex solution spaces, and it is often said that designs that are best for something are often worst for something else.

There may be several different ways of accomplishing the same task. For example, one could use a mouse to select a command by using a pop-up menu, a fixed menu (a palette or command bar), multiple clicking, circling the desired command, or even writing the name of the command with the mouse. Software might even detect patterns of mouse use in the background, such as repeated "surfing" through menus, to automatically suggest commands or help topics (Horvitz et al., 1998). The latter suggests a shift from the classical view of interaction as direct manipulation where the user is responsible for all actions and decisions to one that uses *background sensing* techniques to allow technology to support the user with semiautomatic or implicit actions and services (Buxton, 1995a).

21.3 Composition of Interaction Tasks

Early efforts in HCI sought to identify elemental tasks that appear repeatedly in human–computer dialogues. Foley et al. (1984) proposed that user interface transactions are composed of the following elemental tasks:

Selection: Choosing objects from a set of alternatives.
Position: Specifying a position within a range. This includes picking a screen coordinate with a pointing device.
Orient: Specifying an angle or 3D orientation.
Path: Specifying a series of positions and/or orientations over time.
Quantify: Specifying an exact numeric value.
Text: Entry of symbolic data.

While these are commonly occurring tasks in many direct-manipulation interfaces, a problem with this approach is that the level of analysis at which one specifies "elemental" tasks is not well defined. For example, for *Position* tasks, a screen coordinate could be selected using a pointing device such as a mouse, but might be entered as a pair of numeric values (*Quantify*) using a pair of knobs (like an Etch-A-Sketch) where precision is paramount. But if these represent elemental tasks, why do we find that we must subdivide *Position* into a pair of *Quantify* subtasks for some devices but not for others?

Treating all tasks as hierarchies of subtasks, known as *compound tasks*, is one way to address this. With appropriate design and by using technologies and interaction techniques that parallel the way the user thinks about a task as closely as possible, the designer can phrase together a series of elemental tasks into a single *cognitive chunk*. For example, if the user's task is to draw a rectangle, a device such as an

Etch-A-Sketch is easier to use. For drawing a circle, a pen is far easier to use. Hence the choice of device influences the level at which the user is required to think about the individual actions that must be performed to achieve a goal. See Buxton (1986) for further discussion of this important concept.

The six elemental tasks previously enumerated may be a complete list of "fundamental" low-level tasks that underlie most interaction with computers, but it could be argued that this list is not complete; for example, which of these six tasks does a fingerprint scanner support? Perhaps, if used for password replacement, it could be viewed as supporting the *Text* task; alternatively, one might add "Establishment of Identity" to the list. This points to a problem with the fundamental task approach. While identifying "elemental tasks" can be useful for thinking about interaction techniques in general, a problem with viewing tasks as assemblies of elemental tasks is that it typically only considers explicit input in the classical direct-manipulation paradigm. Where do devices like cameras, microphones, and the fingerprint scanner previously discussed fit in? These support higher-level data types and concepts (e.g., images, audio, and identity). Advances in technology will continue to yield new "elemental" inputs. However, these new technologies also may make increasing demands on systems to move from individual samples to synthesis of meaningful structure from the resulting data (Fitzmaurice et al., 1999).

21.4 Properties of Pointing Devices

The breadth of input devices and displays on the market today can be bewildering. It is important to understand that, when considering these devices, they are not all created equal. We must remember that the traditional mouse-based user interface, the Windows–Icons–Menus–Pointers (WIMP), has as perhaps its most essential component an abstraction of the logical target of user actions. This abstraction has gone by many names. The inventors of the mouse, Engelbart and English, named it the *bug* and later referred to it as the *telepointer.* In Windows, it is the pointer. In OSX, it is alternately the pointer and the cursor. But by whatever name, it remains a focal point for user-generated events in the system. A funny thing has happened with the pointer: a kind of abstraction has grown up around it, where a plethora of hardware can control it, and it is the movement of the pointer, rather than the hardware, around which software designers create their experiences. This has led to widespread misunderstanding that the design of the GUI itself is abstract. It's not. It has been designed over more than 40 years of iteration to be highly optimized for a particular piece of hardware.

Thus, when considering input devices, one must be careful to avoid thinking of them in the overly simplistic terms of how they will control a cursor. Instead, we consider a number of organizing properties and principles in order to make sense of the design space and performance issues. First, we consider continuous, manually operated pointing devices (as opposed to discrete input mechanisms such as buttons or keyboards, or other devices not operated with the hand, which we will discuss briefly later). For further insight readers may also wish to consult complete taxonomies of devices (Buxton, 1983; Card et al., 1991). As we shall see, however, it is nearly impossible to describe properties of input devices without reference to output—especially the resulting feedback on the screen—since after all input devices are only useful insofar as they support interaction techniques that allow the user to accomplish something.

21.4.1 Physical Property Sensed

Traditional pointing devices typically sense position, motion, or force. A tablet senses position, a mouse measures motion (i.e., change in position), and an isometric joystick senses force. An isometric joystick is a unmoving force sensing joystick such as the IBM TrackPoint ("eraser head") found on many laptops. For a rotary device, the corresponding properties are angle, change in angle, and torque. Position-sensing devices are also known as *absolute input devices*, whereas motion-sensing devices are *relative input devices.* An absolute device can fully support relative motion, since it can calculate changes to position, but a relative device cannot fully support absolute positioning and in fact can only emulate "position" at all by introducing a cursor on the screen. Note that it is difficult to move the mouse cursor to a particular area of the screen (other than the edges) without looking at the screen, but with a tablet one can easily point to a region with the stylus using the kinesthetic sense (Balakrishnan and Hinckley, 1999), informally known as "muscle memory."

21.4.2 Transfer Function

A device, in combination with the host operating system, typically modifies its signals using a mathematical transformation that scales the data to provide smooth, efficient, and intuitive operation. An appropriate mapping is a transfer function that matches the physical properties sensed by the input device. Appropriate mappings include force-to-velocity, position-to-position, and velocity-to-velocity functions. For example, an isometric joystick senses force; a nonlinear rate mapping transforms this into a velocity of cursor movement (Rutledge and Selker, 1990; Zhai and Milgram, 1993; Zhai et al., 1997). Ideally, the device should also be self-centering when using a rate mapping, with a spring return to the zero input value, so that the user can stop quickly by releasing the device. A common inappropriate mapping is calculating a speed of scrolling based on the position of the mouse cursor, such as extending a selected region by dragging the mouse close to the edge of the screen. The user has no feedback of when or to what extent scrolling will accelerate, and the resulting interaction can be hard to learn how to use and difficult to control.

Gain entails a simple multiplicative transfer function known as the device gain, which can also be described as a *control-to-display (C:D) ratio*, the ratio between the movement of the input device and the corresponding movement of the object it controls. For example, if a mouse (the control) must be moved 1 cm on the desk to move a cursor 2 cm on the screen (the display), the device has a 1:2 control-display ratio. However, on commercial pointing devices and operating systems, the gain is rarely constant*; an *acceleration function* is often used to modulate the gain depending on velocity. An acceleration function is a transfer function that exhibits an exponential relationship between velocity and gain. Experts believe the primary benefit of acceleration is to reduce the *footprint*, or the physical movement space, required by an input device (Hinckley et al., 2001; Jellinek and Card, 1990). One must also be very careful when studying the possible influence of gain settings on user performance: experts have criticized gain as a fundamental concept, since it confounds two separate concepts (device size and display size) in one arbitrary metric (Accot and Zhai, 2001). Furthermore, user performance may exhibit speed–accuracy tradeoffs, calling into question the assumption that there exists an "optimal" C:D ratio (MacKenzie, 1995).

21.4.3 Number of Dimensions

Devices can measure one or more linear and angular dimensions. For example, a mouse measures two linear dimensions, a knob measures one angular dimension, and a six degree-of-freedom (6 DOF) magnetic tracker measures three linear dimensions and three angular (for examples of 6 DOF input and design issues, see Green and Liang, 1994; Hinckley et al., 1994b; Serra et al., 1997; Ware and Jessome, 1988). If the number of dimensions required by the user's interaction task does not match the number of dimensions provided by the input device, then special handling (e.g., interaction techniques that may require extra buttons, graphical widgets, and mode switching) will need to be introduced. A lack of control degrees-of-freedom is a particular concern for 3D user interfaces and interaction (Hinckley et al., 1994b; Zhai, 1998). Numerous interaction techniques have been proposed to allow standard 2D pointing devices to control 3D positioning or orientation tasks (e.g., Bukowski and Sequin, 1995; Chen et al., 1988; Conner et al., 1992). Well-designed interaction techniques using specialized multiple degree-of-freedom input devices can sometimes offer superior performance (Hinckley et al., 1997; Ware and Rose, 1999), but may be ineffective for standard desktop tasks, so overall performance must be considered (Balakrishnan et al., 1997; Hinckley et al., 1999).

* Direct input devices (see next page) are an exception, since the C:D ratio is typically fixed at 1:1 (but see also Sears and Shneiderman, 1991 and Forlines et al., 2006).

21.4.4 Pointing Speed, Accuracy, and Throughput

The standard way to characterize pointing device performance employs *Fitts' law* paradigm (Fitts, 1954; Soukoreff and MacKenzie, 2004; Wobbrock et al., 2011). Fitts' law relates the *movement time* to point at a target, the *amplitude* of the movement (the distance to the target), and the *width* of the target (i.e., the precision requirement of the pointing movement). The movement time is proportional to the logarithm of the distance divided by the target width, with constant terms that vary from one device to another. While not emphasized in this chapter, Fitts' law is the single most important quantitative analysis, testing, and prediction tool available to input research and device evaluation. For an excellent overview of its application to the problems of HCI, including use of Fitts' law to characterize throughput (a composite measure of both speed and accuracy often informally called bandwidth), see MacKenzie (1992). For discussion of other accuracy metrics, see MacKenzie et al. (2001). Recently Fitts' law testing paradigm has been proposed as an international standard for evaluating pointing devices (Douglas et al., 1999; Keates et al., 2002).

Recent years have seen a number of new insights and new applications for Fitts' law (Guiard, 2009). Fitts' law was originally conceived in the context of rapid, aimed movements, but Fitts' law can also be applied to tasks such as scrolling (Hinckley et al., 2001), multi-scale navigation (Guiard et al., 2001), crossing boundaries (Accot and Zhai, 2002) and predicting pointing error rates instead of movement time (Wobbrock et al., 2008, 2011). Researchers have also recently applied Fitts' law to expanding targets that double in width as the user approaches them. Even if the expansion begins after the user has already covered 90% of the distance from a starting point, the expanding target can be selected as easily as if it had been fully expanded since the movement began (McGuffin and Balakrishnan, 2002); see also Zhai et al. (2003). However, it remains unclear if this can be successfully applied to improve pointing performance for multiple targets that are closely packed together (as typically found in menus and tool palettes). For tasks that exhibit continuous speed–accuracy requirements, such as moving through a hierarchical menu, Fitts' law cannot be applied, but researchers have recently formulated the steering law, which does addresses such tasks. In fact, the steering law was independently discovered outside the computing fields twice before (Accot and Zhai, 1997, 1999, 2001).

21.4.5 Input Device States

To select a single point or region with an input device, users need a way to signal when they are selecting something versus when they are just moving over something to reach a desired target. The need for this fundamental signal of intention is often forgotten by researchers eager to explore new interaction modalities such as empty-handed pointing (e.g., using camera tracking or noncontact proximity sensing of hand position). The *three-state model* of input (Buxton, 1990b) generalizes the states sensed by input devices as *tracking*, which causes the cursor to move; *dragging*, which allows selection of objects by clicking (as well as moving objects by clicking and dragging them); and *out of range*, which occurs when the device moves out of its physical tracking range (e.g., a mouse is lifted from the desk, or a stylus is removed from a tablet). Most pointing devices sense only two of these three states: for example, a mouse senses tracking and dragging, but a touchpad senses tracking and the out-of-range state. Hence, to fully simulate the functionality offered by mice, touchpads need special procedures, such as tapping to click, which are prone to inadvertent activation (e.g., touching the pad by accident causes a click [MacKenzie and Oniszczak, 1998]). For further discussion and examples, see Buxton (1990b) and Hinckley et al. (1998a).

21.4.6 Direct versus Indirect Control

A mouse is an *indirect input device* (one must move the mouse to point to a spot on the screen); a touch screen is a *direct input device* (the display surface is also the input surface). Direct devices raise several unique issues. Designers must consider the possibility of parallax error resulting from a gap between the input and display surfaces, reduced transmissivity of the screen introduced by a sensing layer, or occlusion of the display by the user's hands. Another issue is that touch screens can support a cursor tracking state, or a dragging state, but not both; typically, touch screens move directly from the out-of-range state to the dragging state when the user touches the screen, with no intermediate cursor feedback (Buxton, 1990b). Techniques for touch-screen cursor feedback have been proposed, but typically require that selection occurs on lift-off (Potter et al., 1988; Sears and Shneiderman, 1991; Sears et al., 1992). See Section 21.5.4.

21.4.7 Device Acquisition Time

The average time to pick up or put down an input device is known as **acquisition time** (or sometimes homing time). It is often assumed to be a significant factor for user performance, but Fitts' law throughput of a device tends to dominate human performance time unless switching occurs frequently (Douglas and Mithal, 1994). However, one exception is stylus or pen-based input devices; pens are generally comparable to mice in general pointing performance (Accot and Zhai, 1999), or even superior for some high-precision tasks (Guiard et al., 2001), but these benefits can easily be negated by the much greater time it takes to switch between using a pen and using a keyboard.

21.4.8 Modeling of Repetitive Tasks

The keystroke-level model (KLM) is commonly used to model expert user performance in repetitive tasks such as text editing. The KLM includes standard operators that represent average times required for pointing with an input device, pressing a button, pauses for decision making, and device acquisition time, but the model does not account for errors or nonexpert behaviors such as problem solving (Card et al., 1980). Good examples of research that apply the KLM include Wang et al. (2001) and MacKenzie and Soukoreff (2002). Another approach called GOMS modeling is an extension of the KLM that can handle more complex cases (Olson and Olson, 1990), but many practitioners still use the KLM to evaluate input devices and low-level interaction techniques because of KLM's greater simplicity.

21.4.9 Hardware Criteria

Various other characteristics can distinguish input devices, but are perhaps less important in distinguishing the fundamental types of interaction techniques that can be supported. Engineering parameters of a device's performance such as sampling rate, resolution, accuracy, and linearity can all influence performance. **Latency** is the end-to-end delay between the user's physical movement, sensing this and providing the ultimate system feedback to the user. Latency can be a devious problem as it is impossible to completely eliminate from system performance; latency of more than 75–100 ms significantly impairs user performance for many interactive tasks (MacKenzie and Ware, 1993; Robertson et al., 1989). For vibrotactile or haptic feedback, users may be sensitive to much smaller latencies of just a few milliseconds (Cholewiak and Collins, 1991). The effects of latency can also be magnified in particular interaction techniques; for example, when dragging on a touch screen, users notice latency as a separation between their finger and the dragged object, even for extremely low latencies (Ng et al., 2012).

21.5 Discussion of Common Pointing Devices

Here, we briefly describe commonly available pointing devices and some issues that can arise with them in light of the properties previously discussed.

21.5.1 Mouse

A mouse senses movement relative to a flat surface. Mice exhibit several properties that are well suited to the demands of desktop graphical interfaces (Balakrishnan et al., 1997). The mouse is stable and does not fall over when released (unlike a stylus on a tablet). A mouse can also provide integrated buttons for selection, and since the force required to activate a mouse's buttons is orthogonal to the plane of movement, it helps minimize accidental clicking or interference with motion. Another subtle benefit is the possibility for users to employ a combination of finger, hand, wrist, arm, and even shoulder muscles to span the range of tasks from short precise selections to large, ballistic movements (Balakrishnan and MacKenzie, 1997; Zhai et al., 1996). Finally, Fitts' law studies show that users can point with the mouse about as well as with the hand itself (Card et al., 1978).

21.5.2 Trackball

A trackball is like a roller-ball mouse that has been turned upside down, with a mechanical ball that rolls in place. Trackballs, like mice, are indirect relative pointing devices. The main advantage of trackballs is that they can be used on an inclined surface and they often require a smaller physical footprint than mice. They also employ different muscle groups, which some users find more comfortable, especially users with motor impairments caused by, e.g., spinal cord injuries (Fuhrer and Fridie, 2001; Sperling and Tullis, 1988; Wobbrock and Myers, 2006). However, rolling the trackball while holding down any of its buttons requires significant dexterity, especially when trying to do so with only one hand. Therefore, tasks requiring moving the trackball with a button held down, for example, dragging an icon across the desktop, can be awkward or even impossible for some users (MacKenzie et al., 1991).

21.5.3 Tablets

Touch-sensitive tablets (that are not also touch screens) are indirect pointing devices and may be used in either absolute or relative pointing modes. In absolute mode, each position on the tablet maps directly to a corresponding position on the screen (e.g., the top-right corner). In relative mode, movements on the tablet move a cursor on the screen according to a transfer function, as with a mouse. Most tablets sense the absolute position of a mechanical intermediary such as a stylus or puck on the tablet surface. A puck is a mouse that is used on a tablet; the only difference is that it senses absolute position and it cannot be used on a surface other than the tablet. Absolute mode is generally preferable for tasks such as tracing, digitizing, drawing, freehand inking, and signature capture. Tablets that sense contact of the bare finger are known as touch tablets (Buxton et al., 1985); touchpads are miniature touch tablets, as commonly found on portable computers (MacKenzie and Oniszczak, 1998). A touch screen is a touch-sensitive tablet collocated with an information display, but demands different handling than a tablet (see Section 21.4.6).

21.5.4 Pen Input

Pen-based input for mobile devices is an area of increasing practical concern. Pens effectively support activities such as inking, marking, and gestural input (see Section 21.8.2), but raise a number of problems when supporting graphical interfaces originally designed for mouse input. Pen input raises the concerns of direct input devices as previously described. There is no way to see exactly what position will be selected before selecting it: pen contact with the screen directly enters the dragging state of the three-state model (Buxton, 1990b). There is no true equivalent of a "hover" state for tool tips* nor an extra button for context menus. Pen dwell time on a target can be used to provide one of these two functions. When detecting a double-tap, one must allow for longer interval between the taps (as compared to double-click on a mouse), and one must also accommodate a significant change to the screen position between taps. Finally, users often want to touch the screen of small devices using a bare finger,

* Tool tips are small explanatory labels or balloons that appear next to a button, icon, or other interface widget when the user holds the cursor still over that object. The "hover" state is detected when there is little or no cursor movement for a fixed time-out (many systems use a time-out of approximately 1 s).

so applications should be designed to accommodate imprecise selections. Note that some pen-input devices, such as the Tablet PC, use an inductive sensing technology that can only sense contact from a specially instrumented stylus and thus cannot be used as a touch screen. However, this deficiency is made up for by the ability to track the pen when it is close to (but not touching) the screen, allowing support for a tracking state with cursor feedback (and hence ToolTips as well).

21.5.5 Joysticks

There are many varieties of joysticks. As previously mentioned, an isometric joystick senses force and returns to center when released. Because isometric joysticks can have a tiny footprint, they are often used when space is at a premium, allowing integration with a keyboard and hence rapid switching between typing and pointing (Douglas and Mithal, 1994; Rutledge and Selker, 1990). Isotonic joysticks sense the angle of deflection of the stick; they tend to move more than isometric joysticks, offering better feedback to the user. Such joysticks may or may not have a mechanical spring return to center. Some joysticks even include both force and position sensing and other special features. For a helpful organization of the complex design space of joysticks, see Lipscomb and Pique (1993).

21.5.6 Alternative Means of Pointing

Researchers have explored using the feet (Pearson and Weiser, 1988), head tracking, (Bates and Istance, 2003; LoPresti et al., 2000) and eye tracking (Jacob, 1990; Kumar et al., 2007; Lankford, 2000) as alternative approaches to pointing. Head tracking has much lower pointing bandwidth than the hands and may require the neck to be held in an awkward fixed position, but has useful applications for intuitive coupling of head movements to virtual environments (Brooks, 1988; Sutherland, 1968) and interactive 3D graphics (Hix et al., 1995; Ware et al., 1993). Eye movement-based input, properly used, can provide an unusually fast and natural means of communication, because we move our eyes rapidly and almost unconsciously. The human eye fixates visual targets within the fovea, which fundamentally limits the accuracy of eye gaze tracking to 1 degree of the field of view (Zhai et al., 1999). Eye movements are subconscious and must be interpreted carefully to avoid annoying the user with unwanted responses to his actions, known as the Midas touch problem (Jacob, 1991). Current eye-tracking technology is expensive and has numerous technical limitations, confining its use thus far to research labs and disabled persons with few other options.

21.6 Feedback and Perception-Action Coupling

The ecological approach to human perception (Gibson, 1986) asserts that the organism, the environment, and the tasks the organism performs are inseparable and should not be studied in isolation. Hence, perception and action are intimately linked in a single motor-visual feedback loop, and any separation of the two is an artificial one. The lesson for interaction design is that techniques must consider both the motor control (input) and feedback (output) aspects of the design and how they interact with one another.

From the technology perspective, one can consider feedback as passive or active. Active feedback is under computer control. This can be as simple as presenting a window on a display or as sophisticated as simulating haptic contact forces with virtual objects when the user moves an input device. We will return to discuss active feedback techniques later in this chapter, when we discuss display technologies and techniques.

Passive feedback may come from sensations within the user's own body, as influenced by physical properties of the device, such as the shape, color, and feel of buttons when they are depressed. The industrial design of a device suggests the purpose and use of a device even before a user touches it (Norman, 1990). Mechanical sounds and vibrations that result from using the device provide confirming feedback of the user's action. The shape of the device and the presence of landmarks can help users orient a device without having to look at it (Hinckley et al., 1998b). Proprioceptive and kinesthetic feedback are somewhat imprecise terms, often used interchangeably, that refer to sensations of body posture, motion, and muscle tension (MacKenzie and Iberall, 1994). These senses allow users to feel how they are moving an input device without looking at the device and indeed without looking at the screen in some situations (Balakrishnan and Hinckley, 1999; Kane et al., 2008; Mine et al., 1997). This may be important when the user's attention is divided between multiple tasks and devices (Fitzmaurice and Buxton, 1997). Sellen et al. (1992) report that muscular tension from depressing a foot pedal makes modes more salient to the user than purely visual feedback. Although all of these sensations are passive and not under the direct control of the computer, these examples nonetheless demonstrate that they are relevant to the design of devices, and interaction techniques can consider these qualities and attempt to leverage them.

21.6.1 Impoverished Physicality

Modern input devices such as touch-screens and in-air gesture systems lack some of the tactile and kinesthetic feedback inherently present in traditional input devices. Such feedback is an essential element of the user experience, especially when users are attempting to understand why the system response may not be as expected. Many commercial mobile devices use audio to compensate with phones that beep or click when a virtual key is pressed.

To understand the role that feedback plays, consider the following table (Table 21.1), which describes various states of a system and the feedback that is provided by either the cursor or the hardware itself. As is immediately evident, most touch-based platforms shift a great deal of the feedback burden onto the application developer.

Some work has addressed this issue by providing haptic sensations using piezoelectric actuators, electrovibration, or deformable fluids (Bau et al., 2010; Jansen et al., 2010; Poupyrev and Maruyama, 2003; Poupyrev et al., 2002). Others have supplied physical cutouts as templates to provide passive haptic sensations to a finger (or stylus) (Buxton et al., 1985; Wobbrock et al., 2003). Still others have placed transparent physical plastics, ranging in feel from "squishy" to "rigid," on touch-screen surfaces to provide for haptic sensations (Bilton, 2011; Jansen et al., 2012; Weiss, 2010; Weiss et al., 2009). If haptic possibilities are not available, well-designed visual feedback indicating when and where touches were perceived by the system is crucial for effective interaction (Wigdor et al., 2009).

User performance may be influenced by correspondences between input and output. Some correspondences are obvious, such as the need to present confirming visual feedback in response to the user's

TABLE 21.1 Potential Causes of Unexpected Behavior (Left) and the Source of Feedback That Users Receive to Refute Those Causes in Representative Mouse versus Touch Input Systems

	Feedback Refuting Cause	
Cause of Unexpected Behavior	Mouse	Touch
System is nonresponsive	OS: pointer movement	Application
Hardware failed to detect input	HW: activation of button	Application
Input delivered to wrong location	OS: visible pointer	Application
Input does not map to expected function	Application	Application
System is in a mode	OS + application: pointer icon	Application
Max size reached	OS: pointer moves past edge	Application
Accidental input (arm brushing)	N/A	Application
Over constrained (too many contacts)	N/A	Application
Stolen capture (second user captures control)	N/A	Application

actions. Ideally, feedback should indicate the results of an operation before the user commits to it (e.g., highlighting a button or menu item when the cursor moves over it). Kinesthetic correspondence and perceptual structure, described later, are less obvious.

21.6.1.1 Kinesthetic Correspondence

Kinesthetic correspondence refers to the principle that graphical feedback on the screen should correspond to the direction that the user moves the input device, particularly when 3D rotation is involved (Britton et al., 1978). Users can easily adapt to certain non-correspondences: when the user moves a mouse forward and back, the cursor actually moves up and down on the screen; if the user drags a scrollbar downward, the text on the screen scrolls upward. With long periods of practice, users can adapt to almost anything (e.g., for over 100 years psychologists have known of the phenomena of prism adaptation, where people can eventually adapt to wearing prisms that cause everything to look upside down [Stratton, 1897]). However, one should not force users to adapt to a poor design.

21.6.1.2 Perceptual Structure

Researchers have also found that the interaction of the input dimensions of a device with the control dimensions of a task can exhibit perceptual structure. Jacob et al. (1994) explored two input devices: a 3D position tracker with integral (x, y, z) input dimensions and a standard 2D mouse, with (x, y) input separated from (z) input by holding down a mouse button. For selecting the position and size of a rectangle, the position tracker is most effective: here, the integral 3D position input matches the integral presentation of the feedback on the screen. But for selecting the position and grayscale color of a rectangle, the mouse is most effective: here, the user perceives the position and grayscale color of the rectangle as separate quantities and can more easily perform the task when the input device provides separate controls. Hence neither a 3D integral input nor a 2D (x, y) plus 1D (z) input is uniformly superior; the better performance results when the task and device are both integral or both separable.

21.7 Pointing Facilitation Techniques

In an effort to improve the performance of pointing devices, particularly devices for indirect relative pointing (e.g., mice), researchers have invented numerous *pointing facilitation techniques.* Such techniques attempt to improve on the "raw" pointing performance one would exhibit with one's hand or finger by utilizing various means for making targets easier to acquire. Such techniques are sometimes said to "beat Fitts' law" (Balakrishnan, 2004), which refers to performing better than Fitts' law, the widespread quantitative model of human movement time, would predict (Fitts, 1954; MacKenzie, 1992).

Pointing facilitation techniques can be divided into two flavors: those that are *target-agnostic* and those that are *target-aware* (Balakrishnan, 2004; Wobbrock et al., 2009a). Target-agnostic techniques require no knowledge of target identities, locations, or dimensions, working instead only with information directly available from the input device and on-screen cursor. Such techniques are relatively easy to deploy in real-world systems. Target-aware techniques, on the other hand, require knowledge of targets, often target locations and dimensions. Such techniques may also manipulate targets in some way, such as by expanding them (McGuffin and Balakrishnan, 2005). As a result, target-aware techniques are exceedingly difficult to deploy across commercial systems and, until recently, have mostly been confined to research laboratories. Recent efforts, however, have shown that target-aware techniques *can* be deployed on real-world systems by using an underlying architecture for target identification and interpretation (Dixon et al., 2012). Nevertheless, an ongoing theoretical challenge with target-aware techniques is that what constitutes a target is not easily defined. In a word processor, for example, a target may be every character or every word or perhaps even the space *between* characters. In a calendar program, every half hour slot may be a target. In a paint program, every pixel on the paint canvas is a target. Even on the open desktop, icons are not the only targets; every pixel is a candidate for the start of a drag-rectangle operation or a right-click menu. What constitutes a target is a more complex question than at first it may seem.

Target-agnostic techniques avoid the aforementioned issues but are relatively few in number, as only information from the input device and on-screen cursor is available to techniques. Target-agnostic techniques include conventional pointer acceleration (Casiez et al., 2008), dynamic gain adjustment based on the spread of movement angles (Wobbrock et al., 2009a), dynamic gain adjustment based on velocity changes (Hourcade et al., 2008), freezing the mouse cursor in place when the mouse button is down (Trewin et al., 2006), and visual-and-motor-space magnification of pixels with a lens (Findlater et al., 2010; Jansen et al., 2011). Click histories in the form of "magnetic dust" have also been used as a means for providing target-agnostic gravity (Hurst et al., 2007). Placing targets along impenetrable screen edges is also a target-agnostic pointing facilitation strategy, as edges constrain, guide, and trap on-screen cursors (e.g., in screen corners) (Appert et al., 2008; Farris et al., 2001; Froehlich et al., 2007; Walker and Smelcer, 1990). Related to but not strictly a pointing facilitation technique in itself, *kinematic endpoint prediction* (Lank et al., 2007) predicts the endpoints of mouse movements en route to targets using motion kinematic formulae like the minimum-jerk law (Flash and Hogan, 1985).

Target-aware techniques are more numerous than their impoverished target-agnostic brethren. With target identities, locations, and dimensions available to them, target-aware techniques contain the highest-performing pointing techniques in the world. One strategy for target-aware pointing is to modify the mouse cursor. For general-purpose pointing, the *bubble cursor*, which operates using a simple rule that "the closest target is always selected," still remains unbeaten (Grossman and Balakrishnan, 2005). It is rivaled by the *DynaSpot*, which is a speed-dependent bubble cursor that also transitions to being a point cursor after stopping, making it capable of selecting specific pixels, not just objects (Chapuis et al., 2009). Both are variations on the static *area cursor*, which is target-aware so that when its large area overlaps multiple targets, it can degrade to a point cursor at its center (Kabbash and Buxton, 1995; Worden et al., 1997). Other explorations of area cursors have incorporated *goal crossing* (Accot and Zhai, 1997, 2002) into the cursors themselves (Findlater et al., 2010).

Other target-aware schemes modify the targets themselves. Mouse gain is dropped inside *sticky icons* (Blanch et al., 2004; Worden et al., 1997). Even more aggressive, *gravity wells* and *force fields* actually "pull" cursors into themselves (Ahlström et al., 2006; Hwang et al., 2003). *Bubble targets* and, more generally, *target expansion* enlarge targets as the cursor approaches (Cockburn and Firth, 2003; McGuffin and Balakrishnan, 2002, 2005; Zhai et al., 2003). A more extreme form of target expansion is to simply bring targets close to the mouse cursor as it begins its movement (Baudisch et al., 2003). More radically, the mouse cursor can be made to "jump over" the open space between targets, bypassing nontarget pixels altogether (Guiard et al., 2004). There are a great many ways to "beat Fitts' law" and it is likely that new pointing facilitation techniques will be invented for years to come.

21.8 Keyboards, Text Entry, and Command Input

For over a century, keyboards, whether on typewriters, desktop computers, laptops, or mobile devices, have endured as the mechanism of choice for text entry. The resiliency of the keyboard, in an era of unprecedented technological change, is the result of how keyboards complement human skills and may make keyboards difficult to supplant with new input devices or technologies. We summarize some general issues surrounding text entry later, with a focus on mechanical keyboards; see also Lewis et al. (1997) and MacKenzie and Tanaka-Ishii (2007).

21.8.1 Skill Acquisition and Skill Transfer

Procedural memory is a specific type of memory that encodes repetitive motor acts. Once an activity is encoded in procedural memory, it requires little conscious effort to perform (Anderson, 1980). Because procedural memory automates the physical act of text entry, touch typists can rapidly type words without interfering with the mental composition of text. The process of encoding an activity in procedural memory can be formalized as the *power law of practice*: $T_n = T_1 \times n^{(-\alpha)}$, where T_n is the time to perform the nth task, T_1 is the time to perform the first task, and α reflects the learning rate (Card et al., 1983; De Jong, 1957; Snoddy, 1926). The power law of practice is sometimes recast as $Y = aX^b$, called the power law of learning, as it produces "learning curves" for which Y is an increasing measure of proficiency (e.g., words per minute), X is a measure of the amount of practice (e.g., trial, block, or session), and a and b are regression coefficients. See MacKenzie and Zhang (1999) for an example in text entry. This suggests that changing the keyboard can have a high relearning cost. However, a change to the keyboard can succeed if it does not interfere with existing skills or allows a significant transfer of skill. For example, some ergonomic keyboards preserve the basic key layout, but alter the typing pose to help maintain neutral postures (Honan et al., 1995; Marklin and Simoneau, 1996), whereas the Dvorak key layout may have some small performance advantages, but has not found wide adoption due to high retraining costs (Lewis et al., 1997).

21.8.2 Eyes-Free Operation

With practice, users can memorize the location of commonly used keys relative to the home position of the two hands, allowing typing with little or no visual attention (Lewis et al., 1997). By contrast, soft keyboards (small on-screen virtual keyboards found on many handheld devices) require nearly constant visual monitoring, resulting in a secondary focus-of-attention (FoA), where the primary focus is on the user's work, that is, the text being composed. A third FoA may consist of a source document that the user is transcribing. Furthermore, with stylus-driven soft keyboards, the user can only strike one key at a time. Thus the design issues for soft keyboards differ tremendously from mechanical keyboards (Zhai et al., 2000).

21.8.3 Tactile Feedback

On a mechanical keyboard users can feel the edges and gaps between the keys, and the keys have an activation force profile that provides feedback of the key strike. In the absence of such feedback, as on touch-screen keyboards (Sears, 1993), performance may suffer and users may not be able to achieve eyes-free performance (Lewis et al., 1997).

21.8.4 Combined Text, Command, and Navigation Input

Finally, it is easy to forget that keyboards provide many secondary command and control actions in addition to pure text entry, such as navigation keys (Enter, Home/End, Delete, Backspace, Tab, Esc, Page Up/Down, Arrow Keys, etc.), chord key combinations (such as Ctrl + C for Copy) for frequently used commands, and function keys for miscellaneous functions defined by the current application. Without these keys, frequent interleaving of mouse and keyboard activity may be required to perform these secondary functions.

21.8.5 Ergonomic Issues

Many modern information workers suffer from repetitive strain injury (RSI). Researchers have identified many risk factors for such injuries, such as working under stress or taking inadequate rest breaks. People often casually associate these problems with keyboards, but the potential for RSI is common to many manually operated tools and repetitive activities (Putz-Anderson, 1988). Researchers have advocated themes for ergonomic design of keyboards and other devices (Pekelney and Chu, 1995), including reducing repetition, minimizing force required to hold and move the device or press its buttons, avoiding sharp edges that put pressure on the soft tissues of the hand, and designing for natural and neutral postures of the user's hands and wrists (Honan et al., 1995; Marklin et al., 1997). Communicating a clear orientation for gripping and moving the device through its industrial design also may help to discourage inappropriate, ergonomically unsound grips.

21.8.6 Other Text Entry Mechanisms

One-handed keyboards can be implemented using simultaneous depression of multiple keys; such *chord keyboards* can sometimes allow one to achieve high peak performance (e.g., court stenographers), but take much longer to learn how to use (Buxton, 1990a; Mathias et al., 1996; Noyes, 1983). They are often used in conjunction with wearable computers (Smailagic and Siewiorek, 1996) to keep the hands free as much as possible (but see also Section 21.8.1). With complex written languages, such as Chinese and Japanese, key chording and multiple stages of selection and disambiguation are currently necessary for keyboard-based text entry (Wang et al., 2001). Handwriting and character recognition may ultimately provide a more natural solution, but for Roman languages handwriting (even on paper, with no recognition involved) is much slower than skilled keyboard use. To provide reliable stylus-driven text input, some systems have adopted unistroke (single-stroke) gestural "alphabets" (Goldberg and Richardson, 1993; Wobbrock et al., 2003) that reduce the demands on recognition technology, while remaining relatively easy for users to guess and learn (MacKenzie and Zhang, 1997; Wobbrock et al., 2005). However, small "two-thumb" keyboards (MacKenzie and Soukoreff, 2002) or fold-away peripheral keyboards are becoming increasingly popular for mobile devices. Recently, hybrid stroke-keyboard combinations have emerged that enable users to tap conventionally on keys or, without explicitly switching modes, to make unistroke gestures over the keys that form the desired word (Kristensson and Zhai, 2004; Zhai and Kristensson, 2003). Precise gestures over keys are not required as the stroke is pattern-matched to produce the intended word. Stroke keyboards have even been commercialized (e.g., ShapeWriter and Swype, both acquired in 2012 by Nuance Corporation) and generally have received a welcome response from consumers. Dictation using continuous speech recognition is available on the market today, but the technology still has a long way to go; a recent study found that the corrected words-per-minute rate of text entry using a mouse and keyboard are about twice as fast as dictation input (Karat et al., 1999). We further discuss speech interaction in the next section.

21.9 Modalities of Interaction

Here, we briefly review a number of general strategies and input modalities that have been explored by researchers. These approaches generally transcend a specific type of input device, but rather span a range of devices and applications.

21.9.1 Speech and Voice

Carrying on a full conversation with a computer as one might do with another person is well beyond the state of the art today and, even if possible, may be a naive goal. Yet even without understanding the content of the speech, computers can digitize, store, edit, and replay segments of speech to augment human–human communication (Arons, 1993; Buxton, 1995b; Stifelman, 1996). Conventional voice mail and the availability of MP3 music files on the web are simple examples of this. Computers can also infer information about the user's activity from ambient audio, such as determining if the user is present or perhaps engaging in a conversation with a colleague, allowing more timely delivery of information or suppression of notifications that may interrupt the user (Horvitz et al., 1999; Sawhney and Schmandt, 2000; Schmandt et al., 2000).

Understanding speech as input has been a long-standing area of research. While progress is being made, it is slower than optimists originally predicted, and daunting unsolved problems remain. For limited vocabulary applications with native English speakers, speech recognition can excel at recognizing words that occur in the vocabulary. Error rates can increase substantially when users employ words that are out-of-vocabulary (i.e., words the computer is not "listening" for), when the complexity of the grammar of possible phrases increases or when the microphone is not a high-quality close-talk headset. Even if the computer could recognize all of the user's words, the problem of understanding natural language is a significant and unsolved one. It can be avoided by using an artificial language of special commands or even a fairly restricted subset of natural language. But, given the current state of the art, the closer the user moves toward full unrestricted natural language, the more difficulties will be encountered.

Speech input is not the only form of voice-based input researchers have explored. Nonspeech voice input has been examined as well, particularly for scenarios in which speech commands are not well suited. Such scenarios may involve continuous control responses, as opposed to discrete responses. Since speech utterances themselves are discrete, they are well mapped to discrete commands (e.g., "save document") but poorly mapped to continuous activities (e.g., painting in a voice-driven paint program, maneuvering in a first-person shooter, or panning or zooming into documents or photographs). In addition, nonspeech voice has numerous qualities often ignored in speech: loudness, pitch, intonation, nasality, duration, timbre, prosody, vowel quality, and so on. Some of these qualities have been utilized, for example, in the *Vocal Joystick*, a voice control engine enabling continuous omnidirectional control by making vowel sounds that form a continuous circular phonetic space (Harada et al., 2006). Explorations utilizing the Vocal Joystick include drawing programs (Harada et al., 2007) and video games (Harada et al., 2011). Others also have explored controlling video games using nonspeech voice (Igarashi and Hughes, 2001). Even humming and whistling have been explored as forms of nonspeech input for text entry and gaming (Sporka et al., 2004, 2006a,b).

21.9.2 Pens

Pen-based input can emulate mouse-based input, for example, on a Tablet PC, or can support gesture-based input (see Section 21.9.4). Pens afford acting on documents, for example, crossing out a word to delete it or circling a paragraph and drawing an arrow to move it. Besides commands, pens can also provide *ink*, where strokes are not interpreted as commands but left as annotations. Pens thus constitute a versatile input modality, not just an isolated input device.

Pens have been employed in numerous specific interaction techniques besides just the emulation of mice or general gestural input. One of the most compelling is the *marking menu*. Marking menus use directional pen motion to provide rapid menu selection (Kurtenbach and Buxton, 1993; Kurtenbach et al., 1993). Marking menus have been incorporated into composite pen-based systems for indicating targets of an action and the action itself (Hinckley et al., 2005). Entire pen-specific interfaces for Tablet PCs have been explored that incorporate marking menus among other features (Hinckley et al., 2007). Marking menus have been improved or extended in numerous ways, for example, by enabling access to hierarchical menus with multiple discrete strokes (Zhao and Balakrishnan, 2004), using relative position and orientation (Zhao et al., 2006) or using two hands on a touch surface (Lepinski et al., 2010) (see also Section 21.9.3).

Of course, pen-specific interaction techniques go beyond marking menus. Multiple Tablet PCs can be "stitched" together using pen strokes that begin on the screen of one device, cross the bezel boundary, and end on the screen of an adjacent device (Hinckley et al., 2004). Pen pressure has been used in different ways, for example, to activate a magnifying lens for easier target acquisition (Ramos et al., 2007). Pen-based target acquisition has also been facilitated by slip-resistant bubble cursors (Moffatt and McGrenere, 2010) and target-directional beams (Yin and Ren, 2007).

The need for frequent mode switching in pen interfaces has inspired numerous interaction techniques. Barrel buttons, pen dwell time, pen pressure, and the "eraser end" of pens have been explored for mode switching (Li et al., 2005). Pens themselves have been made into multi-touch devices for grip sensing (Song et al., 2011). Even the tilt angle and azimuth of pens have been studied and used to control modes (Xin et al., 2011, 2012). Researchers have also explored multimodal pen and voice input; this is a powerful combination because pen and voice have complementary strengths and weaknesses and can disambiguate one another (Cohen and Sullivan, 1989; Cohen et al., 1997; Harada et al., 2007; Kurihara et al., 2006; Oviatt, 1997).

An important benefit of pen-based interfaces is their support for *sketching*. Sketching interfaces often raise the difficult issue of how to disambiguate users' marks as ink versus as commands (Kramer, 1994; Moran et al., 1997; Mynatt et al., 1999). One solution to this is *hover widgets* (Grossman et al., 2006), which uses the pen hover state on Tablet PCs for command and property selection, separate from inking on the screen. Pen input, via sketching, has been used to define 3D objects (Igarashi et al., 1999; Zeleznik et al., 1996). Computational support for pen-based sketching has a rich history. For a comprehensive review, readers are directed to a recent survey (Johnson et al., 2009).

21.9.3 Touch Input

With the widespread adoption of mobile touch-screen devices, particularly mobile phones based on Apple's iOS operating system or the Android operating system, touch interaction has become the mainstream. Although touch and gesture are often thought of together—and rightly so, as many gestures employ touch—they are not inherently the same thing, as one may have touch without gesture and gesture without touch. In the former case, incidental contact with no intended meaning such as brushing up against an object involves touch but not gesture. In the latter case, one may perform gestures in midair using a computer vision system without touching anything. Despite these examples, for many of today's interactions, touch and gesture are meaningfully coupled together. We cover touch in this section and gesture in the next (Section 21.9.4).

The earliest explorations of touch interaction were on touch tablets that were input-sensing surfaces separated from their output graphical displays. Such tablets often employed relative pointing with a cursor on the display. Although some tablets supported the use of pens, pucks, or other objects, touch tablets, as their name suggests, required no such apparatuses, enabling fingers to directly act on their surfaces (Buxton et al., 1985). Although touch tablets still exist, they are far outnumbered today by touch screens whose input and output surfaces are collocated.

Direct-touch interfaces on touch-screen devices or interactive tabletops usually operate without a cursor in absolute positioning mode. Although styli and pens have been used with touch-screen devices for many years (see Section 21.9.2), often on resistive touch screens, today's devices often employ capacitive touch-sensing techniques enabling the human finger to interact directly. Using the finger in direct-touch interfaces raises many challenges. One challenge is the "fat finger problem," in which the user's relatively large fingertip proves to be insufficiently precise when selecting small on-screen targets. A related challenge is the "occlusion problem," in which the user's finger or hand occludes any objects beneath it (Vogel and Casiez, 2012). Numerous explorations of these issues have been conducted with a variety of proposals for their amelioration. Such proposals include offset cursors and lift-off selection (Potter et al., 1988; Sears and Shneiderman, 1991), magnifying lenses (Roudaut et al., 2008; Vogel and Baudisch, 2007), zooming (Olwal et al., 2008), precision handles (Albinsson and Zhai, 2003), picture-in-picture-style "radar views" (Karlson and Bederson, 2007), probabilistic hit testing (Schwarz et al., 2010, 2011), and techniques utilizing finger orientation (Wang et al., 2009). In addition, a stream of work has studied and exploited interaction on the *backsides* of devices (or, in the case of interactive tabletops, the *undersides*), where fingers cannot possibly occlude portions of the graphical display (Baudisch and Chu, 2009; Hiraoka et al., 2003; Sugimoto and Hiroki, 2006; Wigdor et al., 2006, 2007; Wobbrock et al., 2007, 2008).

Finally, while the aforementioned work has innovated to ameliorate the challenges of touch, others have studied the properties of human touch itself (Wang and Ren, 2009). Findings indicate that users touch with consistent offsets from their perceived target location depending on their finger angle (Holz and Baudisch, 2010), and users align visible features on the *tops* of their fingers with the underlying target on the screen (Holz and Baudisch, 2011). Extracting fingerprints for modeling users and their touch styles can double touch accuracy (Holz and Baudisch, 2010), and if cameras can be used (e.g., above a tabletop), features atop users' fingers can be correlated with their perceived touch point for improved accuracy (Holz and Baudisch, 2011). It is also possible to distinguish with which *part* of a user's finger he or she touches the screen from the acoustic signature of touches (e.g., fingertip, finger pad, fingernail, or knuckle) (Harrison et al., 2011). Rather than distinguishing single touch points, it can be useful to support whole-hand touches that create entire touch regions (Cao et al., 2008).

21.9.4 Gestural Input

Gesture is a powerful input modality that takes place in a variety of forms: as strokes on 2D surfaces or as motions made in 3D space. Depending on the sensing technologies employed, gestures may require holding or wearing an object (e.g., a stylus or a Nintendo Wiimote), or gestures may be performed with the bare fingers or hands (e.g., on a capacitive touch screen or in the air before a Microsoft Kinect). Gestures are highly expressive, rapid, capable of symbolic or metaphorical association, and often entertaining for performers and observers alike. Although gestures have been part of HCI for years, in the last decade, gestures have become a mainstream input modality on commercialized platforms.

Gestures have a powerful property in that they can designate both the *object* of an action and the *action itself* in one fluid motion (Buxton et al., 1983). Thus, gestures support cognitive chunking by integrating command selection with specification of the command's scope (Kurtenbach and Buxton, 1991a,b).

Gestures may be defined in the simplest case by a single point, as in a *tap gesture* performed by a finger on a capacitive touch screen. At this level, the concepts of "touch" and "gesture" are trivially the same. It does not take long, however, for a touch unfolding over time to become a complex gesture involving *paths* or ordered sequences of points over time. Paths may be *unistrokes* (Goldberg and Richardson, 1993), meaning they begin upon receiving the first input point (e.g., "touch down"), unfold as that point moves (e.g., "touch move"), and end when the input points cease (e.g., "touch up"). Of course, a "touch" is only exemplary; a unistroke may be made with a stylus or in midair with a wand or entire hand. The defining aspect of a unistroke is that it is segmented (separated) from gestures that precede or succeed it by a clear signal equivalent to "begin" and "end" with no intervening segmentation signals. Gestures that comprise multiple *successive* unistrokes are called *multistrokes* (see, e.g., Hinckley et al., 2006). Gestures that comprise multiple *concurrent* unistrokes using fingers on a touch screen are called *multi-touch gestures*. The famous "pinch" gesture for resizing photographs is an example of a multi-touch gesture, where the user's thumb and forefinger each perform a unistroke concurrently.

Besides paths, gestures may be composed of *poses*. In the case of hand gestures, poses involve certain positions of the hand that convey meaning, such as a fist or flat palm. Whole-body gestures may also entail poses, such as raising one's hands above one's head (Cohn et al., 2012). Poses may be preserved over a path, such as a fist pose held while punching in a boxing game. Or poses themselves may be dynamic, changing either in place or over a path. Of course, as devices or other objects are incorporated into gestures, the possibilities grow according to the properties available. Gestures made with a rigid remote may be different than those available when using a bendable computer (Schwesig et al., 2004).

In an attempt to organize the high variation possible in gestures, numerous gesture taxonomies have been erected both within and outside the computing field. Early taxonomies by sociolinguists included categories such as Efron's *physiographics*, *kinetographics*, *ideographics*, *deictics*, and *batons* (Efron, 1941). Similarly, McNeill's taxonomy contained *iconics*, *metaphorics*, *deictics*, and *beats* (McNeill, 1992). Kendon placed his gestures on a spectrum of formality, identifying *sign languages* as the most formal and *gesticulation* as the least (Cadoz, 1994; Kendon, 1988) broadly categorizes hand gestures as *semiotic*, *ergotic*, and *epistemic*. Semiotic gestures are those used to communicate meaningful information, such as "thumbs up." Ergotic gestures are those used to manipulate physical objects. Epistemic gestures are exploratory movements to acquire haptic or tactile information; see also D. Kirsh (1995a,b).

The aforementioned gesture categories are not entirely foreign to those developed with computer input in mind. Deictic gestures in particular have received much attention, with several efforts using pointing, typically captured using instrumented gloves or camera-based recognition to interact with "intelligent" environments (Baudel and Beaudouin-Lafon, 1993; Freeman and Weissman, 1995; Jojic et al., 2000; Maes et al., 1996). Deictic gestures in combination with speech recognition have also been studied (Bolt, 1980; Hauptmann, 1989; Lucente et al., 1998; Wilson and Shafer, 2003). Explorations of tangible interaction techniques (Ullmer and Ishii, 1997) and efforts to sense movements and handling of sensor-enhanced mobile devices may be considered ergotic gestures (Harrison et al., 1998; Hinckley et al., 2000, 2003). Most research in hand gesture recognition focuses on empty-handed semiotic gestures (Cassell, 2003), which (Rime and Schiaratura, 1991) are further classified as follows:

Symbolic—conventional symbolic gestures such as "OK"
Deictic—pointing to fill in a semantic frame, analogous to deixis in natural language
Iconic—illustrating a spatial relationship
Pantomimic—mimicking an invisible tool, such as, pretending to swing a golf club

Wobbrock et al. studied gestures that users make on an interactive tabletop for accomplishing common actions like "move," "copy," and "delete" (Wobbrock et al., 2009b). Over 1000 gestures were collected from 20 people for 27 common commands. Findings indicated that apart from (Cadoz, 1994) *ergotic* gestures in which simulated physical objects are manipulated (e.g., moving a virtual object from one screen location to another), users employed widely varying gestures and had little gestural agreement. This led Wobbrock et al. to formulate the following taxonomy of surface gestures with four dimensions, each with various properties. Note that the "Nature" dimension represents a similar classification to those seen thus far:

Form—static pose, dynamic pose, static pose and path, dynamic pose and path, one-point touch, one-point path
Nature—symbolic, physical, metaphorical, abstract
Binding—object-centric, world-dependent, world-independent, mixed dependencies
Flow—discrete, continuous

Freeman et al. extended Wobbrock et al.'s "Form" dimension with the following categories that capture how a gesture begins and unfolds over time (Freeman et al., 2009):

Registration pose—single finger, multi-finger, single shape, multi-shape
Continuation pose—static, dynamic
Movement—no path, path

Not surprisingly, the use of gestures for computer input raises numerous challenges. With most forms of gestural input, errors of user intent and errors of computer interpretation seem inevitable (Bellotti et al., 2002). The learnability of gestures is another persistent challenge, as unlike visible buttons, menus, or hyperlinks, gestures are not trivially discoverable by users. Rather like speech commands, gestures must be *articulated* by users. Numerous efforts have explored how to teach gestures to users (Bau and Mackay, 2008; Bragdon et al., 2009; Fothergill et al., 2012; Freeman et al., 2009; Martin and Isokoski, 2008; Plimmer et al., 2011). Other work has attempted to assess and improve the guessability, learnability, and "naturalness" of gestures in the first place (Alexander et al., 2012; Grandhi et al., 2011; Kane et al., 2011; Költringer and Grechenig, 2004; MacKenzie and Zhang, 1997; Wobbrock et al., 2005, 2009b). Yet another challenge with gestures has been for system designers to choose appropriate and distinct gestures that avoid collision in the *feature-space* utilized by gesture recognizers (Long et al., 1999, 2000). Given the popularity of gestures, some researchers have worked to make gesture recognizers easier to incorporate into software prototypes (Anthony and Wobbrock, 2010, 2012; Henry et al., 1990; Landay and Myers, 1993; Li, 2010; Myers et al., 1997; Swigart, 2005; Wobbrock et al., 2007).

21.9.5 Bimanual Input

Aside from touch typing, most of the devices and modes of operation discussed thus far and in use today involve only one hand at a time. But people use both hands in a wide variety of the activities associated with daily life. For example, when writing, a right-hander writes with the pen in the right hand, but the left hand also plays a crucial and distinct role. It holds the paper and orients it to a comfortable angle that suits the right hand. In fact, during many skilled manipulative tasks, Guiard observed that the hands take on asymmetric, complementary roles (Guiard, 1987): for right-handers, the role of the left hand precedes the right (the left hand first positions the paper), the left hand sets the frame of reference for the action of the right hand (the left hand orients the paper), and the left hand performs infrequent, large-scale movements compared to the frequent, small-scale movements of the right hand (writing with the pen). Most applications for bimanual input to computers are characterized by asymmetric roles of the hands, including compound navigation/selection tasks such as scrolling a web page and then clicking on a link (Buxton and Myers, 1986), command selection using the nonpreferred hand (Bier et al., 1993; Kabbash et al., 1994), as well as navigation, virtual camera control, and object manipulation in 3D user interfaces (Balakrishnan and Kurtenbach, 1999; Hinckley et al., 1998b; Kurtenbach et al., 1997). Researchers have also applied this approach to keyboard design (MacKenzie and Guiard, 2001; McLoone et al., 2003). For some tasks, such as banging together a pair of cymbals, the hands may take on symmetric roles; for further discussion of bimanual symmetric tasks, see Guiard (1987) and Balakrishnan and Hinckley (2000).

21.9.6 Direct Muscle-Based Input and Brain–Computer Interfaces

Traditional input devices can be thought of as secondary sensors, in that they sense a physical action that is the consequence of cognition and muscle movements. An alternative approach is to attempt primary sensing by detecting brain activity and muscle movements directly. Muscle sensing is accomplished through electromyography, a technique previously employed for measuring muscular activity or controlling prosthetics. Saponas et al. demonstrated its use to enable sensing of muscle activation as fine-grained as detecting and identifying individual fingers (Saponas et al., 2009) and used in combination with touch-screen input to provide a richer data stream (Benko et al., 2009). Brain–computer interfaces (BCI) typically employ electroencephalography or functional near-infrared spectroscopy to detect input. Projects have used the technique to detect workload and user engagement (Hirshfield et al., 2009a) in order to conduct usability studies, as well as to explore the possible dynamic adaptation of user interfaces based on such metrics (Hirshfield et al., 2009b). Such work remains in its infancy, but appears to hold great promise (particularly as assistive technologies for users suffering from devastating injuries or other significant physical limitations) as sensing and signal processing techniques improve.

21.9.7 Passive Measurement: Interaction in the Background

Not all interactions with computers need consist of explicit, intentionally communicated commands. Think about walking into a grocery store with automatic doors. You approach the building; the doors sense this motion and open for you. No explicit communication has occurred, yet a computer has used your action of walking toward the store as an "input" to decide when to open the door. Intentional, explicit interaction takes place in the *foreground*, while implicitly sensed interaction takes place in the *background*, behind the fore of the user's attention (Buxton, 1995a). *Background sensing techniques* will be a major emphasis of future research in automation and sensing systems as users become increasingly mobile and become saturated with information from many sources. Researchers are currently exploring ways of providing context awareness through location sensing; ambient sensing of light, temperature, and other environmental qualities; movement and handling of devices (Goel et al., 2012); detecting the identity of the user and physical objects in the environment; and possibly even physiological measures such as heart-rate variability. This type of information potentially can allow technology to interpret the context of a situation and respond more appropriately (Dey et al., 2001; Hinckley et al., 2003; Schilit et al., 1994; Schmidt, 1999). However, like other recognition-based technologies, there is a risk of errors of user intent or computer interpretation: returning to the automatic door, for example, if you walk by (parallel to) the doors, they may sense your motion and open even if you have no intention of entering the building.

Background interaction can also be applied to explicit input streams through passive behavioral measurements, such as observation of typing speed, manner of moving the cursor (Evans and Wobbrock, 2012), sequence and timing of commands activated in a graphical interface (Horvitz et al., 1998), and other patterns of use. For example, a carefully designed user interface could make intelligent use of such information to modify its dialogue with the user, based on inferences about the user's alertness or expertise. These measures do not require additional input devices, but rather gleaning of additional, typically neglected, information from the existing input stream. These are sometimes known as intelligent or adaptive user interfaces, but mundane examples also exist. For example, cursor control using the mouse or scrolling using a wheel can be optimized by modifying the device response depending on the velocity of movement (Hinckley et al., 2001; Jellinek and Card, 1990).

We must acknowledge the potential for misuse or abuse of information collected in the background. Users should always be made aware of what information is or may potentially be observed as part of a human–computer dialogue. Users should have control and the ability to block any information that they want to remain private (Nguyen and Mynatt, 2001).

21.10 Displays and Perception

We now turn our attention to focus on the fundamental properties of displays and techniques for effective use of displays. We focus on visual displays and visual human perception, since these represent the vast majority of displays, but we also discuss feedback through the haptic and audio channels.

21.10.1 Properties of Displays and Human Visual Perception

Display requirements, such as resolution in time and space, derive from the properties of human vision. Thus, we begin with the basic issues relating to display brightness, uniformity, and spatial and temporal resolution.

21.10.1.1 Dynamic Range

The human eye has an enormous dynamic range. The amount of light reflected from surfaces on a bright day at the beach is about five orders of magnitude higher than the amount available under dim lamplights. Yet the shapes, layouts, and colors of objects look nearly identical to the human eye across much of this range. Most displays in common use are self-luminous CRTs or back-lit liquid crystal displays (LCDs). The best of these devices has a dynamic range (the ratio between the maximum and minimum values produced) of a little more than two orders of magnitude. In practice, under typical room lighting conditions, 15%–40% of the light reaching the user's eye is actually ambient room light reflected by the front surface of the phosphors or off of the screen surface. This means that the effective dynamic range of most devices, unless viewed in dark rooms, is no better than three or four to one. Fortunately the human eye can tolerate extreme variation in the overall level of illumination, as well as the amount of contrast produced by the display.

21.10.1.2 Spatial Frequency

The ability of the human visual system to resolve fine targets is known as *visual acuity.* A standard way of measuring visual acuity is to determine how fine a sinusoidal striped pattern can be discriminated from a uniform gray. Humans are capable of perceiving targets as fine as 50–60 cycles/degree of visual angle

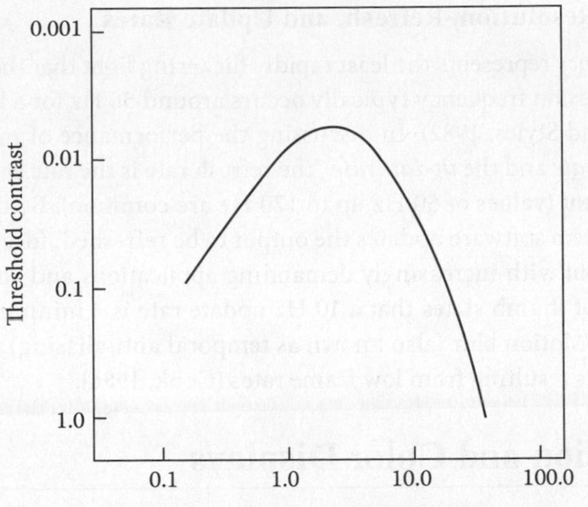

FIGURE 21.1 Spatial contrast sensitivity function of the human visual system. There is a falloff in sensitivity both to detailed patterns (high-spatial frequencies) and to gradually changing gray values (low spatial frequencies).

when the pattern is of very high contrast. Figure 21.1 illustrates the spatial sensitivity of the human eye as a function of spatial frequency. Specifically, it illustrates the degree of contrast required for sinusoidal gratings of different spatial frequencies to be perceived. The function has an inverted U shape with a peak at about two cycles per degree of visual angle. This means that 5 mm stripes at arm's length are optimally visible. The falloff at low spatial frequencies indicates that the human visual system is insensitive to gradual changes in overall screen luminance. Indeed, most CRTs have a brightness falloff toward the edges of as much as 20%, which we barely notice. This nonuniformity is even more pronounced with rear projection systems, due to the construction of screens that project light primarily in a forward direction. This is called the *screen gain*; a gain of 3.0 means that three times as much light is transmitted in the straight through direction compared to a perfect *Lambertian* diffuser. At other angles, less light is transmitted so that at a 45° off-axis viewing angle, only half as much light may be available compared to a perfect diffuser. Screen gain is also available with front projection with similar nonuniformities as a consequence, although the use of curved screens can compensate to some extent.

21.10.1.3 Spatial Resolution

The receptors in the human eye have a visual angle of about 0.8 s of arc. Modern displays provide approximately 40 pixels/cm. A simple calculation reveals that at about a 50 cm viewing distance, pixels will subtend about 1.5 s of arc, about two times the size of cone receptors in the center of vision. Viewed from 100 cm, such a screen has pixels that will be imaged on the retina at about the same size as the receptors. This might suggest that we are in reach of the perfect display in terms of spatial resolution; such a screen would require approximately 80 pixels/cm at normal viewing distances. However, under some conditions the human visual system is capable of producing *superacuities* that imply resolution better than the receptor size. For example, during fusion in stereo vision, disparities smaller than 5 s of arc can be detected (Westheimer, 1979); see also Ware (2000) for a discussion of stereopsis and stereo displays aimed at the practitioner. Another example of superacuity is known as aliasing, resulting from the division of the screen into discrete pixels; for example, a line on the display that is almost (but not quite) horizontal may exhibit a jagged "stairstep" pattern that is very noticeable and unsatisfying. This effect can be diminished by *anti-aliasing*, which computes pixel color values that are averages of all the different objects that contribute to the pixel, weighted by the percentage of the pixel they cover. Similar techniques can be applied to improve the appearance of text, particularly on LCD screens, where individual red, green, and blue display elements can be used for sub-pixel anti-aliasing (Betrisey et al., 2000; Platt, 2000).

21.10.1.4 Temporal Resolution, Refresh, and Update Rates

The *flicker fusion frequency* represents the least rapidly flickering light that the human eye does not perceive as steady. Flicker fusion frequency typically occurs around 50 Hz for a light that turns completely on and off (Wyszecki and Styles, 1982). In discussing the performance of monitors, it is important to differentiate the *refresh rate* and the *update rate*. The refresh rate is the rate at which a screen is redrawn and it is typically constant (values of 60 Hz up to 120 Hz are common). By contrast, the update rate is the rate at which the system software updates the output to be refreshed. Ideally, this should occur at or above the refresh rate, but with increasingly demanding applications and complex data sets, this may not be possible. A rule of thumb states that a 10 Hz update rate is a minimum for smooth animation (Robertson et al., 1989). Motion blur (also known as temporal anti-aliasing) techniques can be applied to reduce the jerky effects resulting from low frame rates (Cook, 1986).

21.11 Color Vision and Color Displays

The single most important fact relating to color displays is that human color vision is trichromatic; our eyes contain three receptors sensitive to different wavelengths. For this reason, it is possible to generate nearly all perceptible colors using only three sets of lights or printing inks. However, it is much more difficult to exactly specify colors using inks than using lights because, whereas lights can be treated as a simple vector space, inks interact in complex nonlinear ways.

21.11.1 Luminance, Color Specification, and Color Gamut

Luminance is the standard term for specifying brightness, that is, how much light is emitted by a self-luminous display. The luminance system in human vision gives us most of our information about the shape and layout of objects in space. The international standard for color measurement is the Commission Internationale de L'Eclairage (CIE) standard. The central function in Figure 21.2 is the CIE $V(\lambda)$ function, which represents the amount that light of different wavelengths contributes to the overall

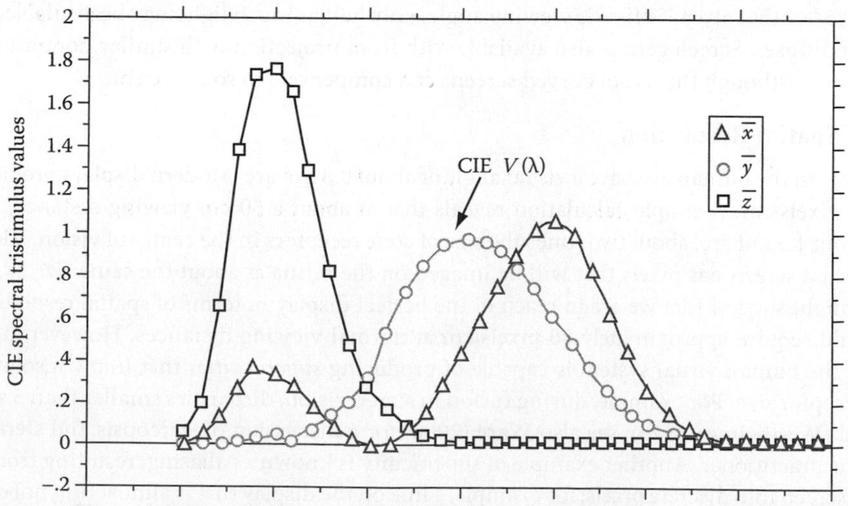

FIGURE 21.2 The CIE tristimulus functions. These are used to represent the standard observer in colorimetry. Short wavelengths at the left-hand side appear blue, in the middle they are green, and to the right they are red. Humans are most sensitive to the green wavelengths around 560 nm.

sensation of brightness. As this curve demonstrates, short wavelengths (blue) and long wavelengths (red) contribute much less than green wavelengths to the sensation of brightness. The CIE tristimulus functions, also shown in Figure 21.2, are a set of color-matching functions that represent the color vision of a typical person. Humans are most sensitive to the green wavelengths around 560 nm. Specifying luminance and specifying a color in CIE tristimulus values are complex technical topics; for further discussion, see Ware (2000) and Wyszecki and Styles (1982).

A chromaticity diagram can be used to map out all possible colors perceptible to the human eye, as illustrated in Figure 21.3. The pure spectral hues are given around the boundary of this diagram in nanometers (10^{-9} m). While the spacing of colors in tristimulus coordinates and on the chromaticity diagram is not perceptually uniform, *uniform color spaces* exist that produce a space in which equal metric distances are closer to matching equal perceptual differences (Wyszecki and Styles, 1982). For example, this can be useful to produce color sequences in map displays (Robertson, 1988).

The gamut of all possible colors is the dark-gray region of the chromaticity diagram, with pure hues at the edge and neutral tones in the center. The triangular region represents the gamut achievable by a particular color monitor, determined by the colors of the phosphors given at the corners of the triangle. Every color within this triangular region is achievable, and every color outside of the triangle is not. This diagram nicely illustrates the trade-off faced by the designer of color displays. A phosphor that produces a very narrow wavelength band will have chromaticity coordinates close to the pure spectral colors, and this will produce more saturated colors (thus enlarging the triangle). However, this narrow band also means that little light is produced.

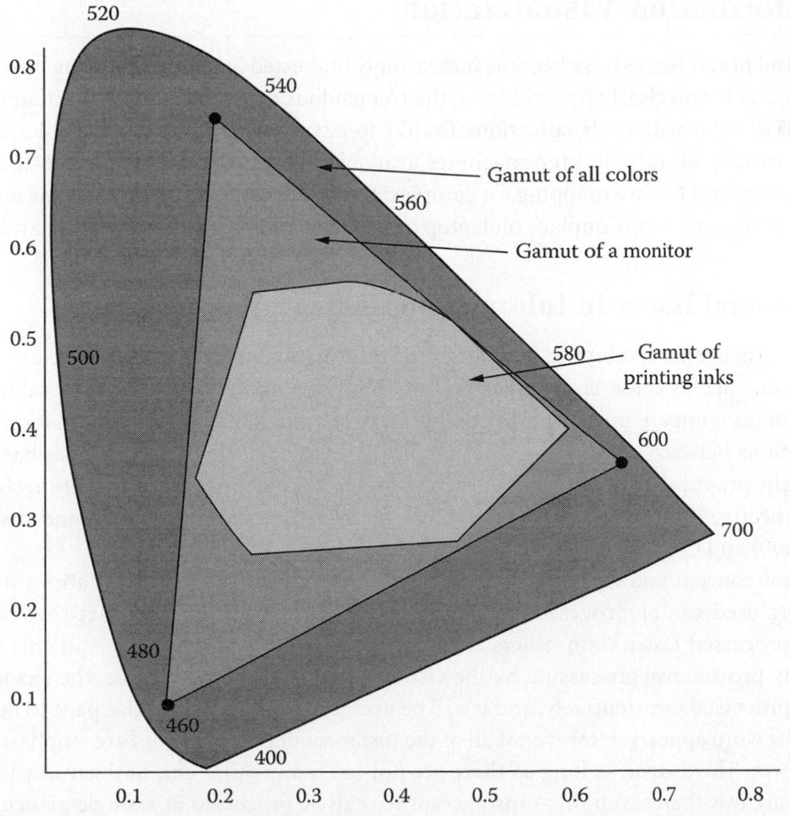

FIGURE 21.3 A CIE chromaticity diagram with a monitor gamut and a printing ink gamut superimposed. The range of available colors with color printing is smaller than that available with a monitor, and both fall short of providing the full range of color that can be seen.

The irregular shape inside the triangle illustrates the gamut of colors obtainable using printing inks. Notice that this set of colors is still smaller, causing difficulties when we try to obtain a hard copy reproduction of the colors on the monitor. Because the eye is relatively insensitive to overall color shifts and overall contrast changes, we can take the gamut from one device and map it into the gamut of the printing inks (or some other device) by compressing and translating it. This is known as gamut mapping, a process designed to preserve the overall color relationships while effectively using the range of a device (Stone et al., 1988). However, it should be noted that the original colors will be lost in this process and that after a succession of gamut mappings, colors may become distorted from their original values.

A process known as chromatic adaptation occurs in the human eye receptors and in the early stages of visual processing: for example, we hardly notice that daylight is much bluer than the yellow cast of tungsten light produced from ordinary light bulbs. The CIE standard does not account for chromatic adaptation nor does it account for color contrast (colors appear differently to the human eye depending on the surrounding visual field). The practical implication is that we can get by with color monitors and printers that are grossly out of calibration. However, accurate color is essential in some applications. It is possible to precisely calibrate a color device so that the particular inputs required to produce a color may be specified in CIE tristimulus values. For monitor calibration, see Cowan (1983); for calibrating print devices, see Stone et al. (1988). It is also possible to correct for the nonlinear response of CRT displays, a process known as *gamma correction*, but keep in mind that CRT designers intentionally insert this nonlinearity to match the human eye's sensitivity to relative changes in light intensity. If one desires a set of perceptually equal gray steps, it is usually best to omit gamma correction. See Ware (2000) for further discussion.

21.12 Information Visualization

Researchers and practitioners have become increasingly interested in communicating large quantities of information quickly and clearly by leveraging the tremendous capabilities of the human visual system, a field known as information visualization. Thanks to advances in computer graphics hardware and algorithms, virtually all new desktop machines available today have sophisticated full-color displays with transparency and texture mapping for complex 2D or 3D scenes, and it now seems inevitable these capabilities will become commonplace on laptop computers, and ultimately even on handheld devices.

21.12.1 General Issues in Information Coding

The greatest challenge in developing guidelines for information coding is that there are usually effective alternatives, such as color, shape, size, texture, blinking, orientation, and gray value. Although a number of studies compare one or more coding methods separately, or in combination, there are so many interactions between the task and the complexity of the display that guidelines based on science are not generally practical. However, Tufte provides excellent guidelines for information coding from an aesthetic perspective (Tufte, 1983, 1990, 1997). For further discussion, examples, and case studies, see also Ware (2000) and Card et al. (1999).

A theoretical concept known as *preattentive processing* has interesting implications for whether or not the coding used can be processed in parallel by the visual system. The fact that certain coding schemes are processed faster than others is called the pop-out phenomenon, and this is thought to be due to early preattentive processing by the visual system. Thus, for example, the shape of the word "bold" is not processed preattentively, and it will be necessary to scan this entire page to determine how many times the word appears. However, if all of the instances of the word *bold* are emphasized, they pop out at the viewer. This is true as long as there are not too many other emphasized words on the same page: if there are less than seven or so instances, they can be processed at a single glance. Preattentive processing is done for color, brightness, certain aspects of texture, stereo disparities, and object orientation and size. Codes that are preattentively discriminable are very useful if rapid search for information is desired (Triesman, 1985). The following visual attributes are known to be preattentive codes and, therefore, useful in differentiating information belonging to different classes:

- *Color*: Use no more than 10 different colors for labeling purposes.
- *Orientation*: Use no more than 10 orientations.
- *Blink coding*: Use no more than 2 blink rates.
- *Texture granularity*: Use no more than 5 grain sizes.
- *Stereo depth*: The number of depths that can be effectively coded is not known.
- *Motion*: Objects moving out of phase with one another are perceptually grouped. The number of usable phases is not known.

However, coding multiple dimensions by combining different pop-out cues is not necessarily effective (Ware, 2000).

21.12.2 Color Information Coding

When considering information display, one of the most important distinctions is between chromatic and luminance information, because these are treated quite differently in human perception. Gray scales are not perceived in the same way as rainbow-colored scales. A purely chromatic difference is one where two colors of identical luminance, such as red and green, are placed adjacent to one another. Research has shown that we are insensitive to a variety of information if it is presented through purely chromatic changes. This includes shape perception, stereo depth information, shape from shading, and motion. However, chromatic information helps us classify the material properties of objects. A number of practical implications arise from the differences in the way luminance and chromatic information is processed in human vision:

- Our spatial sensitivity is lower for chromatic information, allowing image compression techniques to transmit less information about hue relative to luminance.
- To make text visible, it is important to make sure that there is a luminance difference between the color of the text and the color of the background. If the background may vary, it is a good idea to put a contrasting border around the letters (e.g., Harrison and Vicente, 1996).
- When spatial layout is shown either through a stereo display or through motion cues, ensure adequate luminance contrast.
- When fine detail must be shown, for example, with fine lines in a diagram, ensure that there is adequate luminance contrast with the background.
- Chromatic codes are useful for labeling objects belonging to similar classes.
- Color (both chromatic and gray scale) can be used as a quantitative code, such as on maps, where it commonly encodes height and depth. However, simultaneous contrast effects can change the appearance of a patch of color depending on the surrounding colors; careful selection of colors can minimize this (Ware, 1988).

A number of empirical studies have shown color coding to be an effective way of identifying information. It is also effective if used in combination with other cues such as shape. For example, users may respond to targets faster if they can be identified by both shape and color differences (for useful reviews, see Christ [1975]; Stokes et al. [1990]; and Silverstein [1977]). Color codes are also useful in the perceptual grouping of objects. Thus, the relationship between a set of different screen objects can be made more apparent by giving them all the same color. However, it is also the case that only a limited number of color codes can be used effectively. The use of more than about 10 will cause the color categories to become blurred. In general, there are complex relationships between the type of symbols displayed (e.g., point, line, area, or text), the luminance of the display, the luminance and color of the background, and the luminance and color of the symbol (Spiker et al., 1985).

21.12.3 Integrated Control-to-Display Objects

When the purpose of a display is to allow a user to integrate diverse pieces of information, it may make sense to integrate the information into a single visual object or glyph (Wickens, 1992). For example, if the purpose is to represent a pump, the liquid temperature could be shown by changing the color of the pump, the capacity could be shown by the overall size of the pump, and the output pressure might be represented by the changing height of a bar attached to the output pipe, rather than a set of individual dials showing these attributes separately. However, perceptual distortions can result from an ill chosen display mapping, and the object display may introduce visual clutter: if there are 50 pumps to control, then the outlines of all the pumps may interfere with the data of interest (Tufte, 1983). In object displays, input and output can be integrated in a manner analogous to widgets such as the scrollbar, or even more directly by having input devices that resemble the physical object being handled, known as a prop (Hinckley et al., 1994a). For some good examples of the linking of output and input, see Ahlberg and Shneiderman (1994) as well as Ishii and Ullmer (1997).

This style of presentation and interaction can be especially relevant for telepresence or augmented reality applications, where the user needs to interact with actual physical objects that have attributes that must be viewed and controlled (Feiner et al., 1993; Tani et al., 1992). For more abstract data representation tasks, choosing the color, size, orientation, or texture to represent a particular data attribute may be difficult, and there seem to be practical limits on the number of attributes that one can encode simultaneously. Thus, object displays must usually be custom designed for each different display problem. In general, this means that the display and controls should somehow match the user's cognitive model of the task (Cole, 1986; Norman, 1990).

21.12.4 3D Graphics and Virtual Reality

Much research in 3D information visualization and **virtual reality** is motivated by the observation that humans naturally operate in physical space and can intuitively move about and remember where things are (an ability known as spatial memory). However, translating these potential benefits to artificially generated graphical environments is difficult because of limitations in display and interaction technologies. Virtual environment research pushed this to the limit by totally immersing the user in an artificial world of graphics, but this comes at the cost of visibility and awareness of colleagues and objects in the real world. This has led to research in so-called *fish tank virtual reality* displays by using a head tracking system in conjunction with a stereo display (Deering, 1992; Ware et al., 1993) or a mirrored setup, which allows superimposition of graphics onto the volume where the user's hands are located (Schmandt, 1983; Serra et al., 1997). However, much of our ability to navigate without becoming lost depends upon the vestibular system and spatial updating as we physically turn our bodies, neither of which is engaged with stationary displays (Chance et al., 1998; Loomis et al., 1999). For further discussion of navigation in virtual environments, see Darken and Sibert (1993, 1995); for application of spatial memory to 3D environments, see Robertson et al. (1998, 1999).

Outside of virtual reality, *volumetric displays* present imagery in true 3D space, by illuminating "voxels" (volumetric pixels) in midair. Favalora provides a thorough survey of the various technological implementations of volumetric displays (Favalora, 2005). The true 3D imagery in volumetric displays has been shown to improve depth perception (Grossman and Balakrishnan, 2006b) and shape recognition (Rosen et al., 2004). Besides providing true 3D images, the main difference from the other 3D tabletop displays is that volumetric display is generally enclosed by a surface. This means that users cannot directly interact with the 3D imagery. Balakrishnan et al. explored the implications of this unique difference to interaction design by using physical mock-ups (Balakrishnan et al., 2001). More recent working implementations have allowed users to interact with the display by using hand and finger gestures on and above the display surface (Grossman et al., 2004) and by using a handheld six degree-of-freedom input device (Grossman and Balakrishnan, 2006a). Another key differentiator is that volumetric displays better support collocated multiuser interaction, given a shared view of 3D imagery, without the need to wear glasses or head-mounted displays that can interfere with natural collaboration. Grossman and Balakrishnan describe various techniques that can overcome the limitations of multiuser interaction with volumetric displays (Grossman and Balakrishnan, 2008).

21.12.5 Augmented Reality

Augmented reality superimposes information on the surrounding environment rather than blocking it out. For example, the user may wear a semitransparent display that has the effect of projecting labels and diagrams onto objects in the real world. It has been suggested that this may be useful for training people to use complex systems or for fault diagnosis. For example, when repairing an aircraft engine, the names and functions of parts could be made to appear superimposed on the parts seen through the display together with a maintenance record if desired (Caudell and Mizell, 1992; Feiner et al., 1993). The computer must obtain a detailed model of the environment; otherwise it is not possible to match the synthetic objects with the real ones. Even with this information, correct registration of computer graphics with the physical environment is an extremely difficult technical problem due to measurement error and system latency. This technology has been applied to heads-up displays for fighter aircraft, with semitransparent information about flight paths and various threats in the environment projected on the screen in front of the pilot (Stokes et al., 1990), as well as digitally augmented desk surfaces (Wellner, 1993).

21.13 Scale in Displays

It is important to consider the full range of scale for display devices and form factors that may embody an interaction task. Computer displays increasingly span orders of magnitude in size and available computational resources, from watches, handheld personal data assistants (PDAs), tablet computers, and desktop computers, all the way up to multiple-monitor and wall-size displays. A technique that works well on a desktop computer, such as a pull-down menu, may be awkward on a small handheld device or even unusable on a wall-size display (where the top of the display may not even be within the user's reach). Each class of device seems to raise unique challenges, and the best approach may ultimately be to design special-purpose, appliance-like devices (see Want and Borriello [2000] for a survey) that suit specific purposes.

21.13.1 Small Displays

Users increasingly want to do more and more on handheld devices, mobile phones, pagers, and watches that offer less and less screen real estate. Researchers have investigated various strategies for conserving screen real estate. Transparent overlays allow divided attention between foreground and background layers (Harrison and Vicente, 1996; Harrison et al., 1995a,b; Kamba et al., 1996), but some degree of interference seems inevitable. This can be combined with sensing which elements of an interface are being used, such as presenting widgets on the screen only when the user is touching a pointing device (Hinckley and Sinclair, 1999). Researchers have also experimented with replacing graphical interfaces with *graspable* interfaces that respond to tilting, movement, and physical gestures that do not need constant on-screen representations (Fitzmaurice et al., 1995; Harrison et al., 1998; Hinckley et al., 2000; Rekimoto, 1996). Much research in focus plus context techniques, including fish-eye magnification (Bederson, 2000) and zooming metaphors (Bederson et al., 1996; Perlin and Fox, 1993; Smith and Taivalsaari, 1999), has also been motivated by providing more space than the boundaries of the physical screen can provide. Researchers have started to identify principles and quantitative models to analyze the trade-offs between multiple views and zooming techniques (Baudisch et al., 2002; Plumlee and Ware, 2002). There has been considerable effort devoted to supporting web browsing in extremely limited screen space (Baudisch et al., 2004; Buyukkokten et al., 2001; Jones et al., 1999; Trevor et al., 2001; Wobbrock et al., 2002).

21.13.2 Multiple Displays

Researchers have recently recognized that some very interesting design issues arise when multiple displays are considered, rather than the traditional single display of desktop computers. Having multiple monitors for a single computer is not like having one large display (Grudin, 2001). Users employ the boundary between displays to partition their tasks, with one monitor being reserved for a primary task and other monitors being used for secondary tasks. Secondary tasks may support the primary task (e.g., reference material, help files, or floating tool palettes), may provide peripheral awareness of ongoing events (such as an e-mail client), or may provide other background information (to-do lists, calendars, etc.). Switching between applications has a small time penalty (incurred once to switch and again to return), and perhaps more importantly, it may distract the user or force the user to remember information while switching between applications. Having additional screen space "with a dedicated purpose, always accessible with a glance" (Grudin, 2001), reduces these burdens (Czerwinski et al., 2003), and studies suggest that providing multiple, distinct foci for interaction may aid users' memory and recall (Tan et al., 2001, 2002). Finally, small displays can be used in conjunction with larger displays (Myers et al., 1998, 2000; Rekimoto, 1998), with controls and private information on the small device and shared public information on the larger display. This shows how displays of different dimensions support completely different user activities and social conventions. It is also possible to dynamically join multiple displays for collaboration or to create a larger but temporary tiled display (Hinckley, 2003a,b; Tandler et al., 2001).

21.13.3 Large-Format Displays

Trends in display technology suggest that large-format displays will become increasingly affordable and common. A recent journal special issue includes numerous articles on implementing large-format displays using projection, application design for large displays, and specific application domains such as automotive design (Funkhouser and Li, 2000). Large displays often implicitly suggest multiple simultaneous users, with many applications revolving around collaboration (Funkhouser and Li, 2000; Swaminathan and Sato, 1997) and giving a large-scale physical presence to virtual activities (Buxton et al., 2000). To support input directly on whiteboard-size displays, researchers have explored gestural interaction techniques for pens or touch screens (Guimbretiere et al., 2001; Moran et al., 1997). Some technologies cannot handle more than one point of contact, so system developers must check this carefully if simultaneous use by multiple persons is desired. Large displays also seem to lend themselves to interaction at a distance, although using laser pointers to support such interaction (Myers et al., 2002; Olsen and Nielsen, 2001) has met with mixed success due to the lack of separation between tracking versus dragging states (Buxton, 1990b); using small handheld devices to interact with the full area of a large display also is problematic as the ratio of the display size to the control surface size may be very large (Myers et al., 1998). Environmentally situated ambient displays share some properties of large displays, but emphasize subtle presentation of information in the periphery of attention (Ishii and Ullmer, 1997; Wisneski et al., 1998). Large-format displays and virtual realities also share some design issues; see the taxonomy of Buxton and Fitzmaurice for further discussion (Buxton and Fitzmaurice, 1998).

Unless life-size viewing of large objects is necessary (Buxton et al., 2000), in general it is not yet clear what performance benefits a single large display may offer as compared to multiple monitors with the same screen area partitioned by bezels (Czerwinski et al., 2003). One recent study suggests that the increased field of view afforded by large-format displays can lead to improved 3D navigation performance, especially for women (Czerwinski et al., 2002).

21.14 Force and Tactile Displays

Haptic feedback research has sought to provide an additional channel of sensory feedback by synthesizing forces on the skin of the operator. The touch sensation is extraordinarily complex.

In fact, the sense of "touch" is a very imprecise term: it includes an amalgamation of multiple sensory systems, including sensitivity to pressure, small shear forces in the skin, heat and cold, pain, kinesthesis and proprioception, and the vestibular system (Burdea, 1996; MacKenzie and Iberall, 1994).

There appears to be no physical means by which a complex tactile stimulus can be delivered except in a very localized way. As a result, most haptic feedback devices are limited to simulation of a single point of contact, analogous to feeling the world with the tip of a pencil, although a few examples of whole-hand force feedback devices exist (Burdea, 1996; Iwata, 1990). Efforts in haptic feedback include force feedback (active presentation of forces to the user) and tactile feedback (active presentation of vibrotactile stimuli to the user). Haptic feedback is popular for gaming devices, such as force feedback steering wheels and joysticks, but general-purpose pointing devices with force or tactile feedback remain uncommon. For a comprehensive discussion of force and tactile feedback technologies and techniques, as well as perceptual properties of the skin and joints, see Burdea (1996).

Adding force feedback to a mouse or stylus may impose constraints on the mechanical design, since a physical linkage is typically needed to reflect true forces. This may prevent a force feedback mouse from functioning like a traditional mouse, as it may limit range of motion or preclude clutching by lifting the device. Some devices instead increase resistance between the mouse and the pad, but this prevents simulation of hard contact forces. One can also use a vibrotactile stimulus, such as a vibrating pin under the mouse button or a vibrating shaft of an isometric joystick (Campbell et al., 1999). Combination devices have also been explored (Akamatsu and MacKenzie, 1996). Vibrotactile feedback seems especially promising for small mobile devices, for example, to provide the user with feedback of command recognition when the user's attention may not be focused on the screen (Poupyrev et al., 2002). Applications for remote controls and augmented handles also look promising (MacLean et al., 2000; Snibbe and MacLean, 2001).

Using force feedback to provide attractive forces that pull the user toward a target, or tactile feedback to provide additional feedback for the boundaries of the target, has been found to yield modest speed improvements in some target acquisition experiments, although error rates may also increase (Akamatsu and MacKenzie, 1996; MacKenzie, 1995). However, there have been almost no published studies for tasks where multiple targets are present, as on a computer screen with many icons and menus. Haptic feedback for one target may interfere with the selection of another, unless one uses techniques such as reducing the haptic forces during rapid motion (Oakley et al., 2001). Finally, one should also consider whether software constraints, such as snap to grids, are sufficient to support the user's tasks.

The construction of force output devices is extremely technically demanding. They must be stiff in order to be able to create the sensation of solid contact, yet light so that they have little inertia themselves, and there must be a tight loop between input (position) and output (force). Sigoma (1993) has suggested that having this loop iterated at 5 kHz may be necessary for optimal fine motor control. It has been shown that force feedback improves performance in certain telerobotic applications when, for example, inserting a peg into a hole (Sheridan, 1992). The most promising applications of force output seem to appear in domains where simulation of force is essential, such as surgical simulation and telerobotics (Burdea, 1996).

Another fundamental challenge for haptic feedback techniques results from the interaction between the haptic and visual channels. Visual dominance deals with phenomena resulting from the tendency for vision to dominate other modalities (Wickens, 1992). Campbell et al. (1999) show that tactile feedback improves steering through a narrow tunnel, but only if the visual texture matches the tactile texture; otherwise, tactile feedback harms performance.

21.15 Auditory Displays

Here, we consider computer-generated auditory feedback. Speech audio can consist of synthesized or recorded speech. All other audio feedback is known as nonspeech audio. With stereo speakers or a stereo headset, either type of audio can be presented such that it seems to come from a specific 3D location around the user, known as spatialized audio. For speech input and technology-mediated human–human communication applications that treat stored voice as data, see Section 21.8.1.

21.15.1 Nonspeech Audio

Nonspeech auditory feedback is prevalent in video games but largely absent from other interaction with computing devices. Providing an auditory echo of the visual interface has little or no practical utility and may annoy users. Audio should be reserved to communicate simple, short messages that complement visual feedback (if any). Furthermore, one or more of the following conditions should hold: the message should (1) deal with events in time, (2) call for immediate action, or (3) take place when the user's visual attention may be overburdened or directed elsewhere (Buxton, 1995b; Deatherage, 1972). For example, researchers have attempted to enhance scrollbars using audio feedback (Brewster et al., 1994). However, the meaning of such sounds may not be clear. Gaver advocates ecological sounds that resemble real-world events with an analogous meaning. For example, an empty disc drive might sound like a hollow metal container (Gaver, 1989). If a long or complex message must be delivered using audio, it will likely be quicker and clearer to deliver it using speech output. Audio feedback may be crucial to support tasks or functionality on mobile devices that must take place when the user is not looking at the display (for some examples, see Hinckley et al., 2000).

Nonspeech sounds can be especially useful for attracting the attention of the user. Auditory alerting cues have been shown to work well, but only in environments where there is low auditory clutter. However, the number of simple nonspeech alerting signals is limited, and this can easily result in misidentification or cause signals to mask one another. An analysis of sound signals in fighter aircraft (Doll and Folds, 1985) found that the ground proximity warning and the angle-of-attack warning on an F16 were both an 800 Hz tone, a dangerous confound since these conditions require opposite responses from the pilot. It can also be difficult to devise nonspeech audio events that convey information without provoking an alerting response that unnecessarily interrupts the user. For example, this design tension arises when considering nonspeech audio cues that convey various properties of an incoming e-mail message (Hudson and Smith, 1996; Sawhney and Schmandt, 2000).

21.15.2 Speech Output

Speech auditory output is generally delivered through either recorded speech segments or completely synthetic speech (also known as text-to-speech technology). There has been considerable interest, especially for military applications, in the use of speech in providing warnings to the operators of complex systems. Speech can provide information to direct the operator's attention in a way that alarms cannot (since an unfamiliar alarm simply indicates a problem, without telling the user the nature or context of the problem). Synthetic speech is most useful where visual information is not available, for example, in touch-tone phone menu systems or in screen reader software for blind or low-vision users. Although progress is being made, synthetic voices still sound somewhat unnatural and may be more difficult for users to understand. Recorded speech is often used to give applications, particularly games, a more personal feel, but can only be used for a limited number of responses known in advance.

The rate at which words must be produced to sound natural is a narrow range. For warning messages, 178 words per minute is intelligible but hurried, 123 words per minute is distracting and irritatingly slow, and a more natural rate of 156 words per minute is preferred (Simpson and Marchionda-Frost, 1984). The playback rate of speech can be increased by overlapping samples in time such that one sample is presented to one ear and another sample to the other ear. Technologies to correct for pitch distortions and remove pauses have also been developed (Arons, 1993; Sawhney and Schmandt, 2000; Stifelman, 1996). It is recommended by the US Air Force that synthetic speech be 10 dB above ambient noise levels (Stokes et al., 1990).

21.15.3 Spatialized Audio Displays

It is possible to synthesize spatially localized sounds with a quality such that spatial localization in the virtual space is almost as good as localization of sounds in the natural environment (Wenzel, 1992). Auditory localization appears to be primarily a 2D phenomenon, that is, observers can localize in horizontal position (azimuth) and elevation angle to some degree of accuracy. Azimuth and elevation accuracies are of the order of 15°. As a practical consequence this means that sound localization is of little use in identifying sources in conventional screen displays. Where localized sounds are really useful is in providing an orienting cue or warning about events occurring behind the user, outside of the field of vision.

There is also a well-known phenomenon called visual capture of sound. Given a sound and an apparent visual source for the sound, for example, a talking face on a cinema screen, the sound is perceived to come from the source despite the fact that the actual source may be off to one side. Thus, visual localization tends to dominate auditory localization when both kinds of cues are present.

21.16 Future Directions

The future of interaction with computers will both be very different and very much like it is today. Some of our current tools, such as mice and keyboards, have evolved to suit interaction with desktop GUIs and rapid text entry. As long as users' work continues to involve tasks such as calculating budgets, writing reports, looking up citations, exchanging memos, and other knowledge worker tasks that seem to lend themselves to solution using desktop computers, we will continue to see mice and keyboards in use, not only because they are familiar but also because they closely match human skills and the requirements of the tasks. Devising new techniques that provide more efficient pointing at a desktop display than a mouse, for example, is difficult to achieve (Card et al., 1978). Speech recognition will allow new types of interaction and may enable interaction where it previously has been difficult or infeasible. However, even as technical limitations are removed, speech interaction will not replace all forms of interaction: we will continue to interact with computers using our hands and physical intermediaries, not necessarily because our technology requires us to do so but because touching, holding, and moving physical objects is the foundation of the long evolution of tool use in the human species (Wilson, 1998).

But our computers and the tasks they serve are rapidly evolving. Current handheld devices have the display and computational capabilities of common desktop machines from several years ago. What is lacking are new methods of interacting with such devices that uniquely suit mobile interaction, rather than derivatives of the desktop interface. Researchers are still actively exploring and debating the best ways to achieve this. Meanwhile, technology advances and economic trends continue to drive the cost of commodity displays lower and lower, while the limits of the technology continue to increase. Thus, we will continue to see new innovations in both very small and very large displays, and as these become commonplace, new forms of interaction will become prevalent. Very small displays invariably seem to be incorporated into input/output appliances such as watches, pagers, and handheld devices, so interaction techniques for very small form factors will become increasingly important.

The Internet and wireless networking seem to be the main disruptive technologies of the current era. Indeed, it seems likely that 100 years from now the phrase "wireless network" will seem every bit as antiquated as the phrase "horseless carriage" does today. Nobody really understands yet what it will mean for everything and everyone to be connected, but many researchers are working to explore the vision of ubiquitous computing originally laid out by Weiser (1991). Techniques that allow users to communicate and share information will become increasingly important. Biometric sensors or other convenient means for establishing identity will make services such as personalization of the interface and sharing data much simpler (Rekimoto, 1997; Sugiura and Koseki, 1998; Westeyn et al., 2005). Techniques that combine dissimilar input devices and displays in interesting ways also will be important to realize the full potential of these technologies (e.g., Myers et al., 2001; Streitz et al.,

1999). Electronic tagging techniques for identifying objects (Want et al., 1999) may also become commonplace. Such a diversity of locations, users, and task contexts points to the increasing importance of sensors to acquire contextual information, as well as machine learning techniques to interpret them and infer meaningful actions (Bellotti et al., 2002; Buxton, 1995a; Hinckley et al., 2003). This may well lead to an age of ubiquitous sensors (Saffo, 1997) with devices that can see, feel, and hear through digital perceptual mechanisms.

Key Terms

Absolute input device: An input device that reports its actual position, rather than relative movement. A tablet or touch screen typically operates this way (see also relative input device).

Acquisition time: The average time to pick up or put down an input device. It is sometimes known as homing time.

Anti-aliasing: The specification of pixel color values so that they reflect the correct proportions of the colored regions that contribute to that pixel. In temporal anti-aliasing the amount of time a region of a simulated scene contributes to a pixel is also taken into account.

Augmented reality: The superimposition of artificially generated graphical elements on objects in the environment. It is achieved with a see-through head-mounted display.

Background sensing techniques: Implicitly sensed interaction takes place in the background, behind the fore of the user's attention. Background sensing techniques use sensor technology or intelligent algorithms to glean additional, typically neglected, information from the existing input stream, with the goal of supporting the user with semiautomatic or implicit actions and services.

Cognitive chunk: A series of elemental tasks that seems like a single concept to the user. For example, users think of pointing at something as a single chunk, but from a technical perspective it may consist of selecting an (X, Y, Z) coordinate in a 3D environment. By using technologies and interaction metaphors that parallel the way the user thinks about a task as closely as possible, the designer can phrase together a series of elemental tasks into a single cognitive chunk.

Compound tasks: A compound task is a hierarchy of elemental subtasks. For example, the navigate/select compound task consists of scrolling to view an item in a list and then clicking on it to select it. When interacting with a graphical scrollbar, scrolling itself may be a compound task with multiple selection or positioning tasks.

C:D ratio: The ratio between the movement a user must make with an input device and the resulting movement obtained on the display. With a large C:D ratio, a large movement is required to effect a small change on the display, affording greater precision. A low ratio allows more rapid operation and takes less desk space. The C:D ratio is sometimes expressed as a single number, in which case it is referred to as the device gain. Note that many experts have criticized gain as a fundamental concept; one must take great care when manipulating gain in experiments, since it confounds display size and control size in one arbitrary metric.

Direct input device: A device that the user operates directly on the screen or other display to be controlled, such as a touch screen (see also indirect input device).

Fish tank virtual reality: A form of virtual reality display that confines the virtual scene to the vicinity of a monitor screen.

Fitts' law: A model that relates the movement time to point at a target, the amplitude of the movement (the distance to the target), and the width of the target (i.e., the precision requirement of the pointing movement). The movement time is proportional to the logarithm of the distance divided by the target width, with constant terms that vary from one device to another. Fitts' law has found wide application in HCI to evaluating and comparing input devices and transfer functions for pointing at targets.

Flicker fusion frequency: The frequency at which a flickering light is perceived as a steady illumination. It is useful in determining the requirements for a visual display.

Footprint: The physical movement space (area) required to operate an input device.

Fovea: The central part of the retina at which vision is the sharpest, about 2° of visual angle in diameter.

Gamma correction: The correction of nonlinearities of a monitor so that it is possible to specify a color in linear coordinates.

Indirect input device: A device that the user operates by moving a control that is located away from the screen or other display to be controlled, such as a mouse or trackball (see also direct input device).

Input device: A hardware computer peripheral through which the user interacts with the computer.

Interaction task: A low-level primitive input to be obtained from the user, such as entering a text string or choosing a command.

Interaction technique: The fusion of input and output, consisting of all hardware and software elements, that provides a particular way for the user to accomplish a low-level task with a physical input device. For example, the pop-up menu is an interaction technique for choosing a command or other item from a small set, using a mouse and a graphical display.

Lambertian diffuser: A diffuser that spreads incoming light equally in all directions.

Latency: The end-to-end delay between the user's physical movement and the system's ultimate feedback to the user. Latency of more than 75–100 ms significantly impairs user performance for many interactive tasks.

Luminance: The standard way of defining an amount of light. This measure takes into account the relative sensitivities of the human eye to light of different wavelengths.

Preattentive processing: Visual stimuli that are processed at an early stage in the visual system in parallel. This processing is done prior to processing by the mechanisms of visual attention.

Refresh rate: The rate at which a computer monitor is redrawn. It is sometimes different from the update rate.

Relative input device: An input device that reports its distance and direction of movement each time it is moved, but cannot report its absolute position. A mouse operates this way (see absolute input device).

Screen gain: A measure of the amount by which a projection video screen reflects light in a preferred direction. The purpose is to give brighter images if viewed from certain positions. There is a corresponding loss in brightness from other viewing positions.

Superacuities: The ability to perceive visual effects with a resolution that is finer than can be predicted from the spacing of receptors in the human eye.

Three-state model: A model for the discrete states of input devices that models transitions between three states: tracking, dragging, and out of range. Most input devices only sense two of these three states (e.g., a mouse senses tracking and dragging, whereas a touchpad senses tracking and the out-of-range state).

Transfer function. A mathematical transformation that scales the data from an input device to ideally provide smooth, efficient, and intuitive operation. Appropriate mappings are transfer functions that match the physical properties sensed by the input device and include force-to-velocity, position-to-position, and velocity-to-velocity functions.

Uniform color space: A transformation of a color specification such that equal metric differences between colors more closely correspond to equal perceptual differences.

Update rate: The rate at which the image on a computer monitor is changed.

Virtual reality: A method of monitoring a user's head position and creating a perceptive view of an artificial world that changes as the user moves, in such a way as to simulate an illusory 3D scene.

Visual acuity: The ability of the human visual system to resolve fine targets.

References

Accot, J. and S. Zhai (1997). Beyond Fitts' law: Models for trajectory-based HCI tasks. *Proceedings of CHI'97: ACM Conference on Human Factors in Computing Systems*, Atlanta, GA. pp. 295–302. http://portal.acm.org/citation.cfm?id=258760

Accot, J. and S. Zhai (1999). Performance evaluation of input devices in trajectory-based tasks: An application of the Steering law. *Proceedings of CHI'99*, Pittsburgh, PA. pp. 466–472.

Accot, J. and S. Zhai (2001). Scale effects in Steering law tasks. *Proceedings of CHI'2001. ACM Conference on Human Factors in Computing Systems*, Seattle, WA. pp. 1–8.

Accot, J. and S. Zhai (2002). More than dotting the i's—Foundations for crossing-based interfaces. *ACM CHI 2002 Conference on Human Factors in Computing Systems*, Minneapolis, MN. pp. 73–80. http://portal.acm.org/citation.cfm?id=503390

Ahlberg, C. and B. Shneiderman (1994). The alphaslider: A compact and rapid selector. *Proceedings of CHI'94*, Boston, MA. pp. 365–371.

Ahlström, D., M. Hitz, and G. Leitner (2006). An evaluation of sticky and force enhanced targets in multi target situations. Paper presented at the *Proceedings of the Nordic Conference on Human–Computer Interaction (NordiCHI'06)*, Oslo, Norway. http://portal.acm.org/citation.cfm?id=1182482

Akamatsu, M. and I. S. MacKenzie (1996). Movement characteristics using a mouse with tactile and force feedback. *International Journal of Human–Computer Studies* **45**: 483–493.

Albinsson, P.-A. and S. Zhai (2003). High precision touch screen interaction. Paper presented at the *Proceedings of the ACM Conference on Human Factors in Computing Systems (CHI'03)*, Ft. Lauderdale, FL. http://dl.acm.org/citation.cfm?id=642631

Alexander, J., T. Han, W. Judd, P. Irani, and S. Subramanian (2012). Putting your best foot forward: Investigating real-world mappings for foot-based gestures. Paper presented at the *Proceedings of the ACM Conference on Human Factors in Computing Systems (CHI'12)*, Austin, TX. http://dl.acm.org/citation.cfm?id=2208575

Anderson, J. R. (1980). Cognitive skills, Chapter 8. In *Cognitive Psychology and Its Implications*. San Francisco, CA: W. H. Freeman. pp. 222–254.

Anthony, L. and J. O. Wobbrock (2010). A lightweight multistroke recognizer for user interface prototypes. Paper presented at the *Proceedings of Graphics Interface (GI'10)*, Ottawa, Ontario, Canada. http://portal.acm.org/citation.cfm?id=1839214.1839258

Anthony, L. and J. O. Wobbrock (2012). $N-Protractor: A fast and accurate multistroke recognizer. Paper presented at the *Proceedings of Graphics Interface (GI'12)*, Toronto, Ontario, Canada.

Appert, C., O. Chapuis, and M. Beaudouin-Lafon (2008). Evaluation of pointing performance on screen edges. Paper presented at the *Proceedings of the ACM Working Conference on Advanced Visual Interfaces (AVI'08)*, Napoli, Italy. http://portal.acm.org/citation.cfm?id=1385569.1385590

Arons, B. (1993). SpeechSkimmer: Interactively skimming recorded speech. *UIST'93 Symposium on User Interface Software & Technology*, Atlanta, GA. pp. 187–195.

Balakrishnan, R. (2004). "Beating" Fitts' law: Virtual enhancements for pointing facilitation. *International Journal of Human–Computer Studies* **61**(6): 857–874.

Balakrishnan, R., T. Baudel, G. Kurtenbach, and G. Fitzmaurice (1997). The Rockin'Mouse: Integral 3D manipulation on a plane. *CHI'97 Conference on Human Factors in Computing Systems*, Atlanta, GA. pp. 311–318.

Balakrishnan, R., G. Fitzmaurice, and G. Kurtenbach (March 2001). User interfaces for volumetric displays. *IEEE Computer* **34**(3): 37–45.

Balakrishnan, R. and K. Hinckley (1999). The role of kinesthetic reference frames in two-handed input performance. *Proceedings of the ACM UIST'99 Symposium on User Interface Software and Technology*, Asheville, NC. pp. 171–178.

Balakrishnan, R. and K. Hinckley (2000). Symmetric bimanual interaction. *Proceedings of the CHI 2000*, The Hague, the Netherlands. pp. 33–40.

Balakrishnan, R. and G. Kurtenbach (1999). Exploring bimanual camera control and object manipulation in 3D graphics interfaces. *Proceedings of the CHI'99 ACM Conference on Human Factors in Computing Systems*, Pittsburgh, PA. pp. 56–63.

Balakrishnan, R. and I. S. MacKenzie (1997). Performance differences in the fingers, wrist, and forearm in computer input control. *Proceedings of the CHI'97 ACM Conference on Human Factors in Computing Systems*, Atlanta, GA. pp. 303–310.

Bates, R. and H. O. Istance (2003). Why are eye mice unpopular? A detailed comparison of head and eye controlled assistive technology pointing devices. *Universal Access in the Information Society* **2**(3): 280–290.

Bau, O. and W. E. Mackay (2008). OctoPocus: A dynamic guide for learning gesture-based command sets. Paper presented at the *Proceedings of the ACM Symposium on User Interface Software and Technology (UIST'08)*, Monterey, CA. http://portal.acm.org/citation.cfm?id=1449715.1449724

Bau, O., I. Poupyrev, A. Israr, and C. Harrison (2010). TeslaTouch: Electrovibration for touch surfaces. Paper presented at the *Proceedings of the ACM Symposium on User Interface Software and Technology (UIST'10)*, New York. http://dl.acm.org/citation.cfm?id=1866074

Baudel, T. and M. Beaudouin-Lafon (1993). Charade: Remote control of objects using hand gestures. *Communications of the ACM* **36**(7): 28–35.

Baudisch, P. and G. Chu (2009). Back-of-device interaction allows creating very small touch devices. Paper presented at the *Proceedings of the ACM Conference on Human Factors in Computing Systems (CHI'09)*, Boston, MA. http://portal.acm.org/citation.cfm?id=1518701.1518995

Baudisch, P., E. Cutrell, D. Robbins, M. Czerwinski, P. Tandler, B. Bederson, and A. Zierlinger (2003). Drag-and-pop and drag-and-pick: Techniques for accessing remote screen content on touch- and pen-operated systems. Paper presented at the *Proceedings of the 9th IFIP TC13 International Conference on Human–Computer Interaction (INTERACT'03)*, Zurich, Switzerland. http://www.patrickbaudisch.com/publications/2003-Baudisch-Interact03-DragAndPop.pdf

Baudisch, P., N. Good, V. Bellotti and P. Schraedley (2002). Keeping things in context: A comparative evaluation of focus plus context screens, overviews, and zooming. *CHI 2002 Conference on Human Factors in Computing Systems*, Minneapolis, MN. pp. 259–266.

Baudisch, P., X. Xie, C. Wang, and W.-Y. Ma (2004). Collapse-to-zoom: Viewing web pages on small screen devices by interactively removing irrelevant content. *Proceedings of the ACM Symposium on User Interface Software and Technology (UIST'04)*, Santa Fe, NM, October 24–27, 2004. ACM Press, New York. pp. 91–94.

Bederson, B. (2000). Fisheye menus. *Proceedings of the UIST 2000*, San Diego, CA. pp. 217–226.

Bederson, B., J. Hollan, K. Perlin, J. Meyer, D. Bacon, and G. Furnas (1996). Pad++: A zoomable graphical sketchpad for exploring alternate interface physics. *Journal of Visual Languages and Computing* 7: 3–31.

Bellotti, V., M. Back, W. K. Edwards, R. Grinter, C. Lopes, and A. Henderson (2002). Making sense of sensing systems: Five questions for designers and researchers. *Proceedings of the ACM CHI 2002 Conference on Human Factors in Computing Systems*, Minneapolis, MN. pp. 415–422. http://portal.acm.org/citation.cfm?id=503376.503450

Benko, H., T. S. Saponas, D. Morris, and D. Tan (2009). Enhancing input on and above the interactive surface with muscle sensing. Paper presented at the *Proceedings of the ACM International Conference on Interactive Tabletops and Surfaces*, Banff, Alberta, Canada.

Betrisey, C., J. Blinn, B. Dresevic, B. Hill, G. Hitchcock, B. Keely, D. Mitchell, J. Platt, and T. Whitted (2000). Displaced filtering for patterned displays. *Proceedings of the Society for Information Display Symposium*, Redmond, WA. pp. 296–299.

Bier, E., M. Stone, K. Pier, W. Buxton, and T. DeRose (1993). Toolglass and magic lenses: The see-through interface. *Proceedings of SIGGRAPH'93*, Anaheim, CA. pp. 73–80.

Bilton, N. (December 6, 2011). TouchFire adds a physical keyboard to iPad screen. *The New York Times*. Retrieved from http://gadgetwise.blogs.nytimes.com/2011/12/06/touchfire-adds-a-physical-keyboard-to-ipad-screen/

Blanch, R., Y. Guiard, and M. Beaudouin-Lafon (2004). Semantic pointing: Improving target acquisition with control-display ratio adaptation. Paper presented at the *Proceedings of the ACM Conference on Human Factors in Computing Systems (CHI'04)*, Vienna, Austria. http://portal.acm.org/citation.cfm?id=985692.985758

Bolt, R. (1980). Put-that-there: Voice and gesture at the graphics interface. *ACM SIGGRAPH Computer Graphics* **14**(3): 262–270.

Bragdon, A., R. Zeleznik, B. Williamson, T. Miller, and J. J. LaViola (2009). GestureBar: Improving the approachability of gesture-based interfaces. Paper presented at the *Proceedings of the ACM Conference on Human Factors in Computing Systems (CHI'09)*, Boston, MA. http://portal.acm.org/citation.cfm?id=1518701.1519050

Brewster, S. A., P. C. Wright, and A. D. N. Edwards (1994). The design and evaluation of an auditory-enhanced scrollbar. *Conference Proceedings on Human Factors in Computing Systems*, Boston, MA. pp. 173–179.

Britton, E., J. Lipscomb, and M. Pique (1978). Making nested rotations convenient for the user. *Computer Graphics* **12**(3): 222–227.

Brooks, Jr., F. P. (1988). Grasping reality through illusion: Interactive graphics serving science. *Proceedings of CHI'88: ACM Conference on Human Factors in Computing Systems*, Washington, DC. ACM Press, New York. pp. 1–11.

Bukowski, R. and C. Sequin (1995). Object associations: A simple and practical approach to virtual 3D manipulation. *ACM 1995 Symposium on Interactive 3D Graphics*, Monterey, CA. pp. 131–138.

Burdea, G. (1996). *Force and Touch Feedback for Virtual Reality*. New York: John Wiley & Sons.

Buxton, W. (1983). Lexical and pragmatic considerations of input structure. *Computer Graphics* **17**(1): 31–37.

Buxton, W. (1986). Chunking and phrasing and the design of human–computer dialogues. *Information Processing '86, Proceedings of the IFIP 10th World Computer Congress*, Dublin, Ireland. North Holland Publishers, Amsterdam, the Netherlands. pp. 475–480.

Buxton, W. (1990a). The pragmatics of haptic input. *Proceedings of CHI'90: ACM Conference on Human Factors in Computing Systems: Tutorial 26 Notes*, Seattle, WA. ACM Press, New York.

Buxton, W. (1990b). A three-state model of graphical input. *Proceedings of the INTERACT'90*, Cambridge, U.K. Elsevier Science, Amsterdam, the Netherlands. pp. 449–456.

Buxton, W. (1995a). Integrating the periphery and context: A new taxonomy of telematics. *Proceedings of Graphics Interface '95*, Quebec City, Quebec, Canada. pp. 239–246.

Buxton, W. (1995b). Speech, language and audition. In R. Baecker, J. Grudin, W. Buxton, and S. Greenberg (eds.), *Readings in Human–Computer Interaction: Toward the Year 2000*. Los Altos, CA: Morgan Kaufmann Publishers. pp. 525–537.

Buxton, W. and G. Fitzmaurice (1998). HMDs, caves & chameleon: A human-centric analysis of interaction in virtual space. *Computer Graphics* **32**(8): 69–74.

Buxton, W., G. Fitzmaurice, R. Balakrishnan, and G. Kurtenbach (July/August 2000). Large displays in automotive design. *IEEE Computer Graphics and Applications* **20**(4): pp. 68–75.

Buxton, W., E. Fiume, R. Hill, A. Lee, and C. Woo (1983). Continuous hand-gesture driven input. *Proceedings of Graphics Interface (GI'83)*, Edmonton, Alberta, Canada. pp. 191–195. http://www.billbuxton.com/gesture83.html

Buxton, W., R. Hill, and P. Rowley (1985). Issues and techniques in touch-sensitive tablet input. *Computer Graphics* **19**(3): 215–224; *Proceedings of the ACM Conference on Computer Graphics and Interactive Techniques (SIGGRAPH'85)*, San Francisco, CA. http://portal.acm.org/citation.cfm?id=325239

Buxton, W. and B. Myers (1986). A study in two-handed input. *Proceedings of CHI'86: ACM Conference on Human Factors in Computing Systems*, Boston, MA. ACM Press, New York. pp. 321–326.

Buyukkokten, O., H. Garcia-Molina, and A. Paepcke (2001). Accordion summarization for end-game browsing on PDAs and cellular phones. *ACM CHI 2001 Conference on Human Factors in Computing Systems*, Seattle, WA.

Cadoz, C. (1994). *Les realites virtuelles*. Paris, France: Dominos-Flammarion.

Campbell, C., S. Zhai, K. May, and P. Maglio (1999). What you feel must be what you see: Adding tactile feedback to the trackpoint. *Proceedings of INTERACT'99: 7th IFIP Conference on Human Computer Interaction*, Edinburgh, Scotland. pp. 383–390.

Cao, X., A. D. Wilson, R. Balakrishnan, K. Hinckley, and S. E. Hudson (2008). ShapeTouch: Leveraging contact shape on interactive surfaces. Paper presented at the *Proceedings of the 3rd IEEE Workshop on Horizontal Interactive Human–Computer Systems (Tabletop'08)*, Amsterdam, the Netherlands. http://ieeexplore.ieee.org/xpls/abs_all.jsp?arnumber=4660195

Card, S., W. English, and B. Burr (1978). Evaluation of mouse, ratc-controlled isometric joystick, step keys, and text keys for text selection on a CRT. *Ergonomics* **21**: 601–613.

Card, S., J. Mackinlay, and G. Robertson (1991). A morphological analysis of the design space of input devices. *ACM Transactions on Information Systems* **9**(2): 99–122.

Card, S., J. Mackinlay, and B. Shneiderman (1999). *Readings in Information Visualization: Using Vision to Think*. San Francisco, CA: Morgan Kaufmann.

Card, S., T. Moran, and A. Newell (1980). The keystroke-level model for user performance time with interactive systems. *Communications of the ACM* **23**(7): 396–410.

Card, S. K., T. P. Moran, and A. Newell (1983). *The Psychology of Human–Computer Interaction*. Hillsdale, NJ: Lawrence Erlbaum Associates.

Casiez, G., D. Vogel, R. Balakrishnan, and A. Cockburn (2008). The impact of control-display gain on user performance in pointing tasks. *Human–Computer Interaction* **23**(3): 215–250. doi: 10.1080/07370020802278163.

Cassell, J. (2003). A framework for gesture generation and interpretation. In R. Cipolla and A. Pentland (eds.), *Computer Vision in Human–Machine Interaction*. Cambridge, U.K.: Cambridge University Press.

Caudell, T. P. and D. W. Mizell (1992). Augmented reality: An application of heads-up display technology to manual manufacturing processes. *Proceedings of the HICCS'92*, Honolulu, HI.

Chance, S., F. Gaunet, A. Beall, and J. Loomis (1998). Locomotion mode affects the updating of objects encountered during travel: The contribution of vestibular and proprioceptive inputs to path integration. *Presence* 7(2): 168–178.

Chapuis, O., J.-B. Labrune, and E. Pietriga (2009). DynaSpot: Speed-dependent area cursor. Paper presented at the *Proceedings of the ACM Conference on Human Factors in Computing Systems (CHI'09)*, Boston, MA. http://portal.acm.org/citation.cfm?doid=1518701.1518911

Chen, M., S. J. Mountford, and A. Sellen (1988). A study in interactive 3-D rotation using 2-D control devices. *Computer Graphics* **22**(4): 121–129.

Cholewiak, R. and A. Collins (1991). Sensory and physiological bases of touch. In M. Heller and W. Schiff (eds.), *The Psychology of Touch*. Hillsdale, NJ: Lawrence Erlbaum Associates. pp. 23–60.

Christ, R. E. (1975). Review and analysis of color coding research for visual displays. *Human Factors* **25**: 71–84.

Cockburn, A. and A. Firth (2003). Improving the acquisition of small targets. Paper presented at the *Proceedings of the British HCI Group Annual Conference (HCI'03)*, Bath, England. http://citeseer.ist.psu.edu/574125.html

Cohen, P. R., M. Johnston, D. McGee, S. Oviatt, J. Pittman, I. Smith, L. Chen, and J. Clow (1997). QuickSet: Multimodal interaction for distributed applications. *ACM Multimedial 97*, Seattle, WA.

Cohen, P. R. and J. W. Sullivan (1989). Synergistic use of direct manipulation and natural language. *Proceedings of the ACM CHI'89 Conference on Human Factors in Computing Systems*, Austin, TX. pp. 227–233.

Cohn, G., D. Morris, S. N. Patel, and D. S. Tan (2012). Humantenna: Using the body as an antenna for real-time whole-body interaction. Paper presented at the *Proceedings of the ACM Conference on Human Factors in Computing Systems (CHI'12)*, Austin, TX. http://dl.acm.org/citation.cfm?id=2208330

Cole, W. G. (1986). Medical cognitive graphics. *ACM CHI'86 Conference on Human factors in Computing Systems*, Boston, MA. pp. 91–95.

Conner, D., S. Snibbe, K. Herndon, D. Robbins, R. Zeleznik, and A. van Dam (1992). Three-dimensional widgets. *Computer Graphics (Proceedings of the 1992 Symposium on Interactive 3D Graphics)*, Redwood City, CA. pp. 183–188, 230–231.

Cook, R. L. (1986). Stochastic sampling in computer graphics. *ACM Transactions on Graphics* **5**(1): 51–72.

Cowan, W. B. (1983). An inexpensive calibration scheme for calibrations of a color monitor in terms of CIE standard coordinates. *Computer Graphics* **17**(3): 315–321.

Czerwinski, M., G. Smith, T. Regan, B. Meyers, G. Robertson, and G. Starkweather (2003). Toward characterizing the productivity benefits of very large displays. *INTERACT 2003*, Zurich, Switzerland.

Czerwinski, M., D. S. Tan, and G. G. Robertson (2002). Women take a wider view. *Proceedings of the ACM CHI 2002 Conference on Human Factors in Computing Systems*, Minneapolis, MN. pp. 195–202.

Darken, R. P. and J. L. Sibert (1993). A toolset for navigation in virtual environments. *Proceedings of the 6th annual ACM symposium on User interface software and technology*, pp. 157–165. ACM, New York.

Darken, R. P. and J. L. Sibert (October 1995). Navigating large virtual spaces. *International Journal of Human–Computer Interaction* **8**(1): 49–72.

Deatherage, B. H. (1972). Auditory and other sensory forms of information presentation. In H. Van Cott and R. Kinkade (eds.), *Human Engineering Guide to Equipment Design*. Washington, DC: U.S. Government Printing Office.

De Jong, J. R. (1957). The effects of increasing skill on cycle time and its consequences for time standards. *Ergonomics* **1**: 51–60.

Deering, M. (1992). High resolution virtual reality. *Computer Graphics* **26**(2): 195–202.

Dey, A., G. Abowd, and D. Salber (2001). A conceptual framework and a toolkit for supporting the rapid prototyping of context-aware applications. *Journal of Human–Computer Interaction* **16**(2–4): 97–166.

Dixon, M., J. Fogarty, and J. O. Wobbrock (2012). A general-purpose target-aware pointing enhancement using pixel-level analysis of graphical interfaces. Paper presented at the *Proceedings of the ACM Conference on Human Factors in Computing Systems (CHI'12)*, Austin, TX. http://dl.acm.org/citation.cfm?id=2208734

Doll, T. J. and D. J. Folds (1985). Auditory signals in military aircraft: Ergonomic principles versus practice. *Proceedings of the 3rd Symposium on Aviation Psychology*, Ohio State University, Department of Aviation, Colombus, OH. pp. 111–125.

Douglas, S., A. Kirkpatrick, and I. S. MacKenzie (1999). Testing pointing device performance and user assessment with the ISO 9241, Part 9 Standard. *Proceedings of the ACM CHI'99 Conference on Human Factors in Computing Systems*, Pittsburgh, PA. pp. 215–222.

Douglas, S. and A. Mithal (1994). The effect of reducing homing time on the speed of a finger-controlled isometric pointing device. *Proceedings of the ACM CHI'94 Conference on Human Factors in Computing Systems*, Boston, MA. pp. 411–416.

Efron, D. (1941). *Gesture and Environment*. New York: King's Crown Press.

Evans, A. C. and J. O. Wobbrock (2012). Taming wild behavior: The Input Observer for obtaining text entry and mouse pointing measures from everyday computer use. *Proceedings of the ACM Conference on Human Factors in Computing Systems (CHI'12)*, Austin, TX, May 5–10, 2012. ACM Press, New York. pp. 1947–1956.

Farris, J. S., K. S. Jones, and B. A. Anders (2001). Acquisition speed with targets on the edge of the screen: An application of Fitts' Law to commonly used web browser controls. Paper presented at the *Proceedings of the Human Factors and Ergonomics Society 45th Annual Meeting (HFES'01)*, Minneapolis, MN.

Favalora, G. E. (2005). Volumetric 3D displays and application infrastructure. *IEEE Computer* **38**(8): 37–44.

Feiner, S., B. Macintyre, and D. Seligmann (1993). Knowledge-based augmented reality. *Communications of the ACM* **36**(7): 53–61.

Findlater, L., A. Jansen, K. Shinohara, M. Dixon, P. Kamb, J. Rakita, and J. O. Wobbrock (2010). Enhanced area cursors: Reducing fine-pointing demands for people with motor impairments. Paper presented at the *Proceedings of the ACM Symposium on User Interface Software and Technology (UIST'10)*, New York.

Fitts, P. (1954). The information capacity of the human motor system in controlling the amplitude of movement. *Journal of Experimental Psychology* **47**(6): 381–391.

Fitzmaurice, G. W., R. Balakrisnan, and G. Kurtenbach (1999). Sampling, synthesis, and input devices. *Communications of the ACM* **42**(8): 54–63.

Fitzmaurice, G. and W. Buxton (1997). An empirical evaluation of graspable user interfaces: Towards specialized, space-multiplexed input. *Proceedings of CHI'97: ACM Conference on Human Factors in Computing Systems*, Atlanta, GA. ACM Press, New York. pp. 43–50.

Fitzmaurice, G., H. Ishii, and W. Buxton (1995). Bricks: Laying the foundations for graspable user interfaces. *Proceedings of CHI'95: ACM Conference on Human Factors in Computing Systems*, Denver, CO. ACM Press, New York. pp. 442–449.

Flash, T. and N. Hogan (1985). The coordination of arm movements: An experimentally confirmed mathematical model. *The Journal of Neuroscience* **5**(7): 1688–1703.

Foley, J. D., V. Wallace, and P. Chan (November 1984). The human factors of computer graphics interaction techniques. *IEEE Computer Graphics and Applications* **4**(11): 13–48.

Forlines, C., D. Vogel, and R. Balakrishnan (2006). HybridPointing: Fluid switching between absolute and relative pointing with a direct input device. Paper presented at the *Proceedings of the 19th Annual ACM Symposium on User Interface Software and Technology*, Montreux, Switzerland.

Fothergill, S., H. Mentis, P. Kohli, and S. Nowozin (2012). Instructing people for training gestural interactive systems. Paper presented at the *Proceedings of the ACM Conference on Human Factors in Computing Systems (CHI'12)*, Austin, TX. http://dl.acm.org/citation.cfm?id=2208303

Freeman, D., H. Benko, M. R. Morris, and D. Wigdor (2009). ShadowGuides: Visualizations for in-situ learning of multi-touch and whole-hand gestures. Paper presented at the *Proceedings of the ACM Conference on Interactive Tabletops and Surfaces (ITS'09)*, Calgary, Alberta, Canada. http://dl.acm.org/citation.cfm?id=1731903.1731935

Freeman, W. T. and C. Weissman (1995). Television control by hand gestures. *International Workshop on Automatic Face and Gesture Recognition*, Zurich, Switzerland. pp. 179–183.

Froehlich, J., J. O. Wobbrock, and S. K. Kane (2007). Barrier pointing: Using physical edges to assist target acquisition on mobile device touch screens. Paper presented at the *Proceedings of the ACM SIGACCESS Conference on Computers and Accessibility (ASSETS'07)*, Tempe, AZ. http://portal.acm.org/citation.cfm?id=1296843.1296849

Funkhouser, T. and K. Li (July–August 2000). Large format displays. *IEEE Computer Graphics and Applications* **20**(4, special issue): 20–75.

Gaver, W. (1989). The SonicFinder: An interface that uses auditory icons. *Human–Computer Interaction* **4**(1): 67–94.

Gibson, J. (1986). *The Ecological Approach to Visual Perception*. Hillsdale, NJ: Lawrence Erlbaum Associates.

Goel, M., J. O. Wobbrock, and S. N. Patel (2012). GripSense: Using built-in sensors to detect hand posture and pressure on commodity mobile phones. *Proceedings of the ACM Symposium on User Interface Software and Technology (UIST'12)*, Cambridge, MA, October 7–10, 2012. ACM Press, New York.

Goldberg, D. and C. Richardson (1993). Touch-typing with a stylus. *Proceedings of the INTERCHI'93 Conference on Human Factors in Computing Systems*, Amsterdam, the Netherlands. pp. 80–87. http://portal.acm.org/citation.cfm?id=169093

Grandhi, S. A., G. Joue, and I. Mittelberg (2011). Understanding naturalness and intuitiveness in gesture production: Insights for touchless gestural interfaces. Paper presented at the *Proceedings of the ACM Conference on Human Factors in Computing Systems (CHI'11)*, Vancouver, British Columbia, Canada. http://dl.acm.org/citation.cfm?id=1979061

Grossman, T. and R. Balakrishnan (2005). The Bubble Cursor: Enhancing target acquisition by dynamic resizing of the cursor's activation area. Paper presented at the *Proceedings of the ACM Conference on Human Factors in Computing Systems (CHI'05)*, Portland, OR. http://portal.acm.org/citation.cfm?id=1054972.1055012

Grossman, T. and R. Balakrishnan (2006a). The design and evaluation of selection techniques for 3D volumetric displays. Paper presented at the *Proceedings of the 19th Annual ACM Symposium on User Interface Software and Technology*, Montreux, Switzerland.

Grossman, T. and R. Balakrishnan (2006b). An evaluation of depth perception on volumetric displays. Paper presented at the *Proceedings of the Working Conference on Advanced Visual Interfaces*, Venezia, Italy.

Grossman, T. and R. Balakrishnan (2008). Collaborative interaction with volumetric displays. Paper presented at the *Proceedings of the SIGCHI Conference on Human Factors in Computing Systems*, Florence, Italy.

Grossman, T., K. Hinckley, P. Baudisch, M. Agrawala, and R. Balakrishnan (2006). Hover Widgets: Using the tracking state to extend the capabilities of pen-operated devices. Paper presented at the *Proceedings of the ACM Conference on Human Factors in Computing Systems (CHI'06)*, Montréal, Québec, Canada. http://portal.acm.org/citation.cfm?id=1124772.1124898

Grossman, T., D. Wigdor, and R. Balakrishnan (2004). Multi-finger gestural interaction with 3d volumetric displays. Paper presented at the *Proceedings of the 17th Annual ACM Symposium on User Interface Software and Technology*, Santa Fe, NM.

Green, M. and J. Liang (1994). JDCAD: A highly interactive 3D modeling system. *Computers and Graphics* **18**(4): 499–506.

Grudin, J. (2001). Partitioning digital worlds: Focal and peripheral awareness in multiple monitor use. *Proceedings of the CHI 2001*, Seattle, WA. pp. 458–465.

Guiard, Y. (1987). Asymmetric division of labor in human skilled bimanual action: The kinematic chain as a model. *The Journal of Motor Behavior* **19**(4): 486–517.

Guiard, Y. (2009). The problem of consistency in the design of Fitts' law experiments: Consider either target distance and width or movement form and scale. *Proceedings of the ACM Conference on Human Factors in Computing Systems (CHI'09)*, Boston, MA, April 4–9, 2009. ACM Press, New York. pp. 1809–1818.

Guiard, Y., R. Blanch, and M. Beaudouin-Lafon (2004). Object pointing: A complement to bitmap pointing in GUIs. Paper presented at the *Proceedings of Graphics Interface (GI'04)*, London, Ontario, Canada. http://portal.acm.org/citation.cfm?id=1006060

Guiard, Y., F. Buourgeois, D. Mottet, and M. Beaudouin-Lafon (2001). Beyond the 10-bit barrier: Fitts' Law in multi-scale electronic worlds. *IHM-HCI 2001*, Lille, France.

Guimbretiere, F., M. C. Stone, and T. Winograd (2001). Fluid interaction with high-resolution wall-size displays. *Proceedings of the UIST 2001 Symposium on User Interface Software and Technology*, Orlando, FL. pp. 21–30.

Harada, S., J. A. Landay, J. Malkin, X. Li, and J. A. Bilmes (2006). The vocal joystick: Evaluation of voice-based cursor control techniques. Paper presented at the *Proceedings of the ACM SIGACCESS Conference on Computers and Accessibility (ASSETS'06)*, Portland, OR. http://portal.acm.org/citation.cfm?id=1169021

Harada, S., T. S. Saponas, and J. A. Landay (2007). VoicePen: Augmenting pen input with simultaneous non-linguistic vocalization. Paper presented at the *Proceedings of the ACM International Conference on Multimodal Interfaces (ICMI'07)*, Nagoya, Aichi, Japan. http://dl.acm.org/citation.cfm?id=1322192.1322225

Harada, S., J. O. Wobbrock, and J. A. Landay (2007). VoiceDraw: A hands-free voice-driven drawing application for people with motor impairments. Paper presented at the *Proceedings of the ACM SIGACCESS Conference on Computers and Accessibility (ASSETS'07)*, Tempe, AZ. http://portal.acm.org/citation.cfm?id=1296843.1296850

Harada, S., J. O. Wobbrock, and J. A. Landay (2011). VoiceGames: Investigation into the use of non-speech voice input for making computer games more accessible. Paper presented at the *Proceedings of the 13th IFIP TC13 International Conference on Human–Computer Interaction (INTERACT'11)*, Lisbon, Portugal. http://www.springerlink.com/content/p7557762w547064p/?MUD=MP

Harrison, B., K. Fishkin, A. Gujar, C. Mochon, and R. Want (1998). Squeeze me, hold me, tilt me! An exploration of manipulative user interfaces. *Proceedings of the ACM CHI'98 Conference on Human Factors in Computing Systems*, Los Angeles, CA. pp. 17–24.

Harrison, B., H. Ishii, K. Vicente, and W. Buxton (1995a). Transparent layered user interfaces: An evaluation of a display design to enhance focused and divided attention. *Proceedings of CHI'95: ACM Conference on Human Factors in Computing Systems*, Denver, CO. pp. 317–324.

Harrison, B., G. Kurtenbach, and K. Vicente (1995b). An experimental evaluation of transparent user interface tools and information content. *UIST'95*, Pittsburgh, PA. pp. 81–90.

Harrison, C., Schwarz, J., and Hudson, S. E. (2011). TapSense: Enhancing finger interaction on touch surfaces. Paper presented at the *Proceedings of the ACM Symposium on User Interface Software and Technology* (UIST '11), Santa Barbara, CA. http://dl.acm.org/citation.cfm?id=2047279

Harrison, B. and K. Vicente (1996). An experimental evaluation of transparent menu usage. *Proceedings of CHI'96: ACM Conference on Human Factors in Computing Systems*, Vancouver, British Columbia, Canada. pp. 391–398.

Hauptmann, A. (1989). Speech and gestures for graphic image manipulation. *Proceedings of CHI'89: ACM Conference on Human Factors in Computing Systems*, Austin, TX. ACM Press, New York. pp. 241–245.

Henry, T. R., S. E. Hudson, and G. L. Newell (1990). Integrating gesture and snapping into a user interface toolkit. Paper presented at the *Proceedings of the ACM Symposium on User Interface Software and Technology (UIST'90)*, Snowbird, UT. http://portal.acm.org/citation.cfm?id=97938

Hinckley, K. (2003a). Distributed sensing techniques for face-to-face collaboration. *ICMI-PUI'03 Fifth International Conference on Multimodal Interfaces*, Vancouver, British Columbia, Canada.

Hinckley, K. (2003b). Synchronous gestures for multiple users and computers. *UIST'03 Symposium on User Interface Software & Technology*, Vancouver, British Columbia, Canada. 10pp. (full paper).

Hinckley, K., P. Baudisch, G. Ramos, and F. Guimbretière (2005). Design and analysis of delimiters for selection-action pen gesture phrases in Scriboli. Paper presented at the *Proceedings of the ACM Conference on Human Factors in Computing Systems (CHI'05)*, Portland, OR. http://portal.acm.org/citation.cfm?id=1055035

Hinckley, K., E. Cutrell, S. Bathiche, and T. Muss (2001). Quantitative analysis of scrolling techniques. *CHI 2002*, Minneapolis, MN.

Hinckley, K., M. Czerwinski, and M. Sinclair (1998a). Interaction and modeling techniques for desktop two-handed input. *Proceedings of the ACM UIST'98 Symposium on User Interface Software and Technology*, San Francisco, CA. ACM, New York. pp. 49–58.

Hinckley, K., F. Guimbretière, M. Agrawala, G. Apitz, and N. Chen (2006). Phrasing techniques for multi-stroke selection gestures. Paper presented at the *Proceedings of Graphics Interface (GI '06)*, Quebec City, Quebec, Canada. http://dl.acm.org/citation.cfm?id=1143104

Hinckley, K., R. Pausch, J. C. Goble, and N. F. Kassell (1994a). Passive real-world interface props for neurosurgical visualization. *Proceedings of CHI'94: ACM Conference on Human Factors in Computing Systems*, Boston, MA. ACM Press, New York. pp. 452–458.

Hinckley, K., R. Pausch, J. C. Goble, and N. F. Kassell (1994b). A survey of design issues in spatial input. *Proceedings of the ACM UIST'94 Symposium on User Interface Software and Technology*, Marina del Rey, CA. ACM Press, New York. pp. 213–222.

Hinckley, K., R. Pausch, D. Proffitt, and N. Kassell (1998b). Two-handed virtual manipulation. *ACM Transactions on Computer–Human Interaction* 5(3): 260–302.

Hinckley, K., J. Pierce, E. Horvitz, and M. Sinclair (2003). Foreground and background interaction with sensor-enhanced mobile devices. *ACM TOCHI* (submitted for review) (Special Issue on Sensor-Based Interaction).

Hinckley, K., J. Pierce, M. Sinclair, and E. Horvitz (2000). Sensing techniques for mobile interaction. *ACM UIST 2000 Symposium on User Interface Software & Technology*, San Diego, CA. pp. 91–100.

Hinckley, K., G. Ramos, F. Guimbretiere, P. Baudisch, and M. Smith (2004). Stitching: Pen gestures that span multiple displays. Paper presented at the *Proceedings of the ACM Working Conference on Advanced Visual Interfaces (AVI'04)*, Gallipoli, Italy. http://portal.acm.org/citation.cfm?id=989866

Hinckley, K. and M. Sinclair (1999). Touch-sensing input devices. *ACM CHI'99 Conference on Human Factors in Computing Systems*, Pittsburgh, PA. pp. 223–230.

Hinckley, K., M. Sinclair, E. Hanson, R. Szeliski, and M. Conway (1999). The VideoMouse: A camera-based multi-degree-of-freedom input device. *ACM UIST'99 Symposium on User Interface Software & Technology*, Asheville, NC. pp. 103–112.

Hinckley, K., J. Tullio, R. Pausch, D. Proffitt, and N. Kassell (1997). Usability analysis of 3D rotation techniques. *Proceedings of the ACM UIST'97 Symposium on User Interface Software and Technology*, Banff, Alberta, Canada. ACM Press, New York. pp. 1–10.

Hinckley, K., S. Zhao, R. Sarin, P. Baudisch, E. Cutrell, M. Shilman, and D. Tan (2007). InkSeine: In situ search for active note taking. Paper presented at the *Proceedings of the ACM Conference on Human Factors in Computing Systems (CHI'07)*, San Jose, CA.

Hiraoka, S., I. Miyamoto, and K. Tomimatsu (2003). Behind touch: A text input method for mobile phone by the back and tactile sense interface. Paper presented at the *Proceedings of Interaction 2003*, Tokyo, Japan. http://sciencelinks.jp/j-east/article/200405/000020040504A0091046.php

Hirshfield, L. M., K. Chauncey, R. Gulotta, A. Girouard, E. T. Solovey, R. J. Jacob, and S. Fantini (2009a). Combining electroencephalograph and functional near infrared spectroscopy to explore users' mental workload. Paper presented at the *Proceedings of the 5th International Conference on Foundations of Augmented Cognition. Neuroergonomics and Operational Neuroscience: Held as Part of HCI International 2009*, San Diego, CA.

Hirshfield, L. M., E. T. Solovey, A. Girouard, J. Kebinger, R. J. K. Jacob, A. Sassaroli, and S. Fantini (2009b). Brain measurement for usability testing and adaptive interfaces: An example of uncovering syntactic workload with functional near infrared spectroscopy. Paper presented at the *Proceedings of the 27th International Conference on Human Factors in Computing Systems*, Boston, MA.

Hix, D., J. Templeman, and R. Jacob (1995). Pre-screen projection: From concept to testing of a new interaction technique. *Proceedings of the CHI'95*, Denver, CO. pp. 226–233.

Holz, C. and P. Baudisch (2010). The generalized perceived input point model and how to double touch accuracy by extracting fingerprints. Paper presented at the *Proceedings of the ACM Conference on Human Factors in Computing Systems (CHI'10)*, Atlanta, GA. http://dl.acm.org/citation.cfm?id=1753413

Holz, C. and P. Baudisch (2011). Understanding touch. Paper presented at the *Proceedings of the ACM Conference on Human Factors in Computing Systems (CHI'11)*, Vancouver, British Columbia, Canada. http://dl.acm.org/citation.cfm?id=1979308

Honan, M., E. Serina, R. Tal, and D. Rempel (1995). Wrist postures while typing on a standard and split keyboard. *Proceedings of the Human Factors and Ergonomics Society (HFES) 39th Annual Meeting*, Santa Monica, CA. pp. 366–368.

Horvitz, E., J. Breese, D. Heckerman, D. Hovel, and K. Rommelse (July 1998). The Lumiere project: Bayesian user modeling for inferring the goals and needs of software users. *Proceedings of the Fourteenth Conference on Uncertainty in Artificial Intelligence*, Madison, WI. pp. 256–265. Morgan Kaufmann, San Francisco, CA.

Horvitz, E., A. Jacobs, and D. Hovel (1999). Attention-sensitive alerting. *Proceedings of UAI '99, Conference on Uncertainty and Artificial Intelligence*, Stockholm, Sweden. pp. 305–313.

Hourcade, J. P., K. B. Perry, and A. Sharma (2008). PointAssist: Helping four year olds point with ease. Paper presented at the *Proceedings of the ACM Conference on Interaction Design and Children (IDC '08)*, Chicago, IL. http://portal.acm.org/citation.cfm?id=1463689.1463757

Hudson, S. and I. Smith (1996). Electronic mail previews using non-speech audio. *CHI'96 Companion Proceedings*, Vancouver, British Columbia, Canada. pp. 237–238.

Hurst, A., J. Mankoff, A. K. Dey, and S. E. Hudson (2007). Dirty desktops: Using a patina of magnetic mouse dust to make common interactor targets easier to select. Paper presented at the *Proceedings of the ACM Symposium on User Interface Software and Technology (UIST'07)*, Newport, RI. http://portal.acm.org/citation.cfm?id=1294242

Hwang, F., S. Keates, P. Langdon, and P. J. Clarkson (2003). Multiple haptic targets for motion-impaired computer users. Paper presented at the *Proceedings of the ACM Conference on Human Factors in Computing Systems (CHI'03)*, Ft. Lauderdale, FL. http://portal.acm.org/citation.cfm?id=642620

Igarashi, T. and J. F. Hughes (2001). Voice as sound: Using non-verbal voice input for interactive control. Paper presented at the *Proceedings of the ACM Symposium on User Interface Software and Technology (UIST'01)*, Orlando, FL. http://portal.acm.org/citation.cfm?id=502348.502372

Igarashi, T., S. Matsuoka, and H. Tanaka (1999). Teddy: A sketching interface for 3D freeform design. *ACM SIGGRAPH'99*, Los Angeles, CA. pp. 409–416.

Ishii, H. and B. Ullmer (1997). Tangible bits: Towards seamless interfaces between people, bits, and atoms. *Proceedings of CHI'97: ACM Conference on Human Factors in Computing Systems*, Atlanta, GA. ACM Press, New York. pp. 234–241.

Iwata, H. (1990). Artificial reality with force-feedback: Development of desktop virtual space with compact master manipulator. *Computer Graphics* **24**(4): 165–170.

Jacob, R. (1991). The use of eye movements in human–computer interaction techniques: What you look at is what you get. *ACM Transactions on Information Systems* **9**(3): 152–169.

Jacob, R., L. Sibert, D. McFarlane, and M. Mullen, Jr. (1994). Integrality and separability of input devices. *ACM Transactions on Computer–Human Interaction* **1**(1): 3–26.

Jansen, Y., P. Dragicevic, and J.-D. Fekete (2012). Tangible remote controllers for wall-size displays. Paper presented at the *Proceedings of the ACM Conference on Human Factors in Computing Systems (CHI'12)*, Austin, TX. http://dl.acm.org/citation.cfm?id=2208691

Jansen, A., L. Findlater, and J. O. Wobbrock (2011). From the lab to the world: Lessons from extending a pointing technique for real-world use. Paper presented at the *Extended Abstracts of the ACM Conference on Human Factors in Computing Systems (CHI'11)*, Vancouver, British Columbia, Canada. http://portal.acm.org/citation.cfm?id=1979888

Jansen, Y., T. Karrer, and J. Borchers (2010). MudPad: Tactile feedback and haptic texture overlay for touch surfaces. Paper presented at the *Proceedings of the ACM Conference on Interactive Tabletops and Surfaces (ITS '10)*, Saarbrücken, Germany. http://dl.acm.org/citation.cfm?id=1936655

Jellinek, H. and S. Card (1990). Powermice and user performance. *Proceedings of the ACM CHI'90 Conference on Human Factors in Computing Systems*, Seattle, WA. pp. 213–220.

Jacob, R. J. K. (1990). What you look at is what you get: Eye movement-based interaction techniques. *Proceedings of the ACM Conference on Human Factors in Computing Systems (CHI'90)*, Seattle, WA, April 1–5. ACM Press, New York. pp. 11–18.

Jojic, N., B. Brumitt, B. Meyers, and S. Harris (2000). Detecting and estimating pointing gestures in dense disparity maps. *Proceedings of the IEEE International Conference on Automatic Face and Gesture Recognition*, Grenoble, France.

Johnson, G., M. D. Gross, J. Hong, and E. Y.-L. Do (2009). Computational support for sketching in design: A review. *Foundations and Trends in Human–Computer Interaction* **2**(1): 1–93. doi: 10.1561/1100000013.

Jones, M., G. Marsden, N. Mohd-Nasir, K. Boone, and G. Buchanan (1999). Improving Web interaction on small displays. *Computer Networks* **31**(11–16): 1129–1137.

Kabbash, P. and W. Buxton (1995). The "Prince" technique: Fitts' law and selection using area cursors. Paper presented at the *Proceedings of the ACM Conference on Human Factors in Computing Systems (CHI'95)*, Denver, CO. http://portal.acm.org/citation.cfm?id=223939; http://www.billbuxton.com/prince.html

Kabbash, P., W. Buxton, and A. Sellen (1994). Two-handed input in a compound task. *Proceedings of CHI'94: ACM Conference on Human Factors in Computing Systems*, Boston, MA. ACM Press, New York. pp. 417–423.

Kamba, T., S. A. Elson, T. Harpold, T. Stamper, and P. Sukaviriya (1996). Using small screen space more efficiently. *Conference Proceedings on Human Factors in Computing Systems*, Vancouver, British Columbia, Canada. 383p.

Kane, S. K., J. P. Bigham, and J. O. Wobbrock (2008). Slide Rule: Making mobile touch screens accessible to blind people using multi-touch interaction techniques. *Proceedings of the ACM SIGACCESS Conference on Computers and Accessibility (ASSETS '08)*, Halifax, Nova Scotia, Canada, October 13–15. ACM Press, New York. pp. 73–80.

Kane, S. K., J. O. Wobbrock, and R. E. Ladner (2011). Usable gestures for blind people: Understanding preference and performance. Paper presented at the *Proceedings of the ACM Conference on Human Factors in Computing Systems (CHI'11)*, Vancouver, British Columbia, Canada.

Karat, C., C. Halverson, D. Horn, and J. Karat (1999). Patterns of entry and correction in large vocabulary continuous speech recognition systems. *Proceedings of the ACM CHI'99 Conference on Human Factors in Computing Systems*, Pittsburgh, PA. pp. 568–575.

Karlson, A. K. and B. B. Bederson (2007). ThumbSpace: Generalized one-handed input for touchscreen-based mobile devices. Paper presented at the *Proceedings of the 11th IFIP TC13 International Conference on Human–Computer Interaction (INTERACT'07)*, Rio de Janeiro, Brazil. http://www.springerlink.com/content/p85w127j007153k1

Keates, S., F. Hwang, P. Langdon, P. J. Clarkson, and P. Robinson (2002). Cursor measures for motion-impaired computer users. *Proceedings of the ACM SIGCAPH Conference on Assistive Technologies (ASSETS'02)*, July 8–10, 2002, Edinburgh, Scotland. ACM Press, New York. pp. 135–142.

Kendon, A. (1988). How gestures can become like words. In F. Poyatos (ed.), *Cross Cultural Perspectives in Nonverbal Communication*. Toronto, Ontario, Canada: C. J. Hogrefe. pp. 131–141

Kirsh, D. (1995a). Complementary strategies: Why we use our hands when we think. *Proceedings of 7th Annual Conference of the Cognitive Science Society*, Basel, Switzerland. Lawrence Erlbaum Associates, Hillsdale, NJ. pp. 212–217.

Kirsh, D. (1995b). The intelligent use of space. *Artificial Intelligence* **73**: 31–68.

Kirsh, D. and P. Maglio (1994). On distinguishing epistemic from pragmatic action. *Cognitive Science* **18**(4): 513–549.

Költringer, T. and T. Grechenig (2004). Comparing the immediate usability of Graffiti 2 and virtual keyboard. Paper presented at the *Extended Abstracts of the ACM Conference on Human Factors in Computing Systems (CHI'04)*, Vienna, Austria. http://portal.acm.org/citation.cfm?id=986017

Kramer, A. (1994). Translucent patches—Dissolving windows. *Proceedings of the ACM UIST'94 Symposium on User Interface Software & Technology*, Marina del Rey, CA. pp. 121–130.

Kumar, M., A. Paepcke, and T. Winograd (2007). EyePoint: Practical pointing and selection using gaze and keyboard. *Proceedings of the ACM Conference on Human Factors in Computing Systems (CHI'07)*, San Jose, CA, April 28–May 3, 2007. ACM Press, New York. pp. 421–430.

Kurihara, K., Goto, M., Ogata, J., and Igarashi, T. (2006). Speech Pen: Predictive handwriting based on ambient multimodal recognition. *Paper presented at the Proceedings of the ACM Conference on Human Factors in Computing Systems* (CHI'05), Montréal, Québec, Canada. http://portal.acm.org/citation.cfm?id=1124772.1124897

Kurtenbach, G. and W. Buxton (1991a). GEdit: A test bed for editing by contiguous gesture. *SIGCHI Bulletin* **23**(2): 22–26.

Kurtenbach, G. and W. Buxton (1991b). Issues in combining marking and direct manipulation techniques. *Proceedings of UIST'91*, Hilton Head, SC. pp. 137–144.

Kurtenbach, G. and W. Buxton (1993). The limits of expert performance using hierarchic marking menus. *Proceedings of the INTERCHI'93*, Amsterdam, the Netherlands. pp. 482–487.

Kurtenbach, G., G. Fitzmaurice, T. Baudel, and B. Buxton (1997). The design of a GUI paradigm based on tablets, two-hands, and transparency. *Proceedings of CHI'97: ACM Conference on Human Factors in Computing Systems*, Atlanta, GA. ACM, New York. pp. 35–42.

Kurtenbach, G., A. Sellen, and W. Buxton (1993). An empirical evaluation of some articulatory and cognitive aspects of 'marking menus'. *Journal of Human Computer Interaction* **8**(1): 1–23.

Landay, J. and Myers, B. A. (1993). Extending an existing user interface toolkit to support gesture recognition. Paper presented at the *Conference Companion of the ACM Conference on Human Factors in Computing Systems* (INTERCHI'93), Amsterdam, The Netherlands. http://portal.acm.org/citation.cfm?id=260123

Lank, E., Cheng, Y.-C. N., and Ruiz, J. (2007a). Endpoint prediction using motion kinematics. Paper presented at the *Proceedings of the ACM Conference on Human Factors in Computing Systems* (CHI'07), San Jose, CA. http://portal.acm.org/citation.cfm?id=1240724

Lank, E., Cheng, Y.-C. N., and Ruiz, J. (2007b). Endpoint prediction using motion kinematics. Paper presented at the *Proceedings of the ACM Conference on Human Factors in Computing Systems* (CHI'07), San Jose, CA. http://portal.acm.org/citation.cfm?id=1240724

Lankford, C. (2000). Effective eye-gaze input into Windows. *Proceedings of the ACM Symposium on Eye Tracking Research and Applications (ETRA'00)*, Palm Beach Gardens, FL, November 6–8, 2000. ACM Press, New York. pp. 23–27.

Lepinski, G. J., Grossman, T., and Fitzmaurice, G. (2010). The design and evaluation of multitouch marking menus. Paper presented at the *Proceedings of the ACM Conference on Human Factors in Computing Systems* (CHI'10), Atlanta, GA. http://dl.acm.org/citation.cfm?id=1753663

Lewis, J., K. Potosnak, and R. Magyar (1997). Keys and keyboards. In M. Helander, T. Landauer, and P. Prabhu (eds.), *Handbook of Human–Computer Interaction*. Amsterdam, the Netherlands: North-Holland. pp. 1285–1316.

Li, Y., Hinckley, K., Guan, Z., and Landay, J. A. (2005). Experimental analysis of mode switching techniques in pen-based user interfaces. Paper presented at the *Proceedings of the ACM Conference on Human Factors in Computing Systems* (CHI'05), Portland, Oregon. http://dl.acm.org/citation.cfm?id=1055036

Li, Y. (2010). Protractor: A fast and accurate gesture recognizer. Paper presented at the *Proceedings of the ACM Conference on Human Factors in Computing Systems* (CHI'10), Atlanta, GA. http://portal.acm.org/citation.cfm?id=1753326.1753654

Lipscomb, J. and M. Pique (1993). Analog input device physical characteristics. *SIGCHI Bulletin* **25**(3): 40–45.

Long, A. C., Landay, J. A., and Rowe, L. A. (1999). Implications for a gesture design tool. Paper presented at the *Proceedings of the ACM Conference on Human Factors in Computing Systems* (CHI'99), Pittsburgh, Pennsylvania. http://portal.acm.org/citation.cfm?id=302985

Long, A. C., Landay, J. A., Rowe, L. A., and Michiels, J. (2000). Visual similarity of pen gestures. Paper presented at the *Proceedings of the ACM Conference on Human Factors in Computing Systems* (CHI'00), The Hague, The Netherlands. http://portal.acm.org/citation.cfm?id=332458

Loomis, J., R. L. Klatzky, R. G. Golledge, and J. W. Philbeck (1999). Human navigation by path integration. In R. G. Golledge (ed.), *Wayfinding: Cognitive Mapping and Other Spatial Processes*. Baltimore, MD: Johns Hopkins. pp. 125–151.

LoPresti, E. F., D. M. Brienza, J. Angelo, L. Gilbertson, and J. Sakai (2000). Neck range of motion and use of computer head controls. *Proceedings of the ACM SIGCAPH Conference on Assistive Technologies (ASSETS'00)*, Arlington, VA, November 13–15, 2000. ACM Press, New York. pp. 121–128.

Lucente, M., G. Zwart, and A. George (1998). Visualization space: A testbed for deviceless multimodal user interface. *Proceedings of the AAAI '98*, Madison, WI.

MacKenzie, I. S. (1992). Fitts' law as a research and design tool in human–computer interaction. *Human–Computer Interaction* **7**(1): 91–139.

MacKenzie, I. S. (1995). Input devices and interaction techniques for advanced computing. In W. Barfield and T. Furness (eds.), *Virtual Environments and Advanced Interface Design*. Oxford, U.K.: Oxford University Press. pp. 437–470.

MacKenzie, I. S. and Y. Guiard (2001). The two-handed desktop interface: Are we there yet? *Proceedings of the ACM CHI 2001 Conference on Human Factors in Computing Systems: Extended Abstracts*, The Hague, the Netherlands. pp. 351–352.

MacKenzie, C. and T. Iberall (1994). *The Grasping Hand*. Amsterdam, the Netherlands: North Holland.

MacKenzie, I. S., T. Kauppinen, and M. Silfverberg (2001). Accuracy measures for evaluating computer pointing devices. *CHI 2001*, Seattle, WA. pp. 9–16.

MacKenzie, I. S. and A. Oniszczak (1998). A comparison of three selection techniques for touchpads. *Proceedings of the ACM CHI'98 Conference on Human Factors in Computing Systems*, Los Angeles, CA. pp. 336–343.

MacKenzie, I. S., A. Sellen, and W. Buxton (1991). A comparison of input devices in elemental pointing and dragging tasks. *Proceedings of the ACM CHI'91 Conference on Human Factors in Computing Systems*, New Orleans, LA. pp. 161–166.

MacKenzie, I. S. and R. W. Soukoreff (2002). A model of two-thumb text entry. *Proceedings of Graphics Interface*, Canadian Information Processing Society, Toronto, Ontario, Canada. pp. 117–124.

MacKenzie, I. S. and K. Tanaka-Ishii (2007). *Text Entry Systems: Mobility, Accessibility, Universality*. San Francisco, CA: Morgan Kaufmann.

MacKenzie, I. S. and C. Ware (1993). Lag as a determinant of human performance in interactive systems. *Proceedings of the ACM INTERCHI'93 Conference on Human Factors in Computing Systems*, Amsterdam, the Netherlands. pp. 488–493.

MacKenzie, I. S. and S. Zhang (1997). The immediate usability of graffiti. *Proceedings of Graphics Interface (GI'97)*, Kelowna, British Columbia, Canada. pp. 129–137. http://portal.acm.org/citation.cfm?id=266792

MacKenzie, I. S. and S. X. Zhang (1999). The design and evaluation of a high-performance soft keyboard. *Proceedings of the ACM Conference on Human Factors in Computing Systems (CHI'99)*, Pittsburgh, PA, May 15–20, 1999. ACM Press, New York. pp. 25–31.

MacLean, K. E., S. S. Snibbe, and G. Levin (2000). Tagged handles: Merging discrete and continuous control. *Proceedings of the ACM CHI 2000 Conference on Human Factors in Computing Systems*, The Hague, the Netherlands.

Maes, P., T. Darrell, B. Blumberg, and A. Pentland (1997). The ALIVE system: Wireless, full-body interaction with autonomous agents. *ACM Multimedia Systems* (Special Issue on Multimedia and Multisensory Virtual Worlds), **2**(1).

Marklin, R. and G. Simoneau (1996). Upper extremity posture of typists using alternative keyboards. *ErgoCon'96*, Palo Alto, CA. pp. 126–132.

Marklin, R., G. Simoneau, and J. Monroe (1997). The effect of split and vertically-inclined computer keyboards on wrist and forearm posture. *Proceedings of the Human Factors and Ergonomics Society (HFES) 41st Annual Meeting*, Santa Monica, CA. pp. 642–646.

Martin, B., and P. Isokoski (2008). EdgeWrite with integrated corner sequence help. Paper presented at the *Proceedings of the ACM Conference on Human Factors in Computing Systems (CHI'08)*, Florence, Italy. http://portal.acm.org/citation.cfm?id=1357054.1357148

Mathias, E., I. S. MacKenzie, and W. Buxton (1996). One-handed touch typing on a qwerty keyboard. *Human–Computer Interaction* **11**(1): 1–27.

McGuffin, M. and R. Balakrishnan (2002). Acquisition of expanding targets. *CHI Letters* **4**(1): 57–64; *Proceedings of the ACM Conference on Human Factors in Computing Systems (CHI'02)*, Minneapolis, MN. http://portal.acm.org/citation.cfm?id=503388

McGuffin, M. J. and R. Balakrishnan (2005). Fitts' law and expanding targets: Experimental studies and designs for user interfaces. *ACM Transactions on Computer–Human Interaction* **12**(4): 388–422.

McLoone, H., K. Hinckley, and E. Cutrell (2003). Interaction on the Microsoft Office Keyboard. In *Proceedings of IFIP INTERACT03: Human-Computer Interaction*, Zurich, Switzerland, Sept 01–05. IOS Press: Amsterdam, The Netherlands, pp. 49–56.

McNeill, D. (1992). *Hand and Mind: What Gestures Reveal about Thought*. Chicago, IL: University of Chicago Press.

Mine, M., F. Brooks, and C. Sequin (1997). Moving objects in space: Exploiting proprioception in virtual-environment interaction. *Computer Graphics* **31**: 19–26; *Proceedings of the SIGGRAPH'97*, Los Angeles, CA.

Moffatt, K. and J. McGrenere (2010). Steadied-bubbles: Combining techniques to address pen-based pointing errors for younger and older adults. Paper presented at the *Proceedings of the ACM Conference on Human Factors in Computing Systems (CHI'10)*, Atlanta, GA. http://portal.acm.org/citation.cfm?id=1753326.1753495

Moran, T., P. Chiu, and W. van Melle (1997). Pen-based interaction techniques for organizing material on an electronic whiteboard. *Proceedings of the 10th Annual ACM UIST'97 Symposium on User Interface Software & Technology*, Banff, Alberta, Canada. pp. 45–54.

Myers, B. A., R. Bhatnagar, J. Nichols, C. H. Peck, D. Kong, R. Miller, and A. C. Long (2002). Interacting at a distance: Measuring the performance of laser pointers and other devices. *Proceedings of the ACM Conference on Human Factors in Computing Systems (CHI'02)*, Minneapolis, MN, April 2002. ACM Press, New York. pp. 33–40.

Myers, B., K. Lie, and B. Yang (2000). Two-handed input using a PDA and a mouse. *Proceedings of the CHI 2000*, The Hague, the Netherlands. pp. 41–48.

Myers, B. A., R. G. McDaniel, R. C. Miller, A. S. Ferrency, A. Faulring, B. D. Kyle, and P. Doane (1997). The Amulet environment: New models for effective user interface software development. *IEEE Transactions on Software Engineering* **23**(6): 347–365. doi: 10.1109/32.601073.

Myers, B., R. Miller, C. Evankovich, and B. Bostwick (2001). Individual use of hand-held and desktop computers simultaneously (manuscript).

Myers, B., H. Stiel, and R. Gargiulo (1998). Collaboration using multiple PDAs connected to a PC. *Proceedings of the ACM CSCW'98 Conference on Computer Supported Cooperative Work*, Seattle, WA. pp. 285–294.

Mynatt, E. D., T. Igarashi, W. K. Edwards, and A. LaMarca (1999). Flatland: New dimensions in office whiteboards. *ACM SIGCHI Conference on Human Factors in Computing Systems*, Pittsburgh, PA. pp. 346–353.

Ng, A., J. Lepinski, D. Wigdor, S. Sanders, and P. Dietz (2012). Designing for low-latency direct-touch input. Paper presented at the *Proceedings of the 25th Annual ACM Symposium on User Interface Software and Technology*, Cambridge, MA.

Nguyen, D. H. and E. Mynatt (2001). Towards visibility of a Ubicomp environment, private communication.

Norman, D. (1990). *The Design of Everyday Things*. New York: Doubleday.

Noyes, J. (1983). Chord keyboards. *Applied Ergonomics* **14**: 55–59.

Oakley, I., S. Brewster, and P. Gray (2001). Solving multi-target haptic problems in menu interaction. *Proceedings of the ACM CHI 2001 Conference on Human Factors in Computing Systems: Extended Abstracts*, Seattle, WA. pp. 357–358.

Olsen, D. R. and T. Nielsen (2001). Laser pointer interaction. *Proceedings of the ACM CHI 2001 Conference on Human Factors in Computing Systems*, Seattle, WA. pp. 17–22.

Olson, J. R. and G. M. Olson (1990). The growth of cognitive modeling in human–computer interaction since GOMS. *Human–Computer Interaction* **5**(2–3): 221–266.

Olwal, A., S. Feiner, and S. Heyman (2008). Rubbing and tapping for precise and rapid selection on touch-screen displays. Paper presented at the *Proceedings of the ACM Conference on Human Factors in Computing Systems (CHI'08)*, Florence, Italy. http://portal.acm.org/citation.cfm?id=1357105

Oviatt, S. (1997). Multimodal interactive maps: Designing for human performance. *Human–Computer Interaction* **12**: 93–129.

Pearson, G. and M. Weiser (1988). Exploratory evaluation of a planar foot-operated cursor-positioning device. *Proceedings of the ACM CHI'88 Conference on Human Factors in Computing Systems*, Washington, DC. pp. 13–18.

Pekelney, R. and R. Chu (1995). Design criteria of an ergonomic mouse computer input device. *Proceedings of the Human Factors and Ergonomics Society (HFES) 39th Annual Meeting*, Santa Monica, CA. pp. 369–373.

Perlin, K. and D. Fox (1993). Pad: An alternative approach to the computer interface. *SIGGRAPH'93*, Anaheim, CA.

Platt, J. (2000). Optimal filtering for patterned displays. *IEEE Signal Processing Letters* 7(7): 179–183.

Plimmer, B., P. Reid, R. Blagojevic, A. Crossan, and S. Brewster (2011). Signing on the tactile line: A multimodal system for teaching handwriting to blind children. *ACM Transactions on Computer–Human Interaction* **18**(3): 17:11–17:29. doi: 10.1145/1993060.1993067.

Plumlee, M. and C. Ware (2002). Modeling performance for zooming vs multi-window interfaces based on visual working memory. *AVI'02: Advanced Visual Interfaces*, Trento Italy.

Potter, R. L., L. J. Weldon, and B. Shneiderman (1988). Improving the accuracy of touch screens: An experimental evaluation of three strategies. Paper presented at the *Proceedings of the ACM Conference on Human Factors in Computing Systems (CHI'88)*, Washington, DC, May 15–19. ACM Press, New York. http://portal.acm.org/citation.cfm?id=57167.57171

Poupyrev, I. and S. Maruyama (2003). Tactile interfaces for small touch screens. Paper presented at the *Proceedings of the ACM Symposium on User Interface Software and Technology (UIST'03)*, Vancouver, British Columbia, Canada. http://dl.acm.org/citation.cfm?id=964721

Poupyrev, I., S. Maruyama, and J. Rekimoto (2002). Ambient touch: Designing tactile interfaces for handheld devices. *Proceedings of the ACM Symposium on User Interface Software and Technology (UIST 2002)*, Paris, France. pp. 51–60. http://dl.acm.org/citation.cfm?id=571993

Putz-Anderson, V. (1988). *Cumulative Trauma Disorders: A Manual for Musculoskeletal Diseases of the Upper Limbs*. Bristol, PA: Taylor & Francis Group.

Ramos, G., A. Cockburn, R. Balakrishnan, and M. Beaudouin-Lafon (2007). Pointing lenses: Facilitating stylus input through visual-and motor-space magnification. Paper presented at the *Proceedings of the ACM Conference on Human Factors in Computing Systems (CHI'07)*, San Jose, CA. http://portal.acm.org/citation.cfm?id=1240624.1240741

Rekimoto, J. (1996). Tilting operations for small screen interfaces. *Proceedings of the UIST'96*, Seattle, WA. pp. 167–168.

Rekimoto, J. (1997). Pick-and-drop: A direct manipulation technique for multiple computer environments. *Proceedings of the ACM UIST'97 Symposium on User Interface Software & Technology*, Banff, Alberta, Canada. pp. 31–39.

Rekimoto, J. (1998). A multiple device approach for supporting whiteboard-based interactions. *Proceedings of the CHI'98*, Los Angeles, CA. pp. 344–351.

Rime, B. and L. Schiaratura (1991). Gesture and speech. In *Fundamentals of Nonverbal Behavior*, R.S. Feldman and B. Rime' (eds.). New York: Press Syndicate of the University of Cambridge. pp. 239–281.

Robertson, G. G., S. K. Card, and J. D. Mackinlay (1989). The cognitive coprocessor architecture for interactive user interfaces. *Proceedings of the UIST'89 Symposium on User Interface Software and Technology*, Williamsburg, VA. pp. 10–18.

Robertson, G., M. Czerwinski, K. Larson, D. Robbins, D. Thiel, and M. van Dantzich (1998). Data mountain: Using spatial memory for document management. *Proceedings of the UIST'98*, San Francisco, CA.

Robertson, G., M. van Dantzich, D. Robbins, M. Czerwinski, K. Hinckley, K. Risden, V. Gorokhovsky, and D. Thiel (1999). The task gallery: A 3D window manager. *ACM CHI 2000*, pp. 494–501. The Hague, the Netherlands.

Robertson, P. K. (1988). Perceptual color spaces. Visualizing color gamuts: A user interface for the effective use of perceptual color spaces in data display. *IEEE Computer Graphics and Applications* **8**(5): 50–64.

Rosen, P., Z. Pizlo, C. Hoffmann, and V. Popescu (2004). Perception of 3D spatial relations for 3D displays. *Stereoscopic Displays* **XI**: 9–16.

Roudaut, A., S. Huot, and E. Lecolinet (2008). TapTap and MagStick: Improving one-handed target acquisition on small touch-screens. Paper presented at the *Proceedings of the ACM Working Conference on Advanced Visual Interfaces (AVI'08)*, Napoli, Italy. http://dl.acm.org/citation.cfm?id=1385594

Rutledge, J. and T. Selker (1990). Force-to-motion functions for pointing. *Proceedings of INTERACT'90: The IFIP Conference on Human–Computer Interaction*, Cambridge, U.K. pp. 701–706.

Saffo, P. (1997). *Communications of the ACM*, February 1997, **40**(2):93–97.

Saponas, T. S., D. S. Tan, D. Morris, R. Balakrishnan, J. Turner, and J. A. Landay (2009). Enabling always-available input with muscle-computer interfaces. Paper presented at the *Proceedings of the 22nd Annual ACM Symposium on User Interface Software and Technology*, Victoria, British Columbia, Canada.

Sawhney, N. and C. M. Schmandt (2000). Nomadic radio: Speech and audio interaction for contextual messaging in nomadic environments. *ACM Transactions on Computer–Human Interaction* 7(3): 353–383.

Schilit, B. N., N. I. Adams, and R. Want (1994). Context-aware computing applications. *Proceedings of the IEEE Workshop on Mobile Computing Systems and Applications*, IEEE Computer Society, Santa Cruz, CA. pp. 85–90.

Schmandt, C. M. (1983). Spatial input/display correspondence in a stereoscopic computer graphic work station. *Computer Graphics* **17**(3): 253–262; *Proceedings of the ACM SIGGRAPH'83*, Detroit, MI.

Schmandt, C. M., N. Marmasse, S. Marti, N. Sawhney, and S. Wheeler (2000). Everywhere messaging. *IBM Systems Journal* **39**(3–4): 660–677.

Schmidt, A., K. Aidoo, A. Takaluoma, U. Tuomela, K. Van Laerhove, and W. Van de Velde (1999). Advanced interaction in context. *Handheld and Ubiquitous Computing (HUC'99)*, Karlsruhe, Germany. Springer-Verlag, Berlin, Germany. pp. 89–101.

Schwarz, J., S. E. Hudson, J. Mankoff, and A. D. Wilson (2010). A framework for robust and flexible handling of inputs with uncertainty. Paper presented at the *Proceedings of the ACM Symposium on User Interface Software and Technology (UIST'10)*, New York. http://portal.acm.org/citation.cfm?id=1866039

Schwarz, J., J. Mankoff, and S. E. Hudson (2011). Monte Carlo methods for managing interactive state, action and feedback under uncertainty. Paper presented at the *Proceedings of the ACM Symposium on User Interface Software and Technology (UIST'11)*, Santa Barbara, CA. http://dl.acm.org/citation.cfm?id=2047227

Schwesig, C., I. Poupyrev, and E. Mori (2004). Gummi: A bendable computer. Paper presented at the *Proceedings of the ACM Conference on Human Factors in Computing Systems (CHI'04)*, Vienna, Austria. http://portal.acm.org/citation.cfm?id=985726

Sears, A. (1993). Investigating touchscreen typing: The effect of keyboard size on typing speed. *Behavior & Information Technology* **12**(1): 17–22.

Sears, A., C. Plaisant, and B. Shneiderman (1992). A new era for high precision touchscreens. *Advances in Human–Computer Interaction* **3**: 1–33; In R. Hartson and D. Hix (eds.), *Advances in Human–Computer Interaction*. Norwood, NJ: Ablex Publishers.

Sears, A. and B. Shneiderman (1991). High precision touchscreens: Design strategies and comparisons with a mouse. *International Journal of Man-Machine Studies* **34**(4): 593–613.

Sellen, A., G. Kurtenbach, and W. Buxton (1992). The prevention of mode errors through sensory feedback. *Human–Computer Interaction* 7(2): 141–164.

Serra, L., N. Hern, C. Beng Choon, and T. Poston (1997). Interactive vessel tracing in volume data. *ACM/SIGGRAPH Symposium on Interactive 3D Graphics*, Providence, RI. ACM Press, New York. pp. 131–137.

Sheridan, T. B. (1992). *Telerobotics, Automation, and Human Supervisory Control*. Cambridge, MA: MIT Press.

Sigoma, K. B. (1993). A survey of perceptual feedback issues in dexterous telemanipulation: Part I. Finger force feedback. *Proceedings of the IEEE Virtual Reality Annual International Symposium*, San Jose, CA. pp. 263–270.

Silverstein, D. (1977). Human factors for color display systems. In *Color and the Computer: Concepts, Methods and Research*. New York: Academic Press. pp. 27–61.

Simpson, C. A. and K. Marchionda-Frost (1984). Synthesized speech rate and pitch effects on intelligibility of warning messages for pilots. *Human Factors* **26**: 509–517.

Smailagic, A. and D. Siewiorek (February 1996). Modalities of interaction with CMU wearable computers. *IEEE Personal Communications* **3**(1): 14–25.

Smith, R. B. and A. Taivalsaari (1999). Generalized and stationary scrolling. *CHI Letters* **1**(1): 1–9; *Proceedings of UIST'99*, Asheville, NC.

Snibbe, S. and K. MacLean (2001). Haptic techniques for media control. *CHI Letters* **3**(2): 199–208; *Proceedings of the UIST 2001*, Orlando, FL.

Song, H., F. Guimbretière, S. Izadi, X. Cao, and K. Hinckley (2011). Grips and gestures on a multi-touch pen. Paper presented at the *Proceedings of the ACM Conference on Human Factors in Computing Systems (CHI'11)*, Vancouver, British Columbia, Canada. http://dl.acm.org/citation.cfm?id=1979138

Soukoreff, R. W. and I. S. MacKenzie (2004). Towards a standard for pointing device evaluation, perspectives on 27 years of Fitts' law research in HCI. *International Journal of Human-Computer Studies* **61**(6): 751–789.

Sporka, A. J., S. H. Kurniawan, M. Mahmud, and P. Slavík (2006a). Non-speech input and speech recognition for real-time control of computer games. Paper presented at the *Proceedings of the ACM SIGACCESS Conference on Computers and Accessibility (ASSETS'06)*, Portland, OR. http://dl.acm.org/citation.cfm?id=1169023

Sporka, A. J., S. H. Kurniawan, and P. Slavík (2004). Whistling user interface (U^3I). Paper presented at the *Proceedings of the 8th ERCIM Workshop on User Interfaces for All (UI4All'04)*, Vienna, Austria. http://www.springerlink.com/content/dvfetf9xtjj3bv8k/

Sporka, A. J., S. H. Kurniawan, and P. Slavík (2006b). Non-speech operated emulation of keyboard. In J. Clarkson, P. Langdon, and P. Robinson (eds.), *Designing Accessible Technology*. London, U.K.: Springer. pp. 145–154.

Spiker, A., S. Rogers, and J. Cicinelli (1985). Selecting color codes for a computer-generated topographic map based on perception experiments and functional requirements. *Proceedings of the 3rd Symposium on Aviation Psychology*, Ohio State University, Department of Aviation, Columbus, OH. pp. 151–158.

Stifelman, L. (1996). Augmenting real-world objects: A paper-based audio notebook. *CHI'96 Conference Companion*, Vancouver, British Columbia, Canada. pp. 199–200.

Stokes, A., C. Wickens, and K. Kite (1990). *Display Technology: Human Factors Concepts*. Warrendale, PA: SAE.

Stone, M. C., W. B. Cowan, and J. C. Beatty (1988). Color gamut mapping and the printing of digital color images. *ACM Transactions on Graphics* **7**(4): 249–292.

Stratton, G. (1897). Vision without inversion of the retinal image. *Psychological Review* **4**: 360–361.

Streitz, N. A., J. Geißler, T. Holmer, S. Konomi, C. Müller-Tomfelde, W. Reischl, P. Rexroth, P. Seitz, and R. Steinmetz (1999). i-LAND: An interactive landscape for creativity and innovation. *ACM CHI'99 Conference on Human Factors in Computing Systems*, Pittsburgh, PA. pp. 120–127.

Sugimoto, M. and K. Hiroki (2006). HybridTouch: An intuitive manipulation technique for PDAs using their front and rear surfaces. Paper presented at the *Proceedings of the ACM Conference on Human-Computer Interaction with Mobile Devices and Services (MobileHCI'06)*, Helsinki, Finland. http://portal.acm.org/citation.cfm?id=1152243

Sugiura, A. and Y. Koseki (1998). A user interface using fingerprint recognition—Holding commands and data objects on fingers. *UIST'98 Symposium on User Interface Software & Technology*, San Francisco, CA. pp. 71–79.

Sutherland, I. E. (1968). A head-mounted three dimensional display. *Proceedings of the Fall Joint Computer Conference*, San Francisco, CA. pp. 757–764.

Swaminathan, K. and S. Sato (January–February 1997). Interaction design for large displays. *Interactions*, **4**(1):15–24.

Swigart, S. (2005). Easily write custom gesture recognizers for your tablet PC applications. Tablet PC Technical Articles. Retrieved March 19, 2007, from http://msdn.microsoft.com/en-us/library/aa480673.aspx

Tan, D. S., J. K. Stefanucci, D. R. Proffitt, and R. Pausch (2001). The Infocockpit: Providing location and place to aid human memory. *Workshop on Perceptive User Interfaces*, Orlando, FL.

Tan, D. S., J. K. Stefanucci, D. R. Proffitt, and R. Pausch (2002). Kinesthesis aids human memory. *CHI 2002 Extended Abstracts*, Minneapolis, MN.

Tandler, P., T. Prante, C. Müller-Tomfelde, N. A. Streitz, and R. Steinmetz (2001). Connectables: Dynamic coupling of displays for the flexible creation of shared workspaces. *UIST 2001*, Orlando, FL. pp. 11–20.

Tani, M., K. Yamaashi, K. Tanikoshi, M. Futakawa, and S. Tanifuji (1992). Object-oriented video: Interaction with real-world objects through live video. *Proceedings of ACM CHI'92 Conference on Human Factors in Computing Systems*, Monterey, CA. pp. 593–598, 711–712.

Trevor, J., D. M. Hilbert, B. N. Schilit, and T. K. Koh (2001). From desktop to phonetop: A UI for web interaction on very small devices. *Proceedings of the UIST'01 Symposium on User Interface Software and Technology*, Orlando, FL. pp. 121–130.

Trewin, S., S. Keates, and K. Moffatt (2006). Developing steady clicks: A method of cursor assistance for people with motor impairments. Paper presented at the *Proceedings of the ACM SIGACCESS Conference on Computers and Accessibility (ASSETS'06)*, Portland, OR. http://portal.acm.org/citation.cfm?id=1168987.1168993

Triesman, A. (1985). Preattentive processing in vision. *Computer Vision, Graphics, and Image Processing* **31**: 156–177.

Tufte, E. R. (1983). *The Visual Display of Quantitative Information*. Cheshire, CT: Graphics Press.

Tufte, E. R. (1990). *Envisioning Information*. Cheshire, CT: Graphics Press.

Tufte, E. R. (1997). *Visual Explanations: Images and Quantities, Evidence and Narrative*. Cheshire, CT: Graphics Press.

Ullmer, B. and H. Ishii (1997). The metaDESK: Models and prototypes for tangible user interfaces. Paper presented at the *Proceedings of the 10th Annual ACM Symposium on User Interface Software and Technology*, Banff, Alberta, Canada.

Vogel, D. and P. Baudisch (2007). Shift: A technique for operating pen-based interfaces using touch. Paper presented at the *Proceedings of the ACM Conference on Human Factors in Computing Systems (CHI'07)*, San Jose, CA. http://portal.acm.org/citation.cfm?id=1240624.1240727

Vogel, D. and G. Casiez (2012). Hand occlusion on a multi-touch tabletop. Paper presented at the *Proceedings of the ACM Conference on Human Factors in Computing Systems (CHI'12)*, Austin, TX. http://dl.acm.org/citation.cfm?id=2208390

Walker, N. and J. B. Smelcer (1990). A comparison of selection time from walking and bar menus. Paper presented at the *Proceedings of the ACM Conference on Human Factors in Computing Systems (CHI'90)*, Seattle, WA. http://portal.acm.org/citation.cfm?id=97243.97277

Wang, F., X. Cao, X. Ren, and P. Irani (2009). Detecting and leveraging finger orientation for interaction with direct-touch surfaces. Paper presented at the *Proceedings of the ACM Symposium on User Interface Software and Technology (UIST'09)*, Victoria, British Columbia, Canada. http://dl.acm.org/citation.cfm?id=1622182

Wang, F. and X. Ren (2009). Empirical evaluation for finger input properties in multi-touch interaction. Paper presented at the *Proceedings of the ACM Conference on Human Factors in Computing Systems (CHI'09)*, Boston, MA. http://dl.acm.org/citation.cfm?id=1518864

Wang, J., S. Zhai, and H. Su (2001). Chinese input with keyboard and eye-tracking: An anatomical study. *Proceedings of the SIGCHI Conference on Human Factors in Computing Systems*, Seattle, WA.

Want, R. and G. Borriello (May/June 2000). Survey on information appliances. *IEEE Personal Communications* **20**(3): 24–31.

Want, R., K. P. Fishkin, A. Gujar, and B. L. Harrison (1999). Bridging physical and virtual worlds with electronic tags. *Proceedings of the ACM CHI'99 Conference on Human Factors in Computing Systems*, Pittsburgh, PA. pp. 370–377.

Ware, C. (1988). Color sequences for univariate maps: Theory, experiments, and principles. *IEEE Computer Graphics and Applications* **8**(5): 41–49.

Ware, C. (2000). *Information Visualization: Design for Perception*. San Francisco, CA: Morgan Kaufmann.

Ware, C., K. Arthur, and K. S. Booth (1993). Fish tank virtual reality. *Proceedings of ACM INTERCHI'93 Conference on Human Factors in Computing Systems*, Amsterdam, the Netherlands. pp. 37–41.

Ware, C. and D. R. Jessome (November 1988). Using the bat: A six-dimensional mouse for object placement. *IEEE Computer Graphics and Applications* **8**(6): 65–70.

Ware, C. and J. Rose (1999). Rotating virtual objects with real handles. *ACM Transactions on CHI* **6**(2): 162–180.

Weiss, M. (2010). Bringing everyday applications to interactive surfaces. Paper presented at the *Adjunct Proceedings of the ACM Symposium on User Interface Software and Technology (UIST'10)*, New York. http://dl.acm.org/citation.cfm?id=1866227

Weiss, M., R. Jennings, R. Khoshabeh, J. Borchers, J. Wagner, Y. Jansen, and J. D. Hollan (2009). SLAP widgets: Bridging the gap between virtual and physical controls on tabletops. Paper presented at the *Extended Abstracts of the ACM Conference on Human Factors in Computing Systems (CHI'09)*, Boston, MA. http://dl.acm.org/citation.cfm?id=1520462

Weiser, M. (September 1991). The computer for the 21st century. *Scientific American* **265**(3): 94–104.

Wellner, P. (1993). Interacting with paper on the DigitalDesk. *Communications of the ACM* **36**(7): 87–97.

Wenzel, E. M. (1992). Localization in virtual acoustic displays. *Presence* **1**(1): 80–107.

Westeyn, T., P. Pesti, K.-H. Park, and T. Starner (2005). Biometric identification using song-based blink patterns. *Proceedings of the 11th International Conference on Human–Computer Interaction (HCI Int'l'05)*, Las Vegas, NV, July 22–27, 2005. Lawrence Erlbaum Associates, Mahwah, NJ.

Westheimer, G. (1979). Cooperative neural processes involved in stereoscopic acuity. *Experimental Brain Research* **36**: 585–597.

Wickens, C. (1992). *Engineering Psychology and Human Performance*. New York: HarperCollins.

Wigdor, D., C. Forlines, P. Baudisch, J. Barnwell, and C. Shen (2007). LucidTouch: A see-through mobile device. Paper presented at the *Proceedings of the ACM Symposium on User Interface Software and Technology (UIST'07)*, Newport, RI. http://portal.acm.org/citation.cfm?id=1294211.1294259

Wigdor, D., D. Leigh, C. Forlines, S. Shipman, J. Barnwell, R. Balakrishnan, and C. Shen (2006). Under the table interaction. Paper presented at the *Proceedings of the ACM Symposium on User Interface Software and Technology (UIST'06)*, Montreux, Switzerland. http://portal.acm.org/citation.cfm?id=1166253.1166294

Wigdor, D., S. Williams, M. Cronin, R. Levy, K. White, M. Mazeev, and H. Benko (2009). Ripples: Utilizing per-contact visualizations to improve user interaction with touch displays. Paper presented at the *Proceedings of the ACM Symposium on User Interface Software and Technology (UIST'09)*, Victoria, British Columbia, Canada. http://dl.acm.org/citation.cfm?id=1622180

Wilson, A. and S. Shafer (2003). XWand: UI for intelligent spaces. *ACM CHI 2003 Conference on Human Factors in Computing Systems*, Ft. Lauderdale, FL.

Wilson, F. R. (1998). *The Hand: How Its Use Shapes the Brain, Language, and Human Culture*. New York: Pantheon Books.

Wisneski, C., H. Ishii, A. Dahley, M. Gorbet, S. Brave, B. Ullmer, and P. Yarin (1998). Ambient displays: Turning architectural space into an interface between people and digital information. *Lecture Notes in Computer Science* **1370**: 22–32.

Wobbrock, J. O., H. H. Aung, B. Rothrock, and B. A. Myers (2005). Maximizing the guessability of symbolic input. Paper presented at the *Extended Abstracts of the ACM Conference on Human Factors in Computing Systems (CHI'05)*, Portland, OR, April 2–7. ACM Press, New York. pp. 1869–1872. http://portal.acm.org/citation.cfm?doid=1056808.1057043

Wobbrock, J. O., D. H. Chau, and B. A. Myers (2007). An alternative to push, press, and tap-tap-tap: Gesturing on an isometric joystick for mobile phone text entry. Paper presented at the *Proceedings of the ACM Conference on Human Factors in Computing Systems (CHI'07)*, San Jose, CA. http://portal.acm.org/citation.cfm?id=1240624.1240728

Wobbrock, J. O., E. Cutrell, S. Harada, and I. S. MacKenzie (2008). An error model for pointing based on Fitts' law. *Proceedings of the ACM Conference on Human Factors in Computing Systems (CHI'08)*, Florence, Italy, April 5–10, 2008. ACM Press, New York. pp. 1613–1622.

Wobbrock, J. O., J. Fogarty, S.-Y. Liu, S. Kimuro, and S. Harada (2009a). The angle mouse: Target-agnostic dynamic gain adjustment based on angular deviation. Paper presented at the *Proceedings of the ACM Conference on Human Factors in Computing Systems (CHI'09)*, Boston, MA. http://portal2.acm.org/citation.cfm?id=1518701.1518912

Wobbrock, J. O., J. Forlizzi, S. E. Hudson, and B. A. Myers (2002). WebThumb: Interaction techniques for small-screen browsers. *Proceedings of the ACM Symposium on User Interface Software and Technology (UIST'02)*, Paris, France, October 27–30. ACM Press, New York. pp. 205–208.

Wobbrock, J. O., A. Jansen, and K. Shinohara (2011). Modeling and predicting pointing errors in two dimensions. *Proceedings of the ACM Conference on Human Factors in Computing Systems (CHI'11)*, Vancouver, British Columbia, Canada, May 7–12. ACM Press, New York. pp. 1653–1656.

Wobbrock, J. O., M. R. Morris, and A. D. Wilson (2009b). User-defined gestures for surface computing. Paper presented at the *Proceedings of the ACM Conference on Human Factors in Computing Systems (CHI'09)*, Boston, MA. http://portal2.acm.org/citation.cfm?id=1518701.1518866

Wobbrock, J. O., B. A. Myers, and H. H. Aung (2008). The performance of hand postures in front- and back-of-device interaction for mobile computing. *International Journal of Human–Computer Studies* **66**(12): 857–875.

Wobbrock, J. O., B. A. Myers, and J. A. Kembel (2003). EdgeWrite: A stylus-based text entry method designed for high accuracy and stability of motion. Paper presented at the *Proceedings of the ACM Symposium on User Interface Software and Technology (UIST'03)*, Vancouver, British Columbia, Canada, November 2–5. ACM Press, New York. pp. 61–70. http://portal.acm.org/citation.cfm?id=964703

Wobbrock, J.O., K. Shinohara, and A. Jansen (2011). The effects of task dimensionality, endpoint deviation, throughput calculation, and experiment design on pointing measures and models. *Proceedings of the ACM Conference on Human Factors in Computing Systems (CHI '11)*, May 7–12, 2011, Vancouver, British Columbia, Canada. ACM Press, New York. pp. 1639–1648.

Wobbrock, J. O., A. D. Wilson, and Y. Li (2007). Gestures without libraries, toolkits or training: A $1 recognizer for user interface prototypes. Paper presented at the *Proceedings of the ACM Symposium on User Interface Software and Technology (UIST'07)*, Newport, RI. http://portal.acm.org/citation.cfm?id=1294211.1294238

Worden, A., N. Walker, K. Bharat, and S. E. Hudson (1997). Making computers easier for older adults to use: Area cursors and sticky icons. Paper presented at the *Proceedings of the ACM Conference on Human Factors in Computing Systems (CHI'97)*, Atlanta, GA. http://portal.acm.org/citation.cfm?id=258724

Wyszecki, G. and W. S. Styles (1982). *Color Science*, 2nd edn., Wiley, New York.

Xin, Y., X. Bi, and X. Ren (2011). Acquiring and pointing: An empirical study of pen-tilt-based interaction. Paper presented at the *Proceedings of the ACM Conference on Human Factors in Computing Systems (CHI'11)*, Vancouver, British Columbia, Canada. http://dl.acm.org/citation.cfm?id=1978942.1979066

Xin, Y., X. Bi, and X. Ren (2012). Natural use profiles for the pen: An empirical exploration of pressure, tilt, and azimuth. Paper presented at the *Proceedings of the ACM Conference on Human Factors in Computing Systems (CHI'12)*, Austin, TX. http://dl.acm.org/citation.cfm?id=2208518

Yin, J., and X. Ren (2007). The beam cursor: A pen-based technique for enhancing target acquisition. In N. Bryan-Kinns, A. Blanford, P. Curzon, and L. Nigay (eds.), *People and Computers XX: Proceedings of HCI 2006*. London, U.K.: Springer. pp. 119–134.

Zeleznik, R., K. Herndon, and J. Hughes (1996). SKETCH: An interface for sketching 3D scenes. *Proceedings of SIGGRAPH'96*, New Orleans, LA. pp. 163–170.

Zhai, S. (1998). User performance in relation to 3D input device design. *Computer Graphics* **32**(8): 50–54.

Zhai, S., S. Conversy, M. Beaudouin-Lafon, and Y. Guiard (2003). Human on-line response to target expansion. *CHI 2003 Conference on Human Factors in Computing Systems*, Ft. Lauderdale, FL. pp. 177–184. http://portal.acm.org/citation.cfm?id=642644

Zhai, S., M. Hunter, and B. A. Smith (2000). The metropolis keyboard—An exploration of quantitative techniques for virtual keyboard design. *CHI Letters* **2**(2): 119–128.

Zhai, S. and P. Milgram (1993). Human performance evaluation of isometric and elastic rate controllers in a 6DoF tracking task. *Proceedings of the SPIE Telemanipulator Technology*, **2057**:130–141.

Zhai, S., P. Milgram, and W. Buxton (1996). The influence of muscle groups on performance of multiple degree-of-freedom input. *Proceedings of CHI'96: ACM Conference on Human Factors in Computing Systems*, Vancouver, British Columbia, Canada. ACM Press, New York. pp. 308–315.

Zhai, S., C. Morimoto, and S. Ihde (1999). Manual and Gaze Input Cascaded (MAGIC) pointing. *Proceedings of the ACM CHI'99 Conference on Human Factors in Computing Systems*, Pittsburgh, PA. pp. 246–253.

Zhai, S., B. A. Smith, and T. Selker (1997). Improving browsing performance: A study of four input devices for scrolling and pointing tasks. *Proceedings of the INTERACT'97: The Sixth IFIP Conference on Human–Computer Interaction*, Sydney, New South Wales, Australia. pp. 286–292.

Zhao, S., M. Agrawala, and K. Hinckley (2006). Zone and polygon menus: Using relative position to increase the breadth of multi-stroke marking menus. Paper presented at the *Proceedings of the ACM Conference on Human Factors in Computing Systems (CHI'06)*, Montreal, Quebec, Canada. http://dl.acm.org/citation.cfm?id=1124933

Zhao, S. and R. Balakrishnan (2004). Simple vs. compound mark hierarchical marking menus. Paper presented at the *Proceedings of the ACM Symposium on User Interface Software and Technology (UIST'04)*, Santa Fe, NM. http://dl.acm.org/citation.cfm?id=1029639

22 Performance Enhancements

Joseph Dumas
University of Tennessee

22.1 Introduction

In previous chapters, the reader should have become familiar with the basics of instruction set architecture and datapath and control unit design. Knowledge of these principles is sufficient to understand the essentials of central processing unit (CPU) architecture and implementation and, given some effort, to generate the logical design for a complete, usable CPU. Decades ago, purely sequential, von Neumann architecture processors were commercially viable. To succeed in today's marketplace, however, a processor (and the system containing it) must not only work correctly but must perform extremely well on the application(s) of interest. This chapter is devoted to exploring various implementation techniques that computer manufacturers have adopted in order to achieve the goal of making their CPU process information as rapidly as possible. The basis of virtually all of these techniques, which we shall shortly introduce, is known as *pipelining*. Essentially all high-performance computers utilize some form of pipelining.

Implementation technologies used to manufacture computer components change rapidly over time. For decades, transistors (which make up the most basic binary switching elements) have become smaller and faster; circuits that were considered blindingly fast 4 or 5 years ago are hopelessly slow today and will be completely obsolete that far, or less, in the future. Over the history of computing devices, technology has always improved and continues to improve—though at some point, Moore's law must expire as we approach the ultimate physical limits [1]. Where will more speed come from when we have switching elements close to the size of individual atoms, with signal propagation over such tiny distances still limited by the velocity of light? Perhaps the unique approach of quantum computing will allow us to do much more than is currently thought possible within the limits of conventional physics. Only time will tell where new innovations in technology will take us, as designers and users of computer systems.

What is always true is that at any given time, be it 1955 or 1981 or 2013, computer manufacturers could, and can, only make a given piece of hardware so fast using available technology. What if that is not fast enough to suit our purposes? Then we must augment technology with clever design. If a given hardware component is not fast enough to do the work we need to do in the time we have to allow, we build faster hardware if we can, but if we cannot, we build *more* hardware and divide the problem. This approach is known as *parallelism* or *concurrency.* We can achieve concurrency within a single processor core and/or by using multiple processors in a system (either in the form of a multicore chip, multiple CPU chips in a system, or both). The former approach—concurrency within one core—is the subject of the rest of this chapter; the latter will be addressed in Chapters 23 and 24 on multiprocessing.

22.2 Underlying Principles

22.2.1 Limitations of Purely Sequential Processors

The original digital computers, and their successors for a number of years, were all serial (sequential) processors not only in their architecture but also in their implementation. Not only did they *appear* to execute only one instruction at a time, as the von Neumann model suggests, but they actually *did* execute only one instruction at a time. Each machine instruction was processed completely before the next one was started. This sequential execution property allowed the processor's control unit, whether hardwired or microprogrammed, to be kept relatively simple as it only has to generate the control signals for one instruction at a time.

The sequential execution approach that forms the basis of the von Neumann machine cycle is very effective as well as simple, but it has one obvious flaw: it does not make very efficient use of the hardware. In a typical architecture, executing a single machine instruction requires several steps: fetch the instruction from memory, decode it, retrieve its operands, perform the operation it specifies, and store its result. (Somewhat different breakdowns of the machine cycle are possible, but this will suffice for discussion.) If the machine processes one instruction at a time, what is the arithmetic/logic unit (ALU) doing while the instruction (or an operand) is being fetched? Probably nothing. What work are the instruction decoder, memory address register, memory data register, and various other parts of the CPU accomplishing while the ALU is busy performing a computation? Probably none.

For the most efficient use of all the system components in which we have invested design time, chip area, electrical power, and other valuable resources, we would ideally like to keep all of these components as busy as possible as much of the time as possible. A component that is unused part of the time is not giving us our money's worth; designers should search for a way to make more use of it. Conversely, a component that is overused (needed more often than it is available) creates a *structural hazard*; it will often have other components waiting on it and will thus become a bottleneck, slowing down the entire system [2]. Designers may need to replicate such a component to improve throughput. The art of designing a modern processor involves balancing the workload on all the parts of the CPU such that they are kept busy doing useful work as much of the time as possible without any of them "clogging up the works" and making the other parts wait. As the reader might expect, this "balancing act" is not a trivial exercise. Pipelining, which we are about to investigate, is an essential technique for helping bring about this needed balance.

22.2.2 Basic Concepts of Pipelining

Pipelining, in its most basic form, means breaking up a task into smaller subtasks and overlapping the performance of those subtasks for different instances of the task. (The same concept, when applied to

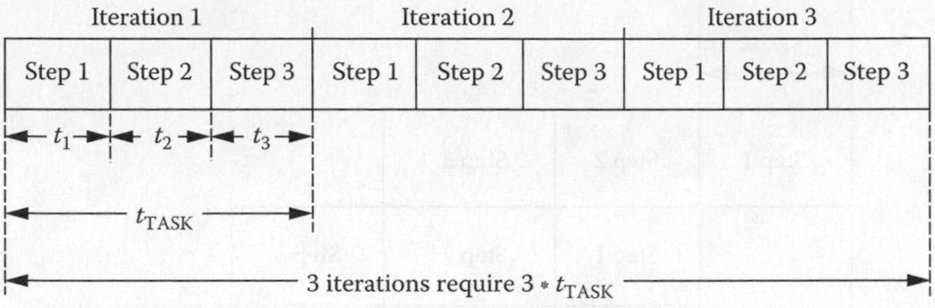

FIGURE 22.1 Subdividing a task into sequential subtasks. (From Dumas II, J.D., *Computer Architecture: Fundamentals and Principles of Computer Design*, CRC Press, Boca Raton, FL, 2006, Figure 4.1.)

the manufacture of automobiles or other objects, is called an assembly line.) Or, to use terms more specifically related to computing, pipelining means dividing a computational operation into steps and overlapping those steps over successive computations. This approach, while much more common in today's computers than it was 20 or 30 years ago, is hardly new. The first use of pipelining in computers dates back to the International Business Machines (IBM) Stretch and Univac LARC machines of the late 1950s [3]. Pipelining, as we shall see, improves the performance of a processor in much the same way that low-order interleaving can improve the performance of main memory while being subject to many of the same considerations and limitations.

To understand how pipelining works, consider a task that can be broken down into three parts, performed sequentially. Let us refer to these parts as step 1, step 2, and step 3 (see Figure 22.1). The time taken to perform step 1 is represented as t_1, while t_2 and t_3 represent the times required to perform steps 2 and 3. Since the three steps (subtasks) are performed sequentially, the total time to perform the task is of course given by

$$t_{TASK} = t_1 + t_2 + t_3$$

Without pipelining, the time to perform three iterations of this task is $3 * t_{TASK}$, while the time to perform fifty iterations is $50 * t_{TASK}$. The time required is directly proportional to the number of iterations to bc performed; there is no advantage to be gained by repeated performance of the task.

Now, suppose that we separate the hardware that performs steps 1, 2, and 3 in such a way that it is possible for them to work independently of each other. (We shall shortly see how this can be accomplished in computer hardware.) Figure 22.2 illustrates this concept. We begin the first iteration of the task by providing its inputs to the hardware that performs step 1 (call this stage 1). After t_1 seconds, step 1 (for the first iteration of the task) is done, and the results are passed along to stage 2 for the performance of step 2. Meanwhile, we provide the second set of inputs to stage 1 and begin the second iteration of the task *before the first iteration is finished*. When stage 2 is finished processing the first iteration and stage 1 is finished processing the second iteration, the outputs from stage 2 are passed to stage 3, and the outputs from stage 1 are passed to stage 2. Stage 1 is then provided a new, third set of inputs—again, before the first and second iterations are finished. At this point, the pipeline is full; all three stages are busy working on something. When all are done this time, iteration 1 is complete, iteration 2 moves to stage 3, iteration 3 moves to stage 2, and the fourth iteration is initiated in stage 1. This process can continue as long as we have more iterations of the task to perform. All stages will remain busy until the last task iteration leaves stage 1 and eventually "drains" from the pipeline (completes through all the remaining stages).

Obviously, this approach works best if the three stage times t_1, t_2, and t_3 are all equal to $t_{TASK}/3$. Let us refer to this as the stage time t_{STAGE}. If t_1, t_2, and t_3 are not all equal, then t_{STAGE} must be set equal to the greatest of the three values; in other words, we cannot advance the results of one stage to the next until the *slowest* stage completes its work. If we try to make t_{STAGE} smaller than that, at least one of the stages will be given new inputs before it finishes processing its current inputs and the process will break down (generate incorrect results).

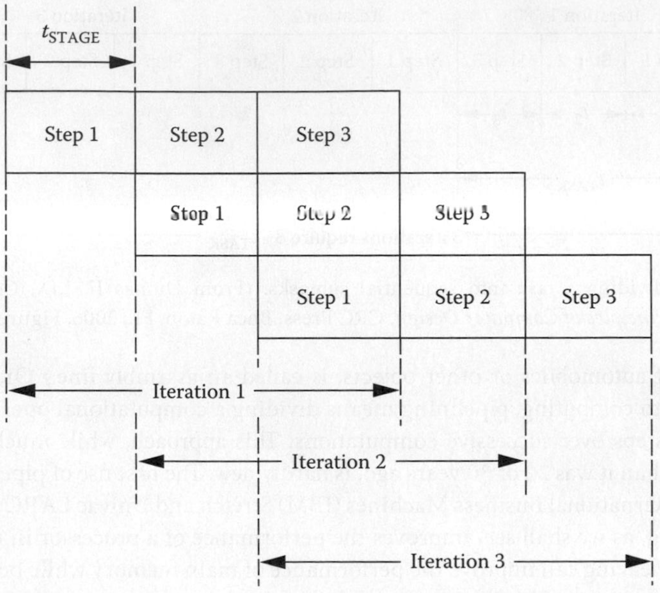

FIGURE 22.2 Basic pipelining concept. (From Dumas II, J.D., *Computer Architecture: Fundamentals and Principles of Computer Design*, CRC Press, Boca Raton, FL, 2006, Figure 4.2.)

22.2.3 Performance Implications of Pipelining

What are the performance implications of this pipelined (or assembly line) approach? It should be obvious that pipelining does nothing to enhance the performance of a *single* computation. A single iteration of the task still requires at least t_{TASK} seconds. In fact, it may take somewhat longer if the stage times are mismatched (we know that $3 * t_{STAGE} \geq t_1 + t_2 + t_3$). The performance advantage occurs only if we perform two or more successive iterations of the task. While the first iteration takes just as long as ever (or perhaps slightly longer), the second and all subsequent iterations are completed in one additional t_{STAGE} each, rather than the $3 * t_{STAGE}$ taken by the first iteration. Two iterations can be completed in $4 * t_{STAGE}$, three iterations in $5 * t_{STAGE}$, four in $6 * t_{STAGE}$, and so on. In general, we can define the time taken to perform n iterations of the task using this three-stage pipeline as

$$t_{TOTAL} = [3 * t_{STAGE}] + [(n - 1) * t_{STAGE}] = [(n + 2) * t_{STAGE}]$$

If t_{STAGE} is equal (or even reasonably close) to $t_{TASK}/3$, then a substantial speedup is possible versus the nonpipelined case; larger values of n lead to a greater advantage for the pipelined implementation. Let us suppose for the sake of simplicity that there is no hardware overhead (we will address this topic later) and that t_{STAGE} equals 1 ns and t_{TASK} equals 3 ns. Five iterations of the task would take (5 * 3) = 15 ns without pipelining but only (7 * 1) = 7 ns using the pipelined approach. The speed ratio in this case is 2.143 to 1 in favor of pipelining. If we consider ten consecutive iterations, the total times required are 30 ns (nonpipelined) versus 12 ns (pipelined), with pipelining yielding a speedup factor of 2.5. For fifty iterations, the numbers are 150 and 52 ns, respectively, for a speedup of 2.885. In the limit, as n grows very large, the speedup factor of the pipelined implementation versus the nonpipelined implementation approaches 3, which is—not coincidentally—the number of stages in the pipe.

Most generally, for a pipeline of s stages processing n iterations of a task, the time taken to complete all the iterations may be expressed as

$$t_{TOTAL} = [s * t_{STAGE}] + [(n - 1) * t_{STAGE}] = [(s + n - 1) * t_{STAGE}]$$

The $[s * t_{STAGE}]$ term represents the *flow-through time*, which is the time for the first result to be completed; the $[(n - 1) * t_{STAGE}]$ term is the time required for the remaining results to emerge from the pipe. The time taken for the same number of iterations without pipelining is $n * t_{TASK}$. In the ideal case of a perfectly balanced pipeline (in which all stages take the same time), $t_{TASK} = s * t_{STAGE}$ and so the total time for the nonpipelined implementation would be $n * s * t_{STAGE}$. The best-case speedup obtainable by using an s-stage pipeline would thus be $(n * s)/(n + s - 1)$, which, as n becomes large, approaches s as a limit.

From this analysis, it would appear that the more stages into which we subdivide a task, the better. This does not turn out to be the case for several reasons. First of all, it is generally only possible to break down a given task so far. (In other words, there is only so much *granularity* inherent in the task.) When each stage of the pipeline represents only a single level of logic gates, how can one further subdivide operations? The amount of logic required to perform a given task thus places a fundamental limitation on the *depth* of (number of stages in) a pipelined implementation of that task.

Another limiting factor is the number of consecutive, uninterrupted iterations of the task that are likely to occur. For example, it makes little sense to build a ten-stage multiplication pipeline if the number of multiplications to be done in sequence rarely exceeds 10. One would spend most of the time filling and draining the pipeline—in other words, with less than the total number of stages doing useful work. Pipelines only improve performance significantly if they can be kept full for a reasonable length of time. Or, mathematically speaking, achieving a speedup factor approaching s (the number of pipeline stages) depends on n (the number of consecutive iterations being processed) being large, where "large" is defined relative to s. "Deep" pipelines, which implement a *fine-grained* decomposition of a task, only perform well on long, uninterrupted sequences of task iterations.

Yet another factor that limits the speedup that can be achieved by subdividing a task into smaller subtasks is the reality of hardware implementation, which always incurs some overhead. Specifically, constructing a pipeline with actual hardware requires the use of a *pipeline register* (a parallel-in, parallel-out storage register comprised of a set of flip-flops or latches) to separate the combinational logic used in each stage from that of the following stage (see Figure 22.3 for an illustration). The pipeline registers effectively isolate the outputs of one stage from the inputs of the next, advancing them only when a clock pulse is received; this prevents one stage from interfering with the operation of those preceding and/or following it. The same clock signal is connected to all the pipeline registers so that the outputs of each stage are transferred to the inputs of the next simultaneously.

Of course, each pipeline register has a cost of implementation: it consumes a certain amount of power, takes up a certain amount of chip area, etc. Also, the pipeline registers have finite propagation delays that add to the propagation delay of each stage; this reduces the performance of the pipeline somewhat, compared to the ideal, theoretical case. The clock cycle time t_C of the pipeline can be no smaller than t_{STAGE} (the longest of the individual stage logic delays) plus t_R (the delay of a pipeline register). The register delays represent an extra cost or "overhead factor" that takes away somewhat from the advantage of a pipelined implementation. If t_R is small compared to t_{STAGE}, as is usually the case, the pipeline will perform close to theoretical limits and certainly much better than a single-stage (purely combinational) implementation of the same task. If we try to divide the task into very small steps, though, t_{STAGE} may become comparable to, or even smaller than, t_R, significantly reducing the advantage of pipelining. We cannot make t_C smaller than t_R no matter how finely we subdivide the task. (Put another way, the maximum clock frequency of the pipeline is limited to $1/t_R$ even if the stages do no work at all!) Thus, there is a point of diminishing returns beyond which it makes little sense to deepen the pipeline. The "best" design would probably be one with a number of stages that maximizes the ratio of performance to cost, where cost may be measured not so much in dollars as in terms of chip area, transistor count, wire length, power dissipation, and/or other factors.

CPU pipelines generally fall into one of two categories: *arithmetic pipelines* or *instruction-unit pipelines*. Arithmetic pipelines are generally found in vector supercomputers, where the same numerical (usually floating-point) computation(s) must be done to many values in succession. Vector machines

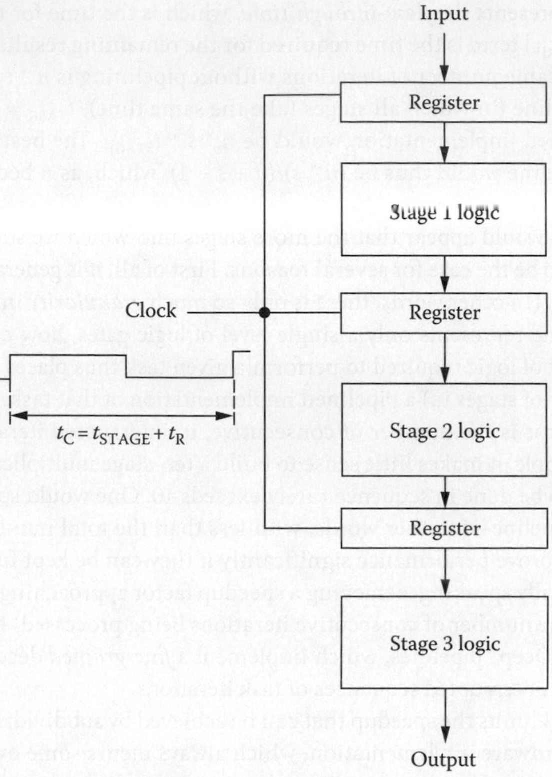

FIGURE 22.3 Pipeline construction showing pipeline registers. (From Dumas II, J.D., *Computer Architecture: Fundamentals and Principles of Computer Design*, CRC Press, Boca Raton, FL, 2006, Figure 4.3.)

are a dying breed, though; while in June 1993, approximately 65% of the world's top 500 supercomputers used a vector architecture, by November 2007, that figure had dropped to less than 1%, where it remains. (The highest-ranked vector supercomputer remaining on the list is Japan's Earth Simulator 2, an NEC SX-9/E system that debuted at number 22 in June 2009 but had dropped to 94th place by November 2011 and all the way to 217th by November 2012 [4].) By contrast, instruction-unit pipelines, which are used to execute a variety of scalar instructions at high speed, are found in practically all modern general-purpose processors—including those used in most high-performance, massively parallel systems. Thus, the following sections will concentrate on techniques used in pipelined instruction execution units.

22.3 Impact on Practice

22.3.1 Instruction-Unit Pipelines

Instruction-unit pipelines are pipelines that are used to execute a machine's scalar instruction set (the instructions that operate on only one or two individual operands to produce a single result—these make up most, if not all, of the instruction set of a typical general-purpose machine). As we discussed in Section 22.2.1, the execution of each machine instruction can be broken into several steps. Often, particularly in a machine that has simple instructions, most of these steps are the same for many (possibly most or all) of the instructions. For example, all of the computational instructions may be done with the same sequence of steps, with the only difference being the operation requested of the ALU once the operands are present. With some effort, the data transfer and control transfer instructions may also be implemented with the same or a very similar sequence of steps. If most or all of the machine's instructions can be accomplished with a similar number and type of steps, it is relatively easy to pipeline the execution of those steps and thus improve the machine's performance considerably versus a completely sequential design.

22.3.2 Basics of an Instruction Pipeline

The basic concept flows from the principles that were introduced at the beginning of Section 22.2.2. Pipelining as we defined it means breaking up a computational task into smaller subtasks and overlapping the performance of those subtasks for different instances of the task. In this case, the basic task is the execution of a generic, scalar instruction, and the subtasks correspond to subdivisions of the von Neumann execution cycle, which the machine must perform in order to execute it. The von Neumann cycle may be broken into more or fewer steps depending on the designer's preference and the amount of logic required for each step. (Remember that for best pipelined performance, the overall logic delay should be divided as equally as possible among the stages.) As an example, let us consider a simple instruction-unit pipeline with four stages, as follows:

F—Fetch instruction from memory.
D—Decode instruction and obtain operand(s).
E—Execute required operation on operand(s).
W—Write result of operation to destination.

A sequential (purely von Neumann) processor would perform these four steps, one at a time, for a given instruction, then go back and perform them one at a time for the next instruction, and so on. If each step required one CPU clock cycle, then each machine instruction would require four clock cycles. Two instructions would require 8 cycles, three would require 12 cycles, etc. As we noted previously, this approach is effective but slow and does not make very efficient use of the hardware.

In an instruction-pipelined processor, we break down the required hardware into smaller pieces, separated by pipeline registers as described previously. Our example divides the hardware into four stages as shown in Figure 22.4. By splitting the hardware into four stages, we can overlap the execution of up to four instructions simultaneously. As Table 22.1 shows, we begin by fetching instruction I_1 during the first time step t_0. During the next time step (t_1), I_1 moves on to the decoding/operand fetch stage, while simultaneously instruction I_2 is being fetched by the first stage. Then during step t_2, I_1 is in the execute (E) stage, while I_2 is in the D stage, and I_3 is being fetched by the F stage. During step t_3, I_1 moves to the final stage (W), while I_2 is in E, I_3 is in D, and I_4 is in F. This process continues indefinitely for subsequent instructions.

As in the nonpipelined case, it still takes four clock cycles to execute the first instruction I_1. However, because of the overlap of operations in the pipeline, I_2 completes one clock cycle later, I_3 completes one cycle after that, and likewise for subsequent instructions. Once the pipeline is full, we achieve a steady-state throughput of one instruction per clock cycle (IPC), rather than one instruction per four cycles as in the nonpipelined case. Thus, in the ideal case, the machine's performance may increase by nearly a factor of four.

Of course, the ideal case is not always the way things work out in the real world. It appears from an initial examination that we can easily achieve a pipeline throughput of one instruction per cycle. (This was the goal of the original reduced instruction set computer [RISC] designers [5] and of course is the goal of any pipelined processor design.) However, a sustained throughput of 1.0 instruction per cycle is never attainable in practice using a single instruction pipeline (although in some cases we can come fairly close). Why not? For the same reason, we can never achieve maximum throughput in an arithmetic or any other type of pipeline: *something* happens to "break the chain" or temporarily interrupt the operation of the pipeline. For a vector arithmetic pipeline, this might result from having to change a dynamic pipeline to another configuration or simply reaching the end of the current vector computation

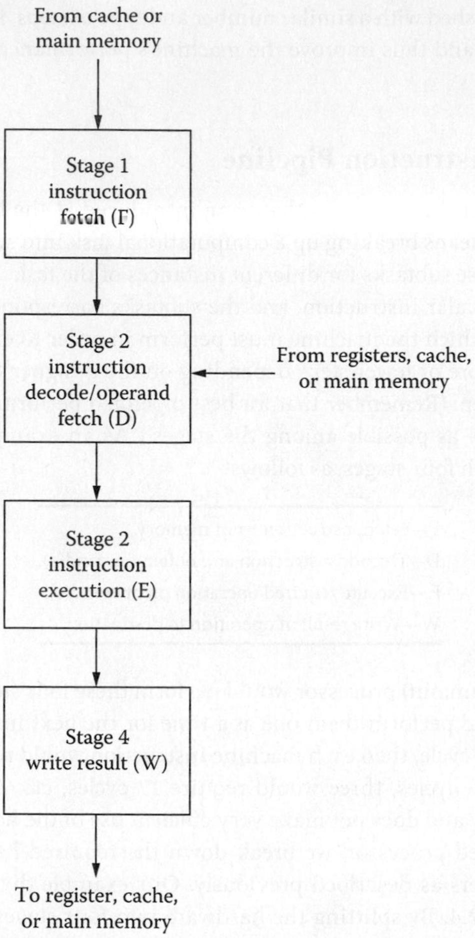

FIGURE 22.4 Stages of a typical instruction pipeline. (From Dumas II, J.D., *Computer Architecture: Fundamentals and Principles of Computer Design*, CRC Press, Boca Raton, FL, 2006, Figure 4.12.)

TABLE 22.1 Execution of Instructions in a Four-Stage Pipeline

	t_0	t_1	t_2	t_3	t_4	t_5	t_6
Stage 1 (F)	I_1	I_2	I_3	I_4	I_5	I_6	I_7
Stage 2 (D)		I_1	I_2	I_3	I_4	I_5	I_6
Stage 3 (E)			I_1	I_2	I_3	I_4	I_5
Stage 4 (W)				I_1	I_2	I_3	I_4

(i.e., running out of operands). Anytime we miss initiating a new operation into the pipe for even a single cycle, we will correspondingly miss completing one operation per cycle at some later time and thus not average performing one operation per cycle.

Neither of the previous scenarios applies directly to an instruction-unit pipeline. Though different instructions are executed, they all use the same sequence of steps; thus, the pipeline structure is not reconfigured. And of course the CPU is always executing *some* instructions! They may be part of a user program or the operating system; they may be computational instructions, data transfers, or even "no operation" (NOP) instructions, but the processor never stops executing instructions as long as it is "up."

(We are ignoring the special case of some embedded processors that have "sleep" or "standby" modes where they halt execution in order to save power.) So if these types of situations are not the problem, what can—and does—happen to hold up the operation of a pipelined instruction unit and keep it from achieving its theoretical throughput of one instruction per cycle? There are several possibilities that we shall explore later, but the most fundamental difficulty is the fact that execution of instructions is not always sequential.

22.3.3 Control Transfers and the Branch Penalty

It is control transfer instructions, which of course are very important to all practical programs, which cause the most obvious problem with respect to pipelined instruction execution. Pipelining instructions that are sequentially stored and sequentially executed is relatively easy, but sooner or later in any useful program, we must make a decision (based on data input or retrieved from memory or on the results of computations performed on data) as to what to do next. This decision-making process is typically done using a comparison and conditional branching technique. The program performs a comparison (or some other arithmetic or logic operations), and then a conditional branch instruction is executed. This branch tests some condition related to the operation just performed and either *succeeds* or *fails* based on whether or not the specified condition is true. A conditional branch that succeeds means that the next instruction executed is the one at the specified target address, while one that fails means that the next instruction executed is the next one in program sequence.

If it were possible to know in advance which of these events would occur, handling the situation would pose fewer problems for the pipelined instruction execution unit. But of course, if it were possible to always know ahead of time that a given branch would be taken or not, it would not have to be encoded as a branch! And in fact the CPU often (particularly in the case of program loops) encounters a particular branch repeatedly over time, where—depending on the particular data being processed—it may succeed on some occasions and fail on others. Conditional transfers of control are an unavoidable fact of life in the logic of useful programs built on the von Neumann execution model, and as we shall see, these control transfers do pose problems, known as *control hazards*, for pipelined execution of instructions [2].

The problems we encounter with branching stem from the obvious fact that a single pipeline can only process one sequence of instructions. There may indeed be only one sequence of instructions leading up to a given conditional branch, but there are always two possible sequences of instructions following it. There is the sequence of instructions immediately following the branch instruction in memory, to be executed if it fails, but there is also the sequence of instructions beginning at the branch target location in memory, to be executed if it succeeds. The pipeline can only process one of these sequences of instructions at a time—what if it is the wrong one?

If the pipeline control logic assumes that a given conditional branch will fail and begins working on the sequential instructions, and it turns out that the branch actually fails, then there is no problem and the pipeline can continue completing instructions at the rate of one per clock cycle. Likewise, if the pipeline logic successfully predicts that the branch will succeed and begins processing at the target location, the pipeline can be kept full or nearly so. (Depending on the pipeline structure and the memory interface, there may be a slight delay as instructions must be obtained from a different part of memory.) But in either of the other two cases, where a branch succeeds while sequential instructions have already started down the pipe or where a branch fails while the processor assumed it would be taken, there is a definite problem that will interrupt the completion of instructions—possibly for several clock cycles.

When any given conditional branch instruction first enters a pipeline, the hardware has no way of knowing whether it will succeed or fail. Indeed, since the first stage of the pipeline always involves fetching the instruction from memory, the control unit would not even know at this point whether or not this instruction is a control transfer! That information is not available until the instruction is decoded

TABLE 22.2 Pipelined Instruction Execution with Conditional Branching

	t_0	t_1	t_2	t_3	t_4	t_5	t_6	t_7	t_8
Stage 1 (F)	I_1	I_2	I_3	I_4	I_5	I_6	I_1	I_2	I_3
Stage 2 (D)		I_1	I_2	I_3	I_4	I_5	—	I_1	I_2
Stage 3 (E)			I_1	I_2	I_3	I_4	—	—	I_1
Stage 4 (W)				I_1	I_2	I_3	I_4	—	—

(in our example, this occurs in the second pipeline stage). And even once the instruction is identified as a conditional branch, it takes some time to check the appropriate condition and make the decision of whether or not the branch will succeed (and then to update the program counter with the target location if the branch does succeed). By this time, one, two, or more subsequent instructions, which may or may not be the correct ones, may have entered the pipe.

Table 22.2 illustrates a possible scenario where a branch that was assumed to fail instead happens to succeed. Assume that instruction I_4 is a conditional branch that implements a small program loop; its target is I_1, the top of the loop, while of course the sequential instruction I_5 will be executed on completion of the loop. We do not know ahead of time how many times the loop will iterate before being exited; either I_1 or I_5 may follow I_4 at any time.

Instruction I_4 is fetched by the first pipeline stage during cycle t_3. During the following cycle, t_4, it is decoded while the pipeline is busy fetching the following instruction I_5. Sometime before the end of cycle t_4, the control unit determines that I_4 is indeed a conditional branch, but it still has to test the branch condition. Let us assume that this does not happen until sometime during cycle t_5, when the fetch of instruction I_6 has begun. At this point, the control unit determines that the branch condition is true and loads the address of I_1 into the program counter to cause the branch to take place. I_1 will thus be fetched during the next clock cycle, t_6, but by this time, two instructions (I_5 and I_6) are in the pipeline where they should not be. Allowing them to continue to completion would cause the program to generate incorrect results, so they are aborted or *nullified* by the control logic (meaning that the results of these two instructions are not written to their destinations). This cancellation of the incorrectly fetched instructions is known as "flushing" the pipeline [6]. While the correctness of program operation can be retained by nullifying the effects of I_5 and I_6, we can never recover the two clock cycles that were wasted in mistakenly attempting to process them. Thus, this branch has prevented the pipeline from achieving its maximum possible throughput of one IPC.

In this example, a successful conditional branch caused a delay of two clock cycles in processing instructions. This delay is known as the *branch penalty* [7]. Depending on the details of the instruction set architecture and the way it is implemented by the pipeline, the branch penalty for a given design might be greater or less than two clock cycles. In the best case, if the branch condition could be tested and the program counter modified in stage 2 (before the end of cycle t_4 in our example), the branch penalty could be as small as one clock cycle. If determining the success of the branch, modifying the PC, and obtaining the first instruction from the target location took longer, the branch penalty might be three or even four cycles (the entire depth of the pipeline, meaning that its entire contents would have to be flushed). The number of lost cycles may vary from implementation to implementation, but branches in a program are never good for a pipelined processor.

It is worth noting that branching in programs can be a major factor limiting the useful depth of an instruction pipeline. In addition to the other reasons mentioned earlier to explain the diminishing returns we can achieve from pipelines with many stages, one can readily see that the deeper an instruction-unit pipeline, the greater the penalty that may be imposed by branching. Rather than one or two instructions, a fine-grained instruction pipe may have to flush several instructions on each successful branch. If branching instructions appear frequently in programs, a pipeline with many stages may perform no better—or even worse—than one with a few stages [8].

It is also worth mentioning that conditional branch instructions are not the only reason why the CPU cannot always initiate a new instruction into the pipeline every cycle. Any type of control transfer instruction, including unconditional jumps and subprogram calls and returns, may cause some delay in processing. Though there is no branch condition to check, these other instructions must still proceed a certain distance into the pipeline before being decoded and recognized as control transfers, during which time one or more subsequent instructions may have entered the pipe. Exception processing (including internally generated traps and external events such as interrupts) also requires the CPU to suspend execution of the sequential instructions in the currently running program, transfer control to a handler located somewhere else in memory, and then return and resume processing the original program. The pipeline must be drained and refilled upon leaving and returning, once again incurring a penalty of one or more clock cycles in addition to other overhead such as saving and restoring register contents.

22.3.4 Branch Prediction

Branch prediction is one approach that can be used to minimize the performance penalty associated with conditional branching in pipelined processors. Consider the example presented in Table 22.2. If the control unit could somehow be made aware that instructions I_1 through I_4 make up a program loop, it might choose to assume that the branch would succeed and fetch I_1 (instead of I_5) after I_4 each time. Using this approach, the full branch penalty would be incurred only once, upon exiting the loop, rather than each time through the loop. Of course, one could equally well envision a scenario where assuming that the branch would fail would be the better course of action. But how, other than by random guessing (which of course is as likely to be wrong as it is to be right), can the control unit predict whether or not a branch will be taken?

Branch prediction, which dates back to the IBM Stretch machine of the late 1950s [9], can be done either *statically* (before the program is run) or *dynamically* (by the control unit at run time) or as a combination of both techniques. The simplest forms of static prediction either assume all branches succeed (assuming they all fail would be equivalent to no prediction at all) or assume that certain types of branches always succeed, while others always fail [10]. As the reader might imagine, these primitive schemes tend not to fare much better than random guessing.

A better way to do static prediction, if the architecture supports it, is to let the compiler do the work. Version 9 of Sun Microsystems' (now Oracle's) scalable processor architecture (SPARC) provides a good example of this approach. Each SPARC V9 conditional branch instruction has two different op codes: one for "branch probably taken" and one for "branch probably not taken [11]." The compiler analyzes the structure of the high-level code and chooses the version of the branch it expects will be correct most of the time; when the program runs, the processor uses the op code as a hint to help it choose which instructions to fetch into the pipeline. If the compiler is right most of the time, this technique will improve performance. However, even this more sophisticated form of static branch prediction has not proven to be especially effective in most applications when used alone (without run-time feedback) [12].

Dynamic branch prediction relies on the control unit keeping track of the behavior of each branch encountered. This may be done by the very simple means of using a single bit to remember the behavior of the branch the last time it was executed. Or, for more accurate prediction, two bits may be used to record the action of the branch for the last two consecutive times it was encountered. If it has been taken, or not taken, twice in succession, that is taken as a strong indication of its likely behavior the next time, and instructions are fetched accordingly. If it was taken once and not taken once, the control unit decides randomly which way to predict the branch. Another, even more sophisticated dynamic prediction technique associates a counter of two or more bits with each branch [10]. The counter is initially set to a threshold value in the middle of its count range. Each time the branch succeeds, the counter is incremented; each time it fails, the counter is decremented. As long as the current count is greater than or equal to the threshold, the branch is predicted to succeed; otherwise, it is predicted to fail. The Intel Pentium processor used a two-bit, four-state counter like this to predict branches [13]. Note that for this technique to work properly, the counter must saturate or "stick" at its upper and lower limits, rather than rolling over from the maximum count to zero, or vice versa. Even more elaborate schemes (including hybrid, adaptive, and two-level mechanisms) are possible [10], but there is a point of diminishing returns beyond which the cost of additional hardware complexity is not justified by significantly better performance on typical compiled code.

While dynamic branch prediction requires more hardware, in general, it has been found to perform better than static prediction (which places greater demands upon the compiler) [12]. Since dynamic prediction is dependent upon the details of a particular implementation rather than the features of the instruction set architecture, it also allows compatibility with previous machines to be more easily maintained. Of course, the two approaches can be used together, with the compiler's prediction being used to initialize the state of the hardware predictor and/or as a tiebreaker when the branch is considered equally likely to go either way.

What are the performance effects of branch prediction? Obviously, performance is very sensitive to the success rate of the prediction scheme(s) used. Highly successful branch prediction can significantly improve performance, while particularly poor prediction may do no better (or possibly even worse) than no prediction at all. While a branch correctly predicted to fail may cost no cycles at all and one correctly predicted to succeed may incur only a minimal penalty (if any), a branch that is mispredicted either way will require the machine to recover to the correct execution path and so may incur a branch penalty as severe (or even more severe) than if the machine had no prediction scheme at all.

For comparison, let us first consider mathematically the throughput of a pipelined processor *without* branch prediction [14]. Any instruction in a program is either a branching instruction or a nonbranching instruction. Let p_b be the probability of any instruction being a branch; then of course $(1 - p_b)$ is the probability of it not being a branch. Also, let C_B be the average number of cycles per branch instruction and C_{NB} be the average number of cycles per nonbranching instruction. The average number of clock cycles per instruction is thus given by

$$C_{AVG} = p_b C_B + (1 - p_b) C_{NB}$$

In a pipelined processor, nonbranching instructions execute in one clock cycle, so $C_{NB} = 1$. (In practice, there may be certain exceptions to this rule, most notably stalls due to data dependencies between instructions, to be discussed in Section 22.3.7, but for now we shall proceed under the simplifying assumption that nonbranching instructions do not stall the pipe.) The average number of cycles per branch C_B depends on the fraction of branches taken (to the target) versus the fraction that are not taken and result in sequential execution. Let p_t be the probability that a branch is taken and $(1 - p_t)$ be the probability it is not taken; also let b be the branch penalty in clock cycles. Since no prediction is employed, failed branches execute in 1 cycle, while successful branches require $(1 + b)$ cycles. The average number of cycles per branch instruction is thus given by

$$C_B = p_t(1 + b) + (1 - p_t)(1) = 1 + p_t b$$

We can substitute this expression for C_B into the previous equation to determine the average number of cycles per instruction as follows:

$$C_{AVG} = p_b(1 + p_t b) + (1 - p_b)(1) = 1 + p_b p_t b$$

The throughput of the pipeline (the average number of instructions completed per clock cycle) is simply the reciprocal of the average number of cycles per instruction:

$$H = \frac{1}{C_{AVG}} = \frac{1}{(1 + p_b p_t b)}$$

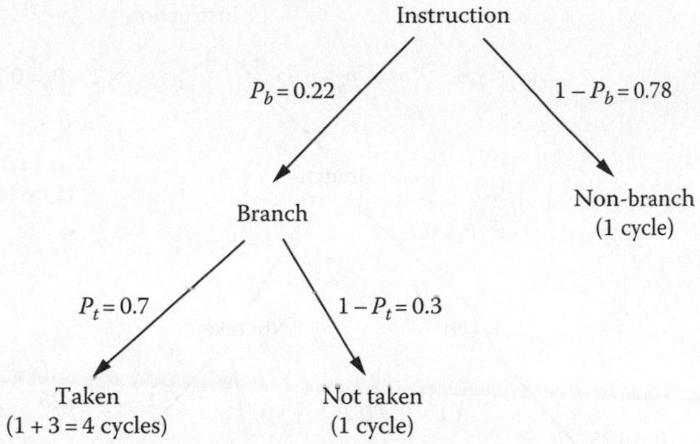

FIGURE 22.5 Probability tree diagram for conditional branching example. (From Dumas II, J.D., *Computer Architecture: Fundamentals and Principles of Computer Design*, CRC Press, Boca Raton, FL, 2006, Figure 4.13.)

The probabilities p_b and p_t obviously will vary for different programs. Typical values for p_b, the fraction of branch instructions, have been found to be in the range of 0.1–0.3 [15]. The probability of a branch succeeding, p_t, may vary widely, but values in the 0.5–0.8 range are reasonable. As a numerical example, suppose for a given program that the branch penalty is three cycles, $p_b = 0.22$, and $p_t = 0.7$. Then the average number of cycles per instruction would be 1 + (0.22)(0.7)(3) = 1.462, and the pipeline throughput would be approximately 0.684 instructions per cycle.

Another way to compute this result without memorizing the formula for C_{AVG} is to construct a simple probability tree diagram as shown in Figure 22.5. To obtain C_{AVG}, it is simply necessary to multiply the number of cycles taken in each case times the product of the probabilities leading to that case and then sum the results. Thus, we obtain $C_{AVG} = (0.22)(0.7)(4) + (0.22)(0.3)(1) + (0.78)(1) = 1.462$ as before, once again giving $H \approx 0.684$.

Now, suppose that the pipelined processor employs a branch prediction scheme to try to improve performance. Let p_c be the probability of a correct prediction, and let c be the reduced penalty associated with a correctly predicted branch. (If correctly predicted branches can execute as quickly as sequential code, c will be equal to zero.) Branches that are incorrectly predicted (either way) incur the full branch penalty of b cycles. In this scenario, the average number of cycles per branch instruction can be shown to be

$$C_B = 1 + b - p_c b + p_t p_c c$$

Substituting this into our original equation $C_{AVG} = p_b C_B + (1 - p_b)(1)$, we find that with branch prediction, the average number of cycles per instruction is given by

$$C_{AVG} = 1 + p_b b - p_b p_c b + p_b p_t p_c c$$

Returning to our numerical example, let us assume that b is still three cycles and p_b and p_t are still 0.22 and 0.7, respectively. Let us further assume that $c = 1$ cycle and the probability of a correct branch prediction, p_c, is 0.75. Substituting these values into the first equation, we find that the average number of cycles per branch instruction is 1 + 3 − (0.75)(3) + (0.7)(0.75)(1) = 2.275. The second equation gives the overall average number of cycles per instruction as 1 + (0.22)(3) − (0.22)(0.75)(3) + (0.22)(0.7)(0.75)(1) = 1.2805. The pipeline throughput H with branch prediction is 1/1.2805 or approximately 0.781, a significant (approximately 14%) improvement over the example without branch prediction.

Again, if one does not wish to memorize formulas, the same result could be obtained using the probability tree diagram shown in Figure 22.6. Multiplying the number of cycles required in each

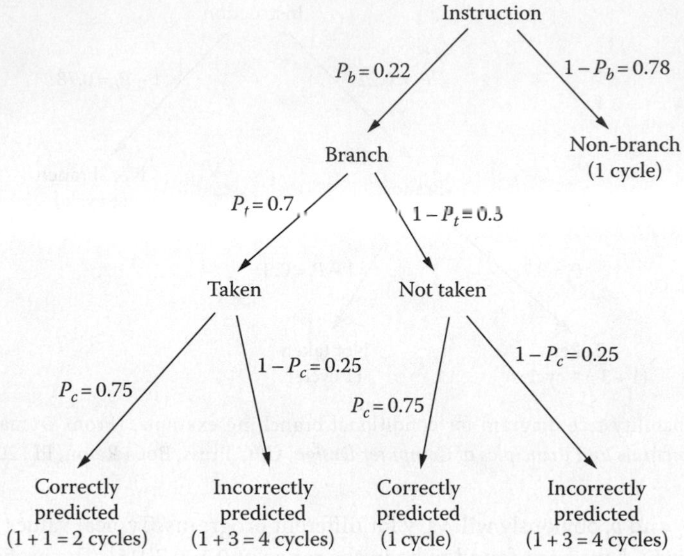

FIGURE 22.6 Probability tree diagram for example with branch prediction. (From Dumas II, J.D., *Computer Architecture: Fundamentals and Principles of Computer Design*, CRC Press, Boca Raton, FL, 2006, Figure 4.14.)

case times the probability of occurrence and then summing the results, we get C_{AVG} = (0.22)(0.7)(0.75)(2) + (0.22)(0.7)(0.25)(4) + (0.22)(0.3)(0.75)(1) + (0.22)(0.3)(0.25)(4) + (0.78)(1) = 1.2805, the same result obtained from our equation.

As we mentioned earlier, the performance benefits derived from branch prediction depend heavily on the success rate of the prediction scheme. To illustrate this, suppose that instead of $p_c = 0.75$, our branch prediction scheme achieved only 50% correct predictions ($p_c = 0.5$). In that case, the average number of cycles per instruction would be 1 + (0.22)(3) − (0.22)(0.5)(3) + (0.22)(0.7)(0.5)(1) = 1.407 for a throughput of approximately 0.711, not much better than the results with no prediction at all. If the prediction scheme performed very poorly, say, with $p_c = 0.3$, the pipeline throughput could be even worse than with no prediction—in this case, 1/1.5082 ≈ 0.663 instructions per cycle.

To facilitate branch prediction, it is common for modern processors to make use of a *branch target buffer* (also known as a *branch target cache* or *target instruction cache*) to hold the addresses of branch instructions, the corresponding target addresses, and the information about the past behavior of the branch [15]. Any time a branch instruction is encountered, this buffer is checked and, if the branch in question is found, the relevant history is obtained and used to make the prediction. A *prefetch queue* may be provided to funnel instructions into the pipeline [16]. To further improve performance, some processors make use of dual instruction prefetch queues to optimize branches [14]. Using this *multiple prefetch* approach, the processor fetches instructions from both possible paths (sequential and target). By replicating the prefetch queue (and possibly even the first few stages of the pipeline itself), the processor can keep both possible execution paths "alive" until the branch decision is made. Instructions from the correct path continue on to execution, while those from the incorrect path are discarded. Of course, for this approach to work, there must be sufficient space on the chip for the prefetch queues and sufficient memory (particularly instruction cache) bandwidth to allow for the simultaneous prefetching of both paths.

22.3.5 Delayed Control Transfers

Delayed control transfers are another approach that has been adopted by some designers to eliminate, or at least minimize, the penalty associated with control transfers (both conditional and unconditional) in pipelined processors. A *delayed branch* instruction is unlike any control transfer instruction the reader has likely encountered before. The branch instructions in most computer architectures take

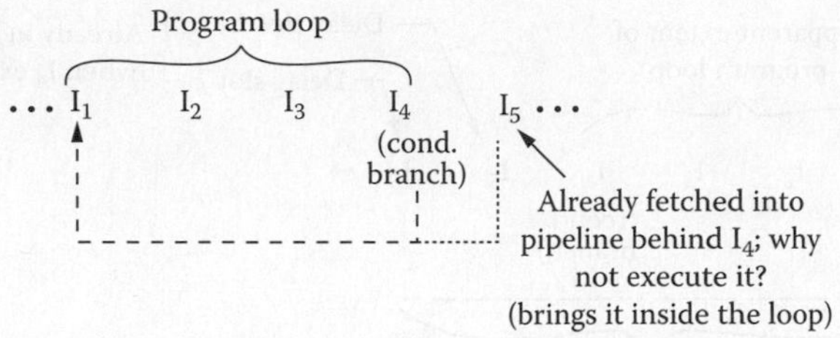

FIGURE 22.7 Delayed branching example. (From Dumas II, J.D., *Computer Architecture: Fundamentals and Principles of Computer Design*, CRC Press, Boca Raton, FL, 2006, Figure 4.15.)

effect immediately. That is to say, if the branch condition is true (or if the instruction is an unconditional branch), the next instruction executed after the branch itself is the one at the target location. A delayed branch, on the other hand, does not take effect immediately. The instruction sequentially following the branch is executed whether or not the branch succeeds [17]. Only after the instruction following the control transfer instruction, which is said to be in the *delay slot* [18], is executed will the target instruction be executed—with execution continuing sequentially from that point.

An example of delayed branching is shown in Figure 22.7. Let us assume that instructions I_1 through I_4 form a program loop, with I_4 being a conditional branch back to I_1. In a "normal" instruction set architecture, I_5 would never be executed until the loop had finished iterating. If this instruction set were implemented with a pipelined instruction unit, I_5 would have to be flushed from the pipeline every time the branch I_4 succeeded. If the number of loop iterations turned out to be large, many clock cycles would be wasted fetching and flushing I_5.

The idea of delayed branching, which only makes sense in an architecture designed for pipelined implementation, comes from the desire not to waste the time required to fetch I_5 each time the branch (I_4) is taken. "Since I_5 is already in the pipeline," the argument goes, "why not go ahead and execute it instead of flushing it?" To programmers used to working in high-level languages or most assembly languages, this approach invariably seems very strange, but it makes sense in terms of efficient hardware implementation. (The delayed branch scheme appears to an uninitiated assembly programmer as a bug, but as all computing professionals should know, any bug that is documented becomes a feature.) I_5 appears to be—and is physically located in memory—outside the loop, but logically, in terms of program flow, it is part of the loop. The instructions are stored in the sequence I_1, I_2, I_3, I_4, I_5—but they are logically executed in the sequence I_1, I_2, I_3, I_5, I_4.

The trick in making use of this feature lies in finding an instruction that logically belongs before the branch, which is independent of the branch decision—neither affecting the branch condition nor being affected by it. (Of course, if the control transfer instruction is unconditional, any instruction that logically goes before it could be placed in its delay slot.) If such an instruction can be identified, it may be placed in the delay slot to make productive use of a clock cycle that would otherwise be wasted. If no such instruction can be found, the delay slot may simply be filled with a time-wasting instruction such as NOP. This is the software equivalent of flushing the delay slot instruction from the pipe.

Delayed control transfer instructions are not found in complex instruction set computer (CISC) architectures, which by and large trace their lineage to a time before pipelining was widely used. However, they are often featured in RISC architectures, which were designed "from the ground up" for a pipelined implementation. CISC architectures were intended to make assembly language programming easier by making the assembly language look more like a high-level language. RISC architectures, on the other hand, were not designed to support assembly language programming at all but rather to support efficient code generation by an optimizing compiler. So the fact that delayed branches make assembly programming awkward is not really significant as long as a compiler can be designed to take advantage of the delay slots when generating code.

There is no reason that an architecture designed for pipelined implementation must have only one delay slot after each control transfer instruction. In fact, the example presented earlier in Table 22.2 is of a

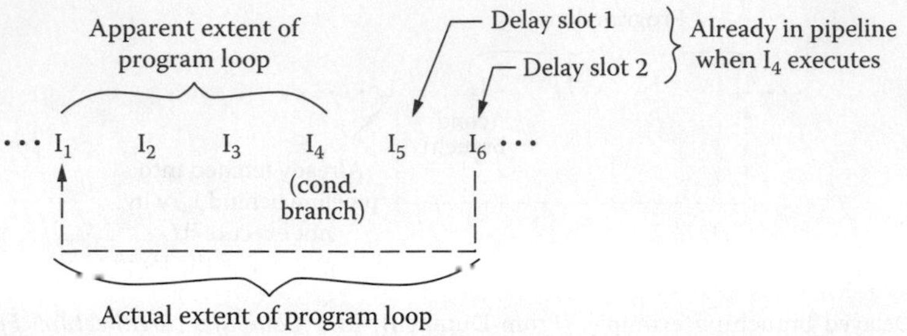

FIGURE 22.8 Example of control transfer with two delay slots. (From Dumas II, J.D., *Computer Architecture: Fundamentals and Principles of Computer Design*, CRC Press, Boca Raton, FL, 2006, Figure 4.16.)

machine that could have two delay slots. (Machines with very deep pipelines might have even more.) Since instructions I_5 and I_6 will both have entered the pipeline before the branch at I_4 can be taken, they could both be executed before the target instruction if the architecture so specifies (see Figure 22.8). Of course, the more delay slots that exist after a conditional branch instruction, the more difficult it will be for the compiler (or masochistic assembly programmer) to find useful, independent instructions with which to fill them. But in the worst case, they can still be filled with NOPs, which keeps the hardware simpler than it would be if it had to be able to recognize this situation and flush the instructions itself. This is just one more example of the hardware/software trade-offs that are made all the time in computer systems design.

22.3.6 Memory Accesses: Delayed Loads and Stores

Control transfers are not the only occurrences that can interrupt and/or slow processing in a pipelined processor. Another potentially costly activity (in terms of performance) is accessing memory for operands. Register operands are generally accessible very quickly, such that they can be used for an arithmetic computation within the same clock cycle. Memory operands, even if they reside in cache, generally require at least one cycle to access before any use can be made of them. RISC architectures, in particular, attempt to minimize this problem by clearly subdividing their instruction sets into computational instructions (which operate only on register contents) and memory access (or load and store) instructions that do not perform computations on data. Even so, a problem may arise when a subsequent instruction tries to perform a computation on a data value being loaded by a previous instruction. For example, suppose the following two instructions appear consecutively in a program:

```
LOAD    VALUE, R5
ADD     R5, R4, R3
```

Given the slower speed of memory relative to most CPUs, the variable VALUE being loaded into register R5 might not be available by the time it was needed by the ADD instruction. One obvious solution to the problem would be to build in some sort of hardware interlock that would "freeze" or stall the pipeline until the data had been retrieved from memory and placed in R5. Only then would the ADD instruction be allowed to execute. The hardware is thus made responsible for the correct operation of the software—a typical, traditional computer design approach.

However, another way of approaching this problem is simply to document the fact that loads from memory always take at least one extra cycle to occur. The compiler, or the assembly language programmer, would be made aware that every load instruction has a "load delay slot [18]." In other words, the instruction immediately following a load must not make use of the value being loaded. (In some architectures, the following instruction may also not be another load.) If the load's destination register is referenced by the following instruction, it is known that the value obtained will be the "old" value, not the new one being loaded from memory. This is another example of documenting what might appear to the casual user to be a bug, thus enshrining it as a feature of the instruction set! Instead of the hardware compensating for delays in loading data and ensuring the correct operation of the software, the software is simply informed of the details of the hardware implementation and forced to ensure correctness on its own. (Note that a hardware interlock will still be required to detect data cache misses and stall the pipeline when they occur, since main memory access may take several clock cycles rather than one.) In the example earlier, correctness of operation is assured by simply inserting an unrelated but useful instruction—or an NOP if no such instruction can be found—between the LOAD and the ADD (or any other computational instruction) that uses the results of the LOAD.

Store operations (writes of data to memory) are less problematic than loads because the CPU is unlikely to need to retrieve the stored information from memory soon enough to pose a problem. (Since stores write data from a register to memory, presumably the CPU can use the copy of the data still in a register if it is needed.) However, back to back memory accesses may pose a problem for some machines because of the time required to complete an access. If two consecutive load or store operations are executed, it may be necessary to stall the pipeline. For that reason, stores in some architectures are also sometimes said to have a "delay slot," which should not contain another load or store instruction. If it does, the pipeline may have to be stalled for one cycle to allow the first store to complete before a second memory access is done.

22.3.7 Data Dependencies and Hazards

Dependency relations among computed results, giving rise to pipeline *data hazards*, may also hold up operations in a pipelined CPU and keep it from approaching its theoretical throughput. In a nonpipelined processor, each instruction completes before execution of the next instruction begins. Thus, values computed by a previous instruction are always available for use in subsequent instructions, and the program always obtains the results anticipated by the von Neumann sequential execution model. Since a pipelined instruction unit overlaps the execution of several instructions, though, it becomes possible for results to become sensitive to the timing of instructions rather than just the order in which they appear in the program. This is obviously not a desirable scenario and must be corrected (or at least accounted for) if we want our pipelined machine to compute the same results as a purely sequential machine.

For an example of a common data dependency problem in a pipelined machine, consider the following situation (illustrated in Figure 22.9). Suppose a CPU with a four-stage pipeline like the one in our previous example executed the following sequence of instructions:

I_1	ADD	R1, R2, R3
I_2	SUB	R3, R4, R6
I_3	XOR	R1, R5, R3

Cycle	Stage 1	Stage 2	Stage 3	Stage 4
1	I_1			
2	I_2	I_1		
3	I_3	I_2	I_1	
4	I_4	I_3	I_2	I_1
5	I_5	I_4	I_3	I_2

(a)

Cycle	Stage 1	Stage 2	Stage 3	Stage 4
1	I_1			
2	I_2	I_1		
3	I_3	I_2	I_1	
4	I_3	I_2	—	I_1
5	I_4	I_3	I_2	—

(Stall)

(b)

FIGURE 22.9 Data dependency problem in pipelined CPU. (a) Incorrect operation stage 3 needs result of I_1 at the beginning of cycle 4; not available until the end of cycle 4. (b) Correct operation due to stalling pipeline for 1 cycle after cycle 3; stage 3 will now have updated value for use in executing I_2. (From Dumas II, J.D., *Computer Architecture: Fundamentals and Principles of Computer Design*, CRC Press, Boca Raton, FL, 2006, Figure 4.17.)

The last operand listed for each instruction is the destination. Thus, it can be seen that instruction I_2 uses the result computed by I_1. But will that result be available in time for I_2 to use it?

In Figure 22.9a, we can see how instruction I_1 proceeds through the pipeline, with I_2 and I_3 following it stage by stage. The result from I_1 is not computed by stage 3 until the end of the third clock cycle and is not stored back into the destination register (R3) by stage 4 until the end of the fourth clock cycle. However, stage 3 needs to read R3 and obtain its new contents at the *beginning* of the fourth clock cycle so they can be used to execute the subtraction operation for I_2. If execution proceeds unimpeded as shown, the previous contents of R3 will be used by I_2 instead of the new contents, and the program will operate incorrectly. This situation, in which there is a danger of incorrect operation due to the fact that the behavior of one instruction in the pipeline depends on that of another, is known as a *data hazard* [2]. This particular hazard is known as a *true data dependence* or (more commonly) a *read after write* (RAW) *hazard* [19] because I_2, which reads the value in R3, comes after I_1, which writes it.

To avoid the hazard and ensure correct operation, the control unit must make sure that I_2 actually reads the data after I_1 writes it. The obvious solution is to stall I_2 (and any following instructions) for one clock cycle in order to give I_1 time to complete. Figure 22.9b shows how this corrects the problem. Instruction I_2 now does not reach stage 3 of the pipeline until the beginning of the fifth clock cycle, and so it is executed with the correct operand value from R3.

RAW hazards are the most common data hazards and in fact the only possible type in a machine such as we have been considering so far, with a single-pipeline instruction unit in which instructions are always begun and completed in the same order and only one stage is capable of writing a result. Other types of data hazards, known as *write after read* (WAR) and *write after write* (WAW) hazards, are only of concern in machines with multiple pipelines (or at least multiple execution units, which might be "fed" by a common pipeline) or in situations where writes can be done by more than one stage of the pipe [20]. In such a machine, instructions may be completed in a different order than they were fetched (*out-of-order execution* or OOE) [21]. OOE introduces new complications to the process of ensuring that the machine obtains the same results as a nonpipelined processor.

Consider again the sequence of three instructions introduced earlier. Notice that I_1 and I_3 both write their results to register R3. In a machine with a single pipeline feeding a single ALU, we know that if I_1 enters the pipeline first (as it must), it will also be completed first. Later, its result will be overwritten by the result calculated by I_3, which of course is the behavior expected under the von Neumann sequential execution model. But suppose the machine has more than one ALU that can execute the required operations. The ADD instruction (I_1) might be executed by one unit, while at the same time the XOR (I_3) is being executed by another. There is no guarantee that I_1 will be completed first or have its result sent to R3 first. Thus, it is possible that due to OOE, R3 could end up containing the wrong value. This situation in which two instructions both write to the same location, and the control unit must make sure that the second instruction's write occurs after the first write, is known as an *output dependence* or WAW hazard [19].

Now, consider the relationship between instructions I_2 and I_3 in the example code. Notice that I_3 writes its result to R3 after the previous value in R3 has been read for use as one of the operands in the subtraction performed by I_2. At least, that is the way things are supposed to happen under the sequential execution model. Once again, however, if multiple execution units are employed, it is possible that I_2 and I_3 may execute out of their programmed order. If this were to happen, I_2 could mistakenly use the "new" value in R3, which had been updated by I_3 rather than the "old" value computed by I_1. This is one more situation, known as an *antidependence* or WAR hazard [19], which must be guarded against in a machine where OOE is allowed.

There is one other possible relationship between instructions that reference the same location for data and might have their execution overlapped in a pipeline(s). This "read after read" situation is the only one that never creates a hazard [19]. In our example, both I_1 and I_3 read R1 for use as a source operand. Since R1 is never modified (written), both instructions are guaranteed to read the correct value regardless of their order of execution. Thus, a simple rule of thumb is that for a data hazard to exist between instructions, at least one of them must modify a commonly used value.

22.3.8 Controlling Instruction Pipelines

Controlling the operation of pipelined processors in order to detect and correct for data hazards is a very important but very complex task, especially in machines with multiple execution units. In a machine with only one pipelined instruction execution unit, RAW hazards are generally the only ones a designer must worry about. (The exception would be in the very unusual case where more than one stage of the pipeline is capable of writing a result.) Control logic must keep track of (or "reserve") the destination register or memory location for each instruction in progress and check it against the source operands of subsequent instructions as they enter the pipe. If a RAW hazard is detected, the most basic approach is to simply stall the instruction that uses the operand being modified (as shown in Figure 22.9b) while allowing the instruction that modifies it to continue. When the location in question has been modified, the reservation placed on it is released and the stalled instruction is allowed to proceed. This approach is straightforward, but forced stalls obviously impair pipeline throughput.

Another approach that can minimize, or in some cases eliminate, the need to stall the pipeline is known as *data forwarding* [22]. By building in additional connectivity within the processor, the result just computed by an instruction in the pipeline can be forwarded to the ALU for use by the subsequent instruction at the same time it is being sent to the reserved destination register. This approach generally saves at least one clock cycle compared to the alternative of writing the data into the first instruction's destination register and then immediately reading it back out for use by the following dependent instruction.

Finally, designers can choose not to build in control logic to detect, and interlocks or forwarding to avoid, these RAW hazards caused by pipelining. Rather, they can document pipeline behavior as an architectural feature and leave it up to the compiler to reorder instructions and/or insert NOPs to artificially stall subsequent instructions, allowing a sufficient number of cycles to elapse so that the value in question will definitely be written before it is read. This solution certainly simplifies the hardware design but is not ideal for designing a family of computers since it ties the instruction set architecture closely to the details of a particular implementation, which may later be superseded. In order to maintain compatibility, more advanced future implementations may have to emulate the behavior of the relatively primitive earlier machines in the family. Still, the approach of handling data dependencies in software is a viable approach for designers who wish to keep the hardware as simple as possible.

Machines with multiple execution units encounter a host of difficulties not faced by simpler implementations. Adding WAW and WAR hazards to the RAW hazards inherent to any pipelined machine makes the design of control logic much more difficult. Two important control strategies were devised back in the 1960s for high-performance, internally parallel machines of that era. Variations of these methods are still used in the microprocessors of today. These design approaches are known as the *scoreboard method* and *Tomasulo's method*.

The scoreboard method for resource scheduling dates back to the Control Data Corporation (CDC) 6600 supercomputer, which was introduced in 1964. James Thornton and Seymour Cray were the lead engineers for the 6600 and contributed substantially to this method, which was used later in the Motorola 88000 and Intel *i*860 microprocessors. The machine had ten functional units that were not pipelined but did operate concurrently (leading to the possibility of OOE). The CDC 6600 designers came up with the idea of a central clearinghouse, or *scoreboard*, to schedule the use of functional units by instructions [23]. As part of this process, the scoreboard had to detect and control interinstruction dependencies to ensure correct operation in an OOE environment.

The scoreboard is a hardware mechanism—a collection of registers and control logic—that monitors the status of all data registers and functional units in the machine. Every instruction is passed through the scoreboard as soon as it is fetched and decoded in order to check for data dependencies and resource conflicts before the instruction is *issued* to a functional unit for execution. The scoreboard checks to make sure the instruction's operands are available and that the appropriate functional unit is also available; it also resolves any write conflicts so that correct results are obtained.

There are three main parts, consisting of *tables* or sets of registers with associated logic, to the scoreboard: *functional unit status*, *instruction status*, and *destination register status* [24]. The functional unit status table contains several pieces of information about the status of each functional unit. This information includes a *busy* flag that indicates whether or not the unit is currently in use, two fields indicating the numbers of its *source registers* and one field indicating its *destination register* number, and *ready* flags for each of the source registers to indicate whether they are ready to be read. (A source register is deemed to be ready if it is not waiting to receive the results of a previous operation.) If a given functional unit can perform more than one operation, there might also be a bit(s) indicating which *operation* it is performing.

The instruction status table contains an entry for each instruction from the time it is first decoded until it completes (until its result is written to the destination). This table indicates the current status of the instruction with respect to four steps: whether or not it has been *issued*, whether or not its *operand(s) has been read*, whether or not its *execution is complete*, and whether or not the *result has been written* to the destination.

The destination register status table is the key to detecting data hazards between instructions. It contains one entry for each CPU register. This entry is set to the number of the functional unit that will produce the result to be written into that register. If the register is not the destination of a currently executing instruction, its entry is set to a null value to indicate that it is not involved in any write dependencies. Tables 22.3 through 22.5 show examples of possible contents of the functional unit status, instruction status, and destination register status tables for a machine with four functional units and eight registers. The actual CDC 6600 was more complex, but this example will serve for illustrative purposes.

TABLE 22.3 Scoreboard Example: Functional Unit Status Table

Unit Name	Unit Number	Busy?	Destination Register	Source Register 1	Ready?	Source Register 2	Ready?
Adder/ Subtractor 1	0	No	—	—	—	—	—
Adder/ Subtractor 2	1	Yes	5	2	Yes	7	Yes
Multiplier	2	Yes	0	1	Yes	3	Yes
Divider	3	No	—	—	—	—	—

TABLE 22.4 Scoreboard Example: Instruction Status Table

Instruction	Instruction Address	Instruction Issued?	Operands Read?	Execution Complete?	Result Written?
ADD R1, R4, R6	1000	Yes	Yes	Yes	Yes
ADD R2, R7, R5	1001	Yes	Yes	Yes	No
MUL R1, R3, R0	1002	Yes	Yes	No	No
MUL R2, R6, R4	1003	No	No	No	No
ADD R0, R5, R7	1004	No	No	No	No
—	—	—	—	—	—

TABLE 22.5 Scoreboard Example: Destination Register Status Table

	R7	R6	R5	R4	R3	R2	R1	R0
Functional unit number	—	—	1	—	—	—	—	2

The tables show a typical situation with several instructions in some stage of completion. Some functional units are busy and some are idle. Unit 0, the first adder/subtractor, has just completed the first ADD operation and sent its result to R6, so it is momentarily idle. Unit 1, the second adder/subtractor, has just completed the second ADD operation. However, the result has yet to be written into R5 (in this case, simply because not enough time has elapsed for the write operation to complete; in some situations, the delay could be a deliberate stall due to a WAR hazard between this and a previous instruction). Therefore, this instruction is not complete, and the reservations on the functional unit and destination register have not been released. Unit 2, the multiplier, has the first MUL operation in progress. Notice that R0 is reserved in the destination register status table as the result register for unit 2. As long as this is true, the control logic will not issue any subsequent instruction that uses R0.

As new instructions are fetched and decoded, the scoreboard checks each for dependencies and hardware availability before issuing it to a functional unit. The second MUL instruction has not yet been issued because the only multiplier, functional unit 2, is busy. Hardware (unit 0) is available such that the third ADD instruction could be issued, but it is stalled due to data dependencies: neither of its operands (in R0 and R5) is yet available. Only when both of these values are available can this instruction be issued to one of the adder/subtractor units. The new R5 value will be available soon, but until the first multiply completes (freeing unit 2 and updating R0), neither of the two instructions following the first MUL can be issued.

As long as scoreboard entries remain, subsequent instructions may be fetched and checked to see if they are ready for issue. (In Table 22.4, the last scoreboard entry is still available and the first one is no longer needed now that the instruction has completed, so two additional instructions could be fetched.) If all scoreboard entries are in use, no more instructions can be fetched until some instruction currently tracked by the scoreboard is completed. The limited number of scoreboard entries, along with the frequent stalls caused by all three types of data hazards and the fact that all results must be written to the register file before use, is the major limitation of the scoreboard approach.

Another important control strategy was first used in the IBM 360/91, a high-performance scientific computer introduced in 1967 [25]. Its multiple pipelined execution units were capable of simultaneously processing up to three floating-point additions or subtractions and two floating-point multiplications or divisions in addition to six loads and three stores. The operation of the floating-point registers and execution units (the heart of the machine's processing capability) was controlled by a hardware scheduling mechanism designed by Robert Tomasulo [26].

Tomasulo's method is essentially a refinement of the scoreboard method with some additional features and capabilities designed to enhance concurrency of operations. One major difference is that in Tomasulo's method, detection of hazards and scheduling of functional units are distributed, not centralized in a single scoreboard. Each data register has a *busy* bit and a *tag* field associated with it. The busy bit is set when an instruction specifies that register as a destination and cleared when the result of that operation is written to the register. The tag field is used to identify which unit will be computing the result for that register. Note that this information is analogous to that kept in the destination status register table of a scoreboard.

Another principal feature of Tomasulo's method that differs from the scoreboard method is the use of *reservation stations* [26] to hold operands (and their tags and busy bits, plus an operation code) for the functional units [24]. Each reservation station is essentially a set of input registers that are used to buffer operations and operands for a functional unit. The details of functional unit construction are not important to the control strategy. For example, if the machine has three reservation stations for addition/subtraction (as the 360/91 did), it does not matter whether it has three nonpipelined adders or one three-stage pipelined adder (as was actually the case). Each reservation station has its own unit number and appears to the rest of the machine as a distinct "virtual" adder. Likewise, if there are two reservation stations for multiplication, the machine appears to have two virtual multipliers regardless of the actual hardware used.

The tags associated with each operand in Tomasulo's method are very important because they specify the origin of each operand independently of the working register set. Though program instructions are written with physical register numbers, by the time an instruction is dispatched to a reservation station, its operands are no longer identified by their original register designations. Instead, they are identified by their tags, which indicate the number of the functional unit (actually the number of a virtual functional unit—one of the reservation stations) that will produce that operand. Operands being loaded from memory are tagged with the number of their *load buffer* [26] (in a modern machine, they might be tagged by cache location, but the 360/91 had no cache). Once an operand has been produced (or loaded from memory) and is actually available in the reservation station, its tag and busy bit are changed to 0 to indicate that the data value is present and ready for use. Any time a functional unit is ready to accept operands, it checks its reservation stations to see if any of them have all operands present; if so, it initiates the requested operation. Thus, although programs for the machine are written sequentially using the von Neumann model, the functional units effectively operate as *dataflow* machines in which execution is driven by the availability of operands.

This use of tags generated "on the fly," rather than the original register numbers generated by the programmer or compiler, to identify operands is known as *register renaming* [27]. The register renaming scheme significantly reduces the number of accesses to data registers; not only do they not have to be read for operands if data are coming directly from a functional unit but also only the last of a series of writes to the same register actually needs to be committed to it. The intermediate values are just sent directly to the reservation stations as necessary. Tomasulo's register renaming scheme is instrumental in avoiding or minimizing stalls due to WAR and WAW hazards and thus achieves a significant advantage over the scoreboard method.

Another important feature of the 360/91 that helped to reduce or eliminate stalls due to RAW hazards was the use of a *common data bus* (CDB) to forward data to reservation stations that need a just-calculated result [26]. Because of this data forwarding mechanism (which goes hand in hand with the register renaming scheme), the reservation stations do not have to wait for data to be written to the register file. If the tag of a value on the CDB matches the tag of an operand needed by any reservation station(s), the operand is captured from the bus and can be used immediately. Meanwhile, the register file also monitors the CDB, loading a new value into any register with a busy bit set whose tag matches the one on the bus. Figure 22.10 is a simplified view of a machine with Tomasulo scheduling and a CDB architecture like the IBM 360/91.

Tomasulo's method has some distinct disadvantages. Not only does it require complex hardware for control but its reliance on a single, shared bus makes it hard to scale up for a machine with many registers

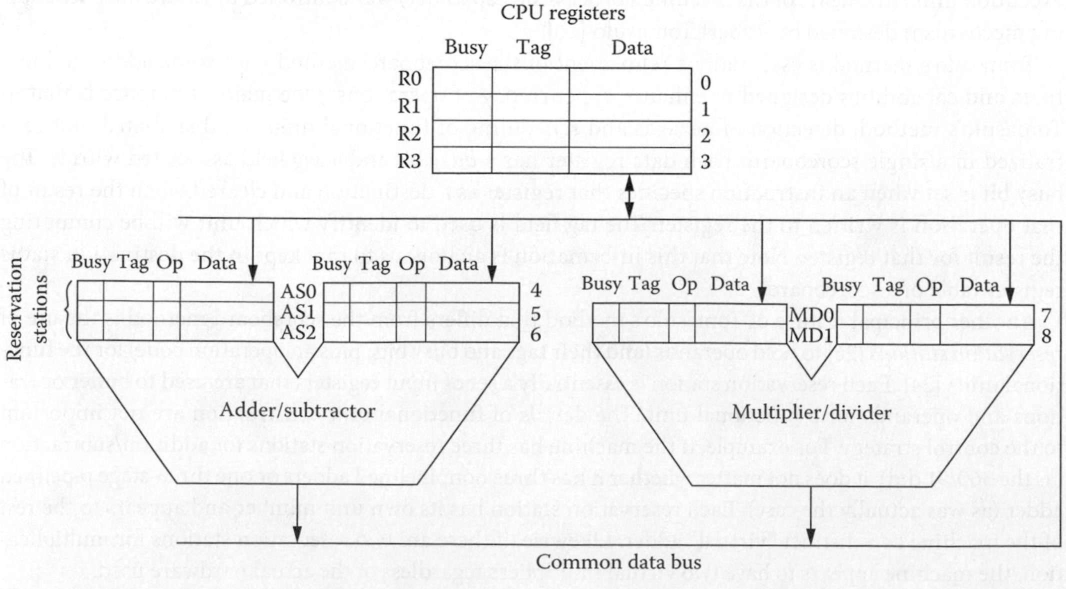

FIGURE 22.10 Example architecture with CDB and reservation stations. (From Dumas II, J.D., *Computer Architecture: Fundamentals and Principles of Computer Design*, CRC Press, Boca Raton, FL, 2006, Figure 4.18.)

and/or functional units (typical of modern CPUs). Of course, given sufficient "real estate" on the chip, additional shared internal buses can be constructed to remove the CDB bottleneck. Likewise, the larger tags (and greater number of tag comparisons) needed in a more complex machine can be accommodated if sufficient space is available. On the other hand, the great selling point for Tomasulo's method is the fact that it helps encourage concurrency of execution (to a greater degree than the scoreboard method) while preserving the dependency relationships inherent to programs written under the von Neumann sequential execution model. Tomasulo's method in particular helps get better performance out of architectures with multiple pipelines. Because of this, approaches based on or very similar to this method are used in many modern, *superscalar* (see Section 22.3.10) microprocessors including the more advanced members of the Alpha, MIPS, Precision Architecture RISC (PA-RISC), Pentium, and PowerPC families.

22.3.9 Superpipelined Architectures

We learned in the preceding sections that pipelining is an effective way to improve the processing performance of computer hardware. However, a simple arithmetic or instruction-unit pipeline can only improve performance by a factor approaching, but never reaching or exceeding, the small number of stages employed. The fundamental limitation on performance improvement using pipelining is the number of stages into which the task is subdivided. A three-stage pipeline can at best yield a speedup approaching a factor of 3; a five-stage pipeline can only approach a 5:1 speed ratio. The simplest and most straightforward approach to achieving further performance improvement, then, is simply to divide the pipeline into a greater number of smaller stages in order to clock it at a higher frequency. There is still only one pipeline, but by increasing the number of stages, we increase its *temporal parallelism*—it is working on more instructions at the same time. This use of a very deep, very high-speed pipeline for instruction processing is called *superpipelining.*

One of the first superpipelined processors was the MIPS R4000 (introduced in 1991), which had an eight-stage pipeline [28] instead of the four- or five-stage design that was common in RISC architectures at the time. The eight stages were instruction fetch (first half), instruction fetch (second half), instruction decode and register fetch, execution, data cache access (first half), data cache access (second half), tag check, and write back. By examining this decomposition of the task, one can see that the speed of the cache had been a limiting factor in CPU performance. In splitting up memory access across two stages, the MIPS designers were able to better match that slower operation with the speed of internal CPU operations and thus balance the workload throughout the pipe.

The MIPS R4000 illustrated both the advantages and disadvantages of a superpipelined approach. On the plus side, it was able to achieve a very high clock frequency for its time. The single-pipeline functional unit was simple to control and took up little space on the chip, leaving room for more cache and other components including a floating-point unit and a memory management unit. However, as with all deeply pipelined instruction execution units, branching presented a major problem. The branch penalty of the R4000 pipeline was "only" three cycles (one might reasonably expect an even greater penalty given an eight-stage implementation). However, the MIPS instruction set architecture had been designed for a shallower pipeline and so only specified one delay slot following a branch. This meant that there were always at least two stall cycles after a branch—three if the delay slot could not be used constructively [28]. This negated much of the advantage gained from the chip's higher clock frequency. The R4000's demonstration of the increased branch delays inherent to a single superpipeline was so convincing that this approach has largely been abandoned. Most current machines that are superpipelined are also *superscalar* (see succeeding text) to allow speculative execution down both paths from a branch.

22.3.10 Superscalar Architectures

A *superscalar* machine is one that uses a standard, von Neumann-type instruction set but can issue (and thus often complete) multiple instructions per clock cycle. Obviously, this cannot be done with a single conventional pipeline; so an alternate definition of superscalar is a machine with a sequential programming model that uses multiple pipelines. Superscalar CPUs attempt to exploit whatever degree of *instruction-level parallelism* (ILP) exists in sequential code by increasing *spatial parallelism* (building multiple execution units) rather than temporal parallelism (building a deeper pipeline). By building multiple pipelines, designers of superscalar machines can get beyond the theoretical limitation of one IPC inherent to any single-pipeline design and achieve higher performance without the problems of superpipelining.

Quite a few microprocessors of the modern generation have been implemented as superscalar designs. Superscalar CPUs are generally classified by the maximum number of instructions they can issue at the same time. (Of course, due to data dependencies, structural hazards, and/or branching, they do not *actually* issue the maximum number of instructions during every cycle.) The maximum number of instructions that can be simultaneously issued depends, as one might expect, on the number of pipelines built into the CPU. Digital Equipment Corporation's (DEC) Alpha 21064 (a RISC processor introduced in 1992) was a "2-issue" chip [29], while its successors the 21164 (1995), 21264 (1998), and 21364 (2003) were all "4-issue" processors [30]. Another way of saying the same thing is to call the 21064 a *two-way superscalar* CPU, while the 21164 and following chips are said to be *four-way superscalar*. The MIPS Technologies R10000 [31], R12000, R14000, and R16000 are also four-way superscalar processors. Even CISCs can be implemented in superscalar fashion: as far back as 1993, Intel's first Pentium chips were based on a two-way superscalar design [32]. The Pentium had essentially the same instruction set architecture as the 486 processor but achieved higher performance by using two execution pipelines instead of one.

In order to issue multiple instructions per cycle from a sequentially written program and still maintain correct execution, the processor must check for dependencies between instructions using an approach like those we discussed previously (the scoreboard method or Tomasulo's method) and issue only those instructions that do not conflict with each other. Since OOE is possible, all types of hazards can occur; register renaming is often used to help solve the problems. Precise handling of exceptions is more difficult in a superscalar environment, too. Of course, the logic required to detect and resolve all of these problems is complex to design and adds significantly to the amount of chip area required. The multiple pipelines also take up more room, making superscalar designs very space sensitive and thus more amenable to implementation technologies with small feature (transistor) sizes. (Superpipelined designs, by contrast, are best implemented with technologies that have short propagation delays.) These many difficulties of building a superscalar CPU are offset by a significant advantage: with multiple pipelines doing the work, clock frequency is not as critical in superscalar machines as it is in superpipelined ones. Since generating and distributing a high-frequency clock signal across a microprocessor are far from a trivial exercise, this is a substantial advantage in favor of the superscalar approach.

Of course, superscalar and superpipelined designs are not mutually exclusive. Some CPUs have been implemented with multiple, deep pipelines, making them both superpipelined and superscalar. Sun's UltraSPARC processor [33], introduced in 1995, was an early example of this hybrid approach: it was both superpipelined (nine stages) and four-way superscalar. A more recent (2003) example is IBM's PowerPC 970 CPU [34]. Even CISC processors such as the AMD Athlon [35] have combined superscalar and superpipelined design in order to maximize performance. Given sufficient chip area, superscalar design is a useful enhancement that makes superpipelining much more practical. When a branch is encountered, a superscalar/superpipelined machine can use one (or more) of its deep pipelines to continue executing sequential code, while another pipeline(s) executes speculatively down the branch target path. Whichever way the branch decision goes, at least one of the pipelines will have correct results; any that took the wrong path can be flushed. Some work is wasted, but processing never comes to a complete halt.

22.3.11 Very Long Instruction Word Architectures

Superscalar CPU designs have many advantages, as we saw in the preceding pages. However, there remains the significant problem of scheduling which operations can be done concurrently and which

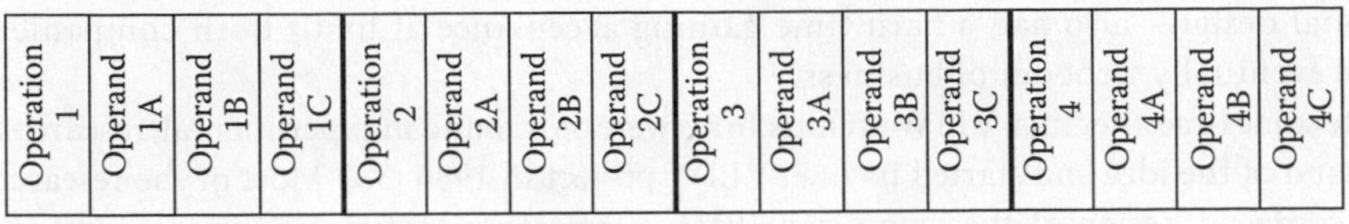

FIGURE 22.11 Example of VLIW format. (From Dumas II, J.D., *Computer Architecture: Fundamentals and Principles of Computer Design*, CRC Press, Boca Raton, FL, 2006, Figure 4.20.)

ones have to wait and be done sequentially after others. Determining the precise amount of ILP present in a program is a complex exercise that is made even more difficult by having to do it "on the fly" within the limitations of hardware design and timing constraints. How can we get the benefits of multiple instruction issue found in superscalar CPUs without having to build in complex, space- and power-consuming circuitry to analyze multiple types of data dependencies and schedule the initiation of operations? A look back at one of the basic RISC design principles provides one possible answer: "let the compiler do the work." This idea is the foundation of a relatively new class of architectures known by the acronym (coined by Josh Fisher of Yale University) *VLIW*, which stands for very long instruction word [36].

The centerpiece of a VLIW architecture is exactly what the name implies: a machine language instruction format that is fixed in length (as in a RISC architecture) but much longer than the 32–64 bit formats common to most conventional CISC or RISC architectures (see Figure 22.11 for an example). Each "very long" instruction contains enough bits to specify not just one machine operation but several to be performed simultaneously. Ideally, the VLIW format includes enough *slots* (or groups of bit fields) to specify operands and operations for every functional unit present in the machine. If sufficient ILP is present in the high-level algorithm, all of these fields can be filled with useful operations, and maximum use will be made of the CPU's internal parallelism. If not enough logically independent operations can be scheduled at one time, some of the fields will be filled with NOPs, and thus some of the hardware will be idle, while other units are working.

The VLIW approach gives essentially the same effect as a superscalar design but with most or all of the scheduling work done statically (offline) by the compiler, rather than dynamically (at run time) by the control unit. It is the compiler's job to analyze the program for data and resource dependencies and pack the "slots" of each VLIW with as many concurrently executable operations as possible. Since the compiler has more time and resources to perform this analysis than would the control unit and since it can see a "bigger picture" of the code at once, it may be able to find and exploit more ILP than would be possible in hardware. Thus, a VLIW architecture may be able to execute more (equivalent) instructions per cycle than a superscalar machine with similar capabilities.

Finding and expressing the parallelism inherent in a high-level program is quite a job for the compiler; however, the process is essentially the same whether it is done in hardware or software. Doing scheduling in software allows the hardware to be kept simpler and hopefully faster. One of the primary characteristics of the original, single-pipeline RISC architectures was reliance on simple, fast hardware with optimization to be done by the compiler. A VLIW architecture is nothing more nor less than the application of that same principle to a system with multiple pipelined functional units. Thus, in its reliance on moving pipeline scheduling from hardware to software (as well as in its fixed-length instruction format), VLIW can be seen as the logical successor to RISC.

The VLIW concept goes back further than most computing professionals are aware. The first VLIW system was the Yale ELI-512 computer (with its Bulldog compiler) built in the early 1980s [36], just about the time that RISC architectures were being born. Josh Fisher and some colleagues from Yale University started a company named Multiflow Computer, Inc., in 1984 [37]. Multiflow produced several TRACE systems, named for the trace scheduling algorithm used in the compiler [36]. Another company, Cydrome Inc., which was founded about the same time by Bob Rau, produced a machine known as the Cydra-5 [37]. However, the VLIW concept was clearly ahead of its time. (Keep in mind that RISC architectures—which were a much less radical departure from conventional design—also had a hard time gaining acceptance at first.) Both companies failed to thrive and eventually went out of business.

For quite some time after this, VLIW architectures were only found in experimental, research machines. IBM got wind of the idea and started its own VLIW project in 1986 [38]. Most of the research involved the (very complex) design of the compilers; IBM's hardware prototype system was not constructed until about 10 years after the project began. This system is based on what IBM calls the Dynamically Architected Instruction Set from Yorktown (DAISY) architecture [39]. Its instructions are 759 bits long; they can simultaneously specify up to 8 ALU operations, 4 loads or stores, and 7 branch conditions. The prototype machine contains only 4 MB of data memory and 64K VLIWs of program memory, so it was obviously intended more as a proof of the VLIW concept (and of the compiler) than as a working system.

Commercially available CPUs based on the VLIW concept entered the market in 2001 with the introduction of the IA-64 architecture [40], which was jointly developed by Intel and Hewlett-Packard (HP) and initially implemented in the Itanium (Merced) processor. Subsequent chips in this family include the Itanium 2 (2002), Itanium 2 9000 series (2006), 9100 series (2007), 9300 series (code name Tukwila, 2010), and 9500 series (code name Poulson), introduced in 2012. Intel does not use the acronym VLIW, preferring to refer to its slightly modified version of the concept as EPIC (which stands for explicitly parallel instruction computing). However, EPIC's lineage is clear: VLIW pioneers Fisher and Rau went to work with HP after their former companies folded and contributed to the development of the new architecture [40]. The IA-64 instruction set incorporates 128-bit "bundles" (see Figure 22.12) containing three 41-bit RISC-type instructions [40]; each bundle provides explicit dependency information. This dependency information (determined by the compiler) is used by the hardware to schedule the execution of the bundled operations. The purpose of this arrangement (with some of the scheduling done in hardware in addition to that done by software) is to allow binary compatibility to be maintained between different generations of chips. Ordinarily, each generation of a given manufacturer's VLIW processors would be likely to have a different number and/or types of execution units and thus a different instruction format. Since hardware is responsible for some of the scheduling and since the instruction format is not directly tied to a particular physical implementation, the Itanium processor family is architecturally somewhere between a "pure VLIW" and a superscalar design (though arguably closer to VLIW).

There are several disadvantages to the VLIW approach, which IBM, HP, and Intel hoped would be outweighed by its significant advantages. Early VLIW machines performed poorly on branch-intensive code; IBM has attempted to address this problem with the tree structure of its system, while the IA-64 architecture addresses it with a technique called *predication* (as opposed to prediction, which has been used in many RISC and superscalar processors) [40]. The predication technique uses a set of *predicate registers*, each of which can be used to hold a true or false condition. Where conditional branches would normally be used in a program to set up a structure such as if/then/else, the operations in each possible sequence are instead *predicated* (made conditional) on the contents of a given predicate register. Operations from both possible paths (the "then" and "else" paths) then flow through the parallel,

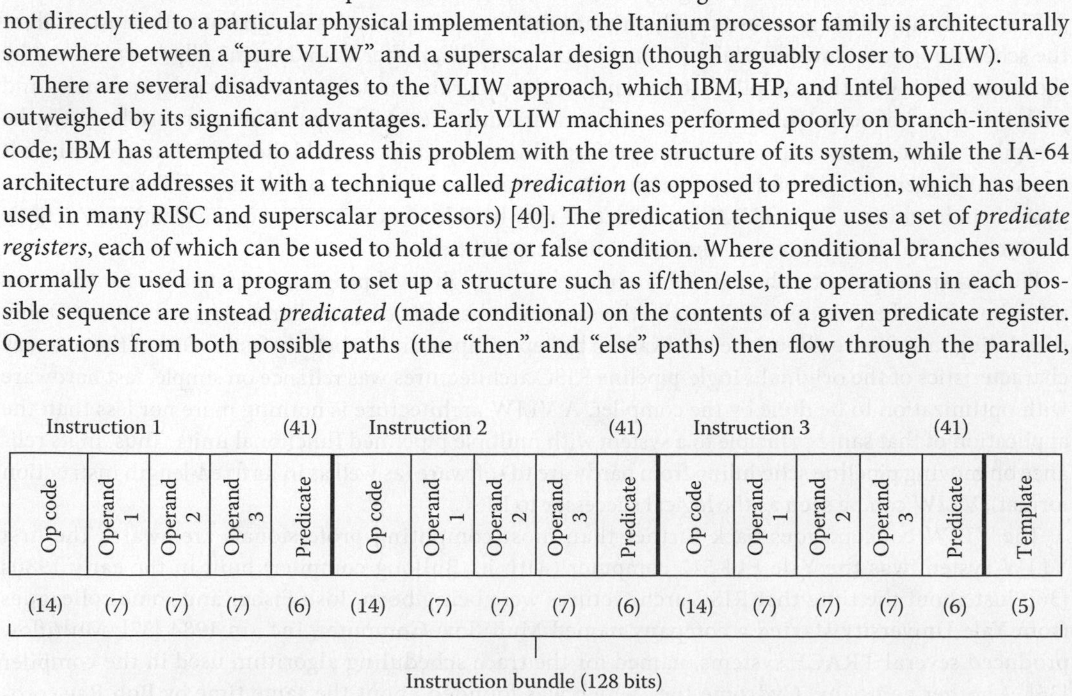

FIGURE 22.12 IA-64 EPIC instruction format. (From Dumas II, J.D., *Computer Architecture: Fundamentals and Principles of Computer Design*, CRC Press, Boca Raton, FL, 2006, Figure 4.21.)

pipelined execution units, but only one set of operations (the ones with the predicate that evaluates to true) are allowed to write their results.

Another problem that negatively affected the commercial success of early VLIW architectures was the lack of compatibility with established architectures. Unless one had access to all the source code and could recompile it, the existing software could not be run on the new architectures. This problem was addressed in IBM's DAISY prototype by performing dynamic translation (essentially, run-time interpretation) of code for the popular PowerPC architecture [39]. Early (pre-2006) EPIC processors also supported the execution of the huge existing Intel IA-32 (x86) code base, as well as code compiled for the HP PA-RISC architecture, through hardware emulation at run time. However, the performance proved to be poor compared to both native IA-64 code running on Itanium series chips and IA-32 code running on existing IA-32 processors. Thus, starting with the Itanium 2 9000 series (Montecito) processors, a software emulator known as the IA-32 Execution Layer [41] was introduced to improve performance on legacy x86 code.

Yet another VLIW disadvantage is poor code density. Since not every VLIW instruction has useful operations in all its slots, some bit fields are inevitably wasted, making compiled programs take up more space in memory than they otherwise would. No real cure has been found for this problem, as the available hardware resources are constant, while the ILP available in software varies. However, this issue has been considerably ameliorated by the fact that memory prices have trended downward over time.

VLIW architectures have several advantages that would seem to foretell their future success. The most obvious is the elimination of most or all scheduling logic from the hardware, freeing up more space for additional functional units, more on-chip cache, multiple cores, etc. Simplifying the control logic may also reduce overall delays and allow the system to operate at a higher clock frequency than a superscalar CPU built with the same technology. Also, as was previously mentioned, the compiler (relatively speaking) has "all day" to examine the entire program and can potentially uncover much more ILP in the algorithm than could a real-time, hardware scheduler. This advantage, coupled with the presence of additional functional units, should allow VLIW machines to execute more operations per cycle than are possible with a superscalar approach. Since the previously steady increase in CPU clock speeds has started to flatten out, the ability of VLIW architectures to exploit ILP would seem to bode well for their success in the high-performance computing market. However, in actual practice, IA-64 chips account for only a small fraction of Intel's overall sales, and Itanium-based systems represent only a fraction of the overall server and high-performance computing market (almost all of it in systems made by HP) [42]. It appears that this trend is likely to continue, and at least in the near future, the vast majority of high-end systems will be based on multicore, and in many cases *multithreaded* (see succeeding text), implementations of more conventional architectures and/or increasingly powerful and popular *graphics processing units* (GPUs).

22.3.12 Multithreaded Architectures

Previously, in Section 22.3.8, we mentioned that Tomasulo's scheduling algorithm for machines using OOE was based on a dataflow approach. Though they outwardly perform as von Neumann machines and are programmed using the conventional, sequential programming paradigm, processors with Tomasulo schedulers operate internally as dataflow machines—a type of "non-von Neumann" architecture. While pure dataflow machines never became popular in their own right, they did inspire this and certain other innovations that have been applied to more traditional CPU designs.

One other way in which dataflow concepts have influenced conventional computer designs is in the area of multithreaded, superthreaded, and hyper-threaded processors. Consider the behavior of programs in conventional machines. We think of programs running on a single instruction stream single data stream (SISD) or multiple instruction stream multiple data stream (MIMD) architecture as being comprised of one or more, relatively coarse-grained, *processes*. Each process is an instruction stream consisting of a number of sequentially programmed instructions. Multiple processes can run on an MIMD system in truly concurrent fashion, while on an SISD machine, they must run one after another in time-sliced fashion. In a dataflow machine, each individual machine operation can be considered to be an extremely "lightweight" process of its own and to be scheduled when it is ready to run (has all its operands) and when hardware (a processing element) is available to run it. This is as fine grained a decomposition of processes as is possible and lends itself well to a highly parallel machine implementation. However, for a number of reasons, massively parallel dataflow machines are not always practical or efficient.

The same concept, to less of a degree (and much less finely grained), is used in *multithreaded* systems. Multithreaded architectures, at least in some aspects, evolved from the dataflow model of computation. The idea of dataflow, and multithreading as well, is to avoid data and control hazards and thus keep multiple hardware functional units busy, particularly when latencies of memory access or communication are high and/or not a great deal of ILP is present within a process. While each *thread* of execution is not nearly as small as an individual instruction (as it would be in a true dataflow machine), having more, "lighter-weight" threads instead of larger, monolithic processes and executing these threads in truly concurrent fashion are in the spirit of a dataflow machine since it increases the ability to exploit replicated hardware. *Time-slice* multithreading, or *superthreading* [43], is a dataflow-inspired implementation technique whereby a single, superscalar processor executes more than one process or thread at a time. In each clock cycle, the control unit's scheduling logic can issue multiple instructions belonging to a single thread, but on a subsequent cycle, it may issue instructions belonging to another thread (Figure 22.13 illustrates this approach). In effect, process/thread execution on a superthreaded machine is still time sliced (as it is in traditional uniprocessor architectures), but the time slice can be as small as one clock cycle.

The superthreaded approach can lead to better use of hardware resources in a superscalar processor due to mitigation of dependency-based hazards, since instructions from one thread are unlikely to depend on results from instructions belonging to other threads. It can also reduce cycles wasted due to access latencies since instructions from other threads can still execute, while a given thread is waiting on data from memory. However, the utilization of hardware resources is still limited by the inherent ILP,

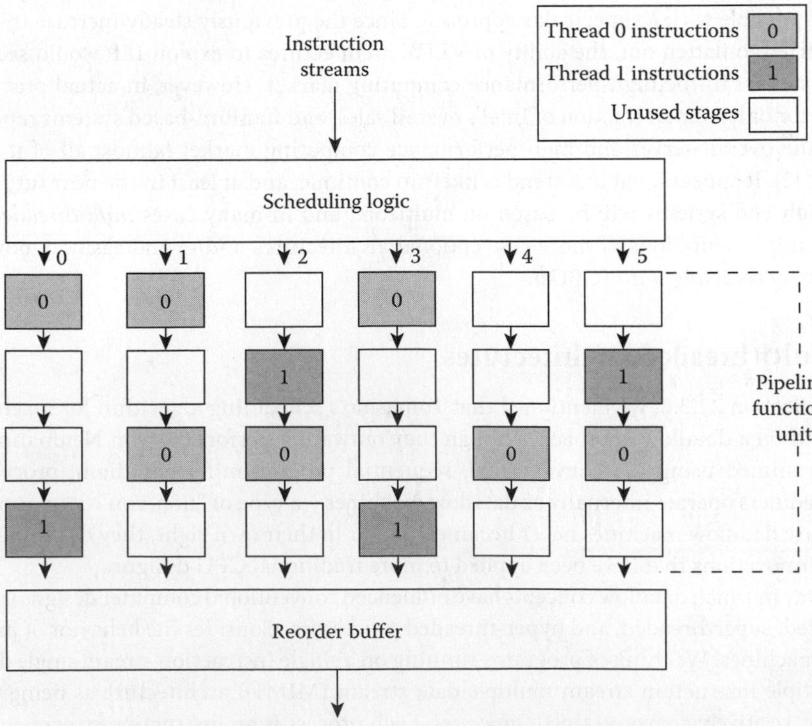

FIGURE 22.13 Instruction execution in superthreaded processor. (From Dumas II, J.D., *Computer Architecture: Fundamentals and Principles of Computer Design*, CRC Press, Boca Raton, FL, 2006, Figure 7.5.)

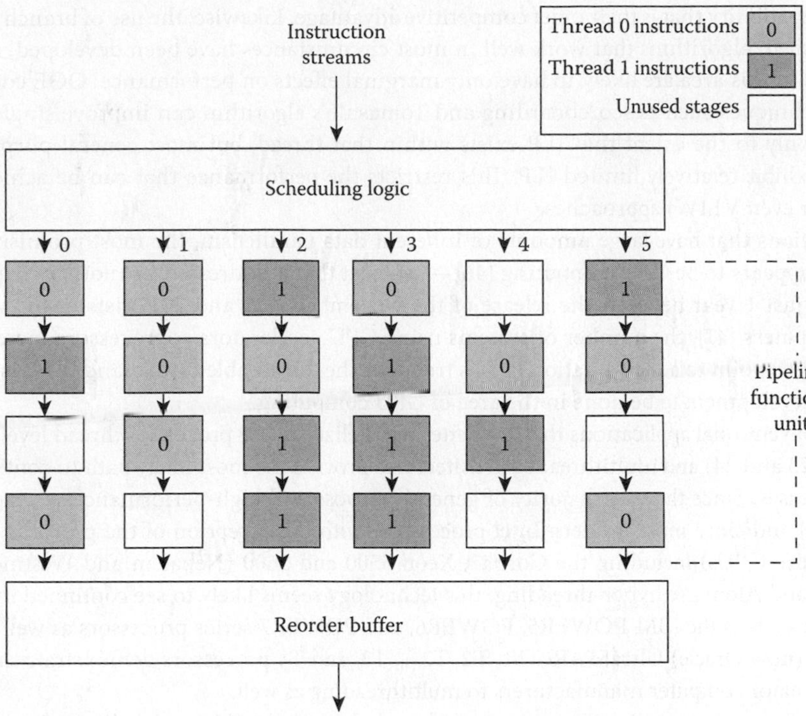

FIGURE 22.14 Instruction execution in hyper-threaded processor. (From Dumas II, J.D., *Computer Architecture: Fundamentals and Principles of Computer Design*, CRC Press, Boca Raton, FL, 2006, Figure 7.6.)

or lack thereof, in the individual threads. (Studies have shown that, on average, a single thread only has sufficient ILP to issue about 2.5 instructions per clock cycle.) To address this limitation, *simultaneous* multithreading (SMT), also known as *hyper-threading* [43], goes a step beyond superthreading.

Compared to superthreading, hyper-threading (which is used in the some of the latest generation CPUs from Intel [44]; the IBM POWER5, POWER6, and POWER7 processors [45]; and others) gives the scheduling logic more flexibility in deciding which instructions to issue at what time. In a hyper-threaded processor, it is not necessary that all the instructions issued during a given clock cycle belong to the same thread [43]. For example, a 6-issue superscalar machine could issue 3 instructions from thread 0 and 3 from thread 1 at the same time, or four from thread 0 and two from thread 1 as illustrated in Figure 22.14.

In effect, SMT internally splits a single, physical superscalar processor into two or more logically separate CPUs, each executing its own instruction stream by sharing processor resources that might otherwise go unused if only a single thread were executing. (Of course, some CPU resources including the program counter and certain other system registers may have to be replicated to allow for hyper-threading.) SMT implementation can significantly increase the efficiency of resource utilization and improve overall system performance as seen by the user when multiple tasks (or a single application coded with multiple threads) are running [43]. Because of its adoption by the largest manufacturer of microprocessors and other industry giants, SMT/hyper-threading could prove to be the most lasting legacy of the dataflow approach.

22.4 Research Issues

Pipelined mainframe processors have been in use since the 1960s, and pipelined microprocessors since the 1980s; so much of the basic research in the field was done decades ago. The use of delay slots to mitigate the performance penalty associated with control transfers is "old hat" among RISC architectures but cannot readily be retrofitted to legacy architectures such as the Intel x86 without destroying the backward compatibility that is their chief competitive advantage. Likewise, the use of branch prediction is well established; algorithms that work well in most circumstances have been developed, and future improvements in this area are likely to have only marginal effects on performance. OOE coupled with scheduling techniques such as scoreboarding and Tomasulo's algorithm can improve single-threaded performance only to the extent that ILP exists within that thread, but most general-purpose, scalar applications exhibit relatively limited ILP. This restricts the performance that can be achieved using superscalar (or even VLIW) approaches.

For applications that have large amounts of inherent data parallelism, the most promising current research area appears to be GPU computing [46]—a subject that is addressed in another chapter of this handbook. In just 1 year between the release of the November 2011 and 2012 lists of the world's top 500 supercomputers [47], the number of systems using GPU accelerators/coprocessors increased from 39 to 62. All signs point to a continuation of this trend for the foreseeable future, and there is still much research and development to be done in the area of GPU computing.

For more conventional applications that are better parallelized at the process or thread level, multicore (see Chapters 23 and 24) and multithreaded architectures provide the most likely path to continuing performance increases. Since the vast majority of general-purpose and high-performance systems use Intel processors [48] and since most modern Intel processors (with the exception of the previous generation Core architecture CPUs) including the Core i7, Xeon 5500 and 5600 (Nehalem and Westmere) series, Itanium 9300, and Atom use hyper-threading, this technology seems likely to see continued use in larger systems. Chips such as the IBM POWER5, POWER6, and POWER7 series processors as well as the Sun Microsystems' (now Oracle) UltraSPARC T1, T2, T2 +, T3, and T4 processors demonstrate the commitment of other major computer manufacturers to multithreading as well.

Some of the issues that are significant to multithreaded architectures include limitations on single-threaded performance when multiple threads are running and disadvantages associated with sharing of memory resources including cache and translation lookaside buffers (TLBs). It is an observed fact that some applications perform noticeably worse on hyper-threaded processors due to contention with other threads for CPU resources such as instruction decoding logic, execution units, and branch predictors [43,49]. With each new design, manufacturers must struggle to balance hardware resources with the anticipated needs of applications. As an alternative, some multithreaded CPUs, for example, the UltraSPARC T4, provide a mechanism for multithreading to be effectively turned off—temporarily dedicating an entire core's hardware to running a single thread that had been suffering from a bottleneck using the Solaris operating system's "Critical Thread Optimization" feature [50]. However, more research and development are needed to provide convenient and efficient means for systems to detect when this is occurring and to use this information to switch back and forth between single-threaded and multithreaded modes for optimum overall performance.

Sharing of memory resources between threads running on a common core creates issues not only for performance but also for security. Obviously, when the instruction and/or data caches belonging to a core are shared between two or more threads, the effective cache size available to each thread—and thus the hit ratio for each—is less than it would be if only one thread were running. Effective cache size may not be cut exactly in half if two running threads share some common code or data, but it is extremely unlikely that they would share everything; thus, some degradation of performance is almost inevitable. Important research questions here involve determining the optimum cache size, line size, and other design parameters for multiple threads to share a common cache with the least disruption. But perhaps an even more important topic is preventing threads from sharing the cache "too much"—in other words, spying on data belonging to other threads. As far back as 2005, researcher Colin Percival discovered that it was possible for a malicious thread running on a Pentium 4 CPU with hyper-threading to use cache access patterns to steal a cryptographic key from another thread running on the same core [51]. While operating systems can be configured to block this attack vector, manufacturers of multithreaded CPUs now know they must keep information security in mind when designing cache line replacement policies; this will surely be an ongoing concern as more processors of this type are designed.

22.5 Summary

No other part of a modern computer system is as complex as its CPU(s). In this chapter, we have examined several advanced design concepts involving internal concurrency of operations within a single CPU core. (Multicore/multiprocessing systems will be addressed in Chapters 23 and 24.) The concept of pipelining was the main focus of our attention, since almost all commonly used methods for enhancing the performance of a CPU are based on pipelining and since, as a result, almost all modern microprocessors are internally pipelined to some extent. As the reader has seen, while the potential performance benefits of pipelining are obvious, "the devil is in the details"—particularly when pipelines become very "deep" and/or when multiple pipelines are used in one processor core.

Of course, the examples provided earlier as illustrations of pipelined, superpipelined, superscalar, VLIW, and multithreaded architectures were kept relatively simple so as not to lose the reader in a myriad of implementation details—details that would have to be addressed in any practical, real-world design. Put another way, we have only scratched the surface of the intricacies of this subject. Computer engineers may spend their entire careers designing high-performance microprocessors and not learn everything there is to know about it—and even comprehensive knowledge of technical details can quickly become obsolete as technology changes. Certainly no single handbook chapter can do full justice to this highly specialized discipline. However, it is hoped that the information in this chapter will be sufficient to provide most computing professionals with a sufficient understanding of these techniques and their implications for performance, to make them better equipped to make wise choices among the many competing computer systems in the constantly changing market of the present and the future.

Key Terms

Arithmetic pipeline: A pipeline designed to perform a given arithmetic operation repetitively. Arithmetic pipelines are often used in vector supercomputers due to the nature of their operands and the operations performed on them but are rarely found in general-purpose machines.

Branch penalty: The time wasted when instructions that sequentially follow a control transfer instruction are brought into the pipeline but must be aborted (nullified) by the control logic because the control transfer causes nonsequential execution. Depending on the specifics of the instruction set and the details of the pipeline's implementation, the branch penalty may be as little as one clock cycle or as great as the time required to drain the entire pipeline of instructions.

Branch prediction: A technique that attempts to improve the performance of an instruction-pipelined processor by guessing whether or not a given conditional control transfer instruction will succeed or fail and fetching subsequent instructions into the pipeline accordingly. Branch prediction can be done either *statically* (before the program is run) or *dynamically* (by the pipeline control logic, while the program is running).

Central Processing Unit (CPU): The portion of a digital computer system that performs computational work by fetching program instructions from memory and executing them. A CPU normally consists of a *control unit*, which controls and sequences hardware operations, plus a *datapath* that consists of storage registers and computational hardware (e.g., an ALU) for performing binary calculations.

Data forwarding: A technique in which the result of a computation, which is needed for use in a subsequent computation, is delivered directly to the hardware functional unit performing the second computation rather than being stored into its destination register and then read back out for use in the second computation. Data forwarding can improve the performance of a pipelined machine by eliminating or reducing pipeline stalls.

Dataflow computer: A computer in which the order of performing operations is determined by the availability of operands and computational hardware, rather than by the sequential programming paradigm associated with von Neumann-type computing architectures.

Delayed branch: A conditional control transfer instruction that attempts to eliminate or reduce the branch penalty (wasted clock cycles) of a successful branch by documenting, as an architectural feature, the execution of one (or more) instruction immediately following the branch, whether or not the branch succeeds. The instruction(s) following the branch is said to be in the branch *delay slot(s)*.

Flow-through time: The time taken for the first iteration of a task to be completed by a pipeline. Assuming that each subtask is performed by one pipeline stage, that a clock signal is used to advance operations to the next stage, and that the pipeline does not stall for any reason, the flow-through time of a pipeline with s stages will be s clock cycles.

ILP: A measure of how many instructions within a computer program can be executed at the same time without altering the results that would be produced by executing them sequentially according to the von Neumann paradigm. The more inherent parallelism a program has at the instruction level, the better it can take advantage of internally parallel processor designs (e.g., superscalar or VLIW).

Instruction-unit pipeline: A pipeline used to process a variety of machine instructions that may perform different functions but can be subdivided into a similar number and type of steps. Instruction-unit pipelines are common in modern scalar, general-purpose microprocessors.

Pipeline register: A parallel-in, parallel-out storage register comprising a set of flip-flops or latches, which is used to separate the logic used in different pipeline stages in order to keep one stage from interfering with the operation of another. The pipeline registers are clocked with a common signal such that each stage's outputs are advanced to the following stage's inputs simultaneously.

Pipeline throughput: The number of task iterations completed by a pipeline per clock cycle. In a pipelined instruction-processing unit, throughput is the number of instructions executed per clock cycle (IPC). The ideal throughput of a single pipeline is one IPC, but in practice it is always less due to control transfers, memory accesses, and/or stalls due to data hazards.

Pipelining: A technique used to improve performance when a task is performed repetitively. A pipeline breaks the task into several subtasks, each performed by a different *stage* of the pipeline, and overlaps the performance of subtasks for subsequent task iterations. Since successive task iterations are started before previous ones have completed, a given number of iterations can be completed in less time as compared to a purely sequential implementation.

RAW hazard: Also known as a *true data dependence*, this is a relationship between instructions that exist when an instruction reads a datum that was produced by a previous instruction. In a purely sequential processor, correct operation would be guaranteed, but in a pipelined processor, it is possible that the subsequent instruction could read the datum before the previous instruction's write was completed, leading to erroneous use of "old" data.

Simultaneous multithreading: Also known as *hyper-threading*, this is a CPU design technique in which the processor executes instructions from multiple threads concurrently. During any given clock cycle, instructions from more than one thread may be issued.

Superpipelined processor: A CPU that uses a deep pipeline structure; in other words, one that has a large number of very simple stages. By finely subdividing the processing of instructions, a superpipelined processor may be able to sustain a higher clock frequency than other processors built using similar technology. Because of the large number of stages, control transfers can cause performance problems.

Superscalar processor: A CPU in which more than one instruction can be issued for execution per clock cycle, and thus, more than one instruction may be completed per clock cycle. This is generally accomplished by having more than one pipeline available for processing instructions. The decision about which instructions can be issued simultaneously without producing incorrect results is made by the pipeline control logic while the program is running.

Time-slice multithreading: Also known as *superthreading*, this is a CPU design technique in which the processor executes instructions from different threads during successive clock cycles. During any given clock cycle, instructions from only one thread are issued.

Very long instruction word (VLIW) processor: A CPU with multiple pipelined functional units, similar in hardware design to a superscalar processor. Because of its internally parallel structure, the CPU can issue (and thus complete) multiple instructions per clock cycle. However, the scheduling of instruction issue is done largely (or completely) by the compiler at program translation time, rather than by the control unit at run time.

Write after read (WAR) hazard: Also known as an *antidependence*, this is a relationship between instructions that exist when an instruction writes a datum that was to have been read by a previous instruction. In a processor with in-order execution, correct operation would be guaranteed, but in a processor with multiple pipelines leading to the possibility of OOE, it is possible that the subsequent instruction could overwrite the datum before the previous instruction's read was completed, leading to erroneous use of "too new" data.

Write after write (WAW) hazard: Also known as an *output dependence*, this is a relationship between instructions that exist when more than one instruction writes a common datum. In a processor with in-order execution, correct operation would be guaranteed, but in a processor with multiple pipelines (and/or multiple write stages) leading to the possibility of OOE, it is possible that a previous instruction could write its result after a subsequent instruction had already done so, leading to an erroneous final value of the datum in question.

References

1. Shankland, S. Moore's law: The rule that really matters in tech. *CNET.com*, October 15, 2012. Retrieved from http://news.cnet.com/8301-11386_3-57526581-76/moores-law-the-rule-that-really-matters-in-tech/.
2. Wikibooks. Hazards, in *Microprocessor Design*. Wikibooks.org, September 17, 2011. Retrieved from http://en.wikibooks.org/wiki/Microprocessor_Design/Hazards.
3. Kent, A. and Williams, J. G. *Encyclopedia of Microcomputers*, Vol. 25, p. 269. Marcel Dekker, New York, 2000. Retrieved from http://books.google.com/books?id=NbaZOeyCb6cC&pg=PA269&lpg=PA269.
4. *Earth Simulator—SX-9/E/1280M160*. Top500.org. November 2012. Retrieved from http://www.top500.org/system/176210.
5. Patterson, D. A. and Sequin, C. H. RISC I: A reduced instruction set VLSI computer, *Proceedings of the 8th Annual Symposium on Computer Architecture*, Los Alamitos, CA, 1981. Retrieved from http://web.cecs.pdx.edu/~alaa/courses/ece587/spring2011/papers/patterson_isca_1981.pdf.
6. Godse, D. A. and Godse, A. P. *Computer Organization, Fourth Revised Edition*, p. 4–40. Technical Publications, Pune, India, 2008. Retrieved from http://books.google.com/books?id=c8_2LMURd3AC&pg=SA4-PA40&lpg=SA4-PA40.
7. Lilja, D. J. Reducing the branch penalty in pipelined processors, *IEEE Computer*, 21(7), 47–55, July 1988. Retrieved from http://www.onversity.net/load/branch-penatly.pdf.
8. Fisher, J. A., Faraboschi, P., and Young, C. *Embedded Computing: A VLIW Approach to Architecture, Compilers, and Tools*, p. 157. Morgan Kaufmann, Saint Louis, MO, 2005. Retrieved from http://books.google.com/books?id=R5UXl6Jo0XYC&pg=PA157&lpg=PA157.
9. Smotherman, M. IBM stretch (7030)—Aggressive uniprocessor parallelism, in *Selected Historical Computer Designs*. Last updated July 2010. Retrieved from http://people.cs.clemson.edu/~mark/stretch.html.
10. Branch Predictor. Wikipedia, the free encyclopedia. Retrieved from http://en.wikipedia.org/wiki/Branch_predictor.
11. Weaver, D. L. and Germond, T. The SPARC Architecture Manual, Version 9, pp. 148–150. SPARC International. Prentice Hall, Englewood Cliffs, NJ, 2000. Retrieved from http://dsc.sun.com/solaris/articles/sparcv9.pdf.
12. Lin, W.-M., Madhavaram, R., and Yang, A.-Y. Improving branch prediction performance with a generalized design for dynamic branch predictors, *Informatica 29*, 365–373, 2005. Retrieved from http://www.informatica.si/PDF/29-3/14_Lin-Improving%20Branch%20Prediction….pdf.

13. Anderson, D. and Shanley, T. *Pentium Processor System Architecture,* 2nd edn., pp. 149–150. MindShare, London, U.K., 1995. Retrieved from http://books.google.com/books?id=TVzjEZg1-YC&pg=PA149&lpg=PA14.
14. Zargham, M. R. *Computer Architecture: Single and Parallel Systems*, pp. 119–124. Prentice Hall, Englewood Cliffs, NJ, 1996.
15. Lee, J. K. F. and Smith, A. J. Branch prediction strategies and branch target buffer design, *IEEE Computer*, 17(1), 6–22, January 1984. Retrieved from http://www.ece.ucdavis.edu/~vojin/CLASSES/EEC272/S2005/Papers/Lee-Smith_Branch-Prediction.pdf.
16. Cragon, H. G. *Memory Systems and Pipelined Processors*, pp. 62–65. Jones & Bartlett, Burlington, MA, 1996. Retrieved from http://books.google.com/books?id=q2w3JSFD7l4C&pg=PA64.
17. Silc, J., Robic, B., and Ungerer, T. *Processor Architecture: From Dataflow to Superscalar and Beyond*, p. 29. Springer-Verlag, Berlin, Germany, 1999. Retrieved from http://books.google.com/books?id=JEYKyfZ3yF0C&pg=PA29&lpg=PA29.
18. Pagetable.com. Having Fun With Branch Delay Slots, November 22, 2009. Retrieved from http://www.pagetable.com/?p=313.
19. Glew, A. Access Ordering Hazards. *Comp-arch.net wiki*. Retrieved from http://semipublic.comp-arch.net/wiki/Access_ordering_hazards.
20. Zargham, M. R. *Computer Architecture: Single and Parallel Systems*, p. 112. Prentice Hall, Englewood Cliffs, NJ, 1996.
21. Shimpi, A. L. Out-of-order architectures, in *Understanding the Cell Microprocessor*, AnandTech.com, March 17, 2005. Retrieved from http://www.anandtech.com/show/1647/7.
22. Prabhu, G. M. Forwarding, in *Computer Architecture Tutorial*. Retrieved from http://www.cs.iastate.edu/~prabhu/Tutorial/PIPELINE/forward.html.
23. Thornton, J. E. Parallel operation in the control data 6600, *Proceedings of the Spring Joint Computer Conference,* Washington, DC, 1964, pp. 33–40. Retrieved from http://dl.acm.org/citation.cfm?doid=1464039.1464045.
24. Govindarajalu, B. *Computer Architecture and Organization: Design Principles and Applications*, 2nd edn., pp. 711–730. Tata McGraw-Hill, Delhi, India, 2010. Retrieved from http://books.google.com/books?id=zzGoVXQ0GzsC&pg=PA712&lpg=PA712.
25. System/360 Model 91, *IBM Archives*, International Business Machines Corporation, Washington, DC. Retrieved from http://www-03.ibm.com/ibm/history/exhibits/mainframe/mainframe_PP2091.html.
26. Tomasulo, R. M. An efficient algorithm for exploiting multiple arithmetic units, *IBM Journal*, 11(1), 25–33, January 1967. Retrieved from http://www.cs.washington.edu/education/courses/cse548/11au/Tomasulo-An-Efficient-Algorithm-for-Exploiting-Multiple-Arithmetic-Units.pdf.
27. Bishop, B., Kelliher, T. P., and Irwin, M.J. The Design of a Register Renaming Unit. *Proceedings of the Ninth Great Lakes Symposium on VLSI*, Ann Arbor, Michigan, March 1999, pp. 34–37. Retrieved from http://cecs.uci.edu/~papers/compendium94-03/papers/1999/glsvlsi99/pdffiles/glsvlsi99_034.pdf.
28. Mirapuri, S., Woodacre, M., and Vasseghi, N. The MIPS R4000 Processor, *IEEE Micro*, 12(4), 10–22, April 1992. Retrieved from http://www.inf.ufpr.br/roberto/ci312/mips4k.pdf.
29. Alpha 21064. *Wikipedia, the Free Encyclopedia*. Retrieved from http://en.wikipedia.org/wiki/Alpha_21064.
30. DEC Alpha. *Wikipedia, the Free Encyclopedia*. Retrieved from http://en.wikipedia.org/wiki/DEC_Alpha.
31. Yeager, K. C. The MIPS R10000 superscalar microprocessor, *IEEE Micro*, 16(2), 28–40, April 1996. Retrieved from http://people.cs.pitt.edu/~cho/cs2410/papers/yeager-micromag96.pdf.
32. Stokes, J. The pentium: An architectural history of the world's most famous desktop processor (part I), *Ars Technica Features*, July 12, 2004. Retrieved from http://arstechnica.com/features/2004/07/pentium-1/.
33. Tremblay, M., Greenley, D., and Normoyle, K. The design of the microarchitecture of UltraSPARC-I, *Proceedings of the IEEE*, 83(12), 1653–1663, December 1995. Retrieved from http://cseweb.ucsd.edu/classes/wi13/cse240a/pdf/03/UltraSparc_I.pdf.

34. Stokes, J. Inside the IBM PowerPC 970—Part I: Design philosophy and front end, *Ars Technica Features*, October 29, 2002. Retrieved from http://arstechnica.com/features/2002/10/ppc970/.
35. Advanced Micro Devices Inc. *AMD Athlon Processor Technical Brief*, Advanced Micro Devices Inc. December 1999. Retrieved from http://support.amd.com/us/Processor_TechDocs/22054.pdf.
36. Colwell, R. P. VLIW: The unlikeliest computer architecture, *IEEE Solid-State Circuits Magazine*, 1(2), 18–22, Spring 2009. Retrieved from http://ieeexplore.ieee.org/stamp/stamp.jsp?tp=&arnumber=5116832.
37. Mathew, B. Very large instruction word architectures (VLIW processors and trace scheduling), 2006. Retrieved from http://www.siliconintelligence.com/people/binu/pubs/vliw/vliw.pdf.
38. Gschwind, M., Kailas, K., and Moreno, J. H. VLIW architecture project page, International Business Machines Corporation. Retrieved from http://researcher.watson.ibm.com/researcher/view_project.php?id=2831.
39. Ebcioglu, K. and Altman, E. R. DAISY: Dynamic compilation for 100% architectural compatibility. IBM Research Report RC 20538, August 5, 1996. Retrieved from http://tinyurl.com/b67ke7q.
40. Smotherman, M. Understanding EPIC architectures and implementations, *Proceedings of the ACM Southeast Conference*, Atlanta, GA, 2002. Retrieved from http://people.cs.clemson.edu/~mark/464/acmse_epic.pdf.
41. Baraz, L., Devor, T., Etzion, O., Goldenberg, S., Skaletsky, A., Wang, Y., and Zemach, Y. IA-32 execution layer: A two-phase dynamic translator designed to support IA-32 applications on Itanium-based systems, *Proceedings of the 36th Annual Symposium on Microarchitecture*, San Diego, CA, 2003. Retrieved from http://www.microarch.org/micro36/html/pdf/goldenberg-IA32ExecutionLayer.pdf.
42. Itanium in Computer Desktop Encyclopedia. Answers.com. Retrieved from http://www.answers.com/topic/itanium.
43. Stokes, J. Introduction to multithreading, superthreading and hyperthreading, *Ars Technica Features*, October 3, 2002. Retrieved from http://arstechnica.com/features/2002/10/hyperthreading/.
44. Cepeda, S. *Intel Hyper-Threading Technology: Your Questions Answered*. Intel Software Library, January 27, 2012. Retrieved from http://software.intel.com/en-us/articles/intel-hyper-threading-technology-your-questions-answered.
45. Extension Media. POWER7 Performance Guide. *EECatalog*, August 4, 2010. Retrieved from http://eecatalog.com/power/2010/08/04/power7-performance-guide/.
46. Owens, J. D., Houston, M., Luebke, D., Green, S., Stone, J. E., and Phillips, J. C. GPU computing, *Proceedings of the IEEE*, 96(5), 879–899, May 2008. Retrieved from http://cs.utsa.edu/~qitian/seminar/Spring11/03_04_11/GPU.pdf.
47. *Highlights—November 2012*. Top500.org. November 2012. Retrieved from http://www.top500.org/lists/2012/11/highlights/.
48. Volpe, J. AMD's market share tiptoes higher, Intel still ruler of the roost. *Engadget*, August 2, 2011. Retrieved from http://www.engadget.com/2011/08/02/amds-market-share-tiptoes-higher-intel-still-ruler-of-the-roos/.
49. Leng, T., Ali, R., Hsieh, J., and Stanton, C. A study of hyper-threading in high-performance computing clusters, *Dell Power Solutions*, November 2002, pp. 33–36. Retrieved from http://ftp.dell.com/app/4q02-Len.pdf.
50. Oracle Corporation. *Oracle's SPARC T4-1, SPARC T4-2, SPARC T4-4, and SPARC T4-1B Server Architecture* (white paper), June 2012. Retrieved from http://www.oracle.com/technetwork/server-storage/sun-sparc-enterprise/documentation/o11–090-sparc-t4-arch-496245.pdf.
51. Percival, C. Cache missing for fun and profit, *Proceedings of BSDCan 2005*. Ottawa, Canada. Retrieved from http://www.daemonology.net/papers/cachemissing.pdf.
52. Dumas II, J.D. *Computer Architecture: Fundamentals and Principles of Computer Design*. CRC Press, Boca Raton, FL, 2006.

23

Parallel Architectures

Michael J. Flynn
Stanford University

Kevin W. Rudd
United States Naval Academy

Much of the computing done today is performed using parallel computers—searching for information (such as used by Google across its applications), analyzing data (such as enabled by the Apache Hadoop map–reduce system for scalable distributed computing), and simulating complex systems (such as done for weather prediction by NOAA). Even the computer present in cell phones is often a parallel processor (handling the user interface, applications, and radio)! In many cases, the exact nature of the parallel computer is hidden from the user—the hardware details are abstracted away from both the user and the programmer. However, understanding the basics of how parallel architectures work will help both users and programmers appreciate, optimize, and improve their ability to use these systems.

Parallel or concurrent execution has many different forms within a computer system. Multiple computers can be executing pieces of the same program in parallel, a single computer can be executing multiple operations or instructions in parallel, or there can be some combination of the two. Parallelism can be exploited at several levels: thread or task level, operation or instruction level, or even some lower machine level (such as done in many AMD and Intel processors). Parallelism may be exhibited in space, with multiple independent computation units (commonly referred to as function units (FU)) or, in time, with a single FU that is many times faster than several instruction or operation issuing units and is shared between them. This chapter attempts to remove some of the complexity regarding parallel architectures but necessarily leaves out significant detail; however, we try to provide suitable references to the reader to follow up for more details both within these volumes and in the open literature. Two recommended introductions to uniprocessor computer architecture are Flynn [7] and Hennessy and Patterson [8]; a recommended introduction to multiprocessor systems is Culler, Singh, and Gupta [4].

With many possible forms of parallelism, we need a framework to describe parallel architectures. One of the oldest and simplest such structures is the stream approach (sometimes called the Flynn taxonomy) [6] that is used here as a basis for describing parallel architectures; it is fundamentally implementation agnostic and can apply to traditional processors as well as to custom computation engines. We will place modern variations and extensions in this framework to show how even what appear to be radically new ideas have their basic characteristics grounded in this fundamental model. These characteristics provide a qualitative feel for the architecture for high-level comparisons between different processors.

23.1 Stream Model

A parallel architecture has, or at least appears to have, multiple interconnected *processor elements* (PE) that operate concurrently to solve a single overall problem. PE may operate independently or may execute in lockstep with other PE depending on the nature of the architecture; a PE that operates independently of the other PE is typically referred to as a *core*. Cores are capable of executing independently of other cores but are distinguished from *processors* in that they do not have the interface and communications logic to interface with the rest of a system; processors are typically packaged in a chip or other self-contained module* that is integrated into the system as a major design element.

Initially, the various parallel architectures can be described using the *stream* concept. A stream is simply one (single; S) or more (multiple; M) actions on data objects or the actual data themselves. There are both actions (instructions; I) and objects (data; D), and the four combinations of these streams describe most familiar parallel architectures (illustrated in Figure 23.1):

1. SISD—single instruction stream, single data stream. This classification is the traditional uniprocessor (Figure 23.1a).
2. SIMD—single instruction stream, multiple data streams. This classification includes both vector processors and massively parallel processors as well as general-purpose graphics processors (Figure 23.1b).
3. MISD—multiple instruction streams, single data stream. This classification includes data streaming processors, data flow machines, and systolic arrays (Figure 23.1c).
4. MIMD—multiple instruction streams, multiple data streams. This classification includes both traditional multiprocessors (including chip multiprocessors) and networks of workstations and cloud-computing environments (Figure 23.1d).

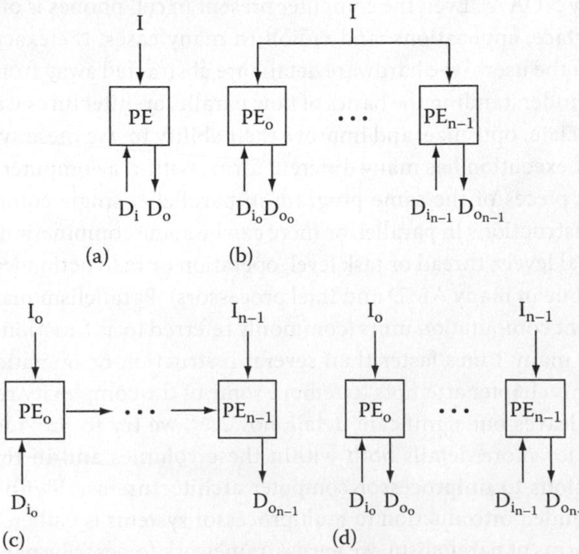

FIGURE 23.1 The stream model. (a) SISD, (b) SIMD, (c) MISD, (d) MIMD.

* In times past, processors were in equipment racks, chassis, boards, chips, etc. Going forward, there is no telling how processors may be packaged or integrated with other features in the future.

The key point to keep in mind is that the stream description of architectures uses the *programmer's* view of the machine (the architecture) as its reference point—the *designer's* view of the machine (the implementation) may be very different and is likely to be much more detailed and complex.* For a programmer to exploit parallelism in their application, they usually need to understand how a parallel architecture is organized; there is the dream of automatic parallelism by either hardware or software, but as of yet that dream is still far from what can be realized in most cases. Note that the stream classification describes only the high-level interaction between instruction and data streams—it does not describe a particular hardware implementation or the means that the various streams are coordinated and synchronized. Thus, a real implementation may be very different than the architecture represented by the stream classification but must ensure that any difference between architecture and implementation is completely transparent to the programmer. Even though an SISD processor can be highly parallel in its execution of operations (as is the case in both superscalar and very long instruction word processors), it still appears to the programmer that the processor executes a single stream of instructions on a single stream of data objects. Understanding the details of a particular implementation may enable the programmer (or compiler) to produce optimized and better-performing code for a particular implementation, but such understanding is not necessary to be able to produce functional (and, typically, reasonably well-performing) code.

There are many factors that determine the overall effectiveness of a parallel processor organization and hence its performance improvement when implemented. What is particularly challenging is that performance can have many different measurements, or metrics, depending on what is desired by the analyst or reviewer of this information. Performance may be the speedup of the measured system compared to the a particular reference system as determined by the observed (wall-clock) time to execute a program, it may be the power consumed to execute the program, or it may be any other definable and measurable value that is of interest. Both Jain [12] and Lilja [15] provide background on performance analysis.

23.2 SISD

The SISD class of processor architecture is the most familiar class to most people—it is found in older video games, home computers, engineering workstations, mainframe computers, cell phones, automobiles, toaster ovens, and even digital watches. From the programmer's view, there is no parallelism in an SISD organization, yet a good deal of real concurrency can be present and can be exploited, resulting in improved performance, or can be abused, resulting in reduced performance. *Pipelining* is an early technique to improve performance that is still used in almost all current processor implementations. Other techniques, such as executing multiple operations or instructions concurrently (in parallel), exploit parallelism using a combination of hardware and software techniques.

23.2.1 Pipelined Processors

Pipelining is a straightforward approach to exploiting parallelism that allows concurrent processing of different phases from multiple instructions in an instruction stream rather than requiring completing the processing of all phases of a single instruction in an instruction stream before starting to process the next instruction. These phases often include fetching an *instruction* from memory (instruction fetch; IF), decoding an instruction to determine its *operation* and *operands* (decode; DE), accessing its operands (register fetch; RF), performing the computation as specified by the operation (execution; EX), and storing the computed value (writeback; WB). Timing diagrams showing the processing of an operation with execution following these phases is shown in Figure 23.2a for a non-pipelined processor and in Figure 23.2b for a pipelined processor.

* Ideally, the *user's* view is much simpler than either the programmer's or the designer's view and is limited to the observable behavior of the application.

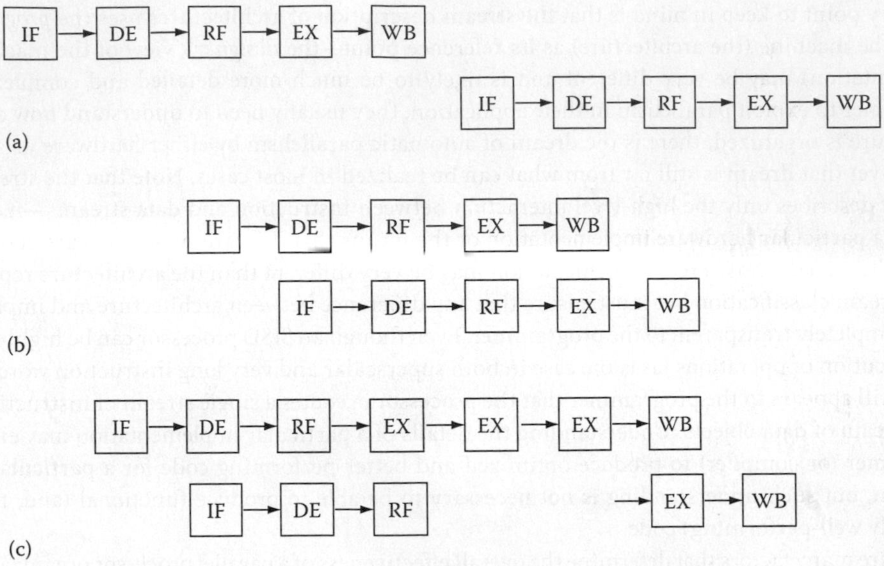

FIGURE 23.2 Instruction issue for scalar processors. (a) Canonical non-pipelined execution. (b) Pipelined execution. (c) Pipelined execution with dependency stall and result bypass.

For both cases, the *latency*, or execution time, of each individual instruction is the same but the overall execution throughput, or *bandwidth*, in terms of cycles per instruction (CPI) or instructions per cycle (IPC) is improved by pipelining; pipelining results in execution requiring fewer CPI (effective execution latency) or completing more IPC (execution bandwidth). The advantage of using pipelining is that multiple instructions have their processing overlapped with each instruction at a different phase of its processing, effectively eliminating all non-computation overhead (where EX is the computationally limiting phase and the rest are overhead). For our hypothetical pipeline, the non-pipelined case has a throughput of five cycles per one instruction or 5 CPI; with the addition of pipelining, we get a throughput of 1 CPI in steady state. This performance is the best that can be achieved by a processor without true concurrent execution—that is, overlapping execution phases for multiple operations in addition to overlapping their overhead phases.

Implementing pipelining assumes that the phases are completely independent between different instructions and therefore can be overlapped—when this assumption does not hold, hardware* must delay processing the dependent later operations or instructions producing a pipeline *stall*; this stall enforces (avoids) the *dependency* as shown in Figure 23.2c; however, it does increase the latency of the dependent instruction and reduces the overall throughput. In this example, the result from the first instruction is delayed and the second instruction must delay the execution phase until the required result becomes available. However, because the result of the first instruction is *bypassed* directly from output of the execution phase of the first instruction to the input of the execution phase of the second instruction the second instruction is then able to begin execution immediately resulting in a two-cycle stall (for the additional two EX phase cycles for the first instruction); without bypassing, execution would have to wait until after the results are available in the register file after WB and then would have to begin executing from the RF phase resulting in a four-cycle stall (now including the WB and and additional two EX phase cycles for the first instruction as well as the RF phase cycle for the second instruction).

* Note that some processors have relied on the programmer, or compiler, to ensure that there are no dependencies between instructions; sometimes this condition required using special no-operation instructions to avoid dependencies. Delayed branches were one instruction that required great care by the programmer (or compiler) to get the correct execution behavior.

Although the first instruction must still perform its register write (during WB), bypassing allows a subsequent instruction to use the value earlier than a simple pipeline would make it available thereby reducing its latency and increasing overall throughput.

23.2.2 Superscalar Processors

To do better than we can with pipelining, we need to use new techniques to discover or to expose the parallelism in the instruction stream. These techniques use some combination of static scheduling (performed by the programmer or the compiler before execution) and dynamic analysis (performed by the hardware during execution) to allow the execution phase of different operations to overlap—potentially yielding an execution rate of greater than one operation every cycle. Because historically most instructions consist of only a single operation (and thus a single computation), this kind of parallelism has been named *instruction-level parallelism* (ILP).

Two common high-performance SISD architectures that exploit ILP are *superscalar* and *VLIW* processors. A superscalar processor uses hardware to analyze the instruction stream and to determine which operations are independent and can be executed concurrently by the hardware; these ready operations are selected by the hardware and routed to the appropriate FU. Figure 23.3a shows the issue of ready operations from a window of available operations. These operations are executed out of order which requires additional hardware to make sure that all operations see the proper results. A larger instruction window makes more operations available for dependency analysis and issue; however, it also increases the hardware cost and complexity used to ensure that all dependencies are correctly satisfied and that all operations execute correctly.

An additional problem with executing operations out of order is how to handle *exceptions* and *interruptions* that occur during execution in an out-of-order processor—there may be many operations before the problem operation which have not executed and many operations after it which have. One solution is for hardware to handle the problem precisely* by determining the affected operation or instruction—then waiting for all previous operations or instructions to complete; at that point, either hardware or software can handle the problem precisely at the affected operation or instruction. Another solution is to handle the problem imprecisely† taking action at the point that the problem is detected and requiring software to fix up any complications resulting from operations or instructions that are incomplete prior to or already executed following the problem operation; the imprecise approach simplifies the hardware dramatically at the expense of software complexity and can result in unrecoverable problems that might have been recoverable with precise exceptions. Some processors, such as the DEC Alpha,

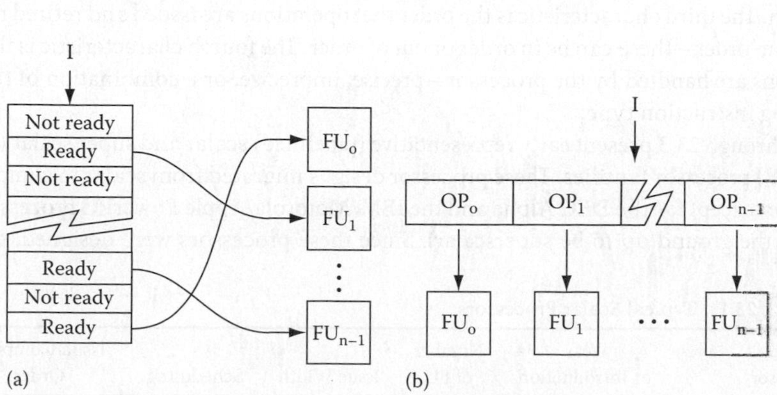

FIGURE 23.3 Instruction issue for superscalar processors. (a) Superscalar. (b) VLIW.

* Precise ⇒ associated with the causing operation or instruction

† Imprecise ⇒ not associated with the causing operation or instruction

use a hybrid precise–imprecise approach that handles problems precisely for most operations (integer and other short-latency operations) but imprecisely for others (long-latency floating-point operations), optimizing the common case and minimizing both hardware and software complexity.

In contrast to superscalar processors, a VLIW processor depends on the programmer or the compiler to analyze the available operations and to schedule independent operations into wide instruction words; the operations in a wide instruction word can be routed without hardware analysis to the appropriate FU. Because operation scheduling is performed statically by the compiler, it is possible to have very wide issue widths without excessive hardware complexity; however, in order to use all of the resources for a wide machine, the compiler must be able to find and schedule enough operations to fill the machine which can be a significant problem. Figure 23.3b shows the direct issue of operations from a wide instruction word to all FU in parallel. VLIW processors have significant challenges associated with exceptions and interrupts as well, but because their execution is always in the order scheduled by the compiler, software has a better chance of success in recovering from the problem than it does with imprecise interrupts in a superscalar processor. Note that there is no reason that VLIW processors could not apply superscalar techniques to execute operations out of order and to handle exceptions and interrupts smoothly (e.g., see Rudd [18] or Wang et. al [21]) however, this approach has not been implemented commercially to the authors' knowledge.

Both superscalar and VLIW use the same compiler techniques to achieve their superscalar performance with the main differences being the complexity of the hardware (superscalar is much more complex) and the tolerance that the processor has to how well the schedule is matched to the processor (VLIW is very intolerant). The practical difference between superscalar and VLIW is that if a program is scheduled inappropriately for the processor a superscalar processor will still execute correctly and may find parallelism from the instruction stream; however, a VLIW processor may fault or, worse, silently produce incorrect answers in the same situation. However, a VLIW processor is likely to have more predictable execution behavior and may prove to be more efficient (in terms of hardware complexity or power consumption); on some applications, a VLIW processor may even have higher performance than a superscalar processor in spite of its execution inflexibility.

23.2.3 SISD Wrap-Up

A SISD processor has four primary characteristics. The first characteristic is whether or not the processor is capable of executing multiple operations concurrently. The second characteristic is the mechanism by which operations are scheduled for execution—statically at compile time, dynamically at execution, or possibly both. The third characteristic is the order that operations are issued and retired relative to the original program order—these can be in order or out of order. The fourth characteristic is the manner in which exceptions are handled by the processor—precise, imprecise, or a combination of the two based on the executing instruction type.

Tables 23.1 through 23.3 present early representative (pipelined) scalar and super scalar (both superscalar and VLIW) processor families. These processor designs migrated from scalar to compatible superscalar processor (except for the DEC Alpha and the IBM/Motorola/Apple PowerPC processors that were designed from the ground up to be superscalar). Since these processors were designed, the trend has

TABLE 23.1 Typical Scalar Processors

Processor	Year of Introduction	Number of FU	Issue Width	Scheduling	Issue/Complete Order
Intel x86	1978	2	1	Dynamic	In order/in order
Stanford MIPS-X	1981	1	1	Dynamic	In order/in order
Sun SPARC	1987	2	1	Dynamic	In order/in order
MIPS R4000	1992	2	1	Dynamic	In order/in order

TABLE 23.2 Typical Superscalar Processors

Processor	Year of Introduction	Number of FU	Issue Width	Scheduling	Issue/Complete Order
DEC 21064	1992	4	2	Dynamic	In order/in order
Sun UltraSPARC	1992	9	4	Dynamic	In order/out of order
MIPS R8000	1994	6	4	Dynamic	In order/in order
HP PA-RISC 8000	1995	10	4	Dynamic	Out of order/out of order
AMD K5	1996	6	4	Dynamic	In-order x86, out of order ROPs/out of order
Intel Pentium 4	2000	7	6[a]	Dynamic	In-order x86, out of order uops/out of order

[a] Although the Pentium 4 can issue 6 uops per cycle, it can only retire 3 uops per cycle.

TABLE 23.3 Typical VLIW Processors

Processor	Year of Introduction	Number of FU	Issue Width	Scheduling	Issue/Complete Order
Cydrome Cydra 5	1987	7	7	Static	In order/in order
Multiflow Trace 28/200	1987	28	28	Static	In order/in order
Philips TM-1	1996	27	5	Static	In order/in order
Intel Itanium 2	2002	9[a]	6	Static	In order/in order

[a] There are actually many more than 9 FU, but there are 9 issue ports that limit the possible combinations of operation issue into the larger pool of physical FU.

been toward improving the resources and capabilities for a processor than their issue width. Common approaches have included increasing the number and type of FU to enable more complex operations to be issued concurrently, increasing the frequency and pipeline depth to increase throughput, and increasing the size and performance of caches to reduce the effective memory latency. More recently, with reaching the "power wall," the trend has been to replicate cores on a processor enabling greater parallelism by executing more threads concurrently but also reducing the performance of an individual thread; these processors exploit MIMD parallelism. Table 23.7 (in Section 23.5) shows the trend for cores in Intel multicore processors. The high-level characteristics for these processors have stabilized on a per-core basis but performance improves each generation from process changes, design optimizations, and new computation capabilities; in addition, die-size reductions enable more cores per die that is not seen at the core level. See Section 23.5 for more information on MIMD processors.

There have been very few VLIW processors to date although advances in compiler technology may cause this to change. VLIW processors have not been very successful as general-purpose processors primarily because of the challenges finding sufficient ILP using only static analysis in the compiler or hand coding and optimizing by experienced programmers. However, there are applications where the VLIW processors are very effective and most current VLIW processors are used as embedded processors, multimedia accelerators, or graphics processors (GPU); for these applications, there can be significant performance advantages and the cost–performance trade-offs are more important than increased software-development costs.

23.3 SIMD

The SIMD class of processor architecture includes both array and vector processors. The SIMD processor is a natural response to the use of certain regular data structures, such as vectors and matrices. From the programmer's view, an SIMD architecture appears very similar to programming a simple SISD processor in that there is a single instruction stream; however, the key difference is that many of these operations perform computations on aggregate data. For example, one instruction might add each pair of elements of two vectors or matrices. Because regular structures are widely used in scientific computing, the SIMD processor has been very successful in these environments.

Two types of SIMD processor will be considered: the *array processor* and the *vector processor* that differ both in their implementations and in their data organizations. An array processor consists of many interconnected PE that are dependent and which share a common instruction stream; these PE operate on the aggregate data elements concurrently across all of the PE. A vector processor consists of a single PE that has both traditional scalar instructions and FU as well as special vector instructions and vector FU; the vector units operate on vectors executing the computation sequentially, one element at a time, but very efficiently, initiating computations rapidly and possibly overlapping computations as the computation on one element is independent of the computations on all other elements.

23.3.1 Array Processors

The array processor is a set of interconnected parallel PE (typically hundreds to tens of thousands). PE operate in lockstep in response to a single broadcast instruction from a control processor and thus all PE synchronously execute exactly the same instruction in each cycle. If supported, when the execution path in different threads diverges, the control processor must issue instructions from each divergent thread to all PE until their paths reconverge; only those PE associated with the execution path for that divergent thread then execute the instructions. When divergence occurs, this method results in a significant slowdown and performance penalty as many of the PE are idle during divergent execution.

Each PE has its own private memory and data are distributed across the elements in a regular fashion that is dependent on both the structure of the data and the structure of the computation to be performed on the data. When allowed, direct access to global memory or to another PE's local memory is expensive; however, constant scalar values can be broadcast by the control processor along with the instruction to all PE. The high remote-data access cost requires that the data be distributed carefully. For performance reasons, it is sometimes easier to duplicate data values and computations than it is to route the intermediate data between PE.

A suitable application for use on an array processor has several key characteristics: a significant amount of data that have a regular structure; computations on the data that are uniformly applied to many or all elements of the data set; simple and regular patterns relating the computations and the data; and minimal communications requirements between computations on individual elements. An example of an application that has these characteristics is the solution of the Naviér–Stokes equations; however, applications that have significant matrix computations and very little execution divergence are likely to benefit from the concurrent capabilities of an array processor.

The programmer's reference point for an array processor is typically the high-level language level—the programmer is concerned with describing the relationships between the data and the the computations on the data but is not directly concerned with the details of scalar and array instruction scheduling. In fact, in many cases, the programmer is not even concerned with the size of the array processor. In general, the programmer specifies the size and structure of the data along with distribution information and the compiler maps the implied virtual processor array (which is sized and shaped based on the size of the data and computation) onto the physical PE (which are actually implemented in hardware); it then generates code to perform the required computations and the management of virtual-to-real PE mappings is performed by some combination of hardware and software methods.

The primary characteristic of an SIMD processor is whether the memory model is shared or distributed. The interconnect network is also important; for more information on interprocessor interconnects, see Chapter ??. Table 23.4 shows several representative SIMD architectures. Note that there have not been a significant number of SIMD architectures developed recently due to a limited application base and market requirement for SIMD processors. In addition, the ready availability of high-performance general-purpose

TABLE 23.4 Typical Array Processors

Processor	Year of Introduction	Memory Model	PE	Number of Processors
Burroughs BSP	1979	Shared	General purpose	16
Thinking Machines CM-1	1985	Distributed	Bit-serial	up to 65,536
Thinking Machines CM-2	1987	Distributed	Bit-serial	4,096–65,536
MasPar MP-1	1990	Distributed	Bit-serial	1,024–16,384

multiprocessor (MIMD) systems and general-purpose graphics processors (GP-GPU) allows exploiting similar parallelism with more flexibility and broader capabilities than available with traditional SIMD processors. Section 23.5 discusses MIMD systems and Section 23.6.1 discusses GP-GPU systems.

23.3.2 Vector Processors

A vector processor is a single processor that resembles a traditional SISD processor except that some of the FU (and registers) operate on vectors—aggregate data comprising sequences of data values—which are operated on as a single entity by the hardware. Vector processor FU typically have a very high clock rate enabling them to be deeply pipelined; in addition, execution phases are overlapped, which is allowed because the calculations are defined on individual data elements and cross element dependencies are not allowed. Although this deep pipeline results in as long or longer latency than a normal scalar FU to compute on the first data element, it produces the results of computations on the remaining elements very rapidly. The vector-processor computation process results in a significant speedup for the overall vector computation that cannot be matched by scalar FU that must explicitly perform the same set of computations on each element in turn.

Early vector processors computed on vectors that were stored in memory. The primary advantage of this approach is that the vectors could be of arbitrary lengths and were not limited by processor resources; however, memory-based vector operations have a high start-up cost due to the long *latency* to get the vector from main memory, limited memory system *bandwidth*, and memory system contention which all proved to be significant limitations. Modern vector processors require that vectors be explicitly loaded from memory into special vector registers and stored back into memory—the same approach that modern scalar processors have taken for similar reasons. However, because vector registers have low latency and preserve intermediate values for reuse without passing back through memory, modern register-based vector processors achieve significantly higher performance than the earlier memory-based vector processors for the same implementation technology.

Vector processors improve performance further using result bypassing (*chaining*) that allows a follow-on vector computation to commence as soon as the first value is available from the preceding computation, pipelining the execution of different vector instructions across different FU. Instead of waiting for the entire vector to be processed, the follow-on computation can be significantly overlapped with the preceding computation that it is dependent on as long as there are available FU. Chaining allows sequential computations to be efficiently compounded and to behave as if they were performed by a single operation with a total latency equal to the latency of the first operation with the pipeline and chaining latencies of the remaining operations but without the per-instruction start-up overhead that would be incurred to perform the same vector computations without chaining. For example, division could be synthesized by chaining a reciprocal operation with a multiply operation at only incrementally greater execution latency than either operation alone (although requiring two FU to perform the chained computation). Chaining typically works for the results of load operations as well as normal computations.

As with the array processor, the programmer's view for a vector machine is the high-level language. In most cases, the programmer sees a traditional SISD machine; however, because vector machines excel on computing with vectors, the programmer typically writes code with simple vectorizable loops that

TABLE 23.5 Typical Vector Processors

Processor	Year of Introduction	Memory or Register Based	Number of Processor Units	Number of Vector Units
Cray 1	1976	Register	1	7
CDC Cyber 205	1981	Memory	1	2–5
Cray T90	1995	Register	4–32	2
NEC SX-9	2008	Register	4–8192	48

can be recognized by the compiler and converted into individual vector instructions. Programmers can also help the compiler by providing hints that help the compiler to locate the vectorizable sections of the code. This situation is purely an artifact of the fact that the programming languages are typically scalar oriented and do not support the treatment of vectors as an aggregate data type but only as a collection of individual values. As new languages are defined that make vectors a fundamental data type, then the programmer is exposed less to the details of the machine and to its SIMD nature; although programs in these languages include more information available to enable automatic parallelization, the compiler may still not be able to exploit all of the performance in a given machine, trading off performance for portability just as we saw previously with superscalar.

The vector processor has two primary characteristics. One characteristic is the location of the vectors—vectors can be memory or register based; the other characteristic is the vector length. Vector processors have developed dramatically from simple memory-based vector processors to modern multiple-processor vector processors that exploit both SIMD vector and MIMD-style processing. Table 23.5 shows some representative vector processors.

There are fewer and fewer big-iron SIMD processors. Currently, SIMD computations are performed on general-purpose SISD processors using instructions that operate on sub-elements of small fixed-length aggregate data structures. These special SIMD instructions look like register-based instructions in a vector processor but instead perform computations in parallel on sub-elements from wide data registers that have far fewer data elements than a vector processor would have in its vector registers. For example, Intel's current version of their Advanced Vector Instructions operate on 256-bit aggregate data structures; the sub-elements of this aggregate data structure can represent four 64-bit values or eight 32-bit values. Other vector extensions may support different aggregate-data sizes and data sub-element sizes. These new SIMD are often programmed explicitly by the programmer using custom libraries or explicit vector primitives inserted manually into the code rather than being directly supported in traditional programming-language; however, over time new programming languages and improved compilers will extend their ability to perform SIMD computations without explicit programmer involvement.

23.4 MISD

Although the term MISD processor is not widely used, conceptually this is an important architectural classification. Here the instructions (or functional operations) are in a fixed configuration specific to an application and the data are streamed across the operation configuration. It is, in many respects, a pipelined processor wherein the programmer can rearrange or create pipeline stages to match the application. Physically, realizing this programmable pipeline has been a challenge.

Abstractly, the MISD can be represented as multiple independently executing FU operating on a single stream of data, forwarding results from one FU to the next. On the microarchitecture level, this is exactly what the vector processor does. However, in the vector pipeline, the operations are simply fragments of an assembly level operation, as distinct from being a complete operation in themselves. Surprisingly, some of the earliest attempts at computers in the 1940s could be seen as the MISD concept. They used plugboards for programs, where data in the form of a punched card were introduced into the first stage of a multistage processor. A sequential series of actions was taken where the intermediate results were forwarded from stage to stage until at the final stage a result would be punched into a new card.

There are more interesting examples of the MISD organization. Data flow machines were pioneered by Prof. Jack Dennis of MIT in the 1970s and 1980s [5]. These machines can use either static or dynamic data flow, the latter allowing limited run time reconfiguration of the data flow pipeline; they can also use either synchronous or asynchronous clocking; finally, they can use either a single data or data streaming. It is only the data-streaming data flow machine that matches the MISD category.

In a more recent development, the Maxeler Technologies Maximum Performance Computing (MPC) products [16] implement static, synchronous data flow machines that are custom designed to match an application's data flow graph. FPGAs are used to realize the data flow machine.

The systolic array [14] is a special form of MISD wherein all nodes of an (usually rectangular) array consist of data processing units (DPU). These are minimally configured PE. Each node is activated by the arrival of data and the result is shared with its neighbors in a mesh topology. It has been effectively used in computations such as matrix multiply.

The programmer's view of an MISD machine depends very heavily on the specific machine organization. It is hard to generalize without this machine knowledge.

23.5 MIMD

The MIMD class of parallel architecture interconnects multiple cores (typically SISD) into a *multicore processor* (when combining multiple cores onto a single die) or *multiprocessor* (when combining multiple processors into a single system). A *core* is a single hardware structure that executes one or more threads but that is not a stand-alone processor, lacking general communications and peripheral support. A *thread* is the combination of an instruction stream with all of its execution state and data—it represents the programmer's view of an SISD processor. Each core supports the execution of one or more threads.

Although there is no requirement that all processor cores or processors be identical, most MIMD configurations are homogeneous with identical cores throughout the system. Homogeneity simplifies the hardware design, operating system implementation, and the programmer's challenge to use the system effectively. Most multicore processors and multiprocessor systems are homogeneous. One well-known example of a heterogeneous MIMD processor is the IBM/Sony Cell processor [9], which has a single conventional RISC main core (SPU) and eight computation cores (SPE).

For simplicity, in this section, we limit ourselves to homogeneous MIMD organizations. However, although each individual core in the system is identical, the notion of homogeneity is complicated by the multiple levels of hierarchy in current multicore and multiprocessor systems. We have multiple threads on a core, multiple cores on a processor, multiple processors on a board, multiple boards in a box, and so forth—so even though each core in such a homogeneous system is the same, different pairs of cores have access to different resources and may also have different communication characteristics resulting in very nonhomogeneous behavior. Because performance may vary depending on where the threads in an application are executing, there are many challenges for the programmer to use such a system effectively. Although it is important to appreciate the kinds of hierarchy associated with multicore or multiprocessor systems (especially if trying to optimize an application on a particular multicore or multiprocessor system), we only discuss the general concept of MIMD here as the variables are many and all are system and configuration dependent.

Both SIMD and MIMD systems are similar in some respects in that they perform many computations concurrently; the greatest difference between SIMD array processors and an MIMD multiprocessor is that the SIMD array processor executes a single instruction stream executing the same instruction synchronously across all PE in the array, while an MIMD multiprocessor executes different instruction streams asynchronously across all cores in the system. Although it may be the case that each core in an MIMD system is running the same program (sometimes referred to as single program multiple data (SPMD)), there is no reason that different cores should not run different programs that are completely unrelated. In fact, most use of multiprocessor systems is to run multiple independent programs concurrently—not to run a single program (or even multiple programs) cooperatively—across all cores in the system; this usage is common across workstations, desktops, laptops, and even mobile devices running multiple applications on mainstream operating systems such as Microsoft Windows, Apple OS-X, Linux, Apple IOS, and Google Android.

The interconnection network in both the SIMD array processor and the MIMD processor passes data between PE. However, in the MIMD processor, the interconnection network is typically much more general and is used to share data as well as to synchronize the independent cooperating threads running across the MIMD processor. Interconnection networks are very diverse including bus interconnects, where all cores share a common bus to communicate; star interconnects, where all cores interconnect via a common hub; ring interconnects, where all cores connect to two other cores and each core forward nonlocal communications to cores further around the ring; and multidimensional interconnects where cores are connected in complex ways requiring careful routing to get the data from the source to the requester. There are many variations on interconnects including multi-level interconnects with different interconnect topologies at each level. For more information on interprocessor interconnects, see Chapter ??.

One key characteristic of a multiprocessor system is the location of memory in the system; the location affects the latency to access the memory and optimizing memory accesses is one of the programmer's key performance challenges. When the memory of the processor is distributed across all processors and only the cores local to that processor have access to it, all sharing and communications between processors (both sharing of data and synchronizing threads) must be done explicitly using messages between processors. Such an architecture is called a *message-passing architecture*. When the memory of the processor is shared across all processors and all cores have access to it, sharing and communicating between threads can be done explicitly or implicitly (and, unfortunately, sometimes inadvertently) through normal memory operations. Such an architecture is called a *shared-memory architecture*. If the shared memory is in a central location and all processors have the same access latency to the shared memory, then it is called a uniform memory access (UMA) latency multiprocessor; bus-based multiprocessors with a centralized main-memory system are a common example of a UMA processor. If the shared memory is distributed across the processors in the system, then it is called a *distributed shared-memory (DSM) architecture* processor; in a DSM architecture, processors see different access latencies to different memory locations and it is thus called a *non-uniform memory access* (NUMA) latency multiprocessor. Because of their hierarchical nature, most multiprocessor systems are NUMA although even a single-processor multicore system locally be an NUMA system depending on its internal architecture of caches and memory channels.

In any multiprocessor system where data are shared, producing the right observable memory behavior is a challenge; in fact, one problem with shared memory is that the "right" behavior for shared memory is not well defined. There are two significant problems that arise when sharing data. One is *memory consistency* describing the ordering constraints for sequences of memory accesses that may be observed by different threads in the system; the other is *memory coherency*, also called *cache coherency*, describing the value constraints on memory accesses to the same location that may be observed by different threads in the system.

Memory consistency is the programmer-visible ordering effects of memory references both within a core and between different cores and processors in a multicore or multiprocessor system. The memory consistency problem is usually solved through a combination of hardware and software techniques. The appearance of perfect memory consistency may be guaranteed for local memory references only—and thus different processors may see different memory content depending on the implemented memory consistency policy. In the most unconstrained case, memory consistency is only guaranteed through explicit synchronization between processors to ensure that all processors see the same memory content.

Memory coherency, or cache coherency, is the programmer invisible mechanism to ensure that all PE see the same value for a given memory location and is normally associated with how caches behave.

The memory coherency problem is usually solved exclusively through hardware techniques. This problem is significant because of the possibility that multiple cores have copies of data from the same address in their local caches and cache coherency describes whether or not different threads accessing this data see the same value as well as the protocol used to ensure this coherency requirement.

There are two primary techniques to maintain cache coherency. The first is to ensure that all affected cores are informed of any change to the shared memory state—these changes are broadcast throughout the MIMD processor and each core monitors the interconnect to track these state changes (commonly by "snooping" communications on the interconnect). The second is to keep track of all cores using a memory address in a directory entry for this address; when there is a change made to the shared memory state for a shared address, each core in the directory for that address is informed. In either case, the result of a change can be one of two things—either the new value is provided to the affected cores, allowing them to update their local value, or other copies of the value are invalidated, forcing these cores to request the data again when it is needed.

Both memory consistency and memory coherency are complex problems and there are many associated trade-offs in hardware and software complexity as well as in performance. There are certainly complexity and scalability advantages to relaxing both memory consistency and coherency, but whether the programmer is able to use such a system effectively is a difficult question to answer. For more information on memory consistency and memory coherency, see Chapter ??.

Scalability for multicore and multiprocessor systems can be challenging as well. For small systems, a few cores are directly connected (typically by a common system bus, but more and more by a point-to-point link such as AMD HyperTransport [10] or Intel QuickPath Interconnect [11]); in these systems, snooping is adequate and minimizes complexity at the cost of coherency traffic on the bus or interconnect traffic. However, as the number of cores in a system increases, it is more likely that there will be a large multi-hop or multidimensional interconnect requiring significant latency to get between cores; there may also be significantly more data sharing that increases the communications bandwidth through the network, further exacerbating the problem. The resulting traffic on the interconnect as well as the long communication latencies associated with these larger systems may make broadcast and snooping undesirable or unusable. In large systems, directory-based memory coherency becomes significantly better because the amount of communications required to maintain coherency is limited to only those processors requesting, owning, and holding copies of the data. Snooping may still be used within a small cluster of PE to track local changes and the cluster would appear to behave as a single memory consumer to the rest of the network.

The hardware implementation of a distributed memory machine is far easier than that of a shared memory machine when memory consistency and cache coherency are taken into account as all information sharing between processors on a distributed memory system is performed through explicit messages rather than through shared memory. However, programming a distributed memory processor can be much more difficult because the applications must be written to exploit, and to not be limited by, the requirement to use of message passing to communicate between threads. On the other hand, despite the problems associated with maintaining consistency and coherency, programming a shared memory processor can take advantage of whatever communications paradigm is appropriate for a given communications requirement and can be much easier to program.

Early multiprocessor systems had few processors; Table 23.6 shows some classic multiprocessor systems. Over time, systems grew larger and the complexity of the systems increased significantly although the basic concepts did not change dramatically through the years. More recently, there have been significant practical changes in multiprocessor systems with single-processor multicore systems becoming common—although these processors are similar to multiprocessor systems through the years, they are often used to run multiple independent programs rather than collaborative programs. In addition, these multicore processors are then used to build larger multiprocessor systems which may then be used for a single application, much like was the case for earlier multiprocessor systems. Table 23.7 shows a few of these processors that can be assembled into much larger multiprocessor systems.

TABLE 23.6 Representative Discrete MIMD Systems

System	Year of Introduction	PE	Number of Processors	Memory Distribution	Programming Paradigm	Interconnection Type
Alliant FX/2800	1990	Intel i860	4–28	Central	Shared memory	Bus + crossbar
Stanford DASH	1992	MIPS R3000	4–64	Distributed	Shared memory	Bus + mesh
MIT Alewife	1994	Sparcle	1–512	Distributed	Message passing	Mesh
Tera Computers MTA	1995	Custom	16–256	Distributed	Shared memory	3D torus
Network of Workstations	1995	Various	Any	Distributed	Message passing	Ethernet
CRAY T3E	1996	DEC Alpha 61164	16–2048	Distributed	Shared memory	3D torus

TABLE 23.7 Intel Multicore Processors for MIMD Server Systems

Processor	Year of Introduction	Number of Cores	Threads per Core	Effective Cache Size
Xeon (Paxville)	2005	2	2	4 MB
Itanium 2 9050 (Montecito)	2006	2	1	24 MB
Xeon X3350 (Yorkfield)	2008	4	1	12 MB
Xeon E7-8870 (Westmere-EX)	2011	10	2	30 MB
Itanium 9560 (Poulson)	2012	8	2	32 MB

TABLE 23.8 TOP500 Supercomputers (The Top Three) as of November 2012

Rank	System	Core	Interconnect	Total Cores
1	Titan – Cray XK7	Opteron 6274 16C, NVIDIA K20x	Cray Gemini	560,640
2	Sequoia – BlueGene/Q	Power BQC 16C	Custom	1,572,864
3	K computer	SPARC64 VIIIfx	Tofu	705,024

Table 23.8 shows the current top three systems in the Supercomputer Top 500 [20]. These systems use large numbers of high-performance general-purpose multicore processors and the top-ranked Titan system is a hybrid system containing general-purpose processors and GP-GPU processors to achieve its performance.

The primary characteristic of an MIMD processor is the nature of the memory address space—it is either local to each processor or shared across all processors. The number of cores in a processor, the number of processors in the system, the memory hierarchy including both caches and memory (and the constraints associated with memory accesses), and the interconnection are also important in characterizing an MIMD processor.

23.6 High-Performance Computing Architectures

Up to this point, we have focused on traditional multiprocessor systems that can be readily described using the Flynn taxonomy. Many of these systems have high performance and fall neatly into the taxonomy. In this section, we discuss three high-performance computing approaches that are becoming ubiquitous but are not readily placed into this simple taxonomy. First, we discuss GPU used for highly parallel general-purpose computing, thus referred to as GP-GPU; we also discuss manycore processors that address similar parallelism but instead use general-purpose CPU cores. Finally, we discuss cloud-computing architectures that are, in a sense, virtual MIMD systems rather than multiprocessor systems themselves but that distinction can be easily transparent to the programmer.

23.6.1 GP-GPU and Manycore Systems

General-purposes use of graphics processors has skyrocketed and resulted in new graphics processors being designed for highly parallel general-purpose use; these processors are called general-purpose GPU (GP-GPU) processors. Parallel applications for GP-GPU are becoming more and more commonplace as the availability of GP-GPU systems (including sharing compatible GPU processors in the graphics subsystem to perform GP-GPU computations in addition to processing graphics) is becoming more common. Unfortunately, GP-GPU have some of the same performance complications of SIMD processors as they are not as flexible as general-purpose processors; thus, they are difficult to program and have a limited range of applications that are suitable for execution on them. In contrast, manycore architectures are derived from general-purpose multiprocessors and although are not as highly parallel as GP-GPU processors are programmed more like traditional multiprocessors and have a broader range of applications that are suitable for execution on them.

NVIDIA calls their Tesla GP-GPU architecture a Single Instruction Multiple Thread (SIMT) architecture; SIMT is a hierarchical hybrid architecture comprising MIMD and SIMD elements that is capable of running many threads across the processor. In this high-level description, we use terms consistent with the rest of the chapter; NVIDIA has their own terminology that is matched to the SIMT architecture and their GP-GPU implementations. For more information on the Tesla GP-GPU architecture, see [13,19].

An SIMT processor contains one or more streaming multiprocessors (the MIMD element) running one or more threads; each streaming multiprocessor comprises a large number of PE running in lockstep (the SIMD element), specialized computing resources, and a memory hierarchy. Both the streaming multiprocessors and the PE are virtualized; thread scheduling is performed using a combination of hardware and software and is very efficient. As with SIMD processors, divergent execution within a streaming multiprocessor still suffers significant performance penalties just as in an SIMD processor under similar execution conditions; however, because each streaming multiprocessor is independent, divergent execution between streaming multiprocessors is not a problem.

The GP-GPU programmer develops an application specifying the number of threads and how those threads are to be distributed across streaming multiprocessors and their PE. By structuring the program carefully, it may be possible (but challenging!) to avoid some of the SIMD complications while achieving the same kind of performance as a SIMD processor.

In contrast to the GP-GPU, the Intel Many Integrated Core (MIC) manycore processor has far fewer cores on the processor than the GP-GPU has PE, but each core is a complete processor that executes independently and can execute the same or different program. Where the flexibility of the manycore processor comes into its own is when each thread is running the same application but follows a different thread of execution—something that the manycore processor is very good at but that the GP-GPU processor is unable to do without suffering significant performance penalties. The manycore processor uses many more cores that are much smaller and simpler than are used in a typical multicore processor; however, in principle, it is very similar to a traditional MIMD processor as discussed in Section 23.5.

What is common across both GP-GPU and manycore architectures is the ability to execute a large number of threads very efficiently. In the case of the GP-GPU processor, these threads tend to be short and execute very quickly along the same execution path; in the case of the manycore processor, these threads tend to be longer and more heavyweight and can execute arbitrary execution paths. Both are used as coprocessors and are controlled by a separate main processor that offloads computations to them and then may be able to continue on with other work (increasing system parallelism) until the coprocessor computations are complete. Table 23.9 shows representative GP-GPU and manycore systems that are currently available.

TABLE 23.9 Massively Parallel GP-GPU and Manycore Coprocessors

Coprocessor	Independent Cores	PE/FU per Core	Threads per Core	Local Memory
NVIDIA Tesla Keppler K20X	14	192 PE	2048	6 GB
AMD Radeon™ HD 7970	32	64 PE	10	3 GB
Intel Xeon Phi 5110P	60	2 FU	4	8 GB

23.6.2 Cloud Computing

A cloud-computing architecture is significantly different from the rest of the multiprocessor architectures discussed in this chapter in that it is essentially a virtual MIMD multiprocessor system (and probably runs on multiple physical multiprocessor systems as well). Early forms of cloud computers were very ad hoc creations, mostly for scientific applications. One well-known community-based virtual cloud computer is the SETI@home project; botnets use similar techniques although, unlike the SETI@home project, they use stolen resources for malicious and criminal purposes. Early workgroup-level cloud computers include the Beowulf computer [17] and Network of Workstations (NOW) [1], but there have been many variations on this theme.

Modern cloud computing is often done in large server farms and uses the notion that computer systems can be aggregated to form a loosely coupled virtual multiprocessor system that can be shared across multiple users. Cloud-computing systems can provide either direct or indirect access to a virtual computer system using SISD, MIMD, or even GP-GPU systems. With direct access, users are provided with their own virtual multiprocessor system and associated administration, application, and development software for their use. With indirect access, users are provided with access to specific services through a service-oriented architecture (SOA) that is an application provided by the cloud-computing infrastructure.

One advantage of cloud computing is that the physical computer systems assigned to a customer can be scaled and configured to adapt to customer demands; they may also be distributed geographically so that the physical computer systems are located near customer centers or low-cost power sources. Another advantage is that physical computer systems are shared among many customers so that the customers do not have to purchase and maintain their own cluster. The individual computer systems may also have diverse functionally (in both performance capabilities and software resources) to provide both cost and performance flexibility and to allow tailoring the compute resources used to each application's computation requirements.

23.7 Afterward

In this chapter, we have reviewed a number of different parallel architectures organized by the stream model. We have described some general characteristics that offer some insight into the qualitative differences between different parallel architectures but, in the general case, provide little quantitative information about the architectures themselves—this would be a much more significant task although there is no reason why a significantly increased level of detail could not be described. Any model, such as the stream model, is merely a way to categorize and characterize processor architectures and if it becomes too detailed and complicated, then it may lose its benefits; the stream model's main benefits are its high-level nature and simplicity. However, there is a place for more detailed models as well; for a more comprehensive characterization of processor characteristics, see Blaauw and Brooks [3] or Bell and Newell [2].

Just as the stream model is incomplete and overlapping (consider that a vector processor can be considered to be a SIMD, MISD, or SISD processor depending on the particular categorization desired), the characteristics for each class of architecture are also incomplete and overlapping. However, the insight gained from considering these general characteristics leads to an understanding of the qualitative differences between gross classes of computer architectures, so the characteristics that we have described provide similar benefits and liabilities.

This is not meant to imply that the aggregate of the stream model along with the relevant characteristics is a complete and formal extension to the original taxonomy—far from it. There are still a wide range of processors that are problematic to describe well in this (and likely in any) framework. Whatever the problems with classifying and characterizing a given architecture, processor architectures, particularly multiprocessor architectures, are developing rapidly. Much of this growth is the result of significant improvements in compiler technology that allow the unique capabilities of an architecture to be efficiently exploited by the programmer with more and more assistance by the compiler.

In many cases, the design (or, perhaps more properly, the success) of a system is based on the ability of a compiler to produce code for it. It may be that a feature is unable to be utilized if a compiler cannot exploit it and thus the feature is wasted (although perhaps the inclusion of such a feature would spur compiler development—there is the chicken-and-egg conundrum). It may also be that an architectural feature is added specifically to support a capability that a compiler can easily support (or already supports in another processor), and thus performance is immediately improved. And in the future, it may be that dynamic code optimization or translation could exploit these new features even in code that was not compiled to take advantage of them; however, that capability is still a research effort. Whether for producing great code or for taking advantage of new features, compiler development is clearly an integral part of system design and architectural effectiveness is no longer limited only to the processor itself. System models must keep up with architecture development and be updated appropriately and new models must be developed to account for these new architectures; these models may include software aspects as well to account for the close coupling between hardware and software that makes these new architectures function to their fullest potential.

Defining Terms

Array processor: An *array processor* is an SIMD processor that executes a single instruction stream on many interconnected that are execute in lockstep. Data are distributed across the array and PE processed in parallel by all the PE.

Bandwidth: *Bandwidth* represents the rate (volume) of computation or data passing through the system or interconnect.

Bypass: A *bypass* is the mechanism used in a pipelined processor to route the result of an operation or instruction to a dependent operation or instruction before the result is available for reading from its register or memory location.

Cache coherency: See *Memory coherency.*

Chaining: *Chaining* is used in vector processors where the results between dependent operations are bypassed element by element from one FU to another. The end result is that compound operations have a latency only incrementally longer than an individual operation resulting in significantly increased execution bandwidth.

Core: A *core* is a logic structure that comprises decode and control logic in addition to PE and provides the ability to execute an instruction stream independently from other cores. As used in this chapter, a core is distinguished from a processor in that a core does not have the communications and peripheral logic to integrate into a system. Cores are the primary logic structures in a processor and include one or more PE.

Dependency: *Dependency* describes the relationship between operation and instruction where the results produced by one instruction are consumed by a subsequent instruction. Dependent instructions must be executed sequentially.

Distributed shared-memory architecture: A *DSM* is a multiprocessor system where the memory is distributed across all processors; although the memory is shared, the distributed nature of the memory results in having a NUMA latency that makes thread and data placement a significant factor in realized performance.

Exception: An *exception* refers to an unexpected event produced internal to the core or processor as a side effect of executing operations or instructions; in contrast, an interrupt is produced external to the core or processor.

Instruction: An *instruction* is the machine-readable representation of one or more independent operations that may be executed concurrently by a core or PE. Although often used indistinguishably with instruction (because most processors use instructions containing only one operation), as used in this chapter, an instruction is distinguished from an operation by being the container of operations rather than the specification of the computation to be performed.

Instruction-level parallelism: ILP is when multiple operations or instructions are extracted from the instruction stream and executed concurrently; it differs from pipelined parallelism in that not only is the overhead overlapped but also the execution as well. Operation or instruction execution may be performed in order or out of order depending on the processor capabilities.

Interrupt: An *interrupt* refers to an unexpected event produced external to the core or processor; it may come from a peripheral or another processor. In contrast, an exception occurs internal to the core or processor as a result of executing operations or instructions.

Latency: *Latency* is the time required to complete an action. For execution of an operation or instruction, latency is the time required from the initial fetch to the successful completion of the operation or instruction; execution results may be available prior to the end of the execution process through bypassing. For memory, latency is the time required to make the request to memory and to receive the result.

Memory coherency: *Memory coherency* is the programmer-invisible mechanism to ensure that all PE see the same value for a given memory location and is normally associated with how caches behave. It is also known as cache coherency.

Memory consistency: *Memory consistency* is the programmer-visible ordering effect of memory references both within a core and between different cores and processors in a multicore or multiprocessor system.

Message-passing architecture: A *message-passing architecture* is a multiprocessor system that uses messages for all communications between processors (or other computer systems) including sharing data as well as synchronization.

Multicore processor: A *multicore processor* comprises a number of cores that share communications and peripheral logic to integrate into a system.

Multiprocessor: A *multiprocessor* comprises multiple interconnected processors.

Nonuniform memory access: An *NUMA* architecture has memory access latencies that are a factor of the location of the processor and referenced data within the system.

Operand: An *operand* specifies the constant value or a storage location—typically either a register or a memory location—that provides data that are used in the execution of the operation or that receives data from the results of an operation.

Operation: An *operation* is the specification of the computation to be performed by a core or PE. It specifies the computation to be performed as well as the operands and other details used to produce the desired computation result. Although often used indistinguishably with operation, as used in this chapter, an instruction is distinguished from an operation by being the container rather than the specification of the computation to be performed.

Pipelining: *Pipelining* is the mechanism that partitions the execution of an instruction into independent phases that be performed in sequence by a core or PE. In addition, these phases can be overlapped with phases from other instructions. Because of overhead, the latency of an individual instruction may be increased, but pipelining improves the resource utilization and execution bandwidth in the core or PE. When there is a dependence or conflict between two instructions, the later instruction is delayed (or stalled) until the dependence or conflict is resolved.

Processor: A *processor* is a stand-alone element in a computer system. As used in this chapter, a processor is distinguished from a core in that a processor has the communications and peripheral logic to integrate into a system with other processors, memory, and other devices. Processors include one or more cores along with other resources that add additional features and functionality to the processor.

Processor element: A *PE* is a logic structure that provides computation functions to a processor. As used in this chapter, a PE is distinguished from a core in that a PE does not have the decode and control logic to give it the ability to execute an instruction stream independently. PE are the primary logic structures in SIMD and GP-GPU processors.

Shared-memory architecture: A *shared-memory architecture* is a multiprocessor system using shared memory for all communications between processors. Memory may be located anywhere in the system and accessed by all interconnected processors.

Stall: A *stall* is the delay inserted during execution so that an operation or instruction is that is dependent on (or conflicts with) a previous operation or instruction is properly executed.

Superscalar: *Superscalar* describes an instruction-level parallel processor that dynamically analyzes the instruction stream and executes multiple operation or instruction as dependencies and operand availability allow. Most processors today are superscalar processors and have one operation per instruction.

Thread: A *thread* is an independent instruction stream that includes its own unique execution context. Although related, hardware threads and software threads are at significantly different abstraction levels. Hardware threads are low-level threads that are managed and executed directly by hardware; software threads are high-level threads that have much more information on how they relate to each other including thread priority, inter-thread dependencies, and data sharing.

Uniform memory access: A *UMA* architecture has memory access latencies that are constant independent of the location of the processor and referenced data within the system.

Vector processor: A *vector processor* is an SIMD processor that executes a single instruction stream on vector objects computing element-by-element on deeply pipelined FU. Operations may be chained to reduce the *latency* of compound operations.

Very long instruction word: *VLIW* describes an instruction-level parallel processor that relies on the compiler to statically analyze the program and to produce an instruction stream where the operations in every instruction are dependence-free. The hardware executes all of the operations in an instruction in parallel without further dependence analysis.

Further Information

There are many good sources of information on different aspects of parallel architectures. The references for this chapter provide a selection of of texts that cover a wide range of issues in this field. There are many professional journals that cover different aspects of this area either specifically or as part of a wider coverage of related areas. Some of these are the following:

- *IEEE Transactions on Computers, Transactions on Parallel and Distributed Systems, Computer, Micro*
- *ACM Transactions on Computer Systems, Computing Surveys*
- *Journal of Supercomputing*
- *Journal of Parallel and Distributed Computing*
- *Journal of Instruction-Level Parallelism*

There are also a number of conferences that deal with various aspects of parallel processing. The proceedings from these conferences provide a current view of research on the topic. Some of these are the following:

- *International Symposium on Computer Architecture (ISCA)*
- *Supercomputing (SC)*
- *International Symposium on Microarchitecture (MICRO)*

- *International Conference on Parallel Processing (ICPP)*
- *International Symposium on High Performance Computer Architecture (HPCA)*
- *Symposium on the Frontiers of Massively Parallel Computation*

References

1. T.E. Anderson, D. E. Culler, D. A. Patterson, and the NOW Team. A case for networks of workstations: NOW. *IEEE Micro*, 15(1):54–64, February 1995.
2. C.G. Bell and A. Newell. *Computer Structures: Readings and Examples*. McGraw-Hill, New York, 1971.
3. G.A. Blaauw and Jr. F. P. Brooks. *Computer Architecture: Concepts and Evolution*. Addison-Wesley Professional Reading, MA, 1997.
4. D.E. Culler, J.P. Singh, and A. Gupta. *Parallel Computer Architecture: A Hardware/Software Approach*. Morgan Kaufmann, San Francisco, CA 1999.
5. J.B. Dennis. Data flow ideas for supercomputers. In *Twenty-Eighth IEEE Computer Society International Conference*, San Francisco, CA, pp. 15–20, February 1984.
6. M.J. Flynn. Very high-speed computing systems. *Proceedings of the IEEE*, 54(12):1901–1909, December 1966.
7. M.J. Flynn. *Computer Architecture: Pipelined and Parallel Processor Design*. Jones and Bartlett, Boston, MA, 1995.
8. J.L. Hennessy and D.A. Patterson. *Computer Architecture A Quantitative Approach*. Morgan Kaufmann, San Francisco, CA, fifth edition, 2011.
9. H.P. Hofstee and A.K. Nanda, eds. *IBM Journal of Research and Development*, 51/5., 1997.
10. HyperTransport Technology Consortium. *HyperTransport™ I/O Link Specification*, 2010. http://www.hypertransport.org/docs/twgdocs/HTC20051222-0046-0035.pdf
11. Intel Corporation. *An Introduction to the Intel® QuickPath interconnect*, 2009. http://www.intel.com/content/dam/doc/white-paper/quick-path-interconnect-introduction-paper.pdf
12. R. Jain. *The Art of Computer Systems Performance Analysis: Techniques for Experimental Design, Measurement, Simulation, and Modeling*. Wiley-Interscience, New York, 1991.
13. D.B. Kirk and W.m.W. Hwu. *Programming Massively Parallel Processors: A Hands-On Approach*. Morgan Kaufmann, San Francisco, CA, 2010.
14. H.T. Kung and C.E. Leiserson. Algorithms for VLSI processor arrays. In *Introduction to VLSI Systems*. Addison-Wesley, Reading, MA, 1979.
15. D.J. Lilja. *Measuring Computer Performance: A Practitioner's Guide*. Cambridge University Press, Cambridge, NY 2000.
16. O. Pell and V. Averbukh. Maximum performance computing with dataflow engines. *Computing in Science Engineering*, 4(4):98–103, July–August 2012.
17. D. Ridge, D. Becker, P. Merkey, and T. Sterling. Beowulf: Harnessing the power of parallelism in a pile-of-PCs. In *IEEE Aerospace Conference Proceedings*, 79–91, 1997.
18. K.W. Rudd. *VLIW Processors: Efficiently Exploiting Instruction Level Parallelism*. PhD thesis, Stanford University, December 1999.
19. J. Sanders and E. Kandrot. *Cuda By Example: An Introduction to General-Purpose GPU Computing*. Addison–Wesley Professional, Upper Sadde River, NJ 2010.
20. TOP500 supercomputer site. http://www.top500.org, accessed on November 27, 2012.
21. P.H. Wang, H. Wang, J.D. Collins, Ed Grochowski, R.M. Kling, and J.P. Shen. Memory latency-tolerance approaches for Itanium processors: Out-of-order execution vs. speculative precomputation. In *HPCA '02 Proceedings, Eighth International Symposium on High Performance Computer Architecture*, pp. 187–196, February 2002.

24

Multicore Architectures and Their Software Landscape

Raphael Poss
University of Amsterdam

24.1 Introduction

In the decade 1990–2000, processor chip architectures have benefited from tremendous advances in manufacturing processes, enabling cheap performance increases from both increasing clock frequencies and decreasing gate size on silicon. These advances in turn enabled an explosive expansion of the software industry, with a large focus on computers based on general-purpose uni-processors. This architecture model, which is of the Von Neumann computer, had emerged at the end of the 1980s as the de facto target of all software developments.

Until the turn of the twenty-first century, system engineers using uni-processors as building blocks could assume ever-increasing performance gains, by just substituting a processor by the next generation in new systems. Then they ran into two obstacles. One was the *memory wall* [45], that is, the increasing divergence between the access time to memory and the execution time of single instructions. To overcome this wall, architects have designed increasingly complex uni-processors using techniques such as branch predictors and out-of-order execution OoOE to automatically find parallelism in single-threaded programs and keep processor pipelines busy during memory accesses. The second is the *sequential performance wall* [1,34], also called "Pollack's rule" [31], that is, the increasing divergence in single processors between performance gains by architectural optimizations and the power-area cost of these optimizations.

To "cut the Gordian knot," in the words of Ronen et al. [34], the industry has since (post-2000) shifted toward multiplying the number of processors on chip, creating increasingly larger chip multiprocessors (CMPs) by processor counts, now called *cores*. The underlying motivation is to *exploit explicit concurrency* in software and distribute workloads across multiple processors to increase performance. The responsibility to find parallelism was pushed again to the software side, where it had been forgotten for 15 years.

During the period 2000–2010, this shift to multicore chips has caused a commotion in those software communities that had gotten used to transparent frequency increases and implicit instruction-level parallelism (ILP) for sequential programs without ever questioning the basic machine model targeted by programming languages and complexity theory. "The free lunch is over" [40], and software ecosystems then had to acknowledge and understand explicit on-chip parallelism and energy constraints to fully utilize current and future hardware.

This transition was disruptive for audiences used to systems where the processor fetches instructions one after another following the control flow of *one program*. Yet the commotion was essentially specific to those traditional audiences of general-purpose uni-processors that had grown in the period 1990–2000. In most application niches, application-specific knowledge about available parallelism had long mandated dedicated support from the hardware and software toward increased performance: scientific and high-performance computing (HPC) have long exploited dedicated single instruction, multiple data (SIMD) units; embedded applications routinely specialize components to program features to reduce logic feature size and power requirements; and server applications in data center have been optimized toward servicing independent network streams, exploiting dedicated I/O channels and hardware multithreading (HMT) for throughput scalability. Moreover, a host of research on parallel systems had been performed in the previous period, up to the late 1980s, and best practices from this period are now surfacing in the software industry again.

In the rest of this chapter, we review the development of multicore processor chips during the last decade and their upcoming challenges.

24.2 Underlying Principles

24.2.1 Multicore Architecture Principles

Two observations from circuit design have motivated the transition to multicore processor chips.

As noted by Bell et al. [4], the scalability of multiple-instruction issue in conventional processors is constrained by the fact that ILP is not improved linearly with the addition of silicon. Scaling up implicit concurrency in superscalar processors gives very large circuit structures. For example, the logic required for out-of-order issue scales quadratically with issue width [22] and would eventually dominate the chip area and power consumption. This situation had been summarized by Pollack [31] who stated that the performance of a single core increased with the square root of its number of transistors.

Meanwhile, the power cost of single-core ILP is disadvantageous. Not only does Pollack's rule suggest more power consumption due to the growing silicon cost per core, to increase the instructions per second (IPS) count, the processor's clock frequency must also increase. As noted in [34], maximum power consumption (*Power*) is increased with the core operating voltage (V) and frequency (F) as follows:

$$Power = C \times V^2 \times F$$

where C is the effective load capacitance of all units and wires on core. Within some voltage range, F may go up with supply voltage V ($F = k \times V^{\alpha-1}$, $\alpha \leq 1$). This is a good way to gain performance, but power is also increased (proportional to $V^{2+\alpha}$). For a given core technology, this entails that a linear increase in IPS via frequency scaling requires at least a quadratic increase in power consumption.

From this circuit perspective, the advantage of explicit parallelism by investing transistor counts toward multiple, simpler cores becomes clear: assuming available concurrency in software, two cores running at half the frequency can perform together the same IPS count at less than half the power usage. Moreover, by keeping the cores simple, more transistors are available to increase the core count on chip and thus maximum IPS scalability.

This perspective also reveals the main challenge of multicore chips: the purported scalability is strongly dependent on the ability of software to exploit the increasing core counts. This is the issue of programmability, covered in the rest of this chapter.

Beyond issues of power and performance, another factor has become visible in the last decade: fault management.

Both transient and permanent faults can be considered. Transient faults are caused mostly by unexpected charged particles traversing the silicon fabric, either emitted by atomic decay in the fabric itself or its surrounding packages, or by cosmic rays, or by impact from atmospheric neutrons; as the density of circuits increases, a single charged particle will impact more circuits. Permanent faults are caused by physical damage to the fabric, for example, via heat-induced stress on the metal interconnect or atomic migration. While further research on energy efficiency will limit heat-induced stress, atomic migration unavoidably entails loss of function of some components over time. This effect increases as the technology density increases because the feature size, that is, the number of atoms per transistor/gate, decreases.

To mitigate the impact of faults, various mechanisms have been used to hide faults from software: redundancy, error correction, etc. However, a fundamental consequence of faults remains: as fault tolerance kicks in, either the *latency changes* (e.g., longer path through the duplicated circuit or error correction logic) or the *throughput changes* (e.g., one circuit used instead of two).

To summarize, the increasing number of faults is a source of unavoidable *dynamic heterogeneity* in larger chips. Either components will appear to enter or leave the system dynamically, for example, when a core must stop due to temporary heat excess, or their characteristics will appear to evolve over time beyond the control of applications. This in turn requires to evolve the abstract model of the chip that programmers use when writing software.

Exposing the chip's structure in abstract models as a network of loosely coupled cores, that is, a *distributed system on chip*, instead of a tightly coupled "central processing unit," will facilitate the management of this dynamic heterogeneity in software.

24.2.2 Multicore Architecture Models

Any computing system today is a composition of memories, caches, processors, I/O interfaces, and data links as "basic blocks." A given architecture is a concrete assembly of these components. In contrast, an *architecture model* is an abstract description of how components are connected together, to capture the general properties common to multiple concrete architectures. Architecture models are in turn characterized by their *typology*, their *topology*, and the *concurrency management primitives* they expose at the hardware/software interface.

The *typology* reports which different types of components are used. In multicore chips, we can distinguish *homogeneous* or "symmetric" designs, where all repeated components have the same type, from *heterogeneous* designs that may use, for example, general-purpose cores in combination with on-chip accelerators, cores with different cache sizes, or cores with different instruction sets. Historically, homogeneous designs have been preferred as they were simpler to program: most multicores prior to 2010 were so-called symmetric multiprocessor (SMP) systems. However, some heterogeneity has since become mandatory: not only as a way to manage faults and energy usage, as highlighted earlier, but also to increase execution efficiency for fixed applications by specializing some functional units to the most demanding software tasks.

The *topology* indicates how components are connected together. In multicores, the topology determines how cores can communicate with each other and the latency and bandwidth of communication. We can distinguish three design spectra for topologies.

The first is the *memory topology*, that is, how cores are connected to main memory. At one end of this spectrum, "UMAs" describe interconnects where a single memory is shared symmetrically by all cores and hence provides homogeneous latency/bandwidth constraints. At the other end, individual or groups of cores have their own local memory, and remote accesses become orders of magnitude more expensive than local accesses. This is where one finds "NUMAs." The position of any specific architecture on this spectrum is a cost trade-off: UMAs are simpler to program but require more silicon and energy to provide increased bandwidth to the shared memory.

The second spectrum is the *cache topology*, that is, how caches are connected to cores, memory, and other caches. At one end of this spectrum, congruent with UMAs, the cache topology can be represented as a *cache tree*, where the common memory forms the root of the tree and the cores form the leaves. This topology fully preserves the notion of "cache hierarchy" from the perspective of individual cores, where there is only one path to main memory and local performance is determined only by the hit and miss rates at each level of requests emitted locally. At the other end, congruent with NUMAs, more diversity exists. With *distributed coherent cache* architectures, a network protocol between caches ensures that updates by one part of the system are consistently visible from all other parts. These are also called cache-coherent NUMAs (ccNUMAs) or distributed shared memory (DSM) architectures. Again, the choice is a cost trade-off: cache trees provide a shorter latency of access to main memory on average but cost more silicon and energy to operate than loosely coupled caches.
Example memory and cache topologies are given in Figure 24.1.

The third spectrum is the *inter-core topology*, that is, what direct links are available between cores for direct point-to-point communication. In most multicores, regardless of the memory architecture, a dedicated *signaling network* is implemented to notify cores asynchronously upon unexpected events. The exchange of notifications across this network is commonly called inter-processor interrupt (IPI) delivery. Inter-core networks-on-chip (NoCs) also exist that offer arbitrary data communication between cores; 2D mesh topologies are most common as they are cheap to implement in tiled designs.

Finally, the *concurrency management primitives* determine how software can exploit the hardware parallelism. There are three aspects to this interface: control, synchronization, and data movement.

In most designs that have emerged from the grouping of cores previously designed for single-core execution, such as most general-purpose SMP chips in use today, the interface for control and synchronization is quite basic. For control, cores execute their flow of instructions until either a "halt" instruction

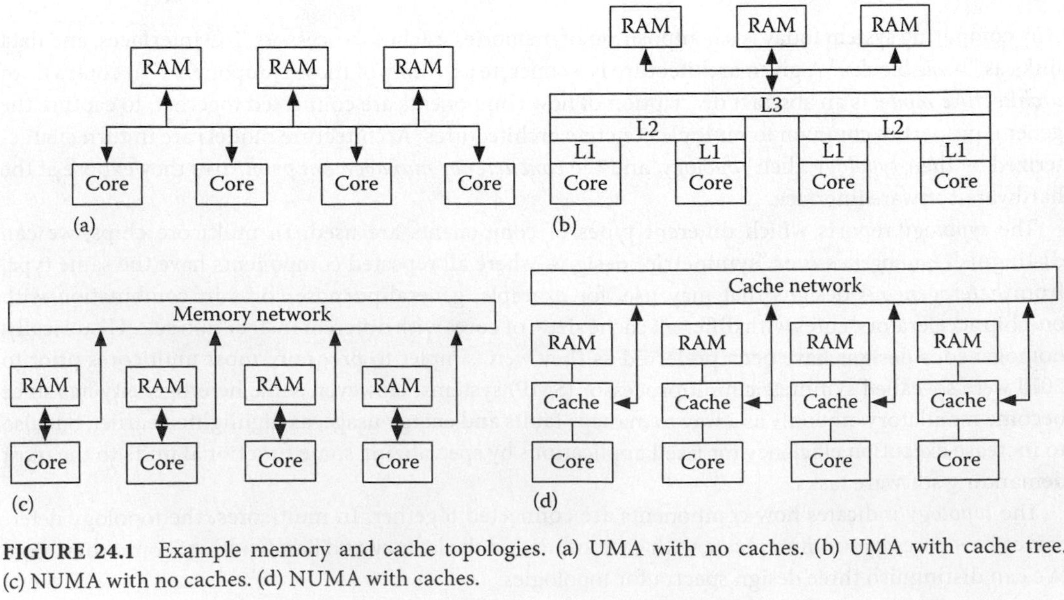

FIGURE 24.1 Example memory and cache topologies. (a) UMA with no caches. (b) UMA with cache tree. (c) NUMA with no caches. (d) NUMA with caches.

is encountered or an IPI is delivered that stops the current instruction flow and starts another. The only primitive for inter-core control is IPI delivery and regular load/store operations to a shared memory. In this context, the only synchronization mechanisms available to software are either *busy loops* that access a memory location until its contents are changed by another core or *passive waiting* that stops the current core, to be awakened by a subsequent IPI from another core. Additional mechanisms may be present but are still unusual. For example, *hardware primitives for synchronization barriers* may be available, whereby two or more cores that execute a barrier will automatically wait for one another.

For data movement, the near universal primitives are still memory load and store instructions: using a shared memory, programs running across multiple cores can *emulate virtual channels by using buffer data structures* at known common locations. In addition to loads and stores, dedicated *messaging primitives* may exist to send a data packet to a named target core or wait upon reception of a packet, although they are still uncommon.

In any case, whichever primitives are available are typically abstracted by the operating systems in software to present a standardized programming interface to applications, such as those described in the remainder of this chapter. Thanks to this abstraction layer, most of the diversity in concurrency management interfaces is hidden to application programmers. However, it is still often necessary to obtain knowledge about which underlying primitives are provided by an architecture to understand its cost/performance trade-offs.

24.2.3 Multicores with Intra-Core Parallelism

Independently and prior to the introduction of multicores, architects had enhanced individual cores to offer internal parallelism. The purpose of internal parallelism is to increase utilization of the processor pipeline, by enabling the overlap of computations with waiting instructions such as I/O or memory accesses. When cores with internal parallelism are combined together to form a multicore chip, two scales of parallelism exist and their interaction must thus be considered.

24.2.3.1 Out-of-Order Execution

In processors with OoOE, instructions wait at the start of the pipeline until their input operands are available, and ready instructions are allowed to enter the pipeline in a different order than program order. Result order is then restored at the end of the pipeline. The key concept is that missing data operands do not prevent the pipeline from executing unrelated, and ready instructions, and *utilization* is increased. For more details, see [12, Chapters 2 and 3] and the chapter on performance enhancements in this volume.

OoOE introduces new challenges for multicore synchronization. For example, a common idiom is to place a computation result into a known memory cell, then write a flag "the result has been computed" into another. If memory stores are performed in order, another core can perform a busy loop, reading the flag until it changes, with the guarantee that the computation result is available afterwards. With OoOE, this pattern is invalidated: although the *program* on the producer core specifies to write the result and only then write the flag, instruction reordering may invert the two stores.

To address this type of situation, new primitives must be introduced to *protect the order of externally visible side effects* in presence of OoOE. The most common are memory barriers, or *fences*. These must be used by programs explicitly between memory operations used for multicore synchronization. When the processor encounters a fence, it will block further instructions until the memory operations prior to the fence have completed. This restores the effect order required by the program, at the expense of less ILP in the pipeline and thus lower utilization.

24.2.3.2 Hardware Multithreading

The key motivation for multithreading in a single core is to exploit the waiting time of blocked threads by running instructions from other threads [33,35]. This is called thread-level parallelism (TLP); it can tolerate longer waiting times than ILP. To enable this benefit of TLP even for small waiting times like individual memory loads or floating-point unit (FPU) operations, multithreading can be implemented in hardware (HMT). With HMT, a processor core will contain multiple program counters (PCs) active simultaneously, together with independent sets of physical registers for each running hardware thread. The fetch unit is then responsible for feeding the core pipeline with instructions from different threads over time, switching as necessary when threads become blocked [23,38,39,41,42].

Because each hardware thread executes an independent instruction stream via its own PC, operating systems in software typically register the hardware threads as independent *virtual processors* in the system. Subsequently, from the perspective of software, care must be taken to distinguish virtual from hardware processors when enumerating hardware resources prior to parallel work distribution. Indeed, when work is distributed to two or more hardware threads sharing the same core pipeline, performance can only increase *until all waiting times in that pipeline are filled with work*. Once a pipeline is fully utilized, no more performance can be gained with hardware threads on that core even though there may be some idle hardware threads available.

24.2.4 Programming Principles

As with all parallel computing systems, multicore programming is ultimately constrained by Amdahl's and Gustafson's laws.

In [2], Amdahl explains that the performance of one program, that is, its time to result or *latency*, will stay fundamentally limited by its longest chain of dependent computations, that is, its *critical path*, regardless of how much platform parallelism is available. The first task of the programmer is thus to shorten the critical path and instead expose more concurrent computations that can be parallelized. When the critical path cannot be shortened, the latency cannot be reduced further with parallelism. However, Gustafson's law [11] in turn suggests that the problem sizes of the parallel sections can be expanded instead, to increase use of the available parallelism and increase *throughput*, that is, computations per second, at constant latency.

Within these boundaries, software design for multicores involves the following concerns:

- Programmers and software frameworks *expose concurrency* in applications. This activity takes two forms. An existing program is *relaxed from ordering constraints* to add concurrency; for example, a sequential loop may be annotated to indicate it can be carried out in parallel. Alternatively, new code can be composed from *concurrent building blocks*, such as primitive map/reduce operators. This activity typically occurs statically, during software development.
- Meanwhile, software frameworks and operating systems *map and schedule program concurrency to the available hardware parallelism*. This activity typically occurs at run-time, to carry out program execution over the available cores.

The connection point between these two activities is the *parallel programming model*. Different languages or software libraries will offer different programming models; each offers both *programming abstractions* toward programmers to specify "what to do" and *operational semantics* that provide an intuition of "how the program will behave" at run-time. Parallel programming models typically diverge from traditional programming models in that they avoid letting the programmer specify "how" to carry out computations, so as to give maximum flexibility to the underlying platform.

Parallel programming models can be categorized along two axes, illustrated in Figure 24.2, depending on how they expose computations and communication.

In one corner, *fork–join parallelism* and *bulk-synchronous parallelism (BSP)* [44] are the most common. With fork–join, a program exposes concurrency by specifying at which points separate threads can start (fork) to compute separate parts of a computation; sequence is then enforced by expressing synchronization on termination (join) of previously created threads. In the higher-level variant BSP, an overall repetitive computation is expressed as a sequence of wide parallel sections, with computations

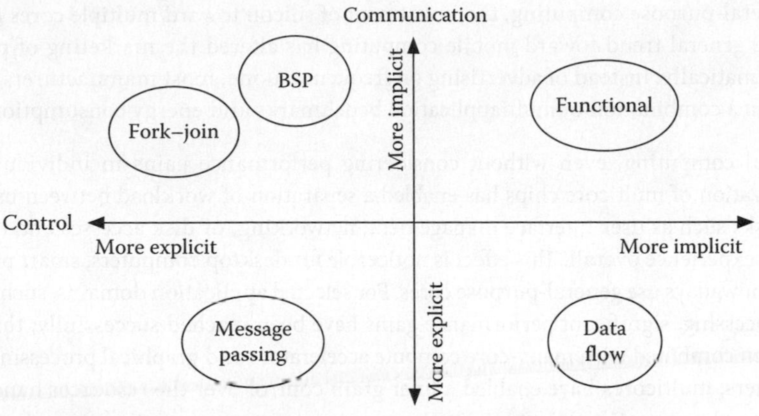

FIGURE 24.2 High-level classification of parallel programming models.

occurring during the parallel section and communication occurring during the synchronization step. With these models, control is specified by the structure of the program, while typically communication is left implicit.

At the corner of explicit control and communication, *message passing* describes a collection of idioms where a program specifies independent processes that act collectively by exchanging data across explicit channels. Message passing stems from decades of research over communicating processes [13,25–27]. It encompasses the most versatile techniques to program multicores, at the expense of a more explicit and error-prone model for programmers.

Diagonally opposed, *functional programming models* encompass the programming style of pure functional languages, where programmers are stimulated to specify only the input–output data relationships of their algorithms using symbolic operations, without introducing or assuming knowledge about the execution environment. These models provide the most flexibility for the platform by removing programmer involvement from mapping and scheduling concurrency, although it thus incurs the expense of a greater technology challenge to fully optimize execution.

Finally, *data-flow programming models* are related to functional models in that the programmer does not specify how to carry out computations. However, whereas with functional programs the data relationships may be implicitly carried by using references (pointers) in data structures, data-flow programs make data edges between computations explicit. This in turn simplifies the mapping of communication to the platform's topology.

A given programming language or software library may expose multiple programming models. The choice of one over another in applications is mostly driven by the trade-off between programmer productivity, technological maturity of the platform, and performance: more implicit models are easier to program with but more difficult to manage for the platform.

24.3 Impact on Practice

After the HPC and general-purpose computing industries stumbled on the sequential performance wall in the last decade, the move to multicore chips has enabled the following breakthroughs:

- For HPC, the grouping of multiple cores on the same chip has enabled an increase of interconnect bandwidths by an order of magnitude, in turn enabling yet higher throughput scalability of supercomputers. As of 2012, nearly all systems from the supercomputer TOP500* are based on multicore chips.

* http://top500.org/.

- For general-purpose computing, the investment of silicon toward multiple cores in combination with the general trend toward mobile computing has altered the marketing of processor products dramatically: instead of advertising on frequency alone, most manufacturers now effectively report on a combination of multiapplication benchmarks and energy consumption.

For personal computing, even without considering performance gains in individual applications, the democratization of multicore chips has enabled a separation of workload between user applications and system tasks such as user interface management, networking, or disk accesses and thus results in a smoother user experience overall. This effect is noticeable for desktop computers, smart phones, and tablets, which all nowadays use general-purpose cores. For selected application domains, such as video games and image processing, significant performance gains have been reached successfully; this effect is even multiplied when combined with many-core compute accelerators and graphical processing units (GPUs).

In data centers, multicores have enabled a finer grain control over the resources handed over to client applications. By separating client applications over separate cores, interference between clients is reduced and resource billing is simplified (resources can be billed per unit of space and time, instead of per actual workload that is more difficult to compute). Some throughput increases have also been possible thanks to multicores, most noticeably for database servers and networked processes such as web servers. However, the preferred way to increase throughput in data centers is still to extend the number of network nodes instead of replacing existing nodes by new nodes with larger core counts.

Meanwhile, multicore processors are a success story for embedded systems, where they are commonly called multiprocessor systems-on-chip (MPSoCs). In these systems, processor chips are typically co-designed with software applications; different cores are implemented on chip to support specific application components. For example, a mobile phone processor chip may contain separate cores for managing wireless networks, encoding/decoding multimedia streams and general-purpose application support. By specializing a processor design to an application, hardware and energy costs are reduced while taking advantage of the parallelism to improve performance. While the embedded landscape is still much focused on application-specific chip designs, it is expected that the field's expertise with mapping and managing application components over heterogeneous resources will propagate to all other uses of multicores during the upcoming decade.

24.3.1 Current Landscape of Multicore Chips

All major technology vendors now have multicore product offerings and continue to invest towards increasing core counts.

After introducing its first mainstream dual-core offerings around 2005 via its Core and Xeon product lines, Intel's processors now nearly all feature a minimum of 2–4 cores. As of this writing, the most popular Core i5 and i7 processors, based on the Sandy Bridge micro-architecture [46], feature 4 cores running at about 3 GHz. These chips use 32 nm silicon technology and nearly a billion transistors. They integrate an on-chip GPU accelerator with 6–12 additional compute cores; the general-purpose cores also optionally feature Hyper-Threading [23] for a maximum of 2 hardware threads per core. The next generation based on the Ivy Bridge micro-architecture is expected to offer similar features at a reduced silicon and power budget.

Meanwhile, AMD has reduced its competition push on single-threaded performance and pushed its multicore road map further. Its Opteron product line currently contains chips with up to 12 general-purpose cores based on the Bulldozer micro-architecture [8]. Frequencies range from 1 to 3 GHz; the chips also use the 32 nm technology. The upcoming Fusion product line, which uses general-purpose cores based on both Bulldozer and the new Bobcat micro-architecture [7], invests more silicon real estate towards on-chip accelerator cores (e.g., 80 in Brazos chips, 160–400 in Lynx/Sabine chips).

On the embedded/mobile computing market, ARM leads the way towards the generalization of multicore platforms. The most licensed architecture towards general-purpose applications is the

Cortex-A design, now available with 2 to 4 cores on chip and frequencies up to 2 GHz in its "MPCore" variant. The most visible user of Cortex-A is currently Apple, which equips its smart phone and tablet offerings with its own A4 and A5 chips based on Cortex-A. The upcoming ARM design Cortex-A15 is planned to feature up to 8 cores on chip, together with optional on-chip accelerators depending on vendor requirements.

On the server market, Oracle (previously Sun Microsystems) has stepped forward with its Niagara [21] micro-architecture. Niagara processors combine multiple cores with HMT, resulting in high core counts per chip: the most recent product, the SPARC T4 [37], exposes 64 hardware threads to software. Although Niagara was previously advertised for throughput due to its lower initial single-thread performance, the latest generations running around 3 GHz with OoOE now compete across all general-purpose workloads.

As can be seen in this overview, the trends suggest a continued increase of core counts on the main processor chip, together with the integration of accelerators. However, separately packaged many-core accelerator chips are still being developed. The two major vendors are nowadays NVidia and AMD, the latter having acquired ATI in 2006. With clock frequencies below 1 GHz but core counts in the hundreds, their accelerator chips deliver orders of magnitude higher peak floating-point performance than general-purpose cores with twice the frequency at the same generation. The main challenge to these designs is bandwidth to memory, where communication between the accelerator and the main processor chip is constrained by the inter-chip system bus. This bottleneck constitutes the main push towards integration with general-purpose cores on the same silicon die.

24.3.2 Shared Memory Multiprogramming for Multicores

For shared memory multiprogramming, the common substrate underlying programming languages and libraries is constituted by *threads*, directly inherited from the era of time sharing on uni-processors. On multicores, threads execute simultaneously instead of interleaved, but these two abstractions remain otherwise identical to their original definition: programs create or *spawn* threads, then the operating system selects cores to execute the workload. The leading low-level application programming interfaces (APIs) to manage threads are the POSIX interface [15] ("pthreads") and the Java virtual machine interface. To enable more fine-grained control over thread-to-core mappings, some operating systems also offer APIs to *pin* threads to specific cores (e.g., `pthread_setaffinity`). These basic interfaces are there to stay: as of 2011, new standards for the C [18] and C++ [17] languages have been published that integrate a native standard threading API similar to POSIX.

Upon these basic interfaces, different programming languages and libraries provide different parallel programming models for application developers. With less than a decade of renewed interest in multiprogramming, these software frameworks for multicores have not yet stabilized and a large diversity of approach still exists across vendors, hardware platforms, and operating systems.

For performance-oriented applications, the leading interfaces are currently OpenMP [29] for C, C++, and FORTRAN and Intel's Threading Building Blocks (TBB) [16,32] for C++. OpenMP is specialized towards the parallelization of existing sequential code by the addition of annotations, or *pragmas*, to indicate which portions of code can be executed concurrently. An example is given in Figure 24.3: A loop is annotated to declare it can be run in parallel when the function "scale" is called; at run-time, the value of "n" is inspected and the workload is distributed across the available cores.

In contrast to OpenMP, TBB is oriented towards the acceleration of new code, where programmers use TBB's control and data structures directly. Primitive constructs are provided for parallel map/reduce, searches, pipelines, sorting algorithm, as well as parallel implementations of container data structures (queues, vectors, hash maps). An example is given in Figure 24.4: The object "`scaler`" is responsible for carrying out the computation over subranges of the array, and TBB ensures that `scaler`'s operator is called in parallel over the available cores.

```
void scale(int n, int a[]) {
    #pragma omp parallel for
    for (i = 0; i < n; i++)
        a[i] = 2 * i;
}
```

FIGURE 24.3 Example OpenMP program fragment.

```
struct scaler {
   vector<int>& a_;
   scaler(vector<int>& a) : a_(a) {}
   void operator()(const blocked_range<size_t>& r) const {
      for (size_t i = r.begin(); i != r.end(); ++i)
         a_[i] = 2 * i;
   }
};
void scale(int n, vector<int>& a) {
   parallel_for(blocked_range<size_t>(0, n), scaler(a));
}
```

FIGURE 24.4 Example TBB program fragment.

Both OpenMP and TBB manage program concurrency in a similar fashion. When reached during execution, the program code generated by the compiler for concurrency constructs causes calls to the language run-time systems to define *tasks*. The run-time system in turn runs a *task scheduler* that spreads the tasks defined by the program over *worker threads*, which it has previously configured to match the number of underlying cores. In both interfaces, primitives are available to control the task scheduler and query the number of worker threads.

When accelerators are involved, typically the accelerator cores cannot run regular application code because they do not support recursion or arbitrary synchronization. To program them, it is still customary to use a different set of APIs. The current leading standards are NVidia's CUDA interface [20], specialized towards its own chips, and OpenCL [19] which intends to provide a unified interface to accelerators. With both interfaces, the application programmer defines *computation kernels*, which can execute on the accelerator cores, and uses the interface's API to trigger data movement and computations using kernels.

An example is given in Figure 24.5. As the example suggests, the main challenge of accelerator-based programming is data movement between the accelerator and main memory. While it is often possible to combine accelerator computations and thus reduce the need for communication, many cases exist where the application structure prevents keeping the data on the accelerator's memory. Again, the industry is moving towards tighter integration of accelerators and general-purpose cores on the same chip, in an effort to alleviate this communication overhead.

24.3.3 Distributed Programming on Chip

It is also possible to consider a multicore chip as a network of single-core nodes sharing a very efficient interconnect. Using this model, a multicore chip can be programmed using *explicitly communicating processes* instead of threads communicating implicitly via shared memory.

Abstractions to program multiple processing units using communicating processes had existed for decades and are now coming back with the advent of multicores. The typical interface for scientific computing, coming from the HPC community, is MPI [24]: it exposes a process management and message passing interface to C, C++, FORTRAN, and Java code. Implementations of MPI are nowadays able to distribute workloads over multiple cores in a single chip as well as over a network of nodes. Meanwhile, the advent of networked applications in the last decade has caused a large diversity of other frameworks for inter-process

```
// the following defines the kernel.
__kernel void scale_kernel(__global int *a) {
    size_t i = get_global_id(0);
    a[i] = 2 * i;
}
// the kernel is used as follows:
void scale(int n, int a[]) {
    /* need to copy the data to GPU memory first */
    void *gpu_mem = gcl_malloc(sizeof(int)*n, a,
                    CL_MEM_COPY_HOST_PTR|CL_MEM_READ_WRITE);
    /* then define a range to operate over the data */
    cl_ndrange r = { 1, {0}, {n,0,0}, {0} };
    /* then call the kernel */
    scale_kernel(&r, gpu_mem);
    /* then copy back the data from GPU to main memory */
    gcl_memcpy(a, gpu_mem, sizeof(int)*n);
    /* then release the GPU memory */
    gcl_free(gpu_mem);
}
```

FIGURE 24.5 Example OpenCL program fragment.

message queuing and brokering in business applications: Java Message Service (JMS), Microsoft Message Queuing (MSMQ), WebSphere Message Broker from IBM, and Apache ActiveMQ are examples. These interfaces are nowadays commonly used to drive processes running over separate cores in the same chip.

24.3.3.1 Post-2010: Era of Multiscale Concurrency

While general-purpose programmers have been struggling to identify, extract, and/or expose concurrency in programs during the last decade, a large amount of untapped higher-level concurrency has also appeared in applications, ready to be exploited. This is a consequence of the increasing number of features or *services* integrated into user-facing applications in the age of the Internet and ever-increasing support of computers for human activities. For example, while a user's focus may be geared towards the decoding of a film, another activity in the system may be dedicated to downloading the next stream, while yet another may be monitoring the user's blood nutrient levels to predict when to order food online, while yet another may be responsible for backing up the day's collection of photographs on an online social platform, etc.

Even programs that are fundamentally sequential are now used in applications with high-level concurrency at scales that were unexpected. For example, the compilation of program source code to machine code is mostly sequential as each pass is dependent on the previous pass' output. However, meanwhile entire applications have become increasingly large in terms of their number of program source files; so even though one individual compilation cannot be accelerated via parallelism, it is possible to massively parallelize an entire application build. While this form of parallelism had been known in large enterprise projects, the advent of multicores makes it accessible to any programmer working with commodity platforms.

24.4 Research Issues

As this book gets published, the multicore programming challenge has taken a new form. Explicit concurrency has appeared in software, both from increased understanding by programmers and by new technology in compilers and software run-time systems to discover concurrency automatically in applications. Meanwhile, CMPs now contain dozens of cores for general-purpose computations and accelerators offer hundreds of smaller cores for specialized computing, and the trends suggest at least a tenfold increase before the end of silicon scaling. The main challenge for software engineers is thus now less to find concurrency, but rather to *express* and *map* it efficiently to the available parallel hardware. Meanwhile, the main challenge for architects is to balance the need of software practitioners to manipulate *simple machine models* while providing *scalable* systems. The architecture and software engineering communities have thus started to work together to overcome the following new challenges:

- Choose which *parallel machine models* to communicate to programmers to give them an intuition of the underlying hardware resources.
- Choose which *programming abstractions* to offer through the software stack to describe concurrency in applications.
- Determine how to *schedule concurrency over parallel resources* in operating software.
- For known application or fields, determine how to *codesign hardware and software* so that the hardware parallelism aligns with application concurrency.

24.4.1 Platform Challenge: Communication Costs

As the example from Figure 24.5 illustrates, the cost of communication is becoming a growing design constraint for algorithms. In upcoming multicore chips, the latency to access memory from cores, or even to communicate between cores, will become large compared to the pipeline cycle time. Any nonlocal data access will become a serious energy and time expenditure in computation. This is a new conceptual development compared to the last decade, where processor speed was still the main limiting factor and memory access latencies were kept under control using ever-growing caches.

From the software architect's perspective, this *communication challenge* takes two forms. For one, either programmers or the concurrency management logic in operating software must become increasingly *aware of the topology* of the platform, so as to match the dependencies between application components to the actual communication links present in hardware. This requirement will require new abstractions and investments in programming languages and operating software, since the current technology landscape still mostly assumes SMPs and UMAs. The second aspect is that the *cost of computations* is no more a function only of the number of "compute steps," correlated with CPU time; it must also involve the "communication steps" correlated with on-chip network latencies. This is a major conceptual shift that will require *advances in complexity theory* before algorithm specifications can be correlated to actual program behavior in massively parallel chips.

From the hardware architect's perspective, the communication challenge takes three forms. One is to develop *new memory architectures* able to serve the bandwidth requirements of growing core numbers. Indeed, the energy and area costs of central caches that serve all cores symmetrically grow quadratically with their capacity and number of clients; caches will thus dominate silicon usage until more distributed cache systems are developed. However, distributed caches in turn require *weaker memory consistency* semantics [28] to be cost advantageous; a generalization of weakly consistent memory architectures will in turn have a dramatic impact on software ecosystems. The second aspect to be covered by hardware architect is *latency tolerance*: as the time cost of nonlocal data accesses grows, individual cores must provide mechanism to overlap computation with communication. A step in this direction is HMT, which will be increasingly complemented with hardware support for point-to-point messaging between cores, such as found in the recent TILE architecture [5]. The third aspect is *dark silicon* [10]: any given design will be fully utilized only by some applications, while most of the silicon will be underutilized by most applications. The role of architects will thus be to determine the best trade-offs between investing silicon real estate towards cores or towards communication links.

24.4.2 Software Challenge: Matching Abstractions to Requirements

A lot of attention has been given on the parallelization of existing software and comparatively less on the improvement of software to better program parallel platforms. There are, in effect, three broad strategies to *optimize performance and cost* on multicore systems.

The first is to provide better abstractions *to programmers* to compose subprograms so that the resulting critical path becomes shorter—that is, decrease the amount of synchronization programmers use—for a given functional specification of the input–output relationship. This is the classical effort towards increasing the amount of concurrency, which must continue as the amount of on-chip parallelism increases; beyond parallelization of individual algorithms, this effort must now also take place at the level of entire applications.

The second strategy is to determine ways to shorten programs to *describe less computations* to be performed at run-time, that is, simplify the input–output relationship. There are two known strategies to do this:

- Let *programmers* use domain-specific knowledge that reduces expectations on program outputs, for example, reduce output "quality" in image processing by introducing non-determinism when the difference is not perceptible. This is the domain of *approximate programming* [36,43] and domain-specific languages (DSLs).
- When *compilers and software run-time systems* transform programs, exploit extra application-level knowledge to remove excess code from the individual subprograms being combined [3].

The third strategy is for programmers to *remove unnecessary constraints* on the execution of algorithms. This strategy stems from the observation that most control and data structures in use today have been designed at a time where computers were predominantly sequential and thus may carry implicit requirements to preserve ordering even when it is not relevant to the application. For example, many programmers use lists and arrays as containers, which implicitly carry a strong ordering guarantee. In contrast, languages like C++ or Haskell provide high-level *type classes* (contracts) that enable the programmers to state their requirements, for example, an non-iterable associative container, and let the implementation choose a suitable parallelizable implementation, for example, a distributed heap or hash table. Research is still ongoing in this direction [6,9,30].

24.5 Summary

The move towards increasing core counts on chip was both an answer to overcome the sequential performance wall [1,34] and to bring higher communication bandwidths to parallel computing. The expected benefits of multicores were both higher performance and lower energy costs, thanks to frequency scaling.

The foremost challenge with multicores is not new, as it was shared by early practitioners with parallel architectures until the 1980s. This "concurrency challenge" requires software engineers to acknowledge platform parallelism and spend extra effort to express concurrency in applications. In response to this, new language and operating software technology has been developed in a short time, resulting in a large diversity of platforms, which have not yet matured nor stabilized. While this diversity creates opportunities in the highly dynamic IT industry, it also means that experience gained by practitioners in the last decade will likely need to be revisited in the coming 10 years.

The move to CMPs, especially with increasing core counts and accelerators on chip, also entails new technological and conceptual issues. More active components in the system imply faults or otherwise resource heterogeneity that must be understood and modeled. Communication links between cores and between cores and memory must be accounted for when mapping application components to the chip's resources. Real parallelism between application components implies that programmers cannot stop an application and observe a consistent global state. The benefits of multiple cores on performance can only be reaped by reducing synchronization, which for some applications means decreasing reliance on determinism. These issue in turn require new abstractions to describe and manipulate the computing system at a high level, and research has barely started to characterize which general aspects of multicore parallelism will be relevant in the next era of growing software concurrency.

Defining Terms

Concurrency vs. parallelism: "Concurrency is non-determinism with regards to the order in which events may occur. Parallelism is the degree to which events occur simultaneously" [14].

Operating software: A software composed of operating systems, compilers, and language run-time systems, in charge of mapping the concurrency expressed in software to the available parallel resources in hardware.

Programming model: A conceptual model available to users of a given programming language. Consists of programming abstractions that allow programmers to specify "what to do" and operational semantics that give programmers an intuition of "how the program will behave." Parallel programming models are special in that they discourage programmers from specifying how to carry out computation, so as to leave operating software maximum flexibility to map and schedule the program's concurrency.

Scalability: The ability of a system to approximate a factor *N* performance improvement for a factor *N* cost investment (e.g., silicon area, number of cores, energy, frequency).

Throughput vs. latency: Throughput is the number of computations achieved by unit of time, whereas latency is the number of seconds necessary to achieve a unit of computation. Parallelism can decrease latency down to a program's critical path, whereas throughput typically remains scalable as the workload on concurrent sections can be arbitrarily increased.

Topology of architecture models: The topology of an architecture consists of how components are connected to each other. On multicore processor chips, one can consider separately the memory topology, that is, how cores are connected to main memory; the cache topology, that is, how caches are connected to cores, main memory and each other; and the inter-core topology.

Typology of architecture models: The typology of an architecture consists of the set of component types that participate in the design. For example, heterogeneous multicore architectures have more than one processor type.

Acronyms

API	application programming interface
BSP	bulk-synchronous parallelism
ccNUMA	cache-coherent NUMA
CMP	chip multiprocessor
DSL	domain-specific language
DSM	distributed shared memory
FPU	floating-point unit
GPU	graphical processing unit
HMT	hardware multithreading
HPC	high-performance computing
ILP	instruction-level parallelism
IPI	inter-processor interrupt
IPS	instructions per second
MIMD	multiple instruction, multiple data
MPSoC	multiprocessor system-on-chip
NoC	network-on-chip
NUMA	nonuniform memory architecture
OoOE	out-of-order execution
PC	program counter
SIMD	single instruction, multiple data
SMP	symmetric multiprocessor

SPMD single program, multiple data
TBB threading building blocks
TLP thread-level parallelism
UMA uniform memory architecture

References

1. V. Agarwal, M. S. Hrishikesh, S. W. Keckler, and D. Burger. Clock rate versus IPC: The end of the road for conventional microarchitectures. *SIGARCH Comput. Archit. News*, 28:248–259, May 2000.
2. G. M. Amdahl. Validity of the single processor approach to achieving large scale computing capabilities. In *Proc. of the April 18–20, 1967, Spring Joint Computer Conference*, AFIPS'67 (Spring), pp. 483–485, ACM, New York, 1967.
3. R. Barik and V. Sarkar. Interprocedural load elimination for dynamic optimization of parallel programs. In *Parallel Architectures and Compilation Techniques, 2009. PACT'09. 18th International Conference on*, pp. 41–52, September 2009.
4. I. Bell, N. Hasasneh, and C. Jesshope. Supporting microthread scheduling and synchronisation in CMPs. *Int. J. Parallel Program.*, 34:343–381, 2006.
5. S. Bell, B. Edwards, J. Amann, R. Conlin, K. Joyce, V. Leung, J. MacKay et al. TILE64 processor: A 64-core SoC with mesh interconnect. In *IEEE International Solid-State Circuits Conference, 2008 (ISSCC 2008)*, San Francisco, CA. *Digest of Technical Papers*, pp. 88–598, IEEE, February 2008.
6. N. Benton, L. Cardelli, and C. Fournet. Modern concurrency abstractions for c#. *ACM Trans. Program. Lang. Syst.*, 26(5):769–804, September 2004.
7. B. Burgess, B. Cohen, M. Denman, J. Dundas, D. Kaplan, and J. Rupley. Bobcat: Amd's low-power x86 processor. *Micro IEEE*, 31(2):16–25, March/April 2011.
8. M. Butler, L. Barnes, D. D. Sarma, and B. Gelinas. Bulldozer: An approach to multithreaded compute performance. *Micro IEEE*, 31(2):6–15, March/April 2011.
9. M. M. T. Chakravarty, R. Leshchinskiy, S. P. Jones, G. Keller, and S. Marlow. Data parallel Haskell: A status report. In *Proceedings of the 2007 Workshop on Declarative Aspects of Multicore Programming*, DAMP'07, pp. 10–18, ACM, New York, 2007.
10. H. Esmaeilzadeh, E. Blem, R. St. Amant, K. Sankaralingam, and D. Burger. Dark silicon and the end of multicore scaling. In *Proceedings of the 38th Annual International Symposium on Computer Architecture, ISCA'11*, pp. 365–376, ACM, New York, 2011.
11. J. L. Gustafson. Reevaluating Amdahl's law. *Commun. ACM*, 31(5):532–533, May 1988.
12. J. L. Henessy and D. A. Patterson. *Computer Architecture—A Quantitative Approach*. Morgan Kaufmann Publishers Inc., San Francisco, CA, 4th edn., 2007.
13. C. A. R. Hoare. *Communicating Sequential Processes*. Prentice Hall, Englewood Cliffs, NJ, 1985.
14. Philip Kaj Ferdinand Hölzenspies. On run-time exploitation of concurrency. PhD thesis, University of Twente, Enschede, the Netherlands, April 2010.
15. IEEE Standards Association. *IEEE Std. 1003.1-2008, Information Technology—Portable Operating System Interface (POSIX®)*. IEEE, 2008.
16. Intel Corporation. *Intel® Threading Building Blocks Reference Manual*, 2011. http://software.intel.com/sites/products/documentation/doclib/tbb_sa/help/index.htm
17. International Standards Organization and International Electrotechnical Commission. *ISO/IEC 14882:2011, Programming Languages—C++*. American National Standards Institute (ANSI), 11 West 42nd Street, New York, 1O036, 1st edn., September 2011.
18. International Standards Organization and International Electrotechnical Commission. *ISO/IEC 9899:2011, Programming Languages—C*. American National Standards Institute (ANSI), 11 West 42nd Street, New York, 1O036, 1st edn., December 2011.
19. Khronos OpenCL Working Group. The OpenCL specification, version 1.0.43, Aaftab Munshi (ed.), 2009. http://www.khronos.org/registry/cl/specs/opencl-1.0.pdf

20. D. Kirk. NVIDIA CUDA software and GPU parallel computing architecture. In *ISMM'07: Proceedings of the Sixth International Symposium on Memory Management*, pp. 103–104, ACM, New York, 2007.
21. P. Kongetira, K. Aingaran, and K. Olukotun. Niagara: A 32-way multithreaded SPARC processor. *IEEE Micro*, 25(2):21–29, March/April 2005.
22. M. H. Lipasti and J. P. Shen. Superspeculative microarchitecture for beyond AD 2000. *Computer*, 30(9):59–66, September 1997.
23. D. T. Marr, F. Binns, D. L. Hill, G. Hinton, D. A. Koufaty, J. A. Miller, and M. Upton. Hyper threading technology architecture and microarchitecture. *Intel Technol. J.*, 6(1):1–12, 2002.
24. Message Passing Interface Forum. *MPI: A Message-Passing Interface Standard, Version 2.2.* High Performance Computing Center Stuttgart (HLRS), September 2009.
25. R. Milner. *A Calculus of Communicating Systems*. Springer-Verlag, New York, 1980.
26. R. Milner, J. Parrow, and D. Walker. A calculus of mobile processes, I. *Informat. and Comput.*, 100(1):1–40, 1992.
27. R. Milner, J. Parrow, and D. Walker. A calculus of mobile processes, II. *Informat. Comput.*, 100(1):41–77, 1992.
28. D. Mosberger. Memory consistency models. *SIGOPS Oper. Syst. Rev.*, 27(1):18–26, 1993.
29. OpenMP Architecture Review Board. OpenMP application program interface, version 3.0, 2008. http://www.openmp.org/mp-documents/spec30.pdf
30. S. P. Jones, A. Gordon, and S. Finne. Concurrent haskell. In *POPL'96: Proceedings of the 23rd ACM SIGPLAN-SIGACT Symposium on Principles of Programming Languages*, pp. 295–308, ACM, New York, 1996.
31. F. J. Pollack. New microarchitecture challenges in the coming generations of CMOS process technologies (keynote address). In *Proceedings of 32nd Annual ACM/IEEE International Symposium on Microarchitecture*, MICRO 32, IEEE Computer Society. Washington, DC, 1999.
32. J. Reinders. *Intel Threading Building Blocks: Outfitting C++ for Multi-Core Processor Parallelism*. O'Reilly Series. O'Reilly, Sebastopol, CA, 2007.
33. D. M. Ritchie and K. Thompson. The UNIX time-sharing system. *Commun. ACM*, 17:365–375, July 1974.
34. R. Ronen, A. Mendelson, K. Lai, S.-L. Lu, F. Pollack, and J. P. Shen. Coming challenges in microarchitecture and architecture. *Proc. IEEE*, 89(3):325–340, March 2001.
35. J. H. Saltzer. CTSS technical notes. Technical Report MAC-TR-16, Massachusetts Institute of Technology, Project MAC, 1965.
36. A. Sampson, W. Dietl, E. Fortuna, D. Gnanapragasam, L. Ceze, and D. Grossman. EnerJ: Approximate data types for safe and general low-power computation. *SIGPLAN Not.*, 46(6):164–174, June 2011.
37. M. Shah, R. Golla, P. Jordan, G. Grohoski, J. Barreh, J. Brooks, M. Greenberg et al. SPARC T4: A dynamically threaded server-on-a-chip. *IEEE Micro*, PP(99):1, 2012.
38. B. J. Smith. Architecture and applications of the HEP multiprocessor computer system. *Proc. SPIE Int. Soc. Opt. Eng. (United States)*, 298:241–248, 1981.
39. A. Snavely, L. Carter, J. Boisseau, A. Majumdar, K. S. Gatlin, N. Mitchell, J. Feo, and B. Koblenz. Multi-processor performance on the Tera MTA. In *Supercomputing '98: Proceedings of the 1998 ACM/IEEE Conference on Supercomputing*, pp. 1–8, IEEE Computer Society, Washington, DC, 1998.
40. H. Sutter. The free lunch is over: A fundamental turn toward concurrency in software. *Dr. Dobb's J.*, 30(3), 2005.
41. J. E. Thornton. Parallel operation in the Control Data 6600. In *Proceedings of the October 27–29, 1964, Fall Joint Computer Conference, Part II: Very High Speed Computer Systems*, AFIPS'64 (Fall, part II), pp. 33–40, ACM, New York, 1965.
42. D. M. Tullsen, S. J. Eggers, and H. M. Levy. Simultaneous multithreading: Maximizing on-chip parallelism. *SIGARCH Comput. Archit. News*, 23:392–403, May 1995.

43. D. Ungar and S. S. Adams. Harnessing emergence for manycore programming: Early experience integrating ensembles, adverbs, and object-based inheritance. In *Proceedings of the ACM International Conference Companion on Object Oriented Programming Systems Languages and Applications Companion*, SPLASH'10, pp. 19–26, ACM, New York, 2010.
44. L. G. Valiant. A bridging model for parallel computation. *Commun. ACM*, 33:103–111, August 1990.
45. Wm. A. Wulf and S. A. McKee. Hitting the memory wall: Implications of the obvious. *SIGARCH Comput. Archit. News*, 23:20–24, March 1995.
46. M. Yuffe, E. Knoll, M. Mehalel, J. Shor, and T. Kurts. A fully integrated multi-CPU, GPU and memory controller 32nm processor. In *Solid-State Circuits Conference Digest of Technical Papers (ISSCC), 2011 IEEE International*, pp. 264–266, February 2011.

25

DNA Computing

Hieu Bui
Duke University

Harish Chandran
Google Corporation

Sudhanshu Garg
Duke University

Nikhil Gopalkrishnan
Duke University

Reem Mokhtar
Duke University

John Reif
Duke University
King Abdulaziz University

Tianqi Song
Duke University

Organization of Chapter

Molecular computing involves computation done at the molecular scale. DNA computing is a class of molecular computing that does computation by the use of reactions involving DNA molecules. DNA computing has been by far the most successful (in scale and complexity of the computations and molecular assemblies done) of all known approaches to molecular computing, perhaps due in part to the very well-established biotechnology and biochemistry on which its experimental demonstration relies, as well as the frequent teaming of scientists in the field with multiple essential disciplines including chemistry, biochemistry, physics, material science, and computer science.

This chapter surveys the field of DNA computing. It begins in Section 25.1 with a discussion of the underlying principles, including motivation for molecular and DNA computations (Section 25.1.1), brief overviews of DNA structures (Section 25.1.2), chemical reaction systems (Section 25.1.3), DNA reactions (Section 25.1.4), and classes of protocols and computations (Section 25.1.5). Then, the chapter discusses potential applications of DNA computing research (Section 25.2). The main section on research issues (Section 25.3) overviews how DNA computation is done, with a discussion of DNA hybridization circuits (Section 25.3.1), including both solution-based and localized hybridization circuits. It also discusses design and simulation software for the same. We discuss DNA detectors in Section 25.3.2, DNA replicators in Section 25.3.3, DNA nanorobotic devices in Section 25.3.4, and DNA dynamical systems in Section 25.3.5. Section 25.3.6 overviews research on tiling assembly computations, including theoretical models and results in Section 25.3.6.1, experimental methods for assembly of tiling lattices in Section 25.3.6.2, design and simulation software in Section 25.3.6.3, assembly in various dimensions in Sections 25.3.6.4 and 25.3.6.5, stepwise tiling assemblies in Section 25.3.6.6, activatable tiles in Section 25.3.6.7, and tiling error-correction methods in Section 25.3.6.8.

25.1 Underlying Principles

25.1.1 Motivation: Why Do Molecular Computation and Why Use DNA for Computation and Self-Assembly?

In an era where electronic computers are powerful and inexpensive, why do we need molecular computation? One response to this question is that conventional electronic devices have been miniaturized to the point where traditional top-down methods for manufacturing these devices are approaching their inherent limits due to constraints in photolithographic techniques and further miniaturization is not cost-effective. On the other hand, bottom-up manufacturing methods such as molecular self-assembly have no such scale size limits. Another response is that molecular computation provides capabilities that traditional computers cannot provide; there are computations that need to be done in environments and at scales where a traditional computer cannot be positioned, for example, within a cell or within a synthetic molecular structure or material.

Why use nucleic acids such as DNA for computation and self-assembly?

DNA and nucleic acids in general are unique in multiple aspects:

1. First of all, they hold and can convey information encoded in their sequences of bases. Most of their key physical properties are well understood.
2. Their tertiary structure is much more predictable, compared to molecules such as proteins.
3. Their hybridization reactions, which allow for addressing of specific subsequences, are also well understood and productively controllable.
4. They allow for a large set of operations to be performed on them. Well-known enzymatic reactions for manipulation of DNA exist.
5. Finally, there is a well-developed set of methods such as gel electrophoresis, Förster Resonance Energy Transfer (FRET), Plasmonics, and Atomic Force Microscopy (AFM) imaging for quantifying the success of experiments involving DNA and DNA nanostructures.

Before we delve into how molecular computation is done, we will discuss DNA structure and function, how information may be stored in it, and what environment it needs to efficiently do computation.

25.1.2 Brief Overview of DNA Structure

DNA is a polymer that can exist in either single- or double-stranded form. Each strand of DNA is made up of a repeating set of monomers called *nucleotides*. A nucleotide consists of three components, a 5-carbon sugar molecule, a nitrogenous base, and a phosphate group. The …-phosphate-sugar-phosphate-… covalent bond forms the backbone of a DNA strand. The phosphate group is attached to the 5th carbon atom (C5) on one end and C3 on another end. This gives the DNA strand directionality, and the two ends of a DNA strand are commonly termed the 5′ (prime) and the 3′ ends. This can be seen in Figure 25.1.

25.1.2.1 DNA Bases

Nitrogenous bases are the component of nucleotides that are not involved in forming the backbone of a strand. There are five types of these bases, named adenine(A), guanine(G), cytosine(C), thymine(T), and uracil(U). Only four of these, A, G, C, and T, are present in DNA, while T is replaced by U in RNA.

Bases A and G belong to a class called *purines*, while C, T, and U fall under the *pyrimidines* class. Figure 25.2 shows the difference in structure of purines and pyrimidines. A purine class has a pyrimidine ring fused to an imidazole ring and contains four nitrogen atoms as opposed to two nitrogen atoms in a pyrimidine class.

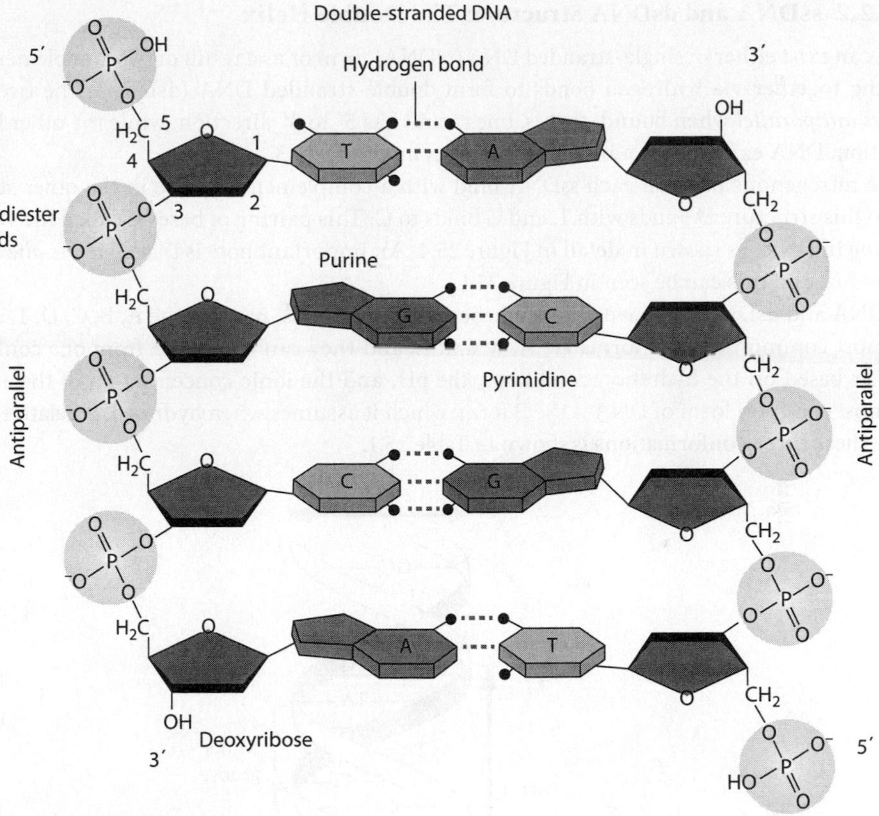

FIGURE 25.1 DNA backbone (on left) and DNA bases involved in hydrogen bonding (middle). (From Campbell, N. et al. *Biology*, 5th edn., Benjamin Cumminges, 1999.)

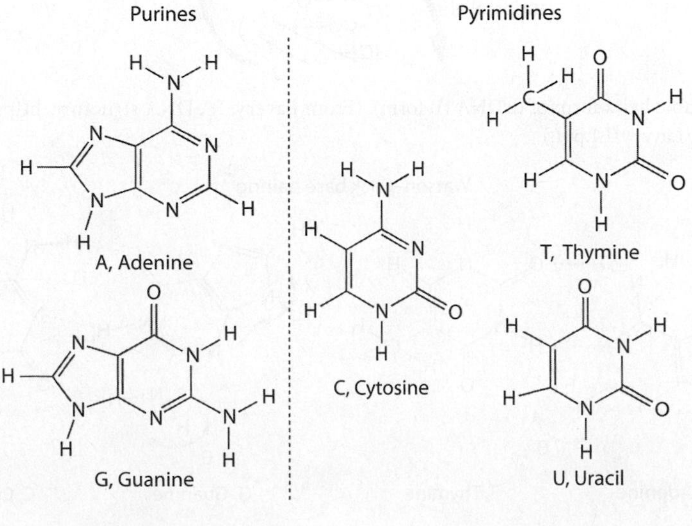

FIGURE 25.2 Structure of DNA/RNA bases.

25.1.2.2 ssDNA and dsDNA Structure: The Double Helix

DNA can exist either in single-stranded DNA (ssDNA) form or as a result of two complementary ssDNA binding together via hydrogen bonds to form double-stranded DNA (dsDNA). The two ssDNA are always *antiparallel* when bound, that is, one strand has 5′ to 3′ direction, while the other has a 3′ to 5′ direction. DNA exists as a double helix, as shown in Figure 25.3.

The nitrogenous bases in each ssDNA bind with a complementary base in the other strand to give rise to this structure: A binds with T, and G binds to C. This pairing of bases is called the *Watson–Crick bonding* in DNA, as shown in detail in Figure 25.4. An important note is that a *purine always binds to a pyrimidine* and this can be seen in Figure 25.1.

ssDNA and dsDNA can have different helical conformations, namely, the A, B, C, D, T, and Z forms. The most common dsDNA forms are A, B, and Z, and they can transform from one conformation to another based on the hydration conditions, the pH, and the ionic concentration of the environment. The most common form of DNA is the B form, which it assumes when hydrated. A relative comparison of the these three conformations is shown in Table 25.1.

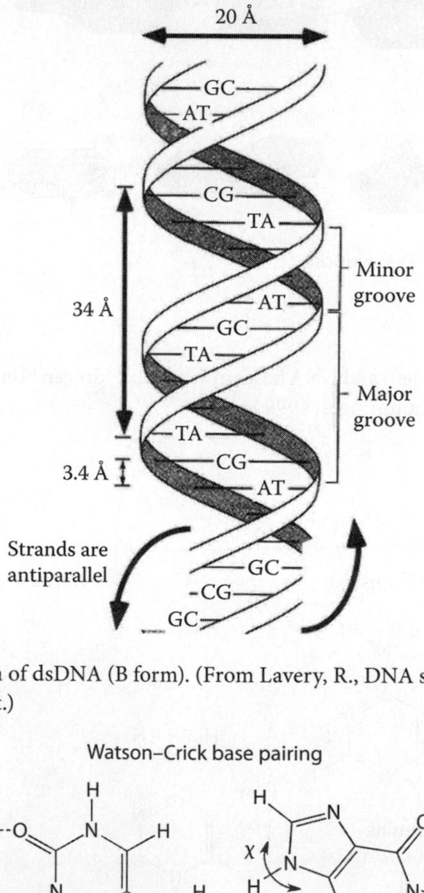

FIGURE 25.3 Double helix form of dsDNA (B form). (From Lavery, R., DNA structure, http://www.phys.ens.fr./monasson/Houches/Lavery/L1.ppt.)

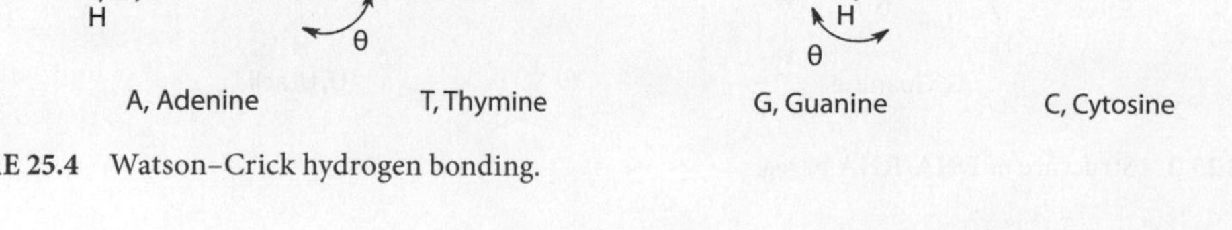

FIGURE 25.4 Watson–Crick hydrogen bonding.

TABLE 25.1 Comparison of A, B, and Z Forms of DNA

Characteristic	A-DNA	B-DNA	Z-DNA
Helix sense	Right-handed	Right-handed	Left-handed
Residues per turn (base pairs)	11	10.4	12
Axial rise	2.55 Å	3.4 Å	3.8 Å
Helix pitch	28 Å	35 Å	45 Å
Base pair tilt	20°	−6°	7°
Helix width (diameter)	23 Å	20 Å	18 Å
Phosphate–phosphate distance	5.9 Å	7.0 Å	5.9 Å
Dimension	Broad	Normal	Narrow

Source: Data compiled from Berg, J. M. et al.,. *Biochemistry*, 7th edn., W.H. Freeman Publishers, December 2010; Saenger, W., *Principles of Nucleic Acid Structure*, Springer-Verlag, New York, p. 556, 1984; Sinden, R.R., *DNA Structure and Function*, Academic Press Inc., San Diego CA, 1994, 398pp., 1994.

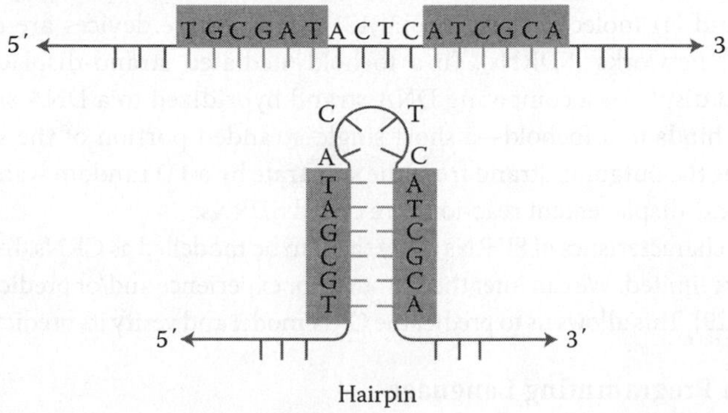

FIGURE 25.5 Hairpin open and closed forms. (From Blicq, D., DNA hairpin, http://xnet.rrc.mb.ca/davidb/dna.htm.)

In its ssDNA form, DNA exists as a long single thread, or in many cases, it forms a *secondary structure*, where the strand loops around itself and forms hydrogen bonds with other bases on itself (called *a random coil*).

25.1.2.3 DNA Hairpins

DNA hairpins are a special *secondary structure* formed by an ssDNA, and contain a stem/neck double-stranded region, and an unhybridized loop region, as seen in Figure 25.5. Hairpins have been recognized as a useful tool in molecular computation because of three reasons: (1) Hairpins store energy in their unhybridized loop, and on hybridization, energy is released driving the reaction forward. (2) In their hairpin form, they are relatively unreactive with other DNA strands and act as excellent monomers until an external entity (usually another DNA strand) causes the stem region to open and react with other DNA complexes. Hence, they can persist with low leaks for a long amount of time. (3) A common way to create DNA complexes is to anneal them. DNA complexes usually contain a large number of strands, and multiple different structures can be formed because of varied interactions between different strands. In low concentrations, DNA hairpins usually form without error and are not involved in spurious structure formation. This is because their formation is not diffusion dependent, that is, the two ends of a hairpin hybridize with each other before two ends of different hairpins hybridize. This property is known as *locality* and is a strong motivation for the use of hairpins.

Hairpins [32] and metastable DNA hairpin complexes [85,102] have been used as fuel in chain reactions to form large polymers [20], in programming pathways in self-assembly [119], and in logic circuits [84]. A common technique to help open a DNA hairpin is via a process known as *toehold-mediated strand displacement*, which we shall discuss in more detail in Section 25.1.4.1.

25.1.3 Brief Overview of Chemical Reaction Networks in DNA Computing

Chemical reaction networks (CRNs) are becoming central tools in the study and practice of DNA computing and molecular programming. Their role is twofold—as a model for analyzing, quantifying, and understanding the behavior of certain DNA computing systems and as a specification/programming language for prescribing information processing (computational) behavior. The first of these roles is traditional and is analogous to the role played in biology by CRNs in describing biochemical processes and genetic reaction networks. The latter role—thinking of CRNs as a programming language—is unique to the field of DNA computing and is a consequence of the ability of DNA to act as an information processing medium and emulate (with certain restrictions) any CRN set down on paper. We will discuss both these roles briefly in the following paragraphs.

25.1.3.1 CRNs Model DNA Strand-Displacement Reaction Networks

Enzyme-free DNA computing devices can execute (1) Boolean circuits and linear threshold gate networks (the latter model neural networks) [68,69,83], (2) nucleic acid amplifiers [20,119,127], (3) finite state automata [30], and (4) molecular walkers [31,121]. All of these devices are examples of strand-displacement reaction networks (SDRNs). In a toehold-mediated strand-displacement reaction, an incoming DNA strand displaces a competing DNA strand hybridized to a DNA substrate strand. The incoming strand first binds to a toehold—a short single-stranded portion of the substrate—and then competitively displaces the outgoing strand from the substrate by a 1D random walk process. A cascade (network) of such strand-displacement reactions are called *SDRNs*.

The modular design characteristics of SDRNs allow them to be modelled as CRNs. In particular, the types and rates of reactions are limited. We can infer them from prior experience and/or predict them from thermodynamic parameters [129]. This allows us to predict the CRN model and verify its predictions experimentally.

25.1.3.2 CRNs as a Programming Language

In theory [95], SDRNs closely approximate the dynamic behavior of any CRN up to a time and concentration scaling. They illustrate how any CRN that we set down upon paper can be translated into a set of DNA molecules that when mixed together in the appropriate concentrations will emulate the behavior of the CRN. Certain CRNs seem hard to emulate in practice, and no successful SDRN implementations currently exist for these, but many others have been successfully implemented.

CRNs are more abstract than SDRNs and can be thought of as a higher-level programming language. The process of translating a CRN into its corresponding SDRN is then analogous to compiling a higher-level programming language down to a lower-level programming language. Programming in the CRN language has the advantages inherent in programming in higher-level languages vs. programming in lower-level languages.

How powerful is the CRN language? Quite powerful, it turns out. It is proven in [94] that a finite CRN obeying stochastic dynamics can achieve efficient Turing universal computation with arbitrarily small (nonzero) error. Error-free computation is impossible in this setting; only semilinear functions can be computed without errors [14].

25.1.4 DNA Reactions

In order to be able to efficiently compute with DNA, we should be aware of its properties and the types of reactions it can undergo. We classify this set of reactions into three types — DNA hybridization reactions, DNA enzyme reactions, and DNAzyme reactions. (1) DNA hybridization reactions are usually enzyme-free and isothermal and encapsulate strand-displacement reactions. (2) DNA enzyme reactions are powerful reactions, which can help cut and join the backbone of DNA strands, as well as synthesize new strands, and are often employed due to this versatility. Enzymatic reactions are often extremely rapid and extremely low error, hence making them attractive to use. (3) More recently, deoxyribozymes (DNAzymes) and aptamers have been discovered and used similar to enzymes to manipulate DNA reactions.

25.1.4.1 DNA Hybridization Reactions

A well-known example of DNA hybridization reactions is the Watson–Crick DNA hybridization between two complimentary ssDNA strands as discussed in Section 25.1.2.2. Two ssDNA strands can attach to each other. However, they can also detach from one another, if the temperature is greater than the melting temperature of the strands (Figure 25.6). The *melting temperature* of a dsDNA is defined as the temperature at which 50% of the dsDNA has converted to single-stranded form.

25.1.4.1.1 Toehold-Mediated Strand Displacement

Yurke et al. reported an interesting DNA hybridization reaction through their DNA tweezer system [124]. As illustrated in Figure 25.7a, two ssDNA strands (B) and $\overline{CB}$) are bound to one another, with one strand called the *incumbent strand* (strand B) completely bound, while the other $\overline{CB}$ has a few unbound bases. These bases ($\overline{C}$) can together be called a *sticky end*, *overhang*, or a *toehold*. A third ssDNA, called

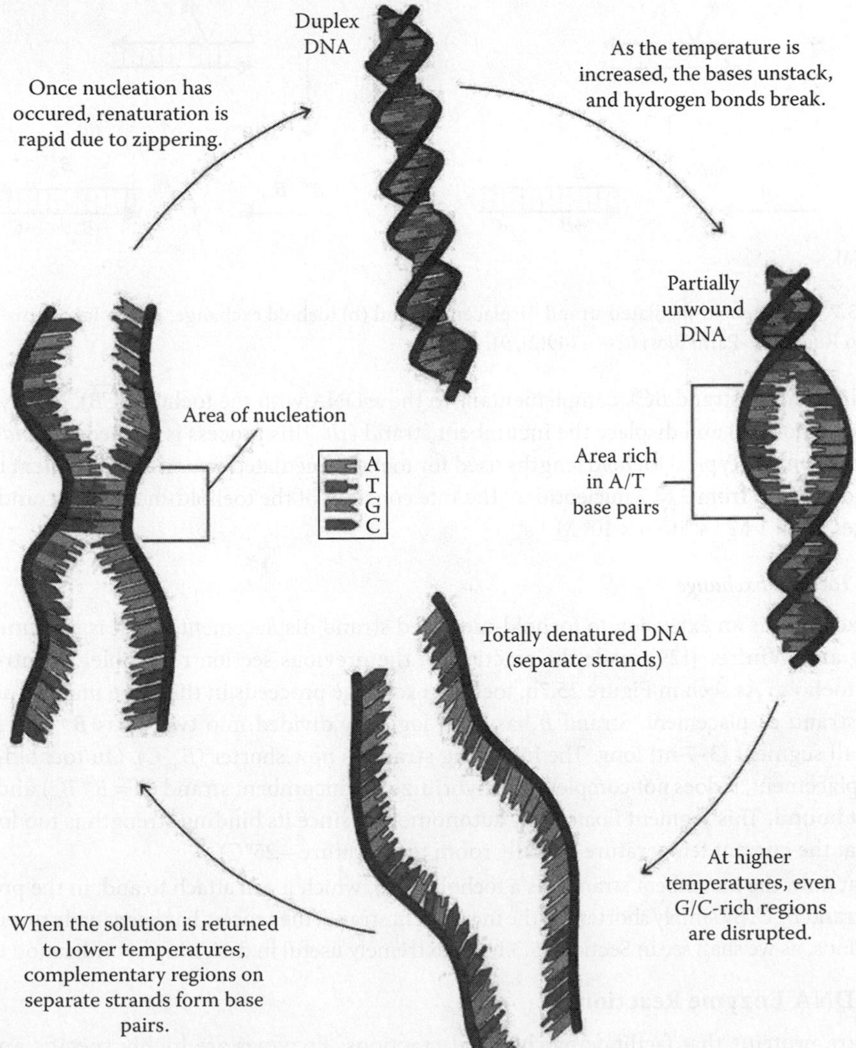
Duplex DNA
As the temperature is increased, the bases unstack, and hydrogen bonds break.
Once nucleation has occured, renaturation is rapid due to zippering.
Partially unwound DNA
Area of nucleation
A
T
G
C
Area rich in A/T base pairs
Totally denatured DNA (separate strands)
At higher temperatures, even G/C-rich regions are disrupted.
When the solution is returned to lower temperatures, complementary regions on separate strands form base pairs.

FIGURE 25.6 DNA denaturation renaturation. (From Moran, L.A. et al., *Biochemistry*, 2nd edn., Neil Patterson publishers, distributed by Prentice Hall Inc., 1994.)

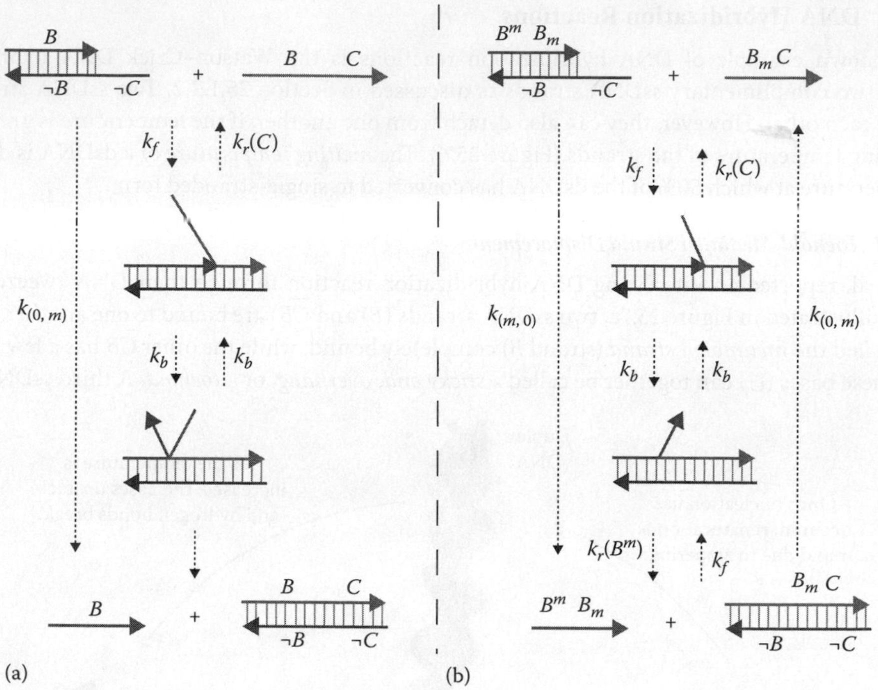

FIGURE 25.7 (a) Toehold-mediated strand displacement and (b) toehold exchange. (From Iaki Sainz de Murieta and Alfonso Rodrguez-Patn, *Biosystems*, 109(2), 91, 2012.)

the *incoming strand* (strand *BC*), complementary to the ssDNA with the toehold ($\overline{CB}$), can hybridize to the toehold region ($\overline{C}$) and displace the incumbent strand (*B*). This process is termed *toehold-mediated strand displacement*. Typical toehold lengths used for toehold-mediated strand-displacement hybridization reactions range from 3 to 7 nucleotides. The rate constant of the toehold-mediated strand displacement ranges from 1 M^{-1} s^{-1} to 6 × 10^6 M^{-1} s^{-1}.

25.1.4.1.2 Toehold Exchange

Toehold exchange is an extension to toehold-mediated strand displacement, but it is extremely powerful. Zhang and Winfree [129] made the reaction in the previous section reversible, by introducing a small exit toehold. As seen in Figure 25.7b, toehold exchange proceeds in the same manner as toehold-mediated strand displacement. Strand *B* has been logically divided into two parts B^m and B_m, where B^m is a small segment (3–7 nt) long. The incoming strand is now shorter (B_m *C*). On toehold-mediated strand displacement, it does not completely dehybridize the incumbent strand ($B = B^m B_m$) and B^m of the incumbent bound. This segment floats away autonomously since its binding strength is too low to keep it in place at the current temperature (usually room temperature—25°C).

Note that now, the incumbent strand has a toehold ($\overline{B^m}$), which it can attach to and, in the process, displace the strand $B_m C$. By simply shortening the incumbent strand, the process has been made reversible. This set of reactions, as we shall see in Section 25.3.1, are extremely useful in designing hybridization circuits.

25.1.4.2 DNA Enzyme Reactions

Enzymes are proteins that facilitate biochemical reactions. Enzymes are highly specific and usually catalyze their desired reactions. Theoretically, enzymes can be reused in a biochemical reaction, without being consumed by the reactants. They also provide high efficiency to their reactions. For example, one enzyme named catalase in liver can break down roughly five million molecules of hydrogen peroxide into oxygen and water in 5 min.

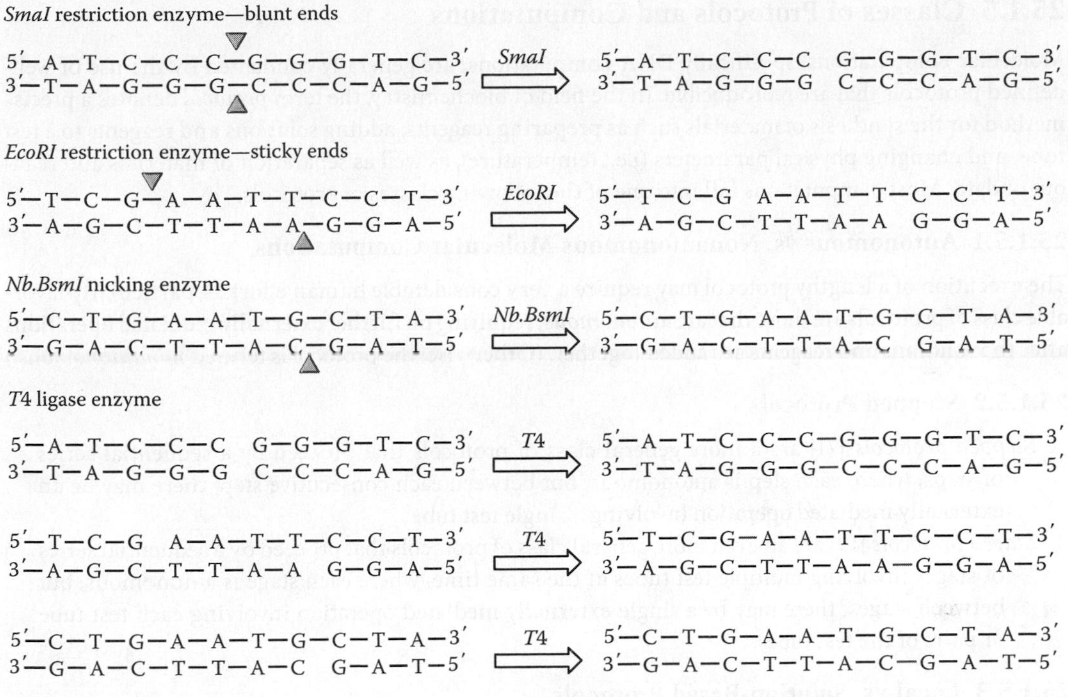

FIGURE 25.8 DNA enzymes.

In the field of DNA-based computation, scientists are currently working with a small subset of enzymes such as restriction enzymes, nicking enzymes, ligase enzymes, and polymerase enzymes. The purpose of restriction enzymes and nicking enzymes is to cleave the phosphodiester bond within a chain of nucleotides, whereas the purpose of ligase enzymes is to repair the phosphodiester bond as illustrated in Figure 25.8. In order for an enzyme to cleave or repair a specific location, the location has to have a recognition site that is designated for that particular enzyme. A recognition site (depicted as triangles in Figure 25.8) is normally a few nucleotides along the DNA double helix. For example, a *SmaI* restriction enzyme cleaves two phosphodiester bonds on both sides of a given DNA double helix to give an end result of two blunted ends of DNA double helices (Figure 25.8). Similarly, an *EcoRI* restriction enzyme cleaves two phosphodiester bonds on both sides of a given DNA double helix to give an end result of two separated overhang DNA helices (Figure 25.8). A *Nb.BsmI* nicking enzyme, on the other hand, cleaves a single phosphodiester bond on one side of a given DNA double helix. Furthermore, a broken phosphodiester bond can be repaired using a T4 ligase enzyme as illustrated in Figure 25.8. To amplify a particular DNA sequence, researchers often employ the polymerase chain reaction (PCR) technique. This technique involves the use of polymerase enzymes to catalyze the polymerization of nucleoside trisphosphates (dNTPs) into a multiple copies of the target DNA sequence.

25.1.4.3 DNAzyme Reactions

Recently, researchers have been using deoxyribozymes (DNAzymes) and aptamers to catalyze DNA hybridization reactions. DNAzymes and aptamers are discovered by *in vitro* evolution search. They are both DNA-based sequences that possess enzymatic activities, which extend beyond Watson–Crick hybridization, and they have highly specific binding to target molecules. An example of the use of DNAzyme for DNA computation was demonstrated by Stojanovic et al. by building a molecular automaton using a network of DNAzymes [96].

25.1.5 Classes of Protocols and Computations

Molecular computations, specifically DNA computations, are generally conducted by the use of well-defined protocols that are reproducible. In the field of biochemistry, the term *protocol* denotes a precise method for the synthesis of materials such as preparing reagents, adding solutions and reagents to a test tube, and changing physical parameters (i.e., temperature), as well as separation of materials and read-out of data. Most computations fall into one of the following classes of protocols.

25.1.5.1 Autonomous vs. Nonautonomous Molecular Computations

The execution of a lengthy protocol may require a very considerable human effort. A particularly favorable class of protocols are those that are *autonomous*, requiring no further externally mediated operations after the solutions and reagents are added together. (Otherwise, the protocol is termed *nonautonomous*.)

25.1.5.2 Stepped Protocols

Stepped protocols [71] are a more general class of protocols that proceed by a sequential series of steps, where each step is autonomous, but between each consecutive step, there may be an externally mediated operation involving a single test tube.

Staged protocols [19] are an even more general class of protocols that proceed by a sequential series of stages involving multiple test tubes at the same time, where each stage is autonomous, but between stages, there may be a single externally mediated operation involving each test tube or pairs of the test tubes.

25.1.5.3 Local vs. Solution-Based Protocols

A reaction is *solution-based* if it occurs within a solution, and materials have to diffuse to each other in order to interact. Likewise, it is *local* if it occurs between materials that are attached to a surface, thereby reducing/eliminating the diffusion time.

25.1.5.4 Activatable Molecular Devices

A molecular device is *activatable* if it has two classes of states: *inactive*, where it generally cannot undergo state transitions until an activating event occurs, and *active*, where it can undergo state transitions. An activatable device usually is initialized inactive and transitions to an active state after an activation event.

25.2 Impact on Practice

Some of the practical applications of DNA computations and DNA nanoassemblies include the following:

1. *Detection of nucleic acid sequences*: For many diseases, the detection of the disease is done via the determination of a characteristic sequence t of DNA or RNA, where t has base length no more than 20 or so base pairs. This detection is conventionally done via the well-known PCR protocol, but this requires somewhat bulky and expensive device for repeated thermocycles and optical detection. The market for these detection protocols is over a billion dollars. DNA computation protocols may provide a much more cost-effective and portable means for detection. Section 25.3.2 describes various isothermal (requiring no thermal cycling) protocols for exquisite sensitive detection of DNA and RNA sequences.
2. *3D DNA nanoassemblies of proteins for x-ray crystallography structure determination*: Almost half of all proteins of interest to medicine cannot be crystallized, and so their 3D structure can be determined from conventional x-ray crystallography studies. An important application of DNA nanoassembly, proposed by Seeman [88], is the assembly of 3D DNA lattices that can hold a given protein at regular positions in the crystalline lattice, allowing for x-ray crystallography studies of an otherwise uncrystallizable protein. For further discussion of this application, see Section 25.3.6.5 discussion on 3D DNA nanoassemblies [130].

3. *3D DNA nanoassemblies for alignment of proteins for NMR structure determination*: Douglas and Shih [22] have proposed and demonstrated a novel method for improved nuclear magnetic resonance (NMR) studies of structure determination that makes use of long DNA nanoassemblies to partially align proteins.
4. *2D DNA nanoassemblies for molecular patterning*: DNA nanoassemblies can be used for programmed patterning of 2D molecular surfaces. This has applications to assembly of molecular electronic devices on molecular surfaces and 2D readout methods [47,76,118].
5. *DNA nanorobotic devices for molecular assembly lines*: Programmed control and sequencing of chemical reactions at the molecular scale have major applications in the chemical industry. DNA nanorobotic devices (see Section 25.3.4) have been demonstrated to execute a programmed sequence of chemical reactions at predetermined positions along a DNA nanostructure [33,57].

25.3 Research Issues

25.3.1 DNA Hybridization Circuits

The construction of circuits made out of DNA has been an area of interest since the advent of molecular computing. Having millions to billions of nanoscale DNA gates working in tandem, to accept an input and produce an unambiguous output, is a challenging task. There have been various designs to create logic gates and Boolean circuits using DNA motifs [1,44,64,65,67,69,80,81,83,84,97,98,113,128]. We examine a few such attempts at creating these circuits and the advantages and pitfalls associated with each. We classify the circuits into two types: DNA circuits contain a large number of DNA molecules, and each molecule reacts with a small subset of other molecules. Circuits in which molecules diffuse through solution, to find a molecule from this subset, are said to be *solution based*. On the other hand, DNA molecules can be attached to a surface, and these molecules only interact with other molecules within their reach. Such circuits are known as localized hybridization circuits.

25.3.1.1 Solution-Based DNA Hybridization Circuits

Qian and Winfree developed a scalable solution-based DNA hybridization circuit system. Their basic motif is a seesaw gate based on the toehold-exchange protocol as seen in Section 25.1.4 developed by [129]. The system consists of a set of seesaw gates. Each gate can accept an input (strand) and produce an output(strand). The input strands act catalytically; hence, a single input strand can help release multiple output strands from one (or more) seesaw gates. The output strands act as inputs for the next seesaw gates downstream in the circuit. Figure 25.9 shows the basic mode of operation of a seesaw gate motif.

As an example, the input strand consists of three domains, S2, T, and S5, and the output strand consists of S5, T, and S6. Note that the 5′ end of the input and the 3′ end of the output need to have the same set of domains, while the converse is not true (S2 and S6 can be different). These domains S2 and S6 can be used to help differentiate different input strands and different output strands that interact with the same gate. We now show how the authors use this difference to construct OR and AND gates.

25.3.1.1.1 Construction of AND–OR Gates

Having constructed the basic motif, the authors then constructed AND and OR gates using the same motif. In order to construct an OR gate, either one of the input strands (x1, x2) should release the output strand. That is, the presence of either x1 OR x2 gives the output. Take the following construction: x1 = (S10 T S5), x2 = (S10 T S5). We can trivially see that either one of x1 or x2 will release the output strand. Hence, this gate is equivalent to an OR gate.

The construction of an AND gate is nontrivial. Only when both x1 and x2 are present should the output strand be released. The authors introduce a gate called the *"threshold gate"* as seen in Figure 25.9. The threshold gate can be thought of as a garbage collector that sucks a strand in and makes it unusable for further reactions. By varying the concentration of the threshold gate, both AND and OR gates can be constructed. This is seen in Figure 25.9.

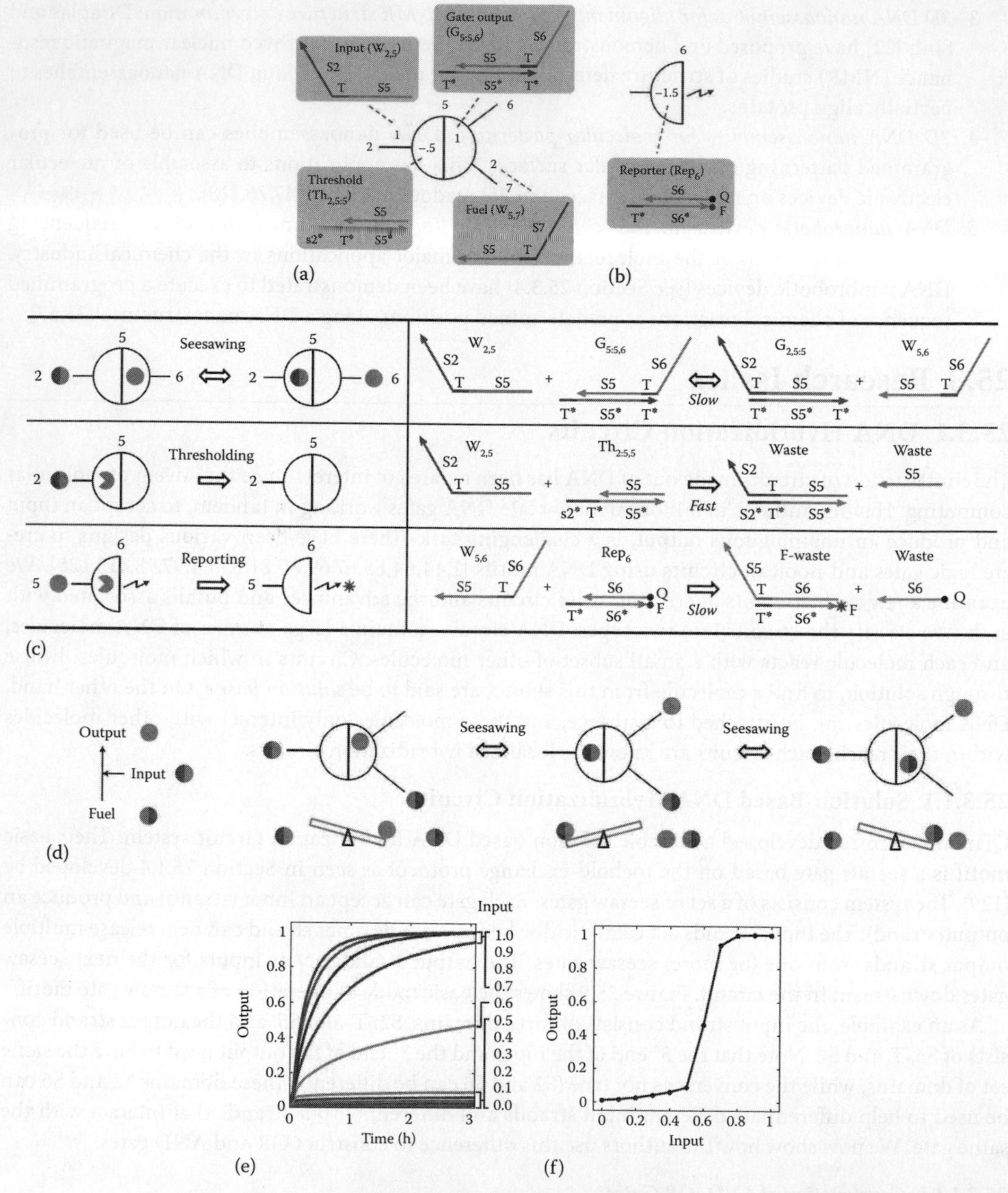

FIGURE 25.9 Components of a seesaw system. (From Qian, L. and Winfree, E., *Science*, 332(6034), 1196, 2011.)

25.3.1.1.2 A DNA Circuit That Computes the Square Root of n ≤ 15

Using the AND and OR gate motifs, we can now construct any circuit by connecting a set of such gates sequentially. The authors construct a circuit that computes the square root of all numbers up to 15. It takes in 8 inputs, a 0 or a 1 for each bit. The output set of strands denote the set of bits that are set to 1 and hence encode the result.

The system described earlier is an example of how DNA can be used to do logical computation *in vitro*, and *detect* any input strands that are present in the vicinity. There are several disadvantages of this approach, with the primary one being that this methodology is slow, and as the depth of the circuit increases, the amount of time taken to perform a reliable computation increases linearly. Also, circuits such as these are susceptible to erroneous reactions taking place between gates and strands that are not intended to happen, and this could lead to faulty outputs. In order to fix these problems, we discuss another set of circuits in the next section.

25.3.1.2 Localized DNA Hybridization Circuits

The solution-based systems described earlier demonstrate the enormous potential of DNA nanosystems. But most of the systems relied on diffusion-based hybridization to perform complex state changes/computation. At low concentrations and temperatures, diffusion can be quite slow and could impede the kinetics of these systems. At higher concentrations and temperature, unintended spurious interactions (often called *leaks*) could hijack the systems from its programmed trajectory. Localized hybridization networks are set of DNA strands attached to an addressable substrate. This localization increases the relative concentration of the reacting DNA strands thereby speeding up the kinetics. Diffusion-based systems possess global states encoded via concentration of various species and hence exhibit limited parallel ability. In contrast, localized hybridization systems allow for each copy of the localized hybridization network to operate independently of each other. Localized hybridization networks also allow one to reuse the same DNA sequence to perform different actions at distinct location on the addressable substrate, increasing the scalability of such systems by exploiting the limited sequence space. An advantage of localized hybridization computational circuit is sharper switching behavior as information is encoded over the state of a single molecule. This also eliminates the need for thresholding as computation is performed locally eliminating the need for a global consensus. These advantages are expounded in greater detail in Chandran et al. (2011). Quite recently, other articles such as Genot et al. (2011) and Aldaye et al. (2011), have also demonstrated that locality can be used successfully.

25.3.1.3 Design and Simulation Software

Computational tools are a growing necessity in order to design DNA strands for use in both simulations and experiments. Visual DSD from [42] is extremely useful in simulating DNA strand-displacement reactions. NUPACK [125] and MFOLD [132] help in design and analysis of sequences at the base level, in order to weed out wrong designs.

25.3.2 Strand-Displacement Reaction Network (SDRN)-Based DNA Detectors

The detection problem is the problem of designing a solution-based chemical molecular circuit (henceforth, the *detector*) for detecting the presence of an input molecular species by producing an output molecule (or molecules) in sufficient quantities so that it can be detected by a measuring instrument. The detector must be sensitive to small amounts of input, specific to the correct type of input, and robust to stochastic variability in the operation of the CRN. We might also ask that the detector function correctly over a wide range of concentrations, temperatures, and buffer conditions.

In this section, we will concentrate on the problem of detecting a short piece of DNA of known sequence. Various other molecules can often be transduced into a short DNA sequence by means of aptamer technology. This problem, interesting in its own right, will not be discussed in this chapter.

25.3.2.1 Advantages of SDRN Detectors

The canonical DNA detector is the PCR, an enzymatic detector that requires temperature cycling. Driven by exquisitely evolved enzymes and technological advances over the past 50 years, real-time PCR protocols are sensitive to as little as 2000 DNA molecules in an mL of solution. However, PCR is not often specific to single base mutations in input, requires temperature cycling, and works only in a narrow range of temperature and buffer conditions. Enzymatic detectors that seek to replace PCR have their own host of issues and rely on special enzymes.

In contrast, SDRN detectors are isothermal, easily modifiable to detect various sequences, can be made specific to single base mutations, and may work correctly over a wide range of temperature, concentration, and buffer conditions [126]. They can also function as a module (subroutine) within a larger SDRN performing a complex computation, for instance, an SDRN circuit that produces an output molecule only in the presence of two types of inputs (AND logic) or one detecting either one of two types of inputs, but not both together (XOR logic) or any arbitrary Boolean predicate. The key limitation with current SDRN detectors is their lack of sensitivity to small amounts of input, but the hope is that this issue can be surmounted [29]. A sensitive SDRN detector could have applications in a variety of settings, including *in vivo*.

25.3.2.2 Leaks Limit Sensitivity of Detectors

A sensitive detector must necessarily amplify small input signal into sufficiently large output signal. This amplification must be conditional; in the presence of input, the amplification must be switched on and switched off in its absence. In terms of CRNs, this implies that detectors must have catalytic reaction pathways. Thermodynamics tells us that if a reaction proceeds in the presence of a catalyst, it must also occur in the absence of that catalyst, typically at a slower rate. These unplanned reactions are termed as *leaks*, and they limit the sensitivity of detectors [28]. Even in the absence of input, the detector produces output due to leak reactions. If the quantity of output produced (as a function of time) in the presence of input is sufficiently larger (detectable by measuring instrument) than the output produced (as a function of time) in the absence of input, the input is detectable. Otherwise, the detector is not sensitive to the input.

Identifying and mitigating leak reactions is the key to implementing sensitive detectors. For simple CRNs, the leak reactions may be easily guessed, since catalytic reactions are easily identified. What are the correct leak reactions when implementing more complex CRNs? Do the leak reactions depend on the particular technology we use to implement the CRN? Ref. [28] hypothesizes that the correct leak model for a CRN is the *saturation* (defined in [27]) of that CRN. Informally, the *saturation* of a CRN introduces some basis set of leak reactions that gets rid of all purely catalytic pathways. There may be many choices for the basis. The *rates* of these reactions depend on the type of technology we use to implement them, but their nature (up to a choice of basis) is determined purely by the connectivity of the CRN.

In SDRN detectors, having identified the leak reactions, knowledge of strand-displacement kinetics [129] allows us to guess the rates of these reactions, and hence we have an accurate model of the dynamic behavior of the SDRN detector before implementing them in the laboratory.

25.3.2.3 Overview of SDRN Detectors

Various successful implementations of SDRN detectors have been demonstrated. The hybridization chain reaction (HCR) of [20] triggers the formation of a linear nicked double-stranded polymer in the presence of an input DNA strand. The polymer units are two DNA strands initially trapped in a metastable hairpin form. The input triggers the opening of one the hairpins, which in turn triggers the other and a linear cascade of triggers results. The driving force for the reaction is the decrease in enthalpy due to the formation of additional hybridization bonds in the nicked double-stranded polymer as compared to the metastable hairpins. The system is described in Figure 25.10.

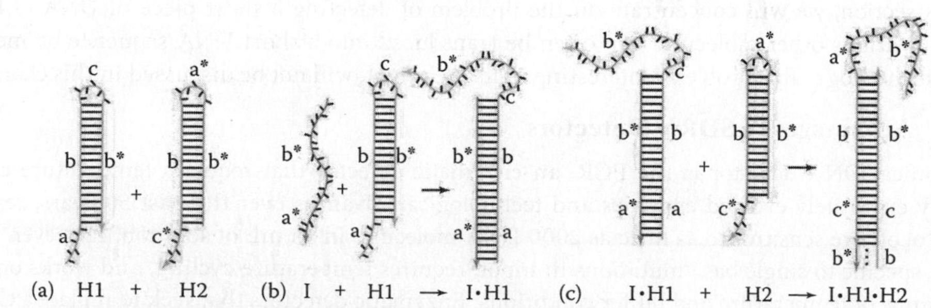

FIGURE 25.10 HCR by [20]. (a) Hairpins H1 and H2 are stable in the absence of initiator I. (b) I nucleates at the sticky end of H1 and undergoes an unbiased strand-displacement interaction to open the hairpin. (c) The newly exposed sticky end of H1 nucleates at the sticky end of H2 and opens the hairpin to expose a sticky end on H2 that is identical in sequence to I. Hence, each copy of I can propagate a chain reaction of hybridization events between alternating H1 and H2 hairpins to form a nicked double helix, amplifying the signal of initiator binding.

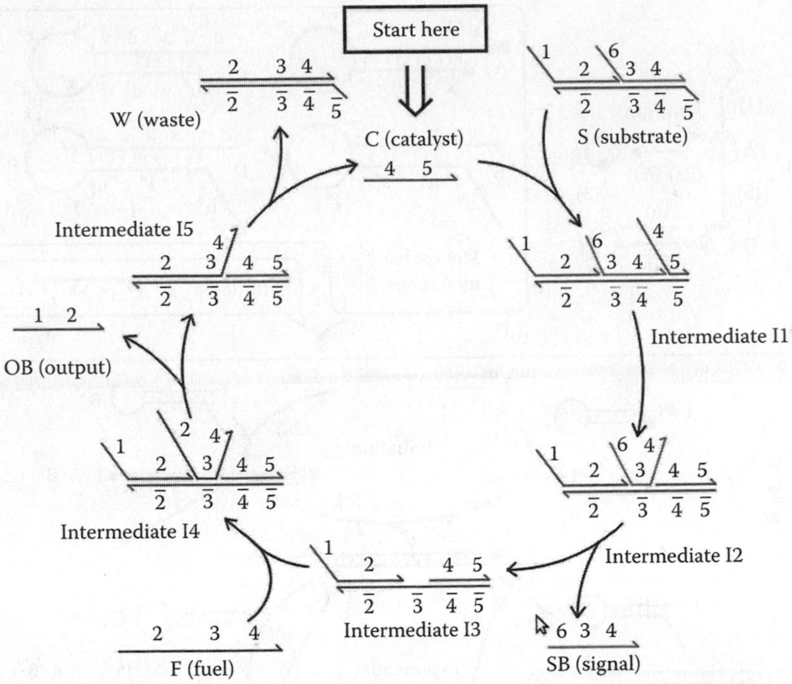

FIGURE 25.11 Entropy-driven catalyst by [127]. C (catalyst) first binds to the single-stranded toehold domain 5 on S to form the four-stranded intermediate I1, which then rearranges (by branch migration) to form I2. The binding between toehold domains 3 and 3 is too weak to keep SB attached, so I2 spontaneously dissociates into SB and I3. Newly exposed 3 then facilitates the binding of F, resulting in I4, which then quickly rearranges to release OB and I5. Finally, I5 rearranges so that C is attached only by the binding of 5 and 5 ,which spontaneously dissociates to leave W (waste) and regenerate C.

Reference [127] demonstrated an SDRN detector driven by entropy increase. The input DNA strand displaces the output from the DNA substrate via a toehold-mediated strand-displacement reaction. The input is released back into the solution by a fuel strand that displaces it from the substrate. The net effect is the fuel replacing an output molecule on the substrate, with the input acting catalytically. The direct displacement of output by fuel, in the absence of the input, is kinetically infeasible due to the lack of a toehold. This simple SDRN detector, like HCR, exhibits linear kinetics. However, when the output from one such SDRN detector is funnelled as input to a second such detector, the overall detector exhibits quadratic kinetics. Even an auto catalytic system where the output feeds back into the detector and a cross-catalytic system were demonstrated. These feedback systems exhibit exponential kinetics and hence detect input quickly. However, they suffer from higher leak rates, limiting sensitivity. The catalytic pathway is shown in Figure 25.11.

Reference [119] adapted a hairpin-based SDRN detector to allow it to exhibit exponential behavior. Their cross-catalytic system consists of two hairpins trapped in a metastable state. The hairpins are triggered by input and form a double-stranded complex with single-stranded overhangs and, in the process, release the input back into the solution. One of the single-stranded overhangs of the complex also acts as a catalytic trigger, resulting in exponential kinetics (Figure 25.12). The sensitivity of this SDRN detector is comparable to that obtained in [127].

25.3.3 DNA Replicators

Scientists have long tried to discover the origin of life. What was the sequence of steps that nature took, in order to create life as we now know it. Given that initially there was just matter, nature used energy from different sources and made autonomous self-replicating cellular machinery. With the discovery of

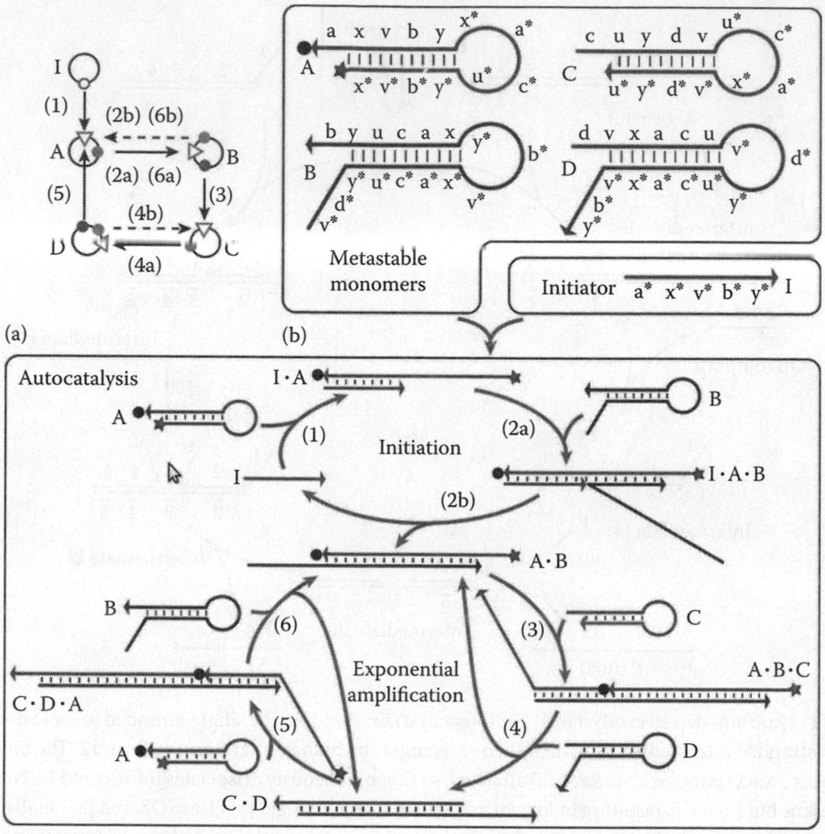

FIGURE 25.12 Hairpin-based cross-catalytic detector by [119] four hairpin species, A, B, C, and D, coexist metastably in the absence of initiator I (In b). The initiator catalyzes the assembly of hairpins A and B to form duplex ANB (steps 1–2, bottom), bringing the system to an exponential amplification stage powered by a cross-catalytic circuit: the duplex ANB has a single-stranded region that catalyzes the assembly of C and D to form CND (steps 3–4); duplex CND in turn has a single-stranded region that is identical to I and can thus catalyze A and B to form ANB (steps 5–6). Hence, ANB and CND form an autocatalytic set capable of catalyzing its own production.

the hereditary material DNA and RNA, it has been speculated that nucleic acids contain all the information required in order to completely define an organism and that the replication of these constituents is sufficient to ensure the replication of the entire organism.

Scientists at the Craig Venter Institute have created synthetic life, by artificially synthesizing a bacterial chromosome, with all the genes necessary for self-replication, and using it to boot up a cell and creating many copies of the same [26]. They ruled out the cause of self-replication being some external contaminant, by placing watermarks within the artificial DNA and re-sequencing the DNA from the cell copies, to prove their claim. However, this cell makes use of various biological molecules (enzymes) already engineered by nature to aid in the replication of an organism. How did organisms first gain the ability to self-replicate in the absence of these enzymes?

Attempts to answer this question have been made by the RNA world, who believe RNA was the first self-sustaining molecule, before DNA was chosen to be the information-bearing constituent of a cell. RNA, unlike DNA, is highly versatile and can perform ligation [24] and polymerization [39], which argue in favor of it being enough to self-replicate an organism. However, the hypothesis that nature accidentally created RNA is highly debated. Many believe that in a pre-RNA world, another molecule(s) was the hereditary material, which could transfer information to and from RNA. Recent research in synthetic molecules such as XNA and TNA is an attempt to resolve questions in this direction.

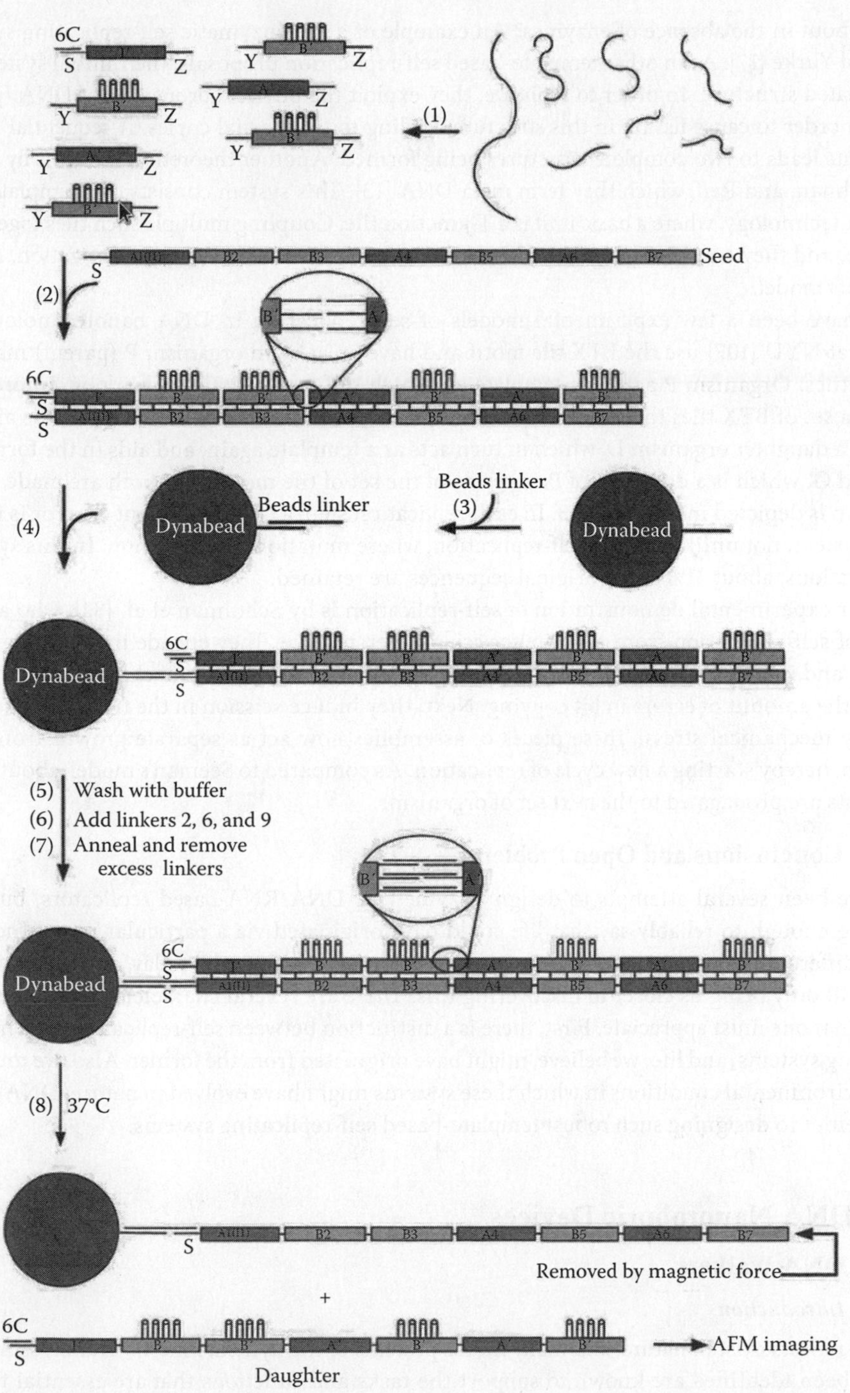

FIGURE 25.13 A seven-tile DNA replicator [107]. Replication of the seven-tile seed pattern in the first generation. In step 1, strands are annealed into tiles, all flanked by the same connectors, designated Y and Z. The initiator tile (I9) contains a protected S-strand, paired with a cover strand, 6C. The B9 tiles contain the 4-hairpin markers for AFM imaging. In the presence of the seed (step 2), the strands assemble into a pattern mimicking the seed pattern. The magnetic dynabead (large gray circle) is prepared in step 3 and attached to the seed (step 4). This is followed by a wash step, the addition of linkers, and their annealing (steps 5–7). Heating the system to 37°C results in the separation of the daughter 7-tile complex and the seed (removed magnetically, using the dynabead).

What about in the absence of enzymes? An example of a nonenzymatic self-replicating system is by Zhang and Yurke [17]. As in other template-based self-replication proposals, their initial system consists of a templated structure. In order to replicate, they exploit the physical properties of DNA (persistence length), in order to cause fission in this structure, leading to two partial copies. A sequential addition of constituents leads to two complete structures being formed. Another theoretical model is by Chandran, Gopalkrishnan, and Reif, which they term meta-DNA [13]. This system consists of a template based on DNA nanotechnology, where a basic unit is a T-junction tile. Coupling multiple such tiles together forms a template, and they emulate a huge set of enzymatic operations, including polymerization, that can be done in this model.

There have been a few experimental models of self-replication in DNA nanotechnology as well. Scientists at NYU [107] use the BTX tile motif and have created an organism P (parent) made up of 7 such BTXtiles. Organism P acts as a template on which self-replication can be done. In order to self-replicate, a set of BTX tiles that represent the complement of P are added to the mix. This aids the formation of a daughter organism D, which in turn acts as a template again, and aids in the formation of a grandchild G, which is a duplicate of P, in terms of the set of tile motifs that both are made up of. This mechanism is depicted in Figure 25.13. In each replication step, a certain amount of error is introduced into the system, not unlike natural self-replication, where mutations are common. In this system, after two replications, about 31% of the original sequences are retained.

Another experimental demonstration of self-replication is by Schulman et al. [82], who attempt the problem of self-replication from a computer science perspective. They encode information in a set of DNA tiles and induce growth via self-assembly. They have a proofreading mechanism built in, in order to reduce the amount of errors in bit copying. Next, they induce scission in the tiles, that is, they break it apart by mechanical stress. These pieces of assemblies now act as separate growth fronts for self-replication, hereby starting a new cycle of replication. As compared to Seeman's model, about 76% of the original bits are propagated to the next set of organisms.

25.3.3.1 Conclusions and Open Problems

There have been several attempts to design enzyme free DNA/RNA-based replicators, but none are convincing enough to reliably say that life could have originated via a particular route. Indeed, there could be different pathways via which life came to exist as we know of it today, and building different systems will only bring us closer to discovering this. There are several characteristics of these systems, however, that one must appreciate. First, there is a distinction between self-replicating systems and self-reproducing systems, and life, we believe, might have originated from the former. Also, we must account for the environmental conditions in which these systems might have evolved in nature. DNA replicators are an attempt to designing such robust template-based self-replicating systems.

25.3.4 DNA Nanorobotic Devices

25.3.4.1 DNA Walkers

25.3.4.1.1 Introduction

Molecular motors exist in nature, and many motor proteins of the dynein, kinesin, and myosin families—that have been identified are known to support the tasks and functions that are essential to life. One of the goals of nanoscience is the ability to synthesize molecular motors that are programmable and capable of operating autonomously. Drawing inspiration from nature, many examples of such synthetic motors have been designed. DNA poses a relatively easy material to use as the building block for these motors, due to the fact that its components are well understood compared to other materials at the nanoscale [72].

Walkers are synthesized DNA systems that are able to move on a substrate. A major ambition for those designing and experimenting with DNA walkers is to be able to design autonomous systems that can be easily controlled and programmed with specific functions. Autonomy in this case is defined as a system's ability to function in an environment without external intervention such that when the source of energy that drives it is depleted, the system halts [123].

Investigators have attempted to alter the method by which the walker moves, the substrate that the walker is moving on, or the fuel source that powers the walker in order to fully realize these two goals. In addition, investigators have also shown that walkers can perform specific tasks as proof of their viability and potential, such as transporting cargo [121], moving along different programmable pathways [52,72], assembling other structures [33], or performing controlled multistep chemical reactions [35]. One can divide the different classes of walkers in terms of (1) their degree of autonomy or (2) the type of fuel that powers these walkers or (3) both. We will start by looking at walkers with minimal autonomy, which respond to changes in their environment. Next, we will cover walkers powered by enzymes, followed by those that are driven by strand-displacement reactions, and finally we will address walkers that are considered autonomous as of this writing. In this section, the terms stators, anchorage sites, or footholds on a track have been used interchangeably, to denote a place where the walker can attach to, on a track or substrate.

25.3.4.1.2 Environmentally Driven Walkers

Nanomachines that function through the introduction of environmental alterations, such as variations to the pH of the solution, salt concentration, or the introduction of agents that cause synthetic DNA to change its conformation or shape, are considered nonautonomous, due to the fact that these nanomachines are heavily dependent on external intervention. Such environmental characteristics have been observed since the early 1970s (see [15,48,66] for some early examples) and later demonstrations of conformational changes based on the properties of the i-motif (see [46,48–51,58]). However, walkers that operate by means of environmental changes have only recently arisen.

Wang et al. [105] demonstrate the operation of a walker that switches back and forth between two sites on a track (Figure 25.14). Each site is made up of a single strand of DNA that is partially hybridized to the track, with a loose, dangling region (15 bp) that acts as the binding site for the walker. The sites are separated by 15 nt, and the first site is made up of an i-motif DNA strand, which collapses onto itself in an acidic environment (pH 5) and switches back to a loose conformation in an alkaline solution (pH 8; for more details, see [45]). Initially, when the walker is introduced, it hybridizes with the first site if the solution is alkaline. Once the solution becomes acidic, the strand of the first site forms a compact i-motif

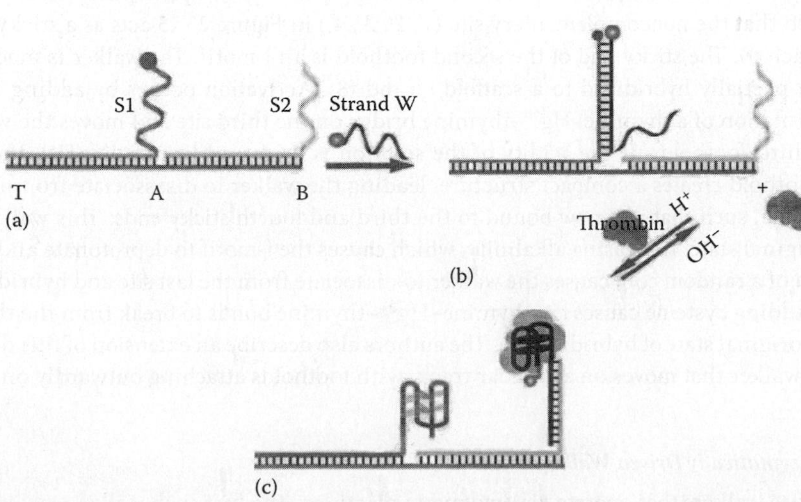

FIGURE 25.14 A walker powered by the pH levels of its environment. (From Wang, C. et al., *Chem. Commun.*, 47(5), 1428, 2011.)

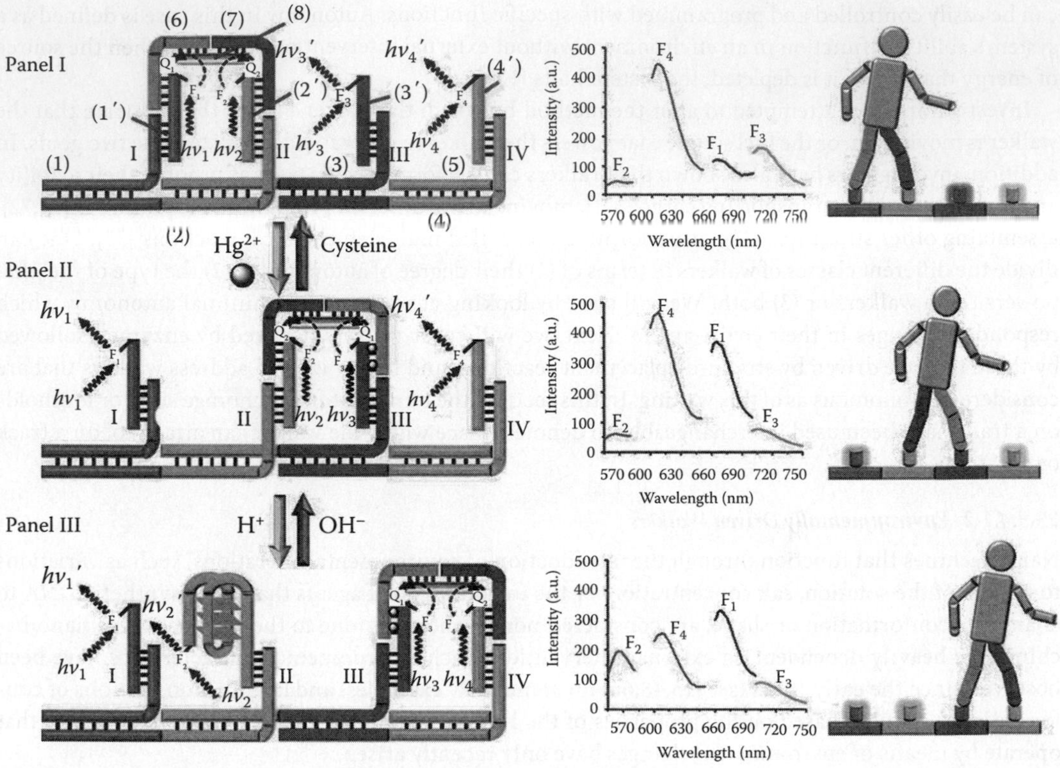

FIGURE 25.15 A walker that moves using both the i-motif, and Hg^{2+} and cysteine. (From Wang, Z.G. et al., *Nano Lett.*, 11(1), 304, January 2011.)

structure, which causes the walker to dissociate from the first site and hybridize with the second site. The limitation of this design is that the walker strand can only move back and forth between two sites, without progressing any further.

Another example is the bipedal walker and stepper by Wang et al. [109] (Figure 25.15). This walker operates using four footholds (I, II, III, IV), each having a partially complementary strand attached to the track, such that the noncomplementary site (1′, 2′, 3′, 4′) in Figure 25.15 acts as a sticky end for the walker to attach to. The sticky end of the second foothold is an i-motif. The walker is made up of two strands, each partially hybridized to a scaffold strand (8). Activation occurs by adding Hg^{2+}, which causes the formation of a thymine–Hg^{2+}–thymine bridge on the third site and moves the walker to the second and third footholds. If the acidity of the solution is increased (increasing H^+), the i-motif on the second foothold creates a compact structure, leading the walker to disassociate from it and attach to the fourth site, such that it is now bound to the third and fourth sticky ends. This walker can cycle back to its original state. Increasing alkalinity, which causes the i-motif to deprotonate and take on the conformation of a random coil, causes the walker to dissociate from the last site and hybridize with the second site. Adding cysteine causes the thymine–Hg^{2+}–thymine bonds to break from the third site and return to the original state of hybridization. The authors also describe an extension of this design to create a bipedal walker that moves on a circular track, with footholds attaching outwardly on the scaffold of the track.

25.3.4.1.3 *Enzymatically Driven Walkers*

Next we discuss walkers that operate via enzymatic reactions. The first such walker was introduced by Reif et al. [121], where a unidirectional walker was able to move from one site to another by the introduction of cleaving and ligating enzymes that targeted specific sites between the walker and the footholds at

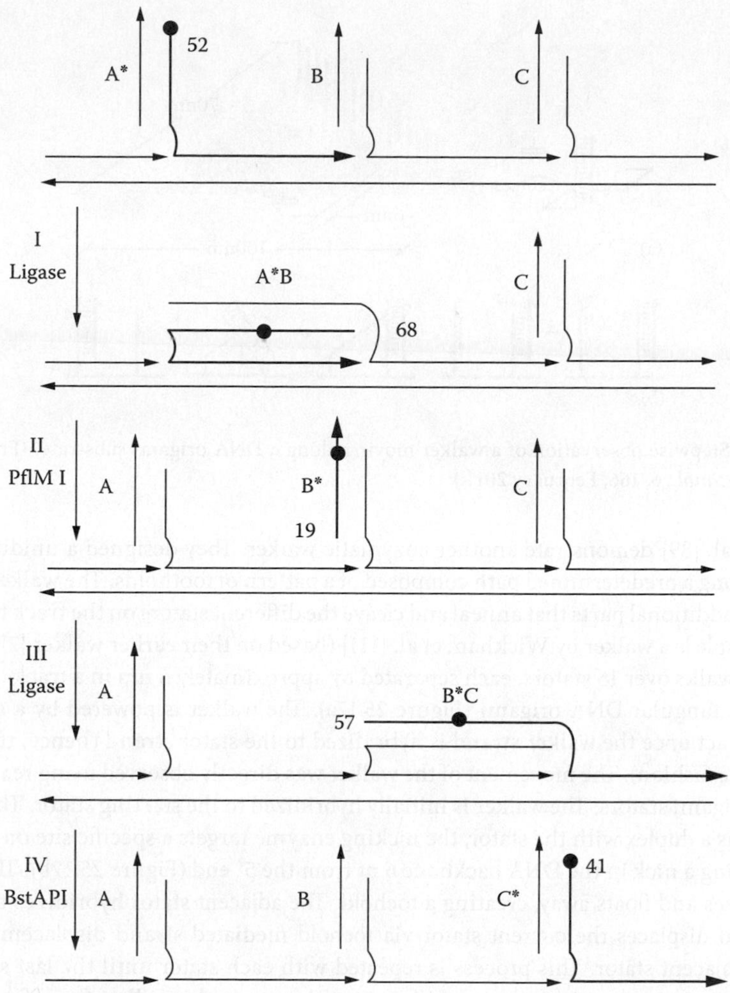

FIGURE 25.16 A unidirectional walker powered by enzymatic reactions. (From Yin, P. et al., *Angew. Chem. Int. Ed.*, 116(37), 5014, 2004.)

each state of movement. The walker is composed of 6 nt that walk over three anchorage sites (A,B,C) on a track (see Figure 25.16; the asterisk denotes the current location of the walker). The anchorage sites are attached via a "hinge," a 4 nt single strand of DNA, which adds flexibility to the sites. The walker starts off at the first anchorage site (A), and when Ligase (step I) is added, it causes the walker to ligate with the second anchorage (B), which creates a recognition site.

When the restriction enzyme (Pf1M I) is added, it cleaves the two anchorages such that the walker ends up on the second site. This cycle gets repeated in steps III (ligase) and IV (restriction enzyme BstAPI). Note that the restriction sites of Pf1M I and BstAPI are different, causing the walker to move in a unique direction. Other notable bidirectional walkers [7,100] are enzymatically driven, such as the walker by Bath et al. [7]. This walker uses a "burnt-bridge" mechanism [54], where random walks are biased in one direction, preventing the walker from moving backwards with a certain probability on a track with weak "bridges" or stators. The walker in this case cleaves the stator that it is hybridized to through a restriction enzyme, which moves the walker forward to hybridize with the next stator. Since the previous stator(s) is "burnt," the walker is much less likely to hybridize with them; then it is to stay with the current stator, until another restriction enzyme cut catalyzes its next move.

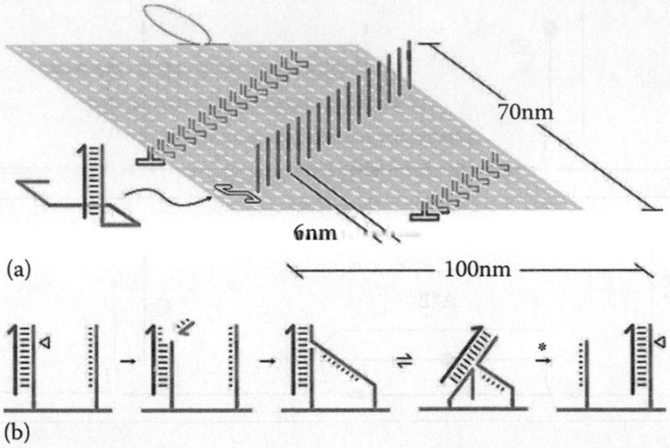

FIGURE 25.17 Stepwise observation of a walker moving along a DNA origami substrate. (From Wictham, S.J. et al., Nat. *Nanotechnol.*, 6, 166, February 2011.)

Sekiguchi et al. [89] demonstrate another enzymatic walker. They designed a unidirectional walker which moves along a predetermined path composed of a pattern of footholds. The walker is composed of three legs, with additional parts that anneal and cleave the different stators on the track to move forward.

The last example is a walker by Wickham et al. [111] (based on their earlier walker [7]) where a walker of length 16 bp walks over 16 stators, each separated by approximately 6 nm in a track of approximately 100 nm on a rectangular DNA origami (Figure 25.17a). The walker is powered by a nicking enzyme, which can only act once the walker strand is hybridized to the stator strand (hence, the enzyme operates in a stepwise fashion). The movement of the walker was directly observed using real-time AFM as it traversed the origami stators. The walker is initially hybridized to the starting stator. The addition Once the walker forms a duplex with the stator, the nicking enzyme targets a specific site on the backbone of the stator, creating a nick in the DNA backbone 6 nt from the 5′ end (Figure 25.17b). The nicked strand (6nt) dehybridizes and floats away, creating a toehold. The adjacent stator hybridizes to the walker via the toehold, and displaces the current stator via toehold mediated strand displacement, moving the walker to the adjacent stator. This process is repeated with each stator until the last stator is reached, which does not contain the nicking recognition site. This stops the walker from further movement. Hairpins were added as an AFM imaging reference parallel to the track. In addition, FRET experiments were run by having the walker carry a quencher with it that reacts with different fluorophores on the track to determine when the walkers reach specific points on their tracks.

25.3.4.1.4 Walkers Driven by Strand-Displacement Reactions

Strand-displacement reactions involve the use of a short, unhybridized portion (called a *toehold*) of a strand in a duplex to mediate the displacement of its complementary hybridized strand. These reactions can be designed to power walkers, as seen in the first demonstration by Seeman and Sherman [91], where a biped walker used set and unset strands to move along a three foothold track. Figure 25.18 shows a cartoon depicting individual steps of the walker. Figure 25.18a shows two sets of strands "Set 1A" and "Set 2B," to keep the walker attached to the footholds. In Figure 25.18b, an unset strand "unset 2B" uses toehold-mediated strand displacement to unbind the "set 2B" strand from both the biped's foot and the foothold strands. This creates a duplex waste product (Figure 25.18c), and causes the biped's foot to dissociate from the foothold. In Figure 25.18d, strand "set 2C" is added, which causes one foot of the walker to hybridize to foothold 2C. In Figure 25.18e, "unset 1A" detaches the left foot, and in Figure 25.18f, "set 1B" attaches the left foot to foothold B. In this manner, the walker is propelled forward.

The walker by Shin and Pierce [92] is similar to Seeman's walker, except that the trailing foot starts the motion forward along the track, while the leading foot keeps the walker attached to the track.

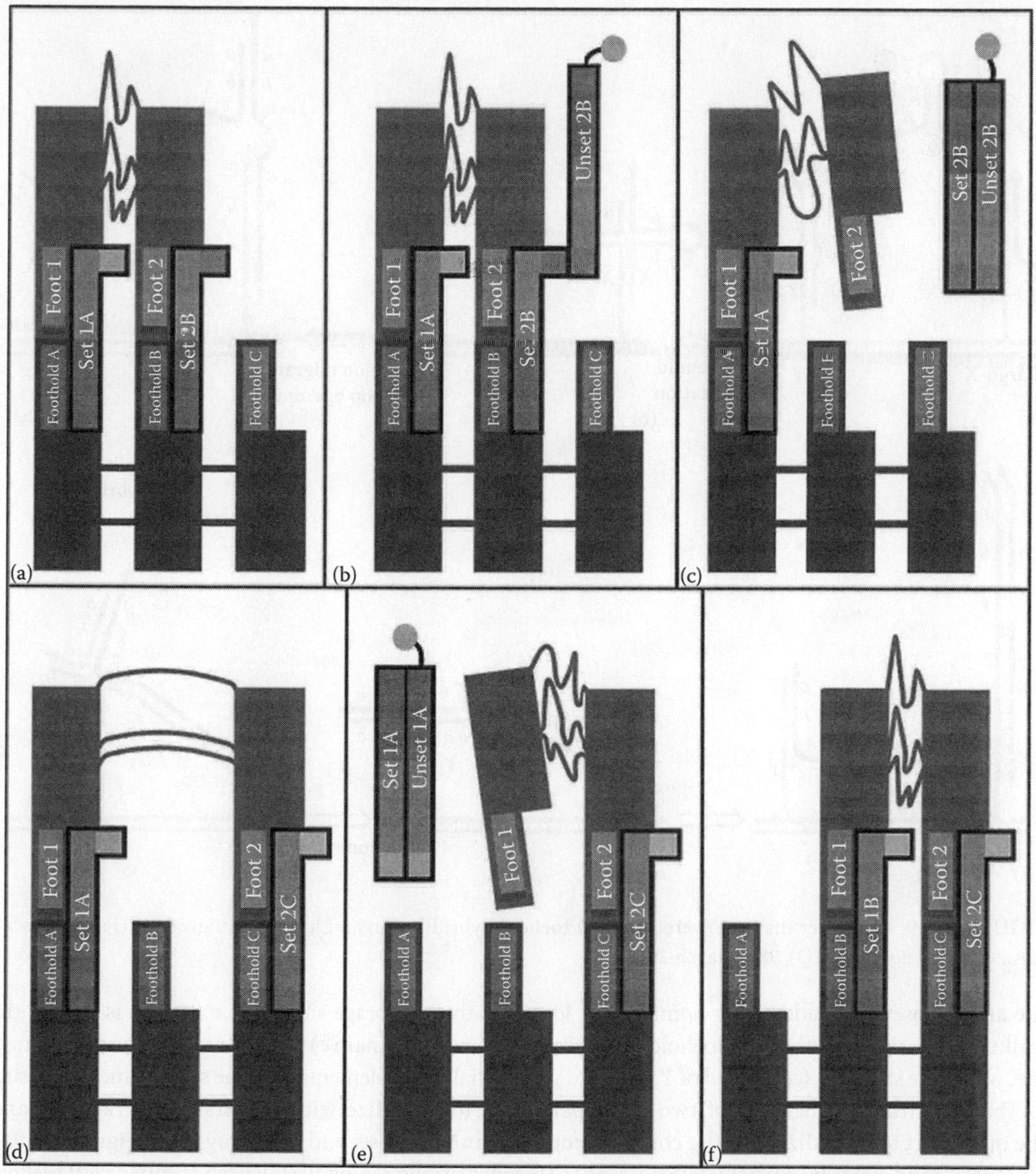

FIGURE 25.18 A biped DNA walker using set and unset strands to move at each step. (a) State 1A, 2B, (b) Starting unset, (c) State 1A, (d) State 1A, 2C, (e) State 2C, (f) State 1B, 2C. (From Sherman, W. and Seeman, N., *Nano Lett.*, 4, 1203, 2004.)

Other similar examples include Mao and Tian's circular biped walker [101], Bath et al. 's Brownian ratchet walker [31], and Dirks et al. 's autonomous cargo transporter [103].

Muscat et al. [61] demonstrated a more recent example, where the walker is composed of a cargo strand, powered forward by the introduction of a fuel strand (Figure 25.19). What distinguishes this walker is its usage of a combination of toehold hybridization and junction migration. In addition, the fuel strands are able to provide the walker with direction instructions.

Anchorage sites attached to a double-stranded track are made up of the domains $\overline{X}\overline{b}\overline{c}$, where $\overline{X}$ is unique to each site, while $\overline{b}\overline{c}$ are common to all. Note that $X = X_1 X_2$. A cargo strand *abc* can bind to anchorage site via hybridization with $\overline{b}\overline{c}$. The domain *a* is left to split (hence is called a split toehold) from the anchorage

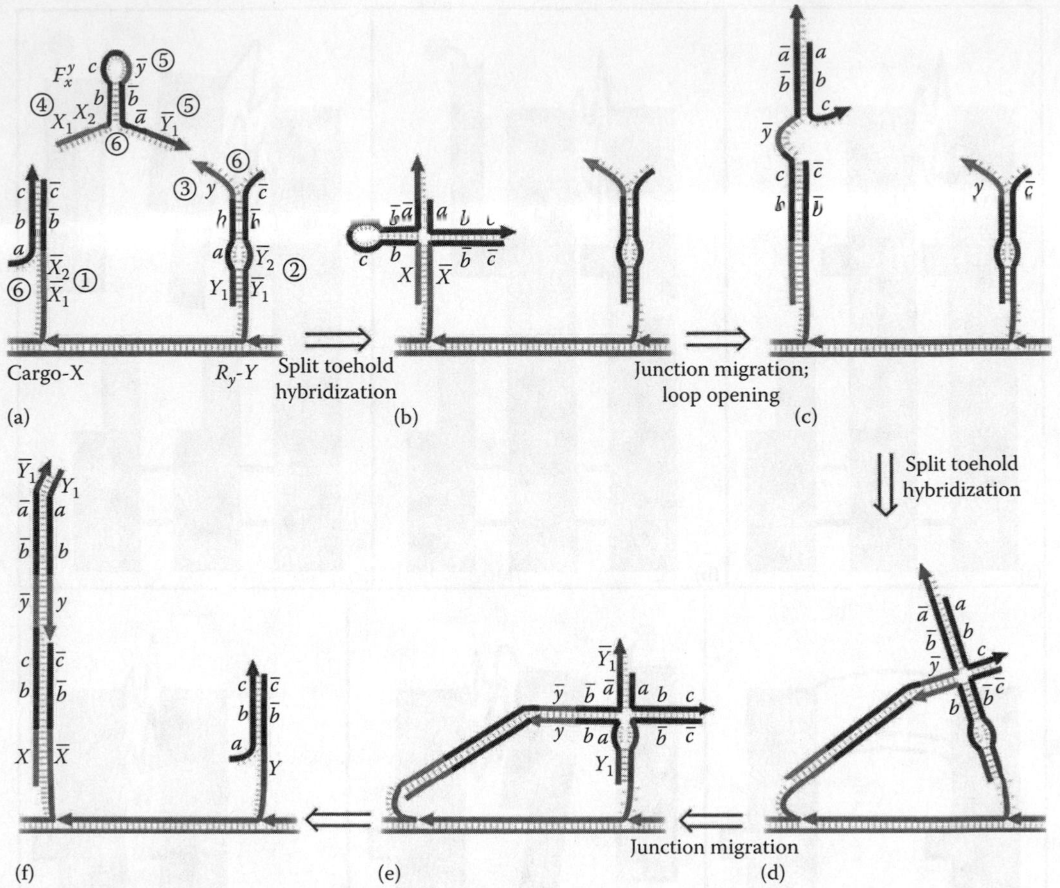

FIGURE 25.19 A walker that is powered by split toehold hybridization and junction migration. (From Muscat, R.A., et al., *Nano Lett.*, 11(3), 982, March 2011.)

site and expose $\overline{X}$ (an addressing domain that identifies the anchorage site). This exposure is key for the walker to function. Another split toehold (between toehold y and domain $\bar{c}$) is made at the second anchorage site, where the strand R_y (composed of Y_1, a, b and y) is partially complementary to the second anchorage site.

The fuel strand is made up of two main parts: one to hybridize with the first anchorage site and the other part to hybridize with the complementary strand of the second anchorage site (Figure 25.19a). The fuel strand starts by partially hybridizing to this identifying region first, which is called *split toehold hybridization* (Figure 25.19b). Afterwards, through junction migration, the fuel strand moves the cargo from the anchorage site onto its common binding domain (Figure 25.19c). Once that occurs, another split toehold hybridization reaction occurs, this time between y and $\bar{c}$ and the now unsequestered complementary domains $\bar{y}$ and the split toehold on the cargo strand c (Figure 25.19d). Through junction migration (Figure 25.19e), the cargo strand is finally moved onto the second anchorage site, and the fuel strand anchors R_y to the first anchorage site (Figure 25.19f).

Another example includes a transporter by Wang et al. [110], which is composed of two parts. The main nanostructure is composed of a 3-arm structure that has four footholds, three on the circumference and one at the center (Figure 25.20a). To each foothold, there is attached one strand each (A, B, C, A_x). This structure is named Module I. The second part of the system (Module II) consists of an arm that can transport a cargo strand (P_x) around the central axis in a circular fashion. Similar to the mechanism of set and unset strand displacement and unset strands seen earlier [91], the cargo strand is detached from the site; it is initially hybridized via the introduction of a fuel strand (Panel I, (Figure 25.20b))

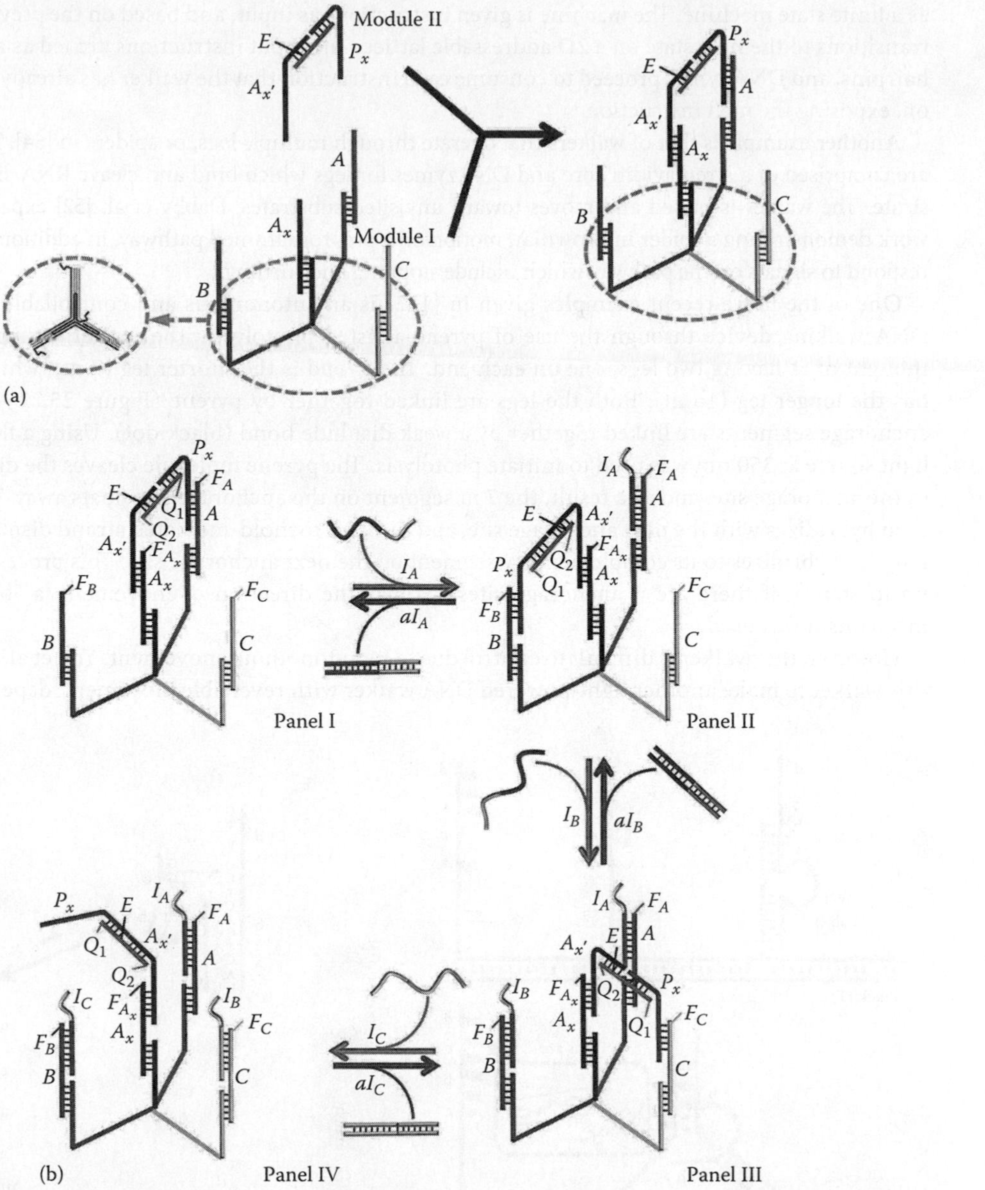

FIGURE 25.20 A transporter that can move cargo along a circular track. (From Wang, Z.-G. et al., *Angew. Chem. Int. Ed.*, 51(18), 4322, 2012.)

complementary to that site. The arm is free to attach to the next site, B, (Panel II) until another fuel strand is introduced that essentially displaces the cargo strand from B to attach to C (Panel III). Anti-fuel strands can be introduced to reverse the sequence of movement. In order to ensure that the walker moves in the order specified, the duplexes that the cargo forms at each site were made energetically less stable than the previous site, starting from the first site A, to the last site C.

25.3.4.1.5 Autonomous, Programmable Walkers

The class of walkers described here are those that are considered programmable and autonomous. Reif and Sahu [72] introduced the theoretical design of an autonomous programmable RNA walker that acts as a finite state machine. The machine is given instructions as input, and based on the previous input, it transitions to the next state on a 2D addressable lattice. The input instructions are fed as a sequence of hairpins, and DNAzymes proceed to consume each instruction that the walker has already transitioned on, exposing the next instruction.

Another example is that of walkers that operate through multiple legs, or spiders in [64]. These spiders are comprised of a streptavidin core and DNAzymes for legs which bind and cleave RNA sites on a substrate. The walker is biased and moves toward unvisited substrates. Dabby et al. [52] expanded on this work demonstrating a spider in Brownian motion on a preprogrammed pathway, in addition to making it respond to signals on the pathway which include stopping and turning.

One of the more recent examples given in [122] is an autonomous and controllable light-driven DNA walking device through the use of pyrene-assisted photolysis. The walker, a single strand, is thought of as having two legs, one on each end. The 3′ end is the shorter leg (7 nt), while the 5′ end has the longer leg (16 nt). Both the legs are linked together by pyrene (Figure 25.22—step 1). The anchorage segments are linked together by a weak disulfide bond (black dot). Using a fluorometer, a light source at 350 nm was used to initiate photolysis. The pyrene molecule cleaves the disulfide bond in the anchorage site, and as a result, the 7 nt segment on the anchorage site floats away. The short leg then hybridizes with the next anchorage site, and through toehold-mediated strand displacement, the long leg hybridizes to its complementary segment on the next anchorage site. This process is repeated up to step n, if there are n anchorage sites. To bias the direction of movement, a "burnt-bridge" mechanism was used.

However, this walker is difficult to control due to its autonomous movement. You et al. [123] altered this walker to make another light-powered DNA walker with reversible movement, depending on the

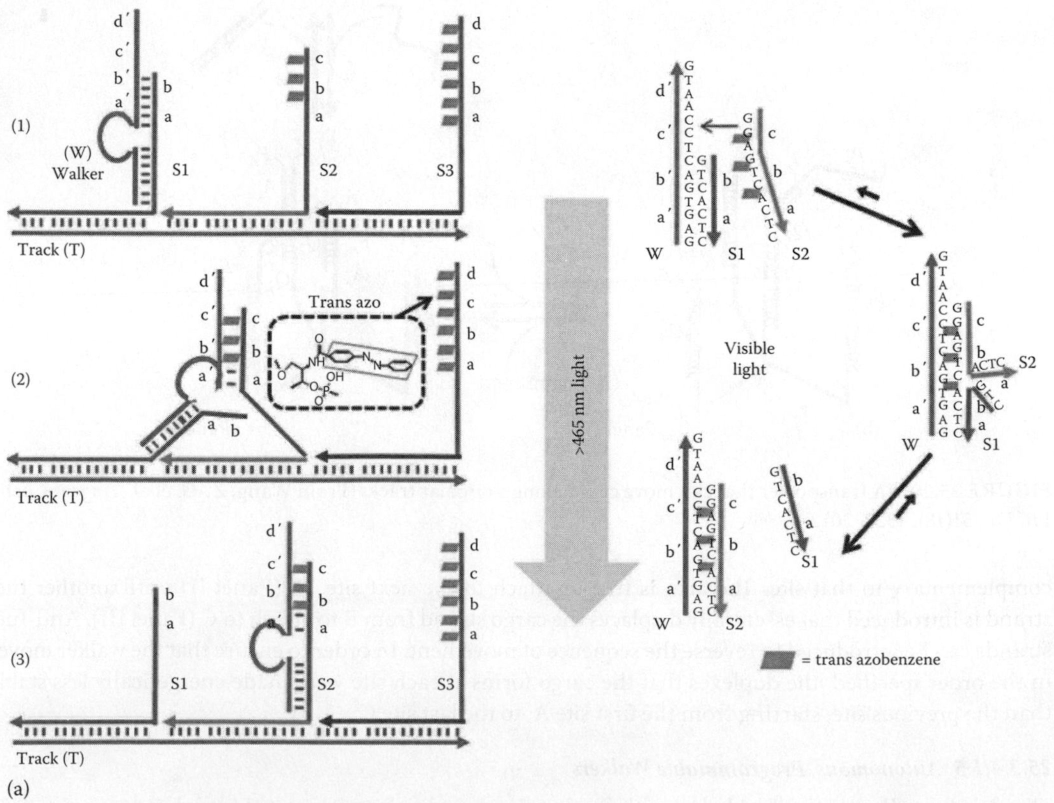

FIGURE 25.21 A reversible walker powered by (a) visible [122].

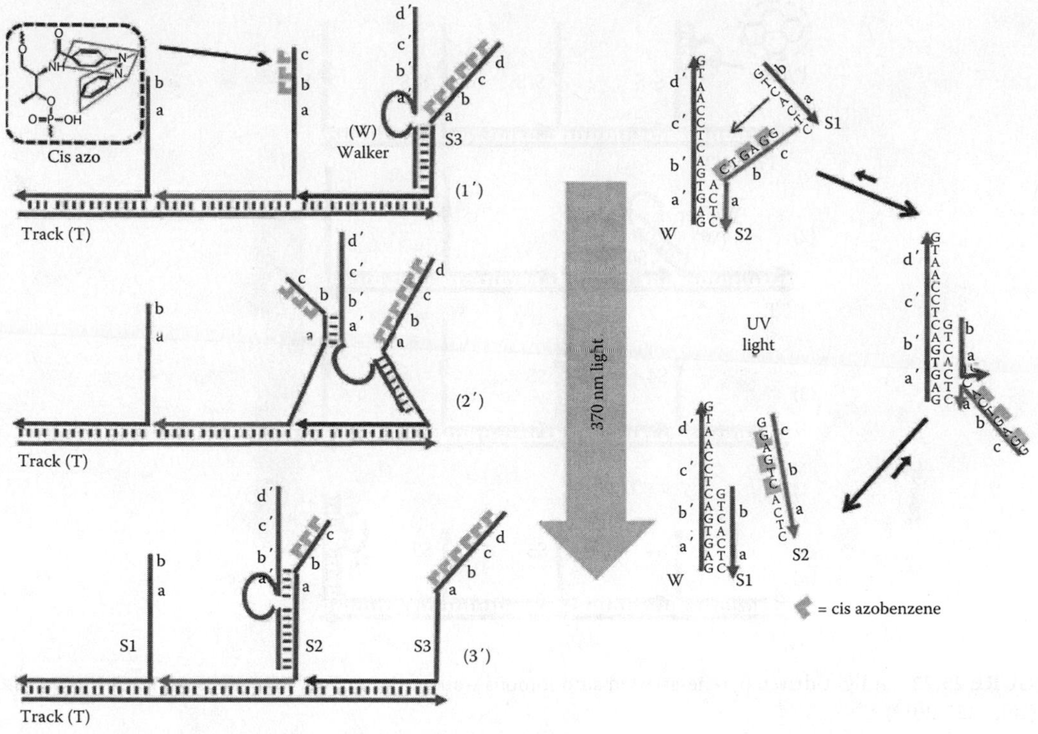

FIGURE 25.21 (continued) A reversible walker powered by (b) UV light irradiation. (From You, M. et al., *Angew. Chem. Int. Ed.*, 51(10), 2457, 2012.)

light's wavelength. In visible light, the walker moves forward, whereas UV light irradiation moves the walker backwards on a track. The anchorage site toeholds, also known as anchorage extender segments, have a different number of azobenzene molecules (one incorporated between every two bases of the toeholds). Azobenzene is a chiral molecule, and it switches its conformation from cis to trans when it is exposed to light wavelengths greater than 465 nm. The reverse is also true, namely, it switches back to cis in the presence of light with wavelength less than 465 nm.

The trans conformation of azobenzene stabilizes the double helix, while the cis conformation destabilizes it. Hence, in visible light, the walker moves from left to right towards the longer strand, which is the more stable duplex (Figure 25.21a). In UV light, however, azobenzene switches from trans to cis, destabilizing the helix. This causes the walker to reverse direction, preferring the more stable hybridization with the first site, again moving via toehold-mediated strand displacement (Figure 25.21b).

25.3.4.1.6 Discussions

DNA walkers are stepping stones to constructing DNA systems that have dynamic behavior, and are able to perform mechanical tasks at the nanoscale. By programming the directed movement of DNA molecules along a track, we can target the walker to deliver cargo reliably, send or receive information, and achieve nanoscale synthesis.

Multistep biosynthesis has been demonstrated by Liu et al. [35]. They designed a DNA walker that performs the synthesis of a linear product, in the order specified by the placement of the anchorage sites on the DNA track. At each step, the walker builds on the product, by attaching the cargo at the current site to that of the prior sites. At the final step a resulting synthesized product is obtained, not unlike mRNA translation.

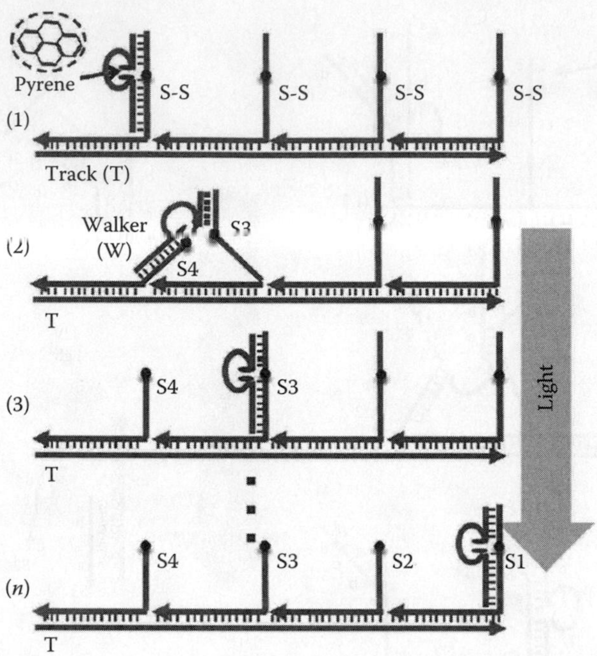

FIGURE 25.22 A light-driven pyrene-assisted autonomous walker. (From You, M. et al., *Angew. Chem. Int. Ed.*, 51(10), 2457, 2012).

25.3.5 DNA Dynamical Systems

The behavior of DNA complexes can be controlled via the action of DNA enzymes that act upon specific sequences of DNA strands [74], competitive DNA hybridization [131], or environmental changes such as pH or temperature [108]. One of the first applications of strand displacement to a DNA nanomachine was the molecular tweezer by [124]. The dynamic behavior of the tweezer is controllable, using strand displacement. Other types of devices include walkers [70,121] and molecular detectors [20], both in vitro and in vivo [59,104]. Delebecque et al. [18] also made nanostructures within a cell. Reference [33] made a molecular assembly line. These and many other devices have been constructed and studied.

25.3.6 DNA Tiling Assembly

25.3.6.1 Models of Tile Assembly

In this section, we give a brief review of the theoretical models of self-assembly. In DNA self-assembly, a tile is the basic unit of assembly, with the initial state consisting of a set of unordered tiles, and the final state consists of an assembly formed by some or all of these tiles sticking to one another. Theoretical models of tile assembly have been developed to study the computational properties of self-assembly. Tile assembly dates back to 1961, when Wang [106] proposed the Wang domino problem. In his tiling model, a tile is a square with a glue on each side. For two tiles to attach, their abutting glues should be the same. Retroflexion and rotation of tiles are not permitted. The problem statement is: "Given a finite set of tiles, can they tile the entire plane." An infinite number of copies of each tile type are provided. A more general problem is: "Does there exist an algorithm that can decide if a finite set of tiles can tile the whole plane." It has been proved that such an algorithm does not exist and that the tiling problem is undecidable [9,75]. The proof shows that any instance of the Turing halting problem can be reduced to an instance of the Wang domino problem. Since the halting problem is not decidable, hence the Wang domino problem is not decidable either.

Rothemund and Winfree proposed the abstract tile assembly model (aTAM) in 2000 [78]. In aTAM, a tiling system is modelled as a quadruple $< T, s_0, \tau, g >$. (1) T here is the tile set. As in the Wang model, a tile in aTAM is a square with a glue on each side. All glues are from a glue set Σ. Retroflexion and rotation of tiles are forbidden. (2) s_0 is a special tile called the *seed tile*. At the beginning of the assembly process, only the seed tile is fixed at a particular position and the other tiles are floating in the plane. Assembly begins from the seed tile. (3) τ is the temperature of the system. When a tile is at a position, it will be fixed at that position if and only if the accumulative glue strength between this tile and tiles around it is not less than τ. (4) g is the glue strength function. It defines the glue strength between two glues. An assembly that grows from the seed tile can grow infinitely large, or it may terminate. A configuration is *terminal* if it cannot grow anymore. This happens when there is no additional site for a tile to attach to at temperature τ. In Winfree's original definition [78], the aTAM is formally defined in 2D, but it can be extended to multidimensions. The 2D aTAM has been proved to be the Turing universal [117].

We begin by introducing some important definitions in aTAM. (1) The tile complexity of shapes [78]. The tile complexity of a shape is the minimum number of tile types that *uniquely* assemble that shape. (2) The time complexity of assembling a shape [3]. This models how fast a shape can be assembled. The assembly process assembly process of an aTAM system can be modelled by a continuous Markov process. The states of the Markov process are the different tile assembly configurations. To proceed from one state to another, a tile(s) attaches or detaches from the growing assembly. The rate of transition from state S_1 to state S_2 is dependent on the concentration of tile x if S_2 is formed by attaching x to S_1. The time taken to reach the terminal state from the initial state is a random variable, and time complexity is defined as the expected value of this random variable.

One important problem in aTAM is the tile complexity and time complexity of assembling a fixed-shaped assembly. The fixed shape used by Winfree and Rothemund is an $N \times N$ square, where N is any positive integer. In other words, "What is the minimum number of tile types needed to construct an $N \times N$ square at temperature τ?" and "What is the minimum expected time needed to construct an $N \times N$ square at temperature τ?"

Rothemund and Winfree give a construction to show that the upper bound of tile complexity of assembling $N \times N$ square in a $\tau = 2$ system is $O(logN)$ [78]. The basic idea of their construction is to use $O(logN)$ tile types to build the frame of an $N \times N$ square and then use a constant number of tile types to fill the frame. The frame has two components: a counter of size $N \times logN$ that grows toward the north and a diagonal of the $N \times N$ square that grows toward the northeast. They also proved a tile complexity lower bound $\Omega(logN/log\ logN)$ for the same, using the Kolmogorov complexity. This lower bound is true for infinitely many N. The question is whether there exists a construction that can reach this lower bound to prove that it is tight.

In 2001, Adleman et al. [3] proved that the lower bound is indeed tight. Adleman et al. proposed a new construction of tile complexity $\Theta(logN/log\ logN)$ that reaches the lower bound proved by Rothemund and Winfree. Their original construction is a $\tau = 3$ construction, but can be generalized to $\tau = 2$. The trick to improve the tile complexity was the technique by which they encoded a number. Rothemund and Winfree encoded N in base 2 (binary), and this technique led to the $O(logN)$ tile complexity. In the new design by Adleman et al., they encoded the number in base $\Theta(logN/log\ logN)$. This strategy reduced the tile complexity to $\Theta(logN/log\ logN)$.

Adleman et al. also generalized Winfree's aTAM by incorporating time complexity to the original model. The construction of tile complexity $O(logN)$ from Rothemund and Winfree was proved to have $\Theta(N\ logN)$ time complexity. Adleman et al. gave a time complexity of $\Theta(N)$, which is optimal for assembling an $N \times N$ square.

Many variants of the aTAM have been developed. Winfree developed the kinetic tile assembly model (kTAM) [114] to model the kinetic properties of crystal growth. This model is based on four assumptions that are essential to it. Aggarwal et al. [4] developed generalized models including the multiple temperature model, flexible glue model, and q-tile model. Sahu et al. [99] proposed the time-dependent glue strength model. In their model, the glue strength between two glues increases with time, until it reaches a maximum. The advantage of this model is that it can model catalysis and self-replication. Reif [71] proposed the stepwise assembly model in 1999. In this model, the assembly process is done by multiple steps. At each step, a separate assembly with a different tile sets and different temperature is created. Only the terminal product of one step is given as input to the next step. The terminal product of the final step is the terminal product of the whole system. We give more details of Reif's model in Section 25.3.6.6. In 2008, Demaine et al. proposed a generalized stepwise assembly model called the *staged assembly model* [19]. They also gave strategies to assemble shapes with $O(1)$ glues under their model. It has been proved that both the stepwise assembly model and the staged assembly model are the Turing universal at temperature 1 [6].

25.3.6.1.1 Theoretical Results of Self-Assembled Shapes

All constructions earlier try to store the information of the target shape in the tile set. This implies the need for a larger tile set to, assemble a target shape of a larger size. However, it is not feasible to synthesize a large number of tile types in practice. There are many constraints to implementing large tile sets such as the length of the pad, spurious reactions among pads of same tile, and sequence design.

An alternate strategy is encoding the information of target shapes in the concentrations of tile types. By this strategy, we can reduce the tile complexity of some self-assembled shapes to $O(1)$ [21,40]. Kao and Robert proposed a random assembly construction for approximate squares [40] by encoding the shape information in the tile concentration. By their construction, an $N \times N$ square can be assembled from a constant number of tile types with high approximation factor and high probability where N should be large enough. The question is whether there exists a construction of $O(1)$ tile complexity that assembles an exact $N \times N$ square with high probability. Doty published such a design in 2009 [21].

The problem of storing shape information in the tile concentration is that it is difficult to control the concentration precisely during the reaction process. Other methods to reduce the tile complexity include attempts by Reif in 1999, the stepwise assembly model [71]. Demaine proposed a generalized version of the stepwise assembly model called the *staged assembly model* [19]. The basic idea of these two models is assembling the target shape by multiple steps. This strategy can reduce the tile complexity efficiently [19,38]. The cost is stage complexity and bin(or tube) complexity [19,38].

25.3.6.2 Tile Assembly Experimental Methods

There are several DNA motifs that are used to design DNA tiles. These include DX and TX motifs [41,116] (Figures 25.23 and 25.24). DNA holliday junctions [25], and DNA origami [76]. These motifs are made stable by the incorporation of *crossover* junctions, where a DNA strand is initially bound to a strand, crosses over, and starts binding to another strand.

A motif attaches to a growing assembly via intermolecular contacts called *sticky ends*. These are complementary single-stranded DNA, which bind to each other once in close proximity. Arbitrarily complex tile

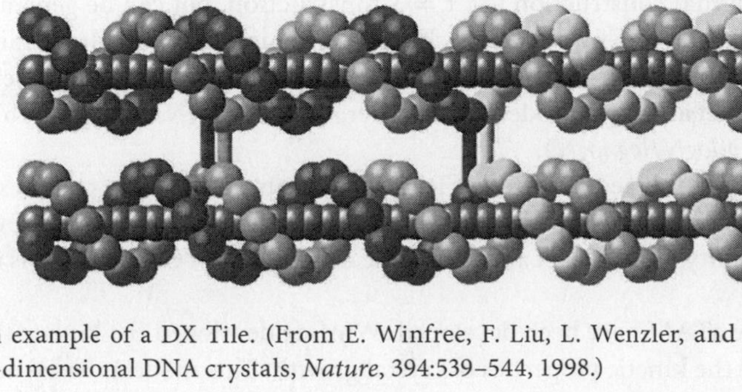

FIGURE 25.23 An example of a DX Tile. (From E. Winfree, F. Liu, L. Wenzler, and N. Seeman, Design and self-assembly of two-dimensional DNA crystals, *Nature*, 394:539–544, 1998.)

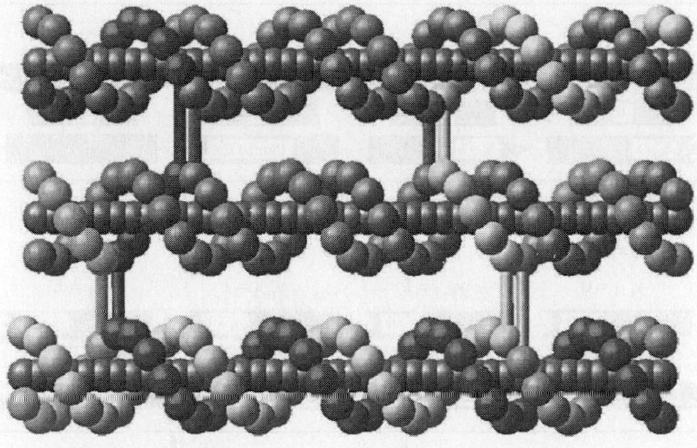

FIGURE 25.24 An example of a TX Tile. (From T. LaBean, H. Yan, J. Kopatsch, F. Liu, E. Winfree, J. Reif, and N. Seeman, Construction, analysis, ligation, and self-assembly of DNA triple crossover complexes, *Journal of the American Chemical Society*, 122(9):1848–1860, 2000.)

assemblies can be made using a small number of component tiles where each tile specifies an individual step of the computation. One of the first 1D experimental demonstrations after Adleman's travelling salesman [2] was the computation of a cumulative XOR function. The inputs were x_1, x_2, x_3, x_4, and the outputs were y_1, y_2, y_3, y_4 where $y_i = y_{i-1}$ XOR x_i and $y_1 = x_1$. A sample computation with input $x_1x_2x_3x_4 = 1110$ and output $y_1y_2y_3y_4 = 1011$ can be seen in Figure 25.25b in four steps by the self-assembly of DNA TX tiles [56]. Binary inputs are encoded in a DNA tile each, and the presence of a unique combination of input tiles results in the formation of a unique set of output tiles, seen in Figure 25.25a.

A more complex construction was the set of 2D DNA Sierpinski triangles made by Rothemund et al. [77]. They designed a 1D cellular automata, whose update rule computes a binary XOR, and the assembly results in a fractal pattern. A detailed design is in Figure 25.26. AFM images can be seen in Figure 25.27. These constructions give evidence that DNA can perform the Turing universal computation and has the ability to implement any algorithmically computable system.

25.3.6.3 Tile Assembly Design and Simulation Software

Computer-aided design (CAD) softwares help molecular architects in the design of DNA-based tiles. To design the DNA sequences of these tiles, sequence symmetry minimization [87] is a common technique used to avoid spurious DNA strand interactions. Tools used in the design of such tiles are NanoEngineer [36], Tiamat [112], TileSoft [120], and GIDEON [10]. Another tool, caDNAno, developed by Douglas et al. [23], primarily designed for the use of DNA origami, can also be used for the same.

Simulating the behavior of DNA-based tiles yields useful theoretical results, which can help predict the kind of errors we should expect when using DNA tiles for computation or construction. Winfree wrote a simulator Xgrow [114] as part of simulating the kTAM (kinetic TAM) model, with extensions to it from [5]. Another simulator is ISU TAS, developed by Patitz at Iowa State University [63].

25.3.6.4 One-Dimensional DNA Tiling Assemblies

In this section, we will give a brief review of probabilistic linear tile assemble model (PLTAM) and results under this model [12]. Assembling linear structure of given length is an important problem in self-assembly both theoretically and experimentally. Many complex structures can be made from linear structures. One question is how big the smallest tile set that assembles linear structure of length N is. If the linear assembly process is deterministic, it can be proved that the smallest tile set to assemble linear structure of length N is of size N.

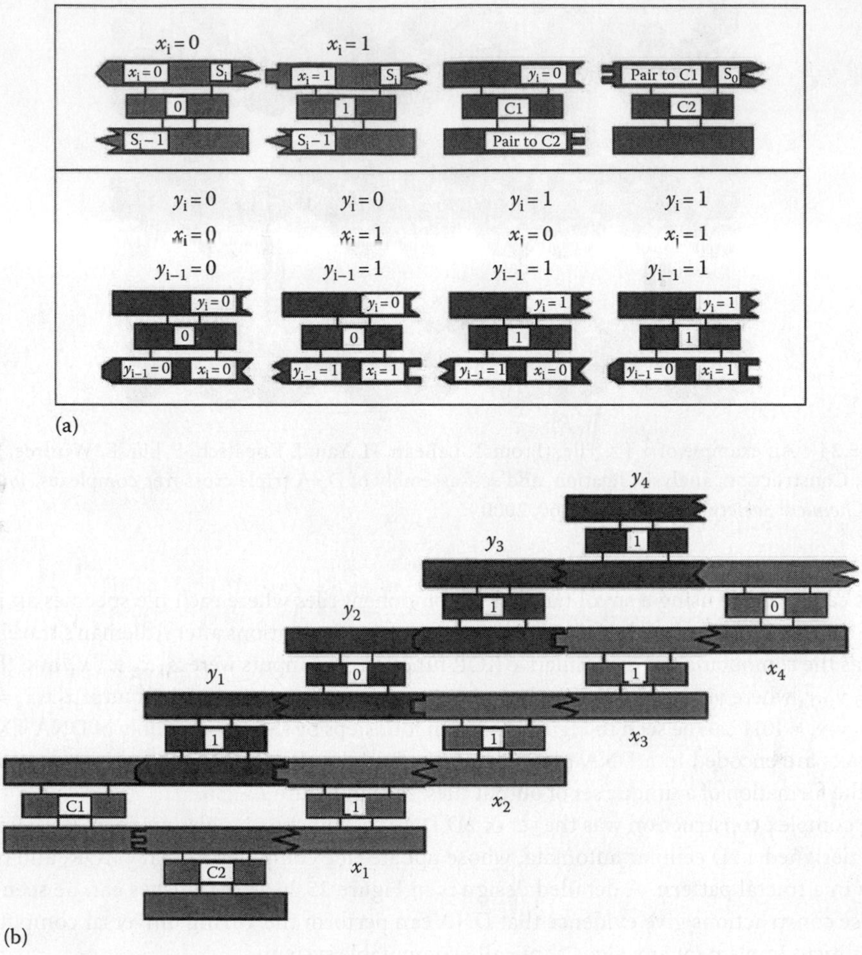

FIGURE 25.25 Calculation of cumulative XOR by self-assembly of DNA tiles. (a) Component tiles. The three helical domains are drawn as rectangles, flanked by sticky ends shown as geometrical shapes. The value of each tile is in the central rectangle. The meaning of each sticky end is also indicated. Shown are the two x tiles (top left), four y tiles (bottom row) and the initialization corner tiles C1 and C2 (top right). The y_i tiles are upside down from the x_i tiles. (b) The values of each tile are the same as in (a), and the sticky ends are the same. Note the complementarity of sticky-ended association at each molecular interface. The operations are designed to proceed from lower left to upper right, because the x_i and C1 and C2 tiles have longer sticky ends than the y_i tiles. (From Mao, C. et al., *Nature*, 407, 493, 2000.)

Chandran et al. proposed a probabilistic linear tile assembly model [12], where "probabilistic" means that given a tile type, there may be multiple tile types that can attach to its western or eastern pad and "linear" means that tiles in this model only have two pads: western pad and eastern pad. They also developed three schemes under PLTAM to assemble linear structure of expected length N with tile complexity $\Theta(logN)$, $O(log^2N)$ and $O(log^3N)$. All schemes are for equimolar standard linear assembly system. A linear assembly system is standard if and only if it is haltable, east growing, diagonal, and uni-seeded where "haltable" means that the assembly process can stop, "east growing" means that the assembly process is from west to east, "diagonal" means that the glue strength between two glues is 0 if they are different and otherwise the strength is 1 if the glue is not null glue, and "uni-seeded" means that there is only one seed tile. It can be proved that standard systems can simulate a larger set of systems. The reason why they focus on equimolar system is that equimolar system is more practical.

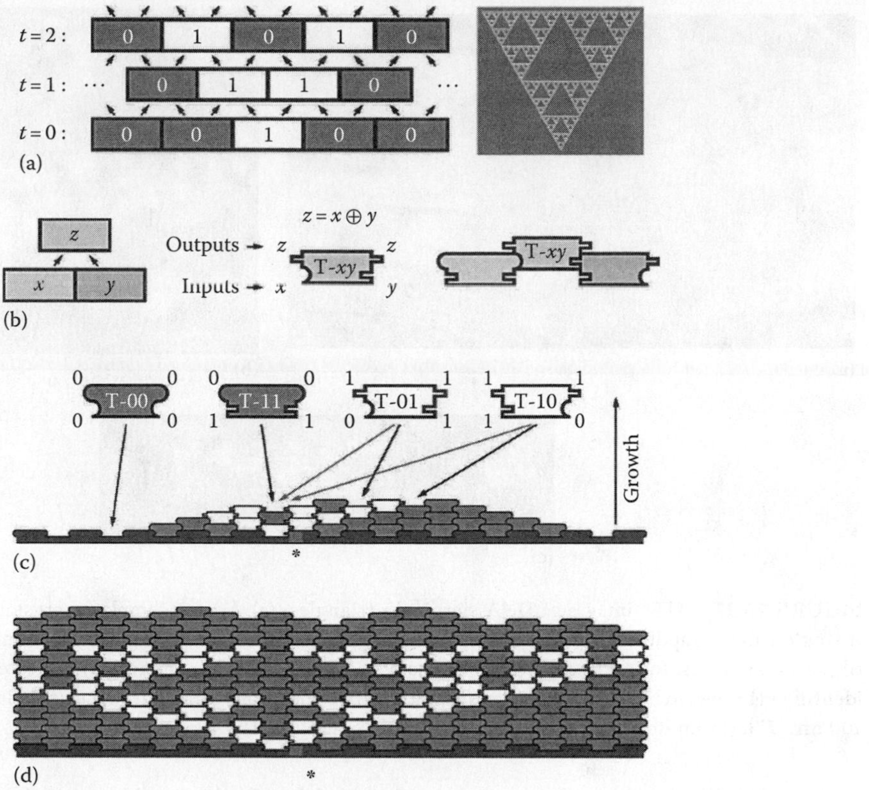

FIGURE 25.26 The XOR cellular automaton and its implementation by tile-based self- assembly. (a) Left: Three time steps of its execution drawn as a spacetime history. Cells update synchronously according to XOR by the equation shown. Cells at even time steps are interleaved with those at odd time steps; arrows show propagation of information. Right: the Sierpinski triangle. (b) Translating the spacetime history into a tiling. For each possible input pair, a tile T-xy that bears the inputs represented as shapes on the lower half of each side and the output as shapes duplicated on the top half of each side. (c) The four Sierpinski rule tiles, T-00, T-11, T-01, and T-10, represent the four entries of the truth table for XOR: 0 XOR 0 → 0, 1 XOR 1 → 0, 0 XOR 1 → 1, and 1 XOR 0 → 1. Lower binding domains on the sides of tiles match input from the layer below; upper binding domains provide output to both neighbors on the layer above. (d) Error-free growth results in the Sierpinski pattern. (From Rothemund, P. et al., *PLoS Biol.*, 2, 424, 2004.)

The basic trick of designing these three schemes is simulating some well-designed stochastic processes by linear tile sets. One issue is that all three schemes do not have a sharp tail bound of assembly length for a general *N*. What has been proved is that the $O(log^3N)$ scheme has sharp tail bound for infinite many *N*, and for large enough *N*, we can get sharp tail bound by dividing *N* into numerous identical chunks. Therefore, the open problem is whether there exists a scheme of $o(N)$ tile complexity with sharp tail bound of assembly length for general *N*.

25.3.6.5 Three-Dimensional DNA Tiling Assemblies

3D assemblies are a natural extension to the construction of 2D assemblies. Cook et al. [16] ask the question: *What is the number of tile types needed to create an N × N 2D square, using a 3D tile system?* In a 3D tile system, a Wang tile is replaced by a Wang cube, each cube having six faces and hence six glues. They give a construction showing that using $O(logn)$ tile types (cubes in 3D) can be used to construct an nxn square, at temperature 1, as opposed to the temperature 2 result by Rothemund and Winfree [78].

Constructions have been done using 3D Wang assembly models that are used to solve NP-complete problems such as maximal clique [53]. The computational power of this system is the Turing universal as

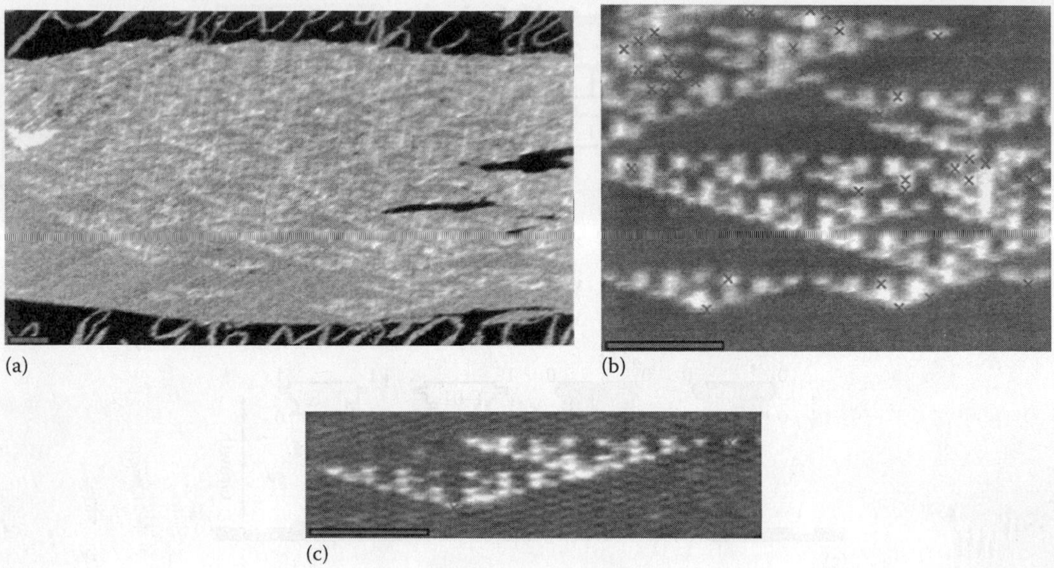

FIGURE 25.27 AFM images of DNA sierpinski triangles. (a) A large templated crystal in a 5-tile reaction. A single 1 in the input row initiates a Sierpinski triangle. (b) A zoomed in region of (a) containing roughly 1000 tiles and 45 errors. (c) A close of up a Sierpinski triangle. Crosses in (b) and (c) indicate tiles that have been identified (by eye) to be incorrect with respect to the two tiles from which they receive their input. Scale bars are 100 nm [77]. (From Rothemund, P. et al., *PLoS Biol.*, 2, 424, 2004).

suggested by Winfree in [117], since it is a superset of the 2D tiling problem, but the additional computing power of this system at different temperatures has still to be ascertained.

25.3.6.6 Stepwise Tiling Assemblies

In this section, we will give a brief review of Reif's stepwise assembly model [71]. Reif's model reduces the tile complexity by dividing the assembly process into steps. It has been proved that this strategy can efficiently reduce tile complexity [19,38]. A tiling system under the stepwise assembly model is a quadruple $< T, s_0, g, \Phi >$, where $T = T_1 \cup T_2 \ldots \cup T_m$ and $\Phi = \{\tau_1, \tau_2, \ldots, \tau_m\}$. The assembly process has m steps. In the first step, tiles in T_1 including s_0 assemble in tube 1 at temperature τ_1 for a long enough time. The terminal product of step 1 is delivered to tube 2. Add T_2 to tube 2 and let tiles (or supertiles) in tube 2 assemble at temperature τ_2 for a long enough time. This process continues until the step m is complete. The terminal product of step m is the terminal product of the whole system. Maňuch et al. proposed schemes of $O(1)$ tile complexity to assemble an $N \times N$ square [38]. Behsaz et al. proved that the stepwise assembly model is the Turing universal at temperature 1 [6]. Demaine et al. proposed the staged assembly model in 2008 [19]. It is a generalized version of the stepwise assembly model.

25.3.6.7 Activatable Tiles: Transition of State across Tiles

Experimental demonstration of the Turing universal tile-based algorithmic DNA self-assembly has been limited by significant assembly errors. An important class of errors, called *co-ordinated growth errors*, occur when an incorrect tile binds to a growing assembly even when some of its pads are mismatched with its neighbors. Activatable DNA tiles, introduced originally by [55], employ a protection/deprotection strategy to strictly enforce the direction of tiling assembly growth that prevents these errors, ensuring the robustness of the assembly process. Tiles are initially inactive, meaning that each tile's pads are protected and cannot bind with other tiles. After an activation event, the tile transitions to an active state and its pads are exposed, allowing further growth.

25.3.6.8 Error Correction in Tiling Assemblies

In this section, we give a brief review of error-correction techniques in self-assembly. Tile assembly has been beset with different types of errors. These are broadly classified into growth errors and nucleation errors. Growth errors happen when a tile incorrectly binds to a growing assembly and propagates this error by causing other incorrect tiles to attach. Nucleation errors happen when an assembly does not grow from the seed tile, but it randomly starts growing using other tiles from the tile set.

Errors in assembly are usually reduced by two approaches. (1) Reducing the errors by optimizing the reaction environment [114]. By this approach, the inherent error rate is reduced. (2) Reducing the errors by optimizing the tile set design [73,115]. However, this approach can only reduce the amount of errors in the assembled pattern and cannot improve the inherent error rate. Winfree proposed an error resilient scheme by the second approach called *proofreading tile sets* [115]. The basic idea of his scheme is using a $k \times k$ block of tiles to replace each tile in the original tile set. Although it does bring down the error from ϵ to ϵ^k, this scheme has two problems. (1) It amplifies the assembly size. (2) It can only reduce the errors in the output, while the errors inside the assembly pattern are dismissed.

Reif et al. proposed a compact error resilient scheme [73]. The basic idea of their scheme is adding proofreading portions to pads. Their scheme does not amplify the assembly size and can reduce the errors inside the assembly pattern. The problem of this scheme is that it is not suitable to general case.

25.4 Summary

DNA computing has matured since Adleman used it to solve a 7-node instance of the Hamiltonian path problem. Adleman's original aim was to leverage the inherent parallelism of DNA for computation and its high information density for storage. DNA computation started with solving a set of "hard" problems interesting to computer scientists but has since evolved to have other applications. Scientists have used DNA devices for a variety of uses, such as in biosensors for molecular detection, diagnostic applications, in nanofabrication, for structure determination, and as potential drug delivery devices.

It is important to analyze the power of any computational device. This is in order to be able to judge it against other devices with similar functionality and to get an estimate of the theoretical limits of any device. This is why CRNs (Section 25.1.3) are important to model the DNA strand-displacement techniques that are inherent to quite a few DNA devices.

DNA can be used to perform complex nanofabrication tasks. Both periodic structures such as lattices [34,86] and nonperiodic structures such as a smiley face and world map [76] have been constructed using DNA. These structures can be used as scaffolds to place gates, transistors, and routing elements at precise locations and thus create circuits from the bottom-up, as opposed to traditional top-down, lithographic techniques. However, as we discussed in Section 25.3.6.8, errors are inherent to scaffold construction, and even without errors, the size of the scaffolds that can be created pales in comparison to that achievable by silicon-based lithography. A parallel line of research is not in using DNA as a scaffold but in using DNA itself to construct the circuit elements. These designs are quite promising as discussed in Section 25.3.1. Using such devices, it would be possible to have medical diagnostic biochemical circuits, with applications similar to Benenson's DNA doctor [90].

DNA devices that are involved in detection have been a steady focus area among scientists. There exists a large set of techniques that has led to the construction of these devices. DNA machines typically respond to external stimuli such as pH, salt concentration, temperature, and other signalling molecules (DNA, proteins) that lead to a change in their state. This mechanical state change is detectable by techniques such as FRET and has led to the development of sensors. In addition, logical operations can be performed on DNA motifs, and coupled with sensors, complex biological systems can be designed as drug delivery systems.

The programmable self-assembly of DNA has also enabled us to construct biochemical circuits that can perform synthesis of complex molecules. For example, DNA-directed synthesis has been shown to be achievable using a variant of DNA walker systems [33,35]. Being programmable, the synthesis of any oligomer can be controlled at every step, not unlike the central dogma that dictates protein synthesis.

One of Seeman's original goals of constructing lattices of DNA was to use it to determine the structure of other foreign (protein) molecules via x-ray crystallography [88]. Most lattices created with DNA have a lot of defects, hence not making DNA very suitable for this use. However, with the latest advances in structural DNA nanotechnology, that goal and others seem to be finally in sight.

Key Terms

Antiparallel: Each DNA strand has a 5′ and 3′ end. When two strands bind to each other, the 5′ end of one strand is adjacent to the 3′ end of the other strand.

Aptamer: DNA molecule that binds to a select target molecule such as other nucleic acids and proteins.

Base pairing: The binding of A (adenine) to T (thymine) or G (guanine) to C (cytosine).

DNAzyme: A molecule made of DNA that possesses enzymatic activity, not unlike that shown by protein enzymes.

Locality: The property of different reactants being in close proximity, due to structural or environmental constraints, causing them to react faster.

Nucleotide: A molecule made up of three components, a 5-carbon (pentose) sugar molecule, a nitrogenous base (one of A, T, G, C), and a phosphate group.

Random walk: A random walk in 1D is the path taken by an object, in either one of the two directions: left or right. At each step, the next direction is determined by an independent fair coin toss.

Secondary structure: The entire set of base pairs formed in a given set of DNA strands, where nucleotides in a pair may lie on either the same strand or on different strands.

Sensitivity: Probability of an actual target (strand) being correctly identified.

Acknowledgments

This work was supported by NSF Grants CCF- 1217457 and CCF-1141847.

References

1. R. Adar, Y. Benenson, G. Linshiz, A. Rosner, N. Tishby, and E. Shapiro. Stochastic computing with biomolecular automata. *Proceedings of the National Academy of Sciences of the United States of America*, 101(27):9960–9965, 2004.
2. L. Adleman. Molecular computation of solutions to combinatorial problems. *Science*, 266(5178):1021–1024, 1994.
3. L. Adleman, Q. Cheng, A. Goel, and M.-D. Huang. Running time and program size for self-assembled squares. *Symposium on Theory of Computing*, 740–748, 2001.
4. G. Aggarwal, M. Goldwasser, M.-Y. Kao, and R. Schweller. Complexities for generalized models of self-assembly. *Symposium on Discrete Algorithms*, 880–889, 2004.
5. A. Chaurasia and S. Dwivedi, and P. Jain and M. K. Gupta. XTile: An error correction package for DNA self-assembly. *Foundations of Nanoscience*, 6, 2009.
6. B. Behsaz, J. Maňuch, and L. Stacho . Turing universality of step-wise and stage assembly at temperature 1. *LNCS*, 7433:1–11, 2012.
7. J. Bath, S. Green, and A. Turberfield. A free-running DNA motor powered by a nicking enzyme. *Angewandte Chemie International Edition*, 44(28):4358–4361, 2005.
8. J. M. Berg, J. L. Tymoczko, and L. Stryer. *Biochemistry*, 7th edn. W.H. Freeman Publishers, New York, December 2010.
9. R. Berger. The undecidability of the domino problem. *Memoirs of American Mathematical Society*, 66:1–72, 1966.

10. J. J. Birac, W. B. Sherman, J. Kopatsch, P. E. Constantinou, and N. C. Seeman. Architecture with GIDEON, a program for design in structural DNA nanotechnology. *Journal of Molecular Graphics & Modelling*, 25(4):470–480, 2006.
11. D. Blicq. DNA Hairpin. http://xnet.rrc.mb.ca/davidb/dna.htm.
12. H. Chandran, N. Gopalkrishnan, and J. Reif. The tile complexity of linear assemblies. *International Colloquium on Automata, Languages and Programming*, 235–253, 2009.
13. H. Chandran, N. Gopalkrishnan, B. Yurke, and J. Reif. Meta-DNA: Synthetic biology via DNA nanostructures and hybridization reactions. *Journal of the Royal Society Interface*, 2012.
14. H.-L. Chen, D. Doty, and D. Soloveichik. Deterministic function computation with chemical reaction networks. *DNA Computing and Molecular Programming*, 7433:25–42, 2012.
15. Y. Chen, S. H. Lee, and C. Mao. A DNA nanomachine based on a duplex-triplex transition. *Angewandte Chemie International Edition*, 43(40):5335–5338, 2004.
16. M. Cook, Y. Fu, and R. T. Schweller. Temperature 1 self-assembly: Deterministic assembly in 3d and probabilistic assembly in 2d. In *SODA*, 570–589, 2011.
17. D. Y. Zhang and B. Yurke. A DNA superstructure-based replicator without product inhibition. *Natural Computing*, 5(2):183–202, 2006.
18. C. J. Delebecque, A. B. Lindner, P. A. Silver, and F. A. Aldaye. Organization of intracellular reactions with rationally designed RNA assemblies. *Science*, 333(6041):470–474, 2011.
19. E. Demaine, M. Demaine, S. Fekete, M. Ishaque, E. Rafalin, R. Schweller, and D. Souvaine. Staged self-assembly: Nanomanufacture of arbitrary shapes with $O(1)$ glues. *DNA Computing*, 4848:1–14, 2008.
20. R. Dirks and N. Pierce. Triggered amplification by hybridization chain reaction. *Proceedings of the National Academy of Sciences of the United States of America*, 101(43):15275–15278, 2004.
21. D. Doty. Randomized self-assembly for exact shapes. *Foundations of Computer Science*, 85–94, 2009.
22. S. Douglas, J. Chou, and W. Shih. DNA-nanotube-induced alignment of membrane proteins for NMR structure determination. *Proceedings of the National Academy of Sciences of the United States of America*, 104(16):6644–6648, 2007.
23. S. M. Douglas, A. H. Marblestone, S. Teerapittayanon, A. Vazquez, G. M. Church, and W. M. Shih. Rapid prototyping of 3D DNA-origami shapes with caDNAno. *Nucleic Acids Research*, 37(15):5001–5006, 2009.
24. E. H. Ekland, J. W. Szostak, and D. P. Bartel. Structurally complex and highly active RNA ligases derived from random RNA sequences. *Science*, 269(5222):364–370, 1995.
25. T.-J. Fu, Y.-C. Tse-Dinh, and N. C. Seeman. Holliday junction crossover topology. *Journal of Molecular Biology*, 236(1):91–105, 1994.
26. D. G. Gibson, J. I. Glass, C. Lartigue, V. N. Noskov, R.-Y. Chuang, M. A. Algire, and G. A. Benders. Creation of a bacterial cell controlled by a chemically synthesized genome. *Science*, 329(5987):52–56, 2010.
27. M. Gopalkrishnan. Catalysis in reaction networks. *Bulletin of Mathematical Biology*, 73(12):2962–2982, 2011.
28. M. Gopalkrishnan and N. Gopalkrishnan. Exquisite detection with chemical circuits. *In preparation*, 2013.
29. N. Gopalkrishnan. *Engineering Exquisite Nanoscale Behavior with DNA*. PhD thesis, Duke University, Durham, NC, 2012.
30. N. Gopalkrishnan, H. Chandran, and J. Reif. High-fidelity DNA hybridization using programmable molecular DNA devices. *DNA Computing and Molecular Programming*, 6518:59–70, 2010.
31. S. Green, J. Bath, and A. Turberfield. Coordinated chemomechanical cycles: A mechanism for autonomous molecular motion. *Physical Review Letters*, 101(23), 2008.
32. S. Green, D. Lubrich, and A. Turberfield. DNA hairpins: Fuel for autonomous DNA devices. *Biophysical Journal*, 91(8):2966–2975, 2006.
33. H. Gu, J. Chao, S.-J. Xiao, and N. Seeman. A Proximity-based programmable DNA nanoscale assembly line. *Nature*, 465(7295):202–205, 2010.

34. Y. He, Y. Chen, H. Liu, A. Ribbe, and C. Mao. Self-assembly of hexagonal DNA two-dimensional (2D) arrays. *Journal of the American Chemical Society*, 127(35):12202–12203, 2005.
35. Y. He and D. Liu. Autonomous multistep organic synthesis in a single isothermal solution mediated by a DNA walker. *Nature Nanotechnology*, 5(11):778782, 2010.
36. Nanorex Inc. NanoEngineer-1. http://www.nanoengineer-1.com/content/, 2006.
37. Iaki Sainz de Murieta and Alfonso Rodrguez-Patn. DNA biosensors that reason. *Biosystems*, 109(2):91–104, 2012.
38. J. Mañuch, L. Stacho and C. Stoll. Step-wise tile assembly with a constant number of tile types. *Natural Computing*, 11(3):535–550, 2012.
39. W. K. Johnston, P. J. Unrau, M. S. Lawrence, M. E. Glasner, and D. P. Bartel. RNA-catalyzed RNA polymerization: Accurate and general RNA-templated primer extension. *Science*, 292(5520):1319–1325, 2001.
40. M.-Y. Kao and R. Schweller. Randomized self-assembly for approximate shapes. *International Colloquium on Automata, Languages and Programming*, pp. 370–384, 2008.
41. T. LaBean, H. Yan, J. Kopatsch, F. Liu, E. Winfree, J. Reif, and N. Seeman. Construction, analysis, ligation, and self-assembly of DNA triple crossover complexes. *Journal of the American Chemical Society*, 122(9):1848–1860, 2000.
42. M. Lakin, S. Youssef, F. Polo, S. Emmott, and A. Phillips. Visual DSD: A design and analysis tool for DNA strand displacement systems. *Bioinformatics*, 27(22):3211–3213, 2012.
43. R. Lavery. DNA structure. www.phys.ens.fr/ monasson/Houches/Lavery/L1.ppt (October 14, 2013).
44. H. Lederman, J. Macdonald, D. Stefanovic, and M. N. Stojanovic. Deoxyribozyme-based three-input logic gates and construction of a molecular full adder. *Biochemistry*, 45(4):1194–1199, 2006. PMID: 16430215.
45. J. L. Leroy, K. Gehring, A. Kettani, and M. Gueron. Acid multimers of oligodeoxycytidine strands: Stoichiometry, base-pair characterization, and proton exchange properties. *Biochemistry*, 32(23):6019–6031, 1993. PMID: 8389586.
46. T. Liedl, M. Olapinski, and F. C. Simmel. A surface-bound DNA switch driven by a chemical oscillator. *Angewandte Chemie International Edition*, 45(30):5007–5010, 2006.
47. C. Lin, R. Jungmann, A. M. Leifer, C. Li, D. Levner, G. M. Church, W. M. Shih, and P. Yin. Submicrometre geometrically encoded fluorescent barcodes self-assembled from DNA. *Nature Chemistry*, 4:832–839, 2012.
48. D. Liu and S. Balasubramanian. A proton-fuelled DNA nanomachine. *Angewandte Chemie International Edition*, 42(46):5734–5736, 2003.
49. D. Liu, A. Bruckbauer, C. Abell, S. Balasubramanian, D.-J. Kang, D. Klenerman, and D. Zhou. A reversible pH-driven DNA nanoswitch array. *Journal of the American Chemical Society*, 128(6):2067–2071, 2006.
50. H. Liu, Y. Xu, F. Li, Y. Yang, W. Wang, Y. Song, and D. Liu. Light-driven conformational switch of i-Motif DNA. *Angewandte Chemie International Edition*, 46(14):2515–2517, 2007.
51. Y. Lu and J. Liu. Functional DNA nanotechnology: Emerging applications of DNAzymes and aptamers. *Current Opinion in Biotechnology*, 17(6):580–588, 2006.
52. K. Lund, A. J. Manzo, N. Dabby, N. Michelotti, A. Johnson-Buck, J. Nangreave, S. Taylor et al. Molecular robots guided by prescriptive landscapes. *Nature*, 465(7295):206–210, May 2010.
53. J. Ma, L. Jia, and Y. Dong. DNA 3D self-assembly algorithmic model to solve maximum clique problem. *International Journal of Image, Graphics and Signal Processing (IJIGSP)*, 3(3):41–48, 2011.
54. J. Mai, I.M. Sokolov, and A. Blumen. Directed particle diffusion under burnt bridges conditions. *Physical Review E*, 64(1):011102, June 2001.
55. U. Majumder, T. LaBean, and J. Reif. Activatable tiles: compact, robust programmable assembly and other applications. *DNA Computing*, 4848:15–25, 2007.
56. C. Mao, T. Labean, J. Reif, and N. Seeman. Logical computation using algorithmic self-assembly of DNA triple-crossover molecules. *Nature*, 407:493–496, 2000.

57. M. L. McKee, P. J. Milnes, J. Bath, E. Stulz, R. K. O'Reilly, and A. J. Turberfield. Programmable one-pot multistep organic synthesis using DNA junctions. *Journal of the American Chemical Society*, 134(3):1446–1449, 2012.
58. S. Modi, M. G Swetha, D. Goswami, G. D. Gupta, S. Mayor, and Y. Krishnan. A DNA nanomachine that maps spatial and temporal pH changes inside living cells. *Nature Nanotechnology*, 4(5):325–330, 2009.
59. S. Modi and Y. Krishnan. A method to map spatiotemporal pH changes inside living cells using a pH-triggered DNA nanoswitch. *Methods in Molecular Biology (Clifton, NJ)*, 749:61–77, May 2011.
60. L.A. Moran, K.G. Scrimageour, H.R. Horton, R.S. Ochs, and J.D. Rawn. *Biochemistry*, 2nd edn. Neil Patterson Publishers, distributed by Prentice Hall Inc., NJ, 1994.
61. R. A. Muscat, J. Bath, and A. J. Turberfield. A programmable molecular robot. *Nano Letters*, 11(3):982–987, March 2011.
62. N. Campbell, J. Reece, and L. Mitchell. *Biology*, 5th edn. Benjamin Cummings, San Francisco, CA, 1999.
63. M. J. Patitz. Simulation of self-assembly in the abstract tile assembly model with isu tas. *CoRR*, abs/1101.5151, 2011.
64. R. Pei and S.K. Taylor. Deoxyribozyme-based autonomous molecular spiders controlled by computing logic gates. *IPCBEE Proceedings*, 2009.
65. R. Pei, E. Matamoros, M. Liu, D. Stefanovic, and M. N. Stojanovic. Training a molecular automaton to play a game. *Nature Nanotechnology*, 5(11):773–777, 2010.
66. F. Pohl and T. Jovin. Salt-induced co-operative conformational change of a synthetic DNA: Equilibrium and kinetic studies with poly(dG-dC). *Angewandte Chemie International Edition*, 67(3):375–396, 1972.
67. L. Qian and E. Winfree. A simple DNA gate motif for synthesizing large-scale circuits. *DNA Computing*, 5347:70–89, 2009.
68. L. Qian and E. Winfree. Scaling up digital circuit computation with DNA strand displacement cascades. *Science*, 332(6034):1196–1201, 2011.
69. L. Qian, E. Winfree, and J. Bruck. Neural network computation with DNA strand displacement cascades. *Nature*, 475(7356):368–372, 2011.
70. J. Reif. The design of autonomous DNA nano-mechanical devices: Walking and Rolling DNA. *DNA Computing*, 2:439–461, 2003.
71. J. Reif. Local parallel biomolecular computation. *International Journal of Unconventional Computing*, 2013. Special Issue: Biomolecular Computing—From Theory to Practical Applications. Invited Paper.
72. J. Reif and S. Sahu. Autonomous programmable nanorobotic devices using DNAzymes. *DNA Computing*, 4848:66–78, 2007.
73. J. Reif, S. Sahu, and P. Yin. Compact error-resilient computational DNA tiling assemblies. *DNA Computing*, 3384:293–307, 2004.
74. R.J. Roberts and K. Murray. Restriction endonucleases. *CRC Critical Reviews in Biochemistry*, 4(2):124–164, 1976.
75. R. Robinson. Undecidability and nonperiodicity for tilings of the plane. *Inventiones Mathematicae*, 12:177–209, 1971.
76. P. Rothemund. Folding DNA to create nanoscale shapes and patterns. *Nature*, 440:297–302, 2006.
77. P. Rothemund, N. Papadakis, and E. Winfree. Algorithmic self-assembly of DNA sierpinski triangles. *PLoS Biology*, 2:424–436, 2004.
78. P. Rothemund and E. Winfree. The program-size complexity of self-assembled squares. *Symposium on Theory of Computing*, 459–468, 2000.
79. W. Saenger. *Principles of Nucleic Acid Structure*. Springer-Verlag, New York, p. 556, 1984.
80. K. Sakamoto, H. Gouzu, K. Komiya, D. Kiga, S. Yokoyama, T. Yokomori, and M. Hagiya. Molecular computation by DNA hairpin formation. *Science*, 288:1223–1226, 2000.
81. K. Sakamoto, D. Kiga, K. Momiya, H. Gouzu, S. Yokoyama, S. Ikeda, H. Sugiyama, and M. Hagiya. State transitions by molecules. *Biosystems*, 52:81–91, 1999.

82. R. Schulman, B. Yurke, and E. Winfree. Robust self-replication of combinatorial information via crystal growth and scission. *Proceedings of the National Academy of Sciences* (PNAS), 109(17):6405–6410, 2012.
83. G. Seelig, D. Soloveichik, D. Y. Zhang, and E. Winfree. Enzyme-free nucleic acid logic circuits. *Science*, 314(5805):1585–1588, 2006.
84. G. Seelig, B. Yurke, and E. Winfree. DNA hybridization catalysts and catalyst circuits. In *DNA*, vol. 3384, pp. 329–343. Springer, 2004.
85. G. Seelig, B. Yurke, and E. Winfree. Catalyzed relaxation of a metastable DNA fuel. *Journal of the American Chemical Society*, 128(37):12211–12220, 2006.
86. N. C. Seeman. Nucleic acid junctions and lattices. *Journal of Theoretical Biology*, 99:237–247, 1982.
87. N. C. Seeman. De novo design of sequences for nucleic acid structural engineering. *Journal of Biomolecular Structure and Dynamics*, 8(3):3211–3213, 1990.
88. N. C. Seeman. Construction of three-dimensional stick figures from branched DNA . *DNA and Cell Biology*, 10(7):475–486, September 1991.
89. H. Sekiguchi, K. Komiya, D. Kiga, and M. Yamamura. A realization of DNA molecular machine that walks autonomously by using a restriction enzyme. In M. Garzon and H. Yan, eds., *DNA Computing*, Vol. 4848, pp. 54–65. Springer, Berlin, Germany 2008.
90. E. Shapiro and Y. Benenson. Bringing DNA computers to life. *Scientific American*, 17(3):40–47, 2006.
91. W. Sherman and N. Seeman. A precisely controlled DNA biped walking device. *Nano Letters*, 4:1203–1207, 2004.
92. J.-S. Shin and N. Pierce. A synthetic DNA walker for molecular transport. *Journal of American Chemical Society*, 126(35):10834–10835, 2004.
93. R. R. Sinden. *DNA Structure and Function*. Academic Press Inc., San Diego, CA, 398pp., 1994.
94. D. Soloveichik, M. Cook, E. Winfree, and J. Bruck. Computation with finite stochastic chemical reaction networks. *Natural Computing*, 7:615–633, 2008.
95. D. Soloveichik, G. Seelig, and E. Winfree. DNA as a universal substrate for chemical kinetics. *Proceedings of the National Academy of Sciences*, 107:5293–5398, 2010.
96. M. Stojanovic and D. Stefanovic. A deoxyribozyme-based molecular automaton. *Nature Biotechnology*, 21(9):1069–1074, August 2003.
97. M. N. Stojanovic, T. E. Mitchell, and D. Stefanovic. Deoxyribozyme-based logic gates. *Journal of the American Chemical Society*, 124(14):3555–3561, 2002.
98. M. N. Stojanovic, S. Semova, D. Kolpashchikov, J. Macdonald, C. Morgan, and D. Stefanovic. Deoxyribozyme-based ligase logic gates and their initial circuits. *Journal of the American Chemical Society*, 127(19):6914–6915, 2005.
99. S. Sahu , P. Yin , and J. H. Reif. A self-assembly model of time-dependent glue strength. *DNA Computing*, 3892:290–304, 2005.
100. Y. Tian, Y. He, Y. Chen, P. Yin, and C. Mao. A DNAzyme that walks processively and autonomously along a one-dimensional track. *Angewandte Chemie International Edition*, 44(28):4355–4358, 2005.
101. Y. Tian and C. Mao. Molecular gears: A pair of DNA circles continuously rolls against each other. *Journal of American Chemical Society*, 126(37):11410–11411, 2004.
102. A. Turberfield, J. Mitchell, B. Yurke, A. Mills, M. Blakey, and F. Simmel. DNA fuel for free-running nanomachines. *Physical Review Letters*, 90(11), 2003.
103. S. Venkataraman, R. Dirks, P. Rothemund, E. Winfree, and N. Pierce. An autonomous polymerization motor powered by DNA hybridization. *Nature Nanotechnology*, 2:490–494, 2007.
104. S. Venkataramana, R. Dirks, C. Uedab, and N. Pierce. Selective cell death mediated by small conditional RNAs. *Proceedings of the National Academy of Sciences*, 107(39):16777–16782, 2010.
105. C. Wang, J. Ren, and X. Qu. A stimuli responsive DNA walking device. *Chemical Communications*, 47(5):1428, 2011.
106. H. Wang. Proving theorems by pattern recognition II. *Bell Systems Technical Journal*, 1–41, 1961.
107. T. Wang, R. Sha, R. Dreyfus, M. E. Leunissen, C. Maass, D. J. Pine, P. M. Chaikin, and N. C. Seeman. Self-replication of information-bearing nanoscale patterns. *Nature*, 478(7368):225–228, 2011.

108. W. Wang, Y. Yang, E. Cheng, M. Zhao, H. Meng, D. Liu, and D. Zhou. A pH-driven, reconfigurable DNA nanotriangle. *Chemical Communications*, 824–826, 2009.
109. Z.-G. Wang, J. Elbaz, and I. Willner. DNA machines: Bipedal walker and stepper. *Nano Letters*, 11(1):304–309, January 2011.
110. Z.-G. Wang, J. Elbaz, and I. Willner. A dynamically programmed DNA transporter. *Angewandte Chemie International Edition*, 51(18):4322–4326, 2012.
111. S. F. J. Wickham, M. Endo, Y. Katsuda, K. Hidaka, J. Bath, H. Sugiyama, and A. J. Turberfield. Direct observation of stepwise movement of a synthetic molecular transporter. *Nature Nanotechnology*, 6:166–169, February 2011.
112. S. Williams, K. Lund, C. Lin, P. Wonka, S. Lindsay, and H. Yan. Tiamat: A Three-Dimensional Editing Tool for Complex DNA Structures. *DNA Computing*, 5347:90–101, 2009.
113. I. Willner, B. Shlyahovsky, M. Zayats, and B. Willner. DNAzymes for sensing, nanobiotechnology and logic gate applications. *Chem. Soc. Rev.*, 37:1153–1165, 2008.
114. E. Winfree. Simulations of computing by self-assembly. Technical report, California Institute of Technology, Pasadena, CA, 1998.
115. E. Winfree and R. Bekbolatov. Proofreading tile sets: Error correction for algorithmic self-assembly. *DNA Computing*, 2943:126–144, 2003.
116. E. Winfree, F. Liu, L. Wenzler, and N. Seeman. Design and self-assembly of two-dimensional DNA crystals. *Nature*, 394:539–544, 1998.
117. E. Winfree, X. Yang, and N. Seeman. Universal computation via self-assembly of DNA: Some theory and experiments. *DNA Based Computers II, DIMACS*, 44:191–213, 1996.
118. H. Yan, T. LaBean, L. Feng, and J. Reif. Directed nucleation assembly of DNA tile complexes for barcode-patterned lattices. *Proceedings of the National Academy of Sciences of the United States of America*, 100(14):8103–8108, 2003.
119. P. Yin, H. Choi, C. Calvert, and N. Pierce. Programming biomolecular self-assembly pathways. *Nature*, 451(7176):318–322, 2008.
120. P. Yin, B. Guo, C. Belmore, W. Palmeri, E. Winfree, T. LaBean, and J. Reif. TileSoft: Sequence optimization software for designing DNA secondary structures. Technical report, Duke and California Institute of Technology, 2004.
121. P. Yin, H. Yan, X. Daniell, A. Turberfield, and J. Reif. A unidirectional DNA walker moving autonomously along a linear track. *Angewandte Chemie International Edition*, 116(37):5014–5019, 2004.
122. M. You, Y. Chen, X. Zhang, H. Liu, R. Wang, K. Wang, K. R. Williams, and W. Tan. An autonomous and controllable light-driven DNA walking device. *Angewandte Chemie International Edition*, 51(10):2457–2460, 2012.
123. M. You, F. Huang, Z. Chen, R.-W. Wang, and W. Tan. Building a nanostructure with reversible motions using photonic energy. *ACS Nano*, 6(9):7935–7941, September 2012.
124. B. Yurke, A. Turberfield, A. Mills, F. Simmel, and J. Neumann. A DNA-fuelled molecular machine made of DNA. *Nature*, 406(6796):605–608, 2000.
125. J. Zadeh, C. Steenberg, J. Bois, B. Wolfe, M. Pierce, A. Khan, R. Dirks, and N. Pierce. NUPACK: analysis and design of nucleic acid systems. *Journal of Computational Chemistry*, 32(1):170–173, 2010.
126. D. Zhang, S. X. Chen, and P. Yin. Optimizing the specificity of nucleic acid hybridization. *Nature Chemistry*, 4:208–214, 2012.
127. D. Zhang, A. Turberfield, B. Yurke, and E. Winfree. Engineering entropy-driven reactions and networks catalyzed by DNA. *Science*, 318:1121–1125, 2007.
128. D. Y. Zhang and G. Seelig. DNA-based fixed gain amplifiers and linear classifier circuits. In *Proceedings of the 16th International Conference on DNA Computing and Molecular Programming*, pp. 176–186, Springer-Verlag, Berlin, Germany, 2011.
129. D. Y. Zhang and E. Winfree. Control of DNA strand displacement kinetics using toehold exchange. *Journal of the American Chemical Society*, 131(48):17303–17314, 2009.

130. J. Zheng, J. Birktoft, Y. Chen, T. Wang, R. Sha, P. Constantinou, S. Ginell, C. Mao, and Nadrian. From molecular to macroscopic via the rational design of a self-assembled 3D DNA crystal. *Nature*, 461(7260):74–78, 2009.
131. S. L. Zipursky and J. Darnell. *Molecular Cell Biology*, 4th edn. Freeman & Co, New York, 2000, pp. 1084, 1993.
132. M. Zuker. MFOLD web server for nucleic acid folding and hybridization prediction. *Nucleic Acids Research*, 31(13):3406–3415, 2003.

Additional Readings

133. J. Bath and A. Turberfield. DNA nanomachines. *Nature Nanotechnology*, 2:275–284, 2007.
134. H. Chandran, N. Gopalkrishnan, S. Garg, and J. Reif. *Biomolecular Computing Systems—From Logic Systems to Smart Sensors and Actuators*. Wiley-VCH, Weinheim, Germany, 2012. Invited Chapter.
135. A. Condon. Designed DNA molecules: Principles and applications of molecular nanotechnology. *Nature Reviews Genetics*, 7:565–575, 2006.
136. Z. Deng, Y. Chen, Y. Tian, and C. Mao. A fresh look at DNA nanotechnology. *Nanotechnology: Science and Computation*, 23–34, June 2006.
137. T. LaBean, K. Gothelf, and J. Reif. Self-assembling DNA nanostructures for patterned molecular assembly. *Nanobiotechnology II*, 79–97, 2007.
138. A. J. Ruben and L. F. Landweber. The past, present and future of molecular computing. *Nature Reviews Molecular Cell Biology*, 1:69–72, 2000.
139. R. Sha, X. Zhang, S. Liao, P. Constantinou, B. Ding, T. Wang, A. Garibotti et al. Structural DNA nanotechnology: Molecular construction and computation. *Unconventional Computing*, 3699:20–31, 2005.
140. E. Winfree. DNA computing by self-assembly. *NAE's The Bridge*, 33:31–38, 2003.
141. H. Yan, P. Yin, S. H. Park, H. Li, L. Feng, X. Guan, D. Liu, J. Reif, and T. LaBean. Self-assembled DNA structures for nanoconstruction. *American Institute of Physics Conference Series*, 725:43–52, 2004.

IV

Computational Science and Graphics

26 Computational Electromagnetics

J.S. Shang
Wright State University

26.1 Introduction

Computational electromagnetics (CEM) is a natural extension of the analytic approach in solving the Maxwell equations. In spite of the fundamental difference between representing the solution in a continuum and in a discretized space, both approaches satisfy all pertaining theorems rigorously. The analytic approach to electromagnetics is elegant, and the results can describe the specific behavior as well as the general patterns of a physical phenomenon in a given regime. However, exact solutions to the Maxwell equations are usually unavailable. Some of the closed-form results that exist have restrictive underlying assumptions that limit their range of validity (Harrington 1961). Solutions of CEM generate only a point value for a specific simulation, but complexity of the physics or of the field configuration is no longer a limiting factor. The numerical accuracy of CEM is an issue to be addressed. Nevertheless, with the advent of high-performance computing systems, CEM is becoming a mainstay for engineering applications (Taflove 1992, Shang 2002).

In the tier-structured CEM discipline, the time-domain method or the frequency domain method is restricted mostly to Rayleigh and resonance regions where the characteristic electric length is of the same order in magnitude as the electromagnetic wavelength. In a high-frequency spectrum or the optical region, all practical applications are supported by the ray tracing of the shooting-and-bouncing (SBR) ray technique. This approximation is based on either physical optics, theory of diffraction, or a combination of both (Bowman et al. 1987). The SBR algorithm is naturally suitable for concurrent computing. Exceptionally high computational efficiency has been consistently demonstrated in the applications for broadband antenna design, ultraband and optical communication, and synthetic aperture radar imaging. In order for the time-domain method to effectively venture into the optical domain, a combination of spectral-like numerical algorithm and concurrent computing is essential.

CEM in the present context is focused on simulation methods for solving the Maxwell equations in the time domain. First of all, time dependence is the most general form of the Maxwell equations, and the dynamic electromagnetic field is not confined to a time-harmonic phenomenon. Therefore, CEM in the time domain has the widest range of engineering applications. In addition, several new numerical algorithms for solving the first-order hyperbolic partial-differential equations, as well as coordinate transformation techniques, were introduced recently to the CEM community. These finite-difference and finite-volume numerical algorithms were devised specifically to mimic the physics involving directional wave propagation. Meanwhile, very complex shapes associated with the field can be easily accommodated by incorporating a coordinate transformation technique. These methodologies have the potential to radically change future research in electromagnetics (Taflove 1992, Shang and Gaitonde 1995, Shang and Fithen 1996).

In order to use CEM effectively, it will be beneficial to understand the fundamentals of numerical simulation and its limitations. The inaccuracy incurred by a numerical simulation is attributable to the mathematical model for the physics, the numerical algorithm, and the computational accuracy. In general, differential equations in CEM consist of two categories: the first-order divergence–curl equations and the second-order curl–curl equations (Harrington 1961, 1968, Elliott 1966). In specific applications, further simplifications into the frequency domain or the Helmholtz equations and the potential formulation have been accomplished. Poor numerical approximations to physical phenomena can result, however, from solving overly simplified governing equations. Under these circumstances, no meaningful quantification of errors for the numerical procedure can be achieved. Physically incorrect values and inappropriate implementation of initial and/or boundary conditions are another major source of error. The placement of the far-field boundary and the type of initial or boundary conditions have also played an important role. These concerns are easily appreciated in the light of the fact that the governing equations are identical, but the different initial or boundary conditions generate different solutions (Shang 2009).

Numerical accuracy is also controlled by the algorithm and computing system adopted. The error induced by the discretization consists of the roundoff and the truncation error. The roundoff error is contributed by the computing system and is problem-size dependent. Since this error is random, it is the most difficult to evaluate. One anticipates that this type of error will be a concern for solution procedures involving large-scale matrix manipulation such as the method of moments and the implicit numerical algorithm for finite-difference or finite-volume methods. The truncation error for time-dependent calculations appears as dissipation and dispersion, which can be assessed and alleviated by mesh-system refinements.

Finally, numerical error can be the consequence of a specific formulation. The error becomes pronounced when a special phenomenon is investigated or when a discontinuous and distinctive stratification of the field is encountered, such as a wave propagating through the interface between media of different characteristic impedances, for which the solution is piecewise continuous. Only in a strongly conservative formulation can the discontinuous phenomenon be adequately resolved. Another example is encountered in radar cross-sectional simulation, where the scattered-field formulation has been shown to be superior to the total-field formulation.

The Maxwell equations in the time domain consist of a first-order divergence–curl system and are difficult to solve by conventional numerical methods. Nevertheless, the pioneering efforts by Yee and others have attained impressive achievements (Yee 1966, Taflove 1992). Recently, numerical techniques in CEM have been further enriched by development in the field of computational fluid dynamics (CFD). In CFD, the Euler equations, which are a subset of the Navier–Stokes equations, have the same classification in the partial-differential system as that of the time-dependent Maxwell equations. Both are hyperbolic systems and constitute initial-value problems (Sommerfeld 1949). For hyperbolic partial-differential equations, the solutions need not be analytic functions. More importantly, the initial values together with any possible discontinuities are continued along a time–space trajectory, which is commonly referred to as the characteristic. A series of numerical schemes have been devised in the CFD community to duplicate the directional information-propagation feature. These numerical procedures are collectively designated as the characteristic-based method, which in its most elementary form is identical to the Riemann problem (Roe 1986).

The characteristic-based method when applied to solve the Maxwell equations in the time domain has exhibited many attractive attributes. A synergism of the new numerical procedures and scalable parallel-computing capability will open up a new frontier in electromagnetics research. For this reason, a major portion of the present chapter will be focused on introducing the characteristic-based finite-volume and finite-difference methods (Shang 1995, Shang and Gaitonde 1995, Shang and Fithen 1996).

26.2 Governing Equations

The time-dependent Maxwell equations for the electromagnetic field can be written as (Harrington 1961, Elliott 1966)

$$\frac{\partial \boldsymbol{B}}{\partial t} + \nabla \times \boldsymbol{E} = 0 \tag{26.1}$$

$$\frac{\partial \boldsymbol{D}}{\partial t} - \nabla \times \boldsymbol{H} = -\boldsymbol{J} \tag{26.2}$$

$$\nabla \cdot \boldsymbol{B} = 0 \tag{26.3}$$

$$\nabla \cdot \boldsymbol{D} = \rho \tag{26.4}$$

The only conservation law for electric charge and current densities is

$$\frac{\partial \rho}{\partial t} + \nabla \cdot \boldsymbol{J} = 0 \tag{26.5}$$

where ρ and $\boldsymbol{J}$ are the charge and current density, respectively, and represent the source of the field. The constitutive relations between the magnetic flux density and intensity and between the electric displacement and field strength are $\boldsymbol{B} = \mu \boldsymbol{H}$ and $\boldsymbol{D} = \epsilon \boldsymbol{E}$. Equation 26.5 is often regarded as a fundamental law of electromagnetics, derived from the generalized Ampere's circuit law and Gauss's law. Since Equation 26.1 and 26.2 contain the information on the propagation of the electromagnetic field, they constitute the basic equations of CEM.

The previous partial-differential equations also can be expressed as a system of integral equations. The following expression is obtained by using Stokes' law and the divergence theorem to reduce the surface and volume integrals to a circuital line and surface integrals, respectively (Elliott 1966). These integral relationships hold only if the first derivatives of the electric displacement $\boldsymbol{D}$ and the magnetic flux density $\boldsymbol{B}$ are continuous throughout the control volume:

$$\oint \boldsymbol{E} \cdot d\boldsymbol{L} = -\iint \frac{\partial \boldsymbol{B}}{\partial t} \cdot d\boldsymbol{S} \tag{26.6}$$

$$\oint \boldsymbol{H} \cdot d\boldsymbol{L} = \iint \left(\boldsymbol{J} + \frac{\partial \boldsymbol{D}}{\partial t} \right) \cdot d\boldsymbol{S} \tag{26.7}$$

$$\iint \boldsymbol{D} \cdot d\boldsymbol{S} = \iiint \rho \, dV \tag{26.8}$$

$$\iint \boldsymbol{B} \cdot d\boldsymbol{S} = 0 \tag{26.9}$$

The integral form of the Maxwell equations is rarely used in CEM. They are, however, invaluable as a validation tool for checking the global behavior of field computations.

The second-order curl–curl form of the Maxwell equations is derived by applying the curl operator to get

$$\nabla \times \nabla \times \boldsymbol{E} + \frac{1}{c^2}\frac{\partial^2 \boldsymbol{E}}{\partial t^2} = -\frac{\partial(\mu \boldsymbol{J})}{\partial t} \tag{26.10}$$

$$\nabla \times \nabla \times \boldsymbol{B} + \frac{1}{c^2}\frac{\partial^2 \boldsymbol{B}}{\partial t^2} = \nabla \times (\mu \boldsymbol{J}) \tag{26.11}$$

The outstanding feature of the curl–curl formulation of the Maxwell equations is that the electric and magnetic fields are decoupled. The second-order equations can be further simplified for harmonic fields. If the time-dependent behavior can be represented by a harmonic function $e^{i\omega t}$, the separation-of-variable technique will transform the Maxwell equations into the frequency domain (Harrington 1961, Elliott 1966). The resultant partial-differential equations in spatial variables become elliptic:

$$\nabla \times \nabla \times \boldsymbol{E} - k^2 \boldsymbol{E} = -i\omega(\mu \boldsymbol{J}) \tag{26.12}$$

$$\nabla \times \nabla \times \boldsymbol{B} - k^2 \boldsymbol{B} = \nabla \times (\mu \boldsymbol{J}) \tag{26.13}$$

where $k = \omega/c$ is the so-called propagation constant or the wave number and is the angular frequency of a component of a Fourier series or a Fourier integral (Harrington 1961, Elliott 1966). The previous equations are frequently the basis for finite-element approaches (Rahman et al. 1991).

In order to complete the description of the differential system, initial and/or boundary values are required. For Maxwell equations, only the source of the field and a few physical boundary conditions at the media interfaces are pertinent (Harrington 1961, Elliott 1966):

$$\boldsymbol{n} \times (\boldsymbol{E}1 - \boldsymbol{E}2) = 0 \tag{26.14}$$

$$\boldsymbol{n} \times (\boldsymbol{H}1 - \boldsymbol{H}2) = \boldsymbol{J}_s \tag{26.15}$$

$$\boldsymbol{n} \times (\boldsymbol{D}1 - \boldsymbol{D}2) = \rho_s \tag{26.16}$$

$$\boldsymbol{n} \times (\boldsymbol{B}1 - \boldsymbol{B}2) = 0 \tag{26.17}$$

where

The subscripts 1 and 2 refer to media on two sides of the interface

$\boldsymbol{J}_s$ and ρ_s are the surface current and charge densities, respectively

Since all computing systems have finite memory, all CEM computations in the time domain must be conducted on a truncated computational domain. This intrinsic constraint requires a numerical far-field condition at the truncated boundary to mimic the behavior of an unbounded field. This numerical boundary unavoidably induces a reflected wave to contaminate the simulated field. In the past, absorbing boundary conditions at the far-field boundary have been developed from the radiation condition (Sommerfeld 1949, Enquist and Majda 1977, Mur 1981, Higdon 1986). In general, a progressive order-of-accuracy procedure can be used to implement the numerical boundary conditions with increasing accuracy (Enquist and Majda 1977, Higdon 1986).

The most popular approaches in the CEM community are the absorbing boundary condition or the perfectly matched-layer scheme (Mur 1981, Berenger 1994). The latter is actually derived from the characteristics or the domain of dependence of the hyperbolic partial-differential equation system. The detailed eigenvector and eigenvalue analyses of the characteristic formulation for the 3D Maxwell equations have been performed (Shang 1995). In this formulation, the flux vector of the governing equations is split according to the signs of the eigenvalue to honor the domain of dependence. The flux-vector splitting formulation is utilized in the present investigation.

On the other hand, the characteristic-based methods that satisfy the physical domain of influence requirement can specify the numerical boundary condition readily. For this formulation, the reflected wave can be suppressed by eliminating the undesirable incoming numerical data. Although the accuracy of the numerical far-field boundary condition depends on the coordinate system, in principle, this formulation under ideal circumstances can effectively suppress artificial wave reflections.

26.3 Characteristic-Based Formulation

The fundamental idea of the characteristic-based method for solving the hyperbolic system of equations is derived from the eigenvalue–eigenvector analyses of the governing equations. For Maxwell equations in the time domain, every eigenvalue is real, but not all of them are distinct (Shang 1995, Shang and Gaitonde 1995, Shang and Fithen 1996). In a time–space plane, the eigenvalue actually defines the slope of the characteristic or the phase velocity of the wave motion. All dependent variables within the time–space domain bounded by two intersecting characteristics are completely determined by the values along these characteristics and by their compatibility relationship. The direction of information propagation is also clearly described by these two characteristics (Sommerfeld 1949). In numerical simulation, the well-posedness requirement on initial or boundary conditions and the stability of a numerical approximation are also ultimately linked to the eigenvalues of the governing equation (Richtmyer and Morton 1967, Anderson et al. 1984). Therefore, characteristic-based methods have demonstrated superior numerical stability and accuracy to other schemes (Roe 1986, Shang 1995). However, characteristic-based algorithms also have an inherent limitation in that the governing equation can be diagonalized only in one spatial dimension at a time. The multidimensional equations are required to split into multiple 1D formulations. This limitation is not unusual for numerical algorithms, such as the approximate factored and the fractional-step schemes (Anderson et al. 1984, Shang 1995). A consequence of this restriction is that solutions of the characteristic-based procedure may exhibit some degree of sensitivity to the orientation of the coordinate selected. This numerical behavior is consistent with the concept of optimal coordinates.

In the characteristic formulation, data on the wave motion are first split according to the direction of phase velocity and then transmitted in each orientation. In each 1D time–space domain, the direction of the phase velocity degenerates into either a positive or a negative orientation. They are commonly referred to as the right-running and the left-running wave components (Sommerfeld 1949, Roe 1986). The sign of the eigenvalue is thus an indicator of the direction of signal transmission. The corresponding eigenvectors are the essential elements for diagonalizing the coefficient matrices and for formulating the approximate Riemann problem (Roe 1986). In essence, knowledge of eigenvalues and eigenvectors of the Maxwell equations in the time domain becomes the first prerequisite of the present formulation.

The system of governing equations cast in the flux-vector form in the Cartesian frame becomes (Shang 1995, Shang and Gaitonde 1995, Shang and Fithen 1996)

$$\frac{\partial U}{\partial t} + \frac{\partial F_x}{\partial x} + \frac{\partial F_y}{\partial y} + \frac{\partial F_z}{\partial z} = -J \tag{26.18}$$

where U is the vector of dependent variables. The flux vectors are formed by the inner product of the coefficient matrix and the dependent variable, $F_x = C_x U$, $F_y = C_y U$, and $F_z = C_z U$, with

$$U = \{B_x, B_y, B_z, D_x, D_y, D_z\}^T \tag{26.19}$$

and

$$F_x = \left\{0, \frac{-D_z}{\varepsilon}, \frac{D_y}{\varepsilon}, 0, \frac{B_z}{\mu}, \frac{-B_y}{\mu}\right\}T$$

$$F_y = \left\{\frac{D_z}{\varepsilon}, 0, \frac{-D_x}{\varepsilon}, \frac{-B_z}{\mu}, 0, \frac{B_x}{\mu}\right\}T \tag{26.20}$$

$$F_z = \left\{\frac{-D_y}{\varepsilon}, \frac{D_x}{\varepsilon}, 0, \frac{B_y}{\mu}, \frac{-B_x}{\mu}, 0\right\}T$$

The coefficient matrices or the Jacobians of the flux vectors C_x, C_y, and C_z are (Shang 1995)

$$C_x = \begin{bmatrix} 0 & 0 & 0 & 0 & 0 & 0 \\ 0 & 0 & 0 & 0 & 0 & -\frac{1}{\varepsilon} \\ 0 & 0 & 0 & 0 & \frac{1}{\varepsilon} & 0 \\ 0 & 0 & 0 & 0 & 0 & 0 \\ 0 & 0 & \frac{1}{\mu} & 0 & 0 & 0 \\ 0 & -\frac{1}{\mu} & 0 & 0 & 0 & 0 \end{bmatrix} \tag{26.21a}$$

$$C_y = \begin{bmatrix} 0 & 0 & 0 & 0 & 0 & \frac{1}{\varepsilon} \\ 0 & 0 & 0 & 0 & 0 & 0 \\ 0 & 0 & 0 & -\frac{1}{\varepsilon} & 0 & 0 \\ 0 & 0 & -\frac{1}{\mu} & 0 & 0 & 0 \\ 0 & 0 & 0 & 0 & 0 & 0 \\ \frac{1}{\mu} & 0 & 0 & 0 & 0 & 0 \end{bmatrix} \tag{26.21b}$$

$$C_z = \begin{bmatrix} 0 & 0 & 0 & 0 & -\frac{1}{\varepsilon} & 0 \\ 0 & 0 & 0 & \frac{1}{\varepsilon} & 0 & 0 \\ 0 & 0 & 0 & 0 & 0 & 0 \\ 0 & \frac{1}{\mu} & 0 & 0 & 0 & 0 \\ -\frac{1}{\mu} & 0 & 0 & 0 & 0 & 0 \\ 0 & 0 & 0 & 0 & 0 & 0 \end{bmatrix} \tag{26.21c}$$

where ε and μ are the permittivity and permeability, which relate the electric displacement to the electric field intensity and the magnetic flux density to the magnetic field intensity, respectively.

The eigenvalues of the coefficient matrices C_x, C_y, and C_z in the Cartesian frame are identical and contain multiplicities (Shang 1995, Shang and Gaitonde 1995). Care must be exercised to ensure that all associated eigenvectors

$$\lambda = \left\{ +\frac{1}{\sqrt{\varepsilon\mu}}, -\frac{1}{\sqrt{\varepsilon\mu}}, 0, +\frac{1}{\sqrt{\varepsilon\mu}}, -\frac{1}{\sqrt{\varepsilon\mu}}, 0 \right\} \tag{26.22a}$$

are linearly independent. The linearly independent eigenvectors associated with each eigenvalue are found by reducing the matrix equation, $(C - I\lambda)U = 0$, to the Jordan normal form (Shang 1995, Shang and Fithen 1996).

Since the coefficient matrices C_x, C_y, and C_z can be diagonalized, there exist nonsingular similar matrices S_x, S_y, and S_z such that

$$\Lambda_x = S_x^{-1} C_x S_x$$

$$\Lambda_y = S_y^{-1} C_y S_y$$

$$\Lambda_z = S_z^{-1} C_z S_z \tag{26.22b}$$

where the Λ_s are the *diagonalized* coefficient matrices. The columns of the similar matrices S_x, S_y, and S_z are simply the linearly independent eigenvectors of the coefficient matrices C_x, C_y, and C_z, respectively.

The fundamental relationship between the characteristic-based formulation and the Riemann problem can be best demonstrated in the Cartesian frame of reference. For the Maxwell equations in this frame of reference and for an isotropic medium, all the similar matrices are invariant with respect to temporal and spatial independent variables. In each time–space plane (*x*–*t*, *y*–*t*, and *z*–*t*), the 1D governing equation can be given just as in the *x*–*t* plane:

$$\frac{\partial U}{\partial t} + C_x \frac{\partial U}{\partial x} = 0 \tag{26.23}$$

Substitute the diagonalized coefficient matrix to get

$$\frac{\partial U}{\partial t} + S_x \Lambda_x S_x^{-1} \frac{\partial U}{\partial x} = 0 \tag{26.24}$$

Since the similar matrix in the present consideration is invariant with respect to time and space, it can be brought into the differential operator. Multiplying the previous equation by the left-hand inverse of S_x, S_x^{-1}, we have

$$\frac{\partial S_x^{-1} U}{\partial t} + \Lambda_x \frac{\partial S_x^{-1} U}{\partial x} = 0 \tag{26.25}$$

One immediately recognizes the group of variables $S_x^{-1}\,U$ as the characteristics, and the system of equations is decoupled (Shang 1995). In scalar-variable form and with appropriate initial values, this is the Riemann problem (Sommerfeld 1949, Courant and Hilbert 1965). This differential system is specialized to study the breakup of a single discontinuity. The piecewise continuous solutions separated by the singular point are also invariant along the characteristics. Equally important, stable numerical operators can now be easily devised to solve the split equations according to the sign of the eigenvalue. In practice, it has been found that if the multidimensional problem can be split into a sequence of 1D equations, this numerical technique is applicable to those 1D equations (Roe 1986, Shang 1995).

The gist of the characteristic-based formulation is also clearly revealed by the decomposition of the flux vector into positive and negative components corresponding to the sign of the eigenvalue:

$$\Lambda = \Lambda^+ + \Lambda^-, \quad F = F^+ + F^- \tag{26.26}$$

$$F^+ = S\Lambda^+ S^{-1}, \quad F^- = S\Lambda^- S^{-1} \tag{26.27}$$

where the superscripts + and − denote the split vectors associated with positive and negative eigenvalues, respectively.

The characteristic-based algorithms have a deep-rooted theoretical basis for describing the wave dynamics. They also have however an inherent limitation in that the diagonalized formulation is achievable only in one dimension at a time. All multidimensional equations are required to be split into multiple 1D formulations. The approach yields accurate results so long as discontinuous waves remain aligned with the computational grid. This limitation is also the state-of-the-art constraint in solving partial-differential equations (Anderson et al. 1984, Roe 1986, Shang 1995).

26.4 Maxwell Equations in a Curvilinear Frame

In order to develop a versatile numerical tool for CEM in a wide range of applications, the Maxwell equations can be cast in a general curvilinear frame of reference (Shang and Gaitonde 1995, Shang and Fithen 1996). For efficient simulation of complex electromagnetic field configurations, the adoption of a general curvilinear mesh system becomes necessary. The system of equations in general curvilinear coordinates is derived by a coordinate transformation (Thompson 1982, Anderson et al. 1984). The mesh system in the transformed space can be obtained by numerous grid generation procedures (Thompson 1982). Computational advantages in the transformed space are also realizable. For a body-oriented coordinate system, the interface between two different media is easily defined by one of the coordinate surfaces. Along this coordinate parametric plane, all discretized nodes on the interface are precisely prescribed without the need for an interpolating procedure. The outward normal to the interface, which is essential for boundary-value implementation, can be computed easily by $n = \nabla S / \|\nabla S\|$. In the transformed space, computations are performed on a uniform mesh space, but the corresponding physical spacing can be highly clustered to enhance the numerical resolution.

As an illustration of the numerical advantage of solving the Maxwell equations on nonorthogonal curvilinear, body-oriented coordinates, a simulation of the scattered electromagnetic field from a reentry vehicle has been performed (Shang and Gaitonde 1995). The aerospace vehicle, X24C- 10D, has a complex geometric shape (Figure 26.1). In addition to a blunt leading-edge spherical nose and a relatively flat delta-shaped underbody, the aft portion of the vehicle consists of five control

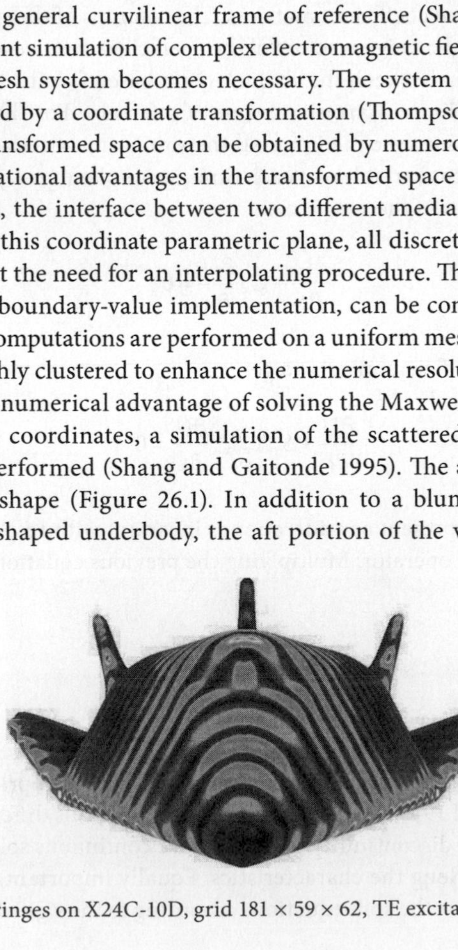

FIGURE 26.1 Radar-wave fringes on X24C-10D, grid 181 × 59 × 62, TE excitation, $L/\lambda = 9.2$.

surfaces—a central fin, two middle fins, and two strakes. A body-oriented, single-block mesh system enveloping the configuration is adopted. The numerical grid system is generated by using a hyperbolic grid generator for the near-field mesh adjacent to the solid surface and a transfinite technique for the far field.

The two mesh systems are merged by the Poisson averaging technique (Thompson 1982, Shang and Gaitonde 1995). In this manner, the composite grid system is orthogonal in the near field but less restrictive in the far field. All solid surfaces of the X24C-10D are mapped onto a parametric surface in the transformed space, defined by $\eta = 0$. The entire computational domain is supported by a 181 × 59 × 162 grid system, where the first coordinate index denotes the number of cross sections in the numerical domain. The second index describes the number of cells between the body surface and the far field boundary, while the third index gives the cells used to circumscribe each cross-sectional plane. The electromagnetic excitation is introduced by a harmonic incident wave traveling along the x-coordinate. The fringe pattern of the scattered electromagnetic waves on the X24C-10D is presented in Figure 26.1 for a characteristic-length-to-wavelength ratio $L/\lambda = 9.2$. A salient feature of the scattered field is brought out by the surface curvature: the smaller the radius of surface curvature, the broader the diffraction pattern. The numerical result exhibits highly concentrated contours at the chine (the line of intersection between upper and lower vehicle surfaces) of the forebody and the leading edges of strakes and fins.

For the most general coordinate transformation of the Maxwell equations in the time domain, a one-to-one relationship between two sets of temporal and spatial independent variables is required. However, for most practical applications, the spatial coordinate transformation is sufficient:

$$\begin{aligned} \xi &= \xi(x, y, z) \\ \eta &= \eta(x, y, z) \\ \zeta &= \zeta(x, y, z) \end{aligned} \tag{26.28}$$

The governing equation in the strong conservation form is obtained by dividing the chain-rule differentiated equations by the Jacobian of coordinate transformation and by invoking metric identities (Anderson et al. 1984, Shang and Gaitonde 1995). The time-dependent Maxwell equations on a general curvilinear frame of reference and in the strong conservative form are

$$\frac{\partial U}{\partial t} + \frac{\partial F_\xi}{\partial \xi} + \frac{\partial F_\eta}{\partial \eta} + \frac{\partial F_\zeta}{\partial \zeta} = -J \tag{26.29}$$

where the dependent variables are now defined as

$$U = U(B_x V, B_y V, B_z V, D_x V, D_y V, D_z V) \tag{26.30}$$

Here, V is the Jacobian of the coordinate transformation and is also the inverse local cell volume. If the Jacobian has nonzero values in the computational domain, the correspondence between the physical and the transformed space is uniquely defined (Thompson 1982, Anderson et al. 1984). Since systematic procedures have been developed to ensure this property of coordinate transformations, detailed information on this point is not repeated here (Thompson 1982, Anderson et al. 1984). We have

$$V = \begin{bmatrix} \xi_x & \eta_x & \zeta_x \\ \xi_y & \eta_y & \zeta_y \\ \xi_z & \eta_z & \zeta_z \end{bmatrix} \tag{26.31}$$

and ξ_x, η_x, and ζ_x are the metrics of coordinate transformation and can be computed easily from the definition given by Equation 26.24. The flux-vector components in the transformed space have the following form:

$$F_\xi = \frac{1}{V}\begin{bmatrix} 0 & 0 & 0 & 0 & -\frac{\xi_z}{\varepsilon} & \frac{\xi_y}{\varepsilon} \\ 0 & 0 & 0 & \frac{\xi_z}{\varepsilon} & 0 & -\frac{\xi_x}{\varepsilon} \\ 0 & 0 & 0 & -\frac{\xi_y}{\varepsilon} & \frac{\xi_x}{\varepsilon} & 0 \\ 0 & \frac{\xi_z}{\mu} & -\frac{\xi_y}{\mu} & 0 & 0 & 0 \\ -\frac{\xi_z}{\mu} & 0 & \frac{\xi_x}{\mu} & 0 & 0 & 0 \\ \frac{\xi_y}{\mu} & -\frac{\xi_x}{\mu} & 0 & 0 & 0 & 0 \end{bmatrix}\begin{Bmatrix} B_x \\ B_y \\ B_z \\ D_x \\ D_y \\ D_z \end{Bmatrix} \tag{26.32}$$

$$F_\eta = \frac{1}{V}\begin{bmatrix} 0 & 0 & 0 & 0 & -\frac{\eta_z}{\varepsilon} & \frac{\eta_y}{\varepsilon} \\ 0 & 0 & 0 & \frac{\eta_z}{\varepsilon} & 0 & -\frac{\eta_x}{\varepsilon} \\ 0 & 0 & 0 & -\frac{\eta_y}{\varepsilon} & \frac{\eta_x}{\varepsilon} & 0 \\ 0 & \frac{\eta_z}{\mu} & -\frac{\eta_y}{\mu} & 0 & 0 & 0 \\ -\frac{\eta_z}{\mu} & 0 & \frac{\eta_x}{\mu} & 0 & 0 & 0 \\ \frac{\eta_y}{\mu} & -\frac{\eta_x}{\mu} & 0 & 0 & 0 & 0 \end{bmatrix}\begin{Bmatrix} B_x \\ B_y \\ B_z \\ D_x \\ D_y \\ D_z \end{Bmatrix} \tag{26.33}$$

$$F_\zeta = \frac{1}{V}\begin{bmatrix} 0 & 0 & 0 & 0 & -\frac{\zeta_z}{\varepsilon} & \frac{\zeta_y}{\varepsilon} \\ 0 & 0 & 0 & \frac{\zeta_z}{\varepsilon} & 0 & -\frac{\zeta_x}{\varepsilon} \\ 0 & 0 & 0 & -\frac{\zeta_y}{\varepsilon} & \frac{\zeta_x}{\varepsilon} & 0 \\ 0 & \frac{\zeta_z}{\mu} & -\frac{\zeta_y}{\mu} & 0 & 0 & 0 \\ -\frac{\zeta_z}{\mu} & 0 & \frac{\zeta_x}{\mu} & 0 & 0 & 0 \\ \frac{\zeta_y}{\mu} & -\frac{\zeta_x}{\mu} & 0 & 0 & 0 & 0 \end{bmatrix}\begin{Bmatrix} B_x \\ B_y \\ B_z \\ D_x \\ D_y \\ D_z \end{Bmatrix} \tag{26.34}$$

After the coordinate transformation, all coefficient matrices now contain metrics that are position dependent, and the system of equations in the most general frame of reference possesses variable coefficients. This added complexity of the characteristic formulation of the Maxwell equations no longer permits the system of 1D equations to be decoupled into six scalar equations and reduced to the true Riemann problem (Shang 1995, Shang and Gaitonde 1995, Shang and Fithen 1996) like that on the Cartesian form.

26.5 Eigenvalues and Eigenvectors

As previously mentioned, eigenvalue and the eigenvector analyses are the prerequisites for characteristic-based algorithms. The analytic process to obtain the eigenvalues and the corresponding eigenvectors of the Maxwell equations in general curvilinear coordinates is identical to that in the Cartesian frame. In each of the temporal–spatial planes t–ξ, t–η, and t–ζ, the eigenvalues are easily found by solving the sixth-degree characteristic equation associated with the coefficient matrices (Sommerfeld 1949, Courant and Hilbert 1965):

$$\lambda_\xi = \left\{ -\frac{\alpha}{V\sqrt{\varepsilon\mu}}, -\frac{\alpha}{V\sqrt{\varepsilon\mu}}, \frac{\alpha}{V\sqrt{\varepsilon\mu}}, \frac{\alpha}{V\sqrt{\varepsilon\mu}}, 0, 0 \right\} \tag{26.35}$$

$$\lambda_\eta = \left\{ -\frac{\beta}{V\sqrt{\varepsilon\mu}}, -\frac{\beta}{V\sqrt{\varepsilon\mu}}, \frac{\beta}{V\sqrt{\varepsilon\mu}}, \frac{\beta}{V\sqrt{\varepsilon\mu}}, 0, 0 \right\} \tag{26.36}$$

$$\lambda_\zeta = \left\{ -\frac{\gamma}{V\sqrt{\varepsilon\mu}}, -\frac{\gamma}{V\sqrt{\varepsilon\mu}}, \frac{\gamma}{V\sqrt{\varepsilon\mu}}, \frac{\gamma}{V\sqrt{\varepsilon\mu}}, 0, 0 \right\} \tag{26.37}$$

where

$$\alpha = \sqrt{\xi_x^2 + \xi_y^2 + \xi_z^2}, \quad \beta = \sqrt{\eta_x^2 + \eta_y^2 + \eta_z^2}, \quad \text{and} \quad \gamma = \sqrt{\zeta_x^2 + \zeta_y^2 + \zeta_z^2}$$

One recognizes that the eigenvalues in each 1D time–space plane contain multiplicities, and hence the eigenvectors do not necessarily have unique elements (Courant and Hilbert 1965, Shang 1995). Nevertheless, linearly independent eigenvectors associated with each eigenvalue still have been found by reducing the coefficient matrix to the Jordan normal form (Shang 1995, Shang and Fithen 1996). For reasons of wide applicability and internal consistency, the eigenvectors are selected in such a fashion that the similar matrices of diagonalization reduce to the same form as in the Cartesian frame. Furthermore, in order to accommodate a wide range of electromagnetic field configurations such as antennas, waveguides, and scatterers, the eigenvalues are no longer identical in the three time–space planes. This complexity of formulation is essential to facilitate boundary-condition implementation on the interfaces of media with different characteristic impedances.

From the eigenvector analysis, the similarity transformation matrices for diagonalization in each time–space plane are formed by using eigenvectors as the column arrays as shown in the following equations. For example, the first column of the similar matrix of diagonalization

$$\left[-\sqrt{\frac{\mu}{\varepsilon}}\frac{\xi_y}{\alpha}, \sqrt{\frac{\mu}{\varepsilon}}\frac{(\xi_x^2 + \xi_z^2)}{\alpha\xi_x}, \sqrt{\frac{\mu}{\varepsilon}}\frac{\xi_y\xi_z}{\alpha\xi_x}, \frac{\xi_y}{\xi_x}, 0, 1 \right]$$

in the t–ξ plane is the eigenvector corresponding to the eigenvalue $\lambda_\xi = -\alpha/V\sqrt{\varepsilon\mu}$. We have

$$S_\xi = \begin{bmatrix} -\sqrt{\frac{\mu}{\varepsilon}}\frac{\xi_y}{\alpha} & \sqrt{\frac{\mu}{\varepsilon}}\frac{\xi_z}{\alpha} & \sqrt{\frac{\mu}{\varepsilon}}\frac{\xi_y}{\alpha} & -\sqrt{\frac{\mu}{\varepsilon}}\frac{\xi_z}{\alpha} & 1 & 0 \\ \sqrt{\frac{\mu}{\varepsilon}}\frac{(\xi_x^2+\xi_z^2)}{\alpha\xi_x} & \sqrt{\frac{\mu}{\varepsilon}}\frac{\xi_y\xi_z}{\alpha\xi_x} & -\sqrt{\frac{\mu}{\varepsilon}}\frac{(\xi_x^2+\xi_z^2)}{\alpha\xi_x} & \sqrt{\frac{\mu}{\varepsilon}}\frac{\xi_y\xi_z}{\alpha\xi_x} & \frac{\zeta_y}{\zeta_x} & 0 \\ -\sqrt{\frac{\mu}{\varepsilon}}\frac{\xi_y\xi_z}{\alpha\xi_x} & -\sqrt{\frac{\mu}{\varepsilon}}\frac{(\xi_x^2+\xi_y^2)}{\alpha\xi_x} & \sqrt{\frac{\mu}{\varepsilon}}\frac{\xi_y\xi_z}{\alpha\xi_x} & \sqrt{\frac{\mu}{\varepsilon}}\frac{(\xi_x^2+\xi_y^2)}{\alpha\xi_x} & \frac{\xi_z}{\xi_x} & 0 \\ -\frac{\xi_z}{\xi_x} & -\frac{\xi_y}{\xi_x} & -\frac{\xi_z}{\xi_x} & -\frac{\xi_y}{\xi_x} & 0 & 1 \\ 0 & 1 & 0 & 1 & 0 & \frac{\xi_y}{\xi_x} \\ 1 & 0 & 1 & 0 & 0 & \frac{\xi_z}{\xi_x} \end{bmatrix} \quad (26.38)$$

$$S_\eta = \begin{bmatrix} -\sqrt{\frac{\mu}{\varepsilon}}\frac{(\eta_y^2+\xi_z^2)}{\beta\eta_y} & -\sqrt{\frac{\mu}{\varepsilon}}\frac{\eta_x\eta_z}{\beta\eta_y} & \sqrt{\frac{\mu}{\varepsilon}}\frac{(\eta_y^2+\xi_z^2)}{\beta\eta_y} & \sqrt{\frac{\mu}{\varepsilon}}\frac{\eta_x\eta_z}{\beta\eta_y} & \frac{\eta_x}{\eta_y} & 0 \\ \sqrt{\frac{\mu}{\varepsilon}}\frac{\eta_x}{\beta} & -\sqrt{\frac{\mu}{\varepsilon}}\frac{\eta_z}{\beta\eta_y} & -\sqrt{\frac{\mu}{\varepsilon}}\frac{\eta_x}{\beta} & \sqrt{\frac{\mu}{\varepsilon}}\frac{\eta_z}{\beta\eta_y} & 1 & 0 \\ -\sqrt{\frac{\mu}{\varepsilon}}\frac{\eta_x\eta_z}{\alpha\eta_y} & \sqrt{\frac{\mu}{\varepsilon}}\frac{(\eta_x^2+\xi_y^2)}{\alpha\eta_y} & \sqrt{\frac{\mu}{\varepsilon}}\frac{\eta_x\eta_z}{\alpha\eta_y} & -\sqrt{\frac{\mu}{\varepsilon}}\frac{(\eta_y^2+\xi_y^2)}{\beta\eta_y} & \frac{\eta_z}{\eta_y} & 0 \\ 0 & 1 & 0 & 1 & 0 & \frac{\eta_x}{\eta_y} \\ -\frac{\eta_z}{\eta_y} & -\frac{\eta_x}{\eta_y} & -\frac{\eta_x}{\eta_y} & -\frac{\eta_x}{\eta_y} & 0 & 1 \\ 1 & 0 & 1 & 0 & 0 & \frac{\eta_z}{\eta_y} \end{bmatrix} \quad (26.39)$$

$$S_\zeta = \begin{bmatrix} \sqrt{\frac{\mu}{\varepsilon}}\frac{(\zeta_y^2+\zeta_z^2)}{\gamma\zeta_z} & \sqrt{\frac{\mu}{\varepsilon}}\frac{\zeta_x\zeta_y}{\gamma\zeta_z} & -\sqrt{\frac{\mu}{\varepsilon}}\frac{(\zeta_y^2+\zeta_z^2)}{\gamma\zeta_z} & -\sqrt{\frac{\mu}{\varepsilon}}\frac{\zeta_x\zeta_y}{\gamma\zeta_z} & \frac{\zeta_x}{\zeta_z} & 0 \\ -\sqrt{\frac{\mu}{\varepsilon}}\frac{\zeta_x\zeta_y}{\gamma\zeta_z} & -\sqrt{\frac{\mu}{\varepsilon}}\frac{(\zeta_x^2+\zeta_z^2)}{\gamma\zeta_z} & \sqrt{\frac{\mu}{\varepsilon}}\frac{\zeta_x\zeta_y}{\gamma\zeta_z} & \sqrt{\frac{\mu}{\varepsilon}}\frac{(\zeta_x^2+\zeta_z^2)}{\gamma\zeta_z} & \frac{\zeta_y}{\zeta_z} & 0 \\ -\sqrt{\frac{\mu}{\varepsilon}}\frac{\zeta_x}{\gamma} & \sqrt{\frac{\mu}{\varepsilon}}\frac{\zeta_y}{\gamma} & \sqrt{\frac{\mu}{\varepsilon}}\frac{\zeta_x}{\gamma} & -\sqrt{\frac{\mu}{\varepsilon}}\frac{\zeta_y}{\gamma} & 1 & 0 \\ 0 & 1 & 0 & 1 & 0 & \frac{\zeta_x}{\zeta_z} \\ 1 & 0 & 1 & 0 & 0 & \frac{\zeta_y}{\zeta_z} \\ -\frac{\zeta_y}{\zeta_z} & -\frac{\zeta_x}{\zeta_z} & -\frac{\zeta_y}{\zeta_z} & -\frac{\zeta_x}{\zeta_z} & 0 & 1 \end{bmatrix} \quad (26.40)$$

Since the similar matrices of diagonalization, S_ξ, S_η, and S_ζ, are nonsingular, the left-hand inverse matrices S_ξ^{-1}, S_η^{-1}, and S_ζ^{-1} are easily found. Although these left-hand inverse matrices are essential to the diagonalization process, they provide little insight for the following flux-vector splitting procedure. The rather involved results are omitted here, but they can be found in Shang and Fithen (1996).

26.6 Flux-Vector Splitting

An efficient flux-vector splitting algorithm for solving the Euler equations was developed by Steger and Warming (1987). The basic concept is equally applicable to any hyperbolic differential system for which the solution need not be analytic (Sommerfeld 1949, Courant and Hilbert 1965). In most CEM applications, discontinuous behavior of the solution is associated only with the wave across an interface between different media, a piecewise continuous solution. Even if a jump condition exists, the magnitude of the finite jump across the interface is much less drastic than the shock waves encountered in supersonic flows. Nevertheless, the salient feature of the piecewise continuous solution domains of the hyperbolic partial-differential equation stands out: the coefficient matrices of the time-dependent, 3D Maxwell equations cast in the general curvilinear frame of reference contain metrics of coordinate transformation. Therefore, the equation system no longer has constant coefficients even in an isotropic and homogeneous medium. Under this circumstance, eigenvalues can change sign at any given field location due to the metric variations of the coordinate transformation. Numerical oscillations have appeared in results calculated using the flux-vector splitting technique when eigenvalues change sign. A refined flux-difference splitting algorithm has been developed to resolve fields with jump conditions (Van Leer 1982, Anderson et al. 1985). The newer flux-difference splitting algorithm is particularly effective at locations where the eigenvalues vanish. Perhaps more crucial for electromagnetics, the polarization of the medium, making the basic equations become nonlinear, occurs only in the extremely high-frequency range (Harrington 1961, Elliott 1966). In general, the governing equations are linear; at most, the coefficients of the differential system are dependent on physical location and phase velocity. For this reason, the difference between the flux-vector splitting (Steger and Warming 1987) and flux-difference splitting (Van Leer 1982, Anderson et al. 1985) schemes, when applied to the time-dependent Maxwell equations, is not of great importance.

The basic idea of the flux-vector splitting of Steger and Warming is to process data according to the direction of information propagation. Since diagonalization is achievable only in each time–space plane, the direction of wave propagation degenerates into either the positive or the negative orientation. This designation is consistent with the notion of the right-running and the left-running wave components. The flux vectors are computed from the point value, including the metrics at the node of interest. This formulation for solving hyperbolic partial-differential equations not only ensures the well posedness of the differential system but also enhances the stability of the numerical procedure (Richtmyer and Morton 1967, Anderson et al. 1984, Roe 1986, Shang 1995). Specifically, the flux vectors F_ξ, F_η, and F_ζ will be split according to the sign of their corresponding eigenvalues. The split fluxes are differenced by an upwind algorithm to allow for the zone of dependence of an initial-value problem (Roe 1986, Shang 1995, Shang and Gaitonde 1995, Shang and Fithen 1996).

From the previous analysis, it is clear that the eigenvalues contain multiplicities, and hence the split flux of the 3D Maxwell equations is not unique (Shang and Gaitonde 1995, Shang and Fithen 1996). All flux vectors in each time–space plane are split according to the signs of the local eigenvalues:

$$
\begin{aligned}
F_\xi &= F_\xi^+ + F_\xi^- \\
F_\eta &= F_\eta^+ + F_\eta^- \\
F_\zeta &= F_\zeta^+ + F_\zeta^-
\end{aligned}
\qquad (26.41)
$$

The flux-vector components associated with the positive and negative eigenvalues are obtainable by a straightforward matrix multiplication:

$$
\begin{aligned}
F_\xi^+ &= S_\xi \lambda_\xi^+ S_\xi^{-1} U \\
F_\xi^- &= S_\xi \lambda_\xi^- S_\xi^{-1} U \\
F_\eta^+ &= S_\eta \lambda_\eta^+ S_\eta^{-1} U \\
F_\eta^- &= S_\eta \lambda_\eta^+ S_\eta^{-1} U \\
F_\zeta^+ &= S_\zeta \lambda_\zeta^+ S_\zeta^{-1} U \\
F_\zeta^- &= S_\zeta \lambda_\zeta^- S_\zeta^{-1} U
\end{aligned} \tag{26.42}
$$

It is also important to recognize that even if the split flux vectors in each time–space plane are no unique, the sum of the split components must be unambiguously identical to the flux vector of the governing Equation 26.29. This fact is easily verifiable by performing the addition of the split matrices to reach the identities in Equation 26.32 through 26.34. In addition, if one sets the diagonal elements of metrics, ξ_x, η_y, and ζ_z, equal to unity and the off-diagonal elements equal to zero, the coefficient matrices will recover the Cartesian form:

$$
F_\varepsilon^+ = \frac{1}{2V}\begin{bmatrix}
\frac{\xi_y^2+\xi_z^2}{\alpha\sqrt{\varepsilon\mu}} & \frac{-\xi_x\xi_y}{\alpha\sqrt{\varepsilon\mu}} & \frac{-\xi_x\xi_z}{\alpha\sqrt{\varepsilon\mu}} & 0 & \frac{-\xi_z}{\varepsilon} & \frac{\xi_y}{\varepsilon} \\
\frac{-\xi_x\xi_y}{\alpha\sqrt{\varepsilon\mu}} & \frac{\xi_x^2+\xi_z^2}{\alpha\sqrt{\varepsilon\mu}} & \frac{-\xi_y\xi_z}{\alpha\sqrt{\varepsilon\mu}} & \frac{\xi_z}{\varepsilon} & 0 & \frac{-\xi_x}{\varepsilon} \\
\frac{-\xi_x\xi_z}{\alpha\sqrt{\varepsilon\mu}} & \frac{-\xi_y\xi_z}{\alpha\sqrt{\varepsilon\mu}} & \frac{\xi_x^2+\xi_y^2}{\alpha\sqrt{\varepsilon\mu}} & \frac{-\xi_y}{\varepsilon} & \frac{\xi_x}{\varepsilon} & 0 \\
0 & \frac{\xi_z}{\mu} & \frac{-\xi_y}{\mu} & \frac{\xi_y^2+\xi_z^2}{\alpha\sqrt{\varepsilon\mu}} & \frac{-\xi_x\xi_y}{\alpha\sqrt{\varepsilon\mu}} & \frac{-\xi_x\xi_z}{\alpha\sqrt{\varepsilon\mu}} \\
\frac{-\xi_z}{\mu} & 0 & \frac{\xi_x}{\mu} & \frac{-\xi_x\xi_y}{\alpha\sqrt{\varepsilon\mu}} & \frac{\xi_x^2+\xi_z^2}{\alpha\sqrt{\varepsilon\mu}} & \frac{-\xi_y\xi_z}{\alpha\sqrt{\varepsilon\mu}} \\
\frac{\xi_y}{\mu} & \frac{-\xi_x}{\mu} & 0 & \frac{-\xi_x\xi_z}{\alpha\sqrt{\varepsilon\mu}} & \frac{-\xi_y\xi_z}{\alpha\sqrt{\varepsilon\mu}} & \frac{\xi_x^2+\xi_y^2}{\alpha\sqrt{\varepsilon\mu}}
\end{bmatrix}
\begin{Bmatrix} B_x \\ B_y \\ B_z \\ D_x \\ D_y \\ D_z \end{Bmatrix} \tag{26.43}
$$

$$
F_\varepsilon^- = \frac{1}{2V}\begin{bmatrix}
\frac{-(\xi_y^2+\xi_z^2)}{\alpha\sqrt{\varepsilon\mu}} & \frac{\xi_x\xi_y}{\alpha\sqrt{\varepsilon\mu}} & \frac{\xi_x\xi_z}{\alpha\sqrt{\varepsilon\mu}} & 0 & \frac{-\xi_z}{\varepsilon} & \frac{\xi_y}{\varepsilon} \\
\frac{\xi_x\xi_y}{\alpha\sqrt{\varepsilon\mu}} & \frac{-(\xi_x^2+\xi_z^2)}{\alpha\sqrt{\varepsilon\mu}} & \frac{\xi_y\xi_z}{\alpha\sqrt{\varepsilon\mu}} & \frac{\xi_z}{\varepsilon} & 0 & \frac{-\xi_x}{\varepsilon} \\
\frac{\xi_x\xi_z}{\alpha\sqrt{\varepsilon\mu}} & \frac{\xi_y\xi_z}{\alpha\sqrt{\varepsilon\mu}} & \frac{-(\xi_x^2+\xi_y^2)}{\alpha\sqrt{\varepsilon\mu}} & \frac{-\xi_y}{\varepsilon} & \frac{\xi_x}{\varepsilon} & 0 \\
0 & \frac{\xi_z}{\mu} & \frac{-\xi_y}{\mu} & \frac{-(\xi_y^2+\xi_z^2)}{\alpha\sqrt{\varepsilon\mu}} & \frac{\xi_x\xi_y}{\alpha\sqrt{\varepsilon\mu}} & \frac{\xi_x\xi_z}{\alpha\sqrt{\varepsilon\mu}} \\
\frac{-\xi_z}{\mu} & 0 & \frac{\xi_x}{\mu} & \frac{\xi_x\xi_y}{\alpha\sqrt{\varepsilon\mu}} & \frac{-(\xi_x^2+\xi_z^2)}{\alpha\sqrt{\varepsilon\mu}} & \frac{\xi_y\xi_z}{\alpha\sqrt{\varepsilon\mu}} \\
\frac{\xi_y}{\mu} & \frac{-\xi_x}{\mu} & 0 & \frac{\xi_x\xi_z}{\alpha\sqrt{\varepsilon\mu}} & \frac{\xi_y\xi_z}{\alpha\sqrt{\varepsilon\mu}} & \frac{-(\xi_y^2+\xi_z^2)}{\alpha\sqrt{\varepsilon\mu}}
\end{bmatrix}
\begin{Bmatrix} B_x \\ B_y \\ B_z \\ D_x \\ D_y \\ D_z \end{Bmatrix} \tag{26.44}
$$

$$F_\eta^+ = \frac{1}{2V}\begin{bmatrix} \frac{\eta_y^2+\eta_z^2}{\alpha\sqrt{\varepsilon\mu}} & \frac{-\eta_x\eta_y}{\alpha\sqrt{\varepsilon\mu}} & \frac{-\eta_x\eta_z}{\alpha\sqrt{\varepsilon\mu}} & 0 & \frac{-\eta_z}{\varepsilon} & \frac{\eta_y}{\varepsilon} \\ \frac{-\eta_x\eta_y}{\alpha\sqrt{\varepsilon\mu}} & \frac{\eta_x^2+\eta_z^2}{\alpha\sqrt{\varepsilon\mu}} & \frac{-\eta_y\eta_z}{\alpha\sqrt{\varepsilon\mu}} & \frac{\eta_z}{\varepsilon} & 0 & \frac{-\eta_x}{\varepsilon} \\ \frac{-\eta_x\eta_z}{\alpha\sqrt{\varepsilon\mu}} & \frac{-\eta_y\eta_z}{\alpha\sqrt{\varepsilon\mu}} & \frac{\eta_x^2+\eta_y^2}{\alpha\sqrt{\varepsilon\mu}} & \frac{-\eta_y}{\varepsilon} & \frac{\eta_x}{\varepsilon} & 0 \\ 0 & \frac{\eta_z}{\mu} & \frac{-\eta_y}{\mu} & \frac{\eta_y^2+\eta_z^2}{\alpha\sqrt{\varepsilon\mu}} & \frac{-\eta_x\eta_y}{\alpha\sqrt{\varepsilon\mu}} & \frac{-\eta_x\eta_z}{\alpha\sqrt{\varepsilon\mu}} \\ \frac{-\eta_z}{\mu} & 0 & \frac{\eta_x}{\mu} & \frac{-\eta_x\eta_y}{\alpha\sqrt{\varepsilon\mu}} & \frac{\eta_x^2+\eta_z^2}{\alpha\sqrt{\varepsilon\mu}} & \frac{-\eta_y\eta_z}{\alpha\sqrt{\varepsilon\mu}} \\ \frac{\eta_y}{\mu} & \frac{-\eta_x}{\mu} & 0 & \frac{-\eta_x\eta_z}{\alpha\sqrt{\varepsilon\mu}} & \frac{-\eta_y\eta_z}{\alpha\sqrt{\varepsilon\mu}} & \frac{\eta_x^2+\eta_y^2}{\alpha\sqrt{\varepsilon\mu}} \end{bmatrix}\begin{Bmatrix} B_x \\ B_y \\ B_z \\ D_x \\ D_y \\ D_z \end{Bmatrix} \tag{26.45}$$

$$F_\eta^- = \frac{1}{2V}\begin{bmatrix} \frac{-(\eta_y^2+\eta_z^2)}{\alpha\sqrt{\varepsilon\mu}} & \frac{\eta_x\eta_y}{\alpha\sqrt{\varepsilon\mu}} & \frac{\eta_x\eta_z}{\alpha\sqrt{\varepsilon\mu}} & 0 & \frac{-\eta_z}{\varepsilon} & \frac{\eta_y}{\varepsilon} \\ \frac{\eta_x\eta_y}{\alpha\sqrt{\varepsilon\mu}} & \frac{-(\eta_x^2+\eta_z^2)}{\alpha\sqrt{\varepsilon\mu}} & \frac{\eta_y\eta_z}{\alpha\sqrt{\varepsilon\mu}} & \frac{\eta_z}{\varepsilon} & 0 & \frac{-\eta_x}{\varepsilon} \\ \frac{\eta_x\eta_z}{\alpha\sqrt{\varepsilon\mu}} & \frac{\eta_y\eta_z}{\alpha\sqrt{\varepsilon\mu}} & \frac{-(\eta_x^2+\eta_y^2)}{\alpha\sqrt{\varepsilon\mu}} & \frac{-\eta_y}{\varepsilon} & \frac{\eta_x}{\varepsilon} & 0 \\ 0 & \frac{\eta_z}{\mu} & \frac{-\eta_y}{\mu} & \frac{-(\eta_y^2+\eta_z^2)}{\alpha\sqrt{\varepsilon\mu}} & \frac{\eta_x\eta_y}{\alpha\sqrt{\varepsilon\mu}} & \frac{\eta_x\eta_z}{\alpha\sqrt{\varepsilon\mu}} \\ \frac{-\eta_z}{\mu} & 0 & \frac{\eta_x}{\mu} & \frac{\eta_x\eta_y}{\alpha\sqrt{\varepsilon\mu}} & \frac{-(\eta_x^2+\eta_z^2)}{\alpha\sqrt{\varepsilon\mu}} & \frac{\eta_y\eta_z}{\alpha\sqrt{\varepsilon\mu}} \\ \frac{\eta_y}{\mu} & \frac{-\eta_x}{\mu} & 0 & \frac{\eta_x\eta_z}{\alpha\sqrt{\varepsilon\mu}} & \frac{\eta_y\eta_z}{\alpha\sqrt{\varepsilon\mu}} & \frac{-(\eta_x^2+\eta_y^2)}{\alpha\sqrt{\varepsilon\mu}} \end{bmatrix}\begin{Bmatrix} B_x \\ B_y \\ B_z \\ D_x \\ D_y \\ D_z \end{Bmatrix} \tag{26.46}$$

$$F_\zeta^+ = \frac{1}{2V}\begin{bmatrix} \frac{\zeta_y^2+\zeta_z^2}{\alpha\sqrt{\varepsilon\mu}} & \frac{-\zeta_x\zeta_y}{\alpha\sqrt{\varepsilon\mu}} & \frac{-\zeta_x\zeta_z}{\alpha\sqrt{\varepsilon\mu}} & 0 & \frac{-\zeta_z}{\varepsilon} & \frac{\zeta_y}{\varepsilon} \\ \frac{-\zeta_x\zeta_y}{\alpha\sqrt{\varepsilon\mu}} & \frac{\zeta_x^2+\zeta_z^2}{\alpha\sqrt{\varepsilon\mu}} & \frac{-\zeta_y\zeta_z}{\alpha\sqrt{\varepsilon\mu}} & \frac{\zeta_z}{\varepsilon} & 0 & \frac{-\zeta_x}{\varepsilon} \\ \frac{-\zeta_x\zeta_z}{\alpha\sqrt{\varepsilon\mu}} & \frac{-\zeta_y\zeta_z}{\alpha\sqrt{\varepsilon\mu}} & \frac{\zeta_x^2+\zeta_y^2}{\alpha\sqrt{\varepsilon\mu}} & \frac{-\zeta_y}{\varepsilon} & \frac{\zeta_x}{\varepsilon} & 0 \\ 0 & \frac{\zeta_z}{\mu} & \frac{-\zeta_y}{\mu} & \frac{\zeta_y^2+\zeta_z^2}{\alpha\sqrt{\varepsilon\mu}} & \frac{-\zeta_x\zeta_y}{\alpha\sqrt{\varepsilon\mu}} & \frac{-\zeta_x\zeta_z}{\alpha\sqrt{\varepsilon\mu}} \\ \frac{-\zeta_z}{\mu} & 0 & \frac{\zeta_x}{\mu} & \frac{-\zeta_x\zeta_y}{\alpha\sqrt{\varepsilon\mu}} & \frac{\zeta_x^2+\zeta_z^2}{\alpha\sqrt{\varepsilon\mu}} & \frac{-\zeta_x\zeta_z}{\alpha\sqrt{\varepsilon\mu}} \\ \frac{\zeta_y}{\mu} & \frac{-\zeta_x}{\mu} & 0 & \frac{-\zeta_x\zeta_z}{\alpha\sqrt{\varepsilon\mu}} & \frac{-\zeta_x\zeta_z}{\alpha\sqrt{\varepsilon\mu}} & \frac{\zeta_x^2+\zeta_y^2}{\alpha\sqrt{\varepsilon\mu}} \end{bmatrix}\begin{Bmatrix} B_x \\ B_y \\ B_z \\ D_x \\ D_y \\ D_z \end{Bmatrix} \tag{26.47}$$

$$F_{\zeta}^{-}=\frac{1}{2V}\begin{bmatrix}\frac{-(\zeta_y^2+\zeta_z^2)}{\alpha\sqrt{\varepsilon\mu}} & \frac{\zeta_x\zeta_y}{\alpha\sqrt{\varepsilon\mu}} & \frac{\zeta_x\zeta_z}{\alpha\sqrt{\varepsilon\mu}} & 0 & \frac{-\zeta_z}{\varepsilon} & \frac{\zeta_y}{\varepsilon}\\ \frac{\zeta_x\zeta_y}{\alpha\sqrt{\varepsilon\mu}} & \frac{-(\zeta_x^2+\zeta_z^2)}{\alpha\sqrt{\varepsilon\mu}} & \frac{\zeta_y\zeta_z}{\alpha\sqrt{\varepsilon\mu}} & \frac{\zeta_z}{\varepsilon} & 0 & \frac{-\zeta_x}{\varepsilon}\\ \frac{\zeta_x\zeta_z}{\alpha\sqrt{\varepsilon\mu}} & \frac{\zeta_y\zeta_z}{\alpha\sqrt{\varepsilon\mu}} & \frac{-(\zeta_x^2+\zeta_y^2)}{\alpha\sqrt{\varepsilon\mu}} & \frac{-\zeta_y}{\varepsilon} & \frac{\zeta_x}{\varepsilon} & 0\\ 0 & \frac{\zeta_z}{\mu} & \frac{-\zeta_y}{\mu} & \frac{-(\zeta_y^2+\zeta_z^2)}{\alpha\sqrt{\varepsilon\mu}} & \frac{\zeta_x\zeta_y}{\alpha\sqrt{\varepsilon\mu}} & \frac{\zeta_x\zeta_z}{\alpha\sqrt{\varepsilon\mu}}\\ \frac{-\zeta_z}{\mu} & 0 & \frac{\zeta_x}{\mu} & \frac{\zeta_x\zeta_y}{\alpha\sqrt{\varepsilon\mu}} & \frac{-(\zeta_x^2+\zeta_z^2)}{\alpha\sqrt{\varepsilon\mu}} & \frac{\zeta_x\zeta_z}{\alpha\sqrt{\varepsilon\mu}}\\ \frac{\zeta_y}{\mu} & \frac{-\zeta_x}{\mu} & 0 & \frac{\zeta_x\zeta_z}{\alpha\sqrt{\varepsilon\mu}} & \frac{\zeta_x\zeta_z}{\alpha\sqrt{\varepsilon\mu}} & \frac{-(\zeta_x^2+\zeta_y^2)}{\alpha\sqrt{\varepsilon\mu}}\end{bmatrix}\begin{Bmatrix}B_x\\B_y\\B_z\\D_x\\D_y\\D_z\end{Bmatrix} \tag{26.48}$$

26.7 Finite-Difference Approximation

Once the detailed split fluxes are known, the formulation of the finite-difference approximation is straightforward. From the sign of an eigenvalue, the stencil of a spatially second- or higher-order-accurate windward differencing can be easily constructed to form multiple 1D difference operators (Richtmyer and Morton 1967, Anderson et al. 1984, Shang 1995). In this regard, the forward difference and the backward difference approximations are used for the negative and the positive eigenvalues, respectively. The split flux vectors are evaluated at each discretized point of the field according to the signs of the eigenvalues.

For the present purpose, a second-order accurate procedure is given:

$$\begin{aligned}\text{If }\lambda<0\quad \Delta U_i&=\frac{-3U_i+4U_{i+1}-U_{i+2}}{2}\\ \text{If }\lambda>0\quad \Delta U_i&=\frac{3U_i-4U_{i+1}+U_{i+2}}{2}\end{aligned} \tag{26.49}$$

The necessary metrics of the coordinate transformation are calculated by central differencing, except at the edges of computational domain, where one-sided differences are used. Although the fractional-step or the time-splitting algorithm (Richtmyer and Morton 1967, Anderson et al. 1984, Shang 1995) has demonstrated greater efficiency in data storage and a higher data-processing rate than predictor–corrector time integration procedures (Shang 1995, Shang and Gaitonde 1995, Shang and Fithen 1996), it is limited to second-order accuracy in time. With respect to the fractional-step method, the temporal second-order result is obtained by a sequence of symmetrically cyclic operators (Richtmyer and Morton 1967, Shang 1995):

$$U_i^{n+1}=L_\xi L_\eta L_\zeta L_\zeta L_\mu L_\xi U_i^n \tag{26.50}$$

where L_ξ, L_η, and L_ζ are the difference operators for 1D equations in the ξ, η, and ζ coordinates, respectively.

In general, second-order and higher temporal resolution is achievable through multiple-time-step schemes (Richtmyer and Morton 1967, Anderson et al. 1984). However, one-step schemes are more attractive because they have less memory requirements and do not need special start-up procedures

(Shang and Gaitonde 1995, Shang and Fithen 1996). For future higher-order accurate solution development potential, the Runge–Kutta family of single-step, multistage procedure is recommended. This choice is also consistent with the accompanying characteristic-based finite-volume method (Shang and Gaitonde 1995).

In the present effort, the two-stage, formally second-order accurate scheme is used:

$$U^0 = U_i^n$$
$$U^1 = U^0 - \Delta U(U^0)$$
$$U^2 = 0.5[\Delta U(U^1 + \Delta U(U^0))]$$
$$U_i^{n+1} = U^2 \tag{26.51}$$

where ΔU comprises the incremental values of dependent variables during each temporal sweep. The resultant characteristic-based finite-difference scheme for solving the 3D Maxwell equations in the time domain is second-order accurate in both time and space.

The most significant feature of the flux-vector splitting scheme lies in its ability to easily suppress reflected waves from the truncated computational domain. In wave motion, the compatibility condition at any point in space is described by the split flux vector (Shang 1995, Shang and Gaitonde 1995, Shang and Fithen 1996). In the present formulation, an approximated no-reflection condition can be achieved by setting the incoming flux component equal to zero:

$$\text{either} \lim_{r\to\infty} F^+(\xi,\eta,\zeta) = 0 \quad \text{or} \quad \lim_{r\to\infty} F^-(\xi,\eta,\zeta) = 0 \tag{26.52}$$

The 1D compatibility condition is exact when the wave motion is aligned with one of the coordinates (Shang 1995). This unique attribute of the characteristic-based numerical procedure in removing a fundamental dilemma in CEM will be demonstrated in detail later.

26.8 Finite-Volume Approximation

The finite-volume approximation solves the governing equation by discretizing the physical space into contiguous cells and balancing the flux vectors on the cell surfaces. Thus, in discretized form, the integration procedure reduces to evaluation of the sum of all fluxes aligned with surface-area vectors:

$$\frac{\Delta U}{\Delta t} + \frac{\Delta F}{\Delta \xi} + \frac{\Delta G}{\Delta \eta} + \frac{\Delta H}{\Delta \zeta} - J = 0 \tag{26.53}$$

In the present approach, the continuous differential operators have been replaced by discrete operators. In essence, the numerical procedure needs only to evaluate the sum of all flux vectors aligned with surface-area vectors (Van Leer 1982, Anderson et al. 1985, Shang and Gaitonde 1995, Shang and Fithen 1996). Only one of the vectors is required to coincide with the outward normal to the cell surface, and the rest of the orthogonal triad can be made to lie on the same surface. The metrics, or more appropriately the direction cosines, on the cell surface are uniquely determined by the nodes and edges of the elementary volume. This feature is distinct from the finite-difference approximation. The shape of the cell under consideration and the stretching ratio of neighbor cells can lead to a significant deterioration of the accuracy of finite-volume schemes (Leonard 1988).

The most outstanding aspect of finite-volume schemes is the elegance of its flux-splitting process. The flux-difference splitting for Equation 26.25 is greatly facilitated by a locally orthogonal system in the transformed space (Van Leer 1982, Anderson et al. 1985). In this new frame of reference, eigenvalues and eigenvectors as well as metrics of the coordinate transformation between two orthogonal systems are well known (Shang 1995, Shang and Gaitonde 1995). The inverse transformation is simply the transpose of the forward mapping. In particular, the flux vectors in the transformed space have the same functional form as that in the Cartesian frame. The difference between the flux vectors in the transformed and the Cartesian coordinates is a known quantity and is given by the product of the surface outward normal and the cell volume, $V(\nabla S/\|\nabla S\|)$ (Shang and Gaitonde 1995). Therefore, the flux vectors can be split in the transformed space according to the signs of the eigenvalues but without detailed knowledge of the associated eigenvectors in the transformed space. This feature of the finite-volume approach provides a tremendous advantage over the finite-difference approximation in solving complex problems in physics.

The present formulation adopts Van Leer's kappa scheme in which solution vectors are reconstructed on the cell surface from the piecewise data of neighboring cells (Van Leer 1982, Anderson et al. 1985). The spatial accuracy of this scheme spans a range from first-order to third-order upwind-biased approximations:

$$U^{+}_{i+(1/2)} = U_i + \frac{\varphi}{4}[(1-\kappa)\nabla + (1+\kappa)\Delta]U_i$$
$$U^{-}_{i+(1/2)} = U_i - \frac{\varphi}{4}[(1+\kappa)\nabla + (1-\kappa)\Delta]U_i \tag{26.54}$$

where $\Delta U_i = U_i - U_{i-1}$ and $\nabla U_i = U_{i+1} - U_i$ are the forward and backward differencing discretizations. The parameters φ and κ control the accuracy of the numerical results. For $\varphi = 1$, $\kappa = -1$, a two-point windward scheme is obtained. This method has an odd-order leading truncation-error term; the dispersive error is expected to dominate. If $\kappa = 1/3$, a third-order upwind-biased scheme will emerge. In fact both upwind procedures have discernible leading phase error. This behavior is a consequence of using the two-stage time integration algorithm, and the dispersive error can be alleviated by increasing the temporal resolution. For $\varphi = 1$, $\kappa = 0$, the formulation recovers the Fromm scheme (Van Leer 1982, Anderson et al. 1985). If $\kappa = 1$, the formulation yields the spatially central scheme. Since the fourth-order dissipative term is suppressed, the central scheme is susceptible to parasitic odd–even point decoupling (Anderson et al. 1984, 1985).

The time integration is carried out by the same two-stage Runge–Kutta method as in the present finite-difference procedure (Shang 1995, Shang and Gaitonde 1995). The finite-volume procedure is therefore second-order accurate in time and up to third-order accurate in space (Shang and Gaitonde 1995, Shang and Fithen 1996). For the present purpose, only the second-order upwinding and the third-order upwind-biased options are exercised. The second-order windward schemes in the form of the flux-vector splitting finite-difference and the flux-difference splitting finite-volume schemes are formally equivalent (Van Leer 1982, Anderson et al. 1985, Leonard 1988, Shang and Gaitonde 1995, Shang and Fithen 1996).

26.9 Summary and Research Issues

The technical merits of the characteristic-based methods for solving the time-dependent, 3D Maxwell equations can best be illustrated by the following two illustrations. In Figure 26.2, the exact electric field of a traveling wave is compared with numerical results. The numerical results were generated at the maximum allowable time-step size defined by the Courant–Friedrichs–Lewy (CFL) number of 2 ($\lambda\Delta x/\Delta t = 2$) (Richtmyer and Morton 1967, Anderson et al. 1984). The numerical solutions presented are at instants when a right-running wave reaches the midpoint of the computational domain and exits the numerical boundary, respectively. For this 1D simulation, the characteristic-based scheme using the

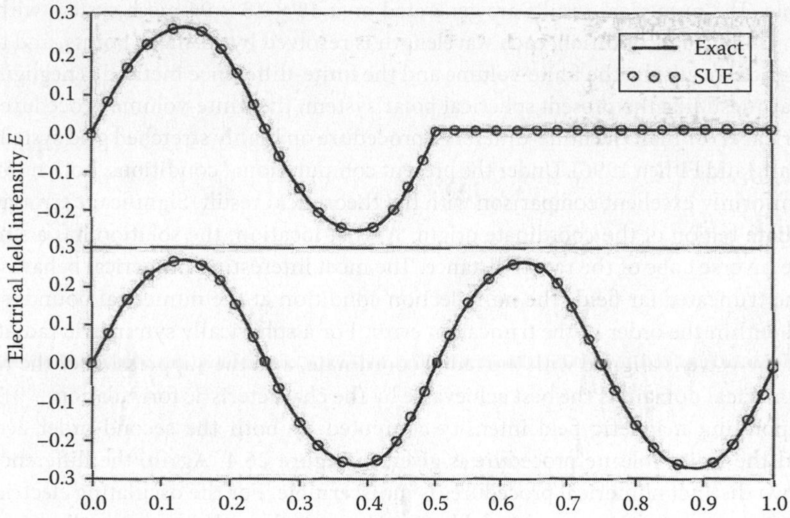

FIGURE 26.2 Perfect-shift property of a 1D wave computation, CFL = 2.

single-step upwind explicit algorithm exhibits the shift property, which indicates a perfect translation of the initial value in space (Anderson et al. 1984). As the impulse wave moves through the initially quiescent environment, the numerical result duplicates the exact solution at each and every discretized point, including the discontinuous incoming wave front. Although this highly desirable property of a numerical solution is achievable only under very restrictive conditions and is not preserved for multi-dimensional problems (Richtmyer and Morton 1967, Anderson et al. 1984), the ability to simulate the nonanalytic solution behavior in the limit is clearly illustrated.

In Figure 26.3, another outstanding feature of the characteristic-based method is highlighted by simulating the oscillating electric dipole. For the radiating electric dipole, the depicted temporal calculations are sampled at the instant when the initial pulse has traveled a distance of 2.24 wavelengths

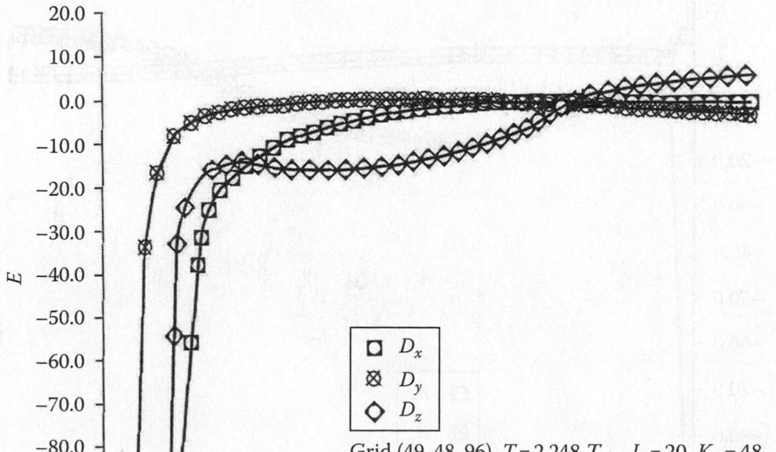

FIGURE 26.3 Instantaneous distributions of oscillating-dipole electric field.

from the dipole. The numerical results are generated on a 48 × 48 × 96 mesh system with the second-order scheme. Under that condition, each wavelength is resolved by 15-mesh points, and the difference between numerical results by the finite-volume and the finite-difference method is negligible. However, on an irregular mesh like the present spherical polar system, the finite-volume procedure has shown a greater numerical error than the finite-difference procedure on highly stretched grid systems (Anderson et al. 1985, Shang and Fithen 1996). Under the present computational conditions, both numerical procedures yield uniformly excellent comparison with the theoretical result. Significant error appeared only in the immediate region of the coordinate origin. At that location, the solution has a singularity that behaves as the inverse cube of the radial distance. The most interesting numerical behavior, however, is revealed in the truncated far field. The no-reflection condition at the numerical boundary is observed to be satisfied within the order of the truncation error. For a spherically symmetric radiating field, the orientation of the wave is aligned with the radial coordinate, and the suppression of the reflected wave within the numerical domain is the best achievable by the characteristic formulation.

The corresponding magnetic field intensity computed by both the second-order accurate finite-difference and the finite-volume procedure is given in Figure 26.4. Again, the difference in solution between the two distinct numerical procedures is indiscernible. For the oscillating electric dipole, only the *x*- and *y*-components of the magnetic field exist. Numerical results attain excellent agreement with theoretical values (Shang and Gaitonde 1995, Shang and Fithen 1996). The third-order accurate finite-volume scheme also produces a similar result on the same mesh but at a greater allowable time-step size (a CFL value of 0.87 is used vs. 0.5 for the second-order method). A numerically more efficient and higher-order accurate simulation is obtainable in theory. However, at present, the third-order windward-biased algorithm cannot reinforce rigorously the zone-of-dependence requirement; therefore, the reflected-wave suppression is incomplete in the truncated numerical domain. For this reason, the third-order accurate results are not included here.

Numerical accuracy and efficiency are closely related issues in CEM. A high-accuracy requirement of a simulation is supportable only by efficient numerical procedures. The inaccuracies incurred by numerical simulations are attributable to the mathematical formulation of the problem, to the algorithm, to the

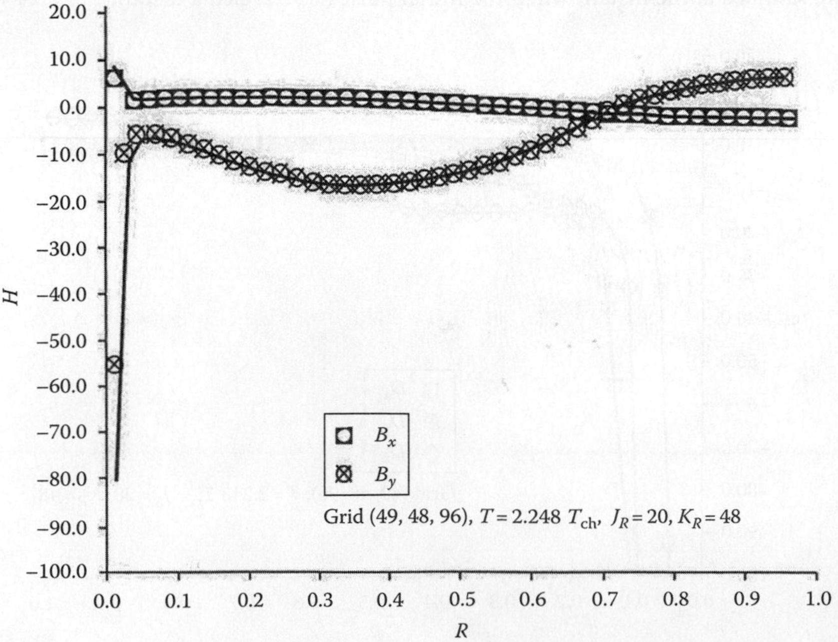

FIGURE 26.4 Instantaneous distributions of oscillating-dipole magnetic field.

numerical procedure, and to computational inaccuracy. A basic approach to relieve the accuracy limitation must be derived from using high-order schemes or spectral methods. The numerical efficiency of CEM can be enhanced substantially by using scalable multicomputers (Shang et al. 1993). The effective use of a distributed-memory, message-passing homogeneous multicomputer still requires a judicious trade-off between a balanced work load and interprocessor communication. A characteristic-based finite-volume computer program has been successfully mapped onto distributed-memory systems by a rudimentary domain decomposition strategy. For example, a square waveguide, at five different frequencies up to the cutoff, was simulated.

Figure 26.5 displays the *x*-component of the magnetic field intensity within the waveguide. The simulated transverse electric mode, $TE_{1,1}$, $E_x = 0$, which has a half period of π along the *x*- and *y*-coordinates, is generated on a (24 × 24 × 128) mesh system. Since the entire field is described by simple harmonic functions, the remaining field components are similar, and only half the solution domain along the *z*-coordinate is presented to minimize repetition. In short, the agreement between the closed-form and numerical solutions is excellent at each frequency. In addition, the numerical simulations duplicate the physical phenomenon at the cutoff frequency, below which there is no phase shift along the waveguide and the wave motion ceases (Harrington 1961, Elliott 1966). For simple harmonic wave motion in an isotropic medium, the numerical accuracy can be quantified. At a grid-point density of 12 nodes per wavelength, the L_2 norm (Richtmyer and Morton 1967) has a nearly uniform magnitude of order 10^{-4}. The numerical results are fully validated by comparison with theory. However, further efforts are still required to substantially improve the parallel and scalable numerical efficiency. In fact, this is the most promising area in CEM research.

The pioneering efforts in CEM usually employed the total-field formulation on staggered-mesh systems (Yee 1966, Taflove 1992). That particular combination of numerical algorithm and procedure has been proven to be very effective. In earlier RCS calculations using the present numerical procedure,

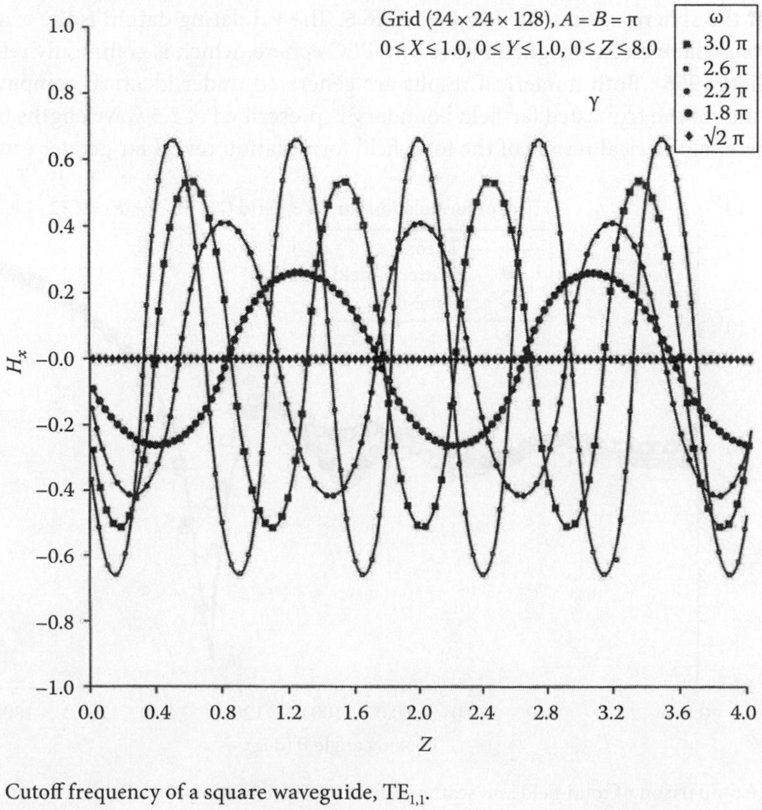

FIGURE 26.5 Cutoff frequency of a square waveguide, $TE_{1,1}$.

the total-field formulation was also utilized (Shang 1995, Shang and Gaitonde 1995). However, for 3D scatterer simulation, the numerical accuracy requirement for RCS evaluations becomes extremely stringent. In the total-field formulation, the dynamic range of the field variables has a substantial difference from the exposed and the shadow region, and the incident wave must also traverse the entire computation domain.

Both requirements impose severe demands on the numerical accuracy of simulation. In addition, the total field often contains only the residual of partial cancelations of the incident and the diffracted waves. The far-field electromagnetic energy distribution becomes a secular problem a small difference between two variables of large magnitude. An alternative approach via the scattered-field formulation for RCS calculations appears to be very attractive. Particularly in this formulation, the numerical dissipation of the incident wave that must propagate from the far-field boundary to the scatterer is completely eliminated from the computations. In short, the incident field can be directly specified on the scatterer's surface. The numerical advantage over the total-field formulation is substantial.

The total-field formulation can be cast in the scattered-field form by replacing the total field with scattered-field variables (Harrington 1961, Elliott 1966):

$$U_s = U_t - U_i \tag{26.55}$$

Since the incident field U_i must satisfy the Maxwell equations identically, the equations of the scattered field remain unaltered from the total-field formulation. Thus, the scattered-field formulation can be considered as a dependent-variable transform of the total-field equations. In the present approach, both formulations are solved by a characteristic-based finite-volume scheme.

The comparison of horizontal polarized RCS of a perfect electrically conducting (PEC) sphere, $\sigma(\theta, 0.0)$, from the total-field and scattered-field formulations at $ka = 5.3$ (where k = wave number and a = diameter of the sphere) is presented in Figure 26.6. The validating datum is the exact solution for the scattering of a plane electromagnetic wave by a PEC sphere, which is commonly referred to as the Mie series (Elliott 1966). Both numerical results are generated under identical computational conditions. The location of the truncated far-field boundary is prescribed at 2.5 wavelengths from the center of the PEC sphere. Numerical results of the total-field formulation reveal far greater error than for the

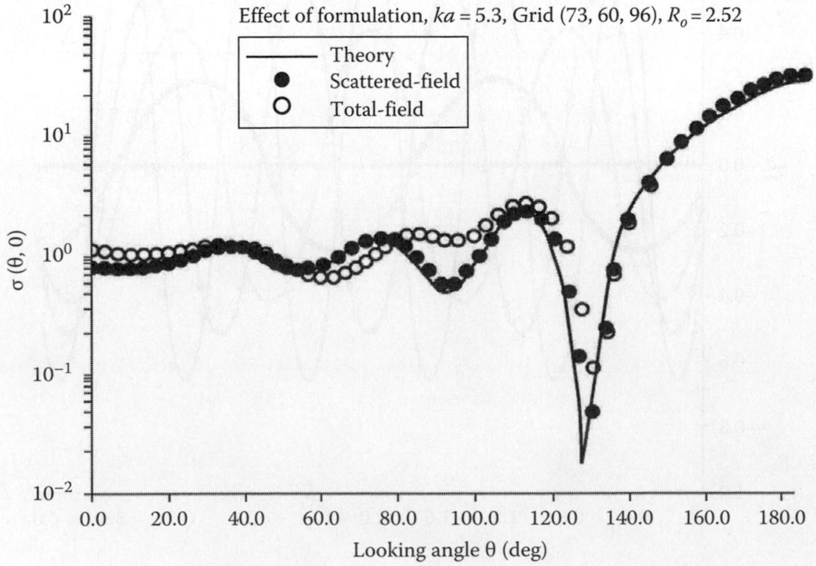

FIGURE 26.6 Comparison of total-field and scattered-field RCS calculations of $\sigma(\theta, 0°)$, $ka = 5.3$.

scattered-field formulation. The additional source of error is incurred when the incident wave must propagate from the far-field boundary to the scatterer. In the scattered-field formulation, the incident field data are described precisely by the boundary condition on the scatterer surface. Since the far-field electromagnetic energy distribution is derived from the near-field parameters (Sommerfeld 1949, Taflove 1992, Shang and Gaitonde 1995), the advantage of describing the data incident on a scatterer without error is tremendous. Numerical errors of the total-field calculations are evident in the exaggerated peaks and troughs over the entire viewing-angle displacement.

In Figure 26.7, the vertically polarized RCS σ(θ, 90.0°) of the *ka* = 5.3 case substantiates the previous observation. In fact, the numerical error of the total-field calculation is excessive in comparison with the result of the scattered-field formulation. Since the results are not obtained for the optimal far-field placement for RCS computation, the results of the scattered-field formulation overpredict the theoretical values by 2.7%. The deviation of the total-field result from the theory, however, exceeds 25.6% and becomes unacceptable. In addition, computations by the total-field formulation exhibit a strong sensitivity to placement of the far-field boundary. A small perturbation of the far-field boundary placement leads to a drastic change in the RCS prediction: a feature resembling the ill-posedness condition, which is highly undesirable for numerical simulation. Since there is very little difference in computer coding for the two formulations, the difference in computing time required for an identical simulation is insignificant. On the Cray C90, 1,505.3 s at a data-processing rate of 528.8 Mflops and an average vector length of 62.9 is needed to complete a sampling period. At present, the most efficient calculation on a distributed-memory system, Cray T3-D, has reduced the processing time to 1204.2 s using 76 computing nodes.

More recent progress of CEM in the time domain can be illustrated in three aspects: the vastly increased data processing on multicomputers, the modeling and simulation of increasingly complex scatters, and the removing the fundamental limitation in conducting wave propagation in a finite computational domain without reflection.

Just a few years ago, an order of tenfold speedup and immense increase in memory size by parallel-computing system permitted an RCS calculation of a PEC sphere at a wave number beyond 100, (*ka* = 50.0). The calculation is carried on a (91 × 241 × 384) mesh system, and the bistatic RCS was sampled for the elapse time of three periods. The RCS computation requires 2887.6 s on an IBM SP 256 node system, which is the first generation commercially available of massive parallel-computing system. The calculated horizontal polarized bistatic RCS distribution σ(θ, 90°) is presented in Figure 26.8 to exhibit an excellent agreement with the theoretical results by the Mie series over a dynamic range of six decades

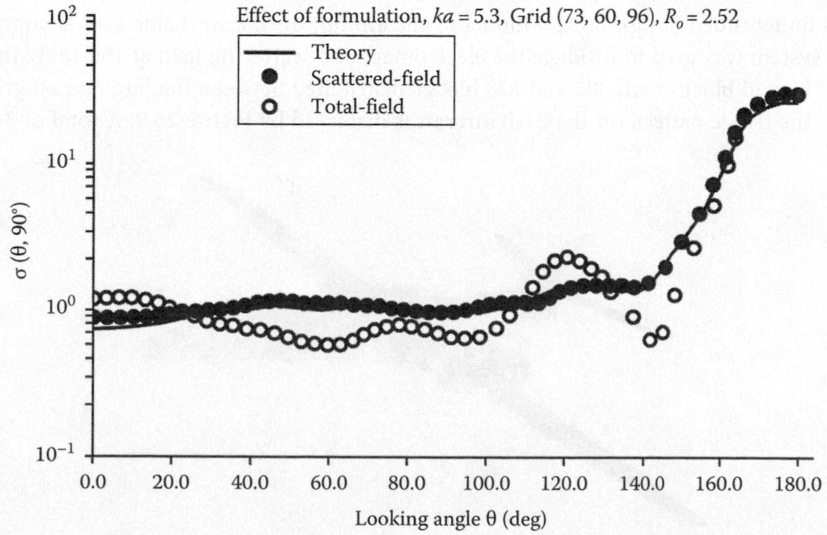

FIGURE 26.7 Comparison of total-field and scattered-field RCS calculations of σ(θ, 90°), *ka* = 5.3.

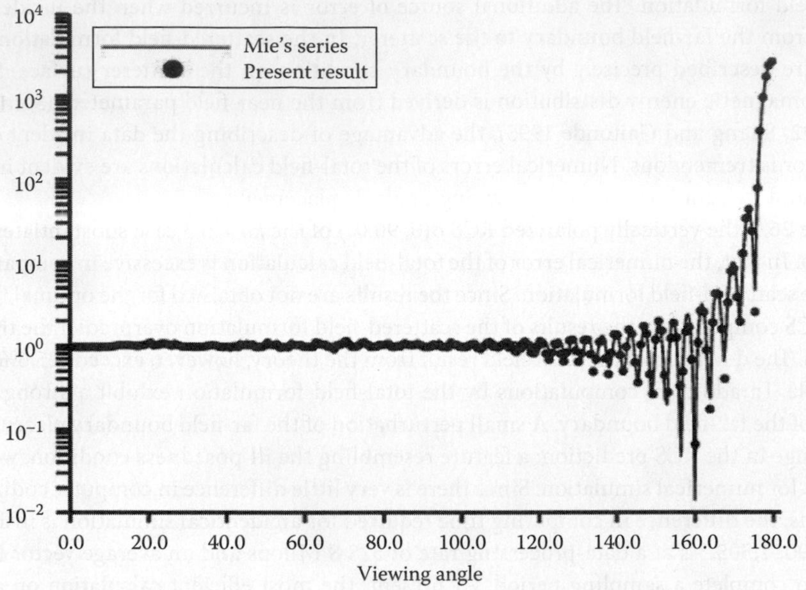

FIGURE 26.8 Comparison bistatic RCS calculations of $\sigma(\theta, 90°)$, $ka = 50$ with Mie series.

(Bowman et al. 1987). Today, the speedup of modern concurrent computers increases almost exponentially to expand the predictable frequency range of incident wave approaching the lower optical regime.

In 1999, the time-domain computational procedure using the structured, overset grid generation process has applied to a scattering simulation of a full-scale B1-B aircraft. This particular simulation is intended to demonstrate the range of practical application of time-domain methods and to seek an alternative grid topology for the discretized field approximation. Two separate multiblock grids for the fore and aft portion of the B1-B are constructed according to the distinct geometric features of the aircraft. A widely used grid generation package from the CFD community is adopted for the process (Shang 2009). The two major multiblock grid systems are overlapped to allow the information exchange between the two systems during the data processing. The level of effort required to generate this overall grid system is substantial; nearly nine months of man-hour has been devoted from the time of the available computer-aided design (CAD) model of the aircraft until a workable grid is completed. The overset grid system was used to produce the electromagnetic scattering field at 100 MHz that consists of a total of 711 grid blocks with 383 and 328 blocks distributed between the fore and aft grid systems, respectively. The fringe pattern on the B1-B aircraft is displayed by Figure 26.9. A total of 34.5 million

FIGURE 26.9 Radar fringe pattern on B1-B aircraft, $f = 100$ MHz.

grid cells were used and with 23.5 million cells are concentrated in describing the fore body of this aircraft including the complex engine nacelles with the air breathing inlet. The aft body configuration is equally challenging with rudder and elevator assembly. The numerical simulation requires then the unprecedented problem size of more than 192,000,000 unknowns. On a 258-node SGI Origin 2000 system, which is another first-generation parallel computer, the data-processing rate is estimated to be about 24.32 Gflops (giga floating point per second). This speedup benchmark in data-processing rate by parallel computing is easily surpassed by the new generation of multicomputer.

Meanwhile, the progress in the technique of grid generation is shifting to the unstructured grid approach, which represents the most general grid connectivity. Basically, the discretized field points are constructed by a group of contiguous tetrahedrons from the solid surface of a complex electric objective extending to far field. This coordinate transforming process requires an explicit connectivity of adjacent grid points. The identical need also exists for the mapping of a numerical procedure to multicomputers by the domain decomposition approach. The explicit connectivity of adjacent grid cells or data blocks for grid generation is also optimal for load balancing and limiting interprocessor communication for concurrent computing to become the most attractive approach for simulating the complex scatter configuration (Shang 2009).

The fundamental challenge for CEM is that the hyperbolic differential system must be solved on a truncated physical domain due to a finite size of computer memory. The limitation was remedied earlier by the absorbing far-field boundary condition of Enquist and Majda (1977) and later by the perfectly matching layer (PML) of Berenge (1992). Both approaches alleviate the wave reflection from the far-field computational domain by numerical dissipation. In the middle 1990s, Shang and Fithen (1996) and Shang and Gaitonde (1995) formally derived the 3D characteristic formulation for the time-dependent Maxwell equation on a generalized curvilinear coordinates. In the characteristic-based formulation, the no-reflection far-field boundary condition is readily achievable by imposed vanishing incoming flux-vector component at the computational domain.

The effectiveness of the characteristic formulation is demonstrated by the numerical simulation of the near-field, microwave radiation from a rectangular pyramidal horn shown in Figure 26.10. In this figure, the transmit wave is the principal transverse electric mode, $TE_{1,0}$, at a frequency of 12.56 GHz with

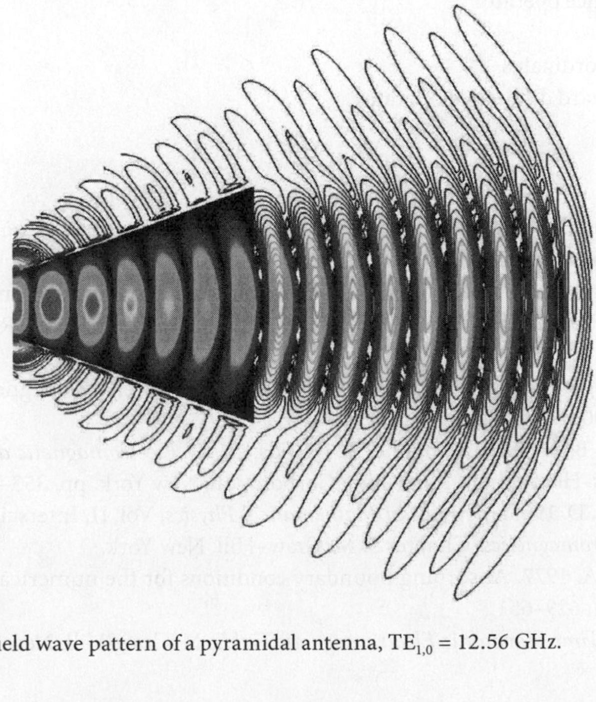

FIGURE 26.10 Near-field wave pattern of a pyramidal antenna, $TE_{1,0}$ = 12.56 GHz.

a wavelength of 2.398 cm, and the aperture of the antenna has a dimension of 3.05 by 3.64 wavelength. The far-field boundary condition for this numerical simulation is placed just at a few wavelengths from the exit of the antenna. The graph is presented at the time frame when the transmitting wave front has just passed through the computational outer domain. The contour of electric intensity exhibits no detectable distortion of any reflected waves from the computational boundary.

In summary, recent progress in solving the 3D Maxwell equations in the time domain has opened a new frontier in electromagnetics, plasmadynamics, and optics, as well as the interface between electrodynamics and quantum mechanics (Taflove 1992). The progress in microchip and interconnect network technology has led to a host of high-performance distributed-memory, message-passing parallel computer systems. The synergism of efficient and accurate numerical algorithms for solving the Maxwell equations in the time domain with high-performance multicomputers will propel the new interdisciplinary simulation technique to practical and productive applications.

Nomenclature

$\mathbf{B}$	magnetic flux density
$\mathbf{C}$	coefficient matrix of flux-vector formulation
$\mathbf{D}$	electric displacement
$\mathbf{E}$	electric field strength
$\mathbf{F}$	flux-vector component
$\mathbf{H}$	magnetic flux intensity
i, j, k	index of discretization
$\mathbf{J}$	electric current density
N	index of temporal level of solution
S	similarity matrix of diagonalization
t	time
U	dependent variables
V	elementary cell volume
x, y, z	Cartesian coordinates
Δ	forward difference operator
λ	eigenvalue
ξ, η, ζ	transformed coordinates
∇	gradient, backward difference operator

References

Anderson, D. A., Tannehill, J. C., and Pletcher, R. H. 1984. *Computational Fluid Mechanics and Heat Transfer*. Hemisphere Publishing Corporation, New York.

Anderson, W. K., Thomas, J. L., and Van Leer, B. 1985. A comparison of finite-volume flux splitting for the Euler equations. *Proceedings of the AIAA 23rd Aerospace Science Meeting*, Reno, NV, AIAA Paper 85-0122.

Berenger, J.-P. 1994. A perfect matched layer for the absorption of electromagnetic waves, *J. Comput. Phys*., 114: 185–200.

Bowman, J. J., Senior, T. B. A., and Uslenghi, P. L. E. (Eds.) 1987. *Electromagnetic and Acoustic Scattering by a Simple Shapes*. Hemisphere Publishing Corporation, New York, pp. 353–415.

Courant, R. and Hilbert, D. 1965. *Methods of Mathematical Physics*, Vol. II. Interscience, New York.

Elliott, R. A. 1966. *Electromagnetics*, Chapter 5. McGraw–Hill, New York.

Enquist, B. and Majda, A. 1977. Absorbing boundary conditions for the numerical simulation of waves. *Math. Comput*., 31: 629–651.

Harrington, R. F. 1961. *Time-Harmonic Electromagnetic Fields*. McGraw-Hill, New York.

Harrington, R. F. 1968. *Field Computation by Moment Methods*, 4th edn., Robert E. Krieger, Malabar, FL.

Higdon, R. 1986. Absorbing boundary conditions for difference approximation to multidimensional wave equation. *Math Comput.*, 47(175): 437–459.

Leonard, B. P. 1988. Simple high-accuracy resolution program for convective modeling of discontinuities. *Int. J. Numer. Methods Fluids*, 8: 1291–1318.

Mur, G. 1981. Absorbing boundary conditions for the finite-difference approximation of the time-domain electromagnetic-field equations. *IEEE Trans. Electromagn. Compat.*, EMC-23(4): 377–382.

Rahman, B. M. A., Fernandez, F. A., and Davies, J. B. 1991. Review of finite element methods for microwave and optical waveguide. *Proc. IEEE*, 79: 1442–1448.

Richtmyer, R. D. and Morton, K. W. 1967. *Difference Methods for Initial-Value Problem*. Interscience, New York.

Roe, P. L. 1986. Characteristic-based schemes for the Euler equations. *Ann. Rev. Fluid Mech.*, 18: 337–365.

Shang, J. S. 1995. A fractional-step method for solving 3-D, time-domain Maxwell equations. *J. Comput. Phys.*, 118(1): 109–119, April 1995.

Shang, J. S. 2002. Shared knowledge in computational fluid dynamics, electromagnetics, and magneto-aerodynamics. *Prog. Aerosp. Sci.*, 38(6–7): 449–467.

Shang, J. S. 2009. Computational fluid dynamics application to aerospace science. *Aeronaut. J.*, 113(1148): 619–632.

Shang, J. S. and Fithen, R. M. 1996. A comparative study of characteristic-based algorithms for the Maxwell equations. *J. Comput. Phys.*, 125: 378–394.

Shang, J. S. and Gaitonde, D. 1995. Scattered electromagnetic field of a reentry vehicle. *J. Spacecraft Rockets*, 32(2): 294–301.

Shang, J. S., Hill, K. C., and Calahan, D. 1993. Performance of a characteristic-based, 3-D, time- domain Maxwell equations solver on a massively parallel computer. *Appl. Comput. Elect. Soc.*, 10(1): 52–62, March 1995.

Sommerfeld, A. 1949. *Partial Differential Equations in Physics*, Chapter 2. Academic Press, New York.

Steger, J. L. and Warming, R. F. 1987. Flux vector splitting of the inviscid gas dynamics equations with application to finite difference methods. *J. Comput. Phys.*, 20(2): 263–293.

Taflove, A. 1992. Re-inventing electromagnetics: Supercomputing solution of Maxwell's equations via direct time integration on space grids. *Comput. Systems Eng.*, 3(1–4): 153–168.

Thompson, J. F. 1982. *Numerical Grid Generation*. Elsevier Science, New York.

Van Leer, B. 1982. Flux-vector splitting for the Euler equations. *Lect. Note. Phys.*, 170: 507–512, ICASE TR 82-30.

Yee, K. S. 1966. Numerical solution of initial boundary value problems involving Maxwell's equations in isotropic media. *IEEE Trans. Antennas Propag.*, 14(3): 302–307.

27

Computational Fluid Dynamics

David A. Caughey
Cornell University

27.1 Introduction

The use of computer-based methods for the prediction of fluid flows has seen tremendous growth in the past half century. Fluid dynamics has been one of the earliest, and most active, fields for the application of numerical techniques. This is due to the essential nonlinearity of most fluid flow problems of practical interest—which makes analytical, or closed-form, solutions virtually impossible to obtain—combined with the geometrical complexity of these problems. In fact, the history of computational fluid dynamics (CFD) can be traced back virtually to the birth of the digital computer itself, with the pioneering work of John von Neumann and others in this area. Von Neumann was interested in using the computer not only to solve engineering problems but to understand the fundamental nature of fluid flows themselves. This is possible because the complexity of fluid flows arises, in many instances, not from complicated or poorly understood formulations but from the nonlinearity of partial differential equations that have been known for more than a century. A famous paragraph, written by von Neumann in 1946, serves to illustrate this point. He wrote [16]:

> "Indeed, to a great extent, experimentation in fluid mechanics is carried out under conditions where the underlying physical principles are not in doubt, where the quantities to be observed are completely determined by known equations. The purpose of the experiment is not to verify a proposed theory but to replace a computation from an unquestioned theory by direct measurements. Thus, wind tunnels are, for example, used at present, at least in part, as computing devices of the so-called analogy type … to integrate the non-linear partial differential equations of fluid dynamics."

This chapter article provides some of the basic background in fluid dynamics required to understand the issues involved in numerical solution of fluid flow problems, then outlines the approaches that have been successful in attacking problems of practical interest, and identifies areas of current research.

27.2 Underlying Principles

In this section, we will provide the background in fluid dynamics required to understand the principles involved in the numerical solution of the governing equations. The formulation of the equations in generalized, curvilinear coordinates and the geometrical issues involved in the construction of suitable grids will also be discussed.

27.2.1 Fluid Dynamical Background

As can be inferred from the quotation of von Neumann presented in the introduction, fluid dynamics is fortunate to have a generally accepted mathematical framework for describing most problems of practical interest. Such diverse problems as the high-speed flow past an aircraft wing, the motions of the atmosphere responsible for our weather, and the unsteady air currents produced by the flapping wings of a housefly, all can be described as solutions to a set of partial differential equations known as the Navier–Stokes equations. These equations express the physical laws corresponding to conservation of mass, Newton's second law of motion relating the acceleration of fluid elements to the imposed forces, and conservation of energy, under the assumption that the stresses in the fluid are linearly related to the local rates of strain of the fluid elements. This latter assumption is generally regarded as an excellent approximation for everyday fluids such as water and air—the two most common fluids of engineering and scientific interest.

We will describe the equations for problems in two space dimensions, for the sake of economy of notation; here, and elsewhere, the extension to problems in three dimensions will be straightforward unless otherwise noted. The Navier–Stokes equations can be written in the form

$$\frac{\partial \mathbf{w}}{\partial t}+\frac{\partial \mathbf{f}}{\partial x}+\frac{\partial \mathbf{g}}{\partial y}=\frac{\partial \mathbf{R}}{\partial x}+\frac{\partial \mathbf{S}}{\partial y} \tag{27.1}$$

where $\mathbf{w}$ is the vector of conserved quantities

$$\mathbf{w}=\{\rho,\ \rho u,\ \rho v,\ e\}^T \tag{27.2}$$

where

ρ is the fluid density
u and v are the fluid velocity components in the x and y directions, respectively
e is the total energy per unit volume

The inviscid flux vectors $\mathbf{f}$ and $\mathbf{g}$ are given by

$$\mathbf{f}=\{\rho u,\ \rho u^2+p,\ \rho uv,\ (e+p)u\}^T \tag{27.3}$$

$$\mathbf{g}=\{\rho v,\ \rho uv,\ \rho v^2+p,\ (e+p)v\}^T \tag{27.4}$$

where p is the fluid pressure, and the flux vectors describing the effects of viscosity are

$$\mathbf{R}=\{0,\ \tau_{xx},\ \tau_{xy},\ u\tau_{xx}+v\tau_{xy}-q_x\}^T \tag{27.5}$$

$$\mathbf{S}=\{0,\ \tau_{xy},\ \tau_{yy},\ u\tau_{xy}+v\tau_{yy}-q_y\}^T \tag{27.6}$$

The viscous stresses appearing in these terms are related to the derivatives of the components of the velocity vector by

$$\tau_{xx} = 2\mu \frac{\partial u}{\partial x} - \frac{2}{3}\mu \left\{ \frac{\partial u}{\partial x} + \frac{\partial v}{\partial y} \right\} \tag{27.7}$$

$$\tau_{xy} = \mu \left\{ \frac{\partial u}{\partial y} + \frac{\partial v}{\partial x} \right\} \tag{27.8}$$

$$\tau_{yy} = 2\mu \frac{\partial v}{\partial y} - \frac{2}{3}\mu \left\{ \frac{\partial u}{\partial x} + \frac{\partial v}{\partial y} \right\} \tag{27.9}$$

and the x and y components of the heat flux vector are

$$q_x = -k \frac{\partial T}{\partial x} \tag{27.10}$$

$$q_y = -k \frac{\partial T}{\partial y} \tag{27.11}$$

according to Fourier's Law. In these equations, μ and k represent the coefficients of viscosity and thermal conductivity, respectively.

If the Navier–Stokes equations are nondimensionalized by normalizing lengths with respect to a representative length L, velocities by a representative velocity V_∞, and the fluid properties (such as density and coefficient of viscosity) by their values in the free stream, an important nondimensional parameter, the *Reynolds number*,

$$\text{Re} = \frac{\rho_\infty V_\infty L}{\mu_\infty} \tag{27.12}$$

appears as a parameter in the equations. In particular, the viscous stress terms on the right-hand side of Equation 27.1 are multiplied by the inverse of the Reynolds number. Physically, the Reynolds number can be interpreted as an inverse measure of the relative importance of the contributions of the viscous stresses to the dynamics of the flow; that is when the Reynolds number is large, the viscous stresses are small almost everywhere in the flow field.

The computational resources required to solve the complete Navier–Stokes equations are enormous, particularly when the Reynolds number is large and regions of *turbulent flow* must be resolved. In 1970, Howard Emmons of Harvard University estimated the computer time required to solve a simple, turbulent pipe-flow problem, including direct computation of all turbulent eddies containing significant energy (Emmons [11]). For a computational domain consisting of approximately 12 diameters of a pipe of circular cross section, the computation of the solution at a Reynolds number based on pipe diameter of $\text{Re}_d = 10^7$ would require approximately 10^{17} s on a 1970s mainframe computer. Of course, much faster computers are now available, but even at a computational speed of 100 gigaflops—that is 10^{11} floating-point operations per second—such a calculation would require more than 3000 *years* to complete.

Because the resources required to solve the complete Navier–Stokes equations are so large, it is common to make approximations to bring the required computational resources to a more modest level for

TABLE 27.1 Hierarchy of Fluid Mechanical Approximations

Stage	Model	Equations	Time–Frame
I	Linear potential	Laplace's equation	1960s
IIa	Nonlinear potential	Full potential equation	1970s
IIb	Nonlinear inviscid	Euler equations	1980s
III	Modeled turbulence	Reynolds-averaged Navier–Stokes (RANS) equations	1990s
IV	Large-eddy simulation (LES)	Navier–Stokes equations with sub-grid turbulence model	1980s–
V	Direct numerical simulation (DNS)	Fully resolved Navier–Stokes equations	1980s–

Source: Chapman, D.R., *AIAA J.*, 17, 1293, 1979.

problems of practical interest. Expanding slightly on a classification introduced by Chapman [9], the sequence of fluid mechanical equations can be organized into the hierarchy shown in Table 27.1.

Table 27.1 summarizes the physical assumptions and time periods of most intense development for each of the stages in the fluid mechanical hierarchy. Stage IV represents an approximation to the Navier–Stokes equations in which only the largest, presumably least isotropic, scales of turbulence are resolved; the sub-grid scales are modeled. Stage III represents an approximation in which the solution is decomposed into time- (or ensemble-) averaged and fluctuating components for each variable. For example, the velocity components and pressure can be decomposed into

$$u = U + u' \tag{27.13}$$

$$v = V + v' \tag{27.14}$$

$$p = P + p' \tag{27.15}$$

where U, V, and P are the average values of u, v, and p, respectively, taken over a time interval that is long compared to the turbulence time scales, but short compared to the time scales of any nonsteadiness of the averaged flow field. If we let $<u>$ denote such a time average of the u component of the velocity, then, for example,

$$<u> = U \tag{27.16}$$

When a decomposition of the form of Equations 27.13 through 27.15 is substituted into the Navier–Stokes equations and the equations are averaged as described previously, the resulting equations describe the evolution of the mean flow quantities (such as U, V, and P). These equations, called the *Reynolds-averaged Navier–Stokes (RANS) equations*, are nearly identical to the original Navier–Stokes equations written for the mean flow variables because terms that are linear in $<u'>$, $<v'>$, $<p'>$, etc., are identically zero. The nonlinearity of the equations, however, introduces terms that depend upon the fluctuating components. In particular, terms proportional to $<\rho u'v'>$, $<\rho u'^2>$, and $<\rho v'^2>$ appear in the averaged equations. Dimensionally, these terms are equivalent to stresses; in fact, these quantities are called the Reynolds stresses. Physically, the Reynolds stresses are the turbulent counterparts of the molecular viscous stresses and appear as a result of the transport of momentum by the turbulent fluctuations. The appearance of these terms in the equations describing the mean flow makes it impossible to determine even the mean flow without some knowledge of these fluctuating components.

In order to solve the RANS equations, the Reynolds stresses must be related to the mean flow at some level of approximation using a phenomenological model. The simplest procedure is to try to relate the

Reynolds stresses to the local mean flow properties. Since the turbulence that is responsible for these stresses is a function not only of the local mean flow state but also of the flow history, such an approximation cannot have broad generality, but such local, or algebraic, turbulence models based on the analogy of mixing length from the kinetic theory of gases are useful for many flows of engineering interest, especially for flows in which boundary-layer separation does not occur.

A more general procedure is to develop partial differential equations for the Reynolds stresses themselves or for quantities that can be used to determine the scales of the turbulence, such as the turbulent kinetic energy k and the dissipation rate ε. The latter, so-called k-ε model, is widely used in engineering analyses of turbulent flows. These differential equations can be derived by taking higher-order moments of the Navier–Stokes equations. For example, if the x-momentum equation is multiplied by v and then averaged, the result will be an equation for the Reynolds stress component $\langle\rho u'v'\rangle$. Again because of the nonlinearity of the equations, however, yet higher moments of the fluctuating components (e.g., terms proportional to $\langle\rho u'^2 v'\rangle$ $\langle\rho u'v'^2\rangle$) will appear in the equations for the Reynolds stresses. This is an example of the problem of closure of the equations for the Reynolds stresses; that is, to solve the equations for the Reynolds stresses, these third-order quantities must be known. Equations for the third-order quantities can be derived by taking yet higher-order moments of the Navier–Stokes equations, but these equations will involve fourth-order moments. Thus, at some point, the taking of moments must be terminated and models must be developed for the unknown higher-order quantities. It is hoped that more nearly universal models may be developed for these higher-order quantities, but the superiority of second-order models, in which the equations for the Reynolds stress components are solved, subject to modeling assumptions for the required third-order quantities, has yet to be established in any generality.

For many design purposes, especially in aerodynamics, it is sufficient to represent the flow as that of an inviscid, or ideal, fluid. This is appropriate for flows at high Reynolds numbers that contain only negligibly small regions of separated flow. The equations describing inviscid flows can be obtained as a simplification of the Navier–Stokes equations in which the viscous terms are neglected altogether. This results in the *Euler equations* of inviscid, compressible flow

$$\frac{\partial \mathbf{w}}{\partial t}+\frac{\partial \mathbf{f}}{\partial x}+\frac{\partial \mathbf{g}}{\partial y}=0 \qquad (27.17)$$

This approximation corresponds to stage IIb in the hierarchy described in Table 27.1.

The Euler equations comprise a hyperbolic system of partial differential equations, and their solutions contain features that are absent from the Navier–Stokes equations. In particular, while the viscous diffusion terms appearing in the Navier–Stokes equations guarantee that solutions will remain smooth, the absence of these dissipative terms from the Euler equations allows them to have solutions that are discontinuous across surfaces in the flow. Solutions to the Euler equations must be interpreted within the context of generalized (or weak) solutions, and this theory provides the mathematical framework for developing the properties of any discontinuities that may appear. In particular, jumps in dependent variables (such as density, pressure, and velocity) across such surfaces must be consistent with the original conservation laws upon which the differential equations are based. For the Euler equations, these jump conditions are called the Rankine–Hugoniot conditions.

Solutions of the Euler equations for flows containing gasdynamic discontinuities, or shock waves, can be computed using either *shock-fitting* or *shock-capturing* methods. In the former, the shock surfaces must be located, and the Rankine–Hugoniot jump conditions enforced explicitly. In shock-capturing methods, *artificial viscosity* terms are added in the numerical approximation to provide enough dissipation to allow the shocks to be captured automatically by the scheme, with no special treatment in the vicinity of the shock waves. The numerical viscosity terms usually act to smear out the discontinuity over several grid cells. Numerical viscosity also is used when solving the Navier–Stokes equations for flows containing shock waves, because usually it is impractical to resolve the shock structure defined by the physical dissipative mechanisms.

In many cases, the flow can further be approximated as steady and irrotational. In these cases, it is possible to define a velocity potential Φ such that the velocity vector **V** is given by

$$\mathbf{V} = \nabla\Phi \tag{27.18}$$

and the fluid density is given by the isentropic relation

$$\rho = \left\{1 + \frac{\gamma - 1}{2}\mathbf{M}_\infty^2[1 - (u^2 + v^2)]\right\}^{\frac{1}{\gamma-1}} \tag{27.19}$$

where

$$u = \frac{\partial \Phi}{\partial x} \tag{27.20}$$

$$v = \frac{\partial \Phi}{\partial y} \tag{27.21}$$

and $\mathbf{M}_\infty$ is a reference Mach number corresponding to the free stream state in which $\rho = 1$ and $u^2 + v^2 = 1$. The steady form of the Euler equations then reduces to the single equation

$$\frac{\partial \rho u}{\partial x} + \frac{\partial \rho v}{\partial y} = 0 \tag{27.22}$$

Equation 27.19 can be used to eliminate the density from Equation 27.22, which then can be expanded to the form

$$(a^2 - u^2)\frac{\partial^2 \Phi}{\partial x^2} - 2uv\frac{\partial^2 \Phi}{\partial x \partial y} + (a^2 - v^2)\frac{\partial^2 \Phi}{\partial y^2} = 0 \tag{27.23}$$

where a is the local speed of sound, which is a function only of the fluid speed $V = \sqrt{u^2 + v^2}$. Thus, Equation 27.23 is a single equation that can be solved for the velocity potential Φ.

Equation 27.23 is a second-order quasi-linear partial differential equation whose type depends on the sign of the discriminant $1 - \mathbf{M}^2$, where $\mathbf{M} = V/a$ is the *local* Mach number. The equation is elliptic or hyperbolic depending upon whether the Mach number is less than, or greater than, unity, respectively. Thus, the nonlinear potential equation contains a mathematical description of the physics necessary to predict the important features of transonic flows. It is capable of changing type, and the conservation form of Equation 27.22 allows surfaces of discontinuity, or shock waves, to be computed. Solutions at this level of approximation, corresponding to stage IIa, are considerably less expensive to compute than solutions of the full Euler equations, since only one dependent variable need be computed and stored. The jump relation corresponding to weak solutions of the potential equation is different than the Rankine–Hugoniot relations, but is a good approximation to the latter when the shocks are not too strong.

Finally, if the flow can be approximated by small perturbations to some uniform reference state, Equation 27.23 can further be simplified to

$$\left(1 - \mathbf{M}_\infty^2\right)\frac{\partial^2 \phi}{\partial x^2} + \frac{\partial^2 \phi}{\partial y^2} = 0 \tag{27.24}$$

where ϕ is the *perturbation* velocity potential defined according to

$$\Phi = x + \phi \tag{27.25}$$

if the uniform velocity in the reference state is assumed to be parallel to the *x*-axis and be normalized to have unit magnitude. Further, if the flow can be approximated as incompressible, that is, in the limit as $M_\infty \to 0$, Equation 27.23 reduces to

$$\frac{\partial^2 \Phi}{\partial x^2} + \frac{\partial^2 \Phi}{\partial y^2} = 0 \tag{27.26}$$

Since Equations 27.24 and 27.26 are linear, superposition of elementary solutions can be used to construct solutions of arbitrary complexity. Numerical methods to determine the singularity strengths for aerodynamic problems are called *Panel methods*, and constitute stage I in the hierarchy of approximations.

It is important to realize that, even though the time periods for development of some of the different models overlap, the applications of the various models may be for problems of significantly differing complexity. For example, DNS calculations were performed as early as the 1970s, but only for the simplest flows—homogeneous turbulence—at very low Reynolds numbers. Flows at higher Reynolds numbers and nonhomogeneous flows are being performed only now, while calculations for 3-D flows with modeled turbulence were performed as early as the mid-1980s.

27.2.2 Treatment of Geometry

For all numerical solutions to partial differential equations, including those of fluid mechanics, it is necessary to discretize the boundaries of the flow domain. For the panel method of stage I, this is all that is required since the problem is linear, and superposition can be used to construct solutions satisfying the boundary conditions. For nonlinear problems, it is necessary to discretize the entire flow domain. This can be done using either structured or unstructured grid systems. *Structured grids* are those in which the grid points can be ordered in a regular Cartesian structure; that is, the points can be given indices (i, j) such that the nearest neighbors of the (i, j) point are identified by the indices $(i \pm 1, j \pm 1)$. The grid cells for these meshes are thus quadrilateral in two dimensions and hexahedral in three dimensions. *Unstructured grids* have no regular ordering of points or cells, and a connectivity table must be maintained to identify which points and edges belong to which cells. Unstructured grids most often consist of triangles (in 2-D), tetrahedra (in 3-D), or combinations of these and quadrilateral and hexahedral cells, respectively. In addition, grids having purely quadrilateral cells may also be unstructured, even though they have a locally Cartesian structure—for example, when multilevel grids are used for adaptive refinement.

Implementations on structured grids are generally more efficient than those on unstructured grids since indirect addressing is required for the latter, and efficient implicit methods often can be constructed that take advantage of the regular ordering of points (or cells) in structured grids. A great deal of effort has been expended to generate both structured and unstructured grid systems, much of which is closely related to the field of computational geometry (see Chapter 32 of this volume).

27.2.2.1 Structured Grids

A variety of techniques are used to generate structured grids for use in fluid mechanical calculations. These include relatively fast algebraic methods, including those based on transfinite interpolation and conformal mapping, as well as more costly methods based on the solution of either elliptic or hyperbolic systems of partial differential equations for the grid coordinates. These techniques are discussed is a review article by Eiseman [10].

For complex geometries, it often is not possible to generate a single grid that conforms to all the boundaries. Even if it is possible to generate such a grid, it may have undesirable properties, such as excessive skewness or a poor distribution of cells, which could lead to poor stability or accuracy in the numerical algorithm. Thus, structured grids for complex geometries are generally constructed as combinations of simpler grid blocks for various parts of the domain. These grids may be allowed to overlap, in which case they are referred to as "Chimera" grids, or be required to share common surfaces of intersection, in which case they are identified as "multi-block" grids. In the latter case, grid lines may have varying degrees of continuity across the interblock boundaries, and these variations have implications for the construction and behavior of the numerical algorithm in the vicinity of these boundaries. Grid generation techniques based on the solution of systems of partial differential equations are described by Thompson et al. [52].

Numerical methods to solve the equations of fluid mechanics on structured grid systems are implemented most efficiently by taking advantage of the ease with which the equations can be transformed to a generalized coordinate system. The expression of the system of conservation laws in the new, body-fitted coordinate system reduces the problem to one effectively of Cartesian geometry. The transformation will be described here for the Euler equations.

Consider the transformation of Equation 27.17 to a new coordinate system (ξ, η). The local properties of the transformation at any point are contained in the Jacobian matrix of the transformation, which can be defined as

$$J = \begin{Bmatrix} x_\xi & x_\eta \\ y_\xi & y_\eta \end{Bmatrix} \tag{27.27}$$

for which the inverse is given by

$$J^{-1} = \begin{Bmatrix} \xi_x & \xi_y \\ \eta_x & \eta_y \end{Bmatrix} = \frac{1}{h} \begin{Bmatrix} y_\eta & -x_\eta \\ -y_\xi & x_\xi \end{Bmatrix} \tag{27.28}$$

where $h = x_\xi y_\eta - y_\xi x_\eta$ is the determinant of the Jacobian matrix. Subscripts in Equation 27.27 and 27.28 denote partial differentiation.

It is natural to express the fluxes in conservation laws in terms of their contravariant components. Thus, if we define

$$\{\mathbf{F}, \mathbf{G}\}^T = J^{-1}\{\mathbf{f}, \mathbf{g}\}^T \tag{27.29}$$

then the transformed Euler equations can be written in the compact form

$$\frac{\partial h\mathbf{w}}{\partial t} + \frac{\partial h\mathbf{F}}{\partial \xi} + \frac{\partial h\mathbf{G}}{\partial \eta} = 0 \tag{27.30}$$

if the transformation is independent of time (i.e., if the grid is not moving). If the grid is moving or deforming with time, the equations can be written in a similar form, but additional terms must be included that account for the fluxes induced by the motion of the mesh.

The Navier–Stokes equations can be transformed in a similar manner, although the transformation of the viscous contributions is somewhat more complicated and will not be included here. Since the nonlinear potential equation is simply the continuity equation (the first of the equations that comprise the Euler equations), the transformed potential equation can be written as

$$\frac{\partial}{\partial \xi}(\rho h U) + \frac{\partial}{\partial \eta}(\rho h V) = 0 \tag{27.31}$$

where

$$\begin{Bmatrix} U \\ V \end{Bmatrix} = J^{-1} \begin{Bmatrix} u \\ v \end{Bmatrix} \tag{27.32}$$

are the contravariant components of the velocity vector.

27.2.2.2 Unstructured Grids

Unstructured grids generally have greater geometric flexibility than structured grids, because of the relative ease of generating triangular or tetrahedral tessellations of 2- and 3-D domains. Advancing-front methods and Delaunay triangulations are the most frequently used techniques for generating triangular/tetrahedral grids. Unstructured grids also can be more efficiently adapted locally to resolve localized features of the solution.

27.3 Best Practices

27.3.1 Panel Methods

The earliest numerical methods used widely for making fluid dynamical computations were developed to solve linear potential problems, described as stage I calculations in the previous section. Mathematically, panel methods are based upon the fact that Equation 27.26 can be recast as an integral equation giving the solution at any point (x, y) in terms of the free stream speed U (here assumed unity) and angle of attack α and line integrals of singularities distributed along the curve C representing the body surface:

$$\Phi(x,y) = x\cos\alpha + y\sin\alpha + \int_C \sigma \ln\left(\frac{r}{2\pi}\right) ds + \int_C \delta\left(\frac{\partial}{\partial n}\right) \ln\left(\frac{r}{2\pi}\right) ds \tag{27.33}$$

In this equation, σ and δ represent the strengths of source and doublet singularities, respectively, distributed along the body contour, r is the distance from the point (x, y) to the boundary point, and n is the direction of the outward normal to the body surface. When the point (x, y) is chosen to lie on the body contour C, Equation 27.33 can be interpreted as giving the solution Φ at any point on the body in terms of the singularities distributed along the surface. This effectively reduces the dimension of the problem by one (i.e., the 2-D problem considered here becomes essentially 1-D, and the analogous procedure applied to a 3-D problem results in a 2-D equation requiring the evaluation only of integrals over the body surface).

Equation 27.33 is approximated numerically by discretizing the boundary C into a collection of panels (line segments in this 2-D example) on which the singularity distribution is assumed to be of some known functional form, but of an as yet unknown magnitude. For example, for a simple non-lifting body, the doublet strength δ might be assumed to be zero, while the source strength σ is assumed to be constant on each segment. The second integral in Equation 27.33 is then zero, while the first integral can be written as a sum over all the elements of integrals that can be evaluated analytically as

$$\Phi(x, y) = x\cos\alpha + y\sin\alpha + \sum_{i=1}^{N} \sigma_i \int_{C_i} \ln r/2\pi \; ds \tag{27.34}$$

where

C_i is the portion of the body surface corresponding to the ith segment

N is the number of segments, or panels, used

The source strengths σ_i must be determined by enforcing the boundary condition

$$\frac{\partial \Phi}{\partial n} = 0 \tag{27.35}$$

which specifies that the component of velocity normal to the surface be zero (i.e., that there be no flux of fluid across the surface). This is implemented by requiring that Equation 27.35 be satisfied at a selected number of control points. For the example of constant-strength source panels, if one control point is chosen on each panel, the requirement that Equation 27.35 be satisfied at each of the control points will result in N equations of the form

$$\sum_{i=1}^{N} A_{i,j}\sigma_i = \mathbf{U} \cdot \hat{\mathbf{n}}, \quad j = 1, 2, \ldots, N \tag{27.36}$$

where

$A_{i,j}$ are the elements of the influence coefficient matrix that give the normal velocity at control point j due to sources of unit strength on panel i

$\mathbf{U}$ is a unit vector in the direction of the free stream

Equation 27.36 constitutes a system of N linear equations that can be solved for the unknown source strengths σ_i. Once the source strengths have been determined, the velocity potential, or the velocity itself, can be computed directly at any point in the flow field using Equation 27.34. A review of the development and application of panel methods is provided by Hess [24].

A major advantage of panel methods, relative to the more advanced methods required to solve the nonlinear problems of stages II – V, is that it is necessary to describe (and to discretize into panels) only the surface of the body. While linearity is a great advantage in this regard, it is not clear that the method is computationally more efficient than the more advanced nonlinear field methods. This results from the fact that the influence coefficient matrix in the system of equations that must be solved to give the source strengths for each panel is not sparse; that is, the velocities at each control point are affected by the sources on all the panels. In contrast, the solution at each mesh point in a finite-difference formulation (or each mesh cell in a finite-volume formulation) is related to the values of the solution at only a few neighbors, resulting in a very sparse matrix of influence coefficients that can be solved very efficiently using iterative methods. Thus, the primary advantage of the panel method is the geometrical one associated with the reduction in dimension of the problem. For nonlinear problems, the use of finite-difference, finite-element, or finite-volume methods requires discretization of the entire flow field, and the associated mesh-generation task has been a major pacing item limiting the application of such methods.

27.3.2 Nonlinear Methods

For nonlinear equations, superposition of elementary solutions is no longer a valid technique for constructing solutions, and it becomes necessary to discretize the solution throughout the entire domain, not just the boundary surface. In addition, since the equations are nonlinear, some sort of iteration must be used to compute successively good approximations to the solution. This iterative process may approximate the physics of an unsteady flow process or may be chosen to provide more rapid convergence to the solution for steady-state problems. Solutions for both the nonlinear potential and Euler equations are generally determined using a finite-difference, finite-volume, or a finite-element method. In any of these techniques, a grid, or network of points and cells, is distributed throughout the flow field. In a finite-difference method, the derivatives appearing in the original differential equation are approximated by discrete differences in the values of the solution at neighboring points (and times, if the solution is unsteady), and substitution of these into the differential equation yields a system of algebraic equations relating the values of the solution at neighboring grid points. In a finite-volume method, the unknowns are taken to be representative of the values of the solution in the control volumes formed by the intersecting grid surfaces, and the equations are constructed by balancing fluxes across the bounding surfaces of each control volume with the rate of change of the solution within the control volume. In a finite-element method, the solution is represented using simple interpolating functions within each mesh cell, or element, and equations for the nodal values are obtained by integrating a variational or residual formulation of the equations over the elements. The algebraic equations relating the values of the solution in neighboring cells can be very similar (or even identical) in appearance for all three methods, and the choice of method often is primarily a matter of taste. Stable and efficient finite-difference and finite-volume methods were developed earlier than finite-element methods for compressible flow problems, but finite-element methods capable of treating very complex compressible flows are now available. A review of recent progress in the development of finite-element methods for compressible flows and remaining issues is given by Glowinski and Pironneau [14].

Since the nonlinear potential, Euler and Navier–Stokes equations all are nonlinear, the algebraic equations resulting from these discretization procedures also are nonlinear, and a scheme of successive approximation usually is required to solve them. As mentioned earlier, however, these equations tend to be highly local in nature and efficient iterative methods have been developed to solve them in many cases.

27.3.2.1 Nonlinear Potential Equation Methods

The primary advantage of solving the potential equation, rather than the Euler (or Navier–Stokes) equations, derives from the fact that the flow field can be described completely in terms of a single scalar function, the velocity potential Φ. The formulation of numerical schemes to solve Equation 27.22 is complicated by the fact that, as noted earlier, the equation changes type according to whether the local Mach number is subsonic or supersonic. Differencing schemes for the potential equation must, therefore, be type dependent—that is, they must change their form depending on whether the local Mach number is less than, or greater than, unity. These methods usually are based upon central, or symmetric, differencing formulas that are appropriate for the elliptic case (corresponding to subsonic flows), which are then modified by adding an upwind bias to maintain stability in hyperbolic regions (where the flow is locally supersonic). This directional bias can be introduced into the difference equations either by adding an artificial viscosity proportional to the third derivative of the velocity potential Φ in the streamwise direction or by replacing the density at each point in supersonic zones with its value at a point slightly upstream of the actual point. Mathematically, these two approaches can be seen to be equivalent, since the upwinding of the density evaluation also effectively introduces a correction proportional to the third derivative of the potential in the flow direction.

It is important to introduce such artificial viscosity (or compressibility) terms in such a way that their effect vanishes in the limit as the mesh is refined. In this way, the numerical approximation will approach the differential equation in the limit of zero mesh spacing, and the method is said to be consistent with the original differential equation. In addition, for flows with shock waves, it is important that the numerical approximation be *conservative;* this guarantees that the properties of discontinuous solutions will be consistent with the jump relations of the original conservation laws. The shock jump relations corresponding to Equation 27.22, however, are different from the Rankine–Hugoniot conditions describing shocks within the framework of the Euler equations. Since entropy is everywhere conserved in the potential theory and since there is a finite entropy jump across a Rankine–Hugoniot shock, it is clear that the jump relations must be different. For weak shocks, however, the differences are small and the economies afforded by the potential formulation make computations based on this approximation attractive for many transonic problems.

Perturbation analysis can be used to demonstrate that the effect of entropy jump across the shocks is more important than the rotationality introduced by these weak shocks. Thus, it makes sense to develop techniques that account for the entropy jump, but still within the potential formulation. Such techniques have been developed (see, e.g., Hafez [20]), but have been relatively little used as developments in techniques to solve the Euler equations have mostly overtaken these approaches.

The nonlinear difference equations resulting from discrete approximations to the potential equation generally are solved using iterative, or relaxation, techniques. The equations are linearized by computing approximations to all but the highest (second) derivatives from the preceding solution in an iterative sequence, and a correction is computed at each mesh point in such a way that the equations are identically satisfied. It is useful in developing these iterative techniques to think of the iterative process as a discrete approximation to a continuous time-dependent process (Garabedian [12]). Thus, the iterative process approximates an equation of the form

$$\beta_0 \frac{\partial \Phi}{\partial t} + \beta_1 \frac{\partial^2 \Phi}{\partial \xi \partial t} + \beta_2 \frac{\partial^2 \Phi}{\partial \eta \partial t} = \frac{a^2}{\rho}\left\{ \frac{\partial}{\partial \xi}(\rho h U) + \frac{\partial}{\partial \eta}(\rho h V) \right\} \tag{27.37}$$

The parameters β_0, β_1, and β_2, which are related to the mix of old and updated values of the solution used in the difference equations, can then be chosen to ensure that the time-dependent process converges to a steady state in both subsonic and supersonic regions. The formulation of transonic potential flow problems and their solution is described by Caughey [7].

Even when the values of the parameters are chosen to provide rapid convergence, many hundreds of iterations may be necessary to achieve convergence to acceptable levels, especially when the mesh is very fine. This slow convergence is a characteristic of virtually all iterative schemes and is a result of the fact that the representation of the difference equations must be highly local if the scheme is to be computationally efficient. As a result of this locality, the reduction of the low-wave-number component of the error to acceptable levels often requires many iterations. A powerful technique for circumventing this difficulty with the iterative solution of numerical approximations to partial differential equations, called the multigrid technique, has been applied with great success to problems in fluid mechanics.

The multigrid method relies for its success on the fact that, after a few cycles of any good iterative technique, the error remaining in the solution is relatively smooth and can be represented accurately on a coarser grid. Application of the same iterative technique on the coarser grid soon makes the error on this grid smooth as well, and the grid coarsening process can be repeated until the grid contains only a few cells in each coordinate direction. The corrections that have been computed on all coarser levels are then interpolated back to the finest grid, and the process is repeated. The accuracy of the converged solution on the fine grid is determined by the accuracy of the approximation on that grid, since the coarser levels are used only to effect a more rapid convergence of the iterative process.

A particularly efficient multigrid technique for steady transonic potential flow problems has been developed by Jameson [26]. It uses a generalized alternating-direction implicit (ADI) smoothing algorithm in conjunction with a full-approximation scheme multigrid algorithm. In theory, for a wide class of problems, the work (per mesh point) required to solve the equations using a multigrid approach is independent of the number of mesh cells. In many practical calculations, 10 or fewer multigrid cycles may be required even on very fine grids.

27.3.2.2 Euler Equation Methods

As noted earlier, the Euler equations constitute a hyperbolic system of partial differential equations, and numerical methods for solving them rely heavily upon the rather well-developed mathematical theory of such systems. As for the nonlinear potential equation, discontinuous solutions corresponding to flows with shock waves play an important role. Shock-capturing methods are much more widely used than shock-fitting methods, and for these methods, it is important to use schemes that are conservative.

As mentioned earlier, it is necessary to add artificial, or numerical, dissipation to stabilize the Euler equations. This can be done by adding dissipative terms explicitly to an otherwise non-dissipative, central difference scheme or by introducing upwind approximations in the flux evaluations. Both approaches are highly developed. The most widely used central difference methods are those modeled after the approach of Jameson et al. [31]. This approach introduces an adaptive blend of second and fourth differences of the solution in each coordinate direction; a local switching function, usually based on a second difference of the pressure, is used to reduce the order of accuracy of the approximation to first order locally in the vicinity of shock waves and to turn off the fourth differences there. The second-order terms are small in smooth regions where the fourth-difference terms are sufficient to stabilize the solution and ensure convergence to the steady state. More recently, Jameson [28] has developed improved symmetric limited positive (SLIP) and upstream limited positive (USLIP) versions of these blended schemes (see also Tatsumi et al. [51]).

Much effort has been directed toward developing numerical approximations for the Euler equations that capture discontinuous solutions as sharply as possible without overshoots in the vicinity of the discontinuity. For purposes of exposition of these methods, we consider the 1-D form of the Euler equations, which can be written as

$$\frac{\partial \mathbf{w}}{\partial t} + \frac{\partial \mathbf{f}}{\partial x} = 0 \tag{27.38}$$

where

$\mathbf{w} = \{\rho, \rho u, e\}^T$

$\mathbf{f} = \{\rho u, \rho u^2 + p, (e + p)u\}^T$

Not only is the exposition clearer for the 1-D form of the equations, but the implementation of these schemes for multidimensional problems also generally is done by dimensional splitting in which 1-D operators are used to treat variations in each of the mesh directions.

For smooth solutions, Equation 27.38 is equivalent to the quasi-linear form

$$\frac{\partial \mathbf{w}}{\partial t} + \mathbf{A}\frac{\partial \mathbf{w}}{\partial x} = 0 \tag{27.39}$$

where $\mathbf{A} = \{\partial \mathbf{f}/\partial \mathbf{w}\}$ is the Jacobian of the flux vector with respect to the solution vector. The eigenvalues of $\mathbf{A}$ are given by $\lambda = u, u + a, u - a$, where $a = \sqrt{\gamma p/\rho}$ is the speed of sound. Thus, for subsonic flows, one of the eigenvalues will have a different sign than the other two. For example, if $0 < u < a$, then $u - a < 0 < u < u + a$. The fact that various eigenvalues have different signs in subsonic flows means that simple one-sided difference methods cannot be stable. One way around this difficulty is to split the flux vector into two parts, the Jacobian of each of which has eigenvalues of only one sign. Such an approach has been developed by Steger and Warming [50]. Steger and Warming used a relatively simple splitting that has discontinuous derivatives whenever an eigenvalue changes sign; an improved splitting has been developed by van Leer [55] that has smooth derivatives at the transition points.

Each of the characteristic speeds can be identified with the propagation of a wave. If a mesh surface is considered to represent a discontinuity between two constant states, these waves constitute the solution to a Riemann (or shock tube) problem. A scheme developed by Godunov [15] assumes the solution to be piecewise constant over each mesh cell and uses the fact that the solution to the Riemann problem can be given in terms of the solution of algebraic (but nonlinear) equations to advance the solution in time. Because of the assumption of piecewise constancy of the solution, Godunov's scheme is only first-order accurate. van Leer [53,54] has shown how it is possible to extend these ideas to a second-order monotonicity-preserving scheme using the so-called monotone upwind scheme for systems of conservation laws (MUSCL) formulation. The efficiency of schemes requiring the solution of Riemann problems at each cell interface for each time step can be improved by the use of approximate solutions to the Riemann problem (Roe [42]).

More recent ideas to control oscillation of the solution in the vicinity of shock waves include the concept of total-variation diminishing (TVD) schemes, first introduced by Harten (see, e.g., Harten [21,22]) and essentially non-oscillatory (ENO) schemes introduced by Osher and his coworkers (see, e.g., Harten et al. [23] and Shu and Osher [45]).

Hyperbolic systems describe the evolution in time of physical systems undergoing unsteady processes governed by the propagation of waves. This feature frequently is used in fluid mechanics, even when the flow to be studied is steady. In this case, the unsteady equations are solved for long enough times that the steady state is reached asymptotically—often to within round-off error. To maintain the hyperbolic character of the equations and to keep the numerical method consistent with the physics it is trying to predict, it is necessary to determine the solution at a number of intermediate time levels between the initial state and the final steady state. Such a sequential process is said to be a time marching of the solution.

The simplest practical methods for solving hyperbolic systems are *explicit* in time. The size of the time step that can be used to solve hyperbolic systems using an explicit method is limited by a constraint known as the Courant–Friedrichs–Lewy *(CFL) condition*. Broadly interpreted, the CFL condition states that the time step must be smaller than the time required for the fastest signal to propagate across a single mesh cell. Thus, if the mesh is very fine, the allowable time step also must be very small, with the result that many time steps must be taken to reach an asymptotic steady state.

Multistage, or Runge–Kutta, methods have become extremely popular for use as explicit time-stepping schemes. After discretization of the spatial operators, using finite-difference, finite-volume, or finite-element approximations, the Euler equations can be written in the form

$$\frac{d\mathbf{w}_i}{dt} + Q(\mathbf{w}_i) + D(\mathbf{w}_i) = 0 \tag{27.40}$$

where

$\mathbf{w}_i$ represents the solution at the ith mesh point, or in the ith mesh cell

Q and D are discrete operators representing the contributions of the Euler fluxes and numerical dissipation, respectively

An m-stage time integration scheme for these equations can be written in the form

$$\mathbf{w}_i^{(k)} = \mathbf{w}_i^{(0)} - \alpha_k \Delta t \left\{ Q(\mathbf{w}_i^{(k-1)}) + D(\mathbf{w}_i^{(k-1)}) \right\}, \quad k = 1, 2, \ldots, m \tag{27.41}$$

with $\mathbf{w}_i^{(0)} = \mathbf{w}_i^n$, $\mathbf{w}_i^{(m)} = \mathbf{w}_i^{n+1}$ and $\alpha_m = 1$. The dissipative and dispersive properties of the scheme can be tailored by the sequence of α_i chosen; note that, for nonlinear problems, this method may be only first-order accurate in time, but this is not necessarily a disadvantage if one is interested only in the steady-state solution. The principal advantage of this formulation, relative to versions that may have better time accuracy, is that only two levels of storage are required regardless of the number of stages used. This approach was first introduced for problems in fluid mechanics by Graves and Johnson [18] and has been further developed by Jameson et al. [31]. In particular, Jameson and his group have shown how to tailor the stage coefficients so that the method is an effective smoothing algorithm for use with multigrid (see, e.g., Jameson and Baker [29]).

Implicit techniques also are highly developed, especially when structured grids are used. Approximate factorization of the implicit operator usually is required to reduce the computational burden of solving a system of linear equations for each time step. Methods based on ADI techniques date back to the pioneering work of Briley and McDonald [6] and Beam and Warming [5]. An efficient diagonalized ADI method has been developed within the context of multigrid by Caughey [8], and a lower–upper symmetric Gauss–Seidel method has been developed by Yoon and Kwak [58]. The multigrid implementations of these methods are based on the work of Jameson [27]. Arguably the fastest convergence for transonic airfoil problems has been achieved by a multigrid implementation of the symmetric Gauss–Seidel algorithm by Jameson and Caughey [30].

An excellent summary of the development of higher-order methods for unstructured grid computations is provided by Vincent and Jameson [56].

27.3.3 Navier–Stokes Equation Methods

As described earlier, the relative importance of viscous effects is characterized by the value of the Reynolds number. If the Reynolds number is not too large, the flow remains smooth, and adjacent layers (or laminae) of fluid slide smoothly past one another. When this is the case, the solution of the Navier–Stokes equations is not too much more difficult than solution of the Euler equations. Greater resolution is required to resolve the large gradients in the boundary layers near solid boundaries, especially as the Reynolds becomes larger, so more mesh cells are required to achieve acceptable accuracy. In most of the flow field, however, the flow behaves as if it were nearly inviscid, so methods developed for the Euler equations are appropriate and effective. The equations must, of course, be modified to include the additional terms resulting from the viscous stresses, and care must be taken to ensure that any artificial dissipative effects are small relative to the physical viscous dissipation in regions where the latter is important. The solution of the Navier–Stokes equations for laminar flows, then, is somewhat more costly in terms of computer resources, but not significantly more difficult from an algorithmic point of view, than solution of the Euler equations. Unfortunately, most flows of engineering interest occur at large enough Reynolds numbers that the flow in the boundary layers near solid boundaries becomes turbulent.

27.3.3.1 Turbulence Models

Solution of the RANS equations requires modeling of the Reynolds stress terms. The simplest models, based on the original mixing-length hypothesis of Prandtl, relate the Reynolds stresses to the local properties of the mean flow. The Baldwin–Lomax model [3] is the most widely used model of this type and gives good correlation with experimental measurements for wall-bounded shear layers so long as there are no significant regions of separated flow. Attempts have also been made to include nonequilibrium effects in algebraic turbulence models, see, for example [1].

The most complete commonly used turbulence models include two additional partial differential equations that determine characteristic length and time scales for the turbulence. The most widely used of these techniques is called the k-ε model, since it is based on equations for the turbulence kinetic energy (usually given the symbol k) and the turbulence dissipation rate (usually given the symbol ε). This method has grown out of work by Launder and Spaulding [34]. Another variant, based on a formulation by Kolmogorov, calculates a turbulence frequency ω instead of the dissipation rate (and hence is called a k-ω model). More complete models that compute all elements of the Reynolds stress tensor have also been developed. The common base for most of these models is the work of Launder, Reese and Rodi [33], with more recent contributions by Lumley [36], Speziale [49], and Reynolds [40]. These models, as well as the k-ε and k-ω models, and their implementation within the context of CFD are described in the book by Wilcox [57].

More limited models, based on a single equation for a turbulence scale, also have been developed. These include the models of Baldwin and Barth [1] and the widely used model of Spalart and Allmaras [48]. These models are applicable principally to boundary layer flows of aerodynamic interest.

27.3.3.2 Large-Eddy Simulations

The difficulty of developing generally applicable phenomenological turbulence models on the one hand and the enormous computational resources required to resolve all scales in turbulent flows at large Reynolds number on the other have led to the development of LES techniques. In this approach, the largest length and time scales of the turbulent motions are resolved in the simulation, but the smaller (sub-grid) scales are modeled. This is an attractive approach because it is the largest scales that contain the preponderance of turbulent kinetic energy and that are responsible for most of the mixing. At the same time, the smaller scales are believed to be more nearly isotropic and independent of the larger scales and thus are less likely to behave in problem-specific ways—that is, it should be easier to develop universal models for these smaller scales.

A filtering technique is applied to the Navier–Stokes equations that results in equations having a form similar to the original equations, but with additional terms representing a sub-grid-scale tensor that describes the effect of the modeled terms on the larger scales. The solution of these equations is not well posed if there is no initial knowledge of the sub-grid scales; the correct statistics are predicted for the flow, but it is impossible to reproduce a specific realization of the flow without this detailed initial condition. Fortunately, for most engineering problems, it is only the statistics that are of interest.

LES techniques date back to the pioneering work of Smagorinsky [46], who developed an eddy-viscosity sub-grid model for use in meteorological problems. The model turned out to be too dissipative for large-scale meteorological problems in which large-scale, predominantly 2-D, motions are affected by 3-D turbulence. Smagorinsky's model and subsequent developments of it find wide application in engineering problems, however. Details of the LES approach are discussed by Rogallo and Moin [43], and more recent developments and applications are described by Lesieur and Métais [35], Meneveau and Katz [37], Sagaut [44], and Garnier et al. [13].

There is a school of thought that believes that the details of the sub-grid model are not important, so long as the kinetic energy of the turbulence is dissipated at the correct rate at the sub-grid scales. Such *implicit*, or ILES, methods rely simply upon the (artificial) dissipation of the numerical scheme to provide this dissipation. A recent summary of these techniques is provided by Grinstein et al. [19].

LES continues to pose problems for wall-bounded flows, for which attached boundary layers require near-DNS resolution in the absence of adequate wall models. An effective approach for many such cases is detached-eddy simulation (DES), which combines the features of RANS turbulence models within the attached boundary layers with an LES approach within regions of massively separated flow. A recent review of this approach is given by Spalart [47].

27.3.3.3 Direct Numerical Simulations

DNS codes generally use spectral or pseudo-spectral approximations for the spatial discretization of the Navier–Stokes equations (see, e.g., Gottleib and Orszag [17] or Hussaini and Zang [25]). The difference between spectral and pseudo-spectral methods is in the way that products are computed; the advantage of spectral methods, which are more expensive computationally, is that aliasing errors are removed exactly (Orszag [39]).

A description of the issues involved and results are given by Rogallo and Moin [43]. DNS results are particularly valuable for the insight that they provide into the fundamental nature of turbulent flows.

The computational resources required for DNS calculations grow so rapidly with increasing Reynolds number that they are unlikely to be directly useful for engineering predictions, but they will remain an invaluable tool providing insight needed to construct better turbulence models for use in LES and with the Reynolds-averaged equations.

27.3.3.4 Verification, Validation, and Uncertainty Quantification

With the ever-widening use of CFD for increasingly complex physical problems, increasing efforts are focused on determining the credibility of predictions based on CFD analyses; these efforts are similar to and, in fact, build upon efforts in related fields, especially operations research.

Because of the inherent difficulty of developing numerical methods to solve highly nonlinear systems of partial differential equations, standards have not yet been developed to certify the accuracy of flow simulations, although some progress has been made in developing a common vocabulary and guidelines for the process; see, for example, the guidelines document developed by the American Institute of Aeronautics and Astronautics (AIAA) [2].

It is generally agreed that *verification* of a code or solution generated by a code refers to the process of determining that the instantiation of a conceptual model as a computer program accurately represents the conceptual model—that is, that the algorithm used is a consistent representation of the conceptual model and that the computer program contains no coding errors. Thus, verification can be carried out without reference to the physical world or to data from experiments; it is an issue purely of the accuracy of the numerical method and its implementation within a particular computer program. Computer codes can be verified for a given range of parameters (e.g., Reynolds number and Mach number):

- By comparing computed solutions with exact solutions of the governing equations. This is possible only in those very limited cases for which exact solutions of the governing equations are known.
- By ascertaining that the convergence rates of solutions over the specified range of parameter values are consistent with those predicted by theory for the numerical methods used. Note that this technique verifies only that the solutions to *some* system of equations converge at the correct rate, but will not detect coding errors that result in an error in the system of equations actually encoded.
- Using the *method of manufactured solutions*, in which source terms are determined that allow a specified function to be a solution to the governing equations. For complex systems of equations, these source terms often are computed using numerical algebra packages to "write" the code representing the source terms, as they can be too complex to derive and code by hand.

The term *validation* refers to the adequacy of the conceptual model to represent the physical phenomena of interest. As such, it requires comparison of the results of a simulation with experimental measurements. The true complexity of validation can be understood when one realizes that the solution(s) one would usually like to verify are those for which no comparable experimental measurements exist; the role of the simulation is, in fact, to replace an (usually expensive and/or time-consuming) experiment with the results of the simulation. Thus, what one would really like to do is to *quantify the uncertainty* in the result of a simulation, where the uncertainty arises from both numerical inaccuracies in the instantiation of the model (e.g., errors associated with truncation or lack of iterative convergence) and inadequacies in the conceptual model. Some information about the uncertainty of simulation results can be provided by sensitivity studies, in which parameters upon which the simulation depends are varied in a systematic way. There are also attempts to develop models *for the inadequacy of the models themselves* within the context of Bayesian inference.

Recent books focusing on these issues in the broader context of scientific simulation include those by Roache [41] and Oberkampf and Roy [38].

27.4 Research Issues and Summary

The field of CFD continues to be a field of intense research activity. The development of accurate algorithms based on unstructured grids for problems involving complex geometries and the increasing application of CFD techniques to unsteady problems, including aeroelasticity and acoustics, are examples of areas of great current interest, as are extensions to multiphase and reacting flows. Algorithmically, the incorporation of adaptive grids that automatically increase resolution in regions where it is required to maintain accuracy and the development of inherently multidimensional high-resolution schemes continue to be areas of fertile development. Finally, the continued expansion of computational capability allows the application of DNS and LES methods to problems of higher Reynolds number and increasingly realistic flow geometries.

Key Terms

Artificial viscosity: Terms added to a numerical approximation that provide artificial—that is, non-physical—dissipative mechanisms to stabilize a solution or to allow shock waves to be captured automatically by the numerical scheme.

CFL condition: The CFL condition is a stability criterion that limits the time step of an explicit time-marching scheme for hyperbolic systems of differential equations. In the simplest 1-D case, if Δx is the physical mesh spacing and $\rho(\mathbf{A})$ is the spectral radius of the Jacobian matrix $\mathbf{A}$, corresponding to the fastest wave speed for the problem, then the time step for explicit schemes must satisfy the constraint $\Delta t \leq K \, \Delta x/\rho(\mathbf{A})$, where K is a constant. For the simplest explicit schemes, $K = 1$, which implies that the time step must be less than that required for the fastest wave to cross the cell.

Conservative: A numerical approximation is said to be conservative if it is based on the conservation (or divergence) form of the differential equations, and the net flux across a cell interface is the same when computed from either direction; in this way the properties of discontinuous solutions are guaranteed to be consistent with the jump relations for the original integral form of the conservation laws.

Direct numerical simulation (DNS): A solution of the complete Navier–Stokes equations in which all length and time scales, down to those describing the smallest eddies containing significant turbulent kinetic energy, are fully resolved.

Euler equations: The equations describing the inviscid flow of a compressible fluid. These equations comprise a hyperbolic system of differential equations; the Euler equations are non-dissipative, and weak solutions containing discontinuities (which can be viewed as approximations to shock waves) must be allowed for many practical problems.

Explicit method: A method in which the solution at each point for the new time level is given explicitly in terms of values of the solution at the previous time level, to be contrasted with an *implicit method* in which the solution at each point at the new time level also depends on the solution at one or more neighboring points at the new time level, so that an algebraic system of equations must be solved at each time step.

Finite-difference method: A numerical method in which the solution is computed at a finite number of points in the domain; the solution is determined from equations that relate the solution at each point to its values at selected neighboring points.

Finite-element method: A numerical method for solving partial differential equations in which the solution is approximated by simple functions within each of a number of small elements into which the domain has been divided.

Finite-volume method: A numerical method for solving partial differential equations, especially those arising from systems of conservation laws, in which the rate of change of quantities within each mesh volume is related to fluxes across the boundaries of the volume.

Inviscid: A model for fluid flow in which the stresses due to internal friction, or viscosity, are neglected; it is also called an ideal fluid. The equations describing the motion of an inviscid, compressible fluid are called the Euler equations.

Large-eddy simulation (LES): A numerical solution of the Navier–Stokes equations in which the largest, energy-carrying eddies are completely resolved, but the effects of the smaller eddies, which are more nearly isotropic, are accounted for by a sub-grid model.

Mach number: The ratio $\mathbf{M} = V/a$ of the fluid velocity V to the speed of sound a. This nondimensional parameter characterizes the importance of compressibility to the dynamics of the fluid motion.

Mesh generation: The field directed at the generation of mesh systems suitable for the accurate representation of solutions to partial differential equations.

Method of manufactured solutions: A technique used in the verification of simulation programs in which source terms are determined that, when added to the right-hand side of the system of equations being solved, allow a specified function to be an exact solution.

Panel method: A numerical method to solve Laplace's equation for the velocity potential of a fluid flow. The boundary of the flow domain is discretized into a set of non-overlapping facets, or panels, on each of which the strength of some elementary solution is assumed constant. Equations for the normal velocity at control points on each panel can be solved for the unknown singularity strengths to give the solution. In some disciplines, this approach is called the boundary integral element method (BIEM).

Reynolds-averaged Navier–Stokes (RANS) equations: Equations for the mean quantities in a turbulent flow obtained by decomposing the fields into mean and fluctuating components and averaging the Navier–Stokes equations. Solution of these equations for the mean properties of the flow requires knowledge of various correlations, the Reynolds stresses, of the fluctuating components.

Shock capturing: Numerical method in which shock waves are treated by smearing them out with artificial dissipative terms in a manner such that no special treatment is required in the vicinity of the shocks; to be contrasted with *shock-fitting* methods in which shock waves are treated as discontinuities with the jump conditions explicitly enforced across them.

Shock wave: Region in a compressible flow across which the flow properties change almost discontinuously; unless the density of the fluid is extremely small, the shock region is so thin relative to other significant dimensions in most practical problems that it is a good approximation to represent the shock as a surface of discontinuity.

Turbulent flow: Flow in which unsteady fluctuations play a major role in determining the effective mean stresses in the field; regions in which turbulent fluctuations are important inevitably appear in fluid flow at large Reynolds numbers.

Upwind method: A numerical method for CFD in which upwinding of the difference stencil is used to introduce dissipation into the approximation, thus stabilizing the scheme. This is a popular mechanism for the Euler equations, which have no natural dissipation, but is also used for Navier–Stokes algorithms, especially those designed to be used at high Reynolds number.

Validation: The process of determining the degree to which a model is an accurate representation of the real world from the perspective of the intended uses of the model.

Verification: The process of determining that an implementation of a model accurately represents the developer's conceptual description of the model and/or that a particular solution is a sufficiently accurate solution to the model equations.

Visualization: The use of computer graphics to display features of solutions to CFD problems.

Further Information

Several organizations sponsor regular conferences devoted completely, or in large part, to CFD. The AIAA sponsors the AIAA Computational Fluid Dynamics Conferences in odd-numbered years, usually in July; the proceedings of this conference are published by AIAA. In addition, there typically are many sessions on CFD and its applications at the AIAA Aerospace Sciences Meeting, held every January, and the AIAA Fluid and Plasma Dynamics Conference, which is held every summer, in conjunction with the AIAA CFD Conference in those years when the latter is held. The Fluids Engineering Conference of the American Society of Mechanical Engineers, held every summer, also contains sessions devoted to CFD. In even-numbered years, the International Conference on Computational Fluid Dynamics is held. This biennial event began in 2000 with the merger of the International Conference on Numerical Methods in Fluid Dynamics and the International Symposium on Computational Fluid Dynamics. The 2012 conference was held in Hawaii.

The *Journal of Computational Physics* contains many articles on CFD, especially covering algorithmic issues. The *AIAA Journal* also has many articles on CFD, including aerospace applications. The *International Journal for Numerical Methods in Fluids* contains articles emphasizing the finite-element method applied to problems in fluid mechanics. The journals *Computers and Fluids* and *Theoretical and Computational Fluid Dynamics* are devoted exclusively to CFD, the latter journal emphasizing the use of CFD to elucidate basic fluid mechanical phenomena. The first issue of the *CFD Review*, which attempts to review important developments in CFD, was published in 1995. The *Annual Review of Fluid Mechanics* also contains a number of review articles on topics in CFD.

The European Research Community on Flow, Turbulence, and Combustion (ERCOFTAC) has held nearly biennial meetings since 1994 on DNS and LES; proceedings of the most recent workshop are contained in the volume edited by Kuerten et al. [32].

The following textbooks provide excellent coverage of many aspects of CFD:

- D. A. Anderson, J. C. Tannehill and R. H. Pletcher, *Computational Fluid Mechanics and Heat Transfer*, Hemisphere, Washington, DC, 1984.
- C. Hirsch, *Numerical Computation of Internal and External Flows, Vol. I: Fundamentals of Numerical Discretization*, 1988; *Vol. II: Computational Methods for Inviscid and Viscous Flows*, 1990; John Wiley & Sons, Chichester, U.K.
- D. C. Wilcox, *Turbulence Modeling for CFD*, DCW Industries, La Cañada, CA, 1993.

Up-to-date summaries on algorithms and applications for high Reynolds number aerodynamics are found in the following references:

- *Frontiers of Computational Fluid Dynamics – 1994*, Caughey, D. A. and Hafez, M. M., Eds., John Wiley & Sons, Chichester, U.K.
- *Frontiers of Computational Fluid Dynamics – 1998*, Caughey, D. A. and Hafez, M. M., Eds., World Scientific Publishing Company, Singapore.
- *Frontiers of Computational Fluid Dynamics – 2002*, Caughey, D. A. and Hafez, M. M., Eds., World Scientific Publishing Company, Singapore.
- *Frontiers of Computational Fluid Dynamics – 2006*, Caughey, D. A. and Hafez, M. M., Eds., World Scientific Publishing Company, Singapore.

References

1. Abid, R., Vatsa, V. N., Johnson, D. A., and Wedan, B. W., 1990. Prediction of separated transonic wing flows with non-equilibrium algebraic turbulence models, *AIAA J.*, 28, 1426–1431.
2. AIAA, 1998. *Guide for the Verification and Validation of Computational Fluid Dynamics Simulations*, American Institute of Aeronautics and Astronautics, Reston, VA.
3. Baldwin, B. S. and Lomax, H., 1978. Thin layer approximation and algebraic model for separated turbulent flows, *AIAA Paper 78-257*, 16th Aerospace Sciences Meeting, Huntsville, AL.
4. Baldwin, B. S. and Barth, T. J., 1991. A one-equation turbulence transport model for high reynolds number wall-bounded flows, *AIAA Paper 91-0610*, 29th Aerospace Sciences Meeting, Reno, NV.

5. Beam, R. M., and Warming, R. F., 1976. An implicit finite-difference algorithm for hyperbolic systems in conservation law form, *J. Comp. Phys.*, 22, 87–110.
6. Briley, W. R. and McDonald, H., 1974. Solution of the three-dimensional compressible Navier-Stokes equations by an implicit technique, *Lecture Notes in Physics*, Vol. 35, Springer-Verlag, New York, pp. 105–110.
7. Caughey, D. A., 1982. The computation of transonic potential flows, *Ann. Rev. Fluid Mech.*, 14, 261–283.
8. Caughey, D. A., 1987. Diagonal implicit multigrid solution of the Euler equations, *AIAA J.*, 26, 841–851.
9. Chapman, D. R., 1979. Computational aerodynamics: review and outlook, *AIAA J.*, 17, 1293–1313.
10. Eiseman, P. R., 1985. Grid generation for fluid mechanics computations, *Ann. Rev. Fluid Mech.*, 17, 487–522.
11. Emmons, H. W., 1970. Critique of numerical modeling of fluid-mechanics phenomena, *Ann. Rev. Fluid Mech.*, 2, 15–36.
12. Garabedian, P. R., 1956. Estimation of the relaxation factor for small mesh size, *Math. Tables Aids Comput.*, 10, 183–185.
13. Garnier, E., Adams, N. and Sagaut, P., 2009. *Large Eddy Simulation of Compressible Flows*, Springer, Berlin, Germany.
14. Glowinski, R. and Pironneau, O., 1992. Finite element methods for Navier-Stokes equations, *Ann. Rev. Fluid Mech.*, 24, 167–204.
15. Godunov, S. K., 1959. A finite-difference method for the numerical computation of discontinuous solutions of the equations of fluid dynamics, *Mat. Sb.*, 47, 357–393.
16. Goldstine, H. H. and von Neumann, J., 1963. On the principles of large scale computing machines, in *John von Neumann, Collected Works*, Vol. 5, p. 4, A. H. Taub, Ed., Pergamon Press, New York.
17. Gottlieb, D. and Orszag, S. A., 1977. Numerical analysis of spectral methods: Theory and application, *CBMS-NSF Reg. Conf. Ser. Appl. Math.*, Vol. 26, SIAM, Philadelphia, PA.
18. Graves, R. A. and Johnson, N. E., 1978. Navier-Stokes solutions using Stetter's method, *AIAA J.*, 16, 1013–1015.
19. Grinstein, F., Margolin, L., and Rider, W., 2007. *Implicit Large Eddy Simulation*, Cambridge University Press, New York.
20. Hafez, M. M. 1985. Numerical algorithms for transonic, inviscid flow calculations, in *Advances in Computational Transonics*, W. G. Habashi, Ed., Pineridge Press, Swansea, UK, pp. 23–58.
21. Harten, A. 1983. High resolution schemes for hyperbolic conservation laws, *J. Comp. Phys.*, 49, 357–393.
22. Harten, A. 1984. On a class of total-variation stable finite-difference schemes, *SIAM J. Numer. Anal.*, 21.
23. Harten, A., Engquist, B., Osher, S., and Chakravarthy, S., 1987. Uniformly high order accurate, essentially non-oscillatory schemes III, *J. Comp. Phys.*, 71, 231–323.
24. Hess, J. L., 1990. Panel methods in computational fluid dynamics, *Ann. Rev. Fluid Mech.*, 22, 255–274.
25. Hussaini, M. Y. and Zang, T. A., 1987. Spectral methods in fluid dynamics, *Ann. Rev. Fluid Mech.*, 19, 339–367.
26. Jameson, A., 1979. A multi-grid scheme for transonic potential calculations on arbitrary grids, *Proc. AIAA 4th Computational Fluid Dynamics Conf.*, pp. 122–146, Williamsburg, VA.
27. Jameson, A., 1983. Solution of the Euler equations by a multigrid method, *MAE Report 1613*, Department of Mechanical and Aerospace Engineering, Princeton University.
28. Jameson, A., 1992. Computational algorithms for aerodynamic analysis and design, *Tech. Rept. INRIA 25th Anniversary Conference on Computer Science and Control*, Paris.
29. Jameson, A. and Baker, T. J., 1983. Solution of the Euler equations for complex configurations, in *Proc. AIAA Computational Fluid Dynamics Conference*, Danvers, MA, pp. 293–302.

30. Jameson, A. and Caughey, D. A. 2001. How many steps are required to solve the Euler equations of compressible flow: In search of a fast solution algorithm. *AIAA Paper 2001-2673 15th Computational Fluid Dynamics Conference*, June 11–14, 2001, Anaheim, CA.
31. Jameson, A., Schmidt, W. and Turkel, E., 1981. Numerical solution of the Euler equations by finite volume methods using Runge-Kutta time stepping schemes, *AIAA Paper 81-1259, AIAA Fluid and Plasma Dynamics Conference*, Palo Alto, CA.
32. Kuerten, H., Geurts, B., Armenio, V., and Frölich, J., Eds., 2011. *Direct and Large-Eddy Simulation VIII*, Springer, Berlin Germany.
33. Launder, B. E., Reese, G. J. and Rodi, W., 1975. Progress in the development of a Reynolds-stress turbulence closure, *J. Fluid Mech.*, 68, (3), pp. 537–566.
34. Launder, B. E. and Spaulding, D. B, 1972. *Mathematical Models of Turbulence*, Academic Press, London, U.K.
35. Lesieur, M. and Métais, 1996. New trends in large-eddy simulations of turbulence, *Ann. Rev. Fluid Mech.*, 28, 45–82.
36. Lumley, J. L., 1978. Computational modeling of turbulent flows, *Adv. Appl. Mech.*, 18, 123–176.
37. Meneveau, C. and Katz, J., 2000. Scale-invariance and turbulence models, *Ann. Rev. Fluid Mech.*, 32, 1–32.
38. Oberkampf, W. L. and Roy, C. J., 2010. *Verification and Validation in Scientific Computing*, Cambridge University Press, Cambridge, U.K.
39. Orszag, S. A., 1972. Comparison of pseudo-spectral and spectral approximation, *Stud. Appl. Math.*, 51, 253–259.
40. Reynolds, W. C., 1987. Fundamentals of turbulence for turbulence modeling and simulation, in lecture notes for von Karman Institute, *AGARD Lecture Series No. 86*, pp. 1–66, NATO, New York.
41. Roache, P. J., 1998. *Verification and Validation in Computational Sciences and Engineering*, Hermosa Publishers, Albuquerque, NM.
42. Roe, P. L., 1986. Characteristic-based schemes for the Euler equations, *Ann. Rev. Fluid Mech.*, 18, 337–365.
43. Rogallo, R. S. and Moin, P., 1984. Numerical simulation of turbulent flows, *Ann. Rev. Fluid. Mech.*, 16, 99–137.
44. Sagaut, P., 2006. *Large Eddy Simulation of Incompressible Flows,* Springer, Berlin, Germany.
45. Shu, C. and Osher, S., 1988. Efficient implementation of essentially non-oscillatory shock-capturing schemes, *J. Comp. Phys.*, 77, 439–471.
46. Smagorinsky, J., 1963. General circulation experiments with the primitive equations, *Mon. Weather Rev*, 91, 99–164.
47. Spalart, P. R., 2009. Detached-eddy simulation, *Ann. Rev. Fluid Mech.*, 41, 181–202.
48. Spalart, P. R. and Allmaras, S. R., 1992. A one-equation turbulence model for aerodynamic flows, *AIAA Paper 92-0439*, 30th Aerospace Sciences Meeting, Reno, NV.
49. Speziale, C. G., 1987. On nonlinear k-ℓ and k-ε models of turbulence, *J. Fluid Mech.*, 178, 459–475.
50. Steger J. L. and Warming, R. F., 1981. Flux vector splitting of the inviscid gasdynamic equations with application to finite-difference methods, *J. Comp. Phys.*, 40, 263–293.
51. Tatsumi, S., Martinelli, L. and Jameson, A., 1995. Flux-limited schemes for the compressible Navier-Stokes equations, *AIAA J.*, 33, 252–261.
52. Thompson, J. F., Warsi, Z. U. A., and Mastin, C. W., 1985. *Numerical Grid Generation*, North Holland, New York.
53. van Leer, B., 1974. Towards the ultimate conservative difference scheme, II. Monotonicity and conservation combined in a second-order accurate scheme, *J. Comp. Phys.*, 14, 361–376.

54. van Leer, B., 1979. Towards the ultimate conservative difference scheme, V. A second-order sequel to Godunov's scheme, *J. Comp. Phys.*, 32, 101–136.
55. van Leer, B., 1982. Flux-vector splitting for the Euler equations, *Lecture Notes in Phys.*, 170, 507–512.
56. Vincent, P. E. and Jameson, A., 2011. Facilitating the adoption of unstructured high-order methods amongst a wider community of fluid dynamicists, *Math. Model. Nat. Phenom.*, 6(3), 97–140.
57. Wilcox, D. C., 1993. *Turbulence Modeling for CFD*, DCW Industries, La Cañada, CA.
58. Yoon, S. and Kwak, D., 1994. Multigrid convergence of an LU scheme, in *Frontiers of Computational Fluid Dynamics–1994*, Caughey, D.A. and Hafez, M.M., Eds., Wiley Interscience, Chichester, U.K. pp. 319–338.

28 Computational Astrophysics

Jon Hakkila
College of Charleston

Derek Buzasi
Florida Gulf Coast University

Robert J. Thacker
St. Mary's University

28.1 Introduction

Modern astronomy and *astrophysics* is a computationally driven discipline. Even during the 1980s, it was said that an astronomer would choose a computer over a telescope if given the choice of only one tool. However, just as it is impossible to separate "astronomy" from astrophysics, almost all astrophysicists would no longer be able to separate the computational components of astrophysics from the processes of data collection, data analysis, and theory. The links between astronomy and computational astrophysics are so close that a discussion of computational astrophysics is essentially a summary of the role of computation in all of astronomy. We have chosen to concentrate on a few specific areas of interest in computational astrophysics rather than attempt the monumental task of summarizing the entire discipline. We further limit the context of this chapter by discussing astronomy rather than related disciplines such as planetary science and the engineering-oriented aspects of space science.

28.2 Astronomical Databases

Physics and astronomy have been leaders among the sciences in the widespread use of and access to online access and data retrieval. This has most likely occurred because the relatively small number of astronomers is broadly distributed so that few astronomers typically reside in any one location; it also perhaps results because astronomy is less protective of data rights than other disciplines, which tend to be more driven by commercial spin-offs. Astronomers regularly enjoy online access to journal articles, astronomical catalogs/databases, and software.

28.2.1 Electronic Information Dissemination

Astronomical journals have been available electronically since 1995 [101]; now all major journals are available electronically. In recent years, several new open access journals have also appeared as alternatives to the highly visible, well-known, yet sometimes costly established journals; astrophysics has not fully moved in the direction of open access but has the possibility of doing so [86]. Meanwhile, astrophysicists tend to make heavy use of electronic preprint servers (e.g., [54]), and these have increased accessibility and citation rates (e.g., [41,65,91]).

The Astrophysical Data Service (ADS) [80] is the electronic index to all major astrophysical publications; it is accessible from a variety of mirror sites worldwide. The ADS is a data retrieval tool allowing users to index journal articles, other publications, and preprints by title, keyword, and author and even by astronomical object. The ADS is also a citation index that can be used to analyze the effectiveness of astronomical dissemination (e.g., [65]). The ADS is capable of accessing the complete text of articles and in many cases original data via electronic files. Although the integrated data management from all astrophysics missions was one of the ADS original purposes, the large amount of astronomical data available, coupled with preprocessing and data analysis tools often needed to analyze these data, has led to the parallel development of astronomical data depositories that will be described later.

Electronic dissemination plays a critical role in the collection of astronomical data. One of the newest frontiers in astronomy has been in the time domain: many variable sources (such as high-energy transients and peculiar *galaxies*) often exhibit changes on extremely short timescales. Follow-up observations require the community to rapidly disseminate source locations, brightnesses, and other pertinent information electronically (notification sometimes must go out to other observatories on timescales as short as seconds). There are a variety of electronic circulars available to the observational community: these include the GRB Coordinates Network (GCN; [12]) and circulars distributed by the International Astronomical Union (http://www.iau.org).

28.2.2 Data Collection

Astronomy is generally an observational rather than an experimental science (although there are laboratory experiments that emulate specific astronomical conditions). Modern astronomical observations are made using computer-controlled telescopes, balloon instruments, and/or satellites. Many of these are controlled remotely, and some are robotic.

All telescopes need electronic guidance because they are placed on moving platforms; even the Earth is a moving platform when studying the heavens. From the perspective of the terrestrial observer, the sky appears to pivot overhead about the Earth's equatorial axis; an electronic drive is needed to rotate the telescope with the sky in order to keep the telescope trained on an object. However, the great weight of large telescopes is easier to mount altazimuthally (perpendicular to the horizon) than equatorially. A processor is needed to make real-time transformations from equatorial to altazimuth coordinates so that the telescope can track objects; it is also needed to accurately and quickly point the telescope. Precession of the equinoxes must also be taken into account; the direction of the Earth's rotational axis slowly changes position relative to the sky. Computers have thus been integrated into modern telescope control systems.

Additional problems are present for telescopes flown on balloons, aircraft, and satellites. The telescopes need to either be pointed remotely or at least have the directions they were pointed while observing accessible after the fact. Flight software (which has now evolved because of increased modern processor speeds to the point where it can be compiled and uploaded) was originally written in machine code to run in real time. Stability in telescope pointing is required because astronomical sources are distant; long observing times are needed to integrate the small amount of light received from sources other than the sun, the moon, and the planets.

As an example, we mention computer guidance on the Hubble Space Telescope (HST). HST has different sensor types that provide feedback in order to maintain a high degree of telescope pointing accuracy; the Fine Guidance Sensors have been key since their installation in 1997. The three Fine Guidance Sensors provide an error signal so that the telescope has the pointing stability needed to produce high-resolution images. The first sensor monitors telescope pitch and yaw, the second monitors roll, and the third serves both as a scientific instrument and as a backup. Each sensor contains lenses, mirrors, prisms, servos, beam splitters, and photomultiplier tubes. The software coordinates the pointing of the sensors onto entries in an extremely large star catalog. The guidance sensors lock onto a star and then deviations in its motion to a 0.0028 arc second accuracy. This provides HST with the ability to point at the target to within 0.007 arc seconds of deviation over extended periods of time. This level of stability and precision is comparable to being able to hold a laser beam constant on a penny 350 miles away.

Robotic telescopes are becoming more common in astronomy [21]. Observations from robotic telescopes can be scheduled in advance (such as observations of variable stars and active galaxies), thus reducing manpower costs because users do not have to travel to an observatory or even be awake when observations are being made. Robotic telescopes also allow real-time follow-up observations to be made, including observations of supernovae, gamma-ray bursts (GRBs), minor planets, and extrasolar planet transits. They are also being used as educational tools to bring astronomy into the classroom [39]. Scheduling and control software are key to the development of a successful robotic telescope design (e.g., [78]).

Many telescopes are designed to observe in *electromagnetic spectral* regimes other than the visible. An accurate computer guidance system is needed to ensure that the telescope is correctly pointing even when a source cannot be identified optically and that the detectors are integrated with the other electronic telescope systems. Many bright sources at nonvisible wavelengths are extremely faint in the visible spectral regime. Furthermore, telescopes must be programmed to avoid pointing at bright objects that can burn out photosensitive equipment and sometimes must avoid collecting data when conditions arise that are dangerous to operation. For example, orbital satellites need to avoid operating when over the South Atlantic Ocean—in a region known as the South Atlantic Anomaly, where the Earth's magnetic field is distorted—because high-energy ions and electrons can interact with the satellite's electronic equipment. This can cause faulty instrument readings and/or introduce additional noise.

There are other examples where computation is necessary to the data collection process. In particular, we mention the developing field of *adaptive optics*. This process requires the fast, real-time inversion of large matrices [14]. *Speckle imaging techniques* (e.g., [119]) also remove atmospheric distortion by simultaneously taking short-exposure images from multiple cameras.

Detectors and electronics often need to be integrated by a single computer system. Many interesting specific data collection problems exist that require modern computer-controlled instrumentation. For example, radio telescopes with large dishes cannot move, and sources pass through the telescope's field of view as the Earth rotates. Image reconstruction is necessary from the data stream, since all sources in the dish's field of view are visible at any given time. Another interesting data collection problem occurs in infrared astronomy. The sky is itself an infrared emitter, and strong source signal to noise can only be obtained by constantly subtracting sky flux from that of the observation. For this reason, infrared telescopes are equipped with an oscillating secondary mirror that wobbles back and forth, and flux measurements alternate between object and sky. A third example concerns the difficulty in observing x-ray and gamma-ray sources. X-ray and gamma-ray sources emit few photons, and these cannot be easily focused due to their short wavelengths. Computational techniques such as Monte Carlo analysis are needed to deconvolve photon energy, flux, and source direction from the instrumental response. Often this analysis needs to be done in real time, requiring a fast processor. The telescope guidance system must also be coordinated with onboard visual observations, since the visual sky still provides the basis for telescope pointing and satellite navigation.

28.2.3 Accessing Astronomical Databases

Database management techniques have allowed astronomers to address the increasing problem of storing and accurately cross-referencing astronomical observations. Historically, bright stars were given catalog labels based on their intensities and on the constellation in which they were found. Subsequent catalogs were sensitive to an increased number of fainter objects and thus became larger while simultaneously assigning additional catalog numbers or identifiers to the same objects. The advent of photographic and photoelectric measurement techniques and the development of larger telescopes dramatically increased catalog sizes. Additional labels were given to stars in specialty catalogs (e.g., those containing bright stars, variable stars, and binary stars). Solar system objects such as asteroids and comets are not stationary and have been placed in catalogs of their own.

28.2.4 Catalogs and Online Databases

In 1781, Charles Messier developed a catalog of fuzzy objects that were often confused with comets. Subsequent observations led to identification of these extended objects as star clusters, galaxies, and gaseous nebulae. Many of these extended astronomical sources (particularly regions of the *interstellar medium*) do not have easily identified boundaries (and the observed boundaries are often functions of the *passband* used); this inherent fuzziness presents a problem in finding unique identifiers as well as a database management problem. Furthermore, sources are often extended in the radial direction (sources are 3-D), which presents additional problems since distances are among the most difficult astrophysical quantities to measure.

As astronomy entered the realm of multi wavelength observations in the 1940s, observers realized the difficulty in directly associating objects observed in different spectral regimes. An x-ray emitter might be undetectable or appear associated with an otherwise unexciting stellar source when observed in the optical. Angular resolution is a function of wavelength, so it is not always easy to directly associate objects observed in different passbands.

Temporal variability further complicates source identification. Some objects detected during one epoch are absent in observations made during other epochs. Signal-to-noise ratios (SNRs) of detectors used in each epoch contribute to the source identification problem. Additionally, gamma-ray and x-ray sources tend to be more intrinsically variable due to the violent, nonthermal nature of their emission. Examples of sources requiring access via their temporal characteristics include supernovae, GRBs, high-energy transient sources, and some variable stars and extragalactic objects.

There are tremendous numbers of catalogs available to astronomers, and many of these are found online. Perhaps the largest single repository of online catalogs and metadata links is VizieR (http://vizier.u-strasbg.fr/viz-bin/VizieR) [99]. Online catalogs also exist at many other sites such as High-Energy Astrophysics Science Archive Research Center (HEASARC) (at http://heasarc.gsfc.nasa.gov), HST (at http://www.stsci.edu/resources/), and NASA/IPAC Extragalactic Database (NED) (at http://nedwww.ipac.caltech.edu).

Due in part to the strong public financing of astronomy in the United States and Europe, significant amounts of astronomical data are publicly available. As a result, large astronomical databases exist for specific ground-based telescopes and orbital satellites. Some of these databases are large enough to present information retrieval problems. Examples of these databases are Two Micron All Sky Survey (2MASS) [77], Digitized Palomar Observatory Sky Survey (DPOSS) [42], Sloan Digital Sky Survey (SDSS) [135], and the National Radio Astronomy Observatory (NRAO) VLA Sky

Survey (NVSS) [36]. Databases span the range of astronomical objects from stars to galaxies, from active galactic nuclei to the interstellar medium (ISM), and from GRBs to the cosmic microwave background (CMB). Databases are often specific to observations made in predefined spectral regimes rather than specific to particular types of sources; this reflects the characteristics of the instrument making the observations.

28.2.5 Virtual Observatories

Perhaps one of the greatest additions to astronomical databases in the recent years has been the introduction and development of *virtual observatories* (*VOs*). VOs are designed to improve accessibility, retrieval, and data analysis from international astronomical catalogs and archives. Currently, twenty international VOs are members of the International Alliance (http://www.ivoa.net): this includes VOs from Europe, North America, South America, Asia, and Australia. Although these cover a wide range of astronomical data, other VOs are more specific (such as the Virtual Solar Observatory [40], the Virtual Wave Observatory, [51]) and the Information and Distributed Information Service, which aims to become the European Virtual Observatory devoted to planetary sciences [32]).

The US Virtual Astronomical Observatory (VAO) was proposed in 2001 and began operations around 2005 [62]. It was established under a cooperative agreement in 2010 with support from the NSF and NASA. The VAO at last count contained over 9000 data collections, tables, and/or databases through interfaces supported at 80 different organizations [44].

28.2.6 Data File Formats

The astronomical community has evolved a standard data format for the transfer of data. The Flexible Image Transport System (FITS) has broad general acceptance within the astronomical community and can be used for transferring images, spectroscopic information, time series, etc. [63]. It consists of an ASCII text header with information describing the data structure that follows. Although originally defined for 9-track half-inch magnetic tape, FITS format has evolved to be generic to different storage media. The general FITS structure has undergone incremental improvements, with acceptance determined by vote of the International Astronomical Union.

Despite the general acceptance of FITS file format, other methods are also used for astronomical data transfer. This is not surprising given the large range of data types and uses. Some data types have been difficult to characterize in FITS formats (such as solar magnetospheric data). Satellite and balloon data formats are often specific to each instrument and/or satellite.

Due to the large need for storing astronomical images, a wide range of image compression techniques have been applied to astronomy. These include fractal, wavelets, pyramidal median, and JPEG (e.g., [84]).

28.3 Data Analysis

Mathematical and statistical analyses, and their applications to large electronic astronomical databases being archived annually, have been among the driving forces behind the use of computation in astrophysics. This has led in the past 10 years to the formal recognition of new astronomy subdisciplines: astrostatistics and astroinformatics [23].

Data analysis and theoretical software can be accessed at a variety of sites worldwide. A few of the most well-known sites are the Astrophysics Source Code Library (http://ascl.net/), the UK Starlink site (http://star-www.rl.ac.uk), and the Astronomical Software and Documentation Service at STScI (http://asds.stsci.edu). Data analysis tools written in RSI's proprietary interactive data language (IDL) programming language are available at the IDL Astronomy User's Library (http://idlastro.gsfc.nasa.gov/homepage.html). IDL has become a computing language of choice by many astronomers because of its design as an image-processing language, because it has been written with mathematical and statistical uses in mind, because it can handle multidimensional arrays easily, and because it has many built-in data visualization tools.

28.3.1 Data Analysis Systems

In the fall of 1981, astronomers at Kitt Peak National Observatory began developing the Image Reduction and Analysis Facility (IRAF), to serve as a general-purpose, flexible, extendable data reduction package. IRAF replaced a myriad of different data analysis systems that had existed within astronomy, where separate data analysis packages were often supported at an institution (or sometimes individual groups within an institution), such that compatibility between these packages was limited or nonexistent. IRAF has grown beyond its original use primarily by ground-based optical astronomers to encompass space-based experiments as well at wavelengths ranging from x-ray to infrared.

IRAF is currently a mature system, with new releases occurring roughly annually, and is operated by about 5000 users at 1500 distinct sites. It is supported under a number of different computer architectures running Unix or Unix-like (e.g., Linux) operating systems. Oversight of IRAF development is formalized, with a Technical Working Group and various User's Committees overseeing evolution of the software. Numerous large astronomical projects have adopted IRAF as their data analysis suite of choice, typically by providing extensions to the basic system. These projects include the x-ray "Great Observatory" Chandra, the HST, PROS (ROSAT X-Ray Data Analysis System), and FTOOLS (a FITS utility package from HEASARC).

The IRAF core system provides data I/O tools, interactive graphics and image display tools, and a variety of image manipulation and statistical tools. Commonly available "packages" that are part of the standard installation include tools for basic CCD data reduction and photometry and support 1-D, 2-D, echelle, and fiber spectroscopy. Most tasks can be operated in either interactive or batch mode.

IRAF supports the FITS file format, as well as its own internal data type. In addition, extensions exist to handle other data types. Currently, those include STF format (used for HST data), QPOE format (used for event list data such as from the x-ray and EUV satellites), and PLIO for pixel lists (used to flag individual pixels in a region of interest). IRAF uses text files as database or configuration files and provides a number of conversion tools to produce images from text-based data. Binary tables can be manipulated directly using the TABLES package.

Other general-purpose data reduction packages coexist with IRAF. These include the Astronomical Image Processing System (AIPS) and its successor, AIPS++, developed at the US NRAO, and still the image analysis suite of choice for radio astronomy. Image processing has been a very important subdiscipline within computational astrophysics, and we mention a number of image reconstruction methods with special applications to astronomy: the maximum entropy method (e.g., [83]), the Pixon method [106], and massive inference [114]. XIMAGE and its relatives, XSPEC and XRONOS, serve a similar function for the x-ray astronomy community. A significant number of optical astronomers use Figaro, developed at the Anglo-Australian telescope and particularly popular throughout Australia and the United Kingdom. Recently, Figaro has been adapted to run within the IRAF system, allowing users to have the best of both worlds. While there is no formal software system operating under IDL, the profusion of programs available in that proprietary language to do astronomy makes it worthy of mention as well, while a small but growing subset of astronomers makes use of Matlab for similar purposes.

One of the most significant drivers for the development of mission-independent data analysis software within astronomy has been NASA's Applied Information Systems Research Program (AISRP). This program has encouraged the development of software to serve community-wide needs and has also fought the recurrent tendency for each project to develop its own software system, but instead to write software "packages" within the IRAF or IDL architecture. Of course, a general data analysis system is not the best solution for all specialized needs, particularly for space-based astronomy. In these cases, many missions have developed their own packages, and not all of these exist within an IRAF/AIPS/XIMAGE/IDL universe.

28.3.2 Data Mining

The topic of data mining in astronomy has become increasingly important in recent years, as databases have become large and more complex [10,75]. These large databases have arisen in both ground-based observatories and orbital ones, and they represent data collected from countries around the world. Most of these archives are available online, though the heterogenous nature of user interfaces and metadata formats—even when constrained by HTML and CGI—can make combining data from multiple sources an unnecessarily involved process. In addition, astronomical archives are growing at rates unheard of in the past: terabytes per year are now typical for most significant archives. Two recently established databases that illustrate the trend are the MACHO database (8 TB) and the SDSS (15 TB). Space-based missions are beginning to approach these data rates as well, with NASA's Kepler mission generating and downlinking approximately 2.5 TB annually [79]. Even these rates will seem small when the Large-Scale Synoptic Telescope (LSST) comes online, as LSST will generate a data rate of roughly 0.5 GB per second [8].

The simplest kinds of questions one might ask of these kinds of data sets involve resource discovery, for example, "find all the sources within a given circle on the sky," or "list all the observations of a given object at x-ray wavelengths." Facilities such as SIMBAD [129], VisieR [99], NED [87], and Astrobrowse [64] permit these kinds of queries, though they are still far from comprehensive in their selection of catalogs and/or data sets to be queried. One problem arises from the nature of the FITS format, which is poorly defined as far as metadata are concerned—XML may provide a solution here, but astronomers have yet to agree on a consistent common set of metadata tags and attributes.

Difficulties due to heterogeneous formats and user interfaces are being addressed by *VO* projects, such as the US VAO [109]. The most basic intent of all of these projects, which are coordinated with one another at some level, is to deliver a form of integrated access to the vast and disparate collection of existing astronomical data. In addition, all are in the process of providing data visualization and mining tools.

In its most encompassing form, astronomical data mining involves combining data from disparate data sets involving multiple sensors, multiple spectral regimes, and multiple spatial and spectral resolutions. Sources within the data are frequently time variable. In addition, the data are contaminated with ill-defined noise and systematic effects caused by varying data sampling rates and gaps and instrumental characteristics. Finally, astronomers typically wish to compare observations with the results of simulations, which may be performed with mesh scales dissimilar to that of the observations and which suffer from systematic effects of their own. It has been observed [103] that data mining in astronomy presently (and in the near future) focuses on the following functional areas:

1. Cross-correlation to find association rules
2. Finding outliers from distributions
3. Sequence analysis
4. Similarity searches
5. Clustering and classification

In spite of the difficulties previously outlined, the nature of astronomical data—it is generally freely available and in computer-accessible form—has led to numerous early applications of data mining techniques in the field. As far back as 1981, Faint Object Classification and Analysis System (FOCAS) [73] was developed for the detection and classification of images on astronomical plates for the automatic assembly of catalogs of faint objects. Neural nets and decision trees have also been applied for the purposes of discriminating between galaxies and stars in image data [100,134] and for morphological classification of galaxies [117].

There has been an increasing need for astronomers to use data mining techniques in recent years, as large astronomical data sets now contain data collected in a variety of detector formats, observed by a range of telescopes and space instruments, and spanning a large range of spectral characteristics. As a result, data mining has progressed from individual tools and techniques developed for specific purposes to toolkits containing a wide range of data mining tools in the tradition of WEKA [61]. Toolkits contain standard data mining tools spanning supervised methods (e.g., artificial neural networks, decision trees, support vector machines, and k nearest neighbor analyses) and unsupervised methods (e.g., kernel density estimation, k-means clustering, mixture models, expectation maximization, Kohonen self-organizing maps, and independent component analysis), as well as other algorithms (including genetic algorithms [GAs], the information bottleneck method, association rule mining, and graphical models) (e.g., [10,27,13,88]).

The greatest successes of data mining are likely to arise from its application to large but coherent data sets. In this case, the data have a common format and noise characteristics, and data mining applications can be planned from the beginning. Recent projects satisfying these conditions have included the SDSS [135] and the 2MASS (e.g., [98]). SDSS uses a dedicated 2.5 m telescope to gather images of approximately 25% of the sky (some 10^8 objects), together with spectra of approximately 10^5 objects of cosmological significance. Pipelines were developed to convert the raw images into astrometric, spectroscopic, and photometric data, which were stored in a common science database that is indexed in a hierarchical manner. The science database is accessible via a specialized query engine. The SDSS required Microsoft to put new features into their SQL server; in particular they added a tree structure into it to allow rapid processing of queries containing massive amounts of data for discrete as well as nebulous sources. Future astronomical observatories will continue this trend; the ground-based 8.4 m LSST, the Chinese Large Sky Area Multi-Object Fiber Spectroscopic Telescope (LAMOST) (with its 6.67 m × 6.05 m mirror), and the orbital James Webb Space Telescope are some examples of this trend.

The key problem in astronomical data mining will most likely continue to revolve around interpretation of discovered classes. Astronomy is ruled by instrumental and sampling biases; these systematic effects can cause data to cluster in ways that mimic true source populations statistically. Since data mining tools are capable of finding classes that are only weakly defined statistically, these tools enhance the user's ability to find "false" classes. Subtle systematic biases are present (but minimized) even in controlled cases when data are collected from only one instrument. The astronomical community must be careful to test the hypothesis that class structures are not produced by instrumental and/or sampling biases before accepting the validity of newly discovered classes.

28.3.3 Multiwavelength Studies

Multiwavelength studies have become of ever-increasing importance in astronomy, as new spectral regimes (x-ray, extreme UV, gamma ray) have opened to practitioners, and astronomers have become aware that most problems cannot be adequately addressed by studies within any one spectral band. Such studies are now recognized as essential to the understanding of objects ranging from the sun and stars through x-ray bursters and classical novae to the ISM and active galactic nuclei. The computational requirements peculiar to multiwavelength astrophysics essentially fall into one of two categories, where the distinction is in the time rather than spectral domain.

In the first case, we have situations where the timescales associated with the phenomena under study are long compared with typical observational timescales. One such case is studies of the ISM, which is important in astrophysics because galactic gas and dust clouds are the source of new generations of stars. The ISM becomes more enriched as generations of stars die and return heavy elements to it. In the study of the ISM, one can identify three characteristic temperature scales, <100 K, $\approx 10^4$ K, and $\approx 10^7$ K—these temperatures necessitate observations at radio/infrared, optical/UV, and x-ray wavelengths, respectively (e.g., [136]). In each case, the strength of its emission or absorption depends on the local density, composition, metallicity, temperature, and distribution of ambient photons. Since this causes the ISM (and to a lesser extent the intergalactic medium) to interfere with observations of other galactic and extragalactic sources, stellar, galactic, and extragalactic astronomers are interested in knowing the radiative properties of the ISM, which is by definition nebulous and 3-D, as well as knowing where this material is located. An example of code used to locate the ISM in the visual spectral regime can be found at http://ascl.net/extinct.html [59]. Since the characteristic evolutionary timescale of the ISM is long by human standards, simultaneous (and even contemporaneous) multiwavelength observations are unnecessary. In this milieu, catalogs and databases come into their own, and archival multiwavelength research is possible, focusing on spatial rather than temporal correlations. An idea of the numerous databases and collections of online data available on the ISM can be found at http://adc.gsfc.nasa.gov/adc/quick_ref/ref_ism.html.

A different situation pertains with variable sources such as gamma-ray bursters, extremely short-duration and high-energy events occurring at cosmological distances. The source of the bursts is as yet unknown, and they may be created by mergers of a pair of neutron stars or black holes or by a hypernova, a type of exceptionally violent exploding star. GRBs have been detected at gamma-ray, x-ray, and optical wavelengths, and further study of these objects clearly calls for multiwavelength studies (e.g., [57]). However, unlike the case obtaining for the ISM, GRBs have timescales ranging from milliseconds up to perhaps $\approx 10^3$ s, and thus, coordinated and simultaneous (or near-simultaneous) multiwavelength observations are essential. In this case, the computational demand is more on rapid deployment of various computer-controlled telescopes (both on the ground and in space) than on correlation analyses of existing databases. Thus, systems such as GCN and Virtual Observatory Events (VOEvents) have been developed [12] to distribute locations of GRBs to observers in real time or near real time and to distribute reports of the resulting follow-up observations.

28.3.4 Specific Examples

28.3.4.1 Cosmic Microwave Background Data Analysis

One of the main goals of CMB data analysis is to derive the power spectrum of the temperature fluctuations that correlates directly to fluctuations in the density of matter in the early universe. The precise spectrum of matter perturbations depends acutely on cosmological parameters, and hence, CMB data are an excellent diagnostic for determining the parameters of cosmological models. However, the task of constructing the power spectrum is daunting since the raw CMB signal has instrument noise, as well as noise from the ISM and other astrophysical objects, imposed upon it. To further complicate matters, all of these sources of noise are often correlated.

Analysis proceeds by first creating a physical map from the time series of instrument pointings. A pixel map is first constructed by dividing up the area of sky surveyed. For the COBE experiment [115], only a few thousand pixels were necessary, while for the PLANCK Surveyor satellite (http://planck.esa.int), tens of millions of pixels are required. The main step in creating the map is then the separation of noise, which can be done under the assumption that the time series of noise signals is drawn from a Gaussian distribution. Linear algebra methods are used to calculate the pixel–pixel noise correlation, which is then used to calculate the map. However, brute force methods on large maps are intractable, and exploiting efficient methods for sparse matrices is the only way to do the calculation. Even so, regardless of the method adopted, the calculation is still sufficiently large that massively parallel computing is necessary.

Having constructed the map, the next step is to calculate the power spectrum. This is a significantly more difficult process than map creation, since each pixel contains information about the signal on all scales within the map. During the early development of the field maximum likelihood, approaches were used that relied upon iterative convergence methods, such as Newton–Raphson schemes. However, for very large maps, the correlation matrices and numerical derivatives that must be calculated make this method too costly. Reducing the resolution of the data is possible, but that limits the resolution of the entire map. More recently Monte Carlo frequentist approaches to calculating the power spectrum have become popular. These methods essentially calculate the harmonics of the power spectrum convolved with all the noise, filters, and mixing (called the pseudo power spectrum) and then attempt to recover the original power spectrum through an inversion of a relationship between the true and pseudo power spectra. Monte Carlo sampling is used in this last step [68]. While such approaches make the calculations more tractable, they require an extremely detailed knowledge of the noise statistical properties that is a considerable challenge in itself. Nonetheless, it seems probably that the Monte Carlo approaches will remain the preferred method for very large data sets for the foreseeable future.

28.3.4.2 Gamma-Ray Burst Data Analysis

GRBs are short bursts of primarily gamma radiation having flux and spectra that evolve on short timescales. Evidence suggests that GRBs result from relativistic shock waves that collide with each other and with the ambient ISM. The source of the shocks is currently presumed to originate from two types of sources: *hypernovae* (supernovae in which a significant portion of the collapsing stellar core's energy is focused into shocked bipolar accelerated particle beams) or colliding binary neutron stars. GRBs exhibit a wide variety of complex behaviors and yet there is evidence of multiple GRB classes. Class identification is as important in astrophysics as it is in other sciences; the properties of distinct GRB classes can lead to a better general understanding of GRB physics as well as to a better understanding of the different sources and/or environments producing the classes. However, class identification is not helpful if the mechanisms responsible for producing the classes cannot be determined.

Data mining techniques have proven useful in the study of GRB classes. Data mining techniques are needed to identify clusters in the GRB attribute space because individual GRB behaviors overlap. Some overlap results from the large intrinsic range of GRB behaviors, some is due to distinctly different properties of GRB classes, and some is due to observational and instrumental bias. Bias can cause phantom classes to appear by creating clustering where no distinct source populations exist.

GRB data mining has been used primarily with the large GRB database collected by the Burst And Transient Source Experiment (BATSE on NASA's Compton Gamma-Ray Observatory) and Swift because the instrumental characteristics of these experiments are well-documented properties [52,102]. Data mining tools identify three distinct GRB classes rather than two known historically (e.g., [71,94]). Interpretation of the data mining data suggests that the third GRB class does not represent a separate source population; it is instead produced by observational biases resulting primarily from low signal-to-noise observations and from the instrumental trigger characteristics [58]. This successful result indicates that scientific data mining can be used not just to determine that classes exist but also to determine *why* they exist. As a further clue behind why these class differences exist, recent observations indicate that the *pulses* making up GRBs have similar, correlated properties regardless of the class to which a GRB belongs [60]. Data mining classification of the pulses, however, does not yield a clear delineation between pulses belonging to different classes [3]; this suggests that it is the distribution of pulses that indicates the class to which a GRB belongs, rather than the pulse properties.

28.3.4.3 Time Series Analysis

Astronomers make use of time series analysis techniques for a variety of purposes, including studies of variable stars and cataclysmic variables [104,133], pulsating or oscillating stars [7,72], asteroid rotation rates [128], active galactic nuclei [47], sunspot number [118], and extrasolar planets [24]. Typically, astronomical time series suffer from relatively low SNRs, uneven sampling, and numerous gaps, at times leading to severe aliasing at the 1 day^{-1} frequency and diffculties in estimating a traditional autocorrelation function. In addition, some applications (Active Galactic Nuclei [AGN] studies and planetary detection) are dominated by nonsinusoidal signals or pulses.

Historically, the primary tools used to support these efforts have been discrete Fourier transforms and periodograms [111]. Both of these estimators of the autocorrelation function can be defined in such a way as to satisfactorily represent unevenly sampled data, and small gaps in the time series can be handled using clever binning or interpolation techniques, but these necessarily distort the data and lead to the loss of information. In some cases, aliasing can be minimized or eliminated by experimental design (e.g., GONG), but more often astronomers are confronted with this fundamental problem and this has driven numerous recent efforts in the area. Recent developments in this area focus on the application of nonlinear, nonstationary techniques and Bayesian methods. The wavelet transformation shows great promise [130], as unlike the DFT it can be used to construct power spectrum estimators that are nearly independent of the signal shape and amplitude in the presence of noise. Such estimators are likely to find increasing use in the planetary detection and AGN modeling communities. Perhaps an even more useful application of wavelets is to the denoising of power spectra, a technique pioneered by the solar physics community. Bayesian methods are also increasing in popularity, as they are well-suited to finding change points in long series of time-tagged data such as is typically obtained from high-energy astrophysics experiments [49,112].

A rapidly growing difficulty is the size of power spectra produced by astronomical experiments. Helioseismological observations can easily give rise to time series with in excess of 10^7 points, and asteroseismological observations are rapidly approaching this level [113]. Achieving the maximum time resolution inherent in data such as time-tagged x-ray photon lists can require estimators with in excess of 10^9 points [108]. Furthermore, upcoming space-based experiments, such as European Space Agency's (ESA) PLATO mission [34], will give rise to data sets that are so large as to mandate the use of automated techniques.

28.4 Theoretical Modeling

Prior to the advent of computers, theoretical modeling consisted largely of solving idealized systems of equations for a given problem. Very often, to make problems tractable, simplifying assumptions like spherical symmetry and linearization of the problem would be necessary. Rapid numerical solutions of equations avoid the need for simplifying assumptions, but at a cost: one can no longer achieve an elegant formula for the solution to the problem, and insight often derived from solving the equations by hand is lost.

28.4.1 Role of Simulation

Although computation is commonplace in theoretical modeling, perhaps the most heavily computationally biased aspect is *simulation*. Simulation can be viewed as an extension of finding numerical solutions to a given equation set; however, the set of equations is often enormous in size (such as that produced by the gravitational interaction problem between a large number of bodies). Simulation also often involves visualizing the "data set" to help understand the phenomenon being studied, and the systems under investigation are almost always cast as an initial value problem.

The roots of simulation in astrophysics can be traced back to at least the 1940s. Driven by a desire to understand the clustering of galaxies, Holmberg built an analog computer consisting of light sources and photocells to simulate the mutual interaction of two galaxies via gravity. Today, almost all simulation is conducted on digital computers. Development of fast efficient algorithms for solving complex equation sets can often lead to programs containing tens of thousands of lines of code. Although it has been the tradition for individual researchers to develop codes in isolation, the last few years have seen the appearance of collaborations of researchers who work together on large coding projects. This trend is likely to continue, and it is probable that in the near future researchers will converge to using a handful of readily available simulation packages (e.g., NEMO, http://bima.astro.umd.edu/nemo/).

28.4.2 Gravitational *n*-Body Problem

Although Newton's law of universal gravitation has been supplanted by *General Relativity*, Newton's law remains highly accurate for a very large number of astrophysical problems. However, solving the interaction problem for any number of bodies ("*n*" bodies) is difficult because at first appearances the number of operations scales as n^2. However, provided small errors in the force calculation are acceptable

(RMS errors typically less than 0.5%), then approximate solutions can be found using order n log n operations. Roughly speaking, the algorithms used by researchers fall into two categories: treecodes [11] and grid (FFT) methods [70]. Treecodes are usually about an order of magnitude slower than grid codes for homogeneous distributions of particles, but are potentially much faster for very inhomogeneous distributions. To date, the largest gravitational simulations conducted contain over 300 billion particles and have been used to coarsely simulate volumes representing as much as 30% of the entire visible universe at high resolution (http://www.hits.org/english/research/tap/projects.php).

28.4.3 Hydrodynamics

Hydrodynamic modeling, or equivalently computational fluid dynamics, plays an extremely important role in astrophysics. Although most astrophysical *plasmas* are not fluids in the everyday sense, the physical description of them is the same. Modern hydrodynamic methods fall into two main groups: Eulerian (fixed) descriptions and Lagrangian (moving) descriptions. Eulerian descriptions can be broadly decomposed into finite difference and finite element methods. In astrophysics, the finite difference method is by far the most common approach. Lagrangian descriptions can be decomposed into "moving mesh" methods where the grid deforms with the flow in the fluid, and particle methods of which smoothed particle hydrodynamics (SPH) is a popular example [53].

Because shocks play an important role in the evolution of stars and the ISM, a significant amount of research has focused on "shock-capturing methods". Most early approaches to shock capturing, and indeed a number of methods still in use today, provide stability by using an artificial viscosity to smooth out flow discontinuities (shocks). Although these methods work well, they often introduce additional, unwanted, dissipation into the simulation. Perhaps the best alternative approach is "Godunov's method" [55], which is a simple example of a first-order method where the Riemann shock tube problem is solved at the interface of each grid cell. More modern algorithms have extended this idea to higher-order integration schemes, such as the piecewise-parabolic method [35].

28.4.4 Magnetohydrodynamics and Radiative Transfer

Magnetic fluid dynamic modeling, or MHD, is the source of a large amount of research in computational astrophysics [116]. The system of equations for MHD is that of hydrodynamics plus the addition of coupling terms corresponding to magnetic and electric forces and Maxwell's equations that constrain and evolve the magnetic and electric fields. Because of severe complexities arising from the divergenceless nature of the magnetic field, most MHD methods are finite difference, and although particle methods have been used, quite often they produce significant integration errors.

Modern methods, as in hydrodynamics, often cast the problem in terms of a system of conservation laws. It is usual to look at variation along a given axis direction and to recast the problem in terms of "characteristic variables" that are constructed from eigenvalues and the primitive variables, such as density, pressure, and flow speed. Such recasting aids development of the numerical method since time stepping can be viewed as propagating the system an infinitesimal amount along a characteristic. This formulation often allows development of stable integration schemes that produce accurate numerical solutions even when large time steps are used.

Radiative Transfer (RT), the study of how radiation interacts with gaseous plasmas, is an extremely difficult problem. It bears parallels to gravity in that all points within a system can usually affect all others, but is further complicated by the possibility of objects along any given direction producing nonisotropic attenuation. The radiation intensity is a function of position, two angles for direction of propagation, and frequency, a total of six independent variables. There any many different approaches to solving RT, ranging from explicit ray tracing to Monte Carlo methods as well as characteristic methods [105]. Much of the modern research effort is focused on deriving useful approximation methods that ease the computational effort required.

28.4.5 Planetary and Solar System Dynamics

Recent advances in telescope instrumentation have lead to a cascade of discoveries of extrasolar planets, and at the time of writing, over 700 extrasolar planets are known, and over 2000 candidates awaiting confirmation. Consequently, there is now a large amount of interest in studying planet and solar system formation. Solarsystem formation occurs during star formation, and the inherent differences in the planets are due to a differentiation process that enables different elements to condense out of the solar nebula at different radii.

Planets form by a hierarchical merging processes within the disk of the solar nebula. Dust grains form the first level of the hierarchy and planets the last, while objects of all mass scales and sizes exist in between. It should be noted that representing this variation of masses and sizes within a simulation is impossible since resolution is always limited by the available computing power and memory. Theoretical models of the agglomeration process must include hydrodynamics and gravity, although currently there is debate about the role of hydrodynamics in gas giant planet formation. At present, theory can be roughly divided into two approaches: the first being the study of stability properties of the solar nebula disk from an analytic perspective and the second being simulation of the process from a first principle perspective. Simulations with a million mass elements, designed to follow the agglomeration process in the inner part of the solar system, were conducted in 1998 [110].

The realization that our solar system contains many small asteroids and meteorites capable of causing severe damage to the Earth has renewed interest in solar system dynamics. Calculating accurate orbits for these systems is hard since very often they may have chaotic orbits. Chaotic systems place great demands on numerical integrations because truncation errors can rapidly pollute the integration. Thus, the integration schemes used need to be highly accurate (often quadruple precision is used), and much effort has been devoted into "long term" integrators (such as "symplectic integrators", see [33]) that preserve numerical stability over long periods of simulation time. The chaos observed in long-term simulations of the solar system inspired a new theory [95] that demonstrates the solar system is chaotic (Uranus could possibly be ejected) but the timescale for this is extremely long (10^{17} years).

28.4.6 Exoplanets

Searches for exoplanets have come of age in astronomy, with 553 confirmed exoplanets as of this writing (the most current count is available at http://www.planetary.org/exoplanets/list.php). Searches generally proceed using Doppler spectroscopy to search for radial velocity variations or high-precision photometry to search for exoplanet transits. NASA's Kepler mission and ESA's Convection Rotation and Planetary Transits (COROT) telescope both adopt the latter approach. Kepler observes some 156,000 stars with 30-min cadence (and a smaller number with 1-min cadence), and data processing and conditioning is hence demanding [74]. In addition to large-scale traditional photometry, the transit search involves application of a wavelet-based, adaptive matched filter to identify transit-like features with durations in the range of 1–16 h. Light curves with transit-like features whose combined (folded) transit detection statistic exceeds 7.1σ for some trial period and epoch are designated as threshold crossing events (TCEs) and subjected to further scrutiny.

One outstanding issue with such programs involves the validation of planet candidates that have been detected photometrically. Space-based missions such as COROT and Kepler, as well as their ground-based counterparts, including OGLE [123], HATNeT [9], and TrES [5], typically give rise to a large number of false positives for each true detection (see, e.g., [13]). The most common form of false positive involves an unresolved eclipsing binary star in the same field of view as the planet candidate; dilution of the binary eclipse can mimic a planetary transit. Generally, differentiating between the two cases involves high-SNR, high-resolution, time-resolved spectroscopy, which is both time- and resource-prohibitive, particularly for the smaller planets that are of the most interest.

Instead, a technique nicknamed BLENDER has been developed that makes use of the photometric light curves themselves [120,121]. BLENDER compares the transit photometry of the planet candidate against a suite of synthetic light curves of eclipsing binaries, attenuated by the light of the planet candidate star. Models account for depth, duration, ingress, and egress time of the candidate, as well as accounting for stellar astrophysical effects including limb darkening, gravity brightening, mutual reflection, oblateness, and differential extinction. The code can also account for more complex systems, including hierarchical triples with multiple planets. The numerous free parameters are varied to map a multidimensional χ^2 surface, with the goal of determining the probability that the observed transit might possess a non-planetary cause.

28.4.7 Stellar Astrophysics

Among the first astrophysical problems to be addressed using modern computational methods were models of stellar interiors [38,66] and atmospheres [81]. One simplifying assumption needed for early stellar codes was that of local thermodynamic equilibrium, which meant that stellar structural variations were not expected to occur on short timescales. Such assumptions are no longer necessary: stellar interior and atmosphere codes have become increasingly complex as computers and computational techniques have evolved. Theoreticians have been able to study rapid evolutionary phases and complex atmospheric processes in stars. Some of the difficult problems currently being addressed include evolution of rotating stars [93], radial and nonradial stellar pulsations [29,37], stellar magnetospheres [124], evolution of stars in binary systems [15], and supernovae [132].

28.4.8 Star Formation and the Interstellar Medium

One of the greatest challenges in astrophysics is understanding the star formation process. Star formation is an enormously difficult problem since it encompasses gravity, hydrodynamics, RT, and magnetic fields. Further, the difference in density between the initial gas cloud from which the star forms and the final star itself is 21 orders of magnitude or, equivalently, a change in physical scale of 7 orders of magnitude.

One of the most significant questions in this field is: Why do most stars form in binary systems? To address this question, numerical simulations have been run that follow the fragmentation of a large cloud of gas. The methods used have been primarily Lagrangian ones (such as SPH) although adaptive mesh refinement (AMR) techniques [17] are becoming more popular. The reason for the growth of interest in AMR is that recent results have demonstrated a severe error in a large body of numerical simulations of the cloud fragmentation process: they lacked resolution to adequately follow the balance between gravitational forces and local pressure forces [122]. Simulations currently suggest that turbulent fragmentation plays a critical role in determining the formation of multiple star systems and that a filamentary structure is the main mechanism for transferring mass to the protostellar disk [76]. A not dissimilar process seems to govern formation of the first stars in the universe [1].

Studying the ISM presents different challenges. Traditionally the ISM is understood as having a series of distinct phases that determine local star formation [89], with regulation of the phases provided by heating and cooling mechanisms. Stellar winds and supernovae are the primary heating mechanism, while radiative cooling is the dominant cooling mechanism for the hot gas phases. The supernovae and winds also constantly stir the ISM, which, in combination with rapid radiative cooling, serves to make it a highly turbulent medium. Turbulent media are difficult to understand because motions on very large scales can quickly couple to motions on much smaller scales, and thus accurate modeling requires resolution of large and small scales [85]. Because of this range of scales, achieving sufficient resolution to be able to accurately model turbulence is difficult, and a number of researchers rely on 2-D models to provide sufficient dynamic range. Recent 3-D simulations have shown that self-gravity alone, without the stirring provided from supernova explosions, is sufficient to produce the spectrum of perturbations expected from analytical descriptions of turbulence [25].

28.4.9 Cosmology

The study of the Big Bang and quantum gravity epoch is still largely conducted analytically, although some aspects of this research lend themselves to computer algebra. Following these earliest moments, the universe undergoes a series of phase transitions (or "symmetry breaking") as the forces of nature separate out of the "unified field" [127]. Computation has been used to examine the nature of the phase transitions that occur as each of the forces separates. For example, the electroweak phase transition has been extensively examined using lattice calculations to explore whether the phase transition is first or second order (e.g., [82]). Current constraints on the measurement of the mass of Higgs boson from the Large Hadron Collider in combination with the lattice models strongly suggest that the phase transition must be second order. Numerical simulations of the early universe have also investigated how nonuniform symmetry breaking can lead to the formation of defects [2].

Computation is used extensively in the study of *Big Bang Nucleosynthesis* (BBN) and the relic CMB. However, at present, CMB data analysis probably represents the biggest challenge computationally. Theoretical modeling of BBN dates back to the 1940s [6], and a very detailed numerical approach to solving the coupled set of equations describing the reaction network was developed comparatively early [125]. Currently there are a number of BBN codes available, and considerable effort has been put into reconciling results from different codes. CMB modeling is comparatively straightforward since the equations describing the evolution of a thermal spectrum of radiation in an expanding universe are not overly complex. However, because the CMB spectrum we measure has foreground effects superimposed upon it (such as clusters of galaxies), a large amount of effort is expended simulating the effect of foreground pollution [22]. Indeed in the past few years, CMB experiments have provided new ways of detecting previously unobserved objects like galaxy clusters (e.g., [107]).

The theoretical modeling of large-scale structure in the universe has relied heavily on computation. Because on large scales "dark matter" dominates dynamics, only gravity need be included, and a Newtonian approximation can be used without significant error. Particle-based algorithms are used to evolve an initially smooth distribution of particles into a clustered state representative of the universe at its current epoch. The first simulations with moderate resolution (3×10^5 mass elements) of the distribution of galaxies were conducted in the early 1980s [45]. Simulations have played a leading role in establishing the accuracy of the *Cold Dark Matter* (CDM) model of structure formation [20]. In this cosmological model, structures are formed via a hierarchical merging process. Simulation has also shown that dark matter tends to form cuspy halos that have a universal core profile [96], while the large-scale distribution of matter is dominated by filamentary structures.

28.4.10 Galaxy Clusters, Galaxy Formation, and Galactic Dynamics

Modeling of clusters of galaxies and galaxy formation relies upon the same codes as the study of large-scale structure, with the addition of hydrodynamics to model the gas that condenses to form stars and nebulae within galaxies. Typically the hydrodynamic methods used are either Eulerian grid-based algorithms or Lagrangian particle-based methods [50] although, as in star formation, AMR methods are being adopted. Modeling of galaxy clusters is comparatively straightforward since the intracluster gas tends toward hydrostatic equilibrium. However, simulations have shown that the gas in galaxy clusters shows evidence of an epoch of preheating [46].

Galaxy formation is an exceptional difficult problem to study numerically because the evolution of the gas is strongly affected by supernova explosions and heating from supermassive black holes in active galactic nuclei, both of which processes occur initially on scales smaller than the best simulations can currently simulate (e.g., [16]). The physics is also technically challenging since galaxy formation occurs in the very nonlinear regime of gravitational collapse while simultaneously being a radiation hydrodynamics problem (although an optically thin approximation for the gas works moderately well, at least in the late time universe). Only within the last 2 years has a sufficient understanding evolved to enable simulations of galaxy formation to produce moderate facsimiles of the galaxies we observe [56]. In the near future, increases in resolution should provide a significant step forward in the accuracy of star formation modeling as the sites of star formation, namely, giant molecular clouds, should be resolved.

The first large-scale numerical studies of galactic dynamics were conducted in the 1960s [69]. At least initially, and to a large extent today, most n-body simulations are used to confirm analytic solutions derived from idealized models of galactic disks [19]. Typically these simulations begin with a model of a given galaxy, which usually consists of a disk of stars and an extended dark halo, which is then perturbed in some fashion to mimic the phenomenon under study. Recently it has been highlighted that accurate modeling of galactic dynamics in CDM universes is exceptionally hard due to coupling between the substructure in the larger galactic halo and the galactic disk [126]. Traditionally, researchers believed that approximately 1 million particles was sufficient to model galaxies reasonably; however, these new results have pushed that estimate at least an order of magnitude higher. Although researchers in this field use codes similar to those in large-scale structure, specialist codes, which are designed to reduce numerical noise, have been developed (i.e., the self consistent field code of [67]).

28.4.11 Numerical Relativity

General relativity calculations are extremely computationally demanding as not only is the theory exceptionally nonlinear, but also there are a number of elliptic constraint equations that must be satisfied at each iteration. Further, the rapid change of scales that can accompany collapse problems often requires adaptive methods to resolve. There are also other subtleties related to the boundary conditions around black holes that present severe intellectual challenges. Currently the strongest science driver behind these calculations is the need to calculate the gravitational wave signal of cataclysmic events (such as binary black hole coalescence) that may be detectable with the LIGO gravitational wave detector (http://www.ligo.caltech.edu). Because of the large amount of computation involved in computing space-time geometry and the comparatively low amount of communication between processors, numerical relativity is an ideal candidate for grid-based computation. The Cactus framework has been developed to aid such calculations (http://www.cactuscode.org).

Relativity calculations are most often grid based (although spectral methods are used occasionally). Before determining the evolution equations for the space-time, a gauge must first be decided upon, and the most common gauge is the so-called "3 + 1 formalism" where the space-time is sliced into a one-parameter family of space-like slices. Other formulations exist, such as the characteristic formalism, and in general, the gauge is chosen to suit the problem being studied. Initial conditions for the space-time are provided, and then the simulation is integrated forward, with suitable boundary conditions being applied. Building upon this body of research, the first calculation of the gravitational waveform from binary black hole coalescence was performed in 2001 [4].

28.4.12 Compact Objects

The study of compact objects such as white dwarf and neutron stars presents a formidable theoretical challenge. These systems exhibit extreme density, in turn requiring detailed nuclear physics as well as relativistic descriptions. Compact objects are widely believed to be the source of energy behind GRBs, with energetic scenarios, such as sudden mergers, driving a highly relativistic "fireball" shock wave that produces an extreme amount of γ-ray radiation during collisions with other shock waves or the ISM [90].

Neutron star collisions have been simulated for similar reasons to black holes: the calculation of their gravitational wave spectrum [31]. Neutron stars are also the beginning point of core collapse (type II) supernovae, and the simulation of the ignition process has attracted much attention. Fully general relativistic models are now beginning to appear [28]. Of particular interest is how the neutrinos drive a wind shortly after collapse begins [30].

Type I supernovae occur when white dwarfs accrete sufficient mass to exceed the Chandrasekhar limit and subsequently undergo collapse. The physics is challenging because the process occurs far from equilibrium and entails RT as well as hydrodynamic instabilities. As in studies of type II supernovae, to date most calculations use a 2-D approximation [97] and Eulerian approaches; however, some explorations have used SPH in 3-dimensions [26]. The push toward full high-resolution 3-D calculations is gathering momentum (e.g., /citeHillebrandt2010). Such simulations are necessary to fully understand instabilities and include more accurate physics. However, the computational challenge is significant and ultimately progress awaits the development of 100 teraflop computers.

28.4.13 Parallel Computation in Astrophysics

Parallel computing in astrophysics is often used to examine problems that simply cannot be addressed on a desktop computer, regardless of how long one could wait. The primary driver in this case is the large amounts of memory available in parallel computers as compared to serial ones: the largest parallel computations just don't fit into a desktop. The secondary use of parallel computation is to speed up data analysis that involves performing the same analysis on many subsets of data. In this case, parallel computers significantly help in lowering the data reduction time. Along similar lines, the significant computational requirements of many astrophysics calculations can utilize special-purpose computational hardware. The rise of general-purpose computing on graphics processors (GPGPU) is attracting much interest within the astrophysics community (e.g., [48]). A number of simulation and analysis codes have already been ported to this specialist hardware.

Around 10% of the total cycles at supercomputing centers are typically devoted to astrophysics. This share is lower than in previous decades as there has been significant growth in parallel computation in other fields, in many cases benefitting directly from new codes and approaches developed within astrophysics. Astrophysicists have a history of developing unique and ingenious parallel algorithms to solve the problems they face. Indeed, a number of research problems in astrophysics, such as binary black hole coalescence and the formation of galaxies, have long been considered computational "grand challenges." These problems have computing demands that are similar to the nuclear ignition codes in the Advanced Simulation and Computing Program, which is an integral element of the US Government Stockpile Stewardship Program.

Parallel codes for distributed memory platforms are most often developed using the Message Passing Interface (MPI). Higher-performance Application Programming Interfaces (APIs), such as the remote direct memory access provided by MPI-2, are yet to receive significant attention primarily because vendors have failed to provide full support for this emerging standard. All of these APIs can lead to a significant increase in the size of a parallel program compared to the serial one. It is not uncommon for parallel codes to be over twice the length of serial ones. Codes written using these APIs have the potential to scale to many thousands of processors.

Shared memory parallel codes are most often developed using the standardized OpenMP API. This API is particular simple to use since it enables simple parallelization of codes by using "pragmas" that are inserted into the code before iterative loops. The iterations within the loop, provided that they meet certain data independence requirements, can then be distributed to different CPUs, thereby speeding up execution speeds. The OpenMP API often does not lead to significantly longer parallel programs, but is limited in terms of scalability by the requirement of running on shared memory computers that typically have a maximum of around 32 processors.

Over the past few years, it has become increasingly apparent that although often the physics being simulated by two codes is quite different, the underlying data structures being used, such as grids or trees, are quite similar. This has lead to the development of skeleton packages in which a researcher needs only to add the numerical implementation of their equations and the communication between processors is handled by the package. Cactus and Paramesh are an example of this type of framework. However, most researchers seem reluctant to rely on these packages and instead develop an optimized communication layers themselves.

The Asteroseismic Modeling Portal (AMP) provides a web-based interface to the MPIKAIA asteroseismology pipeline [92,131], allowing astronomers to run and view simulations that derive the properties of sunlike stars from observations of their pulsation frequencies. AMP represents a simplified orchestration of TeraGrid computational resources; the web-based interface was developed as a traditional stand-alone database-backed web application using the Python-based Django web development framework, leveraging the Django frameworks capabilities while cleanly separating the user interface development from the grid interface development.

The motivation for the interface structure arises from the fact that running a single MPIKAIA simulation requires propagating several independent batches of MPI jobs and can consume 512 processors for over a week of wall-clock time. A complete optimization run to determine the best-fit model for a single Kepler target star consists of four GAs runs executed in parallel, with each GA run modeling a population of 126 stars, using 128 processors, for 200 iterations. Using the science gateway approach allows researchers to run the model without using local resources, disseminates model results to the community, and introduces a uniform analysis structure.

Cloud computing is particularly attractive for a number of problems in analyzing astrophysical data, such as image processing [18]. But it is unclear what role cloud environments will play in the development of theoretical modeling. Large parallel computations remain nontrivial to implement and web-based interfaces to large codes run at supercomputing centers are, at least at present, not widely used.

28.5 Research Issues and Summary

In this section, we first highlight a number of key areas in computational astrophysics that will receive increasing attention in the future. No significance should be attached to the ordering of the items:

1. Data analysis and visualization in the age of VOs: The complexity and large size of VO databases will increase the need for powerful data analysis, visualization, and data mining toolkits for the effective and rapid scientific exploration of the resulting massive data sets. VOs are enabled by technology but driven by science, so data mining tools need to recognize the need to incorporate measurement uncertainty, incomplete sampling, and measurement bias. Visualization tools need to have flexible scaling parameters and resolution, as well as the ability to visualize multidimensional data. Many pattern recognition algorithms currently define classes without regard to data quality; poorly measured data can smear out true clusters that would be easily visible if the data were limited to that of high quality. In some cases, poor-quality measurements can cause data to artificially cluster (e.g., when an upper flux limit is used in lieu of a measurement). Since the scientific method is statistical in nature, computational tools for science are needed that better reflect statistical uncertainties.
2. Planet formation: Current observations are already challenging theoretical models for planet formation and dynamical evolution. Future models need to include radiation, chemistry, hydrodynamics, and dust to better understand both formation and migration processes.
3. Star formation: Following the full collapse of a gas cloud through to nuclear ignition remains an enormously challenging theoretical proposition. A large number of outstanding questions relating to the role of turbulence and magnetic fields remain, and a better understanding of the formation of high-mass stars has a number of implications for other areas of astrophysics, most notably galaxy formation.
4. Galaxy formation: Progress is limited by both a lack of numerical resolution and an incomplete understanding of the physics of the process. It is currently unclear how to model the extremely important effect of heating from supernovae and active galactic nuclei on protogalactic collapse. New discoveries will depend heavily on the growth of understanding of the star formation and black hole growth processes.

5. CMB data analysis: While new approximation algorithms have been developed to reduce the enormity of the analysis calculations for high-resolution data sets, their accuracy is questioned. Brute force and larger computers are not the answer, as their computational requirements exceed the capacity of even the soon to be built exaflop class supercomputers.
6. Black hole coalescence: Despite a vast growth in technical capabilities over the past decade, calculations still have significant room for adding additional physics (incorporating drag from gas disks remains a notable challenge). In addition to the value of calculating gravitational wave forms for future detectors, these calculations have significant implications for galaxy formation scenarios.
7. Supernovae ignition and modeling: Advances in computational power have shown notable differences in the 2-D versus 3-D results, highlighting that hydrodynamic instabilities need much further study. A coherent theory will require developing a clear understanding of the ignition process through to the observational signals (especially transients).
8. Autonomous spacecraft/instrumentation operation: Autonomous spacecraft are being designed to carry out day-to-day operations independent of ground control. Some of these operations include navigation and the scheduling and execution of observations and experiments. Support technologies for autonomous spacecraft include robotics, artificial intelligence, and control theory. This approach will reduce mission operation costs while simultaneously allow orbital satellites to work dynamically rather than passively.

Astrophysics is gradually and irreversibly becoming a computational science. Astronomical data are being stored in and retrieved from progressively large databases; data and metadata are being used by larger audiences. The analyses of these data are aided by pattern recognition algorithms and improved data visualization tools. Theoretical modeling has developed beyond the point where elegant calculations using relatively simple assumptions suffice; detailed models with many physical parameters are often required to adequately explain the detailed observations made by the newest orbital and ground-based telescopes spanning the electromagnetic spectrum. Parallel computation is often used in carrying out time-consuming calculations. High-precision is needed to accurately calculate model parameters, and computationally-intensive statistical tools are required to evaluate theoretical model efficacies. Astrophysics is thus evolving, and the new generation of astrophysicist is increasingly well-versed in the use of computational tools. This can only help us to better understand the structure, evolution, and nature of the universe.

Glossary

Adaptive optics: An active form of observing light (rather than a passive one) in which lasers are fired at the sky from the telescope site, and the reflected laser light is used to differentially correct observed starlight for atmospheric refraction and convection.

Astrophysics: The study of the physical properties, structure, kinematics, dynamics, and evolution of celestial objects.

Big Bang nucleosynthesis: The epoch of the early universe when the nuclei of atoms were formed. Theories of this epoch reproduce observations of elemental abundances exceptionally well, strongly supporting the hot Big Bang theory.

Cold dark matter: Hypothesized subatomic particles that contribute roughly 30% of the mass in the universe. Although in the early universe these particles interacted vigorously, their influence is now seen only via gravitation. Theories that include the effects of these particles match observations exceptionally accurately.

Cosmic microwave background: Relic electromagnetic radiation left over from 300,000 years after the Big Bang. The distribution of radiation is very close to being homogeneous and isotropic, but carries small perturbations on it that correspond to perturbations of the matter density in the early universe.

Cosmology: The branch of astrophysics dealing with the structure and evolution of the universe on the largest scales.

Electromagnetic spectrum: Classification of light by wavelength (which is inversely related to energy). Ranging from light with the shortest wavelength (and largest energy) to that with the longest wavelength (and smallest energy), the electromagnetic spectrum includes gamma-ray, x-ray, ultraviolet, visible, infrared, and radio.

Galaxy: A large, gravitationally-bound ensemble of stars, gas, and dust. Galaxies are traditionally classified as spiral, elliptical, irregular, or peculiar based on their morphological structure. Our *Milky Way* galaxy is a large spiral galaxy containing more than a hundred billion stars.

General relativity: Einstein's theory of gravitation that casts gravity not as a force, but rather a consequence of warping in the local space-time continuum. The theory relies strongly upon geometrical concepts of distance and curvature.

Interstellar Medium (ISM): The material between the stars. It is composed primarily of hydrogen and helium gas, with 1%–2% heavy elements chemically mixed to form larger molecules ("dust"). Dense, cool molecular clouds are the seeds of star formation.

Passband: An electromagnetic regime defined by the spectral response of a particular filter and/or instrument.

Plasma: An ionized gas. A plasma behaves differently than a normal gas (which can often be modeled as a fluid) because a treatment of electromagnetic theory is needed to address the electrical charges found within it.

Supernova: A massive stellar explosion during which a star can briefly brighten to almost one billion times its original luminosity. Such events can be seen at enormous distances, although maximal luminosity usually lasts only for tens of days.

Virtual observatory: Publicly accessible metadatabase of archived ground-based, balloon, and satellite astronomical observations. The database will also be associated with a variety of search engines and data mining tools.

References

1. T. Abel, G. L. Bryan, and M. L. Norman. The formation of the first star in the Universe. *Science*, 295:93–98, January 2002.
2. A. Achucarro, J. Borrill, and A. R. Liddle. Formation rate of semilocal strings. *Phys. Rev. Lett.*, 82:3742–3745, May 1999.
3. J. Adams, J. Hakkila, and C. J. Nettles. Clustering of gamma-ray burst pulse properties: A cleaner delineation of GRB classes. In *American Astronomical Society Meeting Abstracts #215*, volume 42 of *Bulletin of the American Astronomical Society*, Washington, DC, p. 405.07, January 2010.
4. M. Alcubierre, W. Benger, B. Brügmann, G. Lanfermann, L. Nerger, E. Seidel, and R. Takahashi. 3D Grazing collision of two black holes. *Phys. Rev. Lett.*, 87(26):A261103, December 2001.
5. R. Alonso, T. M. Brown, G. Torres, D. W. Latham, A. Sozzetti, G. Mandushev, J. A. Belmonte et al. TrES-1: The transiting planet of a bright K0 V star. *Astrophys. J. Lett.*, 613:L153–L156, October 2004.
6. R. A. Alpher, H. Bethe, and G. Gamow. The origin of chemical elements. *Phys. Rev.*, 73:803–804, April 1948.
7. T. Appourchaux, O. Benomar, M. Gruberbauer, W. J. Chaplin, R. A. García, R. Handberg, G. A. Verner et al. Oscillation mode linewidths of main-sequence and subgiant stars observed by Kepler. *Astron. Astrophys.*, 537:A134, January 2012.
8. T. Axelrod, A. Connolly, Z. Ivezic, J. Kantor, R. Lupton, R. Plante, C. Stubbs, and D. Wittman. The LSST data processing pipeline. In *American Astronomical Society Meeting Abstracts*, volume 36 of *Bulletin of the American Astronomical Society*, Washington, DC, p. 108.11, December 2004.

9. G. Á. Bakos, J. D. Hartman, G. Torres, G. Kovács, R. W. Noyes, D. W. Latham, D. D. Sasselov, and B. Béky. Planets from the HATNet project. *Detection and Dynamics of Transiting Exoplanets, St. Michel l'Observatoire,* France, F. Bouchy, R. Díaz, and C. Moutou, eds., *EPJ Web of Conferences,* Vol 11, id.01002, 11:1002, February 2011.
10. N. M. Ball and R. J. Brunner. Data mining and machine learning in astronomy. *Int. J. Modern Phys. D,* 19:1049–1106, 2010.
11. J. Barnes and P. Hut. A hierarchical O(N log N) force-calculation algorithm. *Nature,* 324:446–449, December 1986.
12. S. Barthelmy. GCN and VOEvent: A status report. *Astronom. Nachrichten,* 329:340, March 2008.
13. N. M. Batalha, J. F. Rowe, R. L. Gilliland, J. J. Jenkins, D. Caldwell, W. J. Borucki, D. G. Koch et al. Pre-spectroscopic false-positive elimination of Kepler planet candidates. *Astrophys. J. Lett.,* 713:L103–L108, April 2010.
14. J. M. Beckers. Adaptive optics for astronomy—Principles, performance, and applications. *Ann. Rev. Astron. Astrophys.,* 31:13–62, 1993.
15. M. E. Beer and P. Podsiadlowski. A general three-dimensional fluid dynamics code for stars in binary systems. *Mon. Not. Royal Astron. Soc.,* 335:358–368, September 2002.
16. A. J. Benson. Galaxy formation theory. *Phys. Rep.,* 495:33–86, October 2010.
17. M. J. Berger and P. Colella. Local adaptive mesh refinement for shock hydrodynamics. *J. Comput. Phys.,* 82:64–84, May 1989.
18. G. B. Berriman, E. Deelman, P. Groth, and G. Juve. The application of cloud computing to the creation of image mosaics and management of their provenance. In *Society of Photo-Optical Instrumentation Engineers (SPIE) Conference Series,* Bellingham, WA, volume 7740 of *Society of Photo-Optical Instrumentation Engineers (SPIE) Conference Series,* July 2010.
19. J. Binney and S. Tremaine. *Galactic Dynamics.* Princeton, NJ: Princeton University Press, 1987.
20. G. R. Blumenthal, S. M. Faber, J. R. Primack, and M. J. Rees. Formation of galaxies and large-scale structure with cold dark matter. *Nature,* 311:517–525, October 1984.
21. M. Boër. Robotic telescopes as science tools. In N. Mebarki and J. Mimouni, eds. *American Institute of Physics Conference Series,* volume 1295 of *American Institute of Physics Conference Series,* Melville, NY, pp. 66–79, October 2010.
22. J. R. Bond, C. R. Contaldi, U.-L. Pen, D. Pogosyan, S. Prunet, M. I. Ruetalo, J. W. Wadsley et al. The Sunyaev-Zel'dovich effect in CMB-calibrated theories applied to the cosmic background imager anisotropy power at l greater than 2000. *Astrophys. J.,* 626:12–30, June 2005.
23. K. Borne. Astroinformatics: A 21st century approach to astronomy research and education. In *American Astronomical Society Meeting Abstracts No. 215,* volume 42 of *Bulletin of the American Astronomical Society,* Washington, DC, p. 230.01, January 2010.
24. W. J. Borucki, D. Koch, G. Basri, N. Batalha, T. Brown, D. Caldwell, J. Caldwell et al. Kepler planet-detection mission: Introduction and first results. *Science,* 327:977, February 2010.
25. F. Bournaud, B. G. Elmegreen, R. Teyssier, D. L. Block, and I. Puerari. ISM properties in hydrodynamic galaxy simulations: Turbulence cascades, cloud formation, role of gravity and feedback. *Mon. Not. Royal Astron. Soc.,* 409:1088–1099, December 2010.
26. E. Bravo and D. Garcia-Senz. Smooth particle hydrodynamics simulations of deflagrations in supernovae. *Astrophys. J. Lett.,* 450:L17, September 1995.
27. M. Brescia, G. Longo, and F. Pasian. Mining knowledge in astrophysical massive data sets. *Nucl. Instrum. Meth. Phys. Res. A,* 623:845–849, November 2010.
28. S. W. Bruenn, K. R. De Nisco, and A. Mezzacappa. General relativistic effects in the core collapse supernova mechanism. *Astrophys. J.,* 560:326–338, October 2001.
29. J. R. Buchler, Z. Kolláth, and A. Marom. An adaptive code for radial stellar model pulsations. *Astrophys. Space Sci.,* 253:139–160, September 1997.
30. A. Burrows, J. Hayes, and B. A. Fryxell. On the nature of core-collapse supernova explosions. *Astrophys. J.,* 450:830, September 1995.

31. A. C. Calder and E. Y. M. Wang. Numerical models of binary neutron star system mergers. II. Coalescing models with post-Newtonian radiation reaction forces. *Astrophys. J.*, 570:303–313, May 2002.
32. M. T. Capria, G. Chanteur, and W. Schmidt. IDIS, the European Planetary Virtual Observatory: Status and perspectives. In EPSC-DPS Joint Meeting 2011, held 2–7 October 2011 in Nantes, France. ¡A href="http://meetings.copernicus.org/epsc-dps2011"¿http://meetings.copernicus.org/epsc-dps2011¡/A¿, p. 947, October 2011.
33. D. A. Clarke and M. J. West, eds. *Theoretical Astrophysics: Computational Astrophysics, 12th Kingston Meeting*, volume 123 of *Astronomical Society of the Pacific Conference Series*, Washington, DC, 1997.
34. R. Claudi. A new opportunity from space: PLATO mission. *Astron. Astrophys. Suppl. Series*, 328:319–323, July 2010.
35. P. Colella and P. R. Woodward. The piecewise parabolic method (PPM) for gas-dynamical simulations. *J. Comput. Phys.*, 54:174–201, September 1984.
36. J. J. Condon, W. D. Cotton, E. W. Greisen, Q. F. Yin, R. A. Perley, G. B. Taylor, and J. J. Broderick. The NRAO VLA sky survey. *Astron. J.*, 115:1693–1716, May 1998.
37. A. H. Corsico and O. G. Benvenuto. A new code for nonradial stellar pulsations and its application to low-mass, helium white dwarfs. *Astrophys. Space Sci.*, 279:281–300, 2002.
38. A. N. Cox, D. L. Bowers, and R. R. Brownlee. A method of computing stellar interior models. *Astron. J.*, 65:486, 1960.
39. L. Cuesta. Robotic telescopes and their use as an educational tool. In L. L. Christensen, M. Zoulias, and I. Robson, eds., *Communicating Astronomy with the Public*, Paris, France, p. 396, June 2008.
40. A. R. Davey and The VSO Team. The virtual solar observatory at eight and a bit! In *AAS/Solar Physics Division Abstracts #42*, Washington, DC, p. 105, May 2011.
41. J. P. Dietrich. The importance of being first: Position dependent citation rates on arXiv:astro-ph. *Pub. Astron. Soc. Pac.*, 120:224–228, February 2008.
42. S. G. Djorgovski, R. R. Carvalho, R. R. Gal, S. C. Odewahn, A. A. Mahabal, R. Brunner, P. A. A. Lopes, and J. L. Kohl Moreira. The digital Palomar observatory sky survey (DPOSS): General description and the public data release. *Bull. Astron. Soc. Brazil*, 23:197–197, August 2003.
43. S. G. Djorgovski, G. Longo, M. Brescia, C. Donalek, S. Cavuoti, M. Paolillo, R. D'Abrusco, O. Laurino, A. Mahabal, and M. Graham. DAta Mining and Exploration (DAME): New tools for knowledge discovery in astronomy. In *American Astronomical Society Meeting Abstracts*, volume 219 of *American Astronomical Society Meeting Abstracts*, Washington, DC, p. 145.12, January 2012.
44. T. Dower, G. Greene, R. Plante, and M. J. Graham. An astronomical data resource registry for the virtual observatory. In D. A. Bohlender, D. Durand, and P. Dowler, eds., *Astronomical Data Analysis Software and Systems XVIII*, volume 411 of *Astronomical Society of the Pacific Conference Series*, Chicago, IL, p. 369, September 2009.
45. G. Efstathiou and J. W. Eastwood. On the clustering of particles in an expanding universe. *Mon. Not. Royal Astron. Soc.*, 194:503–525, February 1981.
46. V. R. Eke, J. F. Navarro, and C. S. Frenk. The evolution of X-ray clusters in a low-density universe. *Astrophys. J.*, 503:569, August 1998.
47. C. Espaillat, J. Bregman, P. Hughes, and E. Lloyd-Davies. Wavelet analysis of AGN x-ray time series: A QPO in 3C 273? *Astrophys. J.*, 679:182–193, May 2008.
48. C. J. Fluke, D. G. Barnes, B. R. Barsdell, and A. H. Hassan. Astrophysical supercomputing with GPUs: Critical decisions for early adopters. *Pub. Astron. Soc. Australia*, 28:15–27, January 2011.
49. E. B. Ford, A. V. Moorhead, and D. Veras. A Bayesian surrogate model for rapid time series analysis and application to exoplanet observations. *ArXiv e-Prints*, July 2011.
50. C. S. Frenk, S. D. M. White, P. Bode, J. R. Bond, G. L. Bryan, R. Cen, H. M. P. Couchman et al. The Santa Barbara cluster comparison project: A comparison of cosmological hydrodynamics solutions. *Astrophys. J.*, 525:554–582, November 1999.

51. L. N. Garcia and S. F. Fung. The virtual wave observatory: A portal for planetary radio and plasma wave data. In *European Planetary Science Congress 2010*, Rome, Italy, p. 394, September 2010.
52. N. Gehrels, G. Chincarini, P. Giommi, K. O. Mason, J. A. Nousek, A. A. Wells, N. E. White et al. The swift gamma-ray burst mission. *Astrophys. J.*, 611:1005–1020, August 2004.
53. R. A. Gingold and J. J. Monaghan. Smoothed particle hydrodynamics—Theory and application to non-spherical stars. *Mon. Not. Royal Astron. Soc.*, 181:375–389, November 1977.
54. P. Ginsparg. ArXiv at 20. *Nature*, 476:145–147, August 2011.
55. S. K. Godunov. A difference method for numerical calculation of discontinuous solutions of the equations of hydrodynamics. *Mat. Sb. (N.S.)*, 47:271–306, 1959.
56. F. Governato, C. Brook, L. Mayer, A. Brooks, G. Rhee, J. Wadsley, P. Jonsson et al. Bulgeless dwarf galaxies and dark matter cores from supernova-driven outflows. *Nature*, 463:203–206, January 2010.
57. J. Greiner. Discoveries enabled by multi-wavelength afterglow observations of gamma-ray bursts. *ArXiv e-Prints*, February 2011.
58. J. Hakkila, T. W. Giblin, R. J. Roiger, D. J. Haglin, W. S. Paciesas, and C. A. Meegan. How sample completeness affects gamma-ray burst classification. *Astrophys. J.*, 582:320–329, January 2003.
59. J. Hakkila, J. M. Myers, B. J. Stidham, and D. H. Hartmann. A computerized model of large-scale visual interstellar extinction. *Astron. J.*, 114:2043, November 1997.
60. J. Hakkila and R. D. Preece. Unification of pulses in long and short gamma-ray bursts: Evidence from pulse properties and their correlations. *Astrophys. J.*, 740:104, October 2011.
61. M. Hall, E. Frank, G. Holmes, B. Pfahringer, P. Reutemann, and I. H. Witten. The weka data mining software: An update. *SIGKDD Explor. Newsl.*, 11(1):10–18, 2009.
62. R. J. Hanisch. The virtual observatory: Retrospective and prospectus. In Y. Mizumoto, K.-I. Morita, and M. Ohishi, eds., *Astronomical Data Analysis Software and Systems XIX*, volume 434 of *Astronomical Society of the Pacific Conference Series*, Chicago, IL, p. 65, December 2010.
63. R. J. Hanisch, A. Farris, E. W. Greisen, W. D. Pence, B. M. Schlesinger, P. J. Teuben, R. W. Thompson, and A. Warnock, III. Definition of the flexible image transport system (FITS). *Astron. Astrophys.*, 376:359–380, September 2001.
64. C. W. Heikkila, T. A. McGlynn, and N. E. White. Astrobrowse: A web agent for querying astronomical databases. In D. M. Mehringer, R. L. Plante, and D. A. Roberts, eds., *Astronomical Data Analysis Software and Systems VIII*, volume 172 of *Astronomical Society of the Pacific Conference Series*, San Francisco, CA, p. 221, 1999.
65. E. A. Henneken, G. Eichhorn, A. Accomazzi, M. J. Kurtz, C. Grant, D. Thompson, E. Bohlen, and S. S. Murray. How the literature is used a view through citation and usage statistics of the ADS. In Haubold, H. J. and Mathai, A. M., eds., *Proceedings of the Third UN/ESA/NASA Workshop on the International Heliophysical Year 2007 and Basic Space Science, Astrophysics and Space Science Proceedings*. Springer-Verlag, Berlin, Germany, p. 141, 2010.
66. L. G. Henyey, L. Wilets, K. H. Böhm, R. Lelevier, and R. D. Levee. A method for automatic computation of stellar evolution. *Astrophys. J.*, 129:628, May 1959.
67. L. Hernquist and J. P. Ostriker. A self-consistent field method for galactic dynamics. *Astrophys. J.*, 386:375–397, February 1992.
68. E. Hivon, K. M. Górski, C. B. Netterfield, B. P. Crill, S. Prunet, and F. Hansen. MASTER of the cosmic microwave background anisotropy power spectrum: A fast method for statistical analysis of large and complex cosmic microwave background data sets. *Astrophys. J.*, 567:2–17, March 2002.
69. R. W. Hockney. Gravitational experiments with a cylindrical galaxy. *Astrophys. J.*, 150:797, December 1967.
70. R. W. Hockney and J. W. Eastwood. *Computer Simulation Using Particles*. McGraw-Hill, New York, 1981.
71. I. Horváth, Z. Bagoly, L. G. Balázs, A. de Ugarte Postigo, P. Veres, and A. Mészáros. Detailed classification of swift 's gamma-ray bursts. *Astrophys. J.*, 713:552–557, April 2010.

72. D. Huber, T. R. Bedding, D. Stello, B. Mosser, S. Mathur, T. Kallinger, S. Hekker et al. Asteroseismology of Red Giants from the first four months of Kepler data: Global oscillation parameters for 800 stars. *Astrophys. J.*, 723:1607–1617, November 2010.
73. J. F. Jarvis and J. A. Tyson. FOCAS—Faint object classification and analysis system. *Astron. J.*, 86:476–495, March 1981.
74. J. M. Jenkins, D. A. Caldwell, H. Chandrasekaran, J. D. Twicken, S. T. Bryson, E. V. Quintana, B. D. Clarke et al. Overview of the Kepler science processing pipeline. *Astrophys. J. Lett.*, 713:L87–L91, April 2010.
75. H. Karimabadi. Intelligent archiving and physics mining of large data sets (Invited). *AGU Fall Meeting Abstracts*, p. B1, December 2009.
76. R. I. Klein, R. Fisher, and C. F. McKee. Fragmentation and star formation in turbulent cores. In H. Zinnecker and R. Mathieu, eds., *The Formation of Binary Stars*, volume 200 of *IAU Symposium*, Dordrecht, The Netherlands, p. 361, 2001.
77. S. G. Kleinmann, M. G. Lysaght, W. L. Pughe, S. E. Schneider, M. F. Skrutskie, M. D. Weinberg, S. D. Price, K. Matthews, B. T. Soifer, and J. P. Huchra. The two micron all sky survey. *Astrophys. Space Sci.*, 217:11–17, July 1994.
78. A. Klotz. Techniques associated with robotic telescopes. In N. Mebarki and J. Mimouni, eds., *American Institute of Physics Conference Series*, volume 1295 of *American Institute of Physics Conference Series*, Melville, NY, pp. 42–45, October 2010.
79. D. G. Koch, W. Borucki, E. Dunham, J. Geary, R. Gilliland, J. Jenkins, D. Latham et al. Overview and status of the Kepler mission. In J. C. Mather, eds., *Society of Photo-Optical Instrumentation Engineers (SPIE) Conference Series*, volume 5487 of *Society of Photo-Optical Instrumentation Engineers (SPIE) Conference Series*, Bellingham, WA, pp. 1491–1500, October 2004.
80. M. Kurtz, A. Accomazzi, and S. S. Murray. The smithsonian/NASA astrophysics data system (ADS) decennial report. In *astro2010: The Astronomy and Astrophysics Decadal Survey*, volume 2010 of *ArXiv Astrophysics e-Prints*, Cornell University, Ithaca, NY, p. 28P, 2009.
81. R. L. Kurucz. A matrix method for calculating the source function, mean intensity, and flux in a model atmosphere. *Astrophys. J.*, 156:235, April 1969.
82. M. Laine and K. Rummukainen. The MSSM electroweak phase transition on the lattice. *Nucl. Phys. B*, 535:423–457, December 1998.
83. A. Lasenby, B. Barreiro, and M. Hobson. Regularization and inverse problems. In A. J. Banday, S. Zaroubi, and M. Bartelmann, eds., *Mining the Sky*, p. 15, 2001.
84. M. Louys, J. L. Starck, S. Mei, F. Bonnarel, and F. Murtagh. Astronomical image compression. *Astron. Astrophys. Suppl.*, 136:579–590, May 1999.
85. M.-M. Mac Low. The dynamical interstellar medium: Insights from numerical models. In D. Alloin, K. Olsen, and G. Galaz, eds., *Stars, Gas and Dust in Galaxies: Exploring the Links*, volume 221 of *Astronomical Society of the Pacific Conference Series*, San Francisco, CA, p. 55, 2000.
86. K. B. Marvel and C. Biemesderfer. Open access: Current status, AAS perspectives. *Future Professional Communication in Astronomy II, Astrophysics and Space Science Proceedings,* Volume 1. Springer Science+Business Media, LLC, New York, 1:91, 2011.
87. J. M. Mazzarella, B. F. Madore, and G. Helou. Capabilities of the NASA/IPAC extragalactic database in the era of a global virtual observatory. In J.-L. Starck and F. D. Murtagh, eds., *Society of Photo-Optical Instrumentation Engineers (SPIE) Conference Series*, volume 4477 of *Society of Photo-Optical Instrumentation Engineers (SPIE) Conference Series*, Bellingham, WA, pp. 20–34, November 2001.
88. S. McConnell, G. Henry, R. Sturgeon, and R. Hurley. A database for data mining applications in astronomy. In I. N. Evans, A. Accomazzi, D. J. Mink, and A. H. Rots, eds., *Astronomical Data Analysis Software and Systems XX*, volume 442 of *Astronomical Society of the Pacific Conference Series*, Chicago, IL, p. 529, July 2011.
89. C. F. McKee and J. P. Ostriker. A theory of the interstellar medium—Three components regulated by supernova explosions in an inhomogeneous substrate. *Astrophys. J.*, 218:148–169, November 1977.

90. P. Mészáros. Gamma-ray bursts. *Rep. Progress Phys.*, 69:2259–2321, August 2006.
91. T. S. Metcalfe. The rise and citation impact of astro-ph in major journals. In *Bulletin of the American Astronomical Society*, volume 37 of *Bulletin of the American Astronomical Society*, Washington, DC, pp. 555–557, May 2005.
92. T. S. Metcalfe and P. Charbonneau. Stellar structure modeling using a parallel genetic algorithm for objective global optimization. *J. Comput. Phs.*, 185:176–193, February 2003.
93. G. Meynet and A. Maeder. Stellar evolution with rotation. VIII. Models at $Z = 10^{-5}$ and CNO yields for early galactic evolution. *Astron. Astrophys.*, 390:561–583, August 2002.
94. S. Mukherjee, E. D. Feigelson, G. Jogesh Babu, F. Murtagh, C. Fraley, and A. Raftery. Three types of gamma-ray bursts. *Astrophys. J.*, 508:314–327, November 1998.
95. N. Murray and M. Holman. Diffusive chaos in the outer asteroid belt. *Astron. J.*, 114:1246–1259, September 1997.
96. J. F. Navarro, C. S. Frenk, and S. D. M. White. A universal density profile from hierarchical clustering. *Astrophys. J.*, 490:493, December 1997.
97. J. C. Niemeyer, W. Hillebrandt, and S. E. Woosley. Off-center deflagrations in chandrasekhar mass type IA supernova models. *Astrophys. J.*, 471:903, November 1996.
98. S. Nikolaev, M. D. Weinberg, M. F. Skrutskie, R. M. Cutri, S. L. Wheelock, J. E. Gizis, and E. M. Howard. A global photometric analysis of 2MASS calibration data. *Astron. J.*, 120:3340–3350, December 2000.
99. F. Ochsenbein, P. Bauer, and J. Marcout. The VizieR database of astronomical catalogues. *Astron. Astrophys. Suppl.*, 143:23–32, April 2000.
100. S. C. Odewahn, E. B. Stockwell, R. L. Pennington, R. M. Humphreys, and W. A. Zumach. Automated star/galaxy discrimination with neural networks. *Astron. J.*, 103:318–331, January 1992.
101. E. Owens. Electronic journal publishing seven years on: Is the revolution over or just beginning? In B. G. Corbin, E. P. Bryson, and M. Wolf, eds., *Library and Information Services in Astronomy IV (LISA IV)*, U.S. Naval Observatory, Washington, DC, p. 83, 2003.
102. W. S. Paciesas, C. A. Meegan, G. N. Pendleton, M. S. Briggs, C. Kouveliotou, T. M. Koshut, J. P. Lestrade et al. The fourth BATSE gamma-ray burst catalog (Revised). *Astrophys. J. Suppl.*, 122:465–495, June 1999.
103. C. Page. Astrogrid and data mining. In J.-L. Starck and F. D. Murtagh, eds., *Society of Photo-Optical Instrumentation Engineers (SPIE) Conference Series*, volume 4477 of *Society of Photo-Optical Instrumentation Engineers (SPIE) Conference Series*, Bellingham, WA, pp. 53–60, November 2001.
104. J. Pelt, N. Olspert, M. J. Mantere, and I. Tuominen. Multiperiodicity, modulations and flip-flops in variable star light curves. I. Carrier fit method. *Astron. Astrophys.*, 535:A23, November 2011.
105. A. Peraiah. *An Introduction to Radiative Transfer*. Cambridge University Press, Cambridge, U.K., December 2001.
106. R. K. Pina and R. C. Puetter. Bayesian image reconstruction—The pixon and optimal image modeling. *Pub. Astron. Soc. Pac.*, 105:630–637, June 1993.
107. Planck Collaboration, P. A. R. Ade, N. Aghanim, M. Arnaud, M. Ashdown, J. Aumont, C. Baccigalupi, A. Balbi et al. Planck early results. VIII. The all-sky early Sunyaev-Zeldovich cluster sample. *Astron. Astrophys.*, 536:A8, December 2011.
108. S. M. Ransom, S. S. Eikenberry, and J. Middleditch. Fourier techniques for very long astrophysical time-series analysis. *Astron. J.*, 124:1788–1809, September 2002.
109. R. O. Redman. Virtual observatory interfaces to CADC data. In R. Kothes, T. L. Landecker, and A. G. Willis, eds., *Astronomical Society of the Pacific Conference Series*, volume 438 of *Astronomical Society of the Pacific Conference Series*, Chicago, IL, p. 409, December 2010.
110. D. C. Richardson, T. Quinn, J. Stadel, and G. Lake. Direct large-scale N-body simulations of planetesimal dynamics. *Icarus*, 143:45–59, January 2000.
111. J. D. Scargle. Studies in astronomical time series analysis. II-Statistical aspects of spectral analysis of unevenly spaced data. *Astrophys. J.*, 263:835–853, December 1982.

112. J. D. Scargle. Studies in astronomical time series analysis. V. Bayesian blocks, a new method to analyze structure in photon counting data. *Astrophys. J.*, 504:405, September 1998.
113. J. Schou and D. L. Buzasi. Observations of p-modes in α Cen. In A. Wilson and P. L. Pallé, eds., *SOHO 10/GONG 2000 Workshop: Helio- and Asteroseismology at the Dawn of the Millennium*, volume 464 of *ESA Special Publication*, Paris, France, pp. 391–394, January 2001.
114. J. Skilling. Massive inference and maximum entropy. In G. J. Erickson, J. T. Rychert, and C. R. Smith, eds., *Maximum Entropy and Bayesian Methods*, Kluwer, Dordrecht, The Netherlands, p. 1, 1998.
115. G. F. Smoot, C. L. Bennett, A. Kogut, E. L. Wright, J. Aymon, N. W. Boggess, E. S. Cheng et al. Structure in the COBE differential microwave radiometer first-year maps. *Astrophys. J. Lett.*, 396:L1–L5, September 1992.
116. J. M. Stone. Astrophysical magnetohydrodynamics. *Bull. Astronom. Soc. India*, 39:129–143, March 2011.
117. M. C. Storrie-Lombardi, O. Lahav, L. Sodre, Jr., and L. J. Storrie-Lombardi. Morphological classification of galaxies by artificial neural networks. *Mon. Not. Royal Astron. Soc.*, 259:8P, November 1992.
118. V. Suyal, A. Prasad, and H. P. Singh. Nonlinear time series analysis of sunspot data. *Solar Phys.*, 260:441–449, December 2009.
119. T. C. Torgersen and D. W. Tyler. Practical considerations in restoring images from phase-diverse speckle data. *Pub. Astron. Soc. Pac.*, 114:671–685, June 2002.
120. G. Torres, F. Fressin, N. M. Batalha, W. J. Borucki, T. M. Brown, S. T. Bryson, L. A. Buchhave et al. Modeling Kepler transit light curves as false positives: Rejection of blend scenarios for Kepler-9, and validation of Kepler-9 d, A super-earth-size planet in a multiple system. *Astrophys. J.*, 727:24, January 2011.
121. G. Torres, M. Konacki, D. D. Sasselov, and S. Jha. Testing blend scenarios for extrasolar transiting planet candidates. II. OGLE-TR-56. *Astrophys. J.*, 619:558–569, January 2005.
122. J. K. Truelove, R. I. Klein, C. F. McKee, J. H. Holliman, II, L. H. Howell, and J. A. Greenough. The jeans condition: A new constraint on spatial resolution in simulations of isothermal self-gravitational hydrodynamics. *Astrophys. J. Lett.*, 489:L179, November 1997.
123. A. Udalski. The optical gravitational lensing experiment. Real time data analysis systems in the OGLE-III survey. *Acta Astronom.*, 53:291–305, December 2003.
124. G. A. Wade, S. Bagnulo, O. Kochukhov, J. D. Landstreet, N. Piskunov, and M. J. Stift. LTE spectrum synthesis in magnetic stellar atmospheres. The interagreement of three independent polarised radiative transfer codes. *Astron. Astrophys.*, 374:265–279, July 2001.
125. R. V. Wagoner, W. A. Fowler, and F. Hoyle. On the synthesis of elements at very high temperatures. *Astrophys. J.*, 148:3, April 1967.
126. M. D. Weinberg and N. Katz. Bar-driven dark halo evolution: A resolution of the cusp-core controversy. *Astrophys. J.*, 580:627–633, December 2002.
127. S. Weinberg. *Cosmology*. Oxford University Press, Oxford, UK, 2008.
128. P. R. Weissman, S. C. Lowry, and Y.-J. Choi. Photometric observations of Rosetta target asteroid 2867 steins. *Astron. Astrophys.*, 466:737–742, May 2007.
129. M. Wenger, A. Oberto, F. Bonnarel, M. Brouty, C. Bruneau, C. Brunet, L. Cambresy et al. The new version of SIMBAD. In S. Ricketts, C. Birdie, and E. Isaksson, eds., *Library and Information Services in Astronomy V*, volume 377 of *Astronomical Society of the Pacific Conference Series*, Chicago, IL, p. 197, October 2007.
130. T. R. White, T. R. Bedding, D. Stello, D. W. Kurtz, M. S. Cunha, and D. O. Gough. Variability in mode amplitudes in the rapidly oscillating Ap star HR 1217. *Monthly Notices Royal Astron. Soc.*, 415:1638–1646, August 2011.
131. M. Woitaszek, T. Metcalfe, and I. Shorrock. AMP: A science-driven web-based application for the TeraGrid. *ArXiv e-Prints*, November 2010.
132. S. E. Woosley, A. Heger, and T. A. Weaver. The evolution and explosion of massive stars. *Rev. Modern Phys.*, 74:1015–1071, November 2002.

133. P. A. Woudt, B. Warner, D. O'Donoghue, D. A. H. Buckley, M. Still, E. Romero-Colemero, and P. Väisänen. Dwarf nova oscillations and quasi-periodic oscillations in cataclysmic variables—VIII. VW Hyi in outburst observed with the Southern African Large Telescope. *Monthly Notices Royal Astron. Soc.*, 401:500–506, January 2010.
134. J. Yoo, A. Gray, J. Roden, U. M. Fayyad, R. R. de Carvalho, and S. G. Djorgovski. Analysis of digital POSS-II catalogs using hierarchical unsupervised learning algorithms. In G. H. Jacoby and J. Barnes, eds., *Astronomical Data Analysis Software and Systems V*, volume 101 of *Astronomical Society of the Pacific Conference Series*, San Francisco, CA, p. 41, 1996.
135. D. G. York, J. Adelman, J. E. Anderson, Jr., S. F. Anderson, J. Annis, N. A. Bahcall, J. A. Bakken, SDSS Collaboration et al. The sloan digital sky survey: Technical summary. *Astron. J.*, 120:1579–1587, September 2000.
136. Q. Zhang, S. M. Fall, and B. C. Whitmore. A multiwavelength study of the young star clusters and interstellar medium in the antennae galaxies. *Astrophys. J.*, 561:727–750, November 2001.

29

Computational Chemistry

J. Phillip Bowen
Mercer University

Frederick J. Heldrich
College of Charleston

Clyde R. Metz
College of Charleston

Shawn C. Sendlinger
North Carolina Central University

Neal E. Tonks
College of Charleston

Robert M. Panoff
Shodor and the National Computational Science Institute

29.1 Introduction

Computational chemistry is now firmly established as an integral part of chemistry at all levels. It has made its way into introductory level textbooks, and it has become a standard tool in many research applications. However, it is still a work in progress. The use of computational programs cannot be expected to provide a perfect solution, but it almost always affords chemists better insights into chemical processes and theoretical issues than would otherwise be available.

In this chapter, we present perspectives from generalists who use computational methods to solve or understand complex problems, and we provide perspectives from specialists who have made careers out of improving or creating new computational methods for their own uses and use by others. This is reflective of the state of the art in the use of computational chemistry. It has progressed from being an area of use only dared by those most intimately aware of the underlying programming and mathematical theory to those who recognize the value and limits of the methods provided in readily available resources.

29.2 Computational Resources

Computational chemistry requires computational resources that are varied and in proportion to the specific problem at hand. The speed, power, and location of the resources needed to conduct computational investigations can also depend on the purpose of the computation, be it education or research. Software licensing issues also impact hardware decisions; hence, decisions require consideration of many factors. Complicating this is the reality that advances in computational chemistry and the state of computational resources are not static; there is a constant need for new upgrades to software and hardware—not just to avoid obsolescence but to ensure the reliability of the computing platform. In most cases, trade-offs must be made to balance time, cost, and accessibility, so there are no absolutes except for "there are no absolutes."

29.2.1 Software Impact on Hardware Decisions

This discussion could just as easily been titled, "Hardware Impact on Software Decisions," but more frequently than not the software decisions are made sooner, based on the desired chemistry application. Since they have been described elsewhere in some detail, for purposes of illustrating the process of choosing hardware based on a prior software decision, consider the trade-offs between Gaussian, MOPAC, and GAMESS, assuming the user is looking for the highest efficiency generated by their computation and communication technology budget. (It should be noted that software providers may argue with the gross generalizations made here, but the characterizations of performance are based on real tests on real hardware in a variety of settings and are therefore considered firm opinion of the authors.)

Given the kinds of problems that one would address with Gaussian, experience shows that this package requires substantial CPU (computing) and RAM (cache and memory) resources, but minimal I/O. In single processor mode then, hardware needed to support a Gaussian installation would be the fastest machine with the most memory that one can afford per processor, without regard for memory or network speed. Choices would range from a stand-alone personal computer or workstation to a loosely coupled cluster of many nodes and cores, each with sufficient memory to support an instance of Gaussian running independently. The licensing details of proprietary software per core or per cpu then will limit the number of copies of Gaussian that can be run simultaneously, but each run of the code will finish in optimal time.

For a single chemist running multiple jobs, or multiple students running a single job, having sufficiently rapid response will require enough computing to handle the "throughput" of many jobs rather than needing to apply multiple processors to accelerate the completion of any given large computation. The total computing infrastructure needed to support a Gaussian installation, therefore, depends on the number of jobs to be run and the average runtime of each job. On a single laptop/desktop/workstation computer, having the faster processor and most memory may keep a small research effort fully productive. For a class of users, a multiprocessor server with substantial core memory to enable multiple jobs to run at the same time would suffice. Faster turnaround times are achieved by getting more processors, assuming the cost of acquiring additional licenses for the software is within the budget. While Gaussian has some parallel capability, it is rather modest at the time this chapter is written, and few users report substantial benefit for single runs on multicore/many-core computer architectures.

Any given run of MOPAC has lower memory requirements along with a light I/O load when compared to Gaussian, but it still requires substantial computing resources. MOPAC, therefore, is best implemented on single machines, or very loosely coupled clusters of single machines. Again, throughput—running as many jobs as possible in the least amount of time rather than getting any single job to run faster—is achieved by having more processors available to distribute the workload of the multiple jobs. As with Gaussian, a well-provisioned individual computer could be fine for a small effort. Large numbers of runs by a single user or a class of students would require many more cores, so a multicore server or massive cluster with sufficient memory per core would be called for.

While the cost may be attractive for GAMESS, among the three exemplars, it has the greatest need for fast communication between the processor and disk memory, since it needs really fast disk access for scratch space. The computing requirements are still substantial, but each run requires less memory than other software packages. If run on a server with slow disks, even for smaller jobs, actual time to completion for multiple jobs running at the same time can be compromised.

29.2.2 Locality and Control

Besides the actual hardware decisions to support a satisfactory experience, there are other issues that impact getting real work done in real time, and whether the chemist must also augment deep knowledge of DFT and molecular orbital theory with substantial system programming, hardware optimization, HVAC, electrical loads, and grant writing.

For a modest research effort, "having your own" computing resource that is managed by the chemist could be sufficient. The user has universal access to the use of the computer as much or little as needed at any hour of the day. There are many chemists who resign themselves to having to run their own cluster. However, for them, the frustration of becoming more of a computing administrator than a chemist must be outweighed by maintaining control over their own computing resources. For a shared resource at the department, lab, or campus level, one has to trade away full-control accessibility for the lower workload and stress of maintaining the computing resources as an operational tool, including acquiring space in a building, system maintenance, storage backups, power, and cooling requirements. Many chemists engaged in education and research adopt a hybrid approach, having one computing infrastructure for their research and another for the class computing laboratories and exercises. Again, this usually requires one or more trade-offs starting with choice of software, size of the problems to be addressed, and expected time to completion.

A growing need is for more and more long-term storage of the results of one's computations. The National Science Foundation (NSF) is now mandating, for instance, a data management plan for all of its grants that requires grantees to assure long-term access to the results of funded research. In similar fashion to the raw computing needs, the need for storage and backup and online repositories is large and growing and has the same local versus control trade-off. The costs of providing this storage, with rare exception, have not been adequately addressed by most individuals, departments, or organizations.

A growing solution to the problem of acquiring substantially more computing resources to implement and support a computational chemistry effort has been to access resources on a network. This solution can range from free resources funded by the NSF for research and class accounts for education (http://www.gridchem.org, http://www.xceded.org) to cloud resources from commercial providers (e.g., http://www.amazon.com/ec2/). While ceding most control to someone else to provide the resource, these remote resources give substantial flexibility to grow or contract computing resources. In the case of NSF resources, many of the software licensing issues have been addressed, but commercial providers have more layers of licensing to go through, depending on the choice of software.

29.3 Computation in Chemical Education

Computational chemistry has transitioned from the developmental stages to become an essential tool used to understand the structure, reactivity, and spectroscopic details of the new compounds. It is fitting that chemistry education should mirror this (Gilbert, 2002). In the chemistry classroom, computation ranges from various forms of modeling (molecular and mathematical modeling, solving complex simulation problems, and molecular visualization/animation) to text and classroom supplements (homework and testing, computational tools, demonstrations and simulations, and interactive figures). Computation appears in all aspects of teaching laboratories: prelaboratory assignments (discussion of theoretical concepts and the proper use of equipment), simulated experiments and instrument use, and the use of computational tools for data analysis. The use of computers and electronic media for data storage/sharing in the academic laboratory as a replacement for the traditional written laboratory notebook is not yet commonplace, but it seems inevitable (especially given the NSF mandate for data management plans and the impending demand to train health-care professionals for mandated use of electronic medical records).

The use of computation helps students understand chemistry, keeps them more engaged, and takes advantage of their accumulated interest and familiarity with technology.

The quality and quantity of computational work by the student continues to increase as the available technology has become more robust. Today, students are using supercomputers, workstation clusters, desktop and laptop computers, classroom "clickers," "smart" phones, and tablet PCs. Common computational tools are readily available with access to classroom software often controlled through a "learning management system" (LMS), a local installation, or a third-party website. The software may be provided by the publisher of a specific textbook or from a more generic commercially available package. Inexpensive software is available in "special issue software collections" from the Journal of Chemical Education's JCE software. Some web-based software is available for use at no charge. Appropriate sites can most readily be found through a hosted collection such as the Computational Science Education Reference Desk (CSERD) or the Chemistry Education Digital Library (ChemEd DL) pathways to the National Science Digital Library (NSDL) (Sendlinger, 2008).

Computational chemistry education involves both pedagogical (chemical education) and disciplinary (computational chemistry) focus. In chemical education, developments include new approaches to the use of existing software, models, apps, visualizations, and other tools to enhance the educational environment. These materials, usually prepared by others, often include visualizations of phenomena occurring at the atomic or molecular level that cannot be easily relayed by other means (Gordon, 1995). In the arena of computational chemistry, the aim is to enable students to learn the process of using software to build models. Workshops and websites are available for this purpose (Sendlinger, 2011). The computational tools used usually include spreadsheets, mathematics software, and molecular modeling programs. The details of model building might require coding cells in a spreadsheet, writing a subroutine or complete program in a programming language, or learning to use advanced graphical user interface (GUI)–enabled molecular modeling software. Students building models also should learn about verification, validation, and accreditation (VVA) of their work. The verification process involves checking that the model is correct, runs properly, uses the correct algorithm, etc. Validation is the process of comparing calculated and experimental data. Accreditation insures that the model achieves the educational purpose and is appropriate for the intended audience.

29.3.1 Computational Chemical Education

A vast number of websites have been created to assist the chemistry educator. To help the educator find quality websites, the NSDL bills itself as the "nation's online portal for education and research on learning in Science, Technology, Engineering, and Mathematics (STEM)" (Zielinski, 2009). A wide variety of Pathway Partners of the NSDL focuses on providing content to specific types of user communities. A key resource for computational chemistry materials is the CSERD (Sendlinger, 2009). The goals of CSERD include assisting students in learning about computational science and helping teachers incorporate effective material into the classroom. CSERD account holders can suggest additions to the computational resources collected and can also sign up to become reviewers. Many of the resources include these reviews, based on the VVA process. The collection can be browsed by several means, and the search functionality offers a quick and convenient method for finding appropriate computational resources.

One example is the molecular model for an ideal gas. The corresponding CSERD site provides a description of the applet along with detailed reviews. In a classroom or laboratory setting, the instructor (with an LCD projector) or individuals/groups of students with computers can be assigned to collect volume (V) data while varying the pressure (P), number of particles (N), particle velocity (v), or the temperature (T). Appropriate graphs of the data reveal $V \propto 1/P$ under conditions of fixed N and v, $V \propto N$ under conditions of fixed P and v, and $V \propto v^2$ (or $V \propto T$) under conditions of fixed P and N. Thus, Boyle's, Avogadro's, and Charles's law can all be deduced. Combining these relationships with the postulates of an ideal gas, a derivation of $PV \propto nT$ can be completed. Instead of being told, the mathematical relationship in each law, the students have had the opportunity of discovering these laws for themselves. Recent work provides evidence for increased student conceptual understanding through the use of this type of inquiry-based approach (Bridle, 2012).

The American Chemical Society (ACS) provides the ChemEd DL as a Pathway Partner of the NSDL (Moore, 2009). This pathway includes resources from the JCE Digital Library, the ACS Education Division, and the ChemCollective project. High school lesson plans that include online resources, a Moodle-based course management system where personalized resources can be added, a question database for tests and homework, and a wide variety of visualization resources are available.

29.3.2 Computational Chemistry in Education: Molecular Modeling

Using available software, students build their own molecular models in order to better understand chemistry in both classroom work and research projects. Students can investigate energy differences of various molecular conformations, calculate and visualize dipole moments, look at molecular orbital energy levels and shapes, visualize electron distributions and electrostatic potentials to investigate reactivity, calculate vibrational and UV–Vis transition frequencies, identify transition structures and reaction pathways, determine NMR chemical shifts and coupling constants, etc. Educational uses of molecular modeling allow the student to visualize key chemical concepts that are difficult to share in another manner. Molecular modeling has become another tool to assist chemists in doing what they do, just as various forms of spectroscopy became important tools in the mid-twentieth century. Compared to spectroscopic techniques (Alexander, 1999), computation is a recent newcomer to the mix (McDougal, 2012). Technological advances such as WebMO software that uses the web browser as a computational interface have made research-grade computational tools readily available.

An example of the educational use of molecular modeling appropriate for general chemistry includes investigating different bond types. Geometry optimization followed by calculation of the electron density for H_2 (covalent bonding), HF (polar covalent bonding), and LiH (ionic bonding) provides the results summarized in Figure 29.1.

After learning about covalent, polar covalent, and ionic bonding, students can easily perform the calculations and visualize results such as those in Figure 29.1 to cement this knowledge and deepen their understanding. Visual learners also greatly benefit from this approach.

Another example of molecular modeling suitable for use in an organic chemistry class would be to calculate potential energy conformational profiles as in an examination of the rotational barrier for the C2–C3 bond in *n*-butane. The calculated results at the density functional theory (DFT) B3LYP/6–31G(d) level are shown in Figure 29.2 (experimental results in parentheses, units of kcal/mol).

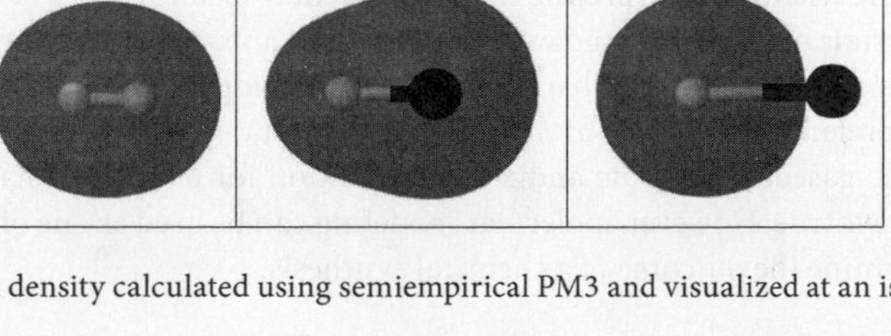

FIGURE 29.1 Electron density calculated using semiempirical PM3 and visualized at an isovalue of 0.01 e⁻/Å³.

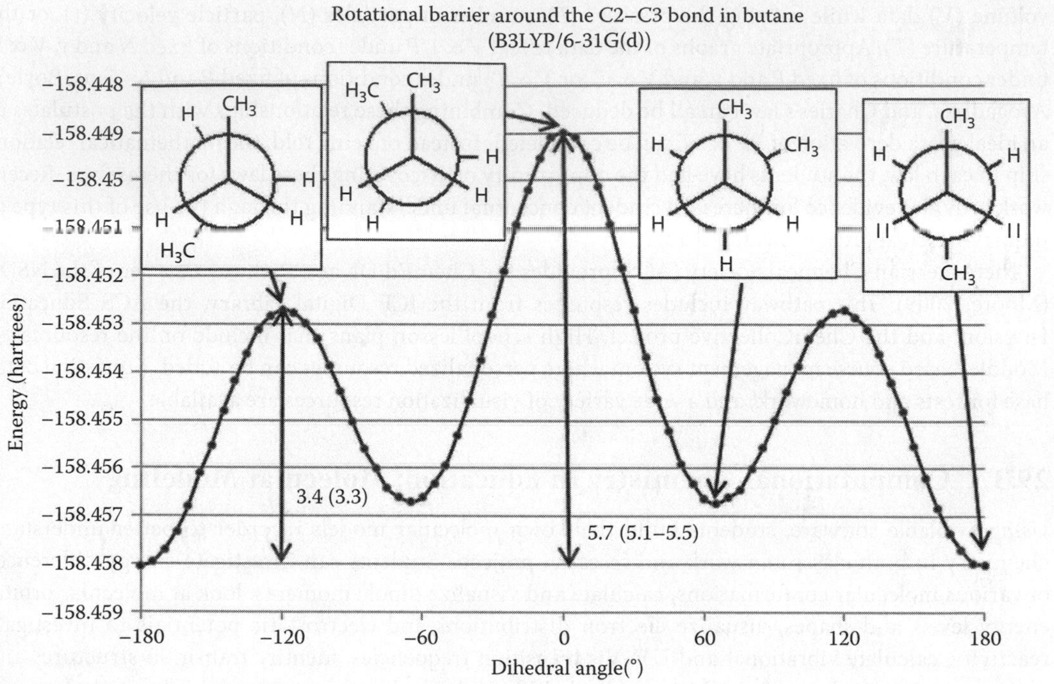

FIGURE 29.2 Energy versus dihedral angle for rotation about the C2–C3 bond in butane, obtained using B3LYP/6-31G(d).

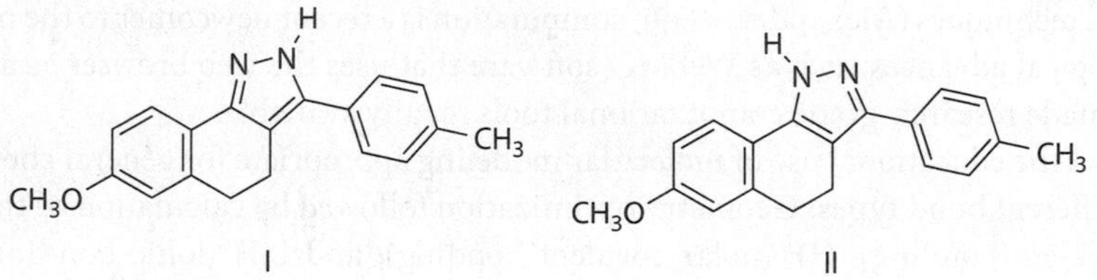

FIGURE 29.3 Tautomeric isomers from research project.

Students at the undergraduate level can also become involved in research projects that utilize computation. For example, an undergraduate research student attempted the synthesis of I but considered that tautomer II might also be produced during the synthesis (Knight, 2007). A clear distinction between these tautomers (see Figure 29.3) is difficult to determine from elemental analysis, nuclear magnetic resonance (NMR) spectroscopy, and infrared (IR) spectroscopy.

A molecular model of each tautomer was constructed and the energy was minimized using B3LYP/6-31G(d). The energy minimization was followed by a vibrational calculation at the same level to assure that a minimum energy structure had been attained. The results showed that I was more stable than II—approximately 0.9 kcal/mol lower in energy than II. Theoretical calculations of NMR and IR spectra for I at the same level were useful in interpreting the experimental spectra.

Suitable single crystals of the compound were obtained and an x-ray diffraction dataset was collected. That structure analysis, available free of charge from the Cambridge Crystallographic Data Centre, indicated that I was the preferred structure. In this case, an assumption was made that results of molecular modeling for a single gaseous molecule and x-ray diffraction for a solid crystal will agree with each other. This is not always true. However, molecular modeling can be used as one of the tools just as NMR, IR, etc., to help determine the outcome of a chemical synthesis.

29.3.3 Computational Chemistry in Education: Mathematical Modeling

Pedagogical applications of mathematical modeling in chemistry include statistical analysis of data, database searching, and reaction mechanisms in chemical kinetics. The software programs associated with these topics range from highly specialized programs to simple spreadsheets. Student involvement with this software can be a straightforward use of preexisting applications to record data and print results, or it can involve having students create an application from "scratch."

A mechanism, sometimes referred to as a chemical reaction sequence (CRS), consists of a theoretical series of elementary steps that describes the interactions of the various species involved in the chemical reaction. Based on the law of mass action, differential rate equations can be written for each species involved in the CRS. To decide whether or not a proposed CRS is an acceptable description of the chemical reaction, either the set of differential rate equations needs to be converted into concentration–time information that can be compared directly with experimental data or the experimental dataset needs to be converted into rate information using a mathematic method or a graphical technique (e.g., a Lineweaver–Burk plot).

There are two major approaches used to convert the differential rate equations from a CRS into the time dependence of concentration. The first approach is to either derive the integrated form of the differential form or locate the solution in a suitable reference (Andraos, 1999, 2008). Robust mathematics programs such as Mathcad, Mathematica, MATLAB®, and Sage have various differential equation solvers that might be successful. Some of these solvers are capable of working with systems of stiff differential systems. The second approach is to solve the differential equations using numerical integration. Numerical integration can be performed using any of these four mathematics programs or for simple systems, using a spreadsheet.

There is another group of programs (e.g., Vensim PLE, Berkeley Madonna, and STELLA) known as system dynamics software that is capable of performing numerical integrations. These programs range from being a free download to being rather expensive. Some of these programs are able to handle sets of stiff differential equations. The numerical integration methods generally range from Euler's method to the Runge–Kutta 4 method and results are available in graphical and tabular formats. The modeling of the differential equations usually uses rectangles (known as box variables, stocks, or reservoirs) to represent quantities of substances, circles (known as variables or constants) to represent time independent quantities, arrows (known as connectors) to represent mathematical connections of various quantities, and a pipe/valve symbol (known as a rate or flow) to represent the flow of substances into or out of box variables. Most mathematical programs, spreadsheets, and system dynamics programs permit adding scroll bars (or sliders) to control the values of the rate constants k_{ij} so that the values can be changed conveniently for instructional purposes or for matching experimental data to the theoretical results.

The system dynamics software can be used to describe complex kinetics systems such as competing reactions, higher-order reactions, Michaelis–Menten systems, and cyclic reaction systems such as the Belousov–Zhabotinskii reaction. In addition, numerical integration calculations can be used to generate acid–base titration curves, perform general and limiting reagent system stoichiometric calculations, and predict various thermodynamic properties (U–U^0, C_V, S, A) using the Debye theory for crystals containing monatomic species. There are many more computational tools available to the researcher in chemical kinetics. Often, these are based on Monte Carlo methods involving probabilities of energy, position, etc. Available as a free download, Chemical Kinetics Simulator is a software package that is based on a stochastic algorithm and is capable of handling stiff systems. This program can model complex reactions such as explosions, systems in which volume changes occur during the reaction, and very small systems.

As one example, consider the use of system dynamics software to model a two-step consecutive, reversible unimolecular reaction. Such a reaction is described as compound A being converted into compound B, then compound B being converted into compound C. This two-step mechanism would be described by the rate equation shown in Figure 29.4, where the equations for the time dependencies of A, B, and C are as described.

$$A \underset{k_{BA}}{\overset{k_{AB}}{\rightleftarrows}} B \underset{k_{CB}}{\overset{k_{BC}}{\rightleftarrows}} C$$

$$\frac{d[A]}{dt} = -k_{AB}[A] + k_{BA}[B]$$

$$\frac{d[B]}{dt} = k_{AB}[A] - k_{BA}[B] - k_{BC}[B] + k_{CB}[C]$$

$$\frac{d[C]}{dt} = k_{BC}[B] - k_{CB}[C]$$

FIGURE 29.4 Mathematical relationships for modeling a simple CRS system. The use of [] indicates the concentration of a value: [A] = concentration of A.

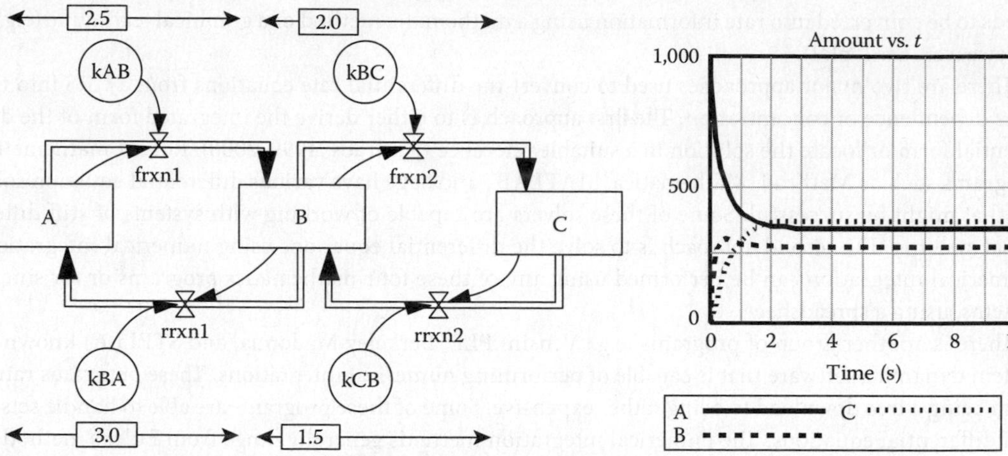

FIGURE 29.5 A system dynamics model to study chemical kinetics.

Although available in the literature, the integrated concentration–time equations are rather complicated. However, the system dynamics model is straightforward, as seen in Figure 29.5. This system dynamics model can be used to describe less complicated systems by setting the values of the respective rate constants equal to zero. For example, setting $k_{BC} = k_{CB} = k_{BA} = 0$ gives the model for a unimolecular forward reaction A to B. Setting $k_{CB} = k_{BA} = 0$ gives the model for a consecutive, nonreversible unimolecular system A to B and B to C. Or setting $k_{BC} = k_{CB} = 0$ gives a reversible CSR system A to B and B to A for which equilibrium studies are possible.

Note that the integrated concentration–time equations become more complicated as the number of steps in the consecutive reaction system increases. Analytical expressions for the integrated rate laws are available for *n* consecutive steps, but the equations cited in the literature are difficult to use and often contain typographical errors. It is easy to add additional sets of reservoirs, constants, and flows to the system dynamics model to describe the exact system under consideration.

29.4 Chemical Principles of Absorption Spectroscopy

The use of ultraviolet–visible (UV–Vis) spectroscopy is one of the oldest forms of characterization of compounds. In the visible range (400–800 nm), the absorption of light gives rise to complimentary colors characteristic of dyes and pigments. The absorption of both visible and ultraviolet light (200–400 nm) provides a means of nondestructive characterization that requires only a very small quantity of a compound to yield both quantitative and qualitative information. Two of the major disadvantages of characterization by UV–Vis spectroscopy are the low resolution of the observed spectrum and the relatively poor fit between experimental observation and theoretical prediction. These limitations are especially evident when compared to the advances made over the past 50 years in matching experimental spectra and theoretical simulations for other common spectroscopic techniques such as IR, Raman, and nuclear magnetic resonance.

The chemical basis of the UV–Vis characterization begins with the chromophore. Organic chromophores possess conjugated pi systems that can readily absorb UV–Vis energy light. Inorganic chromophores may also possess d/f-type electrons that can accomplish the same result. In either case, the general process of the UV–Vis experiment is that the incident light results in promotion of an electron in the chromophore from a position of low energy known as the ground-state electronic configuration into a higher-energy electronic configuration, the excited state. The relaxation of the excited state to the lower-energy ground state is sometimes associated with the emission of UV energy that gives rise to two other spectrophotometric characterizations (fluorescence and phosphorescence). The characteristics of those relaxation processes are associated with many fundamental studies of biological systems.

A particular chromophore is associated with an ability to absorb light of a particular wavelength. But structural alterations to the chromophore can cause relatively large alterations to those initial excitation wavelength expectations. For example, this might be brought about by extending the length of the pi system in an organic system or changes of the conformation of the chromophore.

The chemical problem of understanding the UV–Vis process is further compounded by the fact that the absorption of light is most often measured for compounds in solution. This leads to slight variations in absorption wavelengths due to the interactions of solvent molecules with the solute molecules that are absorbing the light. It also results in the presence of multiple conformational structures of the same compound in solution at the same time, where each conformer can have a different absorption. And a final consideration is that compounds do not absorb light with equal efficiency. Experimentally, the efficiency with which a compound absorbs UV–Vis light is a quantifiable measure called molar absorptivity.

The absorption of light in organic compounds involves molecular orbitals, where electrons from highest energy–filled molecular orbitals (HOMO) or orbitals with energies just below that of the HOMO (designated as HOMO-1 or HOMO-2, etc.) are promoted to the lowest energy–unoccupied molecular orbital (LUMO) or orbitals just above the energy of the LUMO (designated as LUMO+1, LUMO+2, etc.). For organic compounds, this generally involves molecular orbitals associated with pi bonds and lone pairs.

29.4.1 Computational Approaches to Predicting UV–Vis Spectra

Initial approaches to the prediction and correlation of observed UV–Vis spectra to chemical structures were developed in the 1950s and are based entirely on experimental correlations. Once a common feature of intermediate level spectroscopic analysis course texts (Shriner, 1998; Silverstein, 1991), the results of chemists like Woodward, Hoffman, and Smith established nominal excitation values for various structural organic chromophores (e.g., dienes, enones, and benzenoid aromatic compounds) and a set of rules for increases or decreases from those values based on well-defined placements of specific auxochromes, the extension of the conjugated systems, and corrections for use of different solvents. These correlation rule predictions are still the best predictor for simple organic compounds of an absolute value for the key characteristic feature of the UV–Vis spectrum of a compound, the lambda max. However, the use of these rules cannot be extended to new systems.

One of the two most widely used computational methods to predict UV–Vis spectra was developed initially by Zerner and is called the ZINDO method (Ridley, 1973). The ZINDO method is a semiempirical approach based on application of molecular orbital theory. The primary strengths of the ZINDO method for absorption spectroscopy are its speed and reliability. With greater speed comes the ability to predict values for large molecules in reasonable time periods (minutes to hours using common desktop single-core machines). The primary computational limitation of the ZINDO method is that it is applicable only to molecules composed of atoms for which spectroscopic INDO/S parameters are provided (H, Li, B, C, N, O, F, P, S, Cl, Sc, Ti, V, Cr, Mn, Fe, Co, Ni, Cu, and Zn). A recent improvement of the ZINDO/S method, ZINDO-MN (Zerner, 2011) provides enhancements for solvent molecule interactions, partial charges, and inclusion of better parameters for oxygen.

The other commonly used computational method for predicting UV–Vis spectra is the time-dependent DFT, called TD-DFT (Lewars, 2011; Marques, 2003). TD-DFT is often considered to be an *ab initio* method, but that it is not a true *ab initio* method is important in the context of spectroscopic predictions. TD-DFT does make certain assumptions in order to make the computation effort manageable. Key among these assumptions is distinguishing between valence electrons that are involved in bonding and the core electrons that are not involved in bonding. As the principal quantum number of an atom increases, the ratio of valence to core electrons exhibits a significant decline. Since it is generally agreed that the UV–Vis excitation involves the valence electrons, this is not considered to be a significant issue. In most respects however, TD-DFT is considered to be an *ab initio* method. Since TD-DFT is not limited to specific atoms, it is not limited by molecular composition. Its primary limitation is that it is a more expensive computation (requiring hours to days with a desktop computer).

29.4.2 Computational Chemistry Impact on Use of UV–Vis

Despite the elegance and sophistication of the semiempirical and *ab initio* computational methods, the older empirical rules are superior for accurately predicting the lambda max with simple systems, as, for example, with acetophenones (Doig, 2008). The ZINDO and TD-DFT methods are both generally reliable in correctly predicting relative lambda max absorbances of a related series—although the base prediction of wavelength is expected to be shorter than the experimental value in most cases because of a systematic overestimation for the different in energy between the HOMO and LUMO orbitals. However, there are cases where these methods result in predictions of longer wavelength, illustrating the need to calibrate the results to known values for any computational methods.

Both computational methods continue to be used effectively to predict UV–Vis spectral data in a variety of interesting applications ranging from characterizations of potential new dyes (Fleming, 2011) to art history (Fantacci, 2010) to the design of tunable initiators for radical polymerization processes (Jankowiak, 2009). There are numerous studies where either method has been shown to be effective; one such example dealt with substituted coumarins, compounds used as laser dyes (Xu, 2009). One issue in considering the choice of ZINDO or TD-DFT is intramolecular charge transfer (ICT). The successful study of oxazine dyes using TD-DFT is particularly illustrative of the difficulty in dealing with ICT and the need to carefully consider the possibility of failure of the TD-DFT method in such cases (Fleming, 2011). Other systems with large ITC characteristics are dealt with adequately with ZINDO, as in the case of 5-(3-indolyl)oxazoles (Grotkopp, 2011) where the structure shown in Figure 29.6 had a calculated absorbance predicted at 344 nm and an observed absorbance at 306 nm. The HOMO coefficients were most centered the oxazole extending into the indole ring, and the LUMO coefficients were most centered on the aromatic ring attached to the indole at the 2 position extending into the indole ring,

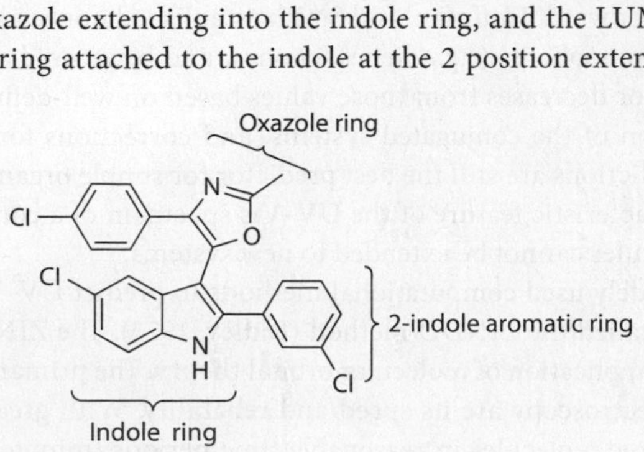

FIGURE 29.6 Substituted 5-(3-indolyl)oxazole.

indicative of ITC. Numerous studies that provide direct comparisons of TD-DFT and ZINDO methods generally have led to the following conclusions. ZINDO is preferred when speed is important, the major expected transitions include n to p* or p to p* transitions, and ITC is a dominant factor. TD-DFT is preferred when economy is not essential, if ITC is not problematic and a close fit to experimental data is desired, or if atom parameterization is not available in ZINDO.

UV–Vis is most useful in structural characterization of species that are found in solution but not easily isolated or stable out of solution. Figure 29.7 illustrates structures associated with two recent examples of this application. The first example is a use of ZINDO to differentiate intermolecular charge transfer alignment of electron donor in a species like pyridine with an electron acceptor like chloranil (Pandey, 2011). In this example, two prominent coordination geometries are proposed, a T-shaped structure where the primary electron donation from pyridine is derived from the lone pair (n type) electrons on nitrogen and the other geometry largely results in electron donation from the aromatic ring electrons (p type). As part of the solution to this differentiation, the authors compared observed and computationally predicted transitions related to excitation of n and p electrons from pyridine to the LUMO π* orbitals of chloranil (Pandey, 2011). These results (observed transitions 374 nm for π to π* and 466 nm for n to π*, T-shaped predicted absorptions 310 and 448 nm, cofacial predictions 380 and 435 nm) pointed, unexpectedly, toward the cofacial geometry. The computational work used self-consistent reaction field (SCRF) to model the influence of solvent on the absorption predictions.

In another application for proof of structure, the ZINDO-MN program available in Gaussian was used in order to identify a transient chromophore generated in the oxidation of *para*-aminophenol (Lerner, 2011). This required predicted absorptions of four proposed structural candidates for comparison to the observed solution chromophore, Chromogen 550. The UV–Vis predictions (550 and 366 nm) were a far better fit for both the longer and shorter wavelengths absorptions (experimentally observed values were 550 and 380 nm; the next closest computational fits were for a different proposed structure, 436 and 343 nm). The problem was compounded by the fact that there are actually two different structures present in solution, both of which have absorbance at 380 nm. The less stable chromophore, Chromogen 550, had absorptions at 550 and 380 nm.

The final research application involves the use of computational ZINDO UV–Vis spectral predictions as a key element in the *de novo* synthesis of new materials designed for use as photovoltaics

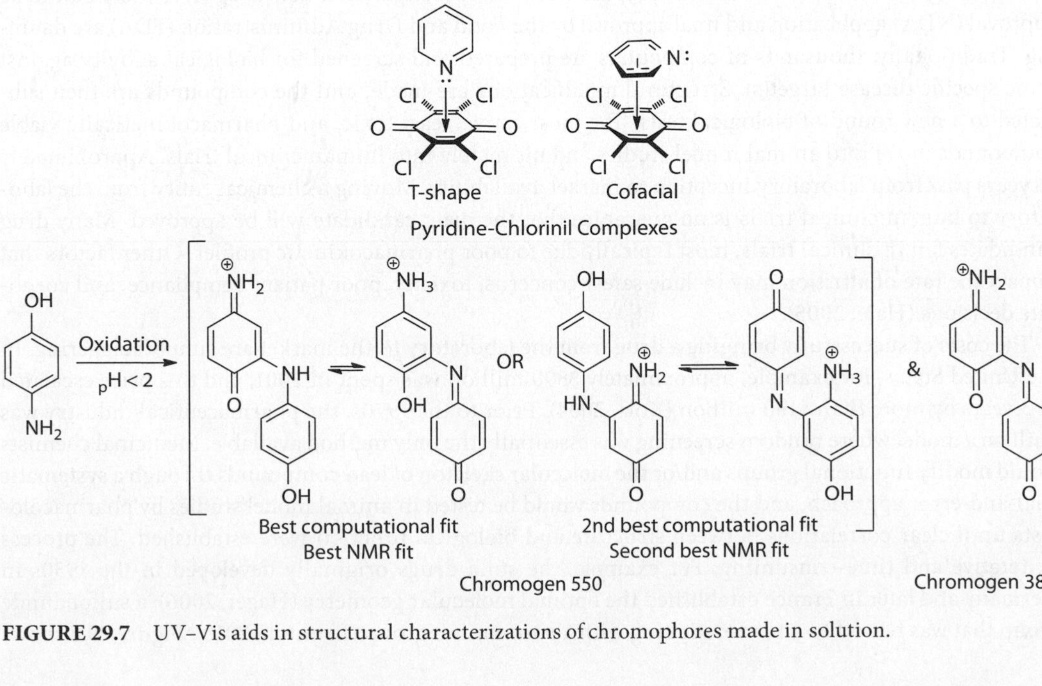

FIGURE 29.7 UV–Vis aids in structural characterizations of chromophores made in solution.

used in solar energy production (Boyle, 2011). These authors performed a computational investigation of 132 monomers that could be cross-coupled to produce over 90,000 oligomeric copolymers with potential application as photovoltaic materials. The compounds were refined geometrically using a fast empirical MMFF94 optimization, then by a PM6 semiempirical method. The authors noted that a more robust geometry optimization might have yielded better UV–Vis predictions because the semiempirical methods generally yield geometries that are underrepresentative of the extent of pi conjugation (the geometries are not as planar as seems evident based on experimental data). However, a subset of 60 oligomers were used to calibrate the UV–Vis spectral predictions to known experimental absorptions (~ +40 to −4 nm variance, with only two predictions of lower energy than the experimental values) indicating that the computational predictions could be used as a starting point to guide selecting the best candidates for synthetic consideration. As interest in materials science grows, the demands for improved computational methods to accurately predict molecular properties such as optical absorption will increase.

29.5 Computational Medicinal Chemistry

The development of new, safe, and effective drugs is one of the most interesting and challenging areas of research today. Scientists involved in drug discovery are at the interface of chemistry and the biomedical sciences. Medicinal chemistry is therefore inherently an interdisciplinary field. Medicinal chemistry includes many of the traditional branches of chemistry (organic, inorganic, physical, analytical, and biochemistry) coupled with pharmacology, physiology, anatomy, statistics, biophysics, and computer sciences. Historically, in collaboration with pharmacologists, medicinal chemists were primarily synthetic organic chemists who prepared novel molecules with desired biological activities. Today, medicinal chemistry may be more broadly defined to include virtually any chemically oriented endeavor related to drug discovery. Pharmaceutical companies, however, still may divide scientists into organizational sections emphasizing their specialties.

29.5.1 Critical Issues in Medicinal Chemistry

The challenges that stand between the preparation of an investigational new drug (IND) to a new drug approval (NDA) application and final approval by the Food and Drug Administration (FDA) are daunting. Traditionally, thousands of compounds are prepared and screened for biological activity against some specific disease target(s). Structural modifications are made, and the compounds are then subjected to a new round of biological tests. The most active, least toxic, and pharmacokinetically viable compounds move into animal model studies and ultimately into human clinical trials. Approximately 14 years pass from laboratory inception to market availability. Moving a chemical entity from the laboratory to human clinical trials is no guarantee that the drug candidate will be approved. Many drug candidates fail in clinical trials, most typically due to poor pharmacokinetic profiles. Other factors that impact the rate of attrition may include safety concerns, toxicity, poor patient compliance, and corporate decisions (Hale, 2005).

The costs of successfully bringing a drug from the laboratory to the market are simply staggering. In the United States, for example, approximately $800 million was spent in 2001, and this price escalated in 3 years by more than $100 million (Kola, 2004). Prior to the 1970s, the pharmaceutical industry was built on a model where random screening was essentially the only method available. Medicinal chemists would modify functional groups and/or the molecular skeleton of lead compounds through a systematic trial-and-error approach, and the compounds would be tested in animal model studies by pharmacologists until clear correlations between structure and biological function were established. The process is iterative and time-consuming. For example, the sulfa drugs originally developed in the 1930s in Germany and later in France established the optimal molecular geometry (Hager, 2006): a sulfonamide group that was bound to an aromatic ring with a *para*-amino group (Brunton, 2010). Figure 29.8 shows

Sulfanilamide   *p*-Aminobenzoic acid (PABA)

FIGURE 29.8 Sulfanilamide, an antimetabolite and the first of the sulfa drugs, is shown in a side-by-side comparison with *p*-aminobenzoic acid.

the structure of sulfanilamide and *p*-aminobenzoic acid (PABA). Sulfanilamide resembles PABA and prevents this metabolite from being used in the synthesis of folic acid. Microorganisms generate their own folic acids, whereas humans get folic acid through the consumption of foods.

29.5.2 Drug Design in Medicinal Chemistry

From the perspective of the practicing medicinal chemist, the combination of the 3D arrangement of functional groups of a molecular structure that is required to produce the desired biological response is known as the pharmacophore. The concept was developed by Paul Ehrlich who many considered to be the father of medicinal chemistry. The knowledge gained about the correlation of biological activity with molecular structure was important. As the biochemical basis of drug action became better understood over time, a more "rational" approach to drug discovery was undertaken.

The objective of rational drug design was to develop a drug by understanding the mechanism of action. A classic example is the discovery and development of captopril (Capoten®), the first FDA-approved angiotensin-converting enzyme (ACE) inhibitor (Cushman, 1977; Ondetti, 1977, 1982). Captopril along with the many fast follow-up drugs revolutionized hypertension therapy. ACE, a Zn(II)-containing enzyme, is responsible for the peptide bond cleavage of two terminal amino acids of the decapeptide angiotensin I to yield the octapeptide angiotensin II. Angiotensin II is a potent vasoconstrictor. In the 1970s, it was hypothesized that inhibition of angiotensin II would be an effective treatment for hypertension and cardiovascular diseases. The story of the development of captopril is an exciting precomputer exercise in rational drug design (Silverman, 2004). At the Royal College of Surgeons, Vane, Ferreira, and Bakhle discovered that the venom of the Brazilian pit viper (*Bothrops jararaca*) was a potent vasodilator (Bakhle, 1968, 1969; Ferreira, 1971). At the University of North Carolina at Chapel Hill, research by Byers and Wolfenden demonstrated that (*R*)-2-benzylsuccinic acid was an inhibitor of carboxypeptidase A (Byers, 1972, 1973). Ondetti, Cushman, and coworkers at Squibb (now Bristol-Myers Squib, BMS) skillfully merged the knowledge of small peptides ACE inhibitors with small molecule inhibitors of carboxypeptidase A to generate captopril (Figure 29.9).

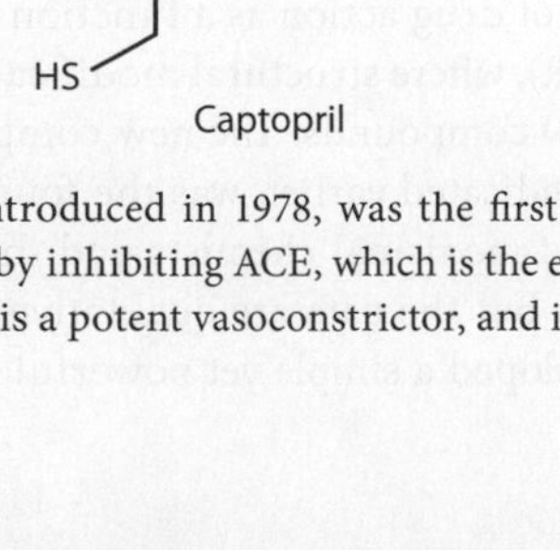

Captopril

FIGURE 29.9 Captopril (Capoten®), introduced in 1978, was the first FDA-approved angiotensin-converting enzyme (ACE) inhibitor. The drug works by inhibiting ACE, which is the enzyme responsible for the conversion of angiotensin I to angiotensin II. The latter is a potent vasoconstrictor, and its use ushered in a new age in hypertension therapy.

29.5.3 Computational Chemistry in Drug Design

Any methodology that would allow pharmaceutical companies to bring a drug to the market more rapidly has huge financial rewards as well as the obvious benefits of a potential new therapeutic treatment for patients. Since the first commercial computers became available in the 1950s, there has been an explosion of computer-based applications in chemistry. Most of the fundamental concepts of quantum mechanics (QM) and molecular mechanics (MM) were developed or outlined in the 1920s and 1940s, respectively. The methodology of treating a system of particles classically with mathematical physics was worked out in the days of Sir Isaac Newton, and this serves as the foundation of modern molecular dynamics (MD) simulations.

The ability to solve complex equations through computations truly has ushered in a new age in science, and the impact in drug discovery has been no less dramatic (Bowen, 2004). The field of computer-assisted drug design (CADD) emerged in the 1970s with the introduction of quantitative structure–activity relationships (QSAR) by Corwin Hansch (Hansch, 1995). Currently, the *in silico*–based methods available to medicinal chemists are far-reaching, but drug discovery is ultimately an experimental science involving *in vitro* biological assays, *in vivo* animal testing, human clinical trials, and long-term postapproval studies. Computer-based methods are clearly reducing experimental efforts, but not replacing them. It is better to view CADD as an exciting and robust set of tools to help medicinal chemists understand drug–receptor interactions at the molecular level and make predictions about biological activity.

29.5.4 Pharmacophore Development and QSAR

The ability to understand interactions at the molecular level with computer-based methodologies is the basis of CADD. There is a full range of potential strategies that may be useful when approaching a drug design problem. The specific methods that may be used range from traditional computational chemistry (e.g., MM, QM, and MD simulations, which are described elsewhere in this chapter) to information-based methods (e.g., QSAR, 3D-QSAR, and database searching that are described herein). In many cases, the methodologies that are used to tackle a problem are a function of the scientists who have been hired, their backgrounds, resources, and available data. The corporate cultural environment of each pharmaceutical company also plays a role. For example, Merck seems to have a greater overall focus on structure-based drug discovery approaches compared to Pfizer, where the focus has been on the development of leads generated from high throughput screening with subsequent synthetic modifications.

Biological activity induced by a drug is a direct result of its molecular structure and ability to interact with the target macromolecule. In addition to the size and shape of the drug, the combination of organic chemistry functional groups and their 3D orientation when interacting with a macromolecular target is responsible for the biological activity of the vast majority of drugs. This simple yet powerful idea of the pharmacophore was first proposed by Paul Ehrlich as the molecular framework that carries (phoros), the essential features responsible for a drug's (pharmacon) biological activity (Ehrlich, 1909). Contrast this original definition to a more modern one as presented by Peter Gund who suggested that the pharmacophore is a set of structural features in a molecule that is recognized at a receptor site and is responsible for that molecule's biological activity (Gund, 1977). The definition of pharmacophore has been generalized in recent years to avoid the use of specific functional groups and highlighting regions of electron density, H-bond donors, H-bond acceptors, etc.

The development of a hypothesis of drug action as a function of molecular structure is the basis of structure–activity relationships (SAR), where structural modifications are made systematically to existing lead compounds to generate new compounds. The new compounds are then tested in animals for biological activity. The process, as indicated earlier, was the foundation of drug discovery, and it was only restricted to the imagination of medicinal chemists and the scientific limitations of the specific times. Knowledge of SAR is critical, but the data are qualitative or semiquantitative at best. It was in the 1960s that Corwin Hansch developed a simple yet powerful mathematical method that correlates biological activity with mathematical equations (Hansch, 1969). His work quantified SAR and is known as QSAR. The QSAR approach introduced the age of CADD in the 1970s.

The basic idea behind QSAR is the recognition that molecular structure determines the physical properties of a compound. Since the biological activity of a compound is a consequence of its molecular structure, it should also be related to the physical properties. For the correlation and prediction of biological activities, some physical properties are more important than others. For example, the narcotic action of a set of compounds on tadpoles was correlated to the oil–water partition coefficients, *P*, by Meyer and Overton over a century ago. There is a thermodynamic tendency of a compound to distribute itself between the two phases of a water–hydrocarbon mixture. The partition coefficient, *P*, (where $P = [\text{drug}]_{\text{organic}}/[\text{drug}]_{\text{aqueous}}$) is the ratio of the drug concentration in an organic phase compared to an aqueous phase. For ionizable compounds, the situation is more complicated and a function of the pH of the medium.

The partition coefficient is an equilibrium constant and therefore related to the differences in the Gibbs energy at standard state conditions for the phase distribution. The relationship between *K* and *P* is given by the famous equation, $\Delta G° = -2.303RT \log K$, where *K* is *P*. The more positive the value of the partition coefficient, the greater is the tendency for a drug to move from the water phase to the hydrocarbon phase. Typically, log *P* is used rather than *P* because of the direct relationship to the Gibbs energy. Octanol has been the organic solvent of choice for many years. Instead of using log *P* values, a substituent hydrophobicity constant π is often used. It is defined as the difference between the log *P* values of the parent compound and a derivative of the parent structure, where a functional group has replaced a hydrogen atom ($\pi_x = \log P_x - \log P_H$). Positive values of π mean that the derivative is more hydrophobic than hydrogen, while negative values of π mean that the derivative is less hydrophobic than hydrogen. In this formulation, it is straightforward to use the π values in an additive scheme along with the original log *P* value of the parent compound to estimate the log *P* of a new compound.

Hansch recognized this term as a physical observable that could be related to the hydrophobic nature of a drug's ability to cross cell membranes and/or interact with hydrophobic regions within the receptor. Other important factors are sterics and electrostatics. The original Hansch approach relied on the data generated by physical organic chemists who examined steric effects (Taft Es parameters) or electrostatic effects (Hammett σ parameters). The Hansch equation describes the biological activity as a function of sterics, electrostatics, and hydrophobicity (biological activity = *f* [sterics, electrostatics, hydrophobicity]). This methodology relies on the use of statistical correlation methods (regression analyses) to develop valid and predictive equations.

The QSAR method has been widely used since the 1970s. Consider the example of hydrochlorothiazide. Hydrochlorothiazide is a specific example of a thiazide diuretic. The drug is administered orally and used extensively in the treatment of high blood pressure, edema, and congestive heart failure. The drug inhibits reabsorption of NaCl in the distal convoluted tubules of the kidney through inhibition of the Na^+/Cl^- cotransporter. The net effect is diuresis, increased excretion of urine. Note the 1,3 relationship between the two sulfonamide groups ($-SO_2NH_2$) on the aromatic ring and shown in Figure 29.10. One of the sulfonamide groups is part of another ring system. The requirement of the locations and types of functional groups necessary for diuresis in this class of diuretics was determined by observation and SAR.

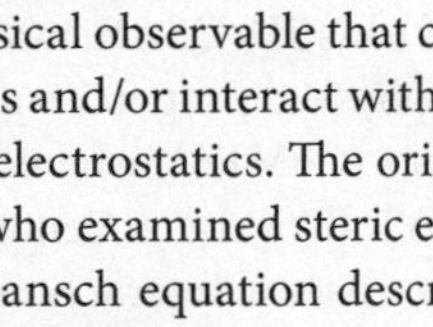

FIGURE 29.10 Hydrochlorothiazide blocks the reabsorption of NaCl, which leads to diuresis. The drug is used primarily in the treatment of hypertension and other disease states affecting the cardiovascular system. In many cases, hydrochlorothiazide is used in combination with other drugs (e.g., Avalide® is a combination of hydrochlorothiazide and irbesartan, an angiotensin II receptor antagonist, which is used for hypertension management).

Like all methods, the Hansch approach has advantages and disadvantages. Overall, traditional QSAR studies are usually (not always) straightforward and predictive. The computing time is minimal. However, the method does require a minimum number of compounds to be prepared and some trial and error in determining the best descriptors. Over the years, scientists have extended the descriptors to include other representative experimental data and/or calculated values (equilibrium A values, dipole moments, charge densities, molecular orbitals, etc.).

Orita et al. published a study that illustrates the QSAR method reasonably well (Orita, 1983) as a working example. They examined thirteen analogs with group variations in the 6- and 7-positions of the core structure. The diuretic activity (A) could be predicted with a formal charge (FC7) calculated for the 7-position, as well as a van der Waals (VW) steric parameter and a hydrophobic parameter (π) at the 6-position. This results in an equation [$\ln A = (0.2695 \times 10^{-3})\, \pi + (7.62535)\mathrm{VW} + (15.7681)\mathrm{FC7} + 0.98502$] generated through a nonlinear regression analysis ($n = 13$, $r = 0.98$). The relationship shows a positive correlation with the diuretic activity and the VW, π, and FC7 descriptors.

The Hansch formulation traditionally has correlated biological activities with the physical properties of the compounds being studied. Some physical properties can be calculated readily. For example, strategies have been developed to generate log *P* values *in silico*–based on the additive scheme of π values and are incorporated into software such as ClogP (Daylight Chemical Information Systems, Inc.; http://www.daylight.com). Perhaps the major complication is that π values are not only additive, but they are also constitutive. This means they are a function of their molecular environment. Such computational approaches are quite sophisticated when taking into account the branching, molecular connectivity, and other structural characteristics that contribute to the overall value of log *P*. This additive scheme is similar to how net dipole moments and thermodynamic properties are generated in MM. Notwithstanding, the QSAR correlation was primarily between experimental biological data and physical properties. More recent methodologies, often referred to as 3D-QSAR, have attempted to make correlations directly with the molecular structure and not the physical properties *per se*.

The comparative molecular field analysis (CoMFA) ushered in a new age for QSAR (Cramer, 1988). CoMFA is available in the program SYBYL (Tripos® a Centara™ Company; http://www.tripos.com). The basic idea is to calculate steric and electrostatic interactions at various positions in space for a set of molecular structures with a probe atom the size of a sp^3 carbon and a positive point charge. The steric and electrostatic interactions are determined by Lennard–Jones and Coulomb's law equations, which are shown in Figure 29.11. Both of these types of equations are found in many MM force fields. The optimal alignment of the structures is critical for CoMFA. The calculated steric and electrostatic information are placed in a table, and a partial least squares (PLS) statistical analysis is carried out to prepare a mathematical model. From the mathematical model, color-coded steric and electrostatic fields are displayed to inform medicinal chemists of the regions in space where molecules may be altered to enhance activity.

3D-QSAR validation is necessary. Unlike the r^2 regression factor that is a positive value ranging from 0.0 to 1.0, where the latter represents a model in perfect agreement with the experimental data, the cross-validated R^2 used in 3D-QSAR is something different. It has been argued that instead of using a cross-validated R^2, it may be better to refer to this result as q^2 to avoid confusion with r^2 because the regression variance may be negative. Therefore, cross-validated numbers range from −1.0 to 1.0. Essentially negative values indicate that random numbers provide a better fit to the experimental data than does the 3D-QSAR model.

Since CoMFA was introduced other methodologies have been developed. An improved method, known as comparative molecular similarity indices analysis (CoMSIA), is now widely used (Klebe, 1994). In this approach, property fields based on similarity indices of drug molecules with a common alignment and fields are calculated. Like CoMFA, a mathematical model is generated and evaluated by a PLS analysis.

Molecular skeleton analysis (MoSA) is a 3D-QSAR program that was developed and presented but never released (Liang, 1996, 2004). Similar to other methodologies, a set of molecular structures are aligned. Physicochemical properties (e.g., partial charge, atomic volume, and H-bonding potential) are assigned to each atom of the set of molecular structures. A 3D grid box is constructed for each property included. Gaussian functions are used in the calculations of molecular property distributions for all the atoms on

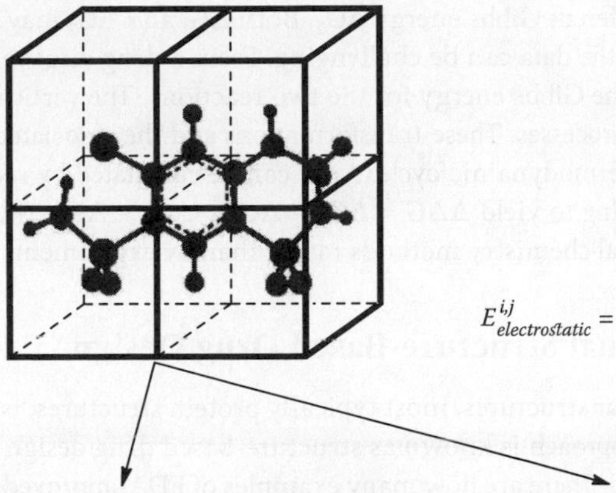

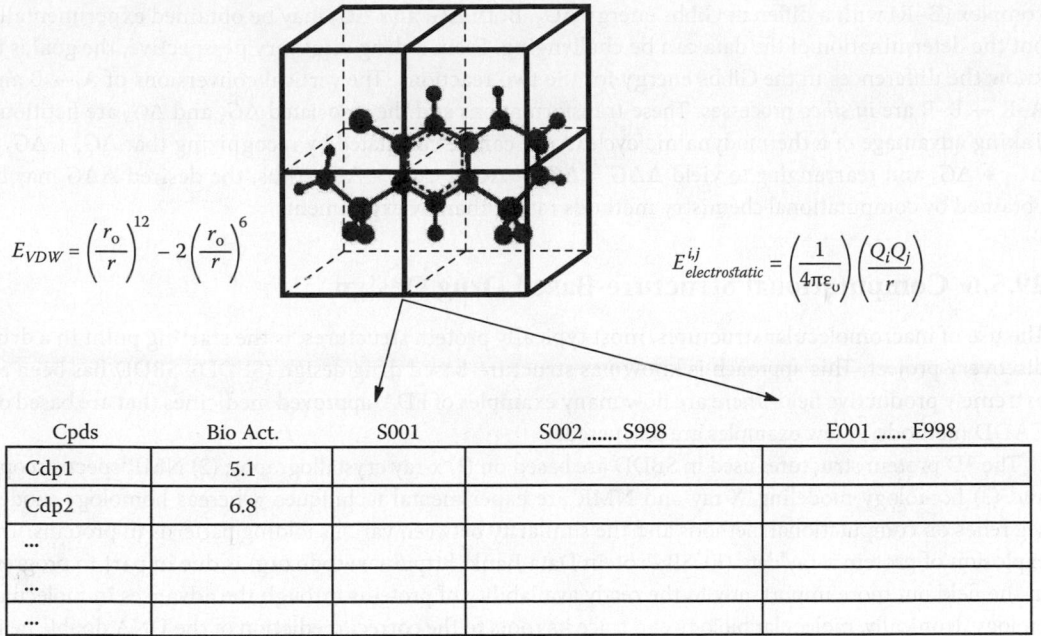

$$E_{VDW} = \left(\frac{r_o}{r}\right)^{12} - 2\left(\frac{r_o}{r}\right)^{6}$$

$$E^{i,j}_{electrostatic} = \left(\frac{1}{4\pi\varepsilon_o}\right)\left(\frac{Q_iQ_j}{r}\right)$$

Cpds	Bio Act.	S001	S002 ...... S998	E001 ...... E998
Cdp1	5.1			
Cdp2	6.8			
...				
...				
...				

FIGURE 29.11 CoMFA creates a 3D grid around molecular structures and uses a probe atom to determine the steric and electrostatic fields at the grid points. The steric energy is based on a simple Lennard–Jones potential and the electrostatic calculations are based on Coulomb's law. Both equations are presented. (Adapted with permission from Cramer III, R. D., Patterson, D. E., Bunce, J. D. Comparative Molecular Field Analysis (CoMFA). 1. Effect of Shape on Binding of Steroids to Carrier Proteins. J. Am. Chem. Soc. 1988, 110, 5959–5967. Copyright 1988 American Chemical Society.)

the grid points. The accumulated property distributions at grid points serve as descriptors. PLS is used to generate the final QSAR model. MoSA has the advantage that it does not use a probe atom, it is force field independent, and multiple conformations with a flexible weighting scheme can be used in the analysis.

29.5.5 Gibbs Free Energy Calculations in Medicinal Chemistry

The free energy perturbation (FEP) method is a computational stratagem used to calculate $\Delta\Delta G$ energy differences (Jorgensen, 1989; McCammon, 1987). Consider a lead compound (A) that binds with a receptor (R) to generate a molecular complex (A–R) in Figure 29.12. The Gibbs energy for this reaction is ΔG_1, whereas a similar reaction with a modified lead compound (B) would yield a different molecular

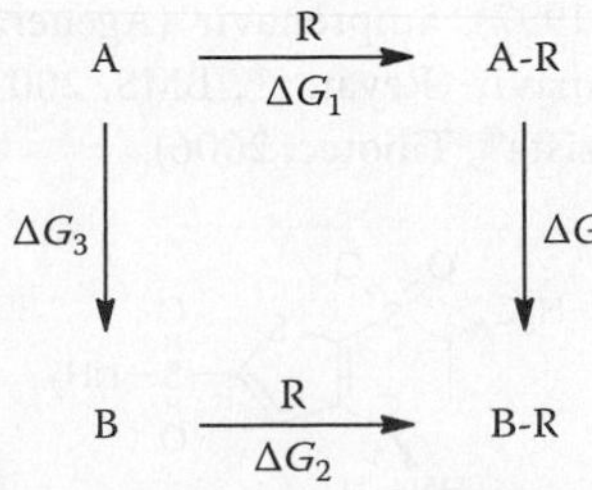

FIGURE 29.12 The thermodynamic cycle depicts the possible formation of the drug–receptor complexes A–R and B–R when drug candidate A reacts with receptor R (A + R→ A–R, ΔG_1) and drug candidate B reacts with receptor R (B + R → B–R, ΔG_2). Both reaction ΔG values can be measured experimentally. The conversion of A → B and A–R → B–R are nonphysical processes, and the associated ΔG_3 and ΔG_4 values may be determined computationally. This process is known as the FEP method, and it is computationally demanding.

complex (B–R) with a different Gibbs energy ΔG_2. Both ΔG_1 and ΔG_2 may be obtained experimentally, but the determination of the data can be challenging. From a drug discovery perspective, the goal is to know the differences in the Gibbs energy for the two reactions. The vertical conversions of A → B and A–R → B–R are *in silico* processes. These transformations and the associated ΔG_3 and ΔG_4 are fictitious. Taking advantage of a thermodynamic cycle, $\Delta\Delta G$ can be calculated by recognizing that $\Delta G_3 + \Delta G_2 = \Delta G_1 + \Delta G_4$ and rearranging to yield $\Delta\Delta G = \Delta G_2 - \Delta G_1 = \Delta G_4 - \Delta G_3$. Thus, the desired $\Delta\Delta G$ may be obtained by computational chemistry methods rather than by experiments.

29.5.6 Computational Structure-Based Drug Design

The use of macromolecular structures, most typically protein structures, is the starting point in a drug discovery project. This approach is known as structure-based drug design (SBDD). SBDD has been an extremely productive field. There are now many examples of FDA-approved medicines that are based on CADD methods. A few examples are discussed.

The 3D protein structures used in SBDD are based on (1) x-ray crystallography, (2) NMR spectroscopy, and (3) homology modeling. X-ray and NMR are experimental techniques, whereas homology modeling relies on computational methods and the similarity between various folding patterns in proteins. The explosion of protein x-ray data (RCSB Protein Data Bank: http://www.pdb.org) is due in part to progress in the field but more importantly to the ready availability of proteins through the advances in molecular biology. Ironically, molecular biology can trace its roots to the correct prediction of the DNA double helix structure by Watson and Crick who used precomputer molecular modeling methods. The development and use of macromolecular structures derived from *in silico* methods (e.g., homology modeling) is important.

Dorzolamide (Trusopt®) is used in the treatment of open-angle glaucoma and hypertension of the eye. The drug, shown in Figure 29.13, reduces intraocular pressures by inhibiting carbonic anhydrase and is administered as an eyedrop. After a decade in development, dorzolamide holds the distinction of being the first FDA-approved drug (1995) that was designed with structure-based methods to reach the market (Greer, 1994). Among the computational chemistry methods used, *ab initio* calculations played a role in the prediction of substitution patterns.

All diseases are cruel, but Alzheimer's disease (AD) seems to strike a particular fear, especially among seniors. There is no cure at this time. It was shown that patients with AD have decreased levels of acetylcholine in their brains. A combination of QSAR, molecular shape analysis, molecular modeling, docking, and related CADD methods resulted in donepezil (Aricept®) being introduced into the market (Kawakami, 1996). This drug, shown in Figures 29.14 and 29.15, is a reversible inhibitor of acetylcholinesterase (AChE).

Another noteworthy example where SBDD played a significant role was in the development of human immunodeficiency virus type 1 (HIV-1) protease inhibitors. The US FDA has approved several HIV protease inhibitors since 1995. They include (in chronological order) saquinavir (Invirase®, Roche, 1995; Fortovase®, Roche, 1997), indinavir (Crixivan®, Merck, 1996), ritonavir (Norvir®, Abbott, 1996), nelfinavir (Viracept®, Pfizer, 1997), amprenavir (Agenerase®, GlaxoSmithKline, 1999), lopinavir (Kaletra®, Abbott, 2000), atazanavir (Reyataz®, BMS, 2003), tipranavir (Aptivus®, Boehringer Ingelheim, 2005), and darunavir (Prezista®, Tibotec, 2006).

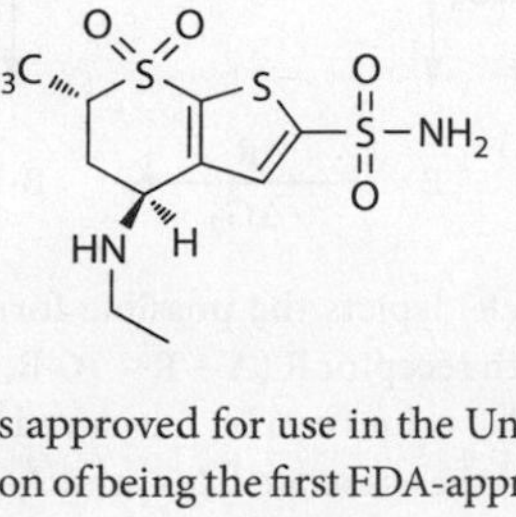

FIGURE 29.13 Dorzolamide (Trusopt®) was approved for use in the United States in 1995. The drug is an eyedrop that is used to treat glaucoma. It has the distinction of being the first FDA-approved drug that was designed with structure-based methods to reach the market. Inhibition of carbonic anhydrase is the mechanism of action for this drug.

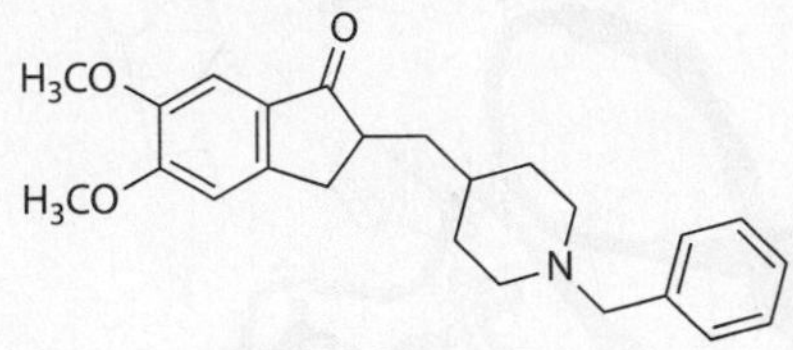

FIGURE 29.14 Donepezil (Aricept®), a reversible inhibitor of AChE, was developed by Eisai and marketed by Pfizer. QSAR, molecular modeling, and SBDD methods were used in its development.

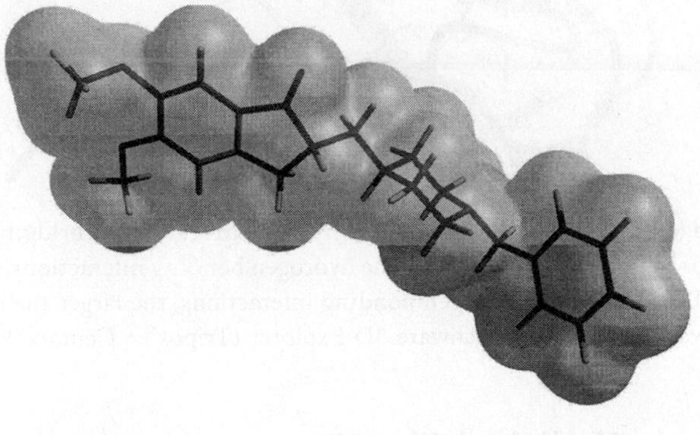

FIGURE 29.15 The energy minimized structure of Donepezil (Aricept®) is shown with a molecular surface displayed.

HIV protease, a member of the aspartyl protease family, is a symmetrical dimer with 99 amino acids in each monomer and has been a target for successful drug discovery (Huff, 1991). The protease is expressed as part of the reproductive cycle of the virus. It is one of several viral enzymes that are initially expressed as a large polyprotein encoded by the HIV-1 pol gene, one of the three major genes of the retroviral genome. The retroviral genome also encodes various small auxiliary proteins. It is known that during the maturation of nascent viral particles, via hydrolysis reactions catalyzed by the catalytic aspartate residues, the protease enzyme cleaves the viral polyproteins at specific sites to generate the structural proteins and enzymes essential for viral replication. A water molecule presumably assists in the peptide bond hydrolysis. X-ray crystal structures show that the NH groups of the protein are bound to the water molecule as well as to the carbonyl groups of the inhibitor. Figures 29.16 and 29.17 illustrate computational modeling of saquinavir.

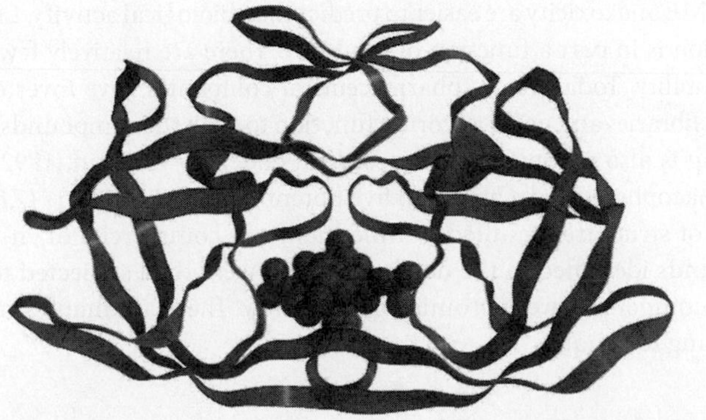

FIGURE 29.16 Saquinavir (Invirase®; Fortovase®) was the first FDA-approved HIV inhibitor developed. It was developed by Roche and, like all drugs in this class, was developed with structure-based CADD methods. Saquinavir, in the center, is shown bound to HIV-1, rendered as a ribbon.

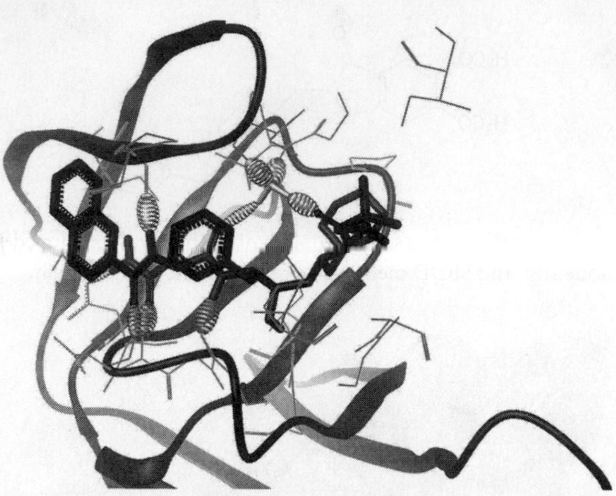

FIGURE 29.17 This image is a close-up view of saquinavir bound to HIV-1. The bridging water molecule between the ligand and the receptor can be seen readily. Note the hydrogen bonding interactions, which are depicted by the football-shaped objects. The greater the hydrogen bonding interactions, the larger the objects. This image of the drug–receptor complex was prepared in Benchware 3D Explorer (Tripos® a Centara™ Company). (Courtesy of Dr. Tom Jones.)

29.5.7 Informatics in Medicinal Chemistry

CADD methods are employed by medicinal and/or computational chemists to enhance understanding of the factors that contribute to the stabilization of the drug–receptor complex. The potency, solubility, and permeability are the physical variables that may be modified to enhance the activity of drug candidates. Maximizing the thermodynamic interactions between a drug and its receptor is not, however, the only aspect involved in successful drug design. The absorption, distribution, metabolism, and excretion properties of a drug are vitally important. It may be more beneficial to sacrifice some favorable thermodynamic interaction energies in a drug–receptor complex in order to minimize some undesired biotransformations or to maximize solubility and permeability. In this vein, Lipinski has championed the idea of predictive ADME. The potential that a drug candidate will have poor absorption or permeation may be expected when at least one of the following four properties is present: (1) more than 5 hydrogen bond donors, (2) more than 10 hydrogen bond acceptors, (3) more than a 500 molecular weight, and (4) more than a 5 computed logP (Lipinski, 1997). There have been other equivalent ways to include these factors in the analysis of small molecule preclinical candidates (e.g., the number of nitrogen centers, the number of oxygen centers).

In many cases, ADME and toxicity are easier to predict than biological activity. Lipinski's rules may work well because absorption is in part a function of solubility. There are relatively few physicochemical principles underlying solubility. Today, many pharmaceutical companies have invested significant resources into screening virtual libraries and using a scoring function to rank the compounds prior to their synthesis.

Database searching is also a popular and successful endeavor (Martin, 1992). For example, Zhong et al. created a pharmacophore model based on five potent MMP-9 inhibitors (Zhong, 2011). A database search of thousands of structures resulted in nine hits from commercial or in-house sources of compounds. The compounds identified in the database search were then subjected to an SVR proliferation assay. Several of the compounds were promising inhibitors. There are many similar examples for successful database mining techniques.

29.5.8 Future Trends in Drug Discovery

Large pharmaceutical companies are undergoing dramatic changes in their approach to research, which includes mergers, downsizing, outsourcing, and even reexamining their research commitments to future drug discovery projects. The use of computational chemistry and computer-based methodologies will be a part of most future drug discovery projects because of the reasonable success that has been demonstrated. Unfortunately, the success rate has been much less than desired, which may in part be due to overexpectations and the inherent difficulty of designing compounds that do not have some hidden flaw that is eventually discovered in human clinical trials. There have been a number of speculations for the problems facing drug approval, and it is beyond the scope of this book chapter to review them. Drug discovery, however, is too important to disappear altogether. Most probably there will be a new emphasis on drug discovery in academic centers and small biopharmaceutical companies. The use of computational chemistry and CADD methods in the right hands can be very effective tools for medicinal chemists.

29.6 Computational Inorganic Chemistry

The improvements in methods and computational power have not yet equaled the challenges presented by large atoms and molecules (Gladysz, 2011). The computational chemistry tools were originally developed for small systems with limited atomic diversity. One of the key decisions in determining how to model an inorganic compound is the trade-off between accuracy and the computational expense. Despite this limitation, computational chemistry has been applied effectively in many areas of inorganic chemistry, including the study of structure and bonding in complex systems; solvation effects on reactive species; chemical catalysis, especially the nature of transient species and energy changes; electrochemical as related to electron transport and storage of organometallic complexes; and the properties of new materials.

29.6.1 Computational Methods for Inorganic Chemistry

Empirical methods, like MM2, do not contain most of the atoms and geometries needed to predict reasonable inorganic structures. Semiempirical methods suffer from the same limitation. Complete *ab initio* calculations using either Hartree-Fock or Møller–Plesset treatments are generally too time-consuming to run for inorganic compounds. Even generally accepted modifications to the complete wave functions like the Born–Oppenheimer approximation for fixed nuclear geometries cannot simplify the calculations enough for these molecules that often contain hundreds to thousands of electrons (Stradi, 2011). It is often necessary to model a portion of the molecule containing just the active metal or other heavy atom center with theories taking into account unpaired delocalized electrons using post self-consistent field models to get an accurate single-point energy calculation after completing a series of modeling trials that use more approximations to gain a good geometric representation. The optimized fragment geometry can then be incorporated into the full structure. Another significant area of research within *ab initio* modeling studies is using plane wave basis sets for the valence electrons with a correction for the effective core potential (ecp) to study conduction in metals instead of Slater- or Gaussian-type orbitals to better match the long-range nature of electron transfer in these materials (Lopez, 2011).

The major computational tools used in all branches of inorganic chemistry utilize DFT models with basis set polarization corrections to take into account d and f orbital contributions and diffuse functions for electronic excited-state calculations. The most frequently used basis sets also take into account the ecp approximations, where the electron interactions for the inner shell electrons are frozen and coupled with the cationic atomic nuclei as a rigid, nonpolarizable core. Even evaluations of ligands like bipyridine, which is among the most studied bidentate organic ligands used in organometallic chemistry, can yield interesting results by utilizing broken symmetry DFT functionals ranging from the oft-referenced B3LYP through more complicated hybrid systems that take into account additional correlation and electron exchange terms in the integrals (Scarborough, 2011). The best fit to the experimental spectral data for the titanium(III) cyclopentadienyl bipyridine diradical complex was seen using B3LYP compared to three other more complex hybrid functionals. Furthermore, for systems that contain a wide range of atom types, a different basis sets need to be used for different atoms within one computational study. Highly complex basis sets that take into account d and f orbitals using polarization and diffusion at long distance from the core are used for the metal atoms, and simpler Pople basis sets are used for the main group atoms and hydrogen.

29.6.2 Main Group Computational Chemistry

A recent study is representative of this type of work (Wittmaack, 2011). A series of derivatives of halogenated structures (X_nA) for the group 4A elements carbon through lead was evaluated to see how the charge transfer from the halides to the group 4A elements was impacted by the number of halides attached to the molecule and by the atomic number of the group 4A central atom. Since these are small, discrete molecules, it was possible to use full *ab initio* Møller–Plesset (MP2) and coupled cluster calculations to incorporate electron correlation effects. For the smaller atoms, full MP2 evaluations were completed; for the largest atoms like lead (Pb) and tin (Sn), approximations to incorporate electron core potential effects were included. Additional calculations using natural bond order rather than molecular orbital calculations were then carried out on the optimized geometries to determine point charges for all of the atoms. These studies have shown that after a certain level of substitution, the charge on the central atom does not change much regardless of the addition of more halides to the structure. This approach of multiple steps in model refinement is again typical of most modeling studies for inorganic compounds.

29.6.3 Computations with Metal-Containing Compounds

New density functionals have been developed that incorporate relativistic effects, ecp's, and electronic correlation effects. For heavier atoms, the near-relativistic speeds of inner shell electrons results in higher effective electron masses that can skew results tremendously. In addition to the hybrid functional B3LYP, many researchers in organometallic chemistry are using the Perdew, Burke, and Ernzerhof (PBE) or the Heyd–Scuseria–Ernzerhof (HSE) generalized gradient approximation hybrid functionals when studying metal atoms as they give good geometrical matches to published crystal structures and available vibrational spectroscopy (Stradi, 2011). Metal phthalocyanine systems have been modeled to determine the source of bands in the emission spectra of these compounds. The major goal of the study was to evaluate the energy minima and vibrational spectra of the materials adsorbed as a thin film on a solid surface to compare the results to the available gas-phase data. The best agreement from the model data in this case came with the HSE hybrid functionals compared to all of the other systems tested.

Sometimes, modeling is a useful method for evaluating chemicals that are difficult to isolate and work with because of stability problems and air sensitivity. While neutral, +2, and +4 oxidation states of palladium have been extensively studied, there are few reports on +3 oxidation state (Pd(III)) complexes. A recent synthetic study on the isolation and x-ray analysis of these compounds used DFT to explain the distorted octahedral geometry for these complexes, and the DFT-derived singly occupied molecular orbital descriptions for these species gave a very accurate prediction for EPR signals that matched the experimental data for these paramagnetic species, showing how complex computational studies are now used by synthetic chemists in conjunction with spectral analysis (Khusnutdinova, 2010).

Molecular modeling using high-level DFT studies have been used in the past few years to study whether assumptions made concerning the non-innocence (lack of involvement in the oxidative process) of homoleptic dithiolene ligands in metal complexes from x-ray, spectral, and simple MO calculations were correct. The models showed that on changing the redox state of the overall organometallic complex the oxidation state changed not just on the metal but on the ligand as well, hence the "non-innocence" sobriquet given this class of ligands. Using modern computational techniques, electron density distributions that matched the spectral data and supported this theory were completed, confirming this long-debated point (Eisenberg, 2011). A recent review highlights many of the uses that have been made of these computational studies in evaluating bonding in very complex compounds (Lin, 2010).

Many inorganic reactions can only be studied effectively by taking into account interactions between the reaction center and its surroundings. A recent study highlights the difficulty in modeling the formation of large clusters that contain highly charged metals, where ion–dipole effects present from solvation change the calculated energy levels of the molecule (López, 2011). The possibility of modeling all of the solvent–solute interactions is very challenging from a computational standpoint, so solvent effects are taken into account using a conductor-like screening model (COSMO) that approximates the electrostatic interactions between the model complex and the solvent by treating the solvent as a set continuum governed by the solvent dielectric constant to give a much more realistic representation of the zero point energies and corresponding vibrational frequencies in highly polarized molecules. These COSMO studies are generally in good agreement with more rigorous molecular dynamic studies using Car–Parrinello simulations where all solvent molecules are treated explicitly and motion of the solvent molecules is allowed during the computational cycle.

29.6.4 Computational Issues Related to Catalysis

Understanding the mechanisms, intermediate structures, and reaction kinetics of catalytic cycles is central to process improvement. Often, these properties can be evaluated by use of molecular models as well as using mathematical models to model rates. In the case of the Suzuki–Miyaura coupling reaction, a system whose discovery was chosen as part of the Nobel Prize in chemistry in 2010, the mechanism is not well understood. The Suzuki–Miyaura coupling reaction uses a palladium catalyst to form new carbon–carbon bonds. The active species in the oxidative addition, transmetallation, and reductive elimination steps in the catalytic cycles are not clearly understood, and experimental characterization does not clarify the structure of these species, so a number of computational studies have been undertaken recently (Jover, 2010). All of the key reaction intermediates were evaluated using very standard DFT functionals and basis sets except for the transition metal palladium and bromine, where basis sets taking into account the ecp as well as triple-zeta orbital descriptions are needed to describe the more complex bonding. The study results were in good agreement with spectral data obtained during actual catalyst trials, and based on variations in the models, the reasons large electron-donating phosphine ligands improve catalytic efficiency was explained, even though the actual energies were not in total agreement with the actual data as fairly simple calculations were attempted. The use of DFT-based molecular modeling is now so pervasive in organometallic catalysis that even groups whose primary goal is the synthetic utility of a catalyst will run DFT studies to confirm proposed reaction cycles.

One of the key parameters in evaluating any catalyst system is the turnover frequency, which measures the number of molecular transformations over a period of time catalyzed by a single active catalytic site; thus, it is a measure of catalytic efficiency. A recent study develops a new computational model for all of the steps in a catalytic cycle using the energetic span model (Kozuch, 2011). The goal of the study was to link experimentally derived rate constants (kinetics) with energy level descriptions of the transitions states (thermodynamics) calculated using modeling by linking the results using the transition state theory developed by Eyring. A computational model developed from this connection between kinetics and thermodynamics was successfully employed with a series of nickel center–catalyzed cross-coupling reactions of anhydrides with organozinc compounds. The resulting transition state measure of turnover frequency was able to accurately predict the experimentally observed turnover frequencies, and the technique is being expanded to other catalyst systems.

29.6.5 Electrochemistry

There has been a tremendous upsurge in interest in electrochemical systems involving inorganic chemistry over the last 5 years, and very often computational methods using DFT analysis are being utilized to evaluate the reduction and oxidation potentials in these systems (Geiger, 2011). Now that there is a wide availability of tools to study these systems, many groups are modeling the changes in the valence-state electrons of the metals using DFT models coupled with basis sets optimized for use with ecp approximations (Roy, 2009). One of the results of these studies on well-understood systems like the one electron oxidation of the iron ferrocene complex is the determination that the choice of the basis sets has less of an impact on the energy and geometry calculations than does the choice of the DFT functionals. Using a wide range of DFT models including generalized gradient approximations and hybridized approximations had a major impact on the overall calculated oxidation potential. By varying the basis set on the iron core atom from all electron triple-zeta models to basis sets with ecp approximations, or with ecp approximations with relativistic effects taken into account, the overall calculated oxidation potentials changed dramatically. Only the basis set taking into account the ecp with relativistic effects incorporated into the approximation matched the experimental data. In the same study, changing the solvent used from acetonitrile to acetone or DMSO resulted in having to change the DFT used from a hybridized DFT to an unhybridized version of the same DFT. Just using one model without evaluating other possibilities can lead to very uncertain results.

29.6.6 Clusters and Nanomaterials

The area of chemistry that results in the most public interest, outside of medicinal chemistry, is the study of nanomaterials. Many metal-containing species are being studied in regard to these nanomaterials; two of the areas of greatest interest involve graphene nanomaterials and metal–organic frameworks (MOFs). These systems are so large and entail so many atoms and electrons that fully modeling them is impossible, but it also increasingly evident that the chemistry of these complexes changes as the size of the clusters grow from localized single metal centers to clusters with thousands to millions of metal centers. The interaction of the first-row transition metal centers with the six-member carbon rings found in graphene nanotubes has now been modeled using computational methods (Valencia, 2010). To be of practical interest, these studies need to take into account enough atoms to allow for measurement of electron transfer across the entire model surface to emulate the band structure for electrical conduction, as well as model the Brillouin zone of the nanomaterial, as that is the representative lattice cell of interest for the determination of the macro properties. The necessity of modeling enough atoms and electrons to take into account these lattice properties is computationally expensive. Since the electrical conduction of these materials is a very important property, the modeling is often done using plane wave basis sets instead of Gaussian-type orbitals.

The study of MOFs is another area of nanomaterials where modeling is being used to predict relevant properties. MOFs are extended 3D cage-like crystal lattices containing multidentate organic ligands that are coordinatively bound to metal centers. These MOFs are being evaluated as catalyst supports, gas storage matrixes, and drug delivery vehicles among many end uses. Since DFT can be used to measure bond strength, it is possible to relate calculated energy profiles and geometries from modeling studies in the MOF to overall mechanical properties such as shear modulus, by determining the individual bond stiffness over the entire unit cell of the lattice in all planes of the cell (Tan, 2011). Actual experimental measurement of these MOF materials is primarily limited to the technique of nanoindentation, since large individual crystals have not yet been created. Nanoindentation only measures penetration distance and applied force, which is of limited utility when comparing to applied shear forces, since many of these compounds exhibit differing levels of dimensional stability. Overall, the comparison to the calculated shear forces is interesting as the shear forces and the force of indentation have a strong linear correlation.

29.6.7 Future Issues for Computational Inorganic Chemistry

Dealing with the lanthanides and actinides, with f orbital electrons, might be the ultimate challenge since the energy levels are nearly degenerate many different electronic orientations are possible. Coordination numbers of as great as 12 are possible and a wide range of oxidation states from 1 to 7 can be seen in some complexes. A recent survey on the use of computational studies in the evaluation of actinide containing molecules summarized these challenges (Schreckenbech, 2010). For bioinorganic structures, the problems are additionally compounded (Ghosh, 2011). Researchers often use combined QM/MM; however, this can lead to large discrepancies in the calculated results. The choice of exchange correlation functionals has a huge impact on the final agreement between computational and experimental results, and occasionally, no DFT models come close to the experimentally determined spectra (Gutten, 2011). Uncertainty in valence electron location between the metal center and the ligand shell makes it difficult to explain electron transfer mechanisms. One group overcomes this problem with DFT methods; some groups are completing complex high-level *ab initio* studies on iron(III) porphyrin studies using post-SCF methods including the complete active space self-consistent field model (CASSCF). By using more rigorous computational models, they are able to predict the spectral properties and give insight into the actual mechanism of electron transfer.

New functional descriptions for DFT to improve the incorporation of long-range electron interactions need to be created. In the absence of a reliable single method, computational chemists must try several different modeling protocols and chose the one that best matches the experimental data. The field of computational inorganic chemistry is just starting to deal with all the complexity inherent in large molecules containing elements with many electrons in a series of atomic orbitals that extend across all available orbitals.

29.7 Summary

There is obviously still room for improvements in computational chemistry. But the role of computation in chemistry is now essential to most areas of chemical inquiry. Perhaps this is most evident in a quick glance at some recent special issues of leading review journals and communications from journal editors. In the opening to the 2006 special issue in *Accounts of Chemical Research*, the editors wrote: "At the outset of the twenty-first century, theoretical and computational chemistry has arrived at a position of central importance not only for theorists but also in the laboratories of most experimentalists and in many disciplines" (Carter, 2006). The December 2011 issue of *Chemical Physics and Physical Chemistry*, dedicated to computational chemistry, includes the following editorial comment: "... the development of new methods and density functionals continues to be a vibrant field of research in a sustained effort to develop reliable approaches for almost any problem in chemistry" (Batista, 2011). Another 2011 editorial attests to the widespread acceptance of computational chemistry: "Approximately one-third of current submissions to the *Journal of Medicinal Chemistry* report computational work of varying weight and complexity." The editors go on to describe a new set of standards for computational chemistry in that journal (Stahl, 2011). The January 2012 issue of *Chemical Reviews* is dedicated to quantum chemistry. In the opening introduction to that section is the following statement: "... quantum chemistry has yielded a very important product: software packages, some of them representing commercial enterprise and others given away freely ... In either case, these programs are being used by an ever growing base of users" (Pyyjkko, 2012). It seems clear that computational chemistry is here to stay and that we can look forward to advances and improvements that will increase the foothold of computation in the myriad disciplines of chemistry.

Key Terms

Acetophenones: A class of organic compounds that have a methyl carbonyl (CH_3–CO–) attached to a benzenoid aromatic ring.

Auxochromes: A term derived from the Greek aux, meaning to grow, and the Greek chroma, meaning color. An auxochrome is any group added to a chromophore that causes a change in the interaction of the group with UV–Vis light.

Belousov–Zhabotinskii: The specific reaction is the oscillating reaction involving BrO_3^-, $HBrO_2$, H^+ to give BrO_2, H_2O, and BrO_2, Ce^{3+}, H^+ to give $HBrO_2$, Ce^{4+}, H_2O. Reactions that oscillate are sometimes referred to as Belousov–Zhabotinskii–type reactions. They are representative of reactions that are not in chemical equilibrium but instead are treated as dynamic systems.

Benzenoid aromatic: A compound that has a six-membered ring with alternating single and double bonds is a benzenoid aromatic compound (e.g., benzene, molecular formula C_6H_6).

Bipyridine: A common organic ligand composed of two pyridine rings, where each ring contains one nitrogen and five carbons in a six-member aromatic structure.

Brillouin zone: The unit cell in a crystal lattice that fully describes the electron distribution and electronic band structure.

Broken symmetry approximations: Single determinant wave functions that are not pure spin states and therefore are not expected to represent the true wave functions of the system. Unsymmetrical determinants are used to represent uneven distribution of electrons in a molecule.

Catalysis: Using a reagent in a chemical reaction that significantly increases the rate at which the reaction proceeds, without using up any of the reagent.

Chromophore: Derived from the Greek *chroma*, for color, and *phoros*, for bearing, the chromophore is the core structural feature of a compound that gives rise to its ability to absorb UV–Vis light.

Conjugated: If atomic orbitals that make up pi bonds can overlap in a side-to-side manner with pi bonds or atomic orbitals on contiguous positions, then the electrons in all of the overlapping orbitals are treated as a single integrated system that is said to be conjugated.

d/f-type electrons: Atoms with a principal quantum number of 3 or higher can have d-type electrons. For each d-type principal quantum number, there are five atomic orbitals with an allowed capacity of 10 d-type electrons. Atoms with a principal quantum number of 4 or higher can also have f-type electrons. For each f-type principal quantum number, there are seven atomic orbitals with an allowed capacity of 14 f-type electrons.

***De novo*:** From Latin *dē*, from, and *novō*, new, *de novo* refers to a process of discovery that makes minimal preliminary assumptions, thus representing a new approach.

Dentate, mono, bi, multi: Binding of atoms to the central atom in a coordination complex is referred to as denticity, where every point of contact is a bond to the central atom.

Dienes: Dienes are compounds that contain two pi bonds between two different sets of carbon atoms. In UV–Vis, the term diene almost always refers to conjugated dienes.

Dyes: Dyes are compounds that both absorb visible light causing the appearance coloration and have intermolecular attractions to compounds/materials to which they are applied, such as textiles.

Electrochemistry: Chemistry occurring at the interface of an electrical conductor, where the flow of electrons either to or from the interface impacts the chemical reaction that occurs.

Enones: Enones refer to a conjugated system composed of a C=C pi bond and a C=O pi bond, as found in the compound methyl vinyl ketone: CH_2=CH–CO–CH_3.

Ferrocene: An organometallic species containing an iron atom bound to two five-member aromatic rings so that the rings are centered perpendicularly above and below the iron atom.

Fluorescence: The emission of light energy of a lower-energy and longer wavelength than the concurrently applied light energy, or other forms of energy, used to generate the excited state, is known as fluorescence.

Functional: A mathematical operation that converts a vector into a scalar field; an example would be the scalar field of temperature gradients around a solid body. A functional is a function that is a function of another function.

Graphene: A form of pure carbon that exists as an extended sheet of six member rings in a very stable form. Sheets of graphene combine to form the mineral graphite.

Halides: Elements in group 17, formerly 7A, on the periodic table that are characterized by high electron affinity (e.g., F, Cl)

Homoleptic: Any organometallic complex in which all of the attached ligands are identical.

INDO/S: It is the acronym for intermediate neglect of differential overlap/screened approximation. The ZINDO method is the Zerner modification of the INDO/S method.

***In silico*:** A process or test or evaluation performed on a computer is termed *in silico*, in deference to the use of silicon chips.

***In vitro*:** A process or test or evaluation performed in glassware is termed *in vitro*.

***In vivo*:** A process or test or evaluation performed in living systems is termed *in vivo*.

Intramolecular charge transfer: The shifting of electron density from an electron-donor region or substance to an electron-accepting region or substance creates a weak attractive force between the two regions. When this occurs between two molecules, it is called a charge transfer complex, when it occurs within one molecule, it is called ICT.

Lambda max: A peak in a UV–Vis spectrum is broad, on the order of 10–100 nm wide, and the identity of the peak is based on the point of highest absorptivity of light in that range, called the lambda (λ) maximum (max), given the notation λ_{max}.

Law of mass action: The law of mass action states that the rate of a chemical reaction at each instant is proportional to the concentrations of the reactants, with each concentration raised to a power equal to the number of molecules of each species participating in the process. If the reaction involves more than one step, the rate of formation or removal of a given substance will contain terms for each step in which the substance is involved. Once overall equilibrium is reached, the law predicts the usual expression for the equilibrium constant K in which the numerator is the product of the concentrations (partial pressures, etc.) of the products raised to the respective stoichiometric coefficient in the overall chemical equation divided by the product of the concentrations of the reactants raised to the respective stoichiometric coefficient.

Ligand: A molecule or ion that donates electrons to the central atom, usually a metal, in an organometallic coordination compound.

Michaelis–Menten: This is the relationship used to describe rates of enzyme-substrate process in biochemical reactions, where the rate of conversion, V, equals the maximum rate of conversion, V_{max}, times the concentration of the substrate, $[S]$, that quantity divided by the sum of the rate constant, K_m, and the concentration of the substrate: $V = (V_{max}[S])/(K_m + [S])$.

Molar absorptivity: Molar absorptivity, also known as extinction coefficient in older literature, describes the quantification of light apportion as a function of the amount of the chromophore molecule present in solution. Beer's law ($A = \varepsilon cb$) is used to define the relationship, where A = measured absorbance intensity, ε is the molar absorptivity, c is the concentration of the molecule, and b is the path length of the sample cell.

Molecular orbitals: As atoms are combined to make molecules, the atomic orbitals and associated electrons from the atoms are combined to produce new orbitals, called molecular orbitals, into which the molecules electrons are placed.

Oxidation potential: A measure in electron volts of the energy required to remove an electron from a compound.

Pharmacokinetic: The description for the rates of the fundamental processes that a drug undergoes in the body is called pharmacokinetics. Those processes include absorption, bioavailability, distribution, metabolism, and excretion.

Phosphine: An organophosphorus compound with an easily donated electron pair on a phosphorus atom that can act as a ligand.

Phosphorescence: The delayed emission of light energy of a lower-energy and longer wavelength than the concurrently applied light energy, or other forms of energy, used to generate the excited state, is known as phosphorescence.

Photovoltaics: The process of capturing sunlight and using that energy to generate electrical energy is known as photovoltaics.

Phthalocyanine: A cyclic organic species with 4 nitrogens on the inside of the cyclic ring that readily binds metals. Structurally related to the porphyrin ring seen in many biological enzymes like hemoglobin.

Pi bond/electrons: If two atomic orbitals on adjacent atom are coplanar so that they can overlap in a side-to-side manner and they share a pair of electrons, the result is covalent bond known as a pi bond, and the electrons in the pi bond are called pi electrons. The pi covalent bond electrons are less tightly held by the two atoms than the initial sigma bond that if formed between the two atoms.

Principal quantum number: The principal quantum number, n, describes the average size of an atomic orbital, where a larger number represents a larger orbital that is located on the average a greater distance away from the nucleus of an atom. In general, as electrons get farther away from a nucleus, they become more susceptible to excitation by UV–Vis light.

Sigma bond/electrons: If two overlapping atomic orbitals on adjacent atom are aligned along the same line, they can overlap in an end-to-end manner and share a pair of electrons to form a strong covalent bond known as a sigma bond. The electrons in the sigma bond are called sigma electrons.

Solvation: The interaction of solvent molecules with a solute, or dissolved species, which leads to an overall stabilization of the solute.

Tautomers: Two compounds are tautomers of each other if they are structural isomers that can be converted one into the other in a rapid equilibrium controlled process. The process generally involves a structural change in connectivity of a hydrogen atom and a pi bond within the molecule.

TD-DFT: The acronym for time-dependent dentistry functional theory. The TD-DFT adds in a time delay to allow for calculation of molecular orbitals and associated properties after a system has been perturbed, such occurs in the excitation.

Valence: The number of electrons in the outer shell of an atom that determines how many bonds an atom can form. The valency of an atom is the number of bonds that atom generally makes.

Zeta: The zeta value is used to describe how diffuse or large an orbital can be. A double-zeta orbital is composed of two separate type orbitals, for instance, an s- and a p-type orbital. Triple zeta is composed of three types of orbitals.

ZINDO: This is the acronym for the computational model developed by Dr. Zerner, of the University of Florida, based on modification of the INSO/S routine.

Further Information

Books

Collett, C.T. and Robson, C.D. 2010. *Handbook of Computational Chemistry Research*. New York: Nova Science.

Heine, T., Joswig, J., and Gelessus, A. 2009. *Computational Chemistry Workbook. Learning through Examples*. Weinheim, Germany: Wiley-VCH.

Rode, B.H., Hofer, T.S., and Kugler, M.D. 2007. *The Basics of Theoretical and Computational Chemistry*. Weinheim, Germany: Wiley-VCH.

Websites (All are accessed on October 4, 2013)

Berkeley Madonna: http://www.berkeleymadonna.com.

CCCE: http://www.computationalscience.org/ccce/

Cambridge Crystallographic Data Centre: http://www.ccdc.cam.ac.uk/data_request/cif (CCDC 615918 contains the supplementary crystallographic data for I).

CSERD: http://www.shodor.org/refdesk/.

Chemical Kinetics Simulator: http://www.almaden.ibm.com/st/computational_science/ck.

JCE software: http://www.jce.divched.org/jce-products

Mathcad: http://www.ptc.com/mathcad.

Mathematica: http://www.wolfram.com.

MATLAB®: http://www.mathworks.com.
Molecular Model for an Ideal Gas: http://www.phy.ntnu.edu.tw/ntnujava/index.php?topic=25
NSDL: http://nsdl.org.
Sage: http://www.sagemath.org.
STELLA: http://www.iseesystems.com.
Vensim PLE: http://www.vensim.com.
WebMO: http://www.webmo.net/

References

Alexander, C.A., Asleson, G.L., Doig, M.T., and Heldrich, F.J. 1999. Spectroscopic instruction in introductory organic chemistry: Results of a national survey. *J. Chem. Educ.* 76: 1294–1296.

Andraos, J. 2008. Corrections: A streamlined approach to solving simple and complex kinetic systems analytically. *J. Chem. Educ.* 85: 1624.

Andraos, J.A. 1999. Streamlined approach to solving simple and complex kinetic systems analytically. *J. Chem. Educ.* 76: 1578–1593.

Bakhle, Y.S. 1968. Conversion of angiotensin I to angiotensin II by cell-free extracts of dog lung. *Nature* 220: 919–921.

Bakhle, Y.S., Reynard, A.M., and Vane, J.R. 1969. Metabolism of the angiotensin in isolated perfused tissues. *Nature* 222: 956–959.

Batista, V.S., Grimme, S., and Reiher, M. 2011. Recent progress in theoretical and computational chemistry. *Chem. Phys. Chem.* 12: 3043–3044.

Bowen, J.P. 2004. Computational chemistry and computer-assisted drug design. In: *Wilson and Gisvold's Textbook of Organic Medicinal and Pharmaceutical Chemistry*, eds. Block J., Beale J.M. Jr., 11th Edn., pp. 919–947. Philadelphia, PA: Lippincott Williams & Wilkins.

Boyle, N.M., Campbell, C.M., and Hutchison, G.R. 2011. Computational design and selection of optimal organic photovoltaic materials. *J. Phys. Chem.* 115: 16200–16210.

Bridle, C.A. and Yezierski, E.J. 2012. Evidence for the effectiveness of inquiry-based, particulate-level instruction on conceptions of the particulate nature of matter. *J. Chem. Educ.* 89: 192–198.

Brunton, L., Chabner, B., and Knollman, B. (Eds.) 2010. *Goodman and Gilman's Pharmacological Basis of Therapeutics*, 12th edn. New York: McGraw-Hill Medical.

Byers, L.D. and Wolfenden, R. 1972. A potent reversible inhibitor of carboxypeptidase A. *J. Biol. Chem.* 247: 606–608.

Byers, L.D. and Wolfenden, R. 1973. Binding of the by-product analog benzylsuccinic acid by carboxypeptidase A. *Biochemistry* 12: 2070–2078.

Carter, M. and Rossky, P.J. 2006. Computational and theoretical chemistry. *Acc. Chem. Res.* 39: 71–72.

Cramer, R.D. III, Patterson, D.E., and Bunce, J.D. 1988. Comparative molecular field analysis (CoMFA). 1. Effect of shape on binding of steroids to carrier proteins. *J. Am. Chem. Soc.* 110: 5959–5967.

Cushman, D.W., Cheung, H.S., Sabo, E.F., and Ondetti, M.A. 1977. Design of potent competitive inhibitors of angiotensin-converting enzyme: Carboxyalkanoyl and mercaptoalkanoyl amino acids. *Biochemistry* 16: 5484–5491.

Doig, M.T., Heldrich, F.J., Meier, P.G., and Metz, C.R. 2008. The use of GC/MS, UV/Vis, and modeling for the limited identification of unknown acetophenones as an introductory characterization experience in an upper-level synthesis and characterization laboratory course. *Chem. Educ.* 13: 148–152.

Ehrlich, P. 1909. Über den jetzigen stand der chemotherapie. *Ber. Dtsch. Chem. Ges.* 42: 17–47.

Eisenberg, R. and Gray, H.B. 2011. Noninnocence in metal complexes: A dithiolene dawn. *Inorg. Chem.* 50: 9741–9751.

Fantacci, S., Amat, A., and Sgamellotti, A. 2010. Computational chemistry meets cultural heritage: Challenges and perspectives. *Acc. Chem. Res.* 43: 802–813.

Ferreira, S.H., Bartelt, D.C., and Greene, L.J. 1971. Isolation of bradykinin-potentiating peptides from *Bothrops jararaca* venom. *Biochem. Pharmacol.* 20: 1557–1567.

Fleming, S., Mills, A., and Tuttle, T. 2011. Predicting the UV-vis spectra of oxazine dyes. *Beilstein. J. Org. Chem.* 7: 432–441.

Geiger, W.E. 2011. Reflections on future directions in organometallic electrochemistry. *Organometallics* 30: 28–31.

Ghosh, A. 2011. *Ab initio* wavefunctions in bioinorganic chemistry: More than a succès d' estime? *J. Biol. Inorg. Chem* 16: 819–820.

Gilbert, J.K., De Jong, O., Justi, R., Treagust, D., and Van Driel, J.H. 2002. *Chemical Education: Towards Research-Based Practice*. London, U.K.: Kluwer Academic Publishers.

Gladysz, J.A. 2011. The future of organometallic chemistry. *Organometallics* 30: 1–4.

Gordin, D. and Pea, R. 1995. Prospects for scientific visualization as an educational technology. *J. Learn. Sci.* 4: 249–279.

Greer, J., Erickson, J.W., Baldwin, J.J., and Varney, M.D. 1994. Application of the three-dimensional structures of protein target molecules in structure-based drug design. *J. Med. Chem.* 37: 1035–1054.

Grotkopp, O., Ahmad, A., Frank, W., and Müller, T.J.J. 2011. Blue-luminescent 5-(3-indoyl)oxazoles via microwave-assisted three component coupling-cycloisomerization-Fischer indole synthesis. *Org. Biomol. Chem.* 9: 8130–8140.

Gund, P. 1977. Three-dimensional pharmacophoric pattern searching. *Prog. Mol. Subcell. Biol.* 11: 117–143.

Gutten, O., Besseová, I., and Rulísek, L. 2011. Interaction of metal ions with biomolecular ligands: How accurate are calculated free energies associated with metal ion complexation? *J. Phys. Chem. A* 115: 11394–11402.

Hager, T. 2006. *The Demon Under the Microscope*. New York: Harmony Books.

Hale, R. 2005. Editorial. *Drug Discov. Today* 10: 377–379.

Hansch, C. 1969. A quantitative approach to biochemical structure-activity. *Acc. Chem. Res.* 2: 232–239.

Hansch, C. and Leo, A. 1995. *Exploring QSAR. Fundamentals and Applications in Chemistry and Biology*. Washington, DC: American Chemical Society.

Huff, J.R. 1991. HIV protease: A novel chemotherapeutic target for AIDS. *J. Med. Chem.* 34: 2305–2314.

Jankowiak, A. and Kaszynski, P. 2009. 4-Substituted 1-acyloxypyridine-2(1*H*)-thiones: Experimental and computational studies of substituent effect on electronic absorption spectra. *J. Org. Chem.* 74: 7441–7448.

Jorgensen, W.L. 1989. Free energy calculations: A breakthrough for modeling organic chemistry in solution. *Acc. Chem. Res.* 22: 184–189.

Jover, J., Fey, N., Purdie, M., Lloyd-Jones, G.C., and Harvey, J.N. 2010. A computational study of phosphine ligand effects in Suzuki-Miyaura coupling. *J. Mol. Catal. A: Chem.* 324: 39–47.

Kawakami, Y., Inoue, A., Kawai, T., Wakita, M., Sugimoto, H., and Hopfinger, A.J. 1996. The rationale for E2020 as a potent acetylcholinesterase inhibitor. *Bioorg. Med. Chem.* 4: 1429–1446.

Khusnutdinova, J.R., Rath, P.R., and Mirica, L.M. 2010. Stable mononuclear organometallic Pd(III) complexes and their C-C bond formation reactivity. *J. Am. Chem. Soc.* 132: 7303–7305.

Kleb, G., Abraham, U., and Mietzner, T. 1994. Molecular similarity indices in a comparative analysis (CoMSIA) of drug molecules to correlate and predict their biological activity. *J. Med. Chem.* 37: 4130–4146.

Knight, J.D., Kramp, C.R., Hilton et al. 2007. Synthesis of *NH-2H*-4,5-dihydrobenz[g]inazoles and related compounds from 1-tetralone carbomethoxyhydrazones and aromatic esters. *Ind. Eng. Chem. Res.* 46: 8959–8967.

Kola, I. and Landis, J. 2004. Can the pharmaceutical industry reduce attrition rates? *Nat. Rev. Drug Discov.* 3: 711–715.

Kozuch, S. and Shaik, S. 2011. How to conceptualize catalytic cycles? The energetic span model. *Acc. Chem. Res.* 44: 101–110.

Lerner, L. 2011. Identity of a purple dye formed by peroxidic oxidation of *p*-aminophenol at low pH. *J. Phys. Chem.* 115: 9901–9910.

Lewars, E.G. 2011. *Computational Chemistry. Introduction to Theory and Application of Molecular Quantum Mechanics*, 2nd Edn. New York: Springer.

Liang, G. and Bowen, J.P. 1996. Molecular Skeleton Analysis (MoSA) and its application for NMDA/glycine/kynurenic binding complex. *Proceedings of the South East Regional Meeting of the American Chemical Society*, Roanoke, VA.

Liang, G., Li, S., Gu, Y., Lewin A., and Bowen, J.P. 2006. Molecular skeleton analysis (MoSA) and its applications for NMDA/glycine/kynurenic binding complexes. *American Chemical Society National Meeting*. Atlanta, GA. March 30, 2006.

Lin, Z. 2010 Interplay between theory and experiment: Computational organometallic and transition metal chemistry *Acc. Chem. Res.* 43: 602–611.

Lipinski, C.A., Lombardo, F., Dominy, B., and Feeney, P. 1997. Experimental and computational approaches to estimate solubility and permeability in drug discovery and development settings. *Adv. Drug Deliv. Rev.* 23: 3–25.

López, X., Miró, P., Carbó, J.J., Rodríguez-Fortea, A., Bo, C., and Poblet, J. 2011. Current trends in modeling of polyoxometalates. *Theor. Chem. Acc.* 128: 393–404.

Marques, M.A.L. and Gross, E.K.U. 2003. Time-dependent density functional theory. Chapter 4. In *A Primer in Density Functional Theory*, eds. Fiolhais, C., Nogueira, F., Marques, M, pp. 144–184. New York: Springer.

Martin, Y.C. 1992. 3D database searching in drug design. *J. Med. Chem.* 35: 2145–2154.

McCammon, J.A. 1987. Computer-aided molecular design. *Science* 238: 486–491.

McDougal, E.E. and Holmes, B.E. 2012. Recent trends in chemistry instrumentation requests by undergraduate institutions to NSF's RUI program. *J. Chem. Educ.* 89: 4–6.

Moore, J.W., Holmes, J.L., and Zielinski, T.J. 2009. Chemical education digital library: Online resources, services, and communities. *J. Chem. Educ.* 86: 122.

Ondetti, M.A. and Cushman, D.W. 1982. Enzymes of the renin-angiotensin system and their inhibitor. *Annu. Rev. Biochem.* 51: 283–308.

Ondetti, M.A., Rubin, B., and Cushman, D.W. 1977. Design of specific inhibitors of angiotensin-converting enzyme: New class of orally active antihypertensive agents. *Science* 196: 441–444.

Orita, Y., Ando, A., Yamabe, S., Nakanishi, T., Arakawa, Y., and Abe, H. 1983. Quantum-chemical and physico-chemical properties of hydrochlorothiazide. *Arzneimm-Forsch/Drug Res.* 33: 688–691.

Pandey, R., Mukhopadhyay, S., Ramasesha, S., and Das, P.K. 2011. Structure of the pyridine-chloranil complex in solution: A surprise from depolarized hyper-Rayleigh scattering measurements. *J. Phys. Chem.* 115: 13842–13846.

Pyykko, P. and Stanton, J.F. 2012. Introduction to quantum chemistry 2012 Issue. *Chem Rev.* 112: 1–3.

Ridley, J. and Zerner, M. 1973. Intermediate neglect of differential overlap (INDO) technique for spectroscopy. Pyrrole and the azines. *Theor. Chem. Acta* 32: 111–134.

Roy, L.E., Jakubikova, E., Guthrie, M.G., and Batista, E.R. 2009. Calculation of one electron redox potentials revisited. Is it possible to calculate accurate potentials with density functional methods? *J. Phys. Chem.* 113: 6745–6750.

Scarborough, C. and Wieghardt, K. 2011. Electronic structure of 2,2′-bipyridine organotransition-metal complexes. Establishing the ligand oxidation level by density functional theory calculations. *Inorg. Chem.* 50: 9773–9793.

Schreckenbech, G. and Shamov, G.A. 2010. Theoretical actinide molecular science. *Acc. Chem. Res.* 43: 19–29.

Sendlinger, S.C., DeCoste, D.J., Dunning, T.H et al. 2008. Transforming chemistry education through computational science. *Comput. Sci. Eng.* 10: 34–39.

Sendlinger, S.C. and Metz, C.R. 2009. CSERD—Another important NSDL pathway for computational science education. *J. Chem. Educ.* 86: 126.

Sendlinger, S.C. and Metz, C.R. 2011. Computational chemistry for chemistry educators. *J. Comput. Sci. Educ.* 1: 28–32.

Shriner, R.A., Hermann, C.K.F., Morrill, T.C., and Curtin, D.Y., and Fuson, R.C. 1998. *The Systematic Identification of Organic Compounds*, 7th Edn. New York: John Wiley & Sons.

Silverman, R.B. 2004. *The Organic Chemistry of Drug Design and Drug Action*, 2nd Edn. New York: Elsevier Academic Press.

Silverstein, R.M., Bassler, G.C., and Morrill, T.C. 1991. *Spectrometric Identification of Organic Compounds*, 5th Edn. New York: John Wiley & Sons.

Stahl, M. and Bajorath, J. 2011. Computational medicinal chemistry. *J. Med. Chem.* 54: 1–2.

Stradi, D., Díaz, C., Martín, F., and Alcamí, M. 2011. A density functional theory study of the manganese phthalocyanine. *Theor. Chem. Acc.* 128: 497–503.

Tan, J.C. and Cheetham, A.K. 2011. Mechanical properties of hybrid inorganic-organic framework materials, establishing fundamental structure-property relationships. *Chem. Soc. Rev.* 40: 1059–1080.

Valencia, H., Gil, A., and Frapper, G. 2010. Trends in the adsorption of 3d transition metal atoms onto graphene and nanotube surfaces: A DFT study and molecular orbital analysis. *J. Phys. Chem.* 114: 14141–14153.

Wittmaack, B.K., Crigger, C., Gaurino, M., and Donald, K.J. 2011. Charge saturation and neutral substitutions in halomethanes and their group 14 analogues. *J. Phys. Chem.* 115: 8743–8753.

Xu, B., Jiang, X.J., Wang, Y., Sun, H., and Yin, J. 2009. Ground and excited states calculations of 7-phenylamino-substituted coumarins. *J. Mol. Struct.* 917: 15–20.

Zerner, M.C., Ridley, J.E., Bacon, A.D et al. 2011. ZINDO-MN version 2011, Quantum Theory Project, University of Florida, Gainesville, and Department of Chemistry, University of Minnesota, Minneapolis, 2011. http://comp.chem.umn.edu/zindo-mn/ (accessed February 24, 2012).

Zhong, H., Wees, M.A., Faure, T.D., Carrillo, C., Arbiser, J., and Bowen, J.P. 2011. The impact of ionization states of matrix metalloproteinase inhibitors on docking-based inhibitor design. *ACS Med. Chem. Lett.* 2: 455–460.

Zielinski, T.J., Holmes, J., and Moore, J. 2009. Chemistry at the national science digital library. *J. Chem. Educ.* 86: 120.

30

Computational Biology: The Fundamentals of Sequence-Based Techniques

Steven M. Thompson
BioInfo 4U

30.1 Introduction

The subdisciplines within computational biology, especially those of bioinformatics, are relatively new, the word "bioinformatics" not being coined until the early to mid-1970s by Hesper and Hogeweg, 2011. With roots in computer science, information theory, biochemistry, and biology, it was and is a natural nexus of the four disciplines. However, before it could really take off, two essential factors had to become a reality, that of readily and freely available sequence and 3D structural databases and the development of powerful and relatively inexpensive computing power. Both of these are now very much a reality, having followed and/or exceeded the 18 month doubling exponential growth rate of Moore's law (e.g., Moore, 1965; Wetterstrand, 2012; and see GenBank growth statistics at ftp://ftp.ncbi.nih.gov/genbank/gbrel.txt) since their development. Within this entire field, one of the most important developments has been the ability for researchers to gain biological understanding of molecules and organismal systems based solely on sequence data.

So, given the nucleotide or amino acid sequence of a biological molecule, what can be known about that molecule? Searching for small relevant patterns in sequences that may reflect some function allows for the inference of biologically relevant information. These can be cataloged motifs ascribed to catalytic activities; restriction enzyme or protease cut sites; regional physical attributes, such as secondary structure predictions, hydrophobicities; or even a sequence's overall content and composition, as is used in some of the gene finding techniques. However, what about comparisons with other sequences? Can one molecule tell another's story? Yes, naturally it can; inference through homology is a fundamental principle to all the biological sciences. A tremendous amount of knowledge can be gained by comparing one sequence against others. When a sequence is found to fall into a preexisting biological gene family, function, enzymatic mechanism, evolution, and possibly even structure can all be inferred based on homology with its neighbors.

However, in order to learn anything by comparing sequences, one needs to know how to compare them. Constrained sequence portions, common patterns between them, could be used as "anchors" to create sequence alignments allowing comparison, but this brings up the alignment problem and "similarity." It is easy to see that sequences are aligned when they have identical symbols at identical positions, but what happens when symbols are not identical and sequences are not the same length? How can anybody know when the most similar portions of sequences are aligned, when is an alignment optimal, and does optimal mean anything as far as biology is concerned?

30.2 Databases

Sequence-based methods more often than not compare similar sequences one way or another. Acquiring the similar sequences usually requires sequence databases. These began with Dayhoff et al.'s hardbound *Atlas of Protein Sequence and Structure* in the mid-1960s (1965–1979). Databases have long since outgrown a hardbound atlas. They have become huge and have evolved considerably. In the United States, the National Center for Biotechnology Information (NCBI) (http://www.ncbi.nlm.nih.gov/), a division of the National Library of Medicine (NLM), at the National Institute of Health (NIH), supports and distributes the GenBank nucleic acid sequence database, which began in 1982 (Bilofsky et al., 1986), and the GenPept coding sequence (CDS) translations database, as well as a host of specialized sequence databases (in particular see the non-redundant RefSeq genomic, complementary DNA (cDNA), and protein databases). The National Biomedical Research Foundation (NBRF) (http://www-nbrf.georgetown.edu/), an affiliate of Georgetown University Medical Center, maintains the protein identification resource (PIR) (George et al., 1986) database of polypeptide sequences. In Europe, the European Molecular Biology Laboratory (EMBL) (http://www.embl.de/) and the European Bioinformatics Institute (EBI) (http://www.ebi.ac.uk/) maintain the EMBL nucleotide sequence database, which began in 1980 (Hamm and Cameron, 1986), and the excellently annotated Swiss-Prot (Bairoch, 1991) protein sequence database (also supported by the Swiss Institute of Bioinformatics [SIB] http://www.isb-sib.ch/), as well as the minimally annotated translations from EMBL (TrEMBL)—those EMBL translations not yet in Swiss-Prot—protein sequence databases, in Heidelberg, Germany; Cambridge, United Kingdom; and Geneva, Switzerland. EBI, SIB, and PIR coordinate to maintain the Universal Protein Resource (UniProt) (http://www.uniprot.org/), which is a single, nearly non-redundant, comprehensive, fully classified, richly and accurately annotated protein sequence knowledge base, with extensive cross-references and querying interfaces, all freely accessible to the scientific community. Additional, less well-known, sequence databases include sites with the military, with private industry, and in Japan (the DNA Data Bank of Japan, DDBJ, http://www.ddbj.nig.ac.jp/). In most cases data are openly exchanged between the databases so that many sites "mirror" one another. This is particularly true with GenBank, EMBL, and DDBJ, organized under the auspices of the International Nucleotide Sequence Databases Collaboration (INSDC). There is never a need to look in more than one of these databases; they all contain the same organized, reliable, comprehensive, and openly available libraries of genetic sequence data.

Additionally, a myriad of specialized sequence databases exist. These include sequence pattern databases such as restriction enzyme (e.g., http://rebase.neb.com/rebase/rebase.html) and protease (e.g., http://merops.sanger.ac.uk/) cleavage sites, promoter sequences and their binding regions (e.g., http://www.gene-regulation.com/pub/databases.html and http://epd.vital-it.ch/), protein motifs (e.g., http://prosite.expasy.org/) and profiles (e.g., http://pfam.janelia.org/), and organism or system-specific databases such as the sequence portions of A C. elegans database (ACeDb) (http://www.acedb.org/), FlyBase (the Drosophila database, http://flybase.org/), the Saccharomyces Genome Database (SGD) (http://www.yeastgenome.org/), and the Ribosomal Database Project (RDP) (http://rdp.cme.msu.edu/). Many sites present their data in the context of a genome map browser, for example, the University of California, Santa Cruz, bioinformatics group's genome browser (http://genome.ucsc.edu/), and the Ensembl project (http://www.ensembl.org/), jointly hosted by the Wellcome Trust Sanger Institute and EBI. Genome map browsers attempt to tie together as many data types as possible using a physical map of a particular genome as a framework. They are a particularly effective way to explore bioinformatics and a great starting point for many analyses. This is but the "tip of the iceberg" when it comes to specialized sequence databases—visit the major sites and explore.

Two other types of databases are commonly accessed in bioinformatics: reference and 3D structure. Reference databases run the gamut from Online Mendelian Inheritance in Man (OMIM) (Pearson et al., 1994), which catalogs human genes and phenotypes, particularly those associated with human disease states, to PubMed (http://www.ncbi.nlm.nih.gov/pubmed), which provides free public access to more than 22 million citations from MedLine (the NLM's citation and author abstract bibliographic database of over 5500 biomedical research and review journals), other life science journals, and online books. In nearly all cases, complete abstracts are supplied, and in many cases, links are provided to full-length text content. Other databases that could be put in this class include things like proprietary medical records databases and population studies databases. Finally, the Research Collaboratory for Structural Bioinformatics (RCSB) (http://home.rcsb.org/), a consortium of three institutions: the State University of New Jersey, Rutgers; the San Diego Supercomputer Center at the University of California, San Diego; and the University of Wisconsin–Madison, supports the 3D structure Protein Data Bank (PDB) (Bernstein et al., 1977). RCSB PDB is now a member of a World Wide PDB organization (http://www.wwpdb.org/, Berman et al., 2003), along with centers in Japan and Europe, whose mission "is to maintain a single Protein Data Bank Archive of macromolecular structural data that is freely and publicly available to the global community." The NIH maintains "Molecules To Go" (http://helixweb.nih.gov/cgi-bin/pdb) as a very easy-to-use interface to PDB. Other 3D structure databases include the Nucleic Acid Databank (NDB) at Rutgers (http://ndbserver.rutgers.edu/) and the proprietary Cambridge small molecule crystallographic structural database (CSD) (http://www.ccdc.cam.ac.uk/products/csd/).

30.3 Primary Algorithms: Just the Basics

Even with the ready availability of sequence databases, the methods of computational biology can be daunting. The bases of the necessary algorithms were already developed within and before the time period when bioinformatics began. Many are discussed elsewhere in this computing handbook series. In particular, various pattern recognition algorithms have been adopted for use in bioinformatics. Computer science concepts used in bioinformatics include dynamic programming, hash tables, edit distances, hierarchical clustering, suffix-trees, hidden Markov models (HMMs), statistical methods of maximum likelihood and Bayesian inference, neural networks, and genetic algorithms. This chapter will concentrate on only those fundamental concepts upon which the entire discipline has grown, that of pair-wise and multiple-sequence alignment, database searching, and their applicability.

Many other realms of computational biology not reviewed in this chapter overlap with the basic concepts of bioinformatics discussed. These include 3D data analysis such as x-ray crystallography, nuclear magnetic spectroscopy, and high-resolution microscopy; "next-generation" DNA sequence assembly; protein and RNA microchip analysis; all the "omics"—genomics, transcriptomics, and proteomics; and, importantly, molecular phylogenetics, population biology, and biogeography. Some of these topics will be discussed elsewhere in this series; all have foundations laid in the fundamental bioinformatics principles discussed here.

30.3.1 Pair-Wise Methods

This category includes all of the programs that perform full length and best segment pair-wise sequence alignment and the numerous database similarity searching algorithms that incorporate alignment. A fundamental aspect of most of these types of programs is a variation of a standard computer science dynamic programming algorithm. The biggest problem is that a "brute force" approach just won't work. Even without considering the introduction of gaps, the computation required to compare all possible alignments between two sequences requires time proportional to the product of the lengths of the two sequences. Therefore, if the two sequences are approximately the same length (N), this is a N^2 problem. To include gaps, one would have to repeat the calculation 2N times to examine the possibility of gaps at each possible position within the sequences, now a N^{4N} problem. Waterman (1989) pointed out that using this naïve approach to align two sequences, each 300 symbols long, would require 10^{88} comparisons, more than the number of elementary particles estimated to exist in the universe and clearly impossible to solve! Part of the solution to this problem is the dynamic programming algorithm, as applied to sequence alignment.

However, first, a very simple method that provides a "gestalt" of all the various ways that two sequences can be compared will be described. This is the dot matrix or "dot plot" technique of sequence analysis, and it can often show things not seen by any other technique, as well as provide a great introduction to all of the pair-wise comparison techniques.

30.3.1.1 Dot Matrix (Plot) Methods

Dot matrix analysis is performed by plotting one sequence on a vertical axis against another on a horizontal axis, upon a grid, using a very simple approach: wherever the two match according to some specified scoring and filtering criteria, a dot is generated in the intersecting cell. So, why use dot matrix analysis? Dot matrices can point out areas of similarity between two sequences that all other methods might miss. This is because most other methods align either the overall length of two sequences or just the "best" parts of each to achieve optimal alignments. Dot matrix methods enable the user to visualize the entirety of both sequences all at the same time; all alignments can be seen, the "less than best" comparisons as well as the main one. Regions of interest can then be "zoomed in" on using more detailed procedures, if desired. The human eye and brain are still better than a computer at discerning complex visual patterns, especially when more than one pattern is being considered. This is what makes the technique such a powerful analytic tool with sequences, from short fragments through complete genes or even chromosomes, all the way to entire genomes (see, e.g., MUMmer, Delcher et al., 1999).

Even comparing a sequence to itself can be informative. A main identity diagonal is obvious; however, other features such as palindromes can also be seen. Perfect palindromes directly cross the main diagonal, and if comparing double-stranded DNA or RNA to itself, these inverted repeat regions could be indicative of potential cruciform pseudoknots at that point. Inverted repeats that are not palindromes show up as perpendicular lines to the diagonals but lie off the main diagonal. Direct internal repeats, duplicated sequence regions, will show up as parallel diagonals off of the main diagonal. Rows or columns of diagonals in any dot plot, whether the sequences are the same or different, clearly point out multiple duplications. Comparing different sequences to one another allows for the immediate recognition of insertions and deletions between the two. It is impossible to tell whether an evolutionary event that caused this sort of discrepancy between two sequences was an insertion or a deletion, and hence this phenomenon is called an "indel." Jumps or shifts in the register of the main diagonal on a dot plot clearly points out the existence of an indel. Diagonals displaced off the center of the plot can also show the occurrence of "transposition." In fact, dot matrix analysis is one of the few ways transpositions can be recognized in sequences.

However, the use and interpretation of dot plots is entirely up to the user—one must know how to successfully filter out extraneous background noise with appropriate parameters and what the plots mean—to identify areas between sequences that may have significant matches that no other method would ever notice. Much of sequence analysis is all about balancing signal to noise, and this is particularly important with dot plots. Often a filtered windowing approach is used—a dot will only be placed on the plot if some "stringency" is met. Within some defined window size, and when some defined criteria is met, then and only then will a dot be placed at the middle of that window. The window is then shifted over one position and the entire process is repeated. This process allows extraneous, random noise, due merely to the composition of the sequences, to be eliminated, while allowing any significant similarity to come through. As a general guide to stringency levels, pick whatever window size is most appropriate for the analysis at hand, that is, about the size of the feature trying to be recognized, if known, and then choose a stringency within that window that produces the cleanest plot, without filtering out too much of the signal.

Dot plots can be particularly helpful in the analysis of small nucleic acid sequences, such as transfer RNAs (tRNAs). Consider the set of examples in Figure 30.1 using the phenylalanine tRNA molecule from yeast, GenBank:K01553, compared against itself. The sequence and structure are both known for this molecule, and the illustration shows how simple dot matrix procedures can quickly lead to functional and structural insights, even without complex folding algorithms, such as Zuker's (1989). Figure 30.1 illustrates many of the previous concepts and how dot plots work using this yeast tRNA molecule as an example.

30.3.1.2 Dynamic Programming

Dynamic programming is a widely applied computer science technique, often used in many disciplines whenever optimal substructure solutions can provide an optimal overall solution. The technique applied to sequence alignment will be illustrated using an overly simplified visualization method. Instead of calculating the "score matrix" on the fly, as is often taught as one proceeds through the graph, a "match matrix" will be completely filled in first, and then points will be added to those positions that produce favorable alignments next. Therefore, the solution will occur in two stages. The first begins very much like the dot matrix methods previously shown; the second is totally different. Furthermore, the process will be illustrated working through the cells, in spite of the fact that many authors prefer to work through the edges; they provide equivalent solutions. Points will be added based on a "looking-back-over-your-left-shoulder" algorithm rule where the only allowable trace-back is diagonally behind and above any particular cell. Matching sequence characters will be worth one point; nonmatching characters will be worth zero points. Subtracting one point for initially creating a gap, and subtracting another point per gap for extending a gap, will penalize the scoring scheme, unless the gaps are at the beginning or end of the sequence. In other words, end gaps will not be penalized; therefore, both sequences do not have to begin or end at the same point in the alignment. This zero penalty end-weighting scheme is the default for most alignment programs but can often be changed with program options, if desired. The gap penalty function described here, and illustrated in Table 30.1, is called an "affine," function, the standard "$y = mx + b$" equation for a line that does not cross the X,Y origin, where "b," the Y intercept, describes how much initial penalty is imposed for creating each new gap (Gotoh, 1982): total penalty = ([length of gap] × [gap extension penalty]) + gap opening penalty. The affine function is the usual gap penalty scheme in most pair-wise alignment programs; however, the gap "creation" or "opening" penalties and gap "extension" penalties used in the example are not typical of most programs. Furthermore, the one/zero match/mismatch scoring scheme is also not typical. These scores are used here for the sake of simplicity. Available program default values are usually quite appropriate and attempt to balance biological feasibility and sensitivity.

The example in Table 30.1 uses two random sequences that happen to fit the TATA promoter region consensus of eukaryotes and of bacteria. The most conserved bases within the consensus are capitalized by convention. The eukaryote promoter sequence is along the X-axis, and the bacterial sequence is along the Y-axis in the example. There may be more than one best path through the matrix. This time, with the relatively simple scoring function used, starting at the top and working down, and then tracing back, only one optimal alignment, with a final score of five, was discovered. This score is the highest, bottom-right value in the trace-back path graph, the sum of five matches minus no interior gaps. Notice that this would not be the optimal alignment, and the matrices would be all wrong, if end gap penalties had been imposed. Also notice that its 62.5% identity score could be improved by sliding the first "c" on the lower sequence to the left but that the gap cost scheme used prevented that from happening. The alignment score, based on the scoring scheme used, is the only number optimized by the algorithm, not any type of a similarity or identity percentage!

FIGURE 30.1 A dot plot analysis of the yeast phenylalanine tRNA. (a) Here the yeast tRNA is compared to itself with a window size of 7 and a stringency value of 5. Several direct repeats are obvious as off-diagonal alignment segments. (b) The yeast tRNA sequence compared to its reverse complement is shown next, using the same 5 out of 7 stringency setting. Now the potential for inverted repeats becomes obvious; these are the well-known stem-loop structures of the tRNA cloverleaf molecular shape. They appear as clearly delineated diagonals. These diagonals are now perpendicular to an imaginary main diagonal running opposite to the previous case, since the orientation of the second sequence was reversed. For example, examine the middle stem; the region of the molecule centered at approximately base number 38 has a clear propensity to base pair with itself without creating a loop, since it crosses that imaginary main diagonal, and then just after a small unpaired gap, another stem is formed between the region from about base number 24 through 30 with approximately 46 through 40. (c) The middle stem described earlier most likely corresponds to the bottom-most stem represented here in a "ribbon" style model of the tRNA. (Drawn from Rotkiewicz, P., iMol molecular visualization program, http://www.pirx.com/iMol, 2007.) This loop is the anticodon region of the yeast phenylalanine tRNA deposited in the 3D PDB under access code 1TRA. (From Sundaralingam, M. et al., *Nucleic Acids Res.*, 10, 2471, 1976.)

TABLE 30.1 Dynamic Programming (One Point for Match, Zero Points for Mismatch, Less One Point for Creating a Gap, Plus Less One Point per Gap for Extending the Gap)

a. First complete the match matrix using whatever match/mismatch scoring scheme used:

	c	T	A	T	A	t	A	a	g	g
c	1	0	0	0	0	0	0	0	0	0
g	0	0	0	0	0	0	0	0	1	1
T	0	1	0	1	0	1	0	0	0	0
A	0	0	1	0	1	0	1	1	0	0
t	0	1	0	1	0	1	0	0	0	0
A	0	0	1	0	1	0	1	1	0	0
a	0	0	1	0	1	0	1	1	0	0
T	0	1	0	1	0	1	0	0	0	0

b. Now add and subtract points based on the best path through the matrix, working diagonally, left to right and top to bottom. However, whenever a box is jumped to make the path, subtract one point for doing so and an additional point per box jumped, except at the beginning or end of the alignment, so that end gaps are not penalized. Fill in all additions and subtractions, calculate the sums and differences along the way, and keep track of all the best paths. The score matrix is shown with all calculations in the following:

	c	T	A	T	A	t	A	a	g	g
c	**1**	**0**	0	0	0	0	0	0	0	0
g	0	**0 + 1 = 1**	**0 + 0 = 0**	0 + 0 = 0	0 + 0 = 0	0 + 0 = 0	0 + 0 = 0	0 + 0 = 0	**1 + 0 = 1**	1 + 0 = 1
T	0	1 + 0 = 1	**0 + 1 = 1**	**1 + 0 = 1**	0 + 0 = 0	1 + 0 = 1	0 + 0 = 0	0 + 0 = 0	0 + 0 = 0	**0 + 1 = 1**
A	0	0 + 0 = 0	**1 + 1 = 2**	**0 + 1 = 1**	**1 + 1 = 2**	0 + 0 = 0	**1 + 1 = 2**	1 + 0 = 1	0 + 0 = 0	0 + 0 = 0
t	0	1 + 0 = 1	0 + 0 = 0	**1 + 2 = 3**	**0 + 1 = 1**	**1 + 2 = 3**	0 + 0 = 0	**0 + 2 = 2**	**0 + 1 = 1**	0 + 0 = 0
A	0	0 + 0 = 0	**1 + 1 = 2**	0 + 0 = 0	**1 + 3 = 4**	**0 + 1 = 1**	**1 + 3 = 4**	**1 + 3 − 2 = 2**	**0 + 2 = 2**	**0 + 1 = 1**
a	0	0 + 0 = 0	1 + 0 = 1	**0 + 2 = 2**	**1 + 3 − 2 = 2**	**0 + 4 = 4**	**1 + 4 − 2 = 3**	**1 + 4 = 5**	**0 + 2 = 2**	**0 + 2 = 2**
T	0	1 + 0 = 1	0 + 0 = 0	1 + 1 = 2	**0 + 2 = 2**	1 + 2 = 3	**0 + 4 = 4**	0 + 3 = 3	**0 + 5 = 5**	**0 + 5 − 2 = 3**

c. Clean up the score matrix next. Just the totals are shown in each cell in the matrix in the following. All best paths are highlighted:

	c	T	A	T	A	t	A	a	g	g
c	**1**	**0**	0	0	0	0	0	0	0	0
g	0	**1**	**0**	0	0	0	0	0	**1**	1
T	0	1	**1**	**1**	0	1	0	0	0	**1**
A	0	0	**2**	**1**	**2**	0	**2**	1	0	0
t	0	1	0	**3**	**1**	**3**	0	**2**	**1**	0
A	0	0	**2**	0	**4**	**1**	**4**	**2**	**2**	**1**
a	0	0	1	**2**	**2**	**4**	**3**	**5**	**2**	**2**
T	0	1	0	**2**	**2**	**3**	**4**	3	**5**	**3**

(*continued*)

TABLE 30.1 (continued) Dynamic Programming (One Point for Match, Zero Points for Mismatch, Less One Point for Creating a Gap, Plus Less One Point per Gap for Extending the Gap)

d. Finally, convert the score matrix into a trace-back path graph by picking the bottom-most, furthest right and highest scoring coordinate. Then choose the trace-back route that created that cell, to connect the cells all the way back to the beginning using the same "over-your-left-shoulder" rule. Only the best trace-back route is now highlighted with an outline font in the trace-back matrix in the following:

	c	T	A	T	A	t	A	a	g	g
c	1	**0**	0	0	0	0	0	0	0	0
g	0	1	**0**	0	0	0	0	0	1	1
T	0	1	1	**1**	0	1	0	0	0	1
A	0	0	2	1	**2**	0	2	1	0	0
t	0	1	0	3	1	**3**	0	2	1	0
A	0	0	2	0	4	1	**4**	2	2	1
a	0	0	1	2	2	4	3	**5**	2	2
T	0	1	0	2	2	3	4	3	**5**	3

e. The final alignment is shown here. It shows the random eukaryotic promoter TATA Box sequence, with a preferred region centered between −36 and −20 of the start codon, versus the random bacterial sequence that fits the consensus from the standard *Escherichia coli* RNA polymerase promoter "Pribnow" box −10 region. (Gap cost: 1, length cost: 1, match score: 1, mismatch: 0. Alignment score: 5, percent identity: 62.5.)

```
cTATAtAagg
  |||||
.cgTAtAaT.
```

Another way to explore the dynamic programming solution space is to reverse the entire process. This can often discover alternative alignments. To recap, and for those people that like mathematics, an optimal pair-wise alignment is defined as an arrangement of two sequences, 1 of length *i* and 2 of length *j*, such that the following holds true:

1. The number of matching symbols between 1 and 2 are maximized.
2. The number of gaps within 1 and 2 are minimized.
3. The number of mismatched symbols between 1 and 2 are minimized. Therefore, the actual solution can be represented by the following recursion:

$$S_{ij} = s_{ij} + \max \begin{Bmatrix} S_{i-1j-1} & \text{or} \\ \max S_{i-xj-1} + w_{x-1} & \text{or} \\ 2 < x < i & \\ \max S_{i-1j-y} + w_{y-1} & \\ 2 < y < i & \end{Bmatrix}$$

where

S_{ij} is the score for the alignment ending at *i* in sequence 1 and *j* in sequence 2
s_{ij} is the score for aligning *i* with *j*
w_x is the score for making a *x* long gap in sequence 1
w_y is the score for making a *y* long gap in sequence 2, allowing gaps to be any length in either sequence

However, just because dynamic programming guarantees an optimal alignment, it is not necessarily the only optimal alignment. Furthermore, the optimal alignment is not necessarily the "right" or biologically relevant alignment! Significance estimators can provide some handle on this, but always question the results of any computerized solution based on what is known about the biology of the system. The example shown in Table 30.1 illustrates the Needleman and Wunsch (1970) global solution, as generalized by Gotoh (1982). Later refinements (Smith and Waterman, 1981) demonstrated how dynamic programming could also be used to find optimal local alignments. To solve dynamic programming using local alignment (without going into all the details) algorithms, do the following:

1. Scoring functions penalize mismatches by assigning negative numbers for them. Therefore, bad paths quickly become very bad. This leads to a trace-back path matrix with many alternative paths, most of which do not extend the full length of the graph.
2. The best trace-back within the overall graph is chosen. This does not have to begin or end at the edges of the matrix—it's the best segment of alignment.

30.3.1.3 Significance

A particularly common misunderstanding in this field regards the concept of homology versus that of similarity: there is a huge difference! Similarity and identity are merely statistical parameters that describe how much two sequences, or portions of them, are alike according to some set scoring criteria. It can be normalized to ascertain statistical significance, as is done in database searching, but it's still just a number. Homology, in contrast and by definition, implies an evolutionary relationship—but more than just the fact that all life on Earth evolved from some common ancestor. One needs to be able to demonstrate an evolutionary lineage between the organisms or genes of interest to claim homology. Better yet, provide experimental evidence, structural, morphological, genetic, or fossil, that corroborates the assertion. There really is no such thing as percent homology; something is either homologous or it's not. Walter Fitch (personal communication) explains with the joke: "Homology is like pregnancy—you can't be 50% pregnant, just like something can't be 50% homologous. You either are or you are not." Do not make the mistake of calling any old sequence similarity homology. Highly significant similarity can argue for homology, but not the other way around.

So, how does one tell if a similarity, in other words, an alignment discovered by some program, means anything? Is it statistically significant, is it truly homologous, and, even more importantly, does it have anything to do with real biology? Many programs generate percent similarity scores; however, as seen earlier, these really don't mean a whole lot. Don't use percent similarities or identities to compare sequences except in the roughest way. They are not optimized or normalized in any manner. Alignment quality scores mean more but are difficult to interpret. At least they take the length of similarity, all of the necessary gaps introduced, and the matching of symbols all into account; however, quality scores are only relevant within the context of a particular comparison or search and are relative to the particular scoring matrix and gap penalties used.

A traditional way of ascertaining alignment significance relies on an old statistics trick—Monte Carlo simulations. This type of significance estimation has implicit statistical problems; however, few practical alternatives exist when just comparing two sequences, and they are fast and easy. Monte Carlo randomization options in dynamic programming alignment algorithms compare an original alignment score against the distribution of scores from alignments where one of the sequences is repeatedly shuffled, at least 100 times after the initial alignment, each with its own alignment score. A standard deviation is calculated based on that distribution. Comparing the mean of the randomized sequence alignment scores to the original score using a "Z-score," calculation can help decide significance. A "rule of thumb" is if the actual score is much more than three standard deviations above the mean of the randomized scores, the analysis may be significant; if it is much more than five, than it probably is significant; and if it is above ten, than it definitely is significant. Many Z-scores measure this distance from the mean using a simplistic Monte Carlo model assuming a normal Gaussian distribution, in spite of the fact that "sequence space" actually follows an "extreme value distribution"; however, this simplistic approximation actually estimates significance quite well:

$$\text{Z-score} = \frac{\left[(\text{Actual score}) - (\text{Mean of randomized scores})\right]}{(\text{Standard deviation of randomized score distribution})}$$

When the two TATA sequences from the previous dynamic programming example are compared to one another using the same scoring parameters as before, but incorporating a Monte Carlo Z-score calculation, their similarity is found not at all significant. The mean score based on 100 randomizations is around 42, with a standard deviation around 8. Therefore, the Z-score is only one, so there is no significance to the match in spite of 62.5% identity! Composition can make a huge difference—the similarity seen here is merely a reflection of the relative abundance of A's and T's in both sequences!

The FastA (Pearson and Lipman, 1988; Pearson, 1990, 1998), Basic Local Alignment Search Tool (BLAST) (Altschul et al., 1990, 1997), Profile (Gribskov et al., 1987, 1989), and HMMer (Eddy, 1996, 1998) database search algorithms and their extensions all use a similar approach but base their statistics on the distance of the query matches from the actual, or a simulated, extreme value distribution of the rest of the, "insignificantly similar," members of the database being searched. The statistics are well characterized for alignments without gaps. The number of ungapped alignments to be expected between any two sequences of length m and n, with a score greater than or equal to a particular score S, is generalized such that the expectation value E relates to S through the function $E = Kmne^{-\lambda s}$ (Karlin and Altschul, 1990, and see http://www.ncbi.nlm.nih.gov/BLAST/tutorial/Altschul-1.html). This is called the E-value for score S. In a database search, m is the length of the query and n is the size of the database in residues, so $N = mn$ is the complete search space size. K and $-\lambda$ are supplied by statistical theory, dependent on the scoring system and the background amino acid or nucleotide frequencies, calculated from actual or simulated database alignment distributions. These two parameters define the statistical significance of an E-value. Expectation values are given in scientific notation; the smaller the number, that is, the closer it is to zero, the more significant the match. In other words, the less likely a particular alignment is due to chance alone. Expectation values provide evidence for the inference of homology, or not, by showing how often to expect a particular alignment to occur as a random phenomenon in a search of that size database. Rough, conservative guidelines to Z-scores and expectation values from a typical protein search are shown in Table 30.2.

Be very careful with guidelines such as these, though, because they are entirely dependent on both the size and content of the database being searched as well as how often the search is performed! Think about it: the odds are very different for rolling dice depending on how many dice are rolled at a time, whether they are "loaded" or not and how many times they are rolled. Another very powerful empirical method of determining significance is to repeat a database search with an entry of questionable significance discovered by the previous search. If the new entry finds more significant "hits" of the same biochemical "family" of sequence as the original search did, then the entry in question is undoubtedly homologous to the original entry. That is, homology is transitive. If it finds entirely different types of sequences, then it probably is not homologous. Modular proteins with distinctly separate domains confuse issues

TABLE 30.2 Rough, Conservative Guidelines to Z-Scores and Expectation Values from a Typical Protein Search

~Z-Score	~*E* Value	Inference
≥3	≤0.1	Little, if any, evidence for homology, but impossible to disprove!
≈5	$\approx 10^{-2}$	Probably homologous, but may be due to convergent evolution
≤10	$\geq 10^{-3}$	Definitely homologous

considerably, but the principles remain the same and can be explained through domain swapping and other examples of non-vertical transmission. And, finally, the "gold standard" of homology is shared structural folds—if two proteins have the same structural fold, then, regardless of similarity, at least that particular domain is homologous between the two.

30.3.1.4 Scoring Matrices

This is all fine, but what about protein sequences—conservative amino acid replacements, as opposed to identities? The scoring functions discussed so are far only allow match or mismatch.

Allowing similarities is certainly an additional complication that would seem important. Particular amino acids are very much alike, structurally, chemically, and genetically. How can the similarity of amino acids be taken advantage of in alignments? People have been struggling with this problem since the late 1960s.

Dayhoff (Schwartz and Dayhoff, 1979) unambiguously aligned closely related protein datasets (no more than 15% difference, and in particular cytochrome c) available at that point in time and noticed that certain residues, if they mutate at all, are prone to change into certain other residues. As it turns out, these propensities for change fell into the same categories that chemists had known for years—those same chemical and structural classes mentioned earlier—conserved through the evolutionary constraints of natural selection. Dayhoff's empirical observation quantified these changes. Based on the multiple-sequence alignments that she created, the empirical amino acid frequencies within those alignments, and the assumption that estimated mutation rates in closely related proteins can be extrapolated to more distant relationships, she was able to empirically specify the relative probabilities at which different residues mutated into other residues through evolutionary history, as appropriate within some level of divergence between the sequences considered. This is the basis of the famous PAM (corrupted acronym of "accepted point mutation") 250 (meaning that the matrix has been multiplied by itself 250 times) log-odds matrix.

Since Dayhoff's time, other biomathematicians (e.g., Henikoff and Henikoff's [1992] BLOSUM [BLOcks SUbstitution Matrix] series of matrices and the Gonnet et al. matrix [1992]) have created matrices regarded more accurate than Dayhoff's original, but the concept remains the same. Plus, the PAM series remains a classic as historically the most widely used. Confusingly these matrices are known variously as symbol comparison, log-odds, substitution, or scoring tables or matrices, but they are fundamental to all sequence comparison techniques. The default amino acid scoring matrix for most protein similarity comparison programs is now the BLOSUM62 table (Henikoff and Henikoff, 1992). The "62" refers to the minimum level of identity within the ungapped sequence blocks that went into the creation of the matrix. Lower BLOSUM numbers are more appropriate for more divergent datasets. The BLOSUM62 matrix is shown in Table 30.3.

30.3.1.5 Database Searching

Database searching programs use elements of all the concepts discussed earlier; however, classic dynamic programming techniques are far too slow when used against an entire sequence database. Therefore, the algorithms use tricks to make things happen faster. These tricks fall into two main categories, that of hashing and that of approximation. Hashing is the process of breaking a sequence into small "words" or "k-tuples" of a set size and creating a "lookup," "hash" table with those words keyed to numbers. Then when any of the words match part of an entry in the database, that match is saved. In general, hashing reduces the complexity of the search problem from N^2 for dynamic programming to N, the length of all the sequences in the database.

Approximation techniques are collectively known as "heuristics." Webster's (1973) defines heuristic as "providing aid or direction in the solution of a problem, but otherwise unjustified or incapable of justification." In database searching techniques, the heuristic usually restricts the necessary search space by calculating some sort of a statistic that allows the program to decide whether further scrutiny of a particular match should be pursued. This statistic may miss things depending on the

TABLE 30.3 BLOSUM62 Amino Acid Scoring Matrix

A	A4	B-2	C0	D-2	E-1	F-2	G0	H-2	I-1	K-1	L-1	M-1	N-2	P-1	Q-1	R-1	S1	T0	V0	W-3	X-1	Y-2	Z-1
B	−2	6	−3	6	2	−3	−1	−1	−3	−1	−4	−3	1	−1	0	−2	0	−1	−3	−4	−1	−3	2
C	0	−3	9	−3	−4	−2	−3	−3	−1	−3	−1	−1	−3	−3	−3	−3	−1	−1	−1	−2	−1	−2	−4
D	−2	6	−3	6	2	−3	−1	−1	−3	−1	−4	−3	1	−1	0	−2	0	−1	−3	−4	−1	−3	2
E	−1	2	−4	2	5	−3	−2	0	−3	1	−3	−2	0	−1	2	0	0	−1	−2	−3	−1	−2	5
F	−2	−3	−2	−3	−3	6	−3	−1	0	−3	0	0	−3	−4	−3	−3	−2	−2	−1	1	−1	3	−3
G	0	−1	−3	−1	−2	−3	6	−2	−4	−2	−4	−3	0	−2	−2	−2	0	−2	−3	−2	−1	−3	−2
H	−2	−1	−3	−1	0	−1	−2	8	−3	−1	−3	−2	1	−2	0	0	−1	−2	−3	−2	−1	2	0
I	−1	−3	−1	−3	−3	0	−4	−3	4	−3	2	1	−3	−3	−3	−3	−2	−1	3	−3	−1	−1	−3
K	−1	−1	−3	−1	1	−3	−2	−1	−3	5	−2	−1	0	−1	1	2	0	−1	−2	−3	−1	−2	1
L	−1	−4	−1	−4	−3	0	−4	−3	2	−2	4	2	−3	−3	−2	−2	−2	−1	1	−2	−1	−1	−3
M	−1	−3	−1	−3	−2	0	−3	−2	1	−1	2	5	−2	−2	0	−1	−1	−1	1	−1	−1	−1	−2
N	−2	1	−3	1	0	−3	0	1	−3	0	−3	−2	6	−2	0	0	1	0	−3	−4	−1	−2	0
P	−1	−1	−3	−1	−1	−4	−2	−2	−3	−1	−3	−2	−2	7	−1	−2	−1	−1	−2	−4	−1	−3	−1
Q	−1	0	−3	0	2	−3	−2	0	−3	1	−2	0	0	−1	5	1	0	−1	−2	−2	−1	−1	2
R	−1	−2	−3	−2	0	−3	−2	0	−3	2	−2	−1	0	−2	1	5	−1	−1	−3	−3	−1	−2	[illegible]
S	1	0	−1	0	0	−2	0	−1	−2	0	−2	−1	1	−1	0	−1	4	1	−2	−3	−1	−2	[illegible]
T	0	−1	−1	−1	−1	−2	−2	−2	−1	−1	−1	−1	0	−1	−1	−1	1	5	0	−2	−1	−2	−[illegible]
V	0	−3	−1	−3	−2	−1	−3	−3	3	−2	1	1	−3	−2	−2	−3	−2	0	4	−3	−1	−1	−2
W	−3	−4	−2	−4	−3	1	−2	−2	−3	−3	−2	−1	−4	−4	−2	−3	−3	−2	−3	11	−1	2	−3
X	−1	−1	−1	−1	−1	−1	−1	−1	−1	−1	−1	−1	−1	−1	−1	−1	−1	−1	−1	−1	−1	−1	−1
Y	−2	−3	−2	−3	−2	3	−3	2	−1	−2	−1	−1	−2	−3	−1	−2	−2	−2	−1	2	−1	7	−2
Z	−1	2	−4	2	5	−3	−2	0	−3	1	−3	−2	0	−1	2	0	0	−1	−2	−3	−1	−2	5

Values whose magnitude is ≥±4 are drawn in shadowed characters to make them easier to recognize.

parameters set—that's what makes it heuristic. The exact implementation varies between the different programs, but the basic ideas follow in all of them.

Two predominant versions exist: the Fast and BLAST programs. Both return local alignments. Both are not a single program but rather a family of programs with implementations designed to compare a sequence to a database every which way imaginable: a DNA sequence against a DNA database (not recommended unless forced to do so because of dealing with a nontranslated genomic region), a translated (where the translation is done "on-the-fly" in all six frames) version of a DNA sequence against a translated ("on-the-fly") version of a DNA database (not available in Fast), a translated ("on-the-fly") version of a DNA sequence against a protein database, a protein sequence against a translated ("on-the-fly") version of a DNA database, or a protein sequence against a protein database. Many implementations allow the recognition of frame shifts in translated comparisons. In more detail:

30.3.1.5.1 FastA and Family (Pearson and Lipman, 1988; Pearson, 1990, 1998)

1. Works well for DNA against DNA searches (within limits of possible sensitivity)
2. Can find only one gapped region of similarity per search
3. Relatively slow, should often be run in the background
4. Does not require specially prepared, preformatted databases

The Fast programs are older than BLAST. They were the first widely used, powerful sequence database searching programs available. Pearson continually refines the algorithms such that they remain a viable alternative to BLAST, especially if one is restricted to searching DNA against DNA without translation. The Fast programs also may be more sensitive than BLAST in situations where BLAST finds no significant alignments.

30.3.1.5.1.1 Algorithm The Fast programs build words of a set k-tuple size, by default two for peptides. The program then identifies all exact word matches between the sequence and the database members. Scores are assigned to each continuous, ungapped, diagonal by adding all of the exact match BLOSUM values. The 10 highest scoring diagonals for each query–database pair are then rescored using BLOSUM similarities as well as identities, and ends are trimmed to maximize the score. The best of each of these is called the *Init1* score.

Next the program "looks" around to see if nearby off-diagonal *Init1* alignments can be combined by incorporating gaps. If so, a new score, *Initn*, is calculated by summing up all the contributing *Init1* scores, penalizing gaps with a penalty for each. The program then constructs an optimal local alignment for all *Initn* pairs with scores better than some set threshold using a variation of dynamic programming "in a band" (Pearson and Lipman, 1988). A sixteen residue band centered at the highest *Init1* region is used by default with peptides. The score generated from this step is called the *opt* score.

Then Fast uses a simple linear regression against the natural log of the search set sequence length to calculate a normalized z-score for the sequence pair (Pearson, 1998). It compares the distribution of these z-scores to the actual extreme value distribution of the search. Using this distribution, the program estimates the number of sequences that would be expected to have, purely by chance, a z-score greater than or equal to the z-score obtained in the search. This is reported as the expectation value. Unfortunately, the z-score used in the Fast programs and the Monte Carlo style Z-score discussed previously are quite different and cannot be directly compared; however, the expectation values between Fast and other programs are equivalent. Finally, the program uses full Smith–Waterman local dynamic programming, not "restricted to a band," to produce its final alignments.

30.3.1.5.2 BLAST: Basic Local Alignment Search Tool (Altschul et al., 1990, 1997)

1. Normally not a good idea to use for DNA against DNA searches (not optimized)
2. Prefilters repeat and "low complexity" sequence regions by default
3. Can find more than one region of gapped similarity
4. Very fast heuristic and parallel implementation
5. Restricted to precompiled, specially formatted databases

30.3.1.5.2.1 Algorithm After BLAST has sorted its lookup table, it tries to find all double-word hits along the same diagonal within some specified distance using a discrete finite automaton (DFA) (NCBI). Word hits of size *W* do not have to be identical; rather, they have to be better than some threshold value *T*. To identify double-word hits, the DFA scans through all strings of words (typically *W* = 3 for peptides) that score at least *T* (usually 11 for peptides). Each double-word hit that passes this step then triggers a process of ungapped extension in both directions, such that each diagonal is extended as far as it can, until the running score starts to drop below a predefined value *X* within a certain range *A*. The result of this pass is called a high-scoring segment pair (HSP). HSPs then, are those stretches of sequence pairs that cannot be further improved by extension or trimming. Those HSPs that pass this step with a score better than *S* then begin a gapped extension step utilizing dynamic programming. Those gapped alignments with expectation values better than the user specified cutoff are reported. The extreme value distribution of BLAST expectation values is pre-computed against each precompiled database available to the program—this is one area that speeds up the algorithm considerably.

30.3.1.5.3 Database Searching Conclusions

In review, both the Fast and BLAST family of programs calculate their Expectation "*E*" values on a more realistic "extreme value distribution," based on either real or simulated "not significantly similar" database alignments, than do most Monte Carlo style Z-scores, since Monte Carlo techniques are often based on the normal distribution, although they do parallel Monte Carlo style Z-scores fairly well. Regardless, the higher an *E*-value, the more probable that the observed match is due to chance in a search of that size database, and the lower its Z-score will be. Conversely, the smaller an *E*-value, that is, the closer it is to zero, the more significant it is, and the higher its Z-score will be. The *E*-value is the number that really matters. A value of 0.01 is usually a decent starting point for significance in most typical searches.

Furthermore, all database searching, regardless of the algorithm, is far more sensitive at the amino acid level than at the DNA level. This is because proteins have twenty match criteria, the 20 naturally occurring amino acids, versus DNA's four nucleotides, and those four DNA nucleotides can only be identical or not, not similar, to each other. Furthermore, many DNA base changes (especially third position changes) do not change the encoded protein at all because of the redundancy of the genetic code. All of these factors drastically increase the "noise" level of a DNA against DNA search and give protein searches a much greater "look-back" time, typically five to ten times longer back in evolutionary history. Therefore, whenever dealing with coding sequence, it is always prudent to search at the protein level. Furthermore, even without a protein sequence query, one can use programs that take a DNA query, translate it in all six frames, and then compare protein databases.

30.3.2 Multiple-Sequence Analysis

The power and sensitivity of sequence-based computational methods dramatically increase with the addition of more data. More data yield stronger analyses—if done carefully! Otherwise, it can confound the issue. The patterns of conservation become ever clearer by comparing the conserved portions of sequences among a larger and larger dataset. Those areas most resistant to change are most important to the molecule. The basic assumption is that those portions of sequence of crucial structural and functional value are most constrained against evolutionary change. They will not tolerate many mutations. Not that mutation doesn't occur in these regions, it is just that most mutation in the area is lethal, so it is never seen. Other areas of sequence are able to drift more readily, being less subject to this evolutionary pressure.

Therefore, sequences end up a mosaic of quickly and slowly changing regions over evolutionary time.

However, a big problem exists. As we've seen, dynamic programming reduces the pair-wise alignment problem's complexity down to order N^2—the solution of a 2D matrix, so the complexity of the solution is equal to the length of the longest sequence squared. But how are more than just two sequences at a time aligned? It becomes much harder. Sequences can be manually aligned with an editor, but some type of an automated solution is desirable, at least as a starting point to manual alignment. Solving the dynamic programming algorithm for more than just two sequences rapidly becomes intractable. Dynamic programming's complexity, and hence its computational requirements, increases exponentially with the number of sequences in the dataset being compared (complexity = $[\text{sequence length}]^{\text{number of sequences}}$), an N-dimensional matrix. So a three-sequence dynamic programming alignment would require the solution of a three-axis matrix, with complexity equal to the length of the longest sequence cubed, and so forth. A 3D matrix can at least be drawn, but more dimensions than that quickly become impossible to even visualize!

Several different heuristics have been employed over the years to simplify the complexity of the problem. One classic program, multiple sequence alignment (MSA) (Gupta et al., 1995), attempts to globally solve the N-dimensional matrix recursion using a bounding box trick. However, the algorithm's complexity precludes its use in most situations, except with very small datasets. Another way to globally solve the algorithm, and yet reduce its complexity, is to restrict the search space to only the most conserved "local" portions of all the sequences involved. This approach is used by the program pattern-induced multi-sequence alignment (PIMA) (Smith and Smith, 1992). Neither program ever really caught on though, as both require relatively small datasets.

30.3.2.1 Heuristic Solutions: How the Algorithms Work

Most multiple-sequence alignment implementations do not attempt to globally solve the algorithm; they modify dynamic programming by establishing a pair-wise order in which to build the alignment. This heuristic modification is known as pair-wise, progressive dynamic programming. Originally attributed to Feng and Doolittle (1987), this variation of the dynamic programming algorithm creates a global alignment but restricts its search space to a local neighborhood of the full length of only two sequences, at any one time. Consider a dataset of sequences. First all are compared to each other, pair-wise, using some quick variation of standard dynamic programming. This establishes an order for the dataset, most to least similar, known as the "guide tree." The algorithm then takes the top two most similar sequences and aligns them. It creates a quasi-consensus of those two and aligns that to the third sequence. Next it creates the same sort of quasi-consensus of the first three sequences and aligns that to the forth most similar. Subgroups are clustered together similarly. The way the various implementations make and use this "consensus" is one of the biggest differences between them. This process, all using standard, pair-wise dynamic programming, continues until it has worked its way through all of the sequences and/or sets of clusters, to complete the full multiple-sequence alignment. The pair-wise, progressive solution is implemented in several programs. Perhaps the most popular is ClustalW of Thompson et al. (1994) and its graphical user interface ClustalX (Thompson et al., 1997). This program achieved the first major advances over the basic Feng and Doolittle algorithm by incorporating variable sequence weighting, dynamically varying gap penalties and substitution matrices, and a neighbor-joining (NJ, Saitou and Nei, 1987) guide tree. Several more variations on the theme have come along since ClustalW. Tree-based consistency objective function for alignment evaluation (T-Coffee) (Notredame et al., 2000) was one of the first and has gained much favor. Its biggest innovation is the use of a preprocessed, weighted library of all the pair-wise global alignments between the sequences in a dataset plus the ten best local alignments associated with each pair of sequences. This helps build both the NJ guide tree and the progressive alignment. Furthermore, the library is used to assure consistency and help prevent errors, by allowing "forward thinking" to see whether the overall alignment will be better one way or another after particular segments are aligned one way or another. Notredame (2006) uses the analogy of school schedules—everybody, students, teachers, and administrators, with some folk being more important than others, that is, the weighting factor, puts the schedule they desire in a big pile, that is, T-Coffee's library, and the trick is to best fit all the schedules to one academic calendar, so that everybody is happiest, that is, T-Coffee's final multiple-sequence alignment. The M-Coffee version (Wallace et al., 2006) can even tie together multiple methods as external modules, making consistency libraries from the results of each. T-Coffee is one of the most accurate multiple-sequence alignment methods available because of this consistency-based rationale, but it is not the fastest.

Muscle (Edgar, 2004; Edgar and Batzoglou, 2006) is another multiple-sequence alignment program that came after ClustalW. It is incredibly fast, yet nearly as accurate as T-Coffee with protein data. Muscle is an iterative method that uses weighted log-expectation profile scoring along with a slew of optimizations. It proceeds in three stages—draft progressive using k-tuple counting, improved progressive using a revised tree from the previous iteration, and refinement by sequential deletion of each tree edge with subsequent profile realignment. Two other programs that claim speed and accuracy beyond standard progressive multiple-sequence alignment are partial order alignment (POA) and profile consistency multiple-sequence alignment (PCMA). POA (Lee et al., 2002) uses graph theory to represent an alignment as a partial order graph and allows for a new edit operator, homologous recombination. PCMA (Pei et al., 2003) combines Clustal-style progressive alignment of very similar sequences with a T-Coffee-like strategy of profile–profile consistency comparison. Perhaps the most accurate multiple-sequence alignment program is ProbCons (Do et al., 2005). It uses HMM techniques and posterior probability matrices that compare random pair-wise alignments to expected pair-wise alignments. Probability consistency transformation is used to reestimate the scores, and a guide tree is then constructed, which is used to compute the alignment, which is then iteratively refined.

The MAFFT (Katoh et al., 2002, 2005) multiple-sequence alignment suite is also quite powerful. It can be run many different ways—a couple of progressive, approximate modes, using a fast Fourier transform (FFT); a couple of iteratively refined methods that add in weighted-sum-of-pairs (WSP) scoring; and several iterative methods that use the WSP scoring combined with a T-Coffee-like consistency-based scoring scheme. Speed and accuracy are inversely proportional for these from fast and rough, to slow and accurate, respectively. MAFFT's FFT provides a huge speedup over most previous methods. Homologous regions are quickly identified by converting amino acid residues to vectors of volume and polarity, thus changing a 20-character alphabet to six, rather than by using an amino acid similarity matrix. Similarly, nucleotide bases are converted to vectors of imaginary and complex numbers. The FFT trick then reduces the complexity of the subsequent comparison to order N logN. FFT identifies potential similarities though, without localizing them; a sliding window step using the BLOSUM62 matrix is used for this. Then MAFFT constructs a distance matrix, and hence a progressive guide tree, on the number of shared six-tuples from this Fourier transform, rather than on a ranking based on full-length, pair-wise sequence similarity. The user can specify how many times a new guide tree is subsequently recalculated from a previous alignment as many times as desired; the alignment is reconstructed using the Needleman–Wunsch algorithm for each pass.

The iterative refinement modes build on this foundation by adding steps that adjust the alignment back and forth until there is either solely no improvement in the WSP score (or the number of cycles has reached a set limit), or it adds both this WSP score and a T-Coffee-like consistency score between pair-wise and multiple alignments to the refinement procedure. Differences in the iterative methods that combine WSP and consistency scores are based on how the pair-wise scores are calculated, globally, locally with affine gap costs, or locally with generalized affine gap costs. Knowing which to choose, especially when dealing with sequences too diverged for the fast methods to work well, depends on the nature of the data. MAFFT's algorithm page (http://mafft.cbrc.jp/alignment/software/algorithms/algorithms.html) explains where and when each mode is most appropriate. If sequences have full-length, but low, similarity, then the global option is appropriate (ginsi); if sequences have one similar domain among a bunch of "junk," then the local affine gap option works best (linsi); whereas if sequence data are composed of multiple, yet alignable domains, then the local generalized affine gap scheme is appropriate (einsi).

MAFFT's capability to handle large datasets and its speed is similar to or greater than Muscle's in its faster modes; its results and capabilities are similar to T-Coffee in its slow, iteratively refined, optimized modes.

30.3.2.2 Coding DNA Issues

As discussed in the database searching section, all alignment algorithms, be they pair-wise, multiple, or database similarity searching, are far more sensitive at the amino acid level than at the DNA level. The signal-to-noise ratio is just so much better with amino acids, for all the reasons previously described. Therefore, database searching and sequence alignment both should always be done at the protein level, unless forced otherwise by dealing with noncoding DNA, or if the sequences are so very similar as to not cause any problems. Therefore, usually, if dealing with coding sequences, translate DNA to its protein counterpart, before performing multiple-sequence alignment. Even if dealing with very similar coding sequences, where the DNA can be directly aligned, it is often best to align the DNA along with its corresponding proteins. In addition to the much more reliable alignment, this also insures that alignment gaps are not placed within codons, which would not make sense biologically. DNA indels in coding regions need to be in multiples of three to avoid frameshift errors. Phylogenetic analysis can then be performed on the DNA rather than on the proteins, if desired. This is especially important when dealing with datasets that are quite similar, since the proteins may not reflect differences hidden in the DNA. Furthermore, many people prefer to run phylogenetic analyses on DNA rather than protein regardless of how similar they are—the multiple substitution models have a long and well-accepted history, and yet are far simpler. However, the more diverged a dataset becomes, the more random third and eventually first codon positions become, which introduces noise (error) into the analysis. Therefore, often third positions and sometimes first positions are masked out of datasets for phylogenetic analysis. Just like in most of computational molecular biology, one is always balancing signal against noise. Too much noise or too little signal, both degrade the analysis to the point of nonsense.

Several scripts and programs, as well as some web servers, can perform this sort of codon-based alignment, but they can be a bit tricky to run. Examples include mrtrans (Pearson, 1990) (also available in EMBOSS [Rice et al., 2000] as tranalign and in BioPerl [Stajich et al., 2002] as aa_to_dna_aln), transAlign (Bininda-Emonds, 2005), RevTrans (Wernersson and Pedersen, 2003), PAL2NAL (Suyama et al., 2006), and TranslatorX (Abascal et al., 2010). Some multiple-sequence alignment editors, such as SeaView (Galtier et al., 1996), can also help with this process.

Multiple-sequence alignment is much more difficult if forced to align nucleotides because a sequence does not code for a protein. Automated methods may be able to help as a starting point, but they are certainly not guaranteed to come up with a biologically correct alignment. The resulting alignment will probably have to be extensively edited, if it works at all. Success will largely depend on the similarity of the nucleotide dataset.

30.3.2.3 Reliability?

One liability of most global, progressive, pair-wise methods is they are entirely dependent on the order in which the sequences are aligned. Fortunately, ordering them from most similar to least similar makes biological sense and works well. However, most of the techniques are very sensitive to the substitution matrix and gap penalties specified. Some programs allow "fine-tuning" areas of an alignment by realignment with different scoring matrices and/or gap penalties; this can be extremely helpful. Regardless, any automated multiple-sequence alignment program should be thought of only as a tool for offering a starting alignment that can be improved upon, not the "end-all-to-meet-all" solution, guaranteed to provide the "one-true" answer. Although, in this post-genomics era, especially when dealing with giga-bases of data, it does make sense to start with the "best" solution possible. This is the premise of using a very accurate multiple-sequence alignment package, such as T-Coffee (Notredame et al., 2000), ProbCons (Do et al., 2005), POA (Lee et al., 2002), PCMA (Pei et al., 2003), or MAFFT (Katoh et al., 2002, 2005).

Whichever program used to create an alignment, always use comparative approaches to help assure its reliability. After the program has offered its best guess, try to improve it further. Think about it—a sequence alignment is a statement of positional homology—it is a hypothesis of evolutionary history. It establishes the explicit homologous correspondence of each individual sequence position, each column in the alignment. Therefore, insure that it is as good as it can be. Be sure that it makes biological sense—align things that make sense to align! Beware of comparing "apples and oranges." Be particularly suspect of sequence datasets found through text-based database searches such as NCBI's Entrez (Schuler et al., 1996) or the EMBL/EBI sequence retrieval system (SRS) (Etzold and Argos, 1993). For example, don't try to align receptors and/or activators with their namesake proteins or with each other. Be wary of trying to align genomic sequences with cDNA when working with DNA; the introns will create big problems, unless using software specifically designed for the task. Similarly, aligning mature and precursor proteins, or alternate splicing forms, from the same organism and locus, doesn't make evolutionary sense, as one is not evolved from the other, rather one is the other. Watch for redundant sequences; there are lots of them in the databases. Some programs can automatically cull them from a dataset. If creating alignments for phylogenetic inference, make either paralogous comparisons (evolution via gene duplication) to ascertain gene phylogenies within one organism or orthologous (within one ancestral loci) comparisons to ascertain gene phylogenies between organisms (which should imply organismal phylogenies). Try not to mix the two without complete data representation. Otherwise, confusion can mislead interpretation, especially if the sequences' nomenclature is inconsistent. These are all easy mistakes to make; try to avoid them.

Devote considerable time and energy toward developing the best alignment possible. Use biological understanding to help guide judgment. Look for conserved functional, enzymatic, regulatory, and structural elements and motifs—they should all line up. Searches of the *PROSITE database of protein families and domains* (Bairoch, 1992) for cataloged structural, regulatory, and enzymatic consensus patterns or "signatures" can help, as can searches of the Pfam protein family database of HMM profiles (Eddy, 1996, 1998), and *de novo* motif discovery tools like MEME (Bailey and Elkan, 1994) and MotifSearch (Bailey and Gribskov, 1998). Look for columns of strongly conserved residues such as tryptophans, cysteines, and histidines; important structural amino acids such as prolines, tyrosines, and phenylalanines; and conserved isoleucine, leucine, and valine substitutions. The conservation of covarying sites in ribosomal and other structural RNA alignments can be very helpful in refining RNA alignments. That is, as one base in a stem structure changes, the corresponding Watson–Crick paired base will change in a corresponding manner. This principle has guided the assembly of ribosomal RNA (rRNA) structural alignments at the Ribosomal Database Project at Michigan State University (RDP, Cole et al., 2007), at the University of Gent, Belgium European Ribosomal RNA database (Wuyts et al., 2004), and at the German SILVA database system (Pruesse et al., 2007).

Editing alignments to insure that all columns are truly homologous should be encouraged, not discouraged. Dedicated sequence alignment editing software such as the genetic data environment (GDE) (Smith et al., 1994), Jalview (Clamp et al., 2004), Se-Al (Rambaut, 1996), and SeaView (Galtier et al., 1996) are great for this, but any editor will do, as long as the sequences end up properly formatted afterward. Structural alignment is the "gold standard," but the luxury of having homologous, experimentally solved structures is often not available. Even with a structural alignment, there'll often be questionable regions of sequence data within it.

These highly saturated regions have the property known as "homoplasy." This is a region of a sequence alignment where so many multiple substitutions have occurred at homologous sites that it is impossible to know if those sites are properly aligned and, thus, impossible to ascertain relationships based on those sites. The primary assumption of all phylogenetic inference algorithms is most violated in these regions, and this phenomenon increasingly confounds evolutionary reconstruction as divergence between the members of a dataset increases.

Therefore, only analyze those sequences and those portions of an alignment that assuredly do align. This often means trimming down or somehow excluding those homoplastic regions, some of the internal gaps, and, minimally, the alignment's extreme amino- and carboxy-termini (5′ and 3′ in DNA), which seldom align well. Sequence editors such as SeaView can do this. Those portions excluded will not be used in subsequent analyses. These decisions are somewhat subjective by nature, experience helps, and some software, such as Gblocks (Talavera and Castresana, 2007), ASaturA (Van de Peer et al., 2002), and T-Coffee (Notredame et al., 2000), has the ability to evaluate the quality of particular regions of an alignment. If, after all else, some region, or an entire sequence, just can't be aligned, if there is any doubt, then do not use it. Cutting an entire sequence out of an alignment may leave columns of gaps across the entire alignment that will need to be removed. Most alignment editors have a function for closing these common gaps. The validity of all subsequent analyses is absolutely dependent upon the quality and accuracy of the input multiple-sequence alignment. Remember the old adage "garbage in—garbage out!" Some general guidelines to remember (Olsen, 1992) include the following:

- If the homology of a region is in doubt, then throw it out
- Avoid the most diverged parts of molecules; they are the greatest source of systematic error
- Do not include sequences that are more diverged than necessary for the analysis at hand

Biocomputing is always a delicate balance—signal against noise—and sometimes it can be quite the balancing act!

30.3.2.4 Applicability

So what's the big deal about multiple-sequence alignment; why would anyone want to bother? Multiple-sequence alignments can be

- Extremely useful in the development of PCR primers and hybridization probes
- Great for producing annotated, publication quality, graphics, and illustrations
- Required for building HMM profiles for remote homology similarity searching and alignment
- Invaluable for structural and functional analyses via homology inference
- Absolutely necessary for molecular evolutionary phylogenetic inference

Multiple-sequence alignments can be very helpful for designing phylogenetic-specific probes and primers by allowing for the clear visualization and localization of the most conserved and the most variable regions within an alignment. Depending on the dataset being analyzed, any level of phylogenetic specificity can be achieved. Areas of high variability in the overall dataset that correspond to areas of high conversation in phylogenetic category subset datasets can differentiate between universal and phylo-specific potential probe sequences. Any of several primer discovery programs, such as MIT's Primer3 (Rozen and Skaletsky, 2000) or the commercial Oligo program (National Biosciences, Inc.), can be used to find and test the best primers within the target areas localized by this method.

Graphics prepared from multiple-sequence alignments can dramatically illustrate functional and structural conservation as it relates to sequence similarity. These can take many forms of all or portions of an alignment—shaded or colored boxes or letters for each residue or base (e.g., BoxShade [Hofmann and Baron] and PrettyPlot in EMBOSS [Rice et al., 2000]), cartoon representations (e.g., WebLogos [Schneider and Stephens, 1990]), running line graphs of overall similarity (as displayed by ClustalX [Thompson et al., 1997] and others), overlays of attributes, various consensus representations, etc.—all can be printed with high-resolution equipment, usually in color or gray tones. These can make a big difference in a poster or manuscript presentation.

Profiles are position-specific scoring matrices (PSSM) that describe a sequence alignment or a portion of an alignment. These powerful tools are created from an existing alignment of similar sequences and can be used to perform the most sensitive database remote similarity search possible. Another advantage of profile techniques is they can be used to build ever-larger multiple-sequence alignments much faster and more accurately than any of the progressive multiple-sequence alignment methods. This ability to easily create larger and larger multiple-sequence alignments is incredibly powerful and much faster than starting all over each time another sequence needs to be added to an alignment. Originally described by Gribskov et al. (1987, 1989), and then automated by NCBI's PSI-BLAST (Altschul et al., 1997), later refinements have added more statistical rigor (see, e.g., Eddy's HMM HMMer profiles and

Pfam profile database [1996, 1998] and Bailey and Elkan's Expectation Maximization [1994]). Briefly described profiles upweight conserved regions such that the more highly conserved a residue is, the more important it becomes, and, in those profiles that allow gaps, gap insertions are penalized more heavily in conserved areas of the alignment than in variable regions.

Conserved sequence really does matter. In addition to the conservation of primary sequence, structure and function are also conserved in these crucial regions. In fact, recognizable structural conservation between true homologues extends way beyond statistically significant sequence similarity. The serine protease superfamily contains a good example. *Streptomyces griseus* protease A shows remarkably little sequence similarity when compared to the rest of the superfamily (expectation values $10^{1.8}$ in a typical protein database search), yet it clearly is a serine protease as its 3D structure can be superimposed over most other members of the family with a root mean square deviation (RMSD) of less than 3 Å (Pearson, W.R., personal communication). These comparative principles are the premise of "homology modeling," which works remarkably well. One of the best applications of these principles is PredictProtein, which uses weighted dynamic programming multiple-sequence alignment methods (MaxHom, Sander and Schneider, 1991), along with neural net technology, to predict protein secondary structure by the profile network method (PHD, Rost and Sander, 1993, 1994) at an expected 70.2% average accuracy for the three states helix, strand, and loop. Furthermore, even 3D modeling without crystal coordinates is possible, if a sequence is similar enough to an experimentally solved structure. In fact, the SWISS-MODEL (Guex and Peitsch, 1997; Guex et al., 1999) system at the ExPASy server in Switzerland, supported by SIB and GlaxoSmithKline, automates the homology modeling process, given similar enough sequence data to at least one solved structure, though as in most cases, a multiple-sequence alignment of solved structure sequences makes the theoretical model even more probable.

Finally, multiple-sequence alignments are required for phylogenetic inference. Many different computational methods can estimate the most reasonable evolutionary tree for a sequence alignment based on the assertion of homologous positions within that alignment. Computational tools that incorporate these methods include Phylogenetic Analysis Using Parsimony (PAUP*) (and other methods) (Swofford, 1989–2012), PHYLogeny Inference Package (PHYLIP) (Felsenstein, 1980–2012), MrBayes (Ronquist and Huelsenbeck, 2003), GARLI (Zwickl, 2006), and RAxML (Stamatakis, 2006). This is a huge and complicated field of study and will not be discussed further here. However, always remember that regardless of the algorithm used—any form of parsimony, all of the distance methods, all maximum likelihood techniques, and even all types of Bayesian phylogenetic inference—all make the absolute validity of the input alignment matrix their first and most critical assumption (but see, e.g., Lunter et al., 2005). The famous Darwinian evolutionist Theodosius Dobzhansky summarized the importance of phylogenetic analyses in 1973, provided as an inscription on the inner cover of the classic organic evolution text *Evolution*: "Nothing in biology makes sense except in the light of evolution" (Dobzhansky et al., 1977). Evolution provides the single, unifying, cohesive force to explain all life.

30.4 Complications

Sequence data format is a huge problem in bioinformatics. The major databases all have their own distinct format, plus many of the different programs and packages require their own. Clustal (Higgins et al., 1992) has a specific format associated with it. The Fast package (Pearson and Lipman, 1988) uses a very basic sequence format that many programs recognize. NCBI uses a library standard called Abstract Syntax Notation One (ASN.1), plus it provides GenBank flat file format for all sequence data. PAUP* (Swofford, 1989–2012), MrBayes (Ronquist and Huelsenbeck, 2003), and many other phylogenetic analysis packages have a required format called the NEXUS file (Maddison et al., 1997). Even PHYLIP (Felsenstein, 1980–2012) has its own unique data format. Standards have been argued over for years, such as using XML for everything, but until everybody agrees, which is not likely to happen, it just won't happen. Fortunately several programs are available to convert formats back and forth between the required standards; however, it can all get quite confusing. BioPerl's SeqIO system (Stajich et al., 2002) and ReadSeq (Gilbert, 1990–2008) are two very helpful tools for format conversion. T-Coffee (Notredame et al., 2000) comes with one built in, and the SeaView (Galtier et al., 1996) editor recognizes NEXUS, Clustal, FastA, PHYLIP, and MASE format. Alignment gaps are still another problem. Different program suites may use different symbols to represent them. Most programs use hyphens "-" but some do not. Furthermore, not all gaps in sequences should be interpreted as deletions. Interior gaps are probably okay to represent this way, as regardless of whether a deletion, insertion, or a duplication event created the gap, logically they are the same. These are the indels mentioned previously. However, end gaps should not be represented as indels, because a lack of information before or beyond the length of any given sequence may not be due to a deletion or insertion event. It may have nothing to do with the particular stretch being analyzed at all. It just may not have been sequenced! These gaps are just placeholders for the sequence. Therefore, it is safest to manually edit an alignment to change leading and trailing gap symbols to "x"'s, which mean "unknown amino acid," or "n"'s, which mean "unknown base," or "?"'s, which is supported by many programs, but not all, and means "unknown residue or indel." This will assure that incorrect assumptions are not made, though most phylogenetic inference algorithms treat indels and missing data equivalently by default.

30.5 Conclusions

The comparative method is a cornerstone of the biological sciences. Multiple-sequence alignment and database searching are the comparative method on a molecular scale and are a vital prerequisite to some of the most powerful computational biology techniques available. Understanding something about the algorithms and the program parameters of each is the only way to rationally know what is appropriate. Knowing and staying well within the limitations of any particular method will avert much frustration. One point that needs to be emphasized is sequence analysis techniques generally have appropriate default parameters. This will usually work just fine, but it is a good idea to think about what these default values imply and adjust them accordingly, especially if the results seem inappropriate after a first pass with the default parameters intact.

Furthermore, the dramatic importance of a multiple-sequence alignment cannot be understated. All subsequent analyses are absolutely dependent upon it, especially phylogenetic inference. Also, if building multiple-sequence alignments for phylogenetic inference, do not base an organism's phylogeny on just one gene. Many complicating factors can produce weird phylogenies: bad alignments, insufficient data, abjectly incorrect models, saturated positions (homoplasy), compositional biases, and/or horizontal gene transfer. Therefore, use several genes—the Ribosomal Database Project (Cole et al., 2007) provides a good, largely accepted alignment and phylogenetic framework with which other phylogenies can be compared. Anytime the orthologous phylogenies of organisms based on two different genes do not agree, there is either some type of problem with the analysis, or lateral gene transfer has occurred. Paralogous gene phylogenies are another story altogether and should be based, if at all possible, on sequences all from the same organism.

Gunnar von Heijne (1987) in his quite readable but very dated treatise, *Sequence Analysis in Molecular Biology; Treasure Trove or Trivial Pursuit*, provides an appropriate conclusion:

> Think about what you're doing; use your knowledge of the molecular system involved to guide both your interpretation of results and your direction of inquiry; use as much information as possible; and do not blindly accept everything the computer offers you.
>
> He continues:
>
> … if any lesson is to be drawn … it surely is that to be able to make a useful contribution one must first and foremost be a biologist, and only second a theoretician …. We have to develop better algorithms, we have to find ways to cope with the massive amounts of data, and above all we have to become better biologists. But that's all it takes.

Key Terms

Affine: An affine function is a linear function, described by the algebraic formula $y = mx + b$.

Expectation value (*E*-value): The likelihood that a particular sequence alignment is due to chance. The value is dependent on sequence and database composition and size and on how often a researcher performs database searches. Most modern sequence database similarity programs such as BLAST and FastA provide this statistic based on the extreme value distribution. The closer the value is to zero, the more significant the match.

Exon: A defined stretch of DNA within a gene that is transcribed into a corresponding complimentary message RNA molecule and that is maintained after RNA processing into mature mRNA, such that only those portions not excised follow the relevant genetic code to be translated into corresponding amino acids.

Gap penalties: In the context of affine gap penalties, the creation or opening penalty is how many points a dynamic programming algorithm is penalized for imposing a gap in an alignment (the *b* in the previous equation).

The extension or lengthening penalty describes how many additional points the algorithm is penalized for each additional gap added to the first one after it is introduced (the *x* in the previous equation).

Gene: A defined stretch of DNA nucleotides that encodes either a protein, with or without introns and exons, or a structural or regulatory RNA (e.g., tRNAs, rRNAs, microRNAs, siRNAs, and snRNAs), which is inherited from one generation to the next.

Global alignment: As opposed to local alignment, which is the alignment of only the best regions within sequences, global alignment is the alignment of the full length of a sequence set and generally applies to the multiple-sequence alignment problem. However, it should be realized that global alignment can be restricted to subsequences within sequences, the distinction being that local alignment "picks" the best regions, whereas global alignment uses the full length of whatever is specified.

HMM: In sequence analysis, HMMs contain the statistical description of a recognized functional unit of some particular biological gene family or superfamily, assembled under the probabilistic model of hidden Markov chains (see profile).

Homology: An organ, trait, gene, or even DNA nucleotide position shared with a common ancestor. Homology, as opposed to sequence identity and similarity, can have no level.

A sequence and, in fact, a position within an alignment, is either demonstrably related via evolution to another or not. Statistically significant similarity can argue for homology; however, a lack of statistically significant similarity cannot be used to argue against homology.

Homology modeling: The secondary and often tertiary, 3D structure of proteins can be inferred by alignment with proteins whose 3D structure has been experimentally determined. Obviously the more similar the sequences are, the more successful the theoretical model will be.

Homoplasy: A region of a sequence alignment where so many multiple substitutions have occurred at homologous sites that it is impossible to know if those sites are properly aligned and, thus, impossible to ascertain relationships based on those sites.

Indel: An indel is a gap introduced into a sequence alignment necessary to reconcile differing lengths due to differing evolutionary histories. It is impossible to ascertain whether an insertion or a deletion event created the discrepancy, hence the term indel.

Intron: A defined stretch of DNA within a gene that is transcribed into a corresponding complimentary message RNA molecule, but that is not retained after RNA processing into a mature mRNA and, therefore, is not translated into amino acids.

Matrices: A match matrix is the first step in dot plot techniques and in the visualization of the dynamic programming algorithm illustrated. Its cells contain the value each position receives for matching (aligning) respective X- and Y-axis characters.

A score matrix also has two meanings. In the context of dynamic programming, it is the matrix in which cell values have received initial match values adjusted by gap penalties and trace-back paths. The alternative meaning describes the values that amino acid residues or nucleotide bases receive for aligning with one another, also known as symbol substitution tables, or log-odds matrices, for example, the PAM and BLOSUM matrix series.

The trace-back path matrix delineates the alignments discovered by the dynamic programming algorithm; it illustrates the optimal paths through the matrix.

Motif: A motif is a described and cataloged region of a sequence, usually shorter than a domain and often, but not always, associated with some biological structural, functional, or regulatory role. Motifs are commonly represented as consensus patterns but are often described by HMM profiles as well.

Orthology and paralogy: Two major classes of sequence homology exist. Orthology describes homologous sequences present in different organisms as a result of speciation processes. Paralogy describes homologous sequences within the same organism as a result of gene duplication. Major confusion can result from mixing paralogues and orthologues in the same analysis.

Profile: A profile is a statistical description of a multiple-sequence alignment, commonly of a region or a motif within a multiple-sequence alignment. Profiles take many forms associated with the particular programs that create them, for example, ProfileBuild, HMMer, MEME, but always increase the importance of conserved residues or bases and decrease the importance of variable areas.

z-score and Z-score: The Z-score is usually based on a normal Gaussian distribution and describes how many standard deviations a particular score is from the distribution's mean, though it may be based on the extreme value distribution. This is confusingly in contrast to Bill Pearson's z-score in the FastA programs that is a linear regression of the opt score against the natural log of the search set sequence length. The two values, Z and z, have entirely different magnitudes and should not be correlated.

References

Abascal, T., Zardoya, R., and Telford, M. 2010. TranslatorX: Multiple alignment of nucleotide sequences guided by amino acid translations. *Nucleic Acids Research* **38**, W7–W13. http://translatorx.co.uk/

Altschul, S.F., Gish, W., Miller, W., Myers, E.W., and Lipman, D.J. 1990. Basic local alignment tool. *Journal of Molecular Biology* **215**, 403–410.

Altschul, S.F., Madden, T.L., Schaffer, A.A., Zhang, J., Zhang, Z., Miller, W., and Lipman, D.J. 1997. Gapped BLAST and PSI-BLAST: A new generation of protein database search programs. *Nucleic Acids Research* **25**, 3389–3402. http://www.ncbi.nlm.nih.gov/BLAST/, and source code at ftp://ftp.ncbi.nih.gov/blast/

Bailey, T.L. and Elkan, C. 1994. Fitting a mixture model by expectation maximization to discover motifs in biopolymers, in *Proceedings of the Second International Conference on Intelligent Systems for Molecular Biology*, AAAI Press, Menlo Park, CA. pp. 28–36.

Bailey, T.L. and Gribskov, M. 1998. Combining evidence using p-values: Application to sequence homology searches. *Bioinformatics* **14**, 48–44.

Bairoch, A. 1991. The Swiss-Prot protein sequence data bank. *Nucleic Acids Research* **19**, 2247–2249.

Bairoch A. 1992. PROSITE: A dictionary of sites and patterns in proteins. *Nucleic Acids Research* **20**, 2013–2018.

Berman, H.M., Henrick, K., and Nakamura, H. 2003. Announcing the worldwide protein data bank. *Nature Structural Biology* **10**, 980.

Bernstein, F.C., Koetzle, T.F., Williams, G.J.B., Meyer Jr. E.F., Brice, M.D., Rodgers, J.R., Kennard, O., Shimanouchi, T., and Tasumi, M. 1977. The protein data bank: A computer-based archival file for macromodel structures. *Journal of Molecular Biology* **112**, 535–542.

Bilofsky, H.S., Burks, C., Fickett, J.W., Goad, W.B., Lewitter, F.I., Rindone, W.P., Swindell, C.D., and Tung, C.S. 1986. The GenBank™ genetic sequence data bank. *Nucleic Acids Research* **14**, 1–4.

Bininda-Emonds, O.R.P. 2005. TransAlign: Using amino acids to facilitate the multiple alignment of protein-coding DNA sequences. *BioMed Central Bioinformatics* **6**, 156. http://www.uni-oldenburg.de/ibu/systematik-evolutionsbiologie/programme/

Clamp, M., Cuff, J., Searle, S.M., and Barton, G.J. 2004. The Jalview java alignment editor. *Bioinformatics* **20**, 426–427. http://www.jalview.org/

Cole, J.R., Chai, B., Farris, R.J., Wang, Q., Kulam-Syed-Mohideen, A.S., McGarrell, D.M., Bandela, A.M., Cardenas, E., Garrity, G.M., and Tiedje, J.M. 2007. The ribosomal database project (RDP-II): Introducing myRDP space and quality controlled public data. *Nucleic Acids Research* **35**, 169–172. http://rdp.cme.msu.edu/

Dayhoff, M.O., Eck, R.V., Chang, M.A., and Sochard, M.R. 1965. in Dayhoff, M.O. (ed.), *Atlas of Protein Sequence and Structure*, Vol. **1**. National Biomedical Research Foundation, Silver Spring, MD.

Delcher, A.L., Kasif, S., Fleischmann, R.D., Peterson, J., White, O., and Salzberg, S.L. 1999. Alignment of whole genomes. *Nucleic Acids Research* **27**, 2369–2376. This is the original paper; see others and download source at http://mummer.sourceforge.net/

Do, C.B., Mahabhashyam, M.S.P., Brudno, M., and Batzoglou, S. 2005. ProbCons: Probabilistic consistency-based multiple sequence alignment. *Genome Research* **15**, 330–340. http://packages.debian.org/squeeze/probcons

Dobzhansky, T., Ayala, F.J., Stebbins, G.L., and Valentine, J.W. 1977. *Evolution*. W.H. Freeman and Co., San Francisco, CA. The source of the original 1973 quote is obscure though it has been cited as being transcribed from the *American Biology Teacher*, March 1973, **35**, 125–129.

Eddy, S.R. 1996. Hidden Markov models. *Current Opinion in Structural Biology* **6**, 361–365.

Eddy, S.R. 1998. Profile hidden Markov models. *Bioinformatics* **14**, 755–763. These are the original papers; see extensive bibliography, software, databases, and Web server at http://hmmer.janelia.org/ and http://pfam.janelia.org/

Edgar, R.C. 2004. MUSCLE: Multiple sequence alignment with high accuracy and high throughput. *Nucleic Acids Research* **32**, 1792–1797. Available at http://www.drive5.com/muscle/

Edgar, R.C. and Batzoglou, S. 2006. Multiple sequence alignment. *Current Opinion in Structural Biology* **16**, 368–373.

Etzold, T. and Argos, P. 1993. SRS—An indexing and retrieval tool for flat file data libraries. *Computer Applications in the Biosciences* **9**, 49–57.

Felsenstein, J. 1980–2012. PHYLIP (Phylogeny Inference Package) version 3.69. Distributed by the author. Department of Genome Sciences, University of Washington, Seattle, WA. http://evolution.genetics.washington.edu/phylip.html

Feng, D.F. and Doolittle, R.F. 1987. Progressive sequence alignment as a prerequisite to correct phylogenetic trees. *Journal of Molecular Evolution* **25**, 351–360.

Galtier, N., Gouy, M., and Gautier, C. 1996. SeaView and Phylo_win, two graphic tools for sequence alignment and molecular phylogeny. *Computer Applications in the Biosciences* **12**, 543–548. http://pbil.univ-lyon1.fr/software/

George, D.G., Barker, W.C., and Hunt, L.T. 1986. The protein identification resource (PIR). *Nucleic Acids Research* **14**, 11–16.

Gilbert, D.G. 1990–1993 (C release) and 1999–2008 (Java release). ReadSeq, public domain software. Distributed by the author. Bioinformatics Group, Biology Department, Indiana University, Bloomington, IN. http://iubio.bio.indiana.edu/soft/molbio/readseq/

Gonnet, G.H., Cohen, M.A., and Benner, S.A. 1992. Exhaustive matching of the entire protein sequence database. *Science* **256**, 1443–1445.

Gotoh, O. 1982. An improved algorithm for matching biological sequences. *Journal of Molecular Biology* **162**, 705–708.

Gribskov, M., Luethy, R., and Eisenberg, D. 1989. Profile analysis, in *Methods in Enzymology*, Russell F. Doolittle (ed.), Vol. **183**, Academic Press, San Diego, CA. pp. 146–159.

Gribskov, M., McLachlan, M., and Eisenberg, D. 1987. Profile analysis: Detection of distantly related proteins. *Proceedings of the National Academy of Sciences of the United States of America* **84**, 4355–4358.

Guex, N., Diemand, A., and Peitsch, M.C. 1999. Protein modelling for all. *Trends in the Biochemical Sciences* **24**, 364–367. http://swissmodel.expasy.org/.

Guex, N. and Peitsch, M.C. 1997. SWISS-MODEL and the Swiss-PdbViewer: An environment for comparative protein modeling. *Electrophoresis* **18**, 2714–2723.

Gupta, S.K., Kececioglu, J.D., and Schaffer, A.A. 1995. Improving the practical space and time efficiency of the shortest-paths approach to sum-of-pairs multiple sequence alignment. *Journal of Computational Biology* **2**, 459–472. MSA available at http://www.ncbi.nlm.nih.gov/CBBresearch/Schaffer/msa.html

Hamm, G.H. and Cameron, G.N. 1986. The EMBL data library. *Nucleic Acids Research* **14**, 5–10.

von Heijne, G. 1987. *Sequence Analysis in Molecular Biology; Treasure Trove or Trivial Pursuit*. Academic Press, San Diego, CA.

Henikoff, S. and Henikoff, J.G. 1992. Amino acid substitution matrices from protein blocks. *Proceedings of the National Academy of Sciences of the United States of America* **89**, 10915–10919.

Higgins, D.G., Bleasby, A.J., and Fuchs, R. 1992. CLUSTALV: Improved software for multiple sequence alignment. *Computer Applications in the Biological Sciences* **8**, 189–191.

Hofmann, K. and Baron, M. Version 3.21 BOXSHADE server at http://www.ch.embnet.org/software/BOX_form.html; software available at http://boxshade.sourceforge.net/ (accessed October 10, 2013).

Hogeweg, P. 2011. The roots of bioinformatics in theoretical biology. *PLoS Computational Biology* 7(3): e1002021.

Karlin, S. and Altschul, S.F. 1990. Methods for assessing the statistical significance of molecular sequence features by using general scoring schemes. *Proceedings of the National Academy of Sciences of the United States of America* **87**, 2264–2268.

Katoh, K., Kuma, K., Toh, H., and Miyata T. 2005. MAFFT version 5: Improvement in accuracy of multiple sequence alignment. *Nucleic Acids Research* **33**, 511–518. http://mafft.cbrc.jp/alignment/software/index.html

Katoh, K., Misawa, K., Kuma, K., and Miyata T. 2002. MAFFT: A novel method for rapid multiple sequence alignment based on fast Fourier transform. *Nucleic Acids Research* **30**, 3059–3066.

Lee, C., Grasso, C., and Sharlow, M. 2002. Multiple sequence alignment using partial order graphs. *Bioinformatics* **18**, 452–464.

Lunter, G., Miklos, I., Drummond, A., Jensen, J.L., and Hein, J. 2005. Bayesian coestimation of phylogeny and sequence alignment. *BioMed Central Bioinformatics* **6**, 83.

Madison, D.R., Swofford, D., and Madison, W.P. 1997. Nexus: An extensible file format for systematic information. *Systematic Biology* **46**, 590–621.

Moore, G.E. 1965. Cramming more components onto integrated circuits. *Electronics Magazine* 38, 114–117.

National Center for Biotechnology Information (NCBI) *Entrez*, public domain software and Webserver. http://www.ncbi.nlm.nih.gov/gquery/ (accessed October 10, 2013). National Library of Medicine, National Institutes of Health, Bethesda, MD.

National Center for Biotechnology Information (NCBI) *PubMed*, Webserver. http://www.ncbi.nlm.nih.gov/pubmed/ National Library of Medicine, National Institutes of Health, Bethesda, MD.

Needleman, S.B. and Wunsch, C.D. 1970. A general method applicable to the search for similarities in the amino acid sequence of two proteins. *Journal of Molecular Biology* **48**, 443–453.

Notredame, C. 2006. *T-Coffee: Tutorial and FAQ* and *Technical Documentation*. Included with the distribution at http://www.tcoffee.org/Projects/tcoffee/

Notredame, C., Higgins, D.G., and Heringa, J. 2000. T-Coffee: A novel method for multiple sequence alignments. *Journal of Molecular Biology* **302**, 205–217.

Olsen, G. 1992. Lecture: *Inference of Molecular Phylogenies*, University of Illinois, Urbana-Champaign, IL. September 3, 1992.

Online Mendelian Inheritance in Man, OMIM™. http://www.ncbi.nlm.nih.gov/omim/ Center for Medical Genetics, Johns Hopkins University, Baltimore, MD, and National Center for Biotechnology Information, National Library of Medicine, Bethesda, MD.

Pearson, W.R. 1990. Rapid and sensitive sequence comparison with FASTP and FASTA. *Methods in Enzymology* **183**, 63–98.

Pearson, W.R. 1998. Empirical statistical estimates for sequence similarity searches. *Journal of Molecular Biology* **276**, 71–84. The complete FastA package is available through http://fasta.bioch.virginia.edu/fasta_www2/fasta_list2.shtml (accessed today, 10-19-2013).

Pearson, P., Francomano, C., Foster, P., Bocchini, C., Li, P., and McKusick, V. 1994. The status of Online Mendelian Inheritance in Man (OMIM) medio 1994. *Nucleic Acids Research* **22**, 3470–3473.

Pearson, W.R. and Lipman, D.J. 1988. Improved tools for biological sequence analysis. *Proceedings of the National Academy of Sciences of the United States of America* **85**, 2444–2448.

Pei, J., Sadreyev, R., and Grishin, N.V. 2003. PCMA: Fast and accurate multiple sequence alignment based on profile consistency. *Bioinformatics* **19**, 427–428. ftp://iole.swmed.edu/pub/PCMA/

Pruesse, E., Quast, C., Knittel, K., Fuchs, B., Ludwig, W., Peplies, J., and Glöckner, F.O. 2007. SILVA: A comprehensive online resource for quality checked and aligned ribosomal RNA sequence data compatible with ARB. *Nucleic Acids Research* **35**, 7188–7196. http://www.arb-silva.de/

Rambaut, A. 1996. Se-Al: Sequence alignment editor. http://tree.bio.ed.ac.uk/software/seal/

Rice, P., Longden, I., and Bleasby, A. 2000. EMBOSS: The European molecular biology open software suite *Trends in Genetics* **16**, 276–277. http://emboss.sourceforge.net/

Ronquist, F. and Huelsenbeck, J.P. 2003. MRBAYES 3: Bayesian phylogenetic inference under mixed models. *Bioinformatics* **19**, 1572–1574. http://mrbayes.sourceforge.net/

Rost, B. and Sander, C. 1993. Prediction of protein secondary structure at better than 70% accuracy. *Journal of Molecular Biology* **232**, 584–599.

Rost, B. and Sander, C. 1994. Combining evolutionary information and neural networks to predict protein secondary structure. *Proteins* **19**, 55–77. https://www.predictprotein.org/

Rotkiewicz, P. 2007. iMol molecular visualization program, http://www.pirx.com/iMol

Rozen, S. and Skaletsky, H. 2000. Primer3 on the WWW for general users and for biologist programmers, in Krawetz, S. and Misener, S. (eds.), *Bioinformatics Methods and Protocols: Methods in Molecular Biology*, Humana Press, Totowa, NJ. pp 365–386. Software available at http://primer3.sourceforge.net/

Saitou, N. and Nei, M. 1987. The neighbor-joining method: A new method of constructing phylogenetic trees. *Molecular Biology and Evolution* **4**, 1406–1425.

Sander, C. and Schneider, R. 1991. Database of homology-derived structures and the structural meaning of sequence alignment. *Proteins* **9**, 56–68.

Schneider, T.D. and Stephens, R.M. 1990. Sequence logos: A new way to display consensus sequences. *Nucleic Acids Research* **18**, 6097–6100. http://weblogo.berkeley.edu/

Schuler, G.D., Epstein, J.A., Ohkawa, H., and Kans, J.A. 1996. Entrez: Molecular biology database and retrieval system. *Methods in Enzymology* **226**, 141–162.

Schwartz, R.M. and Dayhoff, M.O. 1979. Matrices for detecting distant relationships, in Dayhoff, M.O. (ed.), *Atlas of Protein Sequences and Structure*, Vol. **5**, National Biomedical Research Foundation, Washington, DC. pp. 353–358.

Smith, R.F. and Smith, T.F. 1992. Pattern-induced multi-sequence alignment (PIMA) algorithm employing secondary structure-dependent gap penalties for comparative protein modeling. *Protein Engineering* **5**, 35–41. http://genamics.com/software/downloads/pima- 1.40.tar.gz

Smith, S.W., Overbeek, R., Woese, C.R., Gilbert, W., and Gillevet, P.M. 1994. The genetic data environment, an expandable GUI for multiple sequence analysis. *Computer Applications in the Biosciences* **10**, 671–675. Linux and Mac OS X GDE ports at http://www.bioafrica.net/software.php and http://macgde.bio.cmich.edu/

Smith T.F. and Waterman, M.S. 1981. Comparison of bio-sequences. *Advances in Applied Mathematics* **2**, 482–489.

Stajich, J.E., Block, D., Boulez, K., Brenner, S.E., Chervitz, S.A., Dagdigian, C., Fuellen, G. et al. 2002. The Bioperl toolkit: Perl modules for the life sciences. *Genome Research* **12**, 1611–1618. http://www.bioperl.org/

Stamatakis, A. 2006. RAxML-VI-HPC: Maximum likelihood-based phylogenetic analyses with thousands of taxa and mixed models. *Bioinformatics* **22**, 2688–2690. http://www.exelixis-lab.org/ for software and related links

Sundaralingam, M., Mizuno, H., Stout, C.D., Rao, S.T., Liedman, M., and Yathindra, N. 1976. Mechanisms of chain folding in nucleic acids. The Omega plot and its correlation to the nucleotide geometry in Yeast tRNAPhe1. *Nucleic Acids Research* **10**, 2471–2484.

Suyama, M., Torrents, D., and Bork P. 2006. PAL2NAL: Robust conversion of protein sequence alignments into the corresponding codon alignments. *Nucleic Acids Research* **34**, 609–612. http://www.bork.embl.de/pal2nal/

Swofford, D.L. 1989–2012. PAUP* (Phylogenetic Analysis Using Parsimony, and other methods) version 4.0+. Illinois Natural History Survey, 1994; personal copyright, 1997; Smithsonian Institution, Washington, DC, 1998; Florida State University, 2001–2007. Home page at http://paup.scs.fsu.edu/, distributed through Sinaeur Associates, Inc. http://www.sinauer.com/ Sunderland, MA.

Talavera, G. and Castresana, J. 2007. Improvement of phylogenies after removing divergent and ambiguously aligned blocks from protein sequence alignments. *Systematic Biology* **56**, 564–577.

Thompson, J.D., Gibson, T.J., Plewniak, F., Jeanmougin, F., and Higgins, D.G. 1997. The ClustalX windows interface: Flexible strategies for multiple sequence alignment aided by quality analysis tools. *Nucleic Acids Research* **24**, 4876–4882. http://www.clustal.org/

Thompson, J.D., Higgins, D.G., and Gibson, T.J. 1994. CLUSTALW: Improving the sensitivity of progressive multiple sequence alignment through sequence weighting, positions-specific gap penalties and weight matrix choice. *Nucleic Acids Research* **22**, 4673–4680. http://www.clustal.org/

Van de Peer, Y., Frickey, T., Taylor, J.S., and Meyer, A. 2002. Dealing with saturation at the amino acid level: A case study based on anciently duplicated zebrafish genes. *Gene* **295**, 205–211. http://bioinformatics.psb.ugent.be/software/details/ASaturA

Wallace, I.M., O'Sullivan, O., Higgins, D.G., and Notredame, C. 2006. M-Coffee: Combining multiple sequence alignment methods with T-Coffee. *Nucleic Acids Research* **34**, 1692–1699. Included in T-Coffee distribution.

Waterman, M.S. 1989. Sequence alignments, in Waterman, M.S. (ed.), *Mathematical Methods for DNA Sequences*, p. 56, CRC Press, Boca Raton, FL.

Webster's New Collegiate Dictionary (1st edn.) 1973. G. & C. Merriam Co. Springfield, MA.

Wernersson, R. and Pedersen A.G. 2003. RevTrans—Constructing alignments of coding DNA from aligned amino acid sequences. *Nucleic Acids Research* **31**, 3537–3539. http://www.cbs.dtu.dk/services/RevTrans/download.php

Wetterstrand, K.A. 2012. DNA sequencing costs: Data from the NHGRI large-scale genome sequencing program available at: http://www.genome.gov/sequencingcosts/

Wuyts, J., Perriere, G., and Van de Peer, Y. 2004. The European ribosomal RNA database. *Nucleic Acids Research* **32**, 101–103.

Zuker, M. 1989. On finding all suboptimal foldings of an RNA molecule. *Science* **244**, 48–52.

Zwickl, D.J. 2006. Genetic algorithm approaches for the phylogenetic analysis of large biological sequence datasets under the maximum likelihood criterion. PhD dissertation, University of Texas, Austin, TX. https://www.nescent.org/wg_garli/Main_Page

31 Terrain Modeling for the Geosciences

Herman Haverkort
Technische Universiteit Eindhoven

Laura Toma
Bowdoin College

31.1 Introduction

Terrain models are used for many purposes in a variety of hydrological, geomorphological, and biological applications. Examples include estimation of river networks and watersheds, soil moisture and runoff behavior, soil erosion, long-term water availability, delineation of land slide areas, and land cover classification [68]. In the past decades, it has become increasingly easy to collect high-precision elevation samples of the Earth surface and the floors of water bodies. However, many other types of terrain properties are not as easy to collect, because they are not readily observed or measured from above. Examples include the flow of water across the surface or the daily amount of sunlight per square meter. To make maps of such properties across a terrain, scientists use models to compute an estimation of the required data from the available elevation data. Other uses of elevation data include computations for planning purposes, such as computing a least-cost path to construct a road between two given points or to find a location for a building that minimizes or maximizes its visual impact. The increasing availability of ever-higher-resolution elevation data poses an ever-increasing computational challenge, since higher resolutions come with more noise, and the data sets grow too big to be loaded into the main memory of a computer at once.

Terrains have been studied in many different research areas including computational geometry, computational topology, computer graphics, geomatics engineering, civil engineering, hydrology, oceanography, and other earth sciences. It is practically impossible to produce a complete survey of terrain models and applications including results from all these disciplines. On top of that, terrain data are often provided in the form a regular grid of elevation samples, and thus, it is very similar to a large gray-scale image. Therefore, it is not surprising that many of the problems in terrain data processing are similar to problems in image processing. For example, to estimate the slope and aspect at any point on a terrain, one may first have to smoothen the terrain; otherwise, small bumps in the terrain (whether real or the result of errors in the data) would dominate the results; image smoothing techniques can be applied. In this chapter, we highlight algorithmic approaches and challenges that concern some of the most important problems in terrain processing and that are typical to geographic terrain data: removal of noise from light detection and ranging (LiDAR) and sound navigation and ranging (SONAR) data, model construction, the computation of contour maps, water flow analysis, and visibility analysis.

Since terrain data sets acquired with recent technology may be as large as hundreds of gigabytes, they often do not fit in the main memory of a computer at once. Working with such data requires efficient algorithms that scale well and are designed to minimize the I/O: the swapping of data between a fast main memory (or cache) and a larger but slower memory such as a disk. The efficiency of such algorithms is assessed not only by studying the computational complexity (see Chapter 4) and by measuring running times in practical experiments but also by studying how the number of I/O operations grows with the input size. The standard model to do so was defined by Aggarwal and Vitter [5]. In this model, a computer has a memory of size M and a disk of unbounded size. The disk is divided into blocks of size B. Data are transferred between memory and disk by transferring complete blocks: transferring one block is called an I/O operation or, simply, an I/O. Algorithms can only operate on data that are currently in main memory; to access the data in any block that is not in main memory, it first has to be copied from the disk. The I/O efficiency of an algorithm can be assessed by analyzing the number of I/O's it needs as a function of the input size n, the memory size M, and the block size B. Fundamental building blocks for I/O-efficient algorithms are *scanning* and *sorting*: scanning n consecutive records from disk takes $scan(n) = \frac{n}{B}$ I/O's; sorting takes $sort(n) = \frac{n}{B} \log_{M/B} \frac{n}{B}$ I/O's in the worst case [5]. A more restrictive model for algorithms for large data is the *streaming model* [57]. Algorithms that conform to this model make only a small number of sequential passes over the data while processing the data using a small memory buffer.

In Section 31.2, we describe the sources and formats of terrain elevation data, the problem of cleaning and filtering raw elevation data, common interpolation methods used on such data, and how to construct terrain models in the most common formats. In Sections 31.3 through 31.5, we review the state of the art in three applications of terrain models: computing contour maps, computing water flow and watershed hierarchies, and computing viewsheds (visibility maps).

31.2 Data Collection and Model Construction

In this section, we discuss how terrain models are obtained. Section 31.2.1 describes sources of terrain elevation data and in particular LIDAR data, which have recently become the best source of terrain data because of its high resolution and accuracy. We overview cleaning and filtering raw LIDAR data in Section 31.2.2. Next, we describe the most commonly used terrain data formats and interpolation methods (Sections 31.2.3 and 31.2.4). Finally, we overview methods for converting LIDAR data to the main terrain formats (Sections 31.2.5 and 31.2.6).

31.2.1 Digital Elevation Data Sources

Most digital elevation data are obtained from stereoscopic interpretation of aerial photographs and satellite imagery. Elevation data can also be obtained by ground survey, by digitizing the contour lines on existing topographic maps, and, more recently, from technology like LIDAR. The LIDAR technology is one of the most accurate and cost-effective ways to collect elevation data for large areas. The LIDAR technology is based on sending laser signals from an aircraft down to a point on earth and measuring the time it takes for the pulse to reflect back. The reflection time, together with the exact position and orientation of the aircraft, allows us to compute the elevation of that point on the earth. LIDAR measurements can be collected at .5–3 m resolution and are very accurate (15–30 cm) [56]. The resulting product is a densely spaced collection of three-dimensional points, often called a point cloud.

A technology similar to LIDAR, called SONAR, can be used in water to measure the elevation of the seafloor. The SONAR technology operates in a similar way as LIDAR, except that, instead of laser, it uses sound. Seafloor data is used for applications such as searching for oil or maintaining pipes.

In the recent years LIDAR and SONAR data have become increasingly available. Data from LIDAR and SONAR are usually massive, have high resolution, high redundancy and contain noise. To be useful, they need to be cleaned and thinned/simplified, and the algorithms for doing so need to be efficient and able to handle huge inputs.

31.2.2 Cleaning and Filtering LiDAR and SoNAR Data

Raw LIDAR and SONAR data contain a point cloud of elevation measurements representing the surface. In fact, several returns may be collected for each pulse emitted. The last return of a pulse may be associated with reflection of the bare ground (or ocean floor), while the first returns may represent reflections of birds, fish, tree tops, etc. Furthermore, the data may contain noise resulting from, for example, refraction in gas bubbles, influences of the ship's propeller or differences in sound speed.

Applications usually need a digital terrain model, that is, a model of the bare ground. Such a model is obtained from the raw data by *cleaning* the data, that is, detecting and eliminating noise, and *filtering* the data, that is, distinguishing and removing all features such as vegetation and buildings. The criteria and approaches used for cleaning and filtering may differ depending on the morphology of the data (seafloor vs. surface, urban vs. agricultural or forested) and the density of points.

For a review of cleaning and filtering algorithms for LIDAR data we refer to Sithole and Vosselman [61] and Meng et al. [50] and the references therein.

A common method for cleaning SONAR data is through the use of the CUBE algorithm (Combined Uncertainty and Bathymetry Estimator) developed by Calder and Mayer [18]. The CUBE algorithm computes depth estimates at the vertices of a grid laid over the terrain, and in doing so it gives a mechanism to remove noise in the data. The method attributes to each point an estimate of uncertainty, propagates these uncertainties in the local neighborhood, and based on them it estimates the true depth at the points in the grid. The CUBE algorithm is implemented by most hydrographic software. It has a low memory footprint, it is fast on large data sets and is suitable for real-time use. However, according to Arge et al. [11], CUBE works well for isolated outliers, but not so well for clustered noise. Their solution is to define a graph that represents the topology of the data, and to clean the data by removing connected components of the graph that appear to represent noise clusters. To construct the graph they start with the Delaunay triangulation of the points,* for each edge they add its corresponding diagonal, and then they remove all edges of which the elevations of the endpoints differ by more than a user-specified threshold τ. Thus, areas that are delineated by abrupt changes in depth, for example the back of a fish that swims at a height at least τ above the seabed, will be disconnected from the rest. The largest connected component in the resulting graph is assumed to represent the seabed, while the others represent noise. Using I/O-efficient algorithms and under realistic assumptions on the size of the graph ($n = O(M^2)$), the algorithm runs in $O(sort(n))$ I/O's on a SONAR data set of n points. The authors analyse the accuracy of the algorithm and compare it to a manually cleaned data set [11].

31.2.3 Data Models/Representation

Digital elevation data can be structured in several ways: the most common data formats are regular grids and triangulated irregular networks (TINs) of elevation samples.

* The Delaunay triangulation is the triangulation in which the interiors of the circumcircles of the triangles do not contain any vertices, see Chapter 6.

Grid data consists of a set of elevation values in a regular pattern, typically the intersection points of a regular grid of squares formed by equally spaced longitudinal and latitudinal lines. Thus, the data is essentially a matrix of elevation values. When the surface covered is not too large, each pair of neighboring elevation samples can be considered to have equal distance between them; otherwise algorithms will have to include corrections to account for the curvature of the earth surface and the decreasing distance between samples as one moves away from the equator. Grid data is most commonly stored in row-major order: row by row, and in each row, from west to east. This makes it easy to convert between coordinates of a grid point and its position in the file, and it guarantees that neighboring grid points are stored relatively close to each other in the file. Alternatively, Z-order (also known as Morton layout) could be used: the terrain is subdivided into an NW, NE, SW and SE quadrant, which are stored one by one in that order, each ordered recursively in the same way [64]. This order tends to optimize the number of cache misses (accesses to main memory or disk) needed when arbitrary long paths through the grid need to be traversed. Grids with sample points on the vertices of a grid of equilateral triangles have also been considered but do not seem to have found their way into common practice in the geosciences. There does not seem to be an inherent algorithmic reason for this: all algorithms for grid terrains discussed in this survey can easily be adapted to triangular grids.

A more flexible alternative to grid formats is provided by triangulated irregular networks (TINs). Here the elevation samples lie in an irregular pattern, and a triangulation on these points is provided. Typically the triangulation forms a Delaunay triangulation when projected on the horizontal plane (or rather, the reference ellipsoid of the Earth). For high-quality flow and visibility computations it may be advantageous to make sure that known valley and ridge lines are approximated by paths that are composed of edges of the triangulation. This can be done by using a *constrained Delaunay triangulation* (see Section 31.2.5). While for grid formats it suffices to store the elevation data itself, for TINs one also needs to store the horizontal coordinates of the sample points and the topology, that is, what are the edges and the triangles of the triangulation. To allow the TIN to be traversed efficiently, a non-trivial data structure is needed, for example a doubly-connected edge list or half-edge structure [30], or a more efficient data structure specifically designed for triangulations [20,49].

The choice between a grid format or a TIN format is not obvious. On one hand, grid models are very restrictive: they require an exact spacing of sample points, and equal spacing everywhere. Elevations that have actually been measured cannot be used since the measured elevation samples seldom lie exactly on a grid point, so the whole grid has to be constructed by interpolation. Furthermore, if high-resolution data is required or available to model any part of the terrain, then the complete terrain model must be given at high resolution, even where high-resolution data is unavailable or simply not required. Thus, grid formats may carry a substantial storage overhead in the form of interpolated sample points that do not carry any information from actual measurements. On the other hand, algorithms operating on TINs tend to be more complicated than algorithms on grid formats. Furthermore, one should choose the data structure to store the topology information of a TIN carefully. Commonly used data structures store at least seven pointers per vertex on average [49], plus the coordinates of each vertex: thus the total amount of storage needed for a TIN is at least an order of magnitude more than what is needed for the mere elevation data. Recently, Castelli Aleardi and Devillers published a data structure that uses only four pointers and, in our case, three coordinates per vertex [20]—I/O-efficient construction has not been studied yet. Theoretically, data structures exist that use as little as 2 bits per vertex to store the topology [21], but construction and navigation of such data structures is much more difficult and time-consuming and researchers in the field consider them to be of "mainly theoretical interest" [20]. In the end, because of their simple structure, grids are probably the most commonly used and most widely available format in geographic information systems.

Naturally, researchers have been researching the possibilities to overcome the disadvantages of grid and TIN formats. As an alternative to regular grids, grids based on quadtrees could be used: starting from a single square grid cell, we keep subdividing cells into four quadrants recursively until each cell is small enough to allow the morphology of the local terrain to be captured adequately [30]. Thus, the cell size may vary across the terrain. However, algorithms for applications such as those mentioned in the remaining sections of this chapter, have generally not been adapted to work with such formats.

31.2.4 Spatial Interpolation

Terrain elevation data must be considered, either explicitly or implicitly, with an interpolation method that describes how to obtain the elevation values for points other than the sampled points. More precisely, let $P = \{p_1, p_2, \dots p_n\}$ be a set of points, each specified by their latitude and longitude, and denote the measured elevation of p_i by $z[p_i]$. To be able to construct and interpret a terrain model, we need to define an elevation function z that gives an elevation value for *any* two-dimensional point p, also for points p that are not in P. This function z should respect the data, that is, $z(p)$ should be (approximately) equal to $z[p]$ for all $p \in P$. In this section we discuss several common methods to define such a function z. For a more thorough survey, see Mitas and Mitásová [51].

Interpolation by Thiessen polygons or nearest neighbors: With this technique, the interpolated elevation $z(q)$ of any point q is the elevation $z[p]$ of the point $p \in P$ that is nearest to q, when distances are measured in the projection on the horizontal plane. This leads to a partitioning of the terrain into regions called Thiessen polygons. The region $T(p)$ containing a point p of P is exactly the set of points that are closer to p than to any other point of P, and within each region, each point has the same elevation. Interpolation with Thiessen polygons is often used for grids; in that case the Thiessen polygons are simply rectangular cells centered at the grid points. This interpolation method is simple, but has the problem that the (interpolated) values change sharply across polygon boundaries, which is not necessarily plausible. The partitioning in Thiessen polygons is also known as a Voronoi diagram; it is the dual of the Delaunay triangulation that usually underlies a TIN [30].

Interpolation on triangular facets: This interpolation method models the surface as a tessellation of triangular facets. To be able to apply this technique, we first compute a triangulation of the elevation samples P. The elevation of any other point q is then obtained by linear interpolation between the three vertices of the triangular facet that contains q. Although simple, linear TIN interpolation is not accurate; see Mitas and Mitásová [51] and the references therein. Variants of this method use non-linear interpolation functions for each triangle to ensure additional continuity across boundaries and the existence of derivatives.

Inverse-distance-weighting (IDW): This method is reported to be the most commonly used method in geographic information systems [48]. The value $z(q)$ is defined as a weighted average of all points $p \in P$, where the weights depend on q, that is, $z(q) = \Sigma_i w_i(q) z[p_i] / \Sigma_i w_i(q)$. More precisely, we define $w_i(q)$ as $1/|p_i q|^2$, where $|p_i q|$ is the distance between p_i and q. The IDW technique produces a smooth surface but can produce counterintuitive results like fake pits and peaks, that is, peaks that are higher than any elevation that was actually measured in its neighborhood [48]. Applications that use IDW often give the option to ignore points that are further away than a specified distance, or to consider only a specified number of nearest points, or to average only over the closest points in each of a number of directions [48].

Natural-neighbor interpolation: Like IDW, the value of $z(q)$ is defined as a weighted average of all points $p \in P$. Each weight $w_i(q)$ is now defined as the size of the part of the Thiessen polygon $T(p_i)$ that would be lost to the Thiessen polygon $T(q)$ when q would be inserted in P. In other words, $w_i(q)$ is the size of the intersection of $T(p_i)$ in the Voronoi diagram of P and $T(q)$ in the Voronoi diagram of $P \cup \{q\}$. An implementation that exploits the parallel processing power of a graphics processing unit was recently presented by Beutel et al. [16].

Kriging: This is another method in which the value of $z(q)$ is defined as a weighted average of all points $p \in P$. The weights are now derived from a geostatistical analysis of the data. The structural variation in the data is captured by the *variogram*, which is the function that gives an estimate of the average squared elevation difference between p_i and p_j as a function of their horizontal distance. More precisely, if $N(h)$ is set of pairs (i, j) such that $|p_i p_j| = h$, then the variogram is obtained by fitting a certain type of function (for example, a Gaussian, spherical, or exponential function) to the function

$\sigma(h) = \frac{1}{|N(h)|}\sum_{(i,j)\in N(h)} |z[p_i] - z[p_j]|^2$. The variogram is then determined to define the weights. For a detailed discussion of Kriging see Longley et al. [48] and Burrough and McDonnell [17].

Regularized spline with tension (RST): Proposed by Mitásová and Mitas [52], this method consists in defining and minimizing a certain smoothness function. It includes a tension parameter which changes the character of the resulting surface from a thin metal plate to a membrane, which is pinned to the data points. The value of the tension has to be determined empirically in each application. The method enables the computation of local derivatives of all orders, which is useful in various applications. The authors report that RST is more accurate than alternative methods, such as thin plate spline, Hardy quadric, etc. [52]. The authors implemented RST in GRASS [53]; an implementation that can handle larger data sets was recently presented by Danner et al. [27].

31.2.5 From Point Cloud to TIN

As discussed above, terrain data often comes in the form of a point cloud obtained by LIDAR or other means. For further processing, this data is usually transformed to a grid or a TIN. As LIDAR data tend to be very large (hundreds of millions of points), the grid or TIN construction algorithm needs to be I/O-efficient in order to handle such data sets. Below we focus on algorithms that have been proposed for and/or tested on very large data sets. In this subsection we discuss algorithms to construct TINs; in the next subsection we discuss algorithms to construct grid models.

To construct a TIN from a point cloud, one needs to triangulate the point set. Typically a Delaunay triangulation based on the horizontal coordinates of the points—the elevation values are not taken into account—is used. Several $O(n \lg n)$-time, but not necessarily I/O-efficient, algorithms for Delaunay triangulations are known [13]. A state-of-the-art implementation is *Triangle*, developed by Shewchuck [59]. Several I/O-efficient algorithms to compute Delaunay triangulations have been proposed as well [3,26,40].

The first I/O-efficient algorithm accompanied by an experimental evaluation is by Agarwal et al. [3], based on Crauser et al. [26]. The implementation actually solves a more general problem, the *constrained Delaunay triangulation*: given a set V of vertices in the plane together with a set S of predetermined edges, compute the triangulation of V that includes all predetermined edges in S and is "as close as possible" to the Delaunay triangulation (for a more precise definition, see the original publication). This allows us to ensure that the triangulation respects, for example, valley and ridge lines that are known from other sources. The algorithm works in rounds which triangulate increasingly bigger sets of points from the input, in each round using the results of the previous round to divide the problem into subproblems that are solved with the *Triangle* package mentioned above. The algorithm remains efficient if V is huge as long as S is modest: if S does not require more storage than a certain constant fraction of the main memory, the expected number of I/O's required by the algorithm is $O(sort(n))$. The authors report that their algorithm can process 10 GB of LIDAR data (about 500 million points from the Neuse river basin of North Carolina) using 128 MB of main memory in roughly 7.5 h.* This is several orders of magnitude faster than a direct application of Shewchuck's *Triangle*.

A different approach is described by Isenburg et al. [44]. They advocate processing large datasets using the *streaming paradigm*. Streaming algorithms make only a few sequential passes over the data, do not create intermediate files, and perform computations only on the part of the data that is in memory. During each pass, as more data and more partial results are computed, the algorithm needs to decide which previously read data and previously computed results still need to be retained in memory and which data and results can make place for new data and results. The Delaunay triangulation algorithm by Isenburg et al. does this as follows. The algorithm first reads the input point cloud to determine its bounding box. Next the algorithm divides the bounding box in a regular grid of rectangular cells such

* Considering machines with so little RAM is done to illustrate the scalability of the algorithm to data that is much larger than main memory.

that one can easily keep a counter for each cell in memory. The input is then read once more to count the number of points in each cell. Finally the input is read a third time. This time the points of each cell are collected in memory. As soon as all points of a cell have been read, the whole cell is triangulated using the *Triangle* package. Triangles whose circumcircles lie entirely inside the current and previously processed cells are written to the output immediately. The remaining triangles may still change, subject to the contents of the unprocessed cells: only these triangles and their vertices are retained in memory. Isenburg et al. report that the algorithm can process the above mentioned data set of the Neuse river basin in 48 min using 70 MB of memory on a laptop, using a slower processor and slower disk than Agarwal et al. [3]. However, the implementation by Isenburg et al. does not take a set of predetermined edges into account.

Note that the approach taken by Isenburg et al. is very fast in practice, but its worst-case complexity is not. The algorithm exploits the natural spatial coherence of raw LIDAR data sets: points that are close to each other in space tend to be close to each other in the file. Thus the algorithm does not need to keep many points of incomplete cells in memory. Furthermore, the algorithm by Isenburg et al. requires that most triangles do not span many regions: thus the algorithm will not need to keep many triangles that may still change in memory. It seems that this requirement is met in practice. Thus the memory footprint of the algorithm by Isenburg et al. stays small and I/O remains limited to sequentially reading the input and sequentially writing the output.

31.2.6 From Point Cloud to Grid Model

Different approaches to compute a grid model from a point cloud I/O-efficiently were again proposed by Agarwal et al. [1] and Isenburg et al. [45].

The solution proposed by Agarwal et al. [1] is a standard I/O-efficient algorithm, designed and analyzed in the I/O-model. Their approach is based on Mitásová and Mitas [52] and has three steps: first, they process the point cloud into a quadtree (the *segmentation* phase) with at most a user-specified number of points in each leaf; second, for each leaf in the quadtree they find out which cells are its neighbors (the *neighbor finding* phase); and third, they interpolate elevations for the grid points within each cell using the points in the cell and its neighboring cells (the *interpolation* phase). In their implementation, interpolation with regularized splines with tension (RST) is used, but other interpolation methods might be substituted. The complete algorithm runs in $O\left(\frac{N}{B}\frac{h}{\log M/B}\right)$ I/O's, where h is the height of the quadtree. Agarwal et al. implemented and evaluated their approach on some practical examples. A grid of 1300 million points is reported to be constructed from a LIDAR data set of 390 million points (the Neuse river basin data set from http://www.ncfloodmaps.com) in 53 h on an Intel 3.4 GHz Pentium 4 with 1 GB of main memory. This is reported to be faster than two popular geographic information system packages, ArcGIS and GRASS. Of the three phases of the algorithm, interpolation is by far the most time-consuming, accounting for 50%–80% of the running time, with the bottleneck being the required computational effort for RST interpolation, rather than I/O. A new implementation of this approach has been recently described by Danner et al. [27]. This version, designed to take advantage of a modern platform containing multi-core processors and modern graphics processors, is at least an order of magnitude faster than the implementation by Agarwal et al. [1].

Isenburg et al. [45] again adopt the streaming paradigm. Their algorithm is based on the point-cloud-to-TIN algorithm described in Section 31.2.5, which is extended as follows. Instead of writing the triangles produced by the algorithm to an output file, the triangles and the vertices are piped through to another process, the *rasterizer*. The rasterizer receives each triangle and each vertex exactly once and uses the triangles to estimate the elevation at the grid points by linear interpolation. The stream of triangles and vertices that is fed to the rasterizer is interleaved with vertex tags that mark when all the triangles incident on a vertex have appeared in the stream: thus the rasterizer knows when a vertex still needs to be retained in memory and when it can be discarded. More advanced interpolation methods, in which the interpolation within a triangle also depends on the neighboring triangles, may be implemented by keeping triangles in memory until all neighboring triangles of its neighboring triangles have been received. The elevations of grid points as computed by the rasterizer, are immediately written to a file and are not kept in memory. Upon completion of the algorithm, the output is sorted into row-major order to obtain the grid model. The overall process is extremely fast: a grid of 1500 million points is constructed by linear interpolation from a 10 GB LiDAR data set of 500 million points from the Neuse river basin of North Carolina in 67 min on a laptop using less than 100 MB of memory and with a 5400 RPM disk. Although this is much faster than the 53 h of Agarwal et al. [1], a direct comparison cannot be made because Agarwal et al. implement a more complex interpolation method (RST), while Isenburg et al. report running times for linear interpolation.

31.3 Computing Contours and Contour Maps

One of the main applications on terrain models is computing contours and contour maps. A *contour* (or *isoline*) of a terrain T at a given level (elevation) is a connected component of a level set of T, that is, the set of points of T at a given elevation. A *contour map* consists of contours at regular elevation intervals. Contour maps have been used historically in cartography to represent terrains; recently they are used for visualization, where isosurface extraction in general is one of the most effective techniques for the investigation of volume datasets [22].

A straightforward way to compute a contour is to start from a cell intersecting the contour (for example a triangle, if the terrain is represented as a TIN) and trace the contour from cell to cell until coming back to the starting point. The question is how to find a start cell. Traversing the entire dataset to construct a single contour is not efficient, since the size of contours is usually small. Therefore research has focused on finding data structures that can be queried for contours and report them efficiently.

One solution is to associate each cell in the data set (grid cell or triangle) with an interval representing the elevation range of the cell; for example, if the terrain is represented as a TIN, each triangle is associated with the interval bounded by the elevations of its lowest and highest vertices. Computing contours now reduces to an interval search problem, namely finding all intervals that contain a given query elevation h. In the first phase of contour extraction one finds all cells whose elevation range includes h; such cells are called active. In the next phase one can apply an algorithm to actually generate the contours from the active cells. An algorithm based on this idea was given by Cignoni et al. [24], using interval trees to store the intervals corresponding to the dataset. After preprocessing in $O(n \lg n)$ time, contours at a given elevation can be found in optimal $O(\lg n + k)$ time, in internal memory, where k is the size (total number of edges) of the contours. In external memory the same approach was used by Chiang and Silva to propose an I/O-efficient algorithm [22,23]. Using the I/O-efficient interval tree developed by Arge and Vitter [12], Chiang and Silva describe a linear-size data structure to store a TIN on disk such that all edges in the contours at elevation h can be reported in $O(\log_B n + k/B)$ I/O's, where k is the size of the contours. Agarwal et al. [2] note that this algorithm does not sort the edges along a contour. Using standard techniques (list ranking) this would require another $O(sort(k))$ I/O's (possibly with relatively high constant factors hidden in the O-notation) in the worst case, but as long as the contours at any given elevation fit in memory, it is not a problem.

A different approach uses the *contour tree*, which is a data structure that encodes the starting point(s) for each contour and can be used to compute contours and contour maps efficiently. Van Kreveld et al. [67] gave an $O(n \lg n)$ time algorithm for constructing a contour tree of a terrain represented as a TIN. In the I/O-model, an algorithm to compute the contour tree efficiently in $O(sort(n))$ I/O's was proposed by Agarwal et al. [4]. This can be used to find the starting points for a contour efficiently, however it is not clear how to efficiently trace the contour in the I/O-model, since a naive implementation would require one I/O per segment, and thus $O(k)$ I/O's in total.

Agarwal et al. proposed a different I/O-efficient data structure for answering contour queries on TINs [2]. They show how how to construct a linear-size data structure in $O(sort(n))$ I/O's, so that all contours at a given elevation can be reported using $O(\log_B n + k/B)$ I/O's, where k is the size of the contours. Each contour is reported with its segments in sorted order. Their solution is based on exploiting the topology of the contours as analyzed by Edelsbrunner et al. [36], and computing a certain ordering of the triangles in the TIN such that for any contour, the segments in the contour are in order and contours do not interleave.

Unfortunately, none of the I/O-efficient algorithms mentioned above come with experimental results.

When a complete contour map needs to be produced, one may be able to afford to scan the complete input. In this case one can simply scan all cells of the data set to find all edges of the contours. Again, this does not generate the edges of the contours in sorted order. Arge et al. [9] describe an algorithm and implementation to compute and simplify contour maps of massive terrain data sets. As a first step, their algorithm indeed generates the (unsimplified) contours by simply scanning all triangles of a TIN and generating the contour segments that intersect them. For further processing, the contour segments need to be sorted by and along each contour. Arge et al. implement an algorithm that does this I/O-efficiently as long as certain assumptions can be made, such as that each contour by itself is not larger than a certain constant fraction of main memory.

31.4 Flow Analysis

One important application of terrain models is in predicting the flow of water on terrains, for example to predict the risks of floods and the impact of pollution, or, in combination with other parameters, to predict rates of erosion, vegetation cover, or suitability for agriculture. The *catchment area* or *watershed* of a location on a terrain is the region from where water flows to that location. The size of the catchment area of a location (possibly weighted according to variations in rainfall) is also known as the location's *flow accumulation*. The *specific catchment area* or *specific flow accumulation* would be the flow accumulation of a section of a contour, divided by the length of the section. The specific catchment area can be defined at any single point of the terrain by estimating the limit value for a contour section that contains that point and shrinks to zero length. Typically a terrain drains through a network of streams and rivers, which are exactly the locations that have relatively high flow accumulation. The watersheds on a terrain form a hierarchical structure, in which the watershed of each location in the river network is contained in the watershed of all locations downstream of it. In particular, watersheds of rivers are composed of watersheds of tributaries. Efficient computation of a (specific) flow accumulation map of a terrain, possibly together with a representation of the watershed hierarchy, is at the core of many applications of water flow analysis on terrains. Furthermore, many other computational problems related to the drainage structure of a terrain can be solved with similar algorithmic techniques.

In this section we will first give an overview of the state of the art in the computation of flow accumulation maps and watershed hierarchies, assuming a perfect terrain model that is nowhere horizontal. Two main approaches can be distinguished: the continuous-surface approach and the discrete-surface approach. The first approach, discussed in Section 31.4.1, treats the terrain as a continuous surface, of which the exact shape is estimated by interpolating between sample points for which an elevation is given. The second approach, discussed in Section 31.4.2, treats the terrain as a network of cells, in which each cell is treated as an atomic unit, and water that arrives in any cell continues its course down-hill to one or more neighboring cells. Both approaches can produce reasonable results only if water always follows a course that, according to the terrain model, leads strictly downhill. However, in practice this is not always the case, due to the presence of sampling or interpolation errors in the model and the presence of flat areas in the actual terrain. To make a terrain model suitable for flow accumulation computations, it needs to be *hydrologically corrected* and one needs to compute how water may find its way on flat surfaces. This is discussed in Section 31.4.3. Finally, in Section 31.4.4, we discuss the main challenges for further development of flow analysis algorithms: how to deal with imprecision, how to improve hydrological correction, and how to take into account the limited capacities of drainage channels and the passing of time.

31.4.1 Flow in the Continuous-Surface Model

In the continuous-surface model, water is usually assumed to follow the exact direction of steepest descent on the interpolated terrain. Thus, water is not bound to flow from sample point to sample point, but it can also flow across the surface between sample points. This approach allows for a relatively precise representation of valley and ridge lines with relatively few sample points, and it avoids certain artifacts that result from the representation of the terrain. For example, streams are not limited to follow the edges of a triangulation, or to follow strictly the eight cardinal and ordinal directions in a grid. Continuous-surface models can be based on elevation samples in a grid, in a triangulated irregular network, or without structure, and the interpolation may produce a smooth surface (without discontinuities in the gradients) or one with sharp folds (with discontinuities in the gradients).

In the context of geographic information science, research has mostly focussed on the setting where the continuous surface is defined by a triangulation on the sample points and linear interpolation within each triangle—resulting in a surface with sharp folds at the edges of the triangulation. In this setting we can distinguish valley edges, transfluent edges, and ridge edges. Valley edges receive water from both incident triangles. On transfluent edges, water arrives from one incident triangle and continues its course down the other incident triangle. All other edges are ridge edges: on both incident triangles, the gradient is directed away from the edge, so that no water ever flows across or along the edge. In this model, two categories of points in the terrain can be distinguished: those that lie on a steepest-descent path starting from a valley edge, and those that do not. The first category of points form the skeleton of the drainage network. They have non-zero flow accumulation and infinite specific flow accumulation. Each point p of the second category only catches water from an infinitely thin course that can be traced back up from p by following the direction of steepest ascent until reaching a ridge. Thus, such points p all have zero flow accumulation, but they may still have meaningful, non-zero *specific* flow accumulation.

Steepest-ascent paths starting at ridges form a skeleton that encodes the boundaries between watersheds [69]. The drainage network and the watershed boundary skeletons, together with the steepest-ascent and steepest-descent paths from other vertices in the terrain, decompose the terrain into *strips*. This decomposition, called a *strip map*, can be stored in a data structure from which the catchment area, the flow accumulation and the specific flow accumulation of any point on the terrain can be retrieved efficiently [29]. Unfortunately, the whole structure has relatively high complexity. Theoretically, by linear interpolation of a triangulated irregular network on n sample points, one can construct a surface that has a drainage network skeleton and a strip map consisting of $\Theta(n^3)$ line segments—even when the triangulation is a Delaunay triangulation [29]. When all triangles are equilateral triangles of the same size, the complexity is still $\Theta(n^2)$ in the worst case [29]. In practice, it has been reported that the drainage network skeleton has size only $O(n)$ [69]. However, De Berg and Tsirogiannis found that explicit computation and storage of the strip map is a major bottleneck, with the computation requiring at least 16 KB of working memory per input vertex and 8 KB per input vertex in the final result [32].

Flow structures in the continuous-surface model have also been studied for smooth surfaces. Paths of steepest ascent/descent between peaks (local maxima), pits (local minima) and saddle points form a structure known as a *Morse-Smale complex* in computational topology [42], or a *surface network* in geographic information science. On a linearly interpolated surface, paths of steepest ascent or descent do not usually cross many edges before they meet a ridge or valley edge, but in a smooth terrain, such paths may extend a long way, slowly converging towards valley lines without actually meeting them. Thus, a generalization of the approach discussed above for linearly interpolated terrains is unlikely to be practical.

31.4.2 Flow in the Discrete-Surface Model

In practice, usually a discrete-surface model is used. In its simplest form (the D8 model), the model consists of elevation samples arranged in rows and columns. Each sample point represents a cell of unit size. Water arriving in any cell continues to one of the eight neighbors of the cell; more precisely, from a cell c it continues to the lower neighbor c' such that the elevation difference between c and c', divided by the distance between the center points of c and c', is largest. This approach is also called *single-directional flow*. A disadvantage of this approach is that it is prone to artifacts, such as a tendency for streams to follow the ordinal or cardinal directions of the grid model, even on slopes that have a different orientation. More advanced models use *multi-directional flow*: they distribute the water that arrives in a cell among multiple lower neighbors to model dispersion on convex slopes, or to model directions of steepest descent that are not exactly one of the cardinal or ordinal directions [41]. The latter may also be achieved by routing flow over the edges of a TIN that is constructed such that all known valley and ridge lines approximately follow edges of the TIN.

In the discrete-surface model, the difference between flow accumulation and specific flow accumulation is subtle. To compute the specific flow accumulation in a cell from its flow accumulation, one should divide by the length of the contour section that is covered by the cell—but this is not well-defined. Gruber and Peckham [41] discuss some of the subtleties to consider.

In this section, we focus on the algorithmic aspects of flow accumulation computation, rather than the modeling aspects. From an algorithmic point of view, the challenge is not how to distribute flow of a single cell to its multiple lower neighbors, but rather how to visit all cells in the most efficient order. As long as the full terrain model fits in memory, computation of flow accumulation for all cells is relatively easy. Here is just one example of a linear-time algorithm to do so. We start the computation by giving each cell one unit of water and marking each cell as unprocessed. Then we go through all cells one by one, calling a flow distribution algorithm on each cell. When called on a cell that has already been processed, or on a cell that is still to receive flow from unprocessed higher neighbors, the flow distribution algorithm does not process the cell and it returns immediately. Otherwise, it marks the cell as processed, distributes its water to its neighbors, and makes a recursive call on each of the neighbors.

When processing large grids that are stored on a hard disk and do not fit in the main memory of a computer at once, computation becomes more complicated. The simple algorithm sketched above may access the cells of the terrain model in an order that requires many accesses to each block of data on disk and renders caching ineffective, making the computation prohibitively slow. Haverkort and Janssen showed that in practice, the simple algorithm can still be very fast when processing a 28 GB terrain model on a computer with only 1 GB of memory, provided the D8 model is used and a regular grid model. The grid model is best stored and traversed in Z-order, that is, recursively subdivided in square blocks, instead of row by row [43]. Haverkort and Janssen also present a more complicated algorithm that has theoretically optimal performance in terms of disk I/O (even on a worst-case input) and that is equally fast in practice. The algorithm first divides the terrain into square blocks that fit in memory one by one, and computes a small "summary" of each block: this summary captures, for each point on the boundary of the block, how much water flows to that point from the interior of the block, and which other points on the boundary of the block are upstream of it. The summaries are then processed together and used to compute the flow of water between blocks; finally the blocks are processed one by one once more, to distribute the incoming water from the boundary into each block. The amount of disk I/O needed is proportional to the amount needed to scan the input.

When multi-directional flow models or triangulated network models are used, the results of Haverkort and Janssen are not applicable. Instead, one can use algorithms based on *time-forward processing*: here cells are processed in order of decreasing elevation; when a cell is processed, it receives messages that say how much flow comes in from each upstream neighbor, and it sends a message to the downstream neighbors, saying how much flow they will receive. If the message passing is handled by a data structure such as an I/O-efficient priority queue, the total amount of disk I/O needed is proportional to the amount needed to sort the input. The approach has successfully been implemented in the *TerraFlow* [7] and *TerraStream* [28,55] packages. The algorithms appear to be an order of magnitude slower than the algorithms of Haverkort and Janssen [43], but they allow a far wider range of flow models.

When a single-directional flow model is used, it is also possible to compute the hierarchical watershed structure: each cell can be given a decimal label such that, given two cells, one can decide if one cell is in the other cell's watershed by only examining their labels. Under the reasonable assumption that in practice, the number of significant digits of a label is bounded by a small constant, the labeling can be computed efficiently with the algorithm of Arge et al. [8], which has also been implemented in *TerraStream* [28]. Unfortunately, the results are not applicable to multi-directional flow models, because multi-directional flow results in unbounded overlap between watersheds: two cells p and q could have partially overlapping watersheds even when none is inside the watershed of the other.

Thus, while discrete-surface models are attractive because of the availability of efficient algorithms, they also have inherent limitations: single-directional flow is prone to artifacts; multi-directional flow may mitigate the worst artifacts but renders the analysis of watershed hierarchies problematic.

31.4.3 Hydrological Correction and Routing

The flow models and algorithms described above are all based on the assumption that water always flows downhill. However, there are two problems with this assumption. First, due to the presence of sampling or interpolation errors, the course that water follows in the real terrain may not always lead downhill in the model. Problematic areas in the model are characterized by the presence of local minima that one would not expect to find in the real terrain. Second, the real terrain may actually contain flat areas or even local minima from which water still finds a way out, for example because these areas are actually flooded (such as lakes), because they drain slightly below the surface, or because they are the result of man-made structures (such as bridges) under which water actually passes through. The last is especially a problem with high resolution terrain models, where small features become visible; in this case bridges appear as elevated areas which, if not removed, can block the flow of water under the bridge and produce unrealistic river networks.

To make a terrain model suitable for flow accumulation computations, small local minima and bridges have to be removed, and flat areas (such as lakes) have to be preprocessed so that the algorithms of Sections 31.4.1 and 31.4.2 can deal with them. In this section we discuss algorithms that can be used to preprocess terrain models so that they can be used for flow analysis under the discrete-surface model, or under the linearly-interpolated continuous-surface model. The removal of local minima and bridges is known as *hydrological correction* or *hydrological conditioning*. Hydrological correction typically removes local minima by filling them up (known as *flooding*), by carving a way out into the terrain (known as *breaching*), or by a combination of these approaches.

Indiscriminate *flooding* raises the elevation of every point of a terrain until for each point there is a path to the edge of the terrain and that never leads up-hill. In practice, one should not always flood all local minima—typically, larger local minima correspond to real features in the terrain, such as quarries, where water does indeed collect, and only smaller local minima should be removed. *TerraStream* [28,55] implements a method to automatically flood (and thus remove) all local minima smaller than a user-defined threshold. The size of a local minimum is defined based on volume and/or topological persistence [36], that is, the height difference between the lowest point in the local minimum and the lowest point on its boundary. *TerraStream* also works for large triangulated models that do not fit in main memory.

As noted by Carlson and Danner [19], not all spurious local minima have small persistence. In particular, local minima behind bridges in a terrain model may have large persistence—and these are particularly problematic because they tend to arise exactly along locations of water flow. Carlson and Danner [19] describe a system for automatically detecting bridges and bridge-like features using supervised learning. Once identified, the bridges are cut out. Both the detection and the cutting algorithms process the data using small neighborhood windows and the whole process is efficient and scalable. The authors give an extensive analysis of the accuracy of their method including cross-validation. Their method modifies the terrain model significantly less as compared to a flooding approach.

When parts of the drainage network are known from other data sources, one may also correct the terrain model by changing the elevation of points along known streams such that these streams follow an always-descending path in the model. When elevations are only lowered, this approach is known as *stream burning*. Van Kreveld and Silveira discuss how to minimize the total amount of elevation change (up and down), using a linear-programming approach that is fast enough for terrains of moderate size, and they discuss what side-effects of the process to watch out for [66].

After hydrological correction, the terrain model may still contain flat areas: groups of vertices at the same elevation. Typically, only one or some of these vertices have a lower neighbor. Flow from the other vertices in a flat area must somehow find its way across the flat area to one of the vertices that have a lower neighbor. Several heuristics have been developed to direct the edges in a flat area such that from each point in the flat area, all directed paths eventually lead to a point that has a lower neighbor. One approach is to simply compute a breadth-first search tree, starting from a point that has a lower neighbor, and direct all edges of the tree towards the root, but heuristics that result in more natural-looking flow networks have also been proposed [62] and implemented in *TerraStream* [55]. Depending on the algorithms used to compute flow accumulation, the directed edges can be given directly to the flow accumulation algorithm, or flow in the right direction can be enforced by (symbolically) modifying the elevations of the vertices, so that each edge is directed down-hill.

31.4.4 Challenges

31.4.4.1 Imprecision

A major challenge with flow analysis on terrain models, is that small differences in elevation can have large consequences. A small error is sufficient to divert a river that passes close to a natural or artificial saddle point into the wrong valley. One way to deal with this, is to make many (hundred or more) different versions of the terrain model, each constructed from the original model by applying random noise according to some probability distribution. Flow accumulation and watershed hierarchy computations are then performed on each version of the model, and results are considered more reliable when they are consistent across the different versions of the model [58,63]. A disadvantage of this approach is that it is computationally expensive (not only flow accumulation but also hydrological correction needs to be computed for each version of the model), it is sensitive to the distribution of noise and the dependencies between the errors on different vertices, and results come without any guarantees even if bounds on the error of the terrain model are known.

Alternatively, one may try to compute the smallest and the largest possible flow accumulation values (and watersheds) for each point in the terrain. Theoretically, it is NP-hard to do so for a linearly-interpolated continuous-surface model, but it can be done in $O(n \lg n)$ time for single-directional discrete-flow models of n vertices, provided bounds on the elevation error of each vertex are known [35]. The algorithm has not been evaluated in practice and raises several questions for further research. One of its shortcomings may be that is gives extreme lower and upper bounds on flow accumulation values without any probability distribution on the range between these extremes. Furthermore, multi-directional flow has not been investigated.

31.4.4.2 Hydrological Correction

Automated hydrological correction still remains a challenge. To our knowledge, bridge detection algorithms have not been implemented for triangulated irregular networks. A natural optimization criterion for hydrological correction could be to apply a combination of raising vertices (flooding) and lowering vertices (breaching) that minimizes the total change in elevation and removes all spurious local minima. Thus, one would avoid flooding large areas where a breach would be the more natural solution. Unfortunately, hydrological correction under this optimization criterion is NP-hard [60] (for a description of NP-hardness see Chapter XXX). Approximation algorithms (see Chapter 7) and/or (I/O-efficient) algorithms that work well under realistic conditions still remain to be developed.

31.4.4.3 Volume and Time

All the techniques discussed above ignore the fact that water has volume: narrow and/or shallow pits and channels cannot process unlimited amounts of water without flooding and diverting flow onto other paths than the path of steepest descent on the surface. In particular, when the purpose of analyzing the flow of water on a terrain is to predict risks of floods, the limited capacity of pits and channels cannot be ignored. In addition, the time component becomes essential: many floods are not the result of structural undercapacity of a channel over a large time span, but rather the result of too much water from one big shower reaching the same location in the landscape at the same time. Such floods could, for example, be prevented by selectively slowing down or speeding up the drainage from some areas, but the effects of this are not captured by steady-state flow accumulation computations as described in Sections 31.4.1 and 31.4.2. Algorithmic techniques that take volume and time into account and can handle high-resolution terrain models—necessary to capture essential features such as dikes and other barriers adequately—are still in an early stage of development. Arge et al. [10] published an algorithm that can compute the flooding times of all terrain vertices in the discrete-surface model, under the assumption that channels have unlimited capacity but pits fill up as more water arrives. Their algorithm is asymptotically I/O-efficient (running in roughly $O(sort(n))$ I/O's), but there are no reports on implementations and experiments.

31.5 Visibility Analysis

Another application of terrain models is the computation of visibility. Visibility computations are at the core of many applications, such as planning the placement of communication towers or watchtowers, planning of buildings such that they do not spoil anybody's view, finding routes on which you can travel while seeing a lot or without being seen, and computing solar irradiation maps—which can in turn be used in predicting vegetation cover. A variety of problems pertaining to visibility have been researched in computational geometry and computer graphics, as well as in geographic information science and geospatial engineering.

Two points a and b are visible to each other if the interior of their *line of sight* $\overline{ab}$ (the line segment from a to b) lies entirely above the terrain. Given a terrain and an arbitrary (view)point v (not necessarily on the terrain), the *visibility map* or *viewshed* of v is the set of all points of the terrain that are visible from v. As discussed in Section 31.2.3, the standard terrain models are the grid and the TIN. Viewshed algorithms can be grouped in two classes: *discrete* approaches mark each vertex of a grid or TIN as visible or invisible; *continuous* approaches compute the exact boundaries of the viewshed. Both approaches can be used with both terrain models (grid or TIN), although the discrete approach is usually used on grids, while the continuous approach is usually used on TINs.

Since terrain datasets may be as large as tens and even hundreds of gigabytes, they may not fit in the main memory of a computer all at once and most of the data may have to reside on disk during computations. Hence, computing visibility requires I/O-efficient algorithms. In this section we give an overview of I/O-efficient algorithms for computing viewsheds on terrains, with emphasis on algorithms for grids (Section 31.5.1). We also present some results and considerations for TINs in Section 31.5.2, and end with a set of challenges and open problems in Section 31.5.3.

There are other visibility-related problems on terrains, such as answering point-to-point and region-to-region intervisibility queries, computing horizons, computing depth-under-horizon maps, computing a map that shows the size of each point's viewshed, etc. For brevity, in this chapter we only discuss viewsheds. For other visibility problems and results we refer the interested reader to the surveys by De Floriani and Magillo [33,34] and the references therein.

31.5.1 Visibility on Grids

Viewsheds on grids are usually modeled in a discrete way: each point in the grid is marked as visible or invisible, and the viewshed of v is defined as the set of all grid points that are visible from v. To decide whether a point p is visible, one needs to interpolate the elevation of the terrain along the line of sight $\overline{pv}$ between p and v (more precisely, along the projection of the line of sight on the horizontal plane) and check whether the interpolated elevations are below the line of sight. Various algorithms differ in what and how many points they select to interpolate along the line of sight, and in the interpolation method used. These choices crucially affect the efficiency and accuracy of the algorithms. Below we discuss a number of viewshed algorithms. For simplicity, we assume that the grid is square and has size $\sqrt{n}$ by $\sqrt{n}$.

Some of the early algorithms for computing viewsheds have been described by Franklin and Ray [39]. The method called *R*3 calculates the intersections between the horizontal projection of the line of sight and the grid lines, and computes the elevation at these intersection points by linear interpolation. Since a line of sight intersects $O(\sqrt{n})$ grid lines, determining the visibility of a point takes $O(\sqrt{n})$ time, and determining the visibility of all points in the grid takes $O(n\sqrt{n})$ time. Franklin and Ray also describe two algorithms that are faster, running in $O(n)$ time, at the expense of accuracy. The algorithm known as *R*2 examines lines of sight as above but *only* for the $O(\sqrt{n})$ grid points on the boundary of the grid; a grid point that is not on the boundary is considered to be visible if and only if the nearest point of intersection between a grid line and one of the examined lines of sight is determined to be visible. The other algorithm, *XDraw*, computes the visibility of the grid points incrementally in concentric layers around the viewpoint, starting at the viewpoint and working its way outwards. For a grid point v in layer i, the algorithm computes whether v is visible, and what is the maximum height above the horizon along the line of sight to v. To do so, it determines which are the two grid points q and r in layer $i-1$ that are nearest to $\overline{pv}$, and then it estimates the maximum height above the horizon along $\overline{pv}$ by interpolating between the lines of sight to q and r. Thus, the visibility of each point is determined in constant time per point. XDraw is the fastest of the three algorithms, due to the simplicity of the calculations, but it is also the least accurate. Izraelevitz [46] presented a generalization of XDraw that allows to user to set a parameter k, which is the number of previous layers that are taken into account when computing the visibility of a grid point. Thus the user can choose how much speed he is willing to sacrifice for accuracy. For $k = 1$ Izraelevitz's algorithm becomes XDraw, and for $k = \sqrt{n}$ it becomes *R*3.

A different approach for computing viewsheds was described by Van Kreveld [65]. In his model, the terrain is seen as a tessellation of square cells, where each cell is centered around a grid point and has the same view angle as the grid point throughout the cell, that is, the cell appears as a horizontal line segment to the viewer—somewhat similar to *R*3 if nearest-neighbor interpolation is used instead of linear interpolation. Van Kreveld describes how to compute the viewshed by a radar-like algorithm, rotating a ray around the viewpoint while maintaining a data structure that stores the cells that intersect the ray. This data structure supports insertions of cells, deletions of cells, and visibility queries for a point along the ray in $O(\lg n)$ time per operation, and thus the whole viewshed can be computed while rotating the ray in $O(n \lg n)$ time. Unfortunately the data structure functions correctly only under the assumption that each cell appears as a horizontal segment to the viewer, so that it is not readily adapted to use linear interpolation or other interpolation methods.

None of the algorithms mentioned above are I/O-efficient. I/O-efficient viewshed algorithms have been proposed by Fishman et al. [37] and Andrade et al. [6]. The algorithm by Andrade et al. [6], based on Franklin et al. [38], runs in $O(n \lg n)$ time and $O(sort(n))$ I/O's in the worst case. As in *R*2, the idea is to evaluate lines of sight only to the points on the perimeter of the grid. To do this I/O-efficiently, the algorithm first copies all grid points from the input file row by row, annotating each point p with the endpoints of the lines of sight whose evaluation requires the elevation of p. Next, all annotated points are sorted by line of sight. The algorithm then evaluates each line of sight, determining for each point on a line of sight whether it is visible or not, and writes the results to a file, in order of computation. As a result, the file contains the visibility map, ordered by line of sight. The algorithm finishes by sorting this file into row-by-row order. The algorithm is very fast in practice. For example, using at most 1.0 GB of memory and a 7200 RPM hard drive, viewsheds on grids of 1600 million points (6.0 GB, using 4 bytes per elevation value) are computed within 2 h, that is, in 4.4 μs per point.

Fishman et al. describe two I/O-efficient viewshed algorithms based on Van Kreveld's approach [37]. A direct implementation of Van Kreveld's algorithm would result in inefficient I/O-behavior. This can be solved by first sorting all grid cells by the time at which they will be hit by the rotating ray—similar to the sorting by line of sight in the algorithm of Andrade et al. To save on the time needed for sorting, Fishman et al. sort the grid cells only partially, but still enough to make the evaluation phase run I/O-efficiently. Two solutions are considered: sorting into concentric bands around the viewpoint, or sorting into sectors. Both solutions run in $O(n \lg n)$ time and $O(scan(n))$ I/O's on inputs of reasonable size, and can be extended to run in $O(sort(n))$ I/O's on larger inputs. The algorithms are very fast in practice and scale up to terrains that are more than order of magnitude larger than the available memory. For example, on a machine with 512 MB of memory and a laptop-speed (5400 RPM) hard drive, the viewshed of a terrain of 7600 million points (28.4 GB using 4 bytes per elevation value) is computed in 430 min, that is, in 3.4 μs per point.

Fishman et al. also describe a different, faster approach, which runs in $O(n)$ time and $O(scan(n))$ I/Os with a cache-oblivious algorithm, that is, an algorithm that is I/O-efficient even if not tuned explicitly to a specific memory and block size. It is somewhat similar to XDraw: it grows a region around the viewpoint, while maintaining the horizon of the terrain within the region seen so far. The horizon is represented by a grid model itself: the algorithm maintains the maximum elevation angle (the "height") of the horizon for a discrete set of regularly spaced azimuth angle intervals. The horizontal resolution of the horizon model is chosen to be similar to the horizontal resolution of the original terrain model, so that elevation angles are maintained for $\Theta(\sqrt{n})$ azimuth angle intervals. Note that this gives the algorithm the potential for higher accuracy than XDraw, which represents the horizon up to a given layer by only as many grid points as there are in that layer—which can be quite inaccurate close to the viewpoint. Another difference with XDraw is that the algorithm of Fishman et al. does not proceed layer by layer, but instead grows the region in a recursive, more I/O-efficient way. This results in a significant speed-up and makes the algorithm of Fishman et al. the fastest of the I/O-efficient algorithms presented here (provided it is tuned to the available amount of memory, like the algorithms mentioned above). For example, on a machine with 512 MB of memory and a laptop-speed (5400 RPM) hard drive, the viewshed of a terrain of 7600 million points (28.4 GB using 4 bytes per elevation value) is computed in only 200 min, that is, in 1.6 μs per point.

A proper comparison of viewshed algorithms needs to include both efficiency and accuracy. While efficiency is easy to compare, even accounting for different platforms, comparing accuracy is much harder. The straightforward way to assess accuracy is to compare the results of the computations with ground truth data. Ideally one would consider a large sample of viewpoints, compute the viewshed from each one in turn, compare it with the real viewshed at that point, and aggregate the differences. Unfortunately, ground truth viewsheds are hard, if not impossible, to obtain for larger grids. Thus one needs to define accuracy in relative terms, with respect to a base algorithm that is considered exact. An experimental evaluation of the accuracy of the various viewshed algorithms proposed so far remains an open problem.

31.5.2 Visibility on TINs

Visibility on TINs is usually modeled with a continuous model, in which the elevation of any point p on the terrain is defined by linear interpolation between the vertices of the triangle that contains p. Given a point v, the goal of a viewshed computation is to determine, for each triangle τ in the TIN, what part of τ is visible. Several algorithms are known for computing viewsheds on a TIN. The algorithm with the best known theoretical bounds on the number of computational steps is by Katz et al. [47] and runs in $O((n\alpha(n) + k)\lg n)$ time, where n is the number of triangles of the TIN, k is the size of the output and $\alpha()$ is the inverse Ackermann function.*

In the worst case, the size of the output could be quite big. The boundary of the visible part of a triangle is basically the boundary of the shadow cast by other, possibly many, triangles of the TIN when a light source is placed at v. In the worst case, in a TIN of size n there may be $\Theta(n)$ triangles such that the boundary of the visible part of each of them consists of $\Theta(n)$ line segments, and thus a complete visibility map may have size $\Theta(n^2)$ [25]. Such high complexity of the viewshed would make its computation unfeasible, especially on large TINs.

* We note that $\alpha(n) \leq 4$ as long as n is at most the estimated number of atoms in the observable universe; therefore $\alpha(n)$ can be considered constant for all practical purposes.

Fortunately, in practice the complexity of viewsheds seems to be close to linear. Moet et al. [54] and De Berg et al. [31] give a theoretical explanation based on assumptions on the shape of the terrain. Given arbitrary constants $\alpha > 0$, $\theta < \frac{\pi}{2}$, $\sigma \geq 1$, we say that a set of terrain triangles is *fat* if all angles (in the projection on the plane) are at least α; the set is *non-steep* if the angle of each triangle with the horizontal plane is at most θ; the set has *bounded scale factor* if the length ratio of the longest and the shortest edge is at most σ. Furthermore, given an arbitrary constant $\rho \geq 1$, we say a terrain has *bounded aspect ratio* if its projection on the horizontal plane is a rectangle with aspect ratio at most ρ. Moet et al. [54] prove that the worst-case complexity of the viewshed is $\Theta(n\sqrt{n})$ if the terrain has bounded aspect ratio and its triangles are fat and have bounded scale factor (the constants in the asymptotic bound depend on α, σ, and ρ). De Berg et al. [31] prove that the expected complexity of any viewshed on any terrain is $O(n)$ provided all except $O(\sqrt{n})$ triangles are fat, non-steep and of bounded scale factor and the elevation of each TIN vertex is subject to random noise that is proportional to the length of the longest edge.

Many applications need to compute viewsheds from many viewpoints, for example an application that tries to find a small number of locations that can together oversee a complete terrain. Such applications cannot afford to spend linear time on each viewshed. The complexity of the output can be reduced by only reporting the boundary of the complete viewshed and not the visible parts of each triangle, but in the worst case this will not help (even under the assumptions of De Berg et al.). To be sure to obtain a viewshed in sublinear time, one needs to resort to approximations. In many cases the digital terrain model itself is not completely accurate, so an algorithm that gives good approximations fast is a good trade-off.

Several algorithms to compute an approximation of the viewshed on a TIN together with an experimental analysis are described by Ben-Moshe et al. [14]. Their best algorithm works by dividing the area around the viewpoint into a number of slices: each slice is bounded by two rays from the viewpoint. The algorithm computes a one-dimensional visibility map along each ray. If the visibility maps of the two rays bounding a slice are similar enough, the visibility map of the whole slice is constructed by interpolation between the bounding rays; otherwise the slice is subdivided into smaller slices.

Another approach to approximating visibility is to simplify the terrain to reduce its size and compute visibility on the simplified terrain; see, for example, Ben-Moshe et al. [15]. Standard simplification methods typically simplify the terrain such that the vertical distance between the simplified terrain and the original terrain is within a user defined margin ϵ. The idea of Ben-Moshe et al. [15] is that vertical distance does not matter much for visibility: the most important features that need to be preserved are the ridges. Their algorithm computes the ridge network and simplifies it (by keeping only a small number of its vertices); this induces a subdivision of the terrain into "patches," which they simplify, one at a time, using one of the standard methods. To compare visibility on the original terrain T and simplified terrain T', they compute, given a set of points X, the fraction of pairs of points in X that have the same visibility in T and T' (i.e. they are both visible or both invisible). Their experimental analysis shows that this approach preserves visibility better (according to their metric) than standard simplification methods.

31.5.3 Challenges

As already mentioned, a question that remains to be addressed is how various viewshed algorithms proposed so far compare with respect to accuracy. Without such a comparison the relative merits and tradeoffs of the algorithms cannot be judged.

Another question is how to compute viewsheds of sublinear complexity in sublinear time, that is, can we preprocess grids and TINs such that we can compute a viewshed of k grid points, or a viewshed bounded by k line segments, in time linear in k but sublinear in n? In practice viewsheds tend to be very small compared to the size of the terrain, especially on very large terrains. An output-sensitive (and preferably I/O-efficient) algorithm with running time sublinear in n, may significantly improve the overall running time in practice.

Although the I/O-efficient viewshed algorithms proposed by Andrade et al. and by Fishman et al. are fast, their running times are still too high to make them suitable for the computation of many vieswheds of the same terrain, or for online calculations. A challenge is to develop viewshed approximation algorithms that trade off speed and accuracy, and for which the error can be quantified and bounded in some way.

Viewsheds on TINs have been less studied from a practical point of view, probably because continuous viewsheds are more complex and therefore less appealing in practice. Exploring viewsheds on TINs, testing the scalability of the algorithms to very large TINs and developing I/O-efficient algorithms for TINs are problems yet to be considered.

31.6 Concluding Remarks

In this chapter we discussed some basic problems and solutions regarding the analysis of terrain data. We have seen a number of efficient and effective algorithms that have been implemented in commercial and academic geographic information systems. However, there are also many problems for which satisfactory solutions do not exist yet.

Dealing with increasingly high-resolution data means dealing with increasing amounts of noise and errors that cannot be corrected by hand. This requires increasingly better techniques for cleaning, filtering and hydrological correction. Dealing with increasingly high-resolution data also means running into the limits of the simplifying assumptions that were made in the past—such as ignoring the volume of water; new algorithmic techniques need to be brought to the geosciences to deal with this in an efficient way. Last but not least, dealing with increasingly high-resolution data means that we need increasingly fast algorithms. It is therefore important to take full advantage of the parallel computing power that is becoming increasingly available, for example in the form of graphics processing units.

A shortcoming of many of the algorithms presented in this chapter, is that it is not clear what exactly they compute. There is no analysis of how accurate the output of the algorithms is as compared to a ground truth. Algorithms do not take the error margins on the input into account and do not output error margins with their results. This holds for algorithms on grid models as well as for algorithms on (usually linearly interpolated) TIN models, both of which are just approximations of an actual terrain. This leads not only to inherently unreliable results, but also to a wasteful use of computational resources. Lacking a thorough analysis of what errors in the input may cause what errors in the output, we may run hundreds of computations on ever-higher-resolution terrain models, requiring ever-more-sophisticated noise removal techniques, but the extent of uncertainty remains unknown.

Acknowledgments

Laura Toma was supported by NSF award no. 0728780.

References

1. P. K. Agarwal, L. Arge, and A. Danner. From point cloud to grid DEM: A scalable approach. In *Proc. 12th Symp. Spatial Data Handling (SDH 2006)*, Vienna, Austria, pp. 771–788. Springer, 2006.
2. P. K. Agarwal, L. Arge, T. Mølhave, and B. Sadri. I/O-efficient algorithms for computing contours on a terrain. In *Proc. 24th Symp. Computational Geometry (SoCG 2008)*, College Park, MD, pp. 129–138, 2008.
3. P. K. Agarwal, L. Arge, and K. Yi. I/O-efficient construction of constrained delaunay triangulations. In *Proc. 13th Eur. Symp. Algorithms (ESA 2005)*, Palma de Mallorca, Spain, vol. 3669 of Lecture Notes in Computer Science, pp. 355–366, 2005.
4. P. K. Agarwal, L. Arge, and K. Yi. I/O-efficient batched union-find and its applications to terrain analysis. *ACM Transactions on Algorithms*, 7(1):11, 2010.
5. A. Aggarwal and J. S. Vitter. The input/output complexity of sorting and related problems. *Communications of the ACM*, 31(9):1116–1127, 1988.

6. M. V. A. Andrade, S. V. G. Magalhães, M. A. Magalhães, W. R. Franklin, and B. M. Cutler. Efficient viewshed computation on terrain in external memory. *Geoinformatica*, 15(2):381–397, 2011.
7. L. Arge, J. S. Chase, P. N. Halpin, L. Toma, J. S. Vitter, D. Urban, and R. Wickremesinghe. Flow computation on massive grid terrains. *GeoInformatica*, 7(4):104–128, 2003.
8. L. Arge, A. Danner, H. Haverkort, and N. Zeh. I/O-efficient hierarchical watershed decomposition of grid terrain models. In *Proc. 12th Symp. Spatial Data Handling (SDH 2006)*, Vienna, Austria, pp. 825–844. Springer, 2006.
9. L. Arge, L. Deleuran, T. Mølhave, M. Revsbæk, and J. Truelsen. Simplifying massive contour maps. In *Proc. 20th Eur. Symp. Algorithms (ESA 2012)*, Ljubljana, Slovenia, vol. 7501 of Lecture Notes in Computer Science, pp. 96–107, 2012.
10. L. Arge, M. Revsbæk, and N. Zeh. I/O-efficient computation of water flow across a terrain. In *Proc. 26th Symp. Computational Geometry (SoCG 2010)*, Snowbird, UT, pp. 403–412, 2010.
11. L. Arge, K. G. Larsen, T. Mølhave, and F. van Walderveen. Cleaning massive sonar point clouds. In *Proc. 18th ACM SIGSPATIAL Symp. Geographic Information Systems (GIS 2010)*, San Jose, CA, pp. 152–161, 2010.
12. L. Arge and J. S. Vitter. Optimal interval management in external memory. *SIAM Journal on Computing*, 32(6):1488–1508, 2003.
13. F. Aurenhammer and R. Klein. Voronoi diagrams. In *Handbook of Computational Geometry*, J.-R. Sack, J. Urrutia (eds.), pp. 201–290, Elsevier, 2000.
14. B. Ben-Moshe, P. Carmi, and M. J. Katz. Approximating the visible region of a point on a terrain. *Geoinformatica*, 12(1):21–36, 2008.
15. B. Ben-Moshe, M. J. Katz, and I. Zaslavsky. Visibility preserving terrain simplification: An experimental study. *Computational Geometry*, 28(2–3):175–190, 2004.
16. A. Beutel, T. Mølhave, and P. K. Agarwal. Natural neighbor interpolation based grid DEM construction using a GPU. In *Proc. 18th ACM SIGSPATIAL Symp. Geographic Information Systems (GIS 2010)*, San Jose, CA, pp. 172–181, 2010.
17. P. Burrough and R. McDonnell. Optimal interpolation using geostatistics. In *Principles of Geographic Information Systems*, pp. 132–161. Oxford University Press, 1998.
18. B. R. Calder and L. A. Mayer. Automatic processing of high-rate, high-density multibeam echosounder data. *Geochemistry Geophysics Geosystems*, 4(6):1048, 2003.
19. R. Carlson and A. Danner. Bridge detection in grid terrains and improved drainage enforcement. In *Proc. 18th ACM SIGSPATIAL Symp. Geographic Information Systems (GIS 2010)*, San Jose, CA, pp. 250–260, 2010.
20. L. C. Aleardi and O. Devillers. Explicit array-based compact data structures for triangulations. In *Proc. 22nd Int. Symp. Algorithms and Computation (ISAAC 2011)*, Yokohama, Japan, vol. 7074 of Lecture Notes in Computer Science, pp. 312–322, 2011.
21. L. C. Aleardi, O. Devillers, and G. Schaeffer. Succinct representations of triangulations with a boundary. In *Proc. 9th Workshop Algorithms and Data Structures (WADS 2005)*, Waterloo, Canada, vol. 3608 of Lecture Notes in Computer Science, pp. 134–145, 2005.
22. Y.-J. Chiang and C. T. Silva. I/O optimal isosurface extraction. In *Proc. Visualization 1997*, Phoenix, AZ, pp. 293–300, 1997.
23. Y.-J. Chiang, C. T. Silva, and W. J. Schroeder. Interactive out-of-core isosurface extraction. In *Proc. Visualization 1998*, Research Triangle Park, NC, pp. 167–174, 1998.
24. P. Cignoni, C. Montani, E. Puppo, and R. Scopigno. Optimal isosurface extraction from irregular volume data. In *Proc. Symp. Volume Visualization*, San Francisco, CA, pp. 31–38, 1996.
25. R. Cole and M. Sharir. Visibility problems for polyhedral terrains. *Journal of Symbolic Computation*, 7:11–30, 1989.
26. A. Crauser, P. Ferragina, K. Mehlhorn, U. Meyer, and E. A. Ramos. Randomized external-memory algorithms for some geometric problems. *Computational Geometry*, 11(3):305–337, 2001.

27. A. Danner, J. Baskin, A. Breslow, and D. Wilikofsky. Hybrid MPI/GPU interpolation for grid DEM construction. In *Proc. 20th ACM SIGSPATIAL Symp. Geographic Information Systems (GIS 2012)*, Redondo Beach, CA, 2012.
28. A. Danner, T. Mølhave, K. Yi, P. K. Agarwal, L. Arge, and H. Mitásová. Terrastream: From elevation data to watershed hierarchies. In *Proc. 15th ACM SIGSPATIAL Symp. Geographic Information Systems (GIS 2007)*, Seattle, WA, pp. 212–219, 2007.
29. M. de Berg, O. Cheong, H. Haverkort, J.-G. Lim, and L. Toma. The complexity of flow on fat terrains and its I/O-efficient computation. *Computational Geometry*, 43(4):331–356, 2010.
30. M. de Berg, O. Cheong, M. van Kreveld, and M. H. Overmars. *Computational Geometry: Algorithms and Applications*. Springer, 1997, 2000, 2008.
31. M. de Berg, H. Haverkort, and C. P. Tsirogiannis. Visibility maps of realistic terrains have linear smoothed complexity. *Journal of Computational Geometry*, 1(1):57–71, 2010.
32. M. de Berg and C. P. Tsirogiannis. Exact and approximate computations of watersheds on triangulated terrains. In *Proc. 19th ACM SIGSPATIAL Symp. Geographic Information Systems (GIS 2011)*, Chicago, IL, pp. 74–83, 2011.
33. L. de Floriani and P. Magillo. Visibility algorithms on digital terrain models. *International Journal of Geographic Information Systems*, 8(1):13–41, 1994.
34. L. de Floriani and P. Magillo. Algorithms for visibility computation on terrains: A survey. *Environment and Planning B—Planning and Design*, 30(5):709–728, 2003.
35. A. Driemel, H. Haverkort, M. Löffler, and R. I. Silveira. Flow computations on imprecise terrains. *Journal of Computational Geometry*, 41(1):38–78, 2013.
36. H. Edelsbrunner, D. Letscher, and A. Zomorodian. Topological persistence and simplification. *Discrete and Computational Geometry*, 28:511–533, 2002.
37. J. Fishman, H. Haverkort, and L. Toma. Improved visibility computations on massive grid terrains. In *Proc. 17th ACM SIGSPATIAL Symp. Geographic Information Systems (GIS 2009)*, Seattle, WA, pp. 121–130, 2009.
38. W. R. Franklin. Siting observers on terrain. In *Proc. 10th Symp. Spatial Data Handling (SDH 2002)*, Ottawa, Ontario, Canada, pp. 109–120, 2002.
39. W. R. Franklin and C. Ray. Higher isn't necessarily better: Visibility algorithms and experiments. In *Proc. 6th Symp. Spatial Data Handling (SDH 1994)*, Edinburgh, Scotland, pp. 751–763, 1994.
40. M. T. Goodrich, J.-J. Tsay, D. E. Vengroff, and J. S. Vitter. External-memory computational geometry. In *Proc. 34th IEEE Symposium on Foundations of Computer Science (FOCS 1993)*, Palo Alto, CA, pp. 714–723, 1993.
41. S. Gruber and S. D. Peckham. Land-surface parameters and objects in hydrology. In *Developments in Soil Science*, vol. 33, pp. 171–194. Elsevier, 2008.
42. A. Gyulassy, P.-T. Bremer, B. Hamann, and V. Pascucci. A practical approach to Morse-Smale complex computation: Scalability and generality. *IEEE Transactions on Visualization and Computer Graphics*, 14(6):1619–1626, 2008.
43. H. Haverkort and J. Janssen. Simple I/O-efficient flow accumulation on grid terrains. In *Abstracts Workshop Massive Data Algorithms*, Aarhus, Denmark. University of Aarhus, 2009.
44. M. Isenburg, Y. Liu, J. Shewchuk, and J. Snoeyink. Streaming computation of Delaunay triangulations. *ACM Transactions on Graphics*, 25(3):1049–1056, 2006.
45. M. Isenburg, Y. Liu, J. Shewchuk, J. Snoeyink, and T. Thirion. Generating raster DEM from mass points via TIN streaming. In *Proc. GIScience 2006*, Muenster, Germany, pp. 186–198, 2006.
46. D. Izraelevitz. A fast algorithm for approximate viewshed computation. *Photogrammetric Engineering and Remote Sensing*, 69(7):767–774, 2003.
47. M. J. Katz, M. H. Overmars, and M. Sharir. Efficient hidden surface removal for objects with small union size. *Computational Geometry*, 2(4):223–234, 1992.
48. P. Longley, M. Goodchild, D. Maguire, and D. Rhind. Spatial data analysis. In *Geographic Information Systems and Science*, Wiley, New York, pp. 351–379, 2011.

49. A. Mebarki. Implantation de structures de données compactes pour les triangulations. PhD. thesis, Université de Nice-Sophia Antipolis, Nice, France, 2008.
50. X. Meng, N. Currit, and K. Zhao. Ground filtering algorithms for airborne LiDAR data: A review of critical issues. *Remote Sensing*, 2(3):833–860, 2010.
51. L. Mitas and H. Mitásová. Spatial interpolation. In *Geographical Information Systems: Principles, Techniques, Management and Applications*, Wiley, New York, pp. 481–492, 1999.
52. H. Mitásová and L. Mitas. Interpolation by regularized spline with tension: I. Theory and implementation. *Mathematical Geology*, 25(6):641–655, 1993.
53. H. Mitásová, L. Mitas, W. Brown, D. Gerdes, and I. Kosinovsky. Modelling spatially and temporarilly distributed phenomena: New methods and tools for GRASS GIS. *International Journal of Geographic Information Systems*, 9(4):433–446, 1995.
54. E. Moet, M. van Kreveld, and A. F. van der Stappen. On realistic terrains. *Computational Geometry*, 41:48–67, 2008.
55. T. Mølhave, P. K. Agarwal, L. Arge, and M. Revsbæk. Scalable algorithms for large high-resolution terrain data. In *Proc. 1st Int. Conf. and Exhib. on Comp. for Geospatial Research & Application (COM.Geo'10)*, Washington, DC, 2010.
56. National Oceanic and Atmospheric Administration (NOAA) Coastal Services Center. 2012. "Lidar 101: An Introduction to Lidar Technology, Data, and Applications." Revised. Charleston, SC: NOAA Coastal Services Center. http://hurricane.csc.noaa.gov/digitalcoast/_/pdf/lidar101.pdf
57. M. R. Henzinger, P. Raghavan, and S. Rajagopalan. Computing on data streams. In *External Memory Algorithms*, vol. 50, DIMACS series in discrete mathematics and theoretical computer science, pp. 107–118. 1999.
58. H. I. Reuter, T. Hengl, P. Gessler, and P. Soille. *Preparation of DEMs for geomorphometric analysis*, Volume 33 of *Developments in Soil Science*, pp. 112–117. Elsevier, 2008.
59. J. R. Shewchuck. Triangle: engineering a 2D quality mesh generator and Delaunay triangulator. In *Proc. FCRC'96 Workshop Applied Computational Geometry: Towards Geometric Engineering (WACG 1996)*, Volume 1148 of Lecture Notes in Computer Science, pp. 203–222, 1996.
60. R. I. Silveira and R. van Oostrum. Flooding countries and destroying dams. *International Journal of Computational Geometry*, 20(3):361–380, 2010.
61. G. Sithole and G. Vosselman. Experimental comparison of filter algorithms for bare-earth extraction from airborne laser scanning point clouds. *ISPRS Journal Photogrammetry and Remote Sensing*, 59:85–101, 2004.
62. P. Soille, J. Vogt, and R. Colombo. Carving and adaptive drainage enforcement of grid digital elevation models. *Water Resources Research*, 39(12):1366–1375, 2003.
63. A. J. A. M. Temme, G. B. M. Heuvelink, J. M. Schoorl, and L. Claessens. *Geostatistical Simulation and Error Propagation in Geomorphometry*, Volume 33 of *Developments in Soil Science*, pp. 121–140. Elsevier, 2008.
64. J. Thiyagalingam and O. Beckmann. Is Morton layout competitive for large two-dimensional arrays yet? *Concurrency and Computation: Practice and Experience*, 18(11):1509–1539, 2006.
65. M. van Kreveld. Variations on sweep algorithms: efficient computation of extended viewsheds and class intervals. In *Proc. Seventh Symposium of Spatial Data Handling (SDH 1996)*, pp. 15–27, 1996.
66. M. van Kreveld and R. I. Silveira. Embedding rivers in triangulated irregular networks with linear programming. *International Journal of Geographic Information Science*, 25(4):615–631, 2011.
67. M. J. van Kreveld, R. van Oostrum, C. L. Bajaj, V. Pascucci, and D. Schikore. Contour trees and small seed sets for isosurface traversal. In *Proc. 13th Symp. Computational Geometry (SoCG 1997)*, pp. 212–220, 1997.
68. J. Wilson and J. Gallant. *Terrain Analysis: Principles and Applications*, Chapter Digital terrain analysis. Wiley, 2000.
69. S. Yu, M. van Kreveld, and J. Snoeyink. Drainage queries on TINs: From local to global and back again. In *Proc. 7th Symp. Spatial Data Handling (SDH 1996)*, pp. 13A.1–13A.14, 1997.

32

Geometric Primitives

Lee Staff
King Abdullah University of Science and Technology

Gustavo Chávez
King Abdullah University of Science and Technology

Alyn Rockwood
King Abdullah University of Science and Technology

32.1 Introduction

Geometric primitives are the basic building blocks for creating sophisticated objects in computer graphics and geometric design. They provide familiarity, uniformity, and standardization and also enable hardware support.

Initially, the definition of geometric primitives was driven by the capabilities of the hardware. Only simple primitives were available, for example, points, line segments, and triangles. In addition to hardware constraints, other driving forces in the development of geometric primitives have been either general applicability to a broad range of needs or wide applicability in restricted domains. The triangular facet is an example of a primitive that is simple to generate, easy to support in hardware, and widely used to model many graphics objects. The sphere is included in the primitive set of most software libraries and software products, for example, AutoCAD, Maya, Rhinoceros, SketchUP, and CATIA. More specialized primitives have also been developed, for example, in architectural modeling where a door might be considered as a primitive in a building information model (BIM) or in chemical modeling where a particular molecule configuration would be considered as primitive. So, what is a geometric primitive? Here we qualify the definition based on simplicity of the object and frequency of use based on computer-aided design (CAD)/computer-aided manufacturing (CAM), computer-aided engineering (CAE), computer graphics, animation, and industrial styling systems. We are cognizant that other readers may choose a somewhat different set of primitives based on their experiences, for example, editing operations like push/pull, but we have strived to select the most comprehensive base available in current software.

As hardware, central processing unit (CPU), and the number of specialized applications increased in capability, sophistication of primitives grew as well. Primitives became somewhat less dependent on hardware and subsequently software primitives became more common. For raw speed, however, hardware primitives still dominate, for example, it is widespread practice in computer graphics engines to partition any primitive object into triangles.

One direction for the expansion of graphics primitives has been in the geometric order of the primitive. Initially, primitives were discrete or first-order approximations of objects, that is, collections of points, lines, and planar polygons. This has been extended to higher-order primitives represented by polynomial or rational curves and surfaces in any direction. The other direction for the expansion of primitives has been in attributes that are assigned to the geometry. Color, transparency, associated bitmaps and textures, surface normals, and labels are examples of attributes attached to a primitive for use in its display. Since the focus of this chapter is on geometry, attributes are beyond the scope of this discussion.

This summary of graphics primitives will first describe representative examples used in practice. We then draw upon these examples to discuss the fundamental principles underlying primitives. Finally, we highlight some future directions in the field of geometric primitives being pursued in academia and industry.

32.2 Examples in Practice

32.2.1 Screen Specification

Screen specification is closely related to object geometry and is the coordinate frame wherein all primitives lie. The problem of rendering a graphics primitive on a raster screen is called *scan conversion*. It is the scan conversion method that is embedded in hardware to accelerate and improve the display of graphics in a system.

32.2.2 Simple Primitives

The primitives are described here in roughly the chronological order of their development, which basically corresponds to their increase in complexity. This chapter concentrates on common primitives. We will distinguish between *design primitives* and *database primitives*. A *design primitive* is input data that form an integral part of the geometric model, whereas a *database primitive* is a tool that is used to assist in the design process or the way in which it is saved in memory, but does not form part of the model (Figure 32.1).

32.2.2.1 Points

A *point* represents a zero-dimensional object. A point can be part of a bitmap representation of an object, for example, line and text, or used as a construction tool, for example, control vertex (CV). The classic example of point as design primitive is its use in text. There are two standard ways to represent

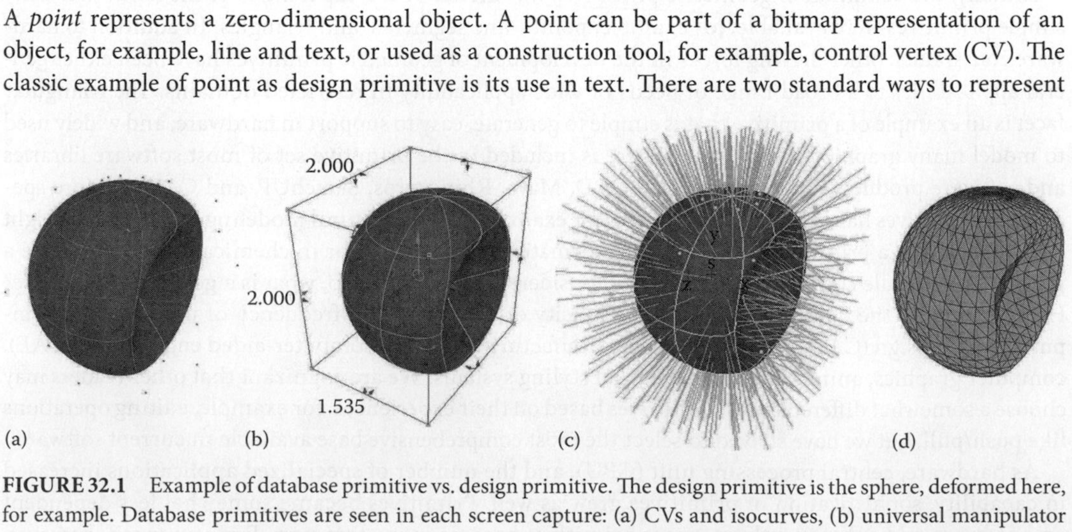

FIGURE 32.1 Example of database primitive vs. design primitive. The design primitive is the sphere, deformed here, for example. Database primitives are seen in each screen capture: (a) CVs and isocurves, (b) universal manipulator frame, (c) normals to surfaces, and (d) underlying triangulated surface tessellation. Model created in Autodesk Maya.

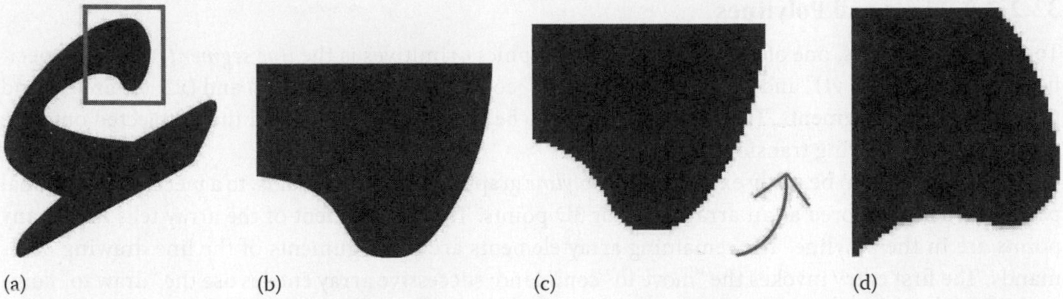

FIGURE 32.2 The drawing in (a) represents the Arabic letter hamza (ء). Details show the effect of (b) anti-aliasing to smooth out jagged edges; (c) bitmap pixilation, jagged edges, as the result of aliasing; and (d) rotation on pixilation; note how transformation changes the pixilation.

textual characters for graphics purposes. The first method is to save a representation of the characters as a bitmap, called a *font cache*. This method allows fast reproduction of the character on-screen. Usually the font cache has more resolution than needed, and the characters are pixelated. Even on high-resolution devices such as a quality laser printer, the discrete nature of the bitmap can be apparent, creating jagged edges. When transformations are applied to bitmaps such as rotations or shearing, aliasing problems can also be apparent. See the examples in Figure 32.2.

To improve the quality of transformed characters and to compress the amount of data needed to transfer text, a second method for representing characters was developed using polygons or curves. When the text is displayed, the curves or polygons are scan-converted; thus, the quality is constant regardless of the transformation. The transformation is applied to the curve or polygon basis before the scan conversion. PostScript is a well-known product based on text transferal (see Adobe Systems Inc., 1999). In a PostScript printer, for instance, the definition of the fonts resides in the printer where it is scan-converted. Transfer across the network requires only a few parameters to describe font size, type, style, etc. Those printers that do not have resident PostScript databases and interpreters must transfer bitmaps with a resulting loss of quality and time. PostScript is based on parametrically defined curves called Bézier curves (see Section 32.3).

An important trend in 3D surface representation is using points instead of polygons, that is, a *point cloud* may be used where a dense set of points defines the surface of an object (Figure 32.3a). Unlike triangular meshes, points do not require connectivity data. For further information see Levoy and Whitted (1985), Rockwood (1986), Rockwood (1987), and Pauly et al. (2003).

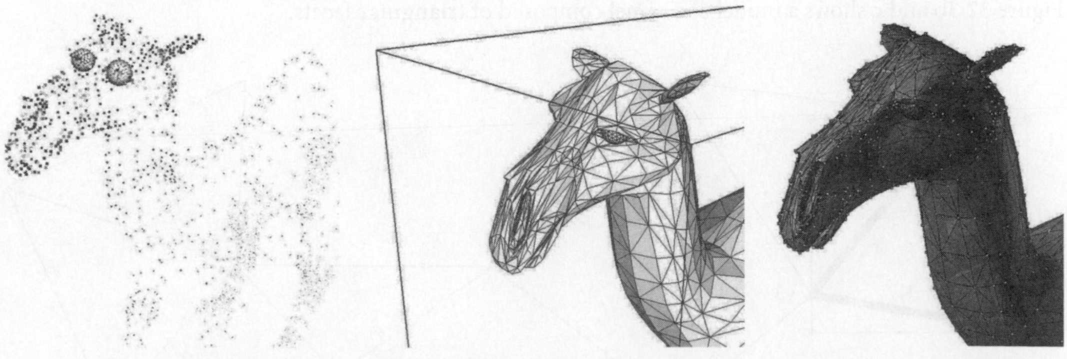

FIGURE 32.3 Three renderings of a camel head model: (a) point cloud model (rendered in MeshLab), (b) triangulated mesh model with bounding box (rendered in SketchUP), and (c) triangulated mesh model with CVs shown (ghosted rendering in Rhinoceros). Open source model "camello" by roberttittto in SketchUP 3D Warehouse (http://sketchup.google.com/3dwarehouse [accessed October 23, 2013]).

32.2.2.2 Lines and Polylines

Together with points, one of the most prevalent graphics primitives is the *line segment*, typically specified by "move to ($x1$, $y1$)" and then "draw to ($x2$, $y2$)" commands, where ($x1$, $y1$) and ($x2$, $y2$) are the end points of the line segments. The end points can also be given as 3D points and then projected onto the screen through viewing transformations.

A line segment can be easily extended to a *polyline* graphics primitive, that is, to a piecewise polygonal path. It is usually stored as an array of 2D or 3D points. The first element of the array tells how many points are in the polyline. The remaining array elements are the arguments of the line drawing commands. The first entry invokes the "move to" command; successive array entries use the "draw to" command as an operator.

Because of the discrete, pixel-based nature of the raster display screen, lines are prone to alias; that is, they may break apart or merge with one another to form distracting *moiré* patterns (Creath and Wyant 1992). They are also susceptible to jagged edges, a signature of older graphics displays. Advanced line drawing systems found it therefore important to add hardware that would *anti-alias* their lines while drawing them (Freeman 1974; Crow 1977) (Figure 32.2b). Where aliasing produces false artifacts by under sampling, anti-aliasing reduces those artifacts either by more intelligent or higher resolution sampling.

32.2.2.3 Polygons

Closing a polyline by matching start and end points creates a polygon, for example, quadrangle and triangle (see Appendices 32.A and 32.B). It is probably the most commonly employed primitive in graphics, because it is easy to define an associated surface. Because of their usefulness for defining surfaces, polygons almost always appear as 3D primitives that subsume the 2D case.

The most sophisticated polygon primitive allows non-convex polygons that contain other polygons, called *holes* or *islands* depending on whether they are filled or not. A complex polygon is often made as a macro out of simple polygon primitives for example, in Open Graphics Library (OpenGL) it would be written as a strip of triangles (Figure 32.4). There are many triangulation routines to reduce polygons to simple triangular facets.

Associated surfaces, or faces, can be created with a polygon where the edges form the boundaries of the surface. Vertices represent the positions where the end points of surface edges meet.

32.2.2.4 Triangular Facet

If the polygon is the most popular primitive, the simple triangle is certainly the most popular polygon. Figure 32.3b and c shows a model of a camel composed of triangular facets.

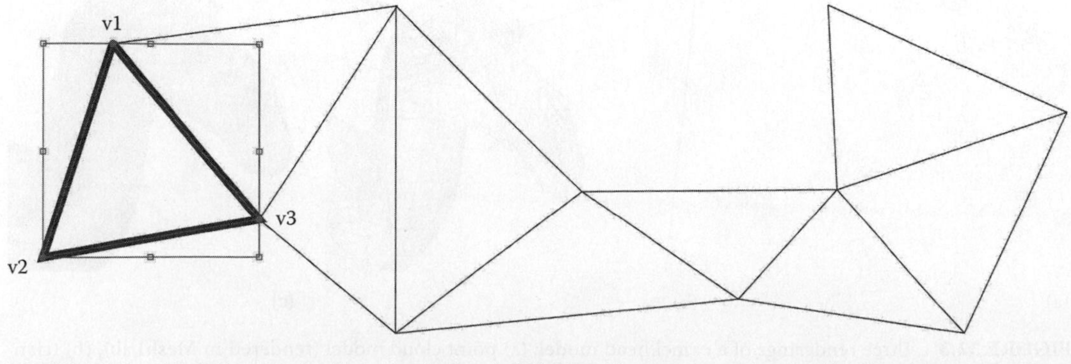

FIGURE 32.4 Strip of triangular facets: lead triangle (thick lines), successive triangles (thin lines); vertices of start triangle indicated as v1, v2, and v3; "path" (lines that connect vertices 1–3); and bounding box of triangle (lines that form a square around the triangle and intersect with vertices).

The triangular facet is fast to draw and supports more complex polygons as well as many more diverse and powerful operations. Another major advantage is that they are always flat; their vertices do not have to leave the plane and avoid numerical problems, data errors, or non-planar preserving transformations. Currently, graphics cards are designed to handle raw triangles in the order of millions, where the limit is given by the memory size of the graphics card and time required for the render process (measured in frames per second); the current limit in memory size is around 6GB.

In order to increase the performance of polygon rendering and decrease the size of the file database, triangular and quadrilateral facets can be stored and processed as meshes, that is, collections of facets that share edges and vertices. The triangular mesh, for instance, is given by defining three vertices for the "lead" triangle and then giving a new vertex for each successive triangle (Figure 32.4). The succeeding triangles use the last two vertices in the list with the newly given one to form the next triangle. Such a mesh contains about one-third the data and uses the shared edges to increase processing speed. Most graphics objects have large contiguous areas that can take advantage of meshing, which turns out to be a compression technique. The quadrilateral mesh requires two additional vertices be used, with the last two given, and therefore has less savings.

Because of its benefits and ubiquity, many other primitives are defined in software as configurations of the triangular facet. One example is the faceted sphere. With enough facets the sphere will look smooth (see Figure 32.9). It is commonly encountered in engineering applications and molecular modeling.

32.2.2.5 Elliptical Arcs

Elliptical arcs in 2D may be specified in many equivalent ways. One way is to give the center position, major and minor axes, and start and end angles. Both angles are given counterclockwise from the major axis. It should be mentioned that arcs can also be represented by a polyline with enough segments; thus, a software primitive for the arc that induces the properly segmented and positioned polyline may be a cost-effective *macro* for elliptical arcs (Figure 32.5). This macro should consider the effects of zoom and perspective to avoid revealing the underlying discretization of the arc. It is surprising how small a number of line segments, properly chosen, can give the impression of a smooth arc. One of the most commonly used elliptical arcs is, of course, the circular arc.

32.2.2.6 Curves: Bézier, Splines, and NURBS

The four most common design primitives for drawing more complex curves are the polynomial and rational *Bézier*, *B-spline*, nonuniform rational B-spline (*NURBS*), and *interpolating splines*, that is, *fit*

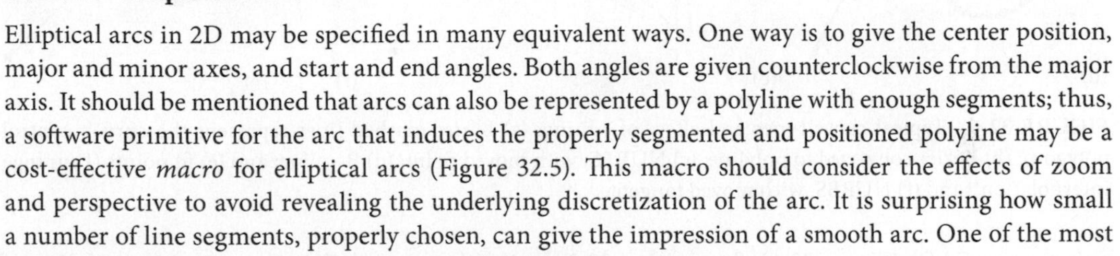

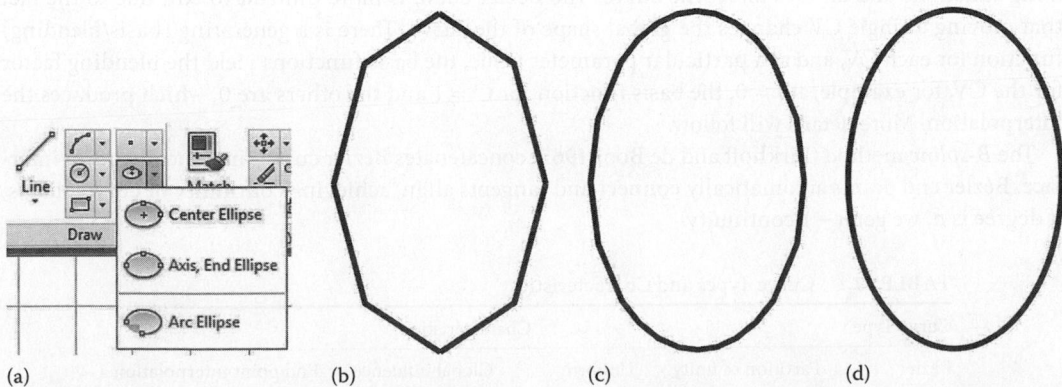

FIGURE 32.5 Ellipse with decreasing discretization of curve, that is, 226, 24, and 8 line segments, from right to left. Inset shows typical user interface for drawing ellipses using alternative input data options, that is, center point, axis direction and end point, or arc length. (Screen capture from AutoCAD.)

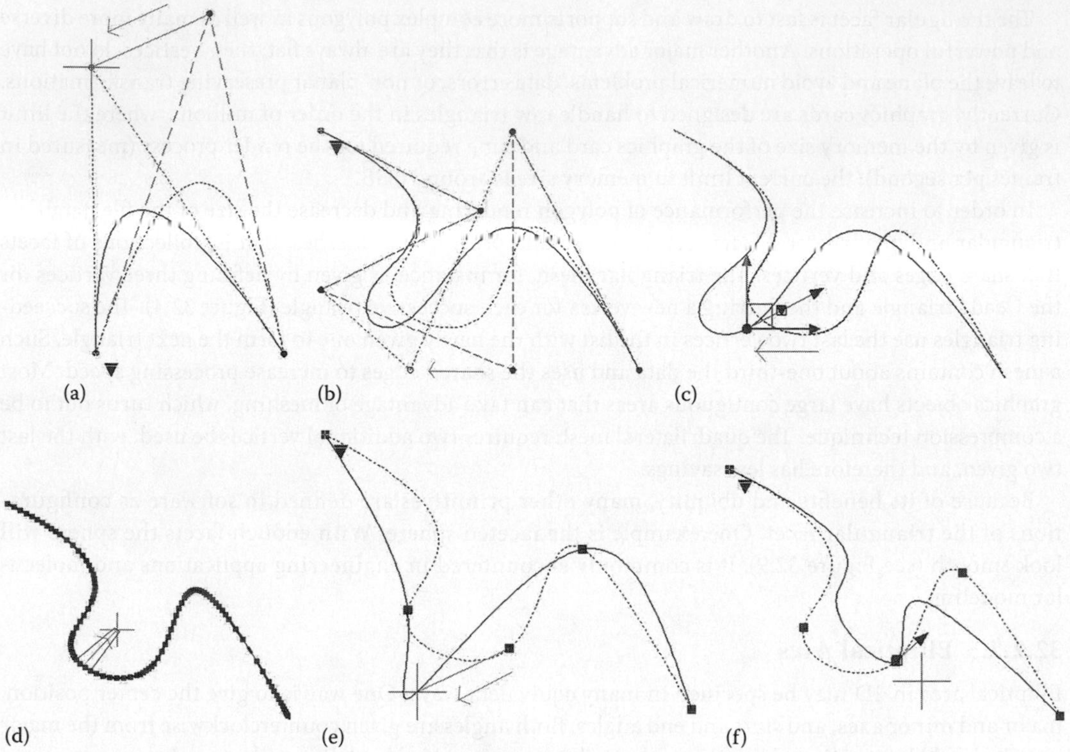

FIGURE 32.6 Typical curve types and effect of editing: (a) Bézier (dashed lines represent the curve's control polygon); (b) Bézier converted to polyline; (c) NURBS, with moved point; (d) B-spline; and (e) fit points (Lagrange interpolation); and (f) NURBS, with moved tangent.

points (AutoCAD) (Figure 32.6). See Section 32.3 for the theory underlying parametric modeling of curves. All polynomial curve types possess a specified degree (the degree implies the number of control points that influence any given point on the curve). Characteristics, such as editability, computational complexity, and ease of use, vary (see Table 32.1).

The *Bézier* method has a strict relation between degree and number of CVs (see Rockwood and Chambers 1996, p.55). End points are fixed and the other CVs lie off the curve. Each curve has generating functions and all CVs affect the curve. The Bézier curve is more difficult to edit due to the fact that moving a single CV changes the global shape of the curve. There is a generating (basis/blending) function for each CV, and at a particular parameter value, the basis functions yield the blending factor for the CV, for example, at $t = 0$, the basis function for $C^0 = 1$ and the others are 0, which produces the interpolation. More details will follow.

The *B-spline* method (Birkhoff and de Boor 1965) concatenates Bézier curves in a succinct user interface. Bézier end points automatically connect and tangents align, achieving continuity of curve, that is, if degree is n, we get $n - 1$ continuity.

TABLE 32.1 Curve Types and Characteristics

Curve Type	Characteristics			
Bézier	Partition of unity	Uniform	Global influence	End point interpolation
B-spline	Partition of Unity	Uniform	Local influence	Noninterpolating
NURBS	Partition of unity	Nonuniform	Local influence	Noninterpolating
Fit points	Non-POU	Nonuniform	Local influence	Interpolates every point

Good editability requires that curves and surfaces allow options of degree and CV weight (influence) together with automatic curvature continuity when concatenating or adding new segments. *NURBS* stands for nonuniform rational B-spline, and the input required are degree, weighted CVs, and knot vectors (a knot vector is a way to specify the weighting factors with a visually convenient sequence of points on the domain line; see Figure 32.12). Nonuniform means that the knots are not evenly spaced. Rational indicates that the CVs are weighted by a rational factor. Adding or removing interior control points, which requires adding or removing a corresponding knot, on a curve allows local refinement. This allows nonuniform subdivision of NURBS curves and surfaces making it a good way to add local complexity. Nonrational splines or Bézier curves may approximate a circle, but they cannot represent it exactly. Rational splines can represent any conic section exactly, including the circle (Piegl and Tiller 1987).

The *fit point* method uses Lagrange interpolation (Berrut and Trefethen 2004) to create an interpolatory B-spline. A special case of the fit point method includes two tangent controls at each point, for example, the "paths" in Adobe (see also Hermite interpolation, e.g., Bartels et al. 1987).

The optimal work flow for modeling curves would be to design and save in B-splines but operate in Bézier form, which provides greater stability and speed, and restore curves in B-spline or NURBS since these are more compact to store.

32.2.3 Solid Geometry Primitives

Before describing each solid geometry primitive, it is worth mentioning the importance of the *Euler characteristic* (see Richeson 2008). In mathematics and computer graphics, one can create objects that could never be built, for example, self-intersecting surfaces. The Euler characteristic is one way of identifying whether or not an object is buildable. The *Euler characteristic,* usually denoted as *chi* χ, was classically defined for the surfaces of polyhedra, according to the formula

$$\chi = V - E + F$$

where V, E, and F are, respectively, the numbers of vertices, edges, and faces in the given polyhedron. Any deformation of the ball without self-intersection, including all convex polyhedra, has Euler characteristic $\chi = 2$, whereas with a torus, for example, $\chi = 0$.

The most widely accepted solid geometry primitives are cube, pyramid, cylinder, cone, sphere, ellipsoid, and torus, as briefly described in the following (Figure 32.7).

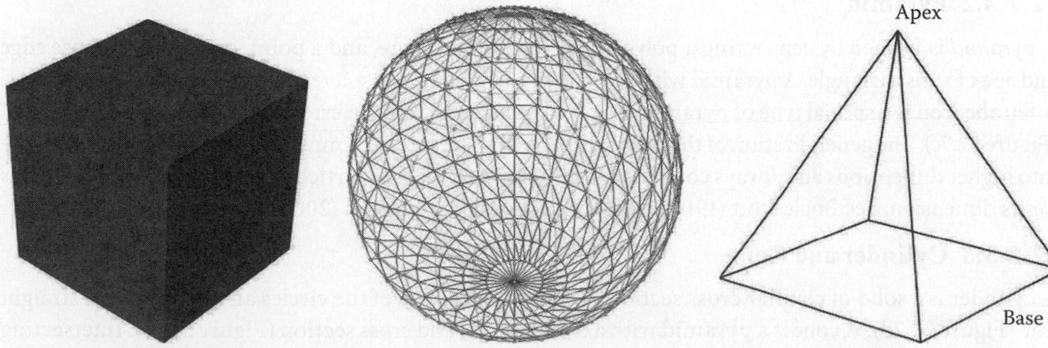

FIGURE 32.7 The most fundamental solid geometry primitives: cube, sphere, pyramid, cylinder, ellipsoid, cone, and torus. The paraboloid and truncated cone are often considered primitives in many modeling software packages. Software used to create these images include Autodesk Maya, AutoCAD, Adobe Photoshop, SketchUP, MeshLab, and Rhinoceros.

(continued)

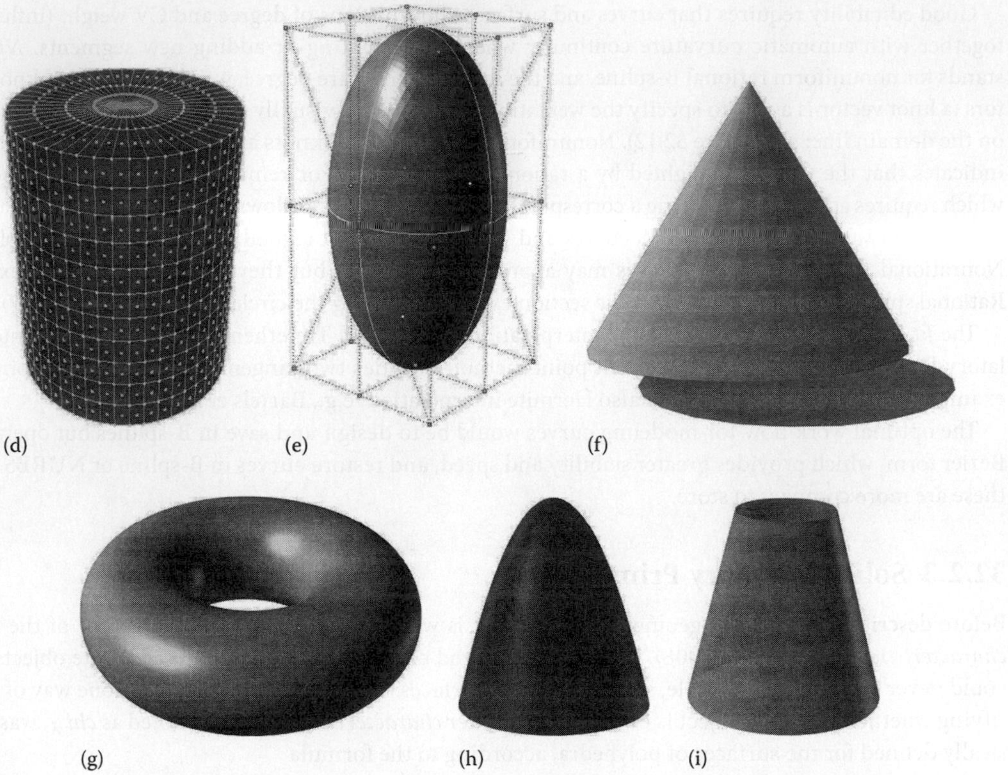

FIGURE 32.7 (continued) The most fundamental solid geometry primitives: cube, sphere, pyramid, cylinder, ellipsoid, cone, and torus. The paraboloid and truncated cone are often considered primitives in many modeling software packages. Software used to create these images include Autodesk Maya, AutoCAD, Adobe Photoshop, SketchUP, MeshLab, and Rhinoceros.

32.2.3.1 Cube

A *cuboid* or *rectangular box* is bounded by six rectangular faces, three of which meet at each vertex. The special case is the cube, a platonic solid, which is bounded by six equal faces (Figure 32.7a).

32.2.3.2 Pyramid

A *pyramid* is formed by connecting a polygonal base, often a square, and a point, or apex. Each base edge and apex forms a triangle. A pyramid with an n-sided base will have $n + 1$ vertices, $n + 1$ faces, and $2n$ edges. A tetrahedron is a special type of pyramid with triangular base, often referred to as a "triangular pyramid" (Figure 32.7c). The generalization of the tetrahedron is the "simplex." A simplex is the extension of a triangle into higher dimensions and always contains the minimum number of vertices, edges, faces, and hyperfaces for its dimension. See Boole Stott (1910), Coxeter (1988), and Grünbaum (2003) for more on polyhedra.

32.2.3.3 Cylinder and Cone

A cylinder is a solid of circular cross section in which the centers of the circles all lie on a single straight line (Figure 32.7d). A cone is a pyramid with a circular base and cross section (Figure 32.7f). Intersecting a double cone with an inclined plane is particularly instructive since the resultant conic sections represent the quadratic curves, that is, circles, ellipses, parabolas, and hyperbolas (Figure 32.8).

32.2.3.4 Sphere, Ellipsoid, and Torus

A sphere is defined as the set of all points in 3D Euclidean space $\mathbb{R}^3$ that are located at a fixed distance, r (radius), from a given center point or origin (see Coxeter 1942) (Figure 32.7b). Figure 32.9d shows a

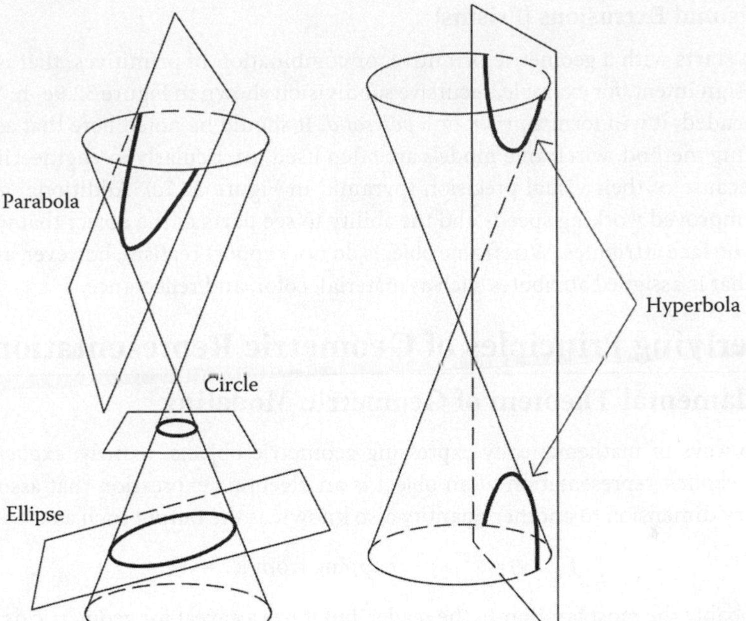

FIGURE 32.8 Conic sections; when a cone is cut through by a plane, depending on the angle of the plane, the resultant section cut is a circle, ellipse, parabola, or hyperbola. (Adapted from Weissten, E.W., MathWorld—A wolfram web resource http://mathworld.wolfram.com/ConicSection.html [accessed October 23, 2011].)

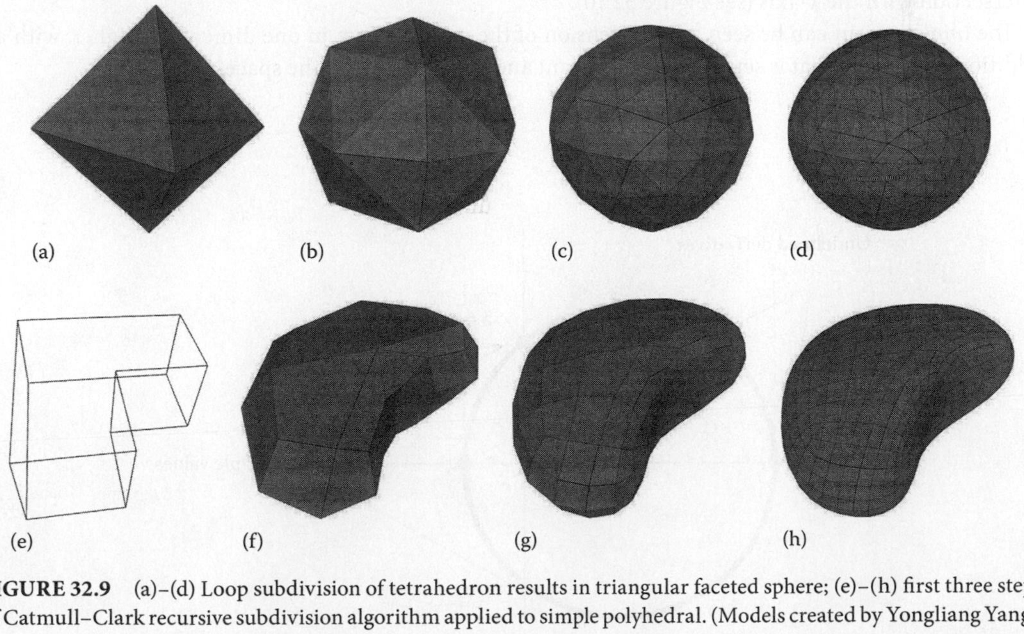

FIGURE 32.9 (a)–(d) Loop subdivision of tetrahedron results in triangular faceted sphere; (e)–(h) first three steps of Catmull–Clark recursive subdivision algorithm applied to simple polyhedral. (Models created by Yongliang Yang.)

faceted sphere resulting from a process of *recursive subdivision* (see the following) that started from a double pyramid. The problem of structuring the surface of a sphere and curved shapes, for example, with triangles, quadrilaterals, combination, or pentagons and hexagons, continues to be a challenging research issue in the field of geometric modeling. A *torus*, a closed surface with a single handle, is the only example thus far of a primitive higher than second order (Figure 32.7g).

32.2.3.5 Polygonal Extrusions (Prisms)

Modeling often starts with a geometric primitive, or combination of primitives, that is then modified according to design intent, for example, recursive subdivision shown in Figure 32.9e–h. When a polygonal plane is extruded, it will form a *prism* or a *polysolid*. It should be noted here that as part of the 3D modeling working method, wireframe models are often used, particularly in engineering and drafting applications, because of their visual precision (pyramid in Figure 32.7c). Additional advantages with wireframe are improved working speeds and the ability to see parts of the object that would otherwise be obscured by surface attributes. Wireframe objects do not support realism, however, and most objects have a surface that is assigned attributes such as material, color, and reflectance.

32.3 Underlying Principles of Geometric Representation

32.3.1 Fundamental Theorem of Geometric Modeling

There are three ways of mathematically expressing geometric objects, namely, *explicit, implicit,* and *parametric*. An *explicit* representation of an object is an algebraic expression that associates an input from an arbitrary dimension to another quantity also known as the output such as

$$f(x, y) = x^2 + y^2, \text{ mapping from } \mathbb{R}^2 \to \mathbb{R}$$

This form is probably the most familiar to the reader, but it has a caveat for geometric design because its shape requires multiple values, but it is a single value function, for example, in the definition of a circle:

$$x = \pm\sqrt{r^2 - y^2}$$

In general, another issue with this form is that derivatives cannot always be defined, for instance, at the intersection with the x-axis (see Figure 32.10).

The *implicit* form can be seen as an extension of the explicit form in one dimension higher, with an additional equation that is set equal to a constant and therefore limits the space, for instance,

$$f(x, y) = z = x^2 + y^2 - r^2$$

$$x^2 + y^2 - r^2 = 0$$

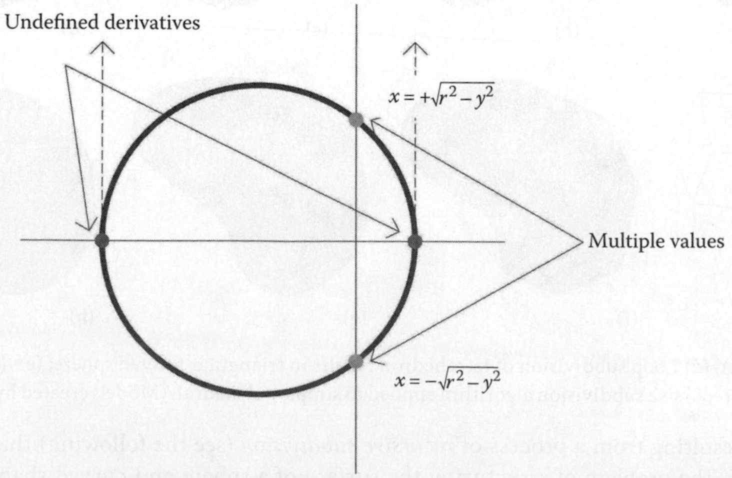

FIGURE 32.10 Diagram showing caveats for explicit representation of geometric design; undefined derivatives (points shown left and right on circle); vertical line test (dotted line through undefined derivative); multiple-valued function (solid vertical line intersecting two points on circle).

This yields a system of 2 equations and 3 unknowns. A major advantage of this representation is that it allows us to determine where a point lies relative to the surface, that is,

$$F(x) > 0, \quad \text{if } x \text{ is "outside" the surface}$$

$$F(x) = 0, \quad \text{if is "on" the surface}$$

$$F(x) < 0, \quad \text{if } x \text{ is "inside" the surface}$$

The *parametric* representation is a system of variables and constraints in which some of the variables, known as *parameters*, are used as generators and the other variables sweep out the object; this representation is helpful when the objective is to easily find points in a plot or to sweep a curve with an additional variable such as time. For example, the parametric equation of a unit circle is

$$x = \cos(t)$$

$$y = \sin(t)$$

with t varying across the interval $[0, 2\pi]$.

In this example x, y specify the object, and t is the generating parameter. This representation also yields a system of equations that has 3 unknowns (x, y, t) and 2 constraints. Note that in order to transform between implicit and explicit forms, the forms must satisfy certain conditions; see implicit and inverse function theorems (Spivak 1965).

32.3.1.1 Degrees of Freedom

If you subtract the number of equations from the number of variables, you can get the intrinsic dimension of the object (Table 32.2). Position in space given by a cubic Bézier curve at a certain time t can be represented as follows:

$$P(t) = (1-t)^3 P_0 + 3(1-t)^2 tP_1 + 3(1-t)^2 t^2 P_2 + t^3 P_3$$

$$P_0 = \begin{pmatrix} x_0 \\ y_0 \\ z_0 \end{pmatrix}, P_1 - \begin{pmatrix} x_1 \\ y_1 \\ z_1 \end{pmatrix}, P_2 - \begin{pmatrix} x_2 \\ y_2 \\ z_2 \end{pmatrix}, P_3 - \begin{pmatrix} x_3 \\ y_3 \\ z_3 \end{pmatrix}$$

P represents a 3-tuple. This parametric form has 4 variables (x, y, z, t), described by 3 equations in x, y, and z, and the degree of freedom is then 1, which implies a curve.

TABLE 32.2 Dimension of Shape Can Be Deduced from Number of Unknowns and Equations

1 unknown	1 equation	0 dimension	Point
3 unknowns	2 equations	1 dimension	Curve
4 unknowns	2 equations	2 dimension	Surface
5 unknowns	2 equations	3 dimension	Volume
6 unknowns	2 equations	4 dimension	Hypervolume

32.3.2 Implicit Modeling

32.3.2.1 Implicit Primitives

The need to model mechanical parts drove the definition of implicitly defined primitives. These modeling primitives naturally became graphics primitives to serve the design industry. They have since been used in other fields such as animation and architecture.

An implicit function f maps a point x in R^n to a real number, that is, $f(x) = r$. Usually, $n = 3$. By absorbing r into the function, we can view the implicit function as mapping to zero, that is, $g(x) = f(x) - r = 0$. The importance of implicit functions in modeling is that the function divides space into three parts: $f(x) < 0$, $f(x) > 0$, and $f(x) = 0$, as mentioned earlier. The last case defines the surface of an *implicit object*. The other two cases define, respectively, the *inside* and *outside* of the implicit object. Hence, implicit objects are useful in *solid modeling* where it is necessary to distinguish the inside and outside of an object. A polygon does not, for instance, distinguish between the inside and outside. It is, in fact, quite possible to describe an object in which inside and outside are ambiguous. This facility to determine inside and outside is further enhanced by using Boolean operations on the implicit objects to define more complex objects (see CSG objects in the following) (Figure 32.11).

Another advantage of implicit objects is that the surface normal of an implicitly defined surface is given simply as the gradient of the function: $N(x) = \nabla f(x)$. Furthermore, many common engineering surfaces are easily given as implicit surfaces; thus, the plane (not the polygon) is defined by $n \cdot x + d = 0$, where n is the surface normal and d is the perpendicular displacement of the plane to the origin. The ellipsoid is defined by $(x/a)^2 + (y/b)^2 + (z/c)^2 - r^2 = 0$, where $x = (x, y, z)$. General *quadratics*, which include ellipsoids, paraboloids, and hyperboloids, are defined by $x^T M x + b \cdot x + d = 0$, where M is a 3-by-3 matrix and b and d are vectors in $\mathbb{R}^3$. The quadrics include such important forms as the cylinder, cone, sphere, paraboloids, and hyperboloids of one and two sheets. Other implicit forms used are the torus, blends (transition surfaces between other surfaces), and *superellipsoids* defined by $(x/a)^k + (y/b)^k + (z/c)^k - \mathbb{R} = 0$ for any integer k.

32.3.2.2 CSG Objects

An important extension to implicit modeling arises from applying set operations such as union, intersection, and difference to the sets defined by implicit objects. The intersection of six half spaces defines

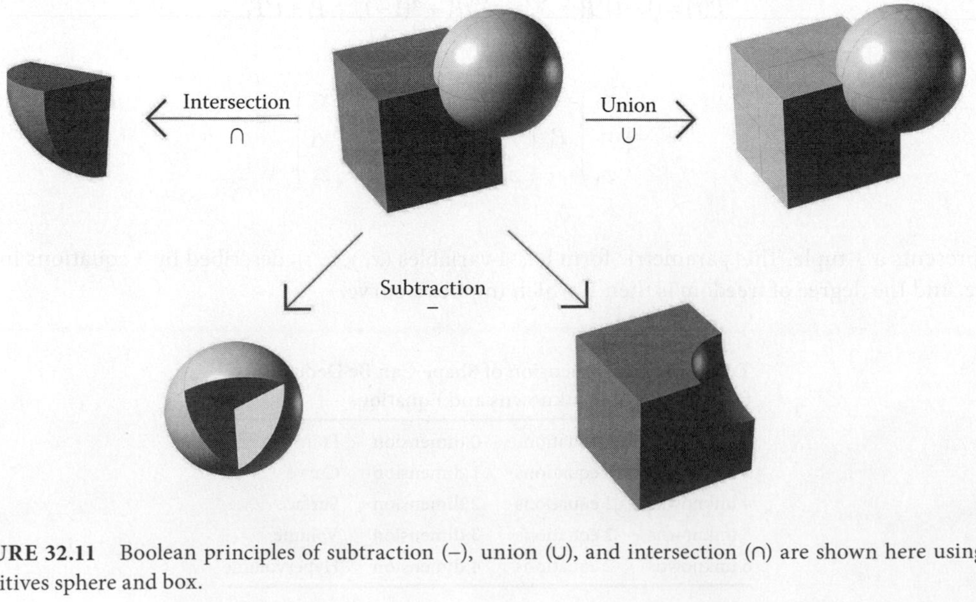

FIGURE 32.11 Boolean principles of subtraction (−), union (∪), and intersection (∩) are shown here using the primitives sphere and box.

a cube, for example. This method of modeling is called *constructive solid geometry* (CSG) (Foley 1990; Foley et al 1994). All set operations can be reduced to some combination of just union, intersection, and complementation (Figure 32.11). Because these create an algebra on the sets that is isomorphic to Boolean algebra, corresponding to multiply, add, and negate, respectively, the operations are often referred to as *Booleans*. Note that this is equivalent to directed acyclic graph (DAG) (Bang-Jensen and Gutin 2009). For example, an object could be made from Boolean parts of plane quadrics, a part torus, and blended transition surfaces. Unfortunately, implicit forms tend to be more expensive to render because algorithms for polygonizing implicit objects and CSG models tend to be quite complex (Bloomenthal 1988). The implicit object gives information about a point relative to the surface in space, but no information is given to tessellate into rendering elements such as triangular facets. In the case of ray tracing, however, the parametric form of a ray $x(t) = (x(t), y(t), z(t))$ composed with implicit function $f(x(t)) = 0$ leads to a root-finding solution of the intersection points on the surface, which are critical points in the ray-tracing algorithm. Determining whether points are part of a CSG model is simply exclusion testing on the CSG tree.

32.3.3 Parametric Modeling

32.3.3.1 Parametric Curves

When modeling with implicit equations, as described earlier, we know whether a point is inside or outside the model. When modeling with parametric equations, on the other hand, we know exactly where a point is on the surface. This makes parametric modeling more convenient for rendering compared to implicit since we can triangulate or texture map.

An important class of geometric primitives is formed by parametrically defined curves and surfaces. These constitute a flexible set of modeling primitives that are locally parameterized; thus in space, the curve is given by $P(t) = (x(t), y(t), z(t))$ and the surface by $S(u, v) = (x(u, v), y(u, v), z(u, v))$. In this section and the next one, we will give examples of only the most popular types of parametric curves and surfaces. There are many variations on the parametric forms (Farin 2002).

32.3.3.1.1 Bézier Curves

A Bézier curve is a parametric curve widely used in computer graphics (De-Casteljau 1959; Bézier 1974). Bézier curves are used to model smooth curves that are not bound by the limits of rasterized images and that are intuitive to modify. In vector graphics software, Bézier curves are linked to form "paths" that are then used in image manipulation.

The general form of a Bézier curve of degree n is

$$f(t) = \sum_{i=0}^{n} b_i B_i^n(t)$$

where b_i are vector coefficients, the *control points*, and

$$B_i(t) = \binom{n}{i} t^i (1 - t^{n-i})$$

where

$\binom{n}{i}$ is the binomial coefficient

$B_i(t)$ are Bernstein *functions*

They form a basis for the set of polynomials of degree n. Bézier curves have a number of characteristics, which are derived from the Bernstein functions and which define their behavior. Figure 32.12 shows examples of B-spline basis functions.

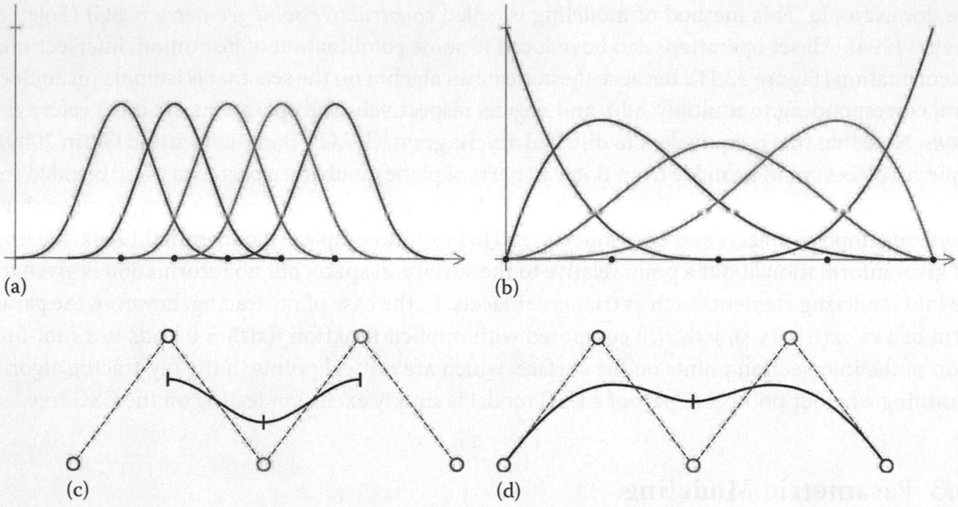

FIGURE 32.12 Comparison of basis functions with uniform and nonuniform knot vectors. (a) The uniform case; each basis function is an identical translated copy of all the others and knots are equally spaced; (b) nonuniform basis functions; the first three knots and the last three knots have zero spacing between them, the remaining knots are equally spaced; (c) and (d) cubic B-splines using the same control points but two different knot vectors, as seen in (a) and (b), respectively. (Adapted from Cashman, T.J., NURBS-compatible subdivision surfaces, Ph.D. University of Cambridge, Computer Laboratory, Queen's College, Cambridge, U.K., p. 21, 2010.)

End point interpolation: The Bézier curve interpolates the first and last control points, b_0 and b_n. In terms of the interpolation parameter t, $f(0) = b_0$ and $f(1) = b_n$.

Tangent conditions: The Bézier curve is cotangent to the first and last segments of the control polygon (defined by connecting the control points) at the first and last control points; specifically $f'(0) = (b_1 - b_0)n$ and $f'(1) = (b_n - b_{n-1})n$.

Convex hull: The Bézier curve is contained in the convex hull of its control points for $0 \leq t \leq 1$ (note: the convex hull can be thought of as enclosing a rubber band around the control points of a curve).

Affine invariance: The Bézier curve is affinely invariant with respect to its control points. This means that any linear transformation or translation of the control points defines a new curve, which is just the transformation or translation of the original curve.

Variation diminishing: The Bézier curve does not undulate any more than its control polygon; it may undulate less.

Linear precision: The Bézier curve has linear precision: If all the control points form a straight line, the curve also forms a line.

32.3.3.1.2 B-Spline Curves

A single *B-spline* curve segment is defined much like a Bézier curve. It is defined as

$$\mathrm{d}(t) = \sum_{i=0}^{n} d_i N_i(t)$$

where d_i are control points, called *de Boor points*. The $N_i(t)$ are the basis functions of the B-spline curve. The degree of the curve is n. Note that the basis functions used here are different from the Bernstein basis polynomials. Schoenberg first introduced the B-spline in 1949. He defined the basis functions using integral convolution (the "B" in B-spline stands for "basis"). Higher-degree basis functions are given by convolving multiple basis functions of one degree lower. Linear basis functions are just "tents"; when convolved together they make piecewise parabolic "bell" curves.

The tent basis function (which has a degree of one) is nonzero over two intervals, the parabola is nonzero over three intervals, and so forth. This gives the region of influence for different degree B-spline control points. Note also that each convolution results in higher-order continuity between segments of the basis function. When the control points (*de Boor* points) are weighted by these basis functions, a B-spline curve results.

32.3.3.1.3 NURBS

NURBS curves and surfaces are generalizations of both B-splines and Bézier curves, the primary difference being the weighting of the control points, which makes NURBS curves rational and admits circles, spheres, cylinders, etc., which are rational. NURBS have become industry standard tools for the representation and design of geometry. Some reasons for the extensive use of NURBS are that they offer one common mathematical form for both standard analytical shapes, for example, conics, and free-form shapes; provide the flexibility to design a large variety of shapes; can be evaluated reasonably fast by numerically stable and accurate algorithms; are invariant under affine as well as perspective transformations such as rotations and scaling; and are generalizations of nonrational B-splines and nonrational and rational Bézier curves and surfaces (Piegl 1991; Rogers and Earnshaw 1991; Farin 2002). One drawback for NURBS is the extra storage required to define traditional shapes, for example, circles. This results from parameters in addition to the control points but finally allows the desired flexibility for defining parametric shapes.

32.3.3.2 Parametric Surfaces

32.3.3.2.1 Bézier and B-Spline Surfaces

Imagine moving the control points of the Bézier curve in three dimensions. As they move in space, new curves are generated. If they are moved smoothly, then the curves formed create a surface, which may be thought of as a bundle of curves (see Figure 32.13). If each of the control points is moved along a Bézier curve of its own, then a Bézier surface patch is created; the result of extruding a B-spline curve is a Bézier surface.

All operations used for B-spline curves carry over to the surface via the nesting scheme implied by its definition. We recall that B-spline curves are especially convenient for obtaining continuity between polynomial segments. This convenience is even stronger in the case of B-spline surfaces; the patches meet with higher-order continuity depending on the degree of the respective basis functions. B-splines define quilts of patches with "local support" design flexibility. Finally, since B-spline curves are more compact to store than Bézier curves, the advantage is "squared" in the case of surfaces. These advantages are tempered by the fact that operations are typically more efficient on Bézier surfaces. Conventional wisdom says that it is best to design and represent surfaces as B-splines and then convert to Bézier form for operations.

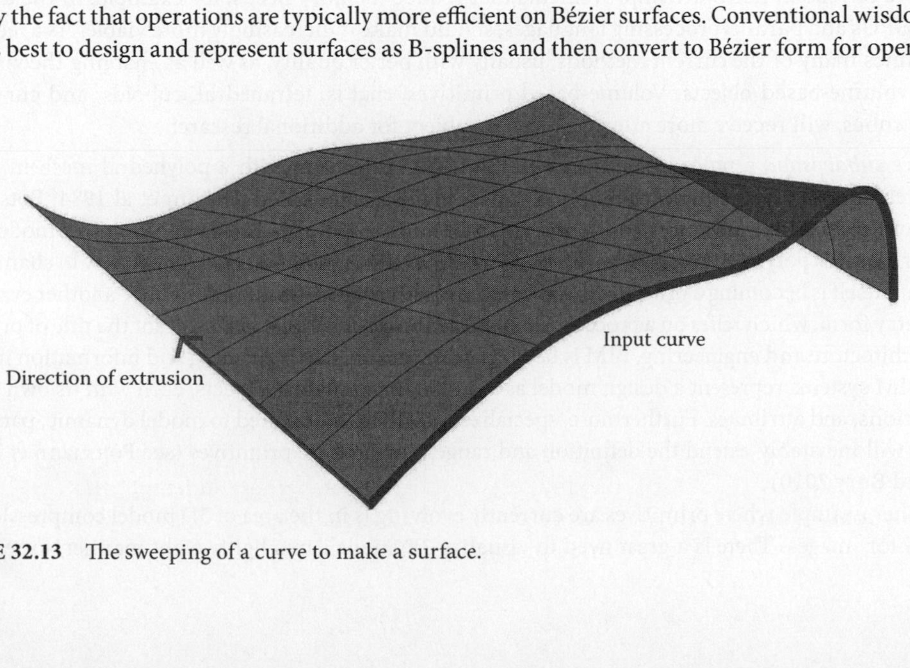

FIGURE 32.13 The sweeping of a curve to make a surface.

32.3.3.2.2 NURBS Surfaces

NURBS surfaces are functions of two parameters mapping to a surface in 3D space. The shape of the surface is determined by control points. A surface under construction, for example, the head of a camel, is usually composed of several NURBS surfaces known as *patches*. Continuity at the patch boundaries will determine the smoothness or otherwise of surface junctions (refer Wiki NURBS; Piegl 1991).

32.4 Standards

Several movements have occurred to standardize primitives. In academia, an early attempt at standardization resulted in Programmer's Hierarchical Interactive Graphics System (PHIGS) + (ISO/IEC 9592–2:1997). Although it has never become the principle standard, many of the ideas were absorbed into later standards. Many standards grew out of the mechanical CAD/CAM industry, because of the early connection with graphics. One such prominent standard is the Initial Graphics Exchange Specification (IGES) (US PRO 1996). Because of its portability, the 3D ACIS format has become a standard in solid modeling (see ACIS Library Reference Documentation, R22). In industry each company has developed and lobbied for its particular set of graphics standards. In the computer graphics development area, the most widely used libraries are OpenGL developed by Silicon Graphics, Inc., and managed by the nonprofit technology consortium Khronos Group, and DirectX, a set of libraries owned and developed by Microsoft. For specific commands used by OpenGL and DirectX, refer to Appendices 32.A and 32.B, respectively.

In this chapter we have concentrated mostly on 3D primitives because they tend to be more complex and demanding. However, there has been great progress standardizing 2D graphics such as that epitomized by Flash, a format developed by Macromedia, now part of the Adobe suite (www.adobe.com/flashplatform/). Indeed, Flash has been so successful that as a standard it dominates 2D graphics.

32.5 Future Directions

There have been efforts to cast higher-order primitives like parametric surfaces into graphics hardware, but the best approach seems to be to convert these into polygons and then render. There may yet be hardware support for higher-order primitives, but the polygon processing remains at the heart of graphics primitives. This trend is likely to continue into the future if for no other reason than its own inertia. Special needs will continue to drive the development of specialized primitives.

One new trend that may affect development is that of volume rendering. Although it is currently quite expensive to render, hardware improvements that reduce memory needs, for example, in the architecture of GPUs and parallel processing languages, should make it increasingly more viable. As a technique it subsumes many of the current methods, usually with better quality, as well as enabling the visualization of volume-based objects. Volume-based primitives, that is, tetrahedral, cuboids, and curvilinear volume cubes, will receive more attention and be subject for additional research.

Recursive subdivision, a procedural modeling technique that begins with a polyhedral mesh in 3D and recursively chamfers the edges until a nicely smoothed object results (see Böhm et al 1984; Botsch et al 2010), continues to be a focus area. This approach is widely used in animation and conceptual modeling. Its input structure is polyhedral mesh and a set of parameters that inform the algorithm of how to chamfer. The algorithm itself is becoming a primitive residing on the graphics card. Fractals would be another example of a geometry form, which relies on a procedure and may be common enough to warrant the title of primitive.

In architecture and engineering, BIM is becoming the standard for modeling and information management. BIM systems represent a design model as a group of interrelated objects, each with its own geometry, relations, and attributes. Furthermore, specialized applications tailored to model dynamic, parametric designs will inevitably extend the definition and range of geometric primitives (see Pottmann et al 2012; Bury and Bury 2010).

Another example where primitives are currently evolving is in the area of 3D model compression akin to JPEG for images. There is a great need to visualize 3D models in a lightweight manner, for example, through increased standardization. Unified Model Language (UML) has been developed jointly by industry and academia to develop 3D standards for model compression and communication of those models.

This cycle of innovative user application, hardware development, and software standardization is never more apparent than now. One of the strongest growing trends that will affect primitives is the further development of digital handheld devices such as smart phones and tablets. A big effort in this direction is to develop OpenGL for their use. This means optimizing OpenGL for the frenetic development in the hardware abilities and imaginative applications. In fact, this trend will only intensify as intelligent computing continues to miniaturize and finds its way into all aspects of our life. We see examples in controlling human implants, for example, digital cochlea, optical implants, and prosthetics. Also it will appear in everything from smart elevators to pilotless flight. Shape detection will play an increasing role, for example, in medicine such as early tumor and aneurism detection, MRI data, and echocardiograms. As in all previous developments, good standards will improve collaboration, efficiency, and further innovation.

Virtual Reality is perhaps the greatest known unknown at present. This is the area of research and development that will push our definition of reality, challenge our physical being, and blur the boundaries between the tangible and the perceptual.

32.6 Summary

Geometric primitives are the most fundamental components of computer visualization. They have been identified here as points, lines, polylines, polygons, triangular facets, elliptical arcs, and curves. Solid geometry primitives include the cube, pyramid, cone, sphere, ellipsoid, torus, and polygonal extrusions. A geometric primitive can be distinguished as either a design primitive or a database primitive, where a design primitive forms part of the final visualized model, for example, the curve that defines the edge of a sphere, and a database primitive is part of the working process and is not visible in the final model, for example, bounding box and CV. Although simple in themselves, geometric primitives can be combined and manipulated, for example, with Boolean operations and recursive subdivision, to form complex and sophisticated objects.

Finding usable implicit forms for quadratic parametric surfaces, for example, sphere, plane, or parabola, is easy and resolved in most software used today. However, the recurring challenge is to find intuitive and mathematically stable solutions for creating cubic patches and more complex shapes.

Key Terms

ACIS: The 3D ACIS Modeler is a geometric modeling kernel developed by Spatial Corporation
BIM: Building information modeling
Boolean: Booleans as used in CAD/CAM (distinguish from Boole's algebra)
CAD: Computer-aided design
CAE: Computer-aided engineering
CAM: Computer-aided manufacturing
CPU: Central processing unit
CSG: Constructive solid geometry
Implicit objects: Defined by implicit functions, they define solid objects of which outside and inside can be distinguished. Common engineering forms such as the plane, cylinder, sphere, and torus are defined simply by implicit functions.
Parametrically defined curves and surfaces: Higher-order surface primitives used widely in industrial design and graphics. Parametric surfaces such as B-spline and Bézier surfaces have flexible shape attributes and convenient mathematical representations.
Platonic solids: A convex solid composed of identical regular polygons. There are five in total.
Polygon: A closed object consisting of vertices, lines, and usually an interior. When pieced together it gives a piecewise (planar) approximation of objects with a surface. Triangular facets are the most common form and form the basis of most graphics primitives.

Polytope: A polytope is a geometric object with flat sides, which exists in any general number of dimensions.
Wireframe: Simplest and earliest form of graphics model, consisting of line segments and possibly elliptical arcs that suggest the shape of an object. It is fast to display and has advantages in precision and "see-through" features.
Zero-dimensional: An object with no volume, area, length, nor any other higher-dimensional analogue.

32.A Appendix A

The table that follows lists the geometric primitive specifications used in OpenGL. "GL" refers to *graphics library* and "v" represents a *vertex*.

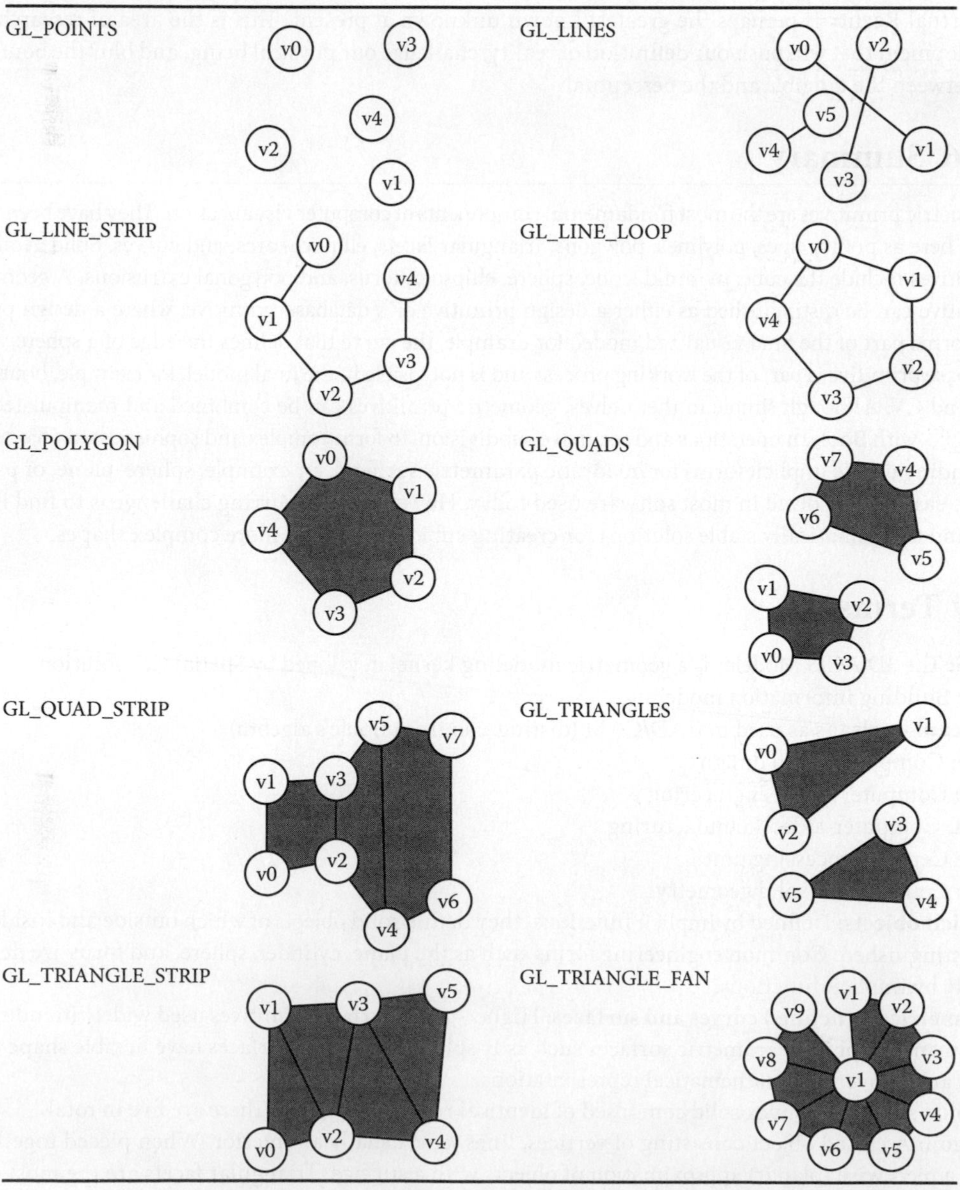

Source: Adapted from http://www.opentk.com/doc/chapter/2/opengl/geometry/primitives.

32.B Appendix B

The table that follows lists the geometric primitive specifications used in DirectX. A *vertex* is represented by either "v" or x, y, z coordinates.

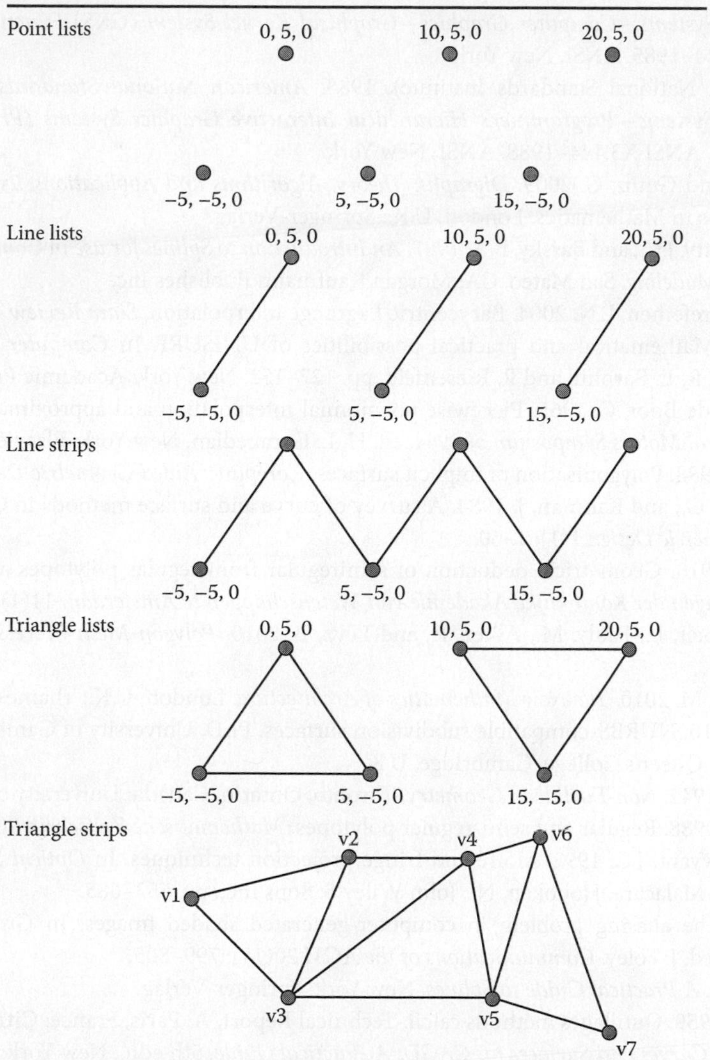

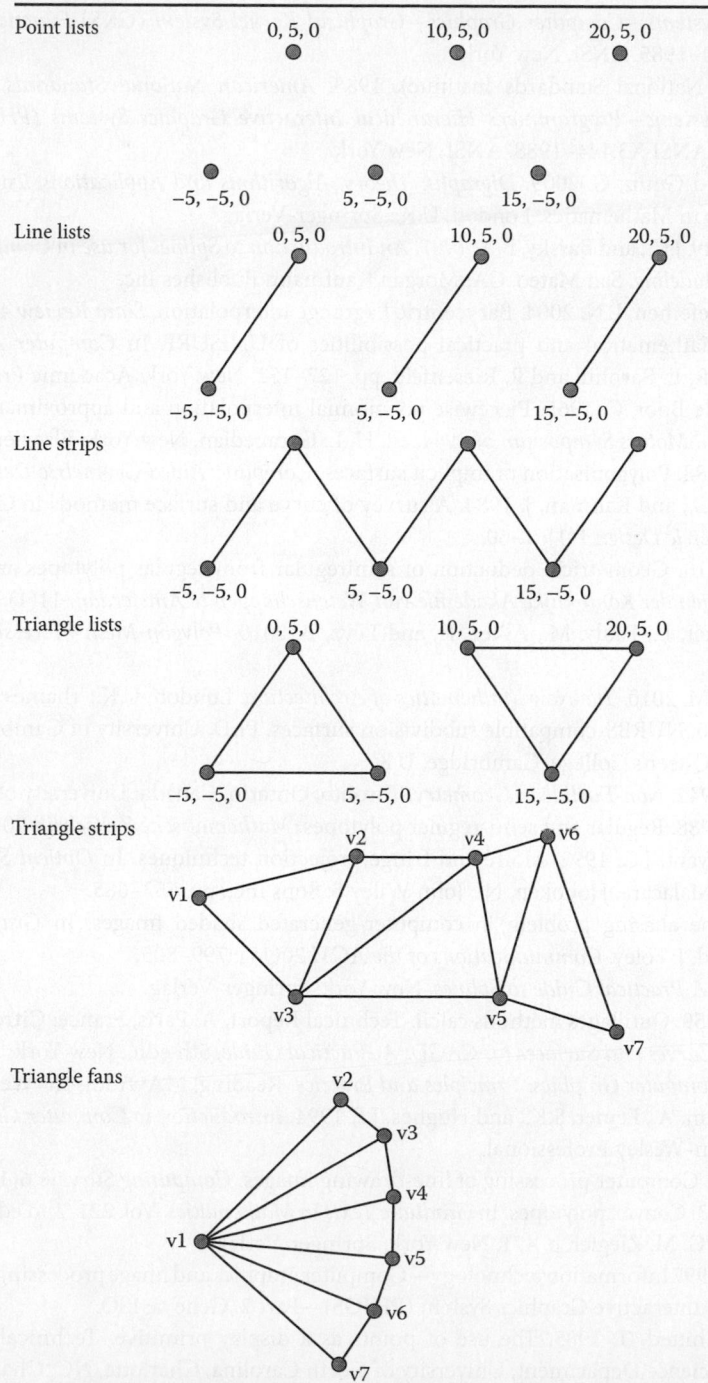

Source: Adapted from http://msdn.microsoft.com/en-us/library/windows/desktop/bb147280(v=VS.85).aspx.

References

Adobe Systems Incorporated. 1999. *PostScript Language Reference*, 3rd edn., Reading, MA: Addison-Wesley Publishing Company.

ANSI (American National Standards Institute). 1985. *American National Standards for Information Processing Systems—Computer Graphics—Graphical Kernel System (GKS) Functional Description*. ANSI X3.124–1985. ANSI, New York.

ANSI (American National Standards Institute) 1988. *American National Standards for Information Processing Systems—Programmer's Hierarchical Interactive Graphics Systems (PHIGS) Functional Description*. ANSI X3.144–1988. ANSI, New York.

Bang-Jensen, J. and Gutin, G. 2009. *Digraphs: Theory, Algorithms and Applications*. 2nd edn., Springer Monographs in Mathematics. London, U.K.: Springer-Verlag.

Bartels, R.H., Beatty, J.C., and Barsky, B.A. 1987. *An Introduction to Splines for use in Computer Graphics & Geometric Modeling*. San Mateo, CA: Morgan Kaufmann Publishes Inc.

Berrut, J-P. and Trefethen, L.N. 2004. Barycentric Lagrange interpolation. *Siam Review* 46(3): 501–517.

Bézier, P. 1974. Mathematical and practical possibilities of UNISURF. In *Computer Aided Geometric Design*, eds. R. E. Barnhill and R. Riesenfeld, pp. 127–152. New York: Academic Press.

Birkhoff, G. and de Boor, C. 1965. Piecewise polynomial interpolation and approximation, *Proceedings of the General Motors Symposium of 1964*, ed. H. L. Garabedian. New York: Elsevier, pp. 164–190.

Bloomenthal, J. 1988. Polygonisation of implicit surfaces. *Computer Aided Geometric Design* 5: 341–345.

Böhm, W., Farin, G., and Kahman, J. 1984. A survey of curve and surface methods in CAGD. *Computer Aided Geometric Design* 1(1): 1–60.

Boole Stott, A. 1910. Geometrical deduction of semiregular from regular polytopes and space fillings. *Verhandelingen der Koninklijke Akademie van Wetenschappen te Amsterdam* 11(1): 3–24.

Botsch, M., Kobbelt, L., Pauly, M., Alliez, P., and Lévz, B. 2010. *Polygon Mesh Processing*. Natick, MA: A.K.Peters.

Bury, J. and Bury, M. 2010. *The New Mathematics of Architecture*. London, U.K.: Thames & Hudson.

Cashman, T.J. 2010. NURBS-compatible subdivision surfaces. PhD, University of Cambridge, Computer Laboratory, Queen's College, Cambridge, U.K.

Coxeter, H.S.M. 1942. *Non-Euclidean Geometry*. Toronto, Ontario, Canada: University of Toronto Press.

Coxeter, H.S.M. 1988. Regular and semi-regular polytopes. *Mathematische Zeitschrift* 200(1): 3–45.

Creath, K. and Wyant, J.C. 1992. Moiré and fringe projection techniques. In *Optical Shop Testing*, 2nd edn., ed. D. Malacara. Hoboken, NJ: John Wiley & Sons Inc., pp. 653–685.

Crow, F. 1977. The aliasing problem in computer-generated shaded images. In *Graphics and Image Processing*, ed. J. Foley. *Communications of the ACM* 20(11): 799–805.

De Boor, C. 1978. *A Practical Guide to Splines*. New York: Springer-Verlag.

De Casteljau, P. 1959. Outillages méthods calcil. Technical Report, A. Paris, France: Citroen.

Farin, G.E. 2002. *Curves and Surfaces for CAGD: A Practical Guide*, 5th edn., New York: Academic Press.

Foley, J.D. 1990. *Computer Graphics: Principles and Practice*. Reading, MA: Addison–Wesley.

Foley, J.D., van Dam, A., Feiner, S.K., and Hughes, J.F. 1994. *Introduction to Computer Graphics*. Reading, MA: Addison-Wesley Professional.

Freeman, H. 1974. Computer processing of line-drawing images. *Computing Surveys* 6(1): 57–97.

Grünbaum, B. 2003. Convex polytopes. In *Graduate Texts in Mathematics*, Vol. 221, 2nd edn., eds. V. Kaibel, V. Klee, and G. M. Ziegler, p. 471. New York: Springer-Verlag.

ISO/IEC 9592–2:1997 Information technology—Computer graphics and image processing—Programmer's Hierarchical Interactive Graphics System (PHIGS)—Part 2. Geneva: ISO.

Levoy, M. and Whitted, T. 1985. The use of points as a display primitive. Technical Report 85–022. Computer Science Department, University of North Carolina, Charlotte, NC: Chapel Hill.

Pauly, M., Keiser, R., Kobbelt, L.P., and Gross, M. 2003. Shape modeling with point-sampled geometry. *ACM Transactions on Graphics (TOG)* 22(3): 641–632; *Proceedings of ACM SIGGRAPH 2003*.

Piegl, L. 1991. On NURBS: A survey. *IEEE Computer Graphics and Applications* 11(1): 55–71.

Piegl, L. and Tiller, W. 1987. Curve and surface constructions using rational B-splines. *Computer Aided Design* 19(9): 485–498.

Pottmann, H., Asperl, A., Hofer, M., and Kilian, A. 2010. *Architectural Geometry*. New York: Bentley Institute Press.

Richeson, D.S. 2008. *Euler's Gem: The Polyhedron Formula and the Birth of Topology*. Princeton, NJ: Princeton University Press.

Rockwood, A. 1986, November 25. US Patent No.4625289: Computer graphics system of general surface rendering by exhaustive sampling. Salt Lake City, Utah: Evans & Sutherland Computer Corporations.

Rockwood, A. 1987. A generalized scanning technique for display of parametrically defined surfaces. *IEEE Computer Graphics and Applications* 7(8): 15–26.

Rockwood, A. and Chambers, P. 1996. *Interactive Curves and Surfaces: A Multimedia Tutorial on CAGD*. San Francisco, CA: Morgan Kauffman Publishers Inc.

Rockwood, A., Davis, T., and Heaton, K. 1989. Real-time rendering of trimmed surfaces. *Computer Graphics* 23(3): 107–116.

Rogers, D.F. and Earnshaw, R.A. (Eds). 1991. *State of the Art in Computer Graphics—Visualization and Modeling*. New York: Springer-Verlag, pp. 225–269.

Spivak, M. 1965. *Calculus on Manifolds*. Reading, MA: Addison-Wesley.

US Product Data Association. 1996. Initial graphics exchange specification IGES 5.3. N. Charleston, SC: IGES/PDES Organization.

Further Resources

http://en.wikipedia.org/wiki/Geometric_primitive (accessed October 9, 2013).

http://en.wikipedia.org/wiki/Non-uniform_rational_B-spline (accessed October 9, 2013).

Wolfram Three Dimensional Graphics Primitives (accessed October 9, 2013).

http://reference.wolfram.com/mathematica/tutorial/ThreeDimensionalGraphicsPrimitives.html (accessed October 9, 2013).

http://www.algorithmist.net/subdivision.html (accessed October 9, 2013).

http://en.wikipedia.org/wiki/Implicit_function_theorem (accessed October 9, 2013).

http://en.wikipedia.org/wiki/Inverse_function_theorem (accessed October 9, 2013).

http://www.cs.clemson.edu/~dhouse/courses/405/notes/implicit-parametric.pdf (accessed October 9, 2013).

http://www.brnt.eu/phd/node11.html (accessed October 9, 2013).

33

Computer Animation

Nadia Magnenat Thalmann
Nanyang Technological University
University of Geneva

Daniel Thalmann
École polytechnique fédérale de Lausanne
Nanyang Technological University

33.1 Introduction

Most virtual worlds are dynamic; they change over time: objects move, rotate, and transform themselves; in inhabited virtual worlds, virtual characters come to life. Object and character motion is an essential part of these virtual worlds. The main goal of computer animation is to synthesize the desired motion effects, which are a mixing of natural phenomena, perception, and imagination. The animator designs the object's dynamic behavior with his or her mental representation of causality. Animators imagine how each object moves, gets out of shape, or reacts when it is pushed, pressed, pulled, or twisted. So, the animation system has to provide the user with motion control tools that enable animators translate their wishes using their own language. Computer animation methods may also aid in the application of physical laws by adding motion control to data in order to show their evolution over time. Formally, any computer animation sequence may be defined as a set of objects characterized by state variables evolving over time. For example, a human character is normally characterized using its joint angles as state variables. To produce a computer animation sequence, the animator has several techniques available depending on the type of object to be animated. Our main focus will be character animation, which corresponds to the animation of articulated bodies (Magnenat-Thalmann 1987, 1991).

We will first explain a few basic animation methods, and then we discuss how to capture motion, which is the most popular technique today used in the production of movies and games. However, its main disadvantage resides in the lack of flexibility of recorded motions, especially when one wants to adapt the raw motion data to another human and/or environment. For this reason, motion editing methods have been developed and a key topic in this chapter. Then we will discuss two fundamental problems for character animation, the deformation of the body and facial animation. Behavioral animation and autonomous characters (ACs) are very important aspects and will also be discussed. Finally, we will address two important topics that have attracted a lot of interest: crowd simulation and cloth simulation.

33.2 Basic Animation Methods

33.2.1 Key Frame

This is an old technique consisting of the automatic generation of intermediate frames, called *in-betweens*, based on a set of key frames supplied by the animator. Originally, the in-betweens were obtained by interpolating the key frame images themselves. As linear interpolation produces undesirable effects such as lack of smoothness in motion, discontinuities in the speed of motion, and distortions in rotations, spline interpolation methods are used. Splines can be described mathematically as piecewise approximations of cubic polynomial functions. Two kinds of splines are very popular: interpolating splines with C1 continuity at knots and approximating splines with C2 continuity at knots. For animation, the most used splines are the interpolating splines: cardinal splines, Catmull–Rom splines, and Kochanek–Bartels (1984) splines.

A way of producing better images is to interpolate parameters of the model of the object itself. This technique is called *parametric key frame animation*, and it is commonly found in most commercial animation systems. In a parametric model, the animator creates key frames by specifying an appropriate set of parameter values. Parameters are then interpolated, and images are finally individually constructed from the interpolated parameters. Spline interpolation is generally used for the interpolation.

33.2.2 Inverse Kinematics

This direct kinematics problem consists in finding the position of end points (e.g., hand, foot) with respect to a fixed-reference coordinate system as a function of time without regard to the forces or the moments that cause the motion. Efficient and numerically well-behaved methods exist for the transformation of position and velocity from joint-space (joint angles) to Cartesian coordinates (end of the limb). Parametric key frame animation is a primitive application of direct kinematics.

Basically, the problem is to determine a joint configuration for which a desired task, usually expressed in Cartesian space, is achieved. For example, the shoulder, elbow, and wrist configurations must be determined so that the hand precisely reaches a position in space. The equations that arise from this problem are generally nonlinear and are difficult to solve in general. In addition, a resolution technique must also deal with the following difficulties. For the positioning and animation of articulated figures, the weighting strategy is the most frequent difficulty: some typical examples are given by Zhao and Badler (1994) for posture manipulation and by Phillips et al. (1990, 1991) to achieve smooth solution blending.

33.2.3 Dynamic Simulation

A great deal of work exists on the dynamics of articulated bodies (Huston 1990), and efficient direct dynamic algorithms have been developed in robotics for structures with many degrees of freedom (Featherstone 1986). In computer animation, these algorithms have been applied to the dynamic simulation of the human body (MacKenna and Zeltzer 1996). Given a set of external forces (like gravity or wind) and internal forces (due to muscles) or joint torques, these algorithms compute the motion of the articulated body according to the laws of rigid body dynamics. Impressive animations of passive structures like falling bodies on stairs can be generated by this technique with little input from the animator. On the other hand, making adjustments is a tedious task as the animator has only indirect control over the animation. Hodgins et al. (1995) achieved simulation of dynamic human activities such as running and jumping. Some controller architectures have also been proposed by Lazlo et al. (1996) to generate a wide range of gaits and to combine elementary behaviors (Faloutsos et al. 2001).

Dynamic approaches aim to describe motion by applying physics laws. For example, it is possible to use control algorithms based on finite state machine to describe a particular motion and proportional derivative servos to compute the forces (Wooten and Hodgins 2000). However, even if these methods produce physically correct animations, configuring their algorithms remains difficult. It is not easy to determine the influence of each parameter on the resulting motions. Many methods based on empirical data and biomechanical observations are able to generate walking (Boulic et al. 2004) or running patterns (Bruderlin and Calvert 1996), reactive to given user parameters. Other similar approaches take into account the environment, to walk on uneven or sloped terrains (Sun and Metaxas 2001) or to climb stairs (Chung and Hahn 1999). Despite their real-time capability, all these methods lack realism, as the legs' motion is considered symmetrical, for example.

33.2.4 Collision Detection and Response

In computer animation, collision detection and response are obviously more important. Hahn (1988) presents bodies in resting contact as a series of frequently occurring collisions. Baraff (1989) presents an analytical method for finding forces between contacting polyhedral bodies, based on linear programming techniques. He also proposed a formulation of the contact forces between curved surfaces that are completely unconstrained in their tangential movement. Bandi et al. (1995) introduced an adaptive spatial subdivision of the object space based on octree structure and presents a technique for efficiently updating this structure periodically during the simulation. Volino and Magnenat Thalmann (1994) described a new algorithm for detecting self-collisions on highly discretized moving polygonal surfaces. It is based on geometrical shape regularity properties that permit avoiding many useless collision tests. Vassilev and Spanlang (2001) propose to use the z-buffer for collision detection to generate depth and normal maps.

33.3 Motion Capture

Motion capture consists of measurement and recording of direct actions of a real person or animal for immediate or delayed analysis and playback. This technique is especially used today in production environments for 3D character animation, like in movies and games. It involves mapping of measurements onto the motion of the digital character.

We may distinguish between the main two kinds of systems: optical and magnetic.

Optical motion capture systems are often based on small reflective sensors called markers attached to an actor's body and on several infrared cameras focused on the performance space. By tracking positions of the markers, one can capture the locations for the key points in an animated model, for example, we attach the markers at joints of a person and record the position of the markers from several different directions. The 3D position of each key point is then reconstructed at each time point. The main advantage of this method is freedom of movement; it does not require any cabling. There is however one main problem: occlusion, that is, lack of data resulting from hidden markers, for example, when the performer lies on his or her back. Another problem is the lack of an automatic way of distinguishing reflectors when they get very close to each other during motion. These problems may be minimized by adding more cameras but at a higher cost, of course. Most optical systems operate with four to six cameras, but for high-end applications like movies, 16 cameras are usual. A good example of optical system is the VICON system (Figure 33.1). In recent years, the motion capture technology has matured with the use of active optical markers (e.g., PhaseSpace), and some recent systems are much more affordable.

Researchers in computer vision have tried to develop similar systems based on simple video cameras without markers. In 2010, Microsoft introduces the Kinect sensor, which is a horizontal bar connected

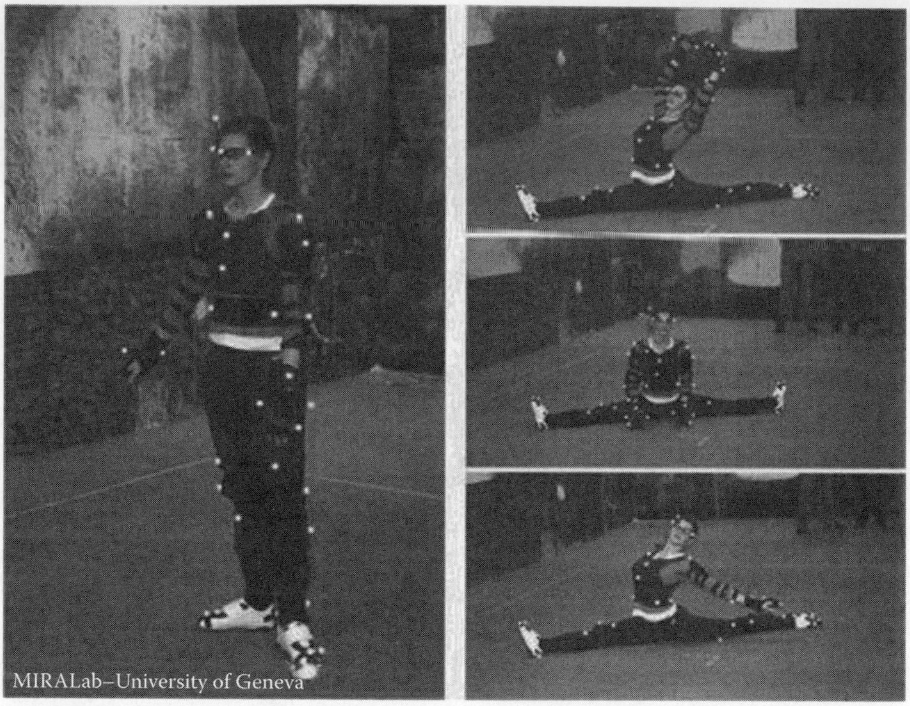

FIGURE 33.1 Optical motion capture based on markers at MIRALab.

to a small base with a motorized pivot, and it is designed to be positioned lengthwise above or below the video display. The device features an RGB camera, depth sensor, and multiarray microphone running proprietary software, which provide full-body 3D motion capture, facial recognition, and voice recognition capabilities. The depth sensor consists of an infrared laser projector combined with a monochrome CMOS sensor, which captures video data in 3D under any ambient light conditions. Although not very accurate, the Kinect (Figure 33.2) provides a cheap way of capturing motion that is becoming very popular.

Magnetic motion capture systems require a real actor to wear a set of sensors, which are capable of measuring their spatial relationship to a centrally located magnetic transmitter. The position and orientation of each sensor is then used to drive an animated character. One problem is the need for synchronizing receivers. The data stream from the receivers to a host computer consists of 3D positions and orientations for each receiver. For human body motion, eleven sensors are generally needed: one on the

FIGURE 33.2 The Kinect sensor.

head, one on each upper arm, one on each hand, one in the center of the chest, one on the lower back, one on each ankle, and one on each foot. To calculate the rest of the necessary information, the most common way is the use of inverse kinematics (IK).

33.4 Motion Editing

Motion retargeting or motion editing is simply editing existing motions to achieve the desired effect. Since motion is difficult to create from scratch using traditional methods, changing existing motions is a quicker way to obtain the desired motion. This means that motion editing is a well-known approach used to produce new motions from existing ones. With space-time constraints (Witkin and Kass 1988), to generate an animation, the user may specify pose constraints that must be satisfied by the resulting motion sequence and also specify an objective function that is a metric of performance such as the total power consumption of all of the character's muscles. Cohen (1992) constructed a more complete system and introduced the concept SpaceTime windows to enable the use of space-time constraints in an interactive framework, providing tools to the animator to examine and thus guide the optimization process. To achieve interactive editing, Gleicher (1997) provided a different formulation to the problem of space-time constraints, by combining displacement maps and space-time constraints. In Popović and Witkin (1999), Newton's laws were applied on a simplified character to minimize computational costs and to preserve the essential physical properties of the motion. Liu and Popović (2002) introduced a method for rapid prototyping of realistic motions. Starting from a simple animation generated using key framing, physical laws are enforced to produce a more realistic one.

Frequency-based methods consider the animation curve as a time-varying signal; by applying signal processing techniques, the input motion is modified. In the field of character animation, Bruderlin and Williams (1995) adapted multiresolution filtering, multitarget motion interpolation with dynamic time warping, wave shaping, and displacement mapping for editing motion data. Witkin and Popović (1995) presented the motion warping method built from the combination of time warping and displacement mapping techniques. Unuma et al. (1995) described a simple technique to represent periodic motions using the so-called rescaled Fourier functional model.

Gleicher (2001) proposed a taxonomy of constraint-based techniques. Constraint-based techniques enable the user to specify constraints over the entire motion or at specific times while editing an animation. An IK solver computes the "best" motion, pose by pose, so that each pose achieves the predefined constraints as much as possible. Boulic and Thalmann (1992) presented the first work in motion editing using IK on a per-frame basis to enforce constraints. They used a numerical IK solver in order to simultaneously handle two tasks: the primary and the secondary. The primary task is responsible for kinematics constraints. The latter is in charge of tracking the original motion so that the corrected motion does not deviate too much from the initial one. Lee and Shin (1999) introduced the first example of per-frame plus filtering class of motion editing techniques. Their interactive motion editing system addresses many common tasks such as retargeting, transitions, and/or editing. Huang et al. (2006) proposed a motion editing technique with collision avoidance to prevent the character's limbs to interpenetrate the body during editing. Callennec and Boulic (2006) proposed an off-line interactive motion editing technique with prioritized constraints. The characteristics of the original motion are preserved by adding the difference between the motion before and after editing as the lowest priority constraint by using an optimization vector.

Data-driven (or model-based) methods focus on the construction of a model from motion capture data and use the model to generate new motions from existing ones. Alexa and Muller (2000) used principal component analysis (PCA) to represent animations as a set of principal animation components, with the aim of decoupling the animation from the underlying geometry. Glardon et al. (2004) generate walking, running, and jumping motions by exploring the low-dimensional space constructed from a hierarchical PCA. This further decomposition of the PCA space was used to extract high-level parameters (e.g., walking speed) to provide user manipulation guiding a character through the virtual environment from a small set of well-understood parameters. Glardon et al. (2006a) also integrated

FIGURE 33.3 PCA-based walking models. (From Glardon, P. et al., *Visual Comput.*, 5(6), 194, 2006a.)

an optimization-based IK solver for preventing foot sliding by exploiting the predictive capability of the PCA motion model (Figure 33.3). Glardon et al. (2006b) developed a motion blending technique in parameter space, such as locomotion speed and jump length, to handle two PCA motion models in order to treat transitions from walking/running to jump motions while guiding a virtual character to avoid obstacles through an environment. Motion graphs and motion interpolation are used to produce new motions from a database (Kovar and Gleicher 2004, Kovar et al. 2002). A number of researches have also developed techniques to synthesize human motion in a low-dimensional space, by using both linear and nonlinear models. Safanova et al. (2004) proposed a motion synthesis framework able to synthesize new motions by optimizing a set of constraints within a low-dimensional space constructed with PCA.

Carvalho et al. (2007) recently proposed a new approach for interactive low-dimensional human motion synthesis, by combining motion models and prioritized IK. They developed a constrained optimization framework to solve the synthesis problem within a low-dimensional latent space. In this space, a movement enforcing a new set of user-specified constraints could be obtained in just one step (Figure 33.4).

FIGURE 33.4 Synthesized motions, from the swing motion shown in the left figure: (a) 7° down slope ground, (b) flat ground, and (c) 7° up slope ground. (From Carvalho, S. et al., *Comput. Anim. Virtual Worlds*, 18, 493, 2007.)

33.5 Character Deformations

Many different approaches have been proposed to connect a deformable skin to its underlying skeleton. They can be roughly subdivided into the following main categories: subspace deformations (SSDs), skin mapping, anatomically based deformations, physics-based approaches, and example-based approaches.

33.5.1 Subspace Deformations

The skeleton-driven deformation, a classical method for basic skin deformation, is probably the most widely used technique in 3D character animation. In the literature, an early version was presented by Magnenat-Thalmann et al. (1988) who introduced the concept of joint-dependent local deformation with operators to deform smoothly the skin surface. This technique has been given various names such as SSD, linear blend skinning, or smooth skinning. This method works by assigning a set of joints with weights to each vertex of the character. The deformation of a vertex is then computed by a weighted combination of the transformations applied to the influencing joints. Mohr et al. (2003) addresses a crucial issue related to linear blending skinning: setting the appropriate vertex weight values. Wang and Phillips (2002) propose an interesting extension of SSD-based approaches: the multiweighting approach assigns each vertex one weight for each coefficient of each influencing joint transformation matrix instead of a single weight per influencing joint. Mohr and Gleicher (2003) have presented an extension to SDD by introducing pseudojoints. The skeleton hierarchy is completed with extra joints inserted between existing ones to reduce the dissimilarity between two consecutive joints. Kavan and Zara (2005) are also aiming at addressing the intrinsic limitations of linear blending such as collapsing joints or twisting problem. To this end, they interpolate the transformations instead of the transformed vertex positions. Merry et al. (2006) proposed a framework called *animation space*, for generalizing SSD by keeping it linear and with the goal to reduce the initial shortcomings of SSD (especially collapses during bending).

33.5.2 Skin Mapping

An example of this approach is proposed by Thalmann et al. (1996) where ellipsoids are attached to the control skeleton to define a simplified musculature. Cani and Desbrun (1997) propose to use an automatic skinning algorithm based on swept surfaces. The skeleton is used as a path for extrusion of interpolating spline surfaces. More complex models include intermediate layers in order to improve and refine the control of the skin. For example, Chadwick et al. (1989) propose an intermediate muscle layer between the flexible skin and the rigid skeleton. This intermediate structure is made of free-form deformation (FFD) control boxes attached to the skeleton. Similar extended and optimized approaches have been proposed by Moccozet and Magnenat-Thalmann (1997) where the deformable skin is embedded inside a control lattice. Singh and Kokkevis (2000) propose an alternate intermediate structure that consists of a low-resolution mesh instead of a control lattice. Kho and Garland (2005) propose a technique for implicitly defining skeleton-based deformation of unstructured polygon meshes. The implicit skeleton is defined by interactively sketching curves onto the 3D shape in the image plane. Zhou et al. (2005) describe a deformation system providing a close interaction scheme: the designer initially specifies a curve on the mesh surface, called the original control curve. The resulting 3D curve is projected onto one or more planes to obtain 2D editable curves. Once edited, the modified 2D curves are projected back to 3D to get the deformed control curve, which is used to control the deformation.

33.5.3 Anatomically Based

Skeleton skinning has also been investigated to simulate anatomy and reproduce as closely as possible the anatomical behavior of characters, such as the work reported by Scheepers et al. (1997) and Wilhems and Vangelder (1997). Aubel and Thalmann (2001) define a two-layered muscle model suitable for computer graphics applications. A muscle is decomposed into two layers: a skeleton and a surface mesh. The skeleton is simply defined by the action lines of the muscle. This idea is based on the work by Nedel and Thalmann (1998). Each vertex of the muscle mesh is parametrized by one or two underlying action lines so that the deformations of the muscle shape are driven by the associated action lines. It therefore reduces the 3D nature of the elasticity problem to 1D for fusiform muscles (one action line) and 2D for flat muscles (several action lines). Pratscher et al. (2005) propose a different approach for anatomical models. Rather than modeling internal tissues and attaching a skin to it, they generate a musculature from a skeleton and a skin mesh at a rest pose, with what they called an outside–in approach.

33.5.4 Physics-Based Approaches

Since Terzopoulos and Fleischer (1988) and Terzopoulos and Witkin (1988), various approaches have been proposed to introduce more realism by using dynamic-based deformations such as those proposed by Faloutsos et al. (1997). Capell et al. (2002) introduce a framework for skeleton-driven animation of elastically deformable characters. They use continuum elasticity and finite element method (FEM) to compute the dynamics of the object being deformed. Huang et al. (2006) presented a method that extends existing gradient domain name mesh deformation approaches. Larboulette et al. (2005) present flesh elements to add secondary dynamic effects to the skinning, while Guo and Wong (2005) propose a pseudoanatomical skinning method. In this chapter, their purpose is to provide a simple intermediate layer between the skeleton and the skin, namely, the chunks, to avoid the tedious design of the anatomical layers.

33.5.5 Example-Based Approaches

The idea of pose space deformation (PSD) (Lewis et al. 2000) approaches is to use artistically sculpted skin surfaces of varying posture that blend them during animation. Each vertex on the skin surface is associated with a linear combination of radial basis functions (RBFs) that compute its position, given the pose of the moving character. Recently, Kry et al. (2002) propose an extension of that technique by using PCA, allowing for optimal reduction of the data and thus faster deformation. These methods have been recently developed because of the full-body 3D scanners. The basic idea of these approaches is to use real data that cover the space of shape variation of the character to train a deformable model. The method proposed by Allen et al. (2002) is based on skin pose examples. Similarly to animation by target morph, they use a set of poses as targets or examples in order to produce the skinning information, whereas traditional approaches rely on much less information, namely, a skeleton and a template skin in a neutral posture. The approach proposed by James and Twigg (2005), named skinning mesh animations (SMAs), presents an original skin example-based scheme. The main difference is that it does not require any predefined skeleton. Sand et al. (2003) propose a method for the acquisition of deformable human geometry from silhouettes. The technique involves a tracking system to capture the motion of the skeleton and estimates skin shape for each bone using constraints provided by the silhouettes from one or more cameras. Anguelov et al. (2005) introduce a data-driven method for building a character shape model. The method is based on a representation that incorporates both skeletal and body deformations. Another recent approach is proposed by Der et al. (2006) extended from Sumner et al. (2005). It transforms a set of unarticulated example shapes into a controllable, articulated model, providing an IK method allowing the interactive animation of detailed shapes using a space learned from a few poses obtained from any source (scanned or hand sculpted).

33.6 Facial Animation

The goal of facial animation systems has always been obtaining a high degree of realism using optimum resolution facial mesh models and effective deformation techniques. Various muscle-based facial models with appropriate parameterized animation systems have been effectively developed for facial animation (Parke 1982, Terzopoulos et al. 1990, Waters 1987). The Facial Action Coding System (Friesen 1978) defines high-level parameters for facial animation, on which several other systems are based. Most facial animation systems typically follow the following steps:

- Define an animation structure on a facial model by parameterization.
- Define "building blocks" or basic units of the animation in terms of these parameters, for example, static expressions and visemes (visual counterparts of phonemes).
- Use these building blocks as key frames and define various interpolation and blending functions on the parameters to generate words and sentences from visemes and emotions from expressions. The interpolation and blending functions contribute to the realism for a desired animation effect.
- Generate the mesh animation from the interpolated or blended key frames. Given the tools of parameterized face modeling and deformation, the most challenging task in facial animation is the design of realistic facial expressions and visemes.

The complexity of the key frame-based facial animation system increases when we incorporate natural effects such as coarticulation for speech animation and blending between a variety of facial expressions during speech. The use of speech synthesis systems and the subsequent application of coarticulation to the available temporized phoneme information are a widely accepted approach (Hill et al. 1988). Coarticulation is a phenomenon observed during fluent speech, in which facial movements corresponding to one phonetic or visemic segment are influenced by those corresponding to the neighboring segments. Two main approaches for coarticulation are the ones by Pelachaud (1991) and Cohen and Massaro (1993). Both these approaches have been based on the classification of phoneme groups and their observed interaction during speech pronunciation. Pelachaud arranged the phoneme groups according to the deformability and context dependence in order to decide the influence of the visemes on each other. Muscle contraction and relaxation times were also considered, and the facial action units were controlled accordingly.

For facial animation, the MPEG-4 standard is particularly important (MPEG-4). The facial definition parameter (FDP) set and the facial animation parameter (FAP) set are designed to encode facial shape, as well as animation of faces, thus reproducing expressions, emotions, and speech pronunciation. The FDPs are defined by the locations of the feature points and are used to customize a given face model to a particular face. They contain 3D feature points such as mouth corners and contours, eye corners, and eyebrow centers. FAPs are based on the study of minimal facial actions and are closely related to muscle actions. Each FAP value is simply the displacement of a particular feature point from its neutral position expressed in terms of the facial animation parameter units (FAPUs). The FAPUs correspond to fractions of distances between key facial features (e.g., the distance between the eyes). For example, the facial animation engine developed at MIRALab uses MPEG-4 facial animation standard (Kshishagar et al. 2000) for animating 3D facial models in real time (see Figure 33.5). This parameterized model is capable of displaying a variety of facial expressions, including speech pronunciation with the help of 66 low-level FAPs.

Recently, the efforts in the field of phoneme extraction have resulted into software systems capable of extracting phonemes from synthetic as well as natural speech and generating lip synchronized speech animation from these phonemes, thus creating a complete talking head system. It is possible to mix emotions with speech in a natural way, thus imparting the virtual character an emotional behavior. The ongoing efforts are concentrated on imparting "emotional" autonomy to the virtual face enabling a dialogue between the real and the virtual humans with natural emotional responses. Kshirsagar and Magnenat-Thalmann (2001) use statistical analysis of the facial feature point movements. As the data are captured for fluent speech, the analysis reflects the dynamics of the facial movements related to speech production. The results of the analysis were successfully applied for a more realistic speech animation. Also, this has enabled us to easily blend between various facial expressions and speech. The use of MPEG-4 feature points for data capture and facial animation enabled us to restrict the quantity

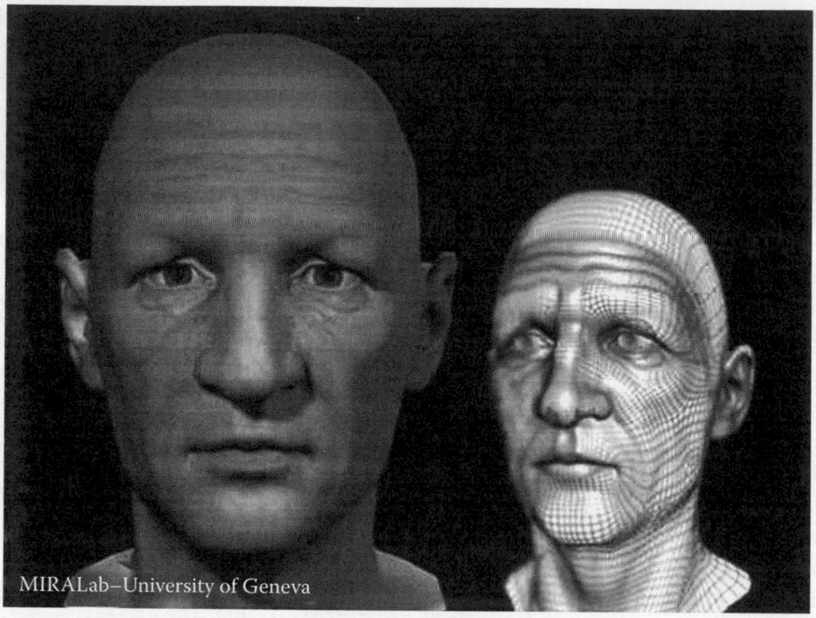

FIGURE 33.5 Facial animation and recreation of expressions of an old man. (Courtesy of MIRALab, University of Geneva, Geneva, Switzerland.)

of data being processed, at the same time offering more flexibility with respect to the facial model. We would like to further improve the effectiveness of the expressive speech using various time envelopes for the expressions that may be linked to the meaning of sentences. Kshirsagar and Magnenat-Thalmann (2002) have also developed a system incorporating a personality model for an emotional autonomous virtual human.

Although MPEG-4 has mainly been used in animating virtual humans, it also proved to be useful for animating humanoid robots. Ben Moussa et al. (2010) and Kasap (2011) present a paradigm for controlling facial expressions using MPEG-4. Through interpolation between the extremes of robot servomotor values and MPEG-4 values, they succeeded in controlling the robot with the same animation mechanisms they use for the animation of virtual humans. This made it possible to perform lip synchronization, emotion expression, and gaze behavior management on the robot in the same way it is done with virtual humans.

In order to achieve realistic facial expressions, one should also take the eye movements into consideration. As has been often expressed, the eyes are a window to the soul. They play an important role in many aspects of human interaction including emotion expression, conversation regulation, attention behavior, and instruction giving. Fukayama et al. (2002) investigated the relationship between gaze parameters (amount and mean duration of gaze at the user and the direction of the gaze when averting the user) and the impression the user gets from the agent's gaze behavior (like/dislike, activity/potency, strong/weak, cold/warm, etc.) and propose a gaze movement model that enables an embodied agent to convey different impressions to users. Another eye movement model is presented by Park Lee et al. (2002), where the eye movement was based on empirical studies of saccades and statistical models of eye-tracking data collected in their study. Based on empirical data and psychological theories, Cig et al. (2010) propose a model where the head and the eyes are moved in accordance with the mood of the embodied agent. Furthermore, another role of gaze is the conversation regulation. Having realistic gaze behavior in a conversation improves the interaction between the user and the embodied agent. Lee et al. (2007) combine different existing psychological theories and present a gaze model guided by turn-taking, discourse, and attention in a conversion. Čereković et al. (2009)

implemented a more advanced multiparty dialogue management system that incorporates gaze tracking and motion detection to recognize the user's conversational cues.

33.7 Behavioral Methods

33.7.1 Autonomous Characters

For games and virtual reality (VR), the animation of characters needs to be dynamically generated in response to the user behavior but not driven by the user. Characters should be able to act on their own; they should be ACs. Autonomy is generally the quality or state of being self-governing. Rather than acting from a script, an AC is aware of its changing virtual environment, making its own decisions in real time in a coherent and effective way. To be autonomous, an AC must be able to perceive its environment and decide what to do to reach an intended goal. The decisions are then transformed into motor control actions, which are animated so that the behavior is believable. Therefore, an AC's behavior often consists of repeating the following sequence: perception of the environment, action selection, and reaction. There are several properties that determine how ACs make their decisions: perception, adaptation and intelligence, memory, emotions, personality, and motivation.

Perception of the elements in the environment is essential for ACs, as it gives them an awareness of what is changing. An AC continuously modifies its environment, which, in turn, influences its perceptions. Therefore, sensorial information drastically influences AC's behavior. This means that one cannot build believable ACs without considering the way they perceive the world and each other. It is tempting to simulate perception by directly retrieving the location of each perceived object straight from the environment. But, to realize believable perception, ACs should have sensors that simulate the functionality of their organic counterparts, mainly for vision, audition, and tactile sensation. These sensors should be used as a basis for implementing everyday behaviors, such as visually directed locomotion, responses to sounds and utterances, and the handling of objects. What is important is the functionality of a sensor and how it filters the information flow coming from the environment. In a typical behavioral animation scene, the actor perceives the objects and the other actors in the environment, which provides information on their nature and position. This information is used by the behavioral model to decide the action to take, which results in a motion procedure. The AC perceives its environment from a small window in which the environment is rendered from its point of view. Synthetic vision was first introduced by Renault et al. (1990) where an actor navigates in a corridor avoiding obstacles by offscreen rendering of the scene from the point of view of the agent. This technique is later applied by Tu and Terzopoulos (1994) in the perception system of artificial fishes. Blumberg et al. (1996) used synthetic vision for his autonomous virtual dog in order to avoid obstacles. Noser et al. (1995) proposed a synthetic vision model based on false coloring and dynamic octrees. Peters and O'Sullivan (2002) extend the synthetic vision model in two different ways. In distinct mode, false coloring is applied to each distant object, and in some cases objects are grouped according to some different criteria that is called *grouped vision mode*. A different approach to synthetic vision is proposed in Bordeux et al. (1999) based on perceptual filters, which receives a perceptible entity from the scene and decides if it should pass or not. A main distinction between virtual human perception models is whether they are based on task perception or biological perception (Kim et al. 2005). Biological perception or bottom-up visual attention is an alerting mechanism to unexpected dangers as explained in Peters and O'Sullivan (2003). Task-based or top-down approaches such as Chopra-Khullar and Badler (1999) and Gillies and Dodgson (2002) consider task-related objects as the focus of attention. This may be extended to gaze attention behaviors for other ACs or even crowds (Grillon and Thalmann 2009). Despite it is rare, aural perception is also modeled and applied by some researchers like Noser and Thalmann (1995) who introduced the concept of synthetic audition. A recent example in Herrero and de Antonio (2003) framework is described to capture the position of the sound source. The concepts are related to human auditory perception such as interaural differences, auditory acuity, and auditory. More recently, an artificial life environment (Conde and Thalmann 2006) framework equips a virtual human with the main virtual sensors, based on an original approach inspired by neuroscience in the form of a small nervous system with a simplified control architecture to optimize the management of its virtual sensors as well as the virtual perception part. The processes of filtering, selection, and simplification are carried out after obtaining the sensorial information.

Adaptation and *intelligence* define how the character is capable of reasoning about what it perceives, especially when unpredictable events happen. An AC should constantly choose the best action so that it can survive in its environment and accomplish its goals. As the environment changes, the AC should be able to react dynamically to new elements, so its beliefs and goals may evolve over time. An AC determines its next action by reasoning about what it knows to be true at specific times. Its knowledge is decomposed into its beliefs and internal states, goals, and plans, which specify a sequence of actions required to achieve a specific goal.

Adding *memory* capabilities to models of human emotions, personality, and behavior traits is a step toward a more natural interaction style. It is necessary for an AC to have a memory so that similar behaviors can be selected when predictable elements reappear. Noser et al. (1995) proposed the use of an octree as the internal representation of the environment seen by an actor because it can represent the visual memory of an actor in a 3D environment with static and dynamic objects. Kasap et al. (2009) developed a memory-based emotion model that uses the memory of past interactions to build long-term relationships between the AC and users.

The believability of an AC is made possible by the emergence of *emotions* clearly expressed at the right moment. An emotion is an emotive reaction to a perception that induces a character to assume a physical response, facial expression, or gesture, or select a specific behavior. The apparent emotions of an AC and the way it reacts are what give it the appearance of a living being with needs and desires. Without them, an actor would just look like an automaton. There is a variety of psychological researches on emotion, and they are classified into four different theoretical perspectives: Darwinian, Jamesian, cognitive, and social constructivist approaches (Cornelius 1996). In order to create computational models of emotions, we need some annotation mechanisms. Ekman (1982) defines six basic labels for emotions: fear, disgust, anger, sadness, surprise, and joy following the Darwinian approach to emotions emphasizing the universality of human emotions. Recently, cognitive appraisal models of emotion are now preferred since they better explain the overall process of how emotions occur and affect our decision-making. Appraisal means a person's assessment of the environment including not only current conditions but past events as well as future prospects (Spindel et al. 1990). The OCC model of Ortony et al. (1988) is widely used for emotion simulation of embodied agents. In the OCC model, the agent's concerns in an environment are divided into goals, standards, and preferences, and twenty-two emotion labels are defined. In addition to the four theoretical approaches mentioned earlier, there are also some studies on emotions based on neurophysiology. The first representative of this approach is Plutchik (1980) who defines eight emotions in an emotion spectrum like the color spectrum, and it is possible to produce new emotions by combining these emotions (e.g., disappointment = sadness + surprise). Other models of emotion such as activation–evaluation (Whissel 1980) define emotions according to abstract and continuous dimensions rather than discrete labels.

Personality influences the way a person perceives their environment, affects their behavior, and distinguishes one from another. Although there is no universally accepted theory of personality, five-factor model or OCEAN model (Mccrae and John 1992) are most widely used ones in the simulation of computer characters. According to this model, the personality of a person can be defined according to five different traits: openness, conscientiousness, extroversion, agreeableness, and neuroticism, and they are explained in Hampson (1999). Openness means being open to experience new things, being imaginative, intelligent, and creative, whereas conscientiousness indicates responsibility, reliability, and tidiness. In other words, conscientious people think about all the outputs of their behaviors before taking action and take the responsibility for their actions. An extravert person is outgoing, sociable, assertive, and energetic to achieve his or her goals. Agreeableness means a person is trustable, kind, and cooperative considering other people's goals and is ready to surrender his or her own goals. Finally, a neurotic person is anxious, nervous, and prone to depression and lacks emotional stability. Usually, a character is represented as a combination of these traits, possibly with emphasis on one of them. Although this static trait-based model of personality does not reflect the complexity of human behavior truly, it has been widely accepted to be used in computational models because of its simplicity.

Motivation is also a key cause of action, and we will consider basic motivations as essential to model life in the virtual world. When looking at the perspective of ACs, an AC can choose whether or not to pay attention and/or react to a given environmental stimulus according to its goals (Caamero 2000). The relation between motivations and emotions of a virtual human is simulated in de Sevin and Thalmann (2005). According to this framework, emotions influence motivations at qualitative levels such as length, perception, activation, and interruption.

33.8 Crowd Simulation

With the advent of crowd simulation (Thalmann and Musse 2007) technology, it becomes possible to create large-scale virtual environments populated by tens of thousands of virtual humans. Crowds have wide applications for both real-time simulation and non-real-time simulation. Virtual crowds are used to simulate epic battles in the movie industry; crowds are exploited to populate virtual worlds in computer games; in the area of safety training, trainees can participate in evacuation scenarios to practice controlling crowds in emergency situations like earthquakes, fires, and floods. Rendering large crowds is usually based on the instantiation of a small set of human templates. McDonnell et al. (2008) showed that introducing color variety was paramount to avoid a clone effect. Maim et al. (2009a) introduced an approach to obtain improved color variety based on segmentation maps using graphics processing unit (GPU) shader programming (deHeras et al. 2005). An accessory is a simple mesh representing any element that can be added to the original mesh of a virtual human (Maim et al. 2009b). Typical accessories include bags, glasses, hats, and mobile phones.

Crowd behavior and motion planning are two topics that have long been studied in fields such as robotics and sociology. More recently, however, and due to the technological improvements, these domains have aroused the interest of the computer graphics community as well.

The first approach studied was agent based and represents a natural way to simulate crowds as independent individuals interacting with each other. Such algorithms usually handle short-distance collision avoidance, in which navigation remains local. Reynolds (1999) proposed to use simple rules to model crowds and groups of interacting agents. Musse and Thalmann (1997, 2001) used sociological concepts in order to simulate relationships among virtual humans. Shao and Terzopoulos (2005) used perceptual, behavioral, and cognitive models to simulate individuals. Kirchner and Shadschneider (2002) used static potential fields to rule a cellular automaton. Metoyer and Hodgins (2003) proposed an avoidance algorithm based on a Bayesian decision process. Nevertheless, the main problem with agent-based algorithms is their low performance. With these methods, simulating thousands of pedestrians in real time requires the use of particular machines supporting parallelizations (Reynolds 2006). Moreover, such approaches forbid the construction of autonomous adaptable behaviors and can only manage crowds of pedestrians with local objectives. To solve the problems inherent in local navigation, some behavioral approaches have been extended with global navigation. Bayazit et al. (2003) stored global information in nodes of a probabilistic road map to handle navigation and introduced simple rules in order to manage groups. Sung et al. (2004) first introduced an architecture for developing situation-based crowd behavior, including groups. They later combined probabilistic road maps with motion graphs to find paths and animations to steer characters to a goal (Sung et al. 2005). Lau and Kuffner (2006) used precomputed search trees of motion clips to accelerate the search for the best paths and motion sequences to reach an objective. Lamarche and Donikian (2004) used automatic topological model extraction of

FIGURE 33.6 Crowd simulation. (From Maim, J. et al., *IEEE Comput. Graphics Appl.*, 29(4), 44, 2009b.)

the environment for individual navigation. Another method, introduced by Kamphuis and Overmars (2004), allowed a group of agents to maintain a given cohesion while trying to reach a goal. Although these approaches offer appealing results, they are not fast enough to simulate thousands of pedestrians in real time. Loscos et al. (2003) presented a behavioral model based on a 2D map of the environment. Their method is suited for simulating wandering crowds but does not provide high-level control on pedestrian goals. Pettré et al. (2006, 2007) presented a novel approach to automatically extract a topology from the scene geometry and handle path planning using a navigation graph (see Figure 33.6). The main advantage of this technique is that it handles uneven and multilayered terrains. Nevertheless, it does not treat interpedestrian collision avoidance. Finally, Helbing et al. (1994, 2000) used agent-based approaches to handle motion planning but mainly focused on emergent crowd behaviors, in particular scenarios.

Another approach to motion planning is inspired by fluid dynamics. Such techniques use a grid to discretize the environment into cells. Hughes (2002, 2003) interpreted crowds as density fields to rule the motion planning of pedestrians. The resulting potential fields are dynamic, guiding pedestrians to their objective while avoiding obstacles. Chenney (2004) developed a model of flow tiles that ensures, under reasonable conditions, that agents do not require any form of collision detection at the expense of precluding any interaction between them. More recently, Treuille et al. (2006) proposed realistic motion planning for crowds. Their method produces a potential field that provides, for each pedestrian, the next suitable position in space (a waypoint) to avoid all obstacles. Compared to agent-based approaches, these techniques allow to simulate thousands of pedestrians in real time and are also able to show emergent behaviors. However, they produce less believable results, because they require assumptions that prevent treating each pedestrian with individual characteristics. For instance, only a limited number of goals can be defined and assigned to sets of pedestrians. The resulting performance depends on the size of the grid cells and the number of sets.

Hybrid approaches have been also proposed. Pelechano et al. (2007) combined psychological, physiological, and geometrical rules with physical forces to simulate dense crowds of autonomous agents. In a recent work, Yersin et al. (2008) introduced a hybrid architecture offering a scalable solution for real-time crowd motion planning. Based on a navigation graph, the environment is divided into regions of varying interest. In regions of high interest, a potential field-based approach is exploited. In other regions, motion planning is ruled by a navigation graph and a short-term collision avoidance algorithm.

Except the topological model of the environment and path planning, collision avoidance is another challenging problem that needs to be addressed. Collision avoidance techniques should be efficient enough to prevent a large number of agents from bumping into each other in real time. Considering the case in which each agent navigates independently without explicit communication with other agents, van den Berg et al. (2008) propose a new concept the "reciprocal velocity obstacle," which takes into account the reactive behavior of the other agents by implicitly assuming that the other agents make a similar collision avoidance reasoning. This concept can be applied to navigation of hundreds of agents in densely populated environments containing both static and moving obstacles for real-time simulation.

Ondřej et al. (2010) explored a novel vision-based approach of collision avoidance between walkers that fits the requirements of interactive crowd simulation. By simulating humans based on cognitive science results, visual stimuli are used to detect future collisions as well as the level of danger.

33.9 Cloth Animation

Weil (1986) pioneered cloth animation using an approximated model based on relaxation of the surface. Kunii and Godota (1990) used a hybrid model incorporating physical and geometrical techniques to model garment wrinkles. Aono (1990) simulated wrinkle propagation on a handkerchief using an elastic model. Terzopoulos and Fleischer (1988) applied their general elastic to a wide range of objects including cloth. Lafleur et al. (1991) and Carignan et al. (1992) have described complex interaction of clothes with synthetic actors in motion, which marked the beginning of a new era in cloth animation (Figure 33.7).

The first particle systems for cloth simulation were grid based (Breen et al. 1994, Eberhardt et al. 1996) and already featured the simulation of nonlinear behavior curves through formulations that made them quite analogous to continuum mechanics models. Their accuracy was however fairly limited for large deformations and required quite long computation times. Faster models, based on mass-spring grids, have become popular since fast implicit numerical integration methods were used (Baraff and Witkin 1998), because they allow a simple expression of the Jacobian of the particle forces while requiring only simple computations (Choi and Ko 2002, Desbrun et al. 1999, Meyer et al. 2001). Combined with advanced implicit integration methods (Eberhardt et al. 2000, Hauth and Etzmuss 2001, Volino and Magnenat-Thalmann 2005b), these simulation schemes have become popular for real-time and interactive applications. Unfortunately, mass-spring systems are unable to model surface elasticity accurately (Wang and Deravajan 2005). Although some techniques have been developed to match their parameters with those of the simulated material, they do not allow full

FIGURE 33.7 3D garment. (Courtesy of MIRALab, University of Geneva, Geneva, Switzerland.)

discrimination among deformation modes (Bianchi et al. 2004), and they remain particularly inaccurate for anisotropic and nonlinear models. Such issues have given particle systems a reputation for inaccuracy.

On the other hand, FEMs have a good maturity for mechanical simulation. Their traditional field of application is elastic solid or shell modeling for mechanical engineering purposes, a context where linear elasticity and small deformations are the underlying assumptions. These formulations have not adapted well to very deformable objects such as cloth, and early attempts to model cloth using high-order elements (Eischen et al. 1996) led to impractically high computation times. Numerous authors have attempted to accelerate the computations required for finite elements. Preinverting the linear system matrix, used by Desbrun et al. (1999) for particle systems, may speed up the computation (Bro-Nielsen and Cotin 1996, Cotin et al. 1999) but is only practical when the mechanical system is small enough. Condensing the dynamics on the boundary of a closed volume can reduce the number of unknowns to solve at each time step (James and Pai 1999). These precomputations are possible when the force displacement relation is linear, which requires both geometrical and material linearity measurements while keeping material linearity. This is called the Saint Venant–Kirchhoff model, and it is used in volume simulation (Barbic and James 2005, Debunne et al. 2001, Hauth et al. 2003, O'Brien and Hodgins 1999, Picinbobo et al. 2003, Zhuang and Canny 2000). Recently, a new approach has been proposed, where the strain tensor is factored as the product of a pure rotation with Cauchy's linear strain tensor aligned along the strain eigendirections (Irving et al. 2004, Muller et al. 2002, Nesme et al. 2005).

Volino et al. (2009) present a model that only addresses tensile viscoelasticity, which deals with in-plane deformations. Meanwhile, bending elasticity deals with out-of-plane deformations (surface curvature), and its main visible effect is to limit fold curvature and wrinkle size. In the context of high-accuracy simulations, this tensile model can easily be complemented by a bending model using the schemes defined by Grinspun et al. (2003) or Volino and Magnenat-Thalmann (2006).

33.10 Research Issues and Summary

Computer animation is no more just a tool to produce movies. It is now the basis for games, simulation, and VR. Moreover, real-time virtual humans are now part of virtual worlds, and people will increasingly communicate with them through broadband multimedia networks. In the future, the application of computer animation to the scientific world will become very common in scientific areas: fluid dynamics, molecular dynamics, thermodynamics, plasma physics, astrophysics, etc. Real-time complex animation systems will be developed taking advantage of VR devices and simulation methods. An integration between simulation methods and VR-based animation will lead to systems allowing the user to interact with complex time-dependent phenomena providing interactive visualization and interactive animation. The future will be intelligent virtual characters with memory and emotion and able to speak, to communicate, and to understand the situation. They will be able to dynamically react to the user and the environment. However, such future systems require real-time spontaneous animation for any situation. This is not possible today, and we are still far from this goal. But the current development is promising, and one will see more and more spectacular dynamic animations.

Further Information

Textbooks on computer

Parent, R. 2008. *Computer Animation: Algorithms and Techniques*, 2nd edn., Morgan-Kaufmann, Burlington, MA.

Kerlow, I. 2004. *The Art of 3-D Computer Animation and Effects*, John Wiley & Sons, Los Angeles, CA.

Magnenat Thalmann, N. and Thalmann, D. (eds.) 2004. *Handbook of Virtual Humans*, John Wiley & Sons, West Sussex, U.K.

Two books could be mentioned for *Crowd Simulation* and Cloth simulation:

Thalmann, D. and Musse, S.R. 2012. *Crowd Simulation*, 2nd edn., Springer, London, U.K.
Volino, P., Magnenat-Thalmann, N. 2000. *Virtual Clothing: Theory and Practice*, Springer, Berlin, Germany.
Journal dedicated to computer animation: *Computer Animation and Virtual Worlds* published by John Wiley & Sons, Chichester, U.K. (since 1990).

Although computer animation is always represented in major computer graphics conferences like SIGGRAPH, Computer Graphics International (CGI), Pacific Graphics, and Eurographics, there are only two annual conferences dedicated to computer animation:

Computer Animation and Social Agents (CASA), organized each year by the Computer Graphics Society. Proceedings are published as a Special Issue of *Computer Animation and Virtual Worlds* (John Wiley & Sons).

Symposium on Computer Animation (SCA) organized by SIGGRAPH and Eurographics.

References

Alexa, M. and Muller, W. 2000. Representing animations by principal components. *Computer Graphics Forum*, 19(3): 411–418.

Allen, B., Curless, B., and Popović, Z. 2002. Articulated body deformation from range scan data. *ACM Transactions on Graphics*, 21(3): 612–619.

Anguelov, D., Srinivasan, P., Koller, D., Thrun, S., Rodgers, J., and Davis, J. 2005. Scape: Shape completion and animation of people. *ACM Transactions on Graphics*, 24(3): 408–416.

Aono, M. 1990. A wrinkle propagation model for cloth. In *Proceedings of the Computer Graphics International '90*, Springer, Tokyo, Japan, pp. 96–115.

Aubel, A. and Thalmann, D. 2001. Interactive modeling of the human musculature. In *Proceedings of the Computer Animation 2001*, pp. 167–174, Seoul, Korea.

Bandi, S. and Thalmann, D. 1995. An adaptive spatial subdivision of the object space for fast collision of animated rigid bodies. In *Proceedings of the Eurographics '95*, Maastricht, the Netherlands, pp. 259–270.

Baraff, D. 1989. Analytical methods for dynamic simulation of non-penetrating rigid bodies. *Computer Graphics*, 23(3): 223–232; *Proceedings of the SIGGRAPH '89*, Boston, MA.

Baraff, D. and Witkin, A. 1998. Large steps in cloth simulation. *Computer Graphics*, (*SIGGRAPH'98*), Orlando, FL, ACM Press, New York, pp. 43–54.

Barbic, J. and James, D.L. 2005. Real-time subspace integration for St.Venant-Kirchhoff deformable models. *ACM Transactions on Graphics*, 24(3): 982–990; *Proceedings of the ACM SIGGRAPH 2005*, Los Angeles, CA.

Bayazit, O.B., Lien, J.M., and Amato, N.M. 2003. Better group behaviors in complex environments using global roadmaps. In *Proceedings of the ICAL 2003*, Cambridge, MA, pp. 362–370.

Ben Moussa, M., Kasap, Z., Magnenat-Thalmann, N., and Hanson, D. 2010. MPEG-4 FAP animation applied to humanoid robot head. In *Proceedings of the Summer School ENGAGE 2010*, Zermatt, Switzerland.

Bianchi, G., Solenthaler, B., Székely, G., and Harders, M. 2004. Simultaneous topology and stiffness identification for mass-spring models based on FEM reference deformations. In *Proceedings of the Medical Image Computing and Computer-Assisted Intervention*, C. Barillot (ed.), Saint Malo, France, Vol. 3, pp. 293–301.

Blumberg, B., Todd, P., and Maes, P. 1996. No BadDogs: Ethological lessons for learning in Hamsterdam. In *Proceedings of the Fourth International Conference on the Simulation of Adaptive Behavior*, pp. 295–304, Cap Cod, MA.

Bordeux, B., Boulic, R., and Thalmann, D. 1999. An efficient and flexible perception pipeline for autonomous agents. *Computer Graphics Forum*, 18(3): 23–30; *Proceedings of the Eurographics '99*, Milan, Italy.

Boulic, R. and Thalmann, D. 1992. Combined direct and inverse kinematic control for articulated figure motion editing. *Computer Graphics Forum*, 11(4): 189–202.

Boulic, R., Ulciny, B., and Thalmann, D. 2004. Versatile walk engine. *Journal of Game Development*, 1: 29–50.

Breen, D.E., House, D.H., and Wozny, M.J. 1994. Predicting the drape of woven cloth using interacting particles. *Computer Graphics*, 4: 365–372; *Proceedings of the SIGGRAPH '94*, Orlando, FL.

Bro-Nielsen, M. and Cotin, S. 1996. Real-time volumetric deformable models for surgery simulation using finite elements and condensation. In *Proceedings of the Eurographics*, Porto, Portugal, ACM Press, New York, pp. 21–30.

Bruderlin, A. and Calvert, T. 1996. Knowledge-driven, interactive animation of human running. In *Proceedings of the Graphics Interface 96*, Toronto, Ontario, Canada, pp. 213–221.

Bruderlin, A. and Williams, L. 1995. Motion signal processing. In *SIGGRAPH '95: Proceedings of the 22nd Annual Conference on Computer Graphics and Interactive Techniques*, Los Angeles, CA, ACM Press, New York, pp. 97–104.

Caamero, L. 2000. Emotions for activity selection, Department of Computer Science Technical Report DAIMI PB 545, University of Aarhus, Denmark.

Callennec, B.L. and Boulic, R. 2006. Interactive motion deformation with prioritized constraints. *Graphical Models*, 68(2): 175–193.

Cani-Gascuel, M.-P. and Desbrun, M. 1997. Animation of deformable models using implicit surfaces. *IEEE Transactions on Visualization and Computer Graphics*, 3(1): 39–50.

Capell, S., Green, S., Curless, B., Duchamp, T., and Popović, Z. 2002. Interactive skeleton-driven dynamic deformations. In *Proceedings of the SIGGRAPH'02*, San Antonio, TX, pp. 41–47.

Carignan, M., Yang, Y., Thalmann, N.M., and Thalmann, D. 1992. Dressing animated synthetic actors with complex deformable clothes. *Computer Graphics*, 26(2): 99–104; *Proceedings of the SIGGRAPH '92*, Chicago, IL.

Carvalho, S., Boulic, R., and Thalmann, D. 2007. Interactive low-dimensional human motion synthesis by combining motion models and PIK. *Computer Animation and Virtual Worlds*, 18: 493–503.

Čereković, A., Huang, H., Furukawa, T., Yamaoka, Y., Pandžić, I.S., Nishida, T.S., and Nakano, Y. 2009. Implementing a multi-user tour guide system with an embodied conversational agent. *Lecture Notes in Computer Science*, 5820: 7–18.

Chadwick, J.E., Haumann, D.R., and Parent, R.E. 1989. Layered construction for deformable animated characters. In *Proceedings of the SIGGRAPH'89*, Boston, MA, pp. 243–252.

Chenney, S. 2004. Flow tiles. In *Proceedings of the Symposium on Computer Animation '04*, New York, pp. 233–242.

Choi, K.J. and Ko, H.S. 2002. Stable but responsive cloth. *Computer Graphics*, 21(3): 604–611; *Proceedings of the SIGGRAPH'02*, San Antonio, TX.

Chopra-Khullar, S. and Badler, N.I. 1999. Where to Look? Automating attending behaviors of virtual human characters. In *AGENTS '99: Proceedings of the Third Annual Conference on Autonomous Agents*, Seattle, WA, pp. 16–23.

Chung, S. and Hahn, J. 1999. Animation of human walking in virtual environments. In *Proceedings of Computer Animation* 1999, Geneva, Switzerland, pp. 4–15.

Cig, C., Kasap, Z., Egges, A., and Magnenat-Thalmann, N. 2010. Realistic emotional gaze and head behavior generation based on arousal and dominance factors. In *Proceedings of the Third International Conference on Motion in Games 2010*, pp. 278–289, Utrecht, the Netherlands, Springer, Berlin, Germany.

Cohen, M.F. 1992. Interactive spacetime control for animation. In *SIGGRAPH '92: Proceedings of the 19th Annual Conference on Computer Graphics and Interactive Techniques*, Chicago, IL, ACM, New York, pp. 293–302.

Cohen, M.M. and Massaro, D.W. 1993. Modelling coarticulation in synthetic visual speech, in N.M. Thalmann and D. Thalmann (eds.), *Models and Techniques in Computer Animation*. Springer-Verlag, Tokyo, Japan, pp. 139–156.

Conde, T. and Thalmann, D. 2006. An integrated perception for autonomous virtual agents: Active and predictive perception. *Computer Animation and Virtual Worlds*, 17(3–4): 457–468.

Cornelius, R. 1996. *The Science of Emotion*. Prentice-Hall, Englewood Cliffs, NJ.

Cotin, S., Delingette, H., and Ayache, N. 1999. Real-time elastic deformations of soft tissues for surgery simulation. *IEEE Transactions on Visualization and Computer Graphics*, 5(1): 62–73.

Debunne, G., Desbrun, M., Cani, M.P., and Barr, A.H. 2001. Dynamic real-time deformations using space and time adaptive sampling. *Computer Graphics*, 31–36; *Proceedings of the SIGGRAPH'01*, Los Angels, CA.

de Heras Ciechomski, P., Schertenleib, S., Maim, J., Maupu, D., and Thalmann, D. 2005. Real-time shader rendering for crowds in virtual heritage. In *Proceedings of the VAST'05*, pp. 1–8, Aire-la-Ville, Switzerland.

de Sevin, E. and Thalmann, D. 2005. An affective model of action selection for virtual humans. In *Proceedings of Agents that Want and Like: Motivational and Emotional Roots of Cognition and Action symposium at the Artificial Intelligence and Social Behaviors Conference*, Hatfield, U.K., pp. 293–297.

Der, K.G., Sumner, R.W., and Popović, J. 2006. Inverse kinematics for reduced deformable models. *ACM Transactions on Graphics*, 25(3): 1174–1179.

Desbrun, M., Schröder, P., and Barr, A.H. 1999. Interactive animation of structured deformable objects. In *Proceedings of Graphics Interface*, Kingston, Ontario, Canada, pp. 1–8.

Eberhardt, B., Etzmuss, O., and Hauth, M. 2000. Implicit-explicit schemes for fast animation with particles systems. In *Proceedings of the Eurographics Workshop on Computer Animation and Simulation*, Springer-Verlag, Interlaken, Switzerland, pp. 137–151.

Eberhardt, B., Weber, A., and Strasser, W. 1996. A fast, flexible, particle-system model for cloth draping. *IEEE Computer Graphics and Application*, 16(5): 52–59, IEEE Press, Washington, DC.

Eischen, J.W., Deng, S., and Clapp, T.G. 1996. Finite element modeling and control of flexible fabric parts. *IEEE Computer Graphics and Application*, 16(5): 71–80, IEEE Press, Washington, DC.

Ekman, P. 1982. *Emotion in the Human Face*. Cambridge University Press, New York.

Faloutsos, P., van de Panne, M., and Terzopoulos, D. 2001. Composable controllers for physics-based character animation. In *Proceedings of ACM SIGGRAPH 2001*, Los Angeles, CA.

Faloutsos, P., Van de panne, M., and Terzopoulos, D. 1997. Dynamic free-form deformations for animation synthesis. *IEEE Transactions on Visualization and Computer Graphics*, 3(3): 201–214.

Featherstone, R. 1986. *Robot Dynamics Algorithms*. Kluwer Academic Publishers, Boston, MA.

Friesen, E.W.V. 1978. *Facial Action Coding System: A Technique for the Measurement of Facial Movement*. Consulting Psychologists Press, Palo Alto, CA.

Fukayama, A., Ohno, T., Mukawa, N., Sawaki, M., and Hagita, N. 2002. Messages embedded in gaze of interface agents—Impression management with agent's gaze. In *Proceedings of the SIGCHI Conference on Human Factors in Computing Systems: Changing Our World, Changing Ourselves*, Minneapolis, MN.

Gillies, M.F.P. and Dodgson, N.A. 2002. Eye movements and attention for behavioural animation. *The Journal of Visualization and Computer Animation*, 13(5): 287–300.

Glardon, P., Boulic, R., and Thalmann, D. 2004. A coherent locomotion engine extrapolating beyond experimental data. In *Proceedings of Computer Animation and Social Agent*, Geneva, Switzerland.

Glardon, P., Boulic, R., and Thalmann, D. 2006a. Robust on-line adaptive footplant detection and enforcement for locomotion. *The Visual Computer*, 5(6): 194–209.

Glardon, P., Boulic, R., and Thalmann, D. 2006b. Dynamic obstacle avoidance for real-time character animation. *The Visual Computer*, 22(6): 399–414.

Gleicher, M. 1997. Motion editing with spacetime constraints. In *Proceedings of SI3D '97*, Providence, RI, ACM Press, New York, pp. 139–ff.

Gleicher, M. 2001. Comparing constraint-based motion editing methods. *Graphical Models*, 63(2): 107–134.

Grinspun, E., Hirani, A.H., Desbrun, M., and Schröder, P. 2003. Discrete shells. In *Proceedings of the Eurographics Symposium on Computer Animation*, San Diego, CA, pp. 62–68.

Grillon, H. and Thalmann, D. 2009. Simulating gaze attention behaviors for crowds. *Computer Animation and Virtual Worlds*, 20: 111–119.

Guo, Z. and Wong, K.C. 2005. Skinning with deformable chunks. *Computer Graphics Forum*, 24(3): 373–382.

Hahn, J.K. 1988. Realistic animation of rigid bodies. *Computer Graphics*, 22(4): 299–308; *Proceedings of the SIGGRAPH '88*, Atlanta, GA.

Hampson, S. 1999. State of the art: Personality. *The Psychologist*, 12(6): 284–290.

Hauth, M. and Etzmuss, O. 2001. A high performance solver for the animation of deformable objects using advanced numerical methods. In *Proceedings of Eurographics*, London, U.K., pp. 137–151.

Hauth, M., Gross, J., and Strasser, W. 2003. Interactive physically based solid dynamics. In *Proceedings of the Eurographics Symposium on Computer Animation*, San Diego, CA, pp. 17–27.

Helbing, D., Farkas, I., and Vicsek, T. 2000. Simulating dynamical features of escape panic. *Nature*, 407(6803): 487–490.

Helbing, D., Molnár, P., and Schweitzer, F. 1994. Computer simulations of pedestrian dynamics and trail formation. In *Evolution of Natural Structures*, pp. 229–234.

Herrero, P. and de Antonio, A. 2003. Introducing human-like hearing perception in intelligent virtual agents. In *AAMAS '03: Proceedings of the Second International Joint Conference on Autonomous Agents and Multiagent Systems*, Melbourne, Victoria, Australia, pp. 733–740.

Hill, D.R., Pearce, A., and Wyvill, B. 1988. Animating speech: An automated approach using speech synthesized by rule. *The Visual Computer*, 3, 277–289.

Hodgins, J., Wooten, W.L., Brogan, D.C., and O'Brien, J.F. 1995. Animating human athletics. In *Proceedings of the Siggraph '95*, Los Angeles, CA, pp. 71–78.

Huang, J., Shi, X., Liu, X., Zhou, K., Wei, L.Y., Teng, S.H., Bao, H., Guo, B., and Shum, H.Y. 2006. Subspace gradient domain mesh deformation. *ACM Transactions on Graphics*, 25(3): 1126–1134.

Hughes, R.L. 2002. A continuum theory for the flow of pedestrians. *Transportation Research Part B Methodological*, 36(29): 507–535.

Hughes, R.L. 2003. The flow of human crowds. *Annual Review of Fluid Mechanics*, 35(1): 169–182.

Huston, R.L. 1990. *Multibody Dynamics*. Butterworth-Heinemann, Stoneham, MA.

Irving, G., Teran, J., and Fedkiw, R. 2004. Invertible finite elements for robust simulation of large deformation. In *Proceedings of the Eurographics Symposium on Computer Animation*, Grenoble, France, pp. 131–140.

James, D. and Pai, D. 1999. ArtDefo–accurate real-time deformable objects. *Computer Graphics*, 65–72; *Proceedings of the SIGGRAPH'99*, Los Angeles, CA, ACM Press, New York.

James, D.L. and Twigg, C.D. 2005. Skinning mesh animations. *ACM Transactions on Graphics*, 24(3): 399–407.

Kamphuis, A. and Overmars, M.H. 2004. Finding paths for coherent groups using clearance. In *Proceedings of the SCA'04*, Grenoble, France, pp. 19–28.

Kasap, Z. 2011. Modeling emotions and memory for virtual characters and social robots, PhD thesis, University of Geneva, Geneva, Switzerland.

Kasap, Z., Ben Moussa, M., Chaudhuri, P., and Magnenat-Thalmann, N. 2009. Making them remember—Emotional virtual characters with memory. *IEEE Computer Graphics and Applications*, 29(2): 20–29.

Kavan, L. and Zara, J. 2005. Spherical blend skinning: A real-time deformation of articulated models. In *Proceedings of the Symposium on Interactive 3D Graphics and Games, SI3D'05*, Washington, DC, pp. 9–16.

Kho, Y. and Garland, M. 2005. Sketching mesh deformations. In *Proceedings of the Symposium on Interactive 3D Graphics and Games, SI3D'05*, Washington, DC, pp. 147–154.

Kim, Y., Hill, R.W., and Traum, D.R. 2005. A computational model of dynamic perceptual attention for virtual humans. In *Proceedings of the 14th Conference on Behavior Representation in Modeling and Simulation*, 2005, Universal City, CA.

Kirchner, A. and Shadschneider, A. 2002. Simulation of evacuation processes using a bionics-inspired cellular automaton model for pedestrian dynamics. *Physica A*, 312(1–2): 260–276.

Kochanek, D.H. and Bartels, R.H. 1984. Interpolating splines with local tension, continuity, and bias control. *Computer Graphics*, 18: 33–41; *Proceedings of the SIGGRAPH '84*, San Francisco, CA.

Kovar, L. and Gleicher, M. 2004. Automated extraction and parameterization of motions in large data sets. *ACM Transactions on Graphics*, 23(3): 559–568.

Kovar, L., Gleicher, M., and Pighin, F. 2002. Motion graphs. *ACM Transactions on Graphics*, 21(3): 473–482.

Kry, P., James, D., and Pai, D. 2002. Eigenskin: Real time large deformation character skinning in hardware. In *Proceedings of the Symposium on Computer Animation, SCA'02*, Osaka, Japan, pp. 153–160.

Kshirsagar, S., Garchery, S., and Thalmann, N. 2000. Feature point based mesh deformation applied to MPEG-4 facial animation. In *Proceedings of the Deform'2000*, Geneva, Switzerland.

Kshirsagar, S. and Magnenat-Thalmann, N. 2001. Principal components of expressive speech animation. In *Proceedings of the Computer Graphics International 2001*, Hong Kong, China, IEEE Computer Society, Washington, DC, pp. 38–44.

Kshirsagar, S. and Magnenat-Thalmann, N. 2002. Virtual humans personified. In *Proceedings of the Autonomous Agents Conference (AAMAS) 2002*, Bologna, Italy, pp. 356–359.

Kunii, T.L. and Gotoda, H. 1990. Modeling and animation of garment wrinkle formation processes. In *Proceedings of the Computer Animation '90*, Tokyo, Japan, Springer, Berlin, Germany, pp. 131–147.

Lafleur, B., Magnenat Thalmann, N., and Thalmann, D. 1991. Cloth animation with self-collision detection. In *Proceedings of the IFIP Conference on Modeling in Computer Graphics*, Tokyo, Japan, Springer, Berlin, Germany, pp. 179–187.

Lamarche, F. and Donikian, S. 2004. Crowd of virtual humans: A new approach for real time navigation in complex and structured environments. *Computer Graphics Forum*, 23(3): 509–518.

Larboulette, C., Cani, M., and Arnaldi, B. 2005. Adding real-time dynamic effects to an existing character animation. In *Proceedings of the Spring Conference on Computer Graphics, SCCG'05*, pp. 87–93, Budmerice, Slovakia.

Lau, M. and Kuffner, J.J. 2006. Precomputed search trees: Planning for interactive goal-driven animation. In *Proceedings of the SCA'06*, pp. 299–308, Vienna, Austria.

Lazlo, J., Van De Panne, M., and Fiume, E. 1996. Limit cycle control and its application to the animation of balancing and walking. In *Proceedings of the SIGGRAPH'96*, New Orleans, LA, pp. 155–162.

Lee, J., Marsella, S., Traum, D., Gratch, J., and Lance, B. 2007b. The rickel gaze model: A window on the mind of a virtual human. *Lecture Notes in Computer Science*, 4722: 296–303; *Proceeding of the Intelligent Virtual Agents*, Paris, France.

Lee, J. and Shin, S.Y. 1999. A hierarchical approach to interactive motion editing for human-like figures. In *Proceedings of the ACM SIGGRAPH*, pp. 39–48, Los Angeles, CA.

Lee, K.H., Choi, M.G., Hong, Q., and Lee, J. 2007a. Group behavior from video: A data-driven approach to crowd simulation. In *Proceedings of the ACMSIGGRAPH/Eurographics Symposium on Computer Animation*, San Diego, CA.

Lewis, J.P., Cordner, M., and Fong, N. 2000. Pose space deformations: A unified approach to shape interpolation and skeleton-driven deformation. In *Proceedings of the SIGGRAPH' 00*, New Orleans, LA, pp. 165–172.

Liu, C.K. and Popović, Z. 2002. Synthesis of complex dynamic character motion from simple animations. In *SIGGRAPH '02: Proceedings of the 29th Annual Conference on Computer Graphics and Interactive Techniques*, San Antonio, TX, ACM, New York, pp. 408–416.

Loscos, C., Marchal, D., and Meyer, A. 2003. Intuitive crowd behaviour in dense urban environments using local laws. In *Proceedings of the TPCG'03*, p. 122, Washington, DC.

Mac Kenna, M. and Zeltzer, D. 1996. Dynamic simulation of a complex human figure model with low level behavior control. *Presence*, 5(4): 431–456.

Magnenat-Thalmann, N., Laperriere, R., and Thalmann, D. 1988. Joint-dependent local deformations for hand animation and object grasping. In *Proceedings of the Graphics Interface GI'88*, Edmonton, Alberta, Canada, pp. 26–33.

Magnenat-Thalmann, N. and Thalmann, D. 1987. The direction of synthetic actors in the film Rendezvous a Montreal. *IEEE CG&A*, 7(12): 9–19.

Magnenat Thalmann, N. and Thalmann, D. 1991. Complex models for animating synthetic actors. *IEEE Computer Graphics and Applications*, 11: 32–44.

Maim, J., Yersin, B., and Thalmann, D. 2009a. Unique character instances for crowds. *IEEE Computer Graphics in Applications*, 29(6): 82–90.

Maim, J., Yersin, B., Thalmann, D., and Pettre, J. 2009b. YaQ: An architecture for real-time navigation and rendering of varied crowds. *IEEE Computer Graphics in Applications*, 29(4): 44–53.

Mccrae, R.R. and John, P.O. 1992. An introduction to the five-factor model and its applications. *Journal of Personality*, 60: 175–215.

McDonnell, R., Larkin, M., Dobbyn, S., Collins, S., and O'Sullivan, C. 2008. Clone attack! Perception of crowd variety. *ACM Transactions on Graphics*, 27(3): 1–8.

Merry, B., Marais, P., and Gain, J. 2006. Animation space: A truly linear framework for character animation. *ACM Transactions on Graphics*, 25(4): 1400–1423.

Metoyer, R.A. and Hodgins, J.K. 2003. Reactive pedestrian path following from examples. In *Proceedings of the CASA'03*, p. 149, Rutgers University, NJ.

Meyer, M., Debunne, G., Desbrun, M., and Barr, A.H. 2001. Interactive animation of cloth-like objects in virtual reality. *The Journal of the Visualization and Computer Animation*, 12(1): 1–12.

Moccozet, L. and Magnenat-Thalmann, N. 1997. Dirichlet free-form deformations and their application to hand simulation. In *Proceedings of the Computer Animation, CA'97*, Washington, DC, pp. 93–102.

Mohr, A. and Gleicher, M. 2003. Building efficient, accurate character skins from examples. *ACM Transactions on Graphics*, 22(3): 562–568.

Mohr, A., Tokheim, L., and Gleicher, M. 2003. Direct manipulation of interactive character skins. In *Proceedings of the Symposium on Interactive 3D Graphics, SI3D'03*, Monterey, CA, pp. 27–30.

Morini, F., Yersin, B., Maïm, J., and Thalmann, D. 2007. Real-time scalable motion planning for crowds. In *Proceedings of the Cyberworlds International Conference*, Hannover, Germany, pp. 144–151.

MPEG-4 standard, Moving Picture Experts Group, http://www.cselt.it/mpeg/.

Muller, M., Dorsey, J., Mcmillan, L., Jagnow, R., and Cutler, B. 2002. Stable real-time deformations. In *Proceedings of the Eurographics Symposium on Computer Animation*, San Antonio, TX, pp. 49–54.

Musse, S.R. and Thalmann, D. 1997. A model of human crowd behavior: Group inter-relationship and collision detection analysis. In *Proceedings of the Eurographics Workshop on Computer Animation and Simulation*, pp. 39–51, Budapest, Hungary.

Musse, S.R. and Thalmann, D. 2001. Hierarchical model for real time simulation of virtual human crowds. *IEEE Transactions on Visualization and Computer Graphics*, 7(2): 152–164.

Nedel, L.P. and Thalmann, D. 1998. Real time muscle deformations using mass-spring systems. In *Proceedings of the Computer Graphics International, CGI'98*, Hannover, Germany, pp. 156–165.

Nesme, M., Payan, Y., and Faure, F. 2005. Efficient, physically plausible finite elements. In *Proceedings of the Eurographics*, pp. 77–80, Dublin, Ireland.

Noser, H., Renault, O., Thalmann, D., and Magnenat Thalmann, N. 1995. Navigation for digital actors based on synthetic vision, memory and learning. *Computers and Graphics*, 19(1): 7–19.

Noser, H. and Thalmann, D. 1995. Synthetic vision and audition for digital actors. In *Proceedings of the Eurographics '95*, Maastricht, the Netherlands, pp. 325–336.

O'Brien, J. and Hodgins, J. 1999. Graphical modeling and animation of brittle fracture. *Computers and Graphics*, 137–146; *Proceedings of the SIGGRAPH'99*, Los Angeles, CA, ACM Press, New York.

Ondrej, J., Pettre, J., Olivier, A-H., and Donikian, S. 2010. A 1 based steering approach for crowd simulation. *ACM Transactions on Graphics*, 29(4): 1–9.

Ortony, A., Collins, A., and Clore, G.L. 1988. *The Cognitive Structure of Emotions*. Cambridge University Press, Cambridge, U.K., 1988.

Paris, S., Pettré, J., and Donikian, S. 2007. Pedestrian steering for crowd simulation: A predictive approach. In *Proceedings of the Eurographics'07*, pp. 665–674, Prague, Czech Republic.

Park, L.S., Badler, J.B., and Badler, N.I. 2002. Eyes alive. *ACM Transactions on Graphics*, 21(3): 637–644.

Parke, F.I. 1982. Parameterized models for facial animation. *IEEE Computer Graphics and Applications*, 2(9): 61–68.

Pelachaud, C. 1991. Communication and coarticulation in facial animation, PhD thesis, University of Pennsylvania, Philadelphia, PA.

Pelechano, N., Allbeck, J.M., and Badler, N.I. 2007. Controlling individual agents in high-density. *Proceedings of the 2007 ACM SIGGRAPH/Eurographics symposium on Computer animation*, pp. 99–108, Eurographics Association, Aire-la-Ville, Switzerland.

Peters, C. and O'Sullivan, C. 2002. Synthetic vision and memory for autonomous virtual humans. *Computer Graphics Forum*, 21(4): 743–752.

Peters, C. and O'Sullivan, C. 2003. Bottom-up visual attention for virtual human animation. In *CASA 03: Proceedings of the 16th International Conference on Computer Animation and Social Agents*, pp. 111–117, Washington, DC.

Pettré, J., de Heras Ciechomski, P., Maïm, J., Yersin, B., Laumond, J.P., and Thalmann, D. 2006. Real-time navigating crowds: Scalable simulation and rendering. *The Journal of Visualization and Computer Animation*, 17(3–4): 445–455.

Pettré, J., Grillon, H., and Thalmann, D. 2007. Crowds of moving objects: Navigation planning and simulation. In *Proceedings of the ICRA'07*, pp. 3062–3067, Rome, Italy.

Philips, C.B., Zhao, J., and Badler, N.I. 1990. Interactive real-time articulated figure manipulation using multiple kinematic constraints. *Computer Graphics*, 24(2): 245–250.

Phillips, C.B. and Badler, N. 1991. Interactive behaviors for bipedal articulated figures. *Computer Graphics*, 25(4): 359–362.

Picinbono, G., Delingette, H., and Ayache, N. 2003. Nonlinear anisotropic elasticity for real-time surgery simulation. *Graphical Models*, 65(5): 305–321.

Plutchik, R. 1980. A general psychoevolutionary theory of emotion, in R. Plutchik and H. Kellerman (eds.), *Emotion: Theory, Research, and Experience*, Vol. 1, pp. 2–33, Academic Press, New York.

Popović, Z. and Witkin, A. 1999. Physically based motion transformation. In *SIGGRAPH '99: Proceedings of the 26th Annual Conference on Computer Graphics and Interactive Techniques*, Los Angeles, CA, ACM, New York, pp. 11–20.

Pratscher, M., Coleman, P., Laszlo, J., and Singh, K. 2005. Outside-in anatomy based character rigging. In *Proceedings of the Symposium on Computer Animation, SCA'05*, New York, pp. 329–338.

Renault, O., Magnenat-Thalmann, N., and Thalmann, D. 1990. A vision-based approach to behavioural animation. *Visualization and Computer Animation Journal*, 1(1): 18–21.

Reynolds, C. 2006. Big fast crowds on ps3. In *Sandbox'06: Proceedings of the 2006 ACM SIGGRAPH Symposium on Videogames*, Boston, MA, pp. 113–121.

Reynolds, C.W. 1999. Steering behaviors for autonomous characters. In *Proceedings of Game Developers Conference*, Miller Freeman Game Group, San Francisco, CA, pp. 763–782.

Safonova, A., Hodgins, J.K., and Pollard, N.S. 2004. Synthesizing physically realistic human motion in low-dimensional, behavior-specific spaces. *ACM Transactions on Graphics*, 23(3): 514–521.

Sand, P., Mcmillan, L., and Popović, J. 2003. Continuous capture of skin deformation. *ACM Transactions on Graphics*, 22(3): 578–586.

Scheepers, F., Parent, R., Carlson, W., and May, S. 1997. Anatomy-based modeling of the human musculature. In *Proceedings of the SIGGRAPH'97*, Los Angels, CA, pp. 163–172.

Shao, W. and Terzopoulos, D. 2005. Autonomous pedestrians. In *Proceedings of the SCA'05*, New York, pp. 19–28.

Singh, K. and Kokkevis, E. 2000. Skinning characters using surface oriented free-form deformations. In *Proceedings of the Graphics Interface, GI'00*, Montreal, Quebec, Canada, pp. 35–42.

Spindel, M.S., Roseman, I., and Jose, P.E. 1990. Appraisals of emotion-eliciting events: Testing a theory of discrete emotions. *Personality and Social Psychology*, 59(5): 899–913.

Sumner, R.W., Zwicker, M., Gotsman, C., and Popović, J. 2005. Mesh-based inverse kinematics. *ACM Transactions on Graphics*, 24(3): 488–495.

Sun, H. and Metaxas, D. 2001. Automating gait generation. In *Proceedings of the SIGGRAPH '01*, Los Angels, CA, pp. 261–270.

Sung, M., Gleicher, M., and Chenney, S. 2004. Scalable behaviors for crowd simulation. *Computer Graphics Forum*, 23(3): 519–528.

Sung, M., Kovar, L., and Gleicher, M. 2005. Fast and accurate goal-directed motion synthesis for crowds. In *Proceedings of the SCA'05*, New York, pp. 291–300.

Terzopoulos, D. and Fleischer, K. 1988. Modeling inelastic deformation: Viscoelasticity, plasticity, fracture. *Computer Graphics*, 22(4): 269–278.

Terzopoulos, D., Platt, J.C., Barr, A.H., and Fleischer, K. 1987. Elastically deformable models. *Computer Graphics*, 21(4): 205–214; *Proceedings of the SIGGRAPH'87*, Anaheim, CA.

Terzopoulos, D. and Waters, K. 1990. Physically based facial modelling, analysis and animation. *Journal of Visualization and Computer Animation*, 1(2): 73–90.

Terzopoulos, D. and Witkin, A. 1988. Physically based models with rigid and deformable components. *IEEE Computer Graphics and Applications*, 8(6): 41–51.

Thalmann, D. and Musse, S.R. 2007. *Crowd Simulation*. Springer, London, U.K.

Thalmann, D., Shen, J., and Chauvineau, E. 1996. Fast realistic human body deformations for animation and vr applications. In *Proceedings of the Computer Graphics International, CGI'96*, Pohang, Korea.

Treuille, A., Cooper, S., and Popović, Z. 2006. Continuum crowds. In *Proceedings of the SIGGRAPH'06*, Pittsburgh, PA, pp. 1160–1168.

Tu, X. and Terzopoulos, D. 1994. Artificial fishes: Physics, locomotion, perception, behavior. *Computer Graphics*, 42–48; *Proceedings of the SIGGRAPH '94*, Orlando, FL.

Van den Berg, J., Patil, S., Sewall, J., Manocha, D., and Lin, M. 2008. Interactive navigation of multiple agents in crowded environments. In *Proceedings of the 2008 Symposium on Interactive 3D Graphics and Games*, pp. 139–147, Redwood City, CA.

Vassilev, T. and Spanlang, B. 2001. *Fast Cloth Animation on Walking Avatars*. The Eurographics Association and Blackwell Publishers.

Volino, P. and Magnenat Thalmann, N. 1994. Efficient self-collision detection on smoothly discretised surface animations using geometrical shape regularity. *Computer Graphics Forum*, 13(3): 155–166; *Proceedings of the Eurographics '94*, Oslo, Norway.

Volino, P. and Magnenat-Thalmann, N. 2005. Implicit midpoint integration and adaptive damping for efficient cloth simulation. *Computer Animation and Virtual Worlds*, 16(3–4): 163–175.

Volino, P. and Magnenat-Thalmann, N. 2006. Simple linear bending stiffness in particle systems. In *Proceedings of the Eurographics Symposium on Computer Animation*, Lausanne, Switzerland, pp. 101–105.

Volino, P., Magnenat-Thalmann, N., and Faure, F. 2009. A simple approach to nonlinear tensile stiffness for accurate cloth simulation. *ACM Transactions on Graphics*, 28(4): 105–116.

Wang, X. and Devarajan, V. 2005. 1D and 2D structured mass-spring models with preload. *The Visual Computer*, 21(7): 429–448.

Wang, X.C. and Phillips, C. 2002. Multi-weight enveloping: least-squares approximation techniques for skin animation. In *Proceedings of the Symposium on Computer Animation, SCA'02*, Osaka, Japan, pp. 129–138.

Waters, K. 1987. A muscle model for animating three-dimensional facial expression. *Computer Graphics*, 21(4): 17–24; *Proceedings of the SIGGRAPH '87*, Anaheim, CA.

Weil, J. 1986. The synthesis of cloth objects. *Computer Graphics*, 20, 49–54; *Proceedings of the Conference on SIGGRAPH 86*, Dallas, TX, *Annual Conference Series, ACM SIGGRAPH*, Addison-Wesley, Boston, MA.

Whissel, C.M. 1980. The dictionary of affect in language, in R. Plutchik and H. Kellerman (eds.), *Emotion: Theory, Research, and Experience, Vol. 4: The Measurement of Emotions*, pp. 113–131, Academic Press, New York.

Wilhelms, J. and Vangelder, A. 1997. Anatomically based modeling. In *Proceedings of the SIGGRAPH'97*, Los Angels, CA, pp. 173–180.

Witkin, A. and Kass, M. 1988. Spacetime constraints. In *SIGGRAPH'88: Proceedings of the 15th Annual Conference on Computer Graphics and Interactive Techniques*, Atlanta, GA, ACM, New York, pp. 159–168.

Witkin, A. and Popović, Z. 1995. Motion warping. In *Proceedings of the 22nd Annual Conference on Computer Graphics and Interactive Techniques SIGGRAPH '95*, Los Angels, CA, ACM, New York, pp. 105–108.

Wooten, W. and Hodgins, J. 2000. Simulating leaping, tumbling, landing and balancing humans. In *Proceedings of the IEEE International Conference on Robotics and Automation*, San Francisco, CA, pp. 656–662.

Yersin, B., Maim, J., Morini, F., and Thalmann, D. 2008. Real-time crowd motion planning. *Visual Computer*, 24(10): 859–870.

Zhao, J. and Badler, N.I. 1994. Inverse kinematics positioning using nonlinear programming for highly articulated figures. *ACM Transactions on Graphics*, 13(4): 313–336.

Zhou, K., Huang, J., Snyder, J., Liu, X., Bao, H., Guo, B., and Shum H.Y. 2005. Large mesh deformation using the volumetric graph laplacian. *ACM Transactions on Graphics*, 24(3): 496–503.

Zhuang, Y. and Canny, J. 2000. Haptic interaction with global deformations. In *Proceedings of the IEEE International Conference on Robotics and Automation*, pp. 2428–2433, San Francisco, CA.

V

Intelligent Systems

34

Paraconsistent Logic-Based Reasoning for Intelligent Systems

Kazumi Nakamatsu
University of Hyogo

34.1 Introduction

Logical systems have been applied to various intelligent systems as their theoretical foundations or reasoning languages such as logic programming. However, some actual systems include a lot of inconsistency such as contradictions, dilemma, and conflicts. It is required to deal with such inconsistency in the same system if we formalize the system in logic aiming at an intelligent system. Ordinal logical systems can not deal with inconsistency such as p and $\neg p$ consistently in the system. A logical system that can deal with inconsistency is known as *paraconsistent logic*. The main purpose of paraconsistent logic is to deal with inconsistency in a framework of consistent logical systems. It has been more than six decades since the first paraconsistent logical system was proposed by S. Jaskowski [8]. It was four decades later than that a family of paraconsistent logic called *annotated logic* was proposed in [3,44]. They can deal with inconsistency with many truth values called annotations that should be attached to each atomic formula, although their formula valuation is two-valued.

The paraconsistent annotated logic was developed from the viewpoint of logic programming [10], aimed at applications to computer science such as semantics for knowledge bases in [2,9,45]. Furthermore, in order to deal with inconsistency and nonmonotonic reasoning in a framework of annotated logic programming, the paraconsistent annotated logic programming was developed to treat ontological (strong) negation and the stable model semantics [7,6] in [13] and named *annotated logic program with strong negation* (ALPSN). It has been applied to some nonmonotonic systems, default logic [42], autoepistemic logic [12], and a nonmonotonic assumption based truth maintenance system (ATMS) [4] as their computational models in [14,31,32]. ALPSN can deal with not only inconsistency such as conflict but also nonmonotonic reasoning such as default reasoning. However, it seems to be more important and useful from a practical viewpoint to deal with resolving conflict in a logical way than just to express conflicts consistently. It is not so adequate for ALPSN to deal with resolving conflict or decision making based on resolving conflict in its logical framework. Defeasible logic [5] is known as one of formalizations for nonmonotonic reasoning called *defeasible reasoning* that can deal with resolving conflict easily in a logical way [1,38,39]. However, defeasible logic cannot treat inconsistency in its syntax and its inference rules are too complicated to implement them easily. Therefore, in order to deal with defeasible reasoning in a framework of paraconsistent annotated logic programming, a new version of ALPSN called *vector annotated logic program with strong negation* (VALPSN) was also proposed and applied to resolving conflict [15–17]. It also has been shown that VALPSN provides a computational model of a defeasible logic [18]. Moreover, VALPSN has been extended to deal with deontic notions (obligation, forbiddance, and permission) [11], and the extended version of VALPSN was named *extended vector annotated logic program with strong negation* (EVALPSN) in [19,20]. EVALPSN can deal with defeasible deontic reasoning [40] and has been applied to various kinds of safety verification-based control. The basic ideas of EVALPSN safety verification are that each control system has norms such as guidelines for safe control. Such norms can be formulated in EVALPSN deontic expressions; then the safety verification for control systems can be carried out by EVALPSN defeasible deontic reasoning. Compared to conventional safety verification methods, EVALPSN safety verification has some advantages such that it can provide a formal safety verification method and a computational framework for control systems as logic programming. Considering the safety verification for process control, there are cases in which the safety verification for process order is significant. EVALPSN has been applied to railway interlocking safety verification [23], robot action control [21,24,25,34], safety verification for air traffic control [22], traffic signal control [26], discrete event control [27–29], and pipeline valve control [30,33].

This chapter is organized as follows: in Section 34.2, the paraconsistent annotated logic program EVALPSN is reviewed; in Section 34.3, the outline of EVALPSN-based safety verification system is introduced and examples of its application to pipeline valve control are also given; in Section 34.4, we conclude this chapter by describing the advantages and disadvantages of applying EVLPSAN to intelligent control systems.

We assume that the readers are familiar with the basic knowledge of logic and logic program.

34.2 EVALPSN

In this section, we review a paraconsistent annotated logic program EVALPSN briefly [20]. Generally, a truth value called an *annotation* is explicitly attached to each literal in annotated logic programs [2]. For example, let p be a literal, μ an annotation, then

$$p:\mu$$

is called an *annotated literal*. The set of annotations constitutes a complete lattice. An annotation in EVALPSN has a form of

$$[(i, j), \mu]$$

called an *extended vector annotation*. The first component (i, j) is called a *vector annotation* and the set of vector annotations constitutes the complete lattice,

$$\mathcal{T}_v(n) = \{(x, y) \mid 0 \le x \le n, 0 \le y \le n, x, y, \text{ and } n \text{ are integers}\},$$

in Figure 34.1. The ordering ($\preceq_v$) of complete lattice $\mathcal{T}_v(n)$ is defined as

$$\text{let } (x_1, y_1), (x_2, y_2) \in \mathcal{T}_v(n),$$

$$(x_1, y_1) \preceq_v (x_2, y_2) \quad \text{iff} \quad x_1 \le x_2 \text{ and } y_1 \le y_2.$$

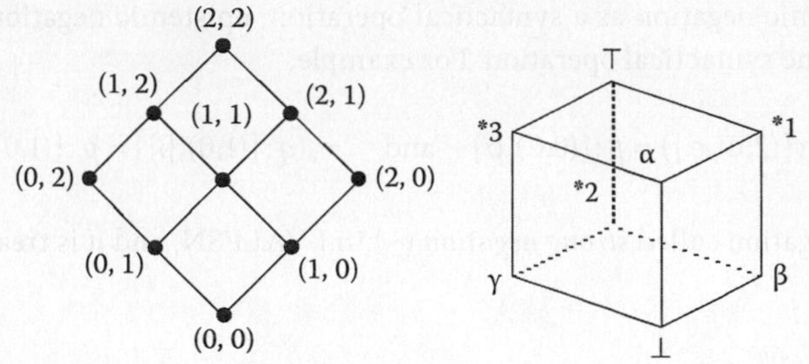

FIGURE 34.1 Lattice $\mathcal{T}_v(2)$ and lattice $\mathcal{T}_d$.

For each extended vector annotated literal $p : [(i, j), \mu]$, integer i denotes the amount of positive information to support literal p and integer j denotes the negative one. The second component μ is an index of fact and deontic notions such as obligation, and the set of the second components constitutes the complete lattice,

$$\mathcal{T}_d = \{\perp, \alpha, \beta, \gamma, *1, *2, *3, \top\}.$$

The ordering ($\preceq_d$) of complete lattice $\mathcal{T}_d$ is described by Hasse's diagram in Figure 34.1. The intuitive meaning of each member of complete lattice $\mathcal{T}_d$ is as follows:

⊥(unknown), ⊤(inconsistency)
α(fact), β(obligation), γ(non-obligation),
*1(fact and obligation), *2(obligation and non-obligation),
*3(fact and non-obligation).

Then the complete lattice $\mathcal{T}_e(n)$ of extended vector annotations is defined as the product $\mathcal{T}_v(n) \times \mathcal{T}_d$ of complete lattices $\mathcal{T}_v(n)$ and $\mathcal{T}_d$. The ordering ($\preceq_e$) of complete lattice $\mathcal{T}_e(n)$ is defined as let $[(i_1, j_1), \mu_1]$ and $[(i_2, j_2), \mu_2] \in \mathcal{T}_e$,

$$[(i_1, j_1), \mu_1] \preceq_e [(i_2, j_2), \mu_2] \quad \text{iff} \quad (i_1, j_1) \preceq_v (i_2, j_2) \text{ and } \mu_1 \preceq_d \mu_2.$$

There are two kinds of *epistemic negation* ($\neg_1$ and $\neg_2$) in EVALPSN, both of which are defined as mappings over complete lattices $\mathcal{T}_v(n)$ and $\mathcal{T}_d$, respectively.

Definition 34.1 (epistemic negations $\neg_1$ and $\neg_2$ in EVALPSN)

$$\neg_1([i, j), \mu]) = [(j, i), \mu], \quad \forall \mu \in \mathcal{T}_d,$$

$$\neg_2([(i, j), \perp]) = [(i, j), \perp], \quad \neg_2([(i, j), \alpha]) = [(i, j), \alpha],$$

$$\neg_2([(i, j), \beta]) = [(i, j), \gamma], \quad \neg_2([(i, j), \gamma]) = [(i, j), \beta],$$

$$\neg_2([(i, j), *_1]) = [(i, j), *_3], \quad \neg_2([(i, j), *_2]) = [(i, j), *_2],$$

$$\neg_2([(i, j), *_3]) = [(i, j), *_1], \quad \neg_2([(i, j), \top]) = [(i, j), \top].$$

If we regard epistemic negation as a syntactical operation, epistemic negations followed by literals can be eliminated by the syntactical operation. For example,

$$\neg_1(p:[(2,0),\alpha]) = p:[(0,2),\alpha] \quad \text{and} \quad \neg_2(q:[(1,0),\beta]) = p:[(1,0),\gamma].$$

There is another negation called *strong negation* (~) in EVALPSN, and it is treated as well as classical negation.

Definition 34.2 (strong negation ~) (see [3])

Let F be any formula and $\neg$ be $\neg_1$ or $\neg_2$:

$$\sim F =_{def} F \rightarrow ((F \rightarrow F) \wedge \neg(F \rightarrow F)).$$

Definition 34.3 (well-extended vector annotated literal)

Let p be a literal.

$$p : [(i, 0), \mu] \quad \text{and} \quad p : [(0, j), \mu]$$

are called *weva (well-extended vector annotated) literals*, where $i, j \in \{1, 2, \ldots, n\}$ and $\mu \in \{\alpha, \beta, \gamma\}$.

Definition 34.4 (EVALPSN)

If $L_0, \cdots, L_n$ are weva literals,

$$L_1 \wedge \cdots \wedge L_i \wedge \sim L_{i+1} \wedge \cdots \wedge \sim L_n \rightarrow L_0$$

is called an *EVALPSN clause*. An *EVALPSN* is a finite set of EVALPSN clauses.

Usually, a deontic operator $\bigcirc$ is used for expressing obligation in deontic logic; we describe the relation between deontic expressions and EVALPSN literals. Let $\bigcirc$ be a deontic operator representing obligation and p a literal and,

p is represented by $p : [(m, 0), \alpha]$ (fact),

$\bigcirc p$ is represented by $p : [(m, 0), \beta]$ (obligation),

$\bigcirc \neg p$ is represented by $p : [(0, m), \beta]$ (forbiddance),

$\neg \bigcirc \neg p$ is represented by $p : [(0, m), \gamma]$ (permission),

where m is a positive integer.

We now address the formal interpretation of EVALPSN literals. Generally, the interpretation of annotated proposition is defined as follows: let ν be the set of all propositional symbols, $\mathcal{F}$ the set of all formulas, and $\mathcal{T}$ the complete lattice of annotations; then, an interpretation I is a function:

$$I : \nu \longrightarrow \mathcal{T}.$$

To each interpretation I, the valuation function such that

$$v_I : \mathcal{F} \longrightarrow \{0,1\}$$

is defined as follows:

(1) Let p be a propositional symbol and μ an annotation:

$$v_I(p:\mu) = 1 \text{ if } \mu \leq I(p), \quad v_I(p:\mu) = 0 \text{ if } \mu \not\leq I(p),$$

where the symbol $\leq$ is the order over complete lattice $\mathcal{T}$.

(2) Let A and B be any formulas, and A not an annotated atom:

$$v_I(\neg A) = 1 \text{ if } v_I(A) = 0, \quad v_I(\sim B) = 1 \text{ if } v_I(B) = 0;$$

other formulas, $A \rightarrow B$, $A \wedge B$, $A \vee B$ are valuated as usual.

We here note that for any literal p, $v_I(p : \perp) = 1$ holds, which intuitively means that annotated literal $p : \perp$ is always satisfied, where $\perp$ is the bottom annotation of the complete lattice of annotations.

Example 34.1 Suppose an EVALPSN

$$P = \{p:[(0,1),\alpha] \wedge \sim q:[(2,0),\beta] \rightarrow r:[(0,1),\gamma], \tag{34.1}$$

$$p:[(2,0),\alpha], \tag{34.2}$$

$$p:[(0,2),\alpha], \tag{34.3}$$

$$q:[(2,0),\gamma], \tag{34.4}$$

$$q:[(1,0),\beta]\}, \tag{34.5}$$

where p, q, and r are literals. We show the derivation of the head literal of EVALPSN clause (34.1), r : [(0, 1), γ] in EVALPSN P. In order to derive the head literal, we need to show that the body of EVALPSN clause (34.1) is satisfied, that is, EVALP literal p : [(0, 1), α] is satisfied and EVALP literal q : [(2, 0), β] is not satisfied in EVALPSN P. If we consider EVALP clause (34.2), as

$$[(0,1),\alpha] \not\preceq_e [(2,0),\alpha],$$

it does not satisfy EVALP literal p : [(0, 1), α]; on the other hand, if we consider EVALP clause (34.3), as

$$[(0,1),\alpha] \preceq_e [(0,2),\alpha],$$

it satisfies EVALP literal p : [(0, 1), α].

On the other hand, if we consider EVALP clause (34.4), as

$$[(2,0),\beta] \not\preceq_e [(2,0),\gamma],$$

it does not satisfy EVALP literal $q : [(2, 0), \beta]$; furthermore, even if we consider another EVALP clause (34.5), as

$$[(2,0),\beta] \not\preceq_e [(1,0),\beta],$$

it does not satisfy EVALP literal $q : [(2, 0), \beta]$. Therefore, the head literal $r : [(0, 1), \gamma]$ of EVALPSN clause (34.1) can be derived.

34.3 EVALPSN Safety Verification for Pipeline Control

This section introduces EVALPSN-based intelligent safety verification for a pipeline valve control with a simple brewery pipeline example.

34.3.1 Pipeline Network

The pipeline network described in Figure 34.2 is taken as an example of brewery pipeline valve control based on EVALPSN safety verification. In Figure 34.2, an arrow represents the direction of liquid flow, a home-plate pentagon does a liquid tank, and a cross does a valve. In the pipeline network, we consider three kinds of physical entities,

$$\text{four tanks } (T_0, T_1, T_2, T_3), \quad \text{two valves } (V_0, V_1), \quad \text{and}$$

$$\text{five pipes } (Pi_0, Pi_1, Pi_2, Pi_3, Pi_4),$$

and three kinds of logical entities,

$$\text{four processes } (Pr_0, Pr_1, Pr_2, Pr_3), \quad \text{two valves } (V_0, V_1), \quad \text{and}$$

$$\text{five subprocesses } (SPr_0, SPr_1, SPr_2, SPr_3, SPr_4),$$

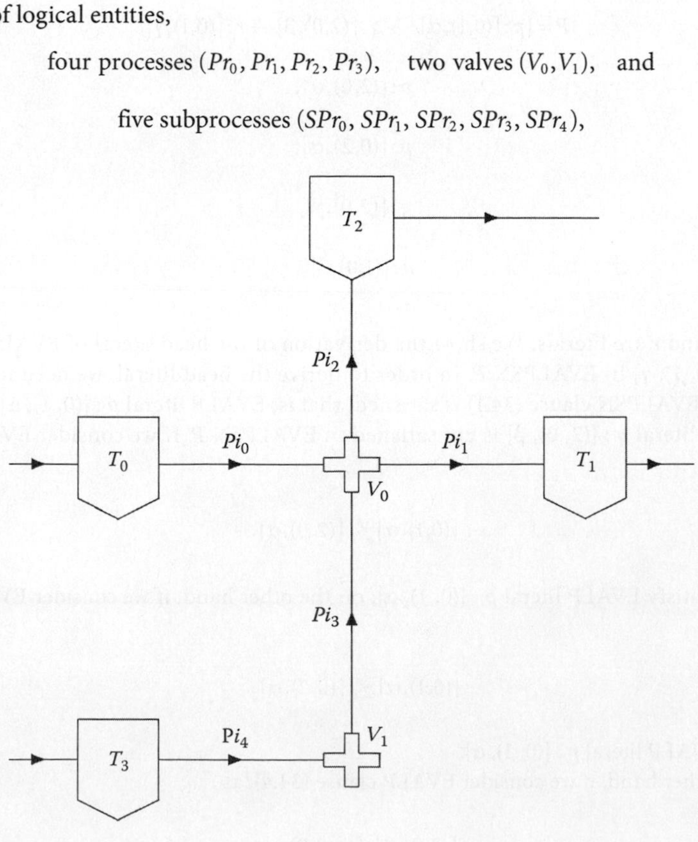

FIGURE 34.2 Pipeline example.

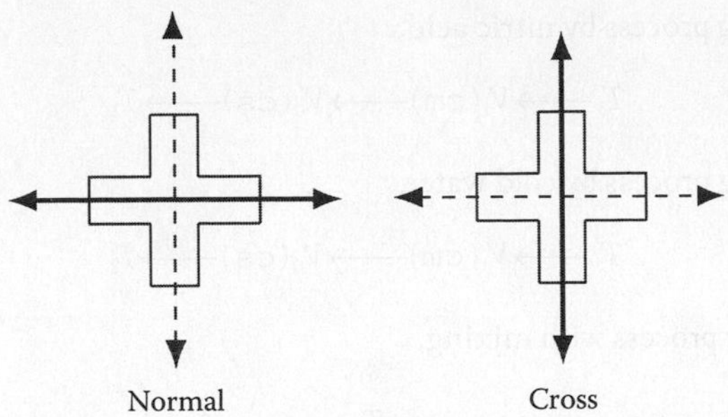

FIGURE 34.3 Normal and cross directions.

where the logical entities are used for EVALPSN-based safety verification. We suppose that a subprocess is defined as a logical process using a pipe, a valve is also defined to be a logically controlled entity, and a process is defined as a set of subprocesses and valves in the verification system. For example, process Pr_0 consists of subprocesses SPr_0 and SPr_1 and valve V_0.

We assume that the valves can physically control two liquid flows in the normal and cross directions as shown in Figure 34.3. Furthermore, each logical entity is supposed to have logical states:

- Subprocesses have two logical states *locked* (l) and *free* (f), where "the subprocess is locked" means that it is logically reserved for one kind of liquid and "free" means that it can be logically reserved for any kind of liquid but unlocked.
- Valves have two logical states, *controlled mixture* (cm), representing that the valve is controlled to mix the liquid flows in normal and cross directions, and *controlled separate* (cs), representing that the valve is controlled to separate the liquid flows in normal and cross directions as shown in Figure 34.4.
- Processes have two states *set* (s) and *unset* (xs), where "the process is set" means that all subprocesses and valves in the process are locked and controlled consistently and "unset" means not set.

We also assume that there are five kinds of cleaning liquid for the pipeline network, cold water (*cw*), warm water (*ww*), hot water (*hw*), nitric acid (*na*), and caustic soda (*cs*). We now consider the following four processes:

Process Pr_0 a beer process,

$$T_0 \longrightarrow V_0(\mathtt{cs}) \longrightarrow T_1$$

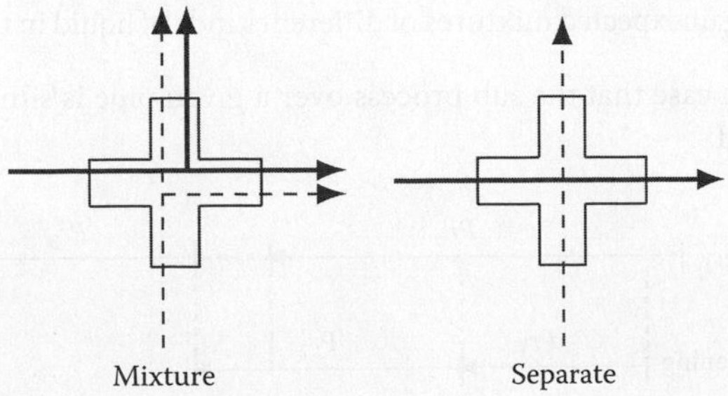

FIGURE 34.4 Controlled mixture and separate.

Process Pr_1 a cleaning process by nitric acid,

$$T_3 \longrightarrow V_1(\mathtt{cm}) \longrightarrow V_0(\mathtt{cs}) \longrightarrow T_2$$

Process Pr_2 a cleaning process by cold water,

$$T_3 \longrightarrow V_1(\mathtt{cm}) \longrightarrow V_0(\mathtt{cs}) \longrightarrow T_2$$

Process Pr_3 a brewery process with mixing,

$$\begin{array}{ccccc} & & T_2 & & \\ & & \uparrow & & \\ T_0 & \longrightarrow & V_0(\mathtt{cm}) & \longrightarrow & T_1 \\ & & \uparrow & & \\ T_3 & \longrightarrow & V_1(\mathtt{cm}) & & \end{array}$$

In order to verify the safety for processes Pr_0, Pr_1, Pr_2, and Pr_3, the pipeline network operator issues process requests to be verified, which consists of an if-part and a then-part. The if-part of a process request describes the current logical state of the pipelines in the process to be safety verified, and the then-part of it describes permission for setting the process. For example, if a process request for verifying the safety of process Pr_1 is described as

if Subprocess SPr_0 is free, subprocess SPr_1 is free,
valve V_0 is controlled separate physically but not locked logically.

then Pr ocess Pr_0 can be set.

We also consider the following process schedule including processes Pr_0, Pr_1, Pr_2, and Pr_3:

PRS-0 process Pr_0 must start before any other processes
PRS-1 process Pr_1 must start immediately before process Pr_0
PRS-2 process Pr_2 must start immediately after process Pr_1
PRS-3 process Pr_3 must start immediately after both processes Pr_0 and Pr_2 finished,

which are shown in the process schedule chart (Figure 34.5).

34.3.2 Pipeline Safety Property

We introduce some safety properties as the criteria of the pipeline process safety verification, *SPr* for subprocesses, *Val* for valves, and *Pr* for processes, which assure the safety for pipeline processes Pr_0, Pr_1, Pr_2, and Pr_3, that is, for avoiding unexpected mixtures of different kinds of liquid in the same pipeline and valve:

- *SPr*: It is a forbidden case that the sub process over a given pipe is simultaneously locked by different kinds of liquid.

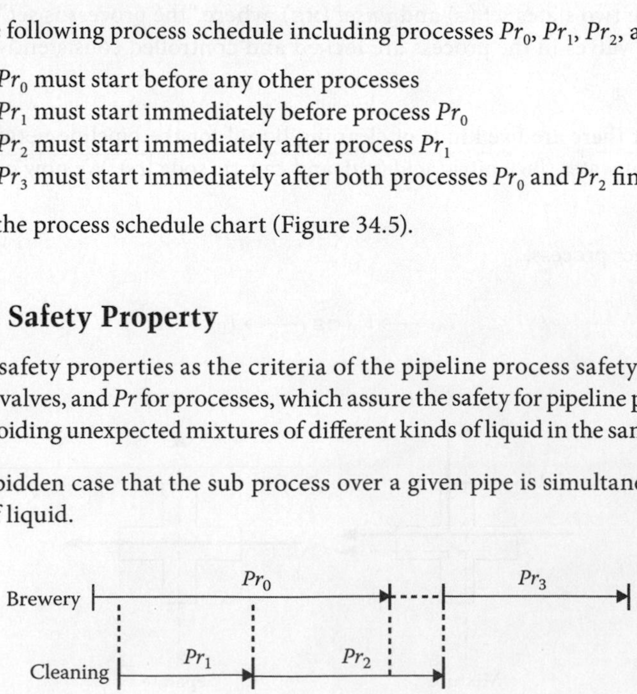

FIGURE 34.5 Process schedule chart.

- *Val*: It is a forbidden case that valves are controlled for unexpected mixture of liquid.
- *Pr*: Whenever a process is set, all its component sub processes are locked and all its component valves are controlled consistently.

34.3.3 Predicates for Safety Verification

The EVALPSN-based safety verification is carried out by verifying whether the process request contradicts the safety properties or not in EVALPSN. Then the following three steps have to be executed:

1. The safety properties and some other control policies are translated into EVALPSN clauses, and they are stored as EVALPSN P_{sc}.
2. The if-part and then-part of the process request are translated into EVALPSN clauses as EVALPSNs P_i and P_t.
3. EVALPSN P_t is inquired from EVALPSN $\{P_{sc} \cup P_i\}$, then if *yes* is returned, the safety for the pipeline process is assured; otherwise it is not assured.

In the safety verification for pipeline processes, the following predicates are used in EVALPSN:

- $Pr(i, l)$ represents that process i for liquid l is set(s) or unset(xs), where each $i \in \{p0, p1, p2, p3\}$ is a process id corresponding to processes Pr_0, Pr_1, Pr_2, and Pr_3, and each $l \in \{b, cw, ww, hw, na, cs\}$ is a liquid id corresponding to beer, cold water, warm water, hot water, nitric acid, and caustic soda, and we have EVALP clause $Pr(i, l) : [\mu_1, \mu_2]$, where

$$\mu_1 \in \mathcal{T}_{v_1} = \{\perp_1, \mathtt{s}, \mathtt{xs}, \top_1\} \quad \text{and} \quad \mu_2 \in \mathcal{T}_d = \{\perp, \alpha, \beta, \gamma, {}^*1, {}^*2, {}^*3, \top\}.$$

Complete lattice $\mathcal{T}_{v_1}$ is a variant of complete lattice $\mathcal{T}_v$ in Figure 34.6 as $n = 1$. Therefore, annotations $\perp_1$, s, xs, and $\top_1$ represent vector annotations (0, 0), (1, 0), (0, 1), and (1, 1), respectively. The epistemic negation $\neg_1$ over $\mathcal{T}_{v_1}$ is defined by the following mapping:

$$\neg_1([\perp_1, \mu_2]) = [\perp_1, \mu_2], \quad \neg_1([\mathtt{s}, \mu_2]) = [\mathtt{xs}, \mu_2],$$

$$\neg_1([\top_1, \mu_2]) = [\top_1, \mu_2], \quad \neg_1([\mathtt{xs}, \mu_2]) = [\mathtt{s}, \mu_2].$$

For example, EVALP clause $Pr(p2, b) : [\mathtt{s}, \alpha]$ can be intuitively interpreted as "it is a fact that beer process Pr_2 is set."

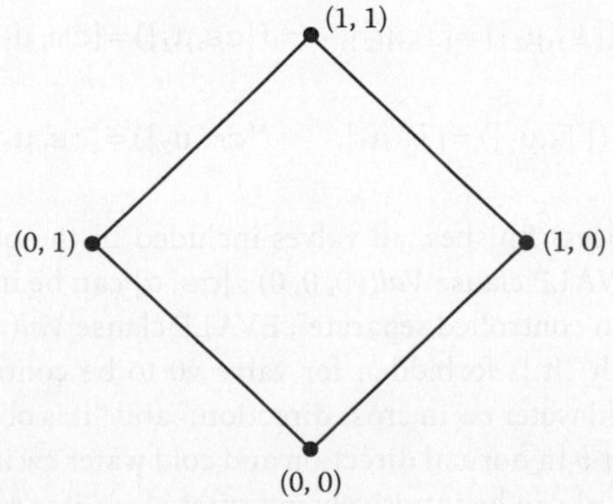

FIGURE 34.6 Complete lattice $\mathcal{T}_v(n = 1)$.

- *SPr(i, j, l)* represents that the sub process from valve *i* (or tank *i*) to valve *j* (or tank *j*) occupied by liquid *l* is locked (`l`) or free (`f`). Moreover, if a sub-process is free, then the kind of liquid in the pipe does not matter, and the liquid is represented by symbol "0" (zero). Therefore, we have

$$l \in \{b, cw, ww, hw, na, cs, 0\} \text{ and } i, j \in \{v0, v1, t0, t1, t2, t3\},$$

which are valve and tank ids corresponding to valves and tanks V_0, V_1, T_0, T_1, T_2, and T_3. Then we have EVALP clause *SPr(i, j, l)* : $[\mu_1, \mu_2]$, where

$$\mu_1 \in \mathcal{T}_{v_2} = \{\bot_2, \mathtt{l}, \mathtt{f}, \top_2\} \quad \text{and} \quad \mu_2 \in \mathcal{T}_d = \{\bot, \alpha, \beta, \gamma, *1, *2, *3, \top\}.$$

Complete lattice $\mathcal{T}_{v_2}$ is a variant of complete lattice $\mathcal{T}_v$ in Figure 34.6 as $n = 1$. Therefore, annotations $\bot_2$, `l`, `f`, and $\top_2$ represent vector annotations (0, 0), (1, 0), (0, 1), and (1, 1), respectively. The epistemic negation $\neg_1$ over $\mathcal{T}_{v_2}$ is defined by the following mapping:

$$\neg_1([\bot_2, \mu_2]) = [\bot_2, \mu_2], \quad \neg_1([\mathtt{l}, \mu_2]) = [\mathtt{f}, \mu_2],$$

$$\neg_1([\top_2, \mu_2]) = [\top_2, \mu_2], \quad \neg_1([\mathtt{f}, \mu_2]) = [\mathtt{l}, \mu_2].$$

For example, EVALP clause *SPr(v0, t1, b)* : [`f`, γ] can be intuitively interpreted as "the subprocess from valve *v0* to tank *t1* is permitted to be locked under the beer process."

- *Val(i, l_n, l_c)* represents that valve *i* occupied by two kinds of liquid l_n, $l_c \in \{b, cw, ww, hw, na, cs, 0\}$ is controlled separate (`cs`) or mixture (`cm`), where $i \in \{v0, v1\}$ is a valve id. We assume that there are two directed liquid flows in *normal* or *cross* directions of valves (Figure 34.3). Therefore, the second argument l_n represents liquid flowing in normal direction and the third argument l_c represents liquid flowing in cross direction. Generally, if a valve is released from the controlled state, liquid flow in the valve is represented by the specific symbol 0 that means "free." We have EVALP clause *Val(i, l_n, l_c)* : $[\mu_1, \mu_2]$, where

$$\mu_1 \in \mathcal{T}_{v_3} = \{\bot_3, \mathtt{cm}, \mathtt{cs}, \top_3\} \quad \text{and} \quad \mu_2 \in \mathcal{T}_d = \{\bot, \alpha, \beta, \gamma, *_1, *_2, *_3, \top\}.$$

Complete lattice $\mathcal{T}_{v_3}$ is a variant of complete lattice $\mathcal{T}_v$ in Figure 34.6 as $n = 1$. Therefore, annotations $\bot_3$, `cm`, `cs`, and $\top_3$ represent vector annotations (0, 0), (1, 0), (0, 1), and (1, 1), respectively. The epistemic negation $\neg_1$ over $\mathcal{T}_{v_3}$ is defined by the following mapping:

$$\neg_1([\bot_3, \mu_2]) = [\bot_3, \mu_2], \quad \neg_1([\mathtt{cs}, \mu_2]) = [\mathtt{cm}, \mu_2],$$

$$\neg_1([\top_3, \mu_2]) = [\top_3, \mu_2], \quad \neg_1([\mathtt{cm}, \mu_2]) = [\mathtt{cs}, \mu_2].$$

We assume that if a process finishes, all valves included in the process are controlled separate (closed). For example, EVALP clause *Val(v0, 0, 0)* : [`cs`, α] can be intuitively interpreted as "valve *v0* has been released from controlled separate"; EVALP clause *Val(v0, b, cw)* : [`cs`, β] can be intuitively interpreted as both "it is forbidden for valve *v0* to be controlled mixture with beer *b* in normal direction and cold water *cw* in cross direction," and "it is obligatory for valve *v0* to be controlled separate with beer *b* in normal direction and cold water *cw* in cross direction"; and EVALP clause *Val(v0, 0, b)* : [`cs`, α] can be intuitively interpreted as "it is a fact that valve *v0* is controlled separate with free flow 0 in normal direction and beer *b* in cross direction."

- $Eql(l_1, l_2)$ represents that the kinds of liquids l_1 and l_2 are the same (`sa`) or different (`di`), where $l_1, l_2 \in \{b, cw, ww, hw, na, cs, 0\}$. We have EVALP clause $Eql(l_1, l_2) : [\mu_1, \mu_2]$, where

$$\mu_1 \in \mathcal{T}_{v_4} = \{\bot_4, \texttt{sa}, \texttt{di}, \top_4\} \quad \text{and} \quad \mu_2 \in \mathcal{T}_d = \{\bot, \alpha, \beta, \gamma, *1, *2, *3, \top\}.$$

Complete lattice $\mathcal{T}_{v_4}$ is a variant of complete lattice $\mathcal{T}_v$ in Figure 34.6 as $n = 1$. Therefore, annotations $\bot_4$, `sa`, `di`, and $\top_4$ represent vector annotations (0, 0), (1, 0), (0, 1), and (1, 1), respectively. The epistemic negation $\neg_1$ is defined by the following mapping.

$$\neg_1([\bot_4, \mu_2]) = [\bot_4, \mu_2], \quad \neg_1([\texttt{di}, \mu_2]) = [\texttt{sa}, \mu_2],$$

$$\neg_1([\top_4, \mu_2]) = [\top_4, \mu_2], \quad \neg_1([\texttt{sa}, \mu_2]) = [\texttt{di}, \mu_2].$$

Now we consider process release conditions when processes have finished and define some more predicates to represent the conditions. We assume that if the terminal tank T_i of process Pr_j is filled with one kind of liquid, the finish signal $Fin(pj)$ of process Pr_j is issued.

- $Tan(ti, l)$ represents that tank T_i has been filled fully (`fu`) with liquid l or empty (`em`). Then we have EVALP clause $Tan(ti, l) : [\mu_1, \mu_2]$, where $i \in \{0, 1, 2, 3\}$ and $l \in \{b, cw, ww, hw, na, cs, 0\}$,

$$\mu_1 \in \mathcal{T}_{v_5} = \{\bot_5, \texttt{fu}, \texttt{em}, \top_5\} \text{ and } \mu_2 \in \mathcal{T}_d = \{\bot, \alpha, \beta, \gamma, *1, *2, *3, \top\}.$$

Complete lattice $\mathcal{T}_{v_5}$ is a variant of complete lattice $\mathcal{T}_v$ in Figure 34.6 as $n = 1$. Therefore, annotations $\bot_5$, `fu`, `em`, and $\top_5$ represent vector annotations (0, 0), (1, 0), (0, 1), and (1, 1), respectively. The epistemic negation $\neg_1$ over $\mathcal{T}_{v_5}$ is defined by the following mapping:

$$\neg_1([\bot_5, \mu_2]) = [\bot_5, \mu_2], \quad \neg_1([\texttt{fu}, \mu_2]) = [\texttt{em}, \mu_2],$$

$$\neg_1([\top_5, \mu_2]) = [\top_5, \mu_2], \quad \neg_1([\texttt{em}, \mu_2]) = [\texttt{fu}, \mu_2].$$

Note that annotation $\bot_5$ can be intuitively interpreted to represent "no information in terms of liquid amount." For example, EVALP clause $Tan(t2, 0) : [\texttt{em}, \alpha]$ can be interpreted as "it is a fact that tank $t2$ is empty."

- $Str(pi)$ represents that the start signal for process Pr_i is issued (`is`) or not (`ni`).
- $Fin(pj)$ represents that the finish signal for process Pr_j has been issued (`is`) or not (`ni`). Then we have EVALP clauses $Str(pi) : [\mu_1, \mu_2]$ and $Fin(pj) : [\mu_1, \mu_2]$, where $i, j \in \{0, 1, 2, 3\}$,

$$\mu_1 \in \mathcal{T}_{v_6} = \{\bot_6, \texttt{ni}, \texttt{is}, \top_6\} \quad \text{and} \quad \mu_2 \in \mathcal{T}_d = \{\bot, \alpha, \beta, \gamma, *1, *2, *3, \top\}.$$

Complete lattice $\mathcal{T}_{v_6}$ is a variant of complete lattice $\mathcal{T}_v$ in Figure 34.6 as $n = 1$. Therefore, annotations $\bot_6$, `is`, `ni`, and $\top_6$ represent vector annotations (0, 0), (1, 0), (0, 1), and (1, 1), respectively. The epistemic negation $\neg_1$ over $\mathcal{T}_{v_6}$ is defined by the following mapping:

$$\neg_1([\bot_6, \mu_2]) = [\bot_6, \mu_2], \quad \neg_1([\texttt{is}, \mu_2]) = [\texttt{ni}, \mu_2],$$

$$\neg_1([\top_6, \mu_2]) = [\top_6, \mu_2], \quad \neg_1([\texttt{ni}, \mu_2]) = [\texttt{is}, \mu_2].$$

For example, EVALP clause $Fin(p3)$: $[\mathtt{ni}, \alpha]$ can be interpreted as "it is a fact that the finish signal for process Pr_3 has not been issued yet."

34.3.4 Safety Property into EVALPSN

Here, we formalize all safety properties *SPr*, *Val*, and *Pr* in EVALPSN. Safety property *SPr* is intuitively interpreted as derivation rules of forbiddance. If a sub process from valve(or tank) *i* to valve(or tank) *j* is locked by one kind of liquid, it is forbidden for the subprocess to be locked by different kinds of liquid processes simultaneously. Thus, we have the following EVALPSN clauses:

$$SPr(i,j,l_1):[\mathtt{l},\alpha]\wedge \sim Eql(l_1,l_2):[\mathtt{sa},\alpha]$$

$$\rightarrow SPr(i,j,l_2):[\mathtt{f},\beta], \tag{34.6}$$

where $l_1, l_2 \in \{b, cw, ww, hw, na, cs\}$. Moreover, in order to derive permission for locking subprocesses, we need the following EVALPSN clauses:

$$\sim SPr(i,j,l):[\mathtt{f},\beta]\rightarrow SPr(i,j,l):[\mathtt{f},\gamma], \tag{34.7}$$

where $l \in \{b, cw, ww, hw, na, cs\}$.

Safety property *Val* is intuitively interpreted as derivation rules of forbiddance. We have to consider two cases: one is for deriving forbiddance from changing the control state of the valve and another one is for deriving forbiddance from mixing different kinds of liquid without changing the control state of the valve.

Case 1

If a valve is controlled separate, it is forbidden for the valve to be controlled mix; conversely, if a valve is controlled mix, it is forbidden for the valve to be controlled separate. Thus, we have the following EVALPSN clauses:

$$Val(i,l_n,l_c):[\mathtt{cs},\alpha]\wedge \sim Eql(l_n,0):[\mathtt{sa},\alpha]\wedge$$

$$\sim Eql(l_c,0):[\mathtt{sa},\alpha]\rightarrow Val(i,l_n,l_c):[\mathtt{cs},\beta], \tag{34.8}$$

$$Val(i,l_n,l_c):[\mathtt{cm},\alpha]\wedge \sim Eql(l_n,0):[\mathtt{sa},\alpha]\wedge$$

$$\sim Eql(l_c,0):[\mathtt{sa},\alpha]\rightarrow Val(i,l_n,l_c):[\mathtt{cm},\beta], \tag{34.9}$$

where $l_n, l_c \in \{b, cw, ww, hw, na, cs, 0\}$.

Case 2

Next, we consider the other forbiddance derivation case in which different kinds of liquid are mixed even if the valve control state is not changed. We have the following EVALPSN clauses:

$$Val(i,l_{n_1},l_{c_1}):[\mathtt{cs},\alpha]\wedge \sim Eql(l_{n1},l_{n_2}):[\mathtt{sa},\alpha]\wedge$$

$$\sim Eql(l_{n_1},0):[\mathtt{sa},\alpha]\rightarrow Val(i,l_{n_2},l_{c_2}):[\mathtt{cm},\beta], \tag{34.10}$$

$$Val(i,l_{n_1},l_{c_1}):[\mathtt{cs},\alpha]\wedge \sim Eql(l_{c1},l_{c_2}):[\mathtt{sa},\alpha]\wedge$$

$$\sim Eql(l_{c_1},0):[\mathtt{sa},\alpha]\rightarrow Val(i,l_{n_2},l_{c_2}):[\mathtt{cm},\beta], \tag{34.11}$$

$$Val(i,l_{m_1},l_{c_1}):[\mathtt{cm},\alpha] \wedge \sim Eql(l_{n_1},l_{n_2}):[\mathtt{sa},\alpha]$$
$$\rightarrow Val(i,l_{n_2},l_{c_2}):[\mathtt{cs},\beta], \tag{34.12}$$

$$Val(i,l_{m_1},l_{c_1}):[\mathtt{cm},\alpha] \wedge \sim Eql(l_{c_1},l_{c_2}):[\mathtt{sa},\alpha]$$
$$\rightarrow Val(i,l_{n_2},l_{c_2}):[\mathtt{cs},\beta], \tag{34.13}$$

where $l_{m_1}, l_{c_1} \in \{b, cw, ww, hw, na, cs, 0\}$ and $l_{n_2}, l_{c_2} \in \{b, cw, ww, hw, na, cs\}$.

Note that the EVALPSN clause $\sim Eql(l_n, 0) : [\mathtt{sa}, \alpha]$ represents that there does not exist information such that the normal direction with liquid l_n in the valve is free (not controlled). As well as the case of subprocesses, in order to derive permission for controlling valves, we need the following EVALPSN clauses:

$$\sim Val(i,l_n,l_c):[\mathtt{cm},\beta] \rightarrow Val(i,l_n,l_c):[\mathtt{cm},\gamma], \tag{34.14}$$

$$\sim Val(i,l_n,l_c):[\mathtt{cs},\beta] \rightarrow Val(i,l_n,l_c):[\mathtt{cs},\gamma], \tag{34.15}$$

where $l_n, l_c \in \{b, cw, ww, hw, na, cs, 0\}$.

Safety property *Pr* is intuitively interpreted as derivation rules of permission and directly translated into EVALPSN clauses as a rule: "if all the components of the process can be locked or controlled consistently, then the process can be set." For example, if beer process Pr_0 consists of the subprocess from tank T_0 to valve V_0, valve V_0 (beer in normal direction), and the subprocess from valve V_0 to tank T_1, then we have the following EVALP clause to obtain permission for setting process Pr_0:

$$SPr(t0,v0,b):[\mathtt{f},\gamma] \wedge SPr(v0,t1,b):[\mathtt{f},\gamma] \wedge$$
$$Val(v0,b,l):[\mathtt{cm},\gamma] \wedge Tan(t0,b):[\mathtt{fu},\alpha] \wedge$$
$$Tan(t1,0):[\mathtt{em},\alpha] \rightarrow Pr(p0,b):[\mathtt{xs},\gamma] \tag{34.16}$$

where $l \in \{b, cw, ww, hw, na, cs, 0\}$. We also have the following EVALP clauses for setting other processes Pr_1, Pr_2, and Pr_3:

$$SPr(t3,v1,na):[\mathtt{f},\gamma] \wedge SPr(v1,v0,na):[\mathtt{f},\gamma] \wedge$$
$$SPr(v0,t2,na):[\mathtt{f},\gamma] \wedge Val(v0,l,na):[\mathtt{cm},\gamma] \wedge$$
$$Val(v1,na,0):[\mathtt{cs},\gamma] \wedge Tan(t3,na):[\mathtt{fu},\alpha] \wedge$$
$$Tan(t2,0):[\mathtt{em},\alpha] \rightarrow Pr(p1,na):[\mathtt{xs},\gamma], \tag{34.17}$$

$$SPr(t3,v1,cw):[\mathtt{f},\gamma] \wedge SPr(v1,v0,cw):[\mathtt{f},\gamma] \wedge$$
$$SPr(v0,t2,cw):[\mathtt{f},\gamma] \wedge Val(v0,l,cw):[\mathtt{cm},\gamma] \wedge$$
$$Val(v1,cw,0):[\mathtt{cs},\gamma] \wedge Tan(t3,cw):[\mathtt{fu},\alpha] \wedge$$
$$Tan(t2,0):[\mathtt{em},\alpha] \rightarrow Pr(p2,cw):[\mathtt{xs},\gamma], \tag{34.18}$$

$$SPr(t0,v0,b):[\mathtt{f},\gamma] \wedge SPr(t3,v1,b):[\mathtt{f},\gamma] \wedge$$

$$SPr(v0,t1,b):[\mathtt{f},\gamma] \wedge SPr(v0,t2,b):[\mathtt{f},\gamma] \wedge$$

$$SPr(v1,v0,b):[\mathtt{f},\gamma] \wedge Val(v0,b,b):[\mathtt{cs},\gamma] \wedge$$

$$Val(v1,b,0):[\mathtt{cs},\gamma] \wedge Tan(t0,b):[\mathtt{fu},\alpha] \wedge$$

$$Tan(t1,0):[\mathtt{em},\alpha] \wedge Tan(t3,b):[\mathtt{fu},\alpha] \wedge$$

$$Tan(t2,0):[\mathtt{em},\alpha] \rightarrow Pr(p3,b):[\mathtt{xs},\gamma], \tag{34.19}$$

where $l \in \{b, cw, ww, hw, na, cs, 0\}$.

34.3.5 Example

Now, let us show an example of EVALPSN safety verification-based pipeline control for processes Pr_0, Pr_1, Pr_2, and Pr_3. According to the process schedule chart (Figure 1.5), we describe each stage of the EVALPSN safety verification.

Initial stage: We assume that all subprocesses and valves in the pipeline network are free (unlocked) and no process has already started. In order to verify the safety for all processes Pr_0, Pr_1, Pr_2, and Pr_3, the following fact EVALP clauses (sensed information) are input to the safety verification EVALPSN:

$$SPr(t0, v0, 0) : [\mathtt{f}, \alpha], \qquad Val(v0, 0, 0) : [\mathtt{cs}, \alpha],$$

$$SPr(v0, t1, 0) : [\mathtt{f}, \alpha], \qquad Val(v1, 0, 0) : [\mathtt{cs}, \alpha],$$

$$SPr(v0, t2, 0) : [\mathtt{f}, \alpha],$$

$$SPr(v1, v0, 0) : [\mathtt{f}, \alpha],$$

$$SPr(t3, v1, 0) : [\mathtt{f}, \alpha],$$

$$Tan(t0, b) : [\mathtt{fu}, \alpha], \qquad Tan(t1, 0) : [\mathtt{em}, \alpha], \tag{34.20}$$

$$Tan(t2, 0) : [\mathtt{em}, \alpha], \qquad Tan(t3, na) : [\mathtt{fu}, \alpha]. \tag{34.21}$$

Then, all the subprocesses and valves are permitted to be locked or controlled. However, the tank states (34.20) and (34.21) do not permit for processes Pr_2 and Pr_3 to be set. Beer process Pr_0 can be verified to be set as follows.

We have neither the forbiddance from locking subprocesses SPr_0 and SPr_1 nor forbiddance from controlling valve V_0 separate with beer in normal direction, by EVALPSN clauses (34.6), (34.9), (34.10), and (34.11) and the input fact EVALP clauses.

Then, we have permission for locking subprocesses SPr_0 and SPr_1 and controlling valve V_0 separate with beer in normal direction and any liquid in cross direction:

$$SPr(t0, v0, b) : [\mathtt{f}, \gamma], \qquad Val(v0, b, l) : [\mathtt{cm}, \gamma],$$

$$SPr(v0, t1, b) : [\mathtt{f}, \gamma],$$

where $l \in \{b, cw, ww, hw, na, cs, 0\}$, by EVALPSN clauses (34.7) and (34.14).

Moreover, we have the following EVALP clauses for the tank state:

$$Tan(t0, b) : [\mathtt{fu}, \alpha], \quad Tan(t1, 0) : [\mathtt{em}, \alpha].$$

Thus, we have permission for setting beer process Pr_0,

$$Pr(p0, b) : [\mathtt{xs}, \gamma],$$

by EVALPSN clause (34.16).

According to the process schedule, beer process Pr_0 has to start first, then nitric acid process Pr_1 has to be verified for safety and processed parallel to beer process Pr_0 as soon as possible. We show the safety verification for process Pr_1 at the following stage.

2nd stage: Beer process Pr_0 has already started but not finished yet, then in order to verify the safety for other processes Pr_1, Pr_2, and Pr_3, the following fact EVALP clauses are input to the safety verification EVALPSN:

$$SPr(t0, v0, b) : [\mathtt{l}, \alpha], \quad Val(v0, b, 0) : [\mathtt{cs}, \alpha],$$

$$SPr(v0, t1, b) : [\mathtt{l}, \alpha], \quad Val(v1, 0, 0) : [\mathtt{cs}, \alpha],$$

$$SPr(v0, t2, 0) : [\mathtt{f}, \alpha],$$

$$SPr(v1, v0, 0) : [\mathtt{f}, \alpha],$$

$$SPr(t3, v1, 0) : [\mathtt{f}, \alpha],$$

$$Tan(t2, 0) : [\mathtt{em}, \alpha], \quad Tan(t3, na) : [\mathtt{fu}, \alpha].$$

The tank information permit neither cold water process Pr_2 nor beer process Pr_3 to be set. We show that only nitric acid process Pr_1 can be verified to be set as follows:

We have neither forbiddance from locking three subprocesses SPr_2, SPr_3, and SPr_4, forbiddance from controlling valves V_0 separate with any liquid in normal direction, and nitric acid in cross direction nor forbiddance from controlling valve V_1 mixture(open) with nitric acid in normal direction and no liquid in cross direction, by EVALPSN clauses (34.6), (34.8), (34.9), (34.10), (34.11), and (34.12), (34.13) and the earlier fact EVALP clauses.

Therefore, we have permission for locking three subprocesses SPr_2, SPr_3, and SPr_4 and controlling valves V_0 and V_1 as described before,

$$SPr(v0, t2, na) : [\mathtt{f}, \gamma], \quad Val(v0, b, na) : [\mathtt{cm}, \gamma],$$

$$SPr(v1, v0, na) : [\mathtt{f}, \gamma], \quad Val(v1, na, 0) : [\mathtt{cs}, \gamma],$$

$$SPr(t3, v1, na) : [\mathtt{f}, \gamma],$$

by EVALPSN clauses (34.7), (34.14) and (34.15).

Moreover, we have the following EVALP clauses for tank information:

$$Tan(t3, na) : [\mathtt{fu}, \alpha], \quad Tan(t2, 0) : [\mathtt{em}, \alpha].$$

Thus, we have permission for setting nitric acid process Pr_1,

$$Pr(p1, na) : [\mathtt{xs}, \gamma],$$

by EVALPSN clause (34.17).

Both beer process Pr_0 and nitric acid process Pr_1 have already started, then processes Pr_2 and Pr_3 have to be verified for safety. We show it at the following stage.

3rd stage: In order to verify the safety for cold water process Pr_2 and beer process Pr_3, the following fact EVALP clauses are input to the safety verification EVALPSN:

$$SPr(t0, v0, b) : [\mathtt{l}, \alpha], \qquad Val(v0, b, na) : [\mathtt{cs}, \alpha],$$

$$SPr(v0, t1, b) : [\mathtt{l}, \alpha], \qquad Val(v1, na, 0) : [\mathtt{cm}, \alpha],$$

$$SPr(v0, t2, na) : [\mathtt{l}, \alpha],$$

$$SPr(v1, v0, na) : [\mathtt{l}, \alpha],$$

$$SPr(t3, v1, na) : [\mathtt{l}, \alpha].$$

Apparently, neither cold water process Pr_2 nor beer process Pr_3 is permitted to be set, since there is no tank information in the input fact EVALP clauses. We show the safety verification for process Pr_2 as an example.

We have forbiddance from locking three subprocesses SPr_2, SPr_3, and SPr_4, forbiddance from controlling valve V_0 under separating with beer in normal direction and cold water in cross direction, and forbiddance from controlling valve V_1 under mixing with cold water in normal direction and no liquid in cross direction,

$$SPr(v0, t2, cw) : [\mathtt{f}, \beta], \qquad Val(v0, b, cw) : [\mathtt{cm}, \beta],$$

$$SPr(v1, v0, cw) : [\mathtt{f}, \beta], \qquad Val(v1, cw, 0) : [\mathtt{cs}, \beta],$$

$$SPr(t3, v1, cw) : [\mathtt{f}, \beta],$$

by EVALPSN clauses (34.6), (34.11), and (34.12), and the input fact EVALP clauses.

The finish condition for nitric acid process Pr_1 is that tank T_2 is fully filled with nitric acid and tank T_3 is empty. If nitric acid process Pr_1 has finished and its finish conditions are satisfied, process Pr_1 is permitted to be released (unset) and has been released. It is also assumed that tank T_3 is filled with cold water and tank T_2 is empty as preparation for cold water process Pr_2 immediately after process Pr_1 finished. Then, cold water process Pr_2 has to be verified and start according to the process schedule. We show the safety verification for cold water process Pr_2 at the following stage.

4th stage: Then, since only beer process Pr_0 is being processed, the other three processes Pr_1, Pr_2, and Pr_3 have to be verified. In order to do that, the following fact EVALP clauses are input to the safety verification EVALPSN:

$$SPr(t0, v0, b) : [\mathtt{l}, \alpha], \qquad Val(v0, b, 0) : [\mathtt{cs}, \alpha],$$

$$SPr(v0, t1, b) : [\mathtt{l}, \alpha], \qquad Val(v1, 0, 0) : [\mathtt{cs}, \alpha],$$

$$SPr(v0, t2, 0) : [\mathtt{f}, \alpha],$$

$$SPr(v1, v0, 0) : [\mathtt{f}, \alpha],$$

$$SPr(t3, v1, 0) : [\mathtt{f}, \alpha],$$

$$Tan(t2, 0) : [\mathtt{em}, \alpha], \qquad Tan(t3, cw) : [\mathtt{fu}, \alpha].$$

Since beer process Pr_0 is still being processed, neither nitric acid process Pr_1 nor beer process Pr_3 is permitted to be set, and only cold water process Pr_2 is permitted to be set as well as process Pr_1 at the 2nd stage. Therefore, we have permission for setting cold water process Pr_2,

$$Pr(p2, cw) : [\mathtt{xs}, \gamma],$$

by EVALPSN clause (34.18).

Now, both beer process Pr_0 and cold water process Pr_2 are being processed. Then apparently any other process is not permitted to be set. Moreover, even if one of processes Pr_0 and Pr_2 has finished, beer process Pr_3 is not permitted to be set until both processes Pr_0 and Pr_2 finished. We show the safety verification for beer process Pr_3 at the following stages.

5th stage: If neither beer process Pr_0 nor cold water process Pr_2 has finished, we have to verify the safety for nitric acid process Pr_1 and beer process Pr_3. The following fact EVALP clauses are input to the safety verification EVALPSN:

$$SPr(t0, v0, b) : [\mathtt{1}, \alpha], \qquad Val(v0, b, cw) : [\mathtt{cs}, \alpha],$$

$$SPr(v0, t1, b) : [\mathtt{1}, \alpha], \qquad Val(v1, cw, 0) : [\mathtt{cm}, \alpha],$$

$$SPr(v0, t2, cw) : [\mathtt{1}, \alpha],$$

$$SPr(v1, v0, cw) : [\mathtt{1}, \alpha],$$

$$SPr(t3, v1, cw) : [\mathtt{1}, \alpha].$$

Then, since all subprocesses and valves are locked and controlled, neither processes Pr_1 nor Pr_3 is permitted to be set. It is shown that beer process Pr_3 is not permitted to be set as follows:

We have forbiddance from locking three subprocesses SPr_2, SPr_3, and SPr_4 in process Pr_3 and controlling both valves V_0 and V_1,

$$SPr(v0, t2, b) : [\mathtt{f}, \beta], \qquad Val(v0, b, b) : [\mathtt{cs}, \beta],$$

$$SPr(t3, v1, b) : [\mathtt{f}, \beta], \qquad Val(v1, b, 0) : [\mathtt{cs}, \beta],$$

$$SPr(v1, v0, b) : [\mathtt{f}, \beta],$$

by EVALPSN clauses (34.6), (34.8), and (34.12), and the input fact EVALP clauses.

Therefore, we cannot have permission for setting process Pr_3.

The finish condition for cold water process Pr_2 is that tank T_2 is fully filled with cold water and tank T_3 is empty. If cold water process Pr_2 has finished and its finish condition is satisfied, process Pr_2 is permitted to be released (unset). It is also assumed that tank T_3 is filled with beer and tank T_2 is empty as preparation for beer process Pr_3 immediately after process Pr_2 finished. Then, beer process Pr_3 has to be verified and start according to the process schedule, but beer process Pr_3 cannot be permitted to be set. We show the safety verification for process Pr_3 at the following stage.

6th stage: Since only beer process Pr_0 is being processed, the other three processes Pr_1, Pr_2, and Pr_3 have to be verified for their safety. In order to do that, the following fact EVALP clauses are input to the safety verification EVALPSN:

$$SPr(t0, v0, b) : [\mathtt{1}, \alpha], \qquad Val(v0, b, 0) : [\mathtt{cs}, \alpha],$$

$$SPr(v0, t1, b) : [\mathtt{1}, \alpha], \qquad Val(v1, 0, 0) : [\mathtt{cs}, \alpha],$$

$$SPr(v0, t2, 0) : [\mathtt{f}, \alpha],$$

$$SPr(v1, v0, 0) : [\mathtt{f}, \alpha],$$

$$SPr(t3, v1, 0) : [\mathtt{f}, \alpha],$$

$$Tan(t2, 0) : [\mathtt{em}, \alpha], \qquad Tan(t3, b) : [\mathtt{fu}, \alpha].$$

Since beer process Pr_0 is still on being processed, beer process Pr_3 is not verified for safety due to the tank conditions and safety property *Val* for valve V_0. The safety verification is carried out as follows:

We have forbiddance from controlling valve V_0 mixture,

$$Val(v0, b, b) : [\mathtt{cs}, \beta],$$

by EVALPSN clause (34.8);

Therefore, we cannot have permission for setting the beer process Pr_3 then.

On the other hand, even if beer process Pr_0 has finished with cold water process Pr_2 being processed, beer process Pr_3 is not permitted to be set. If both processes Pr_0 and Pr_2 have finished, beer process Pr_3 is assured its safety and set. Then, process Pr_3 starts according to the process schedule. For brevity, we omit the remaining of stages.

34.4 Summary

In this chapter, we have introduced a paraconsistent annotated logic program called EVALPSN that can deal with defeasible deontic reasoning and its application to intelligent safety verification-based control with an example of pipeline valve control. EVALPSN can be applied to various kinds of safety verification and intelligent control. We summarize our chapter by showing the advantage and disadvantage of EVALPS safety verification and control.

Advantages

EVALPSN can be easily implemented in common programming languages such as Prolog or C language, moreover, if it is locally stratified [7,43], in a programmable logic controller (PLC) ladder program.

EVALPSN can be implemented as electronic circuits on micro chips [34]. Therefore, if real-time processing is required in the control system, EVALPSN should be very useful.

We have already another EVALPSN called bf(before–after)-EVALPSN, which can deal with before-after relation of processes, although we have not addressed bf-EVALPSN in this chapter. bf-EVALPSN can provide safety verification systems for process order control [35–37]. As far as we know, there seems to be no other efficient computational tool for dealing with safety verification of process order control than bf-EVALPSN. If we use both EVALPSN and bf-EVALPSN, safety verification systems for both process and its order control can be implemented under the same environment.

Disadvantages

Since basically EVALPSN itself is not a tool of logical safety verification, it includes a little complicated and redundant expressions for safety verification systems. Therefore, it might be better to develop safety verification-oriented tools or programming languages based on EVALPSN, if EVALPSN is applied to logical safety verification.

It is difficult to understand how to utilize EVALPSN fully to do practical implementation due to paraconsistent annotated logic.

References

1. D. Billington. Defeasible logic is stable. *J. Logic Computation*, 3(4):379–400, 1993.
2. H.A. Blair and V.S. Subrahmanian. Paraconsistent logic programming. *Theoret. Comp. Sci.*, 68:135–154, 1989.
3. N.C.A. da Costa et al. The paraconsistent logics $P\mathcal{T}$. *Zeitschrift für Mathematische Logic und Grundlangen der Mathematik*, 37:139–148, 1989.
4. O. Dressler. An extended basic ATMS. In *Proceedings of the 2nd Int'l Workshop on Nonmonotonic Reasoning*, Grassau, Germany. *Lecture Notes in Computer Science*, 346:143–163, Springer, 1988.
5. P. Geerts et al. Defeasible logics. In *Handbook of Defeasible Reasoning and Uncertainty Management Systems*, Vol. 2, pp. 175–210, Springer, 1998.
6. M. Gelfond and V. Lifschitz. The stable model semantics for logic programming. In *Proceedings of the 5th Int'l Conf. and Symp. Logic Programming(ICLP/SLP88)*, Seattle, WA, pp. 1070–1080, The MIT Press, 1989.
7. A.V. Gelder et al. The well-founded semantics for general logic programs. *J. ACM*, 38:620–650, 1991.
8. S. Jaskowski. Propositional calculus for contradictory deductive system (English translation of the original Polish paper). *Studia Logica*, 24:143–157, 1948.
9. M. Kifer and V.S. Subrahmanian. Theory of generalized annotated logic programming and its applications. *J. Logic Program.*, 12:335–368, 1992.
10. J.W. Lloyd. *Foundations of Logic Programming* (2nd edn.), Springer, 1987.
11. J-J.C. Meyer and R.J. Wiering (eds.) *Deontic Logic in Computer Science*, John Wiley & Sons, 1993.
12. R. Moore. Semantical considerations on nonmonotonic logic. *Art. Intell.*, 25:75–94, 1985.
13. K. Nakamatsu and A. Suzuki. Annotated semantics for default reasoning. In *Proceedings of the 3rd Pacific Rim Int'l Conf. Artificial Intelligence(PRICAI94)*, Beijing, China, pp. 180–186, International Academic Publishers, 1994.
14. K. Nakamatsu and A. Suzuki. A nonmonotonic ATMS based on annotated logic programs. *Lect. Notes Art. Int.*, 1441:79–93, 1998.
15. K. Nakamatsu and J.M. Abe. Reasonings based on vector annotated logic programs. *Computational Intelligence for Modelling, Control & Automation (CIMCA99)*, Vienna, Austria, Concurrent Systems Engineering Series, Vol. 55, pp. 396–403, IOS Press, 1999.
16. K. Nakamatsu, et al. Defeasible reasoning between conflicting agents based on VALPSN. In *Proceedings of the AAAI Workshop Agents' Conflicts*, Orlando, FL, pp. 20–27, AAAI Press, 1999.
17. K. Nakamatsu et al. Defeasible reasoning based on VALPSN and its application. In *Proceedings of the 3rd Australian Commonsense Reasoning Workshop*, pp. 114–130, 1999.
18. K. Nakamatsu. On the relation between vector annotated logic programs and defeasible theories. *Logic Logical Philos.*, 8:181–205, 2001.
19. K. Nakamatsu et al. A defeasible deontic reasoning system based on annotated logic programming. In *Proceedings of the 4th Int'l. Conf. Computing Anticipatory Systems(CASYS2000), AIP Conference Proceedings*, Liege, Belgium, Vol. 573, pp. 609–620, American Institute of Physics, 2001.
20. K. Nakamatsu et al. Annotated semantics for defeasible deontic reasoning. *Lect. Notes Art. Intell.*, 2005:432–440, 2001.
21. K. Nakamatsu et al. Extended vector annotated logic program and its application to robot action control and safety verification. *Hybrid Information Systems, Advances in Soft Computing Series*, pp. 665–680, IOS Press, 2002.
22. K. Nakamatsu et al. Paraconsistent logic program based safety verification for air traffic control. In *Proceedings of the IEEE Int'l Conf. System, Man and Cybernetics 02(SMC02)*, Yasmine Hammamet, Tunisia, CD-ROM, IEEE SMC, 2002.
23. K. Nakamatsu et al. A railway interlocking safety verification system based on abductive paraconsistent logic programming. *Soft Computing Systems(HIS02), Frontiers in Artificial Intelligence and Applications*, Santiago, Chile, Vol. 7, pp. 775–784, IOS Press, 2002.

24. K. Nakamatsu, J.M. Abe, and A. Suzuki. Defeasible deontic robot control based on extended vector annotated logic programming. In *Proceedings of the 5th International Conference on Computing Anticipatory Systems(CASYS2001), AIP Conference Proceedings*, Liege, Belgium, Vol. 627, pp. 490–500, American Institute of Physics, 2002.
25. K. Nakamatsu, Y. Mita, and T. Shibata. Defeasible deontic action control based on paraconsistent logic program and its hardware application. In *Proceedings of the International Conference on Computational Intelligence for Modelling Control and Automation 2003(CIMCA2003)*, Vienna, Austria, CD-ROM, IOS Press, 2003.
26. K. Nakamatsu, T. Seno, J.M. Abe, and A. Suzuki. Intelligent real time traffic signal control based on a paraconsistent logic program EVALPSN. In *Rough Sets, Fuzzy Sets, Data Mining and Granular Computing(RSFDGrC2003)*, Chongqing, China, *Lecture Notes in Artificial Intelligence*, Vol. 2639, pp. 719–723, Springer, 2003.
27. K. Nakamatsu, H. Komaba, and A. Suzuki. Defeasible deontic control for discrete events based on EVALPSN. In *Proceedings of the 4th International Conference on Rough Sets and Current Trends in Computing(RSCTC2004)*, Uppsala, Sweden, *Lecture Notes in Artificial Intelligence*, Vol. 3066, pp. 310–315, Springer, 2004.
28. K. Nakamatsu, R. Ishikawa, and A. Suzuki. A paraconsistent based control for a discrete event cat and mouse. In *Proceedings of the 8th International Conference on Knowledge-Based Intelligent Information and Engineering Systems(KES2004)*, Wellington, New Zealand, *Lecture Notes in Artificial Intelligence*, Vol. 3214, pp. 954–960, Springer, 2004.
29. K. Nakamatsu, S.-L. Chung, H. Komaba, and A. Suzuki. A discrete event control based on EVALPSN stable model. In *Rough Sets, Fuzzy Sets, Data Mining and Granular Computing(RSFDGrC2005)*, Regina, Canada, *Lecture Notes in Artificial Intelligence*, Vol. 3641, pp. 671–681, Springer, 2005.
30. K. Nakamatsu, K. Kawasumi, and A. Suzuki. Intelligent verification for pipeline based on EVALPSN. In *Advances in Logic Based Intelligent Systems, Frontiers in Artificial Intelligence and Applications*, Himeji, Japan, Vol. 132, pp. 63–70, IOS Press, 2005.
31. K. Nakamatsu and A. Suzuki. Autoepistemic theory and paraconsistent logic program. In *Advances in Logic Based Intelligent Systems, Frontiers in Artificial Intelligence and Applications*, Vol. 132, pp. 177–184, IOS Press, 2005.
32. K. Nakamatsu and A. Suzuki. Annotated semantics for nonmonotonic reasonings in artificial intelligence — I, II, III, IV. In *Advances in Logic Based Intelligent Systems, Frontiers in Artificial Intelligence and Applications*, Vol. 132, pp. 185–215, IOS Press, 2005.
33. K. Nakamatsu. Pipeline valve control based on EVALPSN safety verification. *J. Adv. Comput. Intell. Intelligent Informat.*, 10(3):647–656, 2006.
34. K. Nakamatsu, Y. Mita, and T. Shibata. An intelligent action control system based on extended vector annotated logic program and its hardware implementation. *J. Intell. Automat. Soft Comput.*, 13(3):222–237, 2007.
35. K. Nakamatsu. The paraconsistent annotated logic program EVALPSN and its application. *Computational Intelligence: A Compendium, Studies in Computational Intelligence*, Vol. 115, pp. 233–306, Springer, 2008.
36. K. Nakamatsu and J.M. Abe. The development of paraconsistent annotated logic programs. *J. Reasoning-Based Intelligent Syst.*, 1(1/2):92–112, 2009.
37. K. Nakamatsu, J.M. Abe, and S. Akama. A logical reasoning system of process before-after relation based on a paraconsistent annotated logic program bf-EVALPSN. *J. Knowl. Based Intell. Eng. Syst.*, 15(3):145–163, 2011.
38. D. Nute. Defeasible reasoning. In *Proceedings of the 20th Hawaii International Conference on System Science(HICSS87)*, Honolulu, HI, Vol. 1, pp. 470–477, 1987.
39. D. Nute. Basic defeasible logics. In *Intensional Logics for Programming*, pp. 125–154, Oxford University Press, 1992.

40. D. Nute. Apparent obligatory. *Defeasible Deontic Logic, Synthese Library*, Vol. 263, pp. 287–316, Kluwer Academic Publishers,1997.
41. T.C. Przymusinski. On the declarative semantics of deductive databases and logic programs. *Foundation of Deductive Database and Logic Programs*, pp. 193–216, Morgan Kaufmann, 1989.
42. R. Reiter. A logic for default reasoning. *Art. Intell.*, 13:81–123, 1980.
43. J.C. Shepherdson. Negation as failure, completion and stratification. *Handbook of Logic in Artificial Intelligence and Logic Programming*, Vol. 5, pp. 356–419, Oxford University Press, 1988.
44. V.S. Subrahmanian. On the semantics of qualitative logic programs. In *Proceedings of the 1987 Symposium Logic Programming(SLP87)*, San Francisco, CA, pp. 173–182, IEEE CS Press, 1987.
45. V.S. Subrahmanian. Amalgamating knowledge bases. *ACM Trans. Database Syst.*, 19:291–331, 1994.

35

Qualitative Reasoning

Kenneth Forbus
Northwestern University

35.1 Introduction

Qualitative reasoning is the area of artificial intelligence (AI) that creates representations for continuous aspects of the world, such as space, time, and quantity, which support reasoning with very little information. Typically, it has focused on scientific and engineering domains, hence its other common name, qualitative physics. It is motivated by two observations. First, people draw useful and subtle conclusions about the physical world without equations. In our daily lives, we figure out what is happening around us and how we can affect it, working with far less data, and less precise data, than would be required to use traditional, purely quantitative methods. Creating software for robots that operate in unconstrained environments and modeling human cognition require understanding how this can be done. Second, scientists and engineers appear to use qualitative reasoning when initially understanding a problem, when setting up more formal methods to solve particular problems, and when interpreting the results of quantitative simulations, calculations, or measurements. Thus, advances in qualitative reasoning should lead to the creation of more flexible software that can help engineers and scientists solve problems.

Qualitative reasoning began with de Kleer's investigation on how qualitative and quantitative knowledge interacted in solving a subset of simple textbook mechanics problems (de Kleer, 1977). After roughly a decade of initial explorations, the potential for important industrial applications led to a surge of interest in the mid-1980s, and the area grew steadily, with rapid progress. There have been a variety of successful applications, including supervisory process control, design, failure modes and effects analysis (FMEA), and educational software.

Given its demonstrated utility in applications and its importance in understanding human cognition, work in qualitative modeling will continue to remain an important area in AI.

This chapter first surveys the state of the art in qualitative representations and in qualitative reasoning techniques. The application of these techniques to various problems is discussed subsequently.

35.2 Qualitative Representations

As is the case with many other knowledge representation issues, there is no single best qualitative representation. Instead, there exists a spectrum of choices, each with its own advantages and disadvantages for particular tasks. What all of them have in common is that they provide notations for describing and reasoning about continuous properties of the physical world. Two key issues in qualitative representation are *resolution* and *compositionality*. We discuss each in turn.

Resolution concerns the level of information detail in a representation. Resolution is an issue because one goal of qualitative reasoning is to understand how little information suffices to draw useful conclusions. Low-resolution information is available more often than precise information ("the car heading toward us is slowing down" vs. "the derivative of the car's velocity along the line connecting us is −28 km/h/s"), but conclusions drawn with low-resolution information are often ambiguous. The role of ambiguity is important: the prediction of alternative futures (i.e., "the car will hit us" vs. "the car won't hit us") suggests that we may need to gather more information, analyze the matter more deeply, or take action, depending on what alternatives our qualitative reasoning uncovers. High-resolution information is often needed to draw particular conclusions (i.e., a finite element analysis of heat flow within a tablet computer design to ensure that the CPU will not cook the battery), but qualitative reasoning with low-resolution representations reveals what the interesting questions are. Qualitative representations comprise one form of tacit knowledge that people, ranging from the person on the street to scientists and engineers, use to make sense of the world.

Compositionality concerns the ability to combine representations for different aspects of a phenomenon or system to create a representation of the phenomenon or system as a whole. Compositionality is an issue because one goal of qualitative reasoning is to formalize the modeling process itself. Many of today's AI systems are based on handcrafted knowledge bases that express information about a specific artifact or system needed to carry out a particular narrow range of tasks involving it. By contrast, a substantial component of the knowledge of scientists and engineers consists of principles and laws that are broadly applicable, both with respect to the number of systems they explain and the kinds of tasks they are relevant for. Qualitative reasoning has developed ideas and organizing techniques for knowledge bases with similar expressive and inferential power, called *domain theories*.

The remainder of this section surveys the fundamental representations used in qualitative reasoning for quantity, mathematical relationships, modeling assumptions, causality, space, and time.

35.2.1 Representing Quantity

Qualitative reasoning has explored trade-offs in representations for continuous parameters ranging in resolution from sign algebras to the hyperreals. Most of the research effort has gone into understanding the properties of low-resolution representations because the properties of high-resolution representations tend to already be well understood due to work in mathematics.

The lowest resolution representation for continuous parameters is the status abstraction, which represents a quantity by whether or not it is normal. It is a useful representation for certain diagnosis and monitoring tasks because it is the weakest representation that can express the difference between something working and not working. The next step in resolution is the sign algebra, which represents continuous parameters as either −, +, or 0, according to whether the sign of the underlying continuous parameter is negative, positive, or zero. The sign algebra is surprisingly powerful: because a parameter's derivatives are themselves parameters whose values can be represented as signs, some of the main results of the differential calculus (e.g., the mean value theorem) can be applied to reasoning about sign values (de Kleer and Brown, 1984). This allows sign algebras to be used for qualitative reasoning about dynamics, including expressing properties such as oscillation and stability. The sign algebra is the weakest representation that supports such reasoning.

Representing continuous values via sets of ordinal relations (also known as the *quantity space* representation) is the next step up in resolution (Forbus, 1984). For example, the temperature of a fluid might be represented in terms of its relationship between the freezing point and boiling point of the material that comprises it. Like the sign algebra, quantity spaces are expressive enough to support qualitative reasoning about dynamics. (The sign algebra can be modeled by a quantity space with only a single point of comparison, zero.) Unlike the sign algebra, which draws values from a fixed finite algebraic structure, quantity spaces provide variable resolution because new points of comparison can be added to refine values. The temperature of water in a kettle on a stove, for instance, will likely be defined in terms of its relationship with the temperature of the stove as well as its freezing and boiling points. There are two kinds of comparison points used in defining quantity spaces. *Limit points* are derived from general properties of a domain as applicable to a specific situation. Continuing with the kettle example, the particular ordinal relationships used were chosen because they determine whether or not the physical processes of freezing, boiling, and heat flow occur in that situation. The precise numerical value of limit points can change over time (e.g., the boiling point of a fluid is a function of its pressure). *Landmark values* are constant points of comparison introduced during reasoning to provide additional resolution (Kuipers, 1994). To ascertain whether an oscillating system is overdamped, underdamped, or critically damped, for instance, requires comparing successive peak values. Noting the peak value of a particular cycle as a landmark value, and comparing it to the landmarks generated for successive cycles in the behavior, provides a way of making this inference.

Intervals are a well-known variable-resolution representation for numerical values and have been heavily used in qualitative reasoning. A quantity space can be thought of as partial information about a set of intervals. If we have complete information about the ordinal relationships between limit points and landmark values, these comparisons define a set of intervals that partition a parameter's value. This natural mapping between quantity spaces and intervals has been exploited by a variety of systems that use intervals whose end points are known numerical values to refine predictions produced by purely qualitative reasoning. Fuzzy intervals have also been used in similar ways, for example, in reasoning about control systems.

Order of magnitude representations stratify values according to some notion of scale. They can be important in resolving ambiguities and in simplifying models because they enable reasoning about what phenomena and effects can safely be ignored in a given situation. For instance, heat losses from turbines are generally ignored in the early stages of power plant design, because the energy lost is very small relative to the energy being produced. Several stratification techniques have been used in the literature, including hyperreal numbers, numerical thresholds, and logarithmic scales. Three issues faced by all these formalisms are (1) the conditions under which many small effects can combine to produce a significant effect, (2) the soundness of the reasoning supported by the formalism, and (3) the efficiency of using them (Travé-Massuyès et al., 2003).

Although many qualitative representations of number use the reals as their basis, another important basis for qualitative representations of number is finite algebras. One motivation for using finite algebras is that observations are often naturally categorized into a finite set of labels (i.e., very small, small, normal, large, very large). Research on such algebras is aimed at solving problems such as how to increase the compositionality of such representations (e.g., how to propagate information across different resolution scales).

35.2.2 Representing Mathematical Relationships

Like number, a variety of qualitative representations of mathematical relationships have been developed, often by adopting and adapting systems developed in mathematics. Abstractions of the analytic functions are commonly used to provide the lower resolution and compositionality desired. For example, confluences are differential equations over the sign algebra (de Kleer and Brown, 1984). An equation such as $V = I R$ can be expressed as the confluence

$$[V] = [I] + [R]$$

where $[Q]$ denotes taking the sign of Q. Differential equations can also be expressed in this manner, for instance,

$$[F] = \partial v$$

which is a qualitative version of $F = M A$ (assuming M is always positive). Thus, any system of algebraic and differential equations with respect to time can be described as a set of confluences.

Many of the algebraic operations taken for granted in manipulating analytic functions over the reals are not valid in weak algebras (Struss, 1988). Because qualitative relationships are most often used to propagate information, this is not a serious limitation. In situations where algebraic solutions themselves are desirable, mixed representations that combine algebraic operations over the reals and move to qualitative abstractions when appropriate provide a useful approach (Williams, 1991).

Another low-resolution representation of equations uses monotonic functions over particular ranges; that is,

```
M+(acceleration, force)
```

states that acceleration depends only on the force, and the function relating them is increasing monotonic (Kuipers, 1994). Compositionality is achieved using *qualitative proportionalities* (Forbus, 1984) to express partial information about functional dependency; for example,

```
acceleration αQ+ force
```

states that acceleration depends on force and is increasing monotonic in its dependence on force but may depend on other factors as well. Additional constraints on the function that determines force can be added by additional qualitative proportionalities; for example,

```
acceleration αQ- mass
```

states that acceleration also depends on mass and is decreasing monotonic in this dependence. Qualitative proportionalities must be combined via closed-world assumptions to ascertain all the effects on a quantity. Similar primitives can be defined for expressing relationships involving derivatives, to define a complete language of compositional qualitative mathematics for ordinary differential equations. As with confluences, few algebraic operations are valid for combining monotonic functions. Composition of functions of identical sign is safe; that is,

```
M+(f,g) ∧ M+(g,h) ⇒ M+(f,h)
```

In addition to resolution and compositionality, another issue arising in qualitative representations of mathematical relationships is causality. There are three common views on how mathematical relationships interact with the causal relationships people use in commonsense reasoning. One view is that there is no relationship between them. The second view is that mathematical relationships should be expressed with primitives that also make causal implications. For example, qualitative proportionalities include a causal interpretation, that is, a change in force causes a change in acceleration, but not the other way around. The third view is that acausal mathematical relationships give rise to causal relationships via the particular process of using them. For example, confluences have no built-in causal direction but are used in causal reasoning by identifying the flow of information through them while reasoning with a presumed flow of causality in the physical system they model. One method for imposing causality on a set of acausal constraint equations is by computing a causal ordering (Iwasaki and Simon, 1994) that imposes directionality on a set of equations, starting from variables considered to be exogenous within the system.

Each view of causality has its merits. For tasks where causality is truly irrelevant, ignoring causality can be a reasonable approach. To create software that can span the range of human commonsense reasoning, something like a combination of the second and third views appears necessary because the appropriate notion of causality varies (Forbus and Gentner, 1986). In reasoning about chemical phenomena, for instance, changes in concentration are always caused by changes in the amounts of the constituent parts and never the other way around. In electronics, on the other hand, it is often convenient to consider voltage changes as being caused by changes in current in one part of a circuit and to consider current changes as being caused by changes in voltage in another part of the same circuit.

35.2.3 Ontology

Ontology concerns how to carve up the world, that is, what kinds of things there are and what sorts of relationships can hold between them. Ontology is central to qualitative modeling because one of its main goals is formalizing the art of building models of continuous systems. A key choice in any act of modeling is figuring out how to construe the situation or system to be modeled in terms of the available models for classes of entities and phenomena. No single ontology will suffice for the span of reasoning about continuous systems that people do. What has been developed instead is a catalog of ontologies, describing their properties and interrelationships and specifying conditions under which each is appropriate. While several ontologies are currently well understood, the catalog still contains many gaps.

An example of ontologies will make this point clearer. Consider the representation of liquids. Broadly speaking, the major distinction in reasoning about fluids is whether one individuates fluid according to a particular collection of particles or by location (Hayes, 1985). The former are called Eulerian, or piece of stuff, ontologies. The latter are called Lagrangian, or contained stuff, ontologies. The contained stuff view of liquids is used when one treats a river as a stable entity, although the particular set of molecules that comprises it is changing constantly. The piece of stuff view of liquids is used when one thinks about the changes in a fluid as it flows through a steady-state system, such as a working refrigerator. Ontologies multiply as we try to capture more aspects of human reasoning. For instance, the piece of stuff ontology can be further divided into three cases, each with its own rules of inference: (1) molecular collections, which describe the progress of an arbitrary piece of fluid that is small enough to never split apart but large enough to have extensive properties; (2) slices, which, like molecular collections, never subdivide but unlike them are large enough to interact directly with their surroundings; and (3) pieces of stuff large enough to be split into several pieces (e.g., an oil slick). Similarly, the contained stuff ontology can be further specialized according to whether or not individuation occurs simply by container (abstract-contained stuffs) or by a particular set of containing surfaces (bounded stuffs). Abstract-contained stuffs provide a low-resolution ontology appropriate for reasoning about system-level properties in complex systems (e.g., the changes over time in a lubricating oil subsystem in a propulsion plant), whereas bounded stuffs contain the geometric information needed to reason about the interactions of fluids and shape in systems such as pumps and internal combustion engines.

Cutting across the ontologies for particular physical domains are systems of organization for classes of ontologies. The most commonly used ontologies are the device ontology (de Kleer and Brown, 1984) and the process ontology (Forbus, 1984). The device ontology is inspired by network theory and system dynamics. Like those formalisms, it construes continuous systems as networks of devices whose interactions occur solely through a fixed set of ports. Unlike those formalisms, it provides the ability to write and reason automatically with device models whose governing equations can change over time.

The process ontology is inspired by studies of human mental models and observations of practice in thermodynamics and chemical engineering. It construes continuous systems as consisting of entities whose changes are caused by processes. Process ontologies thus postulate a separate ontological category for causal mechanisms, unlike device ontologies, where causality arises solely from the interaction of the parts. Another difference between the two classes of ontologies is that in the device ontology the system of devices and connections is fixed over time, whereas in the process ontology entities and processes can come into existence and vanish over time. Each is appropriate in different contexts: for most purposes, an electronic circuit is best modeled as a network of devices, whereas a chemical plant is best modeled as a collection of interacting processes.

35.2.4 State, Time, and Behaviors

A qualitative state is a set of propositions that characterize a qualitatively distinct behavior of a system. A qualitative state describing a falling ball, for instance, would include information about what physical processes are occurring (e.g., motion downward, acceleration due to gravity) and how the parameters of the ball are changing (e.g., its position is getting lower and its downward velocity is getting larger). A qualitative state can abstractly represent an infinite number of quantitative states: although the position and velocity of the ball are different at each distinct moment during its fall, until the ball collides with the ground, the qualitative state of its motion is unchanged.

Qualitative representations can be used to partition behavior into natural units. For instance, the time over which the state of the ball falling holds is naturally thought of as an interval, ending when the ball collides with the ground. The collision itself can be described as yet another qualitative state, and the fact that falling leads to a collision with the ground can be represented via a transition between the two states. If the ball has a nonzero horizontal velocity and there is some obstacle in its direction of travel, another possible behavior is that the ball will collide with that object instead of the ground. In general, a qualitative state can have transitions to several next states, reflecting ambiguity in the qualitative representations. Returning to our ball example, and assuming that no collisions with obstacles occur, notice that the qualitative state of the ball falling occurs again once the ball has reached its maximum height after the collision. If continuous values are represented by quantity spaces and the sources of comparisons are limit points, then a finite set of qualitative states is sufficient to describe every possible behavior of a system. A collection of such qualitative states and transitions is called an *envisionment* (de Kleer, 1977). Many interesting dynamical conclusions can be drawn from an envisionment. For instance, oscillations correspond to cycles of states. Unfortunately, the fixed resolution provided by limit points is not sufficient for other dynamical conclusions, such as ascertaining whether or not the ball's oscillation is damped. If comparisons can include landmark values, such conclusions can sometimes be drawn, for example, by comparing the maximum height on one bounce to the maximum height obtained on the next bounce. The cost of introducing landmark values is that the envisionment no longer need be finite; every cycle in a corresponding fixed-resolution envisionment could give rise to an infinite number of qualitative states in an envisionment with landmarks.

A sequence of qualitative states occurring over a particular span of time is called a *behavior*. Behaviors can be described using purely qualitative knowledge, purely quantitative knowledge, or a mixture of both. If every continuous parameter is quantitative, the numerical aspects of behaviors coincide with the notion of trajectory in a state-space model. If qualitative representations of parameters are used, a single behavior can represent a family of trajectories through state space.

An idea closely related to behaviors is *histories* (Hayes, 1985). Histories can be viewed as local behaviors, that is, how a single individual or property varies through time. A behavior is equivalent to a global history, that is, the union of all the histories for the participating individuals. The distinction is important for two reasons. First, histories are the dual of situations in the situation calculus; histories are bounded spatially and extended temporally, whereas situations are bounded temporally and extended spatially. Using histories avoids the frame problem, instead trading it for the more tractable problems of generating histories locally and determining how they interact when they intersect in space and time. The second reason is that history-based simulation algorithms can be more efficient than state-based simulation algorithms because no commitments need to be made concerning irrelevant information.

In a correct envisionment, every possible behavior of the physical system corresponds to some path through the envisionment. Because envisionments reflect only local constraints, the converse is not true; that is, an arbitrary path through an envisionment may not represent a physically possible behavior. When using an envisionment to generate hypotheses about possible behaviors, this can lead to over-generation. Paths must be tested against global constraints, such as energy conservation, to ensure their physical validity, or filtered by some other means. When using an envisionment to explain observed behaviors, this limitation is less serious, since the observed behavior, by its very nature, must be physically possible. A more serious limitation is that envisionments are often exponential in the size of the system being modeled. This means that, in practice, envisionments often are not generated explicitly and, instead, possible behaviors are searched in ways similar to those used in other areas of AI.

Many tasks require integrating qualitative states with other models of time, such as numerical models. Including precise information (e.g., algebraic expressions or floating-point numbers) about the end points of intervals in a history does not change their essential character.

35.3 Space and Shape

Qualitative representations of shape and space play an important role in spatial cognition because they provide a bridge between the perceptual and the conceptual. By discretizing continuous space, they make it amenable to symbolic reasoning. As with qualitative representations of 1D parameters, task constraints govern the choice of qualitative representation. However, problem-independent purely qualitative spatial representations suffice for fewer tasks than in the 1D case, because of the increased ambiguity in higher dimensions (Forbus et al., 1991). Consider, for example, deciding whether a protrusion can fit snugly inside a hole. If we have detailed information about their shapes, we can derive an answer. If we consider a particular set of protrusions and a particular set of holes, we can construct a qualitative representation of these particular protrusions and holes that would allow us to derive whether or not a specific pair would fit, based on their relative sizes. But if we first compute a qualitative representation for each protrusion and hole in isolation, in general the rules of inference that can be derived for this problem will be very weak. Work in qualitative spatial representations thus tends to take two approaches. The first approach is to explore what aspects do lend themselves to purely qualitative representations. The second approach is to use a quantitative representation as a starting point and compute problem-specific qualitative representations to reason with. We summarize each in turn.

An impressive catalog of qualitative spatial representations have been developed that can be used to reason about particular aspects of shape and space (Cohn and Renz, 2008). The general approach is to identify a set of relationships that provide natural decompositions of a spatial property, such as connectivity or orientation, and define calculi over these relationships to support inference. For example, the eight relationship versions of the region connection calculus (RCC) include disjoint, edge connected, partially overlapping, tangential proper part (and its inverse), non-tangential proper part (and its inverse), and equal as relationships. A transitivity table supports inference, that is, if a bird is inside a cage and a cat is outside the cage, then the bird and the cat are disjoint (and unless the connectivity changes, the bird is safe). Note that the sequence of relationships earlier also captures some aspects of the relationships between two objects that are moving: they are apart, they touch, one enters the other, etc. Such physical interpretations have enabled such calculi to be used in reasoning about change and in interpreting visual information (Sridhar et al., 2010). The calculi vary in terms of expressiveness and efficiency: some have been shown to be NP-complete, but often tractable subsets can be found.

The use of quantitative representations to ground qualitative spatial reasoning can be viewed as a model of the ways humans use diagrams and models in spatial reasoning. For this reason such work is also known as *diagrammatic reasoning* (Glasgow et al., 1995). One form of diagram representation is the occupancy array that encodes the location of an object by cells in a (2D or 3D) grid. These representations simplify the calculation of spatial relationships between objects (e.g., whether or not one object is above another), albeit at the cost of making the object's shape implicit. Another form of diagram representation uses symbolic structures with quantitative properties, for example, numerical, algebraic, or interval (e.g., Forbus et al., 1991). These representations simplify calculations involving shape and spatial relationships, without the scaling and resolution problems that sometimes arise in array representations. However, they require a set of primitive shape elements that spans all the possible shapes of interest, and identifying such sets for particular tasks can be difficult. For instance, many intuitively natural sets of shape primitives are not closed with respect to their complement, which can make characterizing free space difficult.

Diagram representations are used for qualitative spatial reasoning in two ways. The first is as a decision procedure for spatial questions. This mimics one of the roles diagrams play in human perception. Often, these operations are combined with domain-specific reasoning procedures to produce an analog style of inference, where, for instance, the effects of perturbations on a structure are mapped into the diagram, the effect on the shapes in the diagram noted, and the results mapped back into a physical interpretation. The second way uses the diagram to construct a problem-specific qualitative vocabulary, imposing new spatial entities representing physical properties, such as the maximum height a ball can reach or regions of free space that can contain a motion. This is the *metric diagram/place vocabulary* model of qualitative spatial reasoning.

Representing and reasoning about kinematic mechanisms was one of the early successes in qualitative spatial reasoning. The possible motions of objects are represented by qualitative regions in the configuration space representing the legitimate positions of parts of mechanisms (Faltings, 1990). Whereas, in principle, a single high-dimensional configuration space could be used to represent a mechanism's possible motions (each dimension corresponding to a degree of freedom of a part of the mechanism), in practice a collection of configuration spaces, one 2D space for each pair of parts that can interact, is used. These techniques suffice to analyze a wide variety of kinematic mechanisms and can support design (Sacks and Joscowicz, 2010).

Another important class of spatial representations concerns qualitative representations of spatially distributed phenomena, such as flow structures and regions in phase space. These models use techniques from computer vision to recognize or impose qualitative structure on a continuous field of information, gleaned from numerical simulation or scientific data. This qualitative structure, combined with domain-specific models of how such structures tie to the underlying physics, enables them to interpret physical phenomena in much the same way that a scientist examining the data would (e.g., Yip, 1991; Huang and Zhao, 2000).

An important recent trend is using rich, real-world data as input for qualitative spatial reasoning. For example, several systems provide some of the naturalness of sketching by performing qualitative reasoning on spatial data input as digital ink, for tasks like mechanical design (Stahovich et al., 1998), reasoning about student's sketches in geoscience and engineering design education (Forbus et al., 2011), and communicating tactics to robot soccer teams (Gspandl et al., 2010). Qualitative representations are being used in computer vision as well, for example, as a means of combining dynamic scenes across time to interpret events (e.g., Sridhar et al., 2010).

35.3.1 Compositional Modeling, Domain Theories, and Modeling Assumptions

There is almost never a single correct model for a complex physical system. Most systems can be modeled in a variety of ways, and different tasks can require different types of models. The creation of a system model for a specific purpose is still something of an art. Qualitative reasoning has developed formalisms that combine logic and mathematics with qualitative representations to help automate the process of creating and refining models. The compositional modeling methodology (Falkenhainer and Forbus, 1991), which has become standard in qualitative reasoning, works like this: models are created from domain theories, which describe the kinds of entities and phenomena that can occur in a physical domain. A domain theory consists of a set of *model fragments*, each describing a particular aspect of the domain. Creating a model is accomplished by instantiating an appropriate subset of model fragments, given some initial specification of the system (e.g., the propositional equivalent of a blueprint) and information about the task to be performed. Reasoning about appropriateness involves the use of modeling assumptions. Modeling assumptions are the control knowledge used to reason about the validity or appropriateness of using model fragments. Modeling assumptions are used to express the relevance of model fragments. Logical constraints between modeling assumptions comprise an important component of a domain theory.

An example of a modeling assumption is assuming that a turbine is isentropic. Here is a model fragment that illustrates how this assumption is used:

```
(defEquation Isentropic-Turbine
   ((turbine ?g ?in ?out)(isentropic ?g))
   (= (spec-s ?in) (spec-s ?out)))
```

In other words, when a turbine is isentropic, the specific entropy of its inlet and outlet are equal. Other knowledge in the domain theory puts constraints on the predicate isentropic,

```
(for-all (?self (turbine ?self))
   (iff (= (nu-isentropic ?self) 1.0) (isentropic ?self)))
```

That is, a turbine is isentropic exactly when its isentropic thermal efficiency is 1. Although no real turbine is isentropic, assuming that turbines are isentropic simplifies early analyses when creating a new design. In later design phases, when tighter performance bounds are required, this assumption is retracted and the impact of particular values for the turbine's isentropic thermal efficiency is explored. The consequences of choosing particular modeling assumptions can be quite complex; the fragments shown here are less than one-fourth the knowledge expressing the consequences of assuming that a turbine is isentropic in a typical knowledge base.

Modeling assumptions can be classified in a variety of ways. An *ontological assumption* describes which ontology should be used in an analysis. For instance, reasoning about the pressure at the bottom of a swimming pool is most simply performed using a contained stuff representation, whereas describing the location of an oil spill is most easily performed using a piece of stuff representation. A *perspective assumption* describes which subset of phenomena operating in a system will be the subject. For example, in analyzing a steam plant, one might focus on a fluid perspective, a thermal perspective, or both at once. A *granularity assumption* describes how much detail is included in an analysis. Ignoring the implementation details of subsystems, for instance, is useful in the conceptual design of an artifact, but the same implementation details may be critical for troubleshooting that artifact. The relationships between these classes of assumptions can be complicated and domain dependent; for instance, it makes no sense to include a model of a heating coil (a choice of granularity) if the analysis does not include thermal properties (a choice of perspective).

Relationships between modeling assumptions provide global structure to domain theories. Assumptions about the nature of this global structure can significantly impact the efficiency of model formulation, as discussed in the following. In principle, any logical constraint could be imposed between modeling assumptions. In practice, two kinds of constraints are the most common. The first are implications, such as one modeling assumption requiring or forbidding another. For example,

```
(for-all (?s (system ?s))
  (implies (consider (black box ?s))
           (for-all (?p (part-of ?p ?s)) (not (consider ?p)))))
```

says that if one is considering a subsystem as a black box, then all of its parts should be ignored. Similarly,

```
(for-all (?l (contained-stuff ?l))
   (implies (consider (pressure ?l))
            (consider (fluid-properties ?l))))
```

states that if an analysis requires considering something's pressure, then its fluid properties are relevant. The second kind of constraint between modeling assumptions is *assumption classes*. An assumption class expresses a choice required to create a coherent model under particular conditions. For example,

```
(defAssumptionClass (turbine ?self)
  (isentropic ?self) (not (isentropic ?self)))
```

states that when something is modeled as a turbine, any coherent model including it must make a choice about whether or not it is modeled as isentropic. The choice may be constrained by the data so far (e.g., different entrance and exit specific entropies), or it may be an assumption that must be made in order to complete the model. The set of choices need not be binary. For each valid assumption class, exactly one of the choices it presents must be included in the model.

35.4 Qualitative Reasoning Techniques

A variety of qualitative reasoning techniques have been developed that use the qualitative representations just outlined.

35.4.1 Model Formulation

Methods for automatically creating models for a specific task are one of the hallmark contributions of qualitative reasoning. These methods formalize knowledge and skills typically left implicit by most of traditional mathematics and engineering.

The simplest model formulation algorithm is to instantiate every possible model fragment from a domain theory, given a propositional representation of the particular scenario to be reasoned about. This algorithm is adequate when the domain theory is very focused and thus does not contain much irrelevant information. It is inadequate for broad domain theories and fails completely for domain theories that include alternative and mutually incompatible perspectives (e.g., viewing a contained liquid as a finite object vs. an infinite source of liquid). It also fails to take task constraints into account. For example, it is possible, in principle, to analyze the cooling of a cup of coffee using quantum mechanics. Even if it were possible in practice to do so, for most tasks simpler models suffice. Just how simple a model can be and remain adequate depends on the task. If I want to know if the cup of coffee will still be drinkable after an hour, a qualitative model suffices to infer that its final temperature will be that of its surroundings. If I want to know its temperature within 5% after 12 min have passed, a macroscopic quantitative model is a better choice. In other words, the goal of model formulation is to create the simplest adequate model of a system for a given task.

More sophisticated model formulation algorithms search the space of modeling assumptions, because they control which aspects of the domain theory will be instantiated. The model formulation algorithm of Falkenhainer and Forbus (1991) instantiated all potentially relevant model fragments and used an assumption-based truth maintenance system to find all legal combinations of modeling assumptions that sufficed to form a model that could answer a given query. The simplicity criterion used was to minimize the number of modeling assumptions. This algorithm is very simple and general but has two major drawbacks: (1) full instantiation can be very expensive, especially if only a small subset of the model fragments is eventually used; and (2) the number of consistent combinations of model fragments tends to be exponential for most problems. The rest of this section describes algorithms that overcome these problems.

Efficiency in model formulation can be gained by imposing additional structure on domain theories. Under at least one set of constraints, model formulation can be carried out in polynomial time (Nayak, 1994). The constraints are that (1) the domain theory can be divided into independent assumption classes and, (2) within each assumption class, the models can be organized by a (perhaps partial) simplicity ordering of a specific nature, forming a lattice of causal approximations. Nayak's algorithm computes a simplest model, in the sense of simplest within each local assumption class, but does not necessarily produce the globally simplest model.

Conditions that ensure the creation of coherent models, that is, models that include sufficient information to produce an answer of the desired form, provide powerful constraints on model formulation. For example, in generating "what-if" explanations of how a change in one parameter might affect particular other properties of the system, a model must include a complete causal chain connecting the changed parameter to the other parameters of interest. This insight can be used to treat model formulation as a best-first search for a set of model fragments providing the simplest complete causal chain (Rickel and Porter, 1994). A novel feature of this algorithm is that it also selects models at an appropriate timescale. It does this by choosing the slowest timescale phenomenon that provides a complete causal model, because this provides accurate answers that minimize extraneous detail.

As with other AI problems, knowledge can reduce search. One kind of knowledge that experienced modelers accumulate concerns the range of applicability of various modeling assumptions and strategies for how to reformulate when a given model proves inappropriate. Dynamically constructing qualitative representations from more detailed models, driven by task requirements, has been explored (Sachenbacher and Struss, 2005). Model formulation often is an iterative process. For instance, an initial qualitative model often is generated to identify the relevant phenomena, followed by the creation of a narrowly focused quantitative model to answer the questions at hand. Similarly, domain-specific error criterion can determine that a particular model's results are internally inconsistent, causing the reasoner to restart the search for a good model. One approach is to formalize model formulation as a dynamic preference constraint satisfaction problem, where more fine-grained criteria for model preference than "simplest" can be formalized and exploited (Keppens and Shen, 2004). Another promising approach is analogical model formulation, since similar models are likely to work in similar circumstances (Klenk et al., 2011).

35.4.2 Causal Reasoning

Causal reasoning explains an aspect of a situation in terms of others in such a way that the aspect being explained can be changed if so desired. For instance, a flat tire is caused by the air inside flowing out, either through the stem or through a leak. To refill the tire, we must both ensure that the stem provides a seal and that there are no leaks. Causal reasoning is thus at the heart of diagnostic reasoning as well as explanation generation.

The techniques used for causal reasoning depend on the particular notion of causality used, but they all share a common structure. First, causality involving factors within a state are identified. Second, how the properties of a state contribute to a transition (or transitions) to another state are identified, to extend the causal account over time. Because causal reasoning often involves qualitative simulation, we turn to simulation next.

35.4.3 Simulation

The new representations of quantity and mathematical relationships of qualitative reasoning expand the space of simulation techniques considerably. We start by considering varieties of purely qualitative simulation and then describe several simulation techniques that integrate qualitative and quantitative information.

Understanding *limit analysis*, the process of finding state transitions, is key to understanding qualitative simulation. Recall that a qualitative state consists of a set of propositions, some of them describing the values of continuous properties in the system. (For simplicity in this discussion, we will assume that these values are described as ordinal relations, although the same method works for sign representations and representations richer than ordinals.) Two observations are critical: (1) the phenomena that cause changes in a situation often depend on ordinal relationships between parameters of the situation, and (2) knowing just the sign of the derivatives of the parameters involved in these ordinal relationships suffices to predict how they might change over time. The effects of these changes, when calculated consistently, describe the possible transitions to other states.

An example will make this clearer. Consider again a pot of water sitting on a stove. Once the stove is turned on, heat begins to flow to the water in the pot because the stove's temperature is higher than that of the water. The causal relationship between the temperature inequality and the flow of heat means that to predict changes in the situation, we should figure out their derivatives and any other relevant ordinal relationships that might change as a result. In this qualitative state, the derivative of the water's temperature is positive, and the derivative of the stove's temperature is zero. Thus, one possible state change is that the water will reach thermal equilibrium with the stove and the flow of heat will stop. That is not the only possibility, of course. We know that boiling can occur if the temperature of the water begins to rise above its boiling temperature. That, too, is a possible transition that would end the state. Which of these transitions occurs depends on the relationship between the temperature of the stove and the boiling temperature of water.

This example illustrates several important features of limit analysis. First, surprisingly weak information (i.e., ordinal relations) suffices to draw important conclusions about broad patterns of physical behavior. Second, limit analysis with purely qualitative information is fundamentally ambiguous: it can identify what transitions might occur, but cannot by itself determine in all cases which transition will occur. Third, like other qualitative ambiguities, higher-resolution information can be brought in to resolve the ambiguities as needed. Returning to our example, any information sufficient to determine the ordinal relationship between the stove temperature and boiling suffices to resolve this ambiguity. If we are designing an electric kettle, for instance, we would use this ambiguity as a signal that we must ensure that the heating element's temperature is well above the boiling point; and if we are designing a drink warmer, its heating element should operate well below the boiling point.

Qualitative simulation algorithms vary along four dimensions: (1) their initial states, (2) what conditions they use to filter states or transitions, (3) whether or not they generate new landmarks, and (4) how much of the space of possible behaviors they explore. Envisioning is the process of generating an envisionment, that is, generating all possible behaviors. Two kinds of envisioning algorithms have been used in practice: attainable envisioners produce all states reachable from a set of initial states, and total envisioners produce a complete envisionment. Behavior generation algorithms start with a single initial state, generate landmark values, and use a variety of task-dependent constraints as filters and termination criteria (e.g., resource bounds, energy constraints).

Higher-resolution information can be integrated with qualitative simulation in several ways. One method for resolving ambiguities in behavior generation is to provide numerical envelopes to bound mathematical relationships. These envelopes can be dynamically refined to provide tighter situation-specific bounds. Such systems are called *semiquantitative* simulators (Kuipers, 1994).

A different approach to integration is to use qualitative reasoning to automatically construct a numerical simulator that has integrated explanation facilities. These *self-explanatory simulators* (Forbus and

Falkenhainer, 1990) use traditional numerical simulation techniques to generate behaviors, which are also tracked qualitatively. The concurrently evolving qualitative description of the behavior is used both in generating explanations and in ensuring that appropriate mathematical models are used when applicability thresholds are crossed. Self-explanatory simulators can be compiled in polynomial time for efficient execution, even on small computers, or created in an interpreted environment. They have been deployed mostly in education, both as a web-based medium and within classroom activities for middle school students.

35.4.4 Comparative Analysis

Comparative analysis answers a specific kind of "what-if" questions, namely, the changes that result from changing the value of a parameter in a situation. Given higher-resolution information, traditional analytic or numerical sensitivity analysis methods can be used to answer these questions; however (1) such reasoning is commonly carried out by people who have neither the data nor the expertise to carry out such analyses, and (2) purely quantitative techniques tend not to provide good explanations. Sometimes, purely qualitative information suffices to carry out such reasoning, using techniques such as exaggeration (Weld, 1990). Consider, for instance, the effect of increasing the mass of a block in a spring-block oscillator. If the mass were infinite, the block would not move at all, corresponding to an infinite period. Thus, we can conclude that increasing the mass of the block will increase the period of the oscillator.

35.4.5 Teleological Reasoning

Teleological reasoning connects the structure and behavior of a system to its goals. (By its goals, we are projecting the intent of its designer or the observer, because purposes often are ascribed to components of evolved systems.) To describe how something works entails ascribing a function to each of its parts and to explain how these functions together achieve the goals. Teleological reasoning is accomplished by a combination of abduction and recognition. Abduction is necessary because most components and behaviors can play several functional roles (de Kleer, 1984). A turbine, for instance, can be used to generate work in a power generation system and to expand a gas in a liquefaction system. Recognition is important because it explains patterns of function in a system in terms of known, commonly used abstractions. A complex power generation system with multiple stages of turbines and reheating and regeneration, for instance, still can be viewed as a Rankine cycle after the appropriate aggregation of physical processes involved in its operation (Everett, 1999).

35.4.6 Data Interpretation

There are two ways that the representations of qualitative reasoning have been used in data interpretation problems. The first is to explain a temporal sequence of measurements in terms of a sequence of qualitative states; the second is to create a qualitative model of spatially distributed numerical data, for example, phase spaces, by interpreting the results of empirical data measurements or successive numerical simulation experiments in terms of regions whose qualitative properties are constant. The underlying commonality in these problems is the use of qualitative descriptions of physical constraints to formulate compatibility constraints that prune the set of possible interpretations. We summarize each in turn.

In measurement interpretation tasks, numerical and symbolic data are partitioned into intervals, each of which can be explained by a qualitative state or sequence of qualitative states. Using precomputed envisionments or performing limit analysis online, possible transitions between states used as interpretations can be found for filtering purposes. Specifically, if a state S_1 is a possible interpretation for interval I_1, then at least one transition from S_1 must lead to a state that is an interpretation for the next interval.

This compatibility constraint, applied in both directions, can provide substantial pruning. Additional constraints that can be applied include the likelihood of particular states occurring, the likelihood of particular transitions occurring, and estimates of durations for particular states. Algorithms have been developed that can use all these constraints to maintain a single best interpretation of a set of incoming measurements that operate in polynomial time (de Coste, 1991).

In spatial data interpretation tasks, a physical experiment (e.g., Huang and Zhao, 2000) or numerical simulation (e.g., Yip, 1991) is used to gather information about the possible behaviors of a system given a set of initial parameters. The geometric patterns these behaviors form are described using vision techniques to create a qualitative characterization of the behavior. For example, initially, simulations are performed on a coarse grid to create an initial description of the system's phase space. This initial description is then used to guide additional numerical simulation experiments, using rules that express physical properties visually.

35.4.7 Spatial Reasoning

Reasoning with purely qualitative representations uses constraint satisfaction techniques to determine possible solutions to networks of relationships. The constraints are generally expressed as transitivity tables. When metric diagrams are used, processing techniques adapted from vision and robotics research are used to extract qualitative descriptions (Bailey-Kellogg and Zhao, 2003). Some reasoning proceeds purely within these new qualitative representations, while other tasks require the coordination of qualitative and diagrammatic representations. There is now a robust flow in the other direction, that is, vision and robotics researchers adopting qualitative representations because they are more robust to compute from the data and are more appropriate for many tasks (e.g., Kuipers and Byun, 1991; Sridhar et al., 2010). For example, qualitative spatial models computed from sensors are being used by teams in RoboCup Soccer (e.g., Schiffer et al., 2006).

35.5 Applications of Qualitative Reasoning

Qualitative reasoning began as a research enterprise in the 1980s, with successful fielded applications starting to appear by the early 1990s. For example, applications in supervisory process control (LeClair et al., 1989) have been successful enough to be embedded in commercial systems. Qualitative reasoning techniques have been used in the design of complex electromechanical systems, such as photocopiers, including the automatic generation of control software. A commercial tool using qualitative reasoning for FMEA in automobile electrical circuits was adopted by a major automobile manufacturer (Price, 2000). Thus, some applications of qualitative reasoning have been in routine use for over a decade, and more applications continue to be developed, with efforts expanding to include education and scientific modeling as well as engineering. Here we briefly summarize some of these research efforts.

35.5.1 Monitoring, Control, and Diagnosis

Monitoring, control, and diagnosis, although often treated as distinct problems, are in many applications deeply intertwined. Because these tasks also have deep theoretical commonalities, they are described together here. Monitoring a system requires summarizing its behavior at a level of description that is useful for taking action. Qualitative representations correspond to descriptions naturally applied by system operators and designers and thus can help provide new opportunities for automation. An important benefit of using qualitative representations is that the concepts the software uses can be made very similar to those of people who interact with software, thus potentially improving human–computer interfaces. Qualitative representations are important for control because qualitative distinctions provide criteria that make different control actions appropriate. Diagnosis tasks impose similar requirements. It is rarely beneficial to spend the resources required to construct a very detailed quantitative model of the way a particular part has failed when the goal is to isolate a problem. Qualitative models often provide sufficient resolution for fault isolation. Qualitative models also provide the framework for organizing fault detection (i.e., noticing that a problem has occurred) and for working around a problem, even when these tasks require quantitative information (e.g., Flad et al., 2010).

An early commercial success of qualitative reasoning was in supervisory process control, where qualitative representations are used to provide more robust control than statistical process control in curing composite parts (LeClair et al., 1989). In the early stage of curing a composite part, the temperature of the furnace needs to be kept relatively low because the part is outgassing. Keeping the furnace low during the entire curing process is inefficient, however, because lower temperatures mean longer cure times. Therefore, it is more productive to keep temperature low until outgassing stops and then increase it to finish the cure process more quickly. Statistical process control methods use a combination of analytic models and empirical tests to figure out an optimal pattern of high/low cooking times. By incorporating a qualitative description of behavior into the controller, it can detect the change in qualitative regime and control the furnace accordingly, providing both faster curing times and higher yield rates than traditional techniques.

Another use of qualitative representations is in describing control strategies used by machine operators, such as unloading cranes on docks. By recording the actions of crane operators, machine learning techniques can be used to reverse-engineer their strategies (Suc and Bratko, 2002).

In some applications a small set of fault models can be pre-enumerated. A set of models, which includes the nominal model of the system plus models representing common faults, can then be used to track the behavior of a system with a qualitative or semiquantitative simulator. Any fault model whose simulation is inconsistent with the observed behavior can thus be ruled out. Relying on a preexisting library of fault models can limit the applicability of automatic monitoring and diagnosis algorithms. One approach to overcoming this limitation is to create algorithms that require only models of normal behavior. Most consistency-based diagnosis algorithms take this approach. For example, in developing onboard diagnostics for Diesel engines, qualitative representations have been found to be useful as a robust level of description for reasoning and for abstracting away from sensor noise (e.g., Sachenbacher et al., 2000).

One limitation with consistency-based diagnosis is that the ways a system can fail are still governed by natural laws, which impose more constraint than logical consistency. This extra constraint can be exploited by using a domain theory to generate explanations that could account for the problem, via abduction. These explanations are useful because they make additional predictions that can be tested and that also can be important for reasoning about safety in operative diagnosis (e.g., if a solvent tank's level is dropping because it is leaking, then where is the solvent going?). However, in many diagnosis tasks, this limitation is not a concern.

35.5.2 Design

Engineering design activities are divided into conceptual design, the initial phase, when the overall goals, constraints, and functioning of the artifact are established, and detailed design, when the results of conceptual design are used to synthesize a constructible artifact or system. Most computer-based design tools, such as computer-aided design (CAD) systems and analysis programs, facilitate detailed design. Yet many of the most costly mistakes occur during the conceptual design phase. The ability to reason with partial information makes qualitative reasoning one of the few technologies that provides substantial leverage during the conceptual design phase (Klenk et al., 2012). Qualitative reasoning can also help automate aspects of detailed design.

One example was Mita Corporation's DC-6090 photocopier (Shimomura et al., 1995), a self-maintenance machine, in which redundant functionality is identified at design time so that the system can dynamically reconfigure itself to temporarily overcome certain faults. An envisionment including fault models, created at design time, was used as the basis for constructing the copier's control software. In operation, the copier keeps track of which qualitative state it is in, so that it produces the best quality copy it can. More recently, fully automatic, real-time composition of control software for "plug and play" high-end printing systems has been fielded by Xerox, using planning techniques over qualitative compositional models to assemble model-based controllers as parts are added to a system in the field (Fromherz et al., 2003).

In some fields experts formulate general design rules and methods expressed in natural language. Qualitative representations can enable these rules and methods can be further formalized, so that they can be automated. In chemical engineering, for instance, several design methods for distillation plants have been formalized using qualitative representations, and designs for binary distillation plants comparable to those in the chemical engineering research literature have been generated automatically.

Automatic analysis and synthesis of kinematic mechanisms have received considerable attention. Complex fixed-axis mechanisms, such as mechanical clocks, can be simulated qualitatively, and a simplified dynamics can be added to produce convincing animations. Initial forays into conceptual design of mechanisms have been made, for example, using configuration space-based techniques to analyze higher-pair mechanisms (Sacks and Joskowicz, 2010). Qualitative representations are also useful in case-based design, because they provide a level of abstraction that simplifies adaptation (e.g., Faltings, 2001).

Qualitative reasoning also is being used to reason about the effects of failures and operating procedures. Such information can be used in FMEA. For example, potential hazards in a chemical plant design can be identified by perturbing a qualitative model of the design with various faults and using qualitative simulation to ascertain the possible indirect consequences of each fault. Commercial FMEA software using qualitative simulation for electrical system design is now being used commercially in automotive design (Price, 2000).

35.5.3 Scientific Modeling

Formalizing the art of modeling is one of the goals of qualitative reasoning, which has led to it being applied to scientific modeling in several ways. One approach is to use modeling ideas from qualitative reasoning to provide languages for organizing mathematical models in ways that support software-assisted model formulation and revision based on comparing numerical data to simulation results (e.g., Bridewell et al., 2006). Another approach is to use spatial aggregation to extract meaningful features from spatially distributed data to provide concise summarizations of structure and function (e.g., Ironi and Tentoni, 2011). Finally, in some domains traditional mathematical models are too fine grained compared to the experimental data available; hence, qualitative models can provide a more natural language for expressing theories. Two examples are ecology, for example, Bredeweg et al. (2008), where qualitative models have been used to help articulate underlying assumptions in modeling, and biology, for example, Batt et al. (2012), where hypotheses about bacterial regulatory networks are being explored via qualitative simulation with model checking. Qualitative models are also useful as an intermediate representation for eliciting and formalizing knowledge. For example, the Agroecological Knowledge Toolkit helps agricultural scientists gather and distill local ecosystem knowledge in field studies (e.g., Cerdan et al., 2012). Another example is the Calvin system (Rassbach et al., 2011), which uses qualitative representations of evidence to help geoscientists construct arguments concerning the interpretation of cosmogenic isotope data in rocks.

35.5.4 Intelligent Tutoring Systems and Learning Environments

One of the original motivations for the development of qualitative reasoning was its potential applications in intelligent tutoring systems (ITSs) and intelligent learning environments (ILEs). Qualitative representations provide a formal language for a student's mental models (Gentner and Stevens, 1983), and thus they facilitate communication between software and student. For example, if a student modeling a block on a table leaves out the force the table exerts on the block, they can be shown a simulation where the block falls through the table, causing them to improve their model (Hirashima et al., 2009). Similarly, in learning organic chemistry, qualitative representations can provide explanations for what happens in reactions (Tang et al., 2010). For younger students, who have not had algebra or differential equations, the science curriculum consists of learning causal mental models that are well captured by the formalisms of qualitative modeling. By using a student-friendly method of expressing models, such as concept maps, software systems have been built that help students learn conceptual models (Forbus et al., 2005; Leelawong et al., 2001), providing software support that scaffolds students in modeling (Bredeweg et al., 2010). In more advanced courses and in industrial training, qualitative reasoning provides explanations that help students make sense of the results of simulators and quantitative analysis tools, for example, to explain why a student's design violates the laws of thermodynamics when teaching about power plants.

35.5.5 Cognitive Modeling

Since qualitative reasoning was inspired by observations of how people reason about the physical world, one natural application of qualitative reasoning is cognitive simulation, that is, the construction of programs whose primary concern is accurately modeling some aspect of human reasoning and learning, as measured by comparison with psychological results. Some research has been concerned with modeling scientific discovery and conceptual change, for example, simulating the trajectories of mental models that students go through when learning intuitive notions of force (Friedman and Forbus, 2010). Several investigations suggest that qualitative representations have major role to play in understanding cognitive processes such as high-level vision. For example, Sridhar et al. (2010) describes how qualitative spatial representations can be used to help analyze video data, and Lovett et al. (2010) describes how analogical reasoning over automatically computed qualitative representations suffices to enable a computational model to outperform most adult Americans on Raven's Progressive Matrices, a visual reasoning test that correlates well with general intelligence. Qualitative order of magnitude representations play a central role in a model of moral decision-making (Dehghani et al., 2008), providing the stratification needed to model the effects of sacred values in decision-making. Commonsense reasoning appears to rely heavily on qualitative representations, although human reasoning may rely more on reasoning from experience than first-principle reasoning (Forbus and Gentner, 1997). Understanding the robustness and flexibility of human commonsense reasoning is an important scientific goal in its own right and will provide clues as to how to build better AI systems. Thus, potential use of qualitative representations by cognitive scientists may ultimately prove to be the most important application of all.

35.6 Research Issues and Summary

Qualitative reasoning is now a mature subfield with a mixture of basic and applied activities, including a number of fielded applications. Its importance is likely to continue to grow, due to the increasing number of successful applications, the continued growth in available computing power, and the rapid digitization of workflows in many human endeavors.

Although there is a substantial research base to draw upon, there are many open problems and areas that require additional research. There are still many unanswered questions about purely qualitative representations (e.g., what is the minimum information that is required to guarantee that all predicted behaviors generated from an envisionment are physically possible?), but the richest vein of research concerns the integration of qualitative knowledge with other kinds of knowledge: numerical, analytic, teleological, etc. The work on modeling to date, although a solid foundation, is still very primitive; better model formulation algorithms, well-tested conventions for structuring domain theories, and robust methods for integrating the results of multiple models are needed. Most domain modeling efforts have been driven by particular application domains and tasks, which is effective in the short run, but a more open development strategy would speed the development of integrated domain theories for a broad range of scientific and engineering knowledge. The growing importance of deep natural language understanding (Ferrucci et al., 2012) is likely to invigorate research in exploring the role of qualitative representations in natural language semantics, since they provide a natural level of description for everyday properties and events (Kuehne and Forbus, 2004; Pustejovsky and Moszkowicz, 2011).

Key Terms

Comparative analysis: A particular form of a "what-if" question, that is, how a physical system changes in response to the perturbation of one of its parameters.

Compositional modeling: A methodology for organizing domain theories so that models for specific systems and tasks can be automatically formulated and reasoned about.

Confluence: An equation involving sign values.

Diagrammatic reasoning: Spatial reasoning, with particular emphasis on how people use diagrams.

Domain theory: A collection of general knowledge about some area of human knowledge, including the kinds of entities involved and the types of relationships that can hold between them, and the mechanisms that cause changes (e.g., physical processes and component laws). Domain theories range from purely qualitative to purely quantitative to mixtures of both.

Envisionment: A description of all possible qualitative states and transitions between them for a system. Attainable envisionments describe all states reachable from a particular initial state; total envisionments describe all possible states.

FMEA (Failure Modes and Effects Analysis): The analysis of the consequences of the failure of one or more components for the operation of an entire system.

Landmark: A comparison point indicating a specific value achieved during a behavior, for example, the successive heights reached by a partially elastic bouncing ball.

Limit point: A comparison point indicating a fundamental physical boundary, such as the boiling point of a fluid. Limit points need not be constant over time, example, boiling points depend on pressure.

Metric diagram: A quantitative representation of shape and space used for spatial reasoning, the computer analog to, or model of, the combination of diagram/visual apparatus used in human spatial reasoning.

Model fragment: A piece of general domain knowledge that is combined with others to create models of specific systems for particular tasks.

Modeling assumption: A proposition expressing control knowledge about modeling, such as when a model fragment is relevant.

Physical process: A mechanism that can cause changes in the physical world, such as heat flow, motion, and boiling.

Place vocabulary: A qualitative description of space or shape that is grounded in a quantitative representation.

Qualitative proportionality: A qualitative relationship expressing partial information about a functional dependency between two parameters.

Qualitative simulation: The generation of predicted behaviors for a system based on qualitative information. Qualitative simulations typically include branching behaviors due to the low resolution of the information involved.

Quantity space: A set of ordinal relationships that describes the value of a continuous parameter.

Semiquantitative simulation: A qualitative simulation that uses quantitative information, such as numerical values or analytic bounds, to constrain its results.

Further Information

Several publically available qualitative reasoning systems that can be downloaded from the web provide an easy way to get started. Those currently being actively maintained and extended include University of Amsterdam's Garp3, a qualitative dynamics simulator; University of Bremen's SparQ; a qualitative spatial reasoning engine; and Northwestern University's CogSketch, a sketch understanding system.

Papers on qualitative reasoning routinely appear in AI, Journal of Artificial Intelligence Research (JAIR), and IEEE Intelligent Systems. Many papers first appear in the proceedings of the American

Association for Artificial Intelligence (AAAI), the International Joint Conferences on Artificial Intelligence (IJCAI), and The European Conference on Artificial Intelligence (ECAI). Every year there is an International Qualitative Reasoning Workshop, which celebrated its 25th anniversary in 2011. Its proceedings, available on the web, document the latest developments in the area.

References

Bailey-Kellogg, C. and Zhao, F. 2003. Qualitative spatial reasoning: Extracting and reasoning with spatial aggregates. *AI Magazine*, 24(4): 47–60.

Batt, G., Besson, B., Ciron, P., de Jong, H., Dumas, E., Geiselmann, J., Monte, R., Monteiro, P., Page, M., Rechenmann, F., and Ropers, D. 2012. Genetic network analyzer: A tool for the qualitative modeling and simulation of bacterial regulatory networks. In van Helden, J., Toussaint, A., and Thieffry, D. (eds.), *Bacterial Molecular Networks*, Humana Press, New York, pp. 439–462.

Bredeweg, B., Liem, J., Beek, W., Salles, P., and Linnebank, F. 2010. Learning spaces as representational scaffolds for learning conceptual knowledge of system behaviour. In Wolpers, M., Kirschner, P.A., Scheffel, M., Lindstaedt, S. and Dimitrova, V. (eds.). *Lecture Notes in Computer Science*, 6383, DOI: 10.1007/978-3-642-16020-2.

Bredeweg, B., Salles, P., Bouwer, A., Liem, J., Nuttle, T., Cioaca, E., Nakova, E. et al. 2008. Towards a structured approach to building qualitative reasoning models and simulations. *Ecological Informatics*, 3(1): 1–12.

Bridewell, W., Nicolas, J., Langley, P., and Billman, D. 2006. An interactive environment for the modeling and discovery of scientific knowledge. *International Journal of Human-Computer Studies*, 64, 1099–1114.

Cerdan, C., Rebolledo, M., Soto, G., Rapidel, B., and Sinclair, F. 2012. Local knowledge of impacts of tree cover on ecosystem services in smallholder coffee production systems. *Agricultural Systems*, 110: 119–130.

Cohn, A. and Renz, J. 2008. Qualitative spatial representation and reasoning. In van Hermelen, F., Lifschitz, V., and Porter, B. (eds.), *Handbook of Knowledge Representation*, Elsevier, Amsterdam, the Netherlands, pp. 551–596.

de Coste, D. 1991. Dynamic across-time measurement interpretation. *Artificial Intelligence*, 51: 273–341.

Dehghani, M., Tomai, E., Forbus, K., and Klenk, M. 2008. An integrated reasoning approach to moral decision-making. *Proceedings of the Twenty-Third AAAI Conference on Artificial Intelligence (AAAI)*, pp. 1280–1286, Chicago, IL.

de Kleer, J. 1977. Multiple representations of knowledge in a mechanics problem solver. *Proceedings of the IJCAI-77*, Cambridge, MA, pp. 299–304.

de Kleer, J. 1984. How circuits work. *Artificial Intelligence*, 24: 205–280.

de Kleer, J. and Brown, J. 1984. A qualitative physics based on confluences. *Artificial Intelligence*, 24: 7–83.

Everett, J. 1999. Topological inference of teleology: Deriving function from structure via evidential reasoning. *Artificial Intelligence*, 113(1–2): 149–202.

Falkenhainer, B. and Forbus, K. 1991. Compositional modeling: Finding the right model for the job. *Artificial Intelligence*, 51: 95–143.

Faltings, B. 1990. Qualitative kinematics in mechanisms. *Artificial Intelligence*, 44(1): 89–119.

Faltings, B. 2001. FAMING: Supporting innovative design using adaptation—A description of the approach, implementation, illustrative example and evaluation. In Chakrabarti, A. (ed.), *Engineering Design Synthesis*, pp. 285–302, Springer-Verlag, New York.

Ferrucci, D. and the Watson Team. 2012. This is Watson. *IBM Journal of Research and Development*, 56(3 and 4).

Flad, S., Struss, P., and Voigt, T. 2010. Automatic detection of critical points in bottling plants with a model-based diagnosis algorithm. *Journal of the Institute of Brewing*, 116(4): 354–359, ISSN 0046-9750.

Forbus, K. 1984. Qualitative process theory. *Artificial Intelligence*, 24: 85–168.

Forbus, K., Carney, K., Sherin, B. and Ureel, L. (2005). VModel: A Visual Qualitative Modeling Environment for Middle-school Students. *AI Magazine*, 26(3), Fall 2005: 63–72.

Forbus, K. and Falkenhainer, B. 1990. Self explanatory simulations: An integration of qualitative and quantitative knowledge, *Proceedings of the AAAI-90*, Boston, MA, pp. 380–387.

Forbus, K. and Gentner, D. 1986. Causal reasoning about quantities. *Proceedings of the Eighth Annual Conference of the Cognitive Science Society*, pp. 196–207, Amherst, MA.

Forbus, K. and Gentner, D. 1997. Qualitative mental models: Simulations or memories? *Proceedings of the Eleventh International Workshop on Qualitative Reasoning*, pp. 97–104, Cortona, Italy.

Forbus, K., Nielsen, P., and Faltings, B. 1991. Qualitative spatial reasoning: The CLOCK project. *Artificial Intelligence*, 51: 417–471.

Forbus, K., Usher, J., Lovett, A., Lockwood, K., and Wetzel, J. 2011. CogSketch: Sketch understanding for Cognitive Science Research and for Education. *Topics in Cognitive Science*, 3(4): 648–666.

Friedman, S.E. and Forbus, K.D. 2010. An integrated systems approach to explanation-based conceptual change. *Proceedings of the 24th AAAI Conference on Artificial Intelligence*, pp. 1523–1529, Atlanta, GA.

Gentner, D. and Stevens, A. (eds.) 1983. *Mental Models*. Erldaum, Hillsdale, NJ.

Glasgow, J., Karan, B., and Narayanan, N. (eds.) 1995. *Diagrammatic Reasoning*. AAAI Press/MIT Press, Cambridge, MA.

Gspandl, S., Reip, M., Steinbauer, G., and Wotawa, F. 2010. From sketch to plan. *Proceeding of QR2010*, Portland, OR.

Hayes, P. 1985. Naive physics 1: Ontology for liquids. In Hobbs, R. and Moore, R. (eds.), *Formal Theories of the Commonsense World*, pp. 71–107, Ablex, Norwood, NJ.

Hirashima, T., Imai, I., Horiguchi, T., and Toumoto, T. 2009. Error-based simulation to promote awareness of errors in elementary mechanics and its evaluation. *Proceedings of the AI & Ed 2009*, Pasadena, CA, pp. 409–416.

Huang, X. and Zhao, F. 2000. Relation based aggregation: Finding objects in large spatial datasets. *Intelligent Data Analysis*, 4: 129–147.

Ironi, L. and Tentoni, S. 2011. Interplay of spatial aggregation and computational geometry in extracting diagnostic features from cardiac activation data. *Computer Methods and Programs in Biomedicine*, 107(3): 456–467, doi:10.1016/j.cmpb.2011.01.009.

Iwasaki, Y. and Simon, H. 1994. Causality and model abstraction. *Artificial Intelligence*, 67(1): 143–194.

Keppens, J. and Shen, Q. 2004. Compositional modeling repositories via dynamic constraint satisfaction with order-of-magnitude preferences. *Journal of AI Research*, 21: 499–550.

Klenk, M., de Kleer, J., Bobrow, D., Yoon, S., Handley, J., and Janssen, B. 2012. DRAFT: Guiding and verifying early design using qualitative simulation. *Proceedings of the ASME IDETC/CIE 2012*, Chicago, IL, pp. 1–7.

Klenk, M., Forbus, K., Tomai, E., and Kim, H. 2011. Using analogical model formulation with sketches to solve Bennett Mechanical Comprehension Test problems. *Journal of Experimental and Theoretical Artificial Intelligence*, 23(3): 299–327.

Kuehne, S. and Forbus, K. 2004. Capturing QP-relevant information from natural language text. *Proceedings of the 18th International Qualitative Reasoning Workshop*, Evanston, IL.

Kuipers, B. 1994. *Qualitative Reasoning: Modeling and Simulation with Incomplete Knowledge*. MIT Press, Cambridge, MA.

Kuipers, B. and Byun, Y. 1991. A robot exploration and mapping strategy based on semantic hierarchy of spatial reasoning. *Journal of Robotics and Autonomous System*, 8: 47–63.

LeClair, S., Abrams, F., and Matejka, R. 1989. Qualitative process automation: Self directed manufacture of composite materials. *Artificial Intelligence for Engineering Design analysis and Manufacturing*, 3(2): 125–136.

Leelawong, K., Wang, Y., Biswas, G., Vye, N., Bransford, J., and Schwartz, D. 2001. Qualitative reasoning techniques to support Learning by Teaching: The Teachable Agents project. *Proceedings of the Fifteenth International Workshop on Qualitative Reasoning*, San Antonio, TX.

Lovett, A., Forbus, K., and Usher, J. 2010. A structure-mapping model of Raven's Progressive Matrices. *Proceedings of CogSci-10*, pp. 276–2766, Portland, OR.

Nayak, P. 1994. Causal approximations. *Artificial Intelligence*, 70: 277–334.

Price, C.J. 2000. AutoSteve: Automated electrical design analysis. *Proceedings ECAI-2000*, Berlin, Germany, pp. 721–725.

Pustejovsky, J. and Moszkowicz, J. 2011. The qualitative spatial dynamics of motion in language. *Spatial Cognition & Computation*, 11(1): 15–44.

Rassbach, L., Bradley, E., and Anderson, K. 2011. Providing decision support for cosmogenic isotope dating. *AI Magazine*, 32: 69–78.

Rickel, J. and Porter, B. 1994. Automated modeling for answering prediction questions: Selecting the time scale and system boundary, *Proceedings of the AAAI-94*, Seattle, WA, pp. 1191–1198.

Sachenbacher, M. and Struss, P. 2005. Task-dependent qualitative domain abstraction. *Artificial Intelligence*, 162: 121–143.

Sachenbacher, M., Struss, P., and Weber, R. 2000. Advances in design and implementation of OBD functions for Diesel injection systems based on a qualitative approach to diagnosis. *SAE World Congress*, pp. 23–32, Detroit, MI.

Sacks, E. and Joscowicz, L. 2010. *The Configuration Space Method for Kinematic Design of Mechanisms*. MIT Press, Cambridge, MA.

Schiffer, S., Ferrein, A., and Lakemeyer, G. 2006. Qualitative world models for soccer robots. *Qualitative Constraint Calculi Workshop at KI 2006*, Bremen, Germany, pp. 3–14.

Shimomura, Y., Tanigawa, S., Umeda, Y., and Tomiyama, T. 1995. Development of self-maintenance photocopiers. *Proceedings of the IAAI-95*, Montreal, Quebec, Canada, pp. 171–180.

Sridhar, M., Cohn, A., and Hogg, D. 2010. Discovering an event taxonomy from video using qualitative spatio-temporal graphs. *Proceedings of the ECAI 2010*, Lisbon, Portugal, doi:10.3233/978-1-60750-606-5-1103.

Stahovich, T.F., David, R., and Shrobe, H. 1998. Generating multiple new designs from a sketch. *Artificial Intelligence*, 104: 211–264.

Struss, P. 1988. Mathematical aspects of qualitative reasoning. *International Journal of Artificial Intelligence Engineering*, 3(3): 156–169.

Suc, D. and Bratko, I. 2002. Qualitative reverse engineering. *Proceedings of the ICML'02* (*International Conference on Machine Learning*), pp. 610–617, New South Wales, Sydney, Australia.

Tang, A., Zain, S., and Abdullah, R. 2010. Development and evaluation of a chemistry educational software for learning organic reactions using qualitative reasoning. *International Journal of Education and Information Technologies*, 4(3): 129–138.

Travé-Massuyès, L., Ironi, L., and Dague, P. 2003. Mathematical foundations of qualitative reasoning. *AI Magazine*, 24(4): 91–106.

Weld, D. 1990. *Theories of Comparative Analysis*. MIT Press, Cambridge, MA.

Williams, B. 1991. A theory of interactions: Unifying qualitative and quantitative algebraic reasoning. *Artificial Intelligence*, 51(1–3): 39–94.

Yip, K. 1991. *KAM: A System for Intelligently Guiding Numerical Experimentation by Computer*. Artificial intelligence series. MIT Press, Cambridge, MA.

36

Machine Learning

Stephen Marsland
Massey University

36.1 Introduction

In essence, machine learning—more completely referred to as *statistical* machine learning—is concerned with identifying and exploiting patterns in data. This has lead to another name for the same area: "pattern recognition." Machine learning is an inherently interdisciplinary field, combining computer science with mathematics and statistics, and it has many application areas such as robotics, bioinformatics, cognitive science, speech recognition, machine vision, and medicine.

At its heart, machine learning is concerned with adaptation: modifying the parameters of a model so that the outputs more closely represent the expected results for given sets of inputs. It is this adaptation that is termed "learning" in the name of the field. Typically, the parameters of the model are adjusted through some form of optimization in order to minimize some error function. The aim of learning a model, rather than using a look-up table of possible outputs or applying predefined rules to the inputs, is to enable "generalization," so that the machine learning algorithm can give sensible predictions on data that were not seen during training.

This chapter provides an overview of the process by which machine learning problems are approached, together with some of the more common algorithms, and then highlights a few areas of current investigation. There has been a lot of research into machine learning over recent years, so this chapter aims to focus on the common features and some differentiating factors between different algorithms before suggesting places where more complete overviews of the field can be found. A set of references for further information on particular algorithms and approaches is also provided.

Examples of the types of problems that machine learning can be usefully applied to include the following:

Predict electricity usage: Based on a set of input factors such as historical electrical usage, weather prediction, season, and what televised events are happening, predict the peak electricity usage for each day into the future. Obviously, electricity providers rely on accurate models forecasting electrical usage, so that they can provide the right amount into the grid without too much waste.

This is an example of a *regression* problem (where a dependent variable, here the electricity usage at a particular time, is predicted based on a set of input variables). It is also an example of time-series prediction, since we have a series of data probably collected at regularly spaced intervals that we wish to propagate into the future.

Predict treatment outcomes: Suppose that for some disease there is a choice of whether or not to provide a particular treatment, for example, because it is not always effective and may have unpleasant side effects. Based on the results of various tests on the patient, and data on their physiological make-up, the algorithm should try to identify those people for whom the treatment is likely to be effective.

This is an example of a *classification* problem, where the output variable identifies which of a set of classes (here just two: treat or don't treat) each input belongs to.

Classify people into different levels of risk for a loan: Using information such as age, credit card debt, and salary, place people into different classes of risk for a loan so that choices of loan limit, interest rate, and so on can be made.

This is written as a classification task, as the input variables are used to separate the individuals into a set of classes based on their loan risk. These different classes could then be used with a set of rules to decide on what loan limit should be applied. Alternatively, the problem could be treated as a regression problem, with the input data used to directly suggest an appropriate loan limit.

Robot control: Based on the positions, velocities, and accelerations of the joints of a robot arm, together with a desired trajectory, compute the electrical current that needs to be applied to the actuators. Again, this is a regression problem, since the torque required at each joint will be a continuous variable.

The four different problems have been described as either regression or classification problems, which describes the dichotomy of statistical machine learning: the desired output for the data is either a set of continuous values for regression or a class membership function. In either case, machine learning can be seen as a form of function approximation, taking a set of inputs and computing a mapping into some appropriate output space.

36.2 Overview of the "Learning from Data" Process

Machine learning algorithms learn from data. The mantra of "garbage in, garbage out" is thus nowhere more true, since algorithms given unsuitable or contradictory data will make useless predictions. However, even with appropriate data, there are still many ways in which errors can be introduced during the data preparation and training process. To assist with avoiding these errors, there is a general approach to data analysis, which can be summarized as follows:

Collect data: While algorithms are often tested, compared, and analyzed on the strength of "toy" (simple) datasets and well-known benchmarks (such as those available at the UCI Machine Learning Repository [5]), for the purposes of real data analysis, original data will need to be chosen and collected. The choice of what data to collect, and how much of it, are important considerations that significantly affect the results of any learning, just as it does any other statistical analysis. As a very basic example, suppose that we wish to predict electricity usage and do not know the weather. This will make it hard to know how much heating people will be using, making the results less accurate. However, knowing the stock market values for the current day is unlikely to help in the prediction, since these variables are presumably largely uncorrelated.

The choice of how much data to collect is also important. We will discuss the need for a variety of different sets of data shortly, but here we are just considering the number of data points required in order to generate a model. Unfortunately, the answer is a rather unhelpful "it depends": the complexity of the underlying model, the features of the learning algorithm, and several other factors all play a role. Notwithstanding computational cost, it is better to have more data than less, and since data are rarely free (there is an inherent cost in collecting, processing, checking, and storing it), the limits on the amount of data are not generally computationally motivated.

Choose features and preprocess: This is often the "art" of machine learning, although some parts of the process can be replaced by experimentation and the requirements of the particular algorithms. In general, it requires a reasonable understanding of the dataset, the application of methods such as *dimensionality reduction* (see Section 36.4.1), and some experimentation. Many of the algorithms make assumptions about how the data will be presented, often requiring that they have been normalized to have zero mean and unit variance.

Choose model(s): There are two parts to the choice of models: the selection of particular algorithms to use and the specification of any parameters required by the model (e.g., the number of neurons in a neural network and the network topology), which often govern the complexity of the model that the algorithm can learn and the amount of training data that needs to be provided.

Train the models on the data: Generally this is relatively simple, being prescribed by the algorithms chosen.

Evaluate on independent data: The model that is trained is based on a set of data known as "training data." These data points are samples from the space of possible data that the model is expected to predict outputs for, but they are certainly not complete, since otherwise there would be no need for any generalization, and so no need for machine learning. One possible problem with the training set is that it may not be representative of the typical data that are seen, leading to a skewed model, but there is a more serious problem that can occur during training, known as *overfitting*. In essence, the amount of training (meaning the number of times that the algorithm runs over the data and modifies its parameters) that is required for a machine learning algorithm is not usually clear. In almost all datasets, the data are not pure representations of the pattern that underlies the dataset—which is what we are trying to find—but also contain variations from an assortment of additional processes. We generally model these as a generic uncorrelated noise term and assume that the model follows a Gaussian distribution. Since the learning algorithm sees only the data points, separating the underlying model from the noise is very difficult. If the learning algorithm is trained for too long, then it will fit the data that have been presented noise and all, meaning that its prediction capabilities are compromized.

Of course, this need for more data puts extra pressure on the data collection, and one method that can be used to make the most of relatively small datasets is to replace data with computation time. The method of *k-fold cross-validation* does this by splitting the data into k folds or subsets, and then training the chosen learning algorithm on $k - 1$ of these subsets, using the missing one for testing. k different learners are trained, each with a different subset missing, and then the most accurate of the learners is selected to be used.

Use the trained model: Having generated and validated the model, it is ready to use on other data.

The most basic way in which machine learning algorithms can be separated into groups is by what data are provided for training. This is a fundamental difference: on one side is *supervised learning* (the vast majority of algorithms; see Section 36.3 for more details), where the training data consist of pairs of input data and target (desired) outputs, while on the other is *unsupervised learning*, where there is no target output provided at all (Section 36.4). Considering the aim of machine learning as either regression or classification, it should be clear that without some form of target data (showing what the correct answer is for some examples) it is impossible to provide a meaningful continuous output, and so regression is not possible. However, there are methods by which similarities between inputs can be used to cluster inputs together, providing an approach to classification that does not require target data.

Sitting in-between supervised and unsupervised learning methods are algorithms where prototype solutions are scored by some form of oracle, and the algorithm aims to maximize its reward by performing search over many possible solutions. The two main classes that fall into this category are *reinforcement learning* (see, e.g., [38]) and *evolutionary learning*, such as genetic algorithms [14,28]. These are different to the other types of machine learning in that what is learnt is not exactly a model of the underlying data generating function, but either a way to act (reinforcement learning considers the actions of an agent interacting with some environment, with the actions of the agent modifying the environment) or the potential solution to a problem. In essence, evolutionary algorithms are stochastic search algorithms that trade off local search around current good solutions with making large-scale changes to the current solutions.

36.3 Supervised Learning

36.3.1 Nearest Neighbors

It seems clear that machine learning needs to be a bit more complicated than just a look-up table of the training data. However, it is useful to start with that simple concept and extend it a little. If the input data cover the space of possible solutions well, then identifying similar inputs to the current input and giving the same output as them seems like a reasonable idea. This is the idea of *nearest neighbor* algorithms, which require that a similarity metric is available so that neighboring inputs can be identified; generally this is the Euclidean metric in the d-dimensional input space created by treating each element of the input vector $\mathbf{x}$ of length d as a separate dimension, so that each $\mathbf{x}$ defines a point in the space.

Using just one neighbor can work for classification tasks but is very prone to errors from mislabelling and other effects of noise, so it is common to use k-nearest neighbors, where k needs to be chosen. For classification tasks, the output is then the majority vote of the k neighbors, while for regression tasks it can be their average value. Whichever is used, choosing an appropriate value for k is generally based on experimentation with the training data.

Figure 36.1 shows an example of a simple classification problem, where the aim is to separate the circles and crosses. The line in each figure is the *decision boundary*, which is the line when the classification changes between the two classes.

Figures like these, which generally show 2D data (for obvious reasons), hide an important consideration. As the number of dimensions in the input vector grows, so does the amount of data that is needed to cover the space. Unfortunately, a linear increase in the number of dimensions leads to an exponential increase in the number of data points required: the sampling density grows as $N^{1/d}$, so 100 samples of a 2D problem have the same coverage as 10^{10} samples of a 10D problem. This is known as the *curse of dimensionality* and it has serious consequences for machine learning, as it means that for high-dimensional spaces the data are always relatively sparse in the input space.

The k-nearest neighbor algorithm produces outputs based on the local inputs, so for accurate predictions in a particular region of input space it requires that the training data represent the data distribution well there. It is also essentially model-free, in that it does not make any assumptions about the underlying statistical distribution of the data. It does, however, assume that within each local region the data are fitted by a constant function. We will see another method that makes this assumption in Section 36.3.3.

36.3.2 Mean-Square Error

Figure 36.1 also shows a feature that is of considerable importance in machine learning: the *decision boundary*. In the figure, it identifies the delineation between crosses and circles, and once this boundary has

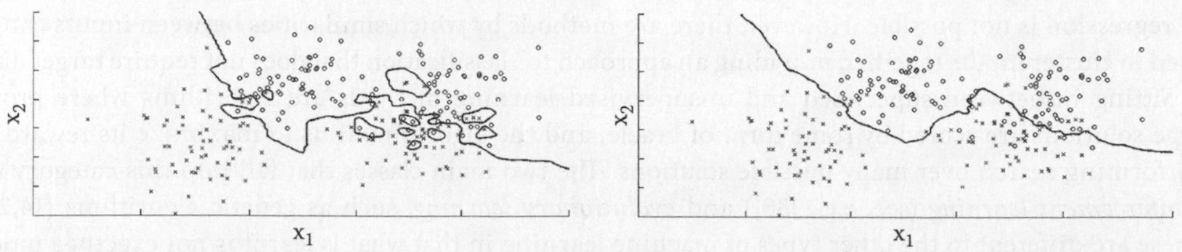

FIGURE 36.1 Decision boundaries for two different values of k (on the left, $k = 2$, and on the right, $k = 20$) for the k-nearest neighbor algorithm.

been identified, the separation between the two classes is complete. For *k*-nearest neighbors, it is typically a very wiggly line, but it doesn't have to be. If we are looking for a model-based solution to machine learning, then, by making assumptions about the number of degrees of freedom that the desired model of the data allows (and hence the complexity of the model), an alternative approach to classification can be found.

Suppose that we wish to fit a straight line between the two classes in the figure. Writing the line as $\hat{y} = \beta_1 x + \beta_0$, the two parameters that need to be found are clear. For regression, the values of $\hat{y}$ should be predictions of the continuous output, but for classification it is less clear what values to use. However, for binary classification, coding the first class (say the circles) as zeros and the second class as ones works fine. The separation between the two classes is then the line $\hat{y} = 0.5$. The generalization of this to higher dimensions fits a hyperplane specified by d parameters, which is again a linear separator of the classes.

There are two observations to make about this linear fit. The first is that for our 2D example there are only two parameters for the line, so the problem is clearly overdetermined, since we have rather more data than that, while the second is that there is no guarantee of a perfect linear separation between the classes. For example, in Figure 36.1, there is no such line. The benefit of having lots of data is that we can fit the line that minimizes some error measure of the training data, such as the mean-square error (MSE). These error measures are known as loss functions $L(\mathbf{y}, f(\mathbf{x}))$; for MSE, it can be written as

$$L(\mathbf{y}, f(\mathbf{x})) = \sum_{i=1}^{n} (\mathbf{y}_i - f(\mathbf{x})_i)^2 \tag{36.1}$$

Suppose that the true function that generates the data is a function g, so that the target data $\mathbf{y} = g(\mathbf{x}) + \varepsilon$ where ε is the noise. The expected value of the fitted MSE on an independent test set is (where $E[\cdot]$ is the statistical expectation)

$$E\left[\frac{1}{N}\sum_{i=1}^{N} (\mathbf{y}_i - \hat{\mathbf{y}}_i)^2\right] \tag{36.2}$$

$$= \frac{1}{N}\sum_{i=1}^{N} E\,[\mathbf{y}_i - g(\mathbf{x}_i) + g(\mathbf{x}_i - \hat{\mathbf{y}}_i)^2] \tag{36.3}$$

$$= \frac{1}{N}\sum_{i=1}^{N} E\left[(\mathbf{y}_i - g(\mathbf{x}_i))^2\right] + E\left[(g(\mathbf{x}_i) - \hat{\mathbf{y}}_i)^2\right] + 2E\left[(g(\mathbf{x}_i) - \hat{\mathbf{y}}_i)\ (\mathbf{y}_i - g(\mathbf{x}_i))\right] \tag{36.4}$$

$$= \frac{1}{N}\sum_{i=1}^{N} E\left[\epsilon^2\right] + E\left[(g(\mathbf{x}_i) - \hat{\mathbf{y}}_i)^2\right] \tag{36.5}$$

$$= \frac{1}{N}\sum_{i=1}^{N} E\left[\epsilon^2\right] + E\left[\ g(\mathbf{x}_i) - E[(\hat{\mathbf{y}}_i] + E[\hat{\mathbf{y}}_i] - \hat{\mathbf{y}}_i)^2\right] \tag{36.6}$$

$$= \frac{1}{N}\sum_{i=1}^{N} E\left[\epsilon^2\right] + E\left[((g(\mathbf{x}_i) - E[\hat{\mathbf{y}}_i])^2\right] + E\left[(E[\hat{\mathbf{y}}_i] - \hat{\mathbf{y}}_i)^2\right] + 2E\left[(E[\hat{\mathbf{y}}_i] - \hat{\mathbf{y}}_i)(g(\mathbf{x}_i) - E[\hat{\mathbf{y}}_i])\right] \tag{36.7}$$

$$= \frac{1}{N}\sum_{i=1}^{N} \mathrm{var(noise)} + \mathrm{bias}(\hat{\mathbf{y}}_i)^2 + \mathrm{var}(\hat{\mathbf{y}}_i) \tag{36.8}$$

In this derivation, we have used the fact that the expectations in the last terms of (36.4) and (36.7) are zero because the noise in the test set is statistically independent of the trained model. Clearly, the variance in the noise is beyond our control. However, the same is not true of the other two terms, and we should favor algorithms that minimize these terms. Unfortunately, a little thought shows that the two terms are in opposition: for a simple example, consider the k-nearest neighbor algorithm of the previous section again. When k is small, the neighbors will all be fairly close to the test point and so the variance in their predictions will be low (in fact, for $k = 1$, the variance will always by 0). However, the bias is high because the points are potentially very noisy. As k increases the variance also increases, since there are more data points to consider and they are further away from each other, but the overall error of the model (the bias) reduces at the same time. This is generally true for any model: the more complex the model, the lower the bias, but the higher the variance. It is known as the *bias–variance trade-off* and means that careful work is typically needed for appropriate model selection; for further details, see [13].

For classification, we can use categorical variables, and the loss function is based on an indicator function, being 0 if the data does not belong to that class, and 1 if it does. We can also compute the posterior probability of each class $P(C_i|\mathbf{x})$ and then pick the *maximum a posteriori (MAP)* estimate, which is the most probable class. While the computations here are relatively simple, using Bayes' theorem, the dependence between the different variables makes them computationally expensive. An assumption that provides a huge simplification is to treat all the variables as independent, an algorithm that goes by the name of *naïve Bayes*. It often works very well, especially in cases where the data are very high dimensional, where it makes the computations tractable.

Figure 36.2 shows three different discriminant functions that separate the (admittedly very limited) data perfectly and therefore have the same MSE value of 0. Without further data to aid in choosing one, it might seem hard to identify a suitable method of doing so, although it should be clear that the middle one is somehow "better" than the others. The concept of the *optimal separating hyperplane*, which is the hyperplane that maximizes the distance to the closest point from either of the two classes that it separates (the margin), provides a solution. This is the basis of the *support vector machine (SVM)* [40], which is one of the most commonly-used machine learning algorithms in practice, and often produces very impressive results.

Up to this stage, we have considered only linear decision boundaries, although it is clear that in general data are not linearly separable. There are two ways to deal with this issue. The first is to use a more complex model that is not limited to hyperplanes, while the second is to consider a transformation of the data into a space where they can be linearly separated. We will consider the first case in Section 36.3.4 and briefly discuss the second case here.

The left of Figure 36.3 shows an example of a simple dataset that cannot be linearly separated using the one parameter (x) provided. However, the addition of a second variable derived from the first (x^2) does allow the data to be linearly separated. Projecting the data into an appropriate set of higher dimensions can thus enable the data to be linearly separated. Methods that make use of these transformations are known as *kernel methods*; there are several potential kernels available. For further details, see [40].

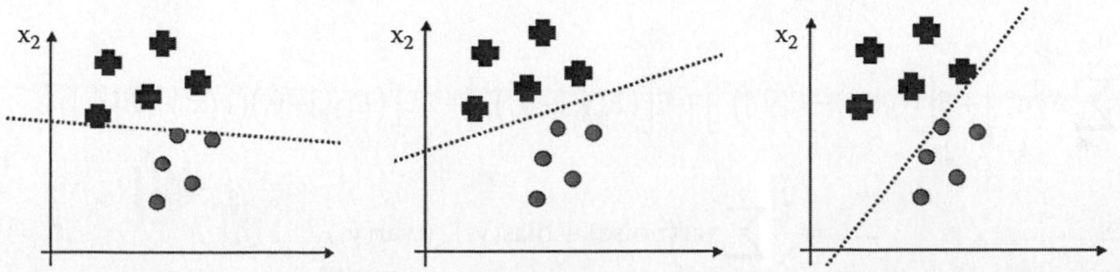

FIGURE 36.2 Three different discrimination functions that all provide linear separation between the two classes of data. The identification of optimal separating hyperplanes is a fairly recent development in machine learning. (From Marsland, S., *Machine Learning: An Algorithmic Perspective*, CRC Press, Lakewood, NJ, 2009.)

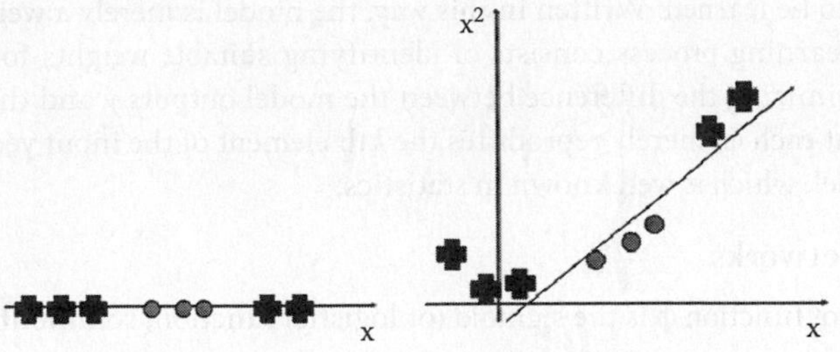

FIGURE 36.3 The data in both figures are based on only one feature variable. However, in the figure on the right, the two dimensions are the variable x and x^2. The addition of this derived variable enables the two classes to be linearly separated. (From Marsland, S., *Machine Learning: An Algorithmic Perspective*, CRC Press, Lakewood, NJ, 2009.)

36.3.3 Decision Trees

It was previously shown that the k-nearest neighbors algorithm effectively fitted a constant value around each data point. It was suggested then that there was another method that takes the same assumption further. This algorithm is the decision tree, which takes the input space and chooses one feature at a time to sequentially split the data, so that eventually the entire space is partitioned into hyper-rectangles, each of which is then represented by a constant value (such as a class membership or a constant regression output), effectively a local basis function. Figure 36.4 shows a dataset with the first two splits of it to demonstrate the process.

Decision trees are built in a greedy fashion, with the feature to be chosen at each level of the tree selected by some criteria (often based on *information gain*, which is the amount by which knowing that feature decreases the *entropy* of the remaining training set) [8].

36.3.4 Kernel Functions

If the aim of the machine learner is to construct a model of the data, then the output $\hat{y}$ (the prediction that it makes) of the learning algorithm for input vector $\mathbf{x}$ can be written as

$$\hat{y} = \sum_{k=1}^{N} w_k \phi_k(\mathbf{x}) \tag{36.9}$$

where the number (N) of functions ϕ_k represents part of the complexity of the model and the choice of functions ϕ represents the remainder. The *weights* w_k are the adjustable parameters of the model, that is,

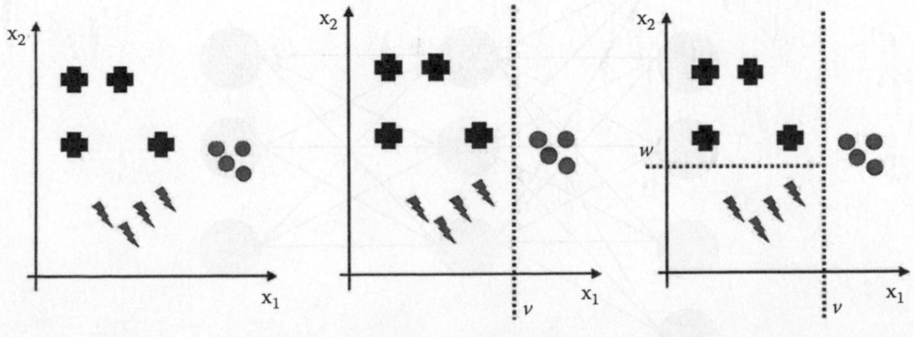

FIGURE 36.4 Potential initial steps for a decision tree learning a dataset. Each choice performs a linear separation based on one of the feature variables. (From Marsland, S., *Machine Learning: An Algorithmic Perspective*, CRC Press, Lakewood, NJ, 2009.)

the things that have to be learned. Written in this way, the model is merely a weighted sum of separate predictors, and the learning process consists of identifying suitable weights for the chosen function model, based on minimizing the difference between the model outputs $\hat{y}$ and the target outputs y. For example, suppose that each ϕ_k merely reproduces the kth element of the input vector. This is simply the *linear regression* model, which is well known in statistics.

36.3.4.1 Neural Networks

One common choice of function ϕ is the sigmoid (or logistic) function, commonly written as (for scalar input value z)

$$\phi(z) = \frac{1}{1 + e^{-\alpha z}} \tag{36.10}$$

where α is some predefined positive parameter. This function has three distinct phases of output: for negative values of z, it produces outputs very close to zero, while for positive values of z it produces outputs near one. Close to zero, however, it varies smoothly between the two saturation values. In the idealized version, where the sigmoid represents a step function, this can be taken as a simple model of a neuron, which accumulates charge on its surface until the potential difference between the surface and the cell body reaches a threshold, whereupon the neuron fires. Any model that is comprised of sets of these simple artificial neurons is known as a *neural network*. In the case where there are just a single set of these artificial neurons, each acting independently of the others, the model is still linear. A particular learning algorithm for this data is the Perceptron learning algorithm, where at each iteration the weights are updated by a learning rule that adapts the weights that are leading to incorrect data.

Suppose now that the functions ϕ can also contain weights within them, so $\phi(\mathbf{x}) = \sum_{k=1}^{L} w_{ik}x_i$, where the index i has been added to identify individual elements of input vector $\mathbf{x}$. This leads to a nonlinear model with two sets of weight vectors that both need to be set by the learning algorithm. The added complexity of such a model enables nonlinear functions to be learned, although at the cost of more training. One way that such networks of neurons can be visualized is as successive layers of neurons, as shown in Figure 36.5, which is an example of the *multilayer Perceptron (MLP)*. The MLP with two layers of neurons is a universal computer, capable of computing arbitrary decision boundaries [10]. The MLP is the most commonly used neural network for supervised learning. Training of the weights of the network are based on a learning rule that propagates the updates back through the network according to the error at the output [34].

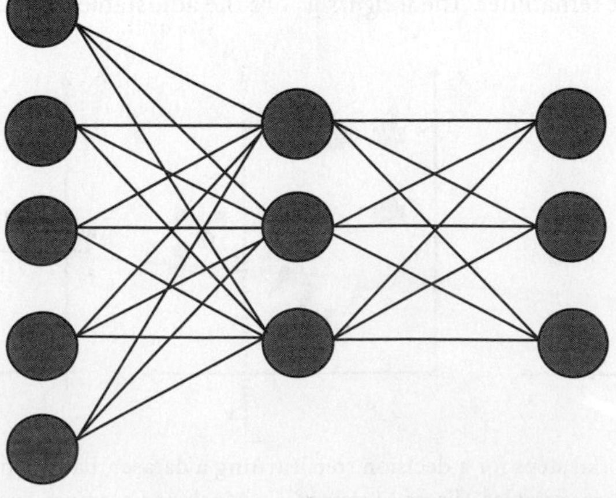

FIGURE 36.5 A schematic showing a feed-forward neural network such as the MLP.

36.3.4.2 Optimization

As with many machine learning algorithms, computation of the weights and other parameters of neural networks are performed by optimization, and many of the algorithms are variations on normal algorithms for gradient descent, particularly Levenberg–Marquardt, since it is based on least squares. For further information on gradient descent algorithms and other work on numerical optimization, see a text such as [31].

One method of optimization that is unique to machine learning is the *expectation–maximization (EM)* algorithm [11], which is used in many areas of machine learning, wherever a *maximum likelihood* solution is needed based on hidden variables. In an EM algorithm, there are two steps. First, the current parameter values are used to compute the posterior distribution of the hidden variables, and then these values are used to find the expectation of the complete data log likelihood for more general parameter values. Then, in the second step, a new estimate of the parameter values is computed by maximizing the function computed in the first step. EM algorithms are guaranteed to converge to a local optimum, but not further.

36.3.4.3 Other Kernels

While the logistic function is the most common choice of function ϕ, there are alternatives. One that is commonly used in function approximation is a *radial basis function (RBF)*, which, as the name implies, is any function whose value depends only on the distance of the input from the cluster center. A common radial basis is e^{-r^2}, which is effectively an unnormalized Gaussian function. RBF networks have the same computational power as feed-forward MLPs, although they are trained rather differently, positioning a set of RBFs in space and then optimizing their parameters independently.

If it is assumed that the weight vectors in Equation 36.9 come from a Gaussian distribution then we generate a simple example of a *Gaussian process*, where the prediction $\hat{y}$ is a linear combination of Gaussian variables and therefore Gaussian itself. The covariance of the outputs is computed as a kernel function between pairs of input (training) data points $\mathbf{x}_i$. For the case of Equation 36.9 the kernel function is

$$k(\mathbf{x}_i, \mathbf{x}_j) = \phi(\mathbf{x}_j)^T \phi(\mathbf{x}_i) \tag{36.11}$$

The kernel defines the covariance matrix of the Gaussian process, which is iteratively updated, since the probability distribution that is being estimated at each training step N is $p(\mathbf{y}_N) = \mathcal{N}(\mathbf{y}_N | \mathbf{0}, C_N)$, where $\mathbf{y}_N$ is the vector of training targets up to step N and C_N is the $N \times N$ covariance matrix whose elements are defined by

$$C(\mathbf{x}_i, \mathbf{x}_j) = k(\mathbf{x}_i, \mathbf{x}_j) + \sigma^2 \delta_{ji} \tag{36.12}$$

where σ is the standard deviation of the noise and $\delta_{ji} = 1$ if $j = i$ and 0 otherwise. For more general problems, the kernel needs to be chosen, and there are several common selections, including the form $e\left(-\|\mathbf{x}_i - \mathbf{x}_j\|\right)$. For more information, see [33].

36.3.5 Statistical Interpretation of Machine Learning

No matter whether or not we are predicting multinomial outputs (classification) or continuous ones (regression), the aim of machine learning is usually to use information from a set of training data $\mathcal{D}$ to create a model $p(\mathbf{x})$. The difference between supervised and unsupervised learning is that for unsupervised learning each element of data in $\mathcal{D}$ consists only of the input vector $\mathbf{x}$, while for supervised learning each input vector is annotated with the target **y**, so that the aim is to learn the mapping between input vector and target, that is, $p(\mathbf{x},\mathbf{y})$. Writing this as

$$p(\mathbf{x},\mathbf{y}) = p(\mathbf{y}|\mathbf{x})\,p(\mathbf{x}) \tag{36.13}$$

It becomes clear that what we are primarily interested in is the conditional probability term on the right: given a set of input data points $\mathbf{x}_i$ we wish to predict the most likely target value $\mathbf{y}_i$ for each of them. In the unsupervised case, where there are no known targets, the entire distribution of interest is $p(\mathbf{x})$ and it is the estimation of this distribution that is the aim of unsupervised learning. For both supervised and unsupervised learning, the value of $p(\mathbf{x})$ is of interest, since low values of it tells us that particular data points are relatively rare. There is a type of learning known as *novelty detection* (or one-class classification) where the training data consists only of examples of what you wish to ignore ("normal" data) and the algorithm identifies inputs with low probability $p(\mathbf{x})$ as "novel." For a summary of this type of learning, see [25].

36.3.6 Graphical Models

As the last section suggests, machine learning can be seen as the meeting point of statistics and computer science. Nowhere is this more clear than in the relatively recent development of graphical models, where graphs (in the computer science sense of a set of nodes and edges) are used to represent probabilistic models. By far the most common approach uses directed acyclic graphs, where the general formulation is known as a *Bayesian network* [20,23]. It represents the joint probability distribution of a set of variables using the nodes of the graph as the variables and the directed edges show some form of dependency between the nodes (more correctly, the absence of an edge between two nodes represents conditional independence of the second node given the first). Many other machine learning algorithms arise as special cases of the Bayesian network, not least the hidden Markov model (HMM), which was originally proposed for speech recognition but has since found many other uses [32], neural networks (including the Boltzmann machine and deep belief network [16]), and particle filters [4]. The second type of network dispenses with the directions on the edges to produce *Markov networks*, which are used in computer vision among other places, particularly the conditional random field [37].

The Bayesian network is specified by the nodes and edges, together with the conditional probability distribution of each node (conditioned on its parents in the graph). For discrete parameters, this is listed as a table, with entries for all the possible values of each of the parents, while for continuous parameters it is specified by the parameters of the chosen probability distribution. Clearly, however, it can be a relatively large collection of numbers to hold and update.

The benefit of Bayesian networks is that the conditional independence relations allow the joint distribution to be represented more compactly than would otherwise be possible. For example, in Figure 36.6, the joint probability of all the nodes in the graph is

$$P(A,B,C,D,E) = P(A) \times P(B) \times P(C|A,B) \times P(D|A) \times P(E|A,B,C,D) \tag{36.14}$$

but by using the fact that E is independent of A given its parents (B, C, and D), this can be reduced to

$$P(A,B,C,D,E) = P(A) \times P(B) \times P(C|A) \times P(D|A) \times P(E|B,C,D) \tag{36.15}$$

While this is not much of a saving in this case, it can often make a substantial difference to the complexity of the calculations in practice.

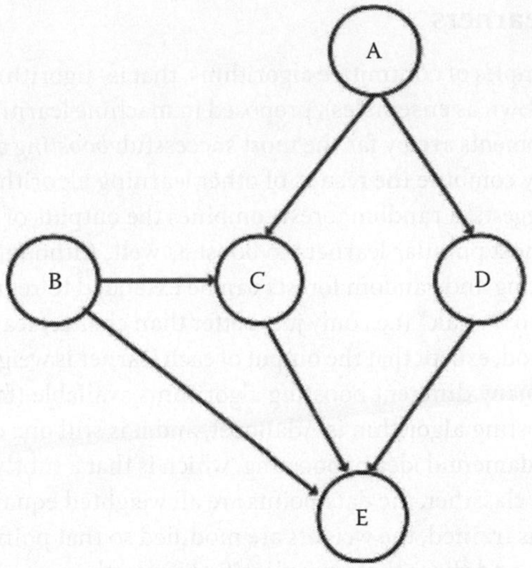

FIGURE 36.6 A possible Bayesian network, consisting of nodes representing random variables in the form of a directed acyclic graph, with the edges implying information about dependence between the variables.

One issue with Bayesian networks is that inference (i.e., computing the posterior distribution of some subset of the nodes given values for the rest) is generally computationally intractable (in fact, it is NP-hard). While several algorithms for approximating the distribution have been developed, typically based on Markov chain Monte Carlo (MCMC) sampling or mean field approximation, these are still computationally expensive.

The training of Bayesian networks is also computationally intractable, since there are a super-exponentially large number of ways to connect up the nodes into a directed acyclic graph. The problem of structure learning (identifying the edges between a set of nodes) is generally predefined based on domain knowledge, although there are algorithms based on greedy heuristics starting from an initial search point [23].

One way around these problems of computational intractability is to impose structure onto the graph, and there are examples where this leads to computationally tractable algorithms, most notably the HMM [32]. In the HMM, a schematic of which is shown in Figure 36.7, a set of unknown states are coupled using an estimate of the transition probabilities between the states, and for each state there are emission probabilities that compute how likely a set of different observations are. As the HMM forms a chain, there are efficient algorithms for both computing the probabilities based on training data and finding the most likely sequence of states for a set of observations [32].

There has been recent interest in combining Bayesian networks with first-order logic, thus obtaining a network that can encode both probabilistic and global constraints and other deterministic relationships. See [18,27] for more information.

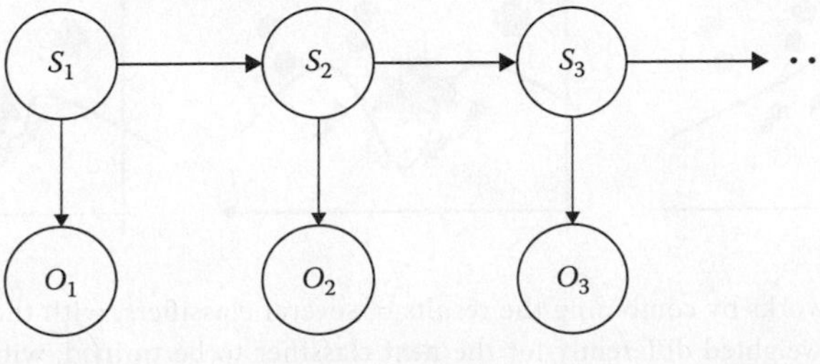

FIGURE 36.7 A schematic of an HMM. Hidden states of the model are inferred, together with the motion between them, based on the observations that are seen.

36.3.7 Combining Learners

There have been several examples of committee algorithms, that is, algorithms that combine the outputs of several classifiers (also known as ensembles), proposed in machine learning over the last 40 years, but two relatively recent developments are by far the most successful: *boosting* and *random forests*. Both are meta-algorithms in that they combine the results of other learning algorithms to produce a more accurate output. As the name suggests, a random forest combines the outputs of many decision trees, and, in fact, decision trees are the most popular learners to boost as well. Although both are easier to describe for classification, both boosting and random forests can be extended to regression problems.

Boosting takes the output of "weak" (i.e., only just better than chance) learners and combines them by using a majority voting method, except that the output of each learner is weighted according to how accurately they work. There are many different boosting algorithms available (for a summary, see, e.g., [35]) but the original adaptive boosting algorithm is AdaBoost, and it is still one of the most commonly used.

Figure 36.8 shows the fundamental idea of boosting, which is that a subtly different dataset is used for each iteration. For the initial classifier, the data points are all weighted equally, and a classifier is trained as normal. However, once it is trained, the weights are modified so that points that were classified incorrectly receive higher weights (and those that were classified correctly lower weights), and a new classifier is trained. In this way, each classifier concentrates particularly on those points that the previous one got wrong. Once training is completed (after a set number of iterations), the outputs of the individual classifiers are combined using a function of the same weights.

Boosting often produces very impressive results. It is functionally different to other learning algorithms in that the loss function that is computed is exponential rather than based on MSE:

$$L(\mathbf{y}, f(\mathbf{x})) = \exp(-\mathbf{y}_i f(\mathbf{x}_i)) \tag{36.16}$$

This means that the algorithm is estimating the probability of getting class 1 as output as a sigmoidal function. For further details, see [35].

The other ensemble algorithm that is increasingly popular is the random forest [3,7]. It is conceptually even simpler than boosting: a collection of *n* bootstrap samples (i.e., samples with replacement) of the training data are made, and a decision tree is trained on each sample. In order to further reduce the potential power of each tree, at each node only some subset of the potential features is searched among, with the subset chosen at random, but with the optimal selection made from that subset. After training, the output of the classifier is the most common prediction among all of the trained trees. Random forests are often reported to be among the most accurate classification methods and work relatively efficiently on large datasets.

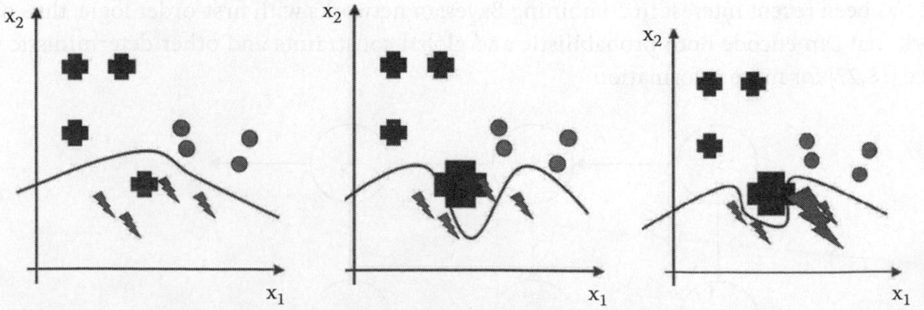

FIGURE 36.8 Boosting works by combining the results of several classifiers, with the classifiers being trained successively. The data are weighted differently for the next classifier to be trained, with points that have previously been classified incorrectly receiving higher weight, shown as getting larger in this figure. (From Marsland, S., *Machine Learning: An Algorithmic Perspective*, CRC Press, Lakewood, NJ, 2009.)

36.4 Unsupervised Learning

Since, by definition, unsupervised learning algorithms do not have the targets that enable problem-specific function approximations to be found, they are rather more limited in what they can achieve. In this section, we will look at algorithms for dimensionality reduction and data clustering.

36.4.1 Dimensionality Reduction

As well as being a tool in its own right, dimensionality reduction is often used as a preprocessing step, before other machine learning algorithms (particularly supervised ones) are used. Underlying the idea of dimensionality reduction is the idea that the *intrinsic dimensionality* of a data manifold is typically substantially lower than that of the original data, which has confounding factors and noise included. For example, suppose that you have a dataset of photographs of one person's face, all taken front on. The only difference is where the face appears within the image, how large it is, and the angle that the head is at. Considering each pixel as a separate element for the input vector will give massively high-dimensional data (even if the images are only 100×100 pixels, this is still a 10,000 element vector). However, the possible variation in the data can be encoded by the changes that are allowed, which is 4D (two for translation in x and y, 1 for rotation, and one for scaling). Thus, the intrinsic dimensionality of the data is 4D rather than 10,000D.

Dimensionality reduction algorithms aim to minimize the reconstruction error of the original data based on a lower-dimensional representation of it, which provides a direct function to optimize. One of the most commonly used dimensionality reduction algorithms, *principal component analysis (PCA)*, selects a linear transformation of the original basis in such a way that the first direction chosen maximizes the variance, the next takes as much as possible of the remaining variance, and so on. In this way, the last few dimensions of the data will have very little variance and can thus be safely ignored [19]. The algorithm for PCA is based on computing eigenvalues and eigenvectors of the covariance matrix of the data.

A related method to PCA is *factor analysis*, which tries to find a smaller set of uncorrelated factors (also known as latent variables) that describe the data. The underlying hypothesis is that there is some unknown set of data sources from which the data are generated, and the method aims to find those. If the sources are assumed to be statistically independent instead of uncorrelated, then the method of *independent component analysis (ICA)* [17] is recovered. This is typically explained in terms of the "cocktail party problem," where even humans can struggle to distinguish the many different speakers and isolate one against the background noise.

All of these methods produce linear basis functions. This typically makes them unsuccessful when applied to data that is nonlinear. However, the use of kernels (as in Section 36.3.4) enables variations of the standard algorithms to be used on such data [36].

36.4.2 Clustering

The second common task in unsupervised learning is to cluster similar data points together, where similar in this case generally means close together in input space according to some distance metric (usually Euclidean distance). Nearest neighbor methods provide a relatively simple way to do this, with each data point in the test data being represented by the point that it is closest to in the training data. However, this is relatively inefficient in that because the training set needs to be large to ensure that enough samples are seen to be confident that the potential test data are covered, and there will often be many samples in the training set that are close together.

Exemplar-based methods attempt to fix this problem by choosing cluster centers that best represent the training data. A simple example is the k-means algorithm, where a predetermined number of cluster centers are positioned at random in the data space and then iteratively moved by being positioned at the mean of a set of data points until they represent the data. The two issues with this approach are that it is hard to pick k appropriately without a lot of knowledge of the dataset and the approach is very subject to local minima, getting stuck in many suboptimal solutions.

The most commonly used algorithm for unsupervised learning of clusters is the *self-organizing map* of Kohoren [22], which has been used for a vast number of different tasks [21]. In this algorithm, a regular lattice of neurons have their weight vectors refined to move successively closer to inputs that they, or their neighbors, represent well. This leads to a *feature map* with nodes that represent similar features being spaced close together in the map space through self-organization.

Further Reading

This chapter has provided a very brief skim over some of the important methods of machine learning. Many of the most important methods of machine learning have been omitted or treated very superficially for reasons of lack of space. In particular, there has been no details of algorithms nor of any research on theoretical developments of machine learning. Examples of this theory include information geometry [2], probably approximately correct (PAC) learning [39], spline fitting [41], and the Vapnik–Chervonenkis dimension [40]. In addition, there is the rich area of model selection, including consideration of various information criteria and description lengths [9].

As well as the specialist books referred to in the text, there are several general textbooks on machine learning for a variety of audiences, from undergraduate students [1,12,26,29,42], to more advanced computer scientists [6,24,30], to statisticians [15]. In addition, there are a wide range of conferences and journals covering both theory and applications.

Glossary

Bayesian network: A form of graphical model based on directed acyclic graphs and conditional probability tables to represent the probability distribution of the whole graph.

Bias–variance trade-off: The two parts of the mean-square error, which means that machine learning algorithms can produce more accurate results only at the cost of more possible variation in result, largely through overfitting.

Boosting: An ensemble algorithm that combines the outputs of weak learners to make more accurate predictions.

Classification: A type of problem in machine learning where the output variable identifies which of a set of classes the data point belongs to.

Decision boundary: The line between classes in a classification problem.

Decision tree: An algorithm that classifies data by asking questions about different elements of the input vector sequentially.

Dimensionality reduction: A form of machine learning, often unsupervised, that tries to find lower dimensional representations of the data manifold.

Evolutionary learning: A form of machine learning that mimics evolution to find better solutions via stochastic search.

Expectation–maximization: An optimization algorithm in machine learning to estimate hidden parameters.

Factor analysis: A form of unsupervised learning where a set of hidden variables are estimated to represent the data, which is assumed to be noisy.

Gaussian process: A supervised learning algorithm that assumes that all random variables are Gaussian distributed.

Graphical model: A representation of machine learning algorithms where random variables are represented as nodes and edges between nodes show some form of conditional dependence.

Hidden Markov model: A graphical model that predicts sequences of hidden states from sets of observations based on two sets of probabilities: transition probabilities between the states and emission probabilities from the states to the observations.

Independent component analysis: A method of dimensionality reduction that creates a basis where all of the basis vectors are statistically independent.

Linearly separable data: Data that can be separated into different classes using a hyperplane (straight line in 2D).

Loss function: A function that computes the cost of getting values wrong.

***k*-fold cross-validation:** A method of training several versions of the same learner on different subsets of the training data and using the one that is most successful.

***k*-means:** An algorithm for unsupervised learning that represents each node by the nearest cluster center.

***k*-nearest neighbors:** An algorithm that predicts outputs based on the k data points from the training set that are closest to the current input.

Kernel: The inner product is some space of a pair of data points.

Markov network: A form of graphical model where the edges are undirected.

Maximum a posteriori: The choice of class in a classification problem as the one with the highest posterior probability from Bayes' theorem.

Multilayer perceptron: A generalization of the perceptron neural network to have multiple layers of neurons, which can thus fit more complex decision boundaries.

Naïve Bayes classifier: A classifier that works by computing the maximum a posteriori estimate under the assumption that all of the features are statistically independent.

Neural network: Any of a set of machine learning algorithms based on a simple model of a neuron, with weighted connections between layers of such neurons.

Novelty detection: A form of machine learning where data are presented only for the normal classes, and anything not seen during training is highlighted as novel.

Optimal separating hyperplane: The hyperplane that separates data by maximizing the distance between the classes.

Overfitting: One cause of poor generalization in machine learning algorithms, where the model fits both the underlying model and the noise in the training data.

Perceptron: A linear neural network consisting of a layer of neurons that are connected to the inputs by a set of adaptive weights.

Principal component analysis: A method of dimensionality reduction where the coordinate axes are linearly transformed so that the first direction has maximal variation, the next direction next most maximal, and so on.

Radial basis function: A function whose output depends only on the radial distance to the point. It is possible to make a machine learning algorithm that linearly combines the outputs of such functions.

Random forest: A supervised learning algorithm that trains a large number of decision trees on different subsets of the data and combines their outputs to get a more accurate answer than any of them individually.

Regression: A problem where a dependent variable is predicted based on a set of input variables.

Reinforcement learning: A form of machine learning where the algorithm is told how well it has performed on the data but not how to improve.

Self-organizing map: An algorithm for unsupervised learning where the weights of neurons are updated to better represent inputs that are close to either than neuron or its neighbors.

Supervised learning: A form of machine learning where each item of training data has an associated target vector showing the correct answer and the aim is to generalize from these examples to all data.

Support vector machine: A machine learning algorithm that maximizes the separation of classes in the data to fit the optimal separating hyperplane.

Unsupervised learning: A form of machine learning where no target data are provided and the algorithm needs to exploit similarities between data points.

References

1. E. Alpaydin. *Introduction to Machine Learning*. MIT Press, Cambridge, MA, 2004.
2. S. Amari and H. Nagaoka. *Methods of Information Geometry*, American Mathematical Society, Providence, RI, 2000.
3. Y. Amit and D. Geman. Shape quantization and recognition with randomized trees. *Neural Computation*, 9(7):1545–1599, 1997.
4. M.S. Arulampalam, S. Maskell, N. Gordon, and T. Clapp. A tutorial on particle filters for online nonlinear/non-Gaussian Bayesian tracking. *IEEE Transactions on Signal Processing*, 50(2):174–188, 2002.
5. K. Bache and M. Lichman. UCI machine learning repository, Irvine, CA: University of California, School of Information and Computer Science, 2013. http://archive.ics.uci.edu/ml
6. C.M. Bishop. *Pattern Recognition and Machine Learning*. Springer, Berlin, Germany, 2006.
7. L. Breiman. Random forests. *Machine Learning*, 45(1):5–32, 2001.
8. L. Breiman, J.H. Friedman, R.A. Olshen, and C.J. Stone. *Classification and Regression Trees*. Wadsworth International Group, Belmont, CA, 1984.
9. K.P. Burnham and D.R. Anderson. *Model Selection and Multimodel Inference: A Practical Information-Theoretic Approach*. Springer-Verlag, New York, 2nd edn., 2002.
10. G. Cybenko. Approximations by superpositions of sigmoidal functions. *Mathematics of Control, Signals, and Systems*, 2(4):303–314, 1989.
11. A.P. Dempster, N.M. Laird, and D.B. Rubin. Maximum-likelihood from incomplete data via the EM algorithm. *Journal of the Royal Statistical Society Part B*, 39:1–38, 1977.
12. R.O. Duda, P.E. Hart, and D.G. Stork. *Pattern Classification*. Wiley-Interscience, New York, 2nd edn., 2001.
13. S. Geman, E. Bienenstock, and R. Doursat. Neural networks and the bias/variance dilemma. *Neural Computation*, 4:1–58, 1992.
14. D.E. Goldberg. *Genetic Algorithms in Search, Optimisation, and Machine Learning*. Addison-Wesley, Reading, MA, 1999.
15. T. Hastie, R. Tibshirani, and J. Friedman. *The Elements of Statistical Learning*. Springer, Berlin, Germany, 2001.
16. G. E. Hinton and T. Sejnowski. Learning and relearning in Boltzmann machines. In D.E. Rumelhart and J.L. McClelland, eds., *Parallel Distributed Processing*, vol. 1, pp. 282–317. The MIT Press, Cambridge, MA, 1986.
17. A. Hyvrinen and E. Oja. Independent component analysis: Algorithms and applications. *Neural Networks*, 13(4–5):411–430, 2000.
18. M. Jaeger. Probabilistic decision graphs—Combining verification and AI techniques for probabilistic inference. *International Journal of Uncertainty, Fuzziness and Knowledge-Based Systems*, 12:19–42, 2004.
19. I.T. Jolliffe. *Principal Component Analysis*. Springer, New York, 2002.
20. M.I. Jordan, ed. *Learning in Graphical Models*. MIT Press, Cambridge, MA, 1999.
21. S. Kaski, J. Kangas, and T. Kohonen. Bibliography of self-organizing map (som) papers: 1981–1997. *Neural Computing Surveys*, 1:102–350, 1998.
22. T. Kohonen. *Self-Organisation and Associative Memory*. Springer, Berlin, Germany, 3rd edn., 1989.
23. D. Koller and N. Friedman. *Probabilistic Graphical Models*. The MIT Press, Cambridge, MA, 2009.

24. D.J.C. MacKay. *Information Theory, Inference and Learning Algorithms*. Cambridge University Press, Cambridge, U.K., 2003.
25. S. Marsland. Novelty detection in learning systems. *Neural Computing Surveys*, 3:157–195, 2003.
26. S. Marsland. *Machine Learning: An Algorithmic Perspective*. CRC Press, Lakewood, NJ, 2009.
27. R. Mateescu and R. Dechter. Mixed deterministic and probabilistic networks. *Annals of Mathematics and Artificial Intelligence*, 54(1–3):3–51, 2008.
28. M. Mitchell. *An Introduction to Genetic Algorithms*. MIT Press, Cambridge, MA, 1996.
29. T. Mitchell. *Machine Learning*. McGraw Hill, New York, 1997.
30. K. Murphy. *Machine Learning: A Probabilistic Perspective*. The MIT Press, Cambridge, MA, 2012.
31. J. Nocedal and S.J. Wright. *Numerical Optimization*. Springer, Berlin, Germany, 2006.
32. L.R. Rabiner. A tutorial on hidden Markov models and selected applications in speech recognition. *Proceedings of the IEEE*, 77(2):257–268, 1989.
33. C.E. Rasmussen and C. Williams. *Gaussian Processes for Machine Learning*. The MIT Press, Cambridge, MA, 2006.
34. D.E. Rumelhart, G.E. Hinton, and R.J. Williams. Learning internal representations by back-propagating errors. *Nature*, 323(99):533–536, 1986.
35. R.E. Schapire and Y. Freund. *Boosting: Foundations and Algorithms*. The MIT Press, Cambridge, MA, 2012.
36. J. Shawe-Taylor and N. Cristianini. *Kernel Methods for Pattern Analysis*. Cambridge University Press, Cambridge, U.K., 2004.
37. C. Sutton and A. McCallum. An introduction to conditional random fields for relational learning. In *Introduction to Statistical Relational Learning*. The MIT Press, Cambridge, MA, 2006.
38. R.S. Sutton and A.G. Barto. *Reinforcement Learning: An Introduction*. MIT Press, Cambridge, MA, 1998.
39. L. Valiant. A theory of the learnable. *Communications of the ACM*, 27(11):1134–1142, 1984.
40. V. Vapnik. *The Nature of Statistical Learning Theory*. Springer, Berlin, Germany, 1995.
41. G. Wahba. *Spline Models for Observational Data*. SIAM, Philadelphia, PA, 1990.
42. I.H. Witten and E. Frank. *Data Mining: Practical Machine Learning Tools and Techniques*. Elsevier, Amsterdam, Holland, 2005.

37

Explanation-Based Learning

Gerald DeJong
University of Illinois at Urbana Champaign

37.1 Introduction

The statistical approach to machine learning has been immensely successful. It often forms the foundation of other AI systems: for example, natural language processing, computer vision, image processing, data mining, intelligent search engines, recommender systems, and bioinformatics all owe much to statistical machine learning. Over the years, it has become apparent that this approach is improved with weaker prior models and greater reliance on data. This has led to the emergence of "big data" and the mantra "Let the data speak for themselves." However, the amount of data required grows exponentially as the gap widens between weaker priors and more sophisticated concepts that require greater subtlety and complexity. Statistical machine learning seems to be reaching a plateau that is well short of human-level concept learning.

Modern explanation-based learning (EBL) addresses this issue by including an additional source of information: a general symbolic inferential domain theory. This domain theory often employs first-order representations that allow much greater expressiveness than the propositional (zeroth-order) expressiveness of statistical methods such as the Bayes or Markov networks. The domain theory can be approximate; it need not be complete, correct, or consistent in the sense required by conventional logic. In EBL, the domain theory serves as *analytic evidence* to complement the empirical statistical evidence of observed examples. EBL's domain theory requires a new semantic interpretation to allow derived expressions to be treated as evidence rather than as truths or constraints on possible worlds. Because of its reliance on a prior domain theory, EBL is not as generally applicable as conventional statistical learning. However, with even modest prior domain knowledge, EBL can learn complex concepts more accurately, more confidently, and with far fewer examples.

The domain theory encodes general understanding discovered by others about the workings of the world. Imagine a conventional "big data" approach to training medical doctors. Instead of medical texts and classroom lectures, each student would be given a long string of specific examples. Each student would follow some idiosyncratic path, perhaps recapitulating the development of medical understanding but culminating finally in inventing his or her own version of the pathogenic theory of disease, physiology, pharmacological biochemistry, and so on. For both humans and computers, this would seem a tall order and unnecessarily complicated. When general prior knowledge of the world is already

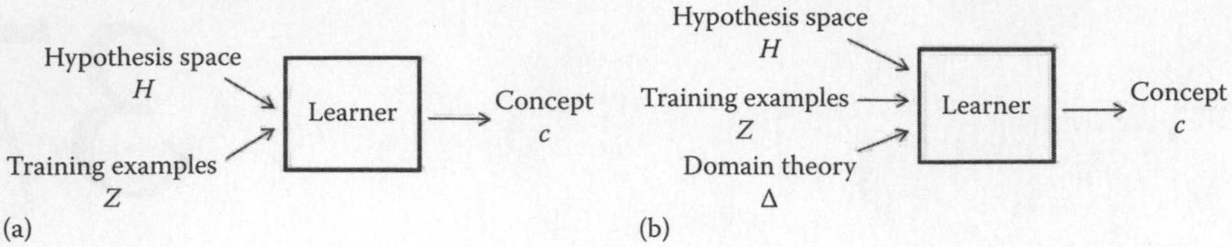

FIGURE 37.1 Machine learners: statistical (a) choosing $c \in H$ based on the empirical evidence of Z, EBL and (b) choosing $c \in H$ based on both the empirical evidence of Z and the analytic evidence of Δ.

available, machine learners, like human learners, should be able to exploit it. Humans seem to learn complex concepts best using a combination of general facts together with a few illustrative examples. Modern EBL adopts this same learning style.

Learning systems (including EBL) can be applied in many paradigms including regression, semi-supervised, unsupervised, and reinforcement learning. Due to space limitations, this chapter focuses on the area of supervised classification learning. Here, a hypothesis space H is given to the learner along with a training set Z composed of examples and their training labels. The learner's task is to choose some concept $c \in H$ that best describes the pattern underlying Z. The concept sought is the one that most accurately classifies future similarly chosen examples.

Figure 37.1 illustrates a conventional statistical learner and an EBL learner. Both choose some hypothesis from the space H to output as the learned concept c. In statistical learning, this choice is based on the empirical evidence embodied by the training set Z. In EBL, the analytic evidence of the domain theory, Δ, is also used.

Modern EBL works by adding new features. These are features that the system believes will be useful in describing the underlying pattern. With such features, the desired pattern will be simpler and easier to find. The new features must be (1) inferable from the domain theory and (2) relevant to the current learning problem. Interaction between the domain theory and the training examples is crucial: a subset of training examples is selected and used to generate explanations from the domain theory. A number of explanations are constructed for the selected subset. Each explanation conjectures why the examples in the subset might merit their teacher-assigned class labels. Each explanation draws intermediate conclusions that are, in effect, new features, latent in the examples but derivable from the domain theory. Given a new example, the value of each feature is computed by a (generally nonlinear) function of the example's raw input features as dictated by the explanation.

A new augmented hypothesis space, Θ, is formed by adding these new derived features to the original raw features. A conventional statistical learner is then applied to the remaining training examples, those not in the subset used to generate explanations. These training examples are embellished with the new derived features, and the learner chooses the best hypothesis from the augmented hypothesis space, some $\theta \in \Theta$. By deriving more appropriate, higher-order features, EBL renders the pattern underlying the examples in the training set clearer and simpler, making this learning task much easier.

As a final step, the EBL system chooses a $c \in H$ that is closest in behavior to the selected θ. In θ, the EBL system has constructed an accurate description of the desired pattern. However, the description is expressed using the vocabulary of its own derived features. This form is not acceptable as an output.

The chosen $c \in H$ may be very complex even though the selected θ is simple. Note, however, that this step does not require labeled training examples. The complexity of c arises from the vocabulary mismatch between the raw input features and the more appropriate explanation-generated features. In a sense, the explanations act as a catalyst for the leaner, providing a low-complexity path to an accurate high-complexity concept in H.

It may be of interest to note that this final step was neglected in early EBL systems. The oversight gave rise to what was known as the *utility problem*, attracting a good deal of research attention [13,21,22,41]. Researchers noticed that newly learned EBL concepts, although accurate, could reduce rather than enhance the performance of the overall system. It is now clear that the utility problem is a consequence of losing control of the hypothesis space. The hypothesis space H specifies the set of acceptable outputs, often via a set of well-formedness constraints. The hypothesis space is an essential input for any machine leaner (EBL or otherwise). But no such input was given to the old EBL systems; there was no guarantee that the resulting pattern was acceptable to the system utilizing the learned concept. Because hard satisfiability problems could be embedded in these unconstrained concepts, their use could be extraordinarily expensive. This could consume processing resources that might have been expended more productively elsewhere if the concept had not been learned. Two approaches were found to the utility problem. The first empirically estimates the cost of utilizing concepts, eliminating overly expensive ones. The second limits the expressiveness of the new concepts themselves to preclude in principle the possibility of expensive patterns. This third, theoretically sound, solution to the utility problem is simply to respect the hypothesis space, H. In practice, however, it can be convenient to employ one of the first two solutions.

The key in EBL is the part played by the domain theory. Modern EBL can be seen as quite similar to conventional statistical learning once the role of the domain theory is properly understood. To see this, consider an analogy between EBL and transfer learning [47] (also multitask learning [5] and domain adaptation [25,39], although with an equivocation on the term "domain"). A learner, after solving one learning problem, will be applied to a second related problem, perhaps followed by a third and fourth. Call the space of related problems a *domain*. Because the problems are related, it is possible to exploit their underlying similarity, utilizing earlier learning experiences to streamline later learning and avoiding the need to relearn the shared structure. Transfer learning is an important area of ongoing research in its own right.

Modern EBL takes this view a step further by generalizing across learners, particularly previous human learners. Instead of one learner encountering a sequence of related learning problems, there are many learners solving many problems from the domain. In this way, far more shared information can be acquired. But now the information must be communicated explicitly. In transfer learning, a single learner builds on its own representations. But different learners cannot be relied on to understand the internal representations of others. Some communication mechanism is required. This is the role of the domain theory. With this newly posited communicative ability, a learner derives the benefit without the need to experience the plethora of previous learning problems or view for itself the myriad of training examples. Rather, it capitalizes on the accumulated knowledge accrued by many other learners. For EBL, the previous learners are humans. They not only solve problems but think about them, postulating and testing new general theories. Importantly, EBL does not attempt to invent such theories. Its goal is the much more modest task of exploiting the understanding developed by others. In so doing, it avoids the difficulty and expense of reinvention.

Thus, the EBL domain theory is seen as a communication medium used by a learner to benefit from the experiences of many other learners that have previously solved many related problems over possibly many generations. The domain theory itself is seen as a cleverly condensed, codified, general summary of the information inherent in an untold number of previous experiences and training examples. Several important properties follow from this view of a domain theory:

- The representation must be sufficiently expressive to capture abstract conceptual domain commonalities. Its function is to highlight significant abstract nonlinearities within the domain. Simple patterns (e.g., linear ones) can be easily recovered via conventional statistical analysis, so their inclusion in a domain theory is of limited value. A good domain theory focuses on subtle, nonlinear, and higher-order covariate structures. This dictates the need for some kind of quantification such as first-order representations. Propositional representations are insufficient.
- The EBL domain theory captures *utility* rather than *correctness* or *truth* as in conventional logic. Like the massive collection of training examples that it summarizes, the EBL domain theory is inherently incomplete and approximate. This dictates the need for a new semantics quite different from the popular "constraints on interpretations of possible worlds" of conventional logic. There can be no notion of a *correct* or *true* domain theory. It follows that analytic inference over the domain theory must be rooted in optimization, rather than satisfiability as used in logic.

- The purpose of the domain theory is to communicate or encode important conceptual distinctions. The better the domain theory, the better this encoding. Better coding schemes are ones that make more important items more easily accessible. Thus, a better domain theory makes its more useful distinctions easier to derive.
- In EBL, a feature's derivational complexity (i.e., the minimal inferential effort sufficient for its construction) provides information on its prior expected utility. This is taken as one definition of a domain theory for EBL:

An EBL domain theory is an inferential mechanism that supports the generation of latent features in which the inverse of derivational complexity encodes expected utility over the space of problems that defines the domain.

37.2 Underlying Principles

Due to space limitations, the principles in this section remain somewhat abstract. Illustrative examples will be given in the next section. It is convenient to consider Δ as following the syntax of first-order logic. Readers unfamiliar with the terms *first-order logic, quantification, symbolic inference, derivation, interpretation, and possible world semantics* may wish to consult a suitable tutorial such as [18,52]. It is not necessary to assume first-order syntax; Δ may be an ad hoc mechanism. Compared to conventional logic, there are two important generalizations and one restriction. The generalizations are as follows: (1) The initial statements (analogs of the axiom set) need not be correct or consistent. By allowing approximations, the domain expert's task of rendering his understanding of the world can be greatly simplified. (2) With little additional penalty, Δ may employ higher-order statements. This is in contrast to conventional logic where including even second-order statements result in the undecidability of most of the interesting questions (such as theoremhood).

The new restriction is that the combining rule must be *paraconsistent* [49]. This avoids explosiveness (known as ECQ for *ex contradictione quodlibet*, from a contradiction everything follows). It is easy to see, for example, that if a refutational inference system is given initial statements that are only approximate, it will likely be able to prove anything. Many contradictions will arise from interactions among approximate statements. Refutational inference can succeed simply by finding one of these contradictions, independent of the negated goal. The phenomenon is not restricted to refutational systems, and the general solution is paraconsistency.

A conventional statistical learner can be viewed as searching for c^+, the most likely hypothesis in H given the training data Z:

$$c^+ \approx \operatorname*{argmax}_{h \in H} Pr(h|Z,H) \tag{37.1}$$

Generally, finding the true *argmax* is intractable. The symbol "≈" is used in place of "=" to denote that it is sufficient for an implemented learner to find a c^+ that with high confidence is a low-loss approximation to the *argmax* expression. Note that the probability of a particular h given Z is independent of the other hypotheses in H. That is, Z is conditionally independent of H given h. Along with Bayes theorem and the fact that the parameter h does not appear in $Pr(Z)$, it follows that

$$c^+ \approx \operatorname*{argmax}_{h \in H} Pr(Z|h) \cdot Pr(h|H) \tag{37.2}$$

where

H is the hypothesis space

Z is a set of examples $\{x_i\}$, drawn independently from some example space X, together with the training label of each

The training labels $\{y_i\}$ are assigned by some true world concept c^* that may be stochastic and in general is not in H.

Expression 37.2 seeks to maximize the likelihood of the training data subject to the probability of h given H. This second factor, $Pr(h|H)$, does not depend on any training examples so with respect to Z it is a prior. It assigns low probability to hypotheses in H that are unlikely. Computationally, it is often treated simply as a regularizer that penalizes overly complex or overly expressive hypotheses. Its role is essential; it prevents overfitting to the training set.

Care must be taken in interpreting the probabilities. In Expression 37.1, for example, $Pr(h|Z,H)$ is *not* the probability given Z and H that h actually *might be* the target concept c^*. In interesting learning problems, c^* is never in H. The patterns embodied by the real world are infinitely subtle and complex. In machine learning, the hypothesis space is chosen to balance expressiveness on the one hand and tractability on the other. It must be sufficiently rich that some element is a good approximation of c^*. Only in simple and artificially clean domains can $c^* \in H$ be legitimately assumed.

By contrast to Expression 37.1, an EBL learner searches for c^+:

$$c^+ \approx \underset{h \in H}{\operatorname{argmax}}\, Pr(h|Z,H,\Delta)$$

which includes the domain theory, Δ, as an additional source of evidence. Since Z is conditionally independent of both H and Δ given h, it follows that

$$c^+ \approx \underset{h \in H}{\operatorname{argmax}}\, Pr(Z|h) \cdot Pr(h|H,\Delta)$$

This is Expression 37.2 adjusted to appreciate the domain theory in the choice of $h \in H$. Rather than influencing h directly, modern EBL employs the domain theory to transform the hypothesis space H into a more effective one, Θ. The domain theory encodes useful expert-supplied distinctions. These distinctions form new features derivable from Δ. Given a subset Φ of these new features (the ones that are likely to be useful for the problem at hand), an enhanced hypothesis space is formed by allowing Φ to participate on equal footing with the original raw features used in H. Let $\Theta(\Phi)$ denote the new hypothesis space enriched with the feature set Φ. An EBL system approximates the underlying pattern by choosing some $\theta^+ \in \Theta(\Phi)$:

$$\theta^+ \approx \underset{\theta \in \Theta(\Phi)}{\operatorname{argmax}}\, Pr(Z|\theta) \cdot Pr(\theta|\Theta(\Phi)) \tag{37.3}$$

This is just Expression 37.2 substituting $\Theta(\Phi)$ for H; $Pr(\theta|\Theta(\Phi))$ is the prior (or regularizer). Expression 37.3 can be realized by a conventional statistical learner once $\Theta(\Phi)$ is defined. While θ^+ may describe the underlying pattern accurately, it is expressed in the vocabulary of the derived features. These are not included in H. Thus, once θ^+ is found, the EBL system chooses some $c^+ \in H$ as an acceptable output:

$$c^+ \approx \underset{h \in H}{\operatorname{argmin}} \left\| h - \theta^+ \right\| \tag{37.4}$$

All three concepts, θ^+ from Θ (Expression 37.3) and both c^+'s from H (Expressions 37.2 and 37.4), approximate the same underlying pattern.

But θ^+ does so using Φ. With a good domain theory and the right selection of derived features, Θ employs a much richer and more focused vocabulary. The underlying pattern will be clearer and its representation simpler.

Generally, θ^+ will be syntactically the simplest. Next simplest will be c^+ from Expression 37.2 because the choice is regularized by $Pr(h|H)$. The c^+ from Expression 37.4 may be quite complex. It may be well beyond the reach of a statistical learner based on the evidence in Z alone. In EBL, regularization has already stabilized the choice of θ^+ in Expression 37.3; direct regularization in H is less of an issue. This underlies EBL's ability to confidently learn complex concepts in an example-efficient manner.

In practice, step (37.4) may be omitted in favor of either classical "utility problem" solution outlined in the previous section. This can be more convenient than the earlier theoretically sound approach, particularly if H is a feature-neutral well-formedness rule. For example, H might employ a linear separator or a Gaussian kernel using all of the example's features without caring what the features actually are. Then θ^+ itself may be returned in place of c^+. But in this case, each new example to be classified must first be explicitly embellished with the new derived features so that the classifier can be properly evaluated.

Let Φ^+ denote the set of derived features sought by EBL. This is the set that yields the best concept from the new hypothesis space. The choice of the feature set is regularized by the derivational complexity of the set, $Pr(\Phi|\Delta)$. The domain theory's analytic evidence for the set balances the empirical fit of Expression 37.3:

$$\Phi^+ \approx \underset{\Phi \in 2^\Delta}{\operatorname{argmax}} \left(Pr(\Phi|\Delta) \cdot \max_{\theta \in \Theta(\Phi)} [Pr(Z|\theta) \cdot Pr(\theta|\Theta(\Phi))] \right) \tag{37.5}$$

Note the approximation; it is not necessary to guarantee that an optimal feature set is found. However, even finding an approximation by directly applying Expression 37.5 would be intractable. Expression 37.5 optimizes over the power set of Δ's derivable expressions which itself may be unboundedly large.

The key is to employ explanations. An explanation for a set of training examples is any derivation from Δ that satisfactorily labels the set of examples. In EBL, a small random subset $Z_E \subset Z$ is selected to guide explanation construction. Guiding explanation construction is similar to classical EBL: a candidate explanation that is unsatisfactory will mislabel some examples. The features of these examples in the light of Δ show the relevant ways that the explanation may be amended; see [4,10,36].

An explanation has the form of a proof tree in a logical deduction. It exposes internal structure and intermediate conclusions, although, of course, now the result is conjectured rather than entailed. Procedurally also, there is a parallel between an EBL domain theory and a theorem prover of conventional logic: (1) Δ is a deterministic inferential system consisting of an inference engine and a number of expert-supplied expressions describing world behavior (the axioms of a conventional system). (2) The inference engine, as with conventional logic, derives new expressions following a prescribed matching and combining process. (3) Also as in conventional logic, Δ can be viewed formally as a finite representation denoting a possibly infinite number of derivable structures.

In EBL, several explanations are generated that cover the subset Z_E. The derived feature set, Φ^+, is the union of derived features from the explanations. An enhanced hypothesis space, $\Theta(\Phi^+)$, is formed that includes these new features. The remaining training examples, $Z_C \equiv Z - Z_E$, are employed conventionally to choose a hypothesis from this augmented hypothesis space:

$$\theta^+ \approx \underset{\theta \in \Theta(\Phi^+)}{\operatorname{argmax}} Pr(Z_C|\theta) \cdot Pr\left(\theta|\Theta,\left(\Phi^+\right)\right) \tag{37.6}$$

How are derived features formed from an explanation? The process starts with the intermediate conclusions (subgoals) that arise in deriving the explanation from Δ. Each intermediate conclusion is a predicate, possibly with arguments. The explanation specifies how the predicate and its arguments are evaluated and the role they play in concluding the classification label. For each predicate and each argument, a new feature is defined. Each predicate and argument is itself derivable from the domain theory. Each is rooted in the example's raw features, and each concludes a property deemed by the explanation to be important in assigning the class label. For each new feature, there is a feature function that follows the dictates of the explanation. When given an example, it computes the feature's value from the example's raw input features.

Thus, explanations sidestep the intractable maximization of Expression 37.5. While features are derivable from Δ, they are not derived directly. Rather explanations are derived, guided by training examples, and the features are extracted from the explanations. Explanation construction must satisfy two requirements:

1. Efficiency demands that only a few explanations be constructed.
2. Accuracy requires that the expected quality of these explanations be high.

A new semantics for Δ is the most important advance in modern EBL and allows these requirements to be met. Unlike conventional logic, the EBL inferential system denotes a *sequence*, rather than a *set*. The ordering of the derivations is by increasing derivational complexity (defined to be the minimal inferential effort sufficient for its derivation). Thus, Δ is viewed formally as the generative denotation of a possibly infinite sequence of derivations:

$$\Delta \equiv \left(\xi_1, \xi_2, \xi_3 \ldots \xi_{521}, \xi_{522}, \xi_{523}, \xi_{524} \ldots \xi_{3,671,495}, \xi_{3,671,496} \ldots\right) \tag{37.7}$$

This apparently small change has far-reaching consequences. Each derivation is associated with some derivational complexity. Derivations with lower complexity appear earlier in the sequence, so the inverse of derivational complexity is a measure of how easily an element of Δ can be derived. This measure induces a probability distribution over Δ with

$$Pr\left(\xi_1|\Delta\right) \geq Pr\left(\xi_2|\Delta\right) \geq Pr\left(\xi_3|\Delta\right) \geq \ldots$$

Consider the full distribution, $Pr(\Xi|\Delta)$, Δ's analytic prior over derivations. What is the nature of this distribution? By definition, Δ embodies a coding scheme that communicates the expert's conceptual understanding of the domain. It follows that $Pr(\Xi|\Delta)$ is the optimal distribution associated with the code words (derivations of Sequence 37.7). Therefore, from coding theory,

$$\log Pr(\xi_i|\Delta) = -\mathcal{D}(\xi_i)$$

where $\mathcal{D}(\xi)$ denotes the derivational complexity of ξ. Let δ be the average difference in derivational complexity between adjacent ξ_i's over Expression 37.7. Note that δ is an intrinsic property of Δ so that $\delta = \delta(\Delta)$. Knowing nothing more about Δ, the best estimate of the derivational complexity of the *i*'th ξ is

$$\mathcal{D}(\xi_i) = \delta \cdot i \text{ which (from earlier) is also} - \log Pr(\xi_i|\Delta)$$

This yields a prior probability over the ξ's given Δ of the form

$$Pr\left(\xi_i|\Delta\right) = \gamma \cdot d^i \tag{37.8}$$

where
d is a more convenient parameter than δ: ($d \equiv e^{-\delta}$)
γ is a normalizing constant

This is precisely a geometric distribution with traditional parameter (1 – *d*). It is a discrete distribution with exponentially decaying probabilities, each multiplied by an additional factor of (1 – *d*). The normalizing constant γ is *d*, and $0 < d < 1$.

Figure 37.2 shows distributions for different Δ's with *d*'s of 0.6, 0.8, and 0.9. As *d* approaches 1, Δ behaves more like a conventional logic. In the limit as $d \to 1$, there is no erosion of confidence in Δ's inferences: Every derivation is firmly believed; the ordering of the sequence carries no information; and the EBL system becomes a conventional logic. It would then inherit all of logic's brittleness: the statements are axiomatic constraints on possible worlds. To work, they must be carefully engineered to avoid unintended pernicious interactions.

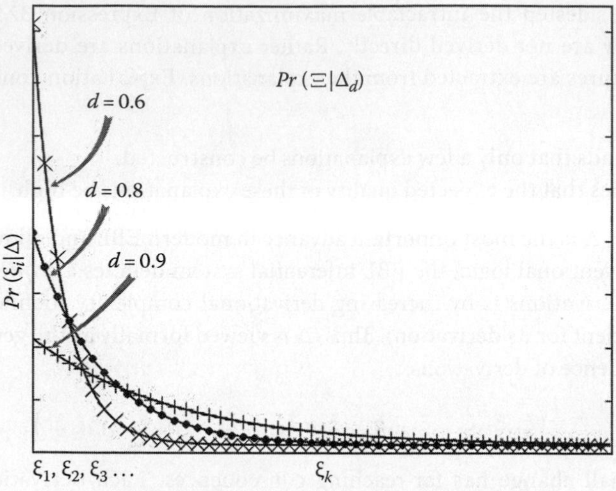

FIGURE 37.2 Three prior distributions over derivations for values 0.6, 0.8, and 0.9 of the domain theory property d.

Conversely, as d decreases toward 0, the EBL system behaves more like a conventional statistical learner. In the limit as $d \rightarrow 0$, each hypothesis stands on its own, and there is no benefit to shorter derivations. The domain theory provides no useful information and cannot serve as analytic evidence.

Now consider what happens as a particular EBL system begins to generate explanations for Z_E, the chosen subset of training examples. Suppose Δ has d of 0.8. Let the first successful explanation of Z_E be ξ_k (the kth derivation in Expression 37.7). The previous $k - 1$ derivations originally merited the highest probability (highest in the prior for $d = 0.8$ as shown in Figure 37.2). But since these derivations cannot explain the training subset Z_E, the explanation generator does not return them.

Finding the first explanation for Z_E at ξ_k changes the distribution considerably. Let $Pr(\Xi\Delta, Z_E)$ denote the posterior distribution over the derivable expressions after the first explanation has been generated that covers Z_E. It follows that the first $k - 1$ derivations cannot cover Z_E, and their posterior falls to 0.

The prior and the posterior for this state of affairs are shown in Figure 37.3. At this point, only the first explanation (the kth derivation in Expression 37.7) is constructed. It is illustrated as the leftmost, tallest

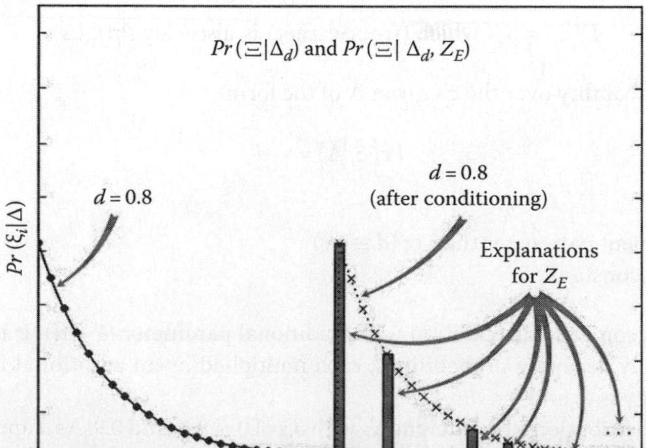

FIGURE 37.3 The prior distribution over derivations for $d = 0.8$, and the posterior distribution given the first explanation is the kth derivation; the explanations that cover the subset of training examples Z_E are shown as gray bars.

vertical gray bar. The other explanations that would cover Z_E if they were constructed are also illustrated as vertical gray bars. The height of each bar indicates its posterior probability.

The posterior distribution $Pr(\Xi|\Delta, Z_E)$ is simply a shifted version of the original $Pr(\Xi|\Delta)$. The change from prior to posterior can be considerable. Clearly, the KL divergence [8] from the prior to the posterior in Figure 37.3 is large. Thus, the information contained in Z_E can be quite large even though the size of Z_E may be quite small (it may even contain just one example). But unlocking this information requires Δ and, of course, sufficient processing resources to construct the first explanation. With these, EBL greatly magnifies the information content of the training examples.

The EBL procedure satisfies the two requirements mentioned earlier: the features with high expected utility are likely to be constructed from just a few of the least expensive explanations covering Z_E. These few explanations, while similar in cost, may conjecture quite different justifications. Each contributes to the set of derived features. As will be illustrated with implementations in Section 37.3, the classifier may ultimately employ relatively few derived features.

Let $\Phi(\Delta, Z_E)$ be the features added from the first few successful explanations of Z_E. Then $\Theta(\Phi(\Delta, Z_E))$ is the resulting enhanced hypothesis space. Expression 37.6 becomes

$$\theta^+ \approx \underset{\theta \in \Theta(\Phi(\Delta, Z_E))}{\operatorname{argmax}} \; Pr\left(Z_C|\theta\right) \cdot Pr\left(\theta|\Theta(\Phi(\Delta, Z_E))\right) \tag{37.9}$$

Thus, the EBL procedure, following Expressions 37.4 and 37.9, is as follows:

1. Employ some of the training examples, Z_E, to construct several explanations, $\{\xi_i\}$.
2. Extract features from the explanations, $\Phi(\Delta,Z_E)$, and assemble them into an extended hypothesis space, $\Theta(\Phi(\Delta, Z_E))$.
3. Use the rest of the training examples, Z_C, to choose a concept from the enhanced space, $\theta^+ \in \Theta(\Phi(\Delta,Z_E))$.
4. Choose a concept from the original space $c^+ \in H$ closest to the enhanced concept, θ^+.

37.3 Practice

The roots of EBL can be traced back a long way. But its formalization is still emerging. The motivation of EBL is to exploit general prior knowledge. Two early AI works served as inspirations for the development of this view. These were the MACROPS learning of the STRIPS system [15] and the notion of goal regression [58]. An important conceptual step was the explicit recognition of the role of *explanation* (i.e., building a structure justifying an unobserved feature of interest such as a class label) [9,43,53]. Seminal papers [11,42] sparked an explosion of interest in EBL.

In modern EBL, the construction of features from an explanation employs similar processing to that of classical EBL [28,44]. But in classical EBL, the acquired fully formed concept was the preserved portion of the explanation, from the "operational" predicates up to the class label conclusion. In modern EBL, there is no operationality criterion; all internal conclusions are used. Furthermore, the important portions of the explanation are from the raw feature leaves up to the internal conclusions. Modern EBL explanations serve to generate features, not the fully formed concept of classical EBL. Several aspects of modern EBL are shared with earlier EBL systems. Multiple explanations were explored in [7,17]. The use of an embedded statistical learner can be seen as a generalization of the research combining EBL with neural networks [56,57].

More recently, EBL has been applied to AI and machine learning problems such as structured prediction [34,36], planning [16,32,26], natural language processing [35], vision and image processing [33,34,38,46], and as a component in larger systems such as [60].

Two implementations illustrate the important EBL principles of the previous sections. The first [33] learns to distinguish ketch from schooner sailing ships in the Pascal VOC data set [14]. It shows how

FIGURE 37.4 (a) Three ketches and (b) three schooners. The EBL-augmented representations are below the raw image representations. Dots denote the highest point and center of geometry. The inferred hull and sail labels are shown light and dark pixels respectively; water and background labels are omitted.

explanation-generated features, although quite approximate, can yield an accurate classifier. The second [59] learns to distinguish similar handwritten Chinese characters. It illustrates feature generation from the interaction of domain theory distinctions and training examples. It also illustrates the ability to grow or shrink the structural complexity of the classifier in response to the difficulty of the particular learning problem.

Both systems employ the acquired EBL feature functions as "virtual sensors." These satisfy the analytic approach to the utility problem as in [13]. In applying the learned concept, these virtual sensors embellish the new example to be classified with derived features.

Consider the right-facing sailing ships in Figure 37.4. The top row shows ketches (Figure 37.4a) and schooners (Figure 37.4b). It is not easy to tell a ketch from a schooner. The key is that the foremast of ketches is taller while in schooners the aft mast is taller. The behavior of an EBL system is compared with two conventional statistical learners.

Figure 37.5 shows learning curves for the three systems. These are error rates on a withheld test set after training on an increasing numbers of examples. The EBL system significantly outperforms the others. The conventional statistical learners are the two top lines. The better conventional learner achieves an error rate of 30% after training on 30 examples. The EBL system is shown as a dashed line with circle markers. It achieves a 21% error rate after 5 training examples and 17% after 30. On the same test images, after being told how to distinguish ketches from schooners, humans turned in an error rate of 11%. This is shown as the bottom horizontal dotted line. The human error rate reflects the relatively poor quality of the VOC data set itself. Several images are misclassified, some are framed so that the top of the masts are beyond the top of the image, and a number of images are neither ketches nor schooners but single-mast sloops.

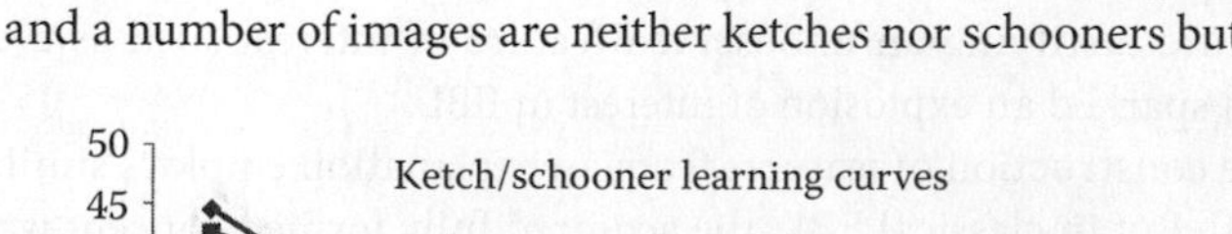

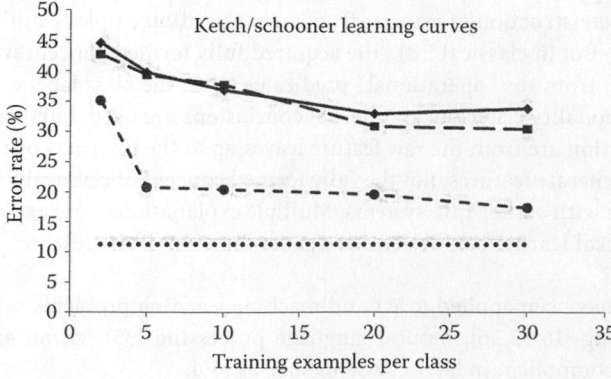

FIGURE 37.5 Ketch vs. schooner learning curves: two conventional learners (diamond and square), the EBL learner (circle), and human performance (horizontal dotted line).

There is no accurate concept within the scope of the conventional systems. It is easy to see why. The conventional state-of-the-art systems extract code words (clustered local patches of pixels) from the training images [37] without regard to class. Two distributions over the code words are constructed. One is the distribution imposed by ketch training images and the other by schooners. Given a new unlabeled image, code words are again extracted from it. The distribution of its code words fits one of the training distributions better than the other and the image is assigned the corresponding label.

Why is this task challenging? The ketch/schooner signal to noise ratio is very low: relatively few pixels encode the class information. Furthermore, those pixels occur in small functionally defined regions that do not correspond to any particular image location. Even worse, they exhibit only high-order systematic covariate structure. The conventional systems perform better than chance only because ketches tend to be used nearer the shore and this results in somewhat different distribution of background pixels. Statistics over code words expose this distinction quite easily. But, as can be seen, their learning rates flatten. Adding many more training examples would serve only to improve the confidence in the sea-to-shore distribution. The ketch/schooner accuracy would not improve much.

Rather than the whole image, the EBL system tries to model distributions for expected conceptual constituents of the ships. These intermediate concepts are introduced by the domain theory: water, hull, sails, and background. It also specifies ways to compute properties of regions such as height, width, and centroid and their geometric relations: above, below, left, right. The chosen EBL classifier employs several derived features: the joint model of the hull and sails, the sail center of geometry, and the highest point of the sails. The learned classifier is as follows: the ketch (schooner) has a sail center of geometry left of (right of) the highest point of the sail. The relevant derived features are shown in Figure 37.4. These include the pixels corresponding to sails (darker gray), the pixels corresponding to the hull (lighter gray), and black dots to indicate the center of geometry and the highest point of the sails. Simpler, more straightforward competitors, such as comparing the positions of the highest sail points, do not perform well. It might at first seem that modeling the hull is an unnecessary complication. But the hull turns out to be more distinctive than the sails that, without the hull, are often confused with sky and clouds.

Of particular note in Figure 37.4 is the relatively poor correspondence between the explained images (lower rows) and the original images (upper rows). This illustrates the power and robustness of combining the analytic and empirical information. One might hope for a more accurate internal representation, modeling, for example, the sails more faithfully. But accuracy has a price in terms of computational efficiency and (especially) in terms of robustness. Accurate sail identification would require a feature function that is much more subtle, derivationally much more complex, and computationally much more expensive. Conventional logic requires such an accurate representation. The brittleness inherent in conventional logic is avoided in EBL precisely because of this "good enough" modeling. Features are constructed so as to be minimally sufficient for their downstream purpose.

A second system applies EBL to the problem of distinguishing offline handwritten Chinese characters. Offline means that the input is a pixel image of the drawn glyph. This is contrasted with online processing in which captured stylus information (such as position, velocity, and acceleration) is available via a touch-sensitive pad. Offline processing is significantly more difficult.

Chinese characters are more complex than digits or Latin characters; each is composed from a few to a few dozen strokes. Because there are 10–20,000 common characters, training data are much more limited. Available databases contain a few tens to a few hundred handwritten examples of each character, compared to the tens of thousands of training images available for handwritten digits and Latin characters. These conditions challenge even the best machine learning algorithms.

A space of binary classification problems can be defined by selecting two Chinese characters to distinguish. This generates problems of widely varying difficulty. Most characters are quite different from each other, resulting in relatively easy problems. But many characters have one or more very similar counterparts. These form extremely challenging classification problems, particularly given the paucity of training

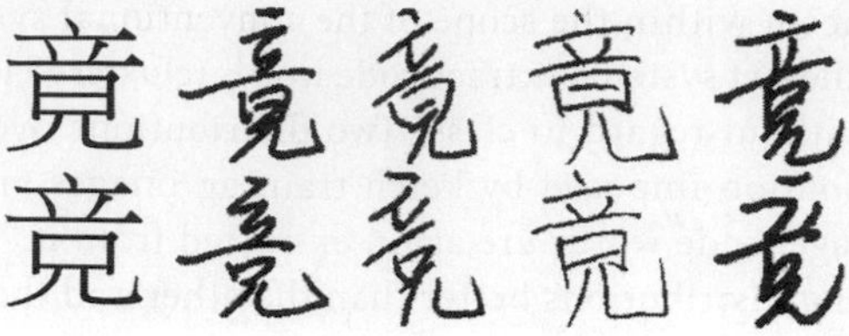

FIGURE 37.6 A printed and four handwritten examples of two similar Chinese characters.

examples and the diversity of writing styles. Figure 37.6 illustrates one such difficult classification problem. The top and bottom rows show two different characters; the left column is the printed glyph, the remaining columns show each character drawn by four different writers. It is clear that the noise induced by the different writing styles is large compared to the systematic class difference between the characters.

Research on handwritten Chinese characters has a long history [24,55]. For Chinese characters, as in most statistical classification problems, the first step in state-of-the-art processing is dimensionality reduction. One of the most effective specialized feature sets is known as WDH (for weighted directional histograms) originally from [30]. First, the image is normalized and then the thousands of pixel features are compressed into several hundred parameters encoding discretized local edge gradients.

The present experiment compares the behavior of an EBL system with a support vector machine (SVM). Chinese characters are input to the SVM as WDH features. The SVM employs a radial basis kernel that preliminary experiments indicate outperforms other standard kernels.

By contrast, the EBL system inputs characters as raw image pixels. Its underlying statistical learner is also an SVM employing a radial basis kernel, but its SVM is applied to a small chosen region rather than the entire character image. Explanation-derived features define this expected high-information region. The domain theory specifies that characters are composed of strokes and that the observable features (the raw input pixels) arise from these strokes. Strokes are modeled (quite imperfectly) as long thin rectangles. Input pixels are "explained" by these rectangular strokes. The domain theory also includes an "ideal" stroke representation for each character class (i.e., a single printed character of the class similar to that of a vector font but composed of the simple rectangular strokes) and the ability to compute WDH features of an image region.

The derived features are problem-specific configurations of strokes. These features are not mentioned explicitly in the domain theory but are constructed automatically during the learning process. Given a different learning problem (i.e., a different pair of difficult characters), quite different features will be constructed.

The features form a sequence. Each feature is evaluated from the image pixels together with the values of sequentially earlier features. The value of a feature is the set of parameters that optimize the local fit of its stroke configuration in the image. The explanation conjectures that with high confidence, each feature function in the sequence can be fit to the image sufficiently accurately to support the next feature function, and that the final target region contains sufficient discriminative evidence to classify the image.

Figure 37.7 illustrates the feature sequence applied to two test instances from a particular difficult-to-distinguish pair of Chinese characters. The first column shows handwritten instances of the two characters. The top row, left to right, shows the sequence of, in this case, five derived features. Each is an expected configuration of idealized (rectangular) strokes. Strokes from the structure common to both characters are black; class-discriminative strokes are reduced in intensity and appear gray.

The value of each feature is the result of an optimization. It is the vector of affine transformation parameters that locally optimizes the fit of the stroke configuration to the pixel image. Sequentially previous values (best fits of earlier features) are used to initialize the optimization search. The correctness of each feature evaluation is contingent on the correctness of earlier feature evaluations. Each must be initialized to within some basin of attraction around the optimal value (a small locally convex optimization region).

The rows of Figure 37.7 directly below the derived features show the application of each feature to each test image. Its value, the optimized affine parameters, is illustrated by superimposing the transformed

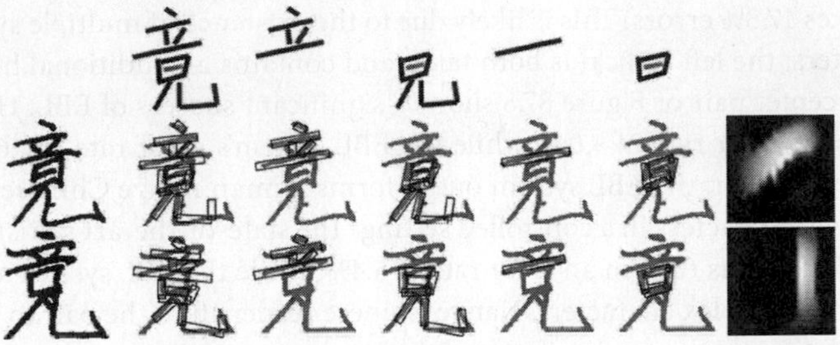

FIGURE 37.7 Application of the learned classifier to instances of a difficult pair showing the intermediate features applied. Left: the handwritten characters; top: the constructed intermediate features; and right: the normalized discriminative feature region given to the SVM.

feature strokes onto the original image. At each step, the feature is initialized at the location predicted by the previous feature values. Then, using the feature-to-pixel fit as a gradient, the locally optimal affine transformation parameters are found. The vector of optimized parameters is the value of the feature.

As can be seen, the first feature is the average full character rendered with rectangular strokes. In this and other features, the gray class-discriminative strokes have less effect on the optimizer because of their diminished intensity. There are no previous steps; the feature is initialized to the global expected location over the training set. The basin of attraction of the feature is large since the feature includes all of the strokes. However, it is clear that the best affine fit is poor for both images. This is due to the limited flexibility afforded by an affine transformation. The fit is far too poor to reliably test for the class-discriminative stroke, but it is sufficient to initialize the second feature, consisting of five strokes, to its basin of attraction, and so on.

Each feature fulfills a role. The roles are (1) to provide information about the target window (the fifth feature performs this role), (2) to provide improved initialization for a downstream features (this is the role of the first and third features), and (3) to expand the basin of attraction of downstream features by matching and thereby removing from later gradient computations, some pixels that empirically proved distracting to the optimizer (the second and fourth features fulfill this role).

The final target region is bounded by the four common strokes of feature five. The region is normalized to a square. The result is treated conventionally: WDH features are computed on this square region and given to the RBF SVM for final classification. The normalized region for each image is shown in the rightmost column of Figure 37.7. The identification of the discriminative window is far from perfect, but it is sufficient for a large margin classifier to detect, in this case, the predominance of horizontal WDH features. Details of the system can be found in [59].

The EBL system significantly improves performance over the state-of-the-art WDH RBF SVM. The average error rate over the ten hardest-to-distinguish pairs is improved by 20%. This is particularly significant since this EBL system is only equipped to consider a single difference. An obvious refinement would be to explain multiple differences. The conventional state-of-the-art classifier, by taking a dot product with support WDH images, integrates all differences over the entire character image. While on average the EBL system outperforms the conventional statistical classifier, one exception is illustrated in the left pair of Figure 37.8. Here the RBF SVM with WDH features achieves an error rate of 15.8%, while

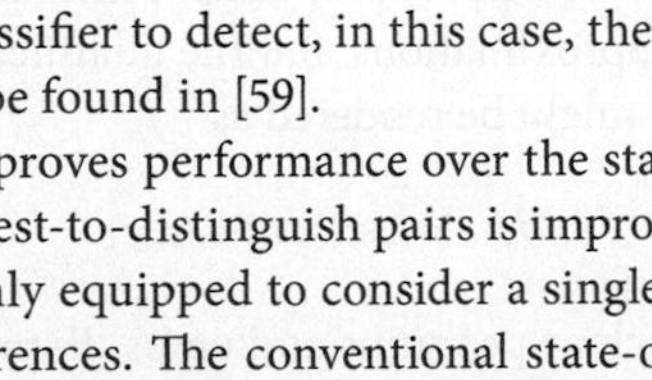

FIGURE 37.8 The performance of EBL compared to the state of the art is (left) slightly worse, (center) significantly better, and (right) better than human native Chinese readers.

the EBL system makes 17.3% errors. This is likely due to the existence of multiple systematic differences between the characters; the left radical is both taller and contains an additional horizontal stroke. On the other hand, the center pair of Figure 37.8 shows a significant success of EBL. Here the state-of-the-art learner achieves an error rate of 9.0%, while the EBL system's error rate is 1.6%. The right pair is also interesting. For this pair, the EBL system outperforms human native Chinese readers tasked with distinguishing the two characters in a controlled setting. The state-of-the-art statistical system achieves an error rate of 7.0%, humans turn in an error rate of 6.1%, while the EBL system achieves 4.0% errors This is one of the more complex characters. Native Chinese readers find the pair to be quite challenging to distinguish in isolation.

37.4 Research Issues

The EBL presented in this chapter represents a significant advance over previous formalisms (e.g., [50]). But it is only the most recent snapshot in a long evolutionary path. Much work remains to be done.

First and foremost is the question of codifying domain theories. Implications for the novel semantics of prior knowledge are far reaching and not yet fully understood. While the syntax of first-order logic is standard and familiar, it evolved to fill a very different need than EBL. Because it is widely known, first-order logic is convenient for defining and analyzing EBL but it seems not to be the representation of choice for practical implementations. Most successful modern EBL systems, including the two discussed in Section 37.3, employ ad hoc inference systems. A more appropriate general representation scheme may emerge from experience with more domains.

Another important need is in the development of a more sophisticated model of analytic uncertainty. In the current presentation, derivational complexity is the measure of confidence. The development of a more refined measure is needed. Some first steps are clear. Information on trustworthiness is encoded within an explanation's internal structure. Consider again the deficiencies of conventional logic. The root cause of logic's brittleness in AI applications is the interaction between its semantics (based on satisfiability) and the qualification problem [40]. The semantics dictates that adopting its axioms means believing just as strongly all of the implications of the axioms. Unintended conclusions arise from any imprecisions or approximations in the statement of the axioms. In conventional logic, if one believes "cars move" and also that "buried things do not move," then one is equally committed to believing that "cars cannot be buried." The qualification problem states that general (universally quantified) statements intended to describe the real world can at best be approximations. One might hope to amend the axioms in order to eliminate such approximations. But the qualification problem assures us that this is impossible. For example, "birds fly" might be rendered as

$$\forall x\ Bird(x) \Rightarrow Flies(x) \tag{37.10}$$

But there are obvious exceptions such as penguins and emus. Expression 37.10 can be easily amended with the known flightless birds. Unfortunately, there are always more exceptions such as cooked Thanksgiving turkeys and birds with their flight feathers clipped. Merely adding more antecedent conditions (qualifications) fails for two reasons: (1) additional (increasingly bizarre but very clear) exceptions continue to arise (birds with their feet cast in cement) and (2) the inference system becomes paralyzed by the need to entertain and defeat a myriad of unlikely but logically necessary conditions.

Let $\{\Gamma_i\}$ be the set of initial statements of EBL's Δ. Perhaps Γ_1 is Expression 37.10, Γ_2 might be a representation of "Penguins do not fly," Γ_3 "Penguins are birds," etc. These are not acceptable in conventional logic as they admit no possible worlds (the set is self-inconsistent). But in an EBL framework, "believed" means something different. To investigate this difference, consider how representations of Δ might be converted to their "closest" logical counterparts:

$$\text{Each } \Gamma_i \text{ must be replaced with } \alpha_i \Rightarrow \Gamma_i$$

where each α_i denotes the weakest sufficient condition that guarantees the truth of the corresponding Γ_i. So α_1 would represent "*x* is not a penguin nor is it an emu nor…nor is it cooked nor is its feet cast in any cement block nor…" It rules out every possible exception to Expression 37.10, including all the bizarre ones. The difference with conventional logic is that in EBL these α's are implicit; they are not represented nor are they reasoned about. Their truth conditions are subtle, depending both on their context within the explanation and on the distribution of examples that will be seen. This dynamic semantics sets EBL apart from past approaches for combining logic and statistics. See [12,19,23,29]. Perhaps the closest in spirit is [29]. Since these α_i's are not reasoned about, their probabilities of holding may be modeled as independent of each other. So an explanation composed of the set of statements $\{\Gamma_i\}$ holds with probability

$$\prod_i Pr(\alpha_i)$$

Like Expression 37.8, this yields an exponentially decaying probability over derivations. Estimates of the $Pr(\alpha_i)$'s (although not the α_i's themselves) may be updated as experience with Δ accrues.

Experience may also be helpful to EBL in exploring more expressive domain theories, such as those with second-order statements. First-order logic represents an attractive trade-off between expressiveness and computability for logic. But the trade-off is quite different for EBL. The undecidability of higher-order logic asserts that no effective procedure can exist that reliably determines whether or not an expression is a theorem. But some, perhaps many, such determinations may be possible. These are precisely the ones with tractable derivations. An undecidable question may not halt. But failing to halt incurs infinite derivational complexity. These are implicitly ignored by an EBL system that constructs explanations in order of Expression 37.7. Furthermore, some higher-order statements can be quite useful. For example, the statement of the qualification problem is itself a higher-order statement. Humans are not paralyzed by such statements so clearly some inferential systems tolerate them well. Additional research should investigate and characterize the utility of such statements.

Another useful direction to explore concerns explanation generation. Instead of a random subset Z_E to guide explanation generation, the system might grow Z_E one example at a time. Early explanations would likely be abandoned quite frequently. But as more examples are taken into account, the explanations grow in complexity, and their expected adequacy improves; it is more likely that important distinctions are already incorporated. Standard online stopping criteria could signal when sufficient evidence has accrued. In this way, Z_E might consume something close to the minimal number of training examples, preserving the largest possible calibration set Z_C for the new concept.

There may be insight to be gained from the parallels with planning. Conventionally, in complex learning domains, some preprocessor performs dimensionality reduction. Only then does a pattern recognizer form the concept. In EBL, this division begins to be blurred. Feature evaluation proceeds sequentially with later features dependent on the choice and values of earlier ones. The enhanced representation also emerges as classification proceeds. This is reminiscent of executing a conditional or disjunctive plan. The learner plays the role of a conditional planner. It more devises a goal-directed procedure than characterizes a probability distribution.

The primary motivation for EBL is to improve example efficiency of learning. This is a clear need. But there is another, less acknowledged but more pernicious problem that EBL may also address. This concerns the representation of input data. Computational learning systems almost universally assume the data are drawn from some example space (X), and each example is represented as a vector of values from some set of features. Sometimes this assumption is warranted. An image is completely defined by a large but finite set of pixels; a natural language sentence is precisely a particular sequence of words and punctuation. Tasks such as identifying pedestrians in an image or performing semantic role labeling on the sentence fit very well into this ubiquitous representational assumption. But these are artificial tasks that, at best, might be subtasks for useful real-world behavior. Suppose, by contrast, we command our household robot "Get me my keys from the bedroom." The robot enters and must deal with the bedroom. Can it be described with a finite set of features? Pick an average room and make a list of all of the "features" that in principle can be noticed. It is impossible. The room is not defined by a finite set of features. Rather, it is an object of great complexity; the closer it is examined, the more features it has. This holds for most real-world objects. Humans almost never completely represent an object. Rather, we possess a suite of sensors through which we perceive the world. We choose to look more closely at some parts of the world than others, building a sufficient representation as we go, enhancing it as necessary. Our representations must consist of only a tiny subset of the possible features. And those features must be adequate for the task at hand. Include too many features and specious patterns abound; include too few and no useful pattern can be constructed. Machine learning research has evolved to avoid tasks where this becomes problematic. Conventional feature selection methods are far from satisfactory. They generally begin by enumerating *all* of the possible features and only then choosing a smaller set or combination. SVM's can effortlessly handle large or infinite inflated feature spaces, but like other machine learners, they are paralyzed by an unboundedly rich primal feature space. These generally result in narrow margins, a high fraction of examples becoming support vectors, and poor predictive generalization. It is clear that most of the attributes of most real-world objects must be *implicitly* ignored. A promising EBL direction might be to include a description of sensors' behaviors in the EBL domain theory, Δ. Then the bias of low derivational complexity, preferring the simplest adequate explanations, might naturally embrace a small but adequate set of object sensing commands.

Another interesting direction is to explore a human in the learning loop. In this regard, a human teacher might collaboratively help to build or to choose explanations. Human interaction has proven to be effective in theorem proving [27], which is similar to explanation building. Alternatively, the human might select or suggest elements for Z_E. In this regard, the approach might be similar to curriculum learning [1] and benefit from similar approaches.

In addition to refinements and extensions, there are several active areas in machine learning that might be explored as complementary to EBL. Perhaps the domain theory, Δ, might itself be learned. The approaches that fall under the rubric of nonparametric Bayesian learning would be relevant. Topic models, for example, allow the learner itself to discover latent structure from examples [3]. It might be fruitful to explore how the structure discovered by these systems might be employed as an EBL domain theory. A complication is that the latent hierarchies produced by nonparametric Bayesian approaches are propositional in expressiveness rather than first order. One interesting alternative is once again to introduce a human in the loop but now using natural language interaction. A number of NLP systems have examined extracting first-order or relational rules from text, essentially parsing text into the sort of semantic knowledge representation needed by EBL. See, for example, [6,20,54].

Finally, the relevance of EBL to psychology demands further exploration. There is ample evidence that humans exploit general prior knowledge during concept learning [45]. A number of studies have shown EBL-like behavior in adults [31,48,51]. There is recent experimental evidence that infants as young as 21 months also employ an EBL-like prior knowledge bias [2], going beyond statistical regularities and exploiting prior knowledge when learning.

37.5 Summary

Modern EBL enhances conventional statistical learning with analytic evidence from a general domain theory. A good domain theory introduces distinctions deemed by experts to be useful. Analytic inference supports the nonlinear interaction of these distinctions giving rise to a potentially huge number of derivable features. The explanation process insures that only a few of these, the ones relevant to the task at hand, are elevated to the status of new features. Some of these new features, perhaps many, will be specious. But others will be effective in describing aspects of the underlying pattern. A conventional statistical learner chooses which derived features are valuable, which to disregard, and how best to combine the good ones. The resulting pattern is then translated back into the target hypothesis space.

Both analytic inference and statistical inference possess strengths and weaknesses as models for the real world. Analytic inference (i.e., symbolic deduction) handles general quantified expressions easily and naturally. It is adept at generating well-formed complex nonlinear combinations of new beliefs from old ones. But in any interesting real-world domain, it unavoidably generates specious incorrect conclusions. All its conclusions are equally believed. Real-world correctness has no place within the formalism. On the other hand, the conclusions of statistical inference are based on observed data and thus are directly connected to the real world. Statistical inference naturally embraces uncertainty and offers formal guarantees of confidence in its conclusions. But it is not easily nor naturally extended beyond propositional representations, and it requires exponentially larger data sets to confidently discover the higher-order covariate structure required by many every-day patterns in the world.

The EBL process exploits the strengths of both analytic inference and statistical inference and uses each to circumvent the inherent weaknesses of the other. The analytic component generates explanations. But these explanations do not have to be "correct" in any sense. An explanation is used only to expose derived features. It is only required that some good features be included within the set of explanations. Each good feature helps to simplify the description of the underlying pattern; each good feature relieves the statistical learner of the need to work out potentially difficult interactions among the raw input features. Bad features, those that are unhelpful in describing the pattern, are easily dismissed or discounted by a statistical learner. By leveraging both analytic and empirical evidence, EBL may help AI to scale beyond its current capabilities.

For further information on specific topics, the reader is referred to the works cited in the text.

Key Terms

Concept or classifier: A function mapping examples to labels; a hypothesis from the hypothesis space chosen by a learner as an acceptable approximation to a target concept.

Derivation: A sequence of statements such that each statement is either an initial statement from the domain theory or the result of applying the domain theory's inference procedure to earlier statements in the sequence.

Derived feature: A property beyond an object's raw features that are likely to be true of the object given its raw features and the domain theory.

Domain: A set or space of related (classification learning) problems.

Domain theory: A set of statements expressing general prior knowledge or beliefs about a domain that are used to drive an inference procedure generating new beliefs from old ones. In EBL it serves as analytic evidence supporting the choice of a concept from the hypothesis space.

Example: An object that can be assigned a label (classified), often represented or individuated by a set of features.

Example space: The set of all classifiable objects.

Explanation: A derivation from the domain theory that, for one or more labeled examples, infers the label from the example's features.

Hypothesis space: The set of all well-formed concepts that in principle could be output by a learner.

Inference: Reasoning from believed statements to new statements that a rational system should also believe. Inference may be empirical (based on observations) or analytic (based on interpreted symbols) and may reflect uncertainty in the prior beliefs.

Overfitting: The selection by a learner of a concept from the hypothesis space that captures a pattern exhibited by the training examples but not exhibited by the underlying population; a learner's concept selection based on an insignificant amount of evidence.

Raw or input features: A set of properties that defines or individuates an example.

Target concept: The veridical function embodied by the world that generates labels for examples; generally it is not contained within the hypothesis space.

Training set: A collection of labeled examples given to a learner serving as empirical evidence supporting the choice of a concept from the hypothesis space.

References

1. Y. Bengio, J. Louradour, R. Collobert, and J. Weston. Curriculum learning. In *Proceedings of the 26th Annual International Conference on Machine Learning, ICML '09*, pp. 41–48, Montreal, Quebec, Canada, 2009, ACM, New York.
2. A. Bernard, R. Baillargeon, and G. DeJong. Beyond statistical regularities: Evidence for causal reasoning in infants. In *Biennial Meeting of the Society for Research in Child Development*, Montreal, Quebec, Canada, April 2011.
3. D. M. Blei, T. L. Griffiths, and M. I. Jordan. The nested Chinese restaurant process and Bayesian nonparametric inference of topic hierarchies. *Journal of ACM*, 57(2):1–30, 2010.
4. M. Brodie and G. DeJong. Iterated phantom induction: A knowledge-based approach to learning control. *Machine Learning*, 45(1):45–76, 2001.
5. R. Caruana. Multitask learning. *Machine Learning*, 28(1):41–75, July 1997.
6. D. L. Chen and R. J. Mooney. Learning to interpret natural language navigation instructions from observations. In *Proceedings of the National Conference on Artificial Intelligence, AAAI*, San Francisco, CA, 2011.
7. W. W. Cohen. Abductive explanation-based learning: A solution to the multiple inconsistent explanation problem. *Machine Learning*, 8:167–219, 1992.
8. T. M. Cover and J. A. Thomas. *Elements of Information Theory*, 2nd edn. John Wiley & Sons, Inc., New York, 2006.
9. G. DeJong. Generalizations based on explanations. In *IJCA181, the Seventh International Joint Conference on Artificial Intelligence*, pp. 67–69, Vancouver, Canada, 1981.
10. G. DeJong. Toward robust real-world inference: A new perspective on explanation-based learning. In *ECML06, the Seventeenth European Conference on Machine Learning*, pp. 102–113, Berlin, Germany, 2006.
11. G. DeJong and R. Mooney. Explanation-based learning: An alternative view. *Machine Learning*, 1(2):145–176, 1986.
12. P. Domingos and D. Lowd. *Markov Logic: An Interface Layer for Artificial Intelligence*. Morgan & Claypool Publishers, San Rafael, CA, 2009.
13. O. Etzioni. Acquiring search-control knowledge via static analysis. *Artificial Intelligence*, 62:255–302, 1993.
14. M. Everingham, L. Van Gool, C. K. I. Williams, J. Winn, and A. Zisserman. The Pascal visual object classes (voc) challenge. *International Journal of Computer Vision*, 88(2):303–338, June 2010.
15. R. Fikes, P. E. Hart, and N. J. Nilsson. Learning and executing generalized robot plans. *Artificial Intelligence*, 3(1–3):251–288, 1972.
16. L. Finkelstein and S. Markovitch. A selective macro-learning algorithm and its application to the N × N sliding-tile puzzle. *Journal of Artificial Intelligence Research (JAIR)*, 8:223–263, 1998.
17. N. S. Flann and T. G. Dietterich. A study of explanation-based methods for inductive learning. *Machine Learning*, 4:187–226, 1989.
18. M. Genesereth and N. Nilsson. *Logical Foundations of Artificial Intelligence*. Kaufmann, Los Altos, CA, 1987.
19. L. Getoor and B. Taskar, eds. *Introduction to Statistical Relational Learning*. MIT Press, Cambridge, MA, 2007.
20. D. Goldwasser and D. Roth. Learning from natural instructions. In *IJCAI*, Barcelona, Spain, pp. 1794–1800, 2011.

21. J. Gratch and G. DeJong. A statistical approach to adaptive problem solving. *Artificial Intelligence*, 88(1–2):101–142, 1996.
22. R. Greiner and I. Jurisica. A statistical approach to solving the EBL utility problem. In *Proceedings of the Tenth National Conference on Artificial Intelligence, AAAI*, pp. 241–248, San Jose, CA, 1992.
23. J. Halpern. *Reasoning about Uncertainty*. MIT Press, Cambridge, MA, 2005.
24. T. Hildebrandt and W. Liu. Optical recognition of handwritten Chinese characters: Advances since 1980. *Pattern Recognition*, 26:205–225, 1993.
25. H. Daumé III and D. Marcu. Domain adaptation for statistical classifiers. *Journal of Artificial Intelligence Research (JAIR)*, 26:101–126, 2006.
26. S. Kambhampati and S. W. Yoon. Explanation-based learning for planning. In *Encyclopedia of Machine Learning* Claude Sammut and Geoffrey Webb (eds.), pp. 392–396, Springer, 2010.
27. M. Kaufmann and R. Boyer. The Boyer-Moore theorem prover and its interactive enhancement. *Computers and Mathematics with Applications*, 29:27–62, 1995.
28. S. Kedar-Cabelli and T. McCarty. Explanation-based generalization as resolution theorem proving. In *Fourth International Workshop on Machine Learning*, pp. 383–398, Irvine, CA, 1987.
29. A. Kimmig, L. De Raedt, and H. Toivonen. Probabilistic explanation based learning. In *ECML07, the Eighteenth European Conference on Machine Learning*, pp. 176–187, Warsaw, Poland, 2007.
30. F. Kimura, T. Wakabyashi, S. Tsurouka, and Y. Miyake, Improvement of handwritten Japanese character recognition using weighted direction code histogram. *Pattern Recognition*, 30:1329–1337, 1997.
31. W. K. Ahn, R. Mooney, W. Brewer, and G. DeJong. Schema acquisition from one example: Psychological evidence for explanation-based learning. In *Proceedings of the Ninth Annual Conference of the Cognitive Science Society*, pp. 50–57, Seattle, WA, 1987.
32. G. Levine and G. DeJong. Explanation-based acquisition of planning operators. In *ICAPS06, the Sixteenth International Conference on Automated Planning and Scheduling*, pp. 152–161, Lake District, U.K., 2006.
33. G. Levine and G. DeJong. Explanation-based object recognition. In *IEEE Workshop on Applications of Computer Vision*, pp. 1–8, Copper Mountain, CO, 2008.
34. G. Levine and G. DeJong. Object detection by estimating and combining high-level features. In *ICIAP*, pp. 161–169, Victri Sul Mare, Italy, 2009.
35. G. Levine and G. DeJong. Automatic topic model adaptation for sentiment analysis in structured domains. In *Fifth International Joint Conference on Natural Language Processing (IJCNLP2011)*, pp. 75–83, Chiang Mai, Thailand, 2011.
36. G. Levine, G. DeJong, L.-L. Wang, R. Samdani, S. Vembu, and D. Roth. Automatic model adaptation for complex structured domains. In *ECML/PKDD (2)*, pp. 243–258, Barcelona, Spain, 2010.
37. F.-F. Li and P. Perona. A Bayesian hierarchical model for learning natural scene categories. In *CVPR (2)*, pp. 524–531, San Diego, CA, 2005.
38. S. H. Lim, L.-L. Wang, and G. DeJong. Integrating prior domain knowledge into discriminative learning using automatic model construction and phantom examples. *Pattern Recognition*, 42(12):3231–3240, 2009.
39. Y. Mansour and M. Schain. Robust domain adaptation. In *International Symposium on Artificial Intelligence and Mathematics*, Fort Lauderdale, FL, 2012.
40. J. McCarthy. Circumscription—A form non-monotonic reasoning. *Artificial Intelligence*, 13:27–39, 1980.
41. S. Minton. Quantitative results concerning the utility of explanation-based learning. *Artificial Intelligence*, 42(2–3):363–391, 1990.
42. T. Mitchell, R. Keller, and S. Kedar-Cabelli. Explanation-based learning: A unifying view. *Machine Learning*, 1(1):47–80, 1986.
43. T. M. Mitchell. Learning and problem solving. In *IJCAI*, pp. 1139–1151, Karlsruhe, West Germany, 1983.

44. R. Mooney and S. Bennett. A domain independent explanation-based generalizer. In *AAAI*, Philadelphia, PA, pp. 551–555, 1986.
45. G. Murphy. *The Big Book of Concepts*. MIT Press, Cambridge, MA, 2002.
46. J. O'Sullivan, T. Mitchell, and S. Thrun. Explanation-based neural network learning for mobile robot perception. In K. Ikeuchi and M. Veloso, eds., *Symbolic Visual Learning*. Oxford University Press, New York, 1997.
47. S. J. Pan and Q. Yang. A survey on transfer learning. *IEEE Transactions on Knowledge and Data Engineering*, 22(10):1345–1359, 2010.
48. M. J. Pazzani and D. F. Kibler. The utility of knowledge in inductive learning. *Machine Learning*, 9:57–94, 1992.
49. G. Priest. Paraconsistent logic. In D. Gabbay and F. Guenthner, eds., *Handbook of Philosophical Logic*, 2nd edn., Vol. 6, pp. 287–393. Kluwer Academic Publishers Dordrecht, the Netherlands, 2002.
50. S. Russell and P. Norvig. *Artificial Intelligence: A Modern Approach*, 3rd edn. Prentice-Hall, Upper Saddle River, NJ, 2010.
51. L. Schulz, T. Kushnir, and A. Gopnik. Learning from doing: Intervention and causal inference. In A. Gopnik and L. Schulz, eds., *Causal Learning: Psychology, Philosophy, and Computation*, pp. 67–85. Oxford University Press, New York, 2007.
52. S. Shapiro. Classical logic. In E. N. Zalta, ed., *The Stanford Encyclopedia of Philosophy*. Winter, 2009. http://plato.stanford.edu/entries/logic-classical
53. B. Silver. Learning equation solving methods from worked examples. In *Second International Machine Learning Workshop*, Monticello, IL, pp. 99–104, 1983.
54. S. Sorower, T. G. Dietterich, J. R. Doppa, W. Orr, P. Tadepalli, and X. Fern. Inverting Grice's maxims to learn rules from natural language extractions. In *NIPS*, pp. 1053–1061, Granada, Spain, 2011.
55. S. Srihari, X. Yang, and G. Ball. Offline Chinese handwriting recognition: An assessment of current technology. *Frontiers of Computer Science in China*, 1:137–155, 2007.
56. S. Thrun and T. M. Mitchell. Integrating inductive neural network learning and explanation-based learning. In *IJCAI*, pp. 930–936, Chambery, France, 1993.
57. G. G. Towell and J. W. Shavlik. Knowledge-based artificial neural networks. *Artificial Intelligence*, 70(1–2):119–165, 1994.
58. R. Waldinger. Achieving several goals simultaneously. In E. Elcock and D. Michie, eds., *Machine Intelligence 8: Machine Representations of Knowledge*, pp. 94 – 136. Ellis Horwood, Chichester, U.K., 1977.
59. L.-L. Wang. Use of prior knowledge in classification of similar and structured objects. PhD thesis, University of Illinois at Urbana, Champaign, IL, 2012.
60. X. (Shelley) Zhang, S. Yoon, P. DiBona, D. S. Appling, L. Ding, J. R. Doppa, D. Green et al. An ensemble architecture for learning complex problem solving techniques from demonstration. *ACM Transactions on Intelligent Systems and Technology (TIST)*, 75(3), 2012.

38 Search

Danny Kopec
Brooklyn College

James L. Cox
Brooklyn College

Stephen Lucci
The City College of New York

38.1 Introduction

Efforts using artificial intelligence (AI) to solve problems with computers—which humans routinely handle by employing innate cognitive abilities, pattern recognition, perception, and experience—invariably must turn to considerations of search. This chapter explores search methods in AI, including both *blind* exhaustive methods and informed heuristic and optimal methods, along with some more recent findings. The search methods covered include (for nonoptimal, uninformed approaches) *state-space search, generate and test, means–ends analysis, problem reduction, AND/OR Trees, depth-first search (DFS), and breadth-first search.* Under the umbrella of heuristic (informed) methods, we discuss *hill climbing, best-first search, bidirectional search, and the A* algorithm.* Tree searching algorithms for games have proved to be a rich source of study and provide empirical data about heuristic methods. Included here are the *SSS* algorithm*, the use of *iterative deepening*, and variations on the *alpha–beta minimax algorithm*, including the *MTD(f)* and *NegaScout* algorithms.

Coincident with the continuing price–performance improvement of small computers is the growing interest in reimplementing some of the heuristic techniques developed for problem-solving and planning programs, to see if they can be enhanced or replaced by other algorithmic methods. Since many of the heuristic methods are computationally intensive, the second half of the chapter focuses on parallel methods, which can exploit the benefits of parallel processing. The importance of parallel search is presented through an assortment of relatively recent algorithms including the parallel iterative deepening algorithm (PIDA*), *principal variation splitting (PVSplit), parallel window search (PWS), and the young brothers wait concept.* In addition, dynamic tree-splitting methods have evolved for both shared-memory parallel machines and networks of distributed computers. Here the issues include load balancing, processor utilization, and communication overhead. For single-agent search problems, we consider not only work-driven dynamic parallelism but also the more recent data-driven parallelism employed in *transposition table-driven (work) scheduling (TDS).* In adversarial games, tree pruning makes work load balancing particularly difficult, and hence we also consider some recent advances in dynamic parallel methods for game-tree search, including the very recent hybrid parallel window/iterative deepening search *(HyPS)*, which combines the strengths of PWS and tree-splitting algorithms such as PVSplit.

Another class of search algorithms is motivated by the desire to model natural processes. We call this class of algorithms *nature-based search.* These methods include *Monte Carlo simulation, simulated annealing (SA), genetic algorithms (GAs) and programming, Tabu search (TS), and ant colony optimization.*

The application of raw computing power, while an anathema to some, often provides better answers than is possible by reasoning or analogy. Thus, brute force techniques form a good basis against which to compare more sophisticated methods designed to mirror the human deductive process.

38.2 Uninformed Search Methods

38.2.1 Search Strategies

All search methods in computer science share in common three characteristics: (1) a world model or database of facts based on a choice of representation providing the current state, as well as other possible states and a goal state; (2) a set of operators that defines possible transformations of states; and (3) a control strategy that determines how transformations among states are to take place by applying operators. *Forward reasoning* is one technique for identifying states that are closer to a goal state. Working backward from a goal to the current state is called *backward reasoning.* As such, it is possible to make distinctions between bottom-up and top-down approaches to problem-solving. Bottom-up is often "goal oriented"—that is, reasoning backward from a goal state to solve intermediary subgoals. Top-down or data-driven reasoning is based on simply being able to reach a state that is defined as closer to a goal. Often the application of operators to a problem state may not lead directly to a goal state, so some *backtracking* may be necessary before a goal state can be found (Barr and Feigenbaum, 1981).

38.2.2 State-Space Search

Exhaustive search of a problem space (or search space) is often not feasible or practical because of the size of the problem space. In some instances, it is, however, necessary. More often, we are able to define a set of legal transformations of a state space (moves in a board game) from which those that are more likely to bring us closer to a goal state are selected, while others are never explored further. This technique in problem-solving is known as *split and prune.* In AI the technique that emulates this approach is called *generate and test.* Good generators are complete and will eventually produce all possible solutions while not proposing redundant ones. They are also informed; that is, they will employ additional information to constrain the solutions they propose. We note that the state space together with the allowable transitions forms a graph. In AI one typically searches the state space from a distinguished starting state and produces a search tree in which some nodes (states) may be repeated on different paths. In this context, AI literature discusses tree search (e.g., game trees) even when the underlying structure is not technically a tree since, for example, the same chess position may occur many times in the game tree and thus technically have multiple parents. In this case, each occurrence of the repeated state can be viewed as a separate node.

Means–ends analysis is another state-space technique whose purpose is to reduce the difference (distance) between a current state and a goal state. Determining "distance" between any state and a goal state can be facilitated by *difference-procedure tables,* which can effectively prescribe what the next state might be. Figure 38.1 illustrates how to use the means–ends analysis.

```
Repeat
        Describe the current state, the goal state,
            and the difference between the two.
        Use the difference between the current state and goal state,
            to select a promising transformation procedure.
        Apply the promising procedure and update the current state.
Until the GOAL is reached or
        no more procedures are available
If the GOAL is reached, announce success;
Otherwise, announce failure.
```

FIGURE 38.1 Means–ends analysis.

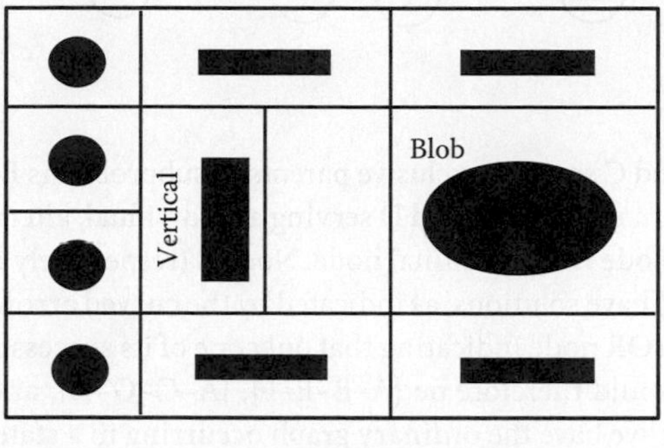

FIGURE 38.2 Problem reduction and the sliding block puzzle.

The technique of *problem reduction* is another important approach in AI. That is, solve a complex or larger problem by identifying smaller manageable problems (or subgoals), which you know can be solved in fewer steps.

For example, Figure 38.2 shows the "donkey" sliding block puzzle, which has been known for over 100 years. Subject to constraints on the movement of "pieces" in the sliding block puzzle, the task is to slide the blob around the vertical bar with the goal of moving it to the other side. The blob occupies four spaces and needs two adjacent vertical or horizontal spaces in order to be able to move, while the vertical bar needs two adjacent empty vertical spaces to move left or right or one empty space above or below it to move up or down. The horizontal bars' movements are complementary to the vertical bar. Likewise, the circles can move to any empty space around them in a horizontal or vertical line. A relatively uninformed state-space search can result in over 800 moves for this problem to be solved, with plenty of backtracking necessary. By problem reduction, resulting in the subgoal of trying the get the blob on the two rows above or below the vertical bar, it is possible to solve this puzzle in just 82 moves!

Another example of a technique for problem reduction is called *AND/OR Trees*. Here, the goal is to find a solution path to a given tree by applying the following rules:
A node is solvable if

1. It is a terminal node (a primitive problem)
2. It is a nonterminal AND node whose successors are all solvable
3. It is a nonterminal OR node and least one of its successors is solvable

Similarly, a node is unsolvable if

1. It is a nonterminal node that has no successors (a nonprimitive problem to which no operator applies)
2. It is a nonterminal AND node and at least one of its successors is unsolvable
3. It is a nonterminal OR node and all of its successors are unsolvable

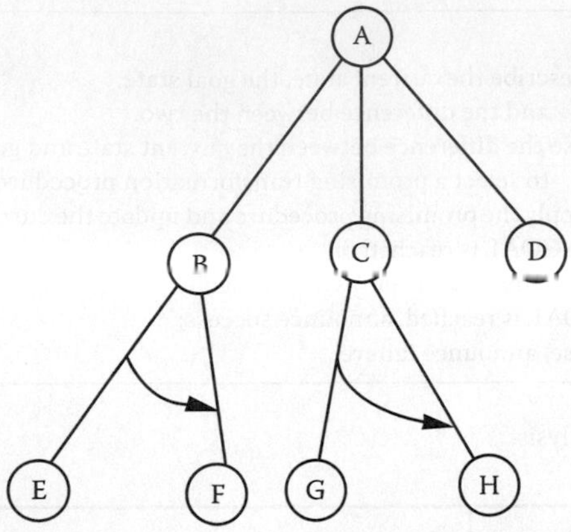

FIGURE 38.3 AND/OR Tree.

In Figure 38.3, nodes B and C serve as exclusive parents to subproblems EF and GH, respectively. One way of viewing the tree is with nodes B, C, and D serving as individual, alternative subproblems. B and C represent AND nodes, and node D is a terminal node. Node B (respectively node C) requires that both E and F (respectively G and H) have solutions, as indicated by the curved arrowheads connecting them. On the other hand, node A is an OR node indicating that only one of its successors (B, C, or D) needs to have a solution. Solution paths would therefore be {A–B–E–F}, {A–C–G–H}, and {A–D}. In the special case where no AND nodes occur, we have the ordinary graph occurring in a state-space search. However, the presence of AND nodes distinguishes AND/OR Trees (or graphs) from ordinary state structures, which call for their own specialized search techniques. Typical problems tackled by AND/OR Trees include games or puzzles and other well-defined state-space goal-oriented problems, such as robot planning, movement through an obstacle course, or setting a robot the task of reorganizing blocks on a flat surface.

38.2.2.1 Breadth-First Search

One way to view search problems is to consider all possible combinations of subgoals, by treating the problem as a tree search. *Breadth-first search* always explores nodes closest to the root node first, thereby visiting all nodes at a given layer first before moving to any longer paths. It pushes uniformly into the search tree. Because of memory requirements, breadth-first search is only practical on shallow trees or those with an extremely low branching factor. It is therefore not much used in practice, except as a basis for such *best-first search* algorithms such as A* and SSS*.

38.2.2.2 Depth-First Search

DFS is one of the most basic and fundamental blind search algorithms. It is used for bushy trees (with high branching factor) where a potential solution does not lie too deep in the tree. That is, "DFS is a good idea when you are confident that all partial paths either reach dead ends or become complete paths after a reasonable number of steps." In contrast, "DFS is a bad idea if there are long paths, particularly indefinitely long paths, that neither reach dead ends nor become complete paths" (Winston, 1992).

DFS always explores the leftmost node to the left first, that is, the one that is farthest down from the root of the tree. When a dead end (terminal node) is reached, the algorithm backtracks one level and then tries to go forward again. To prevent consideration of unacceptably long paths, a depth bound is often employed to limit the depth of search. At each node, immediate successors are generated, and a transition is made to the leftmost node, where the process continues recursively until a dead end or depth limit is reached. In Figure 38.4, DFS explores the tree in the order: I-E-b-F-B-a-G-c-H-C-a-D-A. Here the notation using lowercase letters indicates possible storage of provisional information about the subtree. For example, this could be a lower bound on the value of the tree.

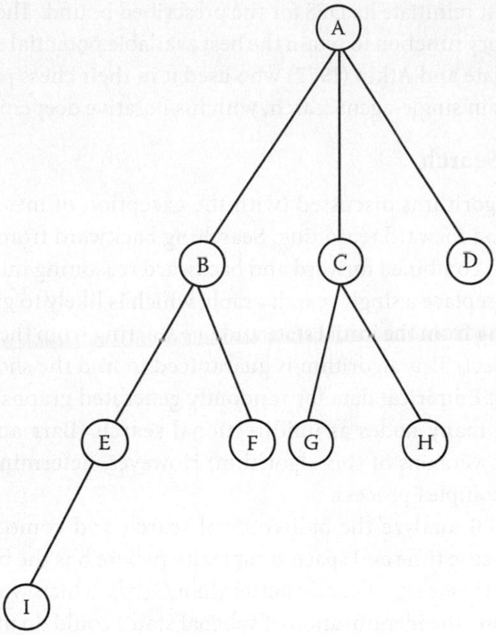

FIGURE 38.4 Tree example for depth-first and breadth-first search.

```
//      The A* (DFS) algorithm expands the N.i successors of node N
//      in best first order. It uses and sets solved, a global indicator.
//      It also uses a heuristic estimate function H(N), and a
//      transition cost C(N,N.i) of moving from N to N.i
//
IDA* (N) → cost
    bound ← H(N)
    while not solved
        bound ← DFS (N, bound)
    return bound                            // optimal cost

DFS (N, bound) → value
    if H(N) ≡ 0                             // leaf node
        solved ← true
        return 0
    new_bound ← ∞
    for each successor N.i of N
        merit ← C(N, N.i) + H(N.i)
        if merit ≤ bound
            merit ← C(N,N.i) + DFS (N.i, bound - C(N,N.i))
            if solved
                return merit
        if merit < new_bound
            new_bound ← merit
    return new_bound
```

FIGURE 38.5 The A* DFS algorithm for use with IDA*.

Figure 38.5 enhances DFS with a form of *iterative deepening* that can be used in a single-agent search like A*. DFS expands an immediate successor of some node N in a tree. The next successor to be expanded is (N.i), the one with lowest cost function. Thus, the expected value of node N.i is the estimated cost C(N,N.i) plus H(N), the known value of node N. The basic idea in iterative deepening is that a DFS is started with a depth bound of 1, and this bound increases by one at each new iteration. With each increase in depth, the algorithm must reinitiate its DFS for the prescribed bound. The idea of iterative deepening, in conjunction with a memory function to retain the best available potential solution paths from iteration to iteration, is credited to Slate and Atkin (1977) who used it in their chess program. Korf (1985) showed how efficient this method is in single-agent search, with his iterative deepening A* (IDA*) algorithm.

38.2.2.3 Bidirectional Search

To this point, all search algorithms discussed (with the exception of means–ends analysis and backtracking) have been based on forward reasoning. Searching backward from goal nodes to predecessors is relatively easy. Pohl (1971) combined forward and backward reasoning into a technique called *bidirectional search*. The idea is to replace a single search graph, which is likely to grow exponentially, with two smaller graphs—one starting from the initial state and one starting from the goal. The search terminates when the two graphs intersect. This algorithm is guaranteed to find the shortest solution path through a general state-space graph. Empirical data for randomly generated graphs show that Pohl's algorithm expands only about 1/4 as many nodes as unidirectional search (Barr and Feigenbaum, 1981). Pohl also implemented heuristic versions of this algorithm. However, determining when and how the two searches will intersect is a complex process.

Russell and Norvig (2003) analyze the bidirectional search and come to the conclusion that it is O ($b^{d/2}$) in terms of average case time and space complexity (where b is the branching factor and d is the depth). They point out that this is significantly better than O(b^d), which would be the cost of searching exhaustively in one direction. The identification of subgoal states could do much to reduce the costs. The large space requirements of the algorithm are considered its weakness.

However, Kaindl and Kainz (1997) have demonstrated that the long held belief that the algorithm is afflicted by the frontiers passing each other is wrong. They developed a new generic approach that dynamically improves heuristic values but is only applicable to bidirectional heuristic search. Their empirical results have found that the bidirectional heuristic search can be performed very efficiently, with limited memory demands. Their research has resulted in a better understanding of an algorithm whose practical usefulness has been long neglected, with the conclusion that it is better suited to certain problems than corresponding unidirectional searches. For more details, the reader should review their paper (Kaindl and Kainz, 1997). The Section 38.3 focuses on heuristic search methods.

38.3 Heuristic Search Methods

George Polya, via his wonderful book *How to Solve It* (1945), may be regarded as the "father of heuristics." Polya's efforts focused on problem-solving, thinking, and learning. He developed a short "heuristic dictionary" of heuristic primitives. Polya's approach was both practical and experimental. He sought to develop commonalties in the problem-solving process through the formalization of observation and experience.

The present-day notions of heuristics are somewhat different from Polya's (Bolc and Cytowski, 1992). Current tendencies seek formal and rigid algorithmic solutions to specific problem domains, rather than the development of general approaches that could be appropriately selected and applied to specific problems.

The goal of a heuristic search is to greatly reduce the number of nodes searched in seeking a goal. In other words, problems whose complexity grows combinatorially large may be tackled. Through knowledge, information, rules, insights, analogies, and simplification, in addition to a host of other techniques, heuristic search aims to reduce the number of objects that must be examined. Heuristics do not guarantee the achievement of a solution, although good heuristics should facilitate this. Over the years, heuristic search has been defined in many different ways:

- It is a practical strategy increasing the effectiveness of complex problem-solving (Feigenbaum and Feldman, 1963).
- It leads to a solution along the most probable path, omitting the least promising ones.
- It should enable one to avoid the examination of dead ends and to use already gathered data.

The points at which heuristic information can be applied in a search include

1. Deciding which node to expand next, instead of doing the expansions in either a strictly breadth-first or depth-first order
2. Deciding which successor or successors to generate when generating a node—instead of blindly generating all possible successors at one time
3. Deciding that certain nodes should be discarded, or pruned, from the search tree

Most modern heuristic search methods are expected to bridge the gap between the completeness of algorithms and their optimal complexity (Romanycia and Pelletier, 1985). Strategies are being modified in order to arrive at a quasi-optimal, instead of optimal, solution with a significant cost reduction (Pearl, 1984). Games, especially two-person, zero-sum games with perfect information, like chess and checkers, have proven to be a very promising domain for studying and testing heuristics.

38.3.1 Hill Climbing

Hill climbing is a DFS with a heuristic measure that orders choices as nodes are expanded. The heuristic measure is the estimated remaining distance to the goal. The effectiveness of hill climbing is completely dependent upon the accuracy of the heuristic measure.

Winston (1992) explains the potential problems affecting hill climbing. They are all related to issue of local "vision" versus global vision of the search space. The *foothills problem* is particularly subject to local maxima where global ones are sought, while the *plateau problem* occurs when the heuristic measure does not hint toward any significant gradient of proximity to a goal. The *ridge problem* illustrates just what it is called: you may get the impression that the search is taking you closer to a goal state, when in fact you are traveling along a ridge that prevents you from actually attaining your goal. SA attempts to combine hill climbing with a random walk in a way that yields both efficiency and completeness (Russell and Norvig, 2003, p. 115). The idea is to "temper" the downhill process of hill climbing in order to avoid some of the pitfalls discussed earlier by increasing the probability of "hitting" important locations to explore. It is akin to intelligent guessing. A more comprehensive discussion of SA can be found in Section 38.6.2.

38.3.2 Best-First Search

Best-first search (Figure 38.6) is a general algorithm for heuristically searching any state-space graph—a graph representation for a problem that includes initial states, intermediate states, and goal states. In this sense, a directed acyclic graph (DAG), for example, is a special case of a state-space graph. Best-first search is equally applicable to data- and goal-driven searchers and supports the use of heuristic evaluation functions. It can be used with a variety of heuristics, ranging from a state's "goodness" to sophisticated measures based on the probability of a state leading to a goal that can be illustrated by examples of Bayesian statistical measures.

Similar to the depth-first and breadth-first search algorithms, best-first search uses lists to maintain states: OPEN to keep track of the current fringe of the search and CLOSED to record states already visited. In addition, the algorithm orders states on OPEN according to some heuristic estimate of their proximity to a goal. Thus, each iteration of the loop considers the most "promising" state on the OPEN list. According to Luger and Stubblefield (1993), best-first search improves at just the point where hill climbing fails with its short-sighted and local vision.

Here is a graph of a hypothetical search space. The problem is to find a shortest path from *d1* to *d5* in this directed and weighted graph (Figure 38.7), which could represent a sequence of local and express subway train stops. The "F" train starts at *d1* and visits stops *d2* (cost 16), *d4* (cost 7), and *d6* (cost 11). The "D" train starts at *d1* and *d3* (cost 7), *d4* (cost 13), and *d5* (cost 12). Other choices involve

```
Procedure Best_First_Search (Start) → pointer
      OPEN ← {Start}                                       // Initialize
      CLOSED ← { }
      While OPEN ≠ { } Do                                  // States Remain
            remove the leftmost state from OPEN, call it X;
            if X ≡ goal then
                   return the path from Start to X
            else
                 generate children of X
                 for each child of X do
                 CASE
                 the child is not on open or CLOSED:
                       assign the child a heuristic value
                       add the child to OPEN
                 the child is already on OPEN:
                       if the child was reached by a shorter path
                       then give the state on OPEN the shorter path
                 the child is already on CLOSED:
                       if the child was reached by a shorter path then
                           remove the state from CLOSED
                           add the child to OPEN
                 end_CASE
                 put X on closed;
                 re-order states on OPEN by heuristic merit (best leftmost)
      return NULL                                          // OPEN is empty
```

FIGURE 38.6 The best-first search algorithm. (From Luger, J. and Stubblefield, W., *Artificial Intelligence: Structures and Strategies for Complex Problem Solving*, 2nd edn., Benjamin/Cummings Publishing Company Inc., Redwood City, CA, p. 121, 1993.)

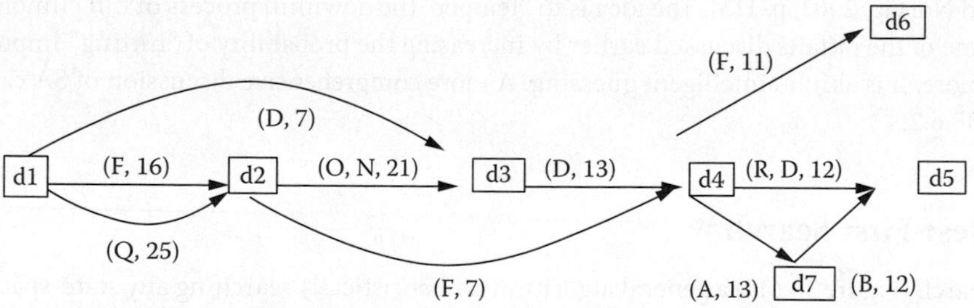

FIGURE 38.7 A state-space graph for a hypothetical subway system.

combinations of "Q," "N," "R," and "A" trains with the F and/or D train. By applying the "best-first search" algorithm, we can find the shortest path from *d1* to *d5*. Figure 38.8 shows a state tree representation of this graph.

The thick arrowed path is the shortest path [d1, d3, d4, d5]. The dashed edges are nodes put on the OPEN node queue, but not further explored. A trace of the execution of procedure best-first search appears in the following:

1. OPEN = [d1]; CLOSED = []
2. Evaluate d1; OPEN = [d3, d2]; CLOSED = [d1]
3. Evaluate d3; OPEN = [d4, d2]; CLOSED = [d3, d1]
4. Evaluate d4; OPEN = [d6, d5, d7, d2]; CLOSED = [d4, d3, d1]
5. Evaluate d6; OPEN = [d5, d7, d2]; CLOSED = [d6, d4, d3, d1]
6. Evaluate d5; a solution is found CLOSED = [d5, d6, d4, d3, d1]

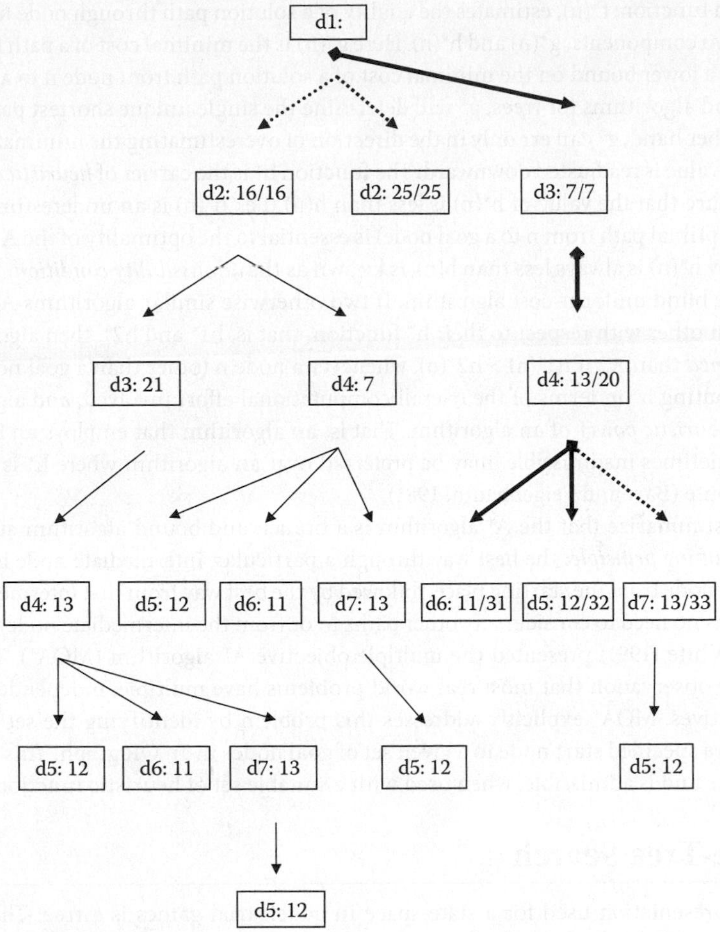

FIGURE 38.8 A search tree for the graph in Figure 38.7.

Note that nodes d6 and d5 are at the same level so we don't take d6 in our search for the shortest path. Hence, the shortest path for this graph is [d1, d3, d4, d5]. After we reach our goal state "d5," we can also find the shortest path from "d5" to "d1" by retracing the tree from d5 to d1.

When the best-first search algorithm is used, the states are sent to the OPEN list in such a way that the most promising one is expanded next. Because the search heuristic being used for measurement of distance from the goal state may prove erroneous, the alternatives to the preferred state are kept on the OPEN list. If the algorithm follows an incorrect path, it will retrieve the "next best" state and shift its focus to another part of the space. In the example earlier, children of node d2 were found to have poorer heuristic evaluations than sibling d3, and so the search shifted there. However, the children of d3 were kept on OPEN and could be returned to later, if other solutions are sought.

38.3.3 A* Algorithm

The A* algorithm, first described by Hart, Nilsson, and Raphael (1968), attempts to find the minimal cost path joining the start node and the goal in a state-space graph. The algorithm employs an ordered state-space search and an estimated heuristic cost to a goal state, f* (known as an evaluation function), as does the best-first search (Section 38.2.2). It uniquely defines f*, so that it can guarantee an optimal solution path. The A* algorithm falls into the *branch and bound* class of algorithms, typically employed in operations research to find the shortest path to a solution node in a graph.

The evaluation function, f*(n), estimates the quality of a solution path through node n, based on values returned from two components, g*(n) and h*(n). Here g*(n) is the minimal cost of a path from a start node to n, and h*(n) is a lower bound on the minimal cost of a solution path from node n to a goal node. As in branch and bound algorithms for trees, g* will determine the single unique shortest path to node n. For graphs, on the other hand, g* can err only in the direction of overestimating the minimal cost; if a shorter path is found, its value is readjusted downward. The function h* is the carrier of *heuristic information*, and the ability to ensure that the value of h*(n) is less than h(n) (i.e., h*(n) is an underestimate of the actual cost, h(n), of an optimal path from n to a goal node) is essential to the optimality of the A* algorithm. This property, whereby h*(n) is always less than h(n), is known as the *admissibility condition*. If h* is zero, then A* reduces to the blind uniform-cost algorithm. If two otherwise similar algorithms A1 and A2 can be compared to each other with respect to their h* function, that is, h1* and h2*, then algorithm A1 is said to be *more informed* than A2 if h1*(n) > h2*(n), whenever a node n (other than a goal node) is evaluated. The cost of computing h* in terms of the overall computational effort involved, and algorithmic utility, determines the *heuristic power* of an algorithm. That is, an algorithm that employs an h* that is usually accurate, but sometimes inadmissible, may be preferred over an algorithm where h* is always minimal but hard to compute (Barr and Feigenbaum, 1981).

Thus, we can summarize that the A* algorithm is a branch and bound algorithm augmented by the *dynamic programming principle*: the best way through a particular, intermediate node is the best way to that intermediate node from the starting place, followed by the best way from that intermediate node to the goal node. There is no need to consider any other paths to or from the intermediate node (Winston, 1992).

Stewart and White (1991) presented the multiple objective A* algorithm (MOA*). Their research is motivated by the observation that most real-world problems have multiple, independent, and possibly conflicting objectives. MOA* explicitly addresses this problem by identifying the set of all nondominated paths from a specified start node to a given set of goal nodes in an OR graph. This work shows that MOA* is complete and is admissible, when used with a suitable set of heuristic functions.

38.4 Game-Tree Search

The standard representation used for a state-space in adversarial games is a tree. The tree is usually explored to a certain depth to determine the best move(s) by each side. Such trees can often grow exponentially, and therefore, heuristics are vital for feasible computation.

38.4.1 Alpha–Beta Algorithms

To the human player of 2-person games, the notion behind the alpha–beta algorithm is understood intuitively as the following:

- If I have determined that a move or a sequence of moves is bad for me (because of a refutation move or variation by my opponent), then I don't need to determine just how bad that move is. Instead I can spend my time exploring other alternatives earlier in the tree.
- Conversely, if I have determined that a variation or sequence of moves is bad for my opponent, then I don't determine exactly how good it is for me.

Figure 38.9 illustrates some of these ideas. Here the thick solid line represents the current solution path. This in turn has replaced a candidate solution, here shown with dotted lines. Everything to the right of the optimal solution path represents alternatives that are simply proved inferior. The path of the current solution is called the principal variation (PV) and nodes on that path are marked as PV nodes. Similarly, the alternatives to PV nodes are CUT nodes, where only a few successors are examined before a proof of inferiority is found. In time the successor to a CUT node will be an ALL node where everything must be examined to prove the cutoff at the CUT node. The number or bound value by each node represents the return to the root of the cost of the solution path.

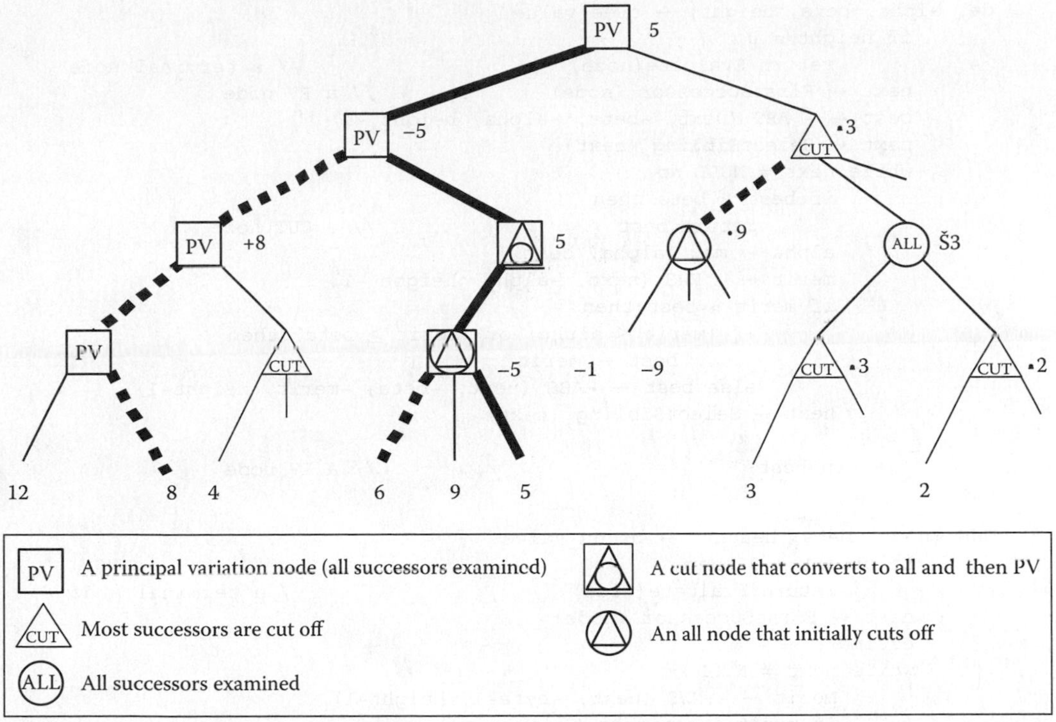

FIGURE 38.9 The PV, CUT, and ALL nodes of a tree, showing its optimal path (bold) and value (5).

In the forty years since its inception (Knuth and Moore, 1975), the alpha–beta minimax algorithm has undergone many revisions and refinements to improve the efficiency of its pruning, and, until the invention of the MTD(f) variant, it has served as the primary search engine for two-person games. There have been many landmarks on the way, including Knuth and Moore's (1975) formulation in a negamax framework, Pearl's (1980) introduction of Scout, and the special formulation for chess with the PV search (Marsland and Campbell, 1982) and NegaScout (Reinefeld, 1983). The essence of the method is that the search seeks a path whose value falls between two bounds called alpha and beta, which form a window. With this approach, one can also incorporate an artificial narrowing of the alpha–beta window, thus encompassing the notion of "aspiration search," with a mandatory research on failure to find a value within the corrected bounds. This leads naturally to the incorporation of null window search (NWS) to improve upon Pearl's test procedure. Here the NWS procedure covers the search at a CUT node (Figure 38.9), where the cutting bound (beta) is negated and increased by 1 in the recursive call. This refinement has some advantage in the parallel search case, but otherwise NWS (Figure 38.10) is entirely equivalent to the minimal window call in NegaScout. Additional improvements include the use of iterative deepening with "transposition tables" and other move-ordering mechanisms to retain a memory of the search from iteration to iteration. A transposition table is a cache of previously generated states that are typically hashed into a table to avoid redundant work. These improvements help ensure that the better subtrees are searched sooner, leading to greater pruning efficiency (more cutoffs) in the later subtrees. Figure 38.10 encapsulates the essence of the algorithm and shows how the first variation from a set of PV nodes, as well as any superior path that emerges later, is given special treatment. Alternates to PV nodes will always be CUT nodes, where a few successors will be examined. In a minimal game tree, only one successor to a CUT node will be examined, and it will be an ALL node where everything is examined. In the general case, the situation is more complex, as Figure 38.9 shows.

```
de, alpha, beta, height) → tree_value
    if height ≡ 0
        return Evaluate(node)                          // a terminal node
    next ← FirstSuccessor (node)                   // a PV node
    best ← - ABS (next, -beta, -alpha, height -1)
    next ← SelectSibling (next)
    while next ≠ NULL do
        if best ≥ beta then
            return best                            // a CUT node
        alpha ← max (alpha, best)
        merit ← - NWS (next, -alpha, height -1)
        if merit > best then
            if (merit ≤ alpha) or (merit ≥ beta) then
                best ← merit
            else best ← -ABS (next, -beta, -merit, height-1)
        next ← SelectSibling (next)
     end
     return best                                   // a PV node
end

NWS (node, beta, height) → bound_value
    if height ≡ 0 then
        return Evaluate(node)                          // a terminal node
    next ← FirstSuccessor (node)
    estimate ← - ∞
    while next ≠ NULL do
        merit ← - NWS (next, -beta+1, height-1)
        if merit > estimate then
           estimate ← merit
        if merit ≥ beta then
           return estimate                             // a CUT node
        next ← SelectSibling (next)
     end
     return estimate                                   // an ALL node
end
```

FIGURE 38.10 Scout/PVS version of ABS in the negamax framework.

38.4.2 SSS* Algorithm

The SSS* algorithm was introduced by Stockman (1979) as a game-searching algorithm that traverses subtrees of the game tree in a best-first fashion similar to the A* algorithm. SSS* was shown to be superior to the original alpha–beta algorithm in the sense that it never looks at more nodes while occasionally examining fewer (Pearl, 1984). Roizen and Pearl (1983), the source of the following description of SSS*, state that

> … the aim of SSS* is the discovery of an optimal solution tree…In accordance with the best-first split-and-prune paradigm, SSS* considers "clusters" of solution trees and splits (or refines) that cluster having the highest upper bound on the merit of its constituents. Every node in the game tree represents a cluster of solution trees defined by the set of all solution trees that share that node.…the merit of a partially developed solution tree in a game is determined solely by the properties of the frontier nodes it contains, not by the cost of the paths leading to these nodes. The value of a frontier node is an upper bound on each solution tree in the cluster it represents,… SSS* establishes upper bounds on the values of partially developed solution trees by seeking the value of terminal nodes, left to right, taking the minimum value of those examined so far. These monotonically nonincreasing bounds are used to order the solution trees so that the tree of highest merit is chosen for development. The development process continues until one solution tree is fully developed, at which point that tree represents the optimal strategy and its value coincides with the minimax value of the root.

...The disadvantage of SSS* lies in the need to keep in storage a record of all contending candidate clusters, which may require large storage space, growing exponentially with search depth. (Pearl, 1984, p. 245)

Heavy space and time overheads have kept SSS* from being much more than an example of a best-first search, but current research seems destined to now relegate SSS* to a historical footnote. Recently, Plaat et al. (1995) formulated the node-efficient SSS* algorithm into the alpha–beta framework using successive NWS search invocations (supported by perfect transposition tables) to achieve a memory-enhanced test procedure that provides a best-first search. With their introduction of the MTD(f) algorithm, Plaat et al. (1995) claim that SSS* can be viewed as a special case of the time-efficient alpha–beta algorithm, instead of the earlier view that alpha–beta is a k-partition variant of SSS*. MTD(f) is an important contribution that has now been widely adopted as the standard two-person game-tree search algorithm. It is described in the succeeding text.

38.4.3 MTD(f) and NegaScout

MTD(f) is usually run in an iterative deepening fashion, and each iteration proceeds by a sequence of minimal or null window alpha–beta calls. The search works by zooming in on the minimax value, as Figure 38.11 shows.

The bounds stored in *upper bound* and *lower bound* form an interval around the true minimax value for a particular search depth d. The interval is initially set to $[-\infty, +\infty]$. Starting with the value f, returned from a previous call to MTD(f), each call to alpha–beta returns a new minimax value g, which is used to adjust the bounding interval and to serve as the pruning value for the next alpha–beta call. For example, if the initial minimax value is 50, alpha–beta will be called with the pruning values 49 and 50. If the new minimax value returned, g, is less than 50, upper bound is set to g. If the minimax value returned, g, is greater than or equal to 50, lower bound is set to g. The next call to alpha–beta will use $g - 1$ and g for the pruning values (or g and $g + 1$, if *g* is equal to the lower bound). This process continues until *upper bound* and *lower bound* converge to a single value, which is returned. MTD(f) will be called again with this newly returned minimax estimate and an increased depth bound until the tree has been searched to a sufficient depth.

As a result of the iterative nature of MTD(f), the use of transposition tables is essential to its efficient implementation. In tests with a number of tournament game-playing programs, MTD(f) outperformed alpha–beta search (ABS) (Scout/PVS, Figure 38.10). It generally produces trees that are 5%–15% smaller than ABS (Plaat et al., 1996). MTD(f) has been shown to examine fewer nodes than any other variants

```
int MTDF ( node_type root, int f, int d)
{
    g = f;
    upperbound = +INFINITY;
    lowerbound = -INFINITY;
    repeat
        if (g == lowerbound)
            beta = g + 1
        else
                         beta = g;
        g = AlphaBetaWithMemory(root, beta - 1, beta, d);
        if (g < beta)
            then upperbound = g
            else lowerbound = g;
    until (lowerbound >= upperbound);
return g;
}
```

FIGURE 38.11 The MTD(f) algorithm pseudo-code.

of the ABS; nevertheless, many chess program developers still prefer NegaScout (Reinefeld, 1983) to MTD(f), as it is free of dependence on transposition tables.

38.4.4 Recent Developments

Sven Koenig has developed Minimax Learning Real-Time A* (Min-Max LRTA*), a real-time heuristic search method that generalizes Korf's (1990) earlier LRTA* to nondeterministic domains. Hence, it can be applied to "robot navigation tasks in mazes, where robots know the maze but do not know their initial position and orientation (pose). These planning tasks can be modeled as planning tasks in non-deterministic domains whose states are sets of poses." Such problems can be solved quickly and efficiently with Min-Max LRTA* requiring only a small amount of memory (Koenig, 2001).

Martin Mueller (2001) introduces the use of partial order bounding (POB) rather than scalar values for the construction of an evaluation function for computer game playing. The propagation of partially ordered values through a search tree has been known to lead to many problems in practice. Instead POB compares values in the leaves of a game tree and backs up Boolean values through the tree. The effectiveness of this method was demonstrated in examples of capture races in the game of GO (Mueller, 2001).

Schaeffer et al. (2001) demonstrate that the distinctions for evaluating heuristic search should not be based on whether the application is for single-agent or two-agent search. Instead, they argue that the search enhancements applied to both single-agent and two-agent problems for creating high-performance applications are the essentials. Focus should be on generality for creating opportunities for reuse. Examples of some of the generic enhancements (as opposed to problem-specific ones) include the alpha–beta algorithm, transposition tables, and IDA*. Efforts should be made to enable more generic application of algorithms.

Hong et al. (2001) present a GA approach that can find a good next move by reserving the board evaluation of new offspring in a partial game-tree search. Experiments have proven promising in terms of speed and accuracy when applied to the game of GO.

The fast-forward (FF) planning system of Hoffman and Nebel (2001) uses a heuristic that estimates goal distances by ignoring delete lists. Facts are not assumed to be independent. It uses a new search strategy that combines hill climbing with systematic search. Powerful heuristic information is extended and used to prune the search space.

38.5 Parallel Search

The proliferation of low-cost processors and, in particular, the advent of multicore computing have stimulated interest in the use of multiple processors for parallel traversals of decision trees. The few theoretical models of parallelism do not accommodate communication and synchronization delays that inevitably impact the performance of working systems. There are several other factors to consider too, including

1. How best to employ the additional memory and I/O resources that become available with the extra processors
2. How best to distribute the work across the available processors
3. How to avoid excessive duplication of computation

Some important combinatorial problems have no difficulty with the third point because every eventuality must be considered, but these tend to be less interesting in an AI context.

One problem of particular interest is game-tree search, where it is necessary to compute the value of the tree while communicating an improved estimate to the other parallel searchers as it becomes available. This can lead to an "acceleration anomaly" when the tree value is found earlier than is possible with a sequential algorithm. Even so, uniprocessor algorithms can have special advantages in that they can be optimized for best pruning efficiency, while a competing parallel system may not have the right information in time to achieve the same degree of pruning and so do more work (suffer from search overhead). Further, the very fact that pruning occurs makes it impossible to determine in advance how big any piece of work (subtree to be searched) will be, leading to a potentially serious work imbalance and heavy synchronization (waiting for more work) delays.

Although the standard basis for comparing the efficiency of parallel methods is simple,

$$Speedup = \frac{\text{Time taken by a sequential single-processor algorithm}}{\text{Time taken by a P-processor system}}$$

This is the practical definition of speedup. Many researchers use a more theoretical definition, in which P-fold speedup is the best possible. However, in practice superlinear speedup can occur for a variety of reasons. This phenomenon happens when the parallel algorithm does less total work than the sequential version. For example, each processor may work on a smaller problem and thus get improved cache performance. In search problems, the parallel version may quickly report a solution on a path that would not be searched until much later by the sequential algorithm, causing other processors to end their search before exploring nodes that the sequential algorithm would have explored.

The exponential growth of the tree size (solution space) with depth of search makes parallel search algorithms especially susceptible to anomalous speedup behavior. Clearly, acceleration anomalies are among the welcome properties, but more commonly anomalously bad performance is seen, unless the algorithm has been designed with care.

In game-playing programs of interest to AI, parallelism is primarily intended not to find the answer more quickly but to get a more reliable result (e.g., based on a deeper search). Here, the emphasis lies on scalability instead of speedup. While speedup holds the problem size constant and increases the system size to get a result sooner, scalability measures the ability to expand the size of both the problem and the system at the same time:
Thus, scale-up close to unity reflects successful parallelism:

$$Scale\text{-}up = \frac{\text{Time taken to solve a problem of size s by a single processor}}{\text{Time taken to solve a } (P \times s) \text{ problem by an P-processor system}}$$

38.5.1 Parallel Single-Agent Search

Single-agent game-tree search is important because it is useful for several robot-planning activities, such as finding the shortest path through a maze of obstacles. It seems to be more amenable to parallelization than the techniques used in adversary games, because a large proportion of the search space must be fully seen—especially when optimal solutions are sought. This traversal can safely be done in parallel, since there are no cutoffs to be missed. Although move ordering can reduce node expansions, it does not play the same crucial role as in dual-agent game-tree search, where significant parts of the search space are often pruned away. For this reason, parallel single-agent search techniques usually achieve better speedups than their counterparts in adversary games.

Most parallel single-agent searches are based on A* or IDA*. As in the sequential case, parallel A* outperforms IDA* on a node-count basis, although parallel IDA* needs only linear storage space and runs faster. In addition, cost-effective methods exist (e.g., PWS described subsequently) that determine nonoptimal solutions with even less computing time.

38.5.1.1 Parallel A*

Given P processors, the simplest way to parallelize A* is to let each machine work on one of the currently best states on a global OPEN list (a place holder for nodes that have not yet been examined). This approach minimizes the search overhead, as confirmed in practice by Kumar et al. (1988). Their relevant experiments were run on a shared-memory BBN Butterfly machine with 100 processors, where a search overhead of less than 5% was observed for the traveling salesperson (TSP) problem.

But elapsed time is more important than the node expansion count, because the global OPEN list is accessed both before and after each node expansion, so that memory contention becomes a serious bottleneck. It turns out that a centralized strategy for managing the OPEN list is only useful in domains where the node expansion time is large compared to the OPEN list access time. In the TSP problem, near-linear time speedups were achieved with up to about 50 processors, when a sophisticated heap data structure was used to significantly reduce the OPEN list access time (Kumar et al., 1988).

Distributed strategies using local OPEN lists reduce the memory contention problem. But again some communication must be provided to allow processors to share the most promising state descriptors, so that no computing resources are wasted in expanding inferior states. For this purpose, a global "Blackboard" table can be used to hold state descriptors of the currently best nodes. After selecting a state from its local OPEN list, each processor compares its f-value (lower bound on the solution cost) to that of the states contained in the Blackboard. If the local state is much better (or much worse) than those stored in the Blackboard, then node descriptors are sent (or received), so that all active processors are exploring states of almost equal heuristic value. With this scheme, a 69-fold speedup was achieved on an 85-processor BBN Butterfly (Kumar et al., 1988).

Although a Blackboard is not accessed as frequently as a global OPEN list, it still causes memory contention with increasing parallelism. To alleviate this problem, Huang and Davis (1989) proposed a distributed heuristic search algorithm called Parallel Iterative A* (PIA*), which works solely on local data structures. On a uniprocessor, PIA* expands the same nodes as A*, while in the multiprocessor case, it performs a parallel best-first node expansion. The search proceeds by repetitive synchronized iterations, in which processors working on inferior nodes are stopped and reassigned to better ones. To avoid unproductive waiting at the synchronization barriers, the processors are allowed to perform speculative processing. Although Huang and Davis (1989) claim that "this algorithm can achieve almost linear speedup on a large number of processors," it has the same disadvantage as the other parallel A* variants, namely, excessive memory requirements.

38.5.1.2 Parallel IDA*

IDA* (Figure 38.5) has proved to be effective, when excessive memory requirements undermine best-first schemes. Not surprisingly it has also been a popular algorithm to parallelize. Rao et al. (1987) proposed PIDA*, an algorithm with almost linear speedup even when solving the 15-puzzle with its trivial node expansion cost. The 15-puzzle is a popular game made up of 15 tiles that slide within a 4×4 matrix. The object is to slide the tiles through the one empty spot until all tiles are aligned in some goal state. An optimal solution to a hard problem might take 66 moves. PIDA* splits the search space into disjoint parts, so that each processor performs a local cost-bounded DFS on its private portion of the state space. When a process has finished its job, it tries to get an unsearched part of the tree from other processors. When no further work can be obtained, all processors detect global termination and compute the minimum of the cost bounds, which is used as a new bound in the next iteration. Note that more than a P-fold speedup is possible when a processor finds a goal node early in the final iteration. In fact, Rao et al. (1987) report an average speedup of 9.24 with 9 processors on the 15-puzzle! Perhaps more relevant is the all-solution case where no superlinear speedup is possible. Here, an average speedup of 0.93P with up to thirty (P) processors on a bus-based multiprocessor architecture (Sequent Balance 21000) was achieved. This suggests that only low multiprocessing overheads (locking, work transfer, termination detection, and synchronization) were experienced.

PIDA* employs a task attraction scheme like that shown in Figure 38.12 for distributing the work among the processors. When a processor becomes idle, it asks a neighbor for a piece of the search space. The donor then splits its DFS stack and transfers to the requester some nodes (subtrees) for parallel expansion. An optimal splitting strategy would depend on the regularity (uniformity of width and height) of the search tree, though short subtrees should never be given away. When the tree is regular

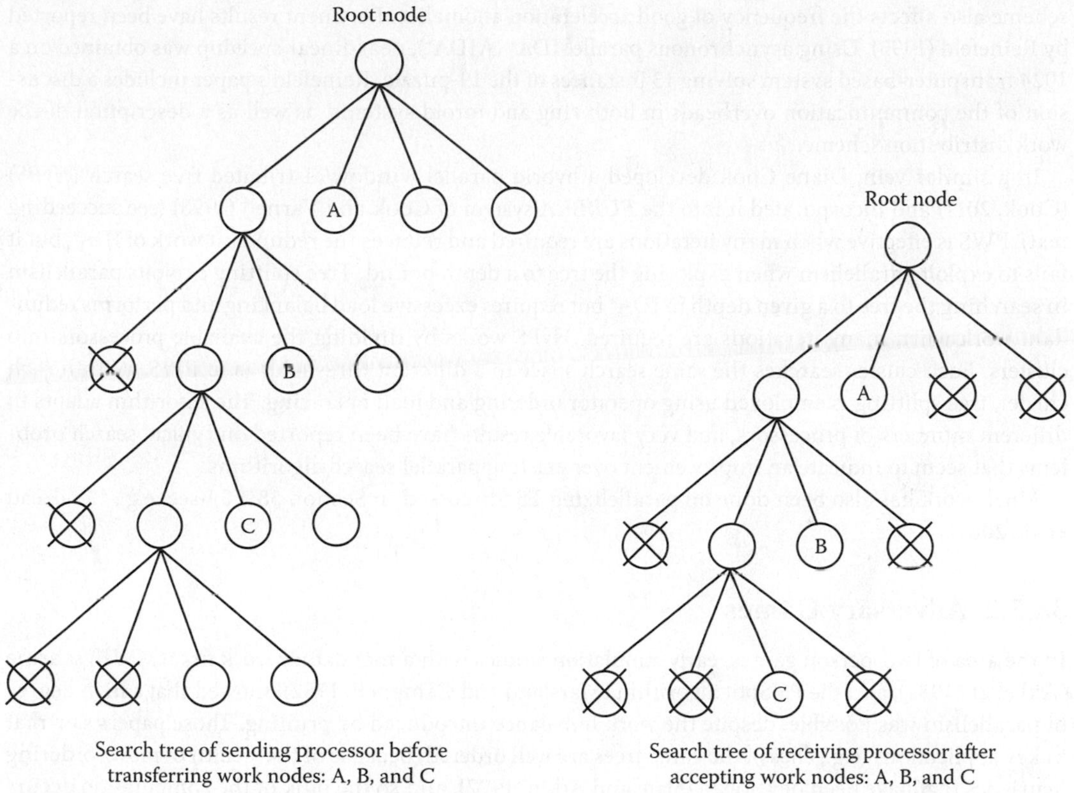

FIGURE 38.12 A work distribution scheme.

(like in the 15-puzzle), a coarse-grained work transfer strategy can be used (e.g., transferring only nodes near the root); otherwise, a slice of nodes (e.g., nodes A, B, and C in Figure 38.12) should be transferred.

38.5.1.3 Comparison with Parallel Window Search

Another parallel IDA* approach borrows from Baudet's (1978) parallel window method for searching adversary games (described later). Powley and Korf (1991) adapted this method to single-agent search, under the title PWS. Their basic idea is to simultaneously start as many iterations as there are processors. This works for a small number of processors, which either expand the tree up to their given thresholds until a solution is found (and the search is stopped) or completely expand their search space. A global administration scheme then determines the next larger search bound and node expansion starts over again.

Note that the first solution found by PWS need not necessarily be optimal. Suboptimal solutions are often found in searches of poorly ordered trees. There, a processor working with a higher cutoff bound finds a goal node in a deeper tree level, while other processors are still expanding shallower tree parts (which may contain cheaper solutions). But according to Powley and Korf (1991), PWS is not primarily meant to compete with IDA*, but it "can be used to find a nearly optimal solution quickly, improve the solution until it is optimal, and then finally guarantee optimality, depending on the amount of time available." Compared to PIDA*, the degree of parallelism is limited, and it remains unclear how to apply PWS in domains where the cost-bound increases are variable.

In summary, PWS and PIDA* complement each other. Hence, it seems natural to combine them to form a single search scheme that runs PIDA* on groups of processors administered by a global PWS algorithm. The amount of communication needed depends on the work distribution scheme. A fine-grained distribution requires more communication, while a coarse-grained work distribution generates fewer messages (but may induce unbalanced work load). Note that the choice of the work distribution scheme also affects the frequency of good acceleration anomalies. Pertinent results have been reported by Reinefeld (1995). Using asynchronous parallel IDA* (AIDA*), near-linear speedup was obtained on a 1024 transputer-based system solving 13 instances of the 19-puzzle. Reinefeld's paper includes a discussion of the communication overheads in both ring and toroid systems, as well as a description of the work distribution scheme.

In a similar vein, Diane Cook developed a hybrid parallel window/distributed tree search (*HyPS*) (Cook, 2011) and incorporated it into the *EUREKA* system of Cook and Varnell (1998) (see succeeding text). PWS is effective when many iterations are required and reduces the redundant work of IDA*, but it fails to exploit parallelism when exploring the tree to a depth bound. Tree splitting exploits parallelism in searching the tree to a given depth in IDA* but requires excessive load balancing and performs redundant work when many iterations are required. HyPS works by dividing the available processors into clusters. Each cluster searches the same search space to a different threshold, as in PWS. Within each cluster, tree splitting is employed using operator ordering and load balancing. The algorithm adapts to different numbers of processors, and very favorable results have been reported on typical search problems that seem to indicate an improvement over existing parallel search algorithms.

Much work has also been done on parallelizing TS (discussed in Section 38.5.2) (see, e.g., Gendreau et al., 2001).

38.5.2 Adversary Games

In the area of two-person games, early simulation studies with a mandatory work first (MWF) scheme (Akl et al., 1982), and the PVSplit algorithm (Marsland and Campbell, 1982), showed that a high degree of parallelism was possible, despite the work imbalance introduced by pruning. Those papers saw that in key applications (e.g., chess), the game trees are well ordered, because of the wealth of move-ordering heuristics that have been developed (Slate and Atkin, 1977), and so the bulk of the computation occurs during the search of the first subtree. The MWF approach uses the shape of the critical tree that must be searched. Since that tree is well defined and has regular properties, it is easy to generate. In their simulation, Akl et al. (1982) consider the merits of searching the critical game tree in parallel, with the balance of the tree being generated algorithmically and searched quickly by simple tree splitting. Marsland and Campbell (1982), on the other hand, recognized that the first subtree of the critical game tree has the same properties as the whole tree, but its maximum height is one less. This so-called PV can be recursively split into parts of about equal size for parallel exploration. PVSplit, an algorithm based on this observation, was tested and analyzed by Marsland and Popowich (1985). Even so, the static processor allocation schemes like MWF and PVSplit cannot achieve high levels of parallelism, although PVSplit does very well with up to half a dozen processors. MWF in particular ignores the true shape of the average game tree, and so is at its best with shallow searches, where the pruning imbalance from the so-called deep cutoffs has less effect. Other working experience includes the first parallel chess program by Newborn, who later presented performance results (Newborn, 1988). For practical reasons Newborn only split the tree down to some prespecified common depth from the root (typically 2), where the greatest benefits from parallelism can be achieved. This use of a common depth has been taken up by Hsu (1990) in his proposal for large-scale parallelism. Depth limits are also an important part of changing search modes and in managing transposition tables.

38.5.2.1 Parallel Aspiration-Window Search

In an early paper on parallel game-tree search, Baudet (1978) suggests partitioning the range of the alpha–beta window rather than the tree. In his algorithm, all processors search the whole tree, but each with a different, nonoverlapping, alpha–beta window. The total range of values is subdivided into P smaller intervals (where P is the number of processors), so that approximately one-third of the range is covered. The advantage of this method is that the processor having the true minimax value inside its narrow window will complete more quickly than a sequential algorithm running with a full window.

Even the unsuccessful processors return a result: They determine whether the true minimax value lies below or above their assigned search window, providing important information for rescheduling idle processors until a solution is found.

Its low communication overhead and lack of synchronization needs are among the positive aspects of Baudet's approach. On the negative side, however, Baudet estimates a maximum speedup of between 5 and 6 even when using infinitely many processors. In practice, PWS can only be effectively employed on systems with two or three processors. This is because even in the best case (when the successful processor uses a minimal window), at least the critical game tree must be expanded. The critical tree has about the square root of the leaf nodes of a uniform tree of the same depth, and it represents the smallest tree that must be searched under any circumstances.

38.5.2.2 Advanced Tree Splitting Methods

Results from fully recursive versions of PVSplit using the Parabelle chess program (Marsland and Popowich, 1985) confirmed the earlier simulations and offered some insight into a major problem: In a P-processor system, P1 processors are often idle for an inordinate amount of time, thus inducing a high synchronization overhead for large systems. Moreover, the synchronization overhead increases as more processors are added, accounting for most of the total losses, because the search overhead (number of unnecessary node expansions) becomes almost constant for the larger systems. This led to the development of variations that dynamically assign processors to the search of the PV. No table is the work of Schaeffer (1989), which uses a loosely coupled network of workstations, and Hyatt et al. (1989) independent implementation for a shared-memory computer. These dynamic splitting works have attracted growing attention through a variety of approaches. For example, the results of Feldmann et al. (1990) show a speedup of 11.8 with 16 processors (far exceeding the performance of earlier systems), and Felten and Otto (1988) measured a 101 speedup on a 256 processor hypercube. This latter achievement is noteworthy because it shows an effective way to exploit the 256 times bigger memory that was not available to the uniprocessor. The use of the extra transposition table memory to hold results of search by other processors provides a significant benefit to the hypercube system, thus identifying clearly one advantage of systems with an extensible address space.

These results show a wide variation not only of methods but also of apparent performance. Part of the improvement is accounted for by the change from a static assignment of processors to the tree search (e.g., from PVSplit) and to the dynamic processor reallocation schemes of Hyatt et al. (1989) and also Schaeffer (1989). These latter systems try to dynamically identify the ALL nodes of Figure 38.9 and search them in parallel, leaving the CUT nodes (where only a few successors might be examined) for serial expansion. In a similar vein, Ferguson and Korf (1988) proposed a "bound-and-branch" method that only assigned processors to the leftmost child of the tree-splitting nodes where no bound (subtree value) exists. Their method is equivalent to the static PVSplit algorithm and realizes a speedup of 12 with 32 processors for alpha–beta trees generated by Othello programs. This speedup result might be attributed to the smaller average branching factor of about 10 for Othello trees, compared to an average branching factor of about 35 for chess. If that uniprocessor solution is inefficient—for example, by omitting an important node-ordering mechanism like the use of transposition tables (Reinefeld and Marsland, 1994)—the speedup figure may look good. For that reason, comparisons with a standard test suite from a widely accepted game are often done and should be encouraged. Most of the working experience with parallel methods for two-person games has centered on the alpha–beta algorithm. Parallel methods for more node-count-efficient sequential methods, like SSS*, have not been successful until recently, when the potential advantages of using heuristic methods like hash tables to replace the OPEN list were exploited (Plaat et al., 1995).

38.5.2.3 Dynamic Distribution of Work

The key to successful large-scale parallelism lies in the dynamic distribution of work. There are four primary issues in dynamic search: (1) *Search overhead* measures the size of the tree searched by the parallel method with respect to the best sequential algorithm. As mentioned earlier, in some cases, superlinear speedup can occur when the parallel algorithm actually visits fewer nodes. (2) *Synchronization overhead* problems occur when processors are idle waiting for the results from other processors, thus reducing the effective use of the parallel computing power (processor utilization). (3) *Load balancing* reflects how evenly the work has been divided among available processors and similarly affects processor utilization. (4) *Communication overhead* in a distributed memory system occurs when results must be communicated between processors via message passing.

Each of these issues must be considered in designing a dynamic parallel algorithm. The distribution of work to processors can either be accomplished in a work-driven fashion, whereby idle processors must acquire new work either from a Blackboard or by requesting work from another processor.

The young brothers wait concept (Feldmann, 1993) is a work-driven scheme in which the parallelism is best described through the help of a definition: The search for a successor N.j of a node N in a game tree must not be started until after the leftmost sibling N.1 of N.j is completely evaluated. Thus, N.j can be given to another processor if and only if it has not yet been started and the search of N.1 is complete. Since this is also the requirement for the PVSplit algorithm, how then do the two methods differ and what are the trade-offs? There are two significant differences. The first is at start-up and the second is in the potential for parallelism. PVSplit starts much more quickly since all the processors traverse the first variation (first path from the root to the search horizon of the tree) and then split the work at the nodes on the path as the processors back up the tree to the root. Thus, all the processors are busy from the beginning, but on the other hand, this method suffers from increasingly large synchronization delays as the processors work their way back to the root of the game tree (Marsland and Popowich, 1985). Thus, good performance is possible only with relatively few processors, because the splitting is purely static. In the work of Feldmann et al. (1990), the start-up time for this system is lengthy, because initially only one processor (or a small group of processors) is used to traverse the first path. When that is complete, the right siblings of the nodes on the path can be distributed for parallel search to the waiting processors. If, for example, in the case of one thousand such processors, possibly less than one percent would initially be busy. Gradually, the idle processors are brought in to help the busy ones, but this takes time. However, and here comes the big advantage, the system is now much more dynamic in the way it distributes work, so it is less prone to serious synchronization loss. Further, although many of the nodes in the tree will be CUT nodes (which are a poor choice for parallelism because they generate high search overhead), others will be ALL nodes, where every successor must be examined and they can simply be done in parallel. Usually CUT nodes generate a cutoff quite quickly, so by being cautious about how much work is initially given away once N.1 has been evaluated, one can keep excellent control over the search overhead while getting full benefit from the dynamic work distribution that Feldmann's method provides.

On the other hand, transposition-driven work scheduling (TDS) for parallel single-agent and game-tree searches, proposed by Romein et al. (1999), is a data-driven technique that in many cases offers considerable improvements over work-driven scheduling on distributed memory architectures. TDS reduces the communication and memory overhead associated with the remote lookups of transposition tables partitioned among distributed memory resources. This permits lookup communication and search computation to be integrated. The use of the transposition tables in TDS, as in IDA*, prevents the repeated searching of previously expanded states.

TDS employs a distributed transposition table that works by assigning to each state a "home processor," where the transposition entry for that state is stored. A signature associated with the state indicates the number of its home processor. When a given processor expands a new state, it evaluates its signature and sends it to its home processor, without having to wait for a response, thus permitting the communication to be carried out asynchronously. In other words, the work is assigned to where the data on a particular state are stored, rather than having to lookup a remote processor's table and wait for the results to be transmitted back. Alternatively, when a processor receives a node, it performs a lookup of its local transposition table to determine whether the node has been searched before. If not, the node is stored in the transposition table and added to the local work queue. Furthermore, since each transposition table entry includes a search bound, this prevents redundant processing of the same subtree by more than one processor.

The resulting reduction in both communication and search overhead yields significant performance benefits. Speedups that surpass IDA* by a factor of more than 100 on 128 processors (Romein et al., 1999) have been reported in selected games.

Cook and Varnell (1998) report that TDS may be led into doing unnecessary work at the goal depth, however, and therefore they favor a hybrid combination of techniques that they term *adaptive parallel iterative deepening search*. They have implemented their ideas in the system called EUREKA. Their system employs machine learning to select the best technique for a given problem domain. They have recently added the HyPS algorithm (Cook, 2011) to their system, as discussed earlier.

38.5.2.4 Recent Developments

Despite advances in parallel single-agent search, significant improvement in methods for game-tree search has remained elusive. Theoretical studies have often focused on showing that linear speedup is possible on worst-order game trees. While not wrong, they make only the trivial point where exhaustive search is necessary and where pruning is impossible, and then even simple work distribution methods may yield excellent results. The true challenge, however, is to consider the case of average game trees or, even better, the strongly ordered model (where extensive pruning can occur), resulting in asymmetric trees with a significant work distribution problem and significant search overhead. The search overhead occurs when a processor examines nodes that would be pruned by the sequential algorithm, but has not yet received the relevant results from another processor.

The intrinsic difficulty of searching game trees under pruning conditions has been widely recognized. Hence, considerable research has been focused on the goal of dynamically identifying when unnecessary search is being performed, thereby freeing processing resources for redeployment. For example, Feldmann et al. (1990) used the concept of making "young brothers wait" to reduce search overhead and developed the "helpful master" scheme to eliminate the idle time of masters waiting for their slaves' results. On the other hand, young brothers wait can still lead to significant synchronization overhead.

Generalized DFSs are fundamental to many AI problems. In this vein, Kumar and Rao (1990) have fully examined a method that is well suited to doing the early iterations of single-agent IDA* search. The unexplored parts of the trees are marked and are dynamically assigned to any idle processor. In principle, this work distribution method (illustrated in Figure 38.12) could also be used for deterministic adversary game trees. Shoham and Toledo (2001) developed a parallel, randomized best-first minimax search (RBFM). RBFM expands terminal nodes randomly chosen, with higher probabilities assigned to nodes with better values. This method seems amenable to parallelism but seems to suffer from speculative search overhead.

The invention of MTD(f) seemed to provide new hope for parallel game-tree search. Romein (2000) developed a parallel version of MTD(f) that achieved speedup over other parallel game-tree algorithms. Kishimoto and Schaeffer (2002) have developed an algorithm called TDSAB that combines MTD(f) with the transposition table-driven scheduling of TDS to achieve significant performance gains in a number of selected games. MIT's parallel chess program Cilkchess employs a parallel version of the algorithm [*supertech.csail.mit.edu/cilk/*]. However, Kishimoto and Schaeffer also report that attempts at parallelizing MTD(f) can suffer from excessive synchronization overhead.

On a massively parallel system, the opportunity arises to effectively employ several techniques simultaneously when searching a game tree. A combination of a PWS search with a distributed tree search (DTS), as embodied by the HyPS algorithm of Cook (2011), is a hopeful approach. Groups of processors are assigned different thresholds or windows, thereby reducing the overhead of the duplicate work in iterative deepening search. Within each group, the threshold search of the tree can be parallelized, as in DTS, using the dynamic workload-balancing techniques discussed earlier. IBM's Blue Gene supercomputers have been used for chess.

Many researchers feel that hardware advances will ultimately increase parallel processing power to the point where we will simply overwhelm the problem of game-tree search with brute force. They feel that this will make the need for continued improvement of parallel game-tree algorithms less critical.

Perhaps in time we will know if massive parallelism solves all our game-tree search problems. However, the results of complexity theory should assure us that unless P = NP, improvements in search techniques and heuristics will remain both useful and desirable.

38.6 Nature-Based Search

This section discusses search techniques that are either inspired by natural processes or derived from computational methods that have been used to simulate natural processes.

38.6.1 Monte Carlo Simulation

Earlier in this work, the speedup of a parallel algorithm was defined as

$$Speedup = \frac{\text{Time taken by a sequential single-processor algorithm}}{\text{Time taken by a P-processor algorithm}}$$

This represents an upper bound on the time savings that can be expected from parallel processing. It is evident that as larger computational tasks are confronted, parallelism alone will fail to provide the requisite computing speeds. All of the algorithms in this section rely upon probability to some extent (even TS described in 6.5 has a probabilistic version).

An early use of probability in search is with Monte Carlo simulation. This approach was widely used by John von Neumann, Stanislaw Ulam, and Nicholas Metropolis in their work on the Manhattan Project during the 1940s. These simulations use random numbers drawn from some prescribed probability distribution. This method has been used in application areas as diverse as portfolio preparation (Wu and Gertner, 2012) to Soduko (Mayorov and Whitlock, 2011). This methodology is explained thoroughly elsewhere in this book.

38.6.2 Simulated Annealing

Previously, we discussed hill climbing, a heuristic search that relies upon our insights into a problem to help make large search trees less daunting. The price we pay is that sometimes hill climbing settles at a local optimum. *SA* is a probabilistic search that allows counterintuitive moves with some prescribed probability. SA gets its name from the physical process of annealing wherein a metal is heated until it liquefies and then is slowly cooled until it resolidifies. Molecules in annealed metals are rearranged and metals that undergo this process are stronger and more pliable.

A successful search algorithm relies upon two strategies: *exploitation* and *exploration*. The intuition behind exploitation is that opportune points in the search space are likely to lie close to one another. Meanwhile, exploration encourages movement to a different portion of the search space, which may yield a superior return. Hill climbing searches rely heavily on exploitation as small incremental jumps are pursued as long as the objective function continues to improve. SA seeks to balance these two seemingly opposed strategies. It does so by employing a temperature parameter T, which is set to a high initial value. As the simulation progresses, T is lowered according to a *cooling schedule*. An SA search will jump from two points in the search space, x_{old} to x_{new}, whenever the objective function F() is improved at x_{new}, that is, whenever $f(x_{new}) \geq f(x_{old})$ (assuming that optimum in this context means maximum). An SA differs from hill climbing, however, in that backward jumps are permitted with a probability P, which is proportional to

$$e^{-[f(x\_old)-f(x\_new)/T]} \quad (38.1)$$

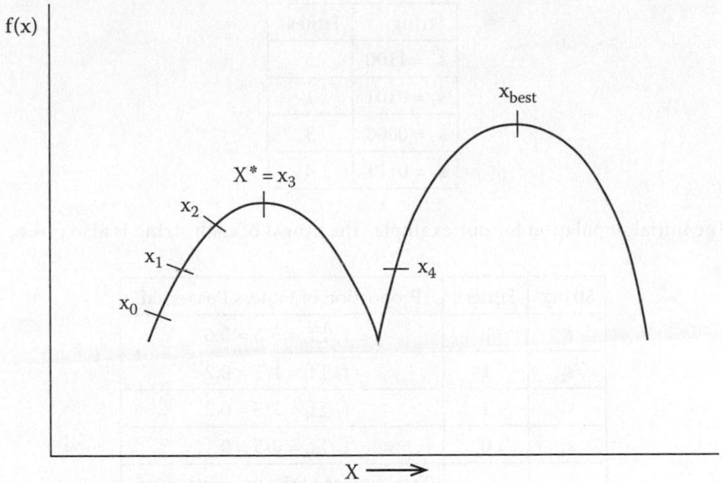

FIGURE 38.13 A search in which SA is likely to outperform hill climbing.

In Figure 38.13, if a hill climbing search were to begin at x_0, exploitation would lead it to the local optimum f(x*). An SA search would likely fare better as a counterintuitive jump from x_3 to x_4 is possible, thereby enabling convergence to the global optimum of $f(x_{best})$. Inspection of Equation 38.1 apprises us that backward jumps in an SA search are more likely when the temperature parameter T is higher. Hence, the early stages of SA favor exploration, whereas later in a search when T is lower, exploitation becomes more probable. We close our discussion by commenting that SA does not guarantee a global optimum. The likelihood of successful termination may be obtained in two ways:

1. Run the simulation longer.
2. Restart the search several times commencing the simulation each time at a different locations in the state space.

S. Kirkpatrick, C.D. Galatt, and M.P. Vecchi discovered SA in 1983 (Kirkpatrick et al., 1983).
This methodology was discovered independently in 1985 by V. Cerny (1985).
The SA search is based upon the Metropolis–Hastings algorithm (1953). SA has been effective in the solution of optimization problems.

38.6.3 Genetic Algorithms

GAs are an uninformed search as no prior knowledge is presumed about the problem domain. GA is similar to the algorithms described in Section 38.5 in that they use parallelism. In GA, a problem's solution is encoded by a string. We consider a trivial problem to help explain the methodology. We wish to learn the two-input NOR function (see Table 38.1).

Observe that the NOR function is the complement of the two-input (inclusive) OR function. The NOR of X_1 and X_2, written $X_1 \downarrow X_2$, equals 1 only when both X_1 and X_2 are equal to zero. We begin with a

TABLE 38.1 2-Input NOR Function

X_1	X_1	$X_1 \downarrow X_2$
0	0	1
0	1	0
1	0	0
1	1	0

String	Fitness
$s_1 = 1100$	3
$s_2 = 0101$	1
$s_3 = 0000$	3
$s_4 = 0111$	0

FIGURE 38.14 The initial population for our example. The fitness of each string is also given.

String	Fitness	Proportion of Fitness Possessed
s_1	3	$f_1/\Sigma f_i = 3/5 = 0.6$
s_2	1	$f_2/\Sigma f_i = 1/5 = 0.2$
s_3	1	$f_3/\Sigma f_i = 1/5 = 0.2$
s_4	0	$f_4/\Sigma f_i = 0/5 = 0$
		Total fitness of the population = 5 Maximum fitness = 3

FIGURE 38.15 The total and maximum fitness for population P_0, together with the proportion of fitness possessed by each string.

String	Mate	Crossover Point	After Crossover	After Mutation
$s_1 = 1100$	s_2	k = 3	1101	1101
$s_2 = 0101$	s_1	k = 3	0100	010$\underline{1}$
$s_3 = 1100$	s_4	k = 2	1000	1000
$s_4 = 0000$	s_3	k = 2	0100	0100

FIGURE 38.16 Selection of mates.

population of strings—population size is arbitrarily set to four. The strings in this initial population (P_0) are formed by using a random number generator (*rng*). GA is an iterative procedure. A *fitness function* is used to obtain better solutions. The fitness of a string (s_i) is a measure of how well a string solves the given problem. We have a choice of representation. We can write the entries in Table 38.1 in row major order, that is, the first row 001 followed by row 010, yielding a string of length 12. Alternately, we can encode solutions via binary strings of length four; the ith bit of a string represents the result of NORing the two inputs in row i. We choose the latter representation in this example.

For example, suppose our *rng* yields the string $s_i = 1100$. It is natural to assign a fitness, $f(s_i) = 3$, as s_i correctly represents the NOR of three of the entries in Table 38.1, that is, rows one, three, and four; however, the second bit of $s_i = 1$, whereas the NOR of 0 with 1 equals 0 and not 1. Assume that our rng yields the four strings $s_1 = 1100$, $s_2 = 0101$, $s_3 = 0000$, and $s_4 = 0111$. These four strings are depicted in Figure 38.14 along with their fitness measures.

We use three so-called *genetic operators* to obtain the next population (or generation) from the current population. Initially we obtain P_1 and so on. We continue this process until a solution is found in some population, or else the algorithm fails to converge to a solution in the allotted computation time. Three popular genetic operators are *selection, crossover* (also recombination), and *mutation*. With selection (more accurately, roulette wheel selection), a string is selected to help form the next population with a probability proportional to the percentage of the population's total fitness that it possesses (see Figure 38.15).

Selection is performed in which two copies of s_1 and one copy each of s_2 and s_3 are chosen. Not surprisingly, string s_4 has not been selected. Next, pairs of strings are randomly chosen to mate with one another (see Figure 38.16). To avoid confusion, strings have been relabeled.

a	b
$Parent_1$ = 1010000	$Offspring_1$ = 1011111
$Parent_2$ = 0011111	$Offspring_2$ = 0010000

FIGURE 38.17 (a) Two parent strings are depicted. Crossover point k = 4 is randomly generated. (b) Offspring after crossover.

String	Fitness
s_1 = 1101	2
s_2 = 0101	1
s_3 = 1000	4
s_4 = 0100	2
	Total fitness = 9 Maximum fitness = 4

FIGURE 38.18 Population P_1 shown with fitness measures.

Crossover is next applied. We use a form of crossover in which two parent strings are used to form two offspring (see Figure 38.17). Other forms of crossover exist (Holland, 1975). A crossover point k is randomly selected for a pair of parent strings. In Figure 38.16, k = 4. Two offspring are shown in Figure 38.16b. $Offspring_1$ is identical to $parent_1$ before the crossover point (bits 1–3) and identical to $parent_2$ after the crossover point (bits 4–7). Similarly, $offspring_2$ is identical to $parent_2$ before the crossover point and identical to $parent_1$ after. The results of crossover are shown in the third and fourth columns of Figure 38.16.

Finally, mutation is applied to form the new population of strings. A bit in a string is inverted with a small probability of perhaps p = 0.001. In our toy example, we let this probability p = 0.1. Observe that the rightmost bit of the second string has been inverted, that is, 0100 has become 0101. The new population (P_1) is shown in Figure 38.18, with updated fitness measures. Once again, to avoid confusion, strings have been relabeled. Both the total and the maximum fitness of this population have increased. Furthermore, string s_3 solves this problem.

Recapitulating, GAs are blind, parallel, iterative, and probabilistic. GAs have been successful in many search applications—ranging from portfolio planning and stock market prediction to scheduling satellite orbits and airport landings (Williams et al., 2001). They have also met success in game playing (Chellapilla and Fogel, 2002).

GA was discovered by John Holland in the late 1960s at the University of Michigan (Holland, 1975). He cites the work of Charles Darwin as providing his inspiration. Darwin's theory of evolution attempts to explain how natural selection enables species of flora and fauna to better adapt to conditions on this planet (Darwin, 1859, 1994, 2001). Goldberg (1989) likens the process of GA search to the process of brainstorming at a business meeting. Additional examples of GA search may be found in Lucci and Kopec (2013). The mathematics that justifies the success of GA may be found in Goldberg (1989) and Holland (1992).

38.6.4 Genetic Programming

GAs are a parallel search wherein a population of strings guided by a fitness function and genetic operators iteratively modifies strings until a solution to a problem is hopefully discovered. With *genetic programming (GP)*, the goal is to design a program whose output is the solution to a given problem. GP is also a parallel algorithm; however, we cannot simply begin with a population of strings as each string must encode a properly constructed program. Traditionally, GP has worked well with LISP and other functional languages, which are based on lists; however, imperative languages such as C++ and Java are currently being used in what is known as linear GP (Brameier and Banzhaf, 2006).

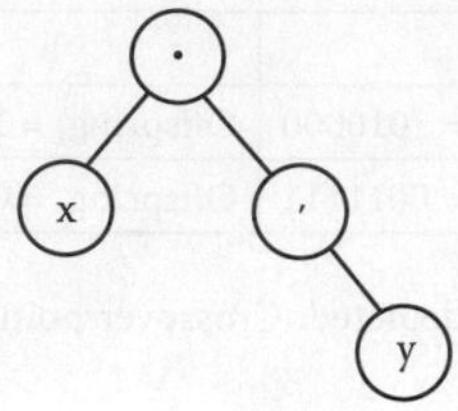

FIGURE 38.19 The list (· x (′y)) represents the tree shown and computes the function x· y′.

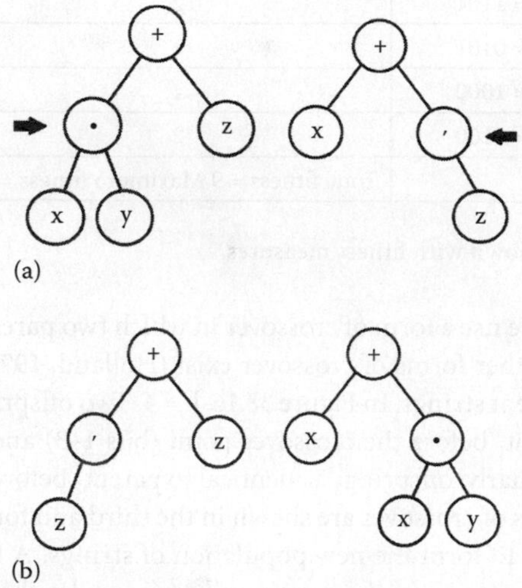

FIGURE 38.20 Crossover in a genetic program. (a) Fracture points (→) are selected in two parents. (b) Offspring shown after crossover.

Programs in LISP may be represented as lists. For example, the function f(x,y) can be represented as (f x y); the function f is followed by its two arguments, x and y. Programs in LISP can consist of functions and terminals. Functions for a problem in Boolean logic might consist of {+,.,′}, that is, the OR function, AND function, and negation. The list (.x (′y)) computes x.y′ as depicted in Figure 38.19.

Three operators that are used in GP are crossover, *inversion*, and mutation. The first step in application of these operators is to identify a fracture point in the list where changes can occur. The beginning of a sublist or at a terminal is permissible fracture points. To perform crossover,

1. Select two programs from the present population
2. Randomly choose two sublists, one from each parent
3. Exchange these sublists in their offspring (see Figure 38.20)

Inversion can be accomplished by

1. Randomly selecting an individual from the present population
2. Randomly choosing two fracture points from the selected individual
3. Exchanging the specified subtrees (see Figure 38.21)

The third operator cited is mutation:

1. Choose an individual program from the present population.
2. Randomly replace a function symbol with some often function symbol OR—replace any terminal symbol with another.

An example of mutation is shown in Figure 38.22.

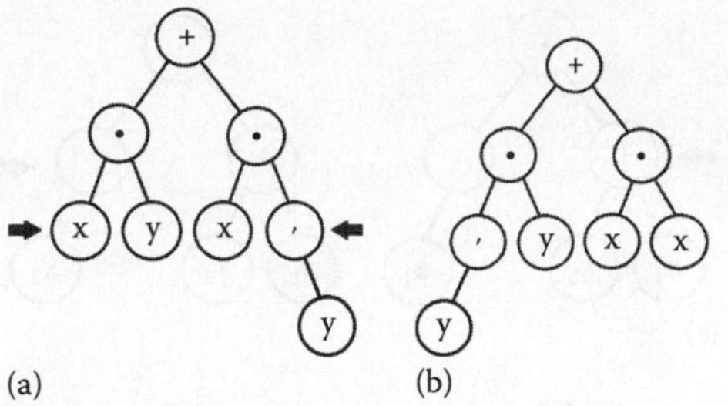

FIGURE 38.21 With inversion, a single parent produces one offspring. (a) Two fracture points (←; →) selected in a program. (b) After inversion, we have a new program.

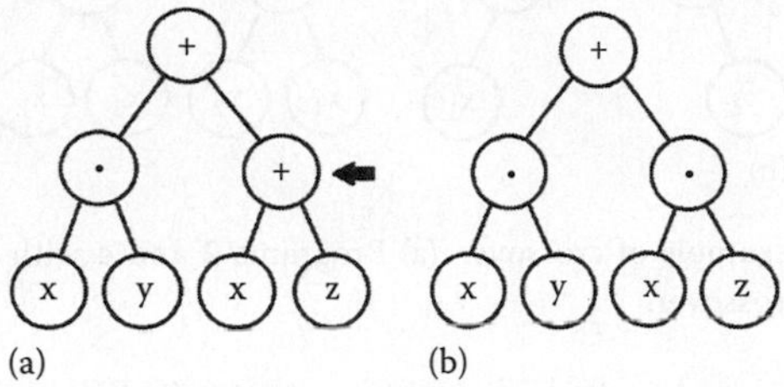

FIGURE 38.22 An individual parent is modified in mutation to produce a single offspring. (a) An individual parent before mutation with its selected node shown. (b) After mutation, Boolean OR in the right subtree has been changed to ANDing.

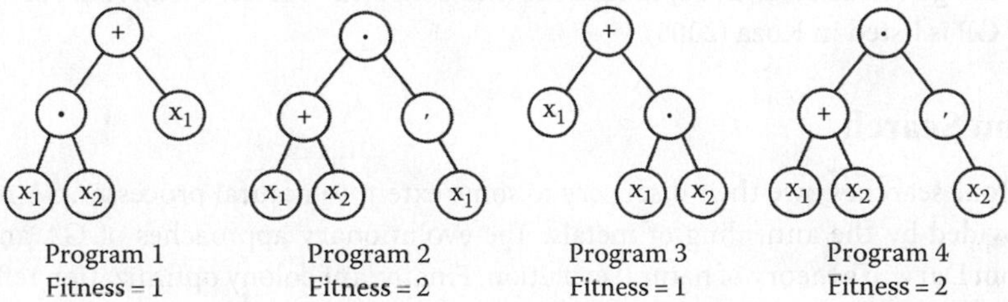

FIGURE 38.23 Four random programs with fitness shown.

With mutation, you need to perform type checking. For example, if Boolean operations are being performed, then 0 and 1 are the only operands permitted.

It is widely held that Koza discovered GP. He advises strategies for the creation of a random population. The interested reader is referred to Koza et al. (2000, 2003) for more details. To provide some insight to this methodology, we revisit our example from the previous section. There, we were using GA to search for a string that mirrored the functionality of the two-input NOR function (see Table 38.1). With a GP we seek to discover a program that when presented with two Boolean values, x_1 and x_2, produces $x_1 \downarrow x_2$ as output. In this toy example, we shall assume that our initial population of programs consists of four trees as shown in Figure 38.23. In realistic applications, an initial population may consist of thousands of programs.

The fitness of each program is also shown in this figure. As a fitness measure, we have used the number of rows of the NOR table that each program correctly computes. For example, when $x_1 = 0$ and $x_2 = 1$, program 1 produces an output of 0, which correctly reflects the second row of Table 38.1; however, for the other three truth assignments for x_1 and x_2, program 1 produces an output that differs from the NOR table.

Suppose that crossover is performed between programs 2 and 4 as shown in Figure 38.23 at the selected fracture points. The left offspring in Figure 38.24b is seen to have a fitness of 4 and hence correctly computes the two-input NOR function. We admit that the initial population in Figure 38.22 was

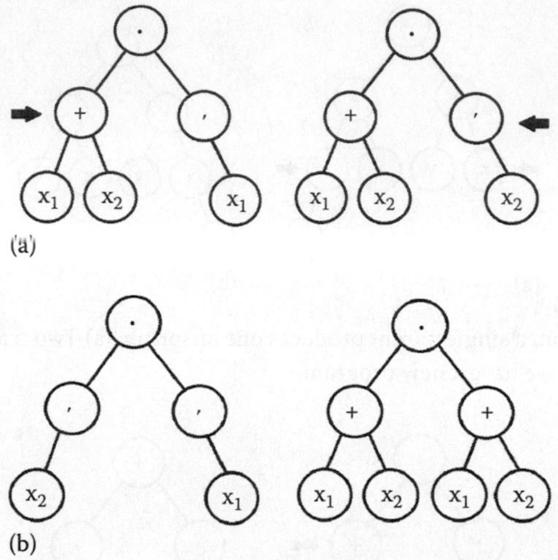

FIGURE 38.24 A fortuitous example of crossover. (a) Programs 2 and 4 with selected fracture points (→; ←) shown. (b) Two offspring after crossover.

not chosen randomly, and the choice of which genetic operator in Figure 38.23 (as well as the selection of population members and fracture points) also was performed with some forethought. It should be apparent, however, that permitted a larger population and allowed ample iterations, a GP would successfully solve this problem. In fact, GPs have obtained remarkable results in a number of areas—ranging from antenna design (Koza et al., 2000) to robotics (Messom and Walker, 2002). A list of new inventions discovered by GP is listed in Koza (2005).

38.6.5 Tabu Search

The previous four searches owe their discovery to some extent to natural processes. SA relies upon the metaphor provided by the annealing of metals. The evolutionary approaches of GA and GP borrow generously from Darwin's theory of natural evolution. Finally, ant colony optimization reflects the intelligence by colonies of ants. *Tabu* (also called *Taboo*) *search* (TS) stresses the importance of memory. To balance both exploitation and exploration, TS maintains two types of lists. A *recency Tabu list* (*RTL*) monitors the moves made that transfer one state of the search into another; these moves cannot be repeated until some period of time (the Tabu tenure that is set by the programmers) has elapsed. The RTL encourages exploration. TS also uses lists to control so-called aspiration criteria.

One such criterion—aspiration by default—will select the oldest move whenever there is a "deadlock" and all moves Tabu. Another aspiration criterion—aspiration by direction—favors moves that have led to superior past values of the objective function. Exploitation is fostered by this heuristic. Finally, there is an aspiration criterion that favors exploration—aspiration by influence—in which moves that lead to unexplored regions of the state space are favored.

TS was developed by Fred Glover in the 1970s (1997). TS has been successfully applied to problems in optimization. This methodology has also been applied to pattern classification, VLSI design, and numerous other problem domains (Emmert et. al., 2004, Glover, 1995, Tung and Chou, 2004).

38.6.6 Ant Colony Optimization

Ant colonies provide the metaphor for our next search paradigm. It is well known by entomologists that ant colonies possess remarkable intelligence. Their intellectual prowess is an example of emergent behavior, behavior at one level (often unexpected) that derives from a lower level. At the microscopic

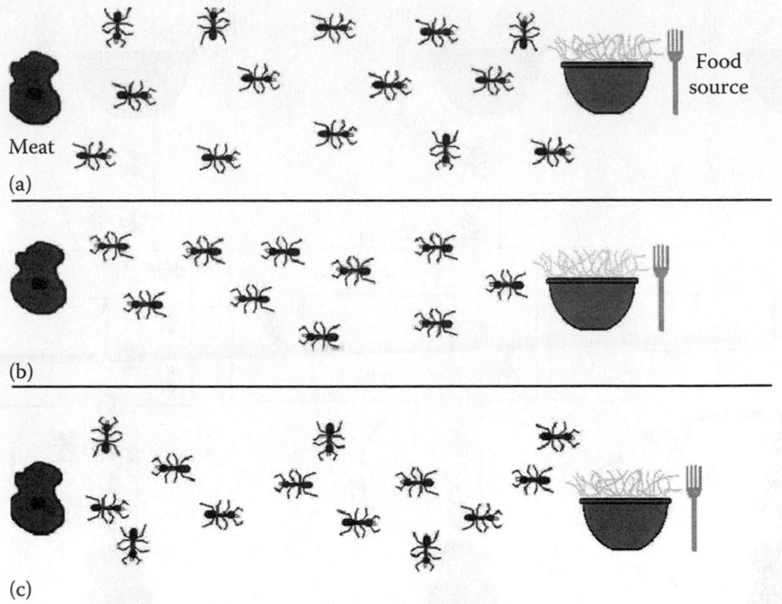

FIGURE 38.25 Ants foraging, depositing, and dispersing from a discovered food source. (a) Ants are moving randomly. (b) Food source discovered. Ants return to nest carrying food and depositing pheromones. (c) Ants follow the pheromone trail from the nest to food source, some still move randomly.

level, however, concepts of temperature and pressure are meaningless as gas molecules have only a location and a kinetic energy level.

The emergent behavior of interest to us here is the intelligence possessed by ant colonies. In the present search, ants are viewed as agents—entities that are capable of sensing the environment, communicating with one another, and responding to the conundrums posed by their environment. M. Dorigo was the first to apply *ant colony optimization* to solve combinatorial problems (Dorigo, 1992).

Ant foraging will prove useful to us in what follows. If you were to introduce a source of food in proximity to an ant colony, ant foragers will undoubtedly form a trail that would enable their brethren to locate and retrieve this nourishment (see Figure 38.25).

In Figure 38.25a, forager ants have discovered the food source. These explorers will communicate their discovery to other ants by a process known as *stigmergy*—this is a form of indirect communication between ants. On the way back to their nest, explorer ants deposit pheromones on their trail (Figure 38.24b); these are chemicals that ants produce. There are two advantages to this form of communication:

1. One ant need not know the location of the other ants that it is communicating with.
2. The communication signal will not perish even if the sender does.

The pheromone trail will encourage exploitation by other ants in the colony.

Consult Figure 38.25c and you observe that not every ant is following the recently discovered trail. The seemingly random behavior of these foragers encourages exploration, that is, they are searching for alternate food sources that are vital for the colony's long-term survival.

We began this section by claiming that ant colony behavior was applicable to combinatorial optimization. The process of trail formation and following is used by some ant species to identify a shortest path between their colony and some food source. This facility for optimization was apparent from experiments conducted by Deneubourg and colleagues (1991).

Figure 38.26 contains a source of food that is separated from an ant colony by two paths of different lengths. In Figure 38.26a there is a barricade separating the ants from the two paths.

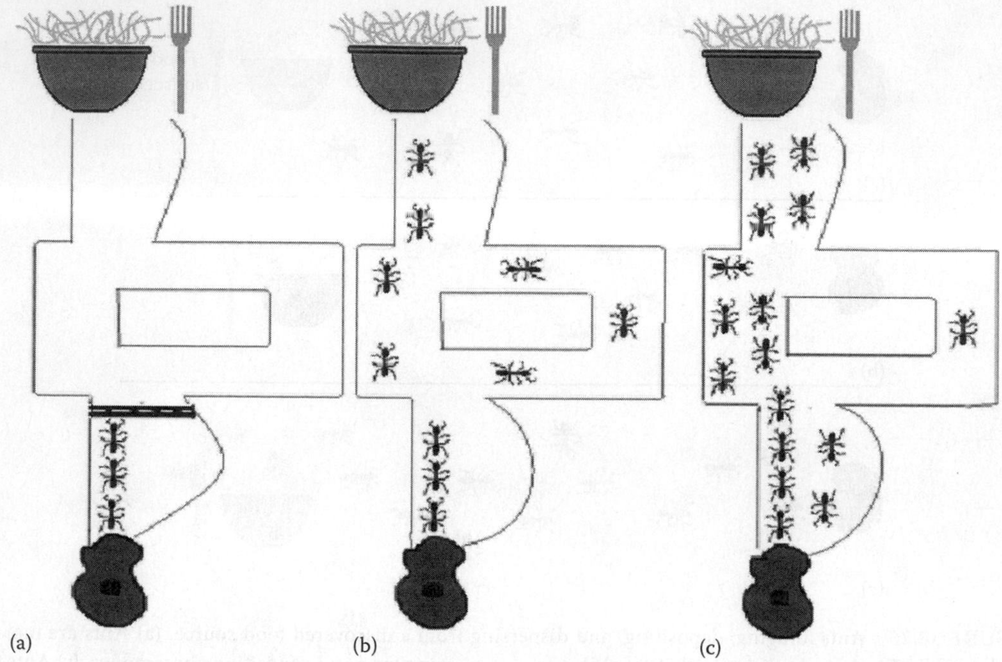

FIGURE 38.26 Ants separated from their colony. (a) Ants blocked by door. (b) Door raised and ants travel arbitrarily to food source, returning to nest to deposit pheromones. (c) More ants at nest follow the left (shorter) path to food source.

In Figure 38.26b the barricade is removed and we observe that ants are as likely to choose either path. The situation after some time has elapsed is shown in Figure 38.26c; ants are favoring the shorter path.

The longer path is not entirely devoid of ant traffic as this is a form of exploration. We mentioned earlier that trails are formed by pheromone deposits from explorer ants. As time passes, this chemical will evaporate. The return trip from the food sources to the ant colony will require more time along the longer path, and hence, more of the pheromone deposited there will have an opportunity to evaporate. Due to this pheromone evaporation, the shorter route will contain more pheromone than the longer one. Carrying food back to the colony will require numerous trips, visited by many ants in these additional journeys.

Choosing a shorter path enables ants to utilize less energy in food collection enabling them to finish faster. Ants thereby avoid competition from nearby colonies and are also less likely to encounter predators.

Dorigo et al. used artificial ants and artificial pheromone trails to solve instances of the TSP problem (Dorigo and Stutzle, 2004).

38.7 Most Recent Developments

The advent of the World Wide Web (WWW) in 1993 naturally led to interest in search techniques for the Internet, especially when intelligence can be applied. Search engines are computer programs that can automatically contact other network resources on the Internet, searching for specific information or key words, and report the results of their search. Intelligent agents are computer programs that help the users to conduct routine tasks, to search and retrieve information, to support decision-making, and to act as domain experts. They do more than just "search and match." There are intelligent agents for consumers that search and filter information. Intelligent agents have been developed for product and vendor finding, for negotiation, and for learning. They can help determine what to buy to satisfy a specific need by looking for specific products' information and then critically evaluating them. Some examples are Firefly, which uses a *collaborative filtering* process that can be described as "word of mouth" to build the profile. It asks a consumer to rate a number of products and then matches his or her ratings with the ratings of other consumers with similar tastes, recommends products that have not yet been rated by the consumer, etc. Intelligent agents for finding products and vendors enabling discovery of bargains.

One example is Jango from NetBot/Excite. It originates requests from the user's site (rather than Jango's). Vendors have no way of determining whether the request is from a real customer or from the agent. Jango provides product reviews. Kasbah, from MIT Lab, is for users who want to sell or buy a product. It assigns the task to an agent who is then sent out to proactively seek buyers or sellers. Intelligent agents for consumers can act as negotiation agents, helping to determine the price and other terms of transactions. Kasbah has multiple agents. Users create agents for the purpose of selling and buying goods. It can exemplify three strategies: anxious, cool headed, and frugal. Tete-a-tete is an intelligent agent for considering a number of different parameters: price, warranty, delivery time, service contracts, return policy, loan options, and other value-added services. Upon request, it can even be argumentative. Finally, there are learning agents capable of learning individuals' preferences and making suitable suggestions based upon these preferences. Memory Agent from IBM and Learn Sesame from Open Sesame use learning theory for monitoring customers' interactions. It learns customers' interests, preferences, and behavior and delivers customized service to them accordingly (Deitel et al., 2001). Expect many interesting developments in this arena combining some of the theoretical findings we have presented with practical situations, to achieve exciting results.

Key Terms

A* algorithm: A best-first procedure that uses an admissible heuristic estimating function to guide the search process to an optimal solution.

Admissibility condition: The necessity that the heuristic measure never overestimates the cost of the remaining search path, thus ensuring that an optimal solution will be found.

Alpha–beta: The conventional name for the bounds on a depth-first minimax procedure that are used to prune away redundant subtrees in two-person games.

AND/OR Tree: A tree that enables the expression of the decomposition of a problem into subproblems; hence, alternate solutions to subproblems through the use of AND/OR node labeling schemes can be found.

Ant colony optimization: An algorithm based on the behavior of natural ant colonies.

Aspiration criteria: Conditions in a TS that encourage both exploitation and exploration. For example, aspiration by influence favors moves that lead to unexplored region of the search space. This condition fosters exploration.

Backtracking: A component process of many search techniques whereby recovery from unfruitful paths is sought by backing up to a juncture where new paths can be explored.

Best-first search: A heuristic search technique that finds the most promising node to explore next by maintaining and exploring an ordered OPEN node list.

Bidirectional search: A search algorithm that replaces a single search graph, which is likely to grow exponentially, with two smaller graphs—one starting from the initial state and one starting from the goal state.

Blind search: A characterization of all search techniques that are heuristically uninformed. Included among these would normally be state-space search, means–ends analysis, generate and test, DFS, and breadth-first search.

Branch and bound algorithm: A potentially optimal search technique that keeps track of all partial paths contending for further consideration, always extending the shortest path one level.

Breadth-first search: An uninformed search technique that proceeds level by level visiting all the nodes at each level (closest to the root node) before proceeding to the next level.

Cooling schedule: Simulating annealing uses a temperature parameter (T). Early in the search, T is high, thus encouraging exploration. As the search continues, T is lowered according to some schedule.

Crossover: A genetic operator that is used in both GA and GP searches. Crossover is used to generate new populations in these searches.

Data-driven parallelism: A load-balancing scheme in which work is assigned to processors based on the characteristics of the data.

Depth-first search: A search technique that first visits each node as deeply and to the left as possible.

Exploitation: A strategy in search algorithms that encourages staying in regions with good objective functions.

Exploration: A search algorithm strategy that encourages movement to unexplored regions of a search space.

Fitness function: A measure of goodness that is assigned to a potential solution to a given problem.

Fracture point: A location in a tree that identifies where a genetic operator is to be applied (used in GP).

Generate and test: A search technique that proposes possible solutions and then tests them for their feasibility.

Genetic algorithm: A blind search that is based on Darwin's theory of evolution. It is a blind search that is both parallel and probabilistic.

Genetic operators: Operators that are used in both GAs and GP to help form a new population.

Genetic programming: A parallel search in which a program is found that solves a given problem. Operators based upon Darwinian evolution drive this search.

Heuristic search: An informed method of searching a state space with the purpose of reducing its size and finding one or more suitable goal states.

HyPS: A hybrid search that combines PWS with DTS.

Inversion: One of the genetic operators used in GP. Inversion will randomly exchange two subtrees in a potential solution program.

Iterative deepening: A successive refinement technique that progressively searches a longer and longer tree until an acceptable solution path is found.

Mandatory work first: A static two-pass process that first traverses the minimal game tree and uses the provisional value found to improve the pruning during the second pass over the remaining tree.

Means–ends analysis: An AI technique that tries to reduce the "difference" between a current state and a goal state.

Monte Carlo simulation: An early search that employs random numbers to model uncertainty.

MTD(f) algorithm: A minimal window minimax search recognized as the most efficient alpha–beta variant.

Mutation: A genetic operator that is used in both GA and GP searches. Mutation fosters genetic diversity in a population.

Parallel window aspiration search: A method where a multitude of processors search the same tree but each with different (nonoverlapping) alpha–beta bounds.

PVSplit: A static parallel search method that takes all the *processors down the first variation to some limiting depth and then splits the* subtrees among the processors as they back up to the root of the tree.

Recency Tabu list: A list maintained in a TS to foster exploration.

Selection: A strategy used in both GA and GP searches. Selection helps to choose those potential solutions that will help to form the next population in a search.

Simulated annealing: A stochastic algorithm that returns optimal solutions when given an appropriate "cooling schedule."

SSS*: A best-first search procedure for two-person games.

Stigmergy: A form of indirect communication used by ants in a colony and ant agents in an ant colony-based search. Ants deposit pheromone on a trail, and this chemical can be detected by other ants.

Tabu search: A search that encourages exploration by forbidding the same moves to be made again until some prescribed time has elapsed. Criteria are also present to encourage exploitation.

Transposition table-driven scheduling (TDS): A data-driven load-balancing scheme for parallel search that assigns a state to a processor based on the characteristics or signature of the given state.

Work-driven parallelism: A load-balancing scheme in which idle processors explicitly request work from other processors

Young brothers wait concept: A dynamic variation of PVSplit in which idle processors wait until the first path of leftmost subtree has been searched before giving work to an idle processor.

Acknowledgments

The authors thank Islam M. Guemey for help with research, Erdal Kose for the best-first example, and David Kopec for technical assistance and assistance with artwork, as well as Andrew Convissar for help with figures in Section 38.6.

Further Information

The most regularly and consistently cited source of information for this article is the Journal of Artificial Intelligence. There are numerous other journals including *AAAI Magazine, CACM, IEEE Expert, ICGA Journal*, and the *International Journal of Computer Human Studies*, which frequently publish articles related to this subject area. Also prominent has been the *Machine Intelligence Series* of Volumes edited by Donald Michie with various others. An excellent reference source is the 3-volume Handbook of Artificial Intelligence by Barr and Feigenbaum (1981).

In addition, there are numerous national and international and national conferences on AI with published proceedings, headed by the *International Joint Conference on AI (IJCAI)*. Classic books on AI methodology include Feigenbaum and Feldman's (1963) *Computers and Thought* and Nils Nilsson's (1971) *Problem-Solving Methods in Artificial Intelligence*. There are a number of popular and thorough textbooks on AI. Two relevant books on the subject of search in are *Heuristics* (Pearl, 1984) and the more recent *Search Methods for* (Bolc and Cytowski). AI texts that have considerable focus on search techniques are George Luger's *Artificial Intelligence* (6th edn., 2008), Russell and Norvig's *Artificial Intelligence: a modern approach* (3rd edn., 2010), and most recently Lucci and Kopec's Artificial Intelligence in the 21st century (2013).

References

Akl, S.G., Barnard, D.T., and Doran, R.J. 1982. Design, analysis and implementation of a parallel tree search machine. *IEEE Transactions on Pattern Analysis and Machine Intelligence*, 4(2): 192–203.

Barr, A. and Feigenbaum, E.A. 1981. *The Handbook of Artificial Intelligence V1*. William Kaufmann Inc., Stanford, CA.

Baudet, G.M. 1978. The design and analysis of algorithms for asynchronous multiprocessors. PhD thesis. Department of Computing Science, Carnegie-Mellon University, Pittsburgh, PA.

Bolc, L. and Cytowski, J. 1992. *Search Methods for Artificial Intelligence*. Academic Press Inc., San Diego, CA.

Brameier, M. and Banzhaf, W. 2006. *Linear Genetic Programming*. Springer-Verlag, New York.

Cerny, V. 1985. Thermodynamical approach to the traveling salesman problem: An efficient simulation algorithm. *Journal of Optimization Theory and Applications*, 45: 44–51.

Chellapilla, K. and Fogel, D.B. 2002. Anaconda's expert rating by computing against Chinook experiments in co-evolving a neural checkers. *Neurocomputing*, 42: 69–86.

Cook, D.J. 2011. A hybrid approach to improving the performance of parallel search. Technical Report, Department of Computer Science Engineering, University of Texas, Austin, TX.

Cook, D.J. and Varnell, R.C. 1998. Adaptive parallel iterative deepening search. *Journal of Artificial Intelligence Research*, 8: 139–166.

Darwin, C. 1859. *Origin of Species*, Bantam.

Darwin, C. 1994. The Correspondence of Charles Darwin. Vol. VI. *The English Historical Review*, 109: 1856–1857.

Darwin, C. 2001. *The Voyage of the Beagle*. New Edition, Random House Inc., New York.

Deitel, H.M., Deitel, P.J., and Steinbuhler, K. 2000. *E-Business and e-Commerce for Managers*. Prentice Hall Publishers, Upper Saddle River, NJ.

Deneubourg, J.L. et al. 1991. *Simulation of Adaptive Behavior: From Animals to Animats*. MIT Press, Cambridge, MA.

Dorigo, M. 1992. Optimization, learning and natural algorithms. PhD thesis. Dipartamento ai Electronics Politechnio di Milano, Milano, Italy.

Dorigo, M. and Stutzle, T. 2004. *Ant Colony Optimization*. MIT Press, Cambridge, MA.

Emmert, J.M., Lodha, S., and Bhatia, D.K. 2004. On using tabu search for design automation of VLSI systems. *Journal of Heuristics*, 9: 75–90.

Feigenbaum, E. and Feldman, J. 1963. *Computers and Thought*. McGraw-Hill, New York.

Feldmann, R. 1993. Game tree search on massively parallel systems, PhD thesis. University of Paderborn, Paderborn, Germany.

Feldmann, R., Monien, B., Mysliwietz, P., and Vornberger, O. 1990. Distributed game tree search. In V. Kumar, P.S. Gopalakrishnan, and L. Kanal (eds.), *Parallel Algorithms for Machine Intelligence and Vision*, pp. 66–101. Springer-Verlag, New York.

Felten, E.W. and Otto, S.W. 1989. A highly parallel chess program. *Proceedings of the International Conference on Fifth Generation Computer Systems*, Tokyo, Japan, pp. 1001–1009.

Ferguson, C. and Korf, R.E. 1988. Distributed tree search and its application to alpha-beta pruning. *Proceedings of the Seventh National Conference on Artificial Intelligence*, Vol. 1, Kaufmann, Saint Paul, LA, pp. 128–132.

Gendreau, M., Laporte, G., and Semet, F. 2001. A dynamic model and parallel Tabu search heuristic for real-time ambulance relocation. *Parallel Computing*, 27(12): 1641–1653.

Glover, F. 1995. Tabu search fundamentals and uses. Technical report, University of California, Davis, CA.

Glover, F. 1997. *Tabu Search*. Springer, New York.

Goldberg, D.E. 1989. *Genetic Algorithms in Search, Optimization, and Machine Learning*. Addison-Wesley, Reading, MA.

Hart, P.E., Nilsson, N.J., and Raphael, B. 1968. A formal basis for the heuristic determination of minimum cost paths. *IEEE Transactions on SSC*, SSC-4: 100–107.

Hoffman, J. and Nebel, B. 2000. The FF planning system: Fast plan generation through heuristic search. *Journal of Artificial Intelligence Research*, 14: 253–302.

Holland, J. 1975. *Adaptation in Natural and Artificial Systems*. University of Michigan Press, Ann Arbor, MI, 183pp.

Holland, J. 1992. *Adaptation in Natural and Artificial Systems*, 2nd edn. MIT Press, Cambridge, MA.

Hong, T.P, Huang, K.Y., and Lin, W.Y. 2001. Adversarial search by evolutionary computation. *Evolutionary Computation*, 9(3): 371–385.

Hsu, F-h. 1990. Large scale parallelization of alpha-beta search: An algorithmic and architectural study with computer chess. Technical report CMU-CS-90-108. Carnegie-Mellon University, Pittsburgh, PA.

Huang, S. and Davis, L.R. 1989. Parallel iterative A* search: An admissible distributed search algorithm. *Proceedings of the 11th International Joint Conference on Artificial Intelligence*, Vol. 1, pp. 23–29, Kaufmann, Detroit, MI.

Hyatt, R.M., Suter, B.W., and Nelson, H.L. 1989. A parallel alpha-beta tree searching algorithm. *Parallel Computing*, 10(3): 299–308.

Kaindl, H. and Kainz, G. 1997. Bidirectional heuristic search reconsidered. *Journal of the Artificial Intelligence Research*, 7: 283–317.

Kalos, M.H. and Whitlock, P. 2008. *Monte Carlo Methods*. Wiley-VCH, Berlin, Germany.

Kirkpatrick, S., Gelatt, C.D., and Vechi, M.P. 1983. Simulated annealing. *Operations Research*, 39: 378–406.

Kishimoto, A. and Schaeffer, J. 2002. Distributed game-tree search using transposition table driven work scheduling, *Proceedings of the International Conference on Parallel Processing (ICPP)*, Vancouver, British Columbia, Canada, pp. 323–330.

Knuth, D. and Moore, R. 1975. An analysis of Alpha-Beta pruning. *Artificial Intelligence*, 6(4): 293–326.

Koenig, S. 2001. Minimax real-time heuristic search. *Artificial Intelligence*, 129: 165–195.

Korf, R.E. 1985. Depth-first iterative-deepening: An optimal admissible tree search. *Artificial Intelligence*, 27(1): 97–109.

Korf, R.E. 1990. Real-time heuristic search. *Artificial Intelligence*, 42(2–3): 189–211.

Koza, J.R., Comisky, W., and Jessen,Y. 2000. Automatic synthesis of a wire antenna using genetic programming, *Proceedings of the Conference on Genetic and Evolutionary Computation*, Las Vegas, NV.

Koza, J.R., Keane, M.A., Streeter, M.J., Adams, T.P., and Jones, L.W. 2005. Invention and creativity in automated design by means of genetic programming. *Artificial Intelligence for Engineering Design Analysis and Manufacturing*, 18(3): 245–269; Cambridge University Press, London, U.K.

Koza, J.R., Keane, M.A., Streeter, M.J., Mydlowee, W., Yu, J., and Lanza, G. 2003. *Genetic Programming: Routine Human-Competitive Machine Intelligence*. Springer, New York.

Kumar, V., Ramesh, K., and Nageshwara-Rao, V. 1988. Parallel best-first search of state-space graphs: A summary of results. *Proceedings of the Seventh National Conference on Artificial Intelligence, AAAI-88*, pp. 122–127, Kaufmann, Los Altos, CA.

Lucci, S. and Kopec, D. 2013. *Artificial Intelligence in the 21st Century*. Mercury Learning and Information, Dulles, VA.

Luger, J. 2008. *Artificial Intelligence: Structures and Strategies for Complex Problem Solving*, 6th edn. Benjamin/Cummings Publishing Company Inc., Redwood City, CA.

Luger, J. and Stubblefield, W. 1993. *Artificial Intelligence: Structures and Strategies for Complex Problem Solving*, 2nd edn. Benjamin/Cummings Publishing Company Inc., Redwood City, CA.

Marsland, T.A. and Campbell, M. 1982. Parallel search of strongly ordered game trees. *ACM Computing Surveys*, 14(4): 533–551.

Marsland, T.A. and Popowich, F. 1985. Parallel game-tree search. *IEEE Transactions on Pattern Analysis and Machine Intelligence*, 7(4): 442–452.

Mayorov, M. and Whitlock, P. Parallelization of algorithms to solve a three dimensional Sudoku puzzle, *Proceedings of the IMACS Seminar on Monte Carlo Methods*, Borovets, Bulgaria, August 28 to September 2, /2011.

Messom, C.H. and Walker, M.G. 2002. Evolving cooperative robotic behavior using distributive genetic programming. *Proceedings of the Seventh International Conference on Control, Automations, Robotics, and Vision ICARCV'02*, Singapore.

Metropolis, N., Rosenbluth, A., Rosenbluth, M., Teller, A., and Teller, E. 1953. Equation of state calculations by fast computing machines. *Journal of Chemical Physics*, 21: 1087–1092.

Mueller, M. 2001. Partial order bounding: A new approach to evaluation in game tree search. *Artificial Intelligence*, 129: 279–311.

Newborn, M.M. 1988. Unsynchronized iteratively deepening parallel alpha-beta search. *IEEE Transactions on Pattern Analysis and Machine Intelligence*, 10(5): 687–694.

Nilsson, N. 1971. *Problem-Solving Methods in Artificial Intelligence*. McGraw-Hill Publishing Co., New York.

Pearl, J. 1980. Asymptotic properties of minimax trees and game-searching procedures. *Artificial Intelligence*, 14(2): 113–138.

Pearl, J. 1984. *Heuristics: Intelligent Search Strategies for Computer Problem Solving*. Addison-Wesley Publishing Co., Reading, MA.

Plaat, A., Schaeffer, J., Pijls, W., and de Bruin, A. 1995. Best-first fixed-depth game-tree search in practice. *Proceedings of the IJCAI-95*, Kaufmann, Montreal, Quebec, Canada, pp. 273–279.

Plaat, A., Schaeffer, J., Pijls, W., and de Bruin, A. 1996. Best-first fixed-depth minimax algorithms. *Artificial Intelligence*, 87(1–2): 1–38.

Pohl, I. 1971. Bi-directional search. In B. Meltzer and D. Michie (eds.), *Machine Intelligence*, Vol. 6, pp. 127–140. American Elsevier, New York.

Polya, G. 1945. *How to Solve It*. Princeton University Press, Princeton, NJ.

Powley, C. and Korf, R.E. 1991. Single-agent parallel window search. *IEEE Transactions on Pattern Analysis and Machine Intelligence*, 13(5): 466–477.

Rao, V.N., Kumar, V., and Ramesh, K. 1987. A parallel implementation of Iterative-Deepening A*. *Proceedings of the Sixth National Conference on Artificial Intelligence*, Seattle, WA, pp. 178–182.

Reinefeld, A. 1983. An improvement to the Scout tree-search algorithm. *International Computer Chess Association Journal*, 6(4): 4–14.

Reinefeld, A. 1995. Scalability of massively parallel depth-first search. In P.M. Pardalos, M.G.C. Resende, and K.G. Ramakrishnan (eds.), *Parallel Processing of Discrete Optimization Problems*, Vol. 22, pp. 305–322. DIMACS Series in Discrete Mathematics and Theoretical Computer Science, American Mathematical Society, Providence, RI.

Reinefeld, A. and Marsland, T.A. 1994. Enhanced iterative-deepening search. *IEEE Transactions on Pattern Analysis and Machine Intelligence*, 6(7): 701–710.

Roizen, I. and Pearl, J. 1983. A minimax algorithm better than alpha-beta?: Yes and no. *Artificial Intelligence*, 21(1–2): 199–220.

Romanycia, M. and Pelletier, F. 1985. What is heuristic? *Computational Intelligence*, 1: 24–36.

Romein, J.W. 2000. Multigame—An environment for distributed game-tree search, PhD thesis. ASCI dissertation series No. 53, Vrije Universiteit, Amsterdam, the Netherlands.

Romein, J., Plaat, A., Bal, H., and Schaeffer, J. 1999. Transposition table driven work scheduling in distributed search. *Proceedings of the AAAI National Conference*, Orlando, FL, pp. 725–731.

Russell, S. and Norvig, P. 2010. *Artificial Intelligence: A Modern Approach*, 3rd edn. Prentice Hall, Englewood Cliffs, NJ.

Schaeffer, J. 1989. Distributed game-tree search. *Journal of the Parallel and Distributed Computing*, 6(2): 90–114.

Schaeffer, J., Plaat, A., and Junghanns, J. 2001. Unifying single-agent and two-player search. *Information Sciences*, 134(3–4): 151–175.

Shoham, Y. and Toledo, S. 2002. Parallel randomized best-first minimax search. *Artificial Intelligence*, 137: 165–196.

Slate, D.J. and Atkin, L.R. 1977. Chess 4.5—The Northwestern University Chess Program. In P. Frey (ed.), *Chess Skill in Man and Machine*, pp. 82–118. Springer-Verlag, New York.

Stewart, B.S. and White, C.C. 1991. Multiobjective A*. *Journal of the ACM*, 38(4): 775–814.

Stockman, G. 1979. A minimax algorithm better than alpha-beta? *Artificial Intelligence*, 12(2): 179–196.

Tung, C. and Chou, C. 2004. Pattern classification using tabu search to identify the spatial distribution of groundwater pumping. *Hydrology Journal*, 12: 488–496.

Williams, E.A., Crossley, W.A., and Lang, T.J. 2001. Average and maximum revisit time trade studies for satellite constellations using a multiobjective genetic algorithm. *The Journal of the Astronomical Sciences*, 49: 385–400.

Winston, P.H. 1992. *Artificial Intelligence*, 3rd edn. Addison-Wesley, Reading, MA.

Wu, W. and Gertner, I. 2011. High performance value at risk (VaR) estimation with monte carlo simulation using general graphics processing unit (GGPU). Technical report. The City College of New York, New York.

39

Planning and Scheduling

Roman Barták
Charles University in Prague

39.1 Introduction

Planning and scheduling are two closely related terms that are frequently interchanged in practical life. In the industrial environment, planning often means long-term scheduling, for example, allocating contract work to departments for the next several months, while scheduling means detailed scheduling, for example, allocating jobs or operations to machines with a minute precision. Hence, it may seem surprising that planning and scheduling are actually two different research areas that originated in different fields and that exploit different problem-solving techniques.

The term *planning* refers to a process of creating an organized set of actions, a *plan*, that can be executed to achieve a desired goal from a given initial situation (state), while the term *scheduling* usually refers to a process of assigning a set of actions to resources over time, the output is called a *schedule*. In complex problems, we typically start with planning, that is, we decide what to do—which actions need to be executed to achieve the goal. After that, we continue with scheduling, that is, we decide how to do it—how to allocate the actions to scarce resources and time to obtain a schedule. Finally, we execute the schedule, which involves monitoring of actions and revoking the planning and scheduling stages when necessary. This chapter covers techniques for planning and scheduling and omits the details of plan/schedule execution.

Planning is an important aspect of rational behavior and it is a fundamental topic of artificial intelligence since its beginning. Planning capabilities are necessary for autonomous controlling of vehicles of many types including space ships (Muscettola et al., 1998) and submarines (McGann et al., 2008), but we can also find planning problems in areas such as manufacturing, games, or even printing machines (Ruml et al., 2005). Planning techniques are frequently domain independent, which means that the planning system gets a formal description of a planning domain in addition to a particular problem instance as its inputs. The planning domains can be as diverse as forming towers from blocks (useful for planning in container terminals), finding paths for robots to transport goods (useful for automated warehouses), and solving puzzles (useful for modeling non-player characters in computer games).

Scheduling has been studied since the 1950s when researchers were faced with problems of efficient management of operations in workshops (Leung, 2004). It became one of the fundamental topics of operations research. Nowadays, scheduling is important in every manufacturing facility to allocate operations to machines and workers as well as, for example, in computing environments to allocate tasks to processors and memory. The research in the area of scheduling focuses mainly on solving specific classes of scheduling problems and on establishing their computational complexity rather than on providing a general scheduling technique applicable to every scheduling problem. The result is that we have a large number of scheduling algorithms for a number of specific classes of problems (Brücker, 2001). This makes scheduling very different from the approach of planning where the focus is on solving general planning problems rather than on developing ad hoc techniques for particular planning domains. In this chapter, we formulate planning and scheduling precisely, we describe the core approaches for solving planning and scheduling problems, and we show how these approaches can be integrated. We also highlight the practical importance of this technology and we overview existing challenges and possible future developments.

39.2 Underlying Principles

As we have already mentioned, despite the close relationship between planning and scheduling, these areas are studied in different fields and use different methodologies. Planning is a typical artificial intelligence topic where it is motivated by general problem solving. Scheduling traditionally belongs to operations research, combinatorial optimization, and computational complexity. It is motivated by everyday industrial requirements for efficient production. These origins are also reflected in approaches used for problem solving. The planning techniques are frequently built around a search algorithm and they cover a broad range of problems, while the scheduling techniques are usually ad hoc algorithms dedicated to a particular class of scheduling problems. Nevertheless, as the real-life problems naturally involve planning and scheduling components, the novel planning techniques are adopting some typical scheduling approaches, such as time and resource allocation, while some scheduling techniques include planning decisions, such as selection of actions to be scheduled. In the following sections, we will describe planning and scheduling approaches separately and we will also show how each approach is related to the other.

39.2.1 Automated Planning

A classical planning task is formulated as the problem of finding a sequence of actions, called a *plan*, that transfer the world from some initial state to a desired state. This task may seem like a path-finding problem in a graph, where nodes describe the world states and arcs are annotated by actions changing the state, which is indeed an abstract view of planning. However, the state space in planning is usually so huge that it is impossible even to store the whole state space (the graph) in a computer memory. Hence, a more compact representation of the state space and the transitions between the states is necessary.

To focus on the core aspects of the planning task, we make several assumptions about the world states and how the actions are changing them. Typically, we assume that the world state is *fully observable*, that is, we know precisely the state of the world where we are. A complementary assumption is a *partially observable* world meaning that we only know a possible set of states where we can be (e.g., we do not know whether the doors that we approached are locked or not and hence we need to assume both states where the doors are locked and unlocked). Another typical assumption is that the world is *deterministic*, that is, the next state is fully determined by the current state and the action that has been applied to the state. In *nondeterministic* worlds, the state after performing an action is not fully known (e.g., after grasping some block, we may hold it or the block may still be on the table, if the grasp is not strong enough and the block fell down). The third major assumption is that the world is *static* meaning that only the planned actions are changing the world states. In *dynamic* worlds, the states may change due to other reasons, for example, nature or other agents may cause such changes. Finally, we assume the actions to be instantaneous so we only deal with action sequencing and we omit time reasoning beyond *causal relations* (actions cause changes of the world state that allow application of other actions). There exist planning systems that can deal with more general assumptions such as durative and parallel actions with uncertain effects and partially known states that may be changed by other entities (Ghallab et al., 2004).

39.2.1.1 Planning Task

As we already mentioned, the state space in planning is usually huge. To overcome this problem when representing the world states, the planning community is frequently using the representation of *world states* as finite sets of predicates. Assume that a predicate describes some particular feature of the world state, for example, hold(blockA) says that we hold some object called blockA, and free(blockB) says that object blockB is free and can be picked up. So each predicate consists of its name describing some feature and some arguments describing the objects. We assume that no functional symbols are used in the predicates. The world state is represented using a set of predicates that are true in that state. In other words, if the predicate is not present in the set representing a given state, then this predicate is not true in that state (a closed world assumption). Assuming a finite number of predicate names and constants representing the objects, we may get only a finite though possibly a huge number of states. In practice, the objects can be grouped to describe a type of the object and the attributes in predicates can be typed to restrict which constants (objects) can be used as a meaningful attribute.

Actions represent a mechanism to describe the change of the world state. We need to describe the states where an action is applicable to and how the action changes the world state. Each *action* in classical planning is specified using a triple (Prec, PosEff, NegEff), where Prec is a set of predicates that must hold for the action to be applicable (*preconditions*), PosEff is a set of predicates that will hold after performing the action (*positive effects*), and NegEff is a set of predicates that will not hold after performing the action (*negative effects*). For example, action grasp(blockA, blockB), describing that we grasp blockA laying on blockB, can be specified as the triple ({free(blockA), on(blockA,blockB), empty_hand}; {hold(blockA), free(blockB)}; {free(blockA), on(blockA,blockB), empty_hand}). In general, we say that action A is applicable to state S if Prec(A) $\subseteq$ S. The result of applying action A to state S is a new state succ(S) = (S\NegEff(A)) $\cup$ PosEff(A). Notice that the action is influencing only a part of the world state. The description of the state after performing the action naturally includes the *frame axiom*—the predicates that are not modified by the action stay valid in the state. In practice, rather than specifying each possible action, we may group similar actions to a so-called *operator*. For example, instead of all possible grasping actions, we can use a single operator grasp(X, Y), where X and Y are variables, saying that any object X can be grasped from any object Y when some particular conditions are satisfied: ({free(X), on(X,Y), empty_hand}, {hold(X), free(Y)}, {free(X), on(X,Y), empty_hand}). An action is obtained from the operator by substituting particular constants for the variables (this is called *grounding*). Again, it is possible to use typing to restrict which constants can be used in which attributes of the operator.

The description of predicate names and operators defines a so-called *planning domain*. For example, we can describe an abstract world where we are moving blocks to build towers of blocks (the famous block world domain). A particular *planning problem* is specified by a list of objects (constants), which we are working with, a full specification of the initial state (typically, this specification defines also the objects), and a goal condition. The goal condition is typically a set of predicates that must hold in the goal state. If G is a goal condition, then any state S such that G $\subseteq$ S is a *goal state*. The planning task is formulated as the problem of finding a sequence of actions that transfer the world (defined via the succ function) from the initial state to some goal state. Notice in particular that the size of the plan (the number of actions) is unknown in advance and that some actions may appear several times in the plan.

The most widely used modeling language to specify planning domains and planning problems is *Planning Domain Definition Language* (PDDL) that was proposed for the International Planning Competition (Koenig, 2012). This language is following the classical logical representation described earlier.

There also exist alternative approaches to the earlier logical formalism. A straightforward extension is based on *multivalued state variables* rather than predicates (Bäckström and Nebel, 1995). In the logical representation, we use a predicate on(blockA, blockB) to specify that blockA is on blockB, but we can also use a function on(blockA) whose value is the name of the object on which the blockA rests or the constant *nothing*. Function on(blockA) is an example of a state variable whose value can change in time by the application of actions. Actions in the multivalued state variable representation are described by the conditions on the values of the state variables and by the assignment of the state variables as the effects. Again, we can use an abstraction to operators so the operator grasp(X,Y) has the precondition {on(X) = Y, free(X) = yes, hand = empty} and the effect {on(X) ← nothing, free(X) ← no, free(Y) ← yes, hand ← blockA}. Notice that we are no more distinguishing the positive and negative effects as the assignment defines explicitly the values of certain state variables and these values "overwrite" the original values.

When using the multivalued state variables, we can see the planning task as finding a synchronized evolution of values of these state variables that starts with given values and reaches some specified values of certain state variables. Briefly speaking, we can describe the possible evolutions of a given state variable to form a so-called *timeline*. Finite state automata can be used to compactly represent these evolutions by specifying which values can follow a given value and how actions are changing these values (Barták, 2011). Then we need to specify how the timelines are synchronized. We described the synchronization of state variables via actions in the previous paragraph, but we can omit the actions and describe the synchronization explicitly. For example, a particular change of value of certain state variable requires a change of value of another state variable. This is a more engineering view for describing system dynamics, which is, for example, used in planning for space applications (Pralet and Verfaillie, 2009). In contrast to PDDL for classical planning, there does not exist yet a standard language for specifying planning problems using timelines.

So far, we discussed a "flat" approach to planning where actions (operators) were not structured and the planning task consisted of connecting the actions via causal relations to form a plan. A different approach to planning is called *hierarchical task networks* (Erol et al., 1994). There are still actions changing the world state (via the preconditions and effects), but the actions are grouped together to form partial plans to perform some tasks. More specifically, in addition to operators, there are compound tasks describing how the operators are connected together to accomplish some job. For example, to go from city X to city Y, we can board the plane in city X, fly from X to Y, and deplane at city Y. Alternatively, we can take a train from city X to city Y. Each of these sequences of operators describes how to accomplish the task of going from city X to city Y. Note that these sequences may also contain tasks so we obtain a hierarchical structure (hence the name hierarchical task networks). A particular way to satisfy the compound task is called a *method*. The planning problem is then given by a set of tasks (goal tasks) and the initial state and the problem is to find a method for each compound task until the primitive actions are obtained and these actions form a valid plan (the preconditions of actions are satisfied).

39.2.1.2 Planning Techniques

A straightforward way to find plans is by searching the state space either from the initial state to some goal state or from the goal states to the initial state. This is called *state-space planning* (Fikes and Nilsson, 1971). Currently, the most widely used planning technique is based on forward state search when the algorithm starts with the initial state, selects some applicable action, and moves to the next state defined by the function succ. This process is repeated until some goal state is found. The A* algorithm is frequently used to implement the search procedure, where the heuristic function estimates the number of actions to the goal. The systems FF (Hoffmann and Nebel, 2011) and Fast Downward (Helmert, 2006) are two examples of successful planners based on this idea.

Backward planning going from the goal to the initial state is also possible though slightly more complicated because it starts from a set of goal states rather than from a single state. The idea of backward planning was first proposed for the pioneering STRIPS system (Fikes and Nilsson, 1971), but it is less frequently used today. The backward search algorithm is exploring the search states defined as sets of world states, which is a major difference from forward search. The search state is defined using a set of predicates and it represents all the states that contain all these predicates. The initial search state is given by the goal condition so it corresponds to all the goal states. We say that an action A is *relevant* for a search state represented by the set of predicates G if $PosEff(A) \cap G \neq \emptyset$ and $NegEff(A) \cap G = \emptyset$. It means that action A contributes to the goal G and does not "destroy" it. Such an action can be the last action in the plan before reaching the goal G. If A is relevant for G, then we can define the new goal $prev(G,A) = (G \setminus PosEff(A)) \cup Prec(A)$. Briefly speaking, if we have a world state satisfying the goal condition prev(G,A), then A is applicable to this state (obviously the precondition of A is satisfied by this state), and after applying A to such a state, we obtain a state satisfying the goal condition G. Hence, the backward search starts with the goal condition, selects an action relevant for this goal, and moves to the next search state (the previous goal) defined by the function prev. The algorithm stops when the initial state satisfies the goal condition given by the search state. The backward search can be modified by a technique called *lifting*. Assume that we want to reach the state where hold(blockA) is true. This can be done by many grasp actions that pick up the blockA from some other block (these actions are relevant to the goal hold(blockA)), for example, grasp(blockA, blockB) and grasp(blockA, blockC). We may decide later from which block we will grasp the blockA so we can use a partially instantiated operator grasp(blockA,Y) instead of the actions in the planning procedure. This approach decreases the branching factor in the search procedure but introduces variables to the search states that must be handled properly. For example, when checking whether the initial state S satisfies the goal condition G, we need to use a substitution σ of variables: $\exists \sigma \; G\sigma \subseteq S$. This substitution instantiates the variables in the operators from the plan.

We defined the plan as a sequence of actions, but primarily we require that causal relations between the actions hold (Sacerdoti, 1990). The causal relation between actions A and B says that action A has some predicate P among its positive effects that is used as a precondition of action B; we write $A \rightarrow^P B$. Hence, the action A must be processed before the action B in the plan and there must not be any action between A and B that deletes the predicate P. To capture the causal relations in the plan, it is enough to specify a partial order of actions rather than the linear order. This is the basic idea behind plan-space planning where we are exploring partial plans until we obtain a complete plan (Penberthy and Weld, 1992). The initial partial plan consists of two actions only: the initial state S is modeled using a special action A0 such that $Prec(A0) = NegEff(A0) = \emptyset$, $PosEff(A0) = S$, the goal G is modeled using a special action A1 such that $Prec(A1) = G$ and $PosEff(A1) = NegEff(A1) = \emptyset$, and action A0 is before A1. Now, the planning process consists of modifying the partial plan by adding new actions, precedence, and causal constraints. The aim is to find a causal relation for each predicate in the preconditions of actions in the partial plan. In the initial plan, the preconditions of action A1 modeling a goal are not provided by any action. Such preconditions are called *open goals*. We can find a causal relation leading from an action that is already in the partial plan (causal relations also define a partial order between the actions) or we can add a new action to the plan that has a given predicate among its effects. The new actions must always be after A0 and before A1 in the partial order of actions. Notice that a causal relation $A \rightarrow^P B$ can be threatened by another action C in the partial plan that deletes P and can be ordered between A and B in the partial order. This *threat* can appear when we add a new causal relation to the partial plan or when we add a new action to the plan. The threat of action C for the causal relation $A \rightarrow^P B$ can be resolved either by ordering C before A or by ordering C after B. If lifting is used in the planning procedure, then we can resolve the threat also by instantiating certain variables so the negative effect of C is no more identical to P. In summary, plan-space planning is based on exploring the space of partial plans starting with the initial plan containing only the actions A0 and A1 and refining the plan by adding causal relations for open goals, which may also add new actions to the plan, and removing threats until we obtain a plan with no open goals and no threats. It is possible to prove that a partial plan with no threats and no open goals is a solution plan; in particular, any linearly ordered sequence of actions that follows the partial order is a plan. Notice that the plan-space approach does not use explicitly the world states;

only the initial state is encoded in the action A0. The systems UCPOP (Penberthy and Weld, 1992) and CPT (Vidal and Geffner, 2004) are two famous partial order planners (partial order planning is another name for plan-space planning).

State-space planning and plan-space planning are two most widely used approaches to solve classical planning problems. In 1995 Blum and Furst proposed a radically new approach to planning based on the notion of a *planning graph*. The original algorithm was called Graphplan (Blum and Furst, 1997). This approach exploits two core principles. First, instead of looking for a sequence of actions, the Graphplan system looks for a parallel plan, which is a sequence of sets of independent actions (we shall define the independence later). The classical plan is then obtained by any ordering of actions in the sets. The second core principle is a reachability analysis realized by the planning graph constructed from the initial state and available actions where the negative (delete) effects are temporally ignored. The *planning graph* is a layered graph where the first layer consists of predicates in the initial state, the second layer consists of actions applicable to the initial state, the third layer consists of predicates from the positive effects of actions in the previous layer (and all predicates from the first layer), and so on until a layer containing all goal predicates is found. Hence, the layers with predicates interleave with the layers with actions. To simplify notation, *no-op actions* are used to transfer predicates between the layers (the no-op action for predicate P has the predicate P as its precondition and also as its sole positive effect). The planning graph defines the search space where we are looking for the parallel plan; the sets of actions are being selected from the action layers. To ensure that actions selected to a single set in the plan can be ordered in any way, the actions must be independent.

Actions A and B are *independent* if Prec(A) ∩ (PosEff(B) ∪ NegEff(B)) = ∅ and Prec(B) ∩ (PosEff(A) ∪ NegEff(A)) = ∅. Selection of the plan from the planning graph is then realized in the backward way by going from the last predicate layer and finding an action for each goal predicate (the no-op actions can also be used). The selected actions must be pairwise independent. After selecting the actions in one layer, the procedure continues with the preconditions of selected actions and finding actions for them in the previous layer and so on until the first layer is reached. If no plan is found, one layer is added to the planning graph and the procedure is repeated. The ideas of planning graphs are nowadays used mainly in heuristics for forward planning.

So far we presented the dedicated algorithms for solving planning problems. There also exists another planning approach based on translating the planning problem to a different formalism, for example, to Boolean satisfiability or constraint satisfaction problems (CSPs). This method was introduced by Kautz and Selman (1992) who showed how the planning problem could be converted to a sequence of Boolean satisfiability (SAT) problems. The idea is that the problem of finding a plan of a given length can be naturally encoded as a SAT formula. If this formula is not satisfiable, then the plan (of a given length) does not exist; otherwise the plan can be decoded back from the model of the formula (from the instantiation of the variables in the formula). However, we do not know the plan length in advance. Hence, the whole planning system works as follows. We encode the problem of finding a plan of length one as a SAT formula. If the plan does not exist, then we continue with the plan of length two and so on until the plan is found (or some fix point is reached and then no plan exists). There exist various encodings of planning problems to SAT; the original approach from Kautz and Selmann (1992) used the encoding of sequential plans, the more recent planners such as SATPLAN (Kautz and Selmann, 1999) used encodings based on a planning graph, and the state-of-the-art SAT-based planners such as SASE (Huang et al., 2010) use multivalued state variables. Several encodings to CSPs were also proposed (Barták, 2011, Barták and Toropila, 2008).

39.2.2 Automated Scheduling

Scheduling deals with the allocation of scarce resources to activities with the objective of optimizing some performance measures. Depending on the situation, resources could be machines, humans, runways, processors etc.; activities could be manufacturing operations, duties, landings and takeoffs, computer programs, etc.; and objectives could be minimization of the schedule length, maximization of resource utilization, minimization of delays, and others.

39.2.2.1 Scheduling Task

As we already mentioned, there are many classes of scheduling tasks. Resources, activities, and the performance measure are three major concepts in each scheduling task so we can define components of the scheduling task as follows. There is a set of n jobs that can run on m machines. Each job i requires some processing time $p_{i,j}$ on a particular machine j. This part of a job is usually called an *operation*; we also used the notions of activity or action in the text. The schedule of each job is an allocation of one or more time intervals (operations) to one or more machines. The scheduling task is to find a schedule for each job satisfying certain restrictions and optimizing a certain given set of objectives. The start time of job j can be restricted by a *release date* r_j, which is the earliest time at which job j can start its processing. The release date models the time when the job arrives to the system. Similarly, a *deadline* d_j can be specified for job j, which is the latest time by which the job j must be completed. The release date and the deadline specify the time window in which the job must be processed. To specify an expected time of job completion, a *due date* δ_j for job j is used. The major difference from the deadline is that the job can complete later or earlier than the due date, but it will incur a cost. Let C_j be the completion time of job j. Then the lateness of job j is defined as $L_j = C_j - \delta_j$ and the tardiness of job j is defined as $T_j = \max(0, L_j)$. The completion time, lateness, and tardiness are three typical participants in the objective functions expressing the quality of the schedule. The most widely used objective function is minimization of the maximal completion time—a so-called *makespan*. The idea is to squeeze production to as early times as possible. A more practically important objective is minimization of the maximal lateness or tardiness that corresponds to on-time delivery. Sometimes, the number of late jobs is used instead of the tardiness and the objective is to minimize the number of late jobs. Other objectives also exist, for example, minimization of the number of used resources.

The jobs (operations) require certain resource(s) for their processing. Operations can be preassigned to particular resources and then we are looking for time allocation only. Or there may be several alternative resources to which the operation can be allocated and then resource allocation is a part of the scheduling task. Such alternative resources can be either identical then we can union them into a so-called *cumulative resource* that can process several operations in parallel until some given capacity, or the alternative resources can be different (e.g., the processing time may depend on the resource). The resource, which can process at most one operation at any time, is called a *unary* or a *disjunctive* resource. If the processing of an operation on a machine can be interrupted by another operation, and then resumed possibly on a different machine, then the job is said to be *preemptible* or *interruptible*. Otherwise, the job is *non-preemptible* (*non-interruptible*).

It is also possible to specify precedence constraints between the jobs, which express the fact that certain jobs must be completed before other jobs can start processing. In the most general case, the precedence constraints are represented by a directed acyclic graph, where each vertex represents a job, and if job i precedes job j, then there is a directed arc from i to j. If each job has at most one predecessor and at most one successor, then we are speaking about *chains*; they correspond to serial production. If each job has at most one successor, then the constraints are referred to as an *in-tree*. This structure is typical for assembly production. Similarly, if each job has at most one predecessor, then the constraint structure is called an *out-tree*. This structure is typical for food or chemical production where from the raw material we obtain several final products with different "flavors."

There exists a specific class of scheduling problems with a particular type of precedence constraints, the so-called *shop problems*. If each job has its own predetermined route to follow, that is, the job consists of a chain of operations where each operation is assigned to a particular machine (the job may visit some machines more than once or never), then we are talking about *job-shop* scheduling (JSS). If the machines are linearly ordered and all the jobs follow the same route (from the first machine to the last machine), then the problem is called a *flow-shop* problem. Finally, if we remove the precedence constraints from the flow-shop problem, that is, the operations of each job can be processed at any order, then we obtain an *open-shop* problem.

To keep track of various classes of scheduling problems, Graham et al. (1979) suggested a classification of scheduling problems using the notation consisting of three components $\alpha \mid \beta \mid \gamma$.

The α field describes the machine environment, for example, whether there is a single machine (1), m parallel identical machines (Pm), job-shop (Jm), flow-shop (Fm), or open-shop (Om) environment with m machines. The β field characterizes the jobs and the scheduling constraints and it may contain no entry or multiple entries. The typical representatives are restriction of the release dates (r_j) and deadlines (d_j) for jobs, assuming the precedence constraints (prec) and allowing preemption of jobs (pmtn). Finally, the γ field specifies the objective function to optimize; usually, a single entry is present but more entries are allowed. The widely used objectives are minimization of the makespan (C_{max}) or minimization of the maximal lateness (L_{max}).

39.2.2.2 Scheduling Techniques

There exists a vast amount of various scheduling techniques proposed for particular scheduling problems (Brucker, 2001, Pinedo, 2002). We will present here just two representatives showing some typical techniques from the scheduling algorithms. In particular, we will present an exact polynomial algorithm for solving the problem $1 \mid r_j, pmtn \mid L_{max}$ to demonstrate a classical scheduling rule and then we will show an exact exponential algorithm for the problem $1 \mid r_j \mid L_{max}$ explaining the branch-and-bound method that uses solutions of the relaxed problems to obtain a good bound.

The scheduling problem $1 \mid r_j \mid L_{max}$ consists of a single machine where jobs have arbitrary release dates and the objective is to minimize the maximal lateness. This problem is important because it appears as a subproblem in heuristic procedures for flow-shop and job-shop problems; unfortunately, the problem is known to be strongly NP-hard (Pinedo, 2002, p. 43). However, if all release dates are identical, or all due dates are identical, or preemption is allowed, then the problem becomes tractable. These tractable problems can be solved in polynomial time by applying the earliest due date (EDD) rule by Jackson (1955) that says "select the job with the smallest due date and allocate it to the resource." We will present the preemptive version of the EDD rule for the problem $1 \mid r_j, pmtn \mid L_{max}$. In the preemptive case, we collect all not-yet finished jobs with the earliest release date and among them we select the job with the EDD for processing and allocate the job to the earliest possible time. Processing of this job continues until the job is finished or another job becomes available (i.e., the next release date is reached). At this point, the next job for processing is selected using the same principle. It could be the current job or a different job in which case the previous job is interrupted (unless it is already finished). Hence, we may interrupt a job when another job with an earlier due date becomes available. Notice that if the schedule constructed does not have any preemptions, then the schedule is optimal also for the problem $1 \mid r_j \mid L_{max}$. Otherwise, the preemptive schedule defines a lower bound of the objective function (L_{max}) for the non-preemptive problem (obviously, if preemption is not allowed, we cannot obtain a better schedule). These two observations can be used for solving the problem $1 \mid r_j \mid L_{max}$. Basically, the algorithm explores possible permutations of jobs that are constructed from left to right (from earlier time to later time). Assume that we already have an upper bound for the maximal lateness (it could be plus infinity at the beginning) and a set of jobs to schedule. We first generate a schedule using the algorithm for the preemptive case. If the maximal lateness of this schedule is greater than a given upper bound, we can stop and report that there does not exist any schedule with the maximum lateness lower than the given bound. Otherwise, we check if the schedule constructed has preemptions. If the schedule is not preemptive, then we are done and we can report that schedule as the optimal schedule. However, if none of the earlier condition holds, we take a job J such that no other job can be completely finished before the release date of job J, we allocate job J to its earliest possible start time, and then we allocate the remaining jobs using the same process as described earlier. Such allocation gives a feasible schedule whose maximal lateness is used as a new upper bound when trying the other jobs J at the first position in the schedule. Note also that if we find a schedule whose maximal lateness equals the lower bound obtained from the preemptive schedule, then the schedule is optimal.

```
solve(Jobs; T; Bound)
// Jobs is a set of jobs to be scheduled
// T is the first possible start time (initialized to 0 for nonnegative release dates)
// Bound is a value of the objective function for the best solution so far (initialized to ∞)
    (Sched; LB) ← solvePreempt(Jobs; T) // find a schedule using the preemptive EDD rule
    if LB ≥ Bound then return (fail, Bound) // no schedule better than Bound
    if Sched is non-preemptive then return (Sched, LB) // optimum found
    First ← {j ∈ Jobs | ¬ ∃i∈Jobs: max(T, r_i) + p_i ≤ r_j}
    Best ← fail
    foreach j ∈ First do
        (Sched; L) ← solve(Jobs – {j}; max(T, r_j) + p_j; Bound)
        if L = LB then return (Sched; L)    // lower-bound schedule is optimal
        if L < Bound then (Best; Bound) ← (Sched; L)
    endfor
    return (Best, Bound)
```

Though we mentioned that scheduling algorithms are typically dedicated to particular classes of scheduling problems, there also exists a general scheduling approach based on constraint satisfaction—this is called *constraint-based scheduling* (Baptiste et al., 2001). The idea is that a scheduling problem is represented as a CSP (Dechter, 2003) and constraint satisfaction techniques are used to solve this problem. This is similar to solving planning problems by converting them to a CSP, but the major difference is that we know the number of actions in the scheduling task so we can encode the task as a single CSP. Recall that the scheduling problem is given by a set of operations that are connected via precedence (temporal) and resource constraints. The task is to allocate these operations to particular times and resources. Hence, the CSP variables can describe the start time of each operation and the resource to which the operation is allocated. So the release date and the deadline define the domain of the start time variable (the set of all possible start times) and the domain of the resource variable consists of identifiers of resources to which the operation can be allocated. Other models are also possible, for example, using a list of 0 1 variables describing if a given operation runs at a given time. This is useful for modeling interruptible (preemptible) operations. The scheduling constraints are directly translated to relations between the variables. For example, the precedence constraint between operations A and B is encoded as $start(A) + p_A \leq start(B)$. Many special constraints have been developed to model resources in scheduling. For example, a unary resource is modeled by a constraint stating that operations are not overlapping in time. The semantic of such a constraint (for non-preemptible operations) can be described by a set of a disjunctions defined between all pairs of operations allocated to that resource: $start(A) + p_A \leq start(B) \lor start(B) + p_B \leq start(A)$. The unary resource constraint itself uses techniques such as edge finding (Baptiste and Le Pape, 1996) and not first/not last (Tores and Lopez, 2000) to filter out values from the domains of variables that violate the constraint. There exist resource constraints for preemptible operations as well as for cumulative resources (Baptiste et al., 2006) and also constraints where the operations are not pre-allocated to resources (Vilím et al., 2005).

39.3 Impact on Practice

There is no doubt that scheduling techniques have a great impact on many areas of human endeavor starting from manufacturing scheduling, going through personal scheduling, and finishing with computer task scheduling. These examples are obviously not exhaustive and there are many others where scheduling plays an important role, for example, scheduling sport tournaments. There are companies such as SAP and i2 Technologies (now owned by JDA Software) that provide complete solutions to solve scheduling problems such as SAP Advanced Planner and Optimizer (APO) and other companies providing tools to solve scheduling problems such as ILOG CP Optimizer by IBM.

The planning techniques are visible mainly in the research areas. The biggest success is probably in space exploration. NASA's Deep Space 1 (DS1) mission was one of the first significant applications of automated planning. The DS1 project successfully tested 12 novel advanced technologies including the autonomous remote agent (RA) (Muscettola et al., 1998). The goal of RA was to plan, execute, and monitor operations of the space probe. Several conclusions were drawn from this experiment including the necessity to validate the formal models, to develop techniques that work in real time, and to provide simple but expressive language to specify goals. The experience from DS1 project was used for developing the EUROPA system (Jonsson et al., 2000) that was applied to do planning for famous Mars Rovers Spirit and Opportunity and later to do planning for underwater autonomous vehicles (McGann et al., 2008). MEXAR and RAXEM (Cesta et al., 2007) are other examples of successful applications of integrated planning and scheduling techniques to plan and schedule communication between the Mars Express orbiter and Earth.

39.4 Research Issues

We mentioned several times that planning and scheduling tasks are quite different problems and different approaches are being used to solve these problems. Nevertheless, complex real-life problems frequently included both types of problems—what to do and how to do it. This demands for approaches that *integrate planning and scheduling.*

Though planning deals mainly with the causal relations between the actions and suppresses the role of time and resources, there already exist planning techniques that involve time and resource reasoning. Plan-space planning can naturally be extended to work with time by using simple temporal constraints (Dechter et al., 1991). The IxTeT planner (Ghallab and Laruelle, 1994) is an example of a planner that uses the so-called *chronicles* to model evolutions of state variables in time. The EUROPA system (Jonsson et al., 2000) is another modern planner that supports time and resources. Both these planners were strongly motivated by practical applications, which justifies that this research direction is highly important for practice.

Similarly, the research in the area of scheduling focuses on time and resource allocation and the question of how to determine the set of scheduled activities is left unsolved. A straightforward approach based on the description of the environment, for example, using the description of the processes to manufacture a given part, is typically used to generate the set of activities for scheduling. If there are some options, for example, alternative processes such as outsourcing, the decision is done by a human operator outside the scheduling process. As the current production environments are becoming more flexible and complicated, the number of options to select the possible activities is increasing and hence the choice of the "right" activities to be scheduled is becoming important. This means that some ideas from planning might be applicable to scheduling as well, for example, in flexible manufacturing environments. There exist attempts to include the so-called *optional activities* in the scheduling process so the scheduling algorithm selects which activities will be scheduled. This direction was pioneered by Beck and Fox (2000) and further extended by Kuster et al. (2007) and Barták and Čepek (2007). The FlowOpt system (Barták et al., 2011) supports optional activities and the recent versions of IBM/ILOG CP Optimizer (Laborie, 2009) provide a strong support for optional activities in the filtering algorithms for resource constraints.

Currently, the research on integrating planning and scheduling is driven mainly by the planning community. This is probably because there is a higher demand from the applications where autonomous agents such as robots are being used.

One of the critical issues when applying planning and scheduling technology is to identify a proper formulation of the problem to be solved. Researchers usually assume that a problem is already given, for example, most planers assume the problem formulation in PDDL is given as the input, and in scheduling the notation of the form $\alpha \mid \beta \mid \gamma$ is assumed for the problem definition. Surprisingly little attention is devoted to the problem of representing real-life problems in a formal model that can be later processed by automated planners and schedulers. The research area that deals with this issue is called *knowledge engineering for planning and scheduling*. This area deals with the acquisition, design, validation and maintenance of domain models, and the selection and optimization of the appropriate machinery to work on them. These processes impact directly on the success of real automated P&S applications. There is also a clear need for modeling tools that bridge the gap between real-life problems and P&S technology. Currently the human modeler must be an expert in the application domain as well as in P&S techniques. This is a rare combination of skills that significantly restricts applicability of the P&S technology. The modeling tools, such as GIPO (Simpson et al., 2001) and itSIMPLE (Vaquero et al., 2007), show a possible way for making the planning technology more accessible; the EUROPA system (Jonsson et al., 2000) also provides many routines for knowledge engineering tasks. The FlowOpt project (Barták et al., 2011) has similar ambitions in the area of scheduling. The main goal of such tools is supporting the modeling process starting with an informal description of the problem and finishing with a precise formal model that can be passed to a planner or a scheduler. The tool should hide the technical complexity of P&S algorithms while providing enough flexibility to describe various real-life problems. Users would greatly benefit from tools that guide the user to design models that are easier to solve.

A closely related topic to knowledge engineering is the gap that is extending between real-life problems and academic solutions. Researchers are going deeper and deeper to understand the problems and proposing techniques to solve them, but the problems to be solved are becoming too far (too artificial) from the real-life problems, which opens the question of practical applicability of current research results. In the previous paragraphs, we pointed out several assumptions made about the problems to be solved and these assumptions are frequently violated in real-life problems. There already exists research that relaxes these assumptions and attempts to solve more realistic problems. For example, solving planning problems under uncertainty is an important topic to make the planning techniques closer to real-life problems. Markov decision processes (MDPs) are currently a typical model to deal with nondeterminism of actions and environment and with partial observability of the world. Instead of plans defined as a sequence of actions, MDPs work with policies that recommend which action should be executed in a given state. Though these techniques are very promising to describe the real-world problems precisely, they do not scale up well to problems of realistic size. Similarly, stochastic scheduling has received attention in the scheduling community where researchers are also interested in online scheduling (not all information is available at the beginning, e.g., new jobs may arrive later). This avenue of research also relates to rescheduling methods where the existing schedule needs to be modified to reflect changes in the problem specification (such as broken machines, late arrival of jobs).

39.5 Summary

Many real-life problems involve planning and scheduling tasks so the knowledge of how to model and solve such problems is becoming increasingly more important. This chapter gave a short overview of classical approaches to formally describe and solve planning and scheduling problems. It is not meant to be exhaustive; for example, we omitted several promising approaches for solving problems such as metaheuristics that are often successful in practical applications (see Chapter 12). The aim was to present the area to a general audience, to introduce the key notions to understand the terminology, and to give several representative examples of techniques involved in solving problems. We also identified three topics, namely, integration of planning and scheduling, knowledge engineering for planning and scheduling, and extensions of existing models toward uncertainty and dynamicity as research areas that are likely to receive more attention in near future due to their relevance for solving practical problems.

Key Terms

A*: A heuristic search algorithm where each search state x is evaluated using the function $g(x) + h(x)$, where $g(x)$ is the length of the path from the initial state to x and h(x) is an estimate of the length of the path from x to some goal state.

Constraint satisfaction problem: A combinatorial optimization problem formulated as a set of variables, each variable has a set of possible values, and constraints, which are relations restricting possible combinations of values; the solution is an assignment of values to variables satisfying the constraints.

Planning: A process of creating an organized set of actions, a plan, that can be executed to achieve a desired goal from a given initial situation (state).

SAT: A Boolean satisfiability problem is the problem of determining if the variables of a given Boolean formula can be assigned in such a way as to make the formula evaluate to *true*.

Scheduling: A process of assigning a set of tasks to resources over time.

Acknowledgment

Roman Barták is supported by the Czech Science Foundation under the contract P202/12/G061.

Further Information

Ghallab M., Nau D., Traverso P. 2004. *Automated Planning: Theory and Practice*. Morgan Kaufmann, San Francisco, CA.

Leung J.Y.T. 2004. *Handbook of Scheduling: Algorithms, Models, and Performance Analysis*. Chapman & Hall, Boca Raton, FL.

Pinedo M. 2012. *Scheduling: Theory, Algorithms, and Systems*, 4th edn. Springer, New York.

The International Conference on Automated Planning and Scheduling (ICAPS): http://www.icaps-conference.org.

The Multidisciplinary International Scheduling Conference (MISTA): http://www.schedulingconference.org.

References

Bäckström Ch. and Nebel B. 1995. Complexity results for SAS+ planning. *Computational Intelligence* 11(4): 625–655.

Baptiste P., Laborie P., Le Pape C., and Nuijten W. 2006. Constraint-based scheduling and planning. In *Handbook of Constraint Programming*, Rossi, F., van Beek, P., Walsh, T. (eds.), pp. 761–799, Elsevier.

Baptiste P. and Le Pape C. 1996. Edge-finding constraint propagation algorithms for disjunctive and cumulative scheduling. *Proceedings of the Fifteenth Workshop of the U.K. Planning Special Interest Group (PLANSIG)*, Liverpool, U.K.

Baptiste P., Le Pape C., and Nuijten W. 2001. *Constraint-Based Scheduling: Applying Constraint Programming to Scheduling*. Kluwer Academic Publishers, Dordrecht, the Netherlands.

Barták R. 2011. On constraint models for parallel planning: The novel transition scheme. *Proceedings of the Eleventh Scandinavian Conference on Artificial Intelligence (SCAI)*, Frontiers of Artificial Intelligence, Vol. 227, IOS Press, Trondheim, Norway, pp. 50–59.

Barták R. and Čepek O. 2007. Temporal networks with alternatives: Complexity and model. *Proceedings of the Twentieth International Florida AI Research Society Conference (FLAIRS)*, AAAI Press, Menlo Park, CA, pp. 641–646.

Barták R., Cully M., Jaška M., Novák L., Rovenský V., Sheahan C., Skalický T., and Thanh-Tung D. 2011. Workflow optimization with flowopt, on modelling, optimizing, visualizing, and analysing production workflows. *Proceedings of Conference on Technologies and Applications of Artificial Intelligence (TAAI)*, IEEE Conference Publishing Services, Taoyuan, Taiwan, pp. 167–172.

Barták R. and Toropila D. 2008. Reformulating constraint models for classical planning. *Proceedings of the Twenty-First International Florida Artificial Intelligence Research Society Conference (FLAIRS)*, AAAI Press, Coconut Grove, FL, pp. 525–530.

Beck J. C. and Fox M. S. 2000. Constraint-directed techniques for scheduling alternative activities. *Artificial Intelligence* 121: 211–250.

Blum A. and Furst M. 1997. Fast Planning Through Planning Graph Analysis. *Artificial Intelligence*, 90: 281–300.

Brucker P. 2001. *Scheduling Algorithms*. Springer Verlag, Berlin, Germany.

Cesta A., Cortellessa G., Denis M., Donati A., Fratini S., Oddi A., Policella N., Rabenau E., and Schulster J. 2007. MEXAR2: AI solves mission planner problems. *IEEE Intelligent Systems* 22(4): 12–19.

Dechter R. 2003. *Constraint Processing*. Morgan Kaufmann, San Francisco, CA.

Dechter R., Meiri I., and Pearl J. 1991. Temporal constraint network. *Artificial Intelligence* 49: 61–95.

Erol K., Hendler J., and Nau D. 1994. HTN planning: Complexity and expressivity. *Proceedings of National Conference on Artificial Intelligence (AAAI)*, AAAI Press, Seattle, WA, pp. 1123–1128.

Fikes R. and Nilsson N. 1971. STRIPS: A new approach to the application of theorem proving to problem solving. *Artificial Intelligence* 2(3–4): 189–208.

Ghallab M. and Laruelle H. 1994. Representation and control in IxTeT, a temporal planner. *Proceedings of the International Conference on AI Planning Systems (AIPS)*, Chicago, IL, pp. 61–67.

Ghallab M., Nau D., and Traverso P. 2004. *Automated Planning: Theory and Practice*. Morgan Kaufmann, Amsterdam, the Netherlands.

Graham R. L., Lawler E. L., Lenstra J. K., and Rinnooy-Kan A. H. G. 1979. Optimization and approximation in deterministic sequencing and scheduling: A survey. *Annals of Discrete Mathematics* 5: 287–326.

Helmert M. 2006. The fast downward planning system. *Journal of Artificial Intelligence Research* 26: 191–246.

Hoffmann J. and Nebel B. 2001. The FF planning system: Fast plan generation through heuristic search. *Journal of Artificial Intelligence Research* 14: 253–302.

Huang R., Chen Y., and Zhang W. 2010. A novel transition based encoding scheme for planning as satisfiability. *Proceedings of National Conference on Artificial Intelligence (AAAI)*, AAAI Press, Vancouver, British Columbia, Canada, pp. 89–94.

Jackson J. R. 1955. Scheduling a production line to minimize maximum tardiness. Research report 43. Management Science Research Project, University of California, San Francisco, CA.

Jonsson A. K., Morris P., Muscettola N., and Rajan K., 2000. Planning in interplanetary space: Theory and practice. *Proceedings of the International Conference on AI Planning Systems (AIPS)*, Breckenridge, CO, pp. 177–186.

Kautz H. and Selman B. 1992. Planning as satisfiability. *Proceedings of the European Conference on Artificial Intelligence (ECAI)*, Vienna, Austria, pp. 359–363.

Kautz H. and Selman B. 1999. Unifying SAT-based and graph-based planning. *Proceedings of the International Joint Conference on Artificial Intelligence (IJCAI)*, Stockholm, Sweden, pp. 318–325.

Koenig S. (ed.) 2012. International planning competition (IPC). http://ipc.icaps-conference.org/, University of Southern California (accessed at September 2012).

Kuster J., Jannach D., and Friedrich G. 2007. Handling alternative activities in resource-constrained project scheduling problems. *Proceedings of Twentieth International Joint Conference on Artificial Intelligence (IJCAI)*, Hyderabad, India, pp. 1960–1965.

Laborie P. 2009. IBM ILOG CP optimizer for detailed scheduling illustrated on three problems. In *Integration of AI and OR Techniques in Constraint Programming for Combinatorial Optimization Problems (CP-AI-OR)*, van Hoeve W.-J., Hooker J.N. (eds.), LNCS 5547, Springer-Verlag, Berlin, Germany, pp. 148–162.

Leung J.Y.T. 2004. Handbook of scheduling: Algorithms, models, and performance analysis. Chapman & Hall, Boca Raton, FL.

McGann C., Py F., Rajan K., Ryan J., and Henthorn, R. 2008. Adaptive control for autonomous underwater vehicles. *Proceedings of the Twenty-Third AAAI Conference on Artificial Intelligence (AAAI 2008)*, Chicago, IL, pp. 1319–1324.

Muscettola N., Nayak P., Pell B., and Williams B. 1998. Remote agent: To boldly go where no AI system has gone before. *Artificial Intelligence* 103: 5–47.

Penberthy J. and Weld D. S. 1992. UCPOP: A sound, complete, partial order planner for ADL. *Proceedings of the International Conference on Knowledge Representation and Reasoning (KR)*, Cambridge, MA, pp. 103–114.

Pinedo M. 2002. *Scheduling: Theory, Algorithms, and Systems*. Prentice Hall, Englewood Cliffs, NJ.

Pralet C. and Verfaillie G. 2009. Forward constraint-based algorithms for anytime planning. *Proceedings of the Nineteenth International Conference on Automated Planning and Scheduling (ICAPS)*, AAAI Press, Thessaloniki, Greece, pp. 265–272.

Ruml W., Do M. B., and Fromherz M. 2005. On-line planning and scheduling for high-speed manufacturing. *Proceedings of the International Conference on Automated Planning and Scheduling (ICAPS 2005)*, Monterey, CA, pp. 30–39.

Sacerdoti E. 1990. The nonlinear nature of plans. *Proceedings of the International Joint Conference on Artificial Intelligence (IJCAI)*, Stanford, CA, pp. 206–214.

Simpson R. M., McCluskey T. L., Zhao W., Aylett R. S., and Doniat C. 2001. GIPO: An integrated graphical tool to support knowledge engineering in AI planning. *Proceedings of the Sixth European Conference on Planning (ECP)*, Toledo, Spain.

Torres P. and Lopez P. 2000. On not-first/not-last conditions in disjunctive scheduling. *European Journal of Operational Research* 127: 332–343.

Vaquero T.S., Romero V., Tonidandel F., and Silva J. R. 2007. itSIMPLE2.0: An Integrated Tool for Designing Planning Domains. *Proceedings of the International Conference on Automated Planning & Scheduling (ICAPS)*, AAAI Press, Menlo Park, CA, pp. 336–343.

Vidal V. and Geffner H. 2004. Branching and pruning: An optimal temporal POCL planner based on constraint programming. *Proceedings of the Nineteenth National Conference on Artificial Intelligence (AAAI)*, San Jose, CA, pp. 570–577.

Vilím P., Barták R., and Čepek P. 2005. Extension of O(n log n) filtering algorithms for the unary resource constraint to optional activities. *Constraints* 10(4): 403–425.

40

Natural Language Processing

Nitin Indurkhya
The University of New South Wales

40.1 Introduction

One of the unique characteristics of human intelligence is the ability to process and understand language. It is a specific human capacity, although the word "language" has been applied more loosely to communication modes used by other animals, for example, dolphins. A key aspect of language is that it is used for communication. However, there is also evidence that it is used for conceptualization. In computing, the word "language" is mostly used within the context of programming, where programs are written in a "computer language" that the computer can process as a set of precise instructions to accept some input and generate some output. In this chapter, when we say "language" we mean human language or "natural language" to distinguish them from artificial languages such as those used in computer programming. Attempts at getting computers to understand natural language began almost as soon as computers were invented. The field of artificial intelligence has its roots in pioneering work in natural language processing (henceforth NLP)—what can be a more convincing demonstration of intelligence in a computer system than its ability to interact with a human being using language? In 2011, when IBM's Watson program convincingly won the Jeopardy tournament, it was seen as a major triumph for NLP. Going back in the past, many of the earliest attempts at building intelligent systems involved some form of language processing. For example, the Turing test required a computer to behave in a manner that is indistinguishable from a human being, and all interactions were to be done in natural language.

The broad interdisciplinary area of NLP has significant intersections with fields such as computational linguistics, speech recognition, phonetics, semantics, cognitive science, logic, and philosophy.

Each field focuses on certain facets of language processing. For example, in speech recognition and synthesis, the input/output is exclusively audio. This leads to an entire class of problems and issues, solutions to which are discussed in Chapter 41. However, in the current chapter, we shall focus on language processing research where the input/output is textual in nature. Such written languages are different from spoken languages (as in speech processing) and visual languages such as sign languages and gestures. A written language is characterized by a script. There are thousands of languages in the world, and almost as many scripts. Many languages tend to share scripts. For example, the romance languages (the major ones are French, Italian, Portuguese, and Spanish) use a script derived from Latin. Most research in NLP is done with English, the lingua franca, but the methods and technologies can often be applied to other languages. When focused on text, NLP tends to overlap somewhat with another field called *text-mining* or *text-analytics*. The main difference is that text-mining deals with text at a very superficial level, whereas NLP goes deeper into the structure of the text and its components.

The ubiquity of word-processors and document preparation systems ensures that most text data originates in digital form. This, and the fact that the Internet and World Wide Web allow easy access and dissemination of textual data, ensures that there is no shortage of data for NLP. As we shall see, these large volumes of digital text have led to the increasing dominance of statistical methods within NLP. However, this does not make the so-called classical approaches to NLP irrelevant or dated. They play an increasingly important role in providing direction for the statistical methods.

Comparing NLP to other fields that are covered in this section, it would seem that NLP could be useful in building a language-processing blackbox. For example, expert systems and robots might be easier to interface with if they were to understand commands and give their responses in natural language. Indeed, one of the earliest applications of NLP was as a front-end to a database where queries are posed in natural language rather than SQL. Analysis of images in computer vision can be benefit from using NLP methods on the text in which the image is embedded. Since many problems in biomedical informatics share similarities to the ones in NLP, it is not surprising that there is significant cross-fertilization of ideas between these two communities. Some other fields within artificial intelligence can be seen as sources for NLP techniques. For example, building a statistical machine translation (henceforth MT) system would require good machine learning algorithms from decision tree and neural network research communities. New research avenues have arisen in the process of solving some of the problems faced in NLP. For example, the programming language PROLOG (which stands for PROgramming LOGic) was developed primarily to serve the NLP community and was later found to be useful in many other scenarios as well. Programming languages are covered in detail later in this book and are not discussed in this chapter.

As mentioned earlier, the field of NLP is as old as computing itself. Research in MT (a subfield of NLP) has been going on since the 1950s. The first meeting of the Association of Computational Linguistics was held in the early 1960s. Hence, one can imagine that there are quite a number of surveys and textbooks and handbooks that describe NLP in great depth. This chapter does not intend to replace them, but rather it gives a quick introduction to the field and its literature, covers the key principles, the notable applications, and outlines future research challenges. The resources section at the end of this chapter will point readers to the most significant ones for further details.

The availability of web-based data and the emergence of mobile computing and cloud-based infrastructure have resulted in a flood of innovative applications of existing NLP techniques and the rapid development of new ones. Tasks in NLP that were considered "too hard" just a decade ago are now done by apps on mobile devices. It is truly an exciting time to be in NLP.

The rest of this chapter is organized as follows: In the next section, we shall discuss the underlying principles upon which the field is based. As mentioned earlier, our focus shall be on the text. Then we shall do a quick survey of the most important applications and the commercial impact of NLP. Following this, we shall identify the major research challenges for the future. Finally, we shall summarize by providing a glossary of terms that may be unfamiliar to the readers and an annotated list of resources (both online and offline) that the reader will find useful in learning more about this field.

40.2 Underlying Principles

A popular view of NLP is that of a series of transducers that decompose the task into a series of stages of increasing abstraction. The eventual goal of NLP is to articulate the text's intended meaning, and the stages break down the complexity of the task with each stage solving a bit of the problem or preparing data for the following stage. These stages are inspired by work in theoretical linguistics where one can analyze the syntax, semantics, and pragmatics of text. These stages are typically understood as a pipeline—the output of one stage fed as input into the next stage with minimal backtracking. At the lowest level, the text is just a sequence of characters, which must be tokenized and segmented into words and sentences. So the first stage, often referred to as text preprocessing, aims to transform the text into words and sentences. The output of this stage is then fed as input to the next stage—the syntax processor. In this stage, the words are subjected to lexical analysis and taken apart to understand them better. This understanding will be useful for later stages. The sentences are also subject to analysis in this stage. This involves putting a structure on top of the words in the sentence such that the relationships between the words are made explicit. This structure is called a *parse* and is often a tree or a directed graph. The parse is based on a grammar of the language, which can be understood as a finite set of rules that govern how words come together to form sentences. Syntactic parsing is a very well-studied topic, and it would be difficult to find a NLP system that did not involve the use of some parsing technique at some stage. At the end of the syntax stage, the text is now transformed into a sequence of structured sentences. The next stage, probably the most difficult one at the moment, attempts to ascribe meaning to the sentences. The notion of meaning in NLP is multifaceted. At a surface level, we are interested in knowing what the sentences say about the various objects/concepts that are the focus of the text. This is the literal meaning of the text and is often described using logic-based formalisms. The techniques are part of semantic analysis. At a more deeper level, we are interested in knowing the intended meaning of the text, which is often more broader than the surface level meaning of the sentence. The context of the sentence, other sentences, is considered, and this sub-field of NLP is called *Pragmatics*. For example, if we consider a text that consists of a conversation between two people, a semantic analysis would tell us the meaning of the individual sentences, but pragmatics (in this example, discourse analysis) would be what tells us whether the text is an argument between the two and whether it ended in an agreement and what was the nature of the agreement or whether it was question and answer session in which one person was asking questions that the other person was answering and whether the session was "satisfactory" from the questioner's perspective.

While such a pipeline of stages is helpful to manage the complexity of NLP, in practice, there is some feedback between stages. For example, a parsing failure may require that sentence segmentation be redone. The availability of faster/distributed computational resources also allow for multiple possibilities to be explored, and the best (most probable) solution be selected. We now explore each of these stages in more detail and describe the underlying principles and issues in each.

40.2.1 Text Preprocessing: Words and Sentences

Digital texts are essentially a stream of bits that must be converted into linguistically meaningful entities—first characters, then words, and sentences. Typically, there are two sub-stages: Document triage, the process of converting a binary file into a well-defined text document, and text segmentation. For the stream of bits to be interpreted as a stream of characters, it is necessary that the character encoding be identified. Optionally, the encoding may need to be changed depending on the kind of algorithms to be used later. There are a number of popular encoding schemes available; and depending on how the digital text was generated, different documents may need to be processed differently. The most common encoding scheme is 7-bit ASCII, which has 128 characters and is usually sufficient for English. Historically, this encoding was so popular that many languages forced their characters to be encoded using ASCII. For example, texts for European languages with accented characters opted to use two ASCII characters to encode the single accented character. Languages such as Arabic required far more elaborate asciification or romanization of their characters so as to be encoded using 7-bit ASCII. Later on, 8-bit character sets arose for virtually most languages. However, since this gives only 256 possibilities, the different character sets for the languages overlap, and prior knowledge of the encoding used is necessary to interpret a given document. The situation is even more complex for languages such as Chinese and Japanese that have several thousand characters. Historically, 2-byte (16 bit) character sets were used to represent them. Further complications happen when multiple encodings are used within the same document. For example, a Chinese document might have an English acronym or a number that is encoded using 1-byte characters. The Unicode standard attempts to eliminate all this complexity by specifying a universal character set that has over 110,000 characters covering over 100 scripts. It also provides a way to consistently deal with issues such as character properties (e.g., lower and upper case), display order (Arabic and Hebrew are right-to-left, unlike English, which is left-to-right), etc. The most common implementation of this standard is in the UTF-8 encoding, which is a variable length character encoding scheme. With no overlap of character sets, it solves the problem of properly interpreting a multilingual document without any outside information. However, despite the rapid adoption of Unicode, the use of different encoding schemes means that an encoding identification algorithm must model the different encoding schemes to figure out which one of them best fits the given file.

A related problem is that of language identification. In some cases, the character set will uniquely identify the language, and if it is a monolingual document, the problem is solved. However, many languages share common characters, and unless one has outside information about the language used, one must look at the distribution of characters in the document and compare it to that of the possible languages to determine which one is most likely. For multilingual documents, this is even more critical since different (language-dependent) word/sentence segmentation algorithms would be applied for different segments of the text.

Texts obtained off the Internet cause additional problems since they are ill-formed and do not follow the traditional language rules. Twitter logs, for instance, use novel spellings and abbreviations. Internet documents also contain a lot of markup and display-specific text that should be filtered out. A non-trivial problem is how to clean the document of the extraneous text.

All these issues become especially important when a corpus of documents is being created for statistical analysis or for model-creation in applications.

Once the text is interpreted as a sequence of characters in a particular language, the next task is to identify the words and sentences. The process of clumping the characters into words is called *tokenization*. The algorithms for doing this are very much language-dependent. Space-delimited languages, such as English, a fixed set of whitespace characters typically delimit words, but there are many special cases and different kinds of texts need custom-built tokenizers. In unsegmented languages such as Chinese and Thai, additional information is needed about what character sequences are legitimate words. The concept of word need not be restricted to a sequence of characters delimited by whitespace. Many tokenization algorithms also extract multi-word expressions such as *New York* or *de facto*, which should to be treated as one token/word.

Once words are extracted, they can be subjected to lexical analysis. This aims to understand the structure of each word. For example, it can help tag the word worked as essentially the word work with a suffix of -ed that signifies the past tense. This is called *inflectional analysis* and can be quite non-trivial. For example, the word ran is really run inflected with a past tense.

A critical unit for NLP is the sentence. Once words are extracted, it is typical to chunk them into sentences. This process is called *sentence segmentation* and is very much language-dependent. Thai, for example, has few punctuation marks and segmentation can be challenging. In a well-formed English document, almost all sentences can be segmented out by applying a set of simple rules. However, for informal English documents, the segmentation can be quite hard as there can be considerable variation in the use of the standard end-of-sentence markers such as periods and exclamation marks. Near perfect results can be obtained by using machine learning techniques on a corpus of documents to build classifiers that decide when a character such as a period is ending a sentence.

Once sentences are obtained, additional interesting word-based processing is possible. One example is part-of-speech (henceforth POS) tagging where every word is labeled with its correct lexical category in the sentence. Unlike lexical analysis, POS tagging deals solely with POS labels. It has been extensively studied and is considered a solved problem for many languages.

With or without lexical analysis, and without further NLP stages, words can be used in many interesting ways. One major focus has been probabilistic word-based models called *N-gram models*. These try to predict a word from the previous N-1 words. These N-grams have been the basis of so-called language models that predict the probability of sequences of words, for example, a sentence. These models rely on counts of words, or sequences of words, in a representative corpus of documents. Models based on N-grams have seen very broad usage in a variety of tasks ranging from spelling correction to MT.

40.2.2 Syntax: Parsing and Structure

A central aspect of NLP is syntactic analysis: determining the structure of a word sequence (typically a sentence) based on a formal grammar. The concept of a grammar for a given language comes from linguistics. It refers to a set of finite rules that define well-formed (valid) strings in the language. The correspondence is typically seen in a generative manner—the grammar rules generate strings in the language and nothing more. The so-called *Chomsky Hierarchy* organizes many of the most well-known grammars into a hierarchy of increasingly powerful (in a generative sense) formalisms. Context-free grammars (henceforth CFG) are particularly interesting and widely studied as useful approximations of natural languages. These are if-then generative rules involving words and constituents (word categories such as noun, verb, etc.) that specify how a constituent might be composed of other constituents and words. In a CFG, the rules do not have a context of application. For example, a rule that says a noun phrase consists of a determiner followed by a noun can be applied regardless of where the noun phrase appears in the string. Given a grammar and a string (sentence), the process of parsing aims to determine how the string was generated by the grammar. Effectively, this reverses the generative process. The result of parsing is a structure that shows exactly how the word string was generated by the grammar. There are many types of parsers available and most are based on the idea of searching through a space of possible structures compatible with the grammar to find one that fits the word string. Two broad classes of such parsers for CFGs are top-down and bottom-up parsers. Efficient parsers such as table-driven parsers combine ideas from both classes using dynamic programming techniques. Another aspect of parsers is how they handle ambiguity such as multiple possible parses of the same string. This leads to probabilistic solutions and specialized methods for processing weighted grammar rules. Increasingly, the field has been moving toward the use of statistical parsing techniques in which samples of text are used to determine various aspects of the underlying grammar.

Grammar theories have led to a wide variety of different grammars for different languages and consequently different parsing algorithms for each. Of particular interest are dependency grammars (and the associated dependency parsing techniques), which have been particularly useful for languages other than English. Unlike phrase-structure grammars, dependency grammars result in flatter structures and considerably ease the eventual goal of most NLP systems—extracting the intended meaning of the text.

40.2.3 Semantics and Meaning

Obtaining the syntactic structure of a word sequence is only a means to the end goal of determining what the word sequence really means. This is the objective of semantic analysis. There are several dimensions to consider. First and foremost, a decision needs to be made as to how should meaning be represented. A number of formalisms, mostly based on first-order logic, have been used for this purpose. They rely on a model-theoretic framework that provides a bridge between linguistic content and knowledge of the real world. The use of first-order logic provides for a computationally tractable mechanism to support inferences and to describe events as state changes. Other representational challenges are how to represent beliefs and common-sense knowledge. One of the most well-studied approaches to semantic analysis is an approach called *compositional semantics*, which is based on the idea that the meaning of a sentence can be obtained (composed) from the meaning of its parts. In this approach, 'parts' refers to both the syntactic components as well as the relationships between them that are the output of they syntactic analysis. With these as inputs, the semantic analyzer will output a meaning representation of the sentence. The way this is done is by augmenting the grammar rules with semantic attachments, which specify how to compute the meaning representation of the construct solely on the basis of the meaning of the constituents. Much of the work in this line of research has focused on correctly representing and reasoning with ambiguity.

Beyond sentences, there has been a rich body of research in the area of discourse representation theory that aims to capture the semantics of coherent sequences of sentences and entire texts. These theories rely on first-order-based structures of discourse and mapping linguistic constructs to operations on these structures so that as the sentences are processed, the structure changes in the way the reader's mental representation of the text referents might change.

Slightly orthogonal to these attempts to understand the meaning of entire sentences and texts is the large body of research called *lexical semantics*, the study of word meanings. Of particular interest is the study of word senses and how to disambiguate among them. A well-known example is the word bass, which has at least eight different interpretations such as a fish or a part of a musical range. Relationships between word senses can be established. Some common relationships are synonymy, antonymy, hypernymy, and hyponymy. Entire databases have been created of such graphs of relationships between words. One of the most well-known examples is WordNet, which is based on the idea that the meaning of a word sense can be understood with reference to synonymous word senses. WordNet exists for multiple languages and is widely used in practical applications. A generalization of the binary concept of synonymy is word similarity. Popular approaches to learning word similarity rely on the notion that similar words have a similar distribution of words around them. WordNet is an instance of an ontology. Early ontologies were painstakingly handcrafted, but more recently, using documents on the Internet, automatic (and semi-automatic) approaches have been used to construct very large general-purpose ontologies as well as domain-specific ones. These ontologies have been important building blocks in the so-called *semantic web*, a term used to describe a network of data elements that can be processed directly by machines. Word sense disambiguation has been a vigorous topic of research in lexical semantics with word-sense classifiers learned from labeled or, lately, unlabeled data. Another important focus of research in lexical semantics has been in modeling events by specifying restrictions and constraints on event predicates. For instance, the agent of an eating event must typically be an animate actor. Such selection restrictions on roles are increasingly being learned from large volumes of text obtained on the Internet.

All the semantic approaches mentioned so far have major problems with processing and modeling pervasive constructs such as idioms and metaphors. For example, to understand the meaning of (or even to accept as valid) the sentence, "Flickr is bleeding to death." the underlying metaphor must be recognized.

The task of recognizing textual entailment refers to determining if one text fragment's truth implies that of another text fragment. This task is at the heart of understanding the meaning of text and making inferences. It has many applications in question answering (henceforth QA), MT, text summarization, etc.

40.2.4 Pragmatics and Discourse Analysis

In the previous section on semantics, we mentioned discourse representation theory as well the problem of recognizing text entailment. These two go beyond words and sentences and focus on understanding an entire block of text. For the purposes of analysis, such a block is referred to as a discourse, and the field is referred to as pragmatics or discourse analysis. Work in this area has been somewhat sparse except for specific applications such as text summarization. Even there, much of the work is applicable only for English-like languages. Part of the reason is that the emergence of statistical techniques in NLP caused a re-examination of many of the problems in syntax and semantics and so discourse analysis was placed on a backburner. However, the problems have not disappeared and remain important to the eventual goal of understanding the intended meaning of the text. Typically, a document does not consist of unrelated sentences but has a coherency to it. In fact, one characteristic of a well-written document is high coherency. A dialogue or conversation between two or more parties is also a form of discourse and can be analyzed in terms of the communication goals of the participants. In such a discourse, there are natural segments as different parties contribute to the text. In a monologue too, even one with high coherency, there is a natural segmentation of the text into subtopics. The movement from topic to topic is governed by coherence relations. These relations might be obtained from a body of research called *rhetorical structure theory*, which identifies a small set of coherence relations normally used in text. Besides looking at coherence, which is a semantic concept dealing with meaning of textual units, one can also examine cohesiveness of text to identify the different segments in it. The intuition is that text within the same segment is more cohesive than across neighboring segments. So if one measures cohesiveness between adjacent sentences, there will be a dip in cohesiveness at topic boundaries.

Besides discourse segmentation, another important task in discourse analysis is co-reference resolution. The idea here is to find the entities that are being talked about irrespective of how they are referred to in individual sentences. For example, a text about Hillary Clinton may refer to her by "Hillary Clinton" in some sentences, by the pronouns "she" or "her" in some others, by just "Clinton" in some others and by "the US Secretary of State" in some others. All these references should be resolved to one and the same person. The simpler task of pronoun resolution, in which one finds the referent to a single pronoun, has been widely studied, and several algorithms exist that do an impressive job. These algorithms rely on parse information of the sentences involved and also various lexical features of potential referents. As mentioned earlier, text summarization has been an important application area for discourse analysis research. In its simplest form, the goal is to decide which sentences are the key ones in the text, and this can be somewhat achieved by examining surface features such as position of sentences and the existence of cue phrases.

40.3 Best Practices: Applications and Outcomes

In this section, we shall do a quick survey of the most prominent practical results and the commercial impact of NLP within computing. Five classes of applications have been identified as the most prominent: MT, information retrieval (IR) and QA, information extraction (henceforth IE), natural language generation (henceforth NLG), and sentiment analysis. For lack of space, many other positive outcomes and applications cannot be covered, and the reader is referred to additional resources at the end of the chapter.

40.3.1 Machine Translation

The most visible and prominent application of NLP is in translation. What would be a better demonstration of a deep understanding of a couple of languages than to accurately translate texts from one into the other? Furthermore, given the multi-billion-dollar size of the translation and interpretation market, it is not surprising that almost every major development in NLP, be it word-level N-grams or sentence-level parsing or semantic analysis, has found its way into MT systems. Early MT systems were rule-based and relied heavily on "deep" knowledge encoded by human experts. The most primitive of these did a word-level translation using dictionaries, while the most sophisticated ones obtained a deep conceptual structure (called *interlingua*) of the text in the source language and then translated the interlingua into the target language. These systems did not do very well on real-world texts outside the narrow domains for which they were designed. The 1990s saw breakthrough techniques in statistical MT and since then both rule-based and statistical approaches have been combined in robust and useful MT systems that are deployed in the real-world. For high-volume translation jobs, such as translations of manuals or regulations, MT systems can be used to assist human translators by quickly providing a draft translation that can then be rapidly fixed manually.

Modern MT systems are heavily statistical in nature and rely on the availability of parallel texts, which are texts available in two languages. Such texts are routinely produced in countries that require government documents be available in all official languages. For example, the proceedings of the Canadian Parliament are kept in both French and English. As another example, Hong Kong legislative records are in both English and Chinese. By aligning such parallel texts, an MT system can obtain enough statistical evidence for now to translate a new piece of text. Early MT systems relied on word-level information, but modern state-of-the-art MT systems extract phrase-level mappings from the parallel texts and use these to translate the new text. Commercial MT systems, such as Google Translate, Systran, etc. have been widely deployed and provide acceptable quality in many scenarios. The availability of open-source MT toolkits, such as Moses, has spurred development of MT systems for new language pairs. Generation of parallel texts too has become easier using low-cost crowd-sourcing techniques. The central issue with parallel texts is how to align them. The so-called *alignment problem* can be addressed at various levels of granularity: paragraphs, sentences, words, and even characters. While many of the alignment algorithms are fairly general and apply for any two language pairs, high performance is obtained by customization to linguistic features. Parsing, especially dependency parsing, can be very useful in aligning texts by matching (i.e., aligning) parse trees/graphs. Parts of the text that cannot be aligned reliably are discarded from the analysis.

A major problem in MT system development has been the issue of evaluation. Ideally, the evaluation must be done by human beings, but this can be very costly and time consuming. Various measures of fidelity, informativeness and intelligibility have been proposed in the past. Modern automatic measures such as bilingual evaluation understudy and TER are useful when comparing similar systems or incremental changes to a system.

Commercial MT systems are not just for end-users. As we shall see in the next section, they can also be effective in multilingual IR as well as many other NLP applications that require collating text in more than one language.

40.3.2 IR and QA

Perhaps the most visible and successful application of NLP is IR and its related cousin QA. Both IR and QA focus on finding information buried within unstructured texts and providing to the end-users in response to a query, which may be posed as a series of keywords (in IR) or a natural language question (in QA). Typically, the texts are pre-processed and auxiliary structures and indices are created in advance to facilitate the task of providing answers in real time. The indices can be extremely large, especially in the case of web-scale search, and pose significant engineering challenges for NLP algorithms. Users of IR systems, such as search engines, tend to write short and ambiguous keywords to describe their needs. There are additional issues of misspellings and the context (search is an iterative process and recent previous queries by the same user can provide valuable additional clues as to what the user is not looking for). Synonyms and multiple word senses pose additional challenges in trying to provide the user with the "correct" set of documents and web pages. Finally, ranking the output by relevance is another challenge that requires the use of NLP methods to assess similarity of queries to texts. At the back-end of IR systems, creating auxiliary structures and indices around the searchable texts requires the use of NLP. A morphological analysis of the text is quite routine so that texts that contain variations of the same word are all indexed in the same manner. The texts need not abide by the same spelling rules and can contain errors as well. Concept ontologies are used to ensure that texts that contain similar concepts are readily fetched in response to user queries. In web-scale IR, an additional consideration is that of multiple languages. The user may pose a query using keywords in English but the most relevant page might well be in French. In such a case, either the page can be automatically translated or else it may be provided to the user without further processing if the user profile suggests that the user is fluent in French. Multilingual IR can be done either by translating the query into multiple languages in real-time or by preprocessing multilingual texts to create a unified index in the language of the user queries.

The use of keywords as a way to query IR systems is ultimately a compromise based on limitations of existing technologies. Ideally, people would like to pose a question and have the system understand the question, do a search of the relevant documents, extract the relevant information, and construct the best possible answer, possibly by consulting multiple documents. A well-known example of this is the Watson system that won the Jeopardy contest. This had the additional complexity of having to reason about its confidence in its answers and also developing strategies to win the overall game. The process of QA is often broken down into three stages: (1) question analysis, (2) text selection, and (3) answer extraction and construction. In the first stage, the most critical part is question classification, which determines the type of answer that should be provided. This can done by something as simple as a set of rules that map question patterns to question (and hence) answer types. In English, this may be flagged by the presence of question words/phrases such as where, what, how many, etc. The question text also serves as a basis for constructing a set of query words that are sent to a search engine to retrieve a set of relevant documents. The documents can then be analyzed to obtain candidate answers. Textual entailment techniques have been explored for ranking the different answers and summarization methods have been used to build an answer from information in multiple documents. Recent trends in QA have also explored approaches such as similar question finding, which exploit the existence of question answer pairs on the web, many of the answers having been provided by human beings and hence probably of high quality.

40.3.3 Information Extraction

We saw in the previous section, how QA systems rely on extracting information from documents in order to construct an answer. The more general task of IE sees applications in a broad spectrum of tasks where structured information is required. A prototypical IE application is database creation by template filling. From unstructured texts of house-for-sale notices, a predefined template's slots might be filled up: location, number of rooms, number of bathrooms, total area, age, price, gas or electric kitchen, parking, etc.

The most common scenario for IE is named entity recognition (NER), which is often seen as the first stage of more advanced relationship extraction. State-of-the-art NER systems are quite good; and many commercial products are deployed in domains such as military intelligence, news analysis, and biomedicine. Generic NER systems extract entities such as people, places, and organizations from newswire articles. More specialized ones for biomedical applications may extract specific genomic entity types from highly technical papers that might refer to these entities using complex and confusing terminology. Most NER systems are designed as a set of classifiers that label the text sequence word by word. In this view, the task is quite similar to POS tagging that we discussed earlier. Practical NER systems use a combination of human-specified patterns, rules, and automatically constructed sequential classifier models.

Once entities are extracted, the next goal of IE is to find the relationships between them. Usually, a predefined list of relationships is considered based on the domain. For example, in intelligence applications, one may be interested in the relationship BELONGS-TO between a PERSON entity and an ORGANIZATION entity. Seed patterns are used to bootstrap the process of extracting additional patterns that specify when the relationship holds between entities. The template-filling application mentioned at the start of this section can be seen as the process of extracting complex *n*-array relationships. For narrow domains, template filling algorithms work very well and are based on sequential labeling approaches.

More recent applications of IE have been in temporal analysis. These involve recognizing temporal expressions (an important goal in QA where a question may start with the word "when") such as absolute dates, relative expressions, and durations. Besides pattern-based rules, chunking approaches that use a set of sequential classifiers have been effective for this task. Another important problem is event detection. Verbs typically flag events, and there is usually a temporal expression to specify when the event happened. Event recognition is domain-specific and begins with a set of seed patterns that feeds into a bootstrapping process that extracts new patterns based on the usage of entities from earlier patterns.

Most of what has been mentioned earlier applies to single-document IE. Emerging multi-document IE systems also take advantage of aspects such as redundancy of information to acquire broadly applicable knowledge. Such open-domain IE systems also provide new challenges for standard NLP tasks. For example, the task of co-reference resolution becomes much harder in the cross-document scenario: when are two documents about the same entities?

Biomedical language processing, also known as BioNLP, is a key IE application. Building NER systems for this domain has been particularly challenging because of the unique and highly technical nature of the texts. Common entities are gene and protein names, and the challenge in the field is that these entities are sometimes referred to by names, other times by symbols, and, to make matters even more confusing, sometimes by the corresponding gene/protein. The names themselves are not unique either and vary based on the organism, function, etc. A unique problem for NER systems in this field is to perform gene normalization—relate different mentions of a gene to some specific gene in a database. The problem is cast as a word sense disambiguation task.

40.3.4 Natural Language Generation

NLG is the process of expressing thoughts using language. Synthesis of language is the flip-side of analysis and shares many of the same concerns and solutions. Just as a simplistic analysis of language can be achieved using patterns, a simple synthesis of text can be achieved by instantiating a parametrized template. In an MT system that translates text from German to English, NLG occurs when English text is generated based on the information/thoughts extracted from an analysis of the German text. In many other scenarios where text is generated, the process is merely one of selecting some fragments (as in text summarization) or instantiating a template (as in QA). It is tempting to view NLG as the analytical process in reverse. In fact, the field of NLG is quite distinct with its own concerns and issues that revolve around speaker intentions, listener models and text planning.

A domain-specific application of NLG is report generation where the process is somewhat the opposite of template filling in IE: take as input some structured data and produce a natural language text that describes/interprets the data in a manner consistent with how a human being would. For example, in a weather report, the input data may contain information about wind speed in km/h; however, a report generator would present this more qualitatively. Furthermore, depending on the strength of the wind speed, it may be necessary to feature it quite prominently at the start of the report, or mention it in passing at the end.

Some of the most exciting developments emerging in NLG are in education and health care. In intelligent tutoring systems, it has been shown that providing more succinct and abstract feedback to students facilitates more learning. Thus, the NLG front-end should take this into account in planning the feedback to be delivered to the student. In health care, the goal of engaging the patient in a dialogue is to persuade them to alter behavior. For example, to adopt a better diet or to quit smoking. Hence, the NLG module must generate text that is based on an argumentation theory that supports this. Patients can be represented as a finite state model of states, and their progress through this model can be tracked as the dialogue proceeds, and the system generating appropriate text based on this analysis.

NLG has also been used to give textual summaries of complex graphs of clinical data. These summaries have been found to improve decision-making and patient care. The reasons should be intuitively clear. A patient's medical history contains so many numeric data from laboratory tests over irregular intervals of time that it becomes difficult to visualize it all. By integrating textual summaries into the visualization process, it is possible to make sense of the data more rapidly and deliver better quality care more effectively.

In generating text, an important consideration is that it be coherent. This is usually ensured by using a top-down planning approach to the construction of the text. At a sentence level, it is important that the text makes use of referring expressions such as pronouns, especially when the reference is unambiguous. Algorithms are available that err on the side of wordiness. It is better to eschew the use of pronouns unless the reference can be identified clearly and unambiguously.

It is extremely hard to evaluate NLG systems. One measure of evaluation is the naturalness of the text. This can be done by human experts. Necessarily subjective, the process would typically be done by multiple experts and inter-annotator agreement recorded. Obviously, such a process is expensive and slow. More recently, there have been attempts at evaluating aspects of NLG instead of treating the entire process as a blackbox. Evaluation of referring expression generation has been explored systematically by setting up a validation set to compare different algorithms. In general though, evaluation and comparison in NLG is very hard to objectify.

40.3.5 Sentiment Analysis

Sentiment analysis or opinion mining is the study of opinions, sentiments, and emotions in text. It has seen extremely rapid growth in recent years and is one of the highest impact applications of NLP.

In general, text can be classified as factual or opinionated. For a long time, people were concerned more with facts and objective information in text. Most search engines, for example, are more concerned with facts than with opinions. We saw earlier how QA systems are primarily concerned with extracting facts from documents and using these fragments of objectivity to construct "correct" answers to questions. With the growth of the web and the rise of social media, however, there has been an increased interest in opinions and feelings. The web has certainly made it easier for people to express their opinions. In addition, social media has made it easy for us to find out what our friends and others we care about and trust have written. When contemplating an expensive purchase, even if we do not care for the opinion of someone we have never met, we just might be interested in what a close friend might have to say about their personal experience using the product. For a company, it is no longer necessary to conduct an expensive survey or organize a focus group to find out how its products are being perceived by consumers. They can simply mine the many blogs, forums, and discussion groups to find out what people are saying. Extracting sentiments is not an easy task. In most cases, opinions are embedded in broader discussions and may even be indirectly expressed. There are no obvious datasources to access—user-generated content is everywhere and anywhere. As people have realized that their opinions are being harvested, many have started to manipulate the system by inserting opinions as spam into popular Internet sources.

Broadly speaking, sentiment analysis is about classification—what are the good things being said, what are the bad things—but before that can be done, a more fundamental question must be answered: which text fragments are objective and which are subjective. This field of research, called *subjectivity analysis*, seeks to separate the facts from the opinions. Typically, it works at the sentence-level, but due to referring expressions and contextual issues, extra information must be included. Once subjective text is identified, there are several aspects of the opinion that must be extracted: who is the opinion holder, what object is the opinion about, is there any particular aspect or feature of the object that the opinion focuses on, is the opinion positive or negative or something in-between, is it a direct opinion about the object, or is it a comparative opinion with reference to another object. In the early days of sentiment analysis, to find the opinion polarity, word-based approaches were popular. Lists of positive and negative words were constructed, and one could simply count the number of words in the text that came from these two lists and use some function of their difference to assign an overall sentiment polarity to the text. More recently though, better results are obtained by using modules from various parts of the

NLP pipeline. For instance, NER systems assist in finding the object of the sentiment and its aspects. Co-reference resolution comes in handy to disambiguate pronouns and collate multiple opinions of the same object. Parsing techniques and POS-tagging help identify the most important indicator words that flag opinions. Classifiers are constructed for combining the multiple sources of opinion within the text fragment to obtain an overall score.

Similar to regular web search, opinion search is another application area of great use. In principle, it should be a straightforward task to restrict regular search to subjective documents only; however, the task of ranking (and maybe aggregating) the different opinionated pages is more complex than ranking factual pages.

Most sentiment analysis research applies to English. However, methods exist to transfer the knowledge gleaned from English texts to perform sentiment analysis in other languages as well.

40.4 Research Issues

We have described the key areas of NLP and commercially significant applications. In this section, we project into the future and identify major challenges that the field will face. Our discussion of research issues is not intended to be comprehensive. For any area of NLP, there will always be an exploration of alternatives and new approaches. The research literature can be accessed at any time to examine current trends. We have used our judgment in selecting themes where NLP will have the maximum impact and hence see the most vigorous research. For some of these themes, techniques will become standardized and mature with routine applications; for others, progress will be made but they will continue to be on research agendas.

40.4.1 Large-Scale NLP

A recent trend in NLP has been the dominance of statistical approaches. The reason for this dominance has been the relatively easy access to text corpora. Instead of relying on humans to encode deep knowledge about language into programs that process text, statistical approaches use corpora to figure out the deep knowledge directly. With corpora sizes growing at an exponential rate, current statistical NLP approaches, primarily based on classical statistics with a small-data view of the world, are increasingly reaching the limits of their usefulness. The routine deployment of distributed systems to handle large-scale data and the emergence of novel computing paradigms are presaging the need for new ways to examine many of the older problems. Beyond the simple application of newer big data analytical tools, this calls for a fundamental new way of looking at problems. For example, approaches to extract ontologies from text need to be self-directed and require minimal human intervention if they are to scale up to the sizes of future text collections. However, these approaches also result in ontologies that are so large and complex that current application scenarios of these ontologies become impractical and these too must be supplanted by new scenarios that are better able to use the ontologies. Algorithms that run on single computers and require all the data to be in one place become irrelevant when the data are necessarily distributed (e.g., due to privacy issues), and the tasks they solve must be rethought at a very fundamental level. In the past, the scaling up of NLP was led by large companies such as Google with the resources and computational horsepower to store and process big data. Many of the published results could not even be replicated or compared by other researchers. Now, with the emergence of cloud-computing, as its use becomes more widespread and costs go down, the same resources will be available to anyone, and we can expect to see an explosion of new ideas and research as big data analytics becomes more widespread in the NLP community.

40.4.2 Multilingual and Cross-Lingual Approaches

The English language is the dominant lingua franca of the world and so, not surprisingly, most NLP research has focused on English. If one examines the amount of available digital text in various languages, English is by far the most common. Part of the reason for this is the so-called digital-divide. Many people from non-English backgrounds simply could not afford computers, and most of the tools and programs that were used to generate digital texts were focused on English. Most population estimates already have Chinese and Spanish as the languages with more native speakers than English. Furthermore, growth projections into the future have estimated that Arabic and Hindi may become more dominant. None of this is to imply that English is going to become insignificant in our lifetimes, but it does imply that as computer usage becomes more widespread, digital text collections in other languages are going to grow rapidly and match in size the collections in English. This is going to result in strong growth of NLP research in other languages. Signs of this are already present. For example, while sentiment analysis used to focus almost exclusively on English text, its application on text in other languages has seen rapid growth. Even in MT, there is evidence of a shift in research from translations to/from English to languages that do not involve English at all. One of the key reasons for the explosion of research in dependency parsing is that dependency grammars are more suited for a wide variety of languages.

A huge focus in multilingual NLP has been in language-independent techniques. The allure of this is that the same technique can then be applied to many different languages. In the future, this may see very rapid growth as English-specific approaches will be found to be less than adequate for other languages. A related approach is language adaptation, where techniques for one language are adapted and modified to work with another language.

The goal of cross-lingual NLP approaches is to be able to work with text in multiple languages simultaneously. For example, in a cross-lingual IE system, the information extracted may come from texts in any of a set of languages. With the rise of user-generated content on the Internet, it is becoming increasingly common to find pages with text in multiple languages without any special tags specifying the switch in languages.

A final issue is the emergence of specialized languages (or sublanguages) in particular domains and scenarios. For example, mobile devices have resulted in the rise of a whole new way of writing text with phonetic abbreviations and special codes. As another example, consider how for-sale postings use a combination of tables, phrases, and visual formatting to convey textual information in an eye-catching manner rather than following the traditional rules of the language. As user-generated content becomes even more widespread, all the familiar steps in a NLP pipeline must be developed for such specialized languages, or existing approaches must be modified to properly process such text.

40.4.3 Semantics

As we saw earlier, semantics is the holy grail of NLP. All the processing that is done to a piece of text has one goal—to understand its meaning. As we saw earlier, most of the current approaches have been limited in scope and scale. Part of the reason has been that there were many issues to solve with regard to syntax. As approaches for that have matured, research has increasingly turned to semantics, and we expect the trend to accelerate in the future. Programs such as IBM's Watson have provided evidence that current approaches can be effective even when competing against human beings. A revival of ambitious projects to map all of human common-sense reasoning knowledge has also added to this trend of vigorous research in semantic processing. As research in general-purpose semantic processing matures, we are likely to see more fielded applications in narrow domains. Fields such as biomedical text processing are indeed ripe for a more semantic analysis. Sentiment analysis too is likely to move from a superficial word-based approach to a deeper understanding using semantics. This trend is already in progress as research is moving from a simplistic positive/negative binary sentiment view to a more fine-grained analysis that examines multiple dimensions such as attitudes and intentions and how they interact with cultural issues and relationships between speakers.

Lexical semantics is another area that is likely to see huge growth. Ontologies such as WordNet are already mature enough to be used in applications as blackboxes. The automated construction and maintenance of such graphs and networks of words is going to become an increasingly important topic of research as demands for larger and more sophisticated versions come from applications.

40.4.4 Discourse Analysis

Discourse analysis has long been the least understood aspect of NLP. Considered the final stage of NLP, research in this area has patiently awaited maturity in earlier stages. Part of the problem has been a lack of tools and proper corpora for testing and evaluation. The more serious problem has been that researchers were still trying to sort out sentence-level semantics before tackling the meaning of an entire discourse. However, lately with some stability in the way semantics is understood, the time is ripe for increased research in this area. The rise of social media has also increased the amount of data available on discourse arising from groups of people. As these data become more prevalent, we can expect more research to be conducted on understanding the many facets of discourse. We mentioned earlier that in education and health care, there is a strong need for automated systems that interact with humans in a situation-appropriate manner. For this, a good understanding of the underlying dialogue structure is critical. Dialogue understanding is also important for building conversational agents embedded in robots and other interfaces that bring humans and computers together. In such settings, the role of several other areas such as NLG and text summarization also could drive research in discourse analysis. With the enormous amounts of text data being generated and recorded, summarizing it in a manner consistent with the needs and abilities of the reader is going to require a much deeper understanding of what makes a piece of text cohesive, coherent, and persuasive. One of the hard problems in NLG and text summarization is evaluation, and we expect that in future there is going to be more vigorous attempts to develop reliable metrics for assessing and comparing quality of the output. Discourse also has social and cultural aspects that we can expect to be explored in more depth in the future. For example, nationality, age, gender, etc. have a strong connection with the overall nature of the discourse. While many of these aspects have been explored by sociolinguists in controlled and somewhat artificial laboratory settings involving mostly students (i.e., people of a narrow age group) in small numbers, we can expect social media and other online forums to provide a wealth of data for a wider and more in-depth analysis. We also expect to see more widespread applications of discourse analysis. Already, there is anecdotal and sporadic evidence of its use in author identification, lie detection, personality analysis, etc. The role of punctuation in a discourse is also poorly understood. For example, if a person writes "SORRY!!" instead of "sorry," it clearly has a different meaning. These issues have been somewhat explored by psycholinguists using heuristic approaches that rely on counts and cues; however, a more thorough statistical analysis is likely in the future with the availability of decent-sized corpus.

40.5 Summary

In this chapter, we have given a birds-eye view of the field of NLP and its role in the discipline of computing. We abstractly defined the key principles upon which the field is based and cast them within a pipeline of stages that takes a piece of text as the input and produces the meaning of the text as the output. These stages process the text from a character-based stream to form words and sentences, which are further structured before being subject to semantic analysis to extract meaning. We then surveyed the most prominent practical results to illustrate the commercial impact of the field. The applications range from MT to sentiment analysis. There have been many notable successes, the most recent one being of IBM's Watson program in the television quiz show *Jeopardy!* where it competed against and defeated two record-holding human champions. Finally, we identified the major challenges and research issues that are likely to shape the field in the future. These challenges deal with scaling up NLP to handle real-world problems with minimal human intervention.

Key Terms

In this section, we give a selection of NLP-specific terms listed in alphabetic order. This selection is by no means complete, and the bibliography will provide additional information on other terms.

Anaphora: An expression whose reference depends on another (usually previous) expression. For example, in the text "John didn't go to work today. He was not feeling well." the word "He" is an anaphor that refers to John. While an anaphora is usually a pronominal expression, it can also be phrase. For example, the reference to "the queen" depends on the context in which it occurs.

BLEU: A widely-used metric for evaluation of MT systems. It stands for bilingual evaluation understudy. The metric assesses the quality of the machine-translated text in terms of closeness to text generated by professional human translators.

Chomsky hierarchy: An organization of formal grammars in terms of their generative powers. Languages higher up in the hierarchy are more expressive than the lower ones. Grammars for natural languages are based on this formal structure.

Coherence: A concept, used in discourse analysis, which refers to how the meaning of different text units can be combined into a meaning of the combined text.

Cohesion: A concept, quite distinct from coherence, which refers to how different text units are linked together using linguistic constructs. Unlike coherence, it does not consider the meaning of the texts or the linking constructs.

Dependency grammar: A kind of grammar based on modern syntactic theories that give primacy to the dependency relationship between words in a sentence. They result in structures that are considered closer to the meaning of the sentence and are quite distinct from phrase structure grammars from the Chomsky hierarchy.

Document triage: Refers to the process of interpreting a set of digital files as text documents. Involves deciding how to decode the bits into characters, how to attribute a language to the character stream, how to segment out the actual textual content from the other elements, such as images and markup, etc.

Interlingua: Deep conceptual structure that captures the meaning of a text independent of the language. Used in MT systems to enable translation between two languages—the text from one language is converted into interlingua from where it is converted into the other language. The conversions to/from interlingua are simpler, and the use of interlingua provides a more scalable approach to translating between multiple languages.

Morphological analysis: Refers to understanding words in terms of their composition from smaller units. It typically involves identifying the morpheme and its relationship to the surface word based on inflection, derivation, compounding, and cliticization.

***N*-grams:** Contiguous blocks of N words. When $N = 1$, they are called *unigrams*. Analysis of a text based on its N-grams has been a popular approach to building language models that have been widely used in applications ranging from spelling correction to machine translation.

Parsing: Process of imposing a structure on a sentence. The structure is usually a tree or a graph that encodes how the sentence could be understood in terms of a grammar for the language in which it is generated.

POS tagging: Refers to tagging each word in a sentence with a part-of-speech tag from a finite set of possibilities. Is usually solved as a classification problem. The tags provide valuable information for other NLP tasks such as parsing, named-entity recognition, etc.

Unicode: An industry standard for consistently encoding and handling text from different languages.

Annotated Bibliography and Additional Resources

In this section, we list sources that can provide readers more detailed information on the issues and topics discussed in this chapter. The list includes websites as well as books. We sought to provide authoritative resources. There is obviously some overlap in the information provided by these resources, but we believe that they cater to slightly different needs and so, between them, most readers will be covered. Since NLP is a fast evolving field, books and articles become outdated rather quickly. Websites too tend to disappear, URL's change or become outdated as their maintainers move on to other organizations. We have tried to list resources that will remain relevant for the lifetime of this handbook. Still, for the latest results, the reader may need to consult other resources. We have tried to minimize this issue by listing the most recent publications at the time of writing. We preferred books and surveys since they provide an overview as well as pointers to the most significant and relevant papers in conferences and journals.

1. http://www.aclweb.org (accessed on October 10, 2013).
   The Association of Computational Linguistics portal is an excellent place to get access to publications in NLP and is a top resource for all professionals and graduate students in NLP. It should be the first stop for anyone serious about working in NLP. Most of the conferences and workshops in the field put their papers in the ACL anthology, which can be browsed as well as searched. Many important journals are accessible here as well. Anything substantive and relevant to NLP usually lands up in the ACL anthology.
2. Indurkhya, N. and Damerau, D. 2010. *Handbook of Natural Language Processing*, 2nd edn. Chapman & Hall/CRC Press, Taylor & Francis Group, Boca Raton, FL.
   This edited collection written by over 40 of the leading experts in NLP provides detailed information of the tools and techniques for implementing NLP systems. Besides surveying classical techniques, it also covers statistical approaches in great detail and discusses emerging applications. The handbook is a good place to delve into subareas of NLP that may interest the readers. The individual chapters are mostly self-contained and each has a very extensive bibliography of the most important papers in the area covered.
3. Jurafsky, D. and Martin, J. 2008. *Speech and Language Processing: An Introduction to Natural Language Processing, Speech Recognition and Computational Linguistics*, 2nd edn. Prentice-Hall, Upper Saddle River, NJ.
   A very comprehensive textbook for NLP that covers the full range of technologies used and relates them to one another. It is the basis of many graduate courses in NLP and is an ideal introduction and reference for NLP.
4. Bird, S., Klein, E., and Loper, E. 2009. *Natural Language Processing with Python*. O'Reilly Media, Sebastapol, CA.
   This is the well-known NLTK book and is a very hands-on introduction to NLP using the open-source freely-available natural language toolkit. Prior knowledge of Python is not required to read this book as it introduces the language features at appropriate places in the book. A nice feature of this book is that it has a very extensive set of exercises, the code is all open-source and there is a lot of online support making it very suitable for self-study. It can be read in conjunction with either the handbook of NLP or the Jurafsky and Martin textbook mentioned earlier. To get NLTK, see the website http://www.nltk.org which is also a useful resource for learning more about NLP.
5. Manning, C. and Schutze, H. 1999. *Foundations of Statistical Natural Language Processing*. MIT Press, Cambridge, MA.
   An excellent in-depth tour of the early work in the field of statistical NLP. It is a bit dated now, but remains an invaluable reference and source for foundational work in this area.
6. https://www.coursera.org/course/nlp (accessed on October 10, 2013).
   This free, online course, taught by Dan Jurafsky and Chris Manning, is a good place to begin learning about NLP.
7. Blackburn, P. and Bos, J. 2005. *Representation and Inference for Natural Language, A First Course in Computational Semantics*. CSLI, Stanford, CA.

One of most accessible discussions of computational semantics. Not only does it clarify, with great elegance, many of the theoretical results in semantics but also provides a very practical perspective with source code for a series of increasingly sophisticated PROLOG programs that implement the ideas discussed in the text.

8. Bunt, H., Merlo, P., and Nivre, J. 2011. *Trends in Parsing Technology*. Springer, New York.
   This is an edited collection written by top researchers in parsing methods. It provides a nice overview of the state-of-the-art in three very important subareas of parsing: dependency parsing, domain adaptation and deep parsing. A must-read for anyone wishing to start active research in this area.
9. Ferucci, D. et al. 2010. Building Watson: An overview of the DeepQA Project. *AI Magazine* 31(3), 59–79. Fall, 2010.
   One of the few in-depth technical descriptions of IBM's high-profile Watson program that won the Jeopardy! Tournament. More publications should appear in the future. Look in the ACL anthology.
10. http://www.ldc.upenn.edu/ (accessed on October 10, 2013).
    Did you say you need data for NLP research? This portal of the Linguistic Data Consortium is the place to visit first. It has a very extensive collection of data and tools. The quality of the data is very high, and it would be difficult to get comparable data elsewhere or to create it yourself.
11. http://wordnet.princeton.edu/ (accessed on October 10, 2013).
    The place to get the latest WordNet, a very popular, and highly regarded lexical database.
12. http://www.eacl.org/ (accessed on October 10, 2013).
    Much of the information on the European chapter of the ACL is available from aclweb. However, for readers interested in European languages, this portal may provide an easier access to the relevant resources.

41

Understanding Spoken Language

Gokhan Tur
Microsoft Research

Ye-Yi Wang
Microsoft

Dilek Hakkani-Tür
Microsoft Research

41.1 Introduction

Spoken language understanding (SLU) is an emerging research area between artificial intelligence, speech, and language processing. The term "SLU" has largely been coined for understanding human speech directed at computers, although understanding human/human conversations is also a vibrant area of SLU. Having a semantic representation for SLU that is both broad in coverage and simple enough to be applicable to several different tasks and domains is challenging, hence most SLU tasks and approaches depend on the application and environment (such as mobile vs. TV) they have been designed for.

Given that speech is the most natural medium people use to communicate with each other, it is hardly used when interacting with computers. Recently, with the advancements in automatic speech recognition (ASR) and natural language processing (NLP) technologies, there is a strong interest—both commercial and academic—in understanding human speech.

SLU applications range from interpreting a small set of commands to complex requests involving a number of steps and reasoning. Most commercial systems employ knowledge-based approaches, such as mainly building hand-crafted grammars or using finite sets of commands, and they have already been deployed in various environments such as smartphones, cars, call centers, and robots. In the near future, we can expect more complex applications, like robots better understanding what we say instead of reacting to a finite number of predefined commands, or we will be able to ask our mobile devices a nontrivial set of requests from a broader range of domains.

Unlike speech recognition, which aims to automatically transcribe the sequence of spoken words, SLU is not a clearly defined task. At the highest level, its goal is to extract "meaning" from natural language. In practice, this may mean any practical application allowing its users to perform some task with speech. As one of the pioneers of statistical spoken language processing, Fred Jelinek,

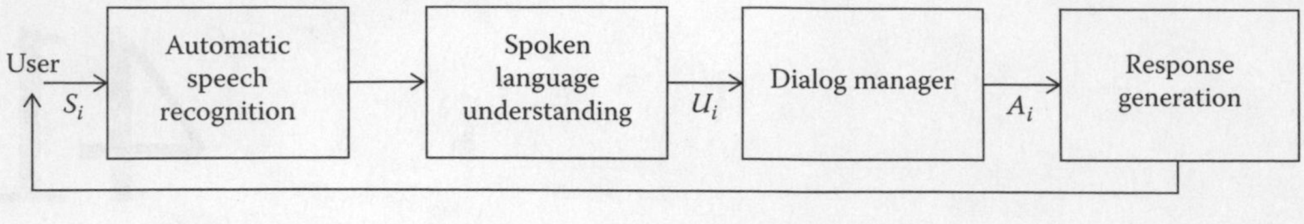

FIGURE 41.1 A conceptual architecture of a human/machine interactive system.

once put it: "The human analogy is misguided because machines do things differently than biological beings. Airplanes don't flap their wings".

The conceptual components of human/machine conversational interaction system are shown in Figure 41.1, following [53]. At each turn, the user's speech, S_i, is recognized and semantically parsed into a task-specific semantic representation of the user's intention, U_i.

For example, if the user asks "*When is the next showing of Hunger Games around here?*" the system should be able to interpret the intent and the associated arguments (slots), such as

Find_Showtimes(movie = Hunger Games, location = (latitude,longitude))

If a statistical model is used, U_i is estimated as

$$\hat{U}_i = \arg\max_{U_i} P(U_i|S_i)$$

Then the dialog manager (DM) updates the belief state using semantic context, C_i, such as the previous belief state of the system (which may include a user goal and actions, as well as dialog history); user-specific meta-information, such as geo-location and personal preferences; and other contextual and lexical information. More formally,

$$\hat{B}_i = \arg\max_{B_i} P(B_i|C_i, \hat{U}_i, S_i)$$

The system decides on the most appropriate system action, A_i. In the statistical approaches, this decision is made based on the expected reward over belief states, which are estimated using the previous machine action and belief state, as well as the observations the machine receives from the user. Then, a response is generated in spoken or multimodal fashion. For instance, if the DM decides to present the information of showtimes with theaters, the generation needs to decide on whether to show a map or list and/or speak this information to the user using text-to-speech synthesis (TTS).

41.2 Impact on Practice

SLU systems can be grouped into three categories based on their high-level approaches. The first category is chatbot systems, like ELIZA, where the system does not really understand what is said but just mimics understanding. ELIZA was built at MIT in the 1960s [49] and has been the most popular artificial intelligence program, even though there was no intelligence but instead the system relied on simple pattern matching. A typical interaction may have such excerpts as shown in Figure 41.2.

There are more complex versions of ELIZA, such as ALICE,* which are moving toward embedding more sophisticated language processing technologies within the same framework.

The understanding systems in the second category are rooted in artificial intelligence. They are using deeper semantics and are demonstrated to be successful for very limited domains. These systems are typically heavily knowledge-based and rely on formal semantic interpretation defined as mapping sentences into their logical forms. In its simplest form, a logical form is a context-independent representation of

* http://alicebot.blogspot.com/

User: I need some help
System: Why do you say you need some help?
User: I am depressed
System: Are you depressed often?

FIGURE 41.2 An example chitchat interaction with ELIZA.

a sentence covering its predicates and arguments. For example, if the sentence is "John loves Mary," the logical form would be "love(john,mary)."

During the 1970s, the first systems for understanding continuous speech were developed with interesting approaches for mapping language features into semantic representations. For this purpose, *case grammars* were proposed for representing sets of semantic concepts with thematic roles such as agent or instrument. The ICSI FrameNet project, for instance, focused on defining semantic frames for each of the concepts [23]. For example, in the "commerce" concept, there is a "buyer" and a "seller" and other arguments such as the "cost" and "good." Therefore, two sentences "A sold X to B" and "B bought X from A" are semantically parsed as the same, with both sentences corresponding to the "commerce" frame. Following these ideas, some researchers worked toward building universal semantic grammars (or interlingua), which assume that all languages have a shared set of semantic features [5]. Such interlingua-based approaches have also heavily influenced machine translation field until the late 1990s before statistical approaches began to dominate the field. [1] may be consulted for more information on the artificial intelligence-based techniques for language understanding.

The last category of understanding systems is the main focus of this chapter, where understanding is reduced to a (mostly statistical) language processing problem. This corresponds to attacking "targeted" speech understanding tasks instead of trying to solve the global machine understanding problem. A good example of targeted understanding is detecting the arguments of an intent given a domain as in the Air Travel Information System (ATIS) project [36]. ATIS was a popular DARPA-sponsored project, focusing on building an understanding system for the airline domain. In this task, users utter queries related to flights such as *I want to fly to Boston from New York next week*. In this case, understanding was reduced to the problem of extracting task-specific arguments in a given frame-based semantic representation involving, for example, *destination* and *departure date*.

In the United States, the study of the frame-based SLU started in the 1970s in the DARPA Speech Understanding Research (SUR) and then the Resource Management (RM) tasks. At this early stage, natural language understanding (NLU) techniques like finite state machine (FSM) and augmented transition networks (ATNs) were applied for SLU [50]. The study of targeted frame-based SLU surged in the 1990, with the DARPA ATIS evaluations [8,21]. Multiple research labs from both academia and industry, including AT&T, BBN Technologies (originally Bolt, Beranek, and Newman), Carnegie Mellon University, MIT, and SRI, developed systems that attempted to understand users' spontaneous spoken queries for air travel information (including flight information, ground transportation information, and airport service information) and then obtain the answers from a standard database. ATIS is an important milestone for the frame-based SLU, largely thanks to its rigorous component-wise and end-to-end evaluation, participated by multiple institutions, with a common test set.

While ATIS focused more or less on the understanding of a single-turn utterance, the more recent DARPA Communicator Program [45] focused on the rapid and cost-effective development of multimodal speech-enabled dialog systems, in which general infrastructures for dialog systems were developed, where different component systems for ASR, SLU, DM, and TTS can be plugged in and evaluated. Naturally, many SLU technologies developed in ATIS were used in the SLU component of the Communicator program. Eight systems from AT&T, BBN Technologies, University of Colorado, Carnegie Mellon University, IBM, Lucent Bell Labs, MIT, and SRI participated in the 2001 evaluation [46]. In the meantime, the AI community had separate efforts in building a conversational planning agent, such as the TRAINS system [2].

Parallel efforts were made on the other side of the Atlantic. The French EVALDA/MEDIA project aimed at designing and testing the evaluation methodology to compare and diagnose the context-dependent and context-independent SLU capability in spoken language dialogs [4]. Participants included both academic organizations (IRIT, LIA, LIMSI, LORIA, VALORIA, CLIPS) and industrial institutions (FRANCE TELECOM R&D, TELIP). Like ATIS, the domain of this study was restricted to database queries for tourist and hotel information.

The more recent LUNA project sponsored by the European Union focused on the problem of real-time understanding of spontaneous speech in the context of advanced telecom services [17]. Its major objective is the development of a robust SLU toolkit for dialog systems, which enhances users experience by allowing natural human/machine interactions via spontaneous and unconstrained speech. One special characteristic of the project, which is absent in similar projects in the United States, is its emphasis on multilingual portability of the SLU components.

While the concept of using semantic frames (templates) is motivated by the case frames of the artificial intelligence area, the slots are very specific to the target domain and finding values of properties from automatically recognized spoken utterances may suffer from ASR errors and poor modeling of natural language variability in expressing the same concept. For these reasons, SLU researchers employed known classification methods for filling frame slots of the application domain using the provided training data set and performed comparative experiments. These approaches used generative models such as hidden Markov models [34], discriminative classification methods [27], knowledge-based methods, and probabilistic context-free grammars [38,47].

Almost simultaneously with the semantic template filling-based SLU approaches, a related SLU task, mainly used for machine-directed dialog in call center interactive voice response (IVR) systems, has emerged. In IVR systems the interaction is completely controlled by the machines. Machine-initiative systems, ask users-specific questions and expect the users-input to be one of predetermined keywords or phrases. For example, a mail delivery system may prompt the user to say *schedule a pick-up, track a package, get rate, or order supply* or a pizza delivery system may ask for the possible toppings (see Figure 41.2). Such IVR systems are typically extended to form a machine-initiative directed dialog in call centers and are now widely implemented using established and standardized platforms such as VoiceXML (VXML) (Figure 41.3).

The success of these IVR systems has triggered more sophisticated versions of this very same idea of *classifying users' utterances into predefined categories* (called as *call types* or *intents*). [14] has published the AT&T *How May I Help You?* system description paper in 1997. HMIHY is essentially a call routing system. The users of the system are greeted by the open-ended prompt *How May I Help You*, encouraging them to talk naturally. This system has also been deployed for AT&T's call center (known as 0300) and used nationwide with great success [13]. The system was capable of categorizing an incoming call into one of the six specific call types (such as *billing* or *rates and plans*) plus an *other* for the remaining intents along with generic intents such as talking to a human agent or greetings. The calls are then routed to corresponding specialized departments. If the initial utterance of the user is not specific (such as *I have a question*), not informative (such as

```
<vxml version="2.0">
<form id= "get_pizza_topping">
<field name= "pizza_topping">
<prompt> What kind of topping would you like to have? </prompt>
<grammar> olive | pepperoni | mushroom </grammar>
<noinput> Please say the topping you wish to have. </noinput>
<nomatch> I didn't understand that. </nomatch>
<filled> Thank you, I added <value expr= "pizza_topping" />
<submit next="next_document.vxml" /> </filled>
</field>
</form>
</vxml>
```

FIGURE 41.3 A VXML example for getting pizza topping information. (From the VXML lecture by François Mairesse.)

HMIHY: How may I help you?
User: Hi, I have a question about my bill (*Billing*)
HMIHY: OK, what is your question?
User: May I talk to a human please? (*CSRepresentative*)
HMIHY: In order to route your call to the most appropriate department can you tell me the specific reason you are calling about?
User: There is an international call I could not recognize (*Unrecognized_Number*)
HMIHY: OK, I am forwarding you to the human agent. Please stay on the line.

FIGURE 41.4 A conceptual example dialog between the user and the AT&T HMIHY system.

Good afternoon), or asks for a human agent, the system reprompts the user accordingly to get some specific intent for their call (Figure 41.4).

Figure 41.2 shows a typical HMIHY dialog for AT&T call centers. The utterance level intents are shown in parentheses next to each user utterance, for a better understanding.

Following the HMIHY system, a number of similar systems have been built, such as the Lucent call routing system [6], the BBN call director [33], AT&T Helpdesk [10], AT&T VoiceTone [15], and the France Telecom 3000 [9] systems.

The biggest difference between the call classification systems and semantic frame filling systems is that the former does not care much about the arguments provided by the user. The main goal is *routing* the call to the appropriate call center department. The arguments provided by the user are important only in the sense that they help make the right classification. For example, if the user specifies the name of a drug in a system used for medical insurance call center, this can improve the confidence for routing to prescription renewal or refill departments. While this is a totally different perspective for the task of SLU, it is actually complementary to frame filling. For example, there are utterances in the ATIS corpus asking about ground transportation or the capacity of planes on a specific flight; hence, the users may have other intents than basically finding flight information.

41.3 Underlying Principles

In this section, we mainly cover the key tasks of SLU as used in human/machine conversational systems. These include utterance classification tasks for domain detection or intent determination and semantic template filling. We will also briefly cover the aspects of a DM interacting with them.

Then, we briefly cover other SLU tasks used in other human/machine applications, like question answering (like IBM Watson) and in human/human speech processing systems like dialog act tagging or summarization.

41.3.1 Domain Detection and Intent Determination

The semantic utterance classification tasks of domain detection and intent determination aim at classifying a given speech utterance X_r into one of M semantic classes, $\hat{C}_r \in C = \{C_1, \ldots, C_M\}$ (where r is the utterance index). Upon the observation of X_r, $\hat{C}_r$ is chosen so that the class-posterior probability given X_r, $P(C_r|X_r)$, is maximized. Formally,

$$\hat{C}_r = \arg\max_{C_r} P(C_r|X_r) \tag{41.1}$$

Semantic classifiers require operation with significant freedom in utterance variations. A user may say *"I want to fly from Boston to New York next week"* and another user may express the same information by saying *"I am looking to fly from JFK to Boston in the coming week."* In spite of this freedom of expression, utterances in such applications have a clear structure that binds the specific pieces of information together. Not only is there no *a priori* constraint on what the user can say, but the system should be able to generalize well from a tractably small amount of training data. For example, the phrase "*Show all flights*" and "*Give me flights*" should be interpreted as variants of a single semantic class "Flight." On the other hand, the command "*Show me fares*" should be interpreted as an instance of another semantic class, "Fare." Typically, binary or weighted n-gram features with n = 1,2,3 to capture the likelihood of the n-grams are generated to express the user intent for the semantic class C. Traditional text categorization techniques devise learning methods to maximize the probability of C_r given the text W_r, that is the *class-posterior* probability $P(C_r|W_r)$. Other semantically motivated features like domain gazetteers (lists of entities), named entities (like organization names or time/date expressions), and contextual features (such as the previous dialog turn) can be used to enrich the feature set.

HMIHY SLU intent classifier heavily depends on portions of input utterances, namely, *salient phrases*, which are salient to some call types. For example, in an input utterance such as "*I would like to change oh about long distance service two one charge nine cents a minute*," the salient phrase "*cents a minute*" is strongly related to the call-type *calling plans*.

The salient phrases are automatically acquired from a corpus of transcribed and labeled training data. The salience of a word n-gram, f, is computed using the formula, that is

$$S(f) = \sum_{c_k} P(c_k|f)\, I(f, c_k)$$

where c_k is a call type and $I(x, y)$ is the mutual information between x and y:

$$I(x, y) = log\frac{P(x, y)}{P(x)}$$

Note that salience fragments are call-type independent. The phrases such as *I would like* still get high salience scores as these are used to improve the language model used for speech recognition.

The selected salient phrases are then clustered into salient grammar fragments (SGFs) [14,51]. The rationale for this clustering is better generalization. Assume two salient phrases *a wrong number* and *the wrong number*; they may be grouped together if they indicate the same call type. Once the SGFs are trained, it is then straightforward to classify an utterance. The SGFs are first compiled in the format of FSMs. Figure 41.5 shows an example simplified SGF. The ASR output (lattice or word confusion network) or input text, s, is then searched for the occurrences of the SGFs, using FSM operations. This is a left-to-right search that searches for the longest matching phrases. This results in a transduction from s to salient phrases, $f_i \in F$. The classification algorithm then chooses the call type with maximum peak value of the a posteriori probability distribution, that is,

$$K(s) = \arg\max_{c_k} P(c_k|f_{i(s)})$$

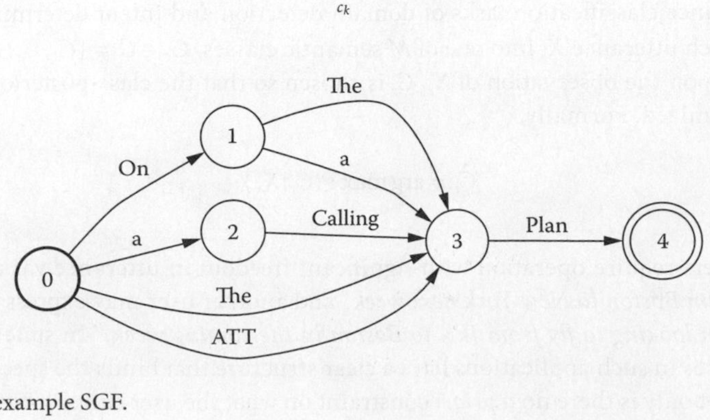

FIGURE 41.5 An example SGF.

where

$$i(s) = \arg\max_{i} \left(\max_{c_k} P(c_k \mid f_i) \right)$$

If this overall peak is less than some threshold, $P(c_k|f_{i(s)}) < \theta$, then the utterance is rejected and classified as *other.*

One of the earliest yet very successful approaches for call classification, adopted by the Bell Labs, is motivated by the information retrieval (IR)-based vector state model [6]. In this approach, first stop-words are removed; stemming is performed on the remaining words, following the usual practice in IR. Then a $F \times K(s)$ term–document matrix is created, where a term can be a word or a word phrase and documents are actually the classes or routes the call is routed to. Each term is weighted according to the well known *tf.idf* (term frequency–inverse document frequency) paradigm. New user requests are then converted and formed as a vector in a similar fashion. The baseline algorithm simply chooses the class where the cosine distance between the user utterance and the route is the minimum as in traditional IR. The original system paper suggests two improvements on top of this baseline system. First, the similarity scores are mapped to between 0 and 1 by fitting a nondecreasing sigmoid function using the least-squares method so as to get confidence scores. The second improvement was employing singular value decomposition (SVD) to improve efficiency especially when the training corpus is not small. In SVD, the routing matrix, W, is decomposed so as to $W = USV^T$, where U and V are orthonormal matrices and S is a diagonal matrix of eigenvalues.

Later [7] have proposed extending this approach via latent semantic analysis (LSA). LSA is also based on using SVD to find a compact subspace for the routing matrix. In LSA, after SVD, top R eigenvalues are retained and others are discarded. Then, all the processing is done in the reduced subspace.

With the advances in machine learning, especially in discriminative classification techniques, in the last decade, researchers have worked on framing intent determination as a classification task. While the traditional solution to intent determination is the bag-of-words approach used in IR, typically word n-grams are used as features after preprocessing with generic entities, such as dates, locations, or phone numbers. Because of the very large dimensions of the input space, large margin classifiers such as SVMs [44] or Adaboost [37] were found to be very good candidates.

Early work has been done by [37] on the AT&T HMIHY system using the Boostexter tool, an implementation of the AdaBoost.MH multiclass multilabel classification algorithm. Boosting is an iterative algorithm; on each iteration, t, a weak classifier, h_t, is trained on a weighted training set, and at the end, the weak classifiers are combined into a single, combined classifier [12]. For example, for text categorization, one can use word n-grams as features, and each weak classifier (e.g., decision stump, which is a single node decision tree) can check the absence or presence of an n-gram. The performance of the classifier closely matched with the salience-based classification approach and beat it especially on high false alarm rates.

Another pioneering work was from the Bell Labs. [28] have proposed the use of discriminative training on the routing matrix, which significantly improved their vector-based call routing system described earlier [6], especially for low rejection rates. Their approach is based on using the minimum classification error (MCE) criterion. To solve the nonlinear optimization problem of achieving MCE, they used the generalized probabilistic descent (GPD) algorithm. The goal is adjusting the routing matrix elements so as to achieve maximum separation of the correct class from competing classes with an iterative approach. Later, they extended this approach so as to cover boosting and automatic relevance feedback (ARF) [56]. They have tested these approaches along with MCE, and proposed a novel combination method of boosting and MCE. While MCE happens to provide the best result among these three discriminative classification approaches, they obtain further gain from the combination of MCE with boosting.

The next category of discriminative call classification systems employed large margin classifiers, such as support vector machines, which were initially demonstrated on binary classification problems. With such classifiers, the definition of the margin is typically unambiguous, and the reasons why this margin leads to good generalization are reasonably well understood. Usually, the performance of SVMs can be reduced to a single number: the binary classification accuracy. Unfortunately, to generalize these binary SVM classifiers to multiclass call classification problems is not straightforward. [16] have proposed a global optimization process based on an optimal channel communication model that allows a combination of *heterogeneous* binary classifiers. As in Markov modeling, computational feasibility is achieved through simplifications and independence assumptions that are easy to interpret. Using this approach, they have managed to decrease the call-type classification error rate for AT&T's HMIHY natural dialog system significantly, especially on the false rejection rates.

41.3.2 Slot Filling

The semantic structure of an application domain is defined in terms of the *semantic frames*. Figure 41.6 shows a simplified example of three semantic frames in the ATIS domain. The semantic frame contains several typed components called "*slots*."

The task of slot filling is then to instantiate the semantic frames. Check Figure 41.7 for a semantic parse for the sentence "Show me flights from Seattle to Boston on Christmas Eve." In this case, the semantic frame is represented as a flat list of *attribute–value pairs*, similar to [35], or *flat concept* representation.

Some SLU systems have adopted a hierarchical representation, as that is more expressive and allows the sharing of substructures. This is mainly motivated by syntactic constituency trees. An example tree for the same sentence is shown in Figure 41.8.

```
<frame name="Flight">
  <slot name="DCity" surface="Departure City">
  <slot name="ACity" surface="Arrival City">
  <slot name="DDate" surface="Departure Date">
  <slot name="Airline" surface="Airline">
  <slot name="NumStops" surface="Number of Stops">
  ...
</frame>
```

FIGURE 41.6 A simplified semantic class schema in the ATIS domain.

"Show me flights from <DCity> Seattle <DCity> to <ACity> Boston <ACity> on <DDate> Christmas eve <DDate>"

FIGURE 41.7 An example flat semantic parse.

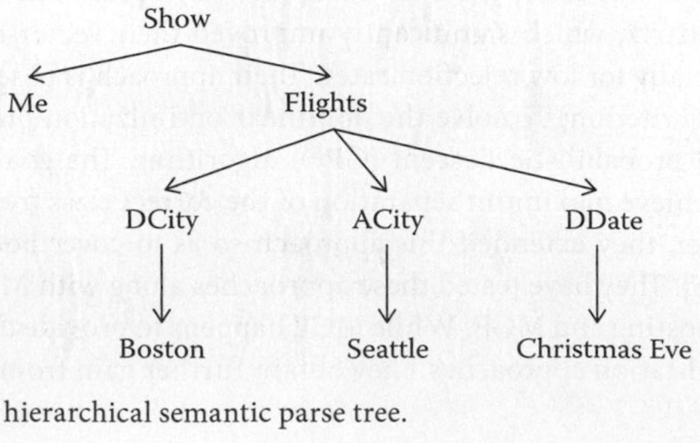

FIGURE 41.8 An example hierarchical semantic parse tree.

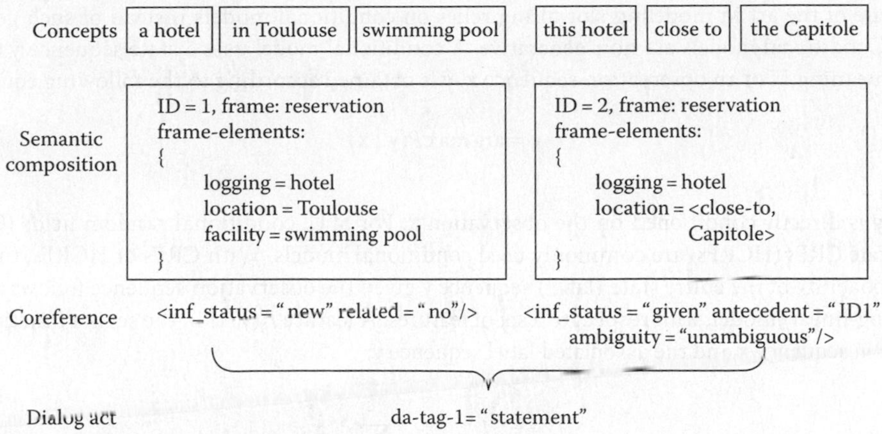

FIGURE 41.9 The multilevel semantic representation for the utterance "a hotel in Toulouse with a swimming pool um this hotel must be close to the Capitole" in the European LUNA project. (Reproduced from Bechet, F., Processing spontaneous speech in SLU systems: A survey, In *Proceedings of the IEEE Spoken Language Technologies (SLT) Workshop*, Goa, India, 2008.)

The semantic representation in the early French MEDIA project adopts an attribute–value list to represent the hierarchical semantic information [4]. This representation, used in the official annotation and evaluation, is ostensibly quite different from the frame-based representation.

The attempt of using this flat attribute–value representation for hierarchical semantics makes the meaning representation quite complicated. In the European LUNA project, a different representation was adopted, in which the MEDIA portion of data was re-annotated according to the new annotation protocol [3]. The semantic annotation protocol adopts a multilevel semantic representation, as illustrated by Figure 41.9. Here the concept level represents the concepts that bear important meanings in the input. The semantic composition level represents the utterance level meaning with semantic frames. A FrameNet-based semantic representation is adopted as an attempt for better generalization beyond a single application domain. The co-reference level annotation depicts the discourse information between the two related frames, while the fourth level models the dialog act (e.g., statement, Y/N question, Wh-question, confirmation).

In the statistical frame-based SLU, the task is often formalized as a pattern recognition problem. Given the word sequence W, the goal of SLU is to find the semantic representation of the meaning M that has the maximum a posteriori probability $P(M \mid W)$. In the generative model framework, the following decision rule is used:

$$\hat{M} = \arg\max_{M} P(M|W) = \arg\max_{M} P(W|M)P(M) \tag{41.2}$$

And the objective function of a generative model is to maximize the joint probability $P(W, M) = P(W \mid M)P(M)$ given a training sample of W and its semantic annotation M.

The first generative model, used by both CHRONUS [35] and the Hidden Understanding Model [32], assumes a deterministic one-to-one correspondence between model states and the segments, that is, there is only one segment per state, and the order of the segments follows that of the states.

As another extension, in the hidden vector state model, the states in the Markov chain represent encode all the structure information about the tree using stacks, so the semantic tree structure (excluding words) can be reconstructed from the hidden vector state sequence. The model imposes a hard limit on the maximum depth of the stack, so the number of the states becomes finite, and the prior model becomes the Markov chain in an HMM [19].

The state of the art in modeling slot filling relies on conditional models instead of such generative models. Conditional models are non-generative. A conditional model uses a state sequence $\hat{\mathbf{y}}$ to represent the meaning M of an observation sequence $\mathbf{x}$. $\hat{\mathbf{y}}$ is obtained according to the following equation:

$$\hat{\mathbf{y}} = \arg\max_{\mathbf{y}} P(\mathbf{y} \mid \mathbf{x}) \tag{41.3}$$

Here, $\mathbf{y}$ is directly conditioned on the observation $\mathbf{x}$. For SLU, conditional random fields (CRFs) or hidden state CRFs (HCRFs) are commonly used conditional models. With CRFs or HCRFs, the conditional probability of the entire state (label) sequence $\mathbf{y}$ given the observation sequence follows an exponential (log-linear) model, with respect to a set of features. A feature $f_k(\mathbf{y}, \mathbf{x})$ in the set is a function of the observation sequence $\mathbf{x}$ and the associated label sequence $\mathbf{y}$:

$$P(\mathbf{y} \mid \mathbf{x}; \Lambda) = \frac{1}{Z(\mathbf{x}; \Lambda)} \exp\left\{ \sum_k \lambda_k f_k(\mathbf{y}, \mathbf{x}) \right\} \tag{41.4}$$

Here $\Lambda = \{\lambda_k\}$ is a set of parameters. The value of λ_k determines the impact of the feature $f_k(\mathbf{y}, \mathbf{x})$ on the conditional probability. $Z(\mathbf{x} : \Lambda) = \sum_{\mathbf{y}} \exp\{\sum_k \lambda_k f_k(\mathbf{y}, \mathbf{x})\}$ is a partition function that normalizes the distribution.

The CRF in Equation 41.4 is unconstrained in the sense that the feature functions are defined on the entire label sequence $\mathbf{y}$. Because the number of all possible label sequences is combinatorial, the model training and inference of an unconstrained CRF is very inefficient. Because of that, it is common to restrict attention to the linear-chain CRFs [29]. The linear-chain CRFs impose a Markov constraint on the model topology and, as a consequence, restrict the feature functions to depend only on the labels assigned to the current and the immediately previous states, in the form $f(y_{t-1}, y_t, \mathbf{x}, t)$. The restrictions enable the application of efficient dynamic programming algorithms in model training and inference yet still support potentially interdependent features defined on the entire observation sequence $\mathbf{x}$.

Figure 41.10 shows the difference between the HMMs and the linear-chain CRFs in graphic model representations. First, since HMMs are generative models, the graph is directed, which points to the direction of the generative process. The state at time t, represented by the random variable $\mathbf{y}_t$, depends on the state at time $t-1$, $\mathbf{y}_{t-1}$. $\mathbf{x}_t$, and the observation at time t is generated with the dependency on $\mathbf{y}_t$. CRFs, on the other hand, are represented by undirected graphs since they are direct, non-generative models. Second, observations in HMMs are generated one at a time in a uniformed way, while in CRFs the observation sequence is given, so at each slice of time it can employ features that depend on the entire observation sequence.

The parameters Λ in CRFs can be optimized according to the objective function with numeric algorithms like stochastic gradient decent or L-BFGS. In a straightforward application of CRFs for the frame-based SLU, $\mathbf{x}$ consists of a transcribed utterance, and $\mathbf{y}$ is the semantic label sequence assigned to $\mathbf{x}$. Non-slot filler words are assigned a null state, as illustrated by Figure 41.11.

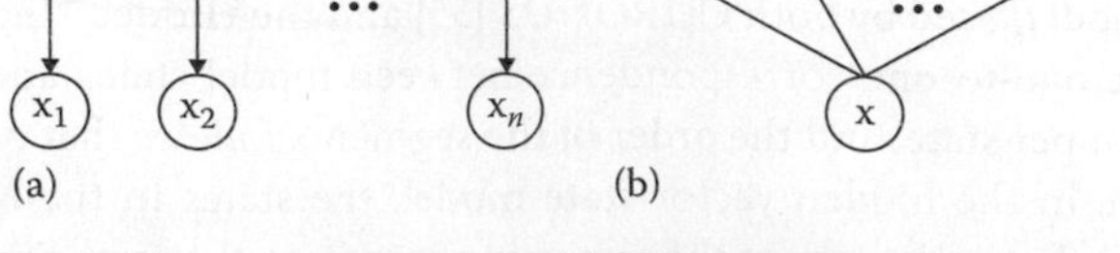

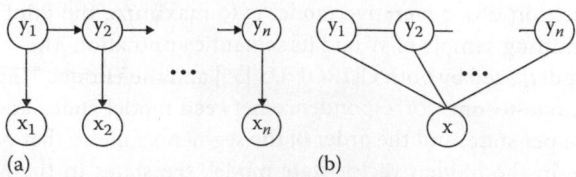

FIGURE 41.10 The graphic model representations for (a) HMMs (generative model) and (b) CRFs (conditional model). A generative model is a directed graph that points to the direction of the generative process, and observations are generated frame by frame in a uniformed way. Conditional models are represented by undirected models, where the entire observation sequence **o** is observable for every state in the model.

list	twa	flights	from	washington	to	philadelphia
|	|	|	|	|	|	|
null	*AirlineCode*	*null*	*null*	*ACity*	*null*	*DCity*

FIGURE 41.11 Frame-based SLU as a sequential labeling problem: words are assigned labels representing their meaning.

Typical features used by the CRFs for SLU include the transition features, the *n*-gram features, and the class membership (aka word list) features. The transition features model the dependency among adjacent states, which capture the influence of context on the tag assigned to a given word. The *n*-gram features often include the lexical features to capture the relation between the current state and the identity of the current (and previous) words. The class member features check if the current word is part of an entry in the lexicon of a syntactic or semantic class.

As an attempt to joint modeling slot filling and intent determination, triangular CRF models have been proposed [24]. These models jointly represent the sequence and meta-sequence labels in a single graphical structure that both explicitly encodes their dependencies and preserves uncertainty between them. While such models jointly optimize both tasks, it requires the training data to be the same, which may be a limiting factor in some applications.

The recent work on semantic parsing heavily relies on mining big data found on the web, web queries, and implicit user feedback, instead of clean transcribed and annotated data for each domain. For example, [18] employed queries mined from search engine query logs to improve domain detection in SLU. Extending the label propagation algorithm, a graph-based semi supervised learning approach is proposed to incorporate noisy domain information estimated from search engine that links the users click following their queries. Furthermore, the semantic web community is building and populating big-scale knowledge bases, with structured and unstructured meta information. By aligning the slots with the nodes and the intents with the edges in the knowledge graph, it is possible to directly leverage the structure of the semantic graphs to automatically specify a semantic representation for a given domain [20,42]. The corresponding natural language surface forms are mined from search queries or web pages and used to automatically train SLU systems in an unsupervised manner that cover the knowledge represented in the semantic graph.

41.3.3 Semantic Parsing for Dialog Management

As shown in Figure 41.1, the output of SLU is fed into the DM. Counterintuitively, the brain of a human/machine conversational system is not the semantic parsing but instead the DM. However, the DM heavily relies on the SLU output given the context (like the previous utterance, dialog context, meta-information about the user and the location, and any other knowledge integrated into the system).

Interestingly, the field of dialog management has emerged somewhat independently from the SLU area. In most cases, the assumption has been a machine-directed dialog and the models are usually trained with simulated data [43].

The key task in a DM is deciding on the next action (A_i) given the SLU output (U_i). The first step for doing this is updating the so-called *belief state*. A belief state, $s \in S$, is an instance of a semantic representation in that context. Usually, the same semantic representation is used as SLU. In most advanced dialog systems, instead of relying on a single belief state, a probability distribution over all possible states is maintained.

The earliest statistical approach relies on a Markov decision process (MDP) [31], focusing on optimizing the dialog strategies. Later this approach is extended to support uncertainty resulting in partially observable MDPs (POMDPs) [54]. Recent efforts focus on using state clustering or lighter-weight discriminative approaches to overcome the problem of POMDPs.

Aside from research labs, statistical dialog management techniques have not been adopted due to their heavy processing cost and complexity. Most commercial systems (as used in call centers, robots, or mobile devices) rely on rule-based DMs for both belief state update and policy determination.

41.3.4 Spoken Language Understanding in Human/Human Conversations

While the term "SLU" is coined for human/machine conversations, it is not hard to think of other interactive understanding tasks, such as spoken question answering, voice search, or other similar human/human conversation understanding tasks such as named entity extraction or topic classification. Each of these tasks are studied extensively, and the progress is fascinating. An example would be the IBM Watson system for question answering [11].

With respect to human/human communication processing, telephone conversations or multiparty meetings are studied heavily. Some open questions include the following. What information from these interactions would be useful for later use? Does this information depend on user/purpose? How could the transcripts of these conversations be used? The reviewed research activities are grouped in the following areas:

- *Dialog act tagging*: Aims to annotate speech acts such as suggestions and questions jointly with conversational linguistic acts such as acknowledgment and agreement. This task is heavily influenced by annotated corpora from human/human and multiparty meetings such as Switchboard and ICSI/AMI data sets. It is mainly treated as an enabling technology for further conversation processing, such as extracting action items or discussions. Dialog act tagging is generally framed as an utterance classification problem [40,41, among others], following the dialog act segmentation. While mostly lexical and contextual features are used, in some cases, such as agreement and disagreement, prosodic features have been shown to help [52].
- *Dialog act segmentation*: Aims to chop spoken utterances into dialog act units as defined by the dialog act tagging schema. While this task is very close to sentence segmentation, there are certain nuances due to the nature of spontaneous conversational speech. Dialog act segmentation is treated as a binary boundary classification problem using lexical, prosodic, and acoustic features [26,39, among others]. There are also few studies performing joint dialog act segmentation and tagging [48,55].
- *Discourse and topic segmentation*: Aims to chop a conversation into topically coherent units. While topic segmentation of text or prewritten speech is a well-established area, there are relatively fewer studies on processing human/human conversations and multiparty meetings.
- *Summarization*: Aims to generate a compact, summary version of meeting discussions. The summaries can be formed by extracting original speaker utterances (extractive summarization) or by formulating new sentences for the summary (abstractive summarization).
- *Action item and decision detection*: Aims to detect task assignments to people and associated deadlines and decision-making subdialogs during a meeting. These can be used to enter such information into the person's calendar or to track the status and progress of these action items or decisions in the following meetings. The decisions made in meetings can be used for indexing meetings, and one can go back and access the content of the meeting where a specific decision was made. They can also be used to track the progress and efficiency of meetings.
- *Speaker role detection*: Aims to classify each of the speakers with respect to their institutional roles. While this is an area deeply rooted in the social sciences, most systems have taken a simplistic view and instead focused on professional roles, such as professor vs. student, boss vs. employee, or project manager vs. software engineer.
- *Addressee detection*: An important enabling task in processing conversations is determining who is talking or who the speaker is referring to. This is useful in downstream understanding of conversations in that it provides essential semantic grounding for the analyzed dialog. This task also covers resolving references (especially pronominal references) in utterances.

41.4 Research Issues

The field of SLU is closely related to NLU, a field that has been studied for more than half a century. NLU focus mainly on the understanding of general domain written texts. Because there is not a specific application domain for the general-purpose NLU, the semantics in NLU have to be defined in a broader sense, such as thematic roles (agents, patients, etc.) In contrast, the SLU has, in the current state of technology, focused only on specific application domains. The semantics are defined very specifically according to the application domain, as illustrated by the earlier examples of semantic frames. Many domain-specific constraints can be included in the understanding model. Ostensibly, this may make the problem easier to solve. Unfortunately, there are many new challenges for SLU, including the following:

- *Extra-grammaticality*: Spoken languages are not as well formed as written languages. People are in general less careful with speech than with writings. They often do not comply with rigid syntactic constraints.
- *Disfluencies*: False starts, repairs, and hesitations are pervasive, especially in conversational speech.
- *Speech recognition errors*: Speech recognition technology is far from perfect. Environment noises, speaker's accent, domain-specific terminologies, all make speech recognition errors inevitable. It is common to see that a generic speech recognizer has over 30% word error rates on domain-specific data.
- *Out-of-domain utterances*: A dialog system can never restrict a user from saying anything out of a specific domain, even in a system-initiated dialog, where users are prompted for answers to specific questions. Because the frame-based SLU focuses on a specific application domain, out-of-domain utterances are not well modeled and can often be confused as an in-domain utterance. Detecting the out-of-domain utterances is not an easy task—it is complicated by the extra-grammaticality, disfluencies, and ASR errors of the in-domain utterances. Furthermore, when there are other humans in the environment, it is nontrivial to separate human-addressed speech from machine-addressed.

In summary, robustness is one of the most important issues in SLU. A system should be able to gracefully handle the unexpected inputs. If an input string is not accepted by a grammar/model, it is still desirable to identify the well-formed concepts in the input that carry important information for a given domain. On the other hand, a robust solution tends to over generalize and introduce ambiguities, leading to the reduction in understanding accuracy. A major challenge to the SLU is thus to strike an optimal balance between the robustness and the constraints that prevent the over generalizations and reduce the ambiguities.

41.5 Summary

We have presented a survey of the state of the art in SLU, focusing on human/computer conversational systems. We mainly described two technical areas of SLU, namely, intent determination and slot filling in detail. The state-of-the-art approaches relying on data-driven methods are being heavily used in academic and industrial research labs, though it started to quickly propagate to commercial applications like personal assistants (e.g., Nuance Nina). This is in contrast to knowledge-based methods that dominate the commercial applications.

Regarding human/machine interactive systems, the candidate applications and environments are growing with tremendous speed. We would like to emphasize on two aspects:

- *Multimodal interaction*: Although there were earlier multimodal interactive systems, such as AT&T MATCH, Microsoft MiPad, or ESPRIT MASK [22,25,30, among others], the boom of smart phones resulted in a completely new array of multimodal applications. All the traditional

SLU and dialog management concepts need to be tailored to multimodal interaction from parsing to grounding to belief state optimization to response generation. Moreover, with the multimodal output, several different actions can be performed simultaneously. This seems to be an exciting area for the near future.

- *Personalization*: Interactive systems are now becoming highly personalized. For example, the mobile device can record vast amounts of private information about the user, ranging from contact lists, calendars, to history of geo-locations coupled with time and previous written or spoken communications. This user-specific metadata is waiting to be exploited for "customized" user interaction, instead of a "one size fits all" speech systems, such as voice search or call routing. Personalized mobile systems also enable longitudinal user studies, since users typically continue to use their mobile multimodal applications over a long period of time. One potential application would be using implicit supervision for user adaptation of models, exploiting user behavior patterns.

Key Term

360° Review: Performance review that includes feedback from superiors, peers, subordinates, and clients.

References

1. J. Allen. *Natural Language Understanding*, Chapter 8. Benjamin/Cummings, Menlo Park, CA, 1995.
2. J. F. Allen, B. W. Miller, E. K. Ringger, and T. Sikorski. A robust system for natural spoken dialogue. In *Proceedings of the Annual Meeting of the Association for Computational Linguistics (ACL)*, Santa Cruz, CA, pp. 62–70, Morristown, NJ, 1996.
3. F. Bechet. Processing spontaneous speech in SLU systems: A survey. In *Proceedings of the IEEE Spoken Language Technologies (SLT) Workshop*, Goa, India, 2008.
4. H. Bonneau-Maynard, S. Rosset, C. Ayache, A. Kuhn, and D. Mostefa. Semantic annotation of the French MEDIA dialog corpus. In *Proceedings of the International Conference on Spoken Language Processing (Interspeech)*, pp. 3457–3460, Lisbon, Portugal, September 2005.
5. N. Chomsky. *Aspects of the Theory of Syntax*. MIT Press, Cambridge, MA, 1965.
6. J. Chu-Carroll and B. Carpenter. Vector-based natural language call routing. *Computational Linguistics*, 25(3):361–388, 1999.
7. S. Cox and B. Shahshahani. A comparison of some different techniques for vector based call-routing. In *Proceedings of the European Conference on Speech Communication and Technology (EUROSPEECH)*, Aalborg, Denmark, September 2001.
8. D. A. Dahl, M. Bates, M. Brown, W. Fisher, K. Hunicke-Smith, D. Pallett, C. Pao, A. Rudnicky, and E. Shriberg. Expanding the scope of the ATIS task: The ATIS-3 corpus. In *Proceedings of the Human Language Technology Workshop*. Morgan Kaufmann, San Francisco, CA, 1994.
9. G. Damnati, F. Bechet, and R. de Mori. Spoken language understanding strategies on the France Telecom 3000 voice agency corpus. In *Proceedings of the International Conference on Acoustics, Speech and Signal Processing (ICASSP)*, Honolulu, HI, 2007.
10. G. Di Fabbrizio, D. Dutton, N. Gupta, B. Hollister, M. Rahim, G. Riccardi, R. Schapire, and J. Schroeter. AT&T help desk. In *Proceedings of the International Conference on Spoken Language Processing (ICSLP)*, Denver, CO, September 2002.
11. D. Ferrucci, E. Brown, J. Chu-Carroll, J. Fan, D. Gondek, A. A. Kalyanpur, A. Lally et al. Building watson: An overview of the DeepQA project. *AI Magazine*, 31(3):59–79, 2010.
12. Y. Freund and R. E. Schapire. A decision-theoretic generalization of on-line learning and an application to boosting. *Journal of Computer and System Sciences*, 55(1):119–139, 1997.

13. A. L. Gorin, A. Abella, T. Alonso, G. Riccardi, and J. H. Wright. Automated natural spoken dialog. *IEEE Computer Magazine*, 35(4):51–56, April 2002.
14. A. L. Gorin, G. Riccardi, and J. H. Wright. How may I help you?. *Speech Communication*, 23:113–127, 1997.
15. N. Gupta, G. Tur, D. Hakkani-Tür, S. Bangalore, G. Riccardi, and M. Rahim. The AT&T spoken language understanding system. *IEEE Transactions on Audio, Speech, and Language Processing*, 14(1):213–222, 2006.
16. P. Haffner, G. Tur, and J. Wright. Optimizing SVMs for complex call classification. In *Proceedings of the International Conference on Acoustics, Speech and Signal Processing (ICASSP)*, Hong Kong, China, April 2003.
17. S. Hahn, M. Dinarelli, C. Raymond, F. Lefevre, P. Lehnen, R. De Mori, A. Moschitti, H. Ney, and G. Riccardi. Comparing stochastic approaches to spoken language understanding in multiple languages. *IEEE Transactions on Audio, Speech, and Language Processing*, 19(6):1569–1583, 2011.
18. D. Hakkani-Tür, G. Tur, L. Heck, A. Celikyilmaz, A. Fidler, D. Hillard, R. Iyer, and S. Parthasarathy. Employing web search query click logs for multi-domain spoken language understanding. In *Proceedings of the IEEE ASRU*, Waikoloa, HI, 2011.
19. Y. He and S. Young. A data-driven spoken language understanding system. In *Proceedings of the IEEE Automatic Speech Recognition and Understanding (ASRU) Workshop*, pp. 583–588, U.S. Virgin Islands, December 2003.
20. L. Heck and D. Hakkani-Tür. Exploiting the semantic web for unsupervised spoken language understanding. In *In Proceedings of the IEEE SLT Workshop*, Miami, FL, December 2012.
21. C. T. Hemphill, J. J. Godfrey, and G. R. Doddington. The ATIS spoken language systems pilot corpus. In *Proceedings of the Workshop on Speech and Natural Language, HLT'90*, pp. 96–101, Morristown, NJ, 1990. Association for Computational Linguistics.
22. X. D. Huang, A. Acero, C. Chelba, L. Deng, J. Droppo, D. Duchene, J. Goodman et al. Mipad: A multimodal interaction prototype. In *Proceedings of the International Conference on Acoustics, Speech and Signal Processing (ICASSP)*, Salt Lake City, Utah, May 2001.
23. C. J. Fillmore J. B. Lowe, and C. F. Baker. A frame-semantic approach to semantic annotation. In *Proceedings of the Annual Meeting of the Association for Computational Linguistics (ACL)-SIGLEX Workshop*, Washington, DC, April 1997.
24. M. Jeong and G. G. Lee. Triangular-chain conditional random fields. *IEEE Transactions on Audio, Speech, and Language Processing*, 16(7):1287–1302, September 2008.
25. M. Johnston, S. Bangalore, G. Vasireddy, A. Stent, P. Ehlen, M. Walker, S. Whittaker, and P. Maloor. MATCH: An architecture for multimodal dialogue systems. In *Proceedings of the Annual Meeting of the Association for Computational Linguistics (ACL)*, Philadelphia, PA, July 2002.
26. J. Kolar, E. Shriberg, and Y. Liu. Using prosody for automatic sentence segmentation of multi-party meetings. In *Proceedings of the International Conference on Text, Speech, and Dialogue (TSD)*, Brno, Czech Republic, 2006.
27. R. Kuhn and R. De Mori. The application of semantic classification trees to natural language understanding. *IEEE Transactions on Pattern Analysis and Machine Intelligence*, 17:449–460, 1995.
28. H.-K. J. Kuo and C.-H. Lee. Discriminative training in natural language call-routing. In *Proceedings of International Conference on Spoken Language Processing (ICSLP)*, Beijing, China, 2000.
29. J. Lafferty, A. McCallum, and F. Pereira. Conditional random fields: Probabilistic models for segmenting and labeling sequence data. In *Proceedings of ICML*, pp. 282–289, San Francisco, CA, 2001.
30. L. Lamel, S. Bennacef, J. Gauvain, H. Dartigues, and J. Temem. User evaluation of the mask kiosk. In *Proceedings of the International Conference on Spoken Language Processing (ICSLP)*, Sydney, New South Wales, Australia, 1998.
31. E. Levin, R. Pieraccini, and W. Eckert. A stochastic model of human-machine interaction for learning dialog strategies. *IEEE Transactions on Speech and Audio Processing*, 8(1):11–23, 2000.

32. S. Miller, R. Bobrow, R. Ingria, and R. Schwartz. Hidden understanding models of natural language. In *Proceedings of the 31st Annual Meeting of the Association for Computational Linguistics*, New Mexico State University, Las Cruces, NM, 1994.
33. P. Natarajan, R. Prasad, B. Suhm, and D. McCarthy. Speech enabled natural language call routing: BBN call director. In *Proceedings of the International Conference on Spoken Language Processing (ICSLP)*, Denver, CO, September 2002.
34. R. Pieraccini, E. Tzoukermann, Z. Gorelov, J.-L. Gauvain, E. Levin, C.-H. Lee, and J. G. Wilpon. A speech understanding system based on statistical representation of semantics. In *Proceedings of the International Conference on Acoustics, Speech and Signal Processing (ICASSP)*, San Francisco, CA, March 1992.
35. R. Pieraccini and E. Levin. A learning approach to natural language understanding. In *1993 NATO ASI Summer School, New Advances and Trends in Speech Recognition and Coding*, Springer-Verlag, Bubion, Spain, 1993.
36. P. J. Price. Evaluation of spoken language systems: The ATIS domain. In *Proceedings of the DARPA Workshop on Speech and Natural Language*, Hidden Valley, PA, June 1990.
37. R. E. Schapire and Y. Singer. Boostexter: A boosting-based system for text categorization. *Machine Learning*, 39(2/3):135–168, 2000.
38. S. Seneff. TINA : A natural language system for spoken language applications. *Computational Linguistics*, 18(1):61–86, 1992.
39. E. Shriberg, A. Stolcke, D. Hakkani-Tür, and G. Tur. Prosody-based automatic segmentation of speech into sentences and topics. *Speech Communication*, 32(1–2):127–154, 2000.
40. A. Stolcke, K. Ries, N. Coccaro, E. Shriberg, R. Bates, D. Jurafsky, P. Taylor, R. Martin, C. van Ess-Dykema, and M. Meteer. Dialogue act modeling for automatic tagging and recognition of conversational speech. *Computational Linguistics*, 26(3):339–373, 2000.
41. G. Tur, U. Guz, and D. Hakkani-Tür. Model adaptation for dialog act tagging. In *Proceedings of the IEEE Spoken Language Technologies (SLT) Workshop*, Aruba, 2006.
42. G. Tur, M. Jeong, Y.-Y. Wang, D. Hakkani-Tür, and L. Heck. Exploiting the semantic web for unsupervised natural language semantic parsing. In *Proceedings of the Interspeech*, Portland, OR, September 2012.
43. G. Tur and R. De Mori, eds. *Spoken Language Understanding: Systems for Extracting Semantic Information from Speech*. John Wiley & Sons, New York, 2011.
44. V. N. Vapnik. *Statistical Learning Theory*. John Wiley & Sons, New York, 1998.
45. M. Walker, J. Aberdeen, J. Boland, E. Bratt, J. Garofolo, L. Hirschman, A. Le et al. DARPA Communicator dialog travel planning systems: The June 2000 data collection. In *Proceedings of the Eurospeech Conference*, Aalborg, Denmark, 2001.
46. M. A. Walker, A. Rudnicky, R. Prasad, J. Aberdeen, E. O. Bratt, J. Garofolo, H. Hastie et al. DARPA Communicator: Cross-system results for the 2001 evaluation. In *Proceedings of the International Conference on Spoken Language Processing (ICSLP)*, Denver, CO, pp. 269–272, 2002.
47. W. Ward and S. Issar. Recent improvements in the CMU spoken language understanding system. In *Proceedings of the ARPA Human Language Technology Conference (HLT) Workshop*, Plainsboro, NJ, pp. 213–216, March 1994.
48. V. Warnke, R. Kompe, H. Niemann, and E. Nöth. Integrated dialog act segmentation and classification using prosodic features and language models. In *Proceedings of the European Conference on Speech Communication and Technology (EUROSPEECH)*, Rhodes, Greece, September 1997.
49. J. Weizenbaum. Eliza–a computer program for the study of natural language communication between man and machine. *Communications of the ACM*, 9(1):36–45, 1966.
50. W. A. Woods. Language processing for speech understanding. In *Computer Speech Processing*. Prentice-Hall International, Englewood Cliffs, NJ, 1983.

51. J. Wright, A. Gorin, and G. Riccardi. Automatic acquisition of salient grammar fragments for call-type classification. In *Proceedings of the European Conference on Speech Communication and Technology (EUROSPEECH)*, Rhodes, Greece, September 1997.
52. F. Yang, G. Tur, and E. Shriberg. Exploiting dialog act tagging and prosodic information for action item identification. In *Proceedings of the International Conference on Acoustics, Speech and Signal Processing (ICASSP)*, Las Vegas, NV, 2008.
53. S. Young. Talking to machines (statistically speaking). In *Proceedings of the International Conference on Spoken Language Processing (ICSLP)*, Denver, CO, September 2002.
54. S. Young. Using POMDPs for dialog management. In *Proceedings of the IEEE Workshop on Spoken Language Technology*, Palm Beach, Aruba, December 2006.
55. M. Zimmermann, Y. Liu, E. Shriberg, and A. Stolcke. Toward joint segmentation and classification of dialog acts in multiparty meetings. In *Proceedings of the Workshop on Multimodal Interaction and Related Machine Learning Algorithms (MLMI)*, Edinburgh, U.K., July 2005.
56. I. Zitouni, H.-K. J. Kuo, and C.-H. Lee. Boosting and combination of classifiers for natural language call routing systems. *Speech Communication*, 41(4):647–661, 2003.

42

Neural Networks

Michael I. Jordan
University of California, Berkeley

Christopher M. Bishop
Microsoft Research

42.1 Introduction

Within the broad scope of the study of artificial intelligence (AI), research in neural networks is characterized by a particular focus on pattern recognition and pattern generation. Many neural network methods can be viewed as generalizations of classical pattern-oriented techniques in statistics and the engineering areas of signal processing, system identification, and control theory. As in these parent disciplines, the notion of "pattern" in neural network research is essentially probabilistic and numerical. Neural network methods have had their greatest impact in problems where statistical issues dominate and where data are easily obtained.

A neural network is first and foremost a graph, with patterns represented in terms of numerical values attached to the nodes of the graph and transformations between patterns achieved via simple message-passing algorithms. Many neural network architectures, however, are also statistical processors, characterized by making particular probabilistic assumptions about data. As we will see, this conjunction of graphical algorithms and probability theory is not unique to neural networks but characterizes a wider family of probabilistic systems in the form of chains, trees, and networks that are currently being studied throughout AI (Spiegelhalter et al. 1993).

Neural networks have found a wide range of applications, the majority of which are associated with problems in pattern recognition and control theory. In this context, neural networks can best be viewed as a class of algorithms for statistical modeling and prediction. Based on a source of *training data*, the aim is to produce a statistical model of the process from which the data are generated, so as to allow the best predictions to be made for new data. We shall find it convenient to distinguish three broad types of statistical modeling problems, which we shall call *density estimation, classification*, and *regression*.

For density estimation problems (also referred to as *unsupervised learning* problems), the goal is to model the unconditional distribution of data described by some vector $\boldsymbol{x}$. A practical example of the application of density estimation involves the interpretation of x-ray images (mammograms) used for breast cancer screening (Tarassenko 1995). In this case, the training vectors $\boldsymbol{x}$ form a sample taken from normal (noncancerous) images, and a network model is used to build a representation of the density $p(\boldsymbol{x})$. When a new input vector x' is presented to the system, a high value for $p(\boldsymbol{x}')$ indicates a normal image, whereas a low value indicates a novel input, which might be characteristic of an abnormality. This is used to label regions of images that are unusual, for further examination by an experienced clinician.

For classification and regression problems (often referred to as *supervised learning* problems), we need to distinguish between *input* variables, which we again denote by $\boldsymbol{x}$, and *target* variables, which we denote by the vector $\boldsymbol{t}$. Classification problems require that each input vector $\boldsymbol{x}$ be assigned to one of C classes $C_1,\dots, C_C$, in which case the target variables represent class labels. As an example, consider the problem of recognizing handwritten digits (LeCun et al. 1989). In this case, the input vector would be some (preprocessed) image of the digit, and the network would have 10 outputs, one for each digit, which can be used to assign input vectors to the appropriate class (as discussed in Section 42.2).

Regression problems involve estimating the values of continuous variables. For example, neural networks have been used as part of the control system for adaptive optics telescopes (Sandler et al. 1991). The network input $\boldsymbol{x}$ consists of one in-focus and one defocused image of a star, and the output $\boldsymbol{t}$ consists of a set of coefficients that describe the phase distortion due to atmospheric turbulence. These output values are then used to make real-time adjustments of the multiple mirror segments to cancel the atmospheric distortion. Classification and regression problems also can be viewed as special cases of density estimation. The most general and complete description of the data is given by the probability distribution function $p(\boldsymbol{x}, \boldsymbol{t})$ in the joint input–target space. However, the usual goal is to be able to make good predictions for the target variables when presented with new values of the inputs. In this case, it is convenient to decompose the joint distribution in the following form:

$$p(\boldsymbol{x},\boldsymbol{t}) = p(\boldsymbol{t} \mid \boldsymbol{x})p(\boldsymbol{x}) \tag{42.1}$$

and to consider only the conditional distribution $p(\boldsymbol{t}|\boldsymbol{x})$, in other words the distribution of $\boldsymbol{t}$ *given* the value of $\boldsymbol{x}$. Thus, classification and regression involve the estimation of *conditional* densities, a problem that has its own idiosyncracies.

The organization of the chapter is as follows. In Section 42.2, we present examples of network representations of unconditional and conditional densities. In Section 42.3, we discuss the problem of adjusting the parameters of these networks to fit them to data. This problem has a number of practical aspects, including the choice of optimization procedure and the method used to control network complexity. We then discuss a broader perspective on probabilistic network models in Section 42.4. The final section presents further information and pointers to the literature.

42.2 Representation

In this section, we describe a selection of neural network architectures that have been proposed as representations for unconditional and conditional densities. After a brief discussion of density estimation, we discuss classification and regression, beginning with simple models that illustrate the fundamental ideas and then progressing to more complex architectures. We focus here on representational issues, postponing the problem of learning from data until the following section.

42.2.1 Density Estimation

We begin with a brief discussion of density estimation, utilizing the Gaussian *mixture model* as an illustrative model. We return to more complex density estimation techniques later in the chapter.

Although density estimation can be the main goal of a learning system, as in the diagnosis example mentioned in the Introduction, density estimation models arise more often as components of the solution to a more general classification or regression problem. To return to Equation 42.1, note that the joint density is composed of $p(\boldsymbol{t}|\boldsymbol{x})$, to be handled by classification or regression models, and $p(\boldsymbol{x})$, the (unconditional) real-life datasets that often have missing components in the input vector. Having a model of the density allows the missing components to be filled in an intelligent way. This can be useful both for training and for prediction (cf. Bishop 1995). Second, as we see in Equation 42.1, a model of $p(\boldsymbol{x})$ makes possible an estimate of the joint probability $p(\boldsymbol{x}, \boldsymbol{t})$. This in turn provides us with the necessary information to estimate the inverse conditional density $p(\boldsymbol{x}|\boldsymbol{t})$. The calculation of such inverses is important for applications in control and optimization.

A general and flexible approach to density estimation is to treat the density as being composed of a set of M simpler densities. This approach involves modeling the observed data as a sample from a *mixture density*,

$$p(\boldsymbol{x} \mid \boldsymbol{w}) = \sum_{i=1}^{M} \pi_i p(\boldsymbol{x} \mid i, \boldsymbol{w}_i) \tag{42.2}$$

where

π_i is constants known as *mixing proportions*

$p(\boldsymbol{x}|i, \boldsymbol{w}_i)$ is the *component densities*, generally taken to be from a simple parametric family

A common choice of component density is the multivariate Gaussian, in which case the parameters $\boldsymbol{w}_i$ are the means and covariance matrices of each of the components. By varying the means and covariances to place and orient the Gaussians appropriately, a wide variety of high-dimensional, multimodal data can be modeled. This approach to density estimation is essentially a probabilistic form of clustering.

Gaussian mixtures have a representation as a network diagram, as shown in Figure 42.1. The utility of such network representations will become clearer as we proceed; for now, it suffices to note that not only mixture models, but also a wide variety of other classical statistical models for density estimation, are representable as simple networks with one or more layers of adaptive weights. These methods include *principal component analysis, canonical correlation analysis, kernel density estimation*, and *factor analysis* (Anderson 1984).

42.2.2 Linear Regression and Linear Discriminants

Regression models and classification models both focus on the conditional density $p(\boldsymbol{t}|\boldsymbol{x})$. They differ in that in regression the target vector $\boldsymbol{t}$ is a real-valued vector, whereas in classification $\boldsymbol{t}$ takes its values from a discrete set representing the class labels.

The simplest probabilistic model for regression is one in which $\boldsymbol{t}$ is viewed as the sum of an underlying deterministic function $f(\boldsymbol{x})$ and a Gaussian random variable ε,

$$t = f(\boldsymbol{x}) + \varepsilon \tag{42.3}$$

where

ε is the *zero* mean, as is commonly assumed

$f(\boldsymbol{x})$ is the *conditional mean* $E(\boldsymbol{t}|\boldsymbol{x})$

It is this function that is the focus of most regression modeling. Of course, the conditional mean describes only the first moment of the conditional distribution, and as we discuss in a later section, a good regression model will also generally report information about the second moment.

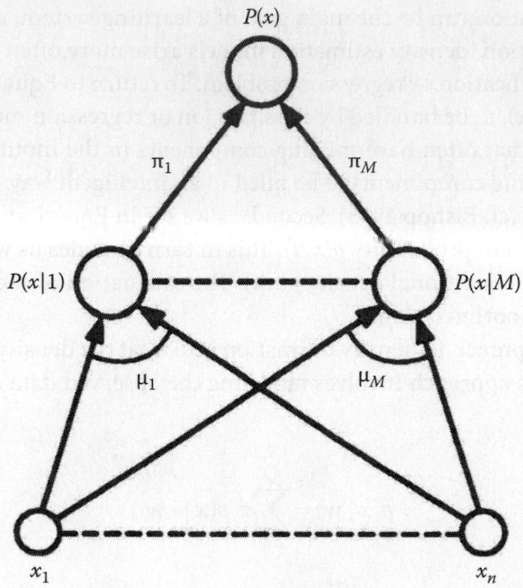

FIGURE 42.1 A network representation of a Gaussian mixture distribution. The input pattern $\boldsymbol{x}$ is represented by numerical values associated with the input nodes in the lower level. Each link has a weight μ_{ij}, which is the jth component of the mean vector for the ith Gaussian. The ith intermediate node contains the covariance matrix Σ_i and calculates the Gaussian conditional probability $p(\boldsymbol{x}|i, \mu_i, \Sigma_i)$. These probabilities are weighted by the mixing proportions π_i, and the output node calculates the weighted sum $p(\boldsymbol{x}) = \sum_i \pi_i p(\boldsymbol{x} \mid i, \mu_i, \Sigma_i)$.

In a linear regression model, the conditional mean is a linear function of $\boldsymbol{x}$: $E(t|\boldsymbol{x}) = W_x$, for a fixed matrix W. Linear regression has a straightforward representation as a network diagram in which the jth input unit represents the jth component of the input vector $\boldsymbol{x}_j$, each output unit i takes the weighted sum of the input values, and the weight w_{ij} is placed on the link between the jth input unit and the ith output unit.

The conditional mean is also an important function in classification problems, but most of the focus in classification is on a different function known as a *discriminant function*. To see how this function arises and to relate it to the conditional mean, we consider a simple two-class problem in which the target is a simple binary scalar that we now denote by t. The conditional mean $E(t|\boldsymbol{x})$ is equal to the probability that t equals one, and this latter probability can be expanded via Bayes rule:

$$p(t=1 \mid \boldsymbol{x}) = \frac{p(\boldsymbol{x} \mid t=1)p(t=1)}{p(\boldsymbol{x})} \tag{42.4}$$

The density $p(t|\boldsymbol{x})$ in this equation is referred to as the *posterior probability* of the class given the input, and the density $p(\boldsymbol{x}|t)$ is referred to as the *class-conditional density*. Continuing the derivation, we expand the denominator and (with some foresight) introduce an exponential:

$$p(t=1 \mid \boldsymbol{x}) = \frac{p(\boldsymbol{x} \mid t=1)p(t=1)}{p(\boldsymbol{x} \mid t=1)p(t=1) + p(\boldsymbol{x} \mid t=0)p(t=0)}$$

$$= \frac{1}{1+\exp\left\{-\ln\left[\frac{p(\boldsymbol{x} \mid t=1)}{p(\boldsymbol{x} \mid t=0)}\right] - \ln\left[\frac{p(t=1)}{p(t=0)}\right]\right\}} \tag{42.5}$$

We see that the posterior probability can be written in the form of the *logistic function*:

$$y = \frac{1}{1+e^{-z}} \tag{42.6}$$

where z is a function of the likelihood ratio $p(x|t=1)/p(x|t=0)$ and the prior ratio $p(t=1)/p(t=0)$. This is a useful representation of the posterior probability if z turns out to be simple.

It is easily verified that if the class-conditional densities are multivariate Gaussians with identical covariance matrices, then z is a linear function of $\boldsymbol{x}$: $z = \boldsymbol{w}^T\boldsymbol{x} + w_0$. Moreover, this representation is appropriate for any distribution in a broad class of densities known as the exponential family (which includes the Gaussian, the Poisson, the gamma, the binomial, and many other densities). All of the densities in this family can be put in the following form:

$$g(\boldsymbol{x};\theta,\phi) = \exp\left\{\frac{(\theta^T\boldsymbol{x} - b(\theta))}{a(\phi) + c(\boldsymbol{x},\phi)}\right\} \tag{42.7}$$

where

θ is the *location parameter*

ϕ is the *scale parameter*

Substituting this general form in Equation 42.5, where θ is allowed to vary between the classes and ϕ is assumed to be constant between classes, we see that z is in all cases a linear function. Thus, the choice of a linear–logistic model is rather robust.

The geometry of the two-class problem is shown in Figure 42.2, which shows Gaussian class-conditional densities and suggests the logistic form of the posterior probability.

The function z in our analysis is an example of a discriminant function. In general, a discriminant function is any function that can be used to decide on class membership (Duda and Hart 1973); our analysis has produced a particular form of discriminant function that is an intermediate step in the

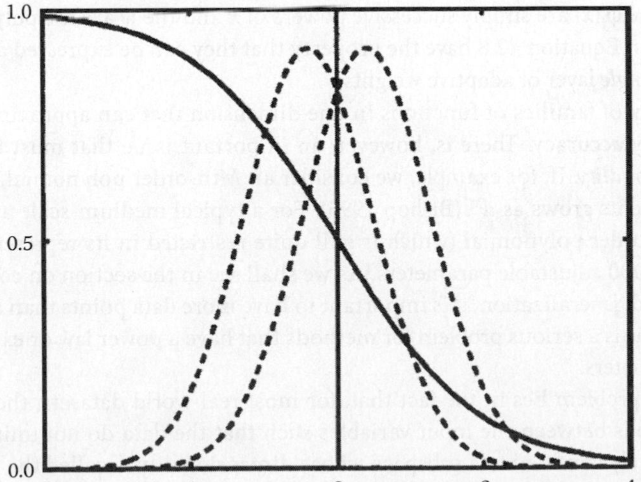

FIGURE 42.2 This shows the Gaussian class-conditional densities $p(x|C_1)$ (dashed curves) for a two-class problem in one dimension, together with the corresponding posterior probability $p(C_1|x)$ (solid curve), which takes the form of a logistic sigmoid. The vertical line shows the decision boundary for $y = 0.5$, which coincides with the point at which the two density curves cross.

calculation of a posterior probability. Note that if we set $z = 0$, from the form of the logistic function, we obtain a probability of 0.5, which shows that $z = 0$ is a *decision boundary* between the two classes.

The discriminant function that we found for exponential family densities is linear under the given conditions on ϕ. In more general situations, in which the class-conditional densities are more complex than a single exponential family density, the posterior probability will not be well characterized by the linear–logistic form. Nonetheless, it still is useful to retain the logistic function and focus on *nonlinear* representations for the function z. This is the approach taken within the neural network field.

To summarize, we have identified two functions that are important for regression and classification, respectively: the conditional mean and the discriminant function. These are the two functions that are of concern for simple linear models and, as we now discuss, for more complex nonlinear models as well.

42.2.3 Nonlinear Regression and Nonlinear Classification

The linear regression and linear discriminant functions introduced in the previous section have the merit of simplicity but are severely restricted in their representational capabilities. A convenient way to see this is to consider the geometrical interpretation of these models. When viewed in the d-dimensional $\boldsymbol{x}$-space, the linear regression function $\boldsymbol{w}^T\boldsymbol{x} + w_0$ is constant on hyperplanes, which are orthogonal to the vector $\boldsymbol{w}$. For many practical applications, we need to consider much more general classes of function. We therefore seek representations for nonlinear mappings, which can approximate any given mapping to arbitrary accuracy. One way to achieve this is to transform the original $\boldsymbol{x}$ using a set of M nonlinear functions $\phi_j(\boldsymbol{x})$ where $j = 1,\ldots, M$ and then to form a linear combination of these functions, so that

$$y_k(\boldsymbol{x}) = \sum_j w_{kj}\phi_j(\boldsymbol{x}) \tag{42.8}$$

For a sufficiently large value of M, and for a suitable choice of the $\phi_j(\boldsymbol{x})$, such a model has the desired universal approximation properties. A familiar example, for the case of 1D input spaces, is the simple polynomial, for which the $\phi_j(\boldsymbol{x})$ are simply successive powers of $\boldsymbol{x}$ and the w are the polynomial coefficients. Models of the form in Equation 42.8 have the property that they can be expressed as network diagrams in which there is a *single* layer of adaptive weights.

There are a variety of families of functions in one dimension that can approximate any continuous function to arbitrary accuracy. There is, however, an important issue that must be addressed, called the *curse of dimensionality*. If, for example, we consider an Mth-order polynomial, then the number of independent coefficients grows as d^M (Bishop 1995). For a typical medium-scale application with, say, 30 inputs, a fourth-order polynomial (which is still quite restricted in its representational capability) would have over 46,000 adjustable parameters. As we shall see in the section on complexity control, in order to achieve good generalization, it is important to have more data points than adaptive parameters in the model, and this is a serious problem for methods that have a power law or exponential growth in the number of parameters.

A solution to the problem lies in the fact that, for most real-world datasets, there are strong (often nonlinear) correlations between the input variables such that the data do not uniformly fill the input space but are effectively confined to a subspace whose dimensionality is called the *intrinsic dimensionality* of the data. We can take advantage of this phenomenon by considering again a model of the form in Equation 42.8 but in which the basis functions $\phi_j(\boldsymbol{x})$ are *adaptive* so that they themselves contain weight parameters whose values can be adjusted in the light of the observed dataset. Different models result from different choices for the basis functions, and here we consider the two most common examples. The first of these is called the *multilayer perceptron* (MLP) and is obtained by choosing the basis

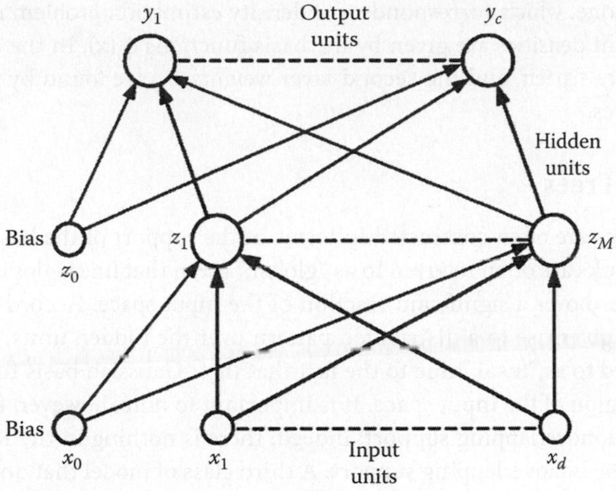

FIGURE 42.3 An example of a feedforward network having two layers of adaptive weights. The bias parameters in the first layer are shown as weights from an extra input having a fixed value of $x_0 = 1$. Similarly, the bias parameters in the second layer are shown as weights from an extra hidden unit, with activation again fixed at $z_0 = 1$.

functions to be given by linear–logistic functions (Equation 42.6). This leads to a multivariate nonlinear function that can be expressed in the form

$$y_k(\boldsymbol{x}) = \sum_{j=1}^{M} w_{kj} g\left(\sum_{i=1}^{d} w_{ji} x_i + w_{j0} \right) + w_{k0} \tag{42.9}$$

Here, w_{j0} and w_{k0} are *bias* parameters, and the basis functions are called *hidden units*. The function $g(\cdot)$ is the logistic sigmoid function of Equation 42.6. This also can be represented as a network diagram as in Figure 42.3. Such a model is able to take account of the intrinsic dimensionality of the data because the first-layer weights w_{ji} can adapt and hence orient the surfaces along which the basis function response is constant. It has been demonstrated that models of this form can approximate to arbitrary accuracy any continuous function, defined on a compact domain, provided the number M of hidden units is sufficiently large. The MLP model can be extended by considering several successive layers of weights. Note that the use of nonlinear activation functions is crucial, because if $g(\cdot)$ in Equation 42.9 was replaced by the identity, the network would reduce to several successive linear transformations, which would itself be linear.

The second common network model is obtained by choosing the basis functions $\phi_j(\boldsymbol{x})$ in Equation 42.8 to be functions of the radial variable $\boldsymbol{x} - \mu_j$ where μ_j is the *center* of the jth basis function, which gives rise to the *radial basis function (RBF) network* model. The most common example uses Gaussians of the following form:

$$\phi_j(\boldsymbol{x}) = \exp\left\{ -\frac{1}{2}(x - \mu_j)^T \sum_j^{-1} (\boldsymbol{x} - \mu_j) \right\} \tag{42.10}$$

Here, both the mean vector μ_j and the covariance matrix Σ_j are considered to be adaptive parameters. The curse of dimensionality is alleviated because the basis functions can be positioned and oriented in input space such as to overlay the regions of high data density and hence to capture the nonlinear correlations between input variables. Indeed, a common approach to training an RBF network is to use a two-stage procedure (Bishop 1995). In the first stage, the basis function parameters are determined using the input data alone, which corresponds to a density estimation problem using a mixture model in which the component densities are given by the basis functions $\phi_j(\boldsymbol{x})$. In the second stage, the basis function parameters are frozen, and the second-layer weights w_{kj} are found by standard least-squares optimization procedures.

42.2.4 Decision Trees

MLP and RBF networks are often contrasted in terms of the support of the basis functions that compose them. MLP networks are often referred to as "global," given that linear–logistic basis functions are bounded away from zero over a significant fraction of the input space. Accordingly, in an MLP, each input vector generally gives rise to a distributed pattern over the hidden units. RBF networks, on the other hand, are referred to as "local," due to the fact that their Gaussian basis functions typically have support over a local region of the input space. It is important to note, however, that local support does not necessarily mean nonoverlapping support; indeed, there is nothing in the RBF model that prefers basis functions that have nonoverlapping support. A third class of model that does focus on basis functions with nonoverlapping support is the *decision tree* model (Breiman et al. 1984). A decision tree is a regression or classification model that can be viewed as asking a sequence of questions about the input vector. Each question is implemented as a linear discriminant, and a sequence of questions can be viewed as a recursive partitioning of the input space. All inputs that arrive at a particular leaf of the tree define a polyhedral region in the input space. The collection of such regions can be viewed as a set of basis functions. Associated with each basis function is an output value, which (ideally) is close to the average value of the conditional mean (for regression) or discriminant function (for classification, a majority vote is also used). Thus, the decision tree output can be written as a weighted sum of basis functions in the same manner as a layered network.

As this discussion suggests, decision trees and MLP/RBF neural networks are best viewed as being different points along the continuum of models having overlapping or nonoverlapping basis functions. Indeed, as we show in the following section, decision trees can be treated probabilistically as mixture models, and in the mixture approach, the sharp discriminant function boundaries of classical decision trees become smoothed, yielding partially overlapping basis functions.

There are trade-offs associated with the continuum of degree-of-overlap; in particular, nonoverlapping basis functions are generally viewed as being easier to interpret and better able to reject noisy input variables that carry little information about the output. Overlapping basis functions often are viewed as yielding lower variance predictions and as being more robust.

42.2.5 General Mixture Models

The use of mixture models is not restricted to density estimation; rather, the mixture approach can be used quite generally to build complex models out of simple parts. To illustrate, let us consider using mixture models to model a conditional density in the context of a regression or classification problem. A mixture model in this setting is referred to as a "mixtures of experts" model (Jacobs et al. 1991).

Suppose that we have at our disposal an elemental conditional model $p(\boldsymbol{t}|\boldsymbol{x}, \boldsymbol{w})$. Consider a situation in which the conditional mean or discriminant exhibits variation on a local scale that is a good match to our elemental model, but the variation differs in different regions of the input space. We could use a more complex network to try to capture this global variation; alternatively, we might wish to combine local variants of our elemental models in some manner. This can be achieved by defining the following probabilistic mixture:

$$p(\boldsymbol{t}\,|\,\boldsymbol{x},\boldsymbol{w}) = \sum_{i=1}^{M} p(i\,|\,\boldsymbol{x},\boldsymbol{v})\,p(\boldsymbol{t}\,|\,\boldsymbol{x},i,\boldsymbol{w}_i) \tag{42.11}$$

Comparing this mixture to the unconditional mixture defined earlier in Equation 42.2, we see that both the mixing proportions and the component densities are now conditional densities dependent on the input vector $\boldsymbol{x}$. The former dependence is particularly important: we now view the mixing proportion $p(i|\boldsymbol{x}, \boldsymbol{v})$ as providing a probabilistic device for choosing different elemental models ("experts") in different regions of the input space. A learning algorithm that chooses values for the parameters $\boldsymbol{v}$ as well as the values for the parameters $\boldsymbol{w}_i$ can be viewed as attempting to find both a good partition of the input space and a good fit to the local models within that partition.

This approach can be extended recursively by considering mixtures of models where each model may itself be a mixture model (Jordan and Jacobs 1994). Such a recursion can be viewed as providing a probabilistic interpretation for the decision trees discussed in the previous section. We view the decisions in the decision tree as forming a recursive set of probabilistic selections among a set of models. The total probability of target $\boldsymbol{t}$ given input $\boldsymbol{x}$ is the sum across all paths down the tree:

$$p(\boldsymbol{t}|\boldsymbol{x},\boldsymbol{w}) = \sum_{i=1}^{M} p(i|\boldsymbol{x},\boldsymbol{u}) \sum_{j=1}^{M} p(j|\boldsymbol{x},i,\boldsymbol{v}_i) \cdots p(\boldsymbol{t}|\boldsymbol{x},i,j,\ldots,\boldsymbol{w}_{ij}\cdots) \quad (42.12)$$

where

i and j are the decisions made at the first level and second level of the tree, respectively

$p(\boldsymbol{t}|\boldsymbol{x}, i, j,\ldots, \boldsymbol{w}_{ij} \ldots)$ is the elemental model at the leaf of the tree defined by the sequence of decisions

This probabilistic model is a conditional hierarchical mixture. Finding parameter values $\boldsymbol{u}$, $\boldsymbol{v}_i$, etc., to fit this model to data can be viewed as finding a nested set of partitions of the input space and fitting a set of local models within the partition.

The mixture model approach can be viewed as a special case of a general methodology known as *learning by committee*. Bishop (1995) provides a discussion of committees; we will also meet them in the section on Bayesian methods later in the chapter.

42.3 Learning from Data

The previous section has provided a selection of models to choose from; we now face the problem of matching these models to data. In principle, the problem is straightforward: given a family of models of interest, we attempt to find out how probable each of these models is in the light of the data. We can then select the most probable model (a selection rule known as *maximum a posteriori* [MAP] estimation), or we can select some highly probable subset of models, weighted by their probability (an approach that we discuss in the section on Bayesian methods). In practice, there are a number of problems to solve, beginning with the specification of the family of models of interest. In the simplest case, in which the family can be described as a fixed structure with varying parameters (e.g., the class of feedforward MLPs with a fixed number of hidden units), the learning problem is essentially one of *parameter estimation*. If, on the other hand, the family is not easily viewed as a fixed parametric family (e.g., feedforward MLPs with a variable number of hidden units), then we must solve the *model selection* problem.

In this section, we discuss the parameter estimation problem. The goal will be to find MAP estimates of the parameters by maximizing the probability of the parameters given the data $\mathcal{D}$. We compute this probability using Bayes rule:

$$p(\boldsymbol{w}|\mathcal{D}) = \frac{p(\mathcal{D}|\boldsymbol{w})p(\boldsymbol{w})}{p(\mathcal{D})} \quad (42.13)$$

where we see that to calculate MAP estimates we must maximize the expression in the numerator (the denominator does not depend on w). Equivalently we can minimize the negative logarithm of the numerator. We thus define the following *cost function* $J(w)$:

$$J(w) = -\ln p(\mathcal{D} \mid w) - \ln p(w) \tag{42.14}$$

which we wish to minimize with respect to the parameters w. The first term in this cost function is a (negative) log *likelihood*. If we assume that the elements in the training set $\mathcal{D}$ are conditionally independent of each other given the parameters, then the likelihood factorizes into a product form. For density estimation, we have

$$p(\mathcal{D}|w) = \prod_{n=1}^{N} p(x_n|w) \tag{42.15}$$

and for classification and regression, we have

$$p(\mathcal{D}|w) = \prod_{n=1}^{N} p(t_n|x_n, w) \tag{42.16}$$

In both cases this yields a log likelihood, which is the sum of the log probabilities for each individual data point. For the remainder of this section, we will assume this additive form; moreover, we will assume that the log prior probability of the parameters is uniform across the parameters and drop the second term. Thus, we focus on *maximum likelihood* (ML) estimation, where we choose parameter values w_{ML} that maximize $\ln p(\mathcal{D}|w)$.

42.3.1 Likelihood-Based Cost Functions

Regression, classification, and density estimation make different probabilistic assumptions about the form of the data and therefore require different cost functions.

Equation 42.3 defines a probabilistic model for regression. The model is a conditional density for the targets t in which the targets are distributed as Gaussian random variables (assuming Gaussian errors ε) with mean values $f(x)$. We now write the conditional mean as $f(x, w)$ to make explicit the dependence on the parameters w. Given the training set $\mathcal{D} = \{x_n, t_n\}_{n=1}^{N}$, and given our assumption that the targets t_n are sampled independently (given the inputs x_n and the parameters w), we obtain

$$J(w) = \frac{1}{2}\sum_{n} \|t_n - f(x_n, w)\|^2 \tag{42.17}$$

where we have assumed an identity covariance matrix and dropped those terms that do not depend on the parameters. This cost function is the standard least-squares cost function, which is traditionally used in neural network training for real-valued targets. Minimization of this cost function is typically achieved via some form of gradient optimization, as we discuss in the following section.

Classification problems differ from regression problems in the use of discrete-valued targets, and the likelihood accordingly takes a different form. For binary classification the Bernoulli probability model $p(t|x, w) = y^t(1 - y)^{1-t}$ is natural, where we use y to denote the probability $p(t = 1|x, w)$. This model yields the following log likelihood:

$$J(w) = -\sum_{n} [t_n \ln y_n + (1 - t_n)\ln(1 - y_n)] \tag{42.18}$$

which is known as the *cross-entropy* function. It can be minimized using the same generic optimization procedures as are used for least squares.

For multiway classification problems in which there are C categories, where $C > 2$, the multinomial distribution is natural. Define $\boldsymbol{t}_n$ such that its elements $t_{n,i}$ are one or zero according to whether the nth data point belongs to the ith category, and define $y_{n,i}$ to be the network's estimate of the posterior probability of category i for data point n; that is, $y_{n,i} \equiv p(t_{n,i} = 1|\boldsymbol{x}_n, \boldsymbol{w})$. Given these definitions, we obtain the following cost function:

$$J(\boldsymbol{w}) = -\sum_n \sum_i t_{n,i} \ln y_{n,i} \tag{42.19}$$

which again has the form of a cross-entropy.

We now turn to density estimation as exemplified by Gaussian mixture modeling. The probabilistic model in this case is that given in Equation 42.2. Assuming Gaussian component densities with arbitrary covariance matrices, we obtain the following cost function:

$$J(\boldsymbol{w}) = -\sum_n \ln \sum_i \pi_i \frac{1}{|\Sigma_i|^{1/2}} \exp\left\{-\frac{1}{2}(\boldsymbol{x}_n - \mu_i)^T \Sigma_i^{-1}(\boldsymbol{x}_n - \mu_i)\right\} \tag{42.20}$$

where the parameters $\boldsymbol{w}$ are the collection of mean vectors μ_i, the covariance matrices Σ_i, and the mixing proportions π_i. A similar cost function arises for the generalized mixture models (cf. Equation 42.12).

42.3.2 Gradients of the Cost Function

Once we have defined a probabilistic model, obtained a cost function, and found an efficient procedure for calculating the gradient of the cost function, the problem can be handed off to an optimization routine. Before discussing optimization procedures, however, it is useful to examine the form that the gradient takes for the examples that we have discussed in the previous two sections.

The ith output unit in a layered network is endowed with a rule for combining the activations of units in earlier layers, yielding a quantity that we denote by z_i and a function that converts z_i into the output y_i. For regression problems, we assume linear output units such that $y_i = z_i$. For binary classification problems, our earlier discussion showed that a natural output function is the logistic: $y_i = 1/(1 + e^{-z_i})$. For multiway classification, it is possible to generalize the derivation of the logistic function to obtain an analogous representation for the multiway posterior probabilities known as the *softmax function* (cf. Bishop 1995):

$$y_i = \frac{e^{z_i}}{\sum_k e^{z_k}} \tag{42.21}$$

where y_i represents the posterior probability of category i.

If we now consider the gradient of $J(\boldsymbol{w})$ with respect to z_i, it turns out that we obtain a single canonical expression of the following form:

$$\frac{\partial J}{\partial \boldsymbol{w}} = \sum_i (t_i - y_i) \frac{\partial z_i}{\partial \boldsymbol{w}} \tag{42.22}$$

As discussed by Rumelhart et al. (1995), this form for the gradient is predicted from the theory of generalized linear models (McCullagh and Nelder 1983), where it is shown that the linear, logistic, and softmax functions are (inverse) *canonical links* for the Gaussian, Bernoulli, and multinomial distributions, respectively. Canonical links can be found for all of the distributions in the exponential family, thus providing a solid statistical foundation for handling a wide variety of data formats at the output layer of a network, including counts, time intervals, and rates.

The gradient of the cost function for mixture models has an interesting interpretation. Taking the partial derivative of $J(\mathbf{w})$ in Equation 42.20 with respect to μ_i, we find

$$\frac{\partial J}{\partial \mu_i} = \sum_n h_{n,i}\, \Sigma_i(\mathbf{x}_n - \mu_i) \tag{42.23}$$

where $h_{n,i}$ is defined as follows:

$$h_{n,i} = \frac{\pi_i\, |\Sigma_i|^{-1/2} \exp\left\{-1/2\,(\mathbf{x}_n - \mu_i)^T\, \Sigma_i^{-1}(\mathbf{x}_n - \mu_i)\right\}}{\sum_k \pi_k\, |\Sigma_k|^{-1/2} \exp\left\{-1/2\,(\mathbf{x}_n - \mu_k)^T\, \Sigma_k^{-1}(\mathbf{x}_n - \mu_k)\right\}} \tag{42.24}$$

When summed over i, the quantity $h_{n,i}$ sums to one and is often viewed as the "responsibility" or "credit" assigned to the ith component for the nth data point. Indeed, interpreting Equation 42.24 using Bayes rule shows that $h_{n,i}$ is the posterior probability that the nth data point is generated by the ith component Gaussian. A learning algorithm based on this gradient will move the ith mean μ_i toward the data point $\mathbf{x}_n$, with the effective step size proportional to $h_{n,i}$.

The gradient for a mixture model will always take the form of a weighted sum of the gradients associated with the component models, where the weights are the posterior probabilities associated with each of the components. The key computational issue is whether these posterior weights can be computed efficiently. For Gaussian mixture models, the calculation (Equation 42.24) is clearly efficient. For decision trees there is a set of posterior weights associated with each of the nodes in the tree, and a recursion is available that computes the posterior probabilities in an upward sweep (Jordan and Jacobs 1994). Mixture models in the form of a chain are known as *hidden Markov models (HMMs)*, and the calculation of the relevant posterior probabilities is performed via an efficient algorithm known as the Baum–Welch algorithm.

For general layered network structures, a generic algorithm known as backpropagation is available to calculate gradient vectors (Rumelhart et al. 1986). Backpropagation is essentially the chain rule of calculus realized as a graphical algorithm. As applied to layered networks, it provides a simple and efficient method that calculates a gradient in $O(W)$ time per training pattern, where W is the number of weights.

42.3.3 Optimization Algorithms

By introducing the principle of ML in Section 42.1, we have expressed the problem of learning in neural networks in terms of the minimization of a cost function, $J(\mathbf{w})$, which depends on a vector, $\mathbf{w}$, of adaptive parameters. An important aspect of this problem is that the gradient vector $\nabla_{\mathbf{w}} J$ can be evaluated efficiently (e.g., by backpropagation). Gradient-based minimization is a standard problem in unconstrained nonlinear optimization for which many powerful techniques have been developed over the years. Such algorithms generally start by making an initial guess for the parameter vector $\mathbf{w}$ and then iteratively updating the vector in a sequence of steps:

$$\mathbf{w}^{(\tau+1)} = \mathbf{w}^{(\tau)} + \Delta \mathbf{w}^{(\tau)} \tag{42.25}$$

where τ denotes the step number. The initial parameter vector $\mathbf{w}^{(0)}$ is often chosen at random, and the final vector represents a minimum of the cost function at which the gradient vanishes. Because of the nonlinear nature of neural network models, the cost function is generally a highly complicated function of the parameters and may possess many such minima. Different algorithms differ in how the update $\Delta w^{(\tau)}$ is computed.

The simplest such algorithm is called *gradient descent* and involves a parameter update which is proportional to the negative of the cost function gradient $\Delta = -\eta \nabla E$ where η is a fixed constant called the learning rate. It should be stressed that gradient descent is a particularly inefficient optimization algorithm. Various modifications have been proposed, such as the inclusion of a *momentum* term, to try to improve its performance. In fact, much more powerful algorithms are readily available, as described in standard textbooks such as Fletcher (1987). Two of the best known are called *conjugate gradients* and *quasi-Newton* (or *variable metric*) methods. For the particular case of a sum-of-squares cost function, the *Levenberg–Marquardt* algorithm can also be very effective. Software implementations of these algorithms are widely available.

The algorithms discussed so far are called *batch* since they involve using the whole dataset for each evaluation of the cost function or its gradient. There is also a *stochastic* or *online* version of gradient descent in which, for each parameter update, the cost function gradient is evaluated using just one of the training vectors at a time (which are then cycled either in order or in a random sequence). Although this approach fails to make use of the power of sophisticated methods such as conjugate gradients, it can prove effective for very large datasets, particularly if there is significant redundancy in the data.

42.3.4 Hessian Matrices, Error Bars, and Pruning

After a set of weights have been found for a neural network using an optimization procedure, it is often useful to examine second-order properties of the fitted network as captured in the Hessian matrix $H = \partial^2 J/\partial w \partial w^T$. Efficient algorithms have been developed to compute the Hessian matrix in time $O(W^2)$ (Bishop 1995). As in the case of the calculation of the gradient by backpropagation, these algorithms are based on recursive message passing in the network.

One important use of the Hessian matrix lies in the calculation of error bars on the outputs of a network. If we approximate the cost function locally as a quadratic function of the weights (an approximation that is equivalent to making a Gaussian approximation for the log likelihood), then the estimated variance of the ith output y_i can be shown to be

$$\hat{\sigma}_{y_i}^2 = \left(\frac{\partial y_i}{\partial w}\right)^T H^{-1} \left(\frac{\partial y_i}{\partial w}\right) \tag{42.26}$$

where the gradient vector $\partial_{y_i}/\partial w$ can be calculated via backpropagation.

The Hessian matrix also is useful in pruning algorithms. A pruning algorithm deletes weights from a fitted network to yield a simpler network that may outperform a more complex, *overfitted* network (discussed subsequently) and may be easier to interpret. In this setting, the Hessian is used to approximate the increase in the cost function due to the deletion of a weight. A variety of such pruning algorithms is available (cf. Bishop 1995).

42.3.5 Complexity Control

In previous sections, we have introduced a variety of models for representing probability distributions, we have shown how the parameters of the models can be optimized by maximizing the likelihood function, and we have outlined a number of powerful algorithms for performing this minimization. Before we can apply this framework in practice, there is one more issue we need to address, which is that of model complexity. Consider the case of a mixture model given by Equation 42.2. The number of input

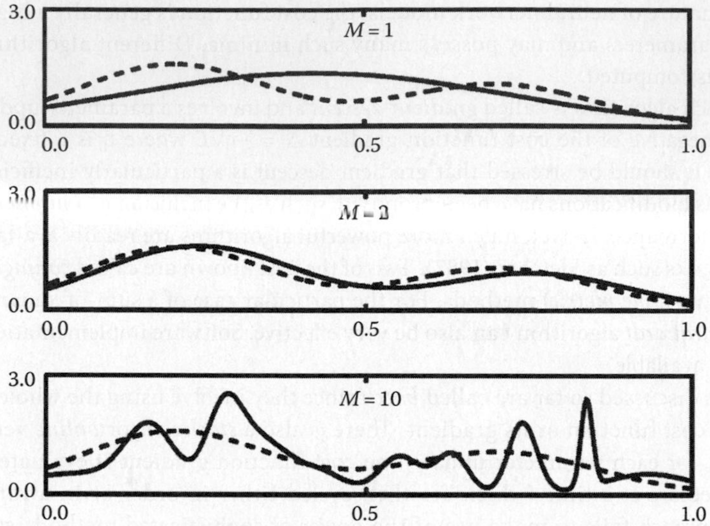

FIGURE 42.4 Effects of model complexity illustrated by modeling a mixture of two Gaussians (shown by the dashed curves) using a mixture of M Gaussians (shown by the solid curves). The results are obtained for 20 cycles of EM.

variables will be determined by the particular problem at hand. However, the number M of component densities has yet to be specified. Clearly, if M is too small, the model will be insufficiently flexible and we will obtain a poor representation of the true density. What is not so obvious is that if M is too large, we can also obtain poor results. This effect is known as *overfitting* and arises because we have a dataset of finite size. It is illustrated using the simple example of mixture density estimation in Figure 42.4. Here a set of 100 data points in one dimension has been generated from a distribution consisting of a mixture of two Gaussians (shown by the dashed curves). This dataset has then been fitted by a mixture of M Gaussians by use of the expectation-maximization (EM) algorithm. We see that a model with 1 component ($M = 1$) gives a poor representation of the true distribution from which the data were generated and in particular is unable to capture the bimodal aspect. For $M = 2$ the model gives a good fit, as we expect since the data were themselves generated from a two-component Gaussian mixture. However, increasing the number of components to $M = 10$ gives a poorer fit, even though this model contains the simpler models as special cases.

The problem is a very fundamental one and is associated with the fact that we are trying to infer an entire distribution function from a finite number of data points, which is necessarily an ill-posed problem. In regression, for example, there are infinitely many functions that will give a perfect fit to the finite number of data points. If the data are noisy, however, the best generalization will be obtained for a function that does not fit the data perfectly but that captures the underlying function from which the data were generated. By increasing the flexibility of the model, we are able to obtain ever better fits to the training data, and this is reflected in a steadily increasing value for the likelihood function at its maximum. Our goal is to model the true underlying density function from which the data were generated since this allows us to make the best predictions for new data. We see that the best approximation to this density occurs for an intermediate value of M.

The same issue arises in connection with nonlinear regression and classification problems. For example, the number M of hidden units in an MLP network controls the model complexity and must be optimized to give the best generalization. In a practical application, we can train a variety of different models having different complexities, compare their generalization performance using an independent validation set, and then select the model with the best generalization. In fact, the process of optimizing the complexity using a validation set can lead to some partial overfitting to the validation data itself, and so the final performance of the selected model should be confirmed using a third independent dataset called a *test* set.

Some theoretical insight into the problem of overfitting can be obtained by decomposing the error into the sum of bias and variance terms (Geman et al. 1992). A model that is too inflexible is unable to represent the true structure in the underlying density function, and this gives rise to a high bias. Conversely, a model that is too flexible becomes tuned to the specific details of the particular dataset and gives a high variance. The best generalization is obtained from the optimum trade-off of bias against variance.

As we have already remarked, the problem of inferring an entire distribution function from a finite dataset is fundamentally ill posed since there are infinitely many solutions. The problem becomes well posed only when some additional constraint is imposed. This constraint might be that we model the data using a network having a limited number of hidden units. Within the range of functions that this model can represent, there is then a unique function that best fits the data. Implicitly, we are assuming that the underlying density function from which the data were drawn is relatively smooth. Instead of limiting the number of parameters in the model, we can encourage smoothness more directly using the technique of *regularization*. This involves adding penalty term Ω to the original cost function J to give the total cost function $\tilde{J}$ of the following form:

$$\tilde{J} = J + \nu\Omega \tag{42.27}$$

where ν is called a regularization coefficient. The network parameters are determined by minimizing $\tilde{J}$, and the value of ν controls the degree of influence of the penalty term Ω. In practice, Ω is typically chosen to encourage smooth functions. The simplest example is called *weight decay* and consists of the sum of the squares of all of the adaptive parameters in the model:

$$\Omega = \sum_i w_i^2 \tag{42.28}$$

Consider the effect of such a term on the MLP function (Equation 42.9). If the weights take very small values, then the network outputs become approximately linear functions of the inputs (since the sigmoidal function is approximately linear for small values of its argument). The value of ν in Equation 42.27 controls the effective complexity of the model, so that for large ν the model is oversmoothed (corresponding to high bias), whereas for small ν the model can overfit (corresponding to high variance). We can therefore consider a network with a relatively large number of hidden units and control the effective complexity by changing ν. In practice, a suitable value for ν can be found by seeking the value that gives the best performance on a validation set.

The weight decay regularizer (Equation 42.28) is simple to implement but suffers from a number of limitations. Regularizers used in practice may be more sophisticated and may contain multiple regularization coefficients (Neal 1994).

Regularization methods can be justified within a general theoretical framework known as *structural risk minimization* (Vapnik 1995). Structural risk minimization provides a quantitative measure of complexity known as the *VC dimension*. The theory shows that the VC dimension predicts the difference between performance on a training set and performance on a test set; thus, the sum of log likelihood and (some function of) VC dimension provides a measure of generalization performance. This motivates regularization methods (Equation 42.27) and provides some insight into possible forms for the regularizer Ω.

42.3.6 Bayesian Viewpoint

In earlier sections we discussed network training in terms of the minimization of a cost function derived from the principle of MAP or ML estimation. This approach can be seen as a particular approximation to a more fundamental, and more powerful, framework based on Bayesian statistics. In the ML approach, the weights $\boldsymbol{w}$ are set to a specific value, $\boldsymbol{w}_{\mathrm{ML}}$, determined by minimization of a cost function. However, we know that there will typically be other minima of the cost function, which might give equally good results. Also, weight values close to $\boldsymbol{w}_{\mathrm{ML}}$ should give results that are not too different from those of the ML weights themselves.

These effects are handled in a natural way in the Bayesian viewpoint, which describes the weights not in terms of a specific set of values but in terms of a probability distribution over all possible values. As discussed earlier (cf. Equation 42.13), once we observe the training dataset $\mathcal{D}$, we can compute the corresponding *posterior* distribution using Bayes' theorem, based on a *prior* distribution function $p(\boldsymbol{w})$ (which will typically be very broad) and a *likelihood* function $p(\mathcal{D}|\boldsymbol{w})$:

$$p(\boldsymbol{w}|\mathcal{D}) = \frac{p(\mathcal{D}|\boldsymbol{w})p(\boldsymbol{w})}{p(\mathcal{D})} \tag{42.29}$$

The likelihood function will typically be very small except for values of $\boldsymbol{w}$ for which the network function is reasonably consistent with the data. Thus, the posterior distribution $p(\boldsymbol{w}|\mathcal{D})$ will be much more sharply peaked than the prior distribution $p(\boldsymbol{w})$ (and will typically have multiple maxima). The quantity we are interested in is the predicted distribution of target values $\boldsymbol{t}$ for a new input vector $\boldsymbol{x}$ once we have observed the dataset $\mathcal{D}$. This can be expressed as an integration over the posterior distribution of weights of the following form:

$$p(\boldsymbol{t}|\boldsymbol{x},\mathcal{D}) = \int p(\boldsymbol{t}|\boldsymbol{x},\boldsymbol{w})p(\boldsymbol{w}|\mathcal{D})d\boldsymbol{w} \tag{42.30}$$

where $p(\boldsymbol{t}|\boldsymbol{x}, \boldsymbol{w})$ is the conditional probability model discussed in the Introduction.

If we suppose that the posterior distribution $p(\boldsymbol{w}|\mathcal{D})$ is sharply peaked around a single most probable value $\boldsymbol{w}_{\mathrm{MP}}$, then we can write Equation 42.30 in the following form:

$$p(\boldsymbol{t}|\boldsymbol{x},\mathcal{D}) \simeq p(\boldsymbol{t}|\boldsymbol{x},\boldsymbol{w}_{\mathrm{MP}}) \int p(\boldsymbol{w}|\mathcal{D})d\boldsymbol{w} \tag{42.31}$$

$$= p(\boldsymbol{t}|\boldsymbol{x},\boldsymbol{w}_{\mathrm{MP}}) \tag{42.32}$$

and so predictions can be made by fixing the weights to their most probable values. We can find the most probable weights by maximizing the posterior distribution or equivalently by minimizing its negative logarithm. Using Equation 42.29, we see that $\boldsymbol{w}_{\mathrm{MP}}$ is determined by minimizing a regularized cost function of the form in Equation 42.27 in which the negative log of the prior–ln $p(\boldsymbol{w})$ represents the regularizer $\nu\Omega$.

For example, if the prior consists of a zero mean Gaussian with variance ν^{-1}, then we obtain the weight decay regularizer of Equation 42.28.

The posterior distribution will become sharply peaked when the size of the dataset is large compared to the number of parameters in the network. For datasets of limited size, however, the posterior distribution has a finite width and this adds to the uncertainty in the predictions for $\boldsymbol{t}$, which can be expressed in terms of error bars. Bayesian error bars can be evaluated using a local Gaussian approximation to the posterior distribution (MacKay 1992). The presence of multiple maxima in the posterior distribution also contributes to the uncertainties in predictions. The capability to assess these uncertainties can play a crucial role in practical applications.

The Bayesian approach can also deal with more general problems in complexity control. This can be done by considering the probabilities of a set of alternative models, given the dataset

$$p(\mathcal{H}_i|\mathcal{D}) = \frac{p(\mathcal{D}|\mathcal{H}_i)p(\mathcal{H}_i)}{p(\mathcal{D})} \tag{42.33}$$

Here, different models can also be interpreted as different values of regularization parameters as these too control model complexity. If the models are given the same prior probabilities $p(\mathcal{H}_i)$, then they can be ranked by considering the *evidence* $p(\mathcal{D}|\mathcal{H}_i)$, which itself can be evaluated by integration over the model parameters $\boldsymbol{w}$. We can simply select the model with the greatest probability. However, a full Bayesian treatment requires that we form a linear combination of the predictions of the models in which the weighting coefficients are given by the model probabilities.

In general, the required integrations, such as that in Equation 42.30, are analytically intractable. One approach is to approximate the posterior distribution by a Gaussian centered on $\boldsymbol{w}_{\text{MP}}$ and then to linearize $p(\boldsymbol{t}|\boldsymbol{x}, \boldsymbol{w})$ about $\boldsymbol{w}_{\text{MP}}$ so that the integration can be performed analytically (MacKay 1992). Alternatively, sophisticated Monte Carlo methods can be employed to evaluate the integrals numerically (Neal 1994). An important aspect of the Bayesian approach is that there is no need to keep data aside in a validation set as is required when using ML. In practical applications for which the quantity of available data is limited, it is found that a Bayesian treatment generally outperforms other approaches.

42.3.7 Preprocessing, Invariances, and Prior Knowledge

We have already seen that neural networks can approximate essentially arbitrary nonlinear functional mappings between sets of variables. In principle, we could therefore use a single network to transform the raw input variables into the required final outputs. However, in practice, for all but the simplest problems, the results of such an approach can be improved upon considerably by incorporating various forms of preprocessing, for reasons we shall outline in the following.

One of the simplest and most common forms of preprocessing consists of a simple normalization of the input, and possibly also target, variables. This may take the form of a linear rescaling of each input variable independently to give it zero mean and unit variance over the training set. For some applications, the original input variables may span widely different ranges. Although a linear rescaling of the inputs is equivalent to a different choice of first-layer weights, in practice the optimization algorithm may have considerable difficulty in finding a satisfactory solution when typical input values are substantially different. Similar rescaling can be applied to the output values, in which case the inverse of the transformation needs to be applied to the network outputs when the network is presented with new inputs. Preprocessing is also used to encode data in a suitable form. For example, if we have categorical variables such as red, green, and blue, these may be encoded using a 1-of-3 binary representation.

Another widely used form of preprocessing involves reducing the dimensionality of the input space. Such transformations may result in loss of information in the data, but the overall effect can be a significant improvement in performance as a consequence of the curse of dimensionality discussed in the complexity control section. The finite dataset is better able to specify the required mapping in the lower dimensional space. Dimensionality reduction may be accomplished by simply selecting a subset of the original variables but more typically involves the construction of new variables consisting of linear or nonlinear combinations of the original variables called *features*. A standard technique for dimensionality reduction is principal component analysis (Anderson 1984). Such methods, however, make use only of the input data and ignore the target values and can sometimes be significantly suboptimal.

Yet another form of preprocessing involves correcting deficiencies in the original data. A common occurrence is that some of the input variables are missing for some of the data points. Correction of this problem in a principled way requires that the probability distribution $p(\boldsymbol{x})$ of input data be modeled.

One of the most important factors determining the performance of real-world applications of neural networks is the use of *prior knowledge*, which is information additional to that present in the data. As an example, consider the problem of classifying handwritten digits discussed in Section 42.1. The most direct approach would be to collect a large training set of digits and to train a feedforward network to map from the input image to a set of 10 output values representing posterior probabilities for the 10 classes. However, we know that the classification of a digit should be independent of its position within the input image. One way of achieving such *translation invariance* is to make use of the technique of *shared weights*. This involves a network architecture having many hidden layers in which each unit takes inputs only from a small patch, called a *receptive field*, of units in the previous layer. By a process of constraining neighboring units to have common weights, it can be arranged that the output of the network is insensitive to translations of the input image. A further benefit of weight sharing is that the number of independent parameters is much smaller than the number of weights, which assists with the problem of model complexity. This approach is the basis for the highly successful US postal code recognition system of LeCun et al. (1989). An alternative to shared weights is to enlarge the training set artificially by generating virtual examples based on applying translations and other transformations to the original training set (Poggio and Vetter 1992).

42.4 Graphical Models

Neural networks express relationships between variables by utilizing the representational language of graph theory. Variables are associated with nodes in a graph, and transformations of variables are based on algorithms that propagate numerical messages along the links of the graph. Moreover, the graphs are often accompanied by probabilistic interpretations of the variables and their interrelationships. As we have seen, such probabilistic interpretations allow a neural network to be understood as a form of a probabilistic model and reduce the problem of learning the weights of a network to a problem in statistics.

Related graphical models have been studied throughout statistics, engineering, and AI in recent years. HMMs, Kalman filters, and path analysis models are all examples of graphical probabilistic models that can be fitted to data and used to make inferences. The relationship between these models and neural networks is rather strong; indeed, it is often possible to reduce one kind of model to the other. In this section, we examine these relationships in some detail and provide a broader characterization of neural networks as members of a general family of graphical probabilistic models.

Many interesting relationships have been discovered between graphs and probability distributions (Pearl 1988, Spiegelhalter et al. 1993). These relationships derive from the use of graphs to represent conditional independencies among random variables. In an undirected graph, there is a direct correspondence between conditional independence and graph separation: random variables X_i and X_k are conditionally independent given X_j if nodes X_i and X_k are separated by node X_j (we use the symbol X_i to represent both a random variable and a node in a graph). This statement remains true for sets of nodes (see Figure 42.5a). Directed graphs have a somewhat different semantics due to the ability of directed graphs to represent induced dependencies. An induced dependency is a situation in which two nodes that are marginally independent become conditionally dependent given the value of a third node. (see Figure 42.5b). Suppose, for example, that X_i and X_k represent independent coin tosses, and X_j represents the sum of X_i and X_k. Then X_i and X_k are marginally independent but are conditionally

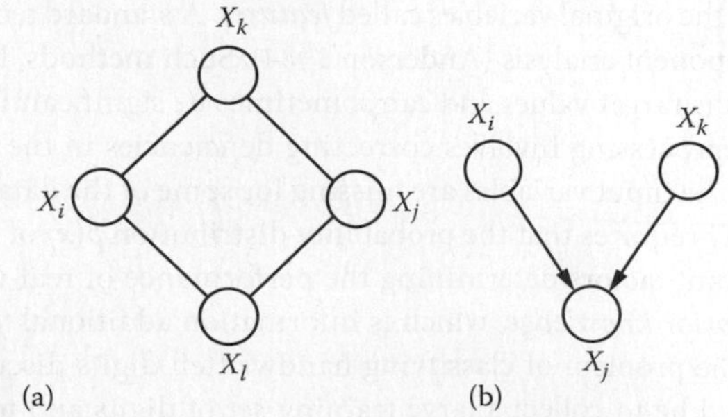

FIGURE 42.5 (a) An undirected graph in which X_i is independent of X_j given X_k and X_l, and X_k is independent of X_l given X_i and X_j. (b) A directed graph in which X_i and X_k are marginally independent but are conditionally dependent given X_j.

dependent given X_j. The semantics of independence in directed graphs is captured by a graphical criterion known as *d-separation* (Pearl 1988), which differs from undirected separation only in those cases in which paths have two arrows arriving at the same node (as in Figure 42.5b).

Although the neural network architectures that we have discussed until now all have been based on directed graphs, undirected graphs also play an important role in neural network research. Constraint satisfaction architectures, including the Hopfield network (Hopfield 1982) and the *Boltzmann machine* (Hinton and Sejnowski 1986), are the most prominent examples. A Boltzmann machine is an undirected probabilistic graph that respects the conditional independency semantics previously described (cf. Figure 42.5a). Each node in a Boltzmann machine is a binary-valued random variable X_i (or, more generally, a discrete-valued random variable). A probability distribution on the 2^N possible configurations of such variables is defined via an *energy function E*. Let J_{ij} be the weight on the link between X_i and X_j, let $J_{ij} = J_{ji}$, let α index the configurations, and define the energy of configuration α as follows:

$$E_\alpha = -\sum_{i<j} J_{ij} X_i^\alpha X_j^\alpha \tag{42.34}$$

The probability of configuration α is then defined via the Boltzmann distribution:

$$P_\alpha = \frac{e^{-E_\alpha/T}}{\sum_\gamma e^{-E_\gamma/T}} \tag{42.35}$$

where the *temperature T* provides a scale for the energy.

An example of a directed probabilistic graph is the HMM. An HMM is defined by a set of *state variables* H_i, where i is generally a time or a space index, a set of output variables O_i, a *probability transition matrix* $A = p(H_i|H_{i-1})$, and an *emission matrix* $B = p(O_i|H_i)$. The directed graph for an HMM is shown in Figure 42.6a. As can be seen from considering the separatory properties of the graph, the conditional independencies of the HMM are defined by the following Markov conditions:

$$H_i \perp \{H_1, O_1, \ldots, H_{i-2}, O_{i-2}, O_{i-1}\} \mid H_{i-1}, \quad 2 \le i \le N \tag{42.36}$$

and

$$O_i \perp \{H_1, O_1, \ldots, H_{i-1}, O_{i-1}\} \mid H_i, \quad 2 \le i \le N \tag{42.37}$$

where the symbol $\perp$ is used to denote independence.

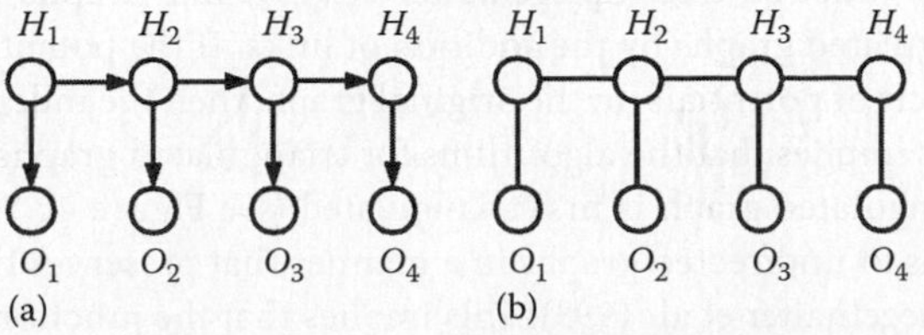

FIGURE 42.6 (a) A directed graph representation of an HMM. Each horizontal link is associated with the transition matrix *A*, and each vertical link is associated with the emission matrix *B*. (b) An HMM as a Boltzmann machine. The parameters on the horizontal links are logarithms of the entries of the *A* matrix, and the parameters on the vertical links are logarithms of the entries of the *B* matrix. The two representations yield the same joint probability distribution.

Figure 42.6b shows that it is possible to treat an HMM as a special case of a Boltzmann machine (Luttrell 1989, Saul and Jordan 1995). The probabilistic structure of the HMM can be captured by defining the weights on the links as the logarithms of the corresponding transition and emission probabilities. The Boltzmann distribution (Equation 42.35) then converts the additive energy into the product form of the standard HMM probability distribution. As we will see, this reduction of a directed graph to an undirected graph is a recurring theme in the graphical model formalism.

General mixture models are readily viewed as graphical models (Buntine 1994). For example, the unconditional mixture model of Equation 42.2 can be represented as a graphical model with two nodes—a multinomial hidden node, which represents the selected component, a visible node representing $\boldsymbol{x}$, with a directed link from the hidden node to the visible node (hidden/visible distinction discussed subsequently). Conditional mixture models (Jacobs et al. 1991) simply require another visible node with directed links to the hidden node and the visible nodes. Hierarchical conditional mixture models (Jordan and Jacobs 1994) require a chain of hidden nodes, one hidden node for each level of the tree.

Within the general framework of probabilistic graphical models, it is possible to tackle general problems of inference and learning. The key problem that arises in this setting is the problem of computing the probabilities of certain nodes, which we will refer to as *hidden nodes*, given the observed values of other nodes, which we will refer to as *visible nodes*. For example, in an HMM, the variables O_i are generally treated as visible, and it is desired to calculate a probability distribution on the hidden states H_i. A similar inferential calculation is required in the mixture models and the Boltzmann machine.

Generic algorithms have been developed to solve the inferential problem of the calculation of posterior probabilities in graphs. Although a variety of inference algorithms have been developed, they can all be viewed as essentially the same underlying algorithm (Shachter et al. 1994). Let us consider undirected graphs. A special case of an undirected graph is a *triangulated graph* (Spiegelhalter et al. 1993), in which any cycle having four or more nodes has a chord. For example, the graph in Figure 42.5a is not triangulated but becomes triangulated when a link is added between nodes X_i and X_j. In a triangulated graph, the cliques of the graph can be arranged in the form of a *junction tree*, which is a tree having the property that any node that appears in two different cliques in the tree also appears in every clique on the path that links the two cliques (the "running intersection property"). This cannot be achieved in nontriangulated graphs. For example, the cliques in Figure 42.5a are $\{X_i, X_k\}$, $\{X_k, X_j\}$, $\{X_j, X_l\}$, and it is not possible to arrange these cliques into a tree that obeys the running intersection property. If a chord is added, the resulting cliques are $\{X_i, X_j, X_k\}$ and $\{X_i, X_j, X_l\}$, and these cliques can be arranged as a simple chain that trivially obeys the running intersection property. In general, it turns out that the probability distributions corresponding to triangulated graphs can be characterized as *decomposable*, which implies that they can be factorized into a product of local functions (potentials) associated with the cliques in the triangulated graph.* The calculation of posterior probabilities in decomposable distributions is straightforward and can be achieved via a local message-passing algorithm on the junction tree (Spiegelhalter et al. 1993). Graphs that are not triangulated can be turned into triangulated graphs by the addition of links. If the potentials on the new graph are defined suitably as products of potentials on the original graph, then the independencies in the original graph are preserved. This implies that the algorithms for triangulated graphs can be used for *all* undirected graphs; an untriangulated graph is first triangulated (see Figure 42.7). Moreover, it is possible to convert *directed* graphs to undirected graphs in a manner that preserves the probabilistic structure of the original graph (Spiegelhalter et al. 1993). This implies that the junction tree algorithm is indeed generic; it can be applied to any graphical model.

---

* An interesting example is a Boltzmann machine on a triangulated graph. The potentials are products of $\exp(J_{ij})$ factors, where the product is taken over all (i, j) pairs in a particular clique. Given that the product across potentials must be the joint probability, this implies that the partition function (the denominator of Equation 42.35) must be unity in this case.

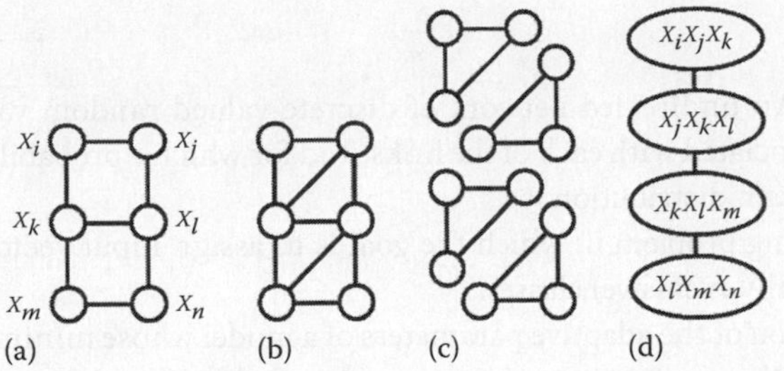

FIGURE 42.7 The basic structure of the junction tree algorithm for undirected graphs. The graph in (a) is first triangulated (b), then the cliques are identified (c), and arranged into a tree (d). Products of potential functions on the nodes in (d) yield probability distributions on the nodes in (a).

The problem of calculating posterior probabilities on graphs is NP-hard; thus, a major issue in the use of the inference algorithms is the identification of cases in which they are efficient. Chain structures such as HMMs yield efficient algorithms, and indeed the classical forward–backward algorithm for HMMs is a special, efficient case of the junction tree algorithm (Smyth et al. 1996). Decision tree structures such as the hierarchical mixture of experts yield efficient algorithms, and the recursive posterior probability calculation of Jordan and Jacobs (1994) described earlier is also a special case of the junction tree algorithm. All of the simpler mixture model calculations described earlier are therefore also special cases. Another interesting special case is the state estimation algorithm of the Kalman filter (Shachter and Kenley 1989). Finally, there are a variety of special cases of the Boltzmann machine, which are amenable to the exact calculations of the junction tree algorithm (Saul and Jordan 1995).

For graphs that are outside of the tractable categories of trees and chains, the junction tree algorithm often performs surprisingly well, but for highly connected graphs, the algorithm can be too slow. In such cases, approximate algorithms such as Gibbs sampling are utilized. A virtue of the graphical framework is that Gibbs sampling has a generic form, which is based on the notion of a *Markov boundary* (Pearl 1988). A special case of this generic form is the stochastic update rule for general Boltzmann machines.

Our discussion has emphasized the unifying framework of graphical models both for expressing probabilistic dependencies in graphs and for describing algorithms that perform the inferential step of calculating posterior probabilities on these graphs. The unification goes further, however, when we consider learning. A generic methodology known as the EM algorithm is available for MAP and Bayesian estimation in graphical models (Dempster et al. 1977). EM is an iterative method, based on two alternating steps: an *E step*, in which the values of hidden variables are estimated, based on the current values of the parameters and the values of visible variables, and an *M step*, in which the parameters are updated, based on the estimated values obtained from the *E* step. Within the framework of the EM algorithm, the junction tree algorithm can readily be viewed as providing a generic *E* step. Moreover, once the estimated values of the hidden nodes are obtained from the *E* step, the graph can be viewed as fully observed, and the *M* step is a standard MAP or ML problem. The standard algorithms for all of the tractable architectures described (mixtures, trees, and chains) are, in fact, instances of this general graphical EM algorithm, and the learning algorithm for general Boltzmann machines is a special case of a generalization of EM known as GEM (Dempster et al. 1977).

What about the case of feedforward neural networks such as the MLP? It is, in fact, possible to associate binary hidden values with the hidden units of such a network (cf. our earlier discussion of the logistic function; see also Amari 1995) and apply the EM algorithm directly. For N hidden units, however, there are patterns whose probabilities must be calculated in the *E* step. For large N, this is an intractable computation, and recent research has therefore begun to focus on fast methods for approximating these distributions (Hinton et al. 1995, Saul et al. 1996).

Key Terms

Boltzmann machine: An undirected network of discrete-valued random variables, where an energy function is associated with each of the links, and for which a probability distribution is defined by the Boltzmann distribution.

Classification: A learning problem in which the goal is to assign input vectors to one of a number of (usually mutually exclusive) classes.

Cost function: A function of the adaptive parameters of a model whose minimum is used to define suitable values for those parameters. It may consist of a likelihood function and additional terms.

Decision tree: A network that performs a sequence of classificatory decisions on an input vector and produces an output vector that is conditional on the outcome of the decision sequence.

Density estimation: The problem of modeling a probability distribution from a finite set of examples drawn from that distribution.

Discriminant function: A function of the input vector that can be used to assign inputs to classes in a classification problem.

Hidden Markov model: A graphical probabilistic model characterized by a state vector, an output vector, a state transition matrix, an emission matrix, and an initial state distribution.

Likelihood function: The probability of observing a particular dataset under the assumption of a given parametrized model, expressed as a function of the adaptive parameters of the model.

Mixture model: A probability model that consists of a linear combination of simpler component probability models.

Multilayer perceptron: The most common form of neural network model, consisting of successive linear transformations followed by processing with nonlinear activation functions.

Overfitting: The problem in which a model that is too complex captures too much of the noise in the data, leading to poor generalization.

Radial basis function network: A common network model consisting of a linear combination of basis functions, each of which is a function of the difference between the input vector and a center vector.

Regression: A learning problem in which the goal is to map each input vector to a real-valued output vector.

Regularization: A technique for controlling model complexity and improving generalization by the addition of a penalty term to the cost function.

VC dimension: A measure of the complexity of a model. Knowledge of the VC dimension permits an estimate to be made of the difference between performance on the training set and performance on a test set.

Further Information

In this chapter, we have emphasized the links between neural networks and statistical pattern recognition. A more extensive treatment from the same perspective can be found in Bishop (1995). For a view of recent research in the field, the *Proceedings of the Annual Neural Information Processing Systems (NIPS)*, MIT Press, and conferences are highly recommended.

Neural computing is now a very broad field, and there are many topics that have not been discussed for lack of space. Here, we aim to provide a brief overview of some of the more significant omissions and to give pointers to the literature.

The resurgence of interest in neural networks during the 1980s was due in large part to work on the statistical mechanics of fully connected networks having symmetric connections (i.e., if unit *i* sends a connection to unit *j*, then there is also a connection from unit *j* back to unit *i* with the same weight value). We have briefly discussed such systems; a more extensive introduction to this area can be found in Hertz et al. (1991).

The implementation of neural networks in specialist very large-scale integrated (VLSI) hardware has been the focus of much research, although by far the majority of work in neural computing is undertaken using software implementations running on standard platforms.

An implicit assumption throughout most of this chapter is that the processes that give rise to the data are stationary in time. The techniques discussed here can readily be applied to problems such as time series forecasting, provided this stationarity assumption is valid. If, however, the generator of the data is itself evolving with time, then more sophisticated techniques must be used, and these are the focus of much current research (see Bengio 1996).

One of the original motivations for neural networks was as models of information processing in biological systems such as the human brain. This remains the subject of considerable research activity, and there is a continuing flow of ideas between the fields of neurobiology and of artificial neural networks. Another historical springboard for neural network concepts was that of adaptive control, and again this remains a subject of great interest.

References

Amari, S. 1995. The EM algorithm and information geometry in neural network learning. *Neural Comput.* 7(1): 13–18.

Anderson, T. W. 1984. *An Introduction to Multivariate Statistical Analysis.* Wiley, New York.

Bengio, Y. 1996. *Neural Networks for Speech and Sequence Recognition.* Thomson Computer Press, London, U.K.

Bishop, C. M. 1995. *Neural Networks for Pattern Recognition.* Oxford University Press, Oxford, U.K.

Breiman, L., Friedman, J. H., Olshen, R. A., and Stone, C. J. 1984. *Classification and Regression Trees.* Wadsworth International Group, Belmont, CA.

Buntine, W. 1994. Operations for learning with graphical models. *J. Artif. Intelligence Res.* 2: 159–225.

Dempster, A. P., Laird, N. M., and Rubin, D. B. 1977. Maximum-likelihood from incomplete data via the EM algorithm. *J. R. Stat. Soc.* B39: 1–38.

Duda, R. O. and Hart, P. E. 1973. *Pattern Classification and Scene Analysis.* Wiley, New York.

Fletcher, R. 1987. *Practical Methods of Optimization*, 2nd edn. Wiley, New York.

Geman, S., Bienenstock, E., and Doursat, R. 1992. Neural networks and the bias/variance dilemma. *Neural Comput.* 4: 1–58.

Hertz, J., Krogh, A., and Palmer, R. G. 1991. *Introduction to the Theory of Neural Computation.* Addison–Wesley, Redwood City, CA.

Hinton, G. E., Dayan, P., Frey, B., and Neal, R. 1995. The wake-sleep algorithm for unsupervised neural networks. *Science* 268: 1158–1161.

Hinton, G. E. and Sejnowski, T. 1986. Learning and relearning in Boltzmann machines. In *Parallel Distributed Processing*, Vol. 1. D. E. Rumelhart and J. L. McClelland (Eds.), pp. 282–317. MIT Press, Cambridge, MA.

Hopfield, J. J. 1982. Neural networks and physical systems with emergent collective computational abilities. *Proc. Nat. Acad. Sci.* 79: 2554–2558.

Jacobs, R. A., Jordan, M. I., Nowlan, S. J., and Hinton, G. E. 1991. Adaptive mixtures of local experts. *Neural Comput.* 3: 79–87.

Jordan, M. I. and Jacobs, R. A. 1994. Hierarchical mixtures of experts and the EM algorithm. *Neural Comput.* 6: 181–214.

LeCun, Y., Boser, B., Denker, J. S., Henderson, D., Howard, R. E., Hubbard, W., and Jackel, L. D. 1989. Backpropagation applied to handwritten zip code recognition. *Neural Comput.* 1(4): 541–551.

Luttrell, S. 1989. The Gibbs machine applied to hidden Markov model problems. Royal Signals and Radar Establishment: SP Research Note 99, Malvern, U.K.

MacKay, D. J. C. 1992. A practical Bayesian framework for back-propagation networks. *Neural Comput.* 4: 448–472.

McCullagh, P. and Nelder, J. A. 1983. *Generalized Linear Models.* Chapman and Hall, London, U.K.

Neal, R. M. 1994. Bayesian learning for neural networks. Unpublished PhD thesis, Department of Computer Science, University of Toronto, Toronto, Ontario, Canada.

Pearl, J. 1988. *Probabilistic Reasoning in Intelligent Systems*. Morgan Kaufmann, San Mateo, CA.

Poggio, T. and Vetter, T. 1992. Recognition and structure from one 2-D model view: Observations on prototypes, object classes and symmetries. Artificial Intelligence Lab., AI Memo 1347, Massachusetts Institute of Technology, Cambridge, MA.

Rabiner, L. R. 1989. A tutorial on hidden Markov models and selected applications in speech recognition. *Proc. IEEE* 77: 257–286.

Rumelhart, D. E., Durbin, R., Golden, R., and Chauvin, Y. 1995. Backpropagation: The basic theory. In *Backpropagation: Theory, Architectures, and Applications*, Y. Chauvin, and D. E. Rumelhart (Eds.), pp. 1–35. Lawrence Erlbaum, Hillsdale, NJ.

Rumelhart, D. E., Hinton, G. E., and Williams, R. J. 1986. Learning internal representations by error propagation. In *Parallel Distributed Processing*, Vol. 1. D. E. Rumelhart and J. L. McClelland (Eds.), pp. 318–363. MIT Press, Cambridge, MA.

Sandler, D. G., Barrett, T. K., Palmer, D. A., Fugate, R. Q., and Wild, W. J. 1991. Use of a neural network to control an adaptive optics system for an astronomical telescope. *Nature* 351: 300–302.

Saul, L. K., Jaakkola, T., and Jordan, M. I. 1996. Mean field learning theory for sigmoid belief networks. *J. Artif. Intelligence Res.* 4: 61–76.

Saul, L. K. and Jordan, M. I. 1995. Boltzmann chains and hidden Markov models. In *Advances in Neural Information Processing Systems* 7, G. Tesauro, D. Touretzky, and T. Leen (Eds.), MIT Press, Cambridge, MA.

Shachter, R., Andersen, S., and Szolovits, P. 1994. Global conditioning for probabilistic inference in belief networks. *Proceedings of the 10th Conference on Uncertainty in Artificial Intelligence*, pp. 514–522, Seattle, WA.

Shachter, R. and Kenley, C. 1989. Gaussian influence diagrams. *Manage. Sci.* 35(5): 527–550.

Smyth, P., Heckerman, D., and Jordan, M. I. 1996. Probabilistic independence networks for hidden Markov probability models. *Neural Comput.* 9(2): 227–269.

Spiegelhalter, D., Dawid, A., Lauritzen, S., and Cowell, R. 1993. Bayesian analysis in expert systems. *Stat. Sci.* 8(3): 219–283.

Tarassenko, L. 1995. Novelty detection for the identification of masses in mammograms. *Proceedings of the 4th IEEE International Conference on Artificial Neural Networks*, Vol. 4, pp. 442–447, Mountain View, CA.

Vapnik, V. N. 1995. *The Nature of Statistical Learning Theory*. Springer–Verlag, New York.

43

Cognitive Modeling

Eric Chown
Bowdoin College

43.1 Introduction

An important goal of cognitive science is to understand human cognition. Good models of cognition can be *predictive*—describing how people are likely to react in different scenarios—as well as *prescriptive*—describing limitations in cognition and potential ways in which the limitations might be overcome. In a sense, the benefits of having cognitive models are similar to the benefits individuals accrue in building their own internal model. To quote Craik (1943):

> If the organism carries a "small-scale model" of external reality and of its own possible actions within its head, it is able to try out various alternatives, conclude which is the best of them, react to future situations before they arise, utilize the knowledge of past events in dealing with the present and future, and in every way to react in a much fuller, safer, and more competent manner to the emergencies which face it. (p. 61)

Among the important questions facing cognitive scientists are how such models are created and how they are represented internally. The Craik quote emphasizes the importance of the predictive power of models, and the ultimate measure of a model's value is the model's predictive accuracy. One important value of computers in cognitive science is that computer simulations provide a means to instantiate theories and to concretely test their predictive power. Furthermore, implementation of a theory in a computer model forces theoreticians to face practical issues that they may never have otherwise considered.

The role of computer science in cognitive modeling is not strictly limited to implementation and testing, however. A core belief of most cognitive scientists is that cognition is a form of computation (otherwise, computer modeling is a doomed enterprise), and the study of computation has long been a source of ideas (and material for debates) for building cognitive models. Computers themselves once served as the dominant metaphor for cognition. In more recent years, the influence of computers has been more in the area of computational paradigms, such as rule-based systems, neural network models, etc.

43.2 Underlying Principles

One of the dangers faced by computer scientists in building cognitive models is falling under the spell of trying to apply computational models directly to cognition. A good example of this is the "mind as computer" metaphor that was once popular but has fallen into disfavor. Computer science offers a range of computational tools designed to solve problems, and it is tempting to apply these tools to psychological data and call the result a cognitive model. As McCloskey (1991) has pointed out, this falls far short of the criteria that could reasonably be used to define a theory of cognition. One replacement for the "mind as computer" metaphor makes this point well. Neurally inspired models of cognition fell out of favor following Minsky and Papert (1969) showed that the dominant neural models at the time were unable to model nonlinear functions (notably exclusive-or). The extension of these models by the parallel distributed processing (PDP) group in 1986 (Rumelhart and McClelland, 1986) is largely responsible for the *connectionist* revolution for the past 25 years. The excitement generated by these models was twofold: (1) they were computationally powerful and simple to use, and (2) as neural-level models, they appeared to be physiologically plausible.

A major difficulty for connectionist theory for the past 25 years has been that, despite the fact that the early PDP-style models (particularly models built upon *feed-forward back-propagation* networks) were proven to be implausible for both physiological and theoretical reasons (e.g., see Lachter and Bever, 1988; Newell, 1990), many cognitive models are still built using such discredited computational engines. The reason for this appears to be simple convenience. Back-propagation networks, for example, can approximate virtually any function and are simple to train. Because any set of psychological data can be viewed as a function that maps an input to a behavior, and because feed-forward back-propagation networks can approximate virtually any function, it is hardly surprising that such networks can "model" an extraordinary range of psychological phenomena. Indeed, because of the versatility of these networks, a running joke in the machine learning community is that "neural networks are the second best way to do anything." To put this in another way, many cognitive models are written in computer languages like C. Although such models may accurately characterize a huge range of data on human cognition, no one would argue that the C programming language is a realistic model of human cognition. Feed-forward neural networks seem to be a better candidate for a cognitive model because of some of their features: they intrinsically learn, they process information in a manner reminiscent of neurons, etc. In any regard, this suggests that the criteria for judging the merits of a cognitive model must include many more constraints than whether or not the model is capable of accounting for a given data set. While issues such as how information is processed are useful for judging models, they are also crucial for constructing models.

There are a number of sources and types of constraints used in cognitive modeling. These break down relatively well by the disciplines that comprise the field. In practice, most cognitive models draw constraints from some, but not all, of these disciplines. In broad terms, the data for cognitive models come from psychology. "Hardware" constraints come from neuroscience. "Software" constraints come from computer science, which also provides methodologies for validation and testing. Two related fields that are relatively new, and therefore tend to provide softer constraints, are evolutionary psychology and environmental psychology. The root idea of each of these fields is that the evolution process, and especially the environmental conditions that took place during evolution, are crucially important to the kind of brain that we now have. We will examine the impact of each field on cognitive modeling in turn.

43.2.1 Psychology

The ultimate test of any cognitive theory is whether or not it can correctly predict human behavior. Psychology as a field is responsible for the vast majority of data on human behavior. Over the last century, the source of this data has evolved from mere introspection on the part of theorists to rigorous laboratory experimentation. Normally, the goal of psychological experiments is to isolate a particular cognitive factor; for example, the number of items a person can hold in short-term memory. In general, this isolation is used as a means of reducing complexity. In principle, this means that cognitive theories can be constructed piecewise instead of out of whole cloth. It would be fair to say that the majority of work in cognitive science proceeds on this principle. A fairly typical paper in a cognitive science conference proceeding, for example, will present a set of psychological experiments on some specific area of cognition, a model to account for the data, and computer simulations of the model.

43.2.2 Neuroscience

The impact of neuroscience on cognitive science has grown dramatically in conjunction with the influence of neural models in the past 25 years. Unfortunately, terms such as "neurally plausible" have been applied fairly haphazardly in order to lend an air of credibility to models. In response, some critics have argued that neurons are not well understood enough to be productively used as part of cognitive theory. Nevertheless, though the low level details are still being studied, neuroscience can provide a rich source of constraints and information for cognitive modelers. Within the field, there are several different types of architectural constraint available. These include the following:

1. *Information flow.* We have learned from neuroscientists, for example, that the visual system is divided into two distinct parts, a "what" system for object identification and a "where" system for determining spatial locations. This suggests computational models of vision should have similar properties. Furthermore, these constraints can be used to drive cognitive theory as with the PLAN model of human cognitive mapping (Chown et al., 1995). In PLAN, it was posited that humans navigate in two distinct ways, each corresponding to one of the visual pathways. Virtually all theories of cognitive mapping had previously included a "what" component based upon topological collections of landmarks, but none had a good theory of how more metric representations are constructed. The split in the visual system led the developers of PLAN to theorize that metric representations would have simple "where" objects as their basic units. This led directly to a representation built out of "scenes," which are roughly akin to snapshots. Models based on scenes or views have since become commonplace in the human navigation literature.
2. *Modularity.* A great deal of work in neuroscience goes toward understanding the kinds of processing done by particular areas of the brain, such as the hippocampus. These studies can range from working with patients with brain damage to intentionally lesioning animal brains. In recent years, noninvasive imaging techniques such as functional magnetic resonance imaging (fMRI) have revolutionized the role of neuroscience in cognitive modeling because they can be used directly on humans. While fMRI cannot provide neuroscientists with information on individual neurons, it can track neural activity across relatively small regions of the brain. This work has provided a picture of the brain far more complex than the simple "right brain-left brain" distinction of popular psychology.

   The usefulness of fMRI, especially in conjunction with computer science, is exemplified by the work of Just et al. (2010) that examines the question of how and where concepts are represented in the brain. Subjects were presented words, generally concrete nouns for three seconds at a time with rest periods in between. During this time, the subjects were monitored by fMRI and were told to generate a set of properties for each item that the participants were free to choose for themselves. Once the data was collected, a variety of techniques were used to analyze it, notably including a number of standard machine learning approaches. The results were remarkable and showed, for example, a commonality of patterns across the individuals tested. The authors concluded that not only do people store concepts in particular places in the brain but that these concepts are represented similarly enough across the population that one person's concept representation can be predicted by using other people's activation patterns.

3. *Mechanisms.* Numerous data are simpler to make sense of in the context of neural processing mechanisms. A good example of this would be the *Necker cube.* From a pure information processing point of view, there is no reason that people would only be able to hold one view of the cube in their mind at a time. From a neural point of view, on the other hand, the perception of the cube can be seen as a competitive process with two mutually inhibitory outcomes. Perceptual theory is an area that has particularly benefited from a neural viewpoint.
4. *Timing.* Perhaps the most famous constraint on cognitive processing offered by neuroscience is the "100 step rule." This rule is based upon looking at timing data of perception and the firing rate of neurons. From these, it has been determined that no perceptual algorithm could be more than 100 steps long (though the algorithm could be massively parallel as the brain itself is).

43.2.3 Computer Science

Aside from providing the means to implement and simulate models of cognition, computer science has also provided constraints on models through limits drawn from the theory of computation and has been a source of algorithms for modelers.

One of the biggest debates in the cognitive modeling community is whether or not computers are even capable of modeling human intelligence. Critics, normally philosophers, point to the limitations on what is computable and have gone as far as suggesting that the mind may not be computational. While some find these debates interesting, they do not actually have a significant impact on the enterprise of modeling. On the other hand, there have been theoretical results from computational theory that have had a huge impact on the development of cognitive models. Probably, the best example of this is the previously mentioned work done by Minsky and Papert on Perceptrons (1969). They showed that perceptrons, which are a simple kind of neural network, are not capable of modeling nonlinear functions (including exclusive-or). This result effectively ended the majority of neural network research for more than a decade until the PDP group developed far more powerful neural network models (Rumelhart and McClelland, 1986).

43.2.4 Evolutionary and Environmental Psychology

In recent years, two branches of psychology have come to prominence as providing alternate sources of constraints based upon evolution, and in particular, the kinds of environments in which humans evolved. Evolutionary psychology is most often associated with the work of Tooby and Cosmides (e.g. Tooby and Cosmides, 1992) while environmental psychology is often associated with the work of Steve and Rachel Kaplan (e.g. Kaplan and Kaplan, 1989). What both of these fields have in common is a belief that the brain should not be studied in a vacuum, that some types of context are extremely meaningful.

In the case of evolutionary psychology, the context is provided by evolution. As has been noted in many places, systems that are evolved rather than designed tend to end up looking like the work of a "tinkerer." The eye is a well-known example of a system that is poorly "designed" but is nonetheless functional and can be understood as a series of successive improvements, each adding functionality to the previous iteration (Dawkins, 1986). This example captures the core tenets of evolutionary psychology—that evolution tends to work in piecemeal fashion with each change adding functionality to what existed previously. This does not tend to be a hard constraint on cognitive models, as it can be argued that the evolutionary story behind any particular theory simply has yet to be found.

Nevertheless, the evolutionary view provides a powerful way to think about how pieces of the cognitive system came about and for what purpose.

As the name would suggest, work in environmental psychology focuses on the environment as a source of constraints. Most environmental psychologists focus their research on how people interact with different kinds of environments and how to use this knowledge to design better spaces. Another branch of the field, however, has noted that the environment adds additional meaningful context to evolutionary history. The evolutionary history of the brain is a story of information processing mechanisms that evolved to address the specific needs of our ancestors. The human ability to represent and reason about large-scale space, for example, allowed our ancestors to forage and hunt over large areas of savanna. In turn, once these spatial abilities were in place, they were available for the greater cognitive system and impact cognition of virtually every type (Chown, 1999). The importance in understanding the evolutionary environments that the brain developed in is highlighted by the work of the Kaplans and their colleagues. The Kaplans have shown, for example, that people will recover more quickly in hospitals with views of nature, perform better in workplaces with views of trees, etc. (for reviews see Kaplan, 1993; Kaplan and Peterson, 1993). This is a clear indicator that people do not treat information neutrally. More recently, this work has been extended to look at the impact of different kinds of information on attention and self-regulation (Kaplan and Berman, 2010). The authors make the case that certain kinds of attention are resource limited and that the depletion of these resources has negative cognitive consequences. Meanwhile, certain kinds of activity, for example, walking in nature, can actually support the replenishment of the lost resources. Work of this sort undercuts theories of cognition based on rationality and can help to explain many of the supposed shortcomings in human reasoning.

43.3 Research Issues

Since so much about cognition is still not well understood, this section will focus on two of the key debates driving research in the field. These include: (1) is the brain a symbol processor, or does it need to be modeled in neural terms? (2) Should the field be working on grand theories of cognition or is it better to proceed on a reductionist path?

43.3.1 Are We Symbol Processors?

The connectionist revolution brought a new way of thinking to cognitive science. The critical idea is rather simple—that since the "hardware" of the brain is neural, then models of the brain should be described in neural terms. Lending credence to this position were a series of neural models that had a number of attractive properties that seemed notably lacking in symbolic models of the time (e.g., *content addressable memory, graceful degradation*, etc. [Rumelhart and McClelland, 1986]). On the face of it, the argument for neural models seems unassailable given that the brain is a neural system. Symbolists, notably Newell (1990) and Fodor and Plyshen (1988) have attacked these models on a number of grounds. Their arguments are based upon the idea that the brain, like any complex system, is hierarchical. Neural models, so the argument goes, provide appropriate computational descriptions for only the lowest levels of the cognitive hierarchy. From the point of view of the symbolists, these levels of cognition are also less well understood and not as interesting from a behavioral point of view as the so-called *cognitive band* (Newell, 1990). From this point of view, the operation of the cognitive band is nothing like a neural network, but is much more like a traditional symbol system. The argument is akin to finding the appropriate level at which to study computers. It is possible, and occasionally necessary, to look at the performance of a computer from the point of view of gates. When trying to understand the performance of a complicated piece of software, however, it is much more appropriate to study the performance at the level of a high level programming language.

The argument for symbolic models comes from computational theory. It is based upon the idea of computational equivalence. Since symbolic models are Turing-complete, they are equivalent computationally to any other Turing-complete model. Along these lines, a number of efforts have been made to implement symbolic models in neural hardware. The case can then be made that this is exactly what the brain does. Symbolists see this equivalence as freeing them from the need to worry about mechanisms. Even so, the impact of connectionism and the capability of neural models for pattern recognition have led even strongly symbolic models like Soar (Laird et al., 1987) to use neural networks as pattern recognizers to obtain the symbols.

The freeing up from the constraints of mechanisms has been something of a double-edged sword for symbolic models. On the one hand, symbolic models are easily implemented on computers and relatively easy to expand, debug, etc. Furthermore, in terms of high-level behavior, symbolic models have been shown to be capable of a much wider range of behaviors than their current connectionist counterparts. For example, Soar agents are capable of modeling the flying behaviors of combat pilots (Jones et al., 1999). On the other hand, critics have complained that systems like Soar are little more than symbolic programming languages. As noted earlier, while a computer running C may be Turing-complete, it would be ridiculous to call it a model of human cognition. Clark (2001) refers to this as "surface mimicry" and points out that the Soar model is far more homogeneous than the "Swiss Army Knife" model suggested by Tooby and Cosmides (1992). The critics argue that symbolic models like Soar and ACT-R are under constrained, and therefore they cannot truly be called cognitive models. The lack of constraints goes even further than just mechanisms; symbolic models have also been attacked on evolutionary grounds. It is difficult to see how symbolic constructs like distal memory access could have evolved in any sort of piecemeal fashion.

There are several arguments that symbolists use against neural models. First is the "levels of modeling" argument. This argument posits that we simply do not understand the behavior of neurons well enough to construct credible models from them. To be fair, the majority of neural models do not model the behavior of individual neurons. Other attacks on connectionist models are based upon what current models cannot do. Popular connectionist models, such as feed-forward back-propagation networks, for example, exhibit a number of problems as memory models including "catastrophic forgetting" of old material when presented with new material (McCloskey and Cohen, 1989). While it is certainly appropriate to attack the individual models on these grounds, it is less so to attack the entire connectionist position based on the failure of even its most notable examples. Similarly, other criticisms of connectionism have attacked the models as being nothing more than new versions of the discredited behaviorist position (Lachter and Bever, 1988). Again, this is certainly the case with many connectionist models, and should rightfully lead to a search for better models, but it cannot undermine the general position.

A more interesting criticism focuses upon the way that connectionists have pursued cognitive modeling. The argument is the same as the one against some of the symbolic programs. As McCloskey (1991) put it, connectionists often pursue a path of "simulation in search of theory" (p. 388). McCloskey argues that many of these simulations are little better than "black boxes." Modelers do not provide insight into what aspects of the network are crucial to its performance with regard to the task. For example, if back-propagation was used to train the network, is that a crucial step, or could another training regime have been used? If back-propagation were crucial, then it would undermine the model if back-propagation were found implausible on other grounds. In other words, connectionist systems are engaged in the same sort of "surface mimicry" as symbolic systems under the guise of "neural plausibility." McCloskey goes on to complain that connectionist models generally fall short in describing how their networks elucidate the cognitive processes they purport to model. To put it simply, connectionist networks are not well understood enough to tell us exactly how they manage to accomplish what they are trained for.

So far in discussing connectionist systems, we have avoided discussing connectionist "symbols." While it may be the case that connectionist units represent collections of neurons, they do not normally represent what most computer scientists would think of as symbols. There are, however, connectionist models that recognize the power of symbols as a basic unit of thought. Many of these models trace their lineage to the work of D.O. Hebb's book *The Organization of Behavior* (1949). Hebb proposed that the "symbols" of thought were cell assemblies, tightly connected groups of neurons capable of functioning as a unit because of their strong interconnections. The problem with cell assemblies, as originally formulated by Hebb, was that all of the connections between neurons were positive and only became stronger through learning. Hebb omitted inhibition because there was still no hard evidence of it at the time of publication. A later simulation of the cell assembly construct (Rochester et al., 1956) showed that without inhibition, activity in the simulated brain quickly grew out of control. Unfortunately, these results were sufficient to essentially stop research on cell assemblies for more than a decade, even though the same paper showed that with the addition of inhibition the cell assembly construct was viable. In recent years, however, the cell assembly idea has undergone a revival as researchers from a number of domains have proposed models based upon Hebb's original idea but modified with modern understandings of neuroscience (Amit, 1995; deVries, 2004; Kaplan et al., 1991). Recent work of the sort done by Just et al. (2010) using fMRI to examine how concepts are represented in the brain has also supported Hebb's position.

Neural models based upon cell assemblies purport to contain the best of both symbolic and connectionist models. It is difficult to study sequence learning, for example, without something approximating a symbol to serve as a unit in the sequence. Furthermore, many connectionist systems do not address temporal issues at all. Conversely, symbolic models do not ground the symbols in any physical mechanism nor are symbolic systems able to take advantage of the properties of neural hardware. Chown (2002), for example, has shown that some learning results that have defied conventional modeling for nearly 40 years can be fairly easily explained when basic neural properties are accounted for in a cell assembly-based model.

43.3.2 Grand Theories?

In his book *Unified Theories of Cognition*, Allen Newell (1990) called for a return to all-encompassing theories of mind called *UTCs* after the title of the book. Newell's reasoning echoes McCloskey's (1991) complaints about connectionism, which by operating at too small a scale, cognitive modelers have worked on under-constrained models. While it may be true, for example, that connectionist system X can model psychological data set Y, such models rarely address questions of how they would or could fit into a larger system. Modeling efforts such as these are sometimes attacked on the grounds that they are "doomed to succeed" in much the same way that models with too many parameters can fit all types of curves. Put another way, if a model is Turing complete, the question of whether or not it can be used to fit some data is not particularly interesting. The interesting question is whether or not there is actually evidence for it. McCloskey and others argue that for these reasons, theory should drive simulation rather than the other way around.

Ironically, the two most notable examples of UTCs, Soar and ACT-R, have been attacked on virtually the same grounds—that they are under-constrained. Both systems are built on the same assumption, which at higher levels of cognition the brain is a rule-based system. At the heart of each system is a production system implementing the rule-base that serves as long-term memory. Since production systems are Turing-complete, they are capable of modeling anything. To be fair, however, Soar does make some key theoretical commitments that can be used to judge its merits. First and foremost, Soar is a symbol manipulation system with all that that implies. Second, in Soar deliberative, thought is equated to a search through a problem space. For example, Soar is equipped with all the basic weak search methods including breadth-first, depth-first, etc. At one time, another key constraint associated with Soar was that all learning came as the result of a single mechanism called "chunking" (Laird et al., 1984). It is not clear, however, from recent work in Soar that this is still held as a central tenet of the system. Furthermore, the Soar group seems to be retreating from positioning Soar as a model of human cognition and is instead focusing on Soar as a system capable of exhibiting "human-level" intelligence.

Soar and ACT-R have both been attacked for their commitment to production systems. This criticism actually pre-dates either system and is most famously associated with the philosopher Hubert Dreyfus (1972). The Dreyfus position is that systems based upon rules are too brittle to account for the richness of human behavior. For example, Dreyfus discusses how knowledge about the health of a jockey's mother might influence how a bettor would make a wager. It seems unlikely that the bettor would have explicit rules dealing with such a situation, and yet humans are capable of dealing with such situations with ease. The response to this criticism has been to test it explicitly, either with Doug Lenat's CYC (Lenat and Feigenbaum, 1992), which aims to capture enough knowledge to perform common sense reasoning, or with the Soar program, which builds more and more complex agents capable of difficult tasks such as flying jet airplanes in combat situations. In part, Dreyfus's criticisms can also be addressed by noting that rules need not all be specified at the same level of generality. For example, while a system for betting on horses might contain many specific rules concerning the records of horses and jockeys, a general cognitive system might reasonably be expected to include rules such as "when something traumatic happens to a person they will not perform at normal levels." Of course, this raises further questions of how such rules are learned, how patterns such as "something traumatic" are recognized, etc. Dreyfus would argue that this leads to an endless cycle for any reasonably complex task.

There are also connectionist programs that work at the level of large theories of cognition. Steven Grossberg, for example, has produced a huge body of work that have never been explicitly put forth as a UTC but which when viewed as a whole have many of the same principles. Probably, the best example of this work is the adaptive resonance theory (ART) model developed in conjunction with Gail Carpenter (see Carpenter and Grossberg, 1987). The SESAME group, operating mainly out of the University of Michigan is also working on a cognitive architecture (Kaplan et al., 1991). The SESAME architecture is based on the cell assembly and is also the only cognitive architecture to include a complete theory of spatial processing (Chown et al., 1995).

43.4 Best Practices

As the previous sections suggest, there are a number of pitfalls involved in putting together a cognitive model. History has shown that there are two problems that crop up again and again. The first is the danger of constructing a simulation without theoretically motivating the details. This is akin to the old saying that "if you have a big enough hammer everything looks like a nail." There is a related danger that once a simulation works (or at least models the data), it is often difficult to say why. Together, these dangers suggest that there should be a close relationship between theory and the simulation process. The goal of a simulation should not be simply to model a dataset, but should also be to elucidate the theory. For example, some connectionist models propose a number of mechanisms as being central to understanding a particular process. These models can be systematically "damaged" by disabling the individual mechanisms. In many cases, the damage to the model can be equated to damage to individuals. This provides a second dataset to model, and provides solid evidence of what the mechanism does in the simulation. Alternatively, models can be built piecewise mechanism by mechanism. Each new piece of the simulation would correspond to a new theoretical mechanism aiming to address some shortcoming of the previous iteration. This motivates each mechanism and helps to clearly delineate its role in the overall simulation. In the SESAME group, this style of simulation has been termed "the systematic exploitation of failure" by one of its members, Abraham Kaplan.

One of the earliest examples of this approach was done by Booker (1982) in an influential work that has helped shape the adaptive systems paradigm. In an adaptive systems paradigm, a simple creature is placed in a microworld where the goal is survival. Creatures are successively altered (and sometimes the environments are as well) by adding and subtracting mechanisms. In each case, the success of the new mechanism can be judged by improvements in the survival rate of the organism. In addition to providing a way to motivate theoretical mechanisms, this paradigm is also essentially the same one used for the development of genetic algorithms.

The Soar architecture is probably the preeminent symbolic cognitive architecture. Soar is based upon a number of crucial premises that constrain all models written in Soar (which can be considered a kind of programming environment). First, Soar, like its major competition ACT-R, is a rule-based system implemented as a production system. In the Soar paradigm, the production rules represent long-term memory and knowledge. One effect of a production firing in Soar can be to put new elements into working memory, Soar's version of short-term memory. For example, a Soar system might contain a number of perceptual productions that aim to identify different types of aircraft. When a production fires, it might create a structure in working memory to represent the aircraft it identifies. This structure in turn might cause further productions to fire. Soar enforces a kind of hierarchy through the use of a subgoaling system. Productions can be written to apply generally or might only match when a certain goal is active. The combination of goals and productions forms a problem space that provides the basic framework for any task. Finally, the Soar architecture contains a single mechanism for learning called "chunking." Essentially, Soar systems learn when they reach an impasse generally created by not being able to match any productions. When impasses occur, Soar can apply weak search methods to the problem space in order to discover what to do. Once a solution is found, a new production or "chunk" is created to apply to the situation.

Here is an example of a Soar production taken from the Soar tutorial (Laird, 2003). In this example, the agent is driving a tank in a battle exercise.

```
sp {wander*propose*move
    (state <s> ^ name wander
               ^io.input-linked-blocked.forward no)
-->
    (<s>      ^operator <o> + =)
    (<o>      ^name move
              ^actions.move.direction forward)
}
```

In the example, "sp" stands for "Soar production" and starts every production. The production is named "wander*propose*move." The elements that come before the arrow represent the "if" part of the production. In this case, the production fires only if the current subgoal is to wander and the forward direction is not blocked (input comes from a specialized structure tagged ^io). The elements that come after the arrow represent the "then" parts of the production. In this case, a new working memory element is created to represent the operator for moving forward. In a typical production cycle, productions are matched in parallel and can propose operators such as the move operator in this case. Then other productions can be used to select among the proposed operators. This selection can be based upon virtually any criteria; for example, cognitive productions may be selected over more reactive productions. Production matching can be done in parallel to simulate the parallelism of the brain.

Both the Soar and ACT-R communities are engaged in programs of simulating more and more human behavior. These simulations can be done at the level of models of simple psychological experiments, or, as is increasingly the case, they can simulate human performance on complex tasks such as flying airplanes. The implicit argument is that if they can simulate anything that humans can do then they must be modeling human cognition. On one level, this argument has merit, if either architecture can accurately simulate human performance, then it can be used in a predictive fashion. Tac-air Soar (Jones et al., 1999), for example, simulates the performance of combat pilots and is used to train new pilots in a more cost-effective way than if experienced pilots had to be used. On the other hand, equivalent functionality is not the same thing as equivalence. As noted previously, critics point out that both Soar and ACT-R are essentially Turing-complete programming environments and therefore are capable of simulating any computable function given clever enough programmers. The fact that both systems still rely heavily on clever programming is still a major limitation with regard to being considered a fully realized model of human cognition. Although Soar's initial success was due in large part to its learning mechanism, for example, little progress has been made within the Soar community in building agents that exhibit any sort of developmental patterns. It is much simpler to build a Soar system that can fly planes than one that can learn to fly planes.

Gail Carpenter, Stephen Grossberg and their associates have also attacked a wide range of problems, but have done so with much more of an eye toward cognitive theory than applications. While Carpenter and Grossberg have not explicitly developed a unified theory of cognition, they have modeled a remarkable range of cognitive processes and have done so using an approach more sympathetic to a systems view of cognition than is typical in connectionist modelers. A good example of this approach can be found in their ART (Carpenter and Grossberg, 1987; Grossberg, 1987). Superficially, ART looks similar to many connectionist learning systems in that it is essentially a classification system, but it was developed to specifically address many of the shortcomings of such models. A feature vector is the input to Art, which uses it to provide a classification of the input. For example, a typical task would be to recognize hand-written numbers. The features would consist of the presence or absence of a pen stroke at different spatial locations.

One of the problems that ART was designed to address was what Grossberg (1987) referred to as the "stability-plasticity" dilemma. This is essentially a problem of how much new knowledge should impact what has been learned before. For example, a system that has been trained to recognize horses might have a problem when confronted with a zebra. The system could either change its representation of horses to include zebras or it could create a separate representation for zebras. This is a significant issue for neural network models because they achieve a great deal of their power by having multiple representations share structure. Such sharing is useful for building compact representations and for automatic abstraction, but it also means that new knowledge tends to constantly overwrite what has come before. Among the problems this raises is "catastrophic forgetting" as mentioned previously.

The stability-plasticity dilemma was addressed, in part, in ART through the introduction of a vigilance parameter that adaptively changed according to how well the system was performing. In some cases, for example, the system would be extremely vigilant and would require an unusually high degree of match before it would recognize an input as being familiar. In cases where inputs were not recognized as familiar, novel structure was created to form a new category or prototype. Such a new category would not share internal structure directly with previously learned categories. Essentially when vigilance is high, the system creates "exemplars" or very specialized categories, whereas when vigilance is low, ART will create "prototypes" that generalize across many instances. This makes ART systems attractive since they do not commit fully either to exemplar or prototype models, but can exhibit properties of both, as seems to be the case with human categorization.

In ART systems, an input vector activates a set of feature cells within an attentional system, essentially storing the vector in short-term memory. These in turn activate corresponding pathways in a bottom-up process. The weights in these pathways represent long-term memory traces and act to pass activity to individual categories. The degree of activation of a category represents an estimate that the input is an example of the category. In the meantime, the categories send top-down information back to the features as a kind of hypothesis test. The vigilance parameter defines the criteria for whether the match is good enough. When a match is established, the bottom-up and top-down signals are locked into a "resonant" state, and this in turn triggers learning, is incorporated into consciousness, etc.

It is important to note that ART, unlike many connectionist learning systems, is unsupervised—it learns the categories without any teaching signals.

A number of extensions have followed the original ART, including ART1, ART2, ART3, and ARTMAP. Grossberg has also tied it to his FACADE model in a system called ARTEX (Grossberg and Williamson, 1999). These models vary in features and complexity but share intrinsic theoretical properties. All of the ART models are self-organizing (i.e., unsupervised, though ARTMAP systems can include supervised learning) and consist of an attentional and an orienting subsystem. A fundamental property of any ART system (and many other connectionist systems) is that perception is a competitive process. Different learned patterns generate expectations that essentially compete against each other. Meanwhile, the orienting system controls whether or not such expectations sufficiently match the input-in other words it acts as a novelty detector.

The ART family of models demonstrate many of the reasons why working with connectionist models can be so attractive. Among them:

- The neural computational medium is natural for many processes including perception. Fundamental ideas such as representations competing against each other (including inhibiting each other) are often difficult to capture in a symbolic model. In a system like ART, on the other hand, a systemic property like the level of activation of a unit can naturally fill many roles from the straightforward transmission of information to providing different measures of the goodness of fit of various representations to input data.
- The architecture of the brain is a source of both constraints and ideas. Parameters, such as ART's vigilance parameter, can be linked directly to real brain mechanisms such as the arousal system. In this way, what is known about the arousal system provides clues as to the necessary effects of the mechanism in the model and provides insight into how the brain handles fundamental issues such as the plasticity-stability dilemma.

43.5 Summary

Unlike many disciplines in computer science, there are no provably correct algorithms for building cognitive models. Progress in the field is made through a process of successive approximation. Models are continually proposed and rejected; and with each iteration of this process, the hope is that the models come closer to a true approximation of the underlying cognitive structure of the brain. It should be clear from the preceding sections that there is no "right" way to do this.

Improvements in cognitive models come from several sources. In many cases, improvements result from an increased understanding of some aspect of cognition. For example, neuroscientists are constantly getting new data on how neurons work, how they are connected, what parts of the brain process what types of information, etc. In the meantime, models are implemented on computers and on robots. These implementations provide direct feedback about model quality and shortcomings. This feedback often will lead to revisions in the models and sometimes may even drive further experimental work. Because of the complexity of cognition and the number of interactions amongst parts of the brain, it is rarely the case that definitive answers can be found; which is not to say that cognitive scientists do not reach consensus on any issues. Over time, for example, evidence has accumulated that there are multiple memory systems operating at different time-scales. While many models have been proposed to account for this, there is general agreement on the kinds of behavior that those models need to be able to display. This represents real progress in the field because it eliminates whole classes of models that could not account for the different time-scales. The constraints provided by data and by testing models work to continually narrow the field of prospective models.

Key Terms

Back-propagation: A method for training neural networks based upon gradient descent. An error signal is propagated backward from output layers toward the input layer through the network.

Cognitive band: In Newell's hierarchy of cognition, the cognitive band is the level at which deliberate thought takes place.

Cognitive map: A mental model. Often, but not exclusively, used for models of large-scale space.

Connectionist: A term used to describe neural network models. The choice of the term is meant to indicate that the power of the models comes from the massive number of connections between units within the model.

Content addressable memory: Memory that can be retrieved by descriptors. For example, people can remember a person when given a general description of the person.

Feed forward: Neural networks are often constructed in a series of layers. In many models, information flows from an input layer toward an output layer in one direction. Models in which the information flows in both directions are called *recurrent*.

Graceful degradation: The principle that small changes in the input to a model, or that result from damage to a model, should result in only small changes to the model's performance. For example, adding noise to a model's input should not break the model.

Necker cube: A three dimensional drawing of a cube drawn in such a way that either of the two main squares that comprise the drawing can be viewed as the face closest to the viewer.

UTC: Unified Theory of Cognition.

Acknowledgment

This work was supported by the National Science Foundation (NSF) under grant number 1017983. The author gratefully acknowledges that support.

Further Information

There are numerous journals and conferences on cognitive modeling. Probably, the best place to start is with the annual conference of the Cognitive Science Society. This conference takes place in a different city each summer. The Society also has an associated journal, *Cognitive Science*. Information on the journal and the conference can be found at the society's homepage at http://www.cognitivesciencesociety.org.

Because of the lag-time in publishing journals, conferences are often the best place to get the latest research. Among other conferences, Neural Information Processing Systems (NIPS) is one of the best for work specializing in neural modeling. The Simulation of Adaptive Behavior conference is excellent for adaptive systems. It has an associated journal as well, *Adaptive Behavior*.

A good place for anyone interested in cognitive modeling to start is Allen Newell's book, *Unified Theories of Cognition*. While a great deal of the book is devoted to Soar, the first several chapters lay out the challenges and issues facing any cognitive modeler. Another excellent starting point is Dana Ballard's 1999 book, *An Introduction to Natural Computation*. Ballard emphasizes neural models, and his book provides good coverage on most of the major models in use. Andy Clark's 2001 book, *Mindware: An Introduction to the Philosophy of Cognitive Science*, covers much of the same ground as this article, but in greater detail, especially with regard to the debate between connectionists and symbolists.

References

Amit, D.J. (1995). The Hebbian paradigm reintegrated: Local reverberations as internal representations. *Behavioral and Brain Sciences*, 18(4), 617–657.

Booker, L.B. (1982). Intelligent behavior as an adaptation to the task environment. PhD dissertation, The University of Michigan.

Carpenter, G.A. and Grossberg, S. (1987). A massively parallel architecture for a self-organizing neural pattern recognition machine. *Computer Vision, Graphics and Image Processing*, 37, 54–115.

Chown, E. (1999). Making predictions in an uncertain world: Environmental structure and cognitive maps. *Adaptive Behavior*, 7(1), 1–17.

Chown, E. (2002). Reminiscence and arousal: A connectionist model. *Proceedings of the Twenty Fourth Annual Meeting of the Cognitive Science Society*, Fairfax, VA, 234–239.

Chown, E., Kaplan, S., and Kortenkamp, D. (1995). Prototypes, location, and associative networks (PLAN): Towards a unified theory of cognitive mapping. *Cognitive Science*, 19, 1–51.

Clark, A. (2001). *Mindware: An Introduction to the Philosophy of Cognitive Science*. New York: Oxford University Press.

Craik, K.J.W. (1943). *The Nature of Exploration*. London, U.K.: Cambridge University Press.

Dawkins, R. (1986). *The Blind Watchmaker*. New York: W.W. Norton & Company.

deVries, P. (2004). Effects of binning in the identification of objects. *Psychological Research*, 69, 41–66.

Dreyfus, H. (1972). *What Computers Can't Do*. New York: Harper & Row.

Fodor, J.A. and Pylyshyn, Z.W. (1988). Connectionism and cognitive architecture: A critical analysis. *Cognition*, 28, 3–71.

Grossberg, S. (1987). Competitive learning: From interactive activation to adaptive resonance. *Cognitive Science*, 11, 23–63.

Grossberg, S. and Williamson, J.R. (1999). A self-organizing neural system for learning to recognize textured scenes. *Vision Research*, 39, 1385–1406.

Hebb, D.O. (1949). *The Organization of Behavior*. New York: John Wiley

Jones, R.M., Laird, J.E., Nielsen, P.E., Coulter, K.J., Kenny, P.G., and Koss, F., (1999). Automated intelligent pilots for combat flight simulation. *AI Magazine*, 20(1), 27–41.

Just, M.A., Cherkassky, V.L., Aryal, S., and Mitchell, T.M. (2010). A neurosemantic theory of concrete noun representation based on the underlying brain codes. *PLoS ONE*, 5(1), e8622.

Kaplan, R. (1993). The role of nature in the context of the workplace. *Landscape and Urban Planning*, 26, 193–201.

Kaplan, R. and Kaplan, S. (1989). *The Experience of Nature: A Psychological Perspective*. New York: Cambridge University Press.

Kaplan, S. and Berman, M.G. (2010). Directed attention as a common resource for executive functioning and self-regulation. *Perspectives on Psychological Science*, 5(1), 43–57.

Kaplan, S. and Peterson, C. (1993). Health and environment: A psychological analysis. *Landscape and Urban Planning*, 26, 17–23.

Kaplan, S., Sonntag, M., and Chown, E. (1991). Tracing recurrent activity in cognitive elements (TRACE): A model of temporal dynamics in a cell assembly. *Connection Science*, 3, 179–206.

Lachter, J. and Bever, T. (1988). The relationship between linguistic structure and associative theories of language learning—A constructive critique of some connectionist teaching models. *Cognition*, 28, 195–247.

Laird, J.E. (2003). The Soar 8 Tutorial. http://ai.eecs.umich.edu/soar/tutorial.html (accessed June 24, 2008).

Laird, J.E., Newell, A., and Rosenbloom, P.S. (1987). Soar: An architecture for general intelligence. *Artificial Intelligence*, 33, 1–64.

Laird, J.E., Rosenbloom, P.S., and Newell, A. (1984). Towards chunking as a general learning mechanism. *Proceedings of the AAAI'84 National Conference on Artificial Intelligence*, Austin, TX. American Association for Artificial Intelligence.

Lenat, D. and Feigenbaum, E. (1992). On the thresholds of knowledge. In D. Kirsh (Ed.), *Foundations of Artificial Intelligence*, 195–250. MIT Press and Elsevier Science, Amsterdam, The Netherlands.

McCloskey, M. (1991). Networks and theories: The place of connectionism in cognitive science. *Psychological Science*, 2(6), 387–395.

McCloskey, M. and Cohen, N.J. (1989). Catastrophic interference in connectionist networks: The sequential learning problem. In G.H. Bower, Ed. *The Psychology of Learning and Motivation*, Vol. 24, pp. 109–123. New York: Academic Press.

Minsky, M. and Papert, S. (1969). *Perceptrons*. MIT Press, Oxford, England.

Newell, A. (1990). *Unified Theories of Cognition*. Cambridge, MA: Harvard University Press.

Rochester, N., Holland, J.H., Haibt, L.H., and Duda, W.L. (1956). Tests on a cell assembly theory of the action of the brain, using a large digital computer. *IRE Transactions on Information Processing Theory*, IT-2, 2(3), 80–93.

Rumelhart, D.E. and McClelland, J.L., Eds. (1986). *Parallel Distributed Processing: Explorations in the Microstructure of Cognition*, Cambridge, MA: The MIT Press.

Tooby, J. and Cosmides, L. (1992). The psychological foundations of culture. In J. Barkow, L. Cosmides, and J. Tooby, Eds., *The Adapted Mind*, New York: Oxford University Press. pp. 19–136.

44

Graphical Models for Probabilistic and Causal Reasoning

Judea Pearl
University of California, Los Angeles

44.1 Introduction

This chapter surveys the development of graphical models known as Bayesian networks, summarizes their semantical basis, and assesses their properties and applications to reasoning and planning.

Bayesian networks are directed acyclic graphs (DAGs) in which the nodes represent variables of interest (e.g., the temperature of a device, the gender of a patient, a feature of an object, the occurrence of an event), and the links represent causal influences among the variables. The strength of an influence is represented by conditional probabilities that are attached to each cluster of parent–child nodes in the network.

Figure 44.1 illustrates a simple yet typical Bayesian network. It describes the causal relationships among the season of the year (X_1), whether rain falls (X_2) during the season, whether the sprinkler is on (X_3) during that season, whether the pavement would get wet (X_4), and whether the pavement would be slippery (X_5). All variables in this figure are binary, taking a value of either true or false, except the root variable X_1 that can take one of four values: spring, summer, fall, or winter. Here, the absence of a direct link between X_1 and X_5, for example, captures our understanding that the influence of seasonal variations on the slipperiness of the pavement is mediated by other conditions (e.g., the wetness of the pavement).

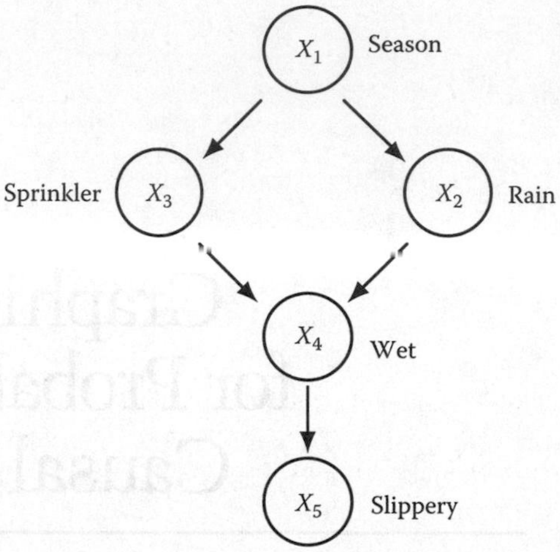

FIGURE 44.1 A Bayesian network representing causal influences among five variables.

As this example illustrates, a Bayesian network constitutes a model of the environment rather than, as in many other knowledge representation schemes (e.g., logic, rule-based systems, and neural networks), a model of the reasoning process. It simulates, in fact, the causal mechanisms that operate in the environment and thus allows the investigator to answer a variety of queries, including associational queries, such as "Having observed *A*, what can we expect of *B*?"; abductive queries, such as "What is the most plausible explanation for a given set of observations?"; and control queries, such as "What will happen if we intervene and act on the environment?" Answers to the first type of query depend only on probabilistic knowledge of the domain, while answers to the second and third types rely on the causal knowledge embedded in the network. Both types of knowledge, associative and causal, can effectively be represented and processed in Bayesian networks.

The associative facility of Bayesian networks may be used to model cognitive tasks such as object recognition, reading comprehension, and temporal projections. For such tasks, the probabilistic basis of Bayesian networks offers a coherent semantics for coordinating top–down and bottom–up inferences, thus bridging information from high-level concepts and low-level percepts. This capability is important for achieving selective attention, that is, selecting the most informative next observation before actually making the observation. In certain structures, the coordination of these two modes of inference can be accomplished by parallel and distributed processes that communicate through the links in the network.

However, the most distinctive feature of Bayesian networks, stemming largely from their causal organization, is their ability to represent and respond to changing configurations. Any local reconfiguration of the mechanisms in the environment can be translated, with only minor modification, into an isomorphic reconfiguration of the network topology. For example, to represent a disabled sprinkler, we simply delete from the network all links incident to the node "sprinkler." To represent a pavement covered by a tent, we simply delete the link between "rain" and "wet." This flexibility is often cited as the ingredient that marks the division between deliberative and reactive agents, and that enables the former to manage novel situations instantaneously, without requiring retraining or adaptation.

44.2 Historical Background

Networks employing DAGs have a long and rich tradition, starting with the geneticist Sewall Wright (1921). Wright (1921) developed a method called *path analysis* (Wright, 1934), which later became an established representation of causal models in economics (Wold, 1964), sociology (Blalock, 1971;

Kenny, 1979), and psychology (Duncan, 1975). Good (1961) used DAGs to represent causal hierarchies of binary variables with disjunctive causes. *Influence diagrams* represent another application of DAG representation (Howard and Matheson, 1981). Developed for decision analysis, they contain both event nodes and decision nodes. *Recursive models* is the name given to such networks by statisticians seeking meaningful and effective decompositions of contingency tables (Lauritzen, 1982; Wermuth and Lauritzen, 1983; Kiiveri et al., 1984).

The role of the network in the applications previously mentioned was primarily to provide an efficient description for probability functions; once the network was configured, all subsequent computations were pursued by symbolic manipulation of probability expressions. The potential for the network to work as a computational architecture, and hence as a model of cognitive activities, was noted in Pearl (1982), where a distributed scheme was demonstrated for probabilistic updating on tree-structured networks. The motivation behind this particular development was the modeling of distributed processing in reading comprehension (Rumelhart, 1976), where both top–down and bottom–up inferences are combined to form a coherent interpretation. This dual mode of reasoning is at the heart of Bayesian updating and in fact motivated Reverend Bayes' original 1763 calculations of posterior probabilities (representing explanations), given prior probabilities (representing causes), and likelihood functions (representing evidence).

Bayesian networks have not attracted much attention in the logic and cognitive modeling circles, but they did in expert systems. The ability to coordinate bidirectional inferences filled a void in expert systems technology in the late 1970s, and it is in this area that Bayesian networks truly flourished. Over the past 10 years, Bayesian networks have become a tool of great versatility and power, and they are now the most common representation scheme for probabilistic knowledge (Neapolitan, 1990; Spiegelhalter et al., 1993; Darwiche, 2009; Koller and Friedman, 2009). They have been used to aid in the diagnosis of medical patients (Andersen et al., 1989; Geiger et al., 1990; Heckerman, 1991; Jiang and Cooper, 2010) and malfunctioning systems (Fenton and Neil, 2012); to understand stories (Charniak and Goldman, 1991); to retrieve documents (de Campos et al., 2004); to interpret pictures (Levitt et al., 1990); to perform filtering, smoothing, and prediction (Weiss and Pearl, 2010); to analyze gene expressions (Friedman et al., 2000), genetic counseling (Uebersax, 2004), semantic search (Koumenides and Shadbolt, 2012), error-correcting codes (McEliece et al., 1998), and speech recognition (Zweig, 1998); to facilitate planning in uncertain environments (Guestrin, 2003; Beaudry et al., 2010); and to study causation, nonmonotonicity, action, change, and attention. Some of these applications are described in Pearl (1988) Russell and Norvig (2003), and Buede (2009).

44.3 Bayesian Networks as Carriers of Probabilistic Information

44.3.1 Formal Semantics

Given a DAG G and a joint distribution P over a set $X = \{X_1,\ldots, X_n\}$ of discrete variables, we say that G *represents* P if there is a one-to-one correspondence between the variables in X and the nodes of G, such that P admits the recursive product decomposition

$$P(x_1,\ldots, x_n) = \prod_i P\left(x_i \middle| \mathbf{pa}_i\right) \tag{44.1}$$

where $\mathbf{pa}_i$ are the direct predecessors (called *parents*) of X_i in G. For example, the DAG in Figure 44.1 induces the decomposition

$$P(x_1, x_2, x_3, x_4, x_5) = P(x_1)\, P\left(x_2 \middle| x_1\right) P\left(x_3 \middle| x_1\right)\, P\left(x_4 \middle| x_2, x_3\right)\, P\left(x_5 \middle| x_4\right) \tag{44.2}$$

The recursive decomposition in Equation 44.1 implies that, given its parent set $\mathbf{pa}_i$, each variable X_i is conditionally independent of all its other predecessors $\{X_1, X_2, \ldots, X_{i-1}\}\backslash\mathbf{pa}_i$. Using Dawid's notation (Dawid, 1979), we can state this set of independencies as

$$X_i \perp\!\!\!\perp \{X_1, X_2, \ldots, X_{i-1}\}\backslash\mathbf{pa}_i \,|\, \mathbf{pa}_i \quad i = 2, \ldots, n \tag{44.3}$$

Such a set of independencies is called *Markovian*, since it reflects the Markovian condition for state transitions: each state is rendered independent of the past, given its immediately preceding state. For example, the DAG of Figure 44.1 implies the following Markovian independencies:

$$X_2 \perp\!\!\!\perp \{0\} \,|\, X_1, \;\; X_3 \perp\!\!\!\perp X_2 \,|\, X_1, \;\; X_4 \perp\!\!\!\perp X_1 \,|\, \{X_2, X_3\}, \;\; X_5 \perp\!\!\!\perp \{X_1, X_2, X_3\} \,|\, X_4 \tag{44.4}$$

In addition to these, the decomposition of Equation 44.1 implies many more independencies, the sum total of which can be identified from the DAG using the graphical criterion of *d*-separation (Pearl, 1988):

Definition 44.1 (*d*-separation)

Let a path in a DAG be a sequence of consecutive edges, of any directionality. A path p is said to be *d*-separated (or blocked) by a set of nodes Z iff

1. p contains a chain $i \to j \to k$ or a fork $i \leftarrow j \to k$ such that the middle node j is in Z, or
2. p contains an inverted fork $i \to j \leftarrow k$ such that neither the middle node j nor any of its descendants (in G) are in Z

If X, Y, and Z are three disjoint subsets of nodes in a DAG G, then Z is said to *d*-separate X from Y, denoted $(X \perp\!\!\!\perp Y \,|\, Z)_G$, iff Z *d*-separates every path from a node in X to a node in Y.

In Figure 44.1, for example, $X = \{X_2\}$ and $Y = \{X_3\}$ are *d*-separated by $Z = \{X_1\}$; the path $X_2 \leftarrow X_1 \to X_3$ is blocked by $X_1 \in Z$, while the path $X_2 \to X_4 \leftarrow X_3$ is blocked because X_4 and all its descendants are outside Z. Thus, $(X_2 \perp\!\!\!\perp X_3 \,|\, X_1)_G$ holds in Figure 44.1. However, X and Y are not *d*-separated by $Z' = \{X_1, X_5\}$, because the path $X_2 \to X_4 \leftarrow X_3$ is rendered active by virtue of X_5, a descendant of X_4, being in Z'. Consequently, $(X_2 \perp\!\!\!\perp X_3 \,|\, \{X_1, X_5\})_G$ does not hold; in words, learning the value of the consequence X_5 renders its causes X_2 and X_3 dependent, as if a pathway were opened along the arrows converging at X_4.

The *d*-separation criterion has been shown to be both necessary and sufficient relative to the set of distributions that are represented by a DAG G (Geiger et al., 1990; Verma and Pearl, 1990). In other words, there is a one-to-one correspondence between the set of independencies implied by the recursive decomposition of Equation 44.1 and the set of triples (X, Z, Y) that satisfy the *d*-separation criterion in G. Furthermore, the *d*-separation criterion can be tested in time linear in the number of edges in G. Thus, a DAG can be viewed as an efficient scheme for representing Markovian independence assumptions and for deducing and displaying all the logical consequences of such assumptions.

An important property that follows from the *d*-separation characterization is a criterion for determining when two dags are observationally equivalent, that is, every probability distribution that is represented by one of the dags is also represented by the other:

Theorem 44.1

(Verma and Pearl, 1990) Two dags are observationally equivalent if and only if they have the same sets of edges and the same sets of v-structures, that is, head-to-head arrows with nonadjacent tails.

The soundness of the *d*-separation criterion holds not only for probabilistic independencies but for any abstract notion of conditional independence that obeys the semi-graphoid axioms (Pearl et al., 1990; Verma and Geiger, 1990). Additional properties of DAGs and their applications to evidential reasoning in expert systems are discussed in Pearl (1988, 1993a), Pearl et al. (1990), Geiger (1990), Lauritzen and Spiegelhalter (1988), Spiegelhalter et al. (1993), Darwiche (2009), and Lauritzen (1996).

44.3.2 Inference Algorithms

The first algorithms proposed for probability updating in Bayesian networks used message-passing architecture and were limited to trees (Pearl, 1982) and singly connected networks (Kim and Pearl, 1983). The idea was to assign each variable a simple processor, forced to communicate only with its neighbors, and to permit asynchronous back-and-forth message-passing until equilibrium was achieved. Coherent equilibrium can indeed be achieved this way, but only in singly connected networks, where an equilibrium state occurs in time proportional to the diameter of the network.

Many techniques have been developed and refined to extend the tree-propagation method to general, multiply connected networks. Among the most popular are Schachter's (1988) method of node elimination, Lauritzen and Spiegelhalter's (1988) method of clique-tree propagation, loop-cut conditioning (Pearl, 1988, Section 44.4.3), and iterative conditioning (Dechter, 1999; Darwiche, 2009).

Clique-tree propagation works as follows. Starting with a directed network representation, the network is transformed into an undirected graph that retains all of its original dependencies. This graph, sometimes called a Markov network (Pearl, 1988, Section 44.3.1), is then triangulated to form local clusters of nodes (cliques) that are tree-structured. Evidence propagates from clique to clique by ensuring that the probability of their intersection set is the same, regardless of which of the two cliques is considered in the computation. Finally, when the propagation process subsides, the posterior probability of an individual variable is computed by projecting (marginalizing) the distribution of the hosting clique onto this variable.

Whereas the task of updating probabilities in general networks is NP-hard (Rosenthal, 1975; Cooper, 1990), the complexity for each of the three methods cited earlier is exponential in the size of the largest clique found in some triangulation of the network. It is fortunate that these complexities can be estimated prior to actual processing; when the estimates exceed reasonable bounds, an approximation method such as stochastic simulation (Pearl, 1987; Henrion, 1988) and belief propagation (Weiss and Pearl, 2010) can be used instead. Learning techniques have also been developed for systematic updating of the conditional probabilities $P(x_i|\mathbf{pa}_i)$ so as to match empirical data (Spiegelhalter and Lauritzen, 1990; Darwiche, 2009).

44.3.3 System's Properties

By providing graphical means for representing and manipulating probabilistic knowledge, Bayesian networks overcome many of the conceptual and computational difficulties of earlier knowledge-based systems (Pearl, 1988). Their basic properties and capabilities can be summarized as follows:

1. Graphical methods make it easy to maintain consistency and completeness in probabilistic knowledge bases. They also define modular procedures of knowledge acquisition that reduce significantly the number of assessments required (Pearl, 1988; Heckerman, 1991).
2. Independencies can be dealt with explicitly. They can be articulated by an expert, encoded graphically, read off the network, and reasoned about, yet they forever remain robust to numerical imprecision (Geiger, 1990; Geiger et al., 1990; Pearl et al., 1990).

3. Graphical representations uncover opportunities for efficient computation. Distributed updating is feasible in knowledge structures that are rich enough to exhibit intercausal interactions (e.g., "explaining away") (Pearl, 1982; Kim and Pearl, 1983). And, when extended by clustering or conditioning, tree-propagation algorithms are capable of updating networks of arbitrary topology (Shachter, 1986; Lauritzen and Spiegelhalter 1988; Pearl, 1988).
4. The combination of predictive and abductive inferences resolves many problems encountered by first-generation expert systems and renders belief networks a viable model for cognitive functions requiring both top–down and bottom–up inferences (Pearl, 1988; Shafer and Pearl, 1990).
5. The causal information encoded in Bayesian networks facilitates the analysis of action sequences, their consequences, their interaction with observations, their expected utilities, and, hence, the synthesis of plans and strategies under uncertainty (Dean and Wellman, 1991; Pearl, 1993b, 1994b).
6. The isomorphism between the topology of Bayesian networks and the stable mechanisms that operate in the environment facilitates modular reconfiguration of the network in response to changing conditions and permits deliberative reasoning about novel situations.

44.3.4 Later Developments

44.3.4.1 Causal Discovery

One of the most exciting prospects in recent years has been the possibility of using the theory of Bayesian networks to discover causal structures in raw statistical data. Several systems have been developed for this purpose (Rebane and Pearl, 1987; Pearl and Verma, 1991; Spirtes et al., 1993, 2000), which systematically search and identify causal structures with hidden variables from empirical data. Technically, because these algorithms rely merely on conditional independence relationships, the structures found are valid only if one is willing to accept weaker forms of guarantees than those obtained through controlled randomized experiments: minimality and stability (Pearl and Verma, 1991; Pearl, 2000, Chapter 2). Minimality guarantees that any other structure compatible with the data is necessarily less specific, and hence less falsifiable and less trustworthy, than the one(s) inferred. Stability ensures that any alternative structure compatible with the data must be less stable than the one(s) inferred; namely, slight fluctuations in experimental conditions will render that structure no longer compatible with the data. With these forms of guarantees, the theory provides criteria and algorithms for identifying genuine and spurious causes, with or without temporal information.

Alternative methods of identifying structure in data assign prior probabilities to the parameters of the network and use Bayesian updating to score the degree to which a given network fits the data (Cooper and Herskovits, 1991; Heckerman et al., 1994). These methods have the advantage of operating well under small sample conditions, but encounter difficulties coping with hidden variables.

Tian and Pearl (2001a,b) developed yet another method of causal discovery based on the detection of "shocks," or spontaneous local changes in the environment, which act like "nature's interventions," and unveil causal directionality toward the consequences of those shocks.

44.3.4.2 Plain Beliefs

In mundane decision making, beliefs are revised not by adjusting numerical probabilities but by tentatively accepting some sentences as "true for all practical purposes." Such sentences, often named *plain beliefs*, exhibit both logical and probabilistic character. As in classical logic, they are propositional and deductively closed; as in probability, they are subject to retraction and to varying degrees of entrenchment (Spohn, 1988; Goldszmidt and Pearl, 1992).

Bayesian networks can be adopted to model the dynamics of plain beliefs by replacing ordinary probabilities with nonstandard probabilities, that is, probabilities that are infinitesimally close to either zero or one. This amounts to taking an "order of magnitude" approximation of empirical frequencies and adopting new combination rules tailored to reflect this approximation. The result is an integer-addition calculus, very similar to probability calculus, with summation replacing multiplication and minimization replacing addition. A plain belief is then identified as a proposition whose negation obtains an infinitesimal probability (i.e., an integer greater than zero). The connection between infinitesimal probabilities, nonmonotonic logic, and iterative belief revision is described in Pearl (1994a), Goldszmidt and Pearl (1996), Darwiche and Pearl (1997), and Spohn (2012).

This combination of infinitesimal probabilities with the causal information encoded by the structure of Bayesian networks facilitates linguistic communication of belief commitments, explanations, actions, goals, and preferences and serves as the basis for qualitative planning under uncertainty (Goldszmidt and Pearl, 1992; Pearl, 1993b; Darwiche and Goldszmidt, 1994; Darwiche and Pearl, 1994). Some of these aspects will be presented in the next section.

44.4 Bayesian Networks as Carriers of Causal Information

The interpretation of DAGs as carriers of independence assumptions does not necessarily imply causation and will in fact be valid for any set of Markovian independencies along any ordering (not necessarily causal or chronological) of the variables. However, the patterns of independencies portrayed in a DAG are typical of causal organizations, and some of these patterns can only be given meaningful interpretation in terms of causation. Consider, for example, two independent events, E_1 and E_2, that have a common effect E_3. This triple represents an intransitive pattern of dependencies: E_1 and E_3 are dependent, E_3 and E_2 are dependent, yet E_1 and E_2 are independent. Such a pattern cannot be represented in undirected graphs because connectivity in undirected graphs is transitive. Likewise, it is not easily represented in neural networks, because E_1 and E_2 should turn dependent once E_3 is known. The DAG representation provides a convenient language for intransitive dependencies via the converging pattern $E_1 \rightarrow E_3 \leftarrow E_2$, which implies the independence of E_1 and E_2 as well as the dependence of E_1 and E_3 and of E_2 and E_3. The distinction between transitive and intransitive dependencies is the basis for the causal discovery systems of Pearl and Verma (1991) and Spirtes et al. (1993) (see Section 44.3.4.1).

However, the Markovian account still leaves open the question of how such intricate patterns of independencies relate to the more basic notions associated with causation, such as influence, manipulation, and control, which reside outside the province of probability theory. The connection is made in the mechanism-based account of causation.

The basic idea behind this account goes back to structural equation models (Wright, 1921; Haavelmo, 1943; Simon, 1953), and it was adapted in Pearl and Verma (1991) for defining probabilistic causal theories, as follows. Each child–parent family in a DAG G represents a deterministic function

$$X_i = f_i(\mathbf{pa}_i, \epsilon_i) \tag{44.5}$$

where $\mathbf{pa}_i$ are the parents of variable X_i in G, and ϵ_i, $0 < i < n$, are mutually independent, arbitrarily distributed random disturbances. Characterizing each child–parent relationship as a deterministic function, instead of the usual conditional probability $P(x_i|\mathbf{pa}_i)$, imposes equivalent independence constraints on the resulting distributions and leads to the same recursive decomposition that characterizes DAG models (see Equation 44.1). However, the functional characterization $X_i = f_i(\mathbf{pa}_i, \epsilon_i)$ also specifies how the resulting distributions would change in response to external interventions, since each function is presumed to represent a stable mechanism in the domain and therefore remains constant unless specifically altered. Thus, once we know the identity of the mechanisms altered by the intervention and the nature of the alteration, the overall effect of an intervention can be predicted by modifying the appropriate equations in the model of Equation 44.5 and using the modified model to compute a new probability function of the observables.

The simplest type of external intervention is one in which a single variable, say X_i, is forced to take on some fixed value x_i'. Such *atomic* intervention amounts to replacing the old functional mechanism $X_i = f_i(\mathbf{pa}_i, \epsilon_i)$ with a new mechanism $X_i = x_i'$ governed by some external force that sets the value x_i'. If we imagine that each variable X_i could potentially be subject to the influence of such an external force, then we can view each Bayesian network as an efficient code for predicting the effects of atomic interventions and of various combinations of such interventions, without representing these interventions explicitly.

44.4.1 Causal Theories, Actions, Causal Effect, and Identifiability

Definition 44.2

A causal theory is a 4-tuple

$$T = \langle V, U, P(\mathbf{u}), \{f_i\} \rangle$$

where

$V = \{X_1, \ldots, X_n\}$ is a set of observed variables
$U = \{U_1, \ldots, U_m\}$ is a set of unobserved variables that represent disturbances, abnormalities, or assumptions
$P(\mathbf{u})$ is a distribution function over $U_1, \ldots, U_n$
$\{f_i\}$ is a set of n deterministic functions, each of the form

$$X_i = f_i(PA_i, u) \quad i = 1, \ldots, n \tag{44.6}$$

where PA_i is a subset of V not containing X_i.

The variables PA_i (connoting parents) are considered the direct causes of X_i, and they define a directed graph G that may, in general, be cyclic. Unlike the probabilistic definition of "parents" in Bayesian networks (Equation 44.1), PA_i is selected from V by considering functional mechanisms in the domain, not by conditional independence considerations. We will assume that the set of equations in (44.6) has a unique solution for $X_i, \ldots, X_n$, given any value of the disturbances $U_1, \ldots, U_n$. Therefore, the distribution $P(\mathbf{u})$ induces a unique distribution on the observables, which we denote by $P_T(\mathbf{v})$.

We will consider concurrent actions of the form $do(X = x)$, where $X \subseteq V$ is a set of variables and x is a set of values from the domain of X. In other words, $do(X = x)$ represents a combination of actions that forces the variables in X to attain the values x.

Definition 44.3 (effect of actions)

The effect of the action $do(X = x)$ on a causal theory T is given by a subtheory T_x of T, where T_x obtains by deleting from T all equations corresponding to variables in X and substituting the equations $X = x$ instead.

The framework provided by Definitions 44.2 and 44.3 permits the coherent formalization of many subtle concepts in causal discourse, such as causal influence, causal effect, causal relevance, average causal effect, identifiability, counterfactuals, and exogeneity. Examples are the following:

- X **influences** Y in context u if there are two values of X, x and x', such that the solution for Y under $U = u$ and $do(X = x)$ is different from the solution under $U = u$ and $do(X = x')$.
- X **can potentially influence** Y if there exist both a subtheory T_z of T and a context $U = u$ in which X influences Y.
- Event $X = x$ **is the (singular) cause of event** $Y = y$ if (1) $X = x$ and $Y = y$ are true, and (2) in every context u compatible with $X = x$ and $Y = y$, and for all $x' \neq x$, the solution of Y under $do(X = x')$ is not equal to y.

The definitions earlier are deterministic. Probabilistic causality emerges when we define a probability distribution $P(u)$ for the U variables, which, under the assumption that the equations have a unique solution, induces a unique distribution on the endogenous variables for each combination of atomic interventions.

Definition 44.4 (causal effect)

Given two disjoint subsets of variables, $X \subseteq V$ and $Y \subseteq V$, the causal effect of X on Y, denoted $P_T(y|\hat{x})$, is a function from the domain of X to the space of probability distributions on Y, such that

$$P_T(y|x) = P_{T_x}(y) \tag{44.7}$$

for each realization x of X. In other words, for each $x \in dom(X)$, the causal effect $P_T(y|\hat{x})$ gives the distribution of Y induced by the action $do(X = x)$.

Note that causal effects are defined relative to a given causal theory T, though the subscript T is often suppressed for brevity.

Definition 44.5 (identifiability)

Let $Q(T)$ be any computable quantity of a theory T; Q is identifiable in a class M of theories if for any pair of theories T_1 and T_2 from M, $Q(T_1) = Q(T_2)$ whenever $P_{T_1}(v) = P_{T_2}(v)$.

Identifiability is essential for estimating quantities Q from P alone, without specifying the details of T, so that the general characteristics of the class M suffice. The question of interest in planning applications is the identifiability of the causal effect $Q = P_T(y|\hat{x})$ in the class M_G of theories that share the same causal graph G. Relative to such classes we now define

Definition 44.6 (causal-effect identifiability)

The causal effect of X on Y is said to be identifiable in M_G if the quantity $P(y|\hat{x})$ can be computed uniquely from the probabilities of the observed variables, that is, if for every pair of theories T_1 and T_2 in M_G such that $P_{T_1}(v) = P_{T_2}(v)$, we have $P_{T_1}(y|\hat{x}) = P_{T_2}(y|\hat{x})$.

The identifiability of $P(y|\hat{x})$ ensures that it is possible to infer the effect of action $do(X = x)$ on Y from two sources of information:

1. Passive observations, as summarized by the probability function $P(v)$
2. The causal graph, G, which specifies, qualitatively, which variables make up the stable mechanisms in the domain or, alternatively, which variables participate in the determination of each variable in the domain

Simple examples of identifiable causal effects will be discussed in the next section.

44.4.2 Acting vs. Observing

Consider the example depicted in Figure 44.1. The corresponding theory consists of five functions, each representing an autonomous mechanism:

$$
\begin{aligned}
X_1 &= U_1 \\
X_2 &= f_2(X_1, U_2) \\
X_3 &= f_3(X_1, U_3) \\
X_4 &= f_4(X_3, X_2, U_4) \\
X_5 &= f_5(X_4, U_5)
\end{aligned}
\tag{44.8}
$$

To represent the action "turning the sprinkler ON," $do(X_3 = \text{ON})$, we delete the equation $X_3 = f_3(x_1, u_3)$ from the theory of Equation 44.8 and replace it with $X_3 = \text{ON}$. The resulting subtheory, $T_{X_3=\text{ON}}$, contains all the information needed for computing the effect of the actions on other variables. It is easy to see from this subtheory that the only variables affected by the action are X_4 and X_5, that is, the descendant of the manipulated variable X_3.

The probabilistic analysis of causal theories becomes particularly simple when two conditions are satisfied:

1. The theory is recursive, that is, there exists an ordering of the variables $V = \{X_1, \ldots, X_n\}$ such that each X_i is a function of a subset PA_i of its predecessors

$$X_i = f_i(PA_i, U_i), \quad PA_i \subseteq \{X_1, \ldots, X_{i-1}\} \tag{44.9}$$

2. The disturbances $U_1, \ldots, U_n$ are mutually independent, that is,

$$P(u) = \prod_i P(u_i) \tag{44.10}$$

These two conditions, also called Markovian, are the bases of the independencies embodied in Bayesian networks (Equation 44.1), and they enable us to compute causal effects directly from the conditional probabilities $P(x_i|\mathbf{pa}_i)$, without specifying the functional form of the functions f_i, or the distributions $P(u_i)$ of the disturbances. This is seen immediately from the following observations: the distribution induced by any Markovian theory T is given by the product in Equation 44.1

$$P_T(x_1, \ldots, x_n) = \prod_i P(x_i|\mathbf{pa}_i) \tag{44.11}$$

where $\mathbf{pa}_i$ are (values of) the parents of X_i in the diagram representing T. At the same time, the subtheory $T_{x'_j}$, representing the action $do(X_j = x'_j)$, is also Markovian, hence it also induces a product-like distribution

$$P_{T_{x'_j}}(x_1, \ldots, x_n) = \begin{cases} \prod_{i \neq j} P(x_i|\mathbf{pa}_i) = \dfrac{P(x_1, \ldots, x_n)}{P(x_j|\mathbf{pa}_j)} & \text{if } x_j = x'_j \\ 0 & \text{if } x_j \neq x'_j \end{cases} \tag{44.12}$$

where the partial product reflects the surgical removal of the

$$X_j = f_j(\mathbf{pa}_j, U_j)$$

from the theory of equation (44.9) (see Pearl, 1993a).

In the example of Figure 44.1, the pre-action distribution is given by the product

$$P_T(x_1,x_2,x_3,x_4,x_5) = P(x_1)P(x_2|x_1)P(x_3|x_1)P(x_4|x_2,x_3)P(x_5|x_4) \tag{44.13}$$

while the surgery corresponding to the action $do(X_3 = \text{ON})$ amounts to deleting the link $X_1 \rightarrow X_3$ from the graph and fixing the value of X_3 to ON, yielding the post-action distribution

$$P_T(x_1,x_2,x_4,x_5|do(X_3=\text{ON})) = P(x_1)\,P(x_2|x_1)\,P(x_4|x_2,x_3=\text{ON})P(x_5|x_4) \tag{44.14}$$

Note the difference between the action $do(X_3 = \text{ON})$ and the observation $X_3 = \text{ON}$. The latter is encoded by ordinary Bayesian conditioning, while the former by conditioning a mutilated graph, with the link $X_1 \rightarrow X_3$ removed. This mirrors indeed the difference between seeing and doing: after observing that the sprinkler is ON, we wish to infer that the season is dry, that it probably did not rain, and so on; no such inferences should be drawn in evaluating the effects of the deliberate action "turning the sprinkler ON." The amputation of $X_3 = f_3(X_1, U_3)$ from (44.8) ensures the suppression of any abductive inferences from X_3, the action's recipient.

Note also that Equations 44.11 through 44.14 are independent of *T*; in other words, the pre-action and post-action distributions depend only on observed conditional probabilities but are independent of the particular functional form of $\{f_i\}$ or the distribution $P(\mathbf{u})$ that generates those probabilities. This is the essence of identifiability as given in Definition 44.6, which stems from the Markovian assumptions (44.9) and (44.10). Section 44.4.3 will demonstrate that certain causal effects, though not all, are identifiable even when the Markovian property is destroyed by introducing dependencies among the disturbance terms.

Generalization to multiple actions and conditional actions is reported in Pearl and Robins (1995). Multiple actions $do(X = x)$, where X is a compound variable, result in a distribution similar to (44.12), except that all factors corresponding to the variables in X are removed from the product in (44.11). Stochastic conditional strategies (Pearl, 1994b) of the form

$$do(X_j = x_j) \text{ with probability } P^*\left(x_j \middle| \mathbf{pa}_j^*\right) \tag{44.15}$$

where $\mathbf{pa}_j^*$ is the support of the decision strategy, also result in a product decomposition similar to (44.11), except that each factor $P(x_j|\mathbf{pa}_j)$ is *replaced* with $P^*\left(x_j \middle| \mathbf{pa}_j^*\right)$.

The surgical procedure described earlier is not limited to probabilistic analysis. The causal knowledge represented in Figure 44.1 can be captured by logical theories as well, for example,

$$\begin{aligned}
x_2 &\Leftrightarrow [(X_1 = \text{Winter}) \vee (X_1 = \text{Fall}) \vee ab_2] \wedge \neg ab_2' \\
x_3 &\Leftrightarrow [(X_1 = \text{Summer}) \vee (X_1 = \text{Spring}) \vee ab_3] \wedge \neg ab_3' \\
x_4 &\Leftrightarrow (x_2 \vee x_3 \vee ab_4) \wedge \neg ab_4' \\
x_5 &\Leftrightarrow (x_4 \vee ab_5) \wedge \neg ab_5'
\end{aligned} \tag{44.16}$$

where x_i stands for $X_i = true$, and ab_i and ab_i' stand, respectively, for trigerring and inhibiting abnormalities. The double arrows represent the assumption that the events on the r.h.s. of each equation are the *only* direct causes for the l.h.s, thus identifying the surgery implied by any action.

It should be emphasized though that the models of a causal theory are not made up merely of truth value assignments that satisfy the equations in the theory. Since each equation represents an autonomous process, the content of each individual equation must be specified in any model of the theory, and this can be encoded using either the graph (as in Figure 44.1) or the generic description of the theory, as in (44.8). Alternatively, we can view a model of a causal theory to consist of a mutually consistent set of submodels, with each submodel being a standard model of a single equation in the theory.

44.4.3 Action Calculus

The identifiability of causal effects demonstrated in Section 44.4.1 relies critically on the Markovian assumptions (44.9) and (44.10). If a variable that has two descendants in the graph is unobserved, the disturbances in the two equations are no longer independent; the Markovian property (44.9) is violated and identifiability may be destroyed. This can be seen easily from Equation 44.12; if any parent of the manipulated variable X_j is unobserved, one cannot estimate the conditional probability $P(x_j|\mathbf{pa}_j)$, and the effect of the action $do(X_j = x_j)$ may not be predictable from the observed distribution $P(x_1,\ldots, x_n)$. Fortunately, certain causal effects are identifiable even in situations where members of $\mathbf{pa}_j$ are unobservable (Pearl, 1993a), and, moreover, polynomial tests are now available for deciding when $P(x_i|\hat{x}_j)$ is identifiable and for deriving closed-form expressions for $P(x_i|\hat{x}_j)$ in terms of observed quantities (Galles and Pearl, 1995).

These tests and derivations are based on a symbolic calculus (Pearl, 1994a, 1995) to be described in the sequel, in which interventions, side by side with observations, are given explicit notation and are permitted to transform probability expressions. The transformation rules of this calculus reflect the understanding that interventions perform "local surgeries" as described in Definition 44.3, that is, they overrule equations that tie the manipulated variables to their pre-intervention causes.

Let X, Y, and Z be arbitrary disjoint sets of nodes in a DAG G. We denote by $G_{\overline{X}}$ the graph obtained by deleting from G all arrows pointing to nodes in X. Likewise, we denote by $G_{\underline{X}}$ the graph obtained by deleting from G all arrows emerging from nodes in X. To represent the deletion of both incoming and outgoing arrows, we use the notation $G_{\overline{X}\underline{Z}}$. Finally, the expression $P\left(y|\hat{x},z\right) \triangleq P\left(y,z|\hat{x}\right)/P\left(z|\hat{x}\right)$ stands for the probability of $Y = y$ given that $Z = z$ is observed and X is held constant at x.

Theorem 44.2

Let G be the directed acyclic graph associated with a Markovian causal theory, and let $P(\cdot)$ stand for the probability distribution induced by that theory. For any disjoint subsets of variables X, Y, Z, and W, we have

Rule 1 (*Insertion/deletion of observations*)

$$P\left(y|\hat{x},z,w\right)=P\left(y,|\hat{x},w\right) \quad \mathit{if} \quad \left(Y \perp\!\!\!\perp Z|X,W\right)_{G_{\overline{X}}} \tag{44.17}$$

Rule 2 (*Action/observation exchange*)

$$P\left(y|\hat{x},\hat{z},w\right)=P\left(y,|\hat{x},z,w\right) \quad \mathit{if} \quad \left(Y \perp\!\!\!\perp Z|X,W\right)_{G_{\overline{X}\underline{Z}}} \tag{44.18}$$

Rule 3 (*Insertion/deletion of actions*)

$$P\left(y|\hat{x},\hat{z},w\right)=P\left(y,|\hat{x},w\right) \ \ if \left(Y \perp\!\!\!\perp Z|X,W\right)_{G_{\overline{X},\overline{Z(W)}}} \tag{44.19}$$

where $Z(W)$ is the set of Z-nodes that are not ancestors of any W-node in $G_{\overline{X}}$.

Each of the inference rules presented earlier follows from the basic interpretation of the "$\hat{x}$" operator as a replacement of the causal mechanism that connects X to its pre-action parents by a new mechanism $X = x$ introduced by the intervening force. The result is a submodel characterized by the subgraph G_X (named "manipulated graph" in Spirtes et al. (1993)) that supports all three rules.

Corollary 44.1

A causal effect q: $P\left(y_1,\ldots,y_k|\hat{x}_1,\ldots,\hat{x}_m\right)$ is identifiable in a model characterized by a graph G if there exists a finite sequence of transformations, each conforming to one of the inference rules in Theorem 44.2, which reduces q into a standard (i.e., hat-free) probability expression involving observed quantities.

Although Theorem 44.2 and Corollary 44.1 require the Markovian property, they can also be applied to non-Markovian, recursive theories because such theories become Markovian if we consider the unobserved variables as part of the analysis and represent them as nodes in the graph. To illustrate, assume that variable X_1 in Figure 44.1 is unobserved, rendering the disturbances U_3 and U_2 dependent since these terms now include the common influence of X_1. Theorem 44.2 tells us that the causal effect $P(x_4|\hat{x}_3)$ is identifiable, because

$$P\left(x_4|\hat{x}_3\right)=\sum_{x_2} P\left(x_4|\hat{x}_3,x_2\right)P\left(x_2|\hat{x}_3\right)$$

Rule 3 permits the deletion

$$P\left(x_2|\hat{x}_3\right)=P(x_2), \ \ \text{because} \left(X_2 \perp\!\!\!\perp X_3\right)_{G_{\overline{X_3}}}$$

while Rule 2 permits the exchange

$$P\left(x_4|\hat{x}_3,x_2\right)=P\left(x_4|x_3,x_2\right), \ \ \text{because} \left(X_4 \perp\!\!\!\perp X_3|X_2\right)_{G_{\underline{X_3}}}$$

This gives

$$P\left(x_4|\hat{x}_3\right)=\sum_{x_2} P\left(x_4|x_3,x_2\right)P(x_2)$$

which is a "hat-free" expression, involving only observed quantities.

In general, it can be shown (Pearl, 1995) that

1. The effect of interventions can often be identified (from nonexperimental data) without resorting to parametric models
2. The conditions under which such nonparametric identification is possible can be determined by simple graphical tests*
3. When the effect of interventions is not identifiable, the causal graph may suggest nontrivial experiments that, if performed, would render the effect identifiable

The ability to assess the effect of interventions from nonexperimental data has immediate applications in the medical and social sciences, since subjects who undergo certain treatments often are not representative of the population as a whole. Such assessments are also important in AI applications where an agent needs to predict the effect of the next action on the basis of past performance records and where that action has never been enacted under controlled experimental conditions, but in response to environmental needs or to other agent's requests.

44.4.4 Recent Developments

44.4.4.1 Complete Identification Results

A key identification condition, which operationalizes Theorem 44.2, has been derived by Jin Tian. It reads:

Theorem 44.3

(Tian and Pearl, 2002) A sufficient condition for identifying the causal effect $P(y|do(x))$ is that there exists no bidirected path (i.e., a path composed entirely of bidirected arcs) between X and any of its children.†

Remarkably, the theorem asserts that, as long as every child of X (on the pathways to Y) is not reachable from X via a bidirected path, then, regardless of how complicated the graph, the causal effect $P(y|do(x))$ is identifiable. In Figure 44.1, for example, adding unobserved confounders for the pairs (X_1, X_3), (X_2, X_4), (X_3, X_5), and (X_2, X_5) will still permit the identification of $P(X_5|\hat{x}_3)$, but adding (X_2, X_4) to this list will prohibit identification.

Tian and Pearl (2002) further showed that the condition is both sufficient and necessary for the identification of $P(v|do(x))$, where V includes all variables except X. A necessary and sufficient condition for identifying $P(w|do(z))$, with W and Z two arbitrary sets, was established by Shpitser and Pearl (2006b). Subsequently, a complete graphical criterion was established for determining the identifiability of *conditional* interventional distributions, namely, expressions of the type $P(y|do(x),z)$ where X,Y, and Z are arbitrary sets of variables (Shpitser and Pearl, 2006a).

These results constitute a complete characterization of causal effects in graphical models. They provide us with polynomial time algorithms for determining whether an arbitrary quantity invoking the

* These graphical tests offer, in fact, a complete formal solution to the "covariate-selection" problem in statistics: finding an appropriate set of variables that need be adjusted for in any study that aims to determine the effect of one factor upon another. This problem has been lingering in the statistical literature since Karl Pearson, the founder of modern statistics, discovered (1899) what in modern terms is called the "Simpson's paradox"; any statistical association between two variables may be reversed or negated by including additional factors in the analysis (Aldrich, 1995).

† Before applying this criterion, one may delete from the causal graphs all nodes that are not ancestors of Y.

do(*x*) operator is identified in a given semi-Markovian model, and if so, what the estimand of that quantity is. Remarkably, one corollary of these results also states that the *do*-calculus is complete, namely, a quantity $Q = P(y|do(x), z)$ is identified if and only if it can be reduced to a *do*-free expression using the three rules of Theorem 44.2.* This turns Corollary 44.1 into an "if-and-only-if" assertion. Tian and Shpitser (2010) provide a comprehensive summary of these results.

44.4.5 Transportability and Transfer Learning

In applications involving identification, the role of the *do*-calculus is to remove the *do*-operator from the query expression. We now discuss a totally different application, to decide if experimental findings can be transported to a new, potentially different environment, where only passive observations can be performed. This problem, labeled "transportability" in Pearl and Bareinboim (2011), is at the heart of every scientific investigation. For example, experiments performed in the laboratory are invariably intended to be used elsewhere, where conditions differ substantially from the laboratory. A robot trained by a simulator should be able to use the knowledge acquired through training to a new environment, in which experiments are costly or infeasible.

To formalize problems of this sort, Pearl and Bareinboim devised a graphical representation called "selection diagrams" that encodes knowledge about differences and commonalities between populations. A selection diagram is a causal diagram annotated with new variables, called *S*-nodes, which point to the mechanisms where discrepancies between the two populations are suspected to take place. The task of deciding if transportability is feasible now reduces to a syntactic problem (using the *do*-calculus) aiming to separate the *do*-operator from the *S*-variables.

Theorem 44.4

(Pearl and Bareinboim, 2011) Let *D* be the selection diagram characterizing two populations, π and π^*, and *S* a set of selection variables in *D*. The relation $R = P^*\left(y|do(x),z\right)$ is transportable from π and π^* if and only if the expression $P(y|do(x), z, s)$ is reducible, using the rules of *do*-calculus, to an expression in which *S* appears only as a conditioning variable in *do*-free terms.

Theorem 44.4 does not specify the sequence of rules leading to the needed reduction when such a sequence exists. Bareinboim and Pearl (2012) established a complete and effective graphical procedure of confirming transportability that also synthesizes the transport formula whenever possible. Each transport formula determines for the investigator what information need to be taken from the experimental and observational studies and how they ought to be combined to yield an unbiased estimate of *R*.

A generalization of transportability theory to multi-environment has led to a principled solution to "meta-analysis." "Meta-analysis" is a data fusion problem aimed at combining results from many experimental and observational studies, each conducted on a different population and under a different set of conditions, so as to synthesize an aggregate measure of effect size that is "better," in some sense, than any one study in isolation. This fusion problem has received enormous attention in the health and social sciences, where data are scarce and experiments are costly (Pearl, 2012a,c).

Using multiple "selection diagrams" to encode commonalities among studies, Bareinboim and Pearl (2013) were able to "synthesize" an estimator that is guaranteed to provide unbiased estimate of the desired quantity based on information that each study shares with the target environment.

* This was independently established by Huang and Valtorta (2006).

44.4.6 Inference with Missing Data

It is commonly assumed that causal knowledge is necessary only when interventions are contemplated and that in purely predictive tasks, probabilistic knowledge suffices. Not so. One predictive task in which causal information is essential and which have thus far been treated in purely statistical terms is the problem of drawing valid inferences from statistical data in which some items are "missing" or failed to be recorded.

The "missing data" problem is pervasive in machine learning and every experimental science; users fail to answer certain items on a questionnaire, sensors may miss certain signal due to bad weather, and so on. The problem is causal in nature, because the mechanism that determines the *reasons* for missingness makes a difference in whether/how we can recover from the data, and that mechanism requires causal language to be properly described, statistics is insufficient.

Mohan and Pearl (2013) have shown that current practices of handling missing data, by relying exclusively on statistical considerations (Rubin, 1976; Little and Rubin, 2002), are deficient in several key areas and could be remedied using causal diagrams. In particular they showed that

1. It is possible to define precisely what relations can be *recovered* from missing data and what causal and statistical knowledge need to be assumed to ensure bias-free recovery
2. Different causal assumptions lead to different routines for recovering a relation from the data
3. Adding auxiliary variables to the data, a popular technique in current practice (Schafer and Graham, 2002) may or may not help the recovery process, depending on conditions that can be read from the causal diagram
4. It is possible to determine when the conditions necessary for bias-free recovery have testable implications

44.4.7 Historical Remarks

An explicit translation of interventions to "wiping out" equations from linear econometric models was first proposed by Strotz and Wold (1960) and later used in Fisher (1970) and Sobel (1990). Extensions to action representation in nonmonotonic systems were reported in Goldszmidt and Pearl (1992) and Pearl (1993a). Graphical ramifications of this translation were explicated first in Spirtes et al. (1993) and later in Pearl (1993b). A related formulation of causal effects, based on event trees and counterfactual analysis, was developed by Robins (1986, pp. 1422–1425). Calculi for actions and counterfactuals based on this interpretation are developed in Pearl (1994b) and Balke and Pearl (1994), respectively.

44.5 Counterfactuals

A counterfactual sentence has the form

If A were true, then C would have been true?

where *A*, the counterfactual antecedent, specifies an event that is contrary to one's real-world observations, and *C*, the counterfactual consequent, specifies a result that is expected to hold in the alternative world where the antecedent is true. A typical example is "If Oswald were not to have shot Kennedy, then Kennedy would still be alive" that presumes the factual knowledge of Oswald's shooting Kennedy, contrary to the antecedent of the sentence.

The majority of the philosophers who have examined the semantics of counterfactual sentences have resorted to some version of Lewis' "closest world" approach; "*C* if it were *A*" is true, if *C* is true in worlds that are "closest" to the real world yet consistent with the counterfactual antecedent *A* (Lewis, 1973). Ginsberg (1986) followed a similar strategy. While the "closest world" approach leaves the precise specification of the closeness measure almost unconstrained, causal knowledge imposes very specific preferences as to which worlds should be considered closest to any given world. For example, considering an array of domino tiles standing close to each other. The manifestly closest world consistent with the antecedent "tile i is tipped to the right" would be a world in which just tile i is tipped, while all the others remain erect. Yet, we all accept the counterfactual sentence "Had tile i been tipped over to the right, tile $i + 1$ would be tipped as well" as plausible and valid. Thus, distances among worlds are not determined merely by surface similarities but require a distinction between disturbed mechanisms and naturally occurring transitions. The local surgery paradigm expounded in Section 44.4.1 offers a concrete explication of the closest world approach that respects causal considerations. A world w_1 is "closer" to w than a world w_2 is, if the set of atomic surgeries needed for transforming w into w_1 is a subset of those needed for transforming w into w_2. In the domino example, finding tile i tipped and $i + 1$ erect requires the breakdown of two mechanism (e.g., by two external actions) compared with one mechanism for the world in which all j-tiles, $j > i$ are tipped. This paradigm conforms to our perception of causal influences and lends itself to economical machine representation.

44.5.1 Formal Underpinning

The structural equation framework offers an ideal setting for counterfactual analysis.

Definition 44.7 (context-based potential response)

Given a causal theory T and two disjoint sets of variables, X and Y, the potential response of Y to X in a context u, denoted $Y(x, u)$ or $Y_x(u)$, is the solution for Y under $U = u$ in the subtheory T_x. $Y(x, u)$ can be taken as the formal definition of the counterfactual English phrase: "the value that Y would take in context u, had X been x,"*

Note that this definition allows for the context $U = u$ and the proposition $X = x$ to be incompatible in T. For example, if T describes a logic circuit with input U, it may well be reasonable to assert the counterfactual: "given $U = u$, Y would be high if X were low," even though the input $U = u$ may preclude X from being low. It is for this reason that one must invoke some notion of intervention (alternatively, a theory change or a "miracle" [Lewis, 1973]) in the definition of counterfactuals.

If U is treated as a random variable, then the value of the counterfactual $Y(x, u)$ becomes a random variable as well, denoted as $Y(x)$ of Y_x. Moreover, the distribution of this random variable is easily seen to coincide with the causal effect $P(y|\hat{x})$, as defined in Equation 44.7, that is,

$$P((Y(x) = y) = P(y|\hat{x})$$

The probability of a counterfactual conditional $x \rightarrow y|o$ may then be evaluated by the following procedure:

- Use the observations o to update $P(u)$ thus forming a causal theory $T^o = < V, U, \{f_i\}, P(u|o) >$.
- Form the mutilated theory T_x^o (by deleting the equation corresponding to variables in X), and compute the probability $P_{T^o}(y|\hat{x})$ that T_x^o induces on Y.

* The term *unit* instead of *context* is often used in the statistical literature Rubin (1974), where it normally stands for the identity of a specific individual in a population, namely, the set of attributes u that characterize that individual. In general, u may include the time of day and the experimental conditions under study. Practitioners of the counterfactual notation do not explicitly mention the notions of "solution" or "intervention" in the definition of $Y(x, u)$. Instead, the phrase "the value that Y would take in unit u, had X been x," viewed as basic, is posited as the definition of $Y(x, u)$.

Unlike causal effect queries, counterfactual queries are not identifiable even in Markovian theories, but require that the functional form of $\{f_i\}$ be specified. In Balke and Pearl (1994), a method is devised for computing sharp bounds on counterfactual probabilities that, under certain circumstances, may collapse to point estimates. This method has been applied to the evaluation of causal effects in studies involving noncompliance and to the determination of legal liability.

44.5.2 Applications to Policy Analysis

Counterfactual reasoning is at the heart of every planning activity, especially real-time planning. When a planner discovers that the current state of affairs deviates from the one expected, a "plan repair" activity need be invoked to determine what went wrong and how it could be rectified. This activity amounts to an exercise of counterfactual thinking, as it calls for rolling back the natural course of events and determining, based on the factual observations at hand, whether the culprit lies in previous decisions or in some unexpected, external eventualities. Moreover, in reasoning forward to determine if things would have been different, a new model of the world must be consulted, one that embodies hypothetical changes in decisions or eventualities, hence, a breakdown of the old model or theory.

The logic-based planning tools used in AI, such as STRIPS and its variants or those based on the situation calculus, do not readily lend themselves to counterfactual analysis; as they are not geared for coherent integration of abduction with prediction, and they do not readily handle theory changes. Remarkably, the formal system developed in economics and social sciences under the rubric "structural equation models" does offer such capabilities, but, as will be discussed in the succeeding text, these capabilities are not well recognized by current practitioners of structural models. The analysis presented in this chapter could serve both to illustrate to AI researchers the basic formal features needed for counterfactual and policy analysis and to call the attention of economists and social scientists to capabilities that are dormant within structural equation models.

Counterfactual thinking dominates reasoning in political science and economics. We say, for example, "If Germany were not punished so severely at the end of World War I, Hitler would not have come to power," or "If Reagan did not lower taxes, our deficit would be lower today." Such thought experiments emphasize an understanding of generic laws in the domain and are aimed toward shaping future policy making, for example, "defeated countries should not be humiliated," or "lowering taxes (contrary to Reaganomics) tends to increase national debt."

Strangely, there is very little formal work on counterfactual reasoning or policy analysis in the behavioral science literature. An examination of a number of econometric journals and textbooks, for example, reveals a glaring imbalance: while an enormous mathematical machinery is brought to bear on problems of estimation and prediction, policy analysis (which is the ultimate goal of economic theories) receives almost no formal treatment. Currently, the most popular methods driving economic policy making are based on so-called *reduced-form* analysis: to find the impact of a policy involving decision variables X on outcome variables Y, one examines past data and estimates the conditional expectation $E(Y|X = x)$, where x is the particular instantiation of X under the policy studied.

The assumption underlying this method is that the data were generated under circumstances in which the decision variables X act as exogenous variables, that is, variables whose values are determined outside the system under analysis. However, while new decisions should indeed be considered exogenous for the purpose of evaluation, past decisions are rarely enacted in an exogenous manner. Almost every realistic policy (e.g., taxation) imposes control over some endogenous variables, that is, variables whose values are determined by other variables in the analysis. Let us take taxation policies as an example. Economic data are generated in a world in which the government is reacting to various indicators and various pressures; hence, taxation is endogenous in the data analysis phase of the study. Taxation becomes exogenous when we wish to predict the impact of a specific decision to raise or lower taxes. The reduced-form method is valid only when past decisions are nonresponsive to other variables in the system, and this, unfortunately, eliminates most of the interesting control variables (e.g., tax rates, interest rates, quotas) from the analysis.

This difficulty is not unique to economic or social policy making; it appears whenever one wishes to evaluate the merit of a plan on the basis of the past performance of other agents. Even when the signals triggering the past actions of those agents are known with certainty, a systematic method must be devised for selectively ignoring the influence of those signals from the evaluation process. In fact, the very essence of *evaluation* is having the freedom to imagine and compare trajectories in various counterfactual worlds, where each world or trajectory is created by a hypothetical implementation of a policy that is free of the very pressures that compelled the implementation of such policies in the past.

Balke and Pearl (1995) demonstrate how linear, nonrecursive structural models with Gaussian noise can be used to compute counterfactual queries of the type: "Given an observation set O, find the probability that Y would have attained a value greater than y, had X been set to x." The task of inferring "causes of effects," that is, of finding the probability that $X = x$ is the cause for effect E, amounts to answering the counterfactual query: "Given effect E and observations O, find the probability that E would not have been realized, had X not been x." The technique developed in Balke and Pearl (1995) is based on probability propagation in dual networks; one representing the actual world, the other the counterfactual world. The method is not limited to linear functions but applies whenever we are willing to assume the functional form of the structural equations. The noisy OR-gate model (Pearl, 1988) is a canonical example where such functional form is normally specified. Likewise, causal theories based on Boolean functions (with exceptions), such as the one described in Equation 44.16, lend themselves to counterfactual analysis in the framework of Definition 44.7.

44.5.3 The Algorithmization of Counterfactuals

Prospective counterfactual expressions of the type $P(Y_x = y)$ are concerned with predicting the average effect of hypothetical actions and policies and can, in principle, be assessed from experimental studies in which X is randomized. Retrospective counterfactuals, on the other hand, consist of variables at different hypothetical worlds (different subscripts), and these may or may not be testable experimentally. In epidemiology, for example, the expression $P\left(Y_{x'} = y' \middle| x, y\right)$ may stand for the fraction of patients who recovered (y) under treatment (x) that would not have recovered (y') had they not been treated (x'). This fraction cannot be assessed in experimental study, for the simple reason that we cannot retest patients twice, with and without treatment. A different question is therefore posed: which counterfactuals can be tested, be it in experimental or observational studies? This question has been given a mathematical solution in Shpitser and Pearl (2007). It has been shown, for example, that in linear systems, $E(Y_x|e)$ is estimable from experimental studies whenever the prospective effect $E(Y_x)$ is estimable in such studies. Likewise, the counterfactual probability $P(Y_{x'}|x)$, also known as the effect of treatment on the treated (ETT), is estimable from observational studies whenever an admissible S exists for $P(Y_x = y)$ (Shpitser and Pearl, 2009).

Retrospective counterfactuals have also been indispensable in conceptualizing direct and indirect effects (Robins and Greenland, 1992; Pearl, 2001), which require nested counterfactuals in their definitions. For example, to evaluate the direct effect of treatment $X = x'$ on individual u, unmediated by a set Z of intermediate variables, we need to construct the nested counterfactual $Y_{x',Z_x(u)}$ where Y is the effect of interest, and $Z_x(u)$ stands for whatever values the intermediate variables Z would take had treatment not been given.* Likewise, the average *indirect effect*, of a transition from x to x', is defined as the expected change in Y affected by holding X constant, at $X = x$, and changing Z, hypothetically, to whatever value it would have attained had X been set to $X = x'$.

* Note that conditioning on the intermediate variables in Z would generally yield the wrong answer, due to unobserved "confounders" affecting both Z and Y. Moreover, in nonlinear systems, the value at which we hold Z constant will affect the result (Pearl, 2000, pp. 126–132).

This counterfactual formulation has enabled researchers to derive conditions under which direct and indirect effects are estimable from empirical data (Pearl, 2001; Petersen et al., 2006) and to answer such questions as "Can data prove an employer guilty of hiring discrimination? " or, phrased counterfactually, "what fraction of employees owes its hiring to sex discrimination?"

These tasks are performed using a general estimator, called the mediation formula (Pearl, 2001, 2009, 2012b), which is applicable to nonlinear models with discrete or continuous variables and permits the evaluation of path-specific effects with minimal assumptions regarding the data-generating process.

Acknowledgments

This research was supported in part by grants from NSF #IIS-1249822 and ONR #N00014-13-1-0153.

References

Aldrich, J. (1995). Correlations genuine and spurious in Pearson and Yule. Forthcoming *Statistical Science* **24(4)** 264–276.

Andersen, S., Olesen, K., Jensen, F., and Jensen, F. (1989). HUGIN—A shell for building Bayesian belief universes for expert systems. In *Eleventh International Joint Conference on Artificial Intelligence*, Detroit, MI.

Balke, A. and Pearl, J. (1994). Counterfactual probabilities: Computational methods, bounds, and applications. In *Uncertainty in Artificial Intelligence 10* (R. L. de Mantaras and D. Poole, eds.). Morgan Kaufmann, San Mateo, CA, pp. 46–54.

Balke, A. and Pearl, J. (1995). Counterfactuals and policy analysis in structural models. In *Uncertainty in Artificial Intelligence 11* (P. Besnard and S. Hanks, eds.). Morgan Kaufmann, San Francisco, CA, pp. 11–18.

Bareinboim, E. and Pearl, J. (2012). Transportability of causal effects: Completeness results. In *Proceedings of the 26th AAAI Conference*. Toronto, ON, Canada.

Bareinboim, E. and Pearl, J. (2013). Meta-transportability of causal effects: A formal approach. Technical Report R-407, Department of Computer Science, University of California, Los Angeles, CA. Submitted. http://ftp.cs.ucla.edu/pub/stat_ser/r407.pdf

Beaudry, E., Kabanza, F., and Michaud, F. (2010). Planning for concurrent action executions under action duration uncertainty using dynamically generated Bayesian networks. In *Proceedings of the Twentieth International Conference on Automated Planning and Scheduling (ICAPS 2010)*. Toronto, Ontario, Canada.

Blalock, JR., H. (1971). *Causal Models in the Social Sciences*. Macmillan, London, U.K.

Buede, D. M. (2009). *The Engineering Design of Systems: Models and Methods*, 2nd edn. Wiley, Hoboken, NJ.

Charniak, E. and Goldman, R. (1991). A probabilistic model of plan recognition. In *Proceedings, AAAI-91*. AAAI Press/The MIT Press, Anaheim, CA, pp. 160–165.

Cooper, G. (1990). Computational complexity of probabilistic inference using Bayesian belief networks. *Artificial Intelligence* **42** 393–405.

Cooper, G. and Herskovits, E. (1991). A Bayesian method for constructing Bayesian belief networks from databases. In *Proceedings of Uncertainty in Artificial Intelligence Conference, 1991* (B. D'Ambrosio, P. Smets, and P. Bonissone, eds.). Morgan Kaufmann, San Mateo, CA, pp. 86–94.

Darwiche, A. (2009). *Modeling and Reasoning with Bayesian Networks*. Cambridge University Press, New York.

Darwiche, A. and Goldszmidt, M. (1994). On the relation between kappa calculus and probabilistic reasoning. In *Uncertainty in Artificial Intelligence* (R. L. de Mantaras and D. Poole, eds.), Vol. 10. Morgan Kaufmann, San Francisco, CA, pp. 145–153.

Darwiche, A. and Pearl, J. (1994). Symbolic causal networks for planning under uncertainty. In *Symposium Notes of the 1994 AAAI Spring Symposium on Decision-Theoretic Planning*. AAAI Press, Stanford, CA, pp. 41–47.

Darwiche, A. and Pearl, J. (1997). On the logic of iterated belief revision. *Artificial Intelligence* **89** 1–29.

Dawid, A. (1979). Conditional independence in statistical theory. *Journal of the Royal Statistical Society, Series B* **41** 1–31.

De Campos, L., Fernandez-Luna, J., and Huete, J. (2004). Bayesian networks and information retrieval: An introduction to the special issue. *Information Processing and Management* **40** 727–733.

Dean, T. and Wellman, M. (1991). *Planning and Control*. Morgan Kaufmann, San Mateo, CA.

Dechter, R. (1999). Bucket elimination: A unifying framework for reasoning. *Artificial Intelligence* **113** 41–85.

Duncan, O. (1975). *Introduction to Structural Equation Models*. Academic Press, New York.

Fenton, N. and Neil, M. (2012). *Risk Assessment and Decision Analysis with Bayesian Networks*. CRC Press, Boca Raton, FL.

Fisher, F. (1970). A correspondence principle for simultaneous equations models. *Econometrica* **38** 73–92.

Friedman, N., Linial, M., Nachman, I., and Pe'er, D. (2000). Using Bayesian networks to analyze expression data. *Journal of Computational Biology* **7** 601–620.

Galles, D. and Pearl, J. (1995). Testing identifiability of causal effects. In *Uncertainty in Artificial Intelligence 11* (P. Besnard and S. Hanks, eds.). Morgan Kaufmann, San Francisco, CA, pp. 185–195.

Geiger, D. (1990). Graphoids: A qualitative framework for probabilistic inference. PhD thesis, Department of Computer Science, University of California, Los Angeles, CA.

Geiger, D., Verma, T., and Pearl, J. (1990). Identifying independence in Bayesian networks. In *Networks*, Vol. 20. John Wiley, Sussex, England, pp. 507–534.

Ginsberg, M. (1986). Counterfactuals. *Artificial Intelligence* **30** 12–20.

Goldszmidt, M. and Pearl, J. (1992). Rank-based systems: A simple approach to belief revision, belief update, and reasoning about evidence and actions. In *Proceedings of the Third International Conference on Knowledge Representation and Reasoning* (B. Nebel, C. Rich, and W. Swartout, eds.). Morgan Kaufmann, San Mateo, CA, pp. 661–672.

Goldszmidt, M. and Pearl, J. (1996). Qualitative probabilities for default reasoning, belief revision, and causal modeling. *Artificial Intelligence* **84** 57–112.

Good, I. (1961). A causal calculus (I). *British Journal for the Philosophy of Science* **11** 305–318.

Guestrin, C. E. (2003). Planning under uncertainty in complex structured environments. PhD thesis, Department of Computer Science, Stanford University, Stanford, CA.

Haavelmo, T. (1943). The statistical implications of a system of simultaneous equations. *Econometrica* **11** 1–12. Reprinted in D. F. Hendry and M. S. Morgan (eds.), *The Foundations of Econometric Analysis*, Cambridge University Press, Cambridge, U.K., pp. 477–490, 1995.

Heckerman, D. (1991). Probabilistic similarity networks. *Networks* **20** 607–636.

Heckerman, D., Geiger, D., and Chickering, D. (1994). Learning Bayesian networks: The combination of knowledge and statistical data. In *Uncertainty in Artificial Intelligence 10* (R. L. de Mantaras and D. Poole, eds.). Morgan Kaufmann, San Mateo, CA, pp. 293–301.

Henrion, M. (1988). Propagation of uncertainty by probabilistic logic sampling in Bayes' networks. In *Uncertainty in Artificial Intelligence 2* (J. Lemmer and L. Kanal, eds.). Elsevier Science Publishers, North-Holland, Amsterdam, the Netherlands, pp. 149–164.

Howard, R. and Matheson, J. (1981). Influence diagrams. *Principles and Applications of Decision Analysis*. Strategic Decisions Group, Menlo Park, CA.

Huang, Y. and Valtorta, M. (2006). Pearl's calculus of intervention is complete. In *Proceedings of the Twenty-Second Conference on Uncertainty in Artificial Intelligence* (R. Dechter and T. Richardson, eds.). AUAI Press, Corvallis, OR, pp. 217–224.

Jiang, X. and Cooper, G. (2010). A Bayesian spatio-temporal method for disease outbreak detection. *Journal of the American Medical Informatics Association* **17** 462–471.

Kenny, D. (1979). *Correlation and Causality*. Wiley, New York.

Kiiveri, H., Speed, T., and Carlin, J. (1984). Recursive causal models. *Journal of Australian Math Society* **36** 30–52.

Kim, J. and Pearl, J. (1983). A computational model for combined causal and diagnostic reasoning in inference systems. In *Proceedings of the Eighth International Joint Conference on Artificial Intelligence (IJCAI-83)*. Karlsruhe, Germany.

Koller, D. and Friedman, N. (2009). *Probabilistic Graphical Models: Principles and Techniques*. MIT Press, Cambridge, MA.

Koumenides, C. L. and Shadbolt, N. R. (2012). Combining link and content-based information in a Bayesian inference model for entity search. In *Proceedings of the 1st Joint International Workshop on Entity-Oriented and Semantic Search (JIWES '12)*, ACM, New York.

Lauritzen, S. (1982). *Lectures on Contingency Tables*, 2nd edn. University of Aalborg Press, Aalborg, Denmark.

Lauritzen, S. (1996). *Graphical Models*. Clarendon Press, Oxford, U.K.

Lauritzen, S. and Spiegelhalter, D. (1988). Local computations with probabilities on graphical structures and their application to expert systems (with discussion). *Journal of the Royal Statistical Society Series B* **50** 157–224.

Levitt, T., Agosta, J., and Binford, T. (1990). Model-based influence diagrams for machine vision. In *Uncertainty in Artificial Intelligence 5* (M. Henrion, R. Shachter, L. Kanal, and J. Lemmer, eds.). North-Holland, Amsterdam, the Netherlands, pp. 371–388.

Lewis, D. (1973). *Counterfactuals*. Harvard University Press, Cambridge, MA.

Little, R. J. and Rubin, D. B. (2002). *Statistical Analysis with Missing Data*, 2nd edn. Wiley, New York.

Mceliece, R., Mackay, D., and Cheng, J. (1998). Turbo decoding as an instance of Pearl's belief propagation algorithm. *IEEE Journal of Selected Areas in Communication* **16** 140–152.

Mohan, K. and Pearl, J. (2013). Recoverability from missing data: A formal approach. Technical Report R-410, http://ftp.cs.ucla.edu/pub/stat_ser/r410.pdf, Department of Computer Science, University of California, Los Angeles, CA. Submitted.

Neapolitan, R. (1990). *Probabilistic Reasoning in Expert Systems: Theory and Algorithms*. Wiley, New York.

Pearl, J. (1982). Reverend Bayes on inference engines: A distributed hierarchical approach. In *Proceedings, AAAI National Conference on AI*. Pittsburgh, PA.

Pearl, J. (1987). Bayes decision methods. In *Encyclopedia of AI*. Wiley Interscience, New York, pp. 48–56.

Pearl, J. (1988). *Probabilistic Reasoning in Intelligent Systems*. Morgan Kaufmann, San Mateo, CA.

Pearl, J. (1993a). From Bayesian networks to causal networks. In *Proceedings of the Adaptive Computing and Information Processing Seminar*. Brunel Conference Centre, London, U.K. See also *Statistical Science* **8**(3) 266–269, 1993.

Pearl, J. (1993b). From conditional oughts to qualitative decision theory. In *Proceedings of the Ninth Conference on Uncertainty in Artificial Intelligence* (D. Heckerman and A. Mamdani, eds.). Morgan Kaufmann, Burlington, MA, pp. 12–20.

Pearl, J. (1994a). From Adams' conditionals to default expressions, causal conditionals, and counterfactuals. In *Probability and Conditionals* (E. Eells and B. Skyrms, eds.). Cambridge University Press, Cambridge, MA, pp. 47–74.

Pearl, J. (1994b). A probabilistic calculus of actions. In *Uncertainty in Artificial Intelligence 10* (R. L. de Mantaras and D. Poole, eds.). Morgan Kaufmann, San Mateo, CA, pp. 454–462.

Pearl, J. (1995). Causal diagrams for empirical research. *Biometrika* **82** 669–710.

Pearl, J. (2000). *Causality: Models, Reasoning, and Inference*, 2nd edn. Cambridge University Press, New York. 2009.

Pearl, J. (2001). Direct and indirect effects. In *Proceedings of the Seventeenth Conference on Uncertainty in Artificial Intelligence*. Morgan Kaufmann, San Francisco, CA, pp. 411–420.

Pearl, J. (2009). Causal inference in statistics: An overview. *Statistics Surveys* **3** 96–146. http://ftp.cs.ucla.edu/pub/stat_ser/r350.pdf

Pearl, J. (2012a). *Do*-calculus revisited. In *Proceedings of the Twenty-Eighth Conference on Uncertainty in Artificial Intelligence* (N. de Freitas and K. Murphy, eds.). AUAI Press, Corvallis, OR.

Pearl, J. (2012b). The mediation formula: A guide to the assessment of causal pathways in nonlinear models. In *Causality: Statistical Perspectives and Applications* (C. Berzuini, P. Dawid, and L. Bernardinelli, eds.). John Wiley & Sons, Ltd, Chichester, U.K., pp. 151–179.

Pearl, J. (2012c). Some thoughts concerning transfer learning, with applications to meta-analysis and data-sharing estimation. Technical Report R-387, http://ftp.cs.ucla.edu/pub/stat_ser/r387.pdf, Department of Computer Science, University of California, Los Angeles, CA.

Pearl, J. and Bareinboim, E. (2011). Transportability of causal and statistical relations: A formal approach. In *Proceedings of the Twenty-Fifth Conference on Artificial Intelligence (AAAI-11)*. Menlo Park, CA. Available at: http://ftp.cs.ucla.edu/pub/stat_ser/r372a.pdf

Pearl, J., Geiger, D., and Verma, T. (1990). The logic and influence diagrams. In *Influence Diagrams, Belief Nets and Decision Analysis* (R. Oliver and J. Smith, eds.). Wiley, New York, pp. 67–87.

Pearl, J. and Robins, J. (1995). Probabilistic evaluation of sequential plans from causal models with hidden variables. In *Uncertainty in Artificial Intelligence 11* (P. Besnard and S. Hanks, eds.). Morgan Kaufmann, San Francisco, CA, pp. 444–453.

Pearl, J. and Verma, T. (1991). A theory of inferred causation. In *Principles of Knowledge Representation and Reasoning: Proceedings of the Second International Conference* (J. Allen, R. Fikes, and E. Sandewall, eds.). Morgan Kaufmann, San Mateo, CA, pp. 441–452.

Petersen, M., Sinisi, S., and van der laan, M. (2006). Estimation of direct causal effects. *Epidemiology* **17** 276–284.

Rebane, G. and Pearl, J. (1987). The recovery of causal poly-trees from statistical data. In *Proceedings of the Third Workshop on Uncertainty in AI*. Seattle, WA.

Robins, J. (1986). A new approach to causal inference in mortality studies with a sustained exposure period—Applications to control of the healthy workers survivor effect. *Mathematical Modeling* 7 1393–1512.

Robins, J. and Greenland, S. (1992). Identifiability and exchangeability for direct and indirect effects. *Epidemiology* **3** 143–155.

Rosenthal, A. (1975). A computer scientist looks at reliability computations. In *Reliability and Fault Tree Analysis* (R. E. Barlow, J. B. Fussell, and N. D. Singpurwalla, eds.). SIAM, Philadelphia, PA, pp. 133–152.

Rubin, D. (1974). Estimating causal effects of treatments in randomized and nonrandomized studies. *Journal of Educational Psychology* **66** 688–701.

Rubin, D. (1976). Inference and missing data. *Biometrika* **63** 581–592.

Rumelhart, D. (1976). Toward an interactive model of reading. Technical Report CHIP-56, University of California, La Jolla, CA.

Russell, S. J. and Norvig, P. (2003). *Artificial Intelligence: A Modern Approach*. Prentice Hall, Upper Saddle River, NJ.

Schafer, J. L. and Graham, J. W. (2002). Missing data: Our view of the state of the art. *Psychological Methods* 7 147–177.

Shachter, R. (1986). Evaluating influence diagrams. *Operations Research* **34** 871–882.

Shachter, R. (1988). Probabilistic inference and influence diagrams. *Operations Research* **36** 589–604.

Shafer, G. and Pearl, J. E. (1990). *Readings in Uncertain Reasoning*. Morgan Kaufmann, San Mateo, CA.

Shpitser, I. and Pearl, J. (2006a). Identification of conditional interventional distributions. In *Proceedings of the Twenty-Second Conference on Uncertainty in Artificial Intelligence* (R. Dechter and T. Richardson, eds.). AUAI Press, Corvallis, OR, pp. 437–444.

Shpitser, I. and Pearl, J. (2006b). Identification of joint interventional distributions in recursive semi-Markovian causal models. In *Proceedings of the Twenty-First National Conference on Artificial Intelligence*. AAAI Press, Menlo Park, CA, pp. 1219–1226.

Shpitser, I. and Pearl, J. (2007). What counterfactuals can be tested. In *Proceedings of the Twenty-Third Conference on Uncertainty in Artificial Intelligence*. AUAI Press, Vancouver, BC, Canada, pp. 352–359. Also, *Journal of Machine Learning Research*, **9** 1941–1979, 2008.

Shpitser, I. and Pearl, J. (2009). Effects of treatment on the treated: Identification and generalization. In *Proceedings of the Twenty-Fifth Conference Annual Conference on Uncertainty in Artificial Intelligence (UAI-09)*. AUAI Press, Corvallis, OR.

Simon, H. (1953). Causal ordering and identifiability. In *Studies in Econometric Method* (W. C. Hood and T. Koopmans, eds.). John Wiley & Sons, Inc., New York, pp. 49–74.

Sobel, M. (1990). Effect analysis and causation in linear structural equation models. *Psychometrika* **55** 495–515.

Spiegelhalter, D. and Lauritzen, S. (1990). Sequential updating of conditional probabilities on directed graphical structures. *Networks* **20** 579–605.

Spiegelhalter, D., Lauritzen, S., Dawid, P., and Cowell, R. (1993). Bayesian analysis in expert systems (with discussion). *Statistical Science* **8** 219–283.

Spirtes, P., Glymour, C., and Scheines, R. (1993). *Causation, Prediction, and Search*. Springer-Verlag, New York.

Spirtes, P., Glymour, C., and Scheines, R. (2000). *Causation, Prediction, and Search*, 2nd edn. MIT Press, Cambridge, MA.

Spohn, W. (1988). A general non-probabilistic theory of inductive reasoning. In *Proceedings of the Fourth Workshop on Uncertainty in Artificial Intelligence*. Minneapolis, MN.

Spohn, W. (2012). *The Laws of Belief: Ranking Theory and its Philosophical Applications*. Oxford University Press, Oxford, U.K.

Strotz, R. and Wold, H. (1960). Recursive versus nonrecursive systems: An attempt at synthesis. *Econometrica* **28** 417–427.

Tian, J. and Pearl, J. (2001a). Causal discovery from changes. In *Proceedings of the Seventeenth Conference on Uncertainty in Artificial Intelligence*. Morgan Kaufmann, San Francisco, CA, pp. 512–521.

Tian, J. and Pearl, J. (2001b). Causal discovery from changes: A Bayesian approach. Technical Report R-285, Computer Science Department, University of California, Los Angeles, CA.

Tian, J. and Pearl, J. (2002). A general identification condition for causal effects. In *Proceedings of the Eighteenth National Conference on Artificial Intelligence*. AAAI Press/The MIT Press, Menlo Park, CA, pp. 567–573.

Tian, J. and Shpitser, I. (2010). On identifying causal effects. In *Heuristics, Probability and Causality: A Tribute to Judea Pearl* (R. Dechter, H. Geffner, and J. Halpern, eds.). College Publications, London, U.K., pp. 415–444.

Uebersax, J. (2004). *Genetic Counseling and Cancer Risk Modeling: An Application of Bayes Nets*. Ravenpack International, Marbella, Spain.

Verma, T. and Pearl, J. (1990). Equivalence and synthesis of causal models. In *Proceedings of the Sixth Conference on Uncertainty in Artificial Intelligence*. Cambridge, MA. Also in P. Bonissone, M. Henrion, L. N. Kanal, and J. F. Lemmer (eds.). *Uncertainty in Artificial Intelligence 6*, Elsevier Science Publishers, B. V., Amsterdam, the Netherlands, pp. 255–268, 1991.

Weiss, Y. and Pearl, J. (2010). Belief propagation—Perspectives. *Communications of the ACM* **53** 1.

Wermuth, N. and Lauritzen, S. (1983). Graphical and recursive models for contingency tables. *Biometrika* **70** 537–552.

Wold, H. (1964). *Econometric Model Building*. North-Holland, Amsterdam, the Netherlands.

Wright, S. (1921). Correlation and causation. *Journal of Agricultural Research* **20** 557–585.

Wright, S. (1934). The method of path coefficients. *The Annals of Mathematical Statistics* **5** 161–215.

Zweig, G. (1998). *Speech Recognition with Dynamic Bayesian Networks*. PhD thesis, Computer Science Division, University of California, Berkeley, CA.

VI

Networking and Communication

45 Network Organization and Topologies

William Stallings
Independent Consultant

In this chapter, we examine the communications software needed to interconnect computers, workstations, servers, and other devices across networks. Then, we look at some of the networks in contemporary use. This chapter focuses exclusively on wired networks. Wireless networks are examined elsewherc, such as in the chapter "Mobile Operating Systems," which has an extensive discussion on wireless networks.

45.1 Transmission Control Protocol/ Internet Protocol Architecture

When communication is desired among computers from different vendors, the software development effort can be a nightmare. Different vendors use different data formats and data exchange protocols. Even within one vendor's product line, different model computers may communicate in unique ways.

As the use of computer communications and computer networking proliferates, a one-at-a-time special-purpose approach to communications software development is too costly to be acceptable. The only alternative is for computer vendors to adopt and implement a common set of conventions. For this to happen, standards are needed. Such standards would have two benefits:

- Vendors feel encouraged to implement the standards because of an expectation that, because of wide usage of the standards, their products will be more marketable with them.
- Customers are in a position to require that the standards be implemented by any vendor wishing to propose equipment to them.

It should become clear from the ensuing discussion that no single standard will suffice. Any distributed application, such as electronic mail or client/server interaction, requires a complex set of communications functions for proper operation. Many of these functions, such as reliability mechanisms, are common across many or even all applications. Thus, the communications task is best viewed as consisting of a modular architecture, in which the various elements of the architecture perform the various required functions. Hence, before one can develop standards, there should be a structure, or *protocol architecture*, that defines the communications tasks.

Transmission control protocol (TCP)/Internet protocol (IP) is a result of protocol research and development conducted on the experimental packet-switched network, ARPANET, funded by the Defense Advanced Research Projects Agency, and is generally referred to as the TCP/IP protocol suite. This protocol suite consists of a large collection of protocols that have been issued as Internet standards by the Internet Activities Board.

45.1.1 TCP/IP Layers

In general terms, communications can be said to involve three agents: applications, computers, and networks. Examples of applications include file transfer and electronic mail. The applications that we are concerned with in this study are distributed applications that involve the exchange of data between two computer systems. These applications, and others, execute on computers that can often support multiple simultaneous applications. Computers are connected to networks, and the data to be exchanged are transferred by the network from one computer to another. Thus, the transfer of data from one application to another involves first getting the data to the computer in which the application resides and then getting it to the intended application within the computer.

There is no official TCP/IP protocol model. However, based on the protocol standards that have been developed, we can organize the communication task for TCP/IP into five relatively independent layers, from bottom to top (Stallings, 2013):

- Physical layer
- Network access layer
- Internet layer
- Host-to-host, or transport layer
- Application layer

The *physical layer* covers the physical interface between a data transmission device (e.g., workstation and computer) and a transmission medium or network. This layer is concerned with specifying the characteristics of the transmission medium, the nature of the signals, the data rate, and related matters.

The *network access layer* is concerned with the exchange of data between an end system (ES) (server, workstation, etc.) and the network to which it is attached. The sending computer must provide the network with the address of the destination computer, so that the network may route the data to the appropriate destination. The sending computer may wish to invoke certain services, such as priority, that might be provided by the network. The specific software used at this layer depends on the type of network to be used; different standards have been developed for circuit switching, packet switching (e.g., frame relay), local-area networks (LANs) (e.g., Ethernet), and others. Thus, it makes sense to separate those functions having to do with network access into a separate layer. By doing this, the remainder of the communications software, above the network access layer, need not be concerned about the specifics of the network to be used. The same higher-layer software should function properly regardless of the particular network to which the computer is attached.

The network access layer is concerned with access to and routing data across a network for two ESs attached to the same network. In those cases where two devices are attached to different networks, procedures are needed to allow data to traverse multiple interconnected networks. This is the function of the *Internet layer.* The IP is used at this layer to provide the routing function across multiple networks. This protocol is implemented not only in the ESs but also in routers. A router is a processor that connects two networks and whose primary function is to relay data from one network to the other on its route from the source to the destination ES.

Regardless of the nature of the applications that are exchanging data, there is usually a requirement that data be exchanged reliably. That is, we would like to be assured that all of the data arrive at the destination application and that the data arrive in the same order in which they were sent. As we shall see, the mechanisms for providing reliability are essentially independent of the nature of the applications. Thus, it makes sense to collect those mechanisms in a common layer shared by all applications; this is referred to as the host-to-host layer or the *transport layer.* The TCP provides this functionality.

Finally, the *application layer* contains the logic needed to support the various user applications. For each different type of application, such as file transfer, a separate module is needed that is peculiar to that application.

45.1.2 Operation of TCP and IP

Figure 45.1 indicates how these protocols are configured for communications. To make clear that the total communications facility may consist of multiple networks, the constituent networks are usually referred to as *subnetworks.* Some sort of network access protocol, such as the Ethernet logic, is used to connect a computer to a subnetwork. This protocol enables the host to send data across the subnetwork to another host or, in the case of a host on another subnetwork, to a router. IP is implemented in all of the ESs and the routers. It acts as a relay to move a block of data from one host, through one or more routers,

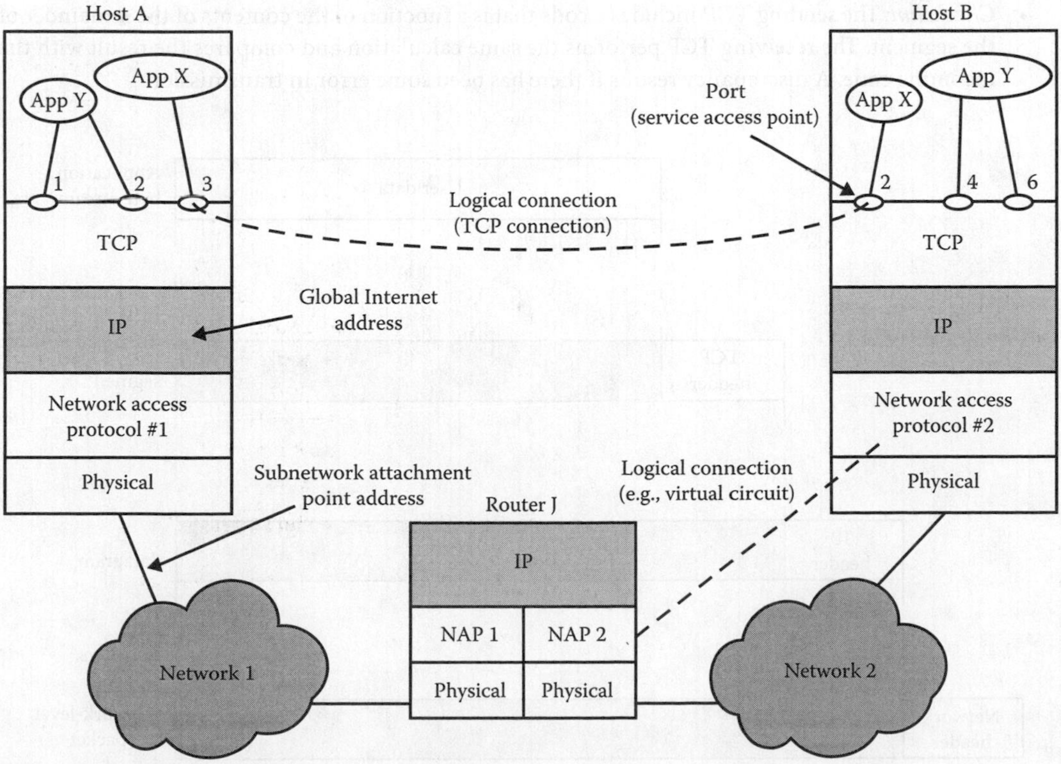

FIGURE 45.1 TCP/IP Concepts.

to another host. TCP is implemented only in the ESs; it keeps track of the blocks of data to assure that all are delivered reliably to the appropriate application.

For successful communication, every entity in the overall system must have a unique address. Actually, two levels of addressing are needed. Each host on a subnetwork must have a unique global Internet address; this allows the data to be delivered to the proper host. Each process with a host must have an address that is unique within the host; this allows the host-to-host protocol (TCP) to deliver data to the proper process. These latter addresses are known as ports.

Let us trace a simple operation. Suppose that a process, associated with port 1 at host *A*, wishes to send a message to another process, associated with port 2 at host *B*. The process at *A* hands the message down to TCP with instructions to send it to host *B*, port 2. TCP hands the message down to IP with instructions to send it to host *B*. Note that IP need not be told the identity of the destination port. All this it needs to know is that the data are intended for host *B*. Next, IP hands the message down to the network access layer (e.g., Ethernet logic) with instructions to send it to router *X* (the first hop on the way to *B*).

To control this operation, control information as well as user data must be transmitted, as suggested in Figure 45.2. Let us say that the sending process generates a block of data and passes this to TCP. TCP may break this block into smaller pieces to make the transmission more manageable. To each of these pieces, TCP appends control information known as the TCP header, forming a *TCP segment*. The control information is to be used by the peer TCP protocol entity at host *B*. Examples of items that are included in this header include the following:

- *Destination port*: When the TCP entity at *B* receives the segment, it must know to whom the data are to be delivered.
- *Sequence number*: TCP numbers the segments that it sends to a particular destination port sequentially, so that if they arrive out of order, the TCP entity at *B* can reorder them.
- *Checksum*: The sending TCP includes a code that is a function of the contents of the remainder of the segment. The receiving TCP performs the same calculation and compares the result with the incoming code. A discrepancy results if there has been some error in transmission.

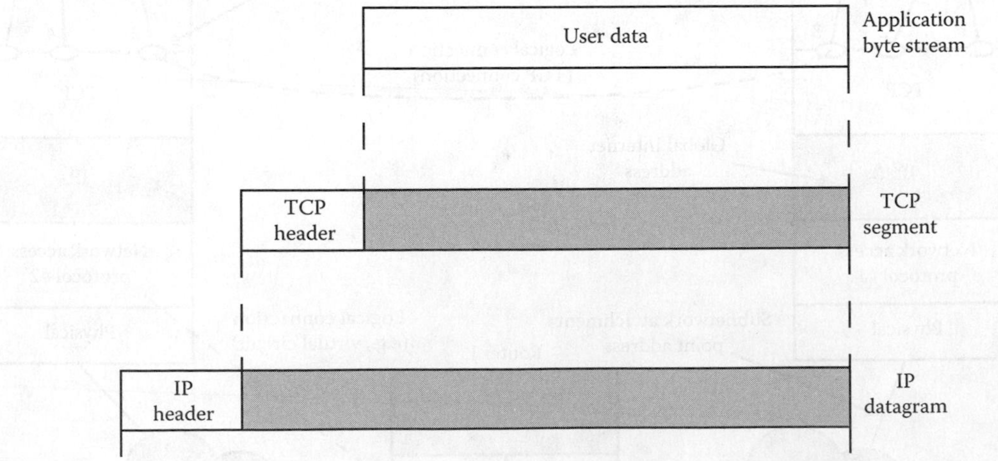

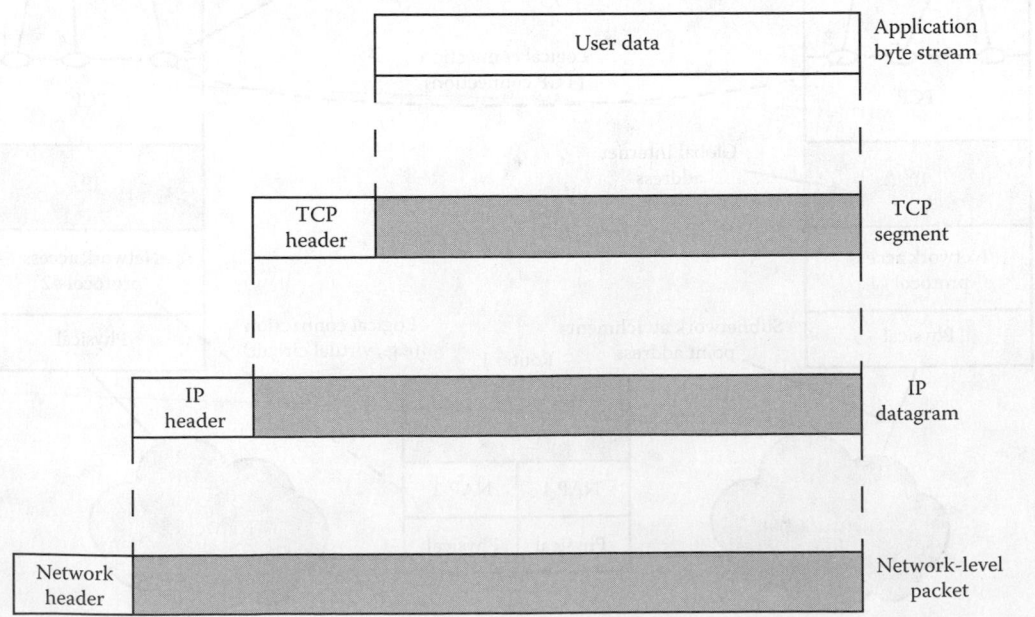

FIGURE 45.2 Protocol data units in the TCP/IP architecture.

Next, TCP hands each segment over to IP, with instructions to transmit it to *B*. These segments must be transmitted across one or more subnetworks and relayed through one or more intermediate routers. This operation, too, requires the use of control information. Thus, IP appends a header of control information to each segment to form an *IP datagram*. An example of an item stored in the IP header is the destination host address (in this example, *B*).

Finally, each IP datagram is presented to the network access layer for transmission across the first subnetwork in its journey to the destination. The network access layer appends its own header, creating a packet, or frame. The packet is transmitted across the subnetwork to router X. The packet header contains the information that the subnetwork needs to transfer the data across the subnetwork. Examples of items that may be contained in this header include the following:

- *Destination subnetwork address*: The subnetwork must know to which attached device the packet is to be delivered.
- *Facilities requests*: The network access protocol might request the use of certain subnetwork facilities, such as priority.

At router X, the packet header is stripped off and the IP header examined. On the basis of the destination address information in the IP header, the IP module in the router directs the datagram out across subnetwork 2 to *B*. To do this, the datagram is again augmented with a network access header.

When the data are received at *B*, the reverse process occurs. At each layer, the corresponding header is removed, and the remainder is passed on to the next higher layer, until the original user data are delivered to the destination process.

45.1.3 TCP/IP Applications

A number of applications have been standardized to operate on top of TCP. We mention three of the most common in this study.

The *simple mail transfer protocol* (*SMTP*) provides a basic electronic mail facility. It provides a mechanism for transferring messages among separate hosts. Features of SMTP include mailing lists, return receipts, and forwarding. The SMTP protocol does not specify the way in which messages are to be created; some local editing or native electronic mail facility is required. Once a message is created, SMTP accepts the message and makes use of TCP to send it to an SMTP module on another host. The target SMTP module will make use of a local electronic mail package to store the incoming message in a user's mailbox.

The *file transfer protocol* (*FTP*) is used to send files from one system to another under user command. Both text and binary files are accommodated, and the protocol provides features for controlling user access. When a user wishes to engage in file transfer, FTP sets up a TCP connection to the target system for the exchange of control messages. These allow user ID and password to be transmitted and allow the user to specify the file and file actions desired. Once a file transfer is approved, a second TCP connection is set up for the data transfer. The file is transferred over the data connection, without the overhead of any headers or control information at the application level. When the transfer is complete, the control connection is used to signal the completion and to accept new file transfer commands.

Secure Shell (*SSH*) provides a secure remote logon capability, which enables a user at a terminal or personal computer to logon to a remote computer and function as if directly connected to that computer. SSH also supports file transfer between the local host and a remote server. SSH enables the user and the remote server to authenticate each other; it also encrypts all traffic in both directions. SSH traffic is carried on a TCP connection.

45.2 Internetworking

In most cases, a LAN or wide-area network (WAN) is not an isolated entity. An organization may have more than one type of LAN at a given site to satisfy a spectrum of needs. An organization may have multiple LANs of the same type at a given site to accommodate performance or security requirements.

Furthermore, an organization may have LANs at various sites and need them to be interconnected via WANs for central control of distributed information exchange.

An interconnected set of networks, from a user's point of view, may appear simply as a larger network. However, if each of the constituent networks retains its identity, and special mechanisms are needed for communicating across multiple networks, then the entire configuration is often referred to as an *Internet*, and each of the constituent networks as a *subnetwork*. The most important example of an Internet is referred to simply as the Internet. As the Internet has evolved from its modest beginnings as a research-oriented packet-switching network, it has served as the basis for the development of internetworking technology and as the model for private Internets within organizations. These latter are also referred to as *intranets*. If an organization extends access to its intranet, over the Internet, to selected customers and suppliers, then the resulting configuration is often referred to as an *extranet*.

Each constituent subnetwork in an Internet supports communication among the devices attached to that subnetwork; these devices are referred to as *hosts* or *ESs*. Within an Internet, subnetworks are connected by devices referred to as *routers*. The differences between them have to do with the types of protocols used for the internetworking logic. The role and functions of routers were introduced in the context of IP earlier in this chapter. However, because of the importance of routers in the overall networking scheme, it is worth providing additional comment in this section.

45.2.1 Routers

Internetworking is achieved by using intermediate systems, or routers, to interconnect a number of independent networks. Essential functions that the router must perform include the following:

1. Provide a link between networks.
2. Provide for the routing and delivery of data between ESs attached to different networks.
3. Provide these functions in such a way as to not require modifications of the networking architecture of any of the attached networks.

Point 3 means that the router must accommodate a number of differences among networks, such as the following:

- *Addressing schemes*: The networks may use different schemes for assigning addresses to devices. For example, an IEEE 802 LAN uses 48 bit binary addresses for each attached device; an ATM network typically uses 15 digit decimal addresses (encoded as 4 bits per digit for a 60 bit address). Some form of global network addressing must be provided, as well as a directory service.
- *Maximum packet sizes*: Packets from one network may have to be broken into smaller pieces to be transmitted on another network, a process known as *fragmentation*. For example, Ethernet imposes a maximum packet size of 1500 bytes; a maximum packet size of 1600 bytes is common on frame relay networks. A packet that is transmitted on a frame relay network and picked up by a router for forwarding on an Ethernet LAN may have to be fragmented into two smaller ones.
- *Interfaces*: The hardware and software interfaces to various networks differ. The concept of a router must be independent of these differences.
- *Reliability*: Various network services may provide anything from a reliable end-to-end virtual circuit to an unreliable service. The operation of the routers should not depend on an assumption of network reliability.

The preceding requirements are best satisfied by an internetworking protocol, such as IP, that is implemented in all ESs and routers.

45.2.2 Internetworking Example

Figure 45.3 depicts a configuration that we will use to illustrate the interactions among protocols for internetworking. In this case, we focus on a server attached to a frame relay WAN and a workstation attached to an IEEE 802 LAN, such as Ethernet, with a router connecting the two networks. The router provides a link between the server and the workstation that enables these ESs to ignore the details of the intervening networks. For the frame relay network, what we have referred to as the network access layer consists of a single frame relay protocol. In the case of the IEEE 802 LAN, the network access layer consists of two sublayers: the logical link control layer and the media access control layer. For purposes of this discussion, we need not describe these layers in any detail.

Consider the typical steps in the transfer of a block of data, such as a file or a Web page, from the server, through an Internet, and ultimately to an application in the workstation. In this example, the message passes through just one router. Before data can be transmitted, the application and transport layers in the server establish, with the corresponding layers in the workstation, the applicable ground rules for a communication session. These include character code to be used, error-checking method, and the like. The protocol at each layer is used for this purpose and then is used in the transmission of the message.

For the actual transmission, the application in the server passes data down to TCP. TCP divides the data into a sequence of data blocks. Each block is assembled as a TCP segment by adding the TCP header, containing protocol control information. Each segment is then passed down to IP, which appends an IP header. A frame relay header and trailer are added to each IP packet. Each frame is then transmitted over the medium as a sequence of bits. The router strips off the frame relay header and trailer and examines the IP header to determine routing information. The router then appends the IEEE 802 headers and trailers and transmits the resulting frame to the destination workstation. At the workstation, all the headers and trailers are stripped off in turn, and the data are delivered to the application.

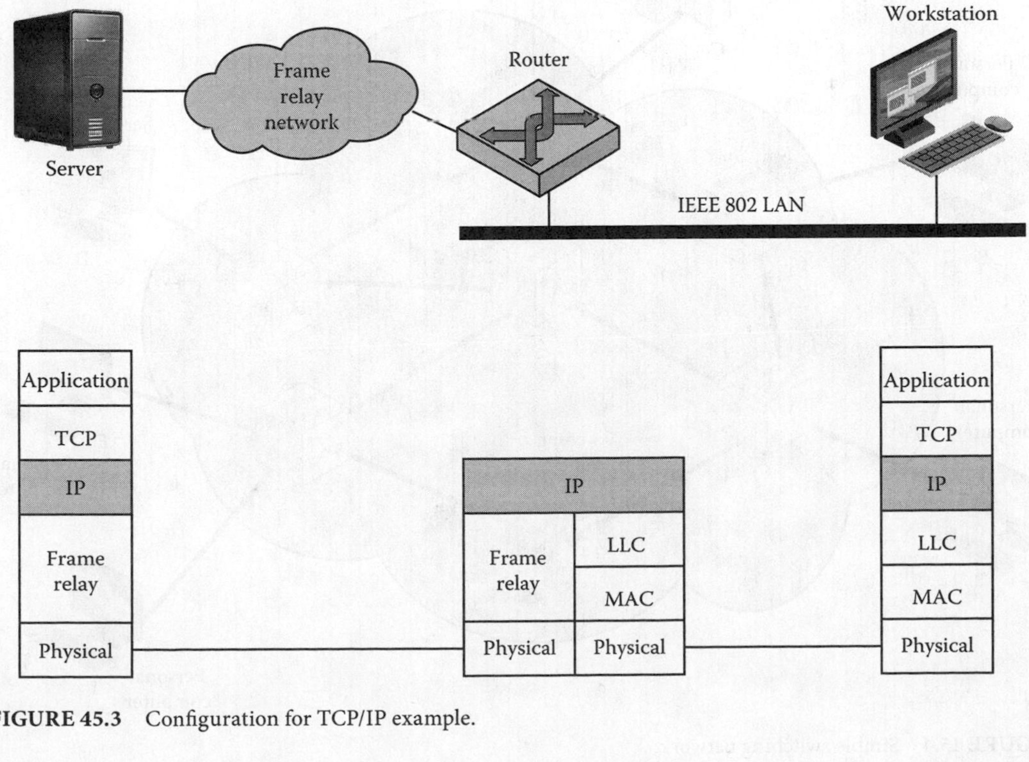

FIGURE 45.3 Configuration for TCP/IP example.

45.3 WAN Organization

Traditionally, data networks have been classified as either WAN or LAN. Although there has been some blurring of this distinction, it is still a useful one. We look first at WANs.

45.3.1 Traditional WANs

Traditional WANs are switched communications networks, consisting of an interconnected collection of nodes, in which information is transmitted from source station to destination station by being routed through the network of nodes. Figure 45.4 is a simplified illustration of the concept. The nodes are connected by transmission paths. Signals entering the network from a station are routed to the destination by being switched from node to node. Two quite different technologies are used in wide-area switched networks: circuit switching and packet switching. These two technologies differ in the way the nodes switch information from one link to another on the way from source to destination.

45.3.1.1 Circuit Switching

Circuit switching is the dominant technology for both voice and data communications today and will remain so for the foreseeable future. Communication via circuit switching implies that there is a dedicated communication path between two stations. That path is a connected sequence of links between network nodes. On each physical link, a channel is dedicated to the connection. The most common example of circuit switching is the telephone network.

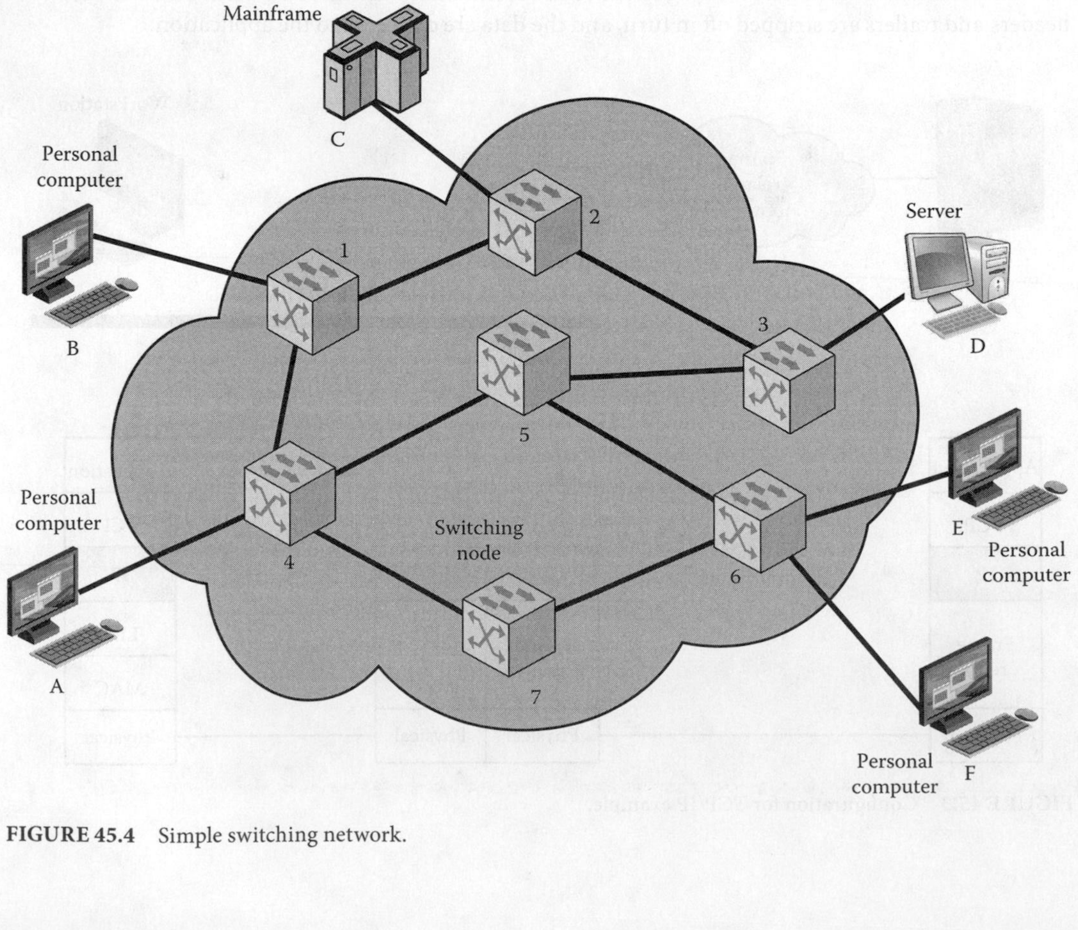

FIGURE 45.4 Simple switching network.

Communication via circuit switching involves three phases, which can be explained with reference to Figure 45.4.

1. *Circuit establishment*: Before any signals can be transmitted, an end-to-end (station-to-station) circuit must be established. For example, station *A* sends a request to node 4 requesting a connection to station *E*. Typically, the link from *A* to 4 is a dedicated line, so that part of the connection already exists. Node 4 must find the next leg in a route leading to node 6. Based on routing information and measures of availability and perhaps cost, node 4 selects the link to node 5, allocates a free channel (using frequency-division multiplexing, FDM, or time-division multiplexing, TDM) on that link and sends a message requesting connection to *E*. So far, a dedicated path has been established from *A* through 4 to 5. Since a number of stations may attach to 4, it must be able to establish internal paths from multiple stations to multiple nodes. The remainder of the process proceeds similarly. Node 5 dedicates a channel to node 6 and internally ties that channel to the channel from node 4. Node 6 completes the connection to *E*. In completing the connection, a test is made to determine if *E* is busy or is prepared to accept the connection.
2. *Information transfer*: Information can now be transmitted from *A* through the network to *E*. The transmission may be analog voice, digitized voice, or binary data, depending on the nature of the network. As the carriers evolve to fully integrated digital networks, the use of digital (binary) transmission for both voice and data is becoming the dominant method. The path is: *A*-4 link, internal switching through 4, 4–5 channel, internal switching through 5, 5–6 channel, internal switching through 6 and 6-*E* link. Generally, the connection is full duplex, and signals may be transmitted in both directions simultaneously.
3. *Circuit disconnect*: After some period of information transfer, the connection is terminated, usually by the action of one of the two stations. Signals must be propagated to nodes 4, 5, and 6 to deallocate the dedicated resources.

Note that the connection path is established before data transmission begins. Thus channel capacity must be reserved between each pair of nodes in the path and each node must have available internal switching capacity to handle the requested connection. The switches must have the intelligence to make these allocations and to devise a route through the network.

Circuit switching can be rather inefficient. Channel capacity is dedicated for the duration of a connection, even if no data are being transferred. For a voice connection, utilization may be rather high, but it still does not approach 100%. For a terminal-to-computer connection, the capacity may be idle during most of the time of the connection. In terms of performance, there is a delay prior to signal transfer for call establishment. However, once the circuit is established, the network is effectively transparent to the users. Information is transmitted at a fixed data rate with no delay other than the propagation delay through the transmission links. The delay at each node is negligible.

Circuit-switching technology has been driven by those applications that handle voice traffic. One of the key requirements for voice traffic is that there must be virtually no transmission delay and certainly no variation in delay. A constant signal transmission rate must be maintained, since transmission and reception occur at the same signal rate. These requirements are necessary to allow normal human conversation. Furthermore, the quality of the received signal must be sufficiently high to provide, at a minimum, intelligibility.

45.3.1.2 Packet Switching

A *packet-switching network* is a switched communications network that transmits data in short blocks called packets. The network consists of a set of interconnected packet-switching nodes. A device attaches to the network at one of these nodes and presents data for transmission in the form of a stream of packets. Each packet is routed through the network. As each node along the route is encountered, the packet is received, stored briefly, and then transmitted along a link to the next node in the route. Two approaches are used to manage the transfer and routing of these streams of packets: datagram and virtual circuit.

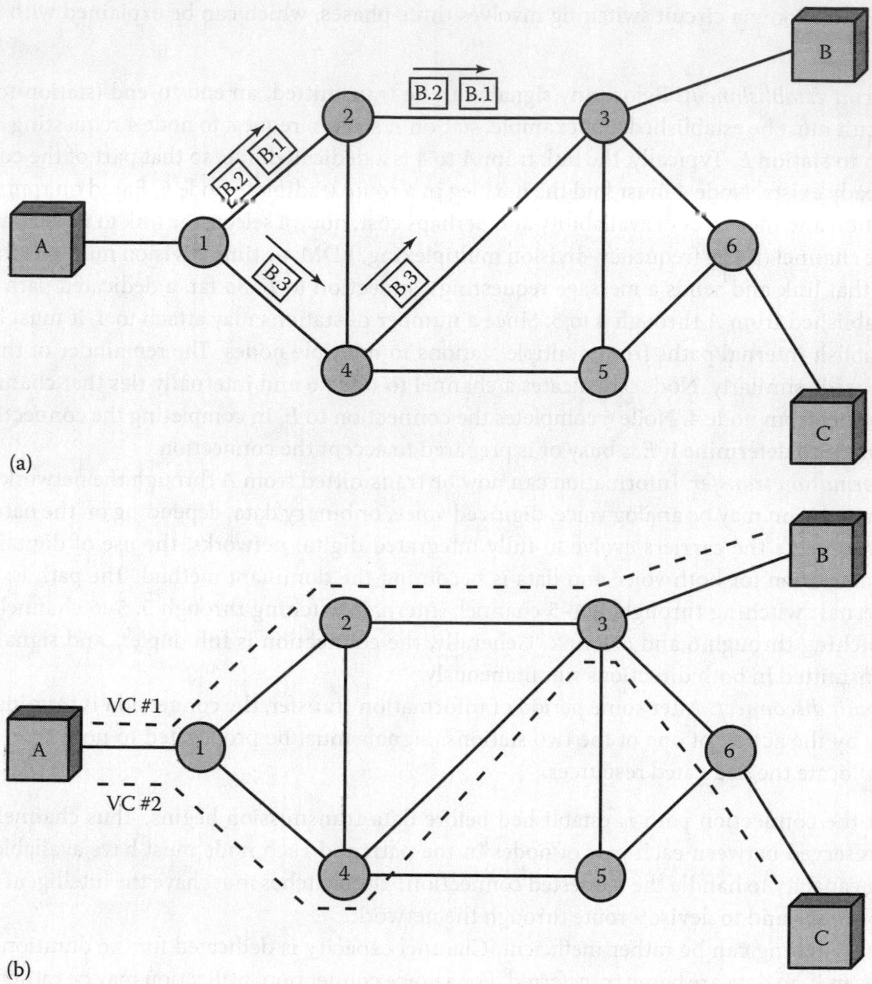

FIGURE 45.5 (a) Internal Datagram and (b) virtual circuit operation.

In the datagram approach, each packet is treated independently, with no reference to packets that have gone before. This approach is illustrated in Figure 45.5a. Each node chooses the next node on a packet's path, taking into account information received from neighboring nodes on traffic, line failures, and so on. So the packets, each with the same destination address, do not all follow the same route, and they may arrive out of sequence at the exit point. In some networks, the exit node restores the packets to their original order before delivering them to the destination. In other datagram networks, it is up to the destination rather than the exit node to do the reordering. Furthermore, it is possible for a packet to be destroyed in the network. For example, if a packet-switching node crashes momentarily, all of its queued packets may be lost. Again, it is up to either the exit node or the destination to detect the loss of a packet and decide how to recover it. In this technique, each packet, treated independently, is referred to as a datagram.

In the virtual circuit approach, a preplanned route is established before any packets are sent. Once the route is established, all of the packets between a pair of communicating parties follow this same route through the network. This is illustrated in Figure 45.5b. Because the route is fixed for the duration of the logical connection, it is somewhat similar to a circuit in a circuit-switching network and is referred to as a virtual circuit. Each packet now contains a virtual circuit identifier as well as data. Each node on the preestablished route knows where to direct such packets; no routing decisions are required. At any time, each station can have more than one virtual circuit to any other station and can have virtual circuits to more than one station.

If two stations wish to exchange data over an extended period of time, there are certain advantages to virtual circuits. First, the network may provide services related to the virtual circuit, including sequencing and error control. *Sequencing* refers to the fact that, since all packets follow the same route, they arrive in the original order. *Error control* is a service that assures not only that packets arrive in proper sequence but also that all packets arrive correctly. For example, if a packet in a sequence from node 4 to node 6 fails to arrive at node 6, or arrives with an error, node 6 can request a retransmission of that packet from node 4. Another advantage is that packets should transit the network more rapidly with a virtual circuit; it is not necessary to make a routing decision for each packet at each node.

One advantage of the datagram approach is that the call setup phase is avoided. Thus, if a station wishes to send only one or a few packets, datagram delivery will be quicker. Another advantage of the datagram service is that, because it is more primitive, it is more flexible. For example, if congestion develops in one part of the network, incoming datagrams can be routed away from the congestion. With the use of virtual circuits, packets follow a predefined route, and thus it is more difficult for the network to adapt to congestion. A third advantage is that datagram delivery is inherently more reliable. With the use of virtual circuits, if a node fails, all virtual circuits that pass through that node are lost. With datagram delivery, if a node fails, subsequent packets may find an alternate route that bypasses that node.

45.3.2 High-Speed WANs

As the speed and number of LANs continue their relentless growth, increasing demand is placed on wide-area packet-switching networks to support the tremendous throughput generated by these LANs. In the early days of wide-area networking, X.25 was designed to support direct connection of terminals and computers over long distances. At speeds up to 64 kbps or so, X.25 copes well with these demands. As LANs have come to play an increasing role in the local environment, X.25, with its substantial overhead, is being recognized as an inadequate tool for wide-area networking. Fortunately, several new generations of high-speed switched services for wide-area networking have moved rapidly from the research laboratory and the draft standard stage to the commercially available, standardized-product stage. We look briefly at the most important alternatives in this section.

45.3.2.1 Frame Relay

Both frame relay and X.25 use a virtual-circuit, packet-switching approach to data transmission. Frame relay provides a streamlined technique for wide-area packet switching, compared to X.25. It provides superior performance by eliminating as much as possible of the overhead of X.25. The key differences of frame relaying from a conventional X.25 packet-switching service are the following:

- Call control signaling (e.g., requesting that a connection be set up) is carried on a logical connection that is separate from the connections used to carry user data. Thus, intermediate nodes need not maintain state tables or process messages relating to call control on an individual per-connection basis.
- There are only physical and link layers of processing for frame relay, compared to physical, link, and packet layers for X.25. Thus, one entire layer of processing is eliminated with frame relay.
- There is no hop-by-hop flow control and error control. End-to-end flow control and error control is the responsibility of a higher layer, if it is employed at all.

Frame relay takes advantage of the reliability and fidelity of modern digital facilities to provide faster packet switching than X.25. Whereas X.25 typically operates only up to speeds of about 64 kbps, frame relay is designed to work at access speeds up to 2 Mbps.

Frame relay at one time enjoyed widespread market share. However, today legacy frame relay systems are gradually being phased out in favor of more efficient techniques.

45.3.2.2 Asynchronous Transfer Mode

Frame relay was designed to support access speeds up to 2 Mbps. Although it has evolved to provide services up to 42 Mbps, frame relay is not able to meet the needs of businesses that require wide area access speeds in the hundreds or thousands of megabits per second. One of the first technologies developed to accommodate such gargantuan requirements was asynchronous transfer mode (ATM), also known as *cell relay.*

ATM is similar in concept to frame relay. Both frame relay and ATM take advantage of the reliability and dependability of modern digital facilities to provide faster packet switching than X.25. ATM is even more streamlined than frame relay in its functionality and can support data rates several orders of magnitude greater than frame relay.

The most noteworthy difference between ATM and frame relay is that ATM uses fixed-size packets, called cells, which are of length 53 octets. There are several advantages to the use of small, fixed-size cells. First, the use of small cells may reduce queuing delay for a high-priority cell, because it waits less if it arrives slightly behind a lower-priority cell that has gained access to a resource (e.g., the transmitter). Second, fixed-size cells can be switched more efficiently, which is important for the very high data rates of ATM. With fixed-size cells, it is easier to implement the switching mechanism in hardware.

An ATM network is designed to be able to transfer many different types of traffic simultaneously, including real-time flows such as voice, video, and bursty TCP flows. Although each such traffic flow is handled as a stream of 53 octet cells traveling through a virtual channel (ATM terminology for virtual circuit), the way in which each data flow is handled within the network depends on the characteristics of the traffic flow and the QoS requirements of the application. For example, real-time video traffic must be delivered within minimum variation in delay.

Although the switching mechanism for ATM transmission is very efficient, and very high data rates can be achieved, ATM as a whole is very complex and expensive to implement. As with frame relay, ATM is gradually losing ground to newer approaches to wide-area networking.

45.3.3 Multiprotocol Label Switching

Multiprotocol Label Switching (*MPLS*) refers to IP-based networking services that are available from a number of carriers. MPLS services are designed to speed up the IP packet-forwarding process while retaining the types of traffic management and connection-oriented QoS mechanisms found in ATM networks.

MPLS is often described as being "protocol agnostic" because it can carry many different kinds of traffic, including ATM cells, IP packets, and Ethernet or SONET frames. Carriers have been able to cost effectively implement MPLS infrastructure because it is possible for MPLS-enabled routers to coexist with ordinary IP routers. MPLS has also been designed to work with ATM and frame relay networks via MPLS-enabled ATM switches and MPLS-enabled frame relay switches.

Because of its ability to provide higher performance network capabilities, MPLS has the potential to completely replace frame relay and ATM. MPLS recognizes that small ATM cells are not needed in the core of modern optical networks with speeds of 40 Gbps or more. In these environments, neither 53-byte cells nor full-length 1500-byte packets experience real-time queuing delays. In addition, MPLS preserves many of the same traffic engineering and out-of-band network control mechanisms that have made frame relay and ATM attractive WAN services for business subscribers.

45.3.3.1 MPLS Operation

An MPLS network or Internet consists of a set of nodes, called *Label Switched Routers* (LSRs) that are capable of switching and routing packets on the basis of a label that has been appended to each packet. Labels define a flow of packets between two endpoints. For each distinct flow, a specific path through the network of LSRs is defined. This path is called a *Forwarding Equivalence Class* (FEC) and each FEC has an associated traffic characterization that defines the QoS requirements for that flow. LSRs forward each packet based on its label value; they do not need to examine or process the packet's IP header. This means that an LSR's forwarding process is simpler and faster than that for an IP router.

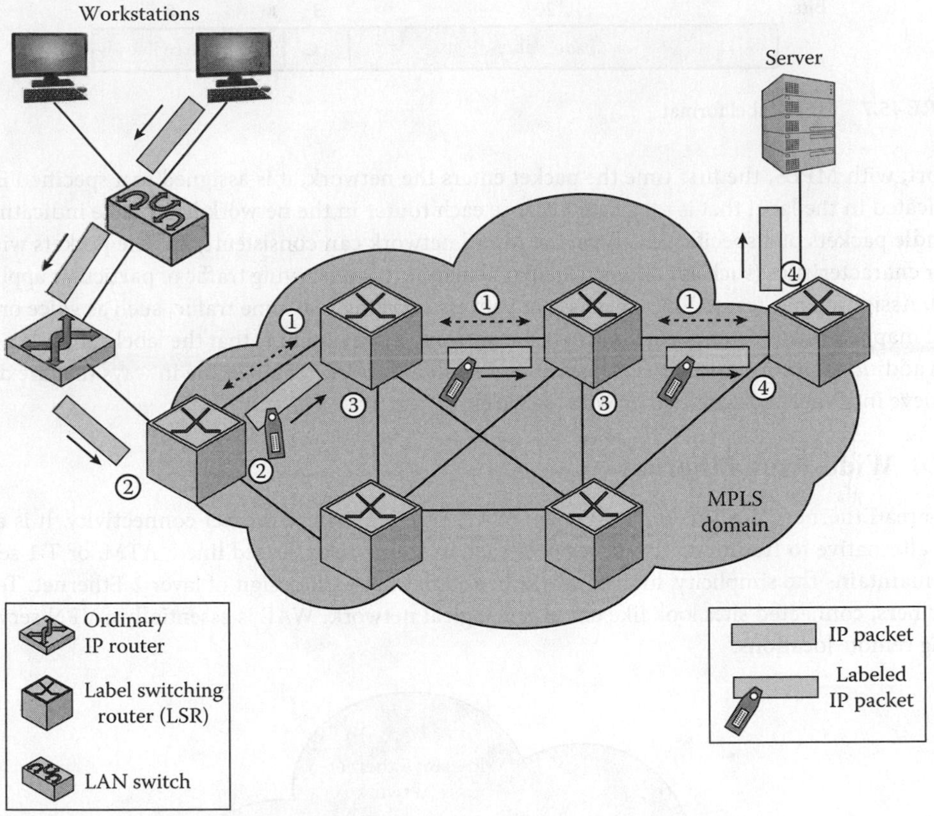

FIGURE 45.6 MPLS operation.

Figure 45.6 illustrates depicts the operation of MPLS within a domain of MPLS-enabled routers. The first step in the process involves establishing a *Label Switched Path* (*LSP*) for the packets that are to be routed and delivered and the QoS parameters that must be established for the LSP. QoS parameters include the queuing and discarding policy for each LSR along the path and the resources that need to be committed to the path. The process of establishing the LSP and its QoS parameters results in the creation of the FEC; this is illustrated in (1) in Figure 45.6. Once the FEC is created, labels can be assigned to the packets for the FEC.

Packets enter a MPLS switching domain through ingress LSR at the edge of the MPLS network. The ingress LSR processes the packet, determines the QoS services that it requires, and assigns the packet to a FEC and LSP. It then appends the appropriate label and forwards the packet to the next LSR along the LSP. This is illustrated in (2) in Figure 45.6. With the MPLS network, each LSR along the LSP receives the labeled packet and forwards it to the next LSR along the LSP, see (3) in Figure 45.6. When the packet arrives at the egress LSR at the edge of the network closest to the packet's destination, the edge LSR strips the label from the packet, reads its IP packet header, and forwards to packet to its final destination. In the example illustrated in (4) in Figure 45.6, the final destination is a server.

As may be observed in Figure 45.7, an MPLS label is a 32 bit field that includes the following elements:

- Label value: A 20 bit label that has local significance
- Traffic class: A three bit label that signifies QoS priority and explicit network congestion
- Bottom of stack bit: If this is set to 1, it indicates that the current label is the last in the stack
- Time to live: 8 bits are used to encode a hop count or time to live value. This is included to avoid looping or having the packet remain too long in the network because of faulty routing

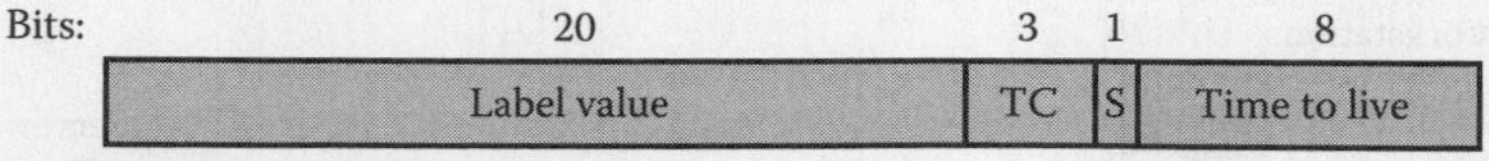

FIGURE 45.7 MPLS label format.

In short, with MPLS, the first time the packet enters the network, it is assigned to a specific FEC that is indicated in the label that is assigned. Because each router in the network has a table indicating how to handle packets of a specific FEC type, the MPLS network can consistently handle packets with particular characteristics (such as coming from particular ports or carrying traffic of particular application types). Assigning packets to FECs means that packets carrying real-time traffic, such as voice or video, can be mapped to low-latency routes across the network. A key point is that the labels provide a way to attach additional information to each packet that facilitates traffic engineering in ways that are difficult to achieve in IP-networks and other WAN services.

45.3.4 Wide-Area Ethernet

Wide-area Ethernet (WAE) is the delivery of WAN services using Ethernet connectivity. It is a high-speed alternative to traditional WAN services such as frame relay, leased lined, ATM, or T-1 services. WAE maintains the simplicity, high bandwidth, and flat network design of layer 2 Ethernet. To WAE subscribers, connected sites look like one single logical network. WAE is essentially a VPN service for linking remote locations.

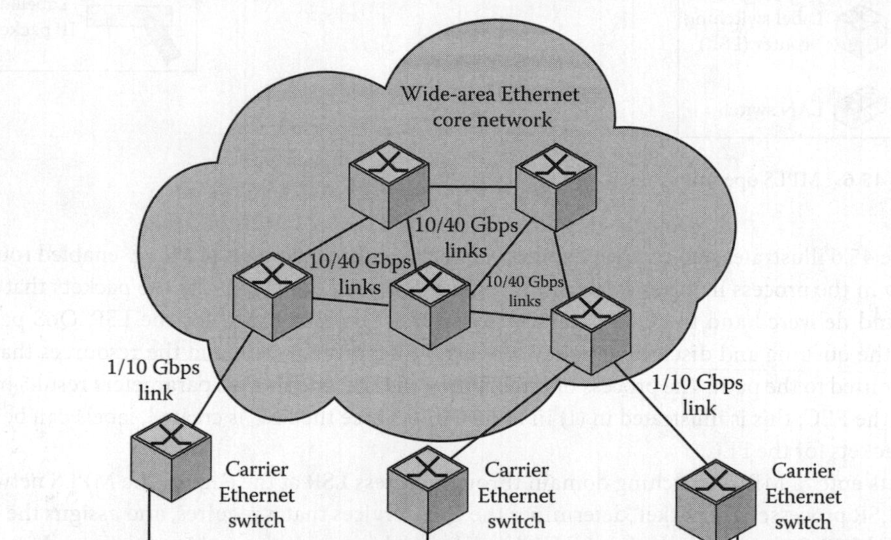

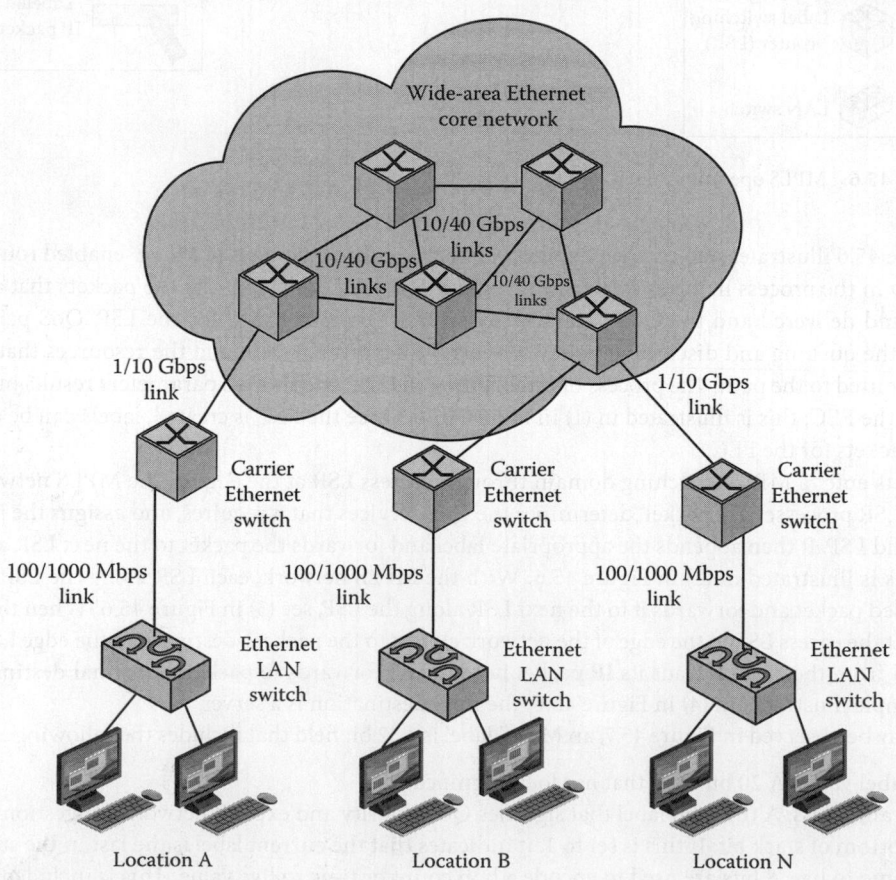

FIGURE 45.8 Wide-Area ethernet.

Most WAE implementations use virtual private LAN services (VPLS) to interconnect network endpoints. VPLS allows carriers to define QoS levels for wide area connections and commit resources, such as sufficient bandwidth, for applications such as video or audio. VPLS also enables carries to create logical Ethernet networks from a variety of WAN services including IP or MPLS networks.

WAE services are sometimes called *Ethernet WAN* or *carrier Ethernet*. Carrier Ethernet includes both WAE and metro Ethernet. Carrier Ethernet services are deployed in several ways, including conventional Ethernet, Ethernet over synchronous digital hierarchy and Ethernet over MPLS. Carrier Ethernet services can accommodate a mixture of residential and business subscribers.

The use of carrier Ethernet technologies to create metropolitan-area networks (MANs) is most commonly called metro Ethernet. Metro Ethernet is often used to provide business LANs and residential subscribers with access to the Internet or other WAN services. Government agencies, educational institutions, and corporations are increasingly using metro Ethernet services to create intranets interconnecting branch offices or campuses.

Carrier Ethernet often capitalizes on the existence of optical fiber and dense wavelength division multiplexing infrastructure to provide WAN and MAN services to subscribers. A high-level example of WAE is provided in Figure 45.8.

45.4 LAN Organization

As with WANs, a LAN is a communications network that interconnects a variety of devices and provides a means for information exchange among those devices. There are several key distinctions between LANs and WANs:

1. The scope of the LAN is small, typically a single building or a cluster of buildings. This difference in geographic scope leads to different technical solutions, as we shall see.
2. It is usually the case that the LAN is owned by the same organization that owns the attached devices. For WANs, this is less often the case, or at least a significant fraction of the network assets is not owned. This has two implications. First, care must be taken in the choice of LAN, because there may be a substantial capital investment (compared to dial-up or leased charges for WANs) for both purchase and maintenance. Second, the network management responsibility for a LAN falls solely on the user.
3. The internal data rates of LANs are typically much greater than those of WANs.

LANs come in a number of different configurations. The most common are switched LANs and wireless LANs. The most common switched LAN is a switched Ethernet LAN, which may consist of a single switch with a number of attached devices, or a number of interconnected switches. Ethernet has come to dominate the wired LAN market, and so this section is devoted to Ethernet systems.

45.4.1 Traditional Ethernet

The Ethernet LAN standard, developed by the IEEE 802.3 standards committee, was originally designed to work over a bus LAN topology. With the bus topology, all stations attach, through appropriate interfacing hardware, directly to a linear transmission medium or bus. A transmission from any station propagates the length of the medium in both directions and can be received by all other stations. A variation on the bus topology is a star topology, with stations attached to a node in such a way that a transmission from any station attached to the node travels to the node and then to all of the other stations attached to the node.

Transmission is in the form of frames containing addresses and user data. Each station monitors the medium and copies frames addressed to itself. Because all stations share a common transmission link, only one station can successfully transmit at a time, and some form of medium access control technique is needed to regulate access. The technique used for traditional Ethernet is carrier sense multiple access with collision detection (CSMA/CD). CSMA/CD, and the types of Ethernet configurations it supports, is now obsolete and is not discussed further in this chapter.

45.4.2 100 Mbps Ethernet

As data rates for Ethernet increased, the nodes were replaced by switches that enabled full-duplex transmission and in which the access control function is regulated by the switch. This is the general arrangement of all contemporary Ethernet LANs.

The first major step in the evolution of Ethernet was the introduction of 100 Mbps Ethernet. This standard specifies the same frame format as the 10 Mbps version and many of the same protocol details. This version of Ethernet quickly crowded out competing schemes, such as ring-based LANs and ATM-based LANs. The reasons for this are instructive. From the vendor's point of view, the Ethernet formatting and protocol are well understood and vendors have experience building the hardware, firmware, and software for such systems. Scaling the system up to 100 Mbps, or more, is easier than implementing an alternative protocol and topology. From the customer's point of view, it is relatively easy to integrate older Ethernet systems running at 10 Mbps with newer systems running at higher speeds if all the systems use the same frame format and similar access protocols. In other words, the continued use of Ethernet-style LANs is attractive because Ethernet is already there. This same situation is encountered in other areas of data communications. Vendors and customers do not always, or even in the majority of cases, choose the technically superior solution. Cost, ease of management, and other factors relating to the already-existing base of equipment are often more important factors in the selection of new LAN equipment than technically superior alternatives. This is the reason that Ethernet-style systems continue to dominate the LAN market and show every sign of continuing to do so in the foreseeable future.

45.4.3 Gigabit Ethernet

The strategy for Gigabit Ethernet is the same as that for 100 Mbps Ethernet. While defining a new medium and transmission specification, Gigabit Ethernet retains the CSMA/CD protocol and frame format of its 10 and 100 Mbps predecessors. It is compatible with both 100BASE-T and 10BASE-T, preserving a smooth migration path. Most business organizations have moved to 100BASE-T and many have jumped to Gigabit Ethernet for at least some of their LANs. These LANs are putting huge traffic loads on backbone networks, which further increase demand for Gigabit Ethernet and 10 Gigabit Ethernet.

Figure 45.9 shows a typical application of Gigabit Ethernet. A 1/10 Gbps LAN switch provides backbone connectivity for central servers and high-speed workgroup switches. Each workgroup LAN switch supports both 1 Gbps links, to connect to the backbone LAN switch and to support high-performance workgroup servers, and 100 Mbps links, to support high-performance workstations, servers, and 100/1000 Mbps LAN switches.

The current 1 Gbps specification for IEEE 802.3 includes the following physical layer alternatives:

- *1000BASE-LX*: This long-wavelength option supports duplex links of up to 550 m of 62.5 μm or 50 μm multimode fiber or up to 5 km of 10 μm single-mode fiber. Wavelengths are in the range of 1270–1355 nm.
- *1000BASE-SX*: This short-wavelength option supports duplex links of up to 275 m using 62.5 μm multimode or up to 550 m using 50 μm multimode fiber. Wavelengths are in the range of 770–860 nm.

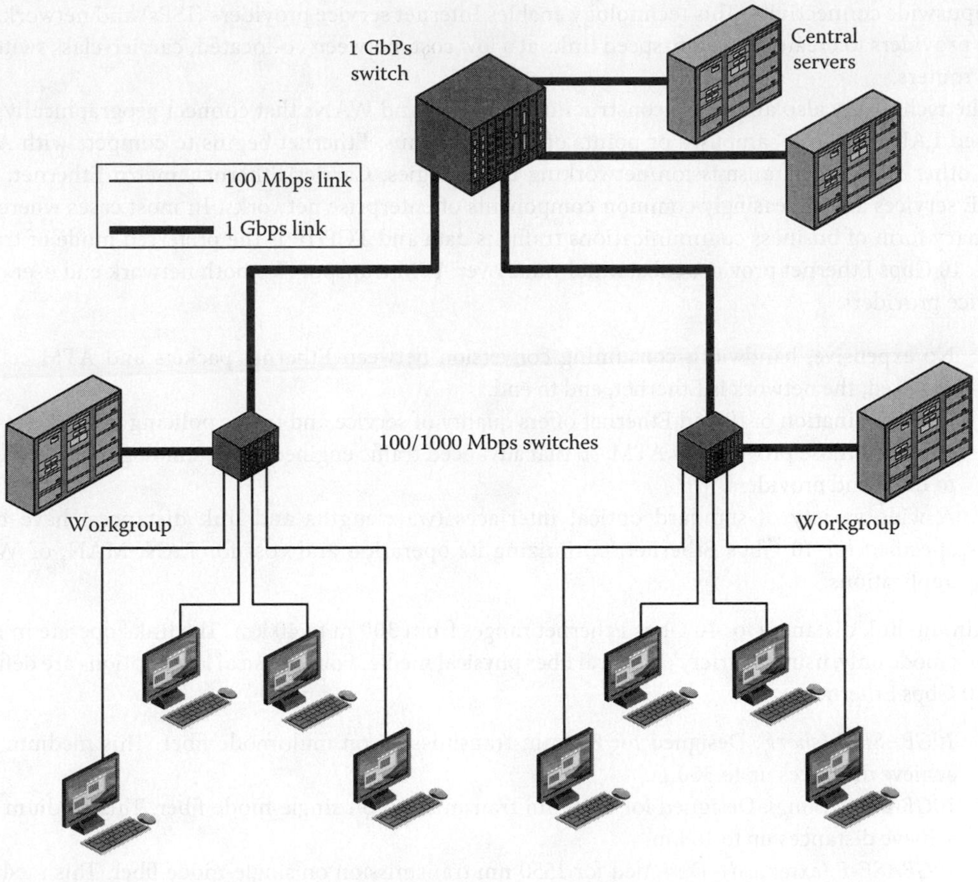

FIGURE 45.9 Example Gigabit ethernet configuration.

- *1000BASE-CX*: This option supports 1 Gbps links among devices located within a single room or equipment rack, using copper jumpers (specialized shielded twisted pair cable that spans no more than 25 m). Each link is composed of a separate shielded twisted pair running in each direction.
- *1000BASE-T*: This option makes use of four pairs of Category 5 unshielded twisted pair to support devices over a range of up to 100 m.

45.4.4 10 Gbps Ethernet

In recent years, 10 Gbps Ethernet switches have made considerable inroads in the LAN market. The principal driving requirement for 10 Gigabit Ethernet is the increase in Internet and intranet traffic. A number of factors contribute to the explosive growth in both Internet and intranet traffic:

- An increase in the number of network connections
- An increase in the connection speed of each end-station (e.g., 10 Mbps users moving to 100 Mbps, analog 56 kbps users moving to DSL and cable modems)
- An increase in the deployment of bandwidth-intensive applications such as high-quality video
- An increase in Web hosting and application hosting traffic

Initially, network managers are using 10 Gbps Ethernet to provide high-speed, local backbone interconnection between large-capacity switches. As the demand for bandwidth increases, 10 Gbps Ethernet will be deployed throughout the entire network and will include server farm, backbone, and campuswide connectivity. This technology enables Internet service providers (ISPs) and network service providers to create very high-speed links at a low cost, between co-located, carrier-class switches and routers.

The technology also allows the construction of MANs and WANs that connect geographically dispersed LANs between campuses or points of presence. Thus, Ethernet begins to compete with ATM and other wide area transmission/networking technologies. Carrier Ethernet, metro Ethernet, and WAE services are increasingly common components of enterprise networks. In most cases where the primary form of business communications traffic is data and TCP/IP is the preferred mode of transport, 10 Gbps Ethernet provides substantial value over ATM transport for both network end users and service providers:

- No expensive, bandwidth-consuming conversion between Ethernet packets and ATM cells is required; the network is Ethernet, end to end.
- The combination of IP and Ethernet offers quality of service and traffic policing capabilities that approach those provided by ATM, so that advanced traffic engineering technologies are available to users and providers.
- A wide variety of standard optical interfaces (wavelengths and link distances) have been specified for 10 Gbps Ethernet, optimizing its operation and cost for LAN, MAN, or WAN applications.

Maximum link distances for 10 Gbps Ethernet ranges from 300 m to 40 km. The links operate in full-duplex mode only, using a variety of optical fiber physical media. Four physical layer options are defined for 10 Gbps Ethernet:

- *10GBASE-S* (*short*): Designed for 850 nm transmission on multimode fiber. This medium can achieve distances up to 300 m.
- *10GBASE-L* (*long*): Designed for 1310 nm transmission on single-mode fiber. This medium can achieve distances up to 10 km.
- *10GBASE-E* (*extended*): Designed for 1550 nm transmission on single-mode fiber. This medium can achieve distances up to 40 km.
- *10GBASE-LX4*: Designed for 1310 nm transmission on single-mode or multimode fiber. This medium can achieve distances up to 10 km. This medium uses wavelength-division multiplexing to multiplex the bit stream across four light waves.

45.4.5 100 Gbps Ethernet

Ethernet is widely deployed and is the preferred technology for wired local-area networking. Ethernet dominates enterprise LANs, broadband access, data center networking, and has also become popular for communication across metropolitan and even WANs. Furthermore, it is now the preferred carrier wire line vehicle for bridging wireless technologies, such as Wi-Fi and WiMAX, into local Ethernet networks.

This popularity of Ethernet technology is due to the availability of cost-effective, reliable, and interoperable networking products from a variety of vendors. Over the years, a number of industry consortiums have participated in the development of ever-faster versions of Ethernet, including the Fast Ethernet Alliance (100 Mbps), the Gigabit Ethernet Alliance, the 10 Gigabit Ethernet alliance, the Ethernet Alliance, and the Road to 100G alliance. As a testament to the continuing evolution of Ethernet, the first three of the alliances just mentioned no longer exist. The Ethernet Alliance is devoted to promoting the development of Ethernet, whatever the speed. The Road to 100G Alliance is focused on the development of standards and technologies for 100 Gbps ethernet.

As this alliance evolution reflects, the development of converged and unified communications, the evolution of massive server farms, and the continuing expansion of VoIP, TVoIP, and Web 2.0

applications have driven the need for ever faster Ethernet switches. The following are market drivers for 100 Gbps Ethernet:

- *Data center/Internet media providers*: To support the growth of Internet multimedia content and Web applications, content providers have been expanding data centers, pushing 10 Gbps Ethernet to its limits. These providers are likely to be high-volume early adopters of 100 Gbps Ethernet.
- *Metro-video/service providers*: Video on demand has been driving a new generation of 10 Gbps Ethernet metropolitan/core network buildouts. These providers are likely to be high-volume adopters in the medium term.
- *Enterprise LANs*: Continuing growth in convergence of voice/video/data and in unified communications is driving up network switch demands. However, most enterprises still rely on 1 Gbps or a mix of 1 and 10 Gbps Ethernet, and adoption of 100 Gbps Ethernet is likely to be slow.
- *Internet exchanges/ISP (ISP) core routing*: With the massive amount of traffic flowing through these nodes, these installations are likely to be early adopters of 100 Gbps Ethernet.

In 2007, the IEEE 802.3 working group authorized the *IEEE P802.3ba 40 Gb/s and 100 Gb/s Ethernet Task Force*. Table 13.4 indicates the physical layer objectives for this task force. As can be seen, these high-speed switches will be standardized to operate at distances from 1 m to 40 km over a variety of physical media.

An example of the application of 100 Gbps Ethernet is shown in Figure 45.10. The trend at large data centers, with substantial banks of blade servers, is the deployment of 10 Gbps ports on individual servers to handle the massive multimedia traffic provided by these servers. Such arrangements are stressing the on-site switches needed to interconnect large numbers of servers. A 100 GbE rate was proposed to provide the bandwidth required to handle the increased traffic load. It is expected that 100 GbE will be deployed in switch uplinks inside the data center as well as providing interbuilding, intercampus, MAN, and WAN connections for enterprise networks.

The success of fast Ethernet, Gigabit Ethernet, and 10 Gbps Ethernet highlights the importance of network management concerns in choosing a network technology. Both ATM and fiber channel, explored subsequently, may be technically superior choices for a high-speed backbone, because of their flexibility

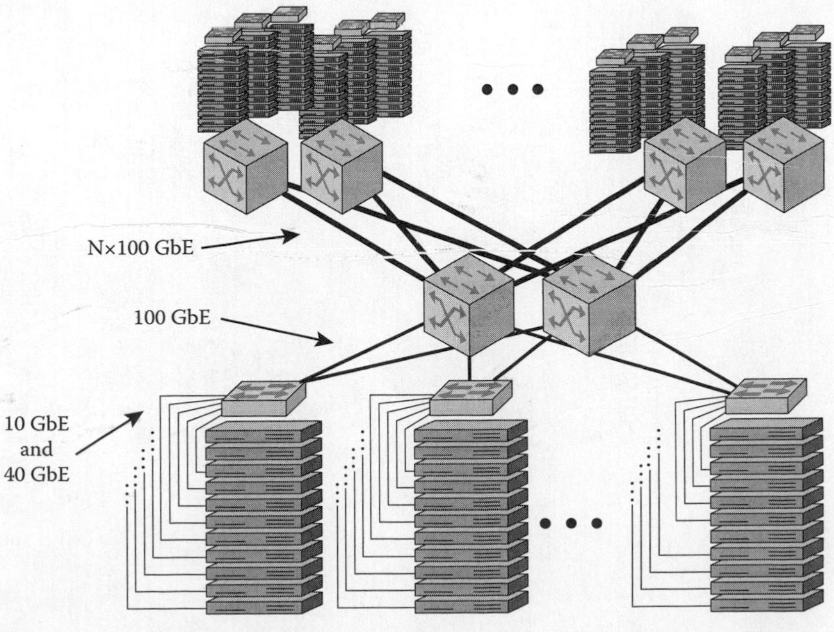

FIGURE 45.10 Example 100-Gbps Ethernet configuration for massive blade server site.

and scalability. However, the Ethernet alternatives offer compatibility with existing installed LANs, network management software, and applications. This compatibility has accounted for the survival of 30-year-old technology in today's fast-evolving network environment.

Key Terms

Asynchronous transfer mode (ATM): A form of packet transmission using fixed-size packets, called cells. Unlike X.25, ATM does not provide error control and flow control mechanisms.

Circuit switching: A method of communicating in which a dedicated communications path is established between two devices through one or more intermediate switching nodes. Unlike packet switching, digital data are sent as a continuous stream of bits. Bandwidth is guaranteed, and delay is essentially limited to propagation time. The telephone system uses circuit switching.

Frame relay: A form of packet switching based on the use of variable-length link-layer frames. There is no network layer and many of the basic functions have been streamlined or eliminated to provide for greater throughput.

Local-area network (LAN): A communication network that provides interconnection of a variety of data communicating devices within a small area.

Packet switching: A method of transmitting messages through a communication network, in which long messages are subdivided into short packets. The packets are then transmitted as in message switching.

Wide-area network (WAN): A communication network that provides interconnection of a variety of communicating devices over a large area, such as a metropolitan area or larger.

Reference

Stallings, W. 2013. *Data and Computer Communications*, 10th edn. Prentice Hall, Upper Saddle River, NJ.

46

Routing Protocols

Radia Perlman
Intel Laboratories

46.1 Introduction

The purpose of this chapter is not as a reference on the details of particular implementations but to understand the variety of routing algorithms and the trade-offs between them.

Computer networking can be a very confusing field. Terms such as network, subnetwork, domain, *local area network (LAN)*, internetwork, *bridge*, *router*, and *switch* are often ill-defined. In the original ISO model of network layering, layer 2, the *data-link layer*, was intended to deliver a packet of information to a neighboring machine, and layer 3, the *network layer*, was intended to relay a packet through a series of switches in order to move data from source to destination, and it was intended to be the job of layer 4, the *transport layer* (a protocol between the source and destination), to recover from lost, duplicated, and out-of-order packets. Although this conceptual model of network layers is a great way to learn about networks, in reality, things have gotten extremely complicated. Ethernet, intended to be a layer 2 technology, includes routing. Various other technologies consider it the task of layer 3 to assure no packet loss or misordering. To make it even more difficult to understand this area, various concepts are reinvented in different standards, usually with different terminology, and little available information about how different technologies relate to each other than marketing hype. In this chapter, we discuss various conceptual techniques for moving data throughout a network, and rather than focusing on the details of a particular standard or deployed technology, we focus on the concepts. We also avoid philosophical questions such as which layer something is.

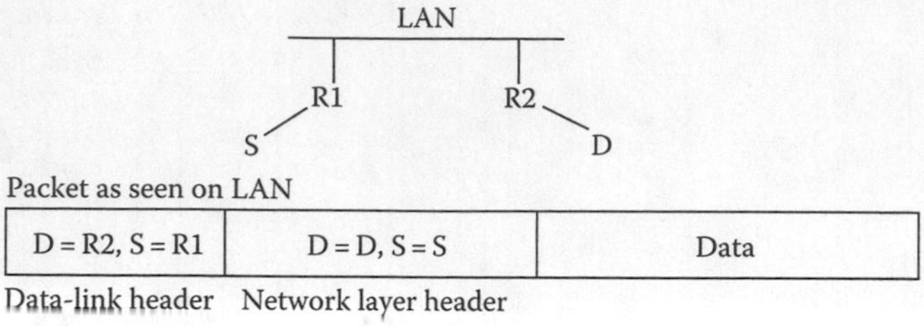

FIGURE 46.1 A multiaccess link and a packet with two headers.

A network consists of several nodes interconnected with various types of links. One type is a *point-to-point link*, which connects exactly two machines. It can either be a dedicated link (e.g., a wire connecting the two machines) or a dial-on-demand link, which can be connected when needed (e.g., when there is traffic to send over the link). Another category of link is a *multiaccess link*. Examples are wireless links, wired LANs, or entire networks that machines can talk to each other across. When one protocol uses a network as a link, it is referred to as a *tunnel*. In such a case, the underlying network can be used to create a point-to-point link between two machines, or a multiaccess link between more than two machines. The underlying network could be something generally assumed to be layer 2 (such as Ethernet) or it could be something generally assumed to be layer 3 (such as IP). A tunneled link presents special challenges to the routing protocol because it has two headers with addressing information. One header (which for simplicity we call the layer 3 header) gives the addresses of the ultimate source and destination. The other header (which for simplicity we call the layer 2 header) gives the addresses of the transmitter and receiver on that particular link (see Figure 46.1, which shows a path incorporating a multiaccess link and a packet with two headers). The terminology cannot be taken too seriously. It is not uncommon to tunnel IP over IP; for example, for an employee to connect to the corporate firewall across the Internet, using an encrypted connection (usually referred to as a "virtual Private Network"). Although the "data link" in that case is an entire network, from the point of view of the protocols using it as a link, it can be considered just a data link.

We define a *switch* to be a device that receives a data packet from the network, uses information in the data packet (e.g., the destination address in the header), plus instructions that are in the switch's *forwarding table*, and then follows those instructions, for example, forwarding the packet onto the indicated output port. We use the term *switch* to include devices that are known as bridges, routers, gateways, and perhaps other names in this chapter. In this chapter, we describe various possibilities, and their implications, for

- Information in the packet, for example, size of address
- How to find instructions for a particular packet in the forwarding table
- How the forwarding table is computed

46.2 Forwarding Table

When a switch receives a packet, it finds relevant information in the packet, and consults its forwarding table, to determine the proper output port onto which to forward the packet. The forwarding table might indicate a single output port, or a set of output ports, any of which are reasonable for the switch to use for forwarding this packet.

46.2.1 Keeping Packets in Order

If the forwarding table gives several output ports rather than a single answer, any of those ports will deliver the packet to the destination; however, many applications that run over a network have come to depend on the network trying really hard to deliver packets for a particular flow in the same order in which they were transmitted. As network protocols are generally taught, it is assumed that layer 3 is allowed to send packets in any order, and it is up to layer 4, the *transport layer* (e.g., the TCP protocol) to put packets back into order. However, applications (including many implementations of TCP) perform badly if the network is allowed to reorder packets within a flow. Therefore, switches tend to be careful to keep packets in order. There are several ways of doing this:

1. If the forwarding table gives just a single output port for a particular destination, then all packets for a flow will travel on the same path, so they will not get out of order. If it is desired that packets to the same destination exploit parallel paths ("load splitting" or "path splitting"), then the destination is given multiple addresses, and occupies two entries in the forwarding table. This is the approach generally used in InfiniBand.
2. If the forwarding table gives a set of output ports, then the switch parses the packet to look at additional information beyond the information in the switch's layer and hashes that information in order to always choose the same output port for a given flow. For instance, if the switch is an IP router, the switch looks beyond the IP header to other information, such as the TCP ports, and hashes fields such as (IP source address, TCP source port, and TCP destination port), and if the forwarding table has indicated, say, five different equally good output ports to get to the destination, the switch might compute the hash mod 5 of those fields. In theory, the switch could choose whichever port was most convenient when it first sees a packet for a given flow and remember that choice, but the hashing procedure allows the switch to accomplish the same thing without keeping state.
3. The packet might contain an "entropy" field, which is specifically for the purpose of choosing a port from a set of ports. There are two reasons for the entropy field instead of doing the hash of fields such as (source IP and TCP ports):
    a. It saves work for the switch, since it can use the single entropy field, right in the header that the switch is supposed to look at, rather than having to parse through other layers to find enough information to allow path diversity between flows.
    b. It allows a single source to exploit path diversity to the same destination, in the case where the application does not require packets to stay in order, or if certain subflows of information between the source–destination pair might need to be in order, but independently of the other subflows. If the source sets the entropy field differently for different subflows, it is likely that at least some switches along the path will make different forwarding choices for the different subflows.

46.2.2 Distributed vs. Centralized Forwarding Table Computation

The forwarding table might be computed with a distributed routing protocol (which we will describe later in this chapter) or by a central node that computes tables for each of the switches and programs the switches. It might be computed only after a topology change (e.g., a link or switch failing or recovering) or each time a new flow of information is initiated between a source S and destination D. We will discuss the tradeoffs between these approaches.

The advantages of a distributed algorithm are as follows:

- A distributed algorithm will react more quickly to topology changes; with a distributed algorithm (as we will discuss later in this chapter), information about the link change travels as quickly as physically possible, from the event, and each switch then adjusts its own forwarding table. With a centralized algorithm, news of the topology change must be conveyed to the central node (somehow—the topology change could affect a switch's configured path to the central node), the central node has to compute new forwarding tables for the switches, and then configure each switch with the new forwarding table.
- There is no central point of failure or bottleneck (as there would be with a centralized algorithm if the central node fails, or is a performance bottleneck).

The reasons some give for a centralized algorithm are as follows:

- It might make the switches simpler because they do not need to implement a routing algorithm. However, a routing algorithm does not make a switch expensive; rather, it is the specialized hardware necessary to move packets quickly through the ports, which needs to be done; however, the forwarding table was computed.
- A centralized algorithm might enable more careful engineering of flows, given other knowledge like which flows will exist and what their bandwidth requirements are. Such things could also be done with a distributed algorithm by changing link costs or creating some paths with reserved bandwidth, but it might be easier with a centralized algorithm.

Most networks today (e.g., the IP-based Internet and Ethernet) use distributed algorithms for computing the forwarding table. Telephony networks of old and InfiniBand (www.infinibandta.org) tend to use a centralized algorithm.

46.2.3 On-Demand vs. Proactive Computation of Paths

The other issue is whether paths should be created "on demand," that is, when a flow starts or whether the switches should always be prepared for any type of packet. Computing the forwarding tables when a new flow starts causes a delay before packets can flow, which, in most cases, will be unacceptable.

In the case of telephony networks of old (e.g., ATM and X.25), the assumptions that made on-demand computation of a path a reasonable choice were as follows:

- The total number of destinations was far greater than the number of actively communicating pairs at any time, so a value in the header indicating the flow, assigned at the time the flow started, could be a smaller field than a field that indicated a destination address.
- It was acceptable to have a delay when a flow started up for the call to be dialed.
- It was acceptable to indicate, for instance, using a "fast busy signal," that there are temporarily no resources to set up the call.

These assumptions are not valid in a data center or the Internet.

46.3 Information in Data Packet

When a switch receives a data packet, what information in that packet does it use to make its forwarding decision? There are various possibilities, and we will discuss the tradeoffs:

- *Flat destination address*: A destination address that is "flat," meaning that the location of a node has no relation to its address; there must be a separate entry for each destination address in the forwarding table.
- *Hierarchical destination address*: A destination address that is hierarchical, meaning that portions of the network can be summarized with a block of addresses, and appear as a single entry in the forwarding table.
- *Path Label*: A path identifier (ID) of significance only on a link; (e.g., as used in X.25, ATM, and MPLS). At each hop, a switch looks up (input port and label 1) and is told by the forwarding table (output port and label 2). The switch not only forwards onto the indicated output port but also replaces label 1 (in the received packet) with label 2.
- *Flow*: The forwarding table gives instructions more specific than simply a destination address (e.g., source address, destination address, source TCP port, destination TCP port, and protocol type).

46.3.1 Flat Destination Address

The forwarding table consists of a set of entries {destination, port(s)}. Each destination occupies a table entry. If a node moves, its address does not change, but the forwarding entry for that node in any particular switch will likely change to be a different port.

Flat addresses are convenient, especially in a data center with virtual machines that might move from one server to another; a virtual machine can move without changing its address. However, the problem with flat addresses is that the forwarding table needs an entry for each destination, so there is a limit to the size of a network that can be created with flat addresses.

For this reason, large networks are based on hierarchical addresses.

46.3.2 Hierarchical Destination Address

Hierarchical addresses are assigned in such a way that a set of addresses can be summarized with a prefix, as in "all addresses that start with 372 are in this circle."

Figure 46.2 shows hierarchical addresses, with multiple levels of hierarchy. The forwarding table of a switch outside the large circle needs only a single forwarding entry for all the nodes inside the circle; the prefix "372." A switch inside the large circle but outside the circle labeled 37279* needs a single forwarding entry for the nodes in the circle 37279*, and likewise, if it is outside the 3725* circle, it only needs a single entry for those nodes.

Hierarchical addresses compress the forwarding table but sometimes lead to suboptimal paths, because forwarding tables will not distinguish the best point to enter a circle.

An address prefix summarizes a set of addresses. The shorter the prefix, the more addresses are in the summary. In other words, the prefix 2* contains more addresses than the prefix 245*.

When a forwarding table contains prefixes, there might be several prefixes that match the destination address, for instance, for a destination 123456789, the prefixes 12345*, 123*, and * all match. Usually, a forwarding table will contain the zero length prefix, which matches anything. The switch needs to find the most specific match, that is, the longest prefix that matches the destination. There are at least two algorithms for doing this, which are described in detail in Perlman (1999). Although these algorithms are certainly implementable, they are more work, and slower, for a switch, than doing a hash of the

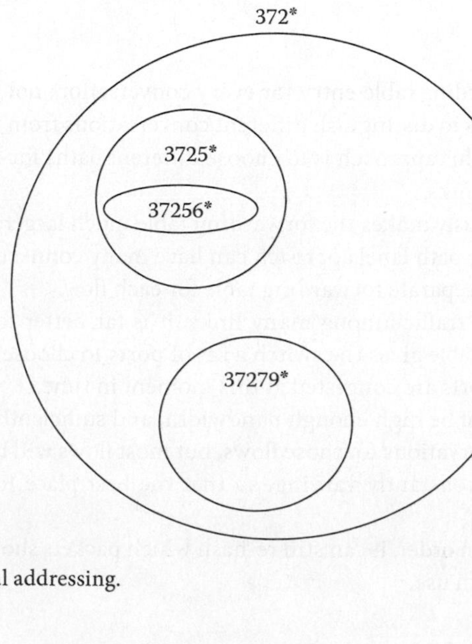

FIGURE 46.2 Hierarchical addressing.

destination address to find the entry (as would be done with, say, a 6 byte Ethernet address) or doing a direct lookup (if the destination is few enough bits so that the switch can have a table directly indexed by the destination address).

46.3.3 Path Label

A path label is a field in a packet that is only significant within the scope of a link. A switch's forwarding table maps (input port and label) to (output port and label). An entry ((p1,x); (p2,y)) means that if a packet is received on port p1 with label x, the switch should forward it onto port p2 and replace the label with y.

A path label represents a source–destination pair. Therefore, if all nodes can talk to all other nodes, the forwarding tables using path labels would have n^2 entries, whereas with destination-based forwarding, the table would only need to have O(n) entries.

The path label approach was invented for telephony networks such as X.25 and ATM. The concept was based on the following three assumptions:

- The total number of addressable nodes attached to the network is much greater than the number of active connections between nodes.
- It is acceptable to incur a delay while a path is set up, before two nodes can communicate.
- It is acceptable if there is currently no available labels on a link, or other state required at a switch to support a particular flow, to allow the connection to fail, with some signal to "try again later." In the U.S. telephone network, this was signaled to the person making a call with a fast busy signal.

A commonly used protocol in the Internet, in backbone ISPs, is MPLS. It uses path labels. A path across the ISP is used for many conversations, and the actual end points are identifiable because the MPLS label encapsulates an IP header. MPLS grew out of a concept known as "tag switching," in which a switch S1 requested that its neighbor S2 put a "label" on a packet for a particular destination, to save S1 from having to do a longest prefix match for that destination. Today, MPLS is primarily used for "traffic engineering," which means configuring a path. The concept of encapsulating an IP (or other) packet with a header that is easier to parse (i.e., not needing to do longest prefix match) is a good one, but MPLS would have been more scalable if it had been based on destinations rather than path labels.

46.3.4 Flow

This approach envisions a forwarding table entry for every conversation; not just a source–destination pair, but using TCP port numbers to distinguish different conversations from the same communicating node pair. The thinking behind this approach is to choose different paths for different flows in order to spread the traffic among a lot of links.

However, this approach obviously makes the forwarding table much larger; even worse, in fact, than the path label approach, since the path label approach can have many connections share the same path, but the flow approach requires a separate forwarding table for each flow.

And if the desire is to spread traffic among many links, it is far better to use a destination-based approach where the forwarding table gives the switch a set of ports to choose between. A switch is in a better position to know which ports are congested at this moment in time.

In some rare cases, flows might be high enough bandwidth, and sufficiently long-lived, that it might be worth creating paths with reservations for those flows, but most flows will be bursty, and the amount of congestion at a port will be constantly varying, so that the best place to make the decision is at the switch.

If a switch must keep packets in order, it can still re-hash which packets should travel on which ports, if some ports are getting too much use.

46.4 Distributed Routing Protocols

There are two basic types of distributed routing protocol:

- Distance vector
- Link state

In both cases, switches communicate with other switches to exchange information, and based on this information, each switch computes its own forwarding table.

An example of a distance vector protocol is RIP (Hedrick, 1988). The spanning tree algorithm used in Ethernet, which will be described later in this chapter, can arguably be called a variant of distance vector as can the Border Gateway protocol (BGP) (Rekhter et al., 2006), which is sometimes called a "path vector" protocol, and which will also be described later in this chapter. EIGRP (http://www.cisco.com/en/US/tech/tk365/technologies_tech_note09186a0080093f07.shtml) is a Cisco proprietary distance vector protocol.

The two best-known link state protocols are IS-IS (Oran, 1990) and OSPF (Moy, 1998).

46.4.1 Distance Vector

The idea behind this class of algorithm is that each router is responsible for keeping a table (known as a distance vector) of distances from itself to each destination. It computes this table based on receipt of distance vectors from its neighbors. For each destination D, router R computes its distance to D as follows:

- 0, if R = D.
- The configured cost, if D is directly connected to R.
- The minimum cost through each of the reported paths through the neighbors.

For example, suppose R has four ports, a, b, c, and d. Suppose also that the cost of each of the links is, respectively, 2, 4, 3, and 5. On port a, R has received the report that D is reachable at a cost of 7. The other (port and cost) pairs R has heard are (b, 6), (c, 10), and (d, 2). Then, the cost to D through port a will be 2 (cost to traverse that link) +7, that is, 9. The cost through b will be 4 + 6, that is, 10. The cost through c will be 3 + 10, that is, 13. The cost through d will be 5 + 2, that is, 7. So the best path to D is through port d, and R will report that it can reach D at a cost of 7 (see Figure 46.3).

The spanning tree algorithm is similar to a distance vector protocol in which each bridge is only computing its cost and path to a single destination, the root. But the spanning tree algorithm does not suffer from the count-to-infinity behavior that distance vector protocols are prone to (see next section).

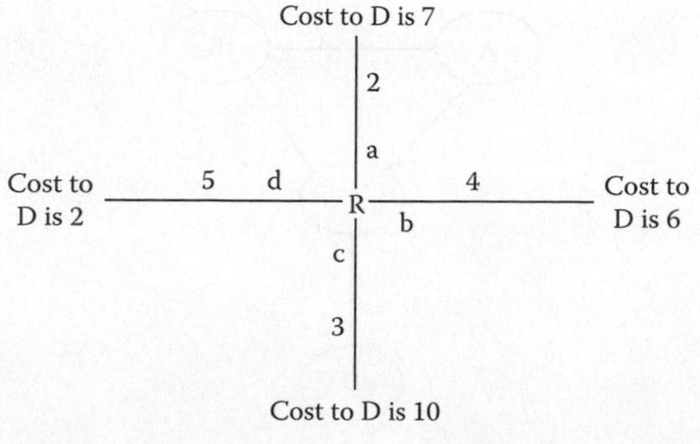

FIGURE 46.3 Example: distance vector protocol.

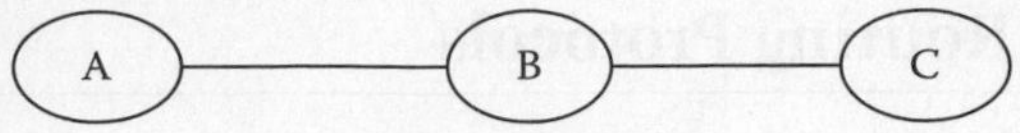

FIGURE 46.4 Example: the count-to-infinity problem.

46.4.1.1 Count to Infinity

One of the problems with distance vector protocols is known as the count-to-infinity problem. Imagine a network with three nodes, A, B, and C (see Figure 46.4).

Let us discuss everyone's distance to C. Node C will be 0 from itself, B will be 1 from C, and A will be 2 from C. When C crashes, B unfortunately does not conclude that C is unreachable, but instead goes to its next best path, which is via neighbor A, who claims to be able to reach C at a cost of 2. So now B concludes it is 3 from C and that it should forward packets for C through A. When B tells A its new distance vector, A does not get too upset. It merely concludes its path (still through B) has gotten a little worse, and now A is 4 from C. Node A will report this to B, which will update its cost to C as 5, and A and B will continue this until they count to infinity. Infinity, in this case, is mercifully not the mathematical definition of infinity, but is instead a parameter (with a definite finite value such as 16). Routers conclude if the cost to a node is greater than this parameter, that node must be unreachable.

A common enhancement that makes the behavior a little better is known as *split horizon*. The split horizon rule is usually implemented as follows: if router R uses neighbor N as its best path to destination D, R should not tell N that R can reach D. This eliminates loops of two routers. For instance, in Figure 46.4, node A would not have told B that A could reach C. Thus, when C crashed, B would conclude B could not reach C at all; and when B reported infinity to A, node A would conclude that A could not reach C either, and everything would work as we would hope it would.

Unfortunately, split horizon does not fix loops of three or more routers. Referring to Figure 46.5, and looking at distances to D, when D crashes, C will conclude C cannot reach D (because of the split horizon rule, A and B are not reporting to C that they can reach D).

Node C will inform A and B that C can no longer reach D. Unfortunately, each of them thinks they have a next-best path through the other. Say A acts first, decides its best path is through B, and that A is now 3 from D. Node A will report infinity to B (because of split horizon), and report 3 to C. B will now (for a moment) think it cannot reach D. It will report infinity to A, but it is too late; A has already reported a finite cost to C. Node C will now report 4 to B, which will now conclude it is 5 from D.

Although split horizon does not solve the problem, it is a simple enhancement, never does any harm, does not add overhead, and helps in many cases.

The simplest distance vector protocols (e.g., RIP) are very slow to converge after a topology change. The idea is that routing information is periodically transmitted quite frequently (on the order of 30 s). Information is discarded if it has not been heard recently (on the order of 2 min). Most implementations only store the best path; and when that path fails, they need to wait for their neighbors' periodic

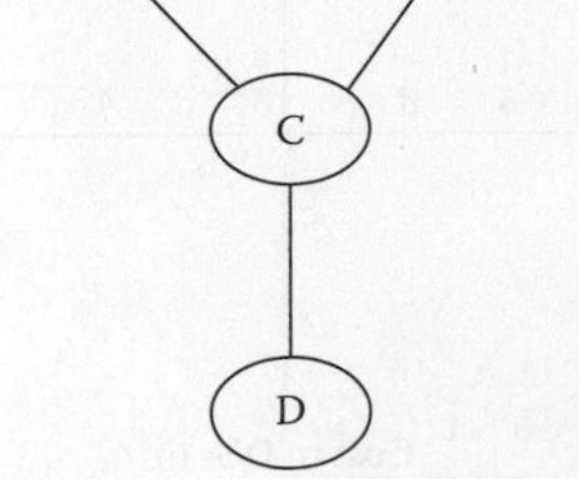

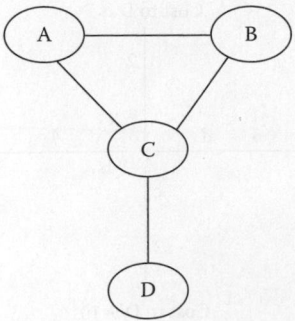

FIGURE 46.5 Example: loops with three or more routers.

transmissions in order to hear about the second best path. Some implementations query their neighbors ("do you know how to reach D?") when the path to D is discarded. In some implementations when R discards its route to D (for instance, it times out), R lets its neighbors know that R has discarded the route, by telling the neighbors R's cost to D is now infinity. In other implementations, after R times out the route, R will merely stop advertising the route, so R's neighbors will need to time out the route (starting from when R timed out the route).

Distance vector protocols need not be periodic. The distance vector protocol used for DECnet Phases 3 and 4 transmits routing information reliably and only sends information that has changed. Information on a LAN is sent periodically (rather than collecting acknowledgments (ack's) from all router neighbors), but the purpose of sending it periodically is solely as an alternative to sending ack's. Distance vector information is not timed out in DECnet, as it is in a RIP-like protocol. Instead, there is a separate protocol in which "Hello" messages are broadcast on the LAN to detect a dead router. If a "Hello" is not received in time, the neighbor router is assumed dead and its distance vector is discarded.

Another variation from the RIP-family of distance vectors is to store the entire received distance vector from each neighbor, rather than only keeping the best report for each destination. Then, when information must be discarded (e.g., due to having that neighbor report infinity for some destination or due to that neighbor being declared dead) information for finding an alternative path is available immediately.

There are variations proposed to solve the count-to-infinity behavior. One variation has been implemented in BGP. Instead of just reporting a cost to destination D, a router reports the entire path from itself to D. This eliminates loops but has high overhead. Another variation proposed by Garcia-Luna-Aceves (1989) and implemented in the proprietary protocol EIGRP involves sending a message in the opposite direction of D, when the path to D gets worse, and not switching over to a next-best path until ack's are received indicating that the information has been received by the downstream subtree. These variations may improve convergence to be comparable to link state protocols, but they also erode the chief advantage of distance vector protocols, which is their simplicity.

46.4.2 Link State Protocols

The idea behind a link state protocol is that each router R is responsible for the following:

- Identifying its neighbors
- Constructing a special packet known as a *link state packet* (LSP) that identifies R and lists R's neighbors (and the cost to each neighbor)
- Cooperating with all of the routers to reliably broadcast LSPs to all the routers
- Keeping a database of the most recently generated LSP from each other router
- Using the LSP database to calculate routes

Identifying neighbors and constructing an LSP is straightforward. Calculating routes using the LSP database is also straightforward. Most implementations use a variation of the shortest path algorithm attributed to Dijkstra. The tricky part is reliably broadcasting the LSP. The original link state algorithm was implemented in the ARPANET. Its LSP distribution mechanism had the unfortunate property that if LSPs from the same source, but with three different sequence numbers, were injected into the network, these LSPs would turn into a virus. Every time a router processed one of them, it would generate more copies, and so the harder the routers worked, the more copies of the LSP would exist in the system. The problem was analyzed and a stable distribution scheme was proposed in Perlman (1983). The protocol was further refined for the IS-IS routing protocol and copied in OSPF (see next section).

One advantage of link state protocols is that they converge quickly. As soon as a router notices one of its links has changed (going up or down), it broadcasts an updated LSP, which propagates in a straight line outwards (in contrast to a distance vector protocol where information might sometimes be ping-ponged back and forth before proceeding further or where propagation of the information is delayed waiting for news from downstream nodes that the current path's demise has been received by all nodes).

Link state protocols have other advantages as well. The LSP database gives complete information, which is useful for managing the network, mapping the network, or constructing custom routes for complex policy reasons (Clark, 1989) or for sabotage-proof routing (Perlman, 1988).

46.4.2.1 Reliable Distribution of LSPs

Each LSP contains

- Identity of the node that generated the LSP
- A sequence number large enough to never wrap around except if errors occur (e.g., 64 bits)
- An age field, estimating time since source generated the LSP
- Other information

Each router keeps a database of the LSP with the largest sequence number seen thus far from each source. The purpose of the age field is to eventually eliminate a LSP from a source that does not exist anymore or that has been down for a very long time. It also serves to get rid of an LSP that is corrupted or for which the sequence number has reached the largest value.

For each LSP, a router R has a table, for each of R's neighbors, as to whether R and the neighbor N are *in sync* with respect to that LSP. The possibilities are as follows:

- R and N are in sync. R does not need to send anything to N about this LSP.
- R thinks N has not yet seen this LSP. R needs to periodically retransmit this LSP to N until N acknowledges it.
- R thinks N does not know R has the LSP. R needs to send N an ack for this LSP.

Router R goes through the list of LSPs round-robin, for each link, and transmits LSPs or acks as indicated. If R sends an ack to N, R changes the state of that LSP for N to be in sync.

The state of an LSP gets set as follows:

- If R receives a new LSP from neighbor N, R overwrites the one in memory (if any) with smaller sequence number, sets send ack for N, and sets send LSP for each of R's other neighbors.
- If R receives an ack for an LSP from neighbor N, R sets the flag for that LSP to be in sync.
- If R receives a duplicate LSP or older LSP from neighbor N, R sets the flag for the LSP in memory (the one with higher sequence number) to send LSP.
- After R transmits an ack for an LSP to N, R changes the state of that LSP to in sync.

If an LSP's age expires, it is important that all of the routers purge the LSP at about the same time. The age is a field that is set to some value by the source and is counted down. In this way, the source can control how long it will last. If R decides that an LSP's age has expired, R refloods it to R's neighbors (by setting the state to send LSP). If R receives an LSP with the same sequence number as one stored, but the received one has zero age, R sets the LSP's age to 0 and floods it to its neighbors. If R does not have an LSP in memory and receives one with zero age, R acks it but does not store it or reflood it.

46.4.2.2 Calculating Routes Based on Link State Information

Given an LSP database, the most popular method of computing routes is to use some variant of an algorithm attributed to Dijkstra. The algorithm involves having each router compute a tree of shortest paths from itself to each destination. Each node on the tree has a value associated with it which is the cost from the root to that node. The algorithm is as follows:

- Step 0: put yourself, with cost 0, on the tree as root.
- Step 1: examine the LSP of the node X just put on the tree. For each neighbor N listed in X's LSP, add X's cost in the LSP to the cost to X to get some number c. If c is smaller than any path to N found so far, place N tentatively in the tree, with cost c.
- Step 2: find the node tentatively in the tree with smallest associated cost c. Place that node permanently in the tree. Go to step 1.

46.5 Evolution of Ethernet from Carrier Sense Multiple Access with Collision Detect to TRansparent Interconnection of Lots of Links

Originally routing was supposed to be done at layer 3; but today, "Ethernet" is a network with multiple links and switches that forward packets along a path. The only way to understand why this is, is to understand the history.

Indeed, originally routing was done at layer 3, and layer 2 was for getting a piece of information from a node to an immediate neighbor. Ethernet was intended to be a single shared link, using a contention protocol known as carrier sense multiple access with collision detect (CSMA/CD). Layer 3 should have been used to interconnect different links.

Unfortunately, many people started implementing protocols directly on Ethernet, omitting layer 3. That meant their protocol could not work beyond the scalability of the original CSMA/CD Ethernet.

This was short-sighted. Indeed, it was desired for those protocols to work beyond a single link. But rather than changing all of the Ethernet endnodes, Digital Equipment Corporation designed (and IEEE eventually adopted) the concept of the "transparent bridge." The goal was to be able to interconnect links without needing to change the endnodes. And there was a fixed maximum size of an Ethernet packet, so nothing could be modified in an Ethernet packet. The transparent bridge concept can work on any sort of link, even though usually today the links are Ethernet.

The basic idea of a transparent bridge is that on each port, the bridge listens promiscuously to all packets, stores them, and forwards each packet onto each other port when given permission by the LAN protocol on that port (e.g., if it were a token ring, when the port received the token, or if it were an Ethernet, when the link was idle). An enhancement is to have the bridge learn, based on addresses in the LAN header of packets received by the bridge, where the stations reside, so that the bridge does not unnecessarily forward packets (see Figure 46.6, where a bridge has learned some of the station addresses). The bridge learns from the source field in the LAN header and forwards based on the destination address. For example, in Figure 46.6, when S transmits a packet with destination address D, the bridge learns which interface S resides on, and then looks to see if it has already learned where D resides. If the bridge does not know where D is, then the bridge forwards the packet onto all interfaces (except the one from which the packet was received). If the bridge does know where D is, then the bridge forwards it only onto the interface where D resides, or if the packet arrived from that interface, the bridge discards the packet).

This created the need for the spanning tree algorithm (Perlman, 1985).

46.5.1 Transparent Bridging

The goal of transparent bridging was to invent a box that would interconnect LANs even though the stations were designed with protocols that only worked on a LAN; that is, they lacked a network layer able to cooperate with routers, devices that were designed for forwarding packets.

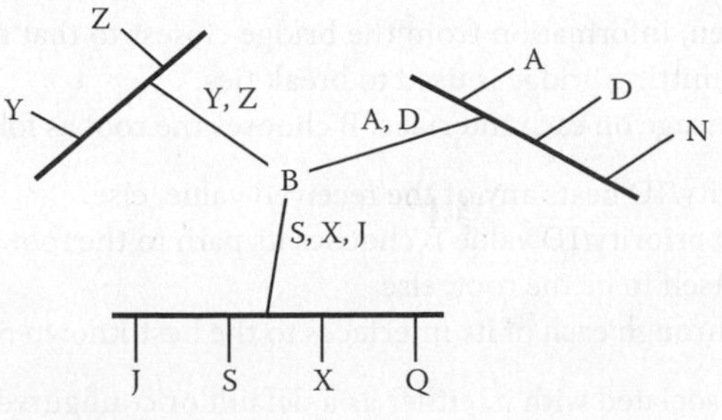

FIGURE 46.6 A bridge learning station addresses.

The basic idea of a transparent bridge is something that is attached to two or more LANs. On each LAN, the bridge listens promiscuously to all packets, stores them, and forwards each packet onto each other LAN when given permission by the LAN protocol on that LAN. An enhancement is to have the bridge learn, based on addresses in the LAN header, of packets received by the bridge, where the stations reside, so that the bridge does not unnecessarily forward packets (see Figure 46.6, where a bridge has learned some of the station addresses). The bridge learns from the source field in the LAN header and forwards based on the destination address. For example, in Figure 46.6, when S transmits a packet with destination address D, the bridge learns which interface S resides on, and then looks to see if it has already learned where D resides. If the bridge does not know where D is, then the bridge forwards the packet onto all interfaces (except the one from which the packet was received). If the bridge does know where D is, then the bridge forwards it only onto the interface where D resides, or if the packet arrived from that interface, the bridge discards the packet.

46.5.1.1 Spanning Tree Algorithm

The spanning tree algorithm is a very simple variant on distance vector, where all bridges calculate their path to a single destination; the root bridge. The tree of shortest paths from that root bridge to each LAN and each bridge is the calculated spanning tree.

46.5.1.1.1 Finding the Tree

How do the bridges decide on the root? Each bridge comes, at manufacture time, with a globally unique 48-bit *IEEE 802 address,* usually one per interface. A bridge chooses one of the 48-bit addresses that it owns as its ID. Because each bridge has a unique number, it is a simple matter to choose the one with the smallest number. However, because some network managers like to easily influence which bridge will be chosen, there is a configurable priority value that acts as a more significant field tacked onto the ID. The concatenated number consisting of priority and ID is used in the election, and the bridge with the smallest value is chosen as the root. The way in which the election proceeds is that each bridge assumes itself to be the root unless it hears, through spanning tree configuration messages, of a bridge with a smaller value for priority/ID. News of other bridges is learned through receipt of spanning tree configuration messages, which we describe shortly.

The next step is for a bridge to determine its best path to the root bridge and its own cost to the root. This information is also discovered through receipt of spanning tree configuration messages.

A spanning tree configuration message contains the following, among other information:

- Priority/ID of best known root
- Cost from transmitting bridge to root
- Priority/ID of transmitting bridge

A bridge keeps the best configuration message received on each of its interfaces. The fields in the message are concatenated together, from most significant to least significant, as root's priority/ID to cost to root to priority/ID of transmitting bridge. This concatenated quantity is used to compare messages. The one with the smaller quantity is considered better. In other words, only information about the best-known root is relevant. Then, information from the bridge closest to that root is considered, and then the priority/ID of the transmitting bridge is used to break ties.

Given a best received message on each interface, B chooses the root as follows:

- Itself, if its own priority/ID beats any of the received value, else
- The smallest received priority/ID value B chooses its path to the root as follows:
- Itself, if it considers itself to be the root, else
- The minimum cost through each of its interfaces to the best-known root

Each interface has a cost associated with it, either as a default or configured. The bridge adds the interface cost to the cost in the received configuration message to determine its cost through that interface.

B chooses its own cost to the root as follows:

- 0, if it considers itself to be the root, else
- The cost of the minimum cost path chosen in the previous step

Node B now knows what it would transmit as a configuration message, because it knows the root's priority/ID, its own cost to that root, and its own priority/ID. If B's configuration message is better than any of the received configuration messages on an interface, then B considers itself the *designated bridge* on that interface and transmits configuration messages on that interface. If B is not the designated bridgc on an interface, then B will not transmit configuration messages on that interface.

Each bridge determines which of its interfaces are in the spanning tree. The interfaces in the spanning tree are as follows:

- The bridge's path to the root: if more than one interface gives the same minimal cost, then exactly one is chosen. Furthermore, if this bridge is the root, then there is no such interface.
- Any interfaces for which the bridge is designated bridge are in the spanning tree.

If an interface is not in the spanning tree, the bridge continues running the spanning tree algorithm but does not transmit any data messages (messages other than spanning tree protocol messages) to that interface and ignores any data messages received on that interface.

If the topology is considered a graph with two types of nodes, bridges, and LANs, the following is the reasoning behind why this yields a tree:

- The root bridge is the root of the tree.
- The unique parent of a LAN is the designated bridge.
- The unique parent of a bridge is the interface that is the best path from that bridge to the root.

46.5.1.1.2 Dealing with Failures

The root bridge periodically transmits configuration messages (with a configurable timer on the order of 1 s). Each bridge transmits a configuration message on each interface for which it is designated, after receiving one on the interface which is that bridge's path to the root. If some time elapses (a configurable value with default on the order of 15 s) in which a bridge does not receive a configuration message on an interface, the configuration message learned on that interface is discarded.

In this way, roughly 15 s after the root or the path to the root has failed, a bridge will discard all information about that root, assume itself to be the root, and the spanning algorithm will compute a new tree.

46.5.1.1.3 Avoiding Temporary Loops

In a routing algorithm, the nodes learn information at different times. During the time after a topology change and before all nodes have adapted to the new topology, there are temporary loops or temporary partitions (no way to get from some place to some other place). Because temporary loops are so disastrous with bridges (because of the packet proliferation problem), bridges are conservative about bringing an interface into the spanning tree. There is a timer (on the order of 30 s, but configurable). If an interface was not in the spanning tree, but new events convince the bridge that the interface should be in the spanning tree, the bridge waits for this timer to expire before forwarding data messages to and from the interface.

IEEE has replaced the timer mechanism of avoiding temporary loops with a complex interlock mechanism of coordinating with neighbors before turning on ports, which allows faster failover. This variant of spanning tree is known as rapid spanning tree protocol.

46.5.2 TRansparent Interconnection of Lots of Links

Once everyone fixed the mistake of omitting layer 3 from network stacks, and having most protocols run directly on top of layer 3, and given that everyone has agreed on a single layer 3 protocol (IPv4, atleast for now), why have not transparent bridges been replaced by IP routers?

A true layer 3 protocol has advantages over spanning tree:

- Spanning tree does not yield shortest paths between any two nodes. For instance, if the topology is a big circle, spanning tree must refrain from using some links. Nodes on either side of that link have to take a long path between each other.
- Not all links can be utilized, so traffic is concentrated on links in the tree and other links are idle.
- For subtle reasons, if spanning tree switches do not have the computational capacity to process every incoming packet, loops can form, which are disastrous because of the lack of a hop count to detect infinitely circulating packets.

However, IP as a layer 3 protocol has disadvantages. In IP, each link has its own block of addresses. If you have a block of IP addresses and want to use them for a network, you have to carve up the address space to make a unique prefix for each link, you have to configure all the routers to know which address blocks are on which of their ports; and if a node moves to a different link, it must change its IP address.

Layer 3 does not need to work that way. In particular, ISO's version of IP, known as ConnectionLess Network Protocol (CLNP), had a 20 byte address. The top 14 bytes were the prefix for an entire cloud of links. The bottom six bytes indicated the node within that cloud.

Routing between clouds was similar to IP, in that the top 14 bytes could be used hierarchically, with routing to the longest prefix match. Within a cloud, endnodes announced their presence to their local router, which announced its endnodes to the other routers in the cloud. If a packet's 14 byte prefix indicated "this cloud," then routers in the cloud routed to the specific endnode.

So with CLNP, there was a large cloud with a flat address space. Routers within a cloud need not be configured; someone in the cloud needed to know the cloud's 14 byte prefix and inform the rest of the routers. Nodes could move within the cloud without changing their address.

But unfortunately, the world does not have CLNP, and instead has IP. Even if the world does migrate to IPv6, with a 16-byte address, IPv6 behaves the same as IPv4, in that each link is intended to have its own prefix.

Therefore, people would like to create large flat Ethernet clouds, with self-configuring switches inside the cloud, and the ability for nodes to migrate within the cloud without changing their address.

TRansparent Interconnection of Lots of Links (TRILL) (RFC 6325; Perlman et al., 2011) accomplishes this, and it does it in an evolutionary way. Within an existing spanning tree-based Ethernet, any subset of the switches can be replaced by TRILL switches; and the more you replace, the better the bandwidth utilization and the more stable the Ethernet appears. Existing endnodes see no change other than that the Ethernet is performing better.

Two TRILL switches are neighbors if they are directly connected (on the same link) or through spanning tree switches. TRILL switches find their neighbors (by sending multicast messages that are not forwarded by TRILL switches) and form a network using a variant of the IS-IS routing protocol, to create forwarding table entries for reaching all the other TRILL switches (but not the endnodes).

Edge TRILL switches (those that are attached to endnodes), create an additional table of (MAC and egress TRILL switch), which is either configured or learned when receiving data packets. Core TRILL switches do not need to have such a table; a core TRILL switch forwards to the last TRILL switch.

The first TRILL switch (the "ingress switch") adds an extra header to an Ethernet packet. This header contains the fields

- Ingress switch
- Egress switch
- Hop count (to detect and discard looping packets)

TRILL switches dynamically acquire 2 byte nicknames. This is done by having TRILL switch T1 choose a nickname N and claim it in T1's LSP. Switch T1 examines all other TRILL switches' LSPs, and if some other TRILL switch T2 has chosen nickname N, then based on priority, one of them gets to keep N and the other chooses a different nickname (one not already claimed in another switch's LSP).

Edge switches learn (or are configured) with an endnode table consisting of Ethernet address of endnode and attached TRILL switch nickname. When an ingress switch S receives an Ethernet packet, S looks up the destination Ethernet address in S's table, finds the proper egress TRILL nickname E, and S adds a TRILL header with TRILL ingress switch nickname = S and TRILL egress switch nickname = E.

The core switches need only look at the TRILL header.

The advantages of the extra TRILL header are as follows:

- Core switches can forward based on an easy lookup in the forwarding table, since a 16-bit switch nickname can be a direct index into the forwarding table.
- Core switches only need a forwarding table, which is as big as the number of TRILL switches; not the number of endnodes.
- There is a hop count in the TRILL header, so it is routable just like layer 3, with all the techniques that are usually associated with layer 3, such as shortest paths, multipathing, etc.

Key Terms

Bridge: A box that forwards information from one link to another but only looks at information in layer 2.

Cloud: An informal representation of a multiaccess link. The purpose of representing it as a cloud is that what goes on inside is irrelevant to what is being discussed. When a system is connected to the cloud, it can communicate with any other system attached to the cloud.

Data-link layer: The layer that gets information from one machine to a neighbor machine (a machine on the same link).

IEEE 802 address: The 48-bit address defined by the IEEE 802 committee as the standard address on 802 LANs.

Hops: The number of times a packet is forwarded by a router.

Local area network (LAN): A multiaccess link with multicast capability.

MAC address: Synonym for IEEE 802 address.

Medium access control (MAC): The layer defined by the IEEE 802 committee that deals with the specifics of each type of LAN (for instance, token passing protocols on token passing LANs).

Multiaccess link: A link on which more than two nodes can reside.

Multicast: The ability to transmit a single packet that is received by multiple recipients.

Network layer: The layer that traditionally forms a path by concatenation of several links.

Router: A box that forwards packets at layer 3.

Switch: Any device that forwards packets (including bridges or routers).

References

Clark, D. 1989. Policy routing in internet protocols. RFC 1102, May.

Garcia-Luna-Aceves, J.J. 1989. A unified approach to loop-free routing using distance vectors or link states. *ACM Sigcomm #89 Symposium*, September.

Griffin, T. and Gordon, W. 1999. An analysis of BGP convergence properties. *ACMSigcomm #99 Symp*.

Hedrick, C. 1988. Routing information protocol. RFC 1058.

Lougheed, K. and Rehkter, Y. 1991. A border gateway protocol 3 (BGP-3). RFC 1267, October.

Moy, J. 1998. OSPF version 2. RFC 2328.

Oran, D. ed. 1990. OSI IS-IS intra-domain routing protocol. RFC 1142.

Perlman, R. 1983. Fault-tolerant broadcast of routing information. *Comput. Networks*, December.

Perlman, R. 1985. A protocol for distributed computation of a spanning tree in an extended LAN, *9th Data Communications Symposium*, Vancouver, British Columbia, Canada.

Perlman, R. 1988. Network layer protocols with byzantine robustness. *MIT Lab. Computer Science Tech. Rep.* #429, October.

Perlman, R. 1999. *Interconnections: Bridges, Routers, Switches, and Internetworking Protocols.* Addison-Wesley, Reading, MA.

Perlman, R. et al., 2011. Routing Bridges (RBridges): Base Protocol Specification.

Rekhter, Y. et al., 2006. RFC 4271, A Border Gateway Protocol 4 (BGP-4).

47

Access Control

Sabrina De Capitani di Vimercati
Università degli Studi di Milano

Pierangela Samarati
Università degli Studi di Milano

Ravi Sandhu
The University of Texas, San Antonio

47.1 Introduction

An important requirement of any information management system is to protect information against improper disclosure or modification (known as *confidentiality* and *integrity*, respectively). *Access control* is a fundamental technology to achieve this goal (e.g., [22,26]). It controls every access request to a system and determines whether the access request should be granted or denied. In access control systems, a distinction is generally made among *policies, models,* and *mechanisms.* Policies are high level guidelines that determine how accesses are controlled and access decisions determined. A policy is then formalized through a security model and is enforced by access control mechanisms. The access control mechanism works as a *reference monitor* that mediates every attempted access by a user (or program executing on behalf of that user) to objects in the system. The reference monitor consults an *authorization database* to determine if the user attempting to do an operation is actually authorized to perform that operation. Authorizations in this database are usually defined with respect to the identity of the users. This implies that access control requires *authentication* as a prerequisite, meaning that the identity of the requesting user has to be correctly verified (e.g., [8,19,25,31]). The reader is surely familiar with the process of signing on to a computer system by providing an identifier and a password. This authentication establishes the identity of a human user to a computer. More generally, authentication can be computer-to-computer or process-to-process and mutual in both directions. Access control is then concerned with limiting the activity of legitimate users who have been successfully authenticated. The set of authorizations are administered and maintained by a security administrator. The administrator sets the authorizations on the basis of the security policy of the organization. Users may also be able to modify some portion of the authorization database, for example, to set permissions for their personal files. The effectiveness of the access control rests on a proper user identification and on the correctness of the authorizations governing the reference monitor.

The variety and complexity of the protection requirements that may need to be imposed in today's systems make the definition of access control policies a far from trivial process. For instance, many services do not need to know the real identity of a user but they may need to know some characteristics/properties of the requesting users (e.g., a user can access a service only if she works in an European country), or different systems may need to collaborate while preserving their autonomy in controlling access to their resources [23,24,26]. The goal of this chapter is therefore to provide an overview of the access control evolution. This overview begins with a discussion on the classical discretionary, mandatory, and role-based access control policies and models (Sections 47.2 through 47.4), and then continues with an illustration of the administration policies (Section 47.5) that determine who can modify the accesses allowed by such policies. We then discuss the most recent advances in access control, focusing on attribute and credential-based access control (Section 47.6). Finally, we present our conclusions in Section 47.7.

47.2 Discretionary Access Control (DAC)

Discretionary access control policies govern the access of users to the information/resources on the basis of the users' identities and authorizations (or rules) that specify, for each user (or group of users) and each object in the system, the access modes the user is allowed on the object. Each request of a user to access an object is checked against the specified authorizations. If there exists an authorization stating that the user can access the object in the specific mode, the access is granted; it is denied, otherwise. In the following, we first describe the access matrix model, which is useful to understand the basic principles behind discretionary access control models and policies, and discuss implementation alternatives. We then illustrate how discretionary policies have been further expanded.

47.2.1 The Access Matrix Model and Its Implementation

A first step in the definition of an access control model is the identification of the set of *objects* to be protected, the set of *subjects* who request access to objects, and the set of *access modes* that can be executed on objects. While subjects typically correspond to users (or groups thereof), objects and access modes may differ depending on the specific system or application. For instance, objects may be files and access modes may be Read, Write, Execute, and Own. The meaning of the first three of these access modes is self evident. Ownership is concerned with controlling who can change the access permissions for the file. An object such as a bank account may have access modes Inquiry, Credit, and Debit corresponding to the basic operations that can be performed on an account. These operations would be implemented by application programs, whereas for a file the operations would typically be provided by the operating system.

A subtle point that is often overlooked is that subjects can themselves be objects. A subject can create additional subjects to accomplish its task. The children subjects may be executing on various computers in a network. The parent subject will usually be able to suspend or terminate its children as appropriate. The fact that subjects can be objects corresponds to the observation that the initiator of one operation can be the target of another. (In network parlance, subjects are often called initiators, and objects called targets.)

The access matrix is a conceptual model which specifies the rights that each subject possesses for each object. There is a row in the matrix for each subject, and a column for each object. Each cell of the matrix specifies the access authorized for the subject in the row to the object in the column. The task of access control is to ensure that only those operations authorized by the access matrix actually get executed. This is achieved by means of a *reference monitor*, which is responsible for mediating all attempted operations by subjects on objects.

TABLE 47.1 An Example of Access Matrix

	File1	File2	File3	File4	Account1	Account2
John	Own R W		Own R W		Inquiry Credit	
Alice	R	Own R W	W	R	Inquiry Debit	Inquiry Credit
Bob	R W	R		Own R W		Inquiry Debit

Table 47.1 shows an example of access matrix where the access modes R and W denote read and write, respectively, and the other access modes are as discussed previously. The subjects shown here are John, Alice, and Bob. There are four files and two accounts. This matrix specifies that, for example, John is the owner of File1 and can read and write that file, but John has no access to File2 and File4. The precise meaning of ownership varies from one system to another. Usually the owner of a file is authorized to grant other users access to the file, as well as revoke access. Since John owns File1, he can give Alice the R right and Bob the R and W rights as shown in Table 47.1. John can later revoke one or more of these rights at his discretion.

The access modes for the accounts illustrate how access can be controlled in terms of abstract operations implemented by application programs. The Inquiry operation is similar to read since it retrieves information but does not change it. Both the Credit and Debit operations will involve reading the previous account balance, adjusting it as appropriate and writing it back. The programs which implement these operations require read and write access to the account data. Users, however, are not allowed to read and write the account object directly. They can manipulate account objects only indirectly via application programs which implement the Debit and Credit operations. Also, note that there is no own right for accounts. Objects such as bank accounts do not really have an owner who can determine the access of other subjects to the account. Clearly, the user who establishes the account at the bank should not be the one to decide who can access the account. Within the bank different officials can access the account depending on their job functions in the organization.

In a large system the access matrix will be enormous in size, and most of its cells are likely to be empty. Accordingly, the access matrix is very rarely implemented as a matrix. We now discuss some common approaches for implementing the access matrix in practical systems.

Access control lists: Each object is associated with an access control list (ACL), indicating for each subject in the system the accesses the subject is authorized to execute on the object. This approach corresponds to storing the matrix by columns. ACLs corresponding to the access matrix of Table 47.1 are shown in Figure 47.1. Essentially the access matrix column for File1 is stored in association with File1, and so on.

By looking at an object's ACL, it is easy to determine which access modes subjects are currently authorized for that object. In other words, ACLs provide for convenient access review with respect to an object. It is also easy to revoke all accesses to an object by replacing the existing ACL with an empty one. On the other hand, determining all the accesses that a subject has is difficult in an ACL-based system. It is necessary to examine the ACL of every object in the system to do access review with respect to a subject. Similarly, if all accesses of a subject need to be revoked, all ACLs must be visited one by one. (In practice revocation of all accesses of a subject is often done by deleting the user account corresponding to that subject. This is acceptable if a user is leaving an organization. However, if a user is reassigned within the organization it would be more convenient to retain the account and change its privileges to reflect the changed assignment of the user.)

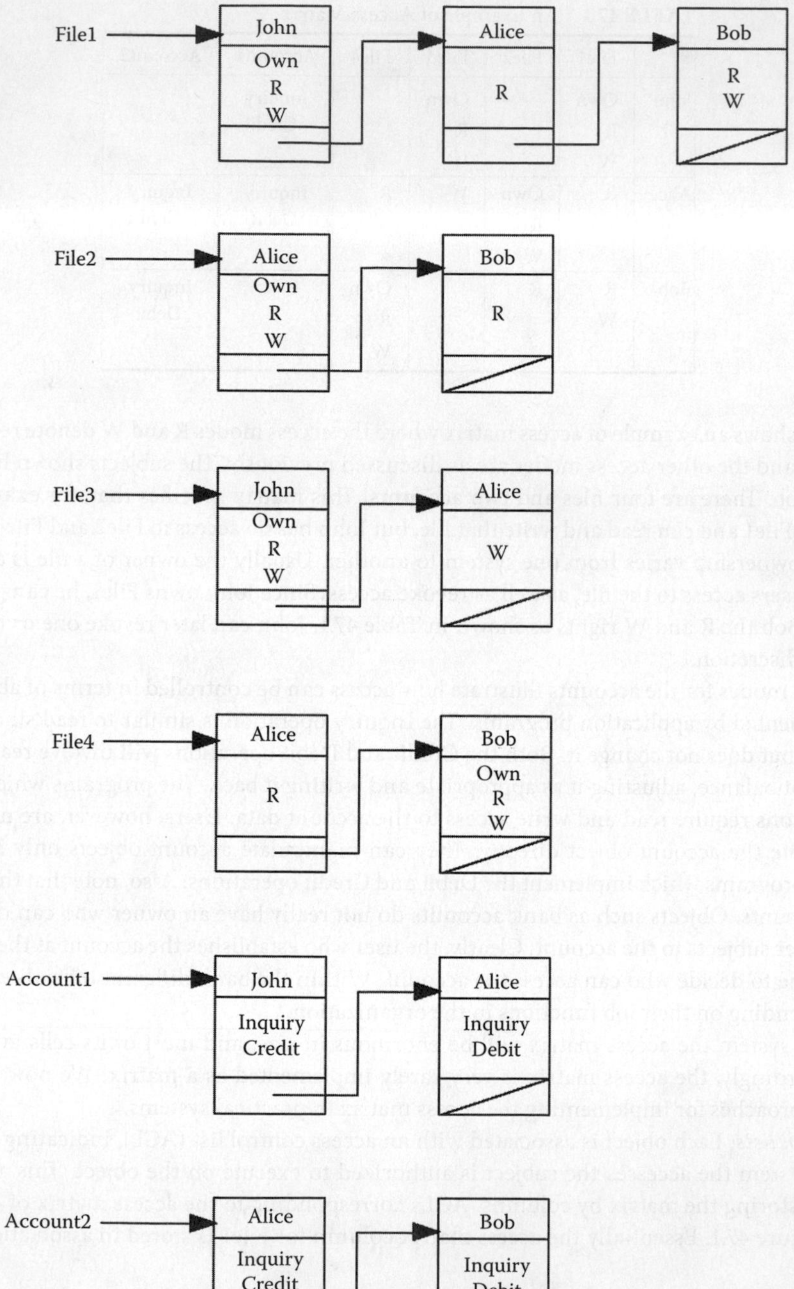

FIGURE 47.1 Access control lists corresponding to the access matrix in Table 47.1.

Many systems allow group names to occur in ACLs (see Section 47.2.2). For instance, an entry such as (ISSE, R) can authorize all members of the ISSE group to read a file. Several popular operating systems (e.g., Unix) implement an abbreviated form of ACLs in which a small number, often only one or two, group names can occur in the ACL. Individual subject names are not allowed. With this approach the ACL has a small fixed size so it can be stored using a few bits associated with the file.

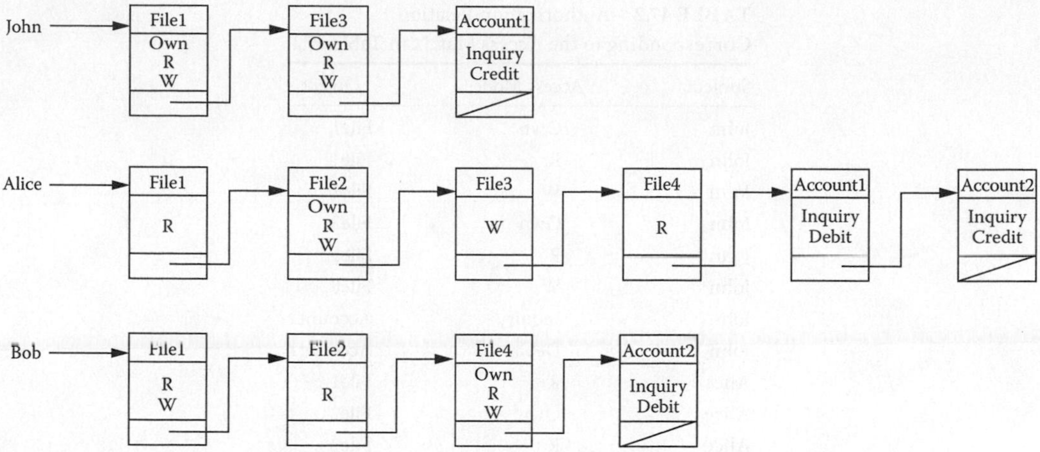

FIGURE 47.2 Capability lists corresponding to the access matrix in Table 47.1.

Capabilities: Capabilities are a dual approach to ACLs. Each subject is associated with a list, called capability list, indicating for each object in the system, the accesses the subject is authorized to execute on the object. This approach corresponds to storing the access matrix by rows. Figure 47.2 shows the capability lists corresponding to the access matrix in Table 47.1. In a capability list approach it is easy to review all accesses that a subject is authorized to perform by simply examining the subject's capability list. However, determination of all subjects who can access a particular object requires examination of each and every subject's capability list. A number of capability-based computer systems were developed in the 1970s, but did not prove to be commercially successful. Modern operating systems typically take the ACL-based approach.

It is possible to combine ACLs and capabilities. Possession of a capability is sufficient for a subject to obtain access authorized by that capability. In a distributed system this approach has the advantage that repeated authentication of the subject is not required. This allows a subject to be authenticated once, obtain its capabilities and then present these capabilities to obtain services from various servers in the system. Each server may further use ACLs to provide finer-grained access control.

Authorization relations: We have seen that ACL- and capability-based approaches have dual advantages and disadvantages with respect to access review. There are representations of the access matrix which do not favor one aspect of access review over the other. For instance, the access matrix can be represented by an authorization relation (or table) as shown in Table 47.2. Each row, or tuple, of this table specifies one access right of a subject to an object. Thus, John's accesses to File1 require three rows. If this table is sorted by subject, we get the effect of capability lists. If it is sorted by object, we get the effect of ACLs. Relational database management systems typically use such a representation.

47.2.2 Expanding Authorizations

Although the access matrix still remains a framework for reasoning about accesses permitted by a discretionary policy, discretionary policies have been developed considerably since the access matrix was proposed. In particular, early approaches to authorization specifications allowed *conditions* to be associated with authorizations to restrict their validity [33]. Conditions may involve some system predicates, and may specify restrictions based on the content of objects or on accesses previously executed. Another important feature supported by current discretionary policies is the definition of abstractions on users and objects. Both users and objects can therefore be hierarchically organized, thus introducing *user groups* and *classes of objects*. Figure 47.3a and b illustrate an example of user group hierarchy and object hierarchy, respectively. The definition of groups of users

TABLE 47.2 Authorization Relation Corresponding to the Access Matrix in Table 47.1

Subject	Access Mode	Object
John	Own	File1
John	R	File1
John	W	File1
John	Own	File3
John	R	File3
John	W	File3
John	Inquiry	Account1
John	Debit	Account1
Alice	R	File1
Alice	Own	File2
Alice	R	File2
Alice	W	File2
Alice	W	File3
Alice	R	File4
Alice	Inquiry	Account1
Alice	Debit	Account1
Alice	Inquiry	Account2
Alice	Credit	Account2
Bob	R	File1
Bob	W	File1
Bob	R	File2
Bob	Own	File4
Bob	R	File4
Bob	W	File4
Bob	Inquiry	Account2
Bob	Debit	Account2

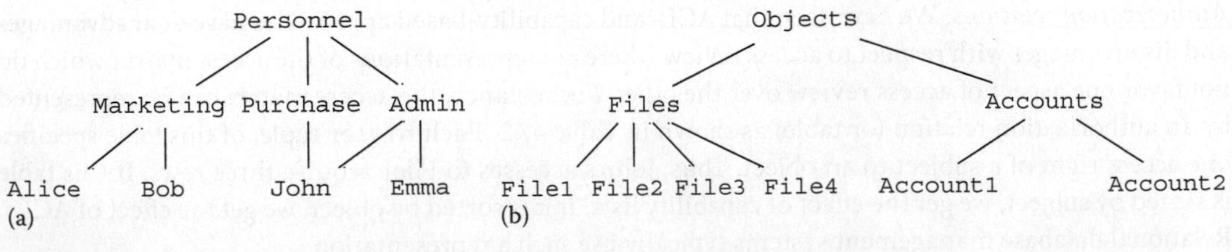

FIGURE 47.3 An example of (a) user group and (b) object hierarchy.

(and classes of objects) requires a technique to easily handle exceptions. For instance, suppose that all users belonging to a group can access a specific object but user *u*. In this case, it is necessary to explicitly associate an authorization with each user in the group but *u*, which is clearly a solution that does not take advantage from the definition of user groups. This observation has been the driving factor supporting the development of access control models that combine *positive* and *negative* authorizations (e.g., [10]). In this way, the previous exception can be easily modeled by the definition of two authorizations: a positive authorization for the group and a negative authorization for user *u*. Hierarchies can also simplify the definition of authorizations because authorizations specified on an abstraction can be propagated to all its members. The propagation of authorizations over a hierarchy may follow different *propagation policies* (e.g., [35,52]). For instance, authorizations associated with an element in the hierarchy may not be propagated, may be propagated to all its descendants, or may be propagated to its descendants if not overridden.

The use of both positive and negative authorizations introduces two problems: (1) *inconsistency*, which happens when for an access there are both a negative and a positive authorization; and (2) *incompleteness*, which happens when some accesses are neither authorized nor denied (i.e., no authorization exists for them). The inconsistency problem is solved by applying a *conflict resolution policy*. There are several conflict resolution policies [35,45] such as: *no conflict*, the presence of a conflict is considered an error; *denials take precedence*, negative authorizations take precedence; *permissions take precedence*, positive authorizations take precedence; *nothing takes precedence*, neither positive nor negative authorizations take precedence and conflicts remain unsolved; *most specific takes precedence*, the authorization that is more specific with respect to a hierarchy wins (e.g., consider the user group hierarchy in Figure 47.3a, the effect on John of a positive authorization for the `Admin` group to read File1, and a negative authorization for reading the same file for `Personnel` is that he is allowed to read File1 since `Admin` is more specific than `Personnel`); and *most specific along a path takes precedence*, the authorization that is more specific with respect to a hierarchy wins only on the paths passing through it (e.g., consider the user group hierarchy in Figure 47.3a, the effect on John of a positive authorization for the `Admin` group to read File1, and a negative authorization for reading the same file for `Personnel` is that there is a conflict for managing John's access to File1 since the negative authorization wins along path ⟨`Personnel`, `Purchase`, John⟩ and the positive authorization wins along path ⟨`Personnel`, `Admin`, John⟩). The incompleteness problem can be solved by adopting a *decision policy*, that is, an *open policy* or a *closed policy*. Open policies are based on explicitly specified authorizations and the default decision of the reference monitor is denial. Open policies are based on the specification of denials instead of permissions. In this case, for each user and each object in the system, the access modes the user is forbidden on the object are specified. Each access request by a user is checked against the specified (negative) authorizations and granted only if no authorization denying the access exists. The combination of a propagation policy, a conflict resolution policy, and a decision policy guarantees a complete and consistent policy for the system.

47.3 Mandatory Access Control (MAC)

The flexibility of discretionary policies makes them suitable for a variety of systems and applications, especially in the commercial and industrial environments. However, discretionary access control policies have the drawback that they do not provide real assurance on the flow of information in a system. It is easy to bypass the access restrictions stated through the authorizations. For instance, a user who is authorized to read data can pass them to other users not authorized to read them without the cognizance of the owner. The reason is that discretionary policies do not impose any restriction on the usage of information by a user once the user has got it (i.e., dissemination of information is not controlled). By contrast, dissemination of information is controlled in mandatory systems by preventing flow of information from high-level objects to low-level objects. Mandatory policies govern access on the basis of a classification of subjects and objects in the system. Note that the concept of subject in discretionary policies is different from the concept of subject in mandatory policies. In fact, authorization subjects correspond to users (or groups thereof) while in mandatory policies users are human beings who can access the system, and subjects are processes operating on behalf of users.

Each subject and each object in a mandatory system is assigned an *access class*. The set of access classes is a partially ordered set and in most cases an access class is composed by a *security level* and a set of *categories*. The security level is an element of a hierarchical ordered set. In the military and civilian government arenas, the hierarchical set generally consists of top secret (TS), secret (S), confidential (C), and unclassified (U), where TS > S > C > U. Each security level is said to dominate itself and all others below it in this hierarchy. The set of categories is a subset of an unordered set, whose elements reflect functional, or competence, areas (e.g., financial, demographic, medical). Given two access classes c_1 and c_2, we say that c_1 dominates c_2, denoted $c_1 \geqslant c_2$, iff the security level of c_1 is greater than or equal to that of c_2 and the categories of c_1 include those of c_2.

The semantics and use of the access classes assigned to objects and subjects within the application of a multilevel mandatory policy depends on whether the access class is intended for a secrecy or an integrity policy. In the following, we illustrate secrecy-based and integrity-based mandatory policies.

47.3.1 Secrecy-Based Model

The main goal of a secrecy-based mandatory policy is to protect the confidentiality of information. In this case, the security level associated with an object reflects the sensitivity of the information contained in the object, that is, the potential damage which could result from unauthorized disclosure of the information. The security level associated with a user, also called *clearance*, reflects the user's trustworthiness not to disclose sensitive information to users not cleared to see it. The set of categories associated with a user reflects the specific areas in which the user operates. The set of categories associated with an object reflects the area to which information contained in the object is referred. Categories enforce restrictions on the basis of the need-to-know principle (i.e., a subject should be only given those accesses which are required to carry out the subject's responsibilities). Users can connect to the system at any access class dominated by their clearance. A user connecting to the system at a given access class originates a subject at that access class.

Access to an object by a subject is granted only if some relationship (depending on the type of access) is satisfied between the access classes associated with the two. In particular, the following two principles are required to hold.

Read-down A subject's access class must dominate the access class of the object being read.
Write-up A subject's access class must be dominated by the access class of the object being written.

Satisfaction of these principles prevents information in high level objects (i.e., more sensitive) to flow to objects at lower levels. Figure 47.4 illustrates the effect of these rules. Here, for simplicity, accesses classes are only composed of a security level. In such a system, information can only flow upwards or within the same security class.

To better understand the rationale behind the read-down and write-up rules, it is important to analyze the relationship between users and subjects in this context. Suppose that the human user Jane is cleared to S (again, for simplicity, access classes are only composed of a security level), and assume she always signs on to the system as an S subject (i.e., a subject with clearance S). Jane's subjects are

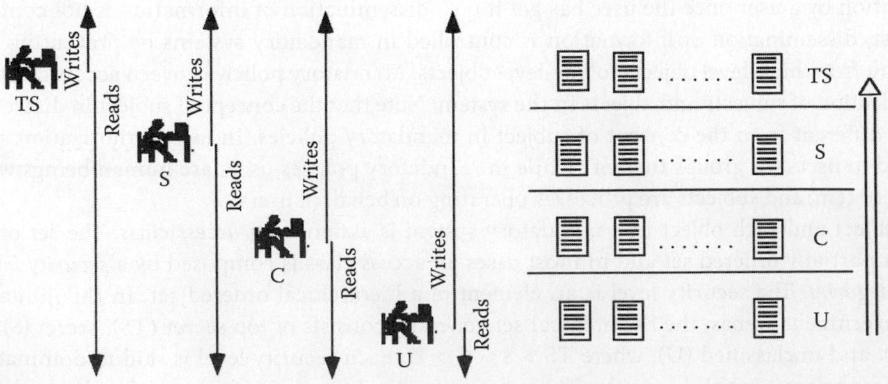

FIGURE 47.4 Controlling information flow for secrecy.

prevented from reading TS objects by the read-down rule. The write-up rule, however, has two aspects that seem at first sight contrary to expectation.

- First, Jane's S subjects can write a TS object (even though they cannot read it). In particular, they can overwrite existing TS data and therefore destroy them. Due to this integrity concern, many systems for mandatory access control do not allow write-up but limit writing to the same level as the subject. At the same time write-up does allow Jane's S subjects to send electronic mail to TS subjects, and can have its benefits.
- Second, Jane's S subjects cannot write C or U data. This means, for example, that Jane can never send electronic mail to C or U users. This is contrary to what happens in the paper world, where S users can write memos to C and U users. This seeming contradiction is easily eliminated by allowing Jane to sign to the system as a C, or U, subject as appropriate. During these sessions she can send electronic mail to C or, U and C, subjects, respectively.

The write-up rule prevents malicious software from leaking secrets downward from S to U. Users are trusted not to leak such information, but the programs they execute do not merit the same degree of trust. For instance, when Jane signs onto the system at U level her subjects cannot read S objects, and thereby cannot leak data from S to U. The write-up rule also prevents users from inadvertently leaking information from high to low.

47.3.2 Integrity-Based Model

Mandatory access control can also be applied for the protection of information integrity. The integrity level associated with an object reflects the degree of trust that can be placed in the information stored in the object, and the potential damage that could result from unauthorized modification of the information. The integrity level associated with a user reflects the user's trustworthiness for inserting, modifying or deleting data and programs at that level. Again, categories define the area of competence of users and data. Principles similar to those stated for secrecy are required to hold, as follows.

Read-up A subject's integrity class must be dominated by the integrity class of the object being read.

Write-down A subject's integrity class must dominate the integrity class of the object being written.

Satisfaction of these principles safeguard integrity by preventing information stored in low objects (and therefore less reliable) to flow to high objects. This is illustrated in Figure 47.5, where, for simplicity, integrity classes are only composed of integrity levels, which can be crucial (C), important (I),

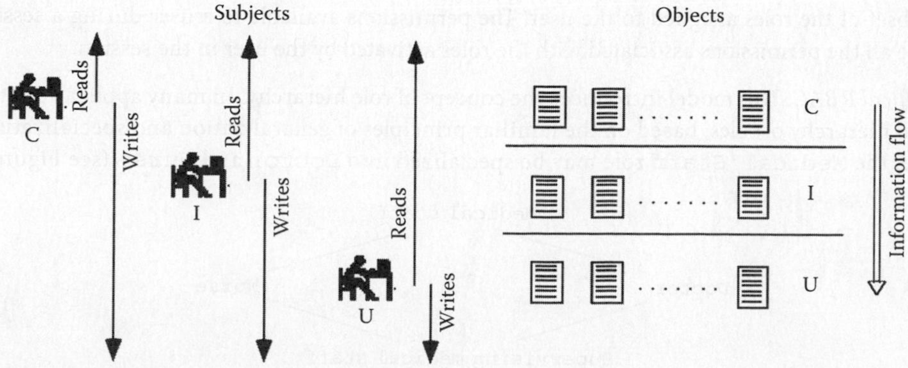

FIGURE 47.5 Controlling information flow for integrity.

and unknown (U) with C > I > U. Controlling information flow in this manner is only one aspect of achieving integrity. Integrity in general requires additional mechanisms, as discussed in [16,53].

Note that the only difference between Figures 47.4 and 47.5 is the direction of information flow, being bottom to top in the former case and top to bottom in the latter. In other words, both cases are concerned with one-directional information flow. The essence of classical mandatory controls is one-directional information flow in a lattice of security classes. For further discussion on this topic see [52].

47.4 Role-Based Access Control (RBAC)

Role-based policies regulate the access of users to the information on the basis of the activities the users execute in the system. Role-based policies require the identification of roles in the system. A role can be defined as a set of actions and responsibilities associated with a particular working activity. Then, instead of specifying all the accesses each user is allowed to execute, access authorizations on objects are specified for roles. Users are given authorizations to play roles. This greatly simplifies the authorization management task. For instance, suppose a user's responsibilities change, say, due to a promotion. The user's current roles can be taken away and new roles assigned as appropriate for the new responsibilities. Another advantage of RBAC is that roles allow a user to sign on with the least privilege required for the particular task at hand. Users authorized to powerful roles do not need to exercise them until those privileges are actually needed. This minimizes the danger of damage due to inadvertent errors or by intruders masquerading as legitimate users. Note that, in general, a user can take on different roles on different occasions. Also, the same role can be played by several users, perhaps simultaneously. Some proposals for role-based access control allow a user to exercise multiple roles at the same time. Other proposals limit the user to only one role at a time, or recognize that some roles can be jointly exercised while others must be adopted in exclusion to one another. In the remainder of this section, we briefly describe the standard RBAC model and then illustrate an extension of RBAC.

47.4.1 Basic RBAC Model

In 2004, the National Institute of Standards and Technology (NIST) proposed a US national standard for role-based access control through the International Committee for Information Technology Standards (ANSI/INCITS) [3,28]. The standard RBAC model is organized in four components, which are briefly described in the following.

Core RBAC: Core RBAC includes five basic elements, that is, users, roles, objects, operations, and permissions. Users are assigned to roles, and permissions (i.e., the association between an object and an operation executable on the object) are assigned to roles. The permission-to-role and user-to-role assignments are both many-to-many, thus providing greater flexibility. In addition, Core RBAC includes the concept of session, where each session is a mapping between a user and a set of activated roles, which are a subset of the roles assigned to the user. The permissions available to a user during a session are therefore all the permissions associated with the roles activated by the user in the session.

Hierarchical RBAC: This model introduces the concept of role hierarchy. In many applications there is a natural hierarchy of roles, based on the familiar principles of generalization and specialization. For instance, the `Medical Staff` role may be specialized into `Doctor` and `Nurse` (see Figure 47.6).

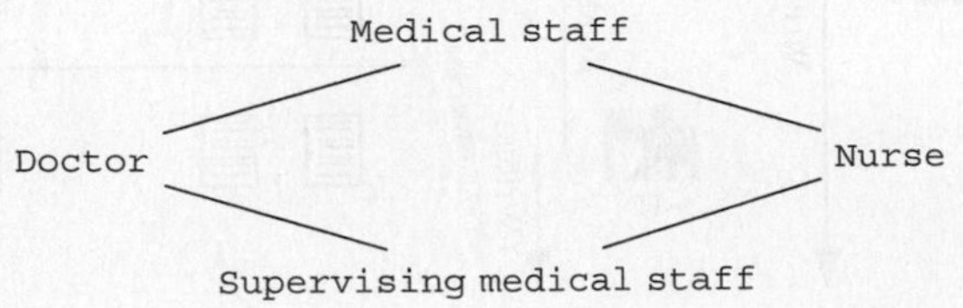

FIGURE 47.6 An example of role hierarchy.

The role hierarchy has implication on role activation and access privileges: a specialized role inherits the authorizations of its generalizations; also, users authorized to activate a role inherit the authorization to activate the generalizations of the role. For instance, with reference to Figure 47.6, role `Nurse` inherits all authorizations of role `Medical Staff`. Also, users authorized to activate role `Nurse` will also be allowed to activate role `Medical Staff`. A role hierarchy can be an arbitrary partial order or it may be possible to impose some restrictions on the type of hierarchy. For instance, it is possible to require that a role hierarchy has to be a tree, meaning that each role may have only one single direct parent.

Static separation of duty: The RBAC model can be enriched by adding separation of duty constraints. Intuitively, static separation of duty imposes restrictions on the assignments of users to roles. For instance, a user assigned to the `Nurse` role may not be assigned to the `Doctor` role. A separation of duty constraint is defined as a pair where the first element is a set *rs* of roles and the second element is an integer *n* greater than or equal to two denoting the number of roles that would constitute a violation. For instance, ⟨{`Doctor`, `Nurse`},2⟩ states that a user may be assigned to one out of the two roles mentioned in the constraint. Since static separation of duty constraints can be also defined in the presence of a role hierarchy, it is important to ensure that role inheritance does not violate the static separation of duty constraints.

Dynamic separation of duty: In the dynamic separation of duty, constraints are enforced at run-time: users can activate any role to which they are assigned but the activation of some roles during a session will rule out their ability to activate another role which is in a separation of duty constraint with the activated roles. An example of dynamic separation of duty is the two-person rule. The first user who executes a two-person operation can be any authorized user, whereas the second user can be any authorized user different from the first. The dynamic separation of duty constraints are still defined as pairs composed by a set *rs* of roles and an integer *n* greater than or equal to two and their enforcement requires that no user is assigned to *n* or more roles from *rs* in a single session.

In addition to the four components described above, the NIST RBAC standard defines administrative functions related to the creation and management of the basic elements and relations of RBAC, and permission review functions.

47.4.2 Expanding RBAC

The RBAC model has been adopted in a variety of commercial systems (e.g., DB2, Oracle), and has been the subject of many research proposals aimed at extending and enriching the model to support particular domain specific security policies (e.g., web services, social networks), administration models, and delegation. Some proposals are also aimed at integrating RBAC with other technologies such as cryptography, trust mechanisms, and XML-based access control languages (e.g., [1,30]). Given the huge amount of work on RBAC, it is clearly not feasible to provide a comprehensive summary of the extensions proposed in the literature. A notable example is the *usage control model* ($UCON_{ABC}$) [51], a framework that encompasses access control, trust management, and digital right management. This proposal is interesting since it supports DAC, MAC, and RBAC and is based on attributes characterizing subjects and objects that are used for specifying authorizations (see Section 47.6 for more details about attribute-based access control). In particular, the $UCON_{ABC}$ model integrates authorizations, obligations, and conditions within a unique framework, and includes the following eight components.

Subjects, objects, and *rights*: Subjects, objects, and rights have the same meaning of the corresponding concepts also used within the DAC, MAC, and RBAC access control models. A subject is therefore an individual who holds some rights on objects.

Subject and *object attributes*: Each subject and object in the system is characterized by a set of attributes that can be used for verifying whether an access request can be granted. Examples of subject attributes include identities, group names, and roles while examples of object attributes include ownerships, security labels, and so on. A peculiarity of the $UCON_{ABC}$ model is that subject and object attributes can be mutable, meaning that their values may change due to an access.

Authorizations: Authorizations are evaluated for usage decision and return whether a subject can perform the required right on an object. Authorizations are based on subject and object attributes, and are distinguished between pre-authorizations and ongoing-authorizations. A pre-authorization is performed before the execution of the requested right while an ongoing-authorization is performed during the access.

Obligations: An obligation is a requirement that a user has to perform before (pre) or during (ongoing) access. For instance, a pre-obligation may require a user to provide her date of birth before accessing a service. The execution of obligations may change the value of mutable attributes and therefore they may affect current or future usage decisions.

Conditions: Conditions evaluate current environmental or system status. Examples are time of the day and system workload. The evaluation of these conditions cannot change the value of any subject or object attributes.

A family of $UCON_{ABC}$ core models is defined according to three criteria: the decision factor, which may be authorizations, obligations, and conditions; the continuity of decision, which may be either pre or ongoing; and the mutability, which can allow changes on subject or object attributes at different times. According to these criteria, the authors in [51] define 16 basic $UCON_{ABC}$ models and show how these models can support traditional DAC, MAC, and RBAC.

47.5 Administration of Authorizations

Administrative policies determine who is authorized to modify the allowed accesses. This is one of the most important, and least understood, aspects of access control.

In mandatory access control the allowed accesses are determined entirely on the basis of the access class of subjects and objects. Access classes are assigned to users by the security administrator. Access classes of objects are determined by the system on the basis of the access classes of the subjects creating them. The security administrator is typically the only one who can change access classes of subjects or objects. The administrative policy is therefore very simple.

Discretionary access control permits a wide range of administrative policies. Some of these are described as follows:

- *Centralized*: A single authorizer (or group) is allowed to grant and revoke authorizations to the users.
- *Hierarchical*: A central authorizer is responsible for assigning administrative responsibilities to other administrators. The administrators can then grant and revoke access authorizations to the users of the system. Hierarchical administration can be applied, for example, according to the organization chart.
- *Cooperative*: Special authorizations on given resources cannot be granted by a single authorizer but need cooperation of several authorizers.
- *Ownership*: A user is considered the owner of the objects he/she creates. The owner can grant and revoke access rights for other users to that object.
- *Decentralized*: In decentralized administration the owner of an object can also grant other users the privilege of administering authorizations on the object.

Within each of these there are many possible variations [52].

Role-based access control has a similar wide range of possible administrative policies. In this case roles can also be used to manage and control the administrative mechanisms.

Delegation of administrative authority is an important aspect in the administration of authorizations. In large distributed systems, centralized administration of access rights is infeasible. Some existing systems allow administrative authority for a specified subset of the objects to be delegated by the central security administrator to other security administrators. For instance, authority to administer objects in a particular region can be granted to the regional security administrator. This allows delegation of administrative authority in a selective piecemeal manner. However, there is a dimension of selectivity that is largely ignored in existing systems. For instance, it may be desirable that the regional security administrator be limited to granting access to these objects only to employees who work in that region. Control over the regional administrators can be centrally administered, but they can have considerable autonomy within their regions. This process of delegation can be repeated within each region to set up sub-regions and so on.

47.6 Attribute and Credential-Based Access Control

Emerging distributed scenarios (e.g., cloud computing and data outsourcing) are typically characterized by several independent servers offering services to anyone who needs them (e.g., [21]). In such a context, traditional assumptions for enforcing access control do not hold anymore. As a matter of fact, an access request may come from unknown users and therefore access control policies based on the identity of the requester cannot be applied. Alternative solutions that have been largely investigated in the last 20 years consist in adopting *attribute-based access control* that uses the attributes associated with the resources/services and requesters to determine whether the access should be granted (e.g., [5]). The basic idea is that not all access control decisions are identity-based. For instance, information about a user's current role (e.g., doctor) or a user's date of birth may be more important than the user's identity for deciding whether an access request should be granted.

In the remainder of this section, we first review the basic concepts about attribute and credential-based access control and then describe some solutions based on such a model.

47.6.1 Basic Elements of the Model

Attribute-based access control differs from traditional discretionary access control since both the subject and the object appearing in an authorization are replaced by a set of *attributes* associated with them. Such attributes may correspond to an identity or a non-identifying characteristic of a user (e.g., date of birth, nationality) and to metadata associated with an object that provide additional context information (e.g., data of creation). In particular, the attributes associated with a user may be specified by the user herself (*declarations*), or may be substantiated by *digital certificates* or *credentials* (e.g., [12,57]). Metadata associated with resources/services can be in different form (e.g., textual or semistructured data). By analyzing previous works in the area, we identify the following main concepts captured by attribute/credential-based access control models.

Authority: An authority is an entity responsible for producing and signing certificates. A party may accept certificates issued by an authority that it trusts or that has been (directly or indirectly) delegated by an authority that it trusts.

Certificate: A certificate is a statement certified by an authority trusted for making such a statement. Each certificate is characterized by the identity of the issuer, the identity of the user for which the certificate has been issued, a validity period, a signature of the issuing authority, and a set of certified attributes. Certificates can be classified according to different dimensions. Since we focus on the use of certificates in access control, we distinguish between *atomic* and *non-atomic* certificates [7]. Atomic certificates (e.g., X.509) can only be released as a whole, meaning that all attributes in a certificate are disclosed. These certificates are the most common type of certificates used today in distributed systems. Although these certificates can only be released as a whole, there is usually the possibility

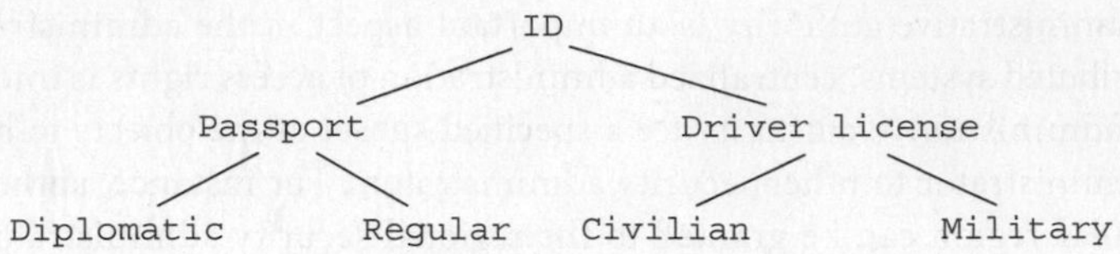

FIGURE 47.7 An example of certificate abstractions.

to refer to specific attributes within a certificate for querying purpose. Given a certificate c including a set $\{a_1,\ldots,a_m\}$ of attributes, we can use the dot notation to refer to a given attribute in c. For instance, given credential `Passport` certifying attributes `name`, `dob`, and `country`, `Passport.name` denotes attribute `name` certified by the `Passport` credential. Non-atomic certificates (e.g., Idemix [15]) allow the selective release of the attributes certified by them. Non-atomic certificates are based on technologies that also permit to certify the possession of a given certificate without disclosing the attributes within it. Abstractions can be defined within the domain of certificates. Figure 47.7 illustrates an example of certificate abstractions. The use of abstractions in the policy specification provides a compact and easy way to refer to complex concepts. For instance, the specification of abstraction `Passport` in a policy states that any kind of passport (diplomatic or regular) can be accepted. Abstractions can be formally modeled through a hierarchy $\mathcal{H} = (\mathcal{C}, \prec)$, with $\mathcal{C}$ a set of certificate abstractions and $\prec$ a partial order relationship over $\mathcal{C}$. Given two certificate abstractions c_i and c_j, $c_i \prec c_j$ if c_j is an abstraction of c_i.

An important concept captured by several credential-based access control models is *delegation*. The delegation is the ability of an authority to produce credentials on behalf of the delegator. Delegation increases flexibility and permits the inexpensive creation of credentials, particularly in an open environment. A *delegation certificate* issued by an authority states that it trusts another authority for issuing certificates that include specific attributes. An authority can delegate other authorities only on given attributes (e.g., a hospital can issue a certificate delegating physicians to certify specific properties of patients) or can give unrestricted delegation to other authorities.

Low-level issues related to the certificate creation, retrieval, validation, and revocation are all usually assumed to be managed by an underlying implementation of the certificate management system (e.g., [38,41,44]) and are therefore outside the scope of this chapter.

Policy: A policy defines the rules (authorizations) regulating access to resources. Such authorizations model access restrictions based on generic properties associated with subjects and objects. In general, these restrictions can refer to attributes within specific credentials or can refer to certificate abstractions. In this latter case, a restriction involving a certificate abstraction applies also to its specialized abstractions/credentials. For instance, with respect to the certificate abstractions in Figure 47.7, `ID.dob` represents attribute `dob` certified by a credential of type `ID`. From an analysis of the current attribute and credential-based access control policies, it is easy to see that at an abstract level the main elements of an authorization that are common to many proposals are the following:

- *Subject expression*: A subject expression identifies a set of subjects having specific attributes. A subject expression can be seen as a boolean formula of basic conditions defined on attributes. Attributes appearing in basic conditions must be certified or can be declared by a user. For instance, expression `ID.dob>01-05-1971` denotes all users with a certificate of type `ID` certifying that the date of birth of the users is after January 05, 1971.
- *Object expression*: An object expression identifies the resources/services to be protected. Like for subject expressions, also an object expression can be seen as a boolean formula of basic conditions defined on the metadata associated with resources/services. For instance, assume that `producer` is a metadata attribute associated with objects. Then, expression `producer="EU"` denotes all objects made in an European country.
- *Action*: An action denotes the operation (or group thereof) to which the authorization refers.

Different languages have been developed for the specification of policies, and each of them supports different features. In the remainder of this section, we first describe the policy communication problem (Section 47.6.2), which is a specific problem of the attribute-based access control systems, and then we present some policy languages (Section 47.6.3). In particular, we present logic-based languages, which are expressive but turn out to be not applicable in practice, and then XACML-based languages, which are easy to use and consistent with consolidated technology.

47.6.2 Policy Communication

A peculiarity of access control systems based on attributes is that the server offering resources/services evaluates the policies without a complete knowledge of users and their properties. The server has then to communicate to users the policies that they should satisfy to have their requests possibly permitted. For instance, consider a service accessible to all people older than 18 and working in an European country. In this case, the server has to communicate to the user that she has to provide her date of birth and the place where she works before accessing the service. This policy communication problem has been under the attention of the research and development communities for more than a decade and several solutions have been proposed, each addressing different issues (e.g., [29,38]). A simple policy communication strategy consists in giving the user a list with all the possible sets of certificate that would allow the access. This solution is not always applicable due to the large number of possible alternatives (e.g., with compound credential requests such as "a passport and one membership certificate from a federated association," there may be a combinatorial explosion of alternatives). Automated trust negotiation strategies have been therefore proposed (e.g., [38,55,56,59,60,61,62]), which are based on the assumption that parties may be unknown a-priori and a multi-step trust negotiation process is necessary for communicating policies and for releasing certificates, which are both considered sensitive. The goal of a trust negotiation process is to gradually establish trust among parties by disclosing credentials and requests for credentials. In successful trust negotiations, credentials eventually are exchanged so that the policy regulating access to the required resource/service is satisfied. In [55] two different strategies are described: *eager* and *parsimonious* credential release strategies. Parties applying the first strategy communicate all their credentials if the release policy for them is satisfied, without waiting for the credentials to be requested. Parsimonious parties only release credentials upon explicit request by the server (avoiding unnecessary releases). In [60] the *PRUdent NEgotiation Strategy* (PRUNES) strategy ensures that a user communicates her credentials to the server only if the access will be granted and the set of certificates communicated to the server is the minimal necessary for granting it. Each party defines a set of credential policies that regulates how and under what conditions the party releases its credentials. The negotiation consists of a series of requests for credentials and counter-requests on the basis of the parties' credential policies. In [61] the authors present a family of trust negotiation strategies, called *disclosure tree strategy* (DTS) family, and show that if two parties use different strategies from the DTS family, they are able to establish a negotiation process. In [59] the authors present a *unified schema for resource protection* (UniPro) for protecting resources and policies in trust negotiation. UniPro gives (opaque) names to policies and allows any named policy P_1 to have its own policy P_2 regulating to what parties policy P_1 can be disclosed. TrustBuilder [62] is a prototype developed to incorporate trust negotiation into standard network technologies. Traust [38] is a third-party authorization service that is based on the TrustBuilder framework for trust negotiation. The Traust service provides a negotiation-based mechanism that allows qualified users to obtain the credentials necessary to access resources provided by the involved server. In [56] the authors introduce a formal framework for automated trust negotiation, and formally define the concept of correct enforcement of policies during a trust negotiation process. In [6] the authors present an expressive and flexible approach for enabling servers to specify, when defining their access control policies, if and how the policy should be communicated to the client. Intuitively, an access control policy is represented through its expression tree and each node of the tree is associated with a disclosure policy (modeled with three colors, namely green, yellow, or red), ensuring a fine-grained support and providing expressiveness and flexibility in establishing disclosure regulations. The disclosure policies state whether an element in the policy can be released as it is or whether it has to be obfuscated. The authors also illustrate how to determine the user's view on the policy (i.e., the policy to be communicated to the user requesting access) according to the specified disclosure policies.

47.6.3 Languages for Access Control

Languages for access control aim to support the expression and the enforcement of policies [52]. Several access control languages have been developed, and most of them rely on concepts and techniques from logic and logic programming (e.g., [13,27,35,36,39,40,42]). Logic languages are particularly attractive due to their clean and unambiguous semantics, suitable for implementation validation, as well as formal policy verification. Logic languages can be expressive enough to formulate all the policies introduced in the literature, and their declarative nature yields a good compromise between expressiveness and simplicity. Nevertheless, many logic-based proposals, while appealing for their expressiveness, are not applicable in practice, where simplicity, efficiency, and consistency with consolidated technology are crucial. Effectively tackling these issues is probably the main motivation for the success of the eXtensible Access Control Markup Language (XACML) [50]. XACML is an OASIS standard that proposes an XML-based language for specifying and exchanging access control policies over the Web. The language can support the most common security policy representation mechanisms and has already found significant support by many players. Moreover, it includes standard extension points for the definition of new functions, data types, and policy combination methods, which provide a great potential for the management of access control requirements in emerging and future scenarios.

In the following, we survey some logic-based access control languages, and then describe the XACML language along with some extensions aiming at including the possibility of using credentials in the XACML policy specification.

47.6.3.1 Logic-Based Languages

The first work investigating logic languages for the specification of authorizations is the work by Woo and Lam [58]. They shown how flexibility and extensibility in access specifications can be achieved by abstracting from the low level authorization triples and adopting a high level authorization language. Their language is a many-sorted first-order language with a rule construct, useful to express authorization derivations and therefore model authorization implications and default decisions (e.g., closed or open policy). Their work has been subsequently refined by several authors (e.g., [35]).

Several logic-based languages have been developed for formulating, for example, dynamic policies, inheritance and overriding, policy composition, and credential-based authorizations (e.g., [9,11,12,35]). In particular, specific solutions have addressed the problem of constraining the validity of authorizations through periodic expressions and appropriate temporal operators. For instance, the proposal in [9] shows a temporal authorization model that supports periodic access authorizations and periodic rules, and allows the derivation of new authorizations based on the presence or absence of other authorizations in specific periods of time. Other logic-based access control languages support inheritance mechanisms and conflict resolution policies. Jajodia et al. [35] present a proposal for a logic-based language that allows the representation of different policies and protection requirements, while at the same time providing understandable specifications, clear semantics (guaranteeing therefore the behavior of the specifications), and bearable data complexity. Authorizations are specified in terms of a locally stratified rule base logic. Such a solution allows the representation of different propagation policies (i.e., policies that specify how to obtain derived authorizations from the explicit authorizations), conflict resolution policies, and decision policies that a security system officer might want to use.

The fact that in open environments there is the need for combining access control restrictions independently stated by different parties motivates the development of solutions specifically targeted to the composition of policies (e.g., [11,54]). These solutions typically do not make any assumption on the language adopted for specifying the given policies and define a set of policy operators used for combining different policies. In particular, in [12] a policy is defined as a set of triples of the form (*s*, *o*, *a*), where *s* is a constant in (or a variable over) the set of subjects $\mathcal{S}$, *o* is a constant in (or a variable over) the set of objects $\mathcal{O}$, and *a* is a constant in (or a variable over) the set of actions $\mathcal{A}$. Here, complex policies can then be obtained by combining policies via specific algebra operators.

Logic-based approaches have been also used for specifying, reasoning about, and communicating protection requirements. In particular, several proposals have addressed the problem of defining and enforcing credential-based authorization policies and trust management (e.g., [12,34,43,48,62]). In particular, in [12] the authors present a framework that includes an access control model, a language for expressing access and release policies, and a policy-filtering mechanism to identify the relevant policies for a negotiation. Access regulations are specified by logical rules, where some predicates are explicitly identified. The system is composed of two entities: the *client* that requests access, and the *server* that exposes a set of services. Abstractions can be defined on services, grouping them in sets, called *classes*. Servers and clients interact via a *negotiation process*, defined as the set of messages exchanges between them. Clients and servers have a *portfolio*, which is a collection of credentials (certified statements) and declarations (unsigned statements). Credentials are modeled as *credential expressions* of the form *credential_name(attribute_list)*, where *credential_name* is the credential name and *attribute_list* is a possibly empty list of elements of the form *attribute_name=value_term*, where *value_term* is either a ground value or a variable. The proposed framework allows a client to communicate the minimal set of certificates to a server, and the server to release the minimal set of conditions required for granting access. For this purpose, the server defines a set of *service accessibility rules*, representing the necessary and sufficient conditions for granting access to a resource. More precisely, this proposal distinguishes two kinds of service accessibility rules: *prerequisites* and *requisites*. Prerequisites are conditions that must be satisfied for a service request to be taken into consideration (they do not guarantee that it will be granted); requisites are conditions that allow the service request to be successfully granted. The basic motivation for this separation is to avoid unnecessary disclosure of information from both parties. Therefore, the server will not disclose a requisite rule until after the client satisfies a corresponding prerequisite rule. Also, both clients and servers can specify a set of *portfolio disclosure rules*, used to define the conditions that govern the release of credentials and declarations.

The rules both in the service accessibility and portfolio disclosure sets are defined through a logic language that includes a set of predicates whose meaning is expressed on the basis of the current *state*. The state indicates the parties' characteristics and the status of the current negotiation process, that is, the certificates already exchanged, the requests made by the two parties, and so on. Predicates evaluate both information stored at the site (persistent state) and acquired during the negotiation (negotiation state). Information related to a specific negotiation is deleted when the negotiation terminates. In contrast, persistent state includes information that spans different negotiations, such as user profiles maintained at web sites.

Since there may exist different policy combinations that may bring the access request to satisfaction, the communication of credentials and/or declarations could be an expensive task. To overcome this issue, the *abbreviation* predicates are used to abbreviate requests. Besides the necessity of abbreviations, it is also necessary for the server, before releasing rules to the client, to evaluate state predicates that involve private information. For instance, the client is not expected to be asked many times the same information during the same session and if the server has to evaluate if the client is considered not trusted, it cannot communicate this request to the client itself.

Communication of requisites to be satisfied by the requester is then based on a filtering and renaming process applied on the server's policy, which exploits partial evaluation techniques in logic programs [12,47]. Access is granted whenever a user satisfies the requirements specified by the filtering rules calculated by means of the original policy and the already released information.

47.6.3.2 XML-Based Languages

With the increasing number of applications that either use XML as their data model, or export relational data as XML data, it becomes critical to investigate the problem of access control for XML. To this purpose, many XML-based access control languages have been proposed (e.g., [14,20,37,50]). As already mentioned, the eXtensible Access Control Markup Language (XACML) is the most relevant XML-based access control language. XACML version 1.0 [49] has been an OASIS standard since 2003. Improvements have been made to the language and incorporated in version 3.0 [50].

XACML supports the definition of policies based on attributes associated with subjects and resources other than their identities. The attributes are assumed to be known during the evaluation time and stored in the XACML evaluation context, or presented by the requester together with the request. While XACML acknowledges that properties can be presented by means of certificates, and in fact, it has been designed to be integrated with the Security Assertion Markup Language (SAML) [2] for exchanging various types of security assertions and for providing protocol mechanisms, it does not provide a real support for expressing and reasoning about digital certificates in the specification of the authorization policies. Intuitively, XACML supports attribute-based access control but does not really support credential-based access control (see Section 47.6.3.3). XACML also supports policies independently specified by multiple authorities on the same resources. When an access request on that resource is submitted, the system has to take into consideration all these policies and their outcomes are combined according to a combining algorithm. Policies defined by different parties may be enforced at different enforcement points. XACML provides a method for specifying some actions, called *obligations*, that must be fulfill in conjunction with the policy enforcement. Figure 47.8 illustrates the XACML data-flow that consists of the following steps:

- The requestor sends an access request to the *Policy Evaluation Point* (PEP) module, which has to enforce the access decision taken by the decision point.
- The PEP module sends the access request to the *Context Handler* that translates the original request in a canonical format, called *XACML request context*, and sends it to the *Policy Decision Point* (PDP).
- The PDP module requests any additional *subject, resource, action, and environment* attributes from the Context Handler.
- The Context Handler requests the attributes required by the PDP module to the *Policy Information Point* (PIP) module. To this purpose, PIP interacts with the *Subjects, Resource, and Environment* modules. The *Environment* module provides a set of attributes that are relevant to take an authorization decision and are independent of a particular *subject, resource, and action.*
- The Contet Handler sends the required attributes (and possibly the resource) to the PDP module.
- The PDP then evaluates the policies made available by the *Policy Administration Point* (PAP) and returns the *XACML response context* to the Context Handler. The context handler translates the XACML response context to the native format of the PEP and returns it to the PEP together with an optional set of obligations.
- The PEP fulfills the obligations and, if the access is permitted, it performs the access. Otherwise, the PEP denies access.

The main concepts of interest in the XACML policy language are *rule*, *policy*, and *policy set.* Each XACML policy has a root element that can be either a `Policy` or a `PolicySet`. A `PolicySet` is a collection of `Policy` or `PolicySet`. A XACML policy consists of a *target*, a set of *rules*, an optional set of *obligations*, an optional set of *advices*, and a *rule combining algorithm*. We now describe these components more in details.

- *Target*: It consists of a simplified set of conditions for the *subject*, *resource*, and *action* that must be satisfied for a policy to be applicable to a given request. Note that the definition of the subjects, resources, and actions in a target are based on attributes. For instance, a physician at an hospital may have the attribute of being a researcher, a specialist in some field, or many other job roles. According to these

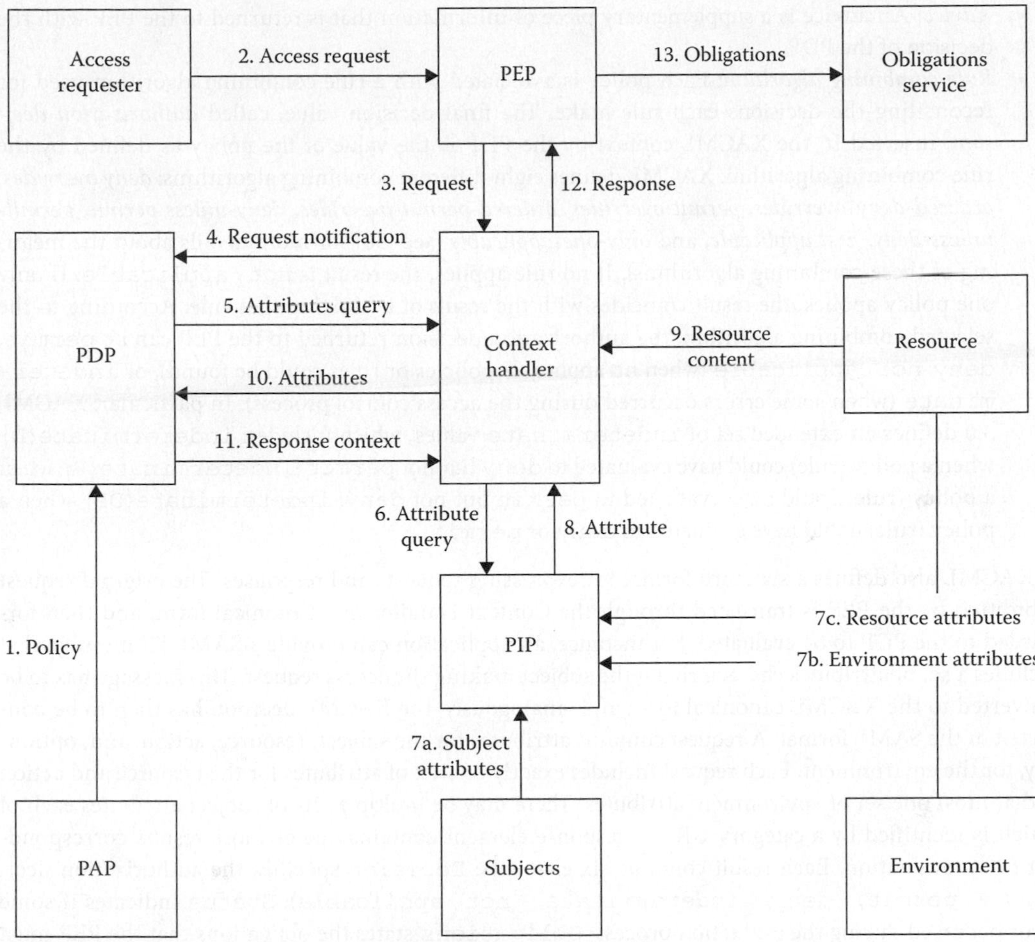

FIGURE 47.8 XACML overview. (From OASIS, *extensible Access Control Markup Language (XACML) Version 3.0*, 2010. http://www.oasis-open.org/committees/xacml.)

attributes, the physician can be able to perform different functions within the hospital. If all the conditions of a `Target` are satisfied, its associated `Policy` (or `Policyset`) applies to the request. If a policy applies to all entities of a given type, that is, all subjects, actions, or resources, an empty element, named `AnySubject`, `AnyAction`, `AnyResource`, respectively, is used.

- *Rule*: The components of a rule are a *target*, an *effect*, a *condition*, *obligation expressions*, and *advice expressions*. The target defines the set of resources, subjects, and actions to which the rule applies. The effect of the rule can be `permit` or `deny`. The condition represents a boolean expression that may further refine the applicability of the rule. Note that the `target` element is an optional element: a rule with no target applies to all possible requests. Obligation and advice expressions are evaluated and they can be returned to the PEP in the response context. Note that while obligations cannot be ignored, advices can be safely ignored by the PEP.
- *Obligation*: An obligation is an operation that has to be performed in conjunction with the enforcement of an authorization decision. For instance, an obligation can state that all accesses on medical data have to be logged. Obligations are returned by the PDP to the PEP along with the response. Note that, only policies that are evaluated and have returned a response of `permit` or `deny` can return obligations. This means that if a policy evaluates to `indeterminate` or `not applicable`, the associated obligations are not returned to the PEP.

- *Advice*: An advice is a supplementary piece of information that is returned to the PEP with the decision of the PDP.
- *Rule combining algorithm*: Each policy is associated with a rule combining algorithm used for reconciling the decisions each rule make. The final decision value, called *authorization decision*, inserted in the XACML context by the PDP is the value of the policy as defined by the rule combining algorithm. XACML defines eight different combining algorithms: *deny overrides*, *ordered-deny-overrides*, *permit overrides*, *ordered-permit-overrides*, *deny-unless-permit*, *permit-unless-deny*, *first applicable*, and *only one-applicable* (see [50] for more details about the meaning of these combining algorithms). If no rule applies, the result is `not applicable`. If only one policy applies, the result coincides with the result of evaluating that rule. According to the selected combining algorithm, the authorization decision returned to the PEP can be `permit`, `deny`, `not applicable` (when no applicable policies or rules could be found), or `indeterminate` (when some errors occurred during the access control process). In particular, XACML 3.0 defines an extended set of `indeterminate` values, which includes: `indeterminate{D}` when a policy (rule) could have evaluated to `deny` but not `permit`; `indeterminate{P}` when a policy (rule) could have evaluated to `permit` but not `deny`; `indeterminate{DP}` when a policy (rule) could have evaluated to `deny` or `permit`.

XACML also defines a standard format for expressing requests and responses. The original request submitted by the PEP is translated through the Context Handler in a canonical form, and then forwarded to the PDP to be evaluated. For instance, an application can provide a SAML [2] message that includes a set of attributes characterizing the subject making the access request. This message has to be converted to the XACML canonical form and, analogously, the XACML decision has then to be converted to the SAML format. A request contains attributes for the subject, resource, action, and, optionally, for the environment. Each request includes exactly one set of attributes for the resource and action and at most one set of environment attributes. There may be multiple sets of subject attributes each of which is identified by a category URI. A response element contains one or more results corresponding to an evaluation. Each result contains six elements: `Decision` specifies the authorization decision (i.e., `permit, deny, indeterminate, not applicable`); `Status` indicates if some error occurred during the evaluation process; `Obligations` states the obligations that the PEP must fulfill; `AssociatedAdvice` is optional and reports a list of advices that provide additional information to the PEP; `Attributes` is optional and contains a list of attributes that were part of the request; `PolicyIdentifierList` is optional and corresponds to a list of policy or policy set identifiers that have been applied to a request.

47.6.3.3 Expanding XACML with Credentials

Although designed to be integrated with the Security Assertion Markup Language (SAML) [2] for exchanging security assertions and providing protocol mechanisms, XACML lacks a real support for considering, reasoning, and expressing conditions on certified properties. Recent proposals have tried to overcome this and other limitations that make XACML not yet suitable for open Web-based systems. In particular, the novel features that should be supported by a practical access control language can be summarized as follows [4]:

- *Certified information*: The attributes used in the XACML policies are assumed to be known during the evaluation time and stored within the XACML context or presented by the user together with the access request. To represent and manage credentials in XACML it is then necessary to express the fact that some attributes should be presented through given certificates, possibly imposing conditions on the value of these attributes and on the certificates themselves.
- *Abstractions*: Intuitively, abstractions represent a shorthand by which a single concept is introduced to represent a more complex one (e.g., a set, a disjunction, or a conjunction of concepts). For instance, `ID` (abstraction head) can be defined as an abstraction for any element in set

{Passport, DriverLicense} of credentials (abstraction tail). A policy specifying that an access requester must provide an ID can then be satisfied by presenting any of the two credentials above.

- *Recursive conditions*: The support for recursive reasoning allows the specification of policies based on chains of credentials and of conditions on data with a recursive structure.
- *Dialog*: The introduction of dialog between the involved parties has the advantages that the server can communicate which information is needed to evaluate an access control policy, and a user can release only the necessary credentials instead of releasing the whole set. A further advantage is that it permits to tackle the issue of the privacy trade-off between providing the whole set of credentials (on the access requester side) and disclosing the whole access control policy (on the server side). The proposal in [4] obtains this result by attaching a disclosure attribute to each condition in an access control policy. This attribute indicates what type of disclosure policy is associated with the condition, and it is enforced by hiding from the access requester the information that cannot be released according to such a disclosure policy. The more (less) of an access control policy is disclosed, the smaller (bigger) is the quantity of information in terms of released credentials that will have to be provided by the user.

The proposal in [4] illustrates how certified information, abstractions, recursive reasoning, and dialog management can be deployed in XACML. In particular, it shows that the integration of XACML with XQuery can be adopted for supporting abstractions and recursive conditions. Credentials and dialog management require a minimal change in the XACML language that consists in the addition of appropriate elements and attributes. Other proposals (e.g., [17,18,32,46]) provide XACML extensions to support trust negotiation.

47.7 Conclusions

In this chapter we introduced the most important concepts related to access control. We first described the discretionary, mandatory, and role-based access control policies, and then we illustrated recent proposals in the area of access control models and languages. In particular, we described novel approaches based on digital certificates, which are more suitable for open scenarios where servers offering services and users requesting such services do not know each other. We also provided an overview of logic-based and XML-based access control languages.

Key Terms

Access control: A process that controls every request to a system and determining, based on specified authorizations, whether the request should be granted or denied.
Access matrix: A matrix representing the set of authorizations defined at a given time in the system.
ACL: Access Control List.
Administrative policy: A policy regulating who can modify the allowed accesses.
Authorization: The right granted to a user to exercise an action (e.g., read, write, create, delete, and execute) on certain objects.
Certificate: A statement certified by an authority trusted for making such a statement.
DAC: Discretionary Access Control.
MAC: Mandatory Access Control.
Obligation: An action that must be performed sometime to allow the execution of a given action.
PDP: Policy Decision Point.
PEP: Policy Evaluation Point.
PIP: Policy Information Point.
RBAC: Role Based Access Control.

Role: A job function within an organization that describes the authority and responsibility related to the execution of an activity.
Security mechanism: Low-level software and/or hardware functions that implement security policies.
Security policy: High-level guidelines establishing rules that regulate access to resources.
XACML: eXtensible Access Control Markup Language.

References

1. G. Ahn and R. Sandhu. Role-based authorization constraints specification. *ACM Transactions on Information and System Security (TISSEC)*, 3(4):207–226, November 2000.
2. E. Rissanen and H. Lockhart (eds.). *SAML 2.0 profile of XACML*. Version 2.0, OASIS, August 2010. http://docs.oasis-open.org/xacml/3.0/xacml-profile-saml2.0-v2-spec-cs-01-en.pdf
3. *ANSI/INCITS 359 American National Standard for Information Technology—Role Based Access Control*, 2004.
4. C. Ardagna, S. De Capitani di Vimercati, S. Paraboschi, E. Pedrini, P. Samarati, and M. Verdicchio. Expressive and deployable access control in open web service applications. *IEEE Transactions on Service Computing (TSC)*, 4(2):96–109, April–June 2011.
5. C.A. Ardagna, M. Cremonini, S. De Capitani di Vimercati, and P. Samarati. A privacy-aware access control system. *Journal of Computer Security (JCS)*, 16(4):369–392, 2008.
6. C.A. Ardagna, S. De Capitani di Vimercati, S. Foresti, G. Neven, S. Paraboschi, F.-S. Preiss, P. Samarati, and M. Verdicchio. Fine-grained disclosure of access policies. In *Proceedings of the 12th International Conference on Information and Communications Security (ICICS 2010)*, Barcelona, Spain, December 2010.
7. C.A. Ardagna, S. De Capitani di Vimercati, S. Foresti, S. Paraboschi, and P. Samarati. Minimising disclosure of client information in credential-based interactions. *International Journal of Information Privacy, Security and Integrity (IJIPSI)*, 1(2/3):205–233, 2012.
8. M. Barni, T. Bianchi, D. Catalano, M. Di Raimondo, R. Donida Labati, P. Failla, R. Lazzeretti, V. Piuri, F. Scotti, and A. Piva. Privacy-preserving fingercode authentication. In *Proceedings of 12th ACM Workshop on Multimedia and Security (MM&Sec 2010)*, Rome, Italy, September 2010.
9. E. Bertino, C. Bettini, E. Ferrari, and P. Samarati. An access control model supporting periodicity constraints and temporal reasoning. *ACM Transactions on Database Systems (TODS)*, 23(3):231–285, September 1998.
10. E. Bertino, P. Samarati, and S. Jajodia. Authorizations in relational database management systems. In *Proceedings of the First ACM Conference on Computer and Communications Security (CCS 93)*, pp. 130–139, Fairfax, VA, November 1993.
11. P. Bonatti, S. De Capitani di Vimercati, and P. Samarati. An algebra for composing access control policies. *ACM Transactions on Information and System Security (TISSEC)*, 5(1):1–35, February 2002.
12. P. Bonatti and P. Samarati. A uniform framework for regulating service access and information release on the web. *Journal of Computer Security (JCS)*, 10(3):241–271, 2002.
13. P. Bonatti and P. Samarati. Logics for authorizations and security. In J. Chomicki, R. van der Meyden, and G. Saake, eds., *Logics for Emerging Applications of Databases*. Springer-Verlag, Berlin, Germany, 2003.
14. S. Bajaj et al. *Web services policy framework (WS-Policy)* Version 1.2, April 2006. http://www.w3.org/Submission/WS-Policy/
15. J. Camenisch and A. Lysyanskaya. An efficient system for non-transferable anonymous credentials with optional anonymity revocation. In *Proceedings of International Conference on the Theory and Application of Cryptographic Techniques (EUROCRYPT 2001)*, Innsbruck, Austria, May 2001.
16. S. Castano, M.G. Fugini, G. Martella, and P. Samarati. *Database Security*. Addison Wesley, Reading, MA, 1994.

17. D.W. Chadwick, S. Otenko, and T.A. Nguyen. Adding support to XACML for dynamic delegation of authority in multiple domains. In *Proceedings of 10th Open Conference on Communications and Multimedia Security (CMS 2006)*, Heraklion, Greece, October 2006.
18. V.S.Y. Cheng, P.C.K. Hung, and D.K.W. Chiu. Enabling web services policy negotiation with privacy preserved using XACML. In *Proceedings of the 40th Annual Hawaii International Conference on System Sciences (HICSS 2007)*, Waikoloa, HI, January 2007.
19. S. Cimato, M. Gamassi, V. Piuri, and F. Scotti. Privacy-aware biometrics: Design and implementation of a multimodal verification system. In *Proceedings of the Annual Computer Security Applications Conference (ACSAC 2008)*, Anaheim, CA, December 2008.
20. E. Damiani, S. De Capitani di Vimercati, S. Paraboschi, and P. Samarati. A fine-grained access control system for XML documents. *ACM Transactions on Information and System Security (TISSEC)*, 5(2):169–202, May 2002.
21. S. De Capitani di Vimercati, S. Foresti, S. Jajodia, S. Paraboschi, and P. Samarati. A data outsourcing architecture combining cryptography and access control. In *Proceedings of the 1st Computer Security Architecture Workshop (CSAW 2007)*, Fairfax, VA, November 2007.
22. S. De Capitani di Vimercati, S. Paraboschi, and P. Samarati. Access control: Principles and solutions. *Software—Practice and Experience*, 33(5):397–421, April 2003.
23. S. De Capitani di Vimercati and P. Samarati. Access control in federated systems. In *Proceedings of the ACM SIGSAC New Security Paradigms Workshop (NSPW 96)*, Lake Arrowhead, CA, September 1996.
24. S. De Capitani di Vimercati and P. Samarati. Authorization specification and enforcement in federated database systems. *Journal of Computer Security (JCS)*, 5(2):155–188, 1997.
25. S. De Capitani di Vimercati, P. Samarati, and S. Jajodia. Hardware and software data security. In D. Kaeli and Z. Navabi, eds., *EOLSS The Encyclopedia of Life Support Systems*. EOLSS Publishers, Oxford, U.K., 2001.
26. S. De Capitani di Vimercati, P. Samarati, and S. Jajodia. Policies, models, and languages for access control. In *Proceedings of the Workshop on Databases in Networked Information Systems (DNIS 2005)*, Aizu-Wakamatsu, Japan, March 2005.
27. J. DeTreville. Binder, a logic-based security language. In *Proceedings of the 2001 IEEE Symposium on Security and Privacy*, Oakland, CA, May 2002.
28. D.F. Ferraiolo and R. Sandhu. Proposed nist standard for role-based access control. *ACM Transactions on Information and System Security (TISSEC)*, 4(3):224–274, August 2001.
29. K. Frikken, M. Atallah, and J. Li. Attribute-based access control with hidden policies and hidden credentials. *IEEE Transactions on Computer (TC)*, 55(10):1259–1270, October 2006.
30. L. Fuchs, G. Pernul, and R. Sandhu. Roles in information security — A survey and classification of the research area. *Computers and Security*, 30(8):748–769, November 2011.
31. M. Gamassi, V. Piuri, D. Sana, and F. Scotti. Robust fingerprint detection for access control. In *Proceedings of the Workshop RoboCare (RoboCare 2005)*, Rome, Italy, May 2005.
32. D.A. Haidar, N. Cuppens, F. Cuppens, and H. Debar. XeNA: An access negotiation framework using XACML. *Annales des télécommunications — Annals of Telecommunications*, 64(1/2):155–169, January 2009.
33. M.H. Harrison, W.L. Ruzzo, and J.D. Ullman. Protection in operating systems. *Communications of the SCM*, 19(8):461–471, August 1976.
34. K. Irwin and T. Yu. Preventing attribute information leakage in automated trust negotiation. In *Proceedings of the 12th ACM Conference on Computer and Communications Security (CCS 2005)*, Alexandria, VA, November 2005.
35. S. Jajodia, P. Samarati, M.L. Sapino, and V.S. Subrahmanian. Flexible support for multiple access control policies. *ACM Transactions on Database Systems (TODS)*, 26(2):214–260, June 2001.
36. T. Jim. Sd3: A trust management system with certified evaluation. In *Proceedings of the 2001 IEEE Symposium on Security and Privacy*, Oakland, CA, May 2001.

37. M. Kudoh, Y. Hirayama, S. Hada, and A. Vollschwitz. Access control specification based on policy evaluation and enforcement model and specification language. In *Proceedings of Symposium on Cryptography and Information Security, (SCIS 2000)*, Okinawa, Japan, January 2000.
38. A.J. Lee, M. Winslett, J. Basney, and V. Welch. The traust authorization service. *ACM Transactions on Information and System Security (TISSEC)*, 11(1):1–3, February 2008.
39. N. Li, B.N. Grosof, and J. Feigenbaum. Delegation logic: A logic-based approach to distributed authorization. *ACM Transactions on Information and System Security (TISSEC)*, 6(1):128–171, February 2003.
40. N. Li and J.C. Mitchell. Datalog with constraints: A foundation for trust-management languages. In *Proceedings of the 5th International Symposium on Practical Aspects of Declarative Languages (PADL 2003)*, New Orleans, LA, January 2003.
41. N. Li and J.C. Mitchell. Understanding SPKI/SDSI using first-order logic. *International Journal of Information Security*, 5(1):48–64, January 2006.
42. N. Li, J.C. Mitchell, and W.H. Winsborough. Design of a role-based trust-management framework. In *Proceedings of the IEEE Symposium on Security and Privacy*, Oakland, CA, May 2002.
43. N. Li, J.C. Mitchell, and W.H. Winsborough. Beyond proof-of-compliance: Security analysis in trust management. *Journal of the ACM*, 52(3):474–514, May 2005.
44. N. Li, W. Winsborough, and J. Mitchell. Distributed credential chain discovery in trust management. *Journal of Computer Security (JCS)*, 11(1):35–86, 2003.
45. T. Lunt. Access control policies: Some unanswered questions. In *Proceedings of the IEEE Computer Security Foundations Workshop (CSFW 88)*, Franconia, NH, June 1988.
46. U.M. Mbanaso, G.S. Cooper, D.W. Chadwick, and S. Proctor. Privacy preserving trust authorization framework using XACML. In *Proceedings of the 2006 International Symposium on World of Wireless, Mobile and Multimedia Networks (WOWMOM 2006)*, Niagara-Falls, NY, June 2006.
47. M. Minoux. Ltur: A simplified linear-time unit resolution algorithm for horn formulae and computer implementation. *Information Processing Letter (IPL)*, 29(1):1–12, 1988.
48. J. Ni, N. Li, and W.H. Winsborough. Automated trust negotiation using cryptographic credentials. In *Proceedings of the 12th ACM Conference on Computer and Communications Security (CCS 2005)*, Alexandria, VA, November 2005.
49. OASIS. *eXtensible Access Control Markup Language (XACML) Version 1.0*, OASIS, February 2003. http://www.oasis-open.org/committees/xacml
50. OASIS. *eXtensible Access Control Markup Language (XACML) Version 3.0*, OASIS, January 2013. http://www.oasis-open.org/committees/xacml
51. J. Park and R. Sandhu. The $ucon_{ABC}$ usage control model. *ACM Transactions on Information and System Security (TISSEC)*, 7(1):128–174, February 2004.
52. P. Samarati and S. De Capitani di Vimercati. Access control: Policies, models, and mechanisms. In R. Focardi and R. Gorrieri, eds., *Foundations of Security Analysis and Design*, volume 2171 of *LNCS*. Springer-Verlag, Berlin, Heidelberg, Germany 2001.
53. R.S. Sandhu. On five definitions of data integrity. In *Proceedings of the IFIP WG11.3 Working Conference on Database Security VII (DBSec 93)*, Lake Guntersville, AL, September 1993.
54. D. Wijesekera and S. Jajodia. A propositional policy algebra for access control. *ACM Transactions on Information and System Security (TISSEC)*, 6(2):286–325, May 2003.
55. W. Winsborough, K.E. Seamons, and V. Jones. Automated trust negotiation. In *Proceedings of the DARPA Information Survivability Conference and Exposition (DISCEX 2000)*, Hilton Head Island, SC, January 2000.
56. W.H. Winsborough and N. Li. Safety in automated trust negotiation. *ACM Transactions on Information and System Security (TISSEC)*, 9(3):352–390, 2006.
57. M. Winslett, N. Ching, V. Jones, and I. Slepchin. Using digital credentials on the World-Wide Web. *Journal of Computer Security (JCS)*, 5(2):255–267, 1997.

58. T.Y.C. Woo and S.S. Lam. Authorizations in distributed systems: A new approach. *Journal of Computer Security (JCS)*, 2(2,3):107–136, 1993.
59. T. Yu and M. Winslett. A unified scheme for resource protection in automated trust negotiation. In *Proceedings of the IEEE Symposium on Security and Privacy*, Berkeley, CA, May 2003.
60. T. Yu, M. Winslett, and K.E. Seamons. Prunes: An efficient and complete strategy for automated trust negotiation over the internet. In *Proceedings of the 7th ACM Conference on Computer and Communications Security (CCS 2000)*, Athens, Greece, November 2000.
61. T. Yu, M. Winslett, and K.E. Seamons. Interoperable strategies in automated trust negotiation. In *Proceedings of the 8th ACM Conference on Computer and Communications Security (CCS 2001)*, Philadelphia, PA, November 2001.
62. T. Yu, M. Winslett, and K.E. Seamons. Supporting structured credentials and sensitive policies trough interoperable strategies for automated trust. *ACM Transactions on Information and System Security (TISSEC)*, 6(1):1–42, February 2003.

48

Data Compression

Ő. Ufuk Nalbantoğlu
University of Nebraska-Lincoln

K. Sayood
University of Nebraska-Lincoln

48.1 Introduction

Data compression involves generating a compact description of information from a representation that may contain redundancies either of exposition or of interpretation. The process of generating a compact description involves understanding how information is organized in data. One way this can be done is by understanding the statistical characteristics of the data. If the data sequence is independent, we could treat the data as a sequence of independent, identically distributed (*iid*) random variables with a particular probability distribution. If the data are correlated, we can develop models for how redundancies could have been introduced into a nonredundant information stream. The study of data compression involves ways to characterize the structure present in the data and then use the characterization to develop algorithms for their compact representation. We will look at a number of different ways in which commonly used sources of data, such as text, speech, images, audio, and video, can be modeled and the compression algorithms that result from these models.

If the reconstruction of the data exactly matches the original data, the compression algorithm is called a *lossless* compression algorithm. If the reconstruction is inexact, the compression algorithm is called a *lossy* compression algorithm. In our study of compression algorithms, we will make use of this natural division to organize the subject.

Before we examine the various approaches to compression, we briefly look at ways of quantifying information. We then look at some of the basic theories behind lossless and lossy compression. We begin our study of compression techniques by describing lossless and lossy compression techniques that assume the source to be from an *iid* distribution. We then examine different ways of modeling structures within the data and the algorithms that are a consequence of the different models. Finally, we conclude with some suggestions for further reading.

48.2 Entropy and Kolmogorov Complexity

In order to realize compression, the code generated by an encoding process must be shorter than the original sequence, something possible only when the original data can be described more concisely than in its original form. Fortunately, many sequences of interest can be described more concisely than with their original representations. What permits this concise description is the organization on different levels within the data. Data compression is generally applied to anthropic or scientific (that could also be considered anthropic) data that are the output of complex systems. Text data are records of natural languages, driven by grammar rules, and containing repetitive structures as a result. Still images usually capture structured objects as observed in the visible band of the electromagnetic spectrum. Motion pictures have a similar configuration as well as containing structured objects moving with bounded velocities. Computer graphics, in general, are generated by rules defined by specific programs. Audio streams are gathered from the instances of acoustic processes. Scientific data sampled from observed physical processes are governed by the laws of physics. When attempting to compress data, it is assumed that the input is not obtained from randomly unorganized media. Clearly, the more structured the data are, the higher the compression the coding system can achieve when given prior knowledge of data organization.

Let's consider the coding problem with two example sequences. Assume two binary sequences S_B and S_π are given as

$$S_B = 10101010100001101111100110000011001011101001110101 \tag{48.1}$$

$$S_\pi = 00100100001111110110101010001000100001011010001100 \tag{48.2}$$

The first sequence S_B is the first 50 outcomes of a Bernoulli trial or the results of a sequence of coin flips where the coin is fair. The second sequence is the first 50 fractional digits of the number π. To describe the first sequence, we do not have many options other than transmitting the entire sequence. However, there is more room for maneuvering when coding the second sequence. We can actually generate the first digits of π with a simple program implementing a series expansion. If $l(S_B)$ is the number of bits needed to describe S_B and $l(S_\pi)$ is the number of bits needed to describe S_π, the description length of these sequences would be

$$l(S_B) = O(n) \tag{48.3}$$

$$l(S_\pi) = l_p + \lceil \log_2 n \rceil \tag{48.4}$$

The sequence S_B is a random sequence and requires around n bits to describe. The sequence S_π can be described using a series expansion program, which can be represented using l_p bits, followed by $\lceil \log_2 n \rceil$ bits to encode the integer n. Clearly, the coding length can differ significantly based on the complexity of a sequence.

Out of all possible descriptions, the one with the shortest length is called the Kolmogorov complexity $K_U(x)$ [1] of a sequence:

$$K_U(x) = \min_{p:U(p)=x} l(p) \tag{48.5}$$

Here, U represents a universal Turing machine; and the output is x when the halting program p is executed.

By definition, a sequence x cannot be compressed to less than $K_U(x)$ bits; therefore, the Kolmogorov complexity determines the theoretical lower bound of data compression. However, there is no guarantee that the Kolmogorov complexity is computable; and even estimating $K_U(x)$ may take infinite time. The estimation of $K_U(x)$ requires a perfect understanding of the system generating a sequence, which is usually not available. Therefore, instead of a physical-model-driven (top-down) description, we use a probabilistic description (bottom-up) to obtain the lower bound for the description length. According to the probabilistic approach, a sequence is a realization of a stationary random process. Therefore, we can model it in terms of the probability distribution of the elements of the sequence. In a landmark paper in 1948 [2], Claude Shannon connected the ideas of information and randomness. Given an event A and its probability of occurrence $p(A)$, he defined the information associated with the occurrence of the event A as *self-information* that is given by

$$i(A) = \log_b \frac{1}{p(A)}$$

where the units of the information depend on the base b of the logarithm. If $b = 10$, the unit is Hartley, after the famous American electrical engineer who laid some of the foundations for what was to become information theory. If $b = e$, the base of the natural logarithm, the unit of information is *nat*. The most common base used in this context is 2; when the base 2 logarithm is used, the unit of information is *bit*.

If we treat each element of a sequence $X_1, X_2, \ldots$ as the independent outcome of an experiment where the possible outcomes come from a set $\mathcal{A}$, then we can define the *average information* as the expected value of the self-information. Shannon called this average information *entropy* and denoted it by H. Thus, if S is a *source* that emits the sequence $X_1, X_2, \ldots$, the entropy of the source is given by

$$H(S) = \sum_{X_i \in \mathcal{A}} p(X_i) \log_2 \frac{1}{p(X_i)}$$

We can show [3] that

$$\frac{1}{n} E\{K_U(X)\} \to \sum_{X_i \in \mathcal{A}} p(X_i) \log_2 \frac{1}{p(X_i)} \tag{48.6}$$

Thus, if we accept the conceit that a given sequence is a realization of a random process, we can see that the lower bound on the number of bits used to represent the sequence is given by the entropy.

We can see from the equation earlier that the expected value of the self-information for an *iid* source is the entropy. For a more general source that emits a sequence of n symbols X_i with a joint probability density function $p(X_1, X_2, \ldots, X_n)$, we can define an entropy per n-tuple as

$$H_n = -\sum_n p(X_1, X_2, \ldots, X_n) \log p(X_1, X_2, \ldots, X_n)$$

The entropy of this general source is defined as the limiting value of the entropy per n-tuple:

$$H(S) = \lim_{n \to \infty} \frac{1}{n} H_n$$

48.2.1 Lossless Compression

Lossless compression refers to a compression process in which the recovered data are identical copies of the input. Erroneous decoding of certain types of data is not tolerable. Transmitting incorrect text to the receiver might result in serious semantic mistakes. Similarly, conveying scientific data with errors might impact the conclusion inferred from the data, such as inferring mutations for DNA or protein sequences that do not actually exist. Because of the need for absolute fidelity in a wide variety of applications, lossless compression algorithms form a significant part of the data compression field.

Data compression involves data processing performed on computers and data transmission via digital communication systems. Therefore, both the data to be encoded and the compressed streams are discrete signals with quantized values. Within this framework, it is possible to view the data to be compressed as a sequence of symbols from a finite set. The set of possible symbols A is called the *alphabet* of the source S, and elements of this set are referred to as *letters*. Alphabets vary according to data types. For text documents with ASCII characters, the alphabet consists of 128 letters (the printable ASCII characters). Grayscale images encoded using 8 bits/pixel have an alphabet size of 256, the values 0–255. A sequence $\{X_1, X_2, \ldots\}$, $X_i \in A$ is generated by the source S. During the encoding process, a sequence is mapped to a binary string called the *code*. Binary strings corresponding to the mapping of each letter in the alphabet are referred to as *codewords*.

Lossless compression is achieved by an invertible coding process bijectively mapping the input sequence to a more compact code stream. Bijective coding imposes the requirement that the encoder generate a unique code for each sequence. In order to attain compression, the code has to have a length shorter than the input sequence. This performance of the compression algorithm is usually measured by the *compression ratio* and the *rate*. The compression ratio is the ratio of the number of bits used to represent the original sequence to the the number of bits used for the compressed representation. For example, if a grayscale image of 800 × 600 pixels, which is encoded using an alphabet of size 256, is stored as a 240,000 byte file, we say that the corresponding compression ratio is 2:1. The rate of the compression algorithm is usually defined as the number of bits per symbol. For the image example described earlier, the average number of bits required to represent a pixel is 4 bits. In other words, the rate is 4 bits/symbol.

48.2.2 Lossy Compression

Lossless compression involves an invertible mapping of data to a shorter compressed string. While this is an essential requirement when information loss cannot be tolerated, a majority of data compression applications do not require perfect reconstruction. The compression techniques where the reconstruction is an approximation of the original data are called *lossy compression* algorithms.

Preference of lossy compression over lossless compression generally depends on the application. One, and perhaps the most significant, motivation for lossy compression is the limited and inferential nature of human perception. Compressing natural images or audio sequences does not require perfect representation of these sequences because they are to be perceived by the limited capabilities of human visual or auditory systems. The human visual system is known to be able to perceive an electromagnetic frequency band between red and violet frequencies. It has higher sensitivity to the intensity of light than to its color components. Moreover, spatial details are usually filtered out; and more attention is paid to the smoother variations. Image and video coding systems are designed to take advantage of all these aspects of human perception, allowing us to discard "unnecessary" data components, focusing resources on those aspects that are important for perception. Similarly, our auditory system can hear the frequency band of 20 Hz–20 kHz. The response in this band is nonlinear, with a greater sensitivity to higher frequencies. Other effects, such as masking frequencies, also play a role in hearing. Using the same rationale as for still or motion picture coding, various "unnecessary" audio components can be discarded.

Beyond that natural reduction that is generally imperceptible, perceptible loss may also need to be tolerated due to economic or physical constraints. Data compression is needed for either storage or transmission over communication systems. Bandwidth and disk spaces are limited that can result in the need for more compression if the demand for these resources is greater than the supply. An example of this is telephone communication in which lower quality, resulting from the low coding rates, may be satisfactory as long as the message is delivered.

A lossy coding scheme in which each codeword can represent more than one element from the original data enables coding at rates lower than the source entropy. The price is the difference between the original data and the reconstructed sequence generally referred to as *distortion*. Obviously, a trade-off between the quality and the rate occurs. However, *quality* is subjective; and it is difficult to define a distortion measure that quantifies the approximations of data based on their quality. On the other hand, simple mathematical definitions of distortion can be useful for studying the limits and opportunities provided by the quality-rate trade-off. *Rate distortion theory* describes the theoretical limits of achievable rates for a given amount of distortion.

48.3 Modeling Data as an *iid* Sequence

We begin with the simplest of models where the data are assumed to be *iid* and are described by a probability model. There has been considerable effort spent on developing compression schemes based on such models. The reason for this is twofold. There are, at times, data that really are *iid* and are well compressed by these algorithms. When the data sequence is not *iid* and can only be modeled by a complex model, it is convenient to use the model to obtain a transformation resulting in a *transformed* or *residual* sequence that is *iid* and then compress the *iid* sequence. In fact most compression systems can be seen as consisting of two stages: (1) a stage of redundancy removal or transformation resulting in the generation of an *iid* sequence followed by (2) a coding of the *iid* sequence thus generated. We look at how to most efficiently represent *iid* sequences both in the case of lossless compression and in the case of lossy compression.

48.3.1 Lossless Compression of *iid* Sequences

We would like to design lossless compression algorithms that promise rates close to the entropy. However, the rate is not the only criterion that needs to be considered. First, a code generated by an encoder must be *uniquely decodable*, so that the decoder will not reconstruct an erroneous sequence. It seems this requirement could be satisfied if every letter in an alphabet was mapped to a unique codeword. However, this is not the case. For example, consider a code in which the letters $\{a, b, c, d\}$ are coded with the codewords {0, 100, 01, 00}, respectively. Each letter has a unique codeword associated with it; however, a code 0100 could be interpreted as either *ab* or *cd* as the sequence of codewords is not uniquely decodable. Mitigation of this problem requires that the code should have only one parsing. One way in which we can guarantee that we have a unique parsing is to make sure that no codeword is a prefix of another codeword. We can check to see if a codeword is a prefix of another codeword by representing the code in the form of a tree. For a code in which no codeword is a prefix of another codeword, called a *prefix code*, the codewords are always assigned to the leaves of the code tree. In Figure 48.1, we show two four-letter codes: {0, 100, 01, 00} and {00, 01, 100, 101}, where the filled-in circles denote codewords. For the first code, we can see that the codeword 0 is a prefix to the codewords 01 and 00. In the case of the second code, all codewords are assigned to the leaves of the tree; and hence, no codeword is a prefix to another codeword. In decoding a sequence encoded using the prefix code, we decode a letter whenever we reach a leaf of the code tree. Clearly, a prefix code will give us a uniquely decodable code. However, while the fact that a code is a prefix code guarantees that the code is uniquely decodable, it is not a necessary condition. Consider the code {0, 01, 11}. This is not a prefix code; however, we can show that this code is uniquely decodable [4]. We are faced with what seems to be a conundrum. If we restrict our search for codes to prefix codes, we can guarantee that the code will be uniquely decodable. However, could we find lower-rate codes if we did not restrict ourselves to prefix code? Happily, the answer is no.

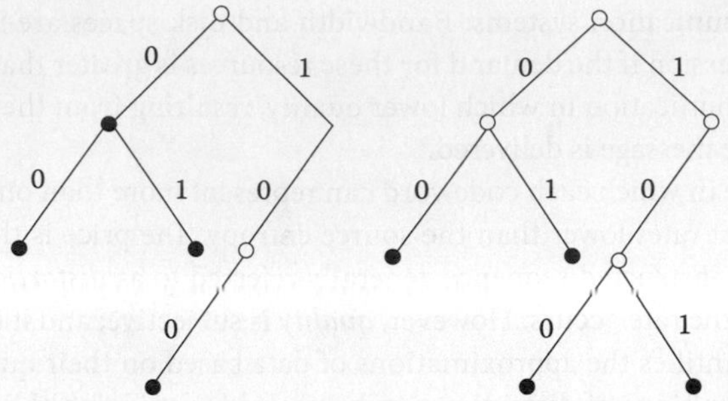

FIGURE 48.1 Code trees for two different codes. The tree on the right corresponds to a prefix code.

To show this, we will use the Kraft–McMillan inequality [4]. Consider a code over an alphabet of m letters, with codeword lengths $l_1, l_2,\ldots, l_m$. If the code is uniquely decodable, the following inequality holds:

$$\sum_i 2^{-l_i} \le 1 \tag{48.7}$$

Furthermore, we can show that given a set of integers $l_1, l_2,\ldots, l_m$ that satisfy this inequality, we can always build a prefix code with codeword lengths $l_1, l_2,\ldots, l_m$. Hence, if a code is uniquely decodable, the lengths of the codewords satisfy the Kraft–McMillan inequality; and if the lengths satisfy the Kraft–McMillan inequality, we can generate a prefix code with those lengths. Thus, for every uniquely decodable code, we can always generate a prefix code with the same average length. This being the case, restricting ourselves to prefix codes does not cost us anything in terms of the average length of the code.

The Kraft–McMillan inequality is also useful for determining bounds on the length of uniquely decodable codes. If the probability of observing the letter i is p_i, then the average codeword length is

$$L = \sum_i p_i l_i \tag{48.8}$$

We want to pick $\{l_i\}$ that minimize L while satisfying the Kraft–McMillan inequality. We can do that by finding $l_1, l_2,\ldots, l_m$ that minimize the Lagrangian

$$J = \sum_i p_i l_i + \lambda \sum_i 2^{-l_i} \tag{48.9}$$

Taking the derivative of J with respect to l_i and setting it equal to 0, we obtain

$$p_i = \lambda 2^{-l_i} \ln 2 \tag{48.10}$$

Substituting this expression in the Kraft–McMillan inequality, we can solve for l_i and get

$$l_i = \log_2 \frac{1}{p_i} \tag{48.11}$$

This length is not restricted to be an integer. If we want integer codeword lengths, we can simply use the ceiling function

$$l_i = \left\lceil \log_2 \frac{1}{p_i} \right\rceil \tag{48.12}$$

(Note that we use the ceiling rather than the floor function in order not to violate the Kraft–McMillan inequality.) Using these codeword lengths, we can find the upper and lower bounds on the average length L:

$$L = \sum_i p_i \left\lceil \log_2 \frac{1}{p_i} \right\rceil \tag{48.13}$$

$$\geq \sum_i p_i \log_2 \frac{1}{p_i} \tag{48.14}$$

$$= H \tag{48.15}$$

Thus, the average length is always greater than or equal to the entropy. We can also obtain an upper bound on the average length:

$$L = \sum_i p_i \left\lceil \log_2 \frac{1}{p_i} \right\rceil \tag{48.16}$$

$$\leq \sum_i p_i \left(\log_2 \frac{1}{p_i} + 1 \right) \tag{48.17}$$

$$= H + 1 \tag{48.18}$$

Therefore, the expected rate of an optimal prefix code is in the 1-bit neighborhood of the source entropy, while it cannot get better than the entropy.

48.3.1.1 Huffman Coding

We have seen that prefix codes can achieve rates close to entropy, in cases where the codeword selection is arranged optimally (i.e., with length close to $-\lfloor \log_2 p_i \rfloor$). A constructive way of building optimal prefix codes is the Huffman coding algorithm [5,6]. This algorithm is based on two observations. First, an optimal code should have longer codewords for less probable symbols. This is a direct result of our previous discussion. Second, the codewords for the two least frequent symbols should have the same length. We can see why this is required by showing the contradictory case. Consider that in an optimal prefix code, the least probable symbol had a longer codeword than the second least probable one. If we deleted the least significant bit of the longer codeword, this shortened codeword would not conflict with another codeword (because no codeword is the prefix of another). Therefore, we obtain another uniquely decodable code with a shorter average codeword length. This is in contradiction with the initial assumption that this prefix code was optimal. Thus, the least probable two codewords must have the same length.

Huffman added a third requirement that, while not necessary, made it easier to obtain the code. The requirement was that the codewords of the two least probable symbols should only differ in the last bit. We can easily show that this requirement does not in any way reduce the optimality of the code.

The Huffman algorithm is a hierarchical codeword construction scheme, starting from the least significant bits of the least probable two codewords, building up adding more probable codewords and significant bits. We can summarize the procedure as follows:

1. Sort the symbol probabilities.
2. Halt if only one symbol exists.
3. Push bits 1 and 0 to the least probable two codewords.
4. Combine the least probable two symbols into one. GOTO 1.

Following this procedure, we allocate bits giving higher priority to the least significant symbols. This is in the spirit of the first property of optimal codes. Since we combine the least probable symbols, they will be treated together in a bit assignment step in any future iteration. As a result, they will have the same length satisfying the second property. Note that combining the symbols that differ in the last bits will not let any codeword be the prefix of another, complying with the definition of prefix codes. Let's perform Huffman coding on an example.

We design a Huffman code for a five-letter alphabet $\mathcal{A} = \{a_1, a_2, a_3, a_4, a_5\}$ with probabilities {.3, .1, .2, .25, .15}. The two letters with the lowest probabilities are a_5 and a_2. We assign a zero as the least significant bit for a_5 and a one as the least significant bit for a_2. We then combine these two letters giving us a letter with a combined probability of .25. We can re-sort this reduced set in one of two ways, $\{a_1, a_4, [a_5a_2], a_3\}$ or $\{a_1, [a_5a_2], a_4, a_3\}$. Depending on which sorting we use, we will get a different code; however, both sortings will result in a code with minimum length, in other words, a Huffman code. In this example, we will use the latter sorting. At this point, the two lowest probability elements in our set are a_3 with a probability of 0.2 and a_4 with the probability of 0.25. We push a zero into the least significant bit for a_4 and a one into the least significant bit for a_3 and continue. The complete process can be represented as in Figure 48.2. Working from the root to the leaves, we obtain the following code:

a_1	10
a_2	111
a_3	01
a_4	00
a_5	110

We have seen that Huffman coding is optimal given the alphabet and the source distribution. However, the Huffman code is not unique. The Huffman coding algorithm finds one of multiple optimal assignments. It is easy to see that Huffman coding is not unique, since other optimal codes can be obtained by inverting bits or swapping the codes of the least probable symbols.

The Huffman coding algorithm requires that the alphabet distribution be known. In practice, the distribution for a source may not be available *a priori* unless the operations are performed offline. For such applications, the algorithm can be made adaptive by using a binary tree update procedure [7]. The leaves represent the codewords and are weighted according to the observing frequency of a symbol, so that the algorithm can keep track of the distribution statistics. The decoder follows the same procedure as the encoder, so that at any instance both the transmitter and the receiver have the same codewords.

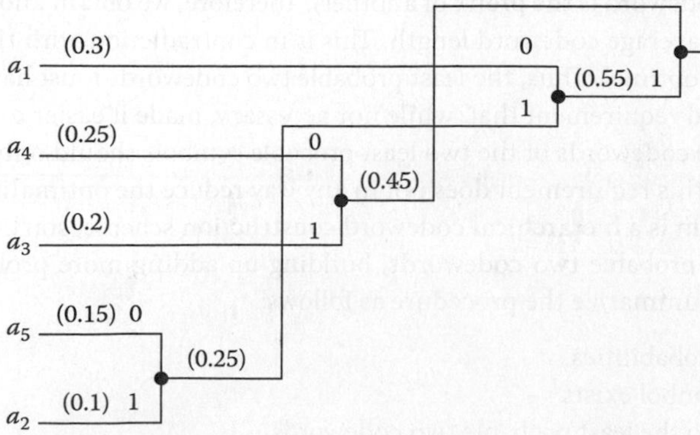

FIGURE 48.2 Construction of a Huffman code tree.

A rate bound tighter than the bound derived for optimal codes can be derived for Huffman codes. It can be shown that the upper rate bound for a Huffman code is $H(X)$ + arg $\max_i P(X_i)$ + 0.086 [7]. This guarantees rates closer to the source entropy for an alphabet distribution in which all letters have small probabilities. However, the code might be inefficient when the probability distribution is highly skewed and the maximum probability is large. This is the case for binary sources such as bilevel documents. Extending the alphabet to blocks of symbols is one way of mitigating this dominant symbol problem. Instead of encoding the source output one symbol at a time, we group the output into blocks and treat the blocks as letters from a larger alphabet. If we block n symbols together, the upper bound on the number of bits needed to encode a block of n symbols is given by $H(X_1, X_2,\ldots, X_n)$ + max $P(X_1, X_2,\ldots, X_n)$ + 0.86. However, for an *iid* sequence,

$$H(X_1, X_2,\ldots, X_n) = nH(X) \tag{48.19}$$

and

$$\max P(X_1, X_2,\ldots, X_n) = \max_i P(X_i)^n \tag{48.20}$$

Therefore, the upper bound on the average number of bits needed to encode a single symbol becomes $H(X) + \max_i P(X_i)^n/n + 0.086/n$. The cost is an increase in the number of codewords required. If the alphabet of the source is of size M, the alphabet of the block of n symbols is M^n. This exponential increase in the number of codewords required makes the approach infeasible in terms of Huffman coding for all but the smallest alphabet sizes. However, if we do not need to generate all the codewords and only need to generate the codeword of the sequence being encoded, this cost is no longer a cost. This is precisely the case for the next coding technique we discuss, namely, arithmetic coding.

48.3.1.2 Arithmetic Coding

The result that a better code (with a rate closer to the source entropy) might be available by extending the alphabet to the blocks of symbols is very promising. This result is especially significant when the source alphabet is small (e.g., bilevel images with binary alphabet). The upper bound of the rate asymptotically converges to the source entropy as the block size increases. Obviously, the best case would be where the block size is equal to the sequence length. However, in practice, this will cause the problem of an exponentially growing alphabet size. Note that the Huffman coding procedure assigns optimal codewords to each letter in the alphabet, whereas in the case of the block length being equal to the sequence length, only one of them, the codeword corresponding to the sequence itself, is observed.

Arithmetic coding offers a solution to the problem of exponentially increasing alphabet size by generating a code only for the observed sequence. In order for this code to be optimal, two fundamental aspects of optimal codes must be preserved. First, the length of the code must be inversely proportional to the observed probability of the symbol (or sequence in this case). This length should be an integer close to $-\log_2 p(S_i)$ where $p(S_i)$ is the probability of observing the sequence. Second, the code must be uniquely decodable. In cases where the entire sequence is mapped into a single codeword, this means different sequences should not be mapped onto the same codeword.

The fundamental idea of arithmetic coding is that every sequence can be mapped to an arbitrary interval on the real number line. The binary code for the sequence is the binary representation of a number in the interval to which the sequence is mapped. In practice, we take this interval to be [0,1). Without loss of generality, we can consider the case of finite sequences of the same length. The number of possible sequences of length n generated from a source with the alphabet A is $m = |A|^n$. All of these sequences can be represented by m nonoverlapping intervals drawn from [0,1). Each interval can be tagged with the binary fractional part of a rational number drawn from it. Therefore, each sequence is tagged by a distinct code; and this satisfies the property of unique decodability. In order to satisfy the optimal coding requirement, a careful allocation of these intervals into the unit interval is needed. Assigning larger intervals to more probable sequences allows them to be represented with shorter codes, because it would take a fewer number of significant bits to locate a rational number in a larger interval. k bits are required to represent an interval of size $(1/2)^k$. Assume we pick the size of the interval to be the probability of the sequence S_i assigned to it. This interval can be tagged with k bits, where k is the smallest integer satisfying $p(S_i) > (1/2)^k$. Therefore, the number of bits required to tag a sequence with probability $p(S_i)$ is

$$k = \lceil -\log_2 p(S_i) \rceil \tag{48.21}$$

which satisfies the optimal code requirement.

Two observations on allocating the intervals representing sequences are important for the arithmetic coding algorithm. First, the nonoverlapping intervals adjacently tile the unit interval. Therefore, since $\sum_i p(S_i) = 1$, the boundaries of these intervals can be determined by the cumulative distribution function (*cdf*). Second, since the symbols are *iid*, the probability of a sequence is equal to the product of the probabilities of the individual elements of the sequence. As the sequence grows, the probability of the sequence decreases because each time a new element is added to the sequence, the probability of the sequence is multiplied by a number less than one. It can be shown that if the sequences are assigned to intervals by lexicographical ordering, each symbol read in shrinks the representative interval to the subset of the corresponding prefix. As a result, the binary codeword corresponding to the sequence can be found by recursively zooming in to the subintervals determined by the *cdf* of symbols. This separability is very useful in practice because it allows the use of an iterative algorithm to generate the code as the symbols are read in, regardless of the length of a sequence.

In practice, we keep track of the upper and lower limits of the interval corresponding to the sequence being encoded. If the upper and lower limits corresponding to the sequence when it consists of k elements are given by $u^{(k)}$ and $l^{(k)}$, then we can show that the recursion equations that generate the upper and lower limits are given by

$$l^{(k)} = l^{(k-1)} + \left(u^{(k-1)} - l^{(k-1)}\right) F\left(X_i^{(k)} - 1\right) \tag{48.22}$$

$$u^{(k)} = l^{(k-1)} + \left(u^{(k-1)} - l^{(k-1)}\right) F\left(X_i^{(k)}\right) \tag{48.23}$$

where

$F\left(X_i^{(k)}\right)$ is the *cdf* evaluated for the letter that is the *kth* element of the sequence S_i

$F\left(X_i^{(k)} - 1\right)$ is the *cdf* evaluated for the letter lexicographically preceding the letter that is the *kth* element of the sequence S_i

Note that after processing the first symbol, the length of the interval is $p\left(X_i^{(1)}\right)$. With the introduction of the second symbol, the length shrinks to $u^{(2)} - l^{(2)} = \left(u^{(1)} - l^{(1)}\right) p\left(X_i^{(2)}\right) = p\left(X_i^{(1)}\right) p\left(X_i^{(2)}\right)$ and for the *kth* symbol to $u^{(k)} - l^{(k)} = \prod_{j=1}^{k} p\left(X_i^{(j)}\right)$. Therefore, the number of bits required to represent a number in the final interval for a sequence of length N is given by

$$\left\lceil -\log_2 \prod_{k=1}^{N} p\left(X_i^{(k)}\right) + 1 \right\rceil \cong NH(X) + 2 \tag{48.24}$$

Although determining the tag using $l^{(N)}$ and $u^{(N)}$ is sufficient for generating the code, some practical issues need to be considered for implementing the algorithm. First, we want to send the bits as we read through the sequence without waiting for the termination of the sequence. More importantly, as we keep multiplying by numbers less than one, the size of the interval becomes smaller and smaller until it falls below the machine precision level. Two strategies are used to provide incremental coding and prevent the upper and lower levels from converging. The first strategy is used when the interval corresponding to the sequence is wholly contained in either the upper or the lower half of the unit interval. In this case, the most significant bit (MSB) of the upper and lower limits is the same, as are the most significant bits of any number in the interval. We can, therefore, release this bit to the communication channel. We can also shift this bit out from the registers holding upper and lower limits and shift in a 1 for the upper and a 0 into the registers holding the upper and lower limits, respectively. By doing this, we effectively double the size of the interval.

When the interval is contained in the middle half of the unit interval, [0.25, 0.75), with the lower limit in [0.25, 0.5) and the upper limit in [0.5, 0.75), we cannot use this strategy. However, we know that in the future, if the interval is contained in either the upper or the lower half of the unit interval there, the size of the interval will be less than 0.25. If this interval is entirely contained in the lower half of the unit interval, the most significant bit of the upper and lower limits, and, hence, all numbers in the interval, will be 0 and the second most significant bit will be 1. Similarly, if the interval is entirely contained in the upper half of the unit interval, the most significant bit of all numbers in the interval will be 1 and the second most significant bit will be 0. This knowledge of the future is used by the algorithm by doubling the interval at the current time without releasing any bits into the communication channel. Instead, a counter, *scale*3, keeps track of the number of times the interval is scaled; and when in the future the upper and lower limits end up with the same most significant bit, the encoder releases that bit followed by *scale*3 bits, which are the complement of the most significant bit. The scaling of the interval is performed by shifting out the most significant bit, complementing the next most significant bit, and shifting in a 1 or a 0 for the upper and lower limits, respectively.

The algorithm for encoding can be summarized as follows:

1. Read the first symbol and initialize l and u and *scale*3 = 0.
2. IF l and u have the same MSB, send MSB and the complement of MSB *scale*3 times; shift-left l, u, reset *scale*3 = 0.
3. IF l and u are in the middle half shift-left&complement MSB of l, u; increment *scale*3.
4. IF u and l are in the lower, middle, or upper half of the unit interval, GOTO 2.
5. Update l and u using the next symbol.
6. IF end of sequence send l; terminate, ELSE GOTO 2.

The encoding algorithm can be implemented using integer buffers containing l and u as described. Floating numbers with corresponding scaling operations can also be employed. The decoder mimics the scaling operations of the encoder. However, instead of sending the MSB, it pushes the new LSB from the code to a tag buffer t. Searching for the *cdf* interval, including the current tag, and representing the corresponding symbol are also performed in the update phase. The updates of l, u, and t are done using the detected interval. Therefore, the sequence can be decoded online as the code is retrieved. The first symbol is decoded after the first m bits are fed into the buffers, where m is determined by the smallest cdf interval.

Let's go through the encoding procedure with an example. Assume a sequence drawn from a three-letter alphabet $\mathcal{A} = \{a_1, a_2, a_3\}$ with probabilities {0.7, 0.2, 0.1}. We encode the sequence $a_1a_3a_2a_2$. The corresponding cdf is $F(1) = 0.7$, $F(2) = 0.9$, $F(3) = 1$. Reading the first symbol a_1, we initialize $l^{(1)} = 0$ and $u^{(1)} = 0.7$. With the second symbol a_3, the updates are performed by Equation 48.22, giving $l^{(2)} = 0.63$ and $u^{(2)} = 0.7$. The MSB of $l^{(2)}$ and $u^{(2)}$ are both 1. Scaling using Step 2 of the algorithm $l^{(2)} = 0.26$ and $u^{(2)} = 0.4$, and 1 is sent to the receiver. The MSB for both are now 0. We can send 0 and scale again using Step 2 once more. Now $l^{(2)} = 0.52$ and $u^{(2)} = 0.8$. One more scaling with Step 2 sends 1 and scales $l^{(2)} = 0.04$ and $u^{(2)} = 0.6$. Now $l^{(2)}$ and $u^{(2)}$ cover more than half of the unit interval. So we proceed to Step 5 and read the third symbol, a_2, and update $l^{(3)} = 0.432$ and $u^{(3)} = 0.5832$. The condition of Step 3 holds, so we increment *scale*3 and scale

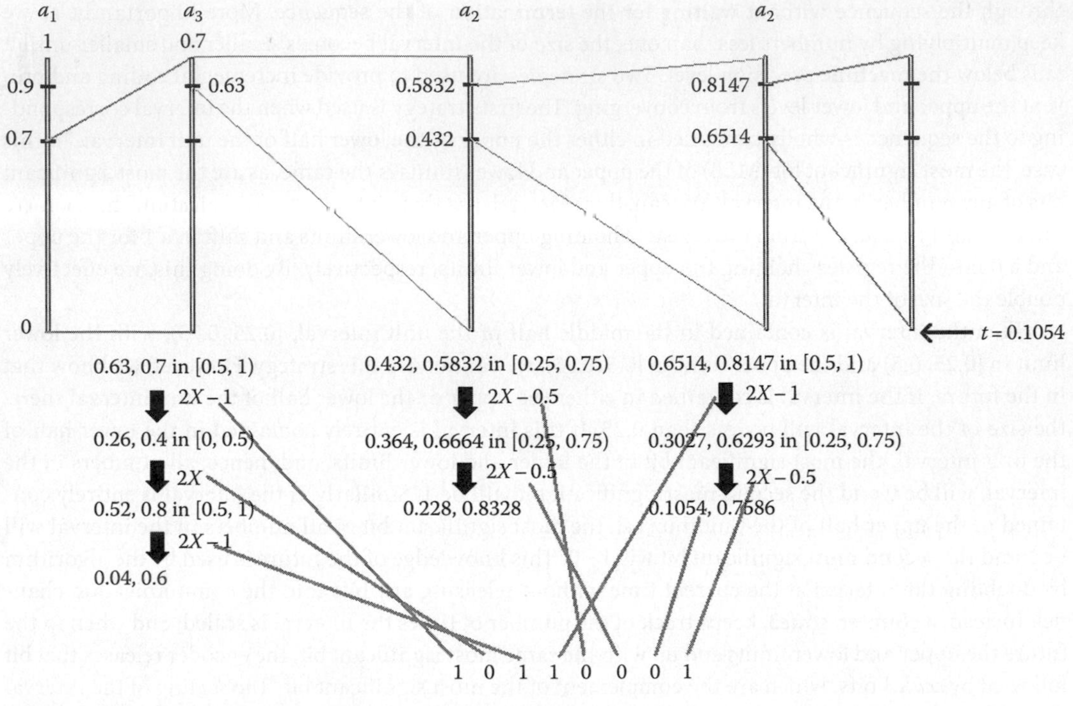

FIGURE 48.3 Example of the arithmetic encoding procedure.

$l^{(3)} = 0.364$ and $u^{(3)} = 0.6664$. Once more $l^{(3)}$ and $u^{(3)}$ lie in the middle half, so repeating Step 3, *scale*3 = 2, $l^{(3)} = 0.228$ and $u^{(3)} = 0.8328$. The interval is greater than half the unit interval, so we read the forth symbol a_2 and update the interval using Step 5, $l^{(4)} = 0.6514$ and $u^{(4)} = 0.8147$. The MSB are both 1, so we send 1 followed by 2 (*scale*3) 0s, and scale $l^{(4)} = 0.3027$ and $u^{(4)} = 0.6293$. The interval is in the middle half, so we follow Step 3, resulting in $l^{(4)} = 0.1054$ and $u^{(4)} = 0.7586$. This is the terminating symbol. So we transmit $l^{(4)}$ using the scaling of Step 2 (it lies in the lower half); a 0 followed by a 1 is transmitted. Finally, the code generated is 10110001. The encoding process can be seen in Figure 48.3.

For unknown distributions, the arithmetic coding algorithm can easily be made adaptive. In this case, initially, equal intervals represent each symbol in the alphabet. As a new symbol is encoded (decoded), the frequency count of the symbol coded is incremented. As the symbols are coded(decoded), the *cdf* is updated accordingly. The compression performance is similar to arithmetic coding with known distributions for large sequences. Since this operation can be performed simultaneously at the encoder and the decoder, no side information needs to be sent.

The performances of Huffman coding and arithmetic coding can be seen on an example of image coding. A 512 × 512 grayscale Lena image is shown in Figure 48.4. Each pixel is coded to 256 levels (8 bits), and the estimated entropy of the image is 7.4455 bits/symbol. Coding the image using a 256-letter alphabet results in a rate of 7.4682 bits/symbol for Huffman coding. The same image can be coded with a rate of 7.4456 bits/symbol by arithmetic coding. While both algorithms behave close to the source entropy, arithmetic coding is even closer since it uses block coding rather than symbol-by-symbol encoding.

48.3.2 Lossy Compression of *iid* Sequences

Lossy compression allows for distortion in the reconstruction hence, the input alphabet and the reconstruction alphabet can have different cardinalities. Lossy compression can be viewed as a mapping from an input space $\aleph$ with a large, possibly infinite, alphabet onto a finite output space $\hat{\aleph}$. In the most

FIGURE 48.4 Lena image. Original size: 262,144 bytes. Entropy: 7.4455 bits/pixel. Huffman coding rate: 7.4682 bits/symbol. Arithmetic coding rate: 7.4456 bits/symbol.

general setting, this map is called *quantization*. Given a distortion measure, which is a function defined on the source and reconstruction alphabet pairs $d: \aleph \times \widehat{\aleph} \rightarrow \Re^{+}$, a quantization method attempts to assign a reconstruction value to the input that results in the smallest possible distortion between the original and reconstruction values. Conversely, given a certain acceptable distortion value, the quantization process picks the particular mapping from the set of acceptable mappings that will result in the minimum rate.

Given a distortion function *d*, the expected distortion due to quantization is defined as

$$D = \sum_{x,x} p(x, x) d(x, x) \tag{48.25}$$

$$= \sum_{x,x} p(x) p(x|x) d(x, x) \tag{48.26}$$

Unlike entropy, the distortion depends not only on the source distribution but also on the conditional distribution of the output.

The rate distortion theorem [8,9] states that for an *iid* source, the distortion *D* for which the minimum rate is achieved is bounded by the average mutual information between the source and the output:

$$R_{\min} = \min I(X; \hat{X}) \tag{48.27}$$

such that

$$= \sum_{x,\hat{x}} p(x) p(\hat{x}|x) d(x, \hat{x}) \leq D \tag{48.28}$$

where $I(X; \hat{X}) = H(X) + H(\hat{X}) - H(X; \hat{X})$ is a measure of the shared information between the source and the output distributions. This theorem provides the theoretical limit on the performance of lossy compression. In this sense, it serves a role similar to the one played by entropy in lossless compression.

48.3.2.1 Scalar Quantization

Scalar quantization maps each symbol of a sequence to a codeword from a finite alphabet. This is a many-to-one mapping and, therefore, not invertible. The goal in quantizer design is to come up with a mapping and a representation set that results in a reconstruction with the minimum average distortion. The input alphabet is partitioned into sets that are each represented by a codeword from the quantizer output alphabet. The design of the partition and the representative codewords turns out to be a Voronoi partition problem. Given a set of reconstruction values, we want to partition the input alphabet such that each partition is represented by the reconstruction value that results in minimum distortion. Furthermore, given a partition of the input alphabet, we want to find the set of reconstruction values, one for each partition, which can be used to represent the input alphabet with the least distortion. For an R bit quantizer, the problem will turn out to be

$$\min_{\hat{x}} \sum_{i=1}^{2^R} \sum_{x \in \mathrm{v}_i} p(x) d(x, \hat{x}_i) \tag{48.29}$$

such that

$$\mathrm{v}_i = \{x : d(x, \hat{x}_i) < d(x, \hat{x}_j), \forall j \neq i\}$$

where $\hat{x}$ represents the reconstruction points. The inner summation changes to an integration for continuous alphabets, and the distribution $p(x)$ turns out to be the *pdf* of the source. This problem is hard even for tractable distortion measures such as the mean squared distance measure. An approximate solution for general distributions is as follows. Starting from a set of arbitrary reconstruction points, the partitions v_i are found by assigning each letter from the input alphabet to the closest letter from the reconstruction; then the new reconstruction points are estimated as the mean of each partition v_i. This procedure is iteratively repeated. The distortion in each iteration can be shown to be nonincreasing, and the total distortion is expected to converge to a local minimum. This algorithm is known as Lloyd's algorithm [10].

The input–output maps for different configurations of a scalar quantizer are shown in Figure 48.5. If the quantization intervals Δ_i are the same for all except the outermost intervals, the quantizer is called

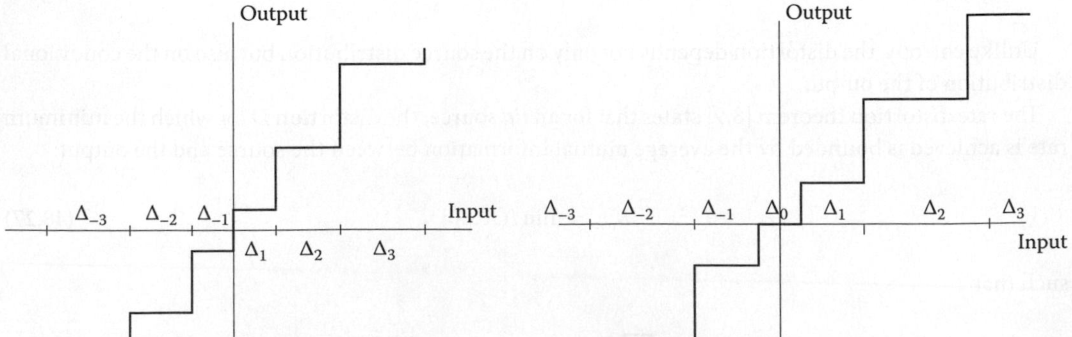

FIGURE 48.5 Two scalar quantizer input–output maps. The one on the left is called a midrise quantizer, while the one on the right is called a midtread quantizer.

a *uniform quantizer*. If not, it is called a *nonuniform quantizer*. One popular implementation of the uniform quantizer is where the quantizer output label l is generated by a division

$$l = \left\lfloor \frac{x}{\Delta} + 0.5 \right\rfloor$$

The reconstruction is given by

$$\hat{x} = l\Delta$$

This results in a quantizer whose innermost interval (corresponding to an output value of zero) is twice the size of the other quantization intervals.

A popular method, widely used in the telecommunication industry, for implementing nonuniform quantizers is the use of companding. Here, the input is passed through a logarithmic compressor function prior to uniform quantization. The reconstruction is obtained by taking the uniform quantizer reconstruction level and mapping it through an expanding function. The companding characteristic used in the United States and Japan is called the μ-law and is given by

$$c(x) = x_{max} \frac{\ln\left(1 + \mu \frac{|x|}{x_{max}}\right)}{\ln(1+\mu)} \operatorname{sgn}(x) \tag{48.30}$$

$$c^{-1}(x) = \frac{x_{max}}{\mu}\left[(1+\mu)^{\frac{|x|}{x_{max}}} - 1\right] \operatorname{sgn}(x) \tag{48.31}$$

where μ = 255. The companding characteristic used in most of the rest of the world is called the *A*-law and is given by

$$c(x) = \begin{cases} \dfrac{A|x|}{1+\ln A} \operatorname{sgn}(x) & 0 \le \dfrac{|x|}{x_{max}} \le \dfrac{1}{A} \\[2ex] x_{max} \dfrac{1 + \ln \dfrac{A|x|}{x_{max}}}{1+\ln A} \operatorname{sgn}(x) & \dfrac{1}{A} \le \dfrac{|x|}{x_{max}} \le 1 \end{cases} \tag{48.32}$$

$$c^{-1}(x) = \begin{cases} \dfrac{|x|}{A}(1+\ln A) & 0 \le \dfrac{|x|}{x_{max}} \le \dfrac{1}{1+\ln A} \\[2ex] \dfrac{x_{max}}{A} \exp\left[\dfrac{|x|}{x_{max}}(1+\ln A) - 1\right] & \dfrac{1}{1+\ln A} \le \dfrac{|x|}{x_{max}} \le 1 \end{cases} \tag{48.33}$$

where $A = 87.6$.

48.3.2.2 Vector Quantization

Recall that employing lossless compression by coding blocks of symbols can result in better performance than symbol-by-symbol encoding even in the case of *iid* sequences. An analogous situation applies for lossy coding. Quantizing blocks of symbols results in better rates for a given distortion for sources with a nonuniform distribution. Using the rate distortion theorem, it is possible to show that for an *iid* source, the achievable rate to quantize blocks of length n with a 2^{nR} level quantizer satisfies

$$nR(D) \leq nR \tag{48.34}$$

where nR is the number of bits allocated for R bit scalar quantizers achieving the same distortion D. Quantizing blocks of input sequences is called *vector quantization* (VQ).

The geometry of multidimensional distributions plays a role in the design of vector quantizers. One class of quantizers indexes the reconstruction points that are distributed regularly, taking the geometry of the distribution into consideration. This class of vector quantizers is called the *lattice quantizers* [11–13].

The vector quantizers can be extended to general distributions by generalizing Lloyd's algorithm for the multidimensional case. If the same procedure is applied for vectors of symbols instead of single symbols, an approximate solution to the VQ problem is provided. This procedure is known as the Linde–Buzo–Gray (LBG) algorithm [14].

48.4 Modeling Data Organization

The rates achieved by entropy coding techniques, such as Huffman coding or arithmetic coding, are close to the first-order entropy of the source. This is only optimal under the assumption that the source is *iid*. However, in general, the data organization cannot be captured only by the alphabet distribution. This organization can stem from various aspects specific to different data sources. For example, in a natural language text, certain words and phrases could be used more frequently than others. Consider the words *dour* and *ordu*. Using entropy coding, the total length of the code allocated for each word would be the same, the sum of the length of the codewords of the letters composing the word. However, *dour* is a much more probable word in English than *ordu* (the reverse is true in Turkish). Therefore, if we knew that the text we were compressing was English, information theory says that we should use fewer bits to encode *dour* than we would use to encode *ordu*. However, such a coding would only be possible if we discarded the assumption that the text sequence was made up of *iid* symbols. The argument against using an *iid* assumption is also true for other types of data. In digital images, the same blocks of pixels can be repeated several times. In motion pictures, adjacent frames have significant parts that remain unchanged, etc. These structures are formed by higher-order organizations that cannot be captured by entropy coding. This structure, if captured, can be used to achieve rates lower than entropy coding of the raw sequence alone can provide.

Similar arguments apply for lossy compression. The quantization schemes assume *iid* distribution of the source. However, correlations in data result in nonzero mutual information between symbols, that is, not utilized by *iid* quantizers. Therefore, it makes sense to first use knowledge of the source model to generate a decorrelated sequence that can then be quantized.

Instead of monolithic coding schemes, a modular approach of modeling and coding is generally adopted for data compression. Since entropy coding is the final phase of compression, the output of the modeling phase should be a sequence with a low first-order entropy. Therefore, we can view the modeling as a data processing step that transforms the data from its original domain to a sequence with uncorrelated elements. Although models can have different objectives, describing data with uncorrelated features is the aim of all models. We can roughly categorize the modeling techniques into three main groups based on their approach to exploring the organization of data.

The first group of algorithms assumes the data are composed of building blocks that are repeatedly used in a sequence. They attempt to parse the data to isolate these building blocks and encode them as units. This can be done by organizing the repeated building blocks into a lookup table or *dictionary*.

The second group attempts to explore the correlations in a sequence by generating a model and encoding the model parameters and the uncorrelated residual. The last group uses the natural structure associated with the physical dynamics of the process generating the data to process the data sequence.

48.4.1 Modeling Building Blocks: Dictionary Methods

In general, a compression algorithm should assign shorter codewords to frequently recurring elements while assigning long codewords to rarely expected elements. Consider a converse approach. Instead of designing variable-length codewords for fixed-size items, we encode frequently occurring variable-length blocks with fixed-size codes. Given an alphabet A, we define a group of letters forming a sequence to be a *word*. A lookup table of words with indices is called a *dictionary*, and compression methods using dictionaries for coding are called *dictionary techniques*.

If the dictionary consists of frequently occurring words within a sequence, which are the building blocks of the corresponding data, the dictionary technique can provide significant compression. It is not difficult to find examples of sequences with recurring items. Texts of natural languages consist of frequent words and phrases. Similarly, computer-generated images have repetitive blocks of the same pixel values. There are many types of data that have similar repetitive structures, which is why dictionary coding techniques are frequently used in general purpose archiving programs.

48.4.1.1 Lempel–Ziv Methods

The most well-known class of dictionary techniques are based on two algorithms proposed by Lempel and Ziv in 1977 [15] and 1978 [16], referred to as the LZ77 and LZ78 algorithms. The LZ77 and LZ78 algorithms and their variations work on the premise that sequences are constructed of repeatedly used building blocks. Therefore, these algorithms attempt to identify and index these blocks. Encoded sequences are streams of indices pointing to entries corresponding to these blocks in a dictionary that is constructed adaptively.

The LZ77 algorithm attempts to represent a variable-sized block of input by an exact match in the history. Starting from the current symbol to be encoded, a variable-sized block is grown one symbol at a time, until the point where the addition of a symbol will result in a pattern that has not been previously encountered. The pointers coding the position and the length of that block in the previously coded sequence and the code for the symbol that terminated the search form the code for the current block. Assume that we are encoding "`that_is_the_block_that_is_repeated`", and we are at the point where we have encoded "`that_is_the_block_`". In the remaining string "`that_is_repeated`", we can grow the block up to the substring "`that_is_`" that can be found in the previously coded sequence. The code for this block is the triplet (18, 7, code[r]). Where 18 is the offset from the beginning of the block "`that_is_`" and its previous occurrence in the sequence, 7 is the length of the block to be copied and code[r] is the code for the symbol r. The decoder will interpret this as "go back 18 symbols, copy 7 symbols, and insert the 7 symbols at the end of the decoded sequence followed by the symbol r." There are numerous variations of this basic algorithm. Most use a variable-length code to encode the various indices. Most also replace the triplet ⟨*offset, length, code*⟩ with a single-bit flag followed by either the code for a symbol or the double ⟨*offset, length*⟩. This latter variation was developed by Storer and Syzmanski [17] and hence is referred to as LZSS. Generally, the distance in the history in which to search for a match is limited for computational efficiency or due to memory requirements.

The LZ78 algorithm differs from the LZ77 algorithm in the way it indexes the repeated sequences. Previously encountered unique blocks are recorded in a lookup table. Similar to LZ77, the block to be encoded grows until there is no match in the lookup table. A double consisting of the index of the entry in the dictionary that matches all but the last element in the block and the last symbol in the block (which made the block unique) are encoded, and the new unique block is recorded in the lookup table.

This can be viewed as parsing the sequence into a minimal number of building blocks constructively. The rate behavior of LZ78 is similar to that of LZ77.

As was the case for LZ77, several variations of LZ78 have been proposed. The most widely used is the LZW algorithm [18]. It eliminates the need for encoding the second element of the double—the last symbol in the block–by priming the dictionary with the source alphabet. The encoder only sends the index of the matching block; however, the dictionary is extended by adding the concatenation of this block and the last symbol. It is easiest to see the process with an example. Consider encoding the sequence *ababbbaa*.... The initial dictionary consists of only two entries:

1	*a*
2	*b*

When we parse the sequence, we encode the index for *a* and begin a new entry in the dictionary:

1	*a*
2	*b*
3	*a*...

The next element in the sequence is *b*. If we extended this, we would get the sequence *ba* that is not in the dictionary; so we encode the index for *b* and add this to the incomplete third element in the dictionary. This results in the third item becoming *ab* that is unique in the dictionary. Thus, the third item is complete; and we bring a fourth item with the symbol *b*:

1	*a*
2	*b*
3	*ab*
4	*b*...

The next symbol to be encoded is *a*. As this exists in the dictionary, we extend this block to include the following symbol; and we get the block *ab*. This also exists in the dictionary so we continue and get the block *abb*. This does not exist in the dictionary so we simply encode the index for *ab*. We now update the dictionary with the symbols that have been encoded. The fourth item in the dictionary is extended to *ba* that is unique so we complete that item and begin the fifth item with *a* followed by *b*. As *ab* exists in the dictionary, this item is incomplete:

1	*a*
2	*b*
3	*ab*
4	*ba*
5	*ab*...

As we continue, we will encode the indices 1,2,3,2,4,1 and the dictionary will become as follows:

1	*a*
2	*b*
3	*ab*
4	*ba*
5	*abb*
6	*bb*
7	*baa*
8	*a*...

All LZ-based algorithms are causal, so that the dictionaries are built both at the encoder and the decoder. That is why no information about the dictionaries other than the indices needs to be sent.

Algorithms based on the LZ algorithms are used in a large number of compression applications. The LZW variant of LZ78 is used in the Unix *compress* command as well as in the graphics interchange format (GIF). The compression algorithms in both portable network graphics (png) [19] as well as *gzip* [20] are based on LZ77.

48.4.2 Modeling Correlated Structures: Predictive Methods

Dictionary methods are well suited to repetitive data. However, the structure might not result in exact repetitions in various situations. Without exact repetition, the use of lookup tables is not useful. However, it is still possible to exploit this imperfect recurrent structure with models capturing the correlations. There are a great variety of approaches for modeling correlated structures. These include prediction filters using the context of a sample, Markov models, and decorrelating data transformations.

Let's consider a simple case where the adjacent symbols are not statistically independent. The entropy of the extended symbols of doubles would be

$$H(x_i x_{i+1}) = H(x_i) + H(x_{i+1}) - I(x_i; x_{i+1}) = 2H(x) - I(x_i; x_{i+1}) \tag{48.35}$$

where $I(x_i; x_{i+1})$ is defined as the average mutual information between adjacent symbols. Since we know that the adjacent symbols are statistically dependent, the mutual information value is greater than zero. Therefore, the achievable rate of this sequence is smaller than $2H(x)$ if we use an extended alphabet of doubles. This example can be generalized by induction. If correlations exist within a sequence, the entropy, and, consequently, the achievable rate of that sequence, is smaller than first-order approximations may suggest. A similar argument can be proposed for the rate distortion function of correlated sequences.

A general strategy to exploit structural redundancy in data compression is the use of a decorrelation phase prior to entropy coding. A model capturing the correlations in the data is used to remove the correlation leaving behind an uncorrelated residual sequence with a first-order entropy considerably less than the original sequence.

This is a *predictive* strategy, and compression algorithms that use this strategy make up an important body of data compression algorithms. The reason behind their importance is that many sources of data of interest for data compression exhibit various types of correlated structures. One form of this structure originates from the dominance of low-frequency components in nature. For example, in daily records of temperature, a measurement is generally correlated with the previous day's measurement. Therefore, given the recent history of temperatures, the current temperature is not very uncertain. Similar time series behavior can be observed in many forms of data from audio streams to medical signals. Data recorded in image format contain spatial correlations. It is very likely that neighboring pixels correspond to the same morphological object resulting in similar intensity and color values. Correlations in each data type might appear in different ways, but they can all be considered as the effect of the low-frequency nature of signals of interest to humans. Another form of correlation is the result of global structure. These are patterns that are repeated in the data though not necessarily in adjacent temporal or spatial locations. Furthermore, unlike the repeated patterns that are captured using dictionary methods, these repeats may be imperfect or approximate. Obviously, models with a more global view of memory are needed to decorrelate data in this context. We can classify predictive models into two categories: models predicting global structure and models predicting local structure.

The predictive model employed at the encoder has to be repeatable at the decoder in order for the input sequence to be recovered. This requirement imposes the use of a causal context in models predicting local structure. For models predicting global structure, we have the option of either encoding model parameters as side information or using causal adaptive models that can mimic the same prediction operation at the receiver.

48.4.2.1 Predicting Local Structure

Predictive coding techniques exploit the local structure using the context of a symbol to predict the current symbol. Since the prediction is deterministic, the uncertainty to be encoded is contained in the residual sequence. If an accurate model is used to predict the current symbol well, the residual should be a white sequence with a low variance. The low variance indicates a sequence with low complexity that can be coded at low rate. In the following, we look at different prediction approaches that use different prediction models.

48.4.2.2 Prediction with Partial Match

Prediction with partial match (ppm) attempts to model the conditional probability of the current symbol given a context of previous symbols. The implicit model is a collection of Markov processes of different orders. The *ppm* algorithm estimates the distribution of symbols in different contexts in the form of frequency tables based on statistics gathered from previously encoded parts of the sequence. The frequency tables associated with a particular context only include symbols that have occurred in that context. As the encoding progresses, the algorithm updates the frequency tables. This aspect is similar to dictionary methods. However, *ppm* differs from dictionary methods in that unlike dictionary coding *ppm* uses the previously occurring words as contexts for a frequency table that is then used to entropy code the symbol under consideration. Instead of continually growing the size of the context as in LZ78-based approaches, words in *ppm* are limited to a maximal length in order to mitigate memory problems and avoid model overfitting.

A maximum context length ℓ is specified as a parameter of the algorithm. To encode a symbol, the algorithm first looks to see if this symbol has been observed before in the context of the ℓ previous symbols. In case such an event exists in the history, the symbol is encoded based on its conditional distribution given those ℓ previous symbols. If this is the first time this symbol has occurred in this context, the algorithm encodes an escape symbol and repeats the process using the context of $(\ell - 1)$ previous symbols. This process is repeated until a context in which the symbol has previously occurred is encountered. If the symbol has never been previously encountered in any context, it is encoded using the first-order distribution of the symbols. The decoder begins by assuming that a context of size ℓ was used. Whenever an escape symbol is decoded, the decoder downsizes the context to look for a suitable match. Because the escape symbol can occur at any context, its frequency of occurrence is included in each frequency table by default.

Assume that we are currently encoding the last letter of the word *prediction* with the maximum context length of 6. We first check for a context *dictio* followed by *n*. If there is none, the contexts of *ictio*, *ctio*, *tio*, *io*, and *o* are checked until one of these contexts followed by an *n* is found. At each step, either the symbol *n* or an escape symbol is encoded using arithmetic coding with the frequency table of the context. Once a symbol has been encoded, the corresponding frequency tables are updated. This makes *ppm* an adaptive arithmetic coding algorithm with variable context length. Since the frequency tables of each context size can be derived causally from previously coded data, the same operations can be conducted both by the encoder and decoder without any further need of side information.

There are several popular variations on the basic *ppm* theme. The modifications deal mostly with the derivation of the frequency tables, especially the escape symbol frequencies. The algorithm *ppm** [21] provides a very creative approach to determining the maximum context length. The algorithm first looks at the longest matching context that has only one symbol (other than the escape symbol) in its frequency table, the idea being that that single symbol is the same as the symbol being encoded with high probability. If this is not the case, the algorithm encodes an escape symbol and defaults to a maximum context length. The algorithm *ppmz*, which can be found at http://www.cbloom.com/src/ppmz.htm, is the best-known variant of the *ppm** approach. Another modification is the disabling of impossible symbols that can be determined by the nonmatches of longer contexts. Assume in encoding *prediction*, we downsized the context to *o*. If the frequency table of context *io* has the symbol *w*, we can discard the symbol *w* in the frequency table of the context *o*, because if the symbol was *w*, an escape would not be encoded.

48.4.2.3 Burrows–Wheeler Transform

The Burrows–Wheeler Transform (BWT) is a powerful technique which is particularly useful when strong correlations between the symbols and their contexts exist. The advantage of this method is that all possible contexts without a size limitation are considered. However, this advantage comes with a cost as it requires an offline phase of BWT.

Given a sequence, the maximal context of each symbol is considered and ordered lexicographically. The sequence of symbols in this order is the BWT. Here, unlike causal methods, the context starts from the next (right-hand) symbol. Since the contexts of the repeated patterns are the same, they are placed adjacent to each other in the transform. For example, the words containing "*predict-*" will be ordered together by the contexts starting with "*redict-*." Therefore, the corresponding part of the transform is expected to have a run of *p*'s. In a correlated sequence, the resulting BWT sequence consists of long runs of the same letter. This provides the opportunity to use coding schemes like run-length coding with low rates.

In order to obtain all possible contexts, the transform takes the original sequence and generates all possible cyclic shifts of the sequence. All cyclic shifts are then ordered lexicographically. The sequence consisting of the last elements of each of the lexicographically ordered cyclic shifts forms the transform of the original sequence. This transform, along with an index pointing to the location of the original sequence in the lexicographically ordered set, can be used to recover the original sequence. The algorithm is described using an example.

Example

As an example, consider the word GALATASARAY, which is the name of a Turkish soccer team. The circular permutations of the word are

G	A	L	A	T	A	S	A	R	A	Y
A	L	A	T	A	S	A	R	A	Y	G
L	A	T	A	S	A	R	A	Y	G	A
A	T	A	S	A	R	A	Y	G	A	L
T	A	S	A	R	A	Y	G	A	L	A
A	S	A	R	A	Y	G	A	L	A	T
S	A	R	A	Y	G	A	L	A	T	A
A	R	A	Y	G	A	L	A	T	A	S
R	A	Y	G	A	L	A	T	A	S	A
A	Y	G	A	L	A	T	A	S	A	R
Y	G	A	L	A	T	A	S	A	R	A

Lexicographically ordering this we get

A	L	A	T	A	S	A	R	A	Y	G
A	R	A	Y	G	A	L	A	T	A	S
A	S	A	R	A	Y	G	A	L	A	T
A	T	A	S	A	R	A	Y	G	A	L
A	Y	G	A	L	A	T	A	S	A	R
G	A	L	A	T	A	S	A	R	A	Y
L	A	T	A	S	A	R	A	Y	G	A
R	A	Y	G	A	L	A	T	A	S	A
S	A	R	A	Y	G	A	L	A	T	A
T	A	S	A	R	A	Y	G	A	L	A
Y	G	A	L	A	T	A	S	A	R	A

The transformed sequence consists of the last letters of the ordered matrix: GSTLRYAAAAA. Notice how all the A's occur together. This means we can easily get a compact representation for this sequence. We encode the transformed sequence and the location of the original sequence in the ordered matrix. In this case, the original sequence is the sixth row of the ordered matrix. At the decoder, we lexicographically order the transformed sequence to obtain AAAAAGLRSTY. The sixth letter is G; therefore, we know that the first letter of the decoded sequence is G. We look for G in the transformed sequence and find that it occurs in the first position, so the second letter of the decoded sequence is the letter in the first position of the lexicographically ordered sequence, namely, A. Now we look for the first A in the transformed sequence. It is in position 7; therefore, the next decoded letter is the letter in position 7 of the lexicographically ordered sequence, namely, L. Continuing in this fashion, we can decode the entire sequence.

48.4.2.4 Differential Encoding

The predictive coding techniques we have discussed attempt to remove the redundancy in correlated sequences using context-based distributions. An adaptive coding scheme based on skewing the symbol probabilities by conditioning over the neighboring context is suitable for data types such as text documents or archiving applications. However, different paradigms of prediction can be more useful based on the data type considered. For example, data sampled from natural waveforms, such as image or audio data, can be viewed as a sequence of correlated numbers. Thus, numerical prediction that uses the correlation works well for this data type.

Lossless image coding is an application of this type of predictive coding. The basic idea behind lossless image coding is that a pixel to be coded can be predicted using the surrounding texture. The difference between the pixel value and the predicted value, or the residual, generally has much lower first-order entropy than the pixels themselves and can, therefore, be encoded efficiently using arithmetic coding. A three-phase procedure is generally followed: (1) initial prediction of the pixel to be encoded, (2) refinement of the initial prediction using the previous events of the same kind, and (3) entropy coding of the residual sequence.

In order to avoid transmitting a large amount of model parameters as side information, the algorithms are causal, or backwards, conducting the same operations on both the transmitter and receiver ends. Therefore, by the surrounding texture, we refer only to the previously encoded neighborhood of the pixel. The initial prediction is obtained using the neighboring pixel values as well as the horizontal and vertical gradients in this neighborhood. The lossless Joint Photographic Expert Group (JPEG) standard uses one of the immediate neighbors (N, W, NW) or the weighted average of these neighbors as the prediction of the current pixel. In JPEG-LS, the prediction is replaced by the median adaptive prediction of the LOCO-I algorithm [22]. Median adaptive prediction selects one of (N, W, W+N-NW) based on the ranking of the three neighboring pixels. If N or W is the median of the neighbors, it is assigned as the prediction. Otherwise, a third assignment is performed.

As the second phase, LOCO-I refines the predicted value considering the previous errors made with different classes of gradient. Consider the pixel labels in Figure 48.6, two local horizontal (NE-N, N-NW) gradients and a vertical (NW-W) local gradient are quantized based on their magnitude into 365 contexts. A correction factor, which is determined by the errors occurring for the corresponding context, is added to the initial prediction. The correction factor is adaptive and is updated at each pixel prediction. This refinement step introduces global capture of the correlations within an image. Therefore, its

NW	N	NE
W	X	

FIGURE 48.6 Current pixel to be encoded X and surrounding texture of causal neighborhood.

effect is to decorrelate the error sequences utilizing the mutual information within the errors. Further whitening of the residual generally results in a Laplacian-like distribution in the encoded symbols. While adaptive Huffman or arithmetic coding can be applied in the final coding phase for lossless JPEG, the JPEG-LS standard prefers Golomb codes that perform well on geometric sources.

An idea similar to that of the *ppm* algorithm can also be used for generating predictions for lossless image compression. Instead of using similar contexts in the history to obtain the frequency counts, an average value of all pixels that have occurred in that particular context can be used as a prediction for the pixel being encoded [23]. This approach has been used to good effect in the lossless compression of hyperspectral images [24].

In the case of lossless compression, the prediction is performed based on the past history of the sequence. In lossy coding, one has to be a bit more cautious. Unlike the case for lossless compression, the reconstruction at the decoder in lossy compression does not exactly match the original sequence. Therefore, if the prediction at the encoder is obtained using the original sequence, the predictions performed by the encoder and decoder will diverge from each other. This is an error accumulation effect caused by the quantization error. At each prediction step, the receiver predictor adds up the previous quantization error with the current one since the prediction is performed using erroneous inputs. To mitigate this effect, the encoder and the decoder are forced to generate the same predictions by generating the reconstructed sequence at the encoder and using this replica of the reconstructed sequence to generate the prediction. A block diagram of this differential encoding scheme is shown in Figure 48.7. The section of the encoder that mimics the decoder is shown in the dotted box.

Data sampled from continuous waveforms, such as speech signals, contain local correlations that are not limited to the immediate neighbors of the symbols. Therefore, the predictive filters should be of a higher order to remove a greater amount of redundancy. Employing autoregressive filters is a well-known solution that has been used successfully in differential pulse code modulation (DPCM) systems. An autoregressive filter uses the linear regression of the previous N samples with minimum variance:

$$\min_{a} E\left[\left(x_n - \sum_{i=1}^{N} a_i x_{n-i}\right)^2\right] \tag{48.36}$$

where

x_n is the nth symbol in the sequence
a_i is the autoregression coefficient

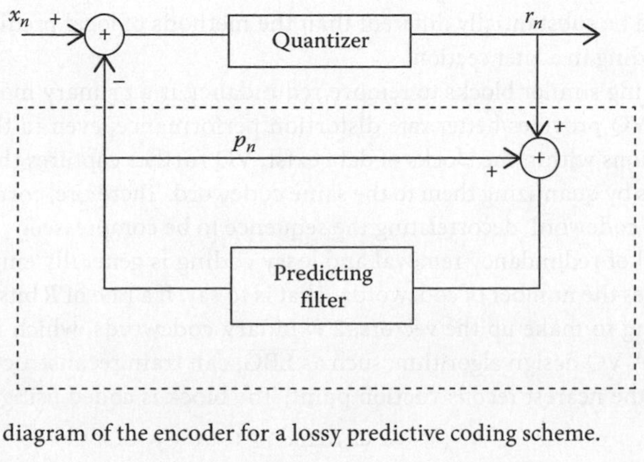

FIGURE 48.7 A block diagram of the encoder for a lossy predictive coding scheme.

The optimal filter coefficients then satisfy the Wiener–Hopf equation, so

$$\sum_{i=1}^{N} a_i R_{xx}(i-j) = R_{xx}(j), \quad \forall j \in \{1, 2, \ldots, N\} \tag{48.37}$$

In this equation system, R_{xx} denotes the autocorrelation function of the input sequence that can be estimated using the sequence samples.

The sequence to be coded can be of a nonstationary nature; thus, the autocorrelation behavior might change temporally or spatially. To capture local correlation structure, DPCM systems are generally adaptive, updating the filter coefficients as the sequence is coded. Adaptation can be offline, estimating the coefficients over a window and transmitting them as side information. For each window, the receiver uses the predetermined coefficients. Blocks of 8 × 8 pixels in images or 16 ms time frames in speech signals have been widely used for offline adaptation. The use of side information and time-delay problems in real-time communication systems illustrates the need for adaptive systems operating online. Updating the filter coefficients using the feedback error provides a solution that can be employed without any need of side information or an offline-pass over the input sequence. The filter coefficients are updated using the gradient descent, least mean squared (LMS), algorithm. The LMS algorithm updates the filter coefficients at each time step using the rule

$$a_i^{(n+1)} = a_i^{(n)} + \alpha r_n x_{n-i}, \quad \forall i \in \{1, 2, \ldots, N\} \tag{48.38}$$

where

r_n is the prediction for the nth symbol

α is the stepsize of the gradient descent

The LMS algorithm can keep track of local correlation structures, which permits the coding system to adapt to nonstationary sources. Quantization, followed by prediction, can also be made adaptive by adjusting the variations in residual range.

48.4.2.5 Predicting Global Structure

Redundancy in data caused by correlated structures may not always be caused by the low variation of symbols within a neighborhood. Similar structures of large context may reoccur due to global organization. An obvious example of this occurs in video sequences. An object captured in a video sequence can exist in different locations in different frames. Although large blocks of pixels corresponding to this object may repeat with some slight variation in the video stream, this redundancy might not be revealed by prediction techniques intended to capture local correlations. So the prediction methods removing global redundancy can be substantially different than the methods of local prediction. We will discuss prediction in video coding in a later section.

The idea of identifying similar blocks to remove redundancy is a primary motivation for the use of VQ for compression. VQ provides better rate distortion performance, even in the case of *iid* sources. Where global correlations within the blocks of data exist, VQ further captures the mutual information between similar blocks by quantizing them to the same codeword. Therefore, correlated contexts can be mapped into the same codeword, decorrelating the sequence to be compressed.

VQ as a joint model of redundancy removal and lossy coding is generally employed as follows. The rate selected determines the number of codewords. That is to say, if a rate of R bits is selected and blocks of M symbols are going to make up the vectors, 2^{MR} binary codewords, which index 2^{MR} reconstruction points, are used. A VQ design algorithm, such as LBG, can train reconstruction points and assign each symbol block to the nearest reconstruction point. The block is coded using MR bits indexing the reconstruction point. During decoding, a lookup table can reconstruct the blocks using the coded indices. Since the lookup table must be available to the decoder, the model has to be sent as side information. This method has two main drawbacks. First, as the bitrate increases, the codebook grows exponentially. This results in a proportional increase in computational complexity to train the codebook and encode the blocks. Also, memory requirements and side information to be transmitted suffer from the same exponential growth. Therefore, VQ has mostly been used in low-rate applications such as low-rate speech coding or image coding. In low-rate speech coding, the vectors being quantized are generally vectors of model parameters, while in image coding applications, the inputs are generally square blocks of image pixels. Variations on VQ have been used to mitigate the exponential complexity and memory increase to enable it to be used in high-rate applications. Structures, such as tree-based and trellis-based quantizers or divide-and-conquer approaches, like multistage quantization, have been used to bring the codeword increase from exponential to linear or logarithmic growth [4,25]. Some adaptive VQ techniques have been used for video coding [26] where the quantization is of such high resolution that the quantized vectors can be used to build and update the vector quantizer codebook without need for side information.

48.5 Modeling Natural Structures

Much of the information in sources such as audio and video is in the form of correlated structures. Consider the way images destined for human perception are structured. Most of the image consists of regions of similar intensity demarcated by sharp transitions. If we think of an image in terms of its spectral characteristics, the regions of similar intensity will contribute lower frequencies, while the demarcations will contribute higher frequencies. As the number of pixels involved in the demarcations, or the edges, is much fewer than the number of pixels in the regions of similar intensities, the amount of low-frequency content is much higher than the amount of high-frequency content. In fact, in many parts of the image, the amount of high-frequency information may be completely negligible. We can take advantage of this fact by decomposing the image into different frequency components, each of which can be treated differently. There are two principal approaches currently being used to do this: *transform coding* and the use of *wavelets* or *subbands* for decomposition. Transform coding has been most widely used in the compression of images and video, for example, the widely used JPEG image coding standard and the Moving Picture Experts Group (MPEG) video coding standard. Subband coding has been used mostly for audio compression, the most well-known application being *mp3* coding. Wavelets have been used in the JPEG 2000 image compression standard [27,28].

Natural structure can also be introduced by the method used to generate the information-bearing signal. Speech is an example of such a signal. We all generate speech using machinery that is very similar. Simply put, air is forced through a vibrating membrane known as the vocal cords; the pressure waves thus generated are modulated by the throat and mouth to generate speech. The physical requirements on the various components in this process impose constraints on the signal generated. Compression algorithms that take advantage of these constraints fall under the general heading of *linear predictive* (LP) *coding*. The ubiquitous cell phones would not be possible without this method of compression.

48.5.1 Transform Coding

In transform coding, the image is divided into blocks, usually 8 × 8 pixels in size, and each block is represented in terms of a set of basis images. For example, the basis images corresponding to the popular discrete cosine transform (DCT) used in JPEG image compression and MPEG video compression are shown in Figure 48.8. The block in the upper left corresponds to the average intensity of the pixels in

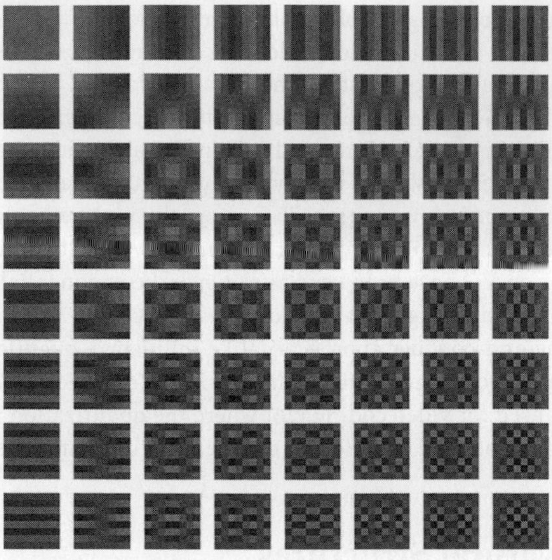

FIGURE 48.8 Basis images for the 8 × 8 DCT.

a block. The basis images increase in spatial frequency with their distance to the basis image in the upper left corner. The *(i, j)th* element of the DCT matrix **C** is given by

$$[\mathbf{C}]_{i,j} = \begin{cases} \sqrt{\dfrac{1}{N}}\cos\dfrac{(2j+1)i\pi}{2N} & i=0,\ j=0,1,\ldots,N-1 \\ \sqrt{\dfrac{2}{N}}\cos\dfrac{(2j+1)i\pi}{2N} & i=1,2,\ldots,N-1,\ j=0,1,\ldots,N-1 \end{cases} \tag{48.39}$$

If we represent an 8 × 8 block of pixels by the matrix **X** where $\mathbf{X}_{i,j}$ is the value of the *(i, j)th* pixel in the block, coefficients corresponding to the various basis images, in other words the transform, are given by

$$\Theta = \mathbf{C}\mathbf{X}\mathbf{C}^T \tag{48.40}$$

where Θ is the matrix of transform coefficients. The pixels can be recovered from the coefficients through the inverse transform

$$\mathbf{X} = \mathbf{C}^T\Theta\mathbf{C} \tag{48.41}$$

The transform coding procedure involves three steps: the transform, quantization of the transform coefficients, and the binary encoding of the quantization labels. We briefly describe these steps for the popular JPEG image coding standard.

The first step in the JPEG encoding is a level shift. Essentially, this means subtracting 2^{b-1} from each pixel where b is the number of bits per pixel. In a grayscale image, the value of b is 8. The coefficients have different perceptual importance with the human visual system being much more sensitive to errors in low-frequency coefficients. The error introduced during transform coding is a result of the quantization process. As there is more sensitivity to errors at lower frequencies, we use coarser quantizers for higher-frequency coefficients than we do for lower-frequency coefficients. The stepsizes for different coefficients that are standard in JPEG compression are shown in Table 48.1. The stepsize for the lowest-frequency coefficient is 16, while the stepsizes for higher-frequency coefficient are

TABLE 48.1 Sample Quantization Table

16	11	10	16	24	40	51	61
12	12	14	19	26	58	60	55
14	13	16	24	40	57	69	56
14	17	22	29	51	87	80	62
18	22	37	56	68	109	103	77
24	35	55	64	81	104	113	92
49	64	78	87	103	121	120	101
72	92	95	98	112	100	103	99

almost an order of magnitude bigger. The quantization process itself consists of dividing the coefficient by the stepsize and then rounding to the nearest integer. The reconstructed value of the coefficient can be obtained by multiplying the quantization label with the appropriate stepsize. For most of the 8×8 blocks in the image, the quantization label corresponding to the higher-frequency coefficients is 0. This is due to two factors. First, as mentioned earlier, the higher-frequency coefficients are generally small. And second, as can be seen from Table 48.1, the quantization stepsizes for the higher-frequency coefficients are much larger than those for the lower-frequency coefficients resulting in a large number of 0 labels. To efficiently code these long runs of 0s, the JPEG algorithm scans the coefficients in a zigzag fashion as shown in Figure 48.9. Notice that the coefficient at the upper left-hand corner is not included in the zigzag scan. As mentioned earlier, this particular coefficient corresponds to the average pixel value in a block. As an 8×8 block is a very small portion of the image, it is reasonable to assume that the average values of neighboring blocks of pixels are going to be similar. The JPEG algorithm takes advantage of this fact and transmits only the difference between these average values. The rest of the coefficients are encoded using a modified Huffman code. Each codeword in this code is indexed by two values: the number of 0 coefficients preceding the coefficient being encoded in the zigzag scan and the value of the coefficient. The last nonzero coefficient in the zigzag scan is followed by an end-of-block code.

Let's work through the JPEG coding process with an example. Consider the 8×8 block shown in Table 48.2. If we subtract 128 from each value in this block and then take the DCT of this level-shifted block, we obtain the coefficients shown in Table 48.3. Quantizing this block of coefficients using the quantization table shown in Table 48.1, we obtain the quantization labels shown in Table 48.4. We have marked the upper left-hand coefficient with an *X* because that particular

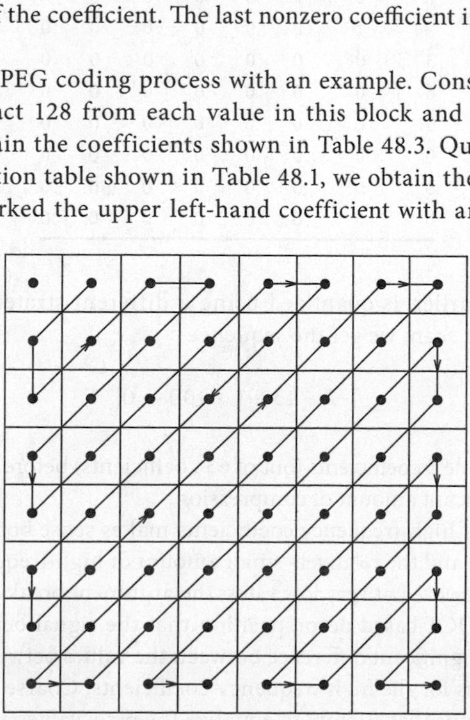

FIGURE 48.9 The zigzag scan used by the JPEG algorithm to scan the quantization labels.

TABLE 48.2 An 8 × 8 Block of Pixels

160	162	164	164	164	166	167	170
158	160	162	164	166	168	170	172
150	154	158	160	162	162	164	164
152	152	156	160	164	166	164	162
162	164	166	164	162	160	161	162
170	172	173	174	168	171	168	169
174	176	178	180	179	181	174	172
176	178	180	182	184	186	188	190

TABLE 48.3 The DCT of the Level-Shifted 8 × 8 Block of Pixels Shown in Table 48.2

317.3750	−18.1082	−7.1088	−3.2756	−1.3750	−0.1463	1.4563	−1.5100
−49.1029	−9.0994	2.5826	−1.2318	0.5473	−0.3857	−0.7265	1.6950
39.7441	−5.3676	3.2312	−0.4727	0.0676	−0.9051	0.1490	−0.5918
−0.1320	15.7378	1.7881	−2.2865	−0.5565	0.4095	1.2199	−2.0727
−1.8750	3.3699	2.7600	−0.1171	0.8750	−0.2135	−0.1961	1.7384
−4.9524	0.0853	−5.7080	4.0486	−0.6103	0.1149	1.1940	−0.9065
−1.7149	−0.7225	2.1490	−3.1358	−0.1633	0.4385	−0.4812	−0.3506
−4.8788	5.8341	−0.5948	−2.4700	−0.4537	−0.7444	−0.6175	1.2710

TABLE 48.4 Quantization Labels for the Coefficients Shown in Table 48.3 Using the Standard Quantization Stepsizes

X	−2	−1	0	0	0	0	0
−4	−1	0	0	0	0	0	0
3	0	0	0	0	0	0	0
0	0	0	0	0	0	0	0
0	0	0	0	0	0	0	0
0	0	0	0	0	0	0	0
0	0	0	0	0	0	0	0
0	0	0	0	0	0	0	0

coefficient, as mentioned earlier, is quantized using a different strategy. If we arrange the other coefficients using the zigzag scan, we get the sequence

$$[-2\ -4\ 3\ -1\ -1\ 0\ 0 \cdots 0]$$

The encoder only has to encode 5 coefficients (out of 63 coefficients) before sending the code for end-of-block. This results in a significant amount of compression.

The coarse quantization of high-frequency coefficients makes sense both in terms of the perceptual quality of the reconstruction and the relatively small amount of high-frequency content in images destined for human viewing. However, at very low rates, the argument breaks down. This is because of an implicit assumption in the DCT-based decomposition that the signal being transformed is periodic. This means that if there is a significant difference between the values between opposite edges of a block, it is reflected in higher values for the high-frequency coefficients. Coarse quantization of these coefficients results in reconstructions that attempt to equalize the pixel values at opposite edges. This means that pixels with large values representing brighter portions of the image will be reconstructed as pixels with smaller values that will be perceived as dark. This effect will be most apparent at the block boundaries leading to a *blocking* artifact. One way around this problem is to use a modified version of the DCT, known, appropriately enough, as the modified discrete cosine transform (MDCT). In this variation, the transform is taken over overlapping blocks, thus mitigating the blocking effect. Another is the use of subband and wavelet decompositions that we describe in the next section.

48.5.2 Wavelet and Subband-Based Coding

Using transforms is one way of separating the source information into spectral components. Another widely used approach relies on filterbanks rather than transforms for decomposing the signal into its spectral component. A typical subband coding system is shown in Figure 48.10. The discrete time input signal is passed through a low-pass filter and a high-pass filter with bandwidth equal to half the bandwidth of the original signal. Thus, the output of the low-pass filter contains the lower half of the frequencies in the input signal, while the output of the high-pass filter contains the upper half of the frequencies contained in the input signal. The Nyquist criterion dictates that the signal be sampled at a rate that is at least twice the highest-frequency component of the signal. Assuming that the original signal was sampled at the Nyquist rate or higher and assuming ideal filters, the output of the filters will be oversampled by at least a factor of two. Therefore, the output of the filters can be downsampled by a factor of two. This downsampling is indicated by the circle with the down-pointing arrow. After downsampling, the total number of samples in the upper and lower branches is equal to the number of samples in the original signal. The low-pass and high-pass signals can be quantized and coded depending on their characteristics and transmitted or stored. In order to reconstruct the signal, the decoder outputs are upsampled by a factor of two, filtered, and summed. Where the filters are not ideal, the downsampling can introduce aliasing distortion. The design of filters that prevent the introduction of aliasing distortion has been a major subject of research in subband coding. The various designs begin with a prototype low-pass filter and derive the other filters from the prototype filter to satisfy *perfect reconstruction* conditions.

In the example earlier, we show the signal being split into two bands. In various applications, the signal can be split into multiple bands. This can be done in a number of ways. One is to design *M*-band filters that split the input into *M* spectral bands followed by a downsampling by a factor of *M*. Another is to use a sequence of two band splits, where generally the low-pass output is split repeatedly. The reason for the latter approach is that most of the information in many natural signals is contained in the lower frequencies; and therefore, there is a benefit to a higher-resolution decomposition of the low-frequency band.

One of the most popular applications of subband coding is the algorithm for the compression of audio signals, which is part of the MPEG standard. There are three audio compression strategies, Layer I, Layer II, and Layer III. The Layer III standard is better known as *mp*3. Each layer is backwards compatible with the previous layer, which means they have substantial commonalities. The input in each case is decomposed into 32 subbands, followed by a 32:1 downsampling. In *mp*3, the output of each subband is further decomposed into 6 or 18 coefficients using an MDCT with 50% overlap. The reason for this level of spectral decomposition is the nature of the hearing process. The human auditory system has a frequency-selective auditory threshold. Signals at frequencies below 300 Hz or above 6 kHz have to be

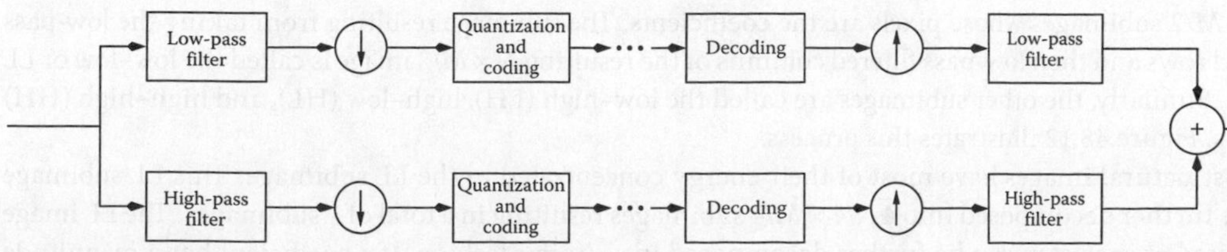

FIGURE 48.10 A basic subband coding system.

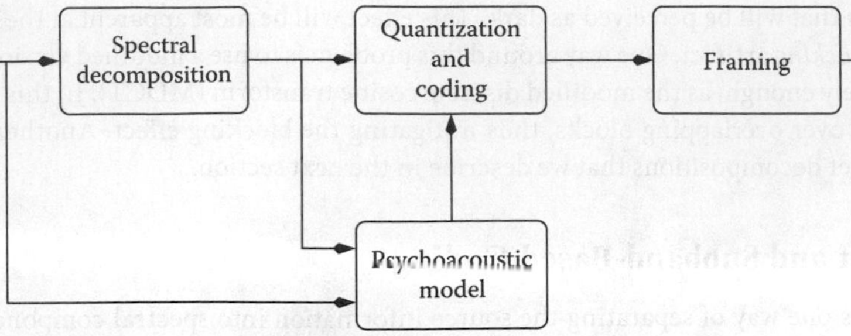

FIGURE 48.11 A conceptual picture of the *mp*3 algorithm.

substantially louder to be heard than signals in the range of 1–5 kHz. Thus, small amplitude signals in the output of some of the subbands can be discarded without any perceptible impact on the audio quality. Furthermore, the auditory system uses several masking strategies that can be exploited to improve the level of compression. If two tones occur within a narrow range of frequencies, known as the critical band, one of the tones will mask the other. This kind of masking is called spectral masking. The auditory system also uses temporal masking; if two tones occur very close to each other, in time, the louder tone will mask out the more quiet one. By not encoding components of the signal that are going to be masked by the human auditory system anyway, we can get significant compression without the loss of perceived fidelity. A conceptual block diagram of this process is shown in Figure 48.11. The option in *mp*3 to switch between an 18 and 6 coefficient MDCT is used to accommodate a wider range of sounds. For a more precise encoding of those portions of the audio signal that include time-limited sharp sounds, such as castanets, the six-coefficient MDCT is used. This permits a more accurate rendition of the sound by allocating more coding resources to the short period in which those sounds occur.

The filters used in subband coding are derived from a Fourier analysis point of view that can be thought of as a way of representing time functions using a basis set made up of complex exponentials that are defined for all time. Because these basis functions have support over the entire time axis, any representation using these basis functions provides no time localization. By looking at the spectral picture, one might be able to say that the signal contained a component of a particular frequency; but it is not possible to determine when that signal occurred. Wavelet analysis is based on translated and dilated versions of a time-limited waveform. This permits localization of components in both time and frequency. From the implementation point of view, however, subband decomposition and wavelet decomposition look very similar. In both cases, the decomposition is implemented using a setup very much like that shown in Figure 48.10.

While the method of decomposition is very similar, compression schemes using wavelets have employed very different coding strategies. The most well-known application of wavelet decomposition is the image compression standard known as JPEG2000 [28]. Unlike speech and audio signals, images are 2D objects; and, thus, the decomposition has to occur in two dimensions. For practical reasons, this decomposition is usually performed as a concatenation of two 1D decompositions. Each row of an $N \times M$ image is treated as a 1D signal and decomposed into a high and low band. After downsampling, we end up with two sets of coefficients of size $N \times M/2$. Each row of these coefficients is decomposed and subsampled to produce four $N/2 \times M/2$ subimages whose pixels are the coefficients. The subimage resulting from taking the low-pass filtered rows and then low-pass filtered columns of the resulting $N \times M/2$ image is called the low–low or LL image. Similarly, the other subimages are called the low–high (LH), high–low (HL), and high–high (HH) images. Figure 48.12 illustrates this process.

Most natural images have most of their energy concentrated in the LL subimage. This LL subimage can be further decomposed into 4 $N/4 \times M/4$ subimages resulting in a total of 7 subimages. The LL image generated at each step can be further decomposed in a similar fashion. If a coefficient has a magnitude greater than a particular value, it is said to be *significant* at that value. As with other wavelet-based image

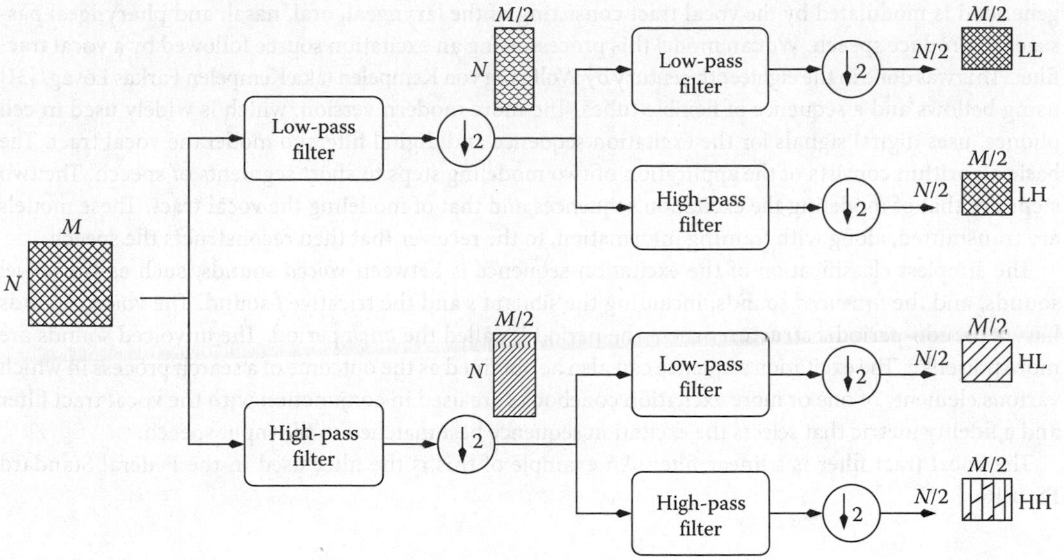

FIGURE 48.12 Decomposition of an $N \times M$ image into four $N/2 \times M/2$ images.

coding schemes [29,30], the JPEG2000 standard uses the concept of significance to improve compression efficiency. As the decomposition is being performed each time for coefficients corresponding to pixels at a particular spatial location, we can group coefficients depending on the pixels they correspond to. In these groups, if the low-frequency coefficient has a magnitude lower than a particular threshold, that is, it is insignificant at that threshold level, then it is highly likely that the other coefficients will also have values lower than that threshold. We can make use of this fact by comparing the low-frequency coefficient against a series of progressively lower threshold and then encoding entire groups of coefficients based on whether the low-frequency coefficient is significant or insignificant at that threshold. If we pick the thresholds to be powers of two, then the encoding of significant coefficients at progressively lower thresholds becomes the encoding of the bitplanes of an image.

The JPEG2000 algorithm divides the image into *tiles* that can be further subdivided into *precincts*. By encoding these subdivisions independently, the algorithm allows for random access to different parts of the image. Within each subdivision, the wavelet coefficients are encoded bitplane by bitplane using a sequence of three passes: a *significance propagation* pass, a *magnitude refinement* pass, and a *cleanup* pass. In the significance propagation pass, all bits in the neighborhood of coefficient locations previously declared significant are examined. Their significance status is encoded. If the coefficient becomes significant, the sign of the coefficient is encoded. In the magnitude refinement pass, the bits corresponding to coefficients declared significant in previous passes are encoded; and in the cleanup pass, all remaining bits are encoded. The bitstream thus generated is an embedded stream, that is, the earlier bits are more important to the reconstruction of the image. As more bits are made available, the decoder can progressively improve the quality of the reconstruction; we do not need to wait for the end of the bitstream to obtain a reconstruction of the image.

Once the bits have been generated and prior to framing, the JPEG2000 algorithm has the option of going through a post-compression rate distortion optimization step where the algorithm organizes the bits into quality layers. This provides the ability to generate the best reconstruction for a given number of bits.

48.5.3 Linear Predictive Coding

While images can be generated in a variety of ways, the mechanism used for generating human speech is relatively standard. Air is forced through a stretched membrane, the vocal cords; and the sound generated is modulated by the vocal tract consisting of the laryngeal, oral, nasal, and pharyngeal passages to produce speech. We can model this process using an excitation source followed by a vocal tract filter. This was done in the eighteenth century by Wolfgang von Kempelen (aka Kempelen Farkas Lovag) [31] using bellows and a sequence of flexible tubes. The more modern version, which is widely used in cell phones, uses digital signals for the excitation sequence and digital filters to model the vocal tract. The basic algorithm consists of the application of two modeling steps to short segments of speech. The two steps are that of modeling the excitation sequences and that of modeling the vocal tract. These models are transmitted, along with framing information, to the receiver that then reconstructs the speech.

The simplest classification of the excitation sequence is between *voiced* sounds, such as the vowel sounds, and the *unvoiced* sounds, including the sibilant *s* and the fricative *f* sound. The voiced sounds have a pseudo-periodic structure where the period is called the *pitch* period. The unvoiced sounds are more noiselike. The excitation sequence can also be obtained as the outcome of a search process in which various elements of one or more excitation codebooks are used in conjunction with the vocal tract filter and a fidelity metric that selects the excitation sequence best matched to the input speech.

The vocal tract filter is a linear filter. An example of this is the filter used in the Federal Standard FS 1016:

$$y_n = \sum_{i=1}^{10} b_i y_{n-i} + \beta y_{n-P} + G\epsilon_n$$

where

ϵ_n is the excitation signal
P is the pitch period
y_n is the synthesized speech

The term βy_{n-P} can be thought of as a correction term. The coefficients $\{b_i\}$ are known as the LP coefficients. These coefficients model the spectral envelope of the voice signal, and it can be shown [4] that the coefficients can be obtained by solving the set of equations:

$$\sum_{i=1}^{M} b_i R_{yy}(|i-j|) = R_{yy}(j); \quad j = 1, 2, \ldots, 10 \tag{48.42}$$

These equations can be efficiently solved using the Levinson–Durbin algorithm [32,33].

To accommodate the rapidly changing sounds that make up the speech signal, this process is repeated every 20–30 ms depending on the particular standard being used.

48.5.4 Video Coding

Video coding has generally been seen as a combination of predictive coding and transform coding. Frames of a video sequence are generally quite similar to each other with the same objects appearing in a succession of frames. Therefore, the approach is to use predictive coding along the temporal axis followed by transform coding of the residual. In order to preserve some random access capability, a frame is periodically encoded without prediction. If we did not do so, we would have to reconstruct all previous frames to reconstruct an individual frame. By periodically encoding a frame without using interframe information, called an I-frame in MPEG parlance, we only need to begin decoding from the nearest I-frame.

The prediction technique used in almost all video coding algorithms is called *motion-compensated prediction*. The frame being encoded is divided into square blocks. For each of the blocks, the algorithm

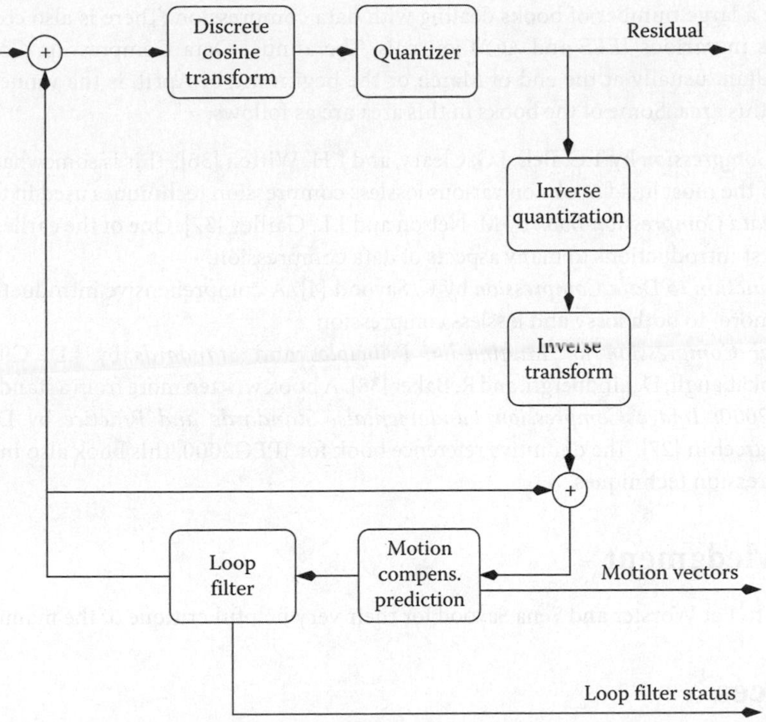

FIGURE 48.13 A block diagram of the H.261 video coding algorithm. (From ITU-T Recommendation H.261, video codec for audiovisual services at $p \times 64$ kbit/s, 1993.)

searches in the previous frame for the closest match in the spatial neighborhood of the frame being encoded. The idea is that as we move from frame to frame, objects will appear in slightly different locations in the frame. The horizontal and vertical offsets of the closest matching block are called the motion vector and can sometimes account for more than 90% of the video bitstream. The difference between the block being encoded and its motion compensated past is generally encoded using a transform coder. A block diagram of one of the earlier video coding schemes, H.261, is shown in Figure 48.13. While there are many differences in details between the H.261 standard and the more modern MPEG-1, MPEG-2, and H.264 (also known as MPEG-4 Part 10) standards, the basic approach has stayed the same. The standards H.261, MPEG-1, and MPEG-2 use an 8×8 DCT, while the H.264 standard uses a 4×4 integer DCT-like transform. While in H.261, the only prediction is based on past frames, the later standards allow for bidirectional prediction, that is, the prediction is formed using an average of the past and future frames. If bidirectional prediction is to be used, the transmission order cannot be causal as the frames used for generating the prediction need to be available to the decoder before the predicted frame can be reconstructed.

48.6 Conclusions and Further Reading

We have barely scratched the surface of the area of data compression in this review. We hope we have convinced the reviewer that in order to design good compression algorithms, one needs to understand the data that are being compressed. If the data can be effectively modeled, it can be compressed very efficiently. There is also a flip side to data compression algorithms. Because the data compression algorithms depend so heavily on discovering structure in the data, they can be used for the purpose of structure discovery rather than data compression. A review of some of the ways data compression algorithms can be used to understand biological sequences can be found in [35].

There are a large number of books dealing with data compression. There is also continued research that appears in various *IEEE* and *ACM* journals. The annual Data Compression Conference held in Snowbird, Utah, usually at the end of March or the beginning of April, is the venue for much of the research in this area. Some of the books in this area are as follows:

- *Text Compression* by T.C. Bell, J.G. Cleary, and I.H. Witten [36]. This is somewhat dated but is still one of the most lucid books on various lossless compression techniques used in text compression.
- *The Data Compression Book* by M. Nelson and J.L. Gailley [37]. One of the earliest and still one of the best introductions to many aspects of data compression.
- *Introduction to Data Compression* by K. Sayood [4]. A comprehensive introduction (and perhaps a bit more) to both lossy and lossless compression.
- *Digital Compression for Multimedia: Principles* and *Standards* by J.D. Gibson, T. Berger, T. Lookabaugh, D. Lindbergh, and R. Baker [38]. A book written more from a standards perspective.
- *JPEG2000: Image Compression Fundamentals, Standards and Practice* by D. Taubman and M. Marcellin [27]. The definitive reference book for JPEG2000, this book also introduces various compression techniques.

Acknowledgment

Our thanks to Pat Worster and Sena Sayood for their very helpful critique of the manuscript.

References

1. W. Li. The study of correlation structures of DNA sequences: A critical review. *Computers & Chemistry*, 21(4):257–271, 1997.
2. C.E. Shannon. A mathematical theory of communication. *Bell System Technical Journal*, 27:379–423, 623–656, 1948.
3. P.D. Grünwald and P.M.B. Vitányi. Kolmogorov complexity and information theory: With an interpretation in terms of questions and answers. *Journal of Logic, Language and Information*, 12:497–529, 2003.
4. K. Sayood. *Introduction to Data Compression*, 4th edn. Morgan Kauffman-Elsevier, San Francisco, CA, 2012.
5. D.A. Huffman. A method for the construction of minimum redundancy codes. *Proceedings of the IRE*, 40:1098–1101, 1951.
6. S. Pigeon. Huffman coding. In K. Sayood, ed., *Lossless Compression Handbook*, pp. 79–100. Academic Press, San Diego, CA, 2003.
7. R.G. Gallager. Variations on a theme by Huffman. *IEEE Transactions on Information Theory*, IT-24(6):668–674, November 1978.
8. C.E. Shannon. Coding theorems for a discrete source with a fidelity criterion. In *IRE International Convention Records*, vol. 7, pp. 142–163. IRE, New York, 1959.
9. T. Berger. *Rate Distortion Theory: A Mathematical Basis for Data Compression*. Prentice Hall, Englewood Cilffs, NJ, 1971.
10. S.P. Lloyd. Least squares quantization in PCM. *IEEE Transactions on Information Theory*, IT-28:127–135, March 1982.
11. K. Sayood, J.D. Gibson, and M.C. Rost. An algorithm for uniform vector quantizer design. *IEEE Transactions on Information Theory*, IT-30:805–814, November 1984.
12. T.R. Fischer. A pyramid vector quantizer. *IEEE Transactions on Information Theory*, IT-32:568–583, July 1986.
13. J.D. Gibson and K. Sayood. Lattice quantization. In P.W. Hawkes, ed., *Advances in Electronics and Electron Physics*, pp. 259–328. Academic Press, Boston, MA, 1990.

14. Y. Linde, A. Buzo, and R.M. Gray. An algorithm for vector quantization design. *IEEE Transactions on Communications*, COM-28:84–95, January 1980.
15. J. Ziv and A. Lempel. A universal algorithm for data compression. *IEEE Transactions on Information Theory*, IT-23(3):337–343, May 1977.
16. J. Ziv and A. Lempel. Compression of individual sequences via variable-rate coding. *IEEE Transactions on Information Theory*, IT-24(5):530–536, September 1978.
17. J.A. Storer and T.G. Syzmanski. Data compression via textual substitution. *Journal of the ACM*, 29:928–951, 1982.
18. T.A. Welch. A technique for high-performance data compression. *IEEE Computer*, 17(6):8–19, June 1984.
19. G. Roelofs. Png lossless compression. In K. Sayood, ed., *Lossless Compression Handbook*, pp. 371–390. Academic Press, San Diego, CA, 2003.
20. J.-L. Gailly and M. Adler. http://www.gzip.org/algorithm.txt (accessed on October 23, 2013).
21. J.G. Cleary and W.J. Teahan. Unbounded length contexts for PPM. *The Computer Journal*, 40:x30–x74, February 1997.
22. M. Weinberger, G. Seroussi, and G. Sapiro. The LOCO-I lossless compression algorithm: Principles and standardization into JPEG-LS. Technical report HPL-98-193, Hewlett-Packard Laboratory, Palo Alto, CA, November 1998.
23. S.D. Babacan and K. Sayood. Predictive image compression using conditional averages. In *Proceedings of the Data Compression Conference, DCC '04*, Snowbird, UT. IEEE, 2004.
24. H. Wang, S.D. Babacan, and K. Sayood. Lossless hyperspectral image compression using context-based conditional averages. *IEEE Transactions on Geoscience and Remote Sensing*, 45:4187–4193, December 2007.
25. A. Gersho and R.M. Gray. *Vector Quantization and Signal Compression*. Kluwer Academic Publishers, Boston, MA, 1991.
26. X. Wang, S.M. Shende, and K. Sayood. Online compression of video sequences using adaptive vector quantization. In *Proceedings Data Compression Conference 1994*, Snowbird, UT. IEEE, 1994.
27. D. Taubman and M. Marcellin. *JPEG2000: Image Compression Fundamentals, Standards and Practice*. Kluwer Academic Press, Boston, MA, 2001.
28. ISO/IEC JTC1/SC29 WG1. JPEG 2000Image coding system: Core coding system, March 2004.
29. J.M. Shapiro. Embedded image coding using zerotrees of wavelet coefficients. *IEEE Transactions on Signal Processing*, SP-41:3445–3462, December 1993.
30. A. Said and W.A. Pearlman. A new fast and efficient coder based on set partitioning in hierarchical trees. *IEEE Transactions on Circuits and Systems for Video Technologies*, 6(3):243–250, June 1996.
31. H. Dudley and T.H. Tarnoczy. Speaking machine of Wolfgang von Kempelen. *Journal of the Acoustical Society of America*, 22:151–166, March 1950.
32. N. Levinson. The Weiner RMS error criterion in filter design and prediction. *Journal of Mathematical Physics*, 25:261–278, 1947.
33. J. Durbin. The fitting of time series models. *Review of the International Statistical Institute*, 28:233–243, 1960.
34. ITU-T Recommendation H.261. Video codec for audiovisual services at $p \times 64$ kbit/s, 1993.
35. O.U. Nalbantoglu, D.J. Russell, and K. Sayood. Data compression concepts and their applications to bioinformatics. *Entropy*, 12:34–52, 2010.
36. T.C. Bell, J.C. Cleary, and I.H. Witten. *Text Compression*. Advanced reference series. Prentice Hall, Englewood Cliffs, NJ, 1990.
37. M. Nelson and J.-L. Gailly. *The Data Compression Book*. M&T Books, California, 1996.
38. J.D. Gibson, T. Berger, T. Lookabaugh, D. Lindbergh, and R. Baker. *Digital Compression for Multimedia: Principles and Standards*. Morgan Kaufmann Publishers, San Francisco, CA, 1998.

49

Localization in Underwater Acoustic Sensor Networks*

Baozhi Chen
Rutgers University

Dario Pompili
Rutgers University

49.1 Introduction

UnderWater Acoustic Sensor Networks (UW-ASNs) [1] consist of a number of sensors and vehicles that interact to collect data and perform tasks in a collaborative manner underwater. They have been deployed to carry out collaborative monitoring tasks including oceanographic data collection, climate monitoring, disaster prevention, and navigation. Autonomous underwater vehicles (AUVs) are widely believed to be revolutionizing oceanography and are enabling research in environments that have typically been impossible or difficult to reach [21]. For example, AUVs have been used for continuous measurement of fresh water exiting the Arctic through the Canadian Arctic Archipelago and Davis Strait in order to study the impact of climate change to the circulation of the world's oceans. The ability to do so under ice is important so that, for example, scientists can measure how much fresh water flows through the strait—and at what times of year—so they have a baseline for comparison in coming years.

For these missions, position information is of vital importance in mobile underwater sensor networks, as the data collected have to be associated with appropriate location in order to be spatially reconstructed onshore. Even though AUVs can surface periodically (e.g., every few hours) to locate themselves using global positioning system (GPS)—which does not work underwater—over time, inaccuracies in models for deriving position estimates, self-localization errors, and drifting due to ocean currents will significantly increase the uncertainty in position of underwater vehicle. Moreover, in

* This chapter is based on our paper in *Proceedings of IEEE Conference on Sensor, Mesh and Ad Hoc Communications and Networks (SECON)*, Seoul, Korea, June 2012 [9].

extreme environments such as under ice, surfacing to get a GPS update is hardly possible and, therefore, position information is highly uncertain. In such environments, relying on standard navigation techniques such as long baseline (LBL) navigation is difficult as the use of static LBL beacons typically limits the operation range to about 10 km [18] and requires great deployment efforts before operation, especially in deep water (more than 100 m deep).

As AUVs are becoming more and more capable and also affordable, deployment of multiple AUVs to finish one mission becomes a widely adopted option. This not only enables new types of missions through cooperation but also allows individual AUVs of the team to benefit from information obtained from other AUVs. Existing localization schemes underwater generally rely on the deployment of transponders or nodes with underwater communication capabilities as reference points, which requires either much deployment effort or much communication overhead. Moreover, these schemes are not able to estimate the uncertainty associated with the calculated position, which is high in under-ice environments, and thus are not able to minimize position uncertainty.

To address this problem, we propose a solution that uses only a subset of AUVs without relying on localization infrastructure. Specifically, a position uncertainty model in [8] is introduced to characterize an AUV's position. This model is extended to estimate the uncertainty associated with the standard distance-based localization technique, resulting in the distance-based localization with uncertainty estimate (DISLU). We further propose a Doppler-based technique with uncertainty estimation capability, which is called Doppler-based localization with uncertainty estimate (DOPLU). The DISLU technique relies on packets (i.e., communication overhead) to measure the inter-vehicle distances (i.e., ranging), which, in conjunction with positions of reference nodes, are utilized to estimate the position. On the other hand, DOPLU, which measures Doppler shifts from ongoing communications and then uses these measurements to calculate velocities for localization, removes the need for ranging packets. As DOPLU only relies on relative measurements, it may not be able to fix displacement errors introduced by the rotation of the AUV group. In this case, DISLU is executed to bound such localization errors. Considering these tradeoffs, using the uncertainty model, the localization error and communication overhead of DISLU and DOPLU can be jointly considered and algorithms are devised to minimize the localization uncertainty and communication overhead while satisfying localization error requirement.

Our solution offers a way to estimate the degree of uncertainty associated with a localization technique and based on this estimation it further minimizes both position uncertainty and communication overhead. The contributions of this work include (1) a probability model to estimate the position uncertainty associated with localization techniques; (2) an algorithm to minimize localization uncertainty by selecting an appropriate subset of reference nodes (in general, other AUVs); (3) an algorithm to optimize the localization interval in order to further minimize the localization overhead; and (4) a Doppler-based localization technique that can exploit ongoing communications for localization.

The remainder of this chapter is organized as follows. In Section 49.2, we introduce the basic knowledge on UW-ASNs. In Section 49.3, we review the related work for localization algorithms in UW-ASNs. We present the motivation and background in Section 49.4 and propose our localization solution in Section 49.5; in Section 49.6, performance evaluation and analysis are carried out, while conclusions are discussed in Section 49.7.

49.2 Basics of Underwater Acoustic Sensor Networks

UW-ASNs are applied in a broad range of applications, including environmental monitoring, undersea exploration, disaster prevention, assisted navigation, and tactical surveillance. Underwater networking is a rather unexplored area although underwater communications have been experimented since World War II, when, in 1945, an underwater telephone was developed in the United States to communicate with submarines [24]. Acoustic communications are the typical physical layer technology in underwater networks. In fact, radio waves propagate at long distances through conductive sea water only at extra low frequencies (30–300:Hz), which require large antennae and high transmission power.

For example, the Berkeley Mica 2 Motes, the most popular experimental platform in the sensor networking community, have been reported to have a transmission range of 120 cm underwater at 433:MHz by experiments performed at the Robotic Embedded Systems Laboratory (RESL) at the University of Southern California. Optical waves do not suffer from such high attenuation but are affected by scattering. Moreover, transmission of optical signals requires high precision in pointing the narrow laser beams. Thus, *acoustic waves* are generally used for underwater communications [26].

The traditional approach for ocean-bottom or ocean-column monitoring is to deploy underwater sensors that record data during the monitoring mission and then recover the instruments [22]. This approach has the following disadvantages:

1. *No real-time monitoring*: The recorded data cannot be accessed until the instruments are recovered, which may happen several months after the beginning of the monitoring mission. This is critical especially in surveillance or in environmental monitoring applications such as seismic monitoring.
2. *No online system reconfiguration*: Interaction between onshore control systems and the monitoring instruments is not possible. This impedes any adaptive tuning of the instruments, nor is it possible to reconfigure the system after particular events occur.
3. *No failure detection*: If *failures* or *misconfigurations* occur, it may not be possible to detect them before the instruments are recovered. This can easily lead to the complete failure of a monitoring mission.
4. *Limited storage capacity*: The amount of data that can be recorded during the monitoring mission by every sensor is limited by the capacity of the onboard storage devices (memories and hard disks).

Therefore, there is a need to deploy underwater networks that will enable real-time monitoring of selected ocean areas, remote configuration, and interaction with onshore human operators. This can be obtained by connecting underwater instruments by means of wireless links based on acoustic communication.

Many researchers are currently engaged in developing networking solutions for terrestrial wireless ad hoc and sensor networks. Although there exist many recently developed network protocols for wireless sensor networks, the unique characteristics of the underwater acoustic communication channel, such as limited bandwidth capacity and variable delays [23], require very efficient and reliable new data communication protocols.

Major challenges in the design of underwater acoustic networks are

- The available bandwidth is severely limited
- The underwater channel is severely impaired, especially due to multi-path and fading problems
- Propagation delay in underwater is five orders of magnitude higher than in radio frequency (RF) terrestrial channels and extremely variable
- High bit error rates and temporary losses of connectivity (shadow zones) can be experienced due to the extreme characteristics of the underwater channel
- Battery power is limited and usually batteries cannot be recharged, also because solar energy cannot be exploited
- Underwater sensors are prone to failures because of fouling and corrosion

Underwater acoustic communications are mainly influenced by *path loss*, *noise*, *multipath*, *Doppler spread*, and *high and variable propagation delay*. All these factors determine the *temporal and spatial variability* of the acoustic channel and make the available bandwidth of the *underwater acoustic channel* limited and dramatically dependent on both range and frequency. Long-range systems that operate over several tens of kilometers may have a bandwidth of only a few kHz, while a short-range system operating over several tens of meters may have more than a hundred kHz of bandwidth. In both cases, these factors lead to low bit rate [5], in the order of tens of kbps for existing devices.

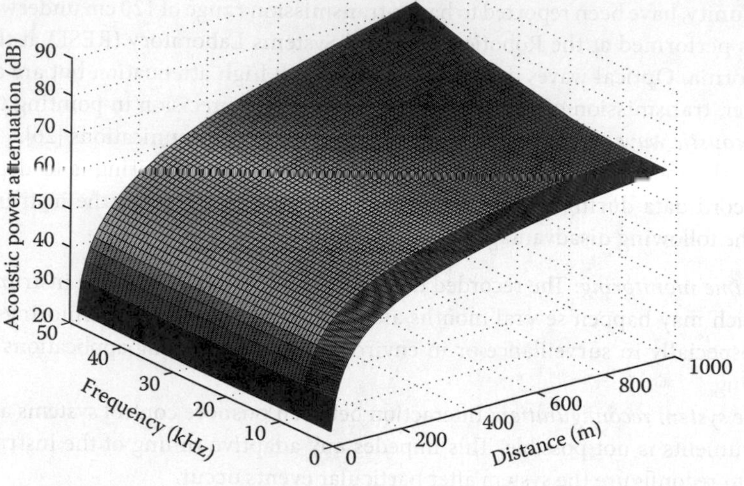

FIGURE 49.1 Path loss of short-range acoustic channel vs. distance and frequency in band 1–50 kHz.

Hereafter we analyze the factors that influence acoustic communications in order to state the challenges posed by the underwater channels for underwater sensor networking. These include the following.

Path loss: *Attenuation* is mainly provoked by absorption due to conversion of acoustic energy into heat. The attenuation increases with distance and frequency. Figure 49.1 shows the acoustic attenuation with varying frequency and distance for a short-range shallow water acoustic channel, according to the propagation model in [27]. The attenuation is also caused by scattering and reverberation (on rough ocean surface and bottom), refraction, and dispersion (due to the displacement of the reflection point caused by wind on the surface). Water depth plays a key role in determining the attenuation. *Geometric spreading* refers to the spreading of sound energy as a result of the expansion of the wavefronts. It increases with the propagation distance and is independent of frequency. There are two common kinds of geometric spreading: *spherical* (omni-directional point source), which characterizes deep water communications, and *cylindrical* (horizontal radiation only), which characterizes shallow water communications.

Noise: *Man-made noise* is mainly caused by machinery noise (pumps, reduction gears, and power plants) and shipping activity (hull fouling, animal life on hull, and cavitation), especially in areas encumbered with heavy vessel traffic. *Ambient noise* is related to hydrodynamics (movement of water including tides, current, storms, wind, and rain) and to seismic and biological phenomena. In [15], boat noise and snapping shrimps have been found to be the primary sources of noise in shallow water by means of measurement experiments on the ocean bottom.

Multipath: Multipath propagation may be responsible for severe degradation of the acoustic communication signal since it generates inter-symbol interference (ISI). The multipath geometry depends on the link configuration. Vertical channels are characterized by little time dispersion, whereas horizontal channels may have extremely long multipath spreads. The extent of the spreading is a strong function of depth and the distance between transmitter and receiver.

High delay and delay variance: The propagation speed in the acoustic channel is five orders of magnitude lower than in the radio channel. This large propagation delay (0.67 s/km) can reduce the throughput of the system considerably. The very high delay variance is even more harmful for efficient protocol design as it prevents from accurately estimating the round trip time (RTT), which is the key parameter for many common communication protocols.

Doppler spread: The Doppler frequency spread can be significant in acoustic channels [26], causing a degradation in the performance of digital communications: transmissions at a high data rate cause many adjacent symbols to interfere at the receiver, requiring sophisticated signal processing to deal with the generated ISI. The Doppler spreading generates a simple frequency translation, which is relatively easy for a receiver to compensate for; and a continuous spreading of frequencies, which constitutes a nonshifted signal, that is more difficult to compensate for. If a channel has a Doppler spread with bandwidth B and a signal has symbol duration T, then there are approximately BT uncorrelated samples of its complex envelope. When BT is much less than unity, the channel is said to be *underspread* and the effects of the Doppler fading can be ignored, while, if greater than unity, it is said to be *overspread* [16].

Most of the described factors are caused by the chemical–physical properties of the water medium such as temperature, salinity, and density, and by their spatio-temporal variations. These variations, together with the wave guide nature of the channel, cause the acoustic channel to be *highly temporally and spatially variable*. In particular, the horizontal channel is by far more rapidly varying than the vertical channel, in both deep and shallow water.

49.3 Related Work

A number of underwater localization schemes, many of which are summarized in [6], have been proposed to date, which take into account a number of factors like the network topology, device capabilities, signal propagation models, and energy requirements. Most of these localization schemes require the positions of some nodes be known. In order to estimate the positions of other nodes, measurements such as distances and angles are made. Range-based schemes, use round-trip time (RTT), time of arrival (ToA), time difference of arrival (TDoA), received-signal-strength (RSS), or angle of arrival (AoA) to estimate their distances to other nodes. Schemes that rely on ToA or TDoA require tight time synchronization between the transmitter and the receiver, whereas RTT does not. However, RTT comes with a higher network cost and associated error. RSS has been implemented in [10], but it comes with an additional network cost.

An hierarchical localization scheme involving surface buoys, anchor nodes, and ordinary nodes is proposed in [30]. Surface buoys are GPS based and used as references for positioning by other nodes. Anchor nodes communicate with surface buoys while ordinary nodes only communicate with the anchor nodes. This distributed localization scheme applied 3D Euclidean distance and recursive location estimation method for the position calculation of ordinary nodes. Mobility of the sensor nodes was not considered in the position estimate.

A recent proposal uses surface-based signal reflection for underwater localization [10]. This approach attempts to overcome limitations imposed by line of sight (LOS) range measurement techniques such as RSS, TOA, and AoA. These limitations are caused by multipath, line of sight attenuation and required reference nodes. The receiver in this approach accepts only signals that have been reflected off the surface. It accomplishes this by applying homomorphic deconvolution to the signal to obtain an impulse response, which contains RSS information. The algorithm then checks the RSS and compares it to calculated reflection coefficients. Mobility is allowed and high localization accuracy can be achieved.

These localization solutions are generally designed for UW-ASNs, which may not include mobile nodes such as AUVs. In the rest of this section, we review the work that is most related, that is, localization in UW-ASNs *using AUVs*. Localization is essential for underwater vehicle navigation, where many localization solutions, as summarized in [18], have been proposed. Short baseline (SBL) and long baseline (LBL) systems [18] are standard ways to localize vehicles underwater, where external transponder arrays are employed to aid localization. In SBL systems, position estimate is determined by measuring the vehicle's distance from three or more transponders that are, for example, lowered over the side of the surface vessel. The LBL systems are similar to SBL, with the difference that an array of transponders is tethered on the ocean bed with fixed locations.

In [12], a localization scheme called AUV aided localization (AAL) is proposed, where position estimation is done using a single AUV. In AAL, an AUV navigates a predefined trajectory, broadcasts its position upon a node's request, and fixes its own position at the surface. Each node estimates the distances to the AUV while the AUV is at different locations, using the RTT between itself and the AUV. Algorithms such as triangulation or bounding box can then be used for position estimate. Another localization solution called dive-and-rise localization (DNRL) is proposed for both static and mobile networks in [11]. The DNRL solution is similar to AAL, with the difference that ocean currents are considered and time synchronization is required between nodes.

In [14], an online algorithm for cooperative localization of submerged AUVs is designed, implemented, and evaluated through experiments. This algorithm relies on a single surface vehicle called communication and navigation aid (CNA) for autonomous navigation. Using the CNA's GPS positions and basic onboard measurements including velocity, heading, and depth, this algorithm can use filtering techniques such as extended Kalman filter (EKF) to bound the error and uncertainty of the onboard position estimates of a low-cost AUV.

Simultaneous localization and mapping (SLAM), also known as concurrent mapping and localization (CML), attempts to merge two traditionally separated concepts, map building and localization. SLAM has been investigated using an imaging sonar and Doppler velocity log (DVL) in combination with dead reckoning [25]. However, despite recent research efforts this technology is still in its infancy and many obstacles still need to be overcome. As of November 2010, there have been several simulations carried out [29], but currently there is no operating solution to the AUV SLAM problem.

Among existing underwater localization techniques (which are generally not suitable for under-ice environments), relatively few under-ice localization techniques have been proposed. Despite these efforts, the technology remains expensive and out of reach for researchers. Current techniques employed in the under-ice environment include combinations of either dead-reckoning using inertial measurements [28], sea-floor acoustic transponder networks such as SBL or LBL [2], and/or a DVL that can be either seafloor or ice relative [17]. These current approaches require external hardware, are cost prohibitive, and suffer from error propagation. For accurate dead reckoning, highly accurate sensors are required because magnetic navigation systems are subject to local magnetic field variations and gyros are subject to drift over time. Quality inertial navigation sensors often cost more than $10,000 [18]. In contrast, our solution is much more economical as it does not require these expensive sensors.

Two solutions for underwater collaborative localization using a probability framework are proposed in [19] and [20], where a sum-product algorithm and a Markov process that are based on the so-called factor graph are used to model the joint distribution of multiple nodes. Both solutions require the global information of the nodes that are involved in localization, which leads to high computation complexity and communication overhead. Our solution offers another probability framework that leverages the self-estimated uncertainty distribution for estimation of other nodes. Therefore, global information is not required, resulting in reduced computation complexity and communication overhead.

49.4 Motivation and Background

In UW-ASNs, inaccuracies in models for position estimation, self-localization errors, and drifting due to ocean currents will significantly increase the uncertainty in position of an underwater vehicle. *Hence, using a deterministic point is not enough to characterize the position of an AUV.* Furthermore, such a deterministic approach underwater may lead to problems such as routing errors in inter-vehicle communications, vehicle collisions, lose of synchronization, and mission failures. In order to address the problems due to position uncertainty, we introduce a probability model to characterize a node's position. In many applications such as geographic routing, AUVs need to estimate the positions of themselves and other AUVs. Therefore, depending on the view of the different nodes, two forms of position uncertainty are defined, that is, *internal uncertainty*, the position uncertainty associated with a particular entity/node (such as an AUV) *as seen by itself*, and *external uncertainty*, the position uncertainty associated with a particular entity/node *as seen by others*.

These two notions introduce a shift in AUV localization: *from a deterministic to a probabilistic view*. This shift can then be leveraged to improve the performance of solutions for problems in a variety of fields.

For example, in *UW-ASNs*, using the external-uncertainty region, routing errors can be decreased and a node can estimate better the transmission power to guarantee the signal-to-noise ratio for correct data reception by taking into account not only channel impairments but also position uncertainty [8]. The notion of external uncertainty can also be used in *underwater robotics* to minimize the risk of vehicle collisions; and in *underwater localization* to decrease the position uncertainty by selecting a proper subset of nodes to be used as "references," as shown later.

Many approaches such as those using Kalman filter (KF) [3] have been proposed to estimate the internal uncertainty assuming that the variables to be estimated have linear relationships between each other and that noise is additive and Gaussian. While simple and quite robust, KF is not optimal when the linearity assumption between variables does not hold. On the other hand, approaches using nonlinear filters such as the extended or unscented KF attempt to minimize the mean squared errors in estimates by jointly considering the navigation location and the sensed states or features such as underwater terrain features, which are nontrivial, especially in an unstructured underwater environment.

Let us denote the internal uncertainty, a 3D region associated with any node $j \in \mathcal{N}$, where $\mathcal{N}$ is the set of network nodes, as $\mathcal{U}_{jj}$, and the external uncertainties, 3D regions associated with j as seen by $i, k \in \mathcal{N}$, as $\mathcal{U}_{ij}$ and $\mathcal{U}_{kj}$, respectively ($i \neq j \neq k$). In general, $\mathcal{U}_{jj}$, $\mathcal{U}_{ij}$, and $\mathcal{U}_{kj}$ are different from each other; also, due to information asymmetry, $\mathcal{U}_{ij}$ is in general different from $\mathcal{U}_{ji}$. External uncertainties may be derived from the broadcast/propagated internal-uncertainty estimates (e.g., using *one-hop* or *multihop neighbor discovery mechanisms*) and, hence, will be affected by end-to-end (e2e) *network latency* and *information loss*.

49.5 Proposed Localization Solution

With the notion of external uncertainty, we can model the uncertainty associated with localization techniques. Based on this uncertainty, optimization problems are formulated to minimize localization uncertainty and communication overhead. In this section, we first show how external uncertainty can be used to estimate the uncertainty with the standard distance-based localization technique (i.e., DISLU). Then we propose a novel Doppler-based localization technique DOPLU that jointly estimates localization uncertainty. The DISLU technique requires ranging packets to measure the distances for position calculation, which introduces communication overhead. This weakness in DISLU can be offset by DOPLU, which exploits ongoing inter-vehicle communications to avoid the need for ranging packets. Such an "opportunistic" approach (i.e., DOPLU) does not guarantee correct absolute locations (as Doppler shifts only characterize *relative* position change) so the team of AUVs needs to go back to DISLU to correct the locations when the error is too large. Based on this idea, we propose algorithms to solve two optimization problems, one for minimization of localization uncertainty and the other for minimization of communication overhead.

The communication protocol for our solution is presented in Figure 49.2. Each AUV first runs DISLU using the distances measured from the round-trip time. Then, DOPLU is run using Doppler-shift

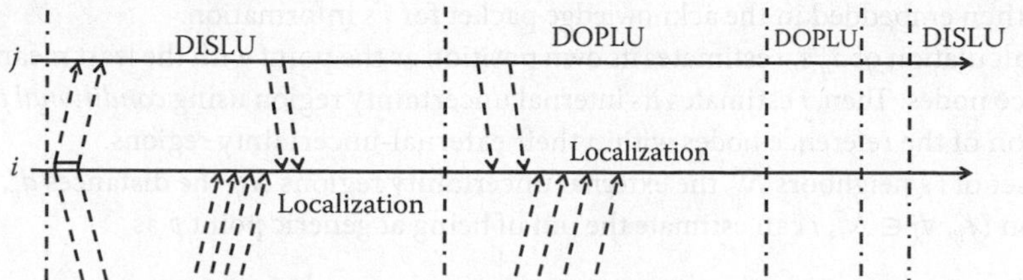

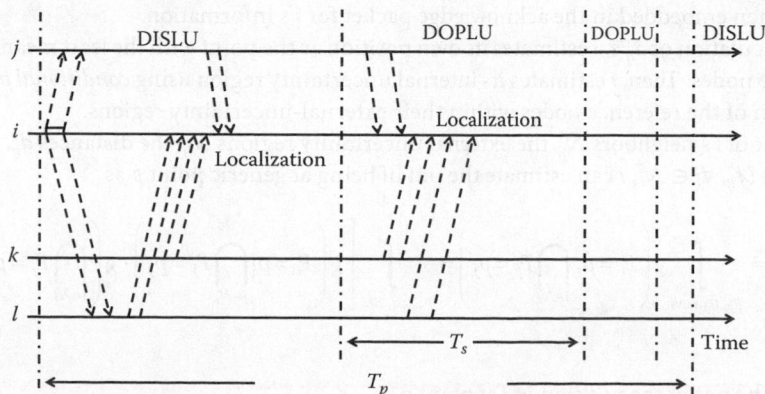

FIGURE 49.2 Overview of the proposed approach.

information extracted from inter-vehicle packets. By overhearing the ongoing packets from the reference nodes, AUV i estimates the Doppler shifts and then extracts the relative velocity, from which the AUVs calculate their absolute velocities. The DISLU technique is run to fix the localization error introduced by DOPLU after T_p, which is the time after the last DISLU is started.

Both DISLU and DOPLU use the external uncertainty and corresponding probability distribution function (pdf) to estimate the uncertainty resulted from the localization technique, that is, the internal uncertainty and pdf of the AUV running the localization algorithm. Then this internal uncertainty information is broadcast for other AUVs to estimate external uncertainties. Our previous work in [8] provided *a statistical solution* to estimate the internal and external uncertainty, which is used for initial estimation here.

49.5.1 Distance-Based Localization with Uncertainty Estimate (DISLU)

We present here the DISLU technique, which is based on the following idea, to estimate its own position, vehicle i needs (1) to *estimate the distances* between itself and its reference vehicles, and (2) to *estimate its own position* based on these distances.

The DISLU technique relies on the round-trip time T_{RTT} to measure the inter-vehicle distance. By extracting the one-way propagation time, i is able to calculate the inter-vehicle distance. That is, the distance between transmitter i and receiver j is $d_{ij} = c\cdot\left(T_{RTT} - T_i^{(TX)} - T_j^{(TX)} - T_j^{(hold)}\right)/2$, where $T_i^{(TX)}$ and $T_j^{(TX)}$ are the duration to transmit the packet at i and the duration to transmit acknowledgment at j (i.e., transmission delays), $T_j^{(hold)}$ is the holdoff time of j to avoid collisions. To reduce the transmission time, we can use the short ping packets. Once j receives the ping packet, it starts a hold-off timer, $T_j^{(hold)}$, which is a uniformly distributed random variable in $\left[0, 2T_{hold}^{mean}\right]$ where T_{hold}^{mean} is given by

$$T_{hold}^{mean} = \left(1 - \frac{d_{ij}}{R}\right)\tau + \frac{\phi_{ij}}{c}, \tag{49.1}$$

where

d_{ij} is the distance from AUV i to j
τ is the estimated transmission time for the current packet
$c = 1500$ m/s is the propagation speed of acoustic waves
R is the transmission radius of the underwater modem
$\phi_{ij} = \max\{0, R - d_{ij}\}$

The first term in (49.1) gives less time to the neighbor that is closer to i, and the second term is the extra delay that a node should wait so that all the nodes receive the packet. This gives fairness by providing synchronization in starting the hold-off timers of all the nodes receiving the data packet. The hold-off time $T_j^{(hold)}$ is then embedded in the acknowledge packet for i's information.

After the calculation of d_{ij}'s, i estimates its own position as the point with the least mean squared error to the reference nodes. Then, i estimates its internal uncertainty region using *conditional probability* and the distribution of the reference nodes within their external-uncertainty regions.

Given the set of i's neighbors $\mathcal{N}_i$, the external uncertainty regions $\mathcal{U}_{ij}$, the distances d_{ij}, and the pdf of j within region $\mathcal{U}_{ij}$, $\forall j \in \mathcal{N}_i$, i can estimate the pdf of being at generic point p as

$$g(P_i = p) = \int_{p_j \in \mathcal{U}_{ij}, j \in \mathcal{N}_i} g\left(P_i = p, \bigcap_{j \in \mathcal{N}_i} P_j = p_j\right) = \int_{p_j \in \mathcal{U}_{ij}, j \in \mathcal{N}_i} \left[g\left(P_i = p \,\middle|\, \bigcap_{j \in \mathcal{N}_i} P_j = p_j\right) \cdot g\left(\bigcap_{j \in \mathcal{N}_i} P_j = p_j\right)\right] \tag{49.2}$$

where

$g(P_i = p)$ is the pdf of the position of i at point p
$g(|)$ denotes conditional probability function

In our solution, p is calculated as the point that has the minimum squared error, that is, $p \in S_i$, where $S_i \equiv \left\{ q = \arg\min \sum_{j \in \mathcal{N}_i} \left\| d(p, p_j) - d_{ij} \right\|^2 \right\}$ (i.e., $\mathcal{U}_{ii} = S_i$). Here, $d(p, p_j)$ is the distance between point p and p_j. We assume point p to be uniformly distributed in S_i. In other words, we have

$$g(P_i = p \mid P_j = p_j, j \in \mathcal{N}_i) = \begin{cases} 1/|S_i|, & p \in S_i \\ 0, & p \notin S_i \end{cases} \tag{49.3}$$

where $|S_i|$ is the number of elements in S_i if S_i is a discrete set, or the area (or volume) of S_i if S_i is a nonempty nondiscrete set.

The joint pdf, $g\left(\bigcap_{j \in \mathcal{N}_i} P_j = p_j\right)$, can be approximated as

$$g\left(\bigcap_{j \in \mathcal{N}_i} P_j = p_j\right) \approx \prod_{j \in \mathcal{N}_j} g\left(P_j = p_j\right) \tag{49.4}$$

as the distributions of these AUVs are independent. Therefore, (49.2) can be expanded as

$$g(P_i = p) \approx \int_{p_j \in \mathcal{U}_{ij}, j \in \mathcal{N}_i} \left[g\left(P_i = p \mid \bigcap_{j \in \mathcal{N}_i} P_j = p_j\right) \cdot \prod_{j \in \mathcal{N}j} g\left(P_j = p_j\right) \right] \tag{49.5}$$

Hence, i's internal uncertainty $\mathcal{U}_{ii}$ with $g()$ being the pdf is estimated, which is then broadcast to other AUVs. The AUVs receiving this information then use $\mathcal{U}_{ii}$ to estimate i's external uncertainty.

49.5.2 Doppler-Based Localization with Uncertainty Estimate (DOPLU)

The DOPLU technique runs between two consecutive runs of DISLU. Obviously, whenever the Doppler shifts from more than three nodes are extracted, DOPLU can be run. The time between two consecutive runs of the DISLU is divided into subslots with appropriate duration T_s (Figure 49.2) so that the DOPLU will be run at an appropriate frequency. Within each subslot, the vehicle that runs DOPLU extracts Doppler shifts from the packet it overhears (even if the packet is not intended to be received by it) from the reference vehicles. With the additional information it obtains from the packet header (such as velocity of the reference node), it computes its own absolute velocity, which is then used to estimate its own position and internal uncertainty. This reduces the communication overhead for sending packets to estimate inter-vehicle distance.

An algorithm is designed so that the duration T_s can be adjusted dynamically according to the frequency of ongoing communication activities. Within T_s, a AUV is expected to collect enough Doppler shifts from its reference neighbors so that the DOPLU algorithm runs efficiently. Note that if T_s is too small, it is very likely that the velocity calculated by DOPLU is close to that obtained from the last calculation, which means waste of computation resources. On the other hand, T_s should not be too large as it would lead to too much localization error. After all, the less frequent a AUV calculates its position, the more position error accumulates.

In the rest of this section, we focus on the main problem, that is, how to estimate the position and internal uncertainty when Doppler shifts are available, and leave the optimization of T_s in Section 49.5.3. Using the Doppler shifts regarding to the reference nodes, i can estimate its own absolute velocity using the projected positions (i.e., by adding history position with history velocity times the time passed) and velocities. Using this relationship for all reference nodes, i obtains an equation group to solve, where absolute velocity $\vec{v}_i$ can be estimated.

To see how to calculate the absolute velocity, assume that at the end of one subslot, AUV i has collected the Doppler shift Δf_{ij} from reference node j. From the definition of Doppler shift, we have $\Delta f_{ij} = -\frac{\mathbf{v}_{ij} \circ \overrightarrow{P_iP_j}}{\left\|\overrightarrow{P_iP_j}\right\|}\frac{f_0}{c}$, where $\mathbf{v}_{ij}$ is the relative velocity of i to j, $\overrightarrow{P_iP_j}$ is the position vector from i to j, f_0 is the carrier frequency, c = 1500 m/s is the speed of sound, and $\circ$ is the inner product operation. From this equation, we have

$$\frac{\overrightarrow{\mathbf{v}_{ij}} \circ \overrightarrow{P_iP_j}}{\left\|\overrightarrow{P_iP_j}\right\|} = -\Delta f_{ij}\frac{c}{f_0}. \tag{49.6}$$

From (49.6), assume that i has collected the Doppler shifts of $N_i^{(ref)}$ reference nodes, we then have an equation group with $N_i^{(ref)}$ equations. We then can derive i's velocity $\vec{\mathbf{v}}_i$. Assume $\vec{\mathbf{v}}_i = \left(v_x^{(i)}, v_y^{(i)}, v_z^{(i)}\right)$ and $\frac{\overrightarrow{P_iP_j}}{\left\|\overrightarrow{P_iP_j}\right\|} = \left(\alpha_x^{(ij)}, \alpha_y^{(ij)}, \alpha_z^{(ij)}\right)$, (49.6) is then $\mathbf{v}_{ij} \circ \frac{\overrightarrow{P_iP_j}}{\left\|\overrightarrow{P_iP_j}\right\|} = \left(\vec{\mathbf{v}}_i - \vec{\mathbf{v}}_j\right) \circ \frac{\overrightarrow{P_iP_j}}{\left\|\overrightarrow{P_iP_j}\right\|} = \left(v_x^{(i)} - v_x^{(j)}\right)\alpha_x^{(ij)} + \left(v_y^{(i)} - v_y^{(j)}\right)\alpha_y^{(ij)} + \left(v_z^{(i)} - v_z^{(j)}\right)\alpha_z^{(ij)} = -\Delta f_{ij}\frac{c}{f_0}$. By manipulating this equation, we have

$$v_x^{(i)}\alpha_x^{(ij)} + v_y^{(i)}\alpha_y^{(ij)} = -\Delta f_{ij}\frac{c}{f_0} - v_z^{(i)}\alpha_z^{(ij)} + v_x^{(j)}\alpha_x^{(ij)} + v_y^{(j)}\alpha_y^{(ij)} + v_z^{(j)}\alpha_z^{(ij)} \tag{49.7}$$

In this equation, $v_x^{(i)}$ and $v_y^{(i)}$ in the left-hand side are variables to be solved, whereas $v_z^{(i)}$ in the right-hand side can be derived from pressure sensor reading, $\left(\alpha_x^{(ij)}, \alpha_y^{(ij)}, \alpha_z^{(ij)}\right)$ is the normalized vector of $\overrightarrow{P_iP_j}$, and $\left(v_x^{(j)}, v_y^{(j)}, v_z^{(j)}\right)$ is obtained from the velocity information embedded in the overheard packet header of j.

Considering all the $N_i^{(ref)}$ reference nodes, we can obtain a linear equation group, which can be expressed in a matrix form as $\mathbf{Ax} = \mathbf{b}$, where

$$\mathbf{A} = \begin{bmatrix} \alpha_x^{i1} & \alpha_y^{i1} \\ \alpha_x^{i2} & \alpha_y^{i2} \\ \vdots & \vdots \\ \alpha_x^{iN_i^{(ref)}} & \alpha_y^{iN_i^{(ref)}} \end{bmatrix}, \quad \mathbf{x} = \begin{bmatrix} v_x \\ v_y \end{bmatrix}, \quad \mathbf{b} = \begin{bmatrix} b_{i1} \\ b_{i2} \\ \vdots \\ b_{iN_i^{(ref)}} \end{bmatrix}. \tag{49.8}$$

Here $b_{ij} = -\Delta f_{ij}\frac{c}{f_0} - v_z^{(i)}\alpha_z^{(ij)} + v_x^{(j)}\alpha_x^{(ij)} + v_y^{(j)}\alpha_y^{(ij)} + v_z^{(j)}\alpha_z^{(ij)}$. We want to find the optimal $\mathbf{x}^*$ such that the sum of squared errors is minimized. That is,

$$\mathbf{x}^* = \arg\min\left\|\mathbf{b} - \mathbf{Ax}\right\|^2. \tag{49.9}$$

From matrix theory, $\mathbf{x}^*$ can be solved as $\mathbf{x}^* = (\mathbf{A}^T\mathbf{A})^{-1}\mathbf{A}^T\mathbf{b}$. Once the velocity is calculated, the position of i is updated as $p_i = p_i' + \vec{\mathbf{v}}$, where $\vec{\mathbf{v}} = \left(v_x^{(i)}, v_y^{(i)}, v_z^{(i)}\right)^T$.

Assume that the uncertainty regions $\mathcal{U}_{ij}$ and the distribution pdf of j within region $\mathcal{U}_{ij}$ are known (by embedding these parameters in the header of the packet), $\forall j \in \mathcal{N}_i$, i can estimate the pdf of being at point p as $g(P_i = p) = \int_{p_j \in \mathcal{U}_{ij}, \forall j \in \mathcal{N}_i}\left[g\left(P_i = p \mid P_j = p_j, j \in \mathcal{N}_i\right) \cdot g\left(P_j = p_j, j \in \mathcal{N}_i\right)\right]$. Similar to the case of DISLU, i can calculate the distribution of its own location and, hence, its internal uncertainty region.

Minimization of location uncertainty: Obviously, localization using different references leads to different estimation of internal uncertainty and corresponding pdf. Our objective is to minimize the estimated internal uncertainty. Using our notions of internal and external uncertainty, this can be achieved by solving an optimization problem. To measure the degree of uncertainty, we use *information entropy* as the metric, that is,

$$H(\mathcal{U}_{ij}, g_{ij}) = -\int_{p\in\mathcal{U}_{ij}} g_{ij}(p)\log(g_{ij}(p))dp \tag{49.10}$$

The bigger $H(\mathcal{U}_{ij}, g_{ij})$ is, the more uncertain $\mathcal{U}_{ij}$ is. The reason to use information entropy instead of simply the size of uncertainty region is that it can better characterize uncertainty.

Example: Assume that an AUV is distributed in [0,10] along x-axis with pdf being 9.9 in [0,0.1] and 0.1/99 in [0.1, 10] (Case 1). Then its entropy is –3.17 bits, which is less than the entropy 3.32 bits when the AUV is uniformly distributed in [0,10] (Case 2) or the entropy 3 bits when the AUV is uniformly distributed in [0,8] (Case 3). Obviously Case 1 is the most certain in these three cases even though Case 2 has the same size and Case 3 has the smallest size of the region. Note that the information flow between AUVs can occur in loops; this may not amplify errors of the positioning algorithm, as our problem can make sure the neighbors that can minimize the uncertainty can always be selected.

With this metric, the problem to minimize localization uncertainty can be formulated as

Given : $\mathcal{N}_i, \mathcal{U}_{ij}, g_{ij}()$;

Find : $\mathcal{A}_i^\star$; **Minimize :** $H(\mathcal{U}_{ii}, g_{ii})$;

S.t. :

$$\mathcal{U}_{ii} \equiv \left\{ q = \arg\min \sum_{j\in\mathcal{A}_i} \left\| d(p, p_j) - d_{ij} \right\|^2 \right\}; \tag{49.11}$$

$$g(P_i = p) = \int_{p_j\in\mathcal{U}_{ij}, j\in\mathcal{A}_i} \left[g(P_i = p \mid P_j = p_j, j\in\mathcal{A}_i) \cdot \prod_{j\in\mathcal{A}_i} g(P_j = p_j) \right]; \tag{49.12}$$

$$|\mathcal{A}_i| \geq 3; \quad \mathcal{A}_i \subset \mathcal{N}_i. \tag{49.13}$$

Here $\mathcal{A}_i$ represents a subset of i's reference nodes, (49.11) and (49.12) estimate the internal uncertainty and corresponding pdf when nodes in $\mathcal{A}_i$ are used as references; and (49.13) are the constraints for $\mathcal{A}_i$ so that enough reference nodes are selected for localization.

To reduce the complexity, we can convert an uncertainty region (internal or external) into discrete counterparts. That is, we divide an uncertainty region into a finite number of equal-size small regions. When the number of small regions is sufficiently large, the pdf of the AUV's position on a point—such as the centroid—in this small region can therefore be approximated by the probability on a small region. Hence, the estimated external-uncertainty region can be approximated as the region contained in the hull of these estimated points. The pdf functions are also be approximated by the probability mass functions on discrete points, which simplifies the pdf estimation. The aforementioned optimization can then be solved using exhaustive search algorithm after the discretization. Depending on the computation capability of the onboard processor, appropriate number of small regions can be used. Further improvement of the solution can be done after converting it into appropriate optimization that can be solved efficiently and we leave this as future work.

49.5.3 Minimization of Communication Overhead

In this section, we discuss how to optimize T_s and T_p so that localization overhead can be minimized while keeping the localization uncertainty low. We first propose an algorithm to dynamically adjust T_s in order to maintain the performance of DOPLU. Then, T_p is optimized to minimize the localization overhead.

As for T_s, it should be large enough so that packets from enough reference nodes are overheard. Suppose $K_{\min}$ is the minimum number of reference nodes (or $|\mathcal{A}_i|$ if the optimization algorithm in Section 49.5.2 is used) so that $\mathbf{x}^*$ can be calculated using DOPLU. In the beginning, T_s is initialized as $T_s = (R/c) + T_{TX} \cdot K_{\min}$, that is, the minimum time to overhear packets from $K_{\min}$ reference nodes. Suppose that during the last T_s' period, Doppler shifts from N' reference nodes with smaller degree of uncertainty (seen by i) than i's are received. On average, it takes T_s'/N' to receive a useful Doppler shift. Then, the expected time to receive $K_{\min}$ useful Doppler shifts is $T_s' \cdot K_{\min}/N'$. We update T_s using a weighted average. That is, $T_s = \beta \cdot T_s' + (1-\beta) \cdot T_s' \cdot K_{\min}/N'$, where $\beta \in (0,1)$ is a weight factor.

Using internal and external uncertainty, we can also optimize the interval T_p running DISLU. By optimizing T_p, we minimize the overhead to use DISLU and hence the overall overhead. The DISLU algorithm is run when the localization error is large. The localization error can be estimated by calculating the distance from the position estimated by DISLU to that estimated by DOPLU. When the localization error is greater than a threshold d_{th}, DISLU is run to correct the error. Since the position is not deterministic, this requirement is expressed in a probabilistic way. That is, DISLU should be run when the probability of the localization error being over d_{th} is above a threshold probability γ. Therefore, to minimize the overhead of running DISLU, T_p should be maximal under the constraint that the probability of the localization error being over d_{th} is below γ. This can be formulated into the following optimization problem:

$$\textbf{Given}: \mathcal{U}_{ij},\ g_{ij}(),\ \gamma;$$

$$\textbf{Find}: T_p^*;\quad \textbf{Maximize}: T_p;$$

$$\textbf{S.t.}: \Pr\left\{\left\|\overline{p_i(T_p)\tilde{p}_i(T_p)}\right\| > d_{th}\right\} < \gamma,$$

where $p_i(T_p)$ and $\tilde{p}_i(T_p)$ are the prediction positions using the DOPLU and DISLU after T_p from the last DISLU run time, respectively. This prediction of future internal uncertainty is based on the current estimated internal uncertainty and AUV's trajectory. A solution has been proposed in [8] for underwater gliders (one type of buoyancy-driven AUVs), which we adopt in this work. As the previous optimization problem, we can also convert it into discrete variable optimization problem and solve it in a similar way.

49.6 Performance Evaluation

The communication solution is implemented and tested on our underwater communication emulator [7]. This emulator is composed of four WHOI micro-modems and a real-time audio processing card to emulate underwater channel propagation. The multiinput multioutput audio interface can process real-time signals to adjust the acoustic signal gains, to introduce propagation delay, to mix the interfering signals, and to add ambient/man-made noise and interference. Our solution is compared against AAL, DNRL, and CNA, as introduced in Section 49.3, under an environment that is described by the Bellhop model [4]. We use the typical Arctic sound speed profile as in [27] and the corresponding Bellhop profile is plotted in Figure 49.3. Note that we use 25 KHz, the sound frequency in use for our WHOI modem. We modify AAL, DNRL, and CNA, as they were originally designed for settings that are quite different from the under-ice environment. Specifically, AAL, DNRL, and CNA all use the AUV that surfaces last as reference node because intuitively the shorter an AUV stays underwater (the less time it stays in an uncertain environment after a GPS fix), the less uncertain its position is. Triangulation is employed for

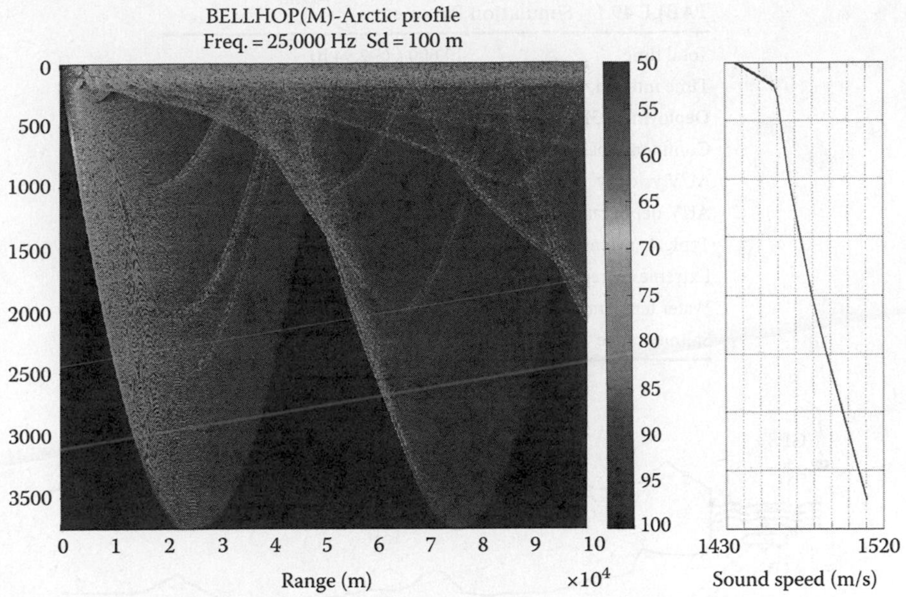

FIGURE 49.3 Bellhop profile for a typical Arctic environment.

position calculation in AAL and DNRL, while EKF filtering is used in CNA. We are also interested in seeing the performance improvement that we get using the external uncertainty notion. Therefore, we implement another version of our proposed solution without using external uncertainty, that is, forcing the position uncertainty to be zero. We denote this modified version and the original version by "Proposed solution w/o EU" and "Proposed solution w/EU," respectively.

In order to evaluate the localization performance, two metrics, the *localization error* and the *deviation of error*, are used. Localization error is defined as the distance between the actual and the estimated AUV position. The deviation of error is the amount the localization error deviating from the total averaged error. The average localization error $\bar{E}$ and deviation of error σ are plotted. The formulae of $\bar{E}$ and σ are expressed as

$$\bar{E} = \frac{1}{L_t}\sum_{j=1}^{L_t}\left(\frac{1}{N}\sum_{i=1}^{N}E_i\right); \ \sigma = \sqrt{\frac{1}{N}\sum_{i=1}^{N}\left(E_i - \bar{E}\right)^2}, \tag{49.14}$$

where

N is the number of AUVs in the UW-ASN

E_i represents the localization error for each AUV operating in the UW-ASN at that particular time

L_t is the number of times the localization is performed, such that $L_t = T_{end}/\Delta T$

49.6.1 Simulation Scenarios

The parameters for our simulations are listed in Table 49.1. We further assume that ongoing communication packets are generated according to the Poisson traffic model with arrival rate being three packets per minute. As shown in Figure 49.4, we utilize the following two specific scenarios.

Scenario 1: This scenario involves a team of AUVs who collaboratively explore an underwater region located under ice. These AUVs remain under-ice for the duration of the mission and do not return to the surface until the mission is completed.

TABLE 49.1 Simulation Parameters

Total time	10,600 s (~2.94 h)
Time interval, ΔT	60 s
Deployment 3D region	2,000(L) × 2,000(W) × 1,000(H) m^3
Confidence parameter, α	0.05
AUV velocity	0.25–0.40 m/s
AUV depth range	[0,1,000] m
Typical currents	0.01–0.03 m/s [13]
Extreme currents	0.04–0.06 m/s [13]
Water temperature range	[−2,2] °C
Salinity range	[32.5,35] ppt

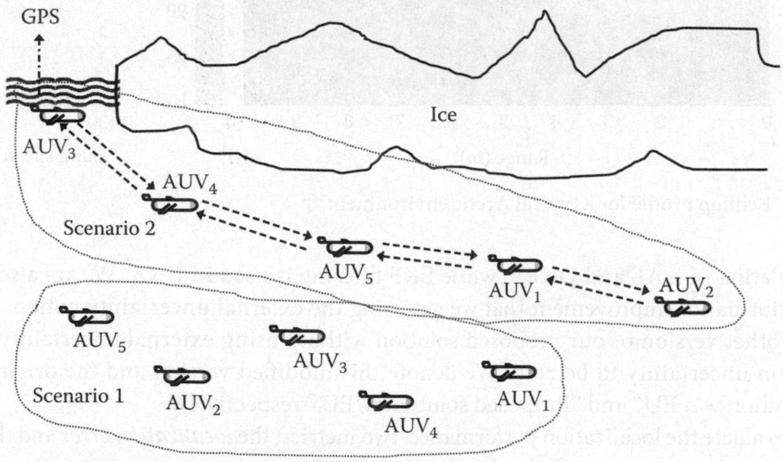

FIGURE 49.4 Two scenarios for simulations: different dotted circles represent different scenarios.

Scenario 2: This scenario is similar to the first except that individual AUVs will periodically surface to update their positioning via GPS. These AUVs take turns to go back to the surface at a predefined interval, which is 4000 s in our simulations. In order to avoid ice cover, these AUVs return to the edge of the ice sheet where they were deployed. Once an AUV surfaces, it acquires a GPS fix and updates its current coordinate position (position uncertainty is also reset).

Both scenarios are tested with typical and extreme currents, whose speed ranges are listed in Table 49.1. A random 3D direction is chosen for the current throughout one round of simulation. The Doppler data are based on the 6-h Doppler speed measurement that we took using WHOI modems on November 20, 2011 in the Bayfront Park bay, Lavallette, NJ, as shown in Figure 49.5. Our measurement shows that most of the Doppler speeds are low, similar to the part we plot here. Note that the right-hand side in (49.6) is replaced with the measured Doppler speed here as there is no need to calculate the Doppler shifts.

49.6.2 Evaluation Results

Real-time (one simulation run) localization errors and deviations of error are plotted in the first two subfigures of Figures 49.6 through 49.9. Moreover, to obtain results of statistical significance, 250 rounds were conducted for varying numbers of AUVs. The average errors for the AUV's predicted location are plotted in Figures 49.6 through 49.9c with 95% confidence intervals.

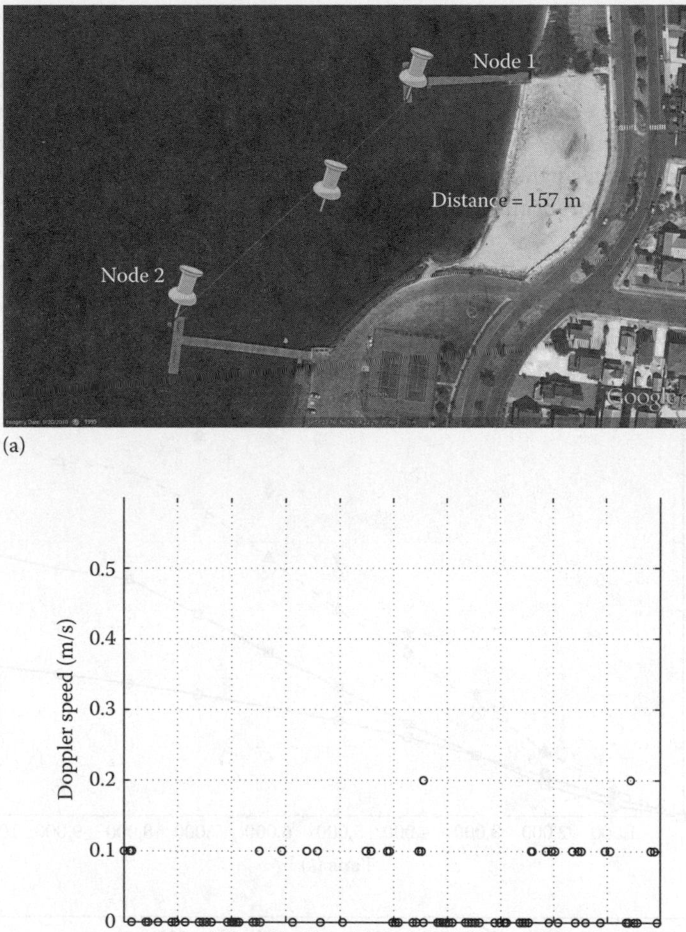

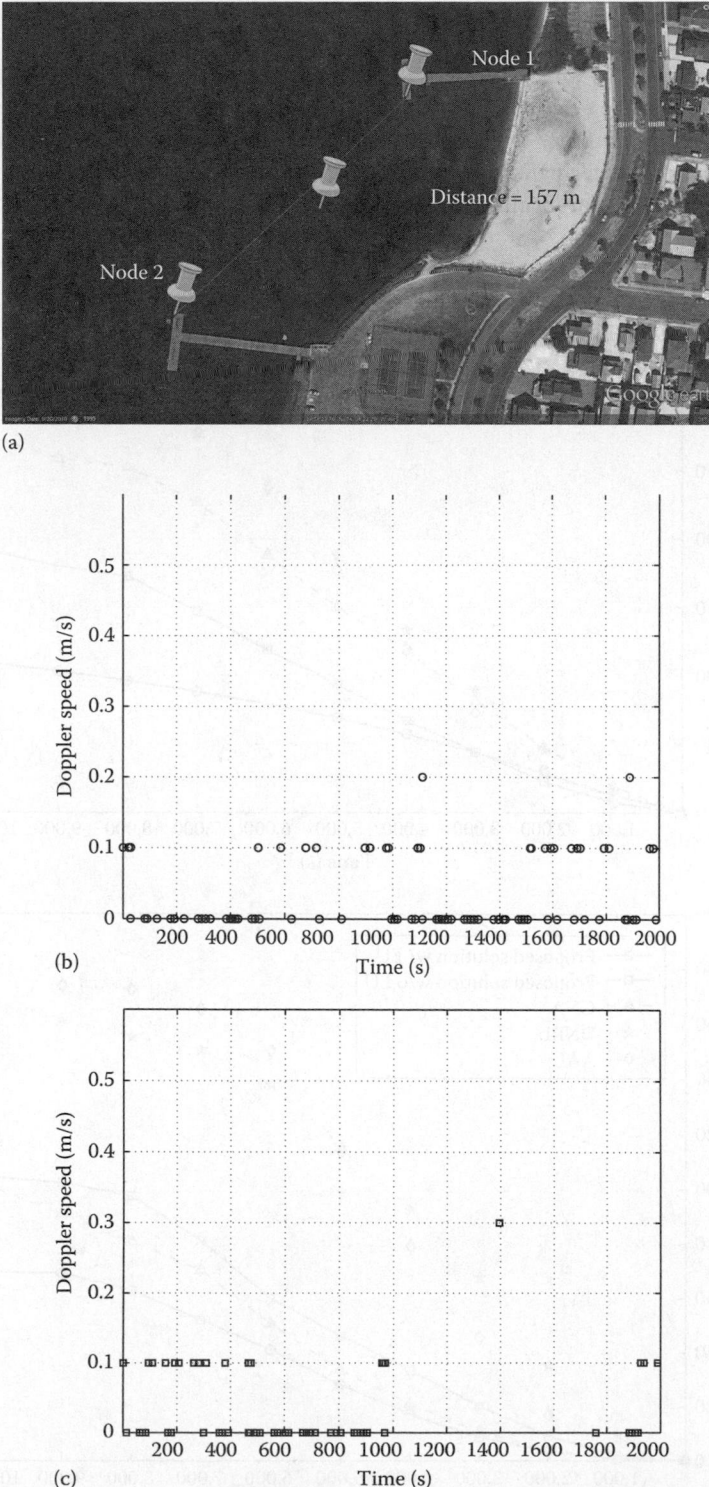

FIGURE 49.5 Doppler speed measurement. Only part of the measurements are plotted for clear visualization. Time coordinates vary due to different reception times. (a) Location: Bayfront Park bay, Lavallette, NJ. (b) Doppler speeds measured at node 1. (c) Doppler speeds measured at node 2.

Scenario 1: As shown in Figures 49.6 and 49.7, our original solution "Proposed solution w/ EU" performs the best. In the typical current setting, "Proposed solution w/ EU" obtains about 74.6% less error than "Proposed solution w/o EU" while it obtains 80.4% less error in the extreme current setting. This is mainly due to the use of the external uncertainty model to predict the position and distribution of the AUVs and the ability to minimize the localization uncertainty. "Proposed solution w/o EU" ranks the second in terms of error performance because an AUV can leverage the ongoing communications and cooperation of other AUVs for localization. Even though CNA uses EKF to predict the positions, its

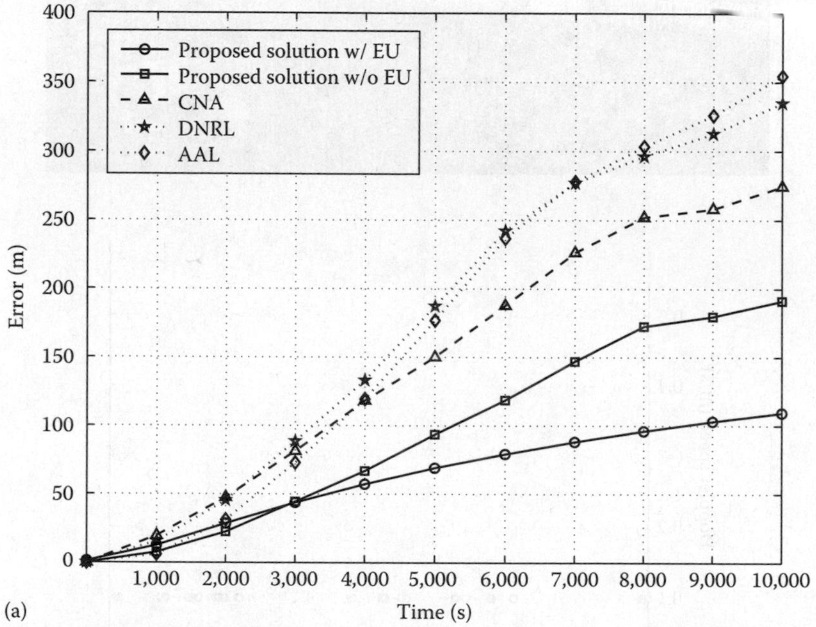

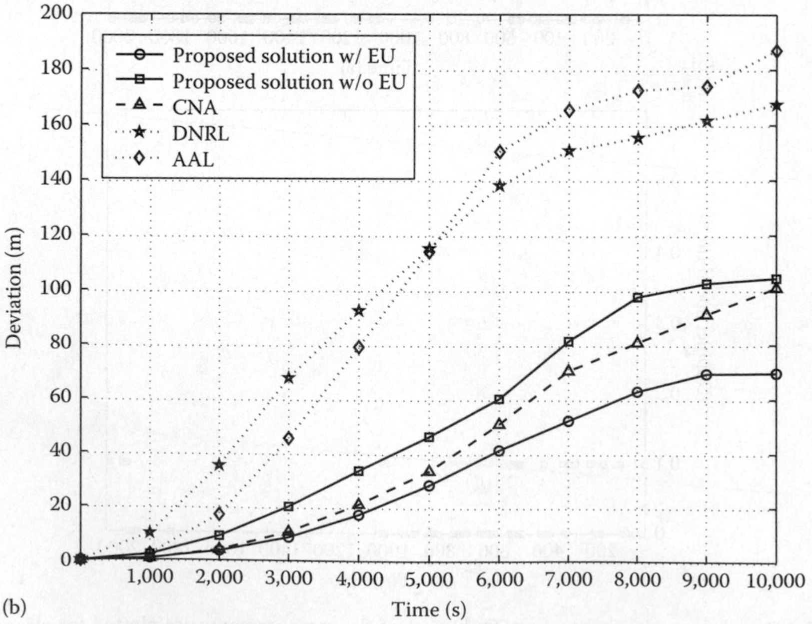

FIGURE 49.6 Scenario 1 with typical currents: under the ice mission with no resurfacing. (a) Localization error Comparison (6 AUVs). (b) Deviation Comparison (6 AUVs).

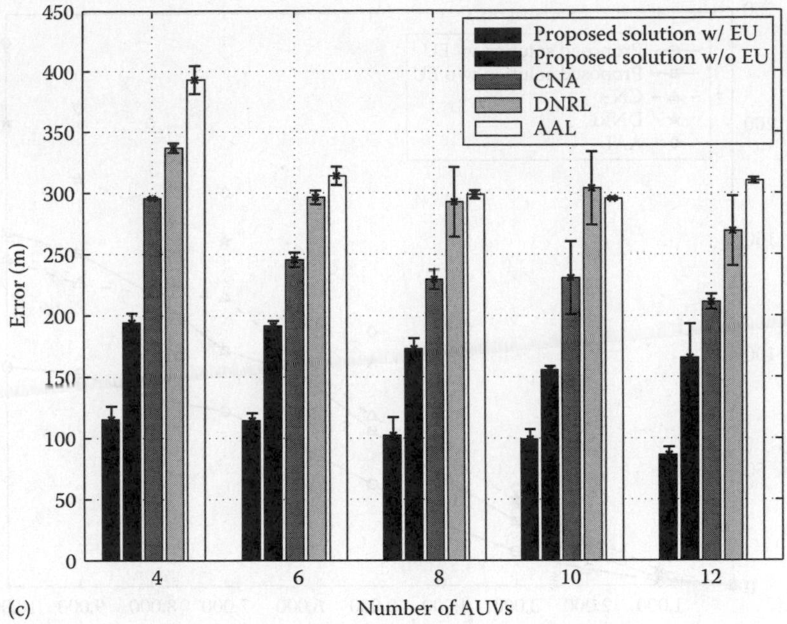

FIGURE 49.6 (continued) Scenario 1 with typical currents: under the ice mission with no resurfacing. (c) Localization error for number of AUVs.

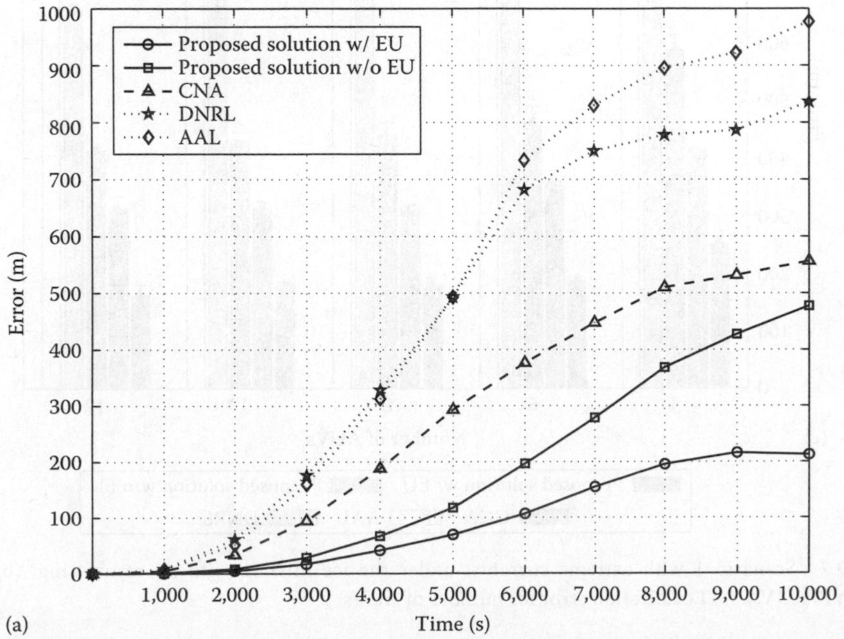

FIGURE 49.7 Scenario 1 with extreme currents: under the ice mission with no resurfacing. (a) Localization error Comparison (6 AUVs).

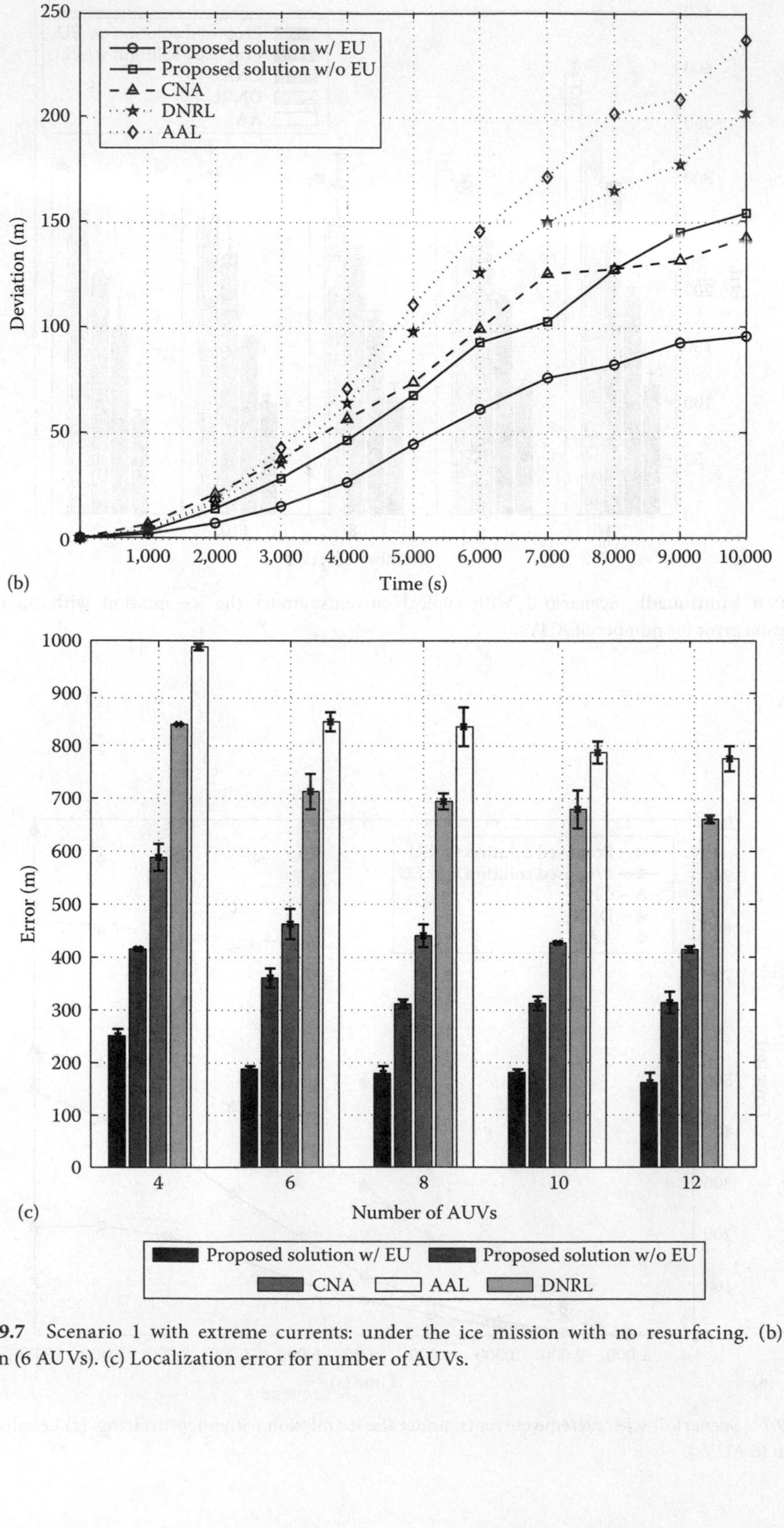

FIGURE 49.7 Scenario 1 with extreme currents: under the ice mission with no resurfacing. (b) Deviation Comparison (6 AUVs). (c) Localization error for number of AUVs.

performance is worse than "Proposed solution w/o EU" since the AUV can only use its own states for position estimation. On the other hand, CNA is still better than DNRL and AAL due to the use of EKF filter, and DNRL performs better than AAL since it takes the current influence into account.

By comparing Figures 49.6 and 49.7, we can see that under extreme conditions, the localization error keeps increasing since more dislocation is incurred by the extreme currents. Interestingly enough, we can see that the performance of our solution without using external uncertainty is not much better than that using CNA. In this case, using Doppler information does not help improve the localization much

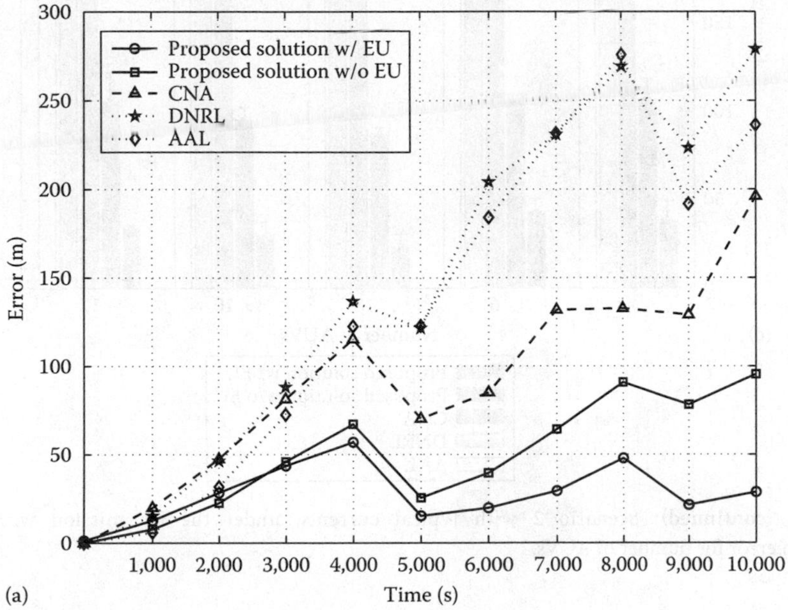

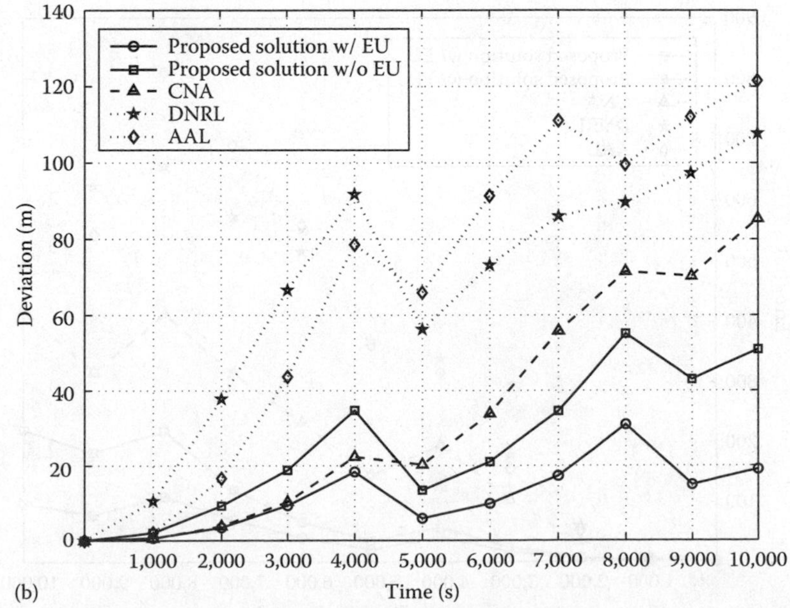

FIGURE 49.8 Scenario 2 with typical currents: under the ice mission with resurfacing. (a) Localization error Comparison (6 AUVs). (b) Deviation Comparison (6 AUVs).

(continued)

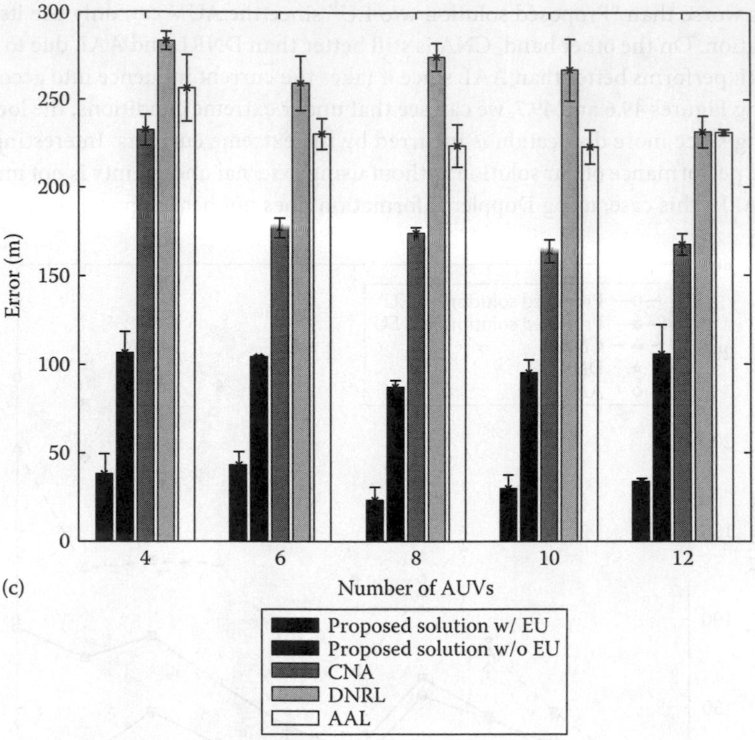

(c)

FIGURE 49.8 (continued) Scenario 2 with typical currents: under the ice mission with resurfacing. (c) Localization error for number of AUVs.

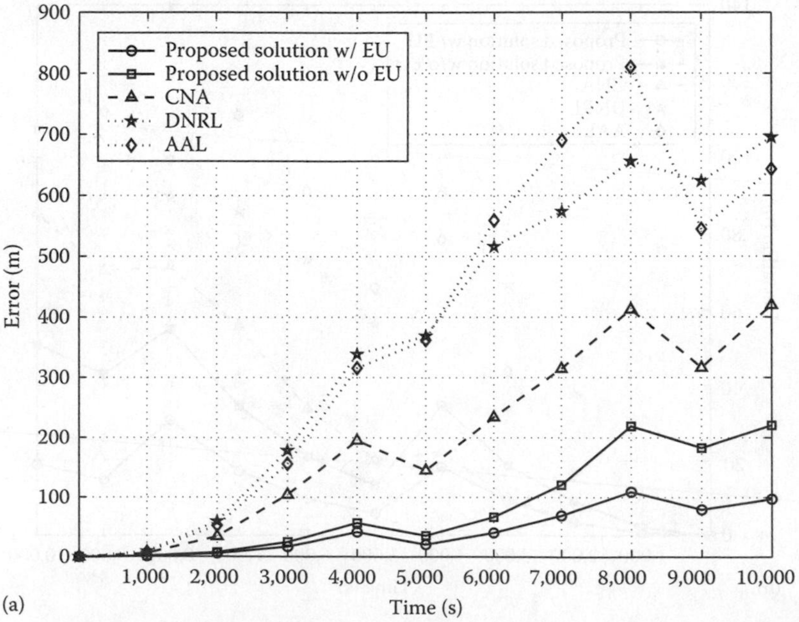

(a)

FIGURE 49.9 Scenario 2 with extreme currents: under the ice mission with resurfacing. (a) Localization error Comparison (6 AUVs).

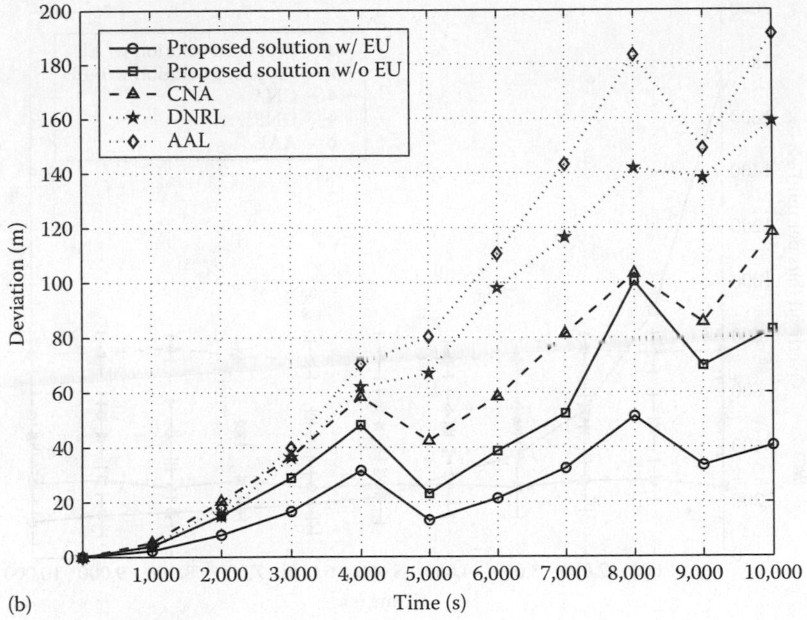

(b)

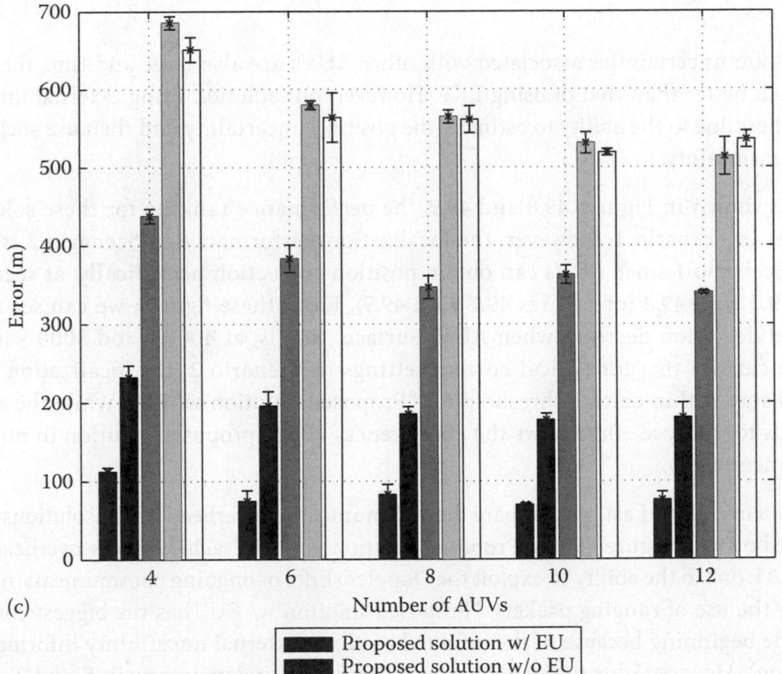

(c)

FIGURE 49.9 (continued) Scenario 2 with extreme currents: under the ice mission with resurfacing. (b) Deviation Comparison (6 AUVs). (c) Localization error for number of AUVs.

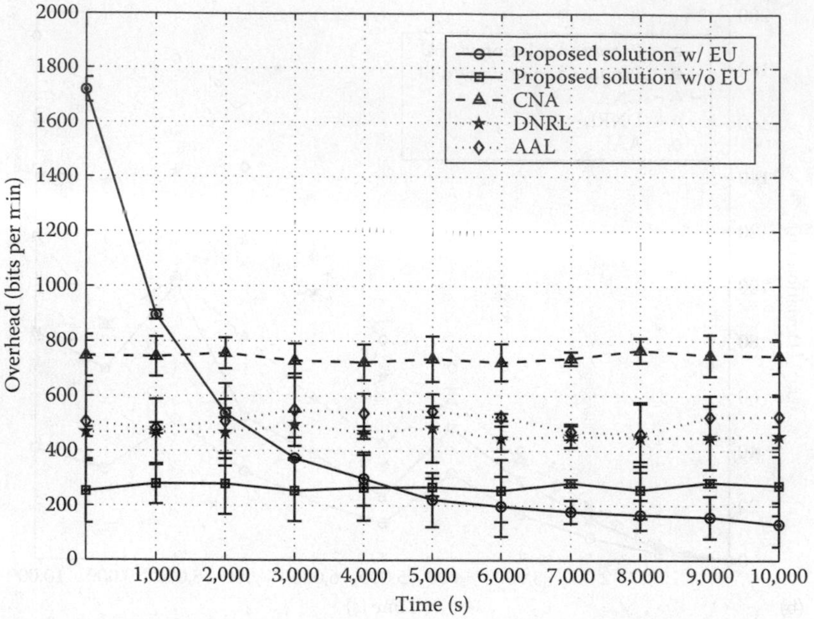

FIGURE 49.10 Comparison of communication overhead.

since the position uncertainties associated with other AUVs are also large and thus the performance is not too much better than that of using EKF. However, our solution using external uncertainty still performs the best due to the ability to estimate the position uncertainty and then use such information to minimize uncertainty.

Scenario 2: As shown in Figures 49.8 and 49.9, the performance ranking for these solutions closely resembles that in Scenario 1. However, the localization performance in Scenario 2 is much better than that in Scenario 1 since AUVs can obtain position correction periodically, as seen by comparing Figures 49.6 with 49.8 (or Figures 49.7 with 49.9). From these figures, we can see that localization error and deviation decrease when AUVs surface, that is, at 4000 s and 8000 s in the results. Moreover, we can see that for typical current settings in Scenario 2, the localization error and its deviation can stay within certain threshold for "Proposed Solution w/ EU," while the error of other solutions tends to increase. This shows the effectiveness of our proposed solution in minimizing the localization uncertainty.

Communication overhead: Last, we compare the communication overhead of our solutions against other solutions. As shown in Figure 49.10, "Proposed solution w/o EU" achieves less overhead than CNA, DNRL, and AAL due to the ability to exploit the Doppler shifts of ongoing communications for localization, reducing the use of ranging packets. "Proposed solution w/ EU" has the biggest communication overhead in the beginning because of the need to broadcast external uncertainty information (such as pdf information). However, due to the ability to optimize the update intervals T_s and T_p as in Section 49.5.3, its communication overhead drops quickly to a level that is lower than CNA, DNRL, and AAL. CNA has higher overhead than DNRL and AAL as CNA needs to broadcast additional information such as velocities and sensor readings for EKF while DNRL and AAL only need to broadcast the position and time information that is embedded in the ranging packet. Note that in "Proposed solution w/ EU," to save the overhead, when the AUVs broadcast the pdf information, they only broadcast the key parameters if the pdf is one of the well-known distributions (e.g., the average and standard deviation for a normal distribution). Otherwise, the point mass function of a finite number of points is broadcast.

49.7 Conclusion and Future Work

We proposed a localization solution that minimizes the position uncertainty and communication overhead of AUVs in the challenging under-ice environments. With the notion of external uncertainty, position uncertainties of the AUV can be modeled in a probabilistic way. This model is further used to estimate the uncertainty resulted from localization techniques, as shown for our proposed Doppler-based localization and the standard distance-based localization. Algorithms are then proposed to minimize the position uncertainty and communication overhead. Our solution is implemented and compared with several existing localization techniques using an acoustic communication emulator. It is shown that our approach achieves excellent localization results with low localization overhead.

Further work will be to implement our proposed localization solution on AUV platforms and evaluate its performance in ocean experiments. We are building our own AUV platform for performance evaluation of our solutions. Our proposed localization can be used as part of the adaptive sampling solution using AUVs to reconstruct the ocean temperature field. Specifically, given a field to sense, the AUVs should coordinate to take measurements with minimal cost such as time or energy in order to reconstruct the field with admissible error. To achieve this, we will propose a framework that jointly considers the online field measurements and multi-AUV trajectory planning. Novel distributed algorithms that can minimize the sampling cost for one-round sampling are proposed, and underwater communication protocols are designed to coordinate the vehicles. In this way, low sampling cost can be achieved while the computation complexity per vehicle can be reduced by using distributed algorithms.

Glossary

In this part, we give a brief introduction to the terms that are used.

Acoustic Transponders: Also called sonar transponders, which operate under water and are used to measure distance and form the basis of underwater location marking, position tracking and navigation.

Autonomous Underwater Vehicles: Programmable robots that, depending on their design, can drift, drive, or glide through the water without the real-time control by human operators. Generally, AUVs are equipped with onboard computers and sensors, which allow them to sense the underwater environment and collect the underwater data as they move through the water, and make decisions on their own. These AUVs, which are divided into two main classes—propeller-less/buoyancy-driven (e.g., gliders) and propeller-driven vehicles (PDVs)—rely on local intelligence with minimal onshore operator dependence.

Communication Overhead: Extra bits of data that does not contribute to the content of the message and that must be sent to convey information about, for example, where the information originated and where it is being sent to, how it is to be routed, time-stamps, or any other information that is not actually the "payload" representing the actual content to be communicated.

Dead Reckoning: The process of calculating the current position by using a previously determined position, and advancing that position based upon known or estimated speeds over elapsed time, and course. Dead reckoning is subject to cumulative errors as each position estimate is relative to the previous one. It is also affected by the accuracy of the velocity and direction estimation.

Deviation of Error: The amount the localization error deviating from the total averaged error. It measures the variation or dispersion of the position error is from the average.

Doppler Velocity Log (DVL): An acoustic device used to measure the speed of a vehicle over the seabed. It can serve as a fundamental component of dead-reckoning (DR) underwater navigation systems and aided-inertial navigation systems (INS). The measurement data comprise speed in three perpendicular directions, calculated from the Doppler shift of the acoustic signal returned from the seabed as the vehicle passes over. Altitude above the seabed is also calculated. Typically the DVL data are used to aid an integrated position calculation incorporating USBL, LBL, or inertial systems.

End-to-end Network Latency (or Delay): The time difference between a packet is sent from the source and the time it is received at the destination.

Extended Kalman Filter (EKF): The nonlinear version of the Kalman filter that is obtained using a linear approximation of the non-linear system based on the Taylor series expansion. In the case of well-defined transition models, the EKF has been proven to be a useful method of obtaining good estimates of the system state.

External Uncertainty: The position uncertainty associated with a particular entity/node as seen by others. External uncertainties may be derived from the broadcast/propagated internal-uncertainty estimates (e.g., using one-hop or multihop neighbor discovery mechanisms) and, hence, will be affected by e2e network latency and information loss. Information loss can be substantial and is attributed to packet losses caused by channel unreliability due to multipath, fading, and ambient noise.

Inertial Navigation System (INS): A self-contained navigation technique in which measurements provided by accelerometers and gyroscopes are used to track the position and orientation of an object relative to a known starting point, orientation and velocity. Inertial measurement units (IMUs) typically contain three orthogonal rate-gyroscopes and three orthogonal accelerometers, measuring angular velocity and linear acceleration respectively. By processing signals from these devices it is possible to track the position and orientation of a device. It is used in a wide range of applications such as the navigation of aircraft, missiles, spacecraft, submarines, and ships.

Information Entropy: A measure of the uncertainty associated with a random variable. It usually refers to the Shannon entropy (or continuous entropy), which quantifies the expected value of the information contained in a discrete random variable (or continues random variable).

Internal Uncertainty: The position uncertainty associated with a particular entity/node (such as an AUV) as seen by itself. Many approaches such as those using Kalman filter (KF) have been used to estimate this uncertainty assuming that the variables to be estimated have linear relationships between each other and that noise is additive and Gaussian. When the linearity assumption does not hold, there is no guarantee of optimality and nonlinear filters such as the extended or unscented KF are used.

Kalman Filter (KF): An algorithm in control theory introduced by Kalman (1960). It operates recursively on streams of noisy input data to produce a statistically optimal estimate of the underlying linear (or nearly linear) system state.

Localization Error: The distance between the actual and the estimated AUV position. It measures the accuracy of the localization algorithm.

Localization Overhead: The communication overhead used for localization. This may include the positions of other nodes, the time-stamps to measure the round-trip time, or information about the time of arrival (ToA), time difference of arrival (TDoA), received-signal-strength (RSS), or angle of arrival (AoA) for distance measurement.

Long Baseline (LBL): One type of underwater acoustic positioning systems. It requires transponders deployed at fixed positions on the sea floor. Localization of a node is done by estimating the distances of this node to the transponders on the sea floor (i.e., acoustically measuring the round-trip time from this node to the transponders). The LBL technique can obtain very high positioning accuracy and position stability that is independent of water depth (generally better than 1-m or a few centimeters accuracy). It generally offer better accuracy than the short baseline systems.

MultiPath Propagation: The propagation phenomenon that results in signals (radio or acoustic) reaching the receiver by two or more paths. Causes of multipath include reflection and refraction, and reflection from boundaries.

Propeller-Driven Vehicles (PDVs): One class of AUVs that are driven by propeller(s). As the propulsion system consumes a lot of energy, the lifespan and operation coverage are limited due to the limited battery capacity. In general, PDVs have a life span ranging from several hours to a few days and can traverse distances of several hundred kilometers. The main benefit of using PDVs is that they can cover long distance in a short time (due to the high speed brought by the propulsion system). In addition, they generally have high maneuverability as the onboard computer can change the course or turn in time.

Simultaneous Localization and Mapping (SLAM): A technique used by robots and autonomous vehicles to update a map within a known environment (with a priori knowledge from a given map) while at the same time estimating their current location, or to build up a map within an unknown environment (without a priori knowledge).

Short Baseline (SBL): One of the major types of underwater acoustic positioning systems. In SBL systems, position estimate is determined by measuring the vehicle's distance from three or more transponders that are, for example, lowered over the side of the surface vessel. The positioning accuracy of SBL improves with transducer spacing. When deployed on larger vessels or a dock, the SBL system can achieve a precision and position robustness that is similar to that of the LBL systems, making the system suitable for high-accuracy survey work.

Underwater Acoustic Sensor Networks: Networks of underwater sensors and vehicles that that interact to collect data and perform tasks in a collaborative manner underwater. They are applied in a broad range of applications, including environmental monitoring, undersea exploration, disaster prevention, assisted navigation and tactical surveillance.

Underwater Gliders: Another class of AUVs that are not driven by propeller. For propulsion, they change their buoyancy in a small amount using a pump and rely on wings to convert vertical velocity into horizontal motion as they rise and fall through the ocean, forming a sawtooth trajectory. As a result, in contrast to PDVs, gliders spend very small amount of energy in propulsion and hence they can operate over long periods of time (weeks or months), even though they are not as fast as PDVs. They travel at a fairly constant horizontal speed, typically 0.25 m/s [1]. The impressive operation time and energy efficiency of these gliders make them a popular choice for oceanographic researchers.

Underwater Localization: The process to determine the 3D position of a sensor, network node, or vehicle underwater. Many localization schemes have been proposed, which can be categorized into range-based schemes and range-free schemes. Range-based schemes use different methods to estimate the distance between nodes and then determine the positions of the nodes. On the other hand, range-free schemes do not use range information to estimate positions of the nodes. Dead reckoning, for example, is such a scheme that relies on velocity measurement to estimate the position.

References

1. I. F. Akyildiz, D. Pompili, and T. Melodia. Underwater acoustic sensor networks: Research challenges. *Ad Hoc Networks (Elsevier)*, 3(3):257–279, May 2005.
2. J. G. Bellingham, C. A. Goudey, T. R. Consi, J. W. Bales, D. K. Atwood, J. J. Leonard, and C. Chryssostomidis. A second generation survey AUV. In *Proceedings of the Symposium on Autonomous Underwater Vehicle Technology (AUV)*, Cambridge, MA, July 1994.
3. M. Blain, S. Lemieux, and R. Houde. Implementation of a ROV navigation system using acoustic/Doppler sensors and Kalman filtering. In *Proceedings of IEEE International Conference on Engineering in the Ocean Environment (OCEANS)*, San Diego, CA, September 2003.

4. W. S. Burdic. Underwater acoustic system analysis. In A. V. Oppenheim, ed., *Prentice-Hall Signal Processing Series*, Chapter 2, p. 49. Prentice-Hall, Englewood Cliffs, NJ, 1984.
5. J. A. Catipovic. Performance limitations in underwater acoustic telemetry. *IEEE Journal of Oceanic Engineering*, 15(3):205–216, July 1990.
6. V. Chandrasekhar, W. K. G Seah, Y. S. Choo, and H. Voon Ee. Localization in underwater sensor networks: Survey and challenges. In *Proceedings of the ACM International Workshop on Underwater Networks (WUWNet)*, New York, September 2006.
7. B. Chen, P. C. Hickey, and D. Pompili. Trajectory-aware communication solution for underwater gliders using WHOI micro-modems. In *Proceedings of IEEE Conference on Sensor, Mesh and Ad Hoc Communications and Networks (SECON)*, Boston, MA, June 2010.
8. B. Chen and D. Pompili. QUO VADIS: QoS-aware underwater optimization framework for inter-vehicle communication using acoustic directional transducers. In *Proceedings of IEEE Conference on Sensor, Mesh and Ad Hoc Communications and Networks (SECON)*, Salt Lake City, UT, June 2011.
9. B. Chen and D. Pompili. Uncertainty-aware localization solution for under-ice autonomous underwater vehicles. In *Proceedings of IEEE Communications Society Conference on Sensor, Mesh and Ad Hoc Communications and Networks (SECON)*, Seoul, Korea, June 2012.
10. R. Emokpae and M. Younis. Surface based anchor-free localization algorithm for underwater sensor networks. In *Proceedings of IEEE International Conference on Communications (ICC)*, Kyoto, Japan, June 2011.
11. M. Erol, L. F. M. Vieira, and M. Gerla. Localization with DiveN'Rise (DNR) beacons for underwater acoustic sensor networks. In *Proceedings of the ACM Workshop on Underwater Networks (WUWNet)*, Montreal, Quebec, Canada, September 14, 2007.
12. M. Erol-Kantarci, L. M. Vieira, and M. Gerla. AUV-aided localization for underwater sensor networks. In *Proceedings of International Conference on Wireless Algorithms, System and Applications (WASA)*, Chicago, IL, August 2007.
13. A. A. Eroshko, V. M. Kushnir, and A. M. Suvorov. Mapping currents in the northern Black Sea using OLT-type profilers. In *Proceedings of the IEEE Working Conference on Current Measurement*, San Diego, CA, March 1999.
14. M. F. Fallon, G. Papadopoulos, J. J. Leonard, and N. M. Patrikalakis. Cooperative AUV navigation using a single maneuvering surface craft. *International Journal of Robotics Research*, 29:1461–1474, October 2010.
15. S. A. L. Glegg, R. Pirie, and A. LaVigne. A study of ambient noise in shallow water. In Florida Atlantic University Technical Report, Boca Raton, FL, 2000.
16. D. B. Kilfoyle and A. B. Baggeroer. The state of the art in underwater acoustic telemetry. *IEEE Journal of Oceanic Engineering*, 25(1):4–27, January 2000.
17. P. Kimball and S. Rock. Sonar-based iceberg-relative AUV navigation. Technical Report, Stanford, CA, February 2008.
18. J. C. Kinsey, R. M. Eustice, and L. L. Whitcomb. A survey of underwater vehicle navigation: Recent advances and new challenges. In *Proceedings of IFAC Conference of Manoeuvering and Control of Marine Craft*, Lisbon, Portugal, September 2006.
19. D. Mirza and C. Schurgers. Motion-aware self-localization for underwater networks. In *Proceedings of ACM International Workshop on UnderWater Networks (WUWNet)*, San Francisco, CA, September 2008.
20. D. Mirza and C. Schurgers. Collaborative tracking in mobile underwater networks. In *Proceedings of ACM International Workshop on UnderWater Networks (WUWNet)*, Berkeley, CA, November 2009.
21. A. Pereira, H. Heidarsson, C. Oberg, D. A. Caron, B. Jones, and G. S. Sukhatme. A communication framework for cost-effective operation of AUVs in coastal regions. In *Proceedings of International Conference on Field and Service Robots (FSR)*, Cambridge, MA, July 2009.
22. J. Proakis, J. Rice, E. Sozer, and M. Stojanovic. Shallow water acoustic networks. In J. G. Proakis, ed., *Encyclopedia of Telecommunications*. John Wiley and Sons, New York, 2003.

23. J. G. Proakis, E. M. Sozer, J. A. Rice, and M. Stojanovic. Shallow water acoustic networks. *IEEE Communications Magazine*, pp. 114–119, November 2001.
24. A. Quazi and W. Konrad. Underwater acoustic communications. *Communications Magazine, IEEE*, 20(2):24–30, March 1982.
25. D. Ribas. Towards Simultaneous Localization and Mapping for an AUV using an Imaging Sonar. PhD thesis, Universitat de Girona Girona, Spain.
26. M. Stojanovic. Acoustic (underwater) communications. In J. G. Proakis, ed., *Encyclopedia of Telecommunications*. John Wiley and Sons, New York, 2003.
27. R. J. Urick. *Principles of Underwater Sound*. McGraw-Hill, New York, 1983.
28. P. Wadhams and D. M. J. Doble. Digital terrain mapping of the underside of sea ice from a small AUV. *Geophysical Research Letters*, 35, 2008.
29. M. E. West and V. L. Syrmos. Navigation of an Autonomous Underwater Vehicle (AUV) using robust SLAM. In *IEEE International Conference on Control Applications*, Munich, Germany, October 2006.
30. Z. Zhou, J. Cui, and S. Zhou. Localization for large-scale underwater sensor networks. NETWORKING 2007, IFIP LNCS 4479, pp. 108–119, 2007.

50

Semantic Web

Pascal Hitzler
Wright State University

Krzysztof Janowicz
University of California, Santa Barbara

50.1 Introduction

The Semantic Web, as an interdisciplinary research field, arose out of the desire to enhance the World Wide Web in such a way that interoperability and integration of multiauthored, multithematic, and multiperspective information and services could be realized seamlessly and on-the-fly [8,14]. Thus, the Semantic Web is concerned with developing methods for information creation, annotation, retrieval, reuse, and integration. It draws from many disciplines, including knowledge representation and reasoning in artificial intelligence, databases, machine learning, natural language processing, software engineering, information visualization, and knowledge management [39,40].

Semantic (Web) technologies are under substantial investigation in many disciplines where information reuse and integration on the web promises significant added value, for example, in the life sciences, in geographic information science, digital humanities research, for data infrastructures and web services, as well as in web publishing. At the same time, semantic technologies are also being picked up to enhance solutions in application areas that are not primarily targeting the World Wide Web but have to access similar challenges, such as enterprise information integration, intelligence data analysis, and expert systems.

The Semantic Web spans from foundational disciplines to application areas. In terms of size and impact of its scientific community, amount of available project funding, and industrial impact, it has emerged as a major field within computer science during the last decade.

50.2 Underlying Principles

As the Semantic Web is driven by an application perspective, namely, to improve information retrieval beyond simple keyword search and foster semantic interoperability on the web, there is a wide variety of proposed and employed methods and principles which differ substantially and are under continuous discussion concerning their suitability for achieving the Semantic Web vision.

Nevertheless, some approaches have become well established and widely accepted. Probably the most commonly accepted principle is the idea of endowing information on the web with *metadata* that shall convey the meaning (or *semantics*) of the information by further describing (or *annotating*) it, and thus enable information retrieval, reuse, and integration. Shared schemas that formally specify the conceptualizations underlying a certain application area or information community are called *ontologies*. In the light of this approach, the development of suitable languages for representing ontologies has been, and still is, a key issue for Semantic Web research.

50.2.1 Languages for Representing Knowledge on the Web

Formal languages that support the representing of information semantics are commonly called *ontology languages*, and among the many proposals for such languages that can be found in the literature, those that have become recommended standards by the World Wide Web Consortium* (W3C) are of primary importance; namely, the *Resource Description Framework RDF* [31], the *web Ontology Language OWL* [22,32], and the *Rule Interchange Format RIF* [10,11]. Syntactically, they are based on XML [12], although their underlying data models are significantly different. During the last years other serializations have gained popularity, for example Notation3 [7] and Turtle [35] for representing RDF graphs. Historically, all ontology languages can be traced back to traditions in the field of knowledge representation and reasoning [23]. We briefly describe each of these languages in turn.

The **Resource Description Framework (RDF)** is essentially a language for representing *triples* of the form `subject predicate object`, where each of the entries is a uniform resource identifier (URI) for web resources. Predicates are usually also called *properties*. An `object` of one triple may be the `subject` of another triple and so forth. An example for such triples would be

```
ex:HoratioNelson ex:diedIn ex:BattleOfTrafalgar and
ex:BattleOfTrafalgar ex:during ex:NapoleonicWars
```

where `ex:` is a namespace identifier so that `ex:HoratioNelson` expands to a proper URI like `http://www.example.org/HoratioNelson`. The RDF specification furthermore includes pre-defined vocabulary, the meaning of which is defined by means of a *formal semantics*, described further later. The most important examples for this are `rdf:type`, `rdfs:subClassOf`, and `rdfs:subPropertyOf`, where `rdf`, respectively `rdfs`, are used as namespace identifiers for the pre-defined namespaces of RDF and RDF Schema, respectively. In fact, RDF Schema is part of the RDF specification, but has been given a separate name to indicate that it has additional expressive features, that is, a richer predefined vocabulary. The informal meaning of this vocabulary examples is as follows. A triple such as

```
ex:BattleOfTrafalgar rdf:type ex:NavalBattle
```

indicates that the individual event `ex:BattleOfTrafalgar` is an instance of the class `ex:NavalBattle` and is interpreted as set membership. A triple such as

```
ex:NavalBattle rdfs:subClassOf ex:Battle
```

* http://www.w3.org/

indicates that every instance of the class `ex:NavalBattle` is also an instance of the class `ex:Battle`. A triple such as

```
ex:diedIn rdfs:subPropertyOf ex:participatedIN
```

indicates that, whenever some *a* is in a `ex:diedIn`-relationship to some *b* (e.g., because a triple *a* `ex:diedIn` *b* has been specified), then *a* and *b* are also in `ex:participatedIn`-relationship.

An RDF document now represents, essentially, a finite set of such triples. It is common practice to think of such a set as representing a finite labeled graph, where each triple gives rise to two vertices (subject and object) and an edge (the predicate). The labels of such subject–predicate–object triples are URIs. Note, however, that it is possible that a URI occurring as a predicate (i.e., an edge) also occurs as a subject or object (i.e., as a vertex) in a different triple. This means that edges can at the same time be vertices, thus breaking what would commonly be considered a graph structure. The W3C recommended standard SPARQL* [37] serves as a language for querying RDF triples that are often stored in a so-called *triplestore*. It is currently undergoing a revision [18], which is expected to make SPARQL also useful for the other ontology languages introduced later.

The **Web Ontology Language (OWL)** has been developed for representing knowledge, that is more complex than what can be represented in RDF. In particular, OWL provides predefined vocabulary to describe complex relationships between classes. For example, in OWL, we can specify that some class is the *union* (logically speaking, the disjunction) or the *intersection* (logically speaking, the conjunction) of two (or more) other classes. Class union and intersection are two examples of *class constructors* that OWL provides. Others also involve properties or even URIs that stand for *individuals* (which should be thought of as instances of classes), such that complex relationships like *Cape Trafalgar is a headland that witnessed at least one battle*, or *Every naval battle has (among other things) two or more involved parties, at least one location, a time span during which it took place, and a reason for engaging in it* can be expressed. Conceptually speaking, OWL is essentially a fragment of first-order predicate logic with equality and can be identified as a so-called *description logic* [6].

The OWL standard defines a number of variants that differ in sophistication and anticipated use cases. OWL EL, OWL QL, and OWL RL correspond to relatively small description logics for which reasoning algorithms (see Section 50.2.2) are relatively easy to implement. The larger OWL DL encompasses the former three and corresponds to a very expressive description logic called $\mathcal{SROIQ}(D)$, while OWL Full is the most general variant and encompasses both OWL DL and RDF—but it is not a description logic, and usually attempts are made to remain within smaller fragments when modeling ontologies.

The **Rule Interchange Format (RIF)** follows a different paradigm than OWL by addressing knowledge modeling by means of rules as those used in logic programming, Datalog, and other rule-based paradigms. Its main purpose is to facilitate the interchange of rules between applications, but it also constitutes a family of knowledge representation languages in its own right. In particular, RIF-Core corresponds to Datalog, while RIF-BLD (from *Basic Logic Dialect*) corresponds to Horn logic (essentially, definite logic programs). Rule paradigms essentially cater for the expression of if-then type relationships such as *If two parties agree on a cease-fire, then a combat operation started by one of these parties constitutes a violation of the agreement.* Rules in the sense of RIF-Core or RIF-BLD also correspond to certain fragments of first-order predicate logic [24].

* "SPARQL" is a recursive acronym that stands for "SPARQL Protocol and RDF Query Language."

50.2.2 Formal Semantics

Ontology languages can be understood as *vocabulary specification languages* in the sense that they can be used to define relationships of vocabulary terms related to a domain of interest. For example, they could be used to define a set of vocabulary terms describing spatial classes and relationships, such as *Headland*, *Coastal Landform*, *Erosion*, or *contributesTo*. Furthermore, terms can be associated via relations to each other, for example, in a taxonomy format: *Headland* is a subclass of *Coastal Landform*, which in turn is a subclass of *Landform*. More complicated relationships are also expressible, such as "every land form has at least one event that contributed to its creation," which, for instance, relates the terms *Erosion* and *contributesTo* to *Headland*, without saying that there is only one erosion per headland or that erosions are the only kind of events that are in a *contributesTo* relation to headlands. Note how we exploit the fact that erosions are subclasses of events and headlands are landforms. The exact kinds of relationships that are expressible depend on the ontology language chosen for the representation. Such vocabulary definitions are often referred to as *terminological* or *schema knowledge*. In the context of OWL, a set of such definitions is often referred to as a *TBox*.

Ontology languages furthermore allow for the expression of *assertional* knowledge or *facts* (often called *ABox* statements in the context of OWL). This refers to statements involving instances of classes, such as "Cape Trafalgar is a headland" or "Nelson is a Person."

Each of the W3C ontology languages presented in Section 50.2.1 comes endowed with a so-called *formal semantics*, which is described in model-theoretic terms borrowed from first-order predicate logic [19,34]. This formal semantics describes in mathematical terms how to draw valid *logical conclusions* from an ontology. For example, if *Headland* is specified as a subclass of *Coastal Landform*, and the ontology furthermore specifies that "every landform was shaped by at least one event," then one such logical consequence would be that "every headland was shaped by at least one event." While this example may seem obvious, in more complex cases it is often difficult to decide intuitively whether a statement is a logical consequence of an ontology. In these cases, the formal semantics serves as a *formal specification* for the valid logical consequences, and it is in this sense in which formal semantics provides a "meaning" to ontologies.

Logical consequences can be understood as knowledge that is *implicit* in an ontology. The formal semantics states that this implicit knowledge constitutes things that are *necessarily true*, given the ontology. As such, formal semantics enables *interoperability*, since it describes consequences which can be drawn from combining previously separated pieces of knowledge. Formal semantics can also be used in various ways when creating ontologies. For example, if an ontology engineer observes that her ontology has some logical consequences which seem to be undesirable from an application perspective, then this is an indication that parts of the ontology may have to be corrected or revised.

50.2.3 Key Issues in Realizing the Semantic Web Vision

A standard template to using ontology languages would thus be as follows. Entities are represented on the web using URIs, for example, *Battle of Trafalgar* could be represented by a URI such as `http://example.org/BattleOfTrafalgar` and the Napoleonic Wars by `http://example.org/NapoleonicWars`. Likewise, vocabulary elements are represented by URIs, such as `http://example.org/during` or `http://example.org/participatedIn`. In a background ontology, terminological information could then express, for example, that "during" is a transitive relation. Now consider, for example, the situation where metadata in a historical knowledge base is provided along with the human-readable content, which states that the Battle of Trafalgar took place during the

Napoleonic Wars, while another, more generic web portal lists facts about historic periods, including the fact that the Napoleonic Wars took place during the age of the Industrial Revolution. By collecting all this information, for example, using a Semantic Web crawler, this information could be combined, and due to the transitivity of "during" we would be able to also obtain the implicit knowledge, through logical deduction, that the Battle of Trafalgar took place during the Industrial Revolution. Even more, by adding semantics to the *participatesIn* relation discussed before, we could infer that Nelson lived during the Industrial Revolution.

The example just given indicates how Semantic Web technologies can in principle be used for information sharing, integration, and reuse across knowledge bases. However, in order to cast this idea into practical approaches, several obstacles need to be overcome. We list some of the most important ones:

- Different websites and knowledge bases may use different URIs for identifying entities on the Semantic Web. For information integration, we need to be able to identify these so-called *co-references.*
- Different websites may use different background ontologies. For information integration, these different ontologies need to be understood in relation to each other and formally related using suitable formats, preferably by using ontology languages or derivatives thereof. This issue is known as *ontology alignment.*
- Information on the web may be *semantically heterogeneous.* For example, on the instance level, different sources may specify other start and end dates for historical periods and, thus, may not agree whether the Napoleonic Wars also fall into the period known as Age of Enlightenment. On the schema level, the meaning assigned to terms such as *Headland*, *Erosion*, or *NavalBattle* may differ to a degree where they may become incompatible. Such issues are addressed by *semantic mediation and translation* research.
- Who is going to endow information on the web with metadata? Can we automate or partly automate this process using data mining or machine learning techniques? What are good interfaces and tools for developers and domain experts to support the creation of metadata?
- Algorithms for deducing logical consequences from ontologies are usually very dependent on input which constitutes ontologies of very high modeling quality. The creation of ontologies of such high quality can currently not be automated, and their creation usually requires experts in both ontology modeling and in the application domain under consideration. Consequently, their creation and maintenance can be very expensive. The creation of methods, workflows, and tools to support ontology modeling (and any other part of the ontology lifecycle) is, thus, of utmost importance for the development of the field and its applications.
- How can we deal with evolving information, for example, when data changes or is revised, or if vocabulary terms change their meaning over time? How can we deal with uncertainties or noise in metadata, which seem to be unavoidable in many realistic settings?

50.2.4 Linked Data

*Linking Data** [9] is a community effort that started in 2007 and is supported by the W3C. Its declared goal is to bootstrap a global graph of interlinked data called the *Web of Data* by converting existing datasets into the format of ontology languages, and interlinking them with meaningful relationships. Linked Data has been a key driver for research and development in the Semantic Web since its inception,

* http://linkeddata.org/

and has in particular created substantial interest in the industry and governments around the world. This Web of Data is currently growing rapidly. As of 2011, it had an estimated size of about 32 billion RDF triples, with contributions coming increasingly from industry and from public sector bodies such as governments and libraries.

In its current form, Linked Data contains mainly assertional knowledge expressed in RDF, and there are only very few light-weight guiding principles for its creation. Methods and tools for making use of Linked Data, and the question how to advance from Linked Data toward the more general Semantic Web vision, are among the most prominent current research questions. Similarly, ontologies that have been developed to annotate and interlink these datasets are often light-weight to a degree where they fail to restrict the interpretation of terms toward their intended meaning [25].

50.3 Impact on Practice

In order to assess the impact of the Semantic Web field in practice, it is important to realize that Semantic Web technologies can also be applied in contexts other than the World Wide Web. They likewise apply to (closed) intranets, but also to many other settings where information creation, retrieval, reuse, and integration are of importance. Typical examples include enterprise information integration and knowledge management systems. In many other cases, for example, search engines, ontologies, and Semantic Web technologies are used in the backend, and, while of major importance, are not visible to the end users.

Indeed, early adoption of Semantic Web technologies in nonweb areas can be traced back to at least the first years of the twenty-first century, when the first spin-off companies left the research realm and applied their methods to enable industrial applications. In the next few years, industry interest kept growing, but slowly, and at the same time other research disciplines that rely on information management and integration adopted Semantic Web technologies for their purposes. A leading role in this latter development was taken by the life sciences, in particular related to health care and bioinformatics.

Usage of Semantic Web technologies on the *web*, however, hardly happened in these years, apart from use-case studies and prototypes. One prominent exception was *Semantic MediaWiki** [30], a semantic enhancement of the MediaWiki software that underlies Wikipedia.† It was presented in 2006 and very quickly found substantial uptake with thousands of known installations worldwide. Indeed, efforts are currently under way to customize the extension for the use in Wikipedia.‡

Since 2007, and substantially driven by the advent of Linked Data, industrial uptake is rising significantly, both on and off the web. At the time of this writing,§ Semantic Web technologies have started to deeply penetrate information technologies and can be expected to become part of the mainstream arsenal for application developers in the near future. This adoption of Semantic Web technologies does not necessarily mean the use of the exact standards described in Section 50.2, but rather the adherence to the general principles of using machine-readable and understandable metadata for information sharing, retrieval, and integration.

Adoption in practice of Semantic Web technologies is, to date, often restricted to the use of simple metadata, and in particular the systematic use of formal semantics is currently only possible in particular application areas and settings, since stronger solutions for some of the key issues listed in Section 50.2.3 need to be developed first (see also Section 50.4).

* http://semantic-mediawiki.org/
† http://www.wikipedia.org/
‡ http://meta.wikimedia.org/wiki/New_Wikidata
§ August 2012.

In the following, we list some of the most visible recent achievements of Semantic Web technologies with respect to practical applications:

- The need for meta-modeling has recently been picked up by some of the most prominent companies on the web. Facebook's Open Graph protocol* is a prominent example, as is schema.org,† a joint effort by Bing, Google, and Yahoo! for improving web search. RDF versions are endorsed by these companies, and the schema.org ontology is officially available in OWL/RDF.‡
- In February 2011, the Watson system by IBM [15] made international headlines for beating the best humans in the quiz show *Jeopardy!*.§ The performance is already being considered as a milestone in the development of artificial intelligence techniques related to general question answering. Semantic Web technologies have been one of the key ingredients in Watson's design.¶
- A significant number of very prominent websites are powered by Semantic Web technologies, including the New York Times,** Thomson Reuters,†† BBC [36], and Google's Freebase, as well as sites using technology such as Yahoo! SearchMonkey [33], or Drupal 7‡‡ [13].
- The speech interpretation and recognition interface *Siri* launched by Apple in 2011 as an intelligent personal assistant for the new generation of IPhone smartphones heavily draws from work on ontologies, knowledge representation, and reasoning [17].
- Oracle Database 11g supports OWL.§§
- Recently, The Wall Street Journal ran an article¶¶ announcing that Google is in the progress of enhancing its web search using Semantic Web technologies.
- The *International Classification of Diseases*, ICD, is the international standard manual for classifying diseases, endorsed by the World Health Organization and in use worldwide. It is currently revised, and ICD-11 (the 11th revision) is scheduled to appear in 2015. The revision is driven by a collaborative platform using OWL as underlying technology [41].
- Many governments and large companies now publish a plethora of governmental and other information as Linked Data, with England and the United States being early adopters.***

50.4 Research Issues

Currently, the Semantic Web field is mainly driven by developments regarding Linked Data (see Section 50.2.4). The amount of information available as Linked Data has shown exponential growth since its inception, and there are no indications that this will slow down soon. The Linked Data cloud indeed serves as an interlinked data source that is available in form of readily processable syntax (namely, RDF), and as such has the potential for widespread usage in data-intensive applications.

However, while it is certainly very helpful that Linked Data is available in RDF, there are very few general principles that would guide its creation.††† Consequently, Linked Data still suffers from many

* http://ogp.me/
† http://schema.org/
‡ http://schema.org/docs/schemaorg.owl
§ http://www.jeopardy.com/
¶ http://semanticweb.com/game-show-circuit-was-just-a-first-step-for-ibms-watson-and-deep-qa_b20431
** http://www.nytimes.com/, see also http://data.nytimes.com/
†† http://www.opencalais.com/
‡‡ http://semantic-drupal.com/
§§ http://www.oracle.com/technetwork/database/options/semantic-tech/index.html
¶¶ http://online.wsj.com/article/SB10001424052702304459804577281842851136290.html
*** See http://www.data.gov/semantic and http://data.gov.uk/linked-data/
††† This is, in principle, a good thing, and, in fact, is in line with the "bottom-up" nature of the World Wide Web itself.

heterogeneity challenges which the Semantic Web field has initially set out to overcome [25,27]. In reference to the issues identified in Section 50.2.3, we note that co-reference identification is still a central and unsolved issue, even for Linked Data [42]. Likewise, efficiently addressing semantic heterogeneity in Linked Data, for example, by means of ontology alignment, has only recently started to be studied [26], in particular for the difficult quest of providing question answering systems based on Linked Data [29]. The development of special techniques for dealing with evolving Linked Data is also in very early stages.

Thus, there currently is a significant disconnect between the factually deployed Linked Data on the web, and the already established research results regarding the use of Semantic Web technologies and ontology languages. It appears to be the foremost current research question, how to enable these strongly and formally semantic approaches for use on and with Linked Data. And this quest is inherently tied with the current lack of useful schema knowledge in Linked Data [25].

The question, how Linked Data can indeed be evolved into a Semantic Web in the stronger sense of the term, is indeed very controversially discussed. And while it is certainly also a question about good underlying conceptual principles, it is foremost also a question of practicability, of finding suitable next steps for development that can find significant adoption in practice.

To address the relative lack of schema knowledge in Linked Data, a foremost requirement is the availability of strong tool support that makes Semantic Web technologies more easily accessible for practitioners. This includes the automated, semi-automated, or manual generation of metadata, and ontologies, powerful ontology alignment systems, lifecycle support for metadata and ontologies such as versioning approaches and revision processes, and tools that enable an easier reuse of metadata and ontologies in applications for which they were not originally developed. In all these aspects, the research community has provided significant research advances which nevertheless remain to be further strengthened before they can find widespread adoption.

At the same time, a significant body of best practices guidelines need to be developed, which includes ontology creation and lifecycle aspects, but also methods and processes for making use of Semantic Web technologies. Of growing importance in this respect is the development of useful ontology design patterns for ontology modeling [16].

Eventually, in order to meet the goals of the Semantic Web vision, strongly semantic approaches—including ontology reasoning—will have to be embraced and brought to bear on realistic web data. In order for this to happen, researchers need to find clear answers on how to establish semantic interoperability without giving up on semantic heterogeneity [28], to scalability issues of reasoning algorithms, and to the question how to deal with metadata which arises from the strongly collaborative web without central control, that is, metadata which substantially varies in modeling quality [21,25].

50.5 Summary

The Semantic Web is an interdisciplinary research field that aims at augmenting the existing World Wide Web with machine-readable and understandable metadata such that information on the web becomes available for processing in intelligent systems. In particular, it shall establish solutions for seamless information creation, retrieval, reuse, and integration on the web.

The key underlying technology is the use of so-called ontology languages for representing metadata information. There exist several such languages endorsed by the World Wide Web Consortium. They support formal semantics that enables automated reasoning using deductive logical methods.

Industry is increasingly adopting Semantic Web technologies, and this also includes adoption for purposes other than for information on the World Wide Web. One of the driving recent developments is the publishing of significant amounts of data in ontology language formats on the World Wide Web—this information is referred to as Linked Data.

Despite its success, many core issues still require further in-depth, and partially foundational, research, such as the systematic use of deep semantics on the web by means of automated reasoning techniques.

Key Terms

Annotation: The attaching of metadata information to entities on the web, such as web pages, text, text snippets, words or terms, images, tables, etc. These annotations are usually not visible in normal web browsers, but can be retrieved from the source code of the web page for further processing. For example, a picture could bear the annotation "Barack Obama," indicating that it is a picture of Barack Obama.

Co-reference: This occurs when two different identifiers are used for one real-world entity. The identification of co-references in ontological data on the Semantic Web is a challenging research problem. The term is borrowed from linguistics.

Datalog: A rule-based knowledge representation language based on function-free Horn logic. It is used in deductive databases.

Description logics: A family of closely related knowledge representation languages. Traditionally, they are strongly based on first-order predicate logic from which they inherit their open-world semantics. Description logics are usually decidable.

Horn logic: A fragment of first-order predicate logic that can be expressed in the form of rules. It is the basis for the logic programming paradigm, and thus also for the logic programming language Prolog.

Linked Data: Data on the World Wide Web that is represented using RDF, following a few simple principles. These datasets are strongly interlinked, thus forming a network of knowledge, often referred to as the *Web of Data*. If the data are available under an open license, it is called Linked *Open* Data, but often this distinction is not made in the literature. Linked Data is considered a major milestone in the development of the Semantic Web.

Logic programming: A knowledge representation and programming paradigm usually based on Horn logic with some modifications and extensions. In particular, it uses a closed-world paradigm. Prolog is the most widely known logic programming language.

Metadata: Data that provides information for other data, often through the process of annotation. Metadata can be of many different forms; in the simplest case it consists only of keywords (often called *tags*), but in a Semantic Web context metadata is usually expressed using some ontology language.

Ontology: A term borrowed from philosophy. In modern computer science, an ontology is a knowledge base that is expressed by describing relationships between concepts in a given domain of interest. The relationships are described using knowledge representation languages—then called ontology languages—which are usually derived from first-order predicate logic or from the logic programming paradigm.

Ontology alignment: The establishing of formal relationships between the concepts and other terms in two different ontologies. This can be done manually, but there is also a significant area of research that develops automated ontology alignment systems.

Ontology design patterns: Schemas for sets of concepts and relations occurring frequently in ontology modeling. They are typically expressed in some ontology language.

Semantics: Sematics, more precisely *formal* semantics, usually refers to the notion of model-theoretic semantics in first-order predicate logic, adapted for an ontology language. The main purpose of such a formal semantics is to define a notion of *logical consequence*. In this sense, a formal semantics is an implicit specification for all the logical consequences that can be derived from an ontology or a knowledge base.

URI: An abbreviation for *uniform resource identifier.* URIs are used to represent resources on the World Wide Web, for example, websites. It is not required that an entity identified by an URI can in fact be located or accessed on the World Wide Web.

Web of Data: See Linked Data.

World Wide Web Consortium W3C: The main international organization that develops standards for the World Wide Web.

XML: XML stands for *Extensible Markup Language.* It is a text-based format for the representation of hierarchically structured data. XML is ubiquitous on the World Wide Web, and often used for the interchange of data on the World Wide Web.

Acknowledgment

Pascal Hitzler acknowledges support by the National Science Foundation under award 1017225 *III: Small: TROn – Tractable Reasoning with Ontologies.* Any opinions, findings, and conclusions or recommendations expressed in this material are those of the authors and do not necessarily reflect the views of the National Science Foundation.

Further Information

Semantic Web as a field of research and applications is still under rapid development. Textbooks on the topic thus tend to be outdated rather quickly, unless they focus on fundamental issues that already find a broad consensus in the community. We recommend the following three books as thorough introductions: *Foundations of Semantic Web Technologies* [23] by Hitzler, Krötzsch, and Rudolph; *Semantic Web for the Working Ontologist* [1] by Allemang and Hendler; and *Semantic Web Programming* [20] by Hebeler, Fisher, Blace, Perez-Lopez and Dean. The first mentioned book [23] focuses on foundations, in particular on a thorough treatment of the standardized ontology languages, including their formal semantics. The other two books [1,20] are written from a more practical perspective—they convey more hands-on knowledge, but are less comprehensive in the formal foundations. All three books are widely acclaimed as excellent introductions.

The Handbook on Ontologies [39] edited by Staab and Studer is a popular resource that introduces many aspects of Semantic Web technologies. The chapters are written by well-known experts in the field. The book is necessarily much less systematic and thorough than a textbook, but its coverage is much wider and includes many topics that are of central importance for the Semantic Web, but that have not been developed far enough yet to be included in a standard textbook.

Primary resources for state-of-the-art developments in the field are the major journals and conferences in the area. Many journals in computer science and adjacent fields in fact publish papers related to Semantic Web, but there are also some prominent ones that are dedicated exclusively to the field. These include the *Journal of Web Semantics,** the *Semantic Web journal,*† and the *International Journal on Semantic Web and Information Systems.*‡ Major conferences in the area are the International Semantic Web Conference§ (see, e.g., [4,5]) and the Extended Semantic Web Conference¶ (see, e.g., [2,3]), which attract primarily researchers, and the Semantic Technology Conference** which targets industry. Many other major conferences also publish research papers in Semantic Web, for example, the comprehensive World Wide Web Conference†† (see, e.g., [38]).

* http://www.websemanticsjournal.org/
† http://www.semantic-web-journal.net/
‡ http://www.ijswis.org/
§ http://swsa.semanticweb.org/
¶ http://eswc-conferences.org/
** See http://semanticweb.com/
†† http://wwwconference.org/

References

1. D. Allemang and J. Hendler. *Semantic Web for the Working Ontologist.* Morgan Kaufmann, Waltham, MA, 2008.
2. G. Antoniou, M. Grobelnik, E. P. B. Simperl, B. Parsia, D. Plexousakis, P. De Leenheer, and J. Z. Pan, eds. *The Semantic Web: Research and Applications—Proceedings of the 8th Extended Semantic Web Conference, ESWC 2011,* Heraklion, Crete, Greece, May 29–June 2, 2011, Part II, volume 6644 of Lecture Notes in Computer Science. Springer, Berlin, Germany, 2011.
3. G. Antoniou, M. Grobelnik, E. P. B. Simperl, B. Parsia, D. Plexousakis, P. De Leenheer, and J. Z. Pan, eds. *The Semantic Web: Research and Applications—Proceedings of the 8th Extended Semantic Web Conference, ESWC 2011,* Heraklion, Crete, Greece, May 29–June 2, 2011, Part I, volume 6643 of Springer, Berlin, Germany, 2011.
4. L. Aroyo, C. Welty, H. Alani, J. Taylor, A. Bernstein, L. Kagal, N. F. Noy, and E. Blomqvist, eds. *The Semantic Web—ISWC 2011—Proceedings of the 10th International Semantic Web Conference,* Bonn, Germany, October 23–27, 2011, Part I, volume 7031 of Lecture Notes in Computer Science. Springer, Berlin, Germany 2011.
5. L. Aroyo, C. Welty, H. Alani, J. Taylor, A. Bernstein, L. Kagal, N. F. Noy, and Eva Blomqvist, eds. *The Semantic Web—ISWC 2011—Proceedings of the 10th International Semantic Web Conference,* Bonn, Germany, October 23–27, 2011, Part II, volume 7032 of Lecture Notes in Computer Science. Springer, Berlin, Germany 2011.
6. F. Baader, D. Calvanese, D. McGuinness, D. Nardi, and P. Patel-Schneider, eds. *The Description Logic Handbook: Theory, Implementation, and Applications.* Cambridge University Press, Cambridge, U.K., 2007.
7. T. Berners-Lee and D. Connolly. Notation3 (N3): A readable RDF syntax. W3C Team Submission March 28, 2011, 2011. Available from http://www.w3.org/TeamSubmission/n3/ (accessed on October, 2013).
8. T. Berners-Lee, J. Hendler, and O. Lassila. The semantic web. *Scientific American,* 284(5):34–43, May 2001.
9. C. Bizer, T. Heath, and T. Berners-Lee. Linked data—The story so far. *International Journal on Semantic Web and Information Systems,* 5(3):1–22, 2009.
10. H. Boley, G. Hallmark, M. Kifer, A. Paschke, A. Polleres, and D. Reynolds, eds. RIF Core Dialect. W3C Recommendation June 22, 2010, 2010. Available from http://www.w3.org/TR/rif-core/ (accessed on October, 2013).
11. H. Boley and M. Kifer, eds. RIF Basic Logic Dialect. W3C Recommendation June 22, 2010, 2010. Available from http://www.w3.org/TR/rif-bld/ (accessed on October, 2013).
12. T. Bray, J. Paoli, C. M. Sperberg-McQueen, E. Maler, and F. Yergeau, eds. *Extensible Markup Language (XML) 1.0,* 5th edn. W3C Recommendation, November 26, 2008. Available at http://www.w3.org/TR/xml (accessed on October, 2013).
13. S. Corlosquet, R. Delbru, T. Clark, A. Polleres, and S. Decker. Produce and consume Linked data with Drupal! In A. Bernstein, D. R. Karger, T. Heath, L. Feigenbaum, D. Maynard, E. Motta, and K. Thirunarayan, eds. *The Semantic Web—ISWC 2009, Proceedings of the 8th International Semantic Web Conference, ISWC 2009,* Chantilly, VA, October 25–29, 2009, volume 5823 of Lecture Notes in Computer Science, pp. 763–778. Springer, Berlin, Germany, 2009.
14. L. Feigenbaum, I. Herman, T. Hongsermeier, E. Neumann, and S. Stephens. The Semantic Web in action. *Scientific American,* 297:90–97, November 2007.
15. D. A. Ferrucci, E. W. Brown, J. Chu-Carroll, J. Fan, D. Gondek, A. Kalyanpur, A. Lally et al. Building Watson: An overview of the DeepQA project. *AI Magazine,* 31(3):59–79, 2010.
16. A. Gangemi. Ontology design patterns for semantic web content. In Y. Gil, E. Motta, V. R. Benjamins, and M. A. Musen, eds. *The Semantic Web—ISWC 2005, Proceedings of the 4th International Semantic Web Conference, ISWC 2005,* Galway, Ireland, November 6–10, 2005, volume 3729 of Lecture Notes in Computer Science, pp. 262–276. Springer, Berlin, Germany, 2005.

17. T. Gruber. Siri: A virtual personal assistant. Keynote Presentation at "*Semantic Technologies Conference*", June 2009. http://tomgruber.org/writing/semtech09.htm (accessed on October, 2013).
18. S. Harris and A. Seaborne. SPARQL 1.1 Query Language. W3C Working Draft, May 12, 2011. Available from http://www.w3.org/TR/sparql11-query/ (accessed on October, 2013).
19. P. Hayes, ed. RDF Semantics. W3C Recommendation, February 10, 2004. Available at http://www.w3.org/TR/rdf-mt/ (accessed on October, 2013).
20. J. Hebeler, M. Fisher, R. Blace, A. Perez-Lopez, and M. Dean. *Semantic Web Programming*. Wiley, Indianapolis, IN, 2009.
21. P. Hitzler. Towards reasoning pragmatics. In K. Janowicz, M. Raubal, and S. Levashkin, eds. *GeoSpatial Semantics Proceedings, Third International Conference, GeoS 2009*, Mexico City, Mexico, December 3–4, 2009. Lecture Notes in Computer Science, pp. 9–25. Springer, Berlin, Germany, 2009.
22. P. Hitzler, M. Krötzsch, B. Parsia, P. F. Patel-Schneider, and S. Rudolph, eds. OWL 2 Web Ontology Language: Primer. W3C Recommendation October 27, 2009, 2009. Available from http://www.w3.org/TR/owl2-primer/ (accessed on October, 2013).
23. P. Hitzler, M. Krötzsch, and S. Rudolph. *Foundations of Semantic Web Technologies*. Chapman & Hall/CRC, Boca Raton, FL, 2009.
24. P. Hitzler and A. K. Seda. *Mathematical Aspects of Logic Programming Semantics*. CRC Press, Boca Raton, FL, 2010.
25. P. Hitzler and F. van Harmelen. A reasonable semantic web. *Semantic Web*, 1(1–2):39–44, 2010.
26. P. Jain, P. Hitzler, A. P. Sheth, K. Verma, and P. Z. Yeh. Ontology alignment for linked open data. In P. Patel-Schneider, Y. Pan, P. Hitzler, P. Mika, L. Zhang, J. Pan, I. Horrocks, and B. Glimm, eds. *Proceedings of the 9th International Semantic Web Conference, ISWC 2010*, Shanghai, China, November 7–11, 2010, volume 6496, pp. 402–417. Springer, Heidelberg, Germany, 2010.
27. P. Jain, P. Hitzler, P. Z. Yeh, K. Verma, and A. P. Sheth. Linked data is merely more data. In D. Brickley, V. K. Chaudhri, H. Halpin, and D. McGuinness, eds. *Linked Data Meets Artificial Intelligence*, pp. 82–86. AAAI Press, Menlo Park, CA, 2010.
28. K. Janowicz. The role of space and time for knowledge organization on the Semantic Web. *Semantic Web*, 1(1–2):25–32, 2010.
29. A. K. Joshi, P. Jain, P. Hitzler, P. Z. Yeh, K. Verma, A. P. Sheth, and M. Damova. Alignment-based querying of linked open data. In R. Meersman et al., eds. *On the Move to Meaningful Internet Systems: OTM 2012 Proceedings. Confederated International Conferences: CoopIS, DOA-SVI, and ODBASE 2012*, Rome, Italy, September 10–14, 2012. Proceedings, Part II, Lecture Notes in Computer Science, Vol. 7566, pp. 807–824, Springer, 2012.
30. M. Krötzsch, D. Vrandecic, M. Völkel, H. Haller, and R. Studer. Semantic wikipedia. *Journal of Web Semantics*, 5(4):251–261, 2007.
31. F. Manola and E. Miller, eds. Resource Description Framework (RDF). Primer. W3C Recommendation, February 10, 2004. Available at http://www.w3.org/TR/rdf-primer/ (accessed on October, 2013).
32. D. L. McGuinness and F. van Harmelen, eds. OWL Web Ontology Language Overview. W3C Recommendation, February 10, 2004. Available at http://www.w3.org/TR/owl-features/ (accessed on October, 2013).
33. P. Mika. Case Study: Improving Web Search Using Metadata. W3C Semantic Web Use Cases and Case Studies, November 2008. Available from http://www.w3.org/2001/sw/sweo/public/UseCases/yahoo/ (accessed on October, 2013).
34. B. Motik, P. F. Patel-Schneider, and B. C. Grau, eds. OWL 2 Web Ontology Language: Direct Semantics. W3C Recommendation, October 27, 2009. Available at http://www.w3.org/TR/owl2-direct-semantics/ (accessed on October, 2013).
35. E. Prud'hommeaux and G. Carothers, eds. Turtle—Terse RDF Triple Language. W3C Working Draft August 09, 2011. Available from http://www.w3.org/TR/turtle/ (accessed on October, 2013).

36. Y. Raimond. Case Study: Use of Semantic Web Technologies on the BBC Web Sites. W3C Semantic Web Use Cases and Case Studies, January 2010. Available from http://www.w3.org/2001/sw/sweo/public/UseCases/BBC/ (accessed on October, 2013).
37. A. Seaborne and E. Prud'hommeaux. SPARQL query language for RDF. W3C Recommendation, January 15, 2008. Available from http://www.w3.org/TR/rdf-sparql-query/ (accessed on October, 2013).
38. S. Srinivasan, K. Ramamritham, A. Kumar, M. P. Ravindra, E. Bertino, and R. Kumar, eds. *Proceedings of the 20th International Conference on World Wide Web, WWW 2011,* Hyderabad, India, March 28–April 1, 2011 (Companion Volume). ACM, New York, 2011.
39. S. Staab and R. Studer, eds. *Handbook on Ontologies*, 2nd edn. International Handbooks on Information Systems. Springer, Berlin, Germany, 2009.
40. R. Studer. The semantic web: Suppliers and customers. In I. F. Cruz, S. Decker, D. Allemang, C. Preist, D. Schwabe, P. Mika, M. Uschold, and L. Aroyo, eds. *The Semantic Web—ISWC 2006, Proceedings of the 5th International Semantic Web Conference, ISWC 2006,* Athens, GA, November 5–9, 2006, volume 4273 of Lecture Notes in Computer Science, pp. 995–996. Springer, Berlin, Germany, 2006.
41. T. Tudorache, S. M. Falconer, C. Nyulas, N. F. Noy, and M. A. Musen. Will semantic web technologies work for the development of ICD-11? In P. F. Patel-Schneider, Y. Pan, P. Hitzler, P. Mika, L. Zhang, J. Z. Pan, I. Horrocks, and B. Glimm, eds. *The Semantic Web—ISWC 2010—9th International Semantic Web Conference, ISWC 2010,* Shanghai, China, November 7–11, 2010, Revised Selected Papers, Part II, volume 6497 of Lecture Notes in Computer Science, pp. 257–272. Springer, Berlin, Germany, 2010.
42. G. Tummarello and R. Delbru. Linked data is merely more data. In D. Brickley, V. K. Chaudhri, H. Halpin, and D. McGuinness, eds. *Linked Data Meets Artificial Intelligence*. AAAI Press, Menlo Park, CA, 2010.

51

Web Search Engines: Practice and Experience

Tao Yang
University of California, Santa Barbara

Apostolos Gerasoulis
Rutgers University

51.1 Introduction

The existence of an abundance of dynamic and heterogeneous information on the web has offered many new opportunities for users to advance their knowledge discovery. As the amount of information on the web has increased substantially in the past decade, it is difficult for users to find information through a simple sequential inspection of web pages or recall previously accessed URLs. Consequently, the service from a search engine becomes indispensable for users to navigate around the web in an effective manner.

General web search engines provide the most popular way of identifying relevant information on the web by taking a horizontal and exhaustive view of the world. Vertical search engines focus on specific segments of content and offer great benefits for finding information on selected topics. Search engine technology incorporates interdisciplinary concepts and techniques from multiple fields in computer science, which include computer systems, information retrieval, machine learning, and computer linguistics. Understanding the computational basis of these technologies is important as search engine technologies have changed the way our society seeks out and interacts with information.

This chapter gives an overview of search techniques to find information on the web relevant to a user query. It describes how billions of web pages are crawled, processed, and ranked, and our experiences in building Ask.com's search engine with over 100 million North American users. Section 51.2

is an overview of search engine components and historical perspective of search engine development. Section 51.3 describes crawling and off-line processing techniques. Section 51.4 describes the online architecture and query-processing techniques. Section 51.5 explains ranking signals and algorithms. Section 51.6 discusses the metrics for evaluating a search engine. Finally, Section 51.7 concludes this chapter.

51.2 Search Engine Components and Historical Perspective

In general, a search engine consists of three main components as shown in Figure 51.1: a crawler, an off-line processing system to accumulate data and produce searchable index, and an online engine for real-time query handling. Their roles are summarized as follows:

- *Crawling and data acquisition.* A crawler discovers and adds the content of the web to the search engine's data repository. Type of content gathered includes text pages, images, videos, and other types of content. Most crawlers find information by beginning at one page and then follow the outgoing URL links on that page. Therefore, if a page is not linked to from another page, this page may never be found by a search engine. In some cases, a search engine may acquire content through private data feeds from other companies. All of the information that a web crawler retrieves is stored in a data repository, which provides a foundation to build an online index searchable by users.
- *The data repository and off-line processing.* Some search engines have extremely large databases with billions of pages while others have comparatively smaller ones. A search engine may aggregate results from its own database and from other search engines hosted in different locations to expand the database coverage. Off-line processing mainly takes a collection of pages and builds appropriate data structure for search. The off-line system also performs duplicate and spam detection and conducts additional data analysis to identify properties of a page as ranking signals.
- *Online search engine and ranking.* A search engine runs as a web service to answer queries from users. The query-processing system uses the index to find relevant documents and rank them. It generates a result page that contains top-ranked documents and serves this page to a user. As users interact with the search results, a search engine logs their browsing behavior and leverages such data for improving search quality.

Information retrieval researchers have studied document search for many years [1] before web search became popular in the early 1990s. An earlier system for locating Internet content was Gopher, used for file name lookup. One of the first "full-text" crawler-based search engines was WebCrawler, which came out in 1994. As the World Wide Web became popular, a number of search engines and portals offered search services, including AltaVista, Excite, Infoseek, Lycos, Inktomi, Northern Light, Yahoo!, and AOL.

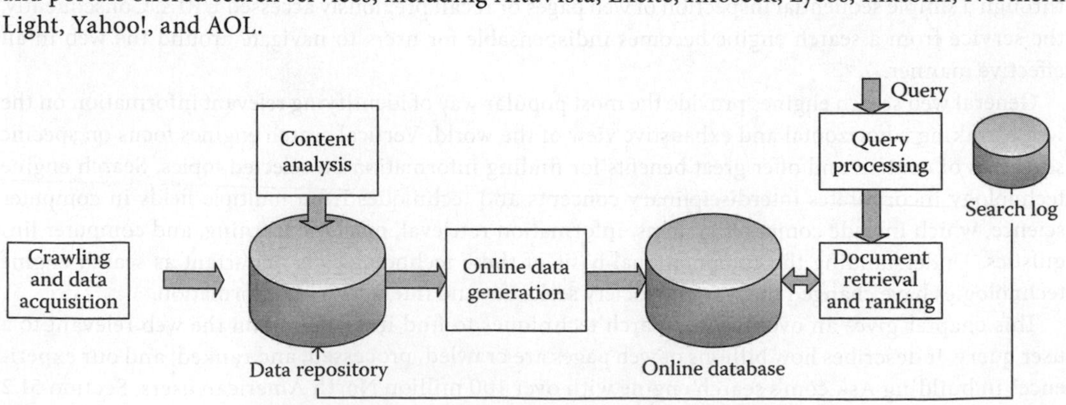

FIGURE 51.1 Components of web search.

In earlier 1999, Google's search engine started to gain popularity with its new link-based ranking paradigm and a simple user interface. In 2002 and 2003, Yahoo! acquired Inktomi search and Overture, which owned AltaVista and All the Web, building its own search technology. Ask.com (formally called Ask Jeeves) acquired Teoma search in 2001 for providing search and question answering. Microsoft initially used search results from Inktomi and has gradually built its own engine: MSN search, Windows Live Search, Live Search, and Bing.

In Europe, Fast search was launched in 1999 to compete with Google and was acquired partially by Overture in 2003 and partially by Microsoft in 2007. In China, Baidu was started in 1999 and has become a dominating search engine. In Russia, Yandex search launched in 1998 and is currently dominating, while in South Korea, Naver is a popular search engine launched in 2000.

Technology behind the major search engines has been seldom published because of fierce competition while some of their techniques have gradually appeared in patents and academic conferences. Academic conferences that cover more on search algorithms and data analysis include SIGIR, WWW, WSDM, and ECIR. Other related conferences for web data analysis include KDD, CIKM, SIGMOD, and VLDB. The system technology covering large-scale Internet systems and data platforms has appeared in various system conferences such as OSDI. Search engines have a major traffic impact to websites, and the search industry activity is closely watched by webmasters and various organizations including searchenginewatch.com. Recent textbooks that describe search algorithms and mining techniques can be found in [2–5].

There exist a number of open-source search engines and related data-processing tools available for search engine research and applications. The Apache Lucene/Solr package [6] is a Java-based search engine in which Lucene provides data indexing and full-text search capability while Solr provides a search user interface built on top of Lucene with additional support such as caching and replication management. Lemur [7] and Xapian [8] are open-source search engines with C++. Various open-source codes are available for underlying system support. Hadoop's MapReduce [9] is a popular clustering package for parallel processing of off-line data. Apache HBase [10] and Cassandra [11] provide the key-value data store support for accessing large datasets.

51.3 Crawling and Data Processing

51.3.1 Crawling

A web crawler is part of the search engine to gather data from the Internet; it can recognize and collect HTML pages and other types of documents including PDF, PowerPoint, Word, and Excel. It extracts text information from these documents through conversion software so that corresponding text information is indexable and searchable. Figure 51.2a illustrates the architecture of a large-scale crawler we have developed in the past for discovering and downloading billions of web pages:

- The URL downloader fetches web pages following the hypertext transfer protocol (HTTP) protocol. Typically, a crawler uses a thread to handle an HTTP connection and maintains many concurrent threads to saturate the available download bandwidth. Since there can be a high latency in downloading a page from a site, a crawler machine can have a large number of HTTP requests outstanding. That increases system resource requirement as each machine needs to maintain information on all open connections.
- A DNS server resolves the IP address of URLs. Since there is a massive amount of DNS lookup of URL hosts, extra caching support can be useful.
- The URL extraction module parses downloaded pages and identifies the outgoing links. It may filter some of those URLs identified as low quality or duplicates. After filtering, it adds the extracted outgoing URLs to a URL queue. Since a crawler typically runs its service on a cluster of machines, extracted URLs from one machine may be distributed to URL queues in multiple machines as illustrated in Figure 51.2b.

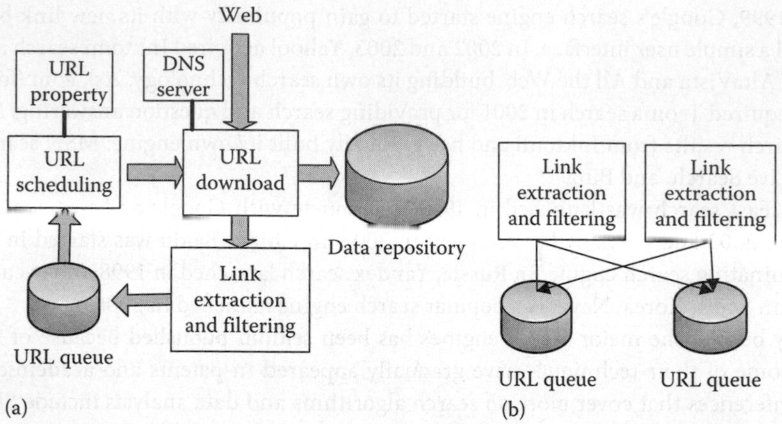

FIGURE 51.2 (a) Architecture of a web crawler. (b) Collaborative outgoing link extraction and URL distribution.

- The URL scheduling decides which URLs are to be visited in the next round of crawling. Given that websites have variable sizes, small sites can be crawled quickly while it takes a long time to visit a large site. A URL scheduler periodically analyzes the properties of the queued URLs, prioritizes them, and selects a subset of URLs to the URL download process, to accomplish the following goals: (1) large sites are crawled completely; (2) URLs are recrawled frequently to ensure freshness; (3) revisiting for recently crawled URLs are excluded to avoid the endless revisit of the same URLs; and (4) low-quality or duplicate pages are not crawled or crawled with a low priority.
- The URL property service maintains properties of URLs. There is a resource contention to accomplish these goals in parallel, and the URL scheduler uses properties of URLs such as the last successful crawling time and unsuccessful data fetch attempts to make a sound decision. Since there is a large number of data accesses in parallel at each second on each crawling machine and there are a large number of URLs to be crawled, accessing URL property information can become a crawling bottleneck. The design of an efficient data structure and crawler architecture is therefore crucial.

There are a number of factors that further complicate crawler management:

- *Politeness.* A website often hosts a large number of pages. To fetch these pages quickly, a crawler can occupy a significant amount of computing and bandwidth resource from this site, and this can affect the normal operation of such a site. To mitigate the negative impact on the site's resources, a crawler must follow a politeness policy to limit its crawl rate. For instance, a crawler may not make more than one request to the same host at the same time, and possibly it should add a time delay between two consecutive requests to the same host.
- *Robots exclusion protocol.* A web administrator can exclude their pages from the index of a search engine by providing the /robots.txt file in the top directory of the website. This file specifies the visit restriction for a search engine's crawler [12]. For example, a line with text "Disallow: /tmp/" in example.com/robots.txt means that visiting example.com/tmp is not allowed. When fetching a page from a site, a crawler must check the robots.txt from the corresponding web host. To avoid repeated access of the robot.txt file from the same host, a crawler may save a copy of such a file in its local cache system with a periodical update.
- *Crawling abnormality.* Some sites respond to a crawler's request slowly or even fail to respond. Some sites may return wrong or broken results to a crawler. A crawler needs to be resilient to the failure or slowness of a website and be able to retry at a later time.

  Another type of crawling abnormality is called *crawling traps.* In this case, a website keeps a crawler busy by dynamically feeding an infinite number of useless web pages. For example,

there can be an infinite number of calendar pages generated dynamically and accessible through "next day" button in a site. Such a trap can be detectable by site-oriented similar content analysis.
- *Hidden web and sitemap.* Crawlers from major search engines can effectively collect surface-level web pages, but there is a vast amount of hidden deep web information invisible to search engines. Any page that is not explicitly pointed by other web pages cannot be found by a crawler through hyperlink analysis. Dynamically generated pages that require HTML-form or Javascript submission are often invisible. Automatic form filling can explore deep web [13], and there is still a challenge to reach high accuracy or coverage. An alternative method is to use the sitemaps protocol [14], which enables webmasters to inform search engines about a list of pages that are available for crawling from their sites. A sitemap is an XML file that lists URLs for a site along with additional metadata about each URL. Thus, webmasters can advertise deep URLs, which are not even hyperlinked from the surface web.

The freshness and content coverage of a data collection crawled are important for the quality of a search engine. The challenge a crawler faces is to discover newly created or updated pages on the web quickly. For known URLs, a crawler can revisit them frequently to detect if their content has been changed. The historical change frequency of a page and its site can guide how often a crawler should revisit this page. While crawling sites frequently can discover new pages, leveraging information published by a site such as RSS feeds can speed up this discovery process. An RSS feed is an XML-based file that provides a structured list of new pages added to a site. The RSS feeds are commonly associated with blogs and other information publishers. For example, CNN provides news RSS feeds under different categories. Amazon publishes RSS feeds on popular products with tags. There are additional signals that can be mined from the web to facilitate the discovery of new URLs, especially for hot or new topics. This includes the detection of new URLs that appear in news, blogs, search logs, and social network sites.

Earlier work that addresses parallel crawling and freshness appears in [15,16]. The order of crawling to fetch important pages under resource constraints was studied in [17]. A research crawler and related techniques were discussed in [18]. Focused crawling was addressed in [19] with content classification. A survey of crawling techniques appears in [20].

51.3.2 Off-Line Data Processing and Content Management

Off-line data processing for a search engine collects and parses web documents to produce a searchable index. The additional information collected in online database generation includes ranking features extracted from page links, anchor text, and user query log data. Content management organizes an online index with classification and partitioning; it also guides the crawling system to discover and refresh web documents. Some search engines have large databases with tens of billions of documents while others have comparatively smaller ones with a few hundred millions of pages, designed to handle a certain type of queries.

51.3.2.1 Document Parsing and Index Compression

Document parsing breaks apart the components of a document to form indexable tokens. Tokenization is straightforward for English documents while it is challenging for documents of other languages such as Chinese, Japanese, or Arabic. In those languages, words are not clearly delineated by white space. Natural language processing techniques can identify boundary of words with a good accuracy. A parser also detects the language of a document and its format because documents do not always clearly or accurately identify their language. Since documents do not always obey the desired format or syntax, a parser has to tolerate syntax errors as much as possible.

In tokenizing a document, there is an option to stem words and remove stop words. Stemming is the process for reducing inflected words to their base form. For example, a stemming algorithm [21,22] reduces the words "fishing," "fished," "fish," and "fisher" to their root word "fish." Stemming can produce an improvement in ranking by matching more relevant documents while it can sometime cause a relevance issue. For example, people search for "fishing" will be less interested in types of "fish." Thus, stemming has to be applied conservatively, and one example is to restrict stemming and only derive plural forms for a selected set of keywords. Stemming is a language-specific technique, and some languages such as Chinese have no or little word variations while stemming for languages such as Arabic and Hebrew is more complex and difficult. Stop words are those words that contribute little on the topic of a document. The examples of stop words in English include "the" and "who." Removing stop words explicitly from an index can impact search relevance because a stop word may carry a meaning in some cases. For example, "the who" is an English rock band. Typically a search engine indexes every word while incorporating removal of some stop words during query-time processing when appropriate.

Given a set of tokens (or called *terms*), a search engine normally represents a searchable database using a data structure called *the inverted index*, and a survey of inverted indexing can be found in [23]. An inverted index consists of a list of terms in the vocabulary of the data collection. For each term, the inverted index maintains a list of documents that contain such a term. This list of documents is often called *a posting* of the corresponding term.

For example, given two documents $d_1 = \{a, b, c, a\}$ and $d_2 = \{c, e, c\}$, the inverted index is

$$a \rightarrow \{d_1\}, b \rightarrow \{d_1\}, c \rightarrow \{d_1, d_2\}, e \rightarrow \{d_2\}$$

In addition to document identifications, a posting can store other information such as frequency of a term and positions of this term in each document. For example, the previous inverted index can be augmented as

$$a \rightarrow \{d_1, 2, (1,4)\}, b \rightarrow \{d_1, 1, (1)\}, c \rightarrow \{d_1, 1, (3), d_2, 2, (1,3)\}, e \rightarrow \{d_2, 1, (1)\}$$

The meaning of the posting for term "a" is interpreted as follows. This term appears in document d_1 with frequency 2 at positions 1 and 4.

The inverted index for a large web dataset takes a huge amount of space, and compression of the index data is necessary to enhance engine's online performance and reduce hosting cost [24,25]. Typically document identifications in a posting are stored in an increasing order, and term positions of each document are also arranged in an increasing order. To compress a sequence of these numbers in a posting, a transformation called *delta encoding* is applied first to represent these numbers using their difference gap. For example, given a list {23, 40, 104, 108, 200}, this list can be reformulated as {23, 17, 64, 4, 92}. The first number of this list is not changed. The second number is the gap between the second number and the first number. The average number in the reformulated list is much smaller than that in the original list. This motivation for using delta encoding is that compressing small numbers is easier.

The next step of compression is to represent these gaps using an encoding method. A bit-aligned encoding method is called *Elias-*γ [26], which encodes each number G using a pair of two numbers (b, r). They satisfy $G = 2^b + r$ and $r = G \bmod 2^b$. Next, we discuss how to store the pair (b, r). The first number is stored as a unary code with b digits of 1s. The second number is stored as a standard binary number with b bits. Bit "0" is inserted to separate these two numbers. For instance, $17 = 2^4 + 1$; the γ code for 17 is 111100001. The leading 4 digits are all 1s, representing 2^4. Offset 0001 is found after the first 0 is encountered. In general, for each number G, this method costs $2 \log G + 1$ bits to encode. Thus, γ encoding is only attractive to compress small numbers. To improve the effectiveness in compressing a larger number, the concept of δ encoding can be applied recursively to the first component b in the previous expression. Namely, $G = 2^{2^b+\delta} + r$. This method is called *Elias-*δ method, which costs approximately $2 \log \log G + \log G$ bits for each number G.

A bit-oriented encoding method has its weakness in performance because computer operations are relatively more efficient toward byte-oriented programs. A simple byte-aligned method is called *v-byte* [27] that uses one byte to represent G if $G < 128$ and two bytes or more bytes for a bigger number. Seven bits of each byte store the content of G while 1 bit of such a byte indicates whether this byte is the last byte or not for the corresponding number. A comparison of compression algorithms can be found in [28,29].

51.3.3 Content Management

Content management plays the following roles in off-line data processing:

- Organize the vast amount of pages crawled to facilitate online search. For example, the online database is typically divided into content groups based on language and country. For each content group, there is a further division of content by tiers based on quality. Such a division also affects the crawling decision to prioritize on crawling targets and frequency.
- Perform in-depth analysis for better understanding of information represented in a document. Examples of such analysis are recognizing document title, section titles, and highlighted words that capture the meaning of a document with different weights; identifying the structure of a page and recognizing site menus, which do not contain primary material; and interpreting Javascript embedded in a page to capture the true content of the page.
- Collect additional content and ranking signals. While on-page text features are important for a search engine, critical ranking information can also be gathered from other sources. For example, user search log provides a feedback on how users interact with search results given their queries. There is a variety of special information available on the web, for instance, movie titles and show schedule. Such information can be organized for answering a class of queries or vertical search. For example, Ask.com has built a large database containing question–answer pairs extracted from the web and used this index for matching and answering live question queries.

An online database is often divided into multiple tiers and multiple partitions. One reason is that the capacity constraint of a data center often restricts the size of a database index that can be hosted. Another reason is the requirement for geographical data distribution. Commonly accessed content hosted close to users' location in a targeted market can reduce the engine's response time. In a multi-tier online database, the first tier usually contains frequently visited content of good quality. Even though there are tens of billions of pages available on the web, a query from a vast majority of users can be answered by using a small percentage of these pages. The lower tiers of an online database contain pages that are less likely to be used to answer queries. To partition content, one can exploit characteristics of pages such as reputation and the historical visit frequency.

Anti-spamming is an important task of content management to detect pages with malicious attempts to influence the outcome of ranking algorithms. For example, some websites present different content to a search engine's crawler compared to what is presented to users' web browsers. These websites differentiate a crawler with a web browser based on the IP addresses or the user–agent header of the HTTP request. This spamming technique, called *web cloaking*, is often used by low-quality sites to deceive search engines. Other spamming tactics include link farms, invisible text with stuffed keywords, and exploitation through user-generated contention such as blogs. Automatic or semiautomatic techniques can be employed to identify spamming pages by examining content features, user browsing patterns, and hyperlink structure.

Search content management has adopted various classification techniques based on document features. The supervised learning techniques [30] fit well for this purpose using a set of labeled data to train a classifier that can label future cases. The examples of automatic or semiautomatic classification tasks include language categorization (e.g., English vs. German), country classification (e.g., United States, United Kingdom, and Canada), and spam detection. Another use of classification is to aid online result ranking for intent matching or result diversification. Page segmentation was conducted in [31] to detect semantic content structure in a web page. While text features represent the important characteristics of a web page, link structure among pages and feedback from a search log dataset also provide critical information in making classification decisions. Surveys of web page classification and text categorization appear in [32,33]. A survey of related techniques can be found in [34].

51.3.3.1 Duplicate Elimination

A large number of pages crawled from the web have near-identical content. For example, several studies [35,36] indicate that there are over 30% of near duplicates in crawled pages. Search services need to return the most relevant answers to a query, and presenting identical or near-identical results leads to bad user experience. By removing duplicates, search result presentation would be more compact and save users' time to browse results. Near duplicate detection can also assist low-quality page removal. For example, by identifying a set of spam pages with certain patterns, other pages that are similar to these pages are the candidates for being classified as spam pages.

There are two types of duplicates among web pages. The first type of duplicates is caused by page redirection when one web page redirects to another, either through permanent or temporary HTTP redirect. Although the source and destination of redirection have identical contents, the crawler may visit them at different times and download different versions. Our experience is that up to 10% of duplicates in a large-scale database can be redirection-based. The second type is content-based when two web pages have near-identical information. This is measured by the percentage of content overlapping between the two pages. A popular way to model content similarity of two documents is to represent each document using shingles [37]. Let each document contain a sequence of terms, and each k-gram shingle contains k consecutive terms in this document. For example, let $D = \{a, b, c, e, f, g\}$, and its 2-gram shingle set is $D^2 = \{ab, bc, ce, ef, fg\}$. Let A^k and B^k denote a set of k-gram shingle signatures derived from pages A and B, respectively. The Jaccard similarity of these two documents is

$$Sim(A, B) = \frac{|A^k \cap B^k|}{|A^k \cup B^k|}.$$

An English web page on average has a few hundreds of words, and computing the Jaccard similarity of web pages is expensive when computation is conducted for every pair. The MinHash technique can be used to speed up with an approximation. The MinHash value with respect to a random permutation function R is

$$MinHash(A^k) = \min_{x \in A^k}(R(x)).$$

It has been shown that $Sim(A, B)$ value is the same as the probability that $MinHash(A^k) = MinHash(B^k)$. Thus, $Sim(A, B)$ can be approximated by using n random permutation functions and by computing the overlapping percentage of the corresponding n *MinHash* values.

In general duplicate removal can be conducted during online query processing or in an off-line processing stage. Removing redundant content in an off-line system requires a very high precision in duplicate judgment. Fast approximation algorithms [37–43] are often not sufficient to simultaneously deliver high precision and high recall. A combination of these algorithms with additional features can improve precision (e.g., [44]). As a result, processing cost increases. Major search engine companies have often taken a compromised approach: while a relatively small percentage of duplicates can be removed conservatively in the off-line process, a large portion of duplicates are kept in a live online database. During online query processing, duplicates are detected among top results matching a query, and only one result from duplicate pages is shown in the final search results. Such an approach simplifies off-line processing but increases the cost of online serving. To develop cost-conscious search applications, duplicates need to be removed in an off-line stage as much as possible [45].

51.3.3.2 Distributed and Incremental Index Update

It is time consuming to process a large web data collection; this affects its freshness because data is continuously updated. Parallel processing using MapReduce [46] or other programming platforms can be used. Handling a large database from scratch still takes too much time, even in parallel on a large computer cluster. It is more effective to perform incremental update of the online database.

Figure 51.3 illustrates the architecture of a large-scale off-line system we have developed in the past for incremental processing of billions of web pages. The key service modules of this system include page repository, update manager, URL property, online content controller, indexer, link manager, and content analysis subsystem. Each of them is running on a cluster of machines:

- Once a set of crawled pages is received, the update manager consults with a URL property service and runs a quick classifier to determine whether these pages are new or have significant content update. If appropriate, these new or updated pages are routed to the online database controller, link manager, and content analysis manager for further data update.
- The online content controller manages the online partition structure. It decides when a new partition should be added to the online service and notifies that some URLs in the existing online partitions are stale.
- The indexer module fetches pages of the desired partition to construct an inverted index; it creates a partition to include new or updated URLs. Noted that the same URLs in the existing partitions will be marked as stale.
- The link manager maintains hyperlinks among web pages and summarizes anchor text from incoming links to a document. The content analysis subsystem detects duplicates incrementally, and it conducts classification and other mining tasks.

Various underlying system support is incorporated for implementing the previous pipeline. That includes a middleware for managing continuously running services on a large cluster of machines, key-value stores with fast disk I/0 [47] for page repository, high-throughput property services with a distributed hash table [48], and distributed message queuing for inter-component coordination.

The previous approach for incremental index has an advantage that online data management is simplified: the modification to the existing partitions only needs to mark staleness when pages are updated. The online content controller makes a decision to remove an online partition when it contains too many stale URLs. Another approach for incremental index update is to merge the old index with new index [49]. The search system support for notification-based data updating is discussed in [50].

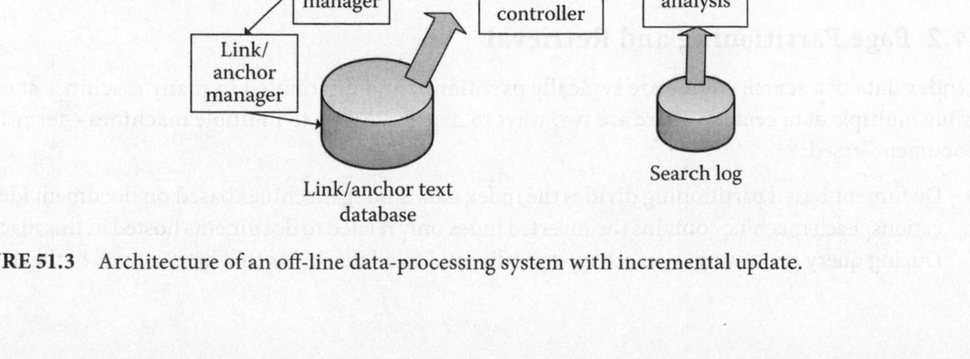

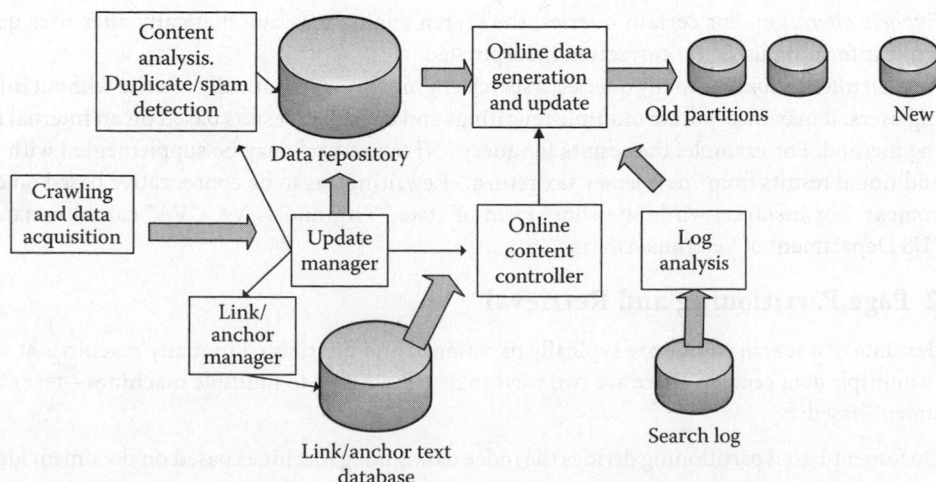

FIGURE 51.3 Architecture of an off-line data-processing system with incremental update.

51.4 Query Processing and Online Architecture

An online system of a search engine typically contains the following components:

1. *Query processing.* This step classifies a query and rewrites it for better matching of relevant results.
2. *Document retrieval.* That is to search the inverted index distributed in a cluster of machines and to collect relevant results.
3. *Result ranking.* Given a large number of matched relevant results, this step filters out low-quality results and ranks top results.
4. *Snippet generation.* This step provides a description for each URL to be shown in a search engine results page (SERP).
5. *Result aggregation and displaying.* Search results may be aggregated from multiple databases including the advertisement engine; this step merges these results and organizes them in a SERP.

51.4.1 Query Processing with Classification

Query processing classifies a query and rewrites it if needed with the goal of closely matching the searcher's intent. Examples of query classification include detecting query language and user location and identify whether this query demands time-sensitive results or entity-oriented information such as a person, an organization, a movie, or a commercial product. The classification of a query helps the search engine to prioritize certain types of results in the final ranking, and it can also gives an indication if highly relevant results from some vertical channels should be included. For example, a query "white house" can trigger the inclusion of news, map, and image search results.

Common query rewriting operations involve stop word elimination, case folding, and spell correction. Stop word elimination detects the use of some popular terms, which may not be able to effectively reflect the intention of a user. For example, terms "the" and "of" may be removed from query "the address of white house." Spell handling involves the examination of popular spell errors to suggest a correction. Case folding commonly used in web search engines does not differentiate cases. Case-insensitive search is more acceptable to users as it includes more relevant results to match users' queries without differentiating cases.

The query rewriting process may be explicit or implicit to users, which can be categorized into the following three situations:

- *Suggestion without query rewriting.* A search engine may not change query terms but recommend a suggestion for correction. That is especially useful when the engine is not certain if the suggested correction is necessary for most of users.
- *Explicit alteration.* For certain queries, the search engine may automatically alter user queries while informing users the correction incorporated.
- *Implicit alternation.* For many queries, a search engine may automatically rewrite without informing users. It may incorporate multiple rewritings and aggregate results based on an internal ranking method. For example, the results for query "NJ tax return" may be supplemented with some additional results from "new jersey tax return." Rewriting has to be conservative based on query context. For instance, while the short form of state "Virginia" is VA, "VA" can also stand for "US Department of Veterans Affairs."

51.4.2 Page Partitioning and Retrieval

The index data of a search engine are typically partitioned and distributed in many machines at one or possibly multiple data centers. There are two ways to distribute data to multiple machines—term-based or document-based:

- Document-based partitioning divides the index data among machines based on document identifications. Each machine contains the inverted index only related to documents hosted in this machine. During query processing, a search query is directed to all machines that host subsets of the data.

- Term-based partitioning divides the index based on terms and assigns postings of selected terms to each machine. This approach fits well for single-word queries, but it requires inter-machine result intersection for multi-word queries.

Document-based partitioning has several advantages. (1) It simplifies the database update process. When documents have a change of their content, only few machines that host these documents are affected. (2) It allows for good load balancing as the term distribution is more skewed. (3) It is more resilient to faults compared to term partitioning. If some machines fail, only the index of documents on these failed machines becomes unavailable, and the search engine may still be able to serve relevant results from other machines. For term-based partitioning, the failure of one machine could lead to more noticeable errors for certain queries, which contain terms hosted in the failed machines. The disadvantage of document-based distribution is that all machines need to participate in query answering. Thus, the serving cost per query is more expensive. A combination of document- and term-based partitioning would be an option to exploit in hosting a large index database.

After the initial query processing, a search engine accesses the index following the semantics of a query. By default, query terms are considered to be conjunctive. Namely, a document is considered to be a match if and only if it contains all query terms. In a disjunctive mode, a document is considered to be a match if only one query term presents in the document. While a search engine often provides an advanced search feature that supports a combination of conjunctive and disjunctive search, the common usage of user queries follows a conjunctive format due to its simplicity.

Since the index is distributed to machines using document- or term-based partitioning, each of these machines returns a set of documents fully or partially matching the processed query; therefore, additional processing is needed. For example, for the query "Britney Spears," the posting of term "Britney" is intersected with the posting of term "Spears." If two postings are hosted in different machines, inter-machine communication is necessary to accomplish this intersection. For document-based partitioning, the posting intersection is done within a machine, and each machine will report a subset of distinct documents that match this query. Once the search engine gathers matched documents, it ranks these matched documents and presents top results to users. We discuss page ranking in the next section.

Retrieval of relevant results during index matching is time consuming when searching a large dataset and various techniques are developed to speed up this process. One technique is to divide the list of documents from the posting of a term into multiple chunks and add skip pointers in the data structure [51,52]. In performing conjunctive search among postings for multiple query terms, the maximum document identification of a chunk can be used to determine if a chunk can be skipped when intersecting with a document identification from another posting. A skipping pointer for a chunk can be inserted into the data structure of inverted index so that the scanning of this chunk can be easily avoided.

51.4.3 Result Caching and Snippet Generation

Caching is a commonly adopted technique to speed up the query answering process because there are often over 50% of queries that can use previously computed search results. Caching provides load smoothing to deal with traffic spikes caused by hot events during which different users often type the same or similar queries again and again to check the latest development. A search engine may adopt a variety of hierarchical caching strategies. For example, cache the top HTML response with snippets, cache the top results without snippets, cache a matched URL list in a local machine before result aggregation, and cache the postings of frequent query terms.

Given a limitation on cache size, results for most frequent queries can be cached, and some results may be pre-cached in anticipation of its popularity. Other replacement policies may consider weighted factors such as recomputing cost [53]. A key issue to consider for caching is freshness of cached results. As the search index is updated continuously, the concern can arise for the staleness of cached results. For example, a query that contains news results may not be cacheable because a later query needs to be answered with the latest news results. One strategy is to cache results only for relatively stable answers.

When top search results are presented to users, each document is described with a summary of its content, which is known as a snippet. There are two categories of techniques for document summarization. One is query-independent static snippet where a document is summarized without knowing queries. The second category is query-dependent dynamic summarization in which a snippet is composed of several short pieces of text, and each short piece often contains query terms. The selection of these text pieces depends on a number of features including the percentage of query terms covered, positions of text pieces in a document, importance of query terms covered, and the readability of the summary. A study on snippet readability was done in [54]. Producing high-quality snippets is time consuming, and efficient techniques for snippet generation were discussed in [55].

51.4.4 An Example of Online Search Architecture

Figure 51.4 illustrates an online search architecture we have developed in the past in hosting billions of web documents and answering queries from tens of millions of queries every day. The database is divided following language and country classification of documents, and the query-processing module can route a query to a proper collection of index based on the targeted user region. For example, English queries are processed using English data partitions. There is a small percentage of queries that contain mixed languages; for those queries, multiple-language content groups are involved in search.

Once a query is received by the online system, this query is distributed to one of query-processing front ends through a load-balancing traffic switch. The query-processing module consults a multi-level cache to see if this query can be answered using the previously computed result with or without snippets. If a query cannot be found from the cache, the page retrieval module searches URLs in Tier 1 of the online index and visits Tier 2 (or other lower priority tiers) only if the results matched from Tier 1 are not satisfactory. Our experience is that up to 70% of queries can be handled by Tier 1 without searching other tiers. There is an internal cache deployed for each partition so that retrieval results for frequent queries or compute-intensive queries are cached within each partition. There is also an aggregation module that combines results from multiple partitions and eliminates duplicates when needed, and this module can be deployed hierarchically also [56].

Ranking is conducted hierarchically. The page retrieval module selects top results from each partition, and these results are combined in an aggregation module, which performs additional filtering. Signals for ranking are discussed in the next section. Once top-ranked results are derived, the snippets for these results can be computed by accessing the text content of these documents from a page repository.

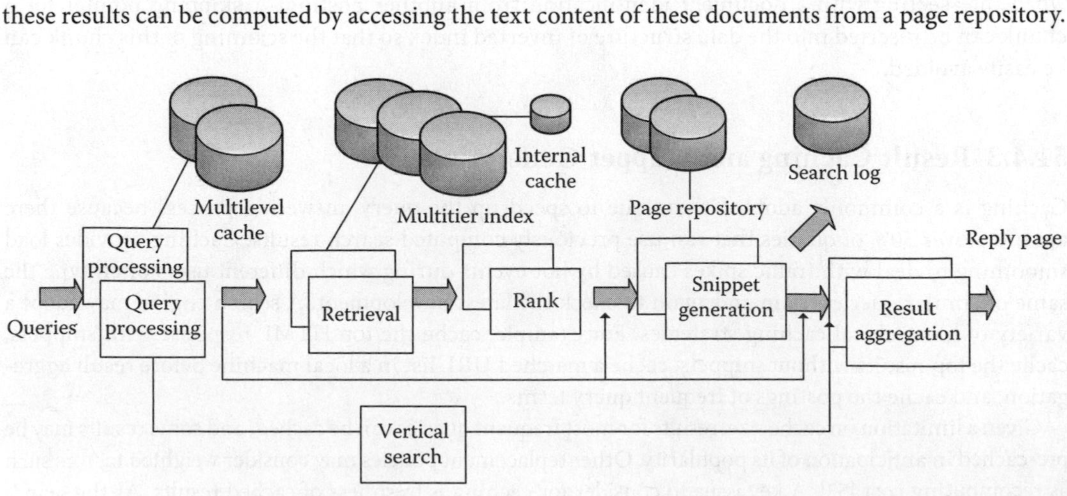

FIGURE 51.4 An online web search architecture.

There is a heavy I/0 activity involved in accessing disk data for snippet generation, and caching snippets for popular queries can be used to improve the performance.

In addition to finding results from the HTML page collection, a query is also routed to vertical search channels including the advertisement engine and other types of popular web collections such as news and blogs, maps, images, and videos. The top results from multiple channels are combined in the SERP. Optimizing the selection and placement of these results from different channels in a SERP is critical for users' experience with a search engine. Once a user browses URLs listed in the SERP, the search engine keeps a log on the click activity of this user. The user behavior extracted from this log provides a feedback for adjustment of ranking of results, triggering and placement of vertical search.

Each logical system component in Figure 51.4 runs on a large-scale computer cluster. As we mainly use a document-based distribution for search index, processing of each search request goes through thousands of machines. System support for the previous engine includes request and internal traffic scheduling, fault tolerance, and replication support; the middleware and related system techniques were discussed in [56–59].

51.5 Ranking

Ranking plays the most visible role in a search engine since most users only look at top results. Ranking can be conducted in a centralized manner or hierarchically so that each subsystem selects the best results first before submitting selected results for the next round of ranking. Relevance scores for documents are computed by exploiting their text and authority features, which are often called *ranking signals.*

We first give an overview of query-dependent and query-independent ranking signals and describe feature analysis based on text, link, and click data. Then we will discuss score aggregation by using these signals.

51.5.1 Ranking Signals

There are four aspects of ranking in matching the intent of a user.

1. *Relevancy.* Top documents need to be relevant to a user query.
2. *Authoritativeness.* High-quality content is normally preferred since users rely on trustful information to learn or make a decision.
3. *Freshness.* Some queries are time sensitive and latest information is more appropriate for such type of queries.
4. *Preference.* Personal or geographical preference can also impact the choices of answers.

Rank signals are designed to reveal those aspects from the query or document point of view, and we summarize two categories of ranking signals as follows:

Query-Dependent Ranking Signals

- *Document text.* This feature measures the closeness between query words and what appears in a document. A web document can be divided into multiple text sections such as document title and body, and query words matched in different sections carry different weights. Additional weighting can be considered if query terms are highlighted in a document. Proximity, positions, and frequency of query words matched in a document area prominent indicator of relevance.

  Some of text features in a web document are not visible to users, for example, description and keyword tags of an HTML document. Spammers often exploit such a feature to present irrelevant keywords; thus, such features have to be used with caution.
- *Anchor text.* Anchor text of a web page appears in the hyperlinks of other web documents referring to this page. There can be many hyperlinks pointing to the same web document, and some of anchor text can be meaningless or of low quality. The commonality in anchor text and popularity of source web pages can be helpful in determining quality of anchor text.

- *URL text.* Text of a URL, especially when appearing in a position within or close to the host name, may sometime reveal the semantic of its content. For example, "download" in http://www.microsoft.com/en-us/download/ is an important keyword for matching query "Microsoft download."
- *Historical click-through data.* A search engine collects the behavior data from users when they search and browse matched results. These URLs clicked by a user following a search query are often relevant to such a query.
- *Query classification and preference.* Implicit features of a query can be extracted during classification to select preferred documents. For example, the language of a query prioritizes the language type of web documents to be matched. The originating geographical location of a query may direct the ranking system to select more content from this region. If a query matches a hot news topic, time-sensitive fresh content is preferred.
- *Link citation.* This feature uses the web hyperlink connectivity among those pages whose text matches a given query. A page cited by many other matched web pages can be a good indicator of authoritativeness for this query.

Query-Independent Signals

- *Link popularity.* This feature analyzes how pages link to each other and then uses this information to determine the importance of each page on the web. If a page is cited from a large number of pages, it is often ranked more prominently.
- *Documentation classification and spam analysis.* A classification of web documents can provide a signal in matching user preference. For example, users are more likely to be looking for content in their own language than in other languages. File types such as HTML and Microsoft Word can also be a preference for users. Since spam analysis leads to a rating of page quality, a spam score is also a useful quality indicator.
- *URL preference.* In many cases, users prefer a URL with a short length and prefer a static URL instead of a dynamically generated URL. Dynamically generated URLs tend to change content from time to time, and their average quality is often lower than that of static URLs.
- *Site or page properties.* Some additional information on a web page or its site can be an indication of web page quality. For example, frequency of user visits, the time duration of visit, and serving speed or reliability of the web host.

There are many methods for modeling above ranking signals, and we next describe some of techniques for analyzing text, links, and click data.

51.5.2 Text Analysis

The goal of text analysis is to quantify the relevance of a document based on how query words appear in this document. Boolean and phrase operators are available in a search engine for advanced retrieval semantics, for example, "Retrieval" or "Search." Since most people use simple and short queries, we focus on text analysis of a search query with conjunctive semantics.

There are three issues in developing a formula measuring the text closeness of document–query matching:

- What are basic units of information to match a query with a document? A document is normally tokenized a sequence of words after excluding HTML tags, and each word is a basic term for matching. Thus, a document can be considered as a bag of words. On the other hand, the order and closeness of terms play an important role. For example, "Paris Hilton" is the actress name while "Hilton Paris" could mean the Hilton hotel at Paris.
- What is the proximity of search words appeared in a document? When search words appear close to each other in a document, this document is often considered to have a high relevancy.

To capture word proximity, an *n*-gram term can be considered, which is defined as a consecutive sequence of *n* words in a document. An n-gram of size 1 is referred as a "unigram," size 2 is a "bigram," and size 3 is a "trigram."

Another model for proximity is to use a text *span* [60,61]. A document is segmented into multiple disjoint spans that cover search terms. Each span may not contain every query term, and it is a minimum text window containing search terms as much as possible. Each span starts with the position of the first query term in a document and contains consecutive query terms in a document. This span ends when the distance between the current query term match position and the next query term match position exceeds a threshold. This span also ends if the current and next matches are the same term, or the previous query term is the same as the next query term.

- How are terms weighted? One method is to use a simple weighting based on where they appear. If a query term appears in the body of a document, it will receive credit. When it appears in the title, it receives weighted credit. Another method that uses the frequency information is known as term frequency and inverse document frequency (TF-IDF) weighting. In this scheme, the importance of a term increases when it appears many times in a document. If a term appears many times in many documents, its weight decreases because it becomes less effective in differentiating a relevant document from a non-relevant document. A common formula for TF-IDF is defined as follows:

$$tfidf(t,d) = tf(t,d) * \log\frac{n}{df_t}$$

where n is the number of documents in the collection, $tf(t, d)$ is the occurrence count of term t in document d, and df_t is the number of documents that term t appears. Because denominator df_t may be zero, it is often adjusted as $1 + df_t$. Additionally, $tf(t, d)$ is sometime normalized against the length of document d. Variations of the TF-IDF weighting have been suggested in the literature, for example, Okapi BM25 [62].

- How can the closeness of matching be represented by aggregating weights from multiple query terms? One model is to use the cosine-based similarity function. Let document d contain a set of terms with weight w_i for each term, and let query q contain a set of query terms with weight qw_i. These weights are normalized, and the aggregated text score is

$$Score(d,q) = \Sigma_{t_i \in q} w_i * qw_i$$

There are other ways to aggregate weights. Since text features can be represented in multiple ways to capture the appearance and proximity of search terms, they can be aggregated using a weighted summation function or a machine learning model. We discuss a weight-learning technique in Section 51.5.5.

51.5.3 Link Analysis

51.5.3.1 PageRank Algorithm

As a query-independent ranking signal, the earlier PageRank algorithm [63] measures the importance of a web document based on its presence in a large web graph. In this graph, each node is a web page and each edge is directed, representing a link from one page to another using the HTML hyperlink notation. The intuition underlying this link analysis is that highly linked pages are more important than pages with just few links, but not all links carry the same importance. A page with a link from a popular site may be ranked higher than other pages with more links but from obscure places.

Let n be the number of pages in a web graph, and let x be a page. A simplified version of the PageRank model can be expressed as

$$R(x) = \frac{\lambda}{n} + (1-\lambda)\Sigma_{y\in P(x)} \frac{R(y)}{|S(y)|}$$

where $S(x)$ is the set of pages that x points to and $P(x)$ is the set of pages that point to x.

PageRank computation can be conducted iteratively. Initially, every page has its initial value as $\frac{1}{n}$. Then all pages will be assigned with a λ-linear combination of a rank value $\frac{1}{n}$ and additional rank credit propagated from its predecessors in the graph. After that, vector $R()$ is normalized. This process repeats until all nodes have a converged rank value.

Computing the in-degree of a web page can approximate its link popularity score, but this score can be less resilient to link spamming. That is because it is relatively easy to create thousands of pages on the web pointing to a page. The PageRank algorithm was initially viewed as the key to Google's differentiation compared to other search engines when Google was launched. Various modifications have been suggested in the literature to improve relevancy or speed up computation. For instance, see [64,65].

51.5.3.2 HITS Algorithm

The HITS algorithm [66] provides query-dependent scoring to determine the authoritativeness of a web page relevant to a query. The key difference compared to PageRank is that HITS computes link scores within a subgraph of a web graph derived based on a user query. The HITS algorithm derives two scores for every page. The first score is called *the authority score*, representing the authoritativeness of a page relevant to the subject of a user query. The second score is called *the hub score* representing its rank value in serving as the compilation of relevant information. Let the authority score of a page x be $A(x)$, and let the hub score of this page be $H(x)$:

$$A(x) = \Sigma_{y\in P(x)} H(y) \quad \text{and} \quad H(x) = \Sigma_{y\in S(x)} A(y)$$

where $P(x)$ contains the predecessors of page x in this graph and $S(x)$ contains the successors of x.

The computation of these scores is performed iteratively. After an initial score assignment, all pages update their hub and authorities based on the previous formula. These values are normalized at each iteration, and this process repeats until all pages have a converged value. Mathematically speaking, the previous process computes the principal eigenvector of matrix C^tC and CC^t where C is the connectivity matrix based on the link structure of the query graph, and C^t is the transpose of C.

The major challenge in using HITS is its online computation cost. There are several extension or follow-up work to improve the seminal work of PageRank and HITS. That includes topic distillation [67] and multi-community hub authority [68,69]. A fast algorithm for computing hub authority scores with sparse matrix approximation was developed at Teoma/Ask [70].

51.5.4 Click Analysis

Users interact with a search engine by clicking search results of a query, browsing them, reformulating their queries, and performing additional browsing. Such interaction is logged in a search engine, and the click data can be thought of as a tuple (t, u, q, c) indicating that user u clicks result c for query q at time t. While click data can be noisy, URLs clicked by users are likely to convey some information for relevance judgment. A document may be ranked high if it has attracted the greatest number of previous users for the same query. A log dataset can be divided into a sequence of search sessions for each user. Each user session contains a sequence of queries and results picked by this user. The following correlations among queries and URL results can be extracted from session analysis:

- *Query-to-pick (Q2P)*. A Q2P correlation associates a query with a pick. When a search engine returns a result in response to a query, and a user picks this result, that constitutes a correlation candidate. When multiple independent users make the same association, that makes this Q2P correlation more plausible.

  As an extended association, after a user enters a query, picks recorded for a different query within the same user session can be associated with the initial query. Another extension is that different pages browsed as a result of a URL pick may also be associated with the original query. The previous two extensions represent the fact that a user often reformulates a query or browses beyond a search link to look for right content. Such an approach carries data errors and additional noise filtering techniques need to be applied. For instance, a restriction can be applied using query similarity from one pick to another in a session, the order and depth of query rewriting or browsing steps, dwelling time, and the first and last choice of browsing.
- *Query-to-query (Q2Q)*. This correlation associates queries issued during a user session. The confidence for such association depends on various factors including query similarity, the issuing time between queries, the number of intervening queries or picks, the order of association, and the number of sessions sharing such an association. The Q2Q correlation can help the generation of query suggestions in search engine response pages.

  For example, "electronic eavesdropping" is related to a query "eavesdropping devices." Misspelled "Hotel Mariot" is related to "Hotel Marriott."
- *Pick-to-pick (P2P)*. This correlation associates picks issued during a user session, which is analogous to the Q2Q correlation described earlier. The P2P correlation can reveal a set of URLs that are similar or related for certain topics. As an example, toyota.com is the top answer for answering "Toyota car," honda.com and ford.com may also be listed as similar results in case users are interested in exploiting similar products.

While click-through data can be very noisy and incomplete with a sparse structure, it still provides valuable information for search quality improvement. Ask.com that bought DirectHit.com in 1999 was the earlier system that ranked search results based on URLs' click-through rates, and the adjustment is applied when URLs with a relatively lower rank may receive more clicks compared to those with a higher rank. Ask.com further adopted session-based click data to improve ranking with Q2P and Q2Q correlations [71]. The use of click data to derive user's relative preference was studied in [72] to train a ranking function. Query reformulation that expands the Q2P correlations in learning to rank was considered in [73]. The relevancy impact of integrating a number of click-based features was demonstrated in [74]. The bipartite graph structure of query–document relationship from a query log for modeling Q2P, Q2Q, and P2P correlations was studied in [75,76]. Integrating browsing activities of a search session for ranking was considered in [77].

51.5.5 Score Aggregation and Learning to Rank

There are many ranking signals that can be extracted or characterized for each matched document in responding to a query. Those signals need to be aggregated, and their weights can be imposed empirically or derived using a machine learning method. Aggregation of these signals can be performed hierarchically. In this way, ranking signals can be divided into several semantically relevant components for fine tuning.

Machine-learned ranking uses supervised or semi-supervised techniques that automatically construct a ranking model from training data. Training data consists of a set of queries, and each query has a list of documents with a relevance label. The example of a relevance label is "relevant," "related," or "irrelevant" for a three-level judgment. There are various ways to model a learning function based on a training dataset, and multiple models can be combined by an ensemble method such as bagging and boosting [78–80] to further improve the learning accuracy.

We illustrate a simple learning function as follows using the RankSVM method [72,81]. Let $\vec{x}_i$ be a feature vector representing a set of ranking signals for document d_i. An aggregated score function $f(x_i)$ is a product of a weight vector $\vec{w}$ with $\vec{x}_i$. Function $f()$ defines ranking as follows: if $f(\vec{x}_i) > f(\vec{x}_j)$, document d_i should be ranked before d_j in answering a query. This condition implies that we need to find $\vec{w}$ so that $\vec{w} \times (\vec{x}_i - \vec{x}_j) > 0$.

Following the previous condition, a training dataset is converted based on the pairwise relationship of documents. Assume that page d_i with a feature vector x_i and another page d_j with a feature vector x_j are labeled for answering the same query. Define a new label $e_{i,j}$ for pair (d_i, d_j) as 1 if d_i should be ranked before d_j and −1 if d_i should be ranked after d_j. The new training data can be thought of as a set of triplets. Each triplet is $(Q, \vec{x}_i - \vec{x}_j, e_{i,j})$, where Q is a test query and d_i and d_j are from the labeled result list of Q. Then the hard-margin linear SVM model is to minimize $\|\vec{w}\|^2$ subject to a constraint for all points in the training data: $\vec{w} \times (\vec{x}_i - \vec{x}_j) \times e_{i,j} \geq 0$.

The previous process demonstrates that a ranking problem can be transformed to a classification problem, and other classification-based algorithms can also be used. Learning-to-rank algorithms have been categorized as pointwise, pairwise, and list-wise [82]. The pointwise approach predicts the relevancy score of each document for a query using a supervised learning method such as regression and classification. The pairwise approach considers learning as a binary classification problem to decide if a document should be ranked prior to another document. Examples of this approach are RankSVM [72,81], RankBoost [80], and RankNet [83]. The list-wise approach tries to directly optimize the cost function, following a relevancy evaluation measure. Examples of such an approach include LambdaMART [84], SVM^{map} [85], and AdaRank [86]. Machine-learned ranking has its advantage for leveraging a large training dataset while interpreting and adjusting its behavior in the failed cases can be hard sometime. In practice, a ranking model that combines machine learning with human tuning may be used.

51.6 Search Engine Evaluation

51.6.1 Strategies in Quality Evaluation

A common evaluation for search engine quality is the relevance of search results as well as the speed in answering a query. It is not easy to define the metric for measuring the relevance of search results. This is because in many cases, it is hard to tell what a user really wants by viewing a query. Precise detection of user's query intent is the holy grail of search. For navigational queries such as Microsoft, most users want to navigate to the respective sites. Queries such "US open" can have multiple answers from time to time. Intent of some queries can be heavily biased by locations or languages used. For instance, CMU can mean Central Michigan University, Carnegie Mellon University, or Central Methodist University, depending users' location.

One way to judge result relevance of a search engine is to employ a team of evaluators, and let them type queries collected from an engine log, and assess if the returned answer matches the typed query. This has a challenge since sometime an evaluator may not be able figure out the right answers. Another complexity is that today's search engine users expect more than just result relevance, and there are multiple aspects of a search engine affecting the satisfactory degree of an end user. For example, are the results from authoritative sources? Are they free of spam? Are their titles and snippets descriptive enough? Are they fresh and timely? Does the answer include additional visual elements such as maps or images that may be helpful? A thorough evaluation needs to cover each of these dimensions when appropriate.

The search log data can be used to assess the quality of an engine and to capture the satisfactory degree of end users. When users engage a search engine and select presented results more frequently, this is a good indicator that users like this engine. The key metrics that can be derived from a search log include the daily or weekly user traffic, the visit frequency of users, result pick rate, and abandon rate when users do not select any result. The metrics derived from a search log carry certain noise. For example, result pick rate is often affected by the placement of results and interface design in a SERP. Site traffic can be also affected by seasonality. The interpretation of these metrics requires a deep data analysis from multiple aspects.

51.6.2 Evaluation Metrics

To measure whether an information retrieval system returns relevant results or not, precision and recall are the two metrics developed in the earlier studies. Precision is the percentage of returned results that are relevant, and recall is the percentage of all relevant results covered in the returned list. For web search, when there are many results relevant to a query, a user normally only browses top few results; thus, precision is more critical than recall in such a setting. When there are only very few results relevant to a query, a recall ratio measures the ranking and crawling capability in finding these documents.

Another measure called *the normalized discounted cumulative gain (NDCG)* [87] that considers the graded relevance judgment and the usefulness of a document based on its position in the ranked result list. In this scheme, the result quality is judged using graded relevance with a multi-level label. For example, a six-level label can be "perfect," "excellent," "good," "fair," "poor," or "bad." The NDCG value starts with a following concept called *the discounted cumulative gain (DCG)* at a particular rank position k:

$$DCG[k] = \sum_{i=1}^{k} \frac{2^{rel_i} - 1}{\log_2(1+i)}$$

where rel_i is the graded relevance value at position i. For example, 0 stands for "bad," 1 stands for "poor," 2 stands for "fair," and so on. The NDCG value normalizes the DCG value as

$$NDCG[k] = \frac{DCG[k]}{DCG'[k]}$$

where $DCG'[k]$ is the ideal DCG at position k. The $DCG'[k]$ value is computed by producing the maximum possible DCG till position k. The NDCG value for all queries is averaged to assess the average ranking quality.

The other evaluation metrics for a search engine includes freshness and coverage. To assess the freshness, one metric can be the time from creation of a document on the web to the time when it appears in the online index. To measure the completeness, the size of database can be used. Since there are a large amount of low-quality URLs on the web, the size alone is not sufficient. An additional strategy is topic-based sampling to measure the satisfactory degree in answering users' queries for selected topics.

The speed of the engine performance can be measured by the round-trip response time of a query. Typically a search engine is required to return a query result within a second, but in practice a search engine is able to deliver a search result within a few hundred milliseconds. During search engine capacity planning, the throughput of an online system is also measured so that a system can be optimized to answer as many queries as possible at every second. When the engine configuration such as the database size changes, its performance metrics may be affected and need to be examined.

51.7 Concluding Remarks

The main challenge of search engine technology is the handling of big data with significant noise. Search engine technology has been improved steadily in the last decade while the expectation of Internet users has also increased considerably. There are still a large number of unsolved technical problems to accurately capture the intent of search users. As more and more ranking signals are introduced to improve result relevance for different categories of queries, the interaction of these signals can sometimes yield to subtle side effects. Thus, further improvement of search quality toward the next level requires more effort. In the next ten years, mobile and multimedia search is a key battleground among search providers. With popularity of social networking, social search is also another hot topic to facilitate information sharing. Search-based information technology will continue to advance and evolve in multiple directions, so Internet and mobile users can find the information that they need quickly.

Acknowledgments

We thank Teofilo Gonzalez for his valuable comments. This work was supported in part by NSF IIS-1118106. Any opinions, findings, and conclusions, or recommendations expressed in this material are those of the authors and do not necessarily reflect the views of the National Science Foundation.

References

1. G. Salton and M. J. McGill. *Introduction to Modern Information Retrieval.* McGraw-Hill, Inc., New York, 1986.
2. C. D. Manning, P. Raghavan, and H. Schütze. *Introduction to Information Retrieval.* Cambridge University Press, Cambridge, U.K., 2008.
3. R. Baeza-Yates and B. Ribeiro-Neto. *Modern Information Retrieval*, 2nd Edn., Addison-Wesley, Wokingham, U.K., 2011.
4. B. Croft, D. Metzler, and T. Strohman. *Search Engines: Information Retrieval in Practice.* Addison-Wesley, Boston, MA, 2010.
5. S. Büttcher, C. L. A. Clarke, and G. V. Cormack. *Information Retrieval: Implementing and Evaluating Search Engines.* The MIT Press, Cambridge, MA, 2010.
6. Apache Software Foundation. Appache Lucene. http://lucene.apache.org
7. The Lemur Project. Lemur Project Home. http://www.lemurproject.org/
8. The Xapian Project. Xapian Project Home. http://xapian.org/
9. Apache Software Foundation. Apache Hadoop Home. http://hadoop.apache.org/
10. Apache Software Foundation. Apache Hbase Home. http://hbase.apache.org/
11. Apache Software Foundation. Apache Cassandra Project Home. http://cassandra.apache.org/
12. robotstxt.org. The Web Robots Pages. http://www.robotstxt.org/
13. J. Madhavan, D. Ko, L. Kot, V. Ganapathy, A. Rasmussen, and A. Y. Halevy. Google's deep web crawl. *PVLDB*, 1(2): 1241–1252, 2008.
14. Sitemaps.org, Sitemaps.org Home. http://www.sitemaps.org
15. J. Cho and H. Garcia-Molina. Parallel crawlers. In *Proceedings of the 11th International Conference on World Wide Web, WWW '02*, pp. 124–135, ACM, New York, 2002.
16. J. Cho and H. Garcia-Molina. Effective page refresh policies for web crawlers. *ACM Transactions on Database System*, 28(4): 390–426, December 2003.
17. R. Baeza-Yates, C. Castillo, M. Marin, and A. Rodriguez. Crawling a country: Better strategies than breadth-first for web page ordering. In *Special Interest Tracks and Posters of the 14th International Conference on World Wide Web, WWW '05*, pp. 864–872, ACM, New York, 2005.
18. H.-T. Lee, D. Leonard, X. Wang, and D. Loguinov. Irlbot: Scaling to 6 billion pages and beyond. In *Proceedings of the 17th International Conference on World Wide Web, WWW '08*, pp. 427–436, ACM, New York, 2008.
19. S. Chakrabarti, M. van den Berg, and B. Dom. Focused crawling: A new approach to topic-specific web resource discovery. *Computer Networks*, 31: 1623–1640, 1999.
20. C. Olston and M. Najork. Web crawling. *Foundations and Trends in Information Retrieval*, 4(3): 175–246, March 2010.

21. M. Porter. An algorithm for suffix stripping. *Program*, 14(3): 130–137, 1980.
22. R. Krovetz. Viewing morphology as an inference process. In *Proceedings of the 16th Annual International ACM- SIGIR Conference on Research and Development in Information Retrieval*, pp. 191–202, Pittsburgh, PA, 1993.
23. J. Zobel and A. Moffat. Inverted files for text search engines. *ACM Computing Surveys*, 38(2): 1–56, July 2006.
24. A. Moffat and L. Stuiver. Binary interpolative coding for effective index compression. *Information Retrieval*, 3(1): 25–47, July 2000.
25. V. N. Anh and A. Moffat. Inverted index compression using word-aligned binary codes. *Information Retrieval*, 8(1): 151–166, January 2005.
26. P. Elias. Universal codeword sets and representations of the integers. *IEEE Transactions on Information Theory*, 21(2): 194–203, March 1975.
27. H. E. Williams and J. Zobel. Compressing integers for fast file access. *The Computer Journal*, 42: 193–201, 1999.
28. S. Büttcher and C. L. A. Clarke. Index compression is good, especially for random access. In *Proceedings of the Conference on Knowledge Management, CIKM*, pp. 761–770, Lisbon, Portugal, 2007.
29. H. Yan, S. Ding, and T. Suel. Inverted index compression and query processing with optimized document ordering. In *Proceedings of the 18th International Conference on World Wide Web, WWW '09*, pp. 401–410, New York, 2009.
30. T. Mitchell. *Machine Learning*. McGraw-Hill, Inc., New York, 1997.
31. S. Yu, D. Cai, J.-R. Wen, and W.-Y. Ma. Improving pseudo-relevance feedback in web information retrieval using web page segmentation. In *Proceedings of the 12th International Conference on World Wide Web, WWW '03*, pp. 11–18, Budapest, Hungary, 2003.
32. F. Sebastiani. Machine learning in automated text categorization. *ACM Computing Surveys*, 34(1): 1–47, March 2002.
33. X. Qi and B. D. Davison. Web page classification: Features and algorithms. *ACM Computing Surveys*, 41(2): 12:1–12:31, February 2009.
34. C. Castillo and B. D. Davison. Adversarial web search. *Foundations and Trends in Information Retrieval*, 4(5): 377–486, 2011.
35. D. Fetterly, M. Manasse, M. Najork, and J. Wiener. A large-scale study of the evolution of web pages. In *Proceedings of the 12th International Conference on World Wide Web, WWW '03*, pp. 669–678, Budapest, Hungary, 2003.
36. A. Z. Broder. Identifying and filtering near-duplicate documents. In *Proceedings of the Combinatorial Pattern Matching, 11th Annual Symposium*, pp. 1–10, Montreal, Quebec, Canada, 2000.
37. A. Z. Broder, S. C. Glassman, M. S. Manasse, and G. Zweig. Syntactic clustering of the web. In *Proceedings of the World Wide Web, WWW 1997*, pp. 1157–1166, Santa Clara, CA, 1997.
38. G. S. Manku, A. Jain, and A. D. Sarma. Detecting near-duplicates for web crawling. In *Proceedings of the World Wide Web, WWW '07*, pp. 141–150, Banff, Alberta, Canada, 2007.
39. A. Gionis, P. Indyk, and R. Motwani. Similarity search in high dimensions via hashing. In *Proceedings of the Conference on Very Large Data Bases, VLDB*, pp. 518–529, Edinburgh, U.K., 1999.
40. M. S. Charikar. Similarity estimation techniques from rounding algorithms. In *Proceedings of the Thirty-Fourth Annual ACM Symposium on Theory of Computing, STOC '02*, pp. 380–388, San Diego, CA, 2002.
41. A. Chowdhury, O. Frieder, D. A. Grossman, and M. C. McCabe. Collection statistics for fast duplicate document detection. *ACM Transactions on Information Systems*, 20(2): 171–191, 2002.
42. A. Kolcz, A. Chowdhury, and J. Alspector. Improved robustness of signature-based near-replica detection via lexicon randomization. In *Proceedings of KDD*, pp. 605–610, Seattle, WA, 2004.
43. M. Theobald, J. Siddharth, and A. Paepcke. Spotsigs: Robust and efficient near duplicate detection in large web collections. In *Proceedings of the 31st Annual International ACM SIGIR Conference on Research and Development in Information Retrieval, SIGIR '08*, pp. 563–570, Singapore, 2008.

44. M. Henzinger. Finding near-duplicate web pages: A large-scale evaluation of algorithms. In *Proceedings of the 29th Annual International ACM SIGIR Conference on Research and Development in Information Retrieval, SIGIR '06*, pp. 284–291, Seattle, WA, 2006.
45. S. Zhu, A. Potapova, M. Alabduljali, X. Liu, and T. Yang. Clustering and load balancing optimization for redundant content removal. In *Proceedings of 22nd International World Wide Web Conference, Industry Track, WWW '12*, Lyon, France, 2012.
46. J. Dean and S. Ghemawat. Mapreduce: Simplified data processing on large clusters. In *Proceedings of the Operating Systems Design and Implementation, OSDI '04*, San Francisco, CA, 2004.
47. F. Chang, J. Dean, S. Ghemawat, W. C. Hsieh, D. A. Wallach, M. Burrows, T. Chandra, A. Fikes, and R. E. Gruber. Bigtable: A distributed storage system for structured data. In *Proceedings of the 7th Symposium on Operating Systems Design and Implementation, OSDI '06*, pp. 205–218, USENIX Association, Seattle, WA, 2006.
48. J. Zhou, C. Zhang, H. Tang, J. Wu, and T. Yang. Programming support and adaptive checkpointing for high-throughput data services with log-based recovery. In *Proceedings of the 2010 IEEE/IFIP International Conference on Dependable Systems and Networks, DSN 2010*, pp. 91–100, Chicago, IL, 2010.
49. N. Lester, A. Moffat, and J. Zobel. Efficient online index construction for text databases. *ACM Transactions on Database Systems*, 33(3): 19:1–19:33, September 2008.
50. D. Peng and F. Dabek. Large-scale incremental processing using distributed transactions and notifications. In *Proceedings of the 9th USENIX Conference on Operating Systems Design and Implementation, OSDI'10*, pp. 1–15, Vancouver, British Columbia, Canada, 2010.
51. A. Moffat and J. Zobel. Self-indexing inverted files for fast text retrieval. *ACM Transactions on Information Systems*, 14(4): 349–379, October 1996.
52. T. Strohman and W. B. Croft. Efficient document retrieval in main memory. In *Proceedings of the 30th Annual International ACM SIGIR Conference on Research and Development in Information Retrieval, SIGIR '07*, pp. 175–182, Amsterdam, the Netherlands, 2007.
53. Q. Gan and T. Suel. Improved techniques for result caching in web search engines. In *Proceedings of the 18th International Conference on World Wide Web, WWW '09*, pp. 431–440, New York, 2009.
54. T. Kanungo and D. Orr. Predicting the readability of short web summaries. In *Proceedings of the Second ACM International Conference on Web Search and Data Mining, WSDM '09*, pp. 202–211, Barcelona, Spain, 2009.
55. A. Turpin, Y. Tsegay, D. Hawking, and H. E. Williams. Fast generation of result snippets in web search. In *Proceedings of the 30th Annual International ACM SIGIR Conference on Research and Development in Information Retrieval, SIGIR '07*, pp. 127–134, Amsterdam, the Netherlands, 2007.
56. L. Chu, H. Tang, T. Yang, and K. Shen. Optimizing data aggregation for cluster-based internet services. In *Principles and Practise of Parellel Program, POPP*, pp. 119–130, San Cisco, CA, 2003.
57. K. Shen, T. Yang, L. Chu, J. Holliday, D. A. Kuschner, and H. Zhu. Neptune: Scalable replication management and programming support for cluster-based network services. In *Proceedings of the 3rd USENIX Symposium on Internet Technologies and Systems*, pp. 197–208, San Francisco, CA, 2001.
58. K. Shen, H. Tang, T. Yang, and L. Chu. Integrated resource management for cluster-based internet services. In *Proceedings of the Fifth USENIX Symposium on Operating Systems Design and Implementation, OSDI'2002*, pp. 225–238, Boston, MA, 2002.
59. J. Zhou, L. Chu, and T. Yang. An efficient topology-adaptive membership protocol for large-scale cluster- based services. In *Proceedings of the 19th International Parallel and Distributed Processing Symposium, IPDPS 2005*. IEEE Computer Society, Denver, CO, 2005.
60. R. Song, M. J. Taylor, J.-R. Wen, H.-W. Hon, and Y. Yu. Viewing term proximity from a different perspective. In *Proceedings of the IR Research, 30th European Conference on Advances in Information Retrieval, ECIR'08*, pp. 346–357, Berlin, Germany, 2008.

61. K. M. Svore, P. H. Kanani, and N. Khan. How good is a span of terms? Exploiting proximity to improve web retrieval. In *Proceedings of the 33rd International ACM SIGIR Conference on Research and Development in Information Retrieval, SIGIR '10*, pp. 154–161, ACM, New York, 2010.
62. K. S. Jones, S. Walker, and S. E. Robertson. A probabilistic model of information retrieval: Development and comparative experiments. *Information Processing and Management*, 36(6): 779–808, November 2000.
63. L. Page, S. Brin, R. Motwani, and T. Winograd. The pagerank citation ranking: Bringing order to the web. Technical report no. 1999–66, Stanford Digital Library project, 1999.
64. S. Kamvar, T. Haveliwala, C. Manning, and G. Golub. Extrapolation methods for accelerating pagerank computations. In *Proceedings of the Twelfth International Conference on World Wide Web, WWW 2003*, pp. 261–270, Budapest, Hungary, 2003.
65. S. D. Kamvar, T. H. Haveliwala, C. D. Manning, and G. H. Golub. Exploiting the block structure of the web for computing pagerank. Stanford University Technical Report no. 2003–17, 2003.
66. J. M. Kleinberg. Authoritative sources in a hyperlinked environment. *Journal of the ACM*, 46: 604–632, 1999.
67. K. Bharat and M. R. Henzinger. Improved algorithms for topic distillation in a hyperlinked environment. In *Proceedings of the 21st Annual International ACM SIGIR Conference on Research and Development in Information Retrieval, SIGIR '98*, pp. 104–111, Melbourne, Victoria, Australia, 1998.
68. B. D. Davison, A. Gerasoulis, K. Kleisouris, Y. Lu, H. J. Seo, W. Wang, and B. Wu. Discoweb: Applying link analysis to web search. In *Proceedings of the Eighth International World Wide Web Conference*, pp. 148–149, Toronto, Ontario, Canada, 1999.
69. A. Gerasoulis, W. Wang, and H.-J. Seo. Retrieval and display of data objects using a cross-group ranking metric. US Patent 7024404, 2006.
70. T. Yang, W. Wang, and A. Gerasoulis. Relevancy-based database retrieval and display techniques. US Patent 7028026, 2006.
71. A. Curtis, A. Levin, and A. Gerasoulis. Methods and systems for providing a response to a query. US Patent 7152061, 2006.
72. T. Joachims. Optimizing search engines using clickthrough data. In *Proceedings of the Eighth ACM SIGKDD International Conference on Knowledge Discovery and Data Mining, KDD '02*, pp. 133–142, ACM, New York, 2002.
73. F. Radlinski and T. Joachims. Query chains: Learning to rank from implicit feedback. In *Proceedings of the Eleventh ACM SIGKDD International Conference on Knowledge Discovery in Data Mining, KDD '05*, pp. 239–248, ACM, New York, 2005.
74. E. Agichtein, E. Brill, S. Dumais, and R. Ragno. Learning user interaction models for predicting web search result preferences. In *Proceedings of the 29th Annual International ACM SIGIR Conference on Research and Development in Information Retrieval, SIGIR '06*, pp. 3–10, ACM, New York, 2006.
75. G.-R. Xue, H.-J. Zeng, Z. Chen, Y. Yu, W.-Y. Ma, W. Xi, and W. Fan. Optimizing web search using web click-through data. In *Proceedings of the Thirteenth ACM International Conference on Information and Knowledge Management, CIKM '04*, pp. 118–126, ACM, New York, 2004.
76. N. Craswell and M. Szummer. Random walks on the click graph. In *Proceedings of the 30th Annual International ACM SIGIR Conference on Research and Development in Information Retrieval, SIGIR '07*, pp. 239–246, ACM, New York, 2007.
77. A. Singla, R. White, and J. Huang. Studying trailfinding algorithms for enhanced web search. In *Proceedings of the 33rd International ACM SIGIR Conference on Research and Development in Information Retrieval, SIGIR '10*, pp. 443–450, ACM, New York, 2010.
78. L. Breiman. Random forests. In *Machine Learning*, pp. 5–32, Kluwer Academic Publisher, Dordrecht, the Netherlands, 2001.

79. J. H. Friedman. Greedy function approximation: A gradient boosting machine. *Annals of Statistics*, 29: 1189–1232, 2000.
80. Y. Freund, R. Iyer, R. E. Schapire, and Y. Singer. An efficient boosting algorithm for combining preferences. *Journal of Machine Learning Research*, 4: 933–969, December 2003.
81. R. Herbrich, T. Graepel, and K. Obermayer. Large margin rank boundaries for ordinal regression. In A. J. Smola et al. eds. *Advances in Large Margin Classifiers*, pp. 115–132, MIT Press, Cambridge, MA, January 2000.
82. T.-Y. Liu. Learning to rank for information retrieval. *Foundations and Trends in Information Retrieval*, 3(3): 225–331, 2009.
83. C. Burges, T. Shaked, E. Renshaw, A. Lazier, M. Deeds, N. Hamilton, and G. Hullender. Learning to rank using gradient descent. In *Proceedings of the 22nd International Conference on Machine Learning, ICML '05*, pp. 89–96, ACM, New York, 2005.
84. Q. Wu, C. J. Burges, K. M. Svore, and J. Gao. Adapting boosting for information retrieval measures. *Informations Retrieval*, 13(3): 254–270, June 2010.
85. Y. Yue, T. Finley, F. Radlinski, and T. Joachims. A support vector method for optimizing average precision. In *Proceedings of the 30th Annual International ACM SIGIR Conference on Research and Development in Information Retrieval, SIGIR '07*, pp. 271–278, Amsterdam, the Netherlands, 2007.
86. J. Xu and H. Li. Adarank: A boosting algorithm for information retrieval. In *Proceedings of the 30th Annual International ACM SIGIR Conference on Research and Development in Information Retrieval, SIGIR '07*, pp. 391–398, Amsterdam, the Netherlands, 2007.
87. K. Järvelin and J. Kekäläinen. Cumulated gain-based evaluation of IR techniques. *ACM Transactions on Information Systems*, 20(4): 422–446, October 2002.

VII

Operating System

52 Process Synchronization and Interprocess Communication

Craig E. Wills
Worcester Polytechnic Institute

52.1 Introduction

Process *synchronization* (also referred to as process coordination) is a fundamental problem in operating system design and implementation. It is a situation when two or more processes coordinate their activities based on a condition. An example is when one process must wait for another process to place a value in a buffer before the first process can proceed. A specific problem of synchronization is *mutual exclusion*, which requires that two or more concurrent activities do not simultaneously access a shared resource. This resource may be shared data among a set of processes where the instructions that access these shared data form a *critical region* (also referred to as a critical section). A solution to the mutual exclusion problem guarantees that among the set of processes only one process is executing in the critical region at a time.

Processes involved in synchronization are indirectly aware of each other by waiting on a condition that is set by another process. A form of this synchronization is waiting on an event, which may occur because of an action of the operating system, of another process or an entity external to the immediate system. Processes can also communicate directly with each other through *interprocess communication* (*IPC*). IPC causes communication to be sent between two (or more) processes. A common form of IPC is message passing.

The origins of process synchronization and IPC are work in concurrent program control by people such as Dijkstra, Hoare, and Brinch Hansen. Dijkstra described and presented a solution to the mutual exclusion problem (Dijkstra 1965) and proposed other fundamental synchronization problems and solutions such as the dining philosophers problem (Dijkstra 1971) and semaphores (Dijkstra 1968). Hansen (1972) and Hoare (1972) suggested the concept of a critical region. Hansen published a classic text with many examples of concurrency in operating systems (Hansen 1973). Hoare (1974) provided a complete description of monitors following work by Hansen (1973).

In discussing process synchronization and communication, it is critical to understand that these are no longer just operating system concerns, but broadly affect system and application design, particularly because multithreaded applications employed on multiprocessor and multicore architectures are now commonplace. Many of the primitives used for synchronizing processes also work for synchronizing threads, which are discussed in the chapter on Thread Management. Investigation is being done on mechanisms that work in multicore/processor environments, discussed in the chapter on Multiprocessing and Multicore Programming, and extend to distributed ones. Concurrent programming using synchronization and IPC in a distributed environment is more complex because there may be a failure by some processes participating in synchronization while others continue, or it is possible for messages to be lost or delayed in an IPC mechanism.

The remainder of this chapter discusses the underlying principles and practices commonly used for synchronization and IPC. Section 52.2 identifies fundamental problems needing solutions and issues that arise in considering various solutions. Section 52.3 discusses prominent results for synchronization and IPC problems and discusses their relative merits in terms of these issues. The chapter concludes with a summary, a glossary of terms that have been defined, references, and sources of further information.

52.2 Underlying Principles

Process synchronization and IPC arose from the need of coordinating concurrent activities in a multiprogrammed operating system. This section defines and illustrates traditional as well as more modern synchronization and IPC problems and characterizes the issues on which to compare the solutions.

52.2.1 Synchronization Problems

A fundamental synchronization problem is mutual exclusion, described by Dijkstra (1965). Lamport (1986) provides a formal treatment of the problem with Anderson (2001) surveying Lamport's contributions on this topic. In this problem, multiple processes wish to coordinate so that only one process is in its critical region of code at any one time. During this critical region, each process accesses a shared resource such as a variable or table in memory. The use of a mutual exclusion primitive for access to a shared resource is illustrated in Figure 52.1, where two processes have been created that each access a shared global variable through the routine `Deposit()`.

The routines `BeginRegion()` and `EndRegion()` define a critical region ensuring that `Deposit()` is not executed simultaneously by both processes. With these routines, the final value of `balance` is always 20, although the execution order of the two processes is not defined.

To illustrate the need for mutual exclusion, consider the same example without the `BeginRegion()` and `EndRegion()` routines. In this case, the execution order of the statements for each process is time dependent. The final value of `balance` may be 20 if `Deposit()` is executed to completion for each process, or the value of `balance` may be 10 if the execution of `Deposit()` is interleaved for the two processes. This example illustrates a *race condition*, where multiple processes access and manipulate the same data with the outcome dependent on the relative timing of these processes. The use of a critical region avoids a race condition.

Many solutions have been proposed for the implementation of these two primitives; the most well known of which are given in the following section. Solutions to the mutual exclusion synchronization

```
int balance = 0;  /* global shared variable */

ProcessA()
{
     Deposit(10);
     print("Balance is %d\n", balance);
}

ProcessB()
{
     Deposit(10);
     print("Balance is %d\n", balance);
}

Deposit(int deposit)
{
     int newbalance; /* local variable */

     BeginRegion();   /* enter critical region */
     newbalance = balance + deposit;
     balance = newbalance;
     EndRegion();    /* exit critical region */
}
```

FIGURE 52.1 Shared variable access handled as a critical region.

problem must meet a number of requirements, which were first set forth in Dijkstra (1965) and are summarized in Stallings (2012). These requirements are as follows:

- Mutual exclusion must be enforced so that at most one process is in its critical region at any point in time.
- A process must spend a finite amount of time in its critical region.
- The solution must make no assumptions about the relative speeds of the processes or the number of processes.
- A process stopped outside of its critical region must not lead to blocking of other processes.
- A process requesting to enter a critical region held by no other process must be permitted to enter without delay.
- A process requesting to enter a critical region must be granted access within a finite amount of time.

Another fundamental synchronization problem is the producer/consumer problem. In this problem, one process produces data to be consumed by another process. Figure 52.2 shows one form of this problem where a *producer* process continually increments a shared global variable and a *consumer* process continually prints out the shared variable. This variable is a fixed-size buffer between the two processes, and hence this specific problem is called the bounded-buffer producer/consumer problem.

The ideal of this example is for the consumer to print each value produced. However, the processes are not synchronized and the output generated is timing dependent. The number 0 is printed 2000 times if the consumer process executes before the producer begins. At the other extreme, the number 2000 is printed 2000 times if the producer process executes before the consumer begins. In general, increasing values of n are printed with some values printed many times and others not at all. This example illustrates the need for the producer and consumer to synchronize with each other.

The producer/consumer problem is a specific type of synchronization that is needed between two processes. In general, many types of synchronization exist between processes. An increasingly important one in the context of parallelization is a *barrier*, which is illustrated in Figure 52.3. A barrier synchronizes the actions of multiple processes (or threads), shown as A, B, etc., in Figure 52.3, such that the execution of each process reaching the barrier is blocked until all processes have reached the barrier.

```
int n = 0; /* shared by all processes */

main()
{
     int producer(), consumer();
     CreateProcess(producer);
     CreateProcess(consumer);
     /* wait until done */
}
producer() // produce values of n
{
     int i;
     for (i=0; i<2000; i++)
         n++; // increment n
}
consumer() // consume and print values of n
{
     int i;
     for (i=0; i<2000; i++)
         printf("n is %d\n", n); /* print value of n */
}
```

FIGURE 52.2 Example of producer/consumer synchronization problem.

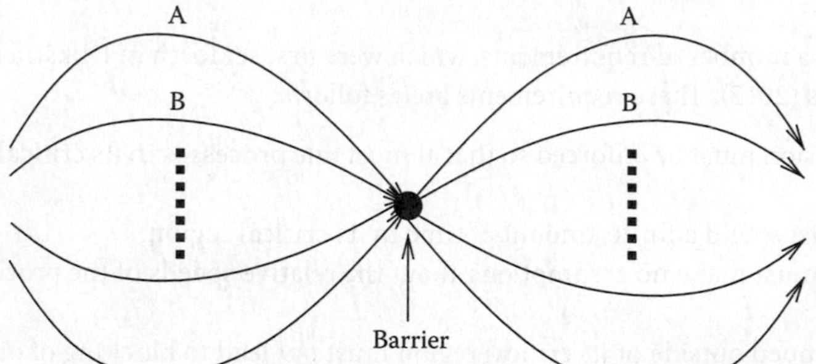

FIGURE 52.3 Barrier synchronization for a set of processes.

This type of synchronization is common for computations that have a mix of serial and parallel phases where a barrier is placed at the end of each parallelized phase.

Another type of synchronization problem that has always existed within operating systems, but has become increasingly important in other contexts is the detection and handling of external events that occur. Solutions to this problem must consider whether such events should be handled asynchronously via event handlers to handle the event when it occurs or synchronously via polling or blocking to stall execution until the event occurs. Solutions may also be needed that allow for one of many events to occur, such as waiting a maximum timeout period for a network packet to arrive.

52.2.2 Synchronization Issues

There are many approaches available for solving mutual exclusion and synchronization problems. There are a number of issues concerning the implementation of solutions to these problems, with the primary ones described in the following:

- *Processor and store synchronicity*: Solutions to synchronization problems that are processor synchronous require that processors can be made uninterruptable. These solutions work for uniprocessor machines but not for multiprocessors where disabling the interrupts on one processor does not affect another. Solutions to synchronization problems that are store synchronous, which assumes individual references to main memory are atomic, are applicable to both uni- and multiprocessor machines.

- *Busy waiting*: Another issue in synchronization solutions is the consumption of CPU resources while a process is waiting for a condition to occur. Some solutions require *busy waiting*, the continued polling of a condition variable. These solutions are less efficient, particularly on a uniprocessor where busy waiting continues until the time slice of the waiting process is complete.
- *Programmer errors*: Potential programming errors inherent in using a particular synchronization approach must also be considered. Ordering of synchronization primitives is necessary for correct solutions with some primitives. Other primitives for synchronization have been specifically designed to minimize the possibility of such a programmer error.
- *Starvation*: It is possible for synchronization solutions to lead to the condition of *starvation*. Starvation occurs when a process is indefinitely denied access to a resource while other processes are granted access to the resource. Starvation is an issue in mutual exclusion if it is possible for one process to be indefinitely denied access to its critical region while access is granted to other processes.
- *Deadlock*: Another fundamental problem confronted by solutions is *deadlock*. Deadlock is the condition when a set of processes using shared resources or communicating with each other are permanently blocked. Coffman et al. (1971) describe three necessary, but not sufficient, conditions that must exist for deadlock to occur among a set of processes sharing resources:
  1. *Mutual exclusion*: A resource may be used by only one process at a time.
  2. *Hold and wait*: Processes holding resources can request new resources.
  3. *No preemption*: A resource given to a process cannot be taken back.
     For deadlock to actually take place, a fourth condition must also exist:
  4. *Circular wait*: A set of processes are in a circular wait; that is, there is a set of processes $\{p_0, p_1, \ldots, p_n\}$ where p_i is waiting for a resource held by p_{i+1}, and p_n is waiting for a resource held by p_0.

  When combined with the three necessary conditions, the presence of the circular wait condition indicates that a deadlock has occurred and that no means exists for breaking the deadlock. When designing solutions to synchronization problems such as mutual exclusion, preventing deadlock involves disallowing one or more of the preceding conditions. Specific solutions to the deadlock problem are given in Section 52.3.

52.2.3 Interprocess Communication Problems

Interprocess communication problems generally involve direct communication between two or more processes, in contrast to synchronization where processes communicate indirectly by waiting on or setting a condition. Thus, communication is indirect between the processes when using synchronization, whereas IPC mechanisms generally communicate by passing messages directly between processes. The primitives for handling messages are

- `send(pid, &message)`
- `pid = receive(&message)`

where `send()` sends the given message to a specific destination process and `receive()` receives the last message sent to it returning the process identifier (pid) of the sender.

Message passing is commonly used in operating systems and user applications for communicating between processes. In a message-based operating system where server processes perform many of the functions associated with an operating system, message passing is used to pass requests to the appropriate server and to receive replies, as is shown in the example of Figure 52.4.

```
int serverpid;           /* well-known pid of the server */

Server()                 /* server process code */
{
    int pid;             /* process id of the requesting process */
    Message request_msg, reply_msg;

    while (TRUE) {
        pid = receive(&request_msg);        /* receive message */
        /* handle request and build reply message */
        send(pid, &reply_msg);              /* send the reply */
    }
}
int ServiceRequest(args)       /* service request function with arguments */
{
    Message request_msg, reply_msg;

    /* load args into request_msg */
    send(serverpid, &request_msg);
    (void)receive(&reply_msg);
    return(reply_msg.status);
}
```

FIGURE 52.4 Service request for a message-based operating system.

52.2.4 Interprocess Communication Issues

As described in the previous example, IPC mechanisms typically involve the exchange of messages from one process to another. There are a number of issues concerning the implementation of a message passing mechanism, with the primary ones described in the following. Implementations of IPC mechanisms, which address these issues, are described in Section 52.3.

- *Direct vs. indirect communication*: A key issue in a message passing system is how messages are addressed to their recipient. In a direct message passing scheme, the delivery address is a process id. This approach is illustrated in Figure 52.4. In an indirect scheme, the address is an intermediate repository, such as a mailbox or port, that is a drop-off point for the message for later pick up by the recipient.
- *Buffering of data*: Another issue is whether the message passing mechanism allows messages to be buffered if the receiving process is currently not ready to receive a message. If buffering is allowed, then another issue is the size of this buffer.
- *Blocking vs. nonblocking operations*: Related to the issue of buffering is the semantics of the `send()` operation when the message cannot be delivered. The operation can either block until the message can be delivered or the operation can immediately return with an error if blocking would occur. Similarly, the `receive()` operation can be defined to either block and wait for message delivery if no message is available or not block and return an error.
- *Fixed- or variable-sized messages*: Fixed-size messages allow for easier implementation but may require more work for the programmer. In contrast, variable-sized messages are easier to program but require more work to implement.
- *Synchronous vs. asynchronous reception*: Message passing mechanisms allow more process control if messages are received only when the `receive()` operation is invoked. However, some mechanisms allow for communication to be received when it arrives through an asynchronous approach using message handlers.

52.3 Impact on Practice

This section discusses prominent results for synchronization and IPC. Each solution contains an example of its use, its relative merits, and problems for which it is useful. More examples of synchronization mechanisms can be found in operating system texts (Tanenbaum 2008, Flynn and McHoes 2011, Silberschatz et al. 2012, Stallings 2012). Comer (2011) and Tanenbaum (1987) present code for actual operating systems to show how synchronization and IPC mechanisms can be implemented. More concurrent programming examples can be found in Ben-Ari (1990), Raynal (1986), and Hansen (1973). Andrews and Schneider (1983) provide a survey of synchronization and IPC techniques. The chapter concludes with more classic synchronization problems and specific solutions to the deadlock problem.

52.3.1 Synchronization Mechanisms

The mechanisms for synchronization are divided into four types based on their level of implementation and support: software only, hardware support, operating system support, and language support. In addition, hybrid solutions exist that combine more than one of these approaches. Each approach shows a solution to the mutual exclusion problem along with other applicable synchronization problems and discusses the relevant synchronization issues.

52.3.1.1 Software Solutions

Software-based synchronization solutions require only that multiple processes can access shared global variables. The solutions use these variables to control access to the critical region. Dekker was the first to devise a software solution that correctly handles the mutual exclusion problem among a set of processes. A discussion of this solution is given in Dijkstra (1965). Peterson (1981) provided a simpler solution of the same problem, which is given in Figure 52.5 for two processes in terms of the `BeginRegion()` and `EndRegion()` routines.

This solution works correctly on any hardware (uni- or multiprocessor) in which references to main memory are atomic. The main disadvantage is that it requires busy waiting, continued polling of a status variable, before gaining access to the critical region if another process is already in the critical region. The solution can generalize to more processes, but better solutions are available.

```
int turn;                            /* whose turn is it? */
int flag[2];                         /* want the mutex? Initially FALSE */
BeginRegion(int pid)                 /* pid is 0 or 1 */
{
    int other;                       /* pid of other process */
    other = 1 - pid;                 /* the opposite of pid */
    flag[pid] = TRUE;                /* express interest in mutex */
    turn = pid;                      /* set flag */
    while ((turn == pid) && (flag[other] == TRUE))
        ;                            /* busy wait */
}
EndRegion(int pid)
{
    flag[pid] = FALSE;          /* drop interest in mutex */
}
```

FIGURE 52.5 Peterson's solution of mutual exclusion for two processes.

52.3.1.2 Solutions Using Hardware Support

Not all problems of synchronization, particularly ones in the operating system itself, can be handled solely with software-based solutions. As in other aspects of computing, hardware support can not only make the task easier but is also necessary for some levels of synchronization within the operating system.

One of the simplest ways to enforce mutual exclusion is to disable hardware interrupts at the start of the critical region, thus ensuring that the process does not give up the CPU (through a context switch) before completing the critical region. When the critical region is done, the process re-enables interrupts. The `BeginRegion()` and `EndRegion()` routines with this approach are shown in Figure 52.6.

This approach is fast and is used for manipulation of shared operating system data structures on a uniprocessor but in general has many disadvantages for general purpose use:

- Programmers must be careful not to disable interrupts for too long; devices that raise interrupts need to be serviced.
- The programmer must be careful about nesting. Activities that disable interrupts must restore them to their previous settings. In particular, if interrupts are already disabled before entering a critical region, they must remain disabled after leaving the critical region. Code in one critical region may call a routine that executes a different critical region.
- Disabling interrupts prevents all other activities, even though many may never execute the same critical region. Disabling interrupts is like using a sledge hammer; it is a powerful tool but bigger than is needed for most jobs.
- The technique is ineffective on multiprocessor architectures, where disabling interrupts on one processor still allows other processes to run on other processors.

Rather than perform mutual exclusion by controlling interrupts, a common approach is to use special instructions provided by the hardware to implement mutual exclusion. One such instruction is `Test_and_Set`, which is functionally defined by the procedure of Figure 52.7. It returns the previous value of a target variable and sets the target to the given value. Most importantly, this instruction is performed in an atomic manner so a context switch cannot occur in the middle of it. In addition, operations performed on two separate processors of a multiprocessor are guaranteed to occur in sequential order because of store synchronicity. Figure 52.7 shows how the `BeginRegion()` and `EndRegion()` primitives are implemented using this instruction. The variable `mutex` is also referred to as a lock variable, and this approach to mutual exclusion is called a *spin lock* because a process spins in an infinite loop waiting for the lock. When the `mutex` variable is set to `FALSE` by a process exiting from its critical region, another process is allowed to gain access to its critical region.

This type of solution is particularly attractive for threaded environments because it allows for *wait-free* or *non-blocking* atomic updates of shared resources. Herlihy (1991) shows that other primitives such as `Compare_and_Swap` are even more computationally strong than `Test_and_Set`. Fraser and Harris (2007) used these primitives as a basis to create practical programming solutions for atomic, nonblocking updates to multiword data structures.

```
BeginRegion()
{
    DisableInterrupts();
}
EndRegion()
{
    EnableInterrupts();
}
```

FIGURE 52.6 Mutual exclusion by disabling/enabling hardware interrupts.

```
int mutex = FALSE;                      /* global variable for mutex */

int Test_and_Set(int *pVar, int value)  /* atomic machine instruction */
{
    int temp;

    temp = *pVar;
    *pVar = value;
    return(temp);
}

BeginRegion()        /* Loop until safe to enter */
{
    while (Test_and_Set(&mutex, TRUE)) ;
        ; /* Loop until return value is FALSE */
}

EndRegion()
{
    mutex = FALSE;
}
```

FIGURE 52.7 Mutual exclusion using the test_and_set machine instruction.

The primary disadvantage of this approach is its use of busy waiting, thus wasting CPU resources. The use of busy waiting also allows for process starvation if multiple processes are contending for a critical region. Finally, deadlock is possible on a uniprocessor machine if a lower priority process gets interrupted in the middle of its critical region and then a higher priority process tries to gain access to the same critical region. The higher priority process busy waits forever because the lower priority process never runs. This is a situation of *priority inversion* where a lower priority process affects access to a resource by a higher priority process. It is typically solved by temporarily raising the priority of the lower priority process.

52.3.1.3 Operating System Support

All of the synchronization approaches shown thus far can be implemented with the bare features of the hardware. The approaches also cause busy waiting if another process already is in its critical region. *Semaphores*, an important synchronization primitive, can be constructed by adding process coordination support to the operating system. Semaphores are data structures consisting of an identifier, a counter, and a queue; where processes waiting on a semaphore are blocked and placed on the queue, processes signaling a semaphore may unblock and remove a process from the queue, and the counter maintains a count of waiting processes.

The concept of a semaphore was first introduced by Dijkstra (1968). Dijkstra defined two atomic semaphore operations: wait and signal, which he termed the P-operation (for wait; from the Dutch word proberen, to test) and the V-operation (for signal; from the Dutch word verhogen, to increment).

A restricted version of a semaphore, called a *binary semaphore*, limits the value of the counter to 0 and 1. However, the more general case is to use a *counting semaphore*, which has the following properties concerning the counter:

- A nonnegative count always means that the queue is empty.
- A count of negative `n` indicates that the queue contains `n` waiting processes.
- A count of positive `n` indicates that `n` resources are available and `n` requests can be granted without delay.

There are four basic operations defined for creating, deleting, waiting on, and signaling a semaphore.

1. `semid = semcreate(val)`: Create a semaphore with the given initial value for the counter.
2. `semdelete(semid)`: Delete a semaphore.

```
int semid;

Initialization()
{
semid = semcreate(1);   /* initialize the semaphore count to 1 */
}

BeginRegion()
{
     semwait(semid);
}

EndRegion()
{
     semsignal(semid);
}
```

FIGURE 52.8 Mutual exclusion using semaphores.

3. `semwait(semid)`: Wait on a semaphore. Decrement the semaphore counter. If the counter is negative, then suspend execution of the process and place it in the semaphore queue.
4. `semsignal(semid)`: Signal a semaphore. Increment the semaphore counter. Make the first process in the semaphore queue ready for execution.

Given these operations, a simple solution to the mutual exclusion problem is shown in Figure 52.8. Unlike previous solutions, a process waiting for its critical region does not busy wait. While it is waiting, it is in a suspended state, allowing the CPU to perform other activities. Because semaphores are provided by the operating system, they work correctly on either uni- or multiprocessor machines.

Semaphores also provide a mechanism to solve the bounded-buffer producer/consumer problem, which was introduced in the example of Figure 52.2. A solution for this problem using semaphores is shown in Figure 52.9. This solution ensures that the consumer process prints all integer values from 1 to 2000 by alternating between the producer and consumer.

Although semaphores provide straightforward solutions to both of these problems, semaphores themselves must be implemented in the operating system with lower level primitives such as the `Test_and_Set` machine instruction. Semaphores work well for processes, which are allowed to block, but must not be used in code that cannot block such as interrupt service routines.

52.3.1.4 Language Constructs

The solutions presented thus far to the mutual exclusion problem each require the programmer to take explicit action to ensure mutual exclusion. To guarantee mutual exclusion some programming languages provide constructs to implicitly guarantee mutual exclusion. One such construct is a *monitor* (Hoare 1974), which permits only one process to be executing in a monitor at a time.

Monitors are a programming language construct, similar to abstract data types in that the programmer defines a set of data types and procedures that can manipulate the data, procedures can be exported to other modules, and the system invokes an initialization routine before execution begins. Monitors differ in that they support guard procedures. When a process invokes a guard procedure, its execution is delayed until no other processes are executing a guard procedure within the monitor. The Java programming language supports monitors using the `synchronized` primitive on methods within a class to prevent threads from simultaneously access of these methods (Arnold et al. 2005). Figure 52.10 revisits the mutual exclusion problem of Figure 52.1 using a solution written in Java. The definition of a monitor guarantees that the variable `balance` is not updated and read simultaneously.

Monitors are a programming language construct that must be implemented using a lower level facility provided by the hardware or operating system, such as semaphores. Although monitors provide

```
int psem, csem; /* semaphores */
int n = 0;      /* shared by all processes */
main()
{
    int producer(), consumer();
    csem = semcreate(0);  /* consumer initially blocks */
    psem = semcreate(1);  /* producer initially allowed to run */
    CreateProcess(producer);
    CreateProcess(consumer);
    /* wait until done */
}
producer()
{
    int i;
    for (i=0; i<2000; i++) {
        wait(psem);
        n++;    /* increment n by 1 */
        signal(csem);
    }
}
consumer()
{
    int i;
    for (i=0; i<2000; i++) {
        wait(csem);
        printf("n is %d\n", n);  /* print value of n */
        signal(psem);
    }
}
```

FIGURE 52.9 Bounded-buffer producer/consumer problem with semaphores.

```
public class Account {
    private int balance;

    public Account() {
        balance = 0;            // initialize balance to zero
    }

    // use synchronized to prohibit concurrent access of balance
    public synchronized void Deposit(int deposit) {
        int newbalance; // local variable

        newbalance = balance + deposit;
        balance = newbalance;
    }

    public synchronized int GetBalance() {
        return balance;         // return current balance
    }
}
```

FIGURE 52.10 Mutual exclusion using Java monitors.

mutual exclusion, they need additional primitives to provide synchronization. To do so, monitors are defined to have condition variables, which are waited on and signaled similar to semaphores. Java provides a single, unnamed condition for a class with the `wait()` method used by a thread to wait on the condition and the `notify()` or `notifyAll()` methods to wake up one or all waiting threads. These primitives allow other synchronization problems such as the producer/consumer problem to be implemented with monitors.

52.3.1.5 Hybrid Solutions

Modern operating systems have migrated from monolithic systems written for a uniprocessor in which critical regions were used to access shared data structures. These critical regions were guarded by setting the interrupt level appropriately. Current operating systems must not only support multiprocessors but also provide real-time capabilities, thus leading to multithreaded designs. In these designs, the use of interrupts to control access to shared data does not work. Rather, these systems have moved to solutions using complex locks, which combine the use of spin locks with the semantics of semaphores.

As one example, the Solaris operating system uses adaptive mutex locks (Eykholt et al. 1992), a type of complex lock, to protect access to shared data among a set of threads. The adaptive lock starts out executing like a standard spin lock. If the lock is currently free, then the issuing thread immediately obtains the lock. However, if the lock is in use, then the operating system checks the status of the thread holding the lock. If this thread is currently in the run state (as could be the case on a multiprocessor), then the issuing thread continues in a spin lock waiting for what is expected to be a short time for the lock to be released. If the holding thread is not in the run state (as would always be the case on a uniprocessor), then the issuing thread is suspended until the lock is released. The rationale is to use a spin lock if the wait for the lock is expected to be short and to actually suspend the thread if the wait is expected to be longer.

This hybrid approach tries to minimize overhead and maximize performance. If the size of a critical region is large (hundreds of instructions), then the adaptive mutex lock is less desirable compared to a lock that simply causes a thread to suspend when the lock is not available.

Another hybrid solution is the Linux construct *futex* (fast userspace mutex) (Franke et al. 2002). This construct allows a task to acquire a lock without making a system call if there are currently no other tasks waiting for the lock, but if it fails to acquire the lock then a system call is made to wait.

Another type of complex lock used in multithreaded operating systems is a read/write lock. These locks allow either a single writer or multiple readers to simultaneously hold the lock, thus increasing the parallelism when reading of a shared data structure predominates. Writers must wait until all readers have released the lock before obtaining the lock, whereas readers are granted immediate access to the lock in the absence of a writer. To prevent starvation of a writer, all read requests after a write request has been issued are queued until the write request has been satisfied. Starvation of readers in the face of multiple writers is similarly avoided.

52.3.1.6 Earlier Solutions

Many other synchronization primitives were initially proposed, but in general can be expressed in terms of the solutions already given. A sampling of these earlier primitives are critical regions (Hansen 1972, Hoare 1972), serializers (Atkinson and Hewitt 1979), path expressions (Campbell and Habermann 1974), and event counts and sequencers (Reed and Kanodia 1979).

52.3.2 Interprocess Communication Mechanisms

As with synchronization, a variety of IPC mechanisms are available. The following discusses a number of these mechanisms and how they handle the IPC issues raised in Section 52.2.

52.3.2.1 Direct Message Passing

The simplest form of message passing is to send messages directly from one process to another. An example of this approach is the low-level message passing facility used in the Xinu operating system (Comer 2011). Where `send()` sends a fixed, integer-sized message to a specific process and `receive()` returns the last message sent to it. A process can buffer only one message. If `send()` detects a message already buffered at the process then it returns immediately with an error, not delivering the message. If no message is buffered, then `send()` buffers the message and readies the receiving process if it is waiting for a message. The `receive()` operation blocks if a message is not available.

Another direct message passing mechanism is implemented in Minix (Tanenbaum 1987) where `send()` sends the given message to a specific destination process and `receive()` receives a message from a particular process. The source process for `receive()` can contain a wildcard value of `ANY`, indicating that messages from any process are accepted. Messages are a fixed size, but there is no buffering. Rather the `send()` and `receive()` operations *rendezvous* so that both operations block until the receiving process has actually copied the message from the sender. The use of rendezvous explicitly synchronizes the execution of the sending and receiving process.

52.3.2.2 Mailboxes/Ports

Rather than send directly to process, a more common approach is to define another operating system abstraction called a *mailbox* (also referred to as a port). Mailboxes are buffers that hold messages sent by one process to be received by another process. Thus, there is indirect communication between the two processes. The primitives for handling messages are

- `send(mailbox, &message)`
- `receive(mailbox, &message)`

where `send()` buffers the message in the given mailbox and `receive()` removes a message from the mailbox.

As an example, the Unix operating system provides ports to allow for intra- and inter-machine communication between processes. Ports often represent well-known services where a server process binds to a port and client processes of the service send requests to the port. The port buffers communication sent to the buffer until it is read by the receiving process. The messages sent to the port can be of variable size.

Message passing can also be implemented by using shared memory and semaphores, illustrating the equivalence of synchronization and IPC primitives. Figure 52.11 shows message passing with shared memory and semaphores to implement a set of mailboxes. The mechanism uses fixed-size messages with each of four mailboxes containing eight message buffers. The `send()` operation blocks if there is no buffer space in the mailbox; similarly, the `receive()` operation blocks if there is no message available in the mailbox. The mutex semaphore is needed to guarantee that no more than one process tries to send to or receive from a mailbox at the same time. This semaphore would not be needed if only one process could send to and receive from a mailbox. The mutex semaphore would also not be needed if a mailbox contained only one buffer slot. This is also an example of a bounded-buffer producer/consumer problem where sending processes produce messages and receiving processes consume them.

52.3.2.3 Pipes

A special case of IPC is the pipe abstraction available, which was first available in the Unix operating system. A *pipe* is a unidirectional, stream communication abstraction. One process writes data to the write end of the pipe, and a second process reads data from the read end of the pipe. The pipe itself is a buffer between the two processes that causes the reader to block if no data are available and the writer to block if the buffer is full. As it implements a stream abstraction, there is no notion of fixed-size messages. A pipe is another example of a solution to the bounded-buffer producer/consumer problem where the writing process is a producer and the reading process is a consumer.

Figure 52.12 shows a simple example of the use of pipes in the Unix operating system. Pipes are typically requested and set up by a Unix command interpreter, but the example shows one process creating another process with `fork()` with a pipe between them. A string of characters is then written to and read from the pipe.

52.3.2.4 Events

Events provide synchronization via a primitive form of IPC. They allow a process to learn of an externally generated event. Environments with events may provide primitives for asynchronously handling

```
#define N 8                    /* number of msgs buffered in a mailbox */
#define M 4                    /* number of mailboxes */

Message mailboxes[M][N];       /* shared memory for mailboxes of messages */

int semidMsg[M];               /* message available */
int semidSlot[M];              /* slot available */
int semidMutex[M];             /* controls access to critical region */

Initialization()
{
    int i;

    for (i = 0; i < M; i++) {
        semidMsg[i] = semcreate(0);   /* no messages are available */
        semidSlot[i] = semcreate(N);  /* N slots are available for messages */
        semidMutex[i] = semcreate(1); /* one process can enter region */
        /* initialize indices for inserting/deleting in mailboxes[i] */
    }
}

send(int m, Message *pmsg)         /* send message to mailbox m */
{
    semwait(semidSlot[m]);         /* is a slot available */
    semwait(semidMutex[m]);        /* enter critical region */
    addmessage(m, pmsg);           /* add msg to circular queue for mailbox m */
    semsignal(semidMutex[m]);      /* exit critical region */
    semsignal(semidMsg[m]);        /* signal message available */
}

receive(int m, Message *pmsg)      /* retrieve message from mailbox m */
{
    semwait(semidMsg[m]);          /* is a message available */
    semwait(semidMutex[m]);        /* enter critical region */
    removemessage(m, pmsg);        /* remove next msg from queue for mailbox m */
    semsignal(semidMutex[m]);      /* exit critical region */
    semsignal(semidSlot[m]);       /* signal slot available */
}
```

FIGURE 52.11 Message passing with semaphores and shared memory.

events, synchronously handling events or synchronously handling one of multiple events. An example for each of these in Unix operating system is provided in the following.

Events for a Unix process are generated as software interrupts. They are similar to hardware interrupts in that when an interrupt of a process occurs, an interrupt handler routine corresponding to the type of interrupt is invoked. Interrupts are asynchronous so, when an interrupt is received, execution of the process stops and is restarted after the interrupt handling routine has been executed. Software interrupts are sent to a process using the process id of the process. Many interrupts are used for well-known functions such as when the user types the interrupt key, a child process completes, or an alarm scheduled by the process has expired.

Two routines are used to send and handle software interrupts:

- `SendInterrupt(pid, num)`: An interrupt of type `num` is sent to process `pid`. In the Unix operating system this routine is `kill()`.
- `HandleInterrupt(num, handler)`: This specifies that user supplied `handler` routine should be invoked when interrupt of type `num` occurs. Typical handlers are to ignore the interrupt, terminate the process, or execute a user supplied interrupt handler. In the Unix operating system, this routine is `signal()`.

```
#define DATA "hello world"
#define BUFFSIZE 1024

int rgfd[2];                    /* file descriptors of pipe ends */

main()
{
    char sbBuf[BUFFSIZE];

    pipe(rgfd);                 /* create a pipe returning two file descriptors */
    if (fork()) {               /* parent, read data from pipe */
        close(rgfd[1]);         /* close write end */
        read(rgfd[0], sbBuf, BUFFSIZE);
        printf("Pipe contents: %snn", sbBuf);
        close(rgfd[0]);
    }
    else {                      /* child, write data to pipe */
        close(rgfd[0]);         /* close read end */
        write(rgfd[1], DATA, sizeof(DATA)); /* write data to pipe */
        close(rgfd[1]);
        exit(0);
    }
}
```

FIGURE 52.12 Pipe example in the Unix operating system.

Figure 52.13 shows a sample program for the Unix operating system with software interrupts. It sets up two interrupt handlers for signals 1 and 2 and then goes into an infinite loop where it updates a counter and sleeps for 1 s. Figure 52.13 also shows a command line script with this program. Invoking the interrupt character from the command interpreter causes interrupt 2 to be sent to the process. The second invocation of the program causes it to be run in the background with a process id of 20822. The `kill` program is then used to send interrupts to the background process.

In some cases, it is desirable to wait for such events in a synchronous manner. Unix provides system calls that block a process until a given interrupt is received. For example, `sleep()` causes a timer to be set with the process blocking until the corresponding alarm software interrupt is generated. Similarly, the `wait()` system call blocks a process until a software interrupts signally the completion of a child process is generated.

In other cases, a process may need to wait for one from a number of events to occur. One example of this situation is a networked server, which may be looking to know when incoming data or connections from multiple network sockets are available while at the same time wanting a timeout event if none occur within a period of time. Figure 52.14 shows a sample function for the Unix operating system using the `select()` system call to multiplex the occurrence of multiple events where activity is detected on a file descriptor (socket) or the given timeout is reached. The function can be easily extended to handle detection of activity on multiple file descriptions.

52.3.3 Classic Problems

Two classic synchronization problems—the critical region and bounded-buffer producer/consumer—have already been discussed. A slight variation of the producer/consumer problem is to use an unbounded buffer, in which case the producer never blocks because the buffer never fills. Many other classic synchronization problems have been proposed and solved. The following describes two such problems.

52.3.3.1 Readers/Writers Problem

The readers/writers problem occurs when multiple readers and writers want access to a shared object such as a database. The problem was introduced in Courtois et al. (1971). In the problem, multiple

```
#include <signal.h>

int n;

main(int argc, char **argv)
{
    void InterruptHandler(), InitHandler();
    n = 0;

    signal(SIGINT, InterruptHandler); /* signal 2 */
    signal(SIGHUP, InitHandler);      /* signal 1 */
    while (1) {
        n++;
        sleep(1);                     /* sleep for one second */
    }
}

void InterruptHandler()
{
    printf("The current value of n is %dnn", n);
    exit(0);
}

void InitHandler()
{
    printf("Resetting the value of n to zeronn");
    n = 0;
}
% cc -o signalex signalex.c
% signalex
^C         (interrupt character)
The current value of n is 3
% signalex &
[1] 20822
% kill -1 20822
Resetting the value of n to zero
% kill -2 20822
The current value of n is 19
[1]     Done                 signalex
```

FIGURE 52.13 Software interrupt program and script in the Unix operating system.

```
int WaitForEvent(int fdIn, int timeout)
{
    fd_set bvfdRead;
    struct timeval tv;

    tv.tv_sec = timeout; /* put timeout secs into timeval struct */
    tv.tv_usec = 0;
    FD_ZERO(&bvfdRead);
    FD_SET(fdIn, &bvfdRead);      /* fd to look for input */
    /* size of bitmap to watch is one more than value of fdIn */
    return(select(fdIn+1, &bvfdRead, 0, 0, &tv));
}
```

FIGURE 52.14 Sample function to wait for multiple events in Unix.

```
int readercount = 0;               /* number of readers currently reading */
int readermutex;                   /* semaphore mutex for reader count */
int dbaccess;                      /* semaphore to control access to database */

main()
{
    readermutex = semcreate(1);    /* mutex for reader count */
    dbaccess = semcreate(1);       /* mutex for database */
    CreateProcess(reader);         /* create a reader process */
    CreateProcess(writer);         /* create a writer process */
}

reader()
{
    while (TRUE) {
        semwait(readermutex);         /* get access to readercount */
        readercount++;                /* increment count */
        if (readercount == 1)         /* if first reader … */
            semwait(dbaccess);        /* gain access to database */
        semsignal(readermutex);       /* done with count */
        /* read database */
        semwait(readermutex);         /* get access to readercount */
        readercount--;                /* decrement count */
        if (readercount == 0)         /* if last reader … */
            semsignal(dbaccess);      /* relinquish access to database */
        semsignal(readermutex);       /* done with count */
        /* use data read */
    }
}
writer()
{
    while (TRUE) {
        /* generate new data */
        semwait(dbaccess);             /* gain access to database */
        /* update database */
        semsignal(dbaccess);           /* relinquish access to database */
    }
}
```

FIGURE 52.15 Solution to the readers/writers problem with semaphores. Based on statements of the algorithm in Tanenbaum (2008). (From Tanenbaum, A., *Modern Operating System*, 3rd edn., Prentice Hall, Upper Saddle River, NJ, 2008.)

readers are allowed to access the database simultaneously, but a writer must have exclusive access to the database before performing any updates for consistency. A practical example of this problem is an airline reservation system with many readers and an occasional update of the information.

Figure 52.15 shows a solution to this problem for multiple reader and writer processes with semaphores. The solution allows multiple readers access to the database at a time. A writer process can gain access only after all reader processes have relinquished the database. The solution gives priority to reader processes, who can gain access to the database even if a writer process is already requesting access to the database. Solutions giving more balanced priority to each type of process can also be constructed as discussed in (Flynn and McHoes 2011).

52.3.3.2 Dining Philosophers Problem

The dining philosophers problem was proposed and solved by Dijkstra (1971). It consists of five philosophers sitting at a round table. Philosophers each have a bowl of rice in front of them and there is a chopstick in between each bowl (alternately, the problem is described using plates of spaghetti and forks).

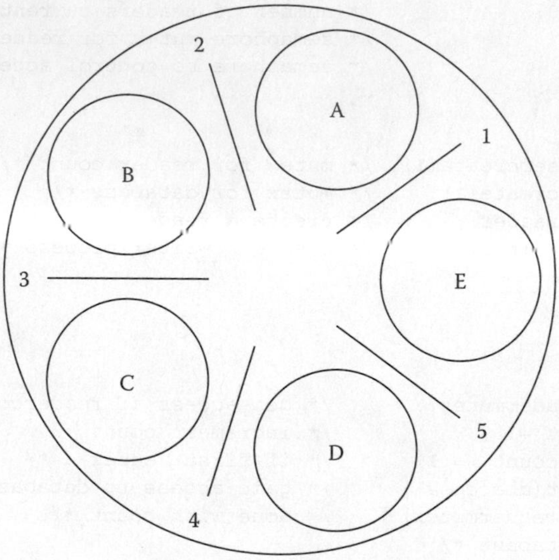

FIGURE 52.16 Illustration of the dining philosophers problem.

The problem is illustrated in Figure 52.16 with the philosophers' bowls labeled A–E and the chopsticks 1–5. These philosophers have two functions in life: think, requiring no interaction with colleagues, and eat, requiring the philosopher to pick up the chopstick on the left and right.

This classic synchronization problem has potential for both deadlock and starvation (literally!). The straightforward solution for a philosopher to eat is to first pick up the left chopstick and then the right chopstick. However, if all philosophers pick up their left chopstick at the same time, they will all deadlock when they go to pick up their right chopstick. A simple modification to this approach is for the philosophers to put down the left chopstick if the right chopstick is not available, wait for some time, and try again. However, there is still a chance that the philosophers will operate in lock step and no philosophers will acquire both chopsticks. This condition is called *livelock* and occurs when attempts by two or more processes (philosophers) to acquire a resource (the left and right chopsticks) run indefinitely without any process succeeding. The dining philosophers problem will be used as a guide in the following discussion on deadlock and starvation.

52.3.4 Deadlock and Starvation

Deadlock occurs when a set of processes using shared resources are permanently blocked trying to gain access to those resources. Classic papers on this topic are Coffman et al. (1971) and Holt (1972). Zobel and Koch (1988) contains an annotated bibliography on the subject. Deadlock can occur with synchronization such as when two processes each need to gain access to two separate critical regions for execution. If one process gains access to the first critical region and the other process gains access to the second critical region, then these two processes will be in deadlock when they attempt to acquire the other needed critical region.

There are four principles used for dealing with the issue of deadlock in operating systems. These principles are prevention, detection, avoidance, and recovery and are summarized in Isloor and Marsland (1980). Solutions to the deadlock problem exemplifying these principles are described in the following. Each of these solutions prevents deadlock by precluding one or more of the four conditions for deadlock given in Section 52.2. The solutions are characterized as being conservative or liberal depending on the degree of concurrency they allow.

The most liberal solution is to allocate requested resources to processes if the resources are available. This approach was given as an initial solution when introducing the dining philosophers problem.

As was shown, it can lead to deadlock and requires that a deadlock detection process be periodically invoked. Deadlock detection involves detecting the circular wait condition through any algorithm for detecting cycles in directed graphs. Once deadlock occurs, a deadlock recovery method must be invoked either to take away resources from a victim process or to terminate the process altogether. In the dining philosophers problem, a recovery method would be to take away a chopstick from one of the philosophers.

At the other extreme, the most conservative deadlock prevention solution is *serialization*. This deadlock prevention approach guarantees no deadlock by allowing only one process to acquire resources at a time. In the dining philosophers problem, this approach means that only one philosopher is allowed to eat at any time. The approach prevents any concurrency and can lead to starvation if a philosopher is constantly passed over in obtaining chopsticks.

Another solution that allows more concurrency but still prevents deadlock is *one-shot allocation*. This approach requires that a process obtain all of its resources at once. Using this approach, a dining philosopher must obtain both chopsticks at the same time. It avoids deadlock by preventing the hold and wait condition. This solution prevents deadlock but not starvation. It also requires that a process obtain all of its resources at the same time even if it does not currently need all of them.

A still more liberal solution that prevents deadlock is *hierarchical allocation*. This solution requires that all resources have a number associated with them as in Figure 52.16. Hierarchical allocation prevents deadlock by requiring that processes can acquire only resources with a higher number than any resource it currently holds. Thus, in Figure 52.16, each philosopher must first acquire the left chopstick and then the right except for Philosopher E, who must first acquire the right chopstick. This solution is still conservative and requires processes to acquire resources not necessarily in the order they are needed but in the order resources are numbered. It avoids deadlock by preventing the circular wait condition.

As opposed to deadlock prevention, deadlock avoidance policies do not prevent deadlock a priori but monitor the allocation requests as they are made so as not to allow deadlock to occur. A solution using this approach is the *bankers algorithm* (also called the advanced claim algorithm) introduced in Dijkstra (1968). This solution is the most liberal that avoids deadlock but requires that each process know the maximum number from a class of resources that it may request at any time during execution. The algorithm performs as a banker giving out loans of money in that any resource requests are granted only if they

1. Do not exceed the total number of resources in a class
2. Do not exceed the maximum number of resources for that process
3. Lead to a *safe state* where a sequence of resource deallocations and allocations can allow all processes to complete without deadlock

52.4 Research Issues and Summary

The research issues in synchronization and IPC correspond to the movement toward multithreaded systems running in multiprocessor/core environments. The adoption of traditional synchronization primitives for these environments and the exploration of better primitives is leading to work on hybrid approaches for synchronization. This work has also extending to distributed environments.

Research work continues on extending synchronization primitives and defining new problems. The issue of performance is a key area as researchers seek to minimize the cost of busy waiting in shared memory multi-processors/cores by using virtual memory to reduce memory access costs. Other research is investigating optimal tradeoff between the use of busy waiting for a lock versus suspending the thread or process. Performance is also a research issue for IPC mechanisms in client/server systems as they execute in intra- and inter-machine environments.

An ongoing research issue is correctness, particularly as the synchronization problems of modern operating systems become more complex. Ad hoc synchronization and its associated problems continue to exist (Xiong et al. 2010). Using language constructs to better encapsulate the synchronization details is being explored, but this approach can lead to tradeoffs with performance. Tools for detecting synchronization errors are also an ongoing area of research.

In summary, synchronization and IPC are fundamental to multiprogrammed operating system design. Many primitives to solve fundamental problems such as mutual exclusion and producer/consumer exist ranging from software only approaches, to special hardware instructions, to primitives constructed by the operating system and programming languages. Many of the primitives are equivalent in terms of their semantics, and one can be implemented in terms of another. Examples include implementing monitors with semaphores or message passing with shared memory and semaphores. Modern operating systems are using hybrid approaches, which adaptively switch between techniques according to the run-time operating environment.

Glossary

Busy waiting: Situation in synchronization when a process continuously polls the status of a condition variable.

Critical region: Set of instructions for a process that access data shared with other processes. Only one process may execute its critical region at a time.

Deadlock: Condition when a set of processes using shared resources or communicating with each other are permanently blocked.

Interprocess communication (IPC): Communication between two (or more) processes directly aware of each other.

Livelock: Condition when attempts by two or more processes to acquire a resource run indefinitely without any process succeeding.

Mailbox: An operating system abstraction containing buffers to hold messages. Messages are sent to and received from the mailbox by processes.

Monitor: Programming language construct providing abstract data types and mutually exclusive access to a set of guard procedures.

Mutual exclusion: A synchronization problem requiring that two or more concurrent activities do not simultaneously access a shared resource.

Priority inversion: Situation where a lower priority process affects access to a resource by a higher priority process.

Race condition: Situation where multiple processes access and manipulate shared data with the outcome dependent on the relative timing of these processes.

Semaphore: Synchronization primitive consisting of an identifier, a counter, and a queue where processes waiting on a semaphore are blocked and placed on the queue; processes signaling a semaphore may unblock and remove a process from the queue, and the counter maintains a count of waiting processes.

Spin lock: Mutual exclusion mechanism where a process spins in an infinite loop waiting for the value of a lock variable to indicate availability.

Starvation: Condition when a process is indefinitely denied access to a resource while other processes are granted access to the resource.

Synchronization: Situation when two or more processes coordinate their activities based upon a condition.

Further Information

Many good text books on operating systems, such as those by Silberschatz et al. (2012), Tanenbaum (2008), and Stallings (2012) exist, which describe problems, issues, and solutions for synchronization and IPC. The book *Principles of Concurrent and Distributed Programming* by M. Ben-Ari (1990) contains a number of problems and worked out solutions for both concurrent and distributed programming. *Algorithms for Mutual Exclusion* by M. Raynal (1986) presents a comprehensive treatment of solutions for the mutual exclusion problem. Andrews and Schneider (1983) provide a survey of processes, synchronization, and interprocess communication.

The Association for Computing Machinery (ACM) Special Interest Group on Operating Systems (SIGOPS) publishes *Operating Systems Review* four times a year. This publication contains work on a variety of operating system topics including synchronization. This group also sponsors the biennial *ACM Symposium on Operating Systems Principles* that cover the latest developments in the field of operating systems. Its proceedings are published in an issue of *Operating Systems Review.* Another ACM publication, the *ACM Transactions on Computer Systems* is good source for relevant work.

The USENIX Association sponsors a number of conferences on operating system-related topics. A general technical conference is sponsored each year. Two conferences, the *Symposium on Operating Systems Design and Implementation (OSDI)* and *Workshop on Hot Topics in Operating Systems (HOTOS)* are specific to operating system issues.

References

Anderson, J. H. 2001. Lamport on mutual exclusion: 27 years of planting seeds. In *Proceedings of the ACM Symposium on Principles of Distributed Computing*, Newport, RI, pp. 3–12.

Andrews, G. R. and Schneider, F. B. 1983. Concepts and notations for concurrent programming. *Comput. Surv.* 15(1):3–43.

Arnold, K., Gosling, J., and Holmes, D. 2005. *The Java Programming Language*, 4th edn. Prentice Hall.

Atkinson, R. and Hewitt, C. 1979. Synchronization and proof techniques for serializers. *IEEE Trans. Software Eng.* 5(1):10–23.

Ben-Ari, M. 1990. *Principles of Concurrent and Distributed Programming*. Prentice Hall, Englewood Cliffs, NJ.

Campbell, R. H. and Habermann, A. N. 1974. The specification of process synchronization by path expressions. In *Operating Systems*, E. Gelenbe and C. Kaiser, eds., pp. 89–102. Springer–Verlag, Berlin, Germany.

Coffman, E., Elphick, M., and Shoshani, A. 1971. System deadlocks. *ACM Comput. Surv.* 3(2):67–78.

Comer, D. 2011. *Operating System Design—The Xinu Approach, Linksys Version*. CRC Press, Taylor & Francis Group, Boca Raton, FL.

Courtois, P., Heymans, F., and Parnas, D. L. 1971. Concurrent control with "readers" and "writers." *Commun. ACM* 14(10):667–668.

Dijkstra, E. W. 1965. Solution of a problem in concurrent programming control. *Commun. ACM* 8(9):569.

Dijkstra, E. W. 1968. Co-operating sequential processes. In *Programming Languages*, F. Genuys, ed., pp. 43–112. Academic Press, New York. Reprint of Technical Report EWD-123, Technological University, Eindhoven, The Netherlands (1965).

Dijkstra, E. W. 1971. Hierarchical ordering of sequential processes. *Acta Inf.* 2(1):115–138.

Eykholt, J. R., Kleiman, S. R., Barton, S., Faulkner, S., Shivalingiah, A., Smith, M., Stein, D., Voll, J., Weeks, M., and Williams, D. 1992. Beyond multiprocessing: Multithreading the SunOS kernel. In *Proceedings of the Summer USENIX Conference*. USENIX Association, pp. 11–18. San Antonio, TX.

Flynn, I. M. and McHoes, A. M. 2011. *Understanding Operating Systems*, 6th edn. Course Technology, Cengage Learning, Boston, MA.

Franke, H., Russell, R., and Kirkwood, M. 2002. Fuss, futexes and furwocks: Fast userlevel locking in Linux. In *Proceedings of the Ottawa Linux Symposium*, Ottawa, Ontario, Canada.

Fraser, K. and Harris, T. 2007. Concurrent programming without locks. *ACM Trans. Comput. Syst.* 25(2):55.

Hansen, P. B. 1972. Structured multiprogramming. *Commun. ACM* 15(7):574–578.

Hansen, P. B. 1973. *Operating Systems Principles*. Prentice Hall, Englewood Cliffs, NJ.

Herlihy, M. 1991. Wait-free synchronization. *ACM Trans. Prog. Lang. Syst.* 11(1):124–149.

Hoare, C. A. R. 1972. Towards a theory of parallel programming. In *Operating Systems Techniques*, C. A. R. Hoare and R. H. Perrott, eds., pp. 61–71. Academic Press, New York.

Hoare, C. A. R. 1974. Monitors: An operating system structuring concept. *Commun. ACM* 17(10):549–557; Erratum 1975. *Commun. ACM* 18(2):95.

Holt, R. 1972. Some deadlock properties of computer systems. *ACM Comput. Surv.* 4(3):179–196.

Isloor, S. S. and Marsland, T. A. 1980. The deadlock problem: An overview. *IEEE Comput.* 13(9):58–78.

Lamport, L. 1986. The mutual exclusion problem. *J. ACM* 33(2):313–348.

Peterson, G. L. 1981. Myths about the mutual exclusion problem. *Inf. Process. Lett.* 12(3):115–116.

Raynal, M. 1986. *Algorithms for Mutual Exclusion*. John Wiley & Sons, New York.

Reed, D. P. and Kanodia, R. K. 1979. Synchronization with eventcounts and sequencers. *Commun. ACM* 22(2):81–92.

Silberschatz, A., Galvin, P. B., and Gagne, G. 2012. *Operating System Concepts*, 8th edn. John Wiley & Sons, Hoboken, NJ.

Stallings, W. 2012. *Operating Systems: Internals and Design Principles*, 7th edn. Prentice Hall, Upper Saddle River, NJ.

Tanenbaum, A. 1987. *Operating Systems: Design and Implementation*. Prentice Hall, Englewood Cliffs, NJ.

Tanenbaum, A. 2008. *Modern Operating Systems*, 3rd edn. Prentice Hall, Upper Saddle River, NJ.

Xiong, W., Park, S., Zhang, J., Zhou, Y., and Ma, Z. 2010. Ad hoc synchronization considered harmful. In *Proceedings of the Symposium on Operating systems design and implementation*, Vancouver, British Columbia, Canada. USENIX Association.

Zobel, D. and Koch, C. 1988. Resolution techniques and complexity results with deadlocks: A classifying and annotated bibliography. *Operating Syst. Rev.* 22(1):52–72.

53

Thread Management for Shared-Memory Multiprocessors

Thomas E. Anderson
University of Washington

Brian N. Bershad
Google

Edward D. Lazowska
University of Washington

Henry M. Levy
University of Washington

53.1 Introduction

Disciplined concurrent programming can improve the structure and performance of computer programs on both uniprocessor and multiprocessor systems. As a result, support for *threads*, or lightweight processes, has become a common element of new operating systems and programming languages.

A thread is a sequential stream of instruction execution. A thread differs from the more traditional notion of a heavyweight process in that it separates the notion of execution from the other state needed to run a program (e.g., an address space). A single thread executes a portion of a program, while cooperating with other threads that are concurrently executing the same program. Much of what is normally kept on a per-heavyweight-process basis can be maintained in common for all threads in a single program, yielding dramatic reductions in the overhead and complexity of a concurrent program.

Concurrent programming has a long history. The operation of programs that must handle real-world concurrency (e.g., operating systems, database systems, and network file servers) can be complex and difficult to understand. Dijkstra (1968) and Hoare (1974, 1978) showed that these programs can be simplified when structured as cooperating sequential threads that communicate at discrete points within the program. The basic idea is to represent a single task, such as fetching a particular file block, within a single thread of control, and to rely on the thread management system to multiplex concurrent activities onto the available processor. In this way, the programmer can consider each function being performed by the system separately and simply rely on automatic scheduling mechanisms to best assign available processing power.

In the uniprocessor world, the principal motivations for concurrent programming have been improved program structure and performance. Multiprocessors offer an opportunity to use concurrency in parallel programs to improve performance, as well as structure. Moderately increasing a uniprocessor's power can require substantial additional design effort, as well as faster and more expensive hardware components. But once a mechanism for interprocessor communication has been added to a uniprocessor design, the system's peak processing power can be increased by simply adding more processors. A shared-memory multiprocessor is one such design in which processors are connected by a bus to a common memory.

Multiprocessors lose their advantage if this processing power is not effectively utilized. If there are enough independent sequential jobs to keep all of the processors busy, then the potential of a multiprocessor can be easily realized: each job can be placed on a separate processor. However, if there are fewer jobs than processors, or if the goal is to execute single applications more quickly, then the machine's potential can only be achieved if individual programs can be parallelized in a cost-effective manner. Three factors contribute to the cost of using parallelism in a program:

- *Thread overhead*: The work, in terms of processor cycles, required to create and control a thread must be appreciably less than the work performed by that thread on behalf of the program. Otherwise, it is more efficient to do the work sequentially, rather than use a separate thread on another processor.
- *Communication overhead*: Again in terms of processor cycles, the cost of sharing information between threads must be less than the cost of simply computing the information in the context of each thread.
- *Programming overhead*: A less tangible metric than the previous two, programming overhead reflects the amount of human effort required to construct an efficient parallel program.

High overhead in any of these areas makes it hard to build efficient parallel programs. Costly threads can only be used infrequently. Similarly, if arranging communication between threads is slow, then the application must be structured so that little interthread communication is required. Finally, if managing parallelism is tedious or difficult, then the programmer may find it wise to sacrifice some speedup for a simpler implementation. Few algorithms parallelize well when constrained by high thread, communication, and programming costs, although many can flourish when these costs are low.

Low overhead in these three areas is the responsibility of the thread management system, which bridges the gap between the physical processors (the suppliers of parallelism) and an application (its consumer). In this chapter, we discuss the issues that arise in designing a thread management system to support low-overhead parallel programming for shared-memory multiprocessors. In the next section, we describe the functionality found in thread management systems. Section 53.3 discusses a number of thread design issues. In Section 53.4, we survey three systems for shared-memory multiprocessors, Windows NT (Custer 1993), Presto (Bershad et al. 1988), and Multilisp (Halstead 1985), focusing our attention on how they have addressed the issues raised in this chapter.

53.2 Thread Management Concepts

53.2.1 Address Spaces, Threads, and Multiprocessing

An address space is the set of memory locations that can be generated and accessed directly by a program. Address space limitations are enforced in hardware to prevent incorrect or malicious programs in one address space from corrupting data structures in others. Threads provide concurrency within a program, while address spaces provide failure isolation between programs. These are orthogonal concepts, but the interaction between thread management and address space management defines the extent to which data sharing and multiprocessing are supported.

The simplest operating systems, generally those for older style personal computers, support only a single thread and a single-address space per machine. A single-address space is simpler and faster since it allows all data in memory to be accessed uniformly. Separate address spaces are not needed on dedicated systems to protect against malicious users; software errors can crash the system but at least are localized to one user, one machine.

Even single-user systems can have concurrency, however. More sophisticated systems, such as Xerox's Pilot (Redell et al. 1980), provide only one address space per machine, but support multiple threads within that single-address space. Because any thread can access any memory location, Pilot provides a compiler with strong type checking to decrease the likelihood that one thread will corrupt the data structures of another.

Other operating systems, such as Unix, provide support for multiple-address spaces per machine, but only one thread per address space. The combination of a Unix address space with one thread is called a Unix *process*; a process is used to execute a program. Since each process is restricted from accessing data that belong to other processes, many different programs can run at the same time on one machine, with errors confined to the address space in which they occur. Processes are able to cooperate by sending messages back and forth via the operating system. Passing data through the operating system is slow, however; only parallel programs that require infrequent communication can be written using threads in disjoint address spaces.

Instead of using messages to share data, processes running on a shared-memory multiprocessor can communicate directly through the shared memory. Some Unix systems allow memory regions to be set up as shared between processes; any data in the shared region can be accessed by more than one process without having to send a message by way of the operating system. The Sequent Symmetry's DYNIX and Encore's UMAX (Encore 1986) are operating systems that provide support for multiprocessing based on shared memory between Unix processes.

More sophisticated operating systems for shared-memory multiprocessors, such as Microsoft's Windows NT and Carnegie Mellon University's Mach operating system (Tevanian et al. 1987), support multiple-address spaces *and* multiple threads within each address space. Threads in the same address space communicate directly with one another using shared memory; threads communicate across address space boundaries using messages. The cost of creating new threads is significantly less than that of creating whole address spaces, since threads in the same address space can share per-program resources. Figure 53.1 illustrates the various ways in which threads and address spaces can be organized by an operating system.

53.2.2 Basic Thread Functionality

At its most basic level, a thread consists of a program counter (PC), a set of registers, and a stack of procedure activation records containing variables local to each procedure. A thread also needs a control block to hold state information used by the thread management system: a thread can be *running* on a processor, *ready-to-run* but waiting for a processor to become available, *blocked* waiting for some other thread to communicate with it, or *finished*. Threads that are ready-to-run are kept on a *ready-list* until they are picked up by an idle processor for execution. There are four basic thread operations:

- *Spawn*: A thread can create or spawn another thread, providing a procedure and arguments to be run in the context of a new thread. The spawning thread allocates and initializes the new thread's control block and places the thread on the ready-list.
- *Block*: When a thread needs to wait for an event, it may block (saving its PC and registers) and relinquish its processor to run another thread.
- *Unblock*: Eventually, the event for which a blocked thread is waiting occurs. The blocked thread is marked as ready-to-run and placed back on the ready-list.
- *Finish*: When a thread completes (usually by returning from its initial procedure), its control block and stack are deallocated, and its processor becomes available to run another thread.

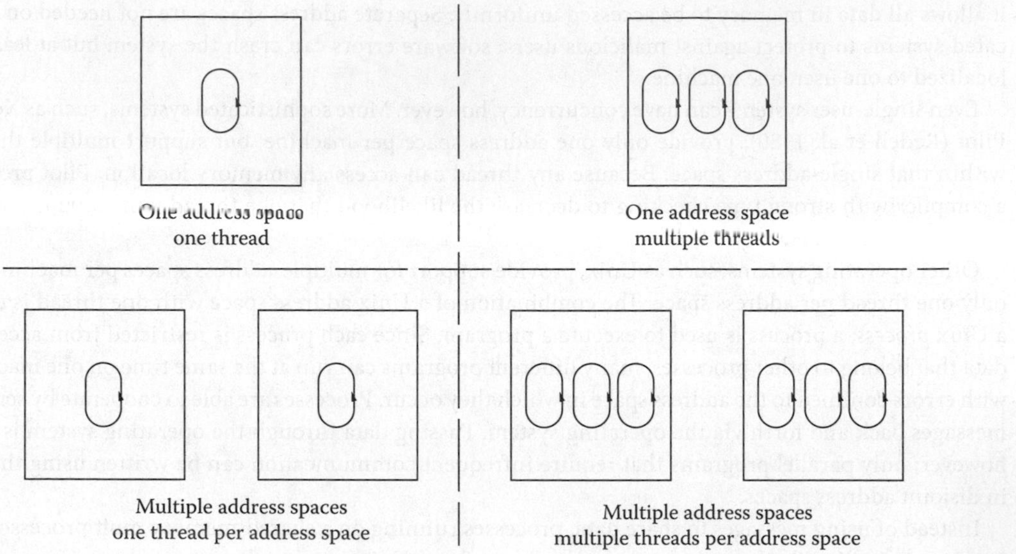

FIGURE 53.1 Threads and address spaces. MS-DOS is an example of a one address space, one thread system. A Java run-time engine is an example of one address space with multiple threads. The Unix operating system is an example of multiple-address spaces, with one thread per address space. Windows NT is an example of a system that has multiple-address spaces and multiple threads per address space.

When threads can communicate with one another through shared memory, *synchronization* is necessary to ensure that threads do not interfere with each other and corrupt common data structures. For example, if two threads each try to add an element to a doubly linked list at the same time, one or the other element may be lost, or the list could be left in an inconsistent state. *Locks* can solve this problem by providing mutually exclusive access to a data structure or region of code. A lock is acquired by a thread before it accesses a shared data structure; if the lock is held by another thread, the requesting thread blocks until the lock is released. (The code that a thread executes while holding a lock is called a *critical section*.) By serializing accesses, the programmer can ensure that threads only see and modify a data structure when it is in a consistent state. Whena program's work is split among multiple threads, one thread may store a result read by another thread. For correctness, the reading thread must block until the result has been written. This data dependency is an example of a more general synchronization object, the *condition variable*, which allows a thread to block until an arbitrary condition has been satisfied. The thread that makes the condition true is responsible for unblocking the waiting thread.

One special form of a condition variable is a *barrier*, which is used to synchronize a set of threads at a specific point in the program. In the case of a barrier, the arbitrary condition is: Have all threads reached the barrier? If not, a thread blocks when it reaches the barrier. When the final thread reaches the barrier, it satisfies the condition and *raises* the barrier, unblocking the other threads.

If a thread needs to compute the result of a procedure in parallel, it can first spawn a thread to execute the procedure. Later, when the result is needed, the thread can perform a *join* to wait for the procedure to finish and return its result. In this case, the condition is: Has a given thread finished? This technique is useful for increasing parallelism, since the synchronization between the caller and the callee takes place when the procedure's result is needed, rather than when the procedure is called.

Locks, barriers, and condition variables can all be built using the basic block and unblock operations. Alternatively, a thread can choose to *spin-wait* by repeatedly polling until an anticipated event occurs, rather than relinquishing the processor to another thread by blocking. Although spin-waiting wastes processor time, it can be an important performance optimization when the expected waiting time is less than the time it takes to block and unblock a thread. For example, spin-waiting is useful for guarding critical sections that contain only a few instructions.

53.3 Issues in Thread Management

This section considers the issues that arise in designing and implementing a thread management system as they relate to the programmer, the operating system, and the performance of parallel programs.

53.3.1 Programmer Issues

53.3.1.1 Programming Models

The flexibility to adapt to different programming models is an important attribute of thread systems. Parallelism can be expressed in many ways, each requiring a different interface to the thread system and making different demands on the performance of the underlying implementation. At the same time, a thread system that strives for generality in handling multiple models is likely to be well suited to none.

One general principle is that the programmer should choose the most restrictive form of synchronization that provides acceptable performance for the problem at hand. For coordinating access to shared data, messages are a more restrictive and, for many kinds of parallel programs, are a more appropriate form of synchronization than locks and condition variables. Threads share information by explicitly sending and receiving messages to one another, as if they were in separate address spaces, except that the thread system uses shared memory to efficiently implement message passing.

There are some cases where explicit control of concurrency may not be necessary for good parallel performance. For instance, some programs can be structured around a single-instruction multiple-data (SIMD) model of parallelism. With SIMD, each processor executes the same instruction in lockstep, but on different data locations. Because there is only one PC, the programmer need not explicitly synchronize the activity of different processors on shared data, thus eliminating a major source of confusion and errors.

Perhaps the simplest programmer interface to the thread system is none at all: the compiler is completely responsible for detecting and exploiting parallelism in the application. The programmer can then write in a sequential language; the compiler will make the transformation into a parallel program. Nevertheless, the compiled program must still use some kind of underlying thread system, even if the programmer does not. Of course, there are many kinds of parallelism that are difficult for a compiler to detect, so automatic transformation has a limited range of use.

53.3.1.2 Language Support

Threads can be integrated into a programming language; they can exist outside the language as a set of subroutines that explicitly manage parallelism; or they can exist both within and outside the language, with the compiler and programmer managing threads together.

Language support for threads is like language support for object-oriented programming or garbage collection: it can be a mixed blessing. On one hand, the compiler can be made responsible for common bookkeeping operations, reducing programming errors. For example, locks can automatically be acquired and released when passing through critical sections. Further, the types of the arguments passed to a spawned procedure can be checked against the expected types for that procedure. This is difficult to do without compiler support.

On the other hand, language support for threads increases the complexity of the compiler, an important factor if a multiprocessor is to support more than one programming language. Further, the concurrency abstractions provided by a single parallel programming language may not do quite what the programmer wants or needs, making it necessary to express solutions in ways that are unnatural or inefficient.

A reasonable way of getting most of the benefits of language support without many of the disadvantages is to define both a language and a procedural interface to the thread management system. Common operations can be handled transparently by the compiler, but the programmer can directly call the basic thread management routines when the standard language support proves insufficient.

53.3.1.3 Granularity of Concurrency

The frequency with which a parallel program invokes thread management operations determines its *granularity*. A *fine-grained* parallel program creates a large number of threads, or uses threads that frequently block and unblock, or both. Thread management cost is the major obstacle to fine-grained parallelism. For a parallel program to be efficient, the ratio of thread management overhead to useful computation must be small. If thread management is expensive, then only *coarse-grained* parallelism can be exploited.

More efficient threads allow programs to be finer grained, which benefits both structure and performance. First, a program can be written to match the structure of the problem at hand, rather than the performance characteristics of the hardware on which the problem is being solved. Just as a single-threaded environment on a uniprocessor can prevent the programmer from composing a program to reflect the problem's logical concurrency, a coarse-grained environment can be similarly restrictive. For example, in a parallel discrete-event simulation, physical objects in the simulated system are most naturally represented by threads that simulate physical interactions by sending messages back and forth to one another; this representation is not feasible if thread operations are too expensive.

Performance is the other advantage of fine-grained parallelism. In general, the greater the length of the ready-list, the more likely it is that a parallel program will be able to keep all of the available processors busy. When a thread blocks, its processor can immediately run another thread provided one is on the ready-list. With few threads though, as in a coarse-grained program, processors idle, while threads do I/O or synchronize with one another.

The performance of a fine-grained parallel program is less sensitive to changes in the number of processors available to an application. For example, consider one phase of a coarse-grained parallel program that does 50 CPU min worth of work. If the program creates five threads on a five-processor machine, the phase finishes in just 10 min. But if the program runs with only four processors, then the execution time of the phase *doubles* to 20 min: 10 min with four processors active followed by 10 min with one processor active. (Preemptive scheduling, which could be used to address this problem, has a number of serious drawbacks, which are discussed subsequently.) If the program had originally been written to use 50 threads, rather than 5, then the phase could have finished in only 13 min, a reasonable degradation in performance.

Of course, one could argue that the programmer erred in writing a program that was dependent on having exactly five processors. The program should have been parameterized by the number of processors available when it starts. But, even so, good performance cannot be ensured if that number can vary, as it can on a multiprogrammed multiprocessor. We consider further the issues of multiprogramming in the next section.

53.3.2 Operating System Issues

53.3.2.1 Multiprogramming

Multiprogramming on a uniprocessor improves system performance by taking advantage of the natural concurrency between computation and I/O. While one program waits for an I/O request, the processor can be running some other program. Because the processor and I/O devices are kept busy simultaneously, more jobs can be completed per unit time than if the system ran only one program at a time.

A multiprogrammed multiprocessor has an analogous advantage. Ideally, periods of low parallelism in one job can be overlapped with periods of high parallelism in another job. Further, multiprogramming allows the power of a multiprocessor to be used by a collection of simultaneously running jobs, none of which by itself has enough parallelism to fully utilize the multiprocessor.

53.3.2.2 Processor Scheduling

Processor scheduling can be characterized by whether physical processors are assigned directly to threads or they are first assigned to jobs and then to threads within those jobs. The first approach, called *one-level* scheduling, makes no distinction between threads in the same job and threads in different jobs. Processors are shared across all runnable threads on the system so that all threads make progress at relatively the same rate. In this case, threads from all jobs are placed on one ready-list that supplies all processors, as shown in Figure 53.2. Although this scheme makes sense for a uniprocessor operating system, it has some unpleasant performance implications on a multiprocessor.

The most serious problem with one-level scheduling occurs when the number of runnable threads exceeds the number of physical processors, because preemptive scheduling is necessary to allocate processor time to threads in a fair manner. With preemption, a processor can be taken away from one thread and given to another at any time. In a sequential program, preemption has a well-defined effect: the program goes from the running state to the nonrunning state as its one thread is preempted. The effect of preemption on the performance of a sequential program is also well defined: if n CPU-intensive jobs are sharing one processor in a preemptive, round-robin fashion, then each job receives $1/n$th of the processor and is slowed down by a factor of n (modulo the preemption and scheduling overhead).

For a parallel program, though, the effects of untimely processor preemption on performance can be more dramatic. In the previous section, we saw how a coarse-grained program can be slowed down by a factor of two when the number of processors is decreased from five to four. That program exemplified a problem that occurs more generally with preemption and barrier-based synchronization. The program had an implicit barrier, which was the final instruction in the phase. Until all threads reached that instruction, the program could not continue. When one processor was removed, it took twice as long to reach the barrier because not all threads within the job could make progress at an equal rate.

Preemptive multiprocessor scheduling also affects program performance when locks are used, but for a different reason than with barriers. Suppose a thread holding a lock while in a critical section is unexpectedly preempted by the operating system. The lock will remain held until the thread is rescheduled. As threads on other processors try to acquire the lock, they will find it held and be forced to block. It is even possible that, as more threads block waiting for the lock to be freed, the number of that job's runnable threads drops to zero and the application can make no progress until

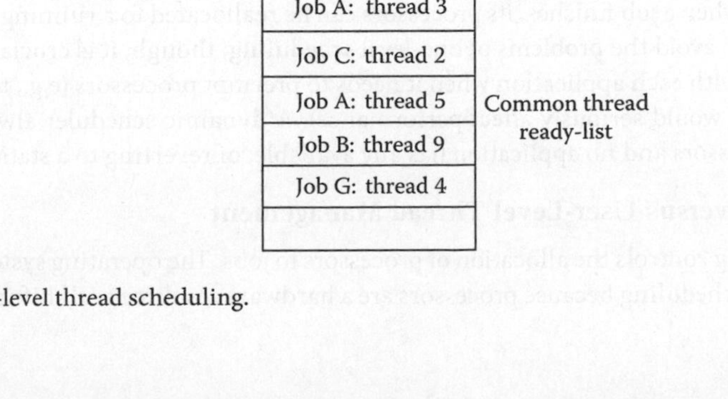

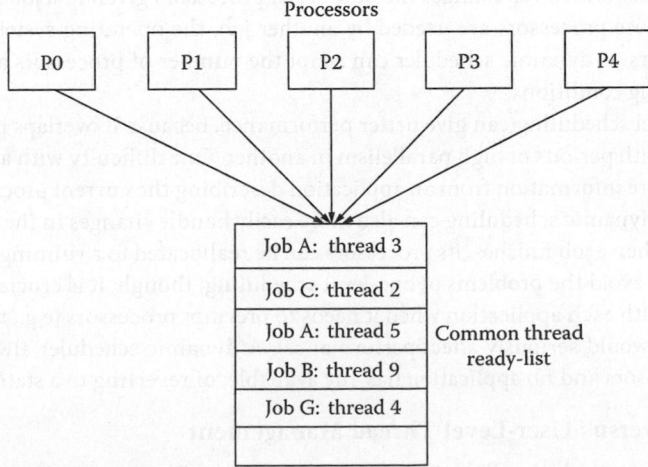

FIGURE 53.2 One-level thread scheduling.

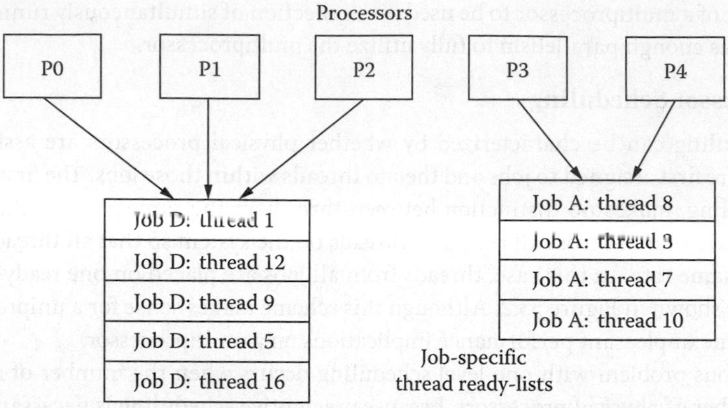

FIGURE 53.3 Two-level thread scheduling.

the preempted thread is rescheduled. The overhead of this unnecessary blocking and unblocking slows down the program's execution.

In the previous section, we saw how fine-grained parallelism can improve a program's performance by increasing the chance that a processor will find another runnable thread when its current thread blocks. Unfortunately, a fine-grained parallel program that packs the ready-list interacts badly with the behavior of a one-level scheduler. In particular, when a program's thread blocks in the kernel on an I/O request, the parallelism of the program can only be maintained if the kernel can schedule another of the program's threads in place of the one that blocked. This benefit, though, comes at the cost of increased preemption activity and diminished overall performance.

The problems of one-level scheduling are addressed by two-level schedulers. With a two-level scheduler, processors are first assigned to a job, and then threads within that job are executed only on the assigned processors. Each job has its own ready-list, which is used only by the job's processors, as shown in Figure 53.3. Thread preemption may no longer be necessary with a two-level scheduler since a preempted thread will only be replaced by another thread from the same job. Further, for long intervals, a processor runs only threads from the same application, and so the cost of switching between threads is kept low.

In a two-level scheduling system, processors can be allocated to jobs either statically or dynamically. A static two-level scheduler never changes the number of processors given to a job from its initial allocation; if some of those processors are needed by another job, the operating system must preempt all of the job's processors. A dynamic scheduler can adapt the number of processors assigned to each job according to changing conditions.

Dynamic two-level scheduling can give better performance, because it overlaps periods of poor parallelism in one job with periods of high parallelism in another. One difficulty with a dynamic scheduler is that it requires more information from an application describing the current processor requirements. As a result, though, dynamic scheduling can also more easily handle changes in the number of running jobs. For example, when a job finishes, its processors can be reallocated to a running job whose parallelism is increasing. To avoid the problems of one-level scheduling, though, it is crucial that the operating system coordinate with each application when it needs to preempt processors (e.g., to avoid preempting a processor when it would seriously affect performance). A dynamic scheduler always has the option, when it needs processors and no application has any available, of reverting to a static policy.

53.3.2.3 Kernel- versus User-Level Thread Management

Processor scheduling controls the allocation of processors to jobs. The operating system must be responsible for processor scheduling because processors are a hardware resource and shifting a processor from one job to another involves updating per-processor address space hardware registers. Spawning a thread so that it runs on an already allocated processor, however, does not require modifying privileged state. Thus, thread management and scheduling within a job can be done entirely by the application instead of by the operating system. In this case, thread management operations can be implemented in an application-level library. The library creates virtual processors using the operating system's processor scheduling interface and schedules the application's threads on top of these virtual processors.

Unlike processor allocation, where a single systemwide scheduling policy can be used, thread scheduling policies benefit from being application specific. Some applications perform well if their threads are scheduled according to some fixed policy, such as first-in–first-out or last-in–first-out, but others need to schedule threads according to fixed or even dynamically changing priorities. For example, consider a parallel simulation where each simulation object is represented by its own thread. Different objects become sequential bottlenecks at different times in the simulation; the amount of parallelism can be increased by preferentially scheduling these objects' threads.

It is difficult to provide sufficient thread scheduling flexibility with kernel-level threads. While the kernel could define an interface that allows each application to select its thread scheduling policy, it is unlikely that the system designer could foresee all possible application needs.

Thread management involves more than scheduling. A trade-off exists between user- and kernel-level thread management. A user-level implementation provides more flexibility and better performance; implementing threads in the kernel guarantees a uniformity that eases the integration of threads with system tools.

The downside of having many custom-built thread management systems is that there is no standard thread. By implication, a kernel-level thread management system defines a single, systemwide thread model that is used by all applications. Operating systems that support only one thread model, like those that support only one programming language, can more easily provide sophisticated utilities, such as debuggers and performance monitors. These utilities must rely on the abstraction and often the implementation of the thread model, and a single model makes it easier to provide complete versions of these tools since their cost can be amortized over a large number of applications. Peripheral support for multiple models is possible, but expensive.

A standard thread model also makes it possible for applications to use libraries or canned software utilities. In the same sense that a standard procedure calling sequence sacrifices speed for the ability to call into separately compiled modules, a standard thread model allows one utility to call into another since they both share the same synchronization and concurrency semantics.

It is important to point out that two-level scheduling does not imply that threads are implemented at the application level; the job-specific ready queues shown in Figure 53.3 could be maintained either within the operating system or within the application. Also, a user-level thread implementation does not imply two-level scheduling, even though threads *are* being scheduled by the application. This implication only holds in the absence of multiprogramming or in cases where processors are explicitly allocated to jobs. For example, a user-level thread implementation built on top of Unix processes that share memory suffers from the same problems relating to preemption and I/O as do one-level kernel threads because both are scheduled in a job-independent fashion.

53.3.3 Performance

The performance of thread operations determines the granularity of parallelism that an application can effectively use. If thread operations are expensive, then applications that have inherently fine-grained parallelism must be restructured (if that is even possible) to reduce the frequency of those operations. As the cost of thread operations begins to approach that of a few procedure calls, several issues become performance critical that, for slower operations, would merely be second-order effects.

Simplicity in the thread system's implementation is crucial to performance (Anderson et al. 1989). There is a performance advantage to building multiple thread systems, each tuned for a single type of application.

Even simple features that are needed by only some applications, such as saving and restoring all floating point registers on a context switch, will markedly affect the performance of applications that do not need the functionality. Each context switch takes only tens of instructions; a feature that adds even a few more instructions must have a large compensating advantage to be worthwhile. For example, the ability to preemptively schedule threads within each job makes the thread management system more sluggish at several levels, because preemption must be disabled (and then re-enabled) whenever scheduling decisions are being made. These scheduling decisions are on the critical path of all thread management operations.

Although kernel-level thread management simplifies the generation and maintenance of system tools, it increases the baseline cost of all thread management operations. Just trapping to the operating system can cost as much as the thread operation itself, making a kernel implementation unattractive for high-performance applications. Further, the generality that must be provided by a kernel-level thread scheduler hurts the performance of those applications needing only basic service. Kernel-level threads are less able to cut corners by exploiting application-specific knowledge. With a user-level thread system, the thread management system can be stripped down to provide exactly the functions needed by an application and no more. User-level thread operations also avoid the cost of trapping to the kernel.

Other performance issues have less to do with what a thread system does than with how it goes about doing it. For example, using a centralized ready-list can limit performance for applications that have extremely fine-grained parallelism. The ready-list is a shared data structure that must be locked to prevent it from being modified by multiple processors simultaneously. Even if the ready-list critical sections consist only of simple enqueue and dequeue operations, they can become a sequential bottleneck, since there is little other work involved in spawning/finishing or blocking/unblocking a thread. For an application with thread overhead of 20% of the total execution time, and half of that overhead is spent accessing the ready-list, then its maximum speedup (the time of the parallel program on P processors divided by the time of the program on one processor) is limited to 10.

The bottleneck at the ready-list can be relieved by giving each processor its own ready-list. In this way, enqueueing and dequeueing of work can occur in parallel, with each processor using a different data structure. When a processor becomes idle, it checks its own list for work, and if that list is empty, it scans other processors' lists so that the workload remains balanced.

Per-processor ready-lists have another nice attribute: threads can be preferentially scheduled on the processor on which they last ran, thereby preserving cache state. Computer systems use caches to take advantage of the principle of *locality*, which says that thread's memory references are directed to or near locations that have been recently referenced. By keeping references close to the processor in fast cache memory, the average time to access a memory location can be kept low. On a multiprocessor, a thread that has been rescheduled on a different processor will initially find fewer of its references in that processor's cache. For some applications, the cost of fetching these references can exceed the processing time of the thread operation that caused the thread to migrate.

The role of spin-waiting as an optimization technique changes in the presence of high-performance thread operations. If a thread needs to wait for an event, it can block, relinquishing its processor, or spin-wait. A thread must spin-wait for low-level scheduler locks, but in application code, a thread should block instead of spin if the event is likely to take longer than the cost of the context switch. Even though context switches can be implemented efficiently, reducing the need to spin-wait, a hidden cost is that context switches also reduce cache locality.

53.4 Three Modern Thread Systems

We now outline three modern thread management systems for multiprocessors: Windows NT, Presto, and Multilisp. The choices made in each system illustrate many of the thread management issues raised in the previous section.

TABLE 53.1 Basic Operations of Thread Management Systems

Basic	Windows NT	Presto	Multilisp
Spawn	Thread_create; thread_resume	Thread::new; Thread::start	(Future…)
Block	Thread_suspend	Thread::sleep	Touch unresolved future
Unblock	Thread_resume	Thread::wakeup	When future is resolved
Finish	Thread_terminate	Thread::terminate	Resolve this future

The thread management primitives for each of these systems are shown in Table 53.1. The table is organized to indicate how the primitives in one system relate to those in the others, as well as those provided by the basic thread interface outlined in the Basic Thread Functionality section.

Windows NT is an operating system designed to support Microsoft Windows applications on uniprocessors, shared-memory multiprocessors, and distributed systems. Windows NT supports multiple threads within an address space. Its thread management functions are implemented in the Windows NT kernel. Since NT's underlying thread implementation is shared by all parallel programs, system services such as debuggers and performance monitors can be economically provided.

Windows NT's scheduler uses a priority-based one-level scheduling discipline. Because Windows NT allocates processors to threads in a job-independent fashion, a parallel program running on top of the Windows NT thread primitives (or even a user-level thread management system based on those primitives) can suffer from anomalous performance profiles due to ill-timed preemptive decisions made by the one-level scheduling system.

Presto is a user-level thread management system originally implemented on top of Sequent's DYNIX operating system, but later ported to DEC workstations. DYNIX provides a Presto program with a fixed number of Unix processes that share memory. The Presto run-time system treats these processes as virtual processors and schedules the user's threads among them. Presto's thread interface is nearly identical to Windows NT's.

Presto is distinguished from most other thread systems in that it is structured for flexibility. Presto is easy to adapt to application-specific needs because it presents a uniform object-oriented interface to threads, synchronization, and scheduling. The object-oriented design of Presto encourages multiple implementations of the thread management functions and so offers the flexibility to efficiently accommodate differing parallel programming needs.

Presto has been tuned to perform well on a multiprocessor; it tries to avoid bottlenecks in the thread management functions through the use of per-processor data structures. Presto does not provide true two-level scheduling, even though the thread management functions (e.g., thread scheduling) are implemented in an application library accessible to the user; DYNIX, the base operating system, schedules the underlying virtual processors (Unix processes) any way that it chooses. Although a Presto program can request that its virtual processors not be preempted, the operating system offers no solid guarantee. As a result, kernel preemption threatens the performance of Presto programs in the same was as it does Windows NT programs.

Although Windows NT and Presto are implemented differently, the interfaces to each represent a similar style of parallel programming in which the programmer is responsible for explicitly spawning new threads of execution *and* for synchronizing their access to shared data. This style is not accidental, but reflects the basic function of the underlying hardware: processors communicating through shared memory. One criticism often made of this style is that it forces the programmer to think about coordinating many concurrent activities, which can be a conceptually difficult task.

Multilisp demonstrates how thread support can be integrated into a programming language in order to simplify writing parallel programs. In Multilisp, a multiprocessor extension to LISP, the basic concurrency mechanism is the *future*, which is a reference to a data value that has not yet been computed. The *future* operator can be included in any Multilisp expression to spawn a new thread that computes the value of the expression in parallel. Once the value has been computed, the future *resolves* to that value.

In the meantime, any thread that tries to use the future's value in an expression automatically blocks until the future is resolved. The language support provided by Multilisp can be implemented on top of a system like Windows NT or Presto using locks and condition variables.

With Multilisp, the programmer does not need to include any synchronization code beyond the future operator; the Multilisp interpreter keeps track of which futures remain unresolved. By contrast, using the Windows NT or Presto thread primitives, the programmer must add calls to the appropriate synchronization primitives wherever the data are needed. Multilisp, like Presto, uses per-processor ready-lists to reduce contention in scheduling operations.

53.5 Summary

This chapter has examined some of the key issues in thread management for shared-memory multiprocessors.

Shared-memory multiprocessors are now commonplace in both commercial and research computing. These systems can easily be used to increase throughput for multiprogrammed sequential jobs. However, their greatest potential—as yet not fully realized—is for accelerating the execution of single, parallelized programs.

As programmers make use of finer-grained parallelism, the design and implementation of the thread management system becomes increasingly crucial. Modern thread management systems must address the programmer interface, the operating system interface, and performance optimizations; language support and scheduling techniques for multiprogrammed multiprocessors are two areas that require further research.

References

Anderson, T. E., Lazowska, E. D., and Levy, H. M. 1989. The performance implications of thread management alternatives for shared memory multiprocessors. In *International Conference on Measurement and Modeling of Computer System ACM SIGMETRICS Performance '89*, Berkeley, CA, pp. 49–60.

Bershad, B., Lazowska, E., and Levy, H. 1988. PRESTO: A system for object-oriented parallel programming. *Software Prac. Exp.* 18(8): 713–732.

Custer, H. 1993. *Inside Windows NT*. Microsoft Press, Redmond, WA.

Dijkstra, E. W. 1968. Cooperating sequential processes. In *Programming Languages*, pp. 43–112. Academic Press, New York.

Encore. 1986. UMAX 4.2 Programmer's Reference Manual. Encore Computer Corporation, Marlborough, MA.

Halstead, R. 1985. Multilisp: A language for concurrent symbolic computation. *ACM Trans. Program. Lang. Syst.* 7(4): 501–538.

Hoare, C. A. R. 1974. Monitors: An operating system structuring concept. *Commun. ACM* 17(10): 549–557.

Hoare, C. A. R. 1978. Communicating sequential processes. *Commun. ACM* 21(8): 666–677.

Redell, D. D., Dalal, Y. K., Horsley, T. R., Lauer, H. C., Lynch, W. C., McJones, P. R., Murray, H. G., and Purcell, S. C. 1980. Pilot: An operating system for a personal computer. *Commun. ACM* 23(2): 81–92.

Tevanian, A., Rashid, R. F., Golub, D. B., Black, D. L., Cooper, E., and Young, M. W. 1987. Mach threads and the Unix kernel: The battle for control. In *Proceedings of the USENIX Summer Conference*, Phoenix, AR, pp. 185–197.

54

Virtual Memory

Peter J. Denning
Naval Postgraduate School in Monterey

54.1 Introduction

Virtual memory is one of the engineering triumphs of the computer age. In the 1960s, it became a standard feature of nearly every operating system and computer chip. In the 1990s, it became also a standard feature of the Internet and the World Wide Web (WWW). It was the subject of intense controversy after its introduction in 1959; today, it is such an ordinary part of infrastructure that few people think much about it.

Early programmers had to solve a memory *overlay problem* as part of their programming work. Early computers had very small amounts of random access memory (RAM) because storage technology was so expensive. Programmers stored a master copy of their programs (and data) on a secondary storage system—then a drum, today a disk—and pulled pieces into the RAM as needed. Deciding which pieces to pull and which parts of RAM to replace was called "overlaying." It was estimated that most programmers spent half to two-thirds of their time planning overlay sequences. As a reliable method of automating, virtual memory had the potential to increase programmer productivity and reduce debugging by severalfold. Virtual memory was invented to automate solutions to the overlay problem.

Virtual memory was taken into the large commercial operating systems of the 1960s to simplify the new features of time-sharing and multiprogramming. The RAMs were considerably larger, sometimes allowing individual programs to be fully loaded. Virtual memory provided three additional benefits in these systems: it isolated users from each other, it allowed dynamic relocation of program pieces within RAM, and it provided read–write access control to individual pieces. These benefits were so for fault tolerance (Denning 1976) that virtual memory was used even when there was sufficient RAM for every program. Thus, the story of virtual memory is not simply a story of progress in automatic storage allocation; it is a story of machines helping programmers to protect information, reuse and share objects, and link software components.

The first operating system with virtual memory was the 1959 Manchester Atlas operating system (Fotheringham 1961, Kilburn et al. 1962). They called their virtual memory a "one-level store." They simulated a large memory—sufficient to hold an entire program—by moving fixed-size pages of a large file on disk in and out of a small RAM. The system appeared as a large, slower RAM to its users. The term "virtual memory" soon superseded the original name because of an analogy to optics: a virtual image in a mirror or lens is an illusion, made from traces of a real object that is actually somewhere else.

54.2 Early Virtual Memory Systems

Electronic digital computers have always had memory hierarchies consisting of at least two levels (Figure 54.1):

1. The main memory, often called RAM, is directly connected to the central processing unit (CPU). RAM is fast and can keep up with the CPU. RAM is volatile; loss of power erases its contents. RAM is expensive.
2. The secondary memory is connected to RAM by a subsystem that implements "up" and "down" moves of blocks of data. It was originally rotating drums, and today it is usually rotating or solid state disks. It is persistent, meaning that data are written as magnetic or optical patterns that do not disappear until explicitly erased. It is much cheaper than RAM.

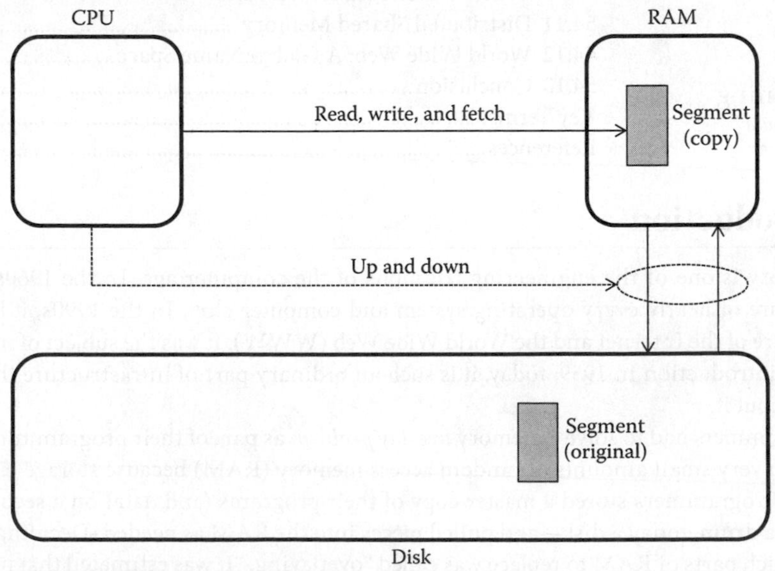

FIGURE 54.1 Two-level memory hierarchy. A processor executes a program from main memory. The entire program is held in the secondary memory (DISK), and segments of it are copied to the main memory (RAM) for processing. The overlay problem is to partition the program into segments and schedule their moves up and down the hierarchy. The manual work to construct a program's overlay sequence doubles or triples the programming time from when no overlays are needed.

Today's disk technology is about a thousand (10^3) times cheaper per bit than RAM and about a million (10^6) times slower. These differentials make it a challenge to find a good trade-off between main and secondary storage. The goal is enough RAM to run programs at close to CPU speed and sufficient permanent storage for all data.

The secondary storage system can include other technologies such as solid state disks, CD/DVD recorders and players, flash memory sticks, tape backups, and remote Internet storage (such as dropbox.com). These non-disk secondary devices are not usually part of the virtual memory system.

Data on RAM or disk are organized into blocks of contiguous bytes. The blocks are units of storage and transfer between the memory levels. Many systems use all blocks of the same fixed size, called *pages*. A few systems allow blocks of different sizes, in which case they are called *segments*.

54.2.1 Address Mapping

The original purpose of virtual memory was to solve the burdensome overlay problem by automating the movement of pages between memory levels. The 1959 Atlas computer system at the University of Manchester was the first working prototype (Fotheringham 1961, Kilburn et al. 1962). The heart of this system was a radical innovation: a distinction between *address* and memory *location*. An address was the name of a byte, and a location was the physical storage holding the byte. This distinction enabled designers to provide programmers a large address space (AS) even though the system had a smaller amount of RAM. Programmers could then design software as if all the code and data fit into a single, contiguous AS.

This distinction led the Atlas designers to three inventions: (1) They built an *address translator*, hardware that automatically converted each address generated by the CPU to its current memory location. (2) They devised *demand paging*, an interrupt mechanism triggered by the address translator that moved a missing page from secondary memory to RAM. (3) They built the first *replacement algorithm*, a procedure to detect and move the least useful pages back to secondary memory.

Despite the success of the Atlas memory system, the literature of 1961 records a spirited debate about the feasibility of automatic storage allocation in general-purpose computers. By that time, Cobol, Algol, Fortran, and Lisp had become the first widely used higher level programming languages. These languages made storage allocation harder because programs were larger, more portable, more modular, and their dynamics more dependent on their input data. Through the 1960s, there were dozens of experimental studies that sought to either affirm or deny that virtual memory systems could do a better job at storage allocation than any compiler or programmer (Denning 1970). The matter was finally laid to rest—in favor of virtual memory—by an extensive study of system performance by an IBM research team led by David Sayre (Sayre 1969).

54.2.2 Multiprogramming

Convinced that virtual memory was the right way to go, the makers of major commercial computers adopted it in the 1960s. Virtual memory was in the IBM 360/67, CDC 7600, Burroughs 5500/6500, RCA Spectra/70, and Multics for the GE 645. By the mid-1970s, the IBM 370, DEC VMS, DEC TENEX, and Unix also had virtual memory.

All these systems used multiprogramming, a mode in which several programs simultaneously reside in RAM. Multiprogramming was intended for better CPU utilization and throughput: if a program stops for input–output, the CPU can switch to another program.

Virtual memory married nicely with multiprogramming. It gave a clean way to logically partition the RAM among multiple programs, preventing them from interfering with one another. This worked because the address translator in the CPU used a distinct mapping table for each program; the mapping table pointed only to the pages of that program's AS. The address translator also recognized access codes, thus protecting read-only pages from being overwritten.

54.2.3 Thrashing

But these design choices exacted an unexpected price. The first multiprogrammed virtual memory systems succumbed to an unexpected and most unwelcome surprise: *thrashing*. Thrashing was a condition of near-total performance collapse when the multiprogramming level became too high (Denning 1968). The performance collapse worked as follows. The system would be operating well, with good throughput. Occasionally, the activation of one additional program pushed the system over an edge into a state with extremely low throughput. Engineers soon determined that in this state, every job was waiting in the disk queue for a page to be moved; and when its page was moved, every job very quickly rejoined the queue with a new request. The engineers called this state "paging to death." They did not understand the conditions that triggered it or how to design a load control that would prevent it.

The solution to thrashing lay with a new model of program behavior, called the *working set (WS) model* (Denning 1968). This model hypothesized that every program had a WS, a dynamic subset of pages that it needed in RAM for efficient execution; and that most of the time, the WS did not change much from one page reference to the next. Thrashing would be impossible as long as every program has its WS loaded in RAM. The operating system would not activate any new program whose WS could not fit into the unused part of RAM.

A series of early experiments confirmed the WS hypothesis and gave engineers insights on how to retrofit their multiprogrammed virtual memory systems to avoid thrashing. It also stimulated a long line of experiments and models seeking to understand why computations would exhibit locality, the principle behind the working-set behavior. By the late 1970s, these investigations had produced deep understanding of the principle of locality and had confirmed that working-set memory controllers would achieve throughput close to the theoretical optimum (Denning 1980).

New studies by Adrian McMenamin in 2011 for Linux reconfirmed the locality behavior of programs in modern systems and the efficacy of WS memory management (McMenamin 2011).

54.3 Cache Systems

The address mapping principle of virtual memory attracted hardware designers as well as software designers. In 1965, Maurice Wilkes proposed the *slave memory*, a small high-speed store included in the CPU to hold a small number of most recently used words from RAM (Wilkes 1965). Like virtual memory, slave memory used address translation, demand loading, and usage-based replacement. Wilkes argued that designing translation, loading, and replacement strategies is easier when the memory hierarchy is two forms of RAM: moving small blocks is efficient because of smaller speed differences. By exploiting the principle of locality, a small slave memory eliminates many data transfers between CPU and main memory, allowing the CPU to run much faster with hardly any increase of memory cost.

This idea became popular among hardware designers. IBM introduced a cache memory in 1968 as part of its 360/85 machine. IBM and others placed small caches in the CPU, enabling the CPU to run faster because it could bypass many accesses to RAM. Modern chips incorporate two levels of cache—L1 operates at register speed and simulates a large set of registers, while L2 operates at a slower speed and bypasses many references to RAM. Cache memory is now a standard principle of computer architecture (Hennessey and Patterson 1990).

The term *caching* is often used to mean that a small set of data are moved to a location close to the CPU, enabling the CPU to run fast on that set, bypassing delays in accessing the data at a larger distance away. Data and instructions cached in the CPU enable faster CPU execution. Files cached in disk controllers enable fast re-access to recently used files. Web pages cached on local servers enable faster re-display and bypass the longer access times in the Internet.

54.4 Object Systems

In the 1960s, virtual memory was seen as a method to make programming easier and more secure:

1. Programmers can write their code without having to worry about overlays, which are done automatically by the system.
2. Programmers can share memory without fear of interference from other programs.

If it ended here, this story would already have guaranteed virtual memory a place in history. But the designers of the 1960s saw even more possibilities to adapt virtual memory to make programming more productive and secure. They added two new programming objectives:

3. Programmers can share, reuse, or recompile any program module without requiring changes to any other program module.
4. Programmers can define abstract data types by specifying a package consisting of a hidden internal data structure to represent an object's state and procedures to access an object's state.

The first new objective was met by a *segmented AS*, an extension of the original linear AS. The second new objective was met by *capability-based architecture*, which later morphed into *object-oriented programming*. Both extended the basic virtual memory in important ways.

In 1965, the designers of Multics at MIT believed that programmer communities develop around interactive, time-shared computing systems, and that members of these communities want to share separately compiled program modules that can be linked together on demand (Dennis 1965, Organick 1972). Link-on-demand was a significant departure from the more common practice of using a linking loader (or makefile program) to bind component files to together into an AS.

In a computer with 32 bit addresses, an AS can contain up to 2^{32} addresses. Multics added a second dimension that selected one of 2^{32} ASs, called segments. Thus, a Multics CPU address consisted of a pair of 32 bit addresses (s, x) denoting segment s, offset x.

In Multics, a program encountering a symbolic reference to a variable X within a segment S would be interrupted by a *linkage fault*. The linkage fault handler would convert the S to a segment number s and the X to a linear offset x. After the conversion, the program would not encounter the same linkage fault again.

The Multics virtual memory demonstrated innovations in sharing, reuse, access control, and protection. Many of its innovations, however, did not find their way beyond Multics; programmers were content with one private, linear AS and a handful of open files. As will be discussed, the WWW (Berners-Lee 2000) changed this: programs and documents contain *hypertext links*: symbolic pointers to other objects that are not linked until the program references them for the first time.

In 1966, Jack Dennis and Earl Van Horn published a landmark paper that initiated a new line of computer architectures: machines that helped programmers create protected managers of classes of objects. They anticipated what is now called object-oriented programming. They were especially concerned that objects be freely reusable and shareable and, at the same time, be protected from unauthorized internal access. They proposed an extension of virtual memory that used two levels of mapping instead of one to get to an object. The first level maps an object name to a persistent global identifier, called a *capability*, for the object. The second level maps the capability to the object's location. Capabilities acted as tickets granting access to their objects. You could share an object by giving another person a copy of your capability for the object. Capability addressing offered an elegant solution to the problem of sharing and reusing modules.

Several commercial and academic capability systems were built in the 1970s: notably the Plessey 250, IBM System 38, Cambridge CAP, Intel 432, SWARD, and Hydra. These systems implemented capabilities as long addresses (for example, 64 bits), which the hardware protected from alteration (Fabry 1974, Myers 1982, Wilkes and Needham 1979). The reduced instruction set computer, coupled with programming languages with type checking, made capability-managing hardware obsolete by the mid-1980s.

But software-managed capabilities, now called *handles*, are indispensable in modern object-oriented programming systems, databases, and distributed operating systems (Chase et al. 1994). The same conceptual structure reappeared in a proposal to manage objects and intellectual property in the Internet (Kahn and Wilensky 1995). It is a powerful structure indeed.

54.5 Virtual Memory in Other Systems

Some people wonder why virtual memory, which was so popular and ultimately successful in the operating systems of the 1960s and 1970s, was not present in initial versions of several new technologies of the 1980s:

- Personal computers (PCs)
- Highly parallel supercomputers
- Distributed memory computers
- WWW

These technologies emerged from new communities that initially had no contact with the earlier generation of operating systems or rebelled against that generation. PC developers, for example, distanced themselves from mainframe companies because they wanted computing affordable to an individual home user. They did not have the memory capacity or knowhow to implement a full operating system and often believed they could do much better with extreme simplicity. After a few years, however, they rediscovered the same programming issues that motivated the early operating system designers. They started to use locality and virtual memory to improve their productivity and build faster machines.

For PCs, the pundits of the microcomputer revolution proclaimed that PCs would not succumb to the diseases of the large commercial operating systems; the PC would be simple, fast, and cheap. Bill Gates once said that no user of a PC would ever need more than 640 K of main memory. His first Microsoft operating system, DOS (1982), did not include most of the common functions, such as multiprogramming and virtual memory. Eventually, the PC makers (Apple, Microsoft, and IBM) added multiprogramming and virtual memory to their operating systems. They were able to do this because the major chip makers had not lost faith; Intel offered virtual memory and cache in its 80386 chip in 1985; Motorola did likewise in its 68020 chip. Apple offered multiprogramming in its MultiFinder and virtual memory in its System 6 operating system. Microsoft offered multiprogramming in Windows 3.1 and virtual memory in Windows 95. IBM offered multiprogramming and virtual memory in OS/2.

A similar pattern appeared in the early development of distributed-memory multicomputers beginning in the mid-1980s. These machines allowed for a large number of computers, sharing a high-speed interconnection network, to work concurrently on a single problem. Around 1985, Intel and N-Cube introduced the first hypercube machines consisting of 128 component microcomputers. Shortly thereafter, Thinking Machines produced the first commercial supercomputer of this genre, the Connection Machine, with as many as 65,536 component computer chips. These machines soon challenged the traditional supercomputer by offering the same aggregate processing speed at a much lower cost (Denning and Tichy 1990). Their designers initially eschewed virtual memory, believing that address translation and page swapping would seriously detract from the machine's performance. But they quickly encountered new programming problems in synchronizing processes on different chips and exchanging data among them. The ordinary method of passing a message to another chip involved three copy operations: first from the sender's local memory to a local buffer, then across the network to a buffer in the receiver, and finally to the receiver's local memory. With virtual memory, the same transfer takes one copy, invoked at the time of reference. Virtual memory significantly decreased communication costs in these machines. Tannenbaum (1995) describes a variety of implementation issues under the topic of distributed shared memory.

The WWW, started in 1989 by Tim Berners-Lee, sought to link together all information in the world. It accomplished this by creating a virtual AS for objects in the WWW. The name of an object is given by its *uniform resource locator* (URL). The Web protocols (such as hypertext transfer protocol [HTTP] and domain name service [DNS]) map URLs to hosts and files in the Internet. The mapping operations are performed only when someone tries to use a URL by clicking a mouse on it. The URL differs from a standard virtual address map because it is possible that a URL points to a nonexistent object or possibly to a new object given the same name as a previously deleted object. To avoid the problem of URLs becoming invalid when the object's owner moves it to a new machine, Kahn and Wilensky proposed that objects be named by global persistent handles; handles are translated with a two-level mapping scheme first into a URL, then into the machine hosting the object (Denning and Kahn 2010, Kahn and Wilensky 1995). The handle scheme recalls the Dennis-Van-Horn capability system of the 1960s, but now with worldwide, decentralized mapping systems. The Java language allowed URLs to address programs as well as documents; when a Java interpreter encounters the URL of another Java program, it brings a copy of that program to the local machine and executes it as an applet. These technologies, now seen as essential for the Internet, vindicate the view of the Multics designers in 1965—that many large-scale computations will consist of many processes roaming a large space of shared objects.

In the Internet, some objects such as Web pages or Web sites become very popular and attract many links. People using the links can create a high demand for a site, causing a queue of backlogged requests and thus a significant delay to a user wanting a fast access to the site. Even though the Internet architecture is relatively flat—the access time ("ping time") to most sites is typically 20–30 ms—queueing at popular sites can cause very slow response times much longer than 30 ms. Caching solves this problem. A cache server can be placed in a cluster of users where it collects their recent web page requests; it can satisfy many repeat requests locally without going to the main site and its long queue. The company Akamai has become a leader in this performance-enhancing technology for the Internet.

From time to time over the past 50 years, various people have argued that virtual memory is not really necessary because advancing memory technology would soon permit us to have all the RAM we could possibly want. Each new generation of users has discovered that its ambitions for processing, memory, and sharing led it to virtual memory. It is unlikely that today's predictions of the passing of virtual memory will prove to be any more reliable than similar predictions made every year since 1960. Virtual memory accommodates essential patterns in the way people use computers to communicate and share information. It will still be used when we are all gone.

54.6 Structure of Virtual Memory

Let us now examine the structure of virtual memory systems. We will begin with the simplest form, a paging system, which was historically the first form.

Virtual memory was originally designed to solve the overlay problem in systems with two levels of memory. In 1959, the time of the first virtual memory, main memory (RAM) access times were about 10^{-6} s (1 μs) and secondary memory (drum) access times were about 10^{-2} s (10 ms), giving a speed ratio of about 10^4. A single-page fault (up-move) would force the CPU be idle for about 10^4 instructions, a stiff penalty. The designers, therefore, sought replacement policies that would minimize the number of page faults. The situation is worse today, with RAM access times around 10^{-9} s and disk around 10^{-3}, for a speed ratio of 10^6. These speed ratios do not make virtual memory very attractive for those who believe that its primary purpose is to move pages up and down the hierarchy.

All virtual memories are based on the principle of distinguishing addresses from locations, and providing a dynamic map that translates an address to its storage location. The map can be updated whenever the location changes.

Every program and its data must fit inside a virtual AS, which is a sequential set of addresses that the CPU can generate. If addresses are k bits long, the AS consists of 2^k bytes, designated 0, 1,…, 2^k–1. Thus, 16 bit addresses can span a space of 65,536 bytes. Today, 32 bit addresses are common, spanning spaces of about 4 gigabytes.

It would be too costly to build a mapping table that mapped individual bytes to their locations. With 32 bit addresses, such a table would require 4 gigabytes of memory. A system with 64 running user programs would require around a terabyte of memory just for the 64 mapping tables. The size of mapping tables is significantly reduced by dividing AS into equal size blocks, called *pages*, and the RAM into same-size blocks, called *frames*. The table maps pages to frames. If the page size were 1024 bytes (10^{10}), the mapping table would be smaller by a factor of 1024. Such tables are feasible.

The mapping is organized as follows. A *page table* contains entries of the form, one for each page:

$$(P, U, M, A, B)$$

where

- *P* is a *presence bit* set to 1 when the page is in RAM
- *U* is a *use bit* set to 1 when the page is read or written
- *M* is a *modified bit* set to 1 when the page is written
- *A* is an *access code* indicating read or write permission
- *B* is the *base address* of the page in RAM

A virtual address of the CPU is encoded as

$$(i,x)$$

meaning page *i*, line *x*. This address is translated to a memory address by adding the displacement *x* to the page's base address *B*.

With the help of Figure 54.2, let us walk through the components of a paged virtual memory. The circled numbers in the figure flag four key elements. Every virtual AS has its own page table; by

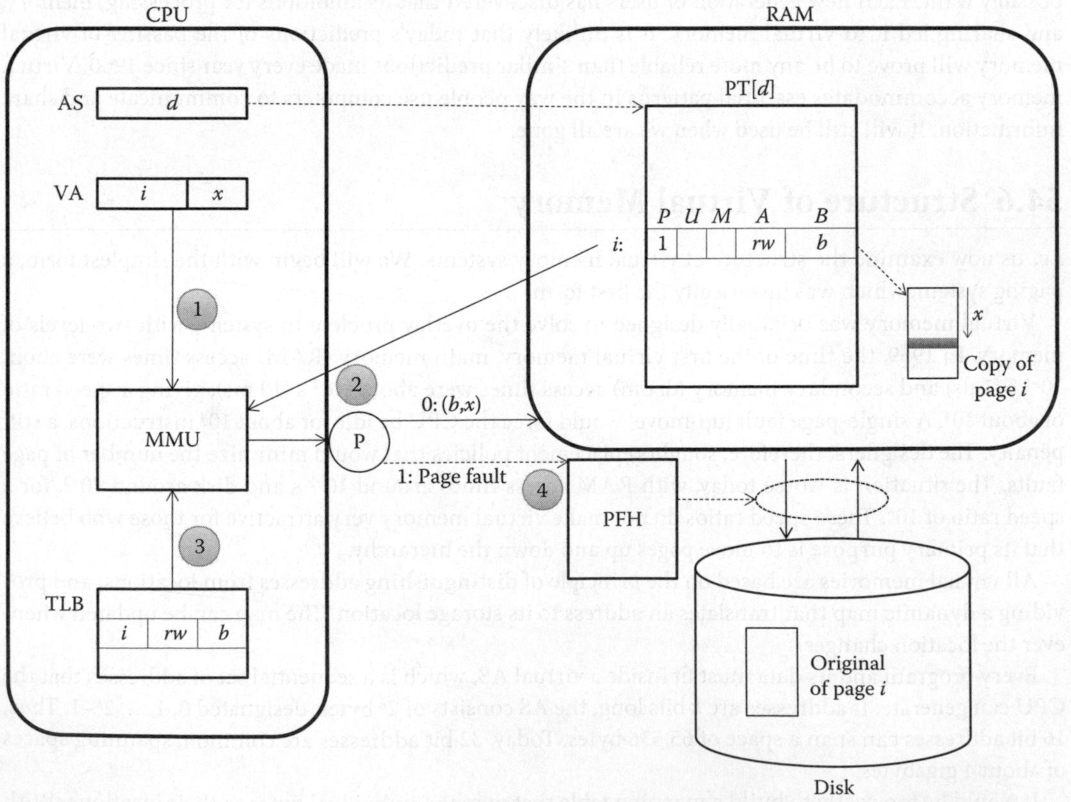

FIGURE 54.2 Structure of paged virtual memory system. The four numbered elements are discussed in the text.

default there are no shared pages. The CPU register AS points to page table the CPU uses for address mapping on behalf of the process it is currently running.

54.6.1 Element 1: Providing the Virtual Address to the Memory Mapping Unit

The memory mapping unit (MMU) is a hardware component that translates virtual addresses to memory locations. A virtual address is a linear offset into the AS. The high order bits of the virtual address in register VA are interpreted as a page number and the low order bits as a line number. An example will show how this works. Assume a page size of $2^{10} = 1024$ bytes in a system with 32-bit addresses. In this case, the virtual AS will contain $2^{22} = 4$ million pages. In a 32-bit address, the 22 high-order bits are page number and 10 low order bits are the line number in the page. Thus, address 8044 will be on page 7, line 876, because $8044 = (7)(2^{10}) + 876$.

54.6.2 Element 2: Mapping the Address

The MMU's job is to map the incoming virtual address to a memory location. The MMU looks up the frame for the current page in the page table. In our example, the page table has one entry per page for a total of 2^{22} entries. If page 7 is in frame 50, the base field *B* of the page table entry for page 7 will contain 50, which we can write as $PT[d,7].B = 50$. The MMU decomposes the target address 8044 into the two components (7,876), retrieves 50 from page table entry 7, and then presents the location address (50,876) to the RAM.

The *P*-bit of a page table entry records whether the page is present in RAM. If page 7 is not present, its *P*-bit will be 0. In that case, the MMU cannot map the virtual address. Instead, it halts with an error condition called *page fault.* The page fault condition triggers an interrupt, which passes control to an operating system routine called *page fault handler* (PFH). The operation of PFH will be discussed shortly at step 4.

54.6.3 Element 3: Checking the Translation Lookaside Buffer

The translation lookaside buffer (TLB) is a small set of high-speed associative registers attached to the MMU. Its purpose is to enable the MMU to bypass the page table lookup as often as possible. Without it, the virtual memory would run at half the rated RAM speed because every virtual address requires two RAM accesses.

A typical register in TLB contains the three components

$$(i,A,B)$$

where

- *i* is the page number (the tag in the TLB)
- *A* is the access code from the page table
- *B* is the frame number from the page table

Before accessing the page table, the MMU checks the TLB for an entry tagged as page *i*. If it finds such an entry, it immediately retrieves the access code *A* and the frame number *B* without having to look them up in the page table. If it does not find an entry for *i*, it proceeds with the normal page table lookup as described earlier and also creates a new TLB entry reflecting that access. The new entry replaces the least recently used (LRU) TLB entry.

Experience shows that with a relatively small TLB—from 32 to 128 registers—the MMU can achieve a sufficiently high hit rate in the TLB that the average slowdown for page table lookups is 1%–3% (Hennessey and Patterson 1990).

54.6.4 Element 4: Managing the RAM Contents

The operating system maintains a complete copy of a program's AS on the disk. The operating system also specifies a RAM allocation as a maximum number of pages from the AS. Thus, RAM contents are a subset of disk contents.

A page fault is an exceptional condition generated by the MMU when it encounters a missing page ($P = 0$). The operating system interrupts the running program and instead runs a special routine called PFH, which follows these steps.

1. Locates the needed page in the secondary memory
2. Uses a replacement policy to select a frame of main memory to put that page in
3. Empties that frame
4. Copies the needed page into that frame
5. Updates the page table entry to reflect these changes
6. Restarts the interrupted program, this time, allowing it to complete its reference

The replacement policy (step 2) has a significant effect on performance of a virtual memory system. Replacement policies generally try to predict which pages are most likely to be reused in the immediate future and protect them from replacement. Because recently used pages tend to be the most likely to be reused soon, the replacement policy identifies the pages with $U = 0$ as candidates for replacement. Among those, it favors one with $M = 0$ because an unmodified page does not need to be copied back to disk.

While no replacement policy can give a perfect prediction of the future, many years of experience and experiment have led to the consensus that the LRU policy is best when RAM allocations are fixed, and the WS policy is best when RAM allocations can vary (Denning 1980, McMenamin 2011).

54.6.5 Summary

Virtual memory makes address translation transparent to the programmer. Since the operating system maintains the contents of the map, it can alter the correspondence between pages and frames dynamically. A program can now be executed on a wide range of system configurations, from small to large main memories, without recompiling it.

54.7 Cache Memories

The caching principle used in the TLB was first proposed by Wilkes (1965) as a direct hardware method for speeding up memory accesses for pages already loaded in RAM (Figure 54.3). The cache contains a set of "slots," each the size of a page. A tag associated with a slot indicates which frame of RAM is copied in that slot. When the MMU of the processor generates location address (b,x)—meaning frame b line x—the cache hardware searches all of the tag registers in parallel for a match on b. If there is a match, it addresses byte x of that slot. If not, the hardware copies frame b into a slot, sets the slot's tag to b, and then addresses byte x of that slot (see Hennessey and Patterson 1990).

Note how the virtual memory mapping principle has been applied to components of the virtual memory system itself. The TLB is a cache of recent (page and frame) paths; a match bypasses the page table lookup. The slot-cache holds copies of recently used RAM frames; a match bypasses the RAM access. The TLB and cache access times are much faster than the corresponding RAM times. The virtual system will demonstrate significant accelerations with these caches, even when they are a small fraction of their maximum potential size. The costs of cache are small compared to full memory size.

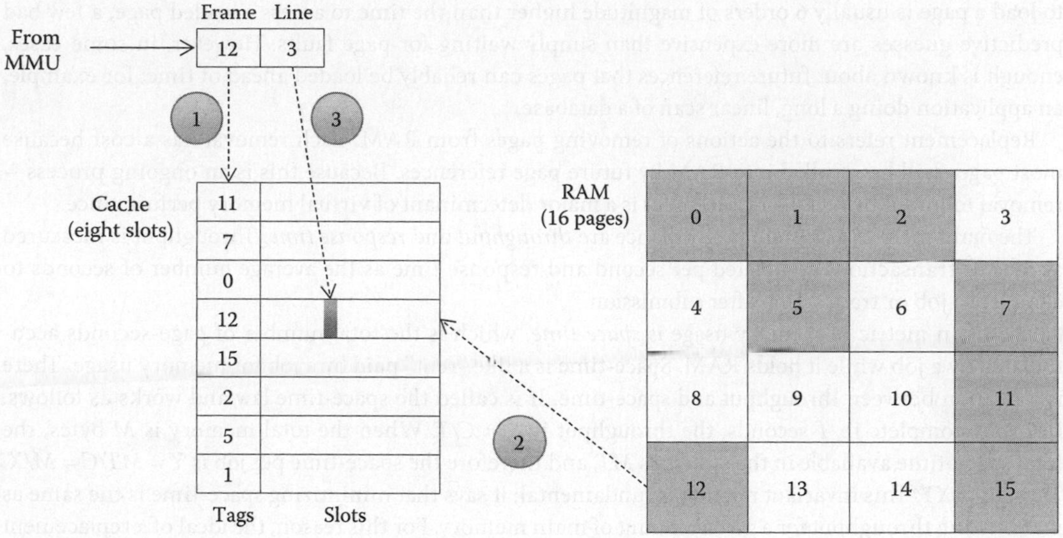

FIGURE 54.3 In a cache memory, the slots hold copies of RAM frames, and the tags indicate which frame is in a slot. The cache hardware searches the tag registers in parallel for a match on the addressed page (1). If no match is found, a page is loaded into the least recently used cache slot (2). Then, the line number selects a byte of that page (3). The search can be accelerated by dividing the tags into 2^m sets, using the m low-order bits of the frame number to select the set, and restricting the parallel search to that set (the figure is drawn for $m = 0$.) This partitions the frames equally among the sets and thereby limits the number of slots into which a given frame may be placed. In the worst case, when 2^m equals the number of cache slots, the set size is 1, and each block can be loaded into one slot only.

54.8 Multiprogramming

Multiprogramming is a mode of operation in which the RAM is shared simultaneously by multiple jobs. The set of pages a job has in RAM is called its resident set (or partition). The resident sets of all loaded jobs partition the RAM into disjoint subsets of frames.

Multiprogramming was originally introduced in the operating systems of the 1960s as a means to improve the utilization of a CPU: When a job stops to wait for an event, such as reading a file from a disk, the CPU can switch to another job ready to go. In the 1990s, multiprogramming was extended to allow each user to start and load multiple jobs in RAM. Users could switch among active programs such as word processor, spreadsheet, mail, and browser. Because virtual memory confines each process to its assigned AS, it provides an elegant and flexible way of partitioning a multiprogrammed memory.

Multiprogramming can be done with fixed or variable partitions. Fixed partitions are easier to implement but variable partitions offer much better performance. With variable partitions, the operating system can adjust the size of the partition so that the rate of page faults stays within acceptable limits. The operating system can transfer space from processes with small memory needs to processes with large memory needs. Variable partitions often improve over fixed even when the variation is random (Denning 1980). System throughput will be near optimal when the virtual memory guarantees each active process just enough space to hold its WS (Denning 1980).

54.9 Performance and the Principle of Locality

We can discuss performance of virtual memory systems in terms of decisions about loading and replacing pages. Loading refers to the actions that bring a page into main memory. Most systems load pages only on demand because they have no reliable method of predicting future page use. Because the time to load a page is usually 6 orders of magnitude higher than the time to access a loaded page, a few bad predictive guesses are more expensive than simply waiting for page faults. However, in some cases, enough is known about future references that pages can reliably be loaded ahead of time; for example, an application doing a long, linear scan of a database.

Replacement refers to the actions of removing pages from RAM. Each removal has a cost because most pages will be recalled into RAM by future page references. Because this is an ongoing process—removal followed by recall—replacement is a major determinant of virtual memory performance.

The main metrics of system performance are *throughput* and *response time*. Throughput is measured as jobs or transactions completed per second and response time as the average number of seconds to complete a job or transaction after submission.

The main metric of memory usage is *space-time*, which is the total number of page-seconds accumulated by a job while it holds RAM. Space-time is a like "rent" paid by a job for memory usage. There is a relation between throughput and space-time. It is called the space-time law and works as follows. If C jobs complete in T seconds, the throughput is $X = C/T$. When the total memory is M bytes, the total space-time available in the system is MT, and therefore the space-time per job is $Y = MT/C = M/X$. Thus, $M = XY$. This invariant relation is fundamental: it says that minimizing space-time is the same as maximizing throughput for a given amount of main memory. For this reason, the ideal of a replacement policy is minimizing the space-time of each job.

When a job holds RAM, it circulates among servers of the system for pieces of service it needs to complete. These servers include disks, directories, network requests, printing devices, and more. Each service delay accumulates some space-time toward the job's total. The delays caused by page faults are only a subset of all the delays contributing to space-time. Still, reducing the number of page faults will reduce a job's total space-time.

The parameters of system usage outside of virtual memory are normally unaffected by paging within the virtual memory. Therefore, the minimum space-time occurs when the page faults are minimum. The ideal policy—let us call it MIN—replaces the page that will not be used again for the longest time (Belady 1966). This policy causes the intervals between page faults to be as long as possible, which minimizes total number of page faults.

MIN requires advance knowledge of the future. Because such knowledge is not usually available, replacement policies use past observations to predict future references. The predictions are necessarily imperfect.

The LRU replacement policy predicts that time until next reference to a page is same as time since last reference. Although not as good as MIN, LRU has been found to be quite robust over a range of programs, typically doing as well or better than other common policies, such as first-in-first-out (Belady 1966). LRU is often used in caches. A simple memory scan approximates LRU well: a pointer cycles through all the pages in the job's resident set, skipping over those with usage bit $U = 1$ (and resetting to $U = 0$), until it finds an unused page.

If we remove the constraints that memory size is fixed and that replacements occur only at page-fault times, we can do better than MIN. The ideal variable-space policy—let us call it VMIN—operates as follows (Prieve and Fabry 1974). After each page reference, VMIN looks ahead to the moment of the next reference to that page: if the time to that reference exceeds a threshold T, VMIN immediately replaces the page; otherwise, it retains the page in memory until that next reference. As T gets larger, VMIN retains more pages and generates fewer page faults. Thus, the parameter T trades off the amount of memory used against the amount of paging generated. The reason no other policy can do as well is that VMIN minimizes space-time at every reference: VMIN retains a page only if the cost of recovering it by page-fault at the next reference exceeds the cost of retaining it.

The WS model gives a good approximation to VMIN (Denning 1968, 1970, 1980). A process's *WS* is defined as the set of pages referenced in a window of size T references looking backwards from the current time. The WS replacement policy guarantees that every page of the WS is in RAM.

Under WS replacement, a page reference causes a page fault only if the time since prior reference to that page exceeds *T*. In other words, WS replacement generates exactly the same faults as VMIN. The difference between WS and VMIN is due solely to the extra space-time generated by WS for retaining pages for *T* seconds when the time until next reference exceeds *T*. The space-time "overshoot" is worst during a change of locality: VMIN unloads the pages of the current locality set prior to the change to the new locality set, while WS keeps the former locality set in memory for a while after the change. Many researchers looked at ways to reduce this overshoot, but none gave much benefit. The WS policy is about as close to optimal as can be found among non-lookahead replacement policies.

The WS policy works is especially well suited for multiprogramming. All jobs use a common window size *T*. The scheduler admits waiting processes to the RAM, one at a time, until the RAM space is filled with WSs. The global window size *T* can be adjusted empirically until it maximizes system throughput. System throughput may improve further by using a customized window size for each running process—but the space-time improvement is at most 5%–10% (Denning 1980).

Working-set memory management became popular because it prevented a system instability called *thrashing* (see Figure 54.4). Many early virtual memory systems attempted to extend the LRU policy to the entire memory by keeping track of the reference times of all pages in RAM. Extended this way, LRU is subject to thrashing because the normal cycles of the scheduler cause a process's pages to look old by before it gets its next time slice. Extended LRU does not have a built-in load limit like the WS policy. Thrashing can be avoided by limiting the multiprogramming level either by a fixed limit or by a working-set policy. The working-set policy generally leads to more stable performance with higher throughput.

These replacement policies all do well because of the *principle of locality*. This principle says that a computation tends to reference the same pages in the immediate future as it referenced in the immediate past.

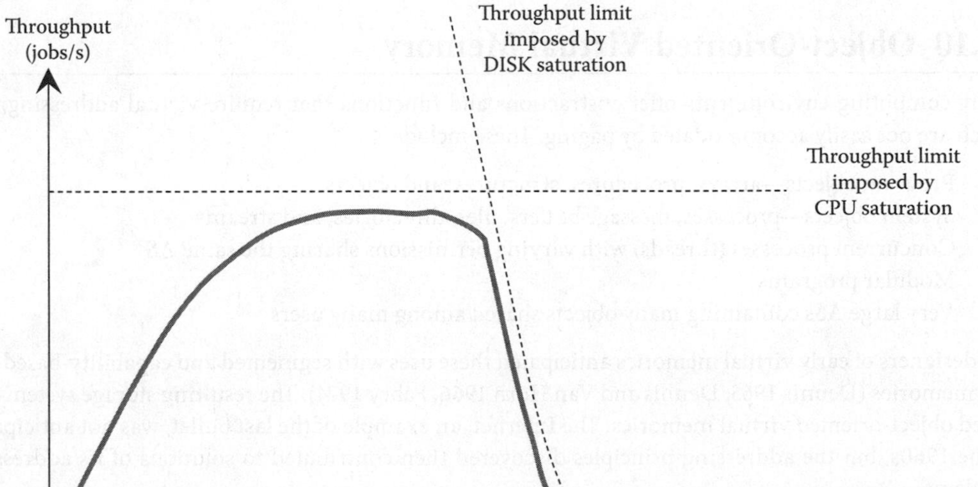

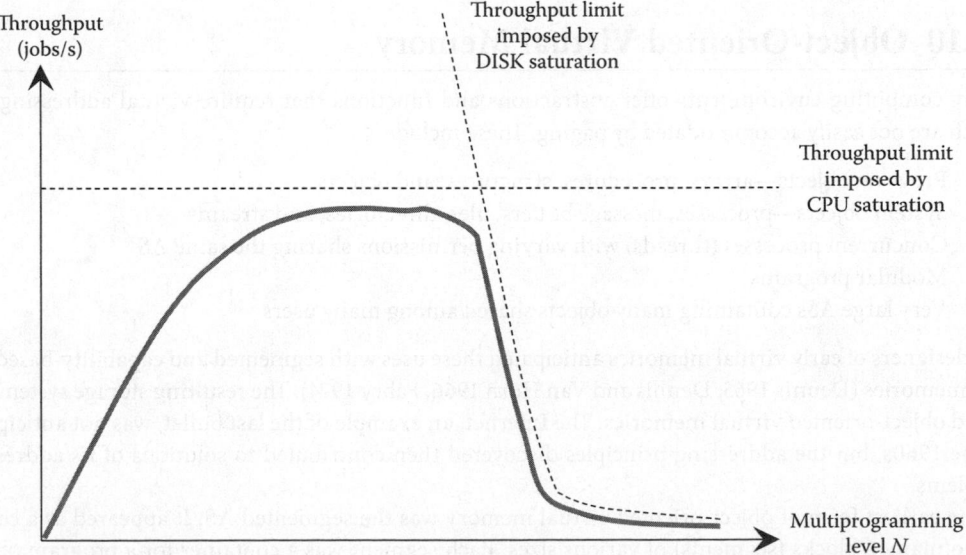

FIGURE 54.4 The system throughput is depicted as a function of the multiprogramming level *N*, which is the number of active programs in RAM. When *N* is too large, programs have too little space for their working sets. This drives the paging rates up, turns the disk into the bottleneck, and dramatically slows down the system. The slowdown is called thrashing. The WS policy is ideal; it holds *N* near the throughput peak but does not allow it to fall off the thrashing cliff.

In fact, computations tend to reference from the same subset of pages over extended intervals. The memory usage of a computation can be represented as a sequence

$$(L_1,P_1), (L_2,P_2),\ldots, (L_k,P_k),\ldots$$

where

L_i is a "locality set"

P_k is the length of time the computation uses only that set (P_k is called the phase time)

Abrupt changes in locality sets at phase transitions are the norm. This principle accounts for the success of WS policies: the backward looking window T is usually much smaller than the phase times, and thus the WS sees the current locality set most of the time. In this way, WS reveals the pages that need to stay in RAM in the immediate future (Denning 1980).

Over the years, it became clear that locality is a deep principle of computing. It accounts for the universal success of caches in every form, whether as part of a virtual memory, a Web browser, or an edge server in the Internet. By 1980, we knew that locality is a product of the way the human mind works—notably its tendencies to pay attention to the same things for a while and to divide big problems into smaller pieces for separate solution (Denning 1980).

In 2010, Moshe Vardi cited some new thinking on the question, "What is an algorithm?" (Vardi 2012). He cites a work of Yuri Gurevich, who defines an algorithm as a description of an abstract state machine, where states can be any data structure, and each operation can cause only a bounded change of state. By limiting the effect of an operation to a local set, this definition pulls the principle of locality into a fundamental definition of computation. Without locality, there is no computation. Locality has long been accepted as a fundamental principle of memory behavior, but it has never been so clearly linked to effective computation. By definition, every computation exhibits locality; some form of backward window can observe the pages of the current locality set and exempt them from replacement.

54.10 Object-Oriented Virtual Memory

Many computing environments offer abstractions and functions that require virtual addressing, but which are not easily accommodated by paging. These include

- Program objects—arrays, procedures, structures, and objects
- System objects—processes, message buffers, files, directories, and streams
- Concurrent processes (threads) with varying permissions sharing the same AS
- Modular programs
- Very large ASs containing many objects shared among many users

The designers of early virtual memories anticipated these uses with segmented and capability-based virtual memories (Dennis 1965, Dennis and Van Horn 1966, Fabry 1974). The resulting storage systems are called object-oriented virtual memories. The Internet, an example of the last bullet, was not anticipated in the 1960s, but the addressing principles discovered then contributed to solutions of its addressing problems.

The earliest form of object-oriented virtual memory was the segmented AS. It appeared as a collection of named blocks (segments) of various sizes. Each segment was a container for a program object. In the Burroughs B5000 and later series, for example, the Algol compiler created program segments containing procedures and data segments containing array rows (Organick 1973). The compiler generated virtual addresses of the form (i,x), meaning segment i, line x. The size of each segment was explicitly recorded in the mapping table, so that the mapper could reject out-of-bounds addresses x.

Multics went further than the Burroughs Algol compiler. It let the programmer define the segments. In Multics PL/I, operands had symbolic two-part names $S.X$; the operating system used a *linkage fault* to invoke a routine that mapped a symbolic name to its corresponding virtual segment number *s* and line number *x* (Organick 1972).

54.10.1 Two-Level Mapping

In recognition that paging led to much simpler virtual memory structures, Multics combined segmentation with paging. It allowed each segment to be divided into pages. The offset into a segment was subdivided into a page and line number. It used a *two-level map* as follows to resolve an address (*s*,*x*):

1. The segment number *s* selects a *segment table* entry, which points to a page table for the segment. There is one segment table for each AS.
2. The offset *x* decomposes into a page number and line number, which are mapped through the page table as before. There is one page table for each segment.

The two-level mapping scheme makes it easy to share segments. Users simply share the segment's page table. Users sharing a segment are likely to get different local segment numbers for their segment tables; but all those segment table entries point to the *same* page table. When it moves a page, the operating system records the new location in the page table. Instantly, all users sharing the page are mapped correctly to the new location. There is no need for the operating system to locate all the users and update their segment mapping tables.

Inspired by Dennis and Van Horn (1966), Robert Fabry formalized the two-level mapping scheme to allow any number of users to share any number of objects in large systems (Fabry 1974). Fabry's addressing principle is summarized in Figure 54.5. The key idea is that the first level of mapping takes a segment number *i* to a system handle *h*; the second level takes a handle to a descriptor of the storage

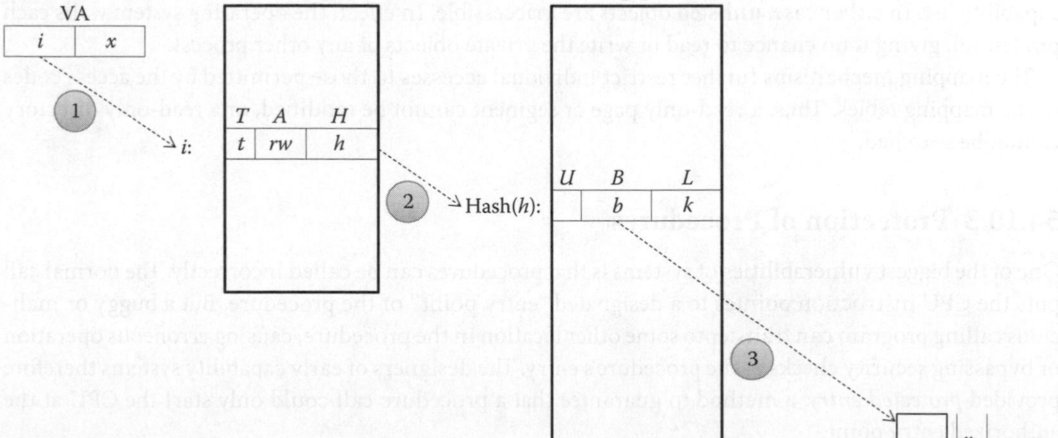

FIGURE 54.5 The two-level mapping scheme replaces a page table with two tables. The capability list CL[*d*] explicitly lists objects accessible in domain *d*. Each entry in CL consists of a tag, an access code, and a handle. The tag identifies the type of object (such as file or directory), the access code the type of access allows (such as read or write), and the handle is the unique global system name for the object. The descriptor table DT is a list of descriptors for every object in the system. This example shows a standard segment with base *b* and length *k*. A handle *h* is passed through a hash function to localize it within the DT for fast mapping. Although the full addressing sequence 1-2-3-4 requires several RAM accesses, the key elements of the path, (T,A,B,L), can be cached in the TLB for fast future lookup.

space holding the object (Fabry used the terms "capability," where we now say "handle"). All information about the location of an object is recorded in the descriptor; any changes instantly affect anyone trying to use the object.

54.10.2 Protection of Handles and Objects

In the two level mapping system, a program that holds a handle can access the object regardless of whether it has permission to do so. In other words, having a handle confers permission to access the object.

Because the mere fact of holding a handle was like having super-user privilege to the designated object, the designers of early capability addressing systems thought that handles had to be heavily protected by the hardware from alteration. They saw no other way to prevent someone from manufacturing a handle pointing to someone else's files or to prevent an erroneous program from corrupting a handle. Some early commercial systems such as Plessey 250 or IBM 360/38 used hardware protection for capabilities. The Cambridge CAP system, a research project, also used hardware protection but concluded that hardware protection led to unwieldy complexity in the operating system (Wilkes and Needham 1979). Eventually, operating system structures were discovered that hid handles from users, thus prevent tampering or alteration, without special hardware protection. For example, file and directory handles are stored in directories, where the user can invoke them by giving their symbolic names; but the handles themselves never enter the user's virtual AS.

One of the fundamental requirements of an operating system is that users cannot interfere with each other. That means, by default, they cannot see each other's ASs. The virtual memory system plays an integral role in meeting this requirement. The images of ASs are always in disjoint regions of main memory. This property of virtual memory is called *logical partitioning*.

With simple virtual memory (Figure 54.2), a processor can address only the pages listed in its page table. With extended virtual memory (Figure 54.5), a processor can address only the objects listed in its capability list. In either case, unlisted objects are inaccessible. In effect, the operating system walls each process off, giving it no chance to read or write the private objects of any other process.

The mapping mechanisms further restrict individual accesses to those permitted by the access codes in the mapping tables. Thus, a read-only page or segment cannot be modified, or a read-only directory cannot be searched.

54.10.3 Protection of Procedures

One of the biggest vulnerabilities of systems is that procedures can be called incorrectly. The normal call puts the CPU instruction pointer to a designated "entry point" of the procedure. But a buggy or malicious calling program can transfer to some other location in the procedure, causing erroneous operation or bypassing security checks at the procedure's entry. The designers of early capability systems therefore provided *protected entry*, a method to guarantee that a procedure call could only start the CPU at the authorized entry point.

It is straightforward to structure a capability system to provide a protected entry operation (Dennis and Van Horn 1966, Wulf et al. 1974). An instruction "ENTER *i*" works only if object *i* (in the caller's domain *d1*) is an enter capability. An enter capability points to the capability list of a new domain *d2*. Object "0" in every domain's capability is the domain's entry procedure. The effect of the enter instruction is to call the entry procedure of the target domain, simultaneously making its capability list the current capability list used by the CPU. The caller's domain and instruction pointer are saved on the stack and restored when the called procedure returns (Figure 54.6).

These structures have important benefits for system fault tolerance. Should a process run amok, it can damage only its own objects: a program crash does not imply a system crash. An untrusted program can be encapsulated in a domain whose capability list contains only the objects it needs to execute;

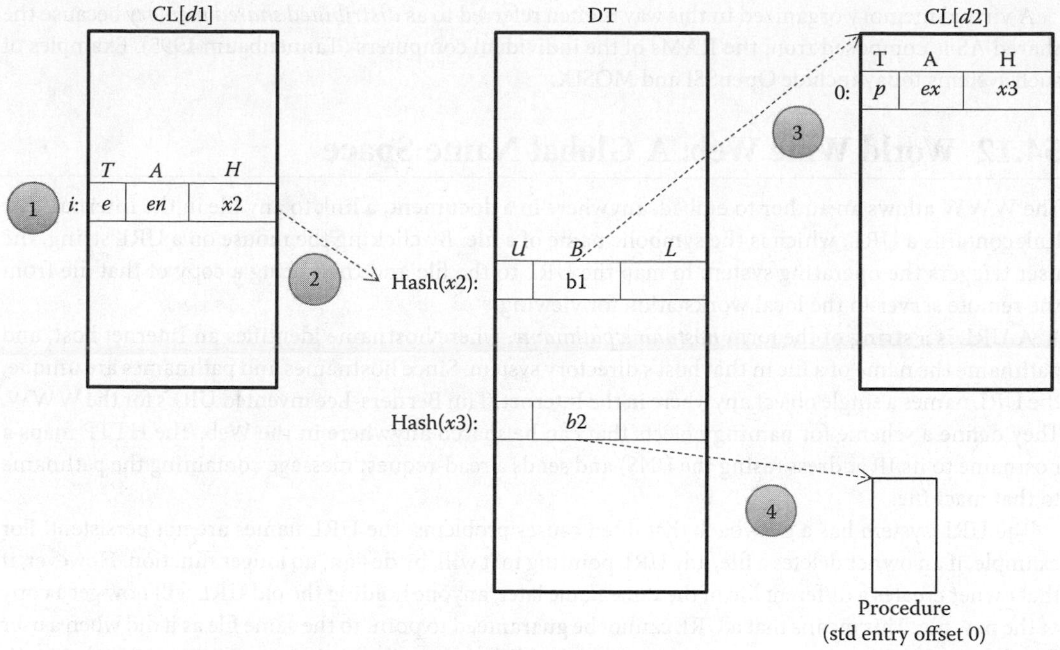

FIGURE 54.6 Protected entry into a protection domain is easily implemented in a capability system. The user in domain *d1* has a domain enter capability (tag *e*) with enter permission (*en*) that points to another domain *d2*. By convention the first entry in every domain (object "0") is a procedure (tag *p*) with execute permission (*ex*) to be automatically invoked when the domain is called. The instruction "ENTER *i*" causes a procedure call on object "0" in the new domain, and the capability list of the new domain becomes the current capability list.

any erroneous or malicious actions cannot access or damage anything outside this encapsulated domain. Systems that support this structure have found it effective against Trojan horses and other malware.

These benefits are so important that many systems use virtual memory even if they have sufficient RAM to eliminate overlaying.

54.11 Distributed Shared Memory

Starting in the mid-1980s, Sequent, Intel, Thinking Machines, N-Cube, and then later IBM, Cray, Kendall Square, and a few others introduced commercial multicomputers. These machines allowed for a large number of computers, sharing a high-speed interconnection network, to work concurrently on a single problem. Multicomputers soon began to challenge the traditional supercomputer by offering the same aggregate processing speed at a much lower cost (Denning and Tichy 1990).

In the mid 1990s, new operating systems like Beowulf provided the means to distribute tasks among ordinary computers registered as a cluster on a network. A large problem can be divided into numerous small problems for the cluster to solve, and their respective answers combined back into an answer to the original problem. Clusters provide supercomputer-grade processing power by mobilizing a large number of ordinary computers.

These new architectures introduced a host of new programming problems having to do with synchronizing the processes on the different computers and exchanging data among them. Because these systems offered no common AS among all of the component computers, their programmers could share data only by copying it between machines. That causes high message overhead.

Much of the overhead can be eliminated if the multicomputers all share an AS. Now data can be passed from one computer to another simply by passing the virtual address. The cost of moving a single pointer is much less than copying the entire object from one AS to another.

A virtual memory organized in this way is often referred to as *distributed shared memory* because the shared AS is composed from the RAMs of the individual computers (Tannenbaum 1995). Examples of such systems today include OpenSSI and MOSIX.

54.12 World Wide Web: A Global Name Space

The WWW allows an author to embed, anywhere in a document, a link to any file in the Internet. The link contains a URL, which is the symbolic name of a file. By clicking the mouse on a URL string, the user triggers the operating system to map the URL to the file and then bring a copy of that file from the remote server to the local workstation for viewing.

A URL is a string of the form *hostname/pathname*, where hostname identifies an Internet host, and pathname the name of a file in that host's directory system. Since hostnames and pathnames are unique, the URL names a single object anywhere in the Internet. Tim Berners-Lee invented URL's for the WWW. They define a scheme for naming objects that can be shared anywhere in the Web. The HTTP maps a hostname to its IP address (using the DNS) and sends a read-request message containing the pathname to that machine.

The URL system has a drawback that often causes problems: the URL names are not persistent. For example, if an owner deletes a file, any URL pointing to it will, by design, no longer function. However, if that owner creates a different file of the same name later, anyone holding the old URL will now get a copy of the new file. This means that a URL cannot be guaranteed to point to the same file as it did when a user first acquired it. Many service providers, notably publishers, want a naming system that uniquely identifies each published object and guarantees that the unique identifier will never point to another object.

The handle system (handle.net) was invented in 1995 by Robert Kahn to provide persistent names for objects. The most well known use of the handle system is with *digital object identifiers* (DOIs) that are of the form *A/B*, where *A* is a numerical string assigned by a DOI registrar to a publisher, and *B* is a unique numerical string assigned by the publisher (Denning and Kahn 2010). For example, the author's paper about WSs (Denning 1968) has DOI 10.1145/363095.363141, where "10.1145" is ACM's unique identifier assigned by the DOI Foundation, and "363095.363141" is a number chosen by ACM to distinguish that paper from every other that ACM has ever published, or ever will. ACM provides a server that translates DOI's to the current URLs of the papers. Thus, invoking the Web address "http://doi.acm.org/10.1145/363095.363141" will get a copy of the paper. ACM can change the URLs, but no matter where ACM actually stores the paper the DOI will always find it.

The handle system (including DOI's) is functionally similar to the handles described earlier in the extended, object-oriented virtual memory. The difference is that no attempt is made to hide the handles from users.

54.13 Conclusion

Virtual memory systems are used to meet one or more of these needs:

- *Automatic storage allocation*: Solving the overlay problem that arises when a program exceeds the size of the computational store available to it. Furthermore, it includes the problems of relocation and partitioning arising with multiprogramming.
- *Logical partitioning*: Each process is given access to a limited set of objects, its protection domain. The operating system enforces the rights granted in a protection domain by restricting references to the memory regions in which objects are stored. Processes cannot access objects beyond those listed in its local mapping tables.
- *Access control*: Within a protection domain, the operating system enforces further restrictions by permitting only the types of reference stated for each object (for example, read, write, or apply a function). These constraints are easily checked by the hardware in parallel with the main computation.

- *Encapsulation and protected entry*: Any untrusted software can be encapsulated in its own protection domain. Any unauthorized action will be automatically blocked by the access controls. It will be impossible to attempt access to any object outside the protection domain.
- *Modular programs*: Programmers should be able to combine separately compiled, reusable, and shareable components into programs without prior arrangements about anything other than interfaces and without having to link the components manually into an AS.
- *Object-oriented programs*: Programmers should be able to define managers of classes of objects and be assured that only the manager can access and modify the internal structures of objects (Myers 1982). Objects should be freely shareable and reusable throughout a distributed system (Chase et al. 1994, Tannenbaum 1995). (This is an extension of the modular programming objective.)
- *Data-centered programming*: Computations in the WWW tend to consist of many processes navigating through a space of shared, mobile objects. Objects are be bound to a computation on demand.
- *Parallel computations on multicomputers*: Scalable algorithms that can be configured at run time for any number of processors are essential to mastery of highly parallel computations on multicomputers. Virtual memory joins the memories of the component machines into a single AS and reduces communication costs by eliminating some of the copying inherent in message passing.

Virtual memory, once the subject of intense controversy, it is now so ordinary that few people think much about it. Its original purpose—automating solutions to the overlay problem—is today less important that it's extended purposes for sharing and access objects in large name spaces. The success of virtual memory is tied directly to the principle of locality, which is a fundamental principle of computation itself. Virtual memory is an enduring technology.

Key Terms

Access control: A means of allowing access to an object based on the type of access sought, the accessor's privileges, and the owner's wishes.

Address fault: An error that halts the mapper when it cannot locate a referenced object in main memory; it invokes an interrupt, whose handler corrects the condition by loading the missing object.

Address map: A table that associates an object (or page) number with the main memory locations containing the object.

Address space: The set of all addresses that a processor can issue while processing a program.

Bounds fault: An error that halts the mapper when it detects that the offset requested into an object exceeds the object's size; it invokes an interrupt that terminates the program.

Capability: A systemwide unique identifier for an object; the bits of a capability are protected from alteration.

Context-switch: An operation that switches the CPU from one process to another by saving all of the CPU registers for the first and replacing them with the CPU registers for the second.

CPU: Central processing unit or processor.

Data-centered view: A view of computing that emphasizes navigation of many concurrent processes within a large space of objects.

Handle: A systemwide unique identifier for an object, like a capability without the system guarantee of integrity.

Location: A memory register with its own address.

Logical partitioning: A property of virtual memory, whereby the address spaces of different jobs are mapped into disjoint regions of memory.

Main memory: The highest level of the memory hierarchy; all CPU memory references are directed to main memory; CPU can access objects only when they are loaded in main memory.

Memory hierarchy: A system of memory devices of different speeds and capacities; allows for trading off between capacity and speed, and between volatility and persistence.

Memory space: The set of all hardware addresses of memory locations in RAM available to a given address space.

Modular programming: Programs are divided into parts that can be shared, reused, and recompiled without affecting other parts of the system as long as the interfaces to modules are unchanged.

Object oriented addressing: A form of virtual addressing in which object numbers are mapped to memory regions and internal object references are mapped to offsets within an object's memory region.

Object-oriented programming: A form of programming in which data are organized into classes of objects, each with a specific set of functions that can be applied to the objects.

Page: a fixed size unit of storage and transfer in a memory hierarchy.

Page frame: A contiguous block of memory locations used to hold a page.

Paging: A method of virtual memory in which address space and memory space are paged.

Partition: A division of memory space into disjoint subsets of pages for each address space.

PC: Personal computer.

Permissions: Access rights granted by an object's owner and represented as bits in the object's access code.

Process: An abstraction of the execution of a program, usually represented as the sequence of values of its CPU state as the program traces through its instruction sequence.

Processor-centered view: A view of computing that emphasizes the work of a processor.

Protection fault: An error condition detected by the address mapper when the type of request is not permitted by the object's access code.

Response time: The time from when a command is submitted to a computer until the computer responds with the result.

RAM: Random access memory.

RISC: Reduced instruction set computer (e.g., PowerPC, Sun SPARC, DEC Alpha, and MIPS).

Secondary memory: Lower, large capacity level of a memory hierarchy, usually a set of disks.

Segmentation: An approach to virtual memory when the mapped objects were variable-size memory regions rather than fixed-size pages.

Slave memory: A hardware cache attached to a CPU, enabling fast access to recently used pages and lowering traffic on the CPU-to-main-memory bus.

Space-time: The accumulated product of the amount of memory and the amount of time used by a process.

Thrashing: A condition of performance collapse in a multiprogramming system when the number of active programs gets too large.

Throughput: The number of jobs (or transactions) per second completed by a computer system.

TLB: Translation lookaside buffer, a cache that holds the most recently followed address paths in the mapper.

Two-level map: A two-tiered mapping scheme; the upper tier converts local object numbers into system unique handles, and the second tier converts handles to the memory regions containing the objects essential for sharing.

URL: Uniform resource locator (in the WWW).

Working set: The smallest subset of a program's pages that must be loaded into main memory to assure acceptable processing efficiency; changes dynamically.

Working-set (WS) policy: A memory allocation strategy that regulates the amount of main memory allocated to a process, so that the process is guaranteed a minimum level of processing efficiency.

World Wide Web (WWW): A set of servers in the Internet and an access protocol that permits fetching documents by following hypertext links on demand.

References

Belady, L. 1966. A study of replacement algorithms for a virtual storage computer. *IBM Systems Journal* 5(2):78–101.

Berners-Lee, T. 2000. *Weaving the Web*. Harper Business, New York.

Chase, J. S., Levy, H. M., Feeley, M. J., and Lazowska, E. D. 1994. Sharing and protection in a single-address-space operating system. *ACM TOCS* 12(4):271–307.

Denning, P. J. 1968. Thrashing: Its causes and prevention. *Proc. AFIPS FJCC* 33:915–922.

Denning, P. J. 1970. Virtual memory. *Computing Survey* 2(3):153–189.

Denning, P. J. 1976. Fault tolerant operating systems. *Computing Survey* 8(3):359–390.

Denning, P. J. 1980. Working sets past and present. *IEEE Transactions on Software Engineering* SE-6(1):64–84.

Denning, P. J. and Kahn, R. E. 2010. The long quest for universal information access. *ACM Communications* 53(12):34–36.

Denning, P. J. and Tichy, W. F. 1990. Highly parallel computation. *Science* 250:1217–1222.

Dennis, J. B. 1965. Segmentation and the design of multiprogrammed computer systems. *Journal of the ACM* 12(4):589–602.

Dennis, J. B. and Van Horn, E. 1966. Programming semantics for multiprogrammed computations. *ACM Communications* 9(3):143–155.

Fabry, R. S. 1974. Capability-based addressing. *ACM Communications* 17(7):403–412.

Fotheringham, J. 1961. Dynamic storage allocation in the Atlas computer, including an automatic use of a backing store. *ACM Communications* 4(10):435–436.

Hennessey, J. and Patterson, D. 1990. *Computer Architecture: A Quantitative Approach*. Morgan-Kaufmann, San Mateo, CA.

Kahn, R. and Wilensky, R. 1995. *A Framework for Distributed Object Services*. Technical Note 95-101, Corporation for National Research Initiatives, Reston, VA. See also www.handle.net.

Kilburn, T., Edwards, D. B. G., Lanigan, M. J., and Sumner, F. H. 1962. One-level storage system. *IRE Transactions* EC-11(2):223–235.

McMenamin, A. 2011. Applying Working Set Heuristics to the Linux Kernel, MSc Project Report, Birbeck College, University of London. Available at: http://cartesianproduct.files.wordpress.com/2011/12/main.pdf, accessed October 6, 2013.

Myers, G. J. 1982. *Advances in Computer Architecture*, 2nd edn. Wiley, New York.

Organick, E. I. 1972. *The Multics System: An Examination of Its Structure*. MIT Press, Cambridge, MA.

Organick, E. I. 1973. *Computer System Organization: The B5700/B6700 System*. Academic Press, New York.

Prieve, B. and Fabry, R. 1974. VMIN: An optimal variable space page replacement algorithm. *ACM Communications* 19(5):295–297.

Sayre, D. 1969. Is automatic folding of programs efficient enough to displace manual? *ACM Communications* 12(12):656–660.

Tannenbaum, A. S. 1995. *Distributed Operating Systems*. Prentice-Hall, Englewood Cliffs, NJ.

Vardi, M. 2012. What is an algorithm? *ACM Communications* 55(3):5.

Wilkes, M.V. 1965. Slave memories and dynamic storage allocation. *IEEE Trans. EC* 14(April):270–271.

Wilkes, M. V. 1975. *Time Sharing Computer Systems*, 3rd edn. Elsevier, Amsterdam, North Holland.

Wilkes, M. V. and Needham, R. 1979. *The Cambridge CAP Computer and Its Operating System*. Elsevier, Amsterdam, North Holland.

Wulf, W., Cohen, E., Corwin, W., Jones, A., Levin, R., Pierson, C., and Pollack, F. 1974. HYDRA: The kernel of a multiprocessor operating system. *ACM Communications* 17(6):337–345.

55

Secondary Storage and Filesystems

Marshall Kirk McKusick
Marshall Kirk McKusick Consultancy

55.1 Introduction

The memory on a computer is organized into a hierarchy of storage. This storage ranges from small and fast to large and slow. Figure 55.1 shows a typical hierarchy. It is composed of two main parts: the primary store and the secondary store.

The main components of this hierarchy are the following:

1. The first level of the primary store is the *cache memory hierarchy*. It is often contained on the same chip as the central processing unit (CPU) or on other nearby chips that can be connected to the CPU with a minimum of delay. Because it must be able to run at close to the speed of the CPU, with access times of as little as a few nanoseconds, cache memory is typically small, rarely exceeding a few megabytes (Mbytes). The cache is never used for permanent storage; it holds values that are actively being processed by the CPU.
2. The second level of the primary store is the *main memory* on the computer. It currently runs with access times of 40–80 ns that may delay a CPU running at 3 GHz by 200–300 instruction cycles. The size of main memory ranges from a few gigabytes (Gbytes) up to several terabytes (Tbytes). Like the cache, main memory is not used for permanent storage; it holds the active part of running programs. Inactive parts of running programs are swapped out of the main memory to secondary store when the main memory becomes full. Thus, the size of a program is not constrained to the size of the main memory.
3. The first level of the secondary store is often built from flash memory. Flash memory is a nonvolatile computer storage chip that can be electrically erased and reprogrammed. NAND flash memory is the type usually used for secondary storage because it is higher density than other types of flash memory. NAND must be read and written in pages much like the sectors on a disk. Unlike disks, it must be erased in large blocks of pages before these pages can be rewritten with new data. Because flash

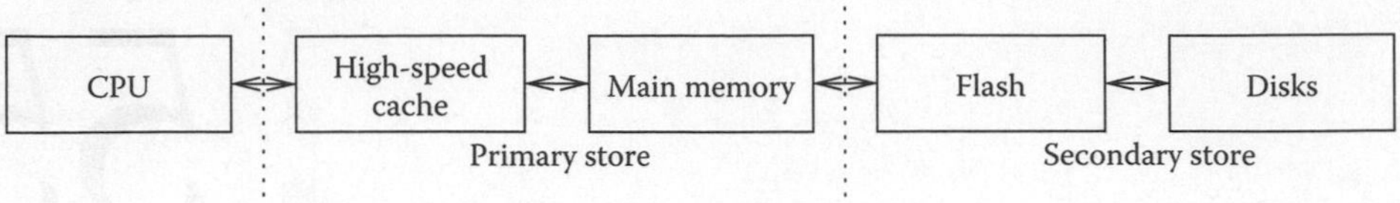

FIGURE 55.1 Computer memory hierarchy.

memory avoids the mechanical delay of disks, it can typically access data about a hundred times faster than a disk (though it is still a hundred to a thousand times slower than main memory).

4. The second level of the secondary store is usually built from one or more disk drives. These disk drives are usually connected directly to the computer, though they may be located across a fast network on a central storage server. Disks are used for intermediate to long-term data storage. Access time for a fast disk is currently about 4–5 ms; thus, a CPU that needs to access data that is on disk will have to wait thousands of instruction cycles. Modern multitasking operating systems will suspend a program that awaits a disk access and run another program. Sometimes after the disk access has completed, the program that requested the data will begin to run again.

This chapter is concerned with the secondary part of the storage hierarchy; Chapters 18 and 54 discuss primary storage and its management. In particular, disks may be used as a temporary store when the main memory becomes full. This chapter considers the secondary storage solely from the perspective of its use as long-term storage media. The first half of this chapter will briefly discuss the hardware used to support secondary storage; Chapter 18 describes the storage hardware in much greater detail. The second half of the chapter will discuss filesystem software used to access and manage secondary storage.

55.2 Secondary Storage Devices

Many types of hardware are used to support secondary storage. This section will describe the most commonly used devices—solid-state disks (SSDs), magnetic disks, digital video disk (DVD) disks, and tape.

55.2.1 Solid-State Disks

A SSD is a data storage device that uses solid-state memory to store persistent data. By contrast, a traditional magnetic disk is an electromechanical device containing spinning disks and movable read/write heads. SSDs use the same interface as magnetic disk drives, thus easily replacing them in most applications. SSDs are typically less susceptible to physical shock, are silent, have lower access time and latency, but are more expensive per Gbyte.

SSDs contain no moving parts and most use NAND-based flash memory that retains memory even without power. NAND flash memory is organized in blocks that are a few Mbytes in size. Each of these blocks is made up of pages that are a few Kbytes in size. Each page can be read or written independently. Once written, a page must be erased before it can be written again. Erasure can only be done at a block level; erasing a block causes all the pages contained in that block to be erased. Flash memory can only be erased and rewritten a few hundred thousand times before it wears out and ceases to work.

Lower priced drives usually use multilevel cell (MLC) flash memory, which is slower and less reliable than single-level cell (SLC) flash memory. For MLC flash memory, a problem called lower page corruption can occur when MLC flash memory loses power while programming an upper page. The result is data written previously and presumed safe can be corrupted if the memory is not supported by a super capacitor covering for a sudden power loss. Another property of MLC flash memory is that the pages within a block must be written sequentially rather than being able to be written in a random order. These problems do not exist with SLC flash memory. Current controllers can mitigate the speed and reliability issues of MLC memory by using more robust error-correcting codes and by using greater over-provisioning (more excess capacity) with which the wear-leveling algorithms can work.

The key components of an SSD are the controller and the flash memory to store the data. The controller incorporates the electronics that bridge the NAND memory components to the host computer. The controller is an embedded processor that executes firmware-level code and is an important factor of SSD performance. Some of the functions performed by the controller include the following:

- *Logical to physical block mapping*: Unlike the magnetic disks that the SSD is emulating, it cannot overwrite an already written block until the page holding that block has been erased. Because erasure must be done in blocks of pages, the only practical way to overwrite a page is to write a new copy elsewhere.

  The SSD needs to maintain a mapping from the logical block numbers being presented to it by the host to the physical block numbers in which those logical blocks are being stored. This mapping table needs to be updated each time a block is over-written and its physical location is changed. The mapping table must be stored in the nonvolatile memory of the drive so that it can be recovered after a crash or power failure as it would otherwise be impossible to recover the contents of the SSD. For performance reasons, the active part of the mapping table is kept in a dynamic-memory cache. The SSD needs either a capacitor or some form of battery to maintain data integrity such that the data in the cache can be flushed to the drive when power is dropped. Some filesystems are beginning to maintain this table for the SSD as part of their operation; this integration of the filesystem with the SSD is described in more detail in the next section.
- *Error correction*: Flash memory is prone to error, especially as it begins to approach the end of its write cycles. Each page needs to have error-correcting codes associated with it so that these errors can be detected and corrected. When the number of errors gets close to the maximum number that the error-correcting code can correct, the page must be retired from use by recording it in a bad-block table.
- *Garbage collection*: When the SSD controller finds that it is running out of unwritten pages, it must create new ones. Because erasure must be done in blocks, it must find a block whose pages have mostly been freed. The remaining valid pages in the block must be copied to free pages in another block. Once the valid pages are copied out, the block can be erased, and all its pages are then available for future write requests.

  This extra writing that must be done to clear out a block before erasing is called write amplification. This multiplying effect increases the number of writes required over the life of the SSD that shortens the time it can reliably operate. The increased writes also consume bandwidth to the flash memory, which mainly reduces random write performance to the SSD. Many factors affect the write amplification of a SSD; some can be controlled by the user and some are a direct result of the data written to and usage of the SSD.

  The SSD knows about the blocks it has freed owing to their being replaced by an over-write. To operate efficiently, it needs to know when blocks are no longer being used by the filesystem such as when an application deletes a file. Most filesystems today provide a TRIM command to inform a SSD that blocks of data are no longer in use by the filesystem and can be erased. While TRIM is frequently spelled in capital letters, it is not an acronym; it is merely a command name.
- *Wear leveling*: Because each page can only be written a finite number of times, pages that are approaching their write limit need to be switched to hold with static or infrequently changing data. Pages with low usage counts, because they have been holding static or infrequently changed data, need to be switched to hold more frequently changing data. The goal is to have as nearly equal write usage as possible. Ideally, the switching of high and low use pages can be done as a side effect of the garbage collection and the new block placement.
- *Read scrubbing and read disturb management*: Pages that are not read for months and pages that are read frequently begin to develop errors. The SSD controller must periodically read every page and rewrite any that have error rates that are close to what the error-correcting code can correct.

- *Encryption*: Eliminating data on a magnetic disk can be done by over-writing it. Eliminating data on an SSD is much more difficult since any attempt to over-write it will leave the old data in place and place the replacement data elsewhere. Unless the drive provides a specific function to erase all the memory in the drive, there is no reliable way to ensure that any particular piece of data has been removed. Thus, many SSD drives provide the ability to encrypt all the data written to the drive and to decrypt it when it is read back. If the encryption is done on a file-by-file basis, a file can be reliably deleted by destroying its associated key.
- *Read and write caching*: As with magnetic disks, SSD drives usually provide cache memory to hold recently read pages and to as a staging area for data to be written to improve their response rates.

The performance of the SSD can scale with the number of parallel NAND flash chips used in the device. A single NAND chip is slow because of a narrow (8/16 bit) asynchronous input/output (I/O) interface and the additional high latency of basic I/O operations. Typically, it takes about 25 ms to fetch a 4 Kbyte page from the array to the I/O buffer on a read, about 250 ms to commit a 4 Kbyte page from the I/O buffer to the array on a write, and about 4 ms to erase a 1 Gbyte block. When multiple NAND devices operate in parallel inside a SSD, the bandwidth scales and the high latencies can be hidden, as long as enough outstanding operations are pending and the load is evenly distributed between devices. The 2012 generation of SSD controllers deliver about 500 Mbyte per second read/write speeds.

55.2.2 Magnetic Disks

Data are transferred between the main memory and the disk by a disk controller. Most disk controllers transfer data between the disk and main memory using direct memory access. The CPU issues a command specifying the range of main memory addresses to be written or read and the disk sectors to or from which they should be transferred. The controller is responsible for issuing a seek command to the disk if necessary, waiting for the desired sectors to rotate under the head, transferring the data between the specified main memory locations, and interrupting the CPU to signal that the transfer has been completed.

To improve performance, disk controllers often provide a *track cache*, which holds the contents of the track containing the currently requested sector. For example, consider a file that fits on one track of a disk. Suppose the disk controller has been requested to read the first part of the file, but when the seek completes, the head is sitting over the middle of the file half a rotation away from its beginning. Instead of waiting for the disk to rotate the beginning of the file into position, the controller immediately begins reading the track into its cache. As the beginning of the file passes under the head, it is transferred into the main memory as requested, and the CPU is notified of its completion. Following the end of the requested data transfer, if there are no further requests awaiting the controller, it reads the remaining quarter of the track into its cache. Since files are often read sequentially, the controller will probably be asked to read the next quarter of the file. Instead of having to wait for the disk to rotate back into position, it can satisfy this request from its cache. Thus, after one disk revolution, the controller can produce any part of the file with no rotational delay. Since the controller can transfer data from its cache to main memory much faster than data are delivered by the disk, the disk cache produces an even greater performance gain than just that measured by eliminating rotational delay.

Using a cache to speed up disk writes is a bit more difficult. If the controller completes a seek half a rotation ahead of the spot that requests the write, there is nothing useful that it can do; it must wait for the requested position to rotate into place. Normally, the controller waits until the write completes before issuing a completion interrupt to the CPU. Typically, a millisecond or so of CPU processing occurs before it issues the next write request. If the next write request contiguously follows the previous write, then the disk head will have just missed the start of the block and will incur nearly an entire rotation's delay reaching the correct starting point. When writing a large contiguous file, the controller will consistently be delayed by a full rotation on each block, which leads to poor throughput. One approach to correcting this problem is to transfer the requested block into the controller cache and then issue the completion interrupt before the block has been placed entirely on the disk. In this study, the CPU can prepare and issue the next request for the controller while the previous block is being written. With the new request in hand, the controller can start its transfer to the disk without loss of a revolution.

This approach has a serious failing if the controller does not have a nonvolatile cache. Many applications, such as databases, depend on the completion interrupt to let them know that critical data such as a transaction log have been stored in a location that will survive a system failure. If the data are only in the volatile controller cache and the system fails before the cache is written to disk, then the database will be unable to recover. Thus, early completion interrupts must be used only if the controller cache uses nonvolatile memory and has software to restart it and write out any incomplete blocks after a system failure.

A much better alternative is to use a technique called tag queuing. In this study, several requests are given to the disk controller, each of which is identified with a unique tag. When a request has been written to the disk, an interrupt is generated that presents the tag of the completed transfer to the system. Tag queuing allows the disk to write contiguously without the need to depend on premature and possibly incomplete I/O. Thus, applications can reliably know that their data are on stable store.

The capacity of disks has been rising steadily; single disks today hold a few Tbytes of data. Unfortunately, disks are still able to transfer data from only one location at a time. As the amount of data on the disk grows, this serialized access has become more and more of a bottleneck. To compensate, some systems deliberately use smaller capacity disks to increase the total number of disks on the system, which allows more parallel data access.

55.2.3 Redundant Array of Independent Disks (RAID)

The biggest problem with magnetic disks is dealing with their inevitable failures. Because the heads fly only microns above the disk surface and because of the need to dissipate the heat that this small gap produces, even the best designed disks have a lifetime of only 5–7 years. On a large file server containing hundreds of disks, a disk failure is a common event. Traditionally, recovery from such failures were handled by use of a tape backup system. Every few weeks, a complete copy of each disk would be made on tape. Each day, a backup would be made of everything that had changed since the last complete copy had been made. When a disk failed, it would be replaced with a new disk; then the contents of the complete copy, followed by all the daily changes, would be made to complete the recovery process. There are three problems with this approach:

1. *Capacity limits*: As the capacity of disks has increased, the amount of data that must be backed up on tape has exploded. The large server above would have many Tbytes of disk space connected to it. Even with the high density tapes available today, it would take hundreds of tapes to back it up. The server would require several tape drives running continuously just to keep up with the backups.
2. *Data loss*: At best, most systems can only schedule backups once per day. If the disk crashes shortly before it is scheduled for its daily backup, everything that was done that day will be lost. For many businesses, losing even a few hours of work is unacceptable.
3. *Recovery delay*: Replacing a disk and recovering its contents from backup tapes takes several hours. During the recovery period, the disk is completely unavailable. This recovery delay is often unacceptable for time critical applications.

Two approaches have been taken to avoid these problems. The first of these approaches is a brute force solution called *mirroring*. The number of disks on the system is doubled. Disks are paired off and each pair keeps a copy of its partner. Thus, each time an application does a write, the changed data are written to both disks in the pair. Reads may be done from either disk since they both contain the same data. If one disk in the pair fails, the other disk can continue servicing requests without interruption.

When the failed disk is replaced, its initial contents are copied in from its operating partner. Although it may take an hour or two to do the replacement and contents copy, users are unaware of the delay since they are running from the remaining good drive. In addition, no data are lost since there is no dependence on backup tapes for recovery. Mirroring has traditionally been used for time or business critical data, such as that handled by banking and airline reservation systems, where the extra cost can be justified.

The second approach to avoiding the tape backup problem is to collect several disks together and use one of them to store a parity of the others. Such an organization is referred to as a redundant array of independent disks (RAID) (Patterson et al., 1988; Chen et al., 1994). A typical RAID cluster will contain five disks. Four of the disks contain data and the fifth contains a parity of the data on the other four. Each time data are written to any of the other four disks, a new parity must be computed and written to the fifth disk. In practice, the parity is not stored entirely on one disk as it would become the bottleneck when trying to write to one of the other four disks. Instead, each disk in a five disk RAID cluster would be divided so that 20% stores parity and 80% stores data that would be covered by parity on the other four disks. A more robust version of RAID maintains two sets of parity that allows it to suffer two disk failures without losing any data.

Recovery from disk failure in a RAID cluster is not as transparent as it is with mirroring. Access to the RAID cluster can continue, but at approximately half of its regular access rate, while the broken disk is replaced and rebuilt. The replacement disk is initialized by reading the other four disks in the cluster and computing what value should be put onto the new disk. In data communications, parity can be used to detect errors, but not to correct them. That is because in data communications, the receiver does not know which bit is in error. Parity can be used for error correction for a RAID cluster because the cluster knows which disk failed. Thus, it can recompute the correct value for the failed disk using the data on the other drives. Failure recovery on a RAID cluster typically takes a few hours.

Once the recovery is complete, the cluster returns to the state it was in just before the disk failed. Thus, RAID clusters solve two of the three tape backup problems. They avoid the need for daily backups and they avoid losing data when they fail. While access to the data is slow during the recovery period, that period is typically about half the time required for a tape recovery. A RAID cluster is considerably cheaper than a mirroring strategy since there is only a 25% redundancy of hardware rather than a 100 percent redundancy. Thus, the RAID cluster usage is increasing in less time-critical environments.

Even with RAID clusters, the need for backups is not completely eliminated. Full backups need to be taken and stored off-site for recovery if a major catastrophe such as a fire destroys the disks in a machine room. Backups are also needed to recover from user errors where an important file is accidentally deleted. Neither of these problems can be handled by RAID clusters.

Mirroring can handle the catastrophic failure if the mirror disks are kept in physically separate locations. For genuine safety, the mirrors should be in different buildings that are at least several miles apart. Buying space on a cloud server is often a less expensive alternative than maintaining a second facility. The communication costs of the high-bandwidth network connection that is required sometimes makes a long distance mirroring solution unacceptably expensive. A less expensive alternative is to make periodic tape backups and transport the tapes to the distant location for archival storage.

Distance mirroring does not provide recovery from user errors where an important file is accidentally deleted. As described in the next section of this chapter, modern filesystems provide the ability to take a *snapshot* of the filesystem. Logically, the snapshot represents a frozen copy of the filesystem as it existed at the time that the snapshot was taken. If a user accidentally deletes a file, they can copy a replacement for it from a recent snapshot.

55.2.4 DVD Disks

The ubiquitous DVD plays a small but important role in computer systems today (Asthana, 1995). Its primary use is as a software distribution medium where it has replaced tapes and floppy disks. It has the benefits of being cheap, reliable, and easily mass produced. It has the random access features of a magnetic disk that makes it much more convenient than a tape. Because the software is often used directly from the DVD rather than being loaded onto the system disk, large software packages can be used on systems that are otherwise short of disk space. Even with all these benefits, the use of DVDs as a software distribution medium is declining as software is increasingly being downloaded directly from the Internet.

The DVD disks are more compact to store than tapes. Like tapes, all user writable DVDs can only hold data reliably for 5–8 years. Factory pressed (read-only) DVDs are expected to hold data reliably for 50–100 years.

55.2.5 Tapes

Tapes are used for archival and backup storage. The access time to get to the start of and begin reading a file stored on a robotically managed tape system is several minutes. If a human operator must get involved, the access time takes longer. Historically, actively running programs directly manipulated data on tapes. Today, most applications arrange to have data read from tapes onto disk before beginning to access that data. Tapes are used primarily to archive data that is not currently accessed.

Tapes remain the most commonly used form of archival storage though they are being increasing replaced by cloud storage solutions. Early tape technology used 12 in. reels of half-inch tape that stored a little over 100 Mbytes of data. Current tape technology uses DLT cartridges that store Tbytes of data. The rule of thumb has been that the largest tapes hold about the same amount of data as the largest disks.

Data transfer to and from tapes tends to be slower than data transfer to and from disks. Random access to tapes is much slower than disks. Even modern DLT drives take about a minute to seek from one end of a tape to the other. The big benefit of tapes is that they are a tenth the cost of disks per Gbyte of storage. Furthermore, by installing a robotic tape system with a capacity of several hundred tapes, it is possible to create a file store capable of storing hundreds of Tbytes of data. While the access time to the data may be a minute or two, it is far cheaper than storing a similar amount of data on disk.

In practice, tape store is generally used as the final repository for data. Recently accessed data are stored on disk where it is more readily available. When the disks become full, the least recently accessed data are copied to tape and deleted from the disk. If it is later needed, it is reloaded onto the disk displacing other less recently accessed data. To maintain reasonable access times, most systems arrange to have enough disk space to keep data disk resident for up to a year.

55.3 Filesystems

Most applications that users run on their computer do not write data directly onto the disk. The operating system provides an interface called a filesystem that organizes the data into files. The filesystem converts the file from the user abstraction as an array of bytes to the structure imposed by the underlying physical medium and is responsible for deciding where the file contents should be placed on the disk. The filesystem provides several important services:

- *Protection*: Most filesystems allow users to control access to their files. At a minimum, they can restrict access to themselves, a defined group of other users, or all other users of the system.
- *Organization*: Data in each file can be manipulated independently of data in other files. Data may be added or deleted from one file without affecting the data contained in other files. In particular, the user need not be concerned that one file might run into another file on the disk as they would if they were directly placing the data on the disk themselves.
- *Multiple access*: Modern filesystems allow multiple files to be accessed at the same time and even allow separate programs to consistently access the same file. Consistent access means that if one application writes a file while another is reading the same part of the file, the reader will see the file either before the write has been started, or after it has finished, but not in a partially written state.

- *Space management*: The filesystem tracks the used and free space on the disk. When any file in the filesystem needs to grow, the filesystem finds an appropriate sized piece of free space and allocates it to the file. As long as there is free space left on the disk, the filesystem is prepared to allocate it to any file within the filesystem that wants to grow or to create a new file and allocate the space to that new file.

In addition, a filesystem is expected to optimize the use of disk bandwidth. Thus, it must not only find space to allocate to files, but it must try to find space that is contiguous or at least rotationally close together to minimize the time that it takes to read and write the file. Issues of space management are discussed in the subsequent sections.

55.3.1 Directory Structure

Most filesystems allow files to be grouped together into directories. These directories may then be further grouped together into other directories. The files and directories are usually grouped together in a tree hierarchy; Figure 55.2 shows a typical filesystem hierarchy. The rounded boxes represent directories; the square boxes represent files. The top of the tree is referred to as the *filesystem root*. Files are accessed by giving the set of names of directories from the root of the tree down to the desired file, separated by slashes; this name is called the *pathname*. Unix-like systems such as Linux, Solaris, and FreeBSD use forward slashes; Windows uses back slashes. For example, access to the file *mydata* in Figure 55.2 would use the pathname */users/myhome/mydata* (IEEE, 1994).

Many filesystems maintain a current working directory for the user. Instead of having to always specify a file by its complete pathname, the system keeps track of which directory the user is currently referencing and does all filename translation relative to that directory. The user initially specifies a current directory using a complete pathname. Using the filesystem shown in Figure 55.2, the user might request that the current directory be set to */users/myhome*. Thereafter, the reference to the file *mydata* can be used without specifying its entire path since it is resident in the current directory.

55.3.2 Describing a File on Disk

To allow both multiple file allocation and random access, most systems use a data structure similar to that shown in Figure 55.3 to describe the contents of a file. This structure includes

- Access permission for the file
- The file's owner
- The time the file was last read and written
- The size of the file in bytes

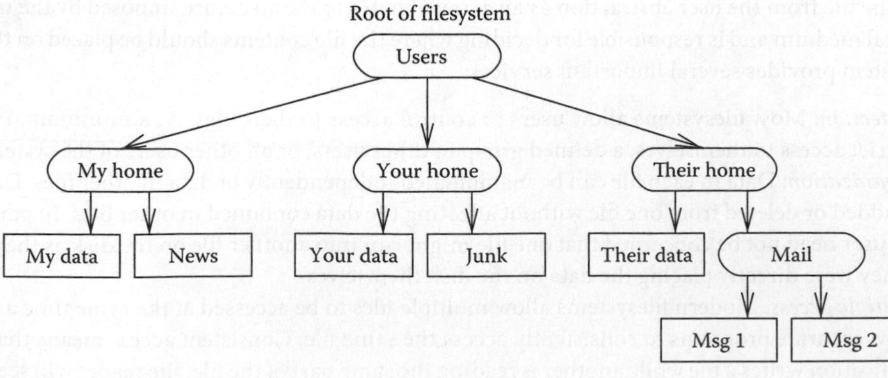

FIGURE 55.2 A set of files and directories.

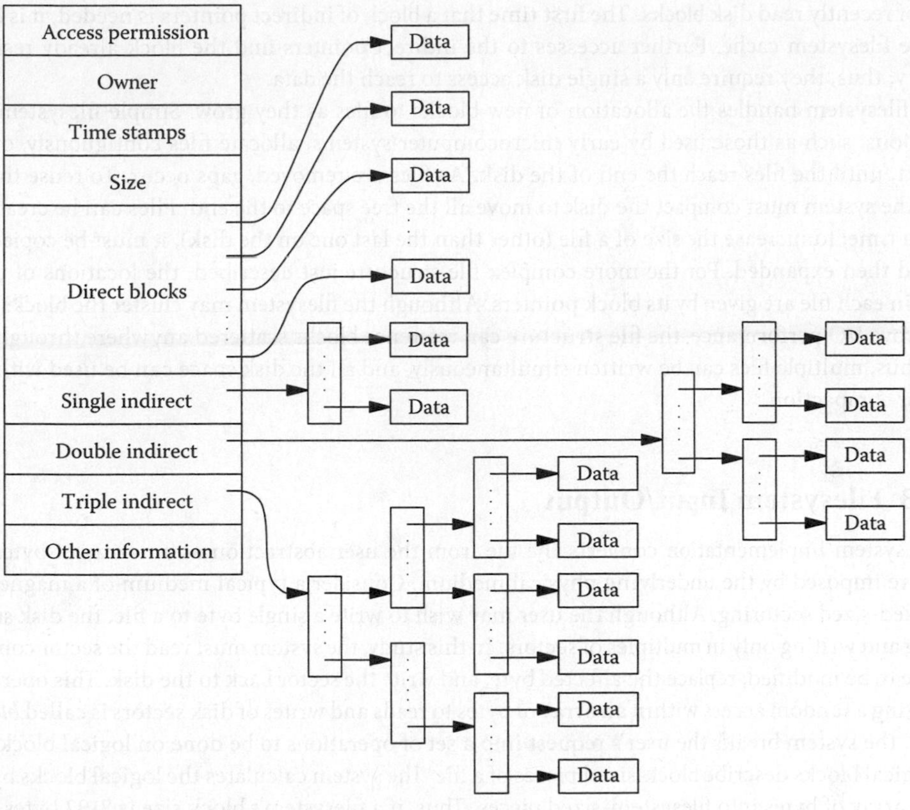

FIGURE 55.3 Extensible data structure used to describe a file.

Notably missing in this structure is the filename. Filenames are usually maintained in directories rather than in this structure because a file may have many names, or links, and the name of a file may be large (often up to 255 bytes in length).

The file structure also contains an array of pointers to the blocks in the file. The system can convert from a *logical block* number to a *physical block* number by indexing into this array using the logical block number. A null array entry shows that no block has been allocated and will cause a block of zeros to be returned on a read. On a write of such an entry, a new block is allocated, the array entry is updated with the new block number, and the data are written on that physical block of the disk.

Since the filesystem must support many files and most files that it stores are small, the file structure has a small array of pointers for efficient use of space. The first few array entries are allocated in the file structure itself. For typical filesystems, these array entries allow the first 400 Kbytes of data to be located directly using a simple indexed lookup.

For somewhat larger files, Figure 55.3 shows how the file structure contains a pointer to a single *indirect block* of pointers to data blocks. To find the hundredth logical block of a file, the system first fetches the block identified by the indirect pointer, then indexes into it by 100 minus the number of direct pointers, and fetches that data block.

For files that are larger than a few Mbytes, this single indirect block is eventually exhausted. These files must resort to using a double indirect block, which is a pointer to a block of pointers to pointers to data blocks. For files of multiple Gbytes or Tbytes, the system uses a triple indirect block, which contains three levels of pointers leading to the data block.

Although indirect blocks appear to increase the number of disk accesses required to reach a block of data, the overhead for this transfer is typically much lower. Most filesystems maintain a memory-based cache of recently read disk blocks. The first time that a block of indirect pointers is needed, it is brought into the filesystem cache. Further accesses to the indirect pointers find the block already resident in memory; thus, they require only a single disk access to reach the data.

The filesystem handles the allocation of new blocks to files as they grow. Simple filesystem implementations, such as those used by early microcomputer systems, allocate files contiguously, one after the next, until the files reach the end of the disk. As files are removed, gaps occur. To reuse this freed space, the system must compact the disk to move all the free space to the end. Files can be created only one at a time; to increase the size of a file (other than the last one on the disk), it must be copied to the end and then expanded. For the more complex file structure just described, the locations of the data blocks in each file are given by its block pointers. Although the filesystem may cluster the blocks of a file to improve I/O performance, the file structure can reference blocks scattered anywhere throughout the disk. Thus, multiple files can be written simultaneously, and all the disk space can be used without the need for compaction.

55.3.3 Filesystem Input/Output

The filesystem implementation converts the file from the user abstraction as an array of bytes to the structure imposed by the underlying physical medium. Consider a typical medium of a magnetic disk with fixed-sized sectoring. Although the user may wish to write a single byte to a file, the disk supports reading and writing only in multiples of sectors. In this study, the system must read the sector containing the byte to be modified, replace the affected byte, and write the sector back to the disk. This operation of converting a random access within an array of bytes to reads and writes of disk sectors is called *block I/O*.

First, the system breaks the user's request into a set of operations to be done on logical blocks of the file. Logical blocks describe block-sized pieces of a file. The system calculates the logical blocks by dividing the array of bytes into filesystem-sized pieces. Thus, if a filesystem's block size is 8192 bytes, logical block 0 would contain bytes 0–8191, logical block 1 would contain bytes 8,192–16,383, and so on.

Figure 55.4 shows the flow of information and work required to access the file on the disk. The abstraction shown to the user is an array of bytes. These bytes are collectively described by a file descriptor that refers to some location in the array. The user can request a write operation on the file by presenting the system with a pointer to a buffer, with a request for some number of bytes to be written. Figure 55.4 shows that the requested data need not be aligned with the beginning or end of a disk sector. Furthermore, the size of the request is not constrained to a single disk sector. In the example shown, the user has requested data to be written to parts of logical blocks 1 and 2.

The data in each logical block are stored in a physical block on the disk. A physical block is the location on the disk to which the system maps a logical block. A physical disk block is constructed from one or more contiguous sectors. For a disk with 4096 byte sectors, an 8192 byte filesystem block would be built from two contiguous sectors. Although the contents of a logical block are contiguous on disk, the logical blocks of the file need not be laid out contiguously.

Returning to our example in Figure 55.4, we now know which logical blocks are to be updated. Since the disk can transfer data only in multiples of sectors, the filesystem must first arrange to read in the data for any part of the block that is to be left unchanged. The system must arrange an intermediate staging area for the transfer. This staging is done through one or more *system-cache buffers*.

In our example, the user wishes to modify data in logical blocks 1 and 2. The operation iterates over five steps:

1. Allocate a system-cache buffer.
2. Determine the location of the corresponding physical block on the disk.
3. Request the disk controller to read the contents of the physical block into the system-cache buffer and wait for the transfer to complete.

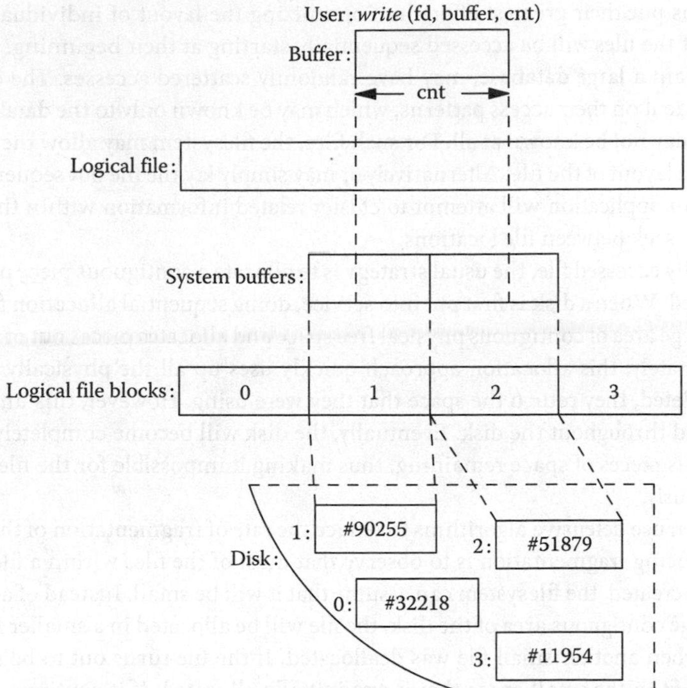

FIGURE 55.4 The block I/O system.

4. Do a memory-to-memory copy of the next piece of the data from the user's I/O buffer to the appropriate portion of the system-cache buffer.
5. Write the block to the disk.

Since the user's request is incomplete, the process is repeated with the next logical block of the file. The system fetches logical block 2 and is able to complete the user's request. Had an entire block been written, the system could have skipped step 3 by simply writing the data to the disk without first reading the old contents. This incremental filling of the write request is transparent to the user's process, because that process is blocked from running during this entire operation. It is also transparent to other processes since the file is locked by this process; any attempted access to the file by any other process will be blocked until this write has completed.

55.3.4 Disk Space Management

The role of the filesystem code is not only to organize the data on a disk but also to minimize the time that it takes to read and write that data. There are two important measurements of access time. The first is the time that it takes to access the data contained within a particular file. The second is the time that it takes to access the data contained within a collection of files that are accessed together.

Examples of collections of files accessed together include files that are collected together in a spool area such as those destined to be sent to a printer or those being batched to be sent as a collection of electronic mail messages. Another example might be a collection of files that make up the components of a spreadsheet, a document, or a program. Directories provide a strong clue that a set of files will be accessed together. Thus, many filesystems will try to allocate all the files contained within a directory in close physical proximity to each other on the disk. If they are accessed together, the disk will not need to make long or multiple seeks to get from one to the next.

Most filesystems put their greatest effort into optimizing the layout of individual files. The default assumption is that the files will be accessed sequentially starting at their beginning. Certain files, such as those that contain a large database, may have randomly scattered accesses. The optimal layout for such files is dependent on their access patterns, which may be known only to the database or application program, or that may not be known at all. For such files, the filesystem may allow the database or application to direct the layout of the file. Alternatively, it may simply lay the file out sequentially and assume that the database or application will attempt to cluster related information within the file to minimize the time it takes to seek between file locations.

For a sequentially accessed file, the usual strategy is to allocate a contiguous piece of space on the disk on which it is stored. When a disk is first put into service, doing sequential allocation for files is easy. The filesystem has a large area of contiguous physical free space and allocates pieces out of that space for each new file. Unfortunately, this allocation approach quickly uses up all the physically contiguous space. As old files are deleted, they return the space that they were using. However, this unused space will be randomly scattered throughout the disk. Eventually, the disk will become completely fragmented with no large contiguous pieces of space remaining, thus making it impossible for the filesystem to allocate new files contiguously.

Filesystems often use defensive algorithms to reduce the rate of fragmentation of the free space on the disk. A key to reducing fragmentation is to observe that most of the files within a filesystem are small. When a new file is created, the filesystem can assume that it will be small. Instead of allocating its initial disk space in a large contiguous area of the disk, the file will be allocated in a smaller fragment that may have been freed when another small file was deallocated. If the file turns out to be small as expected, then it will nicely fit in the small space that it was initially allocated. If it continues to grow, then the filesystem can move it to a fragment left by a somewhat larger file. Only when the file grows large, it is finally relocated to the large contiguous space that the filesystem has now been able to hoard. By only allowing large files to be allocated in the large contiguous space, the filesystem is better able to ensure that some large contiguous space will be available when it is needed.

Moving files around on the disk can be a potentially slow and expensive operation if the file must be read and rewritten every time that the filesystem wants to move it. To reduce the relocation cost, the filesystem will attempt to defer writing the file until it has determined its final location on the disk. The deferral is done by holding blocks of the file in system-cache buffers until its size can be determined (see Figure 55.4). The steps involved in allocating space to a file are as follows:

- When the first block is written, a system-cache buffer is allocated, A single block-sized piece of free disk space is found, the address of that piece of disk space is assigned to the system-cache buffer, but the buffer is not written to the disk.
- If the file continues to grow, a second system-cache buffer is allocated. The filesystem finds a two block-sized piece of free space, frees the single block-sized piece of space originally assigned to the file, and assigns the addresses of the new larger block to the two buffers. As before, neither buffer is written to the disk.
- This process continues until the file has grown to the maximum sized block allowed (typically the size of a disk track) or the file has ceased to grow at which point the buffers are written to their final destination.

If the system-cache buffers are needed for another purpose, or the application explicitly requests that the file be stored on disk (for example, it is a transaction log that must be in stable storage before the application can proceed), the system-cache buffers always have a location to which they can be written. The only implication of doing the write early is the loss of performance that comes from writing the data to disk more than once.

This algorithm has the additional benefit that it allows even slowly growing files to be written contiguously. If a file, such as one holding accounting records, grows at the rate of a few Kbytes per hour, it will be relocated to larger contiguous blocks periodically. In this study, the relocation usually involves reading and rewriting the data, since the turnover in system-cache buffers causes them to be flushed before the new disk addresses for the data are assigned. Since the reallocation occurs only a few times per hour, the added I/O overhead does not adversely affect system performance.

To ensure successful layouts, a filesystem must have a quick method for finding contiguous and nearby free blocks on the disk and then allocating them. The most common data structure used for describing the free disk blocks is an array of bits. The blocks on the disk are sequentially numbered; the corresponding bit in the array is set to 1 if the block is being used and 0 if it is free. Block allocation involves setting the bits corresponding to the blocks being allocated; block deallocation involves clearing the bits corresponding to the blocks being freed. Finding other nearby blocks can be done by looking for 0 bits in the array near the location of the most recently allocated block in the file. Clusters of blocks can be identified by looking for strings of 0 bits in the array.

The free disk block array is large for a big disk. Exhaustive searches of the array (for example, when the disk is nearly full and there are few free blocks remaining) would slow filesystem performance unacceptably. Consequently, most filesystems maintain auxiliary data structures that summarize the contents of subranges of this bit array. Such summaries include the number of free blocks and the maximum sized contiguous piece within each subrange of the bit map. When looking for a block or a cluster of blocks, the filesystem first scans the summary information to find a subrange of the bit map that has the necessary free space. Once it finds the needed space, the filesystem can narrow its search to that subrange of the bit map rather than searching the whole space.

An alternative to using an array of bits is to describe the free space using a more complex data structure such as a B+ tree. For example, a B+ tree is used for this purpose in the XFS filesystem (Sweeney et al., 1996). A B+ tree can also be used to locate the blocks for the file rather than using the indirect block scheme shown in Figure 55.3. B+ trees are described in Chapter 3 on data structures.

55.3.5 Filesystem Crash Recovery

Logging long has been used in database systems to provide recovery after a system failure (Date, 1995). The database periodically does a *checkpoint* of its on-disk data structures to ensure that they are in a consistent state. Following the checkpoint, the database keeps a log of every change that it commits. The log is usually written serially onto a disk that is dedicated to the log. Thus, the log can be written quickly since the disk head is always within no more than a rotational delay from the next location to be written.

If the system fails, the log does a *roll forward* operation on the database to bring it back to consistency. The roll forward works by going through the log and ensuring that all the changes that it lists are reflected in the on-disk database state. The time required to recover the database after a system failure is bounded by the frequency of checkpoints. By checkpointing the database and resetting the log every few minutes, the recovery time for a disk failure can be kept to a few minutes or less.

The idea of using a log to recover a database has also been applied to filesystems. Like a database, the filesystem periodically checkpoints its state. It then writes all changes after that point both to its log and to the filesystem itself. When the log becomes full or a time limit is reached, the filesystem checkpoints itself again and resets its log. As with the database, the log is rolled forward after a system failure to ensure that all the changes made since the checkpoint are reflected in the filesystem. The log only needs to record modifications to the filesystem. While it has no effect on the speed with which data can be read from the filesystem, logging does introduce additional overhead when files are written. However, if the log is stored on a separate disk from the rest of the filesystem, writing to the log is seldom the limiting function on the speed with which data are placed into the filesystem. For systems that require high write performance, the logging disk can be replaced with nonvolatile memory.

Although a log provides fast recovery after a system failure, it does not provide the data recovery of mirroring or RAID that can recover from catastrophic disk failure. The reason that logging does not help with catastrophic disk failure is that data may be lost from parts of the disk that the filesystem thinks are stable and hence did not enter in the log.

55.3.6 Non-Overwriting Filesystems

Traditional filesystem design makes updates by overwriting old versions of the data with new versions. If an application such as an editor writes new contents to a file, the new contents are written into the same space as was previously occupied by the old contents. The benefit of this design is that related files reside near each other on the disk even if they are written at different times. The drawback of this design is that the filesystem may be inconsistent during times when data destined for the disk is still in the main memory waiting to be written. Thus, some mechanism such as logging is necessary ensure that the filesystem can be recovered after a crash.

A more recent filesystem design has taken the approach of never overwriting existing data in the filesystem. Any time an update is made to a file, the changes are accumulated in memory. When a preset interval (usually less than 30 s) or a threshold of changes is reached, all the new data are written to free space on the disk. When all the changes have been written, the filesystem does one last checkpoint write to commit itself to its new state. As part of doing the filesystem checkpoint, the space formerly used by the files that have been updated is released for future use.

The filesystem always moves from one consistent state to the next and is never in an inconsistent state that needs to be recovered after a crash. After a crash, the filesystem can begin operating from its last consistent state. Such filesystems still need a log for use after a crash. Although they do not need it to recover the filesystem, they do need it to recover the changes that occurred since the last checkpoint. After a crash, the filesystem begins from its last checkpoint, all the changes in the log are applied to it, and a new checkpoint is taken to commit those changes.

A drawback of the non-overwriting filesystem design is that the data for related files no longer resides together on the disk. Instead, the files are laid out on the disk in the order in which they were written; if two related files are written at different times, they will likely be far apart from each other on the disk. Thus, reading those files at the same time will take longer than it would on a conventional filesystem. Furthermore, files that are written over time (such as system activity logs) will have their data scattered over the disk. Reading such files will be considerably slower than reading such a file on a traditional filesystem that will put all its data together in one place. These slow access patterns can be mitigated by using a large main-memory cache to hold frequently or commonly accessed data.

The ability to take a snapshot has become a popular filesystem feature. Snapshots are typically compact as they reference the filesystem on a copy-on-write basis. So, the snapshot only needs to retain copies of the disk blocks that are modified. As the contents of most filesystems change slowly, the extra space required to hold a snapshot is small.

A major benefit of non-overwriting filesystems is that they can quickly take snapshots. Because existing data are never overwritten, the filesystem need only to take a checkpoint and then not release the previously used data until the snapshot is deleted. The commit record written to create the checkpoint is saved as the access point for the snapshot. When the snapshot is deleted, the commit record and space associated with it can be released. Because the commit records are small and there is no need to check for over-writing of filesystem blocks, a non-overwriting filesystem can cheaply retain tens or hundreds of snapshots.

Taking a snapshot in a traditional over-writing filesystems is time consuming and requires significant space since it must record every block in use at the time of the snapshot. Every time data are written to the filesystem, it must check every snapshot to see if it will overwrite a block in that snapshot and save a copy of the old contents when such an overwrite is made. Thus over-writing filesystems limit the number of snapshots that they will retain to avoid using up too much space and slowing down filesystem performance.

Non-overwriting filesystems work particularly well with SSDs. Because they never overwrite any existing data, the SSD seldom needs to change the mapping of logical blocks to physical blocks. The only time a remapping is needed is when blocks are moved as part of garbage collection or wear leveling.

Some filesystems are now beginning to integrate more tightly with the SSD. Wear leveling is managed by the filesystem as part of its block allocation strategy. Blocks with low write counts are allocated to files and filesystem metadata structures that are expected to be short-lived, while blocks with high write counts are allocated to files and filesystem metadata structures that are expected to be long-lived.

Garbage collection is also coordinated with the filesystem. When the SSD selects a block for garbage collection, it notifies the filesystem of the blocks that will be moved. The filesystem updates its internal pointers to reference the new locations. Once the filesystem has committed these updates, the SSD can erase the selected block.

When a non-overwriting filesystem can support wear leveling and the block remapping associated with garbage collection, the SSD no longer needs to maintain the logical to physical mapping table. The removal of this extra level of indirection and the need for its maintenance dramatically reduces the cost of the SSD controller as well as improves its performance.

55.3.7 Versioning Systems

Users often want to retain previous versions of files, such as released versions of programs or documents. Usually, they use a revision control utility that maintains a database, which stores and retrieves the selected versions. Most version control systems only keep a complete copy of the original file. Each successive version is then stored as a set of differences from the previous version. Some version control systems operate by storing each version in a separate file, allowing fast and easy access to older versions (since it is just a matter of finding and reading the desired version). However, this method of version control is wasteful of space, especially if the file being versioned is large and the changes between versions are minimal.

The whole file copy version of revision control can be done by the filesystem itself. Each time a file is opened for writing, the contents of the old file are saved instead of being overwritten. The old file is kept around until the filesystem runs out of space. When additional space is needed for new files, the user must either explicitly delete unneeded earlier versions of files or request the filesystem to reclaim the space from the oldest of the earlier file versions.

The user interface for accessing older files may simply provide a list of different versions that are available and allow the user to specify the one that they want. Or it may be more sophisticated, allowing the user travel in time. For example, the user might be able to request all the files making up a particular program as they existed at the release date for the program.

Another variant of versioning involves taking a *snapshot* of the filesystem. When running on a non-overwriting filesystem, it is possible to take a snapshot of a filesystem every few hours during the day. Often, the snapshots are put in a location accessible to the users so that they can go retrieve older versions of their files without the intervention of a system administrator. Thus, if a user wrote a file in the morning and accidentally overwrote it in the afternoon, they could retrieve the original copy from the late-morning snapshot.

Key Terms

Block I/O: The conversion of application reads and writes of records with arbitrary numbers of bytes into reads and writes that can be done based on the block size and alignment required by the underlying hardware.

Checkpoint: The writing of all modified data associated with a filesystem to stable storage (either non-volatile memory or the disk). A checkpoint ensures that all operations completed before the checkpoint will be recovered following a system failure.

Dirty blocks: In computer systems, modified, a system usually tracks whether an object has been modified—is dirty—because it needs to save the object's contents before reusing the space held by the object. For example, in the filesystem, a system-cache buffer is dirty if its contents have been modified. Dirty buffers must be written back to the disk before they are reused.

Indirect block: A filesystem data structure composed of an array of pointers to disk blocks used to locate the data blocks associated with a file.

Logical block: The sequential fixed-sized pieces of a file. The logical block associated with a given byte offset in a file is calculated by dividing the offset by the filesystem block size. For example, byte 20,000 is located in the third logical block of a file residing on a filesystem with 8 kbyte blocks.

No-overwrite policy: The filesystem never rewrites existing data in a file. New data are always written into a new location on the disk.

Physical block: The disk sector addresses associated with a logical block of a file. The filesystem finds the contents of a logical block in a file by using the logical block number as an index into an indirect block to find the disk sector address holding the requested data.

Roll forward: Used to recover after a system failure. The operation of rerunning the update operations stored in a log file against a filesystem or database to bring it to a consistent state as of the last update completed in the log.

Snapshot: A filesystem snapshot is a frozen image of a filesystem at a given instant in time.

System-cache buffers: System memory used to hold recently used data. For example, in the filesystem, system-cache buffers are used to hold recently accessed disk blocks.

Further Information

A good overview of filesystems can be found in Chapter 3 of Silberschatz and Galvin (1994).

Most operating systems today use filesystem designs similar to those found in (McKusick and Neville-Neil, 2005). This chapter summarizes information on file layout on disk described in Section 79.2, filesystem naming described in Section 79.3, and the traditional disk space management described in Section 79.9.

A description of the non-over-writing Open Solaris ZFS filesystem can be found in (Solaris, 2012).

Ars Technica has an excellent in-depth article on SSD technology (Hutchinson, 2012).

The Stanford VLSI Group's CPU database project has a catalog of evolving microprocessor technology (Stanford, 2012).

References

Asthana, P., Superdense optical storage, *IEEE Spectrum* **32**(8), p. 25–31 (August 1995).

Chen, P., E. Lee, G. Garth, R. Katz, and D. Patterson, RAID: High-performance, reliable secondary storage, *ACM Computing Surveys* **26**(2), pp. 145–185 (June 1994).

Date, C. J., *An Introduction to Database Systems* 6 th edn., Addison-Wesley, Reading, MA (1995).

Hutchinson, L., Solid-state revolution: In-depth on how SSDs really work, *Ars Technica*, http://arstechnica.com/information-technology/2012/06/inside-the-ssd-revolution-how-solid-state-disks-really-work/, accessed in June 2012.

IEEE, *POSIX: Part 1: System Application Program Interface*, Institute of Electrical and Electronic Engineers, New York (1994).

McKusick, M. K. and G. V. Neville-Neil, *The Design and Implementation of the FreeBSD Operating System*, Addison-Wesley, Reading, MA (2005).

Patterson, D., G. Garth, and R. Katz, A case for redundant arrays of inexpensive disks (RAID), *SIGMOD Record* **17**(3), pp. 109–116 (June 1988).

Silberschatz, A. and P. Galvin, *Operating System Concepts* 4 th edn., Addison-Wesley, Reading, MA (1994).

Solaris, The ZFS filesystem, http://en.wikipedia.org/wiki/ZFS, accessed in January 2012.

Stanford, Stanford VLSI group CPU database, http://cpudb.stanford.edu/about (Started April 2012).

Sweeney, A., D. Doucette, W. Hu, C. Anderson, M. Nishimoto, and G. Peck, Scalability in the XFS file system, *USENIX Association Conference Proceedings,* San Diego, CA, pp. 1–14 (January 1996).

56

Performance Evaluation of Computer Systems

Alexander Thomasian
Thomasian & Associates

56.1 Introduction

Since the inception of computers, there has been intense interest in their performance. Other dimensions are reliability [143], which is briefly discussed in the chapter on storage systems, power consumption that arises in the broader context of green computing [1], footprint, ease of operation, etc.

Initial attention was paid to the speed of the *central processing unit* (*CPU*), specified in *millions of instructions per second* (*MIPS*) [2]. The MIPS rating of computer systems is specified with respect to the VAX 11/780 marketed as a one MIPS computer [3]. The execution speed of a computer depends on the instruction mix, such as Gibson's, but also the sequence of their execution as affected by data and control dependencies (branches) [73]. There is a slowdown when the instructions and data cannot be accessed in one cycle as a result of cache misses [73].

The slow speed of *arithmetic logic units* (*ALUs*) of early computers led to research in developing and implementing new arithmetic algorithms [63]. Pipelining was introduced to allow a new arithmetic operation to be started per CPU cycle, and this concept was extended to the instruction execution level. For example, the IBM 360/91 decoded one instruction per its 60 nanosecond (ns) cycle, so that at its best it could process 16.7 MIPS [4]. Unlike the IBM 360/85 introduced at about the same time, the 360/91 was not equipped with a cache memory, but rather a sufficient number of instruction registers to hold the inner loop of a program and 16-way memory interleaving for parallel data access.

Early computers were uniprocessors with expensive magnetic core memories. Given the high cost of CPU hardware, multiprogramming was introduced to take advantage of the CPU-I/O overlap in these computers, allowing the CPU to be shared by several jobs. A rule of thumb to determine the *multiprogramming level* (*MPL*) to fully utilize the CPU is discussed in Section 56.3.5. Multiprogramming is also required by *online transaction processing* (*OLTP*) to attain a higher *transaction* (*txn*) throughput at an acceptable mean txn response time.

Multiprocessors and multicomputers have been developed to attain more computing power than uniprocessors. Multiprocessors have been classified as shared everything systems by Stonebraker [107], since resources such as the main memory and disks can be accessed by all processors. There is a performance degradation due to cache interference and also synchronization delays in multiprocessors [70]. In shared disk or data-sharing systems, the processing is shared by two or more computers, which need to coordinate their activities in processing database applications, such as coherency control of database buffers and concurrency control [96]. In the shared-nothing category, there are multicomputers, which share an interconnection network and communicate via message passing or a *distributed shared memory* (*DSM*) [57].

There are numerous applications requiring lengthy execution times if executed sequentially on a uniprocessor. There is also the issue of timeliness, that a computation needs to be completed by a certain deadline, for example, daily weather forecasting (although long-term forecasts are now available). Fortunately, many such computations exhibit a high degree of parallelism [57], so that their processing can be speeded up by parallel computing [5]. Parallelism is also exploited in *database management systems* (*DBMSs*) and web search engines to ensure timeliness of responses.

A semblance of parallelism can be attained by pipelining, which overlaps the processing of successive operations in time. Supercomputers in 1970s were highly pipelined vector computers, with Cray-1 being the world's fastest computer at that time. Flynn classified computers in mid-1960s into four categories based on the number of instruction and data streams [6,73]: *single instruction, single data stream* (*SISD*) [7] computers are uniprocessors. *Single instruction, multiple data stream* (*SIMD*) computers such as the Illiac IV were an early form of parallel computing, where a control unit issued decoded instructions to 64 *processing elements* (*PEs*). This categorization also applies to vector computers, where a single instruction applies to elements of vectors. *Graphics processing units* (*GPUs*) belong to this category [73]. *Multiple instruction, multiple data stream* (*MIMD*)s are multicomputers. Several notable early computers are described in [44].

Amdahl in 1967 enunciated Amdahl's law, as an argument against parallel computing to promote IBM's uniprocessors at that time [8]. Given that a fraction $1 - F$ (of the sequential execution time of a program) cannot be parallelized to run at K processors Speedup $= (1 - F + F/K)^{-1}$, for example, $F = 0.5$ yields at most a speedup of two.

The highest computing power is currently attained by MIMD parallel processors with the *message passing interface* (*MPI*) [9]. IBM's Deep-Blue computer, which allegedly beat world's chess champion Kasparov in 1997 [10], was followed by the Blue-Gene supercomputer for understanding cellular architectures [11]. Another IBM computer called Watson consisting of 90 servers with 16 TB RAM outplayed human contestants of the TV game show Jeopardy [12]. Japan's 10.51 Petaflops/second (10^{15} floating point operations per second) K computer was the world's fastest computer until IBM introduced the Sequoia 16.32 Petaflops computer at the Lawrence Livermore National Laboratory in June 2012 [13]. Changes in supercomputer rankings are listed in [14].

Growth in supercomputer performance is limited by reliability and power consumption considerations. It was said in 1970s that the Illiac IV supercomputer failed every 30 s. When the failure was due to one of its 64 PEs, the system was fixed by manually substituting the failed PE with a spare and restarting. Periodic checkpointing was used to save the state of the computation to ensure that completed work preceding a system failure is not lost [15]. The issue of power consumption is partially addressed by placing the supercomputers in areas where cheap hydroelectric power is available.

Current microprocessor speeds are specified in GigaHertz, which is an upper bound to instruction execution rate, assuming one instruction is decoded per cycle. The growth in microprocessor speeds since late 1970s is shown in Figure 1.1 in [73]. The increase in processor speeds has slowed down in recent years, but the fact that the number of transistors per chip can be increased at a faster rate than its speed is exploited in these systems by providing multiple cores (processors and caches), which provide an increase in raw computing power. An extra effort is, however, required to exploit the parallelism provided by multicore systems [73].

Since the cycle time of main memory is much slower than the processor, cache memories operating at processor speeds are provided to improve performance. Although CPU caches are much smaller than the main memory, they attain small miss rates due to spatial and temporal locality of references [53,73,86,106]. Spatial locality is exploited by prefetching, since it is more advantageous to access consecutive disk (resp. main memory) blocks with one disk (resp. memory) access [73]. An efficient implementation of cache coherence and synchronization in multiprocessor systems is critical for performance [57,73]. Several levels of caches referred to as L1, L2, …, are usually provided. In multiprocessors, L1 caches are dedicated to one processor, while L2 caches are shared by two or more processors [16].

Cache, DRAM (main memories), and (magnetic) disk technologies and their performance are discussed in [75]. *Hard disk drives* (*HDDs*) are slow due to their mechanical nature, and their impact on system performance and reliability is discussed in the chapter on storage systems. Numerous technologies referred to as *storage class memory* (*SCM*) are under development, which in addition to flash memories compete in replacing disks or at least serve as their caches. Performance modeling studies of the processor, caches, main memories, and peripheral devices are discussed in a collection of papers in [81]. The standard computer architecture textbook emphasizing system performance [73].

Benchmarks have been developed over the years for evaluating computer system performance [17]. *Standard Performance Evaluation Corporation*'s SPEC CPU2006 has a floating point and integer version [18]. SPECfp2000 results for several popular microprocessors are given in Figure 1.15 in [73]. Early database benchmarks are discussed in a handbook edited by Gray [19]. He was instrumental in proposing the debit–credit benchmark, which due to its simple nature suited Tandem's shared-nothing organization [20]. The performance goal for the debit–credit OLTP benchmark in 1985 was one *kilo transactions per second* (*KTPS*), which was determined at the point where the mean txn response time stayed below a certain time limit, say 2 s. The metric for batch applications, such as sorting, is elapsed time. Two computer systems are usually involved in benchmarking, with one system generating txns for the *system under test* (*SUT*). This benchmark termed TPC-1 was adopted by the *Transaction Processing Council* (*TPC*). TPC later introduced other benchmarks for processing ad hoc queries in relational DBMSs, decision support, data mining, and web applications [21].

I/O benchmarks were developed by the *Storage Performance Council* (*SPC*) to estimate the performance and energy consumption of disk drives with OLTP and database applications generating random and sequential disk accesses for various RAID levels (see chapter on storage systems) [22].

This chapter deals mainly with methods that can be used to evaluate the performance of computer systems at a macro level, for example, txn response time versus throughput. The emphasis is on queueing theory, which has shown its usefulness in performance evaluation of computer systems, but also computer communication networks. Simulation is a much more flexible tool, which can be used when the system under consideration is analytically intractable. In fact, passive resources such as miss rates in the memory hierarchy can be determined by trace-driven simulation. Finally, performance is evaluated via measurement studies, which can be used for validating analytical and simulation models (see Section 2.2 in [83]).

In what follows, Section 56.2 is a brief introduction to queueing theory. Analysis of Markovian queueing systems with exponential interarrival and service times is discussed in Section 56.3. Non-Markovian queueing systems including M/G/1 and GI/M/1 are discussed in Section 56.4. Several applications of these queueing systems are also provided. Fork-join queueing systems are discussed in Section 56.5. The analysis of *queueing network models* (*QNMs*) is discussed in Section 56.6. Buffer space management in computer systems is discussed in Section 56.7. The performance of *concurrency control* (*CC*) methods is analyzed in Section 56.8. Simulation modeling of computer systems is discussed in Section 56.9. Conclusions are presented in Section 56.10.

56.2 Queueing Theory Basics

Queueing systems consist of a single or multiple servers and a queue that holds incoming requests when all servers are busy. Requests are served as soon as a server becomes available, but forced idleness under some circumstances may reduce the mean overall response time, for example, when the overall queue length is short and one of two servers is considerably faster than the other (see Section 56.3.3). Incoming requests are lost when no queue is provided or the queue is full. Such requests may be resubmitted or are lost as in the case of a caller who does not redial after finding the line busy. The probability of requests being rejected due to a full queue is of interest in the design of multistage switches. Requests wait in the queue until they are served according to a scheduling policy, such as *first-come, first-served* (*FCFS*). Queueing theory mainly deals with estimating the waiting time in queues, which are formed by humans in the more civilized countries of the world.*

According to Kendall's notation, interarrival and service times in queues are denoted by D (deterministic), M (Markovian or exponential), G (general), E_k (Erlang with k exponential stages), H_k (k-way hyperexponential distribution), etc. We use GI/G/m/K/M to represent a queueing system with general independent interarrival time, general service time, m servers, total capacity K to hold requests, and M sources. By default $K = \infty$ and $M = \infty$.

Little's law is a fundamental law in queueing theory, which was well known even before it was proved by him [23]. This law states that the long-term average number of requests in a stable system (L) is equal to the long-term average effective arrival rate (λ) multiplied by the average time requests spend at the system (W) or $L = \lambda W$. A pictorial proof of Little's law is given in [78,80,92].

Arrivals may be from an infinite or finite number of sources. Examples of the former are calls to a telephone switching system, txn arrivals at an OLTP system, and requests to a web server. Quasi-random arrivals apply to the users of a time-sharing system, where a finite number of users generate successive requests following a delay known as think time [80]. The error introduced by approximating a finite number of sources in an M/G/1//N queueing system with M/G/1 is explored in Section 56.4.2.

The Poisson arrival process is associated with exponential interarrival times with *probability distribution function* (*PDF*) $A(t) = 1 - e^{-\lambda t}$, $t > 0$, and *probability density function* (*pdf*) $a(t) = \frac{dA(t)}{dt} = \lambda e^{-\lambda t}$. Poisson arrivals are uniformly distributed over intervals [78], for example, given a time interval (0, 1), n arrivals partition the interval to segments of average length $1/(n + 1)$. This observation was utilized in [134] in determining the mean seek distance according to the scan policy. Generally, we denote the pdf of a continuous *random variable* (*r.v.*) as $f(x)$, so that for positive r.v.'s $\int_0^\infty f(x)dx = 1$ and the PDF is

* John Milton, the English poet, ends his poem "On his blindness" with "They also serve who only stand and wait." At US Congress, money is made by professional "waiters" who are hired by lobbyists to stand in their place for congressional hearings they wish to attend.

$F(t) = \int_0^t f(x)dx$, so that $F(\infty) = 1$. This is not always the case, for example, the pdf of the busy period in M/G/1 is $\int_0^\infty g(y)dy = 1/\rho$, which is smaller than one for $\rho > 1$ (see Section 56.4.1). In the case of discrete r.v.'s instead of the pdf, we have the *probability mass function (pmf)* with $\sum_{k\geq 0} p_k = 1$. The *cumulative distribution function (cdf)* is $P_k = \sum_{j=0}^{k} p_j$, which tends to one as $k \to \infty$ [42,143].

The mean, second moment, the variance, and coefficient of variation squared of the exponential distribution are

$$\overline{t_e} = \int_0^\infty [1 - A(t)]\, dt = \frac{1}{\lambda},\ \overline{t_e^2} = \int_0^\infty t^2 a(t) dt = \frac{2}{\lambda^2},\ \sigma_e^2 = \frac{1}{\lambda^2},\ c_e^2 = \frac{\sigma_e^2}{(\overline{t_e})^2} = 1. \tag{56.1}$$

The *Laplace Stieltjes transform (LST)* of a positive r.v. with pdf $f(x)$ and its ith moment is given as follows [78,143]:

$$\mathcal{F}^*(s) = \int_0^\infty f(x)\, e^{-sx} dx,\quad \overline{x^i} = (-1)^i \frac{d^i \mathcal{F}^*(s)}{ds^i}\Big|_{s=0} = \int_0^\infty x^i f(x) dx. \tag{56.2}$$

For the exponential distribution, the LST. ith moment, the variance, and the coefficient of variation squared are

$$\mathcal{A}^*(s) = \int_{x=0}^\infty (\lambda e^{-\lambda t}) e^{-st} dt = \lambda \int_0^\infty e^{-(s+\lambda)t} dt = \frac{\lambda}{s+\lambda},\quad \overline{t_a^i} = \frac{i!\lambda}{(s+\lambda)^{i+1}}\Big|_{s=0} = \frac{i!}{\lambda^i}.$$

The exponential distribution has the memoryless property, that is, given an elapsed time t_e, the time to the next arrival remains the same:

$$P[t \leq (t+t_e) | t > t_e] = \frac{P[t_e < t \leq (t+t_e)]}{P[t > t_e]} = \frac{1 - e^{-\lambda(t+t_e)} - (1 - e^{-\lambda t_e})}{e^{-\lambda t_e}} = 1 - e^{-\lambda t}.$$

The memoryless property also holds for the discrete geometric distribution. In tossing a coin given that p is the probability of a head and $q = 1 - p$ the probability of a tail, then the pmf of the number of coin tosses leading to a head and the mean number of tosses is [42,143]:

$$p_k = pq^{k-1} \quad \overline{K} = p \sum_{k\geq 1} kq^{k-1} = p \frac{d}{dq} \sum_{k\geq 1} q^k = \frac{1}{p}. \tag{56.3}$$

An alternate method to determine the moments of a discrete r.v. is to compute its z-transform, take its derivatives, and set $z = 1$:

$$P(z) = \sum_{k\geq 1} p_k z^k = pz \sum_{k\geq 1} (qz)^{k-1} = \frac{pz}{1-qz}, \tag{56.4}$$

where the sum converges for $|qz| < 1$.

Taking the ith derivatives of $P(z)$'s, we have

$$\overline{K(K-1)\ldots(K-i+1)} = \frac{d^i P(z)}{dz}\Big|_{z=1} = \sum_{k\geq 0} k(k-1)\ldots(k-i+1) p_k z^{k-i}\Big|_{z=1}. \tag{56.5}$$

In the case of the geometric distribution ($\overline{K}$) and $\overline{K(K-1)}$ are

$$\overline{K} = \frac{dP(z)}{dz}\Big|_{z=1} = \frac{p}{(1-qz)^2}\Big|_{z=1} = \frac{1}{p}, \quad \overline{K(K-1)} = \frac{d^2P(z)}{dz^2}\Big|_{z=1} = \frac{2pq}{(1-qz)^3}\Big|_{z=1} = \frac{2q}{p^2} \tag{56.6}$$

Consider passengers arriving at random instants at a bus station, where the bus interarrival time pdf is $f(x)$. Of interest in the mean residual time to the next bus arrival (r_1). Renewal theory shows that arrivals tend to occur during longer bus interarival times with a pdf $f_X(x) = xf(x)/m_1$, whose mean is $\overline{X}_1 = \int_0^\infty x^2 f(x)dx = m_2/m_1$. The forward and backward recurrence times are identically distributed so that the forward recurrence time or the mean residual lifetimes is $r_1 = m_2/(2m_1)$ [78]. In the case of the exponential distribution for bus interarrival times, $r_1 = 1/\lambda$, which is a consequence of its memoryless property. Given that the passenger arrives $1/\lambda$ time units after the arrival of the previous bus, then the bus interarrival time encountered by passengers is $f_X(x) = \lambda^2 x e^{-\lambda x}$, which is the two-stage Erlang distribution with mean $2/\lambda$ [143].

Generally the k-stage Erlang distribution (E_k) consists of $k > 1$ exponential stages with rate $k\mu$, so that $f_E(x) = k\mu(k\mu x)^{k-1} e^{-k\mu x}/(k-1)!$ for $x \geq 0$. The mean, variance, the coefficient of variation squared, the LST, and the LST with an infinite number of stages ($k \to \infty$) are

$$\overline{x}_E = \frac{1}{\mu}, \ \sigma_E^2 = \frac{1}{k\mu^2}, \ c_E^2 = \frac{1}{k} < 1, \ \mathcal{E}^*(s) = \left(\frac{k\mu}{s+k\mu}\right)^k, \ \lim_{k\to\infty}\left(1+\frac{s}{k\mu}\right)^{-k} = e^{-s/\mu}. \tag{56.7}$$

In fact, $e^{-s/\mu} \Leftrightarrow u_0(x - 1/\mu)$, which is a unit impulse at $x = 1/\mu$. It is shown in Figure 4.5 in [78] that the pdf for E_k tends to a fixed value as $k \to \infty$.

The pdf of the Gamma distribution is given as $f_G(x) = \alpha(\alpha x)^{\beta-1} e^{-\alpha x}/\Gamma(\beta - 1)$ for $x > 0$, where the Gamma function $\Gamma(t) = \int_0^\infty x^{t-1}e^{-x}dx$. When t is an integer $\Gamma(t) = (t-1)!$, which is the factorial function. The Gamma distribution in this case is tantamount to an Erlang distribution with t stages. The mean, variance, coefficient of variation squared, and LST for this distribution are

$$\overline{x}_G = \frac{\beta}{\alpha}, \ \sigma_G^2 = \frac{\beta}{\alpha^2}, \ c_G^2 = \frac{1}{\beta}, \ \mathcal{G}^*(s) = \left(\frac{\alpha}{\alpha+s}\right)^\beta. \tag{56.8}$$

The I-way Hyperexponential distribution (H_2) $f_H(x) = \sum_{i=1}^I p_i\mu_i e^{-\mu_i x}$, $x > 0$. For two-way hyperexponential $p_1 = p$ and $p_2 = 1 - p$ with $p \neq 0$ and $p \neq 1$. The mean, second moment, coefficient of variation squared, and the LST of this distribution are

$$\overline{x}_H = \sum_{i=1}^I \frac{p_i}{\mu_i}, \ \overline{x_H^2} = 2\sum_{i=1}^I \frac{p_i}{\mu_i^2}, \ c_H^2 = \frac{\overline{x_H^2}}{(\overline{x}_H)^2} - 1, \ \mathcal{F}_H(s) = \sum_{i=1}^I p_i \frac{\mu_i}{s+\mu_i}. \tag{56.9}$$

The LST $\mathcal{H}^*(s)$ is a weighted sum of I LSTs for corresponding exponentials. It can be shown easily that $c_H^2 > 1$ for $I = 2$. The CPU service time distribution has been approximated by the hyperexponential distribution as shown in Figure 3.13 in [144].

The exponential and related distributions are popular in that they lend themselves to tractable analytic solutions. The Erlang (resp. hyperexponential) distribution can be used as an approximation to distributions with a coefficient of variation smaller (resp. larger) than one. The Coxian distribution $f_C(x)$ consists of multiple exponentials in tandem similarly to an Erlang distribution, but with a variable number of stages being bypassed. Based on the choice of its parameters, it can attain a coefficient of variation smaller or larger than one. In the case of the Coxian distribution with r stages with services rates μ_i for $1 \le i \le r$, the probability that the ith stage is visited is β_i and $\gamma_i = 1 - \beta_i$ is the probability that an early exit is taken, so that $\gamma_{r+1} = 1$. An Erlang distribution is obtained by setting $\gamma_i = 0$, $\forall i$. The LST for the Coxian distribution can be specified as follows [78,79]:

$$\mathcal{F}_C^*(s) = \int_{x=0^-}^{\infty} f_C(x) e^{-sx}\,dx = \gamma_1 + \sum_{i=1}^{r} \beta_1\beta_2 \ldots \beta_i \gamma_{i+1} \prod_{j=1}^{i} \left(\frac{\mu_j}{s+\mu_j}\right). \tag{56.10}$$

The integration is started at $x = 0^-$ to capture the discontinuity at $x = 0$, which has probability γ_1. Given that $\beta_i = q$, $\gamma_i = 1 - q$, $\forall i$, $\gamma_{r+1} = 1$ and $\mu_i = \mu$, $\forall i$ then the mean, second moment, coefficient of variation squared, and the LST are

$$\bar{x}_C = \sum_{i=1}^{r} \frac{q^i}{\mu}, \quad \overline{x_C^2} = 2\sum_{i=1}^{r} i\frac{q^i}{\mu^2}, \quad c_C^2 = 2\frac{\sum_{i=1}^{r} iq^i}{\left(\sum_{i=1}^{r} q_i\right)^2}, \quad C^*(s) = 1 - q + \sum_{i=1}^{r} q^i \left(\frac{q\mu}{s+\mu}\right)^i. \tag{56.11}$$

The pdf of the uniform distribution can be specified as $f_U(x) = (b - a)^{-1}$ for $a \le x \le b$. Its mean, variance, coefficient of variation squared, and LST are

$$\bar{x}_U = \frac{a+b}{2}, \quad \sigma_U^2 = \frac{(b-a)^2}{12}, \quad c_U^2 = \frac{1}{3}\left(\frac{b-a}{b+a}\right)^2, \quad \mathcal{F}_U^*(s) = \frac{e^{-bs} - e^{-as}}{s(a-b)}. \tag{56.12}$$

Other arrival processes and service time distributions and methods to generate them are discussed in Sections 6.8 and 11.5.2 in [45], Also, see Chapter 5 by Shedler in [82].

We use a disk drive as an example of a server whose service time is its access time. Disks are in fact a complex queueing system in that the service time of requests is affected by the order in which requests are processed and hence the disk scheduling policy. For accesses to randomly placed small disk blocks, the disk access time is dominated by the positioning time of the read/write head, which is the sum of seek time and rotational latency [75]. The transfer time of small disk blocks is negligible for current disks with very high recording densities.

We consider a disk drive with a mean access time $\bar{x}_{disk} = 5$ milliseconds (ms) with FCFS scheduling. It is intuitively obvious that this disk can process at most $\lambda_{max} = 1/\bar{x}_{disk} = 200$ accesses per second. This also follows from the fact that with an arrival rate λ the utilization of a single server, defined as $\rho = \lambda\bar{x}_{disk}$, is at most one. Note that ρ is also the fraction of time that the server is busy. Ordinarily for $\rho < 1$ the server alternates between idle and busy periods, but with an arrival rate $\lambda > \lambda_{max}$ and $\rho > 1$ the length of the queue increases indefinitely. For $\rho > 1$ the probability that the busy period ends is $1/\rho$ (see Equation 56.38 for $s = 0$ and the discussion that follows). When the system load $\lambda\bar{x} > 1$, the number of servers should be increased such that $\rho = \lambda\bar{x}/m$ is less than one. Attention should be paid to maintaining acceptable mean response time, for example, for $\rho = 0.5$ the mean response time in an M/M/1 queueing system is doubled and it increases 10fold for $\rho = 0.9$ (see Equation 56.14). It is interesting to note that this increase is only fivefold in an M/M/2 queueing system at the same utilization (see Equation 56.20).

Disk access time to randomly placed disk blocks is slow, since it is the sum of seek time, rotational latency, and transfer time [139]. This led to the proposal of the SCAN and *shortest seek time first* (*SSTF*) disk scheduling policies as improvements to FCFS scheduling in mid-1960s by Denning in [53]. Both are outperformed by the *shortest access time first* (*SATF*) policy, which was proposed by Jacobson and Wilkes at HP in 1991 to minimize the positioning time, that is, the sum of seek time and latency. SATF's reduction in disk access time with respect to FCFS is $\approx n^{1/5}$ [139], for example, with $n = 32$ enqueued requests $\bar{x}_{SATF} \approx \bar{x}_{FCFS}/2$.

The response time ($\tilde{r}$) of a queueing system is the sum of waiting ($\tilde{w}$) and service time ($\tilde{x}$), that is, $\tilde{r} = \tilde{w} + \tilde{x}$. The mean response time is obtained by taking the expectations of both sides as $R = W + \bar{x}$. The variance of response time ($\tilde{r}$) is given as the sum of the variances of $\tilde{w}$ and $\tilde{x}$

$$\sigma_R^2 = E\left[(\tilde{w}+\tilde{x})^2\right] - (W+\bar{x})^2 = \left(\overline{W^2} - W^2\right) + \left(\overline{x^2} - (\bar{x})^2\right) + 2E[\tilde{w}\tilde{x}] - 2W\bar{x} = \sigma_W^2 + \sigma_X^2.$$

The expression $E[\tilde{w}\tilde{x}] = W\bar{x}$, since the waiting time ($\tilde{w}$) is independent from the service time ($\tilde{x}$) for FCFS, LCFS (nonpreemptive), and *service in random order (SIRO)* scheduling policies.

The mean waiting time W for M/G/1 queues can be obtained by noting that due to *Poisson Arrivals See Time Averages (PASTA)* [45], arrivals encounter the mean queue length $\overline{N}_q = \lambda W_{FCFS}$ and the equality holds due to Little's law. Also due to PASTA a busy server is encountered with probability ρ, where the mean residual service time of the request in service is $m_1 = \overline{x^2}/(2\bar{x})$

$$W_{FCFS} = \overline{N}_q\bar{x} + \rho m_1 \Rightarrow W_{FCFS} = \frac{\lambda\overline{x^2}}{2(1-\rho)} = \frac{\rho\bar{x}\left(1+c_X^2\right)}{2(1-\rho)}. \tag{56.13}$$

This equation is known as the Pollaczek-Khinchine (P. K.) formula for mean waiting time. There are two observations: (1) W_{FCFS} increases in proportion to $\rho/(1-\rho)$ and as $\rho \to 1$ $W \to \infty$, (2) For a fixed ρ, W_{FCFS} increases with c_X^2. With respect to W_{FCFS} for the exponential distribution with $c_X^2 = 1$ is halved for a fixed service time with $c_F^2 = 0$ and doubled with a hyperexponential distribution with $c_H^2 = 3$

$$W_{FCFS} = \frac{\rho\bar{x}}{1-\rho}, \quad R_{FCFS} = W_{FCFS} + \bar{x} = \frac{\bar{x}}{1-\rho}. \tag{56.14}$$

The P. K. formula for mean waiting time (W) also holds for LCFS and SIRO policies, but both policies have higher second moments than FCFS [56], as follows:

$$E\left[w_{LCFS}^2\right] = \frac{E\left[w_{FCFS}^2\right]}{1-\rho}, \quad E\left[w_{SIRO}^2\right] = \frac{E\left[w_{FCFS}^2\right]}{1-(\rho/2)}.$$

$E\left[w_{FCFS}^2\right] = \overline{W^2}$ is given by Equation 56.33 in Section 56.4.1.

A weakness of FCFS scheduling is that all requests, regardless of their service time, encounter the same mean waiting time (W_{FCFS}). *Round-robin* (*RR*) scheduling allows CPU requests to make progress up to a short time quantum q before the request is preempted and another request is scheduled to run. Shorter requests leave the system after receiving a few quanta. This discrete time queueing model is analyzed in [53], and Chapter 6 in [111] is dedicated to discrete time systems. *Processor sharing* (*PS*) is an extreme form of RR with $q \to 0$, which yields $R_{PS}(x) = x/(1-\rho)$ when the service time is x. Given that the service time pdf is $b(x)$, the mean response time for PS is

$$R_{PS} = \int_0^\infty \frac{xb(x)dx}{1-\rho} = \frac{\bar{x}}{1-\rho}, \quad W_{PS} = R_{PS} - \bar{x} = \frac{\rho\bar{x}}{1-\rho}. \tag{56.15}$$

Referring to Equation 56.13 it follows $W_{PS} < W_{FCFS}$ for $c_X^2 > 1$, so that PS is the preferred policy when service times are more variable than the exponential.

Foreground/background (*FB*) scheduling attains even better response times for short requests than RR, since after the time quantum of a CPU expires, longer requests are demoted to lower priority queues, so that shorter requests do not compete with longer requests (see Section 56.4.6 in [79]).

Request priorities can be assigned externally, for example, the processing of OLTP txns at the CPU is given a higher priority than batch jobs. CPU preemptions are possible because they incur a low overhead and that the preemptive resume policy is applicable, so that by saving the state of the computation (the program counter and registers) in main memory, a preempted request may resume from the point it was interrupted. It is inefficient to implement preemptive disk scheduling, since only a preemptive repeat policy is applicable in this case, with the unproductive seek and latency phases being repeated. Preemption is only possible after the completion of a seek and during latency and transfer phases [129]. An efficient implementation of priorities with SATF disk scheduling is attained by multiplying estimated access time of higher priority disk requests by $\alpha \approx 0.625$ [139]. Formulas for M/G/1 queueing systems with priorities are given in Section 56.4.6.

The conditional waiting time for the *shortest processing time* (*SPT*) first scheduling with job service time PDF $B(x)$ is given in [53,56,78] as

$$W_{SPT}(t) = \frac{\lambda\overline{x^2}}{2\left[1-\lambda\int_0^t x\,dB(x)\right]^2}. \tag{56.16}$$

For small t $W_{SPT}(t) = W_0 = \lambda\overline{x^2}/2$, which is the mean residual service time, since the request is processed as soon as the current request is completed. For large t $W_{SPT}(t) = W_0/(1-\rho)^2 = W_{FCFS}/(1-\rho)$.

Table 3.1 in [79] is a summary of optimal scheduling algorithms. One dimension specifies preemptiveness and the second dimension specifies whether (1) the processing time is known, (2) the processing time distribution is known, and (3) no information about processing time is available. In case (1) SPT (also known as *shortest job first* (*SJF*)) and *shortest remaining processing time* (*SRPT*) are applicable depending on preemptiveness. In case (2) *shortest expected processing time* (*SEPT*) and *shortest expected remaining time time* (*SERPT*) are applicable, depending on preemptiveness. Finally in case (3) FCFS, LCFS, SIRO, PS, and FB are applicable.

56.3 Analysis of Markovian Models

Queues with Poisson arrivals and exponential service times, referred to as Markovian are undertaken with *continuous time Markov chains* (*CTMCs*) models [42,45,78,80,143]. Markov chains for queueing systems have discrete states representing the number of requests in the system. Transitions among states are exponentially distributed. The transition rate from each state is the sum of the transitions due to arrivals and completions and the holding time in that state is the inverse of this sum. Systems modeled in this section can be represented by one-dimensional dimensional CTMCs known as *birth-death* (*BD*) processes. Only nearest neighbor transitions are allowed in the more simple queueing systems discussed in Chapter 3 in [78], but this is not so in the case of queueing systems with bulk arrivals and bulk service discussed in Chapter 4 in [78] and Section 56.3.6. The discussions in Chapters 3 and 4 in [78] are referred to as baby queueing theory by Kleinrock [24].

In Section 56.3.1, we use BD models to analyze the performance of M/M/1 queues. In Section 56.3.2, we analyze M/M/1/K queues as an example of a finite capacity system. In Section 56.3.3, we analyze the M/M/m queueing system with $m = 2$ to illustrate the benefits of resource sharing in this context and provide the analytical results for the M/M/m queueing system. In Section 56.3.4, we analyze the performance of an M/M/2 queueing system, whose servers have different service rates. In Section 56.3.5, we analyze the performance of an M/M/1//M queueing system, which has finite sources. In Section 56.3.6, we describe Markovian systems with bulk arrivals and bulk service. The *matrix geometric method* (*MGM*) is briefly introduced in Section 56.3.7. We postulate FCFS scheduling in this section so that this fact is elided in the notation that follows.

56.3.1 M/M/1 Queueing System

The operation of an M/M/1 queue can be represented by a BD process with states denoted by $\mathcal{S}_k$, where $k \geq 0$ is the number of requests at the system. The transition $\mathcal{S}_k \to \mathcal{S}_{k+1}$ with rate λ denotes the arrival of a request or a birth at $\mathcal{S}_k$ and the transition $\mathcal{S}_{k+1} \to \mathcal{S}_k$ with rate μ denotes the completion of a request or a death at $\mathcal{S}_{k+1}$. In summary, the arrival rate is λ and the service rate is μ.

The probability that the system is at state S_k in steady state is denoted by p_k. This also means that for a long time interval T the system is at S_k for time Tp_k. In other words, p_k is the fraction of time that the system is at $\mathcal{S}_k$. In equilibrium, the number of transitions in the two directions with increasing and decreasing number of requests are equal, that is, $\lambda T p_k = \mu T p_{k+1}$, for $k \geq 0$.* Dividing both sides by T we have $\lambda p_k - \mu p_{k+1}$, for $k \geq 0$. Given the utilization factor $\rho = \lambda/\mu$, $p_k = \rho p_{k-1}$ and $p_k = \rho^k p_0$. The sum $S = \sum_{k\geq 0} p_k = p_0(1-\rho)^{-1}$ for $\rho < 1$. Since the probabilities sum to one, setting $S = 1$ yields $p_0 = 1 - \rho$ and hence $p_k = (1 - \rho)\rho^k$ for $k \geq 0$, which is the modified geometric distribution [143]. The mean number of requests at the system is

$$\overline{N} = \sum_{k\geq 1} kp_k = (1-\rho)\sum_{k\geq 1} k\rho^k = (1-\rho)\rho\sum_{k\geq 1} k\rho^{k-1} = (1-\rho)\rho\frac{d}{d\rho}\sum_{k\geq 1}\rho^k = \frac{\rho}{1-\rho}. \tag{56.17}$$

The second moment of N and its variance can be obtained as follows:

$$\overline{N(N-1)} = \sum_{k\geq 0} k(k-1)p_k = (1-\rho)\rho^2\frac{d^2}{d\rho^2}\sum_{k\geq 1}\rho^k = \frac{2\rho^2}{(1-\rho)^2}.$$
$$\sigma_N^2 = \overline{N(N-1)} + \overline{N} - \left(\overline{N}\right)^2 = \frac{\rho}{(1-\rho)^2}. \tag{56.18}$$

The mean response time and mean waiting for M/M/1, which was given as a special case of Equation 56.13 by Equation 56.14 can be obtained by applying Littles law:

$$R_{\text{M/M/1}} = \frac{\overline{N}}{\lambda} = \frac{\overline{x}}{1-\rho}, \quad W_{\text{M/M/1}} = R_{\text{M/M/1}} - \overline{x} = \frac{\rho\overline{x}}{1-\rho}. \tag{56.19}$$

The z-transform for the number in system is

$$P(z) = \sum_{k\geq 0} p_k z^k = (1-\rho)\sum_{k\geq 0}(\rho z)^k = \frac{1-\rho}{1-\rho z},$$

where the sum converges for $|\rho z| < 1$.

56.3.2 M/M/1/K Queueing System with Finite Capacity

An M/M/1/K queue can hold at most K requests, but requests arriving to a full system are lost, so that this queueing system is referred to as *blocked calls cleared* (BCC) in telephony [78]. The state equilibrium probabilities are $p_k = p_0(\lambda/\mu)^k$ for $0 \leq k \leq K$, so that $\sum_{k=0}^{K} p_k = 1$ yields $p_0 = (1-\lambda/\mu)/\left(1-(\lambda/\mu)^{K+1}\right)$. The fraction of arrivals lost is p_K and the throughput of processed requests is $\lambda' = \lambda(1 - p_K)$. It follows that the server utilization is $\rho = \lambda(1 - p_K)/\mu$. In the special case $\lambda = \mu$, applying L'Hospital's rule [95] yields $p_k = 1/(K + 1)$ for $0 \leq k \leq K$. For $K = 1$, $p_0 = \mu/(\lambda + \mu)$ and $p_1 = \lambda/(\lambda + \mu)$.

* The concept of equilibrium is illustrated in [78] by a passenger traveling among three cities: C_1, C_2, and C_3. The probability of ending up in C_j at the end of the day after being in C_i in the morning is given by the the probability matrix $P = (p_{i,j})$ for $1 \leq i, j \leq 3$. This is a discrete time Markov chain and assuming that the passenger is initially at C_i, then he is in one of the cities with probability $\underline{\pi}(1) = \underline{\pi}(0)P$ after 1 day and after n days $\underline{\pi}(n) = \underline{\pi}(0)P^n$ or $\underline{\pi}(n) = \underline{\pi}(n-1)P$. As $n \to \infty$, the time index can be elided and the set of equations can be expressed as $\underline{\pi} = \underline{\pi}P$, which implies the final state is independent of the initial state.

56.3.3 M/M/m Queueing System and Resource Sharing

Consider an M/M/2 queue with arrival rate λ with two servers each with service rate μ, so that the server utilization is $\rho = \lambda/(2\mu)$. The state equilibrium equations are

$$\lambda p_0 = \mu p_1, \quad \lambda p_k = 2\mu p_{k+1} \text{ for } k \geq 1.$$

It follows from $p_k = 2p_0\rho^k$ for $k \geq 1$ and the fact that probabilities sum to one yields $p_k = (1-\rho)/(1+\rho)$ and hence

$$\overline{N}_{M/M/2} = \sum_{k\geq 0} kp_k = \frac{\rho}{1-\rho^2}, \quad R_{M/M/2} = \frac{\overline{N}}{\lambda} = \frac{\overline{x}}{1-\rho^2} \tag{56.20}$$

The M/M/2 queueing system with an arrival rate 2λ, has the same server utilization as two (slow) M/M/1 queues with service rates μ, each with arrival rate λ. It is easy to see that $R_{M/M/2} < R^{slow}_{M/M/1}$, which is because there is the possibility of queueing at one of the servers, while the other server is idle since the jockeying of requests is not provided. We next consider a single fast server with mean service time $\overline{x}_{fast} = 1/(2\mu)$, and arrival rate 2λ, so that $\rho = \lambda/\mu$. We have $R^{fast}_{M/M/1} = \left(1/(2\mu)\right)/(1-\rho)$, which is smaller than $R_{M/M/2}$. A generalization of this discussion appears in Section 5.1 in [79]. It is important to note that for highly variable service times the M/G/m queueing system is preferable to a single fast M/G/1 queue with the same total processing capacity, which is because with highly variable service times a lengthy requests may get in the way of all other requests in a single queue.

The M/M/∞ queueing system has an infinite number of servers so that there is no queueing delay. The state equilibrium equations $\lambda p_k = (k+1)\mu p_{k+1}$ for $k \geq 0$ lead to $p_k = p_0\left(\lambda/\mu\right)^K/k!$ for $k \geq 0$. It follows from $\sum_{k\geq 0} p_k = p_0 \sum_{k\geq 0} \left(\lambda/\mu\right)^k = 1$ that $p_0 = e^{-\lambda/\mu}$ and $p_k = e^{-\lambda/\mu}(\lambda/\mu)^k/k!$, which is the Poisson distribution with mean and variance $\overline{N} = \sigma^2_N = \lambda/\mu$ [42,78,143]. Note that this solution applies to an M/G/∞ queue with mean service time $\overline{x} = 1/\mu$, so that $R = \overline{x}$. The transient probabilty distribution for the M/G/∞ queueing system is given by Equation 56.9 in [143] as $P[X(t) = k] = e^{-\lambda tp}(\lambda tp)^k/k!$, where $p = \int_0^t [(1-G(x))/t]dx$, where $G(x)$ is the PDF for service time. Note that for $t = \infty$, $tp = \int_0^\infty [1-G(x)]\,dx = 1/\mu$ [143], which is previously obtained steady-state probability for M/M/∞.

For the M/M/m queueing system, $\rho = \lambda/mu$ and p_k can be expressed as follows:

$$p_k = p_0\frac{(m\rho)^k}{k!}, \quad k \leq m; \quad p_k = p_0\frac{\rho^k m^m}{m!}, \quad k \geq m; \quad p_0 = \left[\sum_{k=0}^{m-1}\frac{(m\rho)^k}{k!} + \frac{(m\rho)^m}{(1-\rho)m!}\right]^{-1}. \tag{56.21}$$

The probability of queueing $P[k \geq m] = P_m$, the mean number in the system, and mean response time are

$$P_m = \sum_{k=m}^{\infty} p_k = p_0\sum_{k=m}^{\infty}\frac{(m\rho)^k}{m!\times m^{k-m}}, \quad \overline{N} = m\rho + \frac{\rho}{1-\rho}P_m, \quad R = \left[1 + \frac{P_m}{m(1-\rho)}\right]. \tag{56.22}$$

56.3.4 M/M/2 with Heterogeneous Servers

The analysis of a two-server queue with heterogeneous servers, as given by Example 8.29 in [143], is significantly more complex than an M/M/2 queue. The service rate μ_1 at server 1 is assumed to be higher than the service rate μ_2 at server 2, that is, $\mu_1 > \mu_2$. The utilization factor is $\rho = \lambda/(\mu_1 + \mu_2)$.

The states of the CTMC are specified as (n_1, n_2), where n_1 is the number of jobs in the queue including any at the faster server and $n_2 \in \{0, 1\}$ indicates whether the slow server is busy or not. Jobs arriving at an empty system are first served by the fast server. The state equilibrium equations are

$$\lambda p(0,0) = \mu_1 p(1,0) + \mu_2 p(0,1)$$

$$(\lambda + \mu_1) p(1,0) = \mu_2 p(1,1) + \lambda p(0,0)$$

$$(\lambda + \mu_2) p(0,1) = \mu_1 p(1,1)$$

$$(\lambda + \mu_1 + \mu_2) p(1,1) = (\mu_1 + \mu_2) P(2,1) + \lambda \big(P(0,1) + P(1,0) \big)$$

$$(\lambda + \mu_1 + \mu_2) p(n,1) = (\mu_1 + \mu_2) p(n+1,1) + \lambda p(n-1,1) \text{ for } n > 1.$$

The mean number of jobs at the queueing system is

$$\overline{N} = \frac{1}{F(1-\rho)^2}, \quad F = \frac{\mu_1 \mu_2 (1 + 2\rho)}{\lambda(\lambda + \mu_2)} + \frac{1}{1-\rho}. \tag{56.23}$$

This analysis has been utilized to determine the effect of adding a slow computer to fast computer on improving the mean response time. At lower values of λ, the mean response time can be improved by not utilizing the slower computer at all, while at $\lambda > \mu_1$ the slower computer allows higher throughputs up to $\lambda < \mu_1 + \mu_2$ to be attained.

56.3.5 M/M/1//M or Machine Repairman Model

The *machine repairman model* (*MRM*) considers M machines with exponentially distributed times to failure with rate λ. Machines are repaired by a single repairman with exponentially distributed service time with rate μ. Alternatively there are M users with exponentially distributed think times with mean $Z = 1/\lambda$, which share a computer system with exponentially distributed processing times with mean $1/\mu$. The state equilibrium equations with k requests at the computer system are

$$(M-k)\lambda p_k = \mu p_{k+1} \quad \text{for } 0 \le k \le M-1.$$

The probabilities are given as follows:

$$p_k = p_0 \prod_{i=0}^{k-1} \frac{(M-i)\lambda}{\mu} = p_0 \left(\frac{\lambda}{\mu}\right)^k \frac{M!}{(M-k)!}, \quad p_0 = \left[\sum_{k=0}^{M} \left(\frac{\lambda}{\mu}\right)^k \frac{M!}{(M-k)!} \right]^{-1}. \tag{56.24}$$

It follows from Little's result applied to the whole system that

$$M\left(\frac{1}{\lambda} + R\right) = \mu(1 - p_0) \Rightarrow R = \frac{M}{\mu(1-p_0)} - \frac{1}{\lambda}. \tag{56.25}$$

The normalized mean response time $R_n = R\mu$ versus M has two regions: (1) For small M $R_n \approx 1$, that is, no queueing delay. (2) For large M $R_n \approx M - \mu Z$, that is, R_n increases by one for each additional user. The two lines intersect at $M^* = 1 + \mu Z$, which is known as the saturation number (see Figure 4.27 in [79]). Under the best circumstances users do not encounter queuing delays up to M^*. This model can be used to determine the minimum MPL in order to keep the CPU 100% utilized under perfect conditions, that is, fixed disk and CPU service times. Provided 1 millisecond (ms) of CPU processing entails a 10 ms disk access time requires the MPL to be equal to 11.

User productivity at a time-sharing system is affected by the system response time, that is, user think times have been observed to decrease with subsecond response times. The number of programmer interactions per hour is 180 for a mean response time of $R_1 = 3$ s, but this number is doubled (to 371) when the mean response time is $R_2 = 0.3$ s [25]. It follows from $3600 = 180(R_1 + Z_1)$ that $Z_1 = 17$ and similarly $Z_2 = 9.4$, that is, the mean think time is almost halved with the improved R.

This simple QNM was successfully utilized by Scherr in the analysis of the *compatible time sharing system* (*CTSS*) at MIT [26,104]. CTSS is a uniprogrammed system, which follows a round-robin policy in processing one program at a time (see Section 56.2). Once the time quantum is expired, the currently active request is preempted and swapped out to a drum, so that the next pending request can be loaded into main memory.

The single server in MRM can be replaced by a state-dependent server representing the computer system using the hierarchical decomposition approach, so that the service rate of the server is $T(k)$, $k \geq 1$ (see Section 56.6.7).

56.3.6 Bulk Arrival and Bulk Service Systems

These systems analyzed in Chapter 4 in [78] are modeled by BD processes with non-nearest neighbor state transitions. We illustrate the analytic solution method required by considering the bulk arrival system, where the probability of i requests per arrival is g_i with a z-transform $G(z) = \sum_{i \geq 1} g_i z^i$. The state equilibrium equations for the number of requests in the system and the associated z-transform is

$$\lambda p_0 = \mu p_1, \quad (\lambda + \mu) p_k = \mu p_{k+1} + \sum_{i=0}^{k-1} p_i \lambda g_{k-i} \text{ for } k \geq 1.$$

$$P(z) = \frac{\mu p_0 (1-z)}{\mu(1-z) - \lambda z(1 - G(z))} = \frac{p_0}{1 - \frac{\lambda}{\mu} z \frac{1 - G(z)}{1-z}}. \quad (56.26)$$

The unknown p_0 can be obtained by noting that $P(1) = 1$. Since the fraction is indeterminate, we apply L'Hospital's rule [95] by differentiating the numerator and denominator with respect to z. This yields $p_0 = 1 - \rho$ with $\rho = \lambda G'(1)/\mu$, where $G'(1)$ is the average bulk size. The reader is referred to [78] for the solution of M/E_r/1 queueing system.

We next consider bulk service systems, which can serve up to r requests simultaneously, so that the service rate is $r\mu$ and $\rho = \lambda/(r\mu)$. Service is not delayed if there are $k < r$ requests in the systems and are served at service rate $k\mu$. The state equilibrium equations in this case are

$$\lambda p_0 = \mu \sum_{i=1}^{r} p_i, \quad (\lambda + \mu) p_k = \mu p_{k+r} + \lambda p_{k-1} \text{ for } k \geq 1.$$

The z-transform in this case is given as

$$P(z) = \frac{N(z)}{D(z)} = \frac{\sum_{k=0}^{r-1} p_k (z^k - z^r)}{r\rho z^{r+1} - (1 - r\rho) z^r + 1}. \quad (56.27)$$

The denominator $D(z)$ has $r + 1$ roots. Of these $z = 1$ is a root of $N(z)$. It can be shown by Rouche's theorem (see Appendix I in [78]) that $D(z)$ has $r - 1$ roots with $|z| < 1$, which are also roots of $N(z)$, since

$P(z)$ is analytic. There is a single root $|z_0| > 1$, so that $P(z) = [K(1 - z/z_0)]^{-1}$ and $P(1) = 1$ yields $K = 1/(1 - 1/z_0)$. Inverting $P(z)$ we have a modified geometric distribution

$$P(z) = \frac{1-1/z_0}{1-z/z_0} \Leftrightarrow p_k = \left(1-\frac{1}{z_0}\right)\left(\frac{1}{z_0}\right)^k. \tag{56.28}$$

Queueing systems with Erlang interarrival and exponential service times and Poisson arrivals and Erlang service times, respectively, can be dealt with similarly. Note that E_r/M/1 and M/E_r/1 queues are special cases of GI/M/1 and M/G/1 queues and lend themselves to the alternative analysis in Sections 56.4.7 and 56.4.1. A text dedicated to bulk queues is [51].

56.3.7 Brief Note on the Matrix Geometric Method

MGM, developed by Neuts and Lipsky, are a generalization of BD processes with vectors substituting single states. Solution techniques for MGMs are discussed in [45,92]. MGM is used to analyze scheduling policies in mirrored disk systems in [142].

56.4 Analysis of Non-Markovian Queueing Systems

This section is organized as follows. We first provide the analysis of the M/G/1 queueing system in Section 56.4.1. The inadequacy of the two moment approximation for M/G/m queueing system is shown in [71], which also proposes upper and lower bounds for its mean waiting time. In Section 56.4.2, we use the mean response time of the M/G/1//N finite population model to assess the error introduced by the M/G/1 model with a Poisson arrival process. In Section 56.4.3, Lagrange multipliers are used to minimize the mean response time in accessing data blocks from heterogeneous disks. The *vacationing server model* (*VSM*) in the context of M/G/1 queues is discussed in Section 56.4.4. The application of the VSM to the analysis of the scheduling of readers and writers with a threshold policy is discussed in Section 56.4.5. We next proceed to the analysis of priority queueing in Section 56.4.6. The analysis of the GI/M/m queues is discussed in Section 56.4.7. Request routing to minimize response time is discussed in Section 56.4.8. The analysis of polling systems is discussed in Section 56.4.9, followed by load-sharing in homogeneous distributed systems in Section 56.4.10.

56.4.1 M/G/1 and M/G/m Queueing Systems

The queueing system M/G/1 is a single server queue with Poisson arrivals with rate λ and general service time with pdf $b(x)$, whose ith moment is $\overline{x^i}$ and its coefficient of variation squared is $c_X^2 = \overline{x^2}/(\overline{x})^2 - 1$. The LST of the $b(x)$ is $\mathcal{B}^*(s) = \int_o^\infty b(x)e^{-sx}\,dx$. The analysis of the FCFS scheduling policy to determine the mean (W) and second moment ($\overline{W^2}$) of waiting time requires the first two and three moments of service time, respectively.

The distribution of the number in system (p_k) is obtained by specifying the number of requests at the M/G/1 queue at departure instants [78]: $q_{n+1} = q_n - \Delta_{q_n} + v_{n+1}$, where q_{n+1} and q_n is the number of requests left behind by the n + 1*st* and *nth* requests at the system, and v_{n+1} is the number of arrivals while the n + 1*st* request was in service, and $\Delta_{q_n} = 0$ if $q_n = 0$ and is otherwise one. The probabilities at departure instants $d_k = P[q_n]$ can be obtained by solving the recurrence relation for q_n using z-transforms. Given one step transitions, in a nonsaturated M/G/1 system, the number of up and down transitions at arrival and departure instants are equal in the long run. Therefore, the probability of encountering k requests at an arrival instant $r_k = d_k$ and due to PASTA the probabilities at an arrival instant are denoted by r_k, so that $r_k = p_k$ and hence $p_k = d_k$.

It can be shown that the z-transform for the number in the system is

$$Q(z) = V(z)\frac{(1-\rho)(1-z)}{1-V(z)/z}. \tag{56.29}$$

$V(z)$ is the z-transform of number of arrivals during the service time of a single request

$$V(z) = \int_0^\infty e^{-\lambda x} \sum_{k\geq 0} \frac{(\lambda xz)^k}{k!} b(x)dx = \int_0^\infty e^{-\lambda x(1-z)} b(x)dx = \mathcal{B}^*(\lambda(1-z)).$$

The substitution $V(z) = \mathcal{B}^*(\lambda(1-z))$ in Equation 56.29 yields

$$Q(z) = \mathcal{B}^*(\lambda(1-z))\frac{(1-\rho)(1-z)}{\mathcal{B}^*(\lambda(1-z))-z}. \tag{56.30}$$

In the case of exponential service times $\mathcal{B}^*(s) = \mu/(s+\mu)$,

$$Q(z) = \frac{1-\rho}{1-\rho z} \Leftrightarrow p_k = (1-\rho)\rho^k, \ k \geq 0.$$

Inversions of $Q(z)$ to yield p_k for M/H$_2$/1, M/D/1, and M/E$_k$/1 are given in Table 7.2 and graphed for $\rho = 0.75$ in Figure 7.4 in [92] (interestingly $p_1 > p_0 = 0.25$ for M/D/1).

Similarly to $V(z) = \mathcal{B}^*(\lambda(1-z))$, it can be shown that the following relationship holds between the z-transform for the number of arrivals during the response time of a request, which is also the number of requests left behind by FCFS scheduling. Given $d_k = p_k$ this is also the z-transform of the distribution of the number of requests in the system, hence $Q(z) = \mathcal{S}^*(\lambda - \lambda z)$, where $\mathcal{S}^*(s)$ is the LST of response time pdf. Setting $s = (1-\lambda)z$

$$\mathcal{S}^*(s) = \mathcal{B}^*(s)\frac{s(1-\rho)}{s-\lambda+\lambda\mathcal{B}^*(s)}. \tag{56.31}$$

Given that $\tilde{r} = \tilde{w} + \tilde{x}$ and that the mean waiting time $\tilde{x}$ is independent from the service time ($\tilde{x}$) for the FCFS policy, we have $\mathcal{S}^*(s) = \mathcal{W}^*(s)\mathcal{B}^*(s)$. It follows that the LST for the waiting time pdf ($w(t)$) in M/G/1 is given as [78]

$$\mathcal{W}^*(s) = \int_{t=0^-}^{\infty} w(t)e^{-st}\,dt = \frac{s(1-\rho)}{s-\lambda+\lambda\mathcal{B}^*(s)}. \tag{56.32}$$

Integration starting with $t = 0^-$ is necessary, because of the possibility of a unit impulse at the origin, that is, the probability that the waiting time is zero is $1-\rho$. The first and second moment of the waiting time can be obtained by differentiating $\mathcal{W}^*(s)$ once and twice with respect to s and setting $s = 0$

$$\overline{W^i} = (-1)^i \left.\frac{d^i\mathcal{W}^*(s)}{ds^i}\right|_{s=0}, \quad W = \frac{\lambda\overline{x^2}}{2(1-\rho)}, \quad \overline{W^2} = 2W^2 + \frac{\lambda\overline{x^3}}{3(1-\rho)}. \tag{56.33}$$

Takacs recurrence relation can be used to derive higher moments of waiting time as follows [78], where $\overline{W^0} = 1$:

$$\overline{W^k} = \frac{\lambda}{1-\rho}\sum_{i=1}^{k}\binom{k}{i}\frac{\overline{x^i}}{i+1}\overline{W^{k-i}}. \tag{56.34}$$

A generalization of Little's results is $\overline{N(N-1)} = \lambda^2\overline{R^2}$, which can be extended to higher moments [78].

In the case of M/M/1 queueing system, $\mathcal{B}^*(s) = \mu/(s+\mu)$ and using tables for inverting LSTs in [42,78,143] we invert $\mathcal{S}^*(s)$ as follows:

$$\mathcal{S}^*(s) = \frac{\mu(1-\rho)}{s+\mu(1-\rho)} \Leftrightarrow s(t) = \frac{1}{R}e^{-t/R}, \tag{56.35}$$

where $R = [\mu(1-\rho)]^{-1}$ is the mean response time. The LST $\mathcal{S}^*(s)$ for the M/M/1 can be obtained directly by assuming that k requests are encountered by an arrival and unconditioning on p_k

$$\mathcal{S}^*(s) = \sum_{k\geq 0} p_k\left(\frac{\mu}{s+\mu}\right)^{k+1} = \frac{\mu(1-\rho)}{s+\mu}\sum_{\geq 0}\left(\frac{\lambda}{s+\mu}\right)^k = \frac{\mu(1-\rho)}{s+\mu}\frac{1}{1-\dfrac{\lambda}{s+\mu}} = \frac{\mu(1-\rho)}{s+\mu(1-\rho)}.$$

The following approximations for the percentiles of M/G/1 response time, proposed by Martin, are quoted in [42]: $R_{90\%} \approx R + \sigma_R$ and $R_{95\%} \approx R + 2\sigma_R$.

The busy period is the duration of the time required to serve jobs starting with an empty queueing system until it is empty again. For $\rho < 1$ M/G/1 queues alternate between busy and idle periods, with mean durations g_1 and $1/\lambda$, respectively. The mean duration of the busy period can be determined easily by noting that $g_1/(\bar{g}_1 + 1/\lambda) = \rho$, which is the fraction of time the server is busy.

The LST of the busy period can be specified noting that each one of the arrivals during the first request entails subbusy periods with the same distribution as the original distribution, as follows:

$$E\left[e^{-sY}\middle|x_1 = x, \tilde{v} = k\right] = e^{-sx}\left[\mathcal{G}^*(s)\right]^k.$$

$$E\left[e^{-sY}\middle|x_1 = x\right] = \sum_{k\geq 0} e^{-sx}\frac{\left(\lambda\mathcal{G}^*(s)x\right)^k}{k!}e^{-\lambda x} = e^{-x(s+\lambda-\lambda\mathcal{G}^*(s))}.$$

$$\mathcal{G}^*(s) = \int_0^\infty e^{-x(s+\lambda-\lambda\mathcal{G}^*(s))}dB(x) = \mathcal{B}^*(s+\lambda-\lambda\mathcal{G}^*(s)). \tag{56.36}$$

The moments of the busy period can be obtained by taking its derivative

$$g_i = (-1)^i\frac{d^i\mathcal{G}^*(s)}{ds^i}\Big|_{s=0}, \quad g_1 = \frac{\overline{x}}{1-\rho}, \quad g_2 = \frac{\overline{x^2}}{(1-\rho)^3}. \tag{56.37}$$

Note that g_2 increases very rapidly with increasing server utilization. For the M/M/1 queue with $\mathcal{B}^*(s) = \mu/(s+\mu)$, the distribution of the busy period is given by a Bessel function of the first kind of order one, which is obtained by solving $\mathcal{G}^*(s) = \mu/\left(s+\lambda-\lambda\mathcal{G}^*(s)+\mu\right)$ and inverting the negative root of the quadratic equation

$$\mathcal{G}^*(s) = \frac{1}{\lambda}\left[\mu+\lambda+s-\sqrt{(\mu+\lambda+s)^2-4\mu\lambda}\right]. \tag{56.38}$$

The probability that the busy period ends is $\mathcal{G}^*(0) = \int_0^\infty g(y)dy = 1/\rho$. CTMCs have been classified in [78] as positive recurrent, null-recurrent, and transient. In the context of M/G/1 queues, the system is positive recurrent for $\rho < 1$, which implies that the expected time to return to an empty system is finite. For $\rho = 1$, the system is null-recurrent, which means that the probability that the system becomes empty again is one, but the expected time to reach an empty system is infinite. Finally, when $\rho > 1$, the system is transient, which means that with positive probability the system will never empty again.

The *z*-transform for the number of requests served in a busy period is given in [78], which can be used to determine the mean and variance of the number of requests

$$F(z) = z\mathcal{B}^*\big(\lambda(1-F(z)\big), \quad h_1 = (1-\rho)^{-1}, \quad \sigma_h^2 = \frac{\rho(1-\rho)+\lambda^2\overline{x^2}}{(1-\rho)^3}. \tag{56.39}$$

No exact analysis is available for M/G/m queues, but the well-known two moment approximation for M/G/m can be expressed as [45]

$$W \approx \frac{\left(1+c_X^2\right)}{2m\mu(1-\rho)} P_m, \quad P_m = \frac{(m\rho)^m}{m!(1-\rho)} p_0. \tag{56.40}$$

In the corresponding M/M/m queue, $\mu = 1/\bar{x}$ is the service rate and $P_m = P[k \geq m]$. This formula is exact for M/M/m and M/G/1 queues.

This approximation is shown to be highly inaccurate against simulation results for higher values of c_X^2 in [71]. It is also shown that the lognormal and Pareto distribution with the same three moments have very different mean waiting times. Conjecture 1 in [71] provides the following higher and lower bounds for M/G/m with respect to M/D/m for any finite c_X^2 and $\rho < 1$

$$W_h = \left(1+c_X^2\right)E\left[W_{M/D/m}\right], \quad W_\ell = E\left[W_{M/D/m}\right] + \frac{\rho - \frac{m-1}{m}}{1-\rho}\,\frac{c_X^2}{2} \times \left[\rho \geq \frac{m-1}{m}\right].$$

The additional term is added to the first term when the condition in the brackets is true and yields one. A simplified expression for M/D/m is given in [65], where *D* denotes the fixed service time

$$P[W \leq x] = e^{-\lambda(kD-x)} \sum_{j=0}^{km-1} Q_{km-j-1} \frac{\left[-\lambda(kD-x)\right]^j}{j!} \text{ for } (k-1)D \leq x \leq KD. \tag{56.41}$$

Let p_i denote the stationary distribution of *i* requests in the system

$$p_i = \sum_{j=0}^{m} p_j \frac{(\lambda D)^i}{i!} e^{-\lambda D} + \sum_{j=m+1}^{i+m} p_j \frac{(\lambda D)^{i+m-j}}{(i+m-j)!} e^{-\lambda D}, \forall i \in \mathbb{N}_O.$$

Let $q_0 = \sum_{j=0}^{m} p_j$ and $q_i = p_{i+m}$ for $i > 0$, then $Q_j = \sum_{i=0}^{j} p_i$.

The probability Q_n that there are no more than *n* requests is difficult to compute since it requires the solution of an infinite set of linear equations [65,141]. A *fast Fourier transform* (*FFT*) and a geometric tail approach are proposed in [80a].

Several approximations for GI/G/m queues are listed in Section 2.5 in [79] and Section 6.3.6 in [45].

56.4.2 M/G/1 Approximation to M/G/1//N

We repeat the discussion in [49], based on an analysis by Takacs (see [111] Volume 2), to explore the accuracy of approximating its mean response time (R_N) in an M/G/1//N queueing system with N sources with the mean response time of an M/G/1 queue given by Equation 56.13. This is because of the difficulty of calculating R_N as specified next

$$R_N = \frac{N}{\mu} - \frac{1}{\lambda} + \left[\lambda \sum_{n=0}^{N-1} \binom{N-n}{n} \frac{1}{\mathcal{H}_n}\right]^{-1}, \quad \mathcal{H}_n = \prod_{i=1}^{n} \frac{\mathcal{B}^*(i\lambda)}{1-\mathcal{B}^*(i\lambda)}, \quad \mathcal{H}_0 = 1. \tag{56.42}$$

The queueing system M/G/1 overestimates R for the same server utilization ρ. The accuracy of the approximation improves with increasing N as would be expected and is given as $(R/R_N - 1)100\%$. The minimum number of sources to attain a 5% error level increases with the coefficient of variation of the distribution: constant, Erlang-2, exponential, and Hyperexponential-2 (with equal branching probabilities and branch means 0.5 and 1.5) for $\rho = 0.1$ is 2/2/3/4, for $\rho = 0.4$ is 10/16/21/25, and for $\rho = 0.7$ is 61/101/142/188.

The M/G/1//N and M/G/1/K queueing systems are analyzed in Chapters 4 and 5 of Volume 2 in Takagi's three-volume series [111].

56.4.3 Routing to Minimize Mean Response Times

Data allocation to improve response time in the context of heterogeneous servers is considered in this section. Static policies that route requests in proportion to server speeds will balance server utilizations and under Markovian assumptions (exponential service times), the mean response times at server i (R_i) will be proportional to the mean service time at the server $\bar{x}_i = 1/\mu_i$.

The usual criterion is to *minimize (mean) response time (MRP)*, but a more sophisticated policy is to minimize the maximum mean response time policy, which is referred as *minimize maximum (mean response time) policy (MMP)*. The policies considered in this section are static, although dynamic policies such as shortest queue routing are expected to improve performance [91].

Data allocation on disks affects the mean access time. High degrees of access skew are observed among disks in large computer installation, where the utilization of most disks is negligible, see for example, [108]. Data reallocation methods in large file systems are considered in [147]. Striping in RAID (see chapter on storage systems) balances the load in each array, but there is still a need to balance loads across RAID5 arrays [139].

We consider the reallocation of data blocks in a disk array with N heterogeneous disks modeled as M/G/1 queues to minimize the overall mean disk response time. The mean and variance of access time to access data blocks at disk $1 \le i \le N$ is $1/\mu_i$ and σ_i^2, respectively. We assume that disk capacity is not a problem and the data can be reallocated at the level of blocks such that disk access rates λ_i for $1 \le i \le N$ can be treated as continuous variables with a fixed sum $\Lambda = \sum_{i=1}^{N} \lambda_i$ as data blocks are reallocated. The goal of the analysis is to determine the new arrival rates to the N disks also denoted by λ_i for $1 \le i \le N$, which will minimize the mean response time over all disks [48].

The mean overall response time, given the mean responses time R_i for $1 \le i \le N$, is given as

$$R = \sum_{i=1}^{N} \frac{\lambda_i}{\Lambda} R_i(\lambda_i), \quad R_i(\lambda_i) = \frac{1}{\mu_i} + \frac{\lambda_i\left(1/\mu_i^2 + \sigma_i^2\right)}{2(1-\lambda_i/\mu_i)} \text{ for } 1 \le i \le N.$$

The Lagrange multipliers method [95] is applied in [48] to minimize R with respect to λ_i for $1 \le i \le N$ subject to $\Lambda = \sum_{i=1}^{N} \lambda_i$ and $\lambda_i < \mu_i$ for $1 \le i \le N$ [27]. There is no solution if $\Lambda > \sum_{i=1}^{N} \mu_i$. We define F as follows and set its derivatives with respect to λ_i's equal to zero to determine their optimum values

$$F = \sum_{i=1}^{N} \frac{\lambda_i}{\lambda} R_i(\lambda_i) + \alpha\left(\Lambda - \sum_{i=1}^{N} \lambda_i\right).$$

$$\lambda_i = L_i(k) = \mu_i \left[1 - \sqrt{\frac{\mu_i^2\sigma_i^2 + 1}{2k\mu_i + \mu_i^2\sigma_i^2 - 1}}\right] \quad \text{for } 1 \le i \le N \text{ with } k = 2\alpha\Lambda.$$

$L_i(k)$ are increasing functions in k and so is their sum $L(k) = \sum_{i=1}^{N} L_i(k)$. The value of k satisfying $\Lambda = L(k)$ may yield a solution with some of the λ_i's negative at slower disks. This violation is resolved by repeating the optimization step but excluding slower disks from the allocation process.

Nonlinear optimization (Lagrangian functions) is used in [112] to optimize the routing of jobs to heterogeneous host computers, which are interconnected by a communication network whose delays satisfy the triangle inequality.

A static routing policy addressing fairness for user level performance measures, that is, mean response time in a multicomputer system with N heterogeneous processors with different speeds is proposed and evaluated in [68]. The criterion used for optimal workload allocation is the one minimizing the maximum expected response time at a subset of N processors to which jobs are routed. MMP equalizes the expected response times at utilized processors emphasizing fairness. Chapter 6 in [74] deals with the fair resource allocation problem.

The response time characteristic of the N heterogeneous processors $R_n(\lambda)$, for $1 \le n \le N$ known to the router is a nondecreasing function of λ. For an arrival rate Λ, the router sends a fraction of requests $p_n = \lambda(n)/\Lambda$ to the *nth* processor. The mean response time of requests is

$$R(\Lambda) = \sum_{i=n}^{N} R_n\left(\Lambda p_n(\Lambda)\right) p_n(\Lambda).$$

Criterion of Optimality for MMP

For a given $\Lambda > 0$, find the probabilities $p_i(\Lambda)$, for $1 \le i \le N$, so that the maximum average response time incurred on any system is minimized:

$$\min \max_n \{R_n(\Lambda p_n(\Lambda)\} < \infty,$$

where $\sum_{i=n}^{N} p_n(\Lambda) = 1, p_n(\Lambda) \ge 0$ for $1 \le n \le N$.

It is shown in the chapter that provided λ does not exceed the sum of the maximum throughputs at the nodes

Proposition. There is an integer $1 \le K \le N$ and a vector of routing probabilities $\mathbf{p} = (p_1, \ldots, p_K, 0, \ldots, 0)$ such that

$$R_i(\Lambda p_i) = R_{min}(\Lambda) \quad \text{for } 1 \le i \le K$$

where $R_{min}(\Lambda)$ is the minimum mean overall response time.

A simple algorithm to compute the allocation vector is given in the paper. Numerical results show that MMP attains fairness at a tolerable increase in mean response time.

56.4.4 Vacationing Server Model

We consider the *vacationing server model* (*VSM*) in the context of M/G/1 queues [111]. According to the VSM, the server goes on vacation after the busy period ends. In the case of the VSM with multiple vacations, the server repeats its vacation if the queue is empty after it returns from vacation. Otherwise it serves enqueued requests in FCFS order until the queue is emptied. Variations of the VSM within the context of M/G/1 queueing systems are analyzed in detail in [111].

The VSM with multiple vacations of variable durations is utilized in [130] for the analysis of rebuild processing in RAID5 disk arrays (see chapter on storage systems). The variability in service times is due to the fact that the first access for rebuild reading requires a seek, while succeeding rebuild reads do not. This complexity is taken into account in the analysis in [130], but not in the analysis that follows, where all rebuild reads are assumed to have a fixed service time. Given that the ith moment of vacation time is given as $\overline{y^i}$, then the mean residual delay observed by requests arriving while rebuild is in progress is $\overline{n}_1 = \overline{y^2}/(2\overline{y})$ [78]. Based on PASTA, the mean waiting time in M/G/1 queues with vacations is

$$W = \overline{N}_q\overline{x} + \rho m_1 + (1-\rho)n_1 \Rightarrow W = \frac{\lambda \overline{x^2}}{2(1-\rho)} + \frac{\overline{y^2}}{2\overline{y}}. \quad (56.43)$$

That the mean waiting time W in this case is the sum of the mean waiting time in M/G/1 (see Equation 56.13) and the mean mean residual vacation time is attributable to the decomposition property of the VSM [67].

The busy period is elongated by the delay encountered by the first request arriving during the final vacation during which a request arrives. To simplify the discussion, we assume that vacation times have a fixed duration $y = T$, so that mean remaining vacation time after an arrival in a vacation period is given as

$$v_r = T - \frac{\int_0^T te^{-\lambda t}dt}{\int_0^T e^{-\lambda t}dt} = \frac{T}{1-e^{-\lambda T}} - \frac{1}{\lambda}.$$

The mean effective service time of the first request, which starts the next modified busy period, referred to as the delay cycle is given as $\overline{z} = v_r + \overline{x}$ is required to determine its duration [79]

$$\overline{d} = \frac{\overline{z}}{1-\rho}. \quad (56.44)$$

Note similarity to the duration of busy period, which was given as $g_1 = \overline{x}/(1-\rho)$ [78].

The probability that an arrival occurs according to a Poisson process with rate λ in an interval T is $p = 1 - e^{-\lambda T}$. The distribution of the number of vacations taken in the idle period is $P_k = p(1-p)^{k-1}$ with a mean $\overline{k} = 1/p$. The time required per vacation is $(\overline{d} + 1/\lambda)/\overline{k}$.

The VSM can be used to determine the performance of rebuild processing in RAID5 to reconstruct the contents of a failed disk on a spare disk [138]. This requires the reading of consecutive tracks from surviving disks and *eXclusive-ORing* (*XOR*)*ing* them to rebuild missing tracks. The rebuild reading of the track in progress is completed before the service of user requests starts. The response time of user requests is increased by n_1 according to Equation 56.43, which equals $T/2$ in this case. The rebuild time can be estimated using the aforementioned analysis. The discussion can be extended to take into account the effect of read redirection and the update load as rebuild progresses [131,135].

56.4.5 VSM Analysis of Scheduling for Readers and Writers

Readers and writers is a classical paradigm in OS theory and practice. Readers can be processed concurrently, while writers are processed one at a time. The processing of readers and writers in FCFS order can be inefficient compared to a policy that processes read requests uninterrupted as long as possible.

The VSM was applied to the analysis of the *threshold fastest emptying* (*TFE*) policy for scheduling readers and writers in [124]. There is an infinite backlog of readers, which are processed with a degree of concurrency M with rate ν. Writers arrive according to a Poisson process with rate λ and have service time moments $\overline{x^i}$ for $i \geq 1$, so that the utilization of the system due to writers is $\rho = \lambda\bar{x} < 1$. Writers are processed until the writer queue is emptied, after which the processing of readers is resumed. No new readers are processed when the number of writers reaches the threshold $K \geq 1$. The analysis in [124] yields the throughput for readers and the mean number of writers in the queue

$$\gamma(K) = (1-\rho_w)M\nu\frac{K+\lambda/\nu}{K+H_M\lambda/\nu},$$

$$\overline{N_q} = \frac{\lambda^2\overline{x^2}}{2(1-\rho_w)} + \frac{\alpha^{(2)}(1)}{2\alpha^{(1)}(1)}.$$

(note correction to Equation 3.8 in [124]). The z-transform $\alpha(z)$ is for the number of arrivals during a vacation, which is the sum $K+\tilde{J}$, where K is fixed and $\tilde{J}$ is the number of arrivals while the system is being emptied from readers

$$\alpha(z) = M\nu\sum_{m=0}^{M-1}(-1)^m\binom{M-1}{m}\frac{z^K}{(m+1)\nu+\lambda(1-z)}.$$

It is shown in [124] that TFE with threshold $K = 1$ attains a higher throughput for writers than FCFS. The difference between the two policies is that writers arriving in a busy period are served in spite of intervening reader arrivals.

With exponential writer service times, a CTMC, which is infinite in one of its two dimensions, can be used to analyze this system. The more interesting case with both reader and writer arrivals with two thresholds to balance reader and writer response times and to attain a higher throughput was investigated via simulation in [133].

56.4.6 Priority Queueing Systems

The VSM is a special case of priority queues with an infinite backlog of low-priority requests. Consider an M/G/1 queue with P priority classes, where P is the highest priority. The arrival rate for class p requests is λ_p and the first two moments of its service time are $\bar{x}_p$ and $\overline{x_p^2}$, respectively. According to Kleinrock's conservation law [79],

$$\sum_{p=1}^{P}\rho_p W_p = \frac{\rho W_0}{1-\rho}, \tag{56.45}$$

where $\rho_p = \lambda_p\bar{x}_p$, $\rho = \sum_{p=1}^{P}\rho_p$, and $W_0 = \frac{1}{2}\sum_{p=1}^{P}\lambda_p\overline{x_p^2}$. The intuition behind this law is that the reduced response time of some classes is at the cost of increased response time for others (borrow from Peter to pay Paul). In the case of two priority classes,

$$W_1 = \frac{W_0}{(1-\rho_2)(1-\rho)}, \quad W_2 = \frac{W_0}{1-\rho_2}. \tag{56.46}$$

Note that both waiting times are affected by the residual service time of requests in progress (as reflected by W_0), but high-priority requests are only affected by ρ_2.

In the case of preemptive priorities, as far as higher priority classes are concerned. It is as if lower priority classes do not exist. Hence, for class 2, we have an M/G/1 queueing system, the mean response time for class 1 requests according to Equation 3.65 in [82] is

$$E[r_1] = \frac{\bar{x}}{1-\rho_2} + \frac{\lambda_1 \overline{x_1^2} + \lambda_2 \overline{x_2^2}}{2(1-\rho_2)(1-\rho)}, \quad E[r_2] = \bar{x}_2 + \frac{\lambda_2 \overline{x_2^2}}{2(1-\rho_2)}. \tag{56.47}$$

56.4.7 GI/M/m Queueing System

This queueing system has a general interarrival time and an exponential service times. The analysis is based on the embedded Markov chain at arrival instants [78] so that there is no residual interarrival time. The mean waiting time for the GI/M/1 queue can be specified similarly to M/M/1, but with a parameter σ instead of ρ, which in GI/M/1 queues is the probability that an arrival encounters a busy server. This parameter can be obtained by solving the following equation given in [78]

$$\sigma = \mathcal{A}^*(m\mu - m\mu\sigma), \quad m \geq 1, \tag{56.48}$$

where the probability σ that an arrival finds the server busy is obtained by solving the aforementioned equation and $\mathcal{A}^*(.)$ is the LST of interarrival time. The value of σ is different from ρ, so that PASTA does not apply anymore, unless *GI* is *M*. Unlike M/G/m, there is an exact analysis for GI/M/m starting with the solution of Equation 56.48. The mean waiting time encountered by arriving requests at GI/M/1 queues is

$$W_{GI/M/1} = \frac{\sigma/\mu}{1-\sigma}. \tag{56.49}$$

It can be observed from Table 5.3.9 for GI/M/1 queues in [42] that for the four interarrival time distributions, H_2, M, E_3, and D, which have the same mean, the one with the smaller coefficient of variation has a lower σ and hence *W*, but this is not true in general.* That this is so we compute *W* with a uniform (U) and a Gamma (G) interarrival time distribution, whose mean, coefficient of variation, and LST are given by Equations 56.8 and 56.12 with the same mean and coefficient of variation. For the uniform distribution given $a = 0$ and $b = 2$, $\bar{x}_U = 1$ and $c_u^2 = 1/3$. For the Gamma distribution with $\alpha = \beta = 3$ $\bar{x}_G = 1$, $c_G^2 = 1/3$. We set $\bar{x} = 1/\mu = 0.5$, so that $\rho = 0.5$ and $W_{M/M/1} = \rho\bar{x}/(1-\rho) = 0.5$. The solution of Eq. (??) for the two distributions are $\sigma_U \approx 0.361$ and $\sigma_G = 0.334$. The Gamma distribution with $\alpha' = \beta' = 2.9$ with $c_{G'}^2 = 1/2.9 = 0.345 > c_U^2$ has a lower *W* than the uniform distribution since $\sigma_{G'} = 0.345$. We have $W_U = 0.361 \times 0.5/(1-0.361) \approx 0.289$, $W_G = 0.334 \times 0.5/(1-0.334) = 0.25075$ and $W_{G'} = 0.335 \times 0.5/(1-0.335) = 0.252$.†

The probability that an arriving request encounters a queue of length *k* is given by $r_k = (1-\sigma)\sigma^k$. The PDF for the waiting time in GI/M/1 is $W(y) = 1 - \sigma e^{-\mu(1-\sigma)y}$, $y \geq 0$, so that $W(0) = 1 - \sigma$. The pdf and $w(y) = u_0(1-\sigma) + \sigma(1-\sigma)\mu e^{-\mu(1-\sigma)y}$, $y \geq 0$, where $u_0(1-\sigma)$ signifies a unit impulse at the origin.

56.4.8 Routing in Distributed Systems

In the context of mirrored disks classified as RAID level 1, there is the question of which one of the two forms of static routing to two disks is better: (1) uniformly with equal probabilities and (2) round-robin. The service time distribution at the disks is assumed to be exponential with mean service time $\bar{x} = 1/\mu$.

* Private communications from Prof. L. Takacs at Case Western Reserve University, 1978.

† This approach was proposed by Prof. I. Adan at Technical University at Eindhoven, 2012.

Given an arrival rate 2λ, the rate of arrivals to each disk is λ in both cases. In the first case, it follows from the decomposition property of Poisson processes that the arrival process at each disk is Poisson [42,78,80,143], so that the mean waiting time in this case is $W_{M/M/1} = \rho\bar{x}/(1-\rho)$.

With round-robin routing, the interarrival time to each disk follows the Erlang-2 distribution with rate 2λ per stage, so that the interarrival time LST is $\mathcal{A}^*(s) = [2\lambda/(s+2\lambda)]^2$ [78]. This is a GI/M/1 queue, so solving Equation 56.48 we obtain the smaller root $\sigma = \left(1+4\rho-\sqrt{1+8\rho}\right)/2$. It is easy to show that $\sigma < \rho$, so that round-robin routing yields a lower W than uniform routing (at least for exponential service times).

Routing of requests may be dynamic, that is, based on the current state of the system. *Join the shortest queue* (*JSQ*) is an optimal policy for exponential service times [145]. This is not always true, as in the case where most service times are negligible with a high probability, but there is a small probability of encountering a high service time. For nonexponential service times, the expected service times need to be considered and counterintuitive behavior can be observed in this case.

The response time of txns processed in a locally distributed database can be improved by judicious data allocation and txn routing. An integer programming formulation is used in [149] as a first step to assign datasets and txn classes to computer systems (with directly attached disks) to balance processor loads. There are two alternatives dealing with remote accesses: (1) function shipping is a *remote procedure call* (*RPC*) to retrieve remote data. (2) I/O shipping returns referenced disk pages. The former can be implemented in SQL-based databases, which allow transparent distributed query processing, but this is not the case in navigational databases, such as IMS. Optimized static routing results in a further improvement in response time. Dynamic txn routing policies implemented at a frontend, which take into account routing histories and minimize the estimated response time of incoming txns is proposed and found to provide a substantial improvement over the optimal static strategy in [148].

56.4.9 Polling Systems

In the most common case, a single server serves a number of queues in a polling system in cyclic order [109,110]. Queues or buffers holding requests have been classified as single or infinite buffer systems. With infinite buffers the number of requests served per visit is exhaustive or gated. The latter means that requests arriving at a queue after service starts at the queue are not served in that service cycle. The default scheduling discipline from each queue is FCFS. Basic performance measures of interest are the mean cycle time $E[C]$ and the mean waiting time $E[W]$.

We consider a polling system with N queues. The switchover time from the ith node to the next node is r_i and its variance δ_i^2, so that $R = \sum_{i=1}^{N} r_i$ and $\delta^2 = \sum_{i=1}^{N} \delta_i^2$. The arrival rate to node i is λ_i, its mean service time b_i, so that $\rho_i = \lambda b_i$ and $\rho = \sum_{i=1}^{N} \rho_i$.

In the case of the *symmetric* single buffer systems, the mean cycle time and mean waiting time are

$$E[C] = R + bE[Q], \quad E[W] = (N-1)b - \frac{1}{\lambda} + \frac{NR}{E[Q]}, \tag{56.50}$$

where $E[Q]$ is the mean number of messages served in a polling cycle. Given the mean message interdeparture time $E[X] = E[W] + b + 1/\lambda$, the system throughput is $\gamma = 1/E[X]$.

A relatively simple expression for $E[Q]$ is available only for constant service and switchover times

$$E[Q] = \frac{N\sum_{n=0}^{N-1}\binom{N-1}{n}\prod_{j=0}^{n}\left[e^{\lambda(R+jb)}-1\right]}{1+\sum_{n=1}^{N}\binom{N}{n}\prod_{j=0}^{n-1}\left[e^{\lambda(R+jb)}-1\right]}. \tag{56.51}$$

The mean waiting time $E[W]$ of a symmetric exhaustive service system is

$$E[W] = \frac{\delta^2}{2r} + \frac{N\lambda b^{(2)} + r(N-\rho)}{2(1-\rho)}. \tag{56.52}$$

In the case of a symmetric gated service system,

$$E[W] = \frac{\delta}{2r} + \frac{N\lambda b^{(2)} + r(N-\rho)}{2(1-\rho)}. \tag{56.53}$$

Polling analysis has been applied to evaluating the performance of communication network protocols and especially *local area networks* (*LANs*), see for example, [46].

56.4.10 Load Sharing Policies in Distributed Systems

A system with multiple sources and servers is considered by Wang and Morris in [144], which classifies load sharing algorithms as source-initiated and server-initiated. The results of the study based on simulation and analysis show that (1) the choice of load sharing algorithm is a critical design decision. (2) For the same level of scheduling information exchange, server-initiative has the potential of outperforming source-initiative algorithms. (3) The Q-factor (quality of load sharing) performance metric summarizes overall efficiency and fairness of an algorithm and allows them to be ranked. (4) Some previously ignored algorithms may provide effective solutions.

A homogeneous distributed systems with N servers, which can be represented as M/M/1 queues, is considered in [84]. The arrival rate of requests at the N servers is λ. Given that the server service rate is μ, the server utilizations are $\rho = \lambda/\mu$. The probability that i servers are idle is $Q_i = (1-\rho)^i$ and H_j denotes the probability that j servers are not idle and at least one request waits in the queue $H_j = \rho^j - \rho^j(1-\rho)^j$. The probability P_w that at least one request waits, while at least one server is idle is

$$P_w = \sum_{i=1}^{N} \binom{N}{i} Q_i H_{N-i} = \left[1-(1-\rho)^N\right]\left[1-(1-\rho)^N - \rho^N\right].$$

Plotting P_w versus ρ shows that it reaches a peak at $\rho \approx 0.6$, which is close to one at $N \geq 20$, but is smaller for lower values of N. Provided the cost of transferring tasks is low, the system can be approximated by an M/M/N system, so that $P_w = 0$. Tasks are transferred from node to node when (1) A task arrives at a busy server and there are less than N tasks in the system. (2) A server completes at a server whose queue is empty and there are more than N tasks in the system. A lower bound to task transfers to minimize W is given as

$$LT = \sum_{i=1}^{N-1} \left[i\lambda p_i + \mu(N-i)p_{N+i}\right],$$

where p_i are probabilities associated with an M/M/N system. Plotting LT versus N for varying arrival rates, it is observed that LT increases linearly with N and has a higher slope for higher values of λ.

Adaptive load sharing policies in homogeneous distributed systems are discussed in two companion papers by Eager, Lazowska, and Zahorjan [61,62]. The goal of these studies is to improve the performance of systems with N processors (servers) by averting load imbalance as in [84]. Sender initiated policies are explored in [61], while receiver initiated policies are proposed in [62] and compared with sender initiated policies.

In the case of sender-initiated policies, the transfer policy determines whether a task should be processed locally or remotely. The transfer policy is based on a threshold whether the queue length at a node exceeds T. The *location policy* selects the node to which the task should be transferred. The following location policies are considered.

Random policy: Transferred tasks at a receiver node are considered to be as a new arrival, so that they may be transferred again if T is exceeded. This may lead to an instability similar to the Aloha communication system due to packet collisions and retransmissions as discussed in Section 5.11 and shown in Figure 5.39 in [79]. The thrashing phenomenon is alleviated simply by limiting the number of retransmissions. No exchange of information among nodes is required for this policy.

Threshold policy: Random nodes are probed for their queue length. A task is transferred only if the number of tasks at the node is less than L_t, so that useless task transfers are avoided. Probing is stopped when the probe limit L_p is exceeded and the task is processed at the receiving node.

Shortest queue: L_p randomly selected nodes are probed and the one with the shortest queue is selected.

Figure 2 in [62] compares the mean response times (RT) for different policies with $N = 20$ nodes, service time $S = 1$, $L_p = 1$, and $L_t = 3$. The overhead associated with task migration is taken into account as 10% of S. Queueing systems M/M/1 and M/M/20 are upper and lower bounds to RT. It turns out that threshold and shortest outperform the random policy significantly.

In the case of the *receiver initiated* (*RI*) policy, a receiver that has less than T tasks when a task completes probes L_p other nodes to initiate a task transfer [62]. No transfer occurs if the L_p nodes had T or more tasks in their queue (such a policy with $T = 1$ was considered in [84]). The reservation policy is similar to sender policy, but differs in that task transfers are considered upon task arrival. It follows from the graph for RT versus system load in the extended abstract version of [62] that the sender initiated policy for $T = 1$ outperforms others for up $\rho = 0.7$, but is outperformed by a sender initiated policy for $T = 2$ and a receiver initiated policy for $T = 1$. The reservation policy has the highest RT.

56.5 Fork-Join Queueing Systems

In *Fork-Join* (*F/J*) queueing systems arriving jobs spawn K tasks, which are served by K parallel servers. We first consider the case when jobs arriving at a rate λ are processed one at a time. Each jobs spawns K independent tasks to be processed at K servers, but this is done only when all the tasks associated with a previous job are completed. The completion time of all of the tasks of a job is simply the maximum of K service times denoted by R_k^{max}. The maximum throughput is then $\lambda_K^{max} = [R_k^{max}]^{-1}$. When the service times are exponentially distributed with rate μ, then $R_K = H_K/\mu$ and $\lambda_K^{max} = \mu/H_K$, where $H_K = \sum_{k=1}^{K} 1/k$ is the Harmonic sum. The server utilizations are given by $\rho = min(\lambda, \lambda_K^{\max})\bar{x}$, where $\bar{x} = 1/\mu$ is the mean service time.

F/J queueing systems allow overlapped execution of tasks belonging to successive jobs by sending the K tasks of an arriving job to the K queues for FCFS processing. When the arrival rate is λ and the service times are exponentially distributed with mean $\bar{x} = 1/\mu$, then the server utilizations are $\rho = \lambda/\mu$. In this case $\lambda_K^{max} = \mu$, which exceeds the previous throughput by a factor H_K. The distribution of response time at each server is exponential with mean $R(\rho) = \bar{x}/(1 - \rho)$, so that the expected value of the maximum of K response times is $R_K^{max} = H_K \times R(\rho)$. This is an upper bound to $R_K^{\mathrm{F/J}}(\rho)$ for the exponential distribution according to [90]. Under Markovian assumptions, an exact analysis is available for two-way F/J queueing systems

$$R_2^{\mathrm{F/J}}(\rho) = \left(1.5 - \frac{\rho}{8}\right) R(\rho) \le R_2^{max}(\rho), \quad R_2^{max}(\rho) - R_2^{\mathrm{F/J}}(\rho) = \frac{\rho}{8} R(\rho), \tag{56.54}$$

the introduced error increases with ρ. An approximate expression is obtained in [90] for $R_k^{\mathrm{F/J}}(\rho)$ for $2 \le K \le 32$ with exponential service time distribution by curve-fitting against simulation results

$$R_K^{\mathrm{F/J}}(\rho) \approx \left[\frac{H_K}{H_2} + \left(1 - \frac{H_K}{H_2}\right)\frac{4\rho}{11}\right] R_2^{\mathrm{F/J}}(\rho). \tag{56.55}$$

Empirical expressions for $R_K^{\mathrm{F/J}}(\rho)$ are derived in [128] for general service times by curve-fitting.

Fork-join requests may be utilized to estimate the mean response time of accesses to a failed disk in RAID5 disk arrays (see chapter on storage systems). We simplify the discussion by assuming that all disk accesses are due to read requests. The contents of a failed disks in this case are reconstructed by accessing and XORing the corresponding blocks on surviving disks. Note that disks in addition to fork-join requests process requests accessing them directly, so that the access rate and disk utilization on surviving disks are doubled. Simulation results in [128] have shown that for higher disk utilization due to interfering requests $R_K^{F/J}(\rho) \approx R_K^{max}(\rho)$, so that there is no need to use Equation 56.55 as was done in [87].

In order to compute R_K^{max}, the response time distributions at the disks are approximated by the Erlang, Hyperexponential, Coxian, or the extreme-value distribution. The latter is defined as [76] $F_Y(y) = P[Y < y] = \exp\left(-e^{-y-a/b}\right)$, which has a mean $\bar{Y} = a + \gamma b$ and variance $\sigma_Y^2 = \pi^2 b^2/6$. The maximum of K r.v.'s with this distribution is [76]

$$Y_{max}^K = (a + \gamma b) + b\ln(K) = \bar{Y} + \left(\sqrt{6}\ ln(K)/\pi\right)\sigma_Y.$$

Given that the extreme value distribution has two parameters, the mean and variance of response time in RAID5 arrays can be used to estimate them as was done in [136]. A detailed discussion of the queueing analysis of F/J and related systems is given in [140].

56.6 Queueing Network Models

Queueing network models (*QNMs*) comprise a set of nodes, where each node has single or multiple servers. Jobs completing their service at a node are routed to another node according to probabilities that are specified by a first-order or higher-order Markov chain [80] until they depart from the system. Jobs in a QNM may belong to a single class or multiple classes as defined by being open or closed chains, as in the case of OLTP txns and batch jobs, respectively, their routing matrices, and their service demands. Chapter 7 in [83] provides numerical examples for justifying multiple job classes.

Jobs belonging to an open chain arrive at one of its nodes and visit appropriate nodes according to a transition probability matrix $R = [r_{i,j}]$, where $r_{i,j}$ is the probability that a job leaving node i joins node j. Jobs leave the system with a departure rate equal to the arrival rate provided that no node is saturated. The response time of requests in an open chain is the sum of residence times at the nodes visited by the chain.

Closed QNMs with a fixed number of jobs are used to represent batch systems with an infinite backlog of jobs, so that as one job is completed there is another job available for activation. Closed chains have special transitions, signifying the completion of a job. This may be a transition to a delay server, which stands for user think times, or an immediate transition of a job to the same node, indicating its completion and replacement by a new job, so that the number of jobs in the closed QNM remains fixed. Hybrid QNMs allow both open and closed chains.

In what follows, we start our discussion with open QNMs in Section 56.6.1 and then proceed to closed QNMs in Section 56.6.2. Buzen's algorithm for the analysis of closed QNMs is given in Section 56.6.3 and *mean value analysis* (*MVA*) in Section 56.6.4. Bounds applicable to closed QNMs are specified in Section 56.6.5. The analysis of hybrid QNMs with open and closed chains is outlined in Section 56.6.6. Hierarchical decomposition and aggregation in the context of QNMs is discussed in Section 56.6.7. This approach is used to analyze parallel processing based on [122] and serialization delays based on [119].

56.6.1 Open QNMs

In open QNMs, external arrivals to the N nodes of the network are γ_i for $1 \leq i \leq N$. The transition probability from node i to j is given by $r_{i,j}$. The probability that a request leaves the network at node i

is $r_{i,out} = 1 - \sum_{j=1}^{N} r_{i,j}$. The arrival rates to the nodes (λ_i for $1 \le i \le N$) can be obtained by solving a set of linear equations given as

$$\lambda_i = \gamma_i + \sum_{j=1}^{N} \lambda_j r_{j,i} \quad \text{for } 1 \le i \le N \text{ or } \underline{\lambda} = \underline{\gamma} + \underline{\lambda} R, \tag{56.56}$$

where $\underline{\lambda}$ and $\underline{\gamma}$ are row vectors and R is a probabilistic routing matrix. The mean delay encountered by messages arriving at node j and departing at node k is denoted by $Z_{j,k}$ and $\gamma_{j,k}$ is the rate of messages from node j to node k. The mean delay overall messages arc then [79]

$$T = \sum_{j=1}^{N} \sum_{k=1}^{N} \frac{\gamma_{j,k}}{\gamma} Z_{j,k}, \quad \gamma = \sum_{j=1}^{N} \sum_{k=1}^{N} \gamma_{j,k}. \tag{56.57}$$

Consider two queues in tandem, with the output from the first queue being fed to the second queue. The arrivals to the first queue are Poisson with rate λ. The service times at the two queues are exponential with rates μ_1 and μ_2. Provided the first queue is not saturated ($\lambda_1 < \mu_1$), due to conservation of flow the arrival rate to the second queue is also λ. The utilizations of the two servers are $\rho_1 = \lambda/\mu_1$ and $\rho_2 = \lambda/\mu_2$, respectively. The system is not saturated if $\min(\mu_1, \mu_2) > \lambda$. Burke's theorem states that the arrival rate to the second queue is also Poisson with the same rate. A proof of Burke's theorem based on [78] is given by Equation 56.58. The LST for interdeparture time ($\mathcal{D}^*(s)$) depends on whether the queue is busy or not. When the first queue is idle, the interdeparture time is the sum of interarrival and service times (product of the two LSTs), while it is simply the LST of service time when the server is busy

$$\mathcal{D}^*(s) = (1-\rho_1)\frac{\lambda}{s+\lambda}\frac{\mu_1}{s+\mu_1} + \rho_1 \frac{\mu_1}{s+\mu_1} = \frac{\lambda}{s+\lambda} \Leftrightarrow d(t) = \lambda e^{-\lambda t}. \tag{56.58}$$

Both queues can be analyzed simply since they have Poisson arrivals. In case a fraction q of departures from the first queue are fed back, then it can be shown that the arrival process to that queue is not Poisson, but the departure process from the queue remains Poisson [143].

The analysis of tandem queues undertaken in Chapter 3 in [77] demonstrates the difficulty of the problem at hand. This led to the proposal of the *independence assumption* that message lengths are resampled from an exponential distribution in the network. As verified by simulation results this assumption is more accurate for higher fan-ins and fan-outs.

In the case of two queues in tandem with a Poisson arrival rate λ to the first queue and service rates μ_1 and μ_2 at the two servers with $\min(\mu_1, \mu_2) > \lambda$, the state equilibrium equations for the two-dimensional CTMC with states (k_1, k_2) are satisfied by

$$P[k_1, k_2] = (1-\rho_1)\rho_1^{k_1}(1-\rho_2)\rho_2^{k_2} \quad \text{for } k_1 \ge 0, k_2 \ge 0.$$

Even when the transitions are not feedforward and hence the arrival rates to the nodes are not Poisson [143], Jackson's theorem guarantees a product form solution in open QNMs where the external arrivals are Poisson with arbitrary transition probabilities, as if the arrival rates to all nodes are Poisson. Given m_n servers with service rates μ_n at node $1 \le n \le N$, the server utilizations at that node are $\rho_n = \lambda_n/(m_n\mu_n)$. Given that $P_n[k_n]$ is the distribution of the number requests at node n, the probability that the system is at state $\underline{k} = (k_1, k_2, \ldots, k_N)$ is given as [45,78,143]

$$P[\underline{k}] = \prod_{n=1}^{N} P_n[k_n]. \tag{56.59}$$

Analysis of open QNMs with multiple chains is given by Algorithm 7.1 in [83]. The analysis with single servers, which can be easily extended to multiple servers, proceeds as follows: (1) ascertain that the server utilizations at no node exceeds one. (2) Obtain the residence times at various nodes. (3) Obtain queue lengths at various nodes. (4) Obtain residence time in QNM by summing residence times over all nodes constituting a chain (from the arrival node to the departure node). (5) Obtain the mean number of jobs in a chain by using Little's result.

56.6.2 Closed QNMs

We next consider the closed QNM of the *central server model* (*CSM*), which consists of a single CPU (node 1) and $N-1$ disks (nodes $2:N$). The number of jobs at the computer system or its MPL is M. Jobs alternate between processing at the CPU and the disks, returning to the CPU after completing their service at the disks. A job finishing its execution at the CPU completes its processing with probability p_1 or accesses the *nth* disk with probability p_n for $2 \le n \le N$, so that $\sum_{n=1}^{N} p_n = 1$. A completed job is immediately replaced by a new job, so that the MPL remains fixed at M. The number of visits to the CPU is given by the modified geometric distribution $P_k = p_1(1-p_1)^{k-1}$, so that the mean number of visits to the CPU is $v_0 = 1/p_1$ and the disks $v_n = (1/p_1 - 1)(p_n/(1-p_1)) = p_n/p_1$ for $2 \le n \le N$.* Given that $\bar{x}_n$ is the mean service time per visit at node n, the service demands $D_n = v_n \bar{x}_n$ for $1 \le n \le N$. These are the mean times the jobs spend being served at the nodes of the QNM.

The state probabilities of closed QNMs with four types of nodes satisfying the conditions of the BCMP theorem can be expressed in product form [43,45,79,83]:

1. *FCFS*: exponential service time, single or multiple servers
2. *Processor sharing* (*PS*): general service time, single server
3. *Infinite servers* (*IS*): general service time, also a delay server
4. *Last-come, first served* (*LCFS*): preemptive resume, general service time, single server

The service rate at type 1 nodes with m servers equals $k\mu$ for $k \le m$ and $m\mu$ for $k \ge m$ requests at the node. Multiservers and delay servers are examples of state-dependent servers. At type 3 nodes there is no queueing, since there are as many servers as there are requests, so that the mean response time equals the mean service time. The increase in the total service rate in multiprocessors is sublinear in the number of processors, which may be due to cache interference or synchronization delays [70]. For example, in a hypothetical four-way multiprocessor, the processing rate may increase as (1, 1.75, 2.5, 3.25).

Consider a closed QNM with N nodes and M jobs. The state of the QNM can be represented as $\underline{m} = (m_1, m_2, \ldots, m_N)$, so that $\sum_{n=1}^{N} m_i = M$. The probability distribution of the number of jobs at the N nodes is given as†

$$P[\underline{m}] = \frac{1}{G(M)} \prod_{n=1}^{N} F_n(m_n), \quad F_n(m_n) = D_n^{m_n} \quad \text{or} \quad F_n(m_n) = \frac{D_n^{m_n}}{\prod_{i=1}^{m_n} \alpha_n(i)}. \tag{56.60}$$

The first expression for $F_n(m_n)$ is applicable to type 1, 2, and 4 state-independent queueing stations, while the second expression applies to multiservers and delay servers. Given $F_n(0) = 1$, $F_n(i) = (D_n/\alpha_n(i))F_n(i-1)$ for $i \ge 1$. For the aforementioned multiprocessor, $\underline{\alpha} = \{1, 1.75, 2,5, 3.25, 3.25, \ldots\}$.

* The analysis of the fundamental matrix in Section 7.2 in [143] to obtain the mean number of visits to the states of a finite state Markov chain is applicable to this discussion.

† In the case of single servers, the probabilities can be expressed as server utilizations by multiplying them by the system throughput ($T(M)$).

Given the number of jobs M and the service demands $\underline{D} = (D_1, D_2, \ldots, D_N)$, a closed QNM can be solved to determine the throughput for jobs and their mean residence time in the QNM, device queue lengths, and utilizations. The job transition probability matrix should be specified to determine the job response time distribution.

There are several efficient algorithms for analyzing closed QNMs satisfying the BCMP theorem. Buzen's algorithm [28,47] for single job classes was later extended to multiple job classes by Buzen, Reiser and Kobayashi at IBM's T. J. Watson Research Center, and Williams and Bhandiwad using generating functions [45,146]. The analysis in the latter paper was extended by Thomasian and Nadji in [117]. *Mean value analysis* (*MVA*) [29,83,98] and *REcursion by Chain ALgorithm* (*RECAL*) [54,55] are two other algorithms for closed QNMs.

56.6.3 Buzen's Algorithm

It can be shown that in a QNM with N nodes and M jobs there are $S = \binom{M+N-1}{N-1}$ states for the number of jobs at the nodes. To show this is so consider $(N-1)$ bars to delineate the boundaries among the N nodes and M asterisks specifying as many jobs, so that the state of the system can be represented as ***|**|...|*, three jobs at node 1, two jobs at node 2, and one job at node N. There are $(M + N - 1)!$ permutations, which are divided by $M!$ for the M indistinguishable jobs and $(N-1)!$ for the $(N-1)$ indistinguishable bars [59]. Multiple job classes are not considered in this study for the sake of brevity, but it suffices to say that the number of states increases rapidly with K job classes, that is, $S_K = \prod_{k=1}^{K} \binom{M_k+N-1}{N-1}$. The normalization constant in Equation 56.60 is given as

$$G(M) = \sum_{\sum_{n=1}^{N} m_n = M} \prod_{n=1}^{N} F_n(k_n). \tag{56.61}$$

For state-independent nodes with the normalization constant, $G(M)$ is the sum of the S terms of the form $F_n(m_n) = D_n^{m_n}$. A straightforward computation of $G(M)$ requires at most $M - 1$ multiplications for each one of the S terms and $S - 1$ summations.

Arguments based on generating functions [78,143] have been used in [117,146] to derive the recursion for $G(M)$ and other performance metrics. Provided that the dummy variable t is selected to be small enough ($tD_n < 1$), the sum given next converges for state-independent servers

$$x_n(t) = \sum_{m \geq 0} F_n(m)t^m = \sum_{m \geq 0} (tD_n)^m = (1 - tD_n)^{-1}. \tag{56.62}$$

The generating function for the QNM is

$$g(t) = \prod_{n=1}^{N} x_n(t) = \sum_{m \geq 0} G(m)t^m. \tag{56.63}$$

Using Equation 56.62, the following recursion is obtained for the *nth* node

$$g_n(t) = \prod_{i=1}^{n} x_i(t) = g_{n-1}(t)x_n(t) \Rightarrow g_n(t) = g_{n-1}(t) + tD_n g_n(t). \tag{56.64}$$

Setting the coefficients of the same powers of t on both sides equal to each other

$$G_n(m) = G_{n-1}(m) + D_n G_n(m-1) \quad \text{for } 1 \le m \le M, 1 \le n \le N. \tag{56.65}$$

Instead of a two-dimensional array, memory space can be saved by using a one-dimensional array or a vector initialized as $G(0) = 1$ and $G(m) = 0$ for $1 \le m \le M$. The iterative solution proceeds one node at a time, while varying the number of jobs at the node

$$G(m) = G(m) + D_n G(m-1) \quad \text{for } 1 \le m \le M, 1 \le n \le N. \tag{56.66}$$

Generating functions can be used to show that $T(M) = G(M-1)/G(M)$, from which it follows that the mean residence time is $R(M) = M/T(M)$. The utilizations and mean queue lengths at the nodes are $U_n(M) = D_n T(M)$ and $Q_n(M) = \sum_{m=1}^{M} D_n^N G(M-m)/G(M)$ for $1 \le n \le N$, respectively.

Convolution is only applied to state-dependent servers, so that starting with $g_n(t) = g_{n-1}(t)x_n(t)$ for $1 \le n \le N$ and equating the coefficients of m on both sides we have

$$G_n(m) = \sum_{j=0}^{m} G_{n-1}(m-j)Fn(j) \quad \text{for } 1 \le m \le M, 1 \le n \le N. \tag{56.67}$$

If there is only one state-dependent server, the need for convolution is obviated by designating the state-dependent node as node 1 $G(m) = F_1(m)$ for $1 \le m \le M$.

56.6.4 Mean Value Analysis

The MVA method was developed at the IBM Research at Yorktown in late 1970s through a collaboration between Reiser and Schweitzer according to [99], which also states that Lavenberg was instrumental in formalizing the proofs in [98]. The MVA method has the advantage over Buzen's algorithm that it is easy to explain and intuitively appealing. Buzen's algorithm for state-independent station is specified by two nested iterations in Equation 56.66, while the order of iterations is reversed in MVA, as specified next:

MVA algorithm for N single servers nodes and M jobs.
Initialize queue lengths at all nodes $Q_n(0) = 0, \forall n$.
For $1 \le m \le M$ requests repeat steps inside brackets:
{
Compute residence time at all single server nodes: $R_n(m) = D_n[1 + Q_n(m-1)]$.
Compute residence time at all IS nodes: $R_n(m) = D_n$.
Compute system throughput: $T(m) = m \Big/ \sum_{n=1}^{N} R_n(m)$.
Compute queue lengths at all nodes: $Q_n(m) = T(m)R_n(m), \forall n$.
}
Compute device utilizations at all nodes: $U_n(m) = D_n T(m), \forall n$.

The formula for residence times is based on the Arrival Theorem [98], which states that arrivals to a node encounter a queue length, as if there is one less job in the closed QNM [98].

MVA can be easily extended to multiple job classes [45,82,83]. There are numerical stability problems associated with state-dependent servers, since instead of just the mean queue length, MVA computes the distribution of the number of jobs at a node. The probability of zero jobs at a node may turn out to be negative for higher utilizations, since it is computed as one minus the sum of other probabilities (see Chapter 20 in [83]). This does not pose a problem in the case of Buzen's Algorithm developed a decade before MVA. There is also the N-fold increase in memory size requirements for MVA with respect to Buzen's algorithm (see Table 8.5 in [45]). Note that both time and space requirements grow as the product of the number of jobs in different chains.

There are several approximate iterative methods based on MVA that lead to a direct solution for large number of jobs. The Bard–Schweitzer approximate iterative solution method is based on MVA [45,82,83]. It uses the initialization assuming that service demands at the nodes are equal $Q_n(M) = M/N$ and then sets $Q_n(M-1) = [(M-1)/M]Q_n(M)$ for $1 \le n \le N$. The iteration continues until $Q_n(M)$ obtained in successive iterations are approximately equal [45,83]. The *self-correcting approximation technique* (*SCAT*) and the linearizer [45,85] developed by Chandy and Neuse provide more accurate results than the Bard–Schweitzer approximation and allows state-dependent servers, while reducing the computational cost with respect to MVA. A comparison of approximate algorithms for QNMs is given in Table 9.13 in [45]. This is a more important issue in QNMs with multiple job classes. See Table 8.5 in [45], which is extracted from [54].

Priority queueing at the CPU can be incorporated approximately in MVA by inflating the $D_{CPU}^{low} = D_{CPU}^{low-original} \big/ \left(1 - U_{CPU}^{high}\right)$ and since U_{CPU}^{high} is not known a priori, iteration is required (see Algorithm 11.1 in [83]). The so-called shadow CPU technique was proposed by Sevcik to obtain optimistic and pessimistic bounds on the performance of nodes with priorities. Additional approximations to deal with violations of the BCMP theorem, such as different exponential service times across job classes and nonexponential service times, both in the case of FCFS scheduling (see Chapter 11 in [83]).

56.6.5 Bounds to QNM Performance

The *Balanced Job Bounds* (*BJB*) [83] by Zahorjan, Sevcik, Eager, and Galler provides an upper bound to the throughput $T(M)$ of closed QNMs ($T(M)$) with nodes with single servers using the average service demand $D_a = \sum_{n=1}^{N} D_n/N$. In a balanced QNM due to symmetry, the queue lengths encountered at the nodes due to the Arrival Theorem are $Q_n(M-1) = (M-1)/N, \forall n$. The residence time at node n is $R_n(M) = D_a(1 + M - 1)/N), \forall n$, so that

$$R(M) = (N+M-1)D_a \Rightarrow T(M) = \frac{M}{(M+N-1)D_a}. \tag{56.68}$$

It can be observed from Equation 56.68 that in balanced QNMs $T(M)$ increases with M. Generally, according to Theorem 2 in [60] the throughput characteristic of QNMs with state-independent and delay nodes is a concave nondecreasing function. This is shown in Figure 9.1 in [83], that is, $T(M) > [T(M-1) + T(M+1)]/2$.

An alternate analysis using generating functions and the binomial theorem [30] was developed by Thomasian and Nadji to derive Equation 56.68 [118]. The generating function for N single server stations is

$$g(t) = \prod_{n=1}^{N} x_n(t) = (1-Dt)^{-N} = \sum_{m=0}^{\infty} \binom{N+m-1}{m} (Dt)^m, \tag{56.69}$$

hence $G(M) = \binom{N+M-1}{M} D^M$ and $G(M-1) = \binom{N+M-2}{M-1} D^{M-1}$, so that $T(M) = G(M-1)/G(M)$, which is the throughput in Equation 56.68.

The same work develops an approximation based on Taylor series expansion with respect to the mean service demand $D_0 = \sum_{n=1}^{N} D_n/N$. Let $E_j = \sum_{n=1}^{N} D_n^j (D_n/D_0 - 1)^j$

$$R_0(m) = \sum_{j\geq 2} \frac{E_j}{j} \left[\prod_{i=0}^{j-1} \frac{M-i}{N+i} \right].$$

Successively more accurate approximations to $T(M)$ as given by Equation 56.58, which may be considered a 0*th* order approximation, are given as follows

$$T_i(M) = T(M) \frac{1 + R_0(M-1)}{1 + R_0(M)}.$$

Maintaining only one term in the summation for $R_0(M)$ we have

$$T_1(M) = T(M) \left[1 \frac{\frac{M-1}{M(M+1)} E_2}{1 + \frac{M(M-1)}{2M(M+1)} E_2} \right].$$

The analysis has been extended to two job classes in [89].

Nodes with single servers and delay servers are considered in [83]. The service demands at the delay servers can be easily added to the mean residence time in the system $R(M)$ in Equation 56.58.

Asymptotic bounds analysis (*ABA*) developed by Muntz and Wong in 1975 [79] is similar to BJB where the system is obtained using the maximum service demand [83], but in fact the former is more useful since it is applicable to all types of nodes, while the latter applies only to state-independent nodes. Given that the *nth* node has m_n servers with service demand D_n for $1 \leq n \leq N$.

$$T_{max} = min \left\{ \frac{m_1}{D_1}, \frac{m_2}{D_2}, \ldots, \frac{m_N}{D_N} \right\}. \tag{56.70}$$

In the case of state-dependent servers with the implicit assumption that the service rate is an increasing function of the number of requests, the maximum throughput at each node can be used in this calculation. T_{max} is approached as $M \to \infty$. At lower values of M, $T(M)$ is upper bounded by the line passing through (0, 0) and (1, $T(1)$), where $T(1) = 1 \Big/ \left[\sum_{\forall n} D_n \right]$. Chapter 5 in [83] has several examples on utilizing bounds in performance analysis.

56.6.6 Hybrid QNMs with Open and Closed Chains

We conclude the discussion of QNMs, which combine open and closed chains. We summarize Algorithm 7.4 in [83] with multiple open chains and a single closed chain. Only single server state-independent nodes are considered at this point for the sake of brevity.

1. Solve QNM considering open chains alone and determine server utilizations due to them.
2. Expand service demands of closed chains by dividing them by one minus the server utilizations due to open chains.
3. Solve QNM for closed chains to determine queue lengths at the nodes.
4. Determine residence time for open chains and multiply them by one plus the sum of queue lengths for closed chains.

56.6.7 Hierarchical Decomposition and Aggregation for QNMs

Hierarchical decomposition and aggregation can be explained by considering a multiprogrammed computer system represented by a closed QNM with a maximum MPL M_{max}. It is substituted with a single *Flow-Equivalent Service Center (FESC)*, with service rate $T(m)$ for $1 \le m \le M_{max}$ and $T(m) = T(M)$ for $m \ge M_{max}$, which is a flattened throughput characteristic [83].

Given N users with think time $Z = 1/\lambda$ and m jobs at the computer system, the arrival rate to the FESC representing the computer system is $\Lambda(m) = (N - m)\lambda$. The system can be analyzed by solving the BD process for the machine repairman model (MRM) in Section 56.3.5 with variable service rates. Alternatively a graphical method that plots the arrival rate and the completion rate versus the number of the jobs at the computer system can be adopted (see Figure 3.37 in [80] or Figure 20 in [59]). The intersection of the two graphs is the operational point of the system M^*. The distribution of the number of jobs surrounding M^* roughly follows a normal distribution [42,143]. *Equilibrium point analysis (EPA)*, which is based on an iterative *fixed point approximation* (*FPA*), can be used to analyze the system directly without resorting to a detailed solution based on BDs [113].

The problem is more complex in the case of QNMs with external arrivals from multiple job classes and population size constrains. This problem has been studied in the context of multiprogrammed computers processing multiple job streams with Poisson arrivals. The MPL constraint may be per class or over all classes. Solution methods to deal with this problem are given in Chapter 9 in [83], while [121] compares the accuracy of several approximations proposed for this problem.

In a virtual memory environment as the number of activated jobs is increased, the main memory may be overloaded leading to excessive page faults. System resources will be used in handling page faults rather than in processing jobs, so that there is a drop in system throughput, which is known as thrashing [31]. Another reason for the drop in system throughput (in theory) is database lock contention, which is discussed in Section 56.8.1. According to Figure 8 in [127], there is a drop in $T(m)$ for $m \ge 1$ yields three intersection points with $\Lambda(m) = (N - m)\lambda$. The stable intersection points are A and C, where A is desirable and C is in the thrashing region, and B is unstable and the system can drift from it to the other two stable points.

Aggregation and decomposition has been applied in analyzing QNMs that do not lend themselves to a product-form solution. Two such examples are given next.

Serialization delays occur when the processing of certain phases of jobs in a computer system is single-threaded, so that only one of multiple jobs executing concurrently can proceed in the serialized phase, while the other jobs are blocked from entering the serialized phase. The analysis of a single job class with $M = 4$ jobs and $S = 2$ *serialization phases (SPs)* is considered in [119]. The underlying CSM consists of a CPU and disks as in CSM. Jobs completing their service at the CPU access the disks or attempt access to one of two SPs. The job is blocked if there is another job already active at the SP.

A CTMC is required to specify the state of the system, which is represented by the numbers of jobs in the nonserialized phase and two serialized phases $\underline{m} = (m_0, m_1, m_2)$, so that $m_0 + m_1 + m_2 = M$. The analysis of the QNM yields $T_i(\underline{m})$, $0 \le i \le 2$. The explosion in the number of CTMC states can be handled by coalescing states to attain a more compact *state representation* (*SR*). The number of states of the CTMC can be reduced from 15 to 12 by combining states (1,2,1) and (1,1,2) into (1,1,1) and also states (0,3,1), (0,2,2), and (0, 1, 3) into (0,1,1). The reduction in the number of states due to lumping of states is more dramatic when there are multiple job classes. Five different SRs with a large variation in the number of states are considered in [126].

QNMs whose jobs exhibit parallelism are not product-form, but can be analyzed using hierarchical decomposition. An efficient method to analyze such QNMs whose jobs can be specified as task structures was developed by Thomasian and Bay in [122]. We illustrate the analysis via a small example to determine the completion time of two parallel tasks T_1 and T_2, which belong to different job classes. The initial system state is (1, 1), with one job in each class. The throughputs in the two classes are $T_i(1, 1)$ for $i = 1, 2$. Given that the completion rates of the two tasks are exponential, the probability that T_1 completes first is $p_i = T_i(1, 1)/T(1, 1)$, with $T(1, 1) = T_1(1, 1) + T_2(1, 1)$. The holding time in that state is $R(1, 1) = 1/T(1, 1)$. The completion of T_1 and T_2 leads to a transition to states (–, 1) and (1, –) with holding times $R(2)$ and $R(1)$, respectively. It follows that the completion time of the two tasks is given as $C = R(1, 1) + p_1R(2) + p_2R(1)$. The solution method for complex task structures in [122] generates the states of the CTMC and computes their unnormalized probabilities one level at a time, after solving the underlying QNM. The number of states in the CTMC can be trimmed by ignoring states whose occurrence is highly unlikely.

The use of hierarchical decomposition in evaluating the performance of a CSM with higher priorities assigned to one of two classes was proposed in [117]. A similar example is given in Section 8.5.1 in [83].

The effect of lock contention on overall system performance is analyzed using the hierarchical decomposition method in two companion papers. The effect of lock preclaiming on the delay of activating jobs due to lock conflicts with already active jobs is considered in [120], while [102] deals with dynamic locking systems, but uses a more detailed analysis than the one given in [127] (see Section 56.8.1 for details).

Hybrid models combine simulation for higher level activities, such as txn arrivals and job activation, and analysis of QNMs for lower level activities, such as the processing of txns by the CSM. The hybrid modeling of IBM's *System Resource Manager* (*SRM*) for the MVS OS given in [52] is described in Section 15.6 in [83].

Hybrid simulation proposed by Schwetman in [105] counts cycles to determine job progress in a variation of CSM, referred to as the cyclic server model, where jobs leave the system after processing at the disks. This method was adopted in [123] in a simulation study to minimize the mean response time in distributed systems. Job processing demands are given as the service demands per cycle and the number of cycles required by jobs in a cyclic server model, where job completions occur at the disks rather than the CPU. Jobs may be characterized as CPU or I/O bound, so that it is the composition of jobs at the computer system rather than just their MPL that needs to be considered when routing incoming jobs. Jobs are routed to the node that minimizes their mean response time, but this mean response time will only hold if there are no further arrivals to the queue until the job completes. This method works better with high variability in interarrival times and in the extreme case of batch arrivals.

A first-order Markov chain model for routing of jobs implies a geometric distribution for the number of cycles $p_n = (1 - p)p^{n-1}$. Approximating the processing time of jobs per cycle with an exponential distribution (with parameter λ) implies an exponential residence time distribution with mean $[(1 - p)\lambda]^{-1}$

$$\mathcal{Y}^*(s) = \sum_{k\geq 1}(1-p)p^{k-1}\left(\frac{\lambda}{s+\lambda}\right)^k = \frac{(1-p)\lambda}{s+(1-p)\lambda} \Leftrightarrow r(t) = (1-p)\lambda e^{-(1-p)\lambda)t}.$$

Generally for arbitrary distributions for $\tilde{k}$ with pmf p_k and z-transform $K(z)$ and $\tilde{x}$ with LST $\mathcal{X}(s)$, the LST of $\tilde{y} = \sum_{j=1}^{\tilde{k}} x_j$ is

$$\mathcal{Y}^*(s) = \sum_{k\geq 0}[\mathcal{X}^*(s)]^k p_k = K(\mathcal{X}^*(s)), \tag{56.71}$$

where K (.) is the z-transform of the number of cycles.

The mean and variance of this distribution are obtained by integration by parts (in what follows $z = \mathcal{X}^*(s)$)

$$\bar{Y} = -\frac{d\mathcal{Y}^*(s)}{ds} = -\frac{dK(z)}{dz}\Big|_{z=1}\left[-\frac{dz}{ds}\Big|_{s=0}\right] = \bar{K}\times\bar{X}, \quad \sigma_Y^2 = \bar{K}\sigma_X^2 + \sigma_K^2(\bar{X})^2. \tag{56.72}$$

The number of jobs completed in a time interval approaches a Poisson process when the cycling probability $p \to 1$, signifying a large number of cycles, since transitions leading to a job completion are rare. Poisson inter-departure times imply exponentially distributed residence times [103].

Several approximate analyses have been developed to estimate the distribution of response time distribution in QNMs. At a first glance its LST is given as the product of the LSTs at the nodes visited by the chain. Complications such as "overtaking" are discussed in Chapter 9 in [72].

56.7 Page Replacement and Buffer Management

Early studies in this area dealt with the page reference pattern of computer programs residing in a virtual memory environment. The problem becomes more complex when multiple programs share memory space, which is referred to as multiprogrammed memory management. Early discussions of these two topics appear in Chapters 6 and 7 in [53].

Cache memories differ from virtual memories in that they utilize a smaller block size and are more restrictive from the mapping viewpoint, since they rely on a hardware implementation, such as direct or a low-degree set associative mapping for speed, while the more flexible fully associative mapping (with some limitations) is applicable to virtual memories [73]. Main memory access is only required when a CPU cache miss occurs and this requires the translation of virtual addresses generated by processors to physical main memory addresses. This translation is assisted by associative memories known as *translation lookaside buffers* (*TLBs*) [73]. Very low miss rates at the caches and TLBs are required to ensure high performance. System performance at this level can only be determined a priori by trace-driven simulation.

When the miss rates at the two levels are negligibly small, the effective mean access time can be approximated by the access time to the highest level. The following equation for two levels of caching can be generalized to multiple levels [73]

$$\text{Hit\_time}_{L1} + \text{Miss\_rate}_{L1} \times (\text{Hit\_time}_{L2} + \text{Miss\_rate}_{L2} \times \text{Miss\_penalty}_{Main-memory}.$$

The execution of a program is interrupted and context switching occurs when it encounters a page fault in a virtual memory environment. With multiprogramming the processor is reassigned to another program, which is eligible for execution, for example, because its page fault was satisfied. The miss rate, which is the fraction of main memory accesses that result in a page fault, is an appropriate performance metric in this context. When page faults occur, the system throughput is not affected significantly since the CPU is kept busy in multiprogrammed computer systems. With the advent of large main memories, most programs are fully loaded so that the memory miss rate is not an important issue.

A block (or page) replacement policy determines the block is to be overwritten. Better replacement policies minimize the number of misses by not overwriting blocks, which are referenced soon thereafter so that replacing them would result in increasing the execution time of a program. Stack algorithms which include the *least recently used* (*LRU*) policy have the desirable property of reducing the miss rate as the memory space allocated to a program is increased [100]. Policy MIN that minimizes the miss rate by replacing the page referenced furthest in the future, or not at all, is not a feasible policy, but useful in comparing the relative performance improvement. Several other page replacement policies are *first-in, first-out* (*FIFO*), *Last-in, first-out* (*LIFO*), CLOCK, Second-Chance, Generalized CLOCK, *working set* (*WS*), *page fault frequency* (*PFF*), etc. [32]. It is the size of allocated memory that has a first-order effect on the miss rate.

Overextending the main memory underlying the virtual memory may lead to the thrashing phenomenon [58], which may be dealt with by swapping out programs to make more memory space available for active jobs [50]. Programs exhibit a low miss rate when their working set is allocated in main memory so that the number of allocated pages should be increased when the miss rate is high [58]. The WS [58] and the PFF [94] algorithms vary the number of allocated pages in a multiprogrammed computer system to optimize performance. PFF allocates more memory pages when the the miss rate exceeds a certain threshold and otherwise reduces the number of allocated pages.

Buffer management studies estimate the parameters of reference strings for a given reference model and propose and evaluate replacement policies. Two early page reference models are the *independent reference model* (*IRM*) (blocks are accessed with fixed probabilities) and the LRU stack-depth model (stack distances are given by fixed probabilities). McNutt's hierarchical reuse (fractal) model reveals the following relationship between miss rates ($s < s'$) and memory sizes ($m > m'$) $s/s' = (m/m')^{-1/3}$, that is, to halve the miss ratio an eightfold increase in the buffers size is required [86].

The miss rate can be evaluated using random-number-driven or trace-driven simulations. Trace driven simulations are the more credible, but multiple traces should be used for a higher confidence level (86 traces were used in evaluating the cache for the IBM's 360/85, the first commercial computer with a cache in late 1960s). There have been a limited number of analytic models of paging algorithms, for example, the analysis via a Markov chain model of the generalized CLOCK algorithm in [93].

An empirical formula for the number of page faults (n) versus memory size (M) is derived in [114]. Memory sizes M_0 and M^* are the min and max M that are sufficient and necessary. The number of cold memory misses starting with an empty buffer is n_0. The parameters vary with program behavior and replacement policy:

$$n = \frac{1}{2}\left[H + \sqrt{H^2 - 4}\right](n^* + n_0) - n_0, \quad H = 1 + \frac{M^* - M_0}{M - M_0}, \quad M \le M^*.$$

One observation of this study is that policies have different n for smaller memory sizes and otherwise all policies yield about the same n.

The *buffer miss rate* (*BMR*) is applicable to various levels of the memory hierarchy. Misses in the main memory buffer result in accesses to the cache at the *disk array controller* (*DAC*) are accessed following a miss at both levels. BMR is affected by the buffer size, the replacement policy, and the workload. Of interest is improving the performance of shared DAC caches in processing sequential and random streams for large disk arrays. The *sequential prefetching adaptive replacement cache* (*SARC*) improves performance by a judicious combination of caching and prefetching policies [69]. SARC dynamically and adaptively partitions the cache space between sequential and random streams so as to reduce read misses. It outperforms two variants of the LRU policy at peak system throughput in processing a TPC-1 like workload in a RAID5 disk array with an 8 GB cache.

Early articles on page replacement and buffer management policies are summarized in a web page dealing with database implementation [33].

56.8 Analyses of Concurrency Control Methods

Four requirements for txn processing systems are *atomicity, consistency, integrity, and durability* (*ACID*) [97]. *Concurrency control* (*CC*) methods are required to ensure database consistency and integrity and also correctness in txn execution, which is tantamount to serializability, that the execution of a sequence of txns is equivalent to their execution in some serial order [97]. Txn recovery methods, such as write-ahead logging, are used to ensure txn atomicity and durability [97]. Almost all operational database systems utilize the standard locking method and the performance degradation due to this method is analyzed in Section 56.8.1. Two alternative CC methods have been proposed for high data contention environments [64]. Restart-oriented locking methods are discussed and their performance analyzed in Section 56.8.2. The results in the first two sections are due to Thomasian [127,132]. The analysis of the *optimistic CC* (*OCC*) in Section 56.8.3 is based on Morris and Wong in [88]. The extensions of this work by Ryu and Thomasian in [101] are briefly discussed.

56.8.1 Standard Locking and Its Threshing Behavior

Standard locking follows the strict *two-phase locking* (*2PL*) paradigm, where locks are requested on demand and strictness requires that no locks are released until the completion of txn execution known as txn commit. This allows the undoing of txn updates when the txn is aborted [97,131]. Locks are released upon txn abort to resolve deadlocks or to relieve lock congestion in conjunction with restart-oriented policies in Section 56.8.2.

An abstract model of standard locking to analyze the effect of lock contention on the performance of txn processing systems is postulated in [127]. There are M txns in a closed system, so that a completed txn is immediately replaced by a new txn. Txns of size k have $k+1$ steps with mean duration (s), so that the mean txn residence time with no lock contention is $r_k = (k+1)s$. Locks are requested at the end of the first k txn steps and all locks are released at the end of the $k + 1st$ step. The mean duration of each step taking into account txn blocking due to lock conflicts, but not txn restarts due to rare deadlocks is $(s + u)$. We have $u = P_cW$, where P_c is the probability of a lock conflict per lock request and W the mean waiting time per lock conflict [127]. The mean number of locks held by a txn is $\bar{L} \approx k/2$. It is assumed that all locks are exclusive and are distributed uniformly over the D pages in the database, so that the probability of lock conflict $P_c \approx (M-1)\bar{L}/D$ tends to be small. Shared locks whose fraction is f_S can be taken into account by adjusting the database size to $D_{eff} = D/\left(1-f_S^2\right)$ [131]. The probability that a txn encounters a lock conflict is $P_w = 1-(1-P_c)^k \approx kP_c = (M-1)k^2/D$. A two-way deadlock can *only* occur if txn T_A requests a lock held by txn T_B, which is blocked because of its conflict with txn T_A:

$$P_{D(2)}(k) \approx \frac{P[T_A \rightarrow T_B]P[T_B \rightarrow T_A]}{M-1} = \frac{P_w^2}{M-1} \approx \frac{(M-1)k^4}{4D^2}.$$

Multiway deadlocks are not considered since they are rare. The aforementioned analysis was refined to take into account the fact that a deadlock only occurs if T_B was already blocked [125]. We obtain the mean blocking delay of a txn with respect to an active txn holding the requested lock (W_1) by noting that the probability of lock conflict increases with the number of locks held by the txn (j) and that there are $(k-j)$ remaining steps [127]

$$W_1 \approx \sum_{j=1}^{k} \frac{2j}{k(k+1)}[(k-j)(s+u)]+s' = \frac{k-1}{3}[r+u]+s'.$$

where s' is the remaining processing time of a step at which the lock conflict occurred.

Since deadlocks are rare, the mean txn response time only takes into account txn blocking time due to lock conflicts, but not restarts due to deadlocks since they are rare $R_k \approx (k+1)s + kP_cW_1$. The fraction of txn blocking time per lock conflict with an active txn is $A = W_1/R_k \approx 1/3$. A more accurate expression for two-way deadlocks, as verified by simulation, is given by Equation 56.73 in [126] as $P'_{D(2)}(k) \approx AP_{D(2)}(k) = P_{D(2)}(k)/3$. In the case of variable size txns (with a closed model), a completed txn is replaced by another txn of size ℓ with frequency f_ℓ. Given that the ith moment of the number of requested locks is K_i, then $\bar{L} \approx K_2/K_1$ and $A = (K_3 - K_1)(3K_1(K_2 + K_1))$ [128].

Ignoring the difference in the number of locks held by active and blocked txns, the probability that a txn has a lock conflict with an active txn at level $i = 0$ (resp. a blocked txn at level $i = 1$) in the forest of waits-for graph for active and blocks txns is $1 - \beta$ (resp. β) and the mean blocking time is W_1 (resp. $W_2 = 1.5W_1$). The expression for W_2 relies on the fact that at the random instant when the lock conflict occurs, the active txn is halfway to its completion. The probability that a txn is blocked at level i is approximated as

$$P_b(i) \approx \beta^{i-1} \quad \text{for } i>1 \text{ and } P_b(1) = 1-\beta-\beta^2-\beta^3 \ldots.$$

The mean delay at level $i > 1$ is approximated by $W_i \approx (i - 0.5)W_1$. The mean blocking time is a weighted sum of delays incurred by txns blocked at different levels

$$W = \sum_{i\geq 1} P_b(i)W_i = W_1\left[1 - \sum_{i\geq 1}\beta^i + \sum_{i>1}(i-0.5)\beta^{i-1}\right].$$

Let $\overline{M}_A$ and $\overline{M}_B = M - \overline{M}_A$ denote the mean number of active and blocked txns in a closed QNM with M txns. We multiply both sides of the equation by K_1P_c/R and note that $\beta = K_1P_cW/R$ is the fraction of blocked txns. We define the fraction of time txns are blocked by active txns as $\alpha = K_1P_cW_1/R$. It follows $\beta = \alpha(1 + 0.5\beta + 1.5\beta^2 + 2.5\beta^3 + \ldots)$. A closed-form expression for β can be obtained by noting that $\beta < 1$ and assuming that M is infinite yields

$$f(\beta) = \beta^3 - (1.5\alpha + 2)\beta^2 + (1.5\alpha + 1)\beta - \alpha = 0.$$

The function $f(\beta)$ versus β is plotted in Figure 1 in [127]. The cubic equation has three roots $0 \leq \beta_1 < \beta_2 \leq 1$ and $\beta_3 \geq 1$ for $\alpha \leq \alpha^* = 0.226$ and a single root $\beta_3 > 1$ for $\alpha > \alpha^*$, where α^* is the only root of the discriminant function of the cubic [34] (see Appendix B in [127]). The single parameter α determines the level of lock contention for standard locking and its critical value α^* determines the onset of thrashing. The smallest root β_1 for $\alpha \leq \alpha^*$, which determines the mean number of active txns $\overline{M}_A = M(1 - \beta_1)$. $\overline{M}_A$ increases with the MPL M reaching a maximum point before dropping. This is because of a snowball effect that blocked txns cause further txn blocking. The associated thrashing behavior is shown in Figure 2 in [127]. It is also shown in this study that txn size variability has a major effect on lock contention.

56.8.2 Restart-Oriented Locking Methods

Performance degradation in standard locking is mainly caused by txn blocking, since deadlocks are rare and txn aborts to resolve deadlocks have little effect on performance. Restart-oriented locking methods follow the strict 2PL paradigm, but instead of blocking all txns when lock conflicts occur, txns are selectively restarted to increase the degree of txn concurrency. This would allow higher txn throughputs in high data contention systems, such as an early release of DB2 with page level, rather than record level locking [97].

The *running priority* (RP) method aborts txn T_B in the waits-for-graph $T_A \rightarrow T_B \rightarrow T_C$, when T_A encounters a lock conflict with T_B [64]. Similarly according to symmetric RP, T_B, which is already blocking T_A, is aborted when it is blocked requesting a lock held by T_C. In effect RP increases the degree of txn concurrency. The restart of an aborted txn is delayed until the conflicted txn has left the system. The no-waiting and cautious waiting methods abort and restart a txn encountering a lock conflict with an active and blocked txn, respectively. There is no increase in the number of active txns.

The *wait depth limited* (WDL) method measures txn progress by its length, which in [64] is approximated by the number of locks it holds. WDL differs from RP in that it attempts to minimize wasted lock holding time by not restarting txns holding many locks. For example, if T_A, which is blocking another txn, has a lock conflict with T_B, then if $L(T_A) < L(T_B)$ then abort T_A, else abort T_B [64]. The superior performance of WDL over RP and especially standard locking has been shown via random-number-driven simulation in [64], but also trace-driven simulation as discussed in [131]. RP and WDL attempt to attain *essential blocking*, that is, allowing a txn to be blocked only by active txns (bound to complete successfully) and not requesting a lock until required.

Immediately restarting an aborted txn may result in *cyclic restarts* or *livelocks*, which should be prevented to reduce wasted processing and to ensure timely txn completion. *Restart waiting* delays the restart of an aborted txn until *all* conflicting txns are completed [64]. Conflict avoidance delays with random duration are a less reliable method to prevent cyclic restarts.

Restart-oriented locking methods can be analyzed using Markov chain models [131,132]. Active (resp. blocked) txns with j locks are represented by states S_{2j} for $0 \leq j \leq k$ (resp. S_{2j+1} for $0 \leq j \leq k-1$). In the case of the no-waiting or immediate restart method, there is a transition forward when there is no lock conflict: $S_{2j} \rightarrow S_{2j+2}$ for $0 \leq j \leq k-1$, otherwise the txn is aborted $S_{2j} \rightarrow S_0$. The state equilibrium equations for the Markov chain yield the steady-state probabilities π_i for $0 \leq i \leq 2k$. The mean number of visits v_i to S_i can be obtained similarly by noting that $v_{2k} = 1$, which is the state at which the txn commits. Given that h_i is the mean holding time (the inverse of the transition rate) at S_i, then $\pi_i = v_i h_i \Big/ \sum_{j=0}^{2k} v_j h_j$ for $0 \leq i \leq 2k$. The mean number of txn restarts is given by $v_0 - 1$. A shortcoming of this analysis is that locks requested by restarted txns are resampled. Transition rates for the RP and WDL methods are given in [132].

56.8.3 Optimistic Concurrency Control Methods

A txn executing according to OCC copies pages (database blocks) it needs to modify to its private workspace, by noting its timestamp, that is, the last time the page was modified. The validation phase ascertains serializability, which is the universally accepted criterion for correct txn execution, that none of its accessed objects was modified by another txn after it was read. If the validation is successful, the txn commits and otherwise it is immediately restarted. Given that the k objects updated by txns are uniformly distributed over a database of size D, the probability of data conflict between two txns of size k is $\psi = 1-(1-k/D)^k \approx k^2/D$. This is referred to as the *quadratic effect* in [64]. The completion rate of successfully validated txns is $p\mu s(M)$, where p is the probability of successful validation and μ is the completion rate of txns. If there is no hardware resource contention $s(M) \propto M$, but since hardware resource contention is inevitable $s(M)$ is sublinear. A txn with processing time x will require $xM/s(M)$ time units to complete, where $s(M)$ is a nondecreasing function of M. A txn observes the commits of other txns at random times approximated by a Poisson process with rate $\lambda = (1 - 1/M)p\mu s(M)$. Txns encountering data conflict with rate $\gamma = \lambda\psi$ fail their validation with probability $q = 1-e^{-\gamma xM/s(M)}$. The number of txns executing successfully at the system follows a geometric distribution $P_j = (1-q)q^{j-1}$ for $j \geq 1$ with a mean $\bar{J} = 1/(1-q) = e^{\gamma xM/s(M)}$. The system efficiency p can be expressed as the ratio of the mean execution time of a txn without and with data contention, that is, with the possibility of restarts due to failed validation

$$p = \frac{E[xM/s(M)]}{E[(xM/s(M))e^{\gamma xM/s(M)}]} = \frac{\int_0^\infty xe^{-\mu x}dx}{\int_0^\infty xe^{(\gamma M/s(M)-\mu)x}dx} = [1-(M-1)\psi p]^2.$$

The equation yields one acceptable root $p < 1$, provided $\gamma M/s(M) < \mu$ and the integral in the denominator converges:

$$p = \left[1+2(M-1)\psi - \sqrt{1+4(M-1)\psi}\right] \Big/ \left[2(M-1)^2\psi^2\right] < 1.$$

Txn execution time is not resampled in this analysis because resampling processing times results in a mix of completed txns, which is shorter than the original processing time. Shorter txns have a higher probability of successful validation so that resampling would result in overestimating performance.

The aforementioned discussion follows *static* data access considered in [144], which postulates that all objects are accessed at the beginning of txn execution. This work was extended in [101] to allow the more realistic dynamic or on-demand data accesses preceded by lock requests [101]. The analysis is also that of OCC with the *die* method [64], which is inefficient in that a conflicted txn is allowed to execute to the end, at which point it is restarted. The OCC *broadcast* or *kill* method instead aborts a txn as soon as a data conflict is detected, alleviating further wasted processing [64,101].

Although the optimistic die method seems to be less efficient than the kill method, this is not the case when due to database buffer misses txns make disk accesses. It is advantageous then to allow a conflicted txn to run to its completion to prime the database buffer with all the required objects for its re-execution in a second phase, which is expected to have a much shorter processing time, since disk accesses are not required. This is so provided *access invariance* or *buffer retention* prevails, that is, a restarted txn accesses the same set of objects as it did in its first execution phase [64]. The second execution phase may use the optimistic kill method. Standard locking with on demand lock requests, or lock preclaiming. The latter is possible since given access invariance prevails, the identity of pages to be locked by the txn has been determined in the first execution phase. Lock preclaiming ensures that the txn will not be restarted. In general, we have multiphase processing methods with CC methods with increasing strengths, for example, OCC followed by locking, so that a lock request on an object accessed in optimistic mode will result in the abort of the respective txn.

The WDL and RP restart-oriented locking methods outperform two-phase methods and are easier to implement since they only require modifications to the lock manager shared by DBMSs associated with IBM's MVS OS and its descendants so that no modifications are required to the DBMSs.

56.9 Simulation Modeling and Measurement

The performance of computer systems is determined by hardware and software measurement tools. Most systems are also equipped with software to determine the causes of system crashes in operational systems. A description of measurement tools for IBM's MVS operating system (OS) [35], which is a forerunner to the z/390 OS appears in [50] and Chapter 17 in [83]. These measurements are carried out at the level of workloads, for example, response time and throughput, and device level, for example, processors and disk utilizations and mean queue lengths. In the case of BGS's BEST/1 capacity planning tool, complementary software was developed to summarize these results into service demands, which were direct inputs to BEST/1 [50]. For example, when the number of timesharing users is increased, BEST/1 can determine the effect of modifications to the current computer configuration to provide an acceptable response time. It is easier to deal with active rather than passive resources. Before proceeding with capacity planning, there is the important validation phase to ensure that the model accurately predicts performance. Otherwise the model has to be modified (see [59] and Section 2.2 in [83]).

Robust performance models proposed in [115] take into account that the underlying system may be misconfigured or buggy. The IronModel is made part of the system and is aware of all system requests and their performance. It uses localization to determine the sources of mismatches and then uses refinement techniques to make educated guesses. There is reliance on statistical methods and *machine learning (ML)* techniques such as *classification and regression tree (CART)* to modify the model [115]. A preliminary effort to build applications exporting their state and associated performance metrics is described in [116].

Simulation and analytical modeling can be carried out at multiple levels, such as estimating CPU performance as affected by its caches. While there have been analytic studies of caches and memory hierarchies [81], trace-driven simulation is more appropriate in this case. There are also simulation tools for disk subsystems, such as DiskSim from CMU's *Parallel Data Laboratory (PDL)* [36]. Dixtrac extracts disk parameters, which are then used by the DiskSim simulation tool. The ns2 network simulator provides support for simulating *transport control protocol (TCP)*, routing, and multicast over wired and wireless networks [37].

Random number simulations, as opposed to, trace-driven simulations, utilizes the pattern extracted from traces of workload activity, but are less credible than trace-driven simulations since workload characterization may not capture all subtleties of the workload. Trace analysis may show that accesses to a disk are concentrated in two small regions, but given that the accesses occurred in disjoint periods of time implies little seek activity.

While simulations deal with multiple concurrent events, most simulation programs are sequential and use the next event approach processing one event at a time. Several languages developed for simulation programming, but many sophisticated simulators, such as DiskSim and ns2 are mainly based on general purpose programming languages (such as C or C++). Simulator design and programming is described by Markowitz in Chapter 7 in [82].*

As events are generated they are held in an event queue based on their occurrence time in the future. With a large number events (n), efficient data structures such as heaps with an $O(log_2(n))$ insertion and deletion cost may result in a significant reduction in simulation cost in comparison with $O(n)$ for linear lists.

Random number generators (*RNGs*) are based on a pseudo-random sequence of a *linear congruential generator* (*LCG*) of the form $X_n = aX_{n-1} + c(mod\ m)$, where the modulus m is of the form $m = 2^{32}$ [38]. A "built-in radioactive RNG" in addition to a software RNG was used in [77]. X_0 is the seed and determines the sequence of the numbers generated, hence the sequence is pseudo-random. Using the current computer time as the seed has the advantage of generating an independent stream of pseudo-random numbers, which may be used by the batch means method (see below). It is best to use the same seed to reproduce the same events during the debugging phase. Uniform numbers $0 \le u \le 1$ are generated using appropriate transformations, such as $t = -(1/\lambda)ln(1 - u)$ or equivalently $t = (-1/\lambda)ln(u)$ for an exponential distribution with parameter λ. RNG for discrete event simulation is described by Shedler in Chapter 5 in [82].

The simulation of a queueing system, such as a CSM with external arrivals, maintains statistics on response times and mean queue lengths at various devices. Simulations starting with an empty system require statistics accumulated in the warmup period to be discarded. There is the issue of the length of the simulation run. This is usually expressed by the number of completed jobs. The statistical analysis of simulation results is an important aspect of simulation. In the case of an M/M/1 queue with $\lambda = 0.5$ and $\mu = 1$, the mean response times is $R = (\mu - \lambda)^{-1} = 2$, but even very lengthy simulations will only yield approximations to this value.

Thee methods—batch means, regenerative simulation, and spectral—are described in Chapter 6 by Welch in [82]. The batch means method is based on running the simulation n times (with different seeds) to obtain R_i for $1 \le i \le n$. These are then used to obtain the mean and standard deviation: $\bar{R} = \sum_{i=1}^{n} R_i/n$ and $\sigma = \sqrt{\sum_{i=1}^{n} (R_i - \bar{R})^2/(n-1)}$. The confidence interval at a confidence level $1 - \alpha = 90\%, 95\%, 99\%$ is $\bar{R} \pm z_{\alpha/2}\sigma/n$, where $z_{\alpha/2}$ is 1.645, 1.96, and 2.576, respectively.

With the advent of fast inexpensive computers and the possibility of using idle workstation cycles in distributed systems [39], large-scale simulations at a low cost remain a possibility. Understanding the operation of complex systems and developing the simulator and interpreting simulation results remain the chief obstacles to such studies. Simulation tools for Extended QNMs described by Sauer and MacNair in [82] equipped with graphical interfaces are intended to reduce the difficulty of developing simulations.

56.10 Conclusions

A combination of analytic models, simulation, measurement, and statistical methods have been used for designing and operating computer complexes for web servers, e-commerce, etc. With the advent of cloud computing, existing tools may need to be extended for their performance analysis and capacity planning. As noted in [115] there is the issue of understanding system behavior. In a few cases, this has led to fixing implementation bugs, when the predicted performance does not agree with measured performance.

Real systems are rarely amenable to exact and even approximate analytic solutions, although the surprising robustness of QNMs based on *operational analysis* (*OA*) has been noted in [59]. OA proposed

* Markowitz was the winner of a Nobel Prize in economics for his early work on portfolio selection and later got involved in developing the Simscript simulation language.

by Buzen allows "queueing analysis" without resorting to stochastic processes. OA is therefore of great pedagogic value in explaining the behavior of queueing systems and their simulation and has been adopted in several textbooks, for example, [83].

Statistical methods have been used since the early days of computing in computer performance evaluation. An early collection of papers in this area is [66]. Part 3 of [42,80,143] covers the application of statistical methods to computer system modeling. There is a recent interest in applying ML (machine learning) methods to performance analysis.

The use of *Generalized Stochastic Petri Nets (GSPNs)* in the performance analysis of computer systems has been advocated since mid-1980s by researchers at University of Torino in Italy [41]. These models are highly descriptive and can be used to evaluate aspects other than system performance, such as deadlock-freedom. A GSPN has to be translated to a CTMC or a semi-Markov chain for analysis. Related tools have been developed by Trivedi at Duke University [40]. Direct and iterative methods for the analysis of linear equations arising in solving large CTMCs are described in Chapter 3 in [45].

Abbreviations

BD: birth-death process
CC: concurrency control
CPU: central processing unit
CSM: central server model
CTMC: continuous time Markov chain
DBMS: data base management system
FCFS: first-come, first-served
JSQ: join shortest queue
LCFS: last-come, first-served
LST: Laplace–Stieltjes transform
ML: machine learning
MPL: multiprogramming level
MRM: machine repairman model
MVA: mean value analysis
OA: operational analysis
OCC: optimistic concurrency control
OLTP: online transaction processing
pdf: probability density function
PDF: probability distribution function
QNM: queueing network model
RAID: redundant array of independent disks
r.v.: random variable
SATF: shortest access time first
SPT: shortest processing time
txn: transaction
VSM: vacationing server model
XOR: eXclusive-OR

Glossary

Arrival Theorem: The mean number of requests encountered by arrivals at the nodes of a closed QNM with M requests is the queue length of a QNM with $M - 1$ requests.

Asymptotic Bounds Analysis (ABA): The maximum throughput in a closed QNM with N nodes with service demand D_n and m_n servers at node n is $min_{\forall n}(m_n/D_n)$.

Balanced Job Bounds (BJB): Given that the service demands at the N nodes of a QNM with single servers are $\{D_1, D_2, \ldots, D_N\}$, then an upper bound to system throughput at MPL M is $T_{max}(M) = M/[(N + M - 1)D_a]$ with $D_a = \sum_{n=1}^{N} D_n/N$.

BCMP Theorem: Named after the initials of the four authors of [43], who identified the four queueing systems, which lend themselves to a product-form solution. Thees are (1) FCFS with exponential service times, (2) processor sharing (PS), (3) infinite servers (IS), and (4) last-come, first-served (LCFS) preemptive. The last three allow general service times. FCFS with multiple servers and IS are examples of state-dependent servers.

Buzen's Algorithm: The state probabilities and performance metrics in closed QNMs with N nodes and M jobs are costly to calculate using a brute-force method because of the large number of system states. Buzen's algorithm requires a multiplication and an addition per step for state-independent modes. Calculating convolutions is required for state-dependent nodes unless they are designated as the first node.

Central Server Model (CSM): A closed QNM with N nodes, where the central server is the CPU and the peripheral severs are the disks. Job completions are indicated by a self-transition to the CPU, while in the *cyclic server model* jobs leave the system after completing their service at the disks.

Concurrency Control (CC): CC ensures transaction serializability and database integrity. Standard locking utilizes on demand locking and txns encountering a lock conflict are blocked. Restart-oriented locking methods utilize txn restarts to relieve the level of lock contention. Optimistic CC checks whether any data conflicts were encountered during txn execution and if so aborts and restarts it.

Flow Equivalent Service Center (FESC): An FESC replaces multiple nodes in a QNM with a single state-dependent server specified by the throughput characteristic $T(m)$, $1 \leq m \leq M_{max}$. The CTMC for two job classes is specified by its throughputs in processing mixtures of the two job classes $T_i(m_1, m_2)$, $i = 1, 2$.

Fork-Join: Jobs spawn K tasks to K independent servers. A job is considered completed when all K tasks complete, joining at a synchronization point.

Kendall's Notation: A queueing system is specified as A/B/C/D/E. The arrival process is $A \in \{M, GI, \ldots\}$, where M stands for Markovian or Poisson, GI for general independent, etc. The service time distribution is $B \in \{M, G, \ldots\}$, where M is the exponential distribution, G a general distribution, etc. C is the number servers; D is the capacity of the system, which is equal to C if there is no queueing. E is the number of sources. D and E are infinite by default.

Little's Result: The mean number of requests at the queue, the servers, a queueing subnetwork, and the QNM is the product of the arrival rate, which equals the completion rate, and the mean reidence time of requests at that entity.

Mean Value Analysis (MVA): A method to solve a closed QNM using the Arrival Theorem. It has higher computational cost and storage requirements than Buzen's algorithm and is less robust numerically.

Page Replacement Algorithms: FIFO, LRU, Clock, generalized Clock, random, LIFO, OPT, working set (WS), page fault frequency (PFF), etc.

Poisson Arrivals See Time Averages (PASTA): This property is used in obtaining the mean waiting time incurred by incoming requests in the M/G/1 queue.

Queueing Network Model (QNM): In open and closed QNMs, jobs are routed from node to node. In open QNMs, there are external job arrivals and job departures. In closed QNMs, special job transitions signify the completion of the job and its replacement by a new job, so that the number of jobs in the QNM remains fixed.

Readers and Writers Scheduling: Readers can be processed concurrently with other readers, while writers can only be processed singly. A significant improvement in performance can be attained by not interrupting the processing of readers every time a writer arrives, for example, by adopting the *threshold fastest emptying* (*TFE*) policy.

Scheduling Discipline: There are specialized scheduling disciplines at the computer system level for activating jobs, for CPU scheduling, and disk scheduling. Scheduling policies can be nonpreemptive or preemptive and the latter can be preemptive resume or repeat.

Serialization Delays: The processing of certain phases of a job in computer systems is single-threaded so that only one of multiple job executing concurrently can proceed in the serialized phase, while other jobs are blocked.

Thrashing: A severe degradation in system performance, which may be due to unproductive processing, such as excessive paging in a virtual memory environment. In the case of standard locking, trashing occurs when a large fraction of txns are blocked so that there is a significant drop in system throughput.

References to web pages occur in the order of their occurrence in the text, while other references are listed alphabetically according to the last name the first author and the year. Readers should attempt searches of their own with appropriate keywords since web pages may become unavailable over time. It is assumed that the reader has access to digital libraries to check referenced papers.

Further Information

Readers are expected to have access to digital libraries for the journals and conferences referenced in this paper. Given a background in elementary calculus, the interested reader should first study appropriate chapters in one of the following texts [42,143] to gain the necessary background in probability and stochastic processes. The reader can then proceed to read queueing theory texts such as [78,79]. We recommend a comprehensive textbook on performance modeling and analysis by Kobayashi and Mark [80a], which does not however cover some material in Kobayashi's early textbook [80], so that text is given as a reference.

Journals and conferences on systems and performance evaluation and analysis are reviewed in [137]. Conferences and journals sponsored by ACM: http://www.acm.org. IEEE: http://www.ieee.org. and its Computer Society: http://www.computer.org. and USENIX: http://www.usenix.org. There are large number of high quality publications on queueing theory: *Operations Research*, *Queueing Systems*, *Management Science*, *Naval Research Logistics*, etc.

Performance evaluation is covered by numerous conferences. *ACM's SIGMETRICS Conf. on Measurement and Modeling of Computer Systems* is held annually and together with the IFIP Working Group 7.3 Performance Conference every other year. ACM publishes the quarterly *Performance Evaluation Review* (*PER*), which includes the proceedings of the SIGMETRICS Conf. and various workshops, such as *MAthematical performance Modeling and Analysis* (*MAMA*). *Performance Evaluation* is a reviewed journal published since 1981 by Elsevier, which also publishes papers from the *Performance Conference*. The *Computer Performance* journal was published for several years by IPC Science and Technology Press until 1984. *Computer Systems: Science and Engineering* partially replaces *Computer Performance*, since 1986. http://www.crlpublishing.co.uk/.

IEEE's *Trans. on Computers*, *Trans. on Software Engineering*, and *Trans. on Parallel and Distributed Systems* (*TPDS*) publish papers dealing with computer performance analysis. ACM's *Trans. on Computer Systems* (*TOCS*) *Trans. on Storage Systems* (*TOS*), *Trans. on Database Systems* (*TODS*), *Trans. on Modeling and Computer Simulations* (*TOMACS*) are also relevant.

The *IEEE/ACM Int'l Symp. on Modeling, analysis and Simulation of Computer and Telecommunication Systems* (*MASCOTS*) is in its 20th year in 2012. The *Int'l Conf. on Quantitative Evaluation of Systems* (*QEST*) has been held in Europe since 2004 and is sponsored by the IEEE Computer Society. The Winter Simulation Conference http://www.wintersim.org. is held annually with multiple sponsors, such as: *American Statistical Association* (*ASA*), http://www.amstat.org. ACM, IEEE's Man and Cybernetics Society (IEEE/SMC): http://www.ieeesmc.org. Institute for Operations Research and the Management Sciences: Simulation Society (INFORMS-SIM): http://www.informs.org. *National Institute of Standards* (*NIST*): http://www.nist.gov. and the Society for Modeling and Simulation (SCS), which also sponsors the *Int'l Symp. on Performance Evaluation of Computer and Telecommunication Systems* (*SPECTS*). http://www.scs.org. The *Computer Measurement Group* (*CMG*): http://www.cmg.org. holds an annual conference of interest to Information Technology professionals.

ACM's *Int'l Symp. on Computer Architecture* (*ISCA*) and USENIX's *File and Storage Technologies* (*FAST*) are two conferences with papers dealing with the performance of computer systems and storage systems, respectively. Database conferences such as SIGMOD, http://www.sigmod.org. *Very Large Data Bases* (*VLDB*): http://www.vldb.org. and the *IEEE Int'l Conf. on Data Engineering* (*ICDE*) publish papers dealing with database performance. There is also the *VLDB Journal*.

The material in Section 56.8 is complementary to the Chapter on Concurrency Control and Recovery. The material in Section 56.7 on page replacement and buffer management is complementary to the Chapter on Operating Systems. This also applies the analysis of QNMs and measurement and simulation in Section 56.9.

References

1. http://en.wikipedia.org/wiki/Green_computing.
2. http://en.wikipedia.org/wiki/Instructions_per_second.
3. http://en.wikipedia.org/wiki/VAX.
4. http://www-03.ibm.com/ibm/history/exhibits/mainframe/mainframe_PP2091.html.
5. http://en.wikipedia.org/wiki/Parallel_computing.
6. http://en.wikipedia.org/wiki/Cray-1.
7. http://en.wikipedia.org/wiki/SISD.
8. http://en.wikipedia.org/wiki/Amdahl's_law.
9. http://en.wikipedia.org/wiki/Message_Passing_Interface.
10. http://en.wikipedia.org/wiki/Deep_Blue_versus_Kasparov,_1997,_Game_6.
11. http://en.wikipedia.org/wiki/Blue_Gene.
12. http://en.wikipedia.org/wiki/Watson_(computer).
13. http://en.wikipedia.org/wiki/IBM_Sequoia.
14. http://www.top500.org/.
15. http://en.wikipedia.org/wiki/Application_checkpointing.
16. http://en.wikipedia.org/wiki/CPU_cache.
17. http://en.wikipedia.org/wiki/Benchmark_(computing).
18. http://www.spec.org.
19. http://research.microsoft.com/en-us/um/people/gray/BenchmarkHandbook/TOC.htm.
20. http://en.wikipedia.org/wiki/Tandem_Computers.
21. http://www.tpc.org.
22. http://www.storageperformance.org
23. http://en.wikipedia.org/wiki/Little's_law
24. http://www.lk.cs.ucla.edu.
25. http://www.vm.ibm.com/devpages/jelliott/evrrt.html.
26. http://en.wikipedia.org/wiki/Compatible_Time_Sharing_System.
27. http://en.wikipedia.org/wiki/Lagrange_multiplier.
28. http://en.wikipedia.org/Buzen's.algorithm.
29. http://en.wikipedia.org/wiki/Mean_value_analysis.
30. http://en.wikipedia.org/wiki/Binomial_theorem.
31. http://en.wikipedia.org/wiki/thrashing.
32. en.wikipedia.org/wiki/Page_replacement_algorithm.
33. http://www.informatik.uni-trier.de/~ley/db/dbimpl/buffer.html.
34. http://en.wikipedia.org/wiki/Cubic_equation.
35. http://en.wikipedia.org/wiki/Z/OS.

36. http://www.pdl.cmu.edu/DiskSim/.
37. http://www.isi.edu/nsnam/ns/.
38. http://en.wikipedia.org/wiki/Linear_congruential_generator.
39. http://research.cs.wisc.edu/condor/.
40. http://www.ee.duke.edu/~kst.
41. Ajmone Marsan, M., Balbo, G., Conti, G., Donatelli, S., and Franceschinis, G. *Modelling with Generalized Stochastic Petri Nets*, Wiley, New York, 1995.
42. Allen, A. O. *Probability, Statistics, and Queueing Theory with Computer Science Applications*, 2nd edn., Academic Press, Boston, MA, 1990.
43. Baskett, F., Mani Chandy, K., Muntz, R. R., and Palacios, F. G. Open, closed, and mixed networks of queues with different classes of customers, *Journal of ACM* 22(2): 248–260 (April 1975).
44. Bell, C. G., Newell, A., and Siewiorek, D. P. *Computer Structures: Principles and Examples,* 2nd edn., McGraw-Hill, New York, 2001.
45. Bolch, G., Greiner, S., de Meer, H., and Trivedi, K. S. *Queueing Networks and Markov Chains: Modeling and Performance Evaluation with Computer Science Applications,* 2nd edn., Wiley, Hoboken, NJ, 2006.
46. Bux, W. Performance issues in local-area networks, *IBM Systems Journal* 23(4): 351–374 (1984).
47. Buzen, J. P. Computational algorithms for closed queueing networks with exponential servers, *Communications of the ACM* 16(9): 527–531 (September 1973).
48. Buzen, J. P. and Chen, P. P. Optimal load balancing in memory hierarchies, *Proceedings of the IFIP Congress 1974*, Stokholm, Sweden, August 1974, pp. 271–275.
49. Buzen, J. P. and Goldberg, P. S. Guidelines for the use of infinite source queueing models in the analysis of computer system performance, *Proceedings of the 1974 AFIPS National Computer Conference (NCC)*, Vol. 43, AFIPS Press, Montvale, NJ, pp. 371–374.
50. Buzen, J. P. A queueing network model of MVS, *ACM Computing Surveys* 10(3): 319–331 (September 1978).
51. Chaudhry, M. L. and Templeton, J. G. C. *First Course in Bulk Queues*, Wiley, New York, 1993.
52. Chiu, W. W. and Chow, W.-M. A performance model of MVS, *IBM Systems Journal* 17(4): 444–467 (1978).
53. Coffman, E. G. Jr. and Denning, P. J. *Operating Systems Principles*, Prentice-Hall, Englewood Cliffs, NJ, 1973.
54. Conway, A. E. and Georganas, N. D. RECAL—A new efficient algorithm for the exact analysis of multiple-chain closed queuing networks, *Journal of the ACM* 33(4): 768–791 (October 1986).
55. Conway, A. E. and Georganas, N. D. *Queueing Networks: Exact Computational Algorithms*, MIT Press, Cambridge, MA, 1989.
56. Conway, E. W., Maxwell, W. L., and Miller, L. W. *Scheduling Theory*, Dover Publications, New York, 2003 (original Addison-Wesley, Reading, MA, 1967).
57. Culler, D., Singh, J. P., and Gupta, A. *Parallel Computer Architecture: A Hardware/Software Approach*, Morgan-Kauffman, San Francisco, CA, 1998.
58. Denning, P. J. Virtual memory, *ACM Computing Surveys* 2(3): 153–189 (September 1970).
59. Denning, P. J. and Buzen, J. P. The operational analysis of queueing network models, *ACM Computing Surveys* 10(3): 225–261 (September 1978).
60. Dowdy, L. W., Eager, D. L., Gordon, K. D., and Saxton, L. V. Throughput concavity and response time convexity, *Information Processing Letters* 19(4): 209–212 (November 1984).
61. Eager, D. L., Lazowska, E. D., and Zahorjan, J. Load sharing in homogeneous distributed systems, *IEEE Transactions on Software Engineering* 12(5): 662–675 (May 1986).
62. Eager, D. L., Lazowska, E. D., and Zahorjan, J. A comparison of receiver-initiated and sender-initiated adaptive load sharing, *Performance Evaluation* 6(1): 53–68 (1986), also extended Abstract in SIGMETRICS 1985: 1–3.
63. Ercegovac, M. D. and Lang, T. *Digital Arithmetic*, Morgan Kaufmann Publishers, San Francisco, CA, 2004.

64. Franaszek, P. A., Robinson, J. T., and Thomasian, A. Concurrency control for high contention environments, *ACM Transactions on Database Systems* 17(2): 304–345 (June 1992).
65. Franx, G. J. A simple solution for the M/D/c waiting time distribution, *Operations Research Letters* 29, 221–229 (2001).
66. Freiberger, W. *Statistical Methods for the Evaluation of Computer Systems Performance*, Elsevier, 1972.
67. Fuhrmann, S. W. A note on the M/G/1 queue with server vacations, *Operations Research* 33(5): 1368–1373 (November-December 1984).
68. Georgiadis, L., Nikolaou, C., and Thomasian, A. A fair workload allocation policy for heterogeneous systems, *Journal of Parallel and Distributed Computing* 64(4): 507–519 (April 2004).
69. Gill, B. S. and Modha, D. S. SARC: Sequential prefetching in adaptive replacement cache, *Proceedings of 4th USENIX Conference on File and Storage Technologies (FAST)*, San Francisco, CA, December 2005, pp. 293–308.
70. Gunther, N. J. Understanding the MPL effect: Multiprocessing in pictures, *Proceedings of International Computer Measurement Group (CMG) Conference*, San Diego, CA, December 1996, pp. 957–968.
71. Gupta, V., Harchol-Balter, M., Dai, J., and Zwart, B. On the inapproximability of M/G/k: Why two moments of job size distribution are not enough, *Queueing Systems: Theory and Applications* 64(1): 5–48 (January 2010).
72. Harrison, P. G. and Patel, N. M. *Performance Modelling of Communication Networks and Computer Architectures*, Addison-Wesley, Reading, MA, 1993.
73. Hennessy, J. L. and Patterson, D. A. *Computer Architecture: A Quantitative Approach,* 5th edn., Elsevier, Amsterdam, the Netherlands, 2012.
74. Ibaraki, T. and Katoh, N. *Resource Allocation Problems: Algorithmic Approaches*, MIT Press, Cambridge, MA, 1988.
75. Jacob, B. L., Ng, S. W., and Wang, D. T. *Memory Systems: Cache, DRAM, Disk*, Morgan Kaufmann Publishers, Burlington, MA, 2008.
76. Johnson, N. L., Kotz, S., and Balakrishnan, N. *Continuous Univariate Distributions,* Vol. 2. 2nd edn., Wiley Series in Probability and Statistics, New York, May 1995.
77. Kleinrock, L. *Communication Nets: Stochastic Message Flow and Delay*, McGraw-Hill, New York, 1964 (Dover edition 1972).
78. Kleinrock, L. *Queueing Systems, Vol. 1: Theory*, Wiley-Interscience, New York, 1975; Kleinrock, L. and Gail, R. *Queueing Systems: Problems and Solutions*, Wiley, 1996.
79. Kleinrock, L. *Queueing Systems, Vol. 2: Computer Applications*, Wiley-Interscience, New York, 1976; Kleinrock, L. and Gail, R. *Solution Manual for Vol. 2,* Wiley, 1986.
80. Kobayashi, H. *Modeling and Analysis: An Introduction to System Performance Evaluation Methodology*, Addison-Wesley, Reading, MA, 1978.
80a. Kobayashi, H. and Mark, B. L. *System Modeling and Analysis: Foundations of System Performance Evaluation*, Prentice Hall, 2008.
81. Krishna, C. M. *Performance Modeling for Computer Architects*, IEEE Computer Society Press, Los Alamitos, CA, 1996.
82. Lavenberg, S. S. (ed.) *Computer Performance Modeling Handbook*, Academic Press, New York, 1983.
83. Lazowska, E. D., Zahorjan, J., Graham, G. S., and Sevcik, K. C. *Quantitative System Performance Computer System Analysis Using Queueing Network Models*, Englewood Cliffs, NJ, Prentice-Hall, 1984. http://www.cs.washington.edu/homes/lazowska/qsp/.
84. Livny, M. and Melman, M. Load balancing in homogeneous broadcast distributed systems, *Proceedings of ACM SIGMETRICS Conference on Measurement and Modeling of Computer Systems*, College Park, MD, April 1982, pp. 47–56.
85. Mani Chandy, K. and Neuse, D. Linearizer: A heuristic algorithm for queueing network models of computing systems, *Communications of the ACM* 25(2): 126–134 (February 1982).
86. McNutt, B. *The Fractal Structure of Data Reference: Applications to the Memory Hierarchy*, Springer, 2000.

87. Menon, J. Performance of RAID5 disk arrays with read and write caching, *Distributed and Parallel Databases* 2(3): 261–293 (July 1994).
88. Morris, R. J. T. and Wong, W. S. Performance analysis of locking and optimistic concurrency control algorithms, *Performance Evaluation* 5(2): 105–118 (1985).
89. Nadji, B. and Thomasian, A. Throughput estimation in queueing networks, *Computer Performance* 5(4): 197–206 (December 1984).
90. Nelson, R. D. and Tantawi, A. N. Approximate analysis of fork/join synchronization in parallel queues, *IEEE Transactions on Computers* 37(6): 739–743 (June 1988).
91. Nelson, R. D. and Philips, T. K. An approximation for the mean response time for shortest queue routing with general interarrival and service times. *Performance Evaluation* 17(2): 123–139 (1993).
92. Nelson, R. *Probability, Stochastic Processes, and Queueing Theory*, Springer-Verlag, New York, 1995.
93. Nicola, V. F., Dan, A., and Dias, D. M. Analysis of the generalized clock buffer replacement scheme for database transaction processing, *Proceedings of the ACM SIGMETRICS/PERFORMANCE'92 International Conference on Measurement and Modeling of Computer Systems*, Newport, RI, June 1992, pp. 35–46.
94. Opderbeck, H. and Chu, W. W. Performance of the page fault frequency replacement algorithm in a multiprogramming environment, *Proceedings of IFIP Congress 1974*, Stokholm, Sweden, August 1974, pp. 235–241.
95. Pipes, L. A. and Harvill, L. R. *Applied Mathematics for Engineers and Scientists*, McGraw-Hill, New York, 1970.
96. Rahm, E. Empirical performance evaluation of concurrency and coherency control protocols for database sharing systems, *ACM Transactions on Database Systems* 18(2): 333–377 (1993).
97. Ramakrishnan, R. and Gehrke, J. *Database Management Systems*, 3rd edn., McGraw-Hill, Boston, MA, 2003.
98. Reiser, M. and Lavenberg, S. S. Mean-value analysis of closed multichain queuing networks, *Journal of the ACM* 27(2): 313–322 (April 1980); Corrigendum, *Journal of the ACM* 28(3): 629 (July 1981).
99. Reiser, M. Mean value analysis: A personal account, *Performance Evaluation Origins and Directions.* Lecture Notes in Computer Science 1769, Springer, Berlin, Germany, 2000, pp. 491–504.
100. Robinson, J. T. and Devarakonda, M. V. Data cache management using frequency-based replacement, *Proceedings of the ACM SIGMETRICS Conference on Measurement and Modeling of Computer Systems*, Boulder, CO, May 1990, pp. 134–142.
101. Ryu, I. K. and Thomasian, A. Performance analysis of centralized databases with optimistic concurrency control, *Performance Evaluation* 7(3): 195–211 (1987).
102. Ryu, I. K. and Thomasian, A. Analysis of database performance with dynamic locking, *Journal of the ACM* 37(3): 491–523 (July 1990).
103. Salza, S. and Lavenberg, S. S. Approximating response time distributions in closed queueing network models of computer performance, *Proceedings of the PERFORMANCE'81 International Symposium on Computer Performance Modelling, Measurement and Evaluation*, Amsterdam, the Netherlands, November 1981, pp. 133–144.
104. Scherr, A. L. *An Analysis of Time-Shared Computer Systems*, Research Monograph 36, MIT Press, Cambridge, MA, 1967.
105. Schwetman, H. D. Hybrid simulation models of computer systems, *Communications of the ACM* 21(9): 718–723 (September 1878).
106. Spirn, J. R. *Program Behavior: Models and Measurements*, Elsevier, New York, 1977.
107. Stonebraker, M, The case for shared nothing, *IEEE Database Engineering Bulletin* 9(1): 4–9 (March 1986).
108. Sylvester, J. and Thomasian, A. Performance modeling of a large-scale multiprogrammed computer using BEST/1, *Proceedings of the 3rd International Conference on Computer Capacity Management - ICCCM'81*, Chicago, IL, April 1981, pp. 197–211.
109. Takagi, H. *Analysis of Polling Systems*, MIT Press, Cambridge, MA, 1986.

110. Takagi, H. Queuing analysis of polling models, *ACM Computing Surveys* 20(1): 5–28 (March 1988).
111. Takagi, H. *Queueing Analysis: A Foundation of Performance Evaluation, Volume 1: Vacation and Priority Systems, Vol. 2: Finite Systems, Vol. 3: Discrete Time Systems*, North-Holland, Amsterdam, the Netherlands, 1991/93/03.
112. Tantawi, A. N. and Towsley, D. F. Optimal static load balancing in distributed computer systems, *Journal of the ACM* 32(2): 445–465 (April 1985).
113. Tasaka, S. *Performance Analysis of Multiple Access Protocols*, MIT Press, Cambridge, MA, 1986.
114. Tay, Y. C. and Zou, M. A page fault equation for modeling the effect of memory size, *Performance Evaluation* 63(2): 99–130 (February 2006).
115. Thereska, E. and Ganger, G. R. IronModel: Robust performance models in the wild, *Proceedings of the ACM SIGMETRICS Conference on Measurement and Modeling of Computer Systems*, Annapolis, MD, June 2008, pp. 253–264.
116. Thereska, E., Doebel, B., and Zheng, A. X. Practical performance models for complex, popular applications, *Proceedings of the ACM SIGMETRICS Conference on Measurement and Modeling of Computer Systems*, New York, June 2010, pp. 1–12.
117. Thomasian, A. and Nadji, B. Algorithms for queueing network models of multiprogrammed computers using generating functions, *Computer Performance* 2(3): 100–123 (September 1981).
118. Thomasian, A. and Nadji, B. Aggregation in queueing network models of multiprogrammed computer systems, *Proceedings of the 1981 ACM SIGMETRICS Conference on Measurement and Modeling of Computer Systems*, Las Vegas, NV, September 1981, pp. 86–96.
119. Thomasian, A. Queueing network models to estimate serialization delays in computer systems, *Proceedings of the PERFORMANCE'83 Conference on Computer System Modeling and Analysis*, College Park, MD, May 1983, pp. 61–79.
120. Thomasian, A. and Ryu, I. K. A decomposition solution to the queueing network model of the centralized DBMS with static locking, *Proceedings of the ACM SIGMETRICS International Conference on Measurements and Modeling of Computer Systems*, Minneapolis, MN, August 1983, pp. 82–92.
121. Thomasian, A. and Bay, P. F. Analysis of queueing network models with population size constraints and delayed blocked customers, *Proceedings of the ACM SIGMETRICS Conference on Measurement and Modeling of Computer Systems*, Cambridge, MA, August 1984, pp. 202–216.
122. Thomasian, A. and Bay, P. F. Analytic queueing network models for parallel processing of task systems, *IEEE Transactions on Computers* 35(12): 1045–1054 (December 1986).
123. Thomasian, A. A performance study of dynamic load balancing in distributed systems, *Proceedings of the International Conference on Distributed Computing Systems*, Berlin, West Germany, September 1987, pp. 178–184.
124. Thomasian, A. and Nicola, V. F. Performance evaluation of a threshold policy for scheduling readers and writers, *IEEE Transactions on Computers* 42(1): 83–98 (January 1993).
125. Thomasian, A. and Ryu, I. K. Performance analysis of two-phase locking, *IEEE Transactions on Software Engineering* 17(5): 386–402 (May 1991).
126. Thomasian, A. and Nadji, B. State representation trade-offs in Markov chain models for serialization delays in computer systems, *Computer Systems: Science and Engineering (CSSE)* 8(3): 154–165 (July 1993).
127. Thomasian, A. Two-phase locking performance and its thrashing behavior, *ACM Transactions on Database Systems* 18(4): 579–625 (December 1993).
128. Thomasian, A. and Tantawi, A. N. Approximate solutions for M/G/1 fork/join synchronization, *Proceedings of the Winter Simulation Conference (WSC'94)*, Orlando, FL, December 1994, pp. 361–368.
129. Thomasian, A. Rebuild options in RAID5 disk arrays, *Proceedings 7th IEEE Symposium on Parallel and Distributed Systems*, San Antonio, TX, October 1995, pp. 511–518.
130. Thomasian, A. and Menon, J. RAID5 performance with distributed sparing, *IEEE Transactions on Parallel Distributed Systems* 8(6): 640–657 (August 1997).

131. Thomasian, A. Concurrency control: Methods, performance, and analysis, *ACM Computing Surveys* 30(1): 70–119 (March 1998).
132. Thomasian, A. Performance analysis of locking methods with limited wait depth, *Performance Evaluation* 34(2): 69–89 (October 1998).
133. Thomasian, A. A multithreshold scheduling policy for readers and writers, *Information Sciences* 104(3–4): 157–180 (1998).
134. Thomasian, A. and Liu, C. Comment on issues and challenges in the performance analysis of real disk arrays, *IEEE Transactions on Parallel and Distributed Systems* 16(11): 1103–1104 (November 2005).
135. Thomasian, A., Fu, G., and Ng, S. W. Analysis of rebuild processing in RAID5 disk arrays, *Computer Journal* 50(2): 217–231 (March 2007).
136. Thomasian, A., Fu, G., and Han, C. Performance of two-disk failure-tolerant disk arrays, *IEEE Transactions on Computers* 56(6): 799–814 (June 2007).
137. Thomasian, A. Publications on storage and systems research, *ACM SIGARCH Computer Architecture News* 37(4): 1–26 (September 2009).
138. Thomasian, A. and Blaum, M. Higher reliability redundant disk arrays: Organization, operation, and coding, *ACM Transactions on Storage* 5(3): Article 7 (November 2009).
139. Thomasian, A. Survey and analysis of disk scheduling methods, *ACM SIGARCH Computer Architecture News* 39(2): 8–25 (May 2011).
140. Thomasian, A. Analysis of fork-join and related queueing systems. *ACM Computing Surveys*, submitted July 2012.
141. Tijms, H. C. *Stochastic Models: An Algorithmic Approach*, Wiley, Chichester, U.K., 1994.
142. Towsley, D. F., Chen, S.-Z., and Yu, S.-P. Performance analysis of a fault tolerant mirrored disk system, *Proceedings of the 14th IFIP WG 7.3 International Symposium on Computer Performance Modelling, Measurement and Evaluation*, Edinburgh, Scotland, September 1990, pp. 239–253.
143. Trivedi, K. S. *Probability and Statistics with Reliability, Queuing, and Computer Science Applications*, Wiley-Interscience, New York, 2002.
144. Wang, Y.-T. and Morris, R. J. T. Load sharing in distributed systems, *IEEE Transactions on Computers* 34(3): 204–217 (March 1985).
145. Whitt, W. Deciding which queue to join: Some counterexamples, *Operations Research* 34(1): 55–62 (January–February 1986).
146. Williams, A. C. and Bhandiwad, R. A. A generating function approach to queueing network analysis of multiprogrammed computers, *Networks* 6(1): 1–22 (1976).
147. Wolf, J. L. The placement optimization program: A practical solution to the disk file assignment problem, *Proceedings of ACM SIGMETRICS/PERFORMANCE'89 International Conference on Measurement and Modeling of Computer Systems*, Berkeley, CA, May 1989, pp. 1–10.
148. Yu, P. S., Balsamo, S., and Lee. Y.-H. Dynamic transaction routing in distributed database systems, *IEEE Transactions on Software Engineering* 14(9): 1307–1318 (September 1988).
149. Yu, P. S., Cornell, D. W., Dias, D. M., and Thomasian, A. Performance comparison of IO shipping and database call shipping: Schemes in multisystem partitioned databases, *Performance Evaluation* 10(1): 15–33 (1989).

57

Taxonomy of Contention Management in Interconnected Distributed Systems

Mohsen Amini Salehi
The University of Melbourne

Jemal Abawajy
Deakin University

Rajkumar Buyya
The University of Melbourne

57.1 Introduction

Scientists and practitioners are increasingly reliant on large amounts of computational resources to solve complicated problems and obtain results in a timely manner. To satisfy the demand for large computational resources, organizations build or utilize distributed computing systems.

A distributed computing system is essentially a set of computers that share their resources via a computer network and interact with each other toward achieving a common goal [31]. The shared resources in a distributed system include data, computational power, and storage capacity. The common goal can also range from running resource-intensive applications, tolerating faults in a server, and serving scalable Internet applications.

Distributed computing systems such as Clusters, Grids, and recently Clouds have become ubiquitous platforms for supporting resource-intensive and scalable applications. However, surge in demand is still a common problem in distributed systems [26] in a way that no single system (specially Clusters and Grids) can meet the needs of all users. Therefore, the notion of interconnected distributed computing systems has emerged.

In an interconnected distributed computing system, as depicted in Figure 57.1, organizations share their resources over the Internet and consequently are able to access larger resources. In fact,

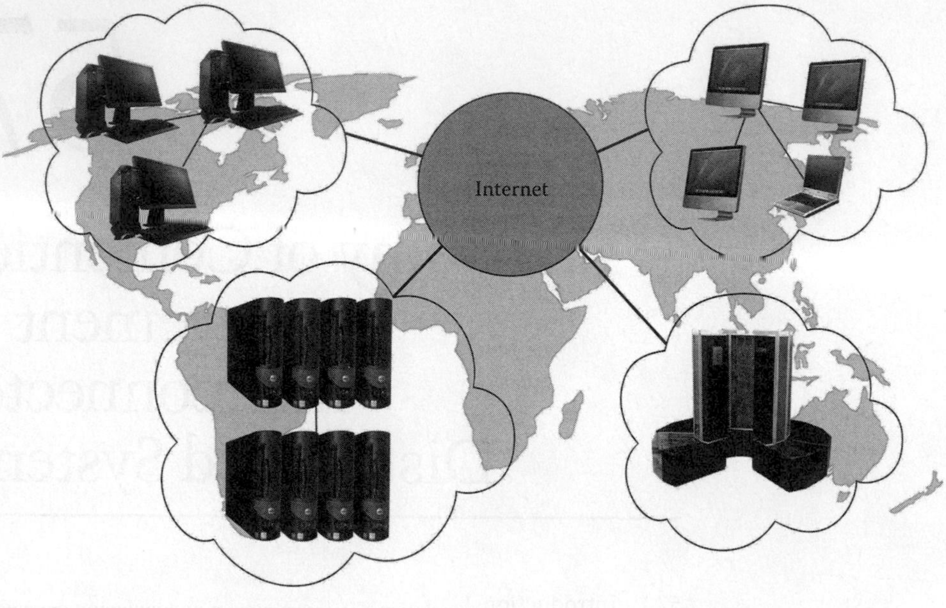

FIGURE 57.1 Interconnected distributed computing systems.

interconnected distributed systems construct an overlay network on top of the Internet to facilitate resource sharing between the constituents.

However, there are concerns in interconnected distributed systems regarding contention between requests to access resources, low access level, security, and reliability. These concerns necessitate a resource management platform that encompasses these aspects. The way current platforms consider these concerns depends on the structure of the interconnected distributed system. In practice, interconnection of distributed systems can be achieved in different levels. These approaches are categorized in Figure 57.2 and explained over the following paragraphs.

- User level (Broker-based): Is useful for creating loosely coupled interconnected distributed systems. In this approach, users/organizations are interconnected through accessing multiple distributed systems. This approach involves repetitive efforts to develop interfaces for different distributed systems and, thus, scaling to many distributed systems is difficult. Gridway [103] and GridBus broker [104] are examples of broker-based interconnection approach. The former, achieves interconnection in organization level, whereas the latter, works in the enduser level.
- Resource level: In this approach, different interfaces are developed on the resource side and consequently the resource can be available to multiple distributed systems. This approach involves administration overhead, since the resource administrator has to be aware of well-known services. This approach is difficult to scale to many distributed systems, hence, it is suggested mostly for large distributed systems. Interconnection of EGEE, NorduGrid, and D-Grid is done based on

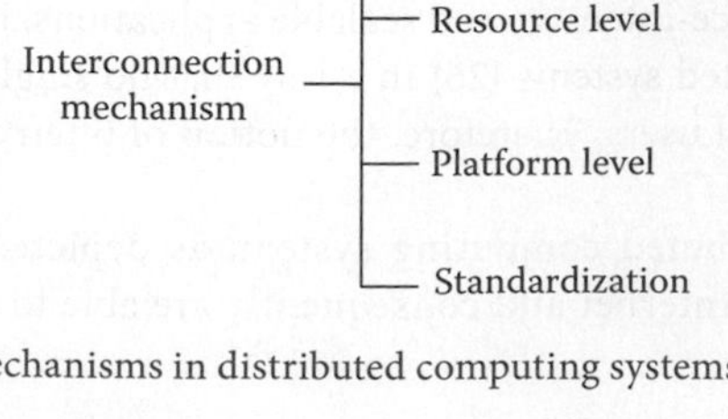

FIGURE 57.2 Interconnection mechanisms in distributed computing systems.

this approach [31]. Particularly, D-Grid [35] leverages interconnectivity via implementing interfaces of UNICORE, gLite, and Globus on each resource provider in a way that resources can be accessed by any of the middlewares.

- Platform level (Gateways): A third platform (usually called a gateway) handles the arrangements between distributed systems. Ideally, the gateway is transparent both from users and resources and makes the illusion of single system for the user. However, in this approach gateways are single point of failure and also a scalability bottleneck. InterGrid [26] and the interconnection of Naregi and EGEE [65] are instances of this approach.
- Standardization: Common and standard interfaces have been accepted as a comprehensive and sustainable solution for interconnecting distributed systems. However, current distributed systems (e.g., current Grid platforms) have already been developed based on different standards and it is a hard and long process to change them to a common standard interface. Issues regarding creating standards for interconnecting distributed systems are also known as interoperability of distributed systems.

  UniGrid [88] is a large-scale interconnected distributed system implemented based on a standard and connects more than 30 sites in Taiwan. It offers a web interface that bridges the user and the lower-level middleware. The core of UniGrid orchestrates different middlewares, including Globus Toolkit [33], Condor [96], and Ganglia [80] transparently from the user. Another project that sought to achieve the idea of World Wide Grid through developing standards and service-oriented architecture is GRIP [25].

Grid computing is a prominent example of interconnected distributed systems. Grids are usually comprised of various organizations that share their resources (e.g., Clusters or SMPs) and form Virtual Organizations (VOs). The concept of Grid has specifically been fascinating for users/organizations that did not have huge resources available or did not have the budget to manage such resources. Nowadays, Grids are utilized predominantly in scientific communities to run high performance computing (HPC) applications. Over the last decade, variety of Grids have emerged based on different interconnection mechanisms. TeraGrid in the United States [102], DAS in the Netherlands [61], and Grid5000 in France [17] are such examples.

Generally, in an interconnected environment, requests from different sources co-exist and, therefore, these systems are prone to contention between different requests competing to access resources. There are various types of contentions that can occur in an interconnected distributed system and, accordingly, there are different ways to cope with these contentions.

The survey will help people in the research community and industry to understand the potential benefits of contention-aware resource management systems in distributed systems. For people unfamiliar with the field, it provides a general overview, as well as detailed case studies.

The rest of this chapter is organized as follows: In Section 57.2, an overview on resource management systems of interconnected distributed systems is presented. Next, in Section 57.3 contention in interconnected distributed systems is discussed which is followed by investigating the architectural models of the contention-aware resource management systems in Section 57.4. In Section 57.5, we discuss about different approaches for contention management in well-known interconnected distributed systems. Finally, conclusion and avenues of future works for researchers are provided in Section 57.6.

57.2 Request Management Systems

Interconnected distributed systems, normally, encounter various users and usage scenarios from users. For instance, the following usage scenarios are expectable:

- Scientists in a research organization run scientific simulations, which are in the form of long running batch jobs without specific deadlines.
- A corporate web site needs to be hosted for a long period of time with a guaranteed availability and low latency.
- A college instructor requires few resources at certain times every week for demonstration purposes.

In response to such diverse demands, interconnected distributed systems offer different service levels (also called multiple quality of service (QoS) levels).

For example, Amazon EC2* supports reserved (availability guaranteed), on-demand, and spot (best-effort) virtual machine (VM) instances. Offering a combination of advance-reservation and best-effort schemes [93], interactive and batch jobs [109], tight-deadline and loose-deadline jobs [37] are common practices in interconnected distributed systems.

These diverse service levels usually imply different prices and priorities for the services that have to be managed by the resource management system. Additionally, interconnected distributed systems can be aware of the origin of the requests and they may discriminate requests based on that. Another challenge in job management of interconnected distributed systems is managing accounting issues of sending/receiving requests to/from peer distributed systems.

There are many approaches for tackling these challenges in resource management systems of interconnected distributed systems. One common approach is prioritizing requests based on criteria, such as service (QoS) or origin. For example, in an interconnected distributed system usually local requests (i.e., local organizations' users) have priority over the requests from external users [5]. Another example is in urgent computing [15] (urgent applications), such as earthquake and bush-fire prediction applications where the applications intend to acquire many resources in an urgent manner. In circumstances that there is surge in demand, requests with different priorities compete to gain access to resources. This condition is generally known as resource contention between requests.

Resource contention is the main challenge in request management of interconnected distributed systems and occurs when a user request cannot be admitted or cannot receive adequate resources, because the resources are occupied by other (higher priority) requests.

In the remainder of this survey, we explore different aspects of resource contention in interconnected distributed systems and also we investigate the possible solutions for them.

57.3 Origins of Resource Contentions

There are various causes for resource contention in interconnected distributed systems. They broadly can be categorized as request-initiated, interdomain-initiated, origin-initiated and hybrid. A taxonomy of different contention types along with their solutions is shown in Figure 57.3.

57.3.1 Request-Initiated Resource Contention

Request-initiated resource contention occurs if any of the requests monopolizes resources to such an extent that deprives other requests from gaining access to them. It is prevalent in all forms of distributed systems, even where there is no interconnection. There are several scenarios that can potentially lead to request-initiated resource contention. One important situation is when there is an imbalance in request sizes, mainly, in terms of required number of nodes or execution time (duration). In this circumstance, small requests may have to wait for a long time behind a long job to access resources.

Another cause for request-initiated resource contention is situation that requests have QoS constraints and they selfishly try to satisfy them. Generally, resource management systems can support three types of QoS requirements for users' requests:

1. Hard QoS: Where the QoS constraints cannot be negotiated. These systems are prone to QoS violation and, hence, managing resource contention is critical [73].
2. Soft QoS: Where the QoS constraints are flexible and can be negotiated based upon the resource availabilities or when there is a surge in demand. The flexibility enables resource management systems to apply diverse resource contention solutions [73].

* http://aws.amazon.com/ec2/

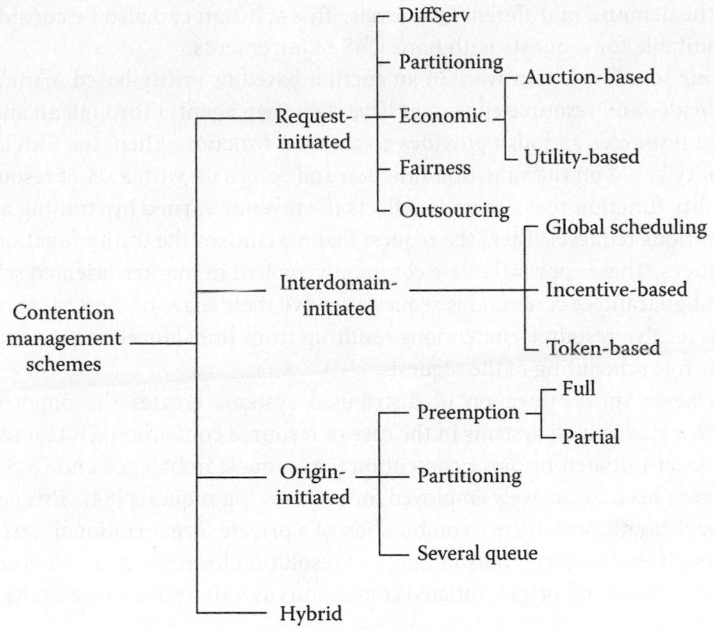

FIGURE 57.3 Taxonomy of different types of resource contentions and possible solutions in interconnected distributed computing systems.

3. Hybrid QoS: Where the resource management system supports a combination of Hard QoS and Soft QoS requirements for the user requests. This fashion is common in commercial resource providers such as Cloud providers. For instance, Amazon EC2 supports services with distinct QoS requirements including reserved (hard QoS), and spot (soft QoS) VM instances. Another example, are the resource management systems that support combination of interactive (hard QoS) and batch requests (usually soft QoS) [109].

Solutions for managing request-initiated contentions are mostly achieved in the context of scheduling and/or admission control units of resource management systems. Over the next paragraphs, we categorize and describe different solutions for resource contention.

Differentiated services (DiffServ) technique that initially was used in Computer Networks and developed to guarantee different QoS levels (with different priorities) for various Internet services, such as VOIP and web. In Computer Networks, DiffServ guarantees different QoSs through dividing the services into distinct QoS levels. According to IETF RFC 2474, each level is supported by dropping TCP packets of lower priority levels.

Similar approach can be taken in the context of request-initiated resource contentions in distributed systems. For this purpose, the resource management system presents different QoS levels for user requests. Then, requests are classified in one of these levels at the admission time. However, in this scheme there is no control on the number of requests assigned to each QoS level. As a result, QoS requirements of request cannot be guaranteed. Therefore, DiffServ scheme is appropriate for soft QoS requirements.

Variations of DiffServ technique can be applied when contention occurs due to imbalanced requests. Silberstein et al. [89] also sought to decrease the response time of short requests in a multigrid environment. For that purpose, they apply a multilevel feedback queue (MLFQ) scheduling. In their policy, Grids are placed in different categories based on their response speed. Requests are all sent to the first queue upon arrival and if they cannot get completed in the time limit of that level, then they are migrated to the lower level queue which is a larger grid. The process continues up until the task finishes or reaches down the hierarchy.

In the *Partitioning scheme*, the resources are reserved for requests with different QoS levels. Unlike DiffServ scheme, in this approach boundaries of the reservations (partitions) can adaptively move, based on the demand in different QoS levels. This solution can also be considered as a type of DiffServ that is suitable for requests with hard QoS requirements.

Economic scheme solutions either work in an auction-based or utility-based manner. In the former, both resource provider and resource consumer have their own agents. Through an auctioneer the consumer bids on the resources and also provides a valuation function. Then, the provider agent tries to maximize the utility based on the valuation function and comes up with a set of resources for the user. In the latter, a utility function that generally reflects the revenue earned by running a request is calculated for all contentious requests. Then, the request that maximizes the utility function has the priority of accessing resources. These approaches are commonly applied in market-oriented scheduling [36].

Fair scheme that guarantees contentious requests receive their share of the system resources [3]. This scheme is used to resolve resource contentions resulting from imbalanced requests in the system and assures starvation-free scheduling of the requests.

Outsourcing scheme: Interconnection of distributed systems creates the opportunity to employ resources from other distributed systems in the case of resource contention. Outsourcing is applied for both causes of request-initiated resource contention (i.e., request imbalance and QoS levels). Specially, Cloud providers have been extensively employed for outsourcing requests [84]. This issue has helped in emergence of hybrid clouds, which are a combination of a private (organizational) resources and public Clouds [12]. Although we categorize outsourcing as a resolution for request-initiated contentions, it can be applied for interdomain and origin initiated contentions as will be discussed in the next parts.

57.3.2 Interdomain-Initiated Resource Contention

Interdomain-initiated resource contention occurs, when the proportion of shared resources to the consumed resources by a constituent distributed system is low. In other words, this resource contention happens when a resource provider contributes few resources while demand more resources from other resource providers in an interconnected distributed system. Unlike request-initiated contention, which merely roots in request characteristics and can take place in any distributed system, interdomain contention is based on the overall consumption and contribution of each resource provider.

There are several approaches for handling interdomain-initiated contentions, namely, global scheduling, incentive, and token-based schemes (see Figure 57.3). These approaches are discussed in detail in what follows .

Global schedulers: In this approach, which is appropriate for large-scale distributed systems, there are local (domain) schedulers and global (meta) schedulers. Global schedulers are in charge of routing user requests to local schedulers and, ultimately, local schedulers, such as Condor [96] or Sun Grid Engine (SGE) [16], allocate resources to the requests.

Global schedulers can manage the interdomain resource contention by admitting requests from different organizations based on the number of requests it has redirected to the resources of each organization. Since global schedulers usually are not aware of the instantaneous load condition in the local schedulers, it is difficult for them to guarantee QoS requirements of users [11]. Thus, this approach is useful for circumstances where requests have soft QoS requirements.

Incentive scheme: In this approach, which is mostly used in peer-to-peer systems [71], resource providers are encouraged to share resources to be able to access more resources. Reputation Index Scheme [58] is a type of incentive-based approach in which the organization cannot submit requests to another organization while it has less reputation than that organization. Therefore, in order to gain reputation, organizations are motivated to contribute more resources to the interdomain sharing environment.

Quality service incentive scheme [70] is a famous type of incentive-based approach. Quality service is an extension of Reputation Index Scheme. The difference is that depending on the number of QoS levels offered by a participant, a set of distinct ratings is presented where each level has its own reputation index.

Token-based scheme: Operates based on the principle where a certain amount of tokens, which are allocated to an organization, is proportional to its resource contribution. If a user wants to get access to another organization resources, its consumer agent must spend amount of tokens to get the access. This scheme encompasses request-initiated and interdomain resource contentions. To address the request-initiated resource contention, valuation functions can be used to translate the QoS demands of user to the number of tokens to be used for a request. The provider agent can then use its own valuation functions to compute the admission price for the request. Finally, the request will be admitted only if the admission price is less or equal to the number of tokens that the requesting organization is willing to pay [73].

57.3.3 Origin-Initiated Resource Contention

In interconnected distributed systems, users' requests originate from distinct organizations. More importantly, these systems are prone to resource contention between local requests of the organization and requests from other organizations (i.e., external requests). Typically, local requests of each organization have priority over external requests [5]. In other words, the organization that owns the resources would like to ensure that its community has priority access to the resources. Under such a circumstance, external requests are welcome to use resources if they are available. Nonetheless, external requests should not delay the execution of local requests.

In fact, origin-initiated resource contention is a specific case of interdomain-initiated and request-initiated resource contentions. Consequently, the approaches of tackling this type of resource contention is similar to the already mentioned approaches. Particularly, partitioning approach both in static and dynamic forms and global scheduling are applicable for origin-initiated resource contentions. There are also other approaches to cope with origin-initiated contentions that we discuss in this part.

Preemption scheme: This mechanism stops the running request and free the resources for another, possibly higher priority, or urgent request. The higher priority request can be a local request or a hard QoS request in an interconnected distributed system. The preempted request may be able to resume its execution from the preempted point. If suspension is not supported in a system, then the preempted request can be killed (canceled) or restarted. For parallel requests, *full preemption* usually is performed, in which whole request leaves the resources. However, some systems support *partial preemption*, in which part of resources allocated to a parallel request is preempted [86].

Although preemption mechanism is a common solution for origin-initiated contentions, it is also widely applied to solve request-initiated resource contentions. Due to the prominent role of preemption in resolving these types of resource contentions, in Section 57.4.5 we explain preemption in details.

Partitioning scheme: Both static and dynamic partitioning of resources, as mentioned in Section 57.3.1, can be applied to tackle origin-initiated contentions.

In dynamic partitioning of resources, the local and external partitions can borrow resources from each other when there is a high demand of local or external requests [11].

Several queues: In this approach when requests arrive [59], they are categorized in distinct queues, based on their origin. Each queue can independently have its own scheduling policy. Then, another scheduling policy determines the appropriate queue that can dispatch a request to the resources.

Combinations of the aforementioned contentions (mentioned as hybrid in Figure 57.3) can occur in an interconnected distributed system. The most common combination is the origin-initiated and request-initiated resource contentions. For instance, in federated Grids and federated Clouds, origin-initiated contention occurs between local and external requests. At the same time, external and local requests can also have distinct QoS levels, which is a request-initiated resource contention [5,6,81]. Generally, Resolution of hybrid resource contentions is a combination of different strategies mentioned earlier.

57.4 Contention Management

Resource management system is the main component of a distributed system that is responsible for resolving resource contentions. Various elements of a resource management system contribute in resolving different types of resource contentions. They apply different approaches in managing contentions. Different components of resource management systems and the way they deal with resource contention is presented in Figure 57.4.

57.4.1 Resource Provisioning

Resource provisioning component of a resource management system is in charge of procuring resources based on user application requirements. Resource provisioning is performed based on a provisioning model that defines the execution unit in a system. In fact, requests are allocated resources based on the resource provisioning model.

Resource provisioning models do not directly deal with resource contentions. However, the way other components of resource management system function strongly depends on the resource provisioning model.

Provisioning resources for users' requests in distributed systems has three dimensions as follows:

1. Hardware resources
2. Software available on the resources
3. Time during which the resources are available (availability)

Satisfying all of these dimensions in a resource provisioning model has been challenging. In practice, past resource provisioning models in distributed systems were unable to fulfill all of these dimensions [93]. Emergence of virtual machine (VM) technology as a resource provisioning model recently has posed an opportunity to address all of these dimensions. Over the next subsections, we discuss common resource provisioning models in current distributed systems.

57.4.1.1 Job Model

In this model, jobs are pushed or pulled across different schedulers in the system to reach the destination node, where they can run. In job-based systems, scheduling a job is the consequence of a request to run

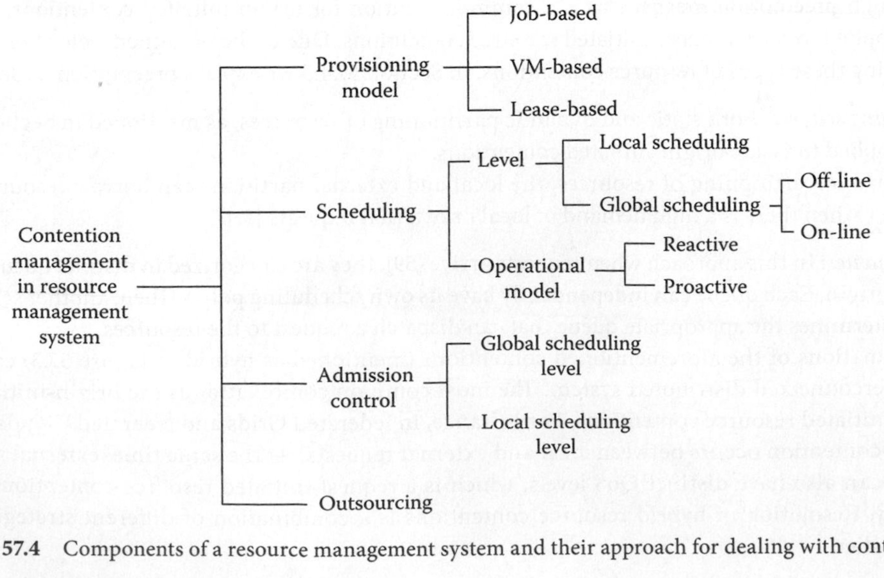

FIGURE 57.4 Components of a resource management system and their approach for dealing with contentions.

the job. Job model resource provisioning has been widely employed in distributed systems. However, this model cannot perfectly support all resource contention solutions.

Job-based systems provision hardware for jobs while they offer a limited support for software availability. In fact, in job-based model users do not have root access, therefore it is difficult to install and use required software packages. Many job-based systems support availability based on queuing theory along with scheduling algorithms. However, queue-based systems usually do not assure specific time availabilities.

To support availability and hardware dimensions, Nurmi et al. [68], present advance reservation (AR) model over the job-based provisioning model. They support AR through predicting waiting time of jobs in the queue. Hovestadt et al. [45] propose plan-based scheduling (opposite to queue-based) that finds the place of each job (instead of waiting in the queue) to be able to support AR model. In this system, on the arrival of each job the whole schedule is re-planned to optimize the resource utilization.

Falkon [75], Condor glidin [34], MyCluster [108], and Virtual Workspace [53] have applied a multilevel/hierarchical scheduling on top of a job-based system to offer other provisioning models (such as lease-based model, which is described in Section 57.4.1.3). In these systems, one scheduler allocates resources to another scheduler and the other scheduler runs the jobs on the allocated resources.

57.4.1.2 Virtual Machine Model

Virtual machines (VMs) are considered as an ideal vehicle for resource provisioning in distributed systems. The reason is that, in VM model, unlike the job model, hardware, software, and availability can be provisioned for user requests. Additionally, VMs' capability in getting suspended, resumed, or migrated without major utilization loss have proved to be useful in resource management. Therefore, VM-based provisioning model is commonly used in current distributed systems.

The VM-based resource provisioning model is used in creating virtual Clusters on top of an existing infrastructure. Virtual clusters (VC) are usually utilized for job-based batch processing. For example, in MOSIX [13], Clusters of VMs are transparently created to run high performance computing (HPC) applications. The Nimbus toolkit [52] provides "one-click virtual Cluster" automatically on heterogeneous sites through contextualizing disk images. Amazon EC2, provides VM-based Cluster instances* that offer supercomputing services to expedite execution of HPC applications, without delaying the user in a queue or acquire expensive hardware. Automatic VM creation and configuration in short time is also considered in In-VIGO [2] and VMplants [56]. An extension of Moab [29] creates VM-based virtual Clusters to run HPC batch applications.

Many commercial datacenters use VM-based provisioning model to provide their services to resource consumers. Such datacenters offer services such as Virtual Cluster, or hosting servers including web, email, and DNS.

Datacenters usually contain large scale computing and storage resources (order of 100s–1000s) and consume so much energy. A remarkable benefit of deploying VM-based provisioning model in datacenters is the consolidation feature of VMs that can potentially saves the energy consumption [105]. However, VM consolidation requires accurate workload prediction in the datacenters. Moreover, the consolidation impact on service level agreements (SLA) needs to be considered. VM consolidation can be performed in a static (also termed cold consolidation) or dynamic (hot consolidation) manner. In the former, VMs needs to be suspended and resumed on another resource that involves time overhead. In the latter approach, live migration [107] of VMs is used, thus, is transparent from the user.

Solutions such as VMware, Orchestrator, Enomalism, and OpenNebula [32] provide resource management for VM-based data centers.

There are also concerns in deploying VM-based provisioning model and Virtual Clusters. Networking and load balancing among physical Clusters is one of the challenges that is considered in Vio-Cluster [79]. Power efficiency aspect and effectively utilizing VMs capability in suspending and migrating are also

* http://aws.amazon.com/hpc-applications/

considered by many researchers [51,66,106]. Overhead and performance issues involved in applying VMs to run compute-intensive and IO-intensive jobs, fault tolerance, and security aspects of VMs are also of special importance in deploying VM-based provisioning model.

57.4.1.3 Lease Model

This model is considered as an abstraction for utility computing in which the user is granted a set of resources for specific interval and agreed quality of service [39]. In this model, job execution is independent from resource allocation, whereas in the job model resource allocation is the consequence of running a job.

Formally, a *lease* is defined by Sotomayor [93] as "a negotiated and renegotiable contract between a resource provider and a resource consumer, where the former agrees to make a set of resources available to the latter, based on a set of lease terms presented by the resource consumer." If lease extension is supported by resource management system, then users would be able to extend their lease for a longer time. This is particularly useful in circumstances that users have inaccurate estimation of required time. Virtual machines are suitable vehicles to implement lease-based model. Depending on the contract, resource procurement for leases can be achieved from a single provider or from multiple providers.

57.4.2 Scheduling Unit

The way user requests are scheduled in an interconnected distributed system affects types of resource contentions occurring. Efficient scheduling decisions can prevent resource contention or reduce its impact whereas poor scheduling decisions can lead to more resource contentions.

In an interconnected distributed system, we can recognize two levels of scheduling, namely, local (domain level) scheduling and global scheduling (meta-scheduling). The global scheduler is generally in charge of assigning incoming requests to resource providers within its domain (e.g., Clusters or sites). In the next step, the local scheduler performs further tuning to run the assigned requests efficiently on resources.

From the resource contention perspective, scheduling methods can either react to resource contention or proactively prevent the resource contention to occur.

57.4.2.1 Local Scheduling

Local scheduler deals with scheduling requests within each distributed system (e.g., Cluster or site). Scheduling policies in this level can mainly deal with request-initiated and origin-initiated contentions. Indeed, there are few local schedulers that handle interdomain-initiated contention.

Backfilling is a common scheduling policy in local resource management systems (LRMS). The aims of backfilling are increasing resource utilization, minimizing average request response time, and reducing queuing fragmentation. In fact, backfilling is an improved version of FCFS in which requests that arrive later, possibly are allocated earlier in the queue, if there is enough space for them. Variations of backfilling policy are applied in local schedulers:

- Conservative: In which a request can be brought forward if it does not delay any other request in the queue.
- Aggressive (EASY): The reservation of the first element in the queue cannot be postponed. However, the arriving request can shift the rest of scheduled requests.
- Selective: If the slowdown of a scheduled request exceeds a threshold, then it is given a reservation, which cannot be altered by other arriving requests.

There are also variations of backfilling method that are specifically designed to resolve request-initiated resource contentions. Snell et al. [91] applied preemption on backfilling policy. They provide policies to select the set of requests for preemption in a way that the requests with higher priority are satisfied and, at the same time, the resource utilization increases. The preempted request is restarted and rescheduled in the next available time slot.

Multiple resource partitioning is another scheduling approach for local schedulers by Lawson and Smirni [59]. In this approach, resources are divided into partitions that potentially can borrow resources from each other. Each partition has its own scheduling scheme. For example, if each partition uses EASY backfilling, then one request from another QoS level can borrow resources, if it does not delay the pivot request of that partition.

In FCFS or backfilling scheduling policies, the start time of a request is not predictable (not determined). Nonetheless, in practice, we need to guarantee timely access to resources for some requests (e.g., deadline-constraint requests in a QoS-based system). Therefore, many local schedulers support Advance Reservation (AR) allocation model that guarantees resource availability for a requested time period. Resource management systems such as LSF, PBSPRO, and MAUI support AR.

Advance Reservation is prone to low-resource utilization specially if the reserved resources are not used by the users. Additionally, it increases the response time of normal requests [63,90]. These side-effects of AR can be minimized by limiting the number of AR, and leveraging flexible AR (in terms of start time, duration, or number of processing elements needed).

57.4.2.2 Global Scheduling (Meta-Scheduling)

Global scheduler in an interconnected distributed system usually has two aspects. On the one hand, the scheduler is in charge of assigning incoming requests to resource providers within its domain (e.g., Clusters). On the other hand, it is responsible to deal with other distributed systems such as schedulers or gateways that delegate other peer distributed systems. This aspect of global schedulers can particularly resolve interdomain-initiated and origin-initiated resource contentions.

The global scheduler either works off-line (i.e., batches incoming requests and assigns each batch to a Cluster), or is online (i.e., assign each request to a local scheduler as it is received). The global schedulers can proactively prevent resource contentions.

57.4.3 Admission Control Unit

Controlling the admission of requests prevents the imbalanced deployment of resources. By employing an appropriate admission control policy different types of resource contentions can be avoided. An example of the situation without admission control in place is when two requests share a resource but one of them demands more time. In this situation, the other request will face low-resource availability and subsequently, high response time. Thus, lack of admission control can potentially lead to request-initiated contention.

Admission control behavior should depend on the workload condition in a resource provider. Applying a strict admission control in a lightly loaded system results in low resource utilization and high rejection of requests. Nonetheless, the consequence of applying less strict admission control in a heavily loaded resource is more QoS violation and less user satisfaction [112].

Admission control can function in different ways. To tackle request-initiated contention, admission control commonly carried out via introducing a valuation function. The valuation function relates the quality constrains of users to a single quantitative value. The value indicates the amount a user is willing to pay for a given quality of service (QoS). Resource management system use the valuation functions to allocate resources with the aim of maximizing aggregate valuation of all users.

Admission control also can be applied in interdomain-initiated contentions to limit the amount of admitted requests of each organization to be proportional to their resource contribution. Similarly, admission control can be applied to avoid origin-initiated resource contention. For this purpose, admission control policy would not admit external requests where there is peak load of local requests.

Placement of admission control component in a resource management system of a interconnected distributed system can be behind the local scheduler and/or behind the global scheduler. In the former, for rejecting a request there should be an alternative policy to manage the rejected request. In fact, rejecting by a local scheduler implies that the request has already been admitted and, hence, has to be taken care. For instance, the rejected request can be redirected to another resource provider or even queued in a separate queue to be scheduled later. Deploying admission control behind the global scheduler is easier in terms of managing the rejected requests. However, the drawback of employing admission control with global scheduler is that the global scheduler may not have updated information about site's workload situation.

57.4.4 Outsourcing Unit

Interconnectivity of distributed systems creates the opportunity to resolve the resource contention via employing resources from other distributed systems. Therefore, resource management systems in interconnected distributed computing systems usually have a unit that decides about details of outsourcing requests (i.e., redirecting arriving requests to other distributed systems) such as when to outsource and which requests should be outsourced. In terms of implementation, in many systems, the outsourcing unit is incorporated into either admission control or scheduling unit. However, it is also possible to have it as an independent unit in the resource management system.

Outsourcing is generally applied when there is a peak demand or there is a resource contention (specially request-initiated contention). In this situation to serve requests without resource contention, some requests (e.g., starved requests) are selected to be redirected to other distributed systems.

Cloud computing providers have been of special interest to be employed for outsourcing (off-loading) requests [84]. This issue has pushed the emergence of hybrid clouds, which are a combination of a private (organizational) Cloud and public Clouds.

57.4.5 Preemption Mechanism

Preemption mechanism in a resource management system makes the resources free and available for another, possibly higher priority, request. Preemption is a useful mechanism to resolve request-initiated and origin-initiated contentions. Preemption of a running process can be performed manually or automatically through the resource management system.

The way preemption mechanism is implemented, depends on the way checkpointing operation is carried out. If the checkpointing is not supported, then the preempted process has to be killed and restarted at a later time. If checkpointing is supported (both by the running process and by the scheduler), then the preempted request can be suspended and resumed at a later time. However, checkpointing is not a trivial task in distributed systems. We will deal with checkpointing hurdles in Section 57.4.5.4.

Due to the critical role of preemption in solving different types resource contentions, in this section, we investigate preemption in distributed systems from different angles. Particularly, we consider various usages of preemption and the way they solve resource contentions. Then, we investigate possible side-effects of preemption. Finally, we discuss how a preempted request (i.e., job/VM/lease) can be resumed in a distributed system.

57.4.5.1 Applications of Preemption Mechanism

Preemption in distributed systems can be applied for reasons that are presented in Figure 57.5. As we can see, preemptions can be used to resolve resource contention. However, there are other usages of preemption in distributed systems that we will discuss them in this part.

Preemption is used to resolve request-initiated resource contentions. One approach is employing preemption in local scheduler along with the scheduling policy (e.g., backfilling) to prevent unfairness. For instance, when a backfilled request exceeds the allocated time slot and interferes with the reservation of other requests preemption mechanism can preempt the backfilled requests and therefore the reservations can be served on time. The preempted request can be allocated another time slot to finish its computation [40].

A preemptive scheduling algorithm is implemented in MOSIX [4] to allocate excess (unclaimed) resources to users that require more resources than their share. However, these resources will be released

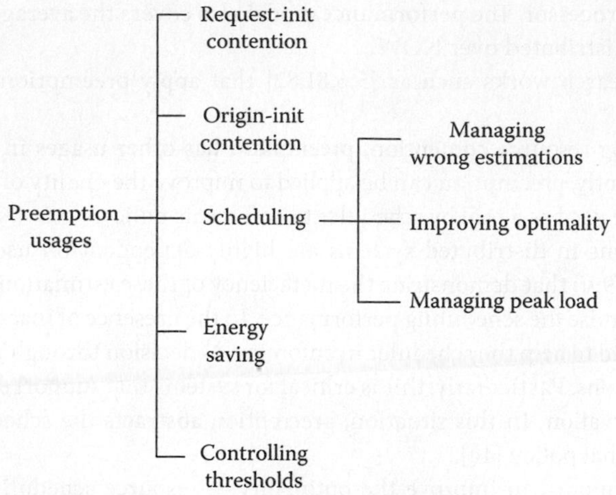

FIGURE 57.5 Different usages of preemption in distributed systems.

as soon as they are reclaimed. MOSIX also support situation that there are local and guest jobs and can consider priority between them (origin-initiated contention).

Scojo-PECT [92] provides a limited response time for several job classes within a virtualized Cluster. It employs DiffServ solution that is implemented via coarse-grained preemption to cope with the request-initiated resource contention. The preemptive scheduler aims at creating a fair-share scheduling between different job classes of a Grid. The scheduler works based on a coarse-grained time sharing and for preemption it suspends VMs on the disk.

Walters et al. [109] introduced a preemption-based scheduling policy for batch and interactive jobs within a virtualized Cluster. In this work, batch jobs are preempted in favor of interactive jobs. The authors introduce different challenges in preempting jobs including selecting a proper job to be preempted, checkpointing the preempted job, VM provisioning, and resuming the preempted job. Their preemption policy is based on weighted summation of factors such as the time spent in the queue.

Haizea [93] is a lease scheduler that schedules a combination of advanced reservation and best effort leases. Haizea preempts best effort leases in favor of advance reservation requests. Haizea also considers the overhead time imposed by preempting a lease (suspending and resuming included VMs).

Preemption of parallel jobs has also been implemented in the Catalina job scheduler [63] in San-Diego Supercomputer Center (SDSC). They have added preemption to conservative backfilling. The job preemption is carried out based on job priorities which is determined based on weighted summation of factors such as the time a request waits in the queue, the size (number of processing elements) required by the request, and expansion factor of the request. In general, the policy tries to preempt jobs that require fewer processing elements because they impose less overhead to the system for preemption. In fact, preempting jobs with larger size (wide jobs) implies more overhead because of the time needed for saving messages between nodes.

Isard et al. [49] have investigated the problem of optimal scheduling for data intensive applications, such as Map-Reduce, on the Clusters where the computing and storage resources are close together. To achieve the optimal resource allocation, their scheduling policy preempts the currently running job in order to maintain data locality for a new job.

Preemption can be applied to resolve the origin-initiated resource contentions. Ren et al. [76] have proposed a prediction method for unavailable periods in fine-grained cycle sharing systems where there are mixture of local jobs and global (guest) jobs. The prediction is used to allocate global requests in a way that do not disturb local requests.

Gong et al. [38] have considered preemption of external tasks in favor of local tasks in a Network of Workstations (NOW) where local tasks have preemptive priority over external tasks. They provided a performance model to work out the run time of an external task that is getting preempted by local tasks in a single processor. The performance model also covers the average runtime of the whole external job which is distributed over NOW.

There are other research works such as [5,6,81,82] that apply preemption for removing origin-initiated contentions.

Apart from removing resource contention, preemption has other usages in resource management systems. More importantly, preemption can be applied to improve the quality of scheduling policies. In fact, preemption can be used as a tool by scheduler to enforce its policy.

Scheduling algorithms in distributed systems are highly dependent on user runtime estimation. There are studies (e.g., [99]) that demonstrate the inefficiency of these estimations and how these wrong estimation can compromise the scheduling performance. In the presence of inaccurate estimations, preemption can be deployed to help the scheduler in enforcing its decision through preempting the process that has wrong estimations. Particularly, this is critical for systems that support strict reservation model such as advanced reservation. In this situation, preemption abstracts the scheduling policy from the obstacles in enforcing that policy [44].

Preemption can be applied to improve the optimality of resource scheduling. Specifically, online scheduling policies are usually not optimal because jobs are constantly arriving over time and the scheduler does not have a perfect knowledge about them [4]. Therefore, preemption can potentially mitigate the nonoptimality of the scheduling policy.

Preemption mechanism can be employed for managing peak load. In these systems, resource-intensive applications or batch applications are preempted to free the resources during the peak time. Accordingly, when the system is not busy and the load is low, the preempted requests can be resumed [69].

Preemption can be employed to improve the system and/or user centric criteria, such as resource utilization and average response time. Kettimuthu et al. [54] have focused on the impact of preempting parallel jobs in supercomputers for improving the average and worst-case slowdown of jobs. They suggest a preemption policy, called *Selective Suspension*, where an idle job can preempt a running job if the suspension factor is adequately more than the running job.

A recent application of preemption is in energy conservation in datacenters. In fact, one prominent approach in energy conservation of virtualized datacenters is VM consolidation, which takes place when resources in the datacenter are not utilized efficiently. In VM consolidation, VMs running on under-utilized resources are preempted (suspended) and resumed on other resources. VM consolidation can also occur through live migration of VMs [107] to minimize the unavailability time of the VMs. When a resource is evacuated, it can be powered off to reduce the energy consumption of the datacenter.

Salehi et al. [83] have applied VM preemption to save energy in a datacenter that supports requests with different SLAs and priorities. They introduce an energy management component for Haizea [93] that determines how resources should be allocated for a high-priority request. The allocation can be carried out through preempting lower-priority requests or reactivating powered off resources. The energy management component can also decide about VM consolidation, in circumstances that powered on resources are not being utilized efficiently.

Preemption can be used for controlling administrative (predetermined) thresholds. The thresholds can be configured on any of the available metrics. For instance, the temperature threshold for CPUs can be established that leads to the system automatically preempts part of the load and reschedule on other available nodes. Bright Cluster Manager [1] is a commercial Cluster resource management system that offers the ability to establish preemption rules by defining metrics and thresholds.

57.4.5.2 Preemption Challenges

Operating systems of single processor computers have been applying preemption mechanism for a long time to offer interactivity to the end-user. However, since interactive requests are not prevalent in distributed systems, there has been less demand for preemption in these systems. More importantly, achieving preemption in distributed systems entails challenges that discourage researchers to investigate deeply on that. This challenges are different based on the resource provisioning model.

TABLE 57.1 Preemption Challenges in Different Resource Provisioning Models

	Resource Provisioning Model		
Challenge	Job-Based	VM-Based	Lease-Based
Coordination	✓	✓	✓
Security	✓	×	×
Checkpointing	✓	×	×
Time overhead	✓	✓	✓
Permission	✓	✓	×
Impact on queue	✓	✓	✓
Starvation	✓	✓	✓
Preemption candidates	✓	✓	✓

In this part, we present the detailed list of challenges that distributed systems encounter in preempting requests in various resource provisioning models. Moreover, a summary of preemption challenges based on different provisioning models is provided in Table 57.1.

- *Coordination*: Distributed requests (jobs/VMs/leases) are scattered on several nodes by nature. Preemption of the distributed requests have to be coordinated between the nodes that are executing them. Lack of such coordination leads to inconsistent situation (e.g., because of message loss) for the running request.
- *Security*: Preemption in job-based systems implies security concerns regarding files that remain open and swapping-in the memory contents before job resumption. In other words, in job-based systems operating system has to provide the security of not accessing files and data of the preempted processes. Since VM- and lease-based systems are self-contained (isolated) by nature, there is not usually security concern in their preemption.
- *Checkpointing*: Lack of checkpointing facilities is a substantial challenge in job-based resource provisioning model. Because of this problem, in job-based systems the preempted job is generally killed, which is a waste of resources [91]. Checkpointing problem is obviated in VM and lease-based resource provisioning models [94]. Due to the fundamental role of checkpointing for preemption mechanism, in Section 57.4.5.4 we discuss it in details.
- *Time overhead*: In VM- and lease-based resource provisioning models, time overhead imposed to the system to perform preemption is a major challenge. If preemption takes place frequently and the time overhead would not be negligible, then the resource utilization will be affected.

  Additionally, disregarding the preemption time overhead in scheduling, prevents requests to start at the scheduled time [94]. In practice, resource management systems that support preemption, must have an accurate estimation of the preemption time overhead. Overestimating the preemption time overhead results in idling resources. However, underestimating the preemption time overhead ends up in starting leases with delay, which subsequently might violate SLA agreements.

  Sotomayor et al. [94] have presented a model to predict the preemption time overhead for VMs. They identified that the size of memory that should be de-allocated, number of VMs mapped to each physical node, local or global memory used for allocating VMs, and the delay related to commands being enacted are effective on the time overhead of preempting VMs. To decrease the preemption overhead, the number of preemptions that take place in the system has to be reduced [87].
- *Permission:* In the lease-based resource provisioning model, preempting leases is not allowed by default. In fact, one difference between lease-based and other resource provisioning models is that jobs and VMs can be preempted without notifying the user (requester), whereas leases require the requester's permission for preemption [39]. Therefore, there must be regulations in the lease

terms to make lease preemption possible. These terms can be in the form of QoS constraints of the requests or can be bound to pricing schemes. For instance, requests with tight deadline, advance reservations, or requests with tight security possibly choose to pay more instead of getting preempted while they are running.

- *Impact on other requests*: Most of the current distributed systems use a variation of backfilling policy as the scheduling policy. In backfilling, future resource availabilities are reserved for other requests that are waiting in the queue. Preempting the running process and allocating resources to a new request affects the running job/lease as well as the reservations waiting in the queue. Re-scheduling of the preempted requests in addition to the affected reservations are side-effects of preemption in distributed systems.
- *Starvation*: Preemption leads to increasing the response time and, in the worst case, starvation for low-priority requests [5]. There is a possibility that low-priority requests get preempted as soon as they start running. This leads to unpredictable waiting time and unstable situation for low-priority requests. Efficient scheduling policies can prevent unstable and long waiting time situation. One approach to cope with the starvation challenge is restricting the number of requests admitted in a distributed system. Salehi et al. [82] have proposed a probabilistic admission control policy that restricts the queue length for low-priority requests in a way that they would not suffer from starvation.
- *Preemption candidates*: By allowing preemption in a distributed system, there is a possibility that several low priority requests have to be preempted to make sufficient vacant resources for the high-priority request. Therefore, there are several sets of candidate requests whose preemption can create adequate space for the high-priority request. As it is expressed in Figure 57.6, there are several candidate sets (Figure 57.6b) that their preemption can vacate resources for the required time interval (i.e., from t_1 to t_2 as indicated in Figure 57.6a).

  Selecting distinct candidate sets affects the amount of unused space (also termed scheduling fragment) appear in the schedule. Furthermore, preempting different candidate sets imposes different time overhead to the system because of the nature of the requests preempted (e.g., being data-intensive). In this situation, choosing the optimal set of requests for preemption is a challenge that needs to be addressed.

  To cope with this challenge, backfilling policy has been extended with preemption ability in Maui scheduler [91] to utilize scheduling fragments. Salehi et al. [5] have proposed a preemption policy that determines the best set of leases to be preempted with the objective of minimizing preemption

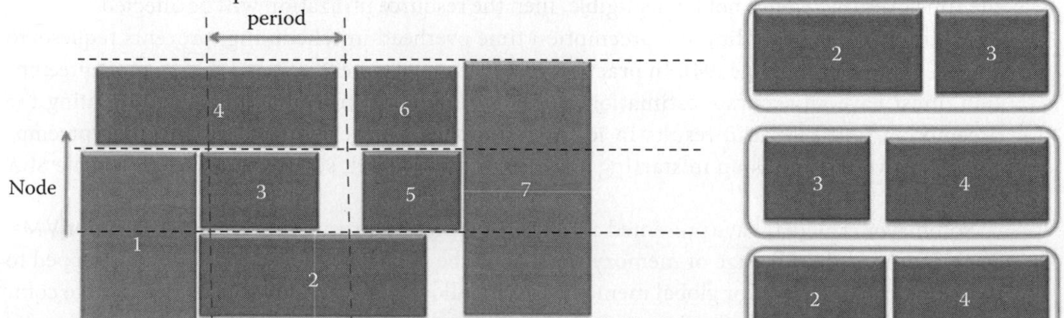

FIGURE 57.6 Preemption candidates for a request that needs two nodes. (a) Shows collision of the requested time interval with running requests within a scheduling queue. (b) Presents three different candidate sets that preempting any of them creates the required space for the new request.

time overhead. A preemption policy is also presented by Walter et al. [109] in a VM-based system with the objective of avoiding starvation for batch requests where a combination of batch and interactive requests co-exist in the system.

57.4.5.3 Possibilities for Preempted Requests

Issues discussed thus far are related to preemption mechanism and its challenges. However, making a proper decision for the preempted request is also important. This decision depends on the facilities provided by the resource management system of a distributed system. For example, migration is one choice that is viable in some distributed systems but not in all of them.

Thanks to the flexibility offered by deploying VM-based resource provisioning models, resource managers are capable of considering various possibilities for the preempted request. Nonetheless, in job-based systems, if preemption is possible, the possible action on the preempted job is usually limited to killing or suspending and resuming of the preempted job. Over the next paragraphs, we introduce various cases that can possibly happen for preempted VMs/leases. Additionally, in Figure 57.7 it is expressed that how different possibilities for the preempted VM affect the VMs' life cycle.

- *Cancelling*: VMs can be canceled (terminated) with/without notifying the request owner. VMs offered in this fashion are suitable for situation that the resource provider does not have to guarantee the availability of the resources for a specific duration. Spot instances offered by Amazon EC2 is an example of cancelling VMs. Isard et al. [49] have used cancelling VMs to execute map-reduce requests. Cancelling VMs imposes the minimum overhead time that is related to the time needed to terminate VMs allocated to the request.

  In job-based systems, cancelling (killing) jobs is a common practice [91] because of the difficulty of performing other possible actions.
- *Restarting*: In both job-based and VM-based systems, the preempted request can be killed (similar to cancelling) and restarted either on the same resource or on another resource. The disadvantage of this choice is losing the preliminary results and wasting the computational power. Restarting can be applied for best-effort and deadline-constraint requests. In the former, restarting can be performed at any time whereas, in the latter, deadline of the request has to be taken into account for restarting.

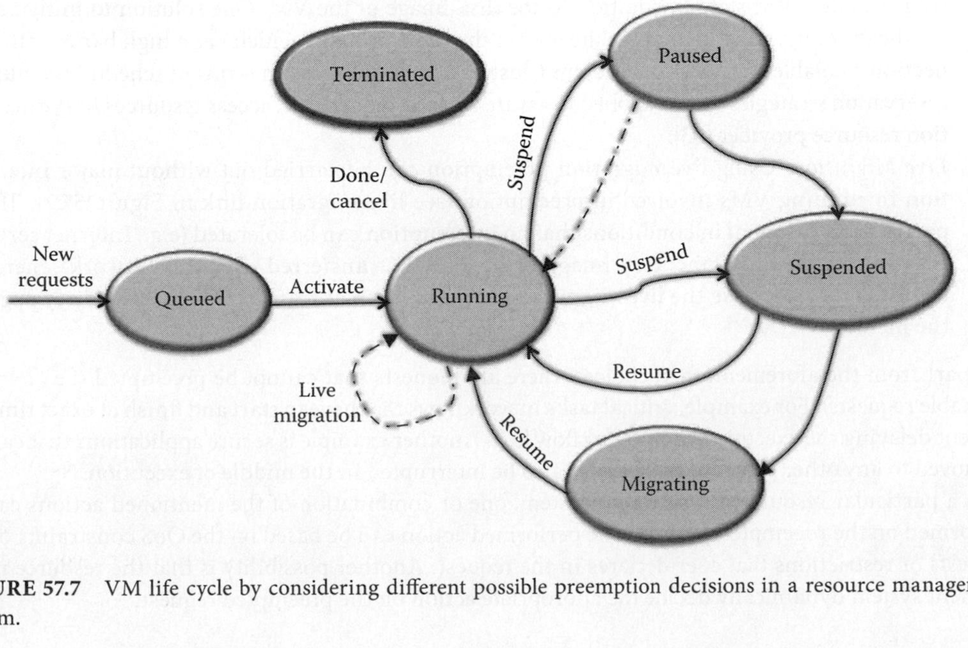

FIGURE 57.7 VM life cycle by considering different possible preemption decisions in a resource management system.

- *Malleability* (*partial preemption*): In this manner, the number of nodes/VMs allocated to a request might be changed while it is executing. In this approach, the request should be designed to adapt dynamically to the changes. This action can be applied on malleable jobs [72] in job-based systems. In VM and lease-based systems, frameworks such as Cluster-on-Demand (COD) [64], support this manner of preemption via cancelling some of the VMs of a lease. Malleability is also known as partial-preemption and can be used to implement dynamic partitioning (see Section 57.3.1).
- *Pausing*: When a VM is paused, it does not get any CPU share, however, it remains in the memory. Resumption of the VM, in this case, happens by getting CPU share and, thus, is very fast (Figure 57.7). Hence, we cannot consider pausing as a complete preemption action.

  Nonetheless, the main usage of pausing is to perform lease-level preemption. In preempting a lease (several correlated VMs), to prevent inconsistency or message loss, first, all VMs are paused and then, suspension takes place [44] (link between pause state and sleep (suspended) state in Figure 57.7). In Section 57.4.5.4, we discuss how pausing VMs helps in preempting leases.
- *Suspending*: When a VM is suspended, the entire state of the VM (including the state of all processes running within the VM) is saved to the disk. At resumption time, the VM continues operating from the suspended point. The suspended request has to be rescheduled to find another free time slot for the remainder of its execution. In job-based systems, the operating system should retain the state of the preempted process and resume the job.

  An important question after suspension is where to resume a VM/lease? Answering this question is crucial particularly for data-intensive applications. A suspended request can be resumed in one of the three following ways:
  1. Resuming on the same resource; this case does not yield to optimal utilization of whole resources.
  2. Resuming on the same site but not essentially on the same resource; In this case, usually data transfer is not required.
  3. Resuming on different site: This case leads to migrating to another site, which entails data transfer. This is, particularly, not recommended for data-intensive requests.
- *Migrating*: VMs of the preempted request are moved to another resource provider to resume the computation (also called cold migration). According to Figure 57.7, migrating involves suspending, transferring, and resuming VMs. Transferring overhead in the worst case includes transferring the latest VM state in addition to the disk image of the VM. One solution to mitigate this overhead is migrating to another site within the same provider which has a high bandwidth connection available (e.g., within different Clusters of a datacenter). In terms of scheduling, multiple reservation strategies can be applied to assure that the request will access resources in the destination resource provider [93].
- *Live Migration:* Using live migration preemption can be carried out without major interruption in running VMs involved in preemption (see live migration link in Figure 57.7). This is particularly essential in conditions that no interruption can be tolerated (e.g., Internet servers). For this purpose, the memory image of the VM is transferred over the network. There are techniques to decrease the live migration overhead, such as transferring just the dirty pages of the memory.

Apart from the aforementioned choices, there are requests that cannot be preempted (i.e., nonpreemptable requests). For example, critical tasks in workflows that have to start and finish at exact times to prevent delaying the execution of the workflow [57]. Another example is secure applications that cannot be moved to any other provider and cannot also be interrupted in the middle of execution.

In a particular resource management system, one or combination of the mentioned actions can be performed on the preempted request. The performed action can be based on the QoS constraints of the requests or restrictions that user declares in the request. Another possibility is that the resource management system dynamically decide the appropriate action on the preempted request.

57.4.5.4 Checkpointing in Distributed Systems

Checkpointing is the function of storing the latest state of a running process (e.g., job, VM, and lease). Checkpointing is an indispensable part of preemption, if the preempted request is going to resume its execution from the preempted point. In fact, checkpointing is the vehicle of implementing preemption. Apart from preemption, checkpointing has other usages including providing fault-tolerance for the requests.

Checkpointed process can be stored on a local storage, or carried over the network to a backup machine for future recovery/resume. Checkpointing has to be achieved in an *Atomic* way, which means either all or none of the modifications are checkpointed (transferred to the backup machine). There are various approaches to achieve checkpointing which are presented briefly in Figure 57.8. In this section, we explain checkpointing strategies for different provisioning models in distributed systems.

57.4.5.4.1 Checkpointing in Job-Based Provisioning Model

Checkpointing approaches are categorized as application-transparent and application-assisted (see Figure 57.8). In application-assisted (user-level) checkpointing, the application defines the necessary information (also called critical data area) that have to be checkpointed. The disadvantage of this approach is that it entails modifying the application by the programmer. However, this approach imposes little overhead to the system because it just checkpoints the necessary parts of the application; additionally, the frequency of performing checkpointing is determined by the user. User-level checkpointing can be further categorized as follows:

- Source-code level: In this manner, checkpointing codes are hard-coded by developers. However, there are some source code analysis tools [21,30] that can help developers to figure out the suitable places that checkpointing codes can be inserted.
- Library level: There are ready-made libraries for checkpointing, such as Libckpt [74] and Condor libraries [62]. To use this kind of checkpointing, developers have to recompile the source code by including the checkpointing library in their program.

As noted in Figure 57.8, checkpointing can also be done in application-transparent manner. This approach is also known as system level, Operating System level, or kernel level in the literature. As the name implies, in this approach the application is not aware of checkpointing process. Therefore, the application does not need to be modified to be checkpointable. Application-transparent checkpointing technique is particularly applied in preemption whereas application-assisted scheme is more used in fault-tolerance techniques. Examples of system level checkpointing in the system level are BLCR [41] and CRAK [111].

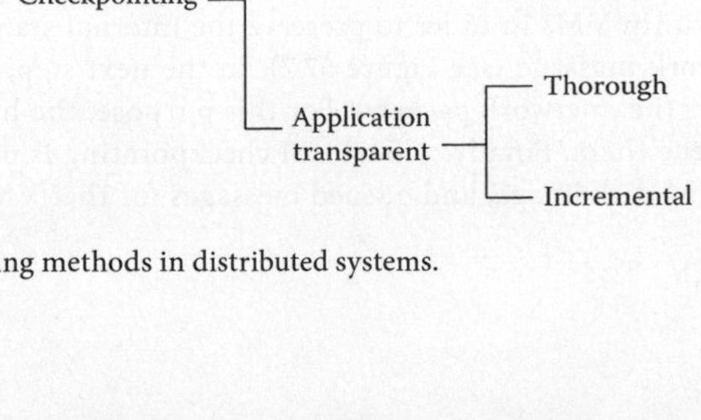

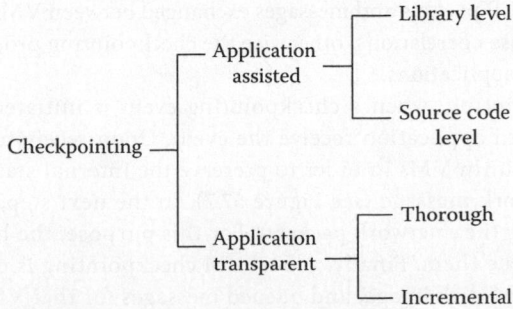

FIGURE 57.8 Checkpointing methods in distributed systems.

Since the application-transparent checkpointing methods have to checkpoint the whole application state, they impose significant time overhead to the system. Another drawback of this approach is that the system-level checkpointing methods are dependent on a specific version of the operating system that they are operating on and, hence, are not entirely portable.

In order to mitigate the checkpointing overhead, incremental checkpointing technique is used [43] in which just the changes since the previous state are checkpointed. Typically, a page-fault technique is used to find the dirty pages and write them to the backup [43,50].

Checkpointing of the distributed applications that run in a distributed system, such as a Cluster, is more complicated. For these applications, not only the state of the application on each running node should be checkpointed, but it has to be assured that the state of the whole application across several nodes remains consistent. Therefore, the checkpointing process across nodes that run the application must be synchronized in a way that there would be neither message loss nor message reordering. Checkpointing of the distributed applications (also termed coordinated checkpointing) traditionally is developed based on the global distributed snapshot concept [22]. These solutions are generally application-level, dependent on a specific version of operating system, and also dependent on the platform implementation (e.g., MPI implementation). Cocheck [95], BLCR [41], and MPICHV [24] are examples of these solutions.

There are various approaches for managing the connections between processes running on different nodes while the checkpointing is performed. In MPICHV [24], the connection among processes has to be disconnected before each process saves its local state to the checkpoint file. In this approach, connections should be re-established before processes can resume their computation. Another approach, which is used in LAM/MPI, uses bookmarking mechanism between sender and receiver processes to guarantee message delivery at the checkpointing time.

57.4.5.4.2 Checkpointing in VM-Based Systems

Virtualization technique provides application-transparent checkpointing as an inherent feature that involves saving (suspending) and restoring (resuming) of the VM state [14,20,55].

In a virtualized platform, hypervisor (also called virtual machine monitor) is an essential component that manages different VMs concurrently running on the same host. Generally, the hypervisor is in charge of VM checkpointing. To checkpoint a VM, its internal state including memory, cache, and data related to the virtual devices have to be stored on the disk. Disk image snapshot also has to be stored, specially when the checkpointed VM is transferred and sharing image is not possible. Current virtual machine monitors, such as VMware, Xen, and KVM, support saving/restoring the state of VMs to/from a file. However, taking a copy of the disk image is not practically possible because of the size of the disk [69]. Therefore, currently, checkpointing is mostly carried out within resources with a shared storage, such as NFS.

Accordingly, distributed applications running on VMs across several nodes within a Cluster can be checkpointed [46]. Checkpointing of such applications is complicated because of the possible correlations between VMs (e.g., TCP packets and messages exchanged between VMs). The checkpointing process should be aware of these correlations, otherwise the checkpointing process leads to inconsistency in running the distributed applications.

To handle the checkpointing, when a checkpointing event is initiated, all the nodes that run a process of the distributed application receive the event. Upon receiving the event, the hypervisor *pauses* computation within VMs in order to preserve the internal state of VM and also to stop submitting any new network message (see Figure 57.7). In the next step, checkpointing protocols save the in-flight messages (i.e., network packets). For this purpose, the hypervisor collects all the incoming packets and queue them. Finally, a local VM checkpointing is performed through which the VM's internal state, VM disk image, and queued messages for that VM are saved in the checkpoint file [44].

57.5 Contention Management in Practice

Various types of distributed systems for resource sharing and aggregation have been developed. They include Clusters, Grids, and Clouds. In this section, we study these systems from the resource contention perspective. We identify and categorize properties of the reviewed systems and summarize them in Table 57.2 for Clusters and in Table 57.3 for Grids and Clouds.

57.5.1 Contention Management in Clusters

Compute Clusters are broadly categorized as dedicated and shared Clusters. In dedicated Clusters a single application exclusively runs on the Cluster's nodes. Mail servers, and web servers are examples of dedicated Clusters.

By contrast, in a shared Cluster the number of requests is significantly higher than the number of Cluster nodes. Therefore, nodes have to be shared between the requests by means of a resource management system [100]. From the resource contention perspective, shared Clusters are generally prone to request-initiated contention.

Virtual Clusters are another variation of Clusters that work based on VMs. Although users of these Clusters are given root access to the VMs, these resources are not dedicated to one user in hardware level (i.e., several VMs on the same node can be allocated to different users).

A multicluster is an interconnected distributed system that consists of several Clusters possibly in different organizations. Multiclusters are prone to origin-initiated contentions as well as request-initiated contention.

TABLE 57.2 Role of Different Components of Cluster Management Systems in Dealing with Resource Contentions

System	Provisioning Model	Operational Model	Context	Contention Initiation	Contention Management	RMS Component
Haizea [93]	Lease	Reactive	Cluster	Request	Preemption	Local sched
VioCluster [79]	VM	Proactive and reactive	MultiCluster	Origin	Preemption	Local sched
Snell et al. [91]	Job	Reactive	Cluster	Request	Preemption	Local sched
Lawson Smirni [59]	Job	Proactive and reactive	Cluster	Request	Partitioning	Local sched
Walters et al. [109]	VM	Reactive	Cluster	Request	Preemption	Local sched
Scojo-PECT [92]	VM	Reactive	Cluster	Request	DiffServ and preemption	Local sched
MOSIX [4]	VM	Reactive	Cluster	Origin	Fairness and preemption	Local sched
Sharc [100]	Job	Reactive	Cluster	Request	Partitioning	Local sched and Admission ctrl
COD [64]	Lease	Reactive	Cluster	Request	Partial preemption	Local sched
Cluster reserves [10]	Job	Proactive	Cluster	Request	Partitioning	Local sched
Muse [23]	Job	Reactive	Cluster	Request	Economic (utility)	Local sched
Shirako [48]	Lease	Reactive	MultiCluster	Interdomain	Token	Local sched
Lee et al. [60]	Job	Proactive and reactive	MultiCluster	Request	Fairness	Global and Local sched
MUSCLE [42]	Job	Proactive	MultiCluster	Request	Utility	Global sched
Percival et al. [73]	Job	Reactive	Cluster	Request	Economic (utility)	Admission ctrl

TABLE 57.3 Role of Different Components of Grid/Cloud Resource Management Systems in Dealing with Resource Contentions

System	Provisioning Model	Operational Model	Context	Contention Initiation	Contention Management	RMS Component
GridWay [103]	Job	Reactive	Grid Federation	Request	Outsourcing	Global sched (outsourcing)
Amazon Spot Market	VM	Reactive	Cloud	Request	Economic (auction)	Global sched
Van et al. [101]	VM	Reactive	Cloud	Request	Economic (utility)	Global sched
OurGrid [9]	Job	Reactive	Grid	Interdomain	Incentive	Global sched
Ren et al. [76]	Job	Proactive	Desktop Grid	Origin	Preemption	Local sched
Salehi et al. [6]	Lease	Proactive	Grid Federation	Origin	Preemption	Global sched
InterGrid [26]	Lease	Proactive	Grid Federation	Origin	Partitioning	Global sched
InterCloud [97]	VM	Reactive	Cloud Federation	Request	Economic (utility)	Admission ctrl
RESERVOIR [78]	VM	Reactive	Cloud Federation	Request	Outsourcing	Outsourcing
Sandholm et al. [86]	VM	Reactive	Grid	Request	Partial preemption	Admission ctrl
InterGrid Peering [26]	Lease	Reactive	Grid Federation	Interdomain	Global scheduling	Global sched
Salehi et al. [82]	Lease	Proactive	Grid Federation	Origin	Preemption	Admission ctrl
NDDE [67]	VM	Reactive	Desktop Grid	Origin	Preemption	Local sched
Gong et al. [38]	Job	Proactive	NOW	Origin	Preemption	Local sched
Delegated-matchmaking [47]	Job	Reactive	Grid Federation	Request	Outsourcing	Outsourcing
Gruber [28]	Job	Reactive	Grid	Interdomain and Request	Global scheduling and outsourcing	Global sched and outsourcing

Shirako [48] is a lease-based platform for on-demand allocation of resources across several Clusters. In Shirako, a broker receives user's application and provides it tickets that are redeemable at the provider Cluster. In fact, Shirako brokers handles interdomain-initiated contentions by coordinating resource allocation across different Clusters. However, the user application should decide how and when to use the resources.

VioCluster [79] is a VM-based platform across several Clusters. It uses lending and borrowing policies to trade VMs between Clusters. VioCluster is equipped with a machine broker that decides when to borrow/lend VMs from/to another Cluster. Machine broker also has policies for reclaiming resources that reacts to origin-initiated contention by preempting a leased VM to another domain. Machine property policy monitors the machine properties that should be allocated to the VMs such as CPU, memory, and storage capacity. Location policy in the VioCluster proactively determines if it is better to borrow VMs from other Cluster or waiting for nodes on a single domain.

Haizea [93] is a lease manager that is able to schedule combination of Advanced Reservation, Best Effort, and Immediate leases. Haizea acts as a scheduling back-end for OpenNebula [32]. The advantage of Haizea is considering and scheduling the preparation overhead of deploying VM disk images. For scheduling Advanced Reservation and Immediate leases, leases with lower priority (i.e., Best Effort) are preempted (i.e., suspended and resumed after the reservation is finished). In fact, Haizea provides a reactive resource contention mechanism for request-initiated contentions.

Sharc [100] is a platform that works in conjunction with nodes' operating system and enables resource sharing within Clusters. Architecturally, Sharc includes two components, namely, *control plane* and *nucleus*. The former is in charge of managing Cluster-wide resources and removing request-initiated contentions, whereas the latter, interacts with the operating system of each node and reserves resources for requests. Control plane uses a tree structure to keep information of resources are currently in use in the Cluster. The root of the tree shows all the resources in the Cluster and each child indicates one job. The nucleus uses a hierarchy that keeps information about what resources are in use on a node and by whom. The root of hierarchy shows all the resources on that node and each child represents a job on that node. In fact, there is a mapping between the control plane hierarchy and the nucleus hierarchy that helps Sharc to tolerate faults.

Cluster-on-Demand [64] (COD) is a resource management system for shared Clusters. COD supports lease-based resource provisioning in the form of virtual Clusters where each Virtual Cluster is an isolated group of hosts inside a shared hardware base. COD is equipped with a protocol that dynamically resizes Virtual Clusters in cooperation with middleware components. COD uses group-based priority and partial preemption scheme to manage request-initiated resource contention. Specifically, when resource contention takes place, COD preempts nodes from a low-priority Virtual Cluster. For preemption the selected Virtual Cluster returns those nodes that create minimal disruption to the Virtual Clusters.

Cluster Reserves [10] is a resource allocation for Clusters that provides services to the clients based on the notion of service class (partitioning). This is performed by allocating resource partitions to parallel applications and dynamically adjusting the partitions on each node based on the user demand. Indeed, Cluster Reserve applies partitioning scheme to cope with the request-initiated contention problems. The resource management problem is considered as a constrained optimization problem where the inputs of the problem are periodically updated based on the resource usage.

Muse [23] is an economy-based architecture for dynamic resource procurement within a job-based Cluster. Muse is prone to request-initiated contention and applies a utility-based, economic solution to resolve that. In the model, each job has a utility function based on its throughput that reflects the revenue earned by running the job. There is a penalty that the job charges the system when its constrains are not met. Resource allocation is worked out through solving an optimization problem that maximizes the overall profit. Muse considers energy as a driving issue in resource management of server Clusters.

MUSCLE [42] is an off-line, global scheduler for multiclusters that batches parallel jobs with high packing potential (i.e., jobs that can be packed into a resource space of a given size) to the same Cluster. In the next step, a local scheduler (called TITAN) performs further tuning to run the assigned jobs with minimized make span and idle times.

Lee et al. [60] have proposed a global and a local scheduler for a multicluster. The local scheduler is a variant of backfilling that grants priority to wide jobs to decrease their waiting time and resolves the request-initiated contention. The global dispatcher assigns requests to the proper Cluster by comparing the proportion of requests with the same size at each participant Cluster. Therefore, a fairly uniform distribution of requests in the Clusters is created which leads to a considerable impact on the performance.

Percival et al. [73] applied an admission control policy for shared Cluster. There is a request-initiated contention because some large jobs takes precedence over many small jobs that are waiting in the queue. Resource providers determine the resource prices based on the degree of contention and instantaneous utilization of resources. Consumers also bid for the resources based on their budget. In general, a job can get a resource if it can compensate the loss of earning resulting from not admitting several small jobs.

57.5.2 Contention Management in Desktop Grids

This form of distributed computing (also known as volunteer computing) inherently relies on participation of resources, mainly Personal Computers. In desktop Grids, participants become available during their idle periods to leverage the execution of long running jobs. They usually use specific events such as screen-saver as an indicator for idle cycles. SETI@home [8] is a famous desktop Grid project that works based on BOINC [7] software platforms and was originally developed to explore the existence of life out of the earth. Desktop Grids are prone to origin-initiated resource contentions that take place between the guest requests (come from the Grid environment) and local requests (initiated by the resource owner) in a node.

In desktop Grids, the guest applications are running in the user (owner) environment. Running the external jobs along with other owner's processes, raised the security concern in desktop Grids and became an obstacle in prevalence of these systems. However, using the emulated platforms, such as Java, and sand-boxing the security concern were mitigated.

Another approach in desktop Grids is rebooting the machine and run an entirely independent operating system for the guest request. As a result, the guest request does not have access to the user environment. Instances of this approach include HP's I-cluster [77] and vCluster [27]. However, this approach can potentially interrupt the interactive user (owner). Therefore, idle cycle prediction has to be done conservatively to avoid interrupting the interactive user (owner). Both of these approaches are heavily dependent on the efficient predicting and harvesting of the idle cycles. Indeed, these approaches function efficiently where there are huge idle cycles.

Recently, VM technology has been used in desktop Grids. The advantages of using VMs in these environments are threefolds. First and foremost is the security that VMs provide through an isolated execution environment. Second, VMs offer more flexibility in terms of the running environment demanded by the guest application. The third benefit is that by using VMs fragmented (unused) idle cycles, such as cycles at the time of typing or other light-weight processes, can be harvested.

NDDE [67] is a platform that utilizes VMs to exploit idle cycles for Grid or Cluster usage in corporations and educational institutions. This system is able to utilize idle cycles that appear even while the user is interacting with the computer. Indeed, in this system the guest and owner applications are run concurrently. This approach increases the harvested idle cycle to as many as possible with minor impact on the interactive user's applications. NDDE has more priority than idle process in the host operating system and, therefore, will be run instead of idle process when the system is idle. At the time the owner has a new request, the VM and all the processes belong to NDDE are preempted and changed to "ready-to-run" state.

Fine-grained cycle sharing system (FGCS) [76] runs a guest request concurrently with the local request whenever the guest process does not degrade the efficiency of the local request. However, FGCS are prone to unavailability because of the following reasons:

1. Guest jobs are killed or migrate off the resource because of a local request
2. Host suddenly discontinue contributing resource to the system

To cope with these problems, they define unavailabilities in the form a state diagram where each state is a condition that resource becomes unavailable (e.g., contention between users, and host unavailability). The authors have applied a Semi-Markov chain Process to predict the availability. The goal of this predictor engine is determining the probabilities of not transferring to unavailable states in a given time period of time in future.

57.5.3 Contention Management in Grids

Grids are initially structured based on the idea of the virtual organizations (VOs). A VO is a set of users from different organizations who collaborate towards a common objective. Several organizations constitute a VO by contributing share of their resources to that and as a result their users gain access to the VO resources. Contributing resources to a VO is carried out via an agreement upon that an organization gets access to the VO resources according to the amount of resources it offers to the VO.

Organizations usually retain part of their resources for their organizational (local) users. In other words, VO (external) requests are welcome to use resources if they are available. However, VO requests should not delay the execution of local requests.

Indeed, Grids are huge interconnected distributed systems that are prone to all kinds of resource contentions [85]. Particularly, interdomain-initiated resource contention arises when organizations need to access VO's resources based on their contributions. Origin-initiated resource contention occurs when there is a conflict between local and external users within the resources of an organization. Finally, request-initiated contention exists between different types of requests (short/long, parallel/serial, etc.).

Gruber/Di-Gruber [28] is a Grid broker that deals with the problem of resource procurement form several VOs and assigns them to different user groups. Gruber provides monitoring facilities that can be used for interdomain-initiated contentions. It also investigates the enforcing of usage policies (SLA) as well as monitoring the enforcement. Another component of Gruber sought to cope with request-initiated resource contention through monitoring resources' loads and outsource jobs to a suitable site (site selector component). Di-Gruber is the distributed version of Gruber which supports multiple decision points.

InterGrid [26] is a federation of Grid systems where each Grid receives lease requests from other Grids based on peering arrangements between InterGrid Gateways (IGG) of the Grids. Each Grid serves its own users (e.g., organizational/local users) as well as users coming from other Grids (external). InterGrid is prone to origin-initiated (between local and external requests) and interdomain-initiated (between different Grids) resource contentions.

Peering arrangements between Grids coordinate exchanging resources and functions based on peer-to-peer relations established among Grids. Each peer is built upon a predefined contract between Grids and handles interdomain-initiated contentions between the two Grids. Outsourcing unit of InterGrid is incorporated in the scheduling and determines when to outsource a request. Salehi et al. [6] have utilized probabilistic methods and proposed contention-aware scheduling that aims at minimizing the number of VM preemptions (and therefore minimizing contention) in a Grid.

They have also come up [82] with an admission control policy to reduce origin-initiated contention in InterGrid. The admission control policy works based on limiting queue length for external requests in a way that their deadline can be met. For that purpose they anticipate the average response time of external requests waiting in the queue by considering characteristics of local requests such as interarrival rate and size. In this situation, the external requests are accepted up until the response time is less than the average deadline.

Delegated-matchmaking [47] proposes an architecture that delegates the ownership of resources to users in a transparent and secure way. More specifically, when a site cannot satisfy its local users, the matchmaking mechanism of Delegated-matchmaking adds remote resources to the local resources. In fact, in Delegated-matchmaking the ownership of resources are delegated in different sites of Grids. From the resource contention perspective, matchmaking mechanism is in charge of dealing with request-initiated contentions through outsourcing scheme.

GridWay [103] is a project that creates loosely coupled connection between Grids via connecting to their meta-schedulers. GridWay is specifically useful when a job does not get the required processing power or the job waiting time is more than an appointed threshold. In these situation, GridWay migrates (outsource) the job to another Grid in order to provides the demanded resources to the job. We can consider GridWay as a global scheduler that deals with request-initiated resource contentions.

OurGrid [9] is a Grid that operates based on a P2P network between sites and share resources based on reciprocity. OurGrid uses network of favors as the resource exchange scheme between participants. According to this network, each favor to a consumer should be reciprocated by the consumer site at a later time. The more favor participants do, the more reward they expect. From the resource contention perspective, OurGrid uses incentive-based approach to figure out the problem of interdomain-initiated contentions in a Grid.

Sandholm et al. [86] investigated how admission control can increase user fulfillment in a computational market. Specifically, they considered the mixture of best effort (to improve resource utilization) and QoS-constrained requests (to improve revenue) within a virtualized Grid. They applied a reactive approach through partial preemption of best-effort requests to resolve request-initiated contentions. However, the admission control proactively accepts a new request if the QoS requirements of the current requests can still be met.

57.5.4 Contention Management in Clouds

Advances in virtual machine and network technologies has led to appearing commercial providers that offer numerous resources to users and charge them in a pay-as-you-go fashion. Since the physical infrastructure is unknown to the users in these providers; they are known as Cloud Computing [18]. There are various fashions for delivering Cloud services, which are generally known as XaaS (X as a Service). Among these services Infrastructure as a Service (IaaS) offers resources in the form of VM to users.

From the availability perspective, Cloud providers are categorized as public, private, and hybrid Clouds [19]. To cope with the shortage of resource availability, particularly in private Clouds, the idea of federated Cloud has been presented [18]. Cloud federation is a possible solution for a Cloud provider in order to access to a larger pool of resources.

Similar to Grid environments, Clouds are also prone to different types of resource contentions. However, as Clouds are more commercialized in comparison with Grids, the resource contentions solutions are also mostly commercially driven.

Recently, Amazon started to offer spot instances to sell the unused capacity of their data centers [110]. Spot instances are priced dynamically based on users' bids. If the bid price is beyond the current spot instance price, the VM instance is created for the user. The spot instance's price fluctuates and if the current price goes beyond the bid price, the VM instance is canceled (terminated) or alternatively suspended up until the current price becomes lower than the bid. Indeed, the spot market presents a request-initiated resource contention where the contention is solved via an auction-based scheme. Kondo and Andryejak [110] have evaluated the dynamic checkpointing schemes, which is adaptive to the current instance price, and achieves cost efficiency and reliability in dealing with spot instances.

Van et al. [101] have proposed a multilayer, contention-aware resource management system for Cloud infrastructure. The resource management system takes into account both request's QoS requirements and energy consumption costs in VM placement. In the request (user) level, a local decision module (LDM) monitors the performance of each request and calculates a utility function that indicates the performance satisfaction of that request. LDM interacts with a global decision module (GDM) which is the decision-making component in the architecture. GDM considers the utility functions of all LDMs along with system-level performance metrics and decides about the appropriate action. In fact, GDM provides a global scheduling solution to resolve request-initiated contentions between requests. The output of the GDM can be management commands to the server hypervisor and notifications for LDMs. The notifications for LDM includes adding a new VM to the application, upgrading or downgrading an existing VM, preempting a VM belonging to a request. Management actions for hypervisors include the starting, stopping, or live migration of a VM.

RESERVOIR [78] is a research initiative that aims at developing the technologies required to address the scalability problems existing in the single provider Cloud computing model. To achieve this goal, Clouds with excess capacity offer their resources, based on an agreed price, to the Clouds that require extra resources. Decision making about where to allocate resources for a given request is carried out through an outsourcing component, which is called placement policy. Therefore, the aim of project is providing an outsourcing solution for request-initiated resource contention.

InterCloud [18] aims to create a computing environment that offers dynamic scaling up and down capabilities (for VMs, services, storage, and database) in response to users' demand variations. The central element in InterCloud architecture is the Cloud Exchange, which is a market that gathers service providers and users' requests. It supports trading of Cloud services based on competitive economic models, such as financial options [98]. Toosi et al. [18,97] consider circumstances that each Cloud offers on-demand and spot VMs. The admission control unit evaluates the cost–benefit of outsourcing an on-demand request to the InterCloud or allocating resource to that via terminating spot VMs (request-initiated contention). Their ultimate objective is to decrease the rejection rate and having access to seemingly unlimited resources for on-demand requests.

57.6 Conclusions and Future Research Directions

Due to resource shortage as well as surge in demand, distributed systems commonly face contention between requests to access resources. Resource contentions are categorized as *request-initiated*, when a user request cannot be admitted or cannot acquire sufficient resources because the resources are occupied by other requests. *Origin-initiated* resource contention refers to circumstances that requests are from different sources with distinct priorities. *Interdomain-initiated* resource contentions take place when the proportion of shared resources to the consumed resources by a resource provider is low.

Resource contention can be handled by different components of resource management system. Therefore, solutions for resource contention depends on the structure of resource management in a distributed system. In this research, we recognized the role of resource provisioning model, local scheduling, global scheduling, and admission control unit in a resource management system on various types of resource contentions. We also realized that the emergence of VM-based resource provisioning model has posed the preemption as a predominant solution for different types of resource contentions. Therefore, in this survey we also investigated the challenges and opportunities of preempting VMs.

We reviewed systems in Clusters, Grids, and Clouds from the contention management perspective and categorized them based on their operational model, the type of contention they deal with, the component of resource management system involved in resolving the contention, and the provisioning model that contention is considered. We also closely investigated preemption mechanism as the substantial resolution for resource contention.

There are avenues of future research works in managing resource contentions that can be pursued by researchers. Proactive resource contention management methods are required specifically for interdomain-initiated contentions. This means that in an interconnected distributed system when resource management system decides to outsource a request, contention probability in the destination provider has to be considered.

Combination of different resource contentions (hybrid contention) requires further investigation. For example, resolving contention where there is a combination of origin-initiated and request-initiated contentions. Moreover, economical solutions can be taken into consideration to resolve the origin-initiated and interdomain-initiated resource contentions.

We also enumerated several options that can be considered for resuming a preempted request. Current systems usually choose one of these options. However, it will be interesting to come up with a mechanism that dynamically (e.g., based on the request condition) chooses one of the available options. A more specific case is when preemption via suspension happens. In this situation, determining the appropriate place to resume the preempted request is a challenge. For instance, if it is data-intensive request, then it might be better to wait in the queue instead of migrating to another resource.

References

1. Bright cluster manager. http://www.brightcomputing.com, accessed October 5, 2013.
2. S. Adabala, V. Chadha, P. Chawla et al. From virtualized resources to virtual computing grids: The in-vigo system. *Future Generation Computer Systems*, 21(6):896–909, 2005.
3. L. Amar, A. Barak, E. Levy, and M. Okun. An on-line algorithm for fair-share node allocations in a cluster. In *Proceedings of the 7th IEEE International Symposium on Cluster Computing and the Grid*, pp. 83–91, Rio de Janeiro, Brazil, 2007.
4. L. Amar, A. Mu'Alem, and J. Stober. The power of preemption in economic online markets. In *Proceedings of the 5th International Workshop on Grid Economics and Business Models (GECON'08)*, pp. 41–57, Berlin, Germany, 2008.
5. M.A. Salehi, B. Javadi, and R. Buyya. Resource provisioning based on leases preemption in InterGrid. In *Proceeding of the 34th Australasian Computer Science Conference (ACSC'11)*, pp. 25–34, Perth, Australia, 2011.
6. M.A. Salehi, B. Javadi, and R. Buyya. QoS and preemption aware scheduling in federated and virtualized grid computing environments. *Journal of Parallel and Distributed Computing (JPDC)*, 72(2):231–245, 2012.
7. D.P. Anderson. Boinc: A system for public-resource computing and storage. In *Proceedings of 5th IEEE/ACM International Workshop on Grid Computing*, pp. 4–10, 2004.

8. D.P. Anderson, J. Cobb, E. Korpela, M. Lebofsky, and D. Werthimer. Seti@ home: An experiment in public-resource computing. *Communications of the ACM*, 45(11):56–61, 2002.
9. N. Andrade, F. Brasileiro, W. Cirne, and M. Mowbray. Automatic grid assembly by promoting collaboration in peer-to-peer grids. *Journal of Parallel and Distributed Computing*, 67(8):957–966, 2007.
10. M. Aron, P. Druschel, and W. Zwaenepoel. Cluster reserves: A mechanism for resource management in cluster-based network servers. In *Proceedings of the International Conference on Measurement and Modelling of Computer Systems (SIGMETRICS'00)*, pp. 90–101, Santa Clara, CA, 2000.
11. M.D. Assunção and R. Buyya. Performance analysis of multiple site resource provisioning: Effects of the precision of availability information. In *Proceedings of the 15th International Conference on High Performance Computing (HiPC'08)*, pp. 157–168, Bangalore, India, 2008.
12. M.D. Assunção, A.D. Costanzo, and R. Buyya. Evaluating the cost-benefit of using cloud computing to extend the capacity of clusters. In *Proceedings of the 19th International Symposium on High Performance Distributed Computing (HPDC09)*, pp. 141–150, Munich, Germany, 2009.
13. A. Barak, A. Shiloh, and L. Amar. An organizational grid of federated mosix clusters. In *IEEE International Symposium on Cluster Computing and the Grid (CCGrid05)*, vol. 1, pp. 350–357, Cardiff, UK, 2005.
14. P. Barham, B. Dragovic, K. Fraser, S. Hand, T. Harris, A. Ho, R. Neugebauer, I. Pratt, and A. Warfield. Xen and the art of virtualization. *ACM SIGOPS Operating Systems Review*, 37(5):164–177, 2003.
15. P. Beckman, S. Nadella, N. Trebon, and I. Beschastnikh. Spruce: A system for supporting urgent high-performance computing. In *Grid-Based Problem Solving Environments*, pp. 295–311, Springer, Boston, MA, 2007.
16. B. Beeson, S. Melniko, S. Venugopal, and D.G. Barnes. A portal for grid-enabled physics. In *Proceeding of the 28th Australasian Computer Science Week (ACSW'05)*, pp. 13–20, 2005.
17. R. Bolze, F. Cappello, E. Caron et al. Grid'5000: A large scale and highly reconfigurable experimental Grid testbed. *International Journal of High Performance Computing Applications*, 20(4):481, 2006.
18. R. Buyya, R. Ranjan, and R.N. Calheiros. InterCloud: Utility-oriented federation of cloud computing environments for scaling of application services. In *Proceedings of the 10th International Conference on Algorithms and Architectures for Parallel Processing-Volume Part I*, pp. 13–31, Busan, South Korea, 2010.
19. R. Buyya, C.S. Yeo, S. Venugopal, J. Broberg, and I. Brandic. Cloud computing and emerging it platforms: Vision, hype, and reality for delivering computing as the 5th utility. *Future Generation Computer Systems*, 25(6):599–616, 2009.
20. K. Chanchio, C. Leangsuksun, H. Ong, V. Ratanasamoot, and A. Shafi. An efficient virtual machine checkpointing mechanism for hypervisor-based HPC systems. In *Proceeding of the High Availability and Performance Computing Workshop (HAPCW)*, pp. 29–35, Denver, CO, 2008.
21. K. Chanchio and X.H. Sun. Data collection and restoration for heterogeneous process migration. *Software: Practice and Experience*, 32(9):845–871, 2002.
22. K.M. Chandy and L. Lamport. Distributed snapshots: Determining global states of distributed systems. *ACM Transactions on Computer Systems (TOCS)*, 3(1):63–75, 1985.
23. J.S. Chase, D.C. Anderson, P.N. Thakar, A.M. Vahdat, and R.P. Doyle. Managing energy and server resources in hosting centers. *ACM SIGOPS Operating Systems Review*, 35(5):103–116, 2001.
24. C. Coti, T. Herault, P. Lemarinier, L. Pilard, A. Rezmerita, E. Rodriguez, and F. Cappello. Blocking vs. non-blocking coordinated checkpointing for large-scale fault tolerant MPI. In *Proceedings of the ACM/IEEE Conference on Supercomputing*, pp. 127–1233, Tampa, FL, 2006.
25. K. Czajkowski, S. Fitzgerald, I. Foster, and C. Kesselman. Grid information services for distributed resource sharing. In *Proceedings of 10th IEEE International Symposium on High Performance Distributed Computing*, pp. 181–194, San Fransisco, CA. IEEE, 2001.
26. M.D. De Assunção, R. Buyya, and S. Venugopal. InterGrid: A case for internetworking islands of Grids. *Concurrency and Computation: Practice and Experience*, 20(8):997–1024, 2008.
27. C. De Rose, F. Blanco, N. Maillard, K. Saikoski, R. Novaes, O. Richard, and B. Richard. The virtual cluster: A dynamic network environment for exploitation of idle resources. In *Proceedings 14th Symposium on Computer Architecture and High Performance Computing*, pp. 141–148 , Pernambuco, Brazil. IEEE, 2002.

28. C. Dumitrescu, I. Raicu, and I. Foster. Di-gruber: A distributed approach to grid resource brokering. In *Proceedings of ACM/IEEE Conference on Supercomputing*, pp. 38–45, Seattle, WA, 2005.
29. W. Emeneker, D. Jackson, J. Butikofer, and D. Stanzione. Dynamic virtual clustering with xen and moab. In *Frontiers of High Performance Computing and Networking–ISPA Workshops*, pp. 440–451, Sorrento, Italy. Springer, 2006.
30. A. Ferrari, S.J. Chapin, and A. Grimshaw. Heterogeneous process state capture and recovery through process introspection. *Cluster Computing*, 3(2):63–73, 2000.
31. L. Field and M. Schulz. Grid interoperability: The interoperations cookbook. *Journal of Physics: Conference Series*, 119:120–139. IOP Publishing, 2008.
32. J. Fontán, T. Vázquez, L. Gonzalez, R.S. Montero, and I.M. Llorente. OpenNebula: The open source virtual machine manager for cluster computing. In *Open Source Grid and Cluster Software Conference*, San Francisco, CA, May 2008.
33. I. Foster and C. Kesselman. Globus: A metacomputing infrastructure toolkit. *International Journal of High Performance Computing Applications*, 11(2):115–128, 1997.
34. J. Frey, T. Tannenbaum, M. Livny, I. Foster, and S. Tuecke. Condor-G: A computation management agent for multi-institutional grids. *Cluster Computing*, 5(3):237–246, 2002.
35. S. Gabriel. Gridka tier1 site management. In *International Conference on Computing in High Energy and Nuclear Physics (CHEP'07)*, 2007.
36. S. Garg, S. Venugopal, and R. Buyya. A meta-scheduler with auction based resource allocation for global grids. In *14th IEEE International Conference on Parallel and Distributed Systems (ICPADS'08)*, pp. 187–194, Melbourne, Australia, 2008.
37. S. Garg, C. Yeo, A. Anandasivam, and R. Buyya. Environment-conscious scheduling of HPC applications on distributed cloud-oriented data centers. *Journal of Parallel and Distributed Computing*, 71(6):732–749, 2011.
38. L. Gong, X. Sun, and E.F. Watson. Performance modeling and prediction of nondedicated network computing. *IEEE Transactions on Computers*, 51(9):1041–1055, 2002.
39. L. Grit, L. Ramakrishnan, and J. Chase. On the duality of jobs and leases. Technical report CS-2007-00, Duke University, Department of Computer Science, Durham, NC, April 2007.
40. I. Grudenić and N. Bogunović. Analysis of scheduling algorithms for computer clusters. In *Proceeding of 31th International Convention on Information and Communication Technology. Electronics and Microelectronics (MIPRO)*, pp. 13–20, Opatija, Croatia, 2008.
41. P.H. Hargrove and J.C. Duell. Berkeley lab checkpoint/restart (BLCR) for linux clusters. *Journal of Physics: Conference Series*, 46:494. IOP Publishing, 2006.
42. L. He, S.A. Jarvis, D.P. Spooner, X. Chen, and G.R. Nudd. Dynamic scheduling of parallel jobs with QoS demands in multiclusters and grids. In *Proceedings of the 5th IEEE/ACM International Workshop on Grid Computing*, pp. 402–409, Pittsburgh, PA. IEEE Computer Society, 2004.
43. J. Heo, S. Yi, Y. Cho, J. Hong, and S.Y. Shin. Space-efficient page-level incremental checkpointing. In *Proceedings of the ACM Symposium on Applied Computing*, pp. 1558–1562, 2005.
44. F. Hermenier, A. Lèbre, and J. Menaud. Cluster-wide context switch of virtualized jobs. In *Proceedings of the 19th ACM International Symposium on High Performance Distributed Computing (HPDC'10)*, New York, pp. 658–666, 2010.
45. M. Hovestadt, O. Kao, A. Keller, and A. Streit. Scheduling in HPC resource management systems: Queuing vs. planning. In *Proceedings of 9th International Workshop on Job Scheduling Strategies for Parallel Processing*, pp. 1–20, New York, 2003.
46. W. Huang, Q. Gao, J. Liu, and D.K. Panda. High performance virtual machine migration with RDMA over modern interconnects. In *Proceedings of the 9th IEEE International Conference on Cluster Computing*, pp. 11–20, Barcelona, Spain. IEEE, 2007.
47. A. Iosup, O. Sonmez, S. Anoep, and D. Epema. The performance of bags-of-tasks in large-scale distributed systems. In *Proceedings of the 17th International Symposium on High Performance Distributed Computing*, pp. 97–108, 2008.

48. D. Irwin, J. Chase, L. Grit, A. Yumerefendi, D. Becker, and K.G. Yocum. Sharing networked resources with brokered leases. In *USENIX Annual Technical Conference*, pp. 199–212, Boston, MA, June 2006.
49. M. Isard, V. Prabhakaran, J. Currey, U. Wieder, K. Talwar, and A. Goldberg. Quincy: Fair scheduling for distributed computing clusters. In *Proceedings of the 22nd Symposium on Operating Systems Principles (SOSP), ACM SIGOPS*, pp. 261–276, Boston, MA, 2009.
50. S.T. Jones, A.C. Arpaci-Dusseau, and R.H. Arpaci-Dusseau. Antfarm: Tracking processes in a virtual machine environment. In *Proceedings of the USENIX Annual Technical Conference*, pp. 1–14, Boston, MA, 2006.
51. A. Kansal, F. Zhao, J. Liu, N. Kothari, and A.A. Bhattacharya. Virtual machine power metering and provisioning. In *Proceedings of the 1st ACM Symposium on Cloud Computing*, pp. 39–50, Indianapolis, IN, 2010.
52. K. Keahey, R. Figueiredo, J. Fortes, T. Freeman, and M. Tsugawa. Science clouds: Early experiences in cloud computing for scientific applications. In *Proceeding of Cloud Computing and Applications*, vol. 1, Chicago, IL, 2008.
53. K. Keahey, I. Foster, T. Freeman, X. Zhang, and D. Galron. Virtual workspaces in the grid. In *Proceeding of the 11th International Euro-Par Conference on Parallel Processing*, pp. 421–431, Lisbon, Portugal, 2005.
54. R. Kettimuthu, V. Subramani, S. Srinivasan, T. Gopalsamy, D.K. Panda, and P. Sadayappan. Selective preemption strategies for parallel job scheduling. *International Journal of High Performance and Networking (IJHPCN)*, 3(2/3):122–152, 2005.
55. A. Kivity, Y. Kamay, D. Laor, U. Lublin, and A. Liguori. KVM: The linux virtual machine monitor. In *Linux Symposium*, p. 225, Ottawa, Canada, 2007.
56. I. Krsul, A. Ganguly, J. Zhang, J.A.B. Fortes, and R.J. Figueiredo. Vmplants: Providing and managing virtual machine execution environments for grid computing. In *Proceedings of the ACM/IEEE Conference Supercomputing*, pp. 7–17, Pittsburgh, PA, 2004.
57. Y. Kwok and I. Ahmad. Dynamic critical-path scheduling: An effective technique for allocating task graphs to multiprocessors. *IEEE Transaction Parallel and Distributed Systems*, 7(5):506–521, 1996.
58. Y. Kwok, S.S. Song, K. Hwang et al. Selfish grid computing: Game-theoretic modeling and nas performance results. In *IEEE International Symposium on Cluster Computing and the Grid (CCGrid'05)*, vol. 2, pp. 1143–1150, 2005.
59. B.G. Lawson and E. Smirni. Multiple-queue backfilling scheduling with priorities and reservations for parallel systems. *ACM SIGMETRICS Performance Evaluation Review*, 29(4):40–47, 2002.
60. H.Y. Lee, D.W. Lee, and R. Ramakrishna. An enhanced grid scheduling with job priority and equitable interval job distribution. In *Advances in Grid and Pervasive Computing*, pp. 53–62, Taichung, Taiwan, 2006.
61. H. Li, D.L. Groep, and L. Wolters. Workload characteristics of a multi-cluster supercomputer. In *Proceedings of 10th International Workshop on Job Scheduling Strategies for Parallel Processing (JSSPP'04)*, pp. 176–193, New York, 2004.
62. M. Litzkow and M. Solomon. The evolution of condor checkpointing. In *Mobility: Processes, Computers, and Agents*, pp. 163–174. ACM Press/Addison-Wesley Publishing Co., New York, 1999.
63. M. Margo, K. Yoshimoto, P. Kovatch, and P. Andrews. Impact of reservations on production job scheduling. In *Proceedings of 13th International Workshop on Job Scheduling Strategies for Parallel Processing*, pp. 116–131, Seattle, WA. Springer, 2008.
64. J. Moore, D. Irwin, L. Grit, S. Sprenkle, and J. Chase. Managing mixed-use clusters with cluster-on-demand. Technical report, Duke University Department of Computer Science, Durham, NC, 2002.
65. H. Nakada, H. Sato, K. Saga, M. Hatanaka, Y. Saeki, and S. Matsuoka. Job invocation interoperability between naregi middleware beta and glite. In *Proceedings of the 9th International Conference on High Performance Computing, Grid and e-Science in Asia Pacific Region (HPC Asia 2007)*, Phoenix, AZ, 2007.
66. R. Nathuji and K. Schwan. Virtualpower: Coordinated power management in virtualized enterprise systems. *ACM SIGOPS Operating Systems Review*, 41(6):265–278, 2007.
67. R.C. Novaes, P. Roisenberg, R. Scheer, C. Northfleet, J.H. Jornada, and W. Cirne. Non-dedicated distributed environment: A solution for safe and continuous exploitation of idle cycles. *Scalable Computing: Practice and Experience*, 6(3):13–22, 2001.

68. D. Nurmi, R. Wolski, C. Grzegorczyk, G. Obertelli, S. Soman, L. Youseff, and D. Zagorodnov. The eucalyptus open-source cloud-computing system. In *Proceedings of 9th IEEE/ACM International Symposium on Cluster Computing and the Grid (CCGRID'09)*, pp. 124–131, Shanghai, China, 2009.
69. H. Ong, N. Saragol, K. Chanchio, and C. Leangsuksun. Vccp: A transparent, coordinated checkpointing system for virtualization-based cluster computing. In *IEEE International Conference on Cluster Computing and Workshops, (CLUSTER'09)*, pp. 1–10, New Orleans, LO. IEEE, 2009.
70. S. Ontañón and E. Plaza. Cooperative case bartering for case-based reasoning agents. In *Topics in Artificial Intelligence*, pp. 294–308, 2002.
71. A. Oram, ed. *Peer-to-Peer: Harnessing the Power of Disruptive Technologies*. O'Reilly & Associates Inc., Sebastopol, CA, 2001.
72. E. Parsons and K. Sevcik. Implementing multiprocessor scheduling disciplines. In *Proceedings of 3rd International Workshop on Job Scheduling Strategies for Parallel Processing*, pp. 166–192, Geneva, Switzerland, 1997.
73. X. Percival, C. Wentong, and L. Bu-Sung. A dynamic admission control scheme to manage contention on shared computing resources. *Concurrency and Computing: Practice and Experience*, 21:133–158, February 2009.
74. J.S. Plank, M. Beck, G. Kingsley, and K. Li. Libckpt: Transparent checkpointing under unix. In *Proceedings of the USENIX Technical Conference*, pp. 18–28, New Orleans, LO. USENIX Association, 1995.
75. I. Raicu, Y. Zhao, C. Dumitrescu, I. Foster, and M. Wilde. Falkon: A fast and light-weight task execution framework. In *Proceedings of the ACM/IEEE Conference on Supercomputing (SC'07)*, pp. 1–12, Reno, NV. IEEE, 2007.
76. X. Ren, S. Lee, R. Eigenmann, and S. Bagchi. Prediction of resource availability in fine-grained cycle sharing systems empirical evaluation. *Journal of Grid Computing*, 5(2):173–195, 2007.
77. B. Richard and P. Augerat. I-cluster: Intense computing with untapped resources. In *Proceeding of 4th International Conference on Massively Parallel Computing Systems*, pp. 127–140, Colorado Spring, CO, 2002.
78. B. Rochwerger, D. Breitgand, A. Epstein et al. Reservoir-when one cloud is not enough. *Computer Journal*, 44(3):44–51, 2011.
79. P. Ruth, P. McGachey, and D. Xu. Viocluster: Virtualization for dynamic computational domain. In *Proceedings of International IEEE Conference on Cluster Computing (Cluster'05)*, pp. 1–10, Boston, MA, September 2005.
80. F.D. Sacerdoti, M.J. Katz, M.L. Massie, and D.E. Culler. Wide area cluster monitoring with ganglia. In *Proceedings of International Conference on Cluster Computing*, pp. 289–298, Hong Kong, China. IEEE, 2003.
81. M.A. Salehi, B. Javadi, and R. Buyya. Performance analysis of preemption-aware scheduling in multi-cluster grid environments. In *Proceedings of the 11th International Conference on Algorithms and Architectures for Parallel Processing (ICA3PP'11)*, pp. 419–432, Melbourne, Australia, 2011.
82. M.A. Salehi, B. Javadi, and R. Buyya. Preemption-aware admission control in a virtualized grid federation. In *Proceeding of 26th International Conference on Advanced Information Networking and Applications (AINA'12)*, pp. 854–861, Fukouka, Japan, 2012.
83. M.A. Salehi, P.R. Krishna, K.S. Deepak, and R. Buyya. Preemption-aware energy management in virtualized datacenters. In *Proceeding of 5th International Conference on Cloud Computing (IEEE Coud'12)*, Honolulu, HI, 2012.
84. M.A. Salehi and R. Buyya. Adapting market-oriented scheduling policies for cloud computing. In *Proceedings of the 10th International Conference on Algorithms and Architectures for Parallel Processing-Volume Part I*, pp. 351–362, Busan, South Korea. Springer-Verlag, 2010.
85. M.A. Salehi, H. Deldari, and B.M. Dorri. Balancing load in a computational grid applying adaptive, intelligent colonies of ants. *Informatica: An International Journal of Computing and Informatics*, 33(2):151–159, 2009.

86. T. Sandholm, K. Lai, and S. Clearwater. Admission control in a computational market. In *Eighth IEEE International Symposium on Cluster Computing and the Grid*, pp. 277–286, Lyon, France, 2008.
87. H. Shachnai, T. Tamir, and G.J. Woeginger. Minimizing makespan and preemption costs on a system of uniform machines. *Algorithmica Journal*, 42(3):309–334, 2005.
88. P.C. Shih, H.M. Chen, Y.C. Chung, C.M. Wang, R.S. Chang, C.H. Hsu, K.C. Huang, and C.T. Yang. Middleware of taiwan unigrid. In *Proceedings of the ACM Symposium on Applied Computing*, pp. 489–493, Ceará, Brazil. ACM, 2008
89. M. Silberstein, D. Geiger, A. Schuster, and M. Livny. Scheduling mixed workloads in multi-grids: The grid execution hierarchy. In *Proceedings of 15th Symposium on High Performance Distributed Computing*, pp. 291–302, Paris, France, 2006.
90. W. Smith, I. Foster, and V. Taylor. Scheduling with advanced reservations. In *Proceedings of the 14th International Parallel and Distributed Processing Symposium*, pp. 127–132, Cancun, Mexico. IEEE, 2000.
91. Q. Snell, M.J. Clement, and D.B. Jackson. Preemption based backfill. In *Proceedings of 8th International Workshop on Job Scheduling Strategies for Parallel Processing (JSSPP)*, pp. 24–37, Edinburgh, Scotland, UK. Springer, 2002.
92. A. Sodan. Service control with the preemptive parallel job scheduler scojo-pect. *Journal of Cluster Computing*, 14(2):1–18, 2010.
93. B. Sotomayor, K. Keahey, and I. Foster. Combining batch execution and leasing using virtual machines. In *Proceedings of the 17th International Symposium on High Performance Distributed Computing*, pp. 87–96, New York, 2008. ACM.
94. B. Sotomayor, R.S. Montero, I.M. Llorente, and I. Foster. Resource leasing and the art of suspending virtual machines. In *Proceedings of the 11th IEEE International Conference on High Performance Computing and Communications*, Washington, DC, pp. 59–68, 2009.
95. G. Stellner. Cocheck: Checkpointing and process migration for mpi. In *Proceedings of the the 10th International Parallel Processing Symposium*, pp. 526–531, Honolulu, HI, 1996.
96. D. Thain, T. Tannenbaum, and M. Livny. Distributed computing in practice: The condor experience. *Concurrency and Computation: Practice and Experience*, 17(2–4):323–356, 2005.
97. A.N. Toosi, R.N. Calheiros, R.K. Thulasiram, and R. Buyya. Resource provisioning policies to increase iaas provider's profit in a federated cloud environment. In *Proceeding of 13th International Conference on High Performance Computing and Communications (HPCC)*, pp. 279–287, Banff, Canada, 2011.
98. A.N. Toosi, R.K. Thulasiram, and R. Buyya. Financial option market model for federated cloud environments. Technical report, Department of Computing and Information System, Melbourne University, Melbourne, Australia, 2012.
99. D. Tsafrir, Y. Etsion, and D.G. Feitelson. Backfilling using system-generated predictions rather than user runtime estimates. *IEEE Transactions on Parallel and Distributed Systems*, 18(6):789–803, 2007.
100. B. Urgaonkar and P. Shenoy. Sharc: Managing cpu and network bandwidth in shared clusters. *IEEE Transactions on Parallel and Distributed Systems*, 15(1):2–17, 2004.
101. H.N. Van, F.D. Tran, and J.M. Menaud. Sla-aware virtual resource management for cloud infrastructures. In *Proceedings of the 9th Conference on Computer and Information Technology, vol. 02*, pp. 357–362, Xiamen, China. IEEE Computer Society, 2009.
102. C. Vázquez, E. Huedo, R.S. Montero, and I.M. Llorente. Federation of TeraGrid, EGEE and OSG infrastructures through a metascheduler. *Future Generation Computing Systems*, 26(7):979–985, 2010.
103. J.L. Vázquez-Poletti, E. Huedo, R.S. Montero, and I.M. Llorente. A comparison between two grid scheduling philosophies: EGEE WMS and grid way. *Multiagent Grid Systems*, 3:429–439, December 2007.
104. S. Venugopal, R. Buyya, and L. Winton. A grid service broker for scheduling e-science applications on global data grids. *Concurrency and Computation: Practice and Experience*, 18(6):685–699, 2006.

105. A. Verma, P. Ahuja, and A. Neogi. pMapper: Power and migration cost aware application placement in virtualized systems. In *Proceedings of the 9th ACM/IFIP/USENIX International Conference on Middleware*, pp. 243–264. Springer-Verlag, New York, Inc., 2008.
106. A. Verma, P. Ahuja, and A. Neogi. Power-aware dynamic placement of HPC applications. In *Proceedings of the 22nd Annual International Conference on Supercomputing*, pp. 175–184, Austin, TX, 2008.
107. W. Voorsluys, J. Broberg, S. Venugopal, and R. Buyya. Cost of virtual machine live migration in clouds: A performance evaluation. *Proceedings of the 1st International Conference on Cloud Computing*, pp. 254–265, 2009.
108. E. Walker, J.P. Gardner, V. Litvin, and E.L. Turner. Creating personal adaptive clusters for managing scientific jobs in a distributed computing environment. In *Proceedings of the Challenges of Large Applications in Distributed Environments (CLADE'06)*, pp. 95–103, Paris, France. IEEE, 2006.
109. J.P. Walters, B. Bantwal, and V. Chaudhary. Enabling interactive jobs in virtualized data centers. *Cloud Computing and Applications*, 2008.
110. S. Yi, D. Kondo, and A. Andrzejak. Reducing costs of spot instances via checkpointing in the amazon elastic compute cloud. In *Proceedings of 3rd IEEE International Conference on Cloud Computing (CLOUD)*, pp. 236–243, Miami, FL, 2010.
111. H. Zhong and J. Nieh. Crak: Linux checkpoint/restart as a kernel module. Technical report, Department of Computer Science, Columbia University, New York, 2001.
112. M. Zwahlen. Managing contention in collaborative resource sharing systems using token-exchange mechanism. PhD thesis, Royal Institute of Technology (KTH), Stockholm, Sweden, 2007.

58

Real-Time Computing

Sanjoy Baruah
The University of North Carolina

58.1 Introduction

The discipline of real-time computing studies computer systems in which certain computations have both a logical and a temporal notion of correctness: the correct result must be produced, at the right time. As computer systems come to play an increasingly important role in monitoring and controlling ever more aspects of the physical world and our interactions with it, principles of real-time computing are fast becoming central to broad areas of computer science and engineering, including embedded systems, cyber-physical systems, electronic commerce, multimedia, communication and networking, and robotics.

This discipline grew out of a need to be able to ensure provably correct behavior in safety-critical embedded systems such as the ones found in aerospace and armaments applications. The focus initially was therefore on the worst-case analysis of very simple systems that are restricted to execute in tightly controlled environments, with the objective of ensuring that all timing constraints are always met. As more and more application domains have come to explicitly recognize the real-time nature of their requirements, the kinds of systems falling within the scope of the discipline of real-time computing have broadened considerably in many ways: systems subject to real-time analysis are no longer simple but display significant complexity and heterogeneity in composition and in requirements, and are expected to operate correctly in relatively unpredictable environments. This expansion in scope has considerably enriched the discipline, rendering it far more challenging and interesting.

58.2 Underlying Principles

A real-time system is specified by specifying a *workload*, and a *platform* upon which this workload is to execute. The workload is often modeled as consisting of a number of *jobs*, each of which has a *release time*, an *execution requirement*, and a *deadline*. The interpretation is that the job needs to execute for an amount equal to its execution requirement within a *scheduling window* that is the interval between its release time and its deadline. If a workload consists of a finite number of jobs that are all known beforehand, it may sometimes be specified explicitly. More commonly, however, real-time systems are *reactive* [26] in the sense that they maintain an ongoing interaction with their environment at a speed determined by the environment: they obtain information about the environment using sensors and may seek to change the environment via actuators. For such reactive systems, *constraints* of various kinds (such as precedence constraints, periodicity constraints, minimum inter-arrival separation constraints, and mutual exclusion constraints) are specified for the kinds of (potentially infinite) collections of jobs that are *legal* for a given system. Such constraints are typically specified within the context of a *formal model* for representing a real-time workload—some commonly used models are described in Section 58.2.4. A *schedule* for a real-time system specifies which jobs are to execute on each part of the platform at each instant in time. A *scheduling algorithm* generates a schedule.

58.2.1 Hard versus Soft Real-Time Systems

For a *hard real-time* (*HRT*) system, it is imperative that all deadlines always be met; in *soft real-time* (*SRT*) systems, it is acceptable to miss an occasional deadline (the consequences of doing so are usually quantitatively specified as part of the system model). A special kind of SRT system is a *firm real-time* system, in which there is a reward associated with completing each job by its deadline (but no penalty for failing to do so), and the objective of a scheduling algorithm is to maximize the cumulative reward obtained.

58.2.2 Pre-Run-Time Analysis

One of the salient features of the discipline of real-time computing is its strong emphasis on the analysis of systems prior to their deployment. This is particularly true for safety-critical systems: such systems must be demonstrated to always exhibit safe behavior during run time before they can be deployed. (In some application domains, including civilian aviation and medical devices, prior certification of such safe behavior may be required from statutory *certification authorities*.) Pre-run-time analysis may take different forms, including the use of theorem-provers, model-checking, and the application of rigorously derived properties of resource-allocation and scheduling algorithms. Most of the analysis techniques considered in this chapter fall into the latter of these possible approaches.

A given schedule is said to be *feasible* for a workload that is specified as a collection of jobs upon a specified platform if it provides sufficient execution for each job to complete within its scheduling window. A specified collection of jobs is feasible upon a specified platform if there exists a feasible schedule for it. A specified collection of jobs is said to be *schedulable* by a given scheduling algorithm if the algorithm produces a feasible schedule.

The aforementioned definitions generalize from workloads that are specified as collections of jobs to workloads that are specified using more general models in the natural way: a workload is feasible if there is a feasible schedule for every legal collection of jobs that is consistent with the specified constraints, and it is schedulable by a given algorithm if the algorithm finds a feasible schedule for every such legal collection of jobs.

With regard to workloads that are specified according to a particular model (see Section 58.2.4 for a listing of some such models) and that are to be implemented upon a specified platform, a *schedulability test* for a given scheduling algorithm is an algorithm that takes as input a workload specification and provides as output an answer as to whether the workload is schedulable by the given scheduling algorithm. A schedulability test is *exact* if it correctly identifies all schedulable and unschedulable task systems. It is *sufficient* if it correctly identifies all unschedulable task systems, but may misidentify some schedulable systems as being unschedulable.

For any scheduling algorithm to be useful for HRT applications, it must have at least a sufficient schedulability test that can verify that a given workload is schedulable. The quality of the scheduling algorithm and the schedulability test are thus inseparable for HRT systems.

58.2.3 Time-Triggered versus Event-Triggered Scheduling

In designing scheduling algorithms, two major alternative paradigms have emerged. In a *time-triggered* (*TT*) schedule, all scheduling activities are initiated by the progress of time. That is, the entire schedule is predetermined before the system is deployed, and all scheduling activities are initiated ("triggered") at prespecified points in time. TT schedules are typically (although not always) stored as lookup tables and are suitable for use in highly deterministic systems, in which the entire real-time workload can be completely characterized as a particular collection of jobs prior to system run time. For systems that are not completely deterministic, the *event-triggered* (*ET*) paradigm is more appropriate. In an ET schedule, it is *events*, rather than the passage of time, that initiate scheduling activities. These events can be external (such as the release of a job or the arrival of an interrupt signal) or internal (e.g., the completion of the execution of a currently executing job). ET scheduling is able to deal with some kinds of incompletely specified workloads in a more resource-efficient manner than TT scheduling.

58.2.4 Models for Recurrent Tasks

The workload of a real-time system is often modeled as consisting of a finite number of recurrent *tasks* or processes. Each such task usually represents code embedded within an infinite loop: each iteration of the loop gives rise to one or more jobs.

The *periodic task model* [39] is the simplest model for representing such recurrent tasks. Each task τ_i in this model is characterized by a *worst-case execution time* (*WCET*) parameter C_i and a *period* T_i, with the interpretation that this task releases a job at each integer multiple of T_i, with this job having a deadline that is T_i time units after its release and needing to execute for up to C_i time units by this deadline. A generalization to the model allows for the specification of an additional parameter O_i called the *initial offset*, with the interpretation that the first job is released at time O_i and subsequent jobs T_i time units apart. A further generalization [36,37] allows for the specification of a *relative deadline* parameter D_i: the deadline of a job is D_i time units after its release (rather than T_i time units). In this most general form, thus, each task $\tau_i = (O_i, C_i, D_i, T_i)$ releases a job with WCET equal to C_i at time instant $O_i + k \cdot T_i$, and deadline at time-instant $O_i + k \cdot T_i + D_i$, for each integer $k \geq 0$.

Recall that the release of a job may be triggered by the occurrence of an external event. In many real-time systems, such external events may not occur periodically (i.e., exactly T_i time units apart); modeling the task that is responsible for processing such events as a periodic task therefore results in a loss of accuracy. If it is, however, possible to provide a *lower bound* on the minimum amount of time that must occur between the occurrence of successive events, the *sporadic task model* [42] is able to model the processing of these events. In this model, a task τ_i is again characterized by the same four-tuple of parameters: $\tau_i = (O_i, C_i, D_i, T_i)$; however, the interpretation of the first and fourth parameters (O_i and T_i) is different. Specifically, such a task generates a sequence of jobs with the release time of the first job $\geq O_i$, and the release times of successive jobs separated by an interval of duration $\geq T_i$; each job has a WCET C_i and a deadline D_i time units after its release time.*

There is a classification of real-time systems consisting entirely of sporadic tasks, depending on the relationship between the D_i and T_i parameters. In this classification,

- Any task system consisting entirely of sporadic tasks is called an *arbitrary-deadline* sporadic task system
- In *constrained-deadline* sporadic task systems, $D_i \leq T_i$ for all tasks τ_i
- In *implicit-deadline* sporadic task systems, $D_i = T_i$ for all tasks τ_i

(It is evident from these definitions that each implicit-deadline sporadic task system is also a constrained-deadline sporadic task system, and each constrained-deadline sporadic task system is also an arbitrary-deadline sporadic task system.)

Several models that are more general than the ones described earlier have also been proposed for representing recurrent real-time task systems; these include the multiframe model [43], the generalized multiframe model [8], the recurring real-time task model [7], and the digraph task model [54]. Models incomparable to the ones discussed earlier include the pinwheel model [28], and various models used for representing firm-real-time task systems such as the imprecise computation model [38] and the *(m, k)*-task model [25].

58.2.5 Priority-Driven Scheduling Algorithms

Most run-time scheduling algorithms are implemented as follows. At every instant during run time, each job that is eligible to execute is assigned a *priority* (either explicitly or implicitly), and the highest-priority jobs are selected for execution. Depending upon the manner in which priorities are assigned, algorithms for scheduling recurrent tasks can be classified [17] into three major categories:

1. In fixed-task-priority (FTP) scheduling algorithms, each recurrent task gets assigned a distinct priority and each job gets the priority of the task that generated it. Common examples of FTP scheduling algorithms include the *rate-monotonic* (*RM*) algorithm, in which tasks are assigned priorities according to their period parameters (tasks with smaller periods get greater priority), and the *deadline monotonic* (*DM*) algorithm, in which priorities are assigned according to the relative deadline parameters (tasks with smaller relative deadline are assigned greater priority).
2. In fixed-job-priority (FJP) scheduling algorithms, different jobs of the same task may be assigned different priorities; however, each job gets assigned exactly one priority upon its release and this priority may not change until the job has completed execution. The *earliest deadline first algorithm* (*EDF*), in which jobs with earlier deadlines are assigned greater priority (with ties broken arbitrarily), is a common example of an FJP scheduling algorithm.
3. There are no restrictions on the priority-assignment procedure for dynamic priority (DP) scheduling algorithms: different jobs of the same task may have different priorities and the priority of a job may change arbitrarily often. The *least laxity* (*LL*) algorithm, in which the priority assigned to a job depends upon the difference between the amount of time remaining to its deadline and its remaining execution, is an example of a DP scheduling algorithm: under LL scheduling, an executing job does not have its priority change, whereas a job that is awaiting execution sees its priority increase.

* In specifying sporadic task systems, the first parameter—the initial offset O_i—is often left unspecified; it seems that knowledge of this parameter does not significantly help in determining scheduling algorithms for systems of sporadic tasks.

58.2.6 Synchronous Real-Time Systems

The synchronous paradigm of real-time computation [24] posits that all actions execute instantaneously. In practice, this requires that *the program reacts rapidly enough to perceive all external events in suitable order* [24, p. 6]. For this paradigm to be applicable, it is therefore necessary that an atomic time-slice be sufficiently large (equivalently, that the clock used be sufficiently coarse-grained to allow any action that begins execution at a clock-tick to complete execution, and have its results communicated to where it may be needed, by the next clock-tick). Synchronous programming languages such as ESTEREL [12] have been developed for implementing real-time systems when these conditions are satisfied.

Although synchronous programming is an extremely powerful programming technique for specifying and proving the correctness of complex real-time systems, it is difficult to obtain implementations of such systems that make efficient use of platform resources; in essence, this is because the time-granularity at which platform resources may be allocated corresponds to the maximum amount of time needed for any action to complete execution. In their purest forms, the synchronous paradigm and the priority-driven one (which is the focus of much of this chapter) can be thought of as representing two alternative approaches: by using the synchronous paradigm it is easier to specify a real-time application but more difficult to obtain a resource-efficient implementation, whereas specifying an application using the task system models discussed in this chapter may be more difficult, but the resulting abstract model can be implemented in a very resource-efficient manner.

58.2.7 Execution Platform

In large part, the discipline of real-time systems has focused on optimizing the use of the processor (often called the CPU); additional platform components (memory, bandwidth, energy, etc.) have received much less attention. This is beginning to change; in Section 58.4, we will see how some of these additional resources give rise to a number of important and interesting research directions.

The execution model can be preemptive or nonpreemptive. In *preemptive* scheduling, an executing job may have its execution interrupted in order to execute some other job, with the interrupted job resuming execution later. In nonpreemptive scheduling, by contrast, an executing job may not be interrupted until it has completed execution.

With regard to the processing resource, both uniprocessor and multiprocessor platforms have been considered. A uniprocessor platform consists of a single CPU upon which the entire real-time workload is required to execute. Real-time systems were initially primarily implemented on uniprocessor platforms, but the advent of multicore technologies prompted many real-time researchers to begin investigating multiprocessor-related resource allocation problems. Given a system of recurrent tasks and a number of processors, it is possible to divide the processors into clusters and assign each task to a cluster. A scheduling algorithm is then applied locally in each cluster to the jobs generated by the tasks that are assigned to the cluster. The following two extremes of clustering have attracted the most attention from the real-time research community:

- *Partitioned scheduling*: Each processor is a cluster of size one. Historically, this has been the first step in extending single-processor analysis techniques to the multiprocessor domain.
- *Global scheduling*: There is only one cluster containing all the processors.

When resources other than the processors are taken into consideration, it is sometimes useful to consider cluster sizes other than these two extremes. For example, many multicore CPUs have on-chip multilevel cache memory that is shared among different sets of processing cores; it may be beneficial to consider cores that share a common cache as a single cluster.

A classification of multiprocessor platforms considers the relationship between the different processors:

- *Identical multiprocessors*: All the processors exhibit the same characteristics.
- *Uniform multiprocessors*: Some processors are "faster" than others, in the sense that a job would require less time to execute on the faster processor. All jobs experience the same speedup upon the faster processor.

  The uniform multiprocessor model is useful for modeling platforms in which all the processors are of the same family of processors, but some may be of a newer generation than others. Alternatively, it may be the case that only a fraction of the computing capacity of each processor is available to the real-time workload (and this fraction is not the same for all the processors).
- *Unrelated multiprocessors*: Different jobs experience different speedups on different processors.

The unrelated multiprocessor model models multiprocessor platforms consisting of different kinds of processors (such as DSPs and GPUs). Depending on what they are doing, different jobs may be expected to have different execution rates on different processors—for example, a graphics-intensive job would probably experience greater speedup on a GPU than some other jobs. As such platforms become ever more common (particularly on multicore CPUs), the unrelated multiprocessor model is expected to become more important for the modeling of real-time application systems.

58.3 Best Practices

58.3.1 Scheduling on Uniprocessors

The scheduling of recurrent task systems on uniprocessor platforms is a well-studied subject. Much of the work has focused on preemptive scheduling; however, there are also several interesting and important results concerning nonpreemptive scheduling.

58.3.1.1 Earliest Deadline First

One of the seminal results in uniprocessor scheduling is that *EDF is optimal for scheduling collections of independent jobs on a preemptive uniprocessor*, in the sense that if a given collection of jobs can be scheduled on a preemptive processor such that each job will complete by its deadline, then scheduling jobs according to their deadlines will also complete each job by its deadline [18,39]. This optimality result also holds for real-time workloads modeled using any of the models described in Section 58.2.4.

58.3.1.2 FTP Scheduling

EDF is an example of a FJP scheduling algorithm (see Section 58.2.5). A couple of FTP algorithms are also very widely used: *rate-monotonic* (*RM*) and *deadline-monotonic* (*DM*). In RM scheduling [39], a task is assigned a priority depending on its period parameter T_i: the smaller the period, the greater the priority (with ties broken arbitrarily). In DM scheduling [37], priorities are assigned depending on the relative deadline parameter D_i: the smaller the relative deadline, the greater the priority (with ties again broken arbitrarily). Observe that for *implicit-deadline* sporadic task systems, RM and DM assign priorities in an identical manner (since $T_i = D_i$ for all tasks in such systems). The following results are known concerning RM and DM scheduling:

1. DM is an optimal FTP algorithm for scheduling constrained-deadline sporadic task systems on a preemptive uniprocessor, in the sense that if a given collection of constrained-deadline sporadic task is schedulable by any FTP algorithm then it is also schedulable by DM [37].
2. DM is also an optimal FTP algorithm for scheduling periodic task systems satisfying the additional properties that (a) all the offset parameters (the O_i's) are equal—periodic task systems satisfying this property are called *synchronous* periodic task systems; and (b) $D_i \leq T_i$ for each task τ_i.

3. A direct corollary of the aforementioned results is that RM is optimal for implicit-deadline sporadic task systems and for synchronous periodic task systems with $D_i = T_i$ for all τ_i; this result was previously proved in [39].
4. DM and RM are *not* optimal for asynchronous periodic task systems, and for sporadic task systems in which D_i may exceed T_i for some tasks τ_i.

58.3.1.3 Shared Nonpreemptable Resources

The uniprocessor platforms upon which real-time systems execute may include nonpreemptable serially reusable resources in addition to the shared processor; these could be software resources such as shared data structures or hardware resources such as sensors and actuators. Access to such shared resources is typically made within critical sections of code that are guarded using mutual exclusion primitives such as semaphores.

Priority-based scheduling algorithms have been extended to handle resource-sharing of this kind. Under *priority inheritance* protocols [49] for resource-sharing, a *blocking job*—that is, a job that holds a nonpreemptable resource and thereby prevents the execution of a higher-priority job—has its scheduling priority temporarily elevated to the priority of the blocked job. A particular kind of a priority inheritance protocol, called the *priority ceiling protocol* (*PCP*), restricts the conditions under which a job is eligible to begin execution. Another version—the *Stack Resource Policy* (*SRP*) [3]—defines a different set of restrictions. It has been shown that PCP is optimal for the FTP scheduling and that SRP is optimal for the FJP scheduling (and, therefore, also for FTP scheduling), of sporadic task systems on a platform consisting of a single preemptive processor and additional nonpreemptive serially reusable resources.

58.3.2 Scheduling on Multiprocessors

There has recently been a considerable amount of effort devoted to obtaining better algorithms for scheduling real-time systems on platforms containing multiple processors. Much of this work studies partitioned and global preemptive scheduling of HRT workloads represented using the sporadic task model (Section 58.2.4) and implemented upon identical multiprocessors; a few results concerning scheduling on uniform and unrelated multiprocessors have also been obtained.

58.3.2.1 Partitioned Scheduling on Identical Multiprocessors

The partitioned scheduling of a collection of recurrent tasks can be thought of as a two-step process. The first step occurs prior to run time: during this step, each task in the collection needs to be assigned to one of the processors. This step reduces the multiprocessor problem to a series of uniprocessor problems, one on each processor—the subsequent step solves these uniprocessor problems using, for example, the techniques discussed in Section 58.3.1.1. In the following, we therefore focus on the partitioning step.

It is evident that the partitioning step is a generalization of the *bin-packing* problem, which is known to be NP-hard in the strong sense. Therefore, most positive results concerning partitioned scheduling concern approximation algorithms. Several such positive results have been obtained for partitioning implicit-deadline sporadic task systems. Lopez et al. [41] have proposed a partitioning algorithm based on the first-fit decreasing bin-packing heuristic and have obtained a sufficient schedulability condition when the subsequent partitioning step is done using EDF; similar sufficient schedulability conditions for RM scheduling have also been obtained [40,44]. In more theoretical work, a polynomial-time approximation scheme (PTAS) for partitioned EDF scheduling is easily derived from the results in [27], while a 3/2-approximation algorithm for partitioned RM scheduling was presented in [29].

The partitioned scheduling of constrained- and arbitrary-deadline sporadic task systems is less well studied. A polynomial-time partitioning algorithm called *first-fit with increasing deadline* (*FFID*) was proposed in [10] for partitioned EDF scheduling; a variant for partitioned DM scheduling was derived in [21].

58.3.2.2 Global Scheduling on Identical Multiprocessors

The global scheduling of implicit-deadline sporadic task systems can be done optimally using a dynamic priority (DP) scheduling algorithm called *pfair* [9] and its variants and refinements.

As stated in Section 58.3.1, upon preemptive uniprocessors EDF is an optimal FJP algorithm, and DM is an optimal FTP algorithm for constrained-deadline task systems. These optimality results do not hold for global multiprocessor scheduling; in fact, a phenomenon commonly called *Dhall's effect* [20] means that such algorithms can perform arbitrarily poorly when used for global scheduling. Phillips et al. [46] studied the global preemptive EDF scheduling of collections of independent jobs; their results have spawned a host of results scheduling algorithms and schedulability tests for the FJP and FTP global scheduling of implicit-deadline, constrained-deadline, and arbitrary-deadline sporadic task systems on identical multiprocessor platforms. These include FTP algorithms in which priorities are assigned using strategies other than RM or DM (see, e.g., [1,2]), as well as *hybrid* scheduling algorithms in which some tasks' jobs are assigned highest priority while remaining tasks' jobs have priorities assigned according to EDF (e.g., [23]). Various sufficient schedulability conditions for EDF and FTP scheduling (with a given priority assignment) have been derived [4,5,13,45].

58.3.2.3 Scheduling on Uniform and Unrelated Multiprocessors

As is evident from the earlier sections, considerable progress has been made on better understanding the scheduling of sporadic task systems on identical multiprocessors. Far less work has been done concerning real-time scheduling in uniform and unrelated multiprocessors. Funk's PhD dissertation [22] has a series of results for global and partitioned scheduling of implicit-deadline sporadic task systems on uniform multiprocessors. On unrelated multiprocessors, results from [33,34,51] can be applied to obtain polynomial-time approximation algorithms for partitioned scheduling of implicit-deadline sporadic task systems on unrelated multiprocessors; results from [31] can be used to obtain a polynomial-time algorithm for global scheduling of implicit-deadline sporadic task systems on unrelated multiprocessors. Global scheduling of arbitrary-deadline sporadic task systems on uniform processors is considered in Ref. [11].

58.3.3 Verification and Certification

As stated in Section 58.2.2, pre run-time analysis is often very important in real-time computing, particularly for safety-critical systems. Considerable effort has been devoted to the design of tools and methodologies for the formal verification of safety-critical real-time systems; these may involve the use of logic-based theorem provers, model-checking, automata-based tools, etc.

In safety-critical systems, *certification* refers to the process of demonstrating the correctness (or at least, safety) of a safety-critical system to a statutory authority. For example, aircraft are generally required to obtain an *airworthiness certificate* from the Federal Aviation Administration in order to be authorized to operate within US civilian airspace. Certification authorities (CAs) are entrusted by governments or other entities with certification responsibility over specified application domains; these CAs establish standards that must be met in order to obtain certification. Coming up with real-time system design processes that ameliorate the effort needed to obtain certification is an active ongoing research area, particularly for *mixed criticality* [6] systems: real-time systems in which different tasks perform functions that are of different degrees of importance (or criticality) to the overall performance of the system and are therefore subject to different certification requirements.

58.3.4 Operating Systems Issues

A real-time operating system (RTOS) provides system support for predictably satisfying real-time constraints. An RTOS differs from a traditional OS in that the time dimension is elevated to a central principle; in contrast, traditional OSes tend to abstract real time away to facilitate specification, design, and verification of systems. An RTOS needs to have a highly integrated time-cognizant resource allocation approach that provides support for meeting deadlines and for predictability; such a resource allocation approach must be built upon the foundational principles described in Sections 58.3.1 and 58.3.2. Some features typically found in current RTOSs* include the following:

- A real-time clock
- A priority scheduling mechanism
- A fast and predictable context switch
- A fast and predictable response to external interrupts
- Provides for special alarms and timeouts
- Permits tasks to invoke primitives to delay by a fixed amount of time and to pause/resume execution

Current RTOSs can be placed in one of two broad categories: those that are specifically built for real-time use and those that are obtained by adding some support for real time to general-purpose operating systems (various flavors of real-time Linux typically fall within this second category). A further classification of purpose-built RTOSs is proposed in [14, Section 2.5.1]:

- *Deeply embedded RTOSs*, for use in applications where resources are at a premium and mostly static, single-purpose workloads are prevalent. Such RTOS kernels are also sometimes called "executives."
- *Full-featured RTOSs*, for applications where UNIX-like OS abstractions and multitasking are desirable, and the system design can afford the additional resources required to support them (e.g., extra memory and faster processors). Such RTOSs may find use in, for example, factory automatization and telecommunication infrastructure.
- *Certification compliant RTOSs*, for use in safety-critical applications that are subject to mandatory certification requirements. RTOSs in this category, also called *separation kernels*, are commonly used in avionics applications, where government-mandated standards such as the DO-178B standard for avionics impose strict isolation of safety-critical subsystems from faults in non- or less-safety-critical subsystems. The RTOS itself may have obtained certification, thereby facilitating its adoption in applications subject to certification.

There are hundreds of small and large RTOS products and projects; see, for example, [53] and [14, Chapter 2.5] for discussions concerning some of them.

58.4 Research Issues

As real-time embedded systems become ever more important, time plays an increasingly more central role in the specification, design, implementation, and validation of computing systems. This phenomenon has led to some prominent researchers in real-time computing making a case (see, e.g., [32,52]) that the entire foundations of computing, starting from the Turing abstraction, need to be revisited to place time in a more central role. Of course, such a radical rethink of the discipline of computing gives rise to a host of deep open questions; these questions, although important, will not be addressed further in this section—the interested reader is encouraged to read the cogent and convincing arguments put forth by Lee [32] and Sifakis [52], and consider their proposed avenues toward solving the problems identified.

Even from a less radical perspective, though, there are many interesting and important unresolved research issues in many areas of real-time computing. We list some of these areas next.

* The term RTOS is often applied to minimal custom-designed run-time environments that are developed for use in specific embedded systems; the discussion during the remainder of this section does exclude such RTOSs.

58.4.1 Scheduling

In *real-time scheduling theory,* researchers are working on developing algorithms and techniques for implementing real-time workloads on *heterogeneous* platforms that consist of different kinds of processing elements. Determining efficient scheduling strategies for meeting specified *end-to-end deadlines* on distributed platforms is also an active research area.

58.4.2 Programming Languages

The design of *specification formalisms* and *programming languages* for specifying and implementing real-time systems is an active research area. This research aims to resolve the ongoing tension between the need for a formalism to be adequately expressive and easy to use, while simultaneously being implementable in a resource-efficient manner. If a programming model is too close to an abstract specification, then it is difficult to obtain an efficient implementation. On the other hand, if a programming model is too close to the execution platform, then programming becomes a challenge. (This tension was highlighted in Section 58.2.6 with respect to the synchronous paradigm of real-time computing, which is easier to program in vis a vis a priorities-based programming model, but typically results in less efficient use of platform resources.)

58.4.3 Real-Time Networking

Since embedded real-time systems are typically implemented upon distributed platforms, it is imperative that communication between the different nodes of the platform be guaranteed to occur in real time. The discipline of *real-time networking* studies the challenges associated in providing such communication.

58.4.4 Energy and Thermal Issues

There was initially an expectation that Moore's law would eventually render resource-efficiency less of an issue, but that expectation has not been realized; instead, resources *other* than processing or memory capacity have become the scarce ones. One particularly important resource in this regard is the *energy* consumed by the real-time system, particularly in mobile embedded systems that are not tethered to the electricity grid. A related concern is ensuring that *temperature* does not rise too much, since the overheating of hardware components can change their behavioral characteristics and compromise their reliability and longevity. *Energy-aware* and *thermal-aware* real-time computing are very active research areas today.

58.4.5 Real-Time Operating Systems

New approaches for implementing RTOSs that are more general (i.e., offer services closer to those offered by general-purpose OSs) and are known to be correct to a high level of assurance (e.g., by being certified correct) are another active research area. There is much work on adapting the server or container abstraction, which allows for multiple applications of different criticalities to coexist on a shared platform, for use in RTOSs.

58.4.6 Soft Real-Time Computing

Much of the discussion in this chapter has been about HRT systems; this reflects the current focus of the community. However, SRT considerations are becoming increasingly more significant, as ever more SRT applications (such as multimedia and gaming) pay increasing attention to temporal issues.

Many different notions of SRT have been explored in some depth; the *time-utility function* approach [48] and the *bounded tardiness* approach [19,35] merit special mention for the comprehensive nature of the research that has been done and the results that have been obtained.

Further Information

The textbooks by Burns and Wellings [15], and Buttazzo [16], provide further information about many of the concepts introduced in this chapter. Kopetz [30] provides in-depth coverage of the concepts and results on time-triggered scheduling; the text by Halbwachs [24] is an introduction to the synchronous paradigm of real-time computing.

Proceedings of research conferences serve to provide an up-to-date snapshot of the current state of knowledge in the discipline and can be a useful source for recent results. The *IEEE Real-Time Systems Symposium* (RTSS), which has been held annually since 1980, is one of the flagship academic research conferences in real-time computing. The *EuroMicro Conference on Real-Time Systems* (ECRTS) and the *ACM International Conference in Embedded Software* (EMSOFT) are other important annual conferences. The archival journal *Real-Time Systems* is, as its name suggests, devoted to publishing research and survey articles on real-time computing principles and applications. The November 2004 issue of this journal is a special anniversary issue containing invited survey papers on a number of important topics in real-time computing, including scheduling theory [50], the issue of predictability in real-time computing [55], real-time databases [47], and real-time operating systems [53].

References

1. B. Andersson, S. Baruah, and J. Jansson. Static-priority scheduling on multiprocessors. In *Proceedings of the IEEE Real-Time Systems Symposium*, pp. 193–202, London, U.K., December 2001. IEEE Computer Society Press, Los Alamitos, CA.
2. B. Andersson and J. Jonsson. Fixed-priority preemptive multiprocessor scheduling: To partition or not to partition. In *Proceedings of the International Conference on Real-Time Computing Systems and Applications*, pp. 337–346, Cheju Island, South Korea, December 2000. IEEE Computer Society Press, Washington, DC.
3. T. Baker. Stack-based scheduling of real-time processes. *Real-Time Systems: The International Journal of Time-Critical Computing*, 3:67–99, 1991.
4. T. Baker. Multiprocessor EDF and deadline monotonic schedulability analysis. In *Proceedings of the IEEE Real-Time Systems Symposium*, pp. 120–129, Cancun, Mexico, December 2003. IEEE Computer Society Press, Los Alamitos, CA.
5. T. Baker. An analysis of EDF schedulability on a multiprocessor. *IEEE Transactions on Parallel and Distributed Systems*, 16(8):760–768, 2005.
6. J. Barhorst, T. Belote, P. Binns, J. Hoffman, J. Paunicka, P. Sarathy, J. S. P. Stanfill, D. Stuart, and R. Urzi. White paper: A research agenda for mixed-criticality systems, April 2009. Available at http://www.cse.wustl.edu/~ cdgill/CPSWEEK09_MCAR (accessed on October 9, 2013).
7. S. Baruah. Dynamic- and static-priority scheduling of recurring real-time tasks. *Real-Time Systems: The International Journal of Time-Critical Computing*, 24(1):99–128, 2003.
8. S. Baruah, D. Chen, S. Gorinsky, and A. Mok. Generalized multiframe tasks. *Real-Time Systems: The International Journal of Time-Critical Computing*, 17(1):5–22, July 1999.
9. S. Baruah, N. Cohen, G. Plaxton, and D. Varvel. Proportionate progress: A notion of fairness in resource allocation. *Algorithmica*, 15(6):600–625, June 1996.
10. S. Baruah and N. Fisher. The partitioned multiprocessor scheduling of sporadic task systems. In *Proceedings of the IEEE Real-Time Systems Symposium*, Miami, FL, pp. 321–329, December 2005. IEEE Computer Society Press, Los Almitos, CA.

11. S. Baruah and J. Goossens. The EDF scheduling of sporadic task systems on uniform multiprocessors. In *Proceedings of the Real-Time Systems Symposium*, pp. 367–374, Barcelona, Spain, December 2008. IEEE Computer Society Press, Los Almitos, CA.
12. G. Berry and G. Gonthier. The ESTEREL synchronous programming language: Design, semantics, implementation. *Science of Computer Programming*, 19:87–152, 1992.
13. M. Bertogna, M. Cirinei, and G. Lipari. Improved schedulability analysis of EDF on multiprocessor platforms. In *Proceedings of the EuroMicro Conference on Real-Time Systems*, pp. 209–218, Palma de Mallorca, Balearic Islands, Spain, July 2005. IEEE Computer Society Press, Washington, DC.
14. B. B. Brandenburg. Scheduling and locking in multiprocessor real-time operating systems. PhD thesis, The University of North Carolina at Chapel Hill, Chapel Hill, NC, 2011.
15. A. Burns and A. Wellings. *Real-Time Systems and Programming Languages*, 4th edn. Addison-Wesley, Boston, MA, 2009.
16. G. C. Buttazzo. *Hard Real-Time Computing Systems: Predictable Scheduling Algorithms and Applications*, 2nd edn. Springer, New York, 2005.
17. J. Carpenter, S. Funk, P. Holman, A. Srinivasan, J. Anderson, and S. Baruah. A categorization of real-time multiprocessor scheduling problems and algorithms. In J. Y.-T. Leung, ed., *Handbook of Scheduling: Algorithms, Models, and Performance Analysis*, pp. 30-1–30-19. CRC Press LLC, Boca Raton, FL, 2003.
18. M. Dertouzos. Control robotics: The procedural control of physical processors. In *Proceedings of the IFIP Congress*, pp. 807–813, Stockholm, Sweden, 1974.
19. U. Devi. Soft real-time scheduling on multiprocessors. PhD thesis, Department of Computer Science, The University of North Carolina at Chapel Hill, Chapel Hill, NC, 2006.
20. S. Dhall. Scheduling periodic time-critical jobs on single processor and multiprocessor systems. PhD thesis, Department of Computer Science, The University of Illinois at Urbana-Champaign, Champaign, IL, 1977.
21. N. Fisher, S. Baruah, and T. Baker. The partitioned scheduling of sporadic tasks according to static priorities. In *Proceedings of the EuroMicro Conference on Real-Time Systems*, Dresden, Germany, pp. 118–127, July 2006. IEEE Computer Society Press Washington, DC.
22. S. Funk. EDF scheduling on heterogeneous multiprocessors. PhD thesis, Department of Computer Science, The University of North Carolina at Chapel Hill, Chapel Hill, NC, 2004.
23. J. Goossens, S. Funk, and S. Baruah. Priority-driven scheduling of periodic task systems on multiprocessors. *Real Time Systems*, 25(2–3):187–205, 2003.
24. N. Halbwachs. *Synchronous Programming of Reactive Systems*. Kluwer Academic Publishers, Dordrecht, the Netherlands, 1993.
25. M. Hamadaoui and P. Ramanathan. A dynamic priority assignment technique for streams with (m, k)-firm deadlines. In *Proceedings of the 23rd Annual International Symposium on Fault-Tolerant Computing* (*FTCS '94*), Austin, TX, pp. 196–205, June 1994. IEEE Computer Society Press, Washington, DC.
26. D. Harel and A. Pnueli. On the development of reactive systems. In Krzysztof R. Apt (ed.), *Logics and Models of Concurrent Systems*, pp. 477–498, Springer-Verlag, New York, 1985.
27. D. Hochbaum and D. Shmoys. Using dual approximation algorithms for scheduling problems: Theoretical and practical results. *Journal of the ACM*, 34(1):144–162, January 1987.
28. R. Holte, A. Mok, L. Rosier, I. Tulchinsky, and D. Varvel. The pinwheel: A real-time scheduling problem. In *Proceedings of the 22nd Hawaii International Conference on System Science*, pp. 693–702, Kailua-Kona, HI, January 1989.
29. A. Karrenbauer and T. Rothvoss. A 3/2-approximation algorithm for rate-monotonic multiprocessor scheduling of implicit-deadline tasks. In *Approximation and Online Algorithms—8th International Workshop*, pp. 166–177, Liverpool, U.K., 2010.
30. H. Kopetz. *Real-Time Systems: Design Principles for Distributed Embedded Applications*. Springer, New York, 2011.

31. E. Lawler and J. Labetoulle. On preemptive scheduling of unrelated parallel processors by linear programming. *Journal of the ACM*, 25(4):612–619, October 1978.
32. E. A. Lee. Computing needs time. *Communications of the ACM*, 52(5):70–79, May 2009.
33. J.-K. Lenstra, D. Shmoys, and E. Tardos. Approximation algorithms for scheduling unrelated parallel machines. In A. K. Chandra, ed., *Proceedings of the 28th Annual Symposium on Foundations of Computer Science*, pp. 217–224, Los Angeles, CA, October 1987. IEEE Computer Society Press, Washington, DC.
34. J.-K. Lenstra, D. Shmoys, and E. Tardos. Approximation algorithms for scheduling unrelated parallel machines. *Mathematical Programming*, 46:259–271, 1990.
35. H. Leontyev and J. Anderson. Generalized tardiness bounds for global multiprocessor scheduling. *Real Time Systems*, 44:26–71, 2010.
36. J. Y.-T. Leung and M. Merrill. A note on the preemptive scheduling of periodic, real-time tasks. *Information Processing Letters*, 11:115–118, 1980.
37. J. Y.-T. Leung and J. Whitehead. On the complexity of fixed-priority scheduling of periodic, real-time tasks. *Performance Evaluation*, 2:237–250, 1982.
38. K.-J. Lin, J. Liu, and S. Natarajan. Scheduling real-time, periodic jobs using imprecise results. In *Proceedings of the Real-Time Systems Symposium*, San Francisco, CA, pp. 210–217, December 1987.
39. C. Liu and J. Layland. Scheduling algorithms for multiprogramming in a hard real-time environment. *Journal of the ACM*, 20(1):46–61, 1973.
40. J. M. Lopez, J. L. Diaz, and D. F. Garcia. Minimum and maximum utilization bounds for multiprocessor rate-monotonic scheduling. *IEEE Transactions on Parallel and Distributed Systems*, 15(7):642–653, 2004.
41. J. M. Lopez, J. L. Diaz, and D. F. Garcia. Utilization bounds for EDF scheduling on real-time multiprocessor systems. *Real-Time Systems: The International Journal of Time-Critical Computing*, 28(1):39–68, 2004.
42. A. Mok. Fundamental design problems of distributed systems for the hard-real-time environment. PhD thesis, Laboratory for Computer Science, Massachusetts Institute of Technology, Cambridge, MA, 1983. Available as Technical Report No. MIT/LCS/TR-297.
43. A. K. Mok and D. Chen. A multiframe model for real-time tasks. *IEEE Transactions on Software Engineering*, 23(10):635–645, October 1997.
44. D.-I. Oh and T. Baker. Utilization bounds for N-processor rate monotone scheduling with static processor assignment. *Real-Time Systems: The International Journal of Time-Critical Computing*, 15:183–192, 1998.
45. R. Pathan and J. Jonsson. Improved schedulability tests for global fixed-priority scheduling. In *Proceedings of the EuroMicro Conference on Real-Time Systems*, Porto, Portugal, pp. 136–147, July 2011. IEEE Computer Society Press, Washington, DC.
46. C. A. Phillips, C. Stein, E. Torng, and J. Wein. Optimal time-critical scheduling via resource augmentation. In *Proceedings of the 29th Annual ACM Symposium on Theory of Computing*, pp. 140–149, El Paso, TX, May 4–6, 1997.
47. K. Ramamritham, S. H. Son, and L. C. DiPippo. Real-time databases and data services. *Real-Time Systems: The International Journal of Time-Critical Computing*, 28(2–3):179–215, 2004.
48. B. Ravindran, E. D. Jensen, and P. Li. On recent advances in time/utility function real-time scheduling and resource management. In *Proceedings of the IEEE International Symposium on Object-Oriented Real-Time Distributed Computing*, Seattle, WA, pp. 55–60, 2005. IEEE Computer Society Press, Washington, DC.
49. L. Sha, R. Rajkumar, and J. P. Lehoczky. Priority inheritance protocols: An approach to real-time synchronization. *IEEE Transactions on Computers*, 39(9):1175–1185, 1990.
50. L. Sha, T. Abdelzaher, K.-E. Årzén, A. Cervin, T. Baker, A. Burns, G. Buttazzo, M. Caccamo, J. Lehoczky, and A. K. Mok. Real-time scheduling theory: A historical perspective. *Real-Time Systems: The International Journal of Time-Critical Computing*, 28(2–3):101–155, 2004.

51. D. Shmoys and E. Tardos. Scheduling unrelated machines with costs. In *Proceedings of the 4th Annual ACM-SIAM Symposium on Discrete Algorithms*, pp. 448–454, Austin, TX, 1993. ACM Press, New York.
52. J. Sifakis. A vision for computer science—The system perspective. *Central European Journal of Computer Science*, 1(1):108–116, 2009.
53. J. Stankovic and R. Rajkumar. Real-time operating systems. *Real-Time Systems: The International Journal of Time-Critical Computing*, 28(2–3):237–253, 2004.
54. M. Stigge, P. Ekberg, N. Guan, and W. Yi. The digraph real-time task model. In *Proceedings of the IEEE Real-Time Technology and Applications Symposium (RTAS)*, pp. 71–80, Chicago, IL, 2011. IEEE Computer Society Press, Washington, DC.
55. L. Thiele and R. Wilhelm. Design for timing predictability. *Real-Time Systems: The International Journal of Time-Critical Computing*, 28(2–3):157–177, 2004.

59

Scheduling for Large-Scale Systems

Anne Benoit
Ecole Normale Supérieure de Lyon

Loris Marchal
Ecole Normale Supérieure de Lyon

Yves Robert
Ecole Normale Supérieure de Lyon
University of Tennessee

Bora Uçar
Ecole Normale Supérieure de Lyon

Frédéric Vivien
Ecole Normale Supérieure de Lyon

59.1 Introduction

In this chapter, *scheduling* is defined as the activity that consists in mapping a task graph onto a target platform. The task graph represents the application, nodes denote computational tasks, and edges model precedence constraints between tasks. For each task, an assignment (choose the processor that will execute the task) and a schedule (decide when to start the execution) are determined. The goal is to obtain an efficient execution of the application, which translates into optimizing some objective function.

The traditional objective function in the scheduling literature is the minimization of the total execution time, or *makespan*. Section 59.2 provides a quick background on task graph scheduling, mostly to recall some definitions, notations, and well-known results.

In contrast to pure makespan minimization, this chapter aims at discussing scheduling problems that arise at large scale and involve one or several different optimization criteria, that depart from, or complement, the classical makespan objective. Section 59.3 assesses the importance of key ingredients of modern scheduling techniques: (1) multicriteria optimization; (2) memory management; and (3) reliability. These concepts are illustrated in the following sections, which cover four case studies. Section 59.4 is devoted to illustrate multicriteria optimization techniques that mix throughput, energy, and reliability. Sections 59.5 and 59.6 illustrate different memory management techniques to schedule large task graphs. Finally, Section 59.7 discusses the impact of checkpointing techniques to schedule parallel jobs at very large scale. We conclude this chapter by stating some research directions in Section 59.8, defining terms in key terms, and pointing to additional sources of information in further information.

59.2 Background on Scheduling

59.2.1 Makespan Minimization

Traditional scheduling assumes that the target platform is a set of p identical processors, and that no communication cost is paid. In that context, a task graph is a directed acyclic vertex-weighted graph $G = (V, E, w)$, where the set V of vertices represents the tasks, the set E of edges represents precedence constraints between tasks $\left(e = (u,v) \in E \text{ if and only if } u \prec v\right)$, and the weight function $w : V \to \mathbb{N}^*$ gives the weight (or duration) of each task. Task weights are assumed to be positive integers. A schedule σ of a task graph is a function that assigns a start time to each task: $\sigma : V \to \mathbb{N}^*$ such that $\sigma(u) + w(u) \leq \sigma(v)$ whenever $e = (u,v) \in E$. In other words, a schedule preserves the *dependence constraints* induced by the precedence relation $\prec$ and embodied by the edges of the dependence graph; if $u \prec v$, then the execution of u begins at time $\sigma(u)$ and requires $w(u)$ units of time, and the execution of v at time $\sigma(v)$ must start after the end of the execution of u. Obviously, if there was a cycle in the task graph, no schedule could exist, hence the restriction to acyclic graphs (DAGs).

There are other constraints that must be met by schedules, namely, *resource constraints*. When there is an infinite number of processors (in fact, when there are as many processors as tasks), the problem is *with unlimited processors*, and denoted $P\infty|prec|Cmax$ in the literature [25]. We use the shorter notation Pb(∞) in this chapter. When there is only a fixed number p of available processors, the problem is *with limited processors*, and denoted Pb(p). In the latter case, an allocation function $\texttt{alloc}: V \to \mathcal{P}$ is required, where $\mathcal{P} = \{1,\ldots, p\}$ denotes the set of available processors. This function assigns a target processor to each task. The resource constraints simply specify that no processor can be allocated more than one task at the same time:

$$\texttt{alloc}(T) = \texttt{alloc}\,(T') \Rightarrow \begin{cases} & \sigma(T) + w(T) \leq \sigma(T') \\ \text{or} & \sigma(T') + w(T') \leq \sigma(T'). \end{cases}$$

This condition expresses the fact that if two tasks T and T' are allocated to the same processor, then their executions cannot overlap in time.

The *makespan* MS(σ, p) of a schedule σ that uses p processors is its total execution time: $\mathrm{MS}(\sigma, p) = \max_{v \in V}\{\sigma(v) + w(v)\}$ (assuming that the first task(s) is (are) scheduled at time 0). The makespan is the total execution time, or finish time, of the schedule. Let $\mathrm{MS}_{opt}(p)$ be the value of the makespan of an optimal schedule with p processors: $\mathrm{MS}_{opt}(p) = \min_\sigma \mathrm{MS}(\sigma, p)$. Because schedules respect dependences, we have $\mathrm{MS}_{opt}(p) \geq w(\Phi)$ for all paths Φ in G (weights extend to paths in G as usual). We also have $\mathrm{Seq} \leq p \times \mathrm{MS}_{opt}(p)$, where $\mathrm{Seq} = \sum_{v \in V} w(v) = \mathrm{MS}_{opt}(1)$ is the sum of all task weights.

While Pb(∞) has polynomial complexity (simply traverse the graph and start each task as soon as possible using a fresh processor), problems with a fixed amount of resources are known to be difficult.

Letting *Dec*(p) be the decision problem associated with Pb(p), and *Indep – tasks*(p) the restriction of *Dec*(p) to independent tasks (no dependence, i.e., when $E = \emptyset$), well-known complexity results are summarized next:

- *Indep – tasks*(2) is NP-complete but can be solved by a pseudo-polynomial algorithm. Moreover, $\forall\, \varepsilon > 0$, *Indep – tasks*(2) admits a $(1 + \varepsilon)$-approximation whose complexity is polynomial in $1/\varepsilon$.
- *Indep – tasks*(p) is NP-complete in the strong sense.
- *Dec*(2) (and hence *Dec*(p)) is NP-complete in the strong sense.

Because Pb(p) is NP-complete, heuristics are used to schedule task graphs with limited processors. The most natural idea is to use greedy strategies: at each instant, try to schedule as many tasks as possible onto available processors. Such strategies deciding *not to deliberately keep a processor idle* are called *list scheduling* algorithms. Of course, there are different possible strategies to decide which tasks are given priority in the (frequent) case where there are more free tasks than available processors. Here a *free* task is a task whose predecessors have all been executed, hence which is available and ready for execution itself. But a key result due to Graham [26] is that any list algorithm can be shown to achieve at most twice the optimal makespan. More precisely, the approximation factor with p processors is $2p-1/p$, and this bound is tight.

A widely used list scheduling technique is *critical path scheduling*. The selection criterion for ready tasks is based on the value of their bottom level, defined as the maximal weight of a path from the task to an exit task of the graph. Intuitively, the larger the bottom level, the more "urgent" the task. The *critical path* of a task is defined as its bottom level and is used to assign priority levels to tasks. Critical path scheduling is list scheduling where the priority level of a task is given by the value of its critical path. Ties are broken arbitrarily.

59.2.2 Dealing with Communication Costs

Thirty years ago, communication costs have been introduced in the scheduling literature. Because the performance of network communication is difficult to model in a way that is both precise and conducive to understanding the performance of algorithms, the vast majority of results hold for a very simple model, which is defined as follows.

The target platform consists of p identical processors that are part of a fully connected clique. All interconnection links have same bandwidth. If a task T communicates data to a successor task T', the cost is modeled as

$$\mathrm{cost}(T, T') = \begin{cases} 0 & \text{if } \texttt{alloc}(T) = \texttt{alloc}(T') \\ c(T, T') & \text{otherwise,} \end{cases}$$

where $\texttt{alloc}(T)$ denotes the processor that executes task T, and $c(T,T')$ is defined by the application specification. The time for communication between two tasks running on the same processor is negligible. A schedule σ must still preserve dependences, a condition that now writes

$$\forall e = (T, T') \in E, \begin{cases} \sigma(T) + w(T) \le \sigma(T') & \text{if } \texttt{alloc}(T) = \texttt{alloc}(T') \\ \sigma(T) + w(T) + c(T, T') \le \sigma(T') & \text{otherwise.} \end{cases}$$

This so-called *macro-dataflow* model makes two main assumptions: (1) communication can occur as soon as data are available and (2) there is no contention for network links. Assumption (1) is reasonable as communication can overlap with (independent) computations in most modern computers. Assumption (2) is much more questionable (although it makes sense in a shared-memory environment). Indeed, there is no physical device capable of sending, say, 1000 messages to 1000 distinct processors, at the same speed as if there were a single message. In the worst case, it would take 1000 times longer (serializing all messages). In the best case, the output bandwidth of the network card of the sender would be a limiting factor. In other words, assumption (2) amounts to assuming infinite network resources. Nevertheless, this assumption is omnipresent in the traditional scheduling literature.

Including communication costs in the model makes everything difficult, including solving Pb(∞), which becomes NP-complete in the strong sense. The intuitive reason is that a trade-off must be found between allocating tasks to either many processors (hence balancing the load but communicating intensively) or few processors (leading to less communication but less parallelism as well). Even the problem in which all task weights and communication costs have the same (unit) value, the so-called UET-UCT problem (unit execution time-unit communication time) is NP-hard [40].

With limited processors, list heuristics can be extended to take communication costs into account, but Graham's bound does not hold any longer. For instance, the *Modified Critical Path (MCP)* algorithm proceeds as follows. First, bottom levels are computed using a pessimistic evaluation of the longest path, accounting for each potential communication (this corresponds to the allocation where there is a different processor per task). These bottom levels are used to determine the priority of free tasks. Then each free task is assigned to the processor that allows its earliest execution, given previous task allocation decisions. It is important to explain further what "previous task allocation decisions" means. Free tasks from the queue are processed one after the other. At any moment, it is known which processors are available and which ones are busy. Moreover, for the busy processors, it is known when they will finish computing their currently allocated tasks. Hence, it is always possible to select the processor that can start the execution of the task under consideration the earliest. It may well be the case that a currently busy processor is selected.

The last extension of the "classical" line of work has been to handle heterogeneous platforms, that is, platforms that consist of processors with different speeds and interconnection links with different bandwidths. The main principle remains the same, namely, to assign the free task with highest priority to the "best" processor, given already given decisions. Now the "best" processor is the one capable to finish the execution of the task at the earliest, rather than starting the execution of the task, because computing speeds are different. These ideas have led to the design of the list heuristic called HEFT, for *Heterogeneous Earliest Finish Time* [51]. We refer the reader to [13] for a complete description of HEFT because we have omitted many technical (but important) implementation details. We stress that HEFT has received considerable acceptance and is the most widely used heuristic to schedule a task graph on a modern computing platform.

59.2.3 Realistic Communication Models

Assuming an application that runs threads on, say, a node that uses multicore technology, the network link could be shared by several incoming and outgoing communications. Therefore, the sum of the bandwidths allotted by the operating system to all communications cannot exceed the bandwidth of the network card. The *bounded multiport model* proposed by Hong and Prasanna [28] assesses that an unbounded number of communications can thus take place simultaneously, provided that they share the total available bandwidth.

A more radical option than limiting the total bandwidth is simply to forbid concurrent communications at each node. In the *one-port model*, a node can either send data or receive data, but not simultaneously. This model is thus very pessimistic as real-world platforms can achieve some concurrency of communication. On the other hand, it is straightforward to design algorithms that follow this model and thus to determine their performance a priori.

There are more complicated models such as those that deal with bandwidth sharing protocols [35,36]. Such models are very interesting for performance evaluation purposes, but they almost always prove too complicated for algorithm design purposes. For this reason, when scheduling task graphs, one prefers to deal with the bounded multiport or the one-port model. As stated earlier, these models represent a good trade-off between realism and tractability, and we use them for all the examples and case studies of this chapter.

59.3 Underlying Principles

As stated in the introduction, the case studies provided in this chapter depart from pure makespan minimization. In this section, we briefly overview the main objectives that can be addressed instead of, or in complement to, the total execution time.

59.3.1 Multicriteria Optimization

The first case study (Section 59.4) deals with energy consumption. Processors can execute tasks at different frequencies, and there is a trade-off to achieve between fast execution (which implies high frequencies) and low-energy consumption (which requests low frequencies). To complicate matters, running at low frequencies increases the probability of a fault during the execution. Altogether, one faces a tri-criteria problem, involving makespan, energy, and reliability.

How to deal with several objective functions? In some approaches, one would form a linear combination of the different objectives and treat the result as the new objective to optimize for. But it is not natural for the user to maximize a quantity like $0.6M + 0.3E + 0.1R$, where M is the makespan, E the energy consumption, and R the reliability. Instead, one is more likely to fix an execution deadline and a reliability threshold, and to search for the execution that minimizes the energy consumption under these makespan and reliability constraints. One single criterion is optimized, under the condition that the constraints on the other criteria are not violated.

59.3.2 Memory Management

The second and third case studies (Sections 59.5 and 59.6) both deal with memory management issues. In Section 59.5, the goal is to traverse a task graph (an out-tree) and execute it with a prescribed amount of main memory. To execute a task, one needs to load the files produced by the parent of the task, the object code of the task itself, and all the files that it produces for its children. If there is not enough main memory available, secondary (out-of-core) memory must be used, at the price of a much slower execution. The ordering of the tasks chosen by the scheduler will have a dramatic impact on the peak amount of memory that is needed throughout the execution.

The problem investigated in Section 59.6 also deals with out-trees and is expressed as follows: How to partition the leaves of the tree in fixed-size parts so that the total memory load is minimized? Here the load of a part if the sum of the weights of the subtree formed by the leaves in the part. Intuitively, the goal is to construct parts whose leaves share many ancestors, so that the cost of loading theses ancestors is paid for only a reduced umber of times. Both case studies arise from scientific applications (such as multifrontal method for sparse linear algebra) that deploy large collections of tasks, and an efficient memory management is the key component to efficiency.

59.3.3 Reliability

The fourth case study (Section 59.7) addresses a completely different problem. Instead of dealing with task graphs, it deals with a single job that must be executed by one or several processors. This (large) job is divisible, meaning that it can be split arbitrarily into chunks of different sizes. Processors are subject to failures, which leads to take a checkpoint after each successful execution of a chunk. Indeed, after a failure, the execution can resume at the state of the last checkpoint. Frequent checkpoints avoid to lose (in fact, re-execute) a big fraction of the work when a failure occurs, but they induce a big overhead.

What is the optimal checkpointing policy? In some way, this problem can be viewed as a trade-off between makespan and reliability, but the framework is totally different from that of Section 59.4. The goal here is to minimize the expected makespan on a platform subject to failures. With the advent of future exascale machines that gather millions of processors, the problem has become central to the community.

59.4 Case Study: Tri-Criteria (Makespan, Energy, and Reliability) Scheduling

In this section, we are interested in energy-aware scheduling: the aim is to minimize the energy consumed during the execution of the target application. Energy-aware scheduling has proven an important issue in the past decade, both for economical and environmental reasons. This holds true for traditional computer systems, not even to speak of battery-powered systems. More precisely, a processor running at frequency f dissipates f^3 watts per unit of time [7,9,15], hence it consumes $f^3 \times d$ joules when operated during d units of time. To help reduce energy dissipation, processors can run at different frequencies, that is, different speeds. A widely used technique to reduce energy consumption is *dynamic voltage and frequency scaling (DVFS)*, also known as speed scaling [7,9,15]. Indeed, by lowering supply voltage, hence processor clock frequency, it is possible to achieve important reductions in power consumption; faster speeds allow for a faster execution, but they also lead to a much higher (supra-linear) power consumption. Processors can have arbitrary speeds, and can vary them continuously in the interval $[f_{\min}, f_{\max}]$. This model of continuous speeds is unrealistic (any possible value of the speed, say $\sqrt{e^{\pi}}$, cannot be obtained), but it is theoretically appealing [9].

Obviously, minimizing the energy consumption makes sense only when coupled with some performance bound to achieve, otherwise, the optimal solution always is to run each processor at the slowest possible speed. We consider a directed acyclic graph (DAG) of n tasks with precedence constraints, and the goal is to schedule such an application onto a fully homogeneous platform consisting of p identical processors. This problem has been widely studied with the objective of minimizing the total execution time, or *makespan*, and it is well known to be NP-complete [12]. Since the introduction of DVFS, many papers have dealt with the optimization of energy consumption while enforcing a deadline, that is, a bound on the makespan [5,7,9,15].

There are many situations in which the mapping of the task graph is given, say by an ordered list of tasks to execute on each processor, and we do not have the freedom to change the assignment of a given task. Such a problem occurs when optimizing for legacy applications, or accounting for affinities between tasks and resources, or even when tasks are pre-allocated [45], for example for security reasons. While it is not possible to change the allocation of a task, it is possible to change its speed. This technique, which consists in exploiting the slack due to workload variations, is called slack reclaiming [32,43]. In our previous work [5], assuming that the mapping and a deadline are given, we have assessed the impact of several speed variation models on the complexity of the problem of minimizing the energy consumption. Rather than using a local approach such as backfilling [43,54], which only reclaims gaps in the schedule, we have considered the problem as a whole.

While energy consumption can be reduced by using speed scaling techniques, it was shown in [17,57] that reducing the speed of a processor increases the number of transient fault rates of the system; the probability of failures increases exponentially, and this probability cannot be neglected in large-scale computing [39]. In order to make up for the loss in *reliability* due to the energy efficiency, different models have been proposed for fault-tolerance. In this section, we consider *re-execution*: a task that does not meet the reliability constraint is executed twice. This technique was also studied in [42,56,57], or in the case study of Section 59.7. The goal is to ensure that each task is reliable enough, that is, either its execution speed is above a threshold, ensuring a given reliability of the task, or the task is executed twice to enhance its reliability. There is a clear trade-off between energy consumption and reliability, since decreasing the execution speed of a task, and hence the corresponding energy consumption, is deteriorating the reliability. This calls for tackling the problem of considering the three criteria (makespan, reliability, and energy) simultaneously. This tri-criteria optimization brings dramatic complications: in addition to choosing the speed of each task, as in the makespan/energy bi-criteria problem, we also need to decide which subset of tasks should be re-executed (and then choose both execution speeds).

Given an application with dependence constraints and a mapping of this application on a homogeneous platform, we present in this section theoretical results for the problem of minimizing the energy consumption under the constraints of both a reliability threshold per task and a deadline bound.

59.4.1 Tri-Criteria Problem

Consider an application task graph $\mathcal{G} = (V, \mathcal{E})$, where $V = \{T_1, T_2, \ldots, T_n\}$ is the set of tasks, $n = |V|$, and where $\mathcal{E}$ is the set of precedence edges between tasks. For $1 \le i \le n$, task T_i has a weight w_i, that corresponds to the computation requirement of the task. We also consider particular class of task graphs, such as *linear chains* where $\mathcal{E} = \bigcup_{i=1}^{n-1}\{T_i \to T_{i+1}\}$, and *forks* with $n+1$ tasks $\{T_0, T_1, T_2, \ldots, T_n\}$ and $\mathcal{E} = \bigcup_{i=1}^{n}\{T_0 \to T_i\}$.

We assume that tasks are mapped onto a parallel platform made up of p identical processors. Each processor has a set of available speeds that is continuous (in the interval $[f_{min}, f_{max}]$). The goal is to minimize the energy consumed during the execution of the graph while enforcing a deadline bound and matching a reliability threshold. To match the reliability threshold, some tasks are executed once at a speed high enough to satisfy the constraint, while some other tasks need to be re-executed. We detail next the conditions that are enforced on the corresponding execution speeds. The problem is therefore to decide which task to re-execute and at which speed to run each execution of a task.

In this section, for the sake of clarity, we assume that a task is executed at the same (unique) speed throughout execution, or at two different speeds in the case of re-execution. In Section 59.4.2, we show that this strategy is indeed optimal. We now detail the three objective criteria (makespan, reliability, and energy), and then define formally the problem.

The *makespan* of a schedule is its total execution time. We consider a *deadline bound D*, which is a constraint on the makespan. Let $\mathcal{E}xe(w_i, f)$ be the execution time of a task T_i of weight w_i at speed f. We assume that the cache size is adapted to the application, therefore ensuring that the execution time is linearly related to the frequency [37]: $\mathcal{E}xe(w_i, f) = w_i/f$. When a task is scheduled to be re-executed at two different speeds $f^{(1)}$ and $f^{(2)}$, we always account for both executions, even when the first execution is successful, and hence $\mathcal{E}xe(w_i, f^{(1)}, f^{(2)}) = \frac{w_i}{f^{(1)}} + \frac{w_i}{f^{(2)}}$. In other words, we consider a worst-case execution scenario, and the deadline D must be matched even in the case where all tasks that are re-executed fail during their first execution.

To define the *reliability*, we use the fault model of Zhu et al. [56,57]. *Transient* failures are faults caused by software errors for example. They invalidate only the execution of the current task and the processor subject to that failure will be able to recover and execute the subsequent task assigned to it (if any). In addition, we use the reliability model introduced by Shatz and Wang [48], which states that the radiation-induced transient faults follow a Poisson distribution. The parameter λ of the Poisson distribution is then

$$\lambda(f) = \tilde{\lambda}_0 \, e^{\tilde{d}\frac{f_{max} - f}{f_{max} - f_{min}}}, \qquad (59.1)$$

where

$f_{min} \le f \le f_{max}$ is the processing speed, the exponent $\tilde{d} \ge 0$ is a constant, indicating the sensitivity of fault rates to DVFS

$\tilde{\lambda}_0$ is the average fault rate corresponding to f_{max}

We see that reducing the speed for energy saving increases the fault rate exponentially. The reliability of a task T_i executed once at speed f is $R_i(f) = e^{-\lambda(f)\times \mathcal{E}xe(w_i,f)}$. Because the fault rate is usually very small, of the order of 10^{-6} per time unit in [8,42], 10^{-5} in [4], we can use the first-order approximation of $R_i(f)$ as

$$R_i(f) = 1 - \lambda(f) \times \mathcal{E}xe(w_i, f) = 1 - \lambda_0 e^{-df} \times \frac{w_i}{f},$$

where

$$d = \frac{\tilde{d}}{f_{\max} - f_{\min}}$$

$$\lambda_0 = \tilde{\lambda}_0 e^{df_{\max}}$$

This equation holds if $\varepsilon_i = \lambda(f) \times w_i / f \ll 1$. With, say, $\lambda(f) = 10^{-5}$, we need $w_i/f \leq 10^3$ to get an accurate approximation with $\varepsilon_i \leq 0.01$: the task should execute within 16 min. In other words, large (computationally demanding) tasks require reasonably high processing speeds with this model (which makes full sense in practice).

We want the reliability R_i of each task T_i to be greater than a given threshold, namely $R_i(f_{\mathrm{rel}})$, hence enforcing a local constraint dependent on the task $R_i \geq R_i(f_{\mathrm{rel}})$. If task T_i is executed only once at speed f, then the reliability of T_i is $R_i = R_i(f)$. Since the reliability increases with speed, we must have $f \geq f_{\mathrm{rel}}$ to match the reliability constraint. If task T_i is re-executed (speeds $f^{(1)}$ and $f^{(2)}$), then the execution of T_i is successful if and only if one of the attempts do not fail, so that the reliability of T_i is $R_i = 1 - \left(1 - R_i(f^{(1)})\right)\left(1 - R_i(f^{(2)})\right)$, and this quantity should be at least equal to $R_i(f_{\mathrm{rel}})$.

The *total energy consumption* corresponds to the sum of the energy consumption of each task. Let E_i be the energy consumed by task T_i. For one execution of task T_i at speed f, the corresponding energy consumption is $E_i(f) = \mathcal{E}xe(w_i, f) \times f^3 = w_i \times f^2$, which corresponds to the dynamic part of the classical energy models of the literature [5,7,9,15]. Note that we do not take static energy into account, because all processors are up and alive during the whole execution.

If task T_i is executed only once at speed f, then $E_i = E_i(f)$. Otherwise, if task T_i is re-executed at speeds $f^{(1)}$ and $f^{(2)}$, it is natural to add up the energy consumed during both executions, just as we add up both execution times when enforcing the makespan deadline. Again, this corresponds to the worst-case execution scenario. We obtain $E_i = E_i\left(f_i^{(1)}\right) + E_i\left(f_i^{(2)}\right)$. In this work, we aim at minimizing the total energy consumed by the schedule in the worst case, assuming that all re-executions do take place. This worst-case energy is $E = \sum_{i=1}^{n} E_i$.

Some authors [56] consider only the energy spent for the first execution, which seems unfair: re-execution comes at a price both in the deadline and in the energy consumption. Another possible approach would be to consider the expected energy consumption, which would require to weight the energy spent in the second execution of a task by the probability of this re-execution to happen. This would lead to a less conservative estimation of the energy consumption by averaging over many execution instances. However, the makespan deadline should be matched in all execution scenarios, and the execution speeds of the tasks have been dimensioned to account for the worst-case scenario, so it seems more important to report for the maximal energy that can be consumed over all possible execution instances.

We are now ready to define the optimization problem: given an application graph $\mathcal{G} = (V, \mathcal{E})$, mapped onto p homogeneous processors with continuous speeds, Tri-Crit is the problem of deciding that tasks should be re-executed and at which speed each execution of a task should be processed, in order to minimize the total energy consumption E, subject to the deadline bound D and to the local reliability constraints $R_i \geq R_i(f_{\mathrm{rel}})$ for each $T_i \in V$.

We also introduce variants of the problems for particular application graphs: Tri-Crit-Chain is the same problem as Tri-Crit when the task graph is a linear chain, mapped on a single processor;

and TRI-CRIT-FORK is the same problem as TRI-CRIT when the task graph is a fork, and each task is mapped on a distinct processor.

59.4.2 Complexity Results

Note that the formal proofs for this section can be found in [6]. As stated in Section 59.4.1, we start by proving that there is always an optimal solution where each task is executed at a unique speed:

Lemma 59.1

It is optimal to execute each task at a unique speed throughout its execution.

The idea is to consider a task whose speed changes during the execution; we exhibit a speed such that the execution time of the task remains the same, but where both energy and reliability are maybe improved, by convexity of the functions.

Next we show that not only a task is executed at a single speed, but that its re-execution (whenever it occurs) is executed at the same speed as its first execution:

Lemma 59.2

It is optimal to re-execute each task (whenever needed) at the same speed as its first execution, and this speed f is such that $f_i^{(inf)} \leq f < \frac{1}{\sqrt{2}} f_{rel}$, where

$$\lambda_0 w_i \frac{e^{-2df_i^{(inf)}}}{\left(f_i^{(inf)}\right)^2} = \frac{e^{-df_{rel}}}{f_{rel}}. \tag{59.2}$$

Similarly to the proof of Lemma 59.1, we exhibit a unique speed for both executions, in case they differ, so that the execution time remains identical but both energy and reliability are improved. If this unique speed is greater than $\frac{1}{\sqrt{2}} f_{rel}$, then it is better to execute the task only once at speed f_{rel}, and if f is lower than $f_i^{(inf)}$, then the reliability constraint is not matched. Note that both lemmas can be applied to any solution of the TRI-CRIT problem, not just optimal solutions. We are now ready to assess the problem complexity:

Theorem 59.1

The TRI-CRIT-CHAIN problem is NP-hard, but not known to be in NP.

Note that the problem is not known to be in NP because speeds could take any real values. The completeness comes from SUBSET-SUM [22]. The problem is NP-hard even for a linear chain application mapped on a single processor (and any general DAG mapped on a single processor becomes a linear chain).

Even if Tri-Crit-Chain is NP-hard, we can characterize an optimal solution of the problem:

Proposition 59.1

If $f_{rel} < f_{\max}$, then in any optimal solution of Tri-Crit-Chain, either all tasks are executed only once, at constant speed $\max\left(\frac{\sum_{i=1}^{n} w_i}{D}, f_{rel}\right)$; or at least one task is re-executed, and then all tasks that are not re-executed are executed at speed f_{rel}.

In essence, Proposition 59.1 states that when dealing with a linear chain, we should first slow down the execution of each task as much as possible. Then, if the deadline is not too tight, that is, if $f_{rel} > \frac{\sum_{i=1}^{n} w_i}{D}$, there remains the possibility to re-execute some of the tasks (and of course it is NP-hard to decide which ones). Still, this general principle *"first slow-down and then re-execute"* can guide the design of efficient heuristics (see [6]).

While the general Tri-Crit problem is NP-hard even with a single processor, the particular variant Tri-Crit-Fork can be solved in polynomial time:

Theorem 59.2

The Tri-Crit-Fork problem can be solved in polynomial time.

The difficulty to provide an optimal algorithm for the Tri-Crit-Fork problem comes from the fact that the total execution time must be shared between the source of the fork, T_0, and the other tasks that all run in parallel. If we know D', the fraction of the deadline allotted for tasks $T_1, \ldots, T_n$ once the source has finished its execution, then we can decide which tasks are re-executed and all execution speeds. Indeed, if task T_i is executed only once, it is executed at speed $f_i^{(once)} = \min\left(\max\left(w_i/D', f_{rel}\right), f_{\max}\right)$. Otherwise, it is executed twice at speed $f_i^{(twice)} = \min\left(\max\left(2w_i/D', f_{\min}, f_i^{(\text{inf})}\right), f_{\max}\right)$, where $f_i^{(\text{inf})}$ is the minimum speed at which task T_i can be executed twice (see Lemma 59.2). The energy consumption for task T_i is finally $E_i = \min\left(w_i \times f_i^{(once)}, 2w_i \times f_i^{(twice)}\right)$, and the case that reaches the minimum determines whether the task is re-executed or not. There remains to find the optimal value of D', which can be obtained by studying the function of the total energy consumption, and bounding the value of D' to a small number of possibilities. Note however that this algorithm does not provide any closed-form formula for the speeds of the tasks, and that there is an intricate case analysis due to the reliability constraints.

If we further assume that the fork is made of identical tasks (i.e., $w_i = w$ for $0 \le i \le n$), then we can provide a closed-form formula. However, Proposition 59.2 illustrates the inherent difficulty of this *simple* problem, with several cases to consider depending on the values of the deadline, and also the bounds on speeds $\left(f_{\min}, f_{\max}, f_{rel}, \text{etc.}\right)$ First, since the tasks all have the same weight $w_i = w$, we get rid of the $f_i^{(\text{inf})}$ introduced earlier, since they are all identical (see Equation 59.2): $f_i^{(\text{inf})} = f^{(\text{inf})}$ for $0 \le i \le n$. Therefore we let $f_{\min} = \max(f_{\min}, f^{(\text{inf})})$ in the following proposition below:

Proposition 59.2

In the optimal solution of Tri-Crit-Fork with at least three identical tasks (and hence $n \ge 2$), there are only three possible scenarios: (1) no task is re-executed; (2) the n successors are all re-executed but not the source; and (3) all tasks are re-executed. In each scenario, the source is executed at speed f_{src} (once or twice), and the n successors are executed at the same speed f_{leaf} (once or twice).

For a deadline $D < \frac{2w}{f_{\max}}$, there is no solution. For a deadline $D \in \left[\frac{2w}{f_{\max}}, \frac{w}{f_{rel}} \frac{\left(1+2n^{\frac{1}{3}}\right)^{\frac{3}{2}}}{\sqrt{1+n}}\right]$, no task is re-executed (scenario (1)) and the values of f_{src} and f_{leaf} are the following:

- If $\frac{2w}{f_{\max}} \leq D \leq \min\left(\frac{w}{f_{\max}}\left(1+n^{\frac{1}{3}}\right), w\left(\frac{1}{f_{rel}}+\frac{1}{f_{max}}\right)\right)$, then $f_{src} = f_{\max}$ and $f_{leaf} = \frac{w}{Df_{\max}-w} f_{\max}$
- If $\frac{w}{f_{\max}}\left(1+n^{\frac{1}{3}}\right) \leq w\left(\frac{1}{f_{rel}}+\frac{1}{f_{\max}}\right)$, then
  - if $\frac{w}{f_{\max}}\left(1+n^{\frac{1}{3}}\right) < D \leq \frac{w}{f_{rel}} \frac{1+n^{\frac{1}{3}}}{n^{\frac{1}{3}}}$, then $f_{src} = \frac{w}{D}\left(1+n^{\frac{1}{3}}\right)$ and $f_{leaf} = \frac{w}{D} \frac{1+n^{\frac{1}{3}}}{n^{\frac{1}{3}}}$
  - if $\frac{w}{f_{rel}} \frac{1+n^{\frac{1}{3}}}{n^{\frac{1}{3}}} < D \leq \frac{2w}{f_{rel}}$, then $f_{src} = \frac{w}{Df_{rel}-w} f_{rel}$ and $f_{leaf} = f_{rel}$
- If $\frac{w}{f_{\max}}\left(1+n^{\frac{1}{3}}\right) > w\left(\frac{1}{f_{rel}}+\frac{1}{f_{\max}}\right)$, then
  - if $w\left(\frac{1}{f_{rel}}+\frac{1}{f_{\max}}\right) < D \leq \frac{2w}{f_{rel}}$, then $f_{src} = \frac{w}{Df_{rel}-w} f_{rel}$ and $f_{leaf} = f_{rel}$
- If $\frac{2w}{f_{rel}} < D \leq \frac{w}{f_{rel}} \frac{\left(1+2n^{\frac{1}{3}}\right)^{\frac{3}{2}}}{\sqrt{1+n}}$, then $f_{src} = f_{leaf} = f_{rel}$

Note that for larger values of D, depending on $f_{\min}$, we can move to scenarios (2) and (3) with partial or total re-execution. The case analysis becomes even more painful, but remains feasible. Intuitively, the property that all tasks have the same weight is the key to obtaining analytical formulas, because all tasks have the same minimum speed $f^{(\text{inf})}$ dictated by Equation 59.2. Beyond the case analysis itself, the result of Proposition 59.2 is interesting: we observe that in all cases the source task is executed faster than the other tasks. This shows that Proposition 59.1 does not hold for general DAGs and suggests that some tasks may be more critical than others. A hierarchical approach that categorizes tasks with different priorities may guide the design of efficient heuristics (see [6]).

59.4.3 Conclusion

In this case study, we have accounted for the energy cost associated to task re-execution in a more realistic and accurate way than the best-case model used in [56]. Coupling this energy model with the classical reliability model used in [48], we have been able to formulate a tri-criteria optimization problem: How to minimize the energy consumed given a deadline bound and a reliability constraint? The "antagonistic" relation between energy and reliability renders this tri-criteria problem much more challenging than the standard bi-criteria (makespan, energy) version. We have assessed the intractability of this tri-criteria problem, even in the case of a single processor. In addition, we have provided several complexity results for particular instances. Note that several polynomial time heuristics are discussed in [6].

Further work could tackle the design of efficient approximation algorithms. However, this is quite a challenging work for arbitrary DAGs, and one may try to focus on special graph structures, for example, series-parallel graphs. However, looking back at the complicated case analysis needed for an elementary fork-graph with identical weights (Proposition 59.2), we cannot underestimate the difficulty of this problem. Also, it could be interesting to study the expected energy consumption, instead of the worst-case energy consumption. However, the complexity of the problems is likely to increase drastically. Finally, we point out that energy reduction and reliability will be even more important objectives with the advent of massively parallel platforms, made of a large number of clusters of multicores. More efficient solutions to the tri-criteria optimization problem (makespan, energy, and reliability) could be achieved through combining replication with re-execution. A promising (and ambitious) research direction would be to search for the best trade-offs that can be achieved between these techniques that both increase reliability, but whose impact on execution time and energy consumption is very different. The comprehensive set of theoretical results presented in this case study provides solid foundations for further studies and constitutes a partial yet important first step for solving the problem at very large scale.

59.5 Case Study: Memory-Aware Scheduling

In this section, we review the latest results in memory-aware scheduling. When processing an application on a computational grid, it is likely that the involved data will have a very high volume. Such large data sizes may prevent the processing of some task on given resources, and more often, may prevent the processing of two data-intensive tasks on the same resource, because of the input and output data needed for these tasks. In this case, some of the data may need to be discarded from the main storage system (such as memory) and to be written to a secondary storage system (such as disk), to be able to proceed with the execution.

In grid computing, workflows are usually modeled as Directed Acyclic Graph (DAG), where nodes represent computational tasks and edges represent the dependencies among tasks. The dependencies are in the form of input and output files: each node accepts a set of (large) files as input, and produces a set of (large) files as output, each of them accepted by different child tasks. The execution scheme of such a workflow corresponds to a traversal of the DAG, where visiting a node translates into reading the associated input files and producing output files. Even on a single machine, finding a way to traverse the DAG to minimize the needed storage is a difficult problem: when restricting the problem to uniform sizes, it is similar to the pebble game problem, which has been proven NP-hard by Sethi [47] (the general problem allowing recomputation, that is, re-pebbling a vertex that has been pebbled before, has been proven PSPACE complete by Gilbert, Lengauer and Tarjan [24]). Latest studies focus on the restricted problem of traversing tree-shaped DAGs to minimize the memory demand on a single machine.

For convenience we refer to the two levels of storage as *main memory* and *secondary memory*, and also as *in-core* and *out-of-core*. Many combinations such as cache and RAM, or RAM and disk, or even disk and tape, lead to the same association of a faster but smaller storage device together with a larger but slower device. The difficulty remains the same for all combinations: find an execution scheme that makes the best use of *main memory*, and minimizes accesses to *secondary memory*.

Throughout the case study we consider *out-trees* where a task can be executed only if its parent has already been executed. However, all results equivalently apply to *in-trees*, where tasks are processed from the leaves up to the root, as explained later. Each task (or node) i in the tree is characterized by the size f_i of its input file (data needed before the execution and received from its parent), and by the size n_i of its execution file. During execution, nonleaf nodes generate several output files, one for each child, which can have different sizes. A task can be processed only in-core; its execution is feasible only if all its files (input, output, and execution) fit in currently available memory. More formally, let M be the size of the main memory, and S the set of files stored in this memory when the scheduler decides to execute task i. Note that S must contain the input file of task i. The processing of task i is possible if we have

$$MemReq(i) = f_i + n_i + \sum_{j \in Children(i)} f_j \leq M - \sum_{j \in S, j \neq i} f_j \qquad (59.3)$$

where *MemReq*(i) denotes the memory requirement of task i (and *Children*(i) its child nodes in the tree). Once i has been executed, its input file and execution file can be discarded, and replaced by other files in main memory; the output files can either be kept in main memory, in order to execute some child of the task, or they can temporarily be stored into secondary memory (and retrieved later when the scheduler decides to execute the corresponding child of i). The volume of accesses (reads or writes) to secondary memory is referred to as the *I/O volume*.

Clearly, the traversal, that is, the order chosen to execute the tasks, plays a key role in determining which amount of main memory and I/O volume are needed for a successful execution of the whole tree. More precisely, there are two main problems that the scheduler must address

MinMemory Determine the minimum amount of main memory that is required to execute the tree without any access to secondary memory.

MinIO Given the size M of the main memory, determine the minimum I/O volume that is required to execute the tree.

Obviously, a necessary condition for the execution to be successful is that the size M of main memory exceeds the largest memory requirement over all tasks:

$$\max_i MemReq(i) \leq M$$

However, this condition is not sufficient, and a much larger main memory size may be needed for the MinMemory problem.

This problem has two main motivations. The first one (chronologically) arose from numerical linear algebra. Tree workflows (assembly or elimination trees) appear during the factorization of sparse matrices, and the huge size of the files involved makes it absolutely necessary to reduce the memory requirement of the factorization. The trees arising in this context are in-trees (as said before, there is no difference between in-trees and out-trees). Liu [33] discusses how to find the best traversal for the MinMemory problem when the traversal is required to correspond to a postorder traversal of the tree. In the follow-up study [34], an exact algorithm is proposed to solve the MinMemory problem, without the postorder constraint on the traversal. Another exact algorithm has been proposed recently [29]. Although its theoretical worst-case complexity is similar to Liu's algorithm, it performs better on elimination trees arising from real sparse matrix factorizations.

The second motivation for this problem comes from computations and simulations involving large objects, such as the accurate modeling of the electronic structure of atoms and molecules in quantum chemistry, or in some computational physics code. In this context, Liu's exact algorithm has been recently rediscovered by Lam et al. [31]. In the following, we first propose a formal model for the problem. Then we review state-of-the-art results and algorithms for the MinMemory and MinIO problems.

59.5.1 Problem Modeling

We consider a tree workflow $\mathcal{T}$ composed of p nodes, or tasks, numbered from 1 to p. Nodes in the tree have an input file, an execution file (or program), and several output files (one per child). More precisely the following conditions hold true:

- Each node i has an input file of size f_i. If i is not the root, its input file is produced by its parent *parent*(i); if i is the root, its input file can be of size zero, or contain input from the outside world.
- Each node i in the tree has an execution file of size n_i.
- Each nonleaf node i in the tree, when executed, produces a file of size f_j for each $j \in$ *Children*(i). Here *Children*(i) denotes the set of the children of i. If i is a leaf node, then *Children*(i) = $\emptyset$ and i produces a file of null size: we then consider that the terminal data produced by leaves are directly written to the secondary memory or sent to the outside world, independently from the I/O mechanism.

The memory requirement *MemReq*(i) of node i is the total amount of main memory that is needed to execute node i, as underlined in Equation 59.3. After i has been processed, its input file and program can be discarded, while its output files can either be kept in main memory (to process the children of i) or be stored in secondary memory temporarily.

59.5.2 In-Core Traversals and the MinMemory Problem

For the MinMemory problem, we are given a tree $\mathcal{T}$ with p nodes and an initial amount of memory M. A traversal is an ordering of the p nodes that specifies at which step they are executed. A traversal must obey precedence constraints (a node is always scheduled after its parent) and must never exceed the available memory. A formal definition of a traversal is given next.

Definition 59.1 (InCoreTraversal)

Given a tree $\mathcal{T}$ and a amount M of available memory, the problem InCoreTraversal($\mathcal{T}$, M) consists in finding a feasible in-core traversal σ described by a permutation of the nodes of a tree $\mathcal{T}$ such that

$$\forall i \neq root, \quad \sigma(parent(i)) < \sigma(i) \tag{59.4}$$

$$\forall i \quad \sum_{\sigma(j)<\sigma(i)} \left(\sum_{k \in Children(j)} f_k - f_j \right) + n_i + \sum_{k \in Children(i)} f_k \leq M \tag{59.5}$$

In this definition, Equation 59.4 accounts for precedence constraints and Equation 59.5 deals with memory constraints. A postorder traversal is a traversal where nodes are visited according to some top-down postorder ordering of the tree nodes. Hence, in a postorder traversal, after processing a vertex i, the whole subtree rooted in i is completely processed.

Definition 59.2 (MinMemory)

Given a tree $\mathcal{T}$, determine the minimum amount of memory M such that InCoreTraversal ($\mathcal{T}$, M) has a solution. MinMemory-PostOrder is the same problem restricted to postorder traversals.

Here, we have chosen to define the problem for out-trees and top-down traversals. However, it is possible to define a similar problem for in-trees and bottom-up traversals. Both problems are equivalent. Consider an in-tree $\mathcal{T}$ and the out-tree $\mathcal{T}'$ obtained when reversing all its edges. Then, from any traversal of $\mathcal{T}$ we can obtain a valid traversal of $\mathcal{T}'$ with same memory footprint simply by reversing the order of visited nodes. Thus, all results mentioned in the following apply both for in-trees and out-trees.

59.5.2.1 Postorder Traversals

Postorder traversals are very natural for the MinMemory problem, and they are widely used in sparse matrix software like MUMPS [2,3]. Liu [33] has characterized the best postorder traversal, leading to a fast but suboptimal solution for MinMemory. In a nutshell, the best postorder is obtained by guaranteeing that in the resulting order the children of a node are listed in the increasing order of the memory requirement of their respective subtrees. This can be done with a recursive algorithm, which computes a memory-optimal schedule for each subtree: at each level subtrees are sorted in nonincreasing order of $\max_{i \in subtree}\left(MemReq(i) - f_{subroot}\right)$ (where *subroot* is the root of the current subtree). In the following, we denote this algorithm by *PostOrder*. Although it has a low time complexity $\left(O\left(p\log(p)\right)\right)$ and it is widely used in practice, it may require arbitrarily more main memory than the optimal traversal.

Theorem 59.3

Given any arbitrarily large integer K, there exist trees for which the best postorder traversal requires at least K times the amount of main memory needed by the optimal traversal for MINMEMORY.

The proof of this results relies on the family of trees illustrated in Figure 59.1, where all branches are identical and all tasks have a zero length execution file. On the single-level tree of Figure 59.1a, any postorder traversal requires an amount of $M + \varepsilon + (b-1)M/b$ main memory, while the optimal traversal (which alternates between branches) only requires $M_{\min} = M + \varepsilon$. Now replace each leaf by a copy of the harpoon graph, as shown in Figure 59.1b. The value of $M_{\min}$ is unchanged, while a postorder traversal requires $M + \varepsilon + 2(b-1)M/b$. Iterating the process K times leads to the desired result.

The tree used to prove that postorder may require an arbitrarily large memory compared from the optimal is very different to usual trees encountered in practice: it has many edges with zero weight, and on a path from the root, the edge weights alternate between large and small values. It is thus interesting to compare the actual performance of *PostOrder* in terms of the memory requirement of the resulting traversal with respect to the optimal value on realistic trees. To this goal, one can use assembly trees that arise from the factorization of matrices obtained from the University of Florida Sparse Matrix Collection (see [29] for details on the experimental dataset). The optimal memory requirement of a traversal is computed using the exact algorithms presented later. The experiments show that in 95.8% of the cases, *PostOrder* is optimal. In the nonoptimal cases, *PostOrder* requires up to 18% more memory than the optimal solution (the average increase is in memory is around 1%).

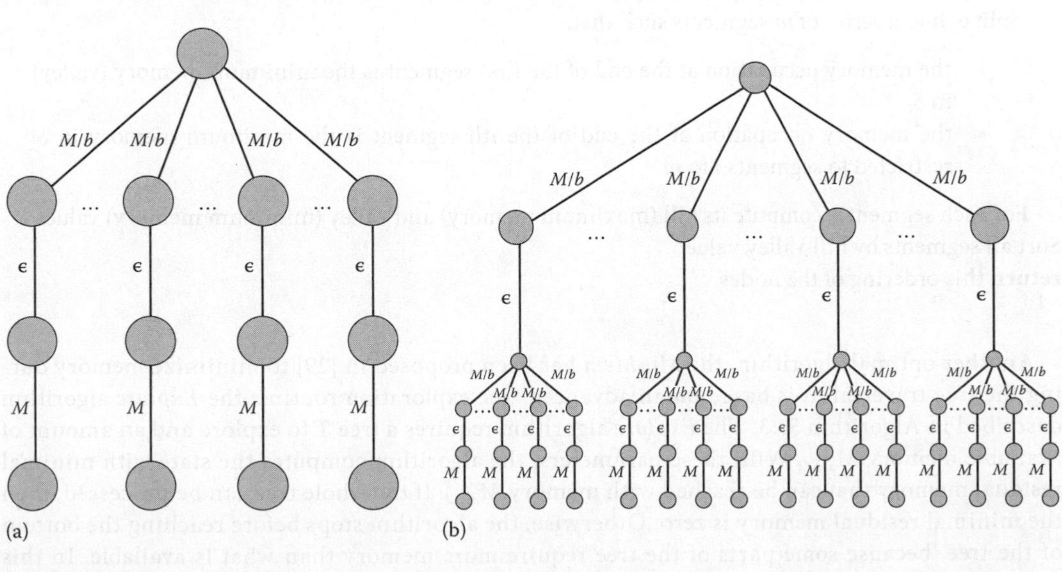

FIGURE 59.1 First levels of the graph for the proof of Theorem 59.3. Here b is the number of children of the nodes with more than one child: (a) One level and (b) two levels.

59.5.2.2 Exact Algorithms

Liu [34] proposes an algorithm for MinMemory that is optimal among all possible traversals, not only postorder ones. It is a recursive bottom-up traversal of the tree that, at each node of the tree, combines the optimal traversals built for all subtrees (see Algorithms 59.1 and 59.2). The combination is based on the notion of *Hill-Valley Segments*: the schedule corresponding to a subtree is split into a sequence of segments. Each segment ends at a (local) minimum memory: to minimize the memory, it is better to totally process a segment before switching to the processing of another subtree. Then, the sequence of segments computed for all subtrees are ordered by nonincreasing *hill – valley* value and merged. The motivation is then to process first segments with high memory peak (hill) but small residual memory requirement (valley).

Liu provides a complex proof for the optimality of the obtained schedule. It also states that the worst-case complexity of this algorithm is $O(p^2)$ using some sophisticated multiway merging algorithm.

Algorithm 59.1 *LiuMinMem(T)*

Input: tree T rooted at r
if *r is a leaf* **then**
  return the sequence consisting of the node r
else
  Let $T_1, \ldots, T_k$ be the subtrees rooted at the children of r **for** $i = 1, \ldots, k$ **do**
  $S_i \leftarrow LiuMinMem(T_i)$
  return $S = Combine(T, S_1, \ldots, S_k)$

Algorithm 59.2 *Combine* $(T, S_1, \ldots, S_k)$

for $i = 1, \ldots, k$ **do**
  Compute the memory occupation at each step of S_i when processing subtree T_i
  Split S_i into a series of m segments such that:

- the memory occupation at the end of the first segment is the minimum memory (valley) in S_i
- the memory occupation at the end of the ith segment is the minimum memory in S_i restricted to segments i to m

  For each segment s, compute its hill (maximum memory) and valley (minimum memory) values
Sort all segments by hill-valley value
return this ordering of the nodes

Another optimal algorithm, the *MinMem* has been proposed in [29] to minimize memory during the tree traversal. It is based on an advanced tree exploration routine: the *Explore* algorithm described in Algorithm 59.3. The *Explore* algorithm requires a tree T to explore and an amount of available memory M_{avail}. With these parameters, the algorithm computes the state with minimal residual memory that can be reached with memory M_{avail}. If the whole tree can be processed, then the minimal residual memory is zero. Otherwise, the algorithm stops before reaching the bottom of the tree, because some parts of the tree require more memory than what is available. In this case, the state with minimal residual memory corresponds to a *cut* in the tree: some subtrees are not yet processed, and the input files of their root nodes are still stored in memory. The *Explore* algorithm outputs the cut with minimal memory occupation, as well as a possible traversal to reach this state with the provided memory.

When called on a tree T, the algorithm first checks if the current node can be executed. If not, the algorithm stops. Otherwise, it recursively proceeds in its subtree. The optimal cut is initialized with its children, and iteratively improved. All the nodes in the cut are explored: if the cut L_j found in the subtree of a child j has a smaller memory occupation than the child itself, the cut is updated by removing child j, and by adding the corresponding cut L_j. When no more nodes in the cut can be improved, then the algorithm outputs the current cut.

The *Explore* algorithm can be used to check whether a given tree can be processed using a given memory. A possible way to compute the minimum memory to process the tree is then to perform a binary search on the available memory. However, there are several ways to speed-up such an algorithm (for details, see [29]):

- We can transform the *Explore* algorithm so that it also returns the minimum increment in the available memory needed to explore more nodes. This helps to reduce the number of calls to *Explore*.

Algorithm 59.3 *Explore*(T, M^{avail})

Input: tree T rooted at r and available memory M^{avail}
Output: $\langle L, m, S\rangle$ where L is a state (described by the input files it contains) with minimum memory m which can be reached with memory M, and S a schedule to reach it.
if *r is a leaf and there is enough memory for its processing* **then**
  return $\langle \emptyset, 0, (r)\rangle$
if *there is not enough memory to process r* **then**
  return $\langle \emptyset, \infty, \emptyset\rangle$
$S \leftarrow (r)$
$CurrentCut \leftarrow \{Children\ of\ r\}$
$stopped \leftarrow 0$
while $stopped \neq 1$ **do**
  $stopped \leftarrow 1$
  foreach $j \in CurrentCut$ **do**
    Compute the available memory for the processing of the subtree at j:
    $M_j^{\text{avail}} \leftarrow M_j^{\text{avail}} - \sum_{k \in CurrentCut \setminus \{j\}} f_k$
    $\langle L_j, m_j, S_j\rangle \leftarrow Explore\left(T, M_j^{\text{avail}}\right)$
    if *exploring the subtree at j leads to state with less memory:* $m_j \leq f_j$ **then**
      $stopped \leftarrow 0$
      $CurrentCut \leftarrow CurrentCut \setminus \{j\} \cup L_j$
      append traversal S_j to the end of S: $S \leftarrow S \oplus S_j$
$m \leftarrow \sum_{j \in CurrentCut} f_j$
return $\langle CurrentCut, m, S\rangle$

- When calling *Explore* several times on the same subtree with an increasing memory, the beginning of the subtree will be explored many times. To avoid this, it is possible to store the current state of exploration, as well a the current schedule. This ensures that nodes are not visited several times.

Using these optimizations, it is possible to prove that the *MinMem* algorithm based on the *Explore* procedures has a similar time complexity with *LiuMinMem*($O(p^2)$). Although both exact algorithms have the same worst-case time complexity, it is interesting to note that *MinMem* is faster than *LiuMinMem* on realistic trees: Experiments on the same set of trees described earlier show that *MinMem* is the fastest algorithm in 80% of the cases. This is explained by the fact that *LiuMinMem* has to sort segments when merging schedules obtained for subtrees. On the contrary, *MinMem* does not use sorting. This can make a significant difference if, for example, for each node, the size of the input file is larger than the size of the output files, *MinMem* will run in $O(n)$ for a tree of n nodes.

59.5.3 Out-of-Core Traversals and the MinIO Problem

Out-of-core processing enables solving large problems, when the size of the data cannot fit into the main memory. In this case, some temporary data are copied into the secondary memory, and unloaded from the main memory, so as to leave room for other computations. Since secondary memory has a smaller access rate, the usual objective is to limit the volume of I/O operations.

Defining traversals that perform I/O operations is more complicated than defining in-core traversals: in addition to determining the ordering of the nodes (the permutation σ), at each step we have to identify which files are written into secondary memory (if necessary). When a task i is scheduled for execution but its input file was moved to secondary memory, that file must be read and loaded back into the main memory before processing task i. Thus, a given file is written at most once in the secondary memory. Moreover, if a file is to be written in the secondary memory, it is always beneficial to write it as soon as it is created. Thus, we can model the I/O operations via a binary function δ_{IO} such that $\delta_{IO}(i) = 1$ if and only if the input file of task i (of size f_i) should be moved to secondary memory. Formally, a valid out-of-core traversal can be defined as follows.

Definition 59.3 (OutOfCoreTraversal)

Given a tree $\mathcal{T}$ and a fixed amount of main memory M, the problem OutOfCoreTraversal($\mathcal{T}$, M) consists in finding an out-of-core traversal, described by a permutation σ of the nodes in $\mathcal{T}$ (corresponding to the schedule of computations), and a binary function δ_{IO} (corresponding to I/O operations), such that:

$$\forall i \neq root, \quad \sigma(parent(i)) < \sigma(i) \tag{59.6}$$

$$\forall i, \quad \sum_{\sigma(j)<\sigma(i)} \left(\sum_{k \in Children(j)} f_k - f_j \right) - \sum_{\substack{\sigma(parent(j)) \\ < \sigma(i) < \sigma(j) \\ \delta_{IO}(j)=1}} f_j + n_i + \sum_{k \in Children(i)} f_k \leq M \tag{59.7}$$

Then the amount of data written in secondary memory is given by

$$IO = \sum_{\delta_{IO}(i)=1} f_i$$

In Equation 59.7, the term $\sum_{\delta_{IO}(j)=1} \sigma(parent(j)) < \sigma(i) < \sigma(j)\ f_j$ corresponds to the files that have been written into secondary memory at step $\sigma(i)$. The MinIO problem, which asks for an out-of-core traversal with the minimum amount of I/O volume, can be defined as follows.

Definition 59.4 (MINIO)

Given a tree T, and a fixed amount of main memory M, determine the minimum *I/O* volume *IO* needed by a solution of OUTOFCORETRAVERSAL (T, M).

The following three variants of the problem are NP-complete.

Theorem 59.4

Given a tree T with p nodes, and a fixed amount of main memory M, consider the following problems:

(i) Given a postorder traversal σ of the tree, determine the *I/O* schedule so that the resulting *I/O* volume is minimized.
(ii) Determine the minimum *I/O* volume needed by any postorder traversal of the tree.
(iii) Determine the minimum *I/O* volume needed by any traversal of the tree.

The (decision version of) each problem is NP-complete.

Note that (iii) is the original MINIO problem. Also note that the NP-completeness of (i) does not a priori imply that of (ii), because the optimal postorder traversal could have a particular structure. The same comment applies for (ii) not implying (iii). The proof of these three results all relies on a reduction from 2-Partition problem [22], using the tree illustrated in Figure 59.2. Scheduling this tree with a memory $M = 2S$ an maximum amount *IO* of $IO = 2/S$ requires to process first task T_{big}, and thus to select a subset of a_is (the input file of tasks T_i), which sums exactly to $S/2$. For more details on the proofs for all variants, see [29].

59.5.4 Perspectives and Open Problems

We have gathered in this section existing results on the problem of scheduling trees under memory constraints. Many problems are still open. The first one concerns splittable files. Indeed, the NP-completeness

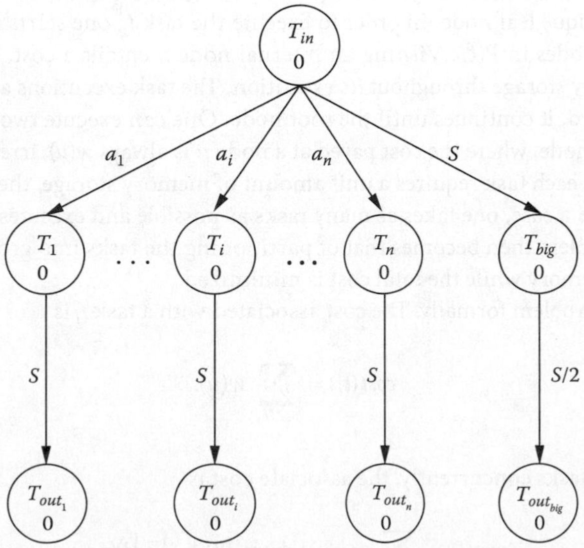

FIGURE 59.2 Reduction of an instance of the 2-Partition problem for the NP-completeness proof of MINIO.

of the MinIO problem comes from the difficulty to find a subset of files whose sizes sum up to a given bound. However, most memory systems nowadays are paginated and allows to write only portions of files to secondary memory. Therefore, studying the complexity of the MinIO problem with splittable files is very relevant. This problem is still open.

The results presented here only concern the processing of trees on a single machine. It is crucial to extend these results for other type of graphs (like series-parallel graphs, or any DAGs), and for parallel machines. It is likely that most problems will be NP-complete, but the results on trees may be useful to derive good heuristics, in the best case with approximation bounds. Even if approximation algorithms seem out of reach for general traversals, there is hope to derive an approximation algorithm for post-order traversals of trees, which are simpler to analyze than arbitrary traversals.

59.6 Case Study: Partitioning Tree Structured Computations

The problem addressed in this case study arises in the solution of linear systems with a direct method, and with an out-of-core environment [1]. In order to make the section self-contained, we investigate the combinatorial problem (in Sections 59.6.1 through 59.6.3) without linear algebra notions and refer the curious reader to the related reference items. We briefly discuss the mentioned application and one more related scenario in Section 59.6.4.

59.6.1 Definitions

Let $T = (V, E)$ be a rooted tree where all the edges are toward the root node r. Each node u, except the root, has a parent. Each node u, except the leaves, has a set of children, which are the nodes whose parent is u. For each node u, the vertices in the unique path from u to r are called the ancestors of u, and they are denoted by $P(u)$. The nodes u and r are included in $P(u)$. A subtree of T rooted at a node u contains all nodes v for which $u \in P(v)$; we use $T(u)$ to denote that subtree and use $L(u)$ to denote the leaf nodes in the subtree $T(u)$. A topological ordering of a tree $T = (V, E)$ is an ordering of the nodes V in such a way that any node is numbered before its parent; the root node is ordered last. A postorder traversal of a tree is a topological ordering where the nodes in each subtree are numbered consecutively.

There is a weight associated with each internal node of T. We use $w(u)$ to denote the weight of the node u. Leaf nodes are special and have zero weight. We are given a set of n tasks $t_1, t_2, \ldots, t_n$. Each task is associated with a unique leaf node. In order to execute the task t_i, one starts from a leaf node ℓ_i and visits all the internal nodes in $P(\ell_i)$. Visiting an internal node u entails a cost, $w(u)$. A task requires a unit amount of memory storage throughout its execution. The task executions are nonpreemptive, that is, when a task is started, it continues until the root node. One can execute two or more tasks concurrently at each internal node, where the cost payed at a node u is always $w(u)$, irrespective of the number of concurrent tasks. As each task requires a unit amount of memory storage, the number of concurrent tasks is limited. In such a case, one takes as many tasks as possible and executes them concurrently till the root node. The problem then becomes that of partitioning the tasks into groups so that each group can be stored in the memory while the total cost is minimized.

We now define the problem formally. The cost associated with a task t_i is

$$cost(t_i) = \sum_{u \in P(\ell_i)} w(u). \tag{59.8}$$

If we execute a set S of tasks concurrently, the associate cost is

$$cost(S) = \sum_{u \in P(S)} w(u) \quad \text{where } P(S) = \bigcup_{t_i \in S} P(t_i).$$

Let B be the maximum number of tasks one can execute concurrently. Then we have the following TreePartitioning problem. Given a tree T, a set of tasks $S = \{t_1, t_2, \ldots, t_n\}$, and an integer $B \le n$, partition S into a number of subsets $S_1, S_2, \ldots, S_K$ such that $|S_k| \le B$, for $k = 1, \ldots, K$ and the total cost

$$\sum_{k=1}^{K} cost(S_k) \tag{59.9}$$

is minimum. The number of parts K is not specified (but for feasibility we have $K \ge \lceil n/B \rceil$).

59.6.2 Complexity Results

Before introducing the complexity results (and algorithms), we present a lower bound for the cost of an optimal partition.

Theorem 59.5

Let T be a node weighted tree, $w(u)$ be the weight of node u, B be the maximum allowed size of a partition, and $L(u)$ be the number of leaf nodes in the subtree rooted at u. Then we have the following lower bound, denoted by η, on the optimal solution c^* of the TreePartitioning problem:

$$\eta = \sum_{u \in T} w(u) \times \left\lceil \frac{L(u)}{B} \right\rceil \le c^*.$$

Proof. Follows easily by noting that the leaf nodes of the subtree $T(u)$ will be partitioned among at least $\left\lceil \frac{L(u)}{B} \right\rceil$ parts.

Each internal node is on a path from (at least) one leaf node, therefore $\lceil L(u)/B \rceil$ is at least 1, and we have $\sum_u w(u) \le c^*$.

Theorem 59.6

The TreePartitioning problem is NP-complete.

Proof. We consider the associated decision problem: given a tree T with n leaves, a value of B, and a cost bound c, does there exist a partitioning S of the n leaves into subsets whose size does not exceed B, and such that $cost(S) \le c$? It is clear that this problem belongs to NP since if we are given the partition S, it is easy to check in polynomial time that it is valid and that its cost meets the bound c. We now have to prove that the problem is in the NP-complete subset.

To establish the completeness, we use a reduction from 3-PARTITION [22], which is NP-complete in the strong sense. Consider an instance $\mathcal{I}_1$ of 3-PARTITION: given a set $\{a_1, \ldots, a_{3p}\}$ of $3p$ integers, and an integer Z such that $\sum_{1 \le j \le 3p} a_j = pZ$, does there exist a partition of $\{1, \ldots, 3p\}$ into p disjoint subsets $K_1, \ldots, K_p$, each with three members, such that for all $1 \le i \le p$, $\sum_{j \in K_i} a_j = Z$?

We build the following instance $\mathcal{I}_2$ of our problem: the tree is a three-level tree composed of $N = 1 + 3p + pZ$ nodes: the root v_r, of cost w_r, has $3p$ children v_i, of the same cost w_v, for $1 \le i \le 3p$. In turn, each v_i has a_i children, each being a leaf node of zero cost. This instance $\mathcal{I}_2$ of the TreePartitioning

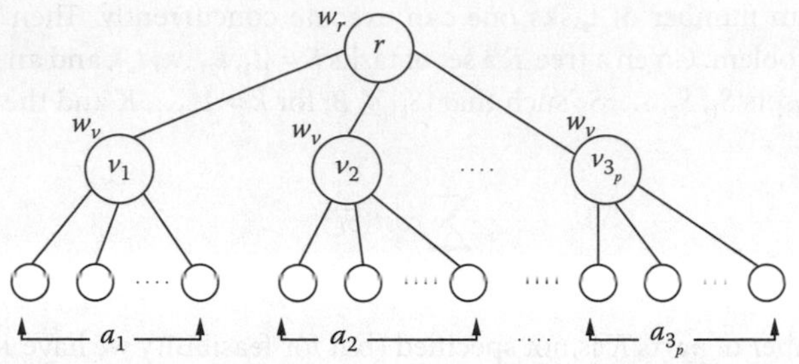

FIGURE 59.3 The instance of the TreePartitioning problem corresponding to a given 3-PARTITION PROBLEM. The weight of each node is shown next to the node. The minimum cost of a solution for $B = Z$ to the TreePartitioning problem is $p \times w_r + 3p \times w_v$, which is only possible when the children of each internal node are all in the same part, and when the children of three different internal nodes, say v_i, v_j, v_k, are put in the same part. This corresponds to putting the numbers a_i, a_j, a_k into a set for the 3-PARTITION problem, which sums up to Z.

problem is shown in Figure 59.3. We let $B = Z$ and ask whether there exists a partition of leaf nodes of cost $c = pw_r + 3pw_v$. Here w_r and w_v are arbitrary values (we can take $w_r = w_v = 1$). We note that the cost c corresponds to the lower bound shown in Theorem 59.5; in this lower bound, each internal node v_i is loaded only once, and the root is loaded p times, since it has $pZ = pB$ leaves below it. Note that the size of $\mathcal{I}_2$ is polynomial in the size of $\mathcal{I}_1$. Indeed, because 3-PARTITION is NP-complete in the strong sense, we can encode $\mathcal{I}_1$ in unary, and the size of the instance is $O(pZ)$.

Now we show that $\mathcal{I}_1$ has a solution if and only if $\mathcal{I}_2$ has a solution. Suppose first that $\mathcal{I}_1$ has a solution $K_1, \ldots, K_p$. The partition of leaf nodes corresponds exactly to the subsets K_i: we build p subsets S_i whose leaves are the children of vertices v_j with $j \in K_i$. Suppose now that $\mathcal{I}_2$ has a solution. To meet the cost bound, each internal node has to be loaded only once, and the root at most p times. This means that the partition involves at most p subsets to cover all leaves. Because there are pZ leaves, each subset is of size exactly Z. Because each internal node is loaded only once, all its leaves belong to the same subset. Altogether, we have found a solution to $\mathcal{I}_1$, which concludes the proof.

We can further show that we cannot get a close approximation to the optimal solution in polynomial time.

Theorem 59.7

Unless P = NP, there is no $1+o\left(\frac{1}{m}\right)$ polynomial approximation for trees with m nodes in the TreePartitioning problem.

Proof. Assume that there exists a polynomial $1+\frac{\varepsilon(m)}{m}$ approximation algorithm for trees with m nodes, where $\lim_{m\to\infty} \varepsilon(m) = 0$. Let $\varepsilon(m) < 1$ for $m \geq m_0$. Consider an arbitrary instance $\mathcal{I}_0$ of 3-PARTITION with a set $\{a_1, \ldots, a_{3p}\}$ of $3p$ integers, and an integer Z such that $\sum_{1\leq j\leq 3p} a_j = pZ$. Without loss of generality, assume that $a_i \geq 2$ for all i (hence $Z \geq 6$). We ask if we can partition the $3p$ integers of $\mathcal{I}_0$ into p triplets of the same sum Z. Now we build an instance $\mathcal{I}_1$ of 3-PARTITION by adding X times the integer $Z - 2$ and $2X$ times the integer 1 to $\mathcal{I}_0$, where $X = \max([(m_0 - 1)/(Z+3)] - p, 1)$. Hence, $\mathcal{I}_1$ has $3p + 3X$ integers and we ask whether these can be partitioned into $p + X$ triplets of the same sum Z. Clearly, $\mathcal{I}_0$ has a solution if and only if $\mathcal{I}_1$ does (the integer $Z - 2$ can only be in a set with two 1s).

We build an instance $\mathcal{I}_2$ of TreePartitioning from $\mathcal{I}_1$ exactly as we did in the proof of Theorem 59.6, with $w_r = w_v = 1$, and $B = Z$. The only difference is that the value p in the proof has been replaced by

$p + X$ here, therefore the three-level tree now has $m = 1 + 3(p + X) + (p + X)Z$ nodes. Note that X has been chosen so that $m \geq m_0$. Just as in the proof of Theorem 59.6, $\mathcal{I}_1$ has a solution if and only if the optimal cost for the tree is $c^* = 4(p + X)$, and otherwise the optimal cost is at least $4(p + X) + 1$.

If $\mathcal{I}_1$ has a solution, and because $m \geq m_0$, the approximation algorithm will return a cost at most

$$\left(1+\frac{\varepsilon(m)}{m}\right)c^* \leq \left(1+\frac{1}{m}\right)4(p+X) = 4(p+X) + \frac{4(p+X)}{m}.$$

But $\frac{4(p+X)}{m} = \frac{4(m-1)}{(Z+3)m} \leq \frac{4}{9} < 1$, so that the approximation algorithm can be used to determine whether $\mathcal{I}_1$, and hence $\mathcal{I}_0$, has a solution. This is a contradiction unless P = NP.

59.6.3 Approximation Algorithm

Consider the heuristic PoPart shown in Algorithm 59.4 for the TreePartitioning problem. As seen in Algorithm 59.4, the PoPart heuristic first orders the leaf nodes according to their rank in the postorder. It then puts the first B leaves in the first part, the next B leaves in the second part, and so on. This simple partitioning approach results in $\lceil n/B \rceil$ parts, for a tree with n leaf nodes, and puts B nodes in each part, except maybe in the last one. We have the following theorem, which states that this simple heuristic obtains results that are at most twice the cost of an optimum solution.

Algorithm 59.4 *PoPart: A postorder-based partitioning*

Input $T = (V, E)$ with n leaves; each requested entry corresponds to a leaf node.
Input B: the maximum allowable size of a part.
Output $\Pi_{PO} = \{S_1, \ldots, S_K\}$ where $K = \lceil n/B \rceil$, a partition on the leaf nodes.
1: compute a postorder of the nodes of T
2: $\mathcal{L} \leftarrow$ sort the leaf nodes according to their rank in the postorder
3: $S_k = \{\mathcal{L}(i) : (k-1) \times B + 1 \leq i \leq \min\{k \times B, n\}$, for $k = 1, \ldots, \lceil n/B \rceil\}$

Theorem 59.8

Let Π_{PO} be the partition obtained by the algorithm PoPart and c^* be the cost of an optimum solution. Then

$$cost(\Pi_{PO}) \leq 2 \times c^*.$$

Proof. Consider node u. Since the leaves of the subtree rooted at u are sorted consecutively in $\mathcal{L}$, the contribution of node u to the cost will be at most $\left\lceil \frac{L(u)}{B} \right\rceil + 1$. Therefore, the overall cost is at most

$$\begin{aligned} \text{cost}(\Pi_{PO}) &\leq \sum_u w(u) \times \left(\left\lceil \frac{L(u)}{B} \right\rceil + 1\right) \\ &\leq \eta + \sum_u w(u) \\ &\leq 2 \times c^* \end{aligned}$$

59.6.4 Application Scenarios

In the application of the TreePartitioning problem described by Amestoy et al. [1], each task corresponds to a sparse triangular solve operation with a right-hand-side (RHS) vector containing a single nonzero. There is a tree that describes the data requirements for each solve operation. Each internal node of the tree corresponds to parts of the triangular matrices stored in a file at disk. Solve operation for a given RHS vector starts from a leaf node and climbs up to the root. At each internal node, the matrix parts associated with that node are read from a file, and their inverse are applied to the RHS vector. All the RHS vectors are of the same size, and there is a huge number of them. Due to the limited memory size, one can hold only a certain number of RHS vectors at once in memory. In this setting, one organizes the solve operations in epochs where at each epoch a set of RHS vectors are solved for. Any internal node is accessed at most once during an epoch, but it can be accessed in multiple epochs. The solution time of an epoch is dominated by the time spent in reading the files containing the parts of the triangular matrices. In order to minimize the completion time of the application, one has to minimize the total size of the files read in. This is an instance of the TreePartitioning problem where the tasks are solve operations; B is the number of unit size RHS vectors that one can hold in memory; the cost associated with an internal node is the size of the corresponding file; and the objective function is the total size of the files read in.

Consider a hypothetical application whose tasks correspond to paths from a leaf node to the root in a given tree. Assume that each internal node is associated with a data. Consider the following parallelization approach. Each task is to be executed by a single processor. This necessitates replicating the nodes of the tree at different processors so that each task can be executed by the owner processor. This can be again described as an instance of the TreePartitioning problem, where this time B is a function of the available processors. Consider again the same application, where we do not want to replicate the data. In a possible parallelization approach, the data should be partitioned, and the tasks should be migrated from one processor to the other whenever needed (e.g., when the parent of a node and the node itself are at different processors). This time, one partitions the internal nodes of a tree so that the leaf-to-root paths are split among as few parts as possible. This seemingly different problem is addressed elsewhere [23] with techniques that are very different than what we saw in this section. However, it is possible to express both problems using hypergraph models and use a common algorithm to obtain effective solutions [52].

59.7 Case Study: Checkpointing Strategies

Any hardware component may be subject to faults. The Mean Time Between Failures (MTBF) of a processor is typically of the order of 100 years. The reader can thus very safely bet that his laptop processor will not fail within the next month. Therefore, she does not need any fault-tolerance technique to safely run a computation on his laptop for, say, 12 h. Things are completely different if the 12-h computation is performed using a large number of processors. Indeed, the probability that one of the processors used by the application will fail during the 12 h of the application execution dramatically increases with the number of processors enrolled in the execution. For instance, the 45,208-processor Jaguar platform is reported to experience on the order of one failure per day [10,38]. In other words, during any execution that would run for 12 h using the entire Jaguar platform, there is one chance out of two that a fault will occur. Faults that cannot be automatically detected and corrected in hardware lead to application failures. Using fault-tolerance techniques is thus mandatory when using a large number of processors for a significant amount of time. Resilience—the ability to resist to failures—is thus a key challenge for modern and future high-performance computing (HPC) systems [18,46].

The most used fault-tolerance technique is rollback recovery: after a failure, job execution is resumed from a previously saved fault-free execution state, or *checkpoint*. Rollback recovery implies frequent (usually periodic) *checkpointing* events at which the job state is saved to resilient storage. More frequent checkpoints lead to higher overhead during fault-free execution, but less frequent checkpoints lead to a larger loss when a failure occurs. The design of efficient *checkpointing strategies,* which specify when checkpoints should be taken, is thus key to high performance.

In this case study, we consider a single job to be executed on a predefined number of processors. The scheduling problem, here, is then to decide when (and how many times) to checkpoint the job in order to minimize its expected execution time.

59.7.1 Problem Statement

We consider an application, or *job*, that executes on *p processors*. The job must complete $\mathcal{W}$ units of (divisible) work: there are no constraints on when the job state can be checkpointed. We assume coordinated checkpointing [53], that is, that all processors checkpoint the job state simultaneously. We call *chunk* the fraction of the whole job that is executed between two consecutive checkpoints. Defining the sequence of chunk sizes is therefore equivalent to defining the checkpointing dates. We use C to denote the time needed to perform a checkpoint. The processors are subject to *failures*, each causing a *downtime* period, of duration D, followed by a *recovery* period, of duration R. The downtime accounts for software rejuvenation (i.e., rebooting [14,30]) or for the replacement of the failed processor by a spare. Regardless, we assume that after a downtime the processor that failed is fault-free and begins a new lifetime at the beginning of the recovery period. This period corresponds to the time needed to restore the last checkpoint. Note that although C and R can fluctuate depending on cluster and network load, as in most works in the field we assume they are constant. Finally, we assume that failures can happen during recovery or checkpointing, but not during a downtime (otherwise, the downtime period could be considered part of the recovery period).

We consider the processors from time t_0 onward. We use $P_{suc}(x|\tau_1, \ldots, \tau_p)$ to denote the probability that none of the processors fail for the next x units of time, knowing that the last failure of processor P_i occurred τ_i units of time ago, for $1 \leq i \leq p$. The failure interarrival times on the different processors are defined by independent and identically distributed random variables. Let X_i be the random variable for the interarrival time on processor P_i between the last failure (that happened τ_i units of time ago) and the next one. Then we have

$$P_{suc}\left(x \mid \tau_1, \ldots, \tau_p\right) = \mathbb{P}\left(X_1 \geq \tau_1 + x, \ldots., X_p \geq \tau_p + x \mid X_1 \geq \tau_1, \ldots, X_p \geq \tau_p\right).$$

Note that we do *not* assume that the failure stochastic process is memoryless.

59.7.2 Minimizing the Expected Makespan

Our aim is to minimize the job execution time. However, as we are trying to execute the job on a failure-prone platform, this execution time cannot be known beforehand: it depends on whether and when failures will impact the job execution. Therefore, our objective will not be the minimization of the makespan, like for instance in Section 59.4, but the *expectation* of the makespan, that is, the average makespan, the average being taken over all possible failure scenarios.

59.7.2.1 Formal Problem Definition

A solution to our scheduling problem is fully defined by a function $f(\omega|\tau_1, \ldots, \tau_p)$ that returns the size of the next chunk to execute given the amount of work ω that has not yet been executed successfully $\left(f(\omega \mid \tau_1, \ldots, \tau_p) \leq \omega \leq \mathcal{W}\right)$ and the amount of time τ_i elapsed since the last failure of processor P_i, $1 \leq i \leq p$. f is invoked at each decision point, that is, after each checkpoint or recovery. Our goal is to determine a function f that defines an optimal solution. Assuming unit-speed processors without loss of generality, the time needed to execute a chunk of size w is $w + C$ if no failure occurs.

For a given amount of work ω and a time elapsed since the last failure τ, we define $T(\omega|\tau_1, \ldots, \tau_p)$ as the random variable that quantifies the time needed for executing ω units of work. Given a solution function f, let $\omega_1 = f(\omega|\tau_1, \ldots, \tau_p)$ denote the size of the first chunk. We can write the following recursion:

$$T(0 \mid \tau_1, \ldots, \tau_p) = 0$$

$$T(\omega \mid \tau_1, \ldots, \tau_p) = \begin{cases} \omega_1 + C + T(\omega - \omega_1 \mid \tau_1 + \omega_1 + C, \ldots, \tau_p + \omega_1 + C) \\ \quad \text{if no processor fails during next } \omega_1 + C \text{ units of time,} \\ T_{wasted}\left(\omega_1 + C \mid \tau_1, \ldots, \tau_p\right) + T\left(\omega \mid \tau'_1, \ldots, \tau'_p\right) \text{ otherwise.} \end{cases} \quad (59.10)$$

These two cases are explained as follows:

- If none of the processors fail during the execution and checkpointing of the first chunk (i.e., for $\omega_1 + C$ time units), there remains to execute a work of size $\omega - \omega_1$ and the time since the last failure is incremented on each processor by $\omega_1 + C$.
- If at least one processor fails before the successful completion of the first chunk and of its checkpoint, then some additional delays are incurred, as captured by the variable $T_{wasted}(\omega_1 + C \mid \tau_1, \ldots, \tau_p)$. The time wasted corresponds to the execution up to the failure, a downtime, and a recovery during which some additional failures (and downtimes) may happen. Regardless, once a successful recovery has been completed, there still remain ω units of work to execute, and the time since the last failure of a processor depends on whether it fails (and when) during the attempt to process the first chunk, or the subsequent downtime and recovery.

We define MAKESPAN formally as find f that minimizes $\mathbb{E}\left(T\left(W \middle| \tau_1^0, \ldots, \tau_p^0\right)\right)$, where $\mathbb{E}(X)$ denotes the expectation of the random variable X, and τ_i^0 the time elapsed since the last failure of processor P_i that happened before t_0.

59.7.2.2 Solving MAKESPAN for the Exponential Distribution

In this section, we assume that the failure inter-arrival times of any processor follow an Exponential distribution with parameter λ, i.e., each $X_i = X$ has probability density $f_X(t) = \lambda e^{-\lambda t}\,dt$ and cumulative distribution $F_X(t) = 1 - e^{-\lambda t}$ for all $t \geq 0$. The advantage of the Exponential distribution, exploited time and again in the literature, is its "memoryless" property: the time at which the next failure occurs does not depend on the time elapsed since the last failure occurred. Therefore, in this section, we simply write $T(\omega)$, $T_{wasted}(\omega)$, and $P_{suc}(\omega)$ instead of $T(\omega|\tau_1, \ldots, \tau_p)$, $T_{wasted}(\omega|\tau_1, \ldots, \tau_p)$, and $P_{suc}(\omega|\tau_1, \ldots, \tau_p)$.

A challenge for solving MAKESPAN is the computation of $T_{wasted}(\omega_1 + C)$. We rely on the following decomposition:

$$T_{wasted}(\omega_1 + C) = T_{lost}(\omega_1 + C) + T_{rec}$$

Where

$T_{lost}(x)$ is the amount of time spent computing before a failure, knowing that the next failure occurs within the next x units of time

T_{rec} is the amount of time needed by the system to recover from the failure (accounting for the fact that other failures may occur during recovery)

The expectation of the makespan can then be expressed as a "simple" recursion:

Lemma 59.3

Let $\mathcal{W}$ be the amount of work to execute on p processors whose failure inter-arrival times follow *iid* Exponential distributions with parameter λ. The MAKESPAN problem is equivalent to finding a function f minimizing the following quantity:

$$\begin{aligned}\mathbb{E}(T(\mathcal{W})) &= P_{suc}(\omega_1 + C)\left(\omega_1 + C + \mathbb{E}\left(T(\mathcal{W} - \omega_1 + \omega_1 + C)\right)\right)\\ &\quad + \left(1 - P_{suc}(\omega_1 + C)\right)\left(\mathbb{E}\left(T_{lost}(\omega_1 + C)\right) + \mathbb{E}(T_{rec}) + \mathbb{E}\left(T(\mathcal{W})\right)\right)\end{aligned}$$

where $\omega_1 = f(\mathcal{W})$ and where $\mathbb{E}\left(T_{lost}(\omega)\right)$ is given by

$$\mathbb{E}\left(T_{lost}(\omega)\right) = \frac{1}{p\lambda} - \frac{\omega}{e^{p\lambda\omega} - 1}.$$

The memoryless property makes it possible to solve the MAKESPAN problem analytically:

Theorem 59.9

Let $\mathcal{W}$ be the amount of work to execute on p processors whose failure inter-arrival times follow *iid* Exponential distributions with parameter λ. Let $K_0 = \frac{p\lambda\mathcal{W}}{1 + \mathbb{L}(-e^{-p\lambda C - 1})}$, where $\mathbb{L}$, the Lambert function, is defined as $\mathbb{L}(z)e^{\mathbb{L}(z)} = z$. Then the optimal strategy to minimize the expected makespan is to split $\mathcal{W}$ into $K^* = \max\left(1, \lfloor K_0 \rfloor\right)$ or $K^* = \lceil K_0 \rceil$ same-size chunks, whichever minimizes $\psi(K^*) = K^*\left(e^{p\lambda\left(\frac{\mathcal{W}}{K^*} + C\right)} - 1\right)$.

The checkpointing strategy in Theorem 59.9 can be shown to be optimal among all deterministic and nondeterministic strategies, as a consequence of Proposition 4.4.3 in [44].

Interestingly, although we know the optimal solution with p processors, we are not able to compute the optimal expected makespan analytically. Indeed, $\mathbb{E}(T_{rec})$, for which there is a closed form in the case of sequential jobs [11], becomes quite intricate in the case of parallel jobs. This is because during the downtime of a given processor another processor may fail. During the downtime of that processor, yet another processor may fail, and so on. We would need to compute the expected duration of these "cascading" failures until all processors are simultaneously available. However, as the value of $\mathbb{E}(T_{rec})$ is independent of the choice of the chunk sizes, not knowing its value does not forbid to find the optimal chunk size function.

59.7.2.3 MAKESPAN and Arbitrary Distributions

Solving the MAKESPAN problem for arbitrary distributions is difficult because, unlike in the memoryless case, there is no reason for the optimal solution to use a single chunk size [50]. In fact, the optimal solution is very likely to use chunk sizes that depend on additional information that becomes available during the execution (i.e., failure occurrences to date). For sequential jobs, approximated dynamic programming solutions have been proposed [11]. They cannot be extended to the parallel case as this would require to memorize the evolution of the time elapsed since the last failure for all possible failure scenarios for each processor, leading to a number of states exponential in p. As we cannot attempt to solve directly MAKESPAN, we introduce a new related optimization problem.

This is especially important as failures in real-world computer systems are recognized not to follow exponential distributions but rather distributions like the Weibull distribution.

59.7.3 Maximizing the Expectation of the Amount of Work Completed

Rather than directly minimizing the job expected makespan, as in the MAKESPAN problem, in the NEXTFAILURE optimization problem we try to maximize the expected amount of work completed before the next failure occurs. NEXTFAILURE amounts to optimizing the makespan on a "failure-by-failure" basis, selecting the next chunk size as if the next failure were to imply termination of the execution. Intuitively, solving NEXTFAILURE should lead to a good approximation of the solution to MAKESPAN, at least for large job sizes $\mathcal{W}$. Therefore, we use the solution of NEXTFAILURE in cases for which we are unable to solve MAKESPAN directly.

59.7.3.1 Formal Problem Definition

For a given amount of work ω and a set of times elapsed since the last failure $\tau_1, \ldots, \tau_p$, we define $W(\omega|\tau_1, \ldots, \tau_p)$ as the random variable that quantifies the amount of work successfully executed before the next failure. Given a solution function f, let $\omega_1 = f(\omega|\tau_1, \ldots, \tau_p)$ denote the size of the first chunk. We can write the following recursion:

$$W(0 \mid \tau_1,\ldots,\tau_p) = 0$$

$$W(\omega \mid \tau_1,\ldots,\tau_p) = \begin{cases} \omega_1 + W(\omega - \omega_1 \mid \tau_1 + \omega_1 + C,\ldots,\tau_p + \omega_1 + C) \\ \quad \text{if no processor fails during the next } \omega_1 + C \text{ units of time,} \\ 0 \text{ otherwise.} \end{cases} \tag{59.11}$$

This recursion is simpler than the one for MAKESPAN because a failure during the computation of the first chunk means that no work (i.e., no fraction of ω) will have been successfully executed before the next failure. We define NEXTFAILURE formally as find f that maximizes $\mathbb{E}\left(W\left(\mathcal{W}|\tau_1^0, \ldots, \tau_p^0\right)\right)$.

59.7.3.2 Solving NEXTFAILURE

Weighting the two cases in Equation 59.11 by their probabilities of occurrence, we obtain the expected amount of work successfully computed before the next failure:

$$\mathbb{E}\left(W(\omega|\tau_1, \ldots, \tau_p)\right) = P_{suc}(\omega_1 + C \mid \tau_1, \ldots \tau_p)\left(\omega_1 + \mathbb{E}\left(W(\omega - \omega_1 \mid \tau_1 + \omega_1 + C, \ldots, \tau_p + \omega_1 + C)\right)\right).$$

Here, unlike for MAKESPAN, the objective function to be maximized can easily be written as a closed form, even for arbitrary distributions. Developing the expression above leads to the following result:

Proposition 59.3

The NEXTFAILURE problem is equivalent to maximizing the following quantity:

$$\mathbb{E}\left(W\left(\mathcal{W} \mid \tau_1^0, \ldots, \tau_p^0\right)\right) = \sum_{i=1}^{K} \omega_i \times \prod_{j=1}^{i} P_{suc}\left(\omega_j + C \mid t_1^j, \ldots, t_p^j\right),$$

where $t_i^j = \tau_i^0 + \sum_{\ell=1}^{j-1} (\omega_\ell + C)$ is the total time elapsed (without failure) before the start of the execution of chunk ω_j and since the last failure of processor P_i prior to the beginning of the application processing, and K is the (unknown) target number of chunks.

Unfortunately, there does not seem to be an exact solution to this problem. However, the recursive definition of $\mathbb{E}\left(W\left(\mathcal{W} \mid \tau_1, \ldots, \tau_p\right)\right)$ lends itself naturally to a dynamic programming algorithm. Formally:

Proposition 59.4

Using a time quantum u, and for any failure inter-arrival time distribution, DPNextFailure (Algorithm 59.5), called with parameters $\left(\mathcal{W}, \tau_1^0, \ldots, \tau_p^0\right)$, computes an optimal solution to NextFailure in time $O\left(\frac{\mathcal{W}^3}{u}\right)$.

Algorithm 59.5: *DPNextFailure* $(W, \tau_1, \ldots, \tau_p)$

if $W = 0$ **then return** 0
best $\leftarrow 0$
chunksize $\leftarrow 0$
for $\omega = \min\{\texttt{quantum}, W\}$ *to* W *step* `quantum` **do**
    (*expected_work*, *1st_chunk*) $\leftarrow$ DPNextFailure$(W - \omega, \tau_1 + \omega + C, \ldots, \tau_p + \omega + C)$
    cur_exp_work $\leftarrow P_{suc}(\tau_1 + \omega + C, \ldots, \tau_p + \omega + C \mid \tau_1, \ldots, \tau_p) \times (\omega + expected\_work)$
    if *cur_exp_work* > *best* **then** *best* $\leftarrow$ *cur_exp_work*
    chunksize $\leftarrow \omega$
return (*best*, *chunksize*)

59.7.4 Conclusion

In this case study, we have considered the problem of deciding when to save the execution state of an application, that is, when to take a checkpoint, in order to minimize the expectation of the execution time of an application in a failure-prone environment. We have showed that the optimal solution was to take periodic checkpoints when failures follow an exponential distribution. In the general case, it is not known how to solve the problem analytically. Instead, we have built an approximation of the optimal solution through a dynamic programming algorithm.

We have considered the problem in a simple setting where all checkpointing and recovery times were constant. Furthermore, we have analyzed the simplest checkpointing protocol, namely, *coordinated checkpointing*. More complex protocols have been designed in order to (try to) alleviate the cost of the coordination [19,20].

Rollback-recovery, as studied here, is only one of several fault-tolerance mechanisms. Another often used mechanism is *replication* (see [21] for a coupling of replication with rollback recovery). The question with replication is to decide which task should be replicated and how many times. The optimization objective could be the minimization of the expectation of the execution time (as in this study), or the maximization of the probability that the computation succeeds (reliability), or a multicriteria optimization mixing reliability and performance.

59.8 Conclusion

In this chapter, we have presented several case studies that we believe to be representative of modern scheduling problems that arise at large scale. Energy consumption, fault-tolerance, and memory management are important objectives that must be taken into account in addition to pure makespan minimization.

The case studies are also representative of the increased level of difficulty of modern scheduling problems, which typically involve multicriteria optimization. A variety of mathematical tools and algorithmic techniques are used to solve such problems, and we hope to have convincingly illustrated many of them in this chapter. In addition, many bibliographical items have been provided within the text, and in further information.

Acknowledgement

This work was supported in part by the ANR RESCUE project. Anne Benoit and Yves Robert are with the Institut Universitaire de France.

Key Terms

Checkpoint and rollback recovery: The technique of saving the current state of the execution at different points in time; when a failure occurs, the application can rollback to, and recover from, the most recently saved state: execution resumes from that state instead of from the very beginning.

Dynamic voltage and frequency scaling (DVFS): By lowering supply voltage, hence processor clock frequency, it is possible to achieve important reductions in power consumption; faster speeds allow for a faster execution, but also lead to a much higher (supra-linear) power consumption.

Heterogeneity: Resources are said to be heterogeneous when they do not have all the exact same characteristics. For instance, the set of processors in a grid is heterogeneous if they do not all have the same processing capabilities or the same amount of memory.

Makespan: The makespan of an application is the time at which the application execution is completed.

Re-execution: A task whose execution is not reliable enough (for instance, because its execution speed is below a threshold) is executed twice (or more) to enhance its reliability.

Resource selection: Is the problem of deciding which, among all available resources, should be used to solve a given problem.

Scheduling: Is the problem of deciding at what time (when), and on which resource (where), to execute each of the (atomic) tasks that must be executed on the given platform.

Further Information

There is a very abundant literature on scheduling and load balancing strategies for parallel and/or distributed systems. The survey [49] is a good starting point to this work. The most famous heuristic to schedule a task graph on an heterogeneous platform is HEFT [51].

A good introduction to multicriteria optimization is the book by Pinedo [41]. Section 59.4 is mainly based on [6]. Section 59.5 is mainly based on [29]. Liu proposes two optimal algorithms for optimal memory traversal of trees, the first one restricted to postorder traversals [33], the second one for any traversal [34]. For more information on the pebble game model such as complexity results for general DAGs, see the seminal work of Sethi [47] and Gilbert, Lengauer and Tarjan [24].

The problem discussed in Section 59.6 is mainly based on a recent work by Amestoy et al. [1]. The paper includes some improved algorithms for the TreePartitioning problem, as well as a hypergraph partitioning based formulation. The postorder-based algorithm discussed in Section 59.6.3 is implemented and incorporated in Mumps [2], a sparse direct solver.

Young is the author of a seminal study of periodic checkpointing policies [55]. His work was later extended by Daly [16]. Section 59.7 is mainly based on [11], which contains references to many studies of checkpointing policies. For noncoordinated checkpointing protocols, [27] by Guermouche et al. is a good starting point.

References

1. P. Amestoy, I. S. Duff, J.-Y. L'Excellent, Y. Robert, F.-H. Rouet, and B. Uçar. On computing inverse entries of a sparse matrix in an out-of-core environment. *SIAM Journal on Scientific Computing (SISC)*, 34(4): A197–A199, 2012.
2. P. R. Amestoy, I. S. Duff, J.-Y. L'Excellent, and J. Koster. A fully asynchronous multifrontal solver using distributed dynamic scheduling. *SIAM Journal on Matrix Analysis and Applications*, 23(1):15–41, 2001.
3. P. R. Amestoy, A. Guermouche, J.-Y. L'Excellent, and S. Pralet. Hybrid scheduling for the parallel solution of linear systems. *Parallel Computing*, 32(2):136–156, 2006.
4. I. Assayad, A. Girault, and H. Kalla. Tradeoff exploration between reliability power consumption and execution time. In *Proceedings of Conference on Computer Safety, Reliability and Security (SAFECOMP)*, Washington, DC, IEEE Computer Society Press, 2011.
5. G. Aupy, A. Benoit, F. Dufossé, and Y. Robert. Reclaiming the energy of a schedule, models and algorithms. *Concurrency and Computation: Practice and Experience*, 2012. Also available as INRIA Research Report 7598. Available at graal.ens-lyon.fr/~abenoit
6. G. Aupy, A. Benoit, and Y. Robert. Energy-aware scheduling under reliability and makespan constraint. Research Report 7757, INRIA, France, February 2012. Available at graal.ens-lyon.fr/~abenoit
7. H. Aydin and Q. Yang. Energy-aware partitioning for multiprocessor real-time systems. In *Proceedings of International Parallel and Distributed Processing Symposium (IPDPS)*, pp. 113–121. Nice, France. IEEE Computer Society Press, 2003.
8. M. Baleani, A. Ferrari, L. Mangeruca, A. Sangiovanni-Vincentelli, M. Peri, and S. Pezzini. Fault-tolerant platforms for automotive safety-critical applications. In *Proceedings of International Confernational on Compilers, Architectures and Synthesis for Embedded Systems*, pp. 170–177 San Jose, CA, ACM Press, 2003.
9. N. Bansal, T. Kimbrel, and K. Pruhs. Speed scaling to manage energy and temperature. *Journal of the ACM*, 54(1):1 – 39, 2007.
10. L. B. Gomez, A. Nukada, N. Maruyama, F. Cappello, and S. Matsuoka. Transparent low-overhead checkpoint for GPU-accelerated clusters. Presentation at *The Fourth Workshop of the INRIA-Illinois Joint Laboratory on Petascale Computing*. Available at https://wiki.ncsa.illinois.edu/display/jointlab/Workshop+Program
11. M. Bougeret, H. Casanova, M. Rabie, Y. Robert, and F. Vivien. Checkpointing strategies for parallel jobs. In *SC'2011, the IEEE/ACM Conference on High Performance Computing Networking, Storage and Analysis*. ACM Press, New York, 2011.
12. P. Brucker. *Scheduling Algorithms*. Springer, Berlin, Germany, 2007.
13. H. Casanova, A. Legrand, and Y. Robert. *Parallel Algorithms*. CRC Press, Boca Raton, FL, 2008.
14. V. Castelli, R. E. Harper, P. Heidelberger, S. W. Hunter, K. S. Trivedi, K. Vaidyanathan, and W. P. Zeggert. Proactive management of software aging. *IBM Journal of Research and Development*, 45(2):311–332, 2001.
15. J.-J. Chen and T.-W. Kuo. Multiprocessor energy-efficient scheduling for real-time tasks. In *Proceedings of International Conference on Parallel Processing (ICPP)*, pp. 13–20, Oslo, Norway. IEEE Computer Society Press, 2005.

16. J. T. Daly. A higher order estimate of the optimum checkpoint interval for restart dumps. *FGCS*, 22(3):303–312, 2004.
17. V. Degalahal, L. Li, V. Narayanan, M. Kandemir, and M. J. Irwin. Soft errors issues in low-power caches. *IEEE Transactions on Very Large Scale Integration Systems*, 13:1157–1166, October 2005.
18. J. Dongarra, P. Beckman, P. Aerts, F. Cappello, T. Lippert, S. Matsuoka, P. Messina et al. The international exascale software project: A call to cooperative action by the global high-performance community. *International Journal of High Performance Computing Application*, 23(4):309–322, 2009.
19. E. Elnozahy and J. Plank. Checkpointing for peta-scale systems: A look into the future of practical rollback-recovery. *IEEE Transactions on Dependable and Secure Computing*, 1(2):97–108, April–June 2004.
20. E. N. M. Elnozahy, L. Alvisi, Y.-M. Wang, and D. B. Johnson. A survey of rollback-recovery protocols in message-passing systems. *ACM Computing Survey*, 34:375–408, 2002.
21. K. Ferreira, J. Stearley, J. H. I. Laros, R. Oldfield, K. Pedretti, R. Brightwell, R. Riesen, P. G. Bridges, and D. Arnold. Evaluating the viability of process replication reliability for exascale systems. In *Proceedings of the 2011 ACM/IEEE Conference on Supercomputing*, New York, 2011.
22. M. R. Garey and D. S. Johnson. *Computers and Intractability, A Guide to the Theory of NP-Completeness*. W.H. Freeman and Company, San Francisco, CA, 1979.
23. J. Gil and A. Itai. How to pack trees. *Journal of Algorithms*, 32(2):108–132, 1999.
24. J. R. Gilbert, T. Lengauer, and R. E. Tarjan. The pebbling problem is complete in polynomial space. *SIAM Journal on Computing*, 9(3):513–524, 1980.
25. R. Graham, E. Lawler, J. Lenstra, and A. R. Kan. Optimization and approximation in deterministic sequencing and scheduling: A survey. *Annals of Discrete Mathematics*, 5:287–326, 1979.
26. R. L. Graham. Bounds for certain multiprocessor anomalies. *Bell System Technical Journal*, 45:1563–1581, 1966.
27. A. Guermouche, T. Ropars, M. Snir, and F. Cappello. HydEE: Failure containment without event logging for large scale send-deterministic MPI applications. In *Proceedings of IPDPS 2012*, Shanghai, China. IEEE Computer Society Press, 2012.
28. B. Hong and V. K. Prasanna. Adaptive allocation of independent tasks to maximize throughput. *IEEE Transactions on Parallel Distributed Systems*, 18(10):1420–1435, 2007.
29. M. Jacquelin, L. Marchal, Y. Robert, and B. Uçar. On optimal tree traversals for sparse matrix factorization. In *IPDPS'2011, the 25th IEEE International Parallel and Distributed Processing Symposium*, Anchorage, AK. IEEE Computer Society Press, 2011.
30. N. Kolettis and N. D. Fulton. Software rejuvenation: Analysis, module and applications. In *FTCS '95*, p. 381, Washington, DC. IEEE Computer Society Press, 1995.
31. C.-C. Lam, T. Rauber, G. Baumgartner, D. Cociorva, and P. Sadayappan. Memory-optimal evaluation of expression trees involving large objects. *Computer Languages, Systems and Structures*, 37(2):63–75, 2011.
32. S. Lee and T. Sakurai. Run-time voltage hopping for low-power real-time systems. In *Proceedings of Annual Design Automation Conference (DAC)*, Los Angeles, CA, pp. 806–809, 2000.
33. J. W. H. Liu. On the storage requirement in the out-of-core multifrontal method for sparse factorization. *ACM Transactions on Mathematical Software*, 12(3):249–264, 1986.
34. J. W. H. Liu. An application of generalized tree pebbling to sparse matrix factorization. *SIAM Journal Algebraic Discrete Methods*, 8(3), 1987.
35. S. H. Low. A Duality model of TCP and queue management algorithms. *IEEE/ACM Transactions on Networking*, 4(11):525–536, 2003.
36. L. Massoulié and J. Roberts. Bandwidth sharing: Objectives and algorithms. *Transactions on Networking*, 10(3):320–328, June 2002.
37. R. Melhem, D. Mosse, and E. Elnozahy. The interplay of power management and fault recovery in real-time systems. *IEEE Transactions on Computers*, 53:2004, 2003.
38. E. Meneses. Clustering parallel applications to enhance message logging protocols. Presentation at *The Fourth Workshop of the INRIA-Illinois Joint Laboratory on Petascale Computing*. Available at http://charm.cs.illinois.edu/talks/ClusteringAppNcsaInriaWorkshop10.pdf

39. A. J. Oliner, R. K. Sahoo, J. E. Moreira, M. Gupta, and A. Sivasubramaniam. Fault-aware job scheduling for BlueGene/L systems. In *Proceedings of International Parallel and Distributed Processing Symposium (IPDPS)*, pp. 64–73, Santa Fe, NM, 2004.
40. C. Picouleau. Task scheduling with interprocessor communication delays. *Discrete Applied Mathematics*, 60(1–3):331–342, 1995.
41. M. L. Pinedo. *Scheduling: Theory, Algorithms, and Systems*. Springer, New York, 2008.
42. P. Pop, K. H. Poulsen, V. Izosimov, and P. Eles. Scheduling and voltage scaling for energy/reliability trade-offs in fault-tolerant time-triggered embedded systems. In *Proceedings of IEEE/ACM International Conference on Hardware/Software Codesign and System Synthesis (CODES+ISSS)*, pp. 233–238, New York, 2007.
43. R. B. Prathipati. Energy efficient scheduling techniques for real-time embedded systems. Master's thesis, Texas A&M University, College Station, TX, May 2004.
44. M. L. Puterman. *Markov Decision Processes: Discrete Stochastic Dynamic Programming*. Wiley, New York, 2005.
45. V. J. Rayward-Smith, F. W. Burton, and G. J. Janacek. Scheduling parallel programs assuming preallocation. In P. Chrétienne, E. G. Coffman Jr., J. K. Lenstra, and Z. Liu, eds., *Scheduling Theory and Its Applications*. John Wiley & Sons, New York, 1995.
46. V. Sarkar et al. Exascale software study: Software challenges in extreme scale systems, 2009. White paper available at: http://users.ece.gatech.edu/mrichard/ExascaleComputingStudyReports/ECSS%20report%20101909.pdf
47. R. Sethi. Complete register allocation problems. In *STOC'73: Proceedings of the Fifth Annual ACM Symposium on Theory of Computing*, pp. 182–195. ACM Press, New York, 1973.
48. S. M. Shatz and J.-P. Wang. Models and algorithms for reliability-oriented task-allocation in redundant distributed-computer systems. *IEEE Transactions on Reliability*, 38:16–27, 1989.
49. B. A. Shirazi, A. R. Hurson, and K. M. Kavi. *Scheduling and Load Balancing in Parallel and Distributed Systems*. IEEE Computer Society Press, Los Alamitos, CA, 1995.
50. A. Tantawi and M. Ruschitzka. Performance analysis of checkpointing strategies. *ACM TOCS*, 2(2):123–144, 1984.
51. H. Topcuoglu, S. Hariri, and M. Y. Wu. Performance-effective and low-complexity task scheduling for heterogeneous computing. *IEEE Transactions on Parallel Distributed Systems*, 13(3):260–274, 2002.
52. B. Uçar. Partitioning problems on trees and simple meshes. Presentation at *15th SIAM Conference on Parallel Processing for Scientific Computing (PP12)*, Savannah, GA, February 2012.
53. L. Wang, P. Karthik, Z. Kalbarczyk, R. Iyer, L. Votta, C. Vick, and A. Wood. Modeling coordinated checkpointing for large-scale supercomputers. In *Proceedings of ICDSN'05*, Yokohama, Japan, pp. 812–821, 2005.
54. L. Wang, G. von Laszewski, J. Dayal, and F. Wang. Towards energy aware scheduling for precedence constrained parallel tasks in a cluster with DVFS. In *Proceedings of CCGrid'2010, the 10th IEEE/ACM International Conferences on Cluster, Cloud and Grid Computing*, pp. 368 –377, Melbourne, Victoria, Australia, May 2010.
55. J. W. Young. A first order approximation to the optimum checkpoint interval. *Communications of the ACM*, 17(9):530–531, 1974.
56. D. Zhu and H. Aydin. Energy management for real-time embedded systems with reliability requirements. In *Proceedings of IEEE/ACM International Conference on Computer-Aided Design (ICCAD)*, pp. 528–534, San Jose, CA, 2006.
57. D. Zhu, R. Melhem, and D. Mossé. The effects of energy management on reliability in real-time embedded systems. In *Proceedings of IEEE/ACM International Conference on Computer-Aided Design (ICCAD)*, pp. 35–40, Washington, DC. IEEE Computer Society Press, 2004.

60

Distributed File Systems*

Thomas W. Doeppner
Brown University

Distributed file systems (DFSs) provide common file access to a collection of computers; all in the collection can access files as if they were local, even though the files do not reside in their local file systems. The need for such a distributed system grew out of the switch in the 1980s from multiuser time-shared systems to networks of workstations (or personal computers). A community of people using a time-sharing system shared a single file system, all files were accessed identically by programs, and files were shared trivially. Furthermore, the file systems were easily administered—they were all attached to one machine and could be easily backed up. A DFS provides all the benefits of the file system on a time-sharing system and opens them up to users of collection of personal computers.

The history of DFSs goes back to the late 1970s and early 1980s. The first commercially successful DFS came as part of the Apollo Domain system, released in 1982, soon followed by Sun's network file system (NFS). What motivated both these systems was the desire not only to share files but also, since disks were fairly expensive, to support relatively cheap, diskless workstations.

To a certain extent, the ideal DFS behaves as if all parties were on the same computer—in other words, as if it really were not a distributed system. Programs designed to work using a local file system should work just as well using a DFS. Coming close to this ideal is possible, but most existing systems do not—they make trade-offs so as to improve performance and reduce complexity.

Things get complicated when we consider failures: client crashes, server crashes, and communication failures. In the single-computer case, either the entire system is up or the entire system is down; we typically do not worry about file-system semantics on "partially up" or "partially down" systems since these cases simply do not arise. In a DFS, though, what should happen if a server crashes but its clients remain up? Or suppose clients and servers are up, but the network connections between them are down. Or suppose a client crashes while the rest of the system remains up.

The most obvious benefit of the DFS is that files are shared. If I want to let you work with one of my files, I simply give you its pathname. You see everything I've done to the file. If you make any modifications to

* Doeppner, Thomas W., *Operating Systems In Depth: Design and Programming* (New Jersey, John Wiley & Sons, 2011). This material is reproduced with permission of John Wiley & Sons, Inc.

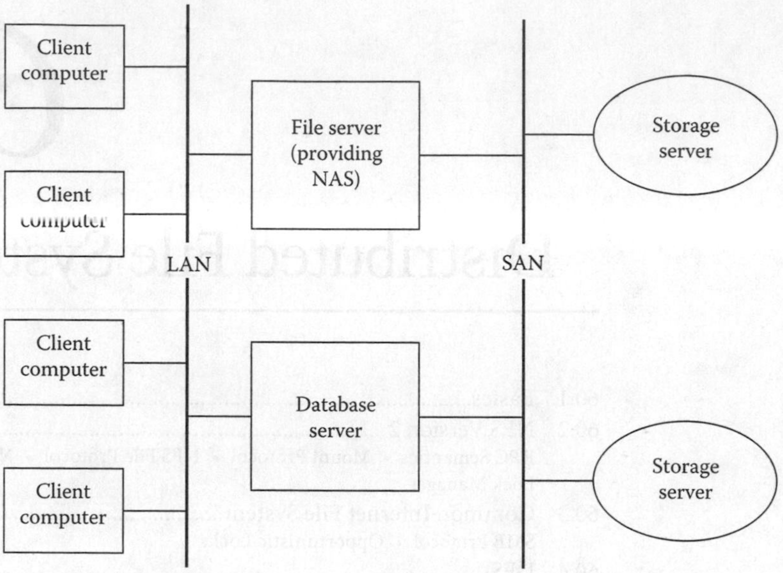

FIGURE 60.1 The relationship between *network attached storage* (NAS) and a *storage area network* (SAN).

it, I see those modifications. Neither of us needs to worry about synchronizing the contents of multiple copies of a file. Without the DFS, we would have to exchange files by e-mail or use applications such as the Internet's file transfer protocol.

A related benefit of a DFS is that my programs can access remote files as if they were local. The remote files have no special application-program interface; they are accessed just like local ones. With this property, known as *access transparency*, programs need not be aware whether a file is remote or local.

Finally, a DFS is easier to administer than multiple local file systems. Typically in such a system, a small number of server machines provide files to a large number of clients. Rather than having to back up files on each of the clients, only the servers' files need to be backed up.

Another term for DFS is *network attached storage*, referring to attaching to the network servers that provide file systems. A similar-sounding term that refers to a very different approach is *storage area network*. Here, the intent is to make available over the network not file systems but storage devices. For example, an organization might own a few large-capacity redundant array of independent disks (RAID) devices, each providing many terabytes of storage. The total capacity might be split into a number of logical storage devices, each assigned to a specific server. Some of the servers might be file servers, which use their logical storage devices to hold a file system. Others might be database systems, which use their logical storage devices to hold databases. The capacity can be shuffled among logical devices as needed to meet new requirements; the redundancy provided by RAID is shared by all. See Figure 60.1.

60.1 Basics

The ideal DFS should not give its users any unpleasant surprises: it should behave as much as possible as if it and its users were all on the same computer. The semantics of file operations should be exactly the same for DFSs as for local file systems. When multiple applications access the same file, the result should be the same on the distributed system as if all parties, clients and server, were on the same system. Failures, such as communication problems and crashes of client and server computers, should be minimally disruptive. And to top it all off, performance should be good.

Designing such a system entails determining how its various components are distributed—what resides on the servers, what resides on the clients. We categorize this information as follows:

- The *data state*: this is the contents of files.
- The *attribute state*: this is information about each file, such as its size, most recent modification time, and access-control information.
- The *open-file state*: this is the information indicating which files are open or otherwise in use as well as describing how files are locked. For example, in both Unix and Windows systems, opening a file adds to its reference count, and a file cannot be deleted unless its reference count is zero. A range of bytes of a file can have a shared (or read) lock or an exclusive (or write) lock applied to it.

As shown in Figure 60.2, each of these components might be kept on the server or on the clients, either in volatile storage (which is lost if a machine crashes) or in non-volatile storage (which is not lost after a crash). We think of the data state as permanently residing on the server's local (disk) file system, but recently accessed or modified information might reside in volatile storage in server or client caches (or both). Some systems even cache data on client disks. Similarly, the attribute state resides permanently on the server's local file system, but might be cached on servers and clients. The open-file state is transitory; it changes as processes open and close files. Thus, it makes sense for this to reside in volatile storage, but, as we discuss in the subsequent sections, for crash-recovery purposes, some systems keep a portion of it in non-volatile storage.

Three basic concerns govern this placement:

1. *Access speed*: Access is slow if the information resides on the server, yet is accessed and modified by clients. Caching such information on clients improves performance considerably. Information stored in volatile storage can be accessed and modified much more quickly than information stored in non-volatile storage.
2. *Consistency*: If clients cache information, do all parties share the same view of it?
3. *Recovery*: If one or more computers crash, to what extent are the others affected? How much information is lost?

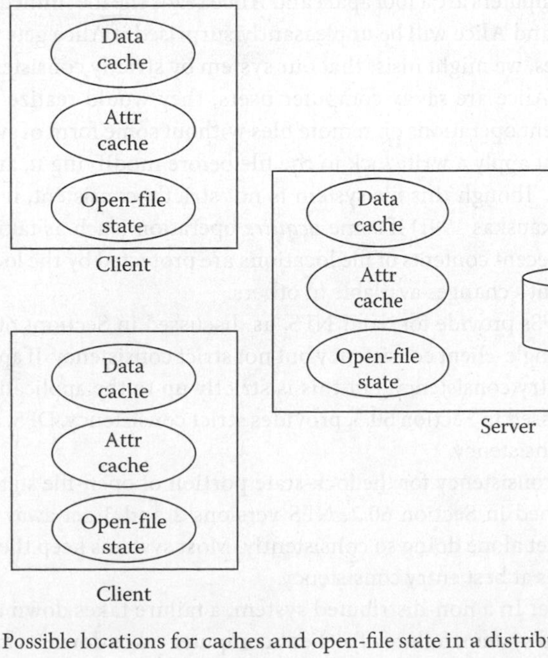

FIGURE 60.2 Possible locations for caches and open-file state in a distributed file system.

Though the factors affecting access speed are straightforward, those affecting consistency and recovery are not. So, before discussing the trade-offs among these concerns, we first spend some time discussing consistency and recovery, giving a quick overview of what is relevant to DFSs and focusing on data consistency.

Suppose a single thread is performing operations on files. We want the order in which these operations occur on the files to be the same as was executed by the thread; that is, we want the operations to be *time-ordered*. If things were otherwise, we would consider the file system broken and would not use it. If no operations are performed by other threads, any read of a particular location of a file by a thread retrieves the result of the thread's most recent write to the location. We call this property *single-thread consistency*.

It is not a big stretch to extend this form of consistency to an entire client computer: the operations on the file system by the computer are time-ordered, and thus, in the absence of operations by other computers, any read of a particular file location by a thread on the computer retrieves the result of the most recent write to the location by any thread of the computer. This we call *single-client consistency*.

It is a bigger step to extend consistency to the threads on all the client machines of a distributed system. But if all operations on a file system are time-ordered, and thus any read of a particular file location by any thread on any computer retrieves the result of the most recent write to that location by any thread on any computer, the file system is said to have *strict consistency*.

It may be difficult if not impossible to time-order the operations coming from many client computers. If operations are executed on different client computers, how do we know which happened first? In general, if this cannot be determined, it probably does not matter. What is important is that all parties agree on whatever order is selected and that the selected order be one that could have happened. Suppose Ted is updating a file on one computer and Alice is reading that file on another computer, if they do these operations at about the same time, and if the clocks on the various computers either are not synchronized well enough or do not have enough resolution to order the two events, then it really is not clear which happened first and either order makes sense—neither Ted nor Alice will be surprised by either outcome. Assuming that the system remains single-client consistent, this weakening of strict consistency is known as *sequential consistency* (Lamport 1979).

But suppose Ted's and Alice's computers are a foot apart and Alice reads the file immediately after Ted tells her he's modified it, both Ted and Alice will be unpleasantly surprised if Alice gets the old version of the file. To rule out such surprises, we might insist that our system by strictly consistent.

On the other hand, if Ted and Alice are savvy computer users, they would realize that expecting deterministic results from concurrent operations on remote files without some form of synchronization might not be realistic. So Ted might apply a write lock to the file before modifying it, and Alice might apply a read lock before reading it. Though this file system is not strictly consistent, it is said to have *entry consistency* (Bershad and Zekauskas 1991) if some *acquire* operation (such as taking a lock) can guarantee the client that the most recent contents of file locations are protected by the lock, and if some *release* operation can make the client's changes available to others.

What sorts of consistency do DFSs provide for data? NFS, as discussed in Sections 60.2 and 60.5 in the subsequent sections, provides single-client consistency but not strict consistency. If applications lock the files they are using, they get entry consistency, but this is strictly up to the applications. Common internet file system (CIFS), as discussed in Section 60.3, provides strict consistency. DFS, as discussed in Section 60.4, also provides strict consistency.

All systems must provide strict consistency for the lock-state portion of open-file state—locks make no sense otherwise. But, as described in Section 60.2, NFS versions 2 and 3 get away with not even sharing the reference state of files, let alone doing so consistently. Most systems keep the attribute state strictly consistent, but NFS provides at best entry consistency.

What happens if there is a failure? In a non-distributed system, a failure takes down the application and everything else. In a distributed system, failure of a client computer takes down its application. But if the server or other clients fail, the application is not necessarily affected at all. Another way of putting this is that for non-distributed systems, either the entire system is up or the entire system is down. When the system comes back up after a crash, people do not expect the cache and the open-file state to be as they were before the crash—all application processes must be restarted anyway.

But with a DFS, partial crashes can happen and must be dealt with. One or more clients might crash, the server might crash, and there might be times during which certain computers cannot communicate with one another due to network problems. In some cases, complete recovery is possible. In others, for instance when a client crashes, problems are restricted to the crashed computer—no other computer is adversely affected. In still other cases, problems can occur that would not have happened if a local file system were being used.

For example, an application that has locked a file might discover that, because of a combination of server crashes and network problems, not only has it lost its lock on the file but also the file has been locked by another application. Ideally, of course, such "surprises" will not happen. But, as explained in the following sections, these surprises, though rare, are unavoidable. The best we can ask for is that applications are notified when they occur.

In the subsequent sections, we look at a few representative DFSs, discussing the trade-offs taken among speed, consistency, and recoverability. We start with NFS version 2, a relatively simple yet successful system. As already noted, it has a fairly relaxed approach to consistency but handles failures well. Next, we cover CIFS, which provides strict consistency but is essentially intolerant of server failures and network problems. We then discuss DFS, which provides strict consistency and is tolerant of server failures and network problems, but requires the cooperation of applications in coping, and finally NFS version 4, which does as good a job with failures as is probably possible.

A final concern, independent of the others, is the file-system name space. If all parties are on one computer, there is probably just a single directory hierarchy and thus one name space. But for the distributed case, the one-name-space approach does not necessarily make sense. Both NFS and CIFS allow each client to set up its own name space, and both also provide a means for all clients to share the same name space. DFS uses a model in which one global name space is shared by all in the distributed system.

60.2 NFS Version 2

NFS was developed by Sun Microsystems in the mid-1980s. Its design is simple,* yet successful; it and its descendants have been widely used ever since, uniquely among DFSs of its time. In this section, we focus on NFS version 2. Version 1 was used strictly within Sun. At the time of this writing, version 3 is in common use, while version 4 (discussed in Section 60.5) is beginning to be adopted.

NFSv2 servers are passive parties that give their clients a simple file store. They respond to client requests, but take no actions on their own. NFS clients actively maintain caches of information obtained from servers, performing read-aheads and write-behinds as necessary. The approach to consistency is rather relaxed but works well in practice. Here is how things work. An NFS server gives its clients access to one or more of its local file systems, providing them with opaque identifiers called *file handles* to refer to files and directories. Clients use a separate protocol, the *mount protocol*, to get a file handle for the root of a server file system. They then use the *NFS file protocol* to follow paths within the file system and to access its files, placing simple remote procedure calls to read them and write them. A third protocol added later, the *network lock manager protocol* (NLM), can be used to synchronize access to files. All communication between client and server is with a remote procedure call (RPC) protocol known as open network computing (ONC) RPC (see Chapter 9 of Doeppner (2011)).

Here, we call the combination of the mount protocol and the NFS file protocol *basic NFS*; basic NFS plus NLM we call *extended NFS*. We begin by discussing basic NFS.

---

* This characterization applies only to version 2 of NFS!

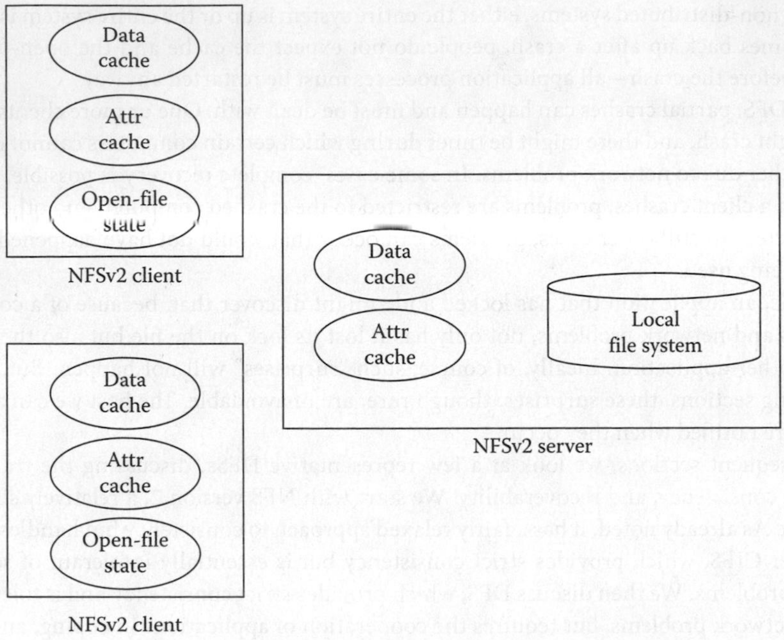

FIGURE 60.3 Distribution of components in NFSv2.

All open-file state information resides on the clients. Servers are thus called *stateless* since they are essentially unaware of what files clients have open, though they necessarily do hold data and communications state.

Figure 60.3 summarizes the information maintained on clients and servers. Clients maintain caches of file data and attributes as well as hold all open-file information (except for locks, which are maintained on the server via the NLM protocol). No other open-file information is held on the server.

To see how NFSv2 operates, let us work through an example. Suppose a Unix thread executes the following instructions, accessing a file system on an NFSv2 server:

```
char buffer[100];
int fd = open("/home/twd/dir/fileX", O_RDWR);
read(fd, buffer, 100);
...
lseek(fd, 0, SEEK_SET);
write(fd, buffer, 100);
```

Our thread opens a file, reads the first 100 bytes, then writes 100 bytes at the beginning of the file, replacing the previous data there. The kernel data structures representing the open file are just what they would be if the file were local. The only difference from the client operating system's point of view is that the actual operations on the file are handled by the NFSv2 client code rather than by a local file system such as FFS.

The NFSv2 server has its own directory hierarchy and exports a number of subtrees of this hierarchy to its clients. The client's operating system has mounted one of them on the local directory /home/twd. It did this by first placing a remote procedure call, using the mount protocol, to the server, and obtaining a file handle for the root of this subtree. As Figure 60.4 shows, any attempt to access the directory /home/twd is now interpreted as an attempt to access the remote directory represented by the file handle.

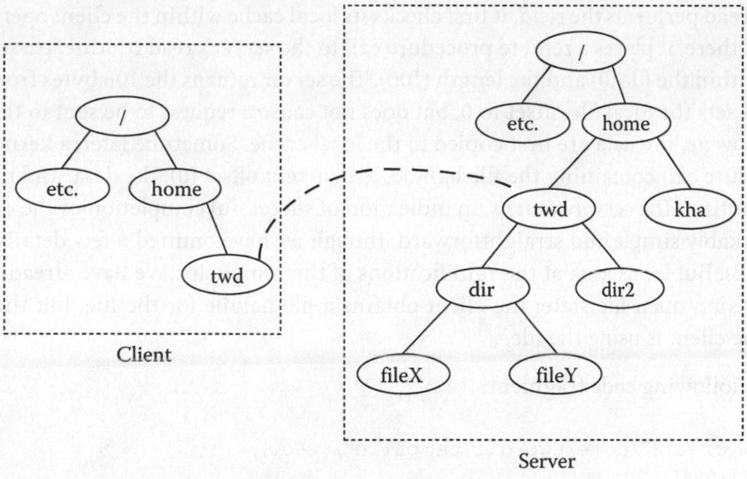

FIGURE 60.4 The client has mounted the server's */home/twd* on its own */home/twd*.

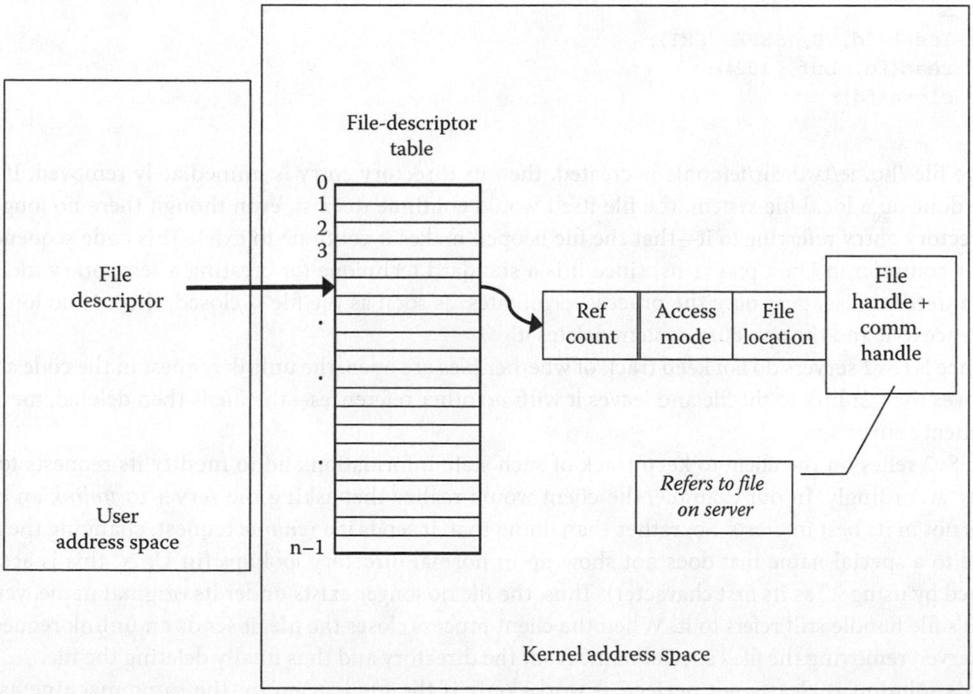

FIGURE 60.5 Open-file data structures on client.

Thus to follow the path /home/twd/dir/fileX, our thread, executing in the client operating system, goes as far as /home/twd and discovers that a remote file system is mounted there. It places a remote procedure call to the NFSv2 server's *lookup* routine, passing it the file handle for the remote file system's root as well as the pathname /dir/fileX. The server returns a file handle for fileX.

The client operating system records the fact that fileX is open, just as it would for a local file. However, rather than representing the file with an inode or equivalent, it uses the file handle together with a communication handle representing the remote server (see Figure 60.5).

When the thread performs the *read*, it first checks its local cache within the client operating system. If the data are not there, it places a remote procedure call to the server's read routine, passing the file handle, the offset within the file (0) and the length (100). The server returns the 100 bytes from that location.

The *lseek* call sets the local file offset to 0, but does not cause a request to be sent to the server. When the thread calls *write*, the data are first copied to the local cache. Sometime later, a kernel thread places a remote procedure call containing the file handle, the current offset (0), the data, and the length to the server's *write* routine. The server returns an indication of successful completion of the operation.

This is remarkably simple and straightforward, though we have omitted a few details (to be covered in Section 60.2.3). But let us look at the ramifications of this simplicity. We have already noted that the server maintains no open-file state: the client obtains a file handle for the file, but the server has no record that some client is using the file.

Consider the following code fragment:

```
int fd = creat("/home/twd/dir/tempfile", 0600);
char buf[1024];
unlink("/home/twd/dir/tempfile");
...
write(fd, buf, 1024);
...
lseek(fd, 0, SEEK_SET);
read(fd, buf, 1024);
close(fd);
```

The file /home/twd/dir/tempfile is created, then its directory entry is immediately removed. If this were done on a local file system, the file itself would continue to exist, even though there no longer is a directory entry referring to it—that the file is open makes it continue to exist. This code sequence is rather common in Unix programs, since it is a standard technique for creating a temporary file, one guaranteed to disappear once the process terminates: as soon as the file is closed, there is no longer a reference to it and the operating system deletes it.

Since NFSv2 servers do not keep track of whether files are open, the unlink request in the code above removes the last link to the file and leaves it with no other references—the file is then deleted, much to the client's surprise.

NFSv2 relies on the client to keep track of such state information and to modify its requests to the server accordingly. In our example, the client would realize that asking the server to *unlink* an open file is not in its best interests. So, rather than doing that, it sends it a *rename* request, changing the file's name to a special name that does not show up in normal directory lookups (in Unix, this is accomplished by using "." as its first character). Thus, the file no longer exists under its original name, yet the client's file handle still refers to it. When the client process closes the file, it sends an unlink request to the server, removing the file's special name from the directory and thus finally deleting the file.

This solution is clearly not perfect: it works only if the file is open on the same machine as the one doing the unlink. In practice, this is the only case that matters, so it does well enough. But what happens if the client renames a file, but crashes before the file is closed? Then the renamed file is not deleted and remains on the server. The server must periodically check for such orphaned files and delete them.

The NFS Mount Protocol, described in Section 60.2.2, lets clients access subtrees of server directory hierarchies. Clients may place these subtrees in their own directory hierarchies wherever they please, but it is usually more convenient to handle this automatically so that all clients have the same file-system name space. Most organizations using NFS define such a name space in an organization-wide database. Each client, using software known as the *automounter*, consults this database and mounts server-provided subtrees into its own directory hierarchy as necessary.

60.2.1 RPC Semantics

The NFSv2 protocols are layered on top of ONC RPC and depend on it for reliable communication between client and server. RPC extends the procedure-call abstraction across different computers. Ideally, when a client calls a remote procedure, it is assured that the procedure is invoked exactly one time, and the results are returned to the client. This assurance is made difficult because of the possibility of network and server failures. For example, if a client invokes a procedure on the server, but does not get a response, is it because of a network failure, a server failure, or both? Perhaps, the remote procedure was actually executed on the server, but due to a network failure, the response did not arrive at the client.

If the effect of executing a remote procedure multiple times is the same as executing it once, that is, the procedure is *idempotent*, then it is safe for the client to repeat a request if it does not get a response. Most of the procedures provided by NFSv2 servers for their clients are idempotent. For example, to modify a file, a client invokes a WRITE procedure providing the file's identity, the data to be written, and the file location that is to be modified. If the procedure is executed twice, the same data are rewritten to the same location, but the effect is as if it has been written only once.

However, some NFSv2 procedures are not idempotent. For example, consider the procedure to unlink a file from a directory (which deletes the file if its only directory link is being removed). If the procedure is executed once, the link is removed. If it is executed again, the request fails, because the link cannot be removed (since it no longer exists). There is no harm done on the server from executing the procedure twice, but the result can harm a client application that inexplicably has a request fail that should have succeeded.

ONC RPC copes with this problem through the use of duplicate request caches (DRCs) on servers, which maintain lists of recently received RPC requests and their responses. Whenever a request is received, the server checks its DRC to determine if the request is a duplicate of one already received (clients assign unique identifiers to their requests; a retransmission of a request carries the same ID as the original transmission). If the request is a duplicate, then, rather than re-execute it, the server simply sends back the original response. This approach works perfectly if items are not removed from the cache unless they will never be received again. NFSv2 and v3 servers remove items from their DRCs after they've been there for a certain amount of time. This works well if clients and servers reside in "reasonably well behaved" networks, but, if the network is not-so-well-behaved, servers may re-execute a duplicate request. NFSv4's DRC does a much better job, as described in the subsequent sections.

ONC RPC handles requests as follows. If user datagram protocol (UDP), an unreliable protocol, is the transport protocol, the client RPC layer repeatedly retransmits each request (with suitable delays in between) until it gets a response. If transmission control protocol (TCP), a reliable protocol, is the transport protocol, the client RPC layer does not retransmit unless there is a loss of connection, in which case, a new connection is made and the client retransmits over it. With both transports, the effect is that requests are received by the server's RPC layer at least once. Most requests are for idempotent procedures, and thus can be safely executed many times in a row. The server uses its DRC to handle duplicate non-idempotent requests.

What is particularly nice about this approach is that client applications need not be aware that they are using a DFS and need do nothing special to cope with crashes or network problems. The client RPC layer eventually makes the desired result happen: if the server crashes while an application is attempting to read from a file, the read system call simply appears to be slow, but it eventually returns with the desired data (assuming the server recovers from the crash).

As an option, a client can set things up so that system calls on files of a particular remote file system eventually time out if the server has crashed or there is a network problem. In practice, this is done only rarely, because few applications do anything intelligent in response to such timeouts.

60.2.2 Mount Protocol

This is by far the simplest of the NFSv2 protocols. Its primary job is to provide clients with file handles for roots of exported file systems. It also provides limited security by giving these file handles only to certain specified clients.

What are file handles? Servers give them to clients to identify files on the server. From the client's point of view, they are completely opaque: their contents have no meaning other than as interpreted on the server.

From the server's point of view, however, file handles play a role similar to that of Unix inodes: they uniquely identify files. While a file might have multiple pathnames, it has only one file handle. A file handle must identify a file even if the file has been renamed; it must continue to be valid until the file is deleted. After a file has been deleted, its file handle must be invalid, so that clients can reliably determine that a file no longer exists.

So, what does a file handle contain? For a Unix server that actually identifies files by their inodes, the file handle contains an identifier for the file-system instance (the server might have many) as well as the inode number. Thus, it identifies a particular file within the file system.

But this is not enough: if a file is deleted, its file handle must become invalid. The problem with using the combination of inode number and file-system identifier as a handle is that, though this combination will be invalid immediately after the file is deleted, the inode number might be reused for a new file. Thus, the handle becomes valid again and refers to a different file from what the holder had intended.

To get around this problem, we need a way to distinguish between different uses of the same inode number (or similar identifiers). What is done is to tag each inode with a generation number, a 32-bit field that takes on a different value each time the inode is reused—a random value works well here. The file handle contains a copy of the value that is current when the handle is created. If the corresponding file is deleted and its inode reused, the generation number in the file handle no longer matches that of the inode, and the handle is rejected—the error message is "stale file handle."

60.2.3 NFS File Protocol

The NFS File Protocol does most of NFS's work. It is responsible for communicating file data between client and server and for maintaining client caches and keeping them reasonably consistent. Its basic operation is remarkably simple. Client application processes make system calls on files. Forget-type operations—fetching data, file attributes, etc.—if the relevant information is in the cache, the call completes immediately. Otherwise, the NFS client code places a RPC to the server both to obtain the needed information for the cache and to return it to the application. For most put-type operations, such as file writes, the information is stored in the local cache, and sometime later NFS client code updates the server via a RPC.

Assuming the communications protocol is providing exactly once semantics for each client's RPC requests, the client caches are single-client consistent. But what happens if more than one client is using a file? The basic premise of NFSv2 is that file sharing takes place primarily for read-only files, such as binaries. Writable files are generally used by only one client at a time. So, with this premise as justification, NSFv2 uses an approximate approach to consistency. For cases in which the premise is unjustified, the separate NLM comes into play, as described in Section 60.2.4.

In the NFSv2, approximate approach to consistency, cached information is assumed to be valid for a certain period of time after it is put in the cache. NFS client code periodically checks with the server for changes to files that are at least partially cached on the client. If so, regardless of the extent of the changes, the cached information for the file is invalidated and subsequent get-type requests result in RPCs to the server.

To aid in doing this, clients maintain a separate cache of file attributes—information obtained from the server about the files they are accessing (see Figure 60.6). This information is essentially anything the server can provide, including access permissions, file size, creation time, access time, and modification time. This cached information is used to satisfy requests by processes for file attributes (via the *stat* and *fstat* system calls in Unix). It is also used in determining whether the cached data from a file is up to date. Cached attributes are valid only for a brief period of time, ranging from a few seconds to a minute, and cached file data are valid only while the file's attributes are valid. If an attempt is made

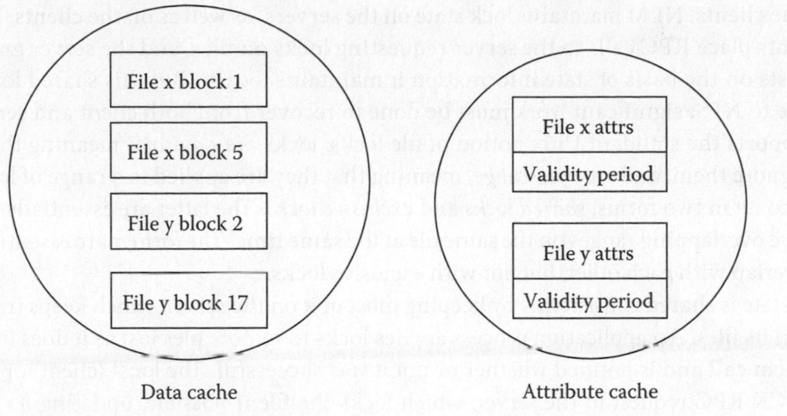

FIGURE 60.6 An NFSv2 client's data and attribute caches.

to access cached data beyond the attributes' validity period, a "GETATTR" RPC request is sent to the server asking for the file's current attributes. On return, the new attributes are cached; if they indicate that the file has not been modified since the previous attributes were cached, the cached data are still used. Otherwise, the file's cached data are flushed—this means that modified file blocks are sent to the server via WRITE RPC requests and unmodified blocks are removed from the cache.

This attribute caching helps keep caches "sort of" consistent but does not give any useful guarantees. Many NFS implementations go a step farther and provide "close-to-open" consistency: when a file is closed, all changes are sent to the server. When a file is opened, the remaining validity time for its cached blocks is ignored, and a GETATTR is sent to the server. The blocks are flushed if the new attributes indicate recent changes. This procedure helps prevent surprises: if a file is shared by multiple clients, but is open on at most one at a time, the clients' caches will be consistent.

Given the near lack of state information the NFS File Protocol keeps on the server, dealing with client and server crashes is straightforward—nothing needs to be done! However, care must be taken with certain operations so that crashes do not cause undue surprises. If a client application performs a synchronous request, such as writing to a file synchronously, the request really needs to be performed synchronously—the application should not be notified of its completion until after the server file system has been updated. Otherwise, in the event of a client or server crash, the user of the application may be under the false impression that the file has been modified.

In fact, all RPC calls that update the server must be synchronous: a client that receives a "successful-completion" response assumes its job is done—the server has been updated and the client can move on. So the modifications really must have been committed on the server. Otherwise, if the server crashes before the updates have been committed, they are lost forever.

This synchronous-update requirement slows file updates on NFSv2 systems, since, for example, all WRITE RPC requests must be synchronous. Some implementations handled this by providing a non-volatile cache into which updates could be committed on the server. The cache survived crashes, and its contents were written to the file system as part of crash recovery.

One of the major improvements in NFSv3 was to weaken the requirement for synchronous updates. Writes in NFSv3 may be asynchronous, meaning that the server may return a response even though the data has not been written to disk. But when the client sends a COMMIT RPC request, the server may not respond until the data from all previous writes to the file are safely on disk.

60.2.4 Network Lock Manager

The final component of NFSv2, forming extended NFS from basic NFS, is its network lock manager (NLM) protocol, which enables clients to lock files. While in the rest of NFSv2, the open-file state resides in the clients, NLM maintains lock state on the servers, as well as on the clients. Its operation is simple: clients place RPC calls to the server requesting locks on files, and the server grants or refuses these requests on the basis of state information it maintains locally. But this shared lock state adds a new wrinkle to NFS: significant work must be done to recover from both client and server crashes.

NLM supports the standard Unix notion of file locks: locks are *advisory*, meaning that applications are free to ignore them, and are *byte-range*, meaning that they are applied to a range of locations within a file. They come in two forms, *shared locks* and *exclusive locks*. The latter are essentially write locks: no two may have overlapping ranges on the same file at the same time. The former are essentially read locks: they may overlap with each other, but not with exclusive locks.

The lock state is shared consistently by keeping most of it on the server, which keeps track of the locks applied to all its files. An application process applies locks to remote files just as it does to local ones—it issues a system call and is notified whether or not it was successful. The local (client) operating system places a LOCK RPC request to the server, which locks the file if possible, updating its own lock-state information, and sends back a success/failure response. If successful, the open-file state on the client is updated to indicate that a portion of the file has a lock applied to it.

The hard part of maintaining lock state is coping with crashes. NLM avoids much of the potential complexity of crash recovery by simply ignoring the difficult cases. In practice, this has worked reasonably well, primarily because NFSv2 is used in well behaved environments that lack Byzantine routers and other networking problems.

The first issue in crash recovery is determining that there has been a crash. This is fraught with difficulty since a false determination causes big problems. NLM avoids these problems by being extremely conservative. A computer is deemed to have crashed only when it comes back up and announces it is restarting. Otherwise, it is assumed to be running fine (despite, say, being unresponsive for the past hour).

Thus, NLM clients keep on disk a list of which servers they are holding locks from, and servers keep on disk a list of which clients are holding locks on their files. When a client restarts, it notifies the servers in its list that it is restarting, and the servers release all the client's locks.

Recovering from a server crash involves recovering the lock state of all the clients. This would be trivial if the state had been stored on disk along with the client list, but for performance reasons, it is not. Instead, the lock state is kept in volatile memory and thus disappears with a server crash. To recover it, servers rely on their clients.

As when clients restart, when a server restarts, it notifies all the clients in its list of lock holders. These clients then have a certain period of time, the *grace period* (typically 45 s), to reclaim the locks they held before the server crashed. During this period, the server accepts no new lock requests.

NLM's approach to recovery is not perfect. We take this up again in Section 60.5.2, when we discuss NFSv4.

60.3 Common Internet File System

CIFS is Microsoft's DFS. Its long history goes back to the mid-1980s, when it was developed by IBM and was known as the server message block (SMB) protocol, running on top of Network Basic Input/Output System (NetBIOS). Microsoft took over further development of SMB and renamed it CIFS in the late 1990s. Though it still can run on top of NetBIOS, NetBIOS is often layered on top of TCP and, with some simple extensions, SMB can run directly on top of TCP (without NetBIOS). The term SMB is still used to refer to the RPC-like communications protocol on which CIFS is layered.

The CIFS approach is to provide strict consistency among clients and servers, allowing client-side caching for efficient handling of the two most common file usages—when a number of clients are sharing a file read-only and when a file is being used exclusively by one client. One way to characterize the differences between CIFS and NFSv2 is that CIFS is always strictly consistent and usually fast, while NFSv2 is always fast and usually strictly consistent. Though keep in mind that by "fast," we mean only that client-side caching is used.

Unlike basic NFS (see Section 60.2), CIFS servers hold open-file state from their clients. Thus if a client crashes, this state must be cleaned up. On the other hand, if a server crashes, clients make no attempt to recover their state when it comes back up: operations on files that were open at the time of a crash will fail.

CIFS requires the transport protocol to be reliable and regards the transport connection as the sole indication of the health of the computer at the other end. So if the connection is lost, perhaps because of a network timeout, CIFS considers the other computer to have crashed.

This approach to failure allows CIFS to be simpler than NFSv2 and other DFSs. A request is not retried if the connection is lost: the application must assume the worst, that the server has crashed and there is no hope of recovery. Whether or not the remote procedure is idempotent does not matter.

The organization of the CIFS namespace is similar to that of NFS. Servers export "shares," subtrees of their own hierarchies. Clients may place them in their local namespaces as desired—Windows clients typically assign drive letters to them (creating a local "forest"). Alternatively, much as NFS defines a namespace shared by all clients by using the automounter and a distributed database, Windows uses a distributed database called "active directory" to define a namespace shared by all clients.

To get a better idea of how CIFS works, consider the code in the following text, which is similar to the code we examined in Section 60.2 when looking at NFSv2:

```
char buffer[100];
HANDLE h = CreateFile(
          "Z:\dir\file",                 // name
          GENERIC_READ|GENERIC_WRITE,    // desired access
          0,                             // share mode
          NULL,                          // security attributes
          OPEN_EXISTING,                 // creation disposition
          0,                             // flags and attributes
          NULL                           // template file
);
ReadFile(h, buffer, 100, NULL, NULL);
…
SetFilePointer(h, 0, NULL, FILE_BEGIN);
WriteFile(h, buffer, 100, NULL, NULL);
CloseHandle(h);
```

In this code, a thread opens an existing file (despite the name of the system call!), requesting both read–write access and that the file not be shared with others until the returned handle is closed (it does this by supplying a "share mode" parameter of 0—we discuss share modes later in this section). It reads 100 bytes from the beginning of the file, then, after resetting the file pointer, writes 100 bytes to the beginning of the file. Finally, it closes the handle, allowing others to open the file.

Here, in outline form, is what takes place for all this to happen. At some point before this code is executed, the client must have created a transport connection with the server. Then, so that the server can do access checking, the client identifies itself to the server (and proves its identity). Next, the client identifies to the server the "disk share" it wants to access. This is a subtree of the server's file system that the server is willing to export. If the client is allowed access to the share, the server returns a "tree ID" (TID) identifying the share. Our client associates this exported file system with the drive letter "Z." Note that this is much as in the NFSv2 mount protocol, via which the client obtains a file handle for the root of a server file system (see Section 60.2.2).

At this point, our client thread can issue the *CreateFile* system call to open the file it wants. It passes to the server the *UID*, identifying the user, the *TID*, identifying the share, and the path (in this case\ *dir\file*) to follow in that share. In addition, it passes the *desired access*, which the server checks to make certain it is allowed, the *share mode*, which the server must honor, and the other parameters (except for the template file, which is ignored when opening an existing file, but may be used as a source for the attributes if creating a new file). If successful, the server returns a "file ID" (FID), which the client subsequently uses to identify the file.

As in NFS, read and write system calls by client threads do not necessarily send requests to the server immediately; they might instead be satisfied by a local cache. But when the client operating system fetches data from the server or sends modified data back to it, it sends read and write requests that identify the file by the FID. The file pointer, used by the application thread to identify file position, is maintained on the client side; read and write requests sent to the server contain an explicit file location.

Unlike NFSv2 servers, CIFS servers must be aware of what their clients are doing. In general, they keep track of not only the reference counts on files, but also the *share modes*. These indicate in what ways a file may have concurrent opens. The default, a share mode of 0, is that concurrent opens are not allowed. But a thread can specify, for example, that others can open the file for reading or that others may open the file for writing, or perhaps for both. Thus, when a thread tries to open a file, the open succeeds only if the desired access does not conflict with the share mode of all the other current opens of that file.

CIFS supports "byte-range" locks that are similar to those supported by NFSv2 in that application threads can lock portions of a file. However, CIFS's locks are mandatory—they must be honored by all threads on all clients, unlike the locks supported by NFSv2 (and NFSv3). From the client application's perspective, locks are associated with the combination of the file handle and the process: all threads of one process using the same file handle share the lock on a byte range of a file. Even if this file handle is shared with threads of a different process, these latter threads do not share the lock. If another handle for the file is created in the original process (by opening the file again), threads using that second handle do not share the lock. Thus, the server must keep track of not only which file handle a lock is associated with but also with which process on which client machine.

60.3.1 SMB Protocol

The SMB protocol plays the role of RPC for CIFS. Like RPC, it is a request-response protocol: clients (and sometime servers) send requests; servers (and sometimes clients) send responses. However, unlike a true RPC protocol, it is not extensible. It consists of a predefined set of requests and responses, all in a fixed format. Nevertheless, it suffices for CIFS (after all, it was designed for CIFS!).*

As the name SMB implies, requests and responses are sent as message blocks, each containing a standard header along with some arguments and data. A client first sets up a session with a server that has already established the client user's identity. The client and server must first negotiate, which version of the SMB protocol they are using and then, via a log-on procedure, establish the user's identity (thus setting the UID) and prove it to the server. After the session is established, the client may request access to one or more server "shares," obtaining a TID for each. With such a TID, the client may open a file, thus obtaining a FID.

Each client operating system must identify which user process is sending a request by supplying a process ID (PID) unique to the client. PIDs are necessary for correctly implementing locks but are also used to match responses to requests. The idea is that a process has only one outstanding request at a time—which made sense before processes were multithreaded, but not since. To allow the client operating

* Microsoft RPC is actually layered on top of SMB, primarily so that it can exploit the authentication and resource (share) identification provided by SMB. However, CIFS does not use Microsoft RPC.

TABLE 60.1 SMB Message Exchange in CIFS

Message	Meaning
SMB_COM_NEGOTIATE -> server	Client gives server list of SMB versions it is willing to use.
client <- SMB_COM_NEGOTIATE	Server chooses one and informs client. It also sends the client a security challenge.
SMB_COM_SESSION_SETUP_ANDX -> server	Client sends encrypted challenge to server, along with user's identity.
client <- SMB_COM_SESSION_SETUP_ANDX	Server verifies identity and fills in UID field.
SMB_COM_TREE_CONNECT_ANDX -> server	Client identifies the share it wants to use.
client <- SMB_COM_TREE_CONNECT_ANDX	Server fills in security fields and TID field.
SMB_COM_OPEN_ANDX -> server	Client sends path name of file.
client <- SMB_COM_OPEN_ANDX	Server performs access checks (based on UID) and, if OK, responds by filling in FID field.
SMB_COM_READ -> server	Client requests data from file.
client <- SMB_COM_READ	Server returns data.
SMB_COM_CLOSE -> server	Client closes file.
client <- SMB_COM_CLOSE	Server returns success.
SMB_COM_TREE_DISCONNECT -> server	Client disconnects from share.
client <- SMB_COM_TREE_DISCONNECT	Server returns success.

system (OS) to sort out responses to multithreaded processes, yet another ID is used: a multiplex ID, which, along with the PID, performs the function of ONC RPC's XID.

The set of IDs form the context for a request and are included in special field in the header, as needed. For many of the details of SMB see Storage Networking Industry Association (2002).

Table 60.1 (adapted from Storage Networking Industry Association (2002)) shows a typical exchange of messages.

A fair number of messages must be exchanged before the client can get any data from a file, and most of these messages simply add one more ID to the context. So, rather than require these messages in sequence, SMB lets them be batched: a client sends a sequence of requests in a single message, and the server responds to all of them in a single response message. This batching ability is given the ungainly name of "andX." Thus in the sample exchange, in Table 60.1, the requests starting with the session setup request through the read request can all be sent in one message. If they all succeed, the server sends back a single response containing all the IDs that have been established and the data requested by the read. If any of the requests fails, the server handles no further requests from the message and returns the cumulative result of all requests up through the failing one. Any request (or response) may be at the end of a batch; only those requests (or responses) whose names end with "_ANDX" may come before other requests (or responses) in a batch.

60.3.2 Opportunistic Locks

One way to achieve strict consistency is to perform all operations at the server without caching data at clients. This is CIFS's default approach, but definitely not the only one possible. Local caching of a file's data at a client is safe, with respect to consistency, either if the client is the only one using the file or if all clients using the file are doing so read-only. The hard part is limiting such caching to just those two cases.

CIFS relies on the server's knowledge of clients' open-file states to determine when it is safe for a client to do local caching. If a client opens a file for reading and writing that is not otherwise open, the server may notify the client that caching is safe. If another client wants to open the file (assuming the share mode permits it), the server must notify the first client that caching is no longer safe and that the client must send all changes it is made to the file back to the server. Then both clients may use the file, but no local caching is permitted on either.

When a client opens a file, as part of the open SMB request, it may ask for an *opportunistic lock* (or *oplock*). There are two categories of such locks: level I and level II. The former allow exclusive access; the latter allow shared access. The server may or may not grant the request depending on whether and how the file is open by others. If the client obtains a level-I oplock, then it is assured that no other client has the file open and thus it can safely cache file data (and metadata) locally. If it obtains a level-II oplock, then it is assured that no other client has the file open in a mode that allows changes, and thus it can safely cache blocks from the file, but any changes it makes must be sent directly to the server, and the server will then revoke all level-II oplocks on the file. Thus, if a file is being used for both reading and writing, a level-I oplock is needed, but if it is being used just for reading, level II suffices. Note that the client application is oblivious of oplocks—it simply opens files and performs operations on them. It is the client's operating system that is responsible for local caching and thus must request oplocks.

To revoke an oplock, the server sends the client an oplock-break SMB request. If a level-I oplock is being revoked, the server will give it a level-II oplock if possible. The client must send back all changes to the server, then signify it is done so by sending a response to the oplock-break request. Since there is nothing to send back when a level-II oplock is being revoked, no response is required.

The CIFS protocol does not allow a client to request a level-II oplock. Instead, it must request a level-I oplock, but it might, at the server's discretion, get a level-II one instead. This might seem a bit strange; the rationale is that though files are often used read-only, rarely are they opened that way. Thus, most opens are read-write, even though writing may well not be done. So, if one client has a file open read-write but does not write, and another client opens the same file read-write but does not write either, then if both clients request level-I oplocks, they will end up with level-II oplocks.

If a client closes a file for which it has an oplock, will its cache still be useful if it opens the file again a short time later? The SMB close request releases all oplocks associated with the file, so there's no guarantee that cached data are valid if the file is reopened—since the client no longer has an oplock, the server is not obliged to notify the client if any other client opens the file. However, the client might cheat a bit: rather than sending an SMB close request to the server when its user application closes the file, it can keep quiet in hopes that some local application opens the file again. If this indeed happens and the client still holds the oplock, then the client is guaranteed that its cache is still valid. On the other hand, if another client opens the file, the server will revoke the original client's oplock, thus telling it that its cache is no longer valid. At this point the client might tell the server that the file is really closed.

For level-II oplocks, such a delayed-close strategy might seem easy: the client holds a level-II oplock when the application closes the file. Shortly thereafter, an application opens the file again. Since the client already has a level-II oplock, not only is the cache still valid but no sort of request needs to be sent to the server. However, as we just mentioned, few applications open files read-only. Thus the second open will probably be for read-write use (even though the file might never be written to). Since the only opportunity the client has to request an oplock is when the file is opened, the client must request a level-I oplock on the off chance that the application might actually write to the file. Thus, an SMB open request must be sent to the server, and the server might grant the level-I oplock or might simply let the client keep the level-II oplock. But the cached file blocks at the client are still valid, so despite the need to send a request to the server, the delayed-close strategy is a win.

The delayed-close strategy for level-I oplocks works a bit differently. If an application closes a file and then it or some other client application opens it again, a level-I oplock need not be requested of the server, since the client already has one. Thus, until the level-I oplock is revoked, there is no need for the client to notify the server of any changes to the open state of the file, nor any need to write back modified file blocks (other than to reduce loss in the event of a client crash). However, if the server does revoke the level-I oplock, the client must not only send back to the server all modified file blocks but also update it to the file's current open state by sending it whatever SMB open and close requests will get it there.

Not all clients are capable of doing this sort of batching, so there are actually two kinds of level-I oplocks—*exclusive* and *batch*. If a client holds an exclusive oplock, it must send the server an SMB close request along with all modified blocks when the application closes the file. It must then obtain a new oplock if the file is reopened, at which time it may not use any previously cached blocks from the file. If a client holds a batch oplock, then it uses the delayed-close strategy mentioned earlier. Only on older systems do clients still use exclusive locks.

60.4 DFS

DFS, the DFS of distributed computing environment (DCE), is no longer in common use but is interesting because of its approach to caching and failure recovery. It is derived from an earlier DFS, Andrew File System (AFS).* Both systems addressed the need for client caching of files in a day when primary storage was expensive: they use client disks to cache files from the server. AFS caches files in their entirety; DFS caches data from files in 64 KB blocks.

Like CIFS, DFS provides strictly consistent sharing of files. However, unlike CIFS, clients always use their caches, even if multiple clients are making changes to the same file. Further unlike CIFS, DFS attempts to recover from server crashes.

The DFS server coordinates its clients' use of their caches by passing out "tokens," each of which grants the client permission to perform a certain operation (such as read or write) on a range of bytes within a file. Since all operations are done in the client's cache, the tokens control the use of the cache.

Suppose an application on client A opens a file for reading and writing, client A's DFS cache manager (running as part of the operating system) sends an open request to the DFS server (using DCE RPC) asking for OPEN_READ and OPEN_WRITE tokens for the file (see Figures 60.7 through 60.12). The server sends back a response containing the tokens, keeping as part of its token state that such tokens are at client A. In response to the application's first read of the file, client A's cache manager requests data from the server, along with a DATA_READ token giving it permission to use this data in its cache. It fetches not just the data needed but 64 kB data blocks, thus performing aggressive read-ahead. If the client application performs a write system call, the client's cache manager first obtains a DATA_WRITE token from the server, granting the client permission to modify a range of bytes in its cache.

Suppose an application on client B opens the same file for reading and writing. Client B's cache manager obtains OPEN_READ and OPEN_WRITE tokens, as did client A's. The application now

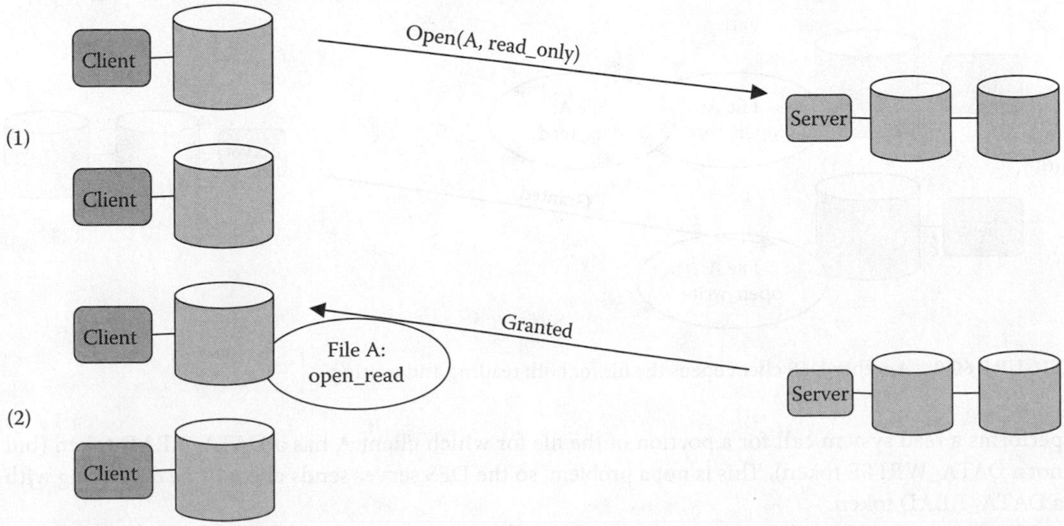

FIGURE 60.7 A DFS client opens a file read-only.

* AFS stands for Andrew File System, so called because it was developed at Carnegie Mellon University: both Mr. Carnegie and Mr. Mellon had the first name Andrew.

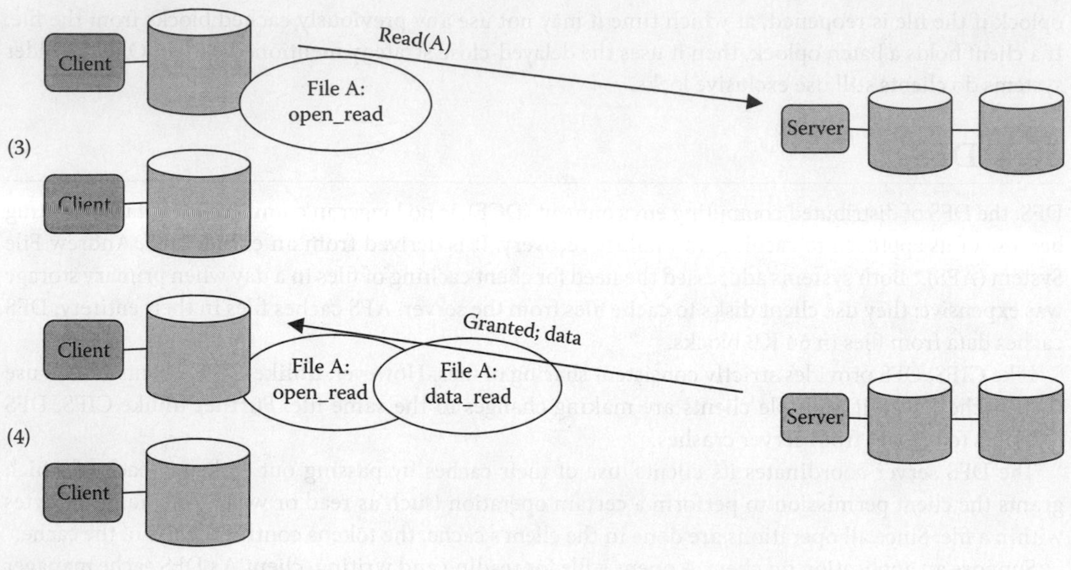

FIGURE 60.8 The DFS client reads data from the file.

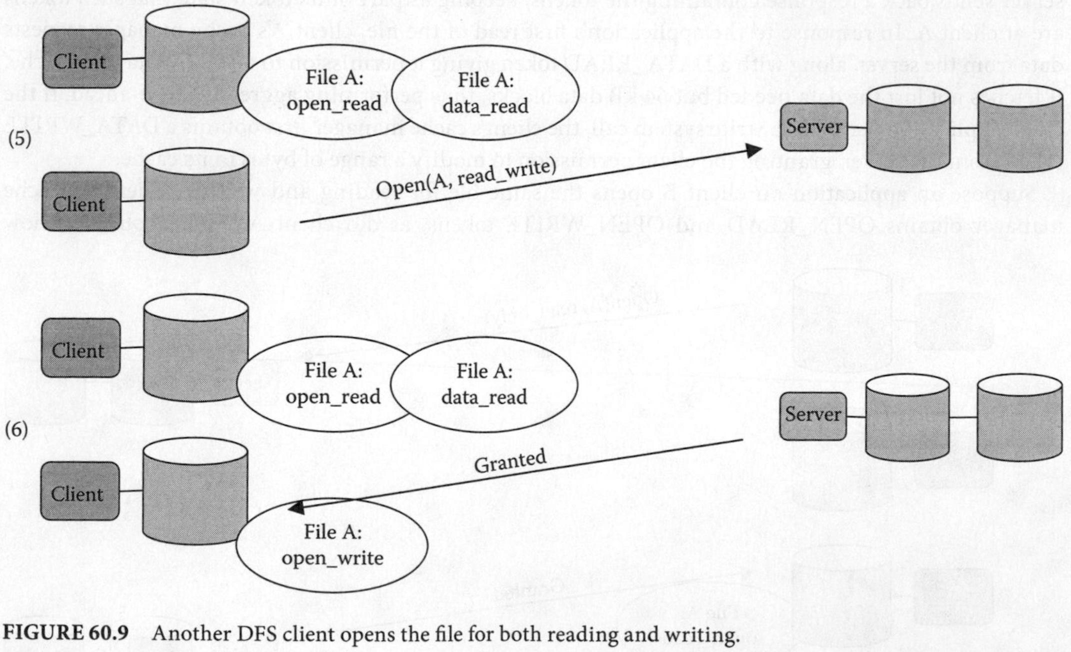

FIGURE 60.9 Another DFS client opens the file for both reading and writing.

performs a read system call for a portion of the file for which client A has a DATA_READ token (but not a DATA_WRITE token). This is not a problem, so the DFS server sends client B the data along with a DATA_READ token.

Now the application performs a read system call for a portion of the file for which client A has a DATA_WRITE token. Before the DFS server can grant B a DATA_READ token, it sends a revoke request to client A revoking its DATA_WRITE token. Client A responds by both giving up the token and sending back the modified file data from its cache. The server now sends client B both a DATA_READ token and the data.

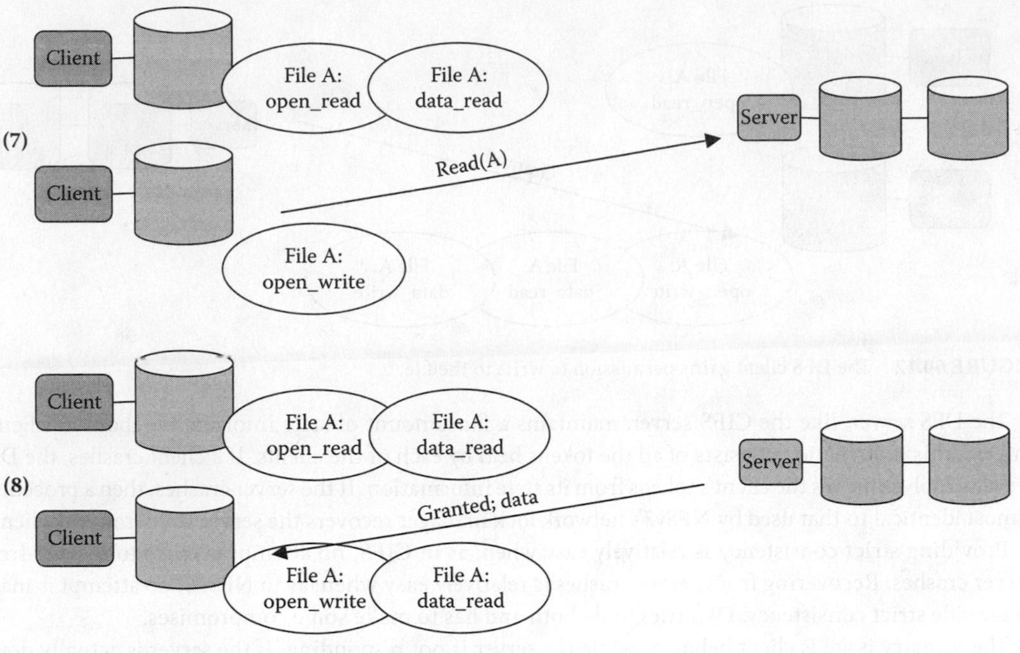

FIGURE 60.10 The second DFS client reads from the file.

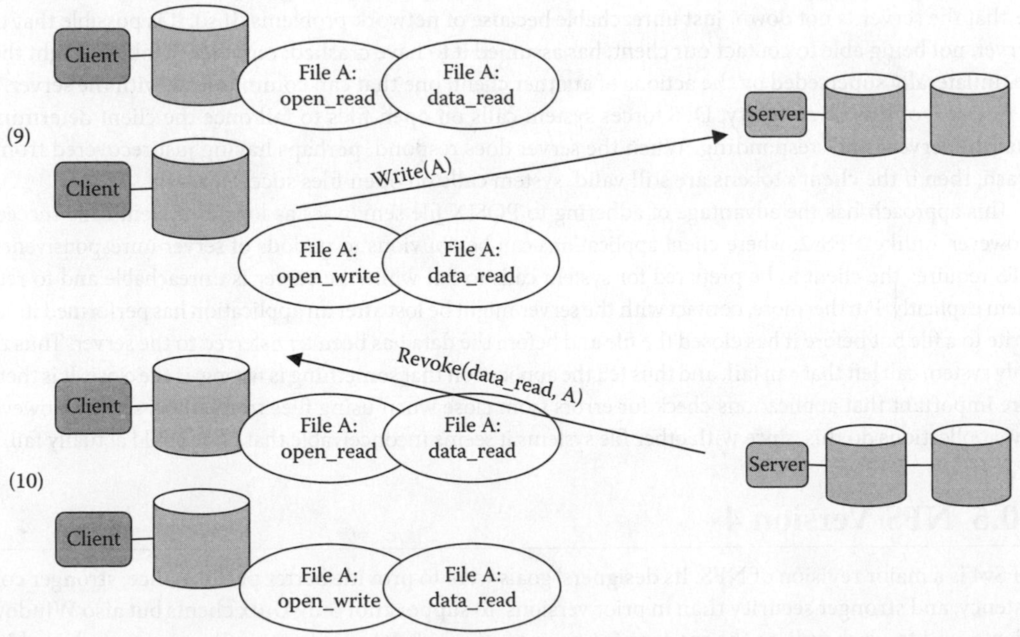

FIGURE 60.11 The server revokes the first client's data_read token.

This is all straightforward, though perhaps laborious. Like both NFSv2 and CIFS, DFS handles the common cases efficiently: if only one client is modifying a file and no clients are reading it, or if a number of clients are reading a file but none are modifying it, all file data resides in client caches and just a minimal number of messages are exchanged. Unlike NFSv2 but like CIFS, all client caches are strictly consistent with the server.

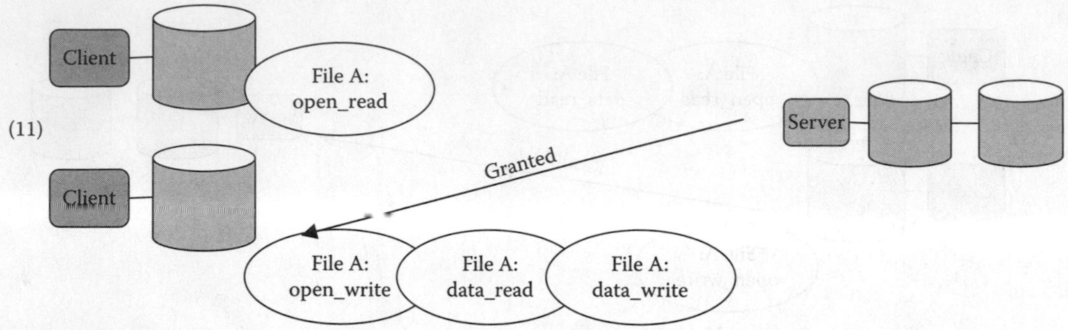

FIGURE 60.12 The DFS client gains permission to write to the file.

The DFS server, like the CIFS server, maintains a fair amount of state information about its clients. In DFS, this information consists of all the tokens held by each of the clients. If a client crashes, the DFS server simply removes the client's tokens from its state information. If the server crashes, then a procedure almost identical to that used by NFSv2's network lock manager recovers the server state from the clients.

Providing strict consistency is relatively easy when, as in CIFS, no attempt is made to recover from server crashes. Recovering from server crashes is relatively easy when, as in NFSv2, no attempt is made to provide strict consistency. DFS tries to do both and has to make some compromises.

The primary issue is client behavior while the server is not responding. If the server is actually down and the client's tokens will not have timed out by the time the server comes back up, then it would be reasonable to let client applications continue to function, where possible, using their local caches. But it could be that the server is not down, just unreachable because of network problems. If so, it is possible that the server, not being able to contact our client, has assumed it to have crashed; our client's tokens might then be unilaterally superseded by the actions of another client, one that can communicate with the server.

Because of this uncertainty, DFS forces system calls on open files to fail once the client determines that the server is not responding. When the server does respond, perhaps having just recovered from a crash, then if the client's tokens are still valid, system calls on open files succeed again.

This approach has the advantage of adhering to POSIX file semantics as long as system calls succeed. However, unlike NFSv2, where client applications can be oblivious to periods of server unresponsiveness, DFS requires the client to be prepared for system calls to fail while the server is unreachable and to retry them explicitly. Furthermore, contact with the server might be lost after an application has performed its last write to a file but before it has closed the file and before the data has been transferred to the server. Thus the only system call left that can fail, and thus tell the application that something is wrong, is the close. It is therefore important that applications check for errors from close when using files from a DFS server. However, few applications do this, since with other file systems it seems inconceivable that close could actually fail.

60.5 NFS Version 4

NFSv4 is a major revision of NFS. Its designers' goals were to provide better performance, stronger consistency, and stronger security than in prior versions, to support not only Unix clients but also Windows clients, and to work well on the modern Internet, particularly where clients and servers may be behind firewalls. A side effect of all this is that the three separate protocols of prior NFS versions (mount protocol, file protocol, and network lock manager) are all combined into a single protocol.

NFSv4 dispenses with the mount protocol by having servers combine their exported file systems into a single "pseudo file system" and supplying a RPC request to obtain the root of that file system. Figure 60.13 shows the file system of a server that exports a few of its subtrees to clients; Figure 60.14 shows the pseudo file system that is made available to clients—note that anything not exported by the server is not seen by clients.

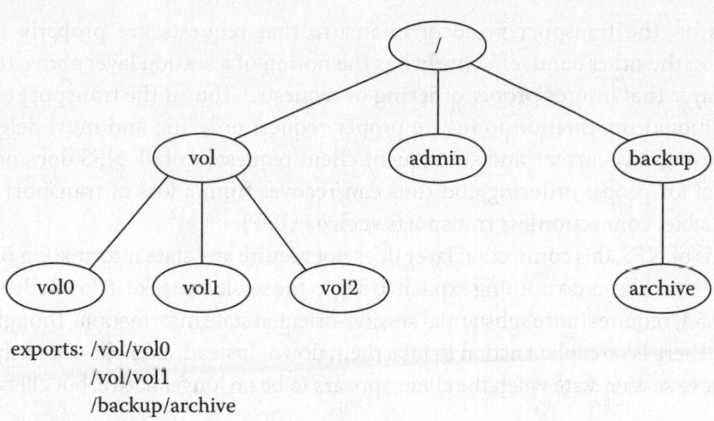

FIGURE 60.13 The directory hierarchy of an NFSv4 server that exports three subtrees.

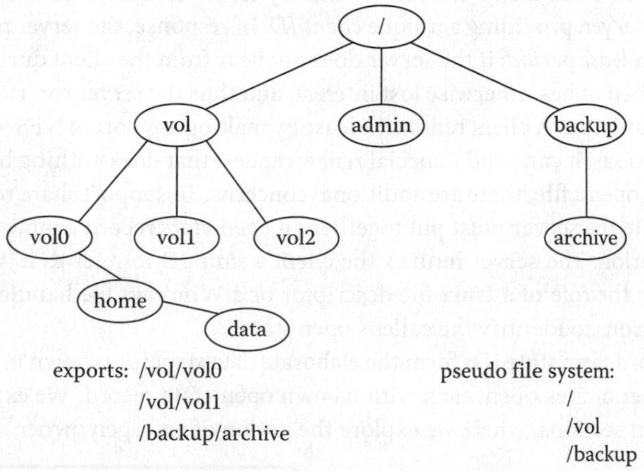

FIGURE 60.14 The pseudo file system that clients of the server of Figures 60.10 through 60.13. (see /admin and /vol/vol2 are not seen.)

NFSv4 reduces network traffic by using *compound RPC*, which allows a number of RPC requests to be encoded in a single message and their responses to be returned in a single message, much as is the case with CIFS's and X facility.

We focus here on consistency and failure handling. Of particular interest is NFSv4's support for *mandatory locks* (it supports advisory locks as well) and *share reservations*—the same thing as share modes in Windows. The network lock manager of prior NFS versions supported only advisory locks. Mandatory locks are expected by Windows applications and supported by some versions of Unix. Share reservations are not really usable by Unix applications—there is no standard API for them—but they are required by Windows applications.

NFSv4's consistency model essentially combines the rather weak consistency of prior NFS versions and the strict consistency of CIFS. If clients use locks or share reservations appropriately when accessing files, NFSv4 guarantees entry consistency. If locks are not used, consistency is the same as that in prior NFS versions. As a performance optimization, NFSv4 uses a technique similar to CIFS' oplocks called *open delegation*. It allows a server to delegate to a client the chores of managing opens, locks, and caching for a file, while no other clients are performing conflicting operations.

The big difference between NFSv4 and CIFS is NFSv4's tolerance for failures. Prior versions of NFS were failure-tolerant as well, but NFSv4 is a step more so, since its servers hold much more state information than those of prior versions.

CIFS depends on the transport protocol to insure that requests are properly ordered, without duplicates. NFS, on the other hand, effectively has the notion of a session layer above the transport, and it is this session layer that insures proper ordering of requests.* Thus if the transport connection is lost, CIFS has no additional information to insure proper request ordering and must delete all client state on the server, causing any current and subsequent client requests to fail. NFS does not depend on the transport protocol for proper ordering and thus can recover from a loss of transport connection (and can run on unreliable, connectionless transports such as UDP).

In prior versions of NFS, this connection layer does not require any state information on the server other than the DRC, and thus clients do nothing explicit to set up the session or take it down. But NFSv4, as we discuss in Section 60.5.1, requires more substantial session-oriented state information. Though clients explicitly establish sessions, there is no explicit action to take them down. Instead, as explained in the subsequent sections, servers remove session state when the client appears to be no longer active, but clients can recover it.

60.5.1 Managing State

NFSv4 servers maintain a hierarchy of state. At the top is the inventory of active clients. Each client identifies itself to the server, providing a unique *client ID*. In response, the server puts together client-ID state, which includes a *lease period*: if the server does not hear from the client during this period, it concludes that it has crashed or has otherwise lost interest, and thus the server can rid itself of the client-ID state and all subordinate state. A client renews its lease by making any sort of NFSv4 request to the server. If it has nothing else to say, it can send a special *renew* request that does nothing but renew the lease.

When a client does open a file, there are additional concerns. To support share reservations, each time some client opens a file the server must put together an *open-state* record containing the access modes and the share reservation. The server returns the client a *state ID* to refer to it, which, along with the NFS file handle, plays the role of a Unix file descriptor or a Windows file handle and is included with subsequent file operations to identify the caller's open state.

The open-state records and state IDs form the elaborate data structure shown in Figure 60.15 in which the client has a number of files open, each with its own open-state record. We explain this figure more fully in the subsequent sections, where we explore the notion of an "open owner."

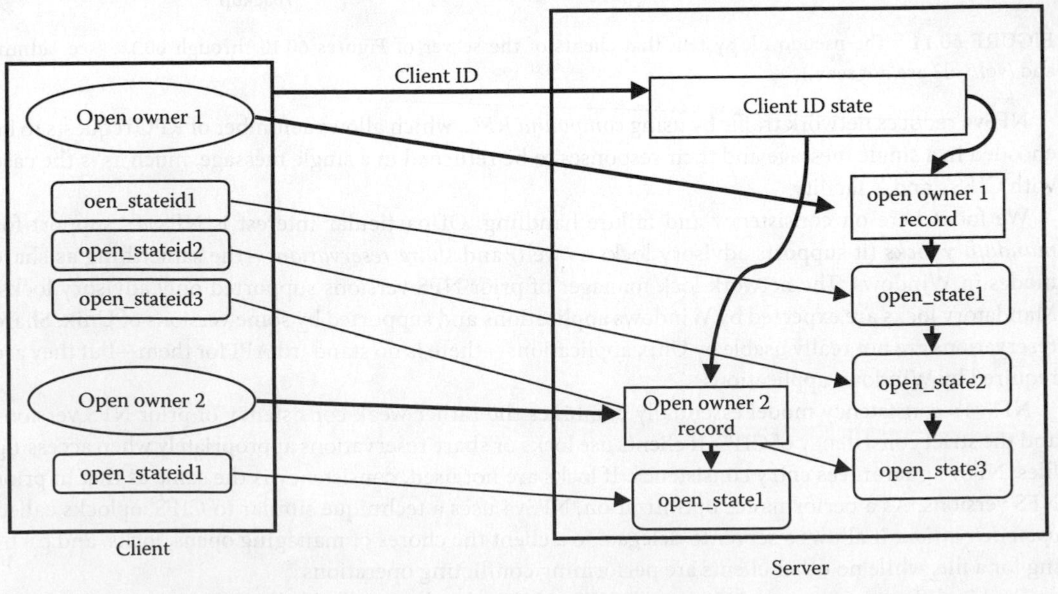

FIGURE 60.15 Open state in NFSv4.

* The session layer is implicit in NFS through version 4.0 and is made explicit starting in version 4.1.

60.5.1.1 Mandatory Locks

NFSv4 supports not only advisory byte-range locks but mandatory ones as well. The latter are a rarely used feature of Unix, partly because it has not been supported by NFS in the past, partly because using it is somewhat obscure (it is enabled by turning on a file's set-group-ID access-permission bit and turning off its group-execute permission bit. On Linux, one must additionally turn on this facility for an entire file system by setting a mount flag). Adding to the confusion is the fact that Windows' semantics for mandatory byte-range locks (it does not support advisory locks) are different from Unix's. In both systems, byte-range locks belong to a process. Thus, for example, in Unix, suppose a process P1 opens file F, producing file descriptor FD1, and then forks a child process P2. If P2 places a mandatory exclusive byte-range lock on byte 0 of the file via FD1, then only P2 may access that byte—not P1, even though they share the same open-file-table entry for the file. Now suppose P2 opens the same file again, this time producing file descriptor FD2. Even though P2 had locked byte 0 via FD1, it can still access that byte via FD2—what matters is that the process has the file locked; it does not matter which file descriptor is used.* However, things are different in Windows, where the lock is associated not just with the process but also with the Windows file handle. Thus in a Windows version of this example, the process is not allowed to access byte 0 via FD2 until the lock placed via FD1 is removed.

When a server receives a read or write request from a client, it needs to know which locks held by the client actually apply to this particular request. For a Unix client, the answer would be all locks held by the client process for the file in question, but for a Windows client, it would be just those locks held by the client process that were placed using the same Windows file handle as was used for the read or write request. But the server does not necessarily know, which requests came from which processes or which requests were made through which Windows file handles.

Rather than give the server this information (which would have to include whether Unix, Windows, or some other lock semantics are being used), the NFSv4 client instead assigns "owners" to locks. When it issues a read or write request, it lets the server know which locks apply by telling it who the owner is. So, for Unix clients, all locks taken by a particular process might be given the same owner, or perhaps all locks taken by a particular process on a particular file might be given the same owner (it does not matter which, since the read or write request identifies the file). For Windows clients, all locks taken by a particular process via a particular Windows file handle are given the same owner. The server does not need to know what the client's lock semantics are; it just needs to keep track of lock owners.

A *lock owner* on a client is simply a unique (to the client) integer. When the client requests a lock requiring a new lock owner, it passes the corresponding integer to the server. In response, the server establishes *lock state* to represent this lock owner and returns to the client a *lock-state ID* referring to the state. Henceforth, the client uses this lock-state ID rather than the lock-owner integer. When the client requests additional locks for the same lock owner, it passes the corresponding lock-state ID to the server. When, for example, the client sends a write request to the server, it includes a lock-state ID to tell the server which locks should apply (a special lock-state ID is sent if no locks apply). Thus the server's lock state represents all the locks held by a particular client lock owner. See Figure 60.16.

60.5.1.2 Maintaining Order

NFSv4's state hierarchy of client ID, open state, and lock state presents challenges. Client crashes must be detected and no-longer-useful state removed from the server. Not only must server crashes be detected, but clients must reestablish their state on the servers. And finally, everything must function correctly in spite of Byzantine routers.

Establishing the client-ID state is far from immune to problems with late-arriving duplicates. If the server receives a request to establish a client ID that already exists, it might decide that the client has

* Another property is that all byte-range locks of a file are removed if any of the process's file descriptors for the file are closed. So suppose P2 in our example was not aware that FD1 and FD2 refer to the same file. If it then locks the file via FD2 and closes FD1, it might be surprised to discover that the file referred to by FD2 no longer has a lock.

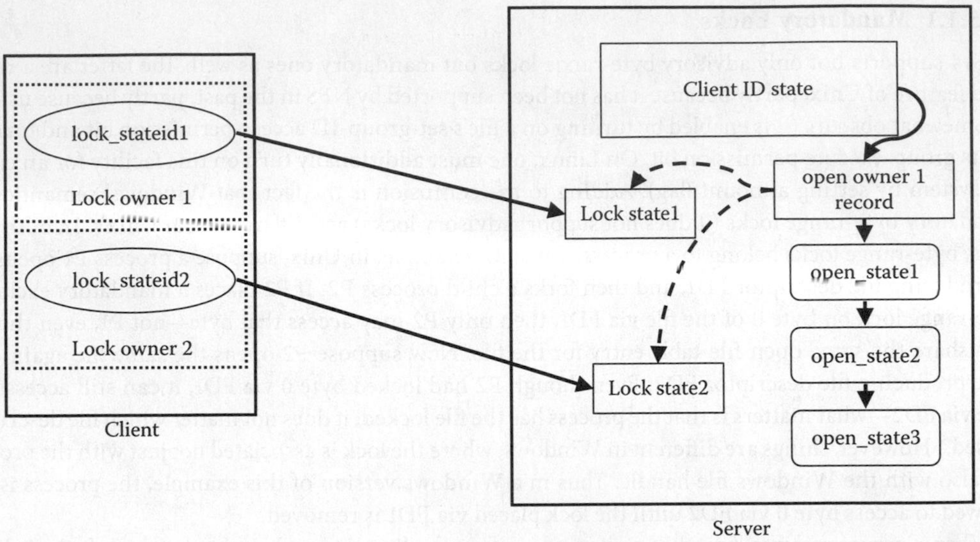

FIGURE 60.16 Lock state in NFSv4.

crashed and has restarted. Even a request to establish a client ID that does not exist might be a long-delayed duplicate from a client that is no longer active. Setting up the state again might tie up resources needlessly.

Open requests (with share reservations) and mandatory lock requests are particularly sensitive to Byzantine routers. Suppose, for example, a portion of a file is locked, then unlocked, and then a much-delayed duplicate of the lock request arrives at the server. If the server does not realize it is a duplicate and locks the file again, then, since the client does not know the file is locked, the file might stay locked for an arbitrarily long period of time.

An explicit session layer, introduced in NFSv4.1, simplifies matters tremendously. Recall that a problem with the DRC of prior NFS versions was determining how long a request must remain in the DRC—the intent is that a late-arriving duplicate request will be recognized as a duplicate, no matter how late it is. Earlier versions of NFS used a simple time-based mechanism for removing requests from the DRC. In NFSv4, clients identify concurrent streams of NFS requests; requests within each stream are assigned sequence numbers, and the streams themselves are assigned slot numbers. The DRC is divided into corresponding slots. When a request is received, it is cached in the associated DRC slot. When a request with a larger sequence number is received, it replaces the previous entry in the DRC slot. If a request with a sequence number equal to the one in the DRC slot is received, the previous response is returned. If a request with a smaller sequence number is received, it must be a very late-arriving duplicate and is discarded (see Chapter 9 of Doeppner (2011) for details). With this approach, *in the absence of server crashes*, open requests, lock requests, data-transfer requests (reads and writes), etc. have exactly once semantics—there are no problems with Byzantine routers and duplicates that cross transport-connection boundaries.

60.5.2 Dealing with Failure

NFSv4 deals with failure issues not handled by its predecessors. Though it uses basically the same strategy for recovery from server failures as the network lock manager of v2 and v3, there is more state to recover and there are more operations that depend on this state. Not only locks but also all open state must be recovered. Because locks may be mandatory and share reservations definitely are mandatory, it is particularly important that crashed clients be detected quickly and restarted servers recover their state quickly. But in addition, there is a new problem: when a server restarts, the client with the most recent lock state might be temporarily unreachable, but a client with older, superseded lock state might try to reclaim that state on the server.

Leases provide the basic defense against client crashes. Clients are expected to renew their leases; if a client ID's lease time expires, the server assumes it has crashed and may dispose of all its state, thus closing all files and breaking all locks. This might be considered pretty drastic if the client has not really crashed but there is merely a temporary network problem. It is also very different behavior from that experienced by NFSv2 and v3 clients: they never had to worry about the server's closing their files, since the server never knew they were open in the first place. However, such behavior is unavoidable when unresponding clients hold lock state or share reservations.

A client might crash, reboot, and recontact the server well before its original lease has expired. The fact that the client is reestablishing its client ID might be a clue to the server that it has restarted, but there are reasons other than crashing that a client might do this. So included with the client-ID state on the server is a 64-bit *verifier* that is supplied by the client when it establishes its client ID and is different each time the client reboots. Thus, if the server receives a set-up-client-ID request from a client for which it still has state, if the verifier supplied with the request is different from what's included with the saved state, then the server knows the client has restarted and the saved state should be discarded. But if the verifiers are the same, then the saved state is still valid.

Clients find out about server crashes when RPC requests that depend on open and lock state return error codes indicating that the state no longer exists. For a period before then (while the server was down), such calls might have timed out, but the only way for a client to distinguish a server problem from a temporary communications problem is for the server to give a positive indication that it has no record of the client. The problem might turn out to be not that the server crashed, but that the client's lease expired; but from the client's point of view, lease expiration is just as bad, if not worse.

As with the network lock manager in v2 and v3, when a server restarts, it establishes a grace period during which clients may recover their states. How long should the grace period be? A client would not discover that the server has crashed until it sends it a request. But the client is not required to send the server a request until just before its lease expires. So, in case a client renewed its lease just before the server crashed, the server's grace period should be at least as long as the lease period.

Suppose a client discovers that its server state has disappeared, its first response is to reestablish its client-ID state, along with a new verifier. Then, as with the network lock manager, it tries to reclaim its previous state.

Now the server has decisions to make. If it is not in its grace period following a restart, then the reason the client's state no longer exists is probably that it was discarded after the client's lease expired. This could be due to incompetence on the client's part, but is most likely the result of a network problem. If the server can determine that no conflicting opens or locks have been given to other clients since the lease expired, then it may allow the client to reclaim its state. But, in general, the server probably cannot determine this, so it must refuse the reclaim request and the client must report an error back to the application (and thus the application must be prepared to cope with the bad news).

If the server is within its grace period, then the reclaim request is probably for state held when the server crashed, and thus it should be granted. However, the client might have been disconnected from the server because of a network problem. In the meantime, the client's lease expired, the server granted a conflicting lock to another client, and then the server crashed. Finally, the network problems experienced by the first client have been solved, and it is now trying to reclaim its (ancient) state. But it really should be the second client whose state is reclaimed.

The general problem for the server when it receives a reclaim request is to make sure that the state being reclaimed has not been superseded by more recent operations by other clients. The NFSv4 specification (RFC 3530*) requires servers to err on the side of safety: if they are not sure, they must refuse the reclaim request.

The only way for a server to be absolutely sure is to record all share reservations and locks in stable storage, so that they can be recovered after a crash. This is probably too time-consuming for most servers.

---

* http://www.faqs.org/rfcs/rfc3530.html.

At the other extreme, a server could simply refuse to accept reclaims on the grounds that it does not have enough information to determine if any of them are for the most recent state.

A reasonable compromise is for the server to keep enough information on stable storage to determine which client IDs hold state. Then, during the grace period, it can refuse all reclaim requests from clients who did not hold state at the time of the most recent crash.

RFC 3530 suggests servers record the following information on stable storage for each client ID:

- Client ID.
- The time of the client's first acquisition of a share reservation or lock after a server reboot or client lease expiration.
- A flag indicating whether the client's most recent state was revoked because of a lease expiration.

In addition, the server stores on stable storage the times of its two most recent restarts. Doing this requires minimal server overhead, yet allows it to determine whether a client definitely held valid server state at the time of the last server crash. We take this up in Exercise 9.

60.6 Conclusions

DFSs are normally made as transparent as possible to users and their applications; as we have seen, much has been done to make accessing remote files indistinguishable from accessing local files. The major problem is coping with failure, both of communication and of machines. NFS has been successful because of its compromises. Prior to version 4, it provided semantics that were "pretty close" to local Unix file semantics, with a response to failure that "pretty well" insulated the client from server and network problems. CIFS, and to a certain extent DFS, were based on the premise that providing semantics that are exactly the same as local semantics is more important than insulating the client from server and network problems. NFSv4 is attempting to come as close as possible to achieving local-file-system semantics while still insulating clients from server problems (and vice versa).

References

Bershad, B. and M. Zekauskas (1991). *Midway: Shared Parallel Programming with Entry Consistency for Distributed Memory Multiprocessors.* Carnegie Mellon University Technical Report Technical Report CMU-CS-91-170.

Doeppner, T. (2011). *Operating Systems in Depth*, Wiley, Hoboken, NJ.

Lamport, L. (1979). How to make a multiprocessor computer that correctly executes multiprocess programs. *IEEE Transactions on Computers* **C-28**(9): 690–691.

Storage Networking Industry Association (2002). Common Internet File System (CIFS) Technical Reference. http://www.thursby.com/sites/default/files/files/CIFS-TR-1p00_FINAL.pdf

61

Mobile Operating Systems

Ralph Morelli
Trinity College

Trishan de Lanerolle
Trinity College

61.1 Introduction

A *mobile operating system*, or *mobile OS*, is an operating system that is designed to run on a mobile device, such as a mobile phone, personal digital assistant (PDA), or a tablet computer. Like traditional operating systems, a mobile OS is the software layer that sits between the device's hardware and its application layers. As such, a mobile OS is an abstraction built in software that hides the details of the hardware layer from the applications that sit atop it.

Its primary role is to manage the device's hardware resources—the processor, memory, input/output (I/O) devices, network, etc.—and to provide a variety of services and easy-to-use software interfaces to support the device's various uses. This would include providing a user interface (UI) as well as providing programming interfaces that allow developers to create applications to extend the device's basic functions.

Although many of the issues and challenges that characterize operating systems research and development are present also in mobile OS development, the field is characterized by distinctive computational issues made important by the size limitations of mobile devices, their reliance on battery power, and by their mobility.

61.1.1 Historical Overview

A *mobile phone* (or cellular phone) can send and receive telephone calls over a radio link. The calls are transmitted over a *cellular network* that is operated by a service provider. The very first mobile telephone call was made in 1946 from a car using Bell Laboratory's Mobile Telephone Service (MTS). MTS became a commercial service provided by AT&T in St. Louis and other metropolitan areas. The system had very few radio channels and relied on a human operator to connect calls. Ten years later, a Swedish company developed the first automatic car phone system, which used vacuum tube technology and weighed 88 pounds [Wikipedia: History of the mobile phone 2012]. Named mobile telephone system A, it allowed calls to be made and received in the car using a rotary dial. The car phone could also be paged.

The first handheld mobile device was developed in 1973 by Motorola. It weighed 2.5 lb and offered a talk time of 30 min and took 10 h to recharge [Wikipedia: History of the mobile phone 2012]. The world's first public cellular call was made in the New York City to a landline by Dr. Martin Cooper, a general manager for the communications systems division at Motorola. "As I walked down the street while talking on the phone," Cooper once told an interviewer, "sophisticated New Yorkers gaped at the sight of someone actually moving around while making a phone call" [Greene 2011].

In 1983, Motorola commercially released the DynaTAC phone—seen sported by Michael Douglas in the 1987 movie *Wall Street*—which is considered one of the world's first mobile phones. Based on the prototype developed in 1973, the DynaTAC weighed almost two pounds, provided 1 h of battery life, and stored 30 phone numbers in its phonebook.

In 1987, Nokia launched the Mobira Cityman 900—the first handheld mobile phone for Nordic Mobile Telephony (NMT) networks. The Mobira Cityman 900 became an iconic device when Soviet leader Mikhail Gorbachev was photographed using a Cityman to make a call from Helsinki to Moscow [Ross Catanzariti n.d.]. In 1989 Motorola MicroTAC 9800X was released, which was one of the first truly portable phones, with its ability to be slipped into a jacket pocket. The Motorola International 3200 followed in 1992 as the first digital hand-sized telephone.

Early mobile phones did not have operating systems in the full sense of the term. Rather, they were controlled by *embedded systems*—that is, a system that is specifically designed to control the device's particular hardware components, such as its battery and radio transmitter, and software. Unlike a true operating system, an embedded operating system does not load and execute applications and is not designed for general purpose computing system.

61.1.2 Cellular Networks

With the advent of the mobile telephony devices, there was a need for cellular network technology that could provide broader coverage and faster and more reliable data transmission capabilities. The introduction of cellular technology made possible the re-use of frequencies by multiple users in small adjacent areas covered by relatively low-powered transmitters, making widespread adoption of mobile telephones economically feasible.

AT&T's MTS, introduced in 1947, now considered zero-generation service, was a precursor to modern day cellular networks. The service relied on an operator to connect both incoming and outgoing calls [Massa 2006]. In the 1970s, AT&T introduced the first-generation (1G) analog mobile phone services, one of the first automatic mobile networks to get Federal Communications Commission approval in the United States [AT&T 2012].

These systems were referred to as *cellular*, due to the method by which the signals were handed off between towers. A *cell* can be visualized as a geographic area with a single *transceiver* known as a base station or cell tower. In order to avoid interference, adjacent cells use different sets of frequencies [Marshall Brain 2011]. Contemporary cellular networks use a system whereby towers are situated at three corners of a hexagonal shaped cell. As shown in Figure 61.1, each tower broadcasts three directional signals, each of a different frequency. The frequencies within a given cell differ from the frequencies in all six adjacent cells. As a device travels through a geographical area, it switches from one frequency to another in a process known as a *handoff* [Wikipedia: Cellular network 2012]. Typically, the base station chooses a frequency as the device enters its area and the device automatically switches to the new frequency. To the cell phone user, the switch is usually imperceptible and, ideally, communication is not interrupted.

In terms of global systems, the Japanese telecommunications company, NTT, built its own network in 1979. Five years later, it was the first 1G network to cover an entire country. The NMT network operated in Denmark, Sweden, Finland, and Norway followed in 1981, becoming the first system to feature international roaming.

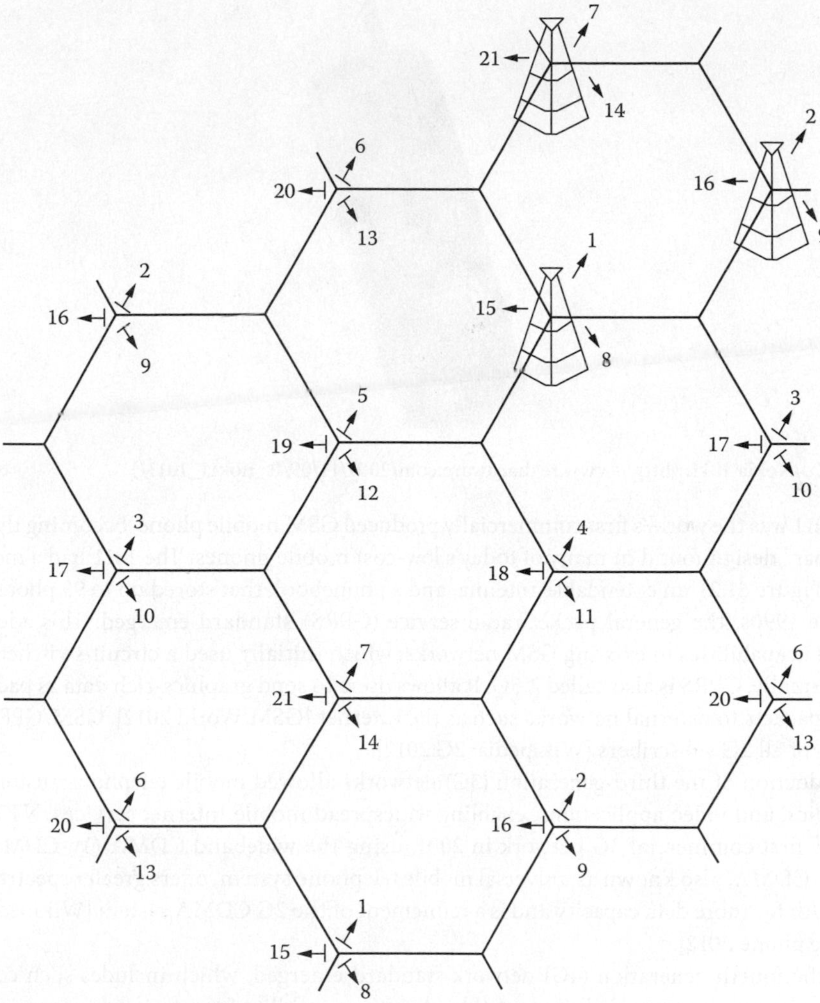

FIGURE 61.1 Cellular network. (http://en.wikipedia.org/wiki/Cellular_network)

The 1990s saw the emergence of two digital networking standards: the European global system for mobile communications (GSM) standard and the North American code division multiple access (CDMA) standard [Wikipedia: 2G 2012]. GSM-based phones use subscriber identity modules—SIM cards—to store user data and are not linked to a particular device. In the United States, CDMA carriers use network-based white lists to verify their subscribers. That means, you can only switch phones with your carrier's permission, and a carrier does not have to accept any particular phone onto its network. Both CDMA and GSM are *multiple access* technologies—that is, they enable users to combine multiple phone calls or Internet connections into one radio channel.

GSM is a *time division* system. Voice signals are transformed into digital data and then given a channel and a time slot. On the other end, the receiver listens only to the assigned time slot and pieces the call back together. CDMA is a *code division* system. A call's data are encoded with a unique key and combined with other calls and all are transmitted in the same time slot. Receivers use the key to divide the combined signals into its individual calls [Segan 2012]. In addition to sound, CDMA and GSM networks are also capable of transmitting small amounts of data, making possible such features as the short message system (SMS), or text messaging, caller ID, and voicemail.

FIGURE 61.2 Nokia 1011. (http://www.reghardware.com/2007/11/09/ft_nokia_1011/)

Nokia's 1011 was the world's first commercially produced GSM mobile phone, becoming the model for the "candy-bar" design found in many of today's low-cost mobile phones. The 1011 had a monochrome display (see Figure 61.2), an extendable antenna, and a phonebook that stored up to 99 phone numbers.

In the late 1990s, the general packet radio service (GPRS) standard emerged. This added packet-switched data capabilities to existing GSM networks, which initially used a circuit-switched telephony system. Informally, GPRS is also called 2.5G. It allows users to send graphics-rich data as packets and to transmit IP packets to external networks such as the Internet [GSM World 2012]. GSM/GPRS accounts for over 80% of all 2G subscribers [Wikipedia: 2G 2012].

The introduction of the third-generation (3G) networks allowed mobile telephone customers to use audio, graphics, and video applications, enabling widespread mobile Internet services. NTT DoCoMo launched the first commercial 3G network in 2001, using the wideband CDMA (W-CDMA) wireless standard. W-CDMA, also known as universal mobile telephone system, offers greater spectral efficiency and bandwidth for more data capacity and is a refinement of the 2G CDMA system [Wikipedia: History of the mobile phone 2012].

In 2009, the fourth-generation (4G) network standard emerged, which includes such communication technologies as long-term evolution, worldwide interoperability for microwave access, and evolved high-speed packet access. These 4G standards are shepherded by the 3G partnership project, a standards governing body for GSM technologies that is responsible for the technical specifications [3rd Generation Partnership Project 2012]. The 4G networks use an IP-based architecture where everything (including voice) is handled as data, similar to the Internet. These data-optimized networks are geared toward an ever expanding demand for bandwidth and mobile capacity.

By the end of 2011, there were 6 billion mobile subscriptions, according to estimates by the International Telecommunication Union, the United Nations Agency responsible for information and communication technologies [International Telecommunication Union 2011]. With each progressive generation of cellular technology, we see improvements in data transfer capabilities, range, and reliability of the cellular networks, as summarized in Table 61.1. The increasing capabilities of services from 1G to 4G has led to the increasing popularity of smartphones.

61.1.3 Smartphone

A *smartphone* is a mobile phone with computer-like features, including a larger screen and an operating system capable of supporting third-party applications and running on different hardware architectures [Oxford University Press 2010]. Phones that typically have smaller screens and the ability to support some advanced functionality, such as web browsing or music and videos, are known as *feature phones*.

TABLE 61.1 Network Generations

Generation	Commercially Available	Dominant Technology	Applications	Speed
1G	Japan, 1979	Analog—AMPS	Voice only and no data services	14.4–28.8 kbps
2G	Finland, 1991	Digital circuit switch—GSM and CDMA	Short messaging service (SMS) and digital encryption	56–115 kbps
3G	Japan, 2001	Digital Packet Switched—W-CDMA (UMTS)	Mobile TV, video on demand, video conferencing, and location-based (GPS) services	2+ to 56 Mbps
4G	South Korea, 2006	IP-based network—LTE and WiMax (3GPP)	Broadband services and HD video streaming	100 Mbps–1 Gbps
5G[a]	Anticipated in 2020	Still in R&D phase	Improved service quality and intelligent networks and devices	1 Gbps +

[a] At present, 5G is not a term officially used for any particular specification.

They usually have a limited operating system and, in general, do not support third-party software. If they do, they usually run on Java or binary runtime environment for wireless and are often standalone items that do not integrate with other features of the phone [Lee 2010].

Smartphones have revolutionized the way people interact with each other and with digital media. The first smartphone can be traced to Bell South's and IBM's Simon, released in 1993, which was a phone, personal data assistant (PDA), and fax machine, built into a single mobile device about the size of brick [Reed 2010]. It included a stylus-operated touchscreen and let you send and receive fax messages. It also included a built-in notepad, an e-mail client, contact list, and a calendar (Table 61.2).

Ericsson was the first brand to coin the term "smartphone" with the release of its GS88 in 1997, which contained a built-in keyboard and touchscreen and personal data assistant application suite [Wikipedia: Smartphone 2012]. Nokia released the Nokia 9000 Communicator in 1996 with a flip open keyboard design. In 2000, the touchscreen-based Ericsson R380 smartphone was the first device to use an open operating system called Symbian OS developed by Nokia. Microsoft released its first operating system targeted at the mobile device market, the Pocket PC, in 2000. This represented the first major push by a software company (as opposed to a phone company) into the mobile space. In 2001, we see Palm OS, BlackBerry OS, and Windows CE coming onto the scene. In 2002, Research in Motion (RIM) entered the mobile phone market with its BlackBerry 5810 device, the first mobile device optimized for e-mail and web browsing. During the 2000s, the Symbian system was the market leader for smartphone operating systems until it was surpassed by the Android operating system in 2011.

In 2007, the introduction of the Apple iPhone's iOS Operating System, the first major mobile OS to come out since the early 2000s, made a notable impact on the mobile OS marketplace. During most of the 2000s, the design of mobile OS, for the most part, did not undergo major change [Brookes 2012]. At the same time, however, hardware devices were making substantial technical leaps in terms of the capabilities and performance. iOS changed the perception of smartphones from that of a device targeted largely toward the business community to that of a commodity that all consumers could use. With this change came an increased emphasis on enhancing and enriching the *user experience*. Consequently, iOS focused more attention on the UI, including high resolution screens with spectacular graphics and such features as *multi-touch*, all of which were designed to appeal to the everyday consumer.

Apple also introduced a software developer kit to enable third-party application developers to write their own programs and distribute them to the end users through a common marketplace. This created a large community of developers writing applications ("apps") that extended the functionality of the devices.

TABLE 61.2 Advanced Devices

Device	Capabilities	Device	Capabilities
Basic mobile phone	Network services, including: Voice telephony and voicemail SMS (short message service) USSD (unstructured supplementary service data)	Smartphone	As featurephone plus: Video camera Web browser GPS (global positioning system) 3G+ internet access Mobile operating systems (such as iOS, android, and blackberry) Ability to download and manage applications VoIP (voice over internet protocol) Mobile TV Removable memory card
Feature phone	As basic mobile phone plus: MMS (multimedia messaging service) Still picture camera MP3 music player 2G+ data access	Tablet	As smartphone plus: Front and rear facing cameras Larger screen and memory capabilities Faster processors External device connectivity ports.

Source: World Bank. Information and Communication for Development 2012: Maximizing Mobile. Case Study, Washington, DC: World Bank, 2012.
Note: The list of capabilities is not exhaustive and not all devices have all features.

In 2008, HTC released the HTC Dream, based on the Android operating system. Android was developed by Google in conjunction with the Open Handset Alliance (OHA), a consortium of 84 technology and mobile companies committed to developing open standards for mobile devices [Open Handset Alliance 2012]. OHA members include mobile operators, handset manufacturers, semiconductor companies, software companies, and commercialization companies. Google also released the Android market place (known today as Google play) for third-party application developers to develop and distribute applications. Unlike Apple's iOS, which is proprietary and not licensed to third parties, Google's Android operating system was released under the Apache open source license and was adopted by many mobile device manufactures. As a result, perhaps of its open licensing approach, Android has become the market leader for mobile OS, capturing over 50% of the market share in 2011 [Gartner Inc. 2012].

61.1.4 Commercial Development

As the preceding discussion highlights, the field of mobile OS is a rapidly growing area of research and development, most of which has been driven by the mobile phone and mobile technology industries. During the first few years of smartphone technology, the market was dominated by one or the other major phone manufacturers such as Nokia and RIM. Since 2007, however, the mobile OS market has come to be dominated by companies known for their software products, mainly Apple and Google as shown in the worldwide smartphone sales in the subsequent sections (Figure 61.3). Similar to the browser wars of the 1990s between Microsoft's Internet Explorer and Mozilla's Firefox and the earlier battles between Macs and IBM PCs, the mobile OS ecosystem has come to be dominated by a couple of players.

The current market leader is the Android operating system with over 56% of the global market as of February 2012, followed by Apple's iOS operating system. The other systems, including RIM's Blackberry, Microsoft's Windows Mobile, and the Symbian OS, trail far behind the two leaders (Figure 61.4).

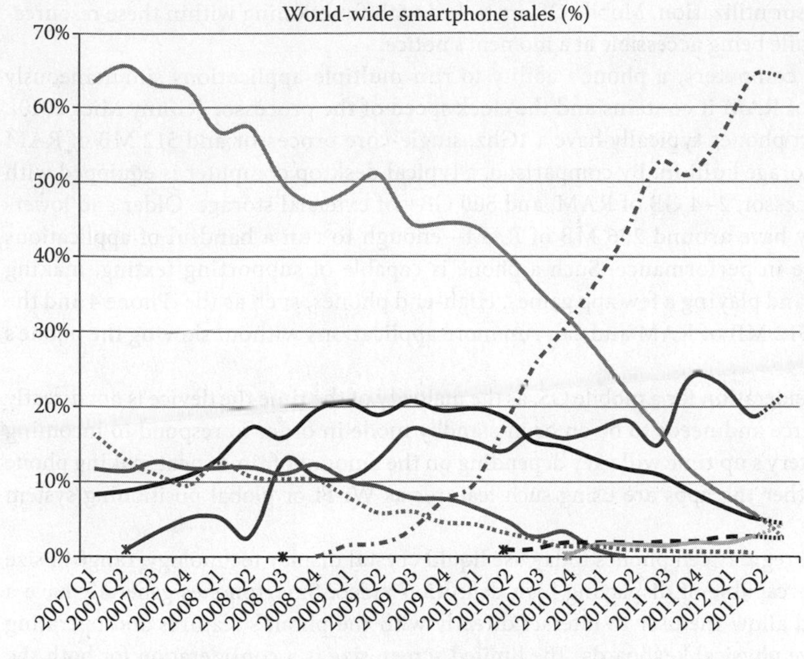

FIGURE 61.3 Smartphone sales. (Smartphone. Wikipedia. 2012. http://en.wikipedia.org/wiki/Smartphone [accessed October 3, 2012].)

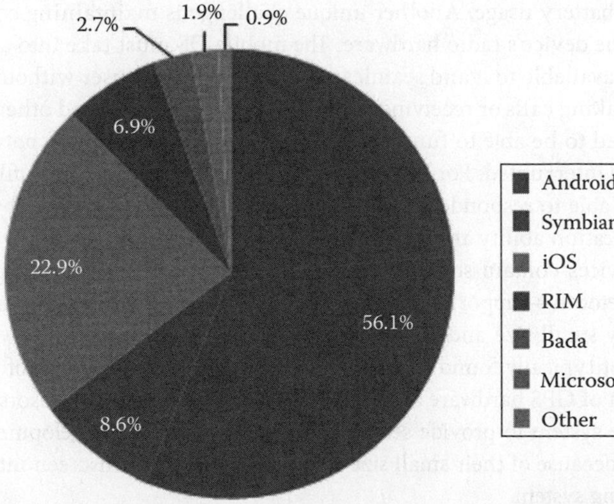

FIGURE 61.4 Mobile operating systems global leaders. (Gartner Inc. "Android Extended Lead While Apple iOS Market Share Growth Paused." Gartner Newsroom. August 12, 2012. http://www.gartner.com/it/page.jsp?id=2120015 [accessed October 3, 2012].)

61.2 Underlying Principles

What distinguishes mobile OS from traditional operating systems are three main characteristics: (1) the relatively small size of the devices, (2) their mobility, and (3) their inclusion of hardware that is not typically found on laptop and desktop computer systems.

The relatively small size of mobile devices and the competitive drive to pack as much functionality as possible into a small space leads to important design considerations regarding power, memory management, and processor utilization. Mobile OS are tasked with functioning within these resource-starved environments, while being accessible at a moment's notice.

As is true of desktop computers, a phone's ability to run multiple applications simultaneously depends on the amount of RAM it contains and the clock speed of the processor [Ginny Mies 2010]. Current generation smart phones typically have a 1Ghz, single-core processor and 512 MB of RAM with about 8–16 GB of storage built in. By comparison, a typical desktop computer is equipped with a 3+ Ghz multi-core processor, 2–4 GB of RAM, and 500 GB+ of external storage. Older and lower-end smartphones usually have around 256 MB of RAM—enough to run a handful of applications with little or no decrease in performance. Such a phone is capable of supporting texting, making calls, browsing the Web, and playing a few app games. High-end phones, such as the iPhone 4 and the Samsung Nexus S, have 512 MB of RAM and can run more applications without slowing the phone's performance.

Battery life is a key consideration for a mobile OS, as the majority of the time the device is not directly plugged into a power source and needs to be on or in-standby mode in order to respond to incoming calls and messages. A battery's up time will vary depending on the amount of time spent making phone calls, running apps, whether the apps are using such features as Wi-Fi or global positioning system (GPS), and other factors.

Most smartphones and regular cell phones today use liquid crystal display technology, range in size from 3 to 4.3 in., and are capable of displaying high definition video. Smartphones generally have a touchscreen interface and allow the user to interact directly with the phone's features and operating system. Some models have physical keyboards. The limited screen size is a consideration for both the mobile OS and application developers.

The phone's mobility poses special challenges for a mobile OS. Perhaps, the primary challenge at the OS level is optimizing battery usage. Another unique challenge is maintaining connectivity with cell phone towers though the device's radio hardware. The mobile OS must take into account the type and nature of a connection available to it and seamlessly transition for the user without interrupting ongoing activities such as making calls or receiving and sending text messages and other data transmissions. Mobile applications need to be able to function in an environment where the network connectivity is intermittent and can be interrupted. For instance, when a user is downloading a file or playing a video, the phone must still be able to respond to an incoming phone call or message. Obviously, phone functionality and communication ability are high priorities for mobile devices.

Although mobile devices contain some of the same features as desktop and laptop computers—for example, cameras, Bluetooth—supporting these features poses unique challenges for the mobile OS because of the relatively small size and power constraints of mobile devices. However, mobile devices also contain features not typically found in laptop and desktop computers. One of the prime examples, perhaps, is the inclusion of GPS hardware and gyroscope and accelerometer sensors. Such features place unique demands on the system to provide services that permit the easy development of location aware applications. Similarly, because of their small size and power needs, touchscreen interfaces place unique demands on the operating system.

Finally, because of the constant connectivity of mobile devices and the increasing consumer demand for sharing all types of information—images, messages, locations—data protection and information security are important considerations for mobile OS designers and app developers alike.

61.2.1 Mobile OS

There are a wide range of mobile OS and releases, far too many to provide a detailed description of each one. In this section, we will provide a brief summary of some of the most influential systems that have emerged, focusing on systems that have played some kind of unique role in the history or development

TABLE 61.3 Worldwide Mobile Device Sales to End Users by Operating System in 2012 (Thousands of Units)

Operating System	2012 Units	2012 Market Share (%)	2011 Units	2011 Market Share (%)
Android	98,529.3	64.1	46,775.9	43.4
iOS	28,935.0	18.8	19,628.8	18.2
Symbian	9,071.5	5.9	23,853.2	22.1
Research in Motion	7,991.2	5.2	12,652.3	11.7
Bada	4,208.8	2.7	2,055.8	1.9
Microsoft	4,087.0	2.7	1,723.8	1.6
Others	863.3	0.6	1,050.6	1.0
Total	153,686.1	100.0	107,740.4	100.0

Source: Gartner Inc. "Android Extended Lead While Apple iOS Market Share Growth Paused." Gartner Newsroom. August 12, 2012. http://www.gartner.com/it/page.jsp?id=2120015 (accessed October 3, 2012).

of this field. According to a 2012 Gartner report, the top mobile OS are Android, Symbian, Apple iOS, RIM BlackBerry, MeeGo, Windows Phone, and Bada [Gartner Inc. 2012] (Table 61.3).

- *Android (Google Inc.)*: Initially released in 2005, programmed in C/C++ and Java and based on the Linux platform, Android has become the dominant operating system for smartphones and tablet devices. As an open source platform, it has been readily adopted and modified by multiple manufacturers and mobile operators and as such there is a proliferation of versions and variants in use today. Android's App Inventor programming language developed by MIT allows users with limited programming ability to write their own apps using their web browser and a block-based development environment [Massachusetts Institute of Technology 2012].
- *Bada (Samsung Electronics)*: A proprietary platform initially released in 2010, programmed in C++. Samsung is using its own Bada operating system, in parallel with Android OS and Windows Phone, for smartphones they develop (Samsung 2012).
- *Blackberry OS (RIM)*: A proprietary platform initially released in 1999, programmed in Java, the BlackBerry platform is best known for its native support for corporate e-mail services including Microsoft Exchange, Lotus Domino, and Novell GroupWise. BlackBerry used to make up 52.4% of the smartphone market in 2008. According to Q2 2012, the present mobile market share held by BlackBerry is 5.2% [Gartner Inc. 2012].
- *iPhone OS/iOS (Apple)*: A proprietary platform initially released in 2007, programmed in C/C++ and Objective-C, the iOS mobile OS is available only on Apple manufactured devices and has not been licensed for third-party hardware. Apple iOS is derived from Apple's Mac OS X Debian Linux-based operating system [Apple Inc: Start Developing iOS Apps Today 2012].
- *Symbian (Accenture)*: Initially released in 1997, programmed in C++ and Qt, Symbian originally developed by Nokia, was transferred to Accenture in 2011 (Accenture 2011). Symbian is one of the oldest and the most successful mobile OS until 2010, when it dramatically lost market share to Android and iOS.
- *Windows Phone (Microsoft)*: Initially released in 2010, the Windows Phone OS is programmed in C/C++ and NET built on the Windows CE platform. With Windows Phone, Microsoft features its design language called Metro, also found in Windows 8. The Windows Phone OS replaces the previous Windows Mobile OS developed in 2000 by Microsoft. Windows Mobile was discontinued in 2011 (Microsoft: Channel 9 2010).

We now provide a more detailed description and discussion of the two most dominant systems at the current time—Google's Android OS and Apple's iOS. We will focus on three main roles that the mobile OS plays: (1) the abstraction it presents in hiding hardware details from the application layer, (2) the approaches it takes to managing the systems resources, and (3) the interface it presents to users and developers.

61.2.2 Android OS

61.2.2.1 Linux Kernel

As shown in Figure 61.5, the Android OS is built on the Linux Kernel. Initial versions of the OS were based on the Linux 2.6 kernel. Version 4.0 and above are based on Linux 3.3 kernel. Although Android runs on Linux, it would be misleading to consider it a Linux distribution, or *distro,* because it leaves so much out of the Linux system, including many standard libraries, shells, editors, and programming frameworks [Bray 2010].

61.2.2.2 System Libraries

On top of the Kernel sits a collection of C/C++ libraries, including, among others, the following [Android: App Framework 2012]:

- Libc—a Berkeley Software Distribution (BSD)-based version of the standard C system library.
- SGL—a 2D graphics library.
- OpenGL—a 3D graphics library.
- SQLite—a lightweight relational database system.
- WebKit—a web browser engine that allows apps to include an embedded Web view.

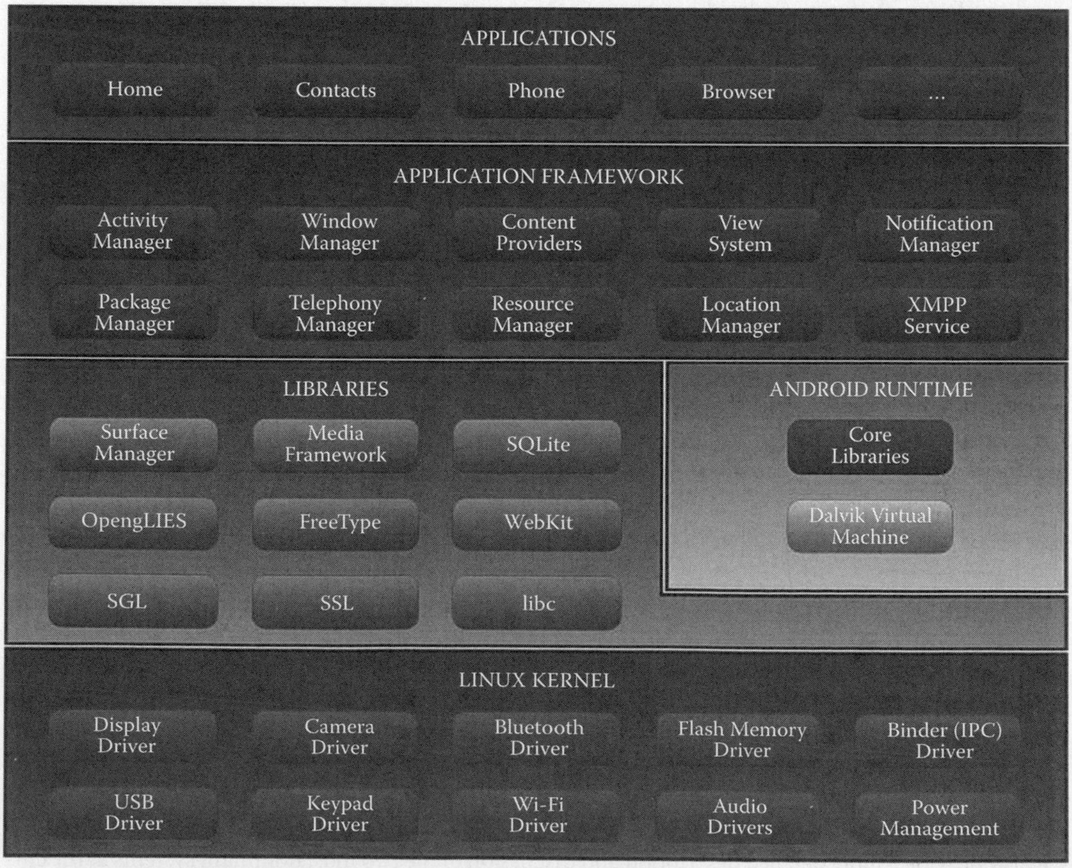

FIGURE 61.5 Android architecture.

61.2.2.3 Dalvik Virtual Machine

Android runs primarily on hardware platforms based on the Advanced RISC Machine (ARM) architecture, a 32 bit reduced instruction set architecture that is widely used in mobile devices [Wikipedia: ARM Architecture 2012]. ARM's dominance among mobile phone architectures—according to Stephen Furber, one of the original designers of the ARM architecture, there are more ARMs than people on the planet [Fitzpatrick 2012]—is due primarily to its power efficiency. It reportedly consumes one-tenth the amount of electricity as other 32 bit processors. Android has also been ported to the ×86 platform [Shah 2011].

Android's runtime environment is based on the open source Dalvik Virtual Machine (VM), originally designed and written by Daniel Boorstein in cooperation with Google [Dalvik Virtual Machine 2008]. The Dalvik VM is designed for machines with memory and speed constraints and is therefore well suited for mobile devices. Unlike the Java VM, which uses a stack-based architecture, the Dalvik VM is register based. It is optimized for low memory requirements and it allows multiple VM instances to run at the same time. It relies on the underlying Android OS to handle process isolation, memory management, threading support, and other system functions, as shown in Figure 61.6.

Android applications are written in Java and compiled first into Java byte code (.class) files. These are then compiled into Dalvik byte code (.dex) files before loading onto the Android device. In early Android versions, the Dalvik VM ran interpreted. The Froyo (2.2) release of Android contains a just-in-time compiler for Dalvik. Although this improves the run-time performance over previous releases, there is still considerable discussion and debate about Android's choice of the Dalvik over the Java VM. For example, one recent comparison of the two systems, by an Oracle engineer, claims that Dalvik runs two to three times more slowly than Oracle's Java SE Embedded Hotspot VM [Vandette 2010]. Of course, this sort of comparison between the Java VM and Dalvik does not take into account other factors that may make Dalvik a more suitable choice for Android,

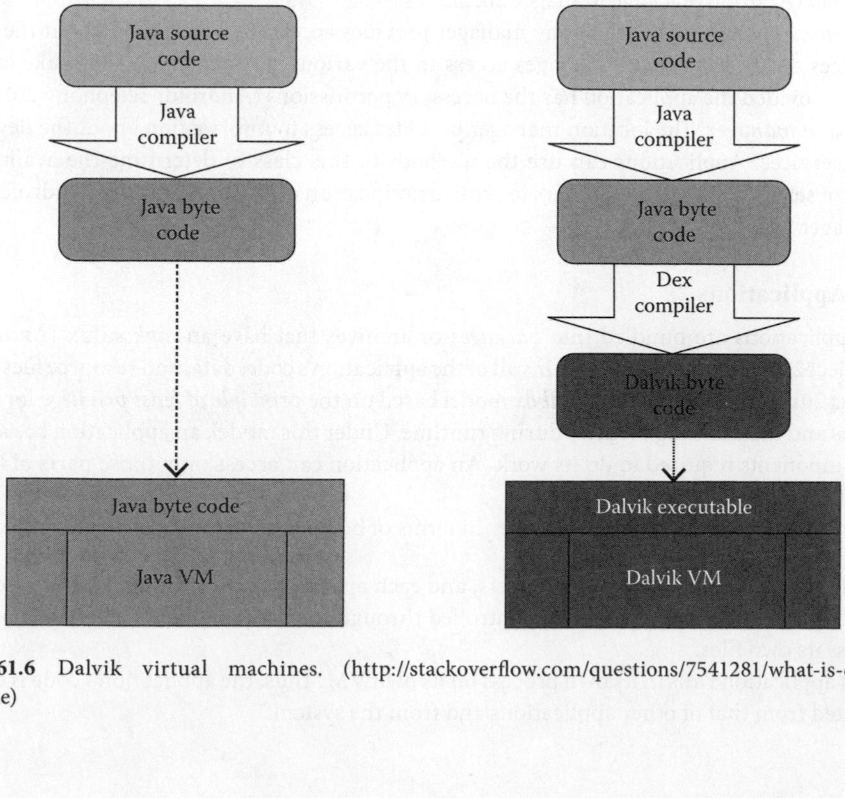

FIGURE 61.6 Dalvik virtual machines. (http://stackoverflow.com/questions/7541281/what-is-dalvik-and-dalvik-cache)

including its open source license and the fact that it can run multiple instances simultaneously. The picture is also clouded by the fact that Google is currently being sued by Oracle over its use of Java in the Android OS [Lowe 2012].

61.2.2.4 Resource Managers

Atop the system libraries sits collection of *managers* that comprise the *application framework*. The framework provides abstractions for controlling and managing the various functions that make up system's operations. Some of the main managers are as follows:

- *Activity manager*: An Android application is comprised of a collection of *activities*, which represent the different tasks or actions that users can do—for example, make a phone call, send an e-mail message, and take a picture. Activities are designed to interact with the user [Android: Activity 2012]. The activity manager interacts with the activities currently running in the system. We discuss the activity life cycle in more detail in the subsequent section.
- *Content providers*: Content providers manage access to structured data. They encapsulate various kinds of data and provide a unified interface (abstraction) for storing, retrieving, and securing data of various kinds. They also provide a mechanism through which data are shared among different processes and applications. Content resolver objects are used as clients that communicate with content providers through a client/server like interface. The provider object receives data requests from clients, performs the requested action, and returns the results. Android provides built-in content providers for audio, video, images, and personal contact information [Android: Content Providers 2012].
- *Package manager*: The package manager installs, removes, and manages all of the application packages that are installed on the system. Among other tasks, the package manager oversees the systems security, granting, and authenticating the various permissions needed by applications to manage systems resources. Android applications run in a sandbox and need explicit permission to perform specific tasks such as making telephone calls, taking pictures, and accessing the Internet (Android: Package Manager 2012).
- *Telephony manager*: The telephony manager provides access to information about the telephony services on the device and manages access to the various services needed to make and receive calls, provided the application has the necessary permissions (Android: Telephony 2012).
- *Location manager*: The location manager provides access to information about the device's location services. Applications can use the methods in this class to determine the availability and type of services to register listeners to receive notification of location changes (Android: Location Manager 2012).

61.2.2.5 Applications

Android applications are bundled into *packages* or archives that have an *.apk* suffix [Android Open Source Project 2012]. The package contains all of the application's code, data, and resource files (Android: Permissions 2012). Android uses a *sandbox* model based on the *principle of least privilege* for managing applications and maintaining security during runtime. Under this model, an application has access only to those components required to do its work. An application can access only those parts of the system for which it is given explicit permission.

Applications are isolated from one another in terms of both their identities and runtime contexts:

- Applications are considered Linux users, and each application has a unique user ID. Permissions to access the application's files are controlled through the user ID. An application is only able to access its own files.
- Each application runs in its own process on its own VM. Thus, the application's code is completely isolated from that of other applications and from the system.

In terms of sharing data among applications, although it is possible for two applications to share the same user ID and inhabit the same process, applications share data or resources primarily through a *permissions* system. For example, access to a user's contacts or SMS messages must be requested. Permissions for such accesses are granted (or not) when the application is installed.

Applications are made up of four basic types of *components* [Android: App Framework 2012]:

1. *Activities*: An *activity* is the basic computational unit that interacts with the user. It consists of a single screen, or *view,* which contains the UI. An application typically consists of several activities that work together. For example, an e-mail application might have an activity for composing e-mail messages and an activity for listing the user's messages.
2. *Services*: A *service* is a component that runs in the background to perform some long-running task. For example, a service might be used to download data from the Internet. Services do not have an UI.
3. *Content providers*: A *content provider* manages a collection of persistent data that can be stored in a database, on a website, in the file system, or in some other form. Applications with the appropriate permissions can query content providers to access or modify data. An example would be saving and retrieving photos in an image gallery.
4. *Broadcast receiver*: A *broadcast receiver* responds to system-wide broadcasts, for example, the announcement that an e-mail or SMS message has been received or that the system's battery is low. Broadcasts may come from the system or from another application. Broadcast receivers do not have a UI, but they can use *notifiers* to display messages on the phone's status bar.

61.2.2.6 Activities and User Interaction

Activities are the basic computational units that interact with users. The *Activity* class takes care of creating a window and defining *the UI* [Android: Activity 2012].

The activity architecture is designed around the principle of *responsiveness*—that is, providing a crisp, responsive UI, with user interactions taking priority over other system operations [Android: Designing for Responsiveness 2012]. The activity manager and the window manager guard against applications that are insufficiently responsive, posting an *Application Not Responding (ANR)* dialog if the application cannot respond to user input or does not respond within a certain time limit. This could happen if the application blocks while waiting for an I/O operation to complete or takes too much time to build a complex data structure in memory. The ANR dialog permits the user to kill the app or to continue waiting, but its presence is usually a sign that something is wrong with the app's design or performance.

Android activities are managed with an *activity stack.* When a new activity is started, it is placed on top of the stack and becomes the *running activity.* The activities below it are, typically, kept in memory, but would not come to the foreground until all the activities above them are popped off the stack. Thus, the user is always interacting with the top activity on the stack.

Activities can be in one of four states [Android: Activity 2012]:

1. *Active* or *running*—it is on the top of the stack and visible in the foreground.
2. *Paused*—it has lost focus but is still visible. It retains its state but the user cannot interact with it. For example, a paused activity may be one that has displayed an interactive dialog that is prompting the user for some action. The system can kill a paused activity if its memory is needed.
3. *Stopped*—it is completely obscured by another activity. Stopped activities retain their state but they can be killed by the system if their memory is needed.
4. *Killed*—if an activity is killed by the system, it is dropped from memory either by asking it to finish or by simply killing its process. When it is displayed again, it must be restarted from scratch.

Figure 61.7 provides an overview of the activity lifecycle. The ovals represent the activity's state, and the rectangles represent callback methods available to the program as the activity progresses through its lifecycle.

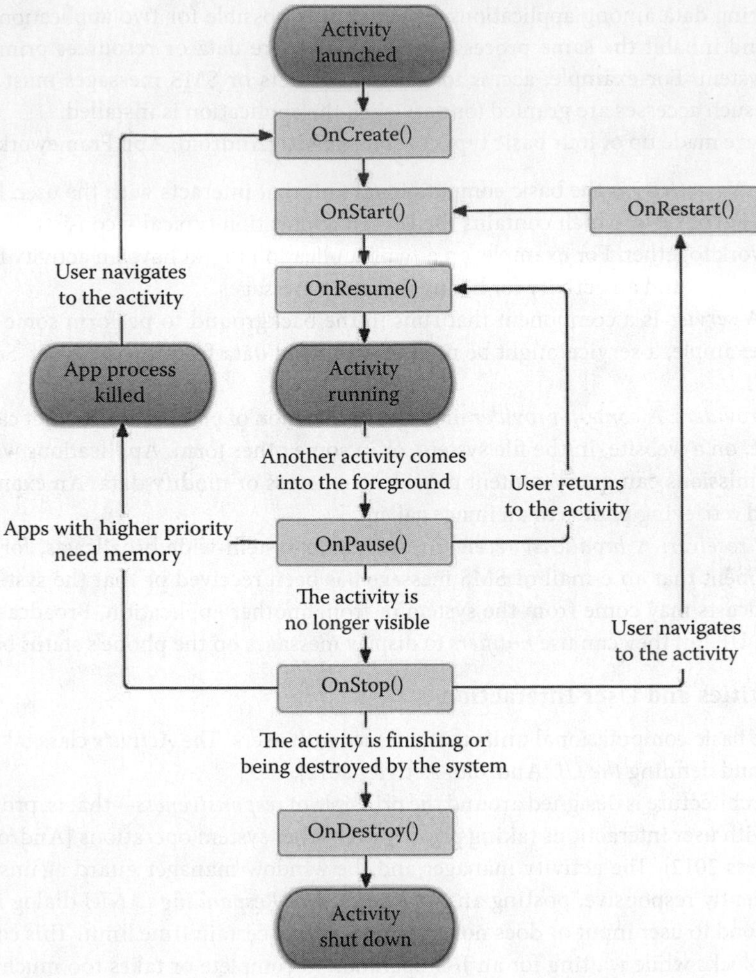

FIGURE 61.7 Activities life cycle. (Android: Activity. "Public Class–Activity." Android Reference. 2012. http://developer.android.com/reference/android/app/Activity.htm [accessed October 9, 2012].)

For the applications programmer, the following interface provides access to all the main points of activity's lifecycle:

```
public class Activity extends ApplicationContext {
        protected void onCreate(Bundle savedInstanceState);
        protected void onStart();
        protected void onRestart();
        protected void onResume();
        protected void onPause();
        protected void onStop();
        protected void onDestroy();
}
```

As the names of these methods suggest, the entire life cycle is *event driven*. When the system finds it necessary to interrupt an activity—for example, because of an incoming phone call—it dispatches an event that can be handled by one or the other of these methods. Because applications can be interrupted at any time, it is important that they be designed for such possibilities. For example, an app that is accepting input from the user—for example, composing an e-mail message—must be able to save the application's state if interrupted so that it can be restored when the application is restarted. Persistent data can be saved in the *onPause()* method and restored in the *onResume()* method. In terms of saving dynamic parts of the application's state, Android provides the *onSaveInstanceState(Bundle)* method, which is invoked just before the activity is placed in the background, and *onRestoreInstanceState(Bundle)*, which is invoked just before the activity is placed in the foreground. Note that the *onCreate()* method can restore such state even if the activity is killed by the OS.

The design of the Activity lifecycle promotes a response UI and the ability of the OS to interrupt an activity at any time in response to incoming phone calls, battery failures, lack of memory, and other events.

61.2.2.7 Intents and Intent Filters

Communication within and among applications and activities take place through messaging objects known as Intents [Android: Intent 2012]. An *Intent* is a passive data structure that represents an operation or task to be performed. Intents can be used to start activities and applications or to communicate with background services. They provide a mechanism for late runtime binding between the code different applications.

The Android system provides both explicit and implicit intents. An *explicit intent* specifies a particular component or activity by its name—for example, an e-mail component or list activity—and is primarily used to communicate among the activities and services that make up a single application. The intent is used to instantiate the component or start the activity and pass information to it—working in a similar fashion to a non-blocking *remote procedure call*. An *implicit intent* does not identify a particular object or activity. Instead, it describes a particular task—such as the "send email" task. The OS then takes responsibility for identifying objects that are available in the system to perform that task. For example, the built-in ACTION_*SENDTO* action expresses a request to send some type of message (e-mail or SMS). The type of message, the destination phone number or e-mail address, and the message itself would be specified in a uniform resource identifier identifying the target of the intent. The system then searches among registered entities for an e-mail or messaging application that handles that type of action and sends a broadcast to all entities that satisfy the target criteria.

Components use *intent filters* to identify to the system which intents they can handle. For example, a broadcast receiver would use an intent filter to identity those applications it will accept text messages from—that is, Google Voice, Android Messaging, or some other application. For system-wide intents, this would typically be done in the *Android manifest*, a static document that describes the application's activities and permissions and informs the system about its various intent filters.

61.2.2.8 User Interface

Android provides a variety of pre-build UI components such as structured layout objects and UI controls that allow you to build the graphical UI for your app. Android also provides other UI modules for special interfaces such as dialogs, notifications, and menus [Android: User Interface 2012].

All UI elements in an Android app are built using `View` and `ViewGroup` objects [Android: User Interface 2012]. A `View` is an object that draws something on the screen that the user can interact with. A `ViewGroup` is an object that holds other `View` (and `ViewGroup`) objects in order to define the layout of the interface. Android provides a collection of both `View` and `ViewGroup` subclasses that offer common input controls (such as buttons and text fields) and various layout models (such as a linear or relative layout).

A typical Android app consists of action bars and the app content area [Android: Design 2012].

1. Main action bar is the command and control center for an app. The main action bar includes elements for navigating the app's hierarchy and views and also surfaces the most important actions.
2. View control allows users to switch between the different views that the app provides. Views typically consist of different arrangements of data or different functional aspects of the app.

3. Content area is the space where the content of the app is displayed.
4. Split action bars provide a way to distribute actions across additional bars located below the main action bar or at the bottom of the screen.

61.2.3 Apple iOS

61.2.3.1 Overview

The mobile OS that runs on the Apple iPhone, iTouch, and iPad devices is iOS. Unlike Android, an open source system running on many hardware platforms, iOS is a proprietary system that is not licensed to third parties and runs exclusively on Apple products.

iOS is derived from Apple's OS X system, with which it shares the Darwin system. Darwin is Apple's open source, POSIX-compliant OS first released in 2000.* Darwin currently includes support for the 64 bit Intel x86 processors as well as for the 32 bit ARM processors used in the iPhone, iTouch, and iPad [Wikipedia 2012]. Darwin does not include many of the defining elements of Mac OS X and iOS, in particular the Cocoa API and thus, Darwin by itself cannot run Mac or iOS applications.

The Cocoa and Cocoa Touch APIs are what give Mac and iOS applications their distinctive look and feel. Cocoa contains several API kits—for example, the *Foundation Kit*, the *Application Kit*, and the *Core Data Frameworks*—and the libraries and frameworks to support those kits. It also includes the Objective-C runtime library [Wikipedia: Cocoa API 2012]. Cocoa applications are typically written in Objective-C using Apple's proprietary development environment, which consists primarily of the Xcode integrated development environment (IDE) and its built-in *Interface Builder*, an application for building UIs (Apple Inc: Cocoa 2012).

Cocoa Touch is an API for building UIs for the iOS (Apple Inc: Cocoa 2012). It includes features and technologies to support multitasking, gesture recognition, and animation and includes several kits and frameworks that extend the functionality of Cocoa. The UI kit, which extends Cocoa's Application Kit, provides the API for supporting graphical user interfaces. The Game Kit framework provides an API for game developers. The *iAd* and *Map Kit* frameworks provide API support for Apple's mobile advertisement platform and its newly unveiled mapping platform. The layer architecture for iOS is depicted in Figure 61.8.

The iOS system interfaces are organized into frameworks. A *framework* is *a directory* that contains a dynamic shared library and the resources (such as header files, images, helper apps, and so on) needed to support that library. Frameworks are linked to the application project.

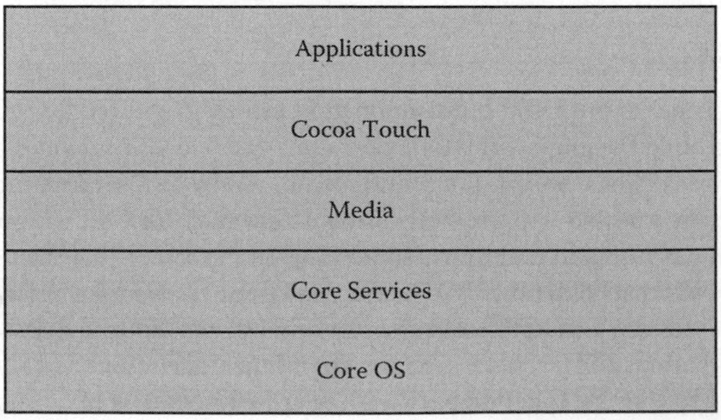

FIGURE 61.8 Layer architecture of iOS.

* POSIX, which stands for portable operating system interface, is a set of IEEE standards for Unix operating systems that define the programming interface (API) and command shells.

61.2.3.2 Core OS

iOS is a Unix system based on the Mach kernel. *Mach* is the operating system kernel developed at Carnegie Mellon University in the mid 1980s. Although sometimes mentioned as one of the first *microkernels*, the version of Mach used in iOS is not a microkernel.*

Darwin, the Mac OS X kernel, is based in turn on the XNU (X is not Unix) kernel, a hybrid kernel that combines features of the Mach kernel with components from BSD, and Driver Kit, an object-oriented API for writing device drivers. The Mach kernel, which was developed at Carnegie Mellon University in the mid 1980s, is a simple kernel that runs core OS functions as separate processes. Darwin incorporates certain features from BSD (Wikipedia 2012)—specifically from the FreeBSD kernel—including the following:

- POSIX API (BSD system calls)
- Unix process model atop Mach tasks
- Basic security policies
- User and group IDs
- Permissions
- The network stack
- Virtual file system code
- Local file systems, including hierarchical file system and network file system
- Cryptographic framework
- Inter-process communication system

From the perspective of the application developer, the core OS layer contains the following low-level system features and frameworks [Apple Inc: The iOS Environment 2012].

- The *accelerator* framework (introduced in iOS 4.0) contains digital signal processing, linear algebra, and image-processing libraries.
- The *core Bluetooth* framework contains libraries for interacting with *Low Energy Bluetooth* accessories.
- The *external accessory framework* provides libraries for communicating with devices that can be connected to an iOS device through its 30 pin dock connector or through its wireless Bluetooth interface.
- The *generic security services* framework based on the IETF RFC 2743 and RFC 4401 standards. In addition to offering the standard interfaces, iOS includes some additions for managing credentials that are not specified by the standard but that are required by many applications.
- The security framework provides interfaces for managing certificates, public and private keys, and trust policies. It supports the generation of cryptographically secure pseudorandom numbers. It also supports the storage of certificates and cryptographic keys in the keychain, which is a secure repository for sensitive user data.

61.2.3.3 Runtime Environment

The iOS system is based on the same technologies used by Mac OS X, namely, the Mach kernel and BSD interfaces. Thus, iOS apps run in a Unix-based system and have full support for threads, sockets, and many of the other technologies typically available at that level. However, there are places where the behavior of iOS differs from that of Mac OS X [Apple Inc: The iOS Environment 2012].

- *The virtual memory system*: In iOS, each program still has its own virtual address space, but unlike Mac OS X, the amount of usable virtual memory is constrained by the amount of physical memory available. iOS does not support paging to disk when memory gets full. Instead, the virtual

* A *microkernel* is the bare minimum of mechanisms needed to implement an operating system. It typically consists of low-level address space management, thread control, and interprocess communication (IPC) functions. These are the software elements that run at the most privileged level (e.g., in kernel mode or supervisor mode). By contrast, a *monolithic kernel*, would include other traditional operating system functions, such as device drivers, schedulers, virtual memory, and file systems.

memory system simply releases read-only memory pages, such as code pages, when it needs more space. If memory continues to be constrained, the system may send notifications to any running applications, asking them to free up additional memory.

- *The automatic sleep timer*: When the system does not detect touch events for an extended period of time, it dims the screen initially and eventually turns it off altogether. For applications that do not use touch inputs, such as a game that relies on the accelerometer, the automatic sleep timer can be disabled to prevent the screen from dimming.
- *Multitasking support*: In iOS 4 and later, *multitasking* allows apps to run in the background even when they are not visible on the screen. Most background apps reside in memory but do not actually execute any code. These apps are suspended by the system shortly after entering the background to preserve battery life.

At install time, iOS places each app (including its preferences and data) in a *sandbox*—that is, a set of fine-grained controls that limit the app's access to files, preferences, network resources, and hardware. As part of the sandboxing process, the system installs each app in its own sandbox directory, which acts as the home for the app and its data. User sensitive data can be stored in *a keychain*, an encrypted container for passwords and other secrets specific to an app. Keychain data for an app are stored outside of the app's sandbox.

61.2.3.4 Programming Model

iOS application development depends on the iOS software development kit (SDK) and Xcode, Apple's IDE. Xcode employs a single window, called the *workspace window* that presents most of the tools needed to develop apps (Apple Inc: Xcode 2012). The iOS SDK extends the Xcode toolset to include the tools, compilers, and frameworks you need specifically for iOS.

iOS apps are written in objective-C, an object-oriented language, that extends the standard ANSI C programming language with syntax for defining classes and methods [Apple Inc: Start Developing iOS Apps Today 2012]. It also promotes dynamic extension of classes and interfaces that any class can adopt.

Figure 61.9 shows the key objects that are most commonly found in an iOS app. iOS apps are organized around the model-view-controller design pattern (Apple Inc: Cocoa 2012). This pattern separates the data objects in the model from the views used to present that data. Models encapsulate application data, views display and edit that data, and controllers mediate the logic between the two. This separation promotes code reuse by making it possible to swap out your views as needed.

61.2.3.5 User Interface

Apple's iOS devices are known for their aesthetically pleasing and intuitive UI that has set Apple devices apart from other mobile OS. The *iOS Human Interface Guidelines* is a high level document that describes the guidelines and principles that developers follow for UI and user experience design. It does not describe how to implement designs in code.

An iOS app has a single window, unless it supports an external display; the user sees one screen at a time. An app's window fills the device's main screen and provides an empty surface that hosts one or more views that present the app's content. An iOS *window* differs in important respects from a window in a typical computer app. For example, an iOS window has no visible components (such as a title bar or a close button) and it cannot be moved to a new location on the device display. Instead, the window simply represents a container for visual content that is mapped to the device's screen. Users experience an iOS app as a sequence of screens through which they navigate. From this perspective, a screen generally corresponds to a distinct visual state or mode in an app.

In iOS, the UIKit framework provides the infrastructure for building graphical applications, managing the event loop, and performing other interface-related tasks [Apple Inc: Start Developing iOS Apps Today 2012].

For a more in-depth look at the iOS programming guidelines and documentation visit the iOS Developer Library (http://developer.apple.com/library/ios).

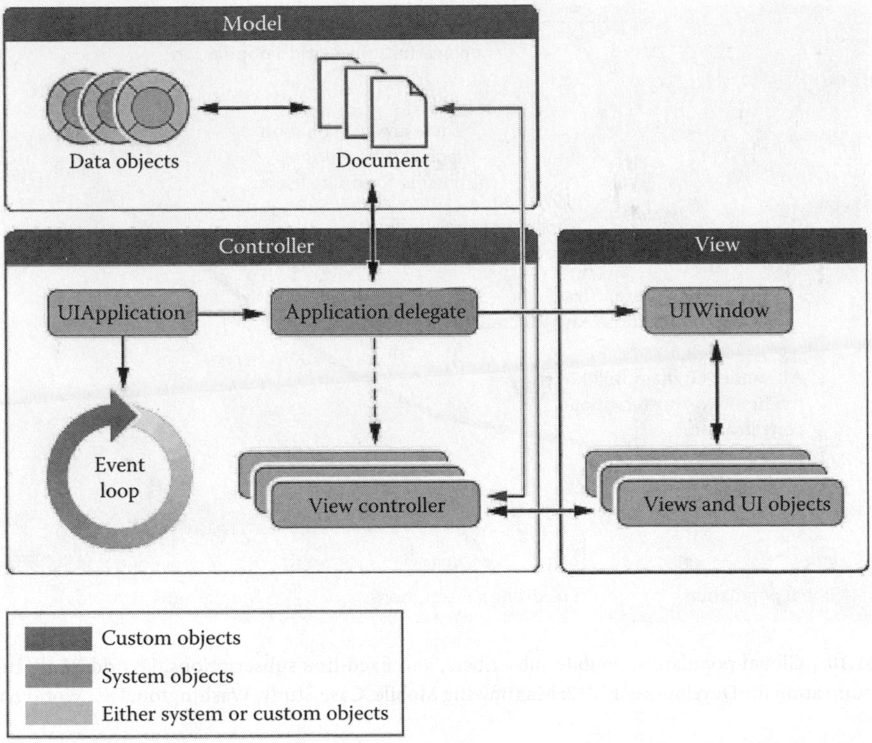

FIGURE 61.9 Key objects found in iOS apps. (http://developer.apple.com/library/ios/#documentation/iPhone/Conceptual/iPhoneOSProgrammingGuide/AppArchitecture/AppArchitecture.html#//apple_ref/doc/uid/TP40007072-CH3-SW1)

61.3 Impact on Practice

In 2012, 75% of the world's population currently has access to a mobile subscription [World Bank 2012]. A total of 13.2% of the world's 6.1 billion cell phones are smartphones, but the rate exceeds 30% in larger markets like the United States, Germany, and Britain [O'Brian 2012]. Mobile computing devices bring computing to the masses in a way not seen since the introduction of the personal computer in the 1980s.

A new wave of "apps," or smartphone applications, and services, driven by high speed networks, social networking, and online crowd sourcing, is helping mobile phones transform the lives of people around the globe. Nearly 40 billion apps for iPhones and Android devices have been downloaded worldwide by mid 2012 [Lunden 2012]. Research reports forecast the worldwide smartphone application market, which it estimates will grow from $1.94 billion in 2009 to $15.65 billion by 2013 [Wauters 2012] (Figure 61.10).

As the mobile landscape develops and smartphones gain market share, content providers are finding new ways to reshape and broadcast information to the public. Regular websites can be reformatted to meet mobile capabilities. Mobile phones are becoming the "third" screen (after television and PCs) used to access content and are reshaping content delivery.

Given technological convergence, smartphones can now function as a digital equivalent of a Swiss army knife a wallet, camera, television, alarm clock, calculator, address book, calendar, newspaper, navigational device, and gaming console combined. The latest smartphones are not just invading the computer space; they are reinventing it by offering so much more in both voice and non-voice services.

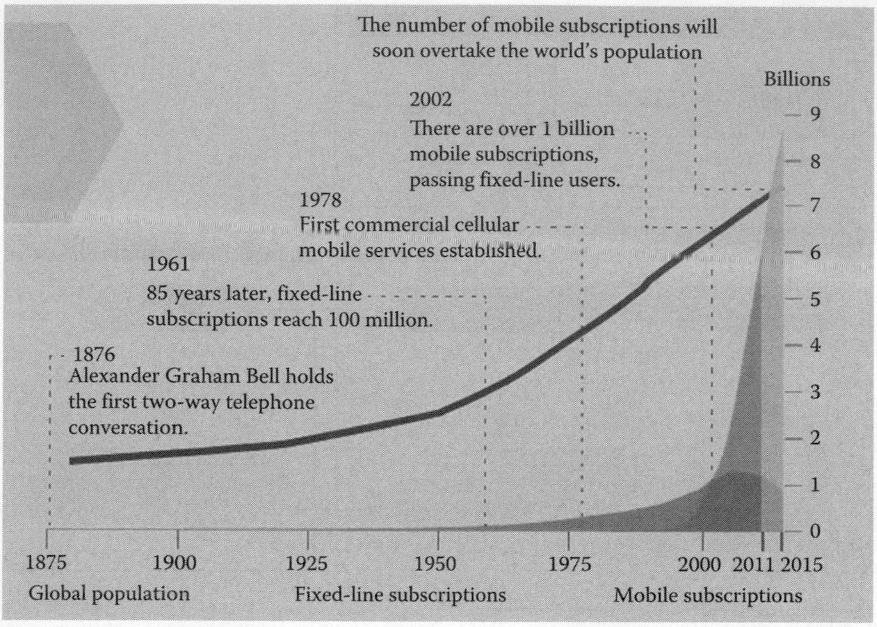
The number of mobile subscriptions will soon overtake the world's population
Billions
2002
There are over 1 billion mobile subscriptions, passing fixed-line users.
1978
First commercial cellular mobile services established.
1961
85 years later, fixed-line subscriptions reach 100 million.
1876
Alexander Graham Bell holds the first two-way telephone conversation.
9
8
7
6
5
4
3
2
1
0
1875
1900
1925
1950
1975
2000
2011
2015
Global population
Fixed-line subscriptions
Mobile subscriptions

FIGURE 61.10 Global population, mobile subscribers, and fixed-line subscriptions. (World Bank. Information and Communication for Development 2012: Maximizing Mobile. Case Study, Washington, DC: World Bank, 2012.)

61.3.1 In Healthcare

According to research conducted by Manhattan Research in 2009 on the professional use of smartphones by physicians, about 64% of the physicians in the United States used smartphones in 2009 compared to only 30% in 2001. This report found a noticeable increase in smartphone adoption and predicted that 81% of physicians would use smartphone in the United States by 2012 [Cooper 2009].

Mobile OS platforms are capable of running third-party applications specifically developed for the healthcare field. For instance, John Hopkins antibiotics guide "Johns Hopkins ABX Guide," which is available for the Android, iOS, Blackberry, and Windows platforms, provides information on the treatment of infectious diseases to help make decisions at the point of care. The guide breaks down details of diagnosis; drug indications, dosing, pharmacokinetics, side effects, and interactions; pathogens; management; and vaccines into easily accessible, frequently-updated, quick-read entries on the mobile device [Unbound Medicine Inc. 2012].

Market research firm Juniper Research believes that some 3 million smartphone users will be using their devices for remote health monitoring by 2016 [Thomas 2012]. Patients and doctor alike can access electronic health records (EHR) through smartphone applications such as MEDITECH EMR, which using a web browser to access clinical data including laboratory results, vital signs, intake and output, allergies, active medications, and documents (reports and notes) (MEDITECH 2012).

There are many health education-based applications available on the App stores that range from teaching children healthy eating habits, blood glucose calculators, telemedicine applications to a fall detector that uses the built-in accelerometer to work out if a person has fallen. Smartphones are poised to change the way we interact with our medical professionals.

The challenges of smartphone-based healthcare include limited battery life, small screen size, potentially erroneous data input, computer viruses including spyware, magnetic interference with medical devices, potentially inefficient patient–physician interactions, loss or theft, and breaches of data privacy and security [Mosa 2012].

61.3.2 In Agriculture

In addition to weather updates, facts and figures that were once the scribbled domain of old receipts and tattered pieces of paper at the farm are now organized on a smartphone [Gerke 2012]. From pasture grazing records to serial numbers and checkbook register, applications are available to help keep track of important information quickly and easily.

Mobile applications for agriculture and rural development have generally been designed locally and for specific target markets, with localized content specific to the languages, crop types, and farming methods in the region.

61.3.3 In Banking

As thc adoption of smartphones continues and speed of access improves, customers across every age group are becoming increasingly comfortable with conducting more and increasingly complex transactions through their mobile phones. EBay Inc.'s PayPal division handled $4 billion in mobile payments in 2011 [Kucera 2012]. Banks now offer customers extended mobile services, which allow them to deposit a check (using a picture of the check), check their balances, transfer money, and locate cash points.

Japan's Jibun Bank, for example, was created in 2008 as a joint venture between the country's largest bank, BTMU and KDDI, a telecommunications operator. Designed from the outset as a mobile-only bank—that is, without being based on a network of physical branch offices—within 2 years, it had more than one million accounts and US $1.7 bn in deposits [KPMG 2011].

61.3.4 Privacy and Security

With the proliferation of mobile apps that provide electronic monetary transactions to accessing medical records on a smartphone, one has to take into consideration the security and privacy of the underlying data. In addition to the obvious risk of losing one's mobile device, smart phones can be vulnerable to third-party information-stealing applications and other malware. Security is intimately bound to the design of the platform for which the mobile application is built.

Smartphone operating systems have both advantages and disadvantages with respect to offered security. On the one hand, they use application sandboxing to contain exploits and limit privileges given to malware. On the other hand, they collect and organize many forms of security and privacy sensitive information—such as location information and phone status—simply as a matter of operation, and make that information easily accessible to downloaded third-party applications [SPSM workshop 2012]. Users and application developers are restricted from accessing and modifying core elements of the OS such as the kernel and system daemons through software and policy controls. These controls are implemented by mobile OS developers, cellular carriers, and smartphone manufactures. But given the fact that consumers can choose, wittingly or unwittingly, to allow third-party applications to have access to that data, such controls can only go so far in protecting sensitive personal data.

For example, when an Android application is downloaded from the market, the user is presented with a list of permissions that the application requires—for example, access to the Internet, to their Google or Twitter accounts, to the phone's camera, etc. But the details and implications of granting such permission is not totally transparent and may lead to unanticipated data sharing.

61.4 Research Issues

The advent of mobile computing has introduced new challenges for the designers of mobile devices and mobile OS. One such challenge is to reduce power consumption in ways that minimize the adverse effect on usage. Savings in power consumption can be "spent" by extending battery life and (or) by reducing the physical size of batteries, thus reducing the physical size of mobile computing devices [Welch 1995].

For software developers, the availability of multiple platforms and the lack of standardization make it difficult to decide what mobile OS to develop on. Processing power, storage, and functionality differ wildly from device to device. It is virtually impossible to build and maintain one application that will run on Blackberry, Symbian, Android, iOS, and other mobile platforms. Supporting multiple operating systems requires one to rewrite the application in each of the different native environments. Needing to build and maintain two or more versions of a smartphone app raises the difficulty level of many development projects significantly.

Indeed, even confining one's focus to the Android platform presents considerable challenges given the large number of manufacturers and phone companies providing Android products and competing with each other for market share. Already, there are significant differences among the various Android versions and among the features promoted by the different vendors. So far, this variability among systems has not proved to be an impediment in Android's adoptability as some had predicted. The situation is somewhat better for iOS developers because of Apple's tight and proprietary control of the platform. But even on the iOS platform, or any single platform, given the rapid development and advancement of device features, an application developed for one version of iOS may quickly become incompatible with newer devices, which require new releases of the OS.

In an attempt to address this challenge, cross-platform development tools for mobile systems are beginning to emerge. Among these tools are Appcelerator Titanium, Phonegap, QuickConnect, Rhomobile, and WidgitPad (Warren 2012). Most use a mix of HTML, CSS, and JavaScript, plus some native "wrapper" code for accessing hardware features like GPS, a camera, or an accelerometer [Dern 2010]. Considerable "tweaking" is required to get applications to function at all across multiple-platforms. One limitation of cross-platform tools is that they may not support particular functionality specific to one operating system or the other. However, in some cases, they do help reduce application development time and costs, while exposing an application to a larger smartphone market segment.

Among the numerous areas of research and development are the following:

- *Haptic technology*, or *haptics*, is a tactile feedback technology that takes advantage of the sense of touch by applying forces, vibrations, or motions to the user [Shubha 2012]. An example would be the use of vibration to confirm that a touchscreen button has been pressed. Haptic devices may incorporate tactile sensors that can measure the force exerted by the user on the interface. These would then be interpreted by the operating system to provide input to the application or feedback to the user. The interface acts as both an input and output mechanism to enhance the user experience. Immersion, a developer of touch feedback technology, announced a new Developer Kit that allows a form of adaptive feedback. For instance, while swiping through Facebook pictures, the force feedback will change from light clicks to longer rumbles according to how many comments it had.
- The emergence of voice activated, *intelligent personal assistants* is an area of applications research. Apple's iPhone has "Siri," the voice-activated virtual assistant that it introduced in October 2011 with the iPhone 4S (Dolcourt and Bennett 2012). The feature has the potential to change how consumers retrieve information and interact with their mobile devices, giving them the ability to find information on the Web with natural voice commands and to perform other tasks. "Google Now" for the Android 4.1 Operating System introduced a similar voice search service for Android-based devices shortly after.
- Google and its mobile partners introduced an alternative to the credit card with a technology called *near-field communications (NFC)* that lets users make payments wirelessly at cash registers [Wingfield 2012]. NFC is a set of standards for smartphones and similar devices to establish radio communication with each other by touching them together or bringing them into close proximity, usually no more than a few centimeters.
- Project Glass is a research and development program by Google to develop an *augmented reality head-mounted display* [Goldman 2012]. The intended purpose of Project Glass products would be

the hands-free displaying of information currently available to most smartphone users. Virtual objects and information could be overlaid on the real world and displayed to the wearer.

- As mobile connectivity and smartphone devices become ubiquitous, we are seeing a shift in the perception and expectations of end users, with access to media rich content virtually anywhere. *Unified communications* (UC) is the integration of real-time communication services with non real-time communication services into a single platform [Vodafone 2010]. As such, UC is not a single product but a portfolio of products that provide a consistent UI and user experience across multiple devices, platforms, and media types. In many cases, these solutions can be hosted remotely, effectively providing a scalable Cloud Service for mobile devices.
- In 2010, Apple filed a patent application for a small *fuel-cell* power supply that could potentially give the iPhone and iPad enough power to last for weeks without the need to plug them in [Wingfield 2012]. Fuel cells use hydrogen and oxygen to produce electricity with no pollution.

61.5 Summary

Mobile devices are arguably the most ubiquitous modern technology: in some developing countries, more people have access to a mobile phone than to a bank account, electricity, or even clean water. Mobile communications now offer major opportunities to advance human development—from providing basic access to education or health information to making cash payments to stimulating citizen involvement in democratic processes [World Bank 2012].

As we have seen in this chapter, the smartphone has evolved over a short period of time and is still at the start of its growth curve. Devices and operating systems are becoming more powerful and less expensive to operate. The App development market has rapidly grown and continues to evolve. Today, mobile development is a multi-disciplinary endeavor requiring familiarity with a host of other areas, including wireless networking, location-based systems and distributed application development [Finkel 2004]. The mobile OS share the common goals to: (1) to hide details of hardware by creating abstractions, (2) to allocate resources to processes, and (3) to provide an effective UI, similar to desktop operating system, while operating within the parameters of a mobile environment with limited connectivity, small footprint, power, and hardware resources.

This chapter has attempted to condense a rich and in-depth field of mobile OS to provide a brief historical overview of the field and the evolution of the present day smartphone and mobile OS. It provided a summary look into the architecture of two of the most popular OS: Google's Android OS and Apple's iOS. For an in-depth study, the reader should refer to the corresponding developer guides.

Key Terms

1G: 1G is short for first-generation wireless telephone technology. This generation of phones and networks is represented by the brick-sized analog phones introduced in the 1980s.

2G: 2G is a second-generation digital mobile phone network allowing for the introduction of digital data services, such as SMS and e-mail.

3G: 3G is a third-generation mobile phone network that supports higher data capacities, enabling features such as mobile Internet access. A 3G network is also capable of video calling.

4G: 4G is like the other generations in that its advantage lies in increased speeds in data transmission. There is currently no formal definition for 4G, but there are objectives. One of these objectives is for 4G to become a fully IP-based system, much like modern computer networks. The supposed speeds for 4G will be between 100 Mbit/s and 1 Gbit/s.

BREW: Binary runtime environment for wireless is an application development platform created by Qualcomm, originally for code division multiple access (CDMA) mobile phones, featuring third-party applications such as mobile games. It is offered in some feature phones but not in smartphones.

GSM: Global system for mobile communications. GSM is the most popular standard for mobile phones in the world.

HSDPA: High-speed downlink packet access. HSDPA is a 3G mobile protocol that offers higher data capacity and transfer speeds. This means that more information—such as video—can be transferred wirelessly to your phone at faster speeds.

Infrared: Infrared (IrDa) is a short-range wireless technology that provides a way to connect and exchange data between mobile phones. It has largely been succeeded by Bluetooth.

MMS: Multimedia messaging service. MMS is the sending of messages that include multimedia such as images, audio, video, and text.

Pre-paid: Pre-paid is a mobile phone account where credit for phone calls is paid up front and not billed by the month.

Push e-mail: When new e-mail is sent to you, it is instantly transferred (or "pushed") to a mobile phone handset.

SIM card: Subscriber identity module card. A SIM card is a removable "smartcard" that is inserted into a mobile phone. SIM cards store the service provider details and identify the customer to the mobile network. You need a SIM to connect to a mobile phone network and make calls.

SMS: Short message service. SMS is the sending of short (160 characters or less) text messages, to and from mobile phones. Receiving SMS messages is free of charge but sending them costs 25 cents across all networks in Australia. Some networks may offer discounted rates for SMS messages sent across their own network.

WAP: Wireless application protocol. WAP is an Internet-like service for a mobile phone. WAP is not the same as using a browser on a computer.

Wi-Fi: Wi-Fi stands for Wireless Fidelity and is a wireless technology that allows connection to the Internet when in range of an "access point." It is normally associated with accessing the Internet from a computer, but many mobile phones—particularly business smartphones—have Wi-Fi as a feature.

References

3rd Generation Partnership Project. About 3rd Generation Partnership Project. 2012. http://www.3gpp.org/About-3GPP (accessed October 3, 2012).

Accenture. Nokia and Accenture Close Symbian Software Development and Support Services Outsourcing Agreement. September 30, 2011. http://newsroom.accenture.com/article_display.cfm?article_id=5297 (accessed October 15, 2012).

Android Open Source Project. "Application Fundamentals". Android Developer Guide. 2012. http://developer.android.com/guide/components/fundamentals.html (accessed October 09, 2012).

Android: Activity. "Public Class – Activity". Android Reference. 2012. http://developer.android.com/reference/android/app/Activity.htm (accessed October 9, 2012).

Android: App Framework. "App Framework." Android Developer. 2012. http://developer.android.com/about/versions/index.html (accessed September 25, 2012).

Android: Content Providers. "Content Providers." Android Developer Guide. 2012. http://developer.android.com/guide/topics/providers/content-providers.html (accessed October 9, 2012).

Android: Design. "Android Design." Android. 2012. http://developer.android.com/design/index.html (accessed October 9, 2012).

Android: Designing for Responsiveness. "Designing for Responsiveness." API Guides. 2012. http://developer.android.com/guide/practices/responsiveness.html (accessed October 9, 2012).

Android: Intent. "Content - Intent." Android Reference. 2012. http://developer.android.com/reference/android/content/Intent.html (accessed October 9, 2012).

Android: Location Manager. "Location Manager." Android Developer Guide. 2012. http://developer.android.com/reference/android/location/LocationManager.html (accessed October 15, 2012).

Android: Package Manager. "Reference - Package Manager." Android Developer Guide. 2012. http://developer.android.com/reference/android/content/pm/PackageManager.html (accessed October 15, 2012).

Android: Permissions. "Security - Permissions." Android Developer Guide. 2012. http://developer.android.com/guide/topics/security/permissions.html (accessed October 15, 2012).

Android: Telephony. "Android.Telephony." Android Developer Guide. 2012. http://developer.android.com/reference/android/telephony/package-summary.html (accessed October 3, 2012).

Android: User Interface. "User Interface Overview." Android Developer. 2012. http://developer.android.com/guide/topics/ui/overview.html (accessed October 9, 2012).

Apple Inc: Cocoa. "Cocoa Overview." Apple Developer Guide. 2012. https://developer.apple.com/technologies/mac/cocoa.html (accessed October 15, 2012).

Apple Inc: Start Developing iOS Apps Today. "Start Developing iOS Apps Today." iOS Developer Library. 2012. http://developer.apple.com/library/ios/#referencelibrary/GettingStarted/RoadMapiOS (accessed October 9, 2012).

Apple Inc: The iOS Environment. "The iOS Environment." iOS Developer Library. 2012. http://developer.apple.com/library/ios/#documentation/iphone/conceptual/iphoneosprogrammingguide/TheiOSEnvironment/TheiOSEnvironment.html (accessed October 9, 2012).

Apple Inc: Xcode. "Xcode Download and Resources." Apple Developer Guide. 2012. https://developer.apple.com/xcode/ (accessed October 3, 2012).

AT&T. "1946: First Mobile Telephone Call." AT&T Corporate. 2012. http://www.corp.att.com/attlabs/reputation/timeline/46mobile.html (accessed October 3, 2012).

Bray, Tim. What Android is? November 24, 2010. http://www.tbray.org/ongoing/When/201x/2010/11/14/What-Android-Is (accessed September 25, 2012).

Brookes, Tim. "A Brief History Of Mobile Phones." Make Use Of. May 17, 2012. http://www.makeuseof.com/tag/history-mobile-phones/ (accessed October 3, 2012).

Cooper, S. Physician Smartphone adoption rate to reach 81% in 2012. Manhattan Research. October 5, 2009. http://manhattanresearch.com/News-and-Events/Press-Releases/physician-smartphones-2012 (accessed October 9, 2012).

Dalvik Virtual Machine. 2008. http://www.dalvikvm.com/ (accessed October 3, 2012).

Dern, D. Cross-Platform Smartphone Apps Still Difficult. *IEEE Spectrum*. June 2010. http://spectrum.ieee.org/geek-life/tools-toys/crossplatform-smartphone-apps-still-difficult/2 (accessed October 09, 2012).

Dolcourt, J, and B Bennett. New Google Voice Search, Siri are closely matched (hands-on). *CNET*. September 21, 2012. http://www.cnet.com/8301-17918_1-57464265-85/new-google-voice-search-siri-are-closely-matched-hands-on/ (accessed October 15, 2012).

Finkel, Raphael. What is an Operating System? In: *Computer Science Handbook*, Second Edition (Hardcover), by Allen B Tucker, 80-1. Chapmen and Hall/CRC, 2004.

Fitzpatrick, J. An interview with Steve Furber. *Communications of the ACM*, 2012: 54.

Gartner Inc. "Android Extended Lead While Apple iOS Market Share Growth Paused." Gartner Newsroom. August 12, 2012. http://www.gartner.com/it/page.jsp?id=2120015 (accessed October 3, 2012).

Gerke, J. "The Rise of Smartphones in Agriculture." Missouri Farm Bureau. 2012. http://www.mofb.org/NewsMedia/ShowMeArticles.aspx?articleID=259 (accessed 2012 йил 9-October).

Ginny Mies, Armando Rodriguez, and Mark Sullivan. "Smartphone Specs Demystified." PC World. December 15, 2010. http://www.pcworld.com/article/213732/ss.html (accessed October 3, 2012).

Goldman, D. "Google Project Glass." CNN Money. April 4, 2012. http://money.cnn.com/2012/04/04/technology/google-project-glass/?source=cnn_bin (accessed October 9, 2012).

Greene, Bob. "38 years ago he made the first cell phone call." CNN. 4 3, 2011. http://edition.cnn.com/2011/OPINION/04/01/greene.first.cellphone.call/index.html (accessed October 3, 2012).

GSM World. "GPRS." GSM World. 2012. http://www.gsmworld.com/aboutus/gsm-technology/gprs (accessed October 3, 2012).

International Telecommunication Union. World in 2011 ICT Facts and Figures from ITU. Statistical Report, Geneva: International Telecommunication Union, 2011.

KPMG. "Emerging role of Smartphones in retail bank." KPMG. 2011. www.kpmg.com/global/.../emerging-role-of-smartphones.aspx (accessed September 03, 2012).

Kucera, D. EBay's PayPal Mobile Payment Volume Rose to $4 Billion Last Year. *Bloomberg Businessweek*. January 11, 2012 (accessed October 03, 2012).

Lee, Nicole. The 411: Feature phones vs. smartphones. *CNET*. March 1, 2010. http://www.cnet.com/8301-17918_1-10461614-85.html (accessed October 3, 2012).

Lowe, A. "Google and Oracle take the fight over Java in Android to court." Hexus. April 16, 2012. http://hexus.net/business/news/legal/37973-google-oracle-take-fight-java-android-court/ (accessed October 4, 2012).

Lunden, I. Google Play About To Pass 15 Billion App Downloads? Pssht! It Did That Weeks Ago. *Tech Crunch*. May 7, 2012. http://techcrunch.com/2012/05/07/google-play-about-to-pass-15-billion-downloads-pssht-it-did-that-weeks-ago (accessed October 9, 2012).

Marshall Brain, Jeff Tyson and Julia Layton. How Cell Phones Work. *How Stuff Works*. 2011. http://electronics.howstuffworks.com/cell-phone3.htm (accessed October 3, 2012).

Massa, Michael Barr and Anthony. Programming Embedded Systems, Second Edition. O'Reilly Media, 2006.

Massachusetts Institute of Technology. MIT AppInventor mobile development language http://appinventor.mit.edu/. 2012. http://appinventor.mit.edu/ (accessed October 3, 2012).

MEDITECH. Meditech Homepage. 2012. http://meditech.com (accessed October 9, 2012).

Microsoft: Channel 9. Windows Phone 7: A New Kind of Phone. Channel 9. June 7, 2010. http://channel9.msdn.com/Events/TechEd/NorthAmerica/2010/WPH201 (accessed October 15, 2012).

Mosa, A.S. A Systematic Review of Healthcare Applications for Smartphones. BMC Medical Informatics and Decision Making 2012. Illhoi Yoo and Lincoln Sheets, 2012. 12–67.

O'Brian, K. Top 1% of Mobile Users Consume Half of World's Bandwidth, and Gap Is Growing. *New York Times*. January 06, 2012. http://www.nytimes.com/2012/01/06/technology/top-1-of-mobile-users-use-half-of-worlds-wireless-bandwidth.html (accessed October 3, 2012).

Open Handset Alliance. "Frequently Asked Questions." Open Handset Alliance. 2012. http://www.openhandsetalliance.com/oha_faq.html (accessed October 3, 2012).

Oxford University Press. smartphone Oxford Dictionaries. April 2010. http://oxforddictionaries.com/definition/american_english/smartphone (accessed September 25, 2012).

Reed, Brad. "A Brief History of Smartphones." Network World. June 18, 2010. http://www.pcworld.com/article/199243/a_brief_history_of_smartphones.html (accessed October 3, 2012).

Ross Catanzariti, Good Gear Guide. The Mobile Phone: A history in pictures. *CIO Magazine*. http://www.cio.com/article/504135/The_Mobile_Phone_A_History_in_Pictures (accessed October 3, 2012).

Samsung. Bada. 2012. http://www.bada.com/ (accessed October 15, 2012).

Segan, Sascha. CDMA vs GSM What's the difference. *PC Magazine*. August 22, 2012. http://www.pcmag.com/article2/0,2817,2407896,00.asp (accessed October 3, 2012).

Shah, Agam. Google's Android 4.0 ported to x86 processors. *Computer World*. December 1, 2011. http://www.computerworld.com/s/article/9222323/Google_s_Android_4.0_ported_to_x86_processors (accessed Feburary 20, 2012).

Shubha. Immersion takes Haptic feedback on smartphones to the next level. Gizmodiva. February 2, 2012. http://www.gizmodiva.com/mobile_phones/immersion-takes-haptic-feedback-on-smartphones-to-the-next-level.php (accessed October 9, 2012).

SPSM workshop. 2nd Annual ACM CCS Workshop on Security and Privacy in Smartphones and Mobile Devices (SPSM). 2012. http://www.spsm-workshop.org/2012/ (accessed October 9, 2012).

Thomas, S. Smartphones to play an increasingly important role in the future of M-Health. February 2, 2012. http://memeburn.com/2012/02/smartphones-to-play-an-increasingly-important-role-in-the-future-of-m-health-report/ (accessed October 9, 2012).

Unbound Medicine Inc. Hopkins Guides. 2012. http://www.hopkinsguides.com/hopkins/ub (accessed October 9, 2012).

Vandette, B. "Java SE Embedded Performance Versus Android 2.2." Oracle Blog. November 22, 2010. https://blogs.oracle.com/javaseembedded/entry/how_does_android_22s_performance_stack_up_against_java_se_embedded (accessed October 9, 2012).

Vodafone. Connecting to the Cloud: Business advantage from Cloud Services. White Paper, Vodafone, 2010.

Warren, C. "The Pros and Cons of Cross-Platform App Design." Mashable Tech. February 16, 2012. http://mashable.com/2012/02/16/cross-platform-app-design-pros-cons/ (accessed October 15, 2012).

Wauters, R. "Global Smartphone App Download Market Could Reach $15 Billion By 2013: Report." Tech Cruch. March 5, 2012. http://techcrunch.com/2010/03/05/global-smartphone-app-download-market-could-reach-15-billion-by-2013-report/ (accessed October 9, 2012).

Welch, Gregory F. A survey of power management techniques in mobile computing operating systems. *SIGOPS Oper. Syst. Rev. (ACM)*, 1995: 47–56.

Wikipedia. "Cocoa Touch." Wikipedia. 2012. http://en.wikipedia.org/wiki/Cocoa_Touch (accessed October 09, 2012).

Wikipedia: "Darwin Operating System." Wikipedia. 2012. http://en.wikipedia.org/wiki/Darwin_%28operating_system%29 (accessed October 9, 2012).

Wikipedia: "XNU." Wikipedia. September 4, 2012. http://en.wikipedia.org/wiki/XNU (accessed October 9, 2012).

Wikipedia: Cocoa API. "Cocoa API." Wikipedia. 2012. http://en.wikipedia.org/wiki/Cocoa_%28API%29 (accessed October 9, 2012).

Wikipedia: 2G. "2G." Wikipedia. 2012. http://en.wikipedia.org/wiki/2G (accessed October 3, 2012).

Wikipedia: ARM Architecture. "ARM Architecture." Wikipedia. 2012. http://en.wikipedia.org/wiki/ARM_architecture (accessed September 25, 2012).

Wikipedia: Cellular Network. "Cellular Network." Wikipedia. 2012. http://en.wikipedia.org/wiki/Cellular_network.

Wikipedia: History of the Mobile phone. History of the Mobile phone. 2012. http://en.wikipedia.org/wiki/History_of_mobile_phones (accessed October 3, 2012).

Wikipedia: Smartphone. "Smartphone." Wikipedia. 2012. http://en.wikipedia.org/wiki/Smartphone (accessed Octobe 3, 2012).

Wingfield, N. "Despite a Slowdown, Smartphone Advances Are Still Ahead." New York Times. September 16, 2012. http://www.nytimes.com/2012/09/17/technology/despite-a-slowdown-smartphone-advances-are-still-ahead.html (accessed October 9, 2012).

World Bank. Information and Communication for Development 2012: Maximizing Mobile. Case Study, Washington, DC: World Bank, 2012.

62

Service-Oriented Operating Systems

Stefan Wesner
University Ulm

Lutz Schubert
University Ulm

Daniel Rubio Bonilla
High Performance Computing Centre Stuttgart

62.1 Background

The original motivation to introduce a layer between the user application and the underlying hardware was based on the assumption that certain tasks are common across certain applications and the application developer needs an abstraction layer from the underlying hardware. Early operating systems had not been much more than a hardware abstraction layer offering basic interfaces to interact with external input and output devices and essential intrinsics of the processing units for computing, graphics, sound, or other special purpose processors.

Over the years operating systems evolved significantly in complexity accordingly with the new hardware functionalities available and the introduction of graphical user interfaces as an essential part of the abstraction layer. However, for performance hungry applications, the layer introduced by the operating system and the libraries on top of it aimed at achieving maximum portability was not appropriate. Using the new features of the most recent graphics cards has been essential for high-end games and key to differentiating them from their competitors. As a result, environments offering a more direct access to the capabilities of their underlying hardware, such as DirectX or OpenGL, offer in principle an alternative operating environment to *normal* applications. As the complexity of the operating system is dependent on the variety of the hardware platforms where it may run, some operating systems have limited such platforms (e.g., Mac OS X has a significantly reduced number of platforms compared to Microsoft Windows or Linux). An extreme case are gaming consoles with a clearly defined hardware profile as well as a significantly reduced footprint.

In this chapter, we argue that with the current trend of exploding complexity of the hardware layer within a computing system, and across computing systems, traditional operating system approaches are not only a threat for the application developers that need to program around their limitations, but prevent the developers from taking advantage of the opportunities of a connected world.

The core part of this chapter is the discussion of a new approach called service-oriented operating system. We conclude the chapter with the application of the model to three scenarios, one from the high-performance computing domain, another one inspired by an ambient intelligence setting, and a third one from intelligent manufacturing systems.

62.1.1 Current and Future Hardware Evolution

While many innovations are integrated into every new generation of computing hardware, systems have appeared quite stable for many years from an application developer viewpoint. Most computers have been using the x86 architecture based microprocessor with increasing frequency and transistor count. Most application developers focus their designs and implementations on a sequential platform considering communication and computing as separate functions. Better performance could be achieved by investing regularly in new hardware with only moderately increasing power costs. The motivation to spent effort in software optimization was limited to special application domains. Legacy applications, even if developed many years ago, would run faster by simply porting them to new architectures.

Recently the situation has dramatically changed, driven by the need to improve the power efficiency. Even low-end computing systems and smartphones include multicore processors. These devices have integrated graphic capabilities which are also used for computing. However, they run into performance bottlenecks when accessing data stored in the memory or communicating with other computing subsystems. The general availability of *parallel* computing capabilities comes with increased heterogeneity of the computing platform and, at the same time, increased demand for optimization. Nowadays a typical computing system is equipped with one or more multicore CPUs with different types of connectivity between each other and to the peripherals (e.g., the core closer to the communication bus can send and receive data over the network faster), and accelerating components such as general purpose graphic processing units. Hierarchical memory approaches with fast and large memory areas and fused processor architectures bringing together CPUs, graphic processors, and communication capabilities on one single chip will soon be available. Moreover, within the family of x86-based CPUs more and more differences between vendors in specific instruction sets and cache architectures are emerging. Competing architectures, such as ARM-based processors, are moving from the embedded systems domain into mainstream computing.

The energy efficiency of applications is now a major concern. An application that is using only five percent of the theoretical peak performance, for example, because it has bad cache hit rates, is wasting much more energy than the one saved through hardware optimization. Consequently, the application developer is facing two major challenges at the same time. On the one hand, the underlying platform is becoming increasingly complex and heterogeneous, and innovation cycles for new hardware is shorter than ever. At the same time, there is a need to handle parallelism and save energy by designing algorithms that exploit data locality and use efficiently the available computing capabilities. Application developers are finding it extremely difficult to cope with these new realities.

62.1.2 Programming Models

Parallel programming is not a new topic, and a couple of programming models have been used on high-performance computing (HPC) systems for over a decade. Besides simple multithread programming approaches, applications are very often parallelized using specific programming models such as OpenMP and/or the Message Passing Interface (MPI). These programming models have been designed for scientific/technical computing intensive applications and are only partially suited for other kinds of applications. In particular, if used in conjunction to realize node parallelism (across the cores of a single server) with OpenMP and using MPI for the communication across nodes, the correct usage is very complex and demanding for programmers.

Partitioned global address space (PGAS) approaches, such as Co-Array Fortran (CAF) or Unified Parallel C (UPC), bring parallelism as a basic element into the programming language to deliver an abstraction layer for distributed data. This approach is more suited to reach the application developer community at large. However, achieving high performance and scalability is complex as it is with other parallelization methods.

For addressing the heterogeneity of the computing platforms, approaches to deliver a higher level of abstraction to the application developer are available. Models, such as OpenAcc or Task-based approaches as StarSs and OpenMP 4.0, are competing with more hardware-oriented models such as OpenCL and CUDA. These models, while conceptually supporting different architectures, are often limited to a few platforms.

62.1.3 Communication Models

With the increasing amount of computing capability in a single node/server, very often applications need to be spread across several computing systems to overcome limitations on memory size or to achieve strong scaling (delivering a solution of fixed size in shorter time, in contrast to weak scaling solving a larger problem in similar time). Consider an extreme case where an application could be distributed across as many nodes as the number of its subproblems, where the subproblems fit in each node's cache. While this would clearly speed up the calculation on each node (as access to the cache is much faster than accessing to the main memory) for most problems, the synchronization across the nodes sharing intermediate results might become the dominating factor in the execution time. While in this scenario the latency will be dominating, in other cases the bandwidth limitations and the time needed to transfer data from main memory to the network and back into the main memory of the receiving server might be the limiting factor.

Consequently, communication and computing cannot be considered to be orthogonal elements of an application, but their balance have to be integral part of the design and implementation.

62.2 Analysis

The trends in hardware development discussed in the previous section have increased, and will continue to increase, the complexity of programming. The fundamental assumption of the authors is that current models will lead to a significant and increasingly, growing gap between the capabilities of the hardware and the ability of the *averagely skilled* application developer to exploit these capabilities. This "parallel divide," where an increasing number of application developers is no longer able to efficiently exploit new hardware capabilities, demands for a new software development model. Such a model has to be able to make use of the wide variety of available computing resource types, taking advantage of its specific characteristics while at the same time reducing the complexity of programming.

Although advanced or specialized application developers will still be able to efficiently use the complex machine's computing resources, it is necessary to provide a solution for the average programmers. The approach must be more abstract, reduce noticeably the development time, and ease portability while having only an acceptable performance penalty compared to handmade optimizations.

The current situation resembles the dilemma between writing code in assembler or in a higher-level language. An experienced programmer with in-depth knowledge of the underlying hardware might extract more performance, but using a high-level language will reduce development time, facilitate debugging, and allow easier portability among different hardware architectures.

62.2.1 Need for a New Programming Model

Existing programming models are not able to deliver this abstraction level as they are typically too tightly coupled with the hardware system and rely on lower layers to control the communication and execution of the program.

Extending or combining existing programming models, for example, to reflect specific user needs or individual application domains (such as OpenMP for parallelization in a cache-coherent common address space machine and MPI for parallelization in a non cache-coherent or noncommon address space machine), will further increase programming complexity. In the current programming models, optimizations have to be done manually by the developer by adapting the source code to the destination platform or by the compiler performing static analysis and specific optimizations for the destination infrastructure. But to obtain an efficient utilization of the infrastructure, the developer must take into consideration as many details as possible of the destination infrastructure. For example, performance is negatively affected by cache misses; organizing and partitioning the data structures according to the processor's cache size to minimize this impact will greatly improve performance. The main problem is that different processors have different cache sizes, access patterns, and hierarchies. Consequently optimizing a program for one platform can lead to performance degradation on another system or even limit program portability if specific instructions are used.

On the other hand, one cannot expect compilers to generate completely optimized code by just using static analysis of the program for different platforms. Typically this leads to parallel branches for different architectures for the same program to overcome limitations of the hardware, the specific compiler capabilities, or even conceptual limitations of the programming language.

A new programming model has to consequently rely on the user knowledge to express optimization hints coming from the problem domain, perform run-time code analysis, and realize a segmentation and distribution across the infrastructure that can change and adapt over time. Language extensions and optimization hints might range from low-level ones (to unroll loops over communication patterns) to data layout assertions to express domain-specific elements.

62.2.2 OS Support

Many of the issues previously exposed could be solved by constant recompilation of an annotated source code or by making the operating system aware of its environment, enabling it to predict memory usage and code dependencies in order to adapt the program to the running environment characteristics. The source code or the programming model, when the application is not written for an specific machine, do not know about the level of parallelism and concurrency of the machine and the infrastructure specifics, while the processor architecture has no view of the overall program and its possible threads. The operating system, being the middle tier between the two perspectives, is hence the ideal place to address this; the only execution "layer" having access to all the information required to resolve these issues is the operating system. The operating system is the interface between the hardware, usually unknown to the software developer, and the program.

Current operating systems cater for threads and processes—its distribution and scheduling—, and to memory management; they control the machine and program interaction with the hardware. But this is one of the main limitations under the current hardware evolution. The execution environment itself should cater for the right functionalities, reflecting the infrastructure specifics to the current deployment. In other words, the developer should have a common interface for all segments/processes for communication, message exchange etc., without having to restrict to the specifics of the current deployment. This would mean, for example, that threads just specify the data they use and access it in a common fashion, regardless whether the underlying platform is a shared-memory, cache-consistent, or nonunified memory access architecture.

Even though the execution environment will know the details of the hardware infrastructure, it is obviously not sufficient to just leave the matching of the access operations to the environment, as it will not know the details about which data is required where and when. In a straight-forward approach, all data access would be matched to the necessary messaging and routing, thereby creating a central memory instance. This in turn would create an immense overhead in communication on the network, downgrading the performance even more than for an implementation that does not match the infrastructure details but takes into account segmentation, distribution, and prefetching of the data.

In fact, information from both sides, that is, the hardware layout and the work/data layout of the program have to be aligned with each other by the operating system. This implies however that our way of specifying the work and data layout of code has to change into a way that allows for the best mapping to different environments. For example, the coder should identify which part of the data is manipulated by which part of the code, so that a compiler or intermediary analyzer can identify how to distribute data and when consistency needs to be maintained. This information can then be used by the execution environment dynamically depending on the destination platform layout, respectively, of the specific architecture of that platform.

Each thread or code instance would therefore need a different way of realizing the same logical functionality, depending on the real data dependencies and access requirements, as well as on the destination platform specifics. With traditional architectures, the operating system would effectively have to cater for all the related functionalities itself and "outsource" the intelligence to finding the right mapping to the developer.

As mentioned earlier, the application developer should not have to worry about the code distribution to nodes, as this depends on the specifics of the destination platform, that is, memory architecture, I/O, available devices, etc. But the developer must provide enough information for the operating system's code analyzer so that it can identify the cross-segment exchange of data, without having to restrict itself to a specific deployment or segmentation.

Even though monolithic operating systems still scale well, they are tightly coupled to a specific resource type, making the integration diverse resource types difficult to manage. Furthermore, monolithic OS assume that the infrastructure is comparatively static and thus cannot provide simple means to cater for fault tolerance response as the corresponding supervision would produce additional management overhead. Monolithic approaches also tend to lose efficiency when cross-process communication reaches a certain threshold as there are too many synchronization steps have to be managed.

62.3 Derived Operating System Model

A primary concern for the efficient execution of distributed programs in future infrastructures consists in providing an execution environment that does not control the process(es) from a central point and that is capable of providing and exploiting the capabilities of each individual resource. A major concern is the degree of messaging overhead created by centralized control and cache misses. To reduce the amount of messaging, it must be ensured that each part of the code can effectively run as a completely independent (i.e., concurrent) process. This means that data exchanges have to be reduced to a minimum, with other processes or even the operating system. The degree of "independence" is thereby not so much defined by the total number of shared data but by the amount of messaging relative to the underlying connectivity speed. If timing is correctly exploited, messaging has no impact on performance since no process has to wait for another due to lack of data (i.e., locking).

As has been noted, most "processes" or segments forming an application require minimal operating system requirements toward the environment they are running on, if they are carefully restructured. Often enough, developers completely depend on integrated frameworks (such as .NET) that are so tightly integrated with the operating system that even simple process calls will lead to system interrupts and therefore to unnecessary overhead.

In principle, any program can be written so that it does not require any operating system support—this implies however that effectively the developer writes his own operating system as part of his intended application logic, which is neither realistic nor helpful. Instead, we require an execution environment (operating system) that can itself be distributed and adjusted in a fashion that it only provides essential capabilities to each executing thread/process, and still maintains consistency with as minimal impact on execution and communication overhead as possible.

62.3.1 Modular OS Architectures

Effectively the architecture should provide minimal OS environments per executing segment that nonetheless still caters for all the additional capabilities and requirements to maintain a distributed execution and therefore all the dependencies between the executing segments. For example, a regular multi-threaded application with one main thread and a hundred worker threads that all read data from a central main memory should implement the functionalities for accessing this memory location, but requiring hardly any other OS support. What is more though, this functionality differs strongly between platforms and their organization, that is, whether memory is shared, whether remote memory access is supported, etc.

This means implicitly that even though only limited functionality is required, the different requirements originating from the environment impose that no single implementation can provide the necessary capabilities. The current approach consists in the operating system providing a variety of strongly related functionalities that can be exploited for a wide set of tasks.

This is where the strength of modular approaches comes in: from the programmer's perspective, only specific functionalities are required that can be exposed in a common way, leaving it to the execution environment to decide on the actual implementation of the respective functionalities. This is not unlike dynamically shared libraries where the right implementation of the library can be chosen on basis of the hardware specifics. Shared libraries thereby are essentially nothing more than programming extensions, built up on the operating system.

With the modular approach the functionalities are tightly integrated with the rest of the execution environment. Effectively, the whole operating system makes use of the same operations, though with different flavors, according to the execution circumstances. This can increase performance drastically and allows for better alignment between execution and platform.

One very essential consequence of this approach is the fact that applications and the operating system are equally structured in a modular/segmented fashion. In other words, the operating system invokes its functionalities in the same way as the applications, putting the applications essentially on the level of control of the operating system.

Modules forming the essential operating system behavior and contributing to the application execution can therefore be deployed in any fashion across the infrastructure. To this end, however, a strong messaging mechanism must be in place that can realize the communication between modules across the deployment—this implicitly puts messaging in a special position compared to all other modules (see section on OS capabilities).

As opposed to classical operating system architectures, this implies that there is not one, common deployment and instance of the whole operating system within the infrastructure, but instead there are different modules instantiated all across the environment that communicate with one another to fulfill the respective functionality. The operating system is effectively spread all over the environment.

In order to allow for coordinated execution, the individual modules need to be carefully controlled and configured so as to compensate and supplement each other, rather than competing against each other as it is currently the case when multiple OS instances run within the same environment.

The modules are not so much higher order extensions, comparable to libraries, but can be compared better to service instances in a distributed environment such as web services. What is more though, the "service" or module does not have to reside at a designated, static location, but can be freely relocated, instantiated, and replicated according to the current requirements, therefore comparing it more with the classical ideas behind grid computing or cloud computing.

62.3.2 Execution Lifecycle and Background Support

Before we take a look at the actual architecture of such a modular, distributed operating system, it is worth examining how the system behaves principally as a whole, if a new application is executed within such an environment. We can thereby identify the following major steps:

1. Analysis
2. Preparation
3. Deployment
4. Execution and maintenance

During *analysis* the code to be executed is analyzed with respect to its specific preconditions and requirements toward the infrastructure and its potential to be executed in parallel/concurrent segments. This means in particular that the indicators provided by the developer with respect to the code–data relationship and the degree of parallelism, and respectively the concurrent code segments, are identified. There are different means to support the automation of this step, such as dependency identification, code segmentation, automated parallelization, etc. For the operating system to work properly, the actual mechanisms to identify these possibilities are secondary. What is of main relevance is how this information relates to the distributed execution of the code.

The main information required for enabling distributed execution consists in bringing out all aspects that create any form of dependency beyond the local code. This implies the following:

1. Remote memory access: Any access to nonlocal memory space, whereby "local" means local to the executing segment (thread, process), which does not necessarily mean local to the function/class/object in the source code.
2. Explicit communication: Any explicit invocation or messaging with nonlocal (see earlier text) code instances, segments, or even devices. Again the actual scope depends more on the executed segmentation than on the source code structure.
3. Work segmentation: Similar to communication and memory access, code execution may jump between threads/processes/segments in order to perform the actual distributed execution—this equally covers intended jumps (invocations, branches), as well as unintended ones (created by segmentation) and ones to exploit concurrency (executing multiple segments at once).

Principally, the developer needs to indicate the data and code segment dependencies and relationships in a form that is understandable by the compiler, analyzer, or interpreter that converts the information into an executable form that the operating system can handle. The most promising approach is a graph representation of the code structure whereby the different types of dependencies are denoted as weighed edges. The weighs can convey different information with respect to amount of data communicated, frequency of communication, etc. Edge direction can represent direct access types (read, write). This graph not only allows for easy identification of concurrent code segments but also of communication points, data dependencies, etc.

In [6], we elaborate about mechanisms that allow for automated characteristics extraction and matching. New models are thereby required to specify the capabilities of hardware in an abstract form.

Preparation of the application code and the computing environments involves conversion of the code to accommodate for the requirements arising from the segmentation and deployment, as well as the selection, instantiation and configuration of the required operating system modules, i.e., enabling provisions with the right functionalities.

Code preparation may include replacing the dependencies with explicit system calls to cater for the distributed execution, that is, in particular, to inject messaging points where according to the original code only foresee memory access and where the dependency graph identifies a cross-segment relationship. Nonetheless, the preparation phase can be regarded as part of the compilation step, if we consider dynamic re-arrangement as a form of re-compilation.

It must be noted that with the growing heterogeneity and principally dynamicity at execution time, the classical concept of completely isolated development, compilation, and execution steps can no longer take place as such—instead, the steps *must* become intertwined to fully exploit the potential of future hardware infrastructures.

Though developers will not and should not write code for a specific destination platform, they will have to give explicit hints that implicitly relate to the mapping of the code to the infrastructure, particularly regarding dependencies and concurrency. As the deployment may change depending on time-sharing and available resources, the operating system must be able to rearrange and re-distribute the code without having to go through an explicit compilation phase. Just-in-time compilers are a clear indicator of this trend.

At the *Deployment* phase, the *selected* segmentation of the code, including all its required operating system modules, will be deployed in the infrastructure according to the application's layout identified during the preceding stages. This step includes all configuration steps required in order to make distributed execution take place, such as registration of the communication endpoints, reservation and registration of the address spaces, etc.

If the analysis and preparation are considered outside the actual operating system lifecycle, the main OS operation starts with the deployment of the code, as well as the deployment of the operating system modules themselves.

Finally, during the *Execution* phase, the application will be executed across the deployment created in the preceding phase. It must be noted that not all code will necessarily be designated to a specific processing unit during the execution initialization. Instead, the application behaves more like a workflow, with individual segments ("services") being deployed at run time upon invocation. This way, concurrency can be exploited even if the exact timing behavior can only be estimated, and thus allow for dynamic reaction to the arising time-sharing constraints.

The operating system must be able to dynamically create and relocate instances of both the application, and—more crucially—of itself; its own modules.

The major task for the operating system during distributed execution consists of ensuring state consistency and communication across the individual instances—both on the OS and the application level. Since the deployment is not fixed and may change dynamically at run time, the operating system *as a whole* has to ensure that all the relevant parts are always reachable from all communicating segments. In other words, the segments can always be executed as if the environment would adhere to the original specification, even as if the platform would be a single-core sequential processor.

In close relationship to the types of dependencies identified earlier, this means, in particular, that the operating system must provide mechanisms to

- Access memory and maintain state
- Access (I/O) devices
- Communication routing between instances/segments
- Event handling across instances

In general, the operating system will provide a dedicated (virtual and managed) address space for each application, allowing the memory manager of the processor to execute relative jumps and calls within the same local address space range and leaving it up to the operating system to handle the page swaps. In most cases (with the exception of applications dedicated to parallel execution), the address space will be continuous, that is, at least within the same physical location. This means that jumps and calls will not leave the machine, unless the developer has explicitly specified it.

In the more dynamic distributed case, this assumption does not generally hold true, as any jump or memory access may imply leaving the physical boundary of the executing entity, and therefore out of the locally assigned address space. The classical way of handling this via a page miss event may not suffice anymore in this case, as the access may imply remote processing. Remote page swapping, that is, copying data from a remote location just into cache, without having another local replica, is generally not advisable, as it implies that every page swap would have to deal with all the communication overhead, even if the cache was not altered, that is, no update is required.

The key factor consists of good data segmentation and identification of the right points for the communication and update of data. Without loss of generality, we assume that the system calls are embedded by the compiler on basis of the annotations provided by the developer and the results from the analysis and matching phase prior to the execution of the code. Memory access in such a distributed environment turns from the explicit access to a specific memory space into requesting the operating system to make the corresponding data set available for accessing. The developer does not have to identify the origin of the corresponding data, nor for distributing it to all linked code segments.

All communications must be executed in the background in order to be nonblocking, and thus allow the execution of other operations before the actual memory access takes place. At the same time, the operating system must offer support to handle communication in different forms, wherever the developer explicitly requests it.

Considering the circumstances, in particular regarding the flexible, potentially dynamic distribution of processes across the infrastructure, a particular task for the operating system consists of maintaining the right communication endpoints. To realize this, such an operating system must support multiple mechanisms:

- Hierarchical lookup tables: Primarily serve the purpose of providing the right communication endpoints for specific address spaces.
- Routing: Due to the potential dynamicity, endpoints may move without all related instances being informed; the operating system must provide means to forward the communication from the old to the new endpoint.
- Fault handling mechanisms: Fault handling mechanisms must ensure that wrong lookups do not automatically lead to failure of the according application, but the study of this issue is beyond the scope of the current document.

62.3.3 Essential OS Capabilities

Based on the preceding discussion, we identify a set of essential functionalities that a distributed operating system must offer over a classical, monolithic and centralized OS architecture. We must thereby explicitly distinguish between exposed functionalities (as visible to the developer or user) and actions arising from the distributed usage of the available functions. As discussed at the beginning of this section, messaging mechanisms must be exposed to the user without adding any further burden to the endpoint. Messaging depends thereby very much on the processor architecture and thus differs between different deployments and execution cycles. Notably, the same level of distinction also applies to the operating system itself: due to its modular organization, many of its functionalities are implicit to the way execution is handled.

On the other hand, the individual functionalities should not hide capabilities and mechanisms that the user may potentially want to exploit for his own benefit that he may want to be able to control at least to a certain degree (such as specifying a security protocol, distinguishing different means for ensuring consistency, etc.).

It is of utmost relevance that the dependencies between functionalities is kept to an absolute minimum, so as to (a) allow easy replacement and adaptation, and (b) not generate management and communication overhead that would exceed the gains of a modular architecture again.

We can identify the following main base capability categories an operating system must provide in order to enable distributed execution:

- Communication: With a messaging-based approach, communication support is implicitly an inherently relevant capability, covering not only the message exchange between the execution segments (code blocks), but also the implicit and explicit communication between operating system modules. Communication is a transparent underlying capability for routing invocations and messages to the necessary endpoint without the respective code having to identify it explicitly. Nonetheless, the developer must be able to explicitly use the communication capabilities to send messages to dedicated endpoints in a thread-/MPI-like fashion.

- Memory management: With the application and its data being distributed all across the infrastructure, memory access is not straight-forward from any executing code; in particular with multiple blocks being executed concurrently, the data space may either be replicated or centralized. In either case, the developer should not have to cater for the specific mechanisms. Instead, the operating system must provide means that grant managed access according to the deployment specifics. This means that memory management is tightly coupled to the communication capabilities in order to enable remote access and state maintenance.
- Scheduling: Multitasking was originally implemented by sharing the same resource between multiple processes and having the scheduler distribute the processing time between them. With large scale, distributed infrastructures, the primary goal of a scheduler is distribution and timely invocation of the blocks. From a single processing unit's point of view, execution ideally becomes a fairly sequential task, thus dedicating most of the unit's performance to the respective task. Nonetheless, some use cases will initiate more processes than resources are available to run them on, and individual processes may be stalled due to communication delays and interdependencies, so that the scheduler will still have to be able to execute classical context switches in order to maximize resource usage.
- Deployment and process management: Strongly related to scheduling, the operating system must be able to *remotely* deploy and trigger processes (or more correctly: code segments) in order to be able to distribute the application across the infrastructure in a way that reflects the parallelism and concurrency of the code. This implies that the designated destination resource provides the base means that allow reception, deployment, and linking of the code. This includes, in particular, communication and memory management functionalities.

In addition to these, we can identify lower-level functionalities that at first glance would have to be considered subsumed by the ones listed earlier. These however may be explicitly used by the application and must therefore be exposed in a way that the developer can invoke them according to his needs:

- File management and I/O management: Devices in a system are generally exposed via a common communication bus, classically denoted as I/O (input and output). Modern systems distinguish between different buses, depending on the type (and location) of the device in question. Most of these devices must be accessible to the user at least in some form, such as the disk drivers, printers, screens, etc. The OS must provide common means to reach the devices and interact with them without having to cater for the protocol details. The location of the according devices may thereby not be known to the user, so that similar routing mechanisms are needed as for memory management.
- Interrupt handling: Handling requests from devices and software instances is an essential feature of all operating systems to control the infrastructure. Some interrupt events are obviously handled by the operating system, but the developer may want to exploit, inspect, and react to such events by capturing them. The corresponding capability must be available to both the operating system itself (infrastructure management) and the developer through an extended forwarding mechanism.

Finally, to enhance the operating system behavior, additional, so-called "supporting" capabilities are required, which provide the essential information for the individual OS components to execute the code in the way described earlier:

- Code analysis and segmentation: The introduction of web services and the cloud gave rise to increased exploitation of remote procedure call like interactions, and thus to a workflow-like segmentation of the application in the first instance. The code analysis must identify external dependencies, that is toward libraries, operating system calls, devices, etc., that effectively define the segmentation and mapping of the code on the infrastructure. Notably, the segmentation identified by the developer does not have to be the most suitable for the destination infrastructure and also the data structure plays an important role in the sensible distribution of code, as it may create dependencies that effectively counter any segmentation. This analysis results in a dependency graph of potential code and data segments, including any form of additional constraint information as given by the developer, that can be exploited for the purpose of distribution and mapping to the destination platform.

- Resource management: In the environment described earlier, resources are no longer treated as concrete processor models, micro-architectures, device types, etc., but instead as abstract entities that provide certain capabilities and impose specific constraints. As heterogeneity increases, devices must be characterized according to their capabilities and must be classified according to how code can make use of these characteristics. Along the same line, resources must be discoverable; changes in the resource setup must be communicated to all related nodes, so as to take potential recovery actions, respectively, to exploit the additional resources. This step is particularly relevant as a precedent to deploying the code.
- Code transformation: In order to match the code against the infrastructure and to exploit its characteristics, it is necessary to transform the code accordingly. This is an extension to the classical compilation step, which involves optimization steps. More important, due to the segmentation and re-arrangement of the memory layout, additional commands for maintaining communication and accessibility to the data have to be injected. Since the layout of the code and infrastructure may change at run time, the mapping and the code transformation should not be a static; it must be possible to adapt the code at run time according to the environmental conditions.

Effectively, the "vertical" supporting capabilities can be regarded as part of the code compilation process, as their main tasks consist in converting the code so as to meet the infrastructure and distribution requirements. It must be noted, however, that many of the steps may have to be executed again at run time. We can thus distinguish between preparational steps and steps to be performed at execution run time under dynamic conditions. This leads to a compilation structure that effectively incorporates the following four steps:

1. Analysis of the code
2. Conversion into a pre-executable code
3. Just-in-time adaptation to infrastructure
4. Just-in-time compilation to destination

62.3.4 OS Layout

The primary concern when building an operating system architecture that meets the capabilities listed earlier consists in identifying a minimal representation of these capabilities in terms of software components. In other words, in identifying the set of minimal components that are sufficient to provide the necessary functionalities, yet do not create performance issues simply because of communication overhead created by segmenting the functionalities into individual components. Essentially, this means that the dependencies between the given set of capabilities in different usage contexts need to be identified.

We can categorize the core functionalities into five major domains (cf. Figure 62.1), namely, capabilities related to executing the processes, handling memory and storage, handling the devices and hardware, supervising the infrastructure, and finally, the messaging support across all these domains.

The primary principle of a service-oriented operating system architecture consists of instantiating only the essential capabilities that are required for executing a specific functionality. Deployment

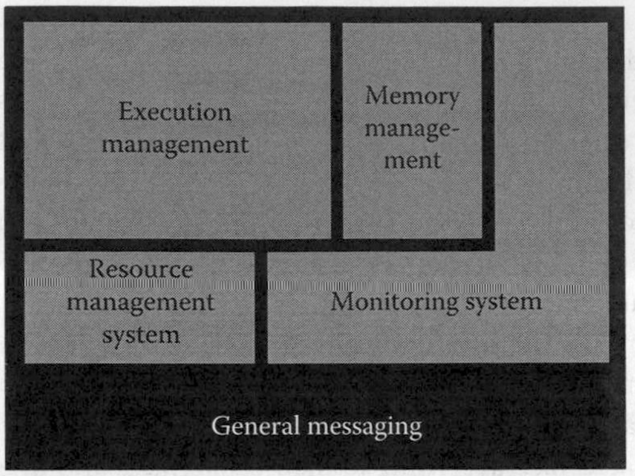

FIGURE 62.1 High-level architecture of the service-oriented OS architecture.

does not have to be co-located with the calling code/function, due to the message-based approach, but remote invocation implicitly leads to higher communication overhead that can stall the process. An optimal segmentation of components keeps the communication overhead and the load on memory (and therefore chances for cache misses) low.

This means that even though the components belonging to the functional groups identified earlier provide the same type of functionality, their individual instantiation can differ quite significantly. We can best exemplify this by a process load operation from a specific designated memory space, local to the full application. In other words, the data space is global to the application but not necessarily to the individual process trying to invoke it. Any memory access can therefore be regarded as a load operation from an unknown source that takes a specific amount of time. If the timing behavior can be predicted and if the usage logic is known, prefetching mechanisms can be enacted to improve processing performance [2].

The developer will not want to decide between server (remote) and client (local) invocations either, but instead the programming convention must follow traditional standards. The compiler or runtime handler must be able to replace the memory accesses with invocations to the memory management system.

A completely managed environment, where all data access is first routed through the operating system, on the other hand, will create additional system management overhead to memory accesses that could be handled directly in cache. The best approach in terms of performance is determined by the type and amount of data, as well as the degree of sharing. In general, the analysis must identify these dependencies and inject the right access means accordingly.

62.4 Applying Service-Oriented Operating Systems

As already discussed, future infrastructures will be characterized by a high degree of mobility, parallelism, and heterogeneity of resources while users expect more and more performance. This can only be achieved by exploiting parallelism and concurrency of the applications. This section will demonstrate how the architecture outlined in the preceding section behaves in the following scenarios.

62.4.1 High-Performance Computing Scenario

High-performance computing (HPC) is designated by a single application requiring a large amount of resources to achieve the necessary performance, with respect to precision, amount of data and/or speed. Performance is primarily defined by the degree of communication taking place during execution, as data consumption/production is in general faster than communication allows. Reducing any form of communication is thereby the primary concern of any developer.

As can be shown, a centralistic operating system creates additional communication overhead that outweighs the overhead generated by a message-based system [1]. What is more, due to the monolithic organization of current operating systems, system invocations may cause a huge amount of cache misses [6]. Finally, the overhead for creating and handling threads ranges in the number of hundreds of thousands of cycles [3,4] as classical operating systems have not been designed for parallel applications, thereby hindering low workload threads in a large-scale programs, and thus optimal exploitation of the environment.

The actual system support required by each thread is comparatively low and relates in particular to handling memory and I/O. With the current structure of OS architectures, each thread is effectively lugging a whole management system with it that obstructs execution and leaves a major footprint on the local memory. The service-oriented approach generates stand-alone (barebone) applications out of each thread that can run without the full management environment of an operating system. As these threads effectively run directly on the chip/core without any intermediary layers, they can be instantiated and executed much faster, very similar to classical DOS applications. Any additional requirements relating to security, reliability, etc., can thereby be dynamically added at the cost of degrading performance.

62.4.2 Connected World Scenario

In the future, we must not expect the user to outsource specific tasks, for example, to a HPC platform anymore, but instead the user will expect to be *integrated* into the resource mesh, that is, that the underlying resources and infrastructure becomes seamlessly available to the running applications, thereby extending capabilities, usage (such as sharing of data), and performance.

In the classical approach for the connected world scenario, the user would own a global ID that enables him to be recognized in a variety of contexts and that allow the respective application to pull information (such as favorite songs) from a distributed or cloud-like data storage environment. This would however only partially address the requirements of the future mobile user, in particular where execution of interactive applications (such as a document editor with shared data) is required. In such cases, some part of the application has to reside locally with the user to provide the necessary interfaces, even if the interface itself is displayed in a browser-like application.

The user's applications may vary widely and will not be restricted to commonly known and deployed applications, such as Open Office, but also include less common applications, and even completely self-developed code. Applications in turn may be composed of a variety of other functionalities (services, applications, etc.) that each in turn has different requirements toward the executing infrastructure, ranging from high performance to shared or social connectivity. Identification of the identity of the user alone is thereby insufficient, for example, if the user wants to run complex simulations on (or from) his mobile device.

Whilst web service like approaches allow integration of remote services, they do not offer the flexibility or performance needed in most cases. For example, communicating the necessary data from the United States to France will lead to significant delays even with the future network developments and increase in performance. This requires a more flexible distribution of code and data than currently considered in any application code. Some global cloud infrastructures may already exhibit indicators of a development in this direction, yet this is far from providing users with the full flexibility of deployment and execution.

With a service-oriented operating system, this capability essentially comes for free with the structure of the OS itself. The principal organization scheme of the OS thereby allows realization of the old grid dream where any resource can be designated host of a dedicated service [5]. The approach goes even further by allowing partial integration of the resource for only limited capabilities and execution, much like sharing cores for multiple tasks in a local desktop. In this case, it is not so much the other device getting access granted to the user data, but hosting temporarily a function on behalf of the user that in turn enables accessing the respective data. In other words, the user will not only be able to play his favorite playlists, but also can use his favorite application for doing so.

Future offices, hotels and conference centers can essentially offer pure resource power to their users who plug in any low power mobile device to run their own applications and perform their individual tasks with extended execution performance. The resources become effectively nothing else but extensions to the mobile device that dynamically move with and adapt to the user's behavior and requirements. The mesh of resources used for this purpose form an isolated environment of their own, with external interfaces only provided and used as specified by the user, respectively developer.

62.5 Conclusions

With the introduction of multicores, specialized infrastructures, Internet provided utility computing, etc., the IT environment is quickly building up to significant changes in our way of thinking of software and hardware. These changes are thereby not matched by our approach for dealing with them, in particular in terms of software development and execution. The average programmer still treats the system like a single core Turing machine with immediate data availability. As the heterogeneity and dynamicity grows, however, solving the problem cannot be put onto the developer alone, but must be matched by the infrastructure's capabilities to adjust itself *and* the code according to the current requirements.

This all builds up toward an environment where the individual resource can never satisfy all the user's requirements without incorporating additional devices and/or services. Already we have the notion of such global infrastructures through web services, social networks, and clouds, yet they are countered by the increasing demand for performance in complex applications.

Classical operating systems are simply not designed for such environments and they lead to increasing overhead both in terms of execution performance and development time/effort. Bringing service orientation not only to the application but also to the underlying execution management framework, that is, the operating system, enables the realization of such scenarios while reducing the inherent performance losses.

The service-oriented approach does not only match the requirements for minimal load on the individual processing unit and for distributed deployment and management of applications spanning multiple devices. It also allows for much easier adaptation toward and integration of new device types. Where classical operating systems effectively had to be rewritten completely even for minor changes (see, e.g., [1]), the service orientation allows for component-wise adaptation and instantiation according to the device properties.

For decades, the monolithic, centralized operating system architecture has prevailed. All the adjustments to meet a specific processor model have made the OS sluggish to adaptation. New infrastructures and usage demand for a much meaner and leaner architecture than monolithic approaches can offer. Service orientation needs to become a natural part of such architectures in order to meet these requirements.

References

1. A. Baumann, P. Barham, P.-E. Dagand, T. Harris, R. Isaacs, S. Peter et al. The multikernel: A new OS architecture for scalable multicore systems. In *Proceedings of the ACM SIGOPS 22nd Symposium on Operating Systems Principles*, pp. 29–44, Big Sky, MT. ACM, New York, 2009.
2. S. Byna, X.-H. Sun, and Y. Chen. A taxonomy of data prefetching mechanisms. In *Proceedings of the International Symposium on Parallel Architectures, Algorithms, and Networks*, pp. 19–24, Sydney, New South Wales, Australia, 2008.

3. V. V. Dimakopoulos, P. E. Hadjidoukas, and G. Ch. Philos. A microbenchmark study of openmp overheads under nested parallelism. In *Proceedings of the 4th International Conference on OpenMP in a New Era of Parallelism, IWOMP'08*, pp. 1–12, West Lafayette, IN. Springer-Verlag, Berlin, Germany, 2008.
4. C. Fleizach and D. Gobera. System measurement project CSE 221: Graduate operating systems, fall 2005.
5. I. Foster and C. Kesselman. *The Grid: Blueprint for a New Computing Infrastructure*. Morgan Kaufmann, San Francisco, CA, 2004.
6. L. Schubert, A. Kipp, and S. Wesner. Above the Clouds: From Grids to Service-oriented Operating Systems., In G. Tselentis, J. Domingue, A. Galis, A. Gavras, D. Hausheer, S. Krco, V. Lotz and T. Zahariadis (eds.), *Future Internet Assembly*, pp. 238–249. IOS Press, Amsterdam, The Netherlands, 2009.

VIII

Programming Languages

63 Imperative Language Paradigm

Michael J. Jipping
Hope College

Kim Bruce
Pomona College

63.1 Introduction

In the 1940s, John von Neumann pioneered the design of basic computer architecture by structuring computers into two major units: a central processing unit (CPU), responsible for computations, and a data storage unit, or memory. This architecture is demand driven, based on a command- and instruction-oriented computing model. The basic unit cycle of execution, typically composed of a single instruction, consists of four steps:

1. Obtain the addresses of the result and operands.
2. Obtain the operand data from the operand location(s).
3. Compute the result data from the operand data.
4. Store the result data in the result location.

Note here how separation of the execution unit from the memory unit has structured the sequence. Data must be located and piped from memory, operated on, and transferred back to memory to be available for the next operation. All operations in a von Neumann machine operate this way, in a stepwise, structured manner. The von Neumann model has been the basis of nearly every computer built since the 1940s.

Imperative programming languages are modeled after the von Neumann model of machine execution and were invented to provide the abstractions of machine components and actions in order to make it easier to program computers. Abstractions such as variables (which model memory cells), assignment statements (which model data transfer), and other language statements are all abstractions of the basic von Neumann approach.

In this chapter, we address the fundamental principles underlying imperative programming languages and examine the way the constructs of imperative languages are represented in several languages. We devote special attention to features of more modern imperative programming languages, among them support for abstract data types (ADTs) and newer control constructs such as iterators and exception handling. Examples in this chapter are given in a variety of imperative programming languages, including FORTRAN, Pascal, C, C++, MODULA-2, and Ada 83. In the Best Practices section, we explore in more detail the languages FORTRAN IV (chosen for historical reasons), C and C++ (its imperative parts), and Ada 83.

63.2 Data Bindings: Variables, Type, Scope, and Lifetime

In this section, we discuss some of the fundamental properties of imperative programming languages. In particular, we address issues related to binding time, the properties of variables, *types*, scope, and lifetime.

63.2.1 Binding Time

We will find it useful to classify many of the differences in programming languages based on the notion of binding time. A *binding* is the association of an attribute to a name. The time at which a binding takes place is an important consideration. There are many times when a binding can occur. Some of these are as follows:

- *Language definition*: when the language is designed. An example is the binding of the constant name true to the corresponding Boolean value.
- *Language implementation*: when a compiler or interpreter is written. An example is the binding of the representation of values of various types.
- *Compile time*: when a program is being translated into machine language. For example, the type of a variable in a statically typed language is bound at compile time. In statically typed languages, overloaded functions are bound at compile time.
- *Load time*: when the executable machine language image of the program is loaded into the memory for execution by the execution unit. The location of global variables is bound at load time.
- *Procedure or function invocation time*: the time a program is being executed. Actual parameters are bound to formal parameters and local variables are bound to locations at procedure invocation time.
- *Run time*: any time during the execution of a program. A new value can be bound to a variable at run time. In dynamically typed languages, overloaded functions are bound at run time.

As we examine fundamental issues in the definition of imperative programming languages, we will keep in mind the distinctions between languages based on differences in binding time.

63.2.2 Variables

Imperative languages support computation by executing commands whose purpose is to change the underlying state of the computer on which they are executed. The *state* of a computer encompasses the contents of memory and also includes both data that are about to be read from outside of the computer and data that have been output.

Variables are central to the definition of imperative languages as they are objects whose values are dependent on the contents of memory. A variable is characterized by its attributes, which generally include its name, location in memory, value, type, scope, and lifetime.

Depending on context, the meaning of a variable may be considered to be either its value or its location. For instance, in the assignment statement, $x := x + 1$, the meaning of the occurrence of the variable x to the left of the assignment symbol is its location (sometimes called the l-value of x), whereas the meaning of the occurrence on the right side is its value, that is, the value stored at the location corresponding to *x* (sometimes called the *r*-value). The location of global variables is bound at load time, whereas the location of local variables and reference parameters is typically bound at procedure entry. The value of the variable can be changed at any point during execution of the program.

63.2.3 Types

Types in programming languages are abstractions that represent sets of values and the operations and relations that are applicable to them. Types can be used to hide the representation of the primitive values of a language, allow type checking at either compile time or run time, help disambiguate overloaded operators, and allow the specification of constraints on the accuracy of computations. Types also can play an important role in compiler optimization.

Types in a programming language include both simple and composite types. The use of *simple types* such as integer, real, Boolean, and character types allows the user to abstract away from the actual computer representation of these values, which may differ from computer to computer. The operations on simple types may or may not be supported directly by the underlying hardware. For instance, many early microprocessors supported only real or floating-point operations in software.

Some languages (e.g., those derived from Pascal) allow the programmer to define their own simple *enumerated* types by simply listing the values of the type. The ordering of elements in this enumeration is significant as these types typically support successor and predecessor functions as well as ordering relations. Later we will discuss mechanisms for supporting *ADTs*, another way of constructing types that can be used as though they were primitive to a language.

Many languages support the creation of *subrange* types, which allows a programmer to define a new type as a copy of a type with a subset of its values. The new type comes equipped with the same operators as its parent type and is usually compatible with the original type.

Composite or structured data types can be created from simple types using *type constructors*. Typical composite types include arrays, records (or structures), variant records (or unions), sets, subranges, pointer types, and, in a few languages, function or procedure types. For instance, arrays are typically constructed from two types: a subrange type that provides the set of indices of the array and another type representing the values stored in the array. Not all languages support all these type constructors. For instance, function and procedure types are provided by MODULA-2 but are not available in Ada 83. Many languages support strings as special types of composite types, for instance, as arrays of characters, but they may also be provided as built-in types.

Most imperative languages bind types to variables statically. These bindings are usually specified in declarations, but some languages, such as FORTRAN, allow implicit declaration of variables, with the type binding determined by the name of the identifier (e.g., in FORTRAN if the name starts with *I* through *N*, then the variable is an integer, otherwise real).

An important issue in type-checking programming languages is type equivalence. When do two terms have equivalent types? The two extremes in the definitions of *type equivalence* are structural and name equivalence:

- *Structural equivalence*: Two types are said to be *structurally (or domain) equivalent* if they have the same structure. That is, they are built from the same type constructors and built-in types in the same way.
- *Name equivalence*: Two types are *name equivalent* if they have the same name.

The language C uses structural equivalence, whereas Ada 83 uses name equivalence. There are also a range of possibilities between these two extremes. For instance, Pascal and MODULA-2 use *declaration equivalence*: two types are declaration equivalent if they are name equivalent or they lead back to the same structure declaration by a series of redeclarations.

Inequivalent types may be compatible in certain situations. For instance, two types are assignment compatible if an expression of one type may be assigned to a variable of another. For instance, in Pascal, a subrange of integer is assignment compatible with integer, even though the types are not equivalent.

An application of these ideas can be found in the rules for determining whether a particular actual parameter may be used in a procedure call for a particular formal parameter. In Pascal, if the formal parameter is a reference parameter, then the actual parameter must be a variable of equivalent type. If the formal parameter is a value parameter, then the actual parameter must be assignment compatible.

As mentioned earlier, some languages support the creation of subrange types. The new subrange type is usually assignment compatible with the original type. Because of this compatibility, the new type is called a *subtype* of the parent in Ada. Another mechanism available in Ada, called *derived typing*, defines a new type by constructing an exact copy of a type that already exists. However, the resulting new type is distinct and is not type equivalent or even assignment compatible with the existing type.

The type equivalence rules are the cause of one of the greatest limitations in the use of Pascal. If a formal parameter has an array type, then the actual parameter must have an equivalent type. In particular, the subscript ranges of the two arrays must be identical. Thus, it is impossible to write a procedure in Pascal that can be used to sort different-sized arrays of real numbers. (Actually, the current ANSI standard Pascal provides a special mechanism to allow exceptions to this rule.)

Ada escapes from this problem by designating some properties of types to be static, while others are dynamic. For example, in a type defined to be a subrange of integers, the underlying static type is integer while the subrange bounds are a dynamic property. Only the static properties of types are considered at compile time by the type checker, whereas restrictions due to dynamic properties are checked at run time.

Consider the following Ada declarations as an example of type bindings:

```
type COINS is (PENNY, NICKEL, DIME, QUARTER);
subtype SILVER is COINS range (NICKEL..QUARTER);
type CHANGE is new COINS;

C1, C2; COINS;
S: SILVER;
CH: CHANGE;
```

`COINS` is an enumerated type, defined by the programmer to allow assignments such as

```
C1 := DIME.
```

`SILVER` is a subrange of `COINS`, which includes only the values `NICKEL`, `DIME`, and `QUARTER`. `CHANGE` is a derived type taken from `COINS`.

Because Ada employs name equivalence, only `C1` and `C2` are equivalent, but `S` is assignment compatible with them. If Ada used structural equivalence, then variables `C1`, `C2`, and `CH` would be equivalent.

63.2.4 Scope

The scope of a binding is the area or section of a program in which that particular binding is effective. The method and extent of *scope rules* that define a binding scope will, to a large degree, affect the usefulness and applicability of a language. If, for instance, the rules allow the scope of a binding to be determined by the execution path of a program, the language might be more flexible, yet the code becomes harder to understand.

Scope rules are tied tightly to concepts of binding time. *Static scope rules* determine the scope of a binding at compile time and are based on the lexical structure of the program. *Dynamic scope rules* determine the scope of a binding at run time. Thus, an occurrence of a variable name in a procedure may refer to one variable the first time it is evaluated yet refer to an entirely different variable the next time, depending on the execution path at run time. Most imperative languages use static scope rules.

As an example of scope rules in Ada, consider the code in Figure 63.1. Static scope rules are determined by the program block structure, which does not change while the program runs. Therefore, the call to procedure `P` prints the variable `J` defined in the outer, main program, no matter where it is called from. Likewise, the assignment in block 1 at reference point A changes `J` from the block and not from

```
with TEXT_IO; use TEXT_IO;
procedure SCOPED is
  package INT_IO is new INTEGER_IO (integer); use INT_IO;
  I, J: integer;
  procedure P is begin put (J); new_line; end P;
  begin
    J := 0;
    I := 10;
    declare -- Block 1
      J: integer;
    begin
      j := I; -- reference point A
      P;
    end;
    put (J); new_line;
    declare -- Block 2
      I: Integer
    begin
      I := 5
      J := I + 1; -- reference point B
      P;
    end;
    put (J); new_line;
  end;
```

FIGURE 63.1 Scoping rules in Ada.

the main program. Dynamic scope rules, on the other hand, typically follow dynamic call paths to determine variable bindings. If Ada used dynamic scope rules, the first call to `P` from block 1 would print the value 10 corresponding to the `J` from block 1, whereas the second call to `P` would print the value 3 corresponding to the `J` from the main program.

63.2.5 Execution Units: Expressions, Statements, Blocks, and Programs

An *expression* is a program phrase that returns a value. Expressions are built up from constants and variables using operators. As described earlier, variables may represent two values, depending on context: their location and the value stored at that location. Operators may be built-in, like the arithmetic and comparison operators, or may be user-defined functions.

Reflecting the sequential order of von Neumann computation, an imperative language specifies the order in which operations are evaluated. Typically, evaluation order is determined by precedence rules. A typical precedence rule set for arithmetic expressions might be the following:

1. Subexpressions inside parentheses are evaluated first (according to the precedence rules).
2. Instances of unary negation are evaluated next.
3. Then, multiplication (*) and division (/) operators are evaluated in left to right order.
4. Finally, addition (+) and subtraction (–) are evaluated left to right.

Although procedure rules are commonly used by imperative languages, some languages use other conventions to avoid precedence rules. For example, PostScript uses postfix notation for expressions, while LISP uses prefix notation. APL evaluates all expressions from right to left without regard to precedence, using only parentheses to change the evaluation order.

The fundamental unit of execution in an imperative programming language is the *statement.* A statement is an abstraction of machine language instructions, grouped together to form a single logical activity.

The simplest and most fundamental statement in imperative programming languages is the assignment statement. This statement, typically written in the form `x := e` or `x = e` with `x` a variable (or other expression representing a location) and e an expression, is usually interpreted by evaluating e and copying its value into the location represented by x. This is known as the *copy semantics* for assignment.

Less common are languages that use the sharing interpretation of assignment. In these languages, variables generally represent references to objects that contain the actual values. The assignment *x*:= *y* would then be interpreted as binding the object referred to by *y* to *x* rather than its value. Since both variables refer to the same object, they share the same value. If the value of one is changed, the value of the other will also change. This is the *sharing semantics* for assignment.

Declarations and statements may be grouped together to form a *block*. Procedure and function bodies are represented as blocks, whereas *control structures* (discussed subsequently) can also be understood as acting on blocks of statements (generally without declarations). The most general form of a block contains a *declarative* section, which contains the declarations that define the bindings that are effective in the block, and an *executable* section, which contains the statements over which the binding is to hold, that is, the scope of the declarations.

In the so-called block-structured languages (including most languages descended from ALGOL 60, e.g., Pascal, Ada, and C), blocks may be nested. Within any block, therefore, there can be two kinds of bindings in force: *local bindings*, which are specified by the declarative sections associated with the block, and *nonlocal bindings* (also known as *global* bindings), which are bindings defined by declarative sections of blocks within which the specific block is nested.

Consider again the code from Figure 63.1. The first two assignments of the main program assign J from the main program the value 0 and I from the main program the value 10. The next assignment assigns the value 10, derived from the global I, to the variable J from the first inner block. When the definition of the second inner block is encountered, the variable I is found in the local scope, while J is found in the *outer* scope, that of the main program. The value 6 will be printed for J at the end of the main program.

63.3 Control Structures

By adopting the semantics of the basic execution cycle of a von Neumann architecture, an imperative language adopts a strict sequential ordering for its statements. By default, the next statement to execute is the next physical statement in the program. Control structures in imperative languages provide ways to alter this strict sequential ordering. The most common control structures are *conditional structures* and *iterative structures. Unconstrained control structures* are also allowed in most languages through the use of goto statements.

63.3.1 Conditional Structures

Conditional control structures (also known as *selection statement*) determine whether or not a block of statements is executed based on the result of one or several tests. These structures fall into one of the two following classes:

63.3.1.1 If Statements

All imperative languages include some form of if statement. This control structure provides a text and a single statement or statement block to be executed if the test evaluates to a true value. Optionally, the programmer may provide another block of statements that can be executed only if the test evaluates to false. The following is a simple example from Ada:

```
if (x = 2) then
  y := 3;
else
  y := 6;
end if;
```

The variable y is set to either 3 or 6 depending on the value of x.

In most languages, if statements can be nested within other control structures, including other if statements. However, nested if statements can result in awkward, deeply nested code. Thus, many languages provide a special construct (e.g., `elsif` in Ada) to represent *else if* constructs without requiring further nesting. The two Ada examples given next are equivalent semantically, though the first, which uses `elsif`, is easier to read than the second, which uses nested conditionals:

```
if (x = 2) then
    y := 3;
elsif (x = 3) then
    y := 15;
elsif (x = 5) then
    y :=18;
else
    y := 6;
end if;
```

```
if (x = 2) then
  y := 3;
else
  if (x = 3) then
    y := 15;
  else
    if (x = 5) then
      y := 18;
    else
      y := 6;
    end if;
  end if;
end if;
```

63.3.1.2 Case Statements

This conditional combines case-by-case expression examination with a restricted multiway conditional. This conditional may be seen to be simply a syntactic convenience, but in many cases its implementation results in a much faster determination at run time of the actual block of code to be executed. Consider the following case statement from Ada:

```
case y is
   when 2 => y := 3;
   when 3 => y := 15;
   when 15 => y := 18;
   when others => y := 6;
end case;
```

An expression (y in this case) of an ordinal type occurs after the keyword `case`. Each `when` clause contains a guard, which is a list of one or more constants of the same type as the expression. Most languages require that there be no overlap between these guards. The expression after the keyword `case` is evaluated, and the resulting value is compared to the guards. The block of statements connected with the first matched alternative is executed. If the value does not correspond to any of the guards, the statements in the `others` clause is executed. Note that the semantics of this example is identical to that of the previous example.

The case statement may be implemented in the same way as a multiway if statement, but in most languages it will be implemented via table lookup, resulting in a constant time determination of which block of code is to be executed.

C's switch statement differs from the case previously described in that if the programmer does not explicitly exit at the end of a particular clause of the switch, program execution will continue with the code in the next clause.

63.3.2 Iterative Structures

One of the most powerful features of an imperative language is the specification of *iteration* or statement repetition. *Iterative structures* can be classified as either definite or indefinite, depending on whether the number of iterations to be executed is known before the execution of the iterative command begins:

- *Indefinite iteration*: The different forms of indefinite iteration control structures differ by where the test for termination is placed and whether the success of the test indicates the continuation or termination of the loop. For instance, in Pascal the *`while-do`* control structure places the test before the beginning of the loop body (a pretest), and a successful test determines that the execution of the loop shall continue (a continuation test). Pascal's `repeat-until` control structure, on the other hand, supports a posttest, which is a termination test. That is, the test is evaluated at the end of the loop and a success results in termination of the loop.

Some languages also provide control structures that allow termination anywhere in the loop. The following example is from Ada:

```
loop
  ...
  exit when test;
  ...
end loop
```

The `exit when test` statement is equivalent to `if test then exit`.

A few languages also provide a construct to allow the programmer to terminate the execution of the body of the loop and proceed to the next iteration (e.g., C's `continue` statement), whereas some provide a construct to allow the user to exit from many levels of nested loop statements (e.g., Ada's named `exit` statements).

- *Definite iteration*: The oldest form of iteration construct is the definite or fixed-count iteration form, whose origins date back to FORTRAN. This type of iteration is appropriate for situations where the number of iterations called for is known in advance. A variable, called the *iteration control variable* (ICV), is initialized with a value and then incremented or decremented by regular intervals for each iteration of the loop. A test is performed before each loop body execution to determine if the ICV has gone over a final, boundary value. Ada provides fixed-count iteration as a for loop; an example is shown next:

```
for i in 1..10 loop
  y := y + i;
  z := z * i;
end loop;
```

Here, `i` is initialized to 1, and incremented by 1 for each iteration of the loop, until it exceeds 10. Note that this type of loop is a pretest iterative structure and is essentially a syntactic sugar for an equivalent `while` loop.

An ambiguity that may arise with a for loop is what value the ICV has after termination of the loop. Most languages specify that the value is formally undetermined after termination of the loop, though in practice it usually contains either the upper limit of the ICV or the value assigned that first passes the boundary value. Ada eliminates this ambiguity by treating the introduction of the control variable as a variable declaration for a block containing only the for loop.

Some modern programming languages have introduced a more general form of for loop called an *iterator* construct. Iterators allow the programmer to control the scheme for providing the ICV with successive values. The following example is from CLU (Liskov et al. 1977). We first define the iterator:

```
string_chars = iter (s : string) yields (char);
  index: Int := 1;
  limit: Int := string$size (s);
  while index <= limit do
    yield (string$fetch(s, index));
    index := index + 1;
  end;
end string_chars;
```

which can be used in a `for` loop as follows:

```
for c: char in string_chars(s) do LoopBody end;
```

When the for loop controlled by an iterator is encountered, control is passed to the iterator, which runs until a `yield` statement is executed. The value associated with the `yield` statement is used as the initial value of the iterator control variable `c`, and the body of the loop is executed. Control is then passed back to the iterator, which resumes execution with the statement following the `yield`. Control is passed to the loop body each time a yield statement is executed and back to the iterator each time the loop body finishes execution. Thus, iterators behave as a restricted form of coroutine, passing control back and forth between the two blocks of code. The loop is terminated when the iterator runs to completion. In the preceding examples, this will occur when `index > limit`.

63.3.3 Unconstrained Control Structures: Goto and Exceptions

Unconstrained control structures, generally known as goto constructs, cause control to be passed to the statement labeled by the *identifier* or line number given in the goto statement. Dijkstra (1968) first questioned the use of goto statements in his famous letter, "Goto statement considered harmful," to the editor of the *Communications of ACM*. The controversy over the goto mostly centers on readability of code and handling of the arbitrary transfer of control into and out of otherwise structured sections of program code.

For example, if a goto statement passes control into the middle of a loop block, how will the loop be initialized, especially if it is a fixed-count loop? Even worse, what happens when a goto statement causes control to enter or exit in the middle of a procedure or function? The problems with readability arise because a program with many goto statements can be very hard to understand if the dynamic (run time) flow of control of the program differs significantly from the static (textual) layout of the program. Programs with undisciplined use of gotos have earned the name of *spaghetti code* for their similarity in structure to a plate of spaghetti.

Although some argue for the continued importance of goto statements, most languages either greatly restrict their use (e.g., do not allow gotos into other blocks) or eliminate them altogether. In order to handle situations where gotos might be called for, other more restrictive language constructs have been introduced to make the resulting code more easily readable. These include the `continue` and `exit` statements (particularly labeled `exit` statements) referred to earlier.

Another construct that has been introduced in some languages in order to replace some uses of the goto statement is the *exception*. An exception is a condition or event that requires immediate action on the part of the program. An exception is *raised* or *signaled* implicitly by an event such as arithmetic overflow or an index out of range error, or it can be explicitly raised by the programmer.

The raising of an exception results in a search for an exception *handler*, a block of code defined to handle the exceptional condition and (hopefully) allow normal processing to resume. The search for an appropriate handler generally starts with the routine that is executing when the exception is raised. If no appropriate handler is found there, the search continues with the routine that called the one that contained the exception. The search continues through the chain of routine calls until an appropriate handler is found, or the end of call chain is passed without finding a handler.

If no handler is found, the program terminates, but if a handler is found, the code associated with the handler is executed. Different languages support different models for resuming execution of the program. The termination model of exception handling results in termination of the routine containing the handler, with execution resuming with the caller of that routine. The continuation model typically resumes execution at the point in the routine containing the handler that occurs immediately after the statement whose execution caused the exception.

The following is an example of the use of exceptions in Ada (which uses the termination model):

```
procedure pop(s: stack) is
  begin
    if empty(s) then raise emptyStack
                else …
  end;
```

```
procedure balance (parens: string) return boolean is
   pStack: stack
  begin
   …
   if … then pop(s) …
  exception
   when emptyStack => return false
  end
```

Many variations on exceptions are found in existing languages. However, the main characteristics of exception mechanisms are the same. When an exception is raised, execution of a statement is abandoned and control is passed to the nearest handler. (Here "nearest" refers to the dynamic execution path of the program, not the static structure.) After the code associated with the handler is executed, normal execution of the program resumes.

The use of exceptions has been criticized by some as introducing the same problems as goto statements. However, it appears that disciplined use of exceptions for truly exceptional conditions (e.g., error handling) can result in much clearer code than other ways of handling these problems.

We complete our discussion of control structures by noting that, although many control structures exist, only very few are actually necessary. At one extreme, simple conditionals and a goto statement are sufficient to replace any control structure. On the other hand, it has been shown (Boehm and Jacopini 1966) that a two-way conditional and a while loop are sufficient to replace any control structure. This result has led some to point out that a language has no need for a goto statement; indeed, there are languages that do not have one.

63.3.4 Procedural Abstraction

Support for abstraction is very useful in programming languages, allowing the programmer to hide details and definitions of objects while focusing on functionality and ease of use. *Procedural abstraction* (Liskov and Guttag 1986) involves separating out the details of an execution unit into a procedure and referencing this abstraction in a program statement or expression. The result is a program that is easier to understand, write, and maintain.

The role of procedural abstraction is best understood by considering the relationships between the four levels of execution units described earlier: expressions, statements, blocks, and programs. A statement can contain several expressions; a block contains several statements; a program may contain several blocks. Following this model, a procedural abstraction replaces one execution unit with another one that is simpler. In practice, it typically replaces a block of statements with a single statement or expression.

The *definition* of a procedure binds the abstraction to a name and to an executable block of statements called the *body*. These bindings are compile-time, declarative bindings. In Ada, such a binding is made by specifying code such as the following:

```
procedure area (height, width: real; result: out real) is
  begin
    result := height * width;
  end;
```

The *invocation* of a procedure creates an activation of that procedure at run time. The *activation record* for a procedure contains data bound to a particular invocation of a procedure. It includes slots for parameters, local variables, other information necessary to access nonlocal variables, and data to enable the return of control to the caller. In languages supporting recursive procedures, more than one activation record can exist at the same time for a given procedure. In those languages, the lifetime of the activation record is the duration of the procedure activation.

Although scoping rules provide access to nonlocal variables, it is generally preferable to access nonlocal information via *parameter* passing. Parameter-passing mechanisms can be classified by the direction in which the information flows: *in parameters*, where the caller passes data to the procedure, but the procedure does not pass data back; *out parameters*, where the procedure returns data values to the caller, but no data are passed in; and *in out parameters*, where data flow in both directions.

Formal parameters are specified in the declaration of a procedure. The *actual parameters* to be used in the procedure activation are specified in the procedural invocation. The procedure passing mechanism creates an association between corresponding formal and actual parameters. The precise information flow that occurs during procedure invocation depends on the parameter-passing mechanism.

The association or mapping of formal to actual parameters can be done in one of three ways. The most common method is *positional parameter association*, where the actual parameters in the invocation are matched, one by one in a left-to-right fashion, to the formal parameters in the procedural definition. *Named parameter association* also can be used, where a name accompanies each actual parameter and determines to which formal parameter it is associated. Using this method, any ordering can be used to specify parameter values. Finally, *default parameter association* can be used, where some actual parameter values are given and some are not. In this case, the unmatched formal parameters are simply given a default value, which is generally specified in the formal parameter declaration.

Note that in a procedural invocation, the actual parameter for an in parameter may be any expression of the appropriate type, since data do not flow back, but the actual parameter for either an out or an in out parameter must be a variable, because the data that are returned from a procedural invocation must have somewhere to go.

Parameter passing is usually implemented as being one of copy, reference, and name. There are two copy parameter-passing mechanisms. The first, labeled *call by value*, copies a value from the actual to the formal parameter before the execution of the procedure's code. This is appropriate for in parameters. A second mode, called *call by result*, copies a value from the formal parameter to the actual parameter after the termination of the procedure. This is appropriate for out parameters. It is also possible to combine these two mechanisms, obtaining *call by value result*, providing a mechanism that is appropriate for in out parameters.

The *call by reference* passes the address of the actual parameter in place of its value. In this way, the transfer of values occurs not by copying but by virtue of the formal parameter and the actual parameter referencing the same location in memory. Call by reference makes the sharing of values between the formal and actual a two-way, immediate transfer, because the formal parameter becomes an alias for the actual parameter.

Call by name was introduced in ALGOL 60 and is the most complex of the parameter-passing mechanisms described here. Although it has some theoretical advantages, it is both harder to implement and generally more difficult for programmers to understand. In call by name, the actual parameter is reevaluated every time the formal parameter is referenced. If any of the constituents of the actual parameter expression has changed in value since the last reference to the formal parameter, a different value may be returned at successive accesses of the formal parameter. This mechanism also allows information to flow back to the main program with an assignment to a formal parameter. Although call by name is no longer used in most imperative languages, a variant is used in functional languages that employ lazy evaluation.

Several issues crop up when we consider parameters and their use. The first is a problem called *aliasing*, where the same memory location is referenced with two or more names. Consider the following Ada code:

```
procedure MAIN is
  a: integer;
  procedure p(x, y: in out integer) is
    begin
      a := 2;
      x := y + a;
    end;
```

```
  begin
    a := 10;
    p(a,a);
    ...
end;
```

During the call of p(a, a), the actual parameter a is bound to both of the formal parameters x and y. Because x and y are in out parameters, the value for a will change after the procedure returns. It is not clear, however, which value a will have after the procedure call. If the parameter-passing mechanism is call by value result, then the semantics of this program depend on the order in which values are copied back to the caller. If they are copied into the parameters from left to right, the value of a will be 10 after the call. The results with call by reference will be unambiguous (though perhaps surprising to the programmer), with the value of a being 4 after the call. In Ada, a parameter specified to be passed as in out may be passed using either call by value result or call by reference. The preceding code provides an example where, because of aliasing, these parameter-passing mechanisms give different answers. Ada terms such programs to be erroneous and considers them not to be legal, even though the compiler may not be able to detect such programs.

Most imperative programming languages support the use of procedures as parameters (Ada is one of the few exceptions). In this case, the parameter declaration must include a specification of the number and types of parameters of the procedure parameter. MODULA-2, for example, supports procedure types that may be used to specify procedural parameters. There are few implementation problems in supporting procedure parameters, though the implementation must ensure that nonlocal variables are accessed properly in the procedure passed as a parameter.

There are two kinds of procedural abstractions. One kind, usually known simply as a *procedure*, is an abstraction of a program statement. Its invocation is like a statement, and control passes to the next statement after the invocation. The other type is called a *value returning procedure* or *function*. Functions are abstractions for an operand in an expression. They return a value when invoked, and, upon return, evaluation of the expression containing the call continues.

Many programming languages restrict the values that may be returned by functions. In most cases, this is simply a convenience for the compiler implementor. However, although common in functional languages, most imperative languages that support nested procedures or functions do not allow functions to return other functions or procedures. The reason has to do with the stack-based implementation of block-structured languages. If allowed, it might be possible to return a nested procedure or function that depends on a nonlocal variable that is no longer available when the procedure is actually invoked.

To avoid confusion, most languages allow a name to be bound to only one procedural abstraction within a particular scope. Some languages, however, permit the *overloading* of names. Overloading permits several procedures to have the same name as long as they can be distinguished in some manner. Distinguishing characteristics may include the number and types of parameters or the data type of the return value for a function. In some circumstances, overloading can increase program readability, whereas in others it can make it difficult to understand which operation is actually being invoked.

Program mechanisms to support concurrent execution of program units are discussed in Chapter 52. However, we mention briefly *coroutines* (Marlin 1980), which can be used to support pseudoparallel execution on a single processor. The normal behavior for procedural invocation is to create the procedural instance and its activation record (runtime environment) upon the call and to destroy the instance and the activation record when the procedure returns. With coroutines, procedural instances are first created and then invoked. Return from a coroutine to the calling unit only suspends its execution; it does not destroy the instance. A resume command from the caller results in the coroutine resuming execution at the statement after the last return.

Coroutines provide an environment much like that of parallel programming; each coroutine unit can be viewed as a process running on a single processor machine, with control passing between processes.

Despite their interesting nature (and clear advantages in writing operating systems), most programming languages do not support coroutines. MODULA-2 is an example of a language that supports coroutines. As mentioned earlier, iterators can be seen as a restricted case of coroutines.

63.3.5 Data Abstraction

Earlier in this chapter, we introduced the idea of data types as specifying a set of values and operations on them. Here we extend that notion of values and operations to ADTs and their definitional structures in imperative languages.

The primitive data types of a language are specified by both a set of values and a collection of operations that may be applied to them. Clearly, the set of integers would be useless without the simultaneous provision of operation on those integers. It is characteristic of primitive data types that the programmer is not allowed access to their representations.

Many modern programming languages provide a mechanism for a programmer to specify a new type that behaves as though it were a primitive type. An ADT is a collection of data objects and operations on those data objects whose representation is hidden in such a way that the new data objects may be manipulated only using the operations provided in the ADT. ADTs abstract away the implementation of a complex data structure in much the same way primitive data types abstract away the details of the underlying hardware implementation.

The *specification* of an ADT presents interface details relevant to the users of the ADT, whereas the *implementation* contains the remaining implementation details that should not be exported to users of the ADT. *Encapsulation* involves the bundling together of all definitions in the specification of the ADT in one place. Because the specification does not depend on any implementation details, the implementation of the ADT may be included in the same program unit with the specification or it may be contained in a separately compiled unit. This encapsulation of the ADT typically supports information hiding so that the user of the ADT (1) need not know the hidden information in order to use the ADT and (2) is forbidden from using the hidden information so that the implementation can be changed without impact on correctness to users of the ADT (at least if the specifications of the operations are still satisfied in the new implementation). Of course, one would expect a change in implementation to affect the efficiency of programs using the ADT. A further advantage of information hiding is that, by forbidding direct access to the implementation, it is also possible to protect the integrity of the data structure.

CLU was among the earliest languages providing explicit language support for ADTs through its clusters. Ada also provides facilities to support ADTs via packages. Let us consider an example from MODULA-2, a successor language to Pascal designed by Niklaus Wirth. The stack ADT provides a stack type as well as operations init, push, pop, top, and empty. A MODULA-2 specification for a stack of integers resembles the following:

```
DEFINITION MODULE StackADT;
    TYPE stack;
    PROCEDURE init (VAR s: stack);
    PROCEDURE push (VAR s: stack; elt: INTEGER);
    PROCEDURE pop (VAR s: stack);
    PROCEDURE top (s: stack): INTEGER;
    PROCEDURE empty (s: stack): BOOLEAN
END StackADT.
```

Note the declaration includes the type name and procedural *headers* only. The type `stack` included in the preceding specification is called an *opaque* type in MODULA-2, because users cannot determine the actual implementation of the type from the specification. This ADT specification uses information hiding to get rid of irrelevant detail. Now, this specification can be placed in a separate file and made available to programmers. By including the following declaration:

```
FROM StackADT IMPORT stack, init, push, pop, top, empty;
```

at the beginning of a module, a programmer could use each of these names as though the complete specification was included in the module. Thus, the user can write

```
var s1, s2: Stack;
begin
  push(s1, 15);
  push(s2, 20);
  if not empty(s1) then pop(s1) …
```

in such a module.

In MODULA-2, the complete definitions of the type `stack` and its associated operations are provided in an implementation module, which typically is stored in a separate file:

```
  IMPLEMENTATION MODULE StackADT;
  TYPE stack = POINTER TO stackRecord;
       stackRecord = RECORD
                       top: 0..100;
                       values: ARRAY[1..100] of INTEGER;
                     END;
  PROCEDURE init(VAR s: stack);
    BEGIN
      ALLOCATE(s, SIZE(stackRecord));
      S^.top :=0
    END
  PROCEDURE push (VAR s: stack; elt: INTEGER);
    BEGIN
      S^.top := S^.top + 1;
      S^.value[S^.top] := elt
    END;
  PROCEDURE pop(VAR s: stack); …
  PROCEDURE top(s: stack): INTEGER; …
  PROCEDURE empty(s: stack) BOOLEAN; …
END StackADT.
```

Notice that the type name `stackRecord` is not exported.

The specification module must be compiled before any module that imports the ADT and before its implementation module, but importing modules and the implementation module of the ADT can be compiled in any order. As previously suggested, the implementation is irrelevant to writing and compiling a program using the ADT, though, of course, the implementation must be compiled and present when the final program is linked and loaded in preparation for execution.

There is one important implementation issue that arises with the use of language mechanisms supporting ADTs. When compiling a module that includes variables of an opaque type imported from an ADT (e.g., `stack`), the compiler must determine how much space to reserve for these variables. Either the language must provide a linguistic mechanism to provide the importing module with enough information to compute the size required for values of each type or there must be a default size that is appropriate for every type defined in an ADT. CLU and MODULA-2 use the latter strategy. Types declared as CLU clusters are represented implicitly as pointers, whereas in MODULA-2 opaque types must be represented explicitly using pointer types as in the `Stack ADT` example just given. In either case, the compiler needs reserve for a variable of these types only an amount of space sufficient to hold a pointer. The memory needed to hold the actual data pointed to is allocated from the heap at run time. As discussed later, Ada uses a language mechanism to provide size information for each type to importing units.

The definition of ADTs can be *parameterized* in several languages, including CLU, Ada, and C++. Consider the definition of the `stack` ADT. Although the preceding example was specifically given for an integer data type, the implementations of the data type and its operations do not depend essentially on the fact that the stack holds integers. It would be more desirable to provide a parameterized definition of `stack` ADT that can be instantiated to create a stack of any type *T*.

Allocating space for these parameterized data types raises the same problems as previously discussed for regular ADTs. C++ and Ada resolve these difficulties by requiring parameterized ADTs to be instantiated at compile time, whereas CLU again resolves the difficulty by implementing types as implicit references.

63.4 Best Practices

In this section, we will examine three quite different imperative languages to evaluate how the features of imperative languages have been implemented in each. The example languages are FORTRAN (FORTRAN IV for illustrative purposes), Ada 83, and C++. We chose FORTRAN to give a historical perspective on early imperative languages. Ada 83 is chosen as one of the most important modern imperative languages that support ADTs. C++ might be considered a controversial choice for the third example language, as it is a hybrid language that supports both ADT-style and object-oriented features. Nevertheless, the more modern feature contained in the C++ language design makes it a better choice than its predecessor, C (though many of the points that will be made about C++ also apply to C). In this discussion, we ignore most of the object-oriented features of C++, as they are covered in more detail in Chapter 64.

63.4.1 Data Bindings: Variables, Types, Scope, and Lifetime

Like most imperative languages, all three of our languages use static binding and static scope rules. FORTRAN is unique, however, because it supports the *implicit declaration* of variables. An identifier whose name begins with any of the letters *I* through *L* is implicitly declared to be of type integer, whereas any other identifier is implicitly declared to be of type real. These implicit declarations can be overridden by explicit declaration. Therefore, in the fragment

```
INTEGER A
I = 0
A = I
B = C + 2.3
```

`A` and `I` are integer variables, while `B` and `C` are of type real. Most other statically typed languages (including Ada and C++) require *explicit declaration* of each identifier before use. Aside from providing better documentation, these declarations lessen the danger of errors due to misspellings of variable names.

FORTRAN has a relatively rich collection of numerical types, including integer, *double precision* (real), and *complex. Logical* (Boolean) is another built-in type, but FORTRAN IV provided no direct support for characters. However, characters could be stored in integer variables. (The *character* data type was added by FORTRAN 77.) FORTRAN IV supported arrays of up to three dimensions but did not support records. Strings were represented as arrays of integers. FORTRAN IV did not provide any facilities to define new named types.

Later languages provided much richer facilities for defining data types. Pascal, C, MODULA-2, Ada, and C++ all provided a full range of primitive types as well as constructors for arrays, records (structures in C and C++), variant records (unions in C and C++), and pointers (access types in Ada). All provided facilities for naming new types and for constructing types hierarchically (constructing nested types).

Variant records or unions opened up holes in the static type systems of most of these languages, but Ada (and CLU before it) provided restrictions on the access to variants and built-in run time checks in order to prevent type insecurities.

The scope rules for each language, though static in nature, differ significantly. In FORTRAN, the rules are the simplest. The unit of scope for an identifier is either the main program or the procedural unit in which it is declared. Declarations of procedures (subroutines) and functions are straightforward as well in FORTRAN, with no nesting and all parameters passed by reference. Whereas FORTRAN does not support access to nonlocal variables through scoping rules, it allows the programmer to explicitly declare that certain variables are to be more globally available. When two subprograms need to share certain variables, they are listed in a *common* statement that is included in each subprogram. If different combinations of subprograms need to share different collections of variables, several distinct common blocks can be set up, with each subprogram specifying which blocks it wishes access to.

Ada (like Pascal) supports *nested declarations* of procedures and functions. As a result, block structure becomes extremely important to scope determination.

Whereas C and C++ do not provide for nested procedures and functions, they do share with Ada the ability to include declaration statements in local blocks of code. In C++, blocks are syntactically enclosed in bracket {...} symbols, and any declarations that occur between the symbols hold for the duration of the block. Consider the following code:

```
for (i = 0; i<20; i++) {
   int i = 1, j;
   j = 0;
   while (i < 25) {
      j += i*2;
      i ++;
   }
}
```

One might think that when the inner loop is done, the outer loop also will be done, because `i` has the value 26. But since the inner block of statements redeclared `i`, the scope rules state that the new, inner `i` was manipulated, leaving the outer `i` untouched and free to correctly manipulate the for loop.

In FORTRAN IV, the lifetime of all variables is the lifetime of the program. As a result, all memory in a program could be statically allocated, including activation records. Because each subprogram has only one activation record, FORTRAN could not support recursion.

Pascal, Ada, C, and C++ all support recursive functions and procedures. In order to support these, implementations generally rely on stack-allocated activation records. Thus, the lifetime of a local variable in a procedure p extends from the call of p to the return to its caller. Each of these languages also supports pointer or access types, generally representing data accessed from the heap (though C and C++ also allow pointers to stack-allocated memory). The lifetime of these variables is generally from the time that the programmer executes a creation instruction until a corresponding destruction statement is executed.

63.4.2 Execution Units

FORTRAN and Ada make a strong distinction between expressions and statements, with expressions simply returning a value, but with statements forming the basic unit for program execution. In C and C++, however, these two units of execution are merged, with statements treated as expressions. The statement `x = 5` assigns the value 5 to the variable x. But, in C++, the = sign is also an operator, and this assignment statement is actually an expression that returns the value being assigned. Thus, the statement `y = x = 5` assigns the value 5 to *both* x and y, because the value 5 is assigned to x and the expression *x* = 5 returns 5, which is assigned to *y*. Although interesting, it can also be very confusing. Because many expressions will have side effects; the order of evaluation will affect the value returned from an expression. Consider the code

```
if ((y = ++x) == (x + 6)) { … }
```

This code actually has two statements embedded in it; first, `++x` increments `x`, then this value is assigned to `y`, then the value assigned is tested against the value of `x + 6`. If the compiler decides to change the order in which the subexpressions are evaluated (a not unheard of occurrence in C++ compilers), it may change whether the guard on the if statement is true or false.

Allowing statements to be part of expressions also means that typographical errors are more likely to give rise to syntactically correct (but logically incorrect) statements. For instance, if one of the = signs in

```
if (x == 6) { … }
```

is omitted, then it will assign of value 6 to *x* and the conditional will always evaluate to true as all non-0 integers in C and C++ are treated as representing true.

63.4.3 Control Structures

Because FORTRAN was one of the earliest high-level languages, it is not surprising that its control structures are much closer to the underlying machine language instructions. Aside from the do loop (which was similar to the for loop in Pascal and Ada), most other control structures were based on the use of goto statements. Thus, the if statement of FORTRAN IV evaluated an integer expression and, depending on whether the result was negative, zero, or positive, resulted in a jump to one of three statement labels included with the statement. Aside from the usual goto statement, FORTRAN IV also included assigned and computed gotos, which provided some of the flexibility of case statements. FORTRAN 77 and the more recent FORTRAN 90 provide more modern control structures such as the if and while statements of other languages.

The control constructs in Ada are similar to those of Pascal, including if, case, while, and for loops, as well as indefinite loops, which are terminated with exit statements. Several of these were described earlier in the general discussion of control structures.

C and C++ include if statements and a switch construct that is similar to the case statement. The while loop is similar to that in Pascal and Ada, but for loops in C and C++ are more general than those in most other languages. For loops have the form

```
for (E₁; E₂; E₃) S;
```

where E_1 is initialization code, E_2 is a test for termination, E_3 contains code to update variables for the next iteration of the loop, and *S* represents the code to be executed each time through the loop. The test for termination is executed before the update code. Thus, a statement of the form

```
for (i = 1; i < 10, i++) S;
```

will result in *S* being executed once for each value of `i` from 1 to 10. (The expression *i* ++ is an expression that increments the value of `i`.) However, much more flexible statements are also possible:

```
for (i = 1; not done and i < 1024; i = 2 * i) S;
```

This statement repeatedly executes `S` while `i` ranges through the powers of 2 from 1 to 1024. If done is ever true, it will terminate early.

63.4.4 Procedural Abstraction

Each of Ada, FORTRAN, C, and C++ provides procedural abstraction. Ada and FORTRAN distinguish between functions and procedures, whereas C and C++ do not since procedures are just functions that return an element of type void. FORTRAN IV also supported single-line statement functions, which could be defined local to a program or subprogram. As noted earlier, Ada, C, and C++ all support recursive functions and procedures, whereas FORTRAN does not.

The languages differ in minor ways in how they return values from functions. FORTRAN, like Pascal, treats the name of the function as a pseudovariable that can be assigned to. An explicit return statement returns control to the calling program unit. When the function returns, the last value stored in the function name is returned as the value of the function. Ada, C, and C++ use return statements of the form return exp to return control to the calling program unit. The value of the expression associated with the return statement is the value returned from the function.

Most programming languages provide system-defined overloaded functions, such as arithmetic operators (+, –, *, etc.) and comparison functions (e.g., =, <). Ada and C++ are relatively unusual, though, in allowing user-defined overloading. In both, the compiler must be able to disambiguate at compile time whichever of the versions of the overloaded operator are called for at each of its occurrences.

C++ determines which version is called for by looking at the number and types of the actual parameters.

Ada goes further and can also use the return type to determine which version works in the particular context in which it is found. Thus, in Ada one may overload the + operator to take two integer parameters and return a user-defined rational type, even though there already exists a built-in version of + that takes two integer parameters and returns an integer. If + occurs in a context in which only an integer result would make sense, the built-in version would be selected. If + occurs in a context in which only a rational value would make sense, the user-defined version would be selected. If the system cannot tell which should be used, then an error will occur at compile time.

Unlike FORTRAN, Pascal, and C, both Ada and C++ provide language support for exceptions. Ada and C++ both use the termination model for program resumption after handling the exception.

63.4.5 Data Abstraction and Separate Compilation

FORTRAN provides support for separate compilation of subroutines and functions but provides no type checking across compilation unit boundaries. Thus, the main program may call a function *F* with two real parameters, but the definition of *F* in a separately compiled unit may have only one formal parameter, and it might be an integer. Because FORTRAN does not support the definition of new named types, it provides no support for ADTs.

C and C++ provide slightly better support for separate compilation by allowing the programmer to put external function and procedure declarations in a header file, which may be included into compilation units that use them. The header files are treated as though they were textually part of the compilation unit into which they are included. C provides no support for ADTs, though C++ does provide strong support through its class facilities. C++ classes give the programmer control over which aspects of a data type the user will be allowed to see and use. Because C++ is described in some depth in Chapter 64, we omit a detailed description here.

Standard Pascal provides no support for separate compilation, though virtually all Pascal implementations provide support for separately compiled units. These units typically include both interface and implementation sections. The interface of a unit may be explicitly imported into another compilation unit with a *uses* statement. This provides for separate, but not independent, compilation without requiring the programmer to create individual header files by hand. These units generally do not provide support for information hiding.

Ada provides both separate compilation and strong support for ADTs. Like MODULA-2's modules described earlier, Ada packages come in two separately compiled units, the specification and body. Only items listed in the nonprivate part of the package specification are accessible at compile time to units that import the package. The following is an Ada package specification for stacks:

```
package StackADT is
     type stack is private;
     procedure push(s: in out stack; elt: in integer);
     procedure pop(s: in out stack);
     procedure top(s: in stack) return integer;
     procedure empty(s: in stack) return boolean;
  private
     type stack is record
        top: integer := 0;
        values: array (1..100) of integer
     end record
  end StackADT;
```

The private section of a package specification is necessary to provide a description of private types. This is necessary so that importing programs know how much space to provide for a variable of that type. This is not as clean as the MODULA-2 solution, since any change of representation of the type will require the recompilation of the specification and hence of any program that imports the package. For this particular representation of stack, no initialization routine is necessary because top is initialized to 0 in the declaration of the type.

If the implementation of the private variable is a pointer, then only partial type information need be provided. That is, if we replace the private part of the preceding example by

```
private
  type stackRecord;
  type stack is access StackRecord;
end StackADT;
```

then this would provide sufficient information for importing programs to determine the memory needs for a variable of this type (i.e., the amount of space necessary to hold a pointer).

An implementation of the original package specification is given next:

```
package body StackADT is
  procedure push (s: in out stack; elt: in integer) is
    begin
      s.top := s.top + 1;
      s.value[s.top] := elt
    end push;
  procedure pop(s: in out stack); …
  function top(s: in stack): integer; …
  function empty(s: in stack) boolean; …
end StackADT;
```

The `StackADT` package specification can be imported into an Ada program unit by including with `StackADT` at the beginning of the unit. Components can be referred to as record components, for example, `StackADT.stack` and `StackADT.push`. The package name prefix can be omitted if use `StackADT` is also included at the beginning of the unit.

Both Ada and C++ provide mechanisms for supporting parameterized packages (or classes in the case of C++). The C++ template mechanism is quite primitive, with template instantiations being treated as being similar to compile-time macroexpansions. The template is never type checked, only its instantiations.

Ada also requires its generic packages to be instantiated at compile time, but the generics are type checked before, rather than after, instantiation. Thus, a generic package can be compiled and later used in another until that does not have access to the implementation.

The following is an example of the header of a generic `BinarySearchTree` package:

```
generic
  type Element is private;
  with function LessThan (x, y: Element) return boolean;
package BinarySearchTree is type BSTree is private;
  ...
end BinarySearchTree;
```

This can be used in another unit by instantiating it with a type and appropriate function, for example,

```
package PeopleDict is new BinarySearchTree(People, PeopleComp)
```

where `PeopleComp` is a function taking pairs of type `People` and returning a Boolean. `PeopleDict` can then be used like any other package. The ability to require generic package instantiations to include necessary functions and values as well as types ensures that they will not be instantiated with types which do not support the appropriate operations.

63.5 Research Issues and Summary

Research issues in imperative languages in recent years have tended to focus on many of the new constructs presented in this chapter. These include support for exceptions, iterators, ADTs, and parameterized or generic types. It is fair to say that most current research in programming languages is devoted to implementation and environment issues or to other programming paradigms. There are not many new concepts currently being introduced into imperative programming languages. Many languages that formerly were purely imperative have recently been extended to include object-oriented concepts (e.g., Object Pascal, Objective C, C++, Ada 95). Another series of extensions has provided features for concurrent and distributed programming.

From our earlier discussion, it is clear that support for abstraction plays an important role in imperative language design and use. Variables abstract away details of memory usage; data types (and in particular ADTs) abstract from the representation of values to provide support for operations that are independent from the actual implementation; execution units abstract away details of machine instruction execution and expression computation while providing clean interfaces for sharing information between caller and callee.

A second major focus in the development of modern imperative programming languages has been the enrichment of type systems, especially static type systems. ADTs can be understood as the enrichment of type systems with the so-called existential types, in which the existence of a type is revealed be instantiated with any type (or in Ada and CLU's case any type that comes supplied with the appropriate operations). These more flexible type systems allow for the construction of safe statically typed programming languages that are more expressive than their predecessors. There is hope that we are moving forward to a time when most programmers will see such secure languages as assisting them in their goal of creating correct and efficient software, rather than getting in the way. (See Chapter 69 for a further discussion of type systems.)

We have surveyed the class of programming languages modeled after the sequential organization of the von Neumann architecture. The imperative programming language paradigm is characterized by its sequential, stepwise statement execution.

As discussed in Section 63.4, the imperative programming constructs are implemented in a variety of ways in different languages. There are many languages to choose from; choosing the right language for the applications at hand is an important first step to software implementation.

It could be argued that the object-oriented paradigm is simply a minor variation on the imperative paradigm in which remote procedure and function calls replace the more familiar imperative calls. However, the object-oriented paradigm requires an entirely different way of thinking about the organization of a program, with the traditional conception of a program as a series of operations being applied to values replaced in the object-oriented view by an organization of more distributed responsibility. In this view, values (typically referred to as objects) are responsible for knowing how to perform their own operations, and the programmer is responsible for bringing together a group of objects with appropriate capabilities and organizing a program that relies on these distributed capabilities to accomplish a task. Subtyping and inheritance provide important organizing tools and promote code reuse in ways unavailable in traditional imperative languages.

Most programmers today are initially taught to program in imperative languages. Thus, these languages reflect the way that most programmers currently think about algorithm construction and program execution. Whether this will continue in the face of the challenge of the object-oriented paradigm will be interesting to see.

Key Terms

Abstract data type: A collection of data type and value definitions and operations on those definitions that behave as a primitive data type. The specifications of these types, values, and operations are generally collected in one place, with the implementations hidden from the user.

Binding: A connection between an abstraction used in the language and a data object as it exists in the computer hardware. The usage, establishment, and number of these bindings characterize the various imperative languages and affect their ease of use and performance.

Control structures: Structures or statements that alter the strict sequential ordering in an imperative program, presenting alternatives to sequential control. Control structures can be conditional, iterative, or unconstrained.

Derived type: A new data type constructed by copying a type that already exists. The resulting new type is distinct and not identified as being copied from the existing type, though operations on the old type are automatically inherited in the new type.

Identifier: The name bound to an abstraction.

Parameters: Data objects passed between the caller and the called procedural abstraction.

Procedural abstraction: Separating out the details of an execution unit in such a way that it may be invoked in a program statement or expression.

Scope rules: Rules in a language that define the area or section of a program in which a particular binding is effective.

Subtype: A new data type defined as a copy of another defined type, typically with a restricted subset of its values. It may generally be used in the same contexts as its parent type.

Type: A collection of values with an associated collection of primitive operations on those values.

Type equivalence: Rules that govern when variables or values from two different data types may be used together.

Variable: An abstraction used in imperative languages for a memory location or cell.

Further Information

A good examination of imperative languages, as well as other paradigms, can be found in the following texts:

Dershem, H. L. and Jipping, M. J. 1995. *Programming Languages: Structures and Models*, 2nd edn. PWS, Boston, MA.

Louden, K. C. 2003. *Programming Languages: Principles and Practice*, 2nd edn. PWS-Kent, Boston, MA.

Pratt, T. W. and Zelkowitz, M. V. 2001. *Programming Languages: Design and Implementation*, 4th edn. Prentice Hall, Englewood Cliffs, NJ.

Sebesta, R. 2013. *Concepts of Programming Languages*, 10th edn. Pearson, New York.

Sethi, R. 1996. *Programming Languages: Concepts and Constructs*, 6th edn. Addison-Wesley, Reading, MA.

Several journals are devoted to programming languages and language design. *ACM Transactions on Programming Languages and Systems* and *Computer Languages* both feature referred papers on programming languages. *ACM SIGPLAN Notices* is a collection of unreferenced papers from the ACM Special Interest Group on Programming Languages. Proceedings of the ACM conferences, *Principles of Programming Languages* (*POPL*), and *Programming Language Design and Implementation*, provide a good presentation of current research in programming languages.

References

Boehm, C. and Jacopini, G. 1966. Flow diagrams, Turing machines, and languages with only two formation rules. *Commun. ACM* 9(5):366–371.

Dijkstra, E. W. 1968. Goto statement considered harmful. *Commun. ACM* 11(3):147–148.

Liskov, B., Snyder, A., Atkinson, R., and Schaffert, C. 1977. Abstraction mechanisms in CLU. *IEEE Trans. Software Eng.* SE-5(6):546–558.

Liskov, B. H. and Guttag, J. V. 1986. *Abstraction and Specification in Program Development*. MIT Press, Cambridge, MA.

Marlin, C. D. 1980. *Coroutines*. Lecture notes in computer science 95. Springer-Verlag, New York.

64

Object-Oriented Language Paradigm

Raimund K. Ege
Northern Illinois University

64.1 Introduction

While invented in the 1960s, it was during the 1990s that *object-oriented programming* (OOP) established itself as the dominant programming paradigm. Although it was first viewed as a revolutionary new programming paradigm, such a characterization is only partly accurate.

OOP is, to be sure, a paradigm in the current sense of that term. It embodies a way of organizing and representing knowledge, "a way of viewing the world" (Budd 2002), that encompasses a wide range of programming activities, including program analysis, design, and implementation. The paradigm derives its power from its view of computation as the simulation of real-world entities. That is, according to Dan Ingalls, "Instead of a bit grinding processor … plundering data structures, we have a universe of well-behaved objects that courteously ask each other to carry out their various desires" (Ingalls 1981).

Central to this view of computation is the notion of self-contained little systems that work together. OOP tools and languages facilitate the description of *objects* as self-contained systems that maintain their own internal state (data), perform actions (*methods*) in their own interest, and interact with other objects (by sending *messages* to one another). Objects can be low-level programming tools, such as lists, stacks, and trees, akin to traditional abstract data types (ADTs). They can also be higher-level abstractions that reflect what a program is intended to model: an automated teller machine, a deck of playing cards, an elevator, or a collection of graphical objects on a screen.

The primary power of OOP derives from the fact that, once defined, objects enjoy a type-like status (as we will explain later, objects are defined via *classes*, which are very much like types). That is,

- An object can be used without knowing the details of its implementation
- An object's implementation detail is hidden and therefore protected
- Objects can be used according to a standard notation, using names, symbols, and operators in conventional ways
- Objects can be combined with other objects and types in expressive and efficient ways (composition and hierarchy) to define new, more complex types

Whereas structured programming languages such as C and PASCAL allowed a programmer to define new types, these were primarily a notational convenience. In such languages, user-defined type names served as shorthand to improve readability and as aids to compilers for recognizing type equivalences. These types, though, did not enjoy the status or flexibility of the built-in types. That is, one could not overload operators to apply to these new types and, more importantly, one could not easily hide the implementation details of a new type.

The seeds for the object-oriented paradigm can be found in the languages that developed in the 1970s and 1980s to provide more direct support for building user-defined ADTs. ADA-83, for example, allows one to build clearly specified, modular software components that effectively hide their implementation details. ADA-83 uses the package construct to describe type specifications and subprograms that can belong to a user-defined ADT. Languages that provide this level of support for ADTs are now called object-based.

In many ways, the evolution of the OOP paradigm extends that of the structured paradigm, which preceded it. We saw the structured paradigm, as manifested in a succession of increasingly high-level programming languages, progress from straight-line code laced with unconditional transfers of control, to block-oriented code that exploited control structures, to finer-grained code built from simple subprograms, to top-down, structured code that relied on parameterized, and separately developed libraries of code which distinguished a subprogram's interface from its implementation. The object-based paradigm followed to enable the capture of ADTs. Finally, OOP combines procedural and data abstraction into the notion of a class that can be part of a class hierarchy.

Hindsight being what it is, we can now see how the seeds of OOP have been nurtured by developments in many subfields of computer science (CS) over the past three decades: the basic idea of computation as simulation was first popularized in the programming language SIMULA 67, which also introduced the concepts of class, object, and message. Hardware architectures were developed in the 1970s using an object-based approach to describe the components of the machine and their interactions at a higher, more natural level. This development was further motivated by operating systems and applications that depended increasingly on graphical interfaces, which lend themselves directly to object-oriented descriptions. In the entity-relationship database model, in which a collection of information is described in terms of entities, attributes, and relations among entities, we see yet another manifestation of objects (at least the information content aspect of them). Even in knowledge representation schemes that supported work in artificial intelligence (frames, scripts, and semantic networks), we can clearly see this object orientation. Most clearly, the first purely object-oriented language, Smalltalk, has been around since the early 1970s.

Still, all of the premonitions of OOP did not coalesce into a recognizable influence on programming language and practice until they were motivated to do so by our collective practical experience in software engineering. As we became increasingly adept at exploiting the procedural paradigm and its languages, we recognized more directly their shortcomings. The more we pushed that paradigm, the more difficult it was to support data abstraction to that same degree that algorithm abstraction could be supported. Furthermore, methods of encapsulation were relatively primitive, thus making it difficult to develop safe, reusable code. Finally, software analysis and design techniques that relied on procedural abstraction as the basis for performing decomposition of complex systems were proving increasingly awkward to apply to domains that were increasingly dependent on data, as opposed to algorithms.

In summary, structured methods flourished in the era of application domains that were either algorithm-centric or information-centric. These same methods prove less useful in an increasingly complex application world, where functionality and information structure must be considered in context. The concept of an object is a simple and elegant combination of all informational and algorithmic aspects of an application entity.

Today, all of these technologies have matured and all of the motivations have been established to the extent that OOP is a viable, recognizable, and popular approach to software development. Programming languages and commercial development environments abound. Smalltalk, as an example of a pure OOP language, is still commercially available; however, Java and C# have become the poster children for pure OOP languages. C++ is also commonly referred to as an object-oriented programming language, while strictly it is a hybrid, that is, it extended an existing language with essential OOP features. Other notable OOP languages are Eiffel, and the hybrids Objective-C and Visual Basic.

Because OOP is rightly referred to as a paradigm, it has spawned the development of many software analysis and design techniques that support the identification and description of objects in a problem specification. As with all paradigms, OOP languages and techniques are best suited for problems that match the paradigm's view of the world. The use and popularity of OOP rises with the increased demand for complex, interface-intensive systems, ones that can be modeled in real-world terms.

Finally, if OOP raises the level of abstraction to bridge the gap between programmer and machine, it may be that novice programmers would stand to benefit the most from its use. Indeed, Smalltalk was developed based on research detailing how young children describe and interact with the world in solving problems. OOP is now influencing significantly how we teach and learn programming. There is a growing recognition that object orientation makes learning to program easier for the novice. Many universities are teaching object orientation as part of the CS introduction. Java is widely used throughout the CS curriculum.

However, it is also recognized that there is a pool of experienced programmers and software engineers that must be retrained significantly: not because object orientation is hard, but because of their experience in traditional, function-centered (or information-centered) problem solving. These programmers must, as Bertrand Meyer puts it, reacquire an object-oriented frame of mind (Meyer 2000).

Proponents of OOP claim that the paradigm represents the state of the art in terms of bridging the language gap between programmer and machine. It offers the prospect for achieving many of the software quality goals to which all programmers aspire: easily designed, safe, efficient, and uniform software.

64.2 Underlying Principles

OOP takes a quite natural, but also quite different, view of programming. In a process-state view of computing (which reflects both the machine's fetch-execute cycle and its data processing nature), one thinks about programs as lists of instructions executed sequentially, and serving to manipulate and change the state of memory is the fundamental strategy of imperative languages. In contrast, OOP sees a software system as set of collaborating objects. Old-style programming involves thinking in terms of algorithms and data structures separately. OOP languages, on the other hand, permit the programmer to take a broader view and think of a program as a collection of cooperating objects, each of which encapsulates both structure and function.

Each object has three aspects: what it is, what it does, and what it is called. We illustrate these aspects by considering a geometric object, a circle. Most programming languages do not have circles as built-in data type, so if we want to write a program that manipulates circles, we must write our own description of circle objects in a program. The descriptive information associated with an object, called its state, includes its properties and the values of these properties. For most objects, the state properties do not change, although the state values may very well be modified during the object's lifetime. For instance, a circle has a radius and a center, and at a given time, these might have the values 4.67 and (0, 0). A circle object will always have a radius and a center, although their values will change if the circle's size or location changes.

The collection of actions an object may perform on itself is called its behavior. In our example, we may want to use a circle as part of a program that draws on a computer screen, and so we would need to be able to set the center and radius of the object and instruct the circle to draw itself on the screen. Notice that we said a circle would draw itself on the screen. This is an important feature of the object-oriented approach. In a language with an algorithmic approach, one could describe circles by radius and center, as we have, but the description of the methods to perform on a circle would not be as tightly associated with the state data as it is here. That is, the drawing method would not be part of the circle object itself.

Encapsulating data and methods together in an object helps us design and understand a program because, for instance, all circle-related data and methods are collected in one place. If we had to modify a circle object's definition, we would know exactly where the modifications would go. Furthermore, we would know that we would not have to modify any other part of the program.

Finally, each object has an identity, which serves to distinguish it from all others. One can refer to an object uniquely via its identity. In most programming languages, we identify an object by giving it a name, so we might call our circle *theCircle—theCircle* is our reference to the object.

It is entirely reasonable to assume that a program that would use one circle object might use several. All of these circle objects would be similar, in that they would have the similar collection of data and methods. It would clearly be a wasteful duplication of effort to describe each separately, since they would differ only in their data values and names. Object-oriented languages allow one to describe a stencil, if you will, for an entire set of objects. Such a stencil is called a *class*.

A class is used to generate a set of objects that share common properties. Continuing our example, we could describe a class, circles, by describing the state properties (but generally not the state values) and behaviors of all objects that belong to that class. Then, we could describe as many circle objects as we needed merely by declaring that they are *instances* of the circles class, that is, they are circle objects. An object can never exist in isolation; every object must be an instance of some class. This means that if we are going to use an object in a program, we must first provide the class to which the object belongs.

One can think of a class as a means for implementing an ADT. That is, to describe a class, we identify the properties that any instance of the class (a particular object) must have. These properties consist of state information (referred to as data fields) and a collection of behaviors (referred to as methods) that such objects are capable of performing. Methods typically involve accessing, setting, or otherwise manipulating the object's data fields.

Methods are invoked by an object in response to a message (i.e., a request of the object to perform one of its methods). Message passing as a means for invoking methods differs subtly from subprogram invocation in most other programming paradigms. In OOP, there is a formal distinction between the subprogram call (the sending of a message) and the invocation of the subprogram (when the message is received). It is the responsibility of the receiver of a message to interpret the message, that is, to determine which of its methods to perform in response. Think of OOP as programming by sending messages to objects.

Taken together, these two mechanisms support the description of full-fledged ADTs in a more general and extensible way than was provided by the first object-based languages. In particular, classes provide an abstraction mechanism that can be used to model a wide range of information structures and information processing agents. Using classes as the primary descriptive vehicle also characterizes an OOP approach to program design that has proved extremely effective for many real-world applications.

Not only are classes a rich descriptive tool they are also safe, in the software engineering sense of that term. That is, classes effectively encapsulate the abstractions they model by controlling (either implicitly or explicitly, depending on the language being used) access to an object's data fields and methods. Some of an object's data fields and methods will be public in nature (visible and accessible to other objects), and some will be private to the object itself (visible and accessible only to other methods). Many OOP languages also support a formal distinction between a class's interface and its implementation.

Classes are also an efficient means of representation by virtue of the fact that they can be related to one another both compositionally (an object of one class can serve as a member field of an object of another class) and hierarchically (to express is-a relationships). This latter feature is unique to OOP, and it affords a means for describing directly any information that is hierarchical in nature.

In cases where classes are arranged in a hierarchy, a *subclass* is said to *inherit* from its *superclass*. That is, instances of the subclass contain (either directly as copies or indirectly via the superclass) both the data fields and methods described by the superclass. These same general properties are associated with all subclasses of the particular superclass. Each subclass extends the descriptions of its superclass by adding specialized data fields and methods that differentiate it from the superclass and from other subclasses.

Going back to our geometric example, we do not have to limit our objects to just circles; we could equally well have objects that are rectangles, triangles, and so on. We could augment our program by including declarations and definitions for the classes rectangles and triangles, just as we did when we defined the circles class. In doing so, we notice that all of these are what we might call geometric objects.

Object-oriented languages allow a class like rectangles to inherit the data fields and methods from a parent class, such as GeometricObjects. For example, because every object in our hierarchy has an extent (a bounding rectangle inside which the object fits as closely as possible), we could declare the method ReportBounds and some appropriate data fields within a superclass named GeometricObjects. These would apply to all the subclasses rectangles, triangles, and circles. Doing this, we would not need to re-declare the method and data fields in any of these subclasses. They would be inherited from the superclass GeometricObjects. Indeed, it is sometimes useful to define a superclass, such as GeometricObjects, knowing full well that we will never be interested in generating direct instances of it. That is, the only instances of GeometricObjects that we will use in our programming are those that are also instances of its subclasses: circles, triangles, or rectangles. In such a case, the superclass is referred to as *abstract class*.

Inheritance is not only distinctive of OOP languages it is also the means by which the paradigm addresses many software engineering concerns. For example, inheritance is an efficient, non-redundant notation for representing information that is hierarchical in nature. It is natural in the sense that it reflects how we humans tend to describe such information (which explains in part why OOP has spawned so many related program design techniques). The efficiency of the notation derives from the ability to reuse code that was used in defining a superclass when defining a subclass. This, in turn, allows the inherited data fields and methods of classes that are derived from one another to project consistent interfaces to their consumers. That is, data and methods can be referred to by the same names in both the super- and subclasses.

There are times, though, when a superclass method or state might not be appropriate for use by a subclass. All of the classes in our hierarchy would have draw methods, but drawing a rectangle might involve different strategies than drawing a circle. OOP languages allow a derived class to redefine, or *override*, a method inherited from a base class. If, for example, we have an object called *theShape* and send it a message to draw itself, how is the computer to know which draw method to use? Whereas procedural languages and object-based ones would resolve this ambiguity statically (i.e., based on the declared type of *theShape*), an OOP language solves this problem at run time, by looking up the class to which *theShape* belongs and finding an appropriate draw method.

This ability to use the same name for methods on objects of different classes is an example of a kind of *polymorphism* (called inclusion polymorphism) that is common to all OOP languages. In the OOP sense of the term, polymorphism describes the fact that a superclass includes all instances of its subclasses. Thus, the notion of a geometric object contains all circle, rectangle, and triangle objects. As a result, a generic entity can take on different types of information during the execution of a program. For example, a geometric object can be a circle, a rectangle, or a triangle at any point during our program.

The concept of interface is related to the polymorphic use of the class concept. An interface is a listing of methods where, for each method, one only lists its name and any potential parameters. OO languages allow classes to implement an interface, that is, the classes provide explicit bodies for all methods listed in the interface.

Other forms of polymorphism often supported in OO languages include the following:

- Overloading: two or more functions or operators can share the same name, so the + operator is allowed to apply to both integers and real numbers.
- Parametric polymorphism: a parameter is used to establish choices: for example, C++ class templates use parameters to allow another level of abstraction.
- Coercion polymorphism: C type casting is an example.

The following characteristics are the essence of OOP:

Objects—As a means for encapsulation of state and behavior
Classes—As stencil for generating objects
Message passing—As the mechanism by which computation takes place
Dynamic method invocation—As the means by which methods are bound and interpreted
Inheritance—As a technique for modifying classes and creating subclasses
Inclusion polymorphism—As a means for allowing expressions of one type to be used in place of expressions of a supertype

Each contributes significantly to the overall utility of the paradigm, and each allows the paradigm to address one of the many software engineering concerns that motivated it. Whereas different programming languages implement them in various combinations and to varying degrees, any language that implements them all is considered object oriented.

64.3 Impact on Practice

To illustrate these characteristics of OOP, let us construct a very simple application to deal with queues of packets as they might appear in a network simulation. Packets in our program maintain their names and priorities and allow their observation and comparison. Different kinds of packets are modeled as subclasses: one for packets that carry protocol information (Ack) and one for packets that carry data (data). The subclasses specialize how packets are observed. The second set of classes models the queue concept. Class FifoQueue represents the algorithmic and data abstraction of a standard first-in-first-out queue. Internally, it employs a doubly linked list—anchored by a head and tail data field—to maintain the packets currently queued. The methods enter and remove implement the standard protocol of such a queue. The FifoQueue class ignores the packet's priority information but also serves as a superclass for two additional subclasses, PriQueue and QueuePri, which use the packet's comparison abilities to handle packets of different degrees of importance.

We will develop the example in terms of Java, C++, C#, and Objective-C. The four major OOP languages are in use today. Our intention here is to provide quick overviews of these languages and to illustrate the different notations and styles for implementing the object-oriented paradigm.

64.3.1 Java

Java is perhaps the most compelling entry into the landscape of OOP languages. Java was developed in the mid-1990s by James Gosling at Sun Microsystems (which is now part of Oracle Corporation) (Gosling et al. 2005). The language was positioned to empower developers to write software independently of specific platforms: "Write once, run anywhere" was a slogan to illustrate the cross-platform benefits of the Java language. Today, Java is one of the most popular programming languages, especially for client-server web applications.

From an object-oriented perspective, Java can be seen as a distillation of many of the good features from Smalltalk and C++. From C++, it inherits its style of syntax, but with great simplifications. From Smalltalk, it inherits its execution model: all objects carry a unique identifier, and all methods are dynamically bound. It is compiled into byte codes that are interpreted within the target environment, and all class information is available at run time, which provides additional type-checking capability and robustness. Java also provides automatic garbage collection, which simplifies a programmer's task significantly and tends to reduce many errors related to memory management. And of course, Java comes with a large class library that contains support for graphical user interfaces (GUIs), networking, web application development, and much more.

Let us develop a Java program for our network/packet/queue example. We start by describing the class *packet*:

```
public class Packet {
    int priority;
    String name;
    public Packet(String n, int p) {
        name = n;
        priority = p;
    }
    public void list() {
        System.out.print(name + " packet: ");
    }
    public boolean more(Packet other) {
        return priority > other.priority;
    }
    public boolean less(Packet other) {
        return priority <= other.priority;
    }
}
```

The basic syntactic style is borrowed from the C programming language. The keyword *class* precedes the name *packet* of the class that is defined here. The keyword *public* denotes that this class is visible beyond the scope of the package in which it is defined. The *packet* class defines two data fields and four methods. Java data fields are explicitly typed: *priority* is of atomic type *int for integer, name is of class String.* The first method is special, it has the same name as the class: *Packet*. It is called a *constructor* and serves to initialize a new instance of the class when it is created. The constructor parameters (n and p) are used to give initial values to the data fields *priority* and *name*. The method *list* prints the name of the packet followed by some text. The boolean methods *more* and *less* implement the comparison capability of packet objects. All four methods use the keyword *public* to denote that they are accessible or visible outside of the packet class. Data fields *priority* and *name* do not have an access specifier, that is, they are only visible within the current package.

Java insists on a closed-class hierarchy: all classes must have a superclass. If a superclass is not specified as in the packet class declaration, then it defaults to class object. In effect, all Java objects can be thought of as instances of that class. This superclass enables broad run-time support, such as automatic garbage collection.

Class *Ack* is defined as a subclass to *packet*. The *extends* clause defines the inheritance relationship among classes. Java allows a class to have a single superclass only.

```
public class Ack extends Packet {
    public Ack() {
        super("Ack",10);
    }

    public void list() {
        System.out.println(" acknowledged");
    }
}
```

It defines two methods: the constructor and *list*. The constructor uses the keyword *super* to invoke the constructor of the superclass. The *list* method redefines the *list* method that would normally be inherited from the *packet* class.

Class *data* is just slightly more complicated: it also extends class *packet*. It defines a data field *body* of class *string*. The two methods are the constructor and a redefined version of *list*.

```
public class Data extends Packet {
     private String body;

     public Data(String b, int p) {
          super("Data", p);
          body = new String(b);
     }

     public void list() {
          super.list();
          System.out.println(body);
     }
}
```

The keyword *super* refers to the method of the same name in the superclass. Method *list* first invokes the superclass version of *list* and then adds the string stored as *body*.

The *packet* class hierarchy is now complete. Before we construct our queue class, we need one more helper class: *node*, which will be used to implement a doubly linked list. The *node* class defines three data fields: *value* to hold a packet object, *next* to refer to the next node in the list, and *previous* to refer to the previous node in the doubly linked list. The *node* constructor sets *value* packet and initializes the *next* and previous *links*.

```
class Node {
     Packet value;
     Node next, previous;
     Node(Packet p) {
          value = p;
          next = previous = null;
     }
}
```

Notable here are the types associated with fields *next* and *previous*. Both are defined as being of type *node*. This does not mean that a *node* object will contain other *node* objects. They will, however, contain the object identifiers of other *node* objects. Thinking of object identifiers as pointers to objects yields the conventional linked list metaphor. Moreover, a true OOP language does not need pointers at all. Since all objects carry their unique identity, it can be used instead. That is why Java need not support pointers.

Now, we can construct the class *FifoQueue* to implement our first-in-first-out queue abstraction. Class *FifoQueue* defines two data fields: *head*, to refer to the beginning of the list of nodes, and *tail*, to hold on to the end of the list. Fields *head* and *tail* are specified as *protected*, which extends their visibility to all subclasses. Protected data fields or methods are accessible in subclasses, whereas private ones are not. The class also defines a constructor and three public methods: *enter*, *remove*, and *list*. The constructor simply initializes the *head* and *tail* to *null* to create an empty queue.

```
public class FifoQueue {
     protected Node head, tail;
     public FifoQueue() {
          head = tail = null;
     }

     public void enter(Packet it) {
          Node tmp = new Node(it);
          tmp.previous = tail;
```

```
        tail = tmp;
        if (head != null)
            tmp.previous.next = tail;
        else
            head = tmp;
    }

    public Packet remove() {
        Packet it = head.value;
        if (head == tail)
            head = tail = null;
        else if (head.next != null) {
            head = head.next;
            head.previous = null;
        }
        return it;
    }
    public void list() {
        for (Node tmp = head; tmp != null; tmp = tmp.next)
            tmp.value.list();
    }
}
```

Method *enter* first creates a new *node* object *tmp* to hold the parameter packet object *it*. Object *tmp* is then added to the end of the existing linked list. If the list is empty, that is, *head* is *null*, then it will become the only object in the list. Method *remove* first captures the packet that is to be removed and stored as *value* of the *head* node. It then removes the first node from the linked list and reconnects the links. Method *list* walks through all nodes in the list and sends message *list* to the packet stored in the node's value.

The next class doubles as the main program. Java does not allow the definition of stand-alone functions. The class has a single static method called main, which creates a few objects and starts execution. The purpose of our main program is to exercise and test our queue class. The main function defines five objects: two data packets, two Ack packets, and one FifoQueue. Objects are entered into the queue in this order: d1, a1, d2, and a2.

```
class Main {
    public static void main(String args[]) {
        Data d1 = new Data("first packet", 5);
        Ack a1 = new Ack(), a2 = new Ack();
        Data d2 = new Data("second packet", 6);

        FifoQueue q = new FifoQueue();

        q.enter(d1);
        q.enter(a1);
        q.enter(d2);
        q.enter(a2);

        System.out.println("The FIFO queue:");
        q.list();

        System.out.println("Order of leaving:");
        q.remove().list();
        q.remove().list();
        q.remove().list();
        q.remove().list();
}
```

Once all four packet objects are entered, sending message *list* to object *q* will produce this output:

```
The FIFO queue:
Data packet: first packet
     acknowledged
Data packet: second packet
     acknowledged
```

which is followed by:

```
Order of leaving:
Data packet: first packet
  acknowledged
Data packet: second packet
  acknowledged
```

Not surprisingly, the packet objects are stored by the queue object in the order in which they were entered, and then they are released in exactly the same order. Since our queue class was meant to implement a first-in-first-out sequence, we notice that our programming effort was successful.

To continue our example, we can now extend our queue abstraction by taking the importance (i.e., priority) of packets into account. Class *PriQueue* is defined as a subclass of *FifoQueue*. It inherits all data fields and methods, but specializes the enter method:

```
public class PriQueue extends FifoQueue {
     public void enter(Packet it) {
          Node tmp = new Node(it);
          if (tail == null) {
               tail = tmp;
               head = tmp;
          } else {
               Node p = tail;
               for (; p != null; p = p.previous)
                    if (it.less(p.value))
                         break;
               if (p == null) {
                    tmp.next = head;
                    head.previous = tmp;
                    head = tmp;
               } else {
                    tmp.previous = p;
                    tmp.next = p.next;
                    if (p == tail)
                         tail = tmp;
                    else
                         p.next.previous = tmp;
                    p.next = tmp;
               }
          }
     }
}
```

The implementation of *enter* for priority queues is significantly more complex. It traverses the existing linked list for each packet that is to be entered to determine its relative importance. The *enter* method uses the *less* method defined for packet objects.

While the *PriQueue* class specializes the *enter* instance method of class *FifoQueue*, another way of providing a queue with priority handling would be to specialize the *remove* method rather than *enter*. Class *QueuePri*, described next, does that:

```
public class QueuePri extends FifoQueue {
    public Packet remove() {
        Node it = head;
        if (head == tail)
            head = tail = null;
        else {
            // find the highest priority
            for (Node p = head; p != null; p = p.next)
                if (p.value.more(it.value))
                    it = p;
            if (it == tail) {
                tail = it.previous;
                tail.next = null;
            } else if (it == head) {
                head = it.next;
                it.next.previous = null;
            } else {
                it.previous.next = it.next;
                it.next.previous = it.previous;
            }
        }
        return it.value;
    }
}
```

The strategy for implementing a priority queue is straight forward. Whenever a packet is to be removed from the queue, we traverse the linked list and determine which of the packets has the highest priority. The good news is that *FifoQueue*, *PriQueue*, and *QueuePri* objects can now be used interchangeably, depending on what kind of queuing strategy is desired. All three classes provide the same protocol; that is, their objects understand the same set of messages.

In summary, Java is a complete and truly OOP language. It also supports a very open and flexible style of encapsulation via the access specifiers *public*, *protected*, and *private* and an unspecified default: package. All data fields and methods are of access right package unless explicitly stated otherwise. All classes in Java belong to packages, and all package-defined fields and methods are accessible from within any class within the package. Java also uses packages to organize its source and compiled code. Java also provides garbage collection as a means to reclaim obsolete objects. Java does not have a delete operator: objects cannot be deleted explicitly.

Java does not support multiple inheritance, where a class can have more than one superclass. Java supports the notion of interface, a specification of the public methods that an object can respond to. The interface does not include any data fields or method bodies. Multiple inheritance is supported for interfaces. Classes can be declared to implement interfaces. Interfaces allow the programmer to establish declared relationships between modules, which in turn can change their underlying implementation as the class changes.

64.3.2 C++

Our next OOP language is C++ developed at Bell Laboratories. C++ is an evolutionary enhancement to the C language to support OOP features. It is most accurately described as a hybrid language (i.e., one that extends a traditional procedural language and remains compatible with that language, to include a variety of OOP features). Basically, this is C with classes, inheritance, and polymorphism. As Bjarne Stroustrup, the designer of the language, likes to point out, C++ is not an OOP language, but a language with which one can write an object-oriented program (Stroustrup 1994).

In C++, a class specification is separated into two files: a header file that contains a class declaration (listing the class name, fields, and method signatures) and a body file that contains all method bodies. C++ uses the C preprocessor to unite the two files upon compilation. It also enables that client code need only to include the header file to be able to use a class.

As our first example, we list the header for class *packet*. The keyword *public* indicates that the methods, now called member functions in C++ terminology are publicly visible. The data fields (or, as referred to in C++, member fields) are private by default. That is, they are not visible outside of this class.

```
class Packet {
     const char *name;
     int priority;
public:
     Packet(const char *n, int p) :
          name(n), priority(p) {
     }
     virtual void list();
     int operator>(Packet &);
     int operator<=(Packet &);
};
```

The class header declares the member fields and member functions, along with their accessibility. *Priority* is defined as an integer number (*int*); *name* is a string (*const char* *). The comparisons are defined as operator functions—the ampersand (&) denotes call by reference. The header shows one other important member function: it has the same name as the class and is referred to as the constructor. The constructor serves to initialize a new instance of the class when it is created. Here, a new packet needs a name and a priority. The keyword *virtual* in the declaration of member function *list* enables dynamic binding in C++. It is needed to allow inclusion polymorphism to work. Without it, C++ will use static binding, as in conventional programming languages.

The class body elaborates the bodies of the member functions:

```
int Packet::operator>(Packet &other) {
     return priority > other.priority;
}

int Packet::operator<=(Packet &other) {
     return priority <= other.priority;
}
void Packet::list() {
     cout << name << " packet: ";
}
```

where *cout* is the predefined output object in C++. It receives message << and prints out its parameters. Subclass *Ack* is very simple:

```
class Ack: public Packet {
public:
     Ack() :
          Packet("Ack", 10) {
     }
```

```
    void list() {
        cout << " acknowledged\n";
    }
};
```

Its constructor uses a special syntax to invoke the constructor of its superclass packet. This class header also shows that a member function can (if it is short enough) be fully defined right away.

Similarly, for class data:

```
class Data: public Packet {
    const char *body;
public:
    Data(const char *b, int p) :
        Packet("Data", p), body(b) {
    }
    void list() {
        Packet::list();
        cout << body << endl;
    }
};
```

The constructor for class *data* is defined with two parameters: one to initialize the *body* field and one for the *priority* field. The member function list refers to *packet::list()*, which is the list function defined in the *packet* class. Java uses the simpler approach of using the keyword *super* to refer to a method defined in a superclass. The word *endl* ensures that the output ends with a new line.

The next class, *FifoQueue*, first needs the definition of class *node*:

```
class Node {
    Packet & value;
    Node *next, *previous;
    Node(Packet &p) :
        value(p) {
        next = previous = NULL;
    }
    friend class FifoQueue;
};
```

The member fields of class *node* default to private. Since class *FifoQueue* needs to be able to access these fields, we include a *friend* clause to extend their visibility to class *FifoQueue*.

Here now is class *FifoQueue*

```
class FifoQueue {
protected:
    Node *head, *tail;
public:
    FifoQueue() :
        head(NULL), tail(NULL) {
    }
    ;
    virtual void enter(Packet &);
    virtual Packet& remove();
    void list();
};
```

The member fields *head* and *tail* are specified as *protected*. C++ allows explicit control whether subclasses have access to inherited members. Protected members are accessible in subclasses, whereas private members are not. The body of class *FifoQueue* details the member functions *enter*, *remove*, and *list*:

```
void FifoQueue::enter(Packet &it) {
    Node *tmp = new Node(it);
    tmp->previous = tail;
    tail = tmp;
    if (head)
        tmp->previous->next = tail;
    else
    head = tmp;
}
Packet& FifoQueue::remove() {
    Packet &it = head->value;
    if (head == tail)
        head = tail = NULL;
    else if (head->next) {
        head = head->next;
        head->previous = NULL;
    }
    return it;
}
void FifoQueue::list() {
    for (Node* tmp = head; tmp; tmp = tmp->next)
        tmp->value.list();
}
```

We can now exercise our objects. This program produces the same output as our previous example written in Java:

```
int main() {
    Data d1("first packet", 5);
    Ack a1, a2;
    Data d2("second packet", 6);
    FifoQueue *q = new FifoQueue();
    q->enter(d1);
    q->enter(a1);
    q->enter(d2);
    q->enter(a2);
    cout << "The FIFO queue:\n";
    q->list();
    cout << "Order of leaving:\n";
    q->remove().list();
    q->remove().list();
    q->remove().list();
    q->remove().list();
}
```

Again, we can use *FifoQueue* as the base class for subclass *PriQueue* to refine the *enter* member function:

```
class PriQueue: public FifoQueue {
public:
    void enter(Packet &) {
```

```
        // logic as before
    }
};
```

In addition, of course, a similar class *QueuePri* is also possible, as follows:

```
class QueuePri: public FifoQueue {
public:
    Packet& remove(){
        // logic as before
    }
};
```

In summary, C++ provides detailed support for specifying the degree of access to its members. C++ goes beyond what we have illustrated so far—it allows one to specify the type of inheritance that is used: public, protected, or private. All our examples use public inheritance, which propagates the accessibility of members to the subclass. Protected and private inheritance allows one to hide the fact that a class is based on a superclass. C++ supports both single and multiple inheritance. It requires that dynamic binding (i.e., the object-oriented behavior of an object to search for a suitable method for a message at run-time) be explicitly requested per member function. C++ uses the keyword *virtual* to request dynamic binding; otherwise, it defaults to static binding. C++ leaves memory management to the programmer, as garbage collection is not supported.

64.3.3 C#

C# (pronounced C sharp) is the latest significant entry into the landscape of OOP languages. C# was developed by Anders Hejlsberg of Microsoft as part of its .NET initiative (Hejlsberg et al. 2010). From an object-oriented perspective, C# can be seen as a distillation of many of the good features of Java and C++, plus influences from ObjectPascal.

Like Java, C# is compiled into byte codes from a common language run-time specification. All class information is available at run time, which provides additional type-checking capability and robustness. C# also provides automatic garbage collection, which simplifies a programmer's task significantly and tends to reduce many errors related to memory management. In addition, of course, C# comes with a large class library, called common language infrastructure, which contains support for common data structures, GUIs, database access, networking, and much more.

Consider this C# version of our *packet* class:

```
public class Packet {
    private int priority;
    private String name;
    public Packet(String n, int p) {
        name = n;
        priority = p;
    }
    virtual public void list() {
        Console.Write(name + " packet: ");
    }
    public bool more(Packet other) {
        return priority > other.priority;
    }
    public bool less(Packet other) {
        return priority < other.priority;
    }
}
```

The resemblance to Java is clear. The basic syntax, including declarations and control structures, use the C++ style. All object handling is done by reference. Access specifiers (like private, protected, or public) are listed per field or method. As in C++, dynamic binding must be requested using the virtual keyword. C# comes with a significant set of predefined classes to allow input and output. The example lists the *console* class, which defines the *write* method. C# insists on a closed-class hierarchy: all classes must have a superclass. If a superclass is not specified in the class declaration, then it defaults to class object.

A C# subclass specification resembles C++ more closely than Java. In the following example, class *Ack* is defined as a subclass of *packet*. The colon is used to designate the superclass. The constructor uses the *base* reference to denote the invocation of the superclass constructor. The *list* method is explicit about redefining the superclass's virtual list method by using the keyword *override*. If the superclass did not have a list method, the compiler would flag an error.

```
public class Ack : Packet {
     public Ack() : base("Ack", 10) { }
     override public void list() {
          Console.WriteLine(" acknowledged");
     }
}
```

Class *data* is just slightly more complicated. It inherits from class *packet* and redefines the *list* method. The keyword *base* is used to refer to methods defined in the superclass.

```
public class Data : Packet {
     private String body;
     public Data(String b, int p): base("Data", p) {
          body = String.Copy(b);
     }
     override public void list() {
          base.list();
          Console.WriteLine(body);
     }
}
```

Again, before we can define the *FifoQueue* class, we need the *node* class. Class *node* is defined here as a simple class with a data field of class *packet*. This limits our queues to contain only *packet* objects. The fields *next* and *previous* embody out doubly-linked list implementation. Like Java, C# does not need pointers.

```
public class Node {
     public Packet value;
     public Node next, previous;
     public Node(Packet p){
          value = p;
          next = previous = null;
     }
}
```

Class *FifoQueue* makes use of these classes to implement our first-in-first-out queue abstraction. The code closely resembles Java. The keyword *virtual* ensures that subclasses can override the methods, and that dynamic binding is used.

```
public class FifoQueue {
     protected Node head, tail;

     public FifoQueue() {
```

```
            head = tail = null;
        }
        public virtual void enter(Packet it) {
            Node tmp = new Node(it);
            tmp.previous = tail;
            tail = tmp;
            if (head != null)
                tmp.previous.next = tail;
            else
                head = tmp;
        }
        public virtual Packet remove() {
            Packet it = head.value;
            if (head == tail)
                head = tail = null;
            else if (head.next != null) {
                head = head.next;
                head.previous = null;
            }
            return it;
        }
        public void list() {
            for (Node tmp = head; tmp != null; tmp = tmp.next)
                tmp.value.list();
        }
}
```

As in the Java version of the example, our main program is a class with a static method *main*, where we assemble a few objects and send messages to test our *packet* and *FifoQueue* classes:

```
class Main {
    public static void Main(String[] args) {
        Data d1 = new Data("first packet", 5);
        Ack a1 = new Ack(), a2 = new Ack();
        Data d2 = new Data("second packet", 6);

        FifoQueue q = new FifoQueue();
        q.enter(d1);
        q.enter(a1);
        q.enter(d2);
        q.enter(a2);

        Console.WriteLine("The FIFO queue:");
        q.list();

        Console.WriteLine("Order of leaving:");
        q.remove().list();
        q.remove().list();
        q.remove().list();
        q.remove().list();
    }
}
```

This program produces the same output as in our Java discussion. Classes *PriQueue* and *QueuePri* can be defined as follows. Both classes inherit from *FifoQueue*, one redefining method *enter* and the other *remove*. Because the superclass did define both methods as virtual, C# needs the keyword *override* here. Dynamic binding is enabled for the *enter* and *remove* methods, that is, given a variable of type *FifoQueue* that holds a reference to a *PriQueue*, it would respond to an incoming message *enter* with the correctly redefined, that is via *override*, method defined in the *PriQueue* class.

```
public class PriQueue : FifoQueue {
     override public void enter(Packet it) {
          // logic as before
     }
}
public class QueuePri : FifoQueue {
     override public Packet remove() {
          // logic as before
     }
}
```

In summary, C# is a complete and truly OOP language. C# also provides garbage collection as a means to reclaim obsolete objects. C# also supports the notion of interface, a specification of the public methods that an object can respond to. The interface does not include any data fields or method bodies. Multiple inheritance is supported for interfaces. Classes can be declared to implement interfaces. Interfaces allow the programmer to establish declared relationships between modules, which in turn can change their underlying implementation as the class changes.

C# supports some additional novel features: it allows programs to treat values of atomic types, such as *int* or *char*, as objects with a process called boxing. Boxing automatically converts an atomic value that is being stored in the execution stack to wrapper objects that are allocated on the heap and referenced from the stack. The reverse process, unboxing, is also done automatically. The advantage of this boxing feature is that values of atomic types can be passed by reference as parameters to methods.

64.3.4 Objective-C

Objective-C is another interesting entry into the landscape of OOP languages. It was developed in the early 1980s by Brad Cox and Tom Love. Like C++, it extends the C language with OOP concepts. However, its style is heavily influenced by Smalltalk. Cox relates the concept of class to that of an integrated circuit (IC), where "Software ICs", that is off-the-shelf classes, could dramatically increase software productivity (Cox et al. 1991).

While Objective-C is an old OOP language, it has only recently come out of obscurity by virtue of being the main programming language in the Apple OS X and iOS operating systems. The Apple Cocoa framework and class library is written in Objective-C.

An Objective-C class is written in two separate parts: a class interface and a class implementation. The interface lists the class name, its superclass, data fields, and method headers. Our *packet* class interface looks like this:

```
@interface Packet : NSObject {
     int priority;
     char *name;
}
+ (Packet *) new: (char *) n pri: (int) p;
- (BOOL) more: (Packet *) other;
- (BOOL) less: (Packet *) other;
- (void) list;
@end
```

Class *packet* is defined as a subclass of *NSObject*. *NSObject* is the typical root of the Objective-C class hierarchy. Most built-in classes feature the "NS" prefix, which is short for NextStep, which is the ancestor of the OS X operating system and framework. The methods are automatically "public." The "+" denotes a class-level method, that is, it is invoked for the class, much like a constructor, whereas "-" methods are regular methods invoked for an instance of a class. Now, consider the *packet* class implementation:

```
@implementation Packet
+ (Packet *) new: (char *) n pri: (int) p {
     Packet *id = [super new];
     id->name = n;
     id->priority = p;
     return id;
}
- (BOOL) more: (Packet *) other {
    return priority > other->priority;
}
- (BOOL) less: (Packet *) other {
    return priority <= other->priority;
}
- (void) list{
    printf("%s packet: ", name);
}
@end
```

It contains the bodies of the methods. Note, how parameters are not enclosed in parenthesis, but rather labeled, such as *new:* and *pri:*. That is Smalltalk style. The phrase *[super new]* is also borrowed from Smalltalk: code enclosed in brackets denotes a message-sending expression. Here, message *new* is sent to object *super*. *super* refers to the superclass of *packet*, here *NSObject*, which implements the basic *new* method to create a new object. The other methods are prefixed with a "-", indicating regular instance methods. Note, in method *print*, good old C standard library function *printf* is used for output. On to subclass *Ack*:

```
@interface Ack : Packet
+ (Ack *) new;
- (void) list;
@end

@implementation Ack

+ (Ack *) new {
    return [ super new: "Ack" pri: 10 ];
}
- (void) list {
    printf(" acknowledged\n");
}
@end
```

The *new* method creates a new instance by sending a message to the *new:pri:* method of the superclass, that is, *packet*. Similarly, for class *data*

```
@interface Data : Packet {
     char * body;
}
+ (Data *) new: (char *) b pri: (int) p;
- (void) list;
@end

@implementation Data
+ new: (char *) b pri: (int) p {
```

```
        Data * id = [ super new: "Data" pri: p ];
        id->body = b;
        return id;
}
- (void) list {
        [ super list ];
        printf("%s\n", body);
}
@end
```

Both the *new:pri:* method and the *list* method take advantage of methods inherited from the *packet* superclass.

The next class, *FifoQueue*, first needs the definition of class *Node*:

```
@interface Node : NSObject {
        @public
        Packet * value;
        Node * next, * previous;
}
@end
```

By default, the data fields of class *node* would not be accessible from the outside of the class. Since class *FifoQueue* needs to be able to access these fields, we include a "@public" clause to make them accessible.

Here now is class *FifoQueue*, first the class interface

```
@interface FifoQueue : NSObject {
        Node * head, *tail;
}
- (void) enter: (Packet *) aPacket;
- (Packet *) remove;
- (void) list;
@end
```

followed by the class implementation

```
@implementation FifoQueue
- (void) enter: (Packet *) aPacket {
        Node * tmp = [ Node new ];
        tmp->value = aPacket;
        tmp->previous = tail;
        tail = tmp;
        if (head)
            tmp->previous->next = tail;
        else
            head = tmp;
}
- (Packet *) remove {
    Packet * it = head->value;
    if (head == tail)
        head = tail = NULL;
    else if (head->next) {
        head = head->next;
        head->previous = NULL;
    }
    return it;
```

```
}
- (void) list {
        for (Node *tmp = head; tmp; tmp = tmp->next)
              [ tmp->value list ];
}
@end
```

Most of the method implementations feature straight C-style code, with the occasional message-send expression enclosed in brackets.

We can now exercise our objects. This program produces the same output as our previous examples written in Java, C++, and C#:

```
int main (int argc, const char * argv[]) {

        Data *d1 = [ Data new: "first packet" pri: 5 ];
        Ack *a1 = [ Ack new ];
        Data *d2 = [ Data new: "second packet" pri: 6 ];
        Ack *a2 = [ Ack new ];

        FifoQueue *q = [ FifoQueue new ];

        [ q enter: d1 ];
        [ q enter: a1 ];
        [ q enter: d2 ];
        [ q enter: a2 ];

        printf("The FIFO queue:\n");
        [ q list ];

        printf("Order of leaving:\n");
        [[ q remove ] list ];
        [[ q remove ] list ];
        [[ q remove ] list ];
        [[ q remove ] list ];
}
```

In summary, Objective-C has a unique style, which traces its Smalltalk heritage. Objective-C supports garbage collection; however, it is at the discretion of the programmer. Instead of garbage collection, a programmer can also opt for explicit reference counting to determine when an object becomes obsolete. Like the other OOP languages, Objective-C can draw upon a large library of classes, which make it easy to add animation, networking, and the native platform appearance and behavior to applications with only a few lines of code.

64.4 Language Implementation Issues

In evolutionary terms, OOP languages can be described as either pure or hybrid. The hybrid approach adds object-oriented features on top of a procedural core: C++ and Objective-C are representatives of this approach. The pure approach, Java and C#, limits the language features to those that are strictly object oriented. C# presents a compromise in that it enforces its object-oriented principles but allows unsafe code, as long as it is clearly labeled.

In theoretical terms, the primary distinction between these kinds of languages lies in their interpretation of the concept of type. Because of their origins, hybrid languages rely on a traditional approach to typing. That is, the type of something resides with its container. Thus, when we declare an integer variable *i* in C++, for example, as *int i;*, variable *i* is of basic type *integer.* In Smalltalk, variables do not carry the type of what they contain. That is, variables are untyped: they can contain any object. Rather, the command *Integer new* returns an integer object. The type in this case is seen to reside with the object, and if we store a reference to it in a variable, the variable could be said to be of type *Integer.* Objective-C retains this Smalltalk approach: it features a type *id*, which can refer to any object. In code, any message can be sent to a variable of type *id*. Only at run time, when the message is actually received by the object, might we notice an error, if the object's class does not implement the corresponding method.

In implementation terms, representing values of basic types as objects carries penalties in storage amount and performance: if an integer value is stored as an object, it will occupy about double the amount of storage, let us say four bytes for its value and four bytes for the object reference. To access the value as an object, we first must dereference the object reference, then retrieve the value from the object. Java stores all values of basic types as simple values and does not treat them as objects. The programmer must explicitly create a wrapper object to treat the value as an object. C# does this implicitly through its boxing and unboxing mechanisms.

Another critical distinction between OOP languages is in how they handle the binding of messages to methods. The purest approach is to perform this binding dynamically (at run time): an object receives a message, and the search for an appropriate method begins at the class of the object. The search continues through superclasses until the message can be processed. Although this approach affords the programmer tremendous flexibility, it can cause instability at run time. A more efficient approach is to leave the binding choice to the system (i.e., the compiler and linker). That is, the system tries to determine at compile time which method should be invoked for each message sent. In cases where inclusion polymorphism is used (and it may be unclear which class to refer to), binding can be performed at run time using techniques varying from simple case statements to a more complex system of virtual method tables. In any case, it is still up to the compiler to detect and indicate the need for run-time binding.

Another important issue to consider in the context of language implementation is the approach one adopts in regards to memory management. OO languages are relatively uniform in their approaches to memory allocation (object creation). Creating objects and all that that entails (determining how much memory to allocate, the types of data fields, etc.) is performed by the system. Initialization, on the other hand, is left to the programmer. Many languages provide direct support for initializing objects: we have constructors in C++, Java, and C#, and *new* methods in Objective-C. There are two common approaches to deleting objects from memory (deallocation). In the programmer-controlled approach, one uses a delete operator. The system-enabled approach to deleting objects uses the concept of garbage collection. All objects that are unknown to other objects are garbage. To check whether an object is referred to by another is quite compute-intensive. To lighten the impact on system performance, garbage collection is typically done either during times of idling or when some threshold of memory usage is reached. On the other hand, memory leaks, that is, memory that is occupied by deleted objects, can occur easily in large bodies of code written without garbage collection.

64.5 Summary

OOP languages have matured dramatically and have gained a significant degree of acceptance over their 40 years of history. Many of the original motivations for the OOP paradigm have been justified by our practical experience with it. OOP has been seen to

- Embody useful analysis and design techniques
- Revise the traditional software life cycle toward analysis and design, instead of coding, testing, and debugging
- Encourage software reuse through the development of useful code libraries of related classes
- Improve the workability of a system so that it is easier to debug, modify, and extend
- Appeal to human instincts in problem solving and description—in particular, to problems that model real-world phenomena

That OOP has become the predominant software development paradigm is evidenced by the multitude of conferences, journals, texts, and websites devoted to both general and language-specific topics. The two most prominent conferences, both of which address a wide range of OOP issues, are the conference on

OOP Systems, Languages, and Applications (www.oopsla.org) and the European Conference on OOP (www.ecoop.org). The *Journal of Object Technology* provides contemporary coverage of OOP languages, applications, and research (www.jot.fm). Perhaps, the most general of the references listed are Budd (2002) and Meyer (2000).

Key Terms

Abstract class: A class that has no direct instances but is used as a base class from which subclasses are derived. These subclasses will add to its structure and behavior, typically by providing implementations for the methods described in the abstract class.

Class: A description of the data and behavior common to a collection of objects. Objects are instances of classes.

Constructor: An operation associated with a class that creates and/or initializes new instances of the class.

Dynamic binding: Binding performed at run time. In OOP, this typically refers to the resolution of a particular name within the scope of a class, so that the method to be invoked in response to a message can be determined by the class to which it belongs at run time.

Inheritance: A relationship among classes, wherein one class shares the structure or behavior defined in an is-a hierarchy. Subclasses are said to inherit both the data and methods from one or more generalized superclasses. The subclass typically specializes its superclasses by adding to its structure and by redefining its behavior.

Instance: A specific example that conforms to a description of a class. An instance of a class is an object.

Interface: A named listing of method headers to be implemented by a class.

Member field: A data item that is associated with (and is local to) each instance of a class.

Message: A means for invoking a subprogram or behavior associated with an object.

Method: A procedure or function that is defined as part of a class and is invoked in a message-passing style. Every instance of a class exhibits the behavior described by the methods of the class.

Object: An object is an instance of a class described by its state, behavior, and identity.

Object-oriented programming (OOP): A method of implementation in which a program is described as a sequence of messages to cooperating collections of objects, each of which represents an instance of some class. Classes can be related through inheritance, and objects can exhibit polymorphic behavior.

Override: The action that occurs when a method in a subclass with the same name as a method in a superclass takes precedence over the method in the superclass.

Polymorphism (or many shapes): That feature of a variable that can take on values of several different types or a feature of a function that can be executed using arguments of a variety of types.

Subclass: A class that inherits data fields and methods from another class (called the superclass).

Virtual function: Most generally, a method of a class that may be overridden by a subclass to the class. In languages in which dynamic binding is not the default, this may also mean that a function is subject to dynamic binding.

References

Budd, T. 2002. *An Introduction to Object-Oriented Programming* (3rd edn.), Addison-Wesley, Reading, MA.

Cox, B. et al. 1991. *Object-Oriented Programming: An Evolutionary Approach* (2nd edn.), Addison-Wesley, Reading, MA.

Gosling, G. et al. 2005. *The Java Language Specification* (3rd edn.), Addison-Wesley, Reading, MA.

Hejlsberg, A. et al. 2010. *C# Programming Language* (4th edn.), Addison-Wesley, Reading, MA.

Ingalls, D. 1981. Design principles behind Smalltalk. *Byte* 6(8), 286–298.

Meyer, B. 2000. *Object-Oriented Software Construction* (2nd edn.), Prentice Hall, Englewood Cliffs, NJ.

Stroustrup, B. 1994. *The Design and Evolution of C++*, Addison-Wesley, Reading, MA.

65

Logic Programming and Constraint Logic Programming

Jacques Cohen
Brandeis University

65.1 Introduction

Logic programming (LP) is a language paradigm based on logic. Its constructs are Boolean *implications* (e.g., *q implies p* meaning that *p* is true if *q* is true), compositions using the Boolean operators *and* (called conjunctions) and *or* (called disjunctions). LP can also be viewed as a procedural language in which the procedures are actually Boolean functions, the result of a program always being either true or false.

In the case of implications, a major restriction applies: when *q* implies *p*, written *p:–q*, then *q* can consist of conjunctions, but *p* has to be a singleton, representing the (sole) Boolean function being defined. The Boolean operator *not* is disallowed, but there is a similar construct that may be used in certain cases.

The Boolean functions in LP may contain parameters, and the parameter matching mechanism is called unification. This type of general pattern matching implies, for example, that a variable representing a formal parameter may be bound to another variable or even to a complex data structure, representing an actual parameter (and vice versa). When an LP program yields a *yes* answer, the bindings of the variables are displayed, indicating that those bindings make the program logically correct and provide a solution to the problem expressed as a logic program.

An important recent extension of LP is *constraint logic programming* (CLP). In this extension, unification can be replaced or complemented by other forms of constraints, depending on the domains of the variables involved. For example, in CLP a relationship such as $X > Y$ can be expressed even in the case where X and Y are unbound real variables. As in LP, a CLP program yields answers expressing that the resulting constraints (e.g., $Z < Y + 4$) must be satisfied for the program to be logically correct.

This chapter includes sections describing the main aspects of LP and CLP. It includes examples, historical remarks, theoretical foundations, implementation techniques, and metalevel interpretation and concludes with the most recent proposed extensions to this language paradigm.

65.2 Introductory Example

Let us consider a simple program written in PROLOG, the main representative among LP languages. The program's objective is to check if a date given by the three parameters, *Month* (a string), *Day*, and *Year* (integers), is valid. For example, *date* (*oct*, 15, 1996) is valid whereas *date* (*june*, 31, 1921) is not. The program can be expressed as follows:

```
date (Month, Day, Year):- member (Month, [jan, march, may, july, aug, oct, dec]),
                          comprised (Day, 1,31).
date (Month, Day, Year):- member (Month, [april, june, oct, sept, nov]),
                          comprised (Day, 1,30).
date (feb, Day, Year):- leap (Year), comprised (Day, 1, 29).
date (feb, Day, Year):- comprised (Day, 1, 28).
comprised (Day, Start, End):- Day >=Start, Day =<End.
leap (Year):- (Year/4) × 4 = Year.
```

The variables in the preceding program start with a capital letter, for example, *Month*; in this simple example, the constants either start with small case letters (strings) or are integers. The symbol:- can be read as is defined by, and it separates the procedure heading from its body, which consists of calls to other procedures. More precisely, it is convenient to view every procedure as a Boolean function yielding *true* or *false*. The body of such a function is made up of conjunctions of calls to functions. Therefore, a comma separating the calls on the right-hand side correspond to the Boolean operator *and*.

The following remarks also apply to the program. There are multiple definitions of *date*, each of which tests for the corresponding month and the compatible number of days. Therefore, the program is nondeterministic in the sense that several possibilities have to be considered either one at a time or in parallel. In many cases there might be several solutions to a given program. Also note that the right-hand side of a rule could be empty. In that case, the rule indicates that the left-hand side is always true.

Let us for the time being assume that *member* (*Element, List*) is a built-in function testing if an *Element* is a member of a *List*. The Boolean function *leap* tests if its parameter is a multiple of 4. One could also add a function *year* (*Year*), which would test if an integer defined by *Year* is positive. Notice that the ordering of the statements in this particular version of the program is important, since a leap year is tested before a nonleap year. Ideally, pure PROLOG programs should be declarative; from a syntactic point of view, that means that the order of the rules and the order of the calls on the right-hand sides should not matter. A new version of the program satisfying this criteria is one in which the Boolean function *nonleap (Year):- (Year/4)* × *4 =/= Year* is called in the definition applicable to the month of February, in the case of a nonleap year.

The preceding program is used with queries (equivalent to main programs) to test the validity of given dates. The answers are either *yes* or *no*. Note that in PROLOG one is allowed to have queries containing variables. In the present example, the query *date* (*X*, 31, 1996) yields as results the successive strings that the variable *X* is bound to for the months that have 31 days. More specifically, the results are as follows:

yes X = jan, More?
yes X = march, More?
...
Yes X = dec, More?
No.

Similarly, the query *date* (*X*, 29, 1996) yields all the months in a year. Note that the built-in functions such as =< and >= require that their arguments be ground; that is, they cannot contain unbound variables.

This is satisfied by the query *date* (*X*, 31, 1996). However, a call *date* (*feb*, *Y*, 1996) entails a problem that provides insight into the desirability of extending PROLOG to handle constraints such as it is done in PROLOG IV. The last query fails in standard PROLOG because the variable *Y* remains unbound. The right answer as provided by a CLP processor consists of the two constraints:

Yes Y >= 1, Y =< 29, More?
No.

Similarly, the correct answer to the query *date* (*feb, Y, Z*) submitted to a CLP processor should yield two pairs of constraints corresponding to leap and nonleap years represented by *Z*. These results are obtained by keeping lists of satisfiable constraints and triggering a failure when the constraints become unsatisfiable. This failure may entail exploring the remaining nondeterministic possibilities.

It is now appropriate to present the recursive Boolean function *member*. It has two parameters: an *Element* and a *List*; the function succeeds if the *Element* appears in the *List*. Lists are represented using the constructor *cons* as in functional languages. This means that a list [*a*, *b*] can be represented by

cons (a, cons (b, nil))

Note that the *cons* in the preceding representation is simply any user-selected identifier describing a data structure (or record) consisting of two fields: (1) a string and (2) a pointer to another such record or to *nil*. In the case of lists there are special built-in features such as [*a*, *b*] simplifying their description. The program for *member* consists of two rules:

member (X, cons (X, T)).
member (X, cons (Y, T)):- X /= Y, member (X, T).

The first rule states that if the head of a list contains the element *X*, then *member* succeeds. The second rule uses recursion to inspect the remaining elements of the list. Note that the query *member (X, cons (a, cons (b, nil))*, equivalent to *member (X, [a, b])*, provides two solutions, namely, *X = a or X = b*. It should be obvious that *member (Y, [a, a])* also yields two (identical) solutions.

Finally, notice that *member* can search for more complex data structures. For example, *member (member (a, Y), Z)* succeeds if the list *Z* contains an element such as *member (a, U)*; if so, the variable *Y* is bound to the variable *U*. The reader should consider the embedded term *member* as the identifier of a record or data structure having two fields. In contrast, the first identifier *member* represents a Boolean function having two parameters. This example illustrates that, in PROLOG, program and data have the same form.

An inquisitive reader will remark that *member* can also be used to place elements in a list if they are not already present in that list. A clue in understanding this property is that the query *member (a, Z)* will bind *Z* to a record *cons (a, W)* in which *W* is a new unbound variable created by the PROLOG processor when applying the first rule defining the Boolean function *member.*

65.3 Features of Logic Programming Languages

Summarized next are some of the features whose combination renders PROLOG unique among languages:

1. Procedures may contain parameters that are both input and output.
2. Procedures may return results containing unbound variables.
3. *Backtracking* is built-in, therefore allowing the determination of multiple solutions to a given problem.
4. General pattern-matching capabilities operate in conjunction with a goal-seeking search mechanism.
5. Program and data are presented in similar forms.

The preceding listing of the features of PROLOG does not fully convey the subjective advantages of the language. There are at least three such advantages:

1. Having its foundations in logic, PROLOG encourages the programmer to describe problems in a logical manner that facilitates checking for correctness and, consequently, reduces the debugging effort.
2. The algorithms needed to interpret PROLOG programs are particularly amenable to parallel processing.
3. The conciseness of PROLOG programs, with the resulting decrease in development time, makes it an ideal language for prototyping.

Another important characteristic of PROLOG that deserves extension, and is now being extended, is the ability to postpone variable bindings as much as is deemed necessary (lazy evaluation). Failure and backtracking are triggered only when the interpreter is confronted with a logically unsatisfiable set of constraints. In this respect, PROLOG's notion of variables approaches that used in mathematics.

The price to be paid for the advantages offered by the language amounts to the increasing demands for larger memories and faster central processing units (CPUs). The history of programming language evolution has demonstrated that, with the consistent trend toward less expensive and faster computers with larger memories, this price becomes not only acceptable but also advantageous because the savings achieved by program conciseness and by a reduced programming effort largely compensate for the space and execution time overheads. Furthermore, the quest for increased efficiency of PROLOG programs encourages new and important research in the areas of optimization and parallelism.

65.4 Historical Remarks

The birth of LP can be viewed as the confluence of two different research endeavors: one in artificial or natural language processing and the other in automatic theorem proving. These endeavors contributed to the genesis of the PROLOG language, the principal representative of LP.

Alain Colmerauer, assisted by Philippe Roussel, is credited as the originator of PROLOG, a language that was first developed in the early 1970s and continues to be substantially extended. Colmerauer's contributions stemmed from his interest in language processing using theorem-proving techniques. Robert Kowalski was also a major contributor to the development of L.P. Kowalski had an interest in logic and theorem proving (Cohen 1988, Bergin 1996). In their collaboration, Kowalski and Colmerauer became interested in problem solving and automated reasoning using resolution theorem proving.

Kowalski's main research was based on the work of Alan Robinson (1965). Robinson had the foresight to distinguish the importance of two components in automatic theorem proving: a single inference rule called *resolution* and the testing for equality of trees called *unification*.

Theorems to be proved using Robinson's approach are placed in a special form consisting of conjunctions of *clauses*. Clauses are disjunctions of positive or negated literals; in the case of Boolean algebra (propositional calculus), the literals correspond to variables. In the more general case of the predicate calculus, literals correspond to a potentially infinite number of Boolean variables, one for each combination of values that the literal has as parameters. A *Horn clause* is one containing (at most) one positive literal; all of the others (if any) are negated.

According to the informal description in the introductory example, a Horn clause corresponds to the definition and body of a Boolean function. The positive literal is the left-hand side of a rule (called *Head*); the negative literals appear on the right-hand side of the rule (called *Body*). A non-Horn clause is one in which (logical) negations can appear qualifying a call in the *Body*. An example of a Horn clause is as follows:

date (feb, Day, Year):- leap (Year), comprised (Day, 1, 29).

because it is equivalent to

leap (Year) and comprised (Day, 1, 29) implies date (feb, Day, Year)

or

date (feb, Day, Year) or not (leap (Year)) or not (comprised (Day, 1, 29))

If one wished to express the clause

date (feb, Day, Year):- not leap (Year), comprisesd (Day, 1, 28).

in which the *not* is the logical Boolean operation, the preceding example would be logically equivalent to

date (feb, Day, Year) and leap (Year) or comprised (Day, 1, 28)

which is not a Horn clause because it has two (positive) elements in the head (and there is no procedural equivalent to it).

Kowalski concentrated his research on reducing the search space in resolution-based theorem proving. With this purpose, he developed with Kuehner a variant of the linear resolution algorithm called SL resolution (for linear resolution with selection function [Kowalski and Kuehner 1970]). Kowalski's view is that, from the automatic theorem-proving perspective, this work paved the way for the development of PROLOG. Having this more efficient (but still general) predicate calculus theorem prover available to them, the Marseilles and Edinburgh groups started using it to experiment with problem-solving tasks. To further increase the efficiency of their prover, the Marseilles group resorted to daring simplifications that would be inadmissible to logicians. These audacious attempts turned out to open new vistas for the future of LP.

Several formulations for solving a given problem were attempted. Almost invariably, the formulations that happened to be written in Horn clause form turned out to be much more natural than those that used non-Horn clauses. A case in which the Horn clause formulation was particularly effective occurred in parsing strings defined by grammar rules. Recall that context-free grammars have a single nonterminal being defined on the left-hand side of each of its rules, therefore establishing a strong similarity with Horn clauses.

The SL inference mechanism applicable to a slight variant of Horn clauses led to the present PROLOG inference mechanism: *selective linear definite* (*SLD*) clause resolution. The word *definite* refers to Horn clauses with exactly one positive literal, whereas general Horn clauses may contain entirely negative clauses. (Because the term "Horn clause" is more widely used than definite clause, the former is often used to denote the latter.)

As a final historical note, one should mention that LP gained renewed impetus by its adoption as the language paradigm for the Japanese fifth-generation program. PROLOG and, in particular, its CLP extensions now count on a significant number of loyal and enthusiastic users.

65.5 Resolution and Unification

Resolution and unification appear in different guises in various algorithms used in computer science. This section first describes these two components separately and then their combination as it is used in LP. In doing so it is useful to consider first the case of the propositional calculus (Boolean algebra) in which unification is immaterial. It is well known that there exist algorithms that can always decide if a system of Boolean formulas is satisfiable or not, albeit with exponential complexity.

In terms of the informal example considered in the introduction, one can view resolution as a (nondeterministic) call of a user-defined Boolean function. Unification is the general operation of matching the formal and actual parameters of a call. Consequently, unification does not occur in the case of parameterless Boolean functions.

The *predicate calculus* includes the quantifiers ∀ and ∃; it can be viewed as a general case of the propositional calculus for which each predicate variable (a literal) can represent a potentially infinite number of Boolean variables. Unification is only used in this latter context. Theorem-proving algorithms for the predicate calculus are not guaranteed to provide a yes-or-no answer because they may not terminate.

65.5.1 Resolution

In the propositional calculus, a simple form of resolution is expressed by the inference rule:

if $a \rightarrow b$ *and* $b \rightarrow c$ *then* $a \rightarrow c$, *or*
$(\neg a \vee b) \wedge (\neg b \vee c) \rightarrow (\neg a \vee c)$

Recall that *a implies b* is equivalent to *not a or b*. The final disjunction $\neg a \vee c$ is called a *resolvant*. In particular, resolving $a \wedge \neg a$ implies the empty clause (i.e., falsity).

To better understand the meaning of the empty clause, consider the implication $a \rightarrow a$, which is equivalent to (*not a*) *or a*. This expression is always true; therefore, its negation *not* ([*not a*] *or a*) equivalent to (*a and* [*not a*]) is always false. If a Boolean expression is always *true*, its negation is always *false*. Resolution theorem proving consists of showing that if the expression is always true, its negation results in contradictions of the type (*a and* [*not a*]), which is always false. The empty clause is simply the resolvant of (*a and* [*not a*]).

Observe the similarity between resolution and the elimination of variables in algebra, for example,

$$a + b = 3 \quad \text{and} \quad -b + c = 5 \text{ imply } a + c = 8$$

Another intriguing example occurs in matching a procedure definition with its call. Consider, for example,

procedure *b*; *a*
...
call *b*; *c*

in which *a* is the body of *b*, and *c* is the code to be executed after the call of *b*. If one views the definition of *b* and its call as complementary, a (pseudo)resolution yields: *a*; *c* in which concatenation is noncommutative, and the resolution corresponds to replacing a procedure call by its body. Actually, the last example provides an intuitive procedural view of resolution as used in PROLOG.

In the case of pure PROLOG programs, only Horn clauses are allowed. For example, if a, b, c, d, and f are Boolean variables (literals), then

$b \wedge c \wedge d \rightarrow a$ and f

are readily transformed in Horn clauses because they correspond, respectively, to

$a \vee \neg b \vee \neg c \vee \neg d$ and f

where the clause f is called unit clause or a fact. The preceding example is written in PROLOG as

a:- *b*, *c*, *d*. and *f*.

where the symbols:- and "," correspond to the logical connectors *only if* and *and*. They are read as follows: a is true only if b and c and d are true. The previous conjunction also requires that f be true; equivalently $b \wedge c \wedge d \rightarrow a$ and f are true.

The resolution mechanism applicable to Horn clauses takes as input a conjunction of Horn clauses $H = h_1 \wedge h_2 \wedge \cdots \wedge h_n$, and a query Q, which is the negation of a theorem to be proved. Q consists of the negation of a conjunction of positive literals or, equivalently, a disjunction of negated literals. Therefore, a query is itself in a Horn clause form in which the head is empty.

A theorem is proved by contradiction, namely, the goal is to prove that $H \wedge Q$ is inconsistent, implying that the result of successive resolutions involving the negated literals of Q inevitably—in the case of the propositional calculus—leads to falsity (i.e., the empty clause). In other words, if H implies the nonnegated Q is always true, then H *and* the negated Q are always false.

Consider, for example, the query *date* (*oct*, 15, 1996) in our introductory example. Its negation is *not date* (*oct*, 15, 1996). This resolves with the first rule yielding the bindings *Month* = *oct*, *Day* = 15, *Year* = 1996. The resolvant is the disjunction *not member* (*oct*, [*jan*, *march*, *may*, *july*, *aug*, *oct.*, *dec*]) or *not comprised* (15, 1, 31).

Although not elaborated here, the reader can easily find out that the successive resolutions using the definition of *member* will fail because *oct* is a member of the list of months containing 31 days. Similarly, day 15 is comprised between 1 and 31. Therefore, the empty (falsity) clause will be reached for both disjuncts of the resolvant.

In what follows, the resolution inference mechanism is applied to Horn clauses representing a PROLOG program. One concrete syntax for PROLOG rules is given by

<rule>::= *<clause>*. |*<unit clause>*.
<clause>::= *<head>*:- *<tail>*
<head>::= *<literal>*
<tail>::= *<literal>* {, *<literal>*}
<unit clause>::= *<literal>*

where the braces { } denote any number of repetitions (including none) of the sequence enclosed by the brackets <>. First consider the simplest case, where a *literal* is a single letter. For example, consider the following PROLOG program in which rules are numbered for future reference:

1. *a*:- *b*, *c*, *d*.
2. *a*:- *e*, *f*.
3. *b*:- *f*.
4. *e*.
5. *f*.
6. *a*:- *f*.

In the first rule, a is the *<head>*, and b, c, d is the *<tail>*, also called the *body*. The fourth and fifth rules are unit clauses; that is, the body is empty. A query Q is syntactically equivalent to a *<tail>*. For example, *a*, *e*. is a query and it corresponds to the Horn clause $\neg a \vee \neg e$. The result of querying the program is one (or multiple) *yes* or a single *no* answer indicating the success or failure of the query. In this particular example, the query *a, e.* yields two *yes* answers. Note that *a* is defined by three rules; the second and the third yield the two solutions; the first fails because *c* is undefined. The successful sequence of the list of goals is as follows:

Solution 1: *a, e* ⇒ *e, f, e* ⇒ *f, e* ⇒ *e* ⇒ *nil;*
Solution 2: *a, e* ⇒ *f, e* ⇒ *e* ⇒ *nil.*

Let us follow in detail the development of the second solution. The negated query *not a or not e* resolves with the sixth rule *a or not f,* yielding the resolvant *not f or not e.* Now the last expression is resolved with the *f* in the fifth rule, yielding *not e* as the resolvant. Finally, *not e* is resolved with the *e* in the fourth rule, yielding the empty clause, which implies the falsity of the negation of the query, using the program rules. The entire search space is shown in Figure 65.1 in the form of a tree. In nondeterministic algorithms, that tree is called the tree of choices. Its leaves are nodes representing failures or successes. The internal nodes are labeled with the list of goals that remain to be satisfied. Note that if the tree of choices is finite, the order of the goals in the list of goals is irrelevant insofar as the presence and number of solutions are concerned. Figure 65.2 shows a proof tree for the first solution of the example. The proof tree is an *and* tree depicting how the proof has been achieved.

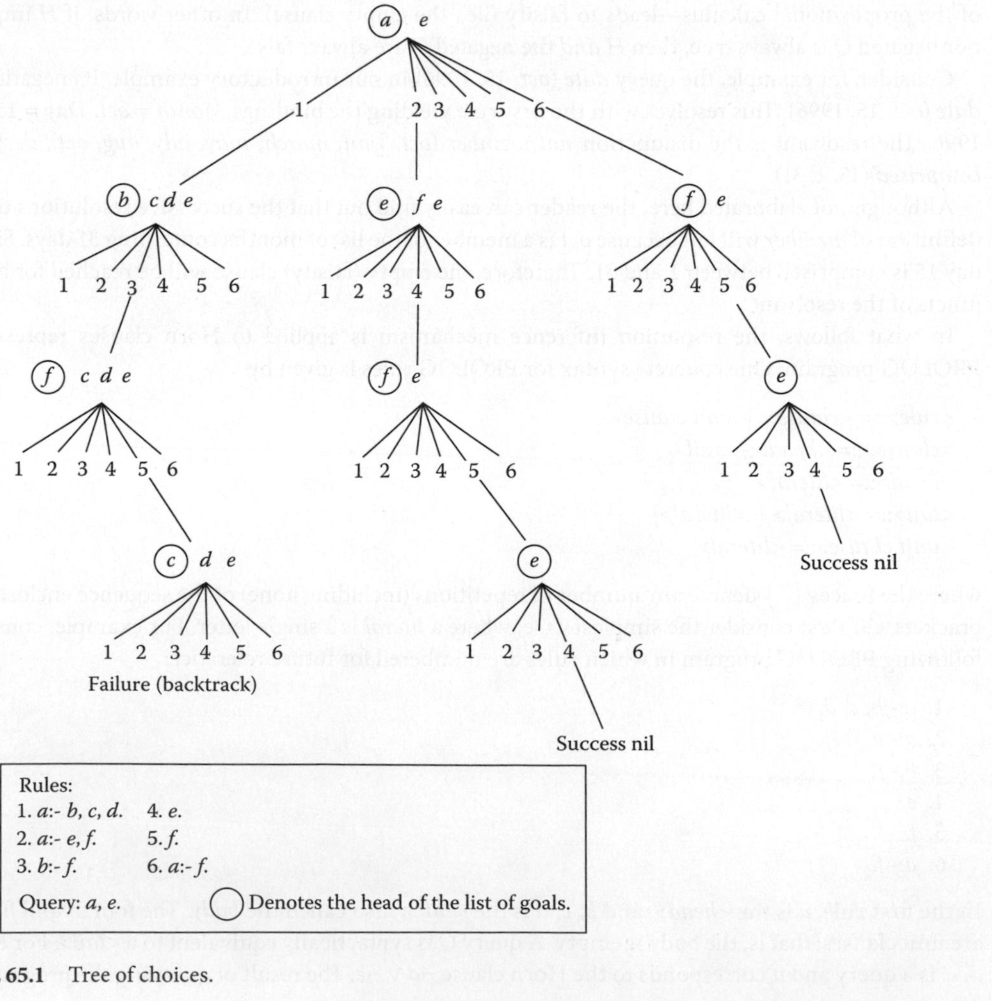

FIGURE 65.1 Tree of choices.

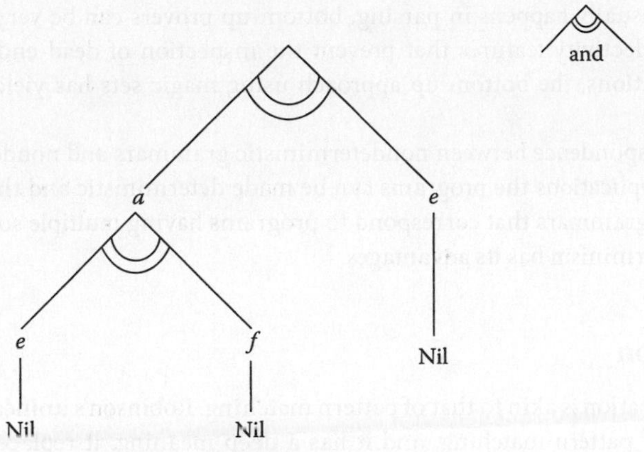

Proof tree for the query *a*, *e*

FIGURE 65.2 Proof tree.

There are three ways of interpreting the semantics of PROLOG rules and queries. The first is based on logic, in this particular case on Boolean algebra. The literals are Boolean variables, and the rules express formulas. The PROLOG program is viewed as the conjunction of formulas it defines. The query *Q* succeeds if it can be implied from the program. In a second (called procedural) interpretation of a PROLOG rule, it is assumed that a <*literal*> is a goal to be satisfied. For example, the first rule states that

goal a can be satisfied if goals b, c, and d can be satisfied.

The unit clause states that the defined goal can be satisfied. The program defines a conjunction of goals to be satisfied. The query succeeds if the goals can be satisfied using the rules of the program. Finally, a third interpretation is based on the similarity between PROLOG rules and context-free grammar rules. A PROLOG program is associated with a context-free grammar in which a <*literal*> is a nonterminal and a <*rule*> corresponds to a grammar rule in which the <*head*> rewrites into the <*tail*>; a unit clause is viewed as a grammar rule in which a nonterminal rewrites into the empty symbol ε. Under this interpretation, a query succeeds if it can be rewritten into the empty string.

Although the preceding three interpretations are all helpful in explaining the semantics of this simplified version of PROLOG, the logic interpretation is the most widely used among theoreticians, and the procedural by language designers and implementors.

The algorithms that test if a query *Q* can be derived from a Horn clause program *P* can be classified in various manners. An analogy with parsing algorithms is relevant: *P* corresponds to a grammar *G*, and *Q* corresponds to the string (of nonterminals) to be parsed (i.e., the sequence of nonterminals that may be rewritten into the empty string using *G*). *Backward-chaining* theorem provers correspond to top-down parsers and are by far the preferred approach presently used in LP. *Forward-chaining* provers correspond to bottom-up parsers. Hybrid algorithms have also been proposed.

In a top-down algorithm, the list of goals in a query is examined from left to right, and the corresponding (recursive) procedures are successively called the equivalent of resolution, until the list of goals becomes empty. Note that the algorithm is essentially nondeterministic because there are usually several choices for the goals (see rules 1 and 2). Another nondeterministic choice occurs when selecting the (next) element of the list of goals to be considered after processing a goal.

Notice that if one had a program consisting of the sole rule *a*:- *a*. and the query *a*, the top-down approach would not terminate. The program states that either *a* or ¬*a* is true. Because the query does not specify *a*, it could be either true or false. A semantically correct interpreter should provide the following constraints as answers *a* = *true* or *a* = *false*. PROLOG programmers have learned to live with these unpalatable characteristics of top-down provers.

Contrary to what usually happens in parsing, bottom-up provers can be very inefficient unless the algorithm contains selectivity features that prevent the inspection of dead-end paths. In the case of database (DB) applications, the bottom-up approach using magic sets has yielded interesting results (Minker 1987).

Also note the correspondence between nondeterministic grammars and nondeterministic PROLOG programs. In many applications the programs can be made deterministic and therefore more efficient. However, ambiguous grammars that correspond to programs having multiple solutions are also useful and therefore nondeterminism has its advantages.

65.5.2 Unification

The operation of unification is akin to that of pattern matching. Robinson's unification algorithm brings rigor to the notion of pattern matching, and it has a deep meaning: it replaces a set of fairly intricate predicate calculus axioms specifying the equality of trees by an efficient algorithm, which is easily implementable in a computer. As mentioned earlier, unification can be viewed as a very general matching of actual and formal parameters.

From the logic point of view, unification is used in theorem proving to equate terms that usually result from the elimination of existential quantifiers when placing a predicate calculus formula in clausal form.

For example,

$\forall X \exists Y$ such that $p(X, Y)$ is always true

is replaced by

$\forall X\, p(X, g(X))$

where $g(x)$ is the Skolem function for y (sometimes referred to as an uninterpreted function symbol). The role of unification is to test the equality of literals containing Skolem functions and, in so doing, bind values to variables.

Consider, for example, the statement "for all positive integer variables X there is always a variable Y representing the successor of X." The predicate p expressing this statement is $p(X, s(X))$:- *integer*(X), where $s(X)$ is the Skolem function representing Y, the successor of X. This representation is commonly used to specify positive integers from a theoretical perspective. (It is called a Peano representation of integers (e.g., $s(s(0))$ represents the integer 2).)

To show the effect of unification, the definition of a literal in the previous subsection on resolution is now generalized to encompass labeled tree structures.

<literal>::= <composite>
<composite>::= <functor>(<term>{, <term>}) | <functor>
<functor>::= <lower case identifiers>
<term>::= <constant> | <variable> | <composite>
<constant>::= <integers and lower case identifiers>
<variable>::= <identifiers starting with an upper case letter or_ >

Examples of terms are *constant*, *Var*, 319, *line*(*point*(*X*, 3), point (4, 3)). It is usual to refer to a single rule in a PROLOG program as a clause. A PROLOG procedure (or predicate) is defined as the set of rules whose head has the same *<functor>* and arity. Unification tests whether two terms $T1$ and $T2$ can be matched by binding some of the variables in $T1$ and $T2$. The simplest algorithm uses the rules summarized in Table 65.1 to match the terms. If both terms are composite and have the same *<functor>*, it recurses on their components. This algorithm only binds variables if it is *absolutely necessary*, so there may be variables that remain unbound. This property is referred to as the most general unifier (mgu).

TABLE 65.1 Summary of Tests for the Unification of Two Terms

Terms 1↓, 2→	*<constant>* C2	*<variable>* X2	*<composite>* T2
<constant> C1	*succeed if* C1 = C2	*succeed with* X2 = C1	*fail*
<variable> X1	*succeed with* X1 = C2	*succeed with* X1 = X2	*succeed with* X1 = T2
<composite> T1	*fail*	*succeed with* X2 = T1	*succeed if* (1) *T1 and T2 have* the same *functor* and arity (2) The matching of corresponding children *succeeds*

One can write a recursive function *unify,* which, given two terms, tests for the result of unification using the contents of Table 65.1. The unification of the two terms $f(X, g(Y), T)$ *and* $f(f(a, b), g(g(a, c), Z))$ succeeds with the following bindings $X = f(a, b)$, $Y = g(a, c)$, and $T = Z$. Note that if one had $g(Y, c)$ instead of $g(a, c)$ in the second term, the binding would have been $Y = g(Y, c)$.

PROLOG interpreters would usually carry out this circular binding but would soon get into difficulties because most implementations of the described unification algorithm cannot handle circular structures (called *infinite trees*). Manipulating these structures (e.g., printing, copying) would result in an infinite loop unless the so-called *occur check* was incorporated to test for circularity. This is an expensive test: unification is linear with the size of the terms unified, and incorporation of the occur check renders unification quadratic.

There are versions of the unification algorithm that use a representation called *solved form*. Essentially, new variables are created whenever necessary to define terms, for example, the system

$$X1 = f(X2, g(X3, X2)) \quad X5 = g(X4, X2)$$

A solved form is immediately known to be satisfiable because $X2$, $X3$, and $X4$ can be bound to any terms formed by using the function symbols f and g and again replacing (*ad infinitum*) their variable arguments by the functions f and g. This is called an element of the *Herbrand universe* for the given set of terms. The solved form version of the unification algorithm is presented by Lassez in Minker (1987).

To further clarify the notion of solved forms, consider the equation in the domain of reals specified by the single constraint $X + Y = 5$. This is equivalent to $X = 5 - Y$, which is in solved form and satisfiable for any real value of Y. Basically a constraint in solved form contains definitions in terms of free variables, that is, those that are not constrained. This form is very useful when unification is extended to be applicable to domains other than trees. For example, linear equations may be expressed in solved form.

65.5.3 Combining Resolution and Unification

Consider now general clauses in which the literals contain arguments that are represented by terms. The result of resolving

$p(\ldots) \vee r(\ldots)$

with

$\neg p(\ldots) \vee s(\ldots)$

is

$r(\ldots) \vee s(\ldots)$

if and only if the arguments of p and $\neg p$ are unifiable and the resulting variables are substituted in the corresponding arguments of r and s.

Using programming language terminology, the combination of resolution and unification corresponds to a Boolean function call in which a variable in the head of a rule is matched (i.e., unified) with its corresponding actual parameter; once this is done, the value to which the variable is bound replaces the instances of that variable appearing in the body. Note that a variable can be bound to another variable or to a complex term. Remark also that when unification fails, another head of a rule has to be tried and so on. For example, consider the query $p(a)$ and the program

$p(b).$
$p(X): - q(X).$
$q(a).$

The first unification of the query $p(a)$ with $p(b)$ fails. The second attempt binds X to a starting a search for $q(a)$ that succeeds. Had the query been $p(Y)$, the two solutions would be $Y = b$ or $Y = a$. The first solution is a consequence of the unit clause $p(b)$; the second follows from binding Y to a (new) variable X in the second clause. A search is then made for $q(X)$, which succeeds with the last clause, therefore binding both Y and X to a.

From a logical point of view, when the result of a resolution is the empty clause (corresponding to the head of a rule with an empty body) and no more literals remain to be examined, the list of bindings is presented as the answer, that is, the bindings (*constraints*) that result from proving that the query is deductible from the program.

65.6 Procedural Interpretation: Examples

It is worthwhile to first understand the importance of the procedure *append* as used in symbolic functional languages. In PROLOG the word procedure is a commonly used misnomer because it always corresponds to a Boolean function, as mentioned in the introductory example; *append* is often used to concatenate two lists. Because one has to avoid destructive assignments, which have very complex semantic meaning, *append* resorts to copying the first list to be appended and making its last element point to the second list. This preserves the original first list. It should be kept in mind that the avoidance of a destructive assignment forces one to produce results that are *cons*tructed from a given data using a LISP-like *cons*. Notice that LP also avoids destructive assignment by having variables bound only once during unification.

This section first demonstrates the transformation of a functional specification of *append* into its PROLOG counterpart corresponding to the previous clauses. Examples of inventive usage of *append* are then presented to further illustrate the procedural interpretation of logic programs. Consider the LISP-like function *append* that concatenates two lists, *L*1 and *L*2:

```
function append(L1, L2: pLIST): pLIST;
    if L1 = nil then append: = L2
                else append: = cons (head (Ll), append (tail (Ll), L2));
```

In the preceding display, *L*1 and *L*2 are pointers to lists; a list is a record containing the two fields *head* and *tail*, which themselves contain pointers to other lists or to atoms. The tail must point to a list or to the special atom *nil*. The constructor *cons* (*H*, *T*) creates the list whose *head* and *tail* are, respectively, *H* and *T*. The function *append* can be rewritten into a procedure having an explicit third parameter *L*3 that will contain the desired result. The local variable *T* is used to store intermediate results:

```
procedure append (Ll, L2: pLIST: var L3: pLIST);
  begin local T: pLIST;
  if L1 = nil then L3: = L2
          else begin append(tail (Ll), L2, T); L3:= cons(head(Ll), T) end
  end
```

The former procedure can be transformed into a Boolean function that, in addition to building *L*3, checks if *append* produces the correct result.

```
function append(Ll, L2: pLIST: var L3: pLIST): Boolean;
  begin local H1, T1, T: pLIST;
    if L1 = nil then begin L3:= L2; append:= true end
        else if {There exists an H1 and a T1 such that
                  H1 = head(L1) and T1 = tail(L1)}
          then begin append:= append(T1, L2, T); L3:= cons(H1, T) end
          else append:= false
  end
```

The Boolean in the conditional surrounded by braces has been presented informally, but it could actually have been programmed in detail. Note that the assignments in the preceding program are executed at most once for each recursive call. The function returns *false* if *L*1 is not a list (e.g., if $L1 = cons(a, b)$ for some atom $b \neq nil$).

One can now transform the last function into a PROLOG-like counterpart in which rules of assignments and conditionals are subsumed by unification and specified by the equality sign. The statement $E1 = E2$ succeeds if *E*1 and *E*2 can be matched. In addition, some of the variables in *E*1 and *E*2 may be bound if necessary:

append(*L*1, *L*2, *L*3) *is true* if *L*1 = *nil* and *L*3 = *L*2
otherwise
append(*L*1, *L*2, *L*3) is true
if *L*1 = *cons*(*H*1, *T*1) and *append*(*T*1, *L*2, *T*) and *L*3 = *cons*(*H*1, *T*)
otherwise *append* is false.

The reader can now compare the preceding results with the previous description of the subset of PROLOG. This comparison yields the PROLOG program

```
append(L1, L2, L3):- L1 = nil, L3 = L2.
append(L1, L2, L3):- L1 = cons(H1, T1), append(T1, L2, T), L3 = cons (H1, T).
```

This version is particularly significative because it clearly separates resolution from unification. In this case, the equality sign is the operator that commands a unification between its left and right operands.

This could also be done using the unit clause *unify* (*X*, *X*) and substituting *L*1 = *nil* by *unify* (*Ll*, *nil*), and so on. Replacing *L*1 and *L*3 with their respective values on the right-hand side of a clause, one obtains the following:

```
append(nil, L2, L2).
append(cons(H1, T1), L2, cons(H1, T)):- append (T1, L2, T).
```

The explicit calls to *unify* have now been replaced by implicit calls that will be triggered by the PROLOG interpreter when it tries to match a goal with the head of a rule. Notice that a <*literal*> now becomes a function name followed by a list of parameters, each of which is syntactically similar to a <*literal*>. In the predicate calculus, the preceding program corresponds to the following:

$\forall L2\ append\ (nil, L2, L2).$
$\forall H1, T1, L2, T\ append(cons(H1, T1), L2, cons(H1, T)) \vee \neg\ append(T1, L2, T)$

A word of caution about types is in order. The preceding version of *append* would produce a perhaps unwanted result if the list *L*2 is a term, say *g*(*a*). For example, *append*(*nil*, *g*(*a*), *Z*) would yield as result *Z* = *g*(*a*). To ensure that this result would not be considered valid, one would have to make explicit the calls to functions that test if the two lists to be appended are indeed lists, namely,

```
is-a-list (nil).
is-a-list (cons(H, T)):- is-a-list (T).
```

The Edinburgh PROLOG representation of *cons(H, T)* is *[H|T]*, and *nil* is [ ]. The Marseilles counterparts are *H.T* and *nil*. In the Edinburgh dialect, *append* is presented by

append([ ], *L2*, *L2*).
append([*H1*|*T1*], *L2*, [*H1*|*T*]):- *append*(*T1*, *L2*, *T*).

The query applicable to the program that uses the term *cons* is *append* (*cons(a, cons(b, nil)), cons(c, nil), Z*), and it yields *Z* = *cons(a, cons(b, cons(c, nil)))*. In the Edinburgh PROLOG, the preceding query is stated as *append*([*a*, *b*], [*c*], *Z*), and the result becomes *Z* = [*a*, *b*, *c*].

A remarkable difference between the original PASCAL-like version and the PROLOG version of *append* is the ability of the latter to determine (unknown) lists that, when appended, yield a given list as a result. For example, the query *append(X, Y, [a])* yields two solutions: *X* = [ ] *Y* = [*a*] and *X* = [*a*] *Y* = [ ].

The preceding capability is due to the generality of the search and pattern-matching mechanism of PROLOG. An often-asked question is: Is the generality useful? The answer is definitely yes! A few examples will provide supporting evidence. The first is a procedure for determining a list *LLL*, which is the concatenation of *L* with *L* the result itself being again concatenated with *L*:

triple(L, LLL):- append(L, LL, LLL), append(L, L, LL).

Note that the first *append* is executed even though *LL* has not yet been bound. This amounts to copying the list *L* and having the variable *LL* as its last element. After the second append is finished, *LL* is bound, and the list *LLL* becomes fully known. This property of postponing binding times can be very useful. For example, a dictionary may contain entries whose values are unknown. Identical entries will have values that are bound among themselves. When a value is actually determined, all communally unbound variables are bound to that value.

Another interesting example is *sublist(X, Y)*, which is true when *X* is a sublist of *Y*. Let *U* and *W* be the lists at the left and right of the sublist *X*. Then the program becomes

sublist(X, Y):- append(Z, W, Y), append(U, X, Z).

where the variables represent the sublists indicated as follows:

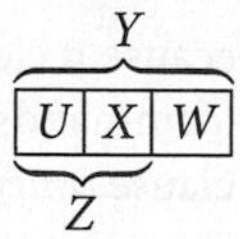

An additional role for *append* is to simulate the behavior of *member* as used in the introductory example. One could state that *member(X, L):- append(Start, [X|End], L).* in which *Start* and *End* are variables representing lists, which, when surrounding *X*, produce *L*. This approach is expensive because it may use a quadratic number of *conses*.

A final example is a bubblesort program. The specification of two adjacent elements *A* and *B* in a list *L* is done by a call:

append(_, [A, B | _], L)

The underscores stand for different variables whose names are irrelevant to the computation, and the notation [*A*, *B* | *C*] stands for *cons(A, cons(B, C))*. The rules to bubblesort then become

bsort(L, S):- append (U, [A, B | X], L), B < A, append (U, [B, A | X], M), bsort(M, S).
bsort(L, L).

The first *append* generates all pairs of adjacent elements in *L*. The literal *B* < *A* is a built-in predicate that tests whether *B* is lexicographically smaller than *A*. (Note that both *A* and *B* must be bound, otherwise a failure occurs! There will be more discussion of this limitation later.) The second *append* reconstructs the modified list, which becomes the argument in a recursive call to *bsort*. If the first clause is no longer applicable, then all pairs of adjacent elements are in order, and the second clause then provides the desired result. This version of bubblesort is space and time inefficient because *U*, the initial segment of the list, is copied twice at each level of recursion. However, the brevity of the program is indicative of the savings that can be accrued in programming and debugging. It should be now clear to the reader that the translation of a functional program into its PROLOG counterpart is easily done, but the reverse translation becomes more difficult because the functional program has to simulate nondeterminism and backtracking.

A unique property of some PROLOG programs is their ability to perform inverse computations. For example, if *p*(*X*, *Y*) defines a procedure *p* taking an input *X* and producing an output *Y*, it can (in certain cases) determine which input *X* produces *Y*. Therefore, if *p* is a differentiation program, it can also perform integration. This is easier said than done, because the use of impure features and simplifications may result in operations that are not correctly backtrackable.

The inverse computation of parsing is string generation, and parsers are now available to perform both operations. A compiler carefully written in PROLOG can also be useful in decompiling. The difficulties encountered in doing the inverse operations are frequently due to the use of impure features.

65.7 Impure Features

In the previous sections, the so-called pure features of PROLOG were described, namely, those that conform with the logic interpretation. A few impure features have been added to the language to make its use more practical. (This situation parallels the introduction of *setq*, *rplaca*, and other impure LISP functions.) However, a word of warning is in order. Some of these impure features vary from one implementation to another (O'Keefe, 1990 is an excellent reference on extralogical features).

The most prominent of these impure features is the *cut*, represented by an exclamation point. Its purpose is to let the programmer change the search control mechanism embodied by the procedure solve in Figure 65.3. Reconsider the example in Section 65.5. Assuming that *a*:- *b*, !, *c*, *d*., then in the forward mode, *b*, *c*, and *d* would, as before, be placed in the list of goals and matched with heads of clauses in the DB. If, however, a goal following the cut fails (e.g., *c*, or *d*), then no further matching of clauses defining *a* or *b* would take place. The cut is often used to increase the efficiency of programs and to prevent the consideration of alternate solutions. PROLOG purists avoid the use of cuts.

Another useful predicate is *fail*, which automatically triggers a failure. To implement it, one simply forbids its presence in the DB. Other built-in predicates that need to be introduced are the input–output commands *read* and *write*.

Once *cut* and *fail* are available, negation by failure is accomplished by the clauses:

not(*X*):- *X*, !, *fail*.
not(*X*).

which fails if *X* succeeds and vice versa. It is important to note that this artifice does not always follow the rules of true negation in logic. This version of negation illustrates that a term in the head of a rule can appear in the body as a call to a function; therefore, in PROLOG, program and data have the same form.

```
procedure solve. L: pLIST/;
 begin local i: integer;
  if L ≠ nil
   then
    for i := 1 to n do
     if match(head(Rule[i]), head(L)) then
       solve(append(tail(Rule[i]), tail(L)));
   else write ('yes')
  end;
```

FIGURE 65.3 An initial version of the interpreter.

The built-in predicates *assert* and *retract* are used to add or remove a clause from the DB; they are often used to simulate assignments. For example, the unit clause *value* (*Variable, Value*). can be asserted using bound parameters (e.g., *value*[*Z*, 37]); it can be subsequently retracted to change the actual *Value*, and then reasserted with a new value. Another use of *assert* and *retract* is associated with the built-in function *setof* that collects the multiple answers of a program and places them in a list.

The assignment is introduced using a built-in binary infix operator such as *is*. For example, Y is $X + 1$ is only valid if X has been bound to a number in which case the right-hand side is evaluated and unified with Y; otherwise the *is* fails. To have a fully backtrackable addition, one would have to use CLP. Note that $I = I + 1$ is invalid in CLP, as it should be. The equality $Z = X + Y$ in CLP is, of course, valid with some or all variables unbound.

65.8 Constraint Logic Programming

Major extensions of the unification component of PROLOG became very significant and resulted in a new area of LP called constraint logic programming or CLP. It may well have had its roots in Colmerauer's approach in generalizing the unification algorithm by making it capable of determining the satisfiability of equalities and disequalities of infinite trees (these are actually graphs containing special loops.) However, the notion of backpropagation (Sussman and Steele) dates back to the 1970s. In the case of PROLOG, the backtracking mode is triggered only in the case of unsatisfiability of a given unification. A logical extension of this approach is to make similar backtracking decisions for other (built-in) predicates, say, inequalities (i.e., ≤, ≥,…) in the domain of rationals.

The first CLP example presented here is the classic program for computing Fibonacci series. Before presenting the program, it is helpful to consider the program's PROLOG counterpart (the annotation *is* corresponds to an assignment and it has been discussed in Section 65.7):

```
fib(0,1).
fib(1,1).
fib(N, R):- N1 is N − 1.
            fib(N1, R1),
            N2 is N − 2,
            fib(N2, R2),
            R is R1 + R2.
```

The *is* predicate prevents the program from being invertible: the query *fib* (10, *X*) succeeds in producing $X = 89$ as a result, but the query *fib* (*Y*, 89) yields an error because N is unbound and the assignment to $N1$ is not performed. Note that if we had placed the predicate $N1 \geq 2$ prior to the first recursive call, the query *fib* (*Y*, 89) would also lead to an error, because the value of N is unbound and the test of inequality cannot be accomplished by the PROLOG interpreter.

The modified CLP version of the program illustrates the invertibility capabilities of CLP interpreters ($N \geq 2$ is a constraint):

```
fib(0,1).
fib(1,1).
fib(N, R1 + R2):- N ≥ 2,
                  fib(N − 1, R1),
                  fib(N − 2, R2).
```

The query ?– *fib*(10, *Fib*) yields $Fib = 89$, and the query ?– *fib*(*N*, 89) yields $N = 10$. The latter result is accomplished by solving systems of linear equations and in equations that are generated when an explicit or implicit constraint must be satisfied. In this example, the matching of actual and formal parameters results in equations. Let us perform an initial determination of those equations. The query ?– *fib* (*N*, 89) matches only the third clause. New variables $R1$ and $R2$ are created as well as the constraint $R1 + R2 = 89$. The further constraint $N >= 2$ is added to the list of satisfiable constraints, which are now $R1 + R2 = 89$ and $N >= 2$.

The recursive calls of *fib* generate further constraints that are added to the previous ones. These are $N1 = N - 1$, $N2 = N - 2$, and so forth. Recall that each recursive call is equivalent to a call by value in which new variables are created and new constraints are added (see the section on Implementation).

Therefore, unification is replaced by testing the satisfiability of systems of equations and inequations. A nontrivial implementation problem is how to determine if the constraints are satisfiable, only resorting to expensive general methods such as Gaussian elimination and the simplex method as a last resource.

The second example presented is a sorting program. For the purposes of this presentation, it is unnecessary to provide the code of this procedure (Sterling and Shapiro 1994); *qsort(L, S)* sorts an input list *L* by constructing the sorted list *S*. When *L* is a list of variables, a PROLOG interpreter would fail because unbound variables cannot be compared using the relational operator ≤. In CLP, the query ?– *qsort*([*X*1, *X*2, *X*3], *S*). yields as result $S = [X1, X2, X3], X1 \leq X2, X2 \leq X3$. When requested to provide all solutions, the interpreter will generate all of the permutations of *L* as well as the applicable constraints.

Jaffar and Lassez (1987) proved that the theoretical foundations of LP languages (see Section 65.11) remain valid for CLP languages. Several CLP languages are now widely used among LP practitioners. Among them one should mention PROLOG III and IV designed by the Marseilles group, CLP(R) designed in Australia and at IBM, CHIP created by members of the European Research Community, and CLP (BNR) designed in Canada at Bell Northern Research.

The subsequent summary describes the main CLP languages and their domains:

PROLOG IV: trees, reals, intervals, linear constraints, rationals, finite domains (including Booleans), strings
CLP(R): trees, linear constraints, floating-point arithmetic
CHIP: trees, linear constraints, floating-point arithmetic, finite domains
CLP(BNR): trees, intervals

The languages considering intervals (defined by their lower and upper bounds) deal with numeric nonlinear constraints; symbolic linear constraints are handled by the first three languages. (Additional information about interval constraints is provided in Section 65.14.)

65.9 Recent Developments in CLP (2002)

The single most important development in CLP is its amalgamation with an area of artificial intelligence (AI) that has been consistently explored since the late 1970s, namely, constraint satisfaction problems or CSPs. This merging has shown to be very fruitful because the two areas CLP and CSP share many common goals, such as concise declarative statements of combinatorial problems and efficient search strategies. In addition, the new combined areas benefit from research that has been done for decades in the field of operations research (OR).

The following subsections summarize the recent developments under the umbrella of the CLP–CSP paradigms. Highly recommended material has appeared in the textbook by Marriot and Stuckey (1998) and in the survey by Rossi (2000). Research articles have been published in the proceedings of the conferences on principles and practice of constraint programming held in 2002 (Van Hentenryck 2002).

65.9.1 CSP and CLP

Consider a set of constraints *C* containing a set of variables x_i; the x_i's are defined in their respective domains D_i. Usually, D_i is a finite set of integers. A CSP involves assigning values to the variables so that *C* is satisfied. CSP is known to be an NP-complete problem (see Chapter 5 of this handbook).

Consider now the cases where the constraints C are (1) unary, that is, they contain a single variable (e.g., $x > 0$), and (2) binary, that is, they contain two variables (e.g., $x^2 + y = 92$) and so forth. It is possible to specify a set of constraints by an undirected graph whose nodes designate each variable and an edge specifies a constraint that is applicable between any two variables. This is often called a constraint network. The notions of node consistency, arc consistency, and path consistency follow from the previous concepts. Furthermore, one can also illustrate the concept of constraint propagation, also called narrowing: it amounts to deleting from a given domain the values that do not satisfy the constraints being specified. More specifically node consistency consists of considering each node of the constraint graph and its associated unary constraints and performing a narrowing: let $c(x_i)$ be the unary constraint applicable to node x_i; then the values satisfying *not* $(c(x_i))$ are deleted from D_i.

Arc consistency consists of considering each edge, the associated binary constraint, and performing a narrowing. In the binary case, the constraints are $c(x_i, x_j)$ and the values satisfying *not* $(c(x_i, x_j))$ are deleted from D_i and D_j.

The previous text can be generalized to define k-consistency by considering constraints involving k variables (the case $k = 3$ is called path consistency, and $k > 3$ hyper-arc consistency).

One should view node and arc consistency as preprocessing the domains of a set of variables in an attempt to eliminate unfeasible values. How much narrowing can be accomplished depends on the constraint problem being considered. However, if after narrowing there remain no feasible values specified for one or more variables, then there is obviously no solution to the given set of constraints.

It is straightforward to write a CLP program that, given the unary and binary constraints pertaining to a problem, would attempt to reduce the domains of its variables. Then one could check if any of the domains becomes empty thus determining that the given constraints are unfeasible. Note that if there is a single value remaining in a domain, then one could replace the corresponding variable by its value and reattempt narrowing using node and arc consistency in the reduced constraint network. This is a typical situation that can occur in many constraint domains, including the case of systems of linear equations and inequations. To further narrow the values of the variables, one could resort to k-consistency. However, it can be shown that for k greater or equal to three, the consistency check is itself exponential. Therefore, one has to ask the following question: Is it worth to attempt further narrowing, known to be computational expensive and possibly fruitless, or should one proceed directly to an exhaustive search and determine if a given CSP is satisfiable or not? The answer is, of course, problem dependent.

Another type of narrowing occurs when dealing with interval constraints, those whose variables are integers or floating-point numbers defined within lower and upper bounds. Whenever an arithmetic expression is evaluated, the variables' bounds may narrow according to the operations being performed. This is called *bound consistency*, and as in the case of other types of consistencies, it is incomplete: failures may be detected (e.g., if the upper bound becomes lower than the lower bound), but bounds may remain unchanged. In those cases, splitting the intervals is the only resource in attempting to solve an interval constraint problem.

Much of the effort in CSP has been done in pruning the search tree by avoiding backtracking as much as possible. The reader is referred to Chapters 63 and 67 for details on how pruning can be accomplished.

65.9.2 Special Constraints

One can distinguish two kinds of special constraints: the first, called global constraints, are very useful in solving OR problems; the second, quadratic constraints, can be viewed as an effort to widen the scope of linear constraints without resorting to methods for handling general nonlinear constraints (e.g., interval constraints in Section 65.14.3).

Global constraints are essentially consistency constraints involving many variables and are implemented using specific and efficient algorithms. They perform the narrowing to each variable given as a parameter. A typical example is the predicate *all_different*, having n variables as parameters; *all_different* only succeeds if all its variables are assigned different values within a given domain.

Another useful global constraint is *cumulative*. Its parameters specify (1) the possible starting times for *n* tasks; (2) their durations; (3) the resources, machines, or personnel needed to perform each task; and (4) the total resources available. This constraint has proven its usefulness in large scheduling problems.

Algorithms for solving linear equations and inequations have been thoroughly studied in algebra and in OR. The latter field has also evidenced significant interest in quadratic programming, whereby one extends the notion of optimality to quadratic objective functions.

A way of handling quadratic constraints is through the use of interval constraints and linearization, followed by multiple applications of the simplex method. More specifically, one can attempt to linearize quadratic terms by introducing new variables and finding their minimal and maximal values using the classic simplex algorithm. This process is repeated until convergence is reached. (See paper by Lebbah et al. in Van Hentenryck (2002).) Therefore, quadratic constraints can be viewed as a variant of global constraints. An earlier interesting algebraic approach to quadratic constraints is described in Pesant and Boyer (1994).

65.9.3 Control Issues

In LP or CLP, one has limited control of control viewed as a component of the pseudo-equations *LP = logic + control* or *CLP = constraints + control*. In the LP–CLP paradigm, control is achieved through (1) ordering of rules, (2) ordering of predicates on the right-hand side of a rule, (3) usage of the *cut* or special predicates like *freeze*, and (4) metalevel interpretation. Most processors for LP–CLP utilize a rather rigid, depth-first, top-down search. (A notable exception is the XSB (Stony Brook) PROLOG processor that uses a tabling approach.)

The issues of backtracking and control have been more thoroughly investigated in CSP than in LP–CLP. This is not surprising because general LP–CLP processors usually utilize a fixed set of strategies. Control in CSP is achieved through (1) ordering of variables, (2) ordering of the values to be assigned to each variable, (3) deciding how much backtracking is performed before resuming a forward execution, and (4) utilizing information gathered at failure states to avoid them in future searches.

In CSP processors, efforts have been made to have language constructs that specify various control strategies available to a user. These have been called *indexicals* and dictate the order in which various alternative narrowing methods are applied.

A trend in control strategies is to use stochastic searches: random numbers and probability criteria are used to select starting nodes of the search tree; information gathered at failure states is also used to redirect searches. Stochastic searches are not complete (i.e., they do not ensure a solution if one exists). Therefore, one must specify a limited time for performing random searches; once that time is reached, the search can restart using new random numbers and information gathered in previous runs. Stochastic searches have proven highly successful in dealing with very large CSP.

Frequently, problems expressed as CSP have multiple successful and unsuccessful situations that are symmetric. The goal of *symmetry breaking* is to bypass searches that, due to the symmetrical nature of a problem, are uninteresting to pursue. A typical example is the *n*-queens problem in which a multitude of equivalent symmetrical solutions are found but have to be discarded; the same occurs with unsuccessful configurations that have already been proven to yield no solutions.

65.9.4 Optimization

CSP may have a large number of valid solutions. This parallels the situation where LP–CLP processors are used to solve combinatorial problems. In those cases, it is important to select one or a few among multiple solutions by specifying an objective function that has to be maximized or minimized. In the case of finite domains or interval constraints, one may have to resort to exhaustive searches.

For interval constraints, this amounts to splitting a given interval into two or more subintervals. When dealing with small finite domains, one may have to explore each possible value within the domain. The reader is referred to Chapters 15 and 63 of this Handbook where searches and combinatorial optimization are covered in greater detail. In particular, the *branch-and-bound* strategies are often used in CSP with objective functions. In the CLP paradigm, it is often up to the user to perform optimization searches by developing specific programs.

65.9.5 Soft Constraints

It is not uncommon that, in trying to establish a set of constraints, one is confronted with an overconstrained configuration that is unsatisfiable. It is then relevant to attempt to modify certain constraints so that the system becomes satisfiable.

There are several criteria for "relaxing" an overconstrained system. In general, those criteria involve some sort of optimization; for example, one may choose to violate the least number of individual constraints. Another possibility is to assign a weight to each constraint, specifying the tolerated "degree of unsatisfaction." In that case, one minimizes some function combining those specified weights.

Alternatively, one can assign probabilities to the desirability that a given constraint should be satisfied and then maximize a function expressing the combined probability for satisfying all the constraints (this is referred to as fuzzy constraints).

The usage of "preferences" is another way of loosening overconstrained systems. Consider, for example, the case of the *n*-queens problem: one may wish to allow solutions that accept attacking queens provided they are far apart. Whenever there are solutions with non-attacking queens, those are preferred over the ones having attacking but distant queens. Preferences are specified by an ordering stipulating that constraint C_1 is preferable to constraint C_2 (e.g., $C_1 > C_2$).

A work in the area of soft constraints provides a general framework for relaxing overconstrained systems (see Bistarelli et al. in Van Hentenryck, 2002). That framework ensures that if certain algebraic (semiring) properties are respected, one can define soft constraints expressing most variants of the previously illustrated constraint relaxation approaches.

65.9.6 Satisfiability Problems

The satisfiability problem can be described as follows: given a Boolean formula *B* with *n* variables, are there *true* or *false* assignments to the variables that render *B* true? (see Chapters 6 and 66 for further details.) It is usual to have *B* expressed in conjunctive normal form (CNF), consisting of conjunctions of disjunctions of variables or their negations.

Boolean formulas in CNF can be further transformed into equivalent ones whose disjunctions contain three variables negated or not or the constants *true* and *false*. That variant of the satisfiability problem is known as 3SAT and enjoys remarkable properties. The class NP-complete (see Chapter 6) congregates combinatorial problems that can be reduced to SAT problems in polynomial time. Therefore, 3SAT can be viewed as a valid standard for studying the practical complexity of hard problems. Furthermore, it is possible to show that 3SAT problems can be transformed into CSP.

Consider a random 3SAT problem with *n* variables and *m* clauses. It is intuitive to check that when *n* is large and *m* is small, the likelihood of satisfiability is high. On the other hand, when *m* is large and *n* is small, that likelihood is low. The curve expressing the ratio m/n versus the probabilities of satisfaction consists of two relatively flat horizontal components indicating high or low probabilities of satisfaction.

It has been empirically determined that in the region around $m/n = 4.5$, the curve decreases sharply; problems in that region have around 50% probability of being satisfiable. These are the computationally intensive problems for 3SAT. The problems where m/n is small or large are in general easily solvable.

The implication of this result is that there are "islands of tractability" within very hard problems. From a theoretical viewpoint, 2SAT is polynomial and the determination of other tractability islands remains of great interest. Unfortunately, the transformation of an NP-complete problem into 3SAT distorts the corresponding values of the m/n ratio. It has become important to find the boundaries of individual NP problems that allow their solutions using inexpensive, moderate, or expensive computational means.

The term *phase transition* establishes an analogy between algorithmic complexity and the physical properties of matter where temperatures and pressures are used to distinguish gaseous, liquid, and solid forms of states. Finding the boundary values delimiting regions of computational easiness and difficulty has become one of the important aims of CSP.

65.9.7 Future Developments

The novel and valuable future developments in CLP–CSP appear to be oriented in melding machine learning and data-mining approaches to the existing models. These developments parallel those that have occurred in the area of inductive logic programming (ILP) vis-a-vis LP. In ILP, Prolog programs are generated from positive and negative examples (Muggleton 1991, Bratko 2001).

In the case of CLP–CSP, the major question becomes: *Can constraints be learned or inferred from data?* If that is the case, then probabilistic or soft constraints are likely to play important roles in providing answers to that quest.

65.10 Applications

The main areas in which LP and CLP have proven successful are summed up in the following:

Symbolic manipulation. Although LISP and PROLOG are currently the main languages in this area, it is probable that a CLP language may replace PROLOG in the next few years. There is a close relationship between the aims of CLP and symbolic languages such as MAPLE, MATHEMATICA, and MACSYMA.

Numerical analysis and OR. The proposed CLP languages allow their users to generate and refine hundreds of equations and inequations having special characteristics (e.g., the generation of linear equations approximating Laplace's differential equations). The possibility of expressing inequations in a computer language has attracted the interest of specialists in OR. Difficult problems in scheduling have been solved using CLP in finite domains.

Combinatorics. Nondeterministic languages such as PROLOG have been successful in the solution of combinatorial problems. The availability of constraints extends the scope of problems that can be expressed by CLP programs.

Artificial intelligence applications. Boolean constraints have been utilized in the design of expert systems. Constraints have also been used in natural language processing and in parsing. The increased potential for invertibility makes CLP languages unique in programming certain applications. For example, the inverse operation of parsing is string generation.

Deductive databases. These applications have attracted a considerable number of researchers and developers who are now extending the DB domains to include constraints. The language DATALOG is the main representative, and its programs contain only variables or constants (no composite terms are allowed).

Engineering applications. The ease with which CLP can be used for generating and refining large numbers of equations and inequations makes it useful in the solution of engineering problems. Ohm's and Kirchhoff's laws can readily be used to generate equations describing the behavior of electrical circuits.

65.11 Theoretical Foundations

This section provides a summary of the fundamental results applicable to logic programs (Apt 1990, Lloyd 1987). It will be shown later that these results remain applicable to a wide class of constraint logic programs.

The semantics of LP and CLP are usually defined using either logic or sets. In the logic approach, one establishes that given both (1) a Horn clause program *P* and (2) a query *Q*, *Q* can be shown to be a consequence of *P*. In other words, $\neg P \vee Q$ is always true, or equivalently, using contradiction, $P \wedge \neg Q$ is always false. This relates the logical and operational meaning of programs; that is, that *Q* is a consequence of *P* can be proved by a resolution-based *breadth-first* theorem prover. (This is because a *depth-first* prover could loop in trying to determine a first solution, being therefore incapable of finding other solutions that may well exist.)

Logic-based semantics are accomplished in two steps. The first considers that a program yields a *yes* answer. In that case, the results (i.e., the bindings of variables to terms as the result of successive unifications) are the constraints that *P* and *Q* must satisfy so that *Q* becomes deducible from *P*. This is accomplished by Horn clause resolutions that render $P \wedge \neg Q$ unsatisfiable.

The second step of the proof is concerned with logic programs that yield a *no* answer and therefore do not specify bindings or constraints. In that case, it becomes important to make a stronger statement about the meaning of a clause.

Recall that in the case of *yes* answers, a Horn clause specifies that the *Head* is a consequence of the *Body* or equivalently that the *Head* is true if the *Body* is true. In the case of a *no* answer, the so-called Clark completion becomes applicable. That means that the *Head* is true *if and only if* the *Body* is true. Then the semantics of programs yielding a *no* answer amount to proving that *not* Q is a consequence of the completion of *P*. This amounts to considering the implication in *Body implies Head* in every clause of *P* as being replaced by *Body equivalent to Head*.

It should be noted that the preceding results are only applicable to queries that do not contain logic negation. For example, the query ¬ *date* (*X*, *Y*, *Z*) in the introductory example is invalid because a negative query is not in Horn clause form. Therefore, only positive queries are allowed and the behavior of the prover satisfies the so-called closed word assumption: only positive queries deducible from the program provide *yes* answers. Recent developments in the so-called nonmonotonic logic extend programs to handle negative queries.

The second approach in defining the semantics of LP and CLP uses sets. Consider the set *S*0 of all unit clauses in a program *P*. This set involves assigning any variables in these unit clauses to elements of the Herbrand universe. Consider then the clauses whose bodies contain elements of that initial set *S*0. Obviously the *Head* of those clauses is now deducible from the program and the new set *S*1 is constructed by taking the union of *S*0 with the heads of clauses that have been found to be true.

This process continues by computing the sets *S*2, *S*3, etc. Notice that *S*(*i*) always contains *S*(*i* – 1). Eventually these sets will not change because all the logical information about a finite program *P* is contained in them. This is called a least fixed point. Then *Q* is a consequence of *P* if and only if each conjunct in *Q* is in the least fixed point of *P*.

Consider, for example, the PROLOG program for adding two positive numbers specified by a successor function *s*(*X*) denoting the successor of *X*:

add(0, *X*, *X*).
add(*s*(*X*), *Y*, *s*(*Z*)):- *add*(*X*, *Y*, *Z*).

First, notice the similarity of the preceding example with *append*. Adding 0 to a number *X* yields *X* (first rule). Adding the successor of *X* to a number *Y* amounts to adding one to the result *Z* obtained by adding *X* to *Y*.

In this simple example, the Herbrand universe consists of 0, *s*(0), *s*(*s*(0)), and so on, namely, the positive natural numbers including the constant 0. The so-called Herbrand base considers all of the literals (in this case *add*) for which a binding of a variable to elements of the Herbrand universe satisfies the program rules. Using the set approach *S*0 consists of all natural numbers because any number can be added to zero. The fixed point corresponds to the infinite set of bindings of *X*, *Y*, and *Z* to elements of the Herbrand universe, which satisfies both rules.

The meaning of a program *P* and query *Q* yielding a *no* answer can also be specified using set theory. In that case, one starts with the set corresponding to the Herbrand universe *H*0. Then this set is reduced to a smaller set by using once the rules in *P*. Call this new set *H*1. By applying again the rules in *P*, one obtains *H*2, and so on. In this case there is not necessarily a fixed point *Hn*. The property pertaining to programs yielding *no* answers then consists of the following statement: *not* Q is a logical consequence of the completion of Q if and only if some conjunct of Q is not a member of *Hn*.

In addition to programs yielding *yes* or *no* answers, there are those that loop. The halting problem tells us that we cannot hope to detect all of the programs that will eventually loop. Let us consider, as an example, the program P_1:

p(*a*).
p(*b*): – *p*(*b*).

As expected, the queries Q_1: $p(a)$ and Q_2: $p(c)$ yield, respectively, *yes* and *no* because $p(a)$ is a consequence of P_1, and $\neg p(c)$ is a consequence of the completion of p_1; that is,

$$p(X) \equiv (X = a) \vee (X = b \wedge p(b))$$

However, the interpreter will loop for the query Q_3: $p(b)$, or when all solutions of Q_4: $p(X)$ are requested.

As mentioned earlier, the preceding theoretical results can be extended to CLP languages. Jaffar and Lassez (1987) established two conditions that a CLP extension to PROLOG must satisfy so that the semantic meaning (using logic or sets) is still applicable. The first is that the replacement of unification by an algorithm that tests the satisfiability of constraints should always yield a *yes* or *no* answer. This property is called *satisfaction-completeness* and it is obviously satisfied by the unification algorithm in the domain of trees. Similarly, the property applies to systems of linear equations in the domain of rationals and even to polynomial equations but with a significantly larger computational cost.

The second of Jaffar and Lassez's conditions is called *solution-compactness*. It basically states that elements of a domain (say, irrational numbers) can be defined by a potentially infinite number of more stringent constraints that bound their actual values by increasingly finer approximations. For example, the real numbers satisfy this requirement. For more detail on Jaffar and Lassez's theory, see Jaffar and Maher (1994) and Cohen (1990). Existing CLP languages satisfy the two conditions established by Jaffar and Lassez.

The beauty of the Jaffar and Lassez metatheory is that they have established conditions under which the basic theorems of LP remain valid, provided that the set of proposed axioms specifying constraints satisfy the described properties.

A convenient (although incomplete) taxonomy for CLP languages is to classify them according to their domains or combinations thereof. One can have CLP (rationals) or CLP (Booleans, reals). PROLOG can be described as CLP (trees) and CLP (R) as CLP (reals, trees). A complete specification of CLP language would also have to include the predicates and operations allowed in establishing valid constraints.

From the language design perspective, the designer would have to demonstrate the correctness of an efficient algorithm implementing the test for constraint satisfiability. This is equivalent to proving the satisfiability of the constraints specified by the axioms.

65.12 Metalevel Interpretation

Metalevel interpretation allows the description of interpreters for the languages (such as LISP or PROLOG), using the languages themselves. In PROLOG, the metalevel interpreter for pure programs consists of a few lines of code. The procedure solve has as a parameter a list of PROLOG goals to be processed. The interpreter assumes that the program rules are stored as unit clauses:

clause(*Head, Body*).

each corresponding to a rule: *Head:- Body.*, where *Head* is a literal and *Body* is a list of literals. Unit clauses are stored as *clause(Head, [ ])*. The interpreter is

solve([ ]).
solve([Goal|Restgoal]):- solve (Goal), solve(Restgoal).
solve(Goal):- clause (Goal, Body), solve(Body).

The first rule states that an empty list of goals is logically correct. (In that case, the interpreter should print the latest bindings of the variables.) The second rule states that when processing (i.e., *solv*ing) a list of goals, the *head* and then the *tail* of the list should be processed. The third rule specifies that when a single *Goal* is to be processed, one has to look up the database containing the clauses and process the *Body* of the applicable clause. In the preceding interpreter, metainterpreter unification is implicit. One could write metainterpreters in which the built-in unification is replaced by an explicit sequence of PROLOG constructs using the impure features.

A very useful extension often incorporated into interpreters is the notion of co-routining or lazy evaluation. The built-in procedure *freeze(X, P)* tests whether the variable *X* has been bound. If so, *P* is executed; otherwise, the pair (*X, P*) is placed in a freezer. As soon as *X* becomes bound, *P* is placed at the head of the list of goals for immediate execution.

The procedure *freeze* can be easily implemented by expressing it as a variant of *solve* also written in PROLOG. Although this metalevel programming will of course considerably slow down the execution, this capability can and has been used for fast prototyping extensions to the language (Sterling and Shapiro 1994, Cohen 1990).

Another important application of metalevel programming is partial evaluation. Its objective is to transform a given program (a set of procedures) into an optimized version in which one of the procedures has one or more parameters that are bound to a known value. An example of partial evaluation is the automatic translation of a simple (inefficient) pattern-matching algorithm, which tests if a given pattern appears in a text. When the pattern is known, a partial evaluator applied to the simple matching algorithm produces the equivalent of the more sophisticated Knuth–Morris–Pratt pattern-matching algorithm.

In a metalevel interpreter for a CLP language, a rule is represented by *clause (Head, Body, Constraints)*. Corresponding to a rule *Head:- Body {Constraints}*.

The modified procedure solve contains three parameters: (1) the list of goals to be processed, (2) the current set of constraints, and (3) the new set of constraints obtained by updating the previous set. The metalevel interpreter for CLP, written in PROLOG becomes

```
solve([], C, C).
solve([Goal | Restgoal], Previous_C, New_C):-
    solve(Goal, Previous_C, Temp_C),
    solve(Restgoal, Temp_C, New_C).
solve(Goal, Previous_C, New_C):-
    clause(Goal, Body, Current_C),
    merge_constraints(Previous_C, Current_C, Temp_C),
    solve(Body, Temp_C, New_C).
```

The heart of the interpreter is the procedure *merge_constraints*, which merges two sets of constraints: (1) the previous constraints, *Previous_C*, and (2) the constraints introduced by the current clause, *Current_C*. If there is no solution to this new set of constraints, the procedure fails; otherwise, it simplifies the resulting constraints, and it binds any variables that have been constrained to take a unique value. For example, the constraints $X \leq 0 \wedge X \geq 0$ simplify to the constraint $X = 0$, which implies that X can now be bound to 0.

The design considerations that influence the implementation of this procedure will be discussed in Section 65.13. Note that the controversial unit logical inference steps per second (LIPS), often used to estimate the speed of PROLOG processors, loses its significance in the case of a constraint language. The number of LIPS is established by counting how many times per second the procedure *clause* is activated; in the case of CLP, this time to process *clause* and *merge_constraints* may vary significantly, depending on the constraints being processed.

65.13 Implementation

It is worthwhile to present the basic LP implementation features by describing an interpreter for the simplified PROLOG of Section 65.5 written in a Pascal or C-like language. The reader should note the similarities between the metalevel interpreter *solve* of the previous section and the one about to be described. The rules will be stored sequentially in a database implemented as a 1D array *Rule*[*l..n*] and containing pointers to a special type of linear list. Such a list is a record with two fields, the first storing a letter and the second being either *nil* or a pointer to a linear list. Let the (pointer) function *cons* be the constructor of a list element, and assume that its fields are accessible via the (pointer) functions *head* and *tail*. The first rule is stored in the database by

Rule[1] := *cons*('*a*', *cons* ('*b*', *cons* ('*c*', *cons* ('*d*', *nil*)))).

The fifth rule defining a unit clause is stored as *Rule*[5] := *cons* ('*e*', *nil*). Similar assignments are used to store the remaining rules.

The procedure *solve* that has as a parameter a pointer to a linear list is capable of determining whether or not a query is successful. The query itself is the list with which *solve* is first called. The procedure uses two auxiliary procedures *match* and *append; match* (*A*, *B*) simply tests if the alphanumeric *A* equals the alphanumeric *B*; *append*(*L1*, *L2*) produces the list representing the concatenation of *L1* with *L2* (this is equivalent to the familiar *append* function in LISP: it basically copies *L1* and makes its last element point to *L2*).

The procedure *solve*, written in a Pascal-like language, appears in Figure 65.3. Recall that the variable *n* represents the number of rules stored in the array *Rule*. The procedure performs a depth-first search of the problem space where the local variable is used for continuing the search in case of a failure. The head of the list of goals *L* is matched with the head of each rule. If a match is found, the procedure is called recursively with a new list of goals formed by adding (through a call of *append*) the elements of the tail of the matching rule to the goals that remain to be satisfied. When the list of goals is *nil*, all goals have been satisfied and a success message is issued. If the attempts to match fail, the search is continued in the previous recursion level until the zeroth level is reached, in which case no more solutions are possible. For example, the query *a, e.* is expressed by *solve* (*cons* ('*a*', *cons* ('*e*', *nil*))) and yields the two solutions presented in Section 65.5 on resolution of Horn clauses.

Note that if the tree of choices is finite, the order of the goals in the list of goals is irrelevant insofar as the presence and number of solutions are concerned. Thus, the order of the parameters of *append* in Figure 65.3 could be switched, and the two existing solutions would still be found. Note that if the last rule were replaced by *a:-f, a.*, the tree of choices would be infinite and solutions similar to the first solution would be found repeatedly. The procedure *solve* in Figure 65.3 can handle these situations by generating an infinite sequence of solutions. However, had the preceding rule appeared as the first one, the procedure *solve* would also loop, but without yielding any solutions. This last example shows how important the ordering of the rules is to the outcome of a query. This explains Kowalski's dictum program = logic + control, in which control stands for the ordering and (impure) control features such as the cut (Kowalski 1979).

It is not difficult to write a recursive function *unify*, which, given two terms, tests for the result of unification using the contents of Table 65.1 (Section 65.5). For this purpose, one has to select a suitable data structure. In a sophisticated version, terms are represented by variable-sized records containing pointers to other records, to constants, or to variables. Remark that the extensive updating of linked data structures inevitably leads to unreferenced structures that can be recovered by a garbage collection. It is frequently used in most PROLOG and CLP processors.

A simpler data structure uses linked lists and the so-called Cambridge Polish notation. For example, the term $f(X, g(Y, c))$ is represented by $(f(var\ x)(g(var\ y)(const\ c)))$, which can be constructed with *conses*.

As mentioned in Section 65.5, if the result of unification results in the binding $Y{:=}\ g(Y, c)$, then (most) PROLOG interpreters would soon get into difficulties because most implementations of the described unification algorithm cannot handle circular structures. Manipulating these structures (e.g., printing, copying) would result in an infinite loop unless the so-called occur check is incorporated to test for circularity.

The additional machinery needed to incorporate unification into the procedure *solve* of Figure 65.3 is described in Cohen (1985). An important remark is in order: when introducing unification, it is necessary to *copy* the clauses in the program and introduce new variables (which correspond to parameters that should be called by value). The frequent copying and updating of lists makes it almost mandatory to use garbage collection, which is often incorporated to LP processors.

65.13.1 Warren Abstract Machine

D.H.D. Warren, a pioneer in the compilation of PROLOG programs, proposed in 1983 a set of primitive instructions that can be generated by a PROLOG compiler, usually written using PROLOG. (Warren's approach parallels that of P-code used in early Pascal compilers.) The *Warren abstract machine* (WAM) primitives can be efficiently interpreted using specific machines.

The main data structures used by the WAM are (1) the recursion stack, (2) the heap, and (3) the trail. The heap is used for storing terms and the trail for backtracking purposes. The WAM uses the copying approach mentioned in the beginning of this section. A local garbage collector takes advantage of the cut by freeing space in the trail.

The WAM has been used extensively by various groups developing PROLOG compilers. Its primitives are of great efficiency in implementing features such as tail-recursion elimination, indexing of the head of the clause to be considered when processing a goal, the cut, and other extralogical features of PROLOG. A useful reference in describing the WAM is the one by Ait-Kaci (1991). A reference on implementation is by Van Roy (1994).

65.13.2 Parallelism

Whereas for most languages it is fairly difficult to write programs that automatically take advantage of operations and instructions that can be executed in parallel, PROLOG offers an abundance of opportunities for parallelization. There are at least three possibilities for performing PROLOG operations in parallel:

1. *Unification*. Because this is one of the most frequent operations in running PROLOG programs, it would seem worthwhile to search for efficient parallel unification algorithms. Some work has already been done in this area (Jaffar et al. 1992). However, the results have not been encouraging.
2. *And parallelism*. This consists of simultaneously executing each procedure in the tail of a clause. For example, in $a(X, Y, U)$:- $b(X, Z)$, $c(X, Y)$, $d(T, U)$, an attempt is made to continue the execution in parallel for the clauses defining *b*, *c*, and *d*. The first two share the common variable *X*; therefore, if unification fails in one but not in the other, or if the unification yields different bindings, then some of the labor done in parallel is lost. However, the last clause in the tail can be executed independently because it does not share variables with the other two.
3. *Or parallelism*. When a given predicate is defined by several rules, it is possible to attempt to apply the rules simultaneously. This is the most common type of parallelism used in PROLOG processors. Kergommeaux and Codognet (1994) is a recommended survey of parallelism in PROLOG.

65.13.3 Design and Implementation Issues in Constraint Logic Programming

There is an important implementation consideration that appears to be fulfilled in both CLP(R) and PROLOG IV: the efficiency of processing PROLOG programs (without constraints) should approach that of current PROLOG interpreters; that is, the overhead for *recognizing* more general constraints should be small.

There are three factors that should be considered when selecting algorithms for testing the satisfiability of systems of constraints used in conjunction with CLP processors. They are (1) incrementality, (2) simplification, and (3) canonical forms. The first is a desirable property that allows an increase in efficiency of multiple tests of satisfiability (by avoiding recomputations). This can be explained in terms of the metalevel interpreter for CLP languages described in Section 65.11: if the current system of constraints *S* is known to be satisfiable, the test of satisfiability should be incremental, minimizing the computational effort required to check if the formula remains satisfiable or not. Classic PROLOG interpreters have this property because previously performed unifications are not recomputed at each inference step. There are modifications of Gaussian methods for solving linear equations that also satisfy this property. This is accomplished by introducing temporary variables and replacing the original system of equations by an equivalent solved form (see Section 65.5): *variable = linear terms involving only the temporary variables*.

The simplex method can also be modified to satisfy incrementality. Similarly, the SL resolution method for testing the satisfiability of Boolean equations and the Gröbner method for testing the satisfiability of polynomial equations have this property.

In nearly all the domains considered in CLP, it may be possible to replace a set of constraints by a simpler set. This simplification can be time-consuming but is sometimes necessary. The implementor of CLP languages may have to make a difficult choice as to what level of simplification should occur at each step verifying constraint satisfaction. It may turn out that a system of constraints eventually becomes unsatisfiable, and all of the work done in simplification is lost. When a final result has to be output, it becomes essential to simplify it and present it to the reader in the clearest, most readable form.

An important function of simplification is to detect the assignment of a variable to a single value (e.g., from $X \geq 1$ *and* $X \leq 1$ *one infers* $X = 1$). This property is essential when implementing a modified simplex method that detects when a variable is assigned to a single value. Note that this detection is necessary when using lazy evaluation.

The incremental algorithms for testing the satisfiability of linear equations and inequations, as well as that used in the Gröbner method for polynomial equations, are capable of discarding redundant equations; therefore, they perform some simplifications (Sato and Aiba 1993).

The canonical (solved) forms referred to earlier in this section can be viewed as (internal) representations of the constraints, which facilitate both the tests of satisfiability and the ensuing simplifications. For example, in the case of the Gröbner method for solving polynomial equations, the input polynomials are internally represented in a normal form, such that variables are lexicographically ordered and the terms of the polynomials are ordered according to their degrees. This ordering is essential in performing the required computations. Also note that if two seemingly different constraints have the same canonical form, only one of them needs to be considered. Therefore, the choice of appropriate canonical forms deserves an important consideration in the implementation of CLP languages (Jaffar and Maher 1994).

65.13.4 Optimization Using Abstract Interpretation

Abstract interpretation is an enticing area of computer science initially developed by Cousot and Cousot (1992); it consists of considering a subdomain of the variables of a program (usually a Boolean variable, e.g., one representing the evenness or oddness of the final result). Program operations and constructs are performed using only the desired subdomain. Cousot proved that if certain conditions are applicable to the subdomains and the operations acting on their variables, the execution is guaranteed to terminate. Dataflow analyses, partial evaluation, detection of safe parallelism, etc., can be viewed as instances of abstract interpretation. The research group at the University of Louvain, Belgium, has been active in exploring the capabilities of abstract interpretation in LP and CLP.

65.14 Research Issues

It is worthwhile to classify the numerous extensions of PROLOG into three main categories, namely, those related to (1) resolution beyond Horn clauses, (2) unification, and (3) others (e.g., concurrency). Major extensions of the unification became very significant and resulted in a new area of LP called constraint logic programming or CLP that was dealt with in Section 65.8; nevertheless, the more recent addition to CLP dealing with the domain of intervals is discussed in this section.

65.14.1 Resolution beyond Horn Clauses

Several researchers have suggested extensions for dealing with more general clauses and for developing semantics for negation that are more general than that of negation by failure (see Section 65.7). Experience has shown that the most general extension, that is, to the general predicate calculus, poses difficult combinatorial search problems. Nevertheless, substantial progress has been made in extending LP beyond pure Horn clauses. Two such extensions deserve mention: stratified programs and generalized predicate calculus formulas in the *body* part of a clause.

Stratified programs are variants of Horn clause programs that are particularly applicable in deductive databases; true negation may appear in the body of clauses, provided that it satisfies certain conditions. These stratified programs have clean semantics based on logic and avoid the undesirable features of negation by failure (see Minker 1987).

To briefly describe the second extension, it is worthwhile to recall that the procedural interpretation of resolution applied to Horn clauses is based on the substitution model: a procedure call consists of replacing the call by the body of the procedure in which the formal parameters are substituted by the actual parameters via unification. The generalization proposed by Ueda and others can use the substitution model to deal with the clauses of the type *head*:- (*a general formula in the predicate calculus containing quantifiers and negation*).

65.14.2 Concurrent Logic Programming and Constraint Logic Programming

A significant extension of LP has been pursued by several groups. A premise of their effort can be stated as follows: a programming language worth its salt should be expressive enough to allow its users to write complex but efficient operating systems software (as is the case of the C language). With that goal in mind, they incorporated into LP the concepts of *don't care nondeterminism* as advocated by Dijkstra. The resulting languages are called concurrent LP languages. The variants proposed by these groups were implemented and refined; they have now converged to a common model that is a specialized version of the original designs.

Most of these concurrent languages use special punctuation marks "?" and "|". The question mark is a shorthand notation for *freezes*. For example, the literal *p*(*X* ?, *Y*) can be viewed as a form of *freeze*(*X*, *p*(*X*, *Y*)). The vertical bar is called *commit* and usually appears once in the tail of clauses defining a given procedure. Consider, for example,

a:- *b*, *c* | *d*, *e*.
a:- *p* | *q*.

The literals *b*, *c*, and *d*, *e* are executed using *and* parallelism. However, the computation using *or* parallelism for the two clauses defining *a* continues only with the clause that first reaches the *commit* sign.

For example, if the computation of *b, c* proceeds faster than *p*, then the second clause is abandoned, and execution continues with *d, e* only (see Saraswat 1993).

65.14.3 Interval Constraints

The domain of interval arithmetic has become a very fruitful area of research in CLP. Older has been a pioneer in this area (Older and Vellino 1993). This domain specifies reals as being defined between lower and upper bounds that can be large integers or rational numbers. The theory of solving most nonlinear and trigonometric equations using intervals guarantees that *if* there is a solution, that solution must lie within the computed intervals. Furthermore, it is also guaranteed that *no* solution exists outside the computed interval or unions of intervals.

The computations involve the operation of *narrowing* that consists of finding new bounds for a quantity denoting the result of an operation (say, +, *, sin, etc.) involving operands, which are also defined by their lower and upper bounds. The narrowing operation also involves intersecting intervals obtained by various computations defining the same variable. The intersection may well fail (e.g., the equality operation applying to operands whose intervals are disjoint). The narrowing is guaranteed to either converge or fail. This, however, may not be sufficient to find possible solutions of interest. One can nevertheless split a given interval into two or more unions of intervals and proceed to find a more precise solution, if one exists. This is akin to enumeration of results in CLP.

The process of splitting is a *don't know nondeterministic* choice, an existing component of LP. The failure of the narrowing operation is analogous to that encountered in CLP when a constraint is unsatisfiable and backtracking occurs. Therefore, there is a natural interaction between LP and the domain of intervals. Interval arithmetic is known to yield valuable results in computing the satisfiability of nonlinear constraints or in the case of finite domains. Its use in linear constraints is an active area of research because results indicate a poor convergence of narrowing. In the case of polynomials, constraints interval arithmetic may well be a strong competitor to Gröebner base techniques.

65.14.4 Constraint Logic Programming Language Design

A current challenge in the design and implementation of CLP is to blend computations in different domains in a harmonious and sound manner. For example, the reals can be represented by intervals whose bounds are floating-point numbers (these have to be carefully implemented to retain soundness due to rounding operations). Actually, floating-point numbers are nothing more than (approximate) very large integers or fractions. This set is, of course, a superset of finite domains, which in turn is a superset of Booleans. Problems in CLP language design that still remain to be solved include how to handle the interaction of these different domains and subdomains. This situation is further complicated by efficiency considerations. Linear inequations, equations, and disequations can be efficiently solved using rational arithmetic, but research remains to be done in adapting simplex-like methods to deal with interval arithmetic.

65.15 Conclusion

As in most sciences, there has always been a valuable symbiosis among the theoretical and experimental practitioners of computer science, including, of course, those working in LP. Three examples come to mind: the elimination of the occur test in unifications, the cut, and the *not* operator as defined in PROLOG. These features were created by practical programmers and are here to stay. They provide a vast amount of food for thought for theoreticians. As mentioned earlier, the elimination of the occur test was instrumental in the development of algorithms for unification of *infinite trees*. Although the concept of the cut has resisted repeated attempts for a clean semantic definition, its use is unavoidable in increasing the efficiency of programs. Finally, PROLOG's *not* operator has played a key role in extending logic programs beyond Horn clauses.

CLP is one of the most promising and stimulating new areas in computer science. It amalgamates the knowledge and experience gained in areas as varied as numerical analysis, OR, artificial languages, symbolic processing, artificial intelligence, logic, and mathematics.

During the past 20 years, LP has followed a creative and productive course. It is not unusual for a fundamental scientific endeavor to branch out into many interesting subfields. An interesting aspect of these developments is that LP's original body of knowledge actually branched into subareas, which joined previously existing research areas. For example, CLP is being merged with the area of CSPs; LP researchers are interested in modal, temporal, intuitionistic, and linear logic; relational database research now includes constraints; and OR and CLP have found previously unexplored similarities.

The several subfields of LP now include research on CLP in various domains, typing, nonmonotonic reasoning, inductive LP, semantics, concurrency, nonstandard logic, abstract interpretation, partial evaluation, and blending with functional and with object-oriented language paradigms. It will not be surprising if each of these subfields will become fairly independent from their LP roots, and the various specialized groups will organize autonomous journals and conferences. The available literature on LP is abundant, and it is likely to be followed by a plentiful number of publications in its autonomous subfields.

Key Terms

Backtracking: A manner to handle (*don't know*) nondeterministic situations by considering *one* choice at a time and storing information that is necessary to restore a given state of the computation. PROLOG interpreters often use backtracking to implement nondeterministic situations.

Breadth first: A method for traversing trees in which all of the children of a node are considered simultaneously. OR parallel PROLOG interpreters use breadth-first traversal.

Clause: A general normal form for expressing predicate calculus formulas. It is a disjunction of literals ($P_1 \vee P_2 \vee \ldots$) whose arguments are terms. The terms are usually introduced by eliminating existential quantifiers.

Constraint logic programming languages: PROLOG-like languages in which unification is replaced or complemented by constraint solving in various domains.

Constraints: Special predicates whose satisfiability can be established for various domains. Unification can be viewed as equality constraints in the domain of trees.

Cut: An annotation used in PROLOG programs to bypass certain nondeterministic computations.

Depth first: A method for traversing trees in which the leftmost branches are considered first. Most sequential PROLOG interpreters use depth-first traversal.

Don't care nondeterminism: The arbitrary choice of one among multiple possible continuations for a computation.

Don't know nondeterminism: Situations in which there are equally valid choices in pursuing a computation.

Herbrand universe: The set of all terms that can be constructed by combining the terms and constants that appear in a logic formula.

Horn clause: A clause containing (at most) one positive literal. The term *definite clause* is used to denote a clause with exactly one positive literal. PROLOG programs can be viewed as a set of definite clauses in which the positive literal is the head of the rule and the negative literals constitute the body or tail of the rule.

Infinite trees: Trees that can be unified by special unification algorithms, which bypass the occur check. These trees constitute a new domain, different from that of usual PROLOG trees.

Metalevel interpreter: An interpreter written in L for the language L.

Occur check: A test performed during unification to ensure that a given variable is not defined in terms of itself (e.g., $X = f(X)$ is detected by an occur check, and unification fails).

Predicate calculus: A calculus for expressing logic statements. Its formulas involve the following:

- *Atoms:* $P(T_1, T_2, \ldots)$ where P is a predicate symbol and the T_i are terms.
- *Boolean connectives:* Conjunction ($\wedge$), disjunction ($\vee$), implication ($\rightarrow$), and negation ($\neg$).
- *Literals:* Atoms or their negations.
- *Quantifiers:* For all ($\forall$), there exists ($\exists$).
- *Terms* (also called *trees*): Constructed from constants, variables, and function symbols.

Resolution: A single inference step used to prove the validity or predicate calculus formulas expressed as clauses. In its simplest version: $P \vee Q$ and $\neg P \vee R$ imply $Q \vee R$, which is called the resolvant.

SLD resolution: Selective linear resolution for definite clauses inference step used in proving the validity of Horn clauses.

Unification: Matching of terms used in a resolution step. It basically consists of testing the satisfiability or the equality of trees whose leaves may contain variables. Unification can also be viewed as a general parameter matching mechanism.

Warren abstract machine (WAM): An intermediate (low-level) language that is often used as an object language for compiling PROLOG programs. Its objective is to allow the compilation of efficient PROLOG code.

Further Information

There are several journals specializing in LP and CLP. Among them we mention the *Journal of Logic Programming* (North-Holland), *New Generation Computing* (Springer-Verlag), and *Constraint* (Kluwer). Most of the conference proceedings have been published by MIT Press. Recent proceedings on constraints have been published in the Lecture Notes in Computer Science (LNCS) series published by Springer-Verlag.

A newsletter is also available (Logic Programming Newsletter, alp@doc.ic.ac.uk). Among the references provided, the following relate to CLP languages: PROLOG III (Colmerauer 1990), CLP(R) (Jaffar et al. 1992), CHIP (Dincbas et al. 1988), CAL (Sato and Aiba 1993), finite domains (Van Hentenryck 1989), and intervals (Older and Vellino 1993). The recommended textbooks include Clocksin and Mellish (1984) and Sterling and Shapiro (1994). The theoretical aspects of LP are well covered in Apt (1990) and Lloyd (1987) and implementation in Ait-Kaci (1991), Kergommeaux and Codognet (1994), Van Roy (1994), and Warren (1983).

References

Ait-Kaci, H. 1991. *The WAM: A (Real) Tutorial.* MIT Press, Cambridge, MA.

Apt, K.R. 1990. Logic programming. In *Handbook of Theoretical Computer Science.* J. van Leewun (ed.), pp. 493–574. MIT Press, Amsterdam, the Netherlands.

Bergin, T.J. 1996. *History of Programming Languages HOPL 2.* Addison-Wesley, Reading, MA.

Borning, A. 1981. The programming language aspects of thing-lab, a constraint-oriented simulation laboratory. *ACM TOPLAS,* 3(4):252–387.

Bratko, I. 2001. *Prolog Programming for Artificial Intelligence,* 3rd edn., Addison-Wesley, Reading, MA.

Clocksin, W.F. and Mellish, C.S. 1984. *Programming in PROLOG,* 2nd edn. Springer-Verlag, New York.

Cohen, J. 1985. Describing PROLOG by its interpretation and compilation. *Commun. ACM,* 28(12):1311–1324.

Cohen, J. 1988. A view of the origins and development of PROLOG. *Commun. ACM,* 31(1):26–36.

Cohen, J. 1990. Constraint logic programming languages. *Commun. ACM,* 33(7):52–68.

Cohen, J. and Hickey, T.J. 1987. Parsing and compiling using PROLOG. *ACM Trans. Prog. Lang. Syst.,* 9(2):125–163.

Colmerauer, A. 1990. An introduction to PROLOG III. *Commun. ACM,* 33(7):69–90.

Cousot, P. and Cousot, R. 1992. Abstract interpretation and applications to logic programs. *J. Logic Program.*, 13(2/3):103–179.

Dincbas, M., Van Hentenryck, P., Simonis, H., Aggoun, A., Graf, T., and Berthier, F. 1988. The constraint logic programming language CHIP, pp. 693–702. In *FGCS'88, Proceedings of the International Conference on Fifth Generation Computer Systems*, Vol. 1. Tokyo, Japan, December.

Jaffar, J. and Lassez, J.-L. 1987. Constraint logic programming, pp. 111–119. In *Proceedings of the 14th ACM Symposium on Principles Programming Language,* Munich, Germany.

Jaffar, J. and Maher, M. 1994. Constraint logic programming, a survey. *J. Logic Program.*, 19(20):503–581.

Jaffar, J., Michaylov, S., and Yap, R.H.C. 1992. The CLP language and system. *ACM Trans. Progr. Lang. Syst.*, 14(3):339–395.

Kergommeaux, J.C. and Codognet, P. 1994. Parallel LP systems. *Comput. Surv.*, 26(3):295–336.

Kowalski, R. and Kuehner, D. 1970. Resolution with selection function. *Artif. Intell.*, 3(3):227–260.

Kowalski, R.A. 1979. Algorithm = logic + control. *Commun. ACM*, 22(7):424–436.

Lloyd, J.W. 1987. *Foundations of Logic Programming*. Springer-Verlag, New York.

Marriot, K. and Stuckey, P.J. 1998. *Programming with Constraints: An Introduction*. MIT Press, Cambridge, MA.

Minker, J. (ed.) 1987. *Foundations of Deductive Databases and Logic Programming*. Morgan Kaufmann, Los Altos, CA.

Muggleton, S. 1991. Inductive logic programming. *New Gener. Comput.*, 8(4):295–318.

O'Keefe, R. A. 1990. *The Craft of PROLOG*. MIT Press, Cambridge, MA.

Older, W. and Vellino, A. 1993. Constraint arithmetic on real intervals. In *Constraint Logic Programming: Selected Research*. F. Benhamou and A. Colmerauer (eds.), MIT Press, Cambridge, MA.

Pesant, G. and Boyer, M. 1994. QUAD-CLP(R): Adding the power of quadratic constraints. In *Principles and Practice of Constraint Programming, Second International Workshop*. A. Borning (ed.), Lecture notes in Computer Science, 1865, pp. 40–74. Springer, Berlin, Germany.

Robinson, J.A. 1965. A machine-oriented logic based on the resolution principle. *J. ACM*, 12(1):23–41.

Rossi, F. 2000. Constraint (logic) programming: A survey on research and applications. In *New Trends in Constraints: Joint ERCIM/Compulog Net Workshop*. K. Apt et al. (eds.), Springer, Berlin, Germany.

Saraswat, V.A. 1993. *Concurrent Constraint Programming Languages*. MIT Press, Cambridge, MA.

Sato, S. and Aiba, A. 1993. An application of CAL to robotics. In *Constraint Logic Programming: Selected Research*. F. Benhamou and A. Colmerauer (eds.), pp. 161–174. MIT Press, Cambridge, MA.

Shapiro, E. 1989. The family of concurrent LP languages. *Comput. Surv.*, 21(3):413–510.

Sterling, L. and Shapiro, E. 1994. *The Art of PROLOG*. MIT Press, Cambridge, MA.

Van Hentenryck, P. 1989. *Constraint Satisfaction in Logic Programming*. Logic programming series, MIT Press, Cambridge, MA.

Van Hentenryck, P. (ed.) 2002. *Principles and Practice of Constraint Programming: Proceedings of the 8th International Conference*, Lecture Notes in Computer Science. Springer, Berlin, Germany.

Van Roy, P. 1994. The wonder years of sequential PROLOG implementation, 1983–1993. *J. Logic Progr.*, 19(20):385–441.

Warren, D.H.D. 1983. *An Abstract PROLOG Instruction Set*. Technical Note 309, SRI International, Menlo Park, CA.

66

Multiparadigm Languages

Michael Hanus
Christian-Albrechts-University of Kiel

66.1 Introduction

High-level programming languages can be classified according to their style of programming, that is, the intended techniques to construct programs from basic elements and composing them to larger units. This is also called a *programming paradigm*. For instance, *imperative programming* is centered around named memory cells (variables) whose contents could be modified by assignments, and assignments are grouped to larger programs using control structures and procedural abstraction. *Functional programming* is based on the concept of defining functions in a mathematical sense and composing them to more complex functions. *Logic and constraint programming* focuses on defining relations between entities and solving goals containing free variables. These are the main programming paradigms.

Other paradigms often occur in combination with these main paradigms. For instance, *object-oriented programming* focuses on objects containing data and operations as basic units and inheritance to structure them. This method of organization can be combined with each of the main paradigms. For instance, Java combines object orientation with imperative programming and OCaml provides object orientation in a functional (and imperative) context. Similarly, *concurrent programming* focuses on threads or processes, which can be composed in a concurrent or sequential manner, and methods for communication and synchronization. Depending on the style of implementing the behavior of threads and processes, concurrent programming is combined with one or more of the main paradigms so that there are imperative concurrent (e.g., Java), functional concurrent (e.g., Alice, Erlang, and Concurrent Haskell), and logic concurrent (e.g., Oz) languages.

When developing software, one should select a programming language that is most appropriate for the specific task to do. If the software system is complex, it is also quite natural to use various programming languages belonging to different paradigms. For instance, database access and manipulation could be done by logic-oriented languages, transformational tasks (like XML transformations) are reasonably implemented by functional languages, and state-oriented user interaction is implemented in imperative (object-oriented) languages. However, the use of different programming languages in one application could be cumbersome since one has to use different programming environments and implement the exchange of data between different parts of the programs. Moreover, a textual data exchange between different systems is a well-known security risk [22].

In order to improve this situation, *multiparadigm languages* have been proposed. They offer more than one programming paradigm in a single language so that "the well-educated computer programmer…should then be able to select a solution technique that best matches the characteristics of the problem to be solved" [10]. Ideally, paradigms are amalgamated in an orthogonal manner so that the programmer could freely mix the different programming techniques. In practice, however, some restrictions, depending on the amalgamated paradigms, are required to keep the advantages of the individual paradigms. For instance, in an imperative functional language, one might not freely use imperative features inside functional parts, otherwise one looses the freedom to choose optimal evaluation strategies for functional computations. Thus, the challenge in designing multiparadigm languages is to provide the features of the individual paradigms in a single language without too many restrictions.

66.2 Principles of Programming Paradigms

To understand the features of multiparadigm languages, we review in this section the basic ideas and features of the main programming paradigms. More details about these paradigms can be found in other chapters of this handbook.

66.2.1 Imperative Languages

Imperative programming is based on the mutation of variables in order to compute an intended result. Thus, they strongly reflect the computer architecture pioneered by John von Neumann where data stored in memory can be updated. In order to abstract from memory addresses, variables have names and the current computation environment maps these names to memory cells. The environment can change in imperative languages with block structures or procedures. The basic operation of an imperative language is the assignment statement where the value of a memory cell associated to a variable is changed. Collections of memory cells can be composed to data structures and collections of assignments can be composed to larger units by control structures. Procedural abstraction is used to organize larger programs. A procedure assigns a name and parameters to some computation unit so that algorithmic entities can be reused as simple statements. Procedures can also have result values (then they are called *functions*) in order to use them inside expressions. There might be a problem if the function has a *side effect*, that is, if it changes some values of nonlocal variables. In this case, the order of evaluating subexpresssions might influence the result of an expression. Since this is difficult to comprehend by a programmer and inhibits or invalidates some compiler optimizations, functions with side effects are considered as a bad programming style. However, this is not forbidden in typical imperative languages.

66.2.2 Functional Languages

Functional programming is based on defining functions and composing them in expressions to more complex functions. A functional computation corresponds to the evaluation of a given expression. Purely functional languages, such as Haskell [26], do not allow side effects so that they are *referentially transparent*: the value of an expression does only depend on the values of the subexpressions and not on the order or time of evaluation. Thus, expressions can be replaced by their values without changing the result of an evaluation, and a functional computation consists of replacing equals by equals w.r.t. the function definitions.

We use the syntax of Haskell in concrete examples, that is, variables and names of defined functions start with lowercase letters and the names of type and data constructors start with an uppercase letter. The application of *f* to *e* is denoted by juxtaposition ("*f e*"). For instance, consider the function

```
square x = x*x
```

This equation expresses that `(square 3)` is equal to `(3*3)` so that the expression `5+(square 3)` can be replaced by `5+(3*3)` without changing its value. The replacement of the left-hand side of a function's rule by its right-hand side in an expression, where the rule's formal parameters (like `x`) are replaced by actual values, is the basic computation step in a functional program and called a *reduction step.*

Referential transparency is important to enable powerful compiler optimizations and to use advanced evaluation strategies. An *evaluation strategy* determines the next subexpression to be reduced in order to evaluate some expression. For instance, subexpressions can be evaluated in parallel or on demand. As an example, consider the function

```
f x y = if x==0 then f y (f x y) else x
```

and the main expression `(f 0 1)`, which is reduced to `f 1 (f 0 1)` (after evaluating the conditional). An *eager strategy* evaluates all arguments before a function application so that the subexpression `(f 0 1)` is reduced next. This leads to an infinite number of computation steps. On the other hand, a *demand-driven strategy* evaluates a subexpression only if its value is demanded (e.g., by a conditional or primitive operation) to proceed the computation. Since the value of the subexpression `(f 0 1)` is not demanded (since it does not occur in the condition), a demand-driven strategy reduces `f 1 (f 0 1)` to `1` (after evaluating the conditional). Hence, demand-driven (also called *lazy*) strategies are computationally more powerful than eager strategies. Demand-driven strategies also support programming with streams and infinite data structures and increase modularity [21]. Moreover, they are the basis of optimal strategies that evaluate expressions with a minimum amount of reduction steps [20].

Functional languages support the definition of data types by enumerating all constructors. For instance, the type of Boolean values is defined in Haskell as

```
data Bool = False | True
```

where `Bool` is the type name and `False` and `True` are the constructors of the type `Bool`. Functions operating on such types can elegantly be defined by *pattern matching*, that is, by several equations, also called *rules*, for different argument values:

```
and False x = False
and True x  = x
```

Thus, if the first argument of a call to `and` has the value `False`, only the first rule is applicable so that such a call reduces to `False` (without evaluating the second argument, if the strategy is demand-driven).

Another important feature of functional languages are *higher-order functions*, that is, functions that have functions as arguments or yield functions as results. Thus, functions are first-class objects that can be passed as any other data structure. The use of higher-order functions provides better code reuse since programming patterns can often be expressed as higher-order functions [21]. To support a safe use of functions, advanced type systems have been developed for functional languages. Although languages like Standard ML [25] or Haskell [26] are strongly typed, the programmer does not have to specify the type of all functions and variables since compilers can *infer* their types.

66.2.3 Logic Languages

Logic programming is based on defining relations between entities by clauses, that is, by facts or implications. For instance, in the logic programming language Prolog [12], `[]` denotes the empty list and `[X|Xs]` a nonempty list with a first element `X` and a tail list `Xs`. Then we can define the relation `member(E,Xs)` with the intended meaning "`E` is a member of the list `Xs`" as follows in Prolog:

```
member(E, [E|Ys]).
member(E, [Y|Ys]):-member(E,Ys).
```

Similarly to functional languages, pattern matching is also supported in logic languages. The first clause states the fact that `E` is a member if it is the first element. The second clause expresses that the membership in the tail list `Ys` implies the membership in the complete list. Logic programming allows to compute with partially known data expressed by free (unbound) variables. For instance, one can compute some members of the list `[1,2]` by solving the query `member(E,[1,2])`, which yields the two *solutions* (i.e., variable bindings for the free variable `E`) `E=1` and `E=2`. Hence, logic programming is based on a ("don't know") nondeterministic computation model where different computation paths are explored in order to find solutions. This principle allows to use free variables at arbitrary argument positions. For instance, the query `member(1,[X,Y])` returns the solutions `X=1` and `Y=1`.

Free variables are also useful inside program clauses. For instance, one can define a predicate that is satisfied if two lists have a common element by

```
overlapping(Xs,Ys):-member(E,Xs),member(E,Ys).
```

Hence, logic programming supports a more abstract programming style where the programmer is not forced to implement all algorithmic details of a problem. For instance, the predicate `overlapping` is implemented without specifying a strategy to find a common element.

In order to compute solutions to a given query, logic language implementations apply program clauses to the query, similarly to functional programming, but use *unification* to equate the query and the clause's left-hand side by binding variables to terms. For instance, the first clause of the definition of `member` can be applied to the query `member(1, [X,Y])` if the variables are bound as `{E↦1, X↦1, Ys↦[Y]}`.

66.3 Multiparadigm Languages in Practice

In the following, we discuss a selection of various programming languages that amalgamate different programming paradigms. Due to numerous approaches to combine different paradigms, we will only discuss a few typical languages in each area.

66.3.1 Imperative Functional Languages

One could consider every modern imperative language as a combination of imperative and functional programming since they support the definition of functional abstractions. However, this viewpoint changes when advanced functional programming concepts are taken into account. For instance, not every imperative language supports higher-order functions, that is, passing defined functions as argument or result values. Since such higher-order features demand for advanced type systems, they are often supported by imperative languages without a static type system. For instance, the languages Smalltalk [15] and Ruby* allow to pass code blocks as first-class objects, and many other imperative languages, like Python† or JavaScript [14], support expressions to denote functional values.

* http://www.ruby-lang.org/.
† http://www.python.org/.

Although the presence of such features appears as a reasonable combination of imperative and functional programming, imperative languages allow the definition of functions with side effects so that the order of evaluation becomes relevant. This is in contrast to functional languages where one wants to abstract from the order of evaluation, for example, in order to optimize the evaluation of expressions. Thus, it is desirable to keep functional evaluation free of side effects, if possible, or to distinguish purely functional computations from imperative computations. In many languages combining imperative and functional programming styles, however, this distinction is not strictly enforced. For instance, the language Scala* extends a typical imperative object-oriented language with functional programming features, like type inference, pattern matching, and anonymous and higher-order functions. Scala supports the separation of imperative and functional programming styles by distinguishing mutable variables and immutable values. Nevertheless, a defined function might depend on or manipulate mutable data so that it is not free of side effects.

Instead of extending imperative languages with functional features, one could also extend functional languages with imperative features. For instance, Lisp and its dialect Scheme† support multiparadigm programming by an assignment operation that allows the mutation of any parameter or locally declared value. The language Standard ML [25] uses the type system to distinguish between mutable values, which are pointers to memory cells, and immutable values. Nevertheless, computations with side effects can be "hidden" inside functional computations. This is different in the purely functional language Haskell [26], which does not allow hiding of side effects. Operations with side effects, also called *I/O actions*, like reading or printing a character, have the result type "`IO` τ", where τ is the type of the value returned by the side effect of the operation. For instance, the operation `getChar`, which reads a character (of type `Char`) from the keyboard as a side effect, has the type signature

```
getChar :: IO Char
```

Similarly, the operation `putChar` to print a character has the type signature

```
putChar :: Char → IO ()
```

that is, it takes a character as argument but the side effect does not return a value, as indicated by the void type "`()`". There are operators like ">>" to compose I/O actions sequentially. Thus, the expression

```
putChar 'a' >> putChar 'a' >> putChar 'a'
```

prints three `'a'`s as a side effect. Since I/O actions are first-class citizens as any other operation, one can also write operations to compose I/O actions. For instance, an operation to print `n` times a character `c` can be defined as follows (`return` is an I/O action that just returns its argument without an additional side effect):

```
putChars c n = if n==0 then return ()
                       else putChar c >> putChars c (n-1)
```

Thus, the effect of the previous expression can also be obtained by the expression `(putChars 'a' 3)`.

It has been shown [28] that typical imperative programming techniques can be implemented using this style of composing basic I/O actions to larger units. The important point is that there are no operations of type `IO` $\tau \rightarrow \tau'$ where τ' is not an `IO` type. Thus, it is not possible to hide an operation with side effects inside another operation, in contrast to most other programming languages with imperative features. As a consequence, the type system of Haskell ensures a clear separation between purely

* http://www.scala-lang.org.
† http://www.schemers.org/.

functional parts and program sections with side effects. Any operation having some side effect has the result type "IO τ". Therefore, purely functional expressions can be evaluated in any "good" order. Actually, Haskell uses a lazy evaluation strategy for expressions. However, I/O actions have to follow the sequential evaluation order as specified by applications of the operator ">>". More details about this kind of integrating imperative features in a purely functional language, also called *monadic I/O*, can be found in [37].

66.3.2 Imperative Logic Languages

The amalgamation of imperative and logic programming paradigms does not seem reasonable due to the very different computation models. Thus, there are only few approaches in this direction. On the one hand, logic programming languages like Prolog offer primitive predicates with side effects to perform I/O or manipulate some global state. Although such impure predicates are often used in application programming, this is a rather ad hoc use of imperative features in logic programming. A conceptually cleaner integration is present in logic languages that offer also functional features, like Mercury [35] or Curry [18], since such languages can use the monadic I/O approach from functional languages sketched earlier. This will be discussed later on.

As an alternative, one could also add logic programming features to an imperative language. This approach has been explored in the multiparadigm language Leda [10] and later in Alma-0 [8], an imperative language with logic programming features. The language Alma-0 is based on a subset of Modula-2 and allows Boolean expressions as statements whose computation can succeed or fail depending on the truth value of the expression. Furthermore, statements have also truth values depending on their successful evaluation, that is, a successful computation has value `TRUE` and a failing one value `FALSE`. Hence, the following statement sequence returns `TRUE` if the (integer) array `a` is smaller than array `b` (where both arrays have `n` elements) w.r.t. the lexicographic ordering [8]:

```
NOT FOR i:=1 TO n DO a[i] = b[i] END;
a[i] < b[i]
```

Since statements, like the `FOR` loop, have also truth values, they can also be used as arguments to Boolean operations like `NOT`. Operationally, the execution of the `FOR` loop fails (and terminates) when it reaches two different elements `a[i]` and `b[i]`. Due to negation (`NOT`), the execution of the first line yields `TRUE` in this case so that the second line is executed that checks the ordering of the different elements.

In order to support the backtracking search mechanism of Prolog, Alma-0 offers an `EITHER-ORELSE` statement to combine alternative computations to be explored by backtracking. Furthermore, there are control structures to commit to one successful alternative or to explore all alternatives (`FORALL`). Imperative features can be used to collect information about all alternatives by mutable variables. For instance, the following program fragment uses a mutable variable `num` to count how many of the variables `a,b,c,d` have the value `x`:

```
num:=0;
FORALL
  EITHER a=x ORELSE b=x ORELSE c=x ORELSE d=x END
DO
  num:=num+1;
END;
```

Furthermore, Alma-0 supports a limited form of unification by distinguishing between uninitialized and initialized variables. The integrated imperative and logic programming features support a compact and comprehensible implementation of typical search problems (e.g., see [8]).

66.3.3 Functional Logic Languages

Functional and logic programming are the main *declarative programming* paradigms. Both are based on the idea to describe logical relationships of a problem rather than the individual steps to compute a solution to a problem. Furthermore, functional as well as logic languages have strong mathematical foundations, namely, the lambda calculus and predicate logic. Thus, one might expect that integrating these paradigms is easily possible. However, there are many different options so that the integration of these paradigms have a long history, starting with the extension of Lisp with logic programming features [30]. Detailed surveys about these developments can be found in [16,17].

A lightweight integration of both paradigms can be obtained by small extensions that implement some programming features of one paradigm in the other. For instance, one can extend a logic programming language by some syntax for functional notations. Since any function can also be considered as a relation between the arguments and the result, functional notations can easily be translated into logic programming constructs [11]. Alternatively, one can use lists and list comprehensions to implement search problems in functional languages [36]. However, such approaches do not exploit the potential offered by the amalgamation of both paradigms. List comprehensions do not offer computation with partial information and constraints, as in logic programming, and the translation of functions into predicates do not exploit functional dependencies to obtain good or optimal evaluation strategies. The main objective of multiparadigm languages is to provide the advantages of the different paradigms in a single language. Thus, a functional logic language should combine efficient, demand-driven evaluation strategies from functional programming with nondeterministic search and computing with partial information from logic programming. If this is carefully done, one gets additional advantages like a demand-driven exploration of the search space. The language Curry [18] is a recent example for such a language and, therefore, discussed in the following in more detail.

Similarly to the nonstrict functional language Haskell, the functional logic language Curry* has been developed by an international initiative to create a standard functional logic language that can be used in research, teaching, and applications of such languages. The syntax of Curry is close to Haskell, as introduced in Section 66.2.2. Hence, operations can be defined by pattern matching:

```
not False = True
not True  = False
```

As in functional programming, computation or evaluation of an expression consists of replacing subexpressions by equal (w.r.t. the rules defining the operations) subexpressions until no more reductions are possible. For instance, `not (not True)` can be evaluated as follows (the subexpression to be replaced is underlined):

```
not (not True) → not False → True
```

So far, these features are identical to functional programming. Logic programming features are supported by allowing computations with partially known values. For instance, in Curry it is allowed to apply an operation to a *free variable* (i.e., a variable not bound to any value). In this case, the free variable will be bound to some value such that the computation can proceed. Consequently, the result of such a computation consists of a binding *and* a value. For instance, the evaluation of `(not x)` yields two results:

```
{x = False} True
{x = True} False
```

* http://www.curry-language.org.

The first result line can be read as "the result of `(not x)` is `True` provided that the variable `x` is bound to `False`." Similarly to logic programming, an expression might have more than one result.

Free variables can also occur inside data structures. Consider the following definition of an infix operator "++" to concatenate two lists (as in Haskell, "`[]`" denotes the empty list and "`x:xs`" a nonempty list with a first element `x` and a tail list `xs`):

```
[]     ++ ys = ys
(x:xs) ++ ys = x : (xs++ys)
```

Then the condition `(zs++[3]) =:= [1,2,3]`, also called an *equational constraint*, is satisfied if the free variable `zs` is bound to the list `[1,2]`, since the concatenation `[1,2]++[3]` evaluates to `[1,2,3]`. Several equational constraints can be combined to one constraint by the infix operation "&" ("and"). For instance, the only solution to the constraint

```
(xs++ys) =:= [1,2,1] & ([2]++xs) =:= ys
```

is {`xs = [1], ys = [2,1]`}. We can use such conditions in the definition of operations to constrain the values to which free variables can be bound. For instance, an operation `last` to compute the last element of a list can be defined as follows (note that free variables must be explicitly declared in program rules):

```
last xs | (zs++[e]) =:= xs
        = e             where zs, e free
```

The equational constraint `(zs++[e]) =:= xs` is satisfied if `e` is bound to the last element of the list `xs`. Such a *conditional rule* can be applied in a computation if the condition is satisfied, for example, by binding free variables. We will later explain how such bindings can be effectively computed.

Note that logic programming features increases the reuse of code. Instead of defining the operation `last` by some recursive rules, we reused the code of "++" to define `last` by a condition on its input list. This condition states that the actual argument `xs` must match a result of evaluating `(zs++[e])`. This can be expressed in Curry by the following alternative definition of `last`:

```
last (zs++[e]) = e
```

Purely functional languages, like Haskell, do not allow such rules since the pattern contains a defined operation rather than constructors and variables. Such *functional patterns* [7] are a powerful feature to match deeply nested subexpressions. Since the evaluation of `(zs++[e])` has infinitely many results (e.g., {`zs=[]`} `[e]`, {`zs=[x1]`} `[x1,e]`, ...), this single rule conceptually abbreviates an infinite number of rules with ordinary patterns. Nevertheless, the overall demand-driven evaluation strategy ensures that the required patterns are evaluated on demand so that a computation space of "`last [1,2]`" is finite.

This example demonstrates the advantage of combining functional and logic programming. By using already defined operations and unknown values, one can define operations as concise specifications. Since these specifications are functional logic *programs*, they are executable (in contrast to formal specifications written in some specification language). The challenge in the design of functional languages is the development of a computation model that avoids an inefficient blind guessing of values for unknowns. Before we discuss such strategies, we proceed with an important further feature.

Logic languages support the search for solutions (they are "don't know nondeterministic") which simplifies many programming tasks. The same is true for functional logic languages but they offer another interesting feature, namely, the definition of *nondeterministic operations*. Such operations deliver more than one result on a given input with the intended meaning that all the results are appropriate. The archetype of a nondeterministic operation is the infix operation "?" that returns one of its arguments:

```
x ? y = x
x ? y = y
```

Since any rule can be used to evaluate an expression (in contrast to functional languages where only the first matching rule is applied), the expression `0?1` has two results: `0` and `1`. Similarly, we can define an operation to insert an element into a list at some indeterminate position:

```
insert x ys     = x : ys
insert x (y:ys) = y : insert x ys
```

Thus, the expression `insert 0 [1,2]` evaluates to any of the results `[0,1,2]`, `[1,0,2]`, and `[1,2,0]`. One can use this operation to define a permutation of some list by inserting the first element at some position of a permutation of the tail list:

```
perm []     = []
perm (x:xs) = insert x (perm xs)
```

Permutations can be used to specify properties of operations. For instance, an operation to sort the elements of a list in ascending order can be defined as a permutation which is sorted. The latter property can be expressed as an identity operation on sorted lists (which fails on lists that are not sorted):

```
sorted []              = []
sorted [x]             = [x]
sorted (x:y:ys) | x<=y = x : sorted (y:ys)
```

Hence, the expression `sorted (perm xs)` evaluates to a sorted version of the input list `xs`. Clearly, the operational behavior of this specification of sorting is inefficient and cannot compete with sophisticated sorting algorithms. However, it should be noted that its behavior is not as bad as it seems at first glance, since the demand-driven strategy has the effect that not all permutations are computed. The evaluation of the argument `(perm xs)` is triggered by the demand of the outer operation `sorted` (which demands at most two elements for the first reduction step). If a computed permutation starts with the first two elements in a wrong order, as in `2:1:(perm ys)`, the operation `sorted` fails on this argument so that the remaining part `(perm ys)` will not be evaluated.

66.3.3.1 Evaluation Strategies

Logic programming frees the programmer from considering the individual computation steps since it is based on strong soundness and completeness results w.r.t. the underlying logic: each computed result is a logical consequence of the program (*soundness*) and each logical consequence is covered by some computed result (*completeness*) [23]. The latter property requires the exploration of all possible computation steps (also called *search space*), in particular, all rules (and not only the first matching one) need to be considered to reduce some subgoal. Similarly, functional logic computations require also the consideration of all rules. For instance, in the definition of the operation `insert` shown above, the second rule must also be considered although the first rule is always applicable. Even worse, it is not obvious to select a subexpression to start a functional logic computation. Therefore, there are many attempts to develop appropriate evaluation strategies for functional logic programming.

In order to enable the evaluation of expressions with free variables, like `not x`, reduction steps from functional programming must be combined with variable instantiation steps used in logic programming. The combination of variable instantiation and reduction is called *narrowing*. It was originally introduced in automated theorem proving [33] and later proposed for using it in programming languages [29]. To ensure completeness in general, all possible rules must be applied to all subexpressions, which results in a huge search space. However, the research in this area has developed much better strategies to restrict the number of possible computation steps without loosing completeness.

Similarly to functional languages, there are strict as well as nonstrict narrowing strategies. Due to the additional instantiation of free variables, strict strategies require sophisticated techniques to restrict the search space. Therefore, contemporary functional logic languages are based on *nonstrict demand-driven narrowing strategies*. In contrast to purely functional languages, the selection of a demanded argument might be influenced by the instantiation of some free variable. For instance, consider an operation to return a prefix of a given length of a list, where we represent the length as a natural number constructed by zero (Z) and successor (S):

```
take Z      xs     = []
take (S n) []      = []
take (S n) (x:xs)  = x : take n xs
```

If we evaluate the expression take x *e*, where x is a free variable, the demand to evaluate the subexpression *e* depends on the instantiation of x: if x is bound to Z, the value of *e* is not demanded, but it is demanded if x is bound to (S *v*). Thus, it is reasonable to instantiate free variables *before* the selection of the demanded subexpression. To avoid unnecessary instantiations, free variables are instantiated only with those constructors that are necessary to determine the demanded subexpression. For instance, the variable x in the expression take x *e* is either instantiated to Z, which yields the result [], or it is instantiated to (S y), where y is a fresh free variable, so that in the next step the subexpression *e* is further evaluated.

This *needed narrowing* strategy [6] is guided by a hierarchical structure called *definitional tree* [2]. A definitional tree of an operation encodes the decisions to be performed in order to select a demanded subexpression. Its structure is determined by the patterns of the rules defining this operation. For instance, all rules defining take require the value of the first argument. The value of the second argument is required only if the first argument is rooted by the constructor S. Hence, the needed narrowing strategy evaluates an expression like take e_1 e_2 by an initial case distinction on e_1:

1. If e_1 is Z, the first rule is applied.
2. If e_1 is rooted by S, the second argument e_2 is examined (resulting in a similar case distinction).
3. If e_1 is a free variable, it is nondeterministically bound to Z or (S y) so that one can proceed with case 1 or 2.
4. If e_1 is a function call, it is evaluated by a needed narrowing step so that we start again with this case distinction.

Definitional trees exist for operations defined by case distinctions on data constructors, as in purely functional programs. In this case, needed narrowing has very strong properties. As other narrowing strategies, it is sound and complete. Moreover, it computes only needed steps, so that each computation has minimal length, and it computes only disjoint solutions, that is, no solution is repeated [6]. These properties also demonstrate the advantages of combining functional and logic programming since similar properties are not known for purely logic programs.

Operations defined by overlapping rules, like insert or "?" earlier, do not have definitional trees in this strong sense. However, if we consider the choice operation "?" as primitive, one can transform any operation into one with a definitional tree and possible occurrences of "?" in the right-hand side. Such programs are also called *overlapping inductively sequential* [3]. For instance, the definition of insert can be transformed into

```
insert x ys = (x:ys) ? (insertTail x ys)
insertTail x (y:ys) = y : insert x ys
```

One can easily extend needed narrowing to this class of programs by introducing a nondeterministic reduction step for the operation "?". This extended strategy enjoys similar optimality properties as needed narrowing [3]. Due to these properties, contemporary implementations of functional logic languages are based on this strategy. Thus, source programs are transformed into overlapping inductively sequential programs that can be implemented by combining functional implementation techniques (pattern matching, graph reduction) with logic implementation techniques (variable bindings, choice points). A good survey on evaluation strategies and classes of functional logic programs can be found in [4].

Backtracking is a technique used in the logic programming language Prolog to explore the search space. If the nondeterministic choices of a logic computation are organized in a tree-like structure (*search tree*), backtracking corresponds to a depth-first search in the search tree. The backtracking strategy is important in Prolog since predicates with side effects require a precisely known evaluation order. A disadvantage of backtracking is its incompleteness, that is, a computation might loop without reporting any solution although solutions exists from a declarative point of view. Since functional logic languages avoid side effects in operations (since they can use the monadic I/O approach from purely functional programming), they do not fix a particular strategy for dealing with nondeterministic computations. Although backtracking is used in some implementations, the definition of Curry [18] does not enforce it so that there are also implementations of Curry supporting complete search strategies, like iterative deepening or breadth-first search. Moreover, functional logic languages also support the encapsulation of nondeterministic computations where the programmer can implement his own search strategies (see later sections).

66.3.3.2 Extensions

The definition of data types, operations, and expressions that are evaluated by narrowing constitutes the kernel of functional logic languages. In order to support features useful for various programming tasks and application areas, functional logic languages offer a variety of extensions. The most important ones are reviewed in the following.

Constraints are an important extension to logic programming. Constraints are based on a specific domain (e.g., truth values, real numbers, or finite sets) and operations and predicates on this domain. Although this could also be expressed as a functional logic program, constraints come also with dedicated constraint solvers, that is, specific algorithms for the considered domain. For instance, an equation like "`3+x =:= 5`" can be naively solved by enumerating values for `x` such that the equation is valid. However, one can solve this equation without any guessing by applying Gaussian elimination and transform the equation into "`x =:= 5-3`," which further evaluates to "`x =:= 2`". Thus, constraints over real numbers and linear equations and inequations can be supported by a constraint solver based on Gaussian elimination and the simplex algorithm. Hence, the integration of constraints is merely a matter of implementing constraint solvers but does not require an extension of the base language. Similarly to equational constraints, further type of constraints can be added. For instance, constraint solvers for finite domains usually offer a constraint `allDifferent`, which has a list of integer values as an argument. It is satisfied if all values are pairwise different. This can be used to compute solutions to SuDoku puzzles in a straightforward way. If we use some higher-order functions that are standard in functional programming, like `map` (applying an operation to each element of a list) or `foldr1` (accumulating a list by applying a binary operation to all elements), we can solve SuDoku puzzles by the following constraint, where `m` is a matrix (list of elements) containing digits and free variables for unknown elements (`transpose` is the standard matrix transposition and `squaresOfNine` computes the list of 3×3 sub-matrices):

```
foldr1 (&) (map allDifferent m)    &
foldr1 (&) (map allDifferent (transpose m))    &
foldr1 (&) (map allDifferent (squaresOfNine m))
```

The first line constrains all rows to contain different digits, the second and the third line constrains the same for all columns and submatrices, where all individual constraints are combined by the conjunction operator `&`. Although the number of potential solutions is quite high, the constraint solver `allDifferent` restricts them so that solutions are found in a few milliseconds. More details about the use of finite domain constraints in functional logic languages are discussed in [13], whereas [31] reviews the general approaches to integrate constraints into functional logic languages.

Encapsulated search is a technique to process the different results of a nondeterministic computation inside a functional logic program. This is often necessary to print all solutions or to select a "best" solution according to some ordering. Whereas logic programming languages like Prolog offer built-in predicates like `findall` with a fixed (backtracking) strategy, functional logic languages generalize this in order to support the possibility to define arbitrary search strategies. For instance, the primitive operation (`getSearchTree` *e*) returns the search tree corresponding to the evaluation of *e*, where a search tree is some computed value, a failure, or a choice between two trees:

```
data SearchTree a = Value a | Fail
                  | Or (SearchTree a) (SearchTree a)
```

Thus, concrete values can be extracted from this structure by operations implementing standard tree traversals. Due to the overall demand-driven strategy, search trees are also constructed on demand so that infinitely many values do not cause the nontermination of the operator `getSearchTree` if one selects only a single value. More details about encapsulated search and a review of different approaches can be found in [9].

Input/output is often neglected or integrated in an ad hoc manner in logic languages. Due to the functional component, functional logic languages often use the monadic I/O concept from purely functional programming (see Section 66.3.1). However, functional logic computations may be nondeterministic so that it might be unclear which I/O action should be finally applied. Therefore, it is required that nondeterministic computations between I/O actions must be encapsulated. This can be done by the methods described above. Using I/O actions, one can access and manipulate global data like databases, networks, etc., so that imperative programming features are also available for programming. However, the type system ensures, as in Haskell, a strict separation of the imperative and declarative programming features, which is essential to support flexible and optimal evaluation strategies.

Concurrency features are available in functional logic languages in various shapes. One approach is based on concurrent logic programming [32] where different predicates are concurrently evaluated and synchronized on a common constraint store. In Curry, the conjunction operator "&" on constraints is actually a concurrent operator: if the evaluation of one argument suspends, the other argument is evaluated. An evaluation might suspend if some information is missing to continue it. For instance, if there is some external operation that should be used in a functional logic program, for example, a multiplication operation implemented in some foreign language like C or Java, then this external operation might not be able to deal with free variables. Instead of raising a run-time failure, one can suspend this computation until the variable is bound in another part of the program. The principle to delay the evaluation of operations applied to free variables is also called *residuation*. This alternative to narrowing is used in some languages integrating functional and logic programming, like Le Fun [1] or Oz [34]. In residuation-based languages, free variables are instantiated only by predicates or disjunctive expressions. Although residuation is a high-level method for concurrent evaluation (no side effects, synchronization on binding constraints), strong results as for needed narrowing are not known.

Another alternative for concurrency features can be found in concurrent functional programming. For instance, Concurrent Haskell [27] uses I/O actions to start concurrent processes. Synchronization and communication is implemented by I/O actions that create, read, and write mutable variables. Since I/O actions are also available in functional logic languages, this concept can also be integrated in such languages.

66.3.3.3 Further Languages

Although Curry is the only functional logic language based on strong foundations (e.g., soundness, completeness, and optimal evaluation) and used to develop larger applications, there are many other languages with similar goals. We briefly discuss some of them and relate them to Curry.

The functional logic language TOY [24] has a different syntax but executes programs also with a demand-driven narrowing strategy. In contrast to Curry, it is purely narrowing-based and does not cover residuation or concurrency. Therefore, the connection with external operations is rather ad hoc in TOY. The language TOY also supports constraints over finite domains or real numbers but does not provide a concept to encapsulate search. In addition to Curry, TOY allows higher-order patterns on the left-hand sides of program rules.

The functional logic language Mercury [35] restricts logic programming features in order to provide a highly efficient implementation. In particular, predicates and functions must have distinct modes so that their arguments are either ground or unbound at call time. This inhibits the application of typical logic programming techniques, like computation with partially instantiated structures. This restriction is dropped in Ciao [19], which is a Prolog implementation but offers various syntactic extensions, in particular, for functional programming. Since Mercury as well as Ciao provide only functional syntax to a logic kernel language, they do not exploit the strong foundations of functional logic programming.

The language Oz [34] is based on a computation model that extends the concurrent constraint programming paradigm [32] with features for distributed programming and stateful computations. Nondeterministic computations must be explicitly represented as disjunctions so that operations used to solve equations require different definitions than operations to rewrite expressions. In contrast to Curry, the base semantics is strict so that optimal evaluations are not directly supported.

66.4 Research Issues

The integration of paradigms has made much progress from early ad hoc integrations to more systematic approaches. In particular, the integration of the functional and logic programming paradigms resulted in languages with advanced programming concepts that are based on strong theoretical foundations developed in the last years. These developments are ongoing and there are many topics for future work. Some of them are exemplified below.

Language concepts: The approaches to combine declarative programming concepts with imperative features ranges from clear separations between both concepts, for example, by type systems as in Haskell, to relaxed combinations, for example, as in Standard ML or Scala. The latter might cause problems when operations have side effects whose evaluation order is important. For instance, languages with functional features allow the implementation of demand-driven or lazy strategies using higher-order functions. Since the detailed evaluation order of a demand-driven strategy is difficult to grasp, operations with side effects should be avoided. Thus, multiparadigm languages with a relaxed combination of declarative and imperative features require techniques to support the programmer to control side effects, for instance, by advanced type and effect systems or program analysis techniques.

Although we have not considered concurrent programming as an own paradigm, the integration of concurrency and distribution features is important for current and future applications. Thus, there are many approaches for this in all main paradigms. However, multiparadigm languages might offer new concepts for concurrent and distributed programming.

Software techniques: Multiparadigm languages allow new programming techniques so that new design patterns need to be developed for such languages. The integration of declarative features allows new ways to develop reliable software. One can start with a declarative formulation and improve its efficiency if necessary. Specific program verification and transformation techniques are required to support such developments, as well as new refactoring methods.

Implementation: Multiparadigm languages can be implemented using known concepts from the individual paradigms, but it might be reasonable to develop new techniques. Sometimes, the support of several paradigms could cause an overhead compared to implementations of the individual paradigms. Thus, program analysis techniques can help to determine the required run-time support at compile time in order to optimize the target programs. New techniques to analyze and optimize multiparadigm programs are required for this purpose.

Debugging: Research in functional programming has shown that traditional trace-oriented debugging tools are often not helpful due to the more complex evaluation strategies, for example, demand-driven evaluation. This provoked the development of declarative debugging methods that visualize the program execution in a different order that is better comprehensible for the programmer. As multiparadigm languages combine features from different paradigms, debugging techniques from different paradigms need to be combined, for example, declarative debugging, trace-oriented debugging, or run-time assertions.

66.5 Summary

This chapter has discussed the main programming paradigms and the approaches to amalgamate them in a single programming language. In principle, all combinations are possible but some combinations do not preserve the properties of the individual paradigms. For instance, the combination of imperative and functional programming can lead to functions with side effects so that referential transparency is destroyed, but there are also approaches to keep the functional parts free of side effects. The most advanced and strongest combination of paradigms are functional logic languages which is due to the fact that both paradigms are based on strong formal foundations. Nevertheless, sophisticated evaluation strategies are necessary in order to combine the efficiency of functional evaluation with the flexibility of logic evaluation in an optimal manner. This combination does not only help to use the nice programming concepts of functional and logic programming in a single language, but it is also useful to improve logic programming by avoiding its impure features (like "cut", arithmetic, I/O) since they can be replaced by functional concepts. Since purely functional languages like Haskell provide clean approaches to deal with imperative computations, they can also be used in functional logic languages leading to a combination of all three main paradigms. The programming language Curry is an example to demonstrate this in practice.

The combination of paradigms is not only useful to provide a variety of programming styles, but they are also helpful to support a smooth software development. For instance, logic programming styles are appropriate to avoid fixing algorithmic details at an early stage of the development. The examples for computing the last element of a list or to sort a list have shown that it is easier to specify what an operation should compute rather than developing an algorithm for it. In a functional logic language, such specifications are executable so that one can test them in order to become confident that they capture the intent. If they are too inefficient, one can use some algorithmic knowledge to improve the implementation, for example, by sophisticated data structures. In this case, the initial specification is valuable since it can be used as an oracle to test the new implementation or to check its execution via run-time assertions.

Finally, side effects cause many complications when parallelizing programs so that they must be carefully analyzed or mainly avoided for parallel programming. Thus, declarative programs have a big potential for parallelization in order to exploit the computational power available by current and future computer hardware. The implementation of these ideas is one of the challenges for the future.

Key Terms

Constraints: Relations on specific domains (real numbers, finite sets) together with specialized constraint solvers for these relations.

Declarative language: A language where the logical relationships of a problem are described rather than the individual steps to compute a solution. Functional and logic languages are the main classes of declarative languages. Declarative languages also occur in dedicated applications, like SQL in database programming.

Demand-driven evaluation: An evaluation strategy that reduces a subexpression only if it is strictly demanded to compute an overall result, also called *lazy evaluation* when subexpressions are evaluated at most once.

Encapsulated search: Collect the results of a nondeterministic computation in some data structure so that they can be processed inside a program.

Evaluation strategy: A technique to determine the next subexpression to be reduced in order to evaluate some expression.

Functional language: A language that is based on the definition of functions without side effects, that is, no mutable variables.

Imperative language: A language that is based on mutable variables and variable assignments as its basic operation.

Logic language: A language that is based on the definition of relations between entities and supports nondeterministic solving of goals containing unknown values.

Multiparadigm language: A programming language that supports more than one of the basic programming paradigms.

Narrowing: The combination of unification and reduction to evaluate expressions containing unknown values.

Pattern matching: The definition of operations by rules having data constructors as formal parameters. A programming style useful to provide specific definitions for the different cases of applying an operation.

Programming paradigm: Style to construct programs from basic elements and composing them to larger units. The basic paradigms are the imperative, functional, and logic programming paradigms.

Reduction: A computation step where the left-hand side of a defining rule is replaced by the corresponding right-hand side after instantiating formal parameters by appropriate expressions.

Referential transparency: A property stating that expressions can be replaced by their values without changing the result of the program. A language allowing functions with side effects is not referentially transparent.

Residuation: Delaying functions calls until all values are known in order to perform a reduction step.

Side effect: A change in the value of some variable when an expression is evaluated, for example, due to assignments in function calls of the expression.

Unification: A method to replace free variables in two expressions so that they become equal.

Further Information

An introduction into multi-paradigm programming together with many programming examples for the language Leda can be found in [10]. A survey on early developments and implementations of functional logic languages can be found in [16]. Later developments are reviewed in [17]. Evaluation strategies for functional logic programs are surveyed in [4]. Reference [5] contains an introduction into functional logic programming techniques.

Research papers on recent developments in the area of multi-paradigm languages can be found in various journals and conferences related to programming languages, in particular, those that are specialized in functional or logic programming, like the *Journal of Functional Programming, Theory and Practice of Logic Programming*, the *International Conference on Functional Programming*, or the *International Conference on Logic Programming*. There are also conference series which covers functional *and* logic programming research, like the *International Symposium on Principles and Practice of Declarative Programming*, the *International Symposium on Functional and Logic Programming*, or the *International Symposium on Practical Aspects of Declarative Languages*.

References

1. H. Aït-Kaci, P. Lincoln, and R. Nasr. Le Fun: Logic, equations, and functions. In *Proceedings of the 4th IEEE International Symposium on Logic Programming*, pp. 17–23, San Francisco, CA, 1987.
2. S. Antoy. Definitional trees. In *Proceedings of the 3rd International Conference on Algebraic and Logic Programming*, pp. 143–157. Springer LNCS 632, Heidelberg, Germany, 1992.

3. S. Antoy. Optimal non-deterministic functional logic computations. In *Proceedings of the International Conference on Algebraic and Logic Programming (ALP'97)*, pp. 16–30. Springer LNCS 1298, Heidelberg, Germany, 1997.
4. S. Antoy. Evaluation strategies for functional logic programming. *Journal of Symbolic Computation*, 40(1):875–903, 2005.
5. S. Antoy. Programming with narrowing: A tutorial. *Journal of Symbolic Computation*, 45(5):501–522, 2010.
6. S. Antoy, R. Echahed, and M. Hanus. A needed narrowing strategy. *Journal of the ACM*, 47(4):776–822, 2000.
7. S. Antoy and M. Hanus. Declarative programming with function patterns. In *Proceedings of the International Symposium on Logic-based Program Synthesis and Transformation (LOPSTR'05)*, pp. 6–22. Springer LNCS 3901, Heidelberg, Germany, 2005.
8. K.R. Apt, J. Brunekreef, V. Partington, and A. Schaerf. Alma-0: An imperative language that supports declarative programming. *ACM Transactions on Programming Languages and Systems*, 20(5):1014–1066, 1998.
9. B. Braßel, M. Hanus, and F. Huch. Encapsulating non-determinism in functional logic computations. *Journal of Functional and Logic Programming*, 2004(6), 2004.
10. T. Budd. *Multiparadigm Programming in Leda*. Addison-Wesley: Reading, MA, 1995.
11. A. Casas, D. Cabeza, and M.V. Hermenegildo. A syntactic approach to combining functional notation, lazy evaluation, and higher-order in lp systems. In *Proceedings of the 8th International Symposium on Functional and Logic Programming (FLOPS 2006)*, pp. 146–162. Springer LNCS 3945, Heidelberg, Germany, 2006.
12. P. Deransart, A. Ed-Dbali, and L. Cervoni. *Prolog—The Standard: Reference Manual*. Springer: Berlin, Germany, 1996.
13. A.J. Fernández, M.T. Hortalá-González, F. Sáenz-Pérez, and R. del Vado-Vírseda. Constraint functional logic programming over finite domains. *Theory and Practice of Logic Programming*, 7(5):537–582, 2007.
14. D. Flanagan. *JavaScript: The Definitive Guide*. 5th edn, O'Reilly: Sebastopol, CA, 2006.
15. A. Goldberg and D. Robson. *Smalltalk-80: The Language and Its Implementation*. Addison-Wesley: Reading, MA, 1983.
16. M. Hanus. The integration of functions into logic programming: From theory to practice. *Journal of Logic Programming*, 19&20:583–628, 1994.
17. M. Hanus. Multi-paradigm declarative languages. In *Proceedings of the International Conference on Logic Programming (ICLP 2007)*, pp. 45–75. Springer LNCS 4670, Heidelberg, Germany, 2007.
18. M. Hanus (ed.). Curry: An integrated functional logic language (vers. 0.8.3). Available at http://www.curry-language.org (accessed on October 4, 2013), 2012.
19. M. Hermenegildo, F. Bueno, M. Carro, P. López-García, E. Mera, J.F. Morales, and G. Puebla. An overview of Ciao and its design philosophy. *Theory and Practice of Logic Programming*, 12(1–2):219–252, 2012.
20. G. Huet and J.-J. Lévy. Computations in orthogonal rewriting systems. In J.-L. Lassez and G. Plotkin, eds., *Computational Logic: Essays in Honor of Alan Robinson*, pp. 395–443. MIT Press: Cambridge, MA, 1991.
21. J. Hughes. Why functional programming matters. In D.A. Turner, ed., *Research Topics in Functional Programming*, pp. 17–42. Addison Wesley: Reading, MA, 1990.
22. S.H. Huseby. *Innocent Code: A Security Wake-Up Call for Web Programmers*. Wiley: New York, 2003.
23. J.W. Lloyd. *Foundations of Logic Programming*. 2nd, extended edn, Springer, Berlin, Heidelberg, New York, 1987.
24. F. López-Fraguas and J. Sánchez-Hernández. TOY: A multiparadigm declarative system. In *Proceedings of RTA'99*, pp. 244–247. Springer LNCS 1631, Heidelberg, Germany, 1999.

25. R. Milner, M. Tofte, and R. Harper. *The Definition of Standard ML*. MIT Press: Cambridge: MA, 1990.
26. S. Peyton Jones, (ed.) *Haskell 98 Language and Libraries—The Revised Report*. Cambridge University Press: Cambridge, U.K., 2003.
27. S.L. Peyton Jones, A. Gordon, and S. Finne. Concurrent Haskell. In *Proceedings of the 23rd ACM Symposium on Principles of Programming Languages (POPL'96)*, pp. 295–308. ACM Press, New York, 1996.
28. S.L. Peyton Jones and P. Wadler. Imperative functional programming. In *Proceedings of the 20th Symposium on Principles of Programming Languages (POPL'93)*, pp. 71–84. ACM Press, New York, 1993.
29. U.S. Reddy. Narrowing as the operational semantics of functional languages. In *Proceedings of the IEEE International Symposium on Logic Programming*, pp. 138–151, IEEE, Boston MA, 1985.
30. J.A. Robinson and E.E. Sibert. LOGLISP: Motivation, design and implementation. In K.L. Clark and S.-A. Tärnlund, eds., *Logic Programming*, pp. 299–313. Academic Press New York, 1982.
31. M. Rodríguez-Artalejo. Functional and constraint logic programming. In *Constraints in Computational Logics: Theory and Applications (CCL'99)*, pp. 202–270. Springer LNCS 2002, Heidelberg, Germany, 2001.
32. V.A. Saraswat. *Concurrent Constraint Programming*. MIT Press: Cambridge, MA, 1993.
33. J.R. Slagle. Automated theorem-proving for theories with simplifiers, commutativity, and associativity. *Journal of the ACM*, 21(4):622–642, 1974.
34. G. Smolka. The Oz programming model. In J. van Leeuwen, ed., *Computer Science Today: Recent Trends and Developments*, pp. 324–343. Springer LNCS 1000, Heidelberg, Germany, 1995.
35. Z. Somogyi, F. Henderson, and T. Conway. The execution algorithm of Mercury, an efficient purely declarative logic programming language. *Journal of Logic Programming*, 29(1–3):17–64, 1996.
36. P. Wadler. How to replace failure by a list of successes. In *Functional Programming and Computer Architecture*, pp. 113–128. Springer LNCS 201, Heidelberg, Germany, 1985.
37. P. Wadler. How to declare an imperative. *ACM Computing Surveys*, 29(3):240–263, 1997.

67

Scripting Languages

Robert E. Noonan
College of William and Mary

67.1 Introduction

According to WebMonkey [15]:

> A scripting language is a simple programming language used to write an executable list of commands, called a script. A scripting language is a high-level command language that is interpreted rather than compiled, and is translated on the fly rather than first translated entirely. JavaScript, Perl, VBscript, and AppleScript are scripting languages rather than general-purpose programming languages.

A major characteristic of scripting languages is that they can serve as *glue* for connecting existing components or applications together. Scripting languages usually have powerful string processing operations since text strings are a fairly universal communication medium.

67.2 History

Scripting languages originated in the early 1960s. At that time, they were used as a device for defining the control sequence for running a series of programs one after the other. Such a series was often known as *batch processing*, and it was the predominant means for accomplishing computing tasks in commercial data processing environments at that time.

67.2.1 Shell Languages

The earliest scripting languages were the job control languages of the 1960s. Their primary use was to instruct the operating system on the sequence of steps required to run a job.

For example, a job to develop a summary sales report for the past month might have the following steps:

1. A program to select all sales records for the previous month from a larger file and copy them to a temporary file
2. A program to sort the temporary file into sequence by sales district, usually using a general-purpose utility
3. A program to calculate subtotals and present the information in a way that is easy for the end-users to read
4. A program to format and display selected pages of the end-user information on a printer or terminal.

However, these languages lacked almost all of the features you might expect, including variables, if and loop statements, expression evaluation, etc. The first true job control languages capable of scripting appeared with the emergence of Unix in the 1970s. The command interpreter in Unix was known as a *command shell*; a file of such shell commands was termed a *shell script*.

The earliest shell was known as the Bourne shell, named after its creator, Steve Bourne. Although later shells, such as the C shell created by Bill Joy, became more popular as a command interpreter, the Bourne shell maintained its popularity for writing shell scripts.

Figure 67.1 shows a typical grading script. This application assumes that all the files needed to grade a class assignment have been collected into a single directory. Each student has his/her own subdirectory. The files in the directory include input test data files, various log files, various shell scripts, etc.

The purpose of the script is to compile each student's program files, and, if successful, execute the resulting program on all of the test cases, producing an output file for each input test case.

Assuming the name of the script was `testem.sh`, it would be invoked as

```
testem.sh Freecell *
```

The first command line argument is the name of the file that contains the main program, minus its suffix.

```
#! /bin/sh
if [$# -lt 2]; then
    echo Usage: testem. sh mainprog dir1 …
    exit
fi

main=$1; shift
# for each student
for d in $*; do
    if [ -d $d -a ! -f $d/DONE. log ]; then
         cd $d
         echo $d

         if [ ! - f $main.class ]; then
              cp ../*.java .
             echo "compiling $d"
             javac $main.java &> compile.log
    fi
         touch DONE.log
    cp ../*. $data.
    for f in ../*. $data; do
             java $main $f < $f & > $f.log
        done
        cd ..
    fi
done
```

FIGURE 67.1 A student test script.

When the script is invoked, it determines whether it has fewer than two arguments: the first is the filename of the file containing the main program. The script must also be invoked with at least one student subdirectory name. If the test fails, the script prints a usage message and then quits.

Otherwise, the first command argument is stored in the variable `$main`, and then deleted from the command line. A for loop is then executed on the remaining command line arguments.

Since each student corresponds to a subdirectory, line 10 tests to see if the argument is a directory and if it is not done. If both conditions are met, then the argument represents an untested student. The next two lines change directories into the student's directory (note the `cd` back to the parent on line??) and echos the student id to the screen.

The script next checks if a class file exists for the java source; if not, it copies any instructor supplied code into the student directory and compiles the result with output diverted to the file `compile.log`.

Next we record the fact that this student has been tested by creating the file `DONE.log` in the student directory. The next four lines test the student program on a set of test data supplied by the instructor, with the output diverted to a separate file for each input test case.

For many class assignments, this is the only script we need to test the student-written programs. It handles a wide variety of cases, including the following:

- Assignments that consist of a single source file, as well as those that include both student and instructor written source code. No assumption is made about who supplied the main program file.
- The student program can obtain the command line argument and then open and read that file. Alternatively, the program can read its input from the keyboard.

Note the use of variables in the script representing the current student and the current test.

67.2.2 Text Processing Languages

As part of the development of Unix, a number of text processing utilities were developed, as well as the troff word processor for writing documentation. The next step in the evolution of scripting languages was the development of specialized languages for processing text, specifically, *grep*, *sed*, and *awk*. The latter was an acronym made up of the first letters of the languages developers: Aho, Weinberger, and Kernighan. *awk* appeared around 1978.

In this section, we present two *awk* scripts: the first is a very small, typical script, while the second is considerably more complex. The first is so small and useful that such scripts are still common today. Scripts like the second served as the impetus for the development of general purpose scripting languages.

The first script was written back in the late 1980s when one of the authors was the systems manager for a network of Unix computers. In the fall semester of each academic year, there were many new students and a few new professors who needed computer accounts. At that time, creating a new user account was a multistep, manual process carried out by student help.

One fall day a new student came to see me because they were unable to login to their new Unix account. My investigation quickly revealed that their record in the user table had one too few fields. It was a simple matter to edit the user table and add the missing field. This quick resolution made the user happy, but not I.

One solution to prevent this problem from recurring was to create a written procedure with a checklist to create a new user account. Although this approach had a certain appeal. I decided to take the extra step of implementing the procedure as a combination of a *Makefile*, a few shell scripts, and a few simple *awk* scripts.

Since editing the user table was a manual process, I decide to implement the checks on this step as a sequence of (largely *awk* scripts). The first of these was to check that each line had the same number of fields:

```
#! /usr/bin/awk -f
NR == 1 { ct = NF }
NF != ct { print NR, $0 }
```

An *awk* program consists of a *guard* followed by a block of statements, repeated as many times as desired; the first line is a pseudo comment used in Unix/Linux computers that identifies the interpreter to use. The *awk* processing cycle consists of reading an input line, then testing each guard in the program. For any guard that is true, its corresponding block is executed, after which the testing of guards resumes. Only after all the guards have been tested and their blocks executed, if applicable, is the next input line read and processed.

Based on a field separator (the default is whitespace), the input line is broken up into fields. The *awk* variable `NR` is the current line (record) number and `NF` the number of fields on the current input line. The variables `$1, $2`, ... hold the contents of each field, whereas the variable `$0` refers to the entire line.

So the aforementioned program a consists of two guarded blocks. The first guard is true only for the first line read; its block stores the number of fields on the first input line into the variable `ct`. The last guarded block prints any lines whose number of fields disagrees with `ct`.

Although the author is no longer a systems manager, this first *awk* script still finds occasional use even today.

The second script is considerably more complex. It was developed to solve the following problem. The authors were writing a text book. In addition to the text, a number of nontrivial programs were developed. The authors needed to include portions of these programs as examples. Most importantly, they wanted the code in the textbook to be identical to the code used in the programs.

So they used a `Makefile` and a set of text filters for extracting the code of interest into a file, which was then automatically included in the book. The *awk* script presented here was given the task of extracting a single function or method in Java. A typical script parameter was the return type and name of the method, which made the starting point easy to find. Having located the starting point, the end point was determined by basically counting braces, based on the authors' coding style:

```
#! /bin/sh
awk "BEGIN { ct = 0; prt = 0 }
/$1/ { prt = 1 }
prt > 0 { print \$0 }
prt > 0 && /{/ { ct++ }
prt > 0 && /}/ { ct-- }
prt >0 && ct == 0 { prt = 0 }
"
```

Before the first input line is read, the script sets a `prt` switch to off (0). It then searches through the input file until it finds the function it is looking for, namely, the second guard, whose block turns the switch on. The third guard prints the input line if the `prt` switch is on. The next two statements increment/decrement the brace count whenever the `prt` switch is on and they detect a left/right brace. The final guarded block turns the switch off when the brace count goes to zero.

There is a very subtle point in this script. The observant reader will have noted that, unlike the previous script, this one is a shell script which then invokes the *awk* interpreter. The parameter to the script is substituted into the second guarded block, while in the statement of the third guarded block, the dollar sign is escaped so that to the *awk* interpreter the statement appears as

```
print $0
```

Quirks like this, together with the limited processing ability of text processing languages, led to the development of more general purpose scripting languages, which are discussed in the next section.

67.2.3 Glue Languages

> Larry Wall … created Perl when he was trying to produce some reports from a Usenet-news-like hierarchy of files for a bug-reporting system, and *awk* ran out of steam. Larry, being the lazy programmer that he is, decided to over-kill the problem with a general purpose tool that he could use in at least one other place. The result was the first version of Perl. [12]

In the 1980s, scripting languages began to move away from simple text processing to more complex tasks with the development of the languages Perl and Tcl/Tk. A hallmark of a *glue* language is its ability to take data output from one application and reformat it so that it could be input to another application. Initiallly, scripting languages were considered much too slow and inefficient to implement an entire application. With the development of faster interpreters and faster computers, this assumption disappeared.

Although Perl and Tcl/Tk had their roots as a Unix scripting languages, all major scripting languages are now available for all major platforms including Unix, Linux, Macintosh, and Windows. These languages include Perl, Tcl/Tk, Python, and Ruby.

For example, the first author has used these scripting languages for the following tasks:

- Systems administration tasks on a network of computers including: setting up new user accounts, running backups, reconfiguring servers, and installing software.
- Class management, including converting electronic class rolls from one form to another, managing email for each course, grading support, and an interactive, GUI-based grading system.
- A textbook support system, which consisted primarily of scripts for including source code in a book.
- Newer versions of programs previously written in conventional programming languages such as C++ and Java. Such programs fall into a variety of subject domains including simple utility programs, computer science research, etc.

To illustrate the robustness of scripting languages, let us consider the academic task of maintaining a grade book of numeric grades earned by students in a class for their assignments, quizzes, tests, and exams. Each grade may have a different weight; for example, an exam may be more heavily weighted than a quiz. The script presented here provides a secure way to deliver grades to students.*

A spreadsheet is used to do the grade management itself, one spreadsheet per class. An example spreadsheet is given in Figure 67.2. In this example, column one is used for student names and column two is used for email addresses. The rightmost two columns used are weighted totals and averages. The intervening columns are used for grades on assignments and tests; at any point in time, any number of grade columns may be blank (empty).

Also, note that the first row is a descriptive title for each grade, and the second row for the maximum points possible for each grade. The last row is used for the average for each assignment or test. The intervening rows contain the students records, one row per student.

The scripting problem here is to move the grade values from the spreadsheet to the email system. This process consists of exporting the data in textual form from the spreadsheet and then using a Python script to generate the email. The spreadsheet output used is comma-separated-values. We have found over the years that each spreadsheet application has its own, often unique definition of the format of

* Specialized web software developed for education, such as Blackboard and WebCt, provide equivalent functionality to instructors through secure login to a dedicated Web server. However, the system described here was developed long before such software became available; it relied on the use of email for conveying grades back to students.

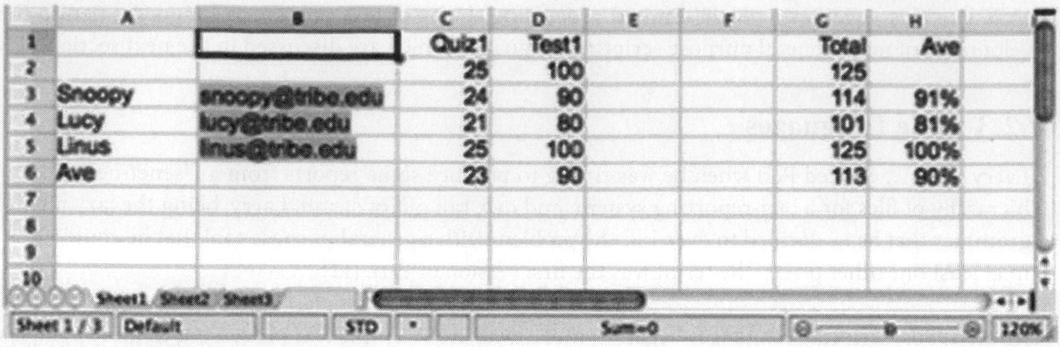

FIGURE 67.2 A spreadsheet gradebook.

comma-separated-values. Fortunately, Python provides a module (or library) which deals with that. For our current spreadsheet the third row would appear as

```
"Snoopy","snoopy@wm.edu",24,90,,,114,91.2
```

In the remainder of this section, we describe the Python script given in Figure 67.3 as a means of examining some of the features of glue languages in general, and Python in particular.

As noted in the previous section, line 1 of the script is a pseudo comment that gives the full path name of the language interpreter. Line 2 imports any needed modules (or libraries).

In lines 4–6, we set any global constants that are likely to change. For this script, these are kept in an dictionary or associative array as a key-value pair. This allows each grade spreadsheet to have its own set of values, which can be read from a file, overriding the defaults. These parameters include the column number of the student's name, his/her email address, etc. For the sake of simplicity, the routine for reading (and checking) these parameters (amounting to 11 lines of code) is omitted here.

The main logic of the program is shown in lines 8–10. First, the CSV input file is read and stored in an array of students. All the student entries must be read because the averages (which are included in each mail message) are last. Then an email is constructed and sent to each student:

All that remains is to examine the functions `readGrades` and `emailGrades`. The first function is passed the name of the file to be read as the first command line argument; a simple GUI alternative would use a dialog box to ask for the file name. The first line of the function sets the variable `student` to be an empty, indexed array. The `filename` is opened and passed to a reader which deals with comma-separated format files. The `for` loop reads one row per iteration, which is appended to the end of the `student` array.

The function named `emailGrades` is passed the `student` array. Next, leading nonstudent entries are copied and trailing ones removed. Then a simple `for` loop that generates an email message per student entry.

Basically, this function iterates over the students, generating a string that contains the student's grades and then mailing them. It counts the lines as it goes so that it can skip the initial lines which contain header information. On a row containing an actual student, the function formats the student's grades in a manner familiar to any programmer, storing the message in the string `outline`.

The function *emailGrades* then creates a subprocess that executes a system command; the `mailx` command works here if you have a static IP address, otherwise it is somewhat more complicated. Note that the subprocess's file `stdin` is set to be a pipe, which is then filled by the `proc.communicate` method call. The function then waits for the subprocess to finish. One essential feature of a *glue* language is the ability to execute system commands (like `mailx`) and send them input or read their output or both.

```
#! /usr/bin/python
import sys, csv, process

param = {“name”:0, “email”:1, “gradecol”:2, “studentrow”:2,
         “titles”:0, “maxgrade”:1}
readPrefs(“ugrades.pref”)

student = readGrades(sys.argv[1])
ct = emailGrades(student)
print ct, “messages sent”

def readGrades(filename):
    student = [ ]
    for row in csv.reader(open(filename, “rU”)):
        student.append(row)
    return student

def emailGrades(student):
    titles = student[prefs[“titles”]]
    maxgrade = student[prefs[“email”]]
    ave = student.pop()
    ct = 0
    for s in student:
        ct += 1
        if ct <= prefs[“studentrow”]:
            continue
        outline = formatGrades(s, titles, maxgrade, ave[s])
        proc = subprocess.Popen(“mailx -sgrades %s@cs.wm.edu”
                 %(student[prefs[“name”]],
                 student[prefs[“email”]]),
                 shell=True, stdin=subprocess.PIPE)
        out = proc.communicate(outline)
        proc.wait( )
    return len(student) - prefs[“studentrow”]
```

FIGURE 67.3 A glue script for emailing grades.

The function `formatGrades` (not shown here) generates a multiline email message; it amounts to 11 lines of code.

Python has some interesting features that are worth noting here. First, it uses the plus symbol as the string concatenation operator. Second, it provides no automatic conversion from a number to a string, even in string concatenation. Third, the percent operator with a string as the left operand and a tuple as the right operand is basically the `sprintf` function of C; the left operand serves as a formatting code for each operand in the tuple. After setting up some column headers, the `for` loop generates one grade per output line, each containing a grade name, a maximum grade, and the grade earned by the student.

As presented here, the entire program has fewer than 40 lines of code. It was written and debugged in less than a day to replace a similar, cryptic, less flexible Perl script. The script accommodates spreadsheet output generated by Excel, OpenOffice Calc, and GoogleDocs using a variety of conventions used by the author and his graders. It has proven to be a very valuable glue script, since it is used about 10 times per semester per class.

This typical Python glue script linking two disparate applications exposes many commonly used features of glue languages such as Python:

- Scripting languages do not require declaration of variables and supports dynamic typing.
- Scripting languages support a wide variety of string operations, including pattern matching.

- Scripting languages provide both dynamic arrays (lists) and dictionaries (also known as associative arrays or hash tables).
- Scripting languages provide a means of executing system commands with the ability to supply input to those commands or to read their output.
- Unlike many languages, Python has a simple syntax. A compound statement such as a function definition (`def`), a `for`, or an `if` end with a colon. The statements forming the loop body (for example) being indented. In Python if a statement looks like it is part of a function definition, loop body, or a then/else body, then it is interpreted as one.
- Python supports a large number of modules, which are very useful language extensions. In this program, we used the `sys` module to access command line arguments, the `csv` module to simplify reading CSV format files, and the `subprocess` module to create a subprocess capable of executing system commands and communicating with its parent.

Finally, we note that many universities now teach Python in their CS1 courses (the first course in a computer science major). One reason for this is that the course is rapidly becoming a service course, taken primarily by physical science and social science majors. Its popularity is partly due to its simple syntax and partly because of this value as a tool for solving computing problems in other fields of study.

67.2.4 Web Languages

According to its developer, Rasmus Lerdorf, the motivation for developing PHP [4] was the following:

> As the Web caught on, the number of non-coders creating Web content grew exponentially. … But soon they were asked to add dynamic content to their sites. … This is where PHP found its niche. … I had written all sorts of CGI [Common Gateway Interface] programs in C, and found that I was writing the same code over and over. What I needed was a simple wrapper that would enable me to separate the HTML portion of my CGI scripts from my C code … This concept became PHP.

Lerdorf initially developed PHP in 1994, but because its usage quickly grew beyond the abilities of a single developer, it became an open-source product. PHP is a server-side scripting language whose scripts can be directly embedded within the HTML code that defines the page layout. In this way, PHP is ideal for use with Web pages that have dynamic content, including forms processing and database access.

A major advance in the development of PHP was the release of PHP 5 in 2004, which added the following major features:

- A new OOP model
- Improved database support for MySQL and SQLite
- A host of new functions to simplify coding of common features, such as date and array handling

These features, along with its basic syntactic similarity with Java, C, and C++, tend to obscure the distinction between PHP and other general purpose languages. They also make PHP easier to learn for programmers who have prior experience with these other languages.

At many sites, PHP is installed on the main Web server. It is often installed together with MySQL, an open source database management language. A popular package used across many server varieties is known as *xAMP*, which combines PHP, MySQL, and an Apache server in an easily installed server configuration that can be run either locally or remotely.*

* The Windows variety is called WAMP, the Mac OS variety is called MAMP, and the Linux variety is called LAMP, appropriately. All three varieties of xAMP are open source and so can be freely downloaded and used by any software developers.

Typical examples of PHP usage on a Web page include

- Dynamic content such as today's date or a news item of the day
- Forms processing, including forms validation
- Database access

The PHP processor takes a document file as input and produces HTML as output. Like JavaScript, the input file consists of a mixture of HTML and PHP, which are marked by special HTML-like tags. The major difference from JavaScript is that a PHP script is executed on the server, not on the client. This provides a level of security that is unachievable with JavaScript, since the PHP script is never downloaded to the client's Web browser.

The PHP processor has two modes of operation. It begins in copy mode in which HTML tags and text are copied directly to the output. When it encounters the special tags:

```
<?php
// one or more lines of PHP code
?>
```

the PHP processor switches to script mode and interprets the script on the fly. The output from the script, whatever is written to stdout, is inserted directly into the HTML page. The double slashes (//) denote a PHP comment, which continues until the end of the line.

Figure 67.4 contains a simple example that illustrates some of the features of PHP. Familiarity with both HTML tags and C programming is presumed. In this example, we use a PHP script in conjunction with a database to produce department directories, one each for the faculty, the staff, and the graduate teaching assistants. The directory desired is specified as a parameter in the URL; for example, http://www.cs.wm.edu/people/index.php?id=Faculty will list the faculty directory.

```
<?php
$title = $_GET['id'];
include "header.inc";
if (! eregi(":Faculty:Staff:GradTA:", ":$title:")
    errorPage($title);
include "dbaccess.inc";
$result = mysql_query ("select * from $title");
print '<center> <table cellspacing=5 cellpadding=5 border=2>
    <tr align=left><th>Name</th><th>Office</th>
    <th>Phone</th><th>Email</th></tr>
';
while (list($name, $url, $office, $phone, $email) =
                mysql_fetch_array ($result)) {
    print "<tr align=left>\n";
    $lname = $name;
    if ($url !="")
        $lname = "<a href=$url>$name</a>";
    print "<td>$lname</td>\n";
    print "<td>$office</td>\n";
    print "<td>$phone</td>\n";
    print "<td><a href=mailto:$email>email</a></td>\n";
    print "</tr>\n";
}
print '</table> </center> ';
include "trailer.inc";
? >
```

FIGURE 67.4 A PHP script for a directory listing.

With this parameter, the script accesses the appropriate database table, in this case the faculty table, and generates the appropriate HTML output. Because these directories are fairly static, it is fair to ask the question: Why not maintain the information as static HTML pages? Here are three reasons:

1. The staff people who maintain this information need only interact with a form that interfaces with the database—they do not need to learn HTML.
2. Using PHP allows the webmaster to more easily maintain a consistent look and feel to the web pages.
3. The information can be accessed from other portions of the web site.

A PHP page begins in HTML mode; this would be used to set up the page in a site-specific standard format, including the title, background color, navigation buttons, etc. Because HTML lacks an include facility, a common use of PHP is to set up the page via parameterized header (line 3) and trailer (line 25) files.

The global `$_GET` is an associative array of variable-value pairs that are passed to the script via the URL parameters. In this case, the parameter `id`, whose value is the string `Faculty`, is passed to the PHP variable `$title`. The `header.inc` script expects the title variable to be set.

Before we begin setting up the body of the page, we want to check to see that the `id` parameter has a valid value. Because this value is used to access a database table, a malicious user could attempt to attack the database by providing an unexpected value. In this case, we test the value supplied for the id parameter using a pattern match against the three legal values. If the pattern match fails, an error routine is called, which generates an error page and exits with no further processing.

Otherwise, database access code is included, which sets the name of the database along with security information such as userid and password. The select SQL statement is shown, which in this example accesses the faculty table in the database. At the current time, each of the three tables contains the following information for each person:

- The person's name
- A URL, if they have a homepage
- Their office address
- Their phone number
- Their email address

The information is to be generated as an HTML table with one row per person. Each column should have an appropriate heading (lines 7–10).

Next comes the main logic of the script. A while loop (line 11 ff.) is used to fetch the rows of the result of the SQL query one person at a time. That result is returned as an array, so the list function is used to assign the array values to conveniently named variables. The body of the loop consists mostly of print statements to write the appropriate columns: As with Perl, variable references may be freely embedded in double-quoted strings. The name field is made into a link if the person has a non-empty URL field.

All that remains is to close the table and center HTML tags and invoke the standard web site trailer:

This script is a typical example of server-side scripting and exposes some of the commonly used features of PHP:

PHP scripts freely alternate between pure HTML and PHP.

PHP does not require the declaration of variables and supports dynamic typing. The same value can be treated both as a string and as a number (provided it can be interpreted as a valid number).

PHP supports a wide variety of string operations. It also supports pattern matching as found in the Unix utilities grep and sed.

PHP provides both dynamically sized arrays and associative arrays (hashtables).

PHP directly supports accessing information from a database.

Besides PHP, all of the glue scripting languages can be used for dynamic web programming, including Perl, Python, and Ruby.

67.2.5 Embedded Languages

Embedded languages provide functionality that can be activated by the user by the user to perform specific tasks, like text editing. One of the earliest examples of an embedded language is GNU Emacs, developed in the 1980s.

Like other versions of Emacs available at the time, GNU Emacs had an embedded Lisp interpreter. Everything from setting startup options to the development of modes was done in Lisp. This made Gnu Emacs customizable by the end user. Most of the customizations themselves were developed by end users writing in Lisp.

For example, a typical customization of Emacs was developed for editing Python programs. This customization automatically indents the program as it is typed, colorizes reserved words, identifiers, strings and other literals, etc. By selecting a region, the user can comment it out, indent it left or right, wrap lines as needed, etc.

The development of computer games, whether on a dedicated console or on a personal computer, has rekindled interest in embedded languages. Many books on writing game programs discuss the use of an embedded scripting language. The choice of language can vary from Lisp/Scheme to Lua to languages similar to Javascript. Factors that are considered in choosing an embedded scripting language include its suitability to the task, familiarity of the development team with the language, and the size and speed of the interpreter for the language.

Tasks performed by the embedded language may include loading options and configuration files at startup, saving and reloading state information, implementing and interpreting finite-state machines, moving objects in the game, etc.

An important consideration in choosing an embedded scripting language is the size and speed of the interpreter. Consider Lua as an example. Compared to Python, Lua has a paucity of data structures and types. Python offers both infinite precision integers, as well as floating point numbers, while Lua offers only the latter.

67.3 Principles

Until recently, scripting languages were designed primarily for programming in the small, as opposed to application programming languages, which were designed for programming in the large. The languages used dynamic typing and were interpreted rather than compiled. The primary application for scripting languages was to glue existing applications and utilities together to make a new application.

The advent of languages like Java and C# began to muddy these waters. Like scripting languages, they were compiled to byte code for a virtual machine and then interpreted. Other than the fact that they used static rather than dynamic typing, there seems little to distinguish them from scripting languages such as Perl and Python.

Entire applications, such as content management systems for web applications, are now being written in scripting languages. Indeed, except for embedded languages, such as Lua, the distinction between scripting languages and application programming languages may soon disappear. In such a case, scripting languages should be held to the same principles as application programming languages. Such principles include

1. A well-defined syntax, expressible in a content-free grammar
2. A clear, well-defined semantics, which is expressible in an efficient implementation
3. A well-defined type system
4. The syntax, semantics and type systems should work together to minimize programming errors

5. The language should support its intended application area with a rich set of constructs that encompass needed functions and idioms
6. The language should be as simple as possible, both to learn and to use
7. The language should support future extensions while maintaining backward compatibility

In the remainder of this section, we will consider each of these in turn. In cases where principles are in conflict with one another, the language designer must make appropriate tradeoffs to achieve a workable implementation for the language.

Further complicating the issue is that, with the exception of Ruby and Lua, most scripting languages began as imperative languages, to which object-oriented features were later added. In some cases this addition resulted in breaking earlier features.

67.3.1 Syntax

A well-designed language starts with a simple syntax that is expressible by a grammar. Most application language definitions present their syntax as a context-free grammar; the size and complexity can vary among languages. For example, see Table 67.1 [13], which shows the number of pages of text required to present the complete grammar for each of four prominent programming languages.

From this table, it can be inferred that the languages Pascal and C are syntactically simpler than Java and C++.

However, a review of the websites for Perl, for example, reveals that Perl may not have a syntactic grammar in the traditional sense.

Moreover, its informal syntactic definition reveals that Perl distinguishes between constant strings using either single or double quotes [14]:

> Besides the backslash escapes listed above, double-quoted strings are subject to *variable interpolation* of scalar and list values. … Variable interpolation may only be done for scalar variables, entire arrays (but not hashes), single elements from an array or hash, or slices (multiple subscripts) of an array or hash.

But "variable interpolation" as described earlier is never given a precise syntax, so beginners often avoid its use altogether.

Secondly, in order to distinguish among scalar variables, arrays, and hashes, Perl requires that each of these start with a unique character. Omitting this character leads to the so-called bare word compile error, which (in our experience) is the most common error in Perl. Languages that have copied this convention, namely, PHP and Ruby, inherit the same weakness.

Since Python does not support variable interpretation of arbitrary strings, variables are not required to start with a special character. This supports Python's reputation as a simpler language than Perl.

Since these languages support the `sprintf` function of C, as well as string concatenation, *variable interpretation* of double quoted strings leads to more problems than it solves.

TABLE 67.1 Grammars for Various Languages

Language	Approximate Pages
Pascal	5
C	6
C++	22
Java	14

67.3.2 Semantics

Languages should have a well-defined semantics based on a formal semantic definition to the extent possible. However, if formal semantic definitions of scripting languages exist, we could not find them. A formal semantic definition allows alternative implementations of language, since it can be proved that the abstract models of various implementations are equivalent. These formal definitions can be based on either operational semantics or denotational semantics.

Again using Perl as an example, the implied subject of most operations is the special variable `$_`. A common error in Perl is to assume that each function has its own copy of this variable but, in fact, there is only a single copy in the entire program. One implication of this rule is that modules or library functions that a programmer develops must avoid using the same special variable.

67.3.3 Type System

Scripting languages are all dynamically typed. A language is defined to be *dynamically typed* if types are associated with values rather than variables; thus, a variable can have a different type at any time during program execution, depending on the value currently assigned to the variable at that time.

Since scripting languages were originally intended for programming in the small, it was felt that a script with only a screenful of code (or two) did not need the extra complexity of having to declare all its variables.

The question then becomes what do you do if one or more operands of an operator has the wrong type. With respect to automatic coercion from one type to another, a permissive language attempts to convert a value from one type to another. A language is more strict with respect to automatic coercions if it allows only those which are guaranteed to be safe.

Permissive languages such as Perl and PHP try to keep the program running if possible. Because of this they tend to allow automatic coercions of values, even when such conversions are more likely to be incorrect rather than correct. In the case of PHP, this attitude eliminates many error messages being written to the browser window. Such a practice can be defended on security grounds.

Because of this permissiveness, scripting languages try to avoid overloading operators based on the types of their operands. Perl and PHP use the dot to denote string concatenation, whereas Python overloads the plus operator for concatenation. Perl even has unique operators for string comparisons, instead of using the mathematical operators already used for numeric types.

Strict languages such as Python do not permit automatic coercions unless they are safe. For example, converting an integer to a string is guaranteed to work, although Python does not permit an automatic conversion. Converting a string to a number would be disallowed, since the string may contain arbitrary text not representing a number. For a glue language, reporting errors tends to be preferable to generating incorrect output.

67.3.4 Other Properties

A good scripting language should support the needed functions, data structures, and idioms of its intended applications. We have seen how PHP supports an HTML mode, which is quite useful in web applications.

Most of the languages we have examined support unbounded indexed lists and dictionaries. They also support iterators for traversing these structures. Whether these features lead to dramatically simpler applications is a subject of debate.

The one area where all of these languages have had problems is in maintaining backward compatibility. As the size and number of scripts written in these languages continues to grow, breaking existing scripts is likely to cause programmers to seek a more stable language.

Consider the language Python as an example. Through versions 1 and 2, Python remained largely backward compatible. However, the design goals for version 3 was that "there should be one—and preferably only one—obvious way to do it." The backward-compatibility problems relative to earlier versions are as follows:

- The `print` statement became a function, necessitating parentheses surrounding the arguments. This broke almost all scripts written for earlier versions of Python.
- The division operator was redefined to always produce a floating point number. A new operator was introduced which produced an integer when one integer was divided by another
- Removing previous backward-compatibility features, including old-style classes, integer-truncating division, string exceptions, and implicit relative imports.
- Many changes were made to key libraries. For example, the glue script in Figure 67.3 will not run under version 3 due to changes in the `process` module. Many user libraries would not work under version 3.

Although Python version 3 came out in December 2008, only in late 2011 and early 2012 have Python textbooks started to appear in large numbers.

To summarize, as the distinction between traditional application programming languages and scripting languages continues to shrink, the principles and qualities of traditional languages should be applied more carefully to scripting languages as well.

67.4 Conclusions

Scripting languages began in the 1960s as a simple device for connecting individual tasks in a sequence of programs that ran in batches.

Throughout the 1970s and 1980s, especially with the rapid emergence of Unix and time sharing systems, scripting languages grew in richness and versatility to accommodate the needs of these systems.

With the widespread emergence of the Internet in the 1990s and recent years, newer scripting languages embodied still more features and functionality to support web-based programming requirements.

In fact, scripting languages of today, like PHP and Python, are not easily distinguishable from other modern programming languages like Java and C++. For that reason, it appears that tomorrow's scripting languages will need to adhere to a more rigorous definitional standard, in order that they maintain such important characteristics as backward compatibility, interoperability, and type safety.

References

Bynum, B. and T. Camp. After you, Alfonse: A mutual exclusion toolkit. *Proceedings of the 28th SIGCSE Technical Symposium on Computer Science Education*, Philadelphia, PA, 1996, pp. 170–174.

Bynum, W. L., R. E. Noonan, and R.H. Prosl. Using a project submission tool across the curriculum. *The Journal of Computing in Small Colleges*, 15 (5), 2000, 96–104.

C Shell, http://en.wikipedia.org/wiki/C_shell, June, 2012.

Dougherty, D. and A. Robbins. *sed & awk*. O'Reilly, 1990.

Hughes, S. *PHP Developer's Cookbook*. SAMS, 2001.

IBM, *Introduction to the New Mainframe: z/OS Basics*. IBM RedBooks, 2012.

Joy, W. *An Introduction to the C Shell*. University of California, Berkeley, CA, 1982.

Kernighan, B. W. and R. Pike. *The UNIX Programming Environment*. Prentice Hall, Englewood Cliffs, NJ, 1984.

Osterhout, J. K. *Tcl and the Tk Toolkit*. Addison-Wesley, 1994.

Osterhout, J. K. Scripting: Higher level programming for the 21st century. *IEEE Computer*, 31 (3), March 1998, 23–30.

Rosenblatt, B. *Learning the Korn Shell*. O'Reilly & Associates, Sebastopol, CA, 1993.
Schwartz, R. L. *Learning Perl*. O'Reilly & Associates, Sebastopol, CA, 1993.
Tucker, A. and R. Noonan. *Programming Languages*, 2nd edn. McGraw-Hill, 2007.
Wall, L., T. Christiansen, and R. L. Schwartz. *Programming Perl*, 2nd edn. O'Reilly, Sebastopol, CA, 1996.
Web Monkey, http://www.webmonkey.com/2010/02/scripting_language.
WikiBooks, *C Shell Scripting*. http://en.wikibooks.org/wiki/C_Shell_Scripting, 2011.

68

Compilers and Interpreters

Kenneth C. Louden
San Jose State University

Ronald Mak
San Jose State University

68.1 Introduction

Compilers and interpreters are language translators that have many functions in common, in that both must read and analyze source code. A compiler, however, produces a program equivalent to the source program in a target language, usually in object or assembly code but also sometimes abstract machine code such as Java virtual machine (JVM) code or even C code, whereas an interpreter directly executes the source program. Any programming language may be either compiled or interpreted, but languages with significant static properties (e.g., FORTRAN, Ada, and C++) are almost always compiled, whereas languages that are more dynamic in nature (e.g., LISP and Smalltalk) are more likely to be interpreted. Languages that differ substantially from the standard von Neumann model of most architectures (e.g., PROLOG) may also be interpreted rather than compiled. A performance penalty is incurred by interpretation over compilation, so in cases where speed is critical, compilation is to be preferred. By mixing compilation and interpretation, this performance penalty can be reduced, usually to well within an order of magnitude. The advantage to interpretation is that the compilation step is avoided (useful during program development), and an interpreter offers greater control over the execution environment (useful for complex run-time environments) and greater flexibility in adapting to different architectures.

The first translators were developed in the 1950s. Prior to the development of high-level languages, a compiler was essentially what is known as a linker today: it "compiled" a collection of machine-language routines from a library to form a single program. A team at IBM under the direction of John Backus is generally credited with developing the first commercial compiler for a high-level language, during the period 1954–1957 (Backus et al., 1957). The language translated by this first compiler was FORTRAN, which was designed simultaneously with the compiler (and is also credited with being the first high-level language). Modern translation techniques were first used in Algol60 compilers a few years later (see, e.g., Randell and Russell, 1964), when the relationship of language translation to the theory of finite automata and context-free grammars became better understood. The study of these subjects was further stimulated by this relationship, and by the early 1970s, many of the standard techniques in use today were known. Since then, general improvements in translators have come in the following areas:

- The automation of a significant portion of the construction of a translator
- Improvements in the speed of the target code, due to the increased application of code-improving (or *optimizing*) algorithms

A greater ability for compilers to be relatively easily *retargeted*, or rewritten for a new target machine language, due in part to automation and in part to a better understanding of the required compiler structure.

Improvements have also come in the implementation of special language features, such as exception handling, generics (parametric polymorphism), object-oriented features (such as dynamic binding), and parallelization, due to increasing understanding of these mechanisms. One area in which significant theoretical advances have taken place in the past few decades is in the translation of functional languages, with new algorithms for type checking, type classes, and interpretation by tree reduction (see, e.g., Peyton Jones, 1987). So far, however, these techniques have remained outside of the mainstream of languages and translators.

An important aspect of the technology of translators is its strong interaction with language design. For example, the introduction of block structure in the Algol language family gave impetus to the development of stack-based translation algorithms. In turn, the stack-based algorithms influenced the development of language semantics specifically designed to take advantage of the algorithms, such as the lexical scope rule and the principle of syntax-directed translation, in which the semantics of a program (i.e., its execution behavior) is directly reflected in its syntax, or structure, so that translation can be guided by this structure. (This rule could equally well have been formulated as *semantics-based syntax*.) Because of its early appearance, FORTRAN was not designed with these principles in mind, and it remains a somewhat more difficult language to translate with the standard techniques than later languages of the Algol family.

Another aspect to the development of language translators has been the tendency of the computing community to constantly rediscover or recreate translation techniques (other than the basic, well-known ones outlined in many texts). Primarily, this is due to the lack of detailed documentation of specific translators in the computer science literature. This, in turn, is largely due to the commercial and/or proprietary nature of most translators. Two translators that historically were reasonably well documented and exerted a corresponding influence on subsequent translator construction were the portable C compiler (Johnson, 1978) and the PASCAL P-compiler (Wirth, 1971; Nori et al., 1981). Lately, the availability of public-domain software from the Internet has greatly improved the opportunity to study existing translators. Particularly well-documented and of high quality are the compilers distributed as part of the GNU software project of the Free Software Foundation (http://gcc.gnu.org).

Common to almost all modern compilers is a conceptual structure in which the tasks performed are divided into *phases*, or logically complete processing steps. The standard phase structure is shown in outline in Figure 68.1.

The first phase, called the scanner or lexical analyzer, is the only phase directly involved in reading the source program. It converts the characters of the source program into tokens, sequences of characters that represent the basic units of program structure. Typical tokens are keywords such as *while*, numeric literals such as 3.14159, and identifiers.

The second phase in a compiler is the parser, or syntactic analyzer, which collects sequences of tokens into complete units, such as expressions, statements, and declarations. The output of the parser (either explicitly or implicitly) is a tree or other equivalent data structure representing the structure of the programming unit just recognized. This tree is called the *syntax tree*.

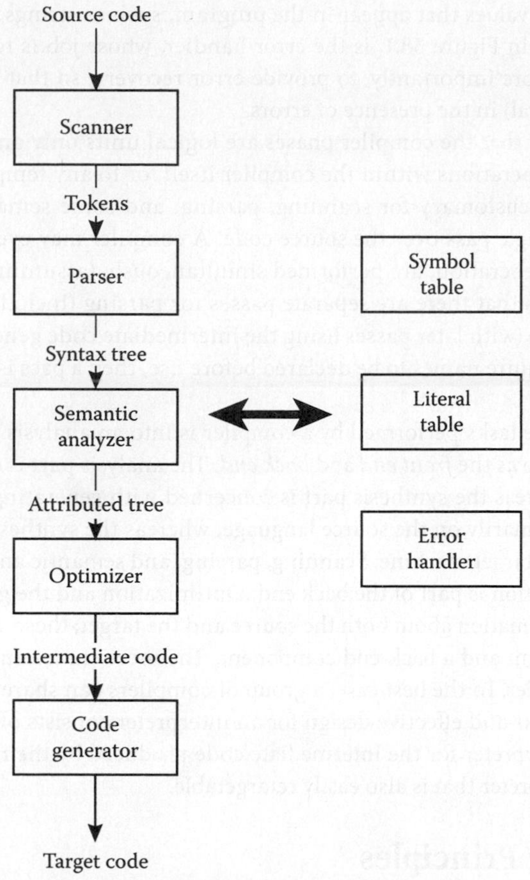

FIGURE 68.1 The phases of a compiler.

The third phase in a compiler is the semantic analyzer, which computes attributes, or properties of each programming construct, as well as its effect on the attributes of other constructs. Typical attributes include data types of expressions and identifiers, memory sizes, and actual or potential memory locations. The semantic analyzer also determines if the construct makes sense according to these attributes. That is, it performs consistency checks, such as type checking or range checking. The output of the semantic analyzer can be represented by an attributed tree representing the original tree modified by the computed attributes. The fourth phase in a compiler is called the optimizer in Figure 68.1. This phase usually generates some form of linear code, called *intermediate code*, from the representation passed to it by the semantic analyzer. It also applies some forms of code-improving algorithms, either before or after generating intermediate code (or both). This phase of compilation is the most variable in different compilers; it may even be absent altogether.

The fifth and final compiler phase shown in Figure 68.1 is the code generator. This phase generates the final target code and may also perform some additional improvements to the code.

All of the phases of a compiler interact with various tables and handlers within the compiler. Figure 68.1 shows three important examples of these other components: the symbol table, the literal table, and the error handler. The figure indicates their interaction with the phases by a large double-headed arrow. The symbol table maintains the names defined by the program under translation, as well as possible predefined names in the language, and associates the names with their attributes, which may include scope, memory location, data type, and memory size. The symbol table may be monolithic, or it may be separated into a tree or graph of smaller tables, representing different scopes in the program. A similar table is needed for literal values that appear in the program, such as strings and numeric literals. The third component, shown in Figure 68.1, is the error handler, whose job is to generate different kinds of error messages but, more importantly, to provide error recovery, so that translation may continue (at least as far as is practical) in the presence of errors.

It must be emphasized that the compiler phases are logical units only and may not correspond to any actual grouping of operations within the compiler itself, or to any temporal sequencing of these operations. Indeed, it is customary for scanning, parsing, and some semantic analysis to be completely integrated in a single pass over the source code. A compiler may even be one-pass, in that all phases, including code generation, are performed simultaneously (assuming that the language itself permits it). More likely is that there are separate passes for parsing (including scanning), optimization, and code generation (with later passes using the intermediate code generated by the first pass). If the language does not require names to be declared before use, then a pass is also necessary to resolve name references.

A useful division of the tasks performed by a compiler is into an analysis part and a synthesis part, sometimes also referred to as the *front end* and *back end*. The analysis part is concerned with analyzing the source program, whereas the synthesis part is concerned with generating the target program. The analysis part depends primarily on the source language, whereas the synthesis part depends primarily on the target language or target machine. Scanning, parsing, and semantic analysis are part of the front end, whereas code generation is part of the back end. Optimization and the generation of intermediate code usually require information about both the source and the target; these are more difficult to divide into a front-end component and a back-end component. The more successfully this is done, the easier it is to retarget the compiler. In the best case, a group of compilers can share front ends and back ends interchangeably. A popular and effective design for an interpreter consists of a compiler front end and a back end that is an interpreter for the intermediate code produced by the front end. This results in a reasonably efficient interpreter that is also easily retargetable.

68.2 Underlying Principles

Algorithms used in translators are based heavily on computation theory and, to a lesser extent, on formal semantics. Scanners are direct implementations of finite automata that solve string recognition problems through nonrecursive pattern matching. Parsers depend on the theory of context-free grammars and pushdown automata, which solve recursive recognition problems through stack-based pattern matching. Semantic analysis depends on solving sets of tree equations called *attribute grammars*. Code generation and interpretation can also be seen as applications of attribute grammars. It is possible to use even more formal semantic specifications, particularly denotational specifications, of the source and target languages to construct semantic analyzers and code generators (see, e.g., Polak, 1981 and Lee, 1989). The advantage to doing so is that the compiler can be proved correct (i.e., the semantics of the source and target programs are guaranteed to be the same). However, these techniques have not become popular, and we do not discuss them further. In the remainder of this section, we will discuss each of the areas mentioned in a little greater detail.

68.2.1 Finite Automata

A finite automaton is an abstract computational machine consisting of a finite number of states and transitions between states based on input symbols. The machine runs by beginning in the starting state and consuming input symbols while entering new states via corresponding transitions, until either an error state or an accepting state is reached. At that point, it may declare success or failure, or possibly continue executing. Each state represents stored knowledge about the computation up to that point. A finite automaton can handle arbitrary repetition by a fixed, finite set of input symbols, but it cannot handle recursive processes, because that would involve an unpredictable number of states.

Finite automata are the basis for the recognition of tokens within a scanner. Based on the well-known correspondence from computation theory between finite automata and regular expressions, tokens are usually given initially by regular expressions, which are specifications for the string patterns, or *lexemes*, that a token represents. The mathematical theory of regular expressions limits itself to the consideration of three matching operations: the choice between two alternatives, indicated by the vertical bar | (similar to the logical OR operation); the concatenation or sequencing of two strings (with no operator symbol); and the repetition of a pattern, indicated by a postfix asterisk* (sometimes called the closure or Kleene closure operation). Parentheses are also used to group subexpressions together. As an example of the use of regular expressions to represent tokens, the following regular expression for a token represents simple, unsigned numbers consisting of a sequence of one or more decimal digits:

```
(0|1|2|3|4|5|6|7|8|9) (0|1|2|3|4|5|6|7|8|9)*
```

Such a regular expression can be converted into a finite automaton by one of several standard algorithms. The basic method is to use Thompson's construction (Aho et al., 2007, p. 159) to derive a nondeterministic automaton (i.e., one with an unpredictable next state) from the regular expression, and then to use the subset construction (Aho et al., 2007, p. 153) to derive an equivalent deterministic automaton from the nondeterministic one. Other algorithms exist that perform this conversion in one step and also construct an automaton with a minimal number of states. Although these algorithms can sometimes be useful for the design of scanners, their primary use is in the construction of scanner generators such as Lex (discussed in Section 68.3.2).

68.2.2 Context-Free Grammars

The theory of context-free grammars extends the ideas of finite automata and regular expressions to recursive situations. A context-free grammar is a collection of named recursive rules of the form $A \rightarrow \alpha_1 | \ldots | \alpha_n$, where A is the name of the rule and the α are strings of tokens and names (including possibly A itself) representing the different possible choices for the structure of A. The names are called *nonterminals* and the tokens are called *terminals*, for technical reasons explained shortly. Grammar rules are also sometimes referred to in the theory as *productions*. Each such rule represents the fact that a structure represented by A may become any one of the structures represented by an α. The absence of context for these potential replacements of A is what makes these grammars context free. It is possible to define grammars that are more general than context-free grammars; these can be arranged according to increasing generality in what is known as the Chomsky hierarchy. However, context-free grammars are the most useful within the computational constraints of language translation.

Every nonterminal in a context-free grammar defines a set of token strings: the strings of tokens that can be legally parsed by the grammar rule for that nonterminal (and associated rules). The most general structure that is defined by the grammar is usually singled out as the structure representing the entire language, and its associated nonterminal is called the *start symbol*. A legal parse is represented by a sequence of replacements of nonterminals by choices of right-hand sides of their associated grammar rules. This sequence of replacements is called a *derivation*. In a derivation, every nonterminal must eventually be eliminated by replacing it with another string, whereas a terminal, once it appears, is never replaced; this may be seen as a justification for the names *terminal* and *nonterminal*. In general, there are many possible derivations for the same string, which can be constructed by varying the order in which the replacements are made. Two kinds of derivations that are important for parsing use fixed orders for the replacements. In the first, called a leftmost derivation, the leftmost nonterminal in the current string is always the one to be replaced at the next step.

Although derivations are useful for expressing exactly which steps are taken in a parse, they do not express the structure of the parsed string very well. A more useful representation of this structure is the parse tree, which represents each terminal and nonterminal by labeled nodes, and each replacement step in a derivation by the construction of a set of children of the node label being replaced. In a parse tree, each leaf node is labeled by a terminal, and each interior node is labeled by a nonterminal. Even parse

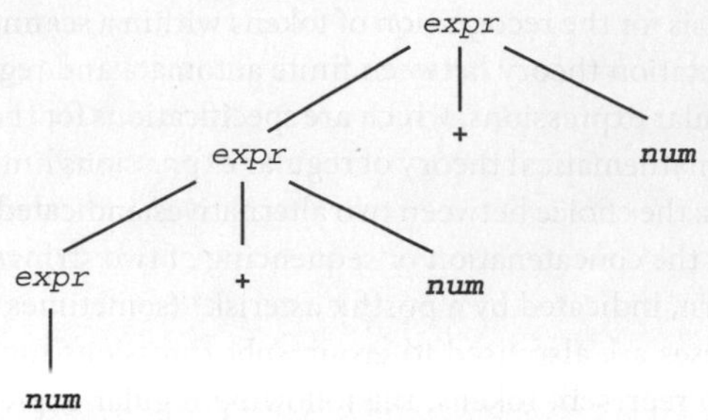

FIGURE 68.2 A parse tree for the string 3 + 4 + 5.

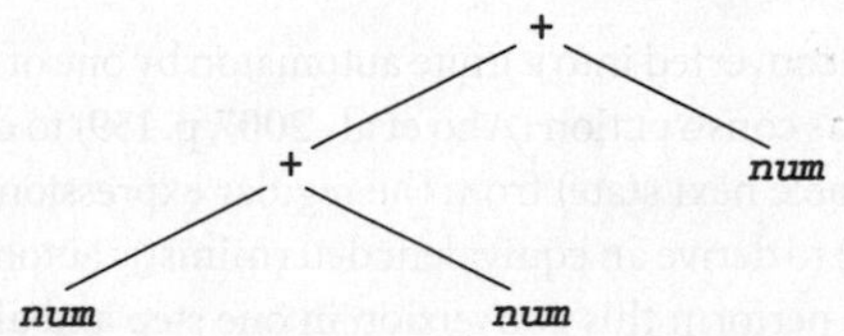

FIGURE 68.3 A syntax tree for the string 3 + 4 + 5.

trees have more detail than is actually needed to determine the meaning of a structure; and a condensed form of a parse tree, called a syntax tree, is the most useful data structure for translation. In a syntax tree, nodes may have a more diverse structure than in a parse tree: they may have more than one label, and both terminals and nonterminals may be used as labels of interior nodes.

As a brief example of the grammar concepts we have summarized, consider the grammar

$$expr \rightarrow expr + \mathbf{num} \mid \mathbf{num}$$

which has one nonterminal *expr*, two terminals **+** and ***num***, and a single production with two choices. (Here, ***num*** represents a number token with lexemes such as 42 or 7.) An example of a legal string for this grammar is 3 + 4 + 5. A leftmost derivation is

$$expr \Rightarrow expr + \mathbf{num} \Rightarrow expr + \mathbf{num} + \mathbf{num} \Rightarrow \mathbf{num} + \mathbf{num} + \mathbf{num}$$

(Note that, because there is only one nonterminal to be replaced at each step, this is also a rightmost derivation.) A parse tree for the same string is given in Figure 68.2. A syntax tree is given in Figure 68.3.

68.2.3 Parsing Algorithms

Algorithms designed to match an input string of tokens based on a grammar and, either implicitly or explicitly, to construct a parse or syntax tree, are called *parsing algorithms*. Parsing algorithms come in two general varieties: *top-down* and *bottom-up*. Top-down algorithms construct a parse tree from the root to the leaves by guessing which structures are about to appear, based on the next part of the input string and the structure expected to be seen. Bottom-up algorithms construct a parse tree from the leaves to the root by consuming the input and forming a set of subtrees until the next structural element can be guessed from the structures seen so far and the next part of the input string. Because of the recursive nature of context-free grammars, both kinds of algorithms must use a stack, either explicitly or implicitly, to hold partial results. When it is constructed explicitly, this stack is referred to as the *parsing stack*, and it will possibly contain terminals, nonterminals, or other symbols representing the state of the parser. Because of the nature of their operation, top-down parsers trace out the steps of a leftmost derivation, and bottom-up parsers trace out in reverse the steps of a rightmost derivation.

There are algorithms of both varieties that will parse any context-free grammar. Early's algorithm (Early, 1970) is the most well-known bottom-up algorithm. A general top-down algorithm may be found in (Graham et al., 1980). General algorithms may run in significantly slower than linear time (Early's algorithm requires cubic time in general), so they are rarely used in practice. Standard algorithms for top-down parsing are the LL(k) algorithms (parsing the input from left to right, giving a leftmost derivation, using k symbols of lookahead) and the LR(k) algorithms for bottom-up parsing (parsing the input from left to right, giving a rightmost derivation, using k symbols of lookahead). Both kinds of algorithms require that a grammar satisfy extra conditions to be parsable. The LL(k) algorithms, in particular, are quite restrictive, although easy to use. The LR(k) algorithms, although less restrictive, are more complex. One top-down parsing method that is more flexible than the LL(k) methods is called *recursive-descent*. A bottom-up algorithm that is simpler than the LR(k) algorithms is called LookAhead LR(1) (LALR(1)) (DeRemer, 1971; DeRemer and Pennello, 1982) and is normally restricted to one symbol of lookahead. Because these algorithms have proved themselves to be the most effective and easiest to use in practice, we discuss them in a little more detail.

In recursive-descent parsing, the grammar rules are viewed as prescriptions for the code of a set of mutually recursive procedures, one for each nonterminal. Recursive-descent, although suffering from some of the same problems as LL(k) parsing, is more flexible and can use simple ad hoc techniques to solve many of the problems of LL(k) parsing (Wirth, 1976). For instance, simple left recursion, which cannot be handled directly by an LL(k) parser, can be handled in recursive-descent by noting that a left-recursive rule $A \to A\alpha \mid \beta$ is equivalent to a parsing procedure that first recognizes β and then a sequence of zero or more α's (because the grammar rule generates strings of the form $\beta\alpha\alpha\ldots$). Thus, a recursive-descent procedure for the grammar *expr* → *expr* + **num** | **num** can be written using a while loop, as follows:

```
void expr(void)
{ match(NUM);
  while (nextToken == PLUS)
  { match(PLUS);
    match(NUM);
  }
}
```

An LALR(1) parser uses an explicit parsing stack instead of recursion. The state of a parse can be expressed by a finite automaton whose states consist of sets of so-called *items*, each item consisting of a production choice, a distinguished position indicated by a period (representing the point of progress in recognizing the rule), and an associated set of lookahead tokens legal at that point in the parse. (In the following discussion, we will use the so-called LR(0) items that lack a lookahead component; although LALR(1) items are more complex, the basics of the LALR(1) algorithm can be understood using these simpler items.)

Consider, for instance, the grammar *expr* → *expr* + **num** | **num**, which we will write for convenience in the form $E \to E + \mathbf{n} \mid \mathbf{n}$. There are six LR(0) items: $E \to .E + \mathbf{n}$, $E \to E. + \mathbf{n}$, $E \to E + .\mathbf{n}$, $E \to E + \mathbf{n}.$, $E \to .\mathbf{n}$, and $E \to \mathbf{n}.$. The start of a parse is indicated by beginning in a state represented by a new rule representing a start symbol that cannot appear elsewhere, and which we will write as $E' \to E$ in the example. The corresponding initial item is $E' \to .E$. Because this rule represents the fact that we may be about to recognize an E, we must also include the items $E \to .E + \mathbf{n}$ and $E \to .\mathbf{n}$ in this state. Transitions to new states are then given by moving the period past the symbols that follow it. For instance, there is a transition on the symbol E from the start state to the state containing the item $E \to E. + \mathbf{n}$, and a transition on the symbol $\mathbf{n}$ from the start state to the state containing the item $E \to \mathbf{n}.$. The complete finite automaton of items is given in Figure 68.4.

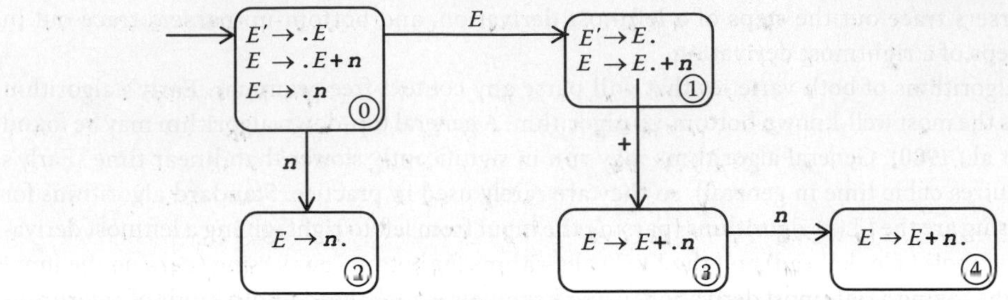

FIGURE 68.4 A DFA of sets of LR(0) items.

This finite automaton has no accepting states and is only used to keep track of the state of the parser. It is used in conjunction with a parser stack that holds the state numbers that the parser has passed through while parsing the input. It is used as follows. First, the initial state is pushed onto the stack. Then, the next input token is consulted. If there is a transition on this token from the current state (on top of the stack), then the token is removed from the input and the new state is pushed onto the stack; this is called a *shift operation*. If, however, there is an item in the current state of the form $A \to \alpha$ (a so-called final item), then this indicates that the string a has already been recognized, and it can be replaced by A; this is called a *reduce operation*.

A reduce operation is performed as follows. The states on the stack corresponding to α are popped from the stack (one state for each symbol in α). The state remaining at the top of the stack must have a transition on A, which is then taken, and the new state is pushed onto the stack.

As an example of this process, consider the automaton of Figure 68.4 and suppose that the input string is $\boldsymbol{n} + \boldsymbol{n}$. We depict the initial state of the parse as follows:

Parsing Stack	Input
\$0	$\boldsymbol{n} + \boldsymbol{n}$\$

The \$ in this representation is used to indicate both the bottom of the stack and the end of the input. The first step in the parse is a shift on $\boldsymbol{n}$ from state 0 to state 2. Then a reduction by $E \to \boldsymbol{n}$ takes place, and the parser moves to state 1. At that point, the + and $\boldsymbol{n}$ are shifted, and the parser is in state 4. Then a reduction by $E \to E + \boldsymbol{n}$ is made, popping states 4, 3, and 1, revealing again state 0. Again, the E transition is followed into state 1. At this point, the end of the input is encountered, and a reduction by $E' \to E$ is made, which corresponds to accepting the input.

The complete set of actions of the parser is given in Table 68.1, in which shift actions also include the new state number. Table 68.2 shows the LALR(1) parsing table for this simple grammar, which is used by the parser to select the actions indicated in Table 68.1. This table is two-dimensional, indexed by state, and lookahead token. Each table entry contains an action; shift entries are indicated by an *s* and the new state number; reduce entries are indicated by an *r* and the rule to be reduced; empty entries are errors. While this table is closely related to the automaton of Figure 68.4, the exact entries can only be inferred from a lookahead component, which we did not compute here.

TABLE 68.1 Actions of an LALR(1) Parser

Parsing Stack	Input	Action
\$0	$\boldsymbol{n} + \boldsymbol{n}$\$	Shift 2
\$02	+$\boldsymbol{n}$\$	Reduce $E \to \boldsymbol{n}$
\$01	+$\boldsymbol{n}$\$	Shift 3
\$013	$\boldsymbol{n}$\$	Shift 4
\$0134	\$	Reduce $E \to E + \boldsymbol{n}$
\$01	\$	Accept

TABLE 68.2 LALR(1) Parsing Table

State	Input			Goto
	n	+	\$	E
0	s2			1
1		s3	Accept	
2		$r\,(E \rightarrow \boldsymbol{n})$	$r\,(E \rightarrow \boldsymbol{n})$	
3	s4			
4		$r\,(E \rightarrow E + \boldsymbol{n})$	$r\,(E \rightarrow E + \boldsymbol{n})$	

An additional area in Table 68.2 is called the goto area. In this area are the transitions on nonterminals that are performed during reductions. These are essentially the same as shift operations, except that no input is consumed. A parser may also choose to condense this table by the use of default entries. For example, because state 2 only has reduce entries by the same rule, this could be made the default, in which case even on input token ***n***, the parser will perform the given reduction. This has the effect of postponing the declaration of error, but it cannot result in an incorrect parse. A similar default can be used for state 4.

68.2.4 Attribute Grammars

Although context-free grammars are generally accepted as the standard way to describe the syntax of a programming language, there is no equivalently accepted method for describing semantics. Formal methods for describing semantics, such as operational semantics or denotational semantics, have not met with universal acceptance, and though translators can be derived from such specifications, this is rarely done. Instead, various *ad hoc* mechanisms are used in translators to perform semantic analysis and code generation or interpretation.

One method that has proven useful for the translator writer is to use a so-called attribute grammar to describe the semantics of a programming language (Knuth, 1968). An attribute grammar associates to each grammar rule a set of equations describing the computational relationships among a set of attributes attached to the symbols in the rule. These attributes can be anything from the data type of a variable to the value of an expression; even the target code generated by a compiler can be represented as a string attribute in an attribute grammar. Most often, attributes are used to represent the static, rather than the dynamic properties of programs, and they are viewed as being fixed values attached to the nodes of a syntax tree. Indeed, attribute values are usually written using a dot notation similar to that of record fields, so that $X.a$ means the value of attribute a of symbol X. Attributes may in fact be implemented as fields in syntax tree nodes, or they may be stored in the symbol table or other data structures elsewhere in the translator.

Given a set of attributes $a_1,\ldots, a_k$ and a grammar rule choice $X_0 \rightarrow X_1\, X_2 \ldots\, X_n$, the jth attribute at the ith symbol is given in an attribute grammar by an equation of the general form

$$X_i.a_j = f_{ij}(X_0.a_1,\ \ldots, X_0.a_k, X_1.a_1,\ \ldots, X_1.a_k,\ \ldots, X_n.a_1,\ \ldots, X_n.a_k) \qquad (68.1)$$

where f_{ij} is a mathematical function. An attribute grammar is thus written in purely functional style without side effects.

As an example, consider the grammar *expr* → *expr* + ***num*** | ***num***, which we may assume expresses the summation of a series of numbers. An attribute grammar for the numeric value of an expression defined by this grammar is given in Table 68.3. Note that the two instances of the nonterminal *expr* in the first grammar rule must be distinguished by subscripting, and that the terminal ***num*** is assumed to have its numeric value (called *lexval* in Table 68.3) precomputed, possibly by the scanner.

Of particular importance to the translator writer are the kinds of dependencies that the attribute equations create among the attributes of different symbols in a parse tree, because these dependencies

TABLE 68.3 Attribute Grammar for a Simple Expression Grammar

Grammar Rule	Attribute Equations
$expr_1 \rightarrow expr_2 + \mathbf{num}$	$expr_1.val = expr_2.val + \mathbf{num}.lexval$
$expr \rightarrow \mathbf{num}$	$expr.val = \mathbf{num}.lexval$

determine when and how—or even if—the attributes can be computed during translation. A primary requirement is that the attribute grammar does not have any circular dependencies. In practical situations, this requirement is virtually guaranteed unless an error has been made. Attributes whose dependencies flow from right to left in the grammar rules (i.e., those whose Equation 68.1 all have only the symbol X_0 on the left) are called *synthesized* attributes; any other attributes are called *inherited*. Synthesized attributes can be computed bottom-up during a parse, or by postorder traversal of the syntax tree, while inherited attributes require a more complex computation scheme. Indeed, as the name implies, inherited attributes are often passed down the syntax tree from parent to child, or from sibling to sibling, and so can be computed by some form of modified preorder traversal of the syntax tree.

A great deal of effort can be expended to ensure that all attribute values are computable during the parsing phase to avoid having to construct the entire syntax tree and to avoid having to make additional passes over the input. The requirements that this places on the attribute grammar vary, depending on the parsing method employed. First, because virtually all parsers read the input from left to right, the attributes must be computable from left to right, and this means that all inherited attributes must depend only on the attribute values of their left siblings. In terms of the Equation 68.1, this means that each equation for an inherited attribute ($X_i.a_j$ is on the left and $i > 0$) must have the form

$$X_i.a_j = f_{ij}(X_0.a_1, \ldots, X_0.a_k, X_1.a_1, \ldots, X_1.a_k, \ldots, X_{i-1}.a_1, \ldots, X_{i-1}.a_k) \tag{68.2}$$

An attribute grammar in which all equations for inherited attributes are of this form is called *L-attributed*.

A further requirement for attribute evaluation during parsing is a form of strong noncircularity, in which an order for attribute evaluation can be fixed in advance without incurring any cycles (naturally occurring attribute grammars satisfy this non-circularity requirement, too).

The particulars of the parsing algorithm can also have a significant effect on which attributes are computable during parsing. Recursive-descent parsers are the most flexible; in the recursive routines, inherited attributes can be implemented as passed parameters, while synthesized attributes become returned values (Katayama, 1984). Bottom-up parsers most naturally compute synthesized attributes. They do this by maintaining a stack of attribute values in parallel with the parsing stack. New synthesized attributes are computed on this stack at each reduction step. It is also possible to evaluate certain inherited attributes during a bottom-up parse, but this often requires that the grammar be rewritten, so that the attribute equations can be converted to a manageable form. Indeed, it is theoretically possible to rewrite a grammar so that all attributes become synthesized (Knuth, 1968). However, the grammar thus produced bears little resemblance to the original. Thus, in difficult situations, it may be preferable to delay an attribute computation until after the parse and avoid rewriting the grammar into an unrecognizable form.

68.3 Best Practices

The theory of scanning, parsing, and attribute analysis implies that, in principle, a compiler or interpreter can be generated automatically from descriptions of the source language, attributes, and target machine. While a number of compilers have been generated in this way (Farrow, 1984), it is not common. This is partially because the necessary tools (particularly for optimization and code generation) are complex and have not become standard, and partially because of the need for efficiency, both in translation and in terms of the generated code, which is more difficult to achieve using general tools. At this writing, it is common to automate only the construction of the front end (i.e., the scanner, parser, and semantic analyzer), although at least partial automation of code generation has become more common with the increasing importance of retargetability.

Although a large number of scanner and parser generators have been written, only a few have gained broad acceptance. We describe here two sets of tools that have been used in many compilers: the Unix tools, Lex and yet another compiler–compiler (Yacc), and their more modern public domain versions, Flex and Bison, and the Java tool, JavaCC. Yacc produces an LALR(1) parser in the C language, whereas Bison can produce a parser using more general algorithms (Denny and Malloy, 2010). JavaCC (Kodaganallur, 2004) produces a recursive-descent parser in the Java language that can be extended to multisymbol lookahead. Another popular tool not discussed here is ANother Tool for Language Recognition (http://www.antlr.org/), which also produces more general recursive-descent parsers (Parr and Fisher, 2011) and can generate source code in a number of different languages. It should be noted that many commercial compilers and interpreters have been written by hand without the use of any tools at all; in such cases, the parsers are usually written using recursive-descent.

68.3.1 Specifying Syntax Using Regular Expressions and Grammars

In order to automate the task of generating a scanner and a parser, the first step is to specify the tokens using regular expressions and the syntax using context-free grammar rules. Although it is essential to understand the mathematical theory of both of these mechanisms, the theory ignores extensions and features of practical importance. We mention a few of these here.

The theory of regular expression relies on only three operations: concatenation, choice, and repetition. While these are enough to match any string recognizable by a finite automaton, most pattern matching systems extend this set of operators in many ways. As a typical example, consider the regular expression

```
[0-9]+(\.[0-9]+)?
```

which is written using standard conventions for the Unix tools Lex and Grep. This expression specifies a pattern for a simple floating point constant without exponential part: the expression (0–9) refers to a choice from the range of characters 0 to 9 (i.e., a digit), the + indicates a repetition of one or more (*a*+ is equivalent to *aa**), the backslash in front of the period *escapes* the metacharacter meaning of the period (otherwise, it would match any character), and the question mark indicates an optional component.

Even with these extensions, it is sometimes difficult to write regular expressions for certain patterns, even when such patterns do exist, and one may choose to apply an *ad hoc* recognition process instead of using a regular expression. A notorious case in point is that of C comments, which can be loosely described as /* (not */) */. The trouble is expressing (not */) as a regular expression. More generally, the non-existence of sequences of more than one character in a string is a difficult property to express as a regular expression. Fortunately, these situations do not occur very often in real languages.

It is also necessary to be aware that the theory of regular expressions can be easily extended to cover simple non-regular situations. For instance, nested comments require recursion, and so cannot be directly expressed as a regular expression. Nevertheless, adding a simple counter variable to a scanner permits it to recognize potentially nested comments.

Similar considerations arise in defining syntax using context-free grammars. Consider the following grammar that specifies a simple four-function floating-point calculator program with a single memory:

```
session → asgn nl session | nl
asgn → = expr | expr
expr → expr addop term | term
addop → + | -
term → term mulop factor | factor
mulop → * | /
factor → - factor | ( expr )| NUMBER | m
```

A `session` consists of a sequence of assignments (`asgn`) followed by newlines (indicated in the grammar by **`nl`**), or just a newline (this is used to end the session). The arithmetic operations have their usual meanings. The optional = sign at the beginning of an asgn indicates assignment of the value of the following expression to the memory. The single letter **`m`** in a factor fetches the value of this memory. The token **`NUMBER`** is the only token with more than one possible lexeme, and it is assumed to be given by the regular expression given earlier in this section.

A sample session with an interpreter for this grammar might look as follows:

```
= 3.14   2
         1.14
= m*3.14 + 3
         6.5796
m*3.14 - 4
         16.6599
```

This represents a computation of the value of the polynomial $x^3 - 2x^2 + 3x - 4$ at the point $x = 3.14$ using Horner's Rule ($x^3 - 2x^2 + 3x - 4 = ((x - 2)x + 3)x - 4$).

This grammar is written in a style called Backus Normal Form or Backus-Naur Form (*BNF*). In it, the only metasymbols are the arrow and the vertical bar (sometimes **::=** or **:** or = is used instead of the arrow, and sometimes nonterminals are distinguished more directly from terminals by surrounding them with angle brackets <…>). Repetition is expressed by recursion, and optional features are expressed by writing separate choices. This grammar also expresses the associativity and precedences of the operators: a left-recursive grammar rule such as `expr → expr addop term` indicates left associativity, and the different precedence levels `expr`, `term`, and `factor`, cause the operators at each level to be given precedences in ascending order (lowest precedence first). When recursion is used only to express repetition, it can be written on the right or the left: the rule for a `session` is written right-recursively for more or less arbitrary reasons.

An alternative to this grammar would be to condense it to the following grammar (still in BNF):

```
session → asgn nl session | nl
asgn → = expr | expr
expr → expr op expr
       | - expr
       | ( expr )
       | NUMBER
       | m
op → + | - | * | /
```

This grammar is ambiguous, however, in that the order in which the operations are applied is not specified, and a legal string may have many different parse trees. This grammar can still be used as the basis for the calculator, as long as separate *disambiguating rules* are stated giving the associativities and precedences of the operators. The advantage to this is that the grammar itself becomes shorter and easier to understand.

A different variety of context-free grammar is obtained by adding metasymbols representing optional and repeated constructs. One standard version of this is called *extended BNF* (*EBNF*), which uses square brackets […] to surround optional constructs and braces {…} to surround repeated constructs (one could just as well use parentheses, ?, and * as in regular expressions). The calculator grammar in EBNF becomes

```
session → { asgn nl } nl
asgn → [=] expr
expr → term { addop term }
addop → + | -
term → factor { mulop factor }
mulop → * | /
factor → - factor | ( expr ) | NUMBER | m
```

In such a grammar, the associativity of the operators is now suppressed in favor of showing the repetition directly. A standard convention is to assume left associativity of operators in any rule using the braces (since a session has no operators, the associativity of its rule is of no consequence).

68.3.2 Lex/Flex and Yacc/Bison

Lex (Lesk, 1975) and Yacc (Johnson, 1975) are scanner and parser generators that are a part of most Unix distributions. Both have modern public domain versions, Flex (Fast Lex; Paxson, 1990, http://flex.sourceforge.net/, based on ideas of Jacobson, 1987) and Bison (http://www.gnu.org/software/bison/) that run under a variety of operating systems. Each of these programs reads a definition file and produces as output a C source code file containing a scanning/parsing procedure. The definition file for each has the same basic format:

```
{definitions}
%%
{rules}
%%
{auxiliary routines}
```

We discuss the contents of the definition files first for Flex and then for Bison, using our running example of a simple calculator program whose tokens and grammar were described previously.

The Flex definition file for the calculator scanner is given in Table 68.4. In this example, the definitions section contains an `#include` directive inside the brackets `%{`and `%}`, and the definitions of the three tokens `digit`, `number`, and `whitespace`, using regular expressions written with previously described metasymbols (e.g., `whitespace` is defined to be a sequence of one or more blanks or tabs, or "return" characters). All code inside the special brackets is inserted directly at the beginning of the C output file, thus allowing the user to provide declarations/definitions that may be used by the rest of the C code. In this case, the only insertion is to indicate the inclusion of the file `calc.tab.h`. This file can be generated by Bison, and it contains the definitions of tokens and other globals that permit communication between the Flex-generated scanner and the Bison-generated

TABLE 68.4 Flex Definition File

```
/*************************************************************
  CALC.L
*************************************************************/
%{
#include "calc.tab.h"
%}
digit       [0-9]
number      {digit}+(\.{digit}+)?
whitespace  [ \t\r]+
%%
{whitespace} { /* skip */ }
{number}     { sscanf(yytext,"%lf",&yylval);
               return NUMBER; }
\n           { return '\n'; }
.            { return yytext[0]; }
```

parser. In our example (and for one particular version of Bison), the essential features contained in this file are as follows:

```
/* file calc.tab.h */
enum yytokentype {
    NUMBER = 258,
    UNARY = 259
  };
extern YYSTYPE yylval;
```

The subsequent Flex code uses only the definition of NUMBER and yylval from this file.

The rules section specifies the actions that the Flex scanner is to take when each token is recognized. These actions are placed inside a C block and are inserted directly at the appropriate places in the scanner. In Table 68.4, the specified actions are as follows. All whitespace is skipped (empty action). A number causes the scanner to compute the floating point value from the yytext string and place it in yylval, and then return the NUMBER token. (The yytext string contains the lexeme matched from the input.) The newline character \n is singled out for special handling, because it is not placed in the yytext string; it is returned directly. Finally, the period indicates a default action (it matches any character), and this causes the character value itself to be returned (yytext[0]).

This concludes the description of the calculator Flex file in Table 68.4. Note that this file contains no auxiliary routines section, and the %% symbol separating this section from the previous rules section is also omitted.

We turn now to a description of the Bison definition file, which is in Table 68.5. The definitions section contains four lines of C code to be inserted in the output file. The first two are library inclusion commands that include standard I/O and library function definitions for use in the code (printf and exit are respective examples). The third line of code defines YYSTYPE to be double; this is the type used to define the Bison value stack, which needs to be double, because expressions compute floating point values. The fourth line defines a static variable mem, which is to be used as the actual memory location for the single calculator memory. The definitions section also contains the token definitions, indicated by the %token directive. In this example, only the NUMBER token need be defined; other tokens are single-character and may be referred to directly. Finally, the definitions section contains a description of the associativity and precedence of the arithmetic operators; these are necessary disambiguating rules, because the rules section uses the ambiguous form of the calculator grammar. The order in which the operators are listed determines their precedence (with lowest precedence listed first). The %left directive indicates that an operator is left associative. Finally, the UNARY token is implicitly defined by the last definition as having the highest precedence. It will be used to give unary minus a higher precedence than any of the binary operators.

The rules section of the Bison specification contains the grammar in a modified BNF format, with actions contained in braces. Since Bison is a bottom-up parser generator, it is easiest to use to compute synthesized attributes. In the calculator grammar, the value attribute is synthesized, and it is this attribute that we use Bison's value stack to compute. The stored memory value, which has an inherited component, is handled directly by using the defined mem variable. The action code refers to the attribute values on the value stack by using symbols beginning with $. The symbol $$ refers to the (synthesized) value to be computed for the nonterminal defined by the rule. Each of the symbols $1, $2, etc., refers to the attribute value computed for each symbol on the right-hand side of the grammar rule. Thus, `$$ = $1 + $3` in the rule `expr: expr + expr` indicates that the value of the first and third symbols (the right hand expressions) are to be added to get the value of the result expression. This convention allows Bison rules to be written in a style very close to the synthesized rules of an attribute grammar.

A few changes have been made to the grammar to make it more usable with Bison. One is that the operators are written directly into the expressions, instead of being listed separately; this eliminates the need for a separate character attribute for an operator rule. Two additional changes have been made to

TABLE 68.5 A Bison Definition File

```
/*************************************************************
  CALC.Y
**************************************************************/
%{
#include <stdio.h>
#include <stdlib.h>
#define YYSTYPE double
static double mem;
%}
%token NUMBER
%left '+' '-'
%left '*' '/'
%left UNARY
%%
session : /*empty*/
        | session '\n' { exit(0); }
        | session asgn '\n' {printf("\t%g\n",$2);}
        | session error '\n' {yyerrok;}
        ;
asgn    : '=' expr { mem = $2; $$ = $2;}
        | expr { $$ = $1; }
        ;
expr    : expr '+' expr { $$ = $1 + $3; }
        | expr '-' expr { $$ = $1 - $3; }
        | expr '*' expr { $$ = $1 * $3; }
        | expr '/' expr { $$ = $1 / $3; }
        | '-' expr %prec UNARY { $$ = - $2; }
        | '(' expr ')' { $$ = $2; }
        | NUMBER { $$ = $1; }
        | 'm' { $$ = mem; }
        ;
%%
int main()
{yyparse(); return 0;}
int yyerror( char* t)
{ fprintf(stderr,"%s\n",t);
  return 0;}
```

the grammar rule for a `session`. First, the rule is written in left recursive instead of right recursive form. A right recursive rule causes the parsing stack to grow without limit; this means that a very long session might cause a stack overflow (a similar situation is caused by tail recursive procedure calls). Thus, all right recursive rules that do not reflect an associativity requirement should be rewritten in left recursive form. The remaining change is the addition of the rule

```
session : session error '\n' {yyerrok;}
```

to the choices for a `session`. This is an error-handling production. It matches any line that is not a legal assignment to the internally defined Bison nonterminal `error`, and the action `yyerrok` resets the parser to accept more input. Bison also automatically calls a `yyerror` procedure that can print an error message or perform some other action.

Finally, the auxiliary routines section of Table 68.5 contains a `main` procedure that just calls the parsing procedure `yyparse` which is generated by Bison. (The scanning procedure generated by Flex is called `yylex` and is automatically called at the appropriate times by `yyparse`.) The `yyerror` procedure is also defined in this section; it simply prints an error message (supplied by Bison, typically `syntax error`).

Assuming that the Flex definition for the calculator is in the file `calc.l`, and the Bison definition is in the file `calc.y`, then a running calculator program can be built in a Unix-like system with the commands

```
flex -I calc.l
bison -d calc.y
gcc calc.tab.c lex.yy.c -lfl -ly
```

The file `calc.tab.c` is the output file produced by Bison, and the file `lex.yy.c` is the output file produced by Flex. The option `-I` causes flex to produce an interactive scanner (i.e., one with lazy lookahead), the option `-d` causes Bison to produce the file `calc.tab.h` automatically, and the options `-lfl` and `-ly` cause the C compiler (in this case the GNU C compiler `gcc`) to consult the Flex and Bison (or Yacc) libraries when linking (if necessary).

One additional Bison feature is the verbose option `-v`. If we give the command

```
bison -v calc.y
```

then a file `calc.output` is produced that contains a description of the parsing table used by the Bison-generated parser, similar to Table 68.2. This can be useful in tracking down exactly the behavior of the parser.

68.3.3 JavaCC

JavaCC (http://javacc.java.net/) is a tool for automatically generating scanners and recursive-descent parsers in Java. While these parsers suffer from some of the same restrictions as other top-down parsers, JavaCC offers some advantages over Flex and Bison. First, the parser and scanner generators are more fully integrated, so that only one definition file needs to be created, and this file contains both the token and grammar definitions. Second, the number of lookahead symbols can be increased and can even vary within a JavaCC-generated parser. (Bison can also handle lookahead problems, but does so by employing multiple parsers, which can substantially decrease execution speeds.) Finally, there are additional tools available for use with JavaCC, such as JJTree, which can assist with the automatic building of syntax trees.

As with Bison and Flex, we give a description of a JavaCC definition file using the calculator grammar as an example. The only file needed for JavaCC is shown in Table 68.6. The general form for this file (in EBNF) is

```
javacc_input → javacc_options
               PARSER_BEGIN ( ID ) java_compilation_unit
               PARSER_END ( ID ) { production } EOF
```

The `javacc_options` section allows for the user to change default properties used by JavaCC to produce the parser, such as lookahead, debug information, and error reporting. The calculator example of Table 68.6 omits this section to accept all defaults; the reader is referred to the JavaCC documentation for further information. The `java_compilation_unit` contains standard Java code to create a runnable program. Note that this code allows the programmer an arbitrary choice of the nonterminal to use as the start symbol (in Table 68.6, it is a `session`). This code also includes the definition of the static variable mem to use as the memory for the calculator.

After the definition of the parser code, the file contains definitions of zero or more production elements. Each of these can be one of four kinds: grammar rules (with embedded Java code), regular expressions defining tokens, Java code separate from other rules, and special declarations to include in

TABLE 68.6 JavaCC Definition File

```
/*********************************************************
  Calc.jj
*********************************************************/
PARSER_BEGIN(Calc)
public class Calc
{
    static double mem;
    public static void main(String args[]) throws ParseException
    {
        Calc parser = new Calc(System.in);
        parser.session();
    }
}
PARSER_END(Calc)
SKIP: { " " | "\t" | "\r" }
TOKEN:
{
  < NUMBER: (<DIGIT>)+ ( "." (<DIGIT>)+ )? >
| < DIGIT: ["0"-"9"] >
}
void session(): { double a; }
{
      (
        a=asgn() "\n" { System.out.println(a); }
      )*
      "\n"
}
double asgn(): { double a; }
{
    "=" a = expr() { mem = a; return a; }
  | a = expr() { return a; }
}
double expr(): { double a; double b; }
{
    a=term()
    (
        "+" b=term() { a += b; }
    |   "-" b=term() { a -= b; }
    )*
                     { return a; }
}
double term(): { double a; double b; }
{
    a=factor()
    (
        "*" b=factor() { a *= b; }
    |   "/" b=factor() { a /= b; }
    )*
```

(continued)

TABLE 68.6 (continued) JavaCC Definition File

```
                              { return a; }
}
double factor(): { Token t; double a; }
{
     "-" a = factor()         { return -a; }
   | t=<NUMBER>               { return Double.parseDouble(t.image); }
   | "(" a=expr() ")"         { return a; }
   | "m"                      { return mem; }
}
```

the scanner. Table 68.6 has examples only of the first two kinds: grammar rules and token definitions. Typically, token definitions are introduced by the keyword TOKEN and a colon. Table 68.6 contains a separate special token definition using the keyword SKIP, which defines tokes that are not passed on to the parser (white space characters in the example).

Finally, the grammar rules are given in a modified EBNF form using regular expression-like symbols; for example, in Table 68.6 the notation (...)* indicates a repetition of 0 or more times. Non-terminal matches are also represented as function calls, with the associated Java code to be executed after the call given in braces immediately afterward, similar to Bison. Unlike Bison, however, there are no automatic variables representing returned values as in Bison. Instead, such variables must be declared and explicitly assigned. Furthermore, in JavaCC these variables must be declared in a separate section surrounded by braces before the production choices are given, and these choices must also be surrounded by braces.

JavaCC, in keeping with Java conventions, requires that the file for the calculator example be named `Calc.jj`. A calculator program can be built and executed using the following command-line commands:

```
javacc Calc.jj
javac *.java
java Calc
```

The first line calls `javacc`, which creates a number of Java code files in the current directory, including a parser in `Calc.java` and a scanner in `CalcTokenManager.java`. The second line compiles these together with the additional Java code files to produce a `Calc.class` file that is executable with the JVM, which is what the last line above performs. The result is an execution similar but not identical to that of the Bison calculator (since different compilers and runtime systems are involved).

It should be noted that no attempt has been made to perform error recovery with this version of the calculator. Indeed, the following execution shows that parse errors end the execution of the calculator with an uncaught Java exception:

```
> java Calc
2 -
Exception in thread "main" ParseException: Encountered " "\n" "\n "" at line
1, column 5.
Was expecting one of:
    <NUMBER> ...
    "-" ...
    "(" ...
    "m" ...
        at Calc.generateParseException(Calc.java:338)
        at Calc.jj_consume_token(Calc.java:276)
        at Calc.factor(Calc.java:153)
        at Calc.term(Calc.java:95)
        at Calc.expr(Calc.java:79)
```

```
        at Calc.asgn(Calc.java:45)
        at Calc.session(Calc.java:26)
        at Calc.main(Calc.java:7)
```

Because of Java's built-in extensive exception handling mechanisms, which JavaCC uses in essential ways, it is possible to perform rigorous error recovery, but doing so involves techniques that are beyond the scope of this chapter.

68.3.4 Dealing with Ambiguity

Although the expectation is that a language grammar should be unambiguous, in that each legal input string will correspond to exactly one syntax tree, there are many practical situations in which it is difficult or impossible to write a grammar unambiguously. In such cases, disambiguating rules must be stated and built into the parser in some way. We have already seen how Yacc/Bison can accommodate precedence and associativity disambiguating rules for operators. We give three additional examples of the use of disambiguating rules.

The dangling-else ambiguity is typical in languages such as C and Java, which allow both simple and compound statements in control structures. In C, the dangling-else ambiguity appears in statements like

```
if(e) if(f) {/*then-part*/} else {/*else-part*/}
```

The question is: should the else-part be executed when `f` is false, or when `e` is false? The standard answer is that it should be executed when `f` is false (and when `e` is true). This is called the most closely nested disambiguating rule for the *if* statement, because it implies that the else-part structure is to be associated in the parse tree with the closest previous *if* statement that does not have an else-part. This also corresponds to always preferring the second choice in the BNF rule

```
if_statement → if ( expression ) statement
             | if ( expression ) statement else statement
```

This disambiguating rule could actually be incorporated directly into the BNF, but it is slightly complicated (and not very illuminating), and the rule as stated is easily implemented in a parser. It is worth noting that languages in which the *if* statement has a closing keyword (such as **`end`** or **`endif`**) do not have the dangling-else ambiguity; Ada is an example of such a language.

A second kind of ambiguity is represented by the availability in C++ of function-style casts, where one can write `int(x)` to mean the same thing as the C cast `(int)x`. Now consider the code fragment

```
typedef int (*F)();
void **x;
void p(void)
{ F (*x)();
  ...
}
```

Is the first code line inside `p` a cast of global `*x` to type `F`, after which this function value is called, or is it a declaration of a local variable `x` of type "pointer to function returning a value of type `F`?" The C++ standard says it is the latter: any statement that may be interpreted as a declaration *is* a declaration. This means that all grammar rules for declarations are to be preferred to all other grammar rules. Note that this is an ambiguity *between* nonterminals, rather than an ambiguity within a single nonterminal, as with the dangling-else ambiguity. It is also impossible to remove this ambiguity directly within the grammar, because the order of rules in a BNF description is immaterial to the language defined.

The final example of an ambiguity that we present here is also one that arises with casts, but this time in C. Consider the C statement

```
(t)-x
```

If `t` is a type name declared in a `typedef`, then this is a cast of the value `-x` to type `t`. On the other hand, if `t` is a variable, then this is the value obtained by subtracting `x` from `t`. In order to parse this correctly, it is necessary to consult the symbol table to see whether `t` is defined as a type. Thus, the symbol table (or at least the `typedef` part of it) must be built as parsing proceeds. Unlike the previous examples of ambiguity, this is a context ambiguity that cannot be resolved by purely syntactic means.

In an LR parser, these ambiguities, or parsing conflicts, can be of two different kinds. The first, more common, situation, is when both a shift and a reduction by a grammar rule choice are called for on the same input token; this is a *shift-reduce conflict*. The other case is when reductions by two different rules are called for on the same input token; this is a *reduce–reduce conflict*. Shift-reduce conflicts are usually resolved by selecting the shift over the reduce. This ensures that the longest possible string will be matched by each rule. The typical example is the dangling-else ambiguity, where preferring the shift corresponds exactly to the most closely nested disambiguating rule. Reduce-reduce conflicts, on the other hand, have no natural disambiguating rule. Parsers generally adopt an *ad hoc* rule, in which the reduction by the lowest-numbered rule is preferred (where the rules are number in the order they are considered by the parser). The C++ ambiguity between cast expressions and declarations (also described previously) can be resolved in this way in favor of the declarations (as the C++ definition requires) by listing the declaration rules before the expression rules. Yacc/Bison adopts both of these disambiguating rules as stated. Unfortunately, the third ambiguity mentioned previously, a cast vs. an arithmetic expression in C, cannot be resolved by either of these means. Typically, this is handled by having the scanner consult the symbol table when recognizing an identifier and returning a different token for a type name than for a variable. Both Flex and JavaCC have mechanisms for introducing code to achieve this into their scanner specifications.

An alternative to these disambiguating rules is to build manual disambiguation mechanisms directly into the parser generator. Indeed, in a top-down parser like that generated by JavaCC, such a capability is a necessity, because ambiguities such as preferring declarations over expressions in C++ cannot be handled by reordering the grammar rules: a top-down parser would simply commit to the first rule possible and never consider a subsequent rule, generating a parse error instead (there are no shift-reduce or reduce-reduce conflicts in LL parsers, only shift–shift conflicts).

In JavaCC, the primary disambiguation mechanism is that of extending the lookahead tests locally to possibly arbitrary lengths, essentially causing backtracking if a grammar rule choice should fail. For example, in order to prefer a C++ declaration over an expression, one might use code such as the following in a JavaCC specification:

```
void statement(): {}
{
    LOOKAHEAD(declaration()) declaration()
  | expression()
}
```

Since this kind of lookahead can be expensive in terms of execution speed, it is also possible in JavaCC to put a bound on the number of tokens to consider, as well as to add tests for specific tokens that would provide clear decision points. For example, writing `LOOKAHEAD(10,declaration())` in the aforementioned code would limit lookahead/backtracking to 10 tokens.

Finally, the `LOOKAHEAD` construct in JavaCC will also accept user-defined Boolean tests (called *semantic predicates*, as opposed to the syntactic tests of previous examples). Details can be found in the JavaCC LOOKAHEAD MiniTutorial at http://javacc.java.net/doc/lookahead.html.

68.3.5 Attribute Analysis

Yacc/Bison and JavaCC restrict the kinds of attributes that can be reasonably computed during a parse, because of the requirements of their parsing algorithms. In cases of difficult attribute computations, these limitations are overcome by either providing ad hoc solutions using external data structures, or by constructing an intermediate representation such as a syntax tree during the parse, and then writing specialized procedures that perform semantic analysis by traversing the intermediate form in one or more passes. Tools for automating the computation of general attributes (other than syntax tree construction) have not been widely used, partially due to the many different varieties of intermediate representations and partially due to the variety and complexity of the semantic attributes of different languages. Some notable systems that do permit the automation of this step include Eli (Gray et al., 1992) (see also http://eli-project.sourceforge.net), JastAdd (Hedin and Magnusson, 2003), and Gentle (Schröer, 1997) (see also http://gentle.compilertools.net). Code generation is a special case of this problem, and special methods for automating the code generation step have been developed, which we describe next.

68.3.6 Intermediate Representations and Code Generation

The abstract syntax tree is a convenient way to represent the source program within a translator, particularly for semantic attribute analysis. However, it is less suited for code generation. Although target code can be generated as a form of attribute analysis, either during parsing or by traversal of the syntax tree, the quality of this code is usually poor, and further processing must be done before an acceptable level of target code is produced. Typically, this is achieved by generating some form of intermediate code that is closer to the code of the target machine but still abstract enough that the compiler can perform code improving transformations on it. Many choices have been used for this intermediate code. Some of the more common are sequences of expression trees, sequences of postfix expressions, an abstract linearized form of syntax tree called *three-address code*, and actual code for a hypothetical target machine, such as P-code for the stack-based P-machine of many Pascal-related compilers (Nori et al., 1981). An example of three-address code corresponding to the source code line `=3.1+m/2` for the calculator language described previously is

```
t1 := m/2
t2 := 3.1 + t1
m := t2
```

Here, the identifiers `t1` and `t2` are temporaries introduced by the compiler that can be thought of as pseudo-registers, and later assigned either to actual registers or to temporary locations in memory. The equivalent (annotated) P-code for this same source code is as follows:

```
ldo r,m     ; load real value onto stack from static location m
ldc r,2.0   ; load constant real value 2.0 onto stack
dvr         ; divide two reals on top of stack, push result
ldc r,3.1   ; load constant real value 3.1 onto stack
adr         ; add two reals on top of stack, push result
sro r,m     ; store real from stack to static location m, pop stack
```

The use of the code of a precisely defined, simple machine, such as the P-machine as the intermediate target language, has several benefits. First, the compiler can be run in interpreter mode, in which the P-code functions as the target code, and a P-machine simulator is used to execute the P-code on the actual target machine. The target code is then kept as P-code, which is generally much more compact than actual executable machine code. Furthermore, in order to retarget the compiler to a new machine, it is only necessary to rewrite the P-machine simulator. Second, in order to obtain a native-code compiler from the P-code compiler, one need only write a translator from P-code to the native code of the target machine. This is usually done by writing individual procedures for each P-code instruction that perform something like a macro expansion of each P-code instruction into target code. By itself, this process will produce very poor code, but it can be combined with a static simulation of the P-machine itself, whose stack can be used to discover opportunities for optimization. An improvement on P-code that adds significant attributes for tracking address calculations and optimizing transformations is U-code (Perkins and Sites, 1979).

An intermediate language that is compact, capable of efficient interpretation, and easily retargeted to different architectures is of special interest for heterogeneous networks, where the platform on which the intermediate code is to be executed or compiled to native code may not be known in advance. If the intermediate code is given a numeric machine code-like encoding (called *bytecode*), and techniques to improve interpretation performance are used (such as on-the-fly compilation or threaded code interpretation [Bell, 1973]), then good performance on a variety of machines across a network can be achieved. This is, for example, the basis for Java implementations (Gosling, 1995). Although one might think that the use of bytecode interpreters in Java implementations should incur a performance penalty, dynamic (*just-in-time* [JIT]) compilation techniques, together with runtime (*hotspot*) monitoring, frequently achieves execution speeds that are the same or even better than those of programs compiled to native code; see Section 68.5 and Lewis and Neumann (2003).

A similar process of macro expansion works for three-address code and its variants, although in this case the underlying abstract machine and its state is not made explicit (as it is for P-code), and the accumulated information during code generation must be stored as attributes in data structures associated with the intermediate code. Typically, these include address descriptors, which record the locations at which named quantities (such as variables) can be found, and register descriptors, which record information about the values that can be statically predicted to be in registers at particular points in the target code. Address descriptors can be kept in the symbol table, while register descriptors can be maintained in an array indexed by the register numbers.

Unfortunately, both of these code generation schemes spread the information about the target machine throughout the code generator. This means that retargeting the compiler is more costly than if the target machine information were collected compactly in one place. Several approaches have been taken towards achieving this.

The first method for improving the retargetability of the code generator involves writing a syntactic description of the macro expansion process for the intermediate code together with code-emitting actions, much as a Yacc/Bison or JavaCC description associates semantic actions with syntactic rules (Glanville and Graham, 1978). This method permits the automatic generation of the code generator from this description, using a tool similar to a parser generator. This method has been expanded in (Ganapathi and Fischer, 1985) to include semantic predicates. As an example, consider generating code for the three-address instruction

```
t1 := m/2
```

A rule that would handle this instruction for a VAX-like architecture might appear as

```
reg(n) → adr(a)/const(i) dead?(n) float?(a) #emit("divd3 #i.,a,rn")
```

This rule establishes that emitting the VAX instruction `divd3 #i.,a,rn` allows the expression consisting of an address `a` divided by a constant `i` to be reduced to a register `n`, provided `n` is dead (i.e., contains a value that is no longer needed), and provided that `a` is the address of a real-valued quantity. Matching this rule to the previous three-address instruction causes `t1` to be identified with `reg(n)`, `m` to be identified with `adr(a)` (thus computing `a` as the address of `m`), and identifying `2` with `const(i)` (thus setting `i` to 2). The resulting line of VAX code generated is

```
divd3 #2.,m,r1
```

In the absence of predicates, such a set of rules describing the target machine could be written as a Yacc/Bison input. The difference between this kind of grammar and a grammar describing the syntax of the source language is that the target machine grammar usually comprises thousands rather than hundreds of rules, and the grammar is highly ambiguous, because there are usually many ways to generate target code to achieve a specific effect. Parser generators such as Yacc/Bison are inefficient and cumbersome when dealing with such grammars, and a code generator specifically tuned to this situation is more appropriately used (Henry, 1984).

The second variety of retargeting mechanism we discuss is that used in the portable C compiler (Johnson, 1978). In this mechanism, a special template language is used to describe the process of matching intermediate code to target code. The code generator then consults a table of these templates during code generation in an attempt to find the optimal match.

The third variety of retargeting mechanism is similar to the second, except that both the intermediate code and the target code templates are written in the template language, and the matching routine runs statically, usually during the installation of the compiler, and selects a match between certain sequences of intermediate code and target code, which is then fixed once and for all. This method was used in the Production-Quality Compiler–Compiler project (Cattell, 1980; Leverett et al., 1980). A related method has also been used in the GNU compiler suite (GCC, 2010), based on ideas of Davidson and Fraser (1984). In the GNU compiler, the template language is called the register transfer language, or RTL. This language is written in LISP-like prefix form. An example of an instruction template is as follows:

```
(define_insn "divdf3"
  [(set (match_operand:DF 0 "register_operand" "=f")
        (over:DF (match_operand:DF 1 "register_operand" "f")
                 (match_operand:DF 2 "register_operand" "f")))]
  "! TARGET_SOFT_FLOAT"
  "divd3 %1,%2,%0"
  [(set_attr "type" "arith")]
)
```

Instruction templates are introduced using the `define_insn` operator. This operator is followed by a symbolic name for the instruction template, a pattern with predicates (such as `register_operand`), constraints (such as `f`, indicating the register may be used for floating point values), and extra conditions (such as !TARGET_SOFT_FLOAT, indicating that there must be a hardware floating point arithmetic unit available in the processor). Following this, there is a pattern for the instruction to be generated, with numbers referring to the operands of the pattern (in this example, it is the VAX instruction `divd3%1,%2,%0`). Finally, there is a specification of attributes for the template.

Such instruction templates are used to generate actual target code in RTL format directly from C code during a parse. Optimizing steps are then applied directly to the RTL intermediate code, and then templates are again used to generate assembly output. Specialized RTL attribute descriptions are also used to guide the code generation process.

One possibility for a compiler writer is to avoid the issue of code generation for a target machine altogether and instead generate code in the C language or other language capable of expressing lower-level code, and then use a widely available C compiler such as the GNU C compiler GCC to generate target code. This represents an elegant and simple solution, but has a number of drawbacks. First, the available C compiler must of course be capable of producing target code for the desired target. Second, it forces the compiler writer to rely on the optimizing capabilities of the C compiler, since many optimizations are not able to be expressed directly in the generated C code. Furthermore, because of its relative lack of constraints, C can be a difficult language to optimize for critical performance. One approach to these difficulties is to define a language similar to C that has better capabilities to be used as an intermediate target language for compilers; this is, for example, the goal of the *C-Minus-Minus* development effort (see http://www.cminusminus.org).

A different approach has been taken by the *LLVM project* (see http://llvm.org; LLVM is now a proper name but earlier was an acronym for *low-level virtual machine*). LLVM consists of a core intermediate language as well as a *compiler infrastructure* consisting of many loosely coupled modules that can be easily manipulated with low coding overhead to achieve desired effects. The core intermediate language itself is an abstract reduced instruction set computer (RISC)-like assembly language, but with calling conventions and register use left unspecified. LLVM has proven to be extremely versatile in reducing coding effort and code size while maintaining high levels of optimization and is the foundation for most of the translation tools in Apple products.

68.3.7 Compiler Output

In general, compilers process source files as input and generate object code files as output. Depending on various compiler directives or command-line options, compilers can also generate metadata about the source program that program development utilities such as debuggers and profilers can use. (See, e.g., http://gcc.gnu.org/onlinedocs/gcc/Debugging-Options.html for Gnu gcc, http://docs.oracle.com/javase/7/docs/technotes/tools/windows/javac.html#options for Java, and http://msdn.microsoft.com/en-US/library/19z1t1wy%28v=vs.80%29.aspx for Microsoft C++.) Compilers can embed this metadata into the object files or they can output the metadata as separate files.

Symbolic, or source-level, debuggers allow developers to interact with their application programs at run time and use identifiers (names of constants, variables, functions, etc.) that appear in the source code and refer to the source program statements by their line numbers. Such symbol table and cross-reference information is often lost by the time a compiler outputs the object files. Compiler directives or command-line options can specify that this information be kept by embedding it into the object code or outputting it as separate files.

Profilers monitor the execution of application programs. Upon completion of the program, the profiler generates a report containing such statistics as how many times each source statement was executed, how many times each variable was accessed and modified, and how much time was spent in each function. Gathering these statistics while an application program is running requires similar metadata to those that debuggers require. Again, compiler directives or command-line options can specify that this information be embedded into the object code or output as separate files.

Another type of output file that some C and C++ compilers can generate is precompiled headers. (See e.g., http://gcc.gnu.org/onlinedocs/gcc/Precompiled-Headers.html for Gnu gcc and http://msdn.microsoft.com/en-us/library/szfdksca%28v=vs.71%29.aspx for Microsoft C++.) A so-called header file (typically named with a `.h` suffix) contains constant and macro definitions that multiple source files include at the beginning of each file. To save the time of rescanning and reparsing a header file each time it is included, the compiler can *precompile* a header file the first time it is included and save it in an intermediate form as a precompiled header file. Then each subsequent time the header file is included, the compiler uses the precompiled header file instead. Compiler directives or command-line options specify whether or not to generate and use precompiled headers.

68.3.8 Code Optimization

A code generator that naively expands intermediate code into target code will produce code that is extremely inefficient, both in terms of execution speed and target code size. A production-quality compiler must include processing steps that improve its ability to generate good target code, so that it more nearly resembles the code that would be produced by an assembly language programmer. Such processing steps are usually referred to as *optimization*, though they almost never produce truly optimal code. Indeed, the production of mathematically optimal code, except in the very simplest cases, is known to be computationally intractable (NP-complete), so to attempt this would result in unacceptably slow compilation speed.

Optimization steps can be built into almost all the phases of a compiler, from parsing to final target code generation. If an optimization pass is performed separately, it usually occurs after intermediate code generation, but before target code generation. Such optimizations are generally *source-level* in that they do not depend heavily on the details of the target machine. Some optimizations, however, are *target-level*, and require that the details of the target code be known. Such optimizations are referred to as *peephole optimizations*, because they were originally (and are sometimes still) performed by examining small sequences of target code and replacing them with more efficient code. This term can be misleading, however, because modern compilers sometimes perform considerably more sophisticated analysis of the target code than the name might imply (an example is the GNU C++ compiler). One complex aspect of the scheduling of optimizations during compilation is that some optimizations may uncover opportunities for other optimizations and vice versa (this is sometimes called the *phase problem*). This leads some compilers to repeat certain optimizations in an attempt to catch such cascaded situations. More adaptable optimizers should in principle be able to solve this problem more effectively in the future, and indeed the versatile LLVM compiler infrastructure (see Section 68.3.6) has been used together with genetic algorithms to achieve better optimization results than hand-chosen sequences (see http://donsbot.wordpress.com/2010/03/01/evolving-faster-haskell-programs-now-with-llvm/).

Optimizations are classified according to the region of the program about which information is gathered in order to perform the code improvement. Local optimizations consider only straight-line segments of code, that is, those not involving jumps or calls. Global optimizations consider the code of a single procedure. Interprocedural optimizations consider the entire program or compilation unit. Most compilers perform some kinds of local optimizations. A more heavily optimizing compiler will perform global optimizations, typically using an information-collection method called *dataflow analysis*, which passes information around a flow graph representing the flow of control through a procedure. It is a rare compiler that performs interprocedural optimization, and for a very good reason: such optimizations are largely ineffective unless they are postponed to link time, because many procedures will not be available in a single compilation unit. This means that the linker must be closely coupled with the compiler; in particular, the system linker cannot be used. This raises the level of complexity of the compilation environment considerably.

In the remainder of this brief discussion of optimizations, we list the principal sources of code improvement over a naive code-generation strategy. While the focus of this list is on speed optimizations, one needs to be aware that there are other possible goals for optimization steps, such as reducing code size, reducing memory or disk usage, or reducing communication bottlenecks. An important optimization for modern portable devices, such as cell phones, is aimed at reducing power consumption; see the Further Information section for a reference. See also the chapter on Scheduling.

Register allocation: This is the most important and pervasive issue in the generation of quality code. Keeping temporaries in registers is indispensable, especially for RISC architectures. In order to extend register allocation to include local variables, parameters, and global variables, a compiler may permanently allocate certain registers to heavily used quantities. An alternative is to build an interference graph and assign registers by graph coloring, permitting noninterfering variables to share the same register. Good global register allocation also requires some form of dataflow analysis in order to identify values that have no further uses in subsequent code and that can be safely overwritten if they are in registers.

Common subexpression elimination: This refers to the identification of expression values that are recomputed one or more times in a program and the avoidance of such recomputation by suitable storing and reuse of the computed value. While one might naively think that a good programmer should not write code that contains common subexpressions, in fact, most common subexpressions are due to address computations of array and record references that cannot be expressed adequately in source code and that therefore require elimination by an optimizer. Common subexpressions are relatively easy to identify in straight line code; the principal difficulty in the general case is to determine what subexpressions may have changed between two computations of the same expression. Thus, a compiler may choose to perform only local common subexpression elimination.

Copy propagation: This optimization tracks code regions in which two or more variables have the same value. After an assignment `x = y` (in C syntax), `x` and `y` have the same value in subsequent expressions until one of them acquires a new value; in this region, any use of `x` can be replaced by a use of `y`, which can lead to better code if, for example, `y` is already in a register. As with common subexpressions, copy statements may exist in the intermediate code even when they are not written by the programmer. A special case of copy propagation is constant propagation: after an assignment such as `x = 2`, uses of `x` can be replaced by the constant 2. As with common subexpressions, copy propagation may be done either as a local or global optimization.

Reduction in strength: This refers to the replacement of arithmetic expressions by equivalent expressions that execute faster. A typical example is the replacement of multiplication or division by a power of 2 with a shift operation. Another example occurs in loop optimization, where a linear combination calculation is replaced by a simple addition (see the subsequent paragraph on loop optimization).

Jump optimization: Opportunities for improving jump code occur when a sequence of jumps can be replaced by a single jump, or when a jump can be eliminated by rearranging the code. For instance, the code sequence

```
   goto lab1
   ...
lab1:  if x = y goto lab2
lab3:  ...
```

can be replaced by the more efficient

```
   if x = y goto lab2
   goto lab3
   ...
lab3:      ...
```

Algebraic laws: An optimizer can look for special cases in expressions such as `x * 1` and `x + 0` (replacing these with `x`). Sometimes, it is also worthwhile to replace a computation `x + y` or `x * y` with its commutative equivalent `y + x` or `y * x`. More difficult is to discover opportunities to use distributive laws, such as replacing `x * y + z * y` with the more efficient `(x + z)*y`.

Loop optimization: Loops are traditionally an area where a great deal of attention has been paid to making code as efficient as possible, because programs tend to spend a lot of time in loops. In programs with goto statements, it is a non-trivial operation to even discover loops, and this usually must be done by building the flow graph. Fortunately, in modern languages, it is reasonable to depend on syntax (such as keywords *while* and *do*) to locate loops. Typical loop optimizations include attempting to discover invariant computations inside loops (i.e., those that always yield the same value, regardless of the loop iteration) and moving their computation to just before the loop entry. Another optimization seeks to discover so-called induction variables, which are linear combinations $a * i + b$, where i is the loop control variable. Since such an expression is incremented by a fixed amount with each iteration, its computation can be reduced in strength to a simple addition. Finally, loop unrolling can be applied: this seeks to reduce the delays (and possible cache misses) caused by the backward jumps in loop iterations by dividing the loop bounds by fixed amounts and inserting the missing iterations as linear code.

Constant folding: This optimization seeks to replace constant expressions (e.g., 3 + 5) by their resulting values (e.g., 8). Such an optimization could, in theory, replace all computations whose results are predictable at compile time by their results alone. In practice, this would involve repeated applications of constant propagation and folding, so this optimal result is rarely achieved.

Dead code elimination: This optimization seeks to skip code generation for those statements that are either never reached during execution or whose actions have no effect on the results of the program.

The first case happens when compile-time constants are set to select certain actions over others (e.g., to suppress the collection of run-time statistics). The second occurs if common subexpression elimination or copy propagation makes a computation or assignment unnecessary.

68.3.9 Error Recovery

An important practical problem in the design of a translator is its response to errors in the input program. Although many translators have been constructed so that they stop at the first error encountered, it is generally considered better to attempt to discover as many errors as possible in the input before halting. Thus, the response to errors includes error recovery, so that translation may continue, as well as the generation of informative error messages. A further step in an error handler may also be error repair, where the translator attempts to correct at least some of the errors. Because it is difficult to infer from an erroneous program what the actual intended program was, error repair is usually not attempted, except in special cases, such as missing punctuation.

Errors may be classified into lexical, syntactic, semantic, and run-time errors. An interpreter must have a reasonable response to runtime errors that involves at least generating an error message and exiting gracefully. A compiler needs only to recover from static errors, although it may need to generate code to report run-time errors. Semantic errors are relatively easy to recover from, because the syntax tree can still be used to guide the remainder of translation. Lexical errors can be simply passed to the parser by the scanner as error tokens. Thus, the principal difficulty in building an error handler occurs in dealing with parsing errors, where the structure of the input is disturbed, and there may be no obvious way to restart the parser.

Several criteria apply to any method for recovery from parse errors. First, it is useful to try to detect errors as early as possible, because continuing to process the input can make it more difficult to determine where the error occurred and to generate an appropriate error message. Second, the translator should attempt to discard as little input as possible when recovering from an error, because discarding input can lead to other errors, as well as mislead the programmer. Third, the translator should avoid generating large numbers of error messages caused by a single error. In particular, a translator must never get into an infinite loop when recovering from an error. This usually requires that at least some input be discarded during error recovery.

A standard technique for error recovery is called *panic mode*, in which tokens are simply discarded, and the parsing stack accordingly adjusted until the parse can resume. This method can be made sophisticated enough so that it becomes better than its name implies. It can be used either as the standard error recovery technique or as a fall-back technique for a more complex method. The important feature of panic mode is the proper computation of a synchronizing set of tokens, which are used to determine when the parser should stop discarding the input and attempt to resume the parse. Panic mode in recursive-descent parsing is discussed in Wirth (1976 [Section 5.9]). A slightly different method is used by Yacc/Bison, in which an error pseudotoken is made available, with which important error recovery locations can be marked in so-called error productions in the input grammar. For example, in the Bison definition file of Table 68.5, the error production

```
session : session error '\n' {yyerrok;}
```

was included in the specification. Yacc/Bison's behavior on encountering an error is as follows. First, it sets a flag to enter an error phase, and then it begins popping the parsing stack until a state is found where the error pseudotoken can be successfully shifted. Then input tokens are discarded until three consecutive tokens can be shifted successfully, whence the error phase is canceled and normal parsing is resumed. The error phase can also be canceled manually by calling the Yacc/Bison procedure yyerrok. For example, the previous error production for a calculator session causes Bison to discard all tokens until a newline is reached, and then to resume the parse. The result is that any incorrect line of input is deleted. (Yacc/Bison also automatically calls `yyerror` whenever the error phase is entered, which can perform other actions and print an error message.)

Other error recovery and repair mechanisms are discussed in Graham (1979), Roehrich (1980), Dion (1982).

68.4 Incremental Compilation

Not only must compilers generate high-quality code they also must do it in a reasonable amount of time. Historically, the speed of compilation was often inversely proportional to the quality of the generated code, with the compilation of a large program under full optimization sometimes taking hours or even days. With modern processors, compilation speed has become less of an issue, but strong optimization still requires significant amounts of processor time. For this reason, compilers usually offer an unoptimized mode that not only avoids code improvements but is also tuned to the speediest possible generation of runnable code; such an option can be very useful for development or instructional work. Even so, large programs can require significant amounts of time to recompile if all of the code must be retranslated during each compilation step. A significant improvement in compiler technology is therefore the ability of a compiler to operate incrementally, that is, to avoid recompilation of code that has not changed since the previous compilation step. Alternatively, compilers that are linked to an editor as part of an interactive development environment (IDE) can compile program text in the background as it is entered by the programmer. Either of these mechanisms is called *incremental compilation*.

One of the original motivations for separate compilation and linking was to allow a form of incremental compilation, where only files that had changed needed to be recompiled, and the results of this recompilation would be combined with previously compiled files by a linker. Of course, this approach is subject to error, because the linker usually has no knowledge of the source language and cannot find incompatibilities between the recompiled code and the unchanged code. Additionally, the programmer has to keep track manually of which files have changed and which have not.

The first major advance over this situation was the Unix *make* utility (Feldman, 1979), varieties of which are part of virtually every IDE today (usually called *build* or *make project*). Make itself is still heavily in use in command-line development environments. Make keeps track automatically of file changes by using file system time stamps: a later time stamp on a source code file over its corresponding object code file indicates the need for recompilation. Additionally, make can keep track of dependencies in a set of source code files (made necessary by the convention of using file inclusion to check incompatibilities in C), although such dependencies must still be maintained manually by a programmer (in a so-called makefile). The makelike features that are built into IDEs usually automate the process of maintaining dependencies even further, particularly in languages with strong interface requirements like Ada and Java (but unlike C or C++).

It is possible, and often even desirable, to go beyond the capabilities of an automated make utility in either or both of the two different approaches to incremental compilation mentioned previously. First, a compiler could implement a finer-grained incremental approach than the file-level increments of make, such as statement-level or declaration-level increments (i.e., recompiling only those statements or function definitions within a file that have changed, rather than the entire file). Such a mechanism requires much more retained information about code structure than just object files and time stamps, however, and can lead to a significant increase in the amount of storage space required for a program or project. Second, the compiler could run in the background behind an editor, recompiling code as it is edited, rather than waiting for a compile command from the user. (Obviously, this approach makes the most sense if finer-grained incremental compilation is also implemented.) However, this approach can also mean more compiler interference in the process of editing or creating code, where an editor may refuse to accept program text that the compiler cannot understand. This can easily create programmer resistance to the use of such systems.

Examples of compilers that offer incremental compilation are the VisualAge Java compiler from IBM (now part of the Java Development Tool Core of the Eclipse Development Environment;

http://www.eclipse.org/jdt/core/) and the Visual Studio C# Compiler from Microsoft. These compilers also provide a contrast in how they mix the two approaches to incremental compilation described in the previous paragraph. The VisualAge compiler is invoked automatically by the IDE whenever a file save operation is performed (or when a file is added or imported into a project). It performs only file-level incremental compilation. However, it immediately reports any compilation problems to the user and will even prevent the completion of a file save operation if syntax errors occur in significant places. In contrast, the C# compiler will perform method-level (function-level) incremental compilation, but only at the direction of the user during a requested build.

68.5 Just-in-Time Compilation

68.5.1 Introduction to Just-in-Time Compilation

Traditionally, application programs were either interpreted or compiled. Interpreters have the advantage of fast start-up; source programs are quickly processed to the point of being executable. This enables highly desired rapid turnaround during program development. However, interpreted code can execute orders of magnitude slower than compiled code. Compilers provide the advantage of fast application program execution by generating native object code that can run at full speed on the target machines. Compilers can also perform optimizations generally not done by interpreters, further improving the execution performance of the application program. But code generation takes time, especially if code optimizations are performed. The generated object code usually requires further processing, such as linking and loading. Therefore, it takes longer for an application program to become executable with a compiler than with an interpreter.

Many compilers for modern programming languages, such as for Java and for Microsoft's .NET family of languages, generate bytecode instead of native code. Bytecode is the "machine language" that implements the instruction set of a virtual machine whose architecture is ideal for the features of a given set of source languages. The virtual machine is simply an interpreter for the bytecode. To execute an application program that has been compiled into bytecode, the virtual machine interprets the bytecode.

The use of bytecode promotes portability. An application program's bytecode can run on any actual machine that implements the virtual machine. A popular example of a virtual machine is the JVM. Once a Java application program has been compiled into bytecode (stored as `.class` files), any actual machine that implements the JVM can execute that program.

Besides interpreting bytecode, it is also possible to translate bytecode into the native object code of an actual machine. This translation can be done much more rapidly than it would take to translate the original source code to native code, since all of the work of scanning and parsing the source code has already been done in order to generate the bytecode.

A *JIT compiler* translates a source application program into bytecode that is interpreted by the virtual machine. While the application is being interpreted, the compiler can step in to dynamically translate blocks of bytecode "on the fly" into native machine code. Such dynamic translation is JIT because a block of bytecode (such as for a function or a method) can be translated into native code just as that code is about to be executed. Then whenever that block of code is executed again, the native code will run at full machine speed. In contrast, a traditional compiler does ahead-of-time (AOT) or static translation.

68.5.2 Advantages of Just-in-Time Compilers

A JIT compiler offers the best of both worlds for the application program developer: fast turnaround by an interpreter and fast program execution by the native code. Since the virtual machine architecture and the bytecode are ideally designed for the source language, a bytecode compiler can translate an application program into bytecode rapidly, and then the virtual machine can execute the bytecode immediately via interpretation. At application run time, the JIT compiler can rapidly translate blocks of bytecode into native machine code that executes at full machine speed.

While dynamic translation from bytecode to native code can be very rapid, it does take time, especially if code optimization is involved. Such a translation, if done serially with the virtual machine, would cause a pause in the interpretation of the application program. Therefore, some JIT compilers perform a runtime analysis of the application program. A block of bytecode is dynamically translated into native machine code only if it is advantageous to do so, such as if that block is executed more than once. Some JIT compilers provide command-line options that allow the developer to specify how many times a block of bytecode must execute before it is translated into native code.

A JIT compiler may be able to perform code optimizations that are not possible by a traditional AOT compiler. Because dynamic translation occurs during the application program's run time, it has the added context of the current state of the underlying actual machine, plus any runtime analysis of the application program. For example, it can generate native code optimized for the how many cores are currently available. Runtime analysis can determine that the native code for a function should be inlined rather than invoked. (See e.g., http://java.sun.com/developer/onlineTraining/Programming/JDCBook/perf2.html for the JVM and http://blogs.msdn.com/b/davidnotario/archive/2004/10/28/248953.aspx for Microsoft.NET.)

To be effective, a JIT compiler must make the tradeoff between time lost by the interpreter to dynamic translation and time gained by faster execution of native code. An application program that does not contain many loops or function calls may not benefit much from a JIT compiler (Li, 2004).

68.6 Compilation for Parallel Execution

An important aspect of modern compiler design is to ensure that the code generated by a compiler makes good use of opportunities provided by hardware for parallel execution. This has become even more important with the recent advent of so-called multicore architectures (see the chapter on Multicore Architectures), where two or more functionally distinct processors are built into the same chip. Virtually all personal computers sold today contain such chips, with the expectation that most programs running on these machines will actually use the hardware to improve execution speeds. Note, however, that the responsibility for assigning execution tasks to different processors is shared between the operating system and the compiler; an operating system can already assign independently running programs to different processors without any compiler intervention. Additionally, the operating system must provide an appropriate set of primitives to the compiler for it to be able to schedule parallel execution effectively. In this section, we will only briefly survey the issues and the current state of the art of compiling for parallel execution.

First, we should note that compiler writers have struggled to find ways for compilers to automatically discover opportunities for parallel execution of code within programs. Reasons for this include the unpredictable interaction of program features with data paths within the hardware, such as bus and cache operations, as well as the difficulty in predicting data dependencies within a program in the presence of control flow and aliasing. As a result, compilers do not usually generate code for multiple processors without programmer intervention, in that the programmer must write source code that explicitly states what code should be executed in parallel and how that code will communicate with the rest of the program. Thus, compilers must not only rely on the operating system to provide adequate mechanisms for parallelism but also the compiler must also rely on the programmer to provide source-level instructions on how to use the operating system mechanisms in a particular program. Further complicating this issue is the fact that many popular programming languages such as C/C++ do not include parallel mechanisms in their standard definitions; Java is a major exception, having included parallel mechanisms in the form of threads (see subsequent section) since its inception. This situation has been partially ameliorated by several widely accepted parallel interface standards, such as POSIX Pthreads (http://www.opengroup.org/onlinepubs/007904975/basedefs/pthread.h.html) and OpenMP (http://openmp.org, supported by GNU, Microsoft, and Intel C/C++ compilers).

One area of parallel execution where compilers play a role without user intervention is in the scheduling of consecutive machine instructions to take advantage of parallel execution within a single processor. Most modern processors have several ways of executing instructions in parallel. Usually, this is expressed as an instruction pipeline that splits the processing of a computer instruction into a series of independent steps. A standard pipeline (called the classic RISC pipeline) consists of five stages: fetch, decode, execute, store, and register writeback; each stage is executed in a single clock cycle, and new instructions are fed at each cycle into the pipeline, so that while one instruction is decoded, the next instruction is simultaneously fetched. In the ideal case, this can result in throughput of one instruction per cycle, even though a typical instruction takes several cycles to complete. Dependencies between successive instructions can inhibit this ideal situation from being achieved, as can jumps, interrupts, and cache misses, which may all stall the pipeline, causing the processor to wait one or more cycles for an action to complete. By constructing a dependency graph, typically during code optimization, a compiler can rearrange target code to substantially reduce pipeline stalls. An additional situation occurs in some architectures (e.g., RISC and SPARC computers) where an explicit delay slot is required in the code immediately after a jump that must either be filled by a *nop* (no operation) or a legal instruction (thus avoiding a pipeline stall caused by a jump). This can cause control flow in a target program to become very counterintuitive (and can lead to debugging and interrupt problems). For example, the simple C function

```
int absdiff(int x, int y)
{ if (x > y) return x - y;
  else return y - x;
}
```

is translated by the GNU C compiler on a SPARC machine as follows:

```
        mov %o0, %g1
        cmp %g1, %o1
        bg   .LL1
        sub %o0, %o1, %o0
        sub %o1, %g1, %o0
.LL1:
        retl
        nop
```

In this code, the first subtraction (in the delay slot after the conditional branch) will always be performed, whether or not the jump is taken, and the second subtraction will overwrite the first only if the jump is not taken. (Note also the `nop` after the return; there is no available instruction to fill this delay slot.) Pipelining and instruction scheduling are discussed in more detail in the Chapter 22 (Performance Enhancements).

A further important situation where parallel execution is frequently scheduled by the compiler (with hints from the programmer) is that of loop optimizations (see Section 68.3.7). Computationally intensive programs, such as those involving matrices, typically spend a lot of execution time within loops, often nested two or three deep. Executing loop iterations in parallel can result in significant speedups. Most Fortran compilers therefore have special instructions for the programmer to mark loops for parallel execution: a `DOALL` Fortran statement indicates that each iteration of a loop is independent of the others, so that each iteration can be executed in parallel with the others. In situations where there are dependencies across iterations, the programmer can still mark a loop for parallel execution with a `DOACROSS` statement, but for this to be useful a compiler must be able to determine how to group successive iterations in ways that admit effective parallel execution. Often, this is combined with loop unrolling operations to achieve speedups.

In order to schedule more general program segments to run concurrently, possibly on multiple cores, the most common mechanism used at present is the *thread*: a program section encapsulated for concurrent execution that does not run as an independent process but shares resources, particularly memory, with the program as a whole. Most modern operating systems, including the latest versions of Linux and Windows, guarantee that code written using thread packages provided by the operating system will use available cores or other parallel hardware when executing a program. In this regard, the Java programming language has a significant advantage, since Java has included threads in its definition from the start. Indeed, a Java compiler transparently generates code that can make use of the latest multicore technology, since a JVM that executes the bytecode programs generated by the compiler encapsulates the required operating system calls, and all modern JVM implementations have been updated to use multicore facilities.

68.7 Research Issues and Summary

Language translators can be decomposed into the phases of scanning (lexical analysis), parsing (syntactic analysis), semantic analysis, optimization, and code generation. A scanner breaks the input program into tokens using the theory of regular expressions and finite automata. A parser constructs, implicitly or explicitly, a representation for the syntactic structure of the program, using the theory of context-free grammars. The construction of both a scanner and a parser can be easily automated, using tools such as Lex/Flex and Yacc/Bison or JavaCC. Yacc/Bison constructs a LALR(1) bottom-up parser; JavaCC constructs a recursive-descent top-down parser. Bottom-up parsers are generally too complex to construct by hand, but recursive-descent can be used to hand-construct a parser. Implementations of scanners as finite automata are also relatively easy to construct by hand.

The semantic analysis and code generation steps of a compiler can be modeled theoretically by an attribute grammar, which expresses in equational form the relationships among the various attributes of language entities. In fact, attribute grammars can be used as a basis for automating the construction of an entire compiler, but this has not become common, possibly because of the complexity of representing the entire semantics of a language as an attribute grammar, and possibly because of the difficulty of producing optimized target code. It may be that other semantic definition mechanisms, such as denotational semantics, will result in better automation techniques, but this remains for future study. Current methods typically construct hand-generated semantic analyzers, which operate during parsing using auxiliary data structures, such as the symbol table, or analyzers that perform recursive traversals of a syntax tree.

Some success has been achieved in automating the code-generation step, with easy retargeting as the primary goal. These methods include the syntax-based approach of Glanville and Graham (1978) and the semantic approach of Ganapathi and Fischer (1985). An alternative is to use a symbolic machine description language to describe the target machine, which is then used by the code generator to produce target code. Effective use of this method has been made in the widely retargeted GNU compiler collection.

An important aspect of the automation and retargetability of a compiler is the choice of an appropriate intermediate code representation for the source code. The best choice appears to be a symbolic code for a hypothetical abstract machine. One may then choose either to interpret the intermediate code using a simulator or to perform code generation based on a static simulation of this machine. The primary requirements of such an intermediate code are flexibility, security, and the availability of enough information to provide good optimization over a wide variety of target architectures. A significant challenge for future translator technology is to develop a standard intermediate code that can be generated by many different language front ends, and which can also be efficiently and safely interpreted and compiled on many different architectures under many different operating systems. Optimized versions of the JVM, such as the Jikes research virtual machine (http://jikesrvm.org), hold the promise that this may happen in the not-too-distant future.

Key Terms

Attribute grammar: A set of equations associated to the grammar rules of a context-free grammar that define a collection of attributes associated to the terminals and nonterminals of the grammar. Attribute equations are written in purely functional form (i.e., without side effects) and may be "solved" for the actual attribute values by different kinds of traversals of the parse or syntax tree. Alternatively, attribute values may be computed by replacing the attribute equations with equivalent side effect-generating code using separate data structures such as the symbol table.

Back end: The part of a compiler that depends only on the target language and is independent of the source language. The back-end receives the intermediate code produced by the front end and translates it into the target language.

BNF: Backus normal form or Backus–Naur form. A notation for context-free grammar rules rst used in the Algol60 report to describe syntax. It comprises only two metasymbols, usually written → or ::= and |.

Bottom-up: A parsing algorithm that constructs the parse tree from the leaves to the root. Bottom-up algorithms include LR and LALR parsers, such as those produced by Yacc/Bison.

Bytecode: Numerically encoded intermediate code generated by a compiler that can be executed by an easily retargeted virtual machine.

Disambiguating rule: A rule stated separately from the rules of a context-free grammar, which specifies the correct choice of syntax tree structure when more than one structure is possible.

EBNF: Extended BNF. Adds bracketing metasymbols [...] and {...} to BNF to indicate optional and repeated structures, respectively. (These can also be written as (...)? and (...)* to remain consistent with standard regular expression notation.)

Front end: The part of a compiler that depends only on the source language and is independent of the target language. The front end translates and analyzes the source program.

Incremental compilation: Compilation or recompilation that occurs in small steps based on incremental changes made during the editing of a program and which encompasses only those parts of the program affected by the changes.

Inherited attribute: An attribute whose value depends on attribute values at syntax tree nodes, which are not descendants. A nonsynthesized attribute.

Intermediate code or representation: An internal data structure or sequence of abstract instructions representing a program as produced by the parser or other phase inside a compiler.

Item: A grammar rule choice with a distinguished position (often indicated by a period), indicating that a parse has reached that position in attempting to recognize the rule. Sets of items are used by a bottom-up parser to record a state reached during a parse. Items may have 0 or more tokens of lookahead attached to them. LR(0) items contain no lookahead, whereas LR(1) items contain one token of lookahead.

Just-in-time (JIT) compiler: A compiler that dynamically translates blocks of bytecode "on the fly" into native machine code during execution in a virtual machine, usually just before the virtual machine would execute the code.

L-attributed grammar: An attribute grammar whose attributes may be computed by a left-to-right traversal of the source program. An attribute grammar must be L-attributed for the attributes to be computable during a parse that processes the input from left to right (as most parsers do). Synthesized attributes are always L-attributed.

LALR(1): The lookahead LR(1) algorithm invented by DeRemer (1971). The algorithm is used in many bottom-up parser generators, including Yacc/Bison. A language is also called LALR(1) if it can be parsed unambiguously by the LALR(1) algorithm.

Lexeme: The actual character string read from the input when recognizing a token. The lexeme of an identifier token is the identifier name.

LL(*k*): A top-down parsing algorithm that processes the input from left to right, producing a leftmost derivation using *k* tokens of lookahead. The term can also be applied to a language that can be unambiguously parsed using this algorithm.

LR(*k*): A bottom-up parsing algorithm that processes the input from left to right, producing a rightmost derivation (in reverse) using *k* tokens of lookahead. The term can also be applied to a language that can be unambiguously parsed using this algorithm.

Nonterminal: A name for a structure defined by a context-free grammar rule. Interior nodes of parse trees are labeled by nonterminals.

Phase: A logical unit of a compiler. Typical phases include scanning, parsing, semantic analysis, and code generation. Phases are distinct from passes, which comprise a complete sequential processing of the input program. Phases may or may not correspond to physical code units within the compiler.

Production: Another term for a context-free grammar rule or grammar rule choice.

Recursive-descent: A top-down parsing algorithm that translates context-free grammar rules into a set of mutually recursive procedures, with each procedure corresponding to a nonterminal. Recursive-descent parsing is usually the method of choice when writing a parser by hand.

Reduce-reduce conflict: In bottom-up parsers, a property of a state in which a parser has a choice of two productions that can be used to reduce the parsing stack, and both are legal for the amount of lookahead allowed. Reduce–reduce conflicts have no natural disambiguating rule.

Retargeting: The process of changing a compiler to produce target code (assembly or machine code) for a different machine. This may involve rewriting the compiler back end or creating a machine definition file for the new machine.

Shift-reduce conflict: In bottom-up parsers, a property of a state in which a parser has a choice of either reducing the parsing stack using a production or shifting a token from the input, and both are legal for the amount of lookahead allowed. A natural disambiguating rule is to prefer the shift, thus allowing the parser to match the longest possible input string at each point.

Synthesized attribute: An attribute whose value depends only on the attribute values of descendants in the parse or syntax tree. Synthesized attributes are the easiest to compute during a parse, requiring no special data structures or techniques. The syntax tree itself is the most important example of a synthesized attribute.

Terminal: Another term for a token in a context-free grammar. Leaf nodes of parse trees are labeled by terminals.

Thread: A program section encapsulated for concurrent execution that does not run as an independent process but shares resources, particularly memory, with the program as a whole.

Top-down: A parsing algorithm that constructs the parse tree from the root to the leaves. Top-down algorithms include LL parsers and recursive-descent parsers, such as those produced by JavaCC.

Virtual machine: An interpreter for intermediate code, such as the Java virtual machine, which executes Java bytecode. A virtual machine provides transparent cross-platform retargetability for compilers and other software.

Further Information

The standard advanced text in compiler design is *Compilers: Principles, Techniques, and Tools*, 2nd edn. by A. V. Aho, M. S. Lam, R. Sethi, and J. D. Ullman (Pearson Addison-Wesley, London, U.K., 2007). Introductory texts include *Writing Compilers and Interpreters: A Software Engineering Approach,* 3rd edn. by R. Mak (Wiley, New York, 2009) and *Compiler Construction: Principles and Practice* by K. C. Louden (Cengage Learning, Stamford, CT, 1997). Somewhat more comprehensive treatment of advanced implementation issues can be found in *Modern Compiler Implementation in Java*, 2nd edn. by A. W. Appel and J. Palsberg (Cambridge University Press, Cambridge, U.K., 2002), which treats object-oriented issues in some detail, for both the implementation language and the target language. For a more detailed study of optimization, see *Advanced Compiler Design and Implementation* by S. S. Muchnick (Morgan Kaufmann, Burlington, MA, 1997). A detailed treatment of parsing algorithms and techniques can be found in *Parsing Techniques: A Practical Guide*, 2nd edn., by D. Grune and C. J. H. Jacobs (Springer, New York, 2008). A book that discusses compiler optimizations for power consumption is *Power Aware Computing* by R. Graybill and R. Melhem, Eds. (Springer, New York, 2002). A book that treats instruction scheduling and parallelization techniques in detail is *Optimizing Compilers for Modern Architectures: A Dependence-Based Approach* by R. Allen and K. Kennedy (Morgan Kaufmann, Burlington, MA, 2002). For those interested in functional languages, *Modern Compiler Implementation in ML* by Andrew W. Appel (Cambridge University Press, Cambridge, U.K., 2004) may be of interest. A book that presents a full C compiler in complete detail is *A Retargetable C Compiler: Design and Implementation* by C. W. Fraser and D. Hanson (Addison-Wesley, Boston, MA, 1995).

Aside from these and many other texts, the best place to locate the latest information on compiler design is the comp.compilers newsgroup on the Internet.

Research papers on language translation can be found in publications by the IEEE and the ACM, particularly the conference proceedings published as part of the ACM SIGPLAN notices (especially the annual Programming Languages Design and Implementation [PLDI] conference), the ACM Annual POPL Conference proceedings, and the ACM TOPLAS journal. For information, contact the ACM via its web site, http://www.acm.org (or by e-mail at acmhelp@acm.org) and the IEEE via its web site http://www.ieee.org.

A brief but useful introduction to the use of Lex and Yacc can be found in *The Unix Programming Environment* by B. W. Kernighan and R. Pike (Prentice Hall, Englewood Cliffs, NJ, 1984). A more detailed study of Flex and Bison is contained in *Flex & Bison: Text Processing Tools* by J. Levine (O'Reilly Media, Sebastopol, CA, 2009). Manuals for the GNU versions Flex and Bison can be found at http://www.gnu.org/manual/manual.html. More detail about the Java parser generator JavaCC is contained in *Generating Parsers with JavaCC*, 2nd edn. by T. Copeland (Centennial Books, Portland, OR, 2012). More information about JavaCC can also be found at http://javacc.java.net. Information about Antlr can be found at http://www.antlr.org.

References

Aho, A.V., Lam, M.S., Sethi, R., and Ullman, J.D. 2007. *Compilers: Principles, Techniques, and Tools*, 2nd edn. Pearson Addison-Wesley, Boston, MA, 2007.

Backus, J. et al. 1957. The FORTRAN automatic coding system. *Western Joint Computer Conference*, pp. 188–198. Reprinted in Rosen, S., *Programming Systems and Languages*, McGraw-Hill, New York, 1967, pp. 29–47.

Bell, J.R. 1973. Threaded code. *Commun. ACM*. 16(6): 370–372.

Cattell, R.G.G. 1980. Automatic derivation of code generators from machine descriptions. *ACM Trans. Prog. Lang. Syst.* 2(2): 173–190.

Davidson, J.W. and Fraser, C.W. 1984. Code selection through object code optimization. *ACM Trans. Prog. Lang. Syst.* 6(4): 505–526.

Denny, J.E. and Malloy, B.A. 2010. The IELR(1) Algorithm for generating minimal LR(1) parser tables for non-LR(1) grammars with conflict resolution. *Sci. Comp. Program.*, 75(11): 943–979. Available at http://dx.doi.org/10.1016/j.scico.2009.08.001 (accessed October 10, 2013).

DeRemer, F.L. 1971. Simple LR(k) grammars. *Commun. ACM* 14(7): 453–460.

DeRemer, F.L. and Pennello, T. 1982. Efficient computation of LALR(1) lookahead sets. *ACM Trans. Prog. Lang. Syst.* 4(4): 615–645.

Dion, B.A. 1982. *Locally Least-Cost Error Correctors for Context-Free and Context-Sensitive Parser*. University of Michigan Research Press, Ann Arbor, MI.

Early, J. 1970. An efficient context-free parsing algorithm. *Commun. ACM* 13(2): 94–102.

Farrow, R. 1984. Generating a production compiler from an attribute grammar. *IEEE Softw.* 1(10): 77–93.

Feldman, S. 1979. Make—A program for maintaining computer programs. *Softw. Pract. Exp.* 9(4): 255–265.

Ganapathi, M.J. and Fischer, C.N. 1985. Affix grammar driven code generation. *ACM Trans. Prog. Lang. Syst.* 7(4): 560–599.

GCC. 2010. Gnu Compiler Collection (GCC) Internals. Available at http://gcc.gnu.org/onlinedocs/gccint/

Glanville, R.S. and Graham, S.L. 1978. A new method for compiler code generation. In *Fifth Annual ACM Symposium on Principles of Programming Languages*, pp. 231–254. ACM Press, Inc., New York.

Gosling, J. 1995. Java intermediate bytecodes. *ACM SIGPLAN Notices* 30(3): 111–118.

Graham, S.L., Haley, C.B., and Joy, W.N. 1979. Practical LR error recover. *SIGPLAN Notices* 14(8): 168–175.

Graham, S.L., Harrison, M.A., and Ruzzo, W. L. 1980. An improved context-free recognizer. *ACM Trans. Prog. Lang. Syst.* 2(3): 415–462.

Gray, R.W. et al. 1992. Eli: A complete, flexible compiler construction system. *Commun. ACM* 35(2): 121–131.

Hedin, G., and Magnusson, E. 2003. JastAdd: An aspect-oriented compiler construction system. *Sci. Comput. Program.* 47(1): 37–58.

Henry, R.R. 1984. Graham-Glanville code generators. PhD thesis, University of California, Berkeley, CA.

Jacobson, V. 1987. Tuning Unix Lex, or it's not true what they say about Lex. *Proceedings of the Winter Usenix Conference*, pp. 163–164. Usenix Association, Washington, D.C.

Johnson, S.C. 1975. Yacc—Yet another compiler-compiler. CS Technical Report #32, Bell Laboratories, Murray Hill, NJ.

Johnson, S.C. 1978. A portable compiler: Theory and practice. In *Fifth Annual ACM Symposium on Principles of Programming Languages*, pp. 97–104. ACM Press, Inc., New York.

Katayama, T. 1984. Translation of attribute grammars into procedures. *ACM Trans. Prog. Lang. Syst.* 6(3): 345–369.

Knuth, D.E. 1968. Semantics of context-free languages. *Math. Syst. Theory* 2(2): 127–145. *Errata* 5(1) (1971): 95–96.

Kodaganallur, V. 2004. Incorporating language processing into Java applications: A JavaCC tutorial. *IEEE Softw.* 21(4): 70–77.

Lee, P. 1989. *Realistic Compiler Generation*. MIT Press, Cambridge, MA.

Lesk, M. 1975. Lex—A lexical analyzer generator. CS Technical Report #39, Bell Laboratories, Murray Hill, NJ.

Leverett, B.W. et al. 1980. An overview of the production-quality compiler-compiler project. *IEEE Comp.* 13(8): 38–40.

Lewis, J.P., and Neumann, U. 2003. Performance of Java versus C++. Available at http://scribblethink.org/Computer/javaCbenchmark.html (accessed October 10, 2013).

Li, Y-H. 2004. *Runtime Performance Evaluation of Just-In-Time Compiler Enabled J9 Virtual Machine*. Master's thesis, Arizona State University, Phoenix, AZ. Available at http://pooh.poly.asu.edu/Lindquist/Students/pubs/YihueyLi.pdf

Nori, K.V. et al. 1981. Pascal P implementation notes. In: *Pascal—The Language and its Implementation*, pp. 125–170, Ed. Barron, D.W. Wiley, Chichester, U.K.

Parr, T.J. and Fisher, K.S. 2011. LL(*): The Foundation of the ANTLR Parser Generator, *Proceedings of the 32nd ACM SIGPLAN Conference on Programming Language Design and Implementation, PLDI 2011*, San Jose, CA, June 4–8, pp. 425–436. Preprint available at http://www.antlr.org/papers/LL-star-PLDI11.pdf

Paxson, V. 1990. Flex users manual. (Part of the GNU ftp distribution.) Available at flex.sourceforge.net/manual/ (accessed October 10, 2013).

Perkins, D.R. and Sites, R.L. 1979. Machine independent Pascal code optimization. *ACM SIGPLAN Notices* 14(8): 201–207.

Peyton Jones, S.L. 1987. *The Implementation of Functional Programming Languages*. Prentice-Hall, Englewood Cliffs, NJ. Available at http://research.microsoft.com/en-us/um/people/simonpj/papers/slpj-book-1987/ (accessed October 10, 2013).

Polak, W. 1981. *Compiler Specification and Verification*. Lecture Notes in Computer Science #124, Springer-Verlag, New York.

Randell, B. and Russell, L.J. 1964. *Algol60 Implementation*. Academic Press, New York.

Roehrich, J. 1980. Methods for the automatic construction of error-correcting parsers. *Acta Inform*. 13(2): 115–139.

Schröer, F.W. 1997. *The GENTLE Compiler Construction System*. R. Oldenbourg Verlag, Munich, Germany.

Wirth, N. 1971. The design of a Pascal compiler. *Softw. Pract. Exp*. 1(4): 309–333.

Wirth, N. 1976. *Algorithms + Data Structures = Programs*. Prentice-Hall, Englewood Cliffs, NJ.

69

Programming Language Semantics

David A. Schmidt
Kansas State University

69.1 Introduction

A programming language possesses *syntax* and *semantics*. Syntax refers to the spelling of the language's programs, and semantics refers to the meanings of the programs. A language's syntax is formalized by a grammar or syntax chart; such formalizations are found in the back of language manuals. A language's semantics should be formalized, too, and this is the topic of this chapter.

Before we begin, we might ask, "What do we gain by formalizing the semantics of a programming language?" Consider the related question, "What was gained when language syntax was formalized with BNF?"

- A language's syntax definition standardizes the official syntax. This is crucial to users, who require a guide to writing syntactically correct programs, and to implementors, who must write a correct parser for the language's compiler.
- The syntax definition permits a formal analysis of its properties, such as whether the definition is *LL(k)*, *LR(k)*, or ambiguous.
- The syntax definition can be used as input to a compiler front-end generating tool, such as YACC; it becomes, in effect, the implementation.

We derive similar benefits from a formal semantics definition:

- The semantics definition standardizes the official semantics of the language. This is crucial to users, who require a guide to understanding the programs that they write, and to implementors, who must write a correct code generator for the language's compiler.
- The semantics definition permits a formal analysis of its properties, such as whether the definition is strongly typed, block structured, uses single-threaded data structures, is parallelizable, etc.
- The semantics definition can be used as input to a compiler back-end generating tool [26,31]; it becomes, in effect, the implementation.

Programming-language syntax was studied intensively in the 1960s and 1970s, and programming language semantics is undergoing similar intensive study. Unlike the acceptance of BNF as the standard for syntax definition, it is unlikely that a single definition method will take hold for semantics—semantics is harder to formalize than syntax, and it has a wider variety of applications.

Semantics-definition methods fall roughly into three groups:

- *Operational*: The meaning of a well-formed program is the trace of computation steps that results from processing the program's input. Operational semantics is also called *intensional* semantics, because the sequence of internal computation steps (the "intension") is most important. For example, two differently coded programs that both compute factorial have different operational semantics.
- *Denotational*: The meaning of a well-formed program is a mathematical function from input data to output data. The steps taken to calculate the output are unimportant; it is the relation of input to output that matters. Denotational semantics is also called *extensional* semantics, because only the "extension"—the visible relation between input and output—matters. Thus, two differently coded versions of factorial have the same denotational semantics.
- *Axiomatic*: A meaning of a well-formed program is a logical proposition (a "specification") that states some property about the input and output. For example, the proposition $\forall x . x \geq 0 \supset \exists y . y = x!$ is an axiomatic semantics of a factorial program.

69.2 Underlying Principles of Semantics Methods

We survey the three semantic methods by applying each in turn to the world's oldest and simplest programming language, arithmetic. The syntax of our arithmetic language is

$$E ::= N \mid E_1+E_2$$

where N stands for the set of numerals {0, 1, 2, …}. Although this language has no notion of input and output, it is computational.

69.2.1 Operational Semantics

There are several versions of operational semantics for arithmetic. The one that you learned as a child is called a *term rewriting system* [6,25]. It uses rule schemes that generate computation steps. There is just one rule scheme for arithmetic:

$$N_1+N_2 \Rightarrow N' \quad \text{where } N' \text{ is the sum of the numerals } N_1 \text{ and } N_2$$

The rule scheme states that adding two numerals is a computation step, for example, $1 + 2 \Rightarrow 3$ is one computation step. An operational semantics of a program is a sequence of such computation steps. For example, an operational semantics of (1 + 2) + (4 + 5) goes as follows:

$$(1+2)+(4+5) \Rightarrow 3+(4+5) \Rightarrow 3+9 \Rightarrow 12$$

Three computation steps led to the answer, 12. An intermediate expression like 3 + (4 + 5) is a "state," so this operational semantics traces the states of the computation.

Another semantics for the example is (1 + 2) + (4 + 5) ⇒ (1 + 2) + 9 ⇒ 3 + 9 ⇒ 12. The outcome is the same, and a set of rules that has this property is called *confluent* [25].

A *structural operational semantics* is a term-rewriting system plus a set of inference rules that state precisely the context in which a computation step can be undertaken. (A structural operational semantics is sometimes called a "small-step semantics," because each computation step is a small step towards the final answer.) Say that we demand left-to-right computation of arithmetic expressions. This is encoded as follows:

$$N_1 + N_2 \Rightarrow N' \quad \text{where } N' \text{ is the sum of } N_1 \text{ and } N_2$$

$$\frac{E_1 \Rightarrow E_1'}{E_1 + E_2 \Rightarrow E_1' + E_2} \qquad \frac{E_2 \Rightarrow E_2'}{N + E_2 \Rightarrow N + E_2'}$$

The first rule goes as before; the second rule states, if the left operand of an addition expression can be rewritten, then do this. The third rule is the crucial one: if the right operand of an addition expression can be rewritten *and* the left operand is already a numeral (completely evaluated), then rewrite the right operand. Working together, the three rules force left-to-right evaluation.

Each computation step must be deduced by the rules. For (1+2)+(4+5), we deduce this initial computation step:

$$\frac{1+2 \Rightarrow 3}{(1+2)+(4+5) \Rightarrow 3+(4+5)}$$

Thus, the first step is (1 + 2) + (4 + 5) ⇒ 3 + (4 + 5); note that we *cannot* deduce that (1 + 2) + (4 + 5) ⇒ (1 + 2) + 9. The next computation step is justified by this deduction:

$$\frac{4+5 \Rightarrow 9}{3+(4+5) \Rightarrow 3+9}$$

The last deduction is simply 3 + 9 ⇒ 12, and we are finished. The example shows why the semantics is "structural": each computation step is explicitly embedded into the structure of the overall program.

Operational semantics is often used to expose implementation concepts, like instruction counters, storage vectors, and stacks. For example, say our semantics of arithmetic must show how a stack holds intermediate results. We use a *state* of form $\langle s, c \rangle$, where s is the stack and c is the arithmetic expression to evaluate. A stack containing n items is written $v_1 :: v_2 :: \ldots :: v_n :: nil$, where v_1 is the topmost item and *nil* marks the bottom of the stack. The c component will be written as a stack as well. The initial state for an arithmetic expression, p, is written $\langle nil, p :: nil \rangle$, and computation proceeds until the state appears as $\langle v :: nil, nil \rangle$; we say that the result is v.

The semantics uses three rewriting rules:

$$\langle s, N :: c \rangle \Rightarrow \langle N :: s, c \rangle$$

$$\langle s, E_1 + E_2 :: c \rangle \Rightarrow \langle s, E_1 :: E_2 :: add :: c \rangle$$

$$\langle N_2 :: N_1 :: s, add :: c \rangle \Rightarrow \langle N' :: s, c \rangle \quad \text{where } N' \text{ is the sum of } N_1 \text{ and } N_2$$

The first rule says that a numeral is evaluated by pushing it on the top of the stack. The second rule decomposes an addition into its operands and operator, which must be computed individually. The third rule removes the top two items from the stack and adds them. Here is the previous example, repeated:

$$\langle nil, (1+2)+(4+5) :: nil\rangle$$

$$\Rightarrow \langle nil, 1+2 :: 4+5 :: add :: nil\rangle$$

$$\Rightarrow \langle nil, 1 :: 2 :: add :: 4+5 :: add :: nil\rangle$$

$$\Rightarrow \langle 1 :: nil, 2 :: add :: 4+5 :: add :: nil\rangle$$

$$\Rightarrow \langle 2 :: 1 :: nil, add :: 4+5 :: add :: nil\rangle$$

$$\Rightarrow \langle 3 :: nil, : 4+5 :: add :: nil\rangle \Rightarrow \ldots \Rightarrow \langle 12 :: nil, nil\rangle$$

This form of operational semantics is sometimes called a *state-transition semantics* because each rewriting rule operates upon the entire state. A state-transition semantics has no need for structural operational-semantics rules.

The formulations presented here are typical of operational semantics. When one wishes to prove properties of an operational-semantics definition, the standard proof technique is *induction on the length of the computation*. That is, to prove that a property, *P*, holds for an operational semantics, one must show that *P* holds for all possible computation sequences that can be generated from the rewriting rules. For an arbitrary computation sequence, it suffices to show that *P* holds no matter how long the computation runs. Therefore, one shows (1) *P* holds after zero computation steps, that is, at the outset, and (2) if *P* holds after *n* computation steps, it holds after *n* + 1 steps. See Nielson and Nielson [34] for examples.

Operational semantics lies close to implementation, and early forms of operational semantics were *definitional interpreters*—implementations that generated the state-transition steps [13,44]. Figure 69.1 displays a Python-coded definitional interpreter that generates the computation steps of the stack-based arithmetic semantics, modeling the stack with a list argument and modeling subexpression evaluation with recursive-function calls. (As an exercise, you should prove the interpreter prints a sequence of stacks that matches the one computed by the state-transition rules.) A definitional interpreter is a powerful, practical tool for language definition and prototyping.

```
def eval (t, stack):
    """eval computes the meaning of t using stack.
       parameters: t - an arithmetic term, parsed into a nested list of form:
                   term ::= N | ["+", term, term], where N is a string of digits
           stack - a list of integers
       returns: stack with the integer meaning of t pushed on its top.
    """
    print "Evaluate", t, "with stack =", stack
    if isinstance(t, str) and t.isdigit(): # is t a string of digits?
       newstack = [int(t)] + stack # push the int onto the stack
    else: # t is a list, ["+", t1, t2]
       stack1 = eval(t[1], stack)
       stack2 = eval(t[2], stack1)
       answer = stack2[1] + stack2[0] # add top two stack values
       newstack = [answer] + stack2[2:] # push answer onto popped stack
    print "Evaluated", t, "Updated stack =", newstack
    return newstack
```

FIGURE 69.1 Definitional interpreter in Python for stack-semantics of arithmetic.

69.2.2 Denotational Semantics

Operational semantics emphasizes internal state transitions. For the arithmetic language, we were distracted by questions about order of evaluation of subphrases, even though this issue is not at all important to arithmetic. Further, the key property that the meaning of an expression is built from the meanings of its subexpressions was obscured.

We use denotational semantics to establish that a program has an underlying *mathematical meaning* that is independent of the computation strategy used to compute it. In the case of arithmetic, an expression like (1 + 2) + (4 + 5) has the meaning, 12. The implementation that computes the 12 is a separate issue, perhaps addressed by an operational semantics.

The assignment of meaning to programs is performed *compositionally*: the meaning of a phrase is built from the meanings of its subphrases. We now see this in the denotational semantics of the arithmetic language. First, we assert that meanings of arithmetic expressions must be taken from the *domain* ("set") of natural numbers, *Nat* = {0, 1, 2,...}, and there is a binary, mathematical function, *plus* : *Nat* × *Nat* → *Nat*, which maps a pair of natural numbers to their sum.

The denotational semantics definition of arithmetic is simple and elegant:

$$\mathcal{E} : \mathit{Expression} \rightarrow \mathit{Nat}$$

$$\mathcal{E}[\![\mathrm{N}]\!] = \mathrm{N}$$

$$\mathcal{E}[\![\mathrm{E}_1 + \mathrm{E}_2]\!] = \mathit{plus}\left(\mathcal{E}[\![\mathrm{E}_1]\!]; \mathcal{E}[\![\mathrm{E}_2]\!]\right)$$

The first line states that $\mathcal{E}$ is the name of the function that maps arithmetic expressions to their meanings. Since there are two BNF constructions for expressions, $\mathcal{E}$ is completely defined by the two equational clauses. (This is a Tarksi-style interpretation, as used in symbolic logic to give meaning to logical propositions [49].) The interesting clause is the one for $\mathrm{E}_1 + \mathrm{E}_2$; it says that the meanings of E_1 and E_2 are combined compositionally by *plus*. Here is the denotational semantics of our example program:

$$\begin{aligned}\mathcal{E}[\![(1+2)+(4+5)]\!] &= \mathit{plus}\left(\mathcal{E}[\![1+2]\!], \mathcal{E}[\![4+5]\!]\right)\\ &= \mathit{plus}\left(\mathit{plus}\left(\mathcal{E}[\![1]\!], \mathcal{E}[\![2]\!]\right), \mathit{plus}\left(\mathcal{E}[\![4]\!], \mathcal{E}[\![5]\!]\right)\right)\\ &= \mathit{plus}\left(\mathit{plus}(1,2), \mathit{plus}(4,5)\right) = \mathit{plus}(3,9) = 12\end{aligned}$$

Read the aforementioned equation as follows: the meaning of (1 + 2) + (4 + 5) equals the meanings of 1 + 2 and 4 + 5 added together. Since the meaning of 1 + 2 is 3, and the meaning of 4 + 5 is 9, the meaning of the overall expression is 12. This reading says nothing about order of evaluation or run-time data structures—it states only mathematical meaning.

Here is an alternative way of understanding the semantics; write a set of simultaneous equations based on the denotational definition:

$$\begin{aligned}\mathcal{E}[\![(1+2)+(4+5)]\!] &= \mathit{plus}\left(\mathcal{E}[\![1+2]\!], \mathcal{E}[\![4+5]\!]\right)\\ \mathcal{E}[\![1+2]\!] &= \mathit{plus}\left(\mathcal{E}[\![1]\!], \mathcal{E}[\![2]\!]\right)\\ \mathcal{E}[\![4+5]\!] &= \mathit{plus}\left(\mathcal{E}[\![4]\!], \mathcal{E}[\![5]\!]\right)\\ \mathcal{E}[\![1]\!] = 1 \quad & \mathcal{E}[\![2]\!] = 2\\ \mathcal{E}[\![4]\!] = 4 \quad & \mathcal{E}[\![5]\!] = 5\end{aligned}$$

Now, solve the equation set to discover that $\mathcal{E}[\![(1+2)+(4+5)]\!]$ is 12.

Since denotational semantics states the meaning of a phrase in terms of the meanings of its subphrases, its associated proof technique is structural induction. That is, to prove that a property, *P*, holds for all programs in the language, one must show that the meaning of each construction in the language has property *P*. Therefore, one must show that each equational clause in the semantic definition produces a meaning with property *P*. In the case that a clause refers to subphrases (e.g., $\mathcal{E}[\![E_1 + E_2]\!]$), one may assume that the meanings of the subphrases have property *P*. Again, see Nielson and Nielson [34] for examples.

69.2.3 Natural Semantics

A semantics method has been proposed that is halfway between operational semantics and denotational semantics; it is called *natural semantics*. Like structural operational semantics, natural semantics shows the context in which a computation step occurs, and like denotational semantics, natural semantics emphasizes that the computation of a phrase is built from the computations of its subphrases.

A natural semantics is a set of inference rules, and a complete computation in natural semantics is a single, large derivation. The natural semantics rules for the arithmetic language are

$$N \Rightarrow N$$

$$\frac{E_1 \Rightarrow n_1 \quad E_2 \Rightarrow n_2}{E_1 + E_2 \Rightarrow m} \quad \text{where } m \text{ is the sum of } n_1 \text{ and } n_2$$

Read a configuration of the form $E \Rightarrow n$ as "E evaluates to *n*." The rules resemble a denotational semantics written in inference rule form; this is no accident—natural semantics can be viewed as a denotational-semantics variant where the internal calculations of meaning are made explicit and the domains are left implicit. The internal calculations are seen in the natural semantics of our example expression:

$$\frac{\dfrac{1 \Rightarrow 1 \quad 2 \Rightarrow 2}{(1+2) \Rightarrow 3} \quad \dfrac{4 \Rightarrow 4 \quad 5 \Rightarrow 5}{(4+5) \Rightarrow 9}}{(1+2)+(4+5) \Rightarrow 12}$$

Unlike denotational semantics, natural semantics does not claim that the meaning of a program is necessarily "mathematical." And unlike structural operational semantics, where a configuration $e \Rightarrow e'$ says that *e* transits to an intermediate state, e', in natural semantics $e \Rightarrow v$ asserts that the final answer for *e* is *v*. For this reason, a natural semantics is sometimes called a "big-step semantics." An interesting drawback of natural semantics is that semantics derivations can be drawn only for terminating programs.

The usual proof technique for proving properties of a natural semantics definition is induction on the height of the derivation trees that are generated from the semantics. Once again, see Nielson and Nielson [34].

69.2.4 Axiomatic Semantics

An axiomatic semantics produces *properties* of programs rather than meanings. The derivation of these properties is done by an inference rule set that looks somewhat like a natural semantics.

As an example, say that we wish to prove even-odd properties of programs in arithmetic and our set of properties is simply {*is_even*, *is_odd*}. For example, 2 : *is_even* and (2 + 3) : *is_odd* both hold true. We can define an axiomatic semantics to do this:

$$N{:}is\_even \text{ if } N \ mod 2 = 0 \qquad N{:}\ is\_odd \text{ if } N \ mod 2 = 1$$

$$\frac{E_1{:}p_1 \quad E_2{:}p_2}{E_1 + E_2{:}p_3} \quad \text{where } p_3 = \begin{cases} is\_even \text{ if } p_1 = p_2 \\ is\_odd \text{ otherwise} \end{cases}$$

The derivation of the even-odd property of our example program is

$$\dfrac{\dfrac{1: is\_odd \quad 2: is\_even}{1+2: is\_odd} \qquad \dfrac{4: is\_even \quad 5: is\_odd}{4+5: is\_odd}}{(1+2)+(4+5): is\_even}$$

In the usual case, the properties to be proved of programs are expressed in the language of predicate logic; see Section 69.3.7. Also, axiomatic semantics has strong ties to the *abstract interpretation* of denotational and natural semantics definitions; see Section 69.4.

69.3 Semantical Principles of Programming Languages

The semantics methods shine when they are applied to programming languages—primary features of a language are made prominent, and subtle features receive proper mention. Ambiguities and anomalies stand out like the proverbial sore thumb. In this section, we use a classic block-structured imperative language to demonstrate. Denotational semantics will be emphasized, but excerpts from the other semantics formalisms will be provided for comparison.

69.3.1 Language Syntax and Informal Semantics

The syntax of the programming language is presented in Figure 69.2. As stated in the figure, there are four "levels" of syntax constructions in the language, and the topmost level, Program, is the primary one. The language is a while-loop language with local, nonrecursive procedure definitions. For simplicity, variables are predeclared and there are just three of them—X, Y, and Z. A program, C., operates as follows: an input number is read and assigned to X's location. Then the body, C, of the program is evaluated, and upon completion, the storage vector holds the results. For example, this program computes n^2 for a positive input n; the result is found in Z's location:

```
begin proc INCR = Z:= Z+X; Y:= Y+1
  in Y:= 0; Z:= 0; while Y not= X do call INCR od end.
```

It is possible to write nonsense programs in the language; an example is: `A:=0; call B`. Such programs have no meaning, and we will not attempt to give semantics to them.

69.3.2 Domains for Denotational Semantics

To give a denotational semantics to the sample language, we must state the sets of meanings, called *domains*, that we use. Our imperative, block-structured language has two primary domains:

P ∈ Program	I ∈ Identifier = alphabetic strings
D ∈ Declaration	V ∈ Variable = {`X, Y, Z`} ⊆ Identifier
C ∈ Command	N ∈ Numeral = {0, 1, 2, ...}
E ∈ Expression	

P ::= C.
D ::= proc I=C
C ::= V:=E | C_1; C_2 | `begin D in C end` | `call I` | `while E do C od`
E ::= N | E_1 + E_2 | E_1 not=E_2 | V

FIGURE 69.2 Language syntax rules.

$$
\begin{aligned}
&\mathit{Store} = \{\langle n_1, n_2, n_3\rangle \mid n_i \in \mathit{Nat},\ i \in 1..3\}\\
&\quad \mathit{lookup} : \{1, 2, 3\} \times \mathit{Store} \to \mathit{Nat}\\
&\qquad \mathit{lookup}(i, \langle n_1, n_2, n_3\rangle) = n_i\\
&\quad \mathit{update} : \{1, 2, 3\} \times \mathit{Nat} \times \mathit{Store} \to \mathit{Store}\\
&\qquad \mathit{update}(1, n, \langle n_1, n_2, n_3\rangle) = \langle n_1, n_2, n_3\rangle\\
&\qquad \mathit{update}(2, n, \langle n_1, n_2, n_3\rangle) = \langle n_1, n, n_3\rangle\\
&\qquad \mathit{update}(3, n, \langle n_1, n_2, n_3\rangle) = \langle n_1, n_2, n\rangle\\
&\quad \mathit{init\ store} : \mathit{Nat} \to \mathit{Store}\\
&\qquad \mathit{init\_store}(n) = \langle n, 0, 0\rangle\\
&\quad \mathit{check} : (\mathit{Store} \to \mathit{Store}_\perp) \times \mathit{Store}_\perp \to \mathit{Store}_\perp \text{ where } \mathit{Store}_\perp = \mathit{Store} \cup \{\perp\}\\
&\qquad \mathit{check}(c, a) = \text{if } (a = \perp) \text{ then } \perp \text{ else } c(a)\\
&\mathit{Environment} = (\text{Identifier} \times \mathit{Denotable})^*\\
&\quad \text{where } A^* \text{ is a list of } A\text{-elements, } a_1 :: a_2 :: \ldots :: a_n :: \mathit{nil},\ n \ge 0\\
&\quad \text{and } \mathit{Denotable} = \{1, 2, 3\} \cup (\mathit{Store} \to \mathit{Store}_\perp)\\
&\qquad \mathit{find} : \text{Identifier} \times \mathit{Environment} \to \mathit{Denotable}\\
&\qquad\quad \mathit{find}(i, \mathit{nil}) = 0\\
&\qquad\quad \mathit{find}(i, (i', d) :: \mathit{rest}) = \text{if } (i = i') \text{ then } d \text{ else } \mathit{find}(i, \mathit{rest})\\
&\qquad \mathit{bind} : \text{Identifier} \times \mathit{Denotable} \times \mathit{Environment} \to \mathit{Environment}\\
&\qquad\quad \mathit{bind}(i, d, e) = (i, d) :: e\\
&\qquad \mathit{init\_env} : \mathit{Environment}\\
&\qquad\quad \mathit{init\_env} = (X, 1) :: (Y, 2) :: (Z, 3) :: \mathit{nil}
\end{aligned}
$$

FIGURE 69.3 Semantic domains.

(1) the domain of storage vectors, called *Store*, and (2) the domain of symbol tables, called *Environment*. There are also secondary domains of booleans and natural numbers. The primary domains and their operations are displayed in Figure 69.3.

The domains and operations deserve study. First, the *Store* domain states that a storage vector is a triple. (Recall that programs have exactly three variables.) The operation *lookup* extracts a value from the store, for example, *lookup*(2,⟨1, 3, 5⟩) = 3, and *update* updates the store, for example, *update*(2, 6, ⟨1, 3, 5⟩) = ⟨1, 6, 5⟩. Operation *init_store* creates a starting store. We examine *check* momentarily.

The environment domain states that a symbol table is a list of identifier-value pairs. For example, if variable X is the name of location 1, and P is the name of a procedure that is a "no-op," then the environment that holds this information would appear (X, 1)::(P, *id*)::*nil*, where *id*(*s*)=*s*. (Procedures will be discussed momentarily.) Operation *find* locates the binding for an identifier in the environment, for example, *find*(X, :(X, 1)::(P, *id*)::*nil*)=1, and *bind* adds a new binding, e.g., *bind*(Y, 2, :(X, 1)::(P, *id*)::*nil*) = (Y, 2)::(X, 1)::(P, *id*)::*nil*. Operation *init_env* creates an environment to start the program.

In the next section, we will see that the job of a command, for example, an assignment, is to update the store—the meaning of a command is a function that maps the current store to the updated one. (That's why a "no-op" command is the identity function, $id(s) = s$, where $s \in Store$.) But sometimes commands "loop," and no updated store appears. We use the symbol, ⊥, read "bottom," to stand for a looping store, and we use $Store_\perp = Store \cup \{\perp\}$ as the possible outputs from commands. Therefore, the meaning of a command is a function of form $Store \to Store_\perp$.

It is impossible to recover from looping, so if there is a command sequence, C_1; C_2, and C_1 is looping, then C_2 cannot proceed. The *check* operation is used to watch for this situation.

Finally, here are two commonly used notations. First, functions like $id(s) = s$ are often reformatted to read $id = \lambda s.s$; in general, for $f(a) = e$, we write $f = \lambda a.e$, that is, we write the argument to the function to the right of the equals sign. This is called *lambda notation*, and stems from the *lambda calculus*, an elegant formal system for functions [4]. The notation $f = \lambda a.e$ emphasizes that (1) the function $\lambda a.e$ is a value in its own right, and (2) the function's name is f.

Second, it is common to revise a function that takes multiple arguments, for example, $f(a,b) = e$, so that it takes the arguments one at a time: $f = \lambda a.\lambda b.e$. So, if the arity of f was $A \times B \rightarrow C$, its new arity is $A \rightarrow (B \rightarrow C)$. This reformatting trick is named *currying*, after Haskell Curry, one of the developers of the lambda calculus.

69.3.3 Denotational Semantics of Programs

Figure 69.4 gives the denotational semantics of the programming language. Since the syntax of the language has four levels, the semantics is organized into four levels of meaning. For each level, we define a *valuation function*, which produces the meanings of constructions at that level. For example, at the Expression level, the constructions are mapped to their meanings by $\mathcal{E}$.

What is the meaning of the expression, say, X + 5? This would be $\mathcal{E}[\![\mathtt{X+5}]\!]$, and the meaning depends on which location is named by X and what number is stored in that location. Therefore, the meaning is dependent on the current value of the environment and the current value of the store. So, if the current environment is $e_0 = (\mathtt{P}, \lambda s.s) :: (\mathtt{X}, 1) :: (\mathtt{Y}, 2) :: (\mathtt{Z}, 3) :: nil$ and the current store is $s_0 = \langle 2, 0, 0\rangle$, then the meaning of X + 5 is 7:

$$\begin{aligned}\mathcal{E}[\![\mathtt{X+5}]\!]e_0 s_0 &= plus\left(\mathcal{E}[\![\mathtt{X}]\!]e_0\, s_0, \mathcal{E}[\![\mathtt{5}]\!]e_0\, s_0\right)\\ &= plus(lookup(find(\mathtt{X}, e_0), s_0), 5)\\ &= plus(lookup(1, s_0), 5) = plus(2, 5) = 7\end{aligned}$$

As this simple derivation shows, data structures like the symbol table and storage vector are modelled by the environment and store arguments. This pattern is used throughout the semantics definition.

$$\begin{aligned}
&\mathcal{P} : Program \rightarrow Nat \rightarrow Nat_\bot\\
&\quad \mathcal{P}[\![\mathtt{C.}]\!] = \lambda n.\mathcal{C}[\![\mathtt{C}]\!]\, init_env\, (init_store\; n)\\
&\mathcal{D} : Declaration \rightarrow Environment \rightarrow Environment\\
&\quad \mathcal{D}[\![\mathtt{Proc\ I = C}]\!] = \lambda e.bind(I, \mathcal{C}[\![\mathtt{C}]\!]e, e)\\
&\mathcal{C} : Command \rightarrow Environment \rightarrow Store \rightarrow Store_\bot\\
&\quad \mathcal{C}[\![\mathtt{V := E}]\!] = \lambda e.\lambda s.update(find(V,e), \mathcal{E}[\![\mathtt{E}]\!]e\, s, s)\\
&\quad \mathcal{C}[\![\mathtt{C_1; C_2}]\!] = \lambda e.\lambda s.check\left(\mathcal{C}[\![\mathtt{C_2}]\!]e, \mathcal{C}[\![\mathtt{C_1}]\!]e\, s\right)\\
&\quad \mathcal{C}[\![\mathtt{begin\ D\ in\ C\ end}]\!] = \lambda e.\lambda s.\mathcal{C}[\![\mathtt{C}]\!](\mathcal{D}[\![\mathtt{D}]\!]e)s\\
&\quad \mathcal{C}[\![\mathtt{call\ I}]\!] = \lambda e.find(\mathtt{I}, e)\\
&\quad \mathcal{C}[\![\mathtt{while\ E\ do\ C\ od}]\!] = \lambda e.\textstyle\bigcup_{i\ge 0} w_i\\
&\quad\quad \text{where } \begin{array}{l} w_0 = \lambda s.\bot \\ w_{i+1} = \lambda s.\text{if } \mathcal{E}[\![\mathtt{E}]\!]e\,s \text{ then } check(w_i, \mathcal{C}[\![\mathtt{C}]\!]e\ s) else\ s\end{array}\\
&\mathcal{E} : Expression \rightarrow Environment \rightarrow Store \rightarrow (Nat \cup Bool)\\
&\quad \mathcal{E}[\![\mathtt{N}]\!] = \lambda e.\lambda s.N\\
&\quad \mathcal{E}[\![\mathtt{E_1 + E_2}]\!] = \lambda e.\lambda s.plus(\mathcal{E}[\![\mathtt{E_1}]\!]e\, s, \mathcal{E}[\![\mathtt{E_2}]\!]e\, s)\\
&\quad \mathcal{E}[\![\mathtt{E_1\ not = E_2}]\!] = \lambda e.\lambda s.notequals(\mathcal{E}[\![\mathtt{E_1}]\!]e\, s, \mathcal{E}[\![\mathtt{E_2}]\!]e\ s)\\
&\quad \mathcal{E}[\![\mathtt{V}]\!] = \lambda e.\lambda s.lookup(find(V,e), s)
\end{aligned}$$

FIGURE 69.4 Denotational semantics.

As noted in the previous section, a command updates the store. Precisely stated, the valuation function for commands is: $\mathcal{C}$:*Command* → *Environment* → *Store* → $Store_\perp$. For example, for e_0 and s_0 given above, we see that

$$\mathcal{C}[\![\texttt{Z:=X+5}]\!]e_0\ s_0 = update(find(\texttt{Z}, e_0), \mathcal{E}[\![\texttt{X+5}]\!]e_0\ s_0, s_0) = update(3, 7, s_0) = \langle 2, 0, 7\rangle$$

But a crucial point about the meaning of the assignment is that it is a function upon stores. That is, if we are uncertain of the current value of store, but we know that the environment for the assignment is e_0, then we can conclude

$$\mathcal{C}[\![\texttt{Z:=X+5}]\!]e_0 = \lambda s.update(3, plus(lookup(1, s), 5), s)$$

That is, the assignment with environment e_0 is a function that updates a store at location 3.

Next, consider this example of a command sequence:

$$\mathcal{C}[\![\texttt{Z:X+5; call P}]\!]e_0\ s_0 = check\big(\mathcal{C}[\![\texttt{call P}]\!]e_0, \mathcal{C}[\![\texttt{Z:X+5}]\!]e_0\ s_0\big)$$

$$= check(find(\texttt{P}, e_0), \langle 2, 0, 7\rangle) = check(\lambda s.s, \langle 2, 0, 70\rangle)$$

$$= (\lambda s.s)\langle 2, 0, 7\rangle = \langle 2, 0, 7\rangle$$

As noted in the earlier section, the *check* operation verifies that the first command in the sequence produces a proper output store; if so, the store is handed to the second command in the sequence. Also, we see that the meaning of `call P` is the store updating function bound to `P` in the environment.

Procedures are placed in the environment by declarations, as we see in this example: let e_1 denote (X, 1)::(Y, 2)::(Z, 3)::*nil*:

$$\mathcal{C}[\![\texttt{begin proc P = Y:=Y in Z:=X+5, call P end}]\!]e_1\ s_0$$

$$= \mathcal{C}[\![\texttt{Z:=X+5, call P}]\!]\big(\mathcal{D}[\![\texttt{proc P = Y:=Y}]\!]\ e_1\big)s_0$$

$$= \mathcal{C}[\![\texttt{Z:=X+5, call P}]\!](bind(\texttt{P}, \mathcal{C}[\![\texttt{Y:=Y}]\!]e_1, e_1))s_0$$

$$= \mathcal{C}[\![\texttt{Z:=X+5, call P}]\!](bind(\texttt{P}, \lambda s.update(2, lookup(2, s), s), e_1))s_0$$

$$= \mathcal{C}[\![\texttt{Z:=X+5, call P}]\!]((\texttt{P}, id) :: e_1)s_0$$

$$\text{where } id = \lambda s.update(2, lookup(2, s), s) = \lambda s.s \qquad (*)$$

$$= \mathcal{C}[\![\texttt{Z:=X+5, call P}]\!]e_0\ s_0 = \langle 2,0,7\rangle$$

The equality marked by (*) is significant; we can assert that the function $\lambda s.update(2, lookup(2, s), s)$ is identical to $\lambda s.s$ by appealing to the *extensionality* law of mathematics: if two functions map identical arguments to identical answers, then the functions are themselves identical. The extensionality law can be used here because in denotational semantics the meanings of program phrases are mathematical—functions. In contrast, the extensionality law cannot be used in operational semantics calculations.

Finally, we can combine our series of little examples into the semantics of a complete program:

$$\mathcal{P}[\![\texttt{begin proc P=Y:=Y in Z:=X+5; call P end.}]\!]2$$

$$=\mathcal{C}[\![\texttt{begin proc P=Y:=Y in Z:=X+5; call P end}]\!]\mathit{init\_env}(\mathit{init\_store}2)$$

$$=\mathcal{C}[\![\texttt{begin proc P=Y:=Y in Z:=X+5; call P end}]\!]e_1\, s_0$$

$$=\langle 2,0,7\rangle$$

69.3.4 Semantics of the While Loop

The most difficult clause in the semantics definition is the one for the while-loop. Here is some intuition: to produce an output store, the loop `while` E `do` C `od` must terminate after some finite number of iterations. To measure this behavior, let $\texttt{while}_i$ E `do` C `od` be a loop that can iterate at most *i* times—if the loop runs more than *i* iterations, it becomes exhausted, and its output is ⊥. For example, for input store ⟨4, 0, 0⟩, the loop $\texttt{while}_k$ `Y not= X do Y:=Y+1 od` can produce the output store ⟨4, 4, 0⟩ only when *k* is greater than 4. (Otherwise, the output is ⊥.)

It is easy to conclude that the family, $\texttt{while}_i$ E `do` C `od`, for $i \geq 0$, can be written equivalently as

$$\texttt{while}_0\ \text{E do C od} = \text{"}\mathit{exhausted}\text{"(that is, its meaning is } \lambda s.\bot)$$

$$\texttt{while}_{i+1}\ \text{E do C od} = \texttt{if E then C; while}_i\ \text{E do C od}\ \texttt{else skip fi}$$

When we refer back to Figure 69.3, we draw these conclusions:

$$\mathcal{C}[\![\texttt{while}_0\ \text{E do C od}]\!]e = w_0$$

$$\mathcal{C}[\![\texttt{while}_{i+1}\ \text{E do C od}]\!]e = w_{i+1}$$

Since the behavior of a while loop must be the "union" of the behaviors of the $\texttt{while}_i$-loops, we conclude that $\mathcal{C}[\![\texttt{while}\ \text{E do C od}]\!]e = \bigcup_{i\geq 0} w_i$. The semantic union operation is well defined because each w_i is a function from the set *Store* → $\text{Store}_\bot$, and a function can be represented as a set of argument-answer pairs. (This is called the *graph of the function*.) So, $\bigcup_{i\geq 0} w_i$ is the union of the graphs of the w_i functions*.

The definition of $\mathcal{C}[\![\texttt{while}\ \text{E do C od}]\!]$ is succinct, but it is awkward to use in practice. An intuitive way of defining the semantics is

$$\mathcal{C}[\![\texttt{while}\ \text{E do C od}]\!]e = w$$

$$\text{where } w = \lambda s.\ \ \text{if } \mathcal{E}[\![\text{E}]\!]e\ s \text{ then } \mathit{check}(w, \mathcal{C}[\![\text{C}]\!]e\ s) \text{ else } s$$

The problem here is that the definition of *w* is circular, and circular definitions can be malformed. Fortunately, this definition of *w* can be claimed to denote the function $\bigcup_{i\geq 0} w_i$ because the following equality holds:

$$\bigcup_{i\geq 0} w_i = \lambda s.\ \ \text{if } \mathcal{E}[\![\text{E}]\!]e\,s\ \ \text{then } \mathit{check}\left(\bigcup_{i\geq 0} w_i, \mathcal{C}[\![\text{C}]\!]e\,s \text{ else}\right)s$$

So, $\bigcup_{i\geq 0} w_i$ is a solution—a *fixed point*—of the circular definition, and in fact it is the smallest function that makes the equality hold. Therefore, it is the *least fixed point*.

* Several important technical details have been glossed over. First, pairs of the form $(s,\bot)$ are ignored when the union of the graphs is performed. Second, for all $i \geq 0$, the graph of w_i is a subset of the graph of w_{i+1}; this ensures the union of the graphs is a function.

Typically, the denotational semantics of the while-loop is presented by the circular definition, and the claim is then made that the circular definition stands for the least fixed point. This is called *fixed-point semantics*. We have omitted many technical details regarding fixed-point semantics; these are available in several texts [15,45,47,51].

69.3.5 Natural Semantics of the Language

We can compare the denotational semantics of the imperative language with a natural semantics formulation. The semantics of several constructions appear in Figure 69.5.

A command configuration has the form $e, s \vdash \mathrm{C} \Rightarrow s'$, where e and s are the "inputs" to command C and s' is the "output." To understand the inference rules, read them "bottom up." For example, the rule for I:=E says, given the inputs e and s, one must first find the location, l, bound to I and then calculate the output, n, for E. Finally, l and n are used to update s, producing the output.

The rules are denotational-like, but differences arise in several key constructions. First, the semantics of a procedure declaration binds I not to a function but to an environment-command pair called a *closure*. When procedure I is called, the closure is disassembled, and its text and environment are executed. Since a natural semantics does not use function arguments, it is called a *first-order semantics*. (Denotational semantics is sometimes called a *higher-order semantics*.)

Second, the while-loop rules are circular. The second rule states, in order to derive a while-loop computation that terminates in s'', one must derive (1) the test, E is true, (2) the body, C, outputs s', and (3) using e and s', one can derive a terminating while-loop computation that outputs s''. The rule makes one feel that the while-loop is "running backwards" from its termination to its starting point, but a complete derivation, like the one shown in Figure 69.6, shows that the iterations of the loop can be read from the root to the leaves of the derivation tree.

$$e \vdash \texttt{proc I = C} \Rightarrow \mathit{bind}(\mathrm{I}, (e, \mathrm{C}), e) \qquad \frac{e \vdash \mathrm{D} \Rightarrow e' \quad e', s \vdash \mathrm{C} \Rightarrow s'}{e, s \vdash \texttt{begin D in C end} \Rightarrow s'}$$

$$\frac{l = \mathit{find}(\mathrm{V}, e) \quad e, s \vdash \mathrm{E} \Rightarrow n}{e, s \vdash \mathrm{V} : \mathrm{E} \Rightarrow \mathit{update}(l, n, s)} \qquad \frac{e, s \vdash \mathrm{C}_1 \Rightarrow s' \quad e, s' \vdash \mathrm{C}_2 \Rightarrow s''}{e, s \vdash \mathrm{C}_1;\ \mathrm{C}_2 \Rightarrow s''}$$

$$\frac{(e', \mathrm{C}') = \mathit{find}(\mathrm{I}, e) \quad e', s \vdash \mathrm{C}' \Rightarrow s'}{e, s \vdash \texttt{call I} \Rightarrow s'} \qquad \frac{e, s \vdash \mathrm{E} \Rightarrow \mathit{false}}{e, s \vdash \texttt{while E do C od} \Rightarrow s}$$

$$\frac{e, s \vdash \mathrm{E} \Rightarrow \mathit{true} \quad e, s \vdash \mathrm{C} \Rightarrow s' \quad e, s' \vdash \texttt{while E do C od} \Rightarrow s''}{e, s \vdash \texttt{while E do C od} \Rightarrow s''}$$

FIGURE 69.5 Natural semantics.

$$\begin{aligned} \text{let } e_0 &= (\mathrm{X}, 1) :: (\mathrm{Y}, 2) :: (\mathrm{Z}, 3) :: \mathit{nil} \\ s_0 &= \langle 2, 0, 0 \rangle, \quad s_1 = \langle 2, 1, 0 \rangle \\ \mathrm{E}_0 &= \texttt{Y not=1}, \quad \mathrm{C}_0 = \texttt{Y: = Y + 1} \\ \mathrm{C}_{00} &= \texttt{while } \mathrm{E}_0 \texttt{ do } \mathrm{C}_0 \texttt{ od} \end{aligned}$$

$$\frac{e_0, s_0 \vdash \mathrm{E}_0 \Rightarrow \mathit{true} \quad \dfrac{2 = \mathit{find}(\mathrm{Y}, e_0) \quad e_0, s_0 \vdash \mathrm{Y} + 1 \Rightarrow 1}{e_0, s_0 \vdash \mathrm{C}_0 \Rightarrow s_1} \quad \dfrac{e_0, s_1 \vdash \mathrm{E}_0 \Rightarrow \mathit{false}}{e_0, s_1 \vdash \mathrm{C}_{00} \Rightarrow s_1}}{e_0, s_0 \vdash \mathrm{C}_{00} \Rightarrow s_1}$$

FIGURE 69.6 Natural semantics derivation.

One important aspect of the natural semantics definition is that derivations can be drawn only for terminating computations. A nonterminating computation is equated with no computation at all.

69.3.6 Operational Semantics of the Language

A fragment of the structural operational semantics of the imperative language is presented in Figure 69.7. For expressions, a computation step takes the form $e \vdash \langle \mathrm{E}, s\rangle \Rightarrow \mathrm{E}'$, where e is the environment, E is the expression that is evaluated, s is the current store, and E′ is E rewritten. In the case of a command, C, a step appears $e \vdash \langle \mathrm{C}, s\rangle \Rightarrow \langle \mathrm{C}', s'\rangle$, because computation on C might also update the store. If the computation step on C "uses up" the command, the step appears $e \vdash \langle \mathrm{C}, s\rangle \Rightarrow s'$.

The rules in the figure are more tedious than those for a natural semantics, because the individual computation steps must be defined, and the order in which the steps are undertaken must also be defined. This complicates the rules for command composition, for example. On the other hand, the rewriting rule for the while-loop merely decodes the loop as a conditional command.

The rules for procedure call are awkward; as with the natural semantics, a procedure, I, is represented as a closure of the form (e', C'). Since C′ must execute with environment, e', which is different from the environment that exists where procedure I is called, the rewriting step for `call` I must retain *two* environments; a new construct, `use` e' `in` C′, remembers that C′ must use e' (and not e). A similar trick is used in `begin` D `in` C `end`.

Unlike a natural semantics definition, a computation can be written for a nonterminating program; the computation is a state sequence of countably infinite length.

69.3.7 Axiomatic Semantics of the Language

An axiomatic semantics computes upon *properties* of stores because programs are understood as "knowledge transformers." For example, the property $\mathrm{X} = 3 \wedge \mathrm{Y} > \mathrm{X}$ asserts knowledge about the values of `X` and `Y` (namely, the store holds 3 at X's location and a number in Y's location that is even larger). The meaning of a command, C, is stated in terms of configurations of form, $\{P\}\ \mathrm{C}\ \{Q\}$, stating that, if predicate (knowledge) P holds true then C generates knowledge Q upon termination (if C does indeed terminate). For example, all of these configurations are true assertions of the command, `X:=X+1`: (1) $\{\mathrm{X} > 2\}$ `X:=X+1` $\{\mathrm{X} > 3\}$; (2) $\{\mathrm{Y} > \mathrm{X}\}$ `X:=X+1` $\{\mathrm{Y} > \mathrm{X} - 1\}$; (3) $\{\mathrm{X} = 3 \wedge \mathrm{Y} > \mathrm{X}\}$ `X:=X+1` $\{\mathrm{X} = 4 \wedge \mathrm{Y} > \mathrm{X} - 1\}$.

Figure 69.8 displays the rules for the primary command constructions. The rule for I: = E states that an assignment transforms knowledge about E into knowledge about E's "new name," I. (To understand

$$e \vdash \langle n_1 + n_2, s\rangle \Rightarrow n_3 \text{ where } n_3 \text{ is the sum of } n_1 \text{ and } n_2$$

$$\frac{e \vdash \langle \mathrm{E}, s\rangle \Rightarrow \mathrm{E}'}{e \vdash \langle \mathrm{I}{:=}\mathrm{E}, s\rangle \Rightarrow \langle \mathrm{I}{:=}\mathrm{E}', s\rangle}$$

$$e \vdash \langle \mathrm{I}{:=}n, s\rangle \Rightarrow \mathit{update}(l, n, s) \text{ where } \mathit{find}(\mathrm{I}, e) = l$$

$$\frac{e \vdash \langle \mathrm{C}_1, s\rangle \Rightarrow \langle \mathrm{C}_1', s'\rangle}{e \vdash \langle \mathrm{C}_1;\ \mathrm{C}_2,\ s\rangle \Rightarrow \langle \mathrm{C}_1';\ \mathrm{C}_2, s'\rangle} \qquad \frac{e \vdash \langle \mathrm{C}_1, s\rangle \Rightarrow s'}{e \vdash \langle \mathrm{C}_1;\ \mathrm{C}_2, s\rangle \Rightarrow \langle \mathrm{C}_2, s'\rangle}$$

$$e \vdash \langle \texttt{while}\ \mathrm{E}\ \texttt{do}\ \mathrm{C}\ \texttt{od}, s\rangle \Rightarrow \langle \texttt{if}\ \mathrm{E}\ \texttt{then}\ \mathrm{C};\ \texttt{while}\ \mathrm{E}\ \texttt{do}\ \mathrm{C}\ \texttt{od else skip fi}, s\rangle$$

$$e \vdash \langle \texttt{call}\ \mathrm{I}, s\rangle \Rightarrow \langle \texttt{use}\ e'\ \texttt{in}\ \mathrm{C}', s\rangle \text{ where } \mathit{find}(\mathrm{I}, e) = (e', \mathrm{C}')$$

$$\frac{e' \vdash \langle \mathrm{C}, s\rangle \Rightarrow \langle \mathrm{C}', s'\rangle}{e \vdash \langle \texttt{use}\ e'\texttt{in}\ \mathrm{C}, s\rangle \Rightarrow \langle \texttt{use}\ e'\texttt{in}\ \mathrm{C}', s'\rangle} \qquad \frac{e' \vdash \langle \mathrm{C}, s\rangle \Rightarrow s'}{e \vdash \langle \texttt{use}\ e'\texttt{in}\ \mathrm{C}, s\rangle \Rightarrow s'}$$

$$e \vdash \texttt{proc}\ \mathrm{I} = \mathrm{C} \Rightarrow \mathit{bind}(\mathrm{I}, (e, \mathrm{C}), e)$$

$$\frac{e \vdash \mathrm{D} \Rightarrow e'}{e \vdash \langle \texttt{begin}\ \mathrm{D}\ \texttt{in}\ \mathrm{C}\ \texttt{end}, s\rangle \Rightarrow \langle \texttt{use}\ e'\ \texttt{in}\ \mathrm{C}, s\rangle}$$

FIGURE 69.7 Structural operational semantics.

$$\{[E/I]P\}\ I{:=}E\ \{P\}$$

$$\frac{P \supset P' \quad \{P'\}\ C\ \{Q'\} \quad Q' \supset Q}{\{P\}\ C\ \{Q\}} \qquad \frac{\{P\}\ C_1\ \{Q\} \quad \{Q\}\ C_2\ \{R\}}{\{P\}\ C_1; C_2\ \{R\}}$$

$$\frac{\{P \wedge E\}\ C_1\ \{Q\} \quad \{P \wedge \neg E\}\ C_2\ \{Q\}}{\{P\}\ \texttt{if}\ E\ \texttt{then}\ C_1\ \texttt{else}\ C_2\ \texttt{fi}\ \{Q\}} \qquad \frac{\{P \wedge E\}\ C\ \{P\}}{\{P\}\texttt{while}\ E\ \texttt{do}\ C\ \texttt{od}\{P \wedge \neg E\}}$$

FIGURE 69.8 Axiomatic semantics inference rules.

this, recall that $[E/I]P$ stands for the substitution of E for all free occurrences of I in P. For example, $[\texttt{Y}+\texttt{1}/\texttt{X}](\texttt{X}>3)$ is $\texttt{Y}+\texttt{1}>3$.) We can use the rule to prove that `X:=X+1` transforms $\texttt{X}+\texttt{1}>3$ into $\texttt{X}>3$, that is, $\{\texttt{X}+\texttt{1}>3\}$ `X:=X+1` $\{\texttt{X}>3\}$. Again, I:=E transforms knowledge about E into knowledge about E's "new name," I.

The second rule in the figure uses logical implication to deduce new knowledge; we deduce that $\{\texttt{X} > 2\}$ `X: = X + 1` $\{\texttt{X} > 3\}$, because $\texttt{X} + \texttt{1} > 3$ implies $\texttt{X} > 2$. The properties of command composition are defined by the third rule: knowlege generated by command C_1 passes to command C_2.

The fourth rule, for the if-command, shows how to make a property hold upon termination regardless of which arm of the conditional is evaluated. For example, the rule proves $\{\texttt{X} \neq 0\}$ `if` $\texttt{X}<0$ `then Y:= - X else Y:=X: fi` $\{\texttt{Y} > 0\}$, because knowledge from the test, $\texttt{X} < 0$, lets us prove both $\{\texttt{X} < 0 \wedge \texttt{X} \neq 0\}$ `Y:= - X` $\{\texttt{Y} > 0\}$ and also $\{\neg(\texttt{X} < 0) \wedge \texttt{X} \neq 0\}$ `Y:=X` $\{\texttt{Y} > 0\}$.

The most fascinating rule is the last one, for `while E do C od`. The loop's body, C, is "repeatable code," which means whatever knowledge, I, is required to *use* C must be regenerated to *reuse* C — $\{E \wedge I\}$ C $\{I\}$! I is called an *invariant* because it remains true for all the repetitions of the loop. We conclude that I holds true if and when the loop terminates.

Another viewpoint is that a loop is kind of "game" played in rounds. If C lists the moves taken in one round, then the invariant, I, is the strategy one uses to "keep the lead" and eventually win. The game ends when the loop's test, E, goes false.

Here is an example: We play a pebble game, where pebbles can be moved only one at a time. There are two boxes of pebbles, named `X` and `Y`. To win the game, all the pebbles must be moved into `Y`'s box without losing any. Here is how we might play one round of the game: `X:= X - 1; Y:= Y + 1`. Our strategy is encoded by the invariant, `TOTAL = X + Y`, which says that we do not lose any pebbles when we move them:

$$\{\text{TOTAL} = \text{X} + \text{Y}\}\text{X} := \text{X} - 1; \text{Y} := \text{Y} + 1\ \{\text{TOTAL} = \text{X} + \text{Y}\}$$

The completed game is this loop program, `while X not=0 do X:=X-1; Y:= Y+1 od`, and the proof that the loop generates a victory is listed in Figure 69.9, which shows at the end that `TOTAL= Y`, that is, all the pebbles rest in `Y`'s box. The derivation is difficult to read, but when reformatted vertically, as seen in the right half of the figure, it shows clearly how the input knowledge, `TOTAL=X + Y`, is transformed into the output knowledge, `TOTAL=Y`.

The key to building the proof is determining a loop invariant. Since the goal was to prove `TOTAL = Y`, this made `TOTAL = X + Y` useful, because `X:= X - 1` eventually lowers `X` to zero.* With the invariant in hand, it is easy to work backwards, from the end of the loop body, to the top, to calculate that the loop body is indeed "repeatable code": The rule for assignment proves both $\{\texttt{TOTAL} = \texttt{X} + (\texttt{Y} + 1)\}$ `Y: = Y + 1` $\{\texttt{TOTAL} = \texttt{X} + \texttt{Y}\}$, and also $\{\texttt{TOTAL} = (\texttt{X} - 1) + (\texttt{Y} + 1)\}$ `X: = X - 1` $\{\texttt{TOTAL} = \texttt{X} + (\texttt{Y} + 1)\}$. Finally, $\texttt{TOTAL} = (\texttt{X} - 1) + (\texttt{Y} + 1)$ implies `TOTAL = X + Y`.

* The loop terminates only if `X`'s value begins as a nonnegative. The laws in Figure 69.8 *do not guarantee termination*. This requires additional logical machinery.

let P_0 be $\texttt{TOTAL = X + Y}$

P_1 be $\texttt{TOTAL = X + (Y + 1)}$, P_2 be $\texttt{TOTAL = (X - 1) + (Y + 1)}$

$E_0 = \texttt{X not = 0}$, $C_0 = \texttt{X:= X - 1; Y:=Y + 1}$

$$\dfrac{\dfrac{(P_0 \wedge E_0) \supset P_2 \quad \dfrac{\{P_2\}\texttt{X:=X-1}\{P_1\} \quad \{P_1\}\texttt{Y:=Y+1}\{P_0\}}{\{P_2\}\ C_0\ \{P_0\}} \quad P_0 \supset P_0}{\{P_0 \wedge E_0\}\ C_0\ \{P_0\}}}{\{P_0\}\texttt{while}\,E_0\ \texttt{do}\ C_0\ \texttt{od}\ \{P0 \wedge \neg E_0\}}$$

```
assert {TOTAL = X+Y}
while X not= 0 do
  invariant {TOTAL = X+Y}
  X:= X - 1
  assert {TOTAL = X+(Y+1)}
  Y:= Y + 1
  assert {TOTAL = X+Y}
od
assert {TOTAL = X+Y and
      not (X not= 0)}
implies {TOTAL = Y}
```

FIGURE 69.9 Axiomatic semantics derivation.

There are three ways of stating the semantics of a command in an axiomatic semantics:

1. Relational semantics: The meaning of C is the set of *P*, *Q* pairs for which {*P*} C {*Q*} holds. Termination is not demanded of C. (This is called "partial correctness.")
2. Postcondition semantics: The meaning of C is a function from an input predicate to an output predicate. We write *slp*(*P*, C)=*Q*; this means that {*P*} C {*Q*} holds, and for all *Q′* such that {*P*} C {*Q′*} holds, it is the case that *Q* implies *Q′*. This is also called *strongest liberal postcondition semantics*. When termination ("total correctness") is demanded also of C, the name becomes *strongest postcondition semantics*.
3. Precondition semantics: The meaning of C is a function from an output predicate to an input predicate. We write *wlp*(C, *Q*)=*P*; this means that {*P*} C {*Q*} holds, and for all *P′* such that {*P′*} C {*Q*} holds, it is the case that *P′* implies *P*. This is also called *weakest liberal precondition semantics*. When termination is demanded also of C, the name becomes *weakest precondition semantics*.

69.4 Practical Impact

Research on programming-language semantics showed how a language can be defined precisely, its correctness properties proved, and its implementation realized. Indeed, the area of *formal methods* (cf. Chapter 116) is an adaptation of the semantics methods from computer languages to computer programs—how a program can be specified precisely, its correctness properties proved, and its specification implemented.

Operational semantics, having the longest life of the semantic approaches, is most entrenched within software development. From operational semantics came virtual machines to which languages can be translated. Definitional interpreters are now the norm for prototyping new languages, especially "little" or "domain-specific" languages, which are designed for problem solving in a limited application domain, for example, file-linking (Make), web-page layout (HTML), spreadsheet layout (Excel), or linear algebra (MATLAB). Scripting languages (see Chapter 110) are also implemented in definitional-interpreter style.

Axiomatic semantics has undergone a significant revival, not only as the starting point for specification writing and verification in formal-methods work, but as the language for description and analysis of secure and safety-critical software systems, where correctness properties, above all, must be stated so that there can validation or justification, whether by testing, monitoring, or theorem proving. Indeed, the object control language (OCL) used in the Unified Modelling Language for software blueprinting is a derivative of the classic axiomatic semantics notation. (See Volume 2 of this handbook.) A wide variety of *proof assistants* and *logical frameworks* are available for helping a designer state and validate crucial properties of software, for example, ACL2 [24], PVS [38], Isabelle/HOL [35], and Coq [41].

The primary impact of denotational semantics has been to the design of modern programmming languages. The modelling of programming constructions as functions that operate on storage motivated many designers to add such functions as primitive constructions to the programming language itself—the results are (1) *higher-order* constructions, such as the anonymous functions in ML and Java; (2) generic and *polymorphic* constructions, such as class templates in C++ and Scala; and (3) *reflective constructions*, like that found in JavaBeans and 3-Lisp. The importance of polymorphic constructions to modern-day software assembly has stimulated extensions of denotational semantics into "higher-order" formats that are expressed within *type theory* [39,36].

Within the subarea of programming-languages research, the semantics methods play a central role. Language designers use semantics definitions to formalize their creations, as was done during the development of Ada [10] and after development for Scheme [43] and Standard ML [27]. Software tools are readily available to prototype a syntax-plus-semantics definition into an implementation [5,8,26,31].

A major success of formal semantics is the analysis and synthesis of data-flow analysis and type-inference algorithms from semantics definitions. This subject area, called *abstract interpretation* [3,7,33], uses algebra techniques to map semantic definitions from their "concrete" (execution) definition to their "abstract" (homomorphic property) definition. The technique can be used to extract properties from the definitions, generate data flow and type inference, and prove program correctness or code-improvement transformations. Most automated methods for program validation use abstract interpretation.

69.5 Research Issues

The techniques in this chapter have proved successful for defining, improving, and implementing sequential programming languages. But new language and software paradigms present new challenges to the semantics methods.

Challenging issues arise in the object-oriented programming paradigm. Not only can objects be arguments ("messages") to other objects' procedures ("methods"), but coercion laws based on *inheritance* allow controlled mismatches between arguments and parameters. For example, an object argument that contains methods for addition, subtraction, and multiplication is bound to a method's parameter that expects an object with just addition and subtraction methods. Carelessly defined coercions lead to unsound programs, so semantics definitions must be extended with inheritance hierarchies [16,40]. See Chapter 107 for details.

Yet another challenging topic is defining communication as it arises in distributed programming, where multiple processes (threads of execution) synchronize through communication. Structural operational semantics has been adapted to formalize such systems. More importantly, the protocol used by a system lies at the center of the system's semantics; semantics approaches based on *process algebra* [12,20,28] have been developed that express a protocol as a family of simultaneous algebra equations that must be "solved" to yield a convergent solution. See Chapter 12.

A current challenge is applying semantics methods to define and explain programs that exploit the architecture of multicore processors, where subcomputations and their storage requirements must be allocated to processor cores and their associated cache hierarchy [18]. See Chapter 34 for background.

Finally, a longstanding crucial topic is the relationship between different forms of semantic definitions: If one has, say, both a denotational semantics and an axiomatic semantics for the same programming language, in what sense do the semantics agree? Agreement is crucial, since a programmer might use the axiomatic semantics to reason about the properties of programs, whereas a compiler writer might use the denotational semantics to implement the language. This question generalizes to the problem of *refinement* of software specifications—see Chapter 116. In mathematical logic, one uses the concepts of *soundness* and *completeness* to relate a logic's proof system to its interpretation, and in semantics there are similar notions of *soundness* and *adequacy* to relate one semantics to another [15,37]. Important as these properties are, they are quite difficult to realize in practice [1].

Key Terms

Axiomatic semantics: The meaning of a program as a property or specification in logic.

Denotational semantics: The meaning of a program as a compositional definition of a mathematical function from the program's input data to its output data.

Fixed-point semantics: A denotational semantics where the meaning of a repetitive structure, like a loop or recursive procedure, is expressed as the smallest mathematical function that satisfies a recursively defined equation.

Loop invariant: In axiomatic semantics, a logical property of a while-loop that holds true no matter how many iterations the loop executes.

Natural semantics: A hybrid of operational and denotational semantics that shows computation steps performed in a compositional manner. Also known as a "big-step semantics."

Operational semantics: The meaning of a program as calculation of a trace of its computation steps on input data.

Strongest postcondition semantics: A variant of axiomatic semantics where a program and an input property are mapped to the strongest proposition that holds true of the program's output.

Structural operational semantics: A variant of operational semantics where computation steps are performed only within prespecified contexts. Also known as a "small-step semantics."

Weakest precondition semantics: A variant of axiomatic semantics where a program and an output property are mapped to the weakest proposition that is necessary of the program's input to make the output property hold true.

Further Information

The best starting point for further reading is the comparative semantics text of Nielson and Nielson [34], which thoroughly develops the topics in this chapter. See also the texts by Slonneger and Kurtz [46] and Watt [50].

Operational semantics has a long history; modern introductions are Hennessey's text [17] and Plotkin's report on structural operational semantics [42]. The principles of natural semantics are documented by Kahn [23]. A good example of the interpreter-based approach to semantics is the text by Friedman et al. [13].

Mosses's paper [32] is a useful introduction to denotational semantics; textbook-length treatments include those by Schmidt [45], Stoy [47], Tennent [48], and Winskel [51]. Gunter's text [15] uses denotational-semantics-based mathematics to compare several of the semantics approaches. Pierce [39] and Mitchell [29] provide foundational supporting material.

Of the many textbooks on axiomatic semantics, one might start with books by Dromey [11] or Gries [14]; both emphasize precondition semantics, which is most effective at deriving correct code. Apt's paper [2] is an excellent description of the formal properties of relational semantics, and Dikstra's text [9] is the standard reference on precondition semantics. Hoare's landmark papers on relational semantics [19,21] are worth reading as well. Many texts have been written on the application of axiomatic semantics to systems development; two samples are by Jones [22] and Morgan [30].

You are also urged to read the other chapters in Section 9 of this Volume.

References

1. S. Abramsky, R. Jagadeesan, and P. Malacaria. Full abstraction for PCF. In *Proceedings of Theoretical Aspects of Computer Software, TACS94*, Sendai, Japan, Lecture Notes in Computer Science 789, pp. 1–15. Springer, New York, 1994.
2. K. Apt. Ten years of Hoare's logic: A survey–part I. *ACM Trans. Programming Languages and Systems*, 3:431–484, 1981.

3. B. Blanchet et al. A static analyzer for large safety-critical software. In *Proceedings of Programming Language Design and Implementation*, San Diego, CA, pp. 196–207. ACM Press, 2003.
4. H. Barendregt. *The Lambda Calculus: Its Syntax and Semantics*. North Holland, Amsterdam, the Netherlands 1984.
5. D. F. Brown, H. Moura, and D. A. Watt. Actress: An action semantics directed compiler generator. In *CC'92, Proceedings of the 4th International Conference on Compiler Construction*, Paderborn, Germany, Lecture Notes in Computer Science 641, pp. 95–109. Springer-Verlag, Berlin, Germany, 1992.
6. M. Clavel et al. *All About Maude—A High-Performance Logical Framework*. Springer-Verlag, Berlin, Germany 2007.
7. P. Cousot and R. Cousot. Abstract interpretation: A unified lattice model for static analysis of programs. In *Proc. 4th ACM Symp. on Principles of Programming Languages*, Los Angeles, CA, pp. 238–252. ACM Press, 1977.
8. Th. Despeyroux. Executable specification of static semantics. In G. Kahn, D.B. MacQueen, and G. Plotkin, eds., *Semantics of Data Types*, pp. 215–234. Lecture Notes in Computer Science 173, Springer-Verlag, New York, 1984.
9. E.W. Dijkstra. *A Discipline of Programming*. Prentice Hall, Englewood Cliffs, NJ, 1976.
10. V. Donzeau-Gouge. On the formal description of Ada. In N.D. Jones, ed., *Semantics-Directed Compiler Generation*. Lecture Notes in Computer Science 94, Springer-Verlag, New York, 1980.
11. G. Dromey. *Program Derivation*. Addison-Wesley, Sydney, New South Wales, Australia, 1989.
12. J. Fokkink. *Introduction to Process Algebra*. Springer-Verlag, Berlin, Germany, 2010.
13. D. Friedman, M. Wand, and C. Haynes. *Essentials of Programming Languages*, 2nd edn. The MIT Press, Cambridge, MA, 2001.
14. D. Gries. *The Science of Programming*. Springer, New York, 1981.
15. C. Gunter. *Semantics of Programming Languages*. MIT Press, Cambridge, MA, 1992.
16. C. Gunter and J. Mitchell, eds. *Theoretical Aspects of Object-Oriented Programming*. The MIT Press, Cambridge, MA, 1994.
17. M. Hennessy. *The Semantics of Programming Languages: An Elementary Introduction Using Structured Operational Semantics*. Wiley, New York, 1991.
18. M. Herlihy and N. Shavit. *Art of Multiprocessor Programming*. Morgan Kaufmann, San Francisco, CA, 2008.
19. C.A.R. Hoare. An axiomatic basis for computer programming. *Communications of the ACM*, 12:576–580, 1969.
20. C.A.R. Hoare. *Communicating Sequential Processes*. Prentice Hall, London, U.K., 1985.
21. C.A.R. Hoare and N. Wirth. An axiomatic definition of the programming language Pascal. *Acta Informatica*, 2:335–355, 1973.
22. C.B. Jones. *Software Development: A Rigorous Approach*. Prentice Hall, London, U.K., 1980.
23. G. Kahn. Natural semantics. In *Proc. STACS '87*, pp. 22–39. Lecture Notes in Computer Science 247, Springer, Berlin, Germany, 1987.
24. M. Kaufmann, P. Manolios, and J.S. Moore. *Computer-Aided Reasoning: ACL2 Case Studies*. Kluwer, Boston, MA, 2000.
25. J.W. Klop. Term rewriting systems. In T. Maibaum S. Abramsky, D. Gabbay, eds., *Handbook of Logic in Computer Science*, Vol. 2, pp. 2–117. Oxford University Press, Oxford, U.K., 1992.
26. P. Lee. *Realistic Compiler Generation*. The MIT Press, Cambridge, MA, 1989.
27. R. Milner, M. Tofte, and R. Harper. *The Definition of Standard ML*. The MIT Press, Cambridge, MA, 1990.
28. R. Milner. *Communication and Concurrency*. Prentice Hall, London, U.K., 1989.
29. J.C. Mitchell. *Foundations for Programming Languages*. The MIT Press, Cambridge, MA, 1996.
30. C. Morgan. *Programming from Specifications*, 2nd. edn. Prentice Hall, London, U.K., 1994.

31. P.D. Mosses. Compiler generation using denotational semantics. In A. Mazurkiewicz, ed., *Mathematical Foundations of Computer Science*, Lecture Notes in Computer Science 45, pp. 436–441. Springer, Berlin, Germany, 1976.
32. P.D. Mosses. Denotational semantics. In J. van Leeuwen, ed., *Handbook of Theoretical Computer Science*, Vol. B, Chapter 11, pp. 575–632. Elsevier, Amsterdam, the Netherlands, 1990.
33. S. Muchnick and N.D. Jones, eds. *Program Flow Analysis: Theory and Applications*. Prentice Hall, Englewood Cliffs, NJ, 1981.
34. H. Riis Nielson and F. Nielson. *Semantics with Applications, a Formal Introduction*. Wiley Professional Computing. John Wiley & Sons, Chichester, U.K., 1992.
35. T. Nipkow, L. Paulson, and M. Wenzel. *Isabelle/HOL: A Proof Assistant for Higher-Order Logic*. Springer-Verlag, New York, 2002.
36. B. Nordström, K. Petersson, and J. Smith. *Programming in Martin-Löf's Type Theory*. Oxford University Press, Oxford, U.K., 1990.
37. C.H.-L. Ong. Correspondence between operational and denotational semantics. In S. Abramsky, D. Gabbay, and T. Maibaum, eds., *Handbook of Computer Science, Vol. 4*. Oxford University Press, Oxford, U.K., 1995.
38. S. Owre et al. PVS: Combining specification, proof checking, and model checking. In *Proc. CAV '87*, pp. 411–414. Springer-Verlag, New York, 1996.
39. B. Pierce. Formal models/calculi of programming languages. In A. Tucker, ed., *CRC Handbook of Computer Science and Engineering*. CRC Press, Boca Raton, FL, 1996.
40. B. Pierce. *Types and Programming Languages*. The MIT Press, Cambridge, MA, 2002.
41. B. Pierce et al. Software Foundations. Technical Report http://www.cis.upenn.edu/~bcpierce/sf, University of Pennsylvania, Philadelphia, PA, 2011.
42. G.D. Plotkin. A structural approach to operational semantics. Technical Report FN-19, DAIMI, Aarhus, Denmark, September 1981.
43. J. Rees and W. Clinger. Revised3 report on the algorithmic language Scheme. *SIGPLAN Notices*, 21:37–79, 1986.
44. J. Reynolds. Definitional interpreters for higher-order programming languages. *Journal of Higher-Order and Symbolic Computation*, 11:363–397, 1998.
45. D.A. Schmidt. *Denotational Semantics: A Methodology for Language Development*. Allyn and Bacon, Inc., 1986.
46. K. Slonneger and B. Kurtz. *Formal Syntax and Semantics of Programming Languages: A Laboratory-Based Approach*. Addison-Wesley, Reading, MA, 1995.
47. J.E. Stoy. *Denotational Semantics*. The MIT Press, Cambridge, MA, 1977.
48. R.D. Tennent. *Semantics of Programming Languages*. Prentice Hall International, Englewood Cliffs, NJ, 1991.
49. D. van Dalen. *Logic and Structure, 3rd ed.* Springer, Berlin, Germany, 1994.
50. D.A. Watt. *Programming Language Syntax and Semantics*. Prentice Hall International, Englewood Cliffs, NJ, 1991.
51. G. Winskel. *Formal Semantics of Programming Languages*. The MIT Press, Cambridge, MA, 1993.

70 Type Systems

Stephanie Weirich
University of Pennsylvania

70.1 Introduction

Many modern programming languages come equipped with *type systems*, which ensure that programs are free from certain classes of execution errors. For example, type systems prevent buffer overruns, invalid code execution, and memory accesses to protected data. In general, type systems describe invariants that all programs must satisfy. They are a valuable component for program development, assisting in the creation of correct, secure, and robust systems.

In this chapter, we introduce research in the design of *safe* programming languages with *static type systems*. In such languages, a tool known as a *type checker* verifies that a program *type checks* before it may be compiled or interpreted. Languages are *safe* when they ensure that no untrapped errors are possible at runtime. Safety is guaranteed in statically typed languages by a *type soundness* theorem that ensures that all *well-typed* programs do not "go wrong." Languages can be statically typed but unsafe when this theorem does not hold, that is, when the process of type checking does not provide firm guarantees about execution. Languages can be safe but dynamically typed when they rule out errors by runtime monitoring instead of compiletime type checking.

The development of safe, statically typed languages is an area of active research. There are many different ways to type programs. The analysis required to claim type soundness requires precision and close attention to detail. Language designers must balance the expressiveness of the type system (which allows it to certify more safe programs) with the complexity of the type soundness proof (any flaw in which could invalidate the entire enterprise).

This chapter provides an overview of that field and examines the following questions: Why do languages check types during compilation? What are the components of a type system? How do language researchers describe and formalize type systems? What are their important properties? What are the limitations of static type checking? What are some of the current areas of type systems research?

This chapter is intended for *students* and *researchers* as an entry point to the study of type systems. Because of the richness of this field and the variety in the design of languages, the goal of this chapter is breadth, not depth. As a result, only the most basic type systems receive the most rigorous mathematical treatment—many topics are covered at an overview level, leaving full discussion to referenced sources.

This chapter is also aimed at *programmers* and *language designers*. Type systems are powerful tools for software development, providing assistance in the challenging task of developing and maintaining correct systems. A secondary goal of this chapter is an appreciation of this power and an understanding of how it may be put to use.

Because the mathematical models that we present here are most similar to the type systems of functional programming languages, this chapter assumes basic experience with languages such as Haskell and ML and draws on examples from those languages. Furthermore, it requires some background in mathematical reasoning, including the knowledge of functions, sets and relations, bound and free variables, the presentation of programming language abstract syntax, and the principle of mathematical induction.

The remainder of this section introduces static type systems by motivating their inclusion in the definitions of programming languages. It is followed by a discussion of what makes a good static type system through a listing of evaluation properties for their design. This description of the concepts and ideas behind static typing is deliberately kept at an informal level.

Section 70.2 introduces the formal notation and mathematical concepts necessary for working with and analyzing type systems. The key components of a type system definition are *judgments*, which assert relationships such as "this program has that type," *derivations*, which provide evidence for the truth of those assertions, and *typing rules*, which are the building blocks of derivations. We introduce the components of modern type systems gradually, building up the vocabulary of structures and relationships piece-by-piece. Section 70.3 uses a core type system, called System F, to demonstrate mathematical reasoning about type systems. This section walks through the type soundness theorem and also discusses the principle of parametricity. Section 70.4 concludes the technical discussion with an overview of type inference, the problem of determining whether a program should type check given limited type annotations. Section 70.5 concludes with references for future study and a list of defining terms.

70.1.1 What Are Type Systems Good For?

A *safe* language is one that does not admit untrapped errors. For example, a buffer overrun in C might not be detected at runtime. If one occurs, the behavior of the program is unspecified. Untrapped errors are an insidious form of execution error because they go unnoticed. As a result, they are a significant source of security vulnerabilities in computer systems. In contrast, trapped errors are permitted in safe languages. For example, most safe languages check that array accesses are in bounds at run time and stop the current computation immediately if the check fails.

A *dynamically typed language* is one that uses only runtime checks to ensure safety. These languages are safe even though they do not have a type system or employ a type checker. The advantage of these languages is simplicity and expressiveness. They have a gentle learning curve because there is no type system that new programmers must learn. Furthermore, they allow some programming patterns that statically typed languages reject because static type checking must approximate program execution and sometimes rules out code that does not trigger run-time failures.

There has been much debate over whether languages should be safe, and if they are, whether they should be dynamically or statically typed. Ultimately, these debates balance trade-offs between expressiveness, performance, development time, maintenance time, and security. Although the choice of implementation language is often determined by economic and social factors, the technical arguments in favor of statically typed languages include the following points.

- *Eliminating bugs and execution errors at compile time*
  The primary purpose of static type systems is compile-time certified safety. Type systems rule out unsafe code by assigning a type to every form of data value and operation in the language, and then ensuring that operations are only applied to the right types of values. In this sense, types are interfaces. They specify the capabilities of data values, and restrict their use to compatible operations.

  For example, boolean values should not be the arguments to numeric operations. A program containing the expression **true** + 3 is sure to be buggy. Furthermore, this bug may not be detected at runtime if after compilation boolean and numeric values have the same machine representation. A sound, static type system can ensure that only numbers are ever added together during the course of execution and detect such bugs when the program is compiled.
- *Making distinctions between values*
  The distinction between boolean and numeric values given earlier is part of the language definition. However, experienced programmers use static type systems to make application-specific data distinctions. For example, a programmer may use integer values as error codes to indicate run-time failures in a system.

0	SUCCESS
1	INVALID_FUNCTION
2	FILE_NOT_FOUND
3	PATH_NOT_FOUND
4	TOO_MANY_OPEN_FILES

  However, not all integer values may have an associated error code. What does error code-41 mean?* What does it mean to add error codes together? When a function or method returns an integer, when should that integer be interpreted as an error code?
  Languages that support enumerations or algebraic datatypes permit users to define new types containing only specified values. That way the type system can prevent error codes from being the argument to arithmetic operations. Functions that return error codes have different types from those that return integers, maintaining this distinction.
- *Information hiding and modularity*
  Static types also provide information hiding through type abstraction. Because types specify the interfaces of data structures, they govern how client code must interact with that data. If that interface includes an abstract type, library implementors are free to make changes in the representation of a data structure as long as the interface is still satisfied.

* See http://xkcd.com/1024/.

For example, a data structure library might export an abstract type of sets of integers, as well as operations to construct and manipulate values of that type. In the OCaml language, this interface could be defined as follows:

```
module type SET =
    type set
    val empty : set
    val insert : int -> int set -> int set
    val contains : int set -> boolean
end
```

This interface does not specify how sets are represented. As a result, the library implementor is free to choose ordered lists, balanced trees, or any other implementation. Furthermore, this implementation can be replaced without invalidating code that uses this interface.

This interface also allows modular reasoning about the correctness of the implementation. Because the type of sets is abstract, the implementor knows that any set that is passed to the `contains` function must have been created by this module. Therefore, preserving invariants about the set implementation requires reasoning only about a single module, not the entire program.

- *Expressive code reuse*

  Although code in dynamically typed languages is often naturally applicable to many situations, *polymorphism* provides a strong mechanism for statically typed languages to create and mark reusable code. A polymorphic function or method is one that works on many different types of arguments—this behavior is safe because the function is required to use those arguments only in ways that do not violate the capabilities specified by their types.

  For example, *subtype polymorphism* allows functions to specify minimum requirements about the capabilities of their arguments. Any argument that provides at least the required functionality may be passed to such a function. Subtyping is common in object-oriented languages, where one object type is a subtype of another if it supports at least the same methods. For example, an object that has both a `reset` and a `move` method may be passed to a function that only calls the `move` method of its argument.

  Similarly, *parametric polymorphism* (also called *generics*) allows the interfaces to functions to include type parameters. For example, the filter function, which preserves only the elements of a list that satisfy a given predicate, can be given a parametric type:

```
val filter : ('a -> bool) -> 'a list -> 'a list
```

  This type indicates that the function can be used with any type of list—the parameter `'a` can be instantiated with any type. However, this type must be instantiated consistently, the supplied predicate must be compatible with the input list, and the type of the resulting list must be same as the input.

  Finally, *ad hoc polymorphism* or *type-directed programming* uses types as metadata to define reusable code. This feature is a form of reflection where execution is determined by type information. A simple, powerful form of type-directed programming is *overloading*—where the types of the arguments to a function determine how the function executes. For example, the Haskell language supports overloading through the use of *type classes*, providing uniform access to operations for serialization, equality, and traversals, as well as algebraic structures such as monoids, functors and monads. Furthermore, Haskell also supports many flavors of *datatype-generic programming* [24], which automates the creation of these overloaded functions by analyzing the structure of datatypes.

- *Code documentation and maintenance*
  Types and interfaces provide documentation for software artifacts. As opposed to comments, they are mechanically checked for correctness, so they can never become out-of-date. Furthermore, in languages that support *type inference*, this documentation can automatically be generated for each program fragment. As such, it is documentation that does not need to be maintained or written, but is available on demand.

  Furthermore, static types assist with code maintenance and refactoring. For example, when programmers change the representation of some data structure, the type system identifies all parts of the program that must be updated with respect to this modification. The programmer need not search by hand nor worry about missing modification sites. Each site will produce a type error when the structure of the data does not match what was provided. Tools such as IDEs can take advantage of type information to support code development and refactoring.

70.1.2 What Are Good Type Systems?

How can we ensure that static type systems live up to their promises? To ensure that type systems provide the benefits discussed earlier, researchers evaluate the quality of type system designs using the following criteria. The following list includes a brief description of each desirable property as well as forward references to discussion of how it may be formally evaluated.

1. Type systems should *eliminate program errors* by ruling out executions that are not consistent with the semantics of the programming language. Formally, the type soundness theorem (Section 70.3.2) provides this assurance.
2. Type systems should be *expressive*, capable of statically describing common data structures and the relationships between program values. Programmers should have a rich vocabulary with which to express the invariants of their code. Type systems should have minimal reliance on runtime checks to ensure safety. Section 70.2 provides an overview of the components of expressive type systems.
3. Type systems should be *modular*. The definitions of values should be checkable independent of their uses and types should provide precise interfaces that govern interaction between different parts of the program. Modular type checking is primitively enabled through the use of typing environments, introduced in Section 70.2.2. Type abstraction, described in Section 70.2.6 enhances this interface by hiding implementation details.
4. Type systems should be *informative* about the semantics of programs. The type soundness property gives some of this information—an expression of type τ evaluates to a value of the same type τ. Going further, Section 70.3.3 uses types as the basis for an equivalence relation between programs, especially in the presence of type abstraction.
5. Type systems should be *effectively checkable* and *predictable*. There should be an algorithm that determines whether a program type checks. Developers should be able to understand why their programs type check and know what program transformations preserve typability. Section 70.2.3 describes the design of annotations that derive simple type checking algorithms. Section 70.4 discusses more sophisticated algorithms that use constraint solving to infer type information that is not explicitly present.

Although we have discussed these properties informally, we can evaluate them rigorously. Such treatment requires a mathematical model of programming languages and their type systems. In the next section, we introduces the notation and tools that are commonly used for this analysis.

70.2 Formal Type Systems

Programming languages are large, complicated artifacts, so it is difficult to design and analyze their type systems wholesale. As a result, programming language researchers have developed tractable mathematical models of languages, and it is the type systems of these formal models that is the topic of study. By defining a model that captures the essence of computation, we can carefully understand its semantics and evaluate its properties.

The language of the formal model of type systems derives from *formal proof systems*. A proof system is a system of *inference rules* that can be used to show that a program has a specific type by constructing a typing *derivation* for a particular typing *judgment*. Learning to understand this language means that type systems can be understood as mathematical specifications, without the need to read the source code of a particular compiler.

Additionally, many type systems specified in this manner are *declarative*, in that they define valid programs without determining a particular type checking algorithm. Having a declarative specification means that formal type systems are part of the language definition: they state what good programs are, in the simplest terms possible, leaving compilers the freedom to implement these specifications in the most appropriate manner.

Furthermore, these formal models provide the basis for the *metatheory* of a programming language—the properties that can be mathematically proven about the language and all its programs. Section 70.3 discusses a few of these properties in detail and describes the mathematical techniques that researchers use for these proofs.

However, the rest of this section discusses what type systems *are* before going into depth about their mathematical properties. It starts with the simplest of languages—arithmetic expressions only—and then introduces new features and type system capabilities incrementally. Understanding these individual features provides insight into their semantics as they occur in the context of full languages.

70.2.1 Arithmetic Expressions

We introduce this mathematical language of type systems with an extremely simple example, that of arithmetic expressions. An arithmetic expression is either a natural number value, such as 0, 1 or 2, a boolean value **true** or **false**, or a binary operator applied to two arguments. For further simplicity, this language only includes operators for addition, subtraction, and equality. Yet, even in this simple language there are expressions, such as 1 + **true**, that the type system should rule out.

The abstract syntax of this language can be described by the following table, presented in Backus-Naur Form (BNF). We use the metavariable n for numeric values, b for boolean values, op for operators, and e for expressions.* We use metavariables consistently—the variable n always stands for a number and e always stands for an expression. If we need to refer to more than one expression at a time, we use primes (e' or e'') and subscripts (e_1 or e_2) to indicate different metavariables.

Numeric values include all natural numbers (not shown in the table but indicated by ...). Boolean values only for equality testing include **true** and **false**. The operator symbols include + for addition – for subtraction and ≡ for equality testing. Finally, we combine these in expressions, which are literal values, an infix operator applied to two arguments, or an **if** expression:

$$n \in \{0, 1, 2, \ldots\}$$

$$b ::= \mathbf{false} \mid \mathbf{true}$$

$$op ::= + \mid - \mid \equiv$$

$$e ::= n \mid b \mid e_1\ op\ e_2 \mid \mathbf{if}\ e_1\ \mathbf{then}\ e_2\ \mathbf{else}\ e_3$$

Note that this grammar does not specify how to parse arithmetic expressions, only how to represent them. In particular, the abstract syntax does not include parentheses, but we often use them in examples for disambiguation. For example, (1 + 2) ≡ 3 is an expression of this language.

* Expressions are also called *terms*.

The syntax of types is defined similarly, again using BNF. All valid expressions in this language have one of two types—they are either numbers or booleans:

$$\tau ::= \mathbf{nat} \mid \mathbf{bool}$$

We can capture the relationship between expressions and their types by defining a logical *judgement*, which asserts that "expression *e* has type τ." This judgment has the following form, starting with a *typing context* Γ and a turnstile* and then putting a colon between the expression and its type:

$$\Gamma \vdash e : \tau$$

Typing contexts, also called *typing environments*, contain assumptions that can be used for type checking. In this simple language, the typing environment plays no role. We include it here to prepare for future extensions (see Section 70.2.2). For now, the syntax of typing contexts includes only the empty context:

$$\Gamma ::= \varepsilon$$

In general, typing judgments often have the form $\Gamma \vdash \mathfrak{J}$, where Γ is a typing environment and $\mathfrak{J}$ is an assertion like "*e*:τ."

Like propositions in logic, particular judgments can be valid or invalid. For example the judgment $\Gamma \vdash 1 + \mathbf{true}{:}\mathbf{bool}$ is invalid. The expression 1 + **true** does not evaluate to a boolean value, so it should not have this (or any other) type. To tell whether a judgment is valid, type systems include *rules* that can be used to *derive* only valid judgments. When a judgment $\Gamma \vdash \mathfrak{J}$ is valid, we say that Γ *entails* $\mathfrak{J}$. The derivation is the justification or proof of the validity of the judgment (and implies the validity of the typing context Γ).

The type system of the arithmetic language includes six *inference rules*, as shown in Figure 70.1. The rules E_NV and E_BV assign the types **nat** and **bool** to natural number and boolean values, respectively. For operator expressions, the rules E_OP and E_EQ require that the two arguments have the appropriate

$$\boxed{\Gamma \vdash e : \tau}$$

$$\frac{\Gamma \vdash \Diamond}{\Gamma \vdash n : \mathbf{nat}}\ \text{E\_NV} \qquad \frac{\Gamma \vdash \Diamond}{\Gamma \vdash b : \mathbf{bool}}\ \text{E\_BV}$$

$$\frac{\Gamma \vdash e_1 : \mathbf{nat} \quad \Gamma \vdash e_2 : \mathbf{nat} \quad op \in \{+,-\}}{\Gamma \vdash e_1\ op\ e_2 : \mathbf{nat}}\ \text{E\_OP}$$

$$\frac{\Gamma \vdash e_1 : \tau \quad \Gamma \vdash e_2 : \tau}{\Gamma \vdash e_1 \equiv e_2 : \mathbf{bool}}\ \text{E\_EQ}$$

$$\frac{\Gamma \vdash e_1 : \mathbf{bool} \quad \Gamma \vdash e_2 : \tau \quad \Gamma \vdash e_3 : \tau}{\Gamma \vdash \mathbf{if}\ e_1\ \mathbf{then}\ e_2\ \mathbf{else}\ e_3 : \tau}\ \text{E\_IF}$$

$$\boxed{\Gamma \vdash \Diamond}$$

$$\frac{}{\varepsilon \vdash \Diamond}\ \text{C\_EMP}$$

FIGURE 70.1 Type system for arithmetic language.

* The "turnstile" symbol (⊢) typically marks a typing judgment and is taken from formal logic. This symbol is derived from notation developed by Frege [14] for formal proofs.

types for the operator. The rule E_IF requires the argument of an **if**-expression to be a boolean expression and the two branches to have equal types.

The last rule of the figure, rule C_EMP, is for an auxiliary judgment that determines only that the typing context itself is *well formed*. Because typing contexts are trivial in this language, this single rule merely states that the empty context is valid. As we extend the type system, we will extend this and other judgments with additional rules.

In general, an inference rule contains two parts divided by a horizontal line. To the right of the rule is its name, for easy reference. Often, a naming convention indicates the form of the judgment in the *conclusion* of the rule (below the line). For example, four rules in Figure 70.1 start with E_ to indicate that they derive judgments about the types of expressions, whereas the single rule for typing contexts starts with C_.

The *premises* of a typing rule appear above the line. These judgments must hold before the rule can be applied. A typing rule can have zero, one or many premises. All must be valid for the rule to apply. The premises of a rule need not have the same form as in the conclusion of the rule. For example, the rules E_NV and E_BV include premises for the judgment $\Gamma \vdash \diamond$ even though they derive the types of expressions. Other premises may be simple logical facts, such as the rule E_OP that explicitly declares the set of valid operators for that rule with the premise $op \in \{+, -\}$.

Rules are schematic templates that can be put together to form *typing derivations*, evidence that a particular program type checks. A derivation is a tree of judgments, where the root of the tree (at the bottom) is the final conclusion of the derivation. Each step of the derivation must be an instance of one of the typing rules, where the metavariables of the rule have been replaced with concrete types, expressions, operators, and contexts.

For example, we can construct the derivation that, using an empty typing context ε, the expression $(1 + 2) \equiv 3$ has type **bool**. We write this judgment as $\varepsilon \vdash (1 + 2) \equiv 3 : \mathbf{bool}$. The last step of the derivation concludes this judgment using the rule E_EQ by satisfying all of the premises of that rule. Each of those premises also satisfy their premises and so on, until in each case reaching the rule C_EMP, which has no premises:

$$\dfrac{\dfrac{\dfrac{\dfrac{}{\varepsilon \vdash \diamond}\,\text{C\_EMP}}{\varepsilon \vdash 1 : \mathbf{nat}}\,\text{E\_NV} \qquad \dfrac{\dfrac{}{\varepsilon \vdash \diamond}\,\text{C\_EMP}}{\varepsilon \vdash 2 : \mathbf{nat}}\,\text{E\_NV}}{\varepsilon \vdash 1 + 2 : \mathbf{nat}}\,\text{E\_OP} \qquad \dfrac{\dfrac{}{\varepsilon \vdash \diamond}\,\text{C\_EMP}}{\varepsilon \vdash 3 : \mathbf{nat}}\,\text{E\_NV}}{\varepsilon \vdash (1 + 2) \equiv 3 : \mathbf{bool}}\,\text{E\_EQ}$$

This derivation is complete because the rule used at the leaves (at the top of the tree) has no premises. Rules with no premises are called *axioms*. Each line in the tree corresponds to the usage of a rule. Note that the rules constrain how the derivation can be formed. The rule E_OP requires that the types of the subexpressions both be **nat**. Likewise, rule E_EQ requires that both subexpressions have the same type, but it does not fix what that type must be.

A *type system* is the entire set of rules necessary for creating derivations. In the arithmetic language, it includes all of the rules that conclude $\Gamma \vdash e{:}\tau$ as well as all of those for $\Gamma \vdash \diamond$.

Programs that have typing derivations are considered *typeable*, or *well formed*. Those that do not are *untypeable*. There may be multiple derivations that assign a type to an expression and an expression may be typeable with more than one type. The rules of a *syntax-directed* type system directly specify an algorithm for determining whether an expression should type check (see Section 70.2.3). In such a type system, each expression has at most one typing derivation.

70.2.2 Assumptions and Functions

The language above does not include many of the most important features of programming languages. In this and the next few subsections, we gradually extend our vocabulary of type systems to develop a more convincing model of computation.

The syntax of the language of this subsection is the same as before, extended with three new expression forms. These forms are variables x, anonymous functions $\lambda x.e$, and applications $e_1\ e_2$. These three forms are listed in the last line of the following syntax table:

$$
\begin{aligned}
n &::= 0, 1, 2, \ldots \\
b &::= \textbf{false} \mid \textbf{true} \\
op &::= + \mid - \mid \equiv \\
e &::= n \mid b \mid e_1\ op\ e_2 \mid \textbf{if } e_1 \textbf{ then } e_2 \textbf{ else } e_3 \\
&\quad\mid x \mid \lambda x.e \mid e_1\ e_2
\end{aligned}
$$

These new additions are taken from the *λ-calculus*, a simple model of computation developed by the logician Alonzo Church in the 1930s to investigate the foundations of mathematics. Because of its foundational nature, researchers often use it for the basis of formal study of programming languages. Here we study a typed version of the λ-calculus, called the *simply-typed lambda calculus* or STLC.

The definition of STLC relies on a number of standard definitions from the theory of the λ-calculus, which appear in many textbooks on programming languages and logic [2]. These definitions include the definition of free and bound variables, the definition of capture-avoiding substitution (where $[e'/x]e$ stands for the replacement of the free variable x with the expression e' in the expression e) and the definition of α-equivalence (where functions are identified up to renaming of their parameters, so that $\lambda x.x$ and $\lambda y.y$ are considered to be the same expressions).

To define a type system for the λ-calculus, we must add a type for functions to the syntax of types.

$$\tau ::= \textbf{nat} \mid \textbf{bool} \mid \tau_1 \rightarrow \tau_2$$

The type $\tau_1 \rightarrow \tau_2$ classifies a function with argument of type τ_1 and result of type τ_2. Using currying, we can define functions that take multiple arguments. The → operator associates to the right, so **nat→bool→nat** means **nat→(bool→nat)**.

The STLC type system uses typing contexts to record assertions about the types of free variables. Therefore, we extend the syntax of typing contexts with assumptions.

$$\Gamma ::= \varepsilon \mid \Gamma, x{:}\tau$$

Typing contexts can be viewed as ordered lists. If a context is nonempty, we often omit parentheses and elide the ε at the beginning, writing a context such as (ε, x:**nat**), y:**nat** more simply as x:**nat**, y:**nat**.

An important condition for typing contexts is *uniqueness*, which means that there is at most one assumption about the type of a particular variable. For example,

$$x{:}\textbf{nat},\ y{:}\textbf{nat}$$

is a valid context but the following are not*:

$$x{:}\textbf{nat}, x{:}\textbf{bool} \qquad x{:}\textbf{nat}, x{:}\textbf{nat}$$

* Although it is possible in some situations to allow later bindings to *shadow* earlier ones, shadowing does not make sense in every type system.

$$\boxed{\Gamma \vdash \tau : \star}$$

$$\frac{\Gamma \vdash \Diamond}{\Gamma \vdash \mathbf{nat} : \star}\ \text{T\_NAT} \qquad \frac{\Gamma \vdash \Diamond}{\Gamma \vdash \mathbf{bool} : \star}\ \text{T\_BOOL}$$

$$\frac{\Gamma \vdash \tau_1 : \star \quad \Gamma \vdash \tau_2 : \star}{\Gamma \vdash \tau_1 \to \tau_2 : \star}\ \text{T\_ARROW}$$

FIGURE 70.2 Well formed types.

When this uniqueness condition holds, we can treat typing contexts as finite maps. In particular, it is straightforward to define a number of operations and predicates on these structures, such as calculating all of the variables with typing assumptions in the context (written $dom(\Gamma)$), stating whether a particular assumption is in the context, written $(x{:}\tau) \in \Gamma$, and appending the assumptions in one context after those in another, written Γ_1, Γ_2.

We ensure uniqueness with a new rule for well formed contexts. An assumption can only be added when the variable does not already appear in the typing context:

$$\frac{\Gamma \vdash \Diamond \quad x \notin dom(\Gamma) \quad \Gamma \vdash \tau : \star}{\Gamma,\ x : \tau \vdash \Diamond}\ \text{C\_EVAR}$$

This rule includes a premise for yet another auxiliary judgment, $\Gamma \vdash \tau : \star$, which asserts that the type τ is a valid type. (The notation ": $\star$" can be read as "is a type.") The rules that derive this judgment appear in Figure 70.2. In this system, the judgment is trivial, as all syntactically well formed types are valid. Again, we introduce this judgment now to prepare for future extension (see Section 70.2.6).

To type check the new expression forms, we need three new rules as shown in Figure 70.3. The first states that variables have the types that the context assumes for them. Next, functions introduce assumptions about the types of their parameters to the typing context when their bodies are checked. There is an implicit requirement that the name of the variable not appear already in the domain of the context. This requirement is enforced when the context is checked for well formedness. The last rule, for function applications, constrains the type of e_1 to be a function type and the type of its argument e_2 to match the domain of the function.

Typing contexts serve several purposes in formal type systems besides typing functions. First, they provide a simple way to add new primitive operations. For example, we can add a primitive boolean negation operation, *not*, to the type system merely by adding an assumption *not*:**bool** → **bool** to the initial typing context.

Furthermore, the ability to type check in the presence of assumptions means that type checking is *modular* and is scalable to large programs. Instead of typing an entire program at once, a program can be broken into smaller parts that can be checked individually. The definitions in a module define an interface that can be described by a typing context. Other modules that use the definitions can be checked using that context. Therefore we do not need to recheck the entire program after each modification, as long as the interfaces are unchanged.

$$\boxed{\Gamma \vdash e : \tau}$$

$$\frac{(x : \tau) \in \Gamma \quad \Gamma \vdash \Diamond}{\Gamma \vdash x : \tau}\ \text{E\_VAR}$$

$$\frac{\Gamma, x : \tau_1 \vdash e : \tau_2}{\Gamma \vdash \lambda x.e : \tau_1 \to \tau_2}\ \text{E\_ABS} \qquad \frac{\Gamma \vdash e_1 : \tau_1 \to \tau_2 \quad \Gamma \vdash e_2 : \tau_1}{\Gamma \vdash e_1\ e_2 : \tau_2}\ \text{E\_APP}$$

FIGURE 70.3 Typing rules for variables, functions, and applications.

(E_BV) $Type(\Gamma, n) = \textbf{nat}$
when $Wf(\Gamma)$

(E_NV) $Type(\Gamma, b) = \textbf{bool}$
when $Wf(\Gamma)$

(E_EQ) $Type(\Gamma, e_1 \equiv e_2) = \textbf{bool}$
when $Type(\Gamma, e_1) = Type(\Gamma, e_2)$

(E_OP) $Type(\Gamma, e_1\ op\ e_2) = \textbf{nat}$
when $Type\ (\Gamma, e_1) = Type(\Gamma, e_2) = \textbf{nat}$ and $op \in \{+, -\}$

(E_IF) $Type(\Gamma, \textbf{if}\ e_1\ \textbf{then}\ e_2\ \textbf{else}\ e_3) = \tau$
when $Type\ (\Gamma, e_1) = \textbf{bool}$ and $Type(\Gamma, e_2) = Type(\Gamma, e_3) = \tau$

(C_EMP) $Wf(\varepsilon) = true$

FIGURE 70.4 Type checking algorithm for arithmetic language.

70.2.3 Syntax-Directed Type Systems

For any type system, it is important that there is an algorithm that decides whether any program type checks. Some type systems are *syntax directed*, meaning that an algorithm can be derived directly from the typing rules. This algorithm works by structural recursion over the form of expressions.

For example, the typing rules of the simple arithmetic language of Section 70.2.1 are syntax directed. For any expression in this language, there is a straightforward algorithm to determine its type. We define this algorithm using the partial function $Type(\Gamma, e)$ shown in Figure 70.4, which is taken almost directly by "reading" the typing rules from bottom to top. The arguments to this function are the typing context Γ and an expression e. If this function returns a type τ, then we know that there is some derivation of the judgment $\Gamma \vdash e : \tau$. The definition of this function pattern matches the syntax of its argument. It then calls itself recursively on any subexpressions of that argument. If the type of any subexpression is undefined, then there is no such typing derivation and the whole expression does not type check. In the base cases, this function uses the auxiliary function $Wf(\Gamma)$ to ensure that the context is well formed.

We can see that this function terminates by noting that each recursive call is to a subterm—a smaller expression. Therefore, this function is partial only when the conditions fail, such as when the types of the arguments to the equality operation do not match. Therefore, for any expression in the arithmetic language, we can use this partial function to either determine its type or determine that it does not type check.

In contrast, STLC is not syntax directed. Although we can extend these partial functions to infer the types of variables and function applications and check the well formedness of contexts with assumptions

(E_VAR) $Type(\Gamma, x) = \tau$

when $(x{:}\tau) \in \Gamma$ and $Wf(\Gamma)$

(E_APP) $Type(\Gamma, e_1\ e_2) = \tau_2$

when $Type(\Gamma, e_1) = \tau_1 \to \tau_2$ and $Type(\Gamma, e_2) = \tau_1$

(C_EVAR) $Wf(\Gamma, x{:}\tau) = Wf(\Gamma)$

when $x \notin dom(\Gamma)$

we run into trouble with anonymous functions. What should the type of $\lambda x.e$ be?

$$(\text{E\_ABS}) \quad Type(\Gamma, \lambda x.e) = ?$$

The problem is that the syntax of functions does not indicate what type should be added to the context when checking the body of the function. Even worse, some functions have many types. For example, $\lambda x.x$ can be typed with **nat** → **nat** and **bool** → **bool** (and many more). In other words, STLC does not have the *uniqueness of types* property, where each expression can be typed with exactly one type.

One way to make this language syntax directed is to annotate functions with the types of their arguments. In other words, we replace $\lambda x.e$ with $\lambda x{:}\tau.e$ and replace the typing rule E_ABS with one for annotated functions:

$$\frac{\Gamma, x{:}\tau_1 \vdash e : \tau_2}{\Gamma \vdash \lambda x{:}\tau_1.e : \tau_1 \rightarrow \tau_2}\ \text{E\_AABS}$$

After these modifications, we can finish the last line of the type checking function as follows:

$$(\text{E\_AABS}) \quad Type(\Gamma, \lambda x{:}\tau.e) = Type((\Gamma, x{:}\tau), e)$$
$$\text{when } x \notin dom(\Gamma)$$

Type checking could fail if the variable x already appears in the domain of the typing context. However, the type checking function is allowed to rename the bound variable in the expression to ensure uniqueness.

Note that this algorithm is not the most efficient way to determine whether an expression is typeable. In particular, it checks that the context is well formed many times. We could replace this algorithm with a more efficient version that checks assumptions as they enter the typing context. This version of the type system is convenient for reasoning about the type system, but even though it derives *an* algorithm for type checking, it does not derive the best algorithm.

In general, the simply-typed λ-calculus without annotations on functions is called a *Curry-style* type system. The system with annotated functions is known as a *Church-style* system [3]. We will also use the words *implicitly typed* and *explicitly typed* to distinguish between the two versions. Even though the Curry-style type system is not syntax directed, there is still an algorithm for type checking it. That algorithm is just not directly determined by the syntax (see Section 70.4).

Many forms besides functions have implicitly-typed and explicitly-typed versions. In the sections that follow, we introduce each new form in its implicit version and then discuss how typing annotations may be added to it afterwards. We order the discussion in this manner to simplify the syntax of each form, draw connections to existing programming languages that often feature these forms without annotation, and focus attention to the important parts of the semantics.

70.2.4 Structured Types

In this section, we extend STLC with unit, product and sum types. These new forms are only the first of many additions that make formal type systems closer to that of actual programming languages such as Haskell and ML.

For each new type construction, we introduce the necessary syntax for the addition and describe the associated rules. In each case, the set of rules follows a basic pattern. These rules declare when the type is well formed, how to *introduce* elements of the type and how to *eliminate* elements of that type.

$$\tau ::= \ldots \mid \textbf{unit}$$
$$e ::= \ldots \mid ()$$

$$\frac{\Gamma \vdash \Diamond}{\Gamma \vdash \textbf{unit} : \star}\ \text{T\_UNIT} \qquad \frac{\Gamma \vdash \Diamond}{\Gamma \vdash () : \textbf{unit}}\ \text{E\_UNIT}$$

FIGURE 70.5 The unit type.

$$\tau ::= \ldots \mid \tau_1 \times \tau_2$$
$$e ::= \ldots \mid \langle e_1, e_2 \rangle \mid \textbf{fst}\ e \mid \textbf{snd}\ e$$

$$\frac{\Gamma \vdash \tau_1 : \star \quad \Gamma \vdash \tau_2 : \star}{\Gamma \vdash \tau_1 \times \tau_2 : \star}\ \text{T\_PROD} \qquad \frac{\Gamma \vdash e_1 : \tau_1 \quad \Gamma \vdash e_2 : \tau_2}{\Gamma \vdash \langle e_1, e_2 \rangle : \tau_2 \times \tau_2}\ \text{E\_PAIR}$$

$$\frac{\Gamma \vdash e : \tau_1 \times \tau_2}{\Gamma \vdash \textbf{fst}\ e : \tau_1}\ \text{E\_FST} \qquad \frac{\Gamma \vdash e : \tau_1 \times \tau_2}{\Gamma \vdash \textbf{snd}\ e : \tau_2}\ \text{E\_SND}$$

FIGURE 70.6 Product types.

For example, in STLC we extended the grammar with variables, functions, and applications. The rule E_ARROW defines when function types are well formed, the introduction rule is T_ABS, and the elimination rule is T_APP.

The **unit** type (Figure 70.5) is a trivial type for an uninteresting value. It is called *void* in some languages. There are no primitive operations available for the single value of this type, written (), so there are only two rules: one for checking that the type is well formed and an introduction rule for showing that the unit value has the unit type.

Product types (Figure 70.6) are the types of composite values. These values contain two components which may be of differing types. The pair expression $\langle e_1, e_2 \rangle$ introduces a product; they are eliminated with **fst** *e* and **snd** *e*, which extract the first and second components of *e*, respectively.

Sum types (Figure 70.7), also called disjoint union types, represent alternatives. A value of type $\tau_1 + \tau_2$, could either be a value of type τ_1 or a value of type τ_2. In either case, the value is tagged so that its type can be determined. The expressions **inl** *e* and **inr** *e* introduce sum types tagged as either the left or the right type of the sum. The elimination form for sum types is case analysis (a simplified form of pattern matching found in functional programming languages). This construct evaluates one of two branches, depending on the tag of the *scrutinee*, the argument of case analysis. Inside each branch, the value that was tagged is available with either the left or the right type depending on the branch.

Note that the inclusion of **unit** and sum types makes the **bool** type unnecessary. We could replace all uses of **bool** with the type **unit** + **unit**, representing **true** as **inl** () and **false** as **inr** (). An **if**-expression,

$$\tau ::= \ldots \mid \tau_1 + \tau_2$$
$$e ::= \ldots \mid \textbf{inl}\ e \mid \textbf{inr}\ e \mid \textbf{case}\ e\ \textbf{of}\ \{\textbf{inl}\ x_1 \to e_1;\ \textbf{inr}\ x_2 \to e_2\}$$

$$\frac{\Gamma \vdash \tau_1 : \star \quad \Gamma \vdash \tau_2 : \star}{\Gamma \vdash \tau_1 + \tau_2 : \star}\ \text{T\_SUM}$$

$$\frac{\Gamma \vdash e : \tau_1}{\Gamma \vdash \textbf{inl}\ e : \tau_1 + \tau_2}\ \text{E\_INL} \qquad \frac{\Gamma \vdash e : \tau_2}{\Gamma \vdash \textbf{inr}\ e : \tau_1 + \tau_2}\ \text{E\_INR}$$

$$\frac{\Gamma \vdash e : \tau_1 + \tau_2 \quad \Gamma, x : \tau_1 \vdash e_1 : \tau \quad \Gamma, y : \tau_2 \vdash e_2 : \tau}{\Gamma \vdash \textbf{case}\ e\ \textbf{of}\ \{\textbf{inl}\ x_1 \to e_1;\ \textbf{inr}\ x_2 \to e_2\} : \tau}\ \text{E\_CASE}$$

FIGURE 70.7 Sum types.

which eliminates the boolean type, can be implemented with case analysis, ignoring the unit values carried by the tags. We formalize this encoding with the following table:

$$\begin{array}{lcl} \textbf{bool} & \triangleq & \textbf{unit} + \textbf{unit} \\ \textbf{true} & \triangleq & \textbf{inl}() \\ \textbf{false} & \triangleq & \textbf{inr}() \\ \textbf{if } e_1 \textbf{ then } e_2 \textbf{ else } e_3 & \triangleq & \textbf{case } e_1 \textbf{ of } \{\textbf{inl } y \rightarrow e_2; \textbf{inr } y \rightarrow e_3\} \\ & & \text{where } y \notin fv(e_2) \cup fv(e_3) \end{array}$$

The additions of this section are intended to give an idea of composite type structure—complex types can be formed from simple components. Programming languages typically generalize products and sums to more convenient versions. For example, n-ary tuple types $\tau_1 \times \tau_2 \times \ldots \tau_n$, generalize products from two components to n components for any $n \geq 2$.

Another generalization is to identify tuple components by name instead of by number—this produces record types, written $\{l_1{:}\tau_1, \ldots l_n{:}\tau_n\}$, where each component of the aggregate data structure is named by some label l. The components of record types can often be accessed using dot notation, $e.l$ accesses the component named l in the record e. Likewise, sums can be generalized from two options to more, and from numbered options to labels. Labelled n-ary sums are called *variant* types.

Although a typing algorithm for unit and product types can be read directly from the rules, the introduction rules for sum types are not syntax directed. To make this language explicitly typed, we replace **inl** e with $\textbf{inl}_\tau\, e$ and **inr** e with $\textbf{inr}_\tau\, e$, where the annotation τ indicates the type of the other alternative:

$$\frac{\Gamma \vdash e : \tau_1 \quad \Gamma \vdash \tau_2 : \star}{\Gamma \vdash \textbf{inl}_{\tau_2}\, e : \tau_1 + \tau_2}\ \text{E\_AINL} \qquad \frac{\Gamma \vdash \tau_1 : \star \quad \Gamma \vdash e : \tau_2}{\Gamma \vdash \textbf{inr}_{\tau_1}\, e : \tau_1 + \tau_2}\ \text{E\_AINR}$$

70.2.5 Recursive Types

Recursive data structures, such as lists and trees, can be typed by recursive types, written $\mu\alpha{:}\star.\tau$. Similar to a λ term, a μ type binds the *type variable* α in the type τ. We use greek letters from the beginning of the alphabet for type variables. As with terms, we can calculate the free type variables that appear in a type with the syntax $fv(\tau)$, substitute types for free type variables (while avoiding capture) with the syntax $[\tau/\alpha]\tau'$, and identify types up to renaming of bound variables. For example, the types $\mu\alpha{:}\star.\alpha \rightarrow \alpha$ and $\mu\beta{:}\star.\beta \rightarrow \beta$ are considered equal.

The new syntax and typing rules are in Figure 70.8. Not only do we add the new forms for variables and recursive type definitions, but we also extend the syntax for typing contexts. Assumptions of the form $\alpha{:}\star$ record the type variables that may appear in types. As with term variables, type variable

$$\begin{array}{rcl} \tau & ::= & \ldots \mid \alpha \mid \mu\alpha{:}\star.\tau \\ \Gamma & ::= & \ldots \mid \Gamma, \alpha{:}\star \end{array}$$

$$\frac{\Gamma \vdash \Diamond \quad \alpha \notin dom(\Gamma)}{\Gamma, \alpha{:}\star \vdash \Diamond}\ \text{C\_TVAR}$$

$$\frac{\Gamma \vdash \Diamond \quad (\alpha{:}\star) \in \Gamma}{\Gamma \vdash \alpha{:}\star}\ \text{T\_VAR} \qquad \frac{\Gamma, \alpha{:}\star \vdash \tau{:}\star}{\Gamma \vdash \mu\alpha{:}\star.\tau{:}\star}\ \text{T\_MU}$$

$$\frac{\Gamma \vdash e : [\mu\alpha{:}\star.\tau/\alpha]\tau}{\Gamma \vdash e : \mu\alpha{:}\star.\tau}\ \text{E\_ROLL} \qquad \frac{\Gamma \vdash e : \mu\alpha{:}\star.\tau}{\Gamma \vdash e : [\mu\alpha{:}\star.\tau/\alpha]\tau}\ \text{E\_UNROLL}$$

FIGURE 70.8 Type variables and equirecursive types.

$$e \ ::= \ \ldots \mathbf{roll}_\tau\ e \mid \mathbf{unroll}\ e$$

$$\frac{\Gamma \vdash e : [\mu\alpha{:}\star.\tau/\alpha]\tau}{\Gamma \vdash \mathbf{roll}_{\mu\alpha:\star\tau}\ e : \mu\alpha{:}\star.\tau}\ \text{E\_RLA} \qquad \frac{\Gamma \vdash e : \mu\alpha{:}\star.\tau}{\Gamma \vdash \mathbf{unroll}\ e : [\mu\alpha{:}\star.\tau/\alpha]\tau}\ \text{E\_URLA}$$

FIGURE 70.9 Isorecursive types.

assumptions in the context must be unique. The rule C_TVAR restricts type variables from occurring multiple times. Rule T_VAR makes the judgment form $\Gamma \vdash \tau{:}\star$ nontrivial. It says that type variables are valid only when they are declared in the context. The formation rule for recursive types (T_MU) adds the bound type variable to the context when it checks the body of the recursive type.

Because the context now contains both type and term variables, the order of assumptions in a typing context matters. Rule C_EVAR (from Section 70.2.2) ensures that the types of assumed term variables are well formed using the prior context. If the context is reordered, these types may no longer be valid.

The last two rules implicitly introduce and eliminate recursive types. In a recursive type $\mu\alpha{:}\star.\tau$, the type variable α stands for the whole type. Therefore these rules equate recursive types $\mu\alpha{:}\star.\tau$ with their unrolling $[\mu\alpha{:}\star.\tau/\alpha]\tau$, allowing terms to have either type as necessary. These type conversions are implicit, so this type system is not syntax directed.

To define a syntax-directed system, we explicitly mark the locations where the type should be "rolled" and "unrolled" in expressions, using the constructs in Figure 70.9. The forms **roll** and **unroll** are explicit coercions between values of type $\mu\alpha{:}\star.\tau$ and $[\mu\alpha{:}\star.\tau/\alpha]\tau$. These forms are witnesses to an isomorphism between a recursive type and its unfolding. Recursive types with implicit conversions are called *equirecursive*; those with **roll** and **unroll** are called *isorecursive types*.

Regular algebraic datatypes, as found in ML and Haskell, can be encoded by recursive types. For example, in Haskell, one might define a tree structure as follows:

```
data Tree = Empty | Node Tree Tree
```

This structure declares the existence of a new data type, called `Tree` and two *data constructors* for the datatype. The data constructor `Empty` constructs a leaf with no children and the data constructor `Node` constructs a tree with left and right children. Recursive functions that work over datatypes use pattern matching, such the case expression in the following function definition:

```
height :: Tree -> Int
height = \ x -> case x of
  Empty -> 0
  Node x y -> 1 + height x + height y
```

We can encode trees with products, sums, unit, and recursive types. The tree type is a recursive type, which is either unit (for the empty tree) or a product of two occurrences of the recursive variable. The constructors `Empty` and `Node` construct values of this type, using the introduction forms for unit, sums, products and (iso)recursion. Finally, case analysis on `Trees` is implemented with the corresponding elimination forms:

$$\begin{aligned}
\texttt{Tree} \quad &\triangleq \quad \mu\beta:\star.\mathbf{unit} + (\beta\times\beta)\\
\texttt{Empty} \quad &\triangleq \quad \mathbf{roll}_{\mu\beta:\star.\mathbf{unit}+(\beta\times\beta)}(\mathbf{inl}())\\
\texttt{Node}\ e_1\ e_2 \quad &\triangleq \quad \mathbf{roll}_{\mu\beta:\star.\mathbf{unit}+(\beta\times\beta)}\left(\mathbf{inr}\langle e_1, e_2\rangle\right)
\end{aligned}$$

$$\texttt{case}\ e\ \texttt{of}\ \{\texttt{Empty} \rightarrow e_1; \texttt{Node}\ x\ y \rightarrow e_2\} \triangleq$$

$$\text{case (unroll } e\text{) of } \{\mathbf{inl}\ x_1 \rightarrow e_1;\ \mathbf{inr}\ x_2 \rightarrow [\mathbf{fst}\ x_2/x]\,[\mathbf{snd}\ x_2/y]e_2\}$$

$$\text{where } x_1 \notin fv\,(e_2), \quad x_2 \notin fv\,(e_2)$$

Furthermore, type-level recursion can be used to add value-level recursion to typed languages. Recursive definitions, such as the height function above, are possible in the λ-calculus with the Y-combinator. The Y-combinator computes the fixed point of its argument, allowing the definition of recursive, anonymous functions. For example, the height function is defined using the Y-combinator as

$$\begin{aligned}&\texttt{height} \triangleq \\ &\quad Y(\lambda h.\,\lambda x.\mathbf{case}(\mathbf{unroll}\;x)\;\mathbf{of}\;\{\\ &\qquad \mathbf{inl}\;y \to 1;\\ &\qquad \mathbf{inr}\;y \to 1 + h\,(\mathbf{fst}\;y) + h(\mathbf{snd}\;y)\})\end{aligned}$$

For call-by-name languages, the Y-combinator is defined as follows:

$$Y \triangleq \lambda f.\big(\lambda x.f(x\;x)\big)\big(\lambda x.f(x\;x)\big)$$

The Y-combinator can be type checked using recursive types. Below, we give the definition of Y in the explicitly-typed language, marking where coercions to and from the recursive type must occur:

$$\begin{aligned}&Y_\tau \triangleq \lambda f : \tau \to \tau.\\ &\quad (\lambda x : (\mu\alpha{:}\star.\alpha \to \tau).f\;((\mathbf{unroll}\;x)x))\\ &\quad (\mathbf{roll}_{\mu\alpha:\star.\alpha\to\tau}(\lambda x : (\mu\alpha{:}\star.\alpha \to \tau).f\;((\mathbf{unroll}\;x)x)))\end{aligned}$$

The implicitly-typed language with equirecursive types can type the definition of Y directly. However, the Y-combinator, like any potentially nonterminating expression, cannot be type checked in STLC [52]. The addition of recursive types is enough to make this language Turing complete.

70.2.6 Parametric Polymorphism

Parametric polymorphism, or type abstraction, enables programs to be generic in the types of data that they work with. This feature increases the flexibility of statically typed languages such as ML, Haskell, Java and Scala. Parametric polymorphism was first studied (independently) by the logician Jean-Yves Girard [17] and the computer scientist John Reynolds [48].

In this subsection, we extend our small language with two different forms of type abstraction: *universally quantified types*, written $\forall\alpha{:}\star.\tau$ and *existentially quantified types*, written $\exists\alpha{:}\star.\tau$. In both cases, the type variable α is bound within the body of the type τ. This type variable α stands for an abstract type, one that is unknown when the term is compiled. Type systems that include quantified types are called *second-order*.

70.2.6.1 Universal Polymorphism

We add universally quantified types to the language with the three rules in Figure 70.10. The introduction form generalizes the type of an expression. If we can type check an expression while treating one part of its type abstractly (i.e., while knowing nothing but its name α), then we can generalize that part

$$\tau \;::=\; \ldots \mid \forall\alpha{:}\star.\tau$$

$$\frac{\Gamma, \alpha{:}\star \vdash \tau : \star}{\Gamma \vdash \forall\alpha{:}\star.\tau : \star}\ \text{T\_ALL}$$

$$\frac{\Gamma, \alpha{:}\star \vdash e:\tau}{\Gamma \vdash e:\forall\alpha{:}\star.\tau}\ \text{E\_TABS} \qquad \frac{\Gamma \vdash e:\forall\alpha{:}\star.\tau \quad \Gamma \vdash \tau' : \star}{\Gamma \vdash e:[\tau'/\alpha]\tau}\ \text{E\_TAPP}$$

FIGURE 70.10 Universally quantified types (implicit rules).

of the type. For example, generalization shows that the identity function $\lambda x.x$ can be given type $\forall\alpha{:}\star.\alpha\to\alpha$ because the type of the argument is unconstrained:

$$\cfrac{\cfrac{\cfrac{\cfrac{\cfrac{}{\varepsilon \vdash \Diamond}\ \text{C\_EMP}}{\varepsilon, \alpha{:}\star \vdash \Diamond}\ \text{C\_TVAR} \quad \cfrac{\cfrac{}{\varepsilon \vdash \Diamond}\ \text{C\_EMP}}{\varepsilon, \alpha{:}\star \vdash \alpha : \star}\ \text{T\_VAR}}{\varepsilon, \alpha{:}\star, x{:}\alpha \vdash \Diamond}\ \text{C\_EVAR}}{\varepsilon, \alpha{:}\star, x{:}\alpha \vdash x : \alpha}\ \text{E\_VAR}}{\varepsilon, \alpha{:}\star \vdash \lambda x.x : \alpha \to \alpha}\ \text{E\_ABS}}{\varepsilon \vdash \lambda x.x : \forall\alpha{:}\star.\alpha \to \alpha}\ \text{E\_TABS}$$

The elimination form allows us to use an expression with a polymorphic type at any instance. Any well formed type can be substituted for the variable. For example, we can apply the identity function to itself by instantiating it with its own type. Part of this derivation is shown as follows:

$$\cfrac{\cfrac{\cfrac{\ldots}{\varepsilon \vdash \lambda x.x{:}\forall\alpha{:}\star.\alpha \to \alpha} \quad \cfrac{\ldots}{\varepsilon \vdash \forall\alpha{:}\star.\alpha \to \alpha{:}\star}}{\varepsilon \vdash \lambda x.x : (\forall\alpha{:}\star.\alpha \to \alpha) \to (\forall\alpha{:}\star.\alpha \to \alpha)} \quad \cfrac{\ldots}{\varepsilon \vdash \lambda x.x : \forall\alpha{:}\star.\alpha \to \alpha}}{\varepsilon \vdash \lambda x.x\ (\lambda x.x) : \forall\alpha{:}\star.\alpha \to \alpha}$$

Languages with parametric polymorphism distinguish between *predicative* and *impredicative* variants. The rules in Figure 70.10 are impredicative. The difference between the two is the meaning of the quantifier. In a type such as $\forall\alpha : \star.\alpha \to \alpha$, can the type variable α really be instantiated by itself as shown in the derivation above? This behavior causes difficulty—if we think of these types as the set of all of their instances, then the meaning of impredicative quantification is circular.

Predicative type systems stratify quantification to escape this circularity. Instantiation is only allowed by "smaller" types. One way to enforce predicativity is to mark each type by a level and allow quantifiers to range only over types from a lower level.

Type systems with predicative polymorphism can be easier to reason about and model semantically because of this stratification. However, impredicative polymorphism is more expressive. It can encode a number of other type forms such as products, sums, existentials (see next section), and some recursive types [5]. Thus, someone studying the semantics of a language with impredicative polymorphism need not include these type forms in the model—they are already available.

Once again, the rules presented in Figure 70.10 are not syntax directed. Explicit universal polymorphism (shown in Figure 70.15) marks the introduction of polymorphic types with the type abstraction form $\Lambda\alpha : \star.e$ and the elimination with a type application form $e\langle\tau\rangle$.

70.2.6.2 Existential Polymorphism

Existentially quantified types, written $\exists\alpha : \star.\tau$, model *abstract data types* (ADT). An ADT is a type that is defined solely by its interface—how it may be used—without revealing its representation.

$$\tau \;::=\; \ldots \mid \exists\alpha{:}\star.\tau$$
$$e \;::=\; \ldots \mid \mathbf{pack}\langle\tau,e\rangle\ \mathbf{as}\ \exists\alpha{:}\star.\tau' \mid \mathbf{let}\langle\alpha,x\rangle = e_1\ \mathbf{in}\ e_2$$

$$\frac{\Gamma,\alpha{:}\star \vdash \tau{:}\star}{\Gamma \vdash \exists\alpha{:}\star.\tau{:}\star}\ \text{T\_EX}$$

$$\frac{\Gamma\vdash\tau:\star \quad \Gamma\vdash e:[\tau/\alpha]\tau'}{\Gamma\vdash \mathbf{pack}\langle\tau,e\rangle\ \mathbf{as}\ \exists\alpha{:}\star.\tau' : \exists\alpha{:}\star.\tau'}\ \text{E\_PACK}$$

$$\frac{\Gamma\vdash e{:}\exists\alpha{:}\star.\tau \quad \Gamma,\alpha{:}\star,x{:}\tau\vdash e'{:}\tau' \quad \Gamma\vdash\tau'{:}\star}{\Gamma\vdash \mathbf{let}\langle\alpha,x\rangle = e\ \mathbf{in}\ e' : \tau'}\ \text{E\_UNPACK}$$

FIGURE 70.11 Existentially quantified types.

Abstract data types are important because they provide encapsulation, localization of change, and flexibility to programmers.

Figure 70.11 contains the necessary definitions and rules for this form of type. Because these types rarely appear in implicitly-typed systems, we describe only the explicitly-typed version. The introduction form **pack** $\langle\tau, e\rangle$ **as** $\exists\alpha:\star.\tau'$ creates an "existential package" that hides part of the structure of a type. The elimination form **let** $\langle\alpha, x\rangle = e$ **in** e' provides access to this value but requires that the hidden type be used abstractly.

For example, we can implement a stack of numbers using an encoding of lists as a recursive sum of products. The *empty* stack is just an empty list. The *push* operation adds a new number to the top of the stack. The *peek* operation returns either unit, when the stack is empty, or the top value on the stack. Finally, the *pop* operation returns the stack without its top value:

$$\begin{aligned}
\mathit{stack} &\triangleq \mu\alpha{:}\star.\mathbf{unit} + (\mathbf{nat}\times\alpha)\\
\mathit{empty} &\triangleq \mathbf{roll}_{\mathit{stack}}(\mathrm{inl}())\\
\mathit{push} &\triangleq \lambda x{:}\mathbf{nat}.\lambda y{:}\mathit{stack}.\mathbf{roll}_{\mathit{stack}}(\mathbf{inr}\langle x,y\rangle)\\
\mathit{peek} &\triangleq \lambda y{:}\mathit{stack}.\mathbf{case}(\mathbf{unroll}\ y)\ \mathbf{of}\ \{\mathbf{inl}\ x\to\mathbf{inl}();\ \mathbf{inr}\ x\to\mathbf{inr}(\mathbf{fst}\,x)\}\\
\mathit{pop} &\triangleq \lambda y{:}\mathit{stack}.\mathbf{case}\ (\mathbf{unroll}\ y)\ \mathbf{of}\ \{\mathbf{inl}\ x\to y;\ \mathbf{inr}\,x\to\mathbf{snd}\,x\}
\end{aligned}$$

Putting these four operations together in a record allows the implementation of the stack type to be hidden. In other words, we can use **pack** to give a record of these operations the following type, where the type *stack* has been hidden by the variable α:

$$\begin{aligned}
\mathit{stkmod} \triangleq \exists\alpha.\{\ \mathit{empty} &: \alpha;\\
\mathit{pusch} &: \mathbf{nat}\to\alpha\to\alpha;\\
\mathit{peek} &: \alpha\to\mathbf{unit}+\mathbf{nat};\\
\mathit{pop} &: \alpha\to\alpha;\\
\}&
\end{aligned}$$

Clients of the stack module use the elimination form for existential types, **let** $\langle\alpha, x\rangle = e_1$ **in** e_2. This form introduces the type variable α and expression variable x with type that mentions α into the context. Clients cannot know the actual type that was used for α when the module was constructed. Instead, they must use this type abstractly, according to the interface specified by the type of x.

For example, once the stack module has been unpacked, the stack operations can be accessed from the record and used with each other. If y is a variable with type as given earlier, then it can be used as follows:

$$\Gamma, y : stkmod \vdash \mathbf{let}\langle \alpha, x\rangle = y \ \mathbf{in}\ x.peek\ (x.push\ 1\ x.empty) : \mathbf{unit} + \mathbf{nat}$$

The third premise of the typing rule for the elimination form for existentials implies that the result type τ' cannot contain the abstract variable α. Because of this premise, the type variable α cannot escape the scope of the **let**-expression.

70.2.7 Parameterized Types and Higher-Order Polymorphism

Parameterized types are those that abstract part of their structure. For example, in Haskell we can define a type of polymorphic lists that can be used with many different types of elements:

```
data List a = Nil | Cons a (List a)
```

This single definition can conveniently be used to construct both lists of integers and lists of boolean values.

```
x :: List Int
x = Cons 1 (Cons 2 (Cons 3 Nil))

y :: List Bool
y = Cons True (Cons False (Cons True Nil))
```

These lists are not heterogeneous; a single list can contain either integers or booleans, but not both. Yet the power of type parameterization means that we can write polymorphic functions that work for all lists, no matter what type of elements they contain. For example, the map function applies its argument to each element in a list in turn, constructing a new list from the results:

```
map :: forall a b. (a -> b) -> List a -> List b
map f Nil = Nil
map f (Cons x xs) = Cons (f x) (map f xs)
```

What exactly is `List` in these definitions? We call it a *parameterized type*. By itself, it does not make sense as a type—it needs a type argument like `Int` and `Bool` to be complete. `List` must be applied to another type to become a type.

Formally, we model parameterized types by adding *type application*, written $\tau_1\ \tau_2$, to the syntax of types as shown in Figure 70.12. That way we can express the application of type constants like `List` to their parameters.

However, with this addition, we must worry about ill-formed types of the form **nat nat**, or **list list**. These types are syntactically correct but meaningless. Allowing them in the type system does not jeopardize type soundness, but they could lead to buggy interfaces as they do not contain any values. We rule such types out using *kinds*, that is, a "type of types." Kind checking ensures that parameterized types are applied to the correct arguments.

There are two forms of kinds. The first, $\star$ is the kind of normal types such as **nat**, **bool**, and **unit + unit**. Only types of this kind contain values. Arrow kinds, written $\kappa_1 \to \kappa_2$, describe parameterized types. For example, `List`, has kind $\star \to \star$ because it requires one type argument to be a complete type.

$$\tau \ ::= \ \ldots \mid \tau_1\,\tau_2 \mid \forall\alpha{:}\kappa.\tau$$
$$\kappa \ ::= \ \star \mid \kappa_1 \rightarrow \kappa_2$$

$\boxed{\Gamma \vdash \tau : k}$

$$\frac{\Gamma \vdash \tau_1 : \kappa_1 \rightarrow \kappa_2 \quad \Gamma \vdash \tau_2 : \kappa_1}{\Gamma \vdash \tau_1\ \tau_2 : \kappa_2}\ \text{T\_APP}$$

$$\frac{\Gamma \vdash \Diamond \quad (\alpha : \kappa) \in \Gamma}{\Gamma \vdash \alpha : \kappa}\ \text{T\_VARK} \qquad \frac{\Gamma, \alpha{:}\kappa \vdash \tau : \star}{\Gamma \vdash \forall\alpha{:}\kappa.\tau : \star}\ \text{T\_ALLK}$$

$\boxed{\Gamma \vdash e : \tau}$

$$\frac{\Gamma, \alpha{:}\kappa \vdash e : \tau}{\Gamma \vdash e : \forall\alpha{:}\kappa.\tau}\ \text{E\_TABSK} \qquad \frac{\Gamma \vdash e : \forall\alpha{:}\kappa.\tau \quad \Gamma \vdash \tau' : k}{\Gamma \vdash e : [\tau'/\alpha]\tau}\ \text{E\_TAPPK}$$

FIGURE 70.12 Higher-order polymorphism.

We also generalize the judgment we have been using to check that types are well formed so that it also tracks their kinds. The form of the judgment is now $\Gamma \vdash \tau : \kappa$. All previous rules are still valid. For a type application, the rule T_APP ensures that the parameterized type can be applied to its argument.

Higher-order polymorphism means that polymorphic functions can quantify over parameterized types. The Haskell programming language includes this feature [25], often using it for generic code that works for arbitrary functors, monads, arrows and idioms. We can add higher-order polymorphism to our formalism by relaxing the rules that introduce and use type variables in the context so that they apply to variables of any kind.

70.2.8 Type-Level Computation

The process of type checking a program is based on comparing types for equivalence. For example, when checking that function applications are well typed, a compiler must compare the type of the argument with the expected type of the formal parameter.

To this point, type equivalence has been unremarkable. In the preceding sections, types are equal only when they are α-equivalent, that is, syntactically equal up to renaming of bound variables. However, some type systems define a richer equality for types and allow them to be compared up to this equivalence. This subsection describes one example of a rich type equality, based on *type-level computation*.

Type-level computation turns the language of types τ into a programming language. However, "programs" in this language are never run, they are only compared for equivalence during type checking. The simplest example of type-level computation is the addition of λ-calculus functions to the language of higher-order polymorphism:

$$\tau ::= \ldots \mid \lambda\alpha{:}\kappa.\tau$$

Such type-level functions have function kinds:

$$\frac{\Gamma, \alpha{:}\kappa_1 \vdash \tau : \kappa_2}{\Gamma \vdash \lambda\alpha{:}\kappa_1.\tau : \kappa_1 \rightarrow \kappa_2}\ \text{T\_ABS}$$

This type system, which combines type-level abstractions with type variables, type application, and kinds (i.e., a simple "type system" for types), is called System F_ω [17].

Languages that include type-level computation *define* what it means for types to be equivalent with the judgment form $\Gamma \vdash \tau_1 \simeq \tau_2 : \kappa$, as shown in Figure 70.13. Derivations of this judgment prove that the

$$\boxed{\Gamma \vdash \tau_1 \simeq \tau_2 : \kappa}$$

$$\frac{\Gamma \vdash (\lambda\alpha{:}\kappa.\tau)\tau' : \kappa}{\Gamma \vdash (\lambda\alpha{:}\kappa.\tau)\tau' \simeq [\tau'/\alpha]\tau : \kappa}\ \text{TE\_BETA}$$

$$\frac{\Gamma \vdash \tau : \kappa_1 \to \kappa_2}{\Gamma \vdash \lambda\alpha{:}\kappa_1.(\tau\ \alpha) \simeq \tau : \kappa_1 \to \kappa_2}\ \text{TE\_ETA}$$

$$\frac{\Gamma \vdash \tau : \kappa}{\Gamma \vdash \tau \simeq \tau : \kappa}\ \text{TE\_REFL} \qquad \frac{\Gamma \vdash \tau' \simeq \tau : \kappa}{\Gamma \vdash \tau \simeq \tau' : \kappa}\ \text{TE\_SYM}$$

$$\frac{\Gamma \vdash \tau_1 \simeq \tau_2 : \kappa \quad \Gamma \vdash \tau_2 \simeq \tau_3 : \kappa}{\Gamma \vdash \tau_1 \simeq \tau_3 : \kappa}\ \text{TE\_TRANS}$$

$$\frac{\Gamma \vdash \tau_1 \simeq \tau_1' : \star \quad \Gamma \vdash \tau_2 \simeq \tau_2' : \star}{\Gamma \vdash \tau_1 \to \tau_2 \simeq \tau_1' \to \tau_2' : \star}\ \text{TE\_ARROW}$$

$$\frac{\Gamma, \alpha{:}\kappa \vdash \tau \simeq \tau' : \star}{\Gamma \vdash \forall\alpha{:}\kappa.\tau \simeq \forall\alpha{:}\kappa.\tau' : \star}\ \text{TE\_ALL}$$

$$\frac{\Gamma, \alpha{:}\kappa_1 \vdash \tau \simeq \tau' : \kappa_2}{\Gamma \vdash \lambda\alpha : \kappa_1.\tau \simeq \lambda\alpha : \kappa_1.\tau' : \kappa_1 \to \kappa_2}\ \text{TE\_ABS}$$

$$\frac{\Gamma \vdash \tau_1 \simeq \tau_1' : \kappa_1 \to \kappa_2 \quad \Gamma \vdash \tau_2 \simeq \tau_2' : \kappa_1}{\Gamma \vdash \tau_1\ \tau_2 \simeq \tau_1'\ \tau_2' : \kappa_2}\ \text{TE\_APP}$$

FIGURE 70.13 Type equivalence.

type τ_1 is equal to the type τ_2. The rules of this judgment capture equivalences derived from computation. For example, the β-equivalence rule (TE_BETA) identifies types such as $(\lambda\alpha : \star.\alpha \times \alpha)$ **nat** and **nat** × **nat**. Type-level functions can also be identified using η-equivalence (TE_ETA). Other rules of the judgment assert that the relation is an equivalence relation (i.e., that it is reflexive, symmetric, and transitive) and that it is compatible with the structure of types.

Once type equality has been *defined*, it can be used with the *conversion* rule:

$$\frac{\Gamma \vdash e : \tau \quad \Gamma \vdash \tau \simeq \tau' : \star}{\Gamma \vdash e : \tau'}\ \text{E\_CONV}$$

This rule permits the type of any expression to be replaced with any equivalent type at any point of a typing derivation. This rule is not syntax directed; it could apply at any point in a typing derivation to convert a type into an equivalent form.

Furthermore, the type equivalence relation $\Gamma \vdash \tau_1 \simeq \tau_2 : \kappa$ is itself not syntax directed because of rules like transitivity. There are several algorithms for determining whether types are equal. If type equivalence does not include η-equivalence, then it is possible to β-reduce both expressions to normal form and compare the results. Crary [8] describes a kind-directed algorithm for η-equivalence.

70.2.9 Subtyping

Subtype polymorphism also increases the expressiveness of type systems. If we view types as collections of values, then there is a natural relationship between them. If all values of a type τ_1 are also contained in the type τ_2, then we consider τ_1 a *subtype* of τ_2, and conversely that τ_2 is a *supertype* of τ_1. This means that any interface that expects an argument of type τ_2 may be supplied a value of type τ_1. Subtyping means that function interfaces can be flexible, specifying only the minimal requirements of their arguments, without fixing a particular representation.

Adding subtyping to a type system involves expressing the relationship between two types as a judgment form:

$$\Gamma \vdash \tau_1 <: \tau_2$$

that states that "in context Γ, the type τ_1 is a subtype of the type τ_2."

The type system uses this judgment in a single typing rule, called *subsumption*, that enables subtyping.

$$\frac{\Gamma \vdash e : \tau \quad \Gamma \vdash \tau <: \tau'}{\Gamma \vdash e : \tau'} \text{ E\_SUB}$$

This rule states that an expression of one type can also be assigned any supertype. Note that this rule is also not syntax directed. It could appear anywhere in a typing derivation to change the type of an expression.

The rules defining the subtyping relation appear in Figure 70.14. The first two rules declare that the subtyping relation is reflexive and transitive. The other rules of the judgment are specific to the individual type forms.

Subtyping is prevalent in object-oriented languages. In such languages, there often is a maximal type that contains all values. For example, the type `Object` is maximal in the Java language and is the supertype of all reference types. Here we extend the syntax of types with a new type **top** as the maximal supertype,

$$\tau ::= \ldots \mid \textbf{top}$$

and declare that all types are subtypes of it with the rule S_TOP.

Products and sums are in the subtype relation when their components are, as in rules S_PROD and S_SUM. These rules are sometimes called *depth subtyping* as they allow subtyping to work deeply within the type. Languages that include records and variants, also permit *width subtyping* in their subtyping rules. With width subtyping, a record type with more components is a subtype of a type with fewer (but compatible) components and a variant type with fewer options is a subtype of one with more cases. For example, a point with a position and mass works in a context that only needs to query the position. Alternatively, a function that has cases for points, lines, and circles can be applied to a value that is known to be either a point or a line.

Function types provide the most confusion in reasoning about subtyping. Note that the rule S_ARROW declares $\tau_1 \to \tau_2$ a subtype of $\tau_1' \to \tau_2'$ when τ_1' is a subtype of τ_1 and τ_2 is a subtype of τ_2'. For the domains, the relationship is inverted with respect to the whole relation: we call this property *contravariance*.

$$\boxed{\Gamma \vdash \tau_1 <: \tau_2}$$

$$\frac{}{\Gamma \vdash \tau <: \tau} \text{ S\_REFL} \qquad \frac{\Gamma \vdash \tau_1 <: \tau_2 \quad \Gamma \vdash \tau_2 <: \tau_3}{\Gamma \vdash \tau_1 <: \tau_3} \text{ S\_TRANS}$$

$$\frac{}{\Gamma \vdash \tau <: \textbf{top}} \text{ S\_TOP} \qquad \frac{\Gamma \vdash \tau_1 <: \tau_1' \quad \Gamma \vdash \tau_2 <: \tau_2'}{\Gamma \vdash \tau_1 \times \tau_2 <: \tau_1' \times \tau_2'} \text{ S\_PROD}$$

$$\frac{\Gamma \vdash \tau_1 <: \tau_1' \quad \Gamma \vdash \tau_2 <: \tau_2'}{\Gamma \vdash \tau_1 \times \tau_2 <: \tau_1' \times \tau_2'} \text{ S\_SUM}$$

$$\frac{\Gamma \vdash \tau_1' <: \tau_1 \quad \Gamma \vdash \tau_2 <: \tau_2'}{\Gamma \vdash \tau_1 \to \tau_2 <: \tau_1' \to \tau_2'} \text{ S\_ARROW}$$

FIGURE 70.14 Subtyping relation.

When the inclusion goes in the same direction, it is called *covariance*. In general, a type variable appears contravariantly within another type if it always occurs on the left of an odd number of arrows and covariantly if it always appears on right. If neither situation is true, then the variable appears invariantly. For example, α is covariant in the types **unit** $\rightarrow \alpha$ and ($\alpha \rightarrow$ **unit**)$\rightarrow\alpha$ and contravariant in the types $\alpha \rightarrow$ **unit** and (**unit** $\rightarrow \alpha$) $\rightarrow$ **unit**. Finally, α is invariant in the type $\alpha \rightarrow \alpha$.

We can extend the notion of subtyping in many ways. Some languages include a minimal subtype, called `bottom`, as an analogue to the maximal supertype `top`. *Bounded polymorphism* combines subtyping with parametric polymorphism. In such systems, type variables are inserted into the typing context with supertype or subtype bounds that reveal partial information about the abstract type. The rules for subtyping quantified types differ in the treatment of these bounds. The combination of subtyping and parameterized types requires reasoning about *variance*.

Relationships like subtyping show up in many contexts, not just in the typing of objects. For example, in Haskell, type class constraints can be satisfied by any subclass [59]. In ML, modules can be given a variety of signatures in a subsignature relationship [31]. The Hindley–Milner type system (see later) orders polymorphic types by a "more-general" relation [10]. Languages with predicativity levels allow types of lower levels to be automatically promoted higher in the hierarchy [29].

As with the conversion rule of the previous section, the subsumption rule in this section is not syntax directed. Furthermore, the presence of a transitivity rule in the subtyping relation also makes it nonalgorithmic. Pierce [41] defines algorithms for subtyping and shows that they are equivalent to these rules.

70.2.10 Other Forms of Types and Type Systems

The preceding sections describe only a few of the many, many features of type systems that have been devised by language researchers. Next, we briefly give pointers to further reading.

Class and object types describe the essential features of object-oriented languages. A starting point is Featherweight Java [23], an essential formalization of classes, objects, methods, and subtyping.

Low-level types describe the structure of machines, including registers, heaps, stacks, and assembly language instructions. Morrisett [32,33] overviews this work.

Dependent types allow types to depend on values. A gentle introduction to dependent types in the context of the Agda programming language is by Bove and Dybjer [6]. Additional information is found in Aspinall and Hofmann [1].

Linear, affine, and relevant types do not support all of the structural properties (see Section 70.3) found by most type systems. They are used to model resources that cannot be forged or forgotten. Walker gives an overview of this area [60].

Session types describe the protocols that govern how channels must be used in distributed systems [53]. Dezani-Ciancaglini and de'Liguoro [11] provide an introduction.

Security types track the flow of information and the integrity of data values through the execution of programs. These types include capabilities for various principals, and ensure, for example, that high-security data is not read by low-security principals [50].

Effect types talk about the properties of computations. Basic variants distinguish between pure and impure code; more precise systems can track memory usage, resource usage, potentially thrown exceptions, and running time. Henglein et al. [21] give an overview.

70.3 Reasoning about Type Systems

The preceding section individually describes the components of formal type systems piecemeal. After these pieces are assembled, it is natural to ask what properties are true of the system as a whole. Now that we have made the semantics of the type system formal, we can return to the properties discussed in Section 70.1.2 and examine them more rigorously.

Syntax

$$\Gamma \ ::= \ \varepsilon \,|\, \Gamma, x{:}\tau \,|\, \Gamma, \alpha{:}\star$$

$$\tau \ ::= \ \mathbf{bool} \,|\, \alpha \,|\, \tau \rightarrow \tau \,|\, \forall \alpha{:}\star.\tau$$

$$e \ ::= \ b \,|\, \mathbf{if}\ e_1\ \mathbf{then}\ e_2\ \mathbf{else}\ e_3 \,|\, x \,|\, \lambda x{:}\tau.e \,|\, e_1\ e_2 \,|\, \Lambda\alpha{:}\star.e \,|\, e\langle\tau\rangle$$

$\boxed{\Gamma \vdash \Diamond}$

$$\frac{}{\varepsilon \vdash \Diamond}\ \text{C\_EMP} \qquad \frac{\Gamma \vdash \Diamond \quad x \notin dom(\Gamma) \quad \Gamma \vdash \tau : \star}{\Gamma, x{:}\tau \vdash \Diamond}\ \text{C\_EVAR}$$

$$\frac{\Gamma \vdash \Diamond \quad \alpha \notin dom(\Gamma)}{\Gamma, \alpha{:}\star \vdash \Diamond}\ \text{C\_TVAR}$$

$\boxed{\Gamma \vdash \tau : \star}$

$$\frac{\Gamma \vdash \Diamond}{\Gamma \vdash \mathbf{bool} : \star}\ \text{T\_BOOL} \qquad \frac{\Gamma \vdash \tau_1 : \star \quad \Gamma \vdash \tau_2 : \star}{\Gamma \vdash \tau_1 \rightarrow \tau_2 : \star}\ \text{T\_ARROW}$$

$$\frac{\Gamma \vdash \Diamond \quad (\alpha{:}\star) \in \Gamma}{\Gamma \vdash \alpha : \star}\ \text{T\_VAR} \qquad \frac{\Gamma, \alpha{:}\star \vdash \tau : \star}{\Gamma \vdash \forall \alpha{:}\star.\tau : \star}\ \text{T\_ALL}$$

$\boxed{\Gamma \vdash e : \tau}$

$$\frac{\Gamma \vdash \Diamond}{\Gamma \vdash b : \mathbf{bool}}\ \text{E\_BV} \qquad \frac{\Gamma \vdash e_1 : \mathbf{bool} \quad \Gamma \vdash e_2 : \tau \quad \Gamma \vdash e_3 : \tau}{\Gamma \vdash \mathbf{if}\ e_1\ \mathbf{then}\ e_2\ \mathbf{else}\ e_3 : \tau}\ \text{E\_IF}$$

$$\frac{(x : \tau) \in \Gamma \quad \Gamma \vdash \Diamond}{\Gamma \vdash x : \tau}\ \text{E\_VAR}$$

$$\frac{\Gamma, x{:}\tau_1 \vdash e : \tau_2}{\Gamma \vdash \lambda x{:}\tau_1.e : \tau_1 \rightarrow \tau_2}\ \text{E\_AABS}$$

$$\frac{\Gamma \vdash e_1 : \tau_1 \rightarrow \tau_2 \quad \Gamma \vdash e_2 : \tau_1}{\Gamma \vdash e_1\ e_2 : \tau_2}\ \text{E\_APP}$$

$$\frac{\Gamma, \alpha{:}\star \vdash e : \tau}{\Gamma \vdash \Lambda\alpha{:}\star.e : \forall \alpha{:}\star.\tau}\ \text{E\_ATABS}$$

$$\frac{\Gamma \vdash e : \forall \alpha{:}\star.\tau \quad \Gamma \vdash \tau' : \star}{\Gamma \vdash e\langle\tau'\rangle : [\tau'/\alpha]\tau}\ \text{E\_ATAPP}$$

FIGURE 70.15 System F with boolean values.

The exact statement of these properties varies from system to system. Therefore, for concreteness, this section states and proves theorems about System F, the explicitly-typed polymorphic language with boolean values listed in Figure 70.15. Working with an explicitly-typed language simplifies some of the metatheory that follows. We discuss type inference, the process of producing explicit type annotations, in Section 70.4.

70.3.1 Structural Properties of Judgments

First, the typing judgments of System F satisfy a number of *structural* properties regarding the treatment of assumptions in the typing context. These properties are not that interesting in their own regard, but they provide the foundation for more significant theorems.

The first property, called *weakening*, states that irrelevant assumptions do not invalidate typing judgments. (Recall that the notation Γ, Γ' refers to the concatenation of two typing contexts.)

Lemma 70.1 (Weakening for typing)

If $\Gamma, \Gamma' \vdash e : \tau$ and $x \notin dom(\Gamma, \Gamma')$ and $\Gamma \vdash \tau' : \star$, then $\Gamma, x{:}\tau', \Gamma' \vdash e : \tau$.

The proof of this theorem, and all of the others in this subsection, is a simple induction on the typing derivation. We do not give the details here but refer readers to other sources [41].

In fact, there are several weakening lemmas that hold for this type system. Type variables may also be added to the context.

Lemma 70.2 (Type variable weakening for typing)

If $\Gamma, \Gamma' \vdash e : \tau$ and $\alpha \notin dom(\Gamma, \Gamma')$, then $\Gamma, \alpha{:}\star, \Gamma' \vdash e : \tau$.

Furthermore, weakening of both type and term variables holds for the other two judgments of the type system, type formation $\Gamma \vdash \tau{:}\star$ and context formation $\Gamma \vdash \diamond$. We can concisely cover all cases at once, including the two lemmas given earlier, by stating a general form of weakening for all judgments.

Lemma 70.3 (Weakening)

1. If $\Gamma, \Gamma' \vdash \mathfrak{J}$ and $\alpha \notin dom(\Gamma, \Gamma')$ then $\Gamma, \alpha : \star, \Gamma' \vdash \mathfrak{J}$
2. If $\Gamma, \Gamma' \vdash \mathfrak{J}$ and $x \notin dom(\Gamma, \Gamma')$ and $\Gamma \vdash \tau : \star$ then $\Gamma, x{:}\tau, \Gamma' \vdash \mathfrak{J}$

The inverse of weakening is *strengthening*, the ability to remove irrelevant assumptions from the context. Variables are irrelevant if they do not appear freely in contexts, expressions, or types. Strengthening is true for all judgments, for both type and term variables. The notation $fv(\Gamma)$ and $fv(\mathfrak{J})$ calculates the set of free variables appearing in the range of typing contexts Γ and in the components of assertions $\mathfrak{J}$, respectively.

Lemma 70.4 (Strengthening)

1. If $\Gamma, \alpha{:}\star, \Gamma' \vdash \tau : \star$ and $\alpha \notin fv(\Gamma') \cup fv(\mathfrak{J})$, then $\Gamma, \Gamma' \vdash \mathfrak{J}$.
2. If $\Gamma, x{:}\tau, \Gamma' \vdash \mathfrak{J}$ and $x \notin fv(\mathfrak{J})$, then $\Gamma, \Gamma \vdash \mathfrak{J}$.

The next structural property, *exchange*, states that the order of assumptions does not matter. Again, this property holds for all orderings of term and type variables, in all judgments. There is one caveat: term variable assumptions can only move in front of type variable assumptions when the latter does not appear in the type of the former.

Lemma 70.5 (Exchange)

1. If $\Gamma, x{:}\tau_1, y{:}\tau_2, \Gamma' \vdash \mathfrak{J}$ then $\Gamma, y{:}\tau_2, x{:}\tau_1, \Gamma' \vdash \mathfrak{J}$.
2. If $\Gamma, \alpha{:}\star, \beta{:}\star, \Gamma' \vdash \mathfrak{J}$ then $\Gamma, \beta{:}\star, \alpha{:}\star, \Gamma' \vdash \mathfrak{J}$.

3. If $\Gamma, x{:}\tau_1, \alpha{:}\star, \Gamma' \vdash \mathfrak{J}$ then $\Gamma, \alpha{:}\star, x : \tau_1, \Gamma' \vdash \mathfrak{J}$.
4. If $\Gamma, \alpha{:}\star, y{:}\tau_2, \Gamma' \vdash \mathfrak{J}$ and $\Gamma \vdash \tau_2 : \star$ then $\Gamma, y{:}\tau_2, \alpha{:}\star, \Gamma' \vdash \mathfrak{J}$.

The *substitution* lemmas justify assumptions. Types can be substituted for type variables in any judgment.

Lemma 70.6 (Type variable substitution)

$\Gamma, \alpha{:}\star, \Gamma' \vdash \mathfrak{J}$ and $\Gamma \vdash \tau : \star$ then $\Gamma, [\tau/\alpha]\Gamma' \vdash [\tau/\alpha]\mathfrak{J}$.

Furthermore, term variables can be replaced with a term of the same type in the typing judgment. (Term variables do not appear freely in types or contexts so there is no need for a general substitution lemma.)

Lemma 70.7 (Term variable substitution)

If $\Gamma, x{:}\tau, \Gamma' \vdash e : \tau'$ and $\Gamma \vdash e' : \tau$, then $\Gamma, \Gamma' \vdash [e'/x]e : \tau'$.

The final structural rule encodes an invariant that we maintain about the type system—all parts of a judgment are well formed. For example, if a term type checks, then the context and its type are also well formed.

Lemma 70.8 (Regularity/generation)

1. If $\Gamma \vdash \tau : \star$ then $\Gamma \vdash \diamond$.
2. If $\Gamma \vdash e : \tau$ then $\Gamma \vdash \diamond$ and $\Gamma \vdash \tau : \star$.

70.3.2 Syntactic Type Soundness

The most fundamental property of a type system is that of *type soundness*. This property captures the slogan (due to Milner [30]) that "well-typed programs don't go wrong." Said another way, a type system can be used to distinguish terms that have well-defined semantics from those that are syntactically correct but meaningless.

There are many ways to give a semantics to the expressions of a programming language. With *operational semantics*, the meaning of a program is given by its computational behavior or operation. A *small step* operational semantics (Figure 70.16) describes the meaning of an expression as a potentially infinite sequence of rewriting steps of the form $e \to e'$. The rules of this semantics come in two flavors: they either directly eliminate various data structures (rules BETA, TBETA, IFTRUE, and IFFALSE) or search through the term looking for a primitive reduction (rules IFCONG, APPCONG, and TCONG). This semantics is deterministic; any expression can step to at most one result. The reflexive transitive closure of this relation, written $e \to\star e'$, models any number of steps of evaluation. We say that an expression e *cannot step* if there is no e' such that $e \to e'$.

There are two sorts of expressions that cannot step. The first sort are the *values*, v, the desired results of a successfully concluded computation. In System F, values include boolean values, functions and type abstraction.

$$v ::= \mathbf{ture} + \mathbf{false} \mid \lambda x{:}\tau.e \mid \Lambda\alpha{:}\kappa.e$$

$$\frac{}{(\lambda x.e)e' \to [e'/x]e}\ \text{BETA} \qquad \frac{}{\textbf{if true then } e_1 \textbf{ else } e_2 \to e_1}\ \text{IFTRUE}$$

$$\frac{}{(\Lambda\alpha{:}\kappa.e)\langle\tau\rangle \to [\tau/\alpha]e}\ \text{TBETA} \qquad \frac{}{\textbf{if false then } e_1 \textbf{ else } e_2 \to e_2}\ \text{IFFALSE}$$

$$\frac{e_1 \to e_1'}{\textbf{if } e_1 \textbf{ then } e_2 \textbf{ else } e_3 \to \textbf{if } e_1' \textbf{ then } e_2 \textbf{ else } e_3}\ \text{IFCONG}$$

$$\frac{e_1 \to e_1'}{e_1\ e_2 \to e_1'\ e_2}\ \text{APPCONG} \qquad \frac{e \to e'}{e\langle\tau\rangle \to e'\langle\tau\rangle}\ \text{TCONG}$$

FIGURE 70.16 Small-step operational semantics for System F.

All other expressions that cannot step are called *stuck* terms. For example, **true** $(\lambda x{:}\textbf{bool}.x)$ is stuck because no rule in Figure 70.16 applies.

The type soundness theorem asserts that well typed expressions must either diverge (i.e., reduce forever) or eventually produce values. This theorem is also called "Type Safety" [62]. They cannot get stuck. By ruling out stuck terms, this theorem asserts that well-typed expressions evaluate using only the behaviors described by the operational semantics.*

We can formally state this property with the following theorem:

Theorem 70.1 (Type soundness)

If $\varepsilon \vdash e : \tau$, then either e diverges or there exists some v such that $e \to\star v$.

The proof of type soundness derives from two separate results, *preservation* and *progress* [62]. Suppose the expression e does not diverge, but steps to some e' which cannot make any more steps. Preservation states that rewriting an expression does not change its type, so we know that e' is also well typed. Progress states that well typed expressions are either values or can step. However, we assumed that e' cannot make any more transitions, so it must be a value.

Theorem 70.2 (Preservation)

If $\Gamma \vdash e : \tau$ and $e \to e'$ then $\Gamma \vdash e' : \tau$

The proof of the preservation theorem is by induction on the typing derivation using the substitution and weakening lemmas. This theorem is sometimes called *Subject Reduction* after a similar result in proof theory.

Theorem 70.3 (Progress)

If $\varepsilon \vdash e : \tau$ then either e is a value, or there exists some e' such that $e \to e'$.

The proof of the progress theorem relies on the following canonical forms lemma. This lemma tells us that the type of a value is all we need to know to discern its form.

* There are no divergent terms in System F. However, this version of the type soundness theorem cannot derive that fact. Consequently, the same proof technique can be used for languages with nontermination.

Lemma 70.9 (Canonical forms)

1. If $\varepsilon \vdash v : \mathbf{bool}$ then v is **true** or **false**.
2. If $\varepsilon \vdash v : \tau_1 \to \tau_2$ then v is some function $\lambda x{:}\tau.e$.
3. If $\varepsilon \vdash v : \forall\alpha{:}\star.\tau$ then v is some type abstraction $\Lambda\alpha{:}\star.e$.

In System F, the proof of the canonical forms lemma is by mere inspection of the typing rules. For each appropriate form of value, there is only one type that it can be assigned. However, the proof of this lemma is not as straightforward in the presence of nontrivial type equality (Section 70.2.8) or subtyping (Section 70.2.9).

The appeal of small-step semantics is extensibility. The language can be extended with many sorts of structure, yet the proof of type soundness is still elementary. In particular, the techniques listed earlier are amenable to constructs such as tuples, disjoint unions, recursive functions, and data structures and objects. Furthermore, mutable state and exceptions require small revisions to the statements of the lemmas and theorems given earlier (e.g., allowing programs to throw exceptions as well as diverge or terminate with values). The reference *Types and Programming Languages* [41] gives the details of these extensions.

70.3.3 Equivalence and Parametricity

An important notion in the semantics of programming language is the definition of program equivalence. This relation is a prerequisite for arguing about the correctness of a compiler optimization, or even trying to prove that a program behaves as expected. But what does it mean for two λ-calculus terms to be equal? What about terms from System F?

The operational semantics presented in the last section gives a starting point for a definition of equivalence but it is not satisfactory. Consider the following naïve version of equivalence: two expressions are equal exactly when they both evaluate to the same value or when they both diverge. This definition makes sense for expressions of type **nat** or **bool**, but it fails to identify some "obviously equal" expressions. For example, $\lambda x{:}\mathbf{nat}.x + 1$ and $\lambda x{:}\mathbf{nat}.1 + x$ are not equal with this definition.

What is a better definition of equivalence? The gold standard is *contextual equivalence*, which identifies the most terms that can be done so in a manner consistent with the operational semantics. Informally, two expressions of type τ are *contextually equal*, written $e \simeq_\tau e'$, if there is no program context that can distinguish them. For base types such as booleans, contextual equivalence coincides with the naïve equivalence defined above—a program context can only observe whether the result of the computation is **true**, **false** or divergence. But for functions, contextual equivalence is much coarser. There is no program context that can distinguish between $\lambda x{:}\mathbf{nat}.x + 1$ and $\lambda x{:}\mathbf{nat}.1 + x$. Furthermore, abstract types can restrict the ability of the context to make distinctions. For example, the terms **pack** $\langle\mathbf{bool}, \mathbf{true}\rangle$ **as** $\exists\alpha : \star.\alpha$ and **pack** $\langle\mathbf{bool}, \mathbf{false}\rangle$ **as** $\exists\alpha : \star.\alpha$ are contextually equivalent because any context must use these terms with type $\exists\alpha : \star.\alpha$.

However, showing that two programs are contextually equivalent can be difficult. Given two expressions, how can we rigorously argue that there is no context that can tell them apart?

Fortunately, we can use types as the basis for a definition of equivalence for System F, written $e_1 \sim_\tau e_2\,[\eta]$. We define this relation by recursion on the structure of the type τ.

This relation takes an argument η as part of its definition. This argument is a finite map from type variables to binary relations R. The empty map is written ε, and the notation $\eta, \alpha \mapsto R$ extends a map η with a new association from the type variable α to some relation $R{:}\tau_1 \Leftrightarrow \tau_2$ between closed expressions of types τ_1 and τ_2. The notation $\eta(\alpha)$ looks up the binary relation associated with the type variable α.

Definition 70.1 (Logical equivalence)

Define the relation between closed expressions $e \sim_\tau e'$ [η] by induction on the structure of τ as follows:*

$$
\begin{aligned}
e \sim_{\mathbf{bool}} e' \ [\eta] \quad &\triangleq \quad e \simeq_{\mathbf{bool}} e' \\
e \sim_{\tau_1 \to \tau_2} e' \ [\eta] \quad &\triangleq \quad e_1 \sim_{\tau_1} e_1' \ [\eta] \text{ implies } e\ e_1 \sim_{\tau_2} e\ e_2 \ [\eta] \\
e \sim_\alpha e' \ [\eta] \quad &\triangleq \quad \eta(\alpha)(e, e') \\
e \sim_{\forall \alpha:\star.\tau} e'[\eta] \quad &\triangleq \quad \text{for every } \varepsilon \vdash \tau_1 : \star, \varepsilon \vdash \tau_2 : \star \text{ and admissible } R : \tau_1 \Leftrightarrow \tau_2, \\
&\qquad e\langle\tau_1\rangle \sim_\tau e'\langle\tau_2\rangle \quad [\eta, \alpha \mapsto R]
\end{aligned}
$$

For expressions of type **bool**, this relation coincides with contextual (and naïve) equivalence. Two expressions of type **bool** are related when they both evaluate to **true**, both evaluate to **false**, or both diverge. Expressions with function type are related *logically*: they are related when they take related arguments to related results. Therefore, we call this relation a *logical relation*.

The last two cases of the definition, for type variables and quantified types, involve the finite map η. This map directly specifies the relation that is used when the type is a variable. Finally, the relation for polymorphic types includes all expressions that, when given any pair of closed types and *any admissible* relation between those types, produces related results. A relation R is admissible if it is one that respects contextual equivalence (i.e., if two terms are related, then all terms contextually equivalent to the first are related to all terms contextually equivalent to the second).

For this relation to make sense as a definition of equivalence, it must be an equivalence relation. Symmetry and transitivity are easy to show by induction on the structure of τ. However, to show that the relation is reflexive, we must extend this definition to open terms—two open terms are equivalent if a closing substitution of related terms produces related results. We can then show reflexivity of the open logical equivalence for all well-typed expressions.

In fact, not only is logical equivalence an equivalence relation, but it is the *same* relation as contextual equivalence. Therefore, to show that two terms are contextually equivalence, we need only show that they are logically equivalent.

For System F, the reflexivity property also goes by the name *parametricity* and is valuable for reasoning about parametric polymorphism. For example, by looking at the definition of logical equivalence, we can derive that the behavior of a program is not affected by the types that are used to instantiate polymorphic functions. If there were such a dependence, then the definition would not be free to choose any pair of types in the last line. This property justifies the type erasure step in the compilation of ML and Haskell programs. Types cannot influence computation so they are not needed at run time.

Furthermore, if we extend the aforementioned definition to include existential types, parametricity means that we can reason about type abstraction. If the type of a data structure is hidden, then it could be anything. Abstract data structures provide *representation independence*—they can be arbitrarily replaced by any other implementation, as long as there is some relation between them.

Another consequence of parametricity is the derivation of "free theorems," properties that can be concluded about polymorphic functions merely by looking at their types [58]. For example, we can use parametricity to show that there is only one expression of type $\forall\alpha:\star.\alpha \to \alpha$ up to contextual equivalence. All expressions of this type must behave like the polymorphic identity function $\Lambda\alpha:\star.\lambda x{:}\alpha.x$.

The structure of the parametricity theorem is similar to Girard's proof that all well-typed System F expressions terminate [17]. In particular, the flexibility of quantifying over all admissible relations in the last case of the definition resembles "Girard's trick" of quantifying over all "reducibility candidates" [15].

* This definition of logical equivalence is adapted from *Practical Foundations for Programming Languages* [19].

However, despite this similarity, parametricity is not limited to terminating languages. Techniques such as syntactic minimal invariance [4,9], top-top closure [45], and step-indexing [12], can be used to extend parametricity results to languages with features such as recursive functions, recursive types, and mutable state.

70.3.4 Machine Assistance for Formal Reasoning

Type systems for real languages are large and the proofs of their fundamental properties, such as preservation and progress, can be tedious. These proofs are not difficult. They often follow similar patterns and use similar lemmas in their development. What is difficult is managing the details to make sure that all interactions between various features are accounted for. Subtle flaws can lead to security holes, especially when type soundness guarantees are used to justify the security of software systems.

Fortunately, a number of reasoning tools exist to help language designers and type system researchers in their work.

One way to find bugs in type system designs is to test them. Tools such as the Twelf System [39] or PLT/Redex [13] assist in the creation of simple reference implementations. Automatic test case generation can help language designers find counterexamples to preservation or progress. The presence of an implementation also assists in the exploration of the expressiveness of the language—demonstrating that desirable programs actually do type check.

Some researchers use logical frameworks and proof assistants to develop and check their formal proofs. There are many suitable tools for this purpose; we mention only a few here. More information on this topic is available from the results of the POPLMARK challenge [55]. The Twelf System [20,39] supports syntactic reasoning. Through using higher-order abstract syntax and hypothetical judgments, many structural rules of the type system, such as substitution and weakening, are automatic. Nominal Isabelle [56] provides a first-order treatment of bound variables, allowing expressive binding patterns in the context of a general purpose logic. Although the Coq proof assistant [54] is a general purpose tool for formal mathematics and does not provide specific support for type system representation, many researchers have used it for this purpose [40,43].

The Ott tool [51] provides assistance for type setting the syntax of programming languages and judgments. It was used to produce this text. The same definitions can also be used to produce input for systems such as OCaml, Twelf, Coq, and Isabelle/HOL so that the paper, reference implementation, and formal system can be synchronized.

70.4 Type Inference

In Section 70.2, we used local annotations to ensure that that it is always possible to type check programs. For several constructs of the language, we introduced both implicit and explicit forms, pointing out sufficient information that ensures that type checking is syntax directed. Of course, the proposed annotations are not the only solution—there are many similar strategies that take advantage of local information during type checking [44].

However, type annotations are not the only way to define an algorithm for type checking. Sometimes ambiguities in the typing process can be resolved through static analysis. This process is called *type reconstruction* or *type inference*. The task is, when given an implicitly-typed program e and a typing environment Γ, to find an explicitly-typed program e' and type such that $\Gamma \vdash e{:}\tau$ if and only if $\Gamma \vdash e'{:}\tau$.

70.4.1 Constraint-Based Type Inference

Constraint-based type inference breaks the problem of type reconstruction into two steps. First, traverse the program collecting constraints that must be satisfied in order for the program to type check. Next, determine if those constraints are indeed satisfiable.

Type reconstruction for implicitly-typed STLC requires the following simple system of constraints. Here, a constraint is an equality between types, the trivial constraint, or a conjunction of constraints:

$$C ::= \{\tau_1 = \tau_2\} \mid \textbf{true} \mid C_1 \cup C_2$$

A constraint is satisfiable if there is some substitution for the type variables such that every equality is between syntactically equal types.

For example, given an expression $\lambda x.x\ 1$, we can elaborate it to an expression $\lambda x{:}\alpha.x\ 1$, with type $\alpha \to \beta$ under the constraint $\{\alpha = \textbf{nat} \to \beta\}$. We can satisfy the constraint with a substitution that maps α to $\textbf{nat} \to \beta$, so we know that $\lambda x.x\ 1$ type checks with type $(\textbf{nat} \to \beta) \to \beta$.

On the other hand, the expression $\lambda x.x\ x$ produces the constraint $\{\alpha = \alpha \to \beta\}$. This constraint is not satisfiable by any substitution, so $\lambda x.x\ x$ does not type check.

For STLC (the language of Section 70.2.2), we can informally define constraint generation with the rules in Figure 70.17. These rules are syntax directed and specify an algorithm that, given a context Γ and expression e, produces a type τ and constraint C. If this constraint is satisfiable, then there is some valid typing derivation for the program e.

The rules in the figure are informal because they include "freshness assumptions," written α **fresh**. This precondition asserts that the type variable α is freshly generated by the algorithm and distinct from all other variables. This new variable may appear in the output types and constraints. This informality can be resolved via bookkeeping, for example, threading a supply of fresh variables through the judgment.

The rules define a total algorithm by structural recursion over the input expression. For variables, this algorithm emits the type specified in the context and a trivial constraint. For functions, constraint generation for the body of the function uses a freshly generated type variable as the type of the argument. The generated constraints may constrain this variable depending on how that argument is used.

Constraint generation for application expressions uses a fresh type variable as the result type of the application. It determines the type of the function and arguments and then constrains the type of the function to have domain equal to the argument type. Likewise, for **if** expressions, the result constrains the type of the scrutinee of the **if** expression to be a boolean and constrains the two branches to have equal types.

$$\boxed{\Gamma \mapsto e : \tau, C}$$

$$\frac{(x:\tau) \in \Gamma}{\Gamma \mapsto e : \tau, \textbf{true}}\ \text{CSVar}$$

$$\frac{\Gamma, x:\alpha \mapsto e : \tau, C \quad \alpha\ \textbf{fresh}}{\Gamma \mapsto \lambda x.e : \alpha \to \tau, C}\ \text{CSAbs}$$

$$\frac{\Gamma \mapsto e_1 : \tau_1, C_1 \quad \Gamma \mapsto e_2 : \tau_2, C_2 \quad \alpha\ \textbf{fresh}}{\Gamma \mapsto e_1\ e_2 : \alpha, C_1 \cup C_2 \cup \{\tau_1 = \tau_2 \to \alpha\}}\ \text{CSApp}$$

$$\frac{}{\Gamma \mapsto n : \textbf{nat}, \textbf{true}}\ \text{CSNat}$$

$$\frac{}{\Gamma \mapsto b : \textbf{bool}, \textbf{true}}\ \text{CSBool}$$

$$\frac{\Gamma \mapsto e_1 : \tau_1, C_1 \quad \Gamma \mapsto e_2 : \tau_2, C_2 \quad \Gamma \mapsto e_3 : \tau_3, C_3}{\Gamma \mapsto \textbf{if}\ e_1\ \textbf{then}\ e_2\ \textbf{else}\ e_3 : \tau_2, C_1 \cup C_2 \cup C_3 \cup \{\tau_1 = \textbf{bool}\} \cup \{\tau_2 \to \tau_3\}}\ \text{CSIf}$$

FIGURE 70.17 Constraint generation for STLC.

For example, we can use these rules to calculate the constraint for the expression $\lambda x.x\ 1$. The produced type is $\alpha \to \beta$ under the constraint $\mathbf{true} \cup \mathbf{true} \cup \{\alpha = \mathbf{nat} \to \beta\}$.

$$\dfrac{\dfrac{\varepsilon, x:\alpha \mapsto x:\alpha, \mathbf{true} \quad \varepsilon, x:\alpha \mapsto 1:\mathbf{nat}, \mathbf{true} \qquad \beta\,\mathbf{fresh}}{\varepsilon, x:\alpha \mapsto x\ 1:\beta, \mathbf{true} \cup \mathbf{true} \cup \{\alpha = \mathbf{nat} \to \beta\} \quad \alpha\,\mathbf{fresh}}}{\varepsilon \mapsto \lambda x.x\ 1:\alpha \to \beta,\ \mathbf{true} \cup \mathbf{true} \cup \{\alpha = \mathbf{nat} \to \beta\}}$$

If the constraint is satisfiable, *unification* [49] can rewrite it into a "solved form," θ. A constraint is in solved form if it can be viewed as a substitution. In other words, the constraint is *solved* when it is either **true** or a sequence of the form $\{\alpha_1 = \tau_1\} \cup \ldots \cup \{\alpha_i = \tau_i\}$, where the variables $\alpha_1 \ldots \alpha_i$ are distinct and do not appear in $\tau_1 \ldots \tau_i$.

Constraint generation is a total operation, but unification only succeeds if the constraint is satisfiable.

Type inference for STLC is sound and complete with respect to the Curry-style type system. Any program that produces a satisfiable constraint type checks without annotations. Likewise, all well-typed programs produce satisfiable constraints.

Theorem 70.4 (Soundness)

If $\Gamma \mapsto e:\tau, C$ and $Unify(C) = \theta$ then $\theta(\Gamma) \vdash e{:}\theta(\tau)$

Theorem 70.5 (Completeness)

If $\Gamma \vdash e:\tau$ and $\Gamma \mapsto e:\tau_1, C$, then there is some θ such that $Unify(C) = \theta$ and $\theta(\tau_1) = \tau$

70.4.2 The Hindley–Milner Type System

The constraint generation process of the previous section directly extends to tuples, records, sums, and variants. We can also adapt the procedure to include recursive types. However, it does not extend to the polymorphic language in Section 70.2.6. In fact, type reconstruction for implicitly-typed System F is known to be undecidable [38,61].

How do languages like Haskell and ML accommodate polymorphism with type inference? The answer is that they restrict the uses of polymorphism to a tractable subset. This subset, called the Hindley–Milner (HM) (or Damas–Hindley–Milner) type system [10,22] restricts polymorphic types to *prenex quantification*. Instead of allowing quantification to appear anywhere in types, it must occur only at the top level.

The HM type system divides types into two sorts: *polytypes* that contain top-level quantification and *monotypes* that do not. Next we use σ for types that may start with a universal quantifier, and τ for types that contain no quantifiers whatsoever:

$$\begin{array}{lcll} \textit{polytypes} & \sigma & ::= & \forall \alpha{:}\star.\sigma \mid \tau \\ \textit{monotypes} & \tau & ::= & \alpha \mid \tau \to \tau \mid \mathbf{nat} \mid \mathbf{bool} \end{array}$$

Monotypes are a subset of the polytypes and a polytype may or may not start with quantification. In other descriptions of this type system, monotypes are called "types" and polytypes are called "type schemes," indicating that polytypes can be thought of as a family of types, parameterized by type variables.

$\boxed{\sigma_1 \leqslant \tau_2}$

$$\frac{}{\tau \leqslant \tau}\ \text{G\_MONO} \qquad \frac{[\tau'/\alpha]\sigma \leqslant \tau}{\forall\alpha{:}\star.\sigma \leqslant \tau}\ \text{G\_INST}$$

$\boxed{\sigma_1 \leqslant \sigma_2}$

$$\frac{\sigma_1 \leqslant \sigma_2 \quad \alpha \notin fv(\sigma_1)}{\sigma_1 \leqslant \forall\alpha{:}\star.\sigma_2}\ \text{G\_SKOL}$$

FIGURE 70.18 More-general relation.

A polytype σ *generalizes* a monotype τ, written $\sigma \leqslant \tau$, when τ is a substitution instance of the quantified variables of σ. This relation is defined at the top of Figure 70.18. Equivalently, if σ is of the form $\forall\alpha_1{:}\star\ldots\forall\alpha_i{:}\star.\tau'$ then τ must be equal to $[\tau_1/\alpha_1]\ldots[\tau_i/\alpha_i]\tau'$. If this relation holds, then conversely, we say that τ is an *instance* or *specialization* of σ.

For example, the polytype $\forall\alpha{:}\star.\alpha \to \alpha$ generalizes the monotypes **nat** → **nat** and **bool** → **bool**.

We can extend this relation to a judgment between polytypes. We say that σ_1 is *more general than* σ_2 when the judgment $\sigma_1 \leqslant \sigma_2$ is derivable using the last rule in Figure 70.18. If σ_1 is more general than σ_2, then any instance of σ_2 is also an instance of σ_1. For example, the polytype $\forall\alpha{:}\star.\alpha \to \alpha$ is more general than the polytype $\forall\alpha{:}\star.\forall\beta{:}\star.(\alpha \to \beta) \to (\alpha \to \beta)$ and the monotype (**nat** → **bool**) → (**nat** → **bool**) is an instance of both polytypes.

The HM type system uses the distinction between monotypes and polytypes to restrict the typing rules of an implicit, polymorphic λ-calculus, as shown in Figure 70.19. The restrictions show up in two places: functions must take monotypes as arguments (rules HM_ABS and HM_APP) and polymorphism can only be instantiated with monotypes (rule HM_INST).

A nonessential difference between this system and the implicit system of Section 70.2.6 is that type variables are not recorded in the typing context. This bookkeeping is unnecessary because all polymorphism in this system is first order (i.e., all variables have kind ⋆). As a result, all types are well formed and there is no need for the $\Gamma \vdash \tau : \star$ judgment. This system ensures that generalized type variables in rule HM_GEN are unique and unconstrained by ensuring that they do not appear free in the typing environment.

If a program type checks using a derivation produced by these rules, then the Damas–Milner type inference *algorithm* will be able to infer its type. As in the previous section, this algorithm

$\boxed{\Gamma \vdash_{HM} e : \tau}$

$$\frac{(x:\sigma) \in \Gamma}{\Gamma \vdash_{HM} x : \sigma}\ \text{HM\_VAR}$$

$$\frac{\Gamma, x:\tau_1 \vdash_{HM} e : \tau_2}{\Gamma \vdash_{HM} \lambda x.e : \tau_1 \mapsto \tau_2}\ \text{HM\_ABS}$$

$$\frac{\Gamma \vdash_{HM} e_1 : \tau_1 \to \tau_2 \quad \Gamma \vdash_{HM} e_2 : \tau_1}{\Gamma \vdash_{HM} e_1\ e_2 : \tau_2}\ \text{HM\_APP}$$

$$\frac{\Gamma \vdash_{HM} e_1 : \sigma \quad \Gamma, x:\sigma \vdash_{HM} e_2 : \sigma'}{\Gamma \vdash_{HM} \mathbf{let}\, x = e_1\, \mathbf{in}\, e_2 : \sigma'}\ \text{HM\_LET}$$

$$\frac{\Gamma \vdash_{HM} e : \sigma \quad \alpha \notin fv(\Gamma)}{\Gamma \vdash_{HM} e : \forall\alpha{:}\star.\sigma}\ \text{HM\_GEN}$$

$$\frac{\Gamma \vdash_{HM} e : \forall\alpha{:}\star.\sigma}{\Gamma \vdash_{HM} e : [\tau/\alpha]\sigma}\ \text{HM\_INST}$$

FIGURE 70.19 Hindley–Milner polymorphism.

generates constraints by traversing an expression and solves them via unification to determine whether the expression is typeable.

Although the HM type system allows polytypes to be instantiated and generalized anywhere in the typing derivation, the algorithm uses instantiation and generalization only at specific points in the expression. Instantiation occurs only at the uses of variables—if a variable has a polymorphic type in the context, then all of its type arguments are immediately instantiated with fresh unification variables. Furthermore, generalization only occurs at **let**-binding. All free type variables in the type of the right-hand-side of a **let**-expression are generalized by the algorithm.

Even though the rules in Figure 70.19 do not have the uniqueness of types property, the Damas–Milner algorithm is able to type check the same programs as the rules. For example, according to the rules, the expression $\lambda x.x$ can be given the types $\mathbf{nat} \to \mathbf{nat}$ and $\forall\alpha{:}\star.\alpha \to \alpha$. The algorithm produces the type $\forall\alpha{:}\star.\alpha \to \alpha$ for this expression because that is the more general type according to the $\preccurlyeq$ relation. In fact, for any typeable expression, there is exactly one most general type for that expression, called the principal type. The algorithm works by always inferring this type.

Theorem 70.6 (Principal types)

If e is typeable, then there is some type σ, such that for all other types σ' that e could be typed with in the same context, $\sigma \preccurlyeq \sigma'$.

The principal type theorem was first shown for a simple type system without let-expressions by Hindley [22] and was then extended to let polymorphism by Damas and Milner [10]. A modern treatment of this result and thorough discussion appears in Pottier and Rémy [46].

HM polymorphism is a "sweet spot" for implicitly-typed functional languages. It has been difficult to extend this type system with greater expressivity while retaining full type inference. One exception is rank two polymorphism, where functions may take polymorphic functions as arguments but restricts quantifiers from appearing to the left of two or more arrows [26].

However, by relaxing the requirement of complete inference, modern functional languages incorporate many extensions of HM polymorphism. Polymorphic recursion [34], higher-rank polymorphism [16,36], and existential types [27] can be adapted using annotations. Annotations also allow the MLF type system [28] to include first-class polymorphism as in System F. The HM(X) type system [37] formalizes an extension with quantification over arbitrary constraints drawn from arbitrary constraint systems. The OCaml compiler uses constraints for subtyping and row types [47], whereas the GHC compiler uses constraints for type classes, GADTs and type-level functions [57].

70.5 Conclusion

This chapter is intended as an entry point to the field of type systems research. It provides a foundation for the mathematical modeling of type systems, defining the prevalent notation, terminology and structures. The type systems featured in recent research papers use the same tools and methods that we have presented here and sometimes use similar notation. In particular, these papers employ inference rules and judgments for the definitions of the type system, base their model of computation on the λ-calculus, and prove preservation and progress theorems.

The study of type systems themselves frame a modern approach to programming language research. The precision that the study of type systems requires extends to a general study of programming language semantics. The formalisms that are the basis of type systems research give benefits to related endeavors, including language design, compiler construction, tool development, language learning, and program understanding. Furthermore, type systems provide several layers of understanding—computation can be modeled not just in terms of what it does at run time, but also what can be known about it at compile time. Finally, type systems maintain the connection between programming languages and logic, continuing a long tradition of cross-fertilization.

Acknowledgments

This work was partially supported by the National Science Foundation under Grant Nos. 0910500 and 1116620. Thanks to Richard Eisenberg for careful reading and feedback.

Key Terms

Abstract type: A type whose representation is hidden, so that the only interactions that it supports are declared by its interface.

α-equivalence: An equivalence relation that identifies terms up to systematic renaming of bound variables.

Axiom: An inference rule with no premises.

Capture-avoiding substitution: The substitution of a term e_1 for a free variable in another term e_2, where the bound variables of e_2 are renamed to avoid capturing the free variables in e_1.

Church-style: An explicitly-typed language.

Compositional type checking: When the type of a compound construct is derived from the types of its subexpressions.

Contravariant: A component of a type where subtyping is defined inversely with respect to the entire type. The primary example is the domain of function types.

Covariant: A component of a type where subtyping is defined analogously with respect to the entire type.

Currying: Transforming a function that takes a tuple of arguments so that it can be partially applied to each argument in turn. For example, currying converts a function of type $(\tau_1 \times \tau_2) \rightarrow \tau$ into a function of type $\tau_1 \rightarrow \tau_2 \rightarrow \tau$.

Curry–Howard Isomorphism: A relationship between type systems and constructive logic, where types correspond to propositions and programs correspond to proofs.

Curry-style: An implicitly-typed language.

Derivation: A tree of judgments obtained by applying the rules of a type system.

First-order type system: A type system that does not include quantified types.

Ill-typed: An expression that cannot be assigned a type using any derivation of a given type system.

Impredicative polymorphism: Polymorphic types, such as $\forall\alpha:\star.\tau$ or $\exists\alpha:\star.\tau$, where the type variable ranges over polymorphic types, including themselves.

Inference rule: A schematic for a typing derivation, declaring the judgment that can be derived (below the line) from premises (the judgments above the line).

Judgment: A proposition about the relationship between the entities of a programming language, including typing context, terms, and types. A derivation is a proof of the validity of a judgment.

Lambda calculus: A foundational model of computation based on functions.

Metatheory: Logical properties that are proven about a type system or programming language.

Predicative polymorphism: Polymorphic types, such as $\forall\alpha:\star.\tau$ or $\exists\alpha:\star.\tau$, where the type variable does not range over all types, but only over some "smaller" collection of types that does not include the types themselves.

Second-order type system: A type system that includes quantified types, either universal or existential.

Static type system: A type system that supports type checking, an algorithm that determines whether a given program is well-typed just by looking at the text of the program at compile time.

Subsumption: A fundamental property of languages with subtyping: any expression of type τ may also be given any supertype of τ.

Subtype and supertype: In a valid judgment $\Gamma \vdash \sigma <: \tau$, σ is a subtype of τ and τ is a supertype of σ. Any context that requires a value of type τ can be safely provided with a value of type σ.

Syntax-directed type system: A set of typing rules that describe a simple algorithm that determines whether an expression is well typed by using the syntactic form of the expression.
Typeable expression: An expression that appears in a derivable typing judgment.
Type checker: The part of the compiler or interpreter that performs type checking.
Type checking: The process of checking a program to make sure that it complies with a given type system.
Type inference: The process of determining the type of a program when it is not directly specified.
Type rule: A component of a type system.
Type system: The collection of rules that describes the types of programs.
Typing environment or typing context: A list of assumptions about the types of variables.
Uniqueness of derivations: For any judgment $\Gamma \vdash e{:}\tau$, there is at most one derivation.
Uniqueness of types: For any expression e and typing context Γ, there is at most one type τ such that the judgment $\Gamma \vdash e{:}\tau$, is derivable.
Valid judgment: A judgment that can be derived by a given type system.
Value: A desired result of computation.
Well-typed: An expression that can be assigned a type and appears in a valid typing judgment.

Further Information

Further information about the study of type systems of programming languages is available from several recent textbooks. Pierce's textbook [41], *Types and Programming Languages* is a standard reference and provides an extensive introduction to type systems research and reference for many topics including subtyping, recursive types, and first-order and higher-order polymorphism. The sequel, *Advanced Topics in Types and Programming Languages*, edited by Pierce [42], contains in-depth articles about logical relations, type inference, and many other topics. Robert Harper's recent book, *Practical Foundations for Programming Languages* [19], provides a modern exposition on the technical issues in this chapter. Cardelli's article [7] in *The Computer Science and Engineering Handbook* is an inspiration for this work and provides additional introductory material.

Many ideas that find their way into programming languages have their roots in Type Theory, a foundational logic grounded in computation. Barendregt's article, "Lambda Calculi with Types" [3], discusses Curry and Church-style type systems as well as their relation to logics via the Curry–Howard isomorphism. Nordstrom, Petersson and Smith [35] summarizes Martin-Löf's work, which introduced the notation for judgments and typing rules. Girard, Lafont and Taylor [18] describe the connections between System F and computation.

The latest research papers about type systems appear in the ACM SIGPLAN conferences *Principles of Programming Languages (POPL), International Conference on Functional Programming (ICFP)*, and *Object-Oriented Programming Languages, Systems and Applications (OOPSLA)*.

References

1. D. Aspinall and M. Hofmann. Dependent types. In B.C. Pierce, ed., *Advanced Topics in Types and Programming Languages*, Chapter 2, pp. 45–86. The MIT Press, Cambridge, MA, 2005.
2. H. P. Barendregt. *The Lambda Calculus: Its Syntax and Semantics*. North Holland, Amsterdam, the Netherlands, 1984.
3. H. P. Barendregt. Lambda calculi with types. In S. Abramsky, D. M. Gabbay, and S. E. Maibaum, eds., *Handbook of Logic in Computer Science* (Vol. 2), pp. 117–309. Oxford University Press, Inc., New York, 1992.
4. L. Birkedal and R. Harper. Relational interpretations of recursive types in an operational setting (summary). In *TACS '97: Proceedings of the Third International Symposium on Theoretical Aspects of Computer Software*, pp. 458–490. Springer-Verlag, London, U.K., 1997.

5. C. Böhm and A. Berarducci. Automatic synthesis of typed Λ-programs on term algebras. *Theoretical Computer Science*, 39(2–3):135–154, August 1985.
6. A. Bove and P. Dybjer. Dependent types at work. In A. Bove, L. Barbosa, A. Pardo, and J. Pinto, eds., *Language Engineering and Rigorous Software Development*, volume 5520 of *Lecture Notes in Computer Science*, pp. 57–99. Springer, Berlin, Germany, 2009.
7. L. Cardelli. Type systems. In A. B. Tucker, ed., *The Computer Science and Engineering Handbook*, Chapter 97. CRC Press, Boca Raton, FL, 2004.
8. K. Crary. Logical relations and a case study in equivalence checking. In B.C. Pierce, ed., *Advanced Topics in Types and Programming Languages*, Chapter 6, pp. 223–244. The MIT Press, Cambridge, MA, 2005.
9. K. Crary and R. Harper. Syntactic logical relations for polymorphic and recursive types. *Electronic Notes on Theoretical Computer Science*, 172:259–299, April 2007.
10. L. Damas and R. Milner. Principal type schemes for functional programs. In *Proceedings 9th ACM Symposium on Principles of Programming Lanuages*, pp. 207–212. New York, 1982.
11. M. Dezani-Ciancaglini and U. De'Liguoro. Sessions and session types: an overview. In *Proceedings of the 6th International Conference on Web Services and Formal Methods, WS-FM '09*, pp. 1–28. Springer-Verlag, Berlin, Heidelberg, 2010.
12. D. Dreyer, A. Ahmed, and L. Birkedal. Logical step-indexed logical relations. In *Proceedings of the 2009 24th Annual IEEE Symposium on Logic In Computer Science*, LICS '09, pp. 71–80. IEEE Computer Society, Washington, DC, 2009.
13. M. Felleisen, R. B. Findler, and M. Flatt. *Semantics Engineering with PLT Redex*. 1st edn., The MIT Press, Cambridge, MA, 2009.
14. G. Frege. *Begriffsschrift: eine der arithmetischen nachgebildete Formelsprache des reinen Denkens*. Verlag von L. Nebert, Halle an der Saale, 1879. Translated as Concept Script, a formal language of pure thought modelled upon that of arithmetic, by S. Bauer-Mengelberg in J. van Heijenoort ed., From *Frege to Gödel: A Source Book in Mathematical Logic*, pp. 1879–1931, Harvard University Press, Cambridge, MA, 1967.
15. J. Gallier. On Girard's "candidates de reductibilité". In P. Odifreddi, ed., *Logic and Computer Science*, pp. 123–203. Academic Press, London, U.K., 1990.
16. J. Garrigue and D. Rémy. Extending ML with semi-explicit higher-order polymorphism. *Information and Computation*, 155(1/2):134–170, 1999.
17. J.-Y. Girard. Une extension de l'interpretation de gödel à l'analyse, et son application à l'élimination des coupures dans l'analyse et la théorie des types. In *Proceedings of the Second Scandinavian Logic Symposium*, p. 6392, Amsterdam, the Netherlands, 1971.
18. J.-Y. Girard, P. Taylor, and Y. Lafont. *Proofs and Types*. Cambridge University Press, New York, 1989.
19. R. Harper. *Practical Foundations for Programming Languages*. Cambridge University Press, New York, 2012.
20. R. Harper and D. R. Licata. Mechanizing metatheory in a logical framework. *Journal of Functional Programming*, 17(4–5):613–673, July 2007.
21. F. Henglein, H. Makholm, and H. Niss. Effect types and region-based memory management. In B.C. Pierce, ed., *Advanced Topics in Types and Programming Languages*, Chapter 3, pp. 87–136. The MIT Press, Cambridge, MA, 2005.
22. J. R. Hindley. The principal type scheme of an object in combinatory logic. *Transactions of the American Mathematical Society*, 146:29–40, 1970.
23. A. Igarashi, B. C. Pierce, and P. Wadler. Featherweight Java: A minimal core calculus for Java and GJ. *ACM Transactions on Programming Languages and Systems*, 23(3):396–450, May 2001.
24. J. Jeuring, S. Leather, J. P. Magalhães, and A. R. Yakushev. Libraries for generic programming in haskell. In P. Koopman, R. Plasmeijer, and D. Swierstra, ed., *Advanced Functional Programming*, volume 5832 of *Lecture Notes in Computer Science*, pages 165–229. Springer, Berlin, Germany, 2009.

25. M. P. Jones. A system of constructor classes: Overloading and implicit higher-order polymorphism. In *Proceedings of the Conference on Functional Programming Languages and Computer Architecture, FPCA '93*, pp. 52–61. ACM, New York, 1993.
26. A. J. Kfoury and J. B. Wells. A direct algorithm for type inference in the rank-2 fragment of the second-order λ-calculus. In *Proceedings of the 1994 ACM Conference on LISP and Functional Programming, LFP '94*, pp. 196–207. ACM, New York, 1994.
27. K. Läufer. Type classes with existential types. *Journal of Functional Programming*, 6(03):485–518, 1996.
28. D. Le Botlan and D. Rémy. MLF: Raising ML to the power of System F. In *Proceedings of the Eighth ACM SIGPLAN International Conference on Functional Programming*, pp. 27–38. ACM, New York, August 2003.
29. Z. Luo. *Computation and Reasoning: A Type Theory for Computer Science*. Oxford University Press, New York, 1994.
30. R. Milner. A theory of type polymorphism in programming. *Journal of Computer and System Sciences*, 17(3):348–375, 1978.
31. R. Milner, M. Tofte, R. Harper, and D. Macqueen. *The Definition of Standard ML (Revised)*. MIT Press, Cambridge, MA, 1997.
32. G. Morrisett. Typed assembly language. In B.C. Pierce, ed., *Advanced Topics in Types and Programming Languages*, Chapter 4, pp. 141–176. The MIT Press, Cambridge, MA, 2005.
33. G. Morrisett, D. Walker, K. Crary, and N. Glew. From system F to typed assembly language. *ACM Transactions on Programming Languages and Systems*, 21(3):527–568, May 1999.
34. A. Mycroft. Polymorphic type schemes and recursive definitions. In M. Paul and B. Robinet, eds., *International Symposium on Programming*, volume 167 of *Lecture Notes in Computer Science*, pp. 217–228. Springer, Berlin, Germany, 1984.
35. B. Nordstrom, K. Petersson, and J. M. Smith. *Programming in Martin-Löf's Type Theory: An Introduction (International Series of Monographs on Computer Science)*. Oxford University Press, New York, 1990.
36. M. Odersky and K. Läufer. Putting type annotations to work. In *Proceedings of the 23rd ACM SIGPLAN-SIGACT Symposium on Principles of Programming Languages*, POPL '96, pp. 54–67. ACM, New York, 1996.
37. M. Odersky, M. Sulzmann, and M. Wehr. Type inference with constrained types. *Theory and Practice of Object Systems*, 5(1):35–55, 1999.
38. F. Pfenning. On the undecidability of partial polymorphic type reconstruction. *Fundamenta Informaticae*, 19(1-2):185–199, September 1993.
39. F. Pfenning and C. Schuermann. Twelf User's Guide Technical Report CMU-CS-98-173, Department of Computer Science, Carnegie Mellon University, November 1998.
40. B. Pierce and S. Weirich. Preface to the issue entitled "Special issue: The POPLmark Challenge". *Journal of Automated Reasoning*, 49:301–302, 2012.
41. B. C. Pierce. *Types and Programming Languages*. MIT Press, Cambridge, MA, 2002.
42. B. C. Pierce. *Advanced Topics in Types and Programming Languages*. The MIT Press, Cambridge, MA, 2004.
43. B. C. Pierce, C. Casinghino, M. Gaboardi, M. Greenberg, C. Hriţcu, V. Sjöberg, and B. Yorgey. *Software Foundations*. Electronic textbook, 2012. http://www.cis.upenn.edu/~bcpierce/sf (accessed October 22, 2013).
44. B. C. Pierce and D. N. Turner. Local type inference. *ACM Transactions on Programming Languages and Systems*, 22(1):1–44, January 2000.
45. A. M. Pitts. Typed operational reasoning. In B. C. Pierce, ed., *Advanced Topics in Types and Programming Languages*, Chapter 7, pp. 245–289. The MIT Press, Cambridge, MA, 2005.
46. F. Pottier and D. Rémy. The essence of ML type inference. In B. C. Pierce, ed., *Advanced Topics in Types and Programming Languages*, Chapter 10, pp. 389–489. MIT Press, Cambridge, MA, 2005.

47. D. Rémy and J. Vouillon. Objective ML: An effective object-oriented extension to ML. *Theory and Practice of Object Systems*, 4(1):27–50, 1998.
48. J. C. Reynolds. Towards a theory of type structure. In *Symposium on Programming '74*, pp. 408–423. New York, 1974.
49. J. A. Robinson. A machine-oriented logic based on the resolution principle. *Journal of the ACM*, 12(1):23–41, January 1965.
50. A. Sabelfeld and A. C. Myers. Language-based information-flow security. *IEEE Journal on Selected Areas in Communications*, 21(1):5–19, IEEE Press Piscataway, NJ, 2003.
51. P. Sewell, F. Z. Nardelli, S. Owens, G. Peskine, T. Ridge, S. Sarkar, and R. Strniša. Ott: Effective tool support for the working semanticist. *Journal of Functional Programming*, 20(1):71–122, January 2010.
52. W. W. Tait. Intensional interpretations of functionals of finite type I. *Journal of Symbolic Logic*, 32(2):198–212, 1967.
53. K. Takeuchi, K. Honda, and M. Kubo. An interaction-based language and its typing system. In *Proceedings of the 6th International PARLE Conference on Parallel Architectures and Languages Europe, PARLE '94*, pp. 398–413. Springer-Verlag, London, U.K. 1994.
54. The Coq Development Team. The Coq proof assistant reference manual: Version 8.4pl2, 2013. http://coq.inria.fr/refman/ (accessed October 22, 2013).
55. The POPLmark team. The POPLmark challenge, 2005. http://plclub.org/poplmark (accessed October 22, 2013).
56. C. Urban. Nominal techniques in Isabelle/HOL. *Journal of Automated Reasoning*, 40(4):327–356, 2008.
57. D. Vytiniotis, S. P. Jones, T. Schrijvers, and M. Sulzmann. OutsideIn(X): Modular type inference with local assumptions. *Journal of Functional Programming*, 21(Special Issue 4-5):333–412, 2011.
58. P. Wadler. Theorems for free! In *Proceedings of the Fourth International Conference on Functional Programming Languages and Computer Architecture, FPCA '89*, pp. 347–359. ACM, New York, 1989.
59. P. Wadler and S. Blott. How to make ad-hoc polymorphism less ad hoc. In *Proceedings of the 16th ACM SIGPLAN-SIGACT Symposium on Principles of Programming Languages, POPL '89*, pp. 60–76. ACM, New York, 1989.
60. D. Walker. Substructural type systems. In B.C. Pierce, ed., *Advanced Topics in Types and Programming Languages*, Chapter 1, pp. 3–43. The MIT Press, Cambridge, MA, 2005.
61. J. B. Wells. Typability and type checking in the second-order λ-calculus are equivalent and undecidable. In *Proceedings of the Ninth Annual IEEE Symposium on Logic in Computer Science (LICS)*, pp. 176–185. New York, 1994.
62. A. K. Wright and M. Felleisen. A syntactic approach to type soundness. *Information and Computation*, 115(1):38–94, November 1994.

71

Formal Methods

Jonathan P. Bowen
Museophile Limited

Michael G. Hinchey
University of Limerick

71.1 Introduction

Computers do not make mistakes, or so we are told. However, computer software is written by, and hardware systems are designed and assembled by, humans, who certainly *do* make mistakes.

Errors in a computer system may be as a result of misunderstood or contradictory requirements, unfamiliarity with the problem, or simply human error during design or coding of the system. Alarmingly, the costs of maintaining software—the costs of rectifying errors and adapting the system to meet changing requirements or changes in the environment of the system—greatly exceed the original implementation costs.

As computer systems are being used increasingly in safety-critical applications—that is, systems where a failure could result in the loss of human life, mass destruction of property, or significant financial loss—both the media and various regulatory bodies involved with standards, especially covering safety-critical applications (Bowen and Stavridou 1993, Gnesi and Margaria 2012) and security applications (Grumberg et al. 2008) have considered formal methods and their role in the specification and design phases of system development and software engineering for many years (Bjørner 2000).

71.2 Underlying Principles

There can be some confusion over what is meant by a "specification" and a "model." David Parnas has differentiated between specification and descriptions or models as follows (Parnas 1995):

- A *description* is a statement of some of the actual attributes of a product, or a set of products.
- A *specification* is a statement of properties required of a product, or a set of products.
- A *model* is a product, neither a description nor a specification. Often it is a product that has some, but not all, of the properties of some "real product."

Others use the terms *specification* and *model* more loosely; a model may sometimes be used as a specification. The process of developing a specification into a final product is one in which a model may be used along the way or even as a starting point.

71.2.1 Formal Methods

Over the last 50 years, computer systems have increased rapidly in terms of both size and complexity. As a result, it is both naive and dangerous to expect a development team to undertake a project without stating clearly and precisely what is required of the system. This is done as part of the *requirements specification* phase of the software life cycle, the aim of which is to describe *what* the system is to do, rather than *how* it will do it.

The use of natural language for the specification of system requirements tends to result in ambiguity and requirements that may be mutually exclusive. Formal methods have evolved as an attempt to overcome such problems by employing discrete mathematics to describe the function and architecture of a hardware or software system, and various forms of *logic* to reason about requirements, their interactions, and validity.

The term *formal methods* is itself misleading; it originates from formal logic but is now used in computing to refer to a plethora of mathematically based activities. For our purposes, a formal method consists of notations and tools with a mathematical basis that are used to unambiguously specify the requirements of a computer system and that support the *proof* of properties of this specification and proofs of correctness of an eventual implementation with respect to the specification.

Indeed, it is true to say that so-called formal methods are not so much methods as formal systems. Although the popular formal methods provide a *formal notation*, or formal specification language, they do not adequately incorporate many of the methodological aspects of traditional development methods.

Even the term *formal specification* is open to misinterpretation by different groups of people. Two alternative definitions for "formal specification" are given in a glossary issued by the IEEE (IEEE 1991):

1. A specification written and approved in accordance with established standards
2. A specification written in a formal notation, often for use in proof of correctness

In this chapter, we adopt the latter definition, which is the meaning assumed by most formal method users.

The notation employed as part of a formal method is "formal" in that it has a mathematical semantics so that it can be used to express specifications in a clear and unambiguous manner and allow us to abstract from actual implementations to consider only the salient issues. This is something that many programmers find difficult to do because they are used to thinking about implementation issues in a very concrete manner.

While programming languages are formal languages, they are generally not used in formal specification, as most languages do not have a full formal semantics, and they force us to address implementation issues before we have a clear description of what we want the system to do. Instead, we use the language of mathematics, which is universally understood, well established in notation, and most importantly enables the generalization of a problem so that it can apply to an unlimited number of different cases. Here we have the key to the success of formal specification—one must *abstract* away from the details of the implementation and consider only the essential relationships of the data, and we can model even the most complex systems using simple mathematical objects: for example, *sets*, *relations*, functions, etc.

At the specification phase, the emphasis is on clarity and precision, rather than efficiency. Eventually, however, one must consider how a system can be implemented in a programming language that, in general, will not support abstract mathematical objects (functional programming languages are an exception) and will be efficient enough to meet agreed requirements and concrete enough to run on the available hardware configuration.

As in structured design methods, the formal specification must be translated to a design—a clear plan for implementation of the system specification—and eventually into its equivalent in a programming language. This approach is known as *refinement* (Back and von Wright 1998).

The process of *data refinement* involves the transition from abstract data types such as sets, sequences, and mappings to more concrete data types such as arrays, pointers, and record structures, and the subsequent verification that the concrete representation can adequately capture all of the data in the formal specification. Then, in a process known as *operation refinement*, each *operation* must be translated so that it operates on the concrete data types. In addition, a number of *proof obligations* must be satisfied, demonstrating that each concrete operation is indeed a "refinement" of the abstract operation—that is, performing at least the same functions as the abstract equivalent, but more concretely, more efficiently, involving less nondeterminism, etc.

Many specification languages have relatively simple underlying mathematical concepts involved. For example, the Z (pronounced "zed") notation (Spivey 2001, ISO 2002) is based on (typed) set theory and first-order predicate logic, both of which *could* be taught at school level. The problem is that many software developers do not currently have the necessary education and training to understand these basic principles, although there have been attempts to integrate suitable courses into university curricula (Bowen and Reeves 2011).

It is important for students who intend to become software developers to learn how to abstract away from implementation detail when producing a system specification. Many find this process of *abstraction* a difficult skill to master. It can be useful for reverse engineering as part of the software maintenance process, to produce a specification of an existing system that requires restructuring. Equally important is the skill of refining an abstract specification toward a concrete implementation, in the form of a program for development purposes (Morgan 1998).

The process of refinement is often carried out informally because of the potentially high cost of fully formal refinement. Given an implementation, it is theoretically possible, although often intractable, to *verify* that it is correct with respect to a specification, if both are mathematically defined. More usefully, it is possible to *validate* a formal specification by formulating required or expected properties and formally proving, or at least informally demonstrating, that these hold. This can reveal omissions or unexpected consequences of a specification. *Verification* and *validation* are complementary techniques, both of which can expose errors.

71.3 Best Practices

Technology transfer (e.g., see Ball et al. 2004) has always been an issue with formal methods, largely because of the significant training and expertise that is necessary for their use. Most engineering disciplines accept mathematics as the underpinning foundations, to allow calculation of design parameters before the implementation of a product. However, software engineers have been somewhat slow to accept such principles in practice, despite the very mathematical nature of all software. This is partly because it is still possible to produce remarkably reliable systems without using formal methods. In any case, the use of mathematics for software engineering is a matter of continuing debate (Tremblay 2000, Hinchey et al. 2008, Parnas 2010).

There are well-documented and well-established industrial examples in which a formal approach has been taken to develop significant systems in a beneficial manner that are easily accessible by professionals (e.g., see Hinchey and Bowen 1999, Ball et al. 2004). Formal methods, including formal specification and modeling, should be considered as one of the possible techniques to improve software quality, where it can be demonstrated to do this cost-effectively.

In fact, just using formal specification within the software development process has been shown to have benefits in reducing the overall development cost (Bowen and Stavridou 1993). Costs tend to be increased early in the life cycle, but reduced later on at the programming, testing, and maintenance stages, where correction of errors is far more expensive. A formal specification can be especially useful in generating test cases rigorously and cost-effectively (Hierons et al. 2009).

An early and widely publicized successful example of the use of formal methods in industry was the IBM CICS (Customer Information Control System) project, where Z was used to specify a portion of this large transaction processing system with an estimated 9% reduction in development costs. There were approximately half the usual number of errors discovered in the software, leading to increased software quality.

Writing a good specification is something that comes only with practice, despite the existence of guidelines. However, there are some good reasons why a mathematical approach may be beneficial in producing a specification:

Precision: Natural language and diagrams can be very ambiguous. A mathematical notation allows the specifier to be very exact about what is specified. It also allows the reader of a specification to identify properties, problems, etc., which may not be obvious otherwise.

Conciseness: A formal specification, although precise, is also very concise compared with an equivalent high-level language program, which is often the first formalization of a system produced if formal methods are not used. Such a specification can be an order of magnitude smaller than the program that implements it, and hence is that much easier to comprehend.

Abstraction: It is all too easy to become mired in detail when producing a specification, making it very confusing and obscure to the reader. A formal notation allows the writer to concentrate on the essential features of a system, ignoring those that are implementation details. However, this is perhaps one of the most difficult skills in producing a specification.

Reasoning: Once a formal specification is available, mathematical reasoning is possible to aid in its validation. This is also useful for discussion implications of features, especially within a team of designers.

A design team that understands a particular formal specification notation can benefit from the above improvements in the specification process. It should be noted that much of the benefit of a formal specification derives from the process of producing the specification, as well as the existence of the formal specification after this (Hall 1990).

71.3.1 Specification Languages

The choice of specification language is likely to be influenced by many factors: previous experience, availability of tools, standards imposed by various regulatory bodies, and the particular aspects that must be addressed by the system in question. Another consideration is the degree to which a specification language is executable. This is the subject of some dispute, and the reader is directed elsewhere for a discussion of this topic (Hayes and Jones 1989, Fuchs 1992, Bowen and Hinchey 1999). Indeed, the development of any complex system is likely to require the use of multiple notations at different stages in the process and to describe different aspects of a system at various levels of abstraction. As a result, over the last several decades, the vast majority of the mainstream formal methods have been extended and re-interpreted to address issues of concurrency (Milner 1999, Abdallah et al. 2005), real-time behavior (Zhou and Hansen 2003), and object orientation (Duke and Rose 2000, Smith 2000).

There is always, necessarily, a certain degree of trade-off between the expressiveness of a specification language and the levels of abstraction that it supports. While certain languages may have wider "vocabularies" and constructs to support the particular situations that need to be handled, they are likely to force us toward particular implementations; while they will shorten a specification, they will make it less abstract and more difficult for reasoning (Bowen and Hinchey, 1995a).

Formal specification languages can be divided into essentially three classes:

Model-oriented approaches as exemplified by ASM (Abstract State Machines) (Börger and Stärk 2003), B-Method (Schneider 2001), Event-B (Abrial 2010), RAISE (Rigorous Approach to Industrial Software Engineering) (Dang Van et al. 2002), VDM (Vienna Development Method) (ISO 1996), and the Z notation (Spivey 2001, ISO 2002). These approaches involve the derivation of an explicit model of the system's desired behavior in terms of abstract mathematical objects.

Property-oriented approaches using *axiomatic semantics* (such as Larch Guttag 1993), which use first-order predicate logic to express *preconditions* and *postconditions* of operations over abstract data types, and *algebraic semantics* (such as the OBJ family of languages including CafeOBJ [Futatsugi et al. 2000]), which are based on multisorted algebras and relate properties of the system in question to equations over the entities of the system.

Process algebras such as CSP (Communicating Sequential Processes) (Abdallah et al. 2005) and the π-calculus (Milner 1999), which have evolved to meet the needs of concurrent, distributed, and real-time systems, and which describe the behavior of such systems by describing their algebras of communicating processes.

It is not always possible to classify a formal specification language in just one of the categories above. LOTOS (Language Of Temporal Ordering Specifications) (ISO 1989, Turner 1993), for example, is a combination of ACT ONE and CCS (Calculus of Communicating Systems); while it can be classified as an algebraic approach, it exhibits many properties of a process algebra too. Similarly, the RAISE development method is based on extending a model-based specification language (specifically, VDM-SL) with concurrent and temporal aspects.

As well as the basic mathematics, a specification language should also include facilities for structuring large specifications. Mathematics alone is all very well in the small, but if a specification is a thousand pages long (and formal specifications of this length exist), there must be aids to organize the inevitable complexity. Z provides the *schema notation* for this purpose, which packages up the mathematics so that it can be reused subsequently in the specification. A number of schema operators, many matching logical connectives, allow recombination in a flexible manner.

A formal specification should also include an informal explanation to put the mathematical description into context and help the reader understand the mathematics. Ideally, the natural language description should be understandable on its own, although the formal text is the final arbiter as to the meaning of the specification. As a rough guide, the formal and informal descriptions should normally be of approximately the same length. The use of mathematical terms should be minimized, unless explanations are being included for didactic purposes.

Formal methods have proved useful in embedded systems and control systems (e.g., see Tretmans et al. 2001, Tiwari et al. 2003, Kordon and Lemoine 2010). Synchronous languages, such as Esterel, Lustre and Signal, have also been developed for *reactive systems* requiring continuous interaction with their environment (Benveniste et al. 2003). Specialist and combined languages may be needed for some systems. *Hybrid systems* (Jones et al. 2007) extend the concept of *real-time systems*. In the latter, time must be considered, possibly as a continuous variable. In hybrid systems, the number of continuous variables may be increased. This is useful in control systems where a digital computer is responding to real-world analog signals.

More visual formalisms, such as the well-established Statecharts approach, are available and are appealing for industrial use, with the associated STATEMENT tool support (Harel and Politi 1998) that has been incorporated into the widely used Unified Modeling Language (UML). However, the reasoning aspects and the exact semantics are less well defined. Some specification languages, such as SDL (Specification and Design Language) (Turner 1993), have provided particularly good commercial tool support, which is very important for industrial use.

There have been many attempts to improve the formality of the various structural design notations in widespread use (Bowen and Hinchey 1999). UML includes the Object Constraint Language (OCL) (Warmer and Kleppe 2003) developed by IBM, an expression language that allows constraints to be formalized, but this part of UML is under-utilized with a lack of tool support and is only a small part of UML in any case.

Object orientation is an important development in programming languages that has also been reflected in specification languages. For example, Object-Z is an object-oriented version of the Z notation that has gained some acceptance (Duke and Rose 2000, Smith 2000, Derrick and Boiten 2001). The Perfect Developer tool has been developed by Escher Technologies to refine formal specifications to object-oriented programming languages such as Java.

71.3.2 Modeling Systems

As previously discussed, the difference between specification and modeling is open to some debate. Different specification languages emphasize and allow modeling to different extents. Algebraic specification eschews the modeling approach, but other specification languages such as Z and VDM actively encourage it.

Some styles of modeling have been formulated for specific purposes. For example, Petri nets may be applied in the modeling of concurrent systems using a specific diagrammatic notation that is quite easily formalizable. The approach is appealing, but the complexity can become overwhelming. Features such as deadlock are detectable, but full analysis can be intractable in practice, since the problem of scaling is not well addressed.

Mathematical modeling allows reasoning about (some parts of) a system of interest. Here, aspects of the system are defined mathematically, allowing the behavior of the system to be predicted. If the prediction is correct this reinforces confidence in the model. This approach is familiar to many scientists and engineers.

Executable models allow rapid prototyping of systems (Fuchs 1992). A very high-level programming language such as a functional program or a logic program (which have mathematical foundations) may be used to check the behavior of the system. Rapid prototyping can be useful in demonstrating a system to a customer before the expensive business of building the actual system is undertaken. Again, scientists and engineers are used to carrying out experiments by using such models.

A branch of formal methods known as *model checking* allows systems to be tested exhaustively (Grumberg et al. 2000, Bérard et al. 2001). Most computer-based systems are far too complicated to test completely because the number of ways the system could be used is far too large. However, a number of techniques, *Binary Decision Diagrams* (BDDs) for example, allow relatively efficient checking of significant systems, especially for hardware (Kropf 2000). An extension of this technique, known as *symbolic model checking*, allows even more generality to be introduced.

Mechanical tools exist to handle BDDs and other model-checking approaches efficiently. SPIN is one of the leading general model-checking tools that is widely used (Holzmann 2003). A more specialist tool based on CSP (Abdallah et al. 2005) known as FDR (Failure Divergence Refinement) allows model checking to be applied to concurrent systems that can be specified in CSP.

71.3.3 Conclusion

Driving forces for best practice include standards, education, training, tools, available staff, certification, accreditation, legal issues, etc. A full discussion of these is out of the scope of this chapter. Aspects of best practice for specification and modeling depend significantly on the selected specification notation. One of the more popular formal notations that has been used in industry is Z. To illustrate the way in which this notation is typically used, a case study using Z is presented in the next section. This demonstrates both some of the underlying principles, and best practice, when employing Z for specification and modeling.

71.4 Case Study

The Z notation (Spivey 2001, ISO 2002) is one of the most widely used formal specification languages and is normally used in a modeling style. An abstract state is first formulated, and then operations on that state are specified. In this section, we present a case study using Z to illustrate this style of specification. The example does not exhaustively present the features of Z, but gives a flavor of the style of presentation of a typical Z specification, with extra informal explanation on the notation and conventions where required. Z constructs are introduced as the example is presented and a glossary of Z notation is provided at the end of the chapter for the convenience of the reader. A basic understanding of set theory and logic will help in understanding the specification.

Window management systems are used extensively for user interfaces to computer systems. The specification given here is of a (fictitious) window system. For more realistic examples of some implemented windows systems presented in Z (see Bowen 2003).

71.4.1 Basic Types

Z is a typed language, which allows a certain amount of consistency checking by a mechanical type-checker. However, the only predefined type is the set of integers, denoted $\mathbb{Z}$. Further types must be defined for a particular specification. These *basic types* (also known as *given sets*) may be introduced as follows:

$$[Position, Value]$$

This provides a set of pixel (picture element) positions (e.g., coordinates on a screen), together with possible pixel values (e.g., colors). Note that we are no more specific than this in the specification presented here. It is important not to introduce irrelevant implementation details into a specification, since this restricts the eventual implementer of the system and clutters the specification with information that is not required at a high level of abstraction.

71.4.2 Abbreviation Definitions

It is often useful to include definitions in a specification for commonly used concepts. This helps to reduce the size of the specification and introduces important concepts to the reader in one place, allowing them to be used later within the specification. Pixel maps, relating pixels to their associated values, are an integral part of most window systems. In fact, each pixel has at most one value (assuming it is defined), so we can model a pixel map as a *partial function* from pixel positions to their values:

$$Pixmap == Position \rightarrowtail\!\!\!\!\!+ Value$$

71.4.3 Generic Definitions

Z has its own library of "tool-kit" operators, formally defined in terms of more basic mathematical concepts, as presented in Spivey (2001). Sometimes it is helpful to extend this library with further *generic definitions* which may be used to define a family of generic constants, applicable to a variety of basic types. Such definitions may be useful for other specifications as well as the one being constructed, allowing reuse of specification components.

For example, a *sequence* of pixel maps may be overlaid in the order given by the sequence to produce a new pixel map. An operator to do this could equally well apply to other partial functions as well as pixel maps, so we can define it generically, using a "distributed overriding" operator:

$$[P, V]$$

$$\oplus / : \mathrm{seq}\ (P \rightarrowtail V) \rightarrow (P \rightarrowtail V)$$

$$\oplus / \langle\rangle = \emptyset$$

$$\forall p : P \rightarrowtail V \cdot \oplus / \langle p \rangle = p$$

$$\forall s,\ t : \mathrm{seq}\ (P \rightarrowtail V) \cdot \oplus / (s \frown t) = (\oplus / s) \oplus (\oplus / t)$$

Here, the base cases for the empty sequence $\langle\rangle$ and a singleton sequence $\langle p \rangle$ are considered, followed by the more general case of two arbitrary sequences concatenated together $s \frown t$. Distributed overriding is particularly useful for the specification presented here in specifying the view on a screen of a display, given a sequence of possibly overlapping pixel maps.

Z tool-kit operators normally have a number of laws associated with them, which are helpful in reasoning about specifications. For example, the following law applies for the distributed overriding operator:

$$p_1, p_2 : Pixmap \vdash \oplus / \langle p_1, p_2 \rangle = p_1 \oplus p_2$$

Such laws must be proved from the original definition. For example, in this case:

$$\oplus / \langle p_1, p_2 \rangle$$

$$= \oplus / (\langle p_1 \rangle \frown \langle p_2 \rangle) \qquad \text{(property of } \frown\text{)}$$

$$= (\oplus / \langle p_1 \rangle) \oplus (\oplus / \langle p_2 \rangle) \qquad \text{(by the general case definition)}$$

$$= p_1 \oplus p_2 \qquad \text{(by the second base case definition, substituting twice)}$$

If the windows in a sequence overlap, it is useful to be able to move selected windows so that their contents may be viewed (or hidden). This is analogous to stuffing a pile of sheets of paper (windows) on a desk (screen). Note that the sheets of paper may be of different sizes and in different positions on the desk.

For example, the following function may be used to move a selected window number in the sequence (if it exists) to the top of the pile (i.e., the end of the sequence). This can also be defined generically:

$$[W]$$

$$top : \mathbb{N} \rightarrow \mathrm{seq}\ W \rightarrow \mathrm{seq}\ W$$

$$\forall n : \mathbb{N};\ s : \mathrm{seq}\ W \cdot$$

$$top\ n\ s = \text{if } n \in \mathrm{dom}\ s \ \textbf{then}\ squash\ (\{n\} \lhd s) \frown \langle s(n) \rangle\ \textbf{else}\ s$$

If the window number n is in the sequence of windows s then it is removed from the sequence (by eliminating that element and squashing the resulting function back into a sequence). This element is then concatenated to the end of the sequence. If the window number is not valid, the sequence of windows is unaffected. The exact technical details require some knowledge of Z, but the above example illustrates the fact that important concepts can be captured formally using relatively short definitions.

In this simple example, we shall ignore the complication of window identifiers. We simply use the position of the window in the sequence to identify it, assuming that the user of the system keeps track of which window is which.

71.4.4 Abstract System State

The window display may be modeled as a sequence of windows against a background "window" which is the size of the display screen itself. The order of the sequence defines which windows are on top in the case of overlapping windows, in ascending order. Only parts of windows that are contained within the background area are displayed.

$$\begin{array}{|l}
\textit{SYS} \\
\hline
\textit{windows} : \text{seq}\ \textit{Pixmap} \\
\textit{screen}, \textit{background} : \textit{Pixmap} \\
\hline
\textit{screen} = \textit{background} \oplus (\text{dom}\ \textit{background} \lhd \oplus/\textit{windows})
\end{array}$$

In the specification of the abstract state earlier, the components *windows* (a sequence of pixel maps), *screen* (as displayed to the user), and *background* (the display if no windows are present), are packaged together in a **schema** box called *SYS*. The **declarations** with their associated type information are above the line and **predicates** defining constraints between these components are (optionally) included below the line. Here, what appears on the display screen is defined in terms of the background pixel map overridden by the sequence of windows in the system, constrained to the background area as defined by its domain of pixel positions.

The screen area is the same as the background area. This can be formalized as follows:

$$SYS \vdash \text{dom}\ \textit{screen} = \text{dom}\textit{background}$$

It is useful to prove such properties correct, either informally, even just mentally, or formally, in order to validate that the specification behaves as expected. Discovering that an expected property does not hold may expose an error in a specification, perhaps in the form of an extra constraint that is required but has been omitted.

Note that the user can see only the display screen. We can specify this view formally by hiding (existentially quantifying) other components in the *SYS* schema to produce a new *View* schema, defined horizontally:

$$\textit{View} \mathrel{\widehat{=}} SYS \setminus (\textit{windows}, \textit{background})$$

Initially there are no windows in the system:

$$\textit{InitSYS} \mathrel{\widehat{=}} [SYS' \mid \textit{windows}' = \langle\rangle]$$

It is important to define the *initial state* and also to ensure that it exists. (Otherwise the system cannot start to operate.) The state at initialization normally consists of the abstract state for the system with some extra constraints. By including *SYS'* in the definition above, all the components in *SYS* are defined, with the extra decoration ′ added to each component name. The prime (or dash) ′ is used by convention in Z to indicate the state *after* an operation. Here we are interested in the state after initialization. Again, we can formulate a property to check that our intuition about the specification is correct:

$$\textit{InitSYS} \vdash \textit{screen}' = \textit{background}'$$

That is, the screen display at initialization should consist of just the background.

Next we can define general properties about the *change of state* for operations on the system. By convention in *Z*, Δ ("delta") is used to indicate a change of state where both an unprimed *before state SYS* (for example) and a primed *after state* (e.g., *SYS′*) are defined:

$$\Delta SYS \mathrel{\widehat{=}} [SYS; SYS' \mid background' = background]$$

Here we have added the extra constraint that the background never changes for any operation. This means that we do not have to consider and define this for each individual operation that uses Δ*SYS* subsequently, since the predicate *background′ = background* is automatically included (conjoined) with any other predicates that are defined.

Note that if the before and after states are not related in an operation scheme, then the after state can take on any value. This is the opposite of most programming languages, where unreferenced variables retain their values by default. However, this style is useful in specifications since it allows *nondeterminism* to be included easily, where more than one outcome of an operation is allowed. Eventually, of course, the implementer will have to choose a particular outcome, but if the choice is not important at the specification level then leaving the options open gives the implementer a greater choice, possibly allowing different optimization strategies in different implementations of the same specification.

Some operations may leave the state of the system unchanged, for example, during a status operation or if an error in the input is detected. Here the Ξ ("xi") convention is used in Z:

$$\Xi\, SYS \mathrel{\widehat{=}} [\Delta SYS \mid \theta SYS' = \theta SYS]$$

The predicate $\theta SYS' = \theta SYS$ is a shorthand way of ensuring that the tuple formed from all the *SYS* components is the same as that formed from all the primed *SYS′* components.

71.4.5 Operations

In order to use the system, we must have the ability to create windows. These are created on top of all the existing windows. We specify that they must fit within the display background.

```
— AddWindow0 ——————————————————————————
  ΔSYS
  window? : Pixmap
  ————————
  dom window? ⊆ dom background
  windows′ = windows ⁀ ⟨window?⟩
————————————————————————————————————————
```

In the aforementioned definition, the Δ*SYS* component automatically includes all the *SYS* and *SYS′* unprimed before and primed after state components, together with the constraint that the background remains unchanged. The *window?* component is an input (as indicated by the Z convention of the added "?"). A ***precondition*** for the operation is that the window area must be contained within (i.e., be a subset of) the background area. If this is so, then the sequence of windows after the operation has the required window concatenated to it. This means that the window appears on top of any other windows already in the system. Note that by default, predicates on separate lines in a schema are conjoined by using "∧."

The ability to update windows is very useful. This may involve changing the size of the window or its contents or moving it about the screen. Again, the updated window must still fit within the display area.

$$
\begin{array}{|l}
\hline
\textit{UpdateWindow0} \\
\Delta SYS \\
which? : \mathbb{N} \\
window? : Pixmap \\
\hline
which? \in \mathrm{dom}\ windows \\
\mathrm{dom}\ window? \subseteq \mathrm{dom}\ background \\
windows' = windows \oplus \{which? \mapsto window?\} \\
\hline
\end{array}
$$

Here two inputs are provided, both which window number is to be updated and the new value for the window. The window to be updated must already be in the system and, as for the *AddWindow* 0 schema, the new window must be within the background for the update to be successful. The sequence of windows is overridden with a new entry for the selected window.

It is desirable to be able to uncover a window that may be partially or even totally obscured by other windows. This can be done by moving the window to the end of the sequence of displayed windows:

$$
\begin{array}{l}
\hline
\textit{ExposeWindow0} \\
\Delta SYS \\
which? : \mathbb{N} \\
\hline
which? \in \mathrm{dom}\ windows \\
windows' = top\ which?\ windows \\
\hline
\end{array}
$$

Sometimes it is useful to simply rotate the order of the displayed windows, one at a time, moving the bottommost window to the top.

$$
\begin{array}{l}
\hline
\textit{RotateWindows0} \\
\Delta SYS \\
\hline
windows \neq \langle\rangle \\
windows' = top\ 1\ windows \\
\hline
\end{array}
$$

Note that sequences are numbered from one updates in Z.

We also wish to be able to delete windows. For instance, we could delete the topmost window (the last window in the sequence):

$$
\begin{array}{|l}
\hline
\textit{RemoveTop0} \\
\Delta SYS \\
\hline
windows \neq \langle\rangle \\
windows' = front\ windows \\
\hline
\end{array}
$$

Alternatively, we may wish to specify which window is to be removed:

$$
\begin{array}{l}
\hline
\textit{RemoveWindow0} \\
\Delta SYS \\
which? : \mathbb{N} \\
\hline
which? \in \mathrm{dom}\ windows \\
windows' = squash(\{which?\} \lhd windows) \\
\hline
\end{array}
$$

The aforementioned schema definitions give a flavor of the way operations are typically presented in a Z specification. They are intended to illustrate that a number of different operations on a system may be specified succinctly by using Z, providing a suitable abstract state has been formulated.

71.4.6 Error Conditions

The operations covered so far detail what should happen in the event of no errors. Normally operations can also handle error conditions in some controlled manner. It is useful to report the status of an operation. For example, the following reports could be issued:

Report ::= "OK"
| "Not a window"
| "No windows"
| "Invalid window"

Here a *free type* definition defines *Report* to be a set with four possible unique values.

It is helpful to report the fact that the operation was successful if this is the case:

—*Success*——————————
rep! : *Report*
——————
rep! = "OK"
————————————————

By convention in Z, "!" indicates an output from an operation.

If errors do occur, then these need to be reported. For example, an invalid window may be specified:

—*NotAWindow*——————————
ΞSYS
which? : $\mathbb{N}$
rep! : *Report*
——————
which? $\notin$ dom *windows*
rep! = "Not a window"
————————————————

In this case, no change of state occurs, as specified by ΞSYS given earlier. As a precondition, a check is made on whether the window number supplied as an input is not a valid existing window in the system, and if this is so an appropriate report is issued as an output.

It is possible that there are no windows displayed when one is required:

—*NoWindows*——————————
ΞSYS
rep! : *Report*
——————
windows = $\langle\rangle$
rep! = "No windows"
————————————————

A specified window may not be within the background area:

$$
\begin{array}{|l}
\hline
\textit{BadWindow} \\
\Xi SYS \\
\textit{window}? : \textit{Pixmap} \\
\textit{rep}! : \textit{Report} \\
\hline
\neg\,(\mathrm{dom}\ \textit{window}? \subseteq \mathrm{dom}\ \textit{background}) \\
\textit{rep}! = \text{“}\texttt{Invalid window}\text{”} \\
\hline
\end{array}
$$

We may include these errors with the previously defined operations that ignored error conditions to produce *total* operations:

$$\textit{AddWindow}1 \mathrel{\widehat{=}} (\textit{AddWindow}0 \wedge \textit{Success}) \vee \textit{BadWindow}$$

$$\textit{UpdateWindow}1 \mathrel{\widehat{=}} (\textit{UpdateWindow}0 \wedge \textit{Success}) \vee \textit{BadWindow} \vee \textit{NotAWindow}$$

$$\textit{ExposeWindow}1 \mathrel{\widehat{=}} (\textit{ExposeWindow}0 \wedge \textit{Success}) \vee \textit{NotAWindow}$$

$$\textit{RotateWindows}1 \mathrel{\widehat{=}} (\textit{RotateWindows}0 \wedge \textit{Success}) \vee \textit{NoWindows}$$

$$\textit{RemoveTop}1 \mathrel{\widehat{=}} (\textit{RemoveTop}0 \wedge \textit{Success}) \vee \textit{NoWindows}$$

$$\textit{RemoveWindow}1 \mathrel{\widehat{=}} (\textit{RemoveWindow}0 \wedge \textit{Success}) \vee \textit{NotAWindow}$$

Here the schema operators of conjunction (∧) and disjunction (∨) are used to combine schemas. For both operators, components are merged. If components have the same name, then they must be type-compatible or the specification becomes meaningless. Using schema conjunction, predicates in each schema are logically conjoined. Similarly, if schema disjunction is used, then the predicates in the two schemas are combined using logical disjunction.

The operations are total in that their preconditions are true. This can be checked by calculation, which is a useful way of ensuring that all error conditions have been handled. This is something that is very easily overlooked if only informal specification using natural language and/or diagrams is used.

71.4.7 Status Operations

The contents of an existing window may be of interest:

$$
\begin{array}{|l}
\hline
\textit{GetWindow}0 \\
\Xi SYS \\
\textit{which}? : \mathbb{N} \\
\textit{window}! : \textit{Pixmap} \\
\hline
\textit{which}? \in \mathrm{dom}\ \textit{windows} \\
\textit{window}! = \textit{windows}\ \textit{which}? \\
\hline
\end{array}
$$

By using ΞSYS, the state of the system does not change during this operation. Status operations normally have one or more outputs returning some aspect of the state of the system. Here a particular window is returned.

We can make this operation total as well:

$$GetWindow \mathrel{\widehat=} (GetWindow0 \land Success) \lor NotAWindow$$

71.4.8 Conclusion

Given the abstract state, initial state, and operation schemas defined in this section, the operation of the system consists of starting in the initial state, followed by an arbitrary sequence of the specified operations on the state, as allowed by the preconditions of the operations. If the preconditions of all the operations are true, then any order of operations is allowed.

This section has presented the use of the Z notation (Spivey 2001, ISO 2002) in a modeling style, as it is widely used for specifying systems. It should be remembered that Z is a general-purpose specification language and can be used in other styles if desired. However, the use of an abstract state and operations on that state has been found to be a style that is easy to understand (once the notation and conventions have been learned), and this is the approach that is often adopted in practice.

For those wishing to learn Z, there are many well-established textbooks available (e.g., see Jacky 1997, Lightfoot 2001). An international standard for Z is available (ISO 2002) and an earlier de facto standard for Z, with a matching type-checker called *fuzz* by the same author, is also widely used (Spivey 2001).

71.5 Technology Transfer and Research Issues

Claims that formal methods can guarantee correct hardware and software, eliminate the need for testing, etc., have previously led some to believe that formal methods are something almost magical (Hall 1990). More significantly, beliefs that formal methods are difficult to use, delay the development process, and raise development costs (Bowen and Hinchey 1995b) have led many to believe that formal methods offer few advantages over traditional development methods. Formal methods are not a panacea; they are just one of a range of techniques that, when correctly applied, have proven themselves to result in systems of the highest integrity in the long term (Bowen and Hinchey 1995a, Bowen and Hinchey 2012).

In the past, the uptake of formal methods has been hindered, at least in part, by a lack of tools. Many of the successful projects discussed in Hinchey and Bowen (1999) required significant investment in tool support. Just as the advent of compiler technology was necessary for the uptake of high-level programming languages, and CASE (Computer Aided Software Engineering) technology provided the impetus for the emergence of structural design methodologies in the 1970s, a significant investment in formal methods tools is required for formal methods to be practical at the level of industrial application. The RODIN tool that supports Event-B (Abrial 2009) is a good example of a robust tool that is now available.

Method integration is one approach that may aid in the acceptance of formal methods and may help in the technology transfer from academic theory to industrial practice. This has the advantage of providing multiple views of a system, for example incorporating a graphical representation that is likely to be more acceptable to nonspecialists, while retaining the ability to propose and prove system properties and to demonstrate that requirements are not contradictory before rather than after implementation.

The Unified Modeling Language (UML) provides a popular software development framework which could benefit from a formal methods approach. For example, amid concerns over the lack of formality (or even uniform interpretation) in UML, the OMG (Object Management Group) issued a request for proposals on re-architecting UML version 2.0.

Cleanroom is a method that provides a middle road between correctness proofs and informal development by stipulating significant checking of programs before they are first run (Prowell et al. 1999). The testing phase then becomes more like a certification phase since the number of errors should be much reduced. *Static analysis* involved rigorous checking of programs without actually executing them. SPARK Ada (Barnes 2003) is a restricted version of the Ada programming language that includes additional comments that facilitate formal tool-assisted analysis, especially worthwhile in high-integrity system development. Such approaches may be more cost-effective than full formal development using refinement techniques.

In any case, formal development is typically not appropriate in all software systems. However, many systems could benefit from selected use of formal methods at some level (perhaps just specification) in their most critical parts (sometimes called "lightweight" formal methods). In particular, many errors are introduced at the requirements stage and some formality at this level could have very beneficial results because the system description is still relatively simple.

Formal methods are complementary to testing in that they aim to avoid the introduction of errors whereas testing aims to remove errors that have been introduced during development. The best balance of effort between these two approaches is a matter for debate (King et al. 2000). In any case, the existence of a formal specification can benefit the testing process by providing an objective and exact description of the system against which to perform subsequent program testing. It can also guide the engineer in deciding which tests are worthwhile (e.g., by considering disjunctive preconditions in operations and ensuring that there is full test coverage of these).

The use of formal methods can and should be used in combination with other software engineering approaches. It is often complementary to many techniques. For example, formal methods could be combined with an agile approach to software development (Black et al. 2009).

In practical industrial use, formal methods have proved to have a niche in high integrity systems such as safety-critical applications where standards may encourage or mandate their use in software at the highest levels of criticality. Formal methods are also being successfully used in security applications such as smart cards where the technology is simple enough to allow fully formal development. They are also useful in discovering errors during cryptographic protocol analysis (Meadows 2003).

Formal methods have largely been used for software development, but they are arguably even more successful in hardware design where engineers may be more open to the use of rigorous approaches because of their background and training. Formal methods can been used for the design of microprocessors where errors can be costly because of the large numbers involved and also because of there possible use in critical applications (Jones et al. 2001). Fully formal verification of significant hardware systems is possible even within the limits of existing proof technology.

Full formal refinement is the ideal but is expensive and can sometimes be impossible to achieve in practice. Retrenchment (Banach and Poppleton 1999) is a suggested liberalization of refinement designed for formal description of applications too demanding for true refinement. Examples are the use of infinite or continuous types or models from classical physics and applications including inconsistent requirements. In retrenchment, the abstraction relation between the models is weakened in the operation postcondition by a *concession* predicate. This weakened relationship allows approximating, inconsistent, or exceptional behaviors to be described in which a *false* concession denotes a refinement.

There are many different formal methods for different purposes, including specification (e.g., the Z notation) and refinement (e.g., the B-Method). There have been moves to develop the B-Method. The simpler Event-B approach has good tool support using RODIN (Abrial 2009). The Alloy tool has been influenced by the Z-style of specification (Jackson 2012). There have also been moves to related different semantic theories like algebraic, denotational and operational semantics using a Unified Theories of Programming (UTP) approach (Hoare and He 1998). A promising approach is that of separation logic, an extension of Hoare logic that allows improved modular reasoning about programs that use dynamically allocated pointers (Reynolds 2002). In any case, there continue to be research and technology transfer challenges in the field of formal methods.

Key Terms

Formal methods: Techniques, notations, and tools with a mathematical basis, used for specification and reasoning in software or hardware system development.

Formal notation: A language with a mathematical semantics, used for formal specification, reasoning, and proof.

Logic: A scheme for reasoning, proof, inference, etc. Two common schemes are *propositional logic* and *predicate logic*, which is propositional logic generalized with quantifiers. Other logics, such modal logics, including *temporal logics* which handle time are also available. Examples include TLA (Temporal Logic of Actions), ITL (Interval Temporal Logic), and Duration Calculus. Schemes may use *first-order logic* or *higher-order logic*. In the former, functions are not allowed on predicates, simplifying matters somewhat, but in the latter they are, providing greater power. Logic includes a calculus which allows reasoning in the logic.

Operation: The performance of some desired action. This may involve the change of state of a system, together with inputs to the operation and outputs resulting from the operation. To specify such an operation, the *before state* (and inputs) and the *after state* (and outputs) must be related with constraining predicates.

Precondition: The predicate which must hold before an operation for it to be successful. Compare ***post-condition***, which is the predicate which must hold after an operation.

Predicate: A constraint between a number of variables which produces a truth value (e.g., *true* of *false*).

Proof: A series of mathematical steps forming an argument of the correctness of a mathematical statement or theorem. For example, the *validation* of a desirable property for a formal specification could be undertaken by proving it correct. Proof may also be used to perform a formal ***verification*** that an implementation meets a specification. A less formal style of reasoning is ***rigorous argument***, where a proof outline is sketched informally, which may be done if the effort of undertaking a fully formal proof is not considered cost-effective.

Refinement: The stepwise transformation of a specification towards an implementation (e.g., as a program). Compare *abstraction*, where unnecessary implementation detail is ignored in a specification.

Relation: A connection or mapping between elements in a number of sets. Often two sets (a *domain* and a *range*) are related in a *binary relation*. A special case of a relation is a function where individual elements in the domain can only be mapped to at most one element in the range of the function. Functions may be further categorized. For example, a *partial function* may not map all possible elements that could be in the domain of the function, whereas a *total function* maps all such elements.

Set: A collection of distinct objects or *elements*, which are also known as *members* of the set. In a typed language, types may consist of maximal sets, as in the Z notation.

Specification: A description of *what* a system is intended to do, as opposed to *how* it does it. A specification may be *formal* (mathematical) or *informal* (natural language, diagrams, etc.). Compare an *implementation* of a specification, such as a program, which actually performs and executes the actions required by a specification.

State: A representation of the possible values which a system may have. In an abstract specification, this may be modeled as a number of sets. By contrast, in a concrete program implementation, the state typically consists of a number of data structures, such as arrays, files, etc. When modeling sequential systems, each operation may include a *before state* and an *after state* which are related by some constraining predicates. The system will also have an *initial state*, normally with some additional constraints, from which the system starts at initialization.

A glossary of the Z mathematical and schema notation is included here for the reader's convenience. For more information on the Z notation, see the *Z Reference Manual* (Spivey 2001) and the ISO international standard (ISO 2002).

Glossary of Z Notation

Names

a, b	Identifiers
d, e	Declarations (e.g., $a : A; b,\ldots: B\ldots$)
f, g	Functions
m, n	Numbers
p, q	Predicates
s, t	Sequences
x, y	Expressions
A, B	Sets
C, D	Bags
Q, R	Relations
S, T	Schemas
X	Schema text (e.g., d, $d \mid p$ or S)

Definitions

$a == x$	Abbreviated definition
$a ::= b \mid \ldots$	Free type definition (or $a ::= b\langle\langle x\rangle\rangle \mid \ldots$)
$[a]$	Introduction of a given set (or $[a,\ldots]$)
$a\_$	Prefix operator
$\_a$	Postfix operator
$\_a\_$	Infix operator

Logic

true	Logical true constant
false	Logical false constant
$\neg p$	Logical negation
$p \land q$	Logical conjunction
$p \lor q$	Logical disjunction
$p \Rightarrow q$	Logical implication ($\neg p \lor q$)
$p \Leftrightarrow q$	Logical equivalence ($p \Rightarrow q \land q \Rightarrow p$)
$\forall X \bullet q$	Universal quantification
$\exists X \bullet q$	Existential quantification
$\exists_1 X \bullet q$	Unique existential quantification
let $a == x; \ldots \bullet p$	Local definition

Sets and expressions

$x = y$	Equality of expressions
$x \neq y$	Inequality ($\neg(x = y)$)
$x \in A$	Set membership
$x \notin A$	Nonmembership ($\neg(x \in A)$)
$\varnothing$	Empty set
$A \subseteq B$	Set inclusion
$A \subset B$	Strict set inclusion ($A \subseteq B \land A \neq B$)
$\{x, y,\ldots\}$	Set of elements
$\{X \bullet x\}$	Set comprehension
$\lambda X \bullet x$	Lambda-expression—function
$\mu X \bullet x$	Mu-expression—unique value
let $a == x;\ldots \bullet y$	Local definition
if p **then** x **else** y	Conditional expression

$(x, y, \ldots)$	Ordered tuple
$A \times B \times \ldots$	Cartesian product
$\mathbb{P}A$	Power set (set of subsets)
$\mathbb{P}_1 A$	Nonempty power set
$\mathbb{F}A$	Set of finite subsets
$\mathbb{F}_1 A$	Nonempty set of finite subsets
$A \cap B$	Set intersection
$A \cup B$	Set union
$A \setminus B$	Set difference
$\bigcup A$	Generalized union of a set of sets
$\bigcap A$	Generalized intersection of a set of sets
first x	First element of an ordered pair
second x	Second element of an ordered pair
$\#A$	Size of a finite set
Relations	
$A \leftrightarrow B$	Relation ($\mathbb{P}(A \times B)$)
$a \mapsto b$	Maplet ((a, b))
$\mathrm{dom} R$	Domain of a relation
$\mathrm{ran} R$	Range of a relation
$\mathrm{id} A$	Identity relation
$Q \mathbin{;} R$	Forward relational composition
$Q \circ R$	Backward relational composition ($R \mathbin{;} Q$)
$A \lhd R$	Domain restriction
$A \ndres R$	Domain anti-restriction
$A \rhd R$	Range restriction
$A \nrres R$	Range anti-restriction
$R(\lvert A \rvert)$	Relational image
iter n R	Relation composed *n* times
R^n	Same as *iter n R*
$R^{\sim}$	Inverse of relation (R^{-1})
R^{*}	Reflexive-transitive closure
R^{+}	Irreflexive-transitive closure
$Q \oplus R$	Relational overriding ($(\mathrm{dom}\, R \ndres Q) \cup R$)
$a \,\underline{R}\, b$	Infix relation
Functions	
$A \pfun B$	Partial functions
$A \to B$	Total functions
$A \pinj B$	Partial injections
$A \rightarrowtail B$	Total injections
$A \psurj B$	Partial surjections
$A \twoheadrightarrow B$	Total surjections
$A \bij B$	Bijective functions
$A \ffun B$	Finite partial functions
$A \finj B$	Finite partial injections
fx	Function application (or $f(x)$)
Numbers	
$\mathbb{Z}$	Set of integers
$\mathbb{N}$	Set of natural numbers $\{0,1,2,\ldots\}$

$\mathbb{N}_1$	Set of nonzero natural numbers ($\mathbb{N}\backslash\{0\}$)
$m + n$	Addition
$m - n$	Subtraction
$m * n$	Multiplication
m div n	Division
m mod n	Modulo arithmetic
$m \leq n$	Less than or equal
$m < n$	Less than
$m \geq n$	Greater than or equal
$m > n$	Greater than
succ n	Successor function $\{0 \mapsto 1, 1 \mapsto 2, \ldots\}$
$m\,..\,n$	Number range
min A	Minimum of a set of numbers
max A	Maximum of a set of numbers

Sequences

seq *A*	Set of finite sequences
$\mathrm{seq}_1\,A$	Set of nonempty finite sequences
iseq *A*	Set of finite injective sequences
$\langle\rangle$	Empty sequence
$\langle x, y, \ldots\rangle$	Sequence $\{1 \mapsto x, 2 \mapsto y, \ldots\}$
$s \frown t$	Sequence concatenation
$\frown/s$	Distributed sequence concatenation
head s	First element of sequence ($s(1)$)
tail s	All but the head element of a sequence
last s	Last element of sequence ($s(\#s)$)
front s	All but the last element of a sequence
rev s	Reverse a sequence
squash f	Compact a function to a sequence
$A \upharpoonleft s$	Sequence extraction (*squash* $(A \lhd s)$)
$s \upharpoonright A$	Sequence filtering (*squash* $(s \rhd A)$)
s prefix *t*	Sequence prefix relation ($s \frown v = t$)
s suffix *t*	Sequence suffix relation ($u \frown s = t$)
s in *t*	Sequence segment relation ($u \frown s \frown v = t$)
disjoint *A*	Disjointness of an indexed family of sets
A partition *B*	Partition an indexed family of sets

Bags

bag *A*	Set of bags or multisets ($A \nrightarrow \mathbb{N}_1$)
$[\![\;]\!]$	Empty bag
$[\![x, y, \ldots]\!]$	Bag $\{x \mapsto 1, y \mapsto 1, \ldots\}$
count C x	Multiplicity of an element in a bag
$C \# x$	Same as *count C x*
$n \otimes C$	Bag scaling of multiplicity
$x \sqsubset C$	Bag membership
$C \sqsubseteq D$	Subbag relation
$C \uplus D$	Bag union
$C \cup\!\!\!\!- D$	Bag difference
items s	Bag of elements in a sequence

Schema notation

Vertical Schema

$$\begin{array}{l} .S\text{———} \\ d \\ \text{—} \\ p \end{array}$$

New lines denote "," and "∧." The schema name and predicate part are optional. The schema may subsequently be referenced by name in the document.

Axiomatic definition

$$\begin{array}{|l} d \\ \hline p \end{array}$$

The definitions may be nonunique. The predicate part is optional. The definitions apply globally in the document.

Generic definition

$$\begin{array}{l} =[a, \ldots]= \\ d \\ \hline p \\ \hline \end{array}$$

The generic parameters are optional. The definitions must be unique. The definitions apply globally in the document.

$s \mathrel{\hat{=}} [X]$	Horizontal schema
$[T;\ldots\mid\ldots]$	Schema inclusion
$z\,.\,a$	Component selection (given $z : S$)
θS	Tuple of components
$\neg S$	Schema negation
pre S	Schema precondition
$S \wedge T$	Schema conjunction
$S \vee T$	Schema disjunction
$S \Rightarrow T$	Schema implication
$S \Leftrightarrow T$	Schema equivalence
$S \backslash (a,\ldots)$	Hiding of component(s)
$S \upharpoonright T$	Projection of components
$S\,;\,T$	Schema composition (S then T)
$S \gg T$	Schema piping (S outputs to T inputs)
$S[a/b,\ldots]$	Schema component renaming (b becomes a, etc.)
$\forall X \bullet S$	Schema universal quantification
$\exists X \bullet S$	Schema existential quantification
$\exists_1 X \bullet S$	Schema unique existential quantification

Conventions

$a?$	Input to an operation
$a!$	Output from an operation
a	State component before an operation
a'	State component after an operation

S	State schema before an operation
S'	State schema after an operation
ΔS	Change of state (normally $S \wedge S'$)
ΞS	No change of state (normally $[S \wedge S' \mid \theta S = \theta S']$)
$d \vdash p$	Theorem

Further Information

A number of organizations have been established to meet the needs of formal methods practitioners; for example:

- Formal Methods Europe (FME) organizes a regular Formal Methods (FM) conference, formerly the VDM symposia and the leading international conference in the field, as well as other activities for users of various formal methods.
- The British Computer Society Specialist Group on Formal Aspects of Computing Science (BCS-FACS) organizes regular talks and periodic workshops on various aspects of formal methods, including being associated with a series of Refinement Workshops.
- The Z User Group (ZUG) has organized a regular international conference, historically known as the Z User Meeting (ZUM), for users of the Z notation. The International B Conference Steering Committee (Association de Pilotage des Conférences B, APCB) organized a similar International B Conference series. Since 2000, these have combined as a single conference that more recently has also covered Abstract State Machines (ASM), Alloy, and VDM, becoming known as the ABZ conference.

There are a number of journals devoted specifically to formal methods. These include *Formal Methods in System Design* and *Formal Aspects of Computing* (FAC). The FAC journal is published by Springer in association with BCS-FACS. Other European-based journals, such as *The Computer Journal*, *IET Software* (formerly *IEE Proceedings—Software* and the *Software Engineering Journal*) and *Information and Software Technology*, publish articles on, or closely related to, formal methods, and they have run special issues on the subject.

While there are no US-based journals that deal specifically with formal methods, they regularly are featured in popular periodicals such as *IEEE Computer*, *IEEE Software*, and the *Communications of the ACM*, as well as in journals such as the *Annals of Software Engineering IEEE Transactions on Software Engineering ACM Transactions on Software Engineering and Methodology* (TOSEM), the *Journal of the ACM*, and *Innovations in Systems and Software Engineering: A NASA Journal* (published by Springer). Survey papers on the state of the art and practice in formal methods have also appeared in the *ACM Computing Surveys* (e.g., see Woodcock et al. 2009).

In addition to the conferences mentioned earlier, The International Conference on Formal Engineering Methods series (ICFEM) and the Software Engineering and Formal Methods (SEFM) conference (Hinchey et al. 2008) also cover formal methods. A number of more specialist conferences have been established. For example, the Integrated Formal Methods (IFM) International Conference concentrates on the use of formal methods with other approaches or in combination. The International Workshop on Formal Methods for Industrial Critical Systems (FMICS) focuses on industrial applications, especially using tools. The IFIP (International Federation of Information Processing) FORTE international conference has concentrated on *Formal Description Techniques* (FDTs). The IFIP WG 6.1 International Conference on Formal Methods for Open Object-Based Distributed Systems (FMOODS) has also been established and more recently combined with FORTE.

Some more wide-ranging conferences give particular attention to formal methods; primary among these are the ICSE (International Conference on Software Engineering) and ICECCS (International Conference on Engineering of Complex Computer Systems) series of conferences. Other specialist conferences in the safety-critical sector, such as SAFECOMP (the International Conference on Computer

Safety, Reliability and Security), and SSS (the Safety-critical Systems Symposium) regularly cover formal methods. The long-established Software Engineering Workshop (SEW) also features formal methods papers.

There have been some collections of case studies on formal methods with various aims and themes. For some industrial applications and approaches, see Hinchey and Bowen (1999), Boulanger (2012), Gnesi and Margaria (2012). The book (Frappier and Habrias 2006) collected together a number of formal specification methods applied to an invoicing case study where the presentations concentrate on the *process* of producing a formal description, including the questions raised along the way. For some research directions in formal methods, see Boca et al. (2010).

The following electronic mailing lists are available, among others:

fm-announcements@lists.nasa.gov	Formal methods announcements (NASA)
fmnet-request@jiscmail.ac.uk	Formal Methods Network (FMnet)
procos-request@jiscmail.ac.uk	Provably Correct Systems (ProCoS)
bforum-request@matrix.inrets.fr	B-Method
zforum-request@jiscmail.ac.uk	Z notation

In addition, some electronic forums are available as Google groups:

comp.specification.misc	Formal specification
comp.specification.z	Z notation

For on-line information on formal methods in general, readers are directed to the following World Wide Web URL (Uniform Resource Locator) that provides formal methods links:

http://formalmethods.wika.com/

References

Abdallah, A. E., Jones, C. B., and Sanders, J. W., eds. 2005. *Communicating Sequential Processes: The first 25 years*. Springer, Berlin, Germany.

Abrial, J.-R. 2010. *Modeling in Event-B: System and Software Engineering*. Cambridge University Press, Cambridge, U.K.

Back, R.-J. and von Wright, J. 1998. *Refinement Calculus: A Systematic Introduction*, Graduate Texts in Computer Science, Springer, New York.

Ball, T., Cook, B., Levin, V., and Rajamani, S. K. 2004. SLAM and static driver verifier: Technology transfer of formal methods inside microsoft. In *Integrated Formal Methods*. E. Boiten, J. Derrick, and G. Smith, eds., pp. 1–20. Lecture Notes in Computer Science 2999. Springer, Berlin, Germany.

Banach, R. and Poppleton, M. 1999. Sharp retrenchment, modulated refinement and simulation, *Formal Aspects of Computing*, 11(5):498–540.

Barnes, J. 2003. *High Integrity Software: The SPARK Approach to Safety and Security*, Addison–Wesley, London, U.K.

Benveniste, A., Caspi, P., Edwards, S. A., Halbwachs, N., Le Guernic, P., and de Simone, R. 2003. The synchronous languages 12 years later. *Proceedings of the IEEE* 91(1):64–83.

Bérard, B., Bidoit, M., Finkel, A., Laroussinie, F., Petit, A., Petrucci, L., Schnoebelen, Ph., and McKenzie, P. 2001. *Systems and Software Verification: Model-Checking Techniques and Tools*. Springer, Berlin, Germany.

Bernardo, M. and Corradini, F., eds. 2004. *Formal Methods for the Design of Real-Time Systems*. Lecture Notes in Computer Science 3185. Springer, Berlin, Germany.

Bjørner, D. 2000. Pinnacles of software engineering: 25 years of formal methods, *Annals of Software Engineering* 10(1–4):11–66.

Black, S., Boca, P. P., Bowen, J. P., Gorman, J., Hinchey, M. G., 2009. Formal versus agile: Survival of the fittest. *IEEE Computer* 42(9):37–45.

Boca, P. P., Bowen, J. P., and Siddiqi, J., eds. 2010. *Formal Methods: State of the Art and New Directions*. Springer, London, U.K.

Börger, E. and Stärk, R. 2003. *Abstract State Machines: A Method for High-Level System Design and Analysis*. Springer, Berlin, Germany.

Boulanger, J.-L., ed. 2012. *Formal Methods: Industrial Use from Model to the Code*. ISTE, London, U.K.

Bowen, J. P. 2003. *Formal Specification and Documentation using Z: A Case Study Approach*. International Thomson Computer Press, London, U.K. Revised online: http://formalmethods.wikia.com/wiki/Z_book.

Bowen, J. P. and Hinchey, M. G. 1995a. Ten commandments of formal methods. *IEEE Computer* 28(4):56–63.

Bowen, J. P. and Hinchey, M. G. 1995b. Seven more myths of formal methods. *IEEE Software* 12(4):34–41.

Bowen, J. P. and Hinchey, M. G. 1999. *High-Integrity System Specification and Design*. FACIT Series, Springer, London, U.K.

Bowen, J. P. and Hinchey, M. G. 2012. Ten commandments of formal methods… ten years on. In *Conquering Complexity*. M. G. Hinchey and L. Coyle, eds., pp. 237–251, Part 3. Springer, London, U.K.

Bowen, J. P. and Reeves, S. 2011. From a community of practice to a body of knowledge: A case study of the formal methods community. In *FM 2011: 17th International Symposium on Formal Methods*. M. Butler and W. Schulte, eds., pp. 308–322. Lecture Notes in Computer Science 6664. Springer, Berlin, Germany.

Bowen, J. P. and Stavridou, V. 1993. Safety-critical systems, formal methods and standards. *IEE/BCS Software Engineering Journal* 8(4):189–209.

Dang Van, H., George, C., Janowski, T., Moore, R., eds. 2002. *Specification Case Studies in RAISE*. FACIT Series, Springer, London, U.K.

Derrick, J. and Boiten, E. A. 2001. *Refinement in Z and Object-Z*. FACIT Series, Springer, London, U.K.

Duke, R. and Rose, G. 2000. *Formal Object-Oriented Specification using Object-Z*. Cornerstones of Computing Series, MacMillan, Basingstoke, U.K.

Frappier, M. and Habrias, H., eds. 2006. *Software Specification Methods: An Overview Using a Case Study*. ISTE, London, U.K.

Fuchs, N. E. 1992. Specifications are (preferably) executable. *IEE/BCS Software Engineering Journal* 7(5):323–334.

Futatsugi, K., Nakagawa, A. T. and Tamai, T., eds. 2000. *CAFE: An Industrial-Strength Algebraic Formal Method*. Elsevier Health Sciences, Amsterdam, the Netherlands.

Gnesi, S. and Margaria, T. 2012. *Formal Methods for Industrial Critical Systems: A Survey of Applications*. IEEE Computer Society Press, Wiley, Hoboken, NJ.

Grumberg, O., Nipkow, T., and Pfaller, C. 2008. *Formal Logical Methods for System Security and Correctness*. IOS Press, Amsterdam, the Netherlands.

Grumberg, O., Peled, D., and Clarke, E. M. 2000. *Model Checking*. MIT Press, Cambridge, MA.

Guttag, J. V. 1993. *Larch: Languages and Tools for Formal Specification*. Springer, New York.

Hall, J. A. 1990. Seven myths of formal methods. *IEEE Software* 7(5):11–19.

Harel, D. and Politi, M. 1998. *Modeling Reactive Systems with Statecharts: The Statemate Approach*. McGraw–Hill, New York.

Hayes, I. J. and Jones, C. B. 1989. Specifications are not (necessarily) executable. *IEE/BCS Software Engineering Journal* 4(6):330–338.

Hierons, R. M., Bogdanov, K., Bowen, J. P., Cleaveland, R., Derrick, J., Dick, J., Gheorghe, M. et al. 2009. Using formal specifications to support testing. *ACM Computing Surveys* 41(2):1–76.

Hinchey, M. G. and Bowen, J. P., eds. 1999. *Industrial-Strength Formal Methods in Practice*. FACIT Series, Springer, London, U.K.

Hinchey, M. G., Jackson, M. Cousot, P., Cook, B., Bowen, J. P., and Margaria, T. 2008. Software engineering and formal methods, *Communications of the ACM* 51(9):54–59.

Hoare, C. A. R. and He, J. 1998. *Unified Theories of Programming*. Prentice Hall International Series in Computer Science. Hemel Hempstead, U.K.

Holzmann, G. 2003. *The Spin Model Checker—Primer and Reference Manual*. Addison-Wesley.

IEEE. 1991. IEEE standard glossary of software engineering terminology. In *IEEE Software Engineering Standards Collection*. Elsevier Applied Science, Amsterdam, the Netherlands.

ISO. 1989. *Information Processing Systems—Open Systems Interconnection—LOTOS*—A formal description technique based on the temporal ordering of observational behaviour. International Standard ISO 8807:1989, International Organization for Standardization, Geneva, Switzerland.

ISO. 1996. *Information Technology—Programming Languages, Their Environments and System Software Interfaces—Vienna Development Method—Specification Language—Part 1: Base Language*. International Standard ISO/IEC 13817-1:1996, International Organization for Standardization, Geneva, Switzerland.

ISO. 2002. *Information Technology—Z Formal Specification Notation—Syntax, Type System and Semantics*. International Standard ISO/IEC 13568:2002. International Organization for Standardization, Geneva, Switzerland.

Jacky, J. 1997. *The Way of Z: Practical Programming with Formal Methods*, Cambridge University Press, Cambridge, U.K.

Jackson, D. 2012. *Software Abstractions: Logic, Language, and Analysis*, revised ed. MIT Press, Cambridge, MA.

Jones, C. B., Liu, Z., and Woodcock, J., eds. 2007. *Formal Methods and Hybrid Real-Time Systems*. Lecture Notes in Computer Science 4700. Springer, Berlin, Germany.

Jones, R. B., O'Leary, J. W., Seger, C.-J. H., Aagaard, M. D., and Melham, T. F. 2001. Practical formal verification in microprocessor design. *IEEE Design and Test of Computers* 18(4):16–25.

King, S., Hammond, J., Chapman, R., Pryor, A. 2000. Is proof more cost-effective than testing? *IEEE Transactions on Software Engineering* 26(8):675–686.

Kordon, F. and Lemoine, M. 2010. *Formal Methods for Embedded Distributed Systems: How to Master the Complexity*. Springer, Berlin, Germany.

Kropf, T. 2000. *Introduction to Formal Hardware Verification*. Springer, New York.

Lightfoot, D. 2001. *Formal Specification using Z*, Grassroots Series, Palgrave, Basingstoke, U.K.

Meadows, C. 2003. Formal methods for cryptographic protocol analysis: Emerging issues and trends. *IEEE Journal on Selected Areas in Communications* 21(1):44–54.

Milner, R. 1999. *Communicating and Mobile Systems: The π-calculus*. Cambridge University Press, Cambridge, U.K.

Morgan, C. 1998. *Programming from Specifications*, 2nd edn. Prentice Hall International Series in Computer Science. Hemel Hempstead, U.K. Revised onlineL: http://www.cs.ox.ac.uk/publications/books/PfS/.

Parnas, D. L. 1995. Using mathematical models in the inspection of critical software. In *Applications of Formal Methods*, M. G. Hinchey and J. P. Bowen, eds., pp. 17–31. Prentice Hall International Series in Computer Science. Hemel Hempstead, U.K.

Parnas, D. L. 2010. Really rethinking 'formal methods'. *IEEE Computer* 43(1):28–34.

Prowell, S. J., Trammell, C. J., Linger, R. C., and Poore, J. H. 1999. *Cleanroom Software Engineering: Technology and Process*. Addison-Wesley, Reading, MA.

Reynolds, J. C. 2002. Separation logic: A logic for shared mutable data structures. *Proc. 17th Annual IEEE Symposium on Logic in Computer Science (LICS)*, Copenhagen, Denmark pp. 55–74.

Schneider, S. 2001. *The B-Method: An Introduction*. Cornerstones of Computing Series, MacMillan, Basingstoke, U.K.

Smith, G. 2000. *The Object-Z Specification Language*. Advances in Formal Methods Series. Kluwer Academic Publishers, Boston, MA.

Spivey, J. M. 2001. *The Z Notation: A Reference Manual*, 2nd edn. Prentice Hall International Series in Computer Science, Hemel Hempstead, U.K. http:/spivey.oriel.ox.ac.uk/mike/zrm/; Revised online: http://spivey.oriel.ox.ac.uk/mike/zrm/

Tiwari, A., Shankar, N., and Rushby, J. 2003. Invisible formal methods for embedded control systems. *Proceedings of the IEEE* 91(1):29–39.

Tremblay, G. 2000. Formal methods: Mathematics, computer science or software engineering? *IEEE Transactions on Education* 43(4):377–382.

Tretmans, J., Wijbrans, K., Chaudron, M. 2001. Software engineering with formal methods: The development of a storm surge barrier control system revisiting seven myths of formal methods. *Formal Methods in System Design* 19(2):195–215.

Turner, K. J., ed. 1993. *Using Formal Description Techniques: An Introduction to Estelle, LOTOS and SDL.* John Wiley & Sons, Chichester, U.K.

Warmer, J. and Kleppe, A. 2003. *The Object Constraint Language: Getting Your Models Ready for MDA*, 2nd edn. Addison–Wesley, Boston, MA.

Woodcock, J., Larsen, P. G., Bicarregui, J., and Fitzgerald, J. 2009. Formal methods: Practice and experience. *ACM Computing Surveys* 41(4):1–40.

Zhou, C. and Hansen, M. R. 2003. *Duration Calculus: A Formal Approach to Real-Time Systems.* Monographs in Theoretical Computer Science, An EATCS Series, Springer, Berlin, Germany.

IX

Discipline of Software Engineering

72

Discipline of Software Engineering: An Overview

Jorge L. Díaz-Herrera
Keuka College

Peter A. Freeman
Georgia Institute of Technology

A handbook is "a book of information, guidance, or information, as for an occupation, travel, or reference,"* a definition that fits this volume and section very well. A "discipline" is "a branch of instruction or learning; behavior in accord with rules of conduct."† More established subjects such as physics and chemistry or disciplines such as the practice of architecture have handbooks that are a compendium of facts, organized for ready reference (such as the specific heat of materials or the standard width of an aisle in an office building) or procedures (such as safety practices).

Software engineering is not yet a fully mature discipline, and it is certainly not a set of facts! Nor are there widely accepted software engineering handbooks.

It is tempting to say that as software engineering matures, it is becoming an engineering discipline. Today, however, software engineering is not widely considered an engineering discipline, at least not in the traditional sense. We will say more about this in the subsequent sections.

* *Random House Dictionary of the English Language*, 1968, New York.
† Op. cit.

It is, however, a branch of instruction and learning that has been around for almost 50 years and about which a good deal is known.* The story of software engineering to date is the story of the development of a practical discipline in response to the demands of the marketplace and the inherent and ever-increasing complexity of software.

The papers in this section of the *Computing Handbook Set* are intended to provide you self-contained descriptions of what is known in specific areas of the discipline. The purpose of this chapter is to provide you a framework and the context in which software engineering has evolved so far, a brief survey of educational efforts, an overview of its elements, and a sketch of how software engineering is practiced.

72.1 Origins of Software Engineering

> Why has this meeting been scheduled? Why have we agreed to participate?
>
> I believe it is because we recognize that a practical problem of considerable difficulty and importance has arisen: The successful design, production and maintenance of useful software systems. The importance is obvious and the more so since we see only greater growth in demands and requirements in the future. The consequences of poor performance, poor design, instability and mismatching of promise and performance are not going to be limited to the computing fraternity, or even their nearest neighbors, but will affect considerable sections of our society whose ability to forgive is inversely proportional to their ignorance of the difficulties we face. The source of difficulty is distributed through the whole problem, easy to identify, and yet its cure is hard to pinpoint so that systematic improvement can be gotten.
>
> **—Keynote speech by A.J. Perlis.**†

The aforementioned quote from the opening speech at a meeting in 1968, generally cited as the start of software engineering, provides the motivation at that time—and it is still true today! Software for general-purpose computers has been created since the earliest days of computers (generally, late 1940s) and the term "software engineering" was used privately by some before 1968.‡ In the mid-1960s, there was an awakening to the fact that creating large, complex software on time, within budget, and that performed as imagined it would by the customers for the software was becoming a huge problem. The dimensions of the problem were multiple, involving not only technical and economic concerns, but also extremely serious military defense concerns—hence, the involvement of the NATO Science Committee.

It was the height of the Cold War during which the Western Democracies felt seriously threatened by the Soviet Union and their allies—and *vice versa*. Computers were already an integral part of defense systems and as their necessary functional complexity grew so did the size and complexity of the software needed. This drove not only the development of military defense systems on both sides of the Iron Curtain but also in the West the commercial development of large, comprehensive operating systems and applications systems.

As intellectually interesting as it may be, our purpose here is not to fully explore the origins and growth of software engineering from its beginnings to today but to give a brief recounting of that history§,¶ in the

* Software engineering was introduced in the early 1970s as a specialization of computer science, and since the mid-1980s, it has been accepted as a formal field of study in academia.

† Naur and Randell (1969), NATO 1968 conference, pp. 135-ff; available at http:/homepages.cs.ncl.ac.uk/brian.randell/NATO/

‡ See http://www.legacy.com/obituaries/bostonglobe/obituary.aspx?n=douglas-t-ross&pid=86341586#fbLoggedOut, accessed on October 6, 2013; also personal communication with Freeman.

§ This should not be read as a definitive history of software engineering, nor should the periods defined by dates be taken as absolutes. There are no histories of software engineering currently, but Freeman (1975) and Zelkowitz (1979), both give overviews of the field at that time and provide multiple references to a variety of papers and books.

¶ Where possible and appropriate, we have used early references to provide entré to the historical literature. These may not be the best for tutorial or modern reference, but may be useful to researchers. Most references, especially to papers, are available in either the ACM Digital Library or IEEE Xplore Digital Library.

belief that it will provide you some necessary context for the effective use of this Handbook section. There is a lot to learn from that history and the efforts to create a software engineering discipline; we encourage you to delve into it. There is no comprehensive history of software engineering, but Wikipedia* and some textbooks will have additional references beyond what are given in the subsequent sections.

72.1.1 Early Efforts to Create a Discipline (1960s)

In the 1960s, the general view in industry was that creating software was purely an adjunct to designing the hardware and the overall system of people, procedures, computers, and other hardware such as sensors and actuators. This view prevailed within the computing community (both industrial and academic) itself, although not as strongly. In addition, creating software was considered, in practice, to be largely a matter of creating the necessary algorithms and then programming them.

Very little attention was paid by anyone to topics such as requirements analysis, specification, design, testing, and maintenance as they pertained to software. (In fairness, all of these issues were mentioned at the NATO meetings.) It is notable that starting in the 1960s and continuing through the 1970s, large programming efforts were usually referred to as "systems programming," hinting at the fact that one was engaged in more than just programming an algorithm.

Thus, it is not surprising that the early efforts at improving the development of software focused on the activity of programming. The small size of available memory and the speed of computers (compared to today) emphasized this concentration of effort. Indeed, it was—and to a somewhat lesser extent, still is—important to make sure that the software that is controlling the hardware is efficient and easy to create. In the academic research community, the creation of languages and the development of mathematically based techniques are better defined problems than the topics of requirements elicitation, design, and maintenance—and they fit better into the mold of what was (is) considered legitimate academic research. In short, a bottom-up approach to what is inherently a system-level problem was largely followed, starting with what people knew and were comfortable with—the programming activity. This phenomenon is still seen today, and new paradigms and ideas often commence at the programming level and then move up to higher levels of abstraction in the development "process," for example, after structured programming, we saw the advent of structured design and structure analysis and design; same for object-oriented programming, design, and analysis and design, and so on; the latest fad is aspect-oriented programming, etc., and feature-oriented software development (Apel and Kästner, 2009) is seeing the light.

72.1.2 Programming Methods (~1968 to 1975 and Beyond)

In software engineering, we speak of a programming method as a set of rules, guidelines, steps, or other ways of organizing the process of creating a program at the programming language level. This is distinct from the use of "method" in object-oriented programming (itself a "method" in the software engineering sense) where "method" denotes a subroutine associated with a class (Bruegge et al., 2000). In at least one programming method, the so-called Booch method (Booch, 1993), the two meanings merge to the extent that the Booch method is based on the use of methods and other devices in object-oriented languages.

The most widely known and influential technique at the time was one pioneered by E.W. Dijkstra called "structured programming" (Dijkstra, 1968). It was described, dissected, and analyzed (not always accurately†) in dozens of books and thousands of papers. It was promoted by some as *the* answer to the

* http://en.wikipedia.org/wiki/History_of_software_engineering, accessed January 17, 2013.

† A review of Dijkstra's papers in (Freeman, 1975, pp. 487ff) points out that he was fundamentally advancing the idea of "constructive programming" (creating programs that are probably correct because of the manner in which they are created) and a set of techniques and program structures (not a hard and fast algorithm nor a set of inviolable structures). The core of his approach was largely ignored at the time and even more so today.

"software problem." Many practitioners in search of a quick fix to their development problem rushed to adopt it. Development of structured programming techniques and tools predominated in many discussions and improvement activities in the late 1960s and much of the 1970s.

Other programming methods you many encounter include iterative programming, functional programming, parallel programming, genetic programming, object-oriented programming, extreme programming, generative programming, and a plethora of X-programming methods, where X = the name of a programming language (e.g., C++ programming method). Almost all of these came about after it was realized that structured programming alone could not solve all software development problems, but that improvements could be made in the way software is created. The technical aspects of some of these are discussed in Section I (Computer Science) programming languages part.

While it turned out to not be a "silver bullet"—a term that Fred Brooks popularized in his 1986 paper (Brooks, 1995)—that so many sought, attention to structured programming did result in some worthwhile improvements. A fundamental result was the realization that the types of language structures that one employed and the way in which one structured the organization of a program could have significant effects on the resulting programs along the dimensions of efficiency and ease of construction and maintenance. A similar, very positive result was the design or redesign of many programming languages and techniques to incorporate some or most of the underlying concepts of structured programming (Wirth, 1971). By the end of the twentieth-century, most of the techniques and structures of structured programming were widely accepted and in use.

Perhaps the most fundamental result was the planting of the seeds necessary for improving the entire method of developing software and the beginning of a focus on modularity (Parnas, 1972) and design methods (Freeman and Wasserman, 1975). Both these topics were discussed at the first NATO conference in 1968, reflecting the fact that the importance and some aspects of both were already known and recognized.

A general definition of a "module" is a "separable component, frequently one that is interchangeable with others, for assembly in units of differing size, complexity, or function,*" with "modularity" and "modularization" being the associated adjective and verb. It is a concept that has long been used in many contexts, including engineering, and when applied to software has the same meaning as elsewhere. The value of modularity in software is not hard to see, but achieving it is much harder and that led to a large amount of attention being paid to it beginning in the 1970s. The seminal paper in this area is the famous one (Parnas, 1972).†

While many of the efforts centered, appropriately, on the mechanisms in programming languages (Wirth, 1974) to enable and encourage modularity some soon became embedded in design methods intended to proceed the writing of the code necessary to implement a design. Like the tenets of structured programming, the ideal of modularizing code is now common throughout software development. A particular form or aspect of modularization is the idea of reusable software exemplified by the notion of separately compiled units, packages, and generics in the Ada programming language (Ada, 2012).

72.1.3 Software Design Methods (~1972 to 1985 and Beyond)

As the limits of programming methods, including modularization, by themselves to "solve" the software problem became evident, the search for the next possibility quickly turned to design techniques and methodologies. By analogy, a design method (or technique) can be considered to be the same sort of construct (i.e., rules, guidelines, steps, etc.) as a programming method, only applied to the problem of creating a design for software that is expressed in something other than executable code (or, its

* Random House Dictionary, op. cit.

† A review of Parnas' papers highlights the fact that "Dave Parnas identified the information hiding principle and showed how to use it to construct workable, reusable modular structures that are stable over time." (Hoffman and Weiss, 2001, preface).

equivalent, a higher-level language)—that is, it is at a higher level of abstraction. A design methodology (or, often, a development methodology) is an organized system of methods and rules of application to guide the overall development of a software-intensive system.

Note that the distinction between programming methods, design methods, and design methodologies is often weak and for the most part unimportant—unless one is trying to compare alternatives. The important factors are the scope of application and the focus of the item under consideration. A programming method based on a specific language may well not be applicable to work in a different language (a scope of application issue). Similarly, any method focused on the specifics of programming will probably not be of any use in organizing the overall workflow of a large development project (a methodology issue).

The origin of many of the concepts of a software design method, especially those employing structure, organized levels of abstraction, and a notation other than a programming language can be traced back to the work of Doug Ross at MIT in the 1960s on numerically controlled machine tools and the techniques for programming them more or less automatically that resulted in the APT language (Ross, 1978). This work led to the development of structured analysis and design technique (SADT) (Marca and MacGowan, 1988) that received significant use in the design of some of the most complicated and sophisticated software-intensive systems in the 1970s and 1980s. There were also attempts to import concepts from other design fields such as engineering design and architecture into the design of software; except at a high level, many of these attempts have not been notably successful in directly changing the practice of software design.

A popular and useful technique for software that is similar to SADT in some respects, structured design, was developed and promoted in the 1970s (Yourdon and Constantine, 1979). A number of other design techniques were developed and used during this period to a varying degree of success. As the importance of data started to grow, along with the realization that collections of data (databases) needed to be designed as well as the programs that accessed them, data design methods—in some cases as distinct from software design methods that tend to focus on the process aspect of software—focused on database design started to appear (Brown, 1975; Chen, 1976).

In the 1980s, as more languages of various sorts appeared (and some older ones such as Cobol started to fade away) as well as new application areas and hardware types, we saw the appearance of design (and programming) methods oriented toward specific language, applications, or hardware. This trend continues today, with many papers and books discussing specific approaches for program or software system or overall system design utilizing specific languages, focused on specific hardware types, or application areas.

The popularity of structured design and other design techniques employing structuring concepts led naturally to the development of structured analysis techniques for use in the stage preceding design (DeMarco, 1979, and others). Some of these bear some resemblance to SADT and did gain a good bit of usage, but in general were not as popular in practice as structured design. At the same time, less technically oriented methods were being developed in the business data processing and information systems communities focused more on the early stages of the development of an overall system.

72.1.4 Software Development Methodologies (~1975 to 1985 and Beyond)

These developments led, in turn, to the creation of more holistic or complete development methodologies intended to provide guidance throughout the lifecycle of a software-intensive system from the time a customer first expresses a need that can be fulfilled by such a system, through complete development, and then on into usage, modification, and de-commissioning phases. These more comprehensive methodologies were generally aimed at and most useful in large organizations dealing with large and complex systems whose lifetimes needed to extend over decades and involve perhaps thousands of people developing and modifying various aspects of the systems.

The largest such organization—and most influential in the early development of software engineering—was the United States Department of Defense (DoD). The Ada programming language was designed specifically to address software engineering issues and was mandated for use by defense contractors once the language was well defined and implemented. The process to define the language was unique in that it went through a series of public iterations on which anyone could offer comments a "straw man proposal." A similar effort was started by the DoD to define a development methodology to accompany the use of Ada, but the effort stalled out after an initial iteration (Freeman and Wasserman, 1982).

Other organizations, however, were more successful in creating and adopting "standard" development methodologies, including the major equipment manufacturers (IBM, Hewlett-Packard, etc.), defense contractors (TRW, Northrup Grumann, Lockheed, etc.), commercial consultancies (Arthur Anderson—now Accenture, Price-Waterhouse, etc.), and large user organizations (banks, transportation companies, manufacturers, etc.). A representative review of the situation as of the early 1980s can be found in (Freeman and Wasserman 1983).

72.1.5 Development Lifecycles (~1980 to 1995 and Beyond)

A *project* is an organized activity with a beginning, end, schedule, budget, and related factors and constraints; *project management* is the discipline of planning, organizing, and controlling a project; all projects typically go through a similar set of *stages* (time periods), depending on the type of project, characterized by the dominant activity in that period; and the organization of the stages is called a *project lifecycle*. These terms have been used in engineering and construction projects for well over a hundred years, but came into much wider use during and after World War II because of the many large projects undertaken in the course of the war.

As the activity of software development became more complex and involved more people over longer time spans, and as it became clear that many of the large systems would be in use and undergo "maintenance" over many years (as of 2012, parts of some software systems have been in use close to 50 years), the concepts and terminology of project management were adapted to software and software-intensive systems. This began in the 1960s in large organizations, but it was not until the mid-1970s or later that much discussion and research attention was focused on the steps involved.

Initially, the "standard" organization of activities was the *waterfall model*, a linear sequence of steps usually defined as requirements analysis and specification, software design, implementation, installation, usage, and maintenance (Royce, 1987). The steps were usually depicted as a set of boxes connected by one-way arrows and arranged in a "waterfall." Many organizations adopted this model as a way of putting structure on the often-chaotic software development process and of controlling it. Adoption and usage of the model, quickly illustrated the simplicity and naiveté of the model, leading to the development (and eventual adoption) of more realistic models. Chief among these was the spiral model (Boehm, 1988) that goes through successive iterations of a basic set of steps. An early comparison of several models can be found in Kerola and Freeman (1981).

As software development has expanded beyond large organizations or one-person, one-off efforts and technical developments including personal computers and the Internet have occurred—to say nothing of the range of applications and the demands of the marketplace—a number of other lifecycle models have been developed and employed. Reviews of the situation today can be found in Part XI, and in the appropriate sections of good software engineering textbooks.

72.1.6 Tool Support for Software Development (1968–1995 and Beyond)

Programs to assist in the writing of other programs (e.g., language syntax checkers) and systems of programs (e.g., linkers, loaders, assemblers, and compilers) have existed almost as long as programs. With the widespread interest in the Unix operating system (which came with a number of tools useful to the programmer) in the late 1960s (e.g., Keringhan and Plauger, 1976), the search for other tools became more prominent. Initially, most tools were focused on programs but as design methods and methodologies were introduced, experimentation with the types of tools broadened. They included drawing tools for graphical representations of design (Díaz-Herrera and Flude, 1980), consistency checkers for specifications and designs, and version control tools.

Testing tools have been in use at least since the 1960s. For example, Pullen and Shuttee (1968) describe "MUSE, a tool for testing and debugging a multi-terminal programming system" became popular.

Work on automated or semi-automated software development dates back at least to the early 1950s (e.g., Carr, 1952) when the term "automatic programming" was used; it usually referred to efforts to automate the translation of higher-level languages into executable code, although the concept of using artificial intelligence in the programming process* was already known (Perlis, 1959). In the early to mid-1960s, research on "compiler compilers" (programs used to generate compilers for higher level languages) was a favorite topic, for example, (Cheatham and Standish, 1970); this work laid the foundation for many of the language tools in common use today.

Using computers to assist or automate parts of the software development process (typically called computer-aided software engineering—CASE) became a hot topic in the 1980s and is still a common topic in software engineering today. Integrated development environments (IDEs) brought together a number of technical and managerial tools to work together in a coordinated way (ECMA, 1993).

72.1.7 Process Maturity Modeling (1989–2000 and Beyond)

The final major component of modern software engineering came into prominence with the publication of (Humphrey, 1989).† This book, and several that followed it along with extensive implementation, training, and standard-setting activities, takes an organizational point of view on the overall activity of creating software (generally understood to be large, complex software systems, although some attempts have been made to apply the concepts to individual work as well).

The fundamental idea is that organizations differ in their ability to create software in predictable, repeatable ways. The "software process," as Humphrey calls it, consists of the set of tools, methods, and practices used to produce a software product and the view is that these are, or should be, actively managed by the organization. He defines five increasing levels of software process maturity (initial, repeatable, defined, managed, and optimizing) and ways of "measuring" an organization's place on the attendant scale. Much of his work then focuses on how organizations can improve their performance.

Government organizations (primarily the US Defense Department, the world's largest buyer of software), some large industrial organizations, and some standards setting bodies (notably the ISO standards process) have adopted in whole or in part this concept.

ISO 9000-3 is a subset of guidelines for applying ISO 9001 to the development, supply, and maintenance of software. ISO 9001 is intended to be used when conformance to specified requirements is to be assured by the supplier during several stages, which may include design, development, production, installation, and servicing.

The ISO/IEC 15504 or Software Process Improvement and Capability Determination (SPICE) is a set of standards for the software development process and related business management functions, initially derived from process lifecycle standard ISO/IEC 12207 and from maturity models like Bootstrap, Trillium, and the Sw-CMM.

---

* Research efforts at applying artificial intelligence concepts and techniques to creating software also started in the 1960s; see Freeman (1974) for an early review.

† Watts Humphrey was awarded the 2003 U.S. National Medal of Technology by President George W. Bush for the impact of his work—the highest award that any software engineering expert has been awarded as of 2012.

These process improvement concepts do not specify anything about the technical details of the methods, tools, and procedures that are used in an organization, only how they should be managed. Some see this as a major failing, but on the whole, application of the concepts has helped organizations improve their ability to create software in predictable, repeatable ways and in so doing they usually adopt one or more development methodologies.

72.1.8 Education and Training (1974–2004 and Beyond)

Education and training are often the drivers in defining the content of a discipline because of their influence on future behavior rather than being a formalization of current and past activity. This was and still is the case in software engineering since, at least in the United States, research is tied to education at the best universities and much of the training in organizations follows their current research directions. Because leaders in the computer science field were prominent among the attendees at the two NATO conferences mentioned earlier and because they carried the message back to their colleagues and students, the importance of finding better approaches to developing software quickly became a major theme in computer science research and hence in the educational efforts of the researchers.* At the second NATO meeting on software engineering (Buxton and Randell, 1969), a session was devoted to discussion of software engineering education.

The first known meeting devoted entirely to software engineering education was an internal IBM meeting in Toronto, Canada, in 1974 that was organized by the late Harlan Mills.† The first broad meeting was organized at the University of California, Irvine, in early 1976 (Freeman et al., 1976) and brought together practitioners and software researchers, as well as representatives of research funding agencies. This meeting served as a baseline of needs expressed by practitioners and served to introduce a number of people with similar interests to each other. This resulted in a number of formative activities over the next few years (Budgem et al., 2003; Ford, 1994).

One of these outcomes was a collaboration that resulted in the first published foundation for software engineering education (Freeman and Wasserman, 1978). For better or worse, the paper served as the basis for essentially all academic programs in software engineering education for the next 10 years or more. As the need for software engineering intensified in industry, especially in defense contracting firms and large computing equipment manufacturers, a number of efforts to define better and support a discipline were undertaken and a number of courses and a few degree programs were designed.

The Software Engineering Institute (SEI) at Carnegie Mellon University founded in 1984, championed educational efforts as one of their main, early activities. A series of conferences on software engineering education and training was started in 1987 and continues annually today. Most universities and colleges today have at least a few courses in software engineering (usually in computer science departments) and an increasing, but not large, number have degree programs (usually at the Master's level). In many basic courses in software engineering and information systems, software engineering concepts are embedded in more general courses on programming and other basic courses (although they are often not identified specifically as software engineering topics). Software engineering training is also a popular topic at community and for-profit colleges, while corporate in-house training programs are perhaps less popular today because of the availability of courses elsewhere—and, also sometimes because companies have turned their attention away from training more generally and in some instances from software engineering in particular.

* One of the authors of this chapter, Freeman, was a graduate student at Carnegie Mellon University in 1968. The week after the first NATO meeting, Alan Perlis who keynoted that meeting and was chairman of the department, met with a group of us to reprise the meeting's discussions and later gave us each a copy of the proceedings (my copy long since purloined). The author was already engaged in research focused on improving software development and that early exposure encouraged him to continue in that pursuit and to develop software engineering courses in the early 1970s and to integrate some of the ideas into other courses as well.

† Harlan Mills was also the chair of the First National Conference on Software Engineering. September 11–12, 1975. Mayflower Hotel, Washington, DC.

A variety of efforts have tried to produce comprehensive "bodies of knowledge" for software engineering, standards for practice, and even certification exams. While useful, these have been of use for a limited time because of the rapid change in technology, the market, and some aspects of the field—to say nothing of the lack of people to develop these artifacts.

Much of the curriculum design for current BS programs was based on the EC2000 accreditation criteria from Accreditation Board for Engineering and Technology (ABET) and the joint IEEE/ACM task force draft accreditation criteria for software engineering (Engel, 1999),* and the curriculum guidelines were approved in 2004 (Lethbridge et al., 2006). "The program must include approximately equal segments in software engineering, in computer science and engineering, in appropriate supporting areas, and in advanced materials. This material should cover about three-quarters of the overall academic program, with the remainder to include institutional requirements and electives."†

The software engineering portion of the program should cover processes and techniques for developing and maintaining software systems. Substantial software design and software architecture work must be included. Engineering responsibility and practice must also be stressed, including ethical, social, legal, economic, and safety issues.

In the supporting areas, we have communications and mathematics. Oral and written communications, including the ability to work in teams, are covered. The mathematics contents focus primarily on discrete mathematics, and probability and statistics (this is particularly useful in process improvement analyses).

The scientific basis for software engineering is primarily computer science, similarly as the natural sciences are for other engineering disciplines.‡ The computer science and engineering portion of a BS program would include foundational topics such as algorithms and data structures, computer architecture, databases, programming languages, operating systems, and networking.

The advanced portion of the program provides depth in one or more areas of study in software engineering and in one or more significant application domains.

The curriculum should also emphasize software practice and development process and these aspects are given a strong hands-on component using the software engineering laboratory. There should be a substantial senior design project (or SwE practicum) component, where a major project is performed in teams. For an earlier historical review of software engineering education refer to (Tomayko, 1999).

72.2 Engineering and Software Engineering

When the term software engineering was introduced 40 years ago at the NATO 1968 conference, it "was deliberately chosen as being provocative, in implying the needs for software manufacture to be based on the types of theoretical foundation and practical disciplines, that are traditional in the established branches of engineering." The term is now widely used in industry, government, and academia: thousands of IT professionals go by the title software engineer; numerous publications, groups and organizations, and professional conferences use the term in their names, as do many educational programs, both at the undergraduate and graduate level (including the Ph.D.).

Is software engineering likely to become a mature engineering discipline in the traditional sense of the word? Could there be a basis for understanding the complexity of software such that we can "engineer" it to have predictable quality and behavior?

As a practical matter, definitions of software engineering in abstract terms are often not overly useful in practice because they are existential in nature and there are wide variations in the importance of specific topics in different situations. In software engineering research endeavors, however,

* G. Engel, "Accreditation criteria for software engineering," *IEEE Software*, 31–34, 1999.

† The majority of these concepts were captured in the design of the curriculum guidelines for software engineering as part of the ACM/IEEE computing curricula effort. One of the authors, Diaz-Herrera, was the co-editor of Volume IV.

‡ Some still consider Computer Science a branch of mathematics.

they are somewhat more useful. Consequently, a general understanding of what software engineering is will be helpful in your use of this handbook.

72.2.1 What Is Engineering?

A dictionary definition is an instructive place to start:

1. *The art or science of making practical application of the knowledge of pure sciences...*
2. *The action, work, or profession of an engineer.*
3. *Skillful or artful contrivance; maneuvering.**

The first definition is probably what you understand informally and is the most general definition. The focus is on "practical," as it should be, especially to satisfy the imperatives placed on the field of software engineering.

The second definition sounds circular, but is also useful in the case of software engineering. As the field expands to new application areas and new elements, practice—that is, what software engineers do—often is ahead of any educational programs or more formal definitions. As an example, software engineers were dealing with human–computer interfaces before that topic was routinely included in software engineering courses.

The third definition is most often found in discussions of non-technical matters—for example, "The customer engineered the procurement so that his company would not be dependent on any one supplier." Of course, it also applies to technical work as well—for example, when a software engineer is faced with a problem for which there is not a straightforward solution.

The second definition in the preceding text begs the question: Who is an engineer?

According to the Model Law (NCEES, 2013) prepared by the National Council of Engineering Examiners, an "engineer" is an

> individual who is qualified to practice engineering by reason of special knowledge and use of the mathematical, physical, and engineering sciences and the principles and methods of engineering analysis and design, acquired by engineering education and engineering experience.

In the same act, the term "practice of engineering means"

> any service or creative work, the adequate performance of which requires engineering education, training, and experience in the application of special knowledge of the mathematical, physical, and engineering sciences to such services and creative work as consultation, investigation, evaluation, planning and design of engineering works and systems, planning the use of land and water, teaching...

The question is whether we can substitute "software engineering" for "engineering" and still have valid statements. For this to work, we would need to replace physical sciences by computer science.

72.2.2 What Is Software Engineering?

The curriculum volume for software engineering (ACM, 2004) listed the following definitions:

1. The establishment and use of sound engineering principles (methods) in order to obtain economically software that is reliable and works on real machines (Bauer, 1972).
2. ... the application of a systematic, disciplined, quantifiable approach to the design, development, operation, and maintenance of software, and the study of these approaches; that is, the application of engineering to software. (Wikipedia, retrieved March 1, 2013)

* Random House Dictionary, op.cit.

3. … concerned with developing and maintaining software systems that behave reliably and efficiently, are affordable to develop and maintain, and satisfy all the requirements that customers have defined for them. (http://computingcareers.acm.org/, retrieved March 1, 2013)
4. Software engineering is that form of engineering that applies the principles of computer science and mathematics to achieving cost-effective solutions to software problems. (CMU/SEI-90-TR-003)
5. … the technological and managerial discipline concerned with systematic production and maintenance of software products that are developed and modified on time and within cost estimates. (Fairley, 1985)
6. The application of a systematic, disciplined, quantifiable approach to the development, operation, and maintenance of software. (IEEE, 1990)

The SE 2004 volume adopted the following definition:

> Software Engineering is about creating high-quality software in a systematic, controlled, and efficient manner. As such, there are important emphases on analysis and evaluation, specification, design, implementation and evolution of software. In addition, there are issues related to management and quality, to novelty and creativity, to individual skills, and to teamwork and professional practice that play a vital role in software engineering.

A definition that one of us has used for many years captures the broad sweep of software engineering:

> Software engineering is the systematic application of methods, tools, and knowledge to achieve stated technical, economic, and human objectives for a software-intensive system.

This definition is purposely general because the specific methods, tools, etc. may vary with the application, the organization, and over time. Note the use of "software-intensive" to signify that in general software engineering is dealing with hardware and human systems as well as software, either directly in the absence of a larger systems methodology or, at least, in conjunction with such an encompassing methodology.

Denning proposes the merging of human-centered design and software engineering process into a profession of software architects, the central practice of which "unites system-oriented engineering and customer-oriented designers" (Denning, 1996). This proposal also focuses on standard practices in a domain of expertise. We should take this recommendation seriously. But, because of its possible misinterpretation with today's use of the term software architecture, the term software systems architecture may be more suitable to describe such a discipline (see also Rechtin, 1997). This definition also risks ignoring the very important implementation, versus architecting, activities that practical software engineering must encompass.

Ian Sommerville (Sommerville, 1999) also concurs with the notion that software engineering is something else, and introduced the idea that at best it is a "soft" systems engineering discipline, "In essence, the type of systems which we are interested in are socio-technical software-intensive systems…" Furthermore, Sommerville claims that "there is no technical solution to software complexity."

Andriole and Freeman (1993) also propose a new discipline of "software systems engineering" that aims to integrate the two disciplines that often operate fairly disconnected from each other. The still developing nature of software engineering as a discipline is underscored by the variety of these definitions.

72.2.3 How Does Software Engineering Differ from Traditional Engineering?

If the set of activities associated with the creation of software is to be considered engineering, we must establish appropriate relationships, in terms of similarities and differences, with conventional engineering disciplines. Software is different from the object of much engineering because of its intangibility and plasticity, of course. There are also important differences between software engineering and traditional engineering (Díaz-Herrera, 2009; Shaw, 1990).

Some specific differences between software engineering and traditional engineering include the following:

- Software engineering builds on a different basis or foundation, because the product it designs and creates is an abstract or logical artifact instead of a concrete or a physically realizable entity.
- Software engineering foundations are primarily in computer science, not in the natural sciences.
- Underpinning all successful engineering is the notion of measurement. This is absolutely fundamental to the engineering process and engineered products, and it is probably the part of software engineering that differs the most from traditional engineering. Measurement as practiced in other engineering disciplines is impossible for software engineering, as DeMarco recently put it (DeMarco, 2009), "Software development is inherently different from a natural science such as physics, and its metrics are accordingly much less precise in capturing the things they set out to describe."

  The three primary things one can measure in software are size, development time, and defects (Humphrey, 1995). These measurements, however, do not address predictability of behavior—a critical aspect of engineered products. There are no universal relationships between software quality and what we can measure in software and, in particular, with predictable behavior (Sommerville, 1999).
- Although some software engineering techniques are based on discrete mathematics, rather than on continuous state systems, there is no software engineering mathematics despite the long quest for a successful axiomatic approach to programming (Shustek, 2009).
- Even if we succeed in finding ways to prove programs correct, software engineering deals with systems that are too complex to be treated by mathematical analysis alone.

In terms of *engineering practice*, software engineering is a different kind of "engineering." The difference that most people think of first is that software is not tangible and is infinitely malleable; most engineering disciplines, in practice, deal with tangible objects (e.g., bridges, mechanical devices, chemicals, and circuits). A major exception is industrial engineering,* which is largely mathematical or abstract, but measurable. Electrical (or electronic) engineering would also appear to be an exception since it deals with physical effects that are not visible—but they can be measured and their effect often seen or otherwise detected—and they are realized in physical circuits.

Another major difference is obvious in the first aforementioned definition; there is no "pure science" on which software engineering is based. This is a major issue in practice and theory that we briefly address in the following text.

There are more practical differences as well—the lack of common standards, a deficiency of case studies and handbooks (which this volume addresses to some extent), the rapidly changing nature of what software engineers do, the lack of much effort spent on developing the structures and artifacts of a traditional engineering discipline, and so on. Many of these are common to any new field as it develops and, especially, when the activities of its practitioners and researchers are so much in demand, leaving little time or motive for most people to devote time to what appear to be long-term, even irrelevant, activities. Over time, this will change.

Fundamentally, however, software engineering will always differ from traditional engineering in its technical details because of the invisibility and malleability of the object of its efforts. It shares this differentiation with computer science as discussed in the subsequent section.

---

* Agresti grouped the collection of Industrial Engineering techniques and tools that appear to offer some prospect for software (Agresti, 1981).

72.2.4 How Does Software Engineering Differ from Computer Science?

Software engineering is more than and less than computer science—put more properly, they share a lot in common, but each addresses issues that the other discipline does not address. This viewpoint was discussed as early as the NATO Conferences (Buxton and Randell, 1969, p. 46).

Chemical engineering, for example, is largely based on chemistry; civil engineering is largely based on mechanical properties known from physics; electrical engineering is largely based on electromagnetic properties known from physics; and so on. While all may have management issues in practice, these tend to be common across many areas of traditional engineering and none in their basic formulation are concerned with human interactions with the systems with which they deal.

Software engineering does not have a well-established science on which most of its tenets can be based. Computer science is also still a developing discipline, although more mature than software engineering, thus making the definition of a software engineering discipline even more difficult.

The major commonality that software engineering and computer science share is, of course, software. Software (programs and systems of programs) is one of two major topics that computer science deals with—the other being the mechanisms (processors) that can execute programs. Software engineering was started by including in the definition of computer science the study of the processes used to create software (the essence of software engineering) but as software engineering has paid more attention to topics such as management, some argue that it needs to be a discipline separate from computer science. Whether or not it should be a separate discipline is an argument that will not be settled any time soon, but is largely an issue of organizational politics and is not terribly relevant to the practical conduct of software engineering.

Differentiating factors include the importance of management, the processes of creating software, technical issues like human–computer interfaces, and so on. This can be explained by noting that computer science, like traditional sciences, is fundamentally focused on *understanding* something, while software engineering, like traditional engineering, is focused on *creating* something. At the same time, the non-traditional nature of computer science blurs this distinction somewhat.

The gap currently is wide between the software engineering knowledge that has been codified, verified, and organized into repeatable, teachable courses and curricula on the one hand and what the practicing software engineer must employ to be successful in many situations on the other (Díaz-Herrera et al., 2007). The objective of software engineering research is to organize and narrow this gap so that future generations can learn how to do software engineering in some way besides making large and costly mistakes!

72.3 Software Engineering Research

Research in software engineering refers mostly to applied research, and a more rigorous, scientific approach is needed (see evidence-based software engineering later in this section of the handbook). We can analyze, in our academic laboratories, existing and proposed software methods and techniques in the light of new software engineering principles. It is a fact that, since the introduction of the term software engineering in 1968, there have been many proposals being put forward to alleviate the software crisis. Rarely, if ever, have these proposals been backed by hard data as evidence of effectiveness. The arrival of Ph.D. programs in software engineering will accelerate the scientific understanding of software.

We need to examine project failures to discover why they fail. Just like ordinary engineering failures (such as a bridge collapsing), major software disasters must be investigated, causes analyzed, principles discovered, and solutions reported and disseminated. This kind of software practice research is difficult to do in a traditional academic setting (Shaw, 2002).

Someday, we may have completely automated ways of producing complex software. While some progress has been made, we are still a long way from reaching that goal. Research aimed at that goal (sometimes called "automatic programming") has been around for longer than software engineering and has taken various forms ranging from artificial intelligence approaches to algorithmic processes. The latter approach has been successful at the level of code generation and program generation for well-structured situations (e.g., I/O drivers for hardware), but progress at higher and broader levels of system generation has not made much progress in unstructured and/or new areas.

72.4 Elements of Software Engineering

The rapidly developing and diverse nature of software engineering, as illustrated earlier, mean that any list of the primary elements in a software engineering curriculum or the primary activities of software engineering will be incomplete and/or soon out of date. Focusing on software, the elements of software engineering typically include *methods* for the various development activities as summarized in Table 72.1 as well as and corresponding verification and validation activities, and *tools* to support and/or automate each of these, including IDEs, plus *skills and knowledge* of software engineering, computer science, engineering management, human perception and behavior, and one or more application domains (see also Díaz-Herrera, 2001).

Notice two characteristics of this set of categories: First, it is focused on methods that are for the most part carried out by human beings and only secondarily (if at all) on the nature of software (in the same way that chemical engineering is influenced by chemical properties of materials). Second, the categorization is still very general.

The great majority of software produced today is still developed following ad-hoc methods and it remains true that "software engineers have not been able to put together a coherent engineering design method for the systematic production of software, specifically for large, complex, ill-defined systems" (Denning, 1996).

The reasons for these problems are many-fold (ACM, 2004):

1. Software products are some of the most complex of man-made systems, and software by its very nature, has intrinsic difficulties (e.g., complexity, visibility, and changeability) that are not easily overcome (Brooks, 1995).
2. Programming techniques and processes that worked effectively in the 1950s and early 1960s, to develop modest-sized programs by an individual or a small team, did not scale-up well to the development of large, complex systems (systems with millions of lines of code, requiring years of work, by hundreds of engineers).
3. The pace of change in computer and communications technology drives the demand for new and evolved software products. Our successes in this area have created customer expectations and competitive forces that stress the quality of software and their development schedules.

The thought of treating software engineering as an engineering discipline is, of course, not new. For several years now, people both in academia and in industry and government have fostered this very idea.

TABLE 72.1 Software Development Activities

Requirements elicitation and analysis
Specification
System or architectural design
Program design
Testing of program modules and subsystems
Integration of system parts into a complete system
Management of software artifacts
Process management

What is new today is the widespread impetus from many fronts to consider software development as engineering. To wit consider the following major milestones:

- A *code of ethics* for software approved by the ACM and the IEEE-CS (Gotterbarn, 1999)
- The processes for outlining the body of knowledge for software engineering has resulted in two widely contributed and distributed versions (Bourque and Dupuis, 2001). SWEBOK V2 (2004 Version) also known as Technical Report ISO/IEC TR 19759 is currently under review to produce V3 under the direction of Dick Fairley.
- *Accreditation criteria* for educational programs in software engineering (see IEEE Spectrum, 1997) accepted by ABET in 1999 (Engel, 1999).
- Texas legislature has being *licensing professionals* in software engineering since July 1999.
- Pioneering undergraduate programs in the United States at Rochester Institute of Technology* in New York, Monmouth University† in New Jersey and Milwaukee School of Engineering‡ led the way to the development of several such programs in the United States and in the European and Australian universities.
- In 1997, we discussed the possibility of creating *Ph.D. programs* in software engineering (Díaz-Herrera, 1997). Two such programs have been put in place, the first at the Naval Postgraduate School§ and the other at Carnegie Mellon University.¶,**
- In 1999, the IEEE-CS started a *professional certification* initiative and the certification exam began in 2001.††

The best way to define the elements of software engineering is by extension. The aforementioned historical material is one example. A good text is another. The content of the curriculum in software engineering from a highly regarded school is a third type of definition. Ultimately, however, the elements of software engineering, at the present state of understanding, are defined by what practicing software engineers need and do in their daily work.

72.5 What Is Used in Practice?

It is impossible to answer the question "What is actually used in practice?" nor is that the purpose of this handbook. We would note that there is no comprehensive, objective survey data on the subject (however, see Lethbridge, 2000). One can find surveys from time to time in the trade press, typically focused on a specific language, tool, or methodology; often times these are little more than promotion pieces by a vendor. We point the reader to later parts of this section discussing empirical and evidence-based software engineering.

Amount of use of "software engineering" or "software engineer" is equally unknown. In some jurisdictions, use of the term "engineer" is illegal unless the jobholder is certified as an engineer in that jurisdiction, for example.

Some engineering based companies that employ engineers of many kinds classify some employees as software engineers while other, similar companies specifically do not unless the employee has a degree in software engineering while yet others have no policy, leaving it to the employee to use whatever title they want. Perhaps the largest usage of "software engineer" is basically as a substitute for and synonymous with "programmer" in terms of what the employee does.

* http://www.rit.edu/~932www/UGrad/UGradCat/colleges/cast/softeng.html
† http://www.monmouth.edu/se/
‡ http://www.msoe.edu/eecs/se/
§ http://www.cs.nps.navy.mil/~se/phd.html
¶ http://www.cs.cmu.edu/afs/cs.cmu.edu/Web/education/doctoral_isri.html
** Graduates have been earning the PhD with software engineering research since the 1970s, usually in CS programs.
†† http://computer.org/certification

72.6 Where Does One Need Software Engineering?

Software engineering originated from difficulties with large, complex software-intensive systems; has been developed with heavy support from industry and government organizations that are usually those that are developing, or contracting for, such systems; and is most often found today in large organizations and/or projects. To some extent, this is because it is perceived (not necessarily accurately) that utilizing software engineering is more expensive, cumbersome, or overly bureaucratic than many organizations feel they can afford. While sometimes true, this need not be the case and often is not when one computes the total lifecycle cost of developing software.

At least until fairly recently and perhaps still today, there are organizations that do not utilize software engineering even though they are building large systems. This tends to be more the case in commercial organizations that are not engineering-oriented overall; in fact, sometimes software engineering is rejected by such organizations precisely because of the term "engineering," that they mistakenly believe refers to something very technical that does not apply to them.

So, does this mean that only large organizations should use software engineering? Absolutely not!

There is nothing inherent in the concepts and elements of software engineering that prevents it from being used in organizations of any size and on projects of any size. It can be used by one person developing a single program, a small team in an organization small or large, teams of any size developing large systems, development spread over geographically distributed groups, and so on. What is needed is to choose carefully the specific elements to be used and the overall process framework to guide the work.

The principles and elements of software engineering are often the embodiment of good, commonsense practice that are just as useful for an individual working alone as they are for a team of thousands in a large organization. Obvious examples are the programming methods that are often part of a software engineering environment—types of language used, structuring principles, review techniques, and so on. These are techniques that are applied by individuals even when they are working in a large group and are just as valuable when they are working alone. Even program review techniques that may call for code reading by another person embody the principle of careful review of work, that is, formal inspections; while it may be more valuable when the review is conducted by someone other than the author of the code, the fundamental principle of review still applies.

Similarly, the concept of following a defined lifecycle with definite stages, explicit entry and exit criteria from each stage, and so on is a concept that is important in small organizations and/or on small projects down to the level of individuals. If one is writing a program alone, it is important to understand the requirements before beginning to program, design the structure of the program if it is more than the simple coding of a straightforward algorithm, then test the program, document it, etc. The depth and formality of most of these steps will certainly be less than if working in a team and/or on a large system, but the structuring of one's activity (workflow) can be beneficial.

Watts Humphrey (1995, 1997) in the two books referenced developed the idea that software engineering is not just a large team, large organization activity.

72.7 Summary

Software is complex and the software industry is large, global, and growing. The increasingly public importance of software-intensive systems in our daily lives, for our safety and security, and for national and global economies makes it imperative to be able to deliver high quality software. We need to find effective ways of evaluating innovative concepts, methods, and tools for developing high quality software.

Over the past 50 years, the act of creating the software that is necessary to implement our visions of automation using a computer has come to be understood as extremely difficult to carry out and, thus, it has been recognized that a discipline akin to more established engineering disciplines might make the task less difficult. The quest outlined earlier has certainly made possible many aspects of modern software systems, but it is not yet a full discipline either in the intellectual or practical sense of the term "discipline."

One overriding theme of the development of software engineering over the past 45 years has been the successive broadening of scope of what must be included in the discipline—from programs to designs to systems to methodologies and beyond. A second theme has been the inclusion of topics such as management and human behavior that are not found in traditional engineering disciplines (civil, chemical, mechanical, and electrical), or are considered separate from the main body of the discipline. A third theme has been the rapid expansion in scope to creating software for countless new applications and situations, usually under increasing pressure to produce the product more rapidly and with less effort.

The practical definition of the discipline has certainly taken shape in the form of a fair number of standards and compilations of knowledge (including this volume). Yet, the still rapidly changing landscape of knowledge and demands keeps the definition fluid, at best. As so often happens, what seems clear in theory is less clear in practice. Perhaps, the most important driver of the disciplinary definition remains the courses and curricula that exist, although at the same time far too many of those do not address the real needs of practice.

As a research discipline, we are probably even further away from a consistent disciplinary definition. Indeed, some would argue that the engineering and management approach will never be completely successful and that the way to address the "software problem" is to automate as much of the software creation process as possible. Even within the engineering paradigm, there are important differences in researcher's beliefs of what the important research questions are and how to attack them—two characteristics that define a discipline of research.

Could there be a basis for understanding the complexity of software such that we can *engineer* it to have predictable quality and behavior?

The answer to this fundamental question is unknown—that is the objective of some software engineering research. Thus, at present, software engineering usually has to address the challenges of architecting complex software-intensive systems, of the kind described earlier, by more heuristic and less analytical methods. As DeMarco (2009) says, "software development is and always will be somewhat experimental." In this way, software development is a different kind of engineering, the kind of engineering associated with computing.

Whatever your point of view and interest in the discipline of software engineering, the history of the past 45 years can be illuminating and helpful to put into context the information that you will find in this handbook. The origins of that history as outlined here will provide you a basis for further exploration of that history in areas that pertain to your needs.

There may be a problem with trying to retrofit this fundamentally computing sub-discipline with the traditional "engineering" paradigm; maybe, what we should be doing is exploring how it can be molded into that paradigm within computing. Computer science is on the way to accomplishing this; it became not a "science" discipline, but rather the "science of computing," and the same can be said of IT—the technology of computing.

We implemented the vision of computing as a distinct branch of knowledge, a discipline in its own right, that culminated with the creation of the computing college—"standing alone, headed by its own dean, with the same stature as traditional colleges," as (Denning, 1998) put it. In addition, here is where software engineering rightfully belongs as the "engineering of computing."

References

[ACM, 2004] *Software Engineering 2004: Curriculum Guidelines for Undergraduate Degree Programs in Software Engineering*. IEEE Computer Society, August 23, 2004.

[ADA 2012] http://www.ada-auth.org/standards/ada12.html. Accessed October 4, 2013.

[Agresti, 1981] Agresti, W. Software Engineering as Industrial Engineering, *ACM SIGSOFT Software Engineering Notes*, 6(5), pp. 11–13, October 1981, New York.

[Andriole and Freeman, 1993]. Andriole, Stephen J. and Freeman, Peter A. Software systems engineering: the case for a new discipline. *Software Engineering Journal*, May 1993.

[Apel and Kästner, 2009] Apel, S. and Kästner, C. An overview of feature-oriented software development. *Journal of Object Technology* 8(5), July–August 2009.

[Bauer, 1972] Bauer, F. L. Software engineering. *Information Processing 71: Proceedings of IFIP Congress 1971*, North-Holland, June 1972, pp. 530–538.

[Boehm, 1988] Boehm, B. A spiral model of software development. *IEEE Computer*, 21(5), pp. 61–72, May 1988.

[Booch, 1993] Booch, G. *Object-Oriented Analysis and Design with Applications* (2nd edn.). Benjamin Cummings, Redwood City, CA.

[Bourque et al., 1999] Bourque, P., Dupuis, R., Abran, A., Moore, J., and Tripp, L. The guide to the software engineering body of knowledge. *IEEE Software* 35–44, November/December 1999.

[Bourque and Dupuis, 2001] Bourque, P. and Dupuis, R. eds. *Guide to the Software Engineering Body of Knowledge*. IEEE CS Press, Los Alamitos, CA, 2001.

[Brooks, 1995] Brooks, F. P. *The Mythical Man-Month, Essays on Software Engineering*, Anniversary Edition. Addison-Wesley, New York, 1995.

[Brown, 1975] Brown, A. P. G. Modelling a real-world system and designing a schema to represent it, in: Douque and Nijssen (eds.), *Data Base Description*. North-Holland, Amsterdam, the Netherlands, 1975.

[Bruegge and Dutoit, 2000] Bruegge, B. and A. Dutoit. *Object-Oriented Software Engineering*. Prentice-Hall, Columbus, OH, 2000.

[Budgen et al, 2003] Budgen, David, and Tomayko, James E., Norm Gibbs and his contribution to software engineering education through the SEI curriculum modules, *Proceedings of the 16th Conference on CSEE&T*, March 2003.

[Buxton, 1970] Buxton, J. N. and Randell, B. (eds.) Software Engineering Techniques, *Report of a Conference Sponsored by NATO Science Committee* (Rome, Italy October 27–31, 1969), 1970.

[Carr, 1952] Carr, J. Progress of the whirlwind computer towards an automatic programming procedure, *ACM'52: Proceedings of the 1952 ACM National Meeting*. ACM, New York. May 1952.

[Cheatham and Standish, 1970] Cheatham, T. and Standish, T. Optimization aspects of compiler-compilers. *ACM SIGPLAN Notices*, 5(10). pp. 10–17, ACM, New York. October 1970.

[Chen, 1976] Chen, P. The entity-relationship model—Toward a unified view of data. *ACM Transactions on Database Systems*, 1(1), 9–36, March 1976.

[DeMarco, 1979] DeMarco, T. *Structured Analysis and System Specification*. Prentice Hall, Columbus, OH, 1979.

[DeMarco, 2009] DeMarco, T. Software engineering: An idea whose time has come and gone? *IEEE Software*, 95–96, July/August 2009.

[Denning and Dargan, 1996] Denning, P. and Dargan, P. Action-centered design, in: T. Winograd (ed.), *Bringing Design to Software*, Addison-Wesley, New York, pp. 105–120, 1996.

[Denning, 1998] Denning, P. Computing the profession: An invitation for computer scientists to cross the chasm. *Educom Review* 33(6), pp. 46–59, 1998.

[Díaz-Herrera, 1997] Diaz-Herrera, J. L. Graduate Software Engineering Education: Issues and opportunities. Panel moderator, *Frontiers in Education Conference*, IEEE. Pittsburgh, PA, November 1997.

[Díaz-Herrera, 2001] Díaz-Herrera, J. L. Engineering design for software: On defining the profession. *IEEE Frontiers in Education Conference*, Reno, NV, October 2001.

[Díaz-Herrera, 2007] Díaz-Herrera, J. L., Makabenta-Ikeda, M., Ardis, M., and Coleman, D. Understanding the University/Industry Gap. *Workshop in Engineering Education; ASIA/Pacific Software Engineering Conference*, Nagoya, Japan, December 2007.

[Díaz-Herrera and Flude, 1980] Díaz-Herrera, J. L. and Flude, R.C. Pascal/HSD: A graphical programming system. *IEEE, COMPSAC Conference*, Chicago, IL, pp. 723–728, 1980.

[Diaz-Herrera, 2009] Díaz-Herrera, J. L. *ACM SIGSOFT Software Engineering Notes*. 34(5), 1, pp. 1–3, September 2009

[Dijkstra. 1968] Dijkstra, E.W. "The Structure of the 'THE' Multiprogramming System." Communications of the ACM, 11(5), 341–346. 1968.

[ECMA, 1993]ECMA TR/55. NIST special publication 500-211. http://www.ecma-international.org/publications/files/ECMA-TR/TR-055.pdf, accessed on 10/6/2013.

[Engel, 1999] Engel, G. Program criteria for software engineering accreditation programs. *IEEE Software*, 31–34 November/December 1999.

[Fairley, 1985] Fairley, R. *Software Engineering Concepts*. McGraw-Hill, New York, 1985.

[Finkelstein, 1993] Finkelstein, A. European computing curricula: A guide and comparative analysis. *Computer Journal*, 36(4), 299–319, 1993.

[Ford, 1994] Ford, G. A Progress Report on Undergraduate Software Engineering Education, CMU/SEI-94-TR-11, Software Engineering Institute, Carnegie Mellon University, May 1994.

[Freeman, 1974] Freeman, P. Automating software design. *Proceedings of 10th IEEE Design Automation Workshop*, IEEE Press, Picataway, NJ, 1974.

[Freeman, 1975] Freeman, P. *Software Systems Principles: A Survey*, Chapter 13. "Methods and Tools for System Creation", pp. 477–505, SRA, Palo Alto, CA, 1975.

[Freeman and Wasserman, 1975] Freeman, P. and Wasserman, A. *Software Design Techniques*, 1st edn., IEEE Computer Society Press, Los Alamitos, CA, 1975.

[Freeman and Wasserman, 1978] Freeman, P. and Wasserman, A. I. A proposed curriculum for software engineering education, *Proceedings of the 3rd International Conference on Software Engineering*, Los Alamitos, CA,1978, pp. 56–62.

[Freeman and Wasserman, 1982] Freeman, P. and Wasserman, A. Software development methodologies and Ada. Ft. Belvoir Technical Information Center, Ft. Belvoir, VA, November 1982.

[Freeman and Wasserman, 1983] Freeman, P. and Wasserman, A. *Software Design Techniques*, 4th edn. IEEE Computer Society Press, New York, 1983.

[Gotterbarn, 1999]. Gotterbarn, D. How the New Software Engineering Code of Ethics. Affects You. *IEEE Software*, 16(6), pp. 58–64,Nov/Dec, 1999.

[Hoffman and Weiss, 2001] Hoffman, D. M. and Weiss , D.M., eds. *Software Fundamentals: Collected Papers by David L. Parnas*. Addison-Wesley, New York, 2001.

[Humphrey, 1989] Humphrey, W. *Managing the Software Process*. Addison-Wesley, New York, 1989.

[Humphrey, 1995] Humphrey, W. *A Discipline for Software Engineering*. Addison-Wesley, New York, 1995.

[Humphrey 1995] Humphrey, W. Managing Technical People: Innovation, Teamwork, and the Software Process. Addison-Wesley, New York, 1997.

[IEEE, 1990] IEEE STD 610.12-1990, IEEE standard glossary of software engineering terminology. IEEE Computer Society, New York, 1990.

[Keringhan and Plauger, 1976] Kernighan, B. and Plauger, P. *Software Tools*. Addison-Wesley, New York, 1976.

[Kerola and Freeman, 1981] Kerola, P. and Freeman, P. A comparison of lifecycle models. *Proceedings of 5th International Conference on Software Engineering*. pp. 90–99. IEEE Press, Piscataway, NJ. 1981.

[Lethbridge, 2000] Lethbridge, T. What knowledge is important to a software engineer? *IEEE Computer*, 33(6), pp. 44–50, May 2000.

[Lethbridge et al., 2006] Lethbridge, T. C., LeBlanc, R. J. Jr., Sobel, A. E. K., Hilburn, T. B., and Díaz-Herrera, J. L. SE2004: Curriculum recommendations for undergraduate software engineering programs. *IEEE Software* 19–25, 2006.

[Marca and MacGowan, 1998] Marca, D. A. and MacGowan, C. L. *SADT: Structured Analysis and Design Technique*. McGraw Hill, New York, 1988.

[NCEES, 2013] http://cdn3.ncees.co/wp-content/uploads/2012/11/Model-Law-2013.pdf. Accessed October 6, 2013.

[Naur, and Randell 1969] Naur, P. and Randell, B., eds. Software engineering: Report on a conference sponsored by the NATO Science Committee (October 7–11, 1968). Scientific Affairs Division, NATO, Brussels, Belgium, 1969.

[Pahl and Beitz, 1996] Pahl, G. and Beitz, W. *Engineering Design: A Systematic Approach*. Springer-Verlag, Berlin, Germany, 1996.

[Parnas, 1972] Parnas, D. On the criteria to be used decomposing systems into modules. *CACM*, 15(12), 1053–1058, 1972.

[Parnas, 1999] Parnas, D. L. Software engineering programs are not computer science programs. *IEEE Software*, 19–30, November/December 1999.

[Perlis, 1959] A.J. *Perlis Papers*, Charles Babbage Institute, University of Minnesota, Minneapolis, MN.

[Pullen and Shuttee, 1968] Pullen, E. and Shuttee, D. MUSE: A tool for testing and debugging a multi-terminal programming system. *Proceedings AFIPS '68 Spring Joint Computer Conference*, May 1968.

[Rechtin, 1997] Rechtin, E. and Maier, M. *The Art of Systems Architecting*. CRC Press LLC, Boca Raton, FL 1997.

[Ross, 1978] Ross, D. T. Origins of the APT language for automatically programmed tools. *ACM SIGPLAN Notices*, 13 (8), pp. 279–338, August 1978.

[Royce 1970] Royce, W. W. Managing the development of large software systems: Concepts and techniques. Originally published in: *Proceedings of WESCON*, August 1970. Reprinted in *Proceedings of the 9th international conference on Software Engineering*, 1987, Pages 328–338, IEEE Computer Society Press Los Alamitos, CA.

[Shaw, 1990] Mary Shaw Prospects for an engineering discipline of software. *IEEE Software*, 7, 6, November 1990, pp.15–24.

[Shaw, 2002] Mary Shaw, What makes good research in software engineering? *International Journal on Software Tools for Technology Transfer*, Vol. 4, DOI 10.1007/s10009-002-0083- 4, June 2002.

[Shustek, 2009] Shustek, L. An interview with C.A.R. Hoare. *CACM*, 52 (3), 38–41, 2009.

[Sommerville, 1999] Sommerville, I. Systems engineering for software engineers. *Annals of Software Engineering*, (1–4), 111–129, April 1999.

[Sommerville, 2012] Sommerville, Ian, D. Cliff, R. Calinescu, J. Keen, T. Kelly, M. Kwiatkowska, J. Mcdermid, and R. Paige. Large-scale complex IT systems. *CACM*, 55 (7), 71–77, July 2012.

[Wasserman and Freeman 1976] Wasserman, A.I. and Freeman, P. (eds.), *Software Engineering Education*. Springer-Verlag, New York, 1976.

[Tomayko, 1999] Tomayko, James E., Forging a discipline: An outline history of software engineering education. *Annuals of Software Engineering*, 6(1–4), pp.3–18, April 1999.

[Yourdon and Constantine, 1979] Yourdon, E. and Constantine, L. *Structured Design*. Prentice-Hall, Upper Saddle River, NJ, 1979. (Earlier versions copyrighted and published in 1975 and 1978 by the authors).

[Zelkowitz, 1979] Zelkowitz, M. Perspectives on software engineering. *Computing Surveys,* 10(2), 197–216, June 1978.

[Wirth, 1971] N. Wirth. The programming language pascal. *Acta Informatica*, 1, 35–63, 1971.

[Wirth, 1974] N. Wirth. *Programming in Modula-2*. Springer, 1974. ISBN 0-387-50150-9.

73 Professionalism and Certification

Stephen B. Seidman
Texas State University

73.1 What Is a Profession?

The term "profession" is used informally to refer to a very wide range of occupations. However, most writers reserve the term for occupations that share a number of characteristics: an explicit body of knowledge, some sort of entrance control, a code of ethics, and a recognized need for continuing education (see, e.g., Davis 2009). Professions are generally governed by collaborations between professional organizations and governmental agencies. Professional credentials may be internationally portable. Occupations that impact individuals' physical and/or financial well-being are most likely to have evolved into professions. Examples include law, health (medicine, dentistry, and nursing), architecture, and accounting. In the United States, each of the aforementioned examples has a body of knowledge that is embedded in a set of entrance examinations. Entrance control is enforced by specialized educational prerequisites for the examinations. All of these professions have explicit codes of ethics. In the United States, members of these professions are required to have government-sanctioned licenses, and continuing education is a requirement for retaining a valid license.

73.2 Engineering as a Profession

Until a few centuries ago, "engineers" were primarily those involved with the design and construction of fortifications and weaponry. However, in the eighteenth century, the occupation "engineer" began to be extended to cover those who designed and built the technological elements (mines, steam engines, canals, bridges, tunnels, locomotives, etc.) that were at the heart of the changes in manufacturing and transportation generally referred to as the Industrial Revolution (Armytage 1961). During this period, the term "civil engineer" was coined to contrast with the more traditional military uses of engineering. This usage is still found in some countries (e.g., Chile). At the same time, the practice of engineering gradually acquired the characteristics described in the preceding section. These characteristics were acquired slowly over time, and the rate and process of acquisition was quite variable across countries. For example, while entrance to the profession was initially achieved by apprenticeship to a practicing engineer, the nineteenth century saw the creation of specialized schools dedicated to the training of engineers. In France, the last decades of the eighteenth century saw the establishment of the École Polytechnique and École des Mines. The first engineering school in the United States was Rensselaer Polytechnic Institute (Troy, New York), established in 1824 "for the application of science to the common purposes of life." During the course of the nineteenth century, other countries followed the example of France and the United States and established engineering schools. In the United Kingdom, this effort began with the establishment of the London Mechanics' Institute in 1823 (Armytage 1961: 144). Engineering entered the UK academic world a bit later; the first courses in civil and mechanical engineering began in 1838, and the first chairs in engineering were created at University College in 1841; both institutions were part of the University of London. In most countries, control of university engineering programs and guaranteeing their quality was the responsibility of ministries of education. This was not possible in the United States, with its federal structure and public/private university system. The gap was eventually filled by the Engineers' Council for Professional Development (now called ABET) established in 1932 as a confederation of professional societies (including the American Institute of Electrical Engineers, a predecessor of the Institute of Electrical and Electronics Engineers [IEEE]). ABET is responsible for the accreditation of US programs in engineering and computing; it also accredits many programs in other countries. ABET accreditation requires that a program in a given engineering discipline meet discipline-specific requirements that are derived from an explicit or implicit body of knowledge for that discipline. Graduates must also recognize the need for lifelong learning in their discipline. Finally, students graduating from an accredited program must have an understanding of professional and ethical responsibility. Corresponding to this requirement, professional societies in the engineering disciplines have developed discipline-specific codes of ethics. Codes of ethics have also been developed by organizations representing engineering more broadly. For example, the US-based National Society for Professional Engineers (NSPE) has developed a code of ethics (NSPE 2007). Governmental regulation of engineering generally involves some sort of collaboration between an arm of government, the organization accrediting engineering education, and one or more professional organizations. In the United States, professional licensure of engineers is regulated by the states, but the requirements and the examinations required for licensure are administered by a professional organization, the National Council of Examiners for Engineering and Surveying (NCEES). Aspiring professional engineers must pass the examinations and have appropriate education and experience; the number of years of experience required is less for those who have completed an ABET-accredited undergraduate degree. Similar structures exist in Canada and Australia; more unitary approaches are used in other countries. It is important to note that (at least in the United States) professional engineers may have their state-issued licenses revoked if they violate the NSPE code of ethics.

73.3 Emergence of Software Engineering

The phrase "software engineering" first saw widespread use in North Atlantic Treaty Organization (NATO) conferences held in 1968 (Naur and Randell 1969) and 1969 (Randell and Buxton 1970). These conferences were a response to a generally perceived "software crisis" that had arisen from experience with the problems associated with designing, developing, and maintaining reliable large-scale software. At the 1968 conference, Friedrich Bauer defined software engineering as "establishment and use of sound engineering principles to obtain economically software that is reliable and works on real machines efficiently" (Naur and Randell 1969). The idea was that just as engineering principles were needed to design and build complex artifacts in the physical domain, they would also be necessary for doing the same in the software domain. During the ensuing decades, the phrase "software engineering" has seen increasingly widespread use. Many scholarly journals, magazines, and research and practitioner conferences have been established to disseminate research and current practice in software engineering and its subdisciplines (see the other chapters). Software engineering research programs are found in universities (most usually in departments of computer science) and in industrial and government laboratories. Courses or modules in software engineering and its subdisciplines are offered by most universities. Many computer science departments offer entire degree programs at undergraduate or Master's level. Some universities have established independent departments of software engineering, usually contained within colleges of engineering. Software engineering clearly has roots and affiliations in computer science, but its name and the associated rhetoric suggest an affiliation with engineering. These alternative affiliations have had an impact on the emergence of the profession of software engineering.

73.4 Software Engineering as a Profession

More than 40 years after the NATO conferences that gave birth to "software engineering" as a concept, the discipline has attained recognition from universities and research funding agencies. At the same time, "software engineer" sees wide use as a job category in industry. While the acceptance of software engineering in academia and industry demonstrates the extent of the change since 1968, it is not so clear whether we can reasonably talk about a profession of software engineering. The following sections will look at this issue.

73.4.1 Body of Knowledge

In late 1998, the Association for Computing Machinery (ACM) and the IEEE Computer Society (IEEE-CS) began a joint effort to develop a body of knowledge for software engineering.* The eventual outcome of this effort was the Software Engineering Body of Knowledge (SWEBOK), first released in 2004 (IEEE-CS 2004b). SWEBOK became an international standard (ISO/IEC TR 19759) in 2005. SWEBOK is built around 10 knowledge areas: requirements, design, construction, testing, maintenance, configuration management, software management, process, tools and methods, and quality. Furthermore, as the discipline of software engineering is subject to rapid change, the SWEBOK project includes an ongoing revision process, described in Appendix B of the SWEBOK document. At about the same time, the two leading professional societies in the general area of computing (ACM and IEEE-CS) began an effort to develop curriculum recommendations for undergraduate programs in software engineering. This effort produced the SE2004 curriculum recommendations (IEEE-CS 2004a). The list of topics included in the curriculum was heavily influenced by the SWEBOK document (IEEE-CS 2004a: 12). More recently, curriculum guidelines for professional Master's programs in software engineering have also been proposed (ISSEC 2009). This Master's curriculum is also based on an explicit body of knowledge; the "core body of knowledge" was derived from the SWEBOK by making relatively minor changes to some of the SWEBOK knowledge areas.

73.4.2 Accreditation

The path to the accreditation of software engineering programs varies by country. Accreditation of US engineering programs by a federation of professional societies (now called ABET) began 80 years ago.† Computing programs are far more recent, and accreditation efforts date back only to the establishment

* Note that ACM withdrew from this effort in 2000. At this time, ACM had taken a strong position opposing licensure of software engineers, and development of a body of knowledge was seen as a step toward licensure. These developments will be discussed further in the subsequent sections.

† Note that ABET now accredits programs outside the United States. These accreditations include software engineering programs.

of the Computer Science Accreditation Board (CSAB) in the 1980s. This entity worked alongside ABET and adopted many of its principles and processes. In the late 1990s, CSAB moved closer to ABET, eventually merging into the larger organization in 2000 when CSAB became an ABET member society. As a result of this merger, ABET now accredits computing programs (computer science, information systems, and information technology) as well as programs in computer engineering and software engineering. ABET accreditation of software engineering programs was first proposed in 1997, and software engineering accreditation criteria were approved in 1998. To understand how all of this works, it is necessary to know a bit more about ABET's organizational structure. ABET is organized into four commissions: the two relevant to this discussion are the Engineering Accreditation Commission (EAC) and the Computing Accreditation Commission (CAC). As might be expected, EAC and CAC are responsible for engineering and computing programs, respectively. The accreditation criteria for a program are set by one or more "sponsoring" societies and must be approved by the relevant commission. For example, electrical engineering criteria are set by IEEE and approved by EAC, while computer science criteria are set by CSAB and approved by CAC. Since software engineering shares aspects of computing and engineering, its criteria are written by CSAB and approved by EAC. The inclusion of CSAB within ABET is one approach to resolving the potential conflict created by software engineering's competing (engineering and computing) affiliations.

Other countries have taken different paths toward resolving the question of who is to accredit software engineering programs. In the United Kingdom, the British Computer Society (BCS) is a chartered engineering organization that is a member of Engineers UK, a federation of professional engineering societies. Since Engineers UK is responsible for accrediting engineering programs in the United Kingdom, BCS can accredit computing programs (including software engineering) within the Engineers UK mandate. In Australia, the Australian Computing Society and Engineers Australia make joint accreditation visits to software engineering programs. However, some countries have found it more difficult to achieve shared responsibility for software engineering accreditation. The Canadian situation is a case in point. Computing programs in Canada are accredited by the Canadian Information Processing Society (CIPS), while engineering programs are accredited by Engineers Canada through the Canadian Engineers Accreditation Board. The situation is exacerbated by the fact that Engineers Canada has trademarked the words "engineer" and "engineering" in Canada and aggressively defends the rights associated with the trademark (Adams 2004). For example, when the computer science department at Memorial University (Newfoundland) attempted to offer a software engineering degree, the university was sued by Engineers Canada. In response, Memorial has trademarked the phrase "software engineering." The CIPS perspective can be found at CIPS (undated). This conflict has never been definitively resolved. It is also worth noting that different countries have different cultural and legal contexts for program accreditation. In the United States, accreditation of engineering programs is part of the culture of engineering, especially because of its link to professional licensure. Accreditation is therefore the norm for US engineering programs. However, the decision to accredit a US computing program involves several factors, including institutional characteristics, academic organization, and local and regional competition. By contrast, other countries (e.g., Germany) have legal requirements for the accreditation of university programs.

73.4.3 Code of Ethics

Most engineering disciplines have developed their own codes of ethics; similar codes have also been developed by national organizations representing engineering more broadly, such as NSPE or Engineers Canada. At the same time, some national computing societies have developed codes of ethics for computing professionals. For example, each of the British, Australian, and Canadian professional societies has a code of ethics (CIPS 2005, BCS 2011, ACS 2012). However, these codes are not specific to software engineering. In the 1990s, ACM and IEEE-CS addressed this need (see Davis [2009], for a detailed history of this effort) by jointly developing a code of ethics for software engineers (ACM 1999); the code was approved by both societies in 1998.

73.5 Professional Qualifications for Software Engineering: Licensure versus Certification

There are two general approaches to providing generally recognized qualifications for a practicing professional. The first approach confers a legal status that carries rights, privileges, and obligations. In the United States, this status is called a "license to practice a profession," and it will be referred to here as "licensure." An alternative approach is for a professional to acquire certificates that demonstrate proficiency in a particular body of knowledge; this general approach will be referred to as "certification." Note that by contrast with licensure, certification is always voluntary and has no legal implications.

73.5.1 Licensure

Many professions require practitioners to hold government-sanctioned licenses or equivalent designations (e.g., "chartered status" in the United Kingdom and Australia). Government supervision of such qualifications is almost always shared with professional societies. It is clear that governments will enforce the regulation of professions that have significant impact on health and safety. The same is true for professions that impact the financial and legal status of individuals and corporations, such as law and accountancy. The situation for engineering is less clear, and there is much cultural variation both within engineering and across countries. In the United States, for example, civil engineers generally hold licenses, but electrical or industrial engineers generally do not. One reason for this difference is that the individual with ultimate responsibility for the design of a project affecting public safety (e.g., a nuclear power plant or oil refinery) must be a licensed engineer; such responsibilities are much more likely for civil engineers than for electrical or industrial engineers. In some Latin American and European countries, the acquisition of professional status as an engineer has social benefits. An engineer with officially recognized professional status is entitled to use the personal title "engineer." Such individuals have higher status in society; they frequently use the title in telephone directory listings and similar places. In these countries, virtually all individuals who are entitled to call themselves "engineer" make use of the title.

What is the situation for software engineering? In countries using the "chartered engineer" licensing model (Canada, Australia, and United Kingdom), the opportunity to attain chartered status has relatively easily been extended to software engineers. In the United States, where the question of licensing or certifying software engineers has been an active topic of discussion for many years, the situation is far less clear. For example, in 1991, the lower house of the New Jersey legislature passed a bill to license "software designers," although the bill eventually died in the upper house (Davis 2009). Shortly thereafter (in early 1993), the IEEE-CS created a Steering Committee for the establishment of software engineering as a profession. At that time, the ACM also expressed interest in assessing professionalism in software engineering. This interest led to collaboration with IEEE-CS, and eventually (in early 1994) to the creation of a joint ACM/IEEE-CS steering committee to deal with this general topic. This committee set up task forces dealing with several aspects of professionalism: code of ethics, body of knowledge, program accreditation, and curriculum recommendations. The question of licensing software engineers frequently arose during these discussions. For example, some software engineering professionals felt that the adoption of a software engineering code of ethics could be seen as one step along the path to licensure (Davis 2009: 211). Furthermore, 1996 saw a forum on the licensing of software professionals; many of the participants were also part of the joint steering committee. The forum attendees were strongly divided on the issue of licensing software professionals. In 1997, the Licensing Committee of the Texas Board of Professional Engineers formally requested ACM and IEEE-CS to work on defining a body of knowledge on which software engineering licensing examinations could be based (Davis 2009: 321). In late 1998, the two societies created a Software Engineering Coordinating Committee (SWECC) to oversee all of these efforts. They agreed to work with the Texas Board of Professional Engineers to "identify a technical basis for recognizing competency in software engineering and [enact] rules that recognized software engineering as a distinct engineering discipline" (Davis 2009: 356). At the same time, the National Council of Examiners for Engineering and Surveying passed a resolution in support of "licensing software engineering throughout the nation." However, the position of the ACM representatives on SWECC was not universal within the ACM software engineering community. This is illustrated by the ACM's creation in 1999 of an Advisory Panel on Professional Licensing in Software Engineering "to assess the current role of ACM as part of a joint effort with IEEE-CS to oversee and actively participate in the development of professional and licensing guidelines through SWECC" (Davis 2009: 359). After receiving a report from this advisory panel, the ACM Council "firmly decided [in May 1999] that it would not endorse licensing software engineers at this time, and that ACM would withdraw from any activity that gave the appearance of condoning the licensing of software engineers" (Allen et al. 2000). However, the question of licensing software engineers was debated in the flagship magazines of the two societies (De Marco 1999, El-Kadi 1999, Frailey 1999a,b); these position papers present strongly opposing views of the suitability of licensure. This situation has persisted over the past decade: IEEE-CS supports professional licensure for software engineers, and ACM continues to oppose it.

At about the same time (in 1999), Texas became the first state to institute professional licensure for software engineers. Since then, several other states have announced their intention to do the same. However, the path to engineering licensure in all US states requires that an applicant take two examinations administered by NCEES: the Fundamentals of Engineering (FE) and Principles and Practice of Engineering (PE) examinations. The FE examination consists of a morning part that is the same for all applicants; it includes mathematics (calculus, differential equations, and linear algebra), basic sciences (chemistry and physics), and engineering sciences (statics, materials, and thermodynamics). For some engineering disciplines (chemical, civil, electrical, environmental, industrial, and mechanical), the FE afternoon examination is discipline-specific; for all others, it deals with additional general engineering topics. The PE exam is specific to each of 24 engineering specialties. For civil, electrical, and mechanical engineers, PE exams are offered for multiple specialties. Since no PE examination was available for software engineering in 1998, Texas allowed practitioners with 12 years of experience and an ABET-accredited degree (or 16 years of experience for those with a related degree) to be licensed (Bagert 2004). In response to requests from engineering licensing boards in 10 states, NCEES announced in August 2009 that it had approved the development of a new PE examination for software engineers. IEEE-USA (an IEEE entity devoted to the career and public policy interests of IEEE's US members) is co-sponsoring the examination development effort, assisted by IEEE-CS, NSPE, and the Texas Society of Professional Engineers. The first step is to conduct a professional activities and knowledge study, which will survey a diverse sample of software engineering professionals to determine the topics to be covered in the exam. However, the creation of a PE examination in software engineering will not do much to increase the number of licensed software engineers, since the FE examination is not a good fit to undergraduate software engineering curricula, and even less so to other computing curricula.

73.5.2 Certification

As stated earlier, certification and licensure are distinct approaches to professional qualifications that carry different implications. While licensure is always associated with legal restrictions and constraints, certification is voluntary and has no legal implications. It is important to distinguish between product-based certifications and broad-based professional certification schemes (Seidman 2009: 354). Product-based certifications are administered by companies (e.g., Microsoft, IBM, and Cisco). A corporate product-based certification requires passing an examination linked closely to products or sets of products produced by or associated with the company offering the examination. For example, Microsoft offers a range of product-based certifications (see, e.g., Microsoft [2012]). According to this site, "The Microsoft Certified Systems Engineer (MCSE) certification helps enable IT professionals to demonstrate their ability to design and implement an infrastructure solution with Windows Server 2003-based business solutions." Note that this is the only Microsoft certification that uses the word "engineer." Interestingly, the appearance of "engineer" in this certification gave rise to legal action in

Canada. In an out-of-court settlement, Microsoft agreed that Canadian holders of this certificate would refer to themselves only by the acronym MCSE. The MCSE certification has now been superseded by a "Microsoft Certified IT Professional" certification. Both Cisco and IBM avoid using the word "engineer" in their certification programs.

By contrast, broad-based certification schemes are intended to recognize professional competence in a recognized body of knowledge. They are offered by societies that represent communities of professionals. Examples include Project Management Professional, Certified Financial Planner, and board certifications for medical professionals. The IEEE-CS's effort to develop a certification scheme for software engineering professionals began in the late 1990s (Seidman 2009: 354–356). The IEEE-CS began by surveying the community and conducting discussions with industry representatives and potential certificate holders. The examination development process followed: lists of task and knowledge statements were created and validated by the community; these materials were used to develop test specifications, which in turn were used to construct test questions. After several rounds of evaluation, the Certified Software Development Professional (CSDP) examination was released in 2002. The original idea had been for the phrase "software engineering" to appear in the name of the designation, but this was dropped in the face of opposition from NSPE and IEEE-USA. Since 2002, nearly 1000 individuals have earned the CSDP designation; more than 460 of these certificates are still valid (Dzerzhinskiy 2012). Over time, the CSDP body of knowledge has been revised to keep it in concordance with the ongoing revisions of SWEBOK. Candidates for CSDP certification must meet the following requirements before they can take the examination; licensed software engineers and senior members of IEEE are automatically eligible. Otherwise, candidates must satisfy the following criteria: (a) hold a bachelor's degree, hold a Certified Software Development Associate (CSDA) certificate (see subsequent sections), be a university-level educator, or be a full member of IEEE and (b) have an advanced degree in software engineering and at least 2 years (about 3500 h) of experience in software engineering/development, or have at least 4 years (about 7000 h) of experience in software engineering/development. The IEEE-CS soon realized that these requirements are too onerous for most candidates from developing countries. This led to the development of an entry-level certification program, the CSDA. While the CSDA program has no formal eligibility requirements, recommended guidelines for candidates indicate that candidates should be recent software or computer engineering graduates, undergraduates in the final year of a bachelor's degree program in software or computer engineering, or non-degree professionals with more than 2 years of programming experience.

Certification of software engineering professionals is also offered by the International Software Quality Institute (iSQI), an independent nonprofit organization in Germany. iSQI offers certification examinations in software architecture, software testing, and project management. The testing examination is also offered at an advanced level.

There is now an international standard (ISO/IEC 24773:2008) for certification schemes for software engineering. This standard was developed by a working group within a joint technical committee of the International Organization for Standardization (ISO), and the International Electrotechnical Commission (IEC). In order to conform to this standard, a certification scheme must demonstrate that it incorporates the certification processes described in a previous international standard ISO/IEC 17024 for certification processes, and that the body of knowledge used by the scheme can be mapped to SWEBOK.

73.6 Conclusions

It has been more than 40 years since the phrase "software engineering" was first used to describe the processes associated with developing and maintaining large-scale software systems. During this period, software engineering has acquired some of the characteristics of a profession. It has a reasonably well-accepted body of knowledge and code of ethics. The pace of change in the computing industry (and in software engineering research) entails the need for ongoing continuing education. However, entrance control is somewhat variable; while there are many university courses of study in software engineering whose quality is governed by governmental or private accreditation agencies, employers far too frequently hire and designate employees as software engineers who lack any formal professional credentials. Furthermore, the terms "software engineer" and "software developer" tend to be conflated. The often careless usage of "software engineer" to designate a software developer tends to weaken the emerging professionalism of the discipline. This situation is exacerbated by the fact that major professional societies (ACM and IEEE-CS) have come to different conclusions as to the pace at which the profession of software engineering is emerging. At the same time, there is increasing pressure to bring software engineering into the professional-credentials sector occupied by the broader community of engineers. As we have seen, this is quite variable by country; the variability has more to do with professional society politics than with substance. In some countries (the United Kingdom and Australia), it has been relatively easy to do so. In others, (Canada and the United States), the pace has been slower. In all cases, though, the direction of movement has been toward government-sanctioned credentials (licensing or chartered status). The recent introduction of broad-based society-sponsored professional certification examinations has opened another way for software engineers to obtain professional credentials. However, the uptake of these credentials has so far been slow driven almost exclusively by practitioner interest; there have been few if any signs of employer interest or involvement. It seems clear that the future of software engineering professionalism will be driven by industry and government. Software quality increasingly impacts public health and safety as well as corporate profits. If either of these drivers leads to an increased emphasis on the competence and professionalism of the software engineers developing the products of the future, we can expect to see a similarly increased emphasis on the professional credentials held by these software engineers.

Key Terms

Accreditation: A defined process to determine whether a university academic program meets specified quality standards.

Body of knowledge: A description of the definition and concepts underlying a discipline. It is usually accompanied by an annotated bibliography.

Certification: A process to determine whether a practitioner has mastered a given body of knowledge.

Chartered engineer: An individual recognized as a professional engineer in certain countries, including the United Kingdom, Australia, and New Zealand.

Code of ethics: A list of ethical expectations for professionals in a specific profession.

CSDA: An entry-level certification examination for software engineering professionals developed by the IEEE Computer Society.

CSDP: A certification examination for experienced software engineering professionals developed by the IEEE Computer Society.

Entrance control: A means of restricting access to a profession, most usually by requiring graduation from an accredited university program.

Licensing: A means of restricting the practice of a profession to those who meet specific requirements, including education, experience, and possibly an examination.

Profession: An occupation characterized by an explicit body of knowledge, entrance control, code of ethics, and a recognized need for continuing education.

Professional: An individual who has received government (and often professional-society) approval to practice a specific profession.

Further Information

Information about professional status for engineers is country-specific. For the United States, the best source is the licensure page of the National Society for Professional Engineers (http://www.nspe.org/Licensure/index.html). The appropriate source for Canada is the Engineers Canada licensure page (http://www.peng.ca). For the United Kingdom, the best source is the Engineering Council's Chartered

Engineer page (http://www.engc.org.uk/professional-qualifications/chartered-engineer/about-chartered-engineer). A relatively recent working paper (Adams 2004) contrasts the development of software engineering as a profession in the United States, the United Kingdom, and Canada.

Curriculum recommendations for computing programs, including software engineering, can be found at http://www.acm.org/education/curricula-recommendations. Additional information can be obtained from the ACM and IEEE-CS computing education pages; see http://www.acm.org/education and http://www.computer.org/portal/web/education.

The ACM/IEEE-CS software engineering code of ethics can be found at http://www.acm.org/about/se-code. A recent monograph (Davis 2009) presents an in-depth discussion of the development of this code.

Accreditation of software engineering programs is also country-specific. ABET accreditation of engineering programs, including software engineering, is described at http://www.abet.org/engineering-criteria-2012-2013. The British Computer Society's accreditation program is described at http://www.bcs.org/category/5844. A good introduction to the German approach to accreditation of engineering and computing programs can be found at http://www.asiin-ev.de/pages/en/asiin-e.-v.php

The best source for the IEEE Computer Society's certification effort is the IEEE-CS certification page: http://www.computer.org/portal/web/getcertified

References

Adams, T., 2004, *Software Engineering in Canada, the U.S., and the U.K.: Inter-Professional Relations and the Emergence of a New Profession*, Working Paper #9, Workforce Aging in the New Economy, University of Western Ontario, London, Ontario, Canada.

Allen, F., Hawthorn, P., and Simons, B., 2000, Not like now, not like this: Why the ACM Council does not support the licensing of software engineers at this time, *Commun. ACM*, 43:29–30.

Armytage, W., 1961, *A Social History of Engineering*, MIT Press, Cambridge, MA.

Association for Computing Machinery, 1999, Software engineering code of ethics and professional practice. http://www.acm.org/about/se-code (retrieved November 15, 2012).

Australian Computer Society, 2012, ACS code of ethics. http://www.acs.org.au/__data/assets/pdf_file/0005/7835/Code-of-Ethics_Final_12.6.12.pdf (retrieved November 15, 2012).

Bagert, D., 2004, Licensing and certification of software professionals, *Ann. Comput.*, 60:1–34.

British Computer Society, 2011, Code of conduct for BCS members. http://www.bcs.org/upload/pdf/conduct.pdf (retrieved November 15, 2012).

Canadian Information Processing Society, 2005, Code of ethics and professional conduct. http://www.cips.ca/?q=system/files/CIPS_COE_final_2007.pdf (retrieved November 15, 2012).

Canadian Information Processing Society, undated, Software engineering. http://www.cips.ca/softeng (retrieved November 15, 2012).

Davis, M., 2009, *Code Making: How Software Engineering Became a Profession*, Center for the Study of Ethics in the Professions, Illinois Institute of Technology, Chicago, IL.

De Marco, T., 1999, It ain't broke, so don't fix it, *IEEE Softw.*, 16:67 ff.

Dzerzhinskiy, F., 2012, How many software engineering professionals hold this certificate. http://arxiv.org/abs/1211 (retrieved November 26, 2012).

El-Kadi, A., 1999, Stop that divorce, *Commun. ACM* 42:27–28.

Frailey, D., 1999a, Software engineering grows up, *IEEE Softw.*, 16:66 ff.

Frailey, D., 1999b, Licensing software engineers, *Commun. ACM*, 42:29–30.

IEEE Computer Society, 2004a, Curriculum guidelines for undergraduate degree programs in software engineering. http://sites.computer.org/ccse (accessed November 15, 2012).

IEEE Computer Society, 2004b, *Guide to the Software Engineering Body of Knowledge*. http://www.computer.org/portal/web/swebok/htmlformat (retrieved November 15, 2012).

Integrated Software and Systems Engineering Curriculum Project, 2009, Curriculum guidelines for graduate degree programs in software engineering. http://www.gswe2009.org/fileadmin/files/GSwE2009_Curriculum_Docs/GSwE2009_version_1.0.pdf (accessed November 15, 2012).

Microsoft, 2012, Microsoft learning: Certifications by name. http://www.microsoft.com/learning/en/us/certification/view-by-name.aspx (retrieved November 15, 2012).

National Society of Professional Engineers, 2007, National Society of Professional Engineers Code of Ethics. http://www.nspe.org/Ethics/CodeofEthics/index.html (retrieved November 15, 2012).

Naur, P. and Randell, B., eds., 1969, *Software engineering techniques: Report on a conference sponsored by the NATO Science Committee*, Garmisch, Germany, October 7–11, 1968, Scientific Affairs Division, NATO, Brussels, Belgium.

Randell, B. and Buxton, J., eds., 1970, Software engineering techniques: Report on a conference sponsored by the NATO Science Committee, Rome, Italy, October 27–31, 1969, Scientific Affairs Division, NATO, Brussels, Belgium.

Seidman, S., 2009, An international perspective on professional software engineering credentials, in: *Software Engineering: Effective Teaching and Learning Approaches and Practices*, H. Ellis, S. Demurjian, and J. F. Naveda, eds., IGI Global, Hershey, PA, pp. 351–361.

74

Software Engineering Code of Ethics and Professional Practice

Don Gotterbarn
East Tennessee State University

74.1 Introduction

In late 1999, the Association for Computing Machinery and the IEEE-Computer Society approved the Software Engineering Code of Ethics and Professional Practice 5.2 (hereafter, the Code). The Code was later translated into Arabic, Croatian, French, German, Hebrew, Italian, Japanese, Mandarin, and Spanish by other computing societies and individuals for their use. The Code encourages software engineers to undertake positive actions and to resist pressures to act unethically. Several major companies have adopted the Code, and its principles have been incorporated into their standards of practice. The Code has also been included as part of employment contracts, which are signed at the time of employment.

The Code describes the ethical and professional obligations against which peers, the public, and legal bodies can measure a software developer's behavior. The development of large complex and safety critical systems is a difficult process. Software development and software developers have not been acclaimed for their successes or professional reputation. There is a long history of highly publicized software failures accompanied by frequent efforts to sidestep the responsibility for those failures by blaming them on "bugs," which somehow creep into the software.

The United States Navy has developed "smart ships" where shipboard applications are being run by software so that fewer sailors are needed to control key ship functions. The smart ships computer system reduced the crew size by 10% and saved more than 2.8 million a year. However, in September of 1997, the software suffered a systems failure, and the propulsion system failed leaving the ship dead in the water. It had to be towed into port (DiGiorgio, 1998).

The failure on the USS Yorktown was described as "a software glitch" implying that the problem was a minor error of little significance. A military ship used to protect populations, a ship, which is vulnerable to attack by opposing forces, was left dead in the water.

A closer look at this "software glitch" illustrates the problem of minimizing responsibility for the final quality of the software. It is common knowledge among computer professionals that computers cannot divide by zero. Most computer systems, including calculators in cell phones, have a basic error check routine, which protects against the results of attempting to divide by zero. The software on the Aegis missile cruiser USS Yorktown was not designed to tolerate such a simple failure. The absence of computer code to protect against a zero divide was considered "a software glitch." This was a failure to include a basic principle of software design. In other professions, similar small professional failures—a physician leaving a small instrument inside a patient after open heart surgery or an engineer "slightly" underestimating the amount of concrete necessary for a dam to hold back a wall of water—would result in legal charges of malpractice or negligence.

This tendency to minimize and shift the responsibility for these events are not acceptable to the practicing professional. It misleads the public about the professional commitment of software engineers and misleads those seeking to enter the profession about their broad responsibilities.

Besides affecting "trust," this tendency to minimize descriptions of software errors is very dangerous because of the impact of computers. This minimizing of "bugs" and "glitches" lends a degree of "normalcy" to software failures. This sense of "normal errors" only re-enforces the lack of ethical sensitivity and professionalism among software developers. Software engineer's complex development issues of coordinating human requirements with computing technical issues are made more difficult by developer's missing the professional and ethical responsibilities involved of their work.

As the impact of the behavior of a group of specialists assumes wider significance, two pressures develop, one from society at large interested in assuring that the behavior of the specialists will not cause any (more) harm to society. This societal concern is manifested in attempts to regulate the specialists by legislation. The other pressure that develops comes from within the ranks of the specialists, in this case software engineers, to organize as a group with a well-defined discipline including a commitment to high-quality practice concerned with the well-being of society at large.

The common elements in all of these efforts are the definition of a domain of practice, which is accomplished in part by circumscribing a body of knowledge and supporting curricula, mandating a level of practical experience (sometimes working with an accepted member of the profession), and a commitment to a level of quality practice generally defined in a code of ethics and professional practice. Many codes of ethics try to educate and inspire the members of the professional group that adopts the code. Codes also inform the public about the responsibilities that are important to a profession.

The Software Engineering Code of Ethics and Professional Conduct 5.2 is the result of one such effort to professionalize software engineering. The Code "instructs practitioners about the standards that society expects them to meet, and what their peers strive for and expect of each other" (Code5.2, 1999). Software engineers need to anticipate unintended consequences, including negative impacts on society, individuals, and the environment. Even developers with the best of intentions can walk into ethical traps. The Code is designed to help software engineers reduce unintended negative professional and ethical consequences of their work.

Additionally, codes are not meant to encourage litigation, and they are not legislation but they do offer practical advice about issues that matter to professionals and their clients, and they do inform policy makers. These concepts have been adopted in the development of this Code.

74.1.1 Brief History and Purpose

The Board of Governors of the IEEE Computer Society established a steering committee in May 1993 for evaluating, planning, and coordinating actions related to establishing software engineering as a profession. In that same year, the ACM Council endorsed the establishment of a Commission on

Software Engineering. By January 1994, both societies formed a joint steering committee "to establish the appropriate set(s) of standards for professional practice of software engineering upon which industrial decisions, professional certification, and educational curricula can be based." To accomplish these tasks, they recommended adopting standard definitions; defining a required body of knowledge and recommended practices; defining ethical standards; and defining educational curricula for undergraduate, graduate, and continuing education. The joint steering committee established a series of task forces to achieve these goals. The initial task forces included: software engineering body of knowledge and recommended practices, software engineering ethics and professional practices (SEEPP), and software engineering curriculum.

Most computing societies already have their own Codes of Ethics; why was there a need for yet another code? Codes of ethics state high-level principles, which cut across all professions; but individual professions need to address those elements, which are unique to a specific profession; thus, a software engineering Code needs to address requirements elicitation and software maintenance—subjects that are not addressed in a chemical engineers code of ethics. Software engineering professionals understand the significance of their work and their ethical obligations to their products' stakeholders. As the Preamble to the full version of the Code states, "Because of their roles in developing software systems, software engineers have significant opportunities to do good or cause harm, to enable others to do good or cause harm, or to influence others to do good or cause harm" (Code5.2, 1999). There was near unanimity that software engineers must behave proactively when they are aware of potential difficulties in a system. Several clauses in the Code require preemptive reporting of potentially dangerous situations and even outline a procedure for whistle blowing.

Software engineers generally agree about their obligations. In reviews, the Code received a consistently high level of agreement about the behavior expected of a professional software engineer. However, various forces pressure software engineers to not always fulfill these obligations and there is something needed to address these pressures.

The development of this Code was unique in that it was designed as a code for a profession with global impact. Ideally "the code instructs practitioners about the standards that society expects them to meet, and what their peers strive for and expect of each other" (Code5.2, 1999). The purpose of the SEEPP task force is to document the ethical and professional responsibilities and obligations of software engineers (Code3.0, 1997).

74.1.1.1 Role of Professional Organizations

The Code addresses both the responsibilities of the practicing professional and of the profession.

Unlike other codes, this Code is not designed to be self-serving to the profession. It requires software engineering professionals to be ethically responsible to all of those who are affected by their products. Several authors (Bayles, 1981; Johnson, 2001) have specified hallmarks of a professional; generally they include: having a code of ethics expressing ideals of service, licensing procedures, and a representative professional association. A profession also requires a monopolistic knowledge that circumscribes a clearly defined territory. The traditional model of a profession is based on the interaction between an individual professional and a recipient of the benefit of the professional's skill. The recipient may be an individual, organization, or the public at large.

The professionalization of any discipline involves the realization that being a professional involves more than the rigid application of formal principles to the artifacts of that discipline (Davis, 1987). The practice of medicine is more than the application of drugs to the human body. Physicians are concerned with the well-being of their patients. In the professionalization of a discipline, there is a realization of the impact on society of the application of the special skills and knowledge of the practitioner of that profession. There is also an acceptance of the responsibility that comes with the privilege of being allowed to apply those principles. The professional's unique understanding of professional situations and how to deal with them places an extra set of ethical obligations on the professional. These obligations are sometimes referred to as professional ethics.

A code of ethics is like a Swiss Army knife, serving many important and useful functions. It is a statement to members about the ethical stand of an organization and profession, a conscience of the profession, an announcement to non-members what the profession stands for (although most often stated in terms of the actions of individuals), and it imposes responsibilities on the professional organizations.

It is not so much that the Code lays down strict rules for the profession to follow, rather that they reflect the maturity of the social contract into which the profession has developed where justice and rights become accepted into the profession. One element that separates professionals (in the philosophical sense) is that they pledge themselves to certain moral responsibilities and a higher order of care. The code leads to that kind of pledge. It is a necessary step to establish a profession.

Professional computing organizations such as the IEEE have been involved in the development standards. Professional organizations have codes of ethics and codes of practice, which require the computing professional to consider a broader range of stakeholders, to consider a range of stakeholders, which includes all those whose lives are affected by the software project and the way it was implemented.

74.2 Code

74.2.1 Purpose and Principles

As Software Engineering developed as a discipline it had considered its primary goal as the "competent creation" of software artifacts, but it has gone beyond that minimalist position to move toward professionalism. The Code goes beyond mere competent creation. Although creation is an important part of software engineering, it is not the primary goal. As an engineering profession, the public good is the primary goal. Although the Code refers to the importance of competence, the Code advocates the primacy of responsibility for public well-being more than others. It reflects the importance of responsibilities to the public that go far beyond competent creation.

The goal of software engineering, as described in the preamble to the Code, is much broader.

> Computers now have a central and growing role in commerce, industry, government, medicine, education, entertainment, social affairs, and ordinary life. Those who contribute, by direct participation or by teaching, to the design and development of software systems have significant opportunities both to do good and to cause harm and to influence and enable others to do good or cause harm. To ensure, as much as possible, that this power will be used for good, software engineers must commit themselves to making the design and development of software a beneficial and respected profession. In accordance with that commitment, software engineers shall adhere to the following code of ethics. (Code5.2, 1999)

As the Code developed from version 3.0 to version 5.2, a shortened version of the Code, including only a brief preamble and the eight principles, was appended to the longer version of the Code containing clauses describing how the principle applies to software engineering, under each principle (Gotterbarn et al. 1997).

The Code contains eight keyword principles related to the behavior of and decisions made by professional software engineers, be they practitioners, educators, managers and supervisors, or policy makers, as well as trainees and students of the profession. The principles identify the various relationships in which individuals, groups, and organizations participate and the primary obligations within these relationships (Preamble, The Code3.0, 1997).

74.2.1.1 Software Engineering Code of Ethics and Professional Practice (Short Version)

The short version of the code summarizes aspirations at a high level of the abstraction; the clauses that are included in the full version give examples and details of how these aspirations change the way we act as software engineering professionals. Without the aspirations, the details can become legalistic and tedious; without the details, the aspirations can become high sounding but empty; together, the aspirations and the details form a cohesive code.

Software engineers shall commit themselves to making the analysis, specification, design, development, testing, and maintenance of software a beneficial and respected profession. In accordance with their commitment to the health, safety, and welfare of the public, software engineers shall adhere to the following eight principles:

1. PUBLIC—Software engineers shall act consistently with the public interest.
2. CLIENT AND EMPLOYER—Software engineers shall act in a manner that is in the best interests of their client and employer consistent with the public interest.
3. PRODUCT—Software engineers shall ensure that their products and related modifications meet the highest professional standards possible.
4. JUDGMENT—Software engineers shall maintain integrity and independence in their professional judgment.
5. MANAGEMENT—Software engineering managers and leaders shall subscribe to and promote an ethical approach to the management of software development and maintenance.
6. PROFESSION—Software engineers shall advance the integrity and reputation of the profession consistent with the public interest.
7. COLLEAGUES—Software engineers shall be fair to and supportive of their colleagues.
8. SELF—Software engineers shall participate in lifelong learning regarding the practice of their profession and shall promote an ethical approach to the practice of the profession.

74.2.2 Full Version: Principles and Illustrative Clauses

The eight principles reflect the order in which software professionals should consider their ethical obligations: The first principle is concerned with the public good. The primacy of well-being and quality of life of the public in all decisions related to software engineering is emphasized throughout the code. This obligation to the public is the final arbiter in all decisions:

> In all these judgments concern for the health, safety and welfare of the public is primary; that is, the 'Public Interest' is central to this Code. (Code5.2, 1999)

These eight general principles have specific clauses describing the goals of good software engineering practice. There has been a transition away from regulatory codes designed to penalize divergent behavior and internal dissent, toward codes which are more normative, giving general guidance (Gotterbarn, 1999a). The intent of the Code was not to impose sanctions on those who violate it but to act as a guide in the relationship between software engineering practice and ethics. It is a declaration of the professional responsibility of those practitioners—licensed or not—who perform the activity, manage, or teach software development. Its statements can be used to help professionals carefully examine alternative actions when they recognize ethically charged situations. Consider the following situations in which software engineers may find themselves (Gotterbarn and Miller, 2004).

74.2.2.1 Good Practice

Long before system testing is complete, scheduled time and money for such testing has run out. The engineer responsible for testing is advised that the company will "let the users do the rest of the testing."

Principle 3 speaks against such an approach and clause 3.10. "Ensure adequate testing, debugging, and review of software and related documents on which they work." asserts the importance of system testing before release.

74.2.2.2 Maintenance

Version 4 of the code added specific language about the importance of ethical behavior during the maintenance phase of software development. The text reflects the amount of time a computer professional spends modifying and improving existing software and also makes clear that we need to treat maintenance with the same professionalism as new development. The quality of maintenance depends upon the professionalism of the software engineer, because maintenance is more likely to be scrutinized only locally, whereas new development is generally reviewed at a broader corporate level.

Knowing that the firm developing a large piece of software for a government agency is unlikely to be awarded the contract to maintain the system after delivery, the company decides to spend less effort on both internal and external documentation. The money saved by this decision is used to pay for additional coders and testers, so that the system can be delivered on time.

This approach violates obligations to the Public in Principle 1 Clause 1.03. "Approve software only if they have a well-founded belief that it is safe, meets specifications, passes appropriate tests, and does not diminish quality of life, diminish privacy or harm the environment. The ultimate effect of the work should be to the public good." and Principle 3.01. "Strive for high quality, acceptable cost, and a reasonable schedule, ensuring significant tradeoffs are clear to and accepted by the employer and the client, and are available for consideration by the user and the public" (Code5.2, 1999).

74.2.2.3 Public Is Paramount

A computer company is working on an experimental fighter. A quality control software engineer suspects that the flight control software is not sufficiently tested, although it has (finally) passed all its contracted test suites. She is being pressured by her employers to sign off on the software. Her employers say they will go out of business if they do not deliver the software on time. She signs off (adapted from McFarland, 1990).

This case is in violation of the product clause "1.03 Approve software only if they have a well-founded belief that it is safe, meets specifications, passes appropriate tests, and does not diminish quality of life, diminish privacy or harm the environment. The ultimate effect of the work should be to the public good." Although Principle 2 seems to say the interests of the employer are relevant, that interest is mitigated by the interest of the public, in this case the pilot and those in her flight path. Principle 2 CLIENT AND EMPLOYER—Software engineers shall act in a manner that is in the best interests of their client and employer, consistent with the public interest.

Most situations faced by software engineers are more complex than these simple examples. This shortened version of the Code is not intended to be a stand-alone abbreviated code. The details of the full version are necessary to provide clear guidance for the practical application of these ethical principles. The software engineering code of ethics and professional practice is significant in that it constitutes the software developer's response to the previously tolerated "misadventures" of software developers as illustrated in the Yorkton example in the preceding text.

74.2.3 Support for Version 5.2

The drafts of the Code, written by volunteers from places like Australia, Canada, Egypt, India, Ireland, the Philippines, the United Kingdom, and the United States, went through several revisions drafts subject to international review by professional societies, large multinational corporations, and small development firms, including the IEEE formal technical standard review process. Because of the way the Code was developed, it is not unreasonable to say that this Code represents a movement toward an international consensus of what software engineers believe to be their professional ethical obligations.

The weight of all the software engineers who reviewed the Code and the support of the professional societies that approved it make it easier to defend whistle blowing, refusal to certify inferior or unsafe software, and support other acts consistent with the moral and professional standards of software engineers. A Code that has been reviewed, voted on, and adopted by professional societies provides ethical direction for new software engineers and gives support to the ethical decisions of other software engineers.

74.2.4 Functions of the Code

Codes can serve many different functions. The choice of function determines the content and style of a code of ethics. It might be designed to be *inspirational* either for positive stimulus for ethical conduct by the practitioner (Martin et al. 1989) or to inspire confidence of the customer or user in the computing artifact and confidence in its creator (Anderson, 1995). Unfortunately, inspirational language tends to be vague, limiting the code's ability to help guide professional behavior.

Historically, there has been a transition away from regulatory codes, designed to penalize divergent behavior and internal dissent, toward more normative codes, which give general *guidance*. Although a professional can use a normative Code to examine alternative actions, such codes are only a partial representation of a profession's ethical standards (Gotterbarn, 1999b). Because the use of normative codes requires moral judgment on the part of the professional, they should not be considered a complete procedure for deciding what is right or wrong (see Section 74.3.3).

- Codes also serve to *educate* both prospective and existing software engineers about their shared commitment to undertake a certain level of quality in their work and their responsibility for the well-being of the customer and user of the developed product.
- Codes also serve to educate managers (see Principle 5 Code5.2, 1999) of software engineers, and to educate those who make rules and laws related to the profession, about expected behavior. Managers' and legislators' expectations will affect what is asked of software engineers and what laws are passed relating to software engineering, respectively. Directly and indirectly, codes also educate management about their responsibility for the effects and impacts of the products developed.
- Codes also indirectly educate the public at large about what professionals consider to be a minimally acceptable ethical practice in that field, even as practiced by nonprofessionals.
- Codes provide a level of *support for the professional* who decides to take positive action. An appeal to the imperatives of a code can be used as counter pressure against others' urging to act in ways inconsistent with the Code.
- Codes can be a means of *deterrence and discipline*. They can serve as a formal basis for action against a professional; for example, some organizations use codes to revoke membership or suspend licenses to practice. Because codes usually define in detail the minimal behavior for all practitioners, the failure to meet this expectation can be used as a reasonable foundation for litigation.
- Codes have been used to *enhance a profession's public image*. They prohibit public criticism of fellow professionals, even if they violate some ethical standard.

The specific functions selected for a code affect its potential impact. This Code, designed for software engineering, emphasizes education, guidance, support, and inspiration. The Code does not specifically address deterrence and discipline or include standards for prohibiting someone from practicing software engineering.

74.2.5 Response to the Code after Approval

The adoption by several computing societies, its translation into multiple languages, and its use by industry as a standard of practice indicate that the international committee that developed the Code captured the ethical ideas of software engineering.

The Code is a hybrid in several ways. Some confusion about codes of ethics arises from a failure to distinguish between closely related concepts about codes, which direct the behavior of practicing professions. Moving from less restrictive to more restrictive codes, there are codes of ethics that are primarily aspirational giving a mission statement for the profession. There are also codes of conduct that describe professional attitudes and some professional behavior. Codes of practice are very specific and tied closely to the practice of the profession. They are the easiest to use as a basis for legal action.

Because practicing professionals deal with human affairs, the underlying ethical principles are the same across professions. Studies have shown that most codes are a hybrid of these three types of code (Berleur and d'Udekem-Gevers, 1994).

The Code is a hybrid focused on software engineering. The Code shows the relation between high level aspirations and the particular practices of software engineering. The international committee made sure the Code focused on what was common to the profession. Even though cultures may differ on the ethics of having 1 wife or having 12 wives, the practice of software engineering is globally consistent in the view that safety critical systems should be tested before they are deployed.

74.3 Foundations and Functions of the Code

The Code is unique in several significant ways. It is the Code for a profession; it is a hybrid of a code of ethics, a code of conduct and a code of practice; its imperatives cross all three levels of ethical obligation—obligations to humanity in general, obligations to professionalism, and obligations to a specific discipline. It is designed to mitigate some philosophical objections to codes of ethics and includes suggestions on how to use the Code in decision making and resolve situations where principles of the Code seem to conflict (see Section 74.2.2).

The Code is organized into a short and a long version; each with its own preamble. The preamble of the short version summarizes aspirations at a high level of abstraction while the specific clause under each principle in the long version ground these aspirations in specific software engineering practices. The Code asserts that "Without the aspirations, the details can become legalistic and tedious; without the details, the aspirations can become high sounding but empty; together, the aspirations and the details form a cohesive code." The clauses under the eight core principles that are included in the full version give examples and details of how these aspirations change the way we act as software engineering professionals.

74.3.1 Levels of Professional Obligation

Many codes are written at a high level of abstraction and only talk about some easily agreed upon human values like "honesty" and "competence" or they are detailed lists of discipline specific practices. Codes that are only out a high level of abstract do not help the individual professional make decisions and codes consisting of detailed practices are not flexible enough to address the rapidly changing computing environment. The detailed list will also be incomplete.

Each principle of this code addresses three levels of ethical obligations owed by professional software engineers. The first level identified is a set of ethical values, which professional software engineers share with all other human beings by virtue of their humanity. The second level obliges software engineering professionals to more challenging obligations than those required at Level one. Level two obligations are required because professionals owe special care to people who may be affected by their work. The third and deeper level comprises several obligations that derive directly from elements unique to the professional practice of software engineering. The clauses of each principle are illustrations of the various levels of obligation included in that principle.

Level one: Aspire (to be human). Statements of aspiration provide vision and objectives and are intended to direct professional behavior. These directives require significant ethical judgment.

Level two: Expect (to be professional). Statements of expectation express the obligations of all professionals and professional attitudes. Again, they do not describe the specific behavior details, but they clearly indicate professional responsibilities in computing.

Level three: Demand (to use good practices). Statements of demand assert more specific behavioral responsibilities within software engineering, which are more closely related to the current state of the art. The range of statements is from the more general aspirational statement to specific measurable requirements.

Although the description of these levels was removed from the Preamble after version 3 of the Code, it is still a philosophical foundation of the Code and underlies the content of the code. The idea is that a complete Code needs to address all three levels. In general, the Code address level one statements in the preambles (see Section 74.2.2), level two issues of professionalism in the eight principles, and software engineering discipline specific issues in the clauses under each principle.

74.3.2 Decision Making

There are several positive elements to this Code, which distinguish it from many other codes of professional ethics. One of the major functions of the Code is to help the software engineer make critical decisions and alert them to potential problems with their work.

Most codes of ethics have some problems. First, they only provide a finite list of principles that are often presented as a complete list; readers might presume that only the things listed should be of ethical concern to the professional. These principles are treated as either rules, which must be followed or as normative guidelines.

Second, many codes provide little (if any) guidance for situations where rules having equal priority appear to conflict. There is not guidance as to which principles should have higher priority. This ambiguity of priority leaves the ethical decision maker confused. The Software Engineering Code addresses both of these limitations.

The Code addresses the first problem in the preceding text. It explicitly rejects the concept of completeness. The problem with an incomplete Code is that it can leave a practitioner without guidance in new situations. To meet this potential difficulty, the software engineering code provides general guidance for ethical decision making, especially in those areas not explicitly mentioned in the Code.

> It is not intended that the individual parts of the Code be used in isolation to justify errors of omission or commission. The list of Principles and Clauses is not exhaustive. The Clauses should not be read as separating the acceptable from the unacceptable in professional conduct in all practical situations. (Code5.2, 1999)

The question of the equal priority of all of the principles and clauses is addressed by stating that in all decisions, the public interest is the primary concern.

Unlike most engineering codes, the Code addresses the second problem about how to make decisions. The Preamble of the Code suggests ways to make decisions when two of its imperatives such as "Honor Contracts" and "Work for the Enhancement of Human Welfare" are in conflict. The Code addresses the problem of choosing between conflicting principles by having a hierarchy of principles, with the health, safety, and welfare of the public primary in all ethical judgments. The Code requires whistle blowing and 6.12, 6.13 suggest a procedure to follow in whistle blowing. This Code advocates a stakeholder analysis that goes beyond the limited concerns of safety, schedule, and budget. The National Society of Professional Engineers' Code requires that an engineer report to a client, " if in the engineer's judgment, a project is likely to fail," whereas this Code requires that a software engineer report "…if in their judgment a project is likely to fail, to prove too expensive, to violate intellectual property law, or otherwise be problematic."

To facilitate decision making the eight principles were reordered to reflect the order in which software professionals should consider their ethical obligations: Version 3.0's first principle concerned the Product, while version 5.2 begins with the Public. The primacy of well-being and quality of life of the public in all decisions related to software engineering is emphasized throughout the code. This obligation is the final arbiter in all decisions: "In all these judgments concern for the health, safety, and welfare of the public is primary; that is, the 'Public Interest' is central to this Code." To reinforce the clear priority of public concern, in several places the code asserts the priority of concern for the public over loyalty to employer or profession. It is a professional obligation to take positive action to address violations of the Code. A sequence of reporting is included in the structure of the Code. Clause 6.12 says that you should first address talk to the person doing the violation. Clause 6.13 says that if following 6.12 does not resolve the matter, carry your concern to the appropriate authorities. Unlike other codes that mandate that one does not criticize fellow professionals, this Code requires that you do criticize those who are not following the Code.

74.3.3 Judgment

An ordered list of principles does not provide an in depth ethical analysis of a complex situation. "The Code is not a simple ethical algorithm that generates ethical decisions. In some situations standards may be in tension with each other or with standards from other sources. These situations require the software engineer to use ethical judgment to act in a manner which is most consistent with the spirit of the Code of Ethics and Professional Practice, given the circumstances" (Code5.2, 1999).

The Code does not leave the reader without support about how to make ethical decisions.

In most cases, using the Code to make decisions will require professional judgment to resolve the ethical tensions.

> "Ethical tensions can best be addressed by thoughtful consideration of fundamental principles, … These Principles should influence software engineers to consider broadly who is affected by their work; to examine if they and their colleagues are treating other human beings with due respect; to consider how the public, if reasonably well informed, would view their decisions; to analyze how the least empowered will be affected by their decisions; and to consider whether their acts would be judged worthy of the ideal professional working as a software engineer. In all these judgments concern for the health, safety and welfare of the public is primary; that is, the 'Public Interest' is central to this Code" (Code5.2, 1999 full preamble).

The first principle—consider who is affected—helps us refocus and the stakeholders beyond the developer and the customer. The second principle—due respect—requires a protection of human values. The third principle points to consideration of extended stakeholders in, those impacted by, our work. The fourth requires the best professional practice, working toward an ideal.

The Code provides specific details about software practitioners' obligations, and if they ignore those obligations, they are not acting in good faith as professionals.

When working on a project, the Code says we must do adequate testing but does not quantify how much testing is adequate. The software engineer should not be working on a system if they lack sufficient judgment to use the various ways in the Code to help make that judgment, including talking to other developers competent in that domain. Judgment is required to extend the principles in the Code to new domain.

74.4 Research

In the following text are a few topics for research related to the development and content of codes of ethics.

74.4.1 How to Handle Enforcement

Many people think that enforcement and deterrence mechanism are necessary elements of a Code of Ethics. Codes of Ethics are generally adopted by organizations and it is up to the organization to establish mechanism to encourage people to follow the adopted Code and deter them from breaking it. In some monopolistic professions, licensing serves both as a credentialing mechanism and a means of deterrence from breaking a code of ethics. In most places, access to working as a software engineer is not controlled by licensing. The global nature of the development and impact of software development means that there would be jurisdictional issues in trying to turn ethical imperatives into laws. Two common problems for codes in computing are that they need to be able to address a rapidly changing environment and there are difficulties in enforcing them. Turning a code into law makes it static and eliminates some of the other important functions of codes of ethics.

Within organizations, there is generally limited enforcement of a code of ethics. Codes get some teeth when they are used by a professional organization to make decisions. Codes of ethics do not define the procedures and sanctions for violating the Code. Codes are not self-referential. Organizations have bylaws, and the Code is a bylaw. The due process and sanctions for violating the code are defined outside the Code. Codes are the mind and conscience of a profession. The profession is what gives the rules nurture and enforcement.

74.4.2 Developing an Ethics Code for a Dynamic Profession

Including a detailed level of software engineering standards in a code would have been a mistake. Specific software engineering standards, which have been proposed by reviewers, such as requiring mutation testing for all mission critical software, are controversial and in flux. Any list of standards included in the Code would freeze the standard, and any list of standards would be incomplete.

The Code, as written does encourage us to exploit standards—but it should not name them. When the profession settles on standards, then we can incorporate them into an ethics code. Codes have different purposes, among them guidance, inspiration, and education. The Code can help lead us to serious discussions about the details of standards; the Code cannot prematurely set those standards.

74.4.3 Approaches to Ethics

Surprisingly, the major tension centered not around technical issues but rather on two distinct approaches to ethics: virtue ethics and rights/obligations ethics (Edgar, 1997). Virtue ethics holds the optimistic view that if people are simply pointed in the right direction, their moral character will guide them through ethical problems. Reviewers from this camp wanted a Code that was mostly inspirational, with minimum detail. They put a heavy emphasis on the autonomy of a professional's judgment. The other position—rights/obligations theory—consists of spelling out precisely one's rights and responsibilities. Believers in this theory wanted a very detailed Code. For example, one reviewer wanted the Code to include a standard of measurement for each imperative—to state exactly how many tests need to be done to ensure adequate testing. The rights/obligations folks used a legalistic model to evaluate each imperative. The problem was that any imperative acceptable to one group was not acceptable to the other. The Code addresses this significant tension in a number of ways.

To address the problem of insufficient guidance for decision making, the Code incorporates some directions for ethical decision making in the preamble and an acknowledgment that the Code's normative premises should not be read as complete descriptions or legalistic statements. It is still an open research question about testing the adequacy of this approach.

The Code addresses the rights—virtue tension in another way: through its structure. The Code is organized around eight broadly based themes. Under each theme or principle is a series of clauses giving examples of how that theme applies to software engineering practice. Thus, under "Principle 1, PUBLIC, Software engineers shall act consistently with the public interest" is the illustrative clause 1.04, "Disclose to appropriate persons or authorities any actual or potential danger to the user, the public, or the environment, that they reasonably believe to be associated with the software or related documents."

74.5 Summary

"The Software Engineering Code of Ethics and Professional Practice is a useful tool. It educates and inspires software engineers. The Code instructs practitioners about the standards that society expects them to meet and what their peers strive for and expect of each other. The Code offers practical advice about issues that matter to professionals and their clients, and it serves to inform policy makers about ethical constraints imposed on software engineers.

The Code encourages the professional to do positive actions. The Code also encourages the professional to resist pressures to act unethically. The professional can appeal to the imperatives of the Code to indicate ethically accepted practice.

The Code indirectly educates the public at large about the responsibilities that are important to and accepted by the profession—what software engineers consider to be minimally acceptable practice—even when a nonprofessional practices it. Thus the code can be a catalyst to simultaneously raising the internal expectations of a profession and the expectations of the society at large.

The Code is a dynamic document, a method for education, inspiration, and continued study and debate. It provokes serious discussion about the software engineering discipline, its responsibilities, and its future. The Code directs us to be a part of that future as we improve our profession and ourselves" (Gotterbarn and Miller, 2009).

References

Anderson, R. 1995. The ACM code of ethics: History, process, and implications, In: *Social Issues in Computing*, eds. C. Huff and T. Finholt, New York: McGraw-Hill, pp. 48–72.

Bayles, M. 1981. *Professional Ethics*. Belmont, CA: Wadsworth Inc.

Berleur, J. and d'Udekem-Gevers. 1994. Codes of ethics, or of conduct, within IFIP and in other computer societies. *13th World Computer Congress 1994*, Hamburg, Germany, pp. 340–348.

Code 5.2. 1999. Association for Computing Machinery, Inc. and the Institute for Electrical and Electronics Engineers, Inc. http://www.acm.org/about/se-code (accessed October 5, 2013).

Davis, M. 1987. The moral authority of a professional code. In: *Nomos XXIX: Authority Revisited*, eds. J.R. Pennock and J.W Chapman, pp. 302–333. New York: New York University Press.

DiGiorgio, A. 1998. Yorktown failure. *US Naval Institute Proceedings Magazine*, June 1998.

Edgar, S. l. 1997. *Morality and Machines: Perspectives in Computer Ethics*, Sudbury, MA; Jones and Bartlett Publication.

Gotterbarn, D., Miller, K., and Rogerson, S. 1997. Software Engineering Code of Ethics, Code 3.2. In *Communications of the ACM*, 40(11):110–118.

Gotterbarn, D. 1992. Software engineering ethics. In: *Encyclopedia of Software Engineering*, ed. J.J. Marciniak, pp. 1422–1428. New York: John Wiley & Sons, Inc.

Gotterbarn, D. 1999a. How the new software engineering code of ethics affects you. *IEEE Software* 16, 6:58–64.

Gotterbarn, D. 1999b. A positive step toward a profession: The software engineering code of ethics and professional practice. *ACM SIGSOFT Software Engineering Notes* 24, 1: 9–14.

Gotterbarn, D. and Miller, K. 2004. Computer ethics in the undergraduate curriculum: Case studies and the joint software engineer's code. *Journal of Computing Sciences in Colleges* 20, 2:156–167.

Gotterbarn, D. and Miller, K. 2009. The public is the priority: Making decisions using the software engineering code of ethics. *Computer*, 42, 6:66–73.

Gotterbarn, D., Miller, K., and Rogerson, S. 1997. Software engineering code of ethics 3.0. *Communications of the ACM* 40, 11:110–118 simultaneously published in *Computer* November 30:11.

Gotterbarn, D., Miller, K., and Rogerson, S. 1999a. Software engineering code of ethics is approved. *Communications of the ACM* 42, 10:102–107 simultaneously published in *Computer* October 1999.

Gotterbarn, D., Miller, K., and Rogerson, S. 1999b. Code reuse: A response. *ACM SIGSOFT Software Engineering Notes* 24, 3:4–6.

Johnson, D. 2001. *Computer Ethics*, 3rd edn. Columbus, OH: Prentice Hall.

Martin, M. et al. 1989. *Ethics in Engineering*, 2nd edn. New York: McGraw-Hill.

McFarland, M.C. 1990. Urgency of ethical standards intensifies in computer community. *IEEE Computer* 23:77–81.

Useful Web Pages

Articles on Code and Ethical Decision Making: http://www.computer.org/cms/Computer.org/Publications/code-of-ethics.pdf and http://dusk.geo.orst.edu/ethics/papers/Quinn_Chapter8.pdf

Code version 3.0: http://www.itk.ilstu.edu/faculty/bllim/itk178/Software%20Engineering%20Code%20of%20Ethics,%20Version%203_0.htm

Code version 5.2 ACM and IEEE-CS sites: http://www.acm.org/about/se-code and http://www.uces.csulb.edu/spin/media/pdf/ethics-ieee-cs.pdf

Davis, M. An electronic book on the development of the Code "Code Making: How Software Engineering Became a Profession" http://ethics.iit.edu/sea/sea.php/100

MacFarland article using the code in making a decision: http://www.scu.edu/ethics/practicing/focusareas/technology/occidental_engineering/occidental_engineering_derived_sources_of_ethical_wisdom.html

Translations of the Code: http://seeri.etsu.edu/Codes/default.shtm

Using the Code to evaluate cases: http://seeri.etsu.edu/Ethics/CodeCases.asp

Versions of the Code: http://library.iit.edu/csep/softwareengineeringarchive/codeversions.html

75

Software Business and Economics

Christof Ebert
Vector Consulting Services GmbH

75.1 Introduction

You are without doubt in the software business, no matter what business you seem to be in. IT and software make the world go round, at ever increasing speeds. Computer-based, software-driven systems are pervasive in today's society. From the software in avionics flight control to the ubiquitous computers in smart phones, automotive, and consumer electronics, software provides features and functions in daily use. Increasingly, the entire system functionality is implemented in software. Where we used to split hardware from software, we see flexible boundaries entirely driven by business cases to determine what is the best package at which level in which component, be it software or silicon. For example, in 1960, only 8% of the functionality of the F-4 fighter aircraft was implemented in software; by 2000, 80% of the F-22's features were provided by software. The same is true for consumer and communication systems. A TV set in the 1970s had no software, while today its competitive advantages are software-driven [1,2].

Software and IT move on a fast pace. When looking in the rear mirror, we see many companies and endeavors, which failed due to overemphasizing technology and not sufficiently implementing a sound business strategy [3]. Take, for example, Netscape. For many of us, it was the first experience of the Internet. In 1995, it had a market share of 80%—more than enough to stay in the pole position forever, as companies such as Google or Amazon show. But already in 1997 it slowed down, lost market share, and in 2003 went into bankruptcy. What went wrong? One of the managers put it in simple words: "We had no product management; it was just a collection of features." [4] More recently and in another domain, a previous Nokia senior manager claimed that the lack of product management and business insight is the primary reason for their loss of market shares in the past years [5]. On the other hand, we all know companies such as Apple and Microsoft and their excelling business management. So, what means software business understanding? We will in this article briefly introduce software business and economics and show success factors toward improving software business understanding. Examples from very different environments (big vs. small and planned vs. agile) will show how we have introduced and improved software business understanding, and what was achieved.

The software business has manifold challenges. They range from the creation process and its inherent risks to direct balance sheet impacts. Many studies reveal that we focus too much on technology and features in the software business and not enough on value and products. Only 52% of the originally allocated requirements appear in the final product release [1]. This is primarily the consequence of this process not having a clear corporate owner with assigned accountability for its success. In a similar study looking at new product development from a broader scope, Cooper et al. [6] found in 105 business units from various industries that the top 20% of enterprises deliver 79% of their new products in time, while the average delivers only 51% in time. The Standish Group's Chaos Report annually surveys information technology and software projects* [1]. They found that only 32% of the projects finished on time and within budget, a staggering 24% were cancelled before delivery, and of the remaining projects, which finished late or over budget, they only delivered a fraction of the planned functionality. Late introduction of a product to market loses market share; cancellation of a product before it ever reaches the market is an even greater loss. Not only is software increasing in size, complexity, and percentage of functionality, it is also increasing in contribution to balance sheet and profit and loss (P&L) statements. The challenge in many IT and software companies lies in better connecting sales and marketing with strategy and product development.

Sounds familiar? Products are pushed to the extreme to be ever more efficient and done at low cost, but when they hit the market, they would not sell as expected. Or customers demand many changes, thus reducing margins dramatically from initial targets. As engineers, we often tend to overhear the voice of the customer. Technology matters, the schedule needs to be kept, features are lined up like a washing list. Poor business management causes insufficient project planning, continuous changes in the requirements and project scope, configuration problems, and defects. The obvious—yet late—symptoms are more delays and overall customer dissatisfaction due to not keeping commitments or not getting the product they expect. Being late with a product in its market has immediate and tremendous business impacts [3,6,7]. In contract business, this often means penalties, and in practically all markets it reduces customer loyalty and the overall returns from sales.

Product development, such as for IT infrastructure, software components, or embedded electronics, traditionally concentrates on the project perspective looking to execute in budget and time. In software, IT, and systems engineering, project execution can be rather easily improved by means of capability maturity model integration (CMMI) and agile principles. Project execution can be improved by means of the project management body of knowledge. Today, there is a lot of exciting results from optimizing projects in terms of cost and cycle time [1,8–10].

Successful product development, though, is more than executing in time and budget. It means to deliver the right products at the right time for the right markets. Naturally the success of a product depends on many factors and stakeholders. However, we realize that it makes a big difference whether many cooks somewhat steer the soup or whether one person is empowered to lead all activities for the product from inception to market and evolution—and hold accountable for the results. This is the product manager.

In many companies and organizations and also in most literature software business, relationships to software development and engineering remain vague. We often see product definition, roadmapping, and marketing decoupled from the engineering project-related processes, which creates deficiencies and overheads, such as heavy changes in requirements and missed market opportunities. While the general principles of business management are known, not much specific guidance is available for software business understanding. With this handbook entry we will provide

* www.standishgroup.com

an introduction and tutorial on software business understanding and provide some concrete experiences, so you as the reader can directly benefit.

From our experiences with many clients in different industries, success comes from anticipating and meeting the relevant customer needs together with being in time and budget. Technical product development, such as for defense systems, automotive components, communication solutions, or IT infrastructure, traditionally focuses on the project perspective and operationally executing a set of given constraints within the triangle of content, budget, and time. Often too late it becomes clear that needs were different from what is built.

This chapter describes software business understanding. It addresses the major activities and provides examples. The target audience is both interested newcomers who wish to get insight into what software business understanding actually means and what techniques are used, and students and industry experts who want to get a fully up-to-date yet concise summary of software business techniques, experiences, and trends.

We will start with the necessary foundations looking into terminology and how the discipline of software business (subsequently abbreviated as SPM) breaks down into different activities. The next section will explain why SPM is necessary. We will then get practical and hands-on and will show how SPM is implemented in the product life cycle. Introducing SPM in an organization is a big challenge for many companies. So we will highlight our experiences from introducing SPM to companies around the world and also provide an overview of SPM-related competencies. Finally, we summarize the article and provide an outlook on trends that impact SPM as a discipline. We conclude with a summary of relevant Internet resources and also a brief overview on good introductory literature.

75.2 Foundations

Software business and economics are about making software-related decisions in a business context. It means aligning software technical decisions with the business goals of the organization. Decisions like "Should we use a specific component?" may look easy from a mere technical perspective but can have serious implications on the business viability of the project and the resulting product.

75.2.1 Finance

Finance is a branch of economics that deals with resource allocation and management, acquisition, and investment. The field of finance deals with the concepts of time, money, and risk and how they are interrelated. It also deals with how money is spent and budgeted. Corporate finance is the task of providing the funds for a company's activities. It generally involves balancing risk and profitability, while attempting to maximize an entity's wealth and the value of its stock. To do this, a company must

- Identify and implement relevant business objectives and constraints, such as organizational or individual goals, time horizon, risk mitigation, and tax considerations
- Identify and implement the appropriate business strategy, such as which portfolio and investment decisions to take, how to manage cash flow, and where to get the funding from
- Measure the financial performance, such as cash flow and return on investment (ROI), and take corrective actions in case of deviation from objectives

75.2.2 Accounting

Accounting is related to finance. It allows people whose money is being used to run the company to know the results of their investment: Did they get the profit they were expecting or not? The primary role of accounting is to measure the company's actual financial performance. Its purpose thus is to communicate financial information about a business entity to users such as shareholders and managers.

The communication is generally in the form of financial statements that show in money terms the economic resources under the control of management; the art lies in selecting the information that is relevant to the user and is reliable. Accounting systems are also a rich source of historical data for estimating.

75.2.3 Controlling

Controlling is part of finance and accounting. Controlling is the measurement and correction of performance in order to make sure that enterprise objectives and the plans devised to attain them are accomplished. Cost controlling is a specialized branch of controlling that is used to detect variances of the actual costs from the planned costs.

75.2.4 Cash Flow

To make a meaningful business decision about any specific proposal, that proposal will need to be evaluated from a business perspective. The concepts of cash flow instances and cash flow streams are used to describe the business perspective of a proposal. A *cash flow instance* is a specific amount of money flowing into or out of the organization at a specific time as a direct result of some proposal.

In a proposal to develop and launch product X, the payment for new development computers, if new hardware is needed, could be an example of an outgoing cash flow instance. Money would need to be spent to carry out that proposal. The sales income from product X in the 11th month after market launch could be an example of an incoming cash flow instance. Money would be coming in because of carrying out the proposal.

The term *cash flow stream* refers to the set of cash flow instances, over time, which would be caused by carrying out some given proposal. The cash flow stream is, in effect, the complete financial picture of that proposal. How much money goes out? When does it go out? How much money comes in? When does it come in? Simply, if the cash flow stream for Proposal A is more desirable than the cash flow stream for Proposal B then—all other things being equal—the organization would be better off carrying out Proposal A than Proposal B. Thus, the cash flow stream is an important input for investment decision-making.

A *cash flow diagram* is a picture of a cash flow stream. In the same sense that a picture is worth a thousand words, the cash flow diagram gives the reader a very quick overview of the financial picture of that subject. Figure 75.1 shows an example cash flow diagram for a proposal.

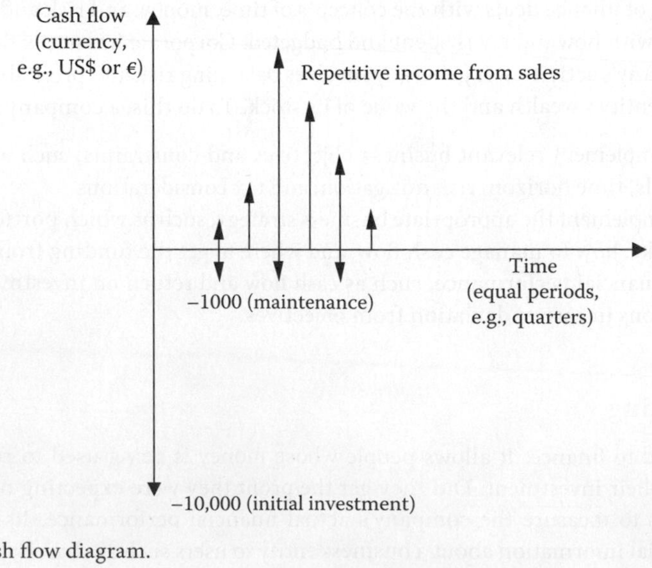

FIGURE 75.1 A cash flow diagram.

A cash flow diagram shows the cash flow stream in two dimensions, time runs from left to right and amounts of money run up and down. Each cash flow instance is drawn on the diagram at a left to right position relative to the timing of that cash flow after the start of the proposal. The horizontal axis is divided into units of time that represent either years, months, weeks, etc. as appropriate for the proposal being studied.

75.2.5 Decision-Making Process

If we assume that candidate solutions solve a given technical problem equally well, why should the business care which one is chosen? The answer is that there is usually a large difference in the costs and incomes from the different solutions. A commercial off the shelf compliant CORBA (object request broker architecture) product might cost a few thousand dollars, but the effort to develop a homegrown service that gave the same functionality could easily cost several hundred times. If the candidate solutions all adequately solve the problem from a technical perspective, then the one that maximizes the return on the organization's investment is the one that should be chosen. To do this, the technical person should follow a systematic process for making decisions. That systematic process is shown in Figure 75.2. It starts with a business challenge at hand and describes the steps to identify alternative solutions, evaluate these solutions, implement one selected solution, and monitor performance of this solution.

Figure 75.2 shows the process as mostly stepwise and serial. The real process is more fluid. Sometimes, the steps can be done in a different order and often several of the steps can be done in parallel. The important thing is to be sure that none of the steps are skipped or shortcut. It is also important to understand that this same process applies at all levels of decision: from a decision as big as should a software project be done at all down to a decision on an algorithm or data structure to use in a software module. The difference is how financially significant the decision is and, therefore, how much effort should be invested in making that decision. The project-level decision is financially significant and probably warrants a relatively high level of effort to make the decision. Selecting an algorithm is much less financially significant and warrants a much lower level of effort to make the decision, even though the same basic decision-making process is being used.

More often than not an organization could carry out more than one proposal if it wanted to. Furthermore, usually there are important relationships between proposals. Maybe Proposal Y can only be carried out if Proposal X is also carried out. Or, maybe Proposal P cannot be carried out if Proposal Q is carried out, nor could Q be carried out if P were. Choices are much easier to make when there are mutually exclusive paths, either A, B, or C, or whatever is chosen. In preparing decisions, it is recommended to turn any given set of proposals, along with their various interrelationships, into a set of mutually exclusive alternatives. The choice can then be made among these alternatives.

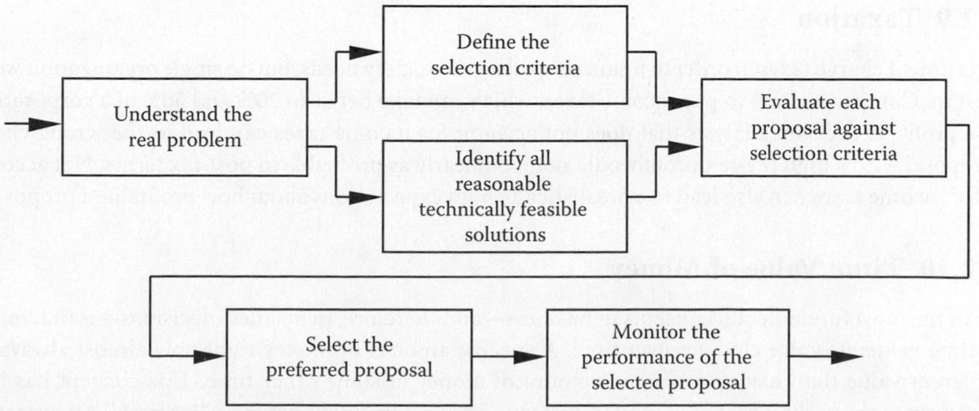

FIGURE 75.2 The basic business decision-making process.

75.2.6 Valuation

In an abstract sense, the decision-making process—be it a financial decision or not—is about maximizing value. The alternative that maximizes total value should always be chosen. A basis for comparison based on value is comparing two or more cash flows. It is a way of using equivalence to meaningfully compare proposals. Several bases of comparison are available, including

- Present worth
- Future worth
- Annual equivalent
- Internal rate of return
- (Discounted) Payback period

Due to the time value of money, two or more cash flows are equivalent only when they equal the same amount of money at a common point in time. Comparing cash flows only makes sense when they are expressed in the same time frame.

Note that value cannot always be expressed in terms of money. For example, whether an item is a brand name or not can significantly affect its perceived value. Relevant values that cannot be expressed in terms of money still need to be expressed in other terms, so that they can be made measurable.

75.2.7 Inflation

Inflation describes long-term trends in prices. Inflation means that the same things cost more than they did before. If the planning horizon of a business decision is longer than a few years, or if the inflation rate is over a couple of percent annually, it can cause noticeable changes in the value of a proposal.

75.2.8 Depreciation

Depreciation addresses how investments in capital assets are charged off against income over several years. Depreciation is an important part of after-tax cash flows, which is critical to accurately addressing income taxes. Software itself typically is not depreciated, but if you are working on proposals with a planning horizon longer than 1 year, the proposals involve capital assets (like buildings and equipment), and you need to accurately reflect the effects of income taxes in the decision analysis, then depreciation will be an important factor to include in the analysis. Another reason to understand depreciation is that your software project proposals will probably be compared against nonsoftware proposals so you should understand how the nonsoftware proposals are being evaluated.

75.2.9 Taxation

Government charge taxes in order to finance expenses the society needs, but no single organization would invest in. Companies have to pay income taxes, which can take between 20% and 50% of a corporation's gross profit. A decision analysis that does not account for income taxes can lead to the wrong choice. A proposal with a high pre-tax profit would not look nearly as profitable in post-tax terms. Not accounting for income taxes can also lead to unrealistically high expectations about how profitable a proposal is.

75.2.10 Time Value of Money

One of the most fundamental concepts in business—and therefore, in business decisions—is that money has time value; its value changes over time. A specific amount of money right now almost always has a different value than having the same amount of money at some other time. This concept has been around since the earliest recorded human history and is commonly known as "interest." Anyone making a business decision needs to understand interest and how it affects that decision.

75.2.11 Earned Value Management

Earned value management is a project management technique for measuring project progress based on created value. At a given moment, the results achieved to date in a project are compared with the projected budget and the planned schedule progress for that date. Progress relates already consumed resources and achieved results at a given point in time with the respective planned values for the same date. It helps to identify possible performance problems at an early stage.

75.2.12 Efficiency

Economic efficiency is the relationship between the result achieved (see Effectiveness) and the resources used to achieve this result. Efficiency means "doing things right." An efficient behavior, like an effective behavior, delivers results but keeps the necessary effort to a minimum.

75.2.13 Effectiveness

Effectiveness is about having impact. It is the relationship between achieved objectives to defined objectives. Effectiveness means "doing the right things." Effectiveness looks only if defined objectives are reached and not how they are reached.

75.2.14 Productivity

Productivity is the ratio of output over input from an economic perspective. Output is the value delivered. Input covers all resources (e.g., effort) spent to generate the output, the influence of environmental factors (e.g., complexity, quality, time, process capability, team distribution, interrupts, feature churn, tools, and language). Productivity combines efficiency and effectiveness from a value-oriented perspective: Productivity is about generating value with lowest resource consumption.

75.2.15 Life-Cycle Economics

75.2.15.1 Product

A product is an economic good (or output), which is created in a process that transforms product factors (or inputs) to an output. When sold, it is characterized by attributes which mean a value to its users. It is a deliverable that creates a value and an experience to its users. A product can be a combination of systems, solutions, materials, and services delivered internally (e.g., inhouse IT solution) or externally (e.g., SW application) as is or as a component for another product (e.g., intellectual property [IP] stack).

75.2.15.2 Project

A project is a temporary endeavor undertaken to create with people a unique product or service. In software engineering, we distinguish different project types (e.g., product development, IT infrastructure, outsourcing, software maintenance, service creation, and so on).

75.2.15.3 Program

A program is a set of related projects. Programs are often used to identify and manage different deliveries to a single customer or market over a time horizon of several years.

75.2.15.4 Portfolio

A portfolio is the sum of all assets and their relationship to the enterprise strategy and its market position. Portfolios are used to group and then manage simultaneously all assets within a business line or company. Looking to an entire portfolio makes sure that impacts of decisions are considered, such as resource allocation to a specific project—which means that the same resources are not available for other projects.

75.2.15.5 Product Life Cycle

The sum of all activities needed to define, develop, implement, build, operate, service, and phase out a product or solution and its related variants. It is subdivided into phases that are separated by dedicated milestones, called decision gates. With the focus on disciplined gate reviews, the product life-cycle (PLC) fosters risk management and providing auditable decision-making information (e.g., complying with product liability needs or Sarbanes-Oxley Act section 404).

75.2.15.6 Project Life Cycle

The set of sequential project phases determined by the control needs of the organizations involved in the project. Typically, the project life cycle can be broken down into at least four phases, namely, initiation, concept/planning, execution, and closure. The project life cycle and the product life cycle are interdependent, that is, a product life cycle can consist of several projects and a project can comprise several products.

75.2.15.7 Proposals

Making a business decision begins with the notion of a *proposal*. Proposals relate to implementing a business objective, either on project, or product, or portfolio level. A proposal is a single, separate option that is being considered, like carrying out a particular software development project or not. Another proposal could be to enhance an existing program and still another might be to redevelop that same software from scratch. Each proposal represents a unit of choice—either you can choose to carry out that proposal or you can choose not to. The whole purpose of business decision-making is to figure out, given the current business circumstances, which proposals should be carried out and which ones should not.

75.2.15.8 Investment Decisions

Investment decisions are made by investors in order to spend money and resources on achieving a target. Investors are either inside (e.g., finance and board) or outside (e.g., banks) the company. The target relates to some economic criteria, such as achieving a high return on the investment, strengthening the capabilities of the company, or improving the value of the company.

75.2.15.9 Planning Horizon

When an organization chooses to invest in a particular proposal, money gets tied up in that proposal—so called "frozen assets." The economic impact of frozen assets ("capital recover with return") tends to start high and decreases over time. On the other hand, operating and maintenance costs of elements associated with the proposal tend to start low but increase over time. The total cost of owning and operating a proposal is the sum of those two costs. Early in time, frozen asset costs dominate and later in time the operating and maintenance costs dominate. There is a point in time where the sum of the costs is minimized or minimum cost lifetime.

To properly compare a proposal with a 4 year life span to a proposal with a 6 year life span, the economic effects of either of the following need to be addressed:

- Cutting the 6 year proposal by 2 years
- Investing the profits from the 4 year proposal for another 2 years

The planning horizon, sometimes known as the study period, is the consistent time frame over which proposals are considered. Effects such as economic life and the time frame over which reasonable estimates can be made will need to be factored into establishing a planning horizon. Once the planning horizon is established, several techniques are available for putting proposals with different life spans into that planning horizon.

75.2.15.10 Price and Pricing

A price is what is paid for in exchange for a good or service. A price is a fundamental aspect of financial modeling and is one of the four Ps of the marketing mix. The other three aspects are product, promotion, and place. Price is the only revenue-generating element amongst the four Ps, the rest being cost centers.

Pricing is part of finance and marketing. It is the process of determining what a company will receive in exchange for its products. Pricing factors are manufacturing cost, market place, competition, market condition, and quality of product. Pricing applies prices to purchase and sales orders, based on factors such as: a fixed amount, quantity break, promotion or sales campaign, specific vendor quote, price prevailing on entry, shipment or invoice date, combination of multiple orders or lines, and many others. The needs of the consumer can be converted into demand only if the consumer has the willingness and capacity to buy the product. Thus pricing is very important in marketing.

75.2.15.11 Cost and Costing

A cost is the value of money that has been used up to produce something, and hence is not available for use anymore. In economics, a cost is an alternative that is given up as a result of a decision. A sunk cost is the expenses before a certain time, typically used to abstract decisions from expenses in the past, which can cause emotional hurdles in forward looking. From a traditional economics point of few, sunk costs should not be considered in decision making.

Costing is part of finance and business management. It is the process to determine the cost based on expenses (e.g., production, software engineering, distribution, and rework) and based on the target cost to be competitive and successful on a market. The target cost can be below the actual estimated cost. Therefore, costing always includes cost management.

75.2.15.12 Performance Measurement

Performance measurement is the process where an organization establishes the parameters within which programs, investments, and acquisitions are measured to control whether they are reaching the desired results. It means to evaluate whether performance objectives are actually achieved, to control budgets, resources, progress and decisions, and to learn and improve performance.

75.2.15.13 Termination Decisions

Termination means to end a project or product. Termination can be planned for a long-time (e.g., when foreseeing that a product will reach its life-time) or can come rather spontaneously (e.g., when performance targets are not achieved). In both cases, the decision must be carefully prepared, considering always the alternatives of continuing versus terminating. Cost of different alternatives must be estimated, covering topics such as replacement, information, suppliers, alternatives, assets, and utilizing resources for other opportunities. Sunk cost must not be considered in such decision-making because they have been spent and will not reappear as a value.

75.2.15.14 Replacement and Retirement Decisions

A replacement decision is a special case of for-profit decision analysis that happens when an organization already has a particular asset and they are considering replacing it with something else, like deciding between keeping a legacy software system and redeveloping it from the ground up. Replacement decisions use the same business decision process as described earlier but there are additional challenges: sunk cost and salvage value. Retirement decisions are about getting out of an activity altogether, such as when a software company considers not selling a software product any more, or a hardware manufacturer thinks about not building and selling a particular model of computer any longer.

75.2.15.15 Friction-Free Economy

Economic friction is everything that keeps markets from having perfect competition. It means distance, cost of delivery, restrictive regulations, or imperfect information. In high-friction markets, customers do not have many suppliers to choose among. Having been in a business for a while or owning a store in a good location determines the economic position. It is hard for new competitors to start business and compete. The marketplace moves slowly and predictably. Friction-free markets are just the reverse. New competitors crop up all over, and customers are quick to respond. The marketplace is anything but predictable. Software and IT clearly are friction-free. New companies can easily create products and often do that at much lower cost than the established companies due to not considering all legacies. Marketing and sales can be done via the Internet that offers with social networks and basically free distribution mechanisms all to ramp up a global business. Software engineering economics aims at providing mechanisms to evaluate different concepts for starting, growing and safe-guarding a software business in such a friction-free economy.

75.2.15.16 Ecosystems

An ecosystem is an environment consisting of all the mutually dependent stakeholders, business units, and companies working in a particular area. In a typical ecosystem, there are producers and consumers, where the consumers add value to the consumed resources. A software ecosystem is, for instance, a supplier of an application working with companies doing the installation and support in different regions. Neither one could exist without the other. Ecosystems can be permanent or temporary. Software engineering economics provides the mechanisms to evaluate alternatives in establishing or extending an ecosystem, for instance assessing whether to work with a specific distributor or have the distribution done by a company doing service in an area.

75.2.15.17 Offshoring and Outsourcing

Offshoring means executing a business activity beyond sales and marketing outside the home country of an enterprise. Enterprises typically either have their offshoring branches in low-cost countries or they ask specialized companies abroad to execute the respective activity. Offshoring should therefore not be confused with outsourcing. Offshoring within the own company is called captive offshoring. Outsourcing is the result-oriented relationship with a supplier, who executes business activities for an enterprise, which traditionally were executed inside the enterprise. Outsourcing is site-independent. The supplier can reside in direct neighborhood of the enterprise or offshore, which is outsourced offshoring. Software engineering economics provides the basic criteria and business tools to evaluate different sourcing mechanisms and control their performance.

75.2.16 Risk and Uncertainty Management

75.2.16.1 Goals, Estimates, and Plans

A goal in software economics is mostly a business goal (or business objective). It is external to the specific software or IT project and sets constraints which a plan has to consider. It relates business needs, such as increasing profitability, to investing resources, such as starting a project or launching a product with a given budget, contents, and timing. Goals are for instance to reach a certain milestone at a given date or to extend testing by some time to achieve a desired quality level.

An estimate is the well-founded evaluation of how much resources and time would be necessary to achieve a stated goal. Estimates are typically internally generated and not necessarily externally visible. They should not be driven by the goals, because this could make the estimate overly optimistic. Of course, the underlying solutions which drive the estimates should be aligned with the goals. Estimates are generated by experts familiar with the product or project. In software projects, this could be the effort necessary to deliver at a given milestone.

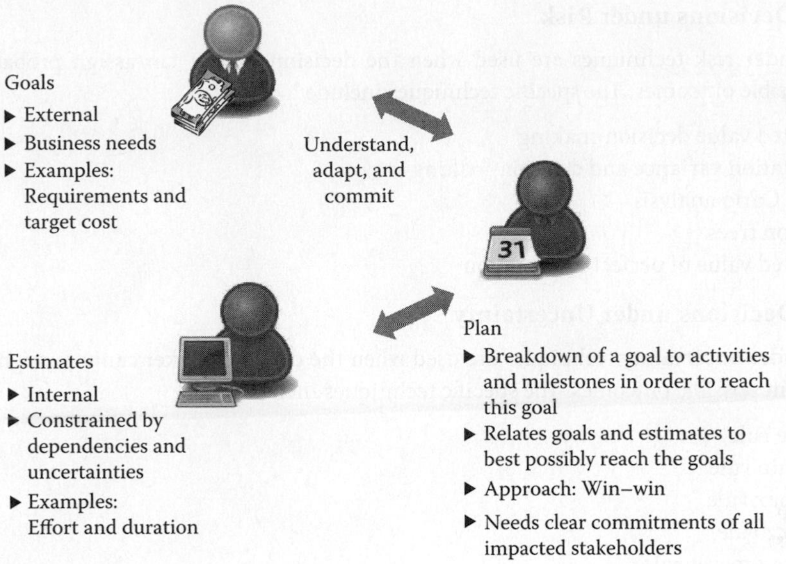
Goals
▶ External
▶ Business needs
▶ Examples: Requirements and target cost
Understand, adapt, and commit
31
Estimates
▶ Internal
▶ Constrained by dependencies and uncertainties
▶ Examples: Effort and duration
Plan
▶ Breakdown of a goal to activities and milestones in order to reach this goal
▶ Relates goals and estimates to best possibly reach the goals
▶ Approach: Win–win
▶ Needs clear commitments of all impacted stakeholders

FIGURE 75.3 Relating goals, estimates, and plans.

A plan is the break-down of a goal to activities and milestones in order to reach this goal. The plan should be in line with the goal and the estimate, which is not necessarily easy and obvious, such as when a software project with given requirements would take longer than the target date foreseen by the client. In such cases, plans demand a review of initial goals as well as estimates and the underlying uncertainties and inaccuracies. Creative solutions with the underlying rationale of achieving a win–win position are applied to resolve conflicts. The plan needs to achieve commitment with impacted stakeholders to be useful (Figure 75.3).

75.2.16.2 Estimation Techniques

Estimations are used to analyze and forecast the necessary resources or time to implement requirements. Four families of estimation techniques exist:

- Expert judgment
- Analogy
- Decomposition
- Statistical (or parametric) methods

75.2.16.3 Addressing Uncertainty

Estimates are inherently uncertain and that uncertainty should be addressed in business decisions. Techniques for addressing uncertainty include

- Consider ranges of estimates
- Sensitivity analysis
- Delay final decisions

75.2.16.4 Prioritization

Prioritization means to compare alternatives based on different criteria and then rank those alternatives to deliver the best possible value. In software projects, often requirements are prioritized in order to deliver most value to the client or to allow for building increments where a first delivery ensures that the client sees a value.

75.2.16.5 Decisions under Risk

Decisions under risk techniques are used when the decision maker can assign probabilities to the different possible outcomes. The specific techniques include

- Expected value decision-making
- Expectation variance and decision-making
- Monte Carlo analysis
- Decision trees
- Expected value of perfect information

75.2.16.6 Decisions under Uncertainty

Decisions under uncertainty techniques are used when the decision maker cannot assign probabilities to the different possible outcomes. The specific techniques include

- Laplace rule
- Maximin rule
- Maximax rule
- Hurwicz rule
- Minimax regret rule

75.2.17 Economic Analysis Methods

75.2.17.1 For-Profit Decision Analysis

Figure 75.4 describes the process for identifying the best alternative from a set of mutually exclusive alternatives. Decision criteria depend on the business objectives and typically include ROI or return on capital employed (ROCE).

The for-profit decision techniques do not apply when the organization's goal is not profit—which is the case in government and in non-profit organizations. In these situations, the organization has a different goal, which means that a different set of decision techniques are needed, such as cost–benefit or cost-effectiveness analysis.

75.2.17.2 Cost–Benefit Analysis

Cost–benefit analysis is one of the most widely used methods for evaluating individual proposals. Any proposal with a benefit–cost ratio of less than 1.0 can usually be rejected without any further analysis because it would cost more than it would benefit. Proposals with a higher ratio need to consider the associated risk of an investment and compare the benefits with the option taking the same money to the bank.

75.2.17.3 Cost-Effectiveness Analysis

Cost-effectiveness analysis shares a lot of the same philosophy and methodology with benefit–cost analysis. There are two versions of cost-effectiveness analysis. The *fixed-cost* version maximizes the benefit given some upper bound on cost. The *fixed-effectiveness* version minimizes the cost needed to achieve a fixed goal.

75.2.17.4 Break-Even Analysis

Given functions describing the costs of two or more proposals, break-even analysis helps in choosing between them by identifying points where the costs are equal. Below a break-even point, one proposal is preferred and above that point the other is preferred.

75.2.17.5 Business Case

The business case is the consolidated information summarizing and explaining a business proposal from different perspectives (cost, benefit, risk, and so on) for a decision maker, often used for assessing the value of a product or requirements of a project, which can be used as a basis in the investment

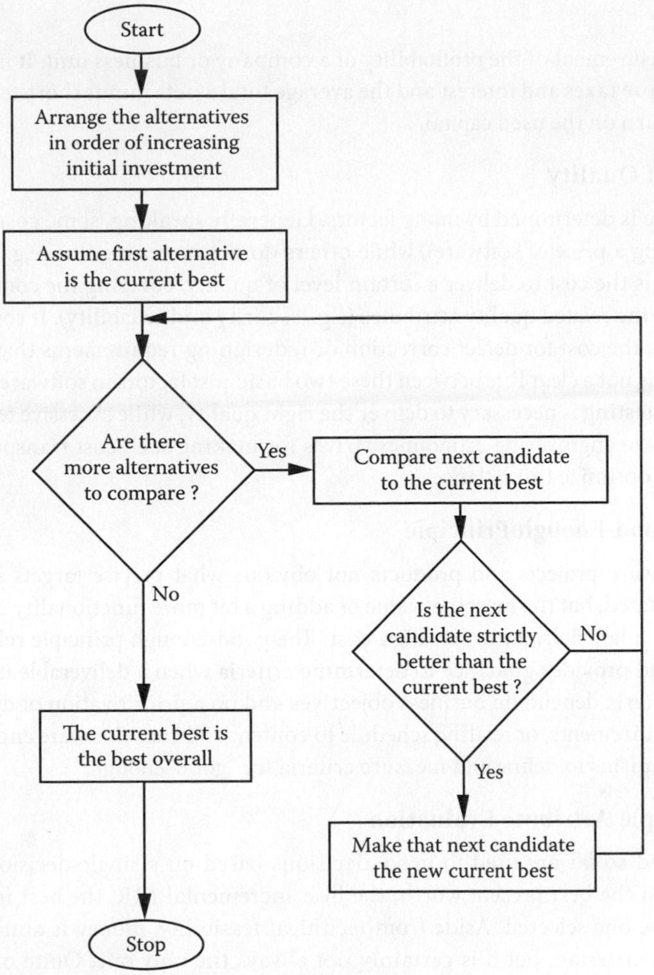

FIGURE 75.4 The for-profit decision-making process.

decision-making process. As opposed to a mere profit-loss calculation, the business case is a "case," which is owned by the product manager and used to support achieving the objectives.

75.2.17.6 ROI

Generally speaking, it would not be smart to invest in an activity with a return of 8% when there's another activity that's known to return 16%. An organization's *minimum attractive rate of return* (MARR) is the lowest internal rate of return the organization would consider to be a good investment. The MARR is a statement that an organization is confident it can achieve at least that rate of return.

Another way of looking at the MARR is that it represents the organization's *opportunity cost* for investments. By choosing to invest in some activity, the organization is explicitly deciding to not invest that same money somewhere else. If the organization is already confident, it can get some known rate of return, other alternatives should be chosen only if their rate of return is at least that high. A simple way to account for that opportunity cost is to use the MARR as the interest rate in business decisions. An alternative's present worth evaluated at the MARR shows how much more or less (in present day cash terms) that alternative is worth than investing at the MARR.

75.2.17.7 ROCE

The ROCE is a measurement of the profitability of a company or business unit. It is defined as the ratio of a gross profit before taxes and interest and the average total assets minus short-term debt minus cash. It describes the return on the used capital.

75.2.17.8 Cost of Quality

The cost of software is determined by many factors. Generally speaking, some cost relate to generating value (e.g., designing a piece of software), while others do not increase value (e.g., rework and errors). The cost of quality is the cost to deliver a certain level of quality, covering the content (e.g., functional requirements), and the related quality attributes (e.g., security and reliability). It contrasts with the cost of non-quality (e.g., the cost for defect correction or redesigning requirements that have been wrongly interpreted). There is not a clear line between these two basic cost factors in software engineering and IT. For example, some testing is necessary to deliver the right quality, while excessive testing means a waste of resources. Software engineering economics strives for making such cost transparent and providing the mechanisms to optimize this balance.

75.2.17.9 The Good-Enough Principle

Often, it is in software projects and products not obvious what precise targets should be achieved. Requirements are stated, but the marginal value of adding a bit more functionality cannot be measured. The result could be a late delivery or too high cost. The good-enough principle relates marginal value to marginal cost and provides guidance to determine criteria when a deliverable is good enough to be delivered. These criteria depend on business objectives and on a prioritization of different alternatives, such as ranking requirements, or relating schedule to content and cost. Software engineering economics provides the mechanisms to define and measure criteria for "good-enough."

75.2.17.10 Multiple Attribute Evaluation

The topics discussed so far are used to make decisions based on a single decision criterion, money. The alternative with the best present worth, the best incremental IRR, the best incremental benefit–cost ratio, etc. is the one selected. Aside from technical feasibility, money is almost always the most important decision criterion, but it is certainly not always the only one. Quite often there are other criteria, other "attributes," that need to be considered and those attributes cannot be cast in terms of money. Multiple attribute decision techniques allow other, non-financial criteria to be factored into the decision.

There are two families of multiple attribute decision techniques that differ in how they use the attributes in the decision. One family is the "compensatory," or single-dimensioned, techniques. This family collapses all of the attributes onto a single figure of merit. The family is called compensatory because, for any given alternative, a lower score in one attribute can be compensated by—or traded off against—a higher score in other attributes. The compensatory techniques include

- Nondimensional scaling
- Additive weighting
- Analytic hierarchy process

In contrast, the other family is the "non-compensatory," or fully dimensioned, techniques. This family does not allow trade-offs among the attributes. Each attribute is treated as a separate entity in the decision process. The non-compensatory techniques include

- Dominance
- Satisficing
- Lexicography

75.2.17.11 Optimization Analysis

The typical use of optimization analysis is to study a cost function over a range of values to find the point where overall performance is best. Software's classic space-time trade-off is an example of optimization; an algorithm that runs faster will often use more memory. Optimization balances the value of the faster run time against the cost of the additional memory.

75.3 Practice

Leadership and teamwork: Install an effective core team. Often, different stakeholders have unaligned agendas, which make the project late and cause lots of overheads and rework. First thing is to formally create a core team with the product, marketing, project, and operations managers for each product (release) and make it fully accountable for the success of a product. These persons represent not only the major internal stakeholders in product or solution development but also sufficiently represent different external perspectives. The core team leads the product development in all its different dimensions. They typically meet once a week to go through all open issues, risks, and relevant aspects of the product. Decisions are taken and implemented by the respective function. We suggest announcing and making this core team operational as early as possible in the product life cycle but certainly when the product or release is defined. The success factor is to give this core team a clear mandate to "own" the project. We see most need for active support in building an effective core team that agrees that they together have to steer the course. Too often, we face silo organizations in marketing, business management, and engineering that do not work together. This means in many cases, not only to build teams, to train, and to coach but also to adjust annual targets and performance management. As we often realize: Culture changes when targets are adjusted.

Managing risks and uncertainty: Enforce the product life cycle. Like the core teams, we urge to make a standardized product life cycle mandatory for all product releases, that is, all engineering projects. Most companies today have such life cycle defined but rarely use it as the pivotal tool to derive and implement decisions. Too often, requirements changes are agreed in sales meetings without checking feasibility, and technical decisions are made without considering business case and downstream impacts. A useful product life cycle has to acknowledge that requirements may never be complete and may indeed be in a "continuum" state. The product life cycle should guide with clear criteria, that is, determining what is good enough or stable enough. This implies that it is sufficiently flexible to handle different types of projects and constraints. This is achieved with basic tailoring techniques and guidance, which elements are mandatory and which should be adjusted to the specific environment. To foster discipline and visibility, the mandatory elements of gate reviews, such as checklists or minutes, must be explicit and auditable. To avoid reporting overheads, we recommend using online workflow support to instrument such product life cycle. Aside ease of execution, such automation facilitates reuse, data quality, and consistency. Today you can find lots of workflow management systems. We suggest to practically evaluating potential solutions versus your own needs, as it otherwise can become a burden.

Mastering stakeholder needs: Evaluate needs and requirements. Requirements must be understood and evaluated by the entire core team to ensure that different perspectives are considered. Each single requirement must be justified to support the business case and to allow managing changes and priorities. We often found requirements simply being "collected" yielding lots of unnecessary features that add to complexity but not to customer perceived value. In fact, almost 50% of all delivered features are rarely used and do not provide any payback [1,8,11]. If a product is developed on such an unjustified basis, it is in trouble because its requirements will continuously change. A product (release) must address a need and must have a strong business vision. This vision (i.e., what will be different with the release of the product) must be coined into a story. The story then translates to business objectives and major requirements. Good business management first understands the customer's needs and business case and then develops necessary features. Requirements are a contract mechanism for the project internally and often for a client externally. They must be documented in a structured and disciplined way, allowing both technical as well as market and business judgment. Their evaluation should specifically look to completeness, consistency, and understandability. Ask a tester to write a test case before processing the requirement. Ask the marketer in the team to check whether he can sell the feature as described. This avoids discussing in the clouds. Requirements should not be perfect or we risk paralysis by analysis. Determine what is good enough and ensure that any further insight is adequately considered. After evaluation, requirements are approved by the core team. Only thereafter they are formally allocated to the project and engineering effort is spent. Requirements and business objectives must be managed (planned, prioritized, agreed, and monitored) throughout the life cycle to assure focus [3,11–14]. We suggest having a project plan that is directly linked with requirements. Work packages within this project plan should show the value they contribute with such links to requirements. This allows not only to focus on what matters but also to also monitor the earned value of the project from begin to end—and to proactively manage risks, such as effort being burned without creating value. Note that your change management needs to be equally formal and disciplined because most issues we face in troubled projects result from creeping requirements and insufficient impact analysis. To ease change management, we suggest installing traceability from requirements upwards to the business case and downwards to test cases.

Business objectives and accountability: Assure a dependable portfolio. Managing release roadmaps and the own portfolio as a mix of resources, projects, and services must be the focus of each product manager. Often, roadmaps are not worth the paper they are printed due to continuous changes and thus lack of buy-in from sales, operations, and service. Projects are started ad-hoc, while necessary reviews and clean-ups in portfolios never happen. With moving targets, sales has no guidance how to influence clients, and engineering will decide on its own which technologies to implement with what resources. When it comes to his own portfolio, the product manager has to show leadership and ensure dependable plans and decisions that are effectively executed. Dependable means that agreed milestones, contents, or quality targets are maintained as committed—unless a change is agreed and documented. Be aware as a product manager, that each ad-hoc content or release change will create the perception that your portfolio is not managed well. Apply adequate risk management techniques to make your portfolio and commitments dependable. Projects may need more resources, suppliers could deliver late or technology would not work as expected. For instance, platform components used by several products might use resource buffers, while application development applies time boxing. If there is a change to committed milestones or contents within your portfolio, they must be approved first by the core team and where necessary by respective steering boards, and then documented and communicated with rationales.

75.4 Business Value

Do aforementioned practices mean better business performance? We performed root cause analyses of hundreds of products that underperformed and found similar causes reappearing. Root causes included business cases that were never re-evaluated, unbalanced portfolios that strangulate new products, insufficient management of new releases and service efforts, and the lack of vision that caused requirements to continuously change. This is underlined by observations such as Cooper's studies indicating that the top 20% of enterprises deliver 79% of new products in time, while the average delivers only 51% of projects in time [6]. The same holds for productivity and performance. Looking, for instance, to how productivity is impacted by good business planning, we found that with a requirements change rate beyond 20% in a project, productivity falls and as such business performance [1,8].

We have been working with hundreds of business managers and achieved reduction of delays of 20% per year [3]. Explanatory factors for this positive impact of business management include leadership and teamwork, managing risks and uncertainty, mastering stakeholder needs, and accountability toward agreed business objectives and accountability—managed by one empowered person across the product life cycle.

Improved product and business management has a profound positive impact on the overall business. For better understanding and knowledge transfer from this handbook entry to your work, we will present a case study from introducing software business understanding in a global information and communication technology (ICT) company with strong focus on corporate consistency. It showed success after ca. 12–18 months of working in the new scheme. This is what should be considered a normal learning curve when strengthening business management and introducing product management.

Strengthening the business focus as sketched earlier, at a major ICT supplier showed that duration (time to market), schedule adherence, and deliverable quality all sustainably improved while introducing and improving business management. The data are drawn from a single business unit over 3 years to allow comparability. The business unit that we selected was amongst the first to introduce consistent business management in a defined way. It operates in North America, Europe, and Asia, thus assuring representative results independent of geography. The products of that business unit are components used in communication networks. Examples of usage include voice over IP and video solutions. They comprise platform products that are developed typically every few years and then customized for contract projects. The approach behind is product-line driven, so that platform products would have the basic functionality and customer products are enhancements (or changes) to those platforms. There are also network management systems included, which help to configure and operate these products.

In order to achieve business success, the product manager has several objectives: Creating a winning product and business case and delivering value to customers. As a control variable, we took the degree of implementation of the product manager role. Naturally, there was a product manager role available since long; however, the responsibilities, competencies, and operational behaviors were very heterogeneous. Only with an orchestrated approach toward defining a competence profile for product managers, aligning their roles and responsibilities versus other roles in the same organization (e.g., marketing manager, project manager, and regional sales) triggered more consistent and effective behaviors. We mapped the implementation degree of the product manager role to three phases [3]:

- Phase 1: Establish foundations. A sense of urgency was created with some critical stakeholders in the organization to strengthen the role of product manager. Leading product managers were brought together and assessed success factors and the elements of the role. We aligned the new role elements to other related roles, such as marketing manager or project manager, on the one hand, and we benchmarked with other companies to formally establishing the role.
- Phase 2: Prepare and pilot. The standard product life cycle was enriched with more templates, self-assessment tools, training materials, and hands-on success stories. Training modules were piloted for key functional competences, such as the writing of a customer business case.
- Phase 3: Deploy. The longest phase dealing with deployment was performed using progress measurements for monitoring and deficiency identification. Incumbent product managers and the newly trained community provided continuous feedback helping in identifying further improvement needs.

The impact of each of these three phases could be deducted from our history database by means of mapping product releases (projects) to dates and organizations. We looked to an overall total of 178 projects without any filtering (Figure 75.5). Each phase has a representative set of projects. As a rule of thumb, one would need at least 10 projects per dependent variable being analyzed. Size is represented from very small projects (few person weeks) up to several hundred person years. The dependent variables of our study are average duration, average delay as percentage of schedule overrun compared to originally committed release dates and quality level in terms of defect detection percentage after handover. We achieved within 3 years of strengthening systematic business and product management in phase 3 (deployment) an improvement of 36% in average duration, 85% in delays, and 82% in quality (Figure 75.6).

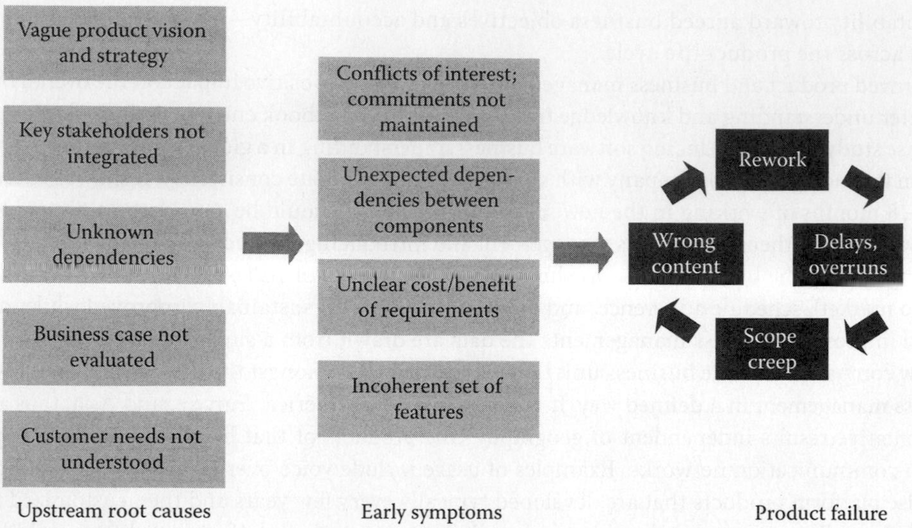

FIGURE 75.5 The results of insufficient business management.

Prod. Mgmt. Introduction Phase	# Projects	Min Size (PY)	Max Size (PY)	Average Delay	Average Duration	Average Defects
1	47	0.1	346	100%	100%	100%
2	55	0.1	84	25%	87%	30%
3	76	0.1	91	15%	64%	18%

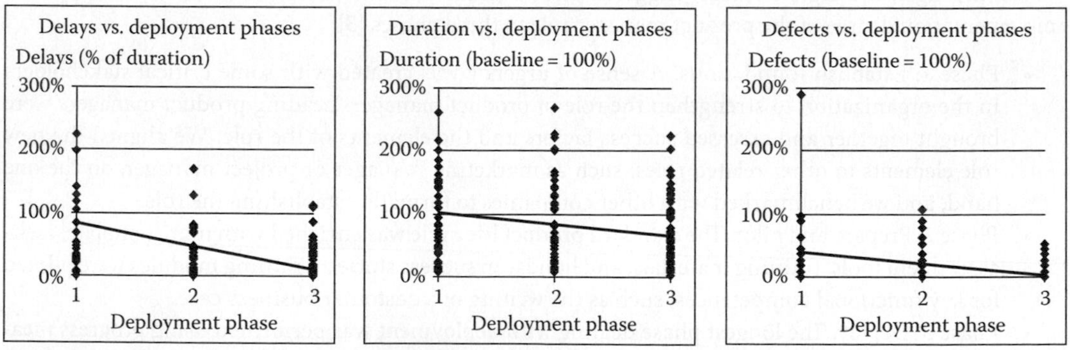

FIGURE 75.6 ICT case study results with 178 projects over 3 years. (Data from 178 industry projects from telecommunication industry over a period of 3 years throughout which the project management role and competency was defined, deployed, and improved. Note that the graphs show regression lines based upon least squares and not averages.)

75.3.1 Outlook: Where Is the Software Business Heading?

We will finally look at trends in software business. These trends indicate close collaboration between industrial needs and research at both universities and enterprises. Business trends naturally are influenced and determined by external trends that impact our society and therefore buyer behaviors as well as individual needs. Such external trends are characterized by: Demand for value, everything is fashion, individualism, ever-changing expectations, demand for ubiquitous services, global competition, economic and ecologic behaviors, and the need for security and stability. From these trends, we can look to what this means for software business.

1. *Innovation with market and solution focus rather than technology-driven*. Product needs and solutions go far beyond those of the software itself. They include service, evolution, integration of business processes, and the like. It is about speed to needs. Software business has to cope with this need and not get cornered by only looking toward what are mere software features. In the end, the customer is not interested in features, but in his needs and how they are best satisfied by products and services.
2. *Value creation with customers*. Value-oriented engineering will grow rapidly, that is, improving the evaluation of requirements within a business case from a portfolio management perspective. Customers are part of the value creation. This implies dynamic segmentation down to the single-buyer segment. Questions include: What is the customer business case behind the requirements? Is the own business case valid and in line with the customer or user business case? What is the contribution of requirements to this business case?
3. *Innovative market rules and business models*. Software has such low entry levels that a new competitor is simply a mouse-click away. Friction-free deliveries further add to this competitive trap. It is therefore important to develop customer and user needs continuously from scratch and innovate how their current and prospective customers can further improve their own business. Examples include community source with networks of stakeholders developing and maintaining software, wikinomics with a global and open access to information and resources, and blue ocean approaches to create new value for customers rather than fighting on price alone.
4. *Quality focus*. Quality starts at the start of a product concept and a project. Quality needs are manifold and can contradict each other. Software business has to deal with these constraints and must offer win–win solutions. Furthermore, it has to develop a quality perspective in the product development, covering for instance the usage of commercial components, including a variety of partners or suppliers, managing the quality delivered by such external partners, and adapting systems quality features as business needs change.
5. *Supplier networks and eco-systems*. Aside from the classic competitive schemes, customers today want to have sustainable networks of suppliers. The traditional concept of supply-chain will be replaced by adaptive supply networks. Suppliers are even more than ever subject to continuous evaluation and replacement where necessary. The success of a supplier depends on how well he is able to create communities and business models together with customers and other suppliers.
6. *Agility to manage uncertainties*. It is increasingly important to predict changes to products and customer or market needs on an individual level. Which requirements are most volatile and at same time exposing the project to highest risk? How can they be addressed by sufficiently flexible solution architecture?
7. *Effective knowledge management*. Knowledge management techniques becomes more and more necessary for capturing the wide-spread knowledge on customers, markets, products, technologies and especially stakeholders, in order to be less people-dependent, and to mature products and business management in an ever more changing environment.

Whether your customer is internal to your company or a traditional external client-user of your product, whether your product is shrink-wrapped and shipped or it is a service or embedded system, creating and addressing customer needs is the primary driver for good business and for good software.

Further Reading

Ebert, C. *Global Software and IT: A Guide to Distributed Development, Projects, and Outsourcing*. Wiley, New York, 2011. Description: Based on the author's first-hand experience and expertise, this book offers a proven framework for global software engineering. Readers will learn best practices for managing a variety of software projects, coordinating the activities of several locations across the globe while accounting for cultural differences. Most importantly, readers will learn how to engineer a first-rate software product as efficiently as possible by fully leveraging global personnel and resources.

Ebert, C. and R. Dumke. *Software Measurement: Establish, Extract, Evaluate, Execute*, Springer, New York, 2007. Description: People who design and develop software like to call themselves software "engineers." Yet few organizations have really institutionalized measurement of their products and processes. This book is bang up-to-date in both fields and packed with practical advice. For every software engineer and manager.

Reifer, D.J. *Making the Software Business Case: Improvement by the Numbers*, Addison Wesley Professional, Boston, MA, September 2001. Description: This practical handbook shows you how to build an effective business case when you need to justify—and persuade management to accept—software change or improvement. Based on real-world scenarios, the book covers the most common situations in which business case analyses are required and explains specific techniques that have proved successful in practice. The book provides examples of successful business cases; along the way, tables, tools, facts, figures, and metrics guide you through the entire analytic process. An excellent book to learn how to prepare and implement a business case and thus make software a successful business.

McGrath, M.E. *Next Generation Product Development*. McGraw-Hill, London, U.K., 2004. Description: A hands-on manager's guide to making the most of today's product development breakthroughs. In-depth explanation of the ways in which companies are able to achieve substantially higher levels of development productivity, while better aligning product development with strategy through new practices and systems.

Condon, D. *Software Product Management: Managing Software Development from Idea to Product to Marketing to Sales*, Aspatore Books, San Francisco, CA, 2002. Description: The book decodes the software product management process with an emphasis on coordinating the needs of stakeholders ranging from engineering, sales, and product support to technical writing and marketing. Based on real-world experience in managing the development of enterprise software, this book details how a team can work together smoothly to achieve their goal of releasing a superior software product on time. While it is not primarily about setting up a business, the book explains hands-on what is necessary in daily operational fights to succeed.

References

1. Ebert, C. and R. Dumke. *Software Measurement*. Springer, New York, 2007.
2. Ebert, C. and T.C. Jones. Embedded software—Facts, figures and future. *IEEE Computer*, 42(4), 42–52, April 2009.
3. Ebert, C. The impacts of software product management. *The Journal of Systems and Software*, 80(6), 850–861, June 2007.
4. Cusumano, M.A. and D.B. Yoffie. *Competing on Internet Time: Lessons from Netscape and its Battle with Microsoft*, Free Press, New York, 1998.
5. Heikkinen, M.P. *Knock, Knock, Nokia's Heavy Fall*. Kuukausiliite monthly supplement for October 2010. Available from: http://www.hs.fi/english/article/Knock+Knock+Nokias+Heavy+Fall+Part+III/1135260623867 (accessed October 13, 2013).
6. Cooper, R.G. et al. Benchmarking best NPD practices: Research—Technology management; Part I: January 2004, p. 31; Part II: May 2004, p. 43; Part III: November 2004, p. 43. Available from: www.apqc.org (accessed October 13, 2013).
7. Gorchels, L. *The Product Manager's Handbook: The Complete Product Management Resource*, McGraw-Hill, New York, 2011.
8. Ebert, C.: *Global Software and IT: A Guide to Distributed Development, Projects, and Outsourcing*. Wiley, New York, 2011.
9. Galin, D. et al. Are CMM program investments beneficial? Analyzing past studies. *IEEE Software*, 23 (6), 81–86, 2006.
10. Reifer, D. J. *Profiles of Level 5 CMMI Organizations*. Crosstalk, 20(1), 24–28, January, 2007. http://www.compaid.com/caiinternet/ezine/reifer-profiles.pdf

11. Davis, A.M. *Just Enough Requirements Management*. Dorset House, New York, 2005.
12. Ebert, C. Understanding the product life cycle: Four key requirements engineering techniques. *IEEE Software*, 23(3), 19–25, 2006.
13. Fricker, S., T. Gorschek, C. Byman, and A. Schmidle. Handshaking with implementation proposals: Negotiating requirements understanding, *IEEE Software*, 27(2), 72–80, March, 2010.
14. Gorschek T., P. Garre, S. Larsson, and C. Wohlin: Industry evaluation of the requirements abstraction model, *Requirements Engineering Journal*, 12(3), 163–190, 2007.

76

Open Source and Governance: COTS*

Brian M. Gaff
McDermott Will and Emery, LLP

Gregory J. Ploussios
Edwards Wildman Palmer, LLP

Brian R. Landry
Saul Ewing, LLP

76.1 Introduction

Open source software is software available in source code form that allows users to view, modify, extend, and, in many cases, further distribute subject to the terms of the applicable license. Use of open source software offers many advantages, including its ready availability and no royalty or license fees. In addition, many open source software applications and modules are continuously being improved, debugged, tested, and updated by a community of users and, therefore, may represent the best solution available from both a quality and cost standpoint.

Open source software rarely comes without some restrictions and conditions. Depending on the license, these restrictions and conditions range from fairly minor to very significant, and may be incompatible with a company's planned use of the open source software.

76.2 Underlying Principles

The availability of open source software at no readily apparent cost to programmers creates much confusion regarding the ownership of rights in the software and what restrictions the owner can impose on use and modification of the code.

* This chapter is an expanded version of an article that first appeared in the June 2012 issue of *Computer*, which is published by the IEEE Computer Society.

Even though open source software may appear to be "free," open source software reflects the distinction between the definition of "free" as meaning "for zero price" (*gratis*) versus "with little or no restriction" (*libre*), with open source software being truly "free" only in the second sense of the word. As the father of the open source movement, Richard Stallman stated, "Think free as in speech, not free beer."

76.2.1 Copyright Law

Like all software (or novels or paintings), open source software is only "free" to the extent that the copyright holders make the software "free."

When distinguishing between, on the one hand, software that a company owns and, on the other hand, open source software that a company obtained from another source, lawyers and business people will often refer to software that a company owns as "proprietary" software. While this distinction is valid from the perspective of the company (i.e., the company does not own the open source software that it downloaded from a website), it is a mistake to infer from this that open source software is not owned by anyone or that it is in the public domain.

The developers of the open source software or their assignees own title to the copyright in the software automatically from its fixation in a tangible medium (e.g., a computer file). While such owners of the open source software may make the open source software available for anyone to use under the terms of the open source license that they have chosen, their ownership of the underlying copyright is what allows them to require that anyone using the software do so in accordance with the license.

76.2.2 Open Source License Terms

Generally speaking, open source software licenses fall into one of three categories: restrictive (also called "copyleft"), moderately restrictive (also called "copycenter"), and permissive. In theory, the number of possible licenses is unlimited because authors of open source software may select any license or create their own. However, the top 10 most common open source software licenses are used by probably the majority of all open source software modules and applications available. With the notable exception of the Affero variant of the GNU General Public License (the "GPL"), in most cases, these restrictions are only applicable or relevant if copies of the software will be provided to third parties and are not applicable or relevant when the software is only being used internally, which includes hosting the software in a software as a service (SaaS) arrangement for customers or using it to operate a customer-facing website. However, these definitions of "distribution" across different versions of the GPL are potentially subject to varying interpretations in different jurisdictions as will be discussed in the subsequent sections.

76.2.2.1 Restrictive

Certain open source software licenses, such as the GNU GPL, are considered "viral" licenses, because they generally require that proprietary software that is distributed with open source software be licensed under terms consistent with the open source software license (hence, like a virus, the license terms are passed on to other software).

Proprietary code distributed with or alongside GPL-licensed code as part of a larger program or application may in many cases be deemed a "covered work" along with the GPL-licensed code, such that the entire covered work (i.e., proprietary code and open source software) may only be distributed on terms compatible with the GPL license, including on the condition that the source code to the entire combined program (including developer's proprietary code) be made available to recipients under the GPL-license terms. Unlike under the GNU Lesser General Public License ("LGPL") and other moderately restrictive licenses discussed in the subsequent sections, maintaining the GPL-licensed code in separate source code and executable files (such as dynamically linked libraries [DLLs]) that only dynamically link to the proprietary code at run time does *not* necessarily prevent the proprietary code distributed alongside GPL-licensed software from becoming a "covered work" under the GPL that may only be distributed subject to the terms of the GPL license.

An exception does exist for a compilation (called an "aggregate") of the open source with other separate and independent works, which are not by their nature extensions of the covered work, and that are not combined with it such as to form a larger program, in or on a volume of a storage or distribution medium, if the compilation and its resulting copyright are not used to limit the access or legal rights of the compilation's users beyond what the individual works permit. Inclusion of a covered work in such an aggregate does not require the license terms to the other parts (such as the company proprietary parts) of the aggregate be also licensed under the GPL. The GPL does not provide much clarity as to the distinction between when multiple components are deemed to be an aggregate (compilation) that is not a covered work and when deemed to be one larger program with multiple parts.* Many companies will not risk including GPL software in distributions of their software products unless the distribution clearly fits into the compilation exception (e.g., where the operating system software is licensed under the GPL and the application running on top of the operating system is commercially licensed).

76.2.2.2 Moderately Restrictive

Moderately restrictive licenses, such as GNU LGPL, may also have a viral effect on proprietary code depending upon the manner and form in which the open source software is used with the proprietary code. For example, if open source software licensed under the LGPL is statically linked into a proprietary software program (e.g., forming an .exe file out of both the LGPL and proprietary code), which is then distributed to customers or third parties, then compliance with the LGPL license would require that the entire executable (including the proprietary code) also be covered by terms compatible with the LGPL license and that, among other things, the source code to the proprietary software be made available to customers (in order to allow the customer to modify the LGPL source code and recompile the executable). On the other hand, distribution of the same LGPL-licensed software with proprietary software where the LGPL software is distributed as a standalone DLL file that is then dynamically linked to by the executable at run time is possible under the terms of the LGPL without requiring the proprietary executable file to be made available in source code form under terms compatible with the LGPL.

76.2.2.3 Permissive

Permissive licenses impose few if any restrictions. The restrictions (if any) are generally not considered to be burdensome (such as a requirement that certain copyright ownership and attribution notices be included with the distributed files), and the licenses are not "viral" as described earlier. Examples of permissive software licenses include the MIT license, the Apache license, and the Berkley Software Distribution license.

---

* The Free Software Foundation (the author of the GPL license) offers only:

"Where's the line between two separate programs, and one program with two parts? This is a legal question, which ultimately judges will decide. We believe that a proper criterion depends both on the mechanism of communication (exec, pipes, rpc, function calls within a shared address space, etc.) and the semantics of the communication (what kinds of information are interchanged)."

"If the modules are included in the same executable file, they are definitely combined in one program. If modules are designed to run linked together in a shared address space, that almost surely means combining them into one program."

"By contrast, pipes, sockets and command-line arguments are communication mechanisms normally used between two separate programs. So when they are used for communication, the modules normally are separate programs. But if the semantics of the communication are intimate enough, exchanging complex internal data structures, that too could be a basis to consider the two parts as combined into a larger program."

"However, the exception would apply, for example, to a CD containing a collection of games where each game was a standalone game that did not rely on the inclusion of any other game to operate and did not link or pass data in any way to the other games. In this example, the CD could contain a mixture of GPL licensed and proprietary licensed games."

76.2.3 Enforcement of Open Source Licenses

Copying, modifying, creating derivative works of, and distributing a work (such as software) that is protected under copyright requires either ownership of the copyright or a valid license from the owner or licensing agent of the copyright. Therefore, if a company distributes software that includes or is derived from open source software, and does not comply with the terms of the applicable license under which such open source software was made available, then such company is infringing the copyright in such software, breaching the terms of the license agreement, or both.

Most open source software modules are available for download and accessible without requiring the recipient to expressly assent to the license terms. For example, usually the recipient is not required by installation software to affirmatively click "I agree" to the terms before the relevant software is installed. However, but for a valid license, the recipient would likely be infringing the software's copyright; therefore, taking the position that the recipient did not agree to the license terms will not usually be a viable option for a recipient.

76.2.3.1 Who Can Enforce an Alleged Open Source Violation?

Authors of open source software, often assisted by organizations formed to advance the enforcement of open source software licensing terms, have been increasingly active in enforcing such licenses. In other cases, commercial software companies that also make a less functional or older version of their proprietary software available under open source licenses, actively seek to enforce the terms of the open source license to ensure that users purchase commercial licenses when appropriate. It is important to note that only the owners of the open source software and their agents acting on their behalf have the right to assert a claim against a user of the open source software for violating the terms of the license or infringing the copyright.

76.2.3.2 Are Open Source Licenses Enforceable?

There is no longer any question that open source licenses are enforceable. The US Court of Appeals for the Federal Circuit, in a rare decision on a copyright matter,* ruled in 2008 in the *Jacobsen v. Katzer and Kamind Industries* case that the terms of the open source "Artistic License" are enforceable. The license covered software used to control model trains. Jacobsen claimed that Katzer/Kamind downloaded the open source software from Jacobsen's website and incorporated it into another product without following the terms of the Artistic license.

Katzer/Kamind admitted that they copied, modified, and distributed the Jacobsen software. The issue for the court was whether these actions were within the scope of the Artistic license. If the actions were outside the scope, then Jacobsen could sue for copyright infringement. Otherwise, the lawsuit would be based on breach of contract.

The court concluded that the terms of the Artistic license that required a copyright notice and tracking of changes to the software were conditions to the license. That is, Katzer/Kamind were required to comply with these conditions to be granted the license to the software. Katzer/Kamind conceded that they did not comply with these conditions.

The court noted that on the facts of the case, Jacobsen made out a prima facie case of copyright infringement. More importantly, however, is that the court acknowledged that the terms of an open source license were enforceable. Therefore, it is likely that noncompliance with a properly drafted open source license, with terms construed as conditions to that license, would result in liability for copyright infringement.

* In the United States, the Court of Appeals for the Federal Circuit has jurisdiction over appeals concerning patent law, while the regional Courts of Appeal typically have jurisdiction over appeals concerning copyright law. The *Jacobsen* case included both copyright and patent law issues, so the Court of Appeals for the Federal Circuit had jurisdiction over the entire case.

76.2.3.3 Available Damages

In the case where, for example, a company distributes a program combining its proprietary software and GPL software and the company does not provide to its customers the source code to its proprietary software (or for that matter license it under the GPL), legal action taken against the company could take the form of a breach of contract claim and/or copyright infringement claim, with the most likely remedies being

- An injunction to prohibit further sale or license of the combined software and require cessation of use of the software by the company and its customers (until the open source software is removed)
- Impoundment and destruction of the combined software
- Monetary damages in the form of direct damages (often measured as copyright holder's lost profits or the price that someone would have been willing to pay) and, if the copyright of the open source software was registered, statutory damages
- Possibly attorney's fees if the copyright to the open source work was registered and the infringement is determined to have been willful

However, a court order forcing the company to "open source" its commercial software to comply with the terms of the GPL is *not* a realistic outcome or liability in this example (or any readily imaginable example). A court order for specific performance of this type is generally *not* available, especially when other remedies, such as monetary damages and court orders enjoining future distribution, afford an adequate equitable remedy.

76.2.4 Interactions between Open Source Licenses and Patent Law

Certain open source software licenses incorporate a provision that purports to cause any person modifying or contributing to the open source software to grant a non-exclusive patent license to recipients of the open source software with respect to any contributions made by such contributor to such open source software. In addition, some of the open source software licenses provide that if a user of the open source software brings a patent infringement claim against another user of the open source software, then the patent license (and, in some cases, also the copyright license) such party had to such open source software terminates. In case of the Mozilla Public License (MPL), a patent infringement claim against a contributor to the MPL licensed software, *regardless whether the patent infringement claim has anything to do with use of such MPL licensed software*, can result in retroactive loss of the MPL license. Because a user of open source software could lose rights to continue using certain open source software modules if it seeks to enforce its patent rights, and because licenses to its patents may be granted if it makes contributions, use of certain open source software licenses may need to be coordinated with decisions on patent strategy.

76.2.5 Commercial Off-the-Shelf

Commercial off-the-Shelf ("COTS") software is generally considered to be software that is widely sold to the general public. In other words, it is software that is not developed in connection with a specific contract or customer. Some advantages of using COTS software as part of a larger software package include decreased overall development time and potentially lower development cost. The reduced time usually means that the software package is delivered to the customer more quickly compared to software packages that consist entirely of custom code.

Although using COTS software may seem attractive, it is important to recognize that significant integration issues may arise, effectively erasing any projected reductions in development cost. Furthermore, the COTS software user has little or no control over the COTS software itself. That is, the manufacturer of the COTS software usually has absolute control over changes and other updates to it. If the changed COTS software is incorporated into the larger software package, it could render the latter inoperable without significant recoding of the package. In an extreme but not uncommon case, the COTS software manufacturer could discontinue the COTS software. In that instance, the COTS software user will likely have no recourse other than to look for an alternative to the obsolete software, which may entail the creation of a custom piece of software as a replacement, retroactively eliminating a primary advantage of using the COTS software.

If a company is evaluating whether to incorporate COTS software into one or more of its released products, it is advisable to consider the inclusion of certain contractual provisions in its purchase orders for the COTS software. Such provisions would operate to protect the company in the event of alterations or obsolescence of the COTS software. For example, the company could require extended notice of planned changes to the COTS software or its availability. This would allow the company to plan effectively for such disruptions and minimize impact to its business and its customers.

76.3 Best Practices

76.3.1 Adopting an Open Source Policy

Having an open source policy that is adhered to is critical. Without such a policy, a company, for example, could find that it lacks the expected ownership and control over software that it just released because the software included certain open source code. The open source code may have been included inadvertently during development. It may have been a remnant of code included only for internal, debugging purposes and unknowingly left in the final, released version of the software. In either case, a company finding itself in such a situation will have significant problem on its hands that it might have avoided if it had a preexisting policy for dealing with open source software.

An open source policy should, at a minimum, include directives on the internal and external use (i.e., redistribution) of open source code, procedures for acquiring open source licenses and monitoring compliance with them, and a formal mechanism that tracks requests to use open source software. The policy should prohibit all use of open source software for which a request to use has not been made or approved.

76.3.1.1 General Philosophy on Open Source

Companies can follow one of two broad paths when setting a policy on the use of open source software. The first path involves an absolute prohibition on the use of open source software. The second path permits the use of open source software, at least to some extent (e.g., for internal use only and/or use in distributed software). Path number one can be easy to manage but will likely add to software developers' workloads and lengthen development time. Path number two may facilitate software development but will require attention to the various open source licenses to ensure compliance.

Irrespective of the path chosen, policing is required. In other words, a company must ensure that it has a complete understanding of its exposure to and use of open source software. Given that software can find its way into a company through unofficial routes, effective policing is necessary even if the company has chosen the first path and prohibited the use of open source material.

76.3.1.2 What Licenses Will the Company Accept?

There are many open source licenses that have been drafted over time and have achieved varying degrees of use by open source developers. As described earlier, these licenses can generally be categorized as restrictive, moderately restrictive, or permissive. One resource that maintains a catalog of these licenses is the open source initiative (opensource.org).

A company's decision as to which open source licenses to accept in connection with using open source software depends on the expected scope of that use. For example, if the expected scope is limited to internal use by the company, only those licenses (if any) that regulate such internal use should be evaluated. On the other hand, if the company intends to couple open source with its proprietary code, then the company needs to consider whether it will agree to license its proprietary code under the same terms as the open source software or seek an alternative license to avoid this requirement.

In view of the significant number of open source licenses in use, albeit in varying degrees, a company should approach a decision on which licenses to accept by first setting a policy regarding its proprietary software. That policy should include a position on the extent to which the company will license that proprietary software. That position will act as a filter and cause the company to reject certain open source licenses that are incompatible with its policy. Consequently, this would foreclose the use of any open source software that is subject to the rejected licenses.

76.3.1.3 Employee Education

An understanding of the benefits and risks of using open source software likely varies widely from employee to employee. Many experienced software developers will understand the relevant issues, while less experienced employees or those having jobs that do not involve coding may be largely unaware of these issues. The solution for dealing with this is having a comprehensive employee education program that covers the open source software landscape.

If a company has already decided to adopt a policy that prohibits the use of all open source software, the education program can focus on how to identify the open source and the risks associated with using it. This should provide the employees with a solid basis for the company's policy. If the employees fully understand that basis, it is more likely that they will follow the policy with minimal resistance.

If the company does not prohibit the use of open source software, it is important that employees be instructed to follow a procedure that ensures they formally request to use the open source. This allows the company to track the use of open source software and evaluate whether the applicable license terms are acceptable so as to authorize the use of the open source. Employees must be educated such that they follow this request-evaluate-approval process without fail.

Irrespective of the choice a company makes about using open source software, it is imperative that employees be able to recognize when they are using open source. This is necessary to guard against the inadvertent use of open source that, if left unchecked, could result in large-scale disruption of the business. If employees are sensitized to recognize open source, they can they take the necessary actions to purge the software or file a formal request to authorize its use.

76.3.1.4 Software Monitoring Tools

Tracking the use of open source software is a critical first step toward ensuring that one does not run afoul of the terms of an open source license. Open source that has worked its way into a company's released software package without the company's knowledge and approval can create major problems, including delaying the release of a software package until the open source is removed or its license terms agreed to.

In view of the complexity of large software engineering projects, it is typically impossible for the developers and their managers to ensure accurate tracking of the use of open source on an individual basis. Accordingly, tools have evolved that monitor the software development process and identify the existence of open source software. These tools are usually largely automated and can include additional features, such as tracking requests to use open source and the decisions made on those requests.

A company should consider the expected complexity of the software that it is developing and the number of developers that it will use and can reasonably monitor in connection with a decision on using an automated tool to police its use of open source.

76.3.2 Deciding Whether to Distribute Software under an Open Source License

76.3.2.1 Balancing Permissiveness with Competitive Concerns

A company that develops its own software for sale to customers should consider whether it will designate that software as open source. In other words, a company should evaluate whether to make its software freely available to customers subject to the appropriate open source license. This contrasts with the company maintaining ownership and control over its software and granting licenses to its customers on terms that it sets.

The trade-off between maintaining software as proprietary versus releasing it as open source is clear: keeping the software proprietary allows a company to preserve the secrecy of its code. Customers who receive and use the software will generally be unable to determine the inner workings of the distributed software, and the company could include license terms that prohibit any type of reverse engineering of the software. In this scenario, the company's competitors will likely be unable to ascertain how the software works. This may be the proper approach when a company decides that it is imperative to protect its technology to the greatest extent possible.

On the other hand, by designating its software as open source, a company can potentially obtain a much larger base of users and customers, many of whom are drawn to the software initially because it would be freely available. Many of the users might exploit the open source nature by creating enhancements and other extensions to the software that could make it more attractive to use. Indeed, a community of user-developers could emerge from the customer base that propels the software toward wider adoption. An important point to keep in mind, however, is that a decision to designate software as open source should be accompanied by a plan to get revenue through alternative means. A different business model is needed because the distribution of the software will no longer generate the expected revenue. This is discussed in the following text.

76.3.2.2 Open Source Business Models

A business model for open source software needs to address how to generate revenue from software that is, by its very nature, distributed freely. Accordingly, business models must incorporate some value added aspect to justify a cost.

Providing proprietary software plugins to open source software is a common way to add value. Companies can offer these plugins, along with subscriptions for training and technical support, for a fee. Another alternative is to offer a stripped-down open source version of a software package for free and at the same time offer an enhanced commercial version for a fee. The theory behind this approach is that customers who are attracted initially to the free version would later want enhanced functionality and would purchase the commercial version.

A key to a successful business model is to identify a service relating to, or other proprietary element of, the product that end users consider sufficiently desirable or valuable such that they are willing to pay a reasonable fee to have access to it. Coming up with that service or other proprietary element is a function of the seller's creativity and knowledge of the marketplace.

76.4 Open Issues

76.4.1 What Constitutes Distribution in the Cloud Computing Environment?

Many websites, and providers of SaaS (where the software is centrally hosted by the provider or in the cloud and not distributed to the customer), make extensive use of open source software. This includes software licensed under the GPL in combination with their own proprietary software. In many cases, if copies of these combinations of open source and proprietary software solutions were distributed to end users as opposed to their functionality being made available under a SaaS model, then under the terms of the GPL or other similar restrictive open source licenses, the party distributing such software would have been required to make available the source code to both the proprietary software and the open source software (including any modifications). However, because the GPL-licensed open source software is hosted by the SaaS provider or website provider (on their own servers or in the cloud) and actual copies are not distributed to customers, the SaaS providers and websites are not required to make available the source code to their own proprietary software or to their modifications to the GPL-licensed open source software (unless Affero GPL-licensed software is used as discussed in the next paragraph).*

The use of GPL-licensed open source software in this manner, without being required to disclose the source code of the modified open source software (and possibly the proprietary software too), was viewed by some to be an unintended or undesirable "loophole" in the GPL, the original versions of which were drafted before the wide adoption of SaaS and cloud computing (though not prior to previous incarnations of such business model, such as application service providers or ASPs). Therefore, in March 2002, the Free Software Foundation (the author of the GPL) published an Affero General Public License ("AGPL"), version 1, as an alternative to the then GPL, version 2. Subsequently, it published the AGPL, version 3, in November 2007, shortly after the publication of GPL, version 3. The AGPL redefines what a distribution is—with express wording that states that interaction over a computer network amounts to a distribution. The use of AGPL-licensed software with proprietary software is generally not compatible with a traditional SaaS model and will likely trigger an obligation to disclose source code. To date, the AGPL has had limited adoption in comparison to the GPL.

76.5 Summary

Open source software is a great resource and option for many companies desiring to supplement and extend the capabilities of their existing software and/or hardware products, allowing companies to focus internal development work on core proprietary functionality. However, as with any third party software that a company desires to embed or include with its own proprietary software, the applicable license terms must first be understood and carefully considered by the company.

The wide availability of open source code presents both great opportunities and risks to software developers. Companies should anticipate that programmers will have access to open source code and be proactive in creating and, more importantly, communicating, and implementing an open source policy so that decisions regarding use of open source software can be quickly made during software development instead of after the fact.

The following is a list of selected open source licenses by category—Permissive, Moderately Restrictive, and Restrictive—as described in Sections 76.2.2.1 through 76.2.2.3:

"Permissive" open source software licenses

- MIT license
- Berkeley Software Distribution (BSD) license
- Apache license

"Moderately restrictive" open source software licenses

- Common Development and Distribution license (CDDL)
- Common Public License (CPL) & Eclipse Public License (EPL)
- Mozilla Public License (MPL)
- Lesser General Public License (LGPL)

"Restrictive" open source software licenses

- General Public License (GPL)
- Affero General Public License (AGPL)

* GPL, version 3, states that mere interaction with a user through a computer network, with no transfer of a copy, is not a distribution.

Key Terms

- "Copyright" is a property right in an original work of authorship fixed in a tangible medium of expression that gives the copyright holder the exclusive right to reproduce, adapt, distribute, perform, and display the work. Black's Law Dictionary 337 (7th edn., 1999). In the United States, copyrights are registered through the US Copyright Office (www.copyright.gov) and can be enforced through suit in the federal courts.
- A "license" is a permission to commit some act (e.g., reproducing, adapting, or distributing software code) that would be otherwise unlawful (e.g., due to a valid copyright). Black's Law Dictionary 931 (7th edn., 1999).
- A "patent" is the exclusive right granted by a government to exclude others from taking certain actions such as making, using, or selling an invention. Patents differ from copyrights in several respects including the potential to protect the basic idea of the invention (as defined the claims of the patent), whereas copyright protection is limited to the particular expression of an idea (e.g., the particular source code used by the programmer). In the United States, patents are granted after examination by the US Patent and Trademark Office (www.uspto.gov) and are enforceable through suit in the federal courts.

Further Information

International free and open source software law review (www.ifosslr.org).
The international free and open source software law book (www.ifosslawbook.org).
Meeker, H.J. *The Open Source Alternative: Understanding Risks and Leveraging Opportunities* (2008).
Open source initiative (www.opensource.org).
Open source resource center (www.osrc.blackducksoftware.com).

X

Software Quality and Measurement

77

Evidence-Informed Software Engineering and the Systematic Literature Review

David Budgen
Durham University

Pearl Brereton
Keele University

77.1 Introduction

In this chapter, we first introduce the evidence-based paradigm, discuss its origins and use in a variety of disciplines, and consider how it can be used in software engineering. We then describe how a secondary study in the form of a systematic literature review (SLR) can be performed in software engineering. After that, we examine the profile of existing secondary studies and what they can tell us, as well as discussing some "gaps" that need to be addressed. Finally, we consider how the use of evidence-based studies might form the basis for *evidence-informed* software engineering practices.

77.1.1 Evidence-Based Paradigm

For centuries, the "traditional" view of *knowledge* in the sciences has been that this should be based on systematic observation and that such experiments should be *repeatable* by other scientists. Observations have led scientists to theorize and produce "models" (such as of gravitational attraction) that have then been tested against further observation. For sciences such as physics where it is usually possible to separate the observer from the experiment, as well as to isolate the key factors, this approach has been highly successful in developing understanding.

Where human beings become *participants* in the experimental situation, rather than simply observers, then issues of repeatability become both more difficult and also more important. Sometimes, the results have been powerfully evident (such as James Lind's demonstration that vitamin C could prevent scurvy or the work of Louis Pasteur), at least to scientists. However, for many studies, the effects have been less pronounced, and because they could be masked by other factors, experiments using humans have not always been repeatable, to the extent that attempts at replication have sometimes produced conflicting results.

Until the last few decades, it was not unheard of for different groups of medical practitioners to advocate quite different treatments for a given problem, each appealing to empirical studies that supported that particular view. In particular, in reviewing existing studies an expert would be likely to "cherry-pick" the data by including only those studies that agreed with their ideas. It was this situation that spurred the noted epidemiologist Archie Cochrane to issue a challenge to clinical medicine that treatment should be based upon the best available evidence—and by implication, how was that evidence to be determined?

The approach that emerged has become known as *evidence-based medicine* (EBM) and with it the use of *secondary studies*, organized largely through the not-for-profit *Cochrane Collaboration*.* A secondary study does not involve conducting any form of "experiment"; instead, it aims to aggregate the outcomes of many primary studies in an *objective* and *unbiased* manner. The process by which this is performed is known as a *systematic review*, although in software engineering we generally use the term *systematic literature review* (SLR) to distinguish it from inspection-style code reviews.

77.1.2 EBM to EBSE

We do not have space here to explore the impact of EBM (an informal but penetrating analysis is provided in [14]). However, the concept rapidly spread from clinical medicine to various other branches of healthcare, through to education and social science. En route there were some adaptations, in particular for the processes used to synthesize the outcomes from different studies. Most clinical studies, where the participants are *recipients* of a treatment, are organized as *randomized controlled trials* (RCTs), and the outcomes from these can be combined using powerful statistical techniques such as meta-analysis, enabling the secondary study to resolve potentially small effects with greater certainty. Education and other branches of healthcare have to deal with more complex forms of treatment and interaction of these with the participants and so have to use other forms for aggregating the outcomes of their studies.

The idea of employing this paradigm in software engineering was advocated by Kitchenham, Dybå, and Jørgensen in 2004 [26]. They suggested that with some adaptation to fit the nature of software engineering studies, this could be a valuable way of consolidating our empirical knowledge about what works, when, and where. From that point, *evidence-based software engineering* (EBSE) has evolved strongly, to the point where by the end of 2011, we could identify around 150 published secondary studies. We will discuss some of these later, but first we examine exactly how an SLR is performed.

77.2 Performing a Systematic Literature Review

We first examine some different reasons for conducting systematic reviews and then the form that an SLR takes in software engineering.

77.2.1 Why Conduct an SLR?

As indicated previously, we are likely to be motivated to undertake an SLR on a software engineering topic if we want to draw together the available knowledge about some phenomenon in some way. This might be to answer a specific research question about a technique, to identify the trends in research on a particular topic, or to find out what primary studies are available that investigate some particular model or technique.

* http://www.cochrane.org

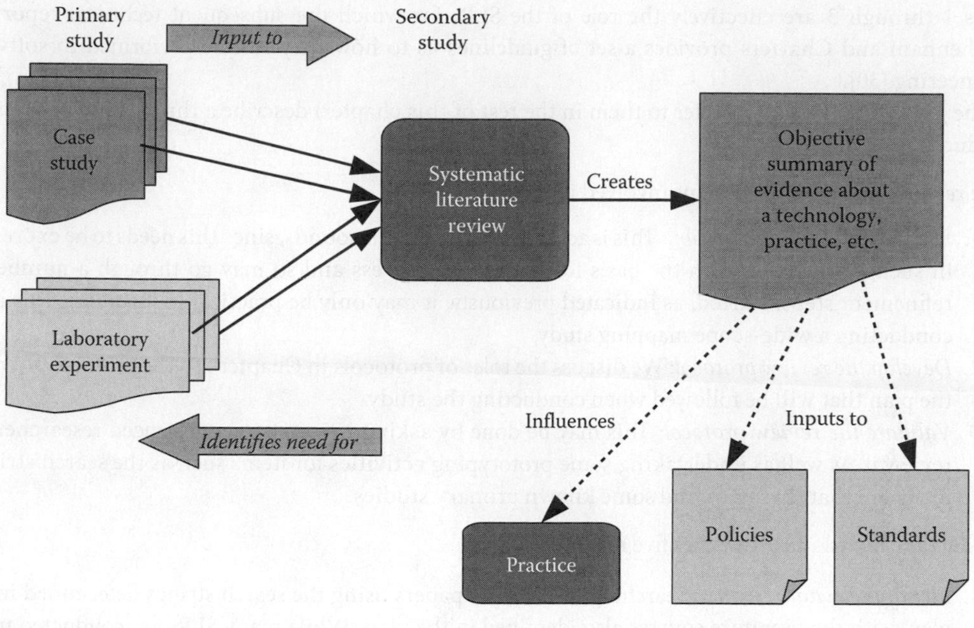

FIGURE 77.1 The role of secondary studies.

We usually term the last of these a *mapping study*, equating this to the *scoping review* described by Petticrew and Roberts [34]. However, we might usefully note that Booth et al. [5] do distinguish between these, characterizing a scoping review as one that identifies *what* primary studies are available, while a mapping study also *categorizes* them. Mapping studies can identify if there is scope to perform a fuller SLR or where new primary studies would be particularly useful [32].

The evidence-based paradigm, through the medium of the secondary study, therefore plays quite an important role in empirical research, as illustrated in Figure 77.1. There is the interaction with primary studies, by which a secondary study both draws upon the data from these and also identifies where new ones are needed (and how reporting standards can be improved). There is also interaction with practice and research, by providing recommendations for practice, for company policies, and for standards bodies. Finally, by identifying research trends (and dead ends), secondary studies may help with identifying where funding for new research might best be directed. Note too that for software engineering, we are potentially interested in a range of forms for primary studies (for simplicity, we have identified two here)—where there is a shortage of useful experiments, we might triangulate their findings with those from other forms of study such as case studies and observational studies [40].

77.2.2 Conducting and Reporting an SLR

In describing how evidence-based practice could be adapted for software engineering, Kitchenham et al. identified the following five steps:

1. Convert the need for information into an answerable question.
2. Find the best evidence with which to answer the question.
3. Critically appraise the evidence for its validity (how close it comes to the truth), its impact (the "size" of the effects observed), and its applicability (how useful it is likely to be).
4. Integrate the critical appraisal with software engineering expertise and stakeholders' values.
5. Evaluate the effectiveness and efficiency in the previous steps 1 through 4, and seek ways to improve them.

Steps 1 through 3 are effectively the role of the SLR, for which the subsequent technical report by Kitchenham and Charters provides a set of guidelines as to how they can be performed in software engineering [30].

The *guidelines* (as we will refer to them in the rest of this chapter) describe a three-phase process for conducting a review.

Plan review. There are three steps involved in this:

1. *Specify the research question*: This is actually harder than it sounds, since this needs to be expressed in such a way as to form the basis for the search process and so may go through a number of refinement steps. Indeed, as indicated previously, it may only be practical to formulate this after conducting a wide-scope mapping study.
2. *Develop the review protocol*: We discuss the roles of protocols in Chapter 79. The protocol provides the plan that will be followed when conducting the study.
3. *Validate the review protocol*: This may be done by asking one or two experienced researchers to review it, as well as undertaking some prototyping activities for items such as the search strings, to ensure that these can find some known primary studies.

Conduct the review. This involves five major activities:

1. *Identify relevant research*: Searching for relevant papers using the search strings determined in the plan, with the literature sources also identified in the plan. While many SLRs are conducted using electronic searches of digital libraries, there are arguments for conducting manual searches [22]. These may be particularly appropriate where it has proved difficult to determine a sound set of search strings, perhaps because of wide variation in the vocabulary being used by researchers.
2. *Select primary studies*: This is quite a labor-intensive task that involves filtering the returns from the search for relevance, and it is common to perform this using two analysts working independently and then reporting and resolving differences, in order to help improve reliability. Common practice is to perform an initial sift based on the titles of the papers found, then on the abstracts, and finally on the full text of the remaining papers. Decisions are based upon the *inclusion/exclusion* criteria that form a key element in the protocol. The SLR on agile methods by Dybå and Dingsøyr provides an excellent description of this process [12].
3. *Assess study quality*: This involves assessing how well the primary study was conducted and reported using a "quality instrument" such as a questionnaire. This is usually conducted with varying degrees of rigor, depending on need—mapping studies often omit this step.
4. *Extract required data*: One of the elements of the research protocol should be a *data extraction form* identifying what should be extracted from a primary study and, where appropriate, how this should be coded or parameterized. As with selection, it is normal to use two analysts for this task, comparing results and resolving differences. This is because primary studies do use very varied reporting styles—and since EBSE is a relatively recent development, few authors of primary studies organize their supporting data with the expectation that anyone will want to extract details from it, although there are indications that this is beginning to change. Again, for a mapping study, this will usually be a rather simpler task, depending chiefly upon how the studies are to be categorized.
5. *Synthesize the data*: This is a potentially challenging task. While in practice there is a range of possible forms for this, software engineers have so far tended to be rather conservative and limit themselves to tabulation and vote counting. Some useful guidance is available from [9].

Document the review. The writing of the methodological element of the final report can be largely based upon the research protocol. However, describing the outcomes, including any recommendations based upon the evidence, should also follow a systematic process.

As a final note on this, we should observe that an SLR might extend an earlier SLR. This may involve repeating the activities described in the protocol for a subsequent period and then synthesizing the augmented set of primary studies. Or it could involve using extended search strings to widen the scope of the original study. An SLR does not necessarily have to begin with an empty sheet.

77.2.3 Realization

We have described some of our early experiences with performing SLRs in [6], and many of the observations made there are still relevant, with many later SLRs identifying very similar issues. So, what challenges do our environment and culture provide?

Specifying the *research question* is a significant issue. This forms the basis for the searching activity and, in particular, for determining the search strings to use. There should be a clear link between research question(s) and search strings, and an iterative process of refinement of these through prototyping of the search process may well be necessary.

Linked to this is the need to write a clear and full *protocol* for the study. SLRs usually involve a team of researchers, not least because we need two analysts for many elements. It is absolutely essential that everyone involved has a shared understanding of the goals of the study and of the criteria that are to be used for the various activities. Developing and discussing the protocol is an essential step in this process.

Another major issue is that of conducting the *search* for papers. Although various search engines are available, most notably the digital libraries from IEEE and ACM, as well as those such as Web of Science and Science Direct, together with indexing services such as Scopus, none have interfaces that are really optimized for our task. In addition, they tend to be organized differently, so that having identified a specific search string, the systematic reviewer then needs to interpret this for the forms used by each search engine.

An additional aspect to note about the searching activity is that electronic searches are rarely able to find all papers. However good our search strings, authors do not always use titles, or even keywords, that are "standard," and so some relevant papers may well get missed. A technique commonly used to alleviate that is termed *snowballing.* This consists of going through the references of the studies found and selected from electronic searches, to identify any potentially relevant papers that they cite. The rule of thumb for this is that snowballing may well find up to around 10% of the final set of papers.

The task of *inclusion/exclusion* likewise offers many challenges, especially for the less experienced researcher. (We should note here that while our experiences suggest that mapping studies often provide a useful starting point for a research student [32], conducting a full SLR is a much more demanding exercise.) Ideally, performing as much as possible of this activity using the title and abstract of papers is far less time-consuming than reading full papers. However, the quality of abstracts in software engineering is often poor (we should observe that we are firm advocates of the use of *structured abstracts* as a means of alleviating this problem [8]), and so decisions have to be deferred. It is not uncommon to begin with 500–2000 "hits" from the searching process, so the scale of this task is also quite substantial.

Both *data extraction* and *synthesis* provide some domain-specific issues, and again it is essential to appreciate that some experience of the domain is really necessary. Indeed, this is one reason why we advocate the use of mapping studies rather than SLRs as a suitable form for postgraduate study.

Fortunately, although all of these are challenging, they can be overcome, and in the following sections, we examine some of the resulting outcomes.

77.3 What SLRs Do We Have?

In this rest of this chapter, we look at the activities that have been undertaken to catalogue the SLRs that address topics within the software engineering domain. We show the growth in numbers over the years since 2004 and summarize the topics that form the focus of studies to date. There are some indications that the cataloguing process has not resulted in the identification and indexing of all of the published studies in software engineering. We discuss the issue of completeness later.

77.3.1 Cataloguing SLRs

As indicated in Section 77.2, the SLR process can be used to map out the research in a particular area of interest. As well as mapping the research undertaken by primary studies, a mapping study can alternatively focus on secondary studies and be used to map out the SLRs undertaken across the whole software engineering domain (or across subareas of the domain). Such *tertiary studies* can provide a thorough overview of the breadth and quality of SLRs and form valuable resources for a range of stakeholders (such as educators, researchers, and practitioners). To date, three broad tertiary studies in software engineering have been published, which we refer to in the rest of this section, as TS1, TS2, and TS3, respectively [10,29,31]. Tertiary studies can also have a narrower focus. For example, a tertiary study of SLRs relating to global software development (GSD) has also been undertaken [39].

TS1, TS2, and TS3 addressed research questions across the broad software engineering domain in order to

- Identify SLR activities over a specific period
- Assess the quality of those studies located
- Identify the individuals and organizations most active in SLR- based research
- Determine the topics covered by the SLRs
- Map the topics covered to the Software Engineering 2004 Curriculum Guidelines for Undergraduate Degree Programs* and the Software Engineering Body of Knowledge (SWEBOK)† (TS2 and TS3 only)

The initial broad tertiary study, TS1, employed a manual search of 10 journals and 4 conference proceedings published between January 2004 and June 2007 (inclusive). From a total of 2506 papers, a set of 18 unique, relevant studies were selected. A further two studies were identified by asking researchers about current work [4] and by searching the Simula Research Laboratory website [21].

The second broad tertiary study, TS2, covered the period between January 2004 and June 2008 (inclusive) and used an automated search of four digital libraries and two indexing services: IEEE Computer Society Digital Library, ACM digital library, CiteSeer, SpringerLink, Web of Science, and Scopus. TS2 identified a further 14 studies in the period covered by TS1 and 19 studies published between July 2007 and June 2008 (inclusive).

The third broad tertiary study, TS3, used, as closely as possible, the protocol that was used for TS2 although it did incorporate manual as well as automated searches. TS3 covered the period July 2008 to December 2009 (inclusive) and identified a total of 67 unique relevant SLRs.

The more focused tertiary study of SLRs in GSD addressed a similar set of questions to those addressed by the three broad tertiary studies although the SLRs located were mapped against Abdullah and Verner's outsourcing risk framework [1] and the ISO 12207 framework [37] rather than the SE curriculum and the SWEBOK [39]. It used an automated search of seven digital libraries/indexing services, a manual search of selected conference proceedings plus snowballing. The study identified 24 SLRs addressing topics relating to GSD and published between 2005 and October 2011. Of these, three were published in the period covered by the broad tertiary studies but were not found by them, and 11 were published outside of the period covered by the broad tertiary studies (i.e., after December 2009).

77.3.2 Results from the Broad Tertiary Studies

Together, the three broad tertiary studies reported 120 SLRs on software engineering topics. The distribution of these reviews across the years from 2004 to 2009 is shown in Figure 77.2 that clearly indicates a growing number of studies through this period.

* http://sites.computer.org/ccse/SE2004Volume.pdf
† http://www.computer.org/portal/web/swebok

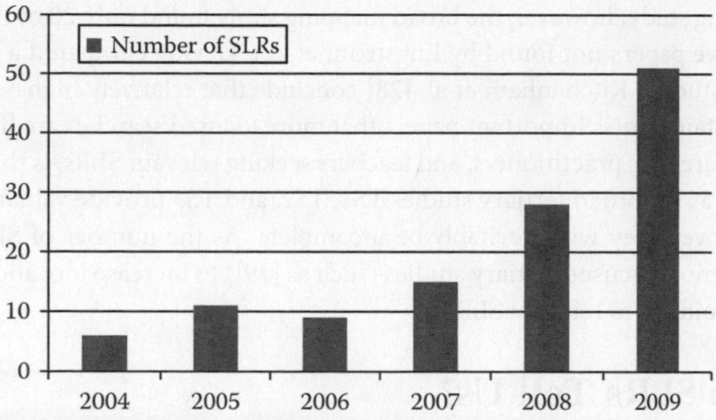

FIGURE 77.2 SLRs published by year. (Data is taken from Table 4 in F.Q.B. da Silva, A.L.M. Santos, S. Soares, A.C′.C. FranÖca, C.V.F. Monteiro, and F.F. Maciel. Six years of systematic literature reviews in software engineering: An updated tertiary study. *Information and Software Technology*, 53(9):899–913, 2011.)

TABLE 77.1 SLRs Mapped to SWEBOK Chapters

SWEBOK Chapter	Number of Studies Reported in TS1, TS2, and TS3
Software requirements	7
Software design	11
Software construction	2
Software testing	6
Software maintenance	6
Software configuration management	0
Software engineering management	14
Software engineering process	9
Software engineering tools and methods	2
Software quality	0

Source: Data is taken from Table 6 in F.Q.B. da Silva, A.L.M. Santos, S. Soares, A.C′.C. FranÖca, C.V.F. Monteiro, and F.F. Maciel. Six years of systematic literature reviews in software engineering: An updated tertiary study. *Information and Software Technology*, 53(9):899–913, 2011.

The topics covered by the studies are shown in Table 77.1, which maps the studies to the SWEBOK chapters.

In TS3, da Silva et al. [10] also reviewed the variation in quality of the SLRs over time and found a steady increase in the mean quality scores in every year except 2007. Over the 6 year period between 2004 and 2009, the increase in quality was 12.5%.

77.3.3 SLRs Not Indexed by the Broad Tertiary Reviews

Not surprisingly, given the year on year increase in published SLRs in software engineering, a substantial number of studies have been published since December 2009. We are not aware of any exhaustive systematic cataloguing of these; however, a manual search of the index pages of five major software engineering journals identified 24 SLRs between January 2010 and mid-2011 [7].

As noted at the beginning of this section, there are some indications that broad mapping studies may not be complete [28]. A study was undertaken to assess the completeness and reliability of a mapping study of research to compare the effectiveness of approaches to unit testing and regression testing [28]. The results indicate that the set of primary studies found by the search undertaken for the (broad) mapping study were less complete than other more focused studies. For example, Ali et al. [2,3] carried out an SLR on search-based test case generation and found 31 papers for the period covered by the mapping study. The mapping study identified only 4 of these papers, although it did find a further 6 papers that were missed by Ali et al. Similarly, in a focused SLR on regression testing, Engström et al. [13] found 24 papers for the period covered by the mapping study; however, the broad mapping study found only 20 of these papers (although it did find a further five papers not found by Engström et al.). Having compared a broad mapping study with 6 more focused studies, Kitchenham et al. [28] conclude that relatively high-level automated search strings will almost certainly miss important papers that more focused searches are likely to find. The message, therefore, to researchers, practitioners, and teachers seeking relevant SLRs, is that although high-level mapping studies such as the broad tertiary studies TS1, TS2, and TS3 provide valuable catalogues of SLRs in the periods they cover, they will inevitably be incomplete. As the number of SLRs increases, we can expect the number of more focused tertiary studies (such as [39]) to increase too, and these will provide an additional source of pointers to relevant SLRs.

77.4 What Do SLRs Tell Us?

In this section, we look at what some of the key stakeholders can learn from existing SLRs. We look particularly at studies that make recommendations to software engineering educators and to practitioners.

77.4.1 Informing Educators

In a study to examine where the outcomes of SLRs can provide support for and inform SE teaching activities, 43 out of 143 candidate SLRs were classified as providing potentially useful teaching material [7]. The study classified the outcomes according to the body of knowledge that is appropriate for an undergraduate program in software engineering (SE2004). The knowledge is designated as the Software Engineering Education Knowledge (SEEK). The study is able to demonstrate that material is available to support all of the major knowledge areas (KAs) within the SEEK.

For the *professional practice* KA, there is evidence about the major challenges facing teams in the collaborative creation of graphical models of systems and also about why IT professionals change their jobs.

In the software *modeling and analysis* KA, evidence relates to the following topics:

- The use of cross-company and within-company data to build and apply management models
- Model-driven software engineering
- The technology acceptance model as a predictor of actual use of a technology
- The use of tools in domain analysis
- Requirements elicitation
- Use of SE models for creating textual requirements specifications

In the *software design* KA, only one study was identified as providing useful material for educators. This provides material about the strengths and weaknesses of aspect-oriented programming compared to other approaches.

Within the software *verification and validation* area, evidence was found across three of the subareas: review, testing, and problem analysis and reporting. Most of the evidence relates to testing and covers the following:

- The selection of unit testing techniques
- Automated acceptance testing
- Use of meta-heuristic search techniques to generate software tests
- Regression test selection

Two of the studies identified covered software evolution activities. These look at whether duplication of code affects system characteristics and how changes at the architecture level map to lower-level system characteristics.

Five studies were found that provide useful material relating to process implementations (falling within the *software process* KA). They focus on the Rational Unified Process, software reuse, pair programming, the use of scrum in GSD, and the challenges facing distributed development teams.

In the *software quality* KA, six studies provide material on topics relating to the following:

- The values of and motivation for adopting CMM for software process improvement (SPI)
- The use of SPI in small organizations
- Analysis of the causes of defects in code to aid product-based process improvement
- Use of coupling metrics as predictors of maintainability for aspect-oriented programming
- Techniques and models for predicting the maintainability of software

The largest proportion of studies (13 out of 43) provides material on topics within the *software management* KA. These cover project planning, project personnel, and organization and project control. For project planning, material was strongly focused on cost estimation and also covered project duration and quality, as well as risks and best practice in global/distributed software projects. For the other subtopics, studies provide material on what motivates and de-motivates developers, the reasons why software productivity varies across different contexts, and the effectiveness of different approaches to measuring and predicting productivity.

77.4.2 Recommendations for Practitioners

Although many of the results from SLRs will be relevant to software developers, projects managers, and other IT decision makers, many reviewers do not make specific recommendations for practitioners. The three broad tertiary studies do indicate which SLRs are or may be useful for practitioners or which specifically include practitioner guidelines. Based on the SLRs identified by the broad tertiary studies (TS1, TS2, and TS3) as potentially providing advice to practitioners, we summarize the recommendations and provide references to publications where more detailed information is available. We note, also, that many other SLRs may provide sufficient evidence from which recommendations for practitioners could be formulated.

From the 33 SLRs identified by TS1, TS2, and TS3 as potentially providing recommendations for practitioners, we were able to extract specific recommendations from 15 studies. The topics covered by these SLRs are shown in Table 77.2.

77.4.2.1 Cost Estimation

Two papers included in TS1 [15,21] use papers selected from a repository that was initially compiled by a mapping study of cost estimation studies [22]. In [15], the following guidelines for practitioners, researchers, and authors of textbooks are provided:

1. Do not mix estimation of most likely effort with planning, budgeting, or pricing.
2. When assessing estimation accuracy, make sure that the estimated and the actual effort are comparable.

Grimstad et al. [15] discuss the implications of these guidelines and indicate that they are based on their own experience as well as the textbook and papers included in the review.

TABLE 77.2 Topics Covered by SLRs That Provide Recommendations for Practitioners

Topic	No. of SLRs
Cost estimation	4
Distributed software development	4
Requirements engineering and elicitation	2
Agile development	2
Software process improvement	2
Testing	1

Jørgensen [21], who compares the use of expert judgement and formal models, provides the following advice to practitioners:

"The use of models, either alone or in combination with expert judgement, may be particularly useful when:

1. there are situational biases that are believed to lead to a strong bias towards over optimism;
2. the amount of contextual information possessed by the experts is low; and
3. the models are calibrated to the organization using them."

Overall, the study concludes that models fail to systematically perform better than the experts when estimating the effort required to complete software development tasks. This is particularly the case when "models are not calibrated to the organization using them" and "experts possess important contextual information not included in the formal models and apply it efficiently".

In an earlier study, Jørgensen [19] presents an evaluation of 12 best practice guidelines for expert cost estimation. He finds that "there is evidence supporting all these principles and, consequently, that software organizations should apply them."

In another study, cost predictions from cross-company models are compared with predictions from within-company models [27]. Ten papers were included in the study and of these, seven present independent results, three of which found no significant difference, and four found cross-company models were significantly worse than within-company models Companies are advised that they should consider how similar project data in the cross-company data set are to the projects undertaken within their own company and should consider the characteristics of their own company.

The focus of a further study on cost estimation is on how to improve the assessment of cost uncertainty of software projects [20]. The reviewer identifies a set of seven guidelines relating to cost uncertainly. Those considered as having the greatest level of supporting evidence are the following:

Do not apply solely unaided, intuition-based processes.
Apply structured and explicit judgement-based processes.

77.4.2.2 Distributed Software Development

The four studies on this topic focus on outsourcing [24,25], the use of Scrum [17], and mechanisms for improving distributed software development (DSD) processes [18].

The studies by Khan et al. [24,25] provide a number of suggestions about the following:

- The problems that outsourcing vendor organizations need to address in order to improve their capabilities
- The factors that influence outsourcing clients in their selection of vendor organizations

The study by Hossain et al. [17] focuses on identifying the challenges of using Scrum in GSD and the strategies used for dealing with those challenges. Hossain et al. provide a set of recommendations but also state that it is difficult to offer specific advice to practitioners based solely on their review and indicate that "the strength of evidence found in the literature about the identified strategies is very low."

The remaining study relating to DSD aimed to identify procedures, models, and strategies for improving DSD processes [18]. Advice is presented as a set of success factors. The reviewers indicate that the evaluation of the results obtained from the proposed improvements is often based on studies in a single organization and sometimes only take into account the developers' subjective perception. Supporting evidence for the recommendations is therefore limited.

77.4.2.3 Requirements Engineering and Elicitation

Guidelines for practitioners on techniques for requirements elicitation are provided by the study reported in [11]. The reviewers include 26 publications reporting 30 empirical studies that compare elicitation techniques. Studies are aggregated according to the techniques that are compared. This results in a set of 35 'aggregations', of the form *techniques A are better than techniques B or techniques A are similar to techniques B*, 17 of which are supported by several pieces of evidence. These 17 'conclusive' aggregations can be considered as guidelines or advice for practitioners. The aggregations supported by more than one study and having no studies that contradict them are (taken from Table 8 in [11]):

- Structured interviews are better than unstructured interviews for eliciting customer needs
- Unstructured interviews are better than sorting techniques for eliciting information
- Hierarchical structuring techniques are better than sorting techniques for eliciting information
- Unstructured interviews are better than sorting technique regarding elicitation time
- Sorting techniques are better than hierarchical structuring techniques regarding total time
- Unstructured interviews are better than introspection and observation regarding completeness
- Introspection and observation are worse than sorting techniques regarding completeness
- Introspection and observation are worse than hierarchical structuring techniques regarding completeness

The reviewers point out that the evidence relates specifically to experienced elicitors and respondents and that for less experienced respondents a number of the techniques are equally effective. They also present a set of more complex guidelines developed through a process of combining the conclusive aggregations.

Also relating to requirements, the study by Nicolás and Toval [33] is concerned with the generation of textual requirements specifications from models. The advice given is embodied in a proposal for a general purpose tool that integrates graphical and textual models of a system.

77.4.2.4 Agile Development

Two studies addressing agile methods provide recommendations for practitioners. One takes a broad perspective with the aims of identifying what is currently known about the benefits and limitations of agile software development and determining the implications of the studies found for the software industry and the research communities [12]; the other focuses more narrowly on pair programming [16].

Dybå & Dingsøyr [12] are concerned with the benefits and limitations of agile methods and discuss the implications of the results of their review for both research and practice. The strongest evidence for practice is the need to focus on human and social factors and in particular on balancing individual autonomy with team autonomy and corporate responsibility. They also highlight the importance of practitioner confidence, good interpersonal skills and trust. Evidence suggests that traditional project management principles, such as state-gate project management models, should be combined with agile project management and that agile methods are not necessarily the best choice for large projects. The reviewers suggest that practitioners "carefully study their projects' characteristics and compare them with the relevant agile methods' required characteristics". However, the reviewers consider that the strength of evidence is very low, making it difficult to offer specific advice to industry. Consequently, they advise practitioners to use the article as a map, to find and investigate relevant studies further and compare the settings in the studies to their own situation.

The study by Hannay et al. [16] offers the following advice about pair programming:

> If you do not know the seniority or skill levels of your programmers, but do have a feeling for task complexity, then employ pair programming either when task complexity is low and time is of the essence, or when task complexity is high and correctness is important.

However, the reviewers note that pair programming is not universally beneficial or effective, inter-study variance is high, and publication bias may be an issue.

77.4.2.5 Software Process Improvement

The two studies on software process improvement (SPI) look at the approaches to SPI that have been applied to small- and medium-sized enterprises [35] and at establishing the state of the art in defect causal analysis (which is a means of product focused SPI) [23]. Pino et al [35] extracted practical advice (success factors) from the primary studies that were included in their review. The factors were analysed and synthesised to provide a set of 11 guidelines however there is no indication of the strength of evidence relating to each guideline.

Kalinowski et al. [23] provide guidance to practitioners in the form of answers to the following seven questions:

1. Is my organization ready for defect causal analysis?
2. What approach should be followed?
3. What metrics should be collected?
4. How should defect causal analysis be integrated with statistical process control?
5. How should defects be categorized?
6. How should cause be categorized?
7. What are the expected costs and results of implementing defect causal analysis?

77.4.2.6 Testing

The study on testing focuses on defect detection, in particular looking at studies that compare testing and inspection techniques [36]. The authors suggest that there are no clear-cut answers about which defect detection method to choose; however, results suggest the following:

- "For requirements defects, no empirical evidence exists at all, but the fact that costs for requirements inspections are low compared to implementing incorrect requirements indicates that reviewers should look for requirements defects through inspection
- For design specification defects, the case studies and one experiment indicate that inspections are both more efficient and more effective than functional testing
- For code, functional or structural testing is ranked more effective or efficient than inspection in most studies. Some studies conclude that testing and inspection find different kinds of defects, so they're complementary. Results differ when studying fault isolation and not just defect detection
- Verification's effectiveness is low; reviewers find only 25%–50% of an artifact's defects using inspection, and testers find 30%–60% using testing. This makes secondary defect detection important. The efficiency is in the magnitude of 1–2.5 defects per hour spent on inspection or testing."

77.5 Conclusion

As we have demonstrated, research based upon EBSE is beginning to provide a better understanding of "what works, when, and where." however, we offer a final word of caution. Employing empirical evidence needs this to be interpreted within a specific context, an activity termed Knowledge Translation (KT), and the outcomes of SLRs are rarely definitive. Hence, for use in teaching and practice, we prefer to use the term *evidence-informed* [7], to emphasize this.

Acknowledgment

The authors would like to thank Prof. Barbara Kitchenham for many useful discussions on this topic.

References

1. L.M. Abdullah and J.M. Verner. Outsourced strategic IT systems development risk. In *IEEE RCIS*, Fes, Morocco, 2009.
2. S. Ali, L.C. Briand, H. Hemmati, and R.K. Panesar-Walawege. A systematic review of the application and empirical investigation of search-based test case generation. Technical report, Simula Technical Report SE 293, Norway, 2008.

3. S. Ali, L.C. Briand, H. Hemmati, and R.K. Panesar-Walawege. A systematic review of the application and empirical investigation of search-based test case generation. *IEEE Transactions on Software Engineering*, 36(6):742–762, 2010.
4. R.F. Barcelos and G.H. Travassos. Evaluation approaches for software architectural documents: A systematic review. In *Proceedings of Ibero-American Workshop on Requirements Engineering and Software Environments (IDEAS)*, La Plata, Argentina, 2006.
5. A. Booth, D. Papaioannou, and A. Sutton. *Systematic Approaches to a Successful Literature Review*. Los Angeles, CA: SAGE Publications Ltd., 2012.
6. O.P. Brereton, B.A. Kitchenham, D. Budgen, M. Turner, and M.A. Khalil. Lessons from applying the systematic literature review process within the software engineering domain. *Journal of Systems and Software*, 80(4):571–583, 2007.
7. D. Budgen, S. Drummond, P. Brereton, and N. Holland. What scope is there for adopting evidence-informed teaching in software engineering? Accepted for ICSE 2012 Education Track (Full Paper), 2012.
8. D. Budgen, B.A. Kitchenham, S. Charters, M. Turner, P. Brereton, and S. Linkman. Presenting software engineering results using structured abstracts: A randomised experiment. *Empirical Software Engineering*, 13(4):435–468, 2008.
9. D.S. Cruzes and T. Dybå. Research synthesis in software engineering: A tertiary study. *Information and Software Technology*, 53(5):440–455, 2011.
10. F.Q.B. da Silva, A.L.M. Santos, S. Soares, A.C.C. Franöca, C.V.F. Monteiro, and F.F. Maciel. Six years of systematic literature reviews in software engineering: An updated tertiary study. *Information and Software Technology*, 53(9):899–913, 2011.
11. O. Dieste and N. Juristo. Systematic review and aggregation of empirical studies on elicitation techniques. *IEEE Transactions on Software Engineering*, 37(2):283–304, 2011.
12. T. Dybå and T. Dingsøyr. Empirical studies of agile software development: A systematic review. *Information & Software Technology*, 50:833–859, 2008.
13. E. Engström, P. Runeson, and M. Skoglund. A systematic review on regression test selection techniques. *Information and Software Technology*, 52(1):14–30, 2010.
14. B. Goldacre. *Bad Science*. London, U.K.: Harper Perennial, 2009.
15. S. Grimstad, M. Jørgensen, and K. Moløkken-Ostvold. Software effort estimation terminology: The tower of babel. *Information and Software Technology*, 48(4):302–310, 2006.
16. J.E. Hannay, T. Dybå, E. Arisholm, and D.I.K. Sjøberg. The effectiveness of pair programming: A meta analysis. *Information and Software Technology*, 51(7):1110–1122, 2009.
17. E. Hossain, M. A. Babar, and H. young Paik. Using scrum in global software development: A systematic literature review. In *Proceedings 4th International Conference on Global Software Engineering (ICGSE)*, Los Alamitos, CA, pp. 175–184, 2009.
18. M. Jiménez, M. Piattini, and A. Vizcaíno. Challenges and improvements in distributed software development: A systematic review. In *Advances in Software Engineering*, New York, pp. 1–15, 2009.
19. M. Jørgensen. A review of studies on expert estimation of software development effort. *Journal of Systems and Software*, 70(1–2):37–60, 2004.
20. M. Jørgensen. Evidence-based guidelines for assessment of software costs uncertainty. *IEEE Transactions on Software Engineering*, 31(11):942–954, 2005.
21. M. Jørgensen. Forecasting of software development work effort: Evidence on expert judgement and formal models. *International Journal of Forecasting*, 23(3):449–462, 2007.
22. M. Jørgensen and M. Shepperd. A systematic review of software development cost estimation studies. *IEEE Transaction on Software Engineering*, 33(1):33–53, 2007.
23. M. Kalinowski, G.H. Travassos, and D.N. Card. Towards a defect prevention based process improvement approach. In *34th Euromicro Conference on Software Engineering and Advanced Applications (SEAA)*, Parma, Italy, pp. 199–206, 2008.

24. S.U. Khan, M. Niazi, and R. Ahmad. Barriers in the selection of offshore software development outsourcing vendors: An exploratory study using a systematic literature review. *Information and Software Technology*, 53(7):693–706, 2011.
25. S.U. Khan, M. Niazi, and R. Ahmad. Factors influencing clients in the selection of offshore software outsourcing vendors: An exploratory study using a systematic literature review. *Journal of Systems and Software*, 84(4):686–699, 2011.
26. B.A. Kitchenham, T. Dybå, and M Jørgensen. Evidence-based software engineering. In *Proceedings of ICSE 2004*, Los Alamitos, CA: IEEE Computer Society Press, pp. 273–281, 2004.
27. B.A. Kitchenham, E. Mendes, and G. Travassos. A systematic review of cross vs within company cost estimation studies. In *Proceedings EASE 2006*, Keele, Staffordshire, UK, pp. 89–98. BCS, 2006.
28. B. Kitchenham, P. Brereton, and D. Budgen. Mapping study completeness and reliability—A case study. In *Proceedings EASE 2012*, Ciadad Real, Spain, pp. 126–135, 2012.
29. B. Kitchenham, P. Brereton, D. Budgen, M. Turner, J. Bailey, and S. Linkman. Systematic literature reviews in software engineering—A systematic literature review. *Information and Software Technology*, 51(1):7–15, 2009.
30. B. Kitchenham and S. Charters. Guidelines for performing systematic literature reviews in software engineering. Technical report EBSE 2007-001, Keele University and Durham University Joint Report, Keele University and University of Durham, UK, 2007.
31. B. Kitchenham, R. Pretorius, D. Budgen, P. Brereton, M. Turner, M. Niazi, and S. Linkman. Systematic literature reviews in software engineering—A tertiary study. *Information and Software Technology*, 52:792–805, 2010.
32. B.A. Kitchenham, D. Budgen, and O. Pearl Brereton. Using mapping studies as the basis for further research—A participant observer case study. *Information and Software Technology*, 53(4):638–651, 2011. Special section from EASE 2010.
33. J. Nicolás and A. Toval. On the generation of requirements specifications from software engineering models: A systematic literature review. *Information and Software Technology*, 51(9):1291–1307, 2009.
34. M. Petticrew and H. Roberts. *Systematic Reviews in the Social Sciences: A Practical Guide*. Malden, MA: Blackwell Publishing, 2006.
35. F.J. Pino, M. Garcia, and M. Piattini. Software process improvement in small and medium enterprises: A review. *Software Quality Journal*, 16:237–261, 2009.
36. P. Runeson, C. Andersson, T. Thelin, A. Andrews, and T. Berling. What do we know about defect detection methods? *IEEE Software*, 23(3):82–86, 2006.
37. R Singh. International Standard ISO/IEC 12207 software lifecycle processes. Technical report, International Organization for Standardization, Washington, DC, 1996.
38. J.M. Verner, O.P. Brereton, B.A. Kitchenham, M. Turner, and M. Niazi. Risk mitigation advice for global software development from systematic literature reviews. Submitted for publication.
39. J.M. Verner, O.P. Brereton, B.A. Kitchenham, M. Turner, and M. Niazi. Systematic literature reviews in global software engineering development: A tertiary study. In *Proceedings EASE 2012*, Ciadad Real, Spain, pp. 2–11, 2012.
40. C. Zhang and D. Budgen. What do we know about the effectiveness of software design patterns? *IEEE Transactions on Software Engineering*, 38(5):1213–1231, 2012.

78

Empirical Software Engineering

David Budgen
Durham University

Barbara Kitchenham
Keele University

78.1 Introduction and History

This chapter provides an overview of how the ideas and practices associated with empirical software engineering have evolved and also aims to supply some pointers toward the ideas that constitute current "best practice." In doing so, we hope to aid the "would-be" empiricist to identify the most suitable form of study for a particular need and to provide enough background for software engineers in general to appreciate the strengths and limitations of empirical knowledge and practice.

78.1.1 Early History

The act of "measurement" is a key element of any empirical study and actually goes back quite a long way in software engineering. Measurement seeks to provide quantitative values for a particular *attribute* of some entity, while a *metric* provides a set of rules for generating those values. Much of the early work in software metrics, stimulated by developments such as Halstead's *Software Science* [9] and McCabe's ideas about control flow through his metric of *cyclomatic complexity* [17], addressed questions of how to measure both the attributes of software itself (usually at the level of code) and also those of the development processes (where prediction was, and is, an important need).

As a result, early empirical work in software engineering, particularly in the period 1970–2000, was concerned with gathering data about attributes related to code and to development practices and studying the effectiveness of these. Research into the extent to which the *lines of code* (LoC) and *cyclomatic complexity* metrics are correlated provided early examples of empirical studies that looked at actual code artifacts, comparing measures with expert assessments where necessary. (The book by Fenton and Pfleeger provides an authoritative review of how ideas about metrics have evolved [6].)

One of the consequences of these activities was the development of a wider realization that the values of these attributes, particularly those related to code, arose at least in part from the processes that led to their creation—and that experimentation was one way of exploring the cause–effect relationships implied by this. This in turn led to attempts to assess the effectiveness of development practices (encompassing the whole spectrum of software development), so that by the mid-1980s, ideas about experimentation (in the broader sense) began to emerge.

78.1.2 Experimentation

Initial thinking about empirical studies in software engineering was very much wedded to the "traditional" notion of the *experiment*—a controlled investigation of the relationship between two measures [3]. This domination of the controlled experiment was to continue until the end of the century [24,26]. This is not to say that other forms were completely neglected, particularly the use of *surveys*, although these were only used to a limited extent—but sometimes very effectively, particularly for process-related aspects [21]. Post-2000, other forms such as the *case study* [27] have come into greater use, as software engineers have realized the need to address deeper issues, especially "in the field," as well as the *systematic literature review*, used to draw together the outcomes of many "primary" studies [13].

One reason for this broadening of the range of empirical forms used in software engineering has been that greater exposure to empirical thinking has made researchers aware of what was happening in other domains. Another influence has been the realization that our heavy dependence upon experiments that use human participants inevitably introduces new factors into our experiments and ones that cannot easily be controlled (or always measured). Indeed, even technology-based experiments have such factors too, such as the choice of software objects used in such studies. Underpinning this has been the nagging question as to how far a controlled experiment investigating the effectiveness of some "treatment" in a laboratory setting can really be a useful indication of its likely value to practice in everyday situations.

In the rest of this chapter, we first discuss a (simplified) model of empirical studies and the role of different forms in exploring questions in an experimental manner. We then go on to examine some of the main empirical forms currently employed in software engineering, offering some suggestions as to when each should be adopted and some examples of successful use. Finally, we suggest some practices that help ensure that these can be managed and used to best effect.

78.2 Scope of Empirical Knowledge

While basing decisions upon empirically based knowledge can confer greater authority upon our choices and actions, we also need to recognize that such knowledge is rarely definitive. In this section, we examine how such "knowledge" may evolve and what influences might limit its scope.

78.2.1 Models of Empirical Knowledge

Figure 78.1 provides a rather simplified description of how we gain knowledge through different forms of empirical interaction with the "real world" (well, simplified by the standards of philosophy).

So, how does this model imply that we gain knowledge about the elements of software engineering practice?

1. The first step is one of *informal observation* that some form of relationship seems to exist between the adoption of some systematic way of doing things and the qualities of the resulting software. As an example, this might be to note (say) that the practice of reading through our code or design with someone else seems to help find the faults more rapidly than working on our own.
2. This raises the question as to whether this might be a useful practice to adopt, so we might then perform some more *systematic observation*, probably involving some informal data collection for

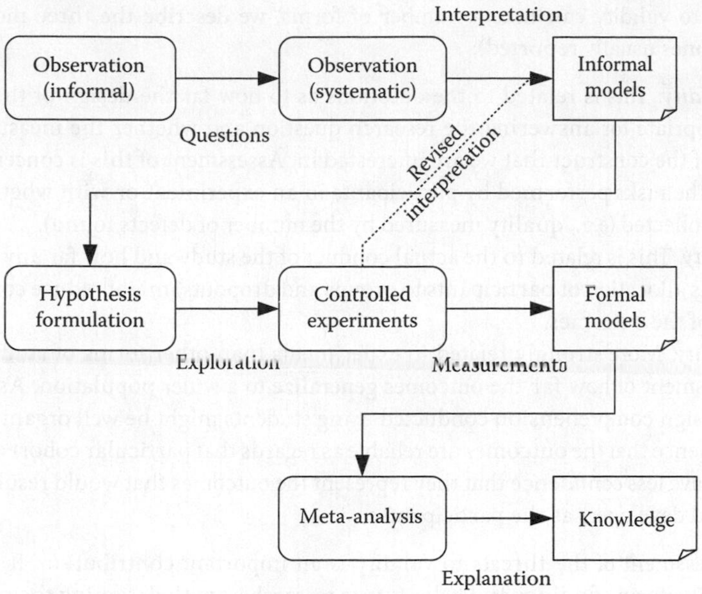

FIGURE 78.1 A simple model of knowledge acquisition.

a number of instances. If we conclude that this does seem to be useful, we can interpret this as meaning that this is a potentially useful practice for others too, and we now also have an informal model of how we can debug our code more effectively.

3. To test this model more rigorously, we can then use it to help formulate a hypothesis ("code inspection is more effective at finding errors than solo debugging") and then perform a controlled laboratory experiment to test this hypothesis. If we conclude that the outcomes reliably support the hypothesis, then we now have a more formal model of how code might be debugged.
4. To give greater confidence in our results, we should repeat our experiment in a number of locations, perhaps involving programmers with different levels of experience, and by aggregating the outcomes, we begin to create some genuine knowledge about how well the technique works and what might limit its effectiveness. Even better, we might arrange that some of the studies we conduct are "field trials" to see if the same effects are still observed when undertaken as part of everyday working practice.

We should note that we have included the possibility that the experiments we perform do not support the hypothesis and that we might go back and revise our informal model. And there are of course many other issues that will *confound* our studies, obscuring the outcomes, since few issues in software engineering are simple cause–effect linkages.

Our purpose in using this rather simple model is to provide some context for discussion of some of the techniques that we will examine in the following sections, rather than to provide a comprehensive explanation of how knowledge is acquired.

78.2.2 Threats to Validity

While often associated with controlled experiments and quasi-experiments [22], perhaps because these lend themselves more to generalization than other forms, the concept of a *threat to validity* is an important issue for all empirical studies. A *threat* is basically a factor that could influence the outcomes of a study in some way. Reporting on the possible threats to validity affecting a study is now a fairly standard element of any empirical paper. Its particular value lies in that it provides an element of reflection from the people who are closest to the study and hence best able to assess the limitations of that study.

While threats to validity can take a number of forms, we describe the three most common ones (as those are the ones usually reported):

Construct validity. This is related to the questions as to how far the design of the study proved to be appropriate for answering the research question and whether the measures were valid measures of the construct that we are interested in. Assessment of this is concerned with such aspects as the tasks performed by participants in an experiment or with whether the "right" data were collected (e.g., quality measured by the number of defects found).

Internal validity. This is related to the actual conduct of the study and how far any problems with this (such as allocation of participants to groups and dropouts) might reduce confidence in the reliability of the outcomes.

External validity. More strongly related to experiments than other forms of study, this consists of an assessment of how far the outcomes generalize to a wider population. As an example, a study of design comprehension conducted using students might be well organized, so that we have confidence that the outcomes are reliable as regards that particular cohort of students, but we might have less confidence that they represent the outcomes that would result if using more experienced developers as the participants.

A realistic assessment of the threats to validity is an important contribution. It indicates possible limitations applying to any findings and helps future researchers with designing their studies. The complex nature of software engineering activities also means that these will always be present, and so they do need to be discussed.

78.3 Four Forms of Primary Study

In this section, we examine four of the major forms of primary study that we use in software engineering. (Anyone needing a rather wider review of empirical practices should note that the book by Oates [19] examines a much wider range of forms in a computing context.)

We should also explain what is meant by a "primary" study in this context. Basically, a primary study is one that directly collects data about the topic of interest, as opposed to a "secondary" study that does not directly collect data, but instead aggregates the data from multiple primary studies. We will examine secondary studies briefly in the next section, and they also form the topic of a separate chapter.

For each form described, we provide the following information:

- A basic description of how it is organized and key references
- A simple outline of the standard process where appropriate or a checklist of key activities
- A brief review of its limitations when used in software engineering
- A discussion of some common pitfalls to be avoided
- Pointers to some good examples of its use

78.3.1 Observational Studies

Sometimes labeled as "experience" or (mislabeled as) "case study" when reported in the literature, an observational study is one where the researcher performs little or no control of any processes, while recording their process in a systematic manner. Indeed, the value of such a study is likely to be proportionate to the degree to which records have been kept systematically, in order to ensure that the outcomes do not simply represent an unsubstantiated degree of "expert opinion." An observational study may well be opportunistic rather than planned, created by the realization that what is being done merits the keeping of records so that others can learn from it. It may be conducted rigorously (such as in an ethnographical study) where the observer keeps firmly "outside" of the situation being observed, avoiding any form of intervention or, more informally, where the observer may be part of the situation.

An observational study may also use "historic" data, employing data mining techniques to analyze large repositories of information related to software. (A good example is the analysis of the records available for open-source software projects.)

Obviously there is no well-defined process for *conducting* such a study, but there are some key activities that relate to the way that the observations are made and recorded. In particular,

1. Be as objective as possible. As an example, if observing the progress of a project within an organization that is the first to use a new tool or technique, the observations are rendered more valuable if the observer can try to avoid being partisan about its success or otherwise.
2. Collect data systematically (e.g., at regular time intervals, or when particular events occur) and record as much as possible, bearing in mind that there is always scope to identify significant factors that were not initially recognized.
3. Wherever possible, collect data that are quantitative (number of faults uncovered, number of classes unchanged, etc.) since this makes it easier to detect trends. Obviously the data have to be relevant to the issue of interest (Basili's *goal–question–metric* (GQM) approach [1,2] may be useful here).
4. Try to provide clear *cause–effect* linkages, in the sense of (say) recording the reasons why the development team believes that using some aspect of a technique might have led to a good (or poor) structure. Simply recording the views/outcomes is not enough without the rationale for these—something that is valuable, but easily forgotten if not recorded at the time.

Many of the *limitations* of this approach are self-evident. It depends upon someone realizing that it would be useful to keep a record of the process of interest and doing so in a systematic manner. The lack of advance planning means that there is a risk that the data itself may be of poor quality and its interpretation may also be open to bias.

The *pitfalls* are linked closely to the limitations. Any study of "experience" papers in the literature quickly reveals how few really provide adequate reasoning to link observations with conclusions. An observational study can easily degenerate into an "opinion piece." Similarly, using historical data carries the risk that the larger the amount of data extracted that is related to an entity, the greater the risk of finding spurious relationships [4].

That said, there is real value in a well-conducted observational study, since it offers the opportunity to study actual practice in a manner that is rarely encountered. An excellent *example* of systematic and thoughtful reporting is provided in Peter Wendorff's report of his experiences with maintaining software created with design patterns [25], not least because where problems were encountered, he probed more deeply to identify why particular patterns had been included in the original system.

78.3.2 Case Studies

A case study can also be considered as a form of observational study, at least inasmuch as it is also noninterventionist, but with the crucial difference that it is *planned*, sets out to answer a defined research question, and aims to collect data that are pertinent to that question in a systematic manner.

While case study research is a well-established practice in other research disciplines (including information systems), we should recognize that there are different views about its use. One camp views it from an *interpretivist* standpoint, whereby the outcomes from a case study can only be considered to be valid within that specific context and that there are multiple realities accessed through social constructs such as language. In contrast, there is also a *positivist* approach to interpretation, based upon the view that the phenomena being studied are subject to some general rules. Scientists and engineers are of course overwhelmingly positivists, so it perhaps not surprising that in embracing case studies, software engineers have tended to adopt this view, usually in the form as described by Robert Yin [27].

According to Yin, a case study is an empirical enquiry that

- Investigates a contemporary phenomenon within its real-life context (as we will see, an experiment removes it from its context into a controlled situation)
- Is particularly appropriate when the boundaries between the phenomenon of interest and its context are not clearly evident (as is often the case in software engineering)

He also observes that a case study

- Copes with the technically distinctive situation where there will be more variables of interest than data points (again, an experiment constrains the set of variables and seeks to measure their relationship across many values)
- Relies upon the use of multiple sources of evidence, seeking to achieve convergence through a process of *triangulation* (whereby we examine the different forms to see where they provide mutual support for a particular interpretation)
- Benefits from the prior development of theoretical *propositions* to guide the processes of data collection and analysis (where a proposition is usually a description of some form of relationship between variables that is less specific than a hypothesis)

As such, case studies are appropriate for situations where we cannot easily separate the object being studied from its context, usually because the *interconnection* of factors is an important element (again, experiments assume that the relationships between pairs of factors can be studied separately). A case study allows for study of a phenomenon in depth, and possibly over a long period of time—but one consequence of this is that it might be hard to generalize from case studies. For example, a case study examining why some development practice is particularly effective will be concerned with looking at the organizational context, the individual software developers, the systems developed with this practice, etc. Any conclusions will be based upon that context and not necessarily be easily translated into the context of a different organization.

Case studies can be used for different purposes. Yin suggests that there are three primary roles: *explanatory* studies (examining how some practice works and why it works successfully), *descriptive* studies providing a "rich and detailed analysis of a phenomenon and its context" but with less interpretation, and *exploratory* studies that help define the questions or scope for a subsequent study. There are also different forms of organization. For this, the primary distinction is between a *single-case* form (maybe involving a representative or extreme case) and a *multiple-case* form, replicating the study across different cases and comparing the outcomes in some way.

Yin identifies five major steps for *conducting* a case study:

1. *Determine the study questions.* These are the high-level concerns that motivate the study. For example, "investigate changes to the development process that arise from adopting the Scrum method in a given organisation."
2. *Identify any propositions.* Examples here might be that "use of the Scrum method will reduce development time" or "use of the Scrum method will increase customer satisfaction."
3. *Select the units of analysis.* This involves determining exactly what will constitute a *case*, such as a "typical" project and also what the *unit* of study will be (a group, division, etc.).
4. *Determine the logic that links the data to the propositions.* Essentially, this involves identifying the data that can be used to evaluate the propositions. An example might be the use of a semistructured interview with customers to assess "satisfaction" (and identification of what that will be considered to be).
5. *Define the criteria to be used for interpreting the findings.* Essentially, this is concerned with how we determine whether or not there is support for the proposition in the absence of being able to employ standard statistical tests. The criteria should be sufficient for the researcher to be able to reject other possible explanations of the outcomes.

A useful checklist approach for using case studies in software engineering is provided in [10].

Perhaps the main *limitation* of case study research is that it is difficult to generalize from one case study to other situations. Indeed, for single-case studies, it may even be difficult to generalize to other situations within the same organization.

What of the *pitfalls*? Perhaps the biggest risk is the relative lack of a well-defined structure, requiring some expertise upon the part of the researcher. Case study research can easily degenerate into something no better than a poorly conducted observational study if not approached in a systematic manner (we discuss the use of research protocols to help overcome this in a later section).

That said, there are a number of *examples* of the effective use of case study research in the software engineering literature. We suggest two contrasting examples here: The first, by Moe et al., examines the use of an agile development method within an organization, demonstrating the benefits of using a case study to study a phenomenon in its context [18]. The second is a study of our own, where we used case study research as an "envelope" to enable us to examine a research process, and hence demonstrates a rather different role for case study research [15].

78.3.3 Surveys

Surveys can also be regarded as a form of observational study. However, whereas a case study aims to probe deeply into a topic, using a small number of cases, a survey aims to do the opposite, probing to only a modest depth (in most cases), but using a large number of respondents. For a survey, the issue of *external validity* (how widely the conclusions can be applied) is particularly important. Essentially, a survey collects information from a sample of the relevant population and then seeks to identify patterns that can be generalized to the wider population.

Like observational studies, surveys can appear deceptively simple to conduct, while concealing some quite difficult design challenges. Perhaps more than the other forms discussed here, they are also difficult to repeat if things go wrong and so depend heavily upon careful planning.

Three common reasons for performing a survey are as follows:

- To use collective experiences in order to *describe* some phenomenon and its attributes. Here the questions being answered tend to be *what* its profile is and what the distribution of the particular attributes is. For example, we might conduct a survey of developers to find out their preferences from a defined list of programming languages, producing a ranking in terms of popularity.
- To make *explanations* about some phenomenon (the question then being *why* it has the form it has). For the example of preferences for programming languages, such a survey might probe into the question of why particular programming languages are preferred.
- To *explore* some issue, perhaps before conducting some other form of investigation (such as a case study), in order to identify the important factors and ensure that these are included in some way.

In software engineering, we tend to make only limited use of surveys. When we do so, it is apt be in a rather focused manner, perhaps because it can be hard to identify and sample from larger groupings (populations) such as "all Java programmers" or "all users of the *Observer* design pattern" in any reliable manner. That said, surveys are used for both technical aspects and also organizational ones, although probably more for the latter (such as investigating the attitude of large software users to the use of open-source components).

A key issue when *conducting* a survey is to determine how the data will be collected. Within software engineering, we tend to make use of *questionnaires*, administered either on paper or online. However, a survey conducted through interviews is sometimes a more appropriate strategy to adopt (especially for organizational issues where the presence of the interviewer can help with interpretation). Having decided on the format, some key decisions that need to be made are as follows:

- Organization of *data generation* in a form most suited to answering the research question. Issues to address will include the use of open or closed questions, whether to use "rating" questions (selecting a value from a list of options) or "ranking" questions (ordering a set of options or identifying the most/least important ones). This is a big decision and it is well beyond the scope of this chapter to discuss it in any detail.
- Identifying the *sampling frame*—the "population of interest" for our question. This might be the users of some software development tool, all Java programmers, managers of agile projects, etc.
- Determining the *sampling technique* by which we seek to obtain a representative sample from the sampling frame. Where possible, we seek to employ a *probabilistic* form for this to aid with later generalization, but sometimes we may not have the necessary control of how we sample and may have to settle for a *non-probabilistic* form (e.g., when we post our questionnaire on the web and invite responses).
- *Organizing* the actual survey process, including ensuring that we obtain an adequate *sample size*, usually by seeking to obtain a good *response rate*, perhaps by following up our original request for those who do not respond.

The general wisdom is that surveys often have quite low response rates (10% is generally regarded as the norm, although some surveys do achieve considerably better rates). As a rule of thumb, a survey needs at least 30 responses to enable reasonable analysis.

There is considerable expertise about the design and conduct of surveys in domains such as social science and psychology. A good reference work is the multivolume *Survey Kit*, edited by Arlene Fink, although for many purposes the first volume may be enough to guide initial design of a survey [7]. For use with software engineering issues, see [14].

When considering the *limitations* for software engineering, we have already previously mentioned a key one, which is that of access to an appropriate sample. For example, a sampling frame consisting of (say) developers who are design pattern users can be difficult to define and almost impossible to sample in a probabilistic manner. In addition, developers and others tend to be multiskilled, so that (to pick another example) the set of Java programmers will intersect with the set of C programmers, the set of Ada programmers, etc. Neither of these examples rules out the use of surveys, but they do have implications for their design and the degree to which are likely to accept the outcomes.

Perhaps the main *pitfall* to beware of with software engineering surveys lies in the sampling process mentioned previously. Many communities of interest may be difficult to identify or to access on a reliable basis, requiring considerable effort on the part of the researcher.

However, there are some good *examples* where surveys help to map out issues of importance. One of these is the survey of project managers to assess different risk factors for software development projects by Ropponen and Lyytinen [21]. This is discussed in the series of columns by Kitchenham and Pfleeger, published in *ACM Software Engineering Notes*, starting with [20] and ending with [16]. A good example of using a survey for exploratory purposes is that by Johnson and Hardgrave that examined which object-oriented design methods were actually being used by developers [11].

78.3.4 Experiments and Quasi-Experiments

For most branches of scientific study, the *experiment* is seen as the most authoritative form of empirical study, as indeed, we indicate in the knowledge acquisition model shown in Figure 78.1. However, one reason why we address this form last is because for a discipline such as software engineering, where the skills and knowledge of the individual play a major role, experiments can rarely be definitive in the way that they are for subjects such as physics. Indeed, building up understanding for software engineering may well need a potpourri of empirical forms.

To address the human element, we usually turn to the use of *randomization* in order to minimize bias, using this to help with such activities as allocating which participants in a study perform particular tasks and the order in which they do them. Where key elements of a study cannot be randomized, perhaps because their specific skills require some people to be allocated to particular groups or tasks or because of constraints in the availability of experimental material, then the experiment becomes a *quasi-experiment* and it becomes necessary to use further techniques to reduce bias [22]. Indeed, a study by Sjøberg et al. of reports of software engineering experiments conducted over a ten-year period suggests that many software engineering studies, despite being labeled as experiments, are actually quasi-experiments [23].

Designing and *conducting* an experiment is quite a complex task, so in this section, we can only discuss a few key aspects—for fuller discussion of empirical forms, we suggest consulting the book by Shadish et al. [22] and for software engineering issues, the book by Wohlin et al. [26].

Any experiment should set out to answer a *research question*. Based upon this, the plan for an experiment (more correctly referred to as the *protocol*—an aspect of study design that we will return to) should address at least the following:

A hypothesis: This is a testable assertion that is related to the research question and refers to the effect that the *treatment*, or the experimental condition, should have. A simple example might be that "the use of code inspection will find more faults than will be found by unit testing." Accompanying this is a *null hypothesis* that effectively states the opposite (that in this case, there is no difference in the number of faults found, or even that unit testing would find more). Here the "treatment" being tested is the use of code inspection.

One problem with software engineering hypotheses is that they are often imprecise about both the treatment and the effect. Here, we might need to clarify the form of code inspection, as well as of unit testing involved, and also what we mean by "more faults." There is also a statistical implication here that any improvement is significant enough to be distinguished from possible random fluctuations.

The independent and dependent variables: The *independent variables* are generally associated with the *cause* element of an experiment (X has the effect of causing Y to occur) and are under the control of the experimenter. While ideally we would only have one independent variable, it is not uncommon to have several. In the previous example, one independent variable is the treatment, that is, whether the participants will use code inspection or unit testing. In this case, where we might want to seed errors in the code to be tested, the forms of error seeded can also be a variable.

The *dependent variable(s)* corresponds to the *effect* element. While again we would like to have one variable, there may actually be several. Here, we might wish to measure the number of errors found within a fixed time (and we might also be interested to know which were found by each form).

The materials: Essentially these are the software objects that are used to test the cause and effect. In the example we are using, this would include the programs to be tested. Ideally these should be as "representative" as possible.

The experimental form: Two widely used ways to organize an experiment are as a *between-subjects* study or one that has a *within-subjects* form.

For a between-subjects study, participants take part in only one treatment. So for our example, they would be assigned to a group that used unit testing or code inspection, but not both. Where one of the treatments is what we might regard as "normal" practice, then we regard that group as a *control group* whose results can be compared with those arising from the treatment being used with the *treatment group*.

For a within-subjects study, participants take part in both treatments (or more than two). Since this may involve learning effects (performing one treatment first might influence the way that the second is performed), one way to organize this is as a *crossover* experiment. We divide the participants into two groups, A and B, and group A performs treatment 1 followed by treatment 2, while group B performs treatment 2 followed by treatment 1. Since this often requires the use of two separate study objects (programs in this case), these also have to be permuted, so that we actually use four groups and orderings.

The allocation of participants to groups is related to this. The aim is to minimize the effect of any factors that might influence the dependent variable (the *confounding factors*), usually by using random allocation to "blocks" in order to avoid or limit any bias arising from having unequal experience or expertise in different groups. (Obviously this may not always be possible in quasi-experiments.)

Measurement plan: This addresses the question of how the dependent variable is to be measured. It may also involve taking measurement before and after the study (pretest and posttest) and address such issues as how time is to be measured, how the number of defects is to be assessed.

Recruitment of participants: Again, this can form a potential source of bias. The participants in a software engineering study are a group of individuals who each bring their own unique past experiences and expertise with them. Some form of pre-study testing may well help assess key aspects of this and assist with such issues as training provision and assignment to groups.

As shown by this (not quite complete) list, conducting an experiment is a complicated exercise. To help assess how well we have planned an experiment, it is also wise to conduct a *dry run* in advance, using one or two "representative" participants in order to check out such aspects as participant training, the tasks allocated, adequacy of time allowed.

A major *limitation* of the controlled experiment is that isolating a relationship (cause–effect) so that we can study it then means that relationship loses its natural context that may well be important. Indeed, a weakness of experiments is that they do not readily allow us to study a phenomenon in truly realistic conditions.

Inevitably for so complex a study form, there are numerous possible *pitfalls* that can render the results of little value. One that should really be avoided is over-complication. A simple study design may well be more worthwhile than a complex one requiring advanced statistical analysis.

What are some good *examples* of well-conducted and well-reported experiments? We suggest two examples that provide particularly clear reporting of what was done and why. The first by Dzidek et al. is a controlled experiment that uses professional developers as participants and studies use of the UML in software maintenance [5]. The second by Güleşir et al. is part of a larger study, but again is particularly well reported in sections 9 through 12 [8]. These two papers provide both well-structured experiments and also excellent examples of how to report an experiment.

78.4 Secondary Studies

Since secondary studies have their own chapter, we will only touch briefly on them here, mainly to discuss their purpose and their role.

Secondary studies are widely used for what is termed *evidence-based* research, an approach pioneered in clinical medicine, but also subsequently adopted in many other disciplines such as education, various other branches of health and social care, and, more recently, software engineering.

A secondary study seeks to answer a research question by systematically locating all of the available primary studies that have relevant data and then *aggregating* that data in some form. While easily stated in this way, there are a number of challenges in performing such a study, not least ensuring that all relevant studies are included and that the analysis is objective. The key tool of evidence-based software engineering (EBSE) is the *systematic literature review* that provides the procedures for conducting an unbiased and objective secondary study [13].

Why have such studies become increasingly important? Well, one reason is that they provide a means of reducing the bias that might come from using expert judgment to select the primary studies. Aggregation also allows us to reduce the effect from local influences upon studies—a confounding factor may have a large influence on one study because of some local issues, but if it is not present in others, then its overall effect will be small. Where statistical forms of aggregation can be used (meta-analysis), then it is also possible to detect smaller effects than can be recognized through a single study.

So why do we mention these studies in this chapter? Well, the reason is mainly because their existence (and effectiveness) depends upon the availability of a number of primary studies addressing relevant issues and also upon the quality of reporting of those studies. Prior to the development of evidence-based practices, each primary study was largely conducted and reported as an isolated item of research. However, this is no longer so, and so when we report our primary studies, we now need to consider how they might contribute to a much wider context.

78.5 Organizing Empirical Studies

For this last section, we examine two important elements for any empirical study: the first is linked to what happens *before* the study is undertaken, while the second is concerned with what we do with the results *after* performing the study.

78.5.1 Experimental Protocol

We have mentioned the *protocol* at various points in this chapter. The role of the protocol is to provide a plan for conducting the study by determining beforehand how all of the key activities will be performed. There is good reason for having a plan, particularly because it avoids the risk that we end up conducting a study that "fishes" for results by adapting the procedures in response to (apparent) observed outcomes. So, before we begin our study, we should have drawn up a complete protocol, which should then have been reviewed by experienced researchers to help identify any oversights (which happen to all of us) or flaws in the design (which unfortunately also tend to happen to all of us).

Studies that make use of human participants should also be subject to ethical approval. The form and procedures for this will vary according to where the study is being undertaken. For software engineering studies, obtaining ethical approval should not form too great a hurdle, since most of our studies involve the participants in performing fairly everyday tasks. They do not usually involve any form of risk nor involve any deceptions, but, particularly when using between-subject forms of study, there is always some possibility that one group will gain extra skills or knowledge simply from being allocated to that group. So, our plan should ensure that all participants should receive equal benefit from having taken part in our study.

Assuming that we have drafted a protocol, had it reviewed, updated it, and had it approved, what comes next? Well, basically the protocol then acts as a guide for conducting the study—and if we have to make changes while conducting the study, we should make careful note of these so that they can be reported as *deviations* from the plan. Finally, the protocol provides a useful starting point for reporting the study since it contains much of the detail about the study design that needs to form part of the final report [12].

The website www.ebse.org.uk provides a number of outline protocols for different forms of primary and secondary study that can be used and adapted by researchers.

78.5.2 Analysis

The analysis of empirical studies can take a wide range of forms, and the details of this are beyond the scope of this chapter. Indeed, this is one aspect where it is often useful to seek expert advice, particularly as regards appropriate statistical forms to employ. However, we do provide a short and simple overview here to indicate what might be appropriate in particular situations.

For any study, it is usually possible to provide some form of *descriptive* statistics. These often use familiar forms such as the *mean*, *median*, and *standard deviation* to describe the "shape" of the resulting data. Visually, data might be presented as a scatter plot (to see if any trends are recognizable) or as box plots (to see how distorted the distribution is). Equally, we might provide counts for different categories in our results—for example, a survey might report how many of the respondents were in particular age brackets.

The descriptive statistics provide an opportunity to see if we can observe patterns (surveys and observational studies in general) or confirm relationships (case studies and experiments). In some cases, they may well be sufficient to enable us to answer our research question.

Where we do seem to observe patterns or trends, then we need to employ *inferential* statistics to confirm that these are truly so and not simply groupings or associations that could easily have occurred by chance. Inferential statistics enable us to answer the original research question with a much greater degree of confidence and are often required in order to decide between the hypothesis and null hypothesis.

78.6 Conclusions

This chapter has addressed a large and complex topic within software engineering and one that draws from many other *reference disciplines*. Indeed, if we are to make software engineering an "engineering" discipline, then its practices do need to be informed by evidence and that evidence needs to come from well-conducted empirical studies. Inevitably we have only been able to skim over many important issues in the space available. However, we hope that the material in this chapter will provide a useful starting point for anyone wanting to find out how empirical studies can be employed within software engineering or to understand their import.

References

1. V. R. Basili and H. D. Rombach. The TAME project: Towards improvement-oriented software environments. *IEEE Transactions on Software Engineering*, 14(6):758–773, 1988.
2. V. R. Basili and D. M. Weiss. A methodology for collecting valid software engineering data. *IEEE Transactions on Software Engineering*, 10(6):728–738, 1984.
3. V. R. Basili and R. W. Reiter. A controlled experiment quantitatively comparing software development approaches. *IEEE Transactions on Software Engineering*, 7(3):299–320, 1981.
4. R. E. Courtney and D. A. Gustafson. Shotgun correlations in software measures. *Software Engineering Journal*, 8(1):5–13, 1992.
5. W. J. Dzidek, E. Arisolm, and L. C. Briand. A realistic empirical evaluation of the costs and benefits of UML in software maintenance. *IEEE Transactions on Software Engineering*, 34(3):407–432, 2008.
6. N. E. Fenton and S. L. Pfleeger. *Software Metrics: A Rigorous and Practical Approach*, 2nd edn. PWS Publishing, Boston, MA, 1997.
7. A. Fink. *The Survey Handbook*, 2nd edn. Sage Publications, Thousand Oaks, CA, 2003. Vol. 1 of the Survey Kit.
8. G. Güleşir, K. van den Berg, L. Bergmans, and M. Akşit. Experimental evaluation of a tool for the verification and transformation of source code in event-driven systems. *Empirical Software Engineering*, 14:720–777, 2009.
9. M. H. Halstead. *Elements of Software Science*. Elsevier, North Holland, the Netherlands, 1977.
10. M. Höst and P. Runeson. Checklists for software engineering case study research. In *1st International Symposium on Empirical Software Engineering and Measurement (ESEM)*, Madrid, Spain, pp. 479–481. IEEE Computer Society Press, 2007.
11. R. A. Johnson and W. C. Hardgrave. Object-oriented methods: Current practices and attitudes. *Journal of Systems and Software*, 48(1):5–12, 1999.
12. B. A. Kitchenham, S. L. Pfleeger, L. M. Pickard, P. W. Jones, D. C. Hoaglin, K. El Emam, and J. Rosenberg. Preliminary guidelines for empirical research in software engineering. *IEEE Transactions on Software Engineering*, 28:721–734, 2002.

13. B. Kitchenham and S. Charters. *Guidelines for Performing Systematic Literature Reviews in Software Engineering*. Technical Report EBSE 2007-001, Keele University and Durham University Joint Report, 2007.
14. B. Kitchenham and S. L. Pfleeger. Personal opinion surveys. In F. Shull, J. Singer, and D. I. K. Sjøberg, eds., *Guide to Advanced Empirical Software Engineering*, pp. 63–92. Springer-Verlag, New York, 2007.
15. B. A. Kitchenham, D. Budgen, and O. P. Brereton. Using mapping studies as the basis for further research—A participant observer case study. *Information and Software Technology*, 53(4):638–651, 2011. Special section from EASE 2010.
16. B. A. Kitchenham and S. L. Pfleeger. Principles of survey research part 6: Data analysis. *ACM Software Engineering Notes*, 28(2):24–27, March 2003.
17. T. J. McCabe. A complexity measure. *IEEE Transactions on Software Engineering*, SE-2:308–320, December 1976.
18. N. B. Moe, T. Dingsøyr, and T. Dybå. A teamwork model for understanding an agile team: A case study of a scrum project. *Information and Software Technology*, 52(5):480–491, 2010.
19. B. J. Oates. *Researching Information Systems and Computing*. Sage Publications, Thousand Oaks, CA, 2006.
20. S. L. Pfleeger and B. A. Kitchenham. Principles of survey research part 1: Turning lemons into lemonade. *ACM Software Engineering Notes*, 26(6):16–18, November 2001.
21. J. Ropponen and K. Lyytinen. Components of software development risk: How to address them. A project manager survey. *IEEE Transactions on Software Engineering*, 26(2):98–111, 2000.
22. W. R. Shadish, T. D. Cook, and D. T. Campbell. *Experimental and Quasi-Experimental Design for Generalized Causal Inference*. Mifflin Co., Houghton, MI, 2002.
23. D. I. K. Sjøberg, J. E. Hannay, O. Hansen, V. B. Kampenes, A Karahasanović, N.-K. Liborg, and A. C. Rekdal. A survey of controlled experiments in software engineering. *IEEE Transactions on Software Engineering*, 31(9):733–753, 2005.
24. W. F. Tichy. Should computer scientists experiment more? *IEEE Computer*, 31(5):32–40, 1998.
25. P. Wendorff. Assessment of design patterns during software reengineering: Lessons learned from a large commercial project. In *Proceedings of 5th European Conference on Software Maintenance and Reengineering (CSMR'01)*, Lisbon, Portugal, pp. 77–84. IEEE Computer Society Press, 2001.
26. C. Wohlin, P. Runeson, M. Host, M. C. Ohlsson, B. Regnell, and A. Wesslen. *Experimentation in Software Engineering: An Introduction*. Kluwer, Norwell, MA, 2000.
27. R. K. Yin. *Case Study Research: Design and Methods*. Sage Publications, Thousand Oaks, CA, 2003.

79

Software Quality and Model-Based Process Improvement

Barış Özkan
Middle East Technical University

Özlem Albayrak
Bilkent University

Onur Demirörs
Middle East Technical University

79.1 Introduction

Software users have been suffering from quality problems for decades (Whittaker and Voas 2002). Quality is more and more often seen as a critical software attribute and a determinant of business success, which is as equally as important as the technology used in software products (Prahalad and Krishnan 1999). The absence of quality in software products and services results in dissatisfied users and financial loss, and may even endanger our lives (Peterson 1996).

Software process improvement (SPI) is a process-oriented approach to solve quality problems. SPI methodologies and concepts are inspired from the heritage of quality movements, such as total quality management (TQM), which have led to significant quality and productivity achievements in other engineering disciplines (Kemp 2006). A large body of software process knowledge has been made available to software organizations via process models and associated methods, such as Software Engineering Institute (SEI) Capability Maturity Model Integrated (CMMI) and ISO/IEC 15504. These have been used by software organizations for assessing and improving their processes systematically.

The basic premise of SPI is that quality of a software product or service is highly dependent on the quality of the process used to develop it. Therefore, an understanding of the concepts of quality, process, and their relationship is fundamental for understanding model-based SPI and evaluating their use. In this chapter, we aim to give an insight into the relation between software quality and model-based SPI. In Section 79.2, we first discuss the definitions for quality and address several characteristics of software products that are important while defining quality for software. Then, we overview the evolution of process-oriented quality improvement paradigm (QIP) and interpret it for software development. We close the section with an analysis of the costs of software quality and relevant models, which are essential tools in managing process improvement programs and evaluating their results. In Section 79.3, we summarize CMMI and ISO/IEC 15504 models for SPI and review the results of their implementations. In the final section, we discuss several challenges in model-based SPI and address relevant future research prospects.

79.2 Underlying Principles

79.2.1 Quality

Quality has always been a difficult concept to define and agree, particularly much more difficult for software quality. As markets, customers and technology, and management paradigms change, the perception of software quality is getting more fluid, complex, and diverse.

Quality experts have brought different definitions of quality, which needs to be reconsidered while using them for software quality. Crosby (1979) defines quality as the "conformance to requirements." The definition is based on the idea that requirements are testable. This brings several questions in mind: "Are software requirements complete and correct? Who determines the requirements?" "Can't they ever change?" According to Juran et al. (1974) "Quality is the fitness for the intended use." This definition accommodates the fact that software requirements may not be fully defined; however, it still needs the determination of who will decide whether it fits or not. Different users may have different definitions for fitness, for example, an expert user may desire a control of fine details on particular software where ease of use will be fit better to a novice user's needs. Table 79.1 gives examples of different meanings of software quality for different stakeholders.

Weinberg (1991) "value to some person" definition of quality emphasizes the relativity of quality from an economic point of view. Weinberg argues that "the definition of quality is always political and emotional, because it always involves a series of decisions about whose opinions count, and how much they count relative to one another."

In light of different definitions that emphasize certain aspects of quality, we can state that quality embodies numerous and various expectations. Garvin (1984) addresses this complexity and gives five approaches to defining quality as perceived from philosophy, economics, marketing, and operations management disciplines. These five perspectives can be outlined as follows:

Transcendental approach: The transcendent definition of quality implies an innate excellence or superiority. Quality cannot be defined precisely nor measured. It is an ideal to be strived for, but never can be completed perfectly, and we can only learn to recognize it through experience. It is this approach when we get the enigmatic answer "I know it when I see" and resort to prototyping to get customer feedback at

TABLE 79.1 Different Viewpoints of Software Quality

Quality Definition	End User	Marketing Department	Project Manager	Software Developers	SW Architect	IT Operations
User friendly	×					
Compatible with competitor	×	×				
Extensible	×	×		×		
High performance	×	×			×	×
Zero defects	×		×		×	×
Rapid development	×	×	×			
Low development cost		×	×			
High modularity				×	×	
Secure	×				×	×

the early stages of development and understand the desired qualities of the software. Pirsig (1974) states that "Quality is neither mind nor matter, but a third entity independent of the two, even though Quality cannot be defined, you know what it is". This philosophical definition is not found useful for modern business management since it is unanalyzable.

Product approach: The product-based definition presupposes that product quality is proportional to the quantity of internal attributes the product possesses and that the internal attributes are measurable. Therefore quality is precise, and the quality level of products is explained by quantifying the attributes and the products can thus be ranked objectively with respect to quality. As an example, consider a polar winter jacket. The grams per square meters of the polar cloth used is directly proportional to the degree of cold resistance the jacket provides. If it is 400 g/m^2, then it will keep us warmer in comparison to a 100 g/m^2 jacket. ISO 9126 software quality standard (ISO/IEC 2001) also supports this view, assuming a positive correlation between internal, external, and in-use attributes, for example, higher the degree of modularity, makes software testable and maintainable, and maintainability increases extensibility. However, there are considerable difficulties of software product measurement in regard to maturity of software measures and measurement models (Pfleeger et al. 1997). This approach has been found mostly useful for economic analysis purposes.

User approach: This approach assumes a highly personalized definition of quality. That is, an individual will call a good or service as high quality if it satisfies the person's needs and preferences. Quality is perceived as highly subjective and in a manufacturing sense, it is "the fitness for use." This definition is subject to open questions challenges when attempting to define quality for a particular market; for example, how should the subset of usability properties of a software product be selected in order to maximize satisfaction in an environment with customers having widely varying preferences and expectations? Uses of this approach have been observed mostly in marketing, economics, and operations management.

Manufacturing approach: Manufacturing perspective looks into the conformance of the products to stated requirements and industrial standard for assessing quality. The focus of quality is the manufacturing and engineering (design and supply) side in contrast to user view where the focus is the consumprion (demand). Any deviation from the specifications is seen as reducing the quality of the product. The concept of process has a key role in the manufacturing view of quality. The approach assumes that organizations have imperfect processes and advocates the improvement of quality by improving processes. A focus on the processes and knowing how to do it, quality improvements are believed to result in prevention of non-conformities (defects) and reductions in rework, scrap, and total manufacturing costs. TQM, zero-defects, six-sigma are the noteworthy quality movements that follow this approach and they have the mantra of "do it right the first time" to express their focus on the processes. This view is found useful in contractual situations and is dominant in standards such as ISO 9000 series and models pertinent to software domain such as CMMI and ISO/IEC 15504. In the following sections, we will mostly refer to quality view of the manufacturing approach.

Value-based approach: This approach is based on the observation that the perception of quality change in relation to the price. That is, a product which is found to be excellent will not have an economical meaning if it is very expensive; it depends on how much a customer is willing to pay for a certain level of excellence. Therefore, the proposal is the definition of quality in terms of costs and prices. A quality product is one that provides a certain degree of performance at an acceptable price or a level of conformance at an acceptable cost.

From the endeavor for defining quality, we can infer that quality is a multifaceted term and it encompasses multiple attributes including value, conformance, fitness, and others. Defining quality is controversial, and it may be exceptionally difficult doing it for software, being an intangible, malleable, and pervasive product.

Recognizing the challenge of defining software quality, ISO published ISO/IEC 9126 (ISO/IEC 2001) series to help organizations define, communicate, and manage quality with respect to their needs.

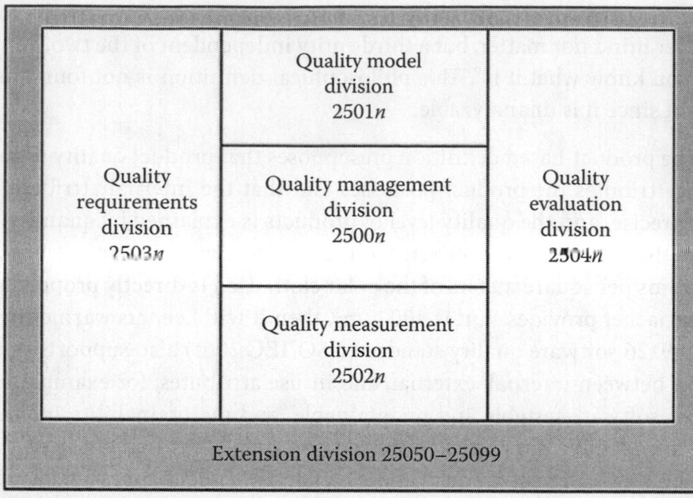

FIGURE 79.1 Organization of the SQuaRE Series. (Adapted from ISO/IEC, *ISO/IEC 25000:2005 Software Engineering—Software Product Quality Requirements and Evaluation (SQuaRE)—Guide to SQuaRE*, International Standards Organization & International Electrotechnical Commission, 2005. With permission.)

The series include characteristics for software quality and a related taxonomy. The standard was inspired by similar software quality models (McCall et al. 1977, Boehm et al. 1979, Dromey 1995) and functionality, usability, reliability, performance, and supportability (Grady and Caswell 1987). The standard has four parts, which address a quality model, external quality metrics, internal quality metrics, and quality in use metrics. Quality model defines characteristics and sub-characteristics common for both external and internal quality aspects. Internal metrics are derived from static properties of the software, which do not rely on software execution. External metrics are derived from the behavior of running software and represent dynamic properties of the product. Quality in-use metrics is available when the final product is in use in a specific context. ISO/IEC 14598 (ISO/IEC 1999) Product evaluation standard series were published to be used in conjunction with ISO 9126 and to provide organizations with methods for measurement, assessment, and evaluation of software product quality.

Starting from 2005, ISO has begun publishing as Software Product Quality Requirements and Evaluation (SQuaRE) series as the next generation software quality standards (ISO/IEC 2005). The motivation for SQuaRE is to extend the scope to three complementary quality processes: requirements specification, measurement, and evaluation (Figure 79.1). Figure 79.2 shows the quality characteristics and their decomposition into sub-characteristics as defined in SQuaRE.

At the end of this section, we can conclude that each software product is unique in its definition of quality. Quality should be properly defined in such a way that it can be managed before, during, and after its development. Only with a good, usable, and communicated definition of quality we will be able to define our strategies, check whether it is built into a product, measure how much of it we have, and predict how much we will have.

79.2.2 Process and Quality

In its simplest sense, a process is what we do to create value for our stakeholders, such as our customers. We deliver value as software products and services. Quality is distinguished through an evaluation of the deliverables and is enjoyed by using them. More formal definitions of processes are as follows:

A series of actions or operations conducing to an end (Merriam-Webster online 2012).

A set of partially ordered steps intended to reach a goal (Feiler and Humphrey 1993).

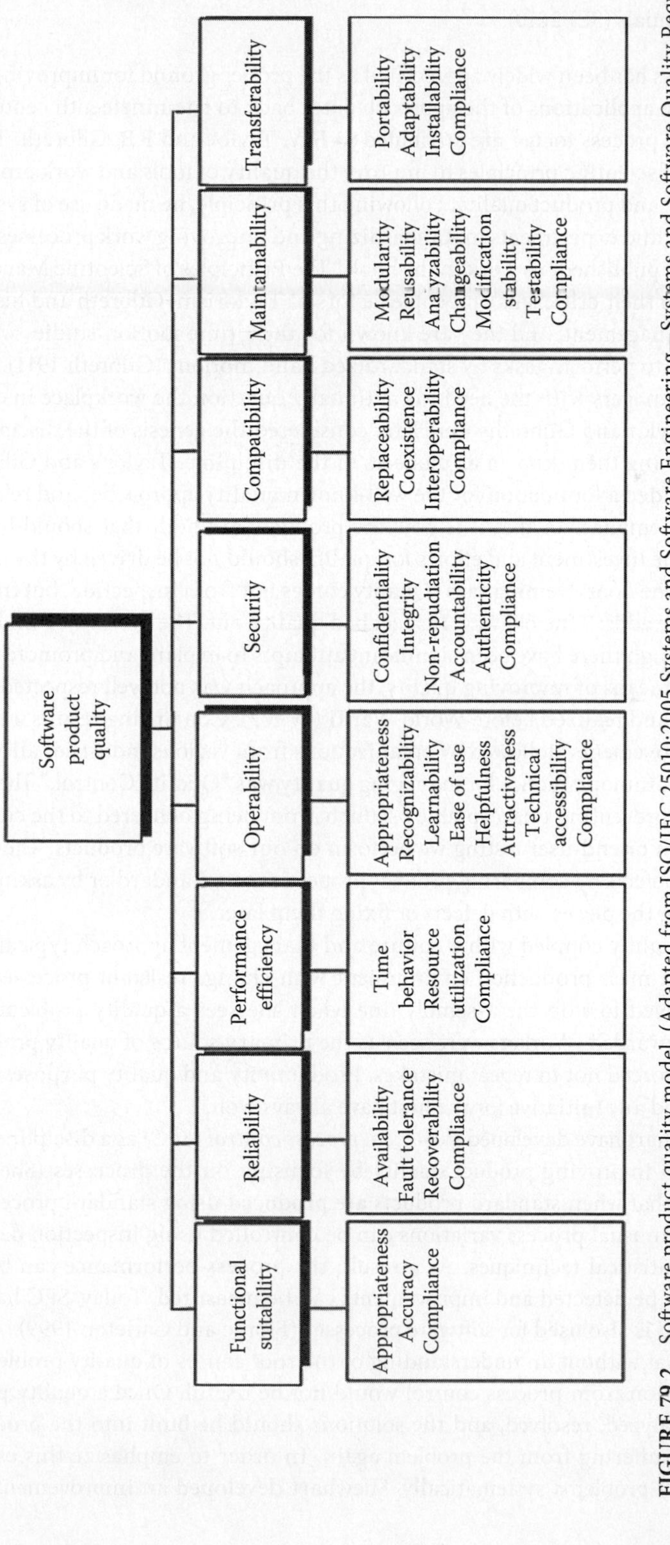

FIGURE 79.2 Software product quality model (Adapted from ISO/IEC 25010:2005 Systems and Software Engineering—Systems and Software Quality Requirements and Evaluation (SQuaRE)—System and Software Quality Models Standards Organization & International Electrotechnical Commission, 2011. With permission.)

> A process is a set of tools, methods and practices used to produce a product (Humphrey 1989).
>
> The set of activities, methods and transformations that people use to Develop and Maintain Software and the Associated Products, for example: product plans, design documents, code, test cases and user manuals (SEI 2010).

The concept of process has been widely appreciated as the proper ground for improving product quality and productivity. The applications of the approach dates back to late nineteenth century and the most influential works on "process focus" are attributed to F.W. Taylor and F.B. Gilbreth. Taylor's approach was the application of scientific principles to improve the quality of tools and work processes in order to improve productivity and product quality. Following this principle, he made use of systematic observations, measurements, and experiments in standardizing and improving work processes and engineering practices. F.W. Taylor published his work in the book "The Principles of Scientific Management" in 1911 (Taylor 1911). Despite their criticisms on people factors of Taylorism, Gilbreth and his wife were advocates of scientific management, and they are known for their time motion studies where they sought for one the best ways to perform tasks by standardized hand motions (Gilbreth 1911). Gilbreths' works included teaching managers with the need to continually question the workplace in order to improve. The works of F.W. Taylor and Gilbreths were later considered the genesis of the discipline of industrial engineering and making them known as pioneers of the discipline. Taylor's and Gilbreths' principles and approaches provided a foundation for the well-known quality approaches and related practices.

The process focus emphasizes that it is how the products are built that should be in question for improving quality; the investment and efforts for quality should not be driven by the motivation in getting the product out the door. Deming says "Quality comes not from inspection, but from improvement of the process" and he adds: "The old way: Inspect bad quality out. The new way: Build good quality in" (Walton 1986). Although there have been significant attempts to explain and promote the role of adhering to processes as a means of improving quality, the approach was not well respected, and the benefits were not recognized and realized before World War II (WW2). Convincing results were obtained later, leading to "quality movements" followed by organizations from various industries all around the world.

Before 1950s, the common practice for managing quality was "Quality Control." The purpose of quality control (QC) was preventing unacceptable products from being delivered to the customer similar to the purpose of system or end-user testing we perform on our software products. The focus was on the finding the product defects by comparing product properties to a standard or by asking an experienced worker and scrapping the pieces with defects or fixing them later.

The practice was tightly coupled with a culture and management approach typical to mass production at the time. In a mass production environment with change resistant processes, no worker was appreciated nor allowed to stop the assembly line when she sees a quality problem. Solving quality problems were not rewarded. Workers were seen as the primary source of quality problems where they were controlled and forced not to repeat mistakes. Productivity and quality purposes were always seen conflicting targets and any initiative for the first have always won.

In 1930s, W. Shewhart have developed *statistical process control* (*SPC*) as a disciplined scientific management approach to improving product quality by focusing on the processes (Shewhart 1931). The premise of SPC was that when standard products are produced using standard processes and the processes are repeated, unusual process variations can be controlled using inspection data on work products and applying statistical techniques. As a result, the process performance can be monitored and quality problems can be detected and improvements can be measured. Today, SPC have evolved into a mature discipline and is also used for software processes (Florac and Carleton 1999).

Shewhart knew that without an understanding of the root causes of quality problems and fostering change, the information from process control would not be useful. Once a quality problem is identified, it should be analyzed, resolved, and the solutions should be built into the processes in order to prevent the process suffering from the problem again. In order to emphasize this essence of the taking actions to quality problems systematically, Shewhart developed an improvement approach called *plan–do–check–act* (*PDCA*). PDCA cycle has been refined and applied successfully by Deming, Juran, Crosby, and many other quality experts after WW2. PDCA improvement cycles have constituted the basis for continuous process improvement in various quality methods and standards including ISO 9000 series. The description of PDCA cycle from ISO 9000 standard is as follows:

- Plan. Establish the objectives and processes necessary to deliver results in accordance with customer requirements and the organization's policies.
- Do. Implement the processes.
- Check. Monitor and measure processes and product against policies, objectives, and requirements for the product and report the results.
- Act. Take actions to continually improve process performance.

SEI IDEAL process improvement program model (McFeeley 1996) and QIP (Basili 1985) are examples that adapt continuous improvement approach to the software engineering domain and provide an outline of the steps necessary to establish a successful improvement program.

During WW2, the concentration on the processes critically increased with an aim to produce standard weapons and equipment without defects. Techniques relying on statistics and operations research were developed and used as the prevailing means of process control and optimizations for production and quality. After WW2, in the western world, the quality related practices have begun evolving from "Quality Control" to "Quality Assurance". *Quality assurance* (*QA*) has become a disciplined approach to quality extending quality practices from controlling the final products to the entire production process. QA primary function was ensuring compliance to organizational process and design standards and improving them. Doing so, defects can be prevented, detected and fixed earlier in the whole process. Although QA includes cross-departmental communication about quality, communication with vendors, it is usually established as an independent organizational function performed by a QA department. It has the potential for adversarial relationships and "Don't tell me my job" hassle between QA teams and employees, for example, between software testers and developers, which create inconsistencies and interruptions in the motivation to improve processes.

TQM was the first quality movement where the business value of improving quality through process improvement was appreciated and continuous improvement was achieved (Kemp 2006). The principles of the approach were established and successfully launched by Deming, Juran, Feigenbaum, Ishikawa, and other quality pioneers, see Hoyer and Hoyer (2001) in postwar Japan. TQM includes elements of improved SPC techniques, extension of quality practices organization-wide, and establishing quality as a continuous function.

The key factor of successful applications of TQM philosophy was managing the cultural change and the achievement of organizational and continuous motivation for quality (Kemp 2006). TQM did not distinguish managers, engineers, and workers; all people working for an organization were empowered with taking initiatives for improving processes and quality. Quality control and improvement was not seen as an isolated and independent function driven by a separate department. Instead, quality improvement was seen a shared responsibility and a cross-functional collaborative work with high involvement from top management. Stopping the assembly line and gathering others to understand the root cause of the problem was truly a desired action when a worker encounters a quality problem. TQM inspired many other quality movements such as Six-Sigma, zero-defects, and ISO 9000 quality standards and SEI CMMI. Today, the accumulated knowledge from the evolution in quality improvement is being studied under "Quality Engineering" (QE) discipline (Borror 2008). Quality engineering combines quality control and assurance techniques and practices whilst maintaining an emphasis on the cultural dimension of process change and continuous improvement. QE have gained good acceptance in software engineering, and "Software Quality Engineering" is now an emerging discipline.

Quality paradigms with a process focus have been effective in manufacturing domains such as automotive and electronics with proven pragmatic methods. They have been influential on software engineering methods and many researchers and organizations have created models and methods that adopt TQM principles. Humphrey, one of the pioneers in SPI, points out "Dr. W.E. Deming, in his work with the Japanese after World War II applied the concepts of statistical process control to many of their industries. While there are important differences these concepts are just as applicable to software as they are to automobiles, cameras, wristwatches and steel" (Humphrey 1988). However, there exist peculiarities of software products and development that should be considered while adapting the methods, techniques, and tools of successful quality paradigms to the software domain.

Software products are intellectual artifacts. Software development is mostly analysis, design, and coding and it requires the integration of knowledge work. This makes software people the most valuable resource and the highest cost item. In contrast to traditional products, the manufacturing step, for example, copying it to CDs, creates an ignorable value. Once you design and implement it, you can create instances of it over and over again almost with no cost, in no time, and without defects. This is also different from large construction projects like stadiums, where you will need large amount of money and years to build every instance of the same design. Another feature that distinguishes software is that it does not deteriorate. It would not be exceptional if our laptop computers start overheating or its screen starts flickering after 3 years of use. However, once the software meets the requirements and defects are removed, an increased use will not cause any faults, and it would be operational as long as requirements do not change. Software products are malleable. When the malleability is exploited by following a disciplined approach, for example, automated regression testing after a change, it is very beneficial and desired since a critical bug can be fixed in hours. It also is the Achilles heel of software since rapid and frequent changes increase control risks and create challenges in avoiding code and fix development habits. The simplicity of software manufacturing increases utilization of software malleability in contrast to physical products. Another distinguishing characteristic of software products is that they change the expectations. Once software is operational, before long, the users start seeing further opportunities, for example, automations to jump to the next level of cost minimization, which can be realized through an enhanced version of the software or an entirely new design.

Adherence to "conformance to requirements" definition may have limitations for software quality. First, software requirements are usually hardly complete and correct and you will not be able to "do it right the first time" if you do not know the requirements. The acquirers and users also make mistakes in their requirements. Second, the requirements are usually considered as a means to an end, and the quality of the eventual product is also evaluated on the basis of how well software helps the users do their work. Although all stated requirements are fulfilled, there can still be a room to increase customer satisfaction and attain their loyalty by delighting them with software features that they did not anticipate but adds significant value to their business. A zero-defect aim can be realistic in a production process where use of automation, robotics, can be quite effective on producing physical products; however, this is not realistic for software, in particular, for the creative parts of software development such as requirements analysis, elicitation, and of software architecture design. A "good enough" software can be considered the optimal for all stakeholders (Yourdon 1996).

79.2.3 Cost of Quality

Customer value is determined by product quality, cost, and cycle time, which are the competitive factors of making business. From a traditional management view, their improvements are seen as conflicting objectives, that is, an improvement in any of them is only possible at the expense of others. For instance, if the objective is to develop software in shorter times, either we will need to work in larger teams of developers or do overtime work or we will deliver fewer features and skip some tests at a risk of low reliability.

From a continuous quality improvement point of view, it is possible to improve them simultaneously through changing processes and organizational culture. This view is based on the fact that not all

TABLE 79.2 Breakdown of Software Development Costs

Total Costs		
	Non–Value Added	
Value Added	Essential	Nonessential
Requirements elicitation	Unit testing	Bug fixes
Design	Stress testing	Unwanted features
Development	Training	Missing features
Documentation	Peer reviews	Bottlenecks

tasks we perform and time and effort we expend in order to develop software generates value for our stakeholders. The premise of continuous improvement is that an improved quality will decrease costs, improve productivity, and shorten delivery times, and in the long term any investment on quality will pay in terms of increased profits and competency.

Total costs of software production can be broken down as shown in Table 79.2. In a strict and ideal sense, the value-added tasks should change the product in some way, make the product more desirable to the stakeholders, and must be done right the first time. These criteria help us identify the proper targets for process improvement, that is, anything that is not value added is a suitable target for removal or improvement. Value streams analysis is also one of the indispensable tools of lean thinking quality approach that originated from Toyota production system and which attracts attention from agile software community (Poppendieck and Poppendieck 2003). Crosby (1979) points out "Quality is free. It's not a gift, but it is free. What costs money are the unquality things—all the actions that involve not doing jobs right the first time." Since our software processes are not perfect with respect to the value-added criteria, every process includes some essential non-value-added parts; they are necessary regarding the current methods of software product development, engineering domain knowledge, and in relation to the peculiarities of software. Non-value-added activities, software verification activities are of this type, that is, if the software design and implementation processes are perfect, verification tasks will not be needed. Quality improvement is also a non-value-added task; nevertheless, we need to invest in improving quality since we benefit from reducing total non-value-added cost.

Any investment decision on improving quality needs an understanding of the cost associated with achieving quality and the economic trade-offs involved in delivering good-quality software. A change to improve processes is justified roughly comparing the returns to the costs of the change. This challenging task involves the analysis, baselining, measurement, and prediction of all internal and external effects of the changes on business of the organization including repercussions on all interacting processes. It is best if the improvement can be represented in monetary units since it is the management language. Cost of Quality (CoQ) analysis and techniques have been developed in order to meet such requirements and have been in use for more than 50 years (Schiffauerova and Thomson 2006). Crosby's CoQ model is one of the popular models in SPI literature. In line with the "conformance to requirements" definition of Crosby, total CoQ is divided into conformance and non-conformance costs (Figure 79.3). The former includes the costs of effort and resources we consume for improving quality (good quality) and the latter includes the costs we pay due to failing to achieve quality (bad quality). Prevention and appraisal costs are incurred to prevent and detect poor quality. Failure costs are further divided into external or internal costs with respect to occurrence of failure before or after product delivery.

Cost of software quality (CoSQ) techniques are the effective tools in feasibility analysis of SPI programs and the measurement and the evaluation of the program performance (Karg et al. 2010, Unterkalmsteiner et al. 2012). They are essential tools in convincing managers and technical staff that poor quality costs more than high quality in the long run, quality problems are serious, hence, it is worth fixing them.

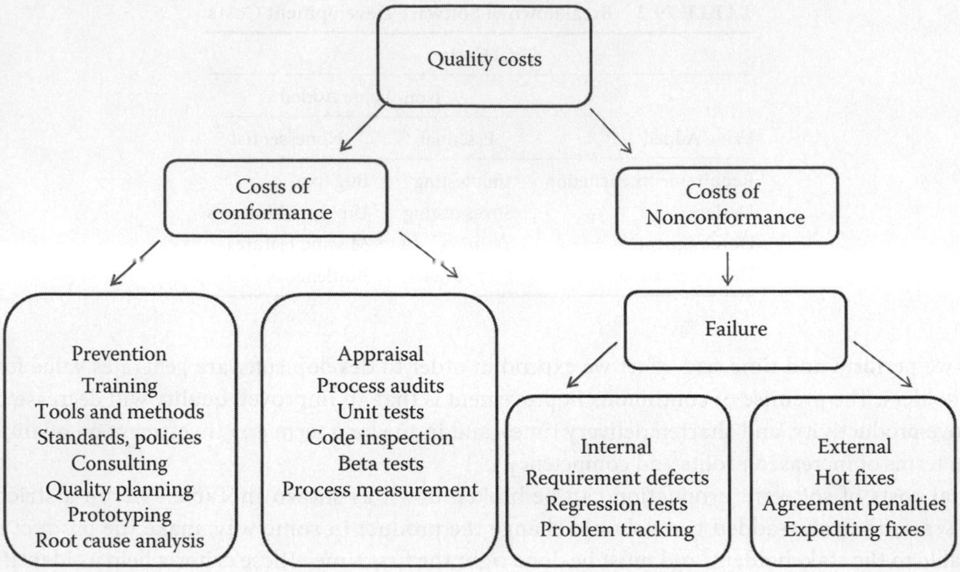

FIGURE 79.3 Costs of software quality.

Both theory and experience advise investing on prevention and appraisal to get the highest returns from the decreased costs of appraisal and failure. If we can prevent a requirements defect from happening, the defect will not cost us. If we detect the defect before design, it may only cost us some effort to update the specification document, but nothing more. If the defect is discovered from the analysis of a failure of the product in use, then a rework of the whole process may be needed. This view is addressed by two dominant views of CoQ theory (Schiffauerova and Thomson 2006). The advocates of the first theorize that there is an economics of quality, and there exists an optimal point where the costs of maintaining a quality above that point will exceed the benefits, where the others challenge the economics of quality and proposes the optimum is the perfect quality-zero defects (Figure 79.4). No matter where the optimum is, this relation is critical for software managers and drives the continuous improvement up-to the optimal, and CoQ analysis helps us decide which non-value-added costs to focus on and where to add non-value-added tasks (cost of conformance) to reduce net costs.

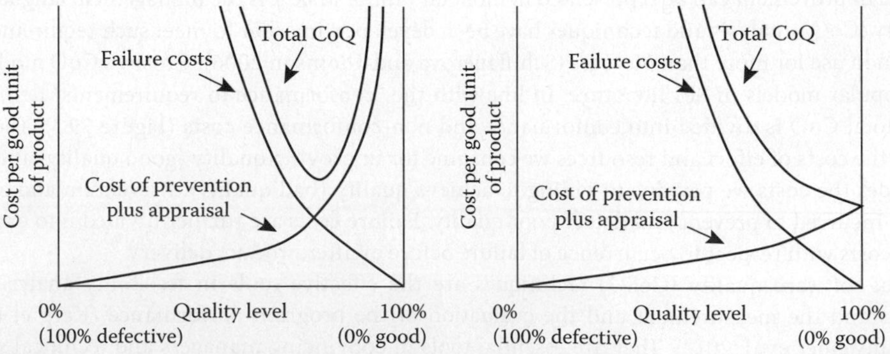

FIGURE 79.4 Two views of economics of quality. (Adapted from Schiffauerova, A. and Thomson, V. *Int. J. Qual. Reliab. Manag.*, 23(6), 647, 2006. With permission.)

79.3 Best Practices

In the previous sections, we have discussed software quality, the relation between process improvement and quality, and explained how value and costs of quality analysis guide us in directing and evaluating our quality improvement efforts.

Today, SPI is being recognized as the core solution behind software development problems. Model-based improvement is a SPI methodology by which the performance of the software development organization's processes is improved on the basis of capability profiles provided by external process assessment models. A number of process models have been developed by people from research and industry that combine theoretical knowledge with practical experience (von Wangenheim et al. 2010). Process models are assumed as the most significant means for transferring process knowledge into process improvement. SPI models contain a set of goals, indicators, and associated best practices that adhere to processes and organized in a maturity context. The term "process quality" is sometimes used for addressing how well a process meets the defined goals that are measured in terms of indicators defined in the models.

In this section, we first discuss patterns of behavior that give us insights into the perception of quality and process by different organizational cultures. We explain how SPI models associate these patterns to the performance of individual processes and the whole process system in a maturity scale. Next, we briefly explain two most wide-spread assessment models used in model-based SPI and discuss the results of their implementations.

79.3.1 Software Process Maturity

During his quality studies in the organizations, Crosby (1979) observed quality management patterns, given as stages of maturity (uncertainty, awakening, enlightenment, wisdom, and certainty) in his quality management maturity grid (QMMG). The QMMG is based on the observation that the attitudes to quality imply certain cohesive characteristics of the elements of an organization. That is, the combination of certain types of product quality, processes, costs of quality, and the problems associated to these types of elements usually do not occur at random. Once the pattern that matches an organization is identified, then the practices, the quality of the products, and rates of CoQ can be predicted in that organization. Furthermore, the direction for improving quality would be the next maturity stage and referring to the target stage in the grid, the actions need to be taken for improvement can be understood. The maturity grid was first adapted to software development by Radice at IBM under the direction of Humphrey (Radice et al. 1985). In their work, they identified 12 software process stages (e.g., requirements, product-level design, and code) and rated them on a five-point ordinal scale. The scale consists of traditional, awareness, knowledge, skill and wisdom, and integrated management system points. Later, Humphrey et al. at SEI expanded the concept to provide the U.S. government with a method for assessing the capability of their software contractors (Humphrey 1988). In his model, Humphrey defined initial, repeatable, defined, managed, and optimized stages of software process maturity at the organizational level with a concentration on types of processes used in software development.

According to the model, at the *initial level*, the software processes are ad hoc, and occasionally chaotic. There is no or few understanding of the concept of process. Quality problems are repeated, deferred, or even forgotten rather than solved. As a consequence to poor planning, problems are discovered late and are usually solved by firefighting (Repenning 2001). The root causes of all problems are in poor management and success depends on individual efforts and heroics. At the *repeatable level*, earlier success on projects can be repeated with similar practices. The process concept is not embodied into engineering but project management practices. The scope, cost, and schedule of the projects are tracked. At the *defined level*, the management and engineering activities are represented as a set of standard software processes for the organization. All processes are documented, standardized, and integrated. The emphasis is on spreading and learning process definition and improvement to the whole organization. At the *managed level*, the emphasis is on measuring the processes, software artifacts,

TABLE 79.3 Comparison Table for Patterns of Software

Maturity Framework	—	Initial	Repeating	Defined	Managed	Optimized
Crosby	—	Uncertainty	Awakening	Enlightenment	Wisdom	Certainty
Weinberg	Oblivious	Variable	Routine	Steering	Anticipating	Congruent
ISO/IEC 15504	Incomplete	Performed	Managed	Established	Predictable	Optimizing
Organizational Competence	Incompetent/ Unaware	Incompetent/ Aware		Competent/ Aware		Competent/ Unaware

and products. They are quantitatively understood and controlled, hence making quantitative management and prediction possible. At the *optimized level*, improvement is internalized and process performance improves incrementally and continually. The focus is on the overall organizational performance. The level was renamed to the optimizing level to correct the misperception that at level optimized, an organization has perfect processes and to emphasize the need that for continuous process improvement. These five progressive stages has constituted the structure of the formalized SEI-Software CMM model, published in 1991, with detailed descriptions of engineering and management practices (Paulk et al. 1991). The use of maturity models and levels has been popularized in software engineering models through SEI CMM.

In response to staged representation of patterns by Crosby and Humphrey, Weinberg gave patterns of subculture with an emphasis on the congruence between what people say and what they do in different parts of a software organization (Weinberg 1991). Although his subculture patterns largely overlap with observations of Crosby and Humphrey, Weinberg avoids the use of word "maturity" and explains that any subculture is not superior to another, thus, any pattern can be successful depending on the circumstances such as right product and customer.

While CMM was gaining popularity in the United States, in 1993, ISO has started an European initiative called Software Process Improvement and Capability dEtermination (SPICE), which has started working on the development of an assessment framework (later formalized as ISO/IEC 15504) and defined six capability levels for software processes. A table that roughly matches the patterns and the levels is given in Table 79.3.

It is also possible to roughly map the organizational competence to progressive patterns using the four stages of competence learning model used in psychology. In the model, the states involved in the process of progressing are expressed in terms of awareness and consciousness. Briefly, at the lowest stage, the organizations are incompetent and they are not aware of that situation. At middle stages, although they are not still considered competitive, the management awareness increases and the management realizes more and more for the need to improve to be competent in the business. As the organization progresses upper stages, the organization is competitive and is aware of it. At the highest stage, despite being competent, the organization gradually loses awareness on its competency since improvement is internalized as a continuous organizational function.

Today, the concept of maturity models and levels is implemented in popular models for SPI (von Wangenheim et al. 2010) and it has been adapted by a number of organizations including international research organizations such as ISO and professional framework governing bodies such as Project Management Institute and Information Systems Audit and Control Association (Grant and Pennypacker 2006).

79.3.2 Models for SPI

Among the models, CMMI and ISO/IEC 15504 framework models are prevalent and have been accepted worldwide (Pino et al. 2008; von Wangenheim et al. 2010). The essence of implementing CMMI and ISO/IEC 15504 models is the adherence to processes and achieving predefined goals set by the models in order to improve business in terms of quality, cost, and schedule. Both models consider quantitative process management and continuous process improvement as the key factors of success in any realm of software business. Since their introduction, the models have evolved in response to theoretical and practical problems and suggestions from software community and emerging needs of the software business. Moreover, the models have influenced each other mutually. The similarities, differences, and compatibility between CMMI and ISO/IEC 15504 are among interesting topics of SPI research (Paulk 1999; Pino et al. 2010; Pardo et al. 2011).

Among many others, important models are BOOTSTRAP, software development capability evaluation (SDCE), and Trillium (von Wangenheim et al. 2010). BOOTSTRAP was developed under European Union (EU) information technologies program (Kuvaja 1999). Initially, it extended SEI Software Capability Maturity Model (SW-CMM) taking ISO 9000 and European Space Agency process standards into account. Later, it was updated to align with ISO 12207 and ISO/IEC 15504. Trillium is a telecommunications extension of SEI SW-CMM that combines it with ISO 9000 and relevant Malcolm Baldrige criteria (Coallier 1995). It was developed by Bell Canadian telecom companies. SDCE is a method for assessing an organization's ability to develop software for mission critical software-intensive systems development. It was developed by the U.S. Air Force to be used in system acquisitions (Babel 1997).

In this section, CMMI and ISO/IEC 15504 frameworks and software process models are summarized. Before proceeding, we would like to draw readers' attention to the following points to clarify possible sources of confusion. CMMI and ISO/IEC 15504 models are not models of software engineering domain knowledge. The models are not intended to define processes, procedures, or work practices for any given activity of an adopting organization. For example, when the software testing process is concerned in ISO/IEC 15504-5 model, defining a model compliant test process will not help if the tester does not know software testing. Having discussed the underlying concepts around process-oriented quality improvement in the preceding sections, we assume that reader has an understanding that the processes are not for the sake of processes. CMMI and ISO/IEC 15504 prescribe norms that processes should possess and a standardized system for measuring their capability. The practices included in the models are intended for improving the existing work practices of an organization.

79.3.2.1 CMMI

SEI published the original SW-CMM model in 1991 and later adapted the model to different domains such as systems engineering, software acquisition, and integrated product development (Paulk 2009). Starting from 2000, SEI combined maturity models into a single CMMI integration framework. CMMI for development (CMMI-DEV), CMMI for acquisition, and CMMI for services are the three flavors of the framework, called as "constellations," developed for the domains of software development, acquisition, and service, respectively.

CMMI-DEV model combines best practices and goals that help software development organizations improve their processes. The model has a process focus for improvement and follows the principle that "the quality of a system or product is highly influenced by the quality of the process used to develop and maintain it" (SEI 2010). The model defines processes as the concept that glues people, procedures and methods, and tools and equipment dimensions of developing software and constitutes the essential means to achieve the organizations business objectives. The conception and development of the capability maturity model was influenced by TQM principles and SPC (Paulk 2009). Humphrey (1988) points that "a truly effective process should be predictable" and explains that process control ensures the same results if we work the same way and when a process is under statistical control, it is possible to anticipate quality cost and schedule within their limits. CMMI model suggests that successive attainment of the goals and the institutionalization of the addressed practices across the software processes will improve software quality, schedule, and costs whilst decreasing costs of software quality. Figure 79.5 illustrates Knox's theoretical model for CoSQ whcrc total CoQ gradually decreases as maturity increases.

CMMI-DEV model consists of process areas where each area encompasses goals and a cohesive set of practices to be fulfilled by organization's processes in order to make improvements in that area (Table 79.4, Figure 79.6). A set of goals, called "generic," apply to multiple process areas and they describe

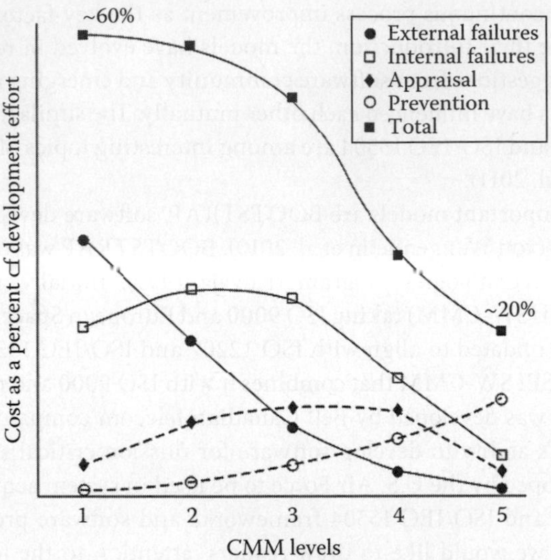

FIGURE 79.5 Knox theoretical CoSQ model. (Adapted from Knox, S.T., *Digit. Tech. J.*, 5, 9, 1993. With permission.)

TABLE 79.4 CMMI-DEV Process Areas

Name (Abbreviation)	Area	Maturity Level
Configuration management (CM)	Support	2
Measurement and analysis (MA)	Support	2
Project monitoring and control (PMC)	Project management	2
Project planning (PP)	Project management	2
Process and product quality assurance (PPQA)	Support	2
Requirements management (REQM)	Project management	2
Supplier agreement management (SAM)	Project management	2
Decision analysis and resolution (DAR)	Support	3
Integrated project management (IPM)	Project management	3
Verification (VER)	Engineering	3
Validation (VAL)	Engineering	3
Product integration (PI)	Engineering	3
Requirements development (RD)	Engineering	3
Technical solution (TS)	Engineering	3
Organizational process definition (OPD)	Process management	3
Organizational process focus (OPF)	Process management	3
Organizational training (OT)	Process management	3
Risk management (RSKM)	Project management	3
Organizational process performance (OPP)	Process management	4
Quantitative project management (QPM)	Project management	4
Causal analysis and resolution (CAR)	Support	5
Organizational performance management (OPM)	Process management	5

the characteristics that must be exhibited for institutionalizing the processes that implement a process area. The goals and related practices that the uniquely characterize a process area called specific goals and specific practices. Subpractices are more concrete and detailed descriptions that help interpreting and implement specific and generic practice. Similarly, work products are supplied to complement practice definitions with typical activity inputs and outputs.

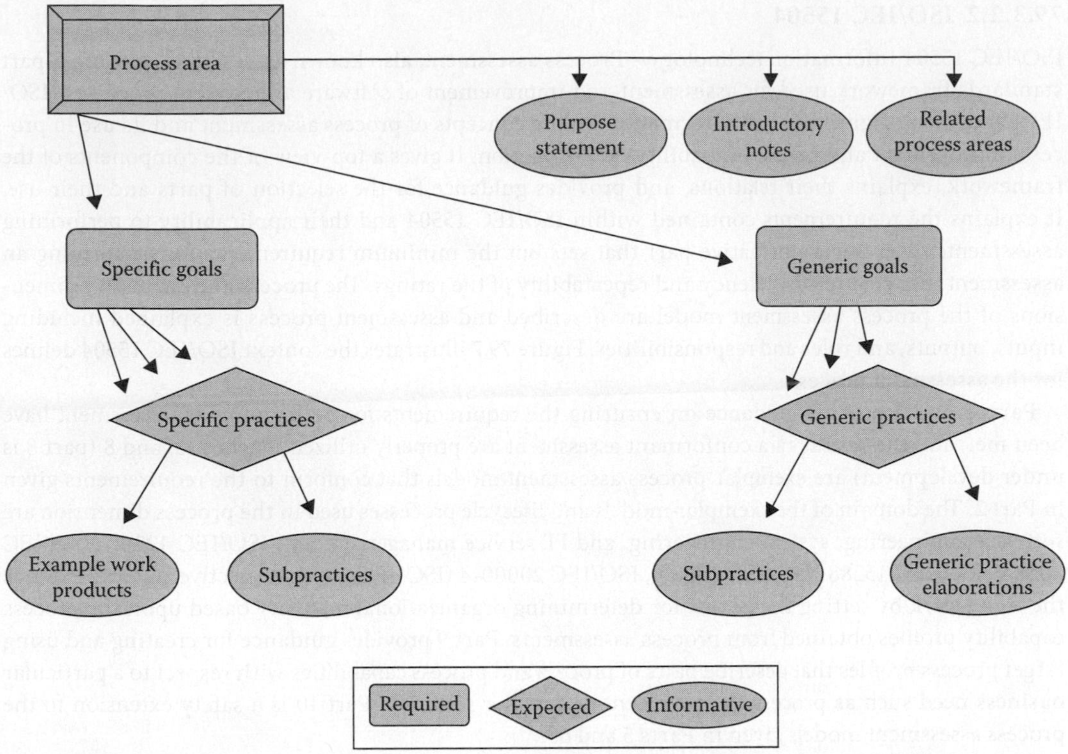

FIGURE 79.6 Process area. (Adapted from SEI, *CMMI for Development* (*CMMI-DEV*), Version 1.3, Software Engineering Institute, Pittsburgh, PA, 2010. With permission.)

CMMI-DEV supports continuous and staged representations of processes and associates them with capability levels and maturity levels, respectively. Capability levels define goals to achieve improvement incrementally in an individually selected process area. Maturity levels address organizations improvement standing with respect to the attainment of goals in successive sets of predefined process areas. The levels can also be used to appraise the corresponding levels of process capability or organizational maturity of an organization. The model defines a "target profile" as the combination of process areas and capability levels selected for improvement. "Achievement profile" represents the actual progress for a combination. Capability evaluations are performed and ratings are given using SEI official appraisal method, standard CMMI appraisal method for process improvement (SCAMPI) (Team 2011). Successful type "A" appraisals are awarded with SEI certificates for the achieved level.

Personal software process and team software process are works of Humphrey that complement CMMI and operationalize the underlying principles of CMMI to individual developer(s) and development teams (Humphrey 2000a,b). These models were developed as a response to the early criticism on SW-CMM that process-orientation ignores human factors. The models have been reported to be successfully used independent from CMMI in companies including Microsoft, IBM, and governmental software organizations (Davis and Mullaney 2003).

CMMI models have been used by more than 5000 companies from over 70 countries. According to a recent maturity profiles report, which is updated regularly (Software Engineering Institute 2012), 4846 appraisals have been reported since 2006. More than 66% of appraised organizational units have a workforce less than 100 employees, and the majority of reported appraisals are maturity level 2 and 3.

79.3.2.2 ISO/IEC 15504

ISO/IEC 15504 Information technology—Process assessment, also known as SPICE is a multiple-part standards framework used for assessment and improvement of software and system processes (ISO/IEC 2004). Part 1 gives general information on the concepts of process assessment and its use in process improvement and process capability determination. It gives a top-view of the components of the framework, explains their relations, and provides guidance for the selection of parts and their use. It explains the requirements contained within ISO/IEC 15504 and their applicability to performing assessments. Part 2 is a normative part that sets out the minimum requirements for performing an assessment that ensure consistency and repeatability of the ratings. The process and capability dimensions of the process assessment model are described and assessment process is explained including inputs, outputs, and roles and responsibilities. Figure 79.7 illustrates the context ISO/IEC 15504 defines for the assessment process.

Parts 3 and 4 provide guidance on ensuring the requirements for performing an assessment have been met, and the results of a conformant assessment are properly utilized. Parts 5, 6, and 8 (part 8 is under development) are exemplar process assessment models that conform to the requirements given in Part 2. The domain of the exemplar models and lifecycle processes used in the process dimension are software engineering, system engineering, and IT service management and ISO/IEC 12207 (ISO/IEC 2008a), ISO/IEC 15288 (ISO/IEC 2008b), ISO/IEC 20000-4 (ISO/IEC 2010), respectively. Part 7 extends the framework by setting the norms for determining organizational maturity based upon the process capability profiles obtained from process assessments. Part 9 provides guidance for creating and using target process profiles that describe pairs of process and process capabilities with respect to a particular business need such as process improvement or supplier selection. Part 10 is a safety extension to the process assessment models given in Parts 5 and 6.

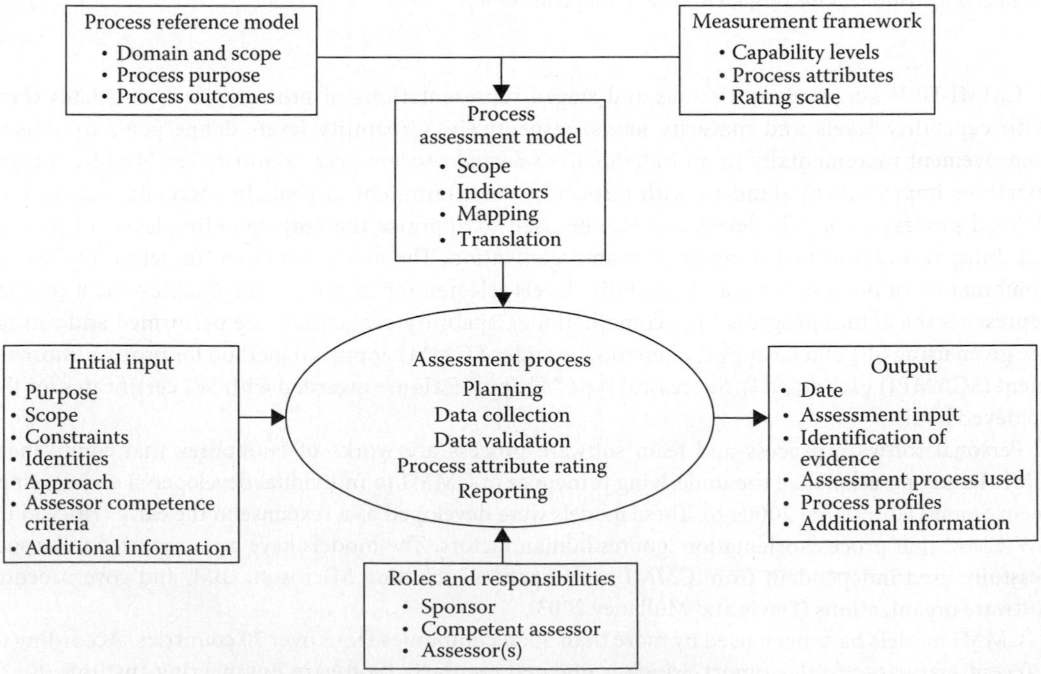

FIGURE 79.7 (Adapted from ISO/IEC 15504 Assessment Model Normative Elements (Adapted from ISO/IEC 15504-2: Information Technology—Process Part 2: Performing an Assessment, International Standards Organization & International Electrotechnical Commission. With permission.)

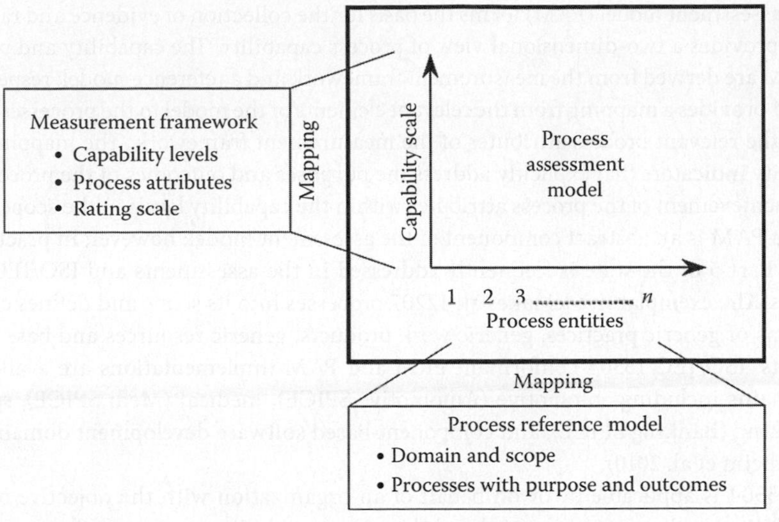

FIGURE 79.8 Process assesment model. (Adapted from ISO/IEC, *ISO/IEC 15504: Information Technology—Process Assessment—Parts [1-10]*, International Standards Organization & International Electrotechnical Commission, 2004. With permission.)

Figure 79.8 illustrates the core components of the framework. The measurement framework comprises six-capability levels and process attributes to be rated on a four-scale rating (Table 79.5). A process reference model (PRM) is a set of processes in a defined domain such as software development. A conformant reference model fulfills the requirements involving process descriptions, process purposes and outcomes, the descriptions of relations between the processes, the relation between the model, and its context of use and community of interest.

ISO/IEC 12207 software life cycle processes standard is a conformant PRM (ISO/IEC 2008a). The standard contains 48 processes and related activities, and tasks that are to be applied during the acquisition, development, operation, and maintenance of software products. It's purpose is to provide a defined set of processes to facilitate communication among acquirers, suppliers, and other stakeholders in the life cycle of a software product.

TABLE 79.5 SPICE Measurement Framework: Capability Levels, Process Attributes, and Rating Scale

Level 5—Optimizing: The process is continuously improved to meet relevant current and projected business goal	PA.5.1: Process innovation PA.5.2: Process optimization
Level 4—Predictable: The process is enacted consistently within defined limits	PA.4.1: Process measurement PA.4.2: Process control
Level 3—Established: A defined process is used based on a standard process	PA.3.1: Process definition PA.3.2: Process deployment
Level 2—Managed: The process is managed and work products are established, controlled, and maintained.	PA.2.1: Performance management PA.2.2: Work product management
Level 1—Performed: The process is implemented and achieves its process purpose	PA.1.1: Process performance
Level 0—Incomplete: The process is not implemented or fails to achieve its purpose	—
N: Not achieved 0%–15% achievement	
P: Partially achieved >15%–50% achievement	
L: Largely achieved >50%–85% achievement	
F: Fully achieved >85%–100% achievement	

A process assessment model (PAM) forms the basis for the collection of evidence and rating of process capability. It provides a two-dimensional view of process capability. The capability and process dimensions of a PAM are derived from the measurement framework and a reference model, respectively. A conformant PAM provides a mapping from the relevant elements of the model to the processes of the selected PRM and to the relevant process attributes of the measurement framework. The mappings are based a set of capability indicators that explicitly address the purposes and outcomes of the processes and demonstrate the achievement of the process attributes within the capability levels in the scope of the PAM.

In Part 2, a PAM is an abstract component of the assessment model; however, in practice, the exemplar PAM in Part 5 of the suite is frequently addressed in the assessments and ISO/IEC 15504-based SPI initiatives. The exemplar model takes all 12207 processes into its scope and defines capability indicators in terms of generic practices, generic work products, generic resources and base practices, and work products. ISO/IEC 15504 conformant PRM and PAM implementations are available for many business domains including automotive (Automotive SPICE), medical (Medi SPICE), space (SPICE 4 SPACE), banking (Banking SPICE), and component-based software development domains (OOSPICE) (von Wangenheim et al. 2010).

ISO/IEC 15504 is applicable by or on behalf of an organization with the objective of understanding any capability gap between the actual and the target capability profiles for the purpose of process improvement or determining the suitability of organization's processes or determining the suitability of another organization's processes for meeting particular requirements. ISO/IEC 15504 framework provide the means to determine process capabilities and can be used for setting improvement targets both at process entity (i.e., process area) and organizational levels.

79.3.3 Results from Implementations

CMMI-DEV and its predecessor SW-CMM have been the dominant models used in model-based SPI investments throughout the world. According to a 2008 literature review (Staples and Niazi 2008), reasons reported for CMM-based SPI includes "software quality" as the most common answer (65%–70%), followed by "development time" and "deployment costs" (around 55%), "productivity" (>40%), and "process visibility" (30%).

Substantial improvements in large companies in terms of software quality, development time, and productivity have been reported by individual and review studies (Agrawal and Chari 2007; Diaz and Sligo 1997; Gibson et al. 2006; Haley 1996; Herbsleb et al. 1997). In a SEI technical report, a survey of 61 assessments from different organizations was conducted, and the results of SEI report in 2006 (Gibson et al. 2006) summarized the empirical results of CMMI-based improvements from 35 organizations including small and mid-sized companies (Table 79.6). The organizations reported significant performance improvements after adopting CMMI, and they attributed their achievement to the model. The reporting organizations mostly defined quality measures as defects, and improvement in product quality is mostly measured by reductions in numbers of defects. Customer satisfaction was measured in terms of award fees and customer survey. Another study explored the relation between process maturity

TABLE 79.6 CMMI Performance Results

Performance Category	Median Improvement	Number of Data Points	Lowest Improvement	Highest Improvement
Cost	34%	29	3%	87%
Schedule	50%	22	2%	95%
Productivity	61%	20	11%	329%
Quality	48%	34	2%	132%
Customer satisfaction	14%	7	−4%	55%
Return on investment	4.0:1	22	1.7:1	27.7:1

and project performance using survey data from 154 experienced software project members (Jiang et al. 2004). The results of the analysis confirmed that software process management maturity is positively associated with project performance.

Evidently, not every organization that has attempted CMMI-based process improvement has succeeded. A group of problems were observed to be general and related to the management of change and underestimated costs and time frames that SPI initiatives typically encounter. Survey results include evidence where SPI efforts are overcome by crisis due political struggles within the organizations or by an underestimated time and resource needs (Herbsleb et al. 1997; Dybå 2003, 2005; Pino et al. 2008). Another group of problems and risks are considered to stem from the characteristics of the CMM itself (Bollinger and McGowan 1991, 2009; Bach 1994; Staples et al. 2007). The model was found to be costly, heavy-weight, bureaucratic, and inherently infeasible for small scale organizations and projects. It was considered to overlook the people (creativity) dimension, not to provide adequate practical guidance on how to improve the process. SCAMPI appraisals were found inflexible with too much emphasis on statistical soundness and fairness. In addition to these causes for failure or inapplicability, the model was criticized to cause over expectations and not to be able to explain high maturity organizations with low performance and low maturity organizations dominating markets.

ISO/IEC 15504 is the second most used model after CMMI. A part of the difference in the usage share has been explained by the emergence of the standard much later than CMMI and continuous marketing of CMMI under the sponsorship of SEI and US Department of Defense. Nevertheless, the utilization of ISO/IEC 15504 standard has begun increasing in a wide-scale and a diversity of uses (Pino et al. 2008, von Wangenheim et al. 2010). This is mainly attributed to the fact that the model was found to be flexible, easy to use and understand. A frequent way of putting ISO/IEC 15504 into action have been through the development and use of conformant process assessment methods, particularly tailored for small and medium sized enterprises (SMEs) (von Wangenheim et al. 2005, 2006). Wangenheim et.al. and Montoni et al. separately reported the results of SPI that are based on ISO/IEC 15504-based conformant models developed to help Brazilian small companies. In von Wangenheim et al. (2006), according to results of a survey performed 1 year after the assessments, they found convincing evidence on the practice of systematic approaches to project management, measurement, customer-relations, software testing, and product installations. Besides, the assessment method for small software companies was found beneficial, because they were able to understand the strengths and weaknesses of their processes and focus on high-priority processes. Similarly, in Montoni et al. (2009), promising improvements were realized at reasonable costs. In the study, it was also reported that the use of a process-centered software engineering environment have catalyzed the improvements, reducing SPI costs up to 60% and amount of project rework up to 36%. Effective uses of ISO/IEC 15504 have been also materialized through the development of domain specific conformant models. For instance, the results from the trail implementations of the S4S model (Cass et al. 2004), which have been developed for space domain, indicate that the model is found useful for effective assessment and addressing paths for improvement. Individual studies on ISO/IEC 15504-based SPI programs also have reported encouraging results where the models were found to enable systematic improvements (Moreau et al. 2003; Truffley et al. 2004).

79.4 Research Issues

We conducted a brief systematic literature review (SLR) to further confirm outstanding research issues. For a more comprehensive SLR, the readers may refer to Unterkalmsteiner et.al. (2012). Our analysis identifies three important topics: SPI for small organizations and agile development, measurement, and automation and tools.

The software processes of organizations are characterized by a vast number of factors, that is, the business goals, the culture, accumulated knowledge and experience, the company size, the market and customer demands, the software domain and environmental and regulatory constraints, etc. Moreover, each execution of an organization's software process is unique as it is dominated by creative human activities.

SPI is challenged by this process diversity issue, and there is no generic reference process model that suits for all software development projects and organizations (Lindvall and Rus 2000). However, SPI models are intended to provide the organizations with a systematic approach and relevant tools for managing the complexity of these factors, increasing their performance; thus, to implement "process-thinking" (Dybå 2005). Organizations increasingly recognize the importance of software processes and model-based SPI is regarded as one of safe paths to improve their business and to survive in the market. In parallel to endeavors for adopting SPI models, SE research has been evaluating the theoretical soundness of SPI models and their validity in practice and proposing improvements to these models. Based on the current trends and analysis of available literature, we address the following issues for future SPI research.

79.4.1 SPI for Small Organizations and Agile Development

SMEs have a growing impact in software development industry. In some countries, the majority of software project development is realized by SMEs. While empirical results have indicated that successful implementation of SPI models were mostly possible in large organizations, SPI research has identified various challenges that make model-based SPI infeasible or risky for SMEs (Emam 2006; Staples et al. 2007; Mishra and Mishra 2008; Pino, García, and Piattini 2008).

A group of factors poses limitations on SMEs such as budget, time, subject-matter expert staff constraints and inadequate guidance. Others are related to structure of the models and their validity in SME business context. Most literature works consensually agree that adopting an SPI model is not just a matter of using the scaled-down versions of the models, and SMEs have highly variable priorities and motivations for doing business when compared to large organizations (Richardson 2001; Mishra and Mishra 2008). Nevertheless, there have been successful SME implementations of the SPI models (Dybå 2003; Gibson et al. 2006). Several factors have been found to be critical in SMEs success with model-based SPI such as model flexibility, improvement paths: continuous or staged, software development team members involvement, involved people's prior experience in the field, existence of tools for self-assessment, and learning strategy (Dybå 2005; Mishra and Mishra 2008; Pino et al. 2008). The problem is increasingly attracting attention from researchers and studies have begun investigating model-based SPI in SME context and collecting evidence on particular requirements and factors of SPI success (Pino et al. 2008). Empirical validation studies and systematic review indicate increasing number of publications on the subject and encourage more empirical results of SPI implementations (Pino et al. 2008; Staples and Niazi 2008).

In addition to proposals for new or tailored models for SMEs, possibilities for harmonizing other software development approaches are being explored (von Wangenheim et al. 2010). Among them, agile methods have been favored in SMEs due to its inherent suitability for small team settings (Dybå and Dingsoyr 2008). Agile methods have also attracted interests of large-scale companies, and attempts have been made to adopt agile practices to large organizations (Auvinen et al. 2005). In agile models, people are in focus, where in SPI models the processes are. Despite this difference, there is a growing discussion on harmonization of model-based SPI and agile practices (Boehm 2002; Nerur et al. 2005; McCaffery et al. 2008; Lami and Falcini 2009). We expect that SPI research will continue to study the context for model-based SPI success for SMEs and agile environments. In the near future, we may expect a separation of SPI models and standards for the SMEs and more emphasis on people and organizational learning factors.

79.4.2 Measurement and Using Statistical Process Control

Clearly, measurement has a crucial role in SPI as it does in any engineering area (Kitchenham 1996; Dybå 2005). There is a vast diversity of measurement related issues for current and future SPI research that seeks to address and eliminate barriers in the effective use of measurement and in making progress through capability levels while implementing process models. However, we believe the following measurement issues will be influential in model-based SPI.

The success and continuity of SPI programs and decisions rely on a well-established measurement program since the benefits from SPI efforts can be justified using proper measures. Software organizations need well-defined and valid SPI measures, measurement methods, performance baselines, and relevant guidance that are applicable even at low maturity levels (Dybå 2005; Unterkalmsteiner et al. 2010). SPI change measurement is a promising topic for future research where means for supporting organizations in measuring, evaluating, benchmarking, and communicating the impact of improvement initiatives can be derived and SPI implementations can be encouraged.

A CMMI compliant organization is required to develop capabilities (e.g., achieve measurement and analysis process area goals) for product, project, and process measurement at starting at lower maturity levels (repeatable and defined). Starting from managed level and measurement becomes a critical capability in order to implement process areas effectively. Similarly, ISO/IEC 15504 defines a measurement process in the assessment model and proposes many other measurement practices in the whole process system. ISO/IEC 15504 model also requires processes to fulfill process measurement process attribute at predictable process capability level (level 5) and higher. SPC is a very powerful tool in process management and improvement, and SPI models direct SPI efforts to apply SPC techniques. However, despite the models exhort organizations to do measurement, the existence of SPC-enabling requirements at high maturity levels (level 4 and above) and the common sense belief that SPC can only be performed after achieving level 4 almost prohibits organizations implementing SPC techniques earlier (Sargut and Demirörs 2006). It normally takes up to 5–10 years to achieve level 4 for an emergent company, which implies that the company will not benefit from SPC techniques before level 4. Several case studies and experience reports demonstrate evidence that SPC can be used at lower levels provided that processes are stabilized and measurements capability and reliable data are available. Another challenge that organizations encounter is the selection of which processes to apply SPC (Sargut and Demirörs 2006; Boffoli et al. 2008) and lack of tools for assessing processes for quantitative management (Tarhan and Demirors 2012). SPC might not be applicable to all software processes and critical processes in a software organization should be considered for SPC, although all SPC techniques seem to be applicable. Therefore, organizations need methods to systematically assess their processes and apply quantitative techniques to understand their potential for improvement. There is limited research on this issue where further research output can increase effective application of techniques such as SPC and Six-Sigma, shorten SPI cycles, and decrease development costs.

Another issue that undermines SPI initiatives, and can be generalized to software engineering, is related to use of software measures and metrics that lack theoretical rigor and validity. Despite the progress in software measurement research and international standardization efforts (e.g., ISO/IEC 14143), there is a lag between research and practice (Pfleeger et al. 1997). Research for defining and validating software measures and developing measurement methods is expected to continue together with further standardization and communication.

79.4.3 Automation/Tools

Process definitions are central for both CMMI defined and ISO/IEC 15504 established levels. In model-based SPI, they also provide the comprehensive and efficient means to describe an organization's processes and compare them to prescriptive process libraries of an SPI model. Therefore, integrating process modeling with process improvement activities can enhance the efficiency of SPI initiatives (Makinen et al. 2007). Tool support is essential for modeling processes enabling formalism in descriptions and other facilities for process automation, error checking, integration, and simulation. However, the relation between process modeling and process improvement has not been examined in the literature thoroughly.

Another challenge in SPI is the use of automated tools at large scale and in integration. For example, a CMM-adopting organization improvement focus extends from individual project performance to multiple projects and then to organizational performance as the organization progresses to upper levels.

This implies the use of integrated tools such as measurement, process modeling, workflow, and document management tools used organization-wide (Sharp and McDermott 2009). Although large companies can afford the burden to develop or acquire custom solutions, the problem challenges the SMEs requiring more resources and subject matter expertise (Montoni et al. 2009). SPI tools are among the important topics of future SPI research.

79.5 Summary

In this chapter, we introduced software quality and model-based process improvement. Quality is more and more often seen as a critical software attribute and a determinant of business success. The absence of quality in software products and services results in dissatisfied users, financial loss, and may even endanger to our lives. SPI is a process-oriented approach to address quality problems.

We presented underlying principles by focusing on quality, process and quality, and the CoQ. We explained quality using different defining approaches, such as transcendental, product, user, manufacturing, and value-based approaches.

We then defined process and quality starting with the concept of process as widely appreciated as the proper ground for improving product quality and productivity. We highlighted the importance of SPC, plan do check act, and TQM.

We also explained CoQ. CoQ analysis and techniques have been in use for more than 50 years and there are multiple models for CoQ. These models are the effective tools in feasibility analysis of SPI programs and the measurement and evaluation of the program performance. Both theory and experience advise investing on prevention and appraisal costs to get the highest returns from the decreased costs of appraisal and failure.

In terms of best practices, we focused on software process maturity, models for SPI, and results from implementations. The use of maturity models has been popularized in software engineering through the SEI software-CMM, which was published in 1991. In 1993, in Europe, ISO started the SPICE initiative. Both these models define capability levels for software processes and corresponding key process areas. Not every organization that has attempted model-based process improvement has succeeded. A group of problems were observed to be general and related to the management of change and to underestimated costs and timeframes. Survey results also included evidence that SPI efforts were overcome by crisis due political struggles within the organizations.

Software processes are characterized by a vast number of factors, that is, business goals, organizational culture, accumulated knowledge and experience, company size, the market, domain and environmental and regulatory constraints, etc. SPI is thus challenged by this process diversity, and there is no generic reference model that suits all software development projects and organizations. Furthermore, our analysis showed that the main areas of future research should focus on SPI for small organizations and agile development, measurement, and using SPC and automation/tools.

Key Terms

Capability maturity model integration (CMMI): An approach to process improvement that provides organizations with models that consist of the essential elements of effective processes. CMMI models define a number of levels of process maturity in terms of the key processes undertaken at these levels.

Cost of quality (CoQ): Any cost that would not have been expended if the quality was perfect such that there were no possibility of substandard service, failure of products, or defects in their manufacture or development.

FURPS: An acronym representing a model for classifying software quality attributes: Functionality, usability, reliability, performance, and supportability. The model was proposed by Robert Grady and Hewlett Packard in 1987 (Grady and Caswell 1987).

ISO/IEC 15504: A generic framework for the assessment of software and system processes published as an international standard. Used most widely by software and systems development organizations in process improvement and capability determination contexts. Also known as SPICE.

Model-based software process improvement: An approach to software process improvement where collections of best practices are organized into reference process models (e.g., CMMI models and ISO 15504-15505) that help organizations improve their processes.)

Plan–do–act–check cycle: An iterative four-step management method used for continuous improvement of processes and products. Also known as Deming or Shewart cycle.

Six-Sigma: Six Sigma is a data-driven quality improvement strategy that values defect prevention over defect detection. It has a focus on the control and the reduction of process variation through the application of statistical methods.

Small and medium sized enterprise (SME): The organizations whose number of employees falls below certain limits. The defined limits vary across countries and organizations.

Software quality requirements and evaluation (SQuaRE): The second-generation software quality standards being published by ISO and aims to harmonize ISO 9106 (product quality) and ISO 14598 (product evaluation).

Total quality management (TQM): An integrated system of principles and procedures whose goal is to improve the quality of an organization's goods and services.

Further Information

As a sequel to book by Humphrey (1989), Software Engineering Institute (1995) provides a description and technical overview of the capability maturity model for software process management, along with guidelines for improving software process management. Hunter et al. (2001) summarizes the best practices for software process improvement (SPI) and related international standards and provides the reader with a collection of republished papers from journals in SPI and papers from international experts in the field. Emam et al. (1997) covers both the theory of SPICE (ISO 15504) and its practical applications, including the lessons learned from the SPICE trials. Kan (2003) gives a thoroughly updated overview and implementation guide for metrics and models used in software quality engineering, which are essential in support of process improvement. In his book, Kemp (2006) gives a general and practical guide to the basic terms, concepts, and tools for defining, measuring, and managing quality as well as giving an insight into history of quality movements. In particular, part 3 of Kemps's book addresses software quality and process improvement. Similarly, two handbooks published by American Society for Quality are two comprehensive guides to Software Quality Engineering discipline, which includes quite valuable information for achieving quality through process improvement (Borror 2008, Westfall 2009).

Research articles on process and quality improvement regularly appear in the following journals, among others: *Journal of Software: Evolution and Process, Software Quality Journal, IEEE Software, IET Software, IEEE Transactions on Software Engineering, Information Technology Journal*, and *Journal of Systems and Software*.

The following annual conferences regularly present research work in software process improvement: *EuroSPI*^{2}*, Product-Focused Software Process Improvement, SPICE Conference on Process Improvement and Capability dEtermination in Software, Systems Engineering and Service Management*, and *International Conference on Software Quality*.

The reader is also referred to the following web sites, which are hosted by organizations that include valuable resources in the field and include many other references to research in software quality and model-based process improvement:

http://www.eurospi.net, European System & Software Process Improvement and Innovation.

http://www.sei.cmu.edu/cmmi, Software Engineering Institute.

http://asq.org, American Society for Quality.

http://www.spiceusergroup.org/, SPICE Users.

References

Agrawal, M. and K. Chari. 2007. Software effort, quality, and cycle time: A study of CMM level 5 projects. *IEEE Transactions on Software Engineering* 33 (3):145–156.

Auvinen, J., R. Back, J. Heidenberg, P. Hirkman, and L. Milovanov. 2005. *Improving the Engineering Process Area at Ericsson with Agile Practices: A Case Study*. Turku Centre for Computer Science.

Babel, P. 1997. Software development capability evaluation: An integrated systems and software approach. *Crosstalk: The Journal of Defense Software Engineering* 10 (4):3–7.

Bach, J. 1994. The Immaturity of the CMM. *American Programmer* (September issue): 13–18.

Basili, V.R. 1985. Quantitative evaluation of software engineering methodology. In *Proceedings of the First Pan Pacific Computer Conference*. Melbourne, Australia.

Boehm, B. 2002. Get ready for agile methods, with care. *Computer* 35 (1):64–69.

Boehm, B.W., J.R. Brown, H. Kaspar, M. Lipow, G.J. MacLeod, and M.J. Merrit. 1979. *Characteristics of Software Quality*, Holland Publishing Company.

Boffoli, N., G. Bruno, D. Caivano, and G. Mastelloni. 2008. Statistical process control for software: A *2nd International Symposium on Empirical Software Engineering and Measurement (ESEM 2008).*

Bollinger, T. and C. McGowan. 2009. A critical look at software capability evaluations: An update. *Software, IEEE* 26 (5):80–83.

Bollinger, T.B. and C. McGowan. 1991. A critical look at software capability evaluations. *Software, IEEE* 8 (4):25–41.

Borror, C.M. 2008. *The Certified Quality Engineer Handbook*. Milwaukee, WI: American Society for Quality Press.

Cass, A., C. Völcker, R. Ouared, A. Dorling, L. Winzer, and J.M. Carranza. 2004. SPICE for SPACE trials, risk analysis, and process improvement. *Software Process: Improvement and Practice* 9 (1):13–21.

Coallier, F. 1995. TRILLIUM: A model for the assessment of telecom product development & support capability. *Software Process Newsletter* 2 (1):3–8.

Crosby, P.B. 1979. *Quality Is Free: The Art of Making Quality Certain*. Vol. 94. New York: McGraw-Hill.

Davis, N. and J. Mullaney. 2003. *The Team Software ProcessSM (TSPSM) in Practice: A Summary of Recent Results*. Pittsburgh, PA: Software Engineering Institute, Carnegie-Mellon University.

Diaz, M. and J. Sligo. 1997. How software process improvement helped Motorola. *Software, IEEE* 14 (5):75–81.

Dromey, R.G. 1995. A model for software product quality. *IEEE Transactions on Software Engineering* 21 (2):146–162.

Dybå, T. 2003. Factors of software process improvement success in small and large organizations: An empirical study in the Scandinavian context. *ACM SIGSOFT Software Engineering Notes* 28 (5):148–157.

Dybå, T. 2005. An empirical investigation of the key factors for success in software process improvement. *IEEE Transactions on Software Engineering* 31 (5):410–424.

Dybå, T. and T. Dingsoyr. 2008. Empirical studies of agile software development: A systematic review. *Information and Software Technology* 50 (9–10):833–859.

Emam, K. 2006. An Overview of Process Improvement in Small Settings. In *Web Engineering*, edited Mosley, pp. 261–275. Springer Berlin, Heidelberg, Germany.

Emam, K. E., Melo, W., and Drouin, J. N. 1997. *SPICE: The Theory and Practice of Software Process Determination*. Los Alamitos, CA: IEEE Computer Society Press.

Feiler, P.H. and W.S. Humphrey. 1993. Software process development and enactment: Concepts and *International Conference on the Software Process*. Berlin: IEEE Computer Society Press.

Florac, W.A. and A.D. Carleton. 1999. *Measuring the Software Process: Statistical Process Control for Software Process Improvement*. Boston, MA: Addison-Wesley Longman Publishing Co., Inc.

Garvin, D.A. 1984. What does" product quality" really mean? *Sloan Management Review* 26 (1):25–43.

Gibson, D., D. Goldenson, and K. Kost. 2006. *Performance Results of CMMI-Based Process Improvement*. Pittsburgh, PA: Software Engineering Institute, Carnegie Mellon University, CMU/SEI-2006-TR-004.

Gilbreth, F.B. 1911. *Motion Study: A Method for Increasing the Efficiency of the Workman*. New York: D. Van Nostrand Company.

Grady, R.B. and D.L. Caswell. 1987. *Software Metrics: Establishing A Company-Wide Program*. Upper Saddle River, NJ: Prentice-Hall, Inc.

Grant, K.P. and J.S. Pennypacker. 2006. Project management maturity: An assessment of project management capabilities among and between selected industries. *IEEE Transactions on Engineering Management*, 53 (1):59–68.

Haley, T.J. 1996. Software process improvement at Raytheon. *Software, IEEE* 13 (6):33–41.

Herbsleb, J., D. Zubrow, D. Goldenson, W. Hayes, and M. Paulk. 1997. Software quality and the capability maturity model. *Communications of the ACM* 40 (6):30–40.

Hoyer, R.W. and B.B.Y. Hoyer. 2001. What is quality. *Quality Progress* 34 (7):53–62.

Humphrey, W.S. 1988. Characterizing the software process: A maturity framework. *Software, IEEE* 5 (2):73–79.

Humphrey, W.S. 1989. *Managing the Software Process*. Vol. 88. Reading, MA: Addison-Wesley.

Humphrey, W.S. 2000a. *The Personal Software Process (sm)(PSP (sm))*. Pittsburgh, PA: Software Engineering Institute, Carnegie Mellon University.

Humphrey, W.S. 2000b. *Team Software Process (TSP)*: Wiley Online Library.

Hunter, R., M. Paulk, and R. Thayer, eds. 2001. *Software Process Improvement*. Hoboken, NJ: Wiley-IEEE Press.

ISO/IEC. 1999. *ISO/IEC 14598:1999—Software Engineering—Product Quality Parts [1–6]*. International Standards Organization & International Electrotechnical Commission.

ISO/IEC. 2001. *ISO/IEC 9126-1:2001 Software Engineering—Product Quality Parts [1–5]*. International Standards Organization & International Electrotechnical Commission.

ISO/IEC. 2004. *ISO/IEC 15504: Information Technology—Process Assessment—Parts [1–10]*. International Standards Organization & International Electrotechnical Commission.

ISO/IEC. 2005. *ISO/IEC 25000:2005 Software Engineering—Software Product Quality Requirements and Evaluation (SQuaRE)—Guide to SQuaRE*. International Standards Organization & International Electrotechnical Commission.

ISO/IEC. 2008a. *ISO/IEC 12207:2008: Systems and Software Engineering—Software Life Cycle Processes*. International Standards Organization & International Electrotechnical Commission.

ISO/IEC. 2008b. *ISO/IEC 15288:2008 Systems and Software Engineering—System Life Cycle Processes*. International Standards Organization & International Electrotechnical Commission.

ISO/IEC. 2010. *ISO/IEC TR 20000-4:2010 Information Technology—Service Management—Part 4: Process Reference Model*. International Standards Organization & International Electrotechnical Commission.

Jiang, J.J., G. Klein, H.G. Hwang, J. Huang, and S.Y. Hung. 2004. An exploration of the relationship between software development process maturity and project performance. *Information & Management* 41 (3):279–288.

Juran, J.M., F.M.J. Gryna, and R.S. Bingham. 1974. *Quality Control Handbook*. London, U.K.: McGraw-Hill Book Company, Chapters 9, p. 22.

Kan, S.H. 2003. *Metrics and Models in Software Quality Engineering*. Uttar Pradesh, India: Pearson Education India.

Karg, L. M., Grottke, M., and Beckhaus, A. (2011). A systematic literature review of software quality cost *Systems and Software*, 84(3):415–427.

Knox, S.T. 1993. Modeling the cost of software quality. *Digital Technical Journal* (Fall issue):9–16.

Kitchenham, B.A. 1996. *Software Metrics: Measurement for Software Process Improvement*. Malden, MA: Blackwell Publishers, Inc.

Knox, S.T. 1993. Modeling the cost of software quality. *Digital Technical Journal* 5:9–9.

Kuvaja, P. 1999. Bootstrap 3.0—A spice1 conformant software process assessment methodology. *Software Quality Journal* 8 (1):7–19.

Lami, G. and F. Falcini. 2009. Is ISO/IEC 15504 Applicable to Agile Methods? In *Agile Processes in Extreme Programming*, Lecture Notes in Business Information Processing Series, Vol. 31, Heidelberg, Germany.

Lindvall, M. and I. Rus. 2000. Process diversity in software development. *IEEE Software* 17 (4):14–18.

Westfall, L. 2009. *The Certified Software Quality Engineer Handbook*. Milwaukee, WI: American Society for Quality Press.

Makinen, T., T. Varkoi, and J. Soini. 2007. Integration of software process assessment and modeling. In *Proceedings of PICMET'07 Conference*, August 5–9, 2007, Portland, OR, pp. 2476–2481.

McCaffery, F., M. Pikkarainen, and I. Richardson. 2008. Ahaa—Agile, hybrid assessment method for automotive, safety critical SMEs. In *30th International Conference on Software Engineering*, Leipzig, Germany, pp. 551–560.

McCall, J.A., P.K. Richards, G.F. Walters, and Rome Air Development Center. 1977. *Factors in Software Quality: Concepts and Definitions of Software Quality*. Rome Air Development Center, Air force Systems Command.

McFeeley, B. 1996. *IDEAL: A User's Guide for Software Process Improvement*. Pittsburgh, PA: Software Engineering Institute, Carnegie Mellon University.

Merriam-Webster online. 2012. Available from www.merriam-webster.com/ (accessed Nov 23, 2012).

Mishra, D. and A. Mishra. 2008. Software Process Improvement Methodologies for Small and Medium *International Conference, PROFES 2008*. Monte Porzio Catone, Italy.

Montoni, M.A., A.R. Rocha, and K.C. Weber. 2009. MPS. BR: A successful program for software process improvement in Brazil. *Software Process: Improvement and Practice* 14 (5):289–300.

Moreau, B., C. Lassudrie, B. Nicolas, O. I'Homme, C. d'Anterroches, and G.L. Gall. 2003. Software quality improvement in France Telecom research center. *Software Process: Improvement and Practice* 8 (3):135–144.

Nerur, S., R.K. Mahapatra, and G. Mangalaraj. 2005. Challenges of migrating to agile methodologies. *Communications of the ACM* 48 (5):72–79.

Pardo, C., F.J. Pino, F. García, M. Piattini Velthius, and M.T. Baldassarre. 2011. Trends in Harmonization Models. In *5th Evaluation of Novel Approaches to Software Engineering*. Athens, Greece.

Paulk, M.C. 1999. Analyzing the Conceptual Relationship Between ISO/IEC 15504 (Software Process Capability Maturity Model for Software). In *1999 International Conference on Software Quality*.

Paulk, M.C. 2009. A history of the capability maturity model for software. *ASQ Software Quality Professional* 12:5–19.

Paulk, M.C., B. Curtis, and M.B. Chrissis. 1991. *Capability Maturity Model for Software (CMU/SEI-91-TR-24)*. Pittsburgh, PA: Software Engineering Institute, Carnegie Mellon University.

Peterson, L., 1996. *Fatal Defect: Chasing Killer Computer Bugs*. New York: Times Books (Random).

Pfleeger, S.L., R. Jeffery, B. Curtis, and B. Kitchenham. 1997. Status report on software measurement. *Software, IEEE* 14 (2):33–43.

Pino, F.J., M.T. Baldassarre, M. Piattini, and G. Visaggio. 2010. Harmonizing maturity levels from CMMI-DEV and ISO/IEC 15504. *Journal of Software Maintenance and Evolution: Research and Practice* 22 (4):279–296.

Pino, F.J., F. García, and M. Piattini. 2008. Software process improvement in small and medium software enterprises: A systematic review. *Software Quality Journal* 16 (2):237–261.

Pirsig, R.M. 1974. *Zen and the Art of Motorcycle Maintenance*. London, U.K.: Bodley Head.

Poppendieck, M. and T. Poppendieck. 2003. *Lean Software Development: An Agile Toolkit*. Reading, MA: Addison-Wesley Professional.

Prahalad, C.K., and M.S. Krishnan. 1999. The new meaning of quality in the information age. *Harvard Business Review* 77 (5):109.

Radice, R.A., J.T. Harding, P.E. Munnis, and R.W. Phillips. 1985. A programming process study. *IBM Systems Journal* 24 (2):91–101.

Repenning, N.P. 2001. Understanding fire fighting in new product development. *Journal of Product Innovation Management* 18 (5):285–300.

Richardson, I. 2001. Software process matrix: A small company SPI model. *Software Process: Improvement and Practice* 6 (3):157–165.

Sargut, K.U. and O. Demirörs. 2006. Utilization of statistical process control (SPC) in emergent software organizations: Pitfalls and suggestions. *Software Quality Journal* 14 (2):135–157.

Schiffauerova, A. and V. Thomson. 2006. A review of research on cost of quality models and best practices. *International Journal of Quality & Reliability Management* 23 (6):647–669.

SEI. 2010. *CMMI for Development (CMMI-DEV)*, Version 1.3. Pittsburgh, PA: Software Engineering Institute.

Software Engineering Institute. 1995. C. Paulk Mark, V. Weber Charles, Curtis Bill, and Chrissis Mary Beth (Eds.). *Maturity Model: Guidelines for Improving the Software Process*. Reading, MA: Addison-Wesley.

Software Engineering Institute. 2012. Company profiles. Available from http://www.sei.cmu.edu/cmmi/why/profiles

Sharp, A. and P. McDermott. 2009. *Workflow Modeling: Tools for Process Improvement and Applications Development*. London, U.K.: Artech House Publishers.

Shewhart, W.A. 1931. *Economic Control of Quality of Manufactured Product*. Vol. 509. Milwaukee, WI: American Society for Quality.

Staples, M., and M. Niazi. 2008. Systematic review of organizational motivations for adopting CMM-based SPI. *Information and Software Technology* 50 (7–8):605–620.

Staples, M., M. Niazi, R. Jeffery, A. Abrahams, P. Byatt, and R. Murphy. 2007. An exploratory study of why organizations do not adopt CMMI. *Journal of Systems and Software* 80 (6):883–895.

Tarhan, A., and Demirors O. Apply Quantitative Management Now. *Software, IEEE* 29.3 (2012): 77.

Taylor, F.W. 1911. *The Principles of Scientific Management*. New York: Harper.

SCAMPI Upgrade Team. 2011. Standard CMMI Appraisal Method for Process Improvement (SCAMPI) A, Version 1.3: Method Definition Document, Software Engineering Institute (SEI).

Truffley, A., B. Grove, and G. McNair. 2004. SPICE for small organizations. *Software Process: Improvement and Practice* 9 (1):23–31.

Unterkalmsteiner, M., T. Gorschek, A. K. M. M. Islam, Cheng Chow Kian, R. B. Permadi, and R. Feldt. Measurement of Software Process Improvement—A Systematic Literature Review. *IEEE Transactions* 38 (2):398–424.

von Wangenheim, C.G., A. Anacleto, and C.F. Salviano. 2006. Helping small companies assess software processes. *Software, IEEE* 23 (1):91–98.

von Wangenheim, C.G., J.C.R. Hauck, C.F. Salviano, and A. von Wangenheim. 2010. Systematic literature review capability/maturity models. *Proceedings of International Conference on Software Process Improvement and (SPICE)*, Pisa, Italy.

Walton, M. 1986. *The Deming Management Method*. New York: Perigee Trade.

Wangenheim, C.G., T. Varkoi, and C.F. Salviano. 2005. Performing ISO 15504 conformant software process assessment in small software companies. In *EUROSPI 2005*, Budapest, Hungary.

Weinberg, G.M. 1991. *Quality Software Management: Systems Thinking*. Vol. 1. New York: Dorset House Publishing.

Whittaker, J.A. and J.M. Voas. 2002. 50 Years of software: Key principles for quality. *IT Professional* 4 (6):28–35.

Yourdon, E. 1996. *Rise & Resurrection of the American Programmer*. Yourdon Press Computing Series Hall PTR, NJ.

80

Software Metrics and Measurements

David Zubrow
Carnegie Mellon University

80.1 Introduction

Software measurement gained momentum as a field of study in the 1970s as researchers began to investigate factors that were thought to influence software development and the performance of software applications. Over time, the scope of software measurement grew, and it took on a critical role in software engineering. In *A Framework of Software Measurement*, Horst Zuse observed, "From a management perspective, measures support up-front estimation of time, resources, and quality, provide true status of projects to permit early insight into potential problems, enable the correction of problems before the consequences consume you, guide decisions on resources, priority adjustments, and schedule stability based on data, not best guesses, and provide hard data on which process can be improved and where to improve them" (Zuse 1998). Software measurement is not a passive field of inquiry; it is an action-oriented activity that supports informed decision making.

80.2 Principles

What are now considered the foundational principles of software measurement have emerged from years of experience and research. The field has evolved as challenges of implementation, adaptation to new methods of software development, and the scale and pervasiveness of software have increased. While some degree of software measurement is frequently employed in practice, overarching theories, systematic applications, and empirically validated results have taken longer to emerge. A 1997 article on the state of software measurement by Shari Pfleeger noted that while measurement theory was receiving considerable attention from researchers, the theoretical underpinnings of software measurement were largely ignored by the practitioners and customers (Pfleeger et al. 1997). Rather, they seemed to be solely concerned with a metric's utility in practice regardless of its scientific grounding. Indeed, Grady Booch has noted regarding his method of measuring architectural complexity, "In practice, I've found these rules of thumb sufficient for distinguishing relative complexity. They're woefully inadequate for measuring absolute complexity; they're also woefully under-tested. And yet, for my purposes they work" (Booch 2008). Shari Pfleeger and her colleagues went on to identify the lack of models in software engineering as symptomatic of a much larger problem: a lack of focus on software engineering as a systematic set of connected activities (Pfleeger et al. 1997).

More recently, Pfleeger highlighted new challenges in her article "Software metrics: Progress after 25 Years?" (Pfleeger 2008). The first challenge, Pfleeger says, is dealing with uncertainty: "[W]hen we use models developed in the research community, we rarely see the assumptions expressed explicitly, let alone evaluate whether they apply to our particular situation. This disconnect is especially true when we apply economic models for cost estimation or trade-off analysis." Another challenge is measuring "soft" characteristics, such as context, experience, and expertise. Involving social scientists in measurement research could be helpful in developing useful measures for how these characteristics affect product and process outcomes. Finally, anticipating change and developing heuristics are identified as additional challenges that need to be addressed to enable progress in software measurement. Together, these articles establish a set of principles around which best practices are defined. These principles include

- Increasing and demonstrating the value provided by measurement
- Consistently capturing product and development context and other metadata
- Understanding variation and explicitly reflecting uncertainty
- Developing validated measures, especially for quality attributes
- Providing predictive models and indicators
- Developing validated benchmarks and heuristics
- Understanding and improving data quality

There is a continuing need for ensuring and demonstrating the value provided by software measurement and making its practice cost effective. For software measures to truly be useful and actionable, they must be based on sound principles, valid, and take the product and development context into account. As noted by Pfleeger, models and analyses also need to explicitly reflect uncertainty.

In recent years, software systems have been expected to possess a growing numbers of qualities such as security, interoperability, and usability. Before we can understand exactly what is meant by these qualities and how well they are implemented, the field of software measurement needs to develop, validate, and agree on measures for these qualities. Additionally, measurement cannot be truly effective if it only reflects what has been. It must provide models and analyses that enable us to foresee problems and take actions to avoid or minimize adverse effects. Finally, given the increasing volume of information that can confront managers and other decision makers, evaluating it and using it to make decisions can be quite difficult. Heuristics and guidelines need to be developed and validated to help people make sense of the information provided.

The area of data quality, although not explicitly identified as an issue in the software measurement literature, also needs attention (Kasunic et al. 2011). Poor-quality data are a problem from multiple perspectives. While the most common understanding relates to the quality of data in a database (such as information about customers), the same principles and techniques can be applied to assess the quality of the data related to software projects and software engineering.

In the following sections, we describe best practices and frameworks related to each of the seven principles noted earlier.

80.3 Best Practices

80.3.1 Increasing and Demonstrating the Value of Software Measurement

80.3.1.1 Software Measurement Standards

The International Organization for Standardization (ISO) defines a standard as "a document that provides requirements, specifications, guidelines, or characteristics that can be used consistently to ensure that materials, products, processes, and services are fit for their purpose" (ISO 2012). Among those standards having an impact in the field of software measurement are ISO/IEC 15939 (ISO 2007a) for software and systems engineering measurement, and the ISO/IEC 25000 series for software quality measurement (ISO 2005). These standards embody principles and best practices as defined and codified by experts in the field.

ISO/IEC 15939 defines a general measurement and analysis set of practices and framework for software. It has made a couple of important contributions to the field in addition to laying out practices related to the measurement and analysis process. First, it standardizes the terminology for the key concepts associated with software measurement. Second, it defines an information model, shown in Figure 80.1, which links the base measure and the method for collecting it to the decision that the measure should inform.

This standard is important because, as Vic Basili noted, the interpretation of a measure and the information that it conveys depends on the context in which it was gathered and the context in which it will be used (Basili et al. 1994). Understanding and sharing the context is vital to the intelligent use of measures from a practical perspective and to validation of the purported utility of the measures by researchers.

To fully understand a measure, it is useful to think about the process that produces the information on which we base our decisions. Various depictions have been used to illustrate the process and the issues associated with producing highly quality data and analysis that supports informed decisions. As depicted in Figure 80.1, the ultimate purpose of the measurement and analysis process is to meet an information need. A measurement product, such as a report, graph, table, or dashboard, is created in this process to represent the interpretation of the analyses performed on the data that have been collected. Continuing to move down the diagram in Figure 80.1, you will notice the concept of a derived measure. This is a measure that is the result of combining two or more base measures, for example, productivity. Common measures for productivity include lines of code or function points per effort hour or effort month. The manner in which the base measures are combined is the measurement function. These functions could include addition, multiplication, division, or a more complex formula.

Further down are the base measures. This is where the raw data collection occurs. As Figure 80.1 indicates, base measures result from applying a measurement method to one attribute of an entity. In software engineering, there are many entities, each with a multitude of attributes, from which to choose. For example, consider an application or file as an entity. Attributes commonly of interest include its size, quality, and cost or effort to produce. Assuming we want to measure its size, we can choose among several measurement methods, such as counting lines of code or function points. Even at this point, there are variants and decisions to make about the specific counting details. For value to be delivered to the organization, choices need to be made and aligned with the information need to be served, not only at this point in the process but also at many other points in the measurement and analysis life cycle.

Several approaches can be taken to implement measurement. The Capability Maturity Model Integration (CMMI) Measurement and Analysis (M&A) process area provides a complete, high-level framework describing what is needed to implement measurement (CMMI 2010). The M&A process area describes important practices needed to institutionalize measurement and analysis, including providing the resources needed to implement it, training those involved, and verifying that the process is being performed. Specific practices for implementation address planning and defining what to measure, how to measure (including instrumentation, tools, and procedures), what types of analyses to perform, and how results will be reported and to whom.

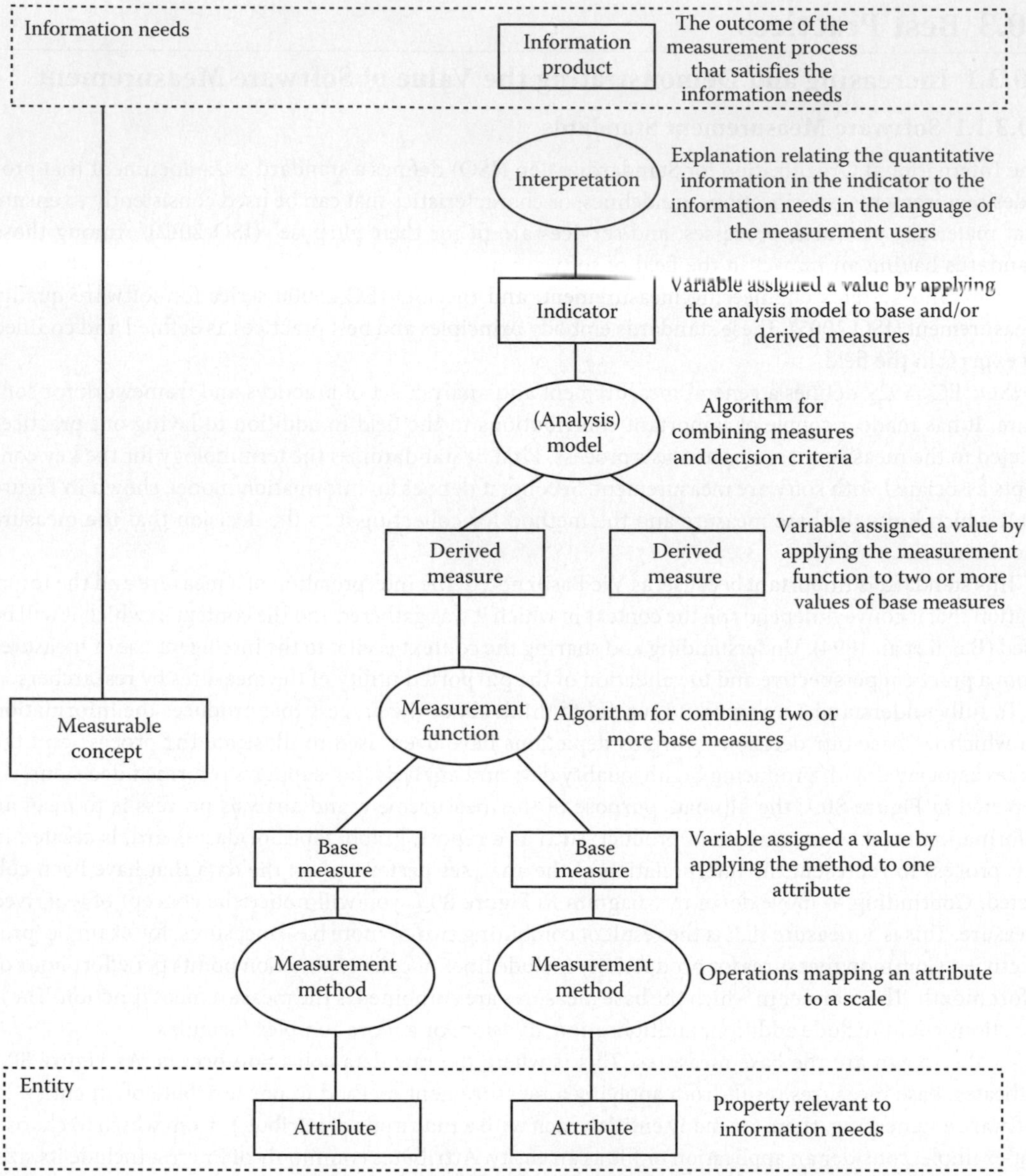

FIGURE 80.1 The ISO 15939 information model.

Organizations can use the Measurement and Analysis Infrastructure Diagnostic (MAID) method to help detect and evaluate shortfalls in the performance of their measurement and analysis process (Kasunic 2010). MAID structures criteria from several sources of best practices into an assessment method. Sources include CMMI, ISO standards, Practical Software and Systems Measurement (McGarry et al. 2001), and the works of leading practitioners in various areas of the measurement and analysis process, including information design, data analysis, and measurement technology (SEMA 2009). Another useful tool for implementing measurement processes is the indicator template (Goethert and Siviy 2004). Building upon the GQM work of Basili et al. (1994), the GQIM approach adds an "indicator" step (Zubrow 1998). In this approach, a sketch of the output of the measurement process, such as a chart, table, or report, serves two purposes. First, it can be analyzed to make sure that it addresses the identified information need prior to implementing the measurement process. Evaluating a sketch of the output helps determine if the data, as analyzed and displayed, will provide the information needed. Second, it specifies the data needed to build the indicator and serves as input for planning its implementation. The rationale and specification for the indicator can be recorded in the indicator template (Goethert and Siviy 2004).

Another notable series of standards, the ISO/IEC 25000 series, relates to the definition and measurement of software qualities. The standard was originally identified as the ISO/IEC 9126 series. The 9126 series and ISO 14598 are being integrated and updated as the ISO/IEC 25000 series (ISO 2005). This series focuses on software quality, with standards addressing requirements development, software evaluation, quality models, and measurement. Within the 25000 series, the measurement set of standards, 25020–25024, includes a general measurement process building on ISO 15939 and catalogs of measures related to the quality characteristics listed in the quality model. The overall software quality model, ISO/IEC 25010, is actually composed of a quality in use model and a product quality model (ISO 2011). Each model lists several quality characteristics and subcharacteristics. The quality in use model is shown in Figure 80.2 and the product quality model is shown in Figure 80.3.

ISO/IEC 25020 builds on the information model from ISO 15939 to show the relationship between the conceptual models depicted in Figures 80.2 and 80.3 and the operationalized measures of those qualities (ISO 2007b). Figure 80.4 shows this relationship.

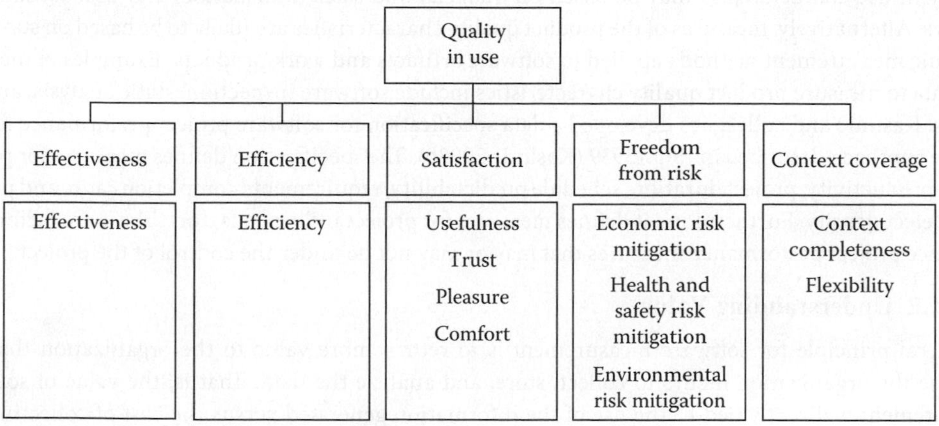

FIGURE 80.2 The ISO 25010 quality in use model.

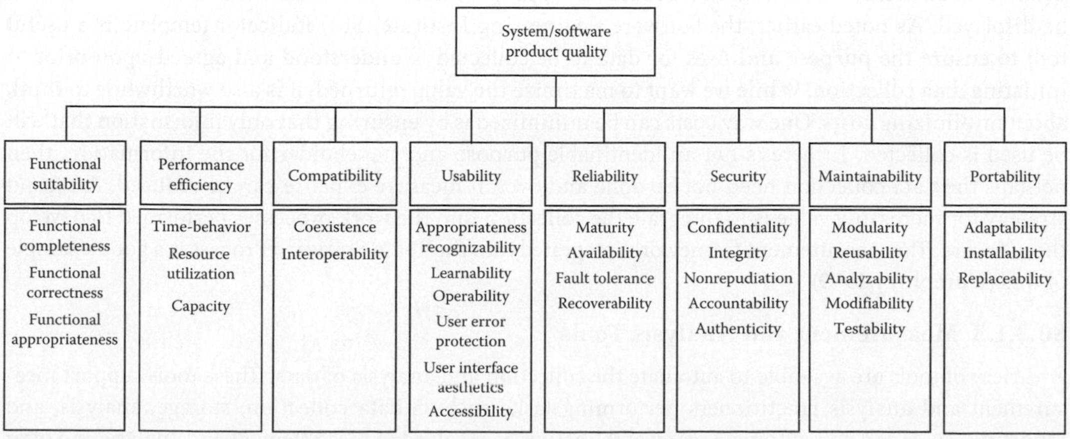

FIGURE 80.3 The ISO 25010 product quality model.

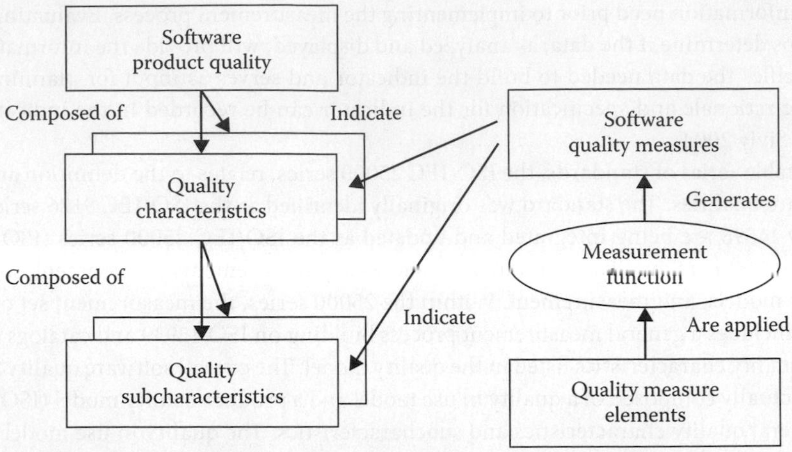

FIGURE 80.4 The ISO 25020 software product quality measurement reference model.

Measures for these quality characteristics can take a number of forms. For example, measures of the quality in use characteristics may be based on financial and sales information, end-user feedback, or surveys. Alternatively, measures of the product quality characteristics are likely to be based on static and dynamic measurement methods applied to software artifacts and work products. Examples of methods and data to measure product quality characteristics include software inspection, static analysis, and test results. Kasunic and colleagues developed a data specification for software project performance following the approach described in ISO 15939 (Kasunic 2008b). The specification defines measures for project effort, productivity, project duration, schedule predictability, requirements completion ratio, and postrelease defect density. Furthermore, it defines measures for project influence factors. These are factors that influence project performance measures that may or may not be under the control of the project.

80.3.1.2 Understanding Value

A general principle for software measurement is to return more value to the organization than the expense the organization incurs to collect, store, and analyze the data. That is, the value of software measurement is directly tied to the use of the information generated versus the cost of collecting and analyzing the software measurement data. To help generate value, the right information needs to be provided to the right decision maker at the right time. To accomplish this, it is important to understand the decision to be made, the information needed to support the decision, and how that information should be displayed. As noted earlier, the Software Engineering Institute (SEI) indicator template is a useful tool to ensure the purpose and uses for data to be collected is understood and agreed upon prior to initiating data collection. While we want to maximize the value returned, it is also worthwhile to think about minimizing costs. One way costs can be minimized is by ensuring that only information that will be used is collected. If there is not an identifiable purpose and stakeholder for the information, then perhaps the data collection need not be done and overall measure expense can be reduced. A second strategy for increasing value is to integrate the collection into the work processes performed to develop the software. The measurement framework integrated into the Team Software Process is a good example of this (Humphrey 1999).

80.3.1.3 Measurement and Analysis Tools

A variety of tools are available to automate the collection and analysis of data. These tools support measurement and analysis practitioners performing tasks such as data collection, storage, analysis, and reporting. Tools can now automate many of the activities involved in measurement and analysis and offer interfaces to engineering and management applications such as requirements management repositories, earned value management systems, and defect tracking systems. These tools ease the burden of collecting data by utilizing the data generated as a natural by-product of performing engineering activities. They can also be used to create a repository where data from multiple sources are stored, which facilitates analyses and reporting for a variety of data types. A common product of this work is a project dashboard or balanced scorecard that reports cost, schedule, requirements, technical progress, risk and dependencies, and quality (Kaplan and Norton 1992, Few 2006).

Organizations today are using a wide variety of analyses to gain insight into and manage their engineering activities and projects. Some of these techniques include modeling and simulation using discrete event simulation, dynamic simulation, and Monte Carlo methods. Discrete event simulation, for instance, can be used to analyze proposed process changes (Raffo et al. 2007), while Monte Carlo simulation can be used to improve schedule analyses by characterizing activities by best, worst, and most likely case distributions and running thousands of combinations to come up with an overall schedule distribution.

General statistics packages and specialized programming languages provide access to an enormous number of parametric and nonparametric multivariate techniques that can prove beneficial, including analysis of variance (ANOVA), regression analysis, correlation analysis, and chi-square analysis. As in any field, formal statistical analysis guards against being fooled by relationships that could be the result of chance and enables the detection of relationships that may be too complex or subtle to perceive through less sophisticated analytical techniques or visual inspections of the data. Software packages now include the ability to conduct more specialized analyses, such as statistical process control (SPC), that were once found in separate, specialized packages. Statistics packages can also provide a full gallery of displays for visualizing raw data. Other special analyses for specific investigations include cost estimation, reliability growth modeling, Bayesian analysis, machine learning, and data mining applications. Finally, there are tools that will perform specified analyses and format and assemble the results in a report or in the form of a dashboard. These tools save much effort and time. However, users should be knowledgeable about their configuration and the business rules they use for aggregating and summarizing results. This is especially true when the dashboard reduces the data to a stoplight chart and the thresholds used to assign red, yellow, and green values are not readily evident.

80.3.2 Capturing Context: The Importance of Metadata

Collecting data about measured values such as size, effort, and quality is useful for analysis and decision making, but metadata and contextual information also need to be captured and included with the analysis to get the most value from the data. Metadata provide context for understanding and interpreting the measured data. Imagine the following scenario: You are asked to develop an estimate for the next project that your team will be performing. You think it would be useful to have a productivity figure to use as input to computing the estimate and look to the web to find an average productivity value. As you think about it, you might decide that you would like a figure based on projects that are similar to the one you will be conducting. Part of this context might include: the project's type, size, when it was executed, the methodology used, characteristics of the team, the customer for the project, whether it was the initial or a subsequent release, the country where is was performed, and so on. This contextual information is considered metadata, although it might be referred to by other names. In the data specification for project performance measures described earlier, the project influence factors are examples of metadata.

The aspects of the context that you capture are driven by your understanding and anticipation of how the measures will be used. Often, the data are being generated and used in a specific project, so much of the context is implicit. A rich set of contextual information, or attributes, about measures enables analyses that span projects, processes, domains, and organizations. It also allows data to be segmented into groups that may exhibit patterns and predictability that are not apparent in the total aggregated data set. One way of identifying useful contextual attributes is by imagining the types of questions you want answered and analyses you might perform. Another technique is to sketch out any charts or reports to be produced and note the breakdowns that are needed.

80.3.3 Understanding Variation and Explicitly Reflecting Uncertainty

Understanding variation in measures is important for making the information as useful as possible for informing decisions (Wheeler 2000). As measures are collected, they combine to form a distribution. For example, the average productivity within an organization might be 3 LoC/h. However, productivity might range from a minimum of 0.5 to a maximum of 5. The distribution reflects the degree to which productivity on projects varies within the organization. Rather than always using the average productivity, decision makers can select a value from the distribution that reflects their risk tolerance. If they want to be sure that the project has adequate resources and time to execute, they can select a value from the lower end of the productivity distribution. The distribution provides much more insight and information for making decisions and allows consideration of uncertainty to be included in the decisions made.

Statistical process control is the most common type of analysis used to understand variation and is often represented by a control chart. The basic elements of a control chart include a center line, upper and lower control limits, and the plotted data in a time-ordered sequence. The center line represents the average performance of the process, while the distance to the control limits represents the variation in process performance and the magnitude of uncertainty associated with future performance. In *Measuring the Software Process*, Florac and Carleton discuss how to appropriately and effectively apply this method of analysis to software engineering processes (Florac and Carleton 1999).

Understanding variation and uncertainty can improve the nature of commitments and management decisions. For instance, when committing to a delivery date or budget, it might be wise to pick a point from the relevant distribution where previous performance suggests a 90% chance of being successful. Note that this is very different than basing the decision on average performance and requires detailed knowledge of the distribution of the relevant performance data.

80.3.4 Validating Measures: The Case of Quality Attributes

Validation in software measurement means the extent to which the measure meets its intended purpose or measures the concept that it is intended to measure. In the field of measurement, validity is not an all-or-nothing concept; rather, measures can have varying degrees of validity that are influenced by the reliability associated with their measurement method. The two concepts, validity and reliability, are inextricably linked. The practice of measurement system evaluation offers techniques for examining the degree to which measures are valid and reliable.

Measurement reliability is defined as the extent to which a measurement method produces the same value when applied to the same object. One way to effectively increase measurement reliability is to have clear and detailed operational definitions of software measures.

The lack of precision in operational definitions of measures was addressed by a series of technical reports by the SEI in 1992 (Carleton et al. 1992). These reports provided a means for communicating exactly what is to be included and excluded in a measure. One example of the lack of common interpretation can be seen in the wide range found when counting the number of lines of code in a small function. Interpretation problems such as this have existed for years, and the problem of definition continues to plague the field as it addresses quality attributes such as security, interoperability, usability, and others. As software becomes more pervasive, an increasing number of these quality attributes are assigned to it. Recently, these attributes have been codified in ISO 25010, which is an update of its predecessor, ISO 9126. The standard is structured into six main characteristics, each of which includes multiple subcharacteristics. These characteristics reflect an increasingly important challenge to the field of software measurement: many measures are proposed but few have been empirically validated. Validating the relationship between the conceptual definition and interpretation of an attribute (the upper part of the 15939 information model) and its operational definition in terms of a specific measurement method (the lower part of the 15939 information model) has rarely been done. Rather, the field has tended to rely on logical rather than empirical validation. This may seem to be a theoretical point, but it is not.

Decisions and actions are taken based on the conceptual model meaning: security is low or maintainability is high, but assignment of the value, low or high, is based on the specific measurement employed.

This bleak assessment is a general, not universal, characterization. Current research is being done in this area and validation studies are being conducted. For instance, many studies in the journal *Empirical Software Engineering* include a description of how their results were validated. However, the generalizability of the results remains questionable. Among the obstacles are the lack of a good reference database against which validation analyses might be conducted, the difficulty developing samples from a reference population, and the general reluctance of organizations to collect and share data. To overcome these obstacles, researchers have become resourceful. They use data from public repositories and open sources, conduct experiments with students, and find industrial organizations with which to collaborate. However, this makes the process of transition into practice slow and uncertain. This type of evidence often does not sway an organization to adopt a new measure; it does not overcome an organization's preference for "what seems to work for them" (Pfleeger et al. 1997, Booch 2008). While the decisions made might be right for individual organizations, they ground the field in local empiricism and slow scientific progression.

A related problem occurs when measures are used without explicit conceptual definitions. This can happen when measures are collected in an ad hoc manner in a repository and then used with assumptions placed on them by the analyst with no real knowledge of the details of measurement and collection. In these instances, there is often misunderstanding and ambiguity as differing interpretations, purposes, and meanings are assumed for a specific measure. This is often the case when measuring lines of code. Lines of code is often considered a needed and important measure, but the reason why is rarely explicitly stated. From a research and community view, the lack of details and metadata diminishes the contribution the measure can make to the community as a whole. Good practice would establish the validity of the measure in four ways corresponding to various forms of validation (Ghiselli et al. 1981). First, the measure would logically relate to what it purports or is intended to measure. Second, the way it is measured—that is, by counting something, by asking experts to provide a rating, or via some testing mechanism—should also make logical sense. Third, the measure should correlate with an independent measure of the same quality attribute. Finally, there should be evidence that the measure is providing value to the user.

80.3.5 Providing Predictive Models and Indicators

Measurement should not be used only to reflect the current state of a product or past performance but also to provide insight into what will be in the future. In order to do this, predictive models must be developed. As an analogy, think of a gas gauge. As a measure, it indicates the amount of fuel remaining in the tank. However, as the fuel level decreases, most drivers and cars can now predict the number of miles they can go before running out of gas. This information can then be used to decide when to go to a fueling station. The number of miles that can be driven before running out of fuel is based on a predictive model.

Predictive models play many roles and inform many decisions in software engineering. For instance, a manager might want to know whether a project will meet its delivery date. The answer is often based on a predictive model, for example, a model that uses productivity to date to predict when the work will be completed. There are a variety of assumptions with such a simple model and they need to be explicitly assessed for their reasonableness and applicability. Furthermore, the model needs to be validated. While there are a variety of ways to demonstrate the validity of a model, the most direct is to measure the difference between the model's predictions and the actual outcomes—that is, its accuracy. Accuracy is not an absolute characteristic but rather a measured amount. Having a measure of accuracy informs those using the model about the level of confidence that can be placed in its predictions. The downside of this is that it requires future data. During the development of the model, there are alternative statistical techniques that can be employed to assess its accuracy or goodness of fit.

Developing a predictive model requires good data representing the quantity, class, or state to be predicted and data representing the variable or variables that the prediction will be based on.

80.3.6 Developing Benchmarks and Heuristics

Although it has been said that "the data speak for themselves," this is rarely the case. There is always some ambiguity and possibility for misinterpretation. The principal purpose for measuring is to use the results to guide decision making and action. This generally requires some expectation about what is good, which often comes from a comparison against past performance, the performance of others, or customer expectations. Useful benchmark values have been published by Caper Jones (2000), the International Software Benchmarking Standards Group (ISBGS) (ISBSG 2012), software cost estimation tool vendors, and others. However, users of this information should take care to understand the provenance and limitations of the underlying data.

80.3.7 Understanding and Improving Data Quality

Software measurement and analysis depends on data, yet there has been relatively little written regarding the methods and practices for assessing and improving the quality of software measurement data. Much more attention in the commercial world has been focused on customer facing data in IT systems (English 2004). The research community has focused its attention on experimental methods, design, and instrumentation, with a relative lack of attention on data quality in practice (Kasunic 2008a). Some progress was made with the recent publication of the ISO/IEC 25012 standard (ISO 2008), a standard similar to ISO/IEC 25010. It contains a data quality model and defines a set of characteristics for data quality. In general, though, data quality is still an area in need of greater attention in terms of research and practice.

It is good practice to first check the data to be analyzed when performing an analysis of organizational, project, or product data (Tabachnick and Fidell 2013). Invariably, checks for data integrity and quality reveal errors and issues in the data. Errors can take the form of missing data, invalid data, and inaccurate data. Missing data occur when a desired piece of data is not available. This is fairly common if the data collection mechanism does not enforce the submission or collection of complete data. There are a number of methods for dealing with missing data that range from ignoring them to estimating the data value. How missing data are best handled depends on the particular circumstance. For instance, a simple solution is to replace missing data with the mean value. This can be reasonable if there are many cases and relatively little missing data. However, it can unduly influence the distribution of a variable if there are many instances of missing data. In those circumstances, it is better to use a method that creates appropriate variation in the missing values. This can be done by building an algorithm to estimate the missing value based on data available for that specific record. Another option is to sample from a distribution based on the present data.

Invalid data can occur when measured data are collected but they are not within the bounds of the valid range for that particular measure or are inaccurate. An example of a problem with measured data would be negative numbers for counts such as lines of code, requirements, and defects. Also, if there is a defined valid range, then numbers outside of the range would be invalid. For example, if a total is supposed to be 100%, then no individual value should be greater than 100% assuming that there can be no negative numbers. This can happen with categorical data in addition to measured values. Examples of problems with categorical data include nonexistent release identifiers, wrong codes for defect types, and status codes for change requests.

While missing data and invalid data are relatively easy to detect and prevent, inaccurate data are not so easy. A typical strategy is to use other information to develop integrity checks that bound the range of likely accurate data. For instance, historical data might inform the likely upper and lower bounds for certain measures thereby identifying outliers that might be errors. Knowing the distribution for defect counts, requirements changes, earned value, and other measures can help detect inaccurate data. Techniques such as use of the Benford distribution can help identify data that have been fabricated or "chunked" (rounded to the nearest 5 or 0 value for instance) when in fact they were supposed to be actual measurements. Also, business rules representing known relationships among the data can be used. For example, it is often expected that the defect density of software will decrease as it progresses through different levels of testing (such as from integration to system testing). Should this expected trend not occur, reviewing the data for any errors is warranted before reviewing the development process.

80.4 Conclusions and Observations

While software measurement has long been a topic of interest and has provided measured benefits, much work remains. Increasing and demonstrating the value of measurement will encourage adoption. As tools and technology mature, they provide an increasing amount of support for measurement activities and can increase the value they provide. Additionally, measures need to be carefully defined, contextual data recorded, information about uncertainty explicitly provided, and greater recognition brought to measurement reliability and validity. Models related to current software engineering methods and products need to be developed.

High-quality data are integral to effective modeling and analysis in support of making decisions related to software engineering. Data quality is directly related to the validity of the data used. Therefore, developing validated measures of quality attributes is also important. Creativity in developing measures of these attributes, as well as measures of progress and the quality of work products, should be encouraged. However, we should also demand evidence of validation and the utility of the measures before adopting and deploying them.

Current trends in software measurement and analysis include a greater focus on analysis, prediction, and integration of measurement with decision making. Better analysis could be done if the data for both dependent and independent variables in models of software engineering are collected. In data mining, this allows us to do supervised learning. While some of this occurred in previous years, it is explicitly included as part of best practice models and standards as well as the design of measurement systems. Software measurement is required to quantify quality and performance and to provide an empirical and objective foundation for decision making—a foundation that is a necessary element if software engineering is truly to be a disciplined field of engineering.

Acknowledgment

I would like to acknowledge Erin Harper who provided her technical writing skills during the preparation of the manuscript.

Glossary

Unless otherwise noted, all definitions are from ISO/IEC 15939.

Attribute: Property or characteristic of an entity that can be distinguished quantitatively or qualitatively by human or automated means

Base measure: Measure defined in terms of an attribute and the method for quantifying it

Derived measure: Measure that is defined as a function of two or more values of base measures

Entity: Object that is to be characterized by measuring its attributes

Indicator: Measure that provides an estimate or evaluation of specified attributes derived from a model with respect to defined information needs

Information model: Structure linking information needs to the relevant entities and attributes of concern software quality

Process: Set of interrelated or interacting activities that transforms inputs into outputs

Software product: Set of computer programs, procedures, and possibly associated documentation and data (ISO/IEC 25000:2005)

Software quality characteristic: Category of software quality attributes that bears on software quality. Note: Software quality characteristics can be refined into multiple levels of subcharacteristics and finally into software quality attributes (ISO/IEC 25000:2005).

Work product: A useful result of a process (CMMI v1.3)

References

Basili, V., Caldiera, G., and Rombach, H.D. The goal question metric approach. In *Encyclopedia of Software Engineering*. New York: John Wiley & Sons, Inc., 1994. http://onlinelibrary.wiley.com/doi/10.1002/0471028959.sof142/full

Booch, G. Measuring architectural complexity. *IEEE Software*, 25(4), 14–15, July/August 2008.

Carleton, A., Park, R., Bailey, E., Goethert, W., Florac, W., and Pfleeger, S. Software measurement for DoD systems: Recommendations for initial core measures (Technical Report CMU/SEI-92-TR-019). Pittsburgh, PA: Software Engineering Institute, Carnegie Mellon University, 1992. http://www.sei.cmu.edu/library/abstracts/reports/92tr019.cfm

CMMI Product Team. CMMI for Development, Version 1.3. (CMU/SEI-2010-TR-033). Pittsburgh, PA: Software Engineering Institute, Carnegie Mellon University, 2010. http://www.sei.cmu.edu/library/abstracts/reports/06tr008.cfm

English, L.P. Information quality: Meeting customer needs. *Information Impact*, 3, 1, 2004.

Few, S. *Information Dashboard Design: The Effective Visual Communication of Data*. Beijing, People's Republic of China: O'Reilly, 2006.

Florac, W.A. and Carleton, A.D. *Measuring the Software Process*. Reading, MA: Addison-Wesley, 1999.

Ghiselli, E.E., Campbell, J.P., and Zedeck, S. *Measurement Theory for the Behavioral Sciences*. San Francisco, CA: W.H. Freeman & Company, 1981.

Goethert, W. and Siviy, J. Applications of the indicator template for measurement and analysis (CMU/SEI-2004-TN-024). Pittsburgh, PA: Software Engineering Institute, Carnegie Mellon University, 2004. http://www.sei.cmu.edu/library/abstracts/reports/04tn024.cfm

Humphrey, W.S. *Introduction to the Team Software Process*. Reading, MA: Addison-Wesley Professional, 1999.

International Organization for Standardization. *ISO/IEC 25000:2005: Software Engineering—Software Product Quality Requirements and Evaluation (SQuaRE)—Guide to SQuaRE*. Geneva, Switzerland: ISO/IEC, 2005. http://www.iso.org/iso/catalogue_detail.htm?csnumber=35683

International Organization for Standardization. *ISO/IEC 15939:2007: Systems and Software Engineering—Measurement Process*. Geneva, Switzerland: ISO/IEC, 2007a. http://www.iso.org/iso/catalogue_detail.htm?csnumber=44344

International Organization for Standardization. *Software Engineering—Software Product Quality Requirements and Evaluation (SQuaRE)—Measurement Reference Model and Guide*. Geneva, Switzerland: ISO/IEC, 2007b. http://www.iso.org/iso/catalogue_detail.htm?csnumber=35744

International Organization for Standardization. *Software Engineering—Software Product Quality Requirements and Evaluation (SQuaRE)—Data Quality Model*. Geneva, Switzerland: ISO/IEC, 2008. http://www.iso.org/iso/catalogue_detail.htm?csnumber=35736

International Organization for Standardization. *ISO/IEC 25010:2011: Systems and Software Engineering—Systems and Software Quality Requirements and Evaluation (SQuaRE)—System and Software Quality Models*. Geneva, Switzerland: ISO/IEC, 2011. http://www.iso.org/iso/catalogue_detail.htm?csnumber=35733

Jones, C. *Software Assessment, Benchmarks, and Best Practices*. Boston, MA: Addison-Wesley, 2000.

Kaplan, R.S. and Norton D.P. The balanced scorecard—Measures that drive performance. *Harvard Business Review*, 70(1), 71–80, January–February, 1992.

Kasunic, M. Can you trust your data? Establishing the need for a measurement and analysis infrastructure diagnostic (CMU/SEI-2008-TN-028). Pittsburgh, PA: Software Engineering Institute, Carnegie Mellon University, 2008a. http://www.sei.cmu.edu/library/abstracts/reports/08tn028.cfm

Kasunic, M. A Data specification for software project performance measures: Results of a collaboration on performance measurement (CMU/SEI-2008-TR-012). Pittsburgh, PA: Software Engineering Institute, Carnegie Mellon University, 2008b. http://www.sei.cmu.edu/library/abstracts/reports/08tr012.cfm

Kasunic, M. Measurement and analysis infrastructure diagnostic, version 1.0: Method definition document (CMU/SEI-2010-TR-035). Pittsburgh, PA: Software Engineering Institute, Carnegie Mellon University, 2010. http://www.sei.cmu.edu/library/abstracts/reports/10tr035.cfm

Kasunic, M., Zubrow, D., and Harper, E. Issues and opportunities for improving the quality and use of data in the Department of Defense (Technical Report CMU/SEI-2011-SR-004). Pittsburgh, PA: Software Engineering Institute, Carnegie Mellon University, 2011. http://www.sei.cmu.edu/library/abstracts/reports/11sr004.cfm

McGarry, J., Card, D., Jones, C., Layman, B., Clark, E., Dean, J., and Hall, F. *Practical Software Measurement: Objective Information for Decision Makers*. Reading, MA: Addison-Wesley Professional, 2001.

Pfleeger, S.L. Software metrics: Progress after 25 Years? *IEEE Software*, 25(6), 32–34, 2008.

Pfleeger, S.L., Jeffery, D.R., Curtis, B., and Kitchenham, B. Status report on software measurement. *IEEE Software*, 14(2), 33–44, 1997.

Raffo, D.M., Ferguson, R., Setamanit, S.-O., and Sethanandha, B.D. Evaluating the impact of the QuARS requirements analysis tool using simulation. In *Proceedings of the 2007 International Conference on Software Process* (*ICSP'07*), Qing W., Dietmar P., and David M.R., (Eds.). Berlin, Germany: Springer-Verlag. pp. 307–319, 2007.

SEMA Group. Software engineering measurement and analysis: Measurement and Analysis Infrastructure Diagnostic (MAID) evaluation criteria, version 1.0 (CMU/SEI-2009-TR-022). Pittsburgh, PA: Software Engineering Institute, Carnegie Mellon University, 2009. http://www.sei.cmu.edu/library/abstracts/reports/09tr022.cfm

Tabachnick, B.G. and Fidell, L.S. *Using Multivariate Statistics*, 6th edn. Boston, MA: Allyn and Bacon, 2013.

Wheeler, D.J. *Understanding Variation: The Key to Managing Chaos*. Knoxville, TN: SPC Press, 2000.

Zubrow, D. Measurement with a focus: Goal-driven software measurement? *CrossTalk*, 11(9), 24–26, September 1998.

Zuse, H. *A Framework of Software Measurement*. Berlin, Germany: Walter de Gruyter, 1998.

Information Resources

Organizations

Practical Software and Systems Measurement
http://www.psmsc.com
Software Engineering Institute
http://www.sei.cmu.edu/measurement
http://www.sei.cmu.edu/tsp
Software Engineering Information Repository
https://seir.sei.cmu.edu
The Cyber Security and Information Systems Information Analysis Center (CSIAC).
http://www.thecsiac.com

The CSIAC is a Department of Defense (DoD) Information Analysis Center (IAC) sponsored by the Defense Technical Information Center (DTIC). The CSIAC is a consolidation of three predecessor IACs: the Data and Analysis Center for Software (DACS), the Information Assurance Technology IAC (IATAC), and the Modeling & Simulation IAC (MSIAC); with the addition of knowledge management and information sharing technical areas.

International Function Point Users Group
http://www.ifpug.org
American Society for Quality Software Division
http://www.asq.org/software
University of Southern California Center for Systems and Software Engineering
http://sunset.usc.edu
Education Programs on Information Quality, University of Arkansas Little Rock, Information Quality Program
http://ualr.edu/informationquality/
This is a graduate program focused on information quality.
MIT Program on Information Quality
http://mitiq.mit.edu/
This is a professional education and research program.

Standards Organizations

International Organization for Standardization (ISO)
http://www.iso.org
Institute of Electrical and Electronics Engineers (IEEE)
ISO/IEC JTC1/SC7
http://www.jtc1-sc7.org/
This subcommittee within this joint committee of ISO and IEC focuses on the "(s)tandardization of processes, supporting tools and supporting technologies for the engineering of software products and systems."
http://www.ieee.org
This site includes and provides access to standards and many publications including *IEEE Software and Transactions on Software Engineering*.

Publications

CrossTalk, The Journal of Defense Software Engineering
http://www.crosstalkonline.org/
Journal of Software Technology
https://journal.thecsiac.com/enews/enews4-measurement.php
General Accounting Office
http://www.gao.gov
The General Accounting Office (GAO) site has many publications addressing aspects of software and software projects within the Federal government.

XI

Software Development Management: Processes and Paradigms

81

Software Development: Management and Business Concepts

Michael A. Cusumano
Sloan School of Management

The persistence of similar problems in software development since the 1960s, as well as common observations that the vast majority of software projects are typically late and overbudget, suggests that the field of software engineering has not made much progress. But any close look at the complexity, quality, and breadth of software systems available today, for devices ranging from supercomputers to smartphones, demonstrates that software engineering has indeed made remarkable advances. In addition to much more sophisticated and easy to use software applications for various devices as well as the Internet, there are now far better programming languages, code libraries and other reusable modules, and support tools available. In addition, we know a lot more today than we did in the past about how to run software projects and software businesses.

Nevertheless, software development remains very much dependent on individual talents for problem solving, design, and creativity, leaving considerable room for managerial discretion. Moreover, not all individual programmers or software project managers incorporate best practices. The development challenge is often further complicated in that many users do not know what they want until they see part of a working system or a prototype. Consequently, software managers and individual engineers still need to think about how to manage the *process* of software development and how software engineering skills can contribute to organizational success.

One reason why software projects do not always apply best practices seems due to ongoing disagreements about what approaches are most effective in different contexts. Some projects require more invention and innovation, as well as trial and error. Invention and innovation are usually more difficult to manage and predict compared to routine work. Some customers have mission-critical reliability requirements, and others do not. But, whatever the situation, in order to be more systematic, software managers and engineers need to avoid treating all software projects as unique events.

At the same time, even though many projects may have much in common, the reality is that software developers still need to adapt their techniques and processes to the needs of different contexts. For example, companies that offer mass-market products or Internet services like Microsoft, Adobe, Google, and Salesforce.com need to anticipate general user needs and release products or features relatively quickly. Professional services companies like IBM, Accenture, and Infosys, whether they are building or enhancing custom systems or hybrid solutions, have to work closely with individual clients and understand these distinct relationships as well as their customers' technical and business requirements. Other kinds of software producers, such as those creating complex, real-time systems for defense or space applications, may have to invent as they go along and try to schedule what truly is rocket science.

In short, this chapter argues that there is no one best way to develop software and manage projects for all kinds of applications and contexts. It is, therefore, natural that the field of software engineering has generated disagreements over what is the best way to manage the development process. Be that as it may, some basic principles are useful to apply to a variety of projects: the need to have strategies for high-level process management, innovation and design, system architecture, team management, project management, and quality assurance. The focus of this chapter is on these general principles, through the lens of my personal experience as a researcher, teacher, and consultant.*

81.1 High-Level Process Concepts

When beginning a software project, rather than trying to plan out schedules and features in detail, managers and engineers should begin by deciding what process is most appropriate for their particular development effort (Cusumano et al. 2009). When making this decision, I have found three observations to be particularly useful. One is the recognition that all projects of more than a handful of people need some well-defined structure to keep everyone in synch (such as how and when to check in components, add features, fix bugs, or ship to the customer). This structure should be repeatable as well as adjustable across different phases or milestones of the same project and across multiple similar projects. Second, the structure should fit the particular technical characteristics and requirements of the product or system being built. This requires some extensive conversations with the customer or customer representatives as well as an understanding of the experience of the team. Third, the structure should fit the market and the business context of the project. This requires understanding the technology and competition as well as customer needs and goals.

81.1.1 No "One Best Process"

Many companies have tried to define a standard development process for their entire organization in an attempt to improve quality, predictability, and perhaps productivity or cost control. One approach should make it easier to introduce a repeatable process, which the Software Engineering Institute (SEI) at Carnegie Mellon University has emphasized since the mid-1980s (Humphrey 1989). While a repeatable process is highly desirable, however, one process for every project or department is usually not. Again, software products and custom (bespoke) projects can differ greatly according to the application and the

* This chapter adapts and updates material originally published in Cusumano (2004, pp. 128–185).

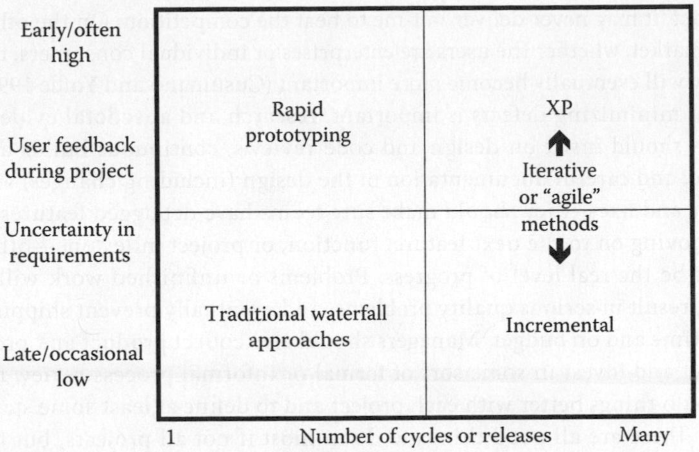

FIGURE 81.1 Spectrum of process approaches in software development. (Reprinted from Cusumano, M.A., *Staying Power*, Oxford University Press, Oxford, U.K., p. 160, 2010. With permission.)

market and due to unique customer requirements. It follows that software development projects even within a single organization need to define different processes or variations for their different needs.

The market and business strategy introduces complexity into process choices, though we can map these out across two dimensions: the level of uncertainty in requirements and thus the need for feedback during development versus the number of releases (i.e., chances to get it right) the team can expect in the future (Figure 81.1). For example, we have new applications where developers know little about what customers want and so feedback from early prototypes or other mechanisms is critical during development. In some cases, rapid prototyping, where engineers evolve the requirements and an actual working system side by side with the customer, may be the best approach. However, if user requirements are well understood, as in the case where a team has built several similar systems before, and managers do not expect there will be multiple releases in the future where engineers can gradually improve the system (such as in embedded satellite software where uploads of new software are difficult), then a carefully documented "waterfall" approach may be best. In a waterfall type of process, development generally proceeds in careful, documented, sequential phases, from concept to detailed design to coding and then to multiple levels and cycles of testing and debugging. In other contexts, again depending on the level of uncertainty in requirements and the number of expected releases in the future, different styles of "agile" or "iterative" development seem most appropriate (Boehm 1988; McConnell 1996; Larman and Basili 2003). In these kinds of projects, developers add functionality in small increments as well as over multiple versions of the system.

81.1.2 What to Emphasize Most

It follows that depending on a company's business objectives (or organizational goals in the case of a noncommercial entity, such as a government department), as well as the familiarity with the application domain and customer requirements, software managers need a different high-level process strategy that leads to different development approaches. The process should vary in terms of how much emphasis to place on practices like writing as complete a spec as possible before coding or applying a "full-court press" in quality assurance (such as intensive design reviews, formal code reviews, and rigorous testing at multiple levels and stages of the project). In mission-critical projects with extremely high reliability requirements, projects generally need extensive specification and architecture work before coding, though still leaving room for feedback and evolution of the spec, such as in the user interface (Fagan 1976; Gilb 1988; Humphrey 1989).

In some cases, such as for mass-market consumer applications, speed to market and innovation may be more important than product reliability. If a firm waits to deploy the latest technology or create a zero-defect product, it may never deliver in time to beat the competition. On the other hand, if a firm targets the mass market, whether the users are enterprises or individual consumers, then quality in the sense of reliability will eventually become more important (Cusumano and Yoffie 1999).

In summary, if minimizing defects is important, research and anecdotal evidence suggests that project managers should insist on design and code reviews, continuous builds and testing of the evolving software, and careful documentation of the design (including changes) as well as the code and code changes and fixes. They should make sure teams have debugged features as thoroughly as possible before moving on to the next feature, function, or project milestone—otherwise, apparent progress will not be the real level of progress. Problems or unfinished work will accumulate and delay the project, result in serious quality problems, and eventually prevent shipping or releasing to the customer on time and on budget. Managers should also collect product and process data during and after a project and invest in some sort of formal or informal process review (postmortems) to figure out how to do things better with each project and to define at least some standards and common procedures. These are all good things to do in most if not all projects, but they are essential practices for firms building mission-critical software or enterprise applications and systems. For less critical projects, it is possible to cut some corners, although a software producer should never want to ship products or release systems with quality levels below the customer requirements or what competitors are offering. This is, of course, easier said than done in competitive markets with time and budget constraints.

81.2 Innovation and Design Strategy

In fast-paced markets, there will be many projects where managers stress creativity and invention and encourage design changes even late in a project and at the risk of producing more defects and delaying the ship date or overrunning the budget. The danger is that loosely managed projects will spin out of control—never ship anything because of "infinite defect loops" (the state where every code change, such as to fix a defect, introduces yet another defect—Cusumano and Selby 1995, pp. 256, 333) or because they are too ambitious, with too little time for experimentation. It follows that managers need a strategy for managing the process of innovation and design, so they do not leave too much to chance.

81.2.1 Level of Control

We have seen distinct and deliberate variations in how much control managers introduce into software projects. For example, at the dawn of the mass adoption of the Internet, a company called Netscape, founded in 1994, built a mass-market browser and the enterprise-class servers and tools. Managers at this firm had a specific philosophy of allowing projects to be "slightly out of control" in order to increase speed to market as well as stimulate innovation and creativity (Cusumano and Yoffie 1998). Microsoft, founded two decades earlier in 1975, pioneered personal computer (PC) software and was more organized than Netscape but less structured than IBM, its initial partner in PCs and the market leader in mainframe computing. Microsoft, like Netscape, wanted to be faster to market and more flexible in experimenting with PC software technology than a highly bureaucratic, waterfall-ish IBM process allowed (Cusumano and Selby 1995).

The downside of less control, taken to the extreme, is that too little structure can lead to bad decisions, failed projects, and missed ship dates. For example, process problems at Netscape helped Microsoft win the browser wars of the 1990s, just as shipping delays for later versions of Windows encouraged users to adopt Apple computers and other alternatives to Microsoft products in the 2000s. Perhaps for this reason, we see a mixture of different approaches at companies such as Google, where mission-critical systems such as for AdWords or back-office operations have been managed much more tightly than smaller projects, and often in more of a waterfall-ish rather than agile style (Striebeck 2006).

81.2.2 Risk Management

One thing we have learned is that iterative or agile development techniques as we know them today—such as building systems in small increments, with frequent or continuous testing, debugging, integration, and synchronization of work under development—are particularly useful for risk management. These types of techniques provide mechanisms to add visibility and assess progress in a software project as well as make frequent adjustments such as in project scope, priorities, schedules, or staffing.

A form of iterative development called "synch-and-stabilize" (a term coined by myself and Richard Selby in our 1995 book *Microsoft Secrets*) relies on techniques such as vision statements, evolving specs, development work broken up into subcycles, daily builds, milestone stabilizations, early integration testing, and various customer feedback mechanisms during development (Cusumano and Selby 1995, pp. 187–326). These practices represent a middle approach to risk management between a highly bureaucratic, waterfall-ish style of software development and a potentially chaotic "hacker" style of development. For example, the vision statement that kicks off a project is essentially a team contract that should scope out, very clearly, what the team hopes to do and what it is *not* going to do. Microsoft vision statements have been as short as one paragraph or one page, and others extend many pages. Then evolving the product spec from an outline and reevaluating it periodically during the project avoids spending time detailing specs for features that the team will never get to or reject. The Microsoft-style iterative process has about a 70% overlap with agile development techniques such as followed in Extreme Programming (XP) projects (Cusumano 2007b).

Also, as part of risk management, project teams should have a "multiversion release" mentality whenever possible. The idea here is that there is no need to try to create a "perfect" product or system that includes every favorite feature from customers, executives, sales, and marketing or that tries to do the equivalent of rocket science in a commercial setting. If a company is making a software product and it is successful, there is probably going to be a second and a third version. Even in custom development, software and services companies usually have a chance to refine their work with maintenance releases or a phased rollout. So there is rarely the need to attempt so much in any one project that the risk of failure becomes higher than the likelihood of success.

81.2.3 Late Changes

Another important observation that comes from experience, and which has been associated with iterative or agile development, is that *late design changes can be good*; they do not necessarily reflect mistakes or a step backward. Some managers and engineers find this hard to understand, no doubt because of their education and experiences with delayed schedules and buggy code. Many software developers and managers once thought the same way, before experiencing the fast-paced markets for PCs and Internet software and services.

My recollection is that software engineering texts and classes in the 1970s and 1980s used to emphasize two particular ideas. One was that most problems occur because projects do not have a good requirements document and a complete specification before engineers start coding. A second was that late changes in the code or the design are too risky because they can destabilize the product, create more bugs, and make the project later, which then creates a destructive dynamic if the project shortchanges additional testing.

Mainframe software producers such as IBM and Japanese software factories got around this problem by giving tremendous authority to their QA departments. For example, at Hitachi, NEC, Toshiba, and Fujitsu, QA managers had to approve any product ship decision, and most used historical data telling them how many bugs they should be finding in design documents or code at different stages of development and how much more testing was necessary before they could consider a product to be of high quality (Cusumano 1991, 1993). But new products for new markets or on new platforms (such as mobile phones compared to mainframes or even PCs) do not have this kind of historical data. Furthermore, fast-paced markets may require different standards and procedures, within certain limits.

The main point is managers and engineers should recognize that any initial specification will be incomplete. Encouraging evolutionary designs through an iterative, agile, or prototype-driven development process allows a team to respond to unforeseen market changes, user feedback, and competitors' moves. Late changes may produce more defects and delay the schedule. But, as some research has shown, a coherent set of countermeasures—such as frequent builds, design and code reviews, and integration testing with each change—can mitigate the level of bugs and lateness (MacCormack et al. 2003). A process that expects and allows projects to accommodate change with a minimal impact on quality and productivity is a great competitive advantage in the business of software.

81.2.4 Economies of Scope

Another strategic aspect of innovation and design is how to increase not simply creativity or structure but also *economies of scope*—efficiency and effectiveness in building multiple products or conducting multiple customer engagements with the same engineering assets. This kind of efficiency and effectiveness is almost always important to an ongoing software business. Simple economies of scale are not generally present, except in replicating a software product or in funding in-house R&D or tools development used by multiple products or projects. Scope economies in general come from reusing system architectures, design frameworks, pieces of working code, support tools, test cases, and historical product and process data (such as how long particular kinds of projects usually take and how much testing resources they normally require).

Code reuse seems to happen most often when companies package components in ways that are easy for developers in other projects or departments to understand and redeploy, like "black box" parts in the auto industry. If programmers have to change a lot of the design or code to use a part, then reuse can become inefficient. For example, Toshiba, before the days of object-oriented design and class libraries, found that its engineers could change up to 20% of a module and still find the reuse effort cost-effective. If engineers had to change more than that, then they were better off writing new code from scratch. Toshiba also kept track of reused modules and gave out awards to encourage programmers to think about reuse and writing popular modules—an interesting way to channel the energies and creativity of the engineers (Cusumano 1991, pp. 264–265).

Another way to achieve scope economies is to buy or license components or whole products and incorporate them into a new system, including open-source "freeware." Cheap or free standardized packages or libraries of components often require some trade-offs in functionality. Nonetheless, from a business point of view, surveying what you can buy or get for free rather than build should be part of every organization's innovation and design strategy.

81.3 Architecture Strategy

An iterative or agile style of development (evolving specs, encouraging lots of feature changes during a project, and doing daily builds and continuous testing) requires a particular type of product or system architecture. With the wrong architecture, too many components and teams will be highly interdependent, leading to difficulties in testing and wasted time when trying to work in parallel.

The best strategy is to encourage a team to think ahead and, from the beginning of the release cycle for a new product or system, devote some engineering effort to designing an architecture that will last at least a few years and accommodate functionality likely to be important in the future. For example, in the early years of development products such as Office and Windows, senior Microsoft managers believed they should devote about 20% of their engineering resources to architectural work and reworking code (Cusumano and Selby 1995, pp. 280–281). They might allocate more for new strategic products or best-selling products that desperately need rearchitecting, though not too much more.

81.3.1 Importance of Modular Designs

Modular designs are important to decouple components and allow at least some coding and testing to proceed independently or in parallel. Modularity also facilitates future design changes and enhancements. There are no specific rules on how large or small a software module should be; it depends on the system. Moreover, there is a sliding scale of modularity for almost any complex product. For example, most automobiles have about 15,000 discrete components but automakers design and build their cars using subsystems. For some companies, the number of subsystems is relatively low, such as 25; for others, it is relatively high, such as 300. The difference is the degree of functionality that each company is designing into the modules or subsystems of its products (Cusumano and Nobeoka 1998, pp. 43–47, 97).

However one defines it, a module in software should be some subset of functionality that is smaller than the whole product and hides information from other modules so that programmers can isolate it from other small chunks of functionality (Parnas 1972). Software companies making products, smartphone apps, or Internet services often think in terms of "features," which contain modules within some larger subsets of functionality understandable to a user.

The opposite of a modular architecture is an *integral* architecture, where components are tightly coupled. It is difficult to change and test pieces of a product with an integral architecture without creating problems in dependent components (Baldwin and Clark 2000; Ulrich and Eppinger 2006). Integral architectures do the equivalent of binding the legs of everybody in a project together—slowing down even fast programmers. Managerial discretion is important here, though, because, in some software systems, an integral architecture is necessary to reduce size or generate superior performance, analogous to a custom-built racing car optimized for speed.

But, as code bases have grown to millions of lines of code for many common products and applications, a modular architecture often becomes essential for development and testing. The architecture needs to lay out what the subsystems of the product are, how the subsystems (collections of modules) relate to each other, perhaps what a module is within a subsystem, and, most important, what the interfaces are so that subsystems and individual modules can exchange data or instructions and work together. Interfaces should be stable for some period of time and not altered without communicating changes carefully to a development team because developers depend on knowing how to get modules to interact with each other.

Modularization also helps a project team prioritize features and build them in order of importance to the product or the business, like a sequential (or "horizontal") list that the team gets to one by one. With prioritization and modularization, a team usually has the option to cut lower priority features if the project falls behind schedule. If the modules are too interdependent, then a project might need a very large team to build all the desired features in parallel. A smaller team might have to build pieces of the product sequentially. With the sequential process, though, the project will usually have to adopt a waterfall type of schedule and not test the pieces in an integrated fashion until the team is mostly done—and when it may be too late to fix major problems or make important changes for the customer.

Again, I can cite examples from my study of Microsoft (Cusumano 2006a, 2007a). This company experienced firsthand the problems of inadequate modularity with the Windows Vista release, which eventually shipped in 2005 after several years of delays and wasted efforts. Microsoft may have adopted a particular strategy in the late 1990s and early 2000s to tie as many functions together into Windows so as not again to appear to violate antitrust law. The Vista (formerly called "Longhorn") project began with the Windows NT code base. The desktop version of the product was supposed to contain many new features and quickly grew to more than 50 million lines of "spaghetti" (i.e., nonmodular) code that proved impossible to build daily or test thoroughly and stabilize. The poor state of the code forced Microsoft to abandon years of work on new features, go back to the 2003 Windows server code base, and make some refinements to its engineering and design approach.

Microsoft eventually decided to break up Windows into different branches and rewrite as much code as possible into smaller, tighter modules. This approach resembled how company engineers had designed

Word, Excel, and PowerPoint and treated these products as "branches" or subsystems of Office, building them separately and then integrating the branches periodically, such as weekly. The new modularization and branching strategy made coding and daily builds for Windows and Office much more manageable than designing and building these as single "monolithic" products. Some teams also used new testing tools from Microsoft Research that helped check automatically for a wider variety of errors (code coverage and correctness, application programming interfaces and component architecture breakage, security, problematic component interdependencies, and memory use) and automatically reject code at desktop builds and branch check-in points (Larus et al. 2004).

It is possible to evolve the architecture of a software system incrementally to make it more modular, even if it did not start out that way because of time pressures or simply a lack of foresight and experience. Microsoft, my example again, gradually rearchitected Office over several years to make the applications within Office able to share features. The company formerly sold Office as a collection of packaged "vertical" applications that were really separate products. Each product (mainly Word, Excel, and PowerPoint) had its own separate features for text processing, file management, table creation, cut and paste, printer drivers, etc. A separate team figured out how to redesign the products and share at least some of these features across the applications. Within a few years, Office became the product, with Word, Excel, and PowerPoint becoming subsystems that shared about half of their code, at least in some versions. This sharing evolved to the point where, for Office 2000, fully 38% of the developers working on the product were creating common features shared by one or more of the applications (MacCormack 2000).

81.3.2 Long-Term versus Short-Term Trade-Offs

Working to get the architecture right or improving it incrementally is really an investment in the future—making it easier to maintain and enhance a system. Not all software companies or producer organizations have the money and time to make such investments, and users may not want to pay higher prices for such work when the trade-off may be less new features. Many new software companies also have to ship products quickly because the window of opportunity for their market may disappear.

It is unreasonable to expect new software businesses to devote too much effort to figuring out how to design a product architecture that will last for years. It is not clear what the right number of staff is to allocate to architecture development, especially for a start-up. But to make zero investment in architecture for the future would seem to be a technical and business mistake for any company that hopes to have a future.

81.4 Team Management

Another piece of common wisdom in the software engineering field is that a small team of very talented programmers works much better than a large team of mediocre people, and that talent is more important than experience (Brooks 1975; DeMarco and Lister 1987). Every experienced software manager has encountered programmers who can write much more and often much better code than other members of the same team. The rule of thumb given by Tom DeMarco and Timothy Lister, the authors of *Peopleware*, is that your best programmer will probably be about 10 times better than your worst programmer and about 2.5 times better than your average programmer (DeMarco and Lister 1987, pp. 44–46).

81.4.1 Problem of Large Teams

But one problem with managing by this philosophy is that "super programmers" are hard to find and maybe harder to keep. No rapidly growing company is likely to find enough programmers at the very upper end of the talent level to develop all the software it needs. So the more common problem in software development is how to get relatively large groups of programmers with varying skills to work together like nimble, efficient small teams.

The synch-and-stabilize techniques, as well as iterative or agile development more broadly, can help managers tackle this problem of how to make large teams work like small teams. Selby and I made this argument in *Microsoft Secrets*, describing how this company tackled the problem with the following approaches: project size and scope limits; modular architectures; project architectures mapped to the product (so that everyone knows why they are building what they are building); projects divided into small relatively autonomous teams (three to eight developers per feature team); rigid rules to force coordination and synchronization, such as through daily builds and periodic milestones; good communications and shared functional responsibilities; and product and process flexibility to handle real-time learning and the appearance of unpredictable problems (Cusumano and Selby 1995, pp. 409–417; Cusumano 1997).

81.4.2 Teamwork Principles

Many researchers have also found it important to have strong project leaders to make sure that even the top programmers follow a few basic rules to improve teamwork. One is the need for overlapping functional responsibilities. In the Microsoft case, for example, product managers take charge of writing vision statements but they are responsible for consulting program managers in order to do this. Program managers write functional specs but they have to consult developers, who generally have de facto veto power because they have to estimate the time and people required to write the code. Developers and testers are paired and jointly responsible for testing code. Good communications and overlapping responsibilities help an organization avoid becoming too functionally oriented, bureaucratic, and compartmentalized, with large separate groups that simply hand off work to each other.

In general, software product companies tend to organize separate groups for each product and then smaller feature teams within these product units. Managers can also scale up this type of structure by creating more product units and more feature teams, as long as the product architectures allow teams to proceed more or less in parallel. Within a product unit that has an effective leader, the right set of development techniques, and a modular product architecture, a company can have a large team of several hundred people or more working together almost like one nimble small team.

Companies such as IBM, Accenture, or Infosys that design or enhance custom systems also rely on small teams to build small chunks of functionality, but their structures are often far more complex. They generally organize at the company level in a matrix, with some managers and personnel assigned to industry or "vertical" specializations, such as manufacturing or banking sectors, and others in a variety of "horizontal" functions and specializations that cut across industry domains, such as experts in SAP, Oracle, Microsoft, or open-source systems. Bespoke projects generally have some team members, such as for the requirements phase or final customer acceptance testing, located at the customer site, while much of the actual software development takes place elsewhere (Carmel and Tjia 2005; Cusumano 2008).

81.5 Project Management

It is worth repeating that the traditional waterfall model, though it may sometimes deliver software on time and meet customer requirements relatively closely and with few bugs, is not a good process for fast-paced markets driven by the need to adapt to continuous innovation, uncertainly in customer requirements, and unpredictable competition. The waterfall model originally came about in fairly stable but complex development projects, like rocket systems, where NASA needed to control requirements and schedules in great detail (Royce 1970). To NASA and contractors such as Lockheed, not making changes that might create bugs has been far more important than being innovative or fast to market. Most commercial software producers, however, whether they make mass-market products or custom-built systems, need a process that lets them evolve designs and incorporate customer feedback in real time, during a project. For these kinds of environments, projects should do some preliminary planning but then more detailed requirements specification, program design, coding, and testing as concurrently as possible and with as much customer involvement as possible.

81.5.1 Divide and Conquer

Tackling any complex task brings up the age-old principle of "divide and conquer." In software, this means that managers should break large projects into multiple subprojects (many firms use the term subcycles or milestones) of no more than a few days, weeks, or months duration. It is much easier to manage several small groups that are doing a focused, small amount of work and that have a deadline not too far in the future than manage a large group building a lot of features scheduled for completion in months or years. Too many things can go wrong or change when a project deadline is too far into the future, when a team is too large, or when the amount of code that needs to be integrated is too extensive.

81.5.2 Individual and Project Discipline

It is also important to get commitments from people to work as a team and deliver on individual promises. Programmers and testers should schedule their own work, rather than have managers dictate schedules. It is often not necessary to press programmers in product companies working in highly competitive markets to shorten their estimates because they tend to be overly optimistic about what they can do. Self-scheduling by developers, therefore, produces aggressive schedules, which managers like, as well as fair schedules because they come from the bottom-up. But historical project data are still useful so that managers can evaluate the realism of individual estimates and schedule some buffer time into a project to accommodate misjudgments as well as unforeseen changes or problems that turn out to be more difficult than anticipated.

81.5.3 Infrastructure Investments

Software producers generally invest heavily in various tools and infrastructure such as build teams or process experts. For example, it is important to make checking in easier and faster for programmers and to automate as much testing as possible. If check-in times take too long, then the frequent build process becomes burdensome and programmers will avoid it. But the real benefit of a smoothly working build process is that, again, a few simple rules can be enough to have discipline and still be subtle. Programmers do not like to rewrite their code.

81.6 Quality Assurance

Finally, we come to the topic of testing and quality assurance. This is an entire subject in itself, as well as essential for any software-producing organization to be successful.

81.6.1 Building in Quality

In automobiles and other industries, we learned from Japanese companies decades ago that it is cheaper ultimately to "build in" quality continuously rather than to test and fix product flaws at the end of a development cycle or production process. This has been especially true in waterfall-style software development projects, where it has long been recognized that fixing a bug at a customer site can cost perhaps a hundred times more than finding and fixing the bug early in a project (Boehm 1976, 1981). Software products with modular architectures and built with iterative or agile techniques appear able to make design changes and fix problems late in a project more easily and with much lower costs than traditional waterfall processes. But fixing problems in the field such as by sending out patches is still expensive and can be harmful to a company's reputation.

Another point here is the importance not only of continuous feature testing—done manually by testers or through automated tests—but continuous *integration* and *system-level* testing. For example, creating a better drawing feature is fine. But if the user cannot print the object then the new feature is not properly integrated and the product has a *bug*. It is better to identify these problems earlier rather than later. Data from process surveys have reinforced this observation—the earlier a project can do integration testing, and the more it does integration testing, the higher the quality and the more likely the project will be closer to the schedule and budget targets (MacCormack et al. 2003).

It is also important for projects to automate as much feature or unit testing as well as component-integration and system-level testing as possible. Automation makes it possible to rerun tests frequently, such as with each code change, and find bugs generated by those changes. However, it is a myth that automation significantly reduces the need for people. The reason is that organizations always need people to update the tests as a project moves forward and incorporate more functionality or changes in user interface designs. Most automated tests run off the user interface and have to change as the user interface changes.

81.6.2 You Can Never Have Too Many Testers

In general, software managers would do well to adopt the philosophy that they *can never do too much testing*. This is especially true for a mass-market software products company. But it is also true for any mission-critical or enterprise-class software system. And, increasingly, consumers are expecting flawless functionality even in inexpensive or free software or functionality accessed over the Internet.

Many people are surprised to learn how many people Microsoft started to allocate to testing in the 1990s—as many as it does to programming, with testers usually assigned as "buddies" to developers in a one-to-one ratio (Cusumano and Selby 1995; Cusumano 2004). Some people are also surprised that, given this enormous investment in testing, Microsoft's quality is not higher. It is important to understand these two observations. First, Microsoft's quality has improved dramatically over time and, in multiple ways, reduced bugs but also produced products that are far more complex yet much easier to install and use compared to the old MS-DOS or early Windows systems and applications. These results directly reflect the enormous investment in testing as well as in process and product improvement more broadly. Second, many of Microsoft's testers, especially in the applications groups, are more like an advance army of beta users. These testers try to use a new product or version under development as a user would and try to detect user types of problems early on. It is a good investment to make because Microsoft sells tens or hundreds of millions of copies of its most popular software products. A few bugs or products that are difficult to install and use can generate millions of customer complaints.

81.6.3 Continuous Process and Product Improvement

Over multiple projects, it is desirable to have a strategy to improve process and product quality on a continuous basis—the now-familiar Japanese notion of *kaizen*. One way is to conduct postmortem analyses and share the results with the team and then act on the conclusions. For example, in the late 1980s and early 1990s, Microsoft adopted a practice of creating postmortem reports. These generally had three parts to them, compiled by the managers for each function on each project—product management, program management, development, testing, customer support, and user education (documentation). Each manager was supposed to interview his or her team members and come up with a summary of (1) what went well on the project, (2) what went poorly, and (3) what should they do differently the next time. Most of the Microsoft teams stayed together for a few years and had an opportunity to apply what they learned in a subsequent project (Cusumano and Selby 1995, pp. 331–339).

Another important source of learning is data from customers through the product support organization. With data-collection tools, it is possible to create detailed lists of bugs and fixes. Good teams will also develop heuristics about how to avoid and fix common bugs. They will create checklists or handbooks or run training sessions for their testers and developers to help avoid common errors.

A phrase common at Microsoft in the 1990s and at other companies such as Netscape and Google—"eat your own dog food"—captures yet another useful practice. That is to use the product internally as you are building it so that you get a firsthand experience of whether or not it is any good. For example, if you are building the next version of Windows, as soon as the team gets to a point where basic functionality works, a developer will start using it to do basic things, like saving files and running the e-mail program. If the product is lousy and crashes, then the developer has to eat the programming equivalent of dog food.

Microsoft and other software product companies have also introduced a variety of mechanisms to get customer feedback during development. Early beta releases can provide important feedback on the quality as well as the design of a product from actual customers. Betas that come too late in the development cycle do not allow the team enough time to make major design changes. Another technique is the usability lab, where companies bring in people to test features or user interfaces under development. In the Microsoft case, programmers watch from behind one-way mirrors to see what percentage of users struggle to understand their new features. Microsoft also has sent developers and testers to staff customer support lines after a new product ships so that, again, they can get firsthand feedback on customer reactions. Product teams complement usability lab data and customer support data (summaries of which teams received on a weekly basis) with customer satisfaction surveys, product usage surveys, and other feedback mechanisms.

Another good practice to monitor and improve the operations of a software development organization is to have each project track a small number of quantifiable metrics covering product quality, the size and performance of the product, and the development process. It is important to understand the major factors driving performance of teams and customer responses to products. Project managers need to be able to measure these factors quantitatively if they are to manage them effectively. This idea of statistical analysis and feedback has also been central to the SEI philosophy, especially for projects aiming to reach CMM Level 5 (Humphrey 1989).

81.7 Results from Project Surveys and Conclusions

During 2001–2003, several colleagues and I became curious about how widespread iterative development techniques were becoming around the world, in contrast to more waterfall-ish approaches, and what, if any, measurable impact they were having on project output measures like quality, productivity, and scheduling. We conducted a pilot study at Hewlett-Packard (HP) and Agilent and then followed this with a survey of 104 projects from a variety of major software producers around the world (Cusumano et al. 2003).

81.7.1 Global Differences in Practices

In the global survey, we first asked about conventional best practices. For example, how many projects wrote architectural and functional specifications as well as detailed designs before coding? How many used code-generation tools? How many implemented design and code reviews? Then we asked about the newer techniques: How many projects divided up into subcycles or milestones, used beta tests, paired programmers with each other and with testers, followed daily builds, and did regression tests on each build? We also divided the sample into regions: India, Japan, the United States, Europe, and others.

About 85% of the sampled projects wrote functional specs, and nearly 70% wrote architectural and detailed design documents, rather than just writing code with minimal planning and documentation. These conventional good practices were especially popular in India, Japan, and Europe. The major difference was that few US projects wrote detailed designs. (I had observed this practice earlier at Microsoft, where projects in general did not write detailed designs but went straight from a functional specification to coding in order to save time and not waste effort writing specs for features that teams might later delete.) Code generation (a technique that uses special software programs to generate code from design frameworks or design tools) was most popular in the Indian sample. Design and code reviews require particular process discipline, as promoted in the SEI recommendations. Not surprisingly, all the Indian and Japanese projects did design reviews, and all but one of the Indian projects did code reviews as well. Most projects in the other regions also followed these good practices, though not universally.

As for the iterative techniques, these were by now popular around the world, but with some variations. Most projects used subcycles, for example, though these were most common in our Indian and European and other samples and least popular in Japan. Projects that did not use subcycles, in our definition, followed a conventional waterfall process. More than half the Japanese projects, therefore, seemed to follow a conventional waterfall schedule. Most projects also used beta releases, which had become a useful testing and feedback tool since the widespread use of the Internet in the mid-1990s. Over 40% of the projects surveyed paired testers with developers—a Microsoft-style practice especially popular in India. Thirty-five percent of the projects used the XP-practice of pairing programmers, and, again, this was especially popular in our Indian sample (58%). More than 80% of the sample used daily builds at some time during the project and about 46% used daily builds at the beginning or middle, which is closer to the Microsoft style of development. More than 83% of the projects also ran regression tests on each build. Again, this good practice was most common in the Indian sample (nearly 92%).

81.7.2 Links between Practices and Performance Metrics

Researchers on software engineering over the past two decades will not be surprised with two of our findings from analyzing the HP and Agilent survey (MacCormack et al. 2003). This set of projects was roughly comparable in techniques and metrics, compared to the global sample, which had a wider variety of projects.

First, the HP and Agilent developers tended to be more productive in terms of code output when they had a more complete design before starting to write code. Second, more complete designs before coding correlated with lower levels of bugs. These results make sense and have led many software managers to insist on having complete specs before programmers start writing code—the old waterfall process. It is logical that programmers can be more productive in a technical sense if they make fewer changes during a project and thus have less rework to do. They also have less chance of introducing errors if they make fewer design and code changes.

In a business sense, however, locking a project early into a particular design may not produce the best result for the customer or enable the firm to compete effectively in a rapidly changing market. We did, in fact, find some evidence that HP and Agilent managers thought their customers were more satisfied with designs that evolved during a project. We also found that use of early betas and prototypes—opportunities for customers to provide early feedback on the design—was associated with higher code productivity and fewer defects, probably because the HP and Agilent projects were able to make early adjustments. In addition, running regression tests with each build, breaking projects into multiple subcycles, and conducting design reviews were associated with fewer bugs.

The most important conclusion from the HP and Agilent data is that, at least within a single development culture, iterative techniques such as described by the synch-and-stabilize philosophy appear to form a coherent, effective set of practices. Software projects can be more flexible in the sense of accommodating design changes with minimal impact on quality and productivity when they use several of these techniques together, rather than just selecting one or two. Our results suggest that there are trade-offs associated with using different techniques, especially with regard to allowing specifications to evolve after the start of coding. Use of particular techniques, however, helps projects overcome these potential trade-offs.

In short, when the HP and Agilent projects used beta releases to get early user feedback, conduct design reviews, and run regression tests on each build of the code (i.e., after each change or addition of new code), then the correlation between having an incomplete design when coding starts and high levels of bugs disappeared. It seems, then, that software producers can have the best of both worlds. With the right set of techniques, they can write high-quality code in a productive manner and quickly adapt to customer feedback and changes in the marketplace *during* rather than *after* a project, as in the old waterfall style.

81.7.3 Regional Differences in Performance

Project performance across firms is difficult to measure and even more difficult to compare regionally from such a small sample, but we used some crude measures and found some noteworthy differences in our global sample. Based on the data we collected, the Japanese had the best quality levels (median of 0.005) in terms of defects reported per 1000 lines of code in the 12 months after implementation at customer sites. The Indian projects (0.033) and US projects (0.030) were quite similar and very good by historical standards but still six times more "buggy" than the Japanese projects. Projects from Europe and other areas (0.05) fell in between the Japanese and the US and Indian levels. In terms of lines of code delivered per programmer per month—a measure of programmer output but not really productivity—the Japanese ranked at the top. They had a median output level of about 469 lines of code per programmer per month, unadjusted for programming language or type of project. This was more than twice the level of the Indian projects and about 70% more than the US projects, though only slightly higher than the European and other projects.

Early research on Japanese software practices found few defects along with high levels of nominal code productivity and reuse, so these results are not surprising (Zelkowitz et al. 1984; Cusumano and Kemerer 1990). US programmers often have different objectives and development styles. They tend to emphasize shorter or more innovative programs and spend much more time thinking about what they are writing and in optimizing code—which reduces lines of code productivity in a gross sense. The Indian companies have mainly US clients and seem to have adopted a US-type programming style, which tends to view shorter programs as better than longer programs. Nonetheless, we expected bug levels in India to be similar to the Japanese, given the emphasis of the Indian companies on achieving high SEI levels.

Overall, the data suggested some technical strengths in India and Japan with regard to software development but ongoing weaknesses from a business perspective. As is common in this type of research, due to the extreme variations in performance from project to project, it is hard to draw any definite conclusions. But it is important to remember as well that no Indian or Japanese company has made any real global mark in software innovation or establishing globally adopted technology platforms, which have long been the province of US and a few European firms. Code productivity is by no means a good measure of business performance and is less valuable than quality numbers in judging a software development organization. Japanese companies still seemed preoccupied with producing close to zero-defect code, and one can only wonder how much this practice constrains their willingness to experiment and innovate in software development. The global survey data also suggest that Indian companies are doing an admirable job of combining conventional best practices with iterative techniques. But they tend to treat software as a custom service business and, like the Japanese, lag behind the United States and some European firms in establishing global platforms for software products (Cusumano 2005, 2006b).

References

Baldwin, C.Y., and K. B. Clark. 2007. Design Rules: The Power of Modularity. Cambridge, MA: MIT Press.
Boehm, B. W. 1976. Software engineering. *IEEE Transactions on Computers* C-25(12): 1226–1241.
Boehm, B. W. 1981. *Software Engineering Economics*. Englewood Cliffs, NJ: Prentice Hall.

Boehm, B. W. 1988. A spiral model of software development and enhancement. *Computer* May: 61–72.

Brooks, Jr., F. P. 1975. *The Mythical Man-Month: Essays on Software Engineering*. Reading, MA: Addison-Wesley.

Carmel, E. and P. Tjia. 2005. *Offshoring Information Technology*. Cambridge, U.K.: Cambridge University Press.

Cusumano, M. A. 1991. *Japan's Software Factories*. New York: Oxford University Press.

Cusumano, M. A. 1993. Objectives and context of software measurement, analysis, and control. In D. Rombach et al. (eds.), *Experimental Software Engineering Issues: Critical Assessment and Future Directions*, Lecture notes in computer science 706, pp. 41–59. London, U.K.: Springer-Verlag.

Cusumano, M. A. 1997. How Microsoft makes large teams work like small teams. *MIT Sloan Management Review* 39(1): 9–20.

Cusumano, M. A. et al. 2003. Software development worldwide: The state of the practice. *IEEE Software* 20(6): 28–34.

Cusumano, M. A. 2004. *The Business of Software*. New York: Free Press.

Cusumano, M. A. 2005. The puzzle of Japanese software. *Communications of the ACM* 48(7): 25–27.

Cusumano, M. A. 2006a. What road ahead for Microsoft and Windows. *Communications of the ACM* 49(7): 21–23.

Cusumano, M. A. 2006b. Envisioning the future of India's software services business. *Communications of the ACM* 49(10): 15–17.

Cusumano, M. A. 2007a. What road ahead for Microsoft the company? *Communications of the ACM* 50(2): 15–18.

Cusumano, M. A. 2007b. Extreme programming compared with Microsoft-style iterative development. *Communications of the ACM* 50(10): 15–18.

Cusumano, M. A. 2008. Managing software development in globally distributed teams. *Communications of the ACM* 51(2): 15–17.

Cusumano, M. A. et al. 2009. Critical decisions in software development: Updating the state of the practice. *IEEE Software* 26(5): 84–87.

Cusumano, M. A. 2010. *Staying Power*. Oxford, U.K.: Oxford University Press.

Cusumano, M. A. and C. F. Kemerer. 1990. A quantitative analysis of U.S. and Japanese practice and performance in software development. *Management Science* 36(11): 1384–1406.

Cusumano, M. A. and K. Nobeoka. 1998. *Thinking Beyond Lean: How Multi-Project Management in Transforming Product Development at Toyota and Other Companies*. New York: Free Press.

Cusumano, M. A. and R. W. Selby. 1995. *Microsoft Secrets*. New York: Free Press.

Cusumano, M. A. and D. B. Yoffie. 1998. *Competing on Internet Time: Lessons from Netscape and Its Battle with Microsoft*. New York: Free Press.

Cusumano, M. A. and D. B. Yoffie. 1999. Software development on Internet time. *IEEE Computer*, Special issue on Software Engineering & Management, October: 2–11.

DeMarco, T. and T. Lister. 1987. *Peopleware: Productive Projects and Teams*. New York: Dorset.

Fagan, M. E. 1976. Design and code inspections to reduce errors in program development. *IBM Systems Journal* 15(3): 182–211.

Gilb, T. 1988. *Principles of Software Engineering Management*. Wokingham, England: Addison-Wesley.

Humphrey, W. S. 1989. *Managing the Software Process*. Reading, MA: Addison-Wesley.

Larman, C. and V. R. Basili. 2003. Iterative and incremental development: A brief history. *IEEE Computer* 36(6): 2–11.

Larus, J. R. et al. 2004. Righting software. *IEEE Software* 21(3): 92–100.

MacCormack, A. 2000. *Microsoft Office 2000*. Boston, MA: Harvard Business School, Multimedia Case #9-600-023.

MacCormack, A. et al. 2003. Trade-offs between productivity and quality in selecting software development practices. *IEEE Software* 20(5): 78–85.

McConnell, S. 1996. *Rapid Development*. Redmond, WA: Microsoft Press.

Parnas, D. L. 1972. On the criteria to be used in decomposing systems into modules. *Communications of the ACM* 5(12): 1053–1058.

Royce, W. 1970. Managing the development of large software systems. *Proceedings of IEEE WESCON* 26 (August): 1–9.

Striebeck, M. 2006. Ssh! We are adding a process. *Proceedings of Agile 2006 Conference*. Los Alamitos, CA: IEEE Computer Society, pp. 1–8.

Ulrich, K. T. and S. D. Eppinger. 2006. *Product Design and Development*. New York: McGraw-Hill.

Zelkowitz, M. et al. 1984. Software engineering practice in the U.S. and Japan. *IEEE Computer* 17(6): 57–66.

82

Project Personnel and Organization

Paul McMahon
PEM Systems

Tom McBride
University of Technology

82.1 Introduction

When planning a project, two critical areas that must be addressed are project personnel and organization. In this chapter we walk the reader through the steps of how to put together a team and explain the options and potential obstacles you will face. We first discuss how you go about deciding the types of personnel that you will need on your team. In this discussion we highlight not only the different types of roles and responsibilities that are common with today's varying methodologies but also emphasize why just assigning a role to an individual is insufficient to ensure project success. How team members acquire the needed skills and competencies to fulfill their assigned role is explained. We also explain the different ways you can organize your personnel along with the advantages and disadvantages of each approach. In the latter part of this chapter, we focus on what makes a team and how a team differs from just a random group of people. This section includes a discussion on how a team is formed and moves through multiple stages. What distinguishes a high-performance team is highlighted.

Throughout this chapter we examine the options you will have, the issues you are likely to face in organizing the personnel, and underlying principles and best practices. We conclude the chapter with a discussion of future research areas with a focus on the social side and the ability of teams to function in whatever circumstances they find themselves in.

82.2 Determining the Types of Personnel Required for Your Project

Before you can decide on the types of personnel, you need to understand the problem you are facing and have some idea about the solution and the technology you will use to solve it. This will lead you to the skills and competencies you will need in the composition of your team. Projects can quickly get into trouble when they prematurely bring people onboard before adequate analysis of the problem. A common best practice many organizations have learned the hard way is "don't add people faster than your understanding of the problem."

82.2.1 Looking Closer at Project Personnel from the Team Perspective

It is easy to state you need to understand the problem and something about your solution to identify your skill needs. However, this oversimplifies the real challenge often faced by project managers. There are usually multiple additional factors the project managers face, such as an aggressive schedule, a limited budget, and unrealistic stakeholder expectations. How well these additional factors are handled often comes down to qualitative aspects related to how well the project personnel perform as a team.

The skills and competencies of the personnel must be deployed and combined in some way to achieve the goals of time, cost, and quality. This is where individual team members must combine with other team members to function as a team and not as a collection of individuals. The topics of what makes a team and what makes a high-performance team are addressed in detail later in this chapter.

82.2.2 Ensuring Project Personnel Understand Their Responsibilities

Once you have decided on the types of personnel needed on your project and have those resources assigned, you will need to ensure your project personnel understand their responsibilities. One common best practice is to provide defined roles and responsibilities.

Clear roles and responsibilities help a team avoid the common problem of too many leaders, support accountability, and minimize the potential problem of excessive work overlap among team members.

There does not exist a single standard set of roles and responsibilities for project personnel. A role includes a set of responsibilities that an individual, or a group of individuals, may take on. The same individual may also take on more than one role at the same time or at different times on a project. Roles are important because they clearly establish the "who" is responsible for certain activities. Without clear roles it is easy for important work to be overlooked since no one is responsible for it or for two team members to take on the same work task. How an organization or project defines its roles depends on many factors including the nature of its product or service and its choice of methodology. The following section provides examples of roles and how their definitions can differ using three common methodologies:

- Scrum
- Team software process (TSP)
- Rational unified process (RUP)

Keep in mind that most organizations tailor their roles and responsibilities based on the specific needs of their project and product. Use the following information as a guide to help you determine appropriate roles in your organizations.

82.2.2.1 Scrum

Scrum (Schwaber, 2004) is a management framework for developing complex products. Scrum defines three roles:

- Product Owner
- Developer
- ScrumMaster

The Product Owner is the person responsible for managing the product backlog.

The product backlog is an ordered list of everything that might be needed in the product and is the single source of requirements for any changes to be made to the product. The Product Owner is responsible for the product backlog, including its content, availability, and ordering.

On small projects the Product Owner may be the actual customer. On large projects where there may exist multiple customers, or the customer is not readily accessible by the development team, the Product Owner may be a representative of the customer or customer community.

By explicitly defining this role as part of the team, Scrum helps to address the common problem observed on many software efforts of scope creep and unclear customer priorities.

Developers are members of the development team, which is a self-organizing team* when using Scrum. The development team is cross-functional with all the skills necessary to accomplish their work. Scrum recognizes no titles other than developer for all development team members. There are no exceptions to this rule as it supports the important Scrum principle that the team as a whole is accountable for their work.

The ScrumMaster is responsible for ensuring the rules of Scrum are understood and followed. The ScrumMaster does not direct the team but rather is a servant–leader. The ScrumMaster coaches the development team and assists the Product Owner in finding effective techniques to manage the product backlog.

82.2.2.2 Team Software Process

TSP (Humphrey, 2006) is a framework and a process for building and guiding development teams. TSP defines 10 standard team roles:

1. Team leader
2. Team member
3. Customer interface manager
4. Design manager
5. Implementation manager
6. Test manager
7. Planning manager
8. Process manager
9. Quality manager
10. Support manager

At first it appears that TSP is more complex in its role definitions than Scrum, but on closer examination the difference may not be as large as one might think. TSP teams, like Scrum teams, are self-directed. This means the team as a whole is responsible for its work and the quality of its products. Software development is a complex and creative activity with many factors that require close attention to ensure the right product is built leading to a satisfied customer. These factors include ensuring the team is listening to its customers, ensuring appropriate design effort is conducted, ensuring the designs are being properly implemented, ensuring appropriate testing is being conducted, ensuring appropriate planning

* Self-organizing teams are discussed in greater detail later in this chapter.

is being conducted, ensuring the team is keeping their processes current and following them, ensuring data are being recorded appropriately, and ensuring team members have appropriate tools and training.

The eight manager roles in TSP are roles that are taken on by team members who essentially become the eyes and ears of their self-directed teammates with respect to these factors. It is the developers on a TSP team who take on these manager roles, not "managers" outside the team. All self-directed software development teams must ensure these factors are being addressed. The difference between Scrum and TSP is the degree to which these responsibilities are documented and formally trained or remain tacit knowledge among team members.

82.2.2.3 Rational Unified Process

The RUP (Kroll et al., 2003) is a software engineering process framework developed by the Rational Software Corporation, a Division of IBM. It is not a single prescriptive process but a framework that is intended to be tailored. A role in RUP goes further than roles in Scrum and TSP in that it breaks out responsibilities at a finer level of granularity, and it also defines the competency required and how the individual should do the work. Examples of roles defined in RUP include

- System analyst
- Business analyst
- Business-model reviewer
- Requirements reviewer
- Requirements specifier
- Test analyst
- User-interface designer
- Software architect
- Developer

For more information on RUP and its role definitions, refer to Kroll et al. (2003).

82.2.3 Additional Roles Dependent on Project Circumstances

As stated previously, there does not exist a single standard set of roles and responsibilities for project personnel. Dependent on your project circumstances, other roles may be required. For example, in organizations that develop large complex systems, there is often a software (or system) architect role. One of the purposes in establishing a Software or System Architect role on a large complex project is to aid communication and coordination across the software development teams (Dikel et al., 2001; McMahon, 2006).

82.2.4 Ensuring Project Personnel Have Appropriate Skills and Competencies

Organizations must compete for the talent required to produce their products or perform their services. One of the major challenges faced by many organizations today is how to acquire and sustain a high-quality work force (Curtis et al., 2001). This includes ensuring your personnel have appropriate skills and competencies to accomplish their assigned responsibilities.

In this section we focus on the skills required of personnel including soft skills, such as communication skills, and listening skills, as well as the technical skills. Group skills will be covered later when we deal with teams.

There are multiple best practices commonly used today by organizations to ensure project personnel have appropriate skills and competencies. Two of those best practices discussed in the following section are

- Training
- Coaching

82.2.5 Training and Coaching

Competencies are not the same as roles. Roles define a set of responsibilities, but when an individual accepts an assigned role, it doesn't guarantee one is competent to perform that role. Helping to ensure project personnel are competent to perform their job is the purpose of training and coaching. Many organizations have developed extensive formal training programs for their personnel and have defined minimum training requirements for their people to fill standard roles defined in their organization. However, not all training needs to be formal classroom training. Over the last few years in an attempt to provide training in a more flexible and efficient way, many organizations have moved toward online web-based training that people can take as a quick refresher when needed or at a time that fits their personal schedule.

While formal training is necessary to ensure competency, it is not sufficient. No prescriptive training course can answer all the questions and address all the situations that arise on the job. This is where coaching is needed to supplement a formal training program.

82.2.5.1 Is the Coach the Same as the Team Leader?

A coach is a distinct role from a team leader although in many small organizations the same individual may take on both roles. The team leader on a self-directed team does not direct the team members but focuses on the project requirements, schedule, and success factors. Basic principles that an effective coach follows include

- Build talent
- Set high standards
- Focus on success
- Focus on improvement
- Improve in steps
- Celebrate every step (Humphrey, 2006)

As projects proceed, project conditions change. It is the responsibility of both the team leader and coach to be on the lookout for conditions that might warrant changes in team personnel to ensure the right people with the right competencies are taking on appropriate roles. While coaching can be effective in many of these situations, sometimes the best approach to address a potential trouble spot is to make a change in the personnel assigned to a project or to change an individual's responsibilities on the project.

82.2.5.2 Relationship of Competency to Methodology and Agility

The competencies that project personnel must acquire to carry out their assigned responsibilities can vary depending on the project's chosen methodology. Boehm and Turner (2004) point out that as you increase the agility of your methodology, the required skill levels of your people also increases. This is due to multiple factors. First, as organizations move increasingly in the agile direction, there is usually less documentation to help the personnel perform. Second, increased agility implies more rapid decisions often with less information. Third, increased agility implies self-direction and self-management, which means personnel need to be competent in planning and estimating and communicating with stakeholders, as well as being technically competent. This is where the "soft" skills often become critical to success.

82.2.6 Need for Measurement, Listening, and Negotiating Skills

Watts Humphrey has stressed that software work is knowledge work (Humphrey, 2011). Because knowledge work takes place in the workers head, rather than with their hands (Drucker, 1996), it is necessary for software professionals to measure and manage their own work. This also implies a need to be able to communicate the results of their measurements to stakeholders, listen to feedback, and negotiate appropriate to the situation.

When using more traditional plan-based methods planning, estimating and communicating with stakeholders often fall under the responsibilities of a manager's role. The manager in traditional functional organizations breaks the work down, estimates the work, and assigns it to project personnel. The manager also assesses progress and maintains the schedule. On self-directed teams all of these tasks fall under the responsibilities of the team that implies a need for greater competency of team members.

82.2.7 Consider All Competency Requirements to Break Down Walls between Interfacing Groups

Personnel in quality assurance groups and systems engineering groups also require communication skills. A common weakness often found in functional hierarchical organizations is a culture that breeds walls between software development groups and interfacing groups, such as quality assurance and systems engineering groups. Similar walls are often found between systems engineering and software engineering. This culture results when systems engineering and quality are viewed as separate functions rather than integrated functions. Too often in these organizations, the view becomes one of quality as being the responsibility of the quality group rather than everyone's responsibility and software being viewed as something distinct from the system it is a part of. Both of these cultures lead to increased bureaucracy and reduced product quality for the customer.

To address these common organizational weaknesses, all required competencies, including the soft communication skills, should be assessed when selecting project personnel to fill key project roles. Too often on software intensive efforts, only technical competencies are considered, which increases the risk of project failure.

82.3 Organizing Project Personnel

Organizations are complex people-based systems constructed to achieve a larger purpose (Rechtin, 1999). In the case of a software organization, that larger purpose is to produce or maintain a software product that may operate stand-alone, embedded within another product or interface to another system. The structure of a software organization can vary depending on the software or service produced. How you go about organizing your project personnel depends on multiple factors including constraints that may be levied upon you by policies in your overall organization, as well as your project's chosen methodology.

In this section we first examine varying organizational structures in use today and how these different organizational structures may affect the decisions you make on your project for organizing your personnel. The organizational structures discussed in this section include

- Functional hierarchical organization
- Integrated product team (IPT) organization
- Self-directed team organization
- Product-line organization
- Hybrid software organization

82.3.1 Functional Hierarchical Organization

Traditional large software organizations are functionally partitioned hierarchical structures with personnel at each level reporting to the level earlier. This organizational structure supports the fundamental principle of solving a large complex problem through "divide and conquer." Typical functional groups may include system architecture, systems engineering, software development, system integration and test, quality assurance, and configuration management. Depending on the product or service, the systems and software groups may be further partitioned based on the product architecture (Figure 82.1).

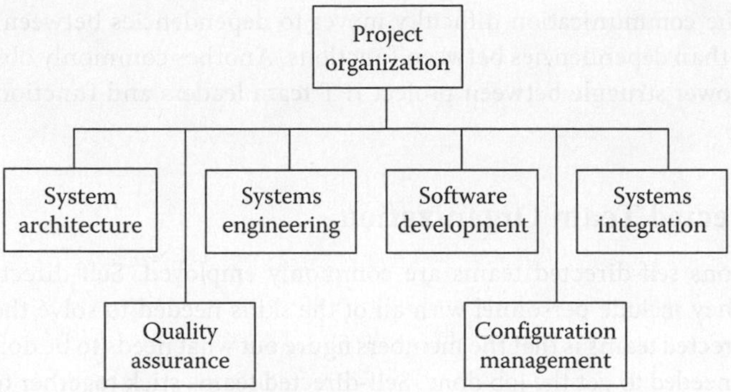

FIGURE 82.1 Functional hierarchical organization.

The primary advantage of a functional organization is that it provides focus on key functional needs and the allocation and management of personnel with the appropriate skills to address each specific function required. There are, however, multiple negative consequences that have been observed when employing a strict functional hierarchical organizational structure.

Common negative consequences observed include too narrow of a focus on the groups specific function causing loss of proper attention to the ultimate end product goal and the customer's needs. Functional organizations often include a system architecture and system integration and test group to help ensure proper attention to dependencies between functional groups and requirements that cross multiple functions. A common side effect often observed especially on large complex projects employing functional hierarchical organizations is long integration schedules due to too poor communication across functional group boundaries. This poor communication is often caused by inefficiencies introduced by a need to communicate up and back down the hierarchical management chain to get things done.

82.3.2 Integrated Product Team Organization

A common approach employed to address the problems observed in strict functional organizations is to use an IPT approach to software development. While this organizational approach still supports the fundamental principle of solving a large complex problem through "divide and conquer," the criteria for division of work are now based on a partitioning of the product architecture rather than a functional partitioning. With IPTs the organizational architecture more closely reflects the product architecture, rather than the generic functions within the product (Figure 82.2).

With IPTs personnel with the required functional skills are assigned to each product team. The primary advantage to the IPT approach is improved communication as it removes the need for communication up and down the chain and therefore supports more rapid decision making. This organizational structure provides more autonomy to individual teams within the overall project organization. However, a common negative side effect of IPTs is that it can still create communication problems,

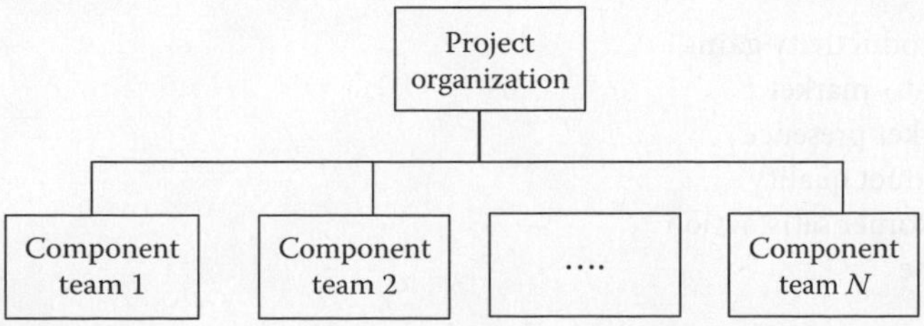

FIGURE 82.2 IPT organization.

but in these cases the communication difficulty moves to dependencies between components of the architecture, rather than dependencies between functions. Another commonly observed negative side effect of IPTs is a power struggle between project IPT team leaders and functional managers in the larger organization.

82.3.3 Self-Directed Team Organization

In small organizations self-directed teams are commonly employed. Self-directed teams are similar to IPTs in that they include personnel with all of the skills needed to solve the software problem. A strength of self-directed teams is that the members figure out what needs to be done on their own, and they do whatever is needed to get the job done. Self-directed teams stick together to the end of the job. Key characteristics of self-directed teams include (Humphrey, 2006)

- A sense of membership and belonging
- Commitment to a common team goal
- Ownership of the process and plan
- The skill to make a plan, the conviction to defend it, and the discipline to follow it
- A shared commitment to honest, truthful, and respectful behavior
- Dedication to excellence

Strengths of self-directed teams include improved communication and more rapid decision making with conflicts handled by the team rather than through escalation to higher level management. One of the weaknesses observed with self-directed teams is scalability and difficulty communicating outside the team. Self-directed teams are typically small. While self-directed teams could in theory be large, in practice self-directed teams that perform well are usually no larger than 15 people. Large projects can be broken down into multiple self-directed teams as long as some kind mechanism is established for communication among the teams.

82.3.4 Product-Line Organization

Projects, regardless of how they are organized, have different motivations than software businesses. Software businesses are motivated by the financial bottom line to the business. Projects are motivated by cost, schedule, quality of the specific software product, and the customer needs (Royce, 1998). Projects do not tend to invest in technology or services that their immediate customer is not asking for. Product-line organizations are structures that are set up specifically to address the broader long-term needs of a product beyond just a single immediate customer.

In organizations with IPTs, team members are usually fully responsible for technical decisions affecting the product. With the product-line approach, technical decisions related to core product assets become the responsibility of the product-line organization. In making technical decisions, this organization must consider the goals of the product and other current or potential customers that use the product-line.

Strengths of a software product-line organization include (Clemens and Northrop, 2001)

- Large-scale productivity gains
- Improve time-to-market
- Maintain market presence
- Improved product quality
- Increased customer satisfaction
- Increased reuse

Because product-line organizations are focused on the broader issues associated with the product, and because they have multiple customer needs to consider when making a change, one weakness often

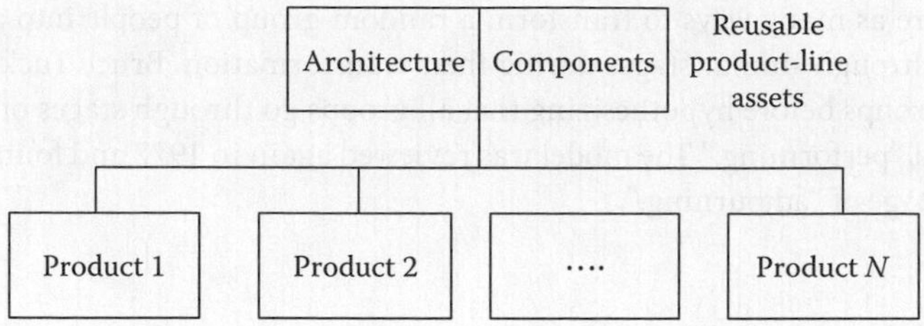

FIGURE 82.3 Product-line organization.

observed is their lack of rapid response to the changing needs of individual customers. Product-line organizations are usually structured around the product architecture with a focus on reuse (Figure 82.3; see Clemens and Northrop, 2001; p. 316, Sections 6–10).

82.3.5 Hybrid Software Organization

Because all software organizational structures have weaknesses, often the best choice is a hybrid approach. An example of a hybrid software organizational structure could be the use of self-directed teams together with an architecture team and an integration and test team to aid coordination and communication.

82.3.6 Where Product, Organization, and People Intersect

It is not uncommon, especially in product-line organizations, to see organizational architecture reflect the product architecture. This can be both good and bad. It can be good because it brings focus to the skills and competencies needed for product development and support. But it can be bad because in today's rapidly changing world, continual product evolution is required to keep pace with changing customer needs. This also implies a need for continual evolution of personnel skills and competencies.

Too often when organizational architecture is structured too close to a product architecture, politics and fear of change get in the way of the product innovation needed to remain competitive. Best in class organizations counter this risk by keeping their personnel challenged continually updating their skills. When organizations provide planned career growth paths for their people, and actively support these plans, organizations and products evolve naturally. When organizations fail to pay proper attention to the growth of their people, fear of change often takes over leading to product stagnation and lost organizational competitiveness.

82.4 What Makes a Team?

To distinguish between a team and a random group of people collected together in the one place for some arbitrary purpose, we must first revisit what a team is. One very useful definition is "A team can be defined as a social system of three or more people, which is embedded in an organization (context), whose members perceive themselves as such and are perceived as members by others (identity), and who collaborate on a common task (teamwork)" (Hoegl and Gemeunden, 2001).

The obvious question is "Why can't the team simply follow the process?" The answer is that processes provide guidance but cannot deal with the uncertainty and variation inherent in software development or, indeed, many of the tasks undertaken by teams. The team needs to understand, frequently to considerable depth, what they are doing, why they are doing it, how to do it, when to know that they have done it among other things. It is more useful to think of the software development as a problem to be solved than as a thing to be produced. A team can be thought of as a distributed cognition system rather than some sort of production system in which each member is a component.

Although there are as many ways to transform a random group of people into a functioning team, they all seem to go through similar stages during their transformation. Bruce Tuckman reviewed published research on groups before hypothesizing that all groups go through stages of "forming," "storming," "norming," and "performing." The model was reviewed again in 1977 and found to be sound other than to add a final stage of "adjourning."

82.4.1 Forming

In software development teams are most often formed anew for each project. It is rare that a team remains intact from one project to the next so forming the team becomes a routine but important activity. During formation the team meets, gathers information about the task and each other, and decides how to approach the task. The team also learns about opportunities and challenges before considering team goals. At this stage the team is likely to be acting independently since they have yet to learn about issues and objectives of the team. Although we may believe that all teams are similar enough that that can simply be brought together and immediately become productive, it does not happen like that.

Watts Humphrey (2001) proposed the following when he considered what it is that a team must agree on:

1. What are our goals?
2. What are the team roles and who will fill them?
3. What are the responsibilities of these roles?
4. How will the team make decisions and settle issues?
5. What standards and procedures does the team need and how do we establish them?
6. What are our quality objectives?
7. How will we track quality performance and what should we do if it falls short?
8. What processes should we use to develop the product?
9. What should be our development strategy?
10. How should we produce the design?
11. How should we integrate and test the product?
12. How do we produce our development plan?
13. How can we minimize the development schedule?
14. How do we assess, track, and manage project risks?
15. How do we determine project status?
16. How do we report status to management and the customer?

Individual circumstances may vary, Humphrey's list will be a good place to start.

A more recent study (Gilley et al., 2010) proposes that teams need to build skills in several areas: conflict resolution, problem solving, communication, organizational understanding, decision making, goal setting and performance management, and planning and task coordination. Certainly a team would need all of those skills to answer Humphrey's questions.

82.4.2 Storming

Every group will next enter the *storming* stage in which different ideas compete for consideration. This is not necessarily a bad thing nor does it need to be antagonistic or acrimonious despite the connotations of the term "storming." Brainstorming is an accepted technique of bringing out different possibilities that each can be considered in turn after the brainstorming session has ended.

This is the time when the team may know "what," but not "how." Somehow the team must identify and consider all the possibilities they can. Premature agreement or forced agreement will usually result in later need to reconsider the matter, usually at greater cost than if time had been taken earlier.

82.4.3 Norming

Having identified all of the decisions the team needed to make in the "forming" stage and explored all the possible answers in the "storming" stage, the team must now decide which of those answers to accept and act upon. Gradually the team builds up a mental model of the way the team will function (Walz et al., 1993). The "norming" stage does not conclude tidily with answers to everything the team may have considered, but enough of them will have been well enough resolved for the team to start performing.

82.4.4 Performing

This stage is reached by some teams who manage to resolve sufficient of their issues to start working smoothly as a unit without inappropriate conflict or the need for external supervision. Team at this level would normally be considered as self-directed teams. Conflicts are expected but are resolved appropriately; new issues arise and are similarly dealt with appropriately.

Teams can, and will, occasionally revert to earlier stages in response to some external changes. Perhaps the team composition changes, or the task changes enough that they need to learn a new domain or new techniques, or their organization changes their supervision arrangements.

82.4.5 Distributed Teams

What must be done differently when the team is distributed?

Distributing the team challenges their ability to communicate with the same immediacy and richness that face to face communication affords. To some extent the diminished communication richness can be compensated by electronic means such as video conferences, electronic chats, or social media. However, the informal coordination that high-performing teams depend on for their effectiveness usually must be augmented by some formal coordination such as project schedules, specifications, backlogs, defect registers, configuration management, and project meetings, among other things. Distance between team members is like any other stress on software development: greater attention is required to ensure small problems do not become large problems. Where a collocated team can rely on informal, and to some extent accidental communication, a distributed team must go to some effort to create communication. Where a collocated team can get by with informal issue tracking, a distributed team usually needs formal issue tracking. That additional formality adds time and slows responsiveness.

Today using physically distributed teams has become an accepted software development approach for many organizations. When the Agile (Cockburn, 2006) movement began around 2000, many of the experts were claiming that agile software development teams needed to be small and collocated. While collocation can aid optimal conditions for software development, today it is recognized that agile approaches, as well as other development approaches, can work effectively with physically distributed teams (refer to Schwaber, 2004).

Nevertheless, one should not assume that the skills required on a distributed development team are no different from that of a collocated team. Experience has shown that many inexperienced project personnel work best when they are placed in an environment where they can easily listen to conversations of their teammates and easily ask questions at the moment the need arises. Collocation usually works best in these situations.

The fundamental problem of global software development is that many of the mechanisms that function to coordinate the work in a colocated setting are absent or disrupted in a distributed project (Herbsleb, 2007). There is much less communication and it is less effective. In many ways it is the problem of "out of sight, out of mind." Communication is less frequent because we must consciously think to communicate instead of being reminded by frequent cues such as their appearance in the hallway. When communication is immediate, there is opportunity to clarify, to respond to puzzled looks, to answer questions, and to change vocabulary to suit the listener. Communication over a distance tends to be less responsive, much more likely to be written and with less opportunity to question and clarify. When people are distant, there is also a lack of awareness of their specific context, which can lead to misunderstandings between sites.

To work effectively as a distributed team requires personnel who are self-starters and have strong self-discipline and self-management skills. Similarly, not all managers are good at managing remote personnel. Some managers are uncomfortable when they can not see and have direct personal interaction with the people they are managing. These factors should always be considered when selecting project personnel for distributed team efforts (McMahon, 2000).

82.4.6 Globally Distributed Teams

Whenever software development is globally distributed, it is usually with a distribution of collocated teams rather than the teams themselves that are distributed. Nevertheless, it is worth reviewing what has been found about distributed software development.

Most attention has been paid to cultural aspects of globally distributed projects possibly because the effects of culture can be subtle and difficult to deal with or can be obvious but requiring attention all the same. Some examples are the attitude to defects where one country could not see the point in classifying or prioritizing defects because they believed that all defects should be rectified before the product was delivered (Borchers, 2003). A more subtle problem is what Hofstede (1991) refers to as power distance. Cultures with small power distances tend to expect subordinates to raise issues with their superiors and openly to discuss matters, whereas cultures with a larger power distance will tend to expect the subordinate to carry out their superior's instructions without question. This can come as a surprise to many western team members who tend to question their superiors but who also can be surprised that a team member has acted on something they believed was offered only as a suggestion to be discussed.

Cultural differences are not the only challenge facing globally distributed teams. Obviously communication can become a challenge when time zones do not overlap by much, even though the IT industry is becoming accustomed to conference calls outside regular working hours.

Technologies are becoming available also to overcome many of the communication challenges facing distributed teams. Video conferencing, cheap voice calls, collaboration websites, and social media sites are all helping not only to reduce the communication costs and barriers but also to help team members become familiar with each other's cultures.

82.4.7 Large Teams

As the problems we try to solve become larger and more complex, the teams we need also become larger. Large problems have been successfully dealt with by decomposition into smaller problems. However, the problem of how to integrate the component solutions into a coherent whole again is less easy to manage. Change is inevitable to the requirements, to the technical solution, to the manner of integration, and to the work that must be done. Solutions to these challenges have emerged in the shape of Microsoft's "synch and stabilize" (Cusumano and Selby, 1997), code check-in gates that provide enough to assure a clean build without becoming burdensome, and interfaces that serve as contracts between components (Brechner, 2005). Coordination is achieved through the common code repository and the rules about how to incorporate code into that repository. It does not do away entirely with the need for some administrative oversight of what gets developed and when it gets integrated, but it reduces dependence on it.

Common too is the idea that there should be just enough process to ensure that people can interact successfully without compromising independence and creativity. So processes will define relationships between people and between components, but not how someone is expected to perform a task. It is assumed that they are competent.

Another approach to increasing team size is "Scrum of Scrums" in which there is a cascade of Scrums. To date this has been reported as reaching a depth of three Scrums but most frequently is only two Scrums. The general idea is that a delegate from each daily Scrum attends a higher level Scrum in which matters of overlap and integration are discussed.

82.5 Challenges and Research Directions

Larger, more complex and more critical projects and the knowledge of what is required to achieve success coevolve. We attempt such projects because we believe we can succeed and we learn what is required to succeed through our attempts. In this area three main challenges seem to be emerging: self-management, governance, and performance measurement. In this section we will first discuss self-management through different approaches to aligning the goals, understanding, and work processes. We will then discuss emerging trends of governance before concluding with a brief discussion of performance measurement.

82.5.1 What Makes a High-Performing Team?

As briefly discussed earlier, a team needs to do more than simply follow instructions. Moreover, with many of the task to be performed in software development, it is not possible to provide a set of instructions that can prescribe how to develop software. Instead there have been various studies of teams and teamwork designed to identify the elements of teamwork.

Work gets done better when there is a common understanding of both the task and the means to accomplish that task. There is less need for over coordination and less need for administrative and organizational support or interference in the task.

82.5.1.1 Mental Models

Team mental models are team members' shared, organized understanding and mental representation of knowledge about key elements of the team's relevant environment (Mohammed and Dumville, 2001). The content of a team mental model includes shared representation of tasks, equipment, working relationships, and situations (Mohammed and Dumville, 2001).

The notion of a team mental model was developed to help account for the differences in performance levels of teams. The mental models of individual team members will be quite different at the start then converge toward a common representation. Both team work and taskwork mental models relate positively to team process and performance. In other words, teams with greater convergence in their mental models will perform better.

The concept of a mental model is quite rich, to the point where it can be useful to separate different aspects of them. For example, team performance could be enhanced if the team shared a model of the team itself, its collective skills, and who had those skills. This could be separate from a mental model of the task, what sequence of activities were needed for its completion and which team member would be responsible for different activities. Separate again from the mental model of the software product to be developed.

Whereas teamwork and taskwork reflect knowledge structures, varying interpretations and perspectives regarding key team issues reflect belief structures (Mohammed and Dumville, 2001). If the belief structures are too dissimilar, miscommunication and misunderstanding will prevail. However, if there is too much similarity, there is a danger of group think.

Information sharing and transactive memory research has been task oriented, concentrating on domains of expertise. Some knowledge will need to be common to all team members, but other, more specialized information is not needed by all team members so can be left to a few individual members who can apply that knowledge when needed.

In contrast, knowledge of how the group will function together, how they will communicate, and how they will resolve their differences should be held by all members (Mohammed and Dumville, 2001).

82.5.1.2 Collective Mind

In a study of flight deck operations, Weick and Roberts (1993) proposed the concept of "collective mind" distinguishing it from "group mind." The term collective refers to individuals who act as if they are a group, taking care to interrelate their activities. Weick and Roberts are interested in the actions and how they are interrelated within the group. An individual can contribute to a collective mind, but a collective mind is distinct from an individual mind because the collective mind is made up from the pattern of interrelated actions within the system.

Weick and Roberts used three characteristics when describing the workings of collective mind: construction, representation, and subordination.

Construction occurs when individuals in the group construct their actions as if there are social forces operating within the group. They decide what actions they believe are expected of them and what actions they believe are expected of others. In a formal organization such actions would be described in formal organization charts and the distribution of roles and responsibilities. To some extent that is also true in teams except that in most teams there are more actions required than are formally described. In some way individuals within the group decide, or construct, their actions and the actions they expect of others. Importantly individuals look for signs that the expected actions have occurred in order to carry out their own actions. For example, if a programmer was scheduled to integrate someone else's code into their own work, they would normally look for cues about the status and stability of the other work before committing their own work to it.

The second characteristic of collective mind is representation in which people envisage a social system of joint actions. Within a team each has some awareness of the totality of actions to be undertaken by the group, who will undertake them and how they interrelate. Again, a certain amount of this is given by the work breakdown structure and allocation of work within a project but this does not capture the detail of all of the actions of software development.

People relate their envisaged actions within the constructed system of actions. They know what they have to do and how it fits into the overall picture. This "subordination" is the third characteristic of collective mind and is clearly necessary in order that the whole collection of interrelated actions functions well.

Software development is not quite as action packed nor as immediately dangerous as flight deck operations of military aircraft carriers, but the lessons of how to form high-performance, high-reliability teams appear to be applicable to a wide range of domains, including software development.

82.5.1.3 Transactive Memory

In tasks of any size, very few people are able to know everything about the task and how to accomplish it. Instead different people know different parts, but the team collectively knows all of it. Transactive memory is defined as the set of knowledge possessed by group members, coupled with an awareness of who knows what (Wegner, 1987).

When teams are a primary mechanism for accomplishing organizational work, effective coordination of teamwork becomes an important organizational issue.

Members of a team come to understand how their individual know-how and skill become linked together within a spatial and technology environment. The team is then able to respond as a complete system to meet situational demands even though the complexity of the task is beyond the cognitive capabilities of individual team members (Faraj and Sproull, 2000).

Need to know where the expertise is. The driving force behind one organization in Sydney, Australia, related how he overcame the problem of developing awareness of team members' knowledge and skills. From time to time, the team members were required to describe the knowledge and skills of another team member to a potential or actual client:

> I think I'm a pretty good technologist but I think my real skill is putting together incredibly high performance teams and people. That's what I think I hang my hat on. The way I do that is I make sure everyone is familiar with the competencies and skills and the value that each other brings to the table. So they can account for that when they are creating or designing or thinking of an idea

> for a customer. How I manifest that within my team is anyone of my team, and I have them do this on a regular basis, can stand up and explain to a group of people what the role, responsibility, history and experience, career path of another member of the team to everyone else. The purpose of that is to ensure that each member of the team understands intrinsically what the skill or the value or the capacity of other members of the team are. The benefit of that, the ability to reflect the knowledge base or the understanding of another team member manifests itself in a very cohesive team and a team that is very open in its communications.
>
> **—Stefan Gillard, CEO Studio Engine**

Faraj and Sproull demonstrated that expertise coordination contribute to team performance above and beyond traditional factors such as group resources and the use of administrative coordination.

82.5.2 High-Reliability Teams*

The level of reliability expected of software systems coupled with the increasing size and complexity of those systems challenges the ability of teams to meet those expectations. We can draw some lessons from high-reliability organizations where the potential for disaster is overwhelming yet the organization has little choice but to function reliably. For software development the equivalent is to produce high-performance software for life critical systems. It can be done with those same humanly fallible people that make up teams in any organization. High-reliability organizations share five characteristics:

82.5.2.1 Preoccupation with Failure

High-reliability organizations pay attention to small problems, treating all of them as symptoms of failure in the system. Small errors could develop into large problems if left undetected and unchecked. Thinking about the potential consequences of each incident counters the tendency to dwell on success and become complacent.

82.5.2.2 Reluctance to Simplify Interpretations

The usual tendency is to simplify in order to pay attention to the few key issues and key indicators. High-reliability organizations fight such a tendency, acknowledging that the environment in which they operate is nuanced, complex, unstable, unknowable, and unpredictable. They encourage diverse experience, skepticism toward conventional wisdom, and ways of negotiating differences of opinion without destroying the nuances that diverse people detect.

82.5.2.3 Sensitivity to Operations

High-reliability organizations are less strategic and more situational in their outlook. When people have well-developed situational awareness, they can make continuous adjustments that prevent errors from accumulating and enlarging.

82.5.2.4 Commitment to Resilience

Resilience is a combination of keeping errors small and improvising work-arounds to keep the system functioning. Both of those avenues of resilience demand deep knowledge of the technology, system, team members, self, and materials. The signature of a high-reliability organization is not that they are error free but that errors do not disable it.

82.5.2.5 Deference to Expertise

High-reliability organizations push decision making down and around. Decisions are made on the front line, authority migrates to those with the most expertise regardless of rank.

* Weick, K. E. and Sutcliffe, K. M., *Managing the Unexpected: Assuring High Performance in an Age of Complexity*, Jossey-Bass, San Francisco, CA, 2001. With permission.

82.5.3 Governance

"Consensus has not yet emerged on a definition of IT governance" (Bannerman, 2009) but can instead be taken as the arrangements to direct, monitor, and control at the level of software development. In the main these arrangements have been the responsibility of project management, quality management, and, to a lesser extent, human resources. What is challenging about software development governance now is that past methods appear to be inadequate for highly complex, highly uncertain volatile project environments that face many teams.

Instead of relying on directives from higher levels of management, teams have to interpret the goals and intent of higher management. Instead of relying on precise specifications from customers and other stakeholders, teams have to understand their goals and intent. This is challenging because there are few reliable methods of knowing if the development team are "on the same page" as their customers, higher management, and other stakeholders. From a corporate perspective, it is insufficient simply to trust that the team will somehow incorporate the goals and intent of others into their own goals.

Software development is not the only organization that operates in highly uncertain, highly volatile and often hazardous environments. Military operations too are faced with many of the same challenges, albeit with higher consequences. The military has the concept of "commander's intent" to guide individual initiatives, actions, and decisions (Augustine et al., 2005).

The military may have significant experience in commander's intent as other domains may have experience with other methods but software development has some catching up to do, both to overcome past experience of development teams pursuing their own agendas and of development teams lacking the tools and methods to elicit and understand the goals and intent of their senior management, customers, and other stakeholders.

82.5.4 Performance Measurement

Measures of performance are used to monitor projects, to determine their state and status, and to determine the need for change. Developing a system for a well-understood problem in a stable environment can rely on long-term planning and associated measures such as budget, earned value, and delivered artifacts. In such an environment, it is reasonable to equate time with progress since actual progress does tend to reflect planned progress.

But in volatile environments and when developing a system for something that is largely unknown, being discovered, and solved on the run, such performance measures prove inadequate. In such volatile and uncertain circumstances, measures of risk or, more specifically, risk reduction have been proposed as more appropriate and more useful (Eisenmann et al., 2011; Ries, 2011). The biggest risk is that a system will be developed that solves the wrong problem or no one wants. So instead of measuring against a plan on the assumption that all will happen as it was planned, projects are measured against risk reduction. What are the biggest risks facing the project and how could those risks be resolved at an acceptable cost?

The challenges of such a shift in perspective are that we tend to revert back to old habits and convert the highly uncertain project into a sequence of prototypes that are assumed to prove what we want them to prove or we simply launch the project with some open-ended commitment to resolve all the risks then watch the escalating cost of doing so rather than stop the project.

While the work of Eric Ries is very appealing and does offer a new direction to pursue, there is much to be investigated. While Ries and others are eager to point out the risks of new product development, there is little guidance about measuring the types and potential impacts of risks in new product development, their interaction with the development, or how best to manage them. Possibly of more importance is knowing when to shift measurement systems from those traditional measures of stable, long-term production projects to those of risk recognition and reduction of highly volatile emergent projects and back again.

82.6 Summary and Conclusion

In this chapter we walked the reader through the steps of how to put together a team and explained the options and potential obstacles you will face. We examined options you have and issues you are likely to face. This included selecting personnel with the right skills and the importance of clearly defined roles and responsibilities to ensure important work is not overlooked and team members are not working on the same task. We identified critical skill needs including both personal and team skills.

Underlying principles and best practices were identified including training and coaching. We explained why coaching is needed to supplement formal training programs. We discussed differing organizational structures including functional hierarchical organizations, IPT organizations, self-directed team organizations, product-line organizations, and hybrid software organizations. Strengths and weaknesses of each were identified.

We explained what distinguishes a team from a random group of people and we discussed the stages groups go through in becoming an effective team including forming, norming, storming, and performing. We discussed what is different when a team is distributed and why distributed teams require personnel with self-starter, and strong self-discipline and self-management skills.

We discussed the management and coordination challenges of large teams and what makes a high-performance team different. Finally we identified and discussed future research areas including mental models, collective mind, transactive memory, and high-reliability teams.

Future changes in the nature of software engineering will require matching changes in the nature of software engineering teams. While some of the responses to those challenges are likely to be technical in nature, more of them are likely to be social in nature placing a premium on our ability to form functioning teams suited to whatever circumstance we find ourselves in.

References

Augustine, S., Payne, B., Sencindiver, F., and Woodcock, S. (2005), Agile project management: Steering from the edges, *Communications of the ACM,* 48(12), 85–89.

Bannerman, P. L. (2009), Software development governance: A meta-management perspective, *Proceedings of the 2009 ICSE Workshop on Software Development Governance*, IEEE Computer Society, Washington, DC, pp. 3–8.

Boehm, B. and Turner, R. (2004), *Balancing Agility and Discipline: A Guide for the Perplexed*, Addison-Wesley, Boston, MA.

Borchers, G. (2003), The software engineering impacts of cultural factors on multi-cultural software development teams, *Proceedings of the 25th International Conference on Software Engineering*, IEEE Computer Society, Portland, OR.

Brechner, E. (2005), Journey of enlightenment: The evolution of development at Microsoft, in *International Conference on Software Engineering*, Anaheim, CA, ACM Press, St Louis, MO.

Clemens, P. and Northrop, L. (2001), *Software Product Lines: Practices and Patterns*, Addison-Wesley Professional, Boston, MA.

Cockburn, A. (2006), *Agile Software Development: The Cooperative Game* (2nd edn.), Addison-Wesley Professional, Upper Saddle River, NJ.

Curtis, B., Hefley, B., and Miller, S. (2001), *People Capability Maturity Model® (P-CMM®) Version 2.0*, Software Engineering Institute, Pittsburgh, PA, CMU/SEI-2001-MM-001.

Cusumano, M. A. and Selby, R. W. (1997), How Microsoft builds software, *Communications of the ACM*, 40(6), 53–61.

Dikel, D. M., Kane, D., and Wilson, J. R. (2001), *Software Architecture: Organizational Principles and Patterns*, Prentice Hall, Upper Saddle River, NJ.

Drucker, P. F. (1996), *Landmarks of Tomorrow: A Report on the New*, Transaction Publishers, New Brunswick, NJ.

Eisenmann, T. R., Ries, E., and Dillard, S. (2011), *Hypothesis-Driven Entrepreneurship: The Lean Startup*, Harvard Business School case study, p. 1.

Faraj, S. and Sproull, L. (2000), Coordinating expertise in software development teams, *Management Science*, 46(12), 1554–1568.

Gemuenden, H. G. and T. Lechler (1997), Success Factors of Project Management: The Critical Few - An Empirical Investigation. *International Conference on Management and Technology*, Portland, OR.

Gilley, J. W., Morris, M. L., Waite, A. M., Coates, T., and Veliquette, A. (2010), Integrated theoretical model for building effective teams, *Advances in Developing Human Resources*, 12(1), 7–28.

Herbsleb, J. D. (2007), Global software engineering: The future of socio-technical coordination, in *Future of Software Engineering, 2007: FOSE '07*, Minneapolis, pp. 188–198.

Hofstede, G. (1991), *Cultures and Organizations: Software of the Mind*, McGraw-Hill, New York.

Humphrey, W. S. (2001), Why don't they practice what we preach? Available: http://www.sei.cmu.edu/publications/articles/practice-preach/practice-preach.html, accessed on September 17, 2002.

Humphrey, W. S. (2006), *TSP: Coaching Development Teams*, Addison-Wesley Professional, Upper Saddle River, NJ.

Humphrey, W. S. (2011), *Leadership, Teamwork, and Trust: Building a Competitive Software Capability*, Addison-Wesley Professional, Upper Saddle River, NJ.

Kroll, P., Kruchten, P., and Booch, G. (2003), *The Rational Unified Process Made Easy: A Practitioner's Guide to the RUP*, Addison-Wesley Professional, Boston, MA.

McMahon, P. E. (2000), *Virtual Project Management: Software Solutions for Today and the Future*, CRC Press, Boca Raton, FL.

McMahon, P. E. (2006), Lessons learned using agile methods on large defense contracts, *Crosstalk: The Journal of Defense Software Engineering*, May 2006, 25–30.

Mohammed, S. and Dumville, B. C. (2001), Team mental models in a team knowledge framework: Expanding theory and measurement across disciplinary boundaries, *Journal of Organizational Behavior*, 22(2), 89–106.

Rechtin, E. (1999), *Systems Architecting of Organizations: Why Eagles Can't Swim*, CRC Press, Boca Raton, FL.

Ries, E. (2011), *The Lean Startup: How Today's Entrepreneurs Use Continuous Innovation to Create Radically Successful Businesses*, Crown Business, New York.

Royce, W. (1998), *Software Project Management: A Unified Framework*, Addison Wesley, Reading, MA.

Schwaber, K. (2004), Scaling projects using scrum, in *Agile Project Management with Scrum*, Chapter 9, pp. 119–132, Microsoft Press, Washington, DC.

The Malcolm Baldrige National Quality Improvement Act 1987, pp. 100–107.

Walz, D. B., Elam, J. J., and Curtis, B. (1993), Inside a software design team: Knowledge acquisition, sharing, and integration, *Communications of the ACM*, 36(10), 63–77.

Wegner, D. M. (1987), Transactive memory: A contemporary analysis of the group mind, in *Theories of Group Behavior* (Eds., Mullen, B. and Goethals, G. R.) Springer-Verlag, New York, pp. 185–205.

Weick, K. E. and Roberts, K. H. (1993), Collective mind in organizations: Heedful interrelating on flight decks, *Administrative Science Quarterly*, 38(3), 357.

Weick, K. E. and Sutcliffe, K. M. (2001), *Managing the Unexpected: Assuring High Performance in an Age of Complexity*, Jossey-Bass, San Francisco, CA.

83

Project and Process Control

James McDonald
Monmouth University

83.1 Introduction

This chapter describes two well-known control methods that can be applied to controlling software development projects and to their underlying software development processes. The first method is closely related to traditional engineering methods for controlling manufacturing and communications systems using feedback. That method is most applicable to the overall management control of entire software development projects. The second method is known as statistical process control (SPC). It can be applied beneficially to portions of the software development processes that can be most easily quantified, such as the inspection process. We will describe both of these methods and their underlying mathematical and engineering foundations, and will then provide some examples of their use in typical software development projects. We will end with a brief discussion of the extensive literature in which other authors have described the application of SPC methods to software projects and the successes and failures that they have reported.

All plan-based software development projects must be planned, organized, monitored, and controlled by their project manager. We will refer to the project manager's control activities in executing the project as project control. We will refer to control of the underlying software development processes as process control. Controlling a project and controlling the underlying processes are very different activities, and consequently, they require very different techniques to make them effective.

83.2 Project Control

A project is usually defined as activities performed by people that produce something unique and that have a limited lifetime. It can be described as a set of tasks that are linked together in a network that shows which of the tasks must be completed before subsequent tasks can begin. Software projects usually require tasks that involve the development of requirements, the development of architectural and design specifications, coding, testing, delivery to the customer, and continuing maintenance. It will also sometimes involve the development of a proposal or a business case, which presents the economic justification for the undertaking, and an agreement, or a contract, with the customer specifying what will be delivered, when it will be delivered, and how much it will cost. At a high level, virtually all large software projects can be defined by these activities. However, because each project produces something unique, the details of tasks that need to be performed to complete the project are usually different from project to project. For most large software projects, the network and schedule relationships usually look similar to Figure 83.1.

For the software development project shown in Figure 83.1, the team needs to develop five different software modules (Mod 1 through Mod 5) and integrate them into two different releases of the product. The project starts with planning and requirements development. The architecture, which defines each of the modules and how they work together, is developed next. Then the design, coding, and unit testing (D, C, and T) of three of the modules is undertaken. When all three modules have been unit tested, they are integrated to produce Release 1 (R1) of the product. Release 1 is then system tested and that release is deployed.

During or shortly after Release 1 has been integrated, two of the resources that developed the first three modules go to work on design, coding, and unit testing of the two additional modules that provide

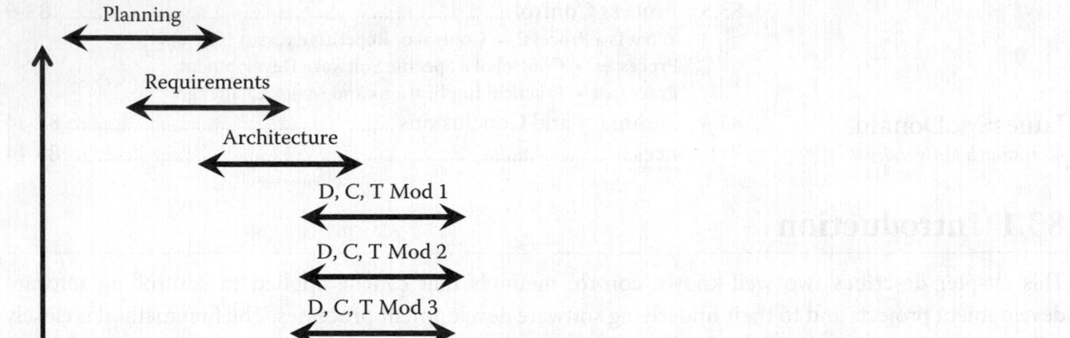

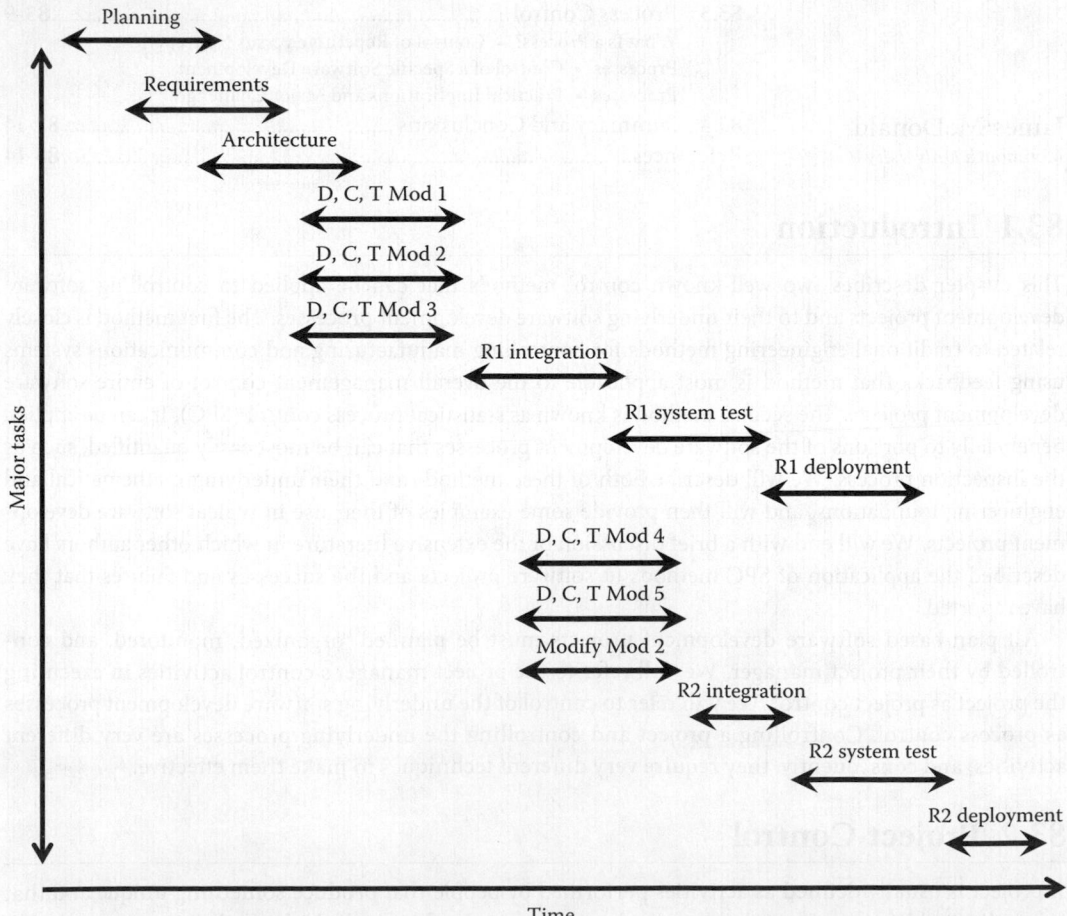

FIGURE 83.1 Typical software development project schedule.

the capabilities included in Release 2. The information in Figure 83.1 also assumes that while Modules 4 and 5 are being designed, coded, and unit tested, someone needs to return to modify Module 2 to accommodate Release 2 (R2) features. When all three modules have been unit tested, they are integrated with the software that was developed for Release 1 to produce Release 2 of the product. Release 2 is then system tested and that release is deployed.

83.2.1 Project Plan

The project plan is a document, prepared by a core team that answers the following questions for the team, for the team's management and for the project's other stakeholders:

What is the project going to produce?
How is the team going to produce the product?
Who is going to do each task?
When will each task, each release, and the whole project be completed?
How much will each task and the entire project cost?

The core team that produces the plan should be a set of people whom we expect to work on the project from beginning to end who can represent all of the functions (such as requirements development, architecture and design, coding) that will be required for completion of the project.

The plan needs to specifically define each of the tasks that must be accomplished, who will be responsible for each task, and visible end points for each task so that the project manager and other members of the team will know when each task has been completed. These visible end points might be things like inspections of requirements, customer sign-off on requirements documents, inspections of detailed designs, unit test results, demonstrations of design, coding and unit testing tasks, demonstrations of test results, and sign-off by team members on installation and user documents.

The reason for specifying very visible end points for each task is so that the project manager does not have to be continually asking each person assigned to a task what the status of work on that task is. Tasks should be defined in such a way that they can be completed by one person working full time on the tasks for no more than a few weeks. By defining relatively small tasks and by having visible end points for each task, the manager no longer needs to ask about the status of the task. The task is either completed or not completed. And the manager can confirm when it is completed by viewing the visible end point. This avoids the challenge of having to use subjective estimates of how much of a given task has been completed. Assigning a percent complete to a project task is a guess on the part of the individual who is responsible for that task. This author has observed that individuals frequently provide overly optimistic estimates when asked this question thereby delaying availability of the manager's knowledge that problems may be developing on the project.

83.2.2 Project Operations

The project plan should also contain some general descriptive materials outlining how the project will be staffed; how it will be kicked off; how and with whom the project will communicate, both internally and with its external stakeholders; and, probably most importantly, how the project will be monitored. Most project managers monitor the status of their projects by arranging periodic project meetings on a weekly, biweekly, or monthly basis. Those project meetings provide a forum in which the project's status and the status of individual tasks can be observed and reported. They also serve as a place where current issues are raised, where responsibilities for addressing issues are assigned, where a variety of project data can be shared with members of the team, and where demonstrations of progress can be conducted.

83.2.3 Project Monitoring and Feedback

The project plan should describe the number of people who are expected to be working on the project as a function of time. It should also describe the cost, or budget, indicating how much the management expects to have to pay for salary and expenses in support of the software development team. Finally, it needs to list all of the tasks that need to be accomplished to complete the project and who will be responsible for each task. This information can be displayed in either tabular or graphical form, that is, in a spreadsheet or in a Gantt chart. As an example, Figure 83.2 shows a Gantt chart for the project that was outlined in Figure 83.1. Figure 83.3 shows the same information in a tabular format.

In addition to the status of the project's tasks, we might also be interested in the resources working on the project and their cost. Figures 83.4 and 83.5 also show, in a graphical format, the number of people we plan to have working on a project and their cost.

83.2.4 Project Control

Now that we have discussed and set up the basics of a project plan, we are ready to start discussing project control. In order to exercise project control, we need to make use of another major project management processes, namely, the monitoring process. Monitoring is the process of tracking the status of the project, comparing the current status with the status contained in the plan for the current point in time, and searching for deviations between the baselined project plan and the actual project status. The deviations are then used as the basis for taking actions to address the deviations, namely, the control process.

Figure 83.6 shows how the monitoring and control processes are used together. Input to the monitoring process includes the project plan, specifically the data shown in Figures 83.2 through 83.5, showing which tasks should be completed at each point in time, the number of resources that should be working on the project at each point in time, and the costs that should have been

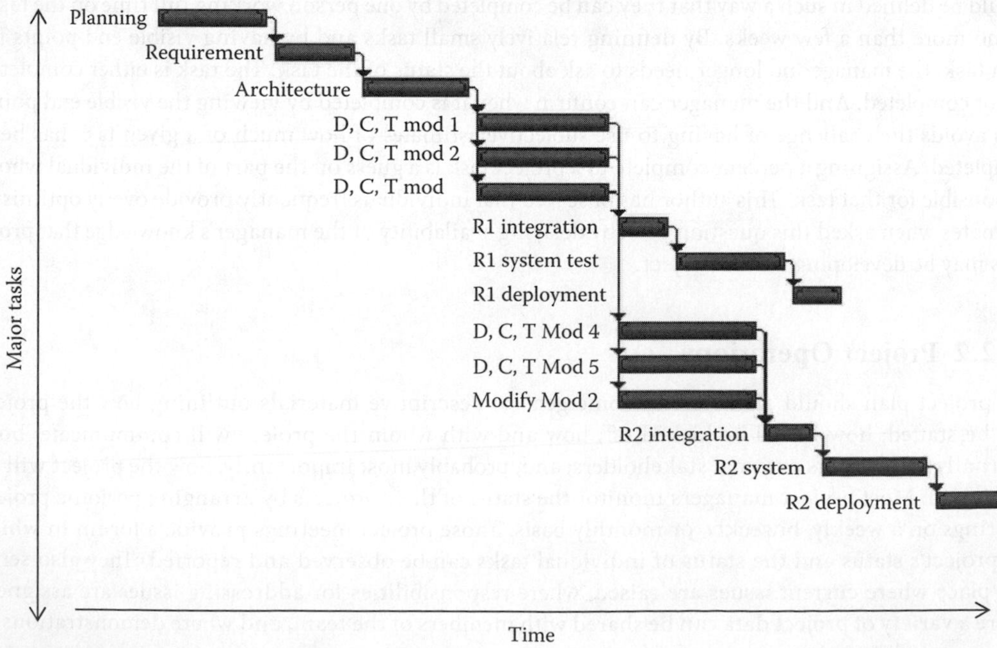

FIGURE 83.2 Gantt chart description of a software development project.

	ℹ	Task Mode	Task Name	Duration	Start	Finish	Predecessors
1			Planning	4 wks	Mon 2/27/12	Fri 3/23/12	
2			Requirements	3 wks	Mon 3/26/12	Fri 4/13/12	1
3			Architecture	4 wks	Mon 4/16/12	Fri 5/11/12	2
4			D,C, T Mod 1	5 wks	Mon 5/14/12	Fri 6/15/12	3
5			D, C, T, Mod 2	5 wks	Mon 5/14/12	Fri 6/15/12	3
6			D, C, T Mod 3	5 wks	Mon 5/14/12	Fri 6/15/12	3
7			R1 Integration	2 wks	Mon 6/18/12	Fri 6/29/12	4,5,6
8			R1 System Test	4 wks	Mon 7/2/12	Fri 7/27/12	7
9			R1 Deployment	2 wks	Mon 7/30/12	Fri 8/10/12	8
10			D, C, T Mod 4	5 wks	Mon 6/18/12	Fri 7/20/12	4,5,6
11			D, C, T Mod 5	5 wks	Mon 6/18/12	Fri 7/20/12	4,5,6
12			Modify Mod 2	5 wks	Mon 6/18/12	Fri 7/20/12	4,5,6
13			R2 Integration	2 wks	Mon 7/23/12	Fri 8/3/12	10,11,12
14			R2 System Test	4 wks	Mon 8/6/12	Fri 8/31/12	13
15			R2 Deployment	3 wks	Mon 9/3/12	Fri 9/21/12	14

FIGURE 83.3 Tabular description of a software development project.

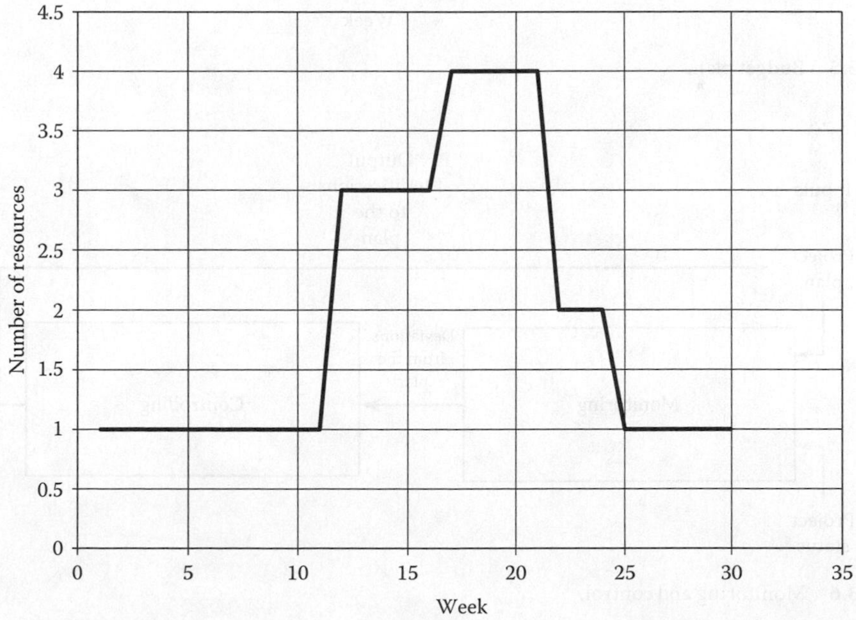

FIGURE 83.4 Staffing plan.

incurred at that time. In preparation for each periodic project meeting, the project manager compiles the status of all tasks, the number of people working on the project, and the cost that has been incurred. When the manager discovers that some of the tasks that should have been completed have not been completed, the number of resources is not at the planned levels or the cost incurred is different from the planned cost the manager needs to take control and do something to address the deviation.

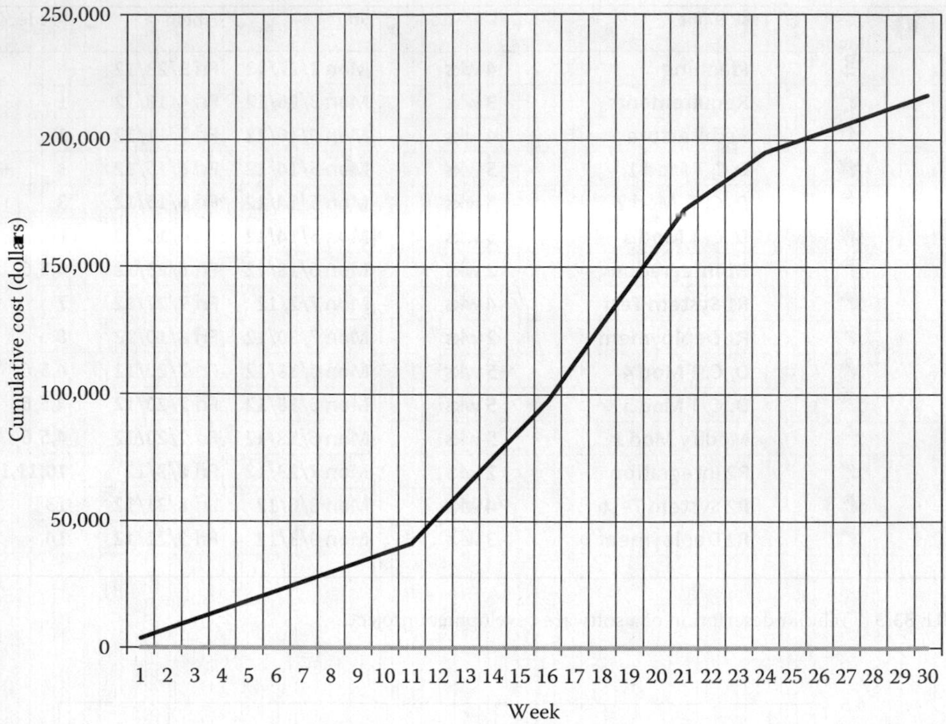

FIGURE 83.5 Budget plan.

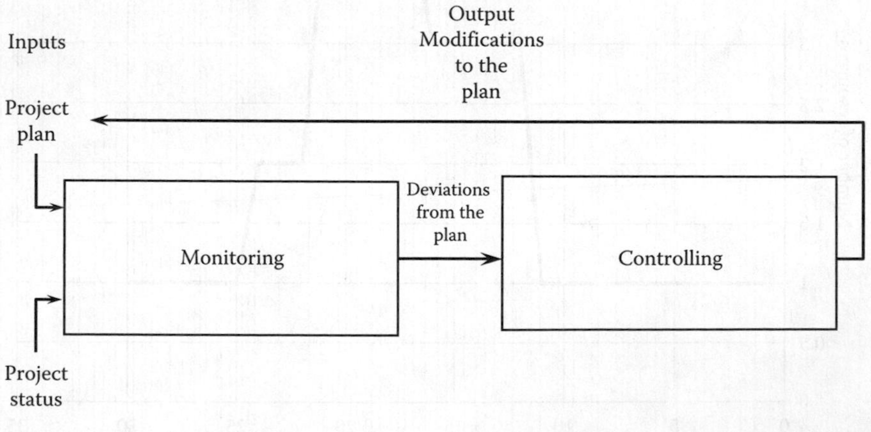

FIGURE 83.6 Monitoring and control.

"Project monitoring and control" is a key process area in the Capability Maturity Model Integrated Development (CMMI-DEV) [6], developed by the Software Engineering Institute. Inclusion of both monitoring and control in the same key process area, and in most other software development best practice descriptions, indicates that monitoring and control are closely related and highly dependent processes.

Figure 83.6 shows that the outputs from the project control process become modifications of the project plan. The purpose of those modifications is to get the project back on track. While these modifications might not literally result in changes to the written project plan, they will certainly result in modifications to the way the project is operating. In practice, when an unacceptable deviation from the project management plan is discovered, the project manager needs to take corrective action. As a first step the project manager may ask the individual who is responsible for the deviation to develop and present the steps needed to get back to the baseline plan as quickly as possible.

The project manager may need to provide support to the responsible person. Corrective actions that do not require additional costs or delays in the project completion date should be considered first. These actions could include nonpaid overtime work or the use of resources assigned to tasks with slack to assist with the delayed work.

If no such modifications can be found, the project manager should consider two additional techniques, "fast tracking" and "crashing." Fast tracking means that tasks that are normally done in sequence are done, instead, partially in parallel. For example, if you would not normally start preparing detailed designs until requirements and an architectural solution were documented, fast tracking might mean starting detailed design or, perhaps, even coding before the requirements and architectural specifications were complete. Fast tracking involves risks that could lead to increased cost and rework later. If the requirements or the architecture changes, this might result in having to redo some of the detailed design or coding work already underway, thereby delaying project completion and/or increasing total project cost.

Crashing the schedule means adding additional resources to tasks on the critical path without necessarily being concerned about the level of efficiency for those tasks. Let's assume that one person was working on a 4-week design, code, and unit test activity and they are having difficulty getting agreement on how their software module will interface with a module designed and coded by another person. If you really need to assure that the task finishes in 4 weeks, you might add a second person to this task. That person might not have all the right skills for this task, and they might have to work for two weeks just to reduce the overall time on this task by 1 week. That trade-off, on the surface, might not make sense. However, if the project's delivery date is of utmost importance, we might be willing to make that trade-off even if it results in an incremental increase of the project's cost. One of the goals of crashing the schedule is to minimize the incremental cost increase. However, in exchange for completing tasks on the critical path on schedule, crashing almost always leads to additional incremental cost. If the project manager is willing and able to spend more to accelerate the schedule, crashing is sometimes the best alternative.

If the schedule will be delayed, the cost increased, or the scope reduced, the project manager will need to consult with the customer and/or other stakeholders to negotiate an increased budget, a project schedule delay, or a reduction in the project's scope (requirements) to accommodate the necessary modifications. Obviously, prevention of schedule delays, cost overruns, or scope reductions is superior to alternatives that require these negotiations.

All of these actions constitute project control.

83.2.5 Theoretical Basis for Project Control

The methods that have been discussed previously are similar to those used in modern control engineering, also called feedback control. Feedback control uses the difference between signals, determined by comparing the values of system's output variables to their desired values, as a method for controlling the system. A very common and well-known use of feedback control is the room thermostat used to control the temperature of a room. The setting provided by the user or occupant of the room is the desired temperature. During periods when heat is required to maintain a comfortable temperature, the occupant will position a lever or, with a digital thermostat, set the desired temperature. Based on that input, the thermostat will determine a temperature slightly lower than the desired temperature at which the thermostat will send a signal to the furnace or other heating device to turn "on" and start delivering heat to the environment. When the temperature rises to slightly more than the desired level, it will send a second signal to the heating device to tell it to turn "off." In the case of

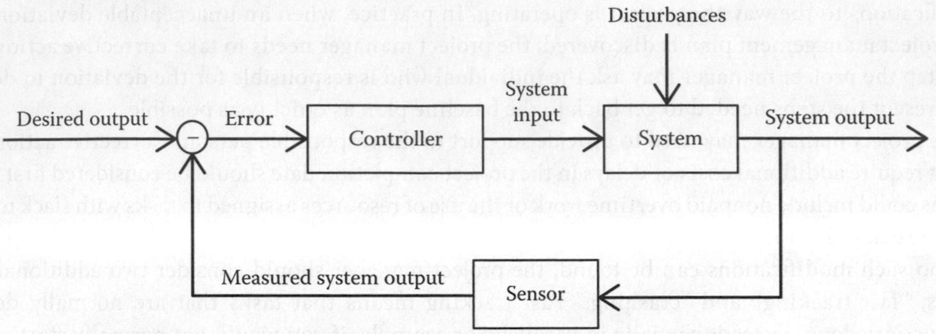

FIGURE 83.7 A formalized control system based on feedback.

room thermostats, turning the switch to "on" or to "off" is usually implemented by a combination of temperature sensitive mechanical devices and electrical circuitry.

This theory of feedback control was first formalized by James Clerk Maxwell who provided the first rigorous analyses of a feedback control system. In 1868, he wrote a famous paper "On Governors," [4] which is widely considered the seminal paper in feedback control theory. In that paper he anticipated many aspects of modern control engineering. Figure 83.7 is a diagram that shows in a more formal way the use of differences between desired results and actual results to control a system.

Note that the relationships shown in this figure are similar to the relationships between the project monitoring process and the project control process shown in Figure 83.6. The "desired output" is the information contained in the project plan. The System represents the project. The Sensor corresponds to the monitoring process, and the Controller represents the actions taken by the project manager to address deviations from the plan. While the theoretical basis for control of many electromechanical systems can frequently be expressed and analyzed quantitatively, that is not usually possible for project control systems that depend very heavily on the experience and capabilities of the project manager and the project team to implement the control actions discussed previously.

83.2.6 Artifact Control

Another aspect of project operations related to control is those activities that have become known as configuration, or artifact, control. Configuration control is the formalized process of keeping track of artifacts such as requirements, architecture, and design documents as well as code and test cases. After a draft of these documents is formalized by being baselined, any further change in those artifacts requires formal approval. That approval is frequently in the hands of managers who are responsible for the project or, more likely for larger projects, a configuration control board, which is usually a subset of team members who take on primary responsibility for controlling what changes are made to the artifacts and how those changes are integrated into the project's formal documentation. Those changes may come from customer trouble reports, from customer requests for changes, or from members of the software development team who are making changes to the system. It is imperative that there be a methodology in place to keep careful track of artifacts included in each release of the project so that all team members know which version(s) of the software is in the hands of customers, which are being tested, and which are in development.

While these activities have the word "control" in their names, they are qualitatively a different kind of control from project control and from process control that will be discussed in later sections of the chapter. The most important characteristic of these artifact control activities is the care with which they must be executed and the cost of exercising this kind of control. There are many software-based commercial configuration management systems available that can be investigated by browsing the web. Information on each is readily available there, so we will not go into any further discussion about artifact control in this chapter.

83.3 Process Control

83.3.1 What Is a Process?

A process can be defined as a set of tasks performed by people, machinery, or nature that transform inputs into outputs using resources [1]. So a process must be considered in the context of the agents carrying out the tasks and the resources that are required [2]. The tasks are usually linked together in such a way that some of the tasks need to be completed before subsequent tasks can begin. A process may also be defined as the workflows and sequence of events inherent in processes such as engineering, business, or manufacturing.

83.3.2 Control of Repetitive versus Nonrepetitive Processes

Several methods have been developed for controlling processes. The continuous feedback control methods that were described briefly in the first half of this chapter are particularly suitable for completely automated electromechanical applications such as the room temperature control system described there and many other repetitive operations, particularly in the chemical manufacturing and electrical communications industries. These are all examples of repetitive processes in which the same processes are executed again and again with the objective of performing them exactly the same way every time.

A second method that has been developed for controlling repetitive processes is SPC. SPC was developed by Walter Shewhart in the 1920s and later applied very successfully by W. Edwards Deming during World War II to improve the manufacturing quality of weapons and other products that were important to the war effort. It has become a very effective method for controlling the variability inherent in many repetitive manufacturing processes. SPC makes use of control charts, as shown in Figure 83.8, to record observations of a numerical variable of interest in the manufacturing process. Those limits are used to raise a flag when the variable goes above or below specified limits indicating that the process is out of control. Such excursions stimulate an investigation to determine the root causes of unusual variability and elimination of those causes.

Based on a sequence of observations, the upper and lower control limits (UCL and LCL) shown in Figure 83.8 were calculated. The control limits are calculated using an assumption that, if the process

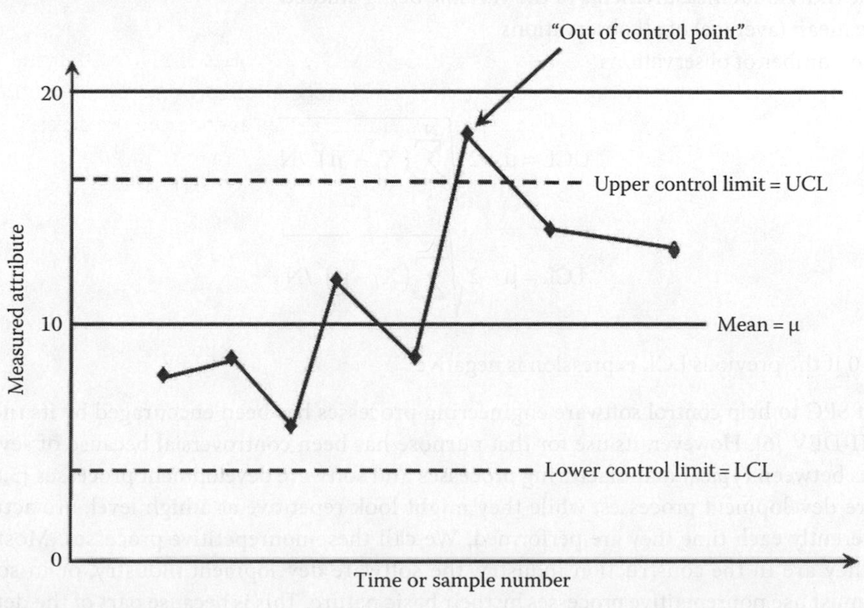

FIGURE 83.8 Control chart.

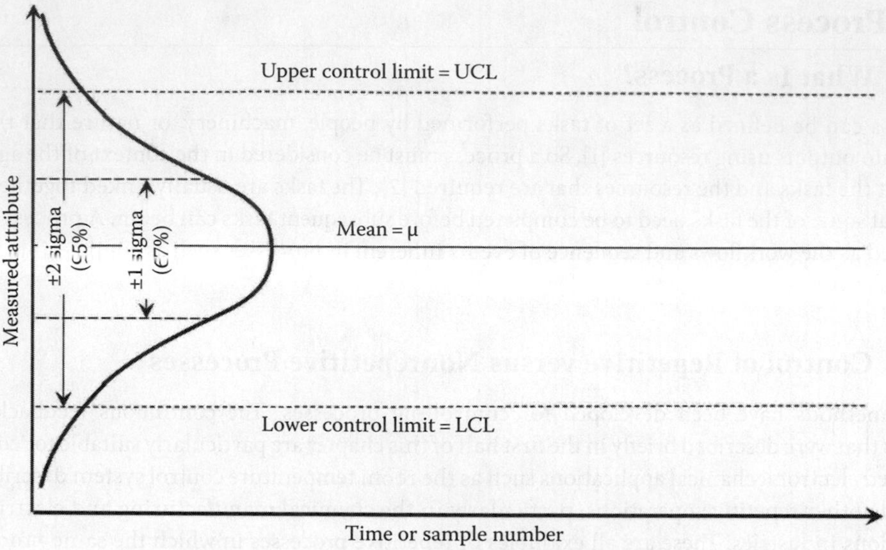

FIGURE 83.9 Assumptions made in calculating UCL and LCL.

is in control, that is, if the mean of the measured attribute and the standard deviation of the attribute are not changing over time, the observed data will fall within plus or minus one standard deviation of the mean approximately 67% of the time and within plus or minus two standard deviations of the mean about 95% of the time (see Figure 83.9). So the observed measurements should appear above the UCL only about once in 40 observations unless there is some systematic cause that results in the observation being out of control more than 5% of the time. Likewise, the same logic could be applied to points appearing below the LCL.

The calculation of control limits is done using the following formulas:
Assume that

X_i is the individual measurements of the variable being studied
μ is the mean (average) of all observations
N is the number of observations

$$\text{UCL} = \mu + 2\sqrt{\sum_{1}^{N} (X_i - \mu)^2 / N}$$

$$\text{LCL} = \mu - 2\sqrt{\sum_{1}^{N} (X_i - \mu)^2 / N}$$

LCL = 0 if the previous LCL expression is negative

The use of SPC to help control software engineering processes has been encouraged by its inclusion in the CMMI-DEV [6]. However, its use for that purpose has been controversial because of several basic differences between typical manufacturing processes and software development processes [5,8].

Software development processes, while they might look repetitive at a high level, are actually executed differently each time they are performed. We call these nonrepetitive processes. Most projects, whether they are in the construction industry, the software development industry, or in some other industry, must use nonrepetitive processes by their basic nature. This is because part of the definition of a project is that it is an activity, performed by people using resources that produce something unique.

If the objective is to produce something unique each time the project's processes are executed, they must be nonrepetitive and must be uniquely planned, organized, monitored, and controlled. While the application of SPC to software development processes is controversial, in a later section of this chapter we will provide an example of an application of this technique to at least one subprocess of the overall software development process, inspections.

Nonrepetitive processes are frequently controlled using a method called management by walking around (MBWA). Using this method a manager would determine if employees are following the desired process by frequently reviewing the work that they are doing, looking over their shoulders while they work, and talking with them about how they are doing the task in which they are engaged. The manager can then provide informal feedback on how the process is being executed and what should be done to improve its execution.

Alternatively, software development processes can be controlled by defining the process in writing, including in that definition a requirement that calls for written artifacts from the process while it is being executed. Documents produced during execution of the process are then reviewed by the manager and/or audited by professional auditors to determine if the process is being executed in compliance with its requirements. This is the method used to achieve International Organization for Standardization (ISO) 9001 certification of a software development process [3]. For that kind of certification, a team of people in the organization seeking certification writes a description of the process. This description is reviewed by the external auditors to ensure that the description complies with general ISO 9001 process description requirements. It becomes part of the organization's ISO 9001 quality manual. Then the development team does the work, following the description while documenting the work along the way. Finally, a team of external auditors reviews the documentation that has been produced, interviews people on the development team about how they have been doing their work, and compares the process that is actually being used with the process described in the quality manual. Any mismatches between the requirements in the quality manual and the process uncovered by the auditors produce nonconformances, which must be addressed and improved before the next external audit to reduce the probability that the nonconformance will occur again.

The CMMI-DEV is a similar model with corresponding processes. That model specifies specific processes for developing software that are generally considered best practices for software development. The model describes an evolutionary improvement path from ad hoc, immature processes to disciplined, mature processes with improved quality and effectiveness. It provides a method for identifying opportunities for improvement and for comparing the maturity of an organization's processes with the model's components to produce a maturity rating from one to five indicating how completely and correctly each of the model's components have been implemented. The model describes the following 22 key process areas listed approximately in the order of maturity with lower levels of maturity using only processes near the top of the list and higher levels of maturity using processes near the bottom of the list in addition to all of the processes above them toward the top of the list:

Configuration management
Measurement and analysis
Project monitoring and control
Project planning
Process and product quality assurance
Supplier agreement management
Requirements management
Decision analysis and resolution
Integrated project management
Organizational process definition
Organizational process focus
Organizational training

Product integration
Requirements development
Risk management
Technical solution
Validation
Verification
Organizational process performance
Quantitative project management
Organizational performance management
Causal analysis and resolution

Organizations may conduct an appraisal and earn a maturity level as specified by the model. Appraisal requirements are specified in CMMI ARC [7]. Appraisals focus on identifying improvement opportunities and comparing the organization's processes to CMMI-DEV best practices.

83.3.3 Control of a Specific Software Development Processes

To achieve high maturity levels of CMMI-DEV, organizations must use quantitative project management. This requirement could conceivably be met by tracking the number of tasks actually completed versus the number planned as a function of time, by tracking trends in numbers of customer trouble reports per unit of time or in number of faults found per line of code during testing. However, the most extensive, and theoretically sound, use of quantitative project management techniques has been the use of SPC in analyzing data produced by inspection processes.

Figure 83.10 shows control charts for a set of date that were produced by 15 requirements inspections. In each chart the number of the inspection is shown on the horizontal axes. The vertical axes show defect density in number of defects per 1000 lines of text, preparation rate in lines of text per hour of preparation, inspection rate in lines of text inspected per hour, and document size in lines of text. The UCLs for each chart were calculated using the formula shown earlier in this chapter. LCL calculations indicated that the LCL for each chart was zero.

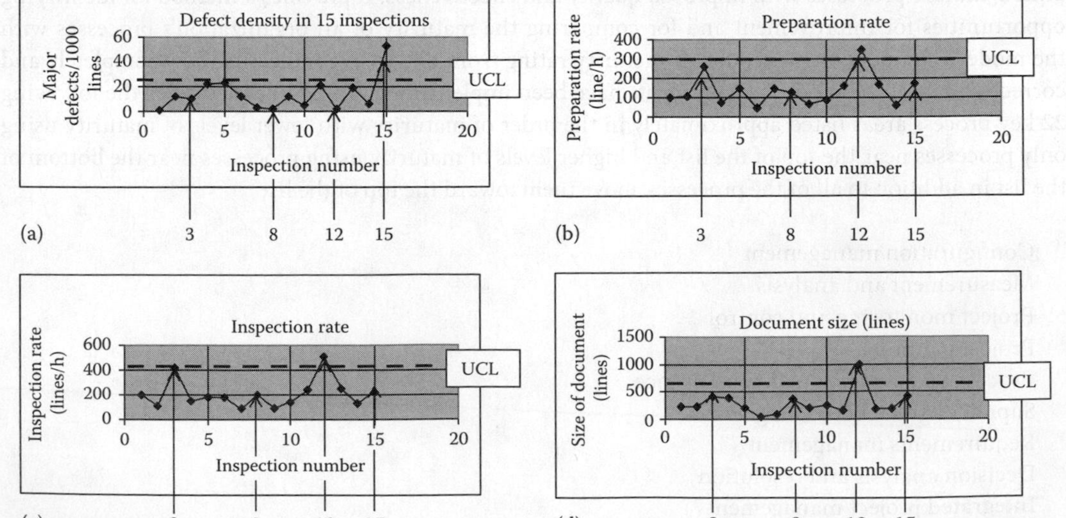

FIGURE 83.10 Control charts for four attributes (a) defect density, (b) preparation rate, (c) inspection rate and (d) document size of a set of 15 requirements inspections.

It is instructive to look at the data for inspections numbered 3, 8, 12, and 15. For inspection three the defect density, the preparation rate, the inspection rate, and the document size are all within the control limits. We could conclude that inspection number three was a normal inspection meeting and that the document was acceptable with a nominal defect rate. For inspection number eight, the defect density was at or very close to zero. The preparation rate, the inspection rate, and the document size were all within the control limits. We could conclude that this was a normal inspection meeting and that, because the defect rate was at the LCL, this was a very-high-quality document. For inspection number twelve, the number of defects was zero. The preparation rate, the inspection rate, and the document size were all high and were all above their UCLs. Therefore, we could conclude that this was likely to have been a poor inspection and that the document should be inspected again. Finally, for inspection number fifteen, the defect density was high and above the UCL. The preparation rate, the inspection rate, and the document size were all within the control limits. Therefore, we could conclude that the inspection was probably a good one but that the quality of the document was poor. The document should probably be reworked and reinspected.

In each of the four control charts shown in Figure 83.10, only one of the points are actually outside the control limits, which is close to the expected number of times such variation would be expected to occur due to random causes, so there is no strong evidence of any systematic cause for these outliers. However, if these excursions continue to occur in the future or if the frequency of occurrence increases, an investigation of the reason why the process is out of control should be undertaken and actions should be taken to prevent these outliers. Such actions might include not starting an inspection meeting if it is determined before the meeting starts that the document size is outside the control limits or stopping an inspection meeting as soon as it is determined that the preparation rate has exceeded the UCLs. These are the kinds of process improvement actions, based on the use of quantitative SPC methods that could be implemented during the execution of the process.

83.3.4 Practical Implications and Major Challenges

Application of the control techniques discussed in this chapter is relatively straightforward in practice. However, the author has observed that they are frequently not applied by project managers. There could be several reasons for this situation. First, and most obvious, is that if a project is not well planned, then the project manager has no basis for comparing the project's status with a baseline plan. A good plan should specify what is going to be done, how it will be done, who is responsible for each task, when each task will be completed, and how much the project will cost.

If the project plan does not specify the necessary tasks in sufficient detail and if there is nothing specified that the project manager can see to determine when each task is completed, the manager needs to rely only on the word of team members that a task has been completed. Anyone who has managed a project has observed that when reporting on task completions, members of software development teams are sometimes overly optimistic, frequently reporting that tasks are completed when there still remains work to be done. This is particularly bothersome during the creation of the project architecture and the design and coding phases of the project. During those phases it is helpful if the project plan calls for formal inspection of the artifacts that are produced during each task so that the project manager can see a short inspection report indicating that the task has been completed. This also provides a basis for implementing an inspection program and controlling its operation as discussed in the second part of this chapter.

When unit and integration testing has begun, the project plan could call for demonstrations of the results of these tests to the team so that completion of the tasks can be visually confirmed. Finally, when system testing begins, the completion of individual test cases, indicating whether they have passed or failed, can and should be reported and tracked.

Developing a plan that includes resources for all of these activities usually requires more work and overcoming more resistance from team members and stakeholders with financial interests than the effort associated with monitoring and control of the project. Therefore, the major challenge faced by project managers is usually the development and implementation of a sound project plan that includes all of these activities.

83.4 Summary and Conclusions

We have reviewed the methods commonly used to control software projects. They involve close monitoring of the project as compared with the project plan and taking appropriate actions when it is found that there are significant deviations from the plan. This method was shown to be closely related to feedback control used for controlling many engineered systems. We then moved to the control of software development processes. Methods discussed there ranged from informally observing how the work was being performed through more formal ISO 9001 certification and CMM assessments to the use of SPC for specific subprocesses of the software development process.

Control of both projects and processes is one of the more difficult activities that must be undertaken by software development project managers and their teams. Monitoring projects to find deviations from the plan is quite straightforward once an appropriate plan has been developed. However, deciding what actions to take in response to deviations requires the exercise of considerable care and judgment. Controlling processes is even more difficult, not only because significant effort and cost are required to do it well but also because there is considerable debate about whether the most soundly based method, SPC, can and should be applied to software development processes.

References

1. ANSI/EIA-63:1998. *Processes for Engineering a System*. Arlington, VA: Electronic Industries Association, p. 66, 1998.
2. Gilb, T. *Competitive Engineering*. Burlington, MA: Elsevier Butterworth-Heinemann, 2005.
3. ISO 9001:2008. *Quality Management Systems Requirements*. International Organization for Standardization, Geneva, Switzerland, 2008.
4. Maxwell, J.C. On governors. *Proceedings of the Royal Society of London*, 16:270–283, 1868.
5. Raczynski, B. and Curtis, B. Software data violate SPC's underlying assumptions. *IEEE Software*, 25(3): 49–51, May/June 2008.
6. Software Engineering Institute. *CMMI®* for Development, Version 1.3, 2010.
7. Software Engineering Institute. *Appraisal Requirements for CMMI®*, Version 1.2 (ARC, V1.3), CMU/SEI-2011-TR-001, 2011.
8. Weller, E. and Card, D. Applying SPC to software development: Where and why. *IEEE Software*, 25(3): 48–50, May/June 2008.

84

Agile

John Favaro
Consulenza Informatica

84.1 Introduction

Agile is an umbrella term used to describe a set of paradigms and approaches to the software development process. The term was popularized with the publication of the Agile Manifesto (Beck et al. 2001) by a group of 17 authors and reflects an intention of the signatories to differentiate agile approaches from a trend in software development methodologies that had gained dominance in previous decades.

That trend, in turn, had been a reaction to the *software crisis* that had been widely publicized by the early 1970s (Dijkstra 1972) and had led to the rise of increasingly elaborate and prescriptive approaches to software development. The publication of *Managing the Software Process* (Humphrey 1989) had culminated a long turn away from a focus on individual developers described in Dijkstra (1972) toward a focus on detailed, defined development processes. The Capability Maturity Model introduced by the Software Engineering Institute (SEI) (Paulk and Curtis 1993) moved the defined software process squarely to the center of attention in software development. Large initiatives such as the Ada Programming Support Environment (Fairley 1980) exemplified the attempt to support defined processes with comprehensive tools. The development of the V-Model in Germany in the mid-1980s enshrined a view of the software development lifecycle as a series of activities based upon a particular sequential interpretation of the waterfall lifecycle model published in 1970 (Royce 1970).

The V-model and similar lifecycle reference models also formed the basis of new software development standards such as IEC 61508 (IEC 2010) with numerous associated workproducts in the form of documentation to be produced by developers. Such documentation workproducts (e.g., user requirements) generally became the object of contractual negotiations that were gated prerequisites to progress in projects. These documentation workproducts were often not associated with functioning software (especially in early phases of projects), which often led to antagonistic relationships between customer and developer as each maneuvered for position regarding functionality whose cost could only be predicted with considerable uncertainty.

By the early 1990s, evidence was accumulating and being reported in publications such as the Chaos reports (Standish 1994) that organizations following established software development doctrine were failing at alarming rates. Those who became the signatories to the Agile Manifesto had come to the conclusion through independent experience that the main source of failure was an unforeseen degree of change in systems over all phases of their lifecycle and an inability of the currently popular defined processes to manage that change. They began to introduce the alternative approaches that became known as agile methods.

The primary differentiating characteristic of the agile methods promoted by the signatories of the Agile Manifesto was an *iterative and incremental* approach to the development process. Although it represented a radical departure from the sequential development processes that had become dominant, it was not unprecedented. As Larman and Basili (2003) recount in detail, iterative and incremental development had been practiced and promoted at numerous times throughout the history of software development. In addition to anecdotally documented experience in IBM Federal Systems Division and NASA Project Mercury, publications explicitly recommending this approach were beginning to appear by the late 1960s (Randell and Zurcher 1968). Gilb was a strong early promoter (Gilb 1989) of iterative and incremental development (Extreme Programming [XP] author Beck has explicitly acknowledged a debt to Gilb's work). The 1980s were also characterized by influential publications promoting an explicit iterative and incremental lifecycle representation, such as the spiral model (Boehm 1988).

Likewise, specific practices that came to be associated with agile methods had important precedents. The practice of *test-first development* was being used in the NASA Project Mercury by 1960 (Larman and Basili 2003). The practice of *continuous integration* was noted by Cusumano at the XP2003 conference on agile processes to be strikingly similar to the *daily build* already being practiced in Microsoft (Cusumano and Selby 1995) in the early 1990s. The agile vision of software being its own best documentation has echoes in *literate programming* (Knuth 1992). *Pair programming* had been introduced by Plauger at Whitesmiths several years earlier than its popular reintroduction in XP (Constantine 1992). In an influential article in the mid-1980s, Brooks (1987) advocated the idea of "growing" a system from simple to more complete functionality, in such a way that a running system was always available to stakeholders. This idea was a precursor to the agile vision of "working software."

Likewise, the refocused attention on individuals and interactions in the Agile Manifesto was not unprecedented, as the landmark book *The Psychology of Computing Programming* (Weinberg 1971) had studied both individual and team software development behavior 30 years earlier, coining terms such as *egoless programming* that became precursors to the group ownership practices seen in agile methods. Even during the heyday of institutional process definition activity, books like *Peopleware* (DeMarco and Lister 1987) were seeking to raise awareness of the human factor in software development.

Thus, the rise of agile software development in the late 1990s and early 2000s can be viewed not so much as innovation in specific software practices as a codification and consolidation of long-existing practices into a number of coherent, usable methodologies and frameworks. This happened within the context of a rebalancing in the nature of large software systems. Whereas the large software systems in the 1970s and 1980s were primarily mission-critical, military, and space systems, a new kind of large software system began to appear in the 1990s with the rise of networks such as the Internet (including the World Wide Web) and financial networks with a global reach. Applications on these new systems were characterized by extremely volatile, rapid feature evolution and high degrees of user interaction and provided the context for the agile methods to take root and thrive.

84.2 Underlying Principles

A set of 12 principles underlying the agile approach to software development was provided by the signatories of the Agile Manifesto (Beck et al. 2001):

1. *Our highest priority is to satisfy the customer through early and continuous delivery of valuable software.*
   A direct link to the surrounding business environment is established by several principles in the Agile Manifesto. Although stakeholder satisfaction had always been at least implicit (e.g., through user requirements) in previous approaches, there was a distinctly technical slant in those approaches. Agile software development places an emphasis on delivery of business value, with an implicit (and often explicit) interpretation as economic value (Beck and Andres 2004). While the link to business value had been made in specific subdisciplines such as software reuse (Favaro et al. 1998) and software product lines, it was relatively unusual to find it in general software development methodologies.
2. *Welcome changing requirements, even late in development. Agile processes harness change for the customer's competitive advantage.*
   Beyond the link to economic value (which could manifest itself simply as cost savings), a strong, explicit link was made by the promoters of agile software development to competitive strategy, going beyond the simpler time-to-market arguments seen previously. Here, the software development process was interwoven with business strategy—in particular, reacting successfully to changing business conditions. Enabling this new interweaving with competitive strategy necessitated a complete reworking of the lifecycle not only in terms of iterative development but also in terms of introducing interaction points between technical development and strategy development. Whereas technically oriented iterative development had been practiced in the past, there had been much less attention paid to its interaction with iterative business strategy development and execution. The principle of harnessing change and uncertainty for competitive business advantage forms the foundation of the *real options* approach, borrowed from mainstream corporate finance (Brealey et al. 2008) and strategy that underlies and is appealed to explicitly in several of the agile methodologies.
3. *Deliver working software frequently, from a couple of weeks to a couple of months, with a preference to the shorter timescale.*
   Aside from enshrining the iterative and incremental development paradigm in agile methods, this principle is also related to promises such as the following (Jeffries et al. 2000):

   > Customer Development Right #1: [...] You may cancel at any time and be left with a useful working system reflecting investment to date.

   Such a promise could not be made with a sequentially defined waterfall lifecycle: a customer wishing to cancel a contract would risk having only a set of requirements and design documents, with no code written. The promise, however, poses serious challenges for implementation techniques, such as how to implement infrastructure incrementally along with the more user-visible features. It also has led to considerable experimentation with the duration of iteration periods (e.g., determining the shortest possible duration of an iteration that can still result in useful working software), as well as the management of that iteration (e.g., fixed timeboxing versus more elastic boundaries). All agile methods confront these issues in various ways.

4. *Business people and developers must work together daily throughout the project.*
   The principle of generating customer satisfaction through frequent delivery of working software implies the need for more frequent interaction between customer and developer than in a sequential process, where customer–developer interaction is primarily at the beginning (requirements definition) and end (acceptance testing) of a project. At a minimum, interaction points must occur between iterations, but agile methods encourage continuous interaction. The most extreme manifestation is the request for a customer representative to be onsite and part of the development team. Not only does this principle redefine the customer–developer relationship in more collaborative terms, but it also places heavy requirements on the customer's resources (i.e., making personnel available on a daily basis) that can be very challenging.
5. *Build projects around motivated individuals. Give them the environment and support they need, and trust them to get the job done.*
   Along with the acceptance of the sequential waterfall process in the 1970s, a view of the cost of changes to systems over lifecycle phases was popularized that embodied the assertion that the cost of changing software grows exponentially over the lifecycle. In particular, it was orders of magnitude more expensive to make a change during programming (the "coding phase") than during the earlier phases of design and requirements analysis.
   The effect of this diagram was that suddenly those earlier phases took over the central importance. The requirements analyst became a "Very Important Person" in the software process, and an entire discipline of requirements analysis grew up around him. The designer (or, more eloquently, the "architect") became another Very Important Person, and an entire discipline of software architecture grew up around him. Conversely, the programmer was suddenly much less important. With all the real effort concentrated in the earlier requirements and design phases, the programmer became a mere assistant, transforming the results of the requirements analyst and architect in a mechanical fashion into code. No special skill was needed, because at least in theory, the requirements analyst and the design architect had already done the important work.
   After the organizational emphasis in the process-oriented approaches of previous decades, with the agile movement individual programmer came back onto center stage, as DeMarco noted:

   > Part of our 20-year-long obsession with process is that we have tried to invest at the organizational level instead of the individual level … If agile means anything to me, it means investing heavily in *individual* skill-building rather than organizational rule sets. (DeMarco and Boehm 2002)

6. *The most efficient and effective method of conveying information to and within a development team is face-to-face conversation.*
   Agile methods reversed a trend toward physical separation of software developers from each other. One source of that trend was psychological: the argument that programming was an intellectually intense activity that suffered greatly from interruptions, and therefore, programmers should ideally have individual offices (DeMarco and Lister 1987). Another source was technical:

increasingly dislocated teams through improved telecommunications, for example, through the Internet. In contrast, agile methods view programming as a social activity involving intense interpersonal interaction. This involves the heavy use of colocated teams, with open spaces favoring face-to-face communication. It also implies a small size of development teams—a departure from the ever-growing size of development teams in previous years, for example, through the *software factory* initiatives in Japan (Cusumano 1991).

7. *Working software is the primary measure of progress.*
   In the agile approach, an expression often heard is, "The code contains all the answers." Although this is a simplification, it is a reflection of a return to what the agile perspective perceives as the most fundamental fact of software development: the ultimate purpose of software development is the production of a working system. The working system embodies the results of all requirements, design, and coding activities and is seen as the only truly reliable source of information about the software. This is in direct opposition to a sequential waterfall lifecycle, where significant requirements and design phases are traversed before any code is produced.
8. *Agile processes promote sustainable development. The sponsors, developers, and users should be able to maintain a constant pace indefinitely.*
   The principle of continuous delivery espoused by agile methods is supported by considerable emphasis on smooth, well-paced production workflow. Much of this emphasis has taken its inspiration from the experience of the authors of agile methodologies with Japanese production workflow management (such as the Toyota Production Management System). For example, the Kanban technique that has been the subject of much recent interest in the agile community is essentially a technique for smoothing the production workflow and rendering it sustainable over long periods. The emphasis on human factors in agile methods also resulted in a set of principles aimed at avoiding overwork, such as the *40 h week* of XP.
9. *Continuous attention to technical excellence and good design enhances agility.*
   In spite of the emphasis on organizational issues in the previous decades, a great deal of progress had also been made in technical areas by the end of the 1990s. In particular, the rise of object orientation had stimulated a kind of Renaissance in software construction techniques. Innovative and highly skilled researchers and programmers had introduced new paradigms such as *architectural styles* and *design patterns*. The agile movement featured small teams of programmers who exploited the technical advances of the past decade (many of which they had authored themselves), to create a highly sophisticated technical environment. New techniques were invented by those in the agile movement to support the concept of "continuous attention to technical excellence": *refactoring* (Fowler et al. 1999) involved continuous improvement of code, a radical departure from the "if it's not broken, don't touch it" philosophy of earlier approaches. Small, targeted tools such as JUnit (Massol and Husted 2003)—codeveloped by Beck, one of the agile signatories—together with a test-first approach supported a sophisticated automated testing environment that provided the immediate feedback necessary to support the resulting continuous interventions in the code base.
10. *Simplicity—the art of maximizing the amount of work not done—is essential.*
   Although the need to master complexity in software development through simplicity has been recognized for a long time, agile methods made it into a fundamental principle. Variations on established principles such as Occam's razor were introduced in precepts such as the The Simplest Thing That Could Possibly Work (i.e., the simplest implementation of a feature) in the XP methodology. New principles were, however, also introduced to support an interpretation of simplicity as the avoidance of unnecessary features. The colorfully named *You Aren't Going to Need It* (YAGNI) principle in XP (Beck 1999) forcefully discourages the gratuitous

implementation of functionality that, while not explicitly required at the present time, is judged not to be difficult to implement at the present time and might be required or considered useful in the future (Favaro 2004). A concept of "design carry" was coined to characterize the added cost of introducing unneeded features in terms of maintenance, understanding, size of the code base, etc., and make explicit the advantages of simplicity in terms of avoided implementation.

11. *The best architectures, requirements, and designs emerge from self-organizing teams.* The agile approach to development team organization is in stark contrast to previous trends, in which development teams were characterized by an increasing number of roles, often defined according to lifecycle phases—requirements analyst, chief architect, and V&V specialist (including suitable organizational independence from the development team), together with management-oriented roles such as project leader. Agile methods encourage the establishment of few roles, which often have no direct counterparts in a defined process, such as facilitating roles (e.g., Scrum master) that reflect the self-organizing nature of agile teams: the facilitator's main duty is to ensure smooth running of the self-organized team, not to impose authority. The self-organizing nature of agile teams introduces a set of issues around the distribution of leadership, authority, responsibility, and accountability—all of which are addressed in various ways in the agile methods—as well as issues with roots in group dynamics.
12. *At regular intervals, the team reflects on how to become more effective, then tunes and adjusts its behavior accordingly.* Although the idea of "process improvement" was already well established in organizational approaches to software development in the past (Paulk and Curtis 1993), agile methods introduce techniques considered to be more lightweight and suitable for the smaller teams. One technique that has gained considerable currency in recent years is *retrospectives* (Derby and Larsen 2006). A retrospective is a meeting held within a project team at certain points in the project. This may be the end of the project, after a certain number of iterations, or even after a single iteration, to reflect upon what happened during the period of time under consideration and how it could be improved upon. The general concept covered by the retrospective may be found in virtually all the major agile methods, including Scrum, XP, and lean development, where it is also known by different names such as Kaizen and continuous improvement.

84.3 Best Practices

Since the publication of the Agile Manifesto, a number of agile methods have entered into regular use in the software industry (Abrahamsson et al. 2002) and have collectively formed a concrete baseline of best practices in agile development. Those that have had the most significant impact on industrial practice are presented in this section, in particular:

1. XP
2. Scrum
3. Lean software development

In addition, the following also have made contributions that are described in a further section:

4. Kanban
5. Rational Unified Process (RUP)
6. Crystal
7. Personal Software Process (PSP) and Team Software Process (TSP)

The following table presents a summary overview of the general relationship between the principles expressed in the Agile Manifesto and the form they take in the three major methods, as well as other characteristics not explicitly appearing in the Agile Manifesto:

Characteristic	XP	Scrum	Lean Software Development
Principles from the agile manifesto			
1. Customer satisfaction first	Economics as a principle, general concern with business value delivery	Product owner determines priorities according to business value.	Optimize the whole, value stream mapping
2. Welcome changing requirements	"Embrace Change"	Continuous renegotiation of product backlog	More implicit than explicit in lean
3. Frequent delivery	Continuous integration	Sprints	Deliver fast
4. Business and developer collaboration	Onsite customer, negotiated scope contracts	Product owner negotiates with implementation team.	Optimize the whole
5. Motivate individuals	Humanity as a principle, accepted responsibility	Scrum master role to facilitate individual excellence	Engage everyone
6. Prefer face-to-face interaction	Collocated teams, communication as a value	Scrum master interacts directly.	
7. Working software as measure of progress	Test-driven development (TDD)	Partial implementation not acceptable in sprints	Implicit in flow-driven paradigm
8. Sustainable pace	40 h week	Extensive planning and velocity estimation to ensure sustainable pace	Attention to flow
9. Technical excellence	Reduced cost of change due to improved technology		Build quality in
10. Simplicity	Direct counterpart in XP		
11. Self-organizing teams	Collective ownership	Direct counterpart in Scrum	
12. Regular reflection for improvement	Daily stand-up meetings	Daily Scrum, sprint retrospective	Learn constantly, Keep getting better
Other principles			
Continuous flow	Flow as a principle	Scrum master facilitates smooth flow.	Comparison with continuous flow system of manufacturing
Eliminate waste	YAGNI principle		Direct origins in lean
Release planning	Planning game	Planning poker	Value stream mapping, Kanban techniques
Options approach	Baby steps, YAGNI, business options, economics principle	Renegotiation of sprint backlog after each sprint	Last responsible moment, learn constantly

84.3.1 Extreme Programming

XP was developed in the mid-1990s and described in an influential book (Beck 1999). It became a catalyst for much of the discourse around agile methods because of its concise formulation, internal coherence, and "…its interesting name" (Larman and Basili 2003). The name is intended to characterize an approach whereby best practices are applied to an extreme degree (e.g., integration is performed far more frequently than in other methods). It received extensive application in the software industry in the decade following its introduction and has stimulated research on various practices that it introduced. The original edition was followed 5 years later by a second edition (Beck and Andres 2004) that provided additional motivation for the approach underlying the method, in addition to reorganizing and extending the set of principles and practices.

To set the context, XP identifies four so-called *variables* that can be manipulated to control (an agile approach to) software development:

- *Money*: The XP method was an early advocate of actively placing constraints on resources available to projects, noting that although underfinanced projects were problematic, overfinanced projects tended to create problems as well, with bloat in both management and technical structures. This point of view may be found also in lean approaches.
- *Time*: Similarly, XP identifies both lower and upper limits in terms of time management. While too little time will clearly cause problems, likewise so will too much time. The preoccupation with time management refers not only to overall project duration but also to the management of iterations.
- *Quality*: The introduction of this variable in XP recognized that delivered quality can be adjusted according to other factors (e.g., time pressure), but the method also explicitly states that quality cannot be lowered beyond a certain point (an aspect of XP that is often misunderstood). The study of *technical debt* that has become popular in recent years is a reflection of this attention to issues in managing the quality of the delivered software over time.
- *Scope*: The most extreme and (initially) least understood variable introduced by XP is scope, which varies along with the set of requirements during execution of the project. In defined sequential lifecycle methods, scope is defined once, at the beginning. Allowing the scope to vary over the course of the project, with requirements being added, modified, and deleted, creates a series of cross-cutting issues ranging from managerial to technical to contractual, each of which has been studied and experimented within the last decade.

84.3.1.1 Values

Although the operational core of XP is formed by its practices, the method was particularly influential in providing a set of four values that summarized an overall approach to software development and set it apart from previous process-oriented methods. Note that these values were published previously (1999) to the Agile Manifesto (2001), and as such, several of them found their way into its principles in a generalized form.

84.3.1.1.1 Communication

This first value reflects the extreme attention to issues of human interaction in the XP method, which is acknowledged in other agile methods but not treated in as much depth. This value is directly associated with several practices that cannot be carried out without high levels of communication. The second edition of the method (Beck and Andres 2004) added a second author with training in psychology, organizational behavior, and decision analysis, further reinforcing the emphasis on human factors. In this second edition, a new and related value was also added: *respect*, both for one's own work (striving for the highest quality) and for the work of others (e.g., by not undertaking activities that may interfere with the work of other team members).

84.3.1.1.2 Simplicity

Another XP value that was absorbed into the Agile Manifesto concerns simplicity, particularly in design. An incremental approach to design is encouraged, allowing the appropriate architecture to emerge over the course of the project. The interpretation of simple design as also the avoidance of implementation that is not explicitly required is justified in the method with a hypothesis on the evolution of technology:

> The software development community has spent enormous resources in recent decades trying to reduce the cost of change—better languages, better database technology, better programming practices, better environments and tools, new notations … It is *the* technical premise of XP. (Beck 1999)

The hypothesis that the cost of change is no longer exponential but rather a flattened curve underlies the recommendation to keep all current implementation as simple as the current requirements justify, because the prospective value of future functionality, the need for which is uncertain, is outweighed by the reduced technological cost of waiting and implementing functionality only when it is explicitly called for by current requirements. This is supported by the practice of refactoring that contributes to minimizing the current complexity of the implementation.

84.3.1.1.3 Feedback

This value enshrines the concept that concrete, continuous feedback about the current state of the system is necessary and invaluable (Williams and Cockburn 2003). Its primary technical support within the XP method is a practice that was developed by the same author and has acquired a significant following on its own: TDD (Beck 2002), in which unit tests are written before feature implementation. Much of the technology for supporting this practice was developed by the same people who developed XP. In addition to JUnit, codeveloped by Beck, the FIT framework (Mugridge and Cunningham 2005) for automated regression testing was developed by Cunningham, often cited by Beck as a primary inspiration for the XP method.

The primary managerial support for this value is the frequent interaction with the customer provided by the short iteration and planning activities, which deliver business level feedback on the state and direction of system implementation.

84.3.1.1.4 Courage

The main interpretation of this value is "a willingness to abandon a failing activity"—in particular, a willingness to discard code that has already been written if it is unsatisfactory in some way (functionality, performance, maintainability, etc.). An additional interpretation in appropriate contexts is a willingness to try something new.

Many consider the *practices* of XP to be its most important contributions to the agile movement, because they give concrete expression to the principles and values. Indeed, they are often combined with other methods (e.g., Scrum), whereby the management of the overall process may be different according to the approach of the other methods, but the practices (with their origins in XP) are maintained. However, wedged between the values and the practices of XP are a set of *principles* that are intended to provide guidelines for putting the practices into place in the local context of the project and implementation team (Fowler 2003). A number of these principles reflect (or conversely, were the basis for) principles expressed in the Agile Manifesto. Certain of these, however, reflect points of view that either originated in XP or at least receive particular emphasis in XP:

- *Accepted responsibility*: From the first edition of the method onward, an emphasis on placing responsibility squarely on the shoulders of the team and not some externally defined role, signaling a major rethinking of the group dynamics of a software development team.
- *Economics*: From the beginning, XP strove to establish a link to established economic theory. The first edition of the method (Beck 1999) included an entire chapter on the economics of software development, and the principles elaborated in that chapter were maintained into the second edition. These rested on the foundation of the time value of money and in particular an approach in terms of real options. As such, XP was one of the earliest agile methods to promote an options-oriented view of agile development, followed subsequently by other methods such as lean software development (Poppendieck and Poppendieck 2003).
- *Baby steps*: Known as "small initial investment" in the first edition, this principle also reflects the economic perspective of *learning options*: staged investments as uncertainty is gradually resolved over the course of the project.
- *Humanity*: A significant emphasis on human factors, leading to several practices that were popularized by XP.

84.3.1.2 Practices

The first edition of XP contained a tightly interwoven set of 12 practices, all of which were characterized as being essential, which were reorganized (Marchesi 2005) in the second edition into a new set of 13 *primary* practices (can be introduced individually) and 11 *corollary* practices (rely on primary practices for their introduction). Among those practices that have been influential on the evolution of agile software development are as follows:

- *TDD*: This practice has spawned an entire subdiscipline of agile development (Jeffries and Melnik 2007).
- *Pair programming*: While it did not originate with XP (Constantine 1992), the practice of *pair programming*—two programmers developing together, at a single keyboard—was popularized through XP and received much subsequent attention in the published literature including extensive tutorials (Williams and Kessler 2002), experience reports (Dybå et al. 2007), and criticism (Stevens and Rosenberg 2003).
- *Planning game*: Iteration planning is a core practice across all agile methods and as practiced in XP has made a significant contribution to the literature on agile planning, including a book coauthored by the author of the method (Beck and Fowler 2000). In the second edition, the planning process was extended and refined with weekly and quarterly planning cycles with characteristics reminiscent of those (e.g., release planning) in the Scrum method.
- *Collocated teams, collective code ownership, whole team*: With practices such as the colocation of personnel in the same open space, cross-functional teams, and the liberty to change any code written by anybody at any time, XP introduced or reinforced the discussion of many of the group dynamic issues that have dominated the discourse in agile methods. For example, the effects of collective code ownership (called Shared Code in the second edition of the method) have been debated vigorously (Levison 2008) since its introduction.
- *Onsite customer*: The practice of having a representative of the customer organization onsite as part of the development team (or at least available to the team for constant consultation) generated much interest and controversy. Detractors point to the increased cost of providing an onsite customer; supporters point to the advantages of being able to resolve questions about requirements quickly. The practice has inspired much experimentation and research (Koskela and Abrahamsson 2004).
- *Negotiated scope contract*: XP entered early into the discussion of legal issues and contractual issues associated with agile software development. As early as the first edition of the method, optional scope contracts were discussed (Beck and Cleal 1999) in which scope was negotiated, in stark contrast to fixed scope contracts associated with sequential lifecycle development processes. Further interest in agile contract mechanisms was stimulated by this practice and continues (Arbogast et al. 2012).

Other practices such as continuous integration were in use in some organizations before the development of XP—indeed, it has been estimated (Cusumano 2007) that there is an overlap of as much as 70% between XP and the iterative development approach previously in use at Microsoft—but without doubt it was the publication of XP that exposed these practices to a broader audience.

84.3.2 Scrum

Where XP has been responsible for much of the success of agile approaches within the IT *development* community, Scrum may be seen as being responsible for much of the success of agile approaches within the IT *management* community. Although XP has disciplined project management practices, they have had less visibility than the development practices and thereby often led to the perception that XP is a "hacker's approach." The project management practices of Scrum, on the contrary, are central in the method. Indeed, Scrum is more properly seen as an agile development *framework*, whereby its project management practices constitute that framework.

Several additional factors have contributed to the relatively broad acceptance of Scrum within the mainstream IT management community:

- Its project management practices have been described in great detail in a number of well-received books (Schwaber and Beedle 2001) that continue to be updated over time (Schwaber 2009).
- The method has been practiced in very visible ways in several leading mainstream IT organizations including Microsoft, IBM, and Nokia.
- A number of commercial and open source tools support the method.
- A number of defined roles, together with programs for *certification* (Scrum Alliance 2012) in those roles, lead to a perception of stability and shared understanding of the method.
- Activities for monitoring progress over the lifetime of a project are codified into the method, leading to an impression of *controllability*, a key concern of project managers.

The method evolved over a period of several years from its origins in Japanese product management in the mid 1980s and a consolidation of multiple independently developed methods by the end of the 1990s. The name *Scrum* is a rugby term reflecting the general idea that the whole team cooperates in the development process.

The method defines a quintessential iterative and incremental lifecycle, serving as a model for an archetypal set of agile project management activities. In addition, the method defines a clear set of roles and their relationships to the activities. The Product Backlog is a prioritized list of *user stories* (essentially equivalent to those of XP) to be implemented by the Scrum team. Much attention has been paid in the Scrum community to the development of guidelines for constructing user stories (Navarro 2010). The Product Backlog is the responsibility of the Product Owner, one of the three core roles. The Product Owner is essentially equivalent to the customer in XP and determines the priorities of the user stories in terms of foreseen business value. The implementation team is self-organizing: there is no "team leader" role.

An implementation iteration is known as a sprint and typically lasts approximately 4 weeks (whereas XP iterations tend to be much shorter). The Sprint Backlog is determined before each sprint in terms of user stories to be implemented during the sprint. The Sprint Backlog is negotiated between the Product Owner and the implementation team and results in a set of stories that the team agrees to implement in that sprint. Note that only entire stories may be implemented—no "partial" implementation of stories is permitted. In addition, the stories are broken down into implementation-related tasks by the team. The Product Owner has no influence on this activity, which is considered wholly the responsibility of the implementation team. Within the sprint cycle, there is a tighter, daily cycle in which the implementation team has *daily Scrum* meetings (equivalent to the *daily stand-up* meetings of XP), discussing current problems and potential issues. Progress is tracked against the scheduled implementation commitments within the sprint.

At the completion of a sprint, a working increment of new functionality is available. A Sprint Review meeting is held in which the results of the sprint are examined. A demonstration of completed functionality is presented to the stakeholders (primarily the Product Owner). The Product Owner decides whether a particular delivered Product Backlog item is acceptable according to specified exit criteria (*Definition of Done* in Scrum terminology). Only completed functionality may be demonstrated. A Sprint Retrospective is held to examine the process and make corrections and improvement.

An important role in Scrum is that of the Scrum master, who is not a team leader (as may be implied by the name) but rather a facilitator, whose job is to ensure that impediments to progress are removed and that the Scrum rules are followed. This role, which is characteristic to Scrum, echoes the ideas of ensuring smooth development flow in other agile methods (such as the "flow" principle in XP second edition and in particular the ideas of Kanban).

In summary, Scrum has had a major role in the penetration of agile approaches into mainstream software development and remains a significant source of agile best practices today. Because of the straightforward, concise definition of the core Scrum framework and a small number of clearly defined core roles, over the span of many years, practitioners have been able to gather concrete implementation experience and then refine and document each of the activities and responsibilities. This has resulted in significant contributions of Scrum to the corpus of established best practices for agile software development, including the following:

- *Sprint planning and estimation*: An essential practice in any iterative and incremental method is iteration planning. As in XP, Scrum practitioners have dedicated many resources to the study and documentation of best practices for sprint planning. Similar to the Planning Game of XP, another "game" (Grenning 2002) known as Planning Poker® (a registered trademark of Mountain Goat Software) provides a technique for estimate consensus building and has been widely documented (Cohn 2005) and discussed in the Scrum community.
- *Sprint management*: There are numerous considerations in managing the actual execution of iterations. The Scrum community has debated the management of sprints at great length including issues such as sprint duration, management of tasks, and dealing with interrupted and unfinished work. Several books (Kniberg 2007) have documented practical experience dealing with such issues and recommended approaches.
- *Tracking*: Much of the popularity of Scrum in the general management community is due to its attention to tracking the results of work in iterations. In addition to recording velocity (the amount of work actually done in a sprint, generally measured in story points or in hours) as an aid to improving estimation, *Burndown Charts*—implemented in many ways ranging from hand-drawn charts to spreadsheets to advanced features of sophisticated Scrum project management tools—track remaining work in a sprint. The practice has been extended to variations deemed useful in overall project management, such as Release Burndown Charts, which track progress against an entire release plan over multiple sprints.
- *Modeling and tool support*: The clear and concise definition of the Scrum process and its associated roles and artifacts (Sutherland and Schwaber 2011) has encouraged the development of more formal models as well as support tools incorporating those models. The overall Scrum framework process and its associated artifacts have been modeled with the Eclipse Process Framework (Eclipse 2012). The clear definition of the process is also an enabler for tool support. Tools supporting Scrum include IBM Rational Concert, Jiri, Microsoft Visual Studio, and Polarion. The Scrum culture has been somewhat more receptive to tool support than the XP culture, leading to broader management acceptance.

Although it emerged out of the object-oriented community, Scrum does not place much emphasis on specific technical practices (indeed, none are listed here). Therefore, it is common to combine the Scrum framework with specific technical practices from XP in industrial projects (Kniberg 2007). It has also been combined with Kanban practices (Ladas 2009).

84.3.3 Lean Software Development

The majority of agile methods in current use were developed by the late 1990s, and most of the authors of those methods were also signatories of the Agile Manifesto in 2001. A later arrival is lean software development, as described in a book (Poppendieck and Poppendieck 2003) whose authors had not been signatories of the Agile Manifesto. In 2003, the concept of "lean" had been in use for nearly 20 years in production management and had resulted primarily from exposure to the Japanese automotive industry and in particular the Toyota Production System, as recounted in Poppendieck and Cusumano, 2012, but its application to software development had not yet been popularized.

Lean software development, as in other methods such as XP, provided a set of *principles* intended to guide the developer in implementing an appropriate lean process in his software development context.

- *Optimize the whole*: This is an appeal to a holistic approach. Consider the whole system, of which software is only a part, when pursuing value creation. Not only the development phase must be emphasized but all of the phases (including design and deployment) that have an impact on value creation.
- *Eliminate waste*: This principle is at the core of a "lean" approach: discover and eliminate anything that does not directly contribute to value creation. In the case of software development, it could be anything from time spent in unproductive meetings to implementation of unused features (as in the YAGNI practice of XP) to time spent debugging. This principle also involves the restriction of work in progress (WIP)—that is, partial, unfinished work.
- *Build quality in*: This principle is essentially an appeal to continuous integration of small development increments and fixing problems as soon as they occur, much like the principle of Single Code Base in the second edition of XP.
- *Learn constantly*: This is an appeal to the *real options* approach seen in agile methods such as XP (Erdogmus and Favaro 2002). Two techniques are recommended: elaboration of multiple alternatives and creating an *option to defer investment* (Sullivan et al. 1999) and creation of a *learning option* through small staged investments (Erdogmus 2002).
- *Deliver fast*: This is an appeal to the metaphor of *flow*, similar to the flow principle in XP second edition. Through seeking an ideal of continuous delivery, software development is no longer seen through a metaphor of a project but rather as a flow system as in manufacturing.
- *Engage everyone*: This principle involves the inclusion of team members reflecting all aspects of value creation, rather than only software development. This may include customer-facing personnel, operations personnel, and appropriate types of support personnel.
- *Keep getting better*: This is an appeal to a general principle of continuous improvement found in many agile methods and the Agile Manifesto itself.

Over the last decade considerable implementation experience has been acquired with lean software development (Kniberg 2011). Lean software development seeks to position itself differently from "mainstream" agile methods such as XP and Scrum (Poppendieck and Cusumano 2012). For example, whereas agile methods define a separate customer role to which responsibility for the overall business success of the product is assigned, lean software development defines no such role but rather considers software development to be integrated in a larger product development under the responsibility of a product manager.

However, Erdogmus argues that the *principles* of lean software development, while often employing different terminology, can be mapped to those of the Agile Manifesto and that the true contribution of lean software development lies rather in its *practices* (Erdogmus 2012).

In this regard, he singles out two main practices. The first is value stream mapping (Locher 2008), which provides a technique for analyzing the information flow across the full product lifecycle and identifying potential sources of problems and waste. The second is Kanban.

84.3.4 Other Methods

The set of methods presented in the previous subsections provide representative coverage of the core characteristics of current agile approaches, but the presented set is by no means complete. In many cases, other methods exhibit different emphases. For example, Adaptive Software Development (Highsmith 2000) and agile project management (Highsmith 2009) focus on management and business-oriented aspects of agile development more than on programming-related issues. The following methods each have special characteristics that merit discussion.

84.3.5 Kanban

Kanban is a system of production control associated with the Toyota Production System (Shingō 1989) and therefore in the software context most closely associated with lean software development. Strictly speaking, it is a practice rather than a method. It represents the "on-demand, pull-oriented" approach of just-in-time manufacturing, in which the signals controlling production propagate backward from customer demand, rather than forward as in more traditional "push-oriented" production control approaches. Kanban (literally "signboard") cards are used to control and limit the amount of WIP in the overall production system. Kanban seeks to achieve an ideal of continuous flow of production, thereby offering an alternative to the time-boxed, iteration-based workflow management found in most mainstream agile methods (Erdogmus 2012). A number of variations exist, such as "Constant Work in Progress" (CONWIP), a simpler single stage variant (Spearman et al. 1990). An economic analogy between WIP and the financial tracking measure of *economic profit* is described in Favaro, 2003. In Humble and Farley (2010), technical practices for continuous release of software in production environments are described in detail.

By its nature as a production control practice, Kanban may be used within different agile methods. It forms an integral part of the lean software development approach, of course, but has also been used together with Scrum (much like the practices of XP are used within the Scrum framework) as described in Kniberg, 2007 and Ladas, 2009. A full method has been developed around Kanban (Anderson 2010) for implementing and managing flow-based software development. It includes a set of core *practices* that are based upon those observed in successful Kanban systems (visualize, limit WIP, manage flow, make policies explicit, implement feedback loops, improve collaboratively, and evolve experimentally).

84.3.6 Crystal

The author of the Crystal software development method was a signatory of the Agile Manifesto. While having much in common with other agile methods, Crystal distinguishes itself in being a "family" of methods (known as Crystal Light) in the sense that it is explicitly tailored in two dimensions: project scale and criticality. In the scale dimension, the Crystal Clear variant is intended for projects of 2–6 persons, Crystal Yellow for 6–20 persons, and Crystal Orange for 20–40 persons (Cockburn 2001). Thus, it was an early method to take into explicit account the problem of scaling agile methods. Even more notable is the second dimension of criticality, which is segmented according to the consequences of system malfunction: loss of comfort, loss of discretionary money, loss of essential money, and loss of life. This scale constitutes an informal version of an *integrity level* scale seen in most modern standards for mission-critical systems, such as the Safety Integrity Levels (SILs) of the IEC 61508 standard (IEC 2010) for electrical and electronic systems. In the Crystal Light family, this informal scale is described as corresponding to "hardening" of the method according to increasing degrees of criticality. This approach corresponds roughly to the way in which standards such as IEC 61508 prescribe increasingly rigorous practices in the development process according to the SIL of the item under development. As such, the Crystal family can be seen as one of the earlier methods to attempt to explicitly address the problem of adapting the agile approach to the development of mission-critical systems.

The best-documented member of the Crystal Light family is Crystal Clear (Cockburn 2004).

84.3.7 Rational Unified Process

The RUP was developed in the 1990s at Rational Software Corporation (Kruchten 1998). Although it was not developed specifically as an "agile method" (none of its authors was a signatory of the Agile Manifesto), it presents an iterative and incremental approach to software development and may be used in a way that is fully compatible with the principles expressed in the Agile Manifesto. Analogous to Scrum, the RUP is more properly a process *framework* (rather than a single, specific process) that is intended to be tailored and instantiated for a specific implementation context, while selecting from a large number of provided practices. In this respect, it has been criticized for being so flexible that anything from an iterative, agile process to a pure sequential, waterfall process can be accommodated. Nonetheless, the intent behind the framework was clearly to support iterative development, as may be seen in the four defined lifecycle phases:

- *Inception*: This is primarily a phase in which the business case and feasibility studies for the project are developed.
- *Elaboration*: In this phase, an architecture is established for the system under development. In some respects this is the most characteristic, distinguishing phase of the method. The *architecture-centric* nature of the RUP has been a stimulus for much discussion about the relationship between agile development and architecture and has also been the source of some criticism that the RUP encourages excessive up-front design.
- *Construction*: In this phase, the system is constructed (following the architecture) in a series of iterations.
- *Transition*: This phase consists essentially of transitioning the system from development to production and deployment.

Within these phases, a number of activities are carried out inside a set of nine predefined disciplines: business modeling, requirements, analysis and design, implementation, test, deployment, project management, configuration and change management, and environment.

Like many agile methods such as XP and lean, the RUP also provides a set of guiding principles, or best practices, to guide the application of the method:

- *Develop software iteratively*: An explicit appeal to iterative and incremental development.
- *Manage requirements*: The authors of the RUP pioneered the use case approach to development that is the heritage of the user story-driven approach of agile methods. This principle has sometimes been misinterpreted to favor pure up-front requirements definition.
- *Visually model software*: The coauthors of the Unified Modeling Language were strongly associated with the RUP and bring a strong bias toward visual modeling into the method.
- *Verify software quality*: The same strong emphasis on continuous assurance of high quality seen elsewhere in agile methods.
- *Control changes to software*: This principle is in the same spirit of "embracing change" as expressed in the Agile Manifesto, whereby constant change is expected and mechanisms are provided to protect the system integrity under change.

Despite criticisms from the agile community regarding its "purity," the RUP has been an important instrument for introducing best practices of iterative and incremental development into the broader software engineering community. It has consolidated tool support such as the Rational Method Composer (IBM 2013) and has been introduced in many large organizations (Sägesser et al. 2013). Its clear definition has made it possible to define variants with a more specifically agile flavor, such as Disciplined Agile Delivery (Ambler and Lines 2012).

84.3.8 Personal Software Process and Team Software Process

The PSP was developed at the SEI in 1993 (Humphrey 2005) and followed by the TSP in 1998 (Humphrey et al. 2010). As such they are precedent to the Agile Manifesto and most current agile methods. Nevertheless, the PSP represents an early attempt to provide a process oriented toward the individual—a fundamental characteristic associated also with agile approaches. Its emphasis on use of concrete data to drive schedule prediction is also reminiscent of the emphasis on feedback in agile methods. Finally, it emphasizes individual accountability as in agile methods. TSP includes PSP-trained individuals in a team process in which product development may be iterative, another fundamental characteristic of agile approaches. It also includes an emphasis on *self-directed teams*, exhibiting respect and responsibility within its membership, reminiscent of the self-organizing teams of the principal agile methods. PSP and TSP thus represent an interesting variant coming from (and trademarked by) a source normally associated with so-called heavyweight processes.

84.3.9 Discussion

Agile methods as exemplified by those presented in this section (in particular the first three methods) have established themselves within the mainstream of software development. Over the last decade, one of the principal issues addressed by practitioners has been their relationship to the methods that preceded them. Shortly after the publication of the Agile Manifesto, analyses began to appear that sought to compare and contrast them to these other methods. In Boehm and Turner, 2004 a distinction between *agile* and *plan-driven* methods was made, and a "home ground" was identified for each in which its application was more suitable, according to four project characteristics:

- *Application*: Rapid value delivery, small teams, and turbulent environment (agile home ground) versus predictability, stability, high assurance, larger teams, and stable environment (plan-driven home ground)
- *Management*: Dedicated onsite customers, internalized plans, and tacit knowledge through interpersonal communication (agile home ground) versus as-needed customer interaction, documented plans, and explicit documented knowledge (plan-driven home ground)
- *Technical*: Prioritized informal stories changing in unforeseeable ways and simple design in short increments with refactoring (agile home ground) versus requirements with foreseeable evolution and extensive design in long cycles (plan-driven home ground)
- *Personnel*: Collocated, high percentage of highly qualified personnel throughout project, and thriving on chaos (agile home ground) versus concentration of highly qualified personnel in early stages with lower-level personnel later, not always collocated, and thriving on order (plan-driven home ground)

A risk-based strategy for balancing agile and plan-driven approaches according to the characteristics of projects is presented in Boehm and Turner, 2003.

For many, the iconic plan-driven approach is represented by the Capability Maturity Model Integration (CMMI) of the SEI (2012), and it has been often remarked that agile methods arose as a reaction against the CMMI. Seven years after the signing of the Agile Manifesto, the SEI published a report (Glazer et al. 2008) that argued forcefully that the perceived incompatibility of CMMI and agile methods is unwarranted and largely due to misunderstandings of the nature of each approach. The report stresses that CMMI is fundamentally a model rather than a process standard, as viewed by many agile proponents, and its rich set of mechanisms for facilitating process improvements has great potential to contribute to a fully compatible agile process. It cites a number of cases (Sutherland et al. 2007) in which CMMI and agile methods have been integrated successfully. The discussion of the relationship between plan-driven approaches and agile approaches is likely to continue in the foreseeable future. In any case, in 2010 the US Department of Defense issued a report (USDOD 2010) recommending that the department's IT procurement policy be modified to promote development practices that are fully compatible with the Agile Manifesto—a strong indication that agile methods are here to stay.

84.4 Research Areas

Since the publication of the Agile Manifesto, there has been vigorous research activity in agile software development. A literature survey in Dingsøyr et al., 2010 reports that from 1997 onward, several hundred papers were published in countries all over the world. Although the single country with the largest number of publications was the United States, Europe constituted the largest overall source when the countries comprising Europe were taken collectively. A systematic review (Dybå and Dingsøyr 2008) of empirical studies of agile software development reported on four broad thematic groups:

- *Introduction and adoption*: Most empirical studies tend to report that agile methods are easy to adopt, although XP was found to be difficult to introduce in organizations of higher complexity.

- *Human and social factors*: Not surprisingly, given the human-centered character of agile approaches, many studies focus on various specific human factors. Interestingly, XP was found to work well in many very different kinds of development environments.
- *Perceptions of agile methods*: Many studies focus on perceptions of agile from the point of view of different stakeholder groups (e.g., architects). Customers in general were found to be satisfied with feedback opportunities but less satisfied with demands for onsite customer involvement. Satisfaction among developers using XP was generally found to be high.
- *Comparative studies*: Studies of agile methods against other approaches (e.g., waterfall) have tended to find that methods such as XP increase productivity.

Despite positive results such as the last one cited earlier, the authors concluded that more empirical studies of agile software development are needed, within a broader and more complete research agenda. Research in specific areas is discussed in the following sections.

84.4.1 Role of Architecture in Agile Software Development

The 1990s were a period of great advances in the discipline of software architecture (Shaw and Garlan 1996), and methods developed during that period such as the RUP were characterized as explicitly *architecture-centric*. Ever since the characterization of architecture as *emergent* within agile software development was coined, however, controversy has surrounded its role within agile methods. The de-emphasis of Big Design Up Front in agile methods seemed to exclude the possibility of architecture-centric approaches in agile methods. In Abrahamsson et al. (2010), it is suggested that much of the controversy may be due to underlying misunderstandings, since authors of several major agile methods have incorporated concepts that reflect a positive attitude toward architecture, for example, "divisible system architecture" of Poppendieck and Poppendieck (lean software development), the "walking skeleton" of Cockburn (Crystal), and the "system metaphor" of Beck (XP).

Some recent research has been oriented toward making the effects of architecture on agile development more explicit (Madison 2010). In Nord et al. (2012), the use of a Kanban board within a lean development process is described as a technique for rendering more visible how architecture either supports or inhibits throughput.

In Falessi et al. (2010), researchers from the University of Rome and the Italian Ministry of Defense collaborated with practitioners from IBM to conduct an exploratory study of attitudes of professional developers who use agile methods toward software architecture. They found that overall, agile developers consider software architecture to be relevant and important to their work; in particular they displayed a favorable attitude toward such practices as the use of architectural design patterns. Conversely, however, they found that those not practicing agile development were more likely to find architecture approaches to be in contrast with agile approaches. Thus, the controversy over their compatibility is likely to continue in the future, both in research and in practice.

84.4.2 Human Factor Research

Given that the very first principle of the Agile Manifesto emphasizes "individuals and interactions over processes and tools," it is not surprising that much research in agile software development has concentrated on human factors. An early ethnographic study (Sharp and Robinson 2004) of XP teams noted that "we seek here to explore the values, beliefs, and assumptions that inform and shape agile practice and … to explore the manner in which a community of agile developers sustains itself." In Robinson and Sharp (2004), several human characteristics of XP teams are identified and discussed: respect, responsibility, quality of working life, faith in abilities, and trust.

Within university environments, courses on human aspects of software engineering began early to present case studies in agile development methods (Hazzan and Tomayko 2004). Interdisciplinary approaches to the study of agile human factors were also introduced. In Falcone and Favaro, 2003 an environment called the Agile Theater is described in which classical techniques of theatrical improvisation were used to teach and explore interactions and roles in agile development teams. Agile Theater sessions with experienced developers revealed that, without exception, the continuous exchange of roles, interaction of participants, and conflict development and resolution always brought out a discussion of the problem of leadership in agile teams. The issue of roles in agile teams was also studied in Dubinsky and Hazzan 2004, which concluded that adding a specific role for each member of an agile team beyond normal technical responsibilities augments the sense of personal accountability and responsibility in the individuals.

Extending earlier interdisciplinary work concerning creativity in requirements engineering (Maiden and Robertson 2005)—in which experts were brought into an air traffic control environment from fields as diverse as railway signaling, oriental textiles, and cooking—Maiden and Hollis (Hollis 2011) performed research on creativity in agile development. Calling into question a common assumption that agile development automatically produces innovative solutions (Oza and Abrahamsson 2011), they embedded creativity techniques into a full agile software development process—specifically, within Scrum, during the envisioning process and at the beginning of selected sprints containing those epics with the most potential for creative outcomes. A case study was performed in a BBC Worldwide project, with results indicating that more novel requirements were generated through the extended epic process.

In XP, collective code ownership is associated with many benefits. A study at Microsoft (Bird et al. 2011) examined the relationship between different measures of ownership and software failures in Windows Vista and Windows 7 development. The study distinguished different levels of ownership (major versus minor contributor to components) and found that strong ownership is indicative of higher quality—which may contradict the usual claims for the benefits of collective ownership.

The colocation of development teams is a cornerstone of agile approaches, but strong business motivations (e.g., offshoring) exist for distributed development; consequently, the effects of distributed development on agile teams have become a subject of research. An overview of the challenges encountered in distributed agile development is provided in (Kajko-Mattsson et al. 2010) in which six classes are identified: culture, time zone, communication, customer collaboration, trust, and training and technical issues. The specific problem of trust in distributed agile projects is also discussed in Dorairaj et al. 2010, while the problem of communication is discussed in Dorairaj et al. 2011. Techniques for supporting multisite and offshore Scrum development are presented in Larman and Vodde 2010.

84.4.3 Agile Development of Mission-Critical and Embedded Systems

Agile software development methods had their first successes within environments exhibiting high interactivity (e.g., web-based systems) and changing often malleable requirements. They were largely ignored in mission-critical and embedded systems environments, due to a perceived lack of compatibility. However, as the benefits of agile development were publicized, practitioners and researchers began to search for suitable adaptations of agile approaches for these environments. An early example is (Fredriksen 2002), in which a mapping of the RUP to the IEC 61508 standard (IEC 2010) for development of safety critical software for electronic systems was made as part of a research initiative and found to be feasible.

In Smith et al. 2009 an adaptation of XP to a lifecycle for the development of embedded medical instrument software is described. It was noted that significant adaptation of XP was required, which may even lead to challenges to its claim to be still an agile approach. A prototype TDD embedded system test framework called Embedded Unit was developed by the researchers. Another example of applying TDD to embedded systems software development is Grenning 2011. A full methodology is developed in Douglass 2009.

A survey of the state of the art in the use of agile methods for embedded systems development (Srinivasan et al. 2009) classified the major challenges into two categories with six areas:

- *Technical*: Requirements management, testing
- *Organizational*: Process tailoring, knowledge sharing and transfer, culture change, support infrastructure

Future research is likely to be concentrated in these areas.

84.4.4 Economics of Agile Development

Given the explicit commitment of agile methods to the delivery of business value, it is natural that there would be a great interest of researchers in the economics of agile software development.

The exhortation in the Agile Manifesto to "welcome change" positions agile methods in the area of economic research known as *decision theory.* In Drury et al. (2012), research is described that analyzes the decision-making process over the iteration cycles of several real-world projects and identifies six obstacles to decision-making (an example is "unwillingness to commit to decisions") in agile development and then maps those obstacles to descriptive decision-making principles to analyze their impact.

Even more narrowly, agile methods are positioned at the heart of decision theory in the area known as *decision-making under uncertainty.* Because of this, the relevance of *real options* was recognized early (Favaro 1999), and a treatment of the YAGNI principle as an option to defer investment appeared in the first book on XP (Beck 1999). Several other real options associated with agile development were subsequently analyzed (Erdogmus and Favaro 2002), and lean software development (Poppendieck and Poppendieck 2003) promoted a real options approach to decision-making, singling out the *option to defer* and introducing an informal characterization of this real option as decision-making at the "last responsible moment." Another discussion of agile decision timing and delay is provided in Favaro (2003) using the "tomato garden" metaphor of Luehrmann. The integration of the concept of the last responsible moment and a general real options approach to architectural decision-making in agile development is described in Blair et al. (2010) as responsibility-driven architecture.

Given the intersection of human factors and economics in agile methods, interest has recently been growing in the relationship of some branches of behavioral economics to agile software development. An early investigation is represented by Erdogmus (2009) in which the application of *diversity* (Page 2007) to the formation of cross-functional teams (as seen in agile methods like Scrum) is discussed. The author draws on diversity theory to conclude that a diverse group of average yet sufficiently competent people will often outperform a homogeneous group of star performers who have overlapping skills. Nonetheless, the caveat is provided that a single top expert will still perform better on average than a not-so-wise group—and thus, competence management in cross-functional agile teams remains relevant and important.

Within behavioral economics, prospect theory (Kahneman 2011) has been perceived as particularly relevant due to its emphasis on the psychological foundations of decision-making under uncertainty. In Makabee (2012), the planning fallacy of prospect theory is studied as a source of insight into why planning estimates in Scrum are often significantly inaccurate. The fourfold pattern of prospect theory is discussed in Favaro and Erdogmus (2012), in the context of the tendency to undertake desperate measures to salvage projects that are failing, rather than adopting the agile values of "courage" and "celebrating failure." The concept of *high-validity environments* is also discussed as a prerequisite for reliable expert intuition, and it is hypothesized that a primary effect of agile development practices such as continuous feedback and integration is to move software development toward a high-validity environment in which expert intuition can develop reliably.

84.4.5 Technical Debt

Technical debt refers to the quick implementation and delivery (e.g., to meet time pressure) of a feature while sacrificing quality, with a promise to remediate at some later point. Although technical debt has become a concern in general software management, it is closely associated with the agile movement both in origin and practice. The expression was coined by agile pioneer Ward Cunningham (Cunningham 1992), and indeed it has been remarked (Bavani 2012) that technical debt is central to the definition of agile itself. Other authors from the agile community (Fowler 2009) have written extensively about technical debt (Gat 2010).

It is studied most frequently in the context of agile project management because of its natural affinity to iteration-oriented development (Sterling 2010). An appeal was recently made (Kruchten et al. 2012) to the research community to establish a theoretical basis for the concept of technical debt that goes beyond the current *ad hoc* definition as a metaphor.

84.4.6 Measuring and Quantifying Agile Development Practices

Agile software development has been presented by proponents as a more effective alternative approach to previous software development methods, but the evidence has often been anecdotal. Likewise, the evidence for comparative effectiveness *among* agile methods themselves has often been mostly anecdotal.

One early exception was a series of empirical investigations of the effectiveness of *pair programming*. An empirical study was undertaken by Williams et al. 2000 in which it was determined that pairs spent on average 15% more total effort than solo programmers to complete the same programming tasks, but that code written by pairs on average passed an additional 15% of the specified acceptance tests. This study was followed by another (Erdogmus and Williams 2003) that concluded that pair programming long-term productivity could achieve a 30% advantage over productivity of solo programmers. Overall, all such studies found that pair programming is costly in the short term but appears to exhibit a quality advantage for complex tasks (Dybå et al. 2007). Notwithstanding, with regard to the effectiveness of pair programming, Wray (2010) points to the need for further experimentation, noting that "we are no longer in the first flush and the pros and cons seem farther apart than ever."

Another subject of numerous empirical studies is TDD. As noted in Turhan et al. (2010), TDD " … is one of the most referenced, yet least used agile practices in industry." An early study was conducted by Erdogmus et al. (2005). A subsequent (Turhan et al. 2010) systematic literature review focused on studies that evaluated TDD with respect to productivity, quality, and test support. Their findings, summarized in Shull et al. (2010), concluded that there is moderate evidence in favor of TDD. More recent studies (Rafique and Misić 2012) report similar conclusions—for example, that TDD appears to support quality improvement with more complex tasks.

With the emergence of *evidence-based software engineering* (Dybå et al. 2005), anecdotal claims regarding effectiveness of agile methods began to come under the closer scrutiny of researchers. For example, one of the key premises underlying claims of the superiority of lean software development approaches is the literature on Japanese automobile manufacturing successes (Womack et al. 1990). But Dybå and Sharp (2012) point out that the evidence for the claims made in that literature is open to valid alternative interpretations that potentially diminish their strength considerably.

As a consequence of the trend toward evidence-based software engineering research, in recent years the research community has undertaken numerous initiatives to measure and quantify the effects of agile practices in particular and agile development in general, along several dimensions. In Sjøberg et al. (2012), an initiative is described in which the comparative effectiveness of the use of the Scrum and the Kanban approaches is investigated using data gathered from more than 12,000 work items produced over a period of 3 years. In the study described, an organization almost halved its lead time after replacing Scrum with Kanban, improved its productivity, and reduced the number of weighted bugs by 10%. The authors provide several caveats regarding interpretation of the results, including the fact that introducing Kanban after Scrum may have introduced a bias toward Kanban because the development professionals were more familiar with agile software development by that time.

One of the largest ongoing initiatives to measure the comparative effectiveness of agile software development methods based upon industrial project data is that of QSM, an early pioneer in software estimation and benchmarking. QSM has collected industrial project data continuously since 1978, compiling a database with validated data from over 10,000 projects suitable for comparative benchmarking. Its Software Lifecycle Model (SLIM) (Putnam 1978) provides input project data parameters of *time*, *size*, and *effort*, which were reinterpreted in an agile context in order to enable data collection for agile projects. For example, size is estimated through stories, story points, and lines of code; development effort is modeled as one estimate with milestones for the iterations, or alternatively each iteration may be an estimate. An initiative of the Central Ohio Agile Association and QSM over the 2012–2013 time frame (QSM 2012) within a specific industrial community (Columbus, Ohio) yielded benchmark data indicating that the agile projects included in the study exhibited defect data that are significantly better than industry averages, with a 30% improvement in completion rate over industry norms.

84.4.7 Conclusions on Agile Research

In Dingsøyr et al. (2012), the observation is made that the research community has been very active ever since the publication of the Agile Manifesto, with over 60 countries contributing to the corpus of research, in numerous special issues of reputable journals, dedicated conferences, and many books, including best sellers in the software engineering market. But the observation is likewise made that researchers are still not paying enough attention to establishing the theoretical underpinnings of agile development and its associated practices. The establishment of these theoretical underpinnings as a basic for solid, effective future research remains of the most urgent themes in agile software development.

84.5 Summary

The so-called agile movement in software development began with the publication of the Agile Manifesto in 2001. Although agile methods were already in use by that time, the Agile Manifesto coined the term and codified its characteristics through twelve principles. The agile movement was widely seen as a reaction to rigid, sequential, defined processes of the previous two decades. The iterative and incremental approach embodied in the agile movement, however, had deep roots in practices that had been proposed at various times throughout the history of software engineering.

Two agile methods in particular are emblematic of industrial best practices in the decade following the publication of the Agile Manifesto. XP offered a concrete set of core practices that exemplified the quintessence of agile software development practice and was well received by programmers. Scrum provided a framework for the overall management of an iterative and iterative development project that was well received by managers. In parallel, methods such as lean software development brought a flow-based character into the movement, later strengthened by the introduction of Kanban. Several analyses appeared over the decade comparing the appropriateness of agile approaches versus more traditional plan-driven approaches in different environments.

Academic research in agile methods began in earnest even before the publication of the Agile Manifesto. Human factors and economic aspects were consistent research topics from the beginning. Agile methods in demanding environments such as critical embedded systems became another research topic, as well as the problem of understanding the role of architecture in agile software development. There has been a steady increase in publication of empirical studies of pre- and post-adoption impact of agile methods.

In February 2011, on the tenth anniversary of the publication of the Agile Manifesto, ten of the original signatories provided commentary (Hunt et al. 2011) on its impact on software development during the intervening decade. Although there was general agreement that the Agile Manifesto and the principles and practices it embraced had significantly affected the software engineering community, there was likewise general disappointment that the agile approaches had not been adopted to the expected degree. In particular, the observation was made that the processes and practices of agile approaches had been implemented by too many practitioners in a slavish and formulaic manner, reflecting a fundamental misunderstanding of the word "agile." Thus, the adoption of agile methodologies as envisioned by the signatories of the Agile Manifesto remains a challenge even today for the software development community.

Key Terms

The agile vocabulary is vast and as varied as the number of agile methods. The following glossary contains terms that are important to understanding agile approaches and/or common to several methods.

Backlog: In Scrum, those tasks or stories that will be worked on in the future.

BDUF: Big Design Up Front. Refers to the practice of designing a system before any implementation has begun, in contrast to the emergent design approach of agile methods.

Burndown chart: A technique in Scrum to track and display work remaining to finish.

Continuous integration: An approach to software development that involves frequent integration of individual work into the whole system. Generally this occurs at least daily.

Daily stand-up: Agile methods often encourage short daily meetings to report on progress and plan future actions. They are often held while standing up in order to discourage a long duration.

Epic: In Scrum, a large user story that eventually is partitioned into smaller stories.

Iteration: A period (ranging from days to 2 months) in which an agile team implements an increment of functionality.

Kanban: Japanese, roughly translates to "signboard." A technique generally associated with lean software development, for visualizing work in progress.

Pair programming: An agile practice in which two programmers share a single workstation, with one acting as coder and the other as reviewer or "navigator."

Refactoring: Rewriting of code to improve its design and maintainability. Generally the refactoring actions are small and local.

Retrospective: A meeting at the end of an iteration (or entire project) in which the team considers what could be improved. A way to implement process improvement in agile methods.

Scrum: Rugby term regarding a method for restarting play after an infringement. Refers to teams huddling together in a group. Used as the name for the agile method.

Spike: An investigation of feasibility outside the normal flow of work.

Sprint: An iteration in the Scrum agile method.

Story: A description (usual textual) of a feature to be implemented. May be seen as a compact version of a use case.

TDD: Test-driven development. An approach to coding that involves first writing a failing test and then writing the code that makes the test succeed.

Timebox: A fixed duration, with reference to the length of an iteration in agile development.

Velocity: A measure of the speed at which an agile team completes its work, usually the number of stories in an iteration.

WIP: Work in progress—a concept in particular from lean software development. Work that has been incurred but not yet delivered to the customer.

YAGNI: You Aren't Going to Need It. Refers to an XP practice of not implementing a feature for which a firm requirement does not yet exist.

References

Abrahamsson, P., Ali Babar, M., and Kruchten, P. 2010. Agility and architecture: Can they coexist? *IEEE Software*. 27(2): 16–22.

Abrahamsson, P., Salo, O., Ronkainen, J., and Warsta, J. 2002. *Agile Software Development Methods: Review and Analysis*. Espoo, Finland: VTT Publications, p. 478.

Ambler, S. and Lines, M. 2012. *Disciplined Agile Delivery: A Practitioner's Guide to Agile Software Delivery in the Enterprise*. Upper Saddle River, NJ: IBM Press.

Anderson, D. 2010. *Kanban: Successful Evolutionary Change for Your Technology Business*. Sequim, WA: Blue Hole Press.

Arbogast, T., Larman, C., and Vodde, B. 2012. Agile Contracts Primer. http://www.agilecontracts.org/. Accessed October 1, 2012.

Bavani, R. 2012. Distributed agile, agile testing, and technical debt. *IEEE Software*. 29(6): 28–33.

Beck, K. 1999. *Extreme Programming Explained: Embrace Change*. Boston, MA: Addison-Wesley.

Beck, K. 2002. *Test Driven Development: By Example*. Boston, MA: Addison-Wesley Professional.

Beck, K. and Andres, C. 2004. *Extreme Programming Explained: Embrace Change*, 2nd edn. Boston, MA: Addison-Wesley.

Beck, K., Beedle, M., and van Bennekum, A. 2001. The Agile Manifesto. http://www.agilemanifesto.org. Retrieved August 8, 2012.

Beck, K. and Cleal, D. 1999. Optional Scope Contracts. http://www.xprogramming.com/ftp/Optional+scope+contracts.pdf. Accessed October 1, 2012.

Beck, K. and Fowler, M. 2000. *Planning Extreme Programming*. Boston, MA: Addison-Wesley Professional.

Bird, C., Nagappan, N., Murphy, B. et al. 2011. Don't touch my code!: examining the effects of ownership on software quality. *Proceedings of the ACM SIGSOFT Symposium of the Foundations of Software Engineering*, Szeged, Hungary, *ESEC/FSE 2011*. ACM, New York. pp. 4–14.

Blair, S., Watt, R., and Cull, T. 2010. Responsibility-driven architecture. *IEEE Software*. 27(2): 26–32.

Boehm, B. 1988. A spiral model of software development and enhancement. *IEEE Computer*. 31(5): 61–72.

Boehm, B. and R. Turner. 2003. Using risk to balance agile and plan-driven methods. *IEEE Computer*. 36(6): 57–66.

Boehm, B. and R. Turner. 2004. *Balancing Agility and Discipline: A Guide for the Perplexed*. Boston, MA: Addison-Wesley.

Brealey, R., Myers, S., and Allen, F. 2008. *Principles of Corporate Finance*. New York: McGraw-Hill.

Brooks, F. 1987. No silver bullet—Essence and accidents of software engineering. *IEEE Computer*. 20(4): 10–19.

Cockburn, A. 2001. Crystal light methods. http://alistair.cockburn.us/Crystal+light+methods. Accessed October 1, 2012.

Cockburn, A. 2004. *Crystal Clear: A Human-Powered Methodology for Small Teams: A Human-Powered Methodology for Small Teams*. Boston, MA: Addison-Wesley Professional.

Cohn, M. 2005. *Agile Estimating and Planning*. Upper Saddle River, NJ: Prentice-Hall.

Constantine, L. 1992. The benefits of visibility. *Computer Language Magazine*, 9(2). Reprinted in *The Peopleware Papers*, Upper Saddle River, NJ: Prentice Hall (2001).

Cunningham, W. 1992. The WyCash Portfolio Management System. OOPSLA 1992 Experience Report. http://c2.com/doc/oopsla92.html. Accessed October 1, 2012.

Cusumano, M. 1991. *Japan's Software Factories: A Challenge to U.S. Management*. New York: Oxford University Press.

Cusumano, M. 2007. Extreme programming compared with microsoft-style iterative development. *Communications of the ACM*. 50(10): 15–18.

Cusumano, M. and Selby, R. 1995. *Microsoft Secrets: How the World's Most Powerful Software Company Creates Technology, Shapes Markets and Manages People*. New York: Free Press.

DeMarco, T. and Lister, T. 1987. *Peopleware*. New York: Dorset House Publishing.

DeMarco, T. and Boehm, B. 2002. The Agile Methods Fray. *Computer*. 35(6): 90–92.

Derby, E. and Larsen, D. 2006. *Agile Retrospectives: Making Good Teams Great*. Raleigh, NC: Pragmatic Bookshelf.

Dijkstra, E. 1972. The humble programmer. *Communications of the ACM*. 15(10): 859–866.

Dingsøyr, T., Dybå, T., and Moe, N. 2010. *Agile Software Development: Current Research and Future Directions*. Berlin, Germany: Springer Verlag.

Dingsøyr, T., Nerur, S., Balijepally, V., and Moe, N. 2012. A decade of agile methodologies: Towards explaining agile software development. *Journal of Systems and Software*. 85(6): 1213–1221.

Dorairaj, S., Noble, J., and Malik, P. 2010. Understanding the importance of trust in distributed agile projects: A practical perspective. In *Agile Processes in Software Engineering and Extreme Programming*. Lecture Notes in Business Information Processing. Vol. 48, pp. 172–177

Dorairaj, S., Noble, J., and Malik, P. 2011. Effective communication in distributed agile software development teams. In *Agile Processes in Software Engineering and Extreme Programming*. Lecture Notes in Business Information Processing. Vol. 77, pp. 102–116.

Douglass, B. 2009. *Real-Time Agility: The Harmony/ESW Method for Real-Time and Embedded Systems Development*. Boston, MA: Addison-Wesley Professional.

Drury, M., Conboy, K., and Power, K. 2012. Obstacles to decision making in Agile software development teams. *Journal of Systems and Software*. 85(6): 1239–1254.

Dubinsky, Y. and Hazzan, O. 2004. Roles in agile software development teams. In *Extreme Programming and Agile Processes in Software Engineering*. Lecture Notes in Computer Science. Vol. 3092, pp. 157–165.

Dybå, T., Arisholm, E., Sjøberg, D. et al. 2007. Are two heads better than one? On the effectiveness of pair programming. *IEEE Software*. 24(6): 12–15.

Dybå, T., Kitchenham, B., and Jørgensen, M. 2005. Evidence-based software engineering for practitioners. *IEEE Software*. 22(1): 58–65.

Dybå, T. and Dingsøyr, T. 2008. Empirical studies of agile software development: A systematic review. *Information and Software Technology*. 50(9–10): 833–859.

Dybå, T. and Sharp, H. 2012. What's the evidence for lean? *IEEE Software*. 29(5): 19–20.

Eclipse Foundation. 2012. Eclipse Process Framework Project. http://www.eclipse.org/epf/. Accessed October 1, 2012.

Erdogmus, H. 2002. Valuation of learning options in software development under private and market risk. *The Engineering Economist*. 47(3): 304–353.

Erdogmus, H. 2009. Diversity and doftware development. *IEEE Software*. 26(3): 2–4.

Erdogmus, H. 2012. Lean is a fad. *IEEE Software*. 29(5): 61–62.

Erdogmus, H. and Favaro, J. 2002. Keep your options open: Extreme programming and the economics of flexibility. In *Extreme Programming Perspectives*. Williams, L. et al., eds. Boston, MA: Addison-Wesley. pp. 503–552.

Erdogmus, H., Morisio, M., and Torchiano, M. 2005. On the effectiveness of the test-first approach to programming. *IEEE Transactions on Software Engineering*. 31(3): 226–237.

Erdogmus, H. and Williams, L. 2003. The economics of software development by pair programmers. *The Engineering Economist*. 48(4): 283–319.

Fairley, R. 1980. Ada debugging and testing support environments. *Proceedings of the ACM-SIGPLAN Symposium on the ADA Programming Language*, New York, pp. 16–25.

Falcone, P. and Favaro, J. 2003. The agile theater. In *Computer Programming*. Ponsacco: Infomedia. CP130. pp. 69–70.

Falessi, D., Cantone, G., Sarcià, S. et al. 2010. Peaceful coexistence: Agile developer perspectives on software architecture. *IEEE Software*. 27(2): 23–25.

Favaro, J. 1999. Managing IT for Value. In: *Proceedings National Polish Software Engineering Conference*, Warsaw, Poland.

Favaro, J. 2003. Value based management and agile methods. In *Extreme Programming and Agile Processes in Software Engineering*. Lecture Notes in Computer Science. Vol. 2675. Berlin, Germany: Springer Verlag. pp. 16–25.

Favaro, J. 2004. Efficient markets, efficient projects, and predicting the future. In *Extreme Programming and Agile Processes in Software Engineering*. Lecture Notes in Computer Science. Vol. 3092. Berlin, Germany: Springer Verlag. pp. 77–84.

Favaro, J. and Erdogmus, H. 2012. The value proposition for agility: A dual perspective. Software Experts Summit 2012. London, U.K. http://www.computer.org/cms/Computer.org/ComputingNow/promo/ses12/Erdogmus.pdf. Accessed October 1, 2012.

Favaro, J., Favaro, K., and Favaro, P. 1998. Value based software reuse investment. *Annals of Software Engineering*. 5: 5–52.

Fowler, M. 2003. Principles of XP. http://www.martinfowler.com/bliki/PrinciplesOfXP.html. Accessed October 1, 2012.

Fowler, M. 2009. Technical Debt. http://martinfowler.com/bliki/TechnicalDebt.html. Accessed October 1, 2012.

Fowler, M., Beck, K., Brant, J., Opdyke, W., and Roberts, D. 1999. *Refactoring: Improving the Design of Existing Code*. Boston, MA: Addison-Wesley Professional.

Fredriksen, R. 2002. Use of the Rational Unified Process for Development of Safety-Related Computer Systems. Østfold University College, Halden, Norway. http://www.hiof.no/neted/upload/attachment/site/group12/Rune_Fredriksen_Use_of_the_Rational_Unified_Process_for_development_of_safety_related_computer_systems.pdf. Accessed October 1, 2013.

Gat, I. 2010. Special issue on technical debt. *Cutter IT Journal*. 23(10). http://www.cutter.com/itjournal/fulltext/2010/10/index.html. Accessed October 1, 2013.

Gilb, T. 1989. *Principles of Software Engineering Management*. Menlo Park, CA: Addison Wesley Longman.

Glazer, H., Dalton, J., Anderson, D., Konrad, M., and Shrum, S. 2008. CMMI or agile: Why not embrace both! Technical Report CMU/SEI-2008-TN-003. Pittsburgh, PA: Software Engineering Institute, Carnegie Mellon University.

Grenning, J. 2002. Planning Poker. http://renaissancesoftware.net/files/articles/PlanningPoker-v1.1.pdf. Accessed October 1, 2012.

Grenning, J. 2011. *Test Driven Development for Embedded C*. Raleigh, NC: Pragmatic Bookshelf.

Hazzan, O. and Tomayko, J. 2004. Human aspects of software engineering: The case of extreme programming. In *Extreme Programming and Agile Processes in Software Engineering*. Lecture Notes in Computer Science. Vol. 3092, pp. 303–311.

Highsmith, J. 2000. *Adaptive Software Development: A Collaborative Approach to Managing Complex Systems*. New York: Dorset House.

Highsmith, J. 2009. *Agile Project Management: Creating Innovative Products*. 2nd edn. Boston, MA: Addison-Wesley Professional.

Hollis, B. 2011. Extending agile methodologies with creativity techniques, Masters Dissertation, School of Informatics, City University London, London, U.K.

Humble, J. and Farley, D. 2010. *Continuous Delivery: Reliable Software Releases through Build, Test, and Deployment Automation*. Boston, MA: Addison-Wesley Professional.

Humphrey, W. 1989. *Managing the Software Process*. Boston, MA: Addison-Wesley Professional.

Humphrey, W. 2005. *PSP: A Self-Improvement Process for Software Engineers*. Reading, MA: Addison-Wesley.

Humphrey, W., Chick, T., Nichols, W. et al. 2010. Team software process body of knowledge. Technical Report CMU/SEI-2010-TR-020.

Hunt, A., Beck, K., Jeffries, R. et al. 2011. Agile @ 10: Ten Authors of The Agile Manifesto Celebrate its Tenth Anniversary. The Pragmatic Bookshelf. http://pragprog.com/magazines/2011-02/agile–. Accessed October 1, 2012.

IEC 2010. IEC 61508: Functional safety of electrical/electronic/programmable electronic safety-related systems. International Electrotechnical Commission.

International Business Machines Corporation. 2013. Rational Method Composer. http://www-03.ibm.com/software/products/us/en/rmc. Accessed October 1, 2013.

Kahneman, D. 2011. *Thinking, Fast and Slow*. New York: Farrar, Straus and Giroux.

Kajko-Mattsson, M., Azizyan, G., and Magarian, M. 2010. Classes of distributed agile development problems. *Proceedings Agile 2010*, Orlando, Florida, pp. 51–58.

Kniberg, H. 2007. *Scrum and XP from the Trenches*. Raleigh, NC: Lulu.com.

Kniberg, H. 2011. *Lean from the Trenches*. Dallas, TX: Pragmatic Bookshelf.

Knuth, D. 1992. Literate programming. *The Computer Journal (British Computer Society)*. 27(2): 97–111.

Koskela, J. and Abrahamsson, P. 2004. On-site customer in an XP project: Empirical results from a case study. In *Software Process Improvement*. Lecture Notes in Computer Science. Vol. 3281, pp. 1–11.

Kruchten, P. 1998. *The Rational Unified Process: An Introduction*. Reading, MA: Addison-Wesley.

Kruchten, P., Nord, R., and Ozkaya, I. 2012. Technical debt: From metaphor to theory and practice. IEEE Software. 29(6): 18–21.

Jeffries, R. and Melnik, G. 2007. TDD: The art of fearless programming. *IEEE Software*. 27(3): 25–30.

Jeffries, R., Anderson, A., and Hendrickson, C. 2000. *Extreme Programming Installed*. Upper Saddle River, NJ: Addison-Wesley Professional.

Ladas, C. 2009. *Scrumban—Essays on Kanban Systems for Lean Software Development*. Seattle, WA: Modus Cooperandi Press.

Larman, C. and V. Basili. 2003. Iterative and incremental development: A brief history. *IEEE Computer*. 36(6): 47–56.

Larman, C. and Vodde, B. 2010. *Practices for Scaling Lean and Agile Development: Large, Multisite, and Offshore Product Development with Large-Scale Scrum*. Upper Saddle River, NJ: Addison-Wesley.

Levison, M. 2008. Are there weaknesses with Collective Code Ownership? InfoQ. http://www.infoq.com/news/2008/05/weaknesses_collective_code. Accessed October 1, 2012.

Locher, D. 2008. *Value Stream Mapping for Lean Development: A How-To Guide for Streamlining Time to Market*. Boca Raton, FL: Productivity Press.

Madison, J. Agile architecture interactions. 2010. *IEEE Software*. 27(2): 41–48.

Maiden, N. and Robertson, S. 2005. Integrating creativity into requirements processes: Experiences with an air traffic management system. In *Proceedings of the 13th IEEE International Requirements Engineering Conference*, Paris, France, pp. 105–114.

Makabee, H. 2012. Planning Poker: Avoiding Fallacies in Effort Estimates. http://effectivesoftwaredesign.com/2012/08/05/planning-poker-avoiding-fallacies-in-effort-estimates/. Accessed October 1, 2012.

Marchesi, M. 2005. The New XP. http://www.agilexp.org/downloads/TheNewXP.pdf. Accessed October 1, 2012.

Massol, V. and Husted, T. 2003. *JUnit in Action*. Greenwich, CT: Manning Publications.

Mugridge, R. and Cunningham, W. 2005. *Fit for Developing Software: Framework for Integrated Tests*. Upper Saddle River, NJ: Prentice-Hall.

Navarro, W. 2010. New to user stories? http://www.scrumalliance.org/articles/169-new-to-user-stories. Accessed October 1, 2012.

Nord, R., Ozkaya, I., and Sangwan, R. 2012. Making architecture visible to improve flow management in lean software development. *IEEE Software*. 29(5): 33–39.

Oza, N. and Abrahamsson, P. 2011. Building blocks of agile innovation. http://www.agileinnovationbook.com. Accessed October 1, 2012.

Page, S. 2007. *The Difference: How the Power of Diversity Creates Better Groups, Firms, Schools, and Societies*. Princeton, NJ: Princeton University Press.

Paulk, M. and Curtis, B. 1993. Capability maturity model for software. Technical Report. CMU/SEI-93-TR-024 ESC-TR-177. Software Engineering Institute, Pittsburgh, Pennsylvania.

Poppendieck, M. and Cusumano, M. 2012. Lean software development: A tutorial. *IEEE Software*. 29(5): 26–32.

Poppendieck, M. and Poppendieck, T. 2003. *Lean Development: A Toolkit for Software Development Managers*. Boston, MA: Addison Wesley Professional. ISBN-10: 0321150783, ISBN-13: 978-0321150783.

Putnam, L. 1978. A general empirical solution to the macro software sizing and estimating problem. *IEEE Transactions on Software Engineering*. 4(4): 345–361.

QSM Associates. 2012. QSM Associates and COHAA to benchmark Columbus, Ohio's Agile Software Development against Industry. http://www.agilesoftwareproject.com/tag/qsm-associates/. Accessed October 1, 2012.

Rafique, Y. and Misić, V. 2012. The effects of test-driven development on external quality and productivity: A meta-analysis. *IEEE Transactions on Software Engineering.*39(6): 835–856.

Randell, B. and Zurcher, F.W. 1968. Iterative Multi-Level Modeling: A Methodology for Computer System Design, In *Proceedings of the IFIP*, Edinburgh, U. K., IEEE CS Press, Yorktown Heights, NY, pp. 867–871.

Robinson, H. and Sharp, H. 2004. The characteristics of XP teams. In *Extreme Programming and Agile Processes in Software Engineering.* Lecture Notes in Computer Science. New York: Springer Verlag. Vol. 3092, pp. 139–147.

Royce, W. 1970. Managing the development of large software systems. In *Proceedings of the Westcon.* IEEE CS Press, Los Angeles, CA, pp. 328–339.

Sägesser, K., Joseph, B., and Grau, R. 2013. Introducing an iterative lifecycle model at credit Suisse IT Switzerland. *IEEE Software.* 30(2).

Schwaber, K. 2009. *Agile Project Management with Scrum.* Redmond, WA: Microsoft Professional.

Schwaber, K. and Beedle, M. 2001. *Agile Software Development with Scrum.* Upper Saddle River, NJ: Prentice-Hall.

Scrum Alliance. 2012. CSP: Become a professional. http://www.scrumalliance.org/pages/certified_scrum_professional. Accessed October 1, 2012.

SEI—Software Engineering Institute. 2012. CMMI Website. http://www.sei.cmu.edu/cmmi/. Accessed October 1, 2012.

Sharp, H. and Robinson, H. 2004. An ethnographic study of XP practice. In *Empirical Software Engineering.* Norwell, MA: Kluwer Academic Publishers. Vol. 9, pp. 353–375.

Shaw, M. and Garlan, D. 1996. *Software Architecture: Perspectives on an Emerging Discipline.* Upper Saddle River, NJ: Prentice-Hall.

Shingō, S. 1989. *A Study of the Toyota Production System from an Industrial Engineering Viewpoint.* Portland, OR: Productivity Press.

Shull, F., Melnik, G., Turhan, B. et al. 2010. What do we know about test-driven development? *IEEE Software.* 27(6): 16–19.

Sjøberg, D., Johnsen, A., and Solberg, J. 2012. Quantifying the effect of using Kanban versus scrum: A case study. *IEEE Software.* 29(5): 47–53.

Smith, M., Miller, J., Huang, L., and Tran, A. 2009. A more agile approach to embedded system development. *IEEE Software.* 26(3): 50–57.

Spearman, M., Woodruff, D., and Hopp, W. 1990. CONWIP: A pull alternative to Kanban. *International Journal of Production Research.* 28: 879–894.

Srinivasan, J., Dobrin, R., and Lundqvist, K. 2009. 'State of the Art' in using agile methods for embedded systems development. *Proceedings of COMPSAC 2009*, Seattle, Washington, Vol. 2, pp. 522–527.

Standish Group International Inc. 1994. Chaos. Technical report. Boston, MA: The Standish Group International Inc. http://www.csus.edu/indiv/v/velianitis/161/ChaosReport.pdf. Accessed October 1, 2013.

Sterling, C. 2010. *Managing Software Debt: Building for Inevitable Change.* Agile Software Development Series. Upper Saddle River, NJ: Addison-Wesley Professional.

Stevens, M. and Rosenberg, D. 2003. *Extreme Programming Refactored: The Case against XP.* Berkeley, CA: Apress.

Sullivan, K., Chalasani, P. Jha, S., and Sazawal. V. 1999. Software design as an investment activity: A real options perspective. In *Real Options and Business Strategy: Applications to Decision Making*, Trigeorgis, L. ed., London, U.K.: Risk Books.

Sutherland, J., Jacobsen, C., and Johnson, K. 2007. Scrum and CMMI Level 5: A Magic Potion for Code Warriors. http://jeffsutherland.com/2007/09/scrum-and-cmmi-level-5-magic-potion-for.html. Accessed October 1, 2012.

Sutherland, J. and Schwaber, K. 2011. The Scrum Guide. http://www.scrum.org/Scrum-Guides. Accessed October 1, 2011.

Turhan, B., Layman, L., Diep, M., Erdogmus, H., and Shull, F. 2010. How effective is test-driven development? In *Making Software What Really Works, and Why We Believe It*, A. Oram and G. Wilson (Eds.). Sebastopol, CA: O'Reilly Media.

USDOD. 2010. United States Department of Defense: A New Approach for Delivering Information Technology Capabilities in the Department of Defense. http://www.afei.org/WorkingGroups/section804tf/Documents/OSD_Sec_804_Report.pdf. Accessed October 1, 2012.

Weinberg, G. 1971. *The Psychology of Computer Programming*. New York: Van Nostrand Reinhold.

Williams, L. and A. Cockburn. 2003. Agile software development: It's about feedback and change. *IEEE Computer*. 36(6): 39–43.

Williams, L. and Kessler, R. 2002. *Pair Programming Illuminated*. Boston, MA: Addison-Wesley.

Williams, L., Kessler, R., Cunningham, W., and Jeffries, R. 2000. Strengthening the case for pair programming. *IEEE Software*. 17(4): 19–25.

Womack, J., Jones, T., and Roos, D. 1990. *The Machine That Changed the World: The Story of Lean Production*. New York: Free Press.

Wray, S. 2010. How pair programming really works. *IEEE Software*. 27(1): 50–55.

Resources

The World Wide Web has vast resources available on agile methods. Some of the most useful and well-known resources are listed as follows:

http://retrospectivewiki.org/index.php?title=Main_Page.
http://www.agilealliance.org.
http://www.scrumprimer.com.
http://www.leanprimer.com.
http:// www.xprogramming.com/index.php.

85

Service-Oriented Development

Andy Wang
Southern Illinois University Carbondale

Guangzhi Zheng
Southern Polytechnic State University

85.1 Introduction

In software engineering, service-oriented development, or service-oriented development of applications (SODA), is a process and activity of developing applications and systems based on service-oriented architectures (SOA). The concept of SOA has received significant attention in the software architecture and development field since the early twenty-first century. SOA is a type of software architecture based on using services available on a network. The fundamental building blocks of the system are services. Services are commonly viewed as self-contained software application modules exposed through interfaces over a distributed environment. The consumers of these services are not end users but rather software applications.

SOA has been emerging as a major integration and architecture framework in today's complex and heterogeneous computing environment (Mahmoud 2005). It represents "a paradigm for organizing and utilizing distributed capabilities that may be under the control of different ownership domains" (MacKenzie and Laskey 2006). Compared with other architecture styles, SOA has the following features and benefits:

- Services have well-defined interfaces that are independent of its implementation. Thus, the clients are primarily concerned with what the services will provide rather than how these services will execute their requests.
- Services are loosely coupled for greater level of flexibility and reuse.
- Services can be dynamically discovered.
- Composite services can be built based on service aggregation.

Service-oriented development has evolved along with a number of computing and software development fields, which have directly or indirectly contributed to the development of SOA and in turn been impacted by SOA.

85.1.1 Component-Based Software Engineering or Component-Based Development

Component-based software engineering (CBSE), or component-based development (CBD), is a branch of software engineering that emphasizes the building of applications based on software components (Wang and Qian 2005). A software component is a reusable piece of code that encapsulates a set of related functions and data and communicates with other parts of the system via interfaces. The concept first became prominent with Douglas McIlroy's NATO conference address on software engineering in 1968 (Mcilroy 1969). The approach is characterized by reusability, development focusing on defining, implementing, and composing loosely coupled independent components into systems. Popular implementations of components include Enterprise Java Beans and web services. Software components are considered to be part of the emerging platform for service orientation. Services can be built from service components and inherit further characteristics that were not part of an ordinary component.

85.1.2 Distributed Computing and Components

Distributed computing is a field that studies distributed systems that consist of multiple autonomous computers that communicate through a computer network. A computer program that runs in a distributed system is called a distributed program or application. Components then can be accessed and used by multiple programs across multiple computers. The significance of the change is the way of communicating with components over the network. The major change of the environment leads to major changes on interfaces and communication protocols and other issues like persistency, state, and synchronization. Microsoft's DCOM and OMG's Common Object Request Broker Architecture (CORBA) are two major architectures supporting the distributed components.

SOA can be viewed as an evolvement of distributed components to a more generic and loosely coupled level. In some cases, they are similar. For example, both services and components could support abstraction (component via an interface, service via a contract) and thus with loose coupling. But component-based architecture (CBA) and SOA are different:

- CBA is based on component models and technologies. There is no universally accepted component standard up to the present (Szyperski 2002).
- SOA is at a higher level and CBA is more dependent of detailed implementation and bind to a framework/environment (Sessions 2004). Both SOA and CBA are similar in principle, but they are different in implementation.
- It is difficult for component-based systems to work over Internet firewalls as they usually use nonstandard ports.

85.1.3 Web Services

A web service is a software program designed to support interoperable machine-to-machine interaction over a network (Haas and Brown 2004). It is also built based on the software components exposed through standard web interface called services. Web services play a pivotal role in a majority of SOA initiatives and are supported by the major development tool and enterprise package vendors (Haines and Rothenberger 2010). However, web services itself are not equal to SOA. It is one implementation that can realize an SOA.

An SOA may be realized with technologies other than existing web services standards. However, web service has been the most promising technology for SOA and for the foreseeable future (Vogels 2003).

85.2 Underlying Principles

Software development involves several stages such as analysis, design, implementation, testing, integration, and maintenance. There are many different paradigms to develop software, each with its strength and limitations. These paradigms could also be called software development styles, approaches, methods, or principles, because each paradigm defines uniquely its own approach to conducting and producing system requirements analysis, system architecture, implementing language and environment, testing methodology, and maintenance activities. For instance, object-oriented software development uses "object" as a fundamental concept during development process. The key of object-oriented development is to identify objects and specify objects during the analysis phase, design a hierarchy of objects with well-defined relationships, and implement those objects with high-performance code. Another example is component-oriented development, which is similar to object-oriented development but more general and more powerful in terms of software reusability and granularity. A component is a self-contained and self-deployable unit, while an object is self-contained but not self-deployable.

Service-oriented development is a new software development paradigm, which has its own system specification language, its own system design approach, and its own implementation and testing methodology. In a service-oriented development process, all development activities evolve around the central concept called "services." A service is a software component, a piece of self-contained, self-deployable computer program with well-defined functionality and can be integrated with other services through its interface. Therefore, service-oriented development can be viewed as a special case of component-oriented development, if we recognize services as reusable components.

With the advances of technologies in computing and networking, software development has evolved from machine-code programming, structured programming, object-oriented programming, to component-oriented programming. Software components are interchangeable units encapsulating functionality and providing services through well-defined interfaces. A service is a software component. Thus, the underlying principles of service-oriented development follow the principles of component-oriented development. This section introduces some general principles of service-oriented development, followed by a detailed discussion of "service infrastructure," including principles and models for service-oriented development.

85.2.1 General Principles for Service-Oriented Development

85.2.1.1 Principle 1: Services Represent Decomposition and Abstraction

The basic and effective strategy for tackling any large and complex problem in software development is "divide and conquer." One major idea in service-oriented development is to create software modules that are themselves self-contained and independently deployable. Thus, different developers will be able to work on different services independently, without needing much communication among developers, and yet the services will work together seamlessly. In addition, during software maintenance phase, it will be possible to modify some of the services without affecting all of the other services. When we decompose a system, we factor it into separable services in such a way that each service is at the same level of details; each service can be solved independently; and the implementations of these services can be integrated to satisfy the original system requirements.

Abstraction is a way to do decomposition productively by changing the level of detail to be considered. Service components hide certain details in an effort to provide only necessary information to the clients through their interface. The strategy of abstracting and then decomposing is typical in software development process.

85.2.1.2 Principle 2: Reusability Should Be Achieved at Various Levels

Software exists in different forms throughout the software development process. At the modeling and analysis phase, the requirements specification is seen as a form of software. In the design phase, architectural design and detailed design documents are part of the software. Source code and executable code deployed to the customer site are certainly software. Therefore, software reusability includes the reuse of any software artifacts in various forms. In service-oriented development, there are five forms of software components, namely, service specification, service interface, service implementation, installed services, and service objects. Each form of the service components could be reused in different stages of software life cycle.

85.2.1.3 Principle 3: Service-Oriented Development Increases the Software Dependability

The dependability of a computing system relies heavily on the trustworthiness of its software part, as opposed to its hardware part. Service-oriented development provides a systematical way to achieve dependable systems. Owing to the abstraction of services and systematic integration of services, it is much easier to validate critical requirements and verify safety for the service-based systems. On the other hand, reusable services have usually been tested through the validation process and real usage for a long time and, therefore, their quality can be assured.

85.2.1.4 Principle 4: Service-Oriented Development Could Increase the Software Productivity

Service-oriented software is constructed by assembling existing reusable components rather than developing from scratch every time—reuse instead of reinventing the wheel. This process is much faster than developing an application from scratch in most cases.

85.2.1.5 Principle 5: Service-Oriented Development Promotes Software Standardization

Service standards must be in place before a mature service market in software industry. Standards can be used for creating an agreement on concrete interface specifications, enabling effective composition, and ensuring service-oriented development to be a new programming paradigm in which "plug and play" becomes a reality in software development just like the hardware counterpart.

85.2.2 Service Infrastructure

A service infrastructure is the basic, underlying framework and facilities for service construction and service management. It consists of three models:

1. A service model
2. A connection model
3. A deployment model

A service model defines what a valid service is and how to create a new service under the service infrastructure. Service developers build reusable services according to the service model. Each service infrastructure has a reusable service library containing building blocks confirming to the service model. The connection model defines a collection of connectors and supporting facilities for service assembling. Thus, the connection model determines how to build an application or a larger component out of existing services. The deployment model describes how to put service components into a working environment. The deployment model deals with all issues related to service distribution and maintenance.

85.3 Best Practices

The concept of SOA has been around for more than 10 years. The field is matured with abundant best practices, design patterns, software products, development methodologies, and standards. In this section, we select and discuss some important practices: using reference architectures, adopting enterprise service bus (ESB), choosing service interface style, and adjusting development activities.

85.3.1 Reference Models and Architectures

SOA is just a general architecture style or paradigm that can lead to various different implementations. In the earlier years, even the understanding of basic SOA concepts and terminologies was inconsistent. Eventually many people realized that it is important to agree on a common understanding of basic concepts and semantics that can be used across different implementations. Reference models and ontologies (MacKenzie and Laskey 2006; The Open Group 2010) have been developed for this purpose. On top of that, reference architectures have been developed to provide high-level architectural that can be used as the basis of many implementations. The reference architectures incorporate the best practices and design patterns of developing SOA systems. Utilizing these reference architectures has a number of benefits:

- They can be used to set the context for a system, to provide a common foundation of understanding for developers and users to map products to and design services for, and to articulate a set of building blocks for solutions (Kreger et al. 2012).
- Reference architectures help produce well-designed system (in both technical and business perspectives).
- Reference architectures can shorten development cycles and increase software productivity.
- Reference architectures can reduce development cost and improve implementation quality.
- Reference architectures serve as the learning guide for junior developers.

There can be more than one reference architecture, and the actual implementation may not necessarily follow exactly one reference architecture. Reference models and architectures have been produced by standards bodies such as the Organization for the Advancement of Structured Information Standards (OASIS) and the Open Group, as well as big IT companies like IBM (Arsanjani et al. 2007), Microsoft,* and Accenture.† In the following sections, two notable efforts by two IT standards organizations are introduced.

85.3.1.1 OASIS SOA Reference Model and Reference Architecture

The OASIS, an IT industry standards body, has been working on the SOA Reference Model and Reference Architecture since 2005. The reference model (SOA-RM) was voted into standards in 2006 while the reference architecture (SOA-RA) is still under development. According to the OASIS, the reference model and reference architecture are different. A reference model aims to provide a higher and more abstract level of common understanding, while a reference architecture provides more detailed and concrete guidelines. A reference model consists of a minimal set of unifying concepts and relationships in a particular problem domain and is not directly tied to any standards, technologies, or other concrete implementation details, such as web services (MacKenzie and Laskey 2006). The OASIS SOA-RM is a conceptual and abstract framework that is used primarily for a common and consistent understanding of key entities and relationships between them. The OASIS SOA-RAF, on the other hand, provides more descriptive solutions to the problem of building an

* See http://msdn.microsoft.com/en-us/library/bb833022.aspx

† See http://www.accenture.com/cn-en/pages/service-technology-soa-reference-architecture-summary.aspx

actual SOA-based system. OASIS-RAF describes a complete SOA ecosystem including the context within which the system functions and the participants involved in making it function (OASIS 2012). It covers both IT and business perspectives in constructing, using, owning, and managing an SOA system.

Following IEEE's recommendation for architecture description (IEEE 2000), the OASIS-RAF describes the SOA development from three views and corresponding viewpoints (OASIS 2012):

1. *Participation in an SOA Ecosystem*: this defines an SOA ecosystem's context, structures, and actions. It identifies roles, responsibilities, and relationships of participants.
2. *Realization of an SOA Ecosystem*: this focuses on the infrastructure elements that are needed to support the construction of SOA-based systems. The viewpoint defines four models: service description, service visibility, service interaction, and policy.
3. *Ownership in an SOA Ecosystem*: this addresses the concerns involved in owning and managing SOA-based systems. Many of these concerns are not easily addressed by automation; instead, they often involve people-oriented processes. Four models are defined: governance, security, management, and testing.

It should be noted that the OASIS SOA-RAF is not a concrete guideline for building a specific SOA system. It is more of a view-based abstract reference architecture foundation that is used to guide other reference architectures and eventual concrete architectures. SOA-RAF models SOA from an ecosystem/paradigm perspective and is the first standard that extends SOA from a pure technical realm into the business world (Poulin 2006).

85.3.1.2 Open Group SOA Standards

The Open Group published a serial of SOA standards and guides including the Open Group SOA Ontology, the Open Group Service Integration Maturity Model (OSIMM), the Open Group SOA Reference Architecture, and others.*

The Open Group SOA Ontology defines SOA-related concepts, explaining what they are and how they relate to each other (The Open Group 2010). Compared to OASIS SOA-RM, it provides a more refined and formal specification of some of the core concepts (Kreger and Estefan 2009). The ontology is formally represented in web ontology language (OWL) to enable processing automation by directly using applications.

The Open Group SOA Integration Maturity Model (OSIMM) provides a way to assess an organization's SOA maturity level. The model consists of seven levels of maturity (silo, integrated, componentized, service, composite services, virtualized services, and dynamically reconfigurable services) based on the evaluation of seven dimensions of consideration that represent significant views of business and IT capabilities (business, organization and governance, method, application, architecture, information, and infrastructure and management). OSIMM defines a process of a staged and incremental adoption and implementation that maximizes business benefits at each stage. It acts as a quantitative model to aid in assessment of current state and desired future state of SOA maturity (The Open Group 2011a).

A little different from OASIS SOA-RA, the Open Group SOA Reference Architecture can represent both abstract enterprise designs as well as concrete SOA implementations (The Open Group 2011b). Originally based on IBM's SOA Solution Stack and Service-Oriented Modeling and Architecture (SOMA) (Kreger et al. 2012), the Open Group SOA-RA includes nine layers or nine key clusters of considerations and responsibilities that typically emerge in the process of designing an SOA solution or defining an enterprise architecture (Figure 85.1).

One key building block is the services layer. It is important to understand different types of services in the development of a service portfolio of an SOA. Categorizing services affects how both business

* These standards and guides can be access through The Open Group SOA Source Book available at http://www.opengroup.org/soa/source-book/intro/

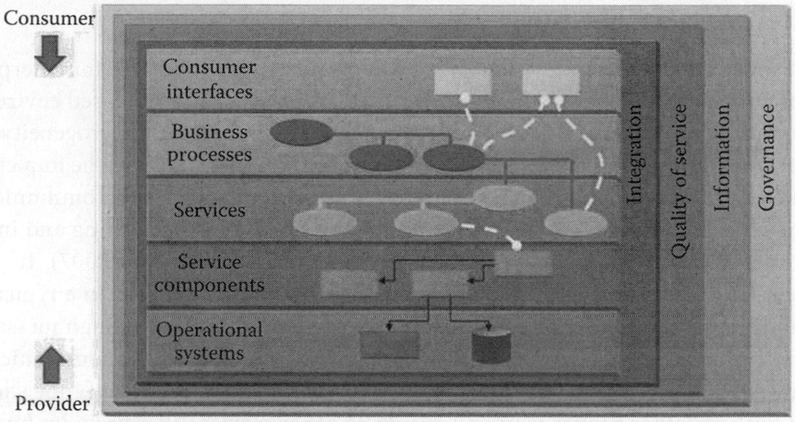

FIGURE 85.1 The Open Group SOA-RA logical solution view. (From The Open Group, SOA reference architecture technical standard, http://www.opengroup.org/soa/source-book/soa_refarch/, 2011b.)

and IT views and understands the architecture and the service portfolio that supports it. A common high-level classification based on logic type puts services into three categories (Erl 2007): entity services (which represents business logic related to business entities), task services (which represent narrower scoped business logic, or steps in a process, that are within a business entity), and utility services (which provide functionality that is common to many functionalities, such as event logging, notification, and exception handling).

The reference architecture provides a finer level categorization scheme for services, based on their functions and purposes (Figure 85.2). One of the significant services is the ESB, which will be described in more detail in the Section 85.3.2.

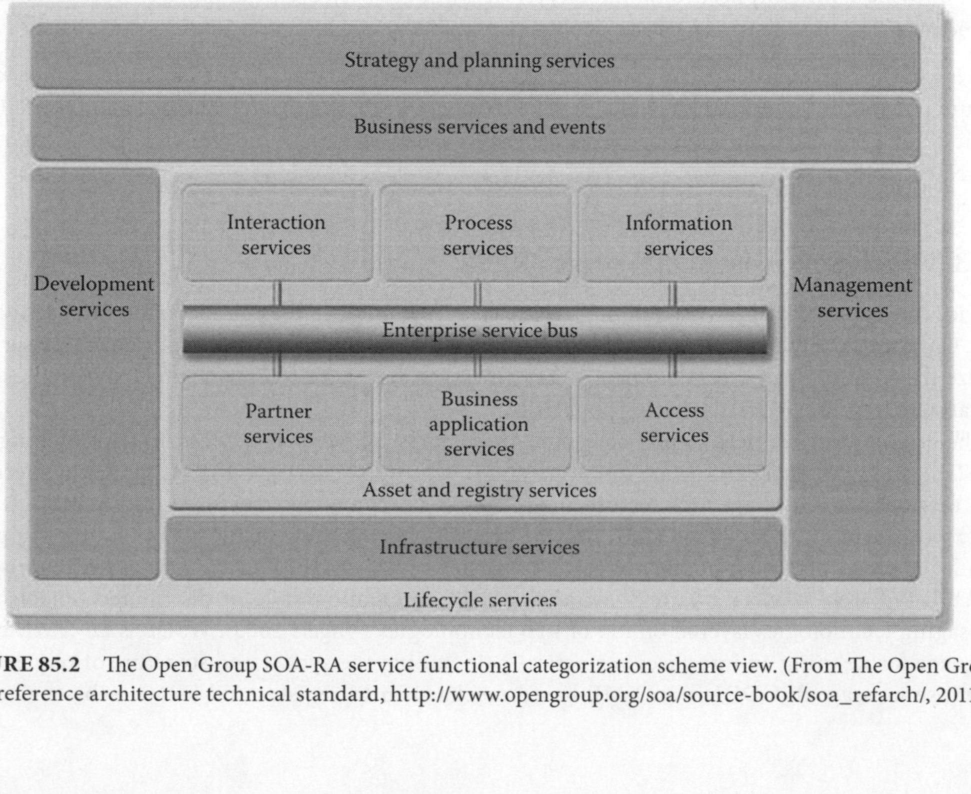

FIGURE 85.2 The Open Group SOA-RA service functional categorization scheme view. (From The Open Group, SOA reference architecture technical standard, http://www.opengroup.org/soa/source-book/soa_refarch/, 2011b.)

85.3.2 Enterprise Service Bus

Because of its loose coupling style and flexibility, SOA is particularly suitable for enterprise systems where various applications need to be integrated. The enterprise is rather a closed environment with management and infrastructure planning consistency, but technological heterogeneity still poses a big challenge. One of the key practices in designing application integration is the implementation of ESB, which is also included in the Open Group Reference Architecture as a communication channel. An ESB is considered as an SOA implementation pattern used for designing and implementing the interaction and communication between services (Flurry and Reinitz 2007). It is one of the key components for messaging in the field of enterprise system integration. In a typical enterprise environment, different applications need to communicate to one another though messages. Instead of having applications to communicate directly end to end, ESB provides a communication bridge for all services. However, ESB should not just be perceived as the SOA itself, nor should ESB be implemented without other proper plan and design of other services that provides business utility (Woolf 2007).

The key concept in ESB is the bus. In ESB-based architecture, the bus acts as the single message path between applications. Every client directs all its requests through the ESB instead of passing it directly to a central broker. Because the communication is usually accomplished using messaging techniques, it is also called a message bus (Hohpe and Woolf 2003). In the traditional enterprise application integration, a centralized message broker is also implemented as a communication bridge. But ESB is different in that the message translation is not centralized to the broker. Each application uses adapters to conform to a canonical message format.

ESB is not a standard and can provide different functions and capabilities in different implantations. The following are usually considered to be the basic set of capabilities of ESB:

- Message queue and routing
- Data transformation and mapping
- Monitor and control routing of messages
- Provide commonly needed commodity services like security, event handling, exception handling, protocol conversion, and service quality monitoring

ESB is usually implemented as a message communication service. It is provided as an enterprise architecture integration tool by vendors like IBM (WebSphere ESB), Microsoft (BizTalk), and Oracle (BEA Logic). There are also open source projects such as Apache ServiceMix and Synapse, JBoss, Open ESB, and NServiceBus.

85.3.3 Web Services: SOAP versus REST

Services are exposed and accessed through interfaces. The interface is a contract of how to use the service. Currently, the most common and reliable implementation is web services. Traditional web services utilize the simple object access protocol (SOAP) message, but more recently another style representational state transfer (REST) is gaining popularity.

The original web services standards include three parts: SOAP, web services description language (WSDL), and universal description discovery and integration (UDDI). SOAP is a protocol specification that defines a message architecture and format for exchanging structured information. It uses an extensible markup language (XML) for its message format and usually relies on other application layer protocols, most notably hypertext transfer protocol (HTTP), for message transmission. SOAP-based web services mirror the traditional IT integration style of distributed objects with messaging technologies layered on top of web technologies (Vogels 2003). WSDL is an XML-based language for defining an interfaces syntactically (including the syntax and structure of request and response messages). UDDI is an XML-based mechanism to register and locate web services on the Internet. All three standards utilize the XML language and can be processed directly by software programs, thus effectively hiding the perceived complexity from the application programmer and integrator (Pautasso et al. 2008).

More recently, another more lightweight style REST has been gaining popularity among many web mash-up applications. REST is an architectural style for distributed resources, introduced by Roy Fielding (Fielding 2000). REST requests and responses between servers and clients are built around the representation (addressing and transferring) of resources (Rodriguez 2008). The World Wide Web is the most well-known example of the REST architecture. A web service implemented based on REST principles is often called a RESTful web service. A concrete RESTful web service follows four basic principles (Rodriguez 2008):

- Explicitly use HTTP methods (GET, POST, PUT, DELETE) for operations.
- Resources are exposed through directory structure-like URIs.
- Resources are usually represented by standard format like XML, JavaScript Object Notation (JSON), or both.
- Be stateless. Stateful interactions are based on the concept of explicit state transfer (Pautasso et al. 2008). A number of techniques exist to make HTTP stateful, including URL rewriting, session cookies, and hidden form fields, all of which embed the state information in response messages for future reference.

Web services are very ambitious, providing extensibility and processing models, with extension of dealing asynchrony, reliability, integrity, confidentiality, etc. It is protocol independent although almost majority is implemented with HTTP. Service interfaces are more complex but more functional. Compared to the heavyweight full stack of web service, REST is a lightweight implementation of services over the web. It is simple in implementation, resource oriented, with simple interfaces. The necessary infrastructure for REST has become pervasive. The effort required to build a client to a RESTful service is also very small. Developers can begin testing such services from a common web browser, without having to develop custom client-side software, and deploying a RESTful web service is very similar to building a dynamic web site (Pautasso et al. 2008). Furthermore, it is possible to discover RESTful web resources without a centralized repository approach that requires registration like UDDI.

However, REST is not a substitute for web services. Rather, it is an alternative style that could be considered as an alternative approach for designing SOA. It has been perceived as a way to provide web services with less dependence on proprietary middleware (e.g., an application server) than the "big" web services. Exposing a system's resources through a RESTful interface is more flexible to meet integration requirements where data need to be combined easily among different kinds of applications (e.g., the mash-ups). Generally, REST is more widely chosen for simple web applications over the Internet where consumer clients are mostly unknown, while SOAP web services are used more within an organization's enterprise system. A quantified method of choosing REST- or SOAP-based service can be found in a tutorial at the 19th International Conference on World Wide Web (Pautasso and Wilde 2010).

85.3.4 SOA Development Activities Adjustment

SOA not only brings technical changes but also business and organizational changes; therefore, the development of SOA needs to make adjustments in development phases and activities. Generally in SOA development, there are three basic tasks: building a service infrastructure, building services, and consuming services in applications. These activities are different in terms of a development methodology. On one hand, the infrastructure is the overall basis upon which the whole system is built. Its changes are more likely to have a greater impact on other applications that use the infrastructure. Services, or at least their interface, are used by multiple applications. They need to be well designed and planned for relatively small changes. Because of this stability requirement, building the service infrastructure and services may require a more rigorous approach with careful planning and design (Haines and Rothenberger 2010). On the other hand, the internal implementation of services is more independent of how their interfaces, and composite applications built on services need to quickly respond to changing requirements. They require a more flexible and shorter development cycles. This may in many circumstances benefit from using a more agile software development approach (Haines and Rothenberger 2010).

Some traditional development activities need to be adjusted because of this duality of SOA development. Certain activities may need to be added, emphasized (or de-emphasized), or rearranged. Figure 85.3 (Haines and Rothenberger 2010) summarizes some key adjustments in the typical development activities in the system development life cycle.

The most radical changes are probably in the design phase due to the paradigm shift. In earlier systems, when the technology and infrastructures were not matured, incorporating too many remotely located services in a composite orchestration would be considered a bad design. Now especially in an internal architecture, service management is more matured and reliable; more loosely coupled system is encouraged to advocate flexibility and agility. One of the important considerations of service design is to determine the granularity of services. Services that are too fine-grained and require many roundtrips to complete a business transaction may cause performance problems, while services that are too coarse-grained limit reusability and may result in the wrong applications being built (Haines and Rothenberger 2010).

Phase Adjustments	Important Characteristics and Changes
Planning *substantial*	• More holistic, enterprise-level • Broader more effective communication channels needed • Higher uncertainty, project estimation more difficult • Assessment of relevant standards and vendor support
Analysis *lowest*	• Requires rigorous, disciplined traditional analytics • Development of common data models and schemas is a key task
Design *most radical*	• Development of reusable enterprise-level components • Interface design must be explicit part early in the lifecycle that requires learning about standards and tool availability • Determining granularity of services • Limiting remote services (based on current remote management capabilities) • Developing effective interfaces descriptions
Implementation *some*	• Need for increased automation with deployment tools • Increased need for coordination with more independent stakeholders • More complex service management • Implementing multiple channels for business partners with different maturity levels of Web services • Awareness of standards maturity and interoperability issues
Testing *substantial*	• Emphasis on integration testing and different environments • Need for increased automation with testing tools, including generation of test clients as well as test services • Compliance testing (e.g., WS-I) • Increased performance and security testing

FIGURE 85.3 Changes needed in the SOA development. (From Haines, M.N. and Rothenberger, M.A., *Commun. ACM*, 53(8), 135, 2010.)

Another phase with substantial changes is the testing phase. Some best practices are (Ahuja 2009) as follows:

- Maintain automated service integration test suites.
- Adopt a service testing framework.
- Use mock services to test consumers ahead of integration with providers.
- Use test suites to adequately test services ahead of consumer integration.
- Always plan for performance testing.

85.4 Research Issues

As we discussed earlier, a service infrastructure consists of three parts: a service model, a connection model, and a deployment model. Each of these three models has research issues and open questions. Research issues related to a service model include the following:

- How do we formally define a service? A service is a reusable component with properties, references, functional and nonfunctional requirements, and implementation details. These are the basic elements of business function units, which are assembled into complete business solutions by the connection model of deployment model. Therefore, it is essential to define these basic services in a precise and concise manner. Many service infrastructures provide both text and graphical representations of a service component. For instance, the picture that follows was used by OASIS (OASIS 2011). A concise text representation of a service component helps to define service infrastructure as a formal system; thus, formal methods could be applied to service oriented architecture (Figure 85.4).

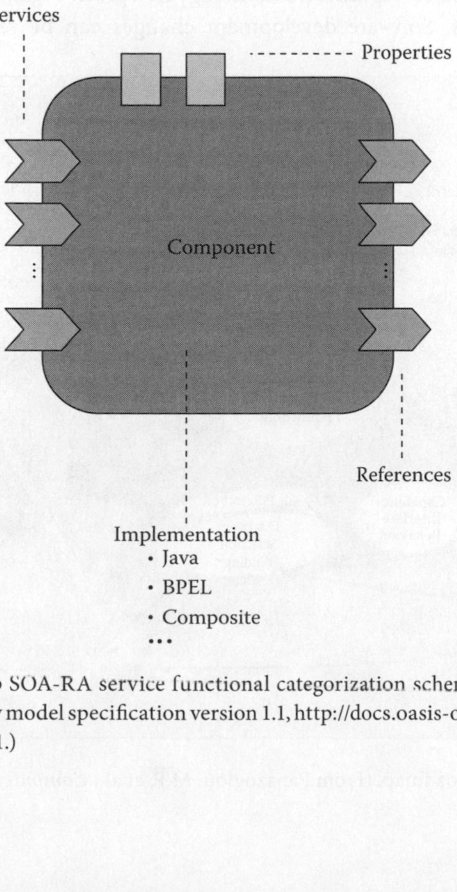

FIGURE 85.4 The Open Group SOA-RA service functional categorization scheme view. (From OASIS, Service component architecture, assembly model specification version 1.1, http://docs.oasis-open.org/opencsa/sca-assembly/sca-assembly-spec-v1.1.html, 2011.)

- Connections as operations. The connection model of a service infrastructure is the most interesting part for further research. The connection model defines ways to combine new and preexisting service components together into a unified system. These rules of assembly could be described as operations of a component algebra (Wang and Qian 2005), and they are the abstractions to represent different composable aspects of the system.
- Deployment as service binding. The deployment model defines the manner in which a service can be made available and how it can be utilized in different applications. It may also define possible methods of access, for instance, through web services, JMS, RMI- IIOP, or EJB. Nonfunctional requirements could be captured by the deployment model as specific bindings such as security, time constraints, transaction, and reliability.

Currently web services represent an active research area. A web service is an online component that can be described by its interface expressed with WSDL, deployed by a deployment tool such as WSDD, published in a registry such as UDDI, located by a SOAP lookup based on UDDI API, and invoked by a SOAP request. Web services are loosely coupled, contracted components that communicate via XML-based WSDL web service interface.

The development and management of SOA (e.g., methodologies) is also a good research area. The software development process is changed due to the adoption of SOA. What is the impact of such changes? How do we assess them? Future research may investigate the productivity and cost reduction by conducting a comparative analysis of traditional development processes and SOA-driven processes. However, beyond the question of whether SOA is good or bad, it is more worthwhile to study how it can be done right in a given context (Haines and Rothenberger 2010).

Papazoglou et al. (2007) envisions an service oriented computing (SOC) roadmap (Figure 85.5) and put the development and management aspects on top of the research pyramid. Several issues were raised in this chapter and we think the following will be the focal point of future research. First, more research could focus on a specific phase or a specific methodology in a given organizational context using detailed case studies or field surveys. Software development changes can be related to underlying technical

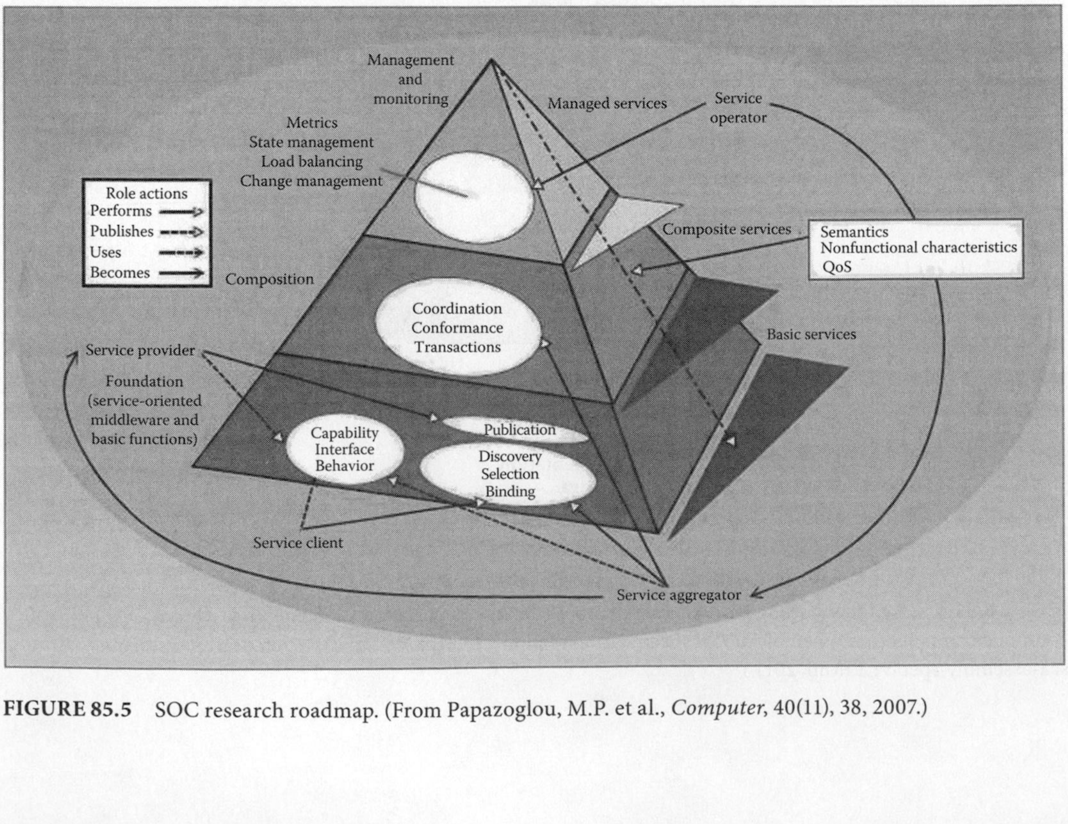

FIGURE 85.5 SOC research roadmap. (From Papazoglou, M.P. et al., *Computer*, 40(11), 38, 2007.)

characteristics to examine if recommendations made for SOA also apply to other development paradigms. Second, the management becomes a critical issue as the number of services grows. Third, it is important to have a highly coordinated process and to consider testing throughout the life cycle. Tools that help automating the process play an important role in effective testing by addressing not only the functional issues but also security and performance.

85.5 Summary

SOA has been around for more than a decade now. It has become a fundamental enabler for many enterprise technologies and systems, such as software as a service (SaaS) and cloud computing. It is also evolving beyond the traditional remote-procedure-call style to encompass more interaction forms as found in REST, web-oriented architecture, and event-driven architecture (Thompson 2009). Reference models, architectures, and patterns have proliferated to guide the service-oriented development. The development also goes beyond simple technical mandate to a high-level conceptual thinking that influences the architectural design. The question nowadays is not just focused on whether or not SOA is a better paradigm but more on how it can be implemented in the right way in a given context.

Key Terms

Enterprise service bus: ESB is an SOA implementation pattern used for designing and implementing the interaction and communication between services.

RESTful web services: Web services implemented based on REST principles.

Service: A service is a software component, a piece of self-contained, self-deployable computer program with well-defined functionality and can be integrated with other services through its interface.

Service-oriented architecture: SOA is a software architecture style utilizing distributed capabilities that may be under the control of different ownership domains. It provides a uniform means to offer, discover, interact with, and use capabilities to produce desired effects consistent with measurable preconditions and expectations. The fundamental building blocks of the system are services.

Service-oriented development: Service-oriented development is a new software development paradigm that develops software systems based on SOAs.

Web services: A web service is a software program designed to support interoperable machine-to-machine interaction over a network.

Further Information

Web Resources

Organizations, standards, and company resources:

- http://www.oasis-ws-i.org
- http://oasis-opencsa.org/sca
- http://www.oasis-open.org/committees/tc_home.php?wg_abbrev=soa-rm
- http://www3.opengroup.org/subjectareas/soa
- http://www.soaspecs.com
- http://www-01.ibm.com/software/solutions/soa
- http://www.oracle.com/us/technologies/soa/index.html
- http://scn.sap.com/community/soa
- http://www.tibco.com/products/resource-centers/service-oriented-architecture

Magazine, community, and reports:

- http://www.servicetechmag.com
- http://soa.sys-con.com

- http://searchsoa.techtarget.com
- http://www.ebizq.net
- http://www.soainstitute.org
- http://www.webservices.org
- http://www.forrester.com/SOA-%26-Web-Services

Journals and conferences:

- http://www.icsoc.org
- http://www.servicetechsymposium.com
- http://www.computer.org/portal/web/tsc

Others:

- http://soapatterns.org
- http://www.soabooks.com
- http://www.soaschool.com
- http://www.thomaserl.com
- http://www.udidahan.com

Books

- Benatallah, B., Casati, F., and Traverso, P. (Eds.), *Service-Oriented Computing—ICSOC 2005*, Lecture Notes in Computer Science, Vol. 3826, Springer, Amsterdam, the Netherlands, 2005, ISBN 978-3-540-30817-1.
- Erl, T., *Service-Oriented Architecture: A Field Guide to Integrating XML and Web Services*, Prentice Hall, Upper Saddle River, NJ, 2004, ISBN 978-0131428980.
- Erl, T., *Service-Oriented Architecture: Concepts, Technology, and Design*, Prentice Hall, Upper Saddle River, NJ, 2005, ISBN 978-0131858589.
- Erl, T., *SOA Principles of Service Design*, Prentice Hall, Upper Saddle River, NJ, 2007, ISBN 978-0132344821.
- Erl, T. et al., *SOA Design Patterns*, January 2009. http://www.amazon.com/dp/0136135161#_
- Marks, E. A. and Bell, M., *Service-Oriented Architecture (SOA): A Planning and Implementation Guide for Business and Technology*, John Wiley & Sons, New York, 2006, ISBN 978-0471768944.
- Singh, M. P. and Huhns, M. N., *Service-Oriented Computing: Semantics, Processes, Agents*, John Wiley & Sons, West Sussex, U.K., 2005, ISBN 978-0470091487.
- Sommerville, I., *Software Engineering*, 6th edn., Addison-Wesley Professional, Reading, MA, 2001.
- Thomas E. et al., *Web Service Contract Design and Versioning for SOA*, Prentice Hall, Upper Saddle River, NJ, September 2008.

References

Ahuja, S. 2009. Adopting SOA best practices and lessons learned. http://www.ibm.com/developerworks/webservices/library/ws-SOAbestpractices/index.html (accessed on October 12, 2012).

Arsanjani, A., L-J. Zhang, M. Ellis, A. Allam, and K. Channabasavaiah. 2007. Design an SOA solution using a reference architecture. http://www.ibm.com/developerworks/library/ar-archtemp/ (accessed on October 12, 2012).

Erl, T. 2007. *SOA Principles of Service Design*, 1st edn. Prentice Hall, Upper Saddle River, NJ. http://www.whatissoa.com/soamethodology/p5.php (accessed on October 12, 2012).

Fielding, R. T. 2000. Architectural styles and the design of network-based software architectures, University of California, Irvine.

Flurry, G. and R. Reinitz. 2007. Exploring the enterprise service bus, part 2: Why the ESB is a fundamental part of SOA. http://www.ibm.com/developerworks/webservices/library/ar-esbpat2/ (accessed on October 12, 2012).

Haas, H. and A. Brown (eds.). 2004. Web services glossary. W3C. http://www.w3.org/TR/ws-gloss/ (accessed on October 12, 2012).

Haines, M. N. and M. A. Rothenberger. 2010. How a service-oriented architecture may change the software development process. *Commun. ACM* 53(8): 135–140. doi:10.1145/1787234.1787269.

Hohpe, G. and B. Woolf. 2003. *Enterprise Integration Patterns: Designing, Building, and Deploying Messaging Solutions*, 1st edn. Addison-Wesley Professional, San Diego, CA.

IEEE. 2000. IEEE recommended practice for architectural description of software-intensive systems. *IEEE Std 1471–2000*: i–23. doi:10.1109/IEEESTD.2000.91944.

Kreger, H., V. Brunssen, R. Sawyer, A. Arsanjani, and R. High. 2012. The IBM advantage for SOA reference architecture standards. http://www.ibm.com/developerworks/webservices/library/ws-soa-ref-arch/ (accessed on October 12, 2012).

Kreger, H. and J. Estefan. 2009. Navigating the SOA open standards landscape around architecture white paper. The Open Group. http://www.opengroup.org/soa/source-book/stds/index.htm (accessed on October 12, 2012).

MacKenzie, C. M. and K. Laskey. 2006. Reference model for service oriented architecture 1.0. Standard. OASIS. http://docs.oasis-open.org/soa-rm/v1.0/ (accessed on October 12, 2012).

Mahmoud, Q. H. 2005. Service-oriented architecture (SOA) and web services: The road to enterprise application integration (EAI). http://www.oracle.com/technetwork/articles/javase/soa-142870.html (accessed on October 12, 2012).

Mcilroy, D. 1969. Mass-produced software components. *Proceedings of Software Engineering Concepts and Techniques*, pp. 138–155. Garmisch, Germany. http://www.cs.dartmouth.edu/~doug/components.txt (accessed on October 12, 2012).

OASIS. 2011. Service component architecture, assembly model specification version 1.1. http://docs.oasis-open.org/opencsa/sca-assembly/sca-assembly-spec-v1.1.html (accessed on October 12, 2012).

OASIS. 2012. Reference architecture foundation for service oriented architecture version 1.0. OASIS Committee Specification 01.04 December 2012. http://docs.oasis-open.org/soa-rm/soa-ra/v1.0/soa-ra.html (accessed on January 10, 2013).

Papazoglou, M. P., P. Traverso, S. Dustdar, and F. Leymann. 2007. Service-oriented computing: State of the art and research challenges. *Computer* 40(11): 38–45. Doi:10.1109/MC.2007.400.

Pautasso, C. and E. Wilde. 2010. RESTful web services: Principles, patterns, emerging technologies. *Proceedings of the 19th International Conference on World Wide Web, WWW '10*, pp. 1359–1360. New York: ACM. doi:10.1145/1772690.1772929. http://doi.acm.org/10.1145/1772690.1772929.

Pautasso, C., O. Zimmermann, and F. Leymann. 2008. Restful web services vs. 'big" web services: Making the right architectural decision. *Proceedings of the 17th International Conference on World Wide Web, WWW '08*, pp. 805–814. New York: ACM. doi:10.1145/1367497.1367606. http://doi.acm.org/10.1145/1367497.1367606.

Poulin, M. 2006. Considering the SOA reference model, part 1: Business grounds. http://java.sys-con.com/node/314124 (accessed on October 12, 2012).

Rodriguez, A. 2008. RESTful web services: The basics. http://www.ibm.com/developerworks/webservices/library/ws-restful/ (accessed on October 12, 2012).

Sessions, R. 2004. Fuzzy boundaries: Objects, components, and web services. *Queue* 2(9): 40–47. doi:10.1145/1039511.1039533.

Szyperski, C. 2002. *Component Software: Beyond Object-Oriented Programming*, 2nd edn., Addison-Wesley Professional, Boston, MA.

The Open Group. 2010. SOA ontology technical standard. The Open Group. http://www.opengroup.org/soa/source-book/ontology/index.htm (accessed on October 12, 2012).

The Open Group. 2011a. OSIMM version 2 technical standard. http://www.opengroup.org/soa/source-book/osimmv2/index.htm (accessed on October 12, 2012).

The Open Group. 2011b. SOA reference architecture technical standard. http://www.opengroup.org/soa/source-book/soa_refarch/ (accessed on October 12, 2012).

Thompson, J. 2009. Q&A: An SOA sanity check in difficult times. Gartner. http://www.gartner.com/it/page.jsp?id=927612 (accessed on October 12, 2012).

Vogels, W. 2003. Web services are not distributed objects. *IEEE Internet Computing* 7(6): 59–66. doi:10.1109/MIC.2003.1250585.

Wang, A. J. A. and K. Qian. 2005. *Component-Oriented Programming*. John Wiley & Sons, Chichester, U.K.

Woolf, B. 2007. ESB-oriented architecture: The wrong approach to adopting SOA. http://www.ibm.com/developerworks/webservices/library/ws-soa-esbarch/ (accessed on October 12, 2012).

86

Software Product Lines

Jorge L. Díaz-Herrera
Keuka College

Melvin Pérez-Cedano
Construx Software

86.1 Introduction

Due to ever-shorter life cycles, maintaining leadership in software-intensive product development organizations depends increasingly on the ability to improve their design and development processes faster than their competitors. There is clearly a need to considerably reduce development effort by increasing productivity and consistently generating higher-quality software.

It is a well-known fact that in more established engineering disciplines, the use of standard parts and known solutions to recurrent design problems is routine and contributes toward the reproducibility and reliability of the solutions produced. By contrast, in the software development universe, there is a strong tendency to start from scratch each time a new project is embarked on, with software continually being reinvented to perform essentially the same function.

A promising approach is to move the focus from building single systems, where one repeatedly custom-writes routine functionality, to orchestrating families of systems in a range of similar products. Focusing on a "range of similar" products can potentially take advantage of *economies of scope*, a benefit that comes from developing one asset used in multiple contexts by identifying "reusable" solutions that support future development of multiple systems, thus, making it possible to derive individual systems in a prescribed manner instead of creating them anew. Moves toward systematic, institutionalized reuse have shown that similar results can be obtained in the software industry as proposed by software product line (SPL) engineering.

A group of related software-intensive systems sharing a *managed* common set of software assets explicitly recognizing *variability* among *similar* user requirements, or *features*, for the same product is considered a SPL. SPL development is attractive because it enables *strategic and systematic software reuse*, that is, software reuse that is planned and aligned with a business strategy [1]. Strategic software reuse increases software development productivity and improves the quality of the products derived while allowing an organization to expand its products portfolio rapidly. The set of common features is used for developing a platform—a set of core reusable assets—that provides mechanisms for introducing individual system variations among the group of related systems by combining mass production with individual customization.

Over the past several years, issues associated with SPL have attracted the attention of a relatively high number of software engineering researchers and practitioners. However, we do know from experience that the SPL paradigm has not been widely adopted in industry. Why is this the case? What are the impediments to adopting SPL in industry? Correspondingly, what software processes are "required" for effective and efficient SPL development?

In this chapter, we answer these and other related questions to the SPL paradigm. In the rest of this chapter we present an overview of SPL development and best practices and a discussion and arguing that to effectively implement SPL, organizations need to achieve a high level of maturity in the process improvement scale. To buttress this argument, we describe the Software Engineering Institute (SEI)-sponsored Framework for Software Product Line Practice and compare its key practices with those found in the Capability Maturity Model Integration for Development (CMMI-DEV). Finally, we close with some of the most relevant contributions in the SPL research over time and highlight unresolved research issues. We conclude with an analysis of SPL adoption and research challenges, outlining the main issues and unsolved problems that impede the widespread adoption of the SPL approach as a way to deliver software products more effectively.

86.2 About Software Product Lines

The overall goal of SPL development is to produce *quality products consistently and predictably* and at the same time greatly *increase productivity.* A key realization is that most software systems developed conventionally, that is, "in isolation, routinely written from scratch" are not completely new or unique [2,3]. That is to say, software-intensive systems satisfying specific needs of a particular market segment or mission share a common set of "features" [4]. Several focused conferences, books, and more general publications have covered SPL [5]. An annual workshop on product family engineering (PFE) started in Europe in 1996 [6], and the international conference on SPLs (SPLC) started in the United States in 2000 [36]. Both events were held separately until 2004 when they merged to form the leading conference in the subject [7].

An underlying assumption is that the benefits from the reuse of components will offset the potential extra cost for any increased organizational and design complexity. The benefits of SPL are clear, provided that assets can be systematically reused several times, including the following:

- Organizations reduce cycle time and cost of new applications by eliminating redundancy.
- Building systems from a common component base reduces risk and improves quality by using trusted, proven components repeatedly.
- An asset-based approach allows the management of legacy systems more efficiently, increasing the likelihood of longer time-IN-market; the organization evolves a common marketing strategy and strengthens its core competency around strategic business interests and goals [8].
- Despite an initial investment on creating a common platform, the approach reduces time to market and costs significantly when compared to developing the corresponding set of systems each independently and in isolation [9].

86.2.1 Software Reuse: An Elusive Goal

Since the dawn of computer programming, different approaches have been developed for exploiting the benefits of software reuse. Programmers have practiced *ad hoc* code reuse since the earliest days of computer programming: they have always reused sections of code, functions, and procedures. As a recognized area of study in software engineering, software reuse dates from 1968 when McIlroy, from Bell Labs, proposed at the now famous NATO conference on software engineering, basing the software industry on reusable components [10].

Earlier definitions of software reuse and reusability include the following:

> Re-use is considered as a means to support the construction of new programs using in a systematical way existing designs, design fragments, program texts, documentation, or other forms of program representation [11].
>
> Reusability is the extent to which a software component can be used (with or without adaptation) in multiple problem solutions [12,13].

Over the decades, more organized forms of software reused, although still not in a systematic way, focused on larger and larger assets: subroutine libraries in the 1960s, modules in the 1970s, objects in the 1980s, components in the 1990s, and web services after that. These approaches represent significant advances in achieving the elusive goal of software reuse; they have indeed increased reusability but only incrementally and in an *ad hoc*, small-scale way—for example, reuse on the basis of a single system. Code from one project can be saved in a "reuse" library in the hope that it will be useful in the future; but unplanned, miscellaneous collections of code components simply fail to achieve high-leverage reuse. They are difficult to locate, understand, and modify, since typically design information is unavailable and adaptability is not designed in. This need for design for commonality and for control of variability results in the introduction of product lines [6,9].

SPL proposes an intentional, systematic, and predictable large-scale software reuse paradigm. Of course, the approach stands on the benefits of the early contributions previously mentioned, which were primarily programming techniques, plus other important ones as discussed later in this chapter.

It is not until systematic reuse is conquered that we will see software reuse widely accepted, and a key differentiator is domain engineering (DE) [14], a process of collecting, organizing, and modeling the knowledge in a domain in the form of reusable assets, "to model software system families and build software modules such that, given particular requirements specs, highly customized and optimized intermediate or end products can be constructed on demand [15]."

Systematic software reuse is the organized creation of common parts or assets with controlled variability that forms the basis for systematically building systems in a domain of expertise by assembling them from these reusable assets, a dual process of DE for building assets and applications engineering for building systems of these assets. When product lines are designed to support an organization's objective or mission, we are institutionalizing reuse [16]. The achievement of design for commonality and control of variability requires the establishment of a reuse infrastructure as discussed next.

86.2.2 Domain Engineering (DE)

In his PhD thesis [17], Neighbors introduces the term domain analysis (DA) and Draco, which is considered the first approach to DE. With Draco he also introduces the idea of domain-specific languages, currently receiving renewed attention [18].

Systematic reuse requires "original engineering design" work at front, done once, followed by "routine practice" [19]. The former known as Domain Engineering (DE) refers to a *development-for-reuse* process effort necessary to build a reusable asset base as *model solutions*. The latter complementary process, known as application engineering (AE), refers to a *development-with-reuse* process to create specific systems from these model solutions and prefabricated assets. Within this duality products are created by instantiating models and by integrating prefabricated artifacts, mapping needs to solutions rather than building from scratch, that is, not building products from first principles. This duality provides a sound basis for industrial-strength software engineering, a necessary condition that naturally leads to systematic reuse [20]. Fundamentally, reuse has the effect of reducing variance and increasing homogeneity, two essential aspects observed from disciplines that have moved into industrialization.

Modeling is an important practice in increasing homogeneity. As Shaw has put it, supporting the modeling first develop second duality (the underlining added):

> Engineering relies on codifying knowledge about a technological problem domain in a form that is directly useful to the practitioner, thereby providing answers for questions that commonly occur in practice.... Engineers of ordinary talent can then apply this knowledge to solve problems far faster that they otherwise could [21].

In this way, development requires original design effort only once while the applications engineering process uses the known solutions repeatedly to create different systems in two ways:

- *Adaptive design*: use known, established solution principles adapting the embodiment to the requirements. It may be necessary to perform original design on some individual parts.
- *Variant design*: arrangements of parts and other product properties (size) vary within the limits set by a previously designed product structure during original engineering (DE).

DE requires careful technical and economic analyses; it involves the discovery of solutions to solve reoccurring problems and goes through all of the development phases of (domain) analysis, (domain) design, (domain) implementation, and testing.

During domain analysis commonality and variability among the systems in the domain are identified and made explicit. Commonality and variability are often traced back to Parnas' work on program families [22], although Parnas himself acknowledges Dijkstra's work on structured programming [23] as the root of his ideas. See [24] for a survey of DA methods.

Commonalities and variabilities of an SPL are typically specified in terms of characteristics visible to the end user called *features*. Kang's work on feature-oriented DA [25] introduces feature models for specifying and modeling relationships among features. Feature models have been further extended and have become the de facto standard for modeling SPL high-level functional requirements. Other important contributions in DA are commonality analysis [26] and SCV (scope, commonality, and variability) [27].

The design activities are supported by system and software architecture technology, most notably architecture description languages (ADLs) [28]. Component design is supported by interface definition languages (IDLs). Jacobson [29] introduced the notion of variation point for identifying locations within software artifacts where variability can occur, including a comprehensive list of variability mechanisms for realizing variation points in the implementation [6,30].

Design-commonality forms the basis for standardizing assets to build products in a domain of expertise by encapsulating common features of related products into a domain model and by defining a common architecture for related products into a domain (or reference) architecture [31]. A specific design is created by instantiating a common design and the product implemented by the identified reusable components.

Control-variability is the basis for providing flexibility in the assets to meet requirements for a variety of products without compromising commonality; it requires careful design to include appropriate levels of parameterization, generalization and specialization, and extension. Like commonality, adaptability must be engineered a priori, and thus analysis must explicitly identify variations that anticipate adaptations specifying optional components, alternate structures, parameterized context dependencies, and the like.

Reuse occurs both within a product line and across product lines, a notion discovered earlier and associated with the concepts of *horizontal* and *vertical* reuse. Assets that cross several systems take advantage of "economies of scope," a benefit that comes from developing one asset used in multiple contexts [32]. Horizontal reuse refers to the use of an asset across several distinct product lines; typically, assets reused in this way tend to be general (i.e., they are application domain independent) and with a very specific set of functionality.

86.2.3 Process Framework for SPL

A framework for practicing SPL has been developed by the SEI [33], and a hall of fame for recognizing representative experiences applying SPL has also been established [34]. The SPL framework provides details of how to perform SPL engineering. It divides the development process into two major categories: core asset development and product development with a management infrastructure.

The SEI SPL framework has 29 practice areas that are divided into three categories: software engineering, technical management, and organizational management as shown in Table 86.1.

Contrast this with CMMI's framework that consists of 25 key process areas (KPAs) and four categories of project management, process management, engineering, and support as shown in Table 86.2. The CMMI-DEV is a software development improvement model that provides an organization with a roadmap for improving their development process. The model is a collection of best practices that help organizations to dramatically improve the effectiveness, efficiency, and quality of their software development practices. CMMI-DEV describes five levels of maturity and provides an assessment instrument to measure an organization's software development capabilities against the five-point scale.

TABLE 86.1 SPL Practice Areas by Category

Software Engineering	Technical Management	Organizational Management
Architecture definition	Configuration management	Building a business case
Architecture evaluation	Data collection, metrics, and tracking	Customer interface management
Component development	Make/buy/mine/commission analysis	Developing an acquisition strategy
COTS utilization	Process definition	Funding
Mining existing assets	Scoping	Launching and institutionalizing
Requirements engineering	Technical planning	Market analysis
Software system integration	Technical risk management	Operations
Testing	Tool support	Organizational planning
Understanding relevant domains		Organizational risk management
		Structuring the organization
		Technology forecasting
		Training

TABLE 86.2 CMMI Process Category and KPAs

Project Management	Process Management	Engineering	Support
1. Project planning	9. Organizational process focus	14. Requirements management	20. Configuration management
2. Project monitoring and control	10. Organizational process definition	15. Requirements development	21. Process and product quality assurance
3. Supplier agreement management	11. Organizational training	16. Technical solution	22. Measurement and analysis
4. Integrated project management (IPPD)	12. Organizational process performance	17. Product integration	23. Causal analysis and resolution
5. Integrated supplier management (SS)	13. Organizational innovation and deployment	18. Verification	24. Decision analysis and resolution
6. Integrated teaming		19. Validation	25. Organizational environment for integration (IPPD)
7. Risk management			
8. Quantitative project management			

A KPA is a body of work or a collection of activities that an organization must master to successfully achieve their goals for the process area. Each KPA has a set of specific goals and specific practices as well as a set of generic goals and generic practices. Specific goals and practices relate "what" must be done to meet the process purpose. The generic goals and practices relate to "how well" the process is carried out. The generic practices guide an organization to move to higher levels of capability.

There are six levels of process capabilities defined in CMMI as follows:

- Level 0: Process performance is incomplete.
- Level 1: Process is performed that means specific practices are performed.
- Level 2: Process is managed, that is, there is a policy that indicates you will perform it. There is a plan for performing it; there are resources provided, responsibilities assigned, training on how to perform it, and selected work products from performing the process area are controlled.
- Level 3: Process is defined, that is, the process is established and maintained organizationally.
- Level 4: Process is quantitatively managed; that is, the defined process is controlled using statistical and other quantitative techniques. A critical distinction between a defined process and a quantitatively managed process is the predictability of the process performance.
- Level 5: Process is optimized with defect prevention, proactive improvement, innovative technology insertion, and deployment.

It is important to distinguish between *process capability* and *organizational maturity*. A single process has capabilities that can be measured on a 1–5 scale (listed earlier). An organization consists of groupings of processes; how well these processes perform their function is a measure of that organization's maturity. If an organization is rated level 4, for example, that implies that all its processes have reached at least capability 4. Maturity levels feed into each other on an upward trajectory. CMMI provides a path to mastering a practice area with the goal of achieving level 4 capability that is quantitatively controlled and continuously improving. We suggest that a CMMI equivalent of level 4 capability is needed to achieve mastery of SPL practices.

Of particular importance to us is level 4 maturity since we believe that is the necessary level of maturity needed to effectively implement product lines. Level 4 maturity (capability) involves use of statistical and other quantitative methods to understand past and predict future quality and process performance. Through a mature measurement system, organizations establish objectives for quality and process performance based on their business objectives. Projects establish their objectives based on those of the organization and the needs of customers and end users. Projects and individuals use statistical and quantitative methods in their activities to plan, monitor, and control progress against their objectives. Organizations use the resulting information to understand process performance, understand variation, target areas for continuing improvement, and evaluate the impact of proposed improvements.

CMMI-DEV and its predecessor CMM have been in use in a variety of organizations globally since 1987. As a result, a wealth of experiential data is available to support the notion that higher maturity results in higher productivity, higher quality, and shorter cycle time. Boeing Australia, as an example, has reported a 33% decrease in the average cost to fix a defect, reductions by half in the amount of time required to turn around releases, and a 60% reduction in work and fewer outstanding actions following pretest and posttest audits [35].

Therefore, CMMI-DEV has a proven migration path guiding organizations toward increasing process capability and organizational maturity. CMMI-DEV, however, contains only "what" to do and is void of "how" to do it. Since the majority of SPL practice areas map into one or more CMMI-KPA, it is prudent to leverage CMMI experience and create a level 4 management infrastructure wrapped around core asset development and product development processes. There are equivalences among the majority of CMMI-KPA's and SPL's practice areas (Table 86.3).

TABLE 86.3 CMMI-KPA versus SPL Practice Areas

SPL Practice Area	CMMI-KPA
Requirements engineering	Requirements management Requirements development
Software system integration	Product integration
Testing	Verification Validation
Configuration management	Configuration management
Process definition	Organizational process definition
Technical planning	Project planning Project monitoring and control
Technical risk management	Risk management
Data collection, metrics, and tracking	Measurement and analysis Organizational process Performance
Architecture definition Architecture evaluation Component development COTS utilization	Technical solution
Make/buy/mine/commission analysis	Decision analysis and resolution

86.3 Research Issues

SPL, although an active research area in software engineering with some recent contributions from the research community, still presents key technical problems that slow down its widespread adoption in industrial settings [36]. Additional work in this direction includes obviously conducting a systematic literature review of the top problems reported and identifying the most influential solutions. What are the main issues and unsolved problems that impede the widespread adoption of the SPL approach as a way to deliver software products more effectively?

Assessing the status of research in this area helps the research community to focus the agenda on the critical problems limiting the advancement of the field. In early studies, including a mapping analysis of the literature, some findings that called our attention were the following:

- *Feature modeling* was ranked high among participants' current research areas (see below). But, according to our study, feature modeling is not one of the top problems.
- *Product derivation* was not considered one of the top problems, but it is the second most active research area.
- *SPL evolution*, the second most important problem we found, has not been featured explicitly in any of the important conferences.

In this section, we report on our findings from a survey directed to understand some of these questions [37]. The main goal was (1) to collect opinions among longtime SPL researchers and practitioners regarding critical problems in SPL adoption, (2) to verify how actively the research community has been working on these problems, and (3) to quantify the SPL research field and find answers to additional questions like the following:

- Which of the previously reported problems remain unsolved?
- How actively has the research community been working on solving these problems?
- What new problems have emerged?

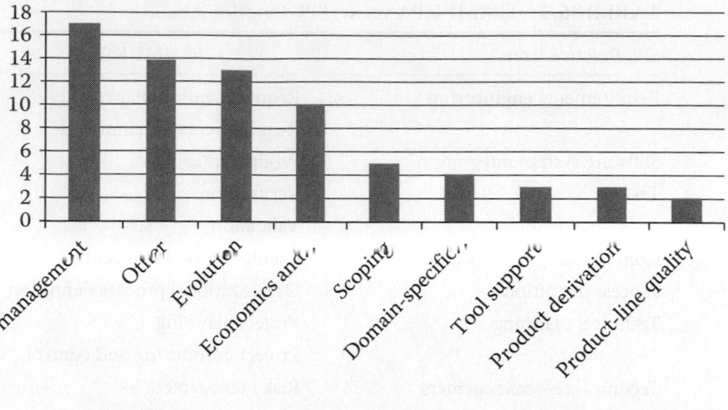

FIGURE 86.1 Top SPL research problems.

To answer these two questions, a mixed empirical study was conducted by combining an expert-focused survey with a systematic mapping study. Expert sampling [38] is useful because knowledgeable researchers and practitioners in SPL are easy to identify but also because data can be collected and coded quantitatively promptly. Researchers are typically ranked based on publications [39–41]. The complementary content analysis and literature review provided more insight on how actively the research community have been working on solving the identified problems.

The findings corroborate previous assessments, citing variability management and evolution among the most important research problems in SPL. Figure 86.1 shows the topmost SPL research problems uncovered by the survey.

Although opinions varied widely, 40 out of 86 reported problems fell into three big groups. These were variability management, evolution, and economic and organizational aspects. Their breakout was as follows:

- *Variability management*, with a frequency of nearly 20% (17 responses), is clearly the top research problem.
- The second most important problem is *evolution* with 15% (13 responses).
- *Economic and organizational* issues were third with 12% (10 responses).
- Others included a plethora of smaller individual problems and counted for 16% of the responses.

86.3.1 Variability Management

In software, variability can be defined as the quality, state, or degree of a software system or artifact to be extendable, changeable, customizable, or configurable for being used in a particular context [42]. In SPL, variability is higher than in conventional software systems, because multiple software systems are derived from a common set of reusable assets. That is to say, reusable assets must anticipate variations for handling differences across the systems in a product line. Furthermore, the driving force behind SPL is that the effort required for introducing these variations is minimal. To achieve this goal, variability needs to be managed following a process for identifying, introducing, constraining, using, implementing, and evolving variations and managing the variants [43,44].

Bosch et al. [45] describe several variability issues that materialize in different phases of the SPL engineering life cycle. The root causes of most of these SPL variability issues are complexity management, visibility of variability, and dependency management.

Complexity management. The value behind an SPL stems from the degree of variability required. The number of possible configurations for an unconstrained feature model is O (2n), where n is the number of variable features; this number of configurations grows polynomially in the number of versions available. Although variability provides flexibility and enriches an SPL, when thousands of options exist, it is difficult to effectively identify the possible configurations that make sense both technically and commercially. Therefore, complexity management is critical for managing an SPL with a large number of features with complex interdependencies, asset variants, and variation points associated with product configurations. As in conventional software, complexity is an essential problem that cannot be completely solved [46]. Typical approaches for taming complexity in SPL include reducing unnecessary variability [47] and clustering variation points into packaged variants.

Visibility of variability. Documenting variability is one of the key properties that characterize SPL engineering. Variability is made explicit from the outset through variation points that are realized in further stages of production. A product line of industrial size can easily incorporate several thousand variation points, which imposes serious issues in modeling, assessing the impact of changes, and in ensuring consistency in such large amount of variability [48]. Although modeling techniques developed in recent years provide first-class representation of variation point, there is no evidence of its scalability in industrial-size SPL. Furthermore, an expected benefit of modeling and managing variability is to be able to derive products at mass production costs—that is, to automate the derivation of products as much as possible. Achieving this goal requires more visibility of the variation mechanisms selected for realizing the variation points.

Dependency management. A critical problem is the lack of efficient mechanisms for maintaining a mapping between the problem space and the solution space. The problem space constitutes a configuration space typically defined using feature models. However, the solution space is comprised of conventional implementation artifacts with no precise traceability to features. This impedance mismatch between features and implementation artifacts occurs because a single feature can be scattered over multiple artifacts (feature scattering) or several features can be tangled into a single component (feature tangling). Without an explicit mapping between features and implementation artifacts, it is problematic to assure feature-wise systems derivation and evolution.

86.3.2 SPL Evolution

Arguably, products and core assets in SPL evolve more often than in other kind of non-software-intensive product lines (e.g., manufacturing). Product lines require an efficient control of evolutionary changes in both the core assets and the derived products. This is due to the higher degree of interdependency between core assets and products. Components in existing instances of the SPL architecture and derived products in use are frequently replaced by new versions. A lack of an effective control of changes deteriorates the SPL and drains its reuse benefits.

Since an SPL is usually aligned with a business strategy, its evolution is driven by changes in the marketed products, for example, features are added, changed, or removed, or new products are added to the product line. It is thus critical to be able to identify accurately the affected implementation artifacts when feature-based changes occur and, conversely, to identify the features affected by changes introduced in implementation artifacts. Having proper dependency management between features and implementation artifacts is key for maintaining an SPL feature-wise consistent. Existing variation management approaches (see above) support SPL evolution on a solution artifact basis but to the best of our knowledge do not provide feature-based evolution.

A textual analysis of the papers published in the SPLCs from 2006 to 2008 identified the most frequently used keywords as shown in Figure 86.2.

This frequency analysis included all the words containing more than five characters that appeared more than 10 times in a paper. Predictably, the keyword "feature" is the topmost frequent word, but it is considered an outlier. Feature modeling is the de facto standard modeling technique used in SPL and it is very likely that any paper about SPL uses this word very often.

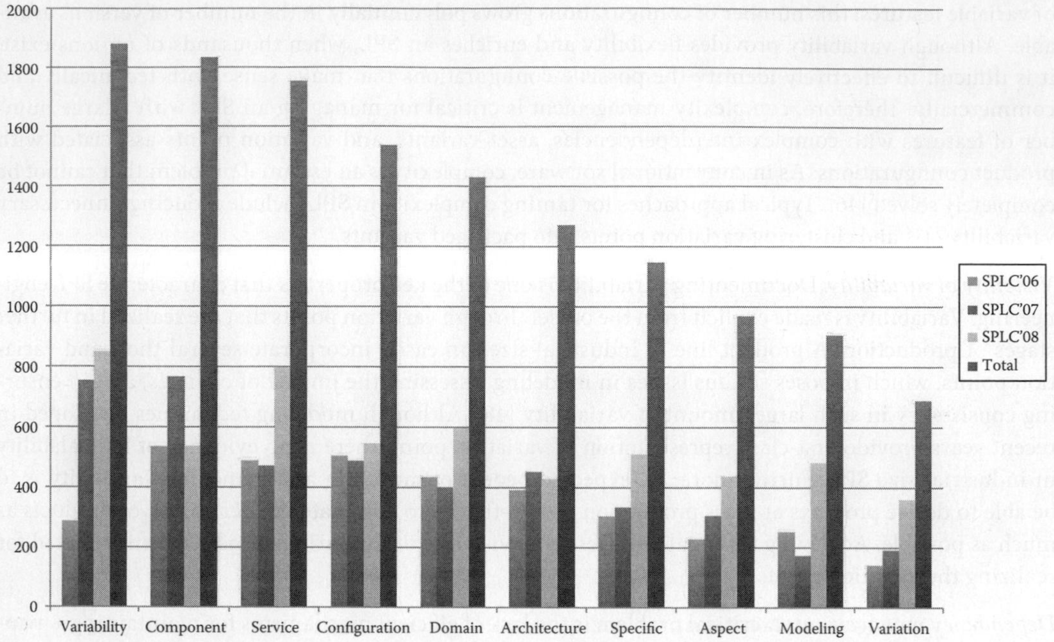

FIGURE 86.2 Top 10 keywords used in SPLC papers.

"Variability" is the most frequent keyword in the last three SPLCs. This result is consistent with its importance in solving SPL adoption problems. "Variation," usually associated with "variability management," is within the top 10 most frequently used keywords we found.

"Derivation" and "evolution" were not explicitly among the most frequently used keywords. However, the word "configuration," the fourth more used keyword found, can be associated with derivation and evolution—that is, product derivation is usually expressed in terms of building configurations, and configuration management usually supports software evolution.

Some other frequent keywords were "component," "service," and "aspect." This reflects the interest of integrating SPL with new technology and paradigms like component-based development, service-oriented architecture, and aspect-oriented approaches.

86.4 Summary and Conclusions

Software gets bigger all the time, becoming very costly to create and modify, and the software industry is very large and growing at unprecedented rates. This enormous demand for software is fundamentally based on two opposing aspects, namely, the critical nature of software for today's well-functioning world and the poor state of the practice in software development.

Plummeting hardware costs are leading to rapid growth in many new application areas requiring more software being built in shorter time frames. In today's post-PC era, new artifacts and systems are increasingly more software-driven. Many artifacts, from consumer electronics (e.g., smart phones, personal digital assistants, set-top boxes), transportation systems (including automotive, avionics, trains, etc.), to industrial plants and corporations, all depend on well-engineered software for their continuing functioning.

The other critical reality is the poor record of accomplishment in software construction: software seldom performs as expected, and cost/schedule overruns are very common even in the most sophisticated organizations. This further increases demand. What is worse, it is quite typical that for many such systems each time a new application is built, routine functionality is custom-written repeatedly from scratch. This also contributes to higher demands for more software. Concomitant with these problems, there is a shortage of qualified software professionals [49].

Software is critical to business' strategic advances, and it has become the bottom line for many organizations, even those who never envisioned themselves to be in the software business. Yet, major software projects failure rates are very high, and, although difficult to quantify the financial cost arising from these low success rates, estimates are in the billion of dollars. There is also a serious impact on all aspects of our lives, national economies, and society in general. Software development organizations must improve both the quality of the software they produce and the productivity of their development groups.

For these software-intensive systems, a reuse-driven approach will potentially reduce time to market while improving product quality and reducing uncertainty on cost and schedule estimates. The approach would allow software professionals to come up with the "right" solution quickly and effectively by assembling applications out of "proven" components. The quality of products is maintained through rigorous analysis and prediction, continuous improvement and refinement, and designed experimentation and feedback from observations of actual systems in a clearly coordinated role in meeting a common need. This represents a main motivation for SPL: a desire to modularize a system in such a way that there is a high probability that significant portions of the system can be built from standard parts.

Quality and budgetary issues with software have their root cause in human error. One proven approach [50] to improving the way software is developed is by knowing what you do and how you do it, as described in the CMMI-DEV published by Carnegie Mellon's SEI. In this chapter we placed the CMMI-DEV in juxtaposition with the also SEI-sponsored Framework for Software Product Line Practice. The former is a proven approach to enhance the quality of software products in general, whereas the latter provides a discipline framework for implementing SPLs, a promising approach that increases productivity multifold yet to be widely adopted by industry.

Our conclusion in this area is that this is a winning combination for the production of quality software predictably and within budget and time constraints.

In the SPL conference held in 2006, SPL research status has been assessed based on the insights of recognized experts in a panel with some of the world's leading SPL researchers. Panelists presented their perspectives regarding lessons learned and outcomes from the prior 10 years as well as directions and potential outcomes for the following 10 years. The presented topics have much in common with the results we obtained in a study presented in this chapter corroborating our findings. In particular, the following were in common:

- Variability management topics included "managing variability in requirements engineering," "introducing variability into software architectures," and the need of novel technologies for setting up "variation mechanisms automatically."
- Evolution included the challenges of "controlling versions of components and architectures" with multiproduct nature and "how to keep products unified overtime."
- Organizational and business issues for adopting SPLs.
- Other topics, including tooling and integration with new technologies such as ubiquitous computing, service-oriented architecture, and model-driven approaches. Participants in the survey also ranked "integration with other technology" as the fourth most cited technical research problem.

Some additional findings that called our attention were as follows:

1. *Feature modeling*, a topic ranked third among participants' current research areas, but feature modeling is not considered one of the top problems.
2. *Product derivation* was not considered one of the top problems, but it is the second most active research area.
3. *SPL evolution*, the second most important problem we found, has not been featured explicitly in any of the SPLCs.

References

1. L. Northrop, Software product lines: Reuse that makes business sense, In *Australian Software Engineering Conference*, Sydney, Australia, 2006, p. 1.
2. W. B. Frakes and K. Kang, Software reuse research: Status and future, *IEEE Transactions on Software Engineering* 31 (2005) 529–536.
3. V. Sugumaran, S. Park, and K. C. Kang, Introduction, *Communications of the ACM* 49(12) (2006) 28–32.
4. P. Clements and L. Northrop, *Software Product Lines: Practices and Patterns,* The SEI Series in Software Engineering, Boston, MA: Addison-Wesley, 2002.
5. ACM, Software Product Line, *Communications of the ACM* 49(12) (2006) 78–81.
6. K. Pohl, G. Bockle, and F. van der Linden, *Software Product Line Engineering: Foundations, Principles, and Techniques*, New York: Springer, 2005.
7. Software product lines conferences, World Wide Web electronic publication. URL http://splc.net
8. J. L. Díaz-Herrera, P. Knauber, and G. Succi, Issues and models in software product lines. *International Journal of Software Engineering and Knowledge Engineering*, 10(4) (2000) 527–539.
9. D. M. Weiss and C. T. R. Lai, *Software Product-Line Engineering: A Family-Based Software Development Process*, Reading, MA: Addison-Wesley, 1999.
10. M. D. McIlroy, (January 1969). Mass produced software components. Software Engineering: Report of a conference sponsored by the NATO Science Committee, Garmisch, Germany, October 7–11, 1968. Scientific Affairs Division, NATO. p. 79.
11. E. M. Dusink and van J. Katwijk, Reflections on reusable software and software components. Ada components: Libraries and tools. In *Proceedings of the Ada-Europe Conference, Stockholm*. Ed. S. Tafvelin, Cambridge, U.K.: Cambridge University Press, 1987, pp. 113–126.
12. J. W. Hooper and R. O. Chester, *Software Reuse, Guidelines and Methods*, New York: Plenum Press, 1991.
13. S. Katz et al., *Glossary of Software Reuse Terms*, Gaithersburg, MD: National Institute of Standards and Technology, 1994.
14. J. L. Díaz-Herrera, Domain engineering. In *Handbook on Software Engineering & Knowledge Engineering*, Vol. I, S. K. Chang (ed.), River Edge, NJ: World Scientific Pub. Co., December 2001.
15. K. Czarnecki and U. W. Eisenecker, *Generative Programming: Methods, Tools, and Applications*, New York: ACM Press/Addison-Wesley Publishing Co., 2000.
16. S. Cohen, Friedman, S., Martin, L., Solderitsch, N., and Webster, R. *Product Line Identification for ESC-Hanscom*. CMU/SEI-95-SR-024, Pittsburgh, PA: Software Engineering Institute, Carnegie Mellon University, 1995.
17. J. M. Neighbors, Software construction using components, PhD thesis, Department of Information and Computer Science, University of California, Oakland, CA, 1980.
18. M. Fowler, *Domain-Specific Languages*, New York: Addison-Wesley, 2011.
19. G. Pahl and W. Beitz, *Engineering Design: A Systematic Approach*. Berlin, Germany: Springer-Verlag, 1996.
20. J. L. Diaz-Herrera, S. Coehn, and J. Withey, Institutionalizing systematic reuse: A model-based approach. In *Proceedings of the Seventh Workshop on Institutionalizing Software Reuse*, Chicago, IL, 1995.
21. M. Shaw and D. Garlan, *Software Architectures: Perspectives on an Emerging Discipline*, Upper Saddle River, NJ: Prentice-Hall, 1996.
22. D. L. Parnas, On the design and development of program families, *IEEE Transactions on Software Engineering*, 2 (1976) 1–9.
23. O. J. Dahl, E. W. Dijkstra, and C. A. R. Hoare, *Structured Programming*, London, U.K.: Academic Press Ltd., 1972.

24. G. Arango, Domain Analysis Methods. In *Software Reusability*. Chichester, U.K.: Ellis Horwood, 1994, pp. 17–49.
25. K. C. Kang, S. G. Cohen, J. A. Hess, W. E. Novak, and A. S. Peterson, Feature-oriented domain analysis (foda) feasibility study, Technical Report CMU/SEI-90-TR-21, Software Engineering Institute, 1990.
26. Ardis, M. A., and Weiss, D. M. Defining families: the commonality analysis (tutorial). *Proceedings of the 19th International Conference on Software Engineering*. ACM, 1997. Boston, MA, May 17–23, 1997.
27. J. Coplien, D. Hoffman, and D. Weiss, Commonality and variability in software engineering, Software, *IEEE* 15(6) (1998) 37–45.
28. F. Hayes-Roth, *Architecture-Based Acquisition and Development of Software: ARPA Domain-Specific Software Architecture Program*, Palo Alto, CA: Te knowledge Federal Systems, 1994.
29. I. Jacobson, M. Griss, and P. Jonsson, *Software Reuse: Architecture, Process and Organization for Business Success*, New York: ACM Press/Addison-Wesley Publishing Co., 1997.
30. J. Bosch, *Design and Use of Software Architectures: Adopting and Evolving a Product-Line Approach*, Reading, MA: Addison-Wesley, 2000.
31. C. A. Gunter, E. L. Gunter, M. Jackson, and P. Zave. A reference model for requirements and specifications. *IEEE Software*, May/June (2000) 37–43.
32. J. Withey, *Investment Analysis of Software Assets for Product Lines*. CMU/SEI-96-TR-010. Pittsburgh, PA: Software Engineering Institute, Carnegie Mellon University, 1996.
33. L. M. Northrop and P. C. Clements, *A Framework for Software Product Line Practice, Version 5.0,* World Wide Web Electronic Publication (2007). URL http://www.sei.cmu.edu/productlines/framework/index.html (accessed October 10, 2013).
34. Product line hall of fame, World Wide Web Electronic Publication (2006). URL http://splc.net/fame.html (accessed October 10, 2013).
35. Michael Campo, "Why CMMI Maturity Level 5?" *CrossTalk*, January/February 2012. World Wide Web Electronic Publication (2012) http://www.crosstalkonline.org/storage/issue-archives/2012/201201/201201-Campo.pdf
36. International Software Product Line Conference (SPLC) 2006. Research panel (2006).
37. M. Pérez-Cedano, RIT, GCCIS research report. 2010. (Internal communication).
38. W. Trochim and J. Donnelly, *The Research Methods Knowledge Base*, 3rd Edn., Cincinnati, OH: Atomic Dog Publishing, 2007.
39. J. Ren and R. N. Taylor, Publications ranking, *Communications of the ACM* 50(6) (2007) 81.
40. W. Wong, T. Tse, R. Glass, V. Basili, and T. Chen, An assessment of systems and software engineering scholars and institutions (2001–2005), *The Journal of Systems & Software* 81(6) (2008) 1059–1062.
41. K. Y. Cai and D. Card, An analysis of research topics in software engineering, *The Journal of Systems & Software* 81(6) (2008) 1051–1058.
42. M. Svahnberg, J. van Gurp, and J. Bosch, A taxonomy of variability realization techniques, *Software—Practice and Experience* 35(8) (2005) 705–754.
43. J. van Gurp, J. Bosch, and M. Svahnberg, On the notion of variability in software product lines. In *Proceedings of the Working IEEE/IFIP Conference on Software Architecture*, Amsterdam, 2012.
44. M. Sinnema and S. Deelstra, Classifying variability modeling techniques, *Information and Software Technology* 49(7) (2007) 717–739.
45. J. Bosch, G. Florijn, D. Greefhorst, J. Kuusela, J. H. Obbink, and K. Pohl, Variability issues in software product lines, *Lecture Notes In Computer Science* 2290 (2001) 13–21.
46. J. Brooks, No silver bullet essence and accidents of software engineering, *Computer* 20(4) (1987) 10–19.
47. F. Loesch, Optimization of variability in software product lines. In *Proceedings of the 11th International Software Product Line Conference*, Kyoto, Japan, 2007, pp. 151–162.

48. C. W. Krueger, Variation management for software production lines. In *Proceedings of the Second International Conference on Software Product Lines*, 2002, pp. 37–48.
49. P. Freeman and W. Aspray, eds. (1999). The Supply of IT workers in the US. Computer Research Association, Washington, D.C., Special report.
50. D. R. Goldenson and D. L. Gibson. Demonstrating the impact and benefits of CMMI®: An update and preliminary results. Special Report: CMU/SEI-2003-SR-009. World Wide Web Electronic Publication, 2003. http://www.ralphyoung.net/articles/PI_Results_SEI.pdf

XII

Software Modeling, Analysis and Design

87

Requirements Elicitation

Daniel M. Berry
University of Waterloo

87.1 Introduction

The *requirements engineering* (RE) task [11,26–28,39,44,49,51] focuses on gathering information from members of a client's organization in order to produce a coherent specification of the requirements for a *computer-based system* (CBS) that the client desires to be built.

Requirements elicitation (RElic) is the subprocess of RE concerned with gathering information about the CBS to be built

- From whatever source it is available, including requests for proposals, organizational policy documents, documentation about the existing system, and videotapes, etc.,
- From whomever might provide information, including the client and the users, etc.,
- By whatever technique or means available, including interviews, questionnaires, focus groups, and observations of the organization at work, etc., even by invention.

Requirements analysis is the process of refining all the elicited information about a CBS into a *requirements specification* (RS) for the CBS by deriving what is possible from the information, validating what is derived with the client and the users, eliciting more information when questions prevent validation, and repeating all as necessary, until the RS can be written.

The more thorough RE and RElic are, the cheaper the subsequent development, because it is well known that fixing a CBS defect during implementation costs an order of magnitude more than fixing it during RE, and fixing the defect after deployment costs two orders of magnitude more than fixing it during RE [9].

This chapter focuses on RElic. Other chapters deal with specifications, including formal, of CBSs (Chapter 88), with formal methods (Chapter 71), and with model checking of formal specifications (Chapter 89).

87.2 Context of RElic and Vocabulary for the Rest of This Chapter

This chapter assumes that a client, *C*, has come to a *requirements analyst (RA)*, *A*, for the purpose of having *A* produce a RS, for a CBS, *S*, that *C* wants built. *S* is intended to solve a problem *P* that *C* and other people, mostly within *C*'s organization, *O*, are having. Almost always, *S*, the new CBS, replaces an existing system, *E*. *E* could be (1) a CBS built earlier or (2) a paper-and-word-of-mouth-driven *manual system*. The problem, *P*, may be that *E* does not do a needed something at all, or that *E* does it poorly. *P* may be that *E* does something that it should not do. Finally, *P* might even be only that *E* should be automated to take advantage of computers, the Internet, the Web, etc.

S is a *bespoke system*, generally developed for one client, *C*, or is a *mass-market system*, developed for many customers. (The distinction between client and customer is coming up.)

The set of people affected by *P*, that want *P* solved, that are using *E*, and that will use the new CBS, *S*, are *S*'s *stakeholders*. Among *S*'s stakeholders are

- *C*, the client, the person or organization paying for *S* to be developed,
- The customer, the person or organization who buys *S* after it is developed,
- Users of both *E* and *S*,
- Domain experts, who understand *P* and *E*,
- Developers, who will implement *S* from its RS,
- Quality assurers, the people who will do inspections and testing of all of *S*'s artifacts, including its RS and artifacts leading to the RS,
- Compliance inspectors, experts on government and safety regulations,
- Market researchers,
- Lawyers, and
- Experts on systems adjacent to *S*.

Note that sometimes the client and customer are the same; sometimes the client, customer, and user are the same; and sometimes the market researchers represent the customers, for example, for a mass-market CBS, in which case the client might be the head of the developing organization's marketing department, who decides that there is a market for *S*. In the rest of this chapter, the term "stakeholder" is used to mean all or much of the set of stakeholders that have been determined as relevant for the development of *S*. When only a specific stakeholder, for example, *C* or the users, is needed in the discussion, only that stakeholder is mentioned.

With the RS produced by *A*, *C* will go to a developer to arrange that *S* will be implemented according to the RS, to *C*'s satisfaction. It is intended that the RS will say enough about *S* that the developer will not have to ask *C* questions about *C*'s requirements for *S*; all the information about *S* that the developer needs to implement *S* is expressed in the RS.

In this context, *A*'s main jobs are

1. To understand *P*, which *C* wants solved by *S*,
2. To extract the essence of the stakeholders' requirements for *S*,
3. To invent a way to solve *P* that is better for the users than the way it is being solved currently by *E*, particularly if *E* is manual,
4. To negotiate with the stakeholders a consistent set of requirements for *S*, and
5. To record the resulting requirements in a RS for *S*.

RElic is used to carry out Jobs 1, 2, 3, and 4. That is, RElic is used to gather information from various sources to inform *A*'s jobs.

In the rest of this chapter, each of "*A*", "*C*", "*E*", "*O*", "*P*", and "*S*" is a proper noun naming a specific RA, CBS, existing system, organization, and problem, namely that in the specific RE task instance at hand. On the other hand, each of "CBS", "MS", "RA", and "RS" is a nonproper noun and is always used with an indefinite or definite article. When one of these nouns is preceded by a definite article and has neither a qualifying adjective nor a specific referent in a previous sentence, then the noun refers to the specific instance of the noun in the specific RE task instance at hand. Each of "RE" and "RElic" is a general noun, used without any article.

87.3 RE and RElic as Part of RE

RE is generally the first step in the development lifecycle of a CBS, especially if the so-called Waterfall Lifecycle Model [45] is followed. Even if an iterative lifecycle model, such as the Spiral Model [10], is followed, then RE is generally the first step in each 360° sweep of the spiral. Even if an agile [1] lifecycle model is followed, then an abbreviated RE process is the first step of each spike.

Two of the problems with the Waterfall Lifecycle Model, shown in Figure 87.1 with the normal flow of information and all possible feedback loops, are the implications that

1. Requirements appear in the starting milestone out of nowhere all ready to use, and
2. One produces a specification, that is, a RS, from requirements just as easily as one produces a design from a RS or a realization from a design.

However, as suggested in Figure 87.2,

1. The starting milestone are the client's ideas,
2. Client ideas are far more nebulous and informal than even the most informally stated natural language RS, which at least *attempts* to be complete and consistent, and
3. The transition from client ideas to the RS is far more torturous, difficult, and haphazard than the relatively systematic transition from the RS to the design milestone.

The wavy line between client ideas and the RS in Figure 87.2 is RE.

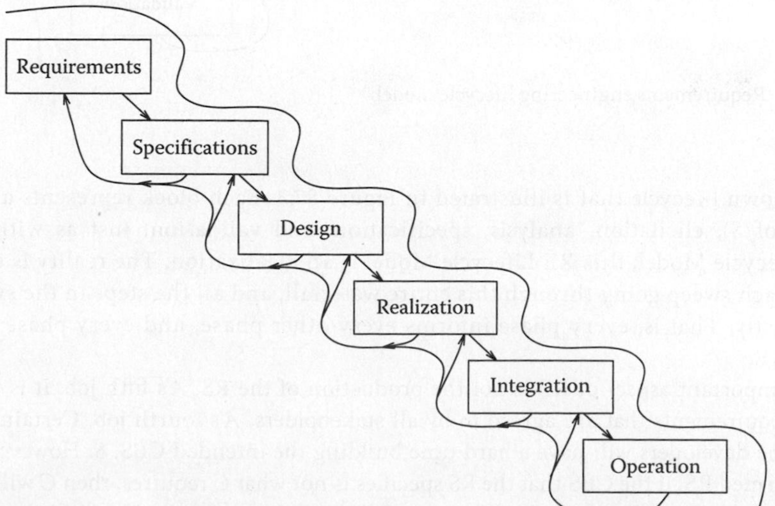

FIGURE 87.1 Waterfall lifecycle model.

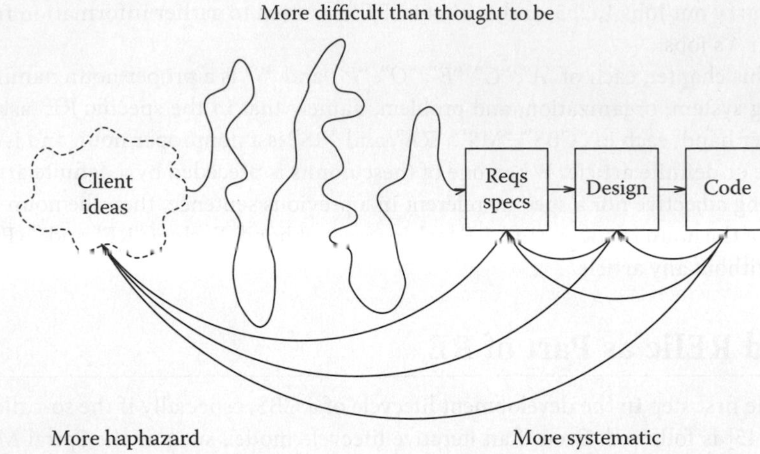

FIGURE 87.2 True lifecycle model.

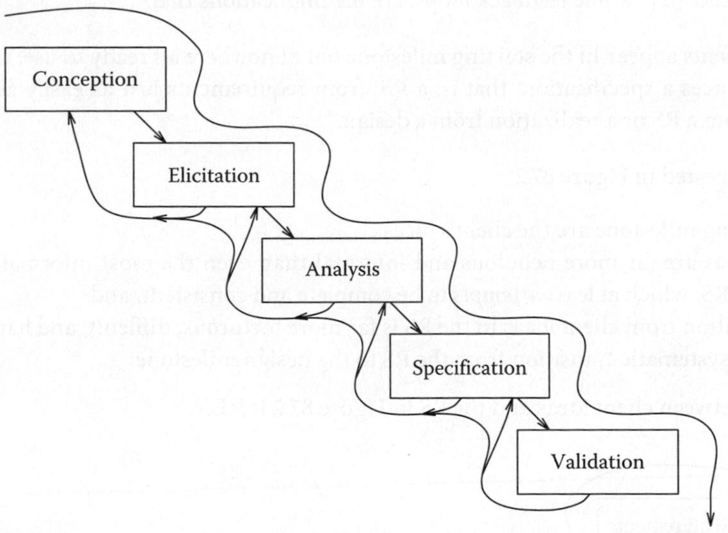

FIGURE 87.3 Requirements engineering lifecycle model.

RE has its own lifecycle that is illustrated in Figure 87.3. Each block represents a phase of RE, conception (of *S*), elicitation, analysis, specification, and validation. Just as with the general Waterfall Lifecycle Model, this RE Lifecycle Model is an idealization. The reality is that there is a spiral, with each sweep going through this entire waterfall, and all the steps in the sweep happening concurrently. That is, every phase informs every other phase, and every phase is happening throughout RE.

The most important aspect of RE is not the production of the RS, *A*'s fifth job. It is the process of negotiating requirements that are agreed to by all stakeholders, *A*'s fourth job. Certainly if the RS is inadequate, the developers will have a hard time building the intended CBS, *S*. However, even with a totally well-formed RS, if the CBS that the RS specifies is not what *C* requires, then *C* will not be happy with the developed CBS. Therefore, the success or failure of the project to develop *S* depends critically on whether *S* satisfies its requirements, that is, whether it solves *P*, *C*'s problem.

87.4 Elicitation and RElic as Human Activities

An amalgamation of several dictionary definitions of "to elicit" is that it means "to bring out, to evoke, to call forth," in the case of RElic, "…, information pertaining to requirements." The main purpose of RElic is to get information about (1) the domain and goals from which the requirements are specified and (2) the requirements from which *S* is developed.

A must get information out of stakeholders' minds without damaging the stakeholders or their minds! Many times, this information does not come out easily. As is discussed later, the stakeholders do not know it themselves, the stakeholders subconsciously do not want to let it out, or both. Additionally, sometimes needed or relevant information is not in *any* stakeholder's mind.

RElic is a human activity involving interaction between human beings [17], the stakeholders (including the nonhuman adjacent systems, represented by human beings). If *A* cannot do the human interaction right, *A* is not going to be able to elicit, no matter what technology and methods he or she uses. Technology and methods might help, but they can also get in the way. Often, success in RElic depends on just being able to talk with the stakeholders about what they really want from *S*, the system being developed.

The skills needed for RElic are (1) identifying contexts, (2) spotting linguistic ambiguities, (3) interviewing, (4) brainstorming, (5) facilitating, (6) getting people to open up, (7) spotting equivocation in people, and (8) inculcating guilt (that the stakeholders have not told the RA everything they know [5]). Only the first two are not human interaction.

Gause and Weinberg's *Exploring Requirements: Quality Before Design* [17] covers a number of techniques, some of which are described later. They are no replacement for the skills, but they do help focus those skills and give ideas on what to do next at an impasse.

A lot of the RE literature talks about eliciting requirements for *S* as if it were simply a matter of asking the right questions of *C* and the users. The reality is not so simple. In practice, requirements for *S* are poorly understood by everyone. It is only by having all stakeholders review and discuss all documents, both provided by *C* and other stakeholders and generated during analysis, that the stakeholders can conduct the negotiations that lead to an agreement about what the requirements *S* really are. If the stakeholders cannot come to an agreement, then the project to develop *S* will probably fail.

87.5 RElic for Each Job of the RA

Recall that *A*'s main jobs are

1. To understand *P*, which *C* wants solved by *S*
2. To extract the essence of the stakeholders' requirements for *S*
3. To invent a way to solve *P* that is better for the users than the way it is being solved currently by *E*, particularly if *E* is manual
4. To negotiate with the stakeholders a consistent set of requirements for *S*
5. To record the resulting requirements in a RS for *S*

Each but the last involves RElic in some way, and each, including the last, is affected by what is learned in RElic.

87.5.1 Understanding the Customer's Problem

RElic is necessary in order for *A* to get the information with which to to understand *P*, which *C* wants solved by *S*. Obtaining this understanding may require analyzing *E*, the existing system that *S* is to replace, if for no other reason that *E* probably does most of what *O* expects of it. It will be necessary to ensure that *S*, the new CBS, performs at least the functionality of *E* that is still needed as well as solve the problem *P* that prompted the initiation of the project to develop *S*. If *S* is radically different from *E*, the users of *E*, that is, the future users of *S*, may be reluctant to use *S*.

If the intent is to build into *S* a new-and-improved version of *E*, then it is necessary for *A*, with the help of the stakeholders, to determine

- What of *E* is used and what is not used
- What of *E* works well and what does not
- How *E* is actually used
- How *E* was intended to be used when it was deployed
- What is missing from *E* that is needed now to solve *P*
- What the stakeholders wish *E* could do

If *A* does not understand *E* well enough to determine these issues, then there is a good chance that *S* will be deficient in at least some of the following ways:

- *S* is so different from *E* and imposes so new a way of working that the users are reluctant to use *S* or even rebel against using *S*.
- *S* fails to have the obvious improvements to *E*.
- *S* fails to implement features of *E* that were not mentioned or are not obvious. In this case, *O* may end up running both *E* and *S*.

The elicitation of the information about *E* is done in one or more of several methods [3,11,26–28,39, 44,49,51]:

1. *A* reviews all available documentation about *E*, including from its development, deployment, and maintenance.
2. *A* observes *O* and its users as they use *E*. This observation should initially be as unobtrusive as possible to guard against the users showing off an unrealistic candy-coated view of the operation of *E*. Later, after the users are used to being observed without showing off, the observers can and should start to ask questions such as "What are you doing?" and "Why are you doing that?" Such questions asked during the actual doing get better answers than if they are asked in questionnaires or interviews. Almost everybody is better at explaining what he or she is doing at the moment than at explaining what he or she has done or will do. Moreover, during the doing of a task, he or she is more likely than other times to be able to explain
   - The steps of the task precisely,
   - Why the task is done the way it is being done, and
   - Most importantly, the exceptions that can occur, even if they do not occur specifically during the present doing of the task.
3. *A* prepares questionnaires that ask stakeholders, particularly users, about features of *E* that they use and do not use.
4. *A* interviews stakeholders, particularly users, about features of *E* that they use and do not use.
5. *A* conducts focus groups with stakeholders, particularly users, about features of *E* that they use and do not use.

Any or more than one of Methods 3, 4, and 5 can be used (1) to elicit information about issues that did not manifest themselves during the observations and (2) to confirm conclusions drawn during any others of Methods 1, 2, 3, 4, and 5.

87.5.2 Extract the Essence of the Stakeholder's Requirements

RElic is necessary for *A* to get all the information to see beyond each stakeholder's goals and requirements to the essence of all stakeholders's common requirements. Generally this job involves building a model of *O*'s business and of the way that *E* supports it. Building the model is itself not part of RElic and is covered partially in Chapter 88. Many textbooks [11,26–28,39,44,49] go in depth into the topic of requirements modeling. Nevertheless, as the model is being built, questions will arise that are not answered by the information that *A* has already elicited. *A* will have to go back to the stakeholders to do more RElic. In this case, *A* will show the stakeholders the current version of the model and ask both:

- Whether the model reflects reality as the stakeholders see it.
- The specific question that *A* had while modeling that prevented him or her from proceeding with the modeling.

87.5.3 Invent New Ways to Do the Users' Work

RElic is important for *A* to be able to invent better ways to do the users' work. Once *A* has an understanding of the work the users of *E* are trying to accomplish, *A* may be able to suggest *new* requirements that would help them because *A* is aware of technology that could help them or *A* can identify patterns in their work of which they are not aware.

Inventing a better way is often overlooked during RElic. Yes, it is important to determine what *C* and users want and to document it. However, if *A* stops with only that determination, then the built system will conform to only the limited notion of what *C* and users think is possible. For real success, *A* must give to *C*, not what he or she wants, but what he or she never even dreamt of having; and when *C* gets it, he or she recognizes it immediately as exactly what he or she wanted all the time—"I'll know it when I see it" (IKIWISI).

One way to invent a better way involves *A*

1. Asking stakeholders *why* the requirements that have been documented thus far are desired, that is, engaging in more RElic, and then thinking about ways to address the goals implied by the offered reasons.
2. Considering ways to give each user more control over his or her activities in *E*.
3. Brainstorming (see Section 87.6.6), with the stakeholders, to invent new requirements.

Thus, *A* needs to go *beyond* RElic to find requirements, to possibly *invent* some.

87.5.4 Help the Stakeholders Negotiate to a Consistent Set of Requirements

RElic is occasionally needed during negotiating with all stakeholders a consistent set of requirements. In general, this negotiation requires that the RA find the inconsistencies among the stakeholders' requirements. Attempting to model all the requirements exposes these inconsistencies as the inconsistencies prevent the formation of the model. The remedy for these inconsistencies is to convene the stakeholders and to show them the inconsistencies. Then ideas on how to resolve the inconsistencies in the form of alternative proposals for requirements are elicited. There are techniques for conducting the negotiations, for example, Win–Win [8] that are outside the scope of this chapter.

87.6 Specific Techniques for RElic

There are a number of specific CBS issues or RElic techniques whose consideration or use help focus RElic or idea invention to be more effective than it might otherwise be [11,26–28,39,44,49,51].

87.6.1 Social and Organizational Factors

In RElic, *A* must keep in mind that any CBS is used within its own *context* that consists of technical *and* social factors. Social factors include organizational factors and these can dominate the technical factors in determining requirements [40,48]. For example, if employees have a particular way of working, *S* should be built to support it. The user interface of *S* must accommodate all of its users, including the most computer illiterate of them. If a future user feels that his or her job is threatened by *S*, he or she may engage in obstructing activities ranging from passive resistance through to active sabotage [16,34,40,48].

87.6.2 Ethnographic Analysis

In ethnographic analysis [22], someone that is trained as a social scientist spends a lot of time observing the members of *O* to observe how they conduct *O*'s work using *E*. The social scientist should try to blend into the background so that the members of the client organization forget that the social scientist is observing them. Normally the social scientist learns strictly by observation. However, variations of ethnographic analysis permit the social scientist to ask questions. Of course, then the members of the client organization are once again aware that they are being observed.

A variation of ethnographic analysis is for *A* to take the role of an apprentice in the client organization, as suggested by Contextual Inquiry [7]. Then, the members of the client organization are aware that they are being observed. However, a person acting as an apprentice is probably perceived more as an insider than a social scientist quietly observing or asking questions because a new employee needs to learn the truth in order to function well within *O*. Thus, the apprentice is less likely than an observing social scientist to be shown abnormal, candy-coated behavior.

87.6.3 PIECES

PIECES [50] can be viewed as a check list for helping to focus RElic. *A* examines *S* from six specific viewpoints and initiates RElic in each viewpoint. The name of the technique is an acronym formed from the first letters of the names of the six viewpoints:

1. Performance, for example, throughput and response time
2. Information and data, for example, relevance, form, timeliness, and accessibility of information and data
3. Economy, for example, trade-offs of costs vs. minimal acceptable performance
4. Control, for example, degrees of automation, of auditing, of robustness, and of security
5. Efficiency, for example, measurement of unintentional waste (differs from economy, which is concerned about acceptable waste)
6. Services, for example, how *S* is to be used by its primary and secondary users

Admittedly, there is some overlap in the viewpoints, but an overlap is better than a gap in trying to ensure that everything that can be relevant is considered. PIECES is oriented toward business information CBSs, but it can be used with other kinds of CBSs. A RA experienced with another kind of CBS can surely think of relevant viewpoints to add to the "PIECES" list for that kind of CBS.

87.6.4 Interviews

In an interview, *A* talks one-on-one with a stakeholder about *P* and *S* and anything else that *A* needs to know. In conducting interviews, it is important to interview *all kinds of* stakeholders if not all stakeholders. It is important for *A* to avoid detailed solution-specific questions and to ask problem-related questions that help understand the problem *P* that *S* is to solve and to extract the essence of the stakeholders' requirements.

A useful technique is for several stakeholders to be interviewed at once to get synergy, as one interviewee amplifies or disputes what another interviewee says.

It is essential that *A* ask calibrating, meta, and unaskable questions. A calibrating question, such as "Is this answer official? ", attempts to determine how much the interviewee's answer can be relied upon. A meta question, such as "Am I asking the right questions?" and "Should I be asking some other questions?" attempts to discover all questions that should be asked. Finally, an unaskable question, such as "Are you opposed to *S*?", "Are you trying to make yourself indispensable?", and "Are you trying to protect your job?" attempts to expose hidden agendas and other hidden information.

An interview should probably start with a list of prepared questions, to at least ensure that *A* does not forget to ask any question. However, *A* should be prepared to deviate from the prepared questions to dig deeper into important issues and to explore emergent issues.

87.6.5 Focus Groups

Basically, a focus group [47] is an interview of a group of group of people conducted by a trained facilitator. In a focus group for RElic, a group of stakeholders are gathered for a short period, for example, about 2 h to ask them specific questions about *E*, *P*, *S*, or some combinations of them.

The technique originated in marketing, in which the focus group is used to ask a group of potential customers their opinions about a planned product, service, advertising, *vel cetera** so that the plans can be changed for maximum marketability. The technique has been used to assess the usability of proposed or prototyped user interfaces; this use of focus groups is most like that in RElic. Indeed, to the extent that determining user interfaces is part of RE, this use of focus groups is a RE technique.

The main advantage of a focus group over the interview of an individual is the potential synergistic effect of one group member's response to a question on the other group members' responses, sort of as in a group brainstorm in which participants build on each other's ideas. Also, a focus group can promote a discussion between stakeholders with differing viewpoints on a question, leading to a resolution of the differences. Thus, a focus group can be used to conduct the negotiations among stakeholders that are one of *A*'s jobs.

The main drawback of a focus group that group mentality may lead it, like a herd of sheep, down one particular idea that is not as good as what would have been exposed if each individual could reply alone. It is important for the facilitator to recognize when this phenomenon has happened, to cut of the wayward discussion, and to refocus the group with other questions.

The advice for conducting interviews concerning the contents of questions, the open-endedness of the planned sequence of questions, and the selection of interviewees for interviews applies also to focus groups. Of course, in a given time period, a focus group will be able to consider many fewer questions than is possible in the interview of one person, if for no other reason than to give everyone a chance to speak, but also to allow any discussion that happens. Also, the advice given in Section 87.6.6, concerning managing the interaction during group brainstorming applies also to focus groups, in order that the participants contribute freely and fully.

The more people in a focus group, the fewer the questions that can be asked in the session. On the other hand, one needs to get a representative set of stakeholders participating. Probably, there should not be more than a dozen people in any focus group.

87.6.6 Brainstorming

Brainstorming [37] is hands' down the most popular method of quickly generating new ideas for any purpose including RElic [12]. A brainstorming session consists of two activities: (1) idea generation and (2) idea pruning.

To begin idea generation for a brainstorming session, the leader of the session throws out a question or problem. People try to think of answers or solutions, never mind how outrageous or outlandish. The goal is to generate as many ideas as possible. An outrageous, outlandish idea could trigger yet other good idea.

The rules are the following:

- Shoot for quantity.
- Mutate and combine ideas.
- Do not criticize anyone's ideas, even your own.
- If you don't like someone's idea, including your own, leave it be, but improve on it.

* "*vel cetera*" means "inclusively or others" as "et cetera" means "and others".

- Nothing anyone says is to be held against him or her after the session—what happens in brainstorming stays in brainstorming.
- Let your imagination soar; wild is good.
- Feel free to be gloriously *wrong*.

The participants in a brainstorming session for RElic are

1. The stakeholders, with an emphasis on those who know most about or are affected most by *E*, *P*, or *S*
2. Some people from outside the project to develop *S*, in order to get out-of-the-project-box ideas

However, the effect of any participant's power in *O* should be minimized. That is, if a person and his or her boss are participants in the same session, the boss's power should be left outside the session, so that the person can contribute freely.

The number of participants should be kept manageable, so that everyone gets a chance to contribute. With too many people, the shy shut up and only the loud are heard. Among the participants, there are two special roles, the scribe and the moderator.

1. The scribe's role is to write down all ideas. He or she also may contribute ideas and may ask clarifying questions, to improve the accuracy of the list of ideas, but not critical questions.
2. The moderator's role can be described as something between two extremes, the traffic cop and the provocative agent.
   a. The traffic cop enforces rules but does not lead.
   b. The provocative agent actively leads, coming prepared with wild ideas to be thrown out at the beginning and whenever idea generation begins to wane. He or she may look for variations and combinations of other ideas, and he or she acts as a traffic cop.

Idea generation should continue until the first of the following happens:

1. The ideas stop coming.
2. Two hours have passed.

Beyond the two-hours limit, efficiency of idea generation and the quality of the ideas seem to drop off significantly.

After the idea generation has been terminated, idea pruning begins. Idea pruning is the removal of *useless* raw ideas and the refining what is left. As *quantity* is the focus of idea generation, *quality* is the focus of idea pruning. The purpose of idea pruning is to make the best ideas possible.

Care should be taken in idea pruning to to avoid to throwing out an idea just because it *looks* useless, hazardous, or uneconomical. In this sense, creativity is needed also in idea pruning. In addition, it may be useful to bring into idea pruning, stakeholders and outsiders who did not participate in idea generation. For example, it might be particularly useful to get the input of marketing and development people for idea pruning.

There are several choices of how to do idea pruning, including the following:

- Voting with threshold: Each person is allowed to vote up to *n* times. Keep those ideas with more than *m* votes. Have multiple rounds thereof with ever smaller *n* and *m*.
- Voting with campaign speeches: Each person is allowed to vote up to *n* times. Keep those ideas with at least one vote. Have someone who did not vote for an idea defend it for the next round. Have multiple rounds thereof with smaller *n*.
- Ranking ideas by priority or quality: Rank ideas by priority, quality, innovativeness, cost-effectiveness, pervasiveness of requirements coverage, speed of implementation, cheapness of implementation, etc., or any combination thereof. The group works to rank ideas by some method, for example, writing each on a card and placing cards into an ordering or a lattice. Find a cut-off point in the ranking, above which ideas are kept and below which ideas are discarded.
- Blending ideas: Consider pairs, triples, etc., of ideas for combining in some way. Then, go into one of the ranking or voting procedures.

Brainstorming is not the only technique for enhancing creativity in idea generation for RElic. There are other techniques that have more systematic or more focused idea generation [6,13,23,32,38,46]. For example, one problem with brainstorming is that it allows a random walk through the space of ideas. EPMcreate makes the walk through the space of ideas more systematic by having *A* lead a group of RAs to consider all 16 combinations of for and against the viewpoints of two stakeholders of *S* and for each combination to think of ideas that meet the requirements for the combination. One such combination would be "for stakeholder 1 and against stakeholder 2" [32].

87.6.7 Gause and Weinberg Techniques

Don Gause and Jerry Weinberg, the authors of two classic books on RElic [17,18], suggest a number of useful RElic techniques that help *A* to find those hard-to-find new requirements ideas:

- Questioning norms is considering things that are done in *E* and asking why they are done the way they are.
- Questioning assumptions is actively searching for assumptions in what is known about *E*, *P*, and *S* and asking if each is *really* true.
- Questioning existence assumptions is a refinement of questioning assumptions in which the assumptions that are sought and questioned are the very existence of things that are needed for *S* as currently conceived to work, for example, the underlying operating system provides an assumed capability.
- Use of the right brain is promoted by getting stakeholders to draw diagrams showing how they see *E* working, where *P* arises, and how they see *S* working and solving *P*.
- Naming is simply brainstorming for cooler or more precise names for things to encourage seeing them in different lights.

87.6.8 Prototyping

Many people, especially the nontechnically oriented, learn only while doing. They have got to see some kind of working version, a prototype, in order to *discover* what they want. There are several kinds of prototypes that can be built to help *C* and users see what is possible and then to comment on it, in order that they begin to discover what they *really* want:

- A functioning, throw-away, quick-and-dirty implementation of at least some of the functionality of *S*
- A mock up, a program that accepts only preplanned input of *S* to produce only preplanned output of *S* and is devoid of all other functionality of *S*
- A wizard of Oz, a user interface operated by a human being that pretends to be *S* to produce the intended results for any input
- A story board that describes, with, say, screen-diagram pictures, anticipated scenarios of the use of *S*
- A draft user's manual for *S* that shows with screen-diagram pictures, anticipated scenarios of the use of *S*

Any of these prototypes can focus on only the main functionality or only the functionality over which there is least understanding or the most contention.

The rationale behind prototyping is IKIWISI. Many clients and users simply cannot describe what they want, but if they see what they want, they can recognize it as what they want immediately. Unfortunately, sometimes AYSTII (anything you show them is it), and *A* will have to judge whether what he or she is observing is IKIWISI or AYSTII.

One problem with the prototyping approach, particularly if it uses a running program, is that it can raise unrealistic expectations in *C* and the users, namely, that the guts of the software is already developed, that development of the production version will go as fast as the production of the prototype, or something else totally unrealistic. *A* must make clear and continually remind *C* and the users that what they see is *not* real software, that it is only a mock up, an empty shell, a wizard of Oz, *vel cetera*.

87.6.9 Joint Application Design

Joint application development (JAD) can be thought of as structured group brainstorming with the emphases on "structure" and "group" or as an extended focus group. The main idea of JAD is get all the stakeholders who matter and who can make binding decisions together in an isolated place to work intensively in a short time period to make the key requirements and design decisions, to which every stakeholder present commits, allowing implementation to proceed straightforwardly from the agreed to RS and design documents [52].

JAD for a CBS is carried out in two steps:

1. JAD/Plan, the RElic step
2. JAD/Design, the software design step

This chapter, being about RElic, focuses on the first step. Each step consists of three phases,

1. Preparing for the session
2. The session itself
3. Reporting on the session

There are six roles for each step:

1. Session leader, a JAD expert who organizes and facilitates the session
2. Analyst–scribe, who understands and develops the big picture and produces the session report
3. Executive sponsor, the manager who has ultimate responsibility for *S*, who understands *O*'s goals for *S* and can make *binding* decisions concerning *S* and its development
4. User representatives, a selection of knowledgeable users and managers
5. Information systems representatives, a selection of technical experts who understand the technology and its trade-offs
6. Specialists, a selection of experts on special topics deemed relevant to *S*, for example, security, the application domain, laws, regulations

Advance preparation is the key to the success of JAD. The executive sponsor picks the participants and invites them to the JAD session with explanations of what is to be done at the session and their roles. The session leader and the executive sponsor familiarize themselves with *P*, the problem at hand; the domain of *S*; and the stakeholders. They try to identify the possible issues and points of contention. They set the scope of the session. They prepare materials both for before the session, for example, reading material for the participants, and for during the session, for example, electronic slides, flip chart pads, markers for the whiteboards, food, etc.

The session is planned for, say, a full day or longer, at some place sufficiently isolated that participants will not be drawn away from the session to deal quickly with short matters. Cell phones should be turned off! Breaks for coffee, meals, and even some recreation are planned.

At the session itself, the session leader welcomes the participants and thanks them for agreeing to isolate themselves for the session period. He or she presents *P* and the goal for what is to be accomplished. He or she establishes the ground rules for the session and delimits the scope. Finally, he or she throws out some initial ideas.

The participants work on solving *P*, that is, doing the RElic that is needed. There will be brainstorming sessions as needed. The goal is to make decisions that everyone present can buy into and that have the power of the executive sponsor behind them.

Toward the end of the session or even very soon afterwards, the scribe writes up what has been decided, in the form of a RS if the session was for RElic. Everyone, including the executive sponsor, reads the write up, corrects it, and signs off on it.

JAD was developed at IBM and was found to be very successful because it is very effective at using the isolation and the short time available to get diverse stakeholders to agree to and buy into concrete decisions that have the power of an organizational executive behind them.

87.7 What Can Go Wrong during RElic?

There are a number of reasons that RElic can fail to lead to an adequate RS for *S*:

1. The existence of an unknown requirement: None of the stakeholders mentioned the requirement because none knew of it. Modeling helps expose the incompleteness of a set of requirements, as the known requirements fail to lead a complete model.
2. The existence of a known but undiscussed requirement: The requirement was known to several stakeholders, but each such stakeholder assumed that it was understood the same way by all and was therefore not worthy of mentioning. The only hope for exposing such requirements is the tendency for a person to say things that depend on unspoken, tacit assumptions, because he or she *knows* that the assumption is true. Having on the RElic team someone who is capable of being puzzled by things that different stakeholders describe differently—perhaps someone ignorant of *S*'s domain [4]—is helpful for exposing these tacit assumptions and undiscussed requirements. To the extent that an undiscussed requirement causes the discussed set of requirements to be incomplete, techniques for finding an unknown requirement can help expose known but undiscussed requirements.
3. The existence of a known, discussed, but undocumented requirement: The requirement was thoroughly discussed, but no one wrote it down, creating the possibility that if it is discovered again later, it will be understood differently from what was agreed to mean when it was first discussed. Having on the RElic team someone who is taking complete minutes and notes helps ensure that everything that is discussed ends up being documented properly.
4. The existence of a wrongly documented requirement. No stakeholder understood the requirement correctly and the incorrect understanding was documented, or it was understood correctly, but a saboteur documented it incorrectly, unbeknown to the other stakeholders.

87.8 Myths about RElic

The biggest myth about RElic is that *A* can just ask *C* and the users what they need and want and can expect to hear accurate answers. The fact is that *S*'s just asking *C* and users what they need and want does not always work.

Sometimes, *C* and the users know what they need and want but cannot describe it, for example, as few people who can ride a bicycle can describe how to ride a bicycle, at least in a way such that anyone hearing the description will be able to just get on a bicycle and start riding without falling. Instead, they have to *show* what they need and want, for example, by showing when, where, and how *E* should deviate from what it does and then showing with pictures what it should do instead.

Other times, they do not know what they need and want; they know only that what *E* does is not what they need and want. They might be able to put their fingers on only what is wrong with what *E* does, and *A* would have to take it from there.

Closely related to not being able to describe what they need and want is not being able to describe in detail what they need and want, the exceptions from the norm, and the responses to the exceptions. When most people are asked what they do, they will quote the official policy, and not what they actually do. Most of what they really do, which is not specified by the policy, is what they do in situations not covered by the policy. This quoting of official policy is not an example of conscious, politically safe mouthing of the policy. Many people simply do not remember the exceptions unless and until they actually come up. The official policy *is* their conscious model of what happens. Domain experts among these people, from their experience with *E* and their understanding of *O*'s goals, instinctively *know* what to do for each exception as it comes up. Ultimately, even with a prototype, *A* has to be watching when any exceptional situation happens during the operation of *E* to see what really happens.

Sometimes requirements information is simply not in the mind of *any* stakeholder. Joseph Goguen [20] observed in 1993 that

> It is not quite accurate to say that requirements are in the minds of clients; it would be more accurate to say that they are in the social system of the client organization. They have to be invented, not captured or elicited, and that invention has to be a cooperative venture involving the client, the users, and the developers. The difficulties are mainly social, political, and cultural, and not technical*.

Finally, many people just do not know why they do something, saying only that it is done this way because the policy says so. They very often do not even know why the policy is the way it is. It could be that the reasons for the policy are lost in history and that they were in response to a situation that is irrelevant today. In other words, the policy once made sense, but the person who formulated the policy, the reasons for it, and the understanding of the reasons are long since gone. For example, many companies that have committed all data to a highly reliable data base continue to print out the summary in quintuplicate. Why? At the time of automation, the five most senior members of the company, who have since retired, refused to learn to use the computer to access the data directly!

87.9 Why Invention and Creativity Are Necessary for RElic

The quotation from Joseph Goguen in Section 87.8 pointed to the need to *invent* requirements. Invention requires creativity. Many others have observed the importance of creativity in RE, particularly for discovering and inventing requirements for *S* [17,20,21,24,29,30,31,33,35,36,43], for solving wicked [42] problems, and for addressing critical business challenges [14,19,41,42].

Creativity is a part of RElic because requirements have to be invented. Requirements have to be invented, because if not, we would never advance beyond what we do now. Anytime *S* is different from *E*, someone had to have invented some new requirement that made the difference. That someone could be *A* or *any* of the stakeholders, including *C* and the users.

87.10 Empirical Studies of RElic Techniques

A lot of what is said about RElic techniques is folkloric. That is, a lot is said without what is accepted as empirical evidence that it is true. A significant fraction of what is said is probably true. That is, experienced practitioners, for example, Gause & Weinberg and Beyer & Holtzblatt, have learned what works and have described the abstracted techniques, suggesting that others try them too.

Nevertheless, there is empirical work testing whether specific techniques work as expected and discovering what really works. For example, for brainstorming and other creativity enhancement techniques,

* Note that this quotation is from a draft of what was later published as a chapter in a book [21]. The quotation did not survive into the book chapter. However, by e-mailed personal communication, Goguen assured the author that he still believed in the contents of the quotation.

there are empirical studies about what works [2,25,32,46]. More empirical research is needed to validate RE methods in general and RElic methods in specific.

Dieste and Juristo have published a systematic review of research empirically validating RElic techniques [15]. Independent of the contents of the review, its bibliography is a gold mine of references to RElic techniques.

87.11 Conclusions

RElic is fundamentally the gathering of information about the CBS *S* to be built, from whatever sources the information is available, beit the client *C*; other stakeholders in *C*'s organization, *O*; the existing system, *E*, to be replaced by *S*, documentation about *E*; and invention. Often times, the information does not come freely from stakeholders in *O* that do not fully understand the problem *P* to be solved by *S* or have a political agenda. This chapter describes techniques that can be used to help start the free flow of this information, so that eventually a good RS can be produced for *S*.

Acknowledgment

The author thanks his colleagues in teaching Software Requirements Analysis and Specification at the University of Waterloo, Joanne Atlee, Michael Godfrey, and Richard Trefler.

References

1. Agile Alliance. Principles: The agile alliance, 2001. http://www.agilealliance.org/ (accessed on December 11, 2012).
2. A. Aurum and E. Martin. Requirements elicitation using solo brainstorming. In *Proceedings of the Third Australian Conference on Requirements Engineering*, pp. 29–37, Deakin University, Geelong, Victoria, Australia, 1998.
3. B. Berenbach, D. Paulish, J. Kazmeier, and A. Rudorfer. *Software & Systems Requirements Engineering: In Practice*. McGraw-Hill Osborne Media, New York, 2009.
4. D. M. Berry. The importance of ignorance in requirements engineering. *Journal of Systems and Software*, 28(2):179–184, 1995.
5. D. M. Berry and O. Berry. The programmer-client interaction in arriving at program specifications: Guidelines and linguistic requirements. In E. Knuth, ed., *Proceedings of IFIP TC2 Working Conference on System Description Methodologies*, Kecskemet, Hungary, pp. 275–292, 1983.
6. C. Berthelson. Patterns of creativity in japan, lessons from successful japanese companies—A resource: The NHK method, viewed, July 2012. http://a-small-lab.com/japanese-creativity/124-the-nhk-method/ (accessed on December 11, 2012).
7. H. Beyer and K. Holtzblatt. *Contextual Design*. Morgan Kaufman, San Francisco, CA, 1998.
8. B. Boehm and R. Ross. Theory W software project management: Principles and examples. *IEEE Transactions on Software Engineering*, SE-15(7):902–916, July 1989.
9. B. W. Boehm. *Software Engineering Economics*. Prentice-Hall, Englewood Cliffs, NJ, 1981.
10. B. W. Boehm. A spiral model of software development and enhancement. In R.H. Thayer, ed., *Software Engineering Project Management*, pp. 128–142. IEEE Computer Society Press, Los Alamitos, CA, 1987.
11. I. K. Bray. *An Introduction to Requirements Engineering*. Addison-Wesley, Harlow, U.K., 2002.
12. J. G. Byrne and T. Barlow. Structured brainstorming: A method for collecting user requirements. In *Proceedings of the Human Factors and Ergonomics Society Thirty-Seventh Annual Meeting*, pp. 427–431, Seattle, WA, 1993.
13. E. de Bono. *Six Thinking Hats*. Viking, London, U.K., 1985.
14. E. de Bono and R. Heller. Can creative management techniques help you survive the recession? August 10, 2010. http://www.thinkingmanagers.com/management/creative-management-techniques.

15. O. Dieste and N. Juristo. Systematic review and aggregation of empirical studies on elicitation techniques. *IEEE Transactions on Software Engineering*, 37(2):283–304, 2011.
16. A. Finkelstein and J. Dowell. A comedy of errors: the London Ambulance Service case study. In *Proceedings of the Eighth International Workshop on Software Specification and Design (IWSSD)*, pp. 2–4, Schloss Velen, Germany, 1996.
17. D. Gause and G. Weinberg. *Exploring Requirements: Quality Before Design*. Dorset House, New York, 1989.
18. D. Gause and G. Weinberg. *Are Your Lights On? How to Figure Out What the Problem REALLY Is*. Dorset House, New York, 1990.
19. H. Geschka. Creativity techniques in product planning and development: A view from West Germany. *R&D Management*, 13(3):169–183, 1983.
20. J. A. Goguen. Requirements engineering as the reconciliation of technical and social issues. Technical report, Centre for Requirements and Foundations, Programming Research Group, Oxford University Computing Lab, U.K., October 1993. modified version later published as [21].
21. J. A. Goguen. Requirements engineering as the reconciliation of technical and social issues. In *Requirements Engineering: Social and Technical Issues*, pp. 165–199. Academic Press, London, U.K., 1994. article in [22].
22. J. A. Goguen and M. Jirotka. *Requirements Engineering: Social and Technical Issues*. Academic Press, London, U.K., 1994.
23. W. J. J. Gordon. *Synectics: The Development of Creative Capacity*. Harper & Row, New York, 1961.
24. O. Hoffmann, D. Cropley, A. Cropley, L. Nguyen, and P. Swatman. Creativity, requirements and perspectives. *Australasian Journal of Information Systems*, 13(1):159–174, 2005.
25. S. Jones, P. Lynch, N. Maiden, and S. Lindstaedt. Use and influence of creative ideas and requirements for a work-integrated learning system. In *Proceedings of the Sixteenth IEEE International Requirements Engineering Conference (RE)*, pp. 289–294, Barcelona, Spain, 2008.
26. G. Kotonya and I. Sommerville. *Requirements Engineering: Processes and Techniques*. Wiley, Chichester, U.K., 1998.
27. S. Laueson. *Software Requirements: Styles and Techniques*. Pearson Education, Harlow, U.K., 2002.
28. L. A. Maciaszek. *Requirements Analysis and System Design*. 2nd edn., Addison-Wesley, Harlow, U.K., 2005.
29. N. Maiden and A. Gizikis. Where do requirements come from? *IEEE Software*, 18(5):10–12, 2001.
30. N. Maiden, A. Gizikis, and S. Robertson. Provoking creativity: Imagine what your requirements could be like. *IEEE Software*, 21(5):68–75, 2004.
31. N. Maiden, S. Robertson, and J. Robertson. Creative requirements: Invention and its role in requirements engineering. In *Proceedings of the Twenty-Eighth International Conference on Software Engineering (ICSE)*, pp. 1073–1074, Shanghai, China, 2006.
32. L. Mich, C. Anesi, and D. M. Berry. Applying a pragmatics-based creativity-fostering technique to requirements elicitation. *Requirements Engineering Journal*, 10(4):262–274, 2005.
33. L. Mich, D. M. Berry, and M. Franch. Classifying web-application requirement ideas generated using creativity fostering techniques according to a quality model for web applications. In *Proceedings of the Twelfth International Workshop Requirements Engineering: Foundation for Software Quality (REFSQ)*, Luxembourg, 2006.
34. A. Milne and N. A. M. Maiden. Power and politics in requirements engineering: Embracing the dark side? *Requirements Engineering*, 17(2):83–98, 2012.
35. L. Nguyen, J. Carroll, and P. A. Swatman. Supporting and monitoring the creativity of IS personnel during the requirements engineering process. In *Proceedings of the Thirty-Third Hawaii International Conference on System Sciences (HICSS)*, Maui, HI, 2000. http://csdl2.computer.org/comp/proceedings/hicss/2000/0493/07/04937008.%pdf.
36. L. Nguyen and G. Shanks. A framework for understanding creativity in requirements engineering. *Journal of Information and Software Technology*, 51(3):655–662, 2009.

37. A. Osborn. *Applied Imagination*. Charles Scribner's, New York, 1953.
38. S. Parnes. *Source Book for Creative Problem Solving*. Creative Foundation, Buffalo, New York, 1992.
39. K. Pohl. *Requirements Engineering, Fundamentals, Principles, and Techniques*. Springer, Heidelberg, Germany, 2010.
40. I. Ramos, D. M. Berry, and J. Á. Carvalho. Requirements engineering for organizational transformation. *Journal of Information and Software Technology*, 47(5):479–495, 2005.
41. T. Rickards. *Creativity and the Management of Change*. Blackwell, Oxford, U.K., 1999.
42. H. W. J. Rittel and M. M. Webber. Dilemmas in a general theory of planning. *Policy Sciences*, 4:155–169, 1973.
43. J. Robertson. Eureka! why analysts should invent requirements. *IEEE Software*, 19:20–22, July 2002.
44. S. Robertson and J. Robertson. *Mastering the Requirements Process*. 2nd edn., Addison-Wesley, Harlow, U.K., 2006.
45. W. W. Royce. Managing the development of large software systems: Concepts and techniques. In *Proceedings of WesCon*, Los Angeles, CA, August 1970.
46. V. Sakhnini, L. Mich, and D. M. Berry. The effectiveness of an optimized EPMcreate as a creativity enhancement technique for website requirements elicitation. *Requirements Engineering Journal*, 17(3): 171–186, 2012.
47. D. W. Stewart and P. N. Shamdasani. *Focus Groups: Theory and Practice*. Sage, Newbury Park, CA, 1990.
48. S. Thew and A. Sutcliffe. Investigating the role of 'soft issues' in the reprocess. In *Proceedings of the Sixteenth IEEE International Requirements Engineering Conference (RE)*, pp. 63–66, Barcelona, Spain, 2008.
49. A. van Lamsweerde. *Requirements Engineering*. Wiley, Chicester, U.K., 2009.
50. J. C. Wetherbe. *Systems Analysis and Design*. 2nd edn., Irwin, Burr Ridge, IL, 1994.
51. K. E. Wiegers. *Software Requirements*. 2nd edn., Microsoft Press, Redmond, WA, 2003.
52. J. Wood and D. Silver. *Joint Application Development*. Wiley, New York, 1999.

88

Specification

Andrew McGettrick
University of Strathclyde

88.1 Underlying Principles

In the discipline of software engineering, the term *specification* has various possible interpretations. Whether applied to a simple routine, to a class, to a small subsystem, or a large system, it relates to the behavior of the entity as seen externally. Thus, how does its operation impact on the external environment, what is the effect of its operation, etc.? In this chapter, the focus will be on specifications for software systems and the term *software requirements specification* (often abbreviated to SRS) is typically used to capture this.

SRSs are relevant to different categories of interested parties. At an early stage, a very informal description can be used to inform users about the system—about what the system will do, what role it will play, and how users can interact with it. Such descriptions are very useful; they are typically accessible to many parties who can then react with helpful insights into providing improvements, amplifications, and clarifications.

Where the software is embedded and forms part of a much larger system, the software requirements are typically derived directly from the specification of the larger system. The functionality of the software is derived in this way together with the manner in which it interacts with the wider system and typically includes constraints and performance issues.

For software system designers and implementers, it is important to have an SRS document that describes the system in unambiguous terms. Such documents have to be written in a rather formal (i.e., precise and unambiguous) manner and are the subject matter of this chapter. These specifications have to be agreed among all interested parties. For large systems, such documents can themselves be relatively large.

For a general overview of different aspects of software specification (see Gehani and McGettrick, 1986 and Dorfman and Thayer, 1997).

88.2 Best Practices

88.2.1 Nature of Specifications

88.2.1.1 Typical Structure of Specifications

Central to the SRS will be two important aspects of specification: typically SRS documents will be couched in terms of

- The *functional specification* that captures the activities that the system has to perform. The various activities have to be identified and their functionality described.

- The *nonfunctional specification* that captures additional aspects such as timing constraints, reliability requirements, safety issues, security and privacy constraints, and recovery from exceptions.

Having said this, an SRS document will have to be far more comprehensive than just this. Such a document will need to capture the context for the software system in all its forms. Much will depend on practices within an organization but typically supplying all the relevant detail can occupy as much as 20%–25% of overall development time. The rationale for such major investment is that errors or omissions at this stage can typically be 10–20 times more expensive than errors at the design stage and 100–200 times more costly than coding errors.

From this perspective, great care should be taken to ensure that as much relevant detail is obtained at this stage and that it is an accurate reflection of what is required. Conversely all irrelevant detail ought to be excluded; for clarity, this refers to detail irrelevant for the explicit purpose of providing an SRS document that is as simple as possible. Having said this, such documents can often be very abstract and the provision of a number of illustrative use cases can be used to add clarity and assist with understanding.

To facilitate the provision of SRS document, guidelines are often supplied to provide a framework for a specification. In this spirit an industrial standard exists—see IEEE (1998). A typical SRS document will include

- An introduction that informally captures the goals and intent of the system.
- *Environmental factors* that capture the characteristics of the environment in which the system is to run, that is, the software, hardware, and communications infrastructure and human interaction issues.
- *Interface considerations* detail how the system will interact with other software and with the various users; this may well be amplified by use cases that capture the class of possible users and the manner in which they have to interact.
- The *functional specification*.
- The *nonfunctional specification*.
- Constraints on design: the languages to be used, standards to be followed, resource issues, policy considerations, etc.

The structuring of these specifications is an important matter. The principles associated with such structuring are similar to the guidance for structuring and organizing programs themselves. Thus, it should be clear where the need for particular parts of the requirements arose and then individual aspects of the specification should be grouped together in some natural and logical fashion, for example, those related to some event, to some data structure, and to some subsystem.

88.2.1.2 Some Fundamental Questions

At this stage there are two fundamental questions to be asked:

1. Is it possible to build this software system? Behind this comment lies the observation that it is possible to specify software systems that can never be built, not because of limitations of resources but because of basic theoretical considerations. For instance, it is just not possible to build a program that tells whether arbitrary programs will terminate or a system that tells whether two arbitrary arithmetic expressions are identical. There are also matters such as the possibility of stating very demanding performance requirements when the reliability of the underlying operating system or hardware makes this impossible.

   A related matter is that, even if it is possible to build a particular system, the task may turn out to be enormously expensive, even far too expensive in retrospect. When this happens it is desirable to know well in advance so that unnecessary development does not occur. The SRS document is the definitive statement and so the input for any discussions or analysis on questions of feasibility. For such purposes, simplicity and clarity are desirable.

2. Are there software systems for which it is not reasonable to think of supplying a specification? Here the issue is not the ability of the person undertaking the task but it relates to the nature of the task itself. It is recognized that providing a specification for a sophisticated search engine falls into this category.

88.2.1.3 Usage of the SRS Document

Generally, the usage of the SRS document falls into several categories:

- It can be viewed almost as a legal document and the basis of a contract between a client and development team; it captures in succinct terms what has to be produced.
- It is a definitive statement that can be used to resolve questions, for example, between members of a development effort, about the final system.

It can be used to address certain (usually limited) questions about the feasibility of building such a system:

- It is the starting point for a serious development effort.
- It can be used as the basis of an evaluation of the resources required for the development effort and this includes costs, manpower, and timing.
- Importantly, it provides the basis for an independent testing effort, that is, an effort that can proceed independently of the work of the development team.
- Finally, it is a sound starting point for considerations of possible future enhancements to the system.

To fully meet these various demands, the SRS document should ideally possess certain desirable characteristics.

88.2.1.4 Desirable Characteristics of SRS Documents

A key characteristic is *simplicity*. This is a somewhat vague concept but in this context includes being easy to read and understand, easy to obtain answers to questions, is well structured and organized, and is free of obscurity and complexity. Ideally also it should be relatively compact and free of extraneous detail. For instance, an argument often given is that an SRS document should be free, for instance, of implementation detail; the latter is the concern of the implementation team who will wish to be free to make implementation decisions and choices.

Another important characteristic is *completeness*. Thus, there should be no areas of omission so that, for instance, all ways in which the system can be accessed by a user and all ways in which it interacts with other software are captured and the relevant functionality explained in detail. Importantly, the issue of completeness is relevant not just to the functional requirements but also to the nonfunctional requirements and the other dimensions of the SRS document. In safety critical systems, for instance, safety requirements need to be carefully clarified and then articulated. Under completeness, an often-neglected area relates to error situations, exception handling, and fault tolerance. It may be desirable that the SRS document should stipulate what happens in the event of exceptions taking place, an upper bound on time to recovery, etc. In safety critical systems, for instance, it can be vital to know how the system responds to exceptions so as to be comfortable that no life will be lost and the system will always be in a safe state.

Requirements need to be captured in such a way that the resulting document is *unambiguous*. All readers must be able to interpret the requirements in the same way; otherwise, there is scope for confusion and inefficiency. An aspect of this, of course, is *consistency* of the requirements; so, for instance, particular terms should be referred to in the same way throughout to avoid possible confusion. When natural languages are used, there is often scope for employing an associated glossary to capture meaning.

Another term often associated with specifications is *correctness*. The term *correctness* usually implies consistency between two entities, for example, in simple mathematics a known answer and a suggested answer. Given this interpretation, if two specifications are supplied, then consistency between these can be addressed and used to produce a definitive SRS document. Otherwise, notions of correctness tend to take the form of comparing an SRS document and some abstract view of the system that resides in someone's mind. If the abstract view is just one person's view that can be problematic, that vision needs to be transmitted to a wider audience and group activity and discussion can facilitate that.

The term *verifiable* is also often used in the context of SRS documents. This is intended to capture the notion that requirements can easily be checked. To illustrate "fast response" is better captured in quantifiable terms so that a developer will know if this is being achieved. Likewise, reliability targets should be quantified.

The concept of *full referential transparency* is also highly relevant. Possessing this property implies that accessing the system for a particular purpose has the stated effect and does not have curious side effects in other parts of the wider system. The reverse is also equally relevant, that is, that activities in other parts of the system (e.g., in other locally resident software systems) do not have an impact on the software under consideration.

To emphasize the earlier text, an additional very important and related aspect relates not to what the final system should do but what it should not do. The possibility of the system producing unexpected or undesirable effects is often ignored. Of course, an implementer may choose to offer additional functionality (hopefully with the approval of the client) that proves to be exceedingly attractive and useful. But as well as implementing the agreed functionality, a devious implementation could include some malicious code that produces highly undesirable side effects. This can happen deliberately or accidentally, for example, through the reuse of code developed elsewhere. It is important to have an eye to such possibilities. Basically, the external impact of the system should not be beyond that captured in the SRS document.

Finally, although every effort has to be made to produce a final definitive SRS document, part of the reality of software systems is that they will need to be modified, perhaps several times. Consideration should be given to ensuring that a SRS document is written, organized, and structured so that it is readily amenable to modification and amplification.

88.2.2 Language of Specifications

An important consideration is the language in which a specification is described. There are various possibilities. Natural languages can be used although normally considerable care has to be taken to ensure precision. However, an important class of formal languages called *specification languages* exists. These are used for capturing what has to be done and not how implementation should proceed.

The motivation here is that a formal structure tends to bring additional rigor and a greater ability to reason about a system. An additional important consideration is that formal languages are amenable to processing by tools of different kinds. Included in these are tools to

- Check such matters as syntax, the proper use of types as appropriate, certain aspects of semantics.
- Assist with ensuring the desirable characteristics such as consistency and completeness mentioned earlier; this may even extend to reasoning about the system.
- Allow a certain amount of simple animation of the specification in the sense that certain inputs can be provided and appropriate outputs produced; this can include error situations.
- Possibly support aspects of the implementation activity; there are systems that aim to carry out correctness—preserving transformations that will ultimately lead to final executable code.

The class of specification languages is subdivided into various categories.

88.2.2.1 Model-Based Specifications

With *model-based* specification languages, a specification captures a system's behavior in terms of an abstract model; this is typically constructed using mathematical structures such as sets, functions, and relations. The functionality of the system is then defined in terms of changes to the model. Examples of these languages include the Z specification language (see Spivey, 1992), the Vienna Development Method (VDM) (see Dawes, 1991), and Milner's Calculus of Communicating Systems (CCS) (see Milner, 1980).

Some insight into the nature of these specifications and the benefits they bring is relevant. Using the formal specification language, an abstract model of a system is developed using the facilities of the language, usually sets relations, variables of particular types, etc. Of course these will be initialized in appropriate ways, for example, initially a particular set may be empty. The various activities that the system has to perform then take the form of functions or routines that change the state of the system in clearly defined ways. So basically, the system is state based. Through the application of functions, the system moves from an initial state through other states by the application of the various functions and routines. Beyond the basic requirements,

- Invariants capture those aspects of the state of the system that must remain unaltered throughout all changes of state, for instance, in a library system invariants may stipulate that all users of the library must be registered and that all books are classified as either on loan or available for lending.
- For each function, there will also be preconditions and post-conditions: the preconditions stipulate state characteristics that must pertain prior to certain state changes being attempted, for example, a book can be borrowed only if it is not already on loan; post-conditions capture the effect of invoking the function.

A key matter is that after the operation of each function, the system invariants must remain intact. This is to reflect the fact that the various states have desirable properties and no undesirable ones. This will typically require proof (for example, see Yang and Hawblitzel, 2011).

There is obvious scope for tools to check syntax, to carry out relevant type checking, to check that variables have been initialized appropriately, and to assist with proof.

88.2.2.2 Property-Based Specification Languages

Here the specification takes the form of system constraints on system behavior; typically the fewer the properties or constraints specified, the richer the range of possible design decisions to be made. This group of languages is typically subdivided as follows:

- The *axiomatic* that uses first-order logic to describe preconditions and post-conditions. The languages OBJ (see Malcolm and Goguen, 1996) and Larch (Guttag et al., 1990) support this form of specification.
- The *algebraic* where data types and processes are captured in the form of an algebra and then axioms in the form of equations are employed.

88.2.2.3 Concurrent and Distributed Systems Specification

Concurrent and distributed systems pose an even greater set of problems for system developers. The timing issues, the potentially complex interactions, the problems of sharing, and the constraints pose a new set of very complex problems. Different approaches, different languages, and different formalisms are needed to address the challenges.

The topics here are not the main focus of this chapter. However, temporal logic has provided the basis for many developments in this area; this is in the form of a logic that permits arguments about happenings over time. Advances in model checking (see Clarke et al., 2009) have made many of the related problems of verifying properties of finite-state systems.

88.2.3 Specification of Nonfunctional Requirements

The specification of nonfunctional requirements is often ignored, and that can lead to serious disruption in project delivery. In reality the nonfunctional specification should be given similar prominence and attention as the functional specification part of an SRS document. It should possess desirable characteristics and be checked for such matters as feasibility and completeness.

88.2.3.1 Classification

Where they are addressed, it is customary to provide a classification of nonfunctional specifications. The IEEE standards (see Kotonya and Sommerville, 1998) use a classification that includes

- Process requirements that cover standards to be used as well as other implementation requirements and importantly delivery dates
- Product requirements that address matters such as efficiency, performance, reliability, and storage requirements as well as safety, security, and usability
- External requirements that address economic considerations, legal matters, etc.

The manner in which these requirements are stated is generally fairly obvious.

88.2.3.2 Reliability and Safety

In the context of reliability requirements, safety requirements, etc., a quantitative approach based on probability can be employed. For instance, in the context of safety, critical systems probability is typically employed to identify the criticality of the software and to capture the required safety levels. Safety integrity levels (SILs) are used to capture the level of criticality of the software and so the required measures of performance of the software. These ideas are captured within the standard IEC 61508—see IEC 61508 (2010).

To illustrate the SILs where the demands on the software are relatively low, the following captures the average probability of failure to perform on demand:

SIL	Maximum Average Probability of Failure (Low Demand)
4	$\geq 10^{-5}$ to $<10^{-4}$
3	$\geq 10^{-4}$ to $<10^{-3}$
2	$\geq 10^{-3}$ to $<10^{-2}$
1	$\geq 10^{-2}$ to $<10^{-1}$

On the other hand, if the demands on the software are very high, even continuous, the following may be more apt:

SIL 1	Maximum Average Probability of Failure (High Demand)
4	$\geq 10^{-9}$ to $<10^{-8}$
3	$\geq 10^{-8}$ to $<10^{-7}$
2	$\geq 10^{-7}$ to $<10^{-6}$
1	$\geq 10^{-6}$ to $<10^{-5}$

The comment should be made that higher SILs are deemed unachievable due to current limits of the technology; there is scope for change here as technology improves.

88.2.3.3 Specification of Security

An increasingly important matter is the specification of security requirements. Such requirements are normally concerned with the careful management and organization of data, but there are huge concerns over such practices as compromising the software that has access to the data.

Of course, this is an ever-changing landscape. Providing nonfunctional requirements in the general area of security is nontrivial. One way of approaching this is to stipulate that steps should be taken to exclude certain common security risks. The latter are known and in Owasp (2013), for instance, there is a list of the 10 most common security risks for 2013.

88.3 Research Issues

Although considerable progress has been made, the field of *SRS* remains one in which there are many challenges. Included in these is the wish to automate to some extent the process of going from an SRS document to code that meets the required objectives. Progress has been made in this regard but much more remains to be done.

The EU funded *Prospectra* project, for instance—see Hoffman and Kreig-Brueckner (1993)—had the ambitious goal of making software production a truly engineering discipline. It sought to support development from a formal specification to executable code through a sequence of correctness preserving transformations and in the process building up knowledge of program construction and the associated proof. There were many interesting developments here but a final usable system did not materialize.

Included among the challenges are

- How can high-quality (meeting the requirements of consistency, completeness, etc.) SRS documents that embrace all aspects of specifications including functional, nonfunctional, and other aspects be produced quickly? What guidance can be provided in this regard?
- Can better approaches be devised to ensure that nonfunctional requirements are captured systematically in a useful and complete manner?
- What guidance can be provided for ensuring that an SRS document can be the basis of the provision of a realistic measure of costs and more generally resources?
- How can formal specification languages and their habitual use in software development process be made ever more accessible to software engineers, with associated proof happening but in a manner that does not require deep insights into logic?
- How can better tools be provided to support the aim of ensuring that software engineering becomes a true engineering discipline, that is, complete with reuse of knowledge about program construction and proof?
- How can correctness preserving program transformations be developed starting with SRS documents, taking account of advances in multi-core?

88.4 Summary

There are differing approaches to the topic of SRS, ranging from the relatively informal through to the very formal. Having high-quality specifications tends to becomes more and more important as the size and complexity of a project increases and also as its significance (in terms of safety, security, etc.) increases. In all cases, high-quality specifications can dramatically enhance the efficiency of the software development process; the more formal approaches open doors for an increased use of support tools and rigorous approaches. Although these are gradually gaining support and favor, some continue to see the lack of skills in this area as a cause for concern and a source of risk.

In the background, there are serious systems issues that need to be addressed. Thus, high-quality specifications are the ultimate goal, but these need to be present in larger systems that are themselves of high quality (thus reliable) and the tools used to process specifications themselves need to be of high quality.

References

Clarke, E.M., Emerson, E.A., and Sifakis, J. Model-checking: Algorithmic verification and debugging. *Communications of the ACM* 52(11), 75–84, 2009.

Dawes, J. *The VDM-SL Reference Guide*, Pitman Publishing, London, U.K., 1991.

Dorfman, M. and Richard H.T. *Software Engineering*, IEEE Computer Society Press, Los Alamitos, CA, 1997.

Gehani, N. and McGettrick, A.D. (eds.) *Software Specification Techniques*, Addison Wesley, Reading, MA, 1986.

Guttag, J.V., Horning, J.J., and A. Modet. Report on the larch shared language, version 2.3, SRC Research Report No. 58, 1990.

Hoffman, B. and Kreig-Brueckner, B. (eds.) *Program Development by Specification and Transformation, the PROSPECTRA Methodology, Language Family, and System*. Lecture Notes in Computer Science 680, Springer, New York, 1993.

IEC 61508. *Functional Safety of Electrical/Electronic/Programmable Electronic Safety-Related Systems*, 2nd edn. International Electrotechnical Commission, Geneva, Switzerland, April 2010.

IEEE, 1998 830-1998. IEEE recommended practice for software requirements specification, sponsored by the IEEE Computer Society, New York.

Kotonya, G. and Sommerville, I. *Requirements Engineering: Processes and Techniques*, John Wiley & Sons, Chichester, NY, 1998.

Malcolm, G. and Goguen, J.A. An executable course in the algebraic semantics of imperative programs, in *Teaching and Learning Formal Methods*, Hinchey, M. and Nevill Dean, C. (eds.), Academic Press, London, U.K., pp. 161–179, 1996.

Milner, R. *A Calculus of Communicating Systems*, Springer Verlag, New York, 1980.

OWASP, 2013 The Owasp Top 10, produced by the Open Web Application Security Project, 2010. See http://www.owasp.org (accessed on October 9, 2013).

Spivey, J.M. *The Z Notation: A Reference Manual*, 2nd edn., Prentice-Hall International, London, U.K., 1992.

Yang, J. and Hawblitzel, C. Safe to the last instruction: Automated verification of a type-safe operating system. *Communications of the ACM*, 54(12), 123–131, 2011.

89

Software Model Checking

Alastair Donaldson
Imperial College

89.1 Introduction

This chapter provides an overview of techniques and tools for analyzing the correctness of software systems using *model checking*. After providing some background on traditional finite-state model checking and related tools (Section 89.2, we give an overview of two classes of techniques that have allowed model checking to be applied directly to the source code of large software systems: counterexample-guided abstraction refinement (CEGAR) using predicate abstraction (Section 89.3), and bounded model checking (Section 89.4). The chapter concludes with a brief survey of tools that leverage traditional model checking methods for the analysis of software systems (Section 89.5). Software model checking techniques based on CEGAR and bounded model checking are also treated in the *Software Verification* chapter of the Handbook of Satisfiability [73]; we believe these chapters complement one another.

89.2 Model Checking

We begin with some background on traditional model checking techniques for finite-state systems. For full coverage of this topic, we refer the reader to several excellent books and tutorials [3,35,66,78,79]).

The model checking problem can be stated as follows:

Given a finite-state model M and a logical property ϕ, does M satisfy ϕ?

The finite-state model M under consideration is usually a *Kripke structure* consisting of a set of states, a transition relation over pairs of states, and a description of the logical formulas that hold in each state, while ϕ is a formula in some temporal logic, such as linear temporal logic (LTL) or computation tree logic (CTL).

A *model checker* is an algorithm for answering the model checking problem with respect to a specific temporal logic. A model checking algorithm either determines that the model satisfies the property, or produces a *counterexample*: a sequence of state transitions through M which demonstrates that the property is violated. Among the many model checking tools that have been designed, perhaps the most notable are SPIN [66] and SMV [77]. Practical model checking tools allow models to be expressed using relatively high-level description languages, such as Promela (the input language for SPIN) or the SMV language.

Model checking can be used in the analysis of any kind of system (e.g., a hardware design, protocol specification or software program) if it is possible to express desired properties of the system as temporal logic formulas, and if it is possible to obtain a finite-state model that accurately describes behaviors of the system that are relevant to these properties. In this case a model checker can be used to prove that the system satisfies the properties of interest, or to yield counterexamples demonstrating that the system is incorrect. The production of counterexamples is a key benefit of model checking over other formal verification techniques: counterexamples can be invaluable aid to engineers in the process of debugging and fixing a system. In practice, model checking tools operate on models provided using high-level description languages which can be expanded automatically into Kripke structures.

The first algorithms for model checking date from the early 1980s [28,83]; see [27] for a retrospective on the birth of the field. The development of *symbolic model checking* [21,77] in the early 1990s led to model checking being used extensively in industrial hardware verification. The application of model checking to software verification, the focus of this chapter, has been more recent.

The main challenge associated with practical model checking is *state space explosion*. Typically a model is described as the composition of a number of components, for example variables or processes, each describing aspects of the system under consideration. The potential state space of the model grows exponentially with the number of components, and can quickly become too large to explore exhaustively using standard algorithms. A significant portion of model checking-related research has been on techniques for ameliorating state space explosion while maintaining soundness guarantees, including symbolic representation [21,77], partial order reduction [51,81], symmetry reduction [31,48,68,91], and abstraction [30,55].

The state space explosion problem is particularly relevant to software model checking because programs exhibit extremely large and often *infinite* state-spaces. This can stem from potentially unbounded allocation of heap data structures, and from recursive procedure calls. In some programming languages, numeric data types such as integers are regarded as having infinite range. In mainstream languages such as C, C++ and Java, all numeric types have finite (though often very large) domains.

89.3 CEGAR-Based Software Model Checking

Counterexample-guided abstraction refinement (CEGAR) [29] aims to leverage the power of traditional finite-state model checking techniques for the analysis of systems with very large or infinite state spaces. Model checking based on CEGAR was originally proposed for the analysis of finite-state transition systems with intractably large state spaces, based on the following idea:

1. Given a description of a finite state transition system C, compute an *abstract* transition system A such that A has a smaller state space than C and A *over-approximates* C: for every path through P there is a corresponding abstract path through A, thus A contains at least the behaviors of P.
2. Apply traditional finite-state model checking techniques to decide whether A is correct. If A is correct then, because A over-approximates C, correctness of C follows.
3. If A is *incorrect*, a *counterexample* π showing incorrectness of A is produced by the model checker. Examine π to determine whether it is also a counterexample to the correctness of C. If it is, incorrectness of C follows directly.

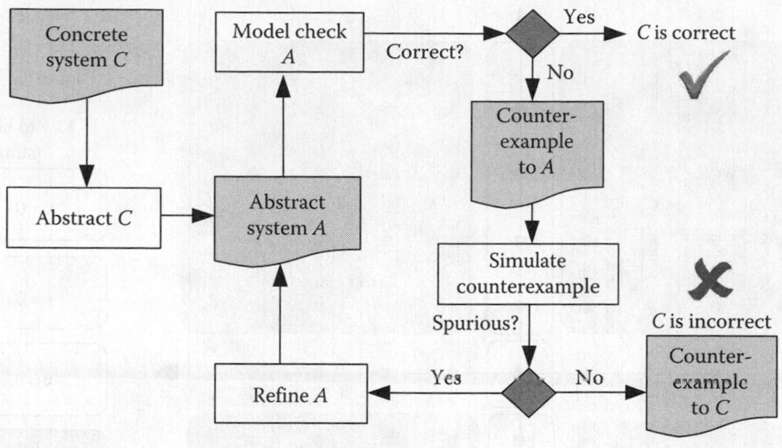

FIGURE 89.1 Counterexample-guided abstraction refinement.

4. If π is deemed *spurious* (it is a counterexample to the correctness of A but not to that of C) then extract information from π to *refine* the abstract program A such that π (and ideally a whole family of counterexamples to which π belongs) is no longer a counterexample to A, and continue from step 2.

The CEGAR process is illustrated in Figure 89.1.

The key motivations for CEGAR are that while there may *exist* an abstraction of a large transition system that facilitates sound, efficient verification, it may not be straightforward to directly pinpoint the required abstraction. CEGAR allows verification to start with a coarse abstraction and gradually refine this abstraction in response to counterexamples, driving the abstraction closer to a point where verification will succeed or, during this process, discovering a genuine counterexample to correctness.

For finite-state transition systems, the CEGAR process is guaranteed to terminate, though in the worst case the state-space of the abstract system may grow as large as that of the concrete system [29].

89.3.1 Predicate Abstraction

CEGAR-based model checking can also be applied to infinite-state systems, including software, but in this case there is no termination guarantee. Each time the abstract system is refined it is possible for new spurious counterexamples to manifest, and there is no limit to how large the abstract system can grow. Nevertheless, CEGAR-based model checking has proven successful in the analysis of infinite state software systems, as demonstrated by several tools including SLAM [10], BLAST [15], SATABS [33], MAGIC [25], and CPACHECKER [16].

The majority of software model checkers that employ CEGAR use *predicate abstraction* [55] as their abstraction method. Predicate abstraction involves choosing a fixed set of predicates over the state of a system and then representing a concrete state abstractly according to the truth values of these predicates when evaluated with respect to the state. For example, consider a simple system with two integer variables x and y. The state space of this system is infinite, consisting of all integer pairs. If we perform predicate abstraction over this state space with respect to predicates $p \triangleq x = 0$ and $q \triangleq y = 0$, then there are four abstract states, as depicted in Figure 89.2, corresponding to each of the four possible valuations of p and q.

A predicate abstraction technique for C programs was proposed in [7]. This method takes a C program together with a set of predicates, and produces a corresponding *Boolean program*. The Boolean

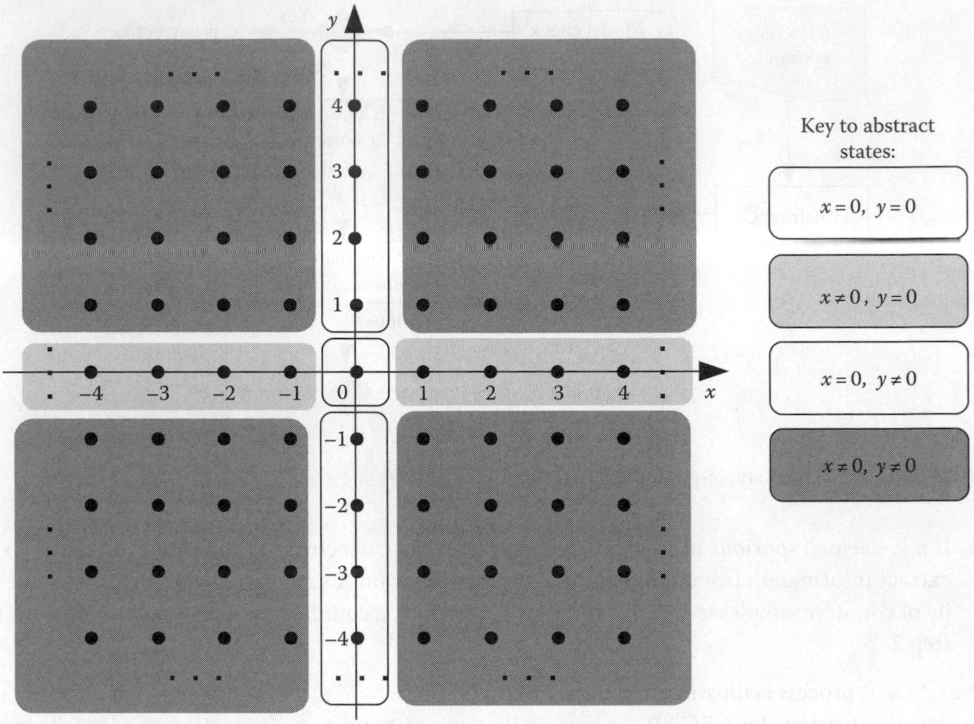

FIGURE 89.2 Predicate abstraction of a system with two integer variables *x* and *y*, with respect to predicates $x = 0$ and $y = 0$. The abstract state space comprises four abstract states, one for each valuation of the predicates.

program has identical control structure to the C program, but instead of operating over the original program it operates over a set of Boolean variables, one for each predicate.

We illustrate predicate abstraction for C programs using an example adapted from [10], and simplified further for presentation here. Consider the fragment of C code shown in Figure 89.3a, which is originally derived from a peripheral component interconnect (PCI) device driver [10]. The method `process` iterates through a queue of requests via global variable `req`. We omit details of the structure of requests, except that the `reqest_t` structure has a `next` field. Before accessing the request queue, the method acquires a lock, shown in Figure 89.3a for simplicity by setting variable `locked` to `true`. The lock is released by setting `locked` to `false`.

We wish to verify that the `process` function obeys correct locking discipline, that is that the lock is not acquired while it is already held, and that the lock not released unless it is held. In Figure 89.3a, we have instrumented the code with assertions about `locked` to check this.

Figure 89.3b is a Boolean program obtained by applying predicate abstraction to the program of Figure 89.3a with respect to the predicate `locked`. The only data tracked by the Boolean program, in addition to the program counter, is the truth of `locked`. This means that statement `locked = true` at line 11 of Figure 89.3a maps to an identical statement in Figure 89.3b, but statement `nPktsOld = nPkts` at line 13 is replaced with `skip` (a no-op), because this statement cannot affect the truth of `locked`. The condition `req` at line 14 is replaced with the *nondeterministic* condition *: the only data available to the Boolean program is the truth of `locked`, so whether `req` is null cannot be precisely determined. The Boolean program soundly over-approximates this condition by evaluating it non-deterministically. The translation of the remaining statements of Figure 89.3a to corresponding statements in Figure 89.3b is similar.

Notice that for every execution trace through the program of Figure 89.3a there is a corresponding trace through the program of Figure 89.3b such that the value of `locked` is identical throughout both traces.

```
 1  unsigned nPkts, nPktsOld;
 2  request_t* req;
 3  bool locked;                    bool locked;                   bool locked, nP_eq_nPO;
 4
 5  // Precondition: !locked       // Precondition: !locked       // Precondition: !locked
 6  void process() {                void process() {               void process() {
 7
 8    do {                            do {                           do {
 9      // Acquire lock
10      assert(!locked);                assert(!locked);               assert(!locked);
11      locked = true;                  locked = true;                 locked = true;
12
13      nPktsOld = nPkts;               skip;                          nP_eq_nPO = true;
14      if(req) {                       if(*) {                        if(*) {
15        req = req->next;                skip                           skip;
16
17        // Release lock
18        assert(locked);                 assert(locked);                assert(locked);
19        locked = false;                 locked = false;                locked = false;
20
21        nPkts++;                        skip;                          nP_eq_nPO =
22                                                                         nP_eq_nPO ?
                                                                           false : *;
23      }                               }                              }
24    } while (nPkts !=               } while (*);                   } while (!nP_eq_nPO);
      nPktsOld);
25
26    // Release lock
27    assert(locked);                 assert(locked);                assert(locked);
28    locked = false;                 locked = false;                locked = false;
29  }                               }                              }
    (a)                             (b)                            (c)
```

FIGURE 89.3 Example programs illustrating CEGAR-based model checking. (a) Device driver code fragment, instrumented for verification of correct locking discipline. (b) Boolean program obtained after abstraction with respect to predicate locked. (c) Boolean program obtained after abstraction with respect to predicates locked and `nPkts==nPktsOld`. (From Ball, T. and Rajamani, S. K., The SLAM toolkit, In Berry, G., Comon, H., and Finkel, A., eds., *CAV*, volume 2102 of *Lecturer Notes in Computer Science*, Springer, New York, pp. 260–264, 2001.)

As we shall see in Section 89.3.3, the reverse is not true: the Boolean program contains additional execution traces that are not feasible in the original program.

Now let us consider applying predicate abstraction to the program of Figure 89.3a with respect to the predicates `locked` and `nPkts==nPktsOld`. The resulting Boolean program is shown in Figure 89.3c. Predicate `locked` is treated as in Figure 89.3b. Boolean variable `nP_eq_nPO` tracks the predicate `nPkts==nPktsOld`. Initially the value of this predicate is unknown. At line 13 the predicate becomes true due to the assignment `nPkts = nPktsOld` in the original program. At line 21 the predicate is updated to model `nPkts` being incremented. It is clear that if `nPkts` and `nPktsOld` are equal then they will no longer be equal after `nPkts` is incremented. However, if the variables are *not* equal then incrementing `nPkts` may or may not lead to equality: if `nPkts==nPktsOld-1` holds then equality will result after the increment, otherwise it will not. In the absence of any such further information about `nPkts` and `nPktsOld`, in this case the best we can say is that after the increment the value of the predicate becomes unknown. The statement:

```
nP_eq_nPO = nP_eq_nPO ? false :*;
```

at line 21 of Figure 89.3c reflects this reasoning.

The predicate `nPkts==nPktsOld` allows the guard of the `do..while` loop to be represented precisely at line 24 of Figure 89.3c. Because the two predicates under consideration say nothing about `req`, testing and updating `req` at lines 14 and 15 of Figure 89.3a still lead to a nondeterministic choice and no-op, respectively, at the corresponding lines of Figure 89.3c.

Boolean and Cartesian Abstractions: We have illustrated the process of predicate abstraction informally through an example. For a complete procedure for this translation, treating pointers and procedure calls, see [7].

Predicate abstraction can be applied with varying degrees of precision. At one extreme, the *Boolean abstraction* represents each program statement as the most precise possible predicate transformer [55]. In the worst case, for a set of *n* predicates, this can involve checking the feasibility of all $2^n \times 2^n$ potential predicate transformations. An alternative is the *Cartesian abstraction*, combined with the *maximum cube length approximation* [7], which allows faster but less precise predicate abstraction. The loss of precision may require additional refinement steps as described in Section 89.3.4. The relationship between the Boolean and Cartesian abstractions has been studied formally [8].

89.3.2 Boolean Program Model Checking

The Boolean program corresponding to a recursion-free C program has a finite state space. If the C program is recursive, then so is the corresponding Boolean program, in which case the Boolean program corresponds to a *pushdown system*, for which the model checking problem has been studied (see, e.g., [20,22,50]). Model checking tools for Boolean programs include Bebop [9], Moped [86], BoPPo [36], and Boom [12] (though BoPPo and Boom do not support recursive Boolean programs). Each of these tools uses symbolic model checking techniques based on binary decision diagrams to curb state explosion.

By construction, the Boolean program obtained from a C program is an over-approximation, thus if Boolean program model checking determines that the Boolean program is correct it follows that the C program is also correct. In the best case, it may be possible to prove correctness of an infinite-state C program by performing predicate abstraction and Boolean program model checking with respect to a small number of predicates. However, if model checking reveals that the Boolean program is not correct, this does not directly reveal anything about the original program; further analysis is required.

89.3.3 Counterexample Simulation

Continuing the example of Section 89.3.1, suppose that we have computed the Boolean program of Figure 89.3b with respect to the predicate `locked`. Model checking this Boolean program will result in a counterexample because the Boolean program is *not* correct. One counterexample that might be returned by a Boolean program model checker is the trace that, starting with `locked = false`, sets `locked` to `true` at line 11, enters the `if` statement at line 14, sets `locked` to `false` at line 19 and leaves the `do..while` loop at line 24. As a result, control reaches line 27 with `locked` set to `false`, and the assertion fails.

It is now necessary to decide whether this counterexample trace corresponds to a feasible trace of the original program. This can be achieved by building a logical formula describing the conditions for the Boolean program counterexample to be followed in the original program and ending in a state where the assertion does not hold. For the counterexample, we have just described the corresponding formula is as follows:

$$!\texttt{locked}_0{}^{5} \wedge \texttt{locked}_1{}^{11} \wedge \texttt{nPktsOld}_1 = \texttt{nPkts}_0{}^{13} \wedge \texttt{req}_0 \neq 0^{14} \wedge \texttt{req}_1 = \texttt{req}_1\texttt{->next}^{15} \wedge$$

$$!\texttt{locked}_2{}^{19} \wedge \texttt{nPkts}_1 = \texttt{nPkts}_0 + 1^{21} \wedge \texttt{nPktsOld}_1 = \texttt{nPkts}_1{}^{24} \wedge !\texttt{locked}_2{}^{27}$$

The superscripts in this formula are for illustration only and indicate from which line of Figure 89.3a each conjunct arises. Multiple assignments to variables are modeled in the formula by giving each variable a fresh index (indicated by a subscript) each time it is assigned, in the style of static single assignment form.

Satisfiability of the formula corresponding to a counterexample can be checked by a state-of-the-art theorem prover or SAT/SMT solver. If the formula is *satisfiable* then abstract counterexample does correspond to a concrete counterexample, and the satisfying assignment provides a test input to the program that will cause the bug to manifest. If the formula is *unsatisfiable*, the abstract counterexample is *spurious*: it does not correspond to a trace of the original program, and the abstraction must be refined to block the counterexample.

The aforementioned formula is easily seen to be unsatisfiable due to the unsatisfiable subformula: $\texttt{nPktsOld}_1 = \texttt{nPkts}_0{}^{13} \wedge \texttt{nPkts}_1 = \texttt{nPkts}_0 + 1^{21} \wedge \texttt{nPktsOld}_1 = \texttt{nPkts}_1{}^{24}$ and thus the counterexample is spurious.

89.3.4 Abstraction Refinement

On discovering a spurious counterexample, it is necessary to *refine* the abstraction, making it sufficiently precise that the counterexample trace is blocked. Eliminating spurious counterexamples clears the way for either finding genuine bugs, or proving correctness.

Two complimentary approaches to abstraction refinement are used by CEGAR-based model checkers: *transition refinement* and *predicate discovery*.

As discussed in Section 89.3.1, it may be infeasible to compute a Boolean program that uses the most precise abstraction, the *Boolean abstraction*, with respect to a set of predicates. Instead, less precise abstractions can be used: the SLAM tool uses *Cartesian abstraction* with *maximum cube length approximation* [7], while SATABS [33] employs an extremely coarse method known in the community as *fast abstraction*. This source of imprecision can lead to spurious counterexamples, which can be refined away using *transition refinement* [5]. Transition refinement considers each abstract transition $a \rightarrow b$ occurring in the counterexample trace. Suppose that the transition $a \rightarrow b$ is associated with the abstract statement at program point i. We can check, using a theorem prover, whether there exists concrete program states s and t such that: s is abstracted by a, t is abstracted by b, and the concrete statement at program point i leads to a transition between s and t. If no such pair of concrete states exists, then the transition $a \rightarrow b$ is *spurious*: it does not belong to the Boolean abstraction of the program, and can thus be safely removed by adding constraints to the Boolean program. The process for obtaining and adding such constraints is described in detail in [5].

If a counterexample is spurious but contains no individually spurious transitions, then even the most precise Boolean abstraction is insufficient for proving the program with respect to the current set of predicates. In this case, the abstraction can be refined by adding further predicates. Some implementations choose to add further predicates before the possibilities for applying transition refinement have been exhausted [5]. Several schemes have been proposed to discover new predicates from counterexamples.

The simplest approach is to symbolically compute the weakest precondition from the failed assertion to the start of the program along the counterexample trace, and to add some or all of the atomic predicates occurring in the resulting formula to the set of predicates to be used for abstraction. This scheme is straightforward to implement and relatively predictable, but can lead to the generation of a large number of irrelevant predicates, and may generate predicates that are too counterexample-specific to facilitate verification. The problem of predicate explosion can be mitigated to some extent by alternating between weakest precondition and strongest postcondition-based analyzes [4], while interpolation techniques have shown success in the generation of relevant and general predicates [63,71].

Returning to our simple running example: transition refinement does not apply to the counterexample trace described in Section 89.3.3. Applying any of the commonly employed predicate discovery techniques identifies the predicate `nPkts==nPktsOld`. Adding this to our set of predicates leads to the Boolean program shown in Figure 89.3c which we already discussed in Section 89.3.1. Model checking for this Boolean program succeeds, thus we can conclude that the program of Figure 89.3a is correct.

89.3.5 Notes on CEGAR

The main bottlenecks for CEGAR-based verification using predicate abstraction are computing and model checking the abstraction. Adding further predicates to refine spurious counterexamples can quickly lead to an abstraction that is expensive to compute and infeasible to check. *Lazy abstraction* [64] aims to address this problem by localizing abstraction refinement. Rather than working with Boolean programs, this technique computes an *abstract reachability tree* (ART) with respect to a set of predicates. When a spurious counterexample through the ART is discovered, yielding new predicates, only portions of the ART relevant to the counterexample are refined. This avoids state explosion in the remainder of the ART.

The scalability of predicate abstraction can be improved by computing predicate transformers with respect to blocks of statements, rather than individual statements. This *large block encoding* technique [14] can dramatically reduce the size of the abstract state space, which is of prime importance for techniques based on ARTs, and facilitates the use of the more precise Boolean abstraction over Cartesian abstraction (see Section 89.3.1). The CPAchecker tool [16] implements a generalization of the large block encoding technique [17].

CEGAR-based model checking for concurrent programs has been considered in several works. The MAGICis designed for analysis of concurrent programs where communication is by message passing [25]. An extension of the BLAST model checker focuses on the detection of data races in shared variable concurrent programs [62]. Two recent techniques and tools aim at verifying freedom from assertion failures in shared variable concurrent programs: THREADER [57,58] and SYMMPA [43,44]. THREADER applies thread modular reasoning to aim to combat state explosion arising due to concurrency, while SYMMPA employs a combination of symmetry reduction and BDD representation [13].

89.4 Bounded Model Checking of Software

Bounded model checking (BMC) is a method for performing depth-bounded verification of transition systems [18]. Given a transition system described by an initial state predicate $I(\mathbf{v})$ and a left-total transition relation* $T(\mathbf{v}, \mathbf{v}')$, and a predicate $P(\mathbf{v})$ describing a property that should hold in all reachable states (where $\mathbf{v}$ and $\mathbf{v}'$ denote vectors of system variables), BMC involves checking the following formula for some $k \geq 0$:

$$I(\mathbf{v}_0) \wedge T(\mathbf{v}_0, \mathbf{v}_1) \wedge \cdots \wedge T(\mathbf{v}_{k-1}, \mathbf{v}_k) \wedge (\neg P(\mathbf{v}_0) \vee \cdots \vee \neg P(\mathbf{v}_k)) \quad (89.1)$$

* A transition relation $T \subseteq S \times S$ is left-total if for every $s \in S$ there exists some $s' \in S$ such that $(s, s') \in T$. If we view T as a directed graph, this means that every node has at least one successor.

This formula encodes all paths of the transition system of length $k + 1$ that commence in an initial state, and along which the property is violated at some point. If the formula is satisfiable, then the satisfying assignment yields a counterexample demonstrating that the property can be violated. Unsatisfiability of the formula establishes that the property holds in all states that are reachable within the given depth bound; no guarantees are provided about deeper execution traces. The technique can also be used for bounded verification of certain temporal logic properties [18].

Bounded model checking has proved very successful as a technique for hardware verification due to the ability of modern SAT solvers to handle large Boolean formulas. We describe two methods by which BMC has been applied to the analysis of software.

89.4.1 Programs as Transition Systems

Traditional bounded model checking can be applied directly to the analysis of nonrecursive programs by inlining all procedure calls and then interpreting the program control flow graph (CFG) as a transition system. We illustrate this using the example program of Figure 89.4a. This simple program fragment initializes variables i and x to zero, then loops 100 times maintaining the invariant $x = 2 * i$. The program then asserts, incorrectly, that x and i are equal.

A possible CFG for this program is depicted in Figure 89.4b. Each block in the CFG is labeled with a unique number.

By using a *program counter* variable, pc, to record which CFG block is due to be executed next, we can encode the semantics of this program via the following formulas:

$$I(pc, x, i) \triangleq pc = 1$$

$$\begin{aligned} T((pc, x, i), (pc', x', i')) \triangleq\ & pc = 1 \wedge x' = 0 \wedge i' = 0 \wedge pc' = (i' < 100\ ?\ 2 : 3) \\ & \vee\ pc = 2 \wedge x' = x + 2 \wedge i' = i + 1 \wedge pc' = (i' < 100\ ?\ 2 : 3) \\ & \vee\ pc = 3 \wedge pc' = end \wedge x' = x \wedge i' = i \\ & \vee\ pc = end \wedge pc' = end \wedge x' = x \wedge i' = i \end{aligned}$$

$$(pc, x, i) \triangleq (pc = 2 \Rightarrow x = 2 * i) \wedge (pc = 3 \Rightarrow x = i)$$

We have used a special program counter value, *end*, to indicate that execution has terminated. If $pc = end$ the transition system exhibits a self-loop; this ensures that the transition relation is left-total.

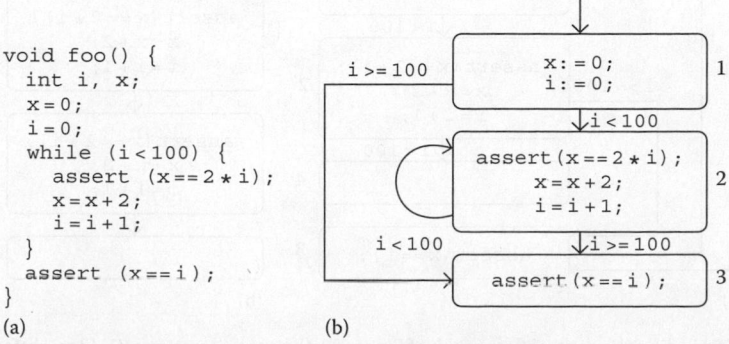

```
void foo() {
  int i, x;
  x=0;
  i=0;
  while (i<100) {
    assert (x==2*i);
    x=x+2;
    i=i+1;
  }
  assert (x==i);
}
```

(a)

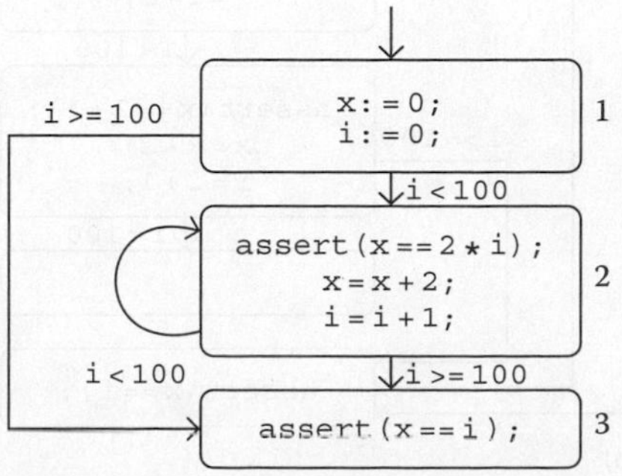

(b)

FIGURE 89.4 A simple program and its control flow graph. (a) Example program. (b) Control flow graph.

Equation 89.1 can now be applied using these formulas to check freedom from assertion violations up to any specified depth. To discover that the assertion at the end of the program can fail, a depth of at least 100 is required.

The F-Soft [69] performs bounded model checking of C programs, treating the program CFG as a transition relation in a manner similar to the above.

89.4.2 Unwinding Loops in Control Flow Graphs

The disadvantage of the method described in Section 89.4.1 is that it does not exploit the structure of the program under consideration. An alternative to unwinding the transition relation associated with a program is to directly unwind loops in the program's CFG.

We illustrate this using the example program of Figure 89.4. Figure 89.5a shows the result of unwinding two iterations of the CFG of Figure 89.4b. Observe that block 2 in Figure 89.4b, the body of the loop, has been duplicated as blocks 2 and 2′ in Figure 89.5a. The edge (2, 3) which leaves the loop in Figure 89.4b results in two edges, (2, 3) and (2′, 3) in Figure 89.5a. The back-edge (2, 2) in Figure 89.4b results in an edge connecting block 2 to block 2′ in Figure 89.5a. Because we have chosen to unwind the loop by just two iterations, the edge leaving block 2′ which corresponds to the back-edge leads to an empty block, 4, with no successors. This causes execution to terminate without error along paths that would exceed the bound of two loop iterations.

The result of this unwinding is a loop-free CFG encoding all program traces that execute at most two loop iterations. Using techniques for computing weakest preconditions [11], this loop-free CFG can be turned into a logical formula whose satisfiability reveals an assertion failure, and whose unsatisfiability implies that the program is correct up to this execution depth. There is clearly room for optimizing this procedure: a cheap static analysis reveals that in the example of Figure 89.5a all the edges guarded by `i >= 100` are infeasible, leading to the simpler CFG of Figure 89.5b in which node 3 is no longer reachable. Such an analysis allows the CFG to be simplified before it is turned into a formula.

The more general case of unwinding k iterations of an arbitrary loop is illustrated in Figure 89.6 (adapted from [41]). Figure 89.6a depicts an arbitrary CFG which contains a natural loop L (see, e.g., [1] for a formal definition of natural loops in control flow graphs). The head of L can be reached from outside

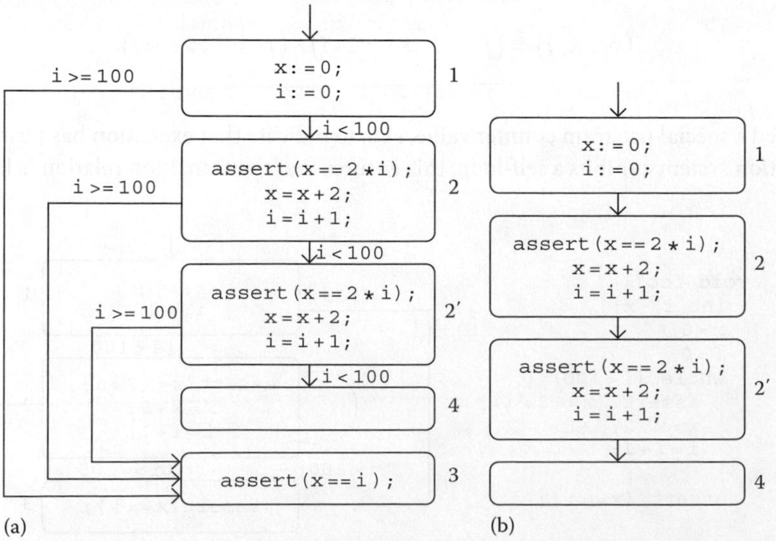

FIGURE 89.5 Unwinding the control flow graph of Figure 89.4b by two iterations. (a) Straightforward unwinding. (b) Unwinding and simplification.

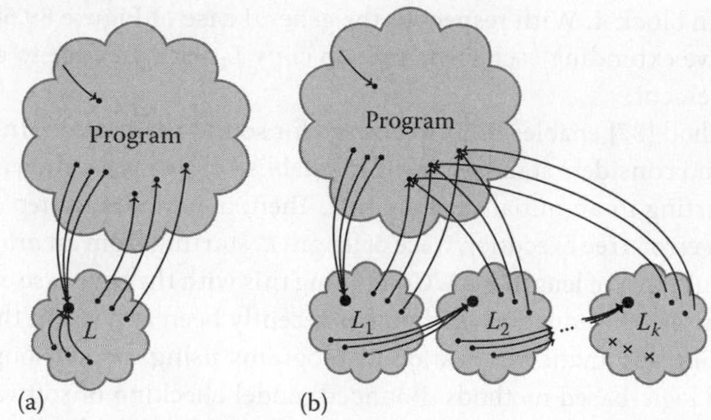

FIGURE 89.6 Illustration of unwinding k iterations of an arbitrary natural loop in a control flow graph. (a) A program containing in a loop L. (b) The program of Figure 89.6a after unwinding k interactions of loop L. Notice that loop back-edges which would lead to a further iteration have been cut.

the loop via a number of CFG edges (edges from *Program* to L). The loop contains a number of *back edges* (edges from nodes in L to the head of L), and a number of *exit edges* (edges from L to *Program*). Figure 89.6b illustrates the effect of unwinding k iterations of loop L. The result is a CFG containing the *Program* part of Figure 89.6a unmodified, together with k copies of the loop body L, denoted $L_1, \ldots, L_k$. Exit edges of L are preserved in each L_i. A back-edge of L leads to an edge from L_i to the head node of L_j, for $1 \leq i < k$. In L_k, edges that would correspond to back edges of L are *not* present; this is indicated by crosses in Figure 89.6b. Omitting these edges restricts program executions to those involving at most k iterations of L, as desired.

Compared with the approach of applying BMC at the transition system level described in Section 89.4.1, loop unwinding has the advantage that different unwinding depths can be applied to different loops in the program. It may be the case that to discover a bug at least k iterations of a certain loop L must be explored, where k is large, while other loops need only be minimally unwound. By unwinding loops individually, it is possible to generate a minimal program unwinding that exposes the bug. Operating at the transition system level it would be necessary to unwind the *entire* transition system up to a bound proportional to k, the unwinding depth for loop L.

Unwinding a single loop by k iterations leads to a linear increase in the size of the CFG. However, if the unwound loop contains a nested loop then the nested loop will be duplicated k times in the unwound CFG. As a result, uniformly unwinding all loops in a loop nest by k iterations leads to $O(k^d)$ growth in the size of the CFG, where d is the depth of the loop nest.

Bounded model checking can be applied to recursive programs by inlining recursive calls up to a given bound.

The CBMC tool [32,34] performs bounded model checking of C programs via loop unwinding and inlining of recursive calls.

89.4.3 Notes on BMC

As discussed earlier, BMC provides no guarantees about program correctness beyond the given execution bound. In embedded software it is common for recursion to be absent and for loops to exhibit fixed upper bounds. In this case, sufficient unwinding of the program transition relation (Section 89.4.1) or loops (Section 89.4.2) is guaranteed to lead to consideration of all possible program behaviors. The loop unwinding method can be combined with the insertion of *unwinding assertions* to determine when sufficient unwinding has been performed to establish program correctness [32,34]: when unwinding k iterations of a loop, we insert an `assert(false);` statement on execution paths that would execute a $(k + 1)$st iteration. In the example of Figure 89.5a, this corresponds to placing an `assert(false);` in block 4. With respect to the general case of Figure 89.6b, adding unwinding assertions would involve extending each cross in loop copy L_k with an edge to a node containing an `assert(false);` statement.

The k-induction method [87] enables BMC to be used for sound verification. In the context of transition systems, the method considers standard BMC to establish a *base case*, showing that all executions of length at most k, starting in an initial state are safe. Then, in addition, a *step case* is checked, in an attempt to show that an error-free execution trace of length k, starting from an *arbitrary* state, cannot be extended to an erroneous trace of length $k + 1$. Combining this with the base case establishes correctness of the transition system. The k-induction method has recently been applied in the context of software verification [41], allowing automatic verification of programs using weaker loop invariants than are required by traditional logic-based methods. Bounded model checking of software using k-induction has also been successful in establishing data race freedom for systems software that manipulates data using direct memory access (DMA) operations [45,46].

Bounded model checking techniques have proven effective in generating test cases for software (see, e.g., [2,61,65]). To generate a test vector that exercises a certain program behavior, one can instrument the program with assertions stating that the desired behavior is *not* possible. Applying BMC to detect violations of this assertion directly yields the desired test vector, if the behavior in question can be exhibited within a reasonable execution bound. Bounded model checking with k-induction has also been used for the detection of equivalent mutants in mutation-based testing [42].

BMC has been applied to the analysis of concurrent programs [39,84]. To limit state explosion arising due to concurrency these works employ the idea of bounding the number of thread context switches that are considered [74,82,88], in addition to bounding execution depth, based on the empirical observation that many concurrency bugs can be exposed using only a small number of thread context switches [76].

A technique closely related to BMC is dynamic symbolic execution, as implemented by tools such as EXE [24] and KLEE [23]. The key difference is that dynamic symbolic execution techniques maintain distinct formulas for different paths through a program, forking execution whenever a branch instruction is encountered such that both branch alternatives are feasible under the current set of constraints. In contrast, a bounded model checker unwinds the entire program, yielding a *single* formula that encodes all execution traces up to a given depth. The disadvantage of exploring execution paths separately, as in dynamic symbolic execution, is *path explosion*: the number of paths that must be considered to guarantee defect freedom up to a given depth k is in the worst-case exponential in d. An advantage is that the formula associated with a specific path remains relatively small and has a conjunctive form for which constraint solvers are highly optimized. The formula generated by program unwinding during bounded model checking can quickly exceed the capacity of state-of-the-art constraint solvers.

89.5 Tools Based on Traditional Model Checking

We conclude the chapter with a brief survey of notable tools that directly employ techniques based on traditional finite-state model checking for the analysis of software.

Bandera and Bogor: The BANDERA tool allows model checking of Java programs through a process of finite-state model extraction [37]. Program slicing is used to fist simplify a Java program with respect to the property under consideration. The program state-space is then restricted through the application of user-provided abstraction functions. This leads to a sound over-approximation, as long as the given abstraction functions are sound. Then, if necessary a *component restriction* phase is used to make the resulting model tractable. This involves, for example, limiting the ranges of variables, or limiting the number of objects that can be allocated in the heap. Component restriction sacrifices soundness, but can allow the detection of subtle errors. The resulting finite-state model is represented in the Bandera Intermediate Language (BIR). A BIR model can then be translated to the input language of any finite-state model checker; translators exist for variety of model checkers, including SPIN [66], dSPIN [40], nuSMV [26], and Java PathFinder [89]. Bandera provides a rich specification language for stating properties to be checked [38] and has been applied in the context of analyzing concurrent Java programs [59].

The BOGOR framework [85] aims to provide an extensible platform for designing software model checkers, motivated by the BANDERA project, during which the often large gap between features of full-blown programming languages and the input languages of finite-state model checking tools proved to be a major source of inefficiency in verification. The core of BOGOR is a model checker for an extended version of the BIR language, where the extensions allow direct encoding of high-level programming constructs.

Java PathFinder: The Java PathFinder (JPF) tool [75,89] supports model checking of Java byte-code, allowing it to be applied to compiled Java programs for which source code is not available. A combination of state compression, partial order and symmetry reduction, abstraction and program slicing is used to combat state space explosion. An earlier version of JPF tool translated Java source code directly to Promela for model checking using SPIN [60].

JPF also provides support for symbolic execution [72], which facilitates test case generation for Java programs [90].

Spin: The SPIN model checker [66] allows fragments of C source code to be included directly in a Promela verification model, and for selected C program variables to be tracked as part of the model state during verification. This allows users of SPIN to extract verification models from C programs, either manually or with the assistance of automated model extraction tools [67].

As discussed earlier, SPIN has been employed as a back-end model checker for a variety of software model checking tools.

VeriSoft: The VeriSoft model checker [53] is geared toward analysis of concurrent C and C++ programs, checking for coordination problems (such as deadlocks), as well as for violations of user-supplied assertions. Based on the observation that the state of C and C++ programs is typically large and complex, VeriSoft employs a *stateless* search approach in which a set of previously visited states is *not* maintained. This dramatically reduces memory consumption during model checking, but means that individual states may be visited many times during exhaustive exploration. VeriSoft employs symmetry and partial order reduction techniques to mitigate this problem to some extent [52]. The VeriSoft approach shares similarities with recent techniques for concurrency testing such as the CHESS tool [80].

89.6 Conclusion

We have surveyed two of the most successful techniques for practical software model checking: predicate abstraction and CEGAR, and bounded model checking, and have provided an overview of other key model checking techniques that are geared towards software analysis.

A common technology shared by many of these techniques is *constraint solving*: the abstraction, counterexample simulation and refinement processes in the CEGAR methods we have studied all involve invoking a constraint solver to decide formulas, and bounded model checking works by unwinding a program into a formula to be checked by a SAT or SMT solver. Thus, a key factor in future scalability of software model checking and other verification techniques will be the extent to which advances in constraint solving continue to improve performance.

Dramatic improvements in the capability of software model checkers have been made over the last 15 years, and there has been some uptake by industry. Arguably the biggest success story for software model checking has been technology transfer from Microsoft's SLAM tool, which is based on predication abstraction and CEGAR, to Static Driver Verifier (SDV) that ships with the Windows Driver Development kit [6].

Despite this success, it is fair to say that software model checking, and verification in general, is *not* employed widely in industry at present. In our view, there are three main barriers to widespread industrial adoption of these techniques, and thus areas for further research:

- *Scalability*: Model checking complex systems is typically expensive; it is not unusual for a verification run to take hours or days to complete.
- *Lack of specifications*: Formal analysis of software depends on the existence of meaningful properties to check. Such specifications are usually not present in general-purpose software.
- *Lack of automation*: Unlike lightweight approaches to improving software reliability, such as regression testing or basic type checking, applying a software model checker is rarely a "push-button" process; significant manual intervention and expertise is usually required.

References

1. A. V. Aho, M. S. Lam, R. Sethi, and J. D. Ullman. *Compilers: Principles, Techniques, and Tools.* Pearson Education, Inc, Boston, MA, 2006.
2. D. Angeletti, E. Giunchiglia, M. Narizzano, A. Puddu, and S. Sabina. Automatic test generation for coverage analysis using cbmc. In R. Moreno-Díaz, F. Pichler, and A. Quesada-Arencibia, eds., *EUROCAST*, volume 5717 of *Lecture Notes in Computer Science*, pp. 287–294. Springer, New York, 2009.
3. C. Baier and J. Katoen. *Principles of Model Checking.* MIT Press, Cambridge, MA, 2008.
4. T. Ball, E. Bounimova, R. Kumar, and V. Levin. SLAM2: Static driver verification with under 4% false alarms. In Bloem and Sharygina, 19, pp. 35–42.
5. T. Ball, B. Cook, S. Das, and S. K. Rajamani. Refining approximations in software predicate abstraction. In Jensen and Podelski, 70, pp. 388–403.
6. T. Ball, B. Cook, V. Levin, and S. K. Rajamani. SLAM and static driver verifier: Technology transfer of formal methods inside microsoft. In E. A. Boiten, J. Derrick, and G. Smith, eds., *Integrated Formal Methods, 4th International Conference, IFM 2004, Canterbury, U.K., April 4–7, 2004, Proceedings*, volume 2999 of *Lecture Notes in Computer Science*, pp. 1–20. Springer, New York, 2004.
7. T. Ball, R. Majumdar, T. D. Millstein, and S. K. Rajamani. Automatic predicate abstraction of C programs. In M. Burke and M. Lou Soffa, eds., *PLDI*, pp. 203–213. ACM, New York, 2001.
8. T. Ball, A. Podelski, and S. K. Rajamani. Boolean and cartesian abstraction for model checking c programs. *STTT*, 5(1):49–58, 2003.
9. T. Ball and S. K. Rajamani. Bebop: A symbolic model checker for boolean programs. In K. Havelund, J. Penix, and W. Visser, eds., *SPIN*, volume 1885 of *Lecture Notes in Computer Science*, pp. 113–130. Springer, New York, 2000.
10. T. Ball and S. K. Rajamani. The SLAM toolkit. In G. Berry, H. Comon, and A. Finkel, eds., *CAV*, volume 2102 of *Lecture Notes in Computer Science*, pp. 260–264. Springer, New York, 2001.
11. M. Barnett and K. R. M. Leino. Weakest-precondition of unstructured programs. In M. D. Ernst and T. P. Jensen, eds., *PASTE*, pp. 82–87. ACM, New York, 2005.
12. G. Basler, M. Hague, D. Kroening, C.-H. Luke Ong, T. Wahl, and H. Zhao. Boom: Taking boolean program model checking one step further. In Esparza and Majumdar, 49, pp. 145–149.
13. G. Basler, M. Mazzucchi, T. Wahl, and D. Kroening. Context-aware counter abstraction. *Formal Methods in System Design*, 36(3):223–245, 2010.
14. D. Beyer, A. Cimatti, A. Griggio, M. E. Keremoglu, and R. Sebastiani. Software model checking via large-block encoding. In *Proceedings of 9th International Conference on Formal Methods in Computer-Aided Design* (FMCAD) November 15–18, Austin, Texas, pp. 25–32. IEEE, 2009.
15. D. Beyer, T. A. Henzinger, R. Jhala, and R. Majumdar. The software model checker blast. *STTT*, 9(5–6):505–525, 2007.
16. D. Beyer and M. E. Keremoglu. CPAchecker: A tool for configurable software verification. In Gopalakrishnan and Qadeer, 54, pp. 184–190.

17. D. Beyer, M. E. Keremoglu, and P. Wendler. Predicate abstraction with adjustable-block encoding. In Bloem and Sharygina, 19, pp. 189–197.
18. A. Biere, A. Cimatti, E. M. Clarke, O. Strichman, and Y. Zhu. Bounded model checking. *Advances in Computers*, 58:117–148, 2003.
19. R. Bloem and N. Sharygina, eds. *Proceedings of 10th International Conference on Formal Methods in Computer-Aided Design, FMCAD 2010*, Lugano, Switzerland, October 20–23. IEEE, 2010.
20. A. Bouajjani, J. Esparza, and O. Maler. Reachability analysis of pushdown automata: Application to model-checking. In A. W. Mazurkiewicz and J. Winkowski, eds., *CONCUR*, volume 1243 of *Lecture Notes in Computer Science*, pp. 135–150. Springer, New York, 1997.
21. J. R. Burch, E. M. Clarke, K. L. McMillan, D. L. Dill, and L. J. Hwang. Symbolic model checking: 10^{20} states and beyond. *Information and Computation*, 98(2):142–170, 1992.
22. O. Burkart and B. Steffen. Model checking for context-free processes. In R. Cleaveland, ed., *CONCUR*, volume 630 of *Lecture Notes in Computer Science*, pp. 123–137. Springer, New York, 1992.
23. C. Cadar, D. Dunbar, and D. R. Engler. KLEE: Unassisted and automatic generation of high-coverage tests for complex systems programs. In Draves and van Renesse, 47, pp. 209–224.
24. C. Cadar, V. Ganesh, P. M. Pawlowski, D. L. Dill, and D. R. Engler. Exe: Automatically generating inputs of death. *ACM Transactions on Information and System Security*, Article No. 10, 12(2), 2008.
25. S. Chaki, E. M. Clarke, J. Ouaknine, N. Sharygina, and N. Sinha. Concurrent software verification with states, events, and deadlocks. *Formal Aspects of Computing*, 17(4):461–483, 2005.
26. A. Cimatti, E. M. Clarke, F. Giunchiglia, and M. Roveri. NUSMV: A new symbolic model checker. *STTT*, 2(4):410–425, 2000.
27. E. M. Clarke. The birth of model checking. In Grumberg and Veith, 56, pp. 1–26.
28. E. M. Clarke and E. A. Emerson. Design and synthesis of synchronization skeletons using branching-time temporal logic. In D. Kozen, ed., *Logic of Programs*, volume 131 of *Lecture Notes in Computer Science*, pp. 52–71. Springer, New York, 1981.
29. E. M. Clarke, O. Grumberg, S. Jha, Y. Lu, and H. Veith. Counterexample-guided abstraction refinement for symbolic model checking. *Journal of ACM*, 50(5):752–794, 2003.
30. E. M. Clarke, O. Grumberg, and D. E. Long. Model checking and abstraction. *ACM Transactions on Programming Languages and Systems*, 16(5):1512–1542, 1994.
31. E. M. Clarke, S. Jha, R. Enders, and T. Filkorn. Exploiting symmetry in temporal logic model checking. *Formal Methods in System Design*, 9(1/2):77–104, 1996.
32. E. M. Clarke, D. Kroening, and F. Lerda. A tool for checking ansi-c programs. In Jensen and Podelski [70], pp. 168–176.
33. E. M. Clarke, D. Kroening, N. Sharygina, and K. Yorav. Predicate abstraction of ANSI-C programs using SAT. *Formal Methods in System Design*, 25(2–3):105–127, 2004.
34. E. M. Clarke, D. Kroening, and K. Yorav. Behavioral consistency of C and Verilog programs using bounded model checking. In *DAC*, pp. 368–371. ACM, New York, 2003.
35. E. M. Clarke, O. Grumberg, and D. A. Peled. *Model Checking*. MIT Press, Cambridge, MA, 2000.
36. B. Cook, D. Kroening, and N. Sharygina. Symbolic model checking for asynchronous boolean programs. In P. Godefroid, ed., *SPIN*, volume 3639 of *Lecture Notes in Computer Science*, pp. 75–90. Springer, New York, 2005.
37. J. C. Corbett, M. B. Dwyer, J. Hatcliff, S. Laubach, C. S. Pasareanu, Robby, and H. Zheng. Bandera: Extracting finite-state models from Java source code. In C. Ghezzi, M. Jazayeri, and A. L. Wolf, eds., *ICSE*, pp. 439–448. ACM, New York, 2000.
38. J. C. Corbett, M. B. Dwyer, J. Hatcliff, and Robby. Expressing checkable properties of dynamic systems: The Bandera specification language. *STTT*, 4(1):34–56, 2002.
39. L. Cordeiro and B. Fischer. Verifying multi-threaded software using SMT-based context-bounded model checking. In R. N. Taylor, H. Gall, and N. Medvidovic, eds., *ICSE*, pp. 331–340. ACM, New York, 2011.

40. C. Demartini, R. Iosif, and R. Sisto. dSPIN: A dynamic extension of SPIN. In D. Dams, R. Gerth, S. Leue, and M. Massink, eds., *SPIN*, volume 1680 of *Lecture Notes in Computer Science*, pp. 261–276. Springer, New York, 1999.
41. A. F. Donaldson, L. Haller, D. Kroening, and P. Rümmer. Software verification using *k*-induction. In E. Yahav, ed., *SAS*, volume 6887 of *Lecture Notes in Computer Science*, pp. 351–368. Springer, New York, 2011.
42. A. F. Donaldson, N. He, D. Kroening, and P. Rümmer. Tightening test coverage metrics: A case study in equivalence checking using *k*-induction. In B. K. Aichernig, F. S. de Boer, and M. M. Bonsangue, eds., *FMCO*, volume 6957 of *Lecture Notes in Computer Science*, pp. 297–315. Springer, New York, 2010.
43. A. F. Donaldson, A. Kaiser, D. Kroening, M. Tautschnig, and T. Wahl. Counterexample-guided abstraction refinement for symmetric concurrent programs. *Formal Methods in System Design*, 41(1):25–44, 2012.
44. A. F. Donaldson, A. Kaiser, D. Kroening, and T. Wahl. Symmetry-aware predicate abstraction for shared-variable concurrent programs. In Gopalakrishnan and Qadeer, 54, pp. 356–371.
45. A. F. Donaldson, D. Kroening, and P. Rümmer. Automatic analysis of scratch-pad memory code for heterogeneous multicore processors. In Esparza and Majumdar, 49, pp. 280–295.
46. A. F. Donaldson, D. Kroening, and P. Rümmer. Automatic analysis of DMA races using model checking and *k*-induction. *Formal Methods in System Design*, 39(1):83–113, 2011.
47. R. Draves and R. van Renesse, eds. *8th USENIX Symposium on Operating Systems Design and Implementation, OSDI 2008*, December 8–10, 2008, San Diego, CA, *Proceedings*. USENIX Association, Berkeley, CA, 2008.
48. E. A. Emerson and A. P. Sistla. Symmetry and model checking. *Formal Methods in System Design*, 9(1/2):105–131, 1996.
49. J. Esparza and R. Majumdar, eds. *Tools and Algorithms for the Construction and Analysis of Systems, 16th International Conference, TACAS 2010, Held as Part of the Joint European Conferences on Theory and Practice of Software, ETAPS 2010, Paphos, Cyprus*, March 20–28, 2010. *Proceedings*, volume 6015 of *Lecture Notes in Computer Science*. Springer, New York, 2010.
50. A. Finkel, B. Willems, and P. Wolper. A direct symbolic approach to model checking pushdown systems. *Electronic Notes in Theoretical Computer Science*, 9:27–37, 1997.
51. P. Godefroid. *Partial-Order Methods for the Verification of Concurrent Systems—An Approach to the State-Explosion Problem*, volume 1032 of *Lecture Notes in Computer Science*. Springer, New York, 1996.
52. P. Godefroid. Exploiting symmetry when model-checking software. In *Proceedings of Formal Methods for Protocol Engineering and Distributed Systems, FORTE XII / PSTV XIX'99, IFIP TC6 WG6.1 Joint International Conference on Formal Description Techniques for Distributed Systems and Communication Protocols (FORTE XII) and Protocol Specification, Testing and Verification (PSTV XIX)*, J. Wu, S.T. Chanson and Q. Gao (eds.) October 5–8, Beijing, China, pp. 257–275. Kluwer, 1999.
53. P. Godefroid. Software model checking: The VeriSoft approach. *Formal Methods in System Design*, 26(2):77–101, 2005.
54. G. Gopalakrishnan and S. Qadeer, eds. *Computer Aided Verification—23rd International Conference, CAV 2011, Snowbird, UT, USA, July 14–20, 2011. Proceedings*, volume 6806 of *Lecture Notes in Computer Science*. Springer, New York, 2011.
55. S. Graf and H. Saïdi. Construction of abstract state graphs with PVS. In O. Grumberg, ed., *CAV*, volume 1254 of *Lecture Notes in Computer Science*, pp. 72–83. Springer, New York, 1997.
56. O. Grumberg and H. Veith, eds. *25 Years of Model Checking - History, Achievements, Perspectives*, volume 5000 of *Lecture Notes in Computer Science*. Springer, New York, 2008.
57. A. Gupta, C. Popeea, and A. Rybalchenko. Predicate abstraction and refinement for verifying multi-threaded programs. In T. Ball and M. Sagiv, eds., *POPL*, pp. 331–344. ACM, New York, 2011.
58. A. Gupta, C. Popeea, and A. Rybalchenko. Threader: A constraint-based verifier for multi-threaded programs. In Gopalakrishnan and Qadeer, 54, pp. 412–417.

59. J. Hatcliff and M. B. Dwyer. Using the Bandera tool set to model-check properties of concurrent Java software. In K. G. Larsen and M. Nielsen, eds., *CONCUR*, volume 2154 of *Lecture Notes in Computer Science*, pp. 39–58. Springer, New York, 2001.
60. K. Havelund and T. Pressburger. Model checking JAVA programs using JAVA pathfinder. *STTT*, 2(4):366–381, 2000.
61. N. He, P. Rümmer, and D. Kroening. Test-case generation for embedded simulink via formal concept analysis. In L. Stok, N. D. Dutt, and S. Hassoun, eds., *DAC*, pp. 224–229. ACM, New York, 2011.
62. T. A. Henzinger, R. Jhala, and R. Majumdar. Race checking by context inference. In W. Pugh and C. Chambers, eds., *PLDI*, pp. 1–13. ACM, New York, 2004.
63. T. A. Henzinger, R. Jhala, R. Majumdar, and K. L. McMillan. Abstractions from proofs. In N. D. Jones and X. Leroy, eds., *POPL*, pp. 232–244. ACM, New York, 2004.
64. T. A. Henzinger, R. Jhala, R. Majumdar, and G. Sutre. Lazy abstraction. In J. Launchbury and J. C. Mitchell, eds., *POPL*, pp. 58–70. ACM, New York, 2002.
65. A. Holzer, C. Schallhart, M. Tautschnig, and H. Veith. Query-driven program testing. In N. D. Jones and M. Müller-Olm, eds., *VMCAI*, volume 5403 of *Lecture Notes in Computer Science*, pp. 151–166. Springer, New York, 2009.
66. G. Holzmann. *The SPIN Model Checker: Primer and Reference Manual.* 1st edn., Addison-Wesley Professional, Boston, MA, 2003.
67. G. J. Holzmann and M. H. Smith. Software model checking: Extracting verification models from source code. *Software Testing, Verification and Reliability*, 11(2):65–79, 2001.
68. C. Norris Ip and D. L. Dill. Better verification through symmetry. *Formal Methods in System Design*, 9(1/2):41–75, 1996.
69. F. Ivancic, I. Shlyakhter, A. Gupta, and M. K. Ganai. Model checking C programs using F-SOFT. In *Proceedings of the 2005 International Conference on Computer Design* (ICCD 2005), October 2–5, San Jose, CA, pp. 297–308. IEEE Computer Society, 2005.
70. K. Jensen and A. Podelski, eds. *Tools and Algorithms for the Construction and Analysis of Systems, 10th International Conference, TACAS 2004, Held as Part of the Joint European Conferences on Theory and Practice of Software, ETAPS 2004, Barcelona, Spain,* March 29–April 2, 2004, *Proceedings*, volume 2988 of *Lecture Notes in Computer Science*. Springer, New York, 2004.
71. R. Jhala and K. L. McMillan. A practical and complete approach to predicate refinement. In H. Hermanns and J. Palsberg, eds., *TACAS*, volume 3920 of *Lecture Notes in Computer Science*, pp. 459–473. Springer, New York, 2006.
72. S. Khurshid, C. S. Pasareanu, and W. Visser. Generalized symbolic execution for model checking and testing. In H. Garavel and J. Hatcliff, eds., *TACAS*, volume 2619 of *Lecture Notes in Computer Science*, pp. 553–568. Springer, New York, 2003.
73. D. Kroening. Software verification. In A. Biere, M. J. H. Heule, H. van Maaren, and T. Walsh, eds., *Handbook of Satisfiability*, Frontiers in Artificial Intelligence and Applications, Chapter 16, pp. 505–532. IOS Press, Amsterdam, The Netherlands, February 2009.
74. A. Lal and T. W. Reps. Reducing concurrent analysis under a context bound to sequential analysis. *Formal Methods in System Design*, 35(1):73–97, 2009.
75. F. Lerda and W. Visser. Addressing dynamic issues of program model checking. In M. B. Dwyer, ed., *SPIN*, volume 2057 of *Lecture Notes in Computer Science*, pp. 80–102. Springer, New York, 2001.
76. S. Lu, S. Park, E. Seo, and Y. Zhou. Learning from mistakes: A comprehensive study on real world concurrency bug characteristics. In S. J. Eggers and J. R. Larus, eds., *ASPLOS*, pp. 329–339. ACM, New York, 2008.
77. K. L. McMillan. Symbolic model checking: An approach to the state explosion problem. PhD thesis, Pittsburgh, PA, 1992. UMI Order No. GAX92-24209.
78. S. Merz. Model checking: A tutorial overview. In F. Cassez, C. Jard, B. Rozoy, and M. D. Ryan, eds., *MOVEP*, volume 2067 of *Lecture Notes in Computer Science*, pp. 3–38. Springer, New York, 2000.

79. M. Müller-Olm, D. A. Schmidt, and B. Steffen. Model-checking: A tutorial introduction. In A. Cortesi and G. Filé, eds., *SAS*, volume 1694 of *Lecture Notes in Computer Science*, pp. 330–354. Springer, New York, 1999.
80. M. Musuvathi, S. Qadeer, T. Ball, G. Basler, P. A. Nainar, and I. Neamtiu. Finding and reproducing heisenbugs in concurrent programs. In Draves and van Renesse, 47, pp. 267–280.
81. D. Peled. Combining partial order reductions with on-the-fly model-checking. *Formal Methods in System Design*, 8(1):39–64, 1996.
82. S. Qadeer and J. Rehof. Context-bounded model checking of concurrent software. In N. Halbwachs and L. D. Zuck, eds., *TACAS*, volume 3440 of *Lecture Notes in Computer Science*, pp. 93–107. Springer, New York, 2005.
83. J.-P. Queille and J. Sifakis. Specification and verification of conurrent systems in Cesar. In Grumberg and Veith, 56, pp. 216–230.
84. I. Rabinovitz and O. Grumberg. Bounded model checking of concurrent programs. In K. Etessami and S. K. Rajamani, eds., *CAV*, volume 3576 of *Lecture Notes in Computer Science*, pp. 82–97. Springer, New York, 2005.
85. Robby, M. B. Dwyer, and J. Hatcliff. Bogor: An extensible and highly-modular software model checking framework. In *ESEC/SIGSOFT FSE*, pp. 267–276. ACM, New York, 2003.
86. S. Schwoon. *Model-Checking Pushdown Systems*. PhD. Thesis, Technische Universität München, Munich, Germany, June 2002.
87. M. Sheeran, S. Singh, and G. Stålmarck. Checking safety properties using induction and a SAT-solver. In W. A. Hunt Jr. and S. D. Johnson, eds., *FMCAD*, volume 1954 of *Lecture Notes in Computer Science*, pp. 108–125. Springer, New York, 2000.
88. S. La Torre, P. Madhusudan, and G. Parlato. Reducing context-bounded concurrent reachability to sequential reachability. In A. Bouajjani and O. Maler, eds., *CAV*, volume 5643 of *Lecture Notes in Computer Science*, pp. 477–492. Springer, New York, 2009.
89. W. Visser, K. Havelund, G. P. Brat, S. Park, and F. Lerda. Model checking programs. *Automated Software Engineering*, 10(2):203–232, 2003.
90. W. Visser, C. S. Pasareanu, and S. Khurshid. Test input generation with Java PathFinder. In G. S. Avrunin and G. Rothermel, eds., *ISSTA*, pp. 97–107. ACM, New York, 2004.
91. T. Wahl and A. F. Donaldson. Replication and abstraction: Symmetry in automated formal verification. *Symmetry*, 2(2):799–847, 2010.

90 Software Design Strategies

Len Bass
National ICT Australia Ltd

90.1 Introduction

The design of a software artifact is a critical step in making the move from the vision of a system to reality. First in the life cycle comes the vision, then the business case for the vision, then the business case is turned into requirements, and then the artifact is designed, implemented, tested, deployed, and released into the environment. These phases are usually not disjoint in time—the business case informs the vision, the requirements inform the business case and so forth, but the design should reflect the preceding phases (the vision, the business case, and the requirements). It is also influenced by the following phases: the design must be implementable, it should be easily testable, and it should be changeable when the artifact must be modified to reflect environmental changes.

One assumption underlying any design is that the design will satisfy some set of requirements. These requirements may be explicit or implicit, known or unknown, but it is the requirements that drive the design strategy. A consequence of this is that a designer is not looking for an optimum design but rather one that is "good enough" in that it satisfies the requirements. A complication is that some requirements may not be known to the designer during the design process but these requirements, nevertheless, must be met by the design.

Requirements can be categorized into three categories:

- The functional requirements—what the artifact should do.
- The quality requirements—how well the artifact will do what it is designed to do. These requirements are frequently called the nonfunctional requirements or the "ilities." Quality requirements can be run time requirements such as performance or availability or they can be development time requirements such as an enumeration of expected modifications or a requirement that the development be distributed across specific teams in specific locations.
- The business requirements—what constraints the various stakeholders add to the other requirements.

We now turn to the principles underlying various design strategies.

90.2 Underlying Principles

A design is a point within a "design space." This is a space formed by considering all possible decisions that might be made within the design process. Since nontrivial systems can have on the orders of thousands of decisions, each with multiple alternatives, arbitrarily searching this space is clearly not feasible. Thus, a systematic method for navigating the design space is critical to produce acceptable designs in a reasonable amount of time.

We begin by discussing two background concepts—views or viewpoints and patterns and tactics. Then we move to the two basic principles of a systematic design strategy—decomposition and generate and test.

90.2.1 Views

Complicated systems such as software systems are difficult to describe because the static version of the system (the code) can have very different structure and properties than the dynamic version of the system (the system in execution). A description of a set of element types in a computer system and their relationships is called a view (Clements et al. 2010). The perspective from a stakeholder who is interested in some properties of the system is called a viewpoint (ISO/IEC/IEEE 2011).

Common types of views are the module views, those that represent the code structures (see Figure 90.1); the component and connector views, those that represent dynamic behavior involving processes,

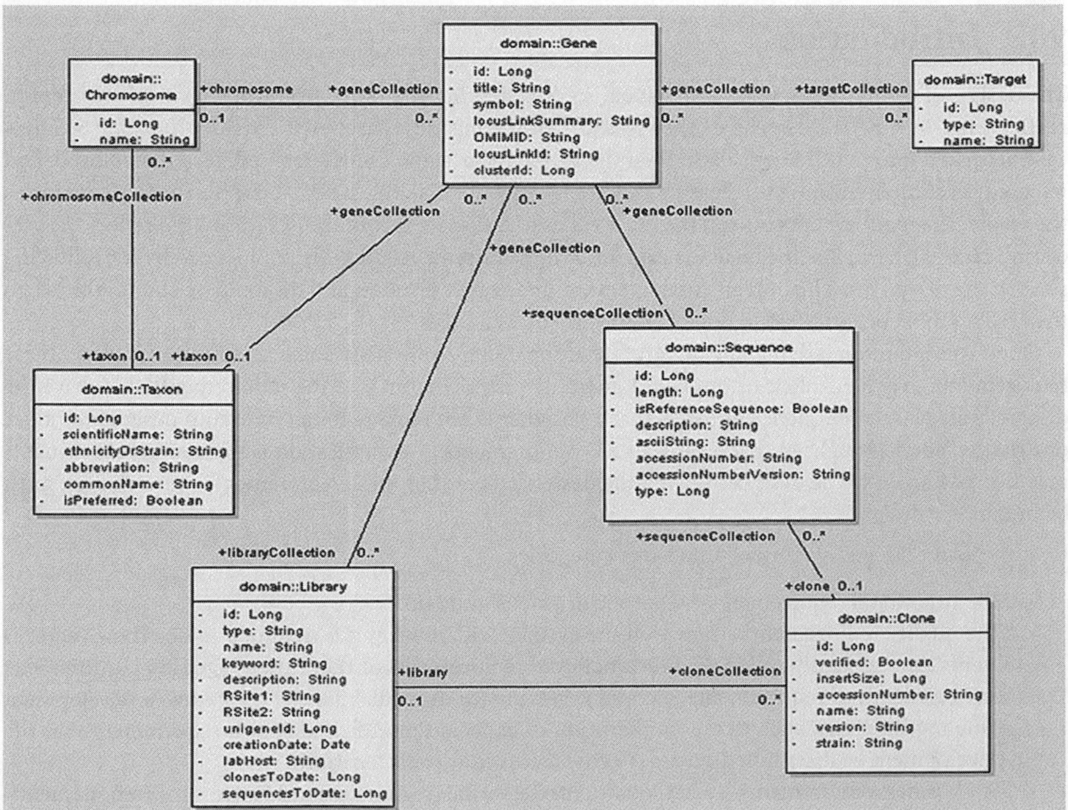

FIGURE 90.1 A sample class diagram.

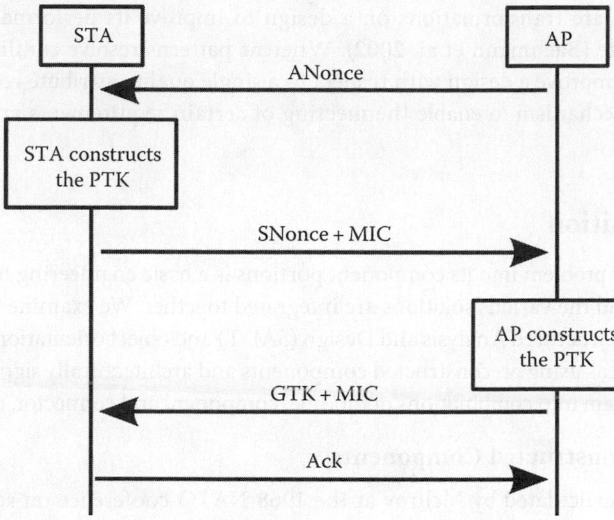

FIGURE 90.2 A sequence diagram is one notation for a component and connector view.

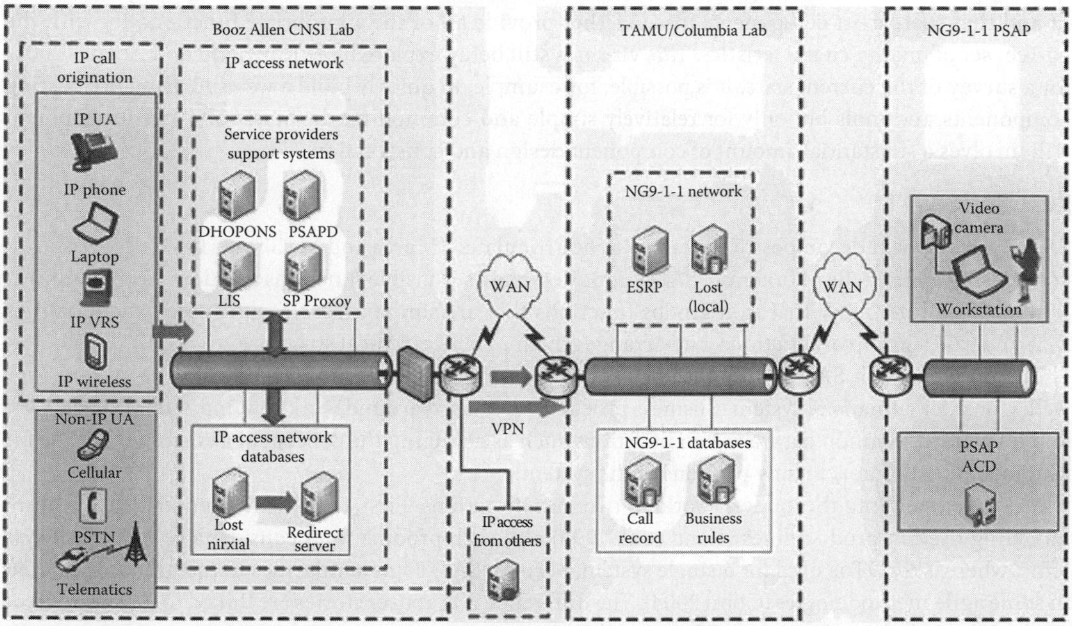

FIGURE 90.3 A sample allocation view.

threads, and components (see Figure 90.2); and the allocation views, those that represent the mapping between software elements and non-software elements such as processors (see Figure 90.3).

90.2.2 Architectural Patterns and Tactics

A pattern is a known solution to resolve conflicting forces (requirements) on a system. Patterns derive from the work of Alexander et al. (1977) and have been widely accepted by the software engineering community (Gamma et al. 1995, Buschmann et al. 1996). It is common to see patterns used in design strategies although the composition of patterns remains an unresolved problem.

Architectural tactics are transformations on a design to improve its performance with respect to a single quality attribute (Bachmann et al. 2002). Whereas patterns resolve conflicting requirements, tactics are intended to improve a design with respect to a single quality attribute requirement. Patterns are a coarse-grained mechanism to enable the meeting of certain requirements and tactics are a fine-grained mechanism.

90.2.3 Decomposition

Decomposing a difficult problem into its component portions is a basic engineering tenet. Each portion is independently solved and the various solutions are integrated together. We examine four different strategies for decomposition. Structured Analysis and Design (SADT) and object orientation decompose the system into modules, whereas using preconstructed components and architecturally significant requirements may decompose the system into combinations of module, component and connector, and allocation views.

90.2.3.1 Using Preconstructed Components

This strategy was first articulated by Mcllroy at the 1968 NATO conference on software engineering (Mcllroy et al. 1968). The strategy is to determine a solution through the composition of preexisting components. The analogy is with the composition of physical components. The assumptions under which this is an effective strategy are that there exists a framework into which all of the components will fit and that there exist components that together provide all of the appropriate functionality with the correct set of quality characteristics. This vision is still being explored; see Kung-Kiu and Zheng (2007) for a survey of the current state. It is possible, for example, to quickly build a web site using preexisting components and tools but only for relatively simple and common needs. Most software development still involves a substantial amount of component design and construction.

90.2.3.2 SADT

SADT was the next decomposition strategy to be articulated (Marca and McGowan 1987). It is based on considering a hierarchy of function. SADT considers the user-visible functions and derives an organization of those functions. That is, it groups functions that are similar, identifies the core functionalities that underlie a group of functions, and arranges them in a hierarchical structure.

The problem with SADT is that it does not accommodate future modifications to the system very well. Consider a financial system. Business processes are based around workflow, individual specialties, and a separate common database. Modifications such as changing the concept of a customer will be far ranging and will impact many portions of the system.

SADT lives on into the modern world in two manifestations. First, it has some similarities to feature modeling used in product lines (Kang et al. 1990) although product lines consider a collection of systems, whereas SADT is used for a single system. Second, SADT has similarities to the user stories used in some agile methodologies (Cohn 2004). The difference is that user stories are linked to a development methodology that involves rapid delivery and future refactoring to accommodate the problems associated with modifiability.

90.2.3.3 Object Orientation

Object-oriented design (Wirfs-Brock and McKean 2002) is a design strategy that builds on the ideas of objects in object-oriented programming (Goldberg and Robson 1989). Object-oriented design's basic decomposition strategy is to decompose the system into objects where the objects represent concepts that are likely to change over time. These objects are determined through examination of scenarios of use (use cases) (Cockburn 2000).

Each object is specified through the allocation of "responsibilities" to the object. A responsibility is an action the object will accomplish, a decision the object will make, or a piece of information the object will retain.

Objects are encapsulated (Parnas 1972) to hide certain data and expose other data. The choice of data to hide is dictated partially by consideration of assumed changes. Any data that are likely to change over time should be hidden so that changing the data will not have ripple effects to the remainder of the system. This prevention of ripple effects is one of the main goals of encapsulation.

Object-oriented design is still a viable design strategy and is taught and written about. The problem with object-oriented design is that it privileges modification among the quality attributes. That is, the decomposition strategy is based on consideration of modification. As we will see in the next section, modifiability is one possible source of a decomposition strategy but there are others.

90.2.3.4 Architecturally Significant Requirements

Object-oriented decomposition makes two assumptions:

1. The primary design drivers are potential changes.
2. Potential changes can be identified through the identification of objects.

Decomposition based on identification of architecturally significant requirements removes those two assumptions. Architecturally significant requirements (Obink et al.) are those requirements that drive the design. Typically, these are quality or business requirements that pertain to either run time or development time.

When one examines a design after it has been completed, one can see the primary requirements that the design is attempting to achieve. This is a *post facto* recognition of the architecturally significant requirements. The assumption behind decomposition based on architecturally significant requirements is that the architecturally significant requirements can be discovered *a priori*. These requirements may be change requirements based on objects but they could also be specific change requirements or other quality attribute requirements such as performance or availability requirements. In this way, decomposition based on architecturally significant requirements can be seen as a generalization of object-oriented decomposition. In Section 90.3.1, we will describe a technique for discovering the architecturally significant requirements.

90.2.4 Generate and Test

Generate and test is a basic strategy of problem solving. An initial hypothesis is generated, it is tested against some set of criteria, and, depending on the results of the test, another hypothesis is generated and tested. In order for this strategy to be effective, successive hypotheses should converge to a solution. Hofmeister et al. 2007 describe this process as one of analysis (understanding the architecturally significant requirements), synthesis (creating a hypothesis), and evaluation (test).

Viewing the design strategy as a process of generate and test leads to three questions:

1. How is the initial hypothesis generated?
2. How is the hypothesis tested and when is the process complete?
3. How are the results of the tests used to generate the next hypothesis?

90.2.4.1 Generating the Initial Hypothesis

The initial hypothesis depends on the context of the particular design. Some possibilities for generating the initial hypothesis are as follows:

- Base the initial hypothesis on an existing system. Many systems are similar to an existing system and basing the initial hypothesis on that existing system will enable the exploitation of the similarities. The differences will come out in the test phase.
- Base the initial hypothesis on an existing framework. A great many software frameworks are available in a wide variety of domains ranging from automotive to web services to content management to gaming. These frameworks provide generic functionality within the domain and must be specialized for the particular system under design.

- Base the initial hypothesis on an object-oriented decomposition. Although the object-oriented decomposition assumes that modifiability is the most important quality attribute, it typically is one of the most important. The testing phase will discover the shortcomings of a purely object-oriented decomposition for the particular system being designed.
- Base the initial hypothesis on patterns or tactics. Basing the initial hypothesis on either patterns or tactics will lead to a very sparse design compared to the final design since patterns consider only a few requirements and tactics even fewer. Patterns and tactics do provide a place to begin, however, as opposed to a strictly random hypothesis.

90.2.4.2 Testing the Hypothesis

Three activities are included in the test phase of generate and test—selecting the test cases, performing the tests, and deciding when the process is complete:

1. Selecting the test cases is a matter of choosing those requirements, both explicit and implicit, that will have the most impact on the design, that is, the architecturally significant requirements. Other test cases will depend on particular usage aspects of the system. For example, a mobile application should have a test case that specifies what happens when the application becomes disconnected and then is reconnected. Start up and shut down are specific usage test cases that are relevant for many large operations. Upgrading the system is relevant for those systems that require 24/7 access, and so forth.
2. The actual process of testing produces one of two outputs. Either the current hypothesis is satisfactory for the test or some problems are determined. These problems are carried forward to the "generate next hypothesis." The tests can be performed through means of increasing fidelity and cost. The choice of which type of test to perform depends on the costs and benefits of the testing type. The cheapest tests have the least fidelity. We present the tests in increasing order of cost and fidelity. The types of tests are as follows:
   a. Thought experiments. These are the most common forms of performing tests on a design. A thought experiment is a walk-through of the design in response to particular stimuli. This is the fastest, cheapest, and the type of test with the least fidelity.
   b. Analytic modeling. Creating a model of the current design will enable a variety of "what-if" types of analyses. Being able to create an analytic model depends on the type of requirement that one is testing. Well-accepted analytic models can be created for performance, for modifiability, and for availability (see, e.g., Smith and Williams 2001). Other quality attributes are more difficult to model and the creation of these models is an active area of research. The fidelity of an analytic model will determine the accuracy of the model's predictions.
   c. Simulation. Creating a simulation of the current design enables more fidelity to be incorporated into the test than through an analytic model. Simulations allow what-ifs to be explored in much the same fashion as do analytic models. Analytic models may be difficult to solve and so creating a simulation will generate the test results through mimicking the operation of the system rather than through generating a closed form solution.
3. Deciding when the generate and test process has completed is the easy portion of the test step. The process is complete when either the requirements are satisfied or the budget for the design portion has been exhausted. In the latter case, the hypothesis that best satisfies the requirements is the hypothesis carried forward to the implementation phase.

90.2.4.3 Generating the Next Hypothesis

The output of the testing phase is essentially a list of requirements that are not met by the current hypothesis. For each requirement not met, there will be either or both of responsibilities that are not covered in the current design and a measurement of how far the current design diverges from the quality attribute requirement.

The next hypothesis should incorporate both responsibilities not covered in the current hypothesis and adjustments to improve the hypothesis with respect to quality attribute requirements:

- Additional responsibilities are either added to modules in the current hypothesis or to newly created modules. If a new responsibility has coherence with the existing responsibilities in a module, it should be added to that module. If the new responsibility does not have coherence with the responsibilities in any existing module, then possibly a new module should be created for the new responsibilities. Since modules also act as work assignments for teams, a new module should only be created when there are sufficient responsibilities within it to assign to a team. Responsibilities that are left dangling because they do not cohere with existing responsibilities and are not large enough to justify the creation of a new module can be assigned arbitrarily.
- Adjusting the current hypothesis to improve its quality attribute performance is the reason for architectural tactics. Each architectural tactic provides additional responsibilities; movement of responsibilities from one module to another; combining, splitting, or creation of a new module; or adjustment of the properties of responsibilities, modules, or components.

One or more tactics are chosen and applied to improve the design, the responsibilities discovered missing from the testing phase are added to the current architecture, and the next test can be performed.

90.3 Best Practices

A major practical problem for a designer is dealing with the masses of detail both in the requirements and in the design. This problem is manifested in three fashions—a designer must abstract the detail to enable the designer to have an overall grasp of the system and its requirements, the designer must produce the design incrementally since designing for all of the requirements at once is not possible for large systems, and design decisions must be recorded, ideally with rationale, since the designer makes too many decisions to remember them all.

Abstracting the masses of detail in the requirements is discussed in Chapter 86. Abstracting the detail from a design perspective is discussed in the next section and appropriate tooling is discussed in Section 90.3.3.

90.3.1 Finding Architecturally Significant Requirements

Architecturally significant requirements are those requirements that drive the design, that is, they are the most important requirements. Deconstructing "important" yields two types of importance—business and structural:

- From a business perspective, a requirement can be essential (the system is unacceptable if the requirement is not met), important (the system is acceptable but not as good as desired if the requirement is not met), and nice to have (the system is fine without it but the meeting of this requirement adds incremental value).
- From a structural perspective, a requirement can be overarching (the design would be substantially different without this requirement), significant (the design would be somewhat different without this requirement), or insignificant (the design would be essentially the same without this requirement).

Using these two types of criteria provides a means for discovering architecturally significant requirements. For each requirement (both explicit and implicit), assign it a business rating and a structural rating as previously mentioned. Those requirements that are both essential and overarching are the architecturally significant requirements. Using a construct called the utility tree (Clements et al. 2002) enables the designer to structure the requirements based on the type of quality attribute that they support. The utility tree is not essential to discovering the architecturally significant requirements, but it provides a structure that is useful for discovering requirements that are essentially the same from a quality attribute perspective. For example, if two requirements have the same required response time and take essentially the same path through the current hypothesis, the utility tree would enable the designer to discover the duplication.

90.3.2 Combining Generate and Test and Decomposition

Both generate and test and the decomposition strategies we discussed as basic principles are motivated by limitations of humans. Generate and test is motivated by the reality that people cannot solve multivariate problems in their head where the number of variables is large. Viewing each requirement as a variable that the design must satisfy means that the design problem is one with thousands of degrees of freedom. Decomposition is also motivated by human limitations. The details of any design are dealt with through the use of abstraction. The initial decomposition is a solution to the most abstract form of the requirements and each subsequent decomposition provides solutions to more detailed forms of the requirements although for a smaller portion of the overall solution.

The two techniques can be combined. Generate and test decomposes the system. Restricting the requirements that drive generate and test to architecturally significant requirements will provide an initial decomposition. Each element of the initial decomposition can then, in turn, be further decomposed by finding the architecturally significant requirements for that element and using generate and test to decompose that element. This leads to the following approach:

1. Determine the architecturally significant requirements.
2. Use generate and test to determine the initial decomposition of the system to satisfy the architecturally significant requirements.
3. For each element of the current decomposition, do the following:
    a. Determine the architecturally significant requirements that pertain to that element.
    b. Use generate and test to develop a solution for the restricted architecturally significant requirements.
4. Iterate until either the requirements are satisfied or the time allocated for designing has elapsed.

The attribute directed design method (Bass et al. 2003) is a method that utilizes this combination of the two techniques.

90.3.3 Tooling

Tooling is essential to support the design of a complicated system. There are three types of functions that a tool or tool chain must support:

1. *The recording of design decisions and their rationale.* Recording of decisions is done in two forms depending on the audience. During the execution of the design process, many decisions are being made. These decisions should be recorded but the audience for this recording is the designers during the design process. The mass of detail in the design will overwhelm any individual's memory, even from day to day. This is especially true when the choice between two decisions is essentially arbitrary. One option must be chosen to move forward but recalling which one was chosen the next day when a similar situation arises may be problematic. Almost any means can be used to record this class of decisions—an engineering notebook, a word processing system, or other ad hoc means. Since the only point of these recording is to supplement the designer's short-term memory, the form does not matter. Tools for recording of design rationale have received renewed attention in recent years (Jansen and Bosch 2005, Kruchten et al. 2009).

    At particular milestones during the design, especially after the completion of the design, decisions will be exposed to a broader audience. The design will be evaluated, it will be presented to developers as a blueprint, and the designers will revisit it after some period of time. In this situation, more formality is needed in the recording of the design and the tooling should support this formality. Different views of the design are presented using different notations and the tool used should support much more formality than needed for the short-term recording of the design decisions. See Clements et al. (2010) for a discussion of various types of notations. Bachmann (2011) proposes using frequent and focused design reviews as a means for encouraging development teams to develop documentation in an incremental fashion.

2. *Traceability to requirements.* A design is intended to provide a solution for a set of requirements. Some requirements will be explicit and others implicit. The explicit requirements will be recorded in some form and a tool should support the tracing of particular sets of design decisions back to the explicit requirements. This will enable review of the design to determine if it satisfies at least those requirements that have been made explicit.
3. *Integration with the development tools.* Creating a design is one step in the development process. The tool chain used in the development should provide a smooth integration from one step to the next and, perhaps more importantly, back. One problem with designs is that they diverge from the code. This happens over time because maintenance activities are performed on the code, frequently without simultaneously updating the design. An integrated tool chain would support the reflection of code modifications in the design (backward integration) as well as the movement from the design to the code (forward integration). How to accomplish this integration is a research question.

90.4 Research Issues

The research issues involved in design revolve around tools and models. Models are necessary for automatic and semiautomatic design. Even with manual design, however, design tools need to be integrated into a coherent tool chain.

90.4.1 Automatic/Semiautomatic Design

One research question is the extent to which a design process can be automated or semiautomated. Several factors make the problem of intelligently supporting the design process a difficult question:

- The size of the design space is huge. Every decision that is made during the design process adds another sub-tree to the set of possible solutions. Since a serious design may encompass thousands of decisions, the tree that represents the design space is huge. Also adding to the size of the design space is the fact that decisions are not commutative. That is, decision to split a module and then deciding to add a particular responsibility to one sub-module may yield a different result than adding that responsibility to the module prior to splitting it. From this, it is clear that any automatic or semiautomatic design engine must have a very sophisticated method for navigating the design space.
- The navigation of the design space must be driven by the requirements. These include the quality and the business requirements as well as the functional requirements. The navigation engine must have models that will guide its decisions in considering its next moves. These models must be able to both deduce the properties of a hypothesis and relate these properties to the satisfaction of requirements and suggest transformations of the hypothesis to improve its behavior with respect to the requirements. For example, a particular hypothesis may be deficient in performance and the design engine must have a menu of options that will improve the performance of the next hypothesis with respect to that particular requirement. See Bachmann and Bass (2003) for one example of a semiautomated tool that relies on human intervention for navigation but provides information to the human based on quality attribute models to guide the navigation.
- Improving the performance of a hypothesis with respect to one requirement will have an impact on how well other requirements are satisfied. This puts any tool for navigating the design space in the realm of multi-attribute optimization. See Makki et al. (2008) for a discussion of how the design process can be cast as a multi-attribute decision process.

90.4.2 Models

As the discussion about automated or semiautomated tool support for design indicated, models are critical to support the design process. The models must relate requirements to the design. The type of requirements that affect the design, as was discussed in the section on architecturally significant requirements, are the quality attribute and the business requirements:

- *Quality attribute models.* The maturity of models for the different quality attributes is very different. Performance has the best models. Queuing models have existed for networked computer systems since the 1970s (Kleinrock 1975) and real-time scheduling models have existed at least since the early 1990s (Harbour et al. 1993). Models for other quality attributes are much less mature and applicable. Security, for example, has no models that relate the design of the system to the probability of a successful attack.
- *Business models.* Models that relate business requirements to design are currently restricted primarily to cost models. Different design choices may be made depending on costs. Many business requirements are manifested as quality attribute requirements, for example, requirements for long-lived systems can usually be translated into modifiability requirements. Other business requirements that impact design decisions, for example, improve the skill set of our employees, have no models that can be used to support various design decisions.

90.4.3 Tool Chain

Design sits between requirements elicitation and implementation in the life cycle. There are a large number of tools to support requirements elicitation and tracing and also a large number of tools to support implementation and testing. Designs are frequently not kept up to date because modifications done to the code are not easily reflected back into the design. What is needed is a tool that supports the design process and integrates well with both the requirements tools and the implementation tools so that the designer and implementer can seamlessly move between their various activities. The design of such a tool is a research problem. IBM has experimented with the Architect's Work Bench (Abrams et al. 2006) but that effort has ended. In addition to the process integrations attacked by the Architect's Work Bench, there is an additional question of the most effective method for reflecting modifications to the code back to the design.

90.5 Summary

Designing a system is a crucial step to realizing the vision articulated for that system. Humans are good at creative activities and computers and processes are good at organizing these activities. Design strategies are built around recognition of human limitations and an understanding of the forces that drive design.

Research issues in design are motivated by increasing the models of important elements of the design process and embedding those models in tools.

References

Abrams, S., B. Bloom, P. Keyser et al. Architectural thinking and modeling with the Architects' Workbench. *IBM Systems Journal* 45(3), 2006, 481–500.

Alexander, C., S. Ishikawa, M. Silverstein et al. *A Pattern Language.* Oxford University Press, Oxford, U.K., 1977.

Bachmann, F. Give the stakeholders what they want: Design peer reviews the ATAM style, CrossTalk. *The Journal of Defense Software Engineering* 24(6), Crosstalk, November–December 2011, pp. 8–10.

Bachmann, F. and L. Bass. *Preliminary Design of ArchE: A Software Architecture Design Assistant.* CMU/SEI-2003-TR-021. Software Engineering Institute, Carnegie Mellon University, Pittsburgh, PA, 2003.

Bachman, F., L. Bass, and M. Klein. *Illuminating the Fundamental Contributors to Software Architecture Quality*. CMU/SEI-2002-TR-025. Software Engineering Institute, Carnegie Mellon University, Pittsburgh, PA, 2002.

Bass, L., P. Clements, and R. Kazman. *Software Architecture in Practice*, 2nd edn. Addison-Wesley, Reading, MA, 2003.

Buschmann, F., R. Meunier, H. Rohnert et al. *Pattern-Oriented Software Architecture: A System of Patterns*. John Wiley, New York, 1996.

Clements, P., F. Bachmann, L. Bass et al. *Documenting Software Architectures: Views and Beyond*, 2nd edn. Addison Wesley, Upper Saddle River, NJ, 2010.

Clements, P., R. Kazman, and M. Klein. *Evaluating Software Architectures: Methods and Case Studies*. Addison-Wesley, Boston, MA, 2002.

Cockburn, A. *Writing Effective Use Cases*. Addison-Wesley, Boston, MA, 2000.

Cohn, M. *User Stories Applied*. Addison Wesley, Boston, MA, 2004.

Gamma, E., R. Helm, R. Johnson et al. *Design Patterns: Elements of Reusable Object-Oriented Software*. Addison Wesley, Reading, MA, 1995.

Goldberg, A. and D. Robson. *Smalltalk 80: The Language*. Addison-Wesley, Reading, MA, 1989.

Harbour, M., M. Klein, R. Obsenza et al. *A Practitioner's Handbook for Real-Time Analysis: Guide to Rate Monotonic Analysis for Real-Time Systems*. Kluwer Academic Publishers, Boston, MA, 1993.

Hofmeister, C., P. Kruchten, R. Nord et al. A general model of software architecture design derived from five industrial approaches. *The Journal of Systems and Software* 80, 2007, 106–126.

ISO/IEC/IEEE 42010:2011, Systems and software engineering—Recommended practice for architectural description of software-intensive systems. International Organization for Standardization, Geneva, Switzerland, 2011.

Jansen, A. and J. Bosch. Software architecture as a set of architectural design decisions. *Proceedings of the 5th Working IEEE/IFIP Conference on Software Architecture*. IEEE, Washington, DC, 2005.

Kang, K., S. Cohen, J. Hess et al. *Feature-Oriented Domain Analysis (FODA) Feasibility Study*. CMU/SEI-90-TR-021. Software Engineering Institute Carnegie Mellon University, Pittsburgh, PA, 1990.

Kleinrock, L. *Queueing Systems*, Vol. 1, Theory. Wiley, New York, 1975.

Kruchten, P., R. Capilla, and J.C. Dueas. The decision view's role in software architecture practice. *IEEE Software* March–April 2009, pp. 36–42.

Kung-Kiu, L. and Zheng, W. Software component models. *IEEE Transactions on Software Engineering* 33(10), October 2007, 709–724.

Makki, M., E. Bagheri, and A. A. Ghorbani. *Automating Architecture Trade-Off Decision Making through a Complex Multi-attribute Decision Process*. Lecture Notes in Computer Science, Springer Verlag, New York, 2008, Vols. 5292/2008, pp. 264–272.

Marca, D. and C. McGowan. *Structured Analysis and Design Technique*. McGraw-Hill, New York, 1987.

Mcllroy, M.D., J.M. Boxton, P. Naur et al. Mass produced software components. *Proceedings of the NATO Software Engineering Conference*, Garmisch, Germany, 1968.

Obink, H., P. Kruchten, W. Kozaczynski et al. Software Architecture Review and Assessment (SARA) Report. http://kruchten.com/philippe/architecture/SARAv1.pdf.

Parnas, D.L. On the criteria to be used in decomposing systems into modules. *Communications of the ACM* 5(12), December 1972, 1053–1058.

Smith, C. and L.G. Williams. *Performance Solutions: A Practical Guide to Creating Responsive, Scalable Software*. Addison-Wesley, Boston, MA, 2001.

Wirfs-Brock, R. and A. McKean. *Object Design: Roles, Responsibilities, and Collaborations*. Addison-Wesley, Boston, MA, 2002.

91 Software Architecture

Bedir Tekinerdogan
Bilkent University

91.1 Underlying Principles

The term *architecture* has been used for centuries to denote the art and science of designing buildings and structures. The term is applied in various different domains scope including building architecture, town planning, urban design, and landscape architecture. One of the earliest written works on the topic of architecture is *De Architectura* (On Architecture), which was published as Ten Books on Architecture, by the Roman architect Vitruvius in the early first century CE [10]. The work is a treatise on architecture and served as a guide for building projects. Currently, the work is one of the most important sources of knowledge of Roman building methods and the planning and design of both large and small structures, including buildings, aqueducts, baths, harbors, and machines. The concept of architecture has further evolved in the history through various traditions and regions.

The software engineering community has also adopted the term architecture to denote the gross-level design of software-intensive systems. The importance of high-level architectural structure was already acknowledged early in the history of software engineering. The first software programs were written for numerical calculations using programming languages that supported mathematical expressions and later algorithms and abstract data types. Programs written at that time served mainly one purpose and were relatively simple compared to the current large-scale diverse software systems. Over time, due to the increasing complexity and the increasing size of the applications, the global structure of the software system became an important issue.

Currently, it is not the algorithms and data structures that constitute the major design problems, but the construction of the systems from many components and the organization of the overall system [16]. Already in 1968, Dijkstra proposed the correct arrangement of the structure of software systems before simply programming [13]. He introduced the notion of layered structure in operating systems, in which related programs were grouped into separate layers, communicating with groups of programs in adjacent layers. Later, Parnas argued that the selected criteria for the decomposition of a system impact the structure of the programs and several design principles must be followed to provide a good structure [27,28]. Within the software engineering community, there is now an increasing consensus that the architectural design of software systems is important to cope with the increased complexity and meet the stakeholder requirements properly. Without loss of generality, we can state that software architecture is recognized as a key determinant of project success. Without the right architecture, the project will likely fail.

91.1.1 Definition

In tandem with the increasing popularity of software architecture design, many definitions of architecture have been introduced over the last decade, though a consensus on a standard definition is still not established. We think that the reason why so many and various definitions on software architectures exist is because every author approaches a different perspective of the same concept of software architecture and likewise provides a definition from that perspective. Notwithstanding the numerous definitions, it appears that the prevailing definitions do not generally conflict with each other and commonly agree that software architecture represents the gross-level design of the software system consisting of components and relations among them.

The ISO/IEC 42010 (formerly IEEE 1471) Recommended Practice for Architectural Description of Software-Intensive Systems provides the following definition for software architecture [19]:

> Architecture is the fundamental organization of a system embodied in its components, their relationships to each other, and to the environment, and the principles guiding its design and evolution.

Another recent definition is given in [8]:

> The software architecture of a program or computing system is the structure or structures of the system, which comprise software elements, the externally visible properties of those elements, and the relationships among them.

There are plenty of other definitions, each focusing on different aspects. Based on the definitions in the literature, we consider the following definition of software architecture that includes the important aspects:

> *Architecture represents the gross level structure of software intensive systems, which consists of software components and relationship among these components, including systemic design decisions and the rationale for these design decisions, to meet the set of concerns for the stakeholders.*

The aforementioned definition has in fact three parts: (1) the structure and behavior of the system, (2) the design decisions that have led to the structure together with the rationale behind these design decisions, and (3) the stakeholder concerns that govern the design decisions and shape the architecture. The term component here is used as an abstraction of different interpretations and likewise may refer to subsystems, processes, software modules, hardware components, or any element that has systemic behavior. In a similar sense, relations may refer to data flows, control flows, call relations, part of relations, etc.

91.1.2 Rationale for Architecture

Software architecture is an abstract representation that serves various purposes. We list the following:

- *Architecture helps to understand the system*
  An important concern in software projects is the understanding of the system and its properties. Software architecture is a high-level abstraction of the system that abstracts away from the irrelevant details and as such can facilitate the understandability.

- *Architecture supports communication among stakeholders*
  Software architecture can be used as a common abstraction that stakeholders can use to communicate and discuss the system properties. A stakeholder is defined as an individual, team, or organization with interests in, or concerns relative to, a system. Each of the stakeholders' concerns impacts the early design decisions that the architect makes. *Architectural drivers* [26] define the concerns of the stakeholders that shape the architecture.
- *Architecture serves as a basis for the subsequent phases of the life cycle process*
  Architecture defines the gross-level structure of the system to which the subsequent life cycle detailed design and code artifacts should conform. Likewise, architecture defines the boundaries of the system and provides the constraints for the implementation. Further, the software architecture is used to guide the subsequent life cycle activities and help to achieve integrity of the system.
- *Architecture supports the quality requirements*
  It is important that the architecture design supports the software system quality requirements that are required by the various stakeholders. Since software architecture represents the earliest design decisions about a system, architecture can be used to define and reason about the quality attributes. In addition, software architecture can be used to analyze the risks, sensitivity points, and trade-offs.
- *Architecture serves as a means for supporting the organization*
  The motivation for architecture-driven development is not just for technical reasons but also for organizational reasons. Architecture defines the gross-level structure of the system that can be used to support work allocation, budget planning, and development structure of the project.

91.2 Architecture Modeling

Every system has an architecture whether it is complex or not. If the architecture is modeled, we can control it and benefit from an architecture-driven development. Initially, software architecture was represented using arbitrary box-and-lines notations leading to ambiguous interpretations. As such, very soon, it was acknowledged to provide more formal support for architectural modeling, both visually and textually. Together with the quest for more formal approaches, a widespread awareness has established stating that architecture should be modeled based on *architecturally significant requirements* (*architecture drivers*). These requirements are systemic in nature and usually concern the overall system or the large part of the system. Modifiability and performance requirements are examples of architecturally significant requirements. This need for modeling the architecture for different stakeholders has led to the notion of *architectural views* that is a representation of a system for a particular concern. Having multiple views of the architecture helps to separate the concerns and as such support the modeling, understanding, communication, and analysis of the software architecture for different stakeholders. Architectural views conform to *viewpoints* that represent the conventions for constructing and using a view. It is now common practice to model and document different architectural views for describing the architecture according to the stakeholders' concerns [4,8,11,17,22,32,33,37]. An architectural view is a representation of a set of system elements and relations associated with them to support a particular concern. Having multiple views helps to separate the concerns and as such support the modeling, understanding, communication, and analysis of the software architecture for different stakeholders. Architectural views conform to viewpoints that represent the conventions for constructing and using a view. An *architecture framework* is defined as the coordinated set of viewpoints that are used to define the views. A more precise definition of architecture framework is given in the ISO standard [21]:

> Conventions, principles, and practices for the description of architectures established within a specific domain of application and/or community of stakeholders

Different architectural frameworks have been proposed in the literature. Examples of architectural frameworks include Kruchten's 4+1 view model [22], the Siemens Four View Model [17], and the Views and Beyond (V&B) approach [8]. An example list of architecture frameworks is shown in Table 91.1.

TABLE 91.1 Example Architecture Frameworks

Architecture Framework	Viewpoints
Rational unified process/ Kruchten's 4+1 [22]	1. The logical view contains the key abstractions from the problem domain
	2. The implementation view represents the system from a programmer's perspective and includes the implementation units
	3. The process view documents the tasks—processes and threads
	4. The deployment view documents the various physical nodes for the most typical platform configurations
	5. The use case view or "plus-one view" contains use cases and scenarios of architecturally significant behavior
V&B approach [8]	Three style categories
	1. Module styles—How it is structured as a set of implementation units
	2. Component-and-connector (C&C) styles—How it is structured as a set of elements that have runtime behavior and interactions
	3. Allocation styles—How it relates to nonsoftware structures in its environment
RM-ODP [33]	1. The enterprise viewpoint describes the business requirements
	2. The information viewpoint describes the information managed by the system and the structure and content type of the supporting data
	3. The computational viewpoint describes the functionality provided by the system and its functional decomposition
	4. The engineering viewpoint describes the distribution of processing performed by the system to manage the information and provide the functionality
	5. The technology viewpoint describes the technologies chosen to provide the processing, functionality and presentation of information

Initially, architectural frameworks were proposed with a fixed set of viewpoints. Because of the different concerns that need to be addressed for different systems, the current trend recognizes that the set of views should not be fixed but multiple viewpoints might be introduced instead. As such, recent architectural frameworks such as the V&B approach provide mechanisms to adapt existing viewpoints or to add new viewpoints. Organizing the system as a set of viewpoints has also been addressed in enterprise application system using the so-called enterprise architecture frameworks. Examples include the early Zachman Framework for Enterprise Architecture [39], The Open Group Architecture Framework (TOGAF) [38], and the ISO (ISO/IEC 10746) Reference Model of Open Distributed Processing (RM-ODP) [30].

So far, most architectural viewpoints seem to have been primarily used either to support the communication among stakeholders or at the best to provide a blueprint for the detailed design. From a historical perspective, it can be observed that viewpoints defined later are more precise and consistent than the earlier approaches but a close analysis shows that even existing viewpoints lack some precision. Moreover, since existing frameworks provide mechanisms to add new viewpoints, the risk of introducing imprecise viewpoints is high. The development of a proper and effective architecture is highly dependent on the corresponding documentation. An incomplete or imprecise viewpoint will impede the understanding and application of the viewpoints to derive the corresponding architectural views and likewise lower the quality of the architectural document.

In Figure 91.1, we show the conceptual model for architectural view modeling that is based on the ISO/IEC 42010 recommended standard for architectural description [19]. The left part of the figure shows basically the definition of the architectural drivers. A system has one or more stakeholders who have interest in the system with respect to one or more concerns. Concerns can be functional concerns or quality concerns. The right part of the figure focuses on the architectural views for the different concerns. Each system has an architecture, which is described by an architectural description. The architectural description consists of a set of views that correspond to their viewpoints. Viewpoints aim to address the stakeholder's concerns. Functional concerns will define the dominant decomposition

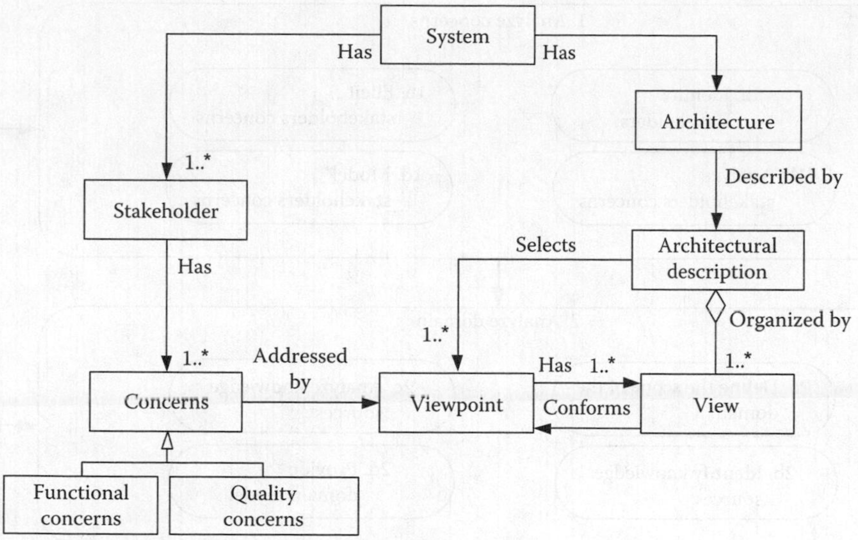

FIGURE 91.1 Conceptual model for architectural views based on ISO/IEC 42010.

along architectural units that are mainly functional in nature. On the other hand, quality concerns will define the dominant decomposition of the architecture along architectural units that explicitly represent quality concerns.

91.3 Architecture Design Methods

Various different software architecture design methods have been published in the literature. In general, the architecture design process is defined early in the software life cycle process. The main activities and a global architecture design process are given in Figure 91.2.

The process is typically defined as an iterative process in which one can iterate to previous activities if needed. In the succeeding text, we describe the activities.

91.3.1 Analyze Concerns

As stated before, architecturally significant requirements define the concerns of the stakeholders that shape the architecture. To analyze the concerns, first, the stakeholders must be identified and their concerns must be elicited. After analyzing and consolidating the elicited concerns, the concerns must be explicitly represented for guiding the subsequent architecture design activities.

91.3.2 Analyze Domain

Domain analysis can be defined as the process of identifying, capturing, and organizing domain knowledge about the problem domain with the purpose of making it reusable when creating new systems [1,9]. The UML glossary provides the following definition of the term domain:

> *Domain*: An area of knowledge or activity characterized by a set of concepts and terminology understood by practitioners in that area.

A survey of domain analysis methods shows that these methods include the similar kind of activities. Figure 91.3 represents the common structure of domain analysis methods as it has been derived from survey studies on domain analysis methods [1,9].

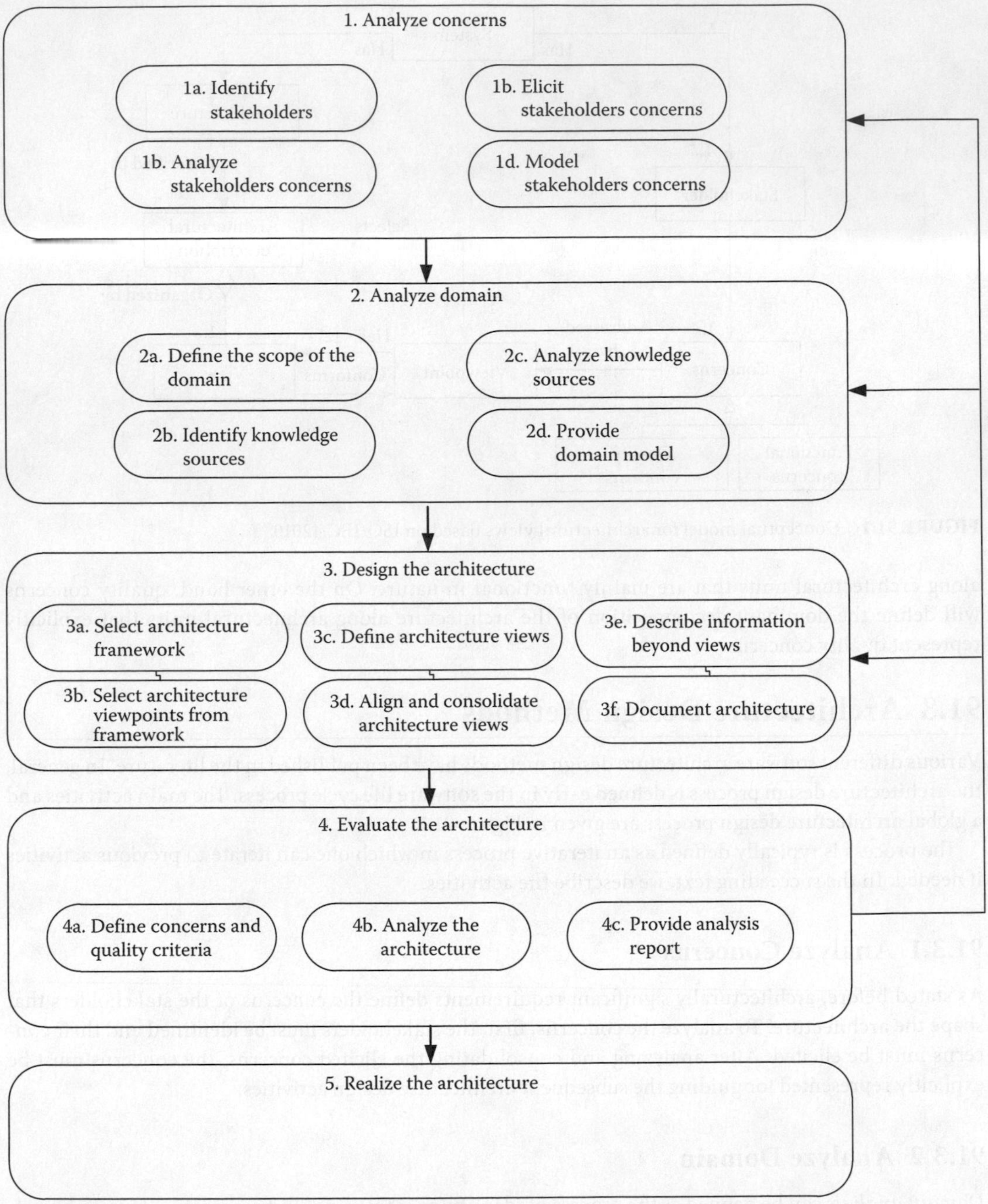

FIGURE 91.2 Example software architecture design process.

Conventional domain analysis methods consist generally of the activities *Domain Scoping* and *Domain Modeling*: *Domain Scoping* identifies the domains of interest, the stakeholders, and their goals and defines the scope of the domain. *Domain Modeling* is the activity for representing the domain, or the *domain model*. The domain model can be represented in different forms such as object-oriented language, algebraic specifications, rules, and conceptual models. Typically, a domain model is formed through a commonality and variability analysis to concepts in the domain. A *domain model* is used as a basis for engineering components intended for use in multiple applications within the domain.

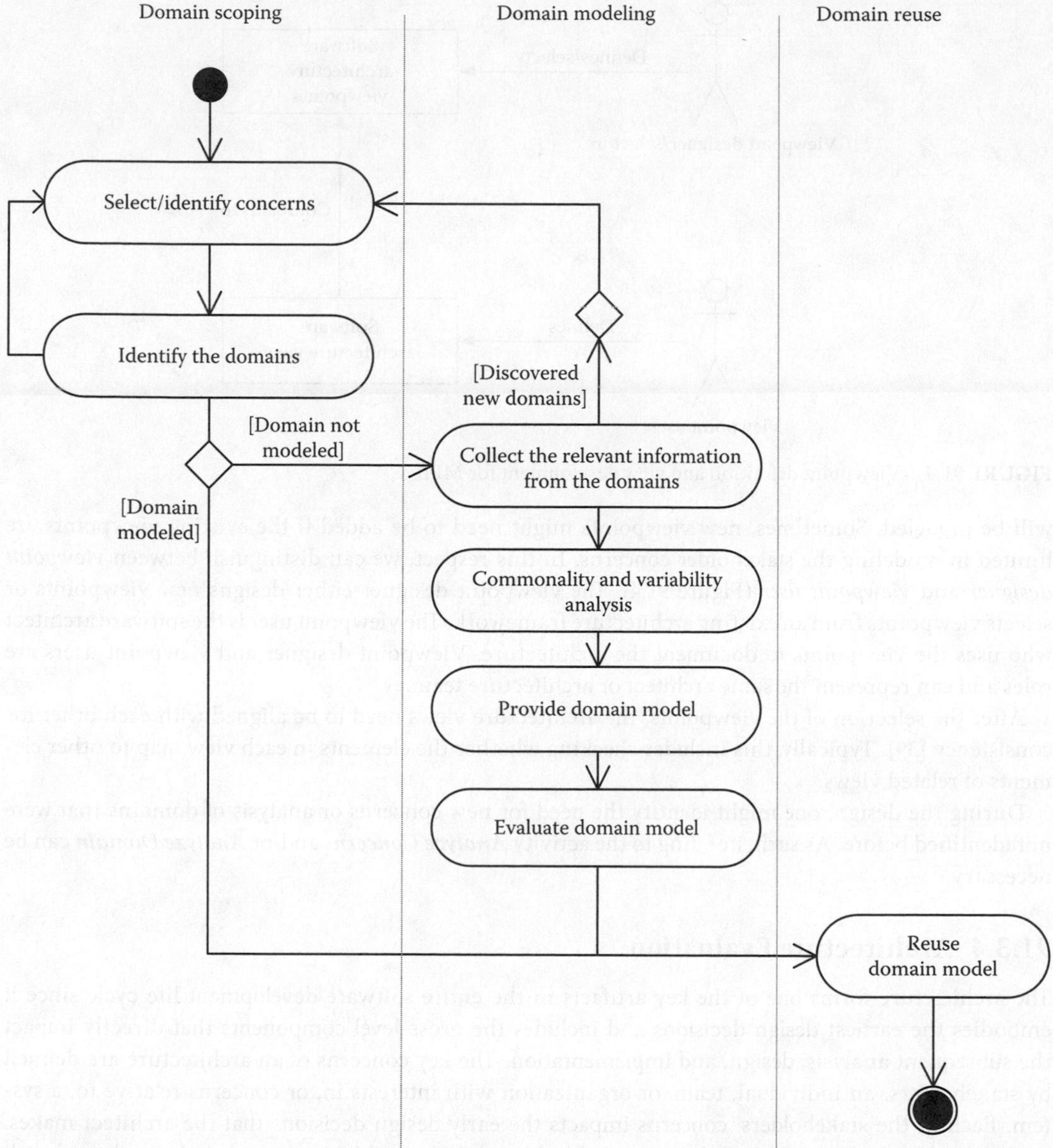

FIGURE 91.3 Common structure of domain analysis methods.

This activity results in a thorough understanding of the domain in which new concerns that were overlooked in the previous activity can be identified. For this, one can iterate back to the previous activity of concern analysis to update the identified concerns.

91.3.3 Architecture Design

Once the stakeholder concerns and the domain has been understood and modeled, the architecture design can be started. In this context, as stated before, the notion of viewpoint plays an important role in modeling and documenting architectures. Before the design process can begin, an architecture framework needs to be selected that contains the viewpoints with which the architecture views

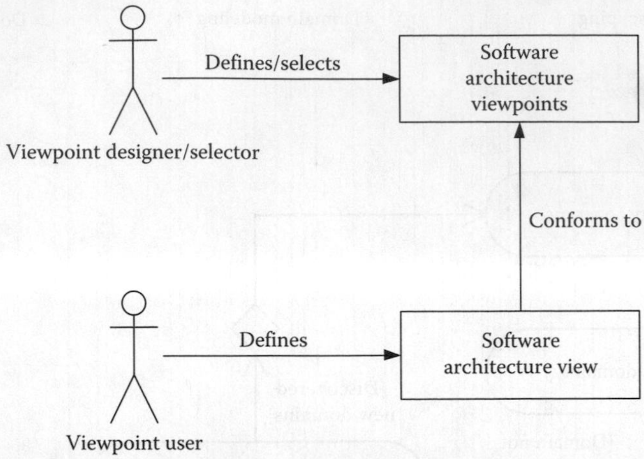

FIGURE 91.4 Viewpoint definition and view development for MPL.

will be modeled. Sometimes, new viewpoints might need to be added if the existing viewpoints are limited in modeling the stakeholder concerns. In this respect, we can distinguish between *viewpoint designer* and *viewpoint user* (Figure 91.4). The viewpoint designer either designs new viewpoints or selects viewpoints from an existing architecture framework. The viewpoint user is the software architect who uses the viewpoints to document the architecture. Viewpoint designer and viewpoint users are roles and can represent the same architect or architecture team.

After the selection of the viewpoints, the architecture views need to be aligned with each other for consistency [39]. Typically, this includes checking whether the elements in each view map to other elements of related views.

During the design, one might identify the need for new concerns or analysis of domains that were not identified before. As such, iterating to the activity *Analyze Concerns* and/or *Analyze Domain* can be necessary.

91.3.4 Architecture Evaluation

The architecture forms one of the key artifacts in the entire software development life cycle since it embodies the earliest design decisions and includes the gross-level components that directly impact the subsequent analysis, design, and implementation. The key concerns of an architecture are defined by stakeholders, an individual, team, or organization with interests in, or concerns relative to, a system. Each of the stakeholders' concerns impacts the early design decisions that the architect makes. To ensure that the concerns have been properly addressed, the architecture needs to be evaluated with respect to the defined concerns. For this, the concerns and the quality criteria are identified. Based on these, the architecture will be evaluated and an impact analysis report will be provided. The result of the evaluation process might require iteration to activities *Analyze Concerns*, *Analyze Domain*, and *Architecture Design*. We elaborate on the evaluation process in Section 91.4.

91.3.5 Architecture Realization

The final step of the architecture development cycle is the realization of the architecture.

To realize the important architecture design concerns, one of the key goals of software architecture design is to serve as a guideline for the subsequent life cycle activities. Based on the architecture design, the detailed design and subsequent activities can be carried out. Figure 91.5 shows the relationships among these concepts. An architecture description is defined using architecture views and addresses a set of concerns. The architecture is followed by the detailed design that realizes the architecture and that

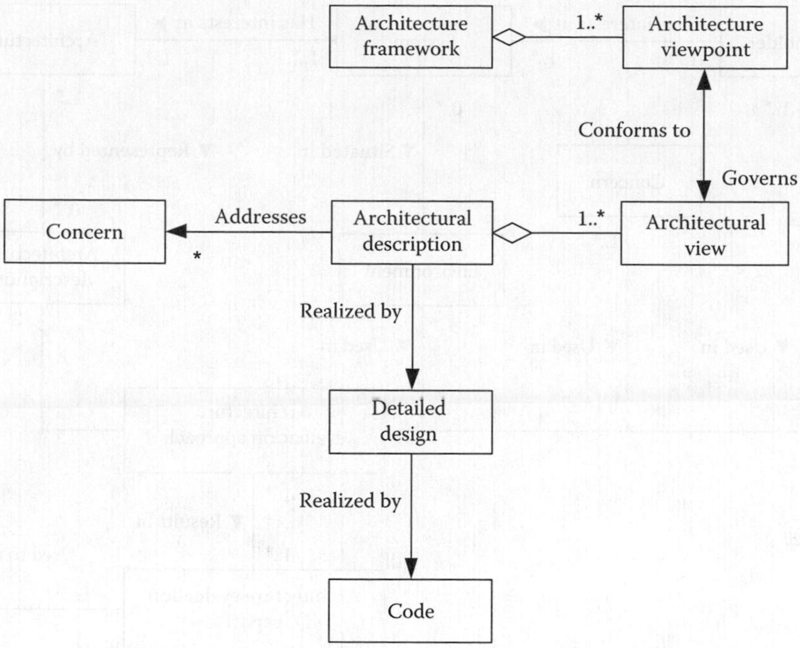

FIGURE 91.5 Relation between architecture viewpoint, architecture view, and detailed design.

is subsequently realized by code. To preserve the benefits of the architecture, it is important to keep the life cycle artifacts consistent with the architecture.

Software systems are rarely static and need to evolve over time due to bug fixes or new requirements. These required changes might lead to a situation in which the realization of the architecture (in the detailed design and code) is different from the architecture description itself. As such, the benefits of an architecture-driven software development approach will be lost. A proper documentation of the architecture is important to minimize the cost of software maintenance. Despite this, the evolution of the system and likewise the discrepancy between the architecture description and the resulting implementation will be highly probable. The discrepancy between the architecture description and the architecture realization is usually referred to as *architectural drift* or *architecture degeneration*. This architecture drift can be even introduced during the initial implementation of the architecture due to lack of knowledge about the architecture or stringent time-to-market constraints. Even if the initial implementation of the architecture is properly aligned with the architecture description, the constant need for new requirements very often leads to a situation that can lead to architecture drift.

The existence of discrepancies among the architecture description and its implementation is not a mere theoretical issue, but it may directly lead to increased maintenance time and cost, because the original design goals have been lost. In the extreme cases, a system's architecture may even deteriorate to a degree where further development is not feasible, leading to a complete system reimplementation. As such, preferably conformance checking of the architecture description and the implementation should be exercised over the architecture of a system during development and maintenance, with the aim of enforcing architecture consistency.

91.4 Architecture Evaluation

In the previous section, we have discussed the generic approach for designing software architecture. Various other architecture design approaches can be found in the literature. In this context, Hofmeister et al. [18] also provide a model of software architecture design approaches. The aim of

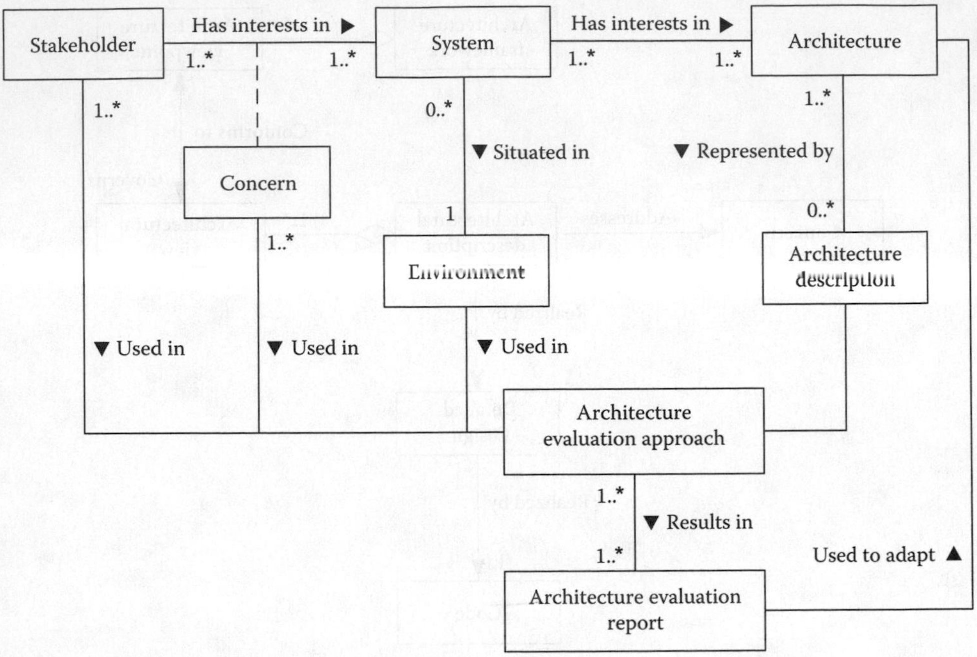

FIGURE 91.6 Conceptual model for software architecture analysis.

all these architecture design approaches is to design a software architecture that meets the stakeholder goals. Since architecture is critical for the success of a project, different architectural evaluation approaches have also been introduced to evaluate the stakeholders' concerns. A comprehensive overview and comparison of architecture analysis methods have been given by, for example, Dobrica et al. [10], and Babar et al. [2]. Kazman et al. [20] have provided a set of criteria for comparing the foundations underlying different methods, the effectiveness and usability of methods. Figure 91.6 provides a conceptual model that relates the architectural concepts with the architecture evaluation approach. Although different architecture evaluation approaches have been proposed in the literature, we can state that most of these follow the model as defined in Figure 91.6. In essence, each architecture evaluation approach takes as input the stakeholder concerns, the environment issues, and the architecture description. Based on these inputs, the evaluation results in an *architecture evaluation report*, which is used to adapt the architecture.

Figure 91.7 shows the generic conceptual process for architecture evaluation that has been based on the proposed architecture evaluation approaches. Each architecture evaluation approach appears to take as input the architecture and the quality criteria that need to be evaluated. If the architecture meets the criteria, then a decision can be made to realize the architecture, if not, then a refactoring of the architecture is required. Architecture evaluation approaches usually differ with respect to, for example, the goal of the approach, the type of inputs, the evaluation techniques, the addressed quality attributes, the stakeholders involvement, the ordering of activities, and the output results [2,20].

91.5 Research Issues

It is now generally accepted that software architecture design plays a fundamental role in coping with the inherent difficulties of the development of large-scale and complex software systems. Research on architecture design in the last two decades has resulted in different useful techniques and approaches. Different architectural modeling approaches for representing multiple views of the architecture have been proposed. Multiple architectural patterns have been introduced in literature to support the quality

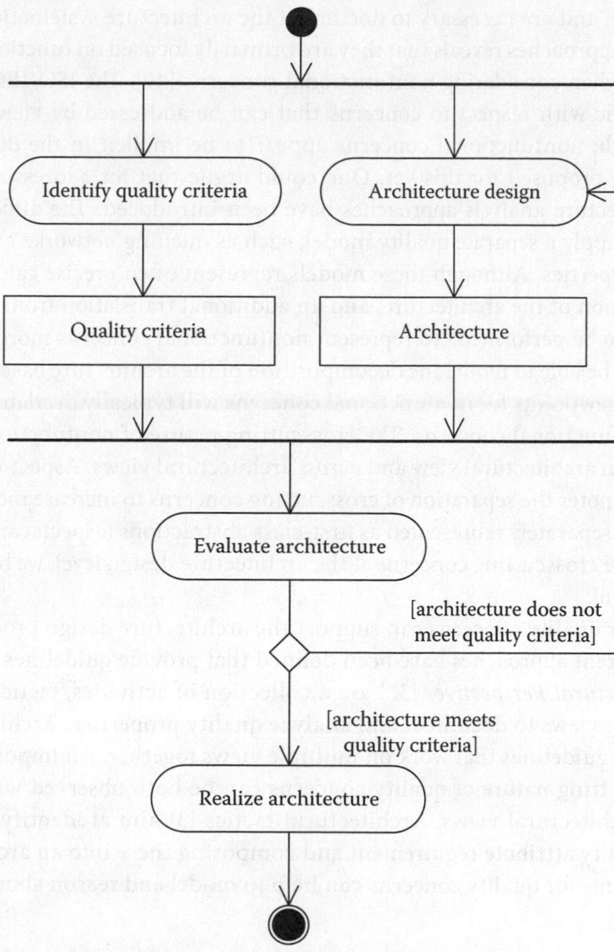

FIGURE 91.7 Generic architecture evaluation process.

of architecture [3,6,22]. A broad range of architecture analysis approaches has been proposed to analyze the architecture before it is implemented [15,34]. We can state that the software architect is now equipped with a broad set of techniques to design a software architecture that aims to meet the stakeholder concerns. Using these approaches, the software architect makes a wide range of design decisions that leads to the selection of a particular design alternative. Yet, there are indeed still challenging issues in the state of the practice and different research questions need to be tackled in the architecture design community [35]. We list some of the important research directions:

91.5.1 Modeling Quality Concerns

In the literature, initially a fixed set of viewpoints has been proposed to model and document the architecture. Because of the different concerns that need to be addressed for various systems, the current trend recognizes that the set of views should not be fixed but *open-ended*. An important example in this context is the V&B approach as proposed by Clements et al. [8]. Hereby, the authors propose the notion of architectural style that is a specialization of architectural element types and relationship types, along with any constraints. The notion of styles makes the V&B approach adaptable and open-ended since the architect is not bound to a fixed set of views but can in principle define any style that is needed. Certainly, existing architectural modeling approaches are important for representing the structure and functionality of the system and are necessary to document the architecture systematically. An analysis of the existing multiview approaches reveals that they are primarily focused on functional concerns and appear to be incomplete when considering nonfunctional concerns [40]. The ISO/IEC 42010 standard is intentionally not specific with respect to concerns that can be addressed by views. In approaches such as the V&B approach, nonfunctional concerns appear to be implicit in the different views but no specific style has been proposed for this yet. One could argue that for addressing nonfunctional concerns, software architecture analysis approaches have been introduced. The difficulty here is that these approaches usually apply a separate quality model, such as queuing networks or process algebra, to analyze the quality properties. Although these models represent often precise calculations, they do not depict the decomposition of the architecture, and an additional translation from the evaluation of the quality model needs to be performed. To represent nonfunctional concerns more explicitly, architectural views should also be able to model the decomposition of the architecture based on the required nonfunctional concern. Viewpoints for nonfunctional concerns will typically overlap or crosscut existing viewpoints based on functional concerns. The crosscutting nature of nonfunctional concerns can be both observed *within* an architectural view and *across* architectural views. Aspect-oriented software development (AOSD) promotes the separation of crosscutting concerns to increase modularity. Hereby, crosscutting concerns are separately represented as first-class abstractions (aspects) and woven into the *base code*. To cope with the crosscutting concerns at the architecture design level, we believe that AOSD techniques are necessary [6].

Appropriate models for quality concerns can support the architecture design process to design for quality. In addition, different approaches have been defined that provide guidelines for designing for various qualities. *Architectural Perspectives* [32] are a collection of activities, tactics, and guidelines to modify a set of existing views to document and analyze quality properties. Architectural perspectives as such are basically guidelines that work on multiple views together. An important observation hereby is that the crosscutting nature of quality concerns can be both observed *within* an architectural view and *across* architectural views. Architectural tactics [3] aim at identifying architectural decisions related to a quality attribute requirement and composing these into an architecture design. Defining explicit viewpoints for quality concerns can help to model and reason about the application of architectural tactics.

91.5.2 Model-Driven Approach for Architecture Design

Architecture design is basically about *modeling* the system from different perspectives. Historically, *models* have had a long tradition in software engineering and have been widely used in software projects. The primary reason for modeling is usually defined as a means for communication, analysis, or guiding the production process. Models are different in nature and quality and different classifications of models have been provided in the literature. Mellor et al. [25] make a distinction between three kinds of models, depending on their level of precision. A model can be considered as a *sketch*, as a *blueprint*, or as an *executable*. According to [25], an executable model is a model that has everything required to produce the desired functionality of a single domain. Executable models are more precise than sketches or blueprints and can be interpreted by model compilers. A similar classification of models is defined by Fowler [15] who suggests a distinction based on three levels of models, namely, *conceptual models*, *specification models*, and *implementation models*.

In model-driven software development, the concept of *models* can be considered as executable models as defined by the aforementioned characterization of Mellor et al. [25,34]. In model-driven software development, models are not mere documentation but become "code" that are executable and that can be used to generate even more refined models or code. This is in contrast to model-based software development in which models are used as blueprints at the most.

The language in which models are expressed is defined by metamodels. As such, a model is said to be an instance of a metamodel, or a model *conforms to* a metamodel. A metamodel itself is a model that conforms to a meta-metamodel, the language for defining metamodels [12,13,23]. Given the different levels in which the models reside in model-driven development, models are usually organized in a four-layered architecture. The top (M3) level in this model is the so-called meta-metamodel and defines the basic concepts from which specific metamodels are created at the meta (M2) level. Normal user models are regarded as residing at the M1 level, whereas real-world concepts reside at level M0.

In fact, we can state that the current architectural modeling practices can be categorized as *model-based development*, rather than *model-driven development*. In the last two to three decades, architectural modeling and the corresponding notations have evolved from simple sketches to more precise models as defined by architectural view concept. Yet, the view models can usually not be considered as executable models. Moreover, the link between architectural models and the link from architectural models are merely implicit and not formal [12].

In architecture modeling literature, the notion of metamodel is not explicitly used. Nevertheless, the concepts related to architectural description are formalized and standardized in ISO/IEC 42010:2007, a fast-track adoption by ISO of IEEE-Std 1471-2000, *Recommended Practice for Architecture Description of Software-Intensive Systems* [21]. The standard holds that an architecture description consists of a set of *views*, each of which conforms to a *viewpoint*, but it has deliberately chosen not to define a particular viewpoint. Here, the concept of view appears to be at the same level as the concept of *model* in the model-driven development approach. The concept of viewpoint, representing the language for expressing views, appears to be on the level of metamodel. Although the ISO/IEC 42010 standard does not really use the terminology of model-driven development, the concepts as described in the standard seem to align with the concepts in the metamodeling framework. In Figure 91.8, we provide a partial view of the standard that has been organized around the metamodeling framework. An *Architecture Description* is a concrete artifact that documents the *Architecture* of a *System of Interest*. The concepts *System of Interest* and *Architecture* reside at layer M0. *System of Interest* defines a system for which an *Architecture* is defined. *Architecture* is described using *Architectural Description* that resides at level M1. *Architectural Description* includes one or more *Architectural Views* that represent the system from particular stakeholder concern's perspective. Architectural views are described based on *Architectural Viewpoint*, the language for the corresponding view. *Architectural Viewpoints* are organized in *Architectural Framework*. The latter two reside at level M2. The standard does not provide a concept that we could consider at level M3, and as such we have omitted this in Figure 91.8.

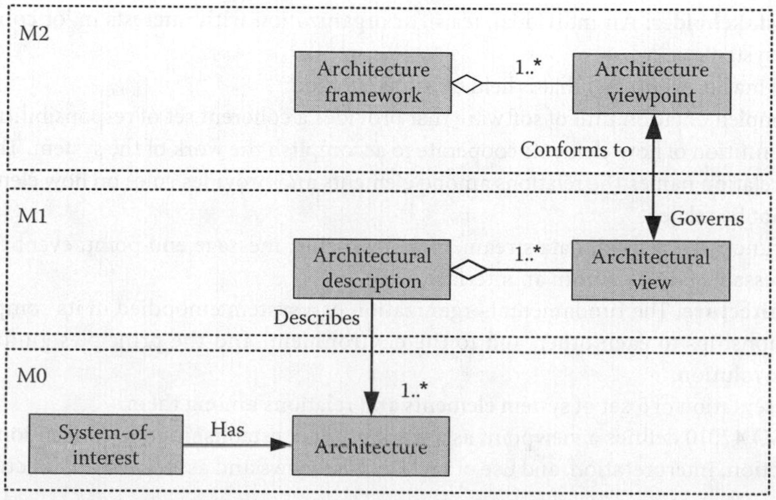

FIGURE 91.8 Architectural description concepts from a metamodeling perspective.

91.6 Summary

In this chapter, we have defined an overview of software architecture design. We have focused on three basic topics including software architecture modeling, software architecture design methods, and software architecture evaluation.

Software architecture modeling has evolved from informal box-and-line drawings to extensive architecture frameworks that include various viewpoints for modeling the architecture based on the concerns of stakeholders. Different architecture frameworks have been introduced in the literature, each with its own set of viewpoints. Earlier frameworks assumed a fixed set of viewpoints from which the architect can select the required ones to model the architecture. Recent approaches such as the V&B approach adopt an open-ended approach in which new viewpoints can be designed if deemed necessary.

The process of software architecture design usually starts in the early phases of the software development life cycle. Various architecture design methods have been introduced. We have identified five key activities that can be observed in most architecture design methods. The activities include *Analyzing Concerns, Analyze Domain, Architecture Design, Evaluate Architecture,* and *Architecture Realization.*

Once the architecture is designed, it can be evaluated based on quality criteria. The evaluation results in impact analysis report that might require refactoring the architecture to align it with the defined quality criteria.

Key Terms

Architectural (architecture) pattern: "An architectural pattern expresses a fundamental structural organization schema for software systems. It provides a set of predefined subsystems, specifies their responsibilities, and includes rules and guidelines for organizing the relationships between them" [5, p. 12].

Architecture description language (ADL): A language for representing a software and/or system architecture. ADLs are usually graphical languages that provide semantics that enable analysis and reasoning about architectures, often using associated tools.

Architecture framework: "Conventions and common practices for architecture description established within a specific domain or stakeholder community" (ISO/IEC 42010:2007).TOGAF and DoDAF are examples of architecture frameworks.

Architecture stakeholder: An individual, team, or organization with interests in, or concerns relative to, a system.

Concern: Any matter of interest that is held by a stakeholder.

Module: An implementation unit of software that provides a coherent set of responsibilities.

Relation: A definition of how elements cooperate to accomplish the work of the system. The description of a relation names the relations among elements and provides rules on how elements can and cannot be related.

Resource: A function, method, data stream, global variable, message end point, event trigger, or any addressable facility within an interface.

Software architecture: The fundamental organization of a system embodied in its components, their relationships to each other, and to the environment, and the principles guiding its design and evolution.

View: A representation of a set of system elements and relations among them.

Viewpoint: ISO 42010 defines a viewpoint as a work product establishing the conventions for the construction, interpretation, and use of architecture views and associated architecture models.

References

1. G. Arrango. Domain analysis methods. In *Software Reusability*, W. Schäfer, R. Prieto-Díaz, and M. Matsumoto (Eds.), Ellis Horwood, New York, pp. 17–49, 1994.
2. M.A. Babar, L. Zhu, and R. Jeffrey. A framework for classifying and comparing software architecture evaluation methods, In *Proceedings of 5th Australian Software Engineering Conference*, Melbourne, Australia, pp. 309–319, April 2004.
3. F. Bachmann, L. Bass, and M. Klein. *Deriving Architectural Tactics: A Step Toward Methodical Architectural Design*, CMU/SEI-2003-TR-004, ADA413644, Pittsburgh, PA, March 2003.
4. G. Booch, J. Rumbaugh, and I. Jacobson. *The Unified Modeling Language User Guide*, Addison-Wesley, Reading, MA, 1999.
5. F. Buschmann, R. Meunier, H. Rohnert, P. Sommerlad, and M. Stal, *Pattern-Oriented Software Architecture, Volume 1: A System of Patterns*, Wiley, New York, 1996.
6. R. Chitchyan, A. Rashid, P. Sawyer, A. Garcia, J. Bakker, M. Pinto Alarcon, B. Tekinerdogan, S. Clarke, and A. Jackson. Survey of aspect-oriented analysis and design, AOSD Europe network of excellence, no. AOSD–Europe-ULANC-9, pp. 1–259, 2005.
7. P. Clements and L. Northrop. *Software Product Lines: Practices and Patterns*, Addison-Wesley, Boston, MA, 2002.
8. P. Clements et al. *Documenting Software Architectures: Views and Beyond*, Addison-Wesley, Boston, MA, September 2011.
9. C. Czarnecki and U. Eisenecker. *Generative Programming: Methods, Tools, and Applications*, Addison-Wesley, Boston, MA, 2000.
10. D. Rowland and T.N. Howe. *Vitruvius*. Ten Books on Architecture. Cambridge University Press, Cambridge 1999, ISBN 0-521-00292-3.
11. E. Demirli and B. Tekinerdogan. Software language engineering of architectural viewpoints, in *Proceedings of the 5th European Conference on Software Architecture (ECSA 2011), LNCS 6903*, Essen, Germany, pp. 336–343, 2011.
12. E. Demirli and B. Tekinerdogan. SAVE: Software architecture environment for modeling views, in *Proceedings of the WICSA 2011: 9th Working IEEE/IFIP Conference on Software Architecture*, Boulder, CO, pp. 355–358, June 20–24, 2011.
13. E.W. Dijkstra. The structure of the "THE" multiprogramming system. *Communications of the ACM* 18(8), 453–457, 1968.
14. L.F. Dobrica and E. Niemela. A survey on software architecture analysis methods. *IEEE Transactions on Software Engineering* 28(7), 638–653, 2002.
15. M. Fowler, S. Scott, and G. Booch. *UML Distilled, Object Oriented Series*, Addison-Wesley, Reading, MA, p. 179, 1999.
16. D. Garlan and M. Shaw. *An Introduction to Software Architecture*, Carnegie Mellon University, Pittsburgh, PA, 1994.
17. C. Hofmeister, R. Nord, and D. Soni. *Applied Software Architecture*, Addison-Wesley, NJ, 2000.
18. C. Hofmeister, P. Kruchten, R.L. Nord, H. Obbink, A. Ran, and P. America. Generalizing a model of software architecture design from five industrial approaches. In *Proceedings of the 5th Working IEEE/IFIP Conference on Software Architecture (WICSA5)*, Pittsburgh, PA, pp. 77–86, November 6–9, 2005
19. ISO/IEC, ISO/IEC 42010. Systems and Software Engineering—Recommended Practice for Architectural Description of Software-Intensive Systems, 2007.
20. R. Kazman, L. Bass, M. Klein, T. Lattanze, and L. Northrop. A basis for analyzing software architecture analysis methods. *Software Quality Journal* 13(4), 329–355, 2005.
21. A. Kleppe. *Software Language Engineering: Creating Domain-Specific Languages Using Metamodels*, Addison-Wesley Longman Publishing Co., Inc., Boston, MA, 2009.

22. P. Kruchten. The 4+1 view model of architecture. *IEEE Software* 12(6), 42–50, 1995.
23. A.J. Lattanze. *Architecting Software Intensive Systems: A Practitioner's Guide*, Auerbach Publications, Boca Raton, FL, 2009.
24. N. Medvidovic and R. N. Taylor. A classification and comparison framework for software architecture description languages. *IEEE Transactions on Software Engineering* 26(1), 70–93, 2000.
25. S.J. Mellor, K. Scott, A. Uhl, and D. Weise. *MDA Distilled: Principle of Model Driven Architecture*, Addison-Wesley, Reading, MA, 2004.
26. H. Obbink, P. Kruchten, W. Kozaczynski, R. Hilliard, A. Ran, H. Postema, D. Lutz, R. Kazman, W. Tracz, and E. Kahane. Report on software architecture review and assessment (SARA), Version 1.0. At http://kruchten.com/philippe/architecture/SARAv1.pdf. February 2002.
27. D. Parnas. On the criteria for decomposing systems into modules. *Communications of the ACM* 15(12), 1053–1058, December 1972.
28. D. Parnas. On the design and development of program families. *IEEE Transactions on Software Engineering* SE-2(1), 1–9, 1976.
29. C. Peña and J. Villalobos. An MDE approach to design enterprise architecture viewpoints, in *IEEE 12th Conference on Commerce and Enterprise Computing (CEC)*, Shanghai, China, pp. 80–87, November 10–12, 2010.
30. J. R. Romero, J. M. Troya, and A. Vallecillo. Modeling ODP computational specifications using UML. *The Computer Journal* 51, 435–450, 2008.
31. B. Roy and T. C. N. Graham. Methods for evaluating software architecture: A survey. *Computing* 545(2008-545), 82, 2008.
32. N. Rozanski and E.Woods. *Software Systems Architecture—Working with Stakeholders using Viewpoints and Perspectives*, Addison-Wesley, Upper Saddle River, NJ, 2005.
33. H. Sozer and B. Tekinerdogan. Introducing recovery style for modeling and analyzing system recovery, in *Seventh Working IEEE/IFIP Working Conference on Software Architecture*, Vancouver, BC, pp. 167–176, 2008.
34. T. Stahl and M. Voelter. *Model-Driven Software Development*, Addison-Wesley, New York, 2006.
35. B. Tekinerdogan. Exploring research directions in software architecture modeling. *International Journal of Software Architecture* 1(1), 7–9, July 2010.
36. B. Tekinerdogan, C. Hofmann, and M. Aksit. Modeling traceability of concerns for synchronizing architectural views. *Journal of Object Technology* 6(7), Special Issue: Aspect-Oriented Modeling, pp. 7–25, August 2007.
37. B. Tekinerdogan and H. Sözer. Defining architectural viewpoints for quality concerns, in *Proceedings of the 5th European Conference on Software Architecture (ECSA 2011), LNCS 6903*, Essen, Germany, pp. 26–34, 2011.
38. TOGAF. The Open Group Architecture Framework, Version 8.1.1., 1995. http://www.opengroup.org/architecture/togaf8-doc/arch/
39. J.A. Zachman. A framework for information systems architecture. *IBM Systems Journal* 26(3), 276–292, 1987.

92

Human–Computer Interfaces for Speech Applications

Shelby S. Darnell
Clemson University

Naja Mack
Clemson University

France Jackson
Clemson University

Hanan Alnizami
Clemson University

Melva James
Clemson University

Josh Ekandem
Clemson University

Ignacio Alvarez
Clemson University

Marvin Andujar
Clemson University

Dekita Moon
Clemson University

Juan E. Gilbert
Clemson University

92.1 Speech Interface Design

A speech application is considered effective if it simulates the core aspects of human conversation. Language is deeply entrenched in human behavior; therefore, a successful speech interface should be based on the different ways that people use speech to communicate. Language conventions that help people know what they should say next should be adopted when creating speech applications (Weinschenk, 2003).

92.1.1 Introduction to Speech Application Development

Human–computer interaction (HCI) focuses on the interaction between humans and computers. This can be done by studying the current design of technology and how it could be redesigned to make technology more user-friendly, more interesting, and engaging. One of the best ways to achieve this is through speech. Integrating speech in an application is a vast benefit because speech is natural. Speaking is a skill humans acquire at an early age and is something we as humans practice quite frequently. The naturalness of a conversation refers to the way that people cooperate with each other in order to ensure successful communication. A speech application is considered effective if it simulates the core aspects of human–human conversation. Language is deeply entrenched in human behavior; therefore, a successful speech interface should be based on the different ways that people use speech to communicate. Language conventions that help people know what they should say next should be adopted when creating speech applications (Weinschenk, 2003).

92.1.1.1 When Is Speech Used?

When determining the success of a speech application, whether or not using speech input is beneficial is a crucial factor. When using a speech application, the user's expectations are high; therefore, speech only needs to be used when the need is clear. When the users are enthusiastic about cooperating, a speech application is most successful (Weinschenk, 2003). For example, airline companies successfully use speech recognition to automate checking flight statuses. People checking flight statuses want to do this as quickly as possible, so they listen and answer the prompts very carefully. People checking flight statuses are also motivated to cooperate, because they do not want to miss any important information, such as a gate change. Automated systems assist the users and save companies' money as well. Speech is suitable for some tasks, but not for others. Table 92.1 lists when speech input and output are appropriate to use (Weinschenk, 2003).

92.1.1.1.1 Speech Challenges

According to Weinschenk et al., using a speech application poses many substantial challenges, and understanding these challenges and the various trade-offs help to produce an effective interface. Six challenges were identified, which are as follows:

- *What did you say?* Once it has been heard or said, it's gone. Since speech is transient, only a limited number of items in a list can be remembered and a user may forget important information that were given at the beginning of a sentence. This is also true when speaking to a dictation system. The users sometimes forget what they have just spoken. The ability to remember transient information has substantial implication for speech interface design.

TABLE 92.1 When Is Speech Input Appropriate?

Use When	Avoid When
Users have a physical disability (e.g., seizures)	Users work in noisy environment
No keyboard is available (e.g., phone)	Task can be accomplished more efficiently using keyboard and mouse
Users are unable to type or are not comfortable with typing	Task requires users to engage in conversation with others while using the application
Task requires the user's hands to be occupied (e.g., driving)	Information is personal or confidential
Task requires user's eyes to be looking at something other than the screen	Large amounts of information need be presented
Situation requires getting the user's attention	Task requires the user to do a comparison

- *What can I say?* In a speech application, the lack of visibility makes it challenging to communicate the functional boundaries of the application to the user. Furthermore, it is challenging to tell the user what actions can be performed and what words or phrases they can use to perform these actions.
- *Asymmetry.* People are able to speak easily and quickly, but cannot listen as easy and quickly. This means that we as humans can speak faster than we type but listen more slowly than we can read. Asymmetry produces an implication for "the type and the amount of information to speak" when designing speech interfaces.
- *Feedback and Latency.* Timing is very important when having a conversation. People read into the meaning behind pauses. Pauses in places where they naturally do not belong causes unfortunate processing delays in speech applications. For example, a user may reply to a prompt and because the system may take a long time to respond, then the user will think the system did not hear them and repeat the response again. This will cause the system to return the wrong response or the user to miss the systems response.
- *Prompting.* Well-designed prompts in a speech application provide the user with a smooth interaction. When designing prompts, many things have to be considered. Good design is accomplished by assessing the trade-off between performance and flexibility. The more the user is constrained on what they can say to an application, the chances of recognition errors are decreased. However, allowing the users to answer flexibly can speed up the interaction and makes the conversation feel more natural.
- *Handling Errors.* The quality of users experiences depend on how a system is able to handle recognition errors. If the users or the application notices an error, the speech interface should be able to provide a way for correcting the error. The most important thing not to do when handling these errors is to repeat the same error message if the users encounter the same rejection error in a row.

92.1.1.2 Speech Recognition

Today, when we call most companies, a person does not answer the phone, an automated machine does. The automated machine instructs the user how to move through the menu of options using touchtone keys or voice, allowing the user to say certain commands or keywords. This is done by speech recognition, the technology that enables computers or other electronic systems to understand human speech (Sharma, 2002). Speech recognition systems have two main functions: to understand the words spoken by the user and convert the spoken words to text for future use and to convert text to speech (TTS) for information access (Weinschenk, 2003). Speech recognition can be divided into two different categories: continuous and discrete.

92.1.1.2.1 Continuous versus Discrete

Continuous recognition allows the user to engage with the system in an everyday manner without using specific commands. Using this type of recognition makes the system error prone and extremely expensive. Discrete recognition allows the user to use a limited vocabulary of single words and phrases. Discrete vocabulary is easy to learn and produces higher accuracy rates; however, it is not as natural (Weinschenk, 2003).

92.1.1.2.2 Recognition Challenges

Speech application design, with its concerns for understanding human language, interpreting different dialects, intonations, and speaking speed, makes speech application design a great research area. The difficulties just listed are a few of the challenges faced for understanding human utterances. The following sections explain the challenges of recognition performance and flexibility versus accuracy.

92.1.1.2.2.1 Performance When designing a speech application, a big challenge is working with an imperfect speech recognizer. The biggest challenges for recognition performance are background noise, accents, and ambiguity. Background noises and utterances can interfere or distort the input that user gives to the system. The variations of different users' voices such as accents can also decrease the accuracy of the speech recognition system (Carroll, 2009). The larger the vocabulary or grammar of a speech recognition system the chances of ambiguity increases. Although the strings may be different when spoken by the user, they might sound alike and therefore the wrong output will be given (Cohen, 2004). Even though speech recognizers are improving consistently, it is very unlikely that in the foreseeable future it will be as robust as they are in science fiction movies. There are some recognition systems that over time adapt to their users, but in order to have good recognition performance, the system needs users who are willing to adapt their speech needs to the recognition system (Weinschenk, 2003).

92.1.1.2.2.2 Flexibility versus Accuracy The more flexible an application is, the more likely errors are to occur. Flexible systems allow the users to speak a command in various ways. When designing a speech application, there must be a balance between flexibility and recognition accuracy (Weinschenk, 2003). For example, a car application may allow the user to change the radio station as follows:

- Change to one oh one point three f m please.
- Can you please go to one hundred one point three?
- Go to station one oh one point three f m.
- Please change the station to one hundred one point three.

In theory, this may be the natural but if the recognition performance is poor, users will not be in favor of using the application. Accepting a variety of ways to complete a task can offer flexibility without causing the recognition performance to decrease (Weinschenk, 2003).

92.1.1.3 Speech Synthesis

A speech synthesizer, synonymous with TTS, converts written TTS (Sharma, 2002). Speech synthesis is complementary to speech recognition (Carroll, 2009). The history of speech synthesis surpasses that of speech recognition but is still progressing today. It is used as a screen reader and provides verbal instruction, feedback, or assistance (Weinschenk, 2003). Speech synthesis can be broken into two categories: concatenated and formant.

92.1.1.3.1 Concatenated versus Formant

Concatenated synthesis uses segments of recorded human speech and pieces them together to produce meaningful speech output (Sharma, 2002). Using concatenated synthesis produces a more natural sound. Having a human reader read segments of speech and storing the segment of speech is the process for creating concatenated synthesizers. This type of synthesis is used for systems that require a small vocabulary (Weinschenk, 2003). Formant synthesis uses a set of guidelines that control audio waveforms that simulate human speech. Formant synthesis creates true machine-generated speech that often sounds like a robot. This form of synthesis is used in most TTS applications (Sharma, 2002).

92.1.1.3.2 Synthesis Limitations

Speech synthesizers can make errors at any given time. Here is a list of some of the sources of error in producing TTS output (Weinschenk, 2003).

- *Structure analysis*: The beginning and end of paragraphs and sentences are inconsistently indicated due to punctuation and formatting.
- *Text preprocessing*: The speech synthesizer cannot know all the possible abbreviation and acronyms. It also cannot determine all the ways things such as time and dates can be written.
- *Text-to-phoneme conversion*: Although speech synthesizers can pronounce a lot of words correctly, there are still some words that it has to guess.
- *Prosody analysis*: Speech synthesizers try to guess the emphasis that a human uses on correctly phrasing a sentence and sometimes the output sounds artificial and unnatural.
- *Waveform production*: Due to the fact that speech synthesizers do not have lips, mouths, lungs, and other things used in human speech, they often produce speech that is mechanical.

92.1.1.3.3 Evaluating Synthesis

Being able to evaluate speech synthesis quality and understanding the factors that can affect the output are important in the utilization of speech synthesis (Weinschenk, 2003). The measures of the quality of synthesized speech are (Cohen, 2004) as follows:

- Intelligibility: An indication of how well the listener understands the words and sentences spoken by the synthesizer.
- Naturalness: An indication of the extent of how much the synthesized speech sounds like a human.
- Accuracy: An indication of how correct the speech synthesizer is at converting TTS.
- Listenability: An indication of how long and well the users can listen to the speech synthesizer for an extended amount of time without becoming tired.

92.1.2 Application Development Process

A voice–user interface (VUI) is an application with which a person interacts using her or his voice (Guang-li, 2011). These interfaces are evaluated based on how well they perform when interacting with various people. A system is said to perform well when it returns the correct responses to a user without lagging, making mistakes, and/or causing frustration. Other vital factors to consider are solving the problem, communicating naturally, and user satisfaction (Cohen, 2004). The accuracy of VUIs depends heavily on how well the application was designed and on how well the speech recognizer can parse through the input. Because most people speak, and speak often, designing an application to respond well to natural speech is often severely underestimated. Designing a machine that understands and communicates using natural language to provide a seamless user experience is a major undertaking. Responding well to natural speech is one of the most substantial challenges a VUI designer faces. Approaching VUI design in a similar way as a software engineer approaches designing software enables quicker development time, and upon application, completion results in a better user experience.

92.1.2.1 Think about the Problem: Requirements Analysis

In order to develop an application, the requirements must first be understood. Requirements Engineering (RE) is the systematic process of developing requirements through an iterative cooperative process. The process entails analyzing the problem, documenting the resulting observations in a variety of representation formats, and checking the accuracy of the understanding gained (Loucopoulos, 1995).

This section of the development process often begins with a mission statement and becomes more elaborate through studies and meetings with prospective clients (Rosson, 2002). The result of this effort is a full list of all functions and features the system must satisfy. Requirements analysis is an ongoing process but many users cannot declare or understand their real needs until they see what kind of options are available. According to (Rosson, 2002), "Usability engineers participate in requirements analysis by studying how work currently takes place to see if they can identify problems or opportunities that might be addressed by new technology."

Scenarios or narratives are also often used during requirements analysis because they evoke reflection and conversation. Additionally, scenarios foster mutual understanding among the different groups who participate in the analysis (Rosson, 2002).

92.1.2.2 VUI Design

Once the problem is clear, designers can begin to envisage solutions. "The field of human-computer interaction (HCI) is concerned with the joined performance of tasks by humans and computational machines" (Hewett, 1996). Some would say that HCI's major contribution to the field of computer science is the design, evaluation, and implementation of interactive computing systems for human use (Hewett, 1996).

Speech technology has improved significantly within the last decade where it has become possible and rather preferable to integrate voice into applications. Using voice as a means of interaction with a technology has its advantages over the traditional input/output modalities. Certain environments such as applications developed for in-vehicle use encourage the use of voice over the traditional input/output modalities. Designing VUIs differs from graphical user interface (GUI) design (Yankelovich, 1995).

While visual sketches are fundamental in GUI design to mock up user experience with an application, designers of VUIs find it vital to understand people's natural conversational styles and communications. A well-designed voice interface dialogue does not violate natural conversation behaviors. A natural dialogue would be one that simulates a natural conversation with the user without influencing the voice input. For example, a system would say, "what would you like to eat?" instead of prompting the user with "do you want an apple or an orange?" While the latter is restricting, the first example simulates more natural interaction. It is vital that VUI designers put together a dialogue that prompts and allows for natural navigation. The design of such a dialogue would consist of scenarios represented by a flowchart of in/output prompts.

As with any other input modality, using voice to place information into an application isn't perfectly reliable. Errors in speech input are encountered in myriad ways. Noise is one of the more troublesome challenges for speech application input. Another major challenge is the limitations of the applications grammar. A well-designed VUI allows a seamless workaround for these major challenges that does not frustrate the user (Yankelovich, 1998).

A VUI should be intuitive and complement human communication. A user would be easily frustrated if miscommunication and misunderstanding were too frequent while interacting with a VUI. Humans are wired to use speech to communication needs and desires among each other. A voice application should be able to attend to that need and deal with miscommunications smoothly. A VUI design fails if the user could detect the application's inefficiency in handling errors and completing tasks for which it was designed.

As aforementioned, errors occur when interacting with voice applications. A designer should incorporate prompts in the application to sympathetically aid the user. The user, with the systems help, should be able to backtrack to the point before miscommunication and continue interactions.

92.1.2.3 Implementation

One of the main aspects of speech application design is the grammar. The grammar tells your speech recognition application that to which it should listen. Once the system knows that to which it should listen, you can program an application to respond to utterances. Because humans generally communicate through speech, a good grammar is designed to process natural language. In order to implement a speech design, you need a speech recognition application and be able to build a grammar.

92.1.2.3.1 Grammar

A grammar consists of an organized list of rules specifying the words or phrases that can be used for that particular speech recognition system. The list of rules gives the guidelines for the applications spoken input. These guidelines are used to predict the possible speech input that the user will give and also restricts the words or phrases that the system can recognize. Grammars can be written in one of two forms: a grammar file or an inline script. A grammar file is an external file that holds the words and phrases for your speech recognition system. An inline script is within the actual code of the speech recognition application (Microsoft, 2004a).

A grammar limits vocabulary, filters response recognition, matches speech, and identifies rules. The grammar only contains the words or phrases that the system needs to successfully match the user's speech input. An application might only need to recognize a small set of words and the grammar prevents the system from having to search an entire dictionary. By specifying a group of words in

TABLE 92.2 Grammar Specification Language Operators

Operator	Expressions	Meaning
() concatenation	(A B C D)	A and B and C and D
[] disjunction	[A B C D]	One of A or B or C or D
? optional	?B	B is optional
+ positive closure	+C	One or more repetitions of C
* Kleene closure	*D	Zero or more repetitions of D

a grammar, the recognition accuracy may also increase. Using a grammar also filters the results by only returning a successful recognition event only if the grammar is matched. Even though grammar structures need to be flexible for the user, it also needs to restrict the users input for a specific situation or task. Grammars have rules or entities that define and give order to the possible words, phrases, or utterances of the user. The rules can be referenced repeatedly within the particular grammar or rules in other grammars. Grammars can define all combinations of how to say a phrase in a single rule (Microsoft, 2004b).

Grammars can be written in various formats such as GSL, XML, JSGF, and ABNF. All grammars are constructed by using five basic operators. Table 92.2 describes how the operators are used in GSL (Nuance, 2001). The symbols A, B, C, and D denote a grammar or word name in the table.

Here are some GSL expressions and some of the phrase they describe: [coffee juice tea] *matches* "coffee," "juice," or "tea;" (I would like [coffee juice tea]) *matches* "I would like coffee," "I would like juice," or "I would like tea;" (?(I would like) [coffee juice tea]) *matches* "I would like coffee," "I would like juice," "I would like tea," or simply "coffee," "juice," or "tea;" and (thanks + very much) *matches* "thanks very much," "thanks very very much," etc.

92.1.2.3.2 Highlighted Speech Platforms and APIs

Some easily available tools that enable speech application design for the web and desktop are Web-Accessible Multimodal Applications (WAMI) by MIT and Dragon NaturallySpeaking by Nuance. These tools are highlighted because of their good reputation and their ability to aid in the design of speech applications. WAMI is a speech platform that is available to use for free online. Dragon NaturallySpeaking is considered the state of the art for personal speech recognition. The following paragraphs further explain how WAMI and Dragon NaturallySpeaking can aid speech application design.

92.1.2.3.2.1 Web WAMI is a JavaScript API that allows a user to add speech recognition capabilities to a website. WAMI was developed by MIT's Computer Science and Artificial Intelligence (CSAIL) Spoken Language Systems group. WAMI provides developers with a sufficient set of components to build a web-based multimodal interface. WAMI is designed to ease the development of multimodal interfaces with modest natural language understanding needs. Grammars are written using the Java Speech Grammar Format (JSGF) standard. WAMI can be used to produce applications on recent versions of Firefox, Opera, Safari, or Internet Explorer on a Windows, Linux, or OS X system (McGraw, 2008).

92.1.2.3.2.2 Desktop Dragon NaturallySpeaking is a speech platform created by Nuance. It allows the user to converse with the computer instead of typing. While the user is talking, whatever they are saying is transcribed to the screen and into documents or e-mail messages. Dragon can be used to compose letters and memos and send e-mail message. Dragon can also be used to do things such as enter data in forms or spreadsheets, work on the web, start programs, and open menus. Using Dragon boosts the productivity of your application. Personalizing Dragon's vocabulary is easy and very important for productivity because it preempts recognition errors. Correcting Dragon's errors will also help it learn. The more Dragon is used, the more it learns and adapts to the user in terms of acoustics and vocabulary (Nuance, 2012).

92.1.2.4 VUI Evaluation

A VUI designer should have a clear understanding of what the purpose of the application is and what needs to be accomplished when accessing the application. Any interruption that stops a user from accomplishing the task would negatively affect the experience. VUI designers should always aim to achieve higher user satisfaction. User satisfaction could be evaluated and measured via questionnaires, interviews, and focus groups. There are two ways to evaluate a VUI, a summative and a formative evaluation. A summative evaluation is one that takes place once after the application has been developed. Data collected from this evaluation are used to improve the next version of the VUI. A formative evaluation is one that takes place several times while the system is being developed and appears as part of the prototyping application. There are several ways to measure task succession. One is running a number of pilot studies to smooth out any wrinkles in interaction flow. A designer should calculate task completion time and evaluate the system's ability to fix errors.

92.1.2.5 Wizard of Oz

The Wizard of Oz (WOZ) is a method of system prototyping where a human experimenter sits behind the scenes and interacts with the user on behalf of the system. This method is useful in trying to understand the interaction that the user would expect under near perfect conditions. Since system prompts are simulated via the human experimenter, the interaction behavior between both the participant and the experimenter depends heavily on the believability of the wizard. Therefore, the user would have different interactions based upon the believability of the wizard. The two major ways of evaluating a VUI are summative and formative. Summative evaluation is an informal and is performed at the end of the interface development phase. The data collected are usually used at the next major release. Formative evaluations are more formal procedures that occur during prototyping.

92.1.2.6 Usability Measures

The best method for measuring the usability of a system is through observation. This method requires analyzers to observe the system during broad situations that closely match the predicted usage patterns. One issue with this method is the amount of effort needed to construct a robust prototype (Campos, 2003).

Obviously, analytical methods do not solve usability issues, but they do help generate early feedback about interactive systems. It is valuable to attain as much feedback as possible before expensive decisions are made. Analytical techniques can be used with informal representations of the design, such as a storyboards or wireframes. It should be noted that the feedback attained from this method alone is seldom enough to provide adequate specifications for programming the system (Campos, 2003). There are a few issues associated with informal methods. Informal methods are difficult to perform systematically and should only be used by someone who has sufficient expertise.

Two common methods for measuring usability include usability inspection and cognitive walkthrough. The inspection involves an examination of the design representation using a sequence of standard questions. Although both methods involve a systematic set of questions, the walkthrough actually requires an initial representation of how the system will be used. This is due to the fact that the walkthrough questions are related to the task representations (Table 92.3).

TABLE 92.3 Sample Questions for Assessing Usability

Sample Usability Questions				
Is the system easy to use (circle the most appropriate response)?				
Strongly agree	Agree	Neutral	Disagree	Strongly disagree
How satisfied are you with this system (circle the most appropriate response)?				
Very satisfied	Satisfied	Neutral	Unsatisfied	Very unsatisfied

TABLE 92.4 Microsoft's Lessons in the Art of Automated Conversation

Selected Best Practice Guidelines for Speech Interfaces
1. Make It Real
Users come with their own set of terminologies (?), metaphors, and organizational structures in mind. This is known as their *mental model*, or expectation of the interaction. Usable applications tap into this knowledge to give users a head start in understanding how to interact with the application.
2. Clearly and Consistently Communicate System Capabilities
VUIs must be carefully designed to help users understand the capabilities of the system. Interfaces need to unobtrusively guide users to speak predictable utterances and avoid the unconstrained conversational speech that we use talking to another person. The goal is to achieve a natural conversation within the technological boundaries of speech recognition.

The usability inspection method uses heuristic evaluation. This approach systematically inspects the system using a set of best practice guidelines. A team of evaluators, preferably not those involved in the design process, is assembled to perform the heuristic evaluation (Table 92.4). A sample of best practice guidelines used by Microsoft can be found in the succeeding text. Once the team of expert evaluators is assembled, they perform their assessment based on the design guidelines and aggregate the results. By aggregating results from multiple evaluators, a more comprehensive evaluation is created.

Because the usability inspection method does not explicitly mention how the system is to be used, a cognitive walkthrough is sometimes useful (Campos and Harrison, 2003). The purpose of a cognitive walkthrough is to assess how well an interface will guide users in performing a task. In order for this to be effective, the expected tasks must be identified and there must be a model of the interface that allows these tasks to be performed. According to Campos and Harrison (2003), the assessment is achieved through answering the following three key questions at each stage of interaction:

- Will the correct action be made sufficiently evident to users?
- Will users connect the correct actin's description with that they are trying to achieve?
- Will users interpret the system's response to the chosen action correctly?

They also state usage models documenting predicted courses of action for the user is a prerequisite. If "no" is an answer to any of these questions, it exposes a problem that will be addressed.

92.2 Research Issues and Future Development

Previous sections of this chapter have introduced design strategies and tools targeted toward developers. In this section, however, user-focused applications take center stage. Here, we present a general overview of the speech technology application landscape with signs pointing toward more specialized readings. Broad classes of real-world applications of VUIs in the past and present are explored, and based on the trajectory defined by these developments, a brief glimpse into the future of speech applications is described.

92.2.1 Speech Technology Applications: Then

When referring to speech applications "then," we refer to applications that accomplish the most fundamental tasks that allow for the existence of speech systems. Stored voice applications allow for the recording and playback of recognizable human language. Synthesized speech applications are enabling applications that allow for the mechanic or digital production of human-like speech. The final type of fundamental speech application, or speech application "then," is those that are able to listen to human speech and in some form understand utterances. Speech applications "then" consist of systems that store voice information for later listening, synthesize speech, and recognize words or interactive voice response (IVR) systems.

92.2.1.1 Stored Voice and Interactive Voice Response Systems

Among the most widely used speech technology systems are those that permit both the storage and the playback of human (or human-*like*) voices. Systems of this kind, in fact, have been in existence for at least 150 years (Feaster, 2009). Stored voice applications use limited speech collections as "read only" data in which individual sounds, or responses, are mapped to specific user actions, or input (Schmandt, 1993). Stored voice applications are most appropriately used when the set of acceptable responses is finite. Example categories of stored voice applications, which may use either synthesized speech or digitally recorded human speech as output, are the following: playback-only applications, interactive record and playback applications, dictation applications, and multimedia/multimodal applications (Rodman, 1999; Schmandt, 1993, 1999).

92.2.1.1.1 Interactive Voice Response Systems

The deployment of VUIs, especially for use over the telephone, has grown rapidly over the past decade, and incentives for both corporations and their clients have stimulated this growth (Cohen, 2004). The quintessential example of this general observation is the IVR system, or call center.

92.2.1.2 Speech Synthesis Applications

Speech synthesis is one of the oldest branches of speech technology research. Speech synthesis (i.e., TTS) systems have applications in many different domains, including telecommunications (see Section 92.1.1), public safety, entertainment, and education (Rodman, 1999). In addition, aid for the disabled is a small, but very important, application niche. Synthetic speech systems can be used by the blind or visually impaired. These systems provide computer access, to read, and to manage personal productivity (Schmandt, 1993). Those who have lost the ability to speak by traditional means can also use synthetic speech systems. Theoretical physicist Stephen Hawkins, who lives with a form of amyotrophic lateral sclerosis (ALS), or "Lou Gehrig's disease," is one of the most famous users of speech synthesis technology (Lange, 2012).

92.2.2 Speech in the Automobile: Speech Applications "Now"

Speech technologies "now" refer to speech applications that take the fundamental speech applications and allow a person to interface with devices. One of the main thrusts of a lot of speech application design research for speech technologies "now" is for speech applications in the automobile.

The increase in the electronics within many vehicles can be attributed to the fact that automobiles have become more than just a means of transportation and for many people have become a multifunctional living space (Kern, 2009). Reducing distraction and cognitive load is the main thrust for designing speech application in the automobile.

92.2.2.1 Reducing Distraction

Any task whether cognitively, physically, or visually demanding can have a significant influence on driver distraction (Young, 2007). However, unlike other input modalities, like direct touch or gestures, the speech input does not require drivers to take their hands off the steering wheel and eliminates much of the overall difficulty of manual device interactions (Gruenstein, 2009). Even more, researchers have found speech to perform better than other input modalities for text input (Ablassmeier, 2006; Camilli, 2011; Maciej, 2009). Ford motor company has done a considerable amount of work in developing a VUI for their vehicles (Rana, 2010).

92.2.2.2 Design to Reduce Cognitive Load

Speech as an input modality is very effective; however, if improperly designed, it can increase cognitive loads. When used as an output modality, Christiansen et al. demonstrated that even though auditory feedback had the fewest eye glances, it had the longest completion times, and the authors state that "listening to audio output while driving causes an increase in the cognitive load of the driver, thereby drawing mental resources away from the task of driving" (Christiansen, 2011).

92.2.2.2.1 Evaluating Speech Interfaces in the Vehicle

Usability measures should be collected using standardized tools such as the system usability scale (Lewis, 2009) or the technology acceptance model (Venkatesh, 2003). These questionnaires assess the overall perception of the system from the user's perspective and are decisive indicator of the possible acceptance of the system in real-life scenarios.

92.2.2.3 Toward Tomorrow

It has also been noted that many auditory and speech alerts are clearly useful when the driver is not looking at an interface. However, additional research will be necessary to ensure that these alerts are not masked by other auditory information (Kramer, 2007).

Currently, handling speech in the automobile is a rich area of research and with increased developments in natural language processing and automotive user interfaces, speech research and developments in the automobile will flourish.

92.2.3 Future of Speech Application Design

Like most technologies, speech application design can be improved by incorporating aspects of complementary technologies. Current speech application design is limited. A limit is the inability of current speech applications to distinguish one person from another and the inability to determine a speaker's mood in order to manage their affect. Integrating biometrics and affective technologies, speech application design can be improved.

In order to make speech applications more acceptable for users of all kinds, it is necessary to use them in conjunction with other technologies. As HCI designers, we want to use technologies that improve and enhance the usability of our applications. Adding security features and basing responses on user affect are two fantastic ways to improve the efficacy of speech applications. Voice biometrics is the modality that uses spoken utterances, as do speech applications, and affective technologies can use voice, facial expressions, body language, and many other human responses and physiological factors. One of the main things that biometrics and affective computing have in common is they take physiological aspects of humans and find patterns from these aspects that help recognize or understand people. The next couple of sections will speak to the usage of voice biometrics and affective computing in conjunction with speech interfaces.

92.2.3.1 Biometrics

Biometrics is the science of measuring the patterns of life. Biometrics is quantitatively taking measurements of physiological and behavioral aspects of humans and using them for tasks of identification or classification. Biometrics can be physiological or behavioral. A single type of biometric reading, physiological or behavioral, is referred to as a modality. Some of the physical modalities are fingerprint, iris, face, periocular, hand, finger geometry, odor, ear, and DNA (Jain, 2004). Some of the behavioral modalities (those included in direct and indirect HCI biometrics) are keystroke dynamics, mouse dynamics, haptic, programming style, audit logs, GUI interaction, and call stack. Yampolskiy and Govindaraju have come up with a definition for direct and indirect HCI biometrics and give examples of several modalities. Direct HCI biometrics are modalities that evaluate a person based on abilities, style, preference, knowledge, and strategy. Indirect HCI biometrics are those deduced from behaviors recorded when a person interacts with technology (Yampolskiy, 2007).

92.2.3.1.1 Biometrics HCI Concerns

Biometrics is a science that can make people uncomfortable because it learns them on a very intimate level. For example, fingerprint is a modality that makes people uncomfortable because law enforcement uses it when registering people upon arrest. Hence, making people submit their fingerprint in order to access a location or service has negative connotations.

Considering HCI concerns with respect to biometrics allows us choose modalities whose connotations are not inherently negative and helps find uses of negative modalities in a manner that are less abhorrent to users (Wertheim, 2010). Gamboa and Fred also use HCI techniques to create behavioral biometric modalities. Given the previous work with respect to HCI and biometrics, how do we leverage the research to improve speech application design (Gamboa, 2003)? The most obvious way is incorporate voice biometrics into speech application design practices.

92.2.3.1.2 *Voice Biometrics*

The biometric modality that makes the most sense to use with speech applications is voice. Voice biometrics is a science that leverages the intricacies of the voice for the purposes of recognition and classification.

Voice biometrics, like most biometric modalities, comes in two recognition-based flavors: speaker verification and speaker identification. Voice has the four main requirements to be a biometric modality: distinctiveness, universality, permanence, and collectability. The systems employing this biometric modality operate similarly to all biometrics systems, which have four major components: sensor, feature extractor, matcher, and system database (Jain, 2004). In order to use voice biometrics with speech application design, these components must be included.

92.2.3.1.3 *Speech Application Design Using Voice Biometrics*

A speech application can prompt the user for their name and after making sure this name is in the system database prompt for other information, but the system may also be able to verify the speaker with the identity claim alone. Once the speaker is verified, they will be privy to using the system further. Markowitz writes about a case study involving a company *Girl Tech, Inc.* that makes products *Door Pass* and Password Journal (Markowitz, 2000). Both *Door Pass* and *Password Journal* use speaker verification to keep a door locked or a journal closed.

Obtaining useful information from a biometric signature that is or isn't adequate for authentication or identification is referred to as soft biometrics (Ross, 2007). Soft biometrics deals with the classification of people into groups based on similar attributes. Incorporating voice biometrics in speech applications takes speech applications to the next level.

92.2.3.1.4 *A Look at the State of the Art*

Voice biometrics has been steadily improving for over a decade (González-Rodríguez, 2008; Markowitz, 2008; Miller, 2012). Recently, a company has improved speaker verification using state-of-the-art recognition technology in conjunction with a smartphone (Miller, 2012). This new technology boasts that it accepts the incorrect user (false accept rate—FAR) once out of 10,000 times and rejects the correct user (false rejection rate—FRR) five times out of one hundred. Affective computing isn't a biometric modality, but using it in conjunction with voice biometrics can provide more secure and usable applications than voice biometrics alone.

92.2.3.2 Affective Computing

Affective computing attempts to recognize emotions and uses what it recognizes to improve a user's computing experience (Wu, 2010). Affective computing uses sensors to capture information. From the sensors, algorithms analyze the information in the context of the situation in which the data were captured (Healey, 1998). Because the data are analyzed in context, it can be labeled to match emotions and feelings a user has when his or her data are recorded. A system is created that uses the information learned from previous sensor readings and is able to understand what a user is feeling. Facial expression is an important and often used indicator of mood. Galvanic skin response (GSR) is another popular indicator of mood. Facial expression information can be captured by camera, whereas specialized sensors must be attached or worn to capture GSR data.

92.2.3.2.1 Affect Improving Speech Applications

Affect can improve speech applications in a couple of ways. There are sensors that can be worn similar to wristwatches or gloves that record GSR data (Affectiva, 2012; Strauss, 2005). Using a wireless version of one of these sensors in conjunction with a speech application, the system would be able to cause the voice to respond to changes in a user's mood. Affective technology can improve speech applications and design by keeping track of user engagement or alertness. There are many more ways affective technology can improve and complement speech applications but that is for future research.

References

Ablassmeier, M. (2006). A new approach of a context-adaptive search agent for automotive environments. In *CHI '06 Extended Abstracts on Human Factors in Computing Systems* (pp. 1613–1618), Montreal, Canada. New York: ACM.

Affectiva. (2012). *Affectiva Q Sensor*. Retrieved August 1, 2012, from http://www.affectiva.com/q-sensor/.

Camilli, M. (2011). Searching digital audio broadcasting radio stations: Usability and safety issues for in-vehicle devices. In *Proceedings of the 9th ACM SIGCHI Italian Chapter International Conference on Computer-Human Interaction: Facing Complexity* (pp. 143–147), Alghero, Italy. New York: ACM.

Campos, J. C. (2003). From HCI to Software Engineering and back. In *International Conference on Software Engineering (ICSE)* (pp. 49–56), Portland, Oregon.

Campos, J. and D. Harrison. (2003). In bridging the gaps between software engineering and human computer interaction. ICSE 2003.

Carroll, J. M. (2009). *Human Computer Interaction (HCI)*. (M. A. Soegaard, Ed.) Aarhus, Denmark: The Interaction Design Foundation.

Christiansen, L. H. (2011). Don't look at me, I'm talking to you: Investigating input and output modalities for in-vehicle systems. In *Proceedings of the 13th IFIP TC 13 International Conference on Human-Computer Interaction—Volume Part II* (pp. 675–691), Lisbon, Portugal. Berlin, Germany: Springer-Verlag.

Cohen, M. H. (2004). *Voice User Interface Design*. Boston, MA: Addison-Wesley Longman Publishing Company Incorporated.

Feaster, P. (2009). *The Phonautographic Manuscripts of Edouard-Leon Scott de Martinville*. (P. Feaster, Ed.). http://www.firstsounds.org/public/Phonautographic-Manuscripts.pdf. PDF with sound at http://archive.org/details/ThePhonautographicManuscriptsOfEdouard-leonScottDeMartinville

Gamboa, H. (2003). An identity authentication system based on human computer interaction behavior. In: Ogier, J.M and Trupin, E (Eds.), Pattern Recognition in Information Systems, Proceedings of the 3rd International Workshop on Pattern Recognition in Information Systems, PRIS 2003, In conjunction with ICEIS 2003, Angers, France, April 2003. ICEIS Press (pp. 46–55).

González-Rodriguez, J. A.-G. (2008). Voice biometrics. In *Handbook of Biometrics*, A. K. Jain, (Ed.) (pp. 151–170). New York: Springer.

Gruenstein, A. (2009). City browser: Developing a conversational automotive HMI. In *Computer Human Interaction* (pp. 4291–4296). Boston, MA: ACM Press.

Guang-li, S. (2011). Improved VUI system based on maintenance device. In *2011 International Conference on Computational Problem-Solving (ICCP)* (pp. 510–513), Chengdu, China.

Healey, J. (1998). Digital processing of affective signals. In *Proceedings of the 1998 IEEE International Conference on Acoustics, Speech and Signal Processing*, Vol. 6 (pp. 3749–3752), Seattle, WA.

Hewett, T. (1996). *ACM SIGCHI Curricula for Human-Computer Interaction*. Retrieved from The ACM Special Interest Group on Computer Human Interaction: http://www.sigchi.org (accessed on September 10, 2013).

Jain, A. (2004). An introduction to biometric recognition. *IEEE Transactions on Circuits and Systems for Video Technology*, 14(1), 4–20.

Kern, D. (2009). Design space for driver-based automotive user interfaces. In *Proceedings of 1st International Conference on Automotive User Interfaces and Interactive Vehicular Applications—Automotive UI 2009* (pp. 3–10), Essen, Germany.

Kramer, A. F. (2007). Influence of age and proximity warning devices on collision avoidance in simulated driving. *Human Factors: The Journal of the Human Factors and Ergonomics Society*, 49(5), 935–949.

Lange, C. d. (2012). The Man Who Saves Stephen Hawking's Voice. *New Scientist*, p. 27.

Lewis, J. R. (2009). The factor structure of the system usability scale. In *First International Conference on Human Centered Design* (pp. 70–80), San Diego, CA. https://www.springer.com/computer/hci/book/978-3-642-02805-2

Loucopoulos, P. (1995). *System Requirements Engineering*. New York: McGraw-Hill.

Maciej, J. (2009). Comparison of manual vs. speech-based interaction with in-vehicle information systems. *Accident Analysis and Prevention*, 41(5), 924–930.

Markowitz, J. A. (2000). Voice biometrics. *Communications of the ACM*, 43(9), 66–73.

McGraw, I. (2008, October). *WAMI a Javascript Api for Speech Recognition*. Retrieved August 1, 2012, from The WAMI toolkit: http://wami.csail.mit.edu/.

Microsoft, S. S. (2004a). *Grammars Overview*. Retrieved August 1, 2012, from http://msdn.microsoft.com/en-us/library/ms870145.aspx.

Microsoft, S. S. (2004b). *Grammars: Purpose and Structure*. Retrieved August 1, 2012, from http://msdn.microsoft.com/en-us/library/ms870145.aspx.

Miller, D. (2012). *VoiceVault Achieving Impressive Levels of Accuracy in Real World Implementations*. Retrieved August 1, 2012, from voicebiocon.com: http://voicebiocon.com/2012/08/02/voicevault-achieving-impressive-levels-of-accuracy-in-real-world-implementations/.

Nuance, C. (2001). *Grammar Developer's Guide*. Retrieved December 1, 2012, from voxeo.com: http://evolution.voxeo.com/library/grammar/grammar-gsl.pdf.

Nuance, C. (2012). *Dragon Naturally Speaking Version 12 End User Workbook*. Retrieved August 1, 2012, from http://www.nuance.com/for-individuals/by-product/dragon-for-pc/home-version/index.htm.

Rana, O. (2010). *New SYNC with MyFord Touch Allows more than 10,000 Voice Commands*. Retrieved August 1, 2012, from egmcartech.com: http://www.egmcartech.com/2010/07/15/new-generation-sync-with-myford-touch-allows-up-to-10000-voice-commands/.

Rodman, R. (1999). *Computer Speech Technology*. Boston, MA: Artech House.

Ross, A. (2007). An Introduction to Multibiometrics. In *Proceedings of the 15th European Signal Processing Conference (EUSIPCO)*, Poznan, Poland.

Rosson, M. B. (2002). *Usability Engineering: Scenario-Based Development of Human-Computer Interaction*. San Francisco, CA: Morgan Kaufmann Publishers Inc.

Schmandt, C. (1993). *Voice Communication with Computers: Conversational Systems*. New York: Van Nostrand Reinhold.

Sharma, C. a. (2002). *VoiceXML: Strategies and Techniques for Effective Voice Application Development with VoiceXML 2.0*. New York: Wiley.

Strauss, M. D. (2005). *HandWave: Design and Manufacture of a Wearable Wireless Skin Conductance Sensor and Housing*. Cambridge, MA: MIT.

Venkatesh, V. (2003). User acceptance of information technology: Towards a unified view. *MIS Quarterly*, 27(3), 425–478.

Weinschenk, S. (2003). *Designing Effective Speech Interfaces*. New York: Wiley Computer Publishing.

Wertheim, K. (2010). Human Factors in Large-Scale Biometric Systems: A Study of the Human Factors Related to Errors in Semiautomatic Fingerprint Biometrics. *IEEE Systems Journal*, 4, 138–146.

Wu, D. (2010). Optimal arousal identification and classification for affective computing using physiological signals: Virtual reality stroop task. *IEEE Transactions on Affective Computing*, 1(2), 109–118.

Yampolskiy, R. V. (2007). Direct and indirect human computer interaction based biometrics. *Journal of Computers*, 2(10), 76–88.

Yankelovich, N. (1995). Designing SpeechActs: Issues in speech user interfaces. In *Proceedings of the SIGCHI Conference on Human Factors in Computing Systems*, (pp. 369–376), Denver, CO. New York: ACM Press/Addison-Wesley Publishing Co.

Yankelovich, N. (1998). Designing speech user interfaces. In *CHI 98 conference Summary on Human factors in computing systems* (pp. 131–132), Los Angeles, CA. New York: ACM.

Young, K. (2007). Driver distraction: A review of the literature. In *Distracted Driving*, I. J. Faulks (Ed.) (pp. 379–405).

93 Software Assurance

Nancy R. Mead
Software Engineering Institute

Dan Shoemaker
University of Detroit Mercy and International Cyber Security Education Coalition

Carol Woody
Software Engineering Institute

93.1 Introduction

In this chapter, we introduce software assurance, discussing the impact of software on our lives, the risks associated with software vulnerabilities, and some basic definitions of software assurance. We then introduce modern principles of software assurance, along with historical principles of software assurance. With this background in hand, we were able to identify a number of process models, frameworks, and best practices that were relevant to software assurance. Finally, we identify a research framework that could be used to support current research and to identify gap areas for future research.

93.2 What Is Software Assurance?

Formal practices to ensure correct software date back at least 40 years (Royce 1970). America's official interest in secure software dates back 9 years to Action Recommendation 2–14 in the *National Strategy to Secure Cyberspace* (DHS 2003). So if the aim of the prior 40-year practices was not to ensure that our code was secure, what *were* we trying to do? That work was to ensure the *quality* of thc software product. To serve that aim, practices to validate product correctness have become an integral part of the industry. Likewise, because of the long history of software quality assurance, there is a seemingly endless list of assurance practices, which have always been considered part of the software life cycle. Activities such as design practices (adopted in 1972), inspections (1976), change management (1979), incremental integration (1979), branch coverage testing (1979), and source code control (1980), which predate today's interest in security by decades, are often included in the steps for developing secure code (McConnell 2005).

What has happened over the past decade to change the emphasis from quality assurance to security assurance? The answer to that question lies in the appearance of the adversary. Prior to 2000, developers' only concern was to develop products that worked right under ordinary circumstances. Therefore, the aim of the software engineering process was to ensure properly functioning code. The fact that defects might be present did not matter as long as the product satisfied user requirements and functioned properly. However, when every bad actor on the planet began using the Internet as an avenue of attack, it became insufficient that the code only works right. Now software has to be free of all exploitable defects. Defects that might not have been worth fixing in the past if they did not affect operation may now represent a pathway for trouble. That changes the focus of the software assurance process from quality to security. This chapter will center on the shift in the definition of terms that results from this change in focus as well as the newest models and methods for finding and fixing exploitable defects in software.

93.2.1 Changing the Assurance Focus to Security

According to the US Committee on National Security Systems' (CNSS) *National Information Assurance (IA) Glossary*, software assurance is "the level of confidence that software is free from vulnerabilities, either intentionally designed into the software or accidentally inserted at any time during its life cycle, and that the software functions in the intended manner" (CNSS 2010, p. 56). In addition, a report from the USC Center for Systems and Software Engineering (CSSE) says that software assurance amounts to the addition of concrete functionality to secure the product (Colbert and Yu 2006). That includes the complete set of additional development activities from specification through design and on to code that are aimed at ensuring security functionality. Finally, a Department of Homeland Security (DHS) report suggests that the purpose of the assurance process is to build software-based systems that "limit the damage resulting from any failures caused by attack-triggered faults, ensuring that the effects of any attack are not propagated, and recover as quickly as possible from those failures as well as ensuring that the software will continue to operate correctly in the presence of most attacks by resisting the exploitation of weaknesses in the software by the attacker, or by tolerating the failures that result from such exploits" (Redwine et al. 2007). For the purpose of developing a curriculum model for software assurance in the *Master of Software Assurance Reference Curriculum* report, the CNSS definition was expanded as follows (Mead et al. 2010):

> Application of technologies and processes to achieve a required level of confidence that software systems and services function in the intended manner, are free from accidental or intentional vulnerabilities, provide security capabilities appropriate to the threat environment, and recover from intrusions and failures.

The expanded definition emphasizes the importance of both technologies and processes in software assurance, observes that computing capabilities may be acquired through services as well as new development, recognizes that security capabilities must be appropriate to the expected threat environment, and identifies recovery from intrusions and failures as an important capability for organizational continuity and survival.

These definitions all suggest that software assurance involves the execution of a well-defined process to ensure that all vulnerabilities are identified and actively mitigated and that the everyday software development, sustainment, and acquisition processes always follow good security practices. However, in addition to finding and fixing faults in the product and process, another important software assurance role is to field software-based systems that limit the damage resulting from any failures caused by attack-triggered faults. The objective of this activity is to ensure that defects that represent vulnerabilities are identified, that the effects of any subsequent attack are not propagated, and that the system recovers as quickly as possible from those attacks. The aim of this type of assurance is to ensure that the system continues to operate correctly in the presence of most attacks. This correct operation can be achieved by either establishing functions to resist the exploitation of weaknesses in the software or by building software functions that will tolerate the failures that result from such exploits. This shifts the purpose of the software assurance process from the simple identification and remediation of identifiable vulnerabilities to the establishment of real-time survivability and recovery in the product. Thus, for our purposes, the goal of software assurance is to ensure comprehensive organizational capability in the specification, design, and coding of the software, as well as in the development of specific security functionality to minimize damage from any unforeseen attacks and ensure a systematic recovery.

93.3 Why Does Software Assurance Matter?

The General Accounting Office (GAO) summarized the security concerns associated with software organizations in a March 2012 report. Software issues fall into five categories. Each category has slightly different implications for product integrity: "Installation of malicious logic on hardware or software, installation of counterfeit hardware or software, failure or disruption in the production or distribution of a critical product or service, reliance upon a malicious or unqualified service provider for the performance of a technical service and installation of unintentional vulnerabilities on software or hardware" (GAO 2012, p.1).

Embedding a malicious object in a product is always a hostile act that fulfills some specific purpose. Malicious objects are by definition not part of the intended functionality. Therefore, in order to find and eliminate any instance of malicious objects, which should be a high priority for any software customer, rigorous testing and inspection is required. Because it is hard enough to ensure the quality and security of the functions that should be present in a piece of software, it is asking a lot to expect that functions that should *not* be present are also identified and eliminated. Therefore, it is almost impossible to estimate how much malicious code currently resides in software products. Because the decision to embed a piece of malicious logic in a product is intentional, one of the most effective ways to ensure against the presence of such objects is to maintain strict oversight and control over the software development, sustainment, and acquisition work.

Some suppliers use counterfeit parts to save money or supply a feature that the maker is otherwise incapable of providing. Counterfeit parts can appear at any stage in the development and sustainment of any information and communication technology product. While counterfeits execute product functions as intended, they threaten product security and integrity because they are not the same as the actual part. As a result, counterfeits embody shortcuts in product quality or security that can fail in many ways. Because they function like the original part, it is often hard to spot a counterfeit in a large array of legitimate components. Consequently, it is critical that the customer for the part fully and completely understands their suppliers' business and technical practices prior to engaging in any use of their products. A capability model is particularly helpful in enforcing that understanding since it establishes a common and auditable basis among both organizations.

The problems caused by breakdowns in the supply chain mirror the problems encountered in conventional manufacturing, in that the failure lies in the inability to do the work because a component is missing. From the standpoint of product security, a failure to deliver a critical part prevents the software product from being used, which is the equivalent of a denial of service in conventional security terms. So efforts to mitigate security risks or risks to product integrity tend to concentrate on identifying and managing single points of failure. Capability models help in that respect because they establish common management functions designed to monitor and control the overall process of construction or maintenance.

From a technical service standpoint, the focus is on learning whether the supplier's operation is capable of delivering the product as specified. Since supplier capability is at the center of any acquisition or outsourcing decision, it is important to find out in advance if the contractors that comprise the supply chain possess all of the capabilities required to do the work. Specifically, suppliers have to prove that they are capable of developing and integrating a secure product. Overall capability is usually demonstrated by the suppliers' past history with similar projects as well as their documented ability to adopt good software engineering practices. The presence of a commonly accepted model of best practice that is shared between customer and supplier and that is fully auditable helps to cement that assurance.

The issue of unintentional vulnerabilities is a result of the overall development and sustainment problem in that defects in software and hardware occur because of failure in the process. By definition, the installation of unintentional flaws is not a hostile act. However, because the problem is so pervasive, the sheer number of exploitable vulnerabilities placed in software products makes it a major concern. There is an extensive body of knowledge (BoK) in software product assurance; however, because the steps necessary to ensure product integrity have to be instituted, managed, and sustained in a logical way, best practices are often not followed or are performed halfheartedly. The result is that common defects in software products are exploited by a growing array of criminal and other bad actors. The installation and sustainment of a commonly accepted capability model addresses this concern directly. However, it is critical that the activities in that model be executed in a continuous and disciplined fashion.

93.4 Software Assurance Principles

93.4.1 Historical Principles for Information Protection

Much of the information protection in place today is based on principles established by Saltzer and Schroeder (1974) in their paper titled "The Protection of Information in Computer Systems," which appeared in *Communications of the ACM* in 1974. They defined security as "techniques that control who may use or modify the computer or the information contained in it" and described the three main categories of concern: confidentiality, integrity, and availability (Saltzer and Schroeder 1974). Their proposed design principles that focus on protection mechanisms to "guide the design and contribute to an implementation without security flaws" (Saltzer and Schroeder 1974) are still taught in today's classrooms. They established eight principles for security in software design and development (Saltzer and Schroeder 1974):

1. *Economy of mechanism: Keep the design as simple and small as possible.*
2. *Fail-safe defaults: Base access decisions on permission rather than exclusion.*
3. *Complete mediation: Every access to every object must be checked for authority.*
4. *Open design: The design should not be secret. The mechanisms should not depend on the ignorance of potential attackers, but rather on the possession of specific, and more easily protected, keys or passwords.*
5. *Separation of privilege: Where feasible, a protection mechanism that requires two keys to unlock it is more robust and flexible than one that allows access to the presenter of only a single key.*
6. *Least privilege: Every program and every user of the system should operate using the least set of privileges necessary to complete the job.*
7. *Least common mechanism: Minimize the amount of mechanism common to more than one user and depended on by all users.*
8. *Psychological acceptability: It is essential that the human interface be designed for ease of use, so that users routinely and automatically apply the protection mechanisms correctly.*

Time has shown the value and utility in these principles; however, it is appropriate to consider that these were developed prior to the Morris worm that generated a massive denial of service by infecting more than 6000 UNIX machines on November 2, 1988 (Wikipedia 2012a). To provide a technology context, consider that the IBM System 360 was in use from 1964 through 1978 and the IBM System 370 came on the market in 1972. The advanced operating system Multiple Virtual Storage (MVS) was released in March 1974 (Wikipedia 2012b).

These principles were assembled prior to the identification of the more than 46,500 software vulnerabilities and exposures that are currently exploitable in today's software products as described in the Common Vulnerabilities and Exposures (CVE) database at http://cve.mitre.org/ (MITRE 2012). When these principles were developed, "buffer overflow," "malicious code," "cross-site scripting," and "zero-day vulnerabilities" were not part of the everyday vocabulary of operational software support personnel. Patches were carefully tested and scheduled to minimize operational disruption instead of pushed into operation to minimize attack vectors.

While these principles are still usable today to address security within an individual piece of technology, they are no longer sufficient to address the complexity and sophistication of the environment within which that component must operate. We must broaden our horizon to consider the large-scale, highly networked, software-dependent systems upon which our entire critical infrastructure depends, from phones, power, and water to industries such as banking, medicine, and retail.

93.4.2 Modern Principles for Software Assurance

There are vast lists of practices and procedures that describe what should be done to address software assurance.* There are also an equal number of complaints that effective assurance is not being addressed in today's software. We posit that some of the inaction stems from a general lack of understanding about why this additional work is needed. In our scrutiny of the wide range of materials published, the case for why to focus on software assurance has not yet been addressed. We proposed the following seven principles in response in our paper "Foundations for Software Assurance" (Woody 2012):

93.4.2.1 Risk

Risk drives assurance decisions: A perception of risk drives assurance decisions. Organizations without effective software assurance perceive risks based on successful attacks to software and systems and usually respond reactively. They may implement assurance choices such as policies, practices, tools, and restrictions based on their perception of the threat of a similar attack and the expected impact should that threat be realized. Organizations can incorrectly perceive risk when they do not understand their threats and impacts. Effective software assurance requires that risk knowledge be shared among all stakeholders and technology participants; however, too frequently, risk information is considered highly sensitive and is not shared, resulting in uninformed organizations making poor risk choices.

93.4.2.2 Interactions

Risk is aligned across all stakeholders and all interconnected technology elements: Highly connected systems like the Internet require alignment of risk across all stakeholders and all interconnected technology elements; otherwise, critical threats will be missed or ignored at different points in the interactions. It is no longer sufficient only to consider highly critical components when everything is highly interconnected. Interactions occur at many technology levels (e.g., network, security appliances, architecture, applications, and data storage) and are supported by a wide range of roles. Protections can be applied at each of these points and may conflict if not well orchestrated. Because of interactions, effective assurance requires that all levels and roles consistently recognize and respond to risk.

93.4.2.3 Trusted Dependencies

Dependencies are trusted: Because of the wide use of supply chains for software, assurance of an integrated product depends on other people's assurance decisions and the level of trust placed on these dependencies. The integrated software inherits all of the assurance limitations of each interacting component. In addition, unless specific restrictions and controls are in place, every operational component including infrastructure, security software, and other applications can be affected by the assurance of

* For a starting point see https://buildsecurityin.us-cert.gov/swa/ (DHS 2012a).

every other component. There is a risk each time an organization must depend on others' assurance decisions. Organizations should decide how much trust they place in dependencies based on a realistic assessment of the threats, impacts, and opportunities represented by an interaction. Dependencies are not static, and trust relationships should be regularly reviewed to identify changes that warrant reconsideration. The following examples describe assurance losses resulting from dependencies:

- Defects in standardized pieces of infrastructure (such as operating systems, development platforms, firewalls, and routers) can serve as widely available threat entry points for applications.
- Using many standardized software tools to build technology establishes a dependency for the assurance of the resulting software product. Vulnerabilities can be introduced into software products by the tool builders.

93.4.2.4 Attacker

There are no perfect protections against attacks: A broad community of attackers with growing technology capabilities are able to compromise the confidentiality, integrity, and availability of an organization's technology assets. There are no perfect protections against attacks, and the attacker profile is constantly changing. The attacker will use technology, processes, standards, and practices to craft a compromise (known as a socio-technical response). Attacks are crafted to take advantage of the ways we normally use technology or designed to contrive exceptional situations where defenses are circumvented.

93.4.2.5 Coordination and Education

Assurance is effectively coordinated among all technology participants: Protection must be applied broadly across the people, processes, and technology in an organization because the attacker will take advantage of all possible entry points. Authority and responsibility for assurance must be clearly established at an appropriate level in the organization to ensure the organization effectively participates in software assurance. This assumes that all participants know about assurance and that is not usually a reality. There is much to be done to educate people on software assurance.

93.4.2.6 Well Planned and Dynamic

Assurance is well planned and dynamic: Assurance must represent a balance among governance, construction, and operation of software and systems and is highly sensitive to changes in each of these areas. An adaptive response is required for assurance because the applications, interconnections, operational usage, and threats are always changing. Assurance is not a once-and-done activity. It must continue beyond the initial operational implementation through operational sustainment. Assurance cannot be added later; it must be built to the level of acceptable assurance that organizations need. No one has resources to redesign systems every time the threats change, and assurance cannot be readily adjusted upward after the fact.

93.4.2.7 Measurable

A means to measure and audit overall assurance is built in: That which is not measured cannot be managed. Each stakeholder or technology user will address only the assurance for which they are held accountable. Assurance will not compete successfully with other competing needs unless results are monitored and measured. All elements of the socio-technical environment, including practices, processes, and procedures, must be tied together to evaluate operational assurance. Organizations with more successful assurance measures react and recover faster, learn from their reactive responses and that of others, and are more vigilant in anticipating and detecting attacks. Defects per lines of code, a common development measure, may be useful for code quality but are not sufficient evidence for overall assurance because they provide no perspective on how that code behaves in an operational context. Both focused and systemic measures are needed to ensure that the components are engineered with sound security and that the interaction among components establishes effective assurance.

93.5 Software Assurance Process Models and Practices

93.5.1 Putting Capability into Practice

The GAO report presents these three common sense principles based on process capability: "control the development and sustainment work using common best practice," "adopt rigorous assurance practice at the component level," and "rationally plan for failure" (GAO 2012). A very large percentage of the counterfeiting, supply-chain-critical point-of-failure breakdowns and capability concerns can be mitigated by simply ensuring that every one of the entities up and down the supply chain is under strict management control. In addition, unwanted functionality and development failures should be addressed by rigorous product assurance from the time of inception to the time of acceptance. Then, when the inevitable failure does occur, there is a well-defined strategy in place to ensure that the problem is resolved in a rational fashion.

Control processes that satisfy these three principles are specified in a range of capability models, which we will discuss in the next section. These are called "capability models" because they provide a complete classification structure of best practices and standards for every aspect of software product development and sustainment work. Because those standards are highly authoritative, their recommendations provide a coherent and logical high-level framework to ensure both the logic and relevance of all the elements of fundamental best practices.

93.5.2 Why Use a Process Model?

The role of software assurance is to ensure that faults do not occur in the first place. However, since managers don't actually do the software development work, they have to use some sort of generic management process to ensure product integrity. The aim of that management process is to establish order, or system, to the way the organization conducts its software work. System, or orderliness, is important because at present, there is no explicit, standard set of best practices for software development work in most technology organizations. Instead, there is some sort of "customary" approach, which is probably not even written down. And whether that approach is agile development, extreme programming, or the waterfall, there is extensive evidence to indicate that software products in general continue to have defects.

A capability-based process directly addresses the problem with defects. Increasing an organization's capability ensures that reliability and integrity are designed and built into the products in the first place rather than added on at the end. Process capability improvement standards provide a given organization with a template for setting up and running an effective management system. The assumption is that a software management system that is based on and follows a commonly accepted model of best practice, or conforms to a formal best practice standard, ensures successful management. Most leading-edge corporations have had such a management system in place for years. Now, with the current set of generic software best-practice standards, any organization can implement the successful practices that more advanced corporations employ.

93.5.3 Best Practices: Software Security Frameworks, Models, and Roadmaps

In addition to considering process models for software development and acquisition, a framework for building assured systems needs to build upon and reflect known, accepted, common practice for software security. There are a growing number of promising frameworks and models for building more secure software. For example, Microsoft has defined their security development life cycle (SDL) (Howard 2006) and made it publicly available (Microsoft 2010a). In version 2, the authors of the Building Security In Maturity Model (BSIMM) (McGraw, Chess, and Migues) have collected and analyzed software security practices in 30 organizations (McGraw et al. 2010).

In this section, we summarize seven models, frameworks, and roadmaps, excerpting descriptive information from publicly available websites and reports. This section summarizes each effort to show the current state of the practice in building secure software and to aid in identifying promising research opportunities to fill gaps.

93.5.3.1 Building Security in Maturity Model

The BSIMM introduction states the following (McGraw et al. 2010):

> The Building Security In Maturity Model (BSIMM) is designed to help an organization understand, measure, and plan a software security initiative. The BSIMM was created through a process of understanding and analyzing real-world data from nine leading software security initiatives and then validated and adjusted with data from twenty-one additional leading software security initiatives. Altogether, the BSIMM collectively represents the wisdom and knowledge of thirty firms with active and successful software security initiatives. Though particular methodologies differ, such as the Open Web Application Security Project (OWASP) Comprehensive, Lightweight Application Security Process (CLASP), the Microsoft SDL, or the Cigital Touchpoints, many initiatives share common ground. This common ground is captured and described in the BSIMM.
>
> BSIMM is appropriate for an organization whose overall business goals for software security include
>
> - Informed risk management decisions
> - Clarity on what is "the right thing to do" for everyone involved in software security
> - Cost reduction through standard, repeatable processes
> - Increased code quality
>
> BSIMM is not a complete "how to" guide for software security, nor is it a one-size-fits-all model. Instead, BSIMM is a collection of good practices and activities that are in use today.

93.5.3.2 CMMI Assurance Process Reference Model

The DHS Software Assurance (SwA) Processes and Practices Working Group developed a draft process reference model (PRM) for assurance in July 2008 (DHS 2008). This PRM recommends additions to CMMI for Development (CMMI-DEV) v1.2 to address software assurance. The "assurance thread" description includes Figure 93.1 (DHS 2012b).

The DHS SwA Processes and Practices Working Group's additions and updates to CMMI-DEV v1.2 are focused at the specific practices level for the following (Table 93.1) CMMI-DEV process areas (CMMI Product Team 2006).

93.5.3.3 Open Web Application Security Project Software Assurance Maturity Model

The following discussion of OWASP SAMM is from *SAMM v1.0* (OWASP 2009):

> The Software Assurance Maturity Model (SAMM) is an open framework to help organizations formulate and implement a strategy for software security that is tailored to the specific risks facing the organization. The resources provided by SAMM will aid in
>
> - Evaluating an organization's existing software security practices
> - Building a balanced software security assurance program in well-defined iterations
> - Demonstrating concrete improvements to a security assurance program
> - Defining and measuring security-related activities throughout an organization

SAMM was defined to be flexible so that small, medium, and large organizations can use it with any style of development. Additionally, this model can be applied organization wide, for a single line of business, or even for an individual project. Beyond these traits, SAMM was built on the following principles (OWASP 2009):

- *An organization's behavior changes slowly over time*—A successful software security program should be specified in small iterations that deliver tangible assurance gains while incrementally working toward long-term goals.
- *There is no single recipe that works for all organizations*—A software security framework must be flexible and allow organizations to tailor their choices based on their risk tolerance and the way in which they build and use software.

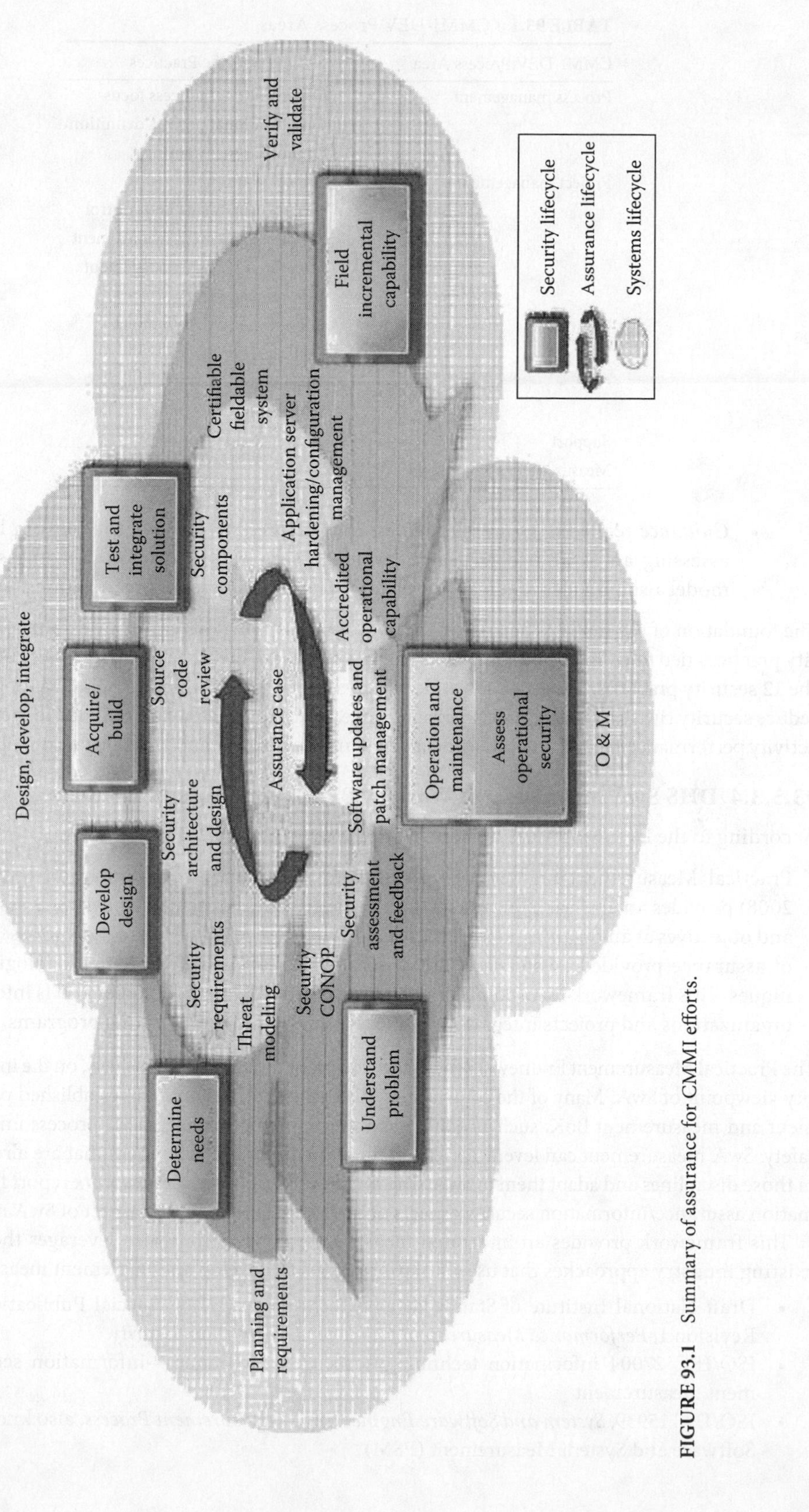

FIGURE 93.1 Summary of assurance for CMMI efforts.

TABLE 93.1 CMMI-DEV Process Areas

CMMI-DEV Process Area	Specific Practices
Process management	Organizational process focus
	Organizational process definition
	Organizational training
Project management	Project planning
	Project monitoring and control
	Supplier agreement management
	Integrated project management
	Risk management
Engineering	Requirements development
	Technical solution
	Verification
	Validation
Support	
Measurement and analysis	

- *Guidance related to security activities must be prescriptive*—All the steps in building and assessing an assurance program should be simple, well-defined, and measurable. This model also provides roadmap templates for common types of organizations.

The foundation of the model is built upon the core business functions of software development with security practices tied to each. The building blocks of the model are the three maturity levels defined for each of the 12 security practices. These define a wide variety of activities in which an organization could engage to reduce security risks and increase software assurance. Additional details are included to measure successful activity performance, understand the associated assurance benefits, and estimate personnel and other costs.

93.5.3.4 DHS SwA Measurement Work by Bartol and Moss

According to the DHS SwA Measurement Working Group (DHS 2010),

> Practical Measurement Framework for Software Assurance and Information Security (Bartol 2008) provides an approach for measuring the effectiveness of achieving software assurance goals and objectives at an organizational, program, or project level. It addresses how to assess the degree of assurance provided by software, using quantitative and qualitative methodologies and techniques. This framework incorporates existing measurement methodologies and is intended to help organizations and projects integrate SwA measurement into their existing programs.

The Practical Measurement Framework focuses principally, though not exclusively, on the information security viewpoint of SwA. Many of the contributing disciplines of SwA enjoy an established process improvement and measurement BoK, such as quality assurance, project management, process improvement, and safety. SwA measurement can leverage measurement methods and techniques that are already established in those disciplines and adapt them to SwA. The *Practical Measurement Framework* report focuses on information assurance/information security aspects of SwA to help mature that aspect of SwA measurement.

This framework provides an integrated measurement approach, which leverages the following five existing industry approaches that use similar processes to develop and implement measurement:

- Draft National Institute of Standards and Technology (NIST) Special Publication (SP) 800-55, Revision 1, *Performance Measurement Guide for Information Security*
- ISO/IEC 27004 information technology—security techniques—information security management measurement
- ISO/IEC 15939, *System and Software Engineering—Measurement Process*, also known as Practical Software and System Measurement (PSM)

- CMMI measurement and analysis process area
- CMMI GQ(I)M—Capability Maturity Model Integration Goal Question Indicator Measure

The framework describes candidate goals and information needs for each stakeholder group. The framework then presents example supplier measures as a table, with project activity, measures, information needs, and benefits as the columns. Example measures for acquirers are similarly presented in the framework and are intended to answer the following questions:

- Have SwA activities been adequately integrated into the organization's acquisition process?
- Have SwA considerations been integrated into the SDLC and resulting product by the supplier?

93.5.3.5 Microsoft Security Development Life Cycle

The Microsoft SDL* is an industry-leading software security process. A Microsoft-wide initiative and a mandatory policy since 2004, the SDL has played a critical role in embedding security and privacy in Microsoft software and culture. Combining a holistic and practical approach, the SDL introduces security and privacy early and throughout all phases of the development process.

The reliable delivery of more secure software requires a comprehensive process, so Microsoft defined *Secure by Design, Secure by Default, Secure in Deployment, and Communications* (SD3+C) to help determine where security efforts are needed. The guiding principles for SD3+C are identified in the following subsections, which are excerpted from *Microsoft SDL version 5.0* (Microsoft 2010b):

Secure by Design

- *Secure architecture, design, and structure. Developers consider security issues part of the basic architectural design of software development. They review detailed designs for possible security issues, and they design and develop mitigations for all threats.*
- *Threat modeling and mitigation. Threat models are created, and threat mitigations are present in all design and functional specifications.*
- *Elimination of vulnerabilities. No known security vulnerabilities that would present a significant risk to the anticipated use of the software remain in the code after review. This review includes the use of analysis and testing tools to eliminate classes of vulnerabilities.*
- *Improvements in security. Less secure legacy protocols and code are deprecated, and, where possible, users are provided with secure alternatives that are consistent with industry standards.*

Secure by Default

- *Least privilege. All components run with the fewest possible permissions.*
- *Defense in depth. Components do not rely on a single threat mitigation solution that leaves users exposed if it fails.*
- *Conservative default settings. The development team is aware of the attack surface for the product and minimizes it in the default configuration.*
- *Avoidance of risky default changes. Applications do not make any default changes to the operating system or security settings that reduce security for the host computer. In some cases, such as for security products, it is acceptable for a software program to strengthen (increase) security settings for the host computer. The most common violations of this principle are games that either open firewall ports without informing the user or instruct users to open firewall ports without informing users of possible risks.*
- *Less commonly used services off by default. If fewer than 80 percent of a program's users use a feature, that feature should not be activated by default. Measuring 80 percent usage in a product is often difficult because programs are designed for many different personas. It can be useful to consider whether a feature addresses a core/primary use scenario for all personas. If it does, the feature is sometimes referred to as a P1 feature.*

* More information is available in *The Security Development Lifecycle* (Howard 2006), at the Microsoft Security Development Lifecycle website (Microsoft 2010a), and in the document *Microsoft Security Development Lifecycle version 5.0* (Microsoft 2010b).

Secure in Deployment

- *Deployment guides. Prescriptive deployment guides outline how to deploy each feature of a program securely, including providing users with information that enables them to assess the security risk of activating non-default options (and thereby increasing the attack surface).*
- *Analysis and management tools. Security analysis and management tools enable administrators to determine and configure the optimal security level for a software release.*
- *Patch deployment tools. Deployment tools aid in patch deployment.*

Communications

- *Security response. Development teams respond promptly to reports of security vulnerabilities and communicate information about security updates.*
- *Community engagement. Development teams proactively engage with users to answer questions about security vulnerabilities, security updates, or changes in the security landscape.*

93.5.3.6 CERT® Resilience Management Model (RMM) Resilient Technical Solution Engineering Process Area

Like software security and software assurance, resilience is a property of software and systems. Developing and acquiring resilient* software and systems requires a resilience-focused dedicated process that encompasses the software and system life cycle. As described in the CERT-RMM (CERT 2012) resilient technical solution engineering (RTSE) process area (Caralli et al. 2010), the process defines what is required to develop resilient software and systems and is as follows (Caralli et al. 2010):

- *Establish a plan for addressing resiliency as part of the organization's (or supplier's) regular development life cycle and integrate the plan into the organization's corresponding development process. Plan development and execution includes identifying and mitigating risks to the success of the project.*
- *Identify practice-based guidelines that apply to all phases such as threat analysis and modeling as well as those that apply to a specific life cycle phase.*
- *Elicit, identify, develop, and validate assurance and resiliency requirements (using methods for representing attacker and defender perspectives, for example). Such processes, methods, and tools are performed alongside similar processes for functional requirements.*
- *Use architectures as the basis for design that reflect a resiliency and assurance focus, including security, sustainability, and operations controls.*
- *Develop assured and resilient software and systems through processes that include secure coding of software, software defect detection and removal, and the development of resiliency and assurance controls based on design specifications.*
- *Test assurance and resiliency controls for software and systems and refer issues back to the design and development cycle for resolution.*
- *Conduct reviews throughout the development life cycle to ensure that resiliency (as one aspect of assurance) is kept in the forefront and given adequate attention and consideration.*
- *Perform system-specific continuity planning and integrate related service continuity plans to ensure that software, systems, hardware, networks, telecommunications, and other technical assets that depend on one another are sustainable.*
- *Perform a post-implementation review of deployed systems to ensure that resiliency (as well as assurance) requirements are being satisfied as intended.*

---

* There is substantial overlap in the definitions of assured software (or software assurance) and resilient software (or software resilience). Resilient software is software that continues to operate as intended (including recovering to a known operational state) in the face of a disruptive event (satisfying business continuity requirements) so as to satisfy its confidentiality, availability, and integrity requirements (reflecting operational and security requirements) (Caralli et al. 2010).

- *In operations, monitor software and systems to determine if there is variability that could indicate the effects of threats or vulnerabilities and to ensure that controls are functioning properly.*
- *Implement configuration management and change control processes to ensure software and systems are kept up to date to address newly discovered vulnerabilities and weaknesses (particularly in vendor-acquired products and components) and to prevent the intentional or inadvertent introduction of malicious code or other exploitable vulnerabilities.*

In addition to RTSE, the following are goals and practices in other CERT-RMM process areas that organizations should consider when developing and acquiring software and systems that need to meet assurance and resiliency requirements (Caralli et al. 2010):

- *Resiliency requirements for software and system technology assets in operation, including those that may influence quality attribute requirements in the development process, are developed and managed in the Resiliency Requirements Development (RRD) and Resiliency Requirements Management (RRM) process areas respectively.*
- *Identifying and adding newly developed and acquired software and system assets to the organization's asset inventory is addressed in the Asset Definition and Management (ADM) process area.*
- *The management of resiliency for technology assets as a whole, particularly for deployed, operational assets, is addressed in the Technology Management (TM) process area. This includes, for example, asset fail-over, backup, recovery, and restoration.*
- *Acquiring software and systems from external entities and ensuring that such assets meet their resiliency requirements throughout the asset life cycle is addressed in the External Dependencies Management process area. That said, RTSE specific goals and practices should be used to aid in evaluating and selecting external entities that are developing software and systems (EXD:SG3.SP3), formalizing relationships with such external entities (EXD:SG3.SP4), and managing an external entity's performance when developing software and systems (EXD:SG4).*
- *Monitoring for events, incidents, and vulnerabilities that may affect software and systems in operation is addressed in the Monitoring (MON) process area.*
- *Service continuity plans are identified and created in the Service Continuity (SC) process area. These plans may be inclusive of software and systems that support the services for which planning is performed.*

The RTSE process area assumes that the organization has one or more existing, defined processes for software and system development into which resiliency controls and activities can be integrated. If this is not the case, the organization should not attempt to implement the goals and practices identified in RTSE or other CERT-RMM process areas as previously described.

93.6 Research Framework

93.6.1 Building Assured Systems Framework

In an earlier research project, a broad framework based on the Master of Software Assurance Reference Curriculum BoK (Mead et al. 2010) was developed and named the Building Assured Systems Framework (BASF).

We developed maturity levels to support our work in software security engineering (refer to *Software Security Engineering* [Allen et al. 2008]). We then applied them to the software assurance BoK in the Master of Software Assurance Reference Curriculum (Mead et al. 2010). The association of BoK elements and maturity levels was accomplished by evaluating the extent to which relevant sources, practices, curricula, and courseware exist for a particular BoK element and the extent to which the element was observed in practice in organizations.

The maturity levels can serve as a litmus test for whether research is needed in a specific area. Presumably we want to focus software assurance research on topic areas that are less mature. The mapping of the maturity levels to the Master of Software Assurance BoK knowledge areas is in the Appendix 93.A.

93.6.1.1 Promising Software Assurance Research Areas

Using the software assurance curriculum model BoK (Mead et al. 2010), we performed an analysis to identify some promising software assurance research areas. In some cases, there has been a fair amount of research work, but more is needed; Assured Software Development is an example. With this in mind, here is our initial list of candidate research areas:

- Baseline level of assurance; allowable tolerances, if quantitative
- Product and process measures by life-cycle phase
- Other performance indicators that test for the baseline, by life-cycle phase
- Making the Business Case for Assurance
- Assured Software Development
- Assured Software Analytics
- Assurance in Acquisition

93.7 Summary

In this chapter we introduced software assurance, discussing the impact of software on our lives, the risks associated with software vulnerabilities, and some basic definitions of software assurance. We then introduced modern principles of software assurance, along with historical principles of software assurance. With this background in hand, we were able to identify a number of process models, frameworks, and best practices that were relevant to software assurance. Finally, we identified a research framework that could be used to support current research and to identify gap areas for future research.

We then performed a gap analysis to identify areas where additional research would be needed. The benefit of this approach is that it establishes a desirable linkage between software assurance research and the associated educational curriculum research. We also wanted to address a more generic problem, one that we had seen in our own work and elsewhere, namely that various research projects in building assured systems appear unrelated to one another; we and other research entities consequently don't have a good way to prioritize and select new research.

From a research perspective, researchers could periodically consider rating the maturity levels of their methods. This would assist users in deciding which methods to use. It would also be helpful if researchers could collect and/or provide available cost/benefit data and encourage users to assist in such data collection. All too often users decide on a particular method but do not collect enough information to determine whether the benefit justified the cost. At the same time, the smaller projects that researchers conduct on their own do not usually result in enough cost/benefit data to be sufficiently compelling.

The gap analysis that we have done could be used to help select, and to some extent prioritize, new research. For example, if research is proposed for an area where there are a number of mature approaches, it would be helpful to understand why that research would be considered a good investment, compared to areas where there are no mature approaches. Since there is a lot of research aimed at building assured systems, we anticipate that this framework would need regular review and revision in order to stay current.

93.A Appendix: Master of Software Assurance Knowledge Areas Mapped to Maturity Levels in the Building Assured Systems Framework

Maturity levels are defined as follows (Table 93.A.1):

- L1—The area provides guidance for how to think about a topic for which there is no proven or widely accepted approach. The intent of the area is to raise awareness and aid the reader in thinking about the problem and candidate solutions. The area may also describe promising research results that may have been demonstrated in a constrained setting.

TABLE 93.A.1 Master of Software Assurance Knowledge Areas Mapped to Maturity Levels in the Building Assured Systems Framework

Knowledge Area	Knowledge Units	Maturity Level
Assurance across life cycles	Software life-cycle processes	
	• New development	[L4]
	• Integration, assembly, and deployment	[L4]
	• Operation and evolution	[L4]
	• Acquisition, supply, and service	[L3]
	Software assurance processes and practices	
	• Process and practice assessment	[L3]
	• Software assurance integration into SDLC phases	[L2/3]
Risk management	Risk management concepts	
	• Types and classification	[L4]
	• Probability, impact, severity	[L4]
	• Models, processes, metrics	[L4] [L3—metrics]
	Risk management process	
	• Identification	[L4]
	• Analysis	[L4]
	• Planning	[L4]
	• Monitoring and management	[L4]
	Software assurance risk management	
	• Vulnerability and threat identification	[L3]
	• Analysis of software assurance risks	[L3]
	• Software assurance risk mitigation	[L3]
	• Assessment of software assurance processes and practices	[L2/3]
Assurance assessment	Assurance assessment concepts	
	• Baseline level of assurance; allowable tolerances, if quantitative	[L1]
	• Assessment methods	[L2/3]
	Measurement for assessing assurance	
	• Product and process measures by life-cycle phase	[L1/2]
	• Other performance indicators that test for the defined assurance baseline by life-cycle phase	[L1/2]
	• Measurement processes and frameworks	[L2/3]
	• Business survivability and operational continuity	[L2]
	Assurance assessment process (collect and report measures that demonstrate the baseline as defined in 3.1.1.)	
	• Comparison of selected measurements to the established baseline	[L3]
	• Identification of out-of-tolerance variances	[L3]
Assurance management	Making the business case for assurance	
	• Valuation and cost/benefit models, cost and loss avoidance, return on investment	[L3]
	• Risk analysis	[L3]
	• Compliance justification	[L3]
	• Business impact/needs analysis	[L3]
	Managing assurance	
	• Project management across the life cycle	[L3]
	• Integration of other knowledge units	[L2/3]

(continued)

TABLE 93.A.1 (continued) Master of Software Assurance Knowledge Areas Mapped to Maturity Levels in the Building Assured Systems Framework

Knowledge Area	Knowledge Units	Maturity Level
	Compliance considerations for assurance	
	• Laws and regulations	[L3]
	• Standards	[L3]
	• Policies	[L2/3]
System security assurance	For newly developed and acquired software for diverse systems	
	• Security and safety aspects of computer-intensive critical infrastructure	[L2]
	• Potential attack methods	[L3]
	• Analysis of threats to software	[L3]
	• Methods of defense	[L3]
	For diverse operational (existing) systems	
	• Historic and potential operational attack methods	[L4]
	• Analysis of threats to operational environments	[L3]
	• Designing of and plan for access control, privileges, and authentication	[L3]
	• Security methods for physical and personnel environments	[L4]
	Ethics and integrity in creation, acquisition, and operation of software systems	
	• Overview of ethics, code of ethics, and legal constraints	[L4]
	• Computer attack case studies	[L3]
System functionality assurance	Assurance technology	
	• Technology evaluation	[L3]
	• Technology improvement	[L3]
	Assured software development	
	• Development methods	[L2/3]
	• Quality attributes	[L3—depends on the property]
	• Maintenance methods	[L3]
	Assured software analytics	
	• Systems analysis	(L2 architectures; L3/4 networks, databases [identity management, access control])
	• Structural analysis	[L3]
	• Functional analysis	[L2/3]
	• Analysis of methods and tools	[L3]
	• Testing for assurance	[L3]
	• Assurance evidence	[L2]
	Assurance in acquisition	
	• Assurance of acquired software	[L2]
	• Assurance of software services	[L3]

TABLE 93.A.1 (continued) Master of Software Assurance Knowledge Areas Mapped to Maturity Levels in the Building Assured Systems Framework

Knowledge Area	Knowledge Units	Maturity Level
System operational assurance	Operational procedures	
	• Business objectives	[L3]
	• Assurance procedures	[L3]
	• Assurance training	[L4]
	Operational monitoring	
	• Monitoring technology	[L4]
	• Operational evaluation	[L4]
	• Operational maintenance	[L3]
	• Malware analysis	[L2/3]
	System control	
	• Responses to adverse events	[L3/4]
	• Business survivability	[L3]

- L2—The area describes practices that are in early pilot use and are demonstrating some successful results.
- L3—The area describes practices that have been successfully deployed (mature) but are in limited use in industry or government organizations. They may be more broadly deployed in a particular market sector.
- L4—The area describes practices that have been successfully deployed and are in widespread use. Readers can start using these practices today with confidence. Experience reports and case studies are typically available.

Glossary

Software Assurance Definitions

The following are software assurance definitions we reviewed while developing the BASF.

The following is the US Department of Defense's (DoD) definition of systems assurance taken from *A DoD-Oriented Introduction to the NDIA's System Assurance Guidebook* (Popick et al. 2010):

> The justified measures of confidence that a system functions as intended and is free of exploitable vulnerabilities, either intentionally or unintentionally designed or inserted as part of the system at any time during the life cycle (NDIA 2008).

The following definition is taken from the NDIA conference paper *Engineering Improvement in Software Assurance: A Landscape Framework* (Brownsword et al. 2009):

> Environment of use:
> Actual environment of use (not just the expected environment of use)
> Means evaluating robustness against unexpected use, threats, and changes in the environment

The following is from the SEI webinar *Engineering Improvement in Software Assurance: A Landscape Framework* (Brownsword and Woody 2010):

> Software assurance: a justified level of confidence that software-reliant systems function as intended within their operational environment

The following is the CNSS definition (CNSS 2010) used in the DHS SwA website (DHS 2010b) and *Software Security Engineering* book (Allen et al. 2008):

> Software assurance (SwA) is the level of confidence that software is free from vulnerabilities, either intentionally designed into the software or accidentally inserted at any time during its life cycle, and that the software functions in the intended manner (from CNSS 4009 IA Glossary—see Wikipedia for definitions and descriptions).

The following is the SAFECode software assurance definition (SAFECode 2008):

> Confidence that software, hardware and services are free from intentional and unintentional vulnerabilities and that the software functions as intended.

The following excerpt is from the software security assurance state-of-the-art report (SOAR) (Goertzel et al. 2007).

> 2.1 Definition 1: Software Assurance
>
> Until recently, the term *software assurance* was most commonly relating two software properties: *quality* (i.e., "software assurance" as the short form of "software quality assurance"), and *reliability* (along with reliability's most stringent quality—safety). Only in the past 5 years or so has the term software assurance been adopted to express the idea of the assured security of software (comparable to the assured security of information that is expressed by the term "information assurance").
>
> The discipline of software assurance can be defined in many ways. The most common definitions complement each other but differ slightly in terms of emphasis and approach to the problem of assuring the security of software.
>
> In all cases, all definitions of software assurance convey the thought that software assurance must provide a reasonable level of *justifiable confidence* that the software will function correctly and predictably in a manner consistent with its documented requirements. Additionally, the function of software cannot be compromised either through direct attack or through sabotage by maliciously implanted code to be considered assured. Some definitions of software assurance characterize that assurance in terms of the software's trustworthiness or "high confidence."
>
> Several leading definitions of software assurance are discussed below.
>
> Instead of choosing a single definition of software assurance for this report, we synthesized them into a definition that most closely reflects software security assurance as we wanted it to be understood in the context of this report—*Software security assurance: The basis for gaining justifiable confidence that software will consistently exhibit all properties required to ensure that the software, in operation, will continue to operate dependably despite the presence of sponsored (intentional) faults. In practical terms, such software must be able to resist most attacks, tolerate as many as possible of those attacks it cannot resist, and contain the damage and recover to a normal level of operation as soon as possible after any attacks it is unable to resist or tolerate.*
>
> 2.1.1 CNSS Definition
>
> The ability to establish confidence in the *security* as well as the predictability of software is the focus of the Committee on National Security Systems (CNSS) definitions of software assurance in its National Information Assurance Glossary [8]. The glossary defines software assurance as—
>
> The level of confidence that software is free from vulnerabilities, regardless of whether they are intentionally designed into the software or accidentally inserted later in its life cycle, and that the software functions in the intended manner.

This understanding of software assurance is consistent with the use of the term in connection with information, i.e., information assurance (IA). By adding the term software assurance to its IA glossary, CNSS has acknowledged that software is directly relevant to the ability to achieve information assurance.

The CNSS definition is purely descriptive: it describes what software must *be* to achieve the level of confidence at which its desired characteristics—lack of vulnerabilities and predictable execution—can be said to be assured. The definition does not attempt to prescribe the means by which that assurance can, should, or must be achieved.

2.1.2 DoD Definition

The Department of Defense's (DoD) Software Assurance Initiative's definition is identical in meaning to that of the CNSS, although more succinct—

The level of confidence that software functions as intended and is free of vulnerabilities, either intentionally or unintentionally designed or inserted as part of the software [9].

2.1.3 NASA Definition

The National Aeronautics and Space Administration (NASA) defines software assurance as—

The planned and systematic set of activities that ensure that software processes and products conform to requirements, standards, and procedures.

The "planned and systematic set of activities" envisioned by NASA include—

Requirements specification

Testing

Validation

Reporting

The application of these functions "during a software development life cycle is called software assurance" [10].

The NASA software assurance definition predates the CNSS definition but similarly reflects the primary concern of its community—in this case, safety. Unlike the CNSS definition, NASA's definition is both descriptive and prescriptive in its emphasis on the importance of a "planned and systematic set of activities." Furthermore, NASA's definition states that assurance must be achieved not only for the software itself but also the processes by which it is developed, operated, and maintained. To be assured, both software *and* processes must "conform to requirements, standards, and procedures."

2.1.3 DHS Definition

Like CNSS, the Department of Homeland Security (DHS) definition of software assurance emphasizes the properties that must be present in the software for it to be considered "assured," i.e.—

Trustworthiness, which DHS defines, like CNSS, in terms of the absence of exploitable vulnerabilities whether maliciously or unintentionally inserted

Predictable execution, which "provides justifiable confidence that the software, when executed, will function as intended [11].

Like NASA, DHS's definition explicitly states that "a planned and systematic set of multidisciplinary activities" must be applied to ensure the conformance of both software and processes to "requirements, standards, and procedures" [12].

2.1.4 NIST Definition

The National Institute of Standards and Technology (NIST) defines software assurance in the same terms as NASA, whereas the required properties to be achieved are those included in the DHS definition: trustworthiness and predictable execution. NIST essentially fuses the NASA and DHS definitions into a single definition, thereby clarifying the cause-and-effect relationship between "the planned and systematic set of activities" and the expectation that such activities will achieve software that is trustworthy and predictable in its execution [13].

2.2 Definition 2: Secure Software

DHS's *Security in the Software Life Cycle* defines secure software in terms that have attempted to incorporate concepts from all of the software assurance definitions discussed in Section 93.1 as well as reflect both narrow-focused and holistic views of what constitutes secure software. The document attempts to provide a "consensus" definition that has, in fact, been vetted across the software security assurance community (or at least that part that participates in meetings of the DHS Software Assurance Working Groups [WG] and DoD/DHS Software Assurance Forums). According to *Security in the Software Life Cycle—Secure software cannot be intentionally subverted or forced to fail. It is, in short, software that remains correct and predictable in spite of intentional efforts to compromise that dependability. Security in the Software Life Cycle* elaborates on this definition—

Secure software is designed, implemented, configured, and supported in ways that enable it to:

Continue operating correctly in the presence of most attacks by either resisting the exploitation of faults or other weaknesses in the software by the attacker, or tolerating the errors and failures that result from such exploits

Isolate, contain, and limit the damage resulting from any failures caused by attack-triggered faults that the software was unable to resist or tolerate, and recover as quickly as possible from those failures

The document then enumerates the different security properties that characterize secure software and clearly associates the means by which software has been developed with its security:

Secure software has been developed such that—

Exploitable faults and other weaknesses are avoided by well-intentioned developers.

The likelihood is greatly reduced or eliminated that malicious developers can intentionally implant exploitable faults and weaknesses or malicious logic into the software.

The software will be attack-resistant or attack-tolerant, and attack-resilient.

The interactions among components within the software-intensive system, and between the system and external entities, do not contain exploitable weaknesses.

Primary Sources

Mead, N. R. and J. H. Allen. September 2010. *Building Assured Systems Framework* (CMU/SEI-2010-TR-025). Pittsburgh, PA: Software Engineering Institute, Carnegie Mellon University.

Woody, C., N. R. Mead, and D. Shoemaker. 2012. Foundations for software assurance. *Proceedings of the Hawaii International Conference on System Sciences*, Maui, HI.

Further Information

Babylon, Ltd. 2009. *Definition of Framework*. http://dictionary.babylon.com/framework/.

Bishop, M. and S. Engle. June 2006. The Software Assurance CBK and University Curricula. *Proceedings of the 10th Colloquium for Information Systems Security Education*. Adelphi, MD: University of Maryland, University College.

CERT Program. 2010. 2009 CERT Research Annual Report. Pittsburgh, PA: Software Engineering Institute, Carnegie Mellon University. http://www.cert.org/research/2009research-report.pdf.

CMMI Product Team. 2007. CMMI® for Acquisition, Version 1.2 (CMU/SEI-2007-TR-017). Pittsburgh, PA: Software Engineering Institute, Carnegie Mellon University. http://www.sei.cmu.edu/library/abstracts/reports/07tr017.cfm.

Department of Homeland Security (DHS). November 2009. *A Roadmap for Cybersecurity Research.* Washington, DC: The White House. http://www.cyber.st.dhs.gov/docs/DHS-Cybersecurity-Roadmap.pdf.

Department of Homeland Security (DHS) Software Assurance (SwA). 2010. *Software Assurance Community Resources and Information Clearinghouse.* Washington, DC: The White House. https://buildsecurityin.us-cert.gov/swa/.

Devanbu, P. T. and S. Stubblebine. June 2000. Software engineering for security: A roadmap. *ICSE 2000, 22nd International Conference on Software Engineering, Future of Software Engineering Track,* Limerick, Ireland.

Dorofee, A., J. Walker, C. Albert, R. Higuera, R. Murphy, and R. Williams. 1996. *Continuous Risk Management Guidebook.* Pittsburgh, PA: Carnegie Mellon University, pp. 7–9.

Drew, C. December 2009. Wanted: 'Cyber Ninjas.' *New York Times.* http://www.nytimes.com/2010/01/03/education/edlife/03cybersecurity.html?emc=eta1.

Ellison, R. J., A. P. Moore, A. P. Bass, M. H. Klein, and A. P. Bachmann. 2004. *Security and Survivability Reasoning Frameworks and Architectural Design Tactics* (CMU/SEI-2004-TN-022). Pittsburgh, PA: Software Engineering Institute, Carnegie Mellon University.

European Research Consortium for Informatics and Mathematics (ERCIM) and European Commission. 2008. *Strategic Seminar: Engineering Secure Complex Software Systems and Services.* Brussels, Belgium: ERCIM. http://www.ercim.eu/activity/strategic-seminar.

Farlex, Inc. 2010. *The Free Dictionary: Definition of Model.* Huntingdon Valley, PA: Farlex, Inc. http://www.thefreedictionary.com/model.

Goertzel, K. M. Updated January 2009; Accessed June 2011. *Introduction to Software Security. Build Security In.* Washington, DC: Department of Homeland Security. https://buildsecurityin.us-cert.gov/bsi/547.html.

Integrated Software & Systems Engineering Curriculum (iSSEc) Project. 2009. *Graduate Software Engineering 2009 (GSwE2009) Curriculum Guidelines for Graduate Degree Programs in Software Engineering, Version 1.0.* Hoboken, NJ: Stevens Institute of Technology.

International Process Research Consortium and F. Eileen, ed. 2006. *A Process Research Framework.* Pittsburgh, PA: Software Engineering Institute, Carnegie Mellon University.

Jones, C. 2005. *Software Quality in 2005: A Survey of the State of the Art.* Marlborough, MA: Software Productivity Research.

Jones, N. A. 2009. Building In … information security, privacy and assurance—A high-level roadmap. *Cyber Security Knowledge Transfer Network,* Paris, France.

Leaders in Security. March 30, 2009. Building In… information security, privacy and assurance. Paper presented at the *Knowledge Transfer Network Paris Information Security Workshop,* Paris, France.

Leiwo, J. August 1999. Observations on information security crisis (computer science and information systems reports technical reports TR-21). *Proceedings of the 22nd Information Systems Research Seminar in Scandinavia (IRIS22): "Enterprise Architectures for Virtual Organizations,"* Vol. 2, ed. Kakola, T. K., Keuruu, Finland, pp. 313–324.

Lipner, S. and M. Howard. 2005. *The Trustworthy Computing Security Development Lifecycle.* http://msdn.microsoft.com/en-us/library/ms995349.aspx.

Maughan, D. January 2010. The need for a National Cybersecurity Research and Development Agenda. *Communications of the ACM* 32, 2. http://www.csl.sri.com/users/neumann/insiderisks08.html#220.

McGraw, G. and B. Chess. October 15, 2008. *Software [In]security: A Software Security Framework: Working Towards a Realistic Maturity Model. InformIT.* http://www.informit.com/articles/article.aspx?p=1271382.

Mead, N. R., E. K. Hawthorne, and M. Ardis. 2011. *Software Assurance Curriculum Project Volume IV: Community College Education* (CMU/SEI-2011-TR-017). Pittsburgh, PA: Software Engineering Institute, Carnegie Mellon University. http://www.sei.cmu.edu/library/abstracts/reports/11tr017.cfm.

Mead, N. R., T. B. Hilburn, and R. Linger. 2010. *Software Assurance Curriculum Project Volume II: Undergraduate Course Outlines* (CMU/SEI-2010-TR-019). Pittsburgh, PA: Software Engineering Institute, Carnegie Mellon University. http://www.sei.cmu.edu/library/abstracts/reports/10tr019.cfm.

National Association of State Chief Information Officers (NASCIO). 2009. *Desperately Seeking Security Frameworks—A Roadmap for State CIOs*. Washington, DC: NASCIO. http://www.nascio.org/publications/documents/NASCIO-SecurityFrameworks.pdf.

Newman, M. 2002. *Software Errors Cost U.S. Economy $59.5 Billion Annually*. Gaithersburg, MD: National Institute of Standards and Technology (NIST).

Okubo, T. and H. Tanaka. 2007. Secure software development through coding conventions and frameworks. *Second International Conference on Availability, Reliability and Security (ARES'07)*. Fujitsu Laboratories Ltd., Institute of Information Security.

Partnership for Public Service and B. A. Hamilton. Accessed July 2009. *Cyber IN-Security: Strengthening the Federal Cybersecurity Workforce*. Partnership for Public Service. http://ourpublicservice.org/OPS/publications/viewcontentdetails.php?id=135.

Pavlich-Mariscal, J. A., S. A. Demurjian, and L. D. Michel. 2006. *A Framework for Composable Security Definition, Assurance, and Enforcement*. Berlin, Germany: Springer.

President's Information Technology Advisory Committee. 2005. *Cybersecurity: A Crisis of Prioritization*. Arlington, TX: Executive Office of the President, National Coordination Office for Information Technology Research and Development.

Redwine, S. T., Ed. 2006. *Software Assurance: A Guide to the Common Body of Knowledge to Produce, Acquire and Sustain Secure Software, Version 1.1*. Washington, DC: U.S. Department of Homeland Security.

Redwine, S. T. 2008. *Toward an Organization for Software System Security Principles and Guidelines*. Harrisonburg, VA: James Madison University.

Royce, W. W. 1970. Managing the development of large software systems: Concepts and techniques. *Proceedings WESCON*. Los Alamitos, CA: IEEE Computer Society Press.

Software Assurance Forum for Excellence in Code (SAFECode). 2010. *SAFECode*. http://www.safecode.org.

Software Engineering Institute (SEI). 2010a. *Capability Model Integration (CMMI)*. Pittsburgh, PA: Software Engineering Institute, Carnegie Mellon University. http://www.sei.cmu.edu/cmmi/.

Software Engineering Institute (SEI). 2010b. *CMMI for Development (CMMI-DEV)*. Pittsburgh, PA: Software Engineering Institute, Carnegie Mellon University. http://www.sei.cmu.edu/cmmi/tools/dev/index.cfm.

Software Engineering Institute (SEI). 2010c. *CMMI for Acquisition*. Pittsburgh, PA: Software Engineering Institute, Carnegie Mellon University. http://www.sei.cmu.edu/cmmi/tools/acq/index.cfm.

Tsipenyuk O'Neil, Y., B. Chess, and J. West. November 14, 2008. JavaScript hijacking: Only 1 out 12 popular AJAX frameworks prevents it. *AjaxWorld Magazine*. http://ajax.sys-con.com/node/747965.

References

Allen, J. H., S. Barnum, R. J. Ellison, G. McGraw, and N. R. Mead. 2008. *Software Security Engineering: A Guide for Project Managers*. Boston, MA: Addison-Wesley Professional.

Bartol, N. 2008. *Practical Measurement Framework for Software Assurance and Information Security, Version 1.0*. Practical Software & Systems Measurement (PSM), Picatinny Arsenal, NJ. http://www.psmsc.com/Prod_TechPapers.asp.

Brownsword, L. and C. Woody. May 2010. *Engineering Improvement in Software Assurance: A Landscape Framework* (SEI Webinar). Pittsburgh, PA: Software Engineering Institute, Carnegie Mellon University. http://www.sei.cmu.edu/library/abstracts/webinars/Engineering-Improvement-in-Software-Assurance-A-Landscape-Framework.cfm.

Brownsword, L., C. Woody, C. Alberts, and A. Moore. 2009. Achieving acquisition excellence via effective systems engineering. *12th Annual Systems Engineering Conference, NDIA*, October 26–29, 2009, San Diego, CA. http://www.dtic.mil/ndia/2009systemengr/8996ThursdayTrack8Brownsword.pdf.

Caralli, R. A., J. H. Allen, P. D. Curtis, D. W. White, and L. R. Young. 2010. *CERT®* Resilience Management Model, Version 1.0: Resilient Technical Solution Engineering (RTSE). Pittsburgh, PA: Software Engineering Institute, Carnegie Mellon University. http://www.cert.org/resilience/rmm.html.

CERT Program. 2012. *Resilience Management*. Pittsburgh, PA: Software Engineering Institute. http://www.cert.org/resilience/.

CMMI Product Team. 2006. *CMMI®* for Development, Version 1.2 (CMU/SEI-2006-TR-008). Pittsburgh, PA: Software Engineering Institute, Carnegie Mellon University. http://www.sei.cmu.edu/library/abstracts/reports/06tr008.cfm.

Committee on National Security Systems (CNSS). April 2010. *National Information Assurance (IA) Glossary: CNSS Instruction No. 4009*. http://www.cnss.gov/Assets/pdf/cnssi_4009.pdf.

Colbert, E. and D. Yu. 2006. Costing secure systems workshop report. *21st International Forum on COCOMO and Software Cost Modeling*, Herndon, VA, October 29–November 2, 2006. Los Angeles, CA: Center for Systems and Software Engineering.

Department of Homeland Security (DHS). February 2003. *A National Strategy to Secure Cyberspace*. Washington, DC: The White House. http://www.dhs.gov/xlibrary/assets/National_Cyberspace_Strategy.pdf.

Department of Homeland Security (DHS) Software Assurance (SwA). 2010. *Measurement Working Group*. Washington, DC: The White House. https://buildsecurityin.us-cert.gov/swa/measwg.html.

Department of Homeland Security (DHS) Software Assurance (SwA). 2012a. *Software Assurance Community Resource and Information Clearinghouse*. Washington, DC: The White House. https://buildsecurityin.us-cert.gov/swa.

Department of Homeland Security (DHS) Software Assurance (SwA). 2012b. *Processes and Practices Working Group*. Washington, DC: The White House. https://buildsecurityin.us-cert.gov/swa/procwg.html.

Department of Homeland Security (DHS) Software Assurance (SwA) Processes and Practices Working Group. 2008. *Process Reference Model for Assurance Mapping to CMMI-DEV V1.2*. Washington, DC: The White House. https://buildsecurityin.us-cert.gov/swa/procwg.html.

Goertzel, K. M., T. Winograd, H. L. McKinley, L. Oh, M. Colon, T. McGibbon, E. Fedchak, and R. Vienneau. July 2007. *Software Security Assurance State-of-the-Art Report (SOAR)*. Joint endeavor by IATAC (Information Assurance Technology Analysis Center) with DACS (Data and Analysis Center for Software). http://iac.dtic.mil/iatac/download/security.pdf.

Howard, M. and S. Lipner. 2006. *The Security Development Lifecycle*. Microsoft Press, Redmond, WA.

McConnell, S. 2005. *The Business Case for Software Development*. Construx Software Builders Inc., Bellevue, WA. http://www.igda.org/qol/IGDA_2005_QoLSummit_Business-Case.pdf.

McGraw, G., B. Chess, and S. Migues. Accessed July 2010. *Building Security In Maturity Model BSIMM v2.0*. http://www.bsimm2.com/.

Mead, N. R., J. H. Allen, M. Ardis, T. B. Hilburn, A. J. Kornecki, R. Linger, and J. McDonald. 2010. *Software Assurance Curriculum Project Volume I: Master of Software Assurance Reference Curriculum* (CMU/SEI-2010-TR-005, ESC-TR-2010-005). Pittsburgh, PA: Software Engineering Institute, Carnegie Mellon University. http://www.sei.cmu.edu/library/abstracts/reports/10tr005.cfm.

Microsoft. 2010a. *Microsoft Security Development Lifecycle*. Redmond, WA: Microsoft. http://www.microsoft.com/security/sdl/about/process.aspx.

Microsoft. Updated March 31, 2010b. *Microsoft Security Development Lifecycle Version 5.0*. Redmond, WA: Microsoft. http://download.microsoft.com/download/F/2/0/F205C451-C59C-4DC7-8377-9535D0A208EC/Microsoft%20SDL_Version%205.0.docx.

MITRE. June 2012. *Common Vulnerabilities and Exposures*. Bedford, MA: MITRE. http://cve.mitre.org/.

NDIA—Systems Assurance Committee. October 2008. *Engineering for System Assurance*. NDIA, Arlington, VA. www.acq.osd.mil/sse/docs/SA-Guidebook-v1-Oct2008.pdf.

Okubo, T. and H. Tanaka. 2007. Secure software development through coding conventions and frameworks. *Second International Conference on Availability, Reliability and Security* (ARES'07). Institute of Information Security, Vienna, Austria.

Open Web Application Security Project (OWASP). 2009. *Software Assurance Maturity Model (SAMM) v1.0.* OWASP, Bel Air, MD. http://www.owasp.org/index.php/Category:Software_Assurance_Maturity_Model.

Orman, H. 2003. The Morris worm: A fifteen-year perspective. *Security and Privacy, IEEE*, 1(5), 35–43, September-October, doi: 10.1109/MSECP.2003.1236233. http://ieeexplore.ieee.org/stamp/stamp.jsp?tp=&arnumber=1236233&isnumber=27717

Partnership for Public Service and B. A. Hamilton. Cyber IN-Security: Strengthening the Federal Cybersecurity Workforce. Partnership for Public Service, Washington, DC. http://ourpublicservice.org/OPS/publications/viewcontentdetails.php?id=135 (accessed July, 2009).

Popick, P., Dr. T. E. Devine, and R. Moorthy. Mar/April 2010. *A DoD-Oriented Introduction to the NDIA's System Assurance Guidebook. CrossTalk 23*, 2, Hill AFB, UT. http://www.stsc.hill.af.mil/crosstalk/2010/03/1003PopickDevineMoorthy.html.

Pugh, E., Johnson, L., and Palmer, J. 2003. *IBM's 360 and Early 370 Systems*, MIT Press. http://en.wikipedia.org/wiki/System/370.

Redwine, S. T., R. O. Baldwin, M. L. Polydys, D. P. Shoemaker, J. A. Ingalsbe, and L. D. Wagoner. October 2007. *Software Assurance: A Curriculum Guide to the Common Body of Knowledge to Produce, Acquire, and Sustain Secure Software*. Washington, DC: Department of Homeland Security.

Saltzer, J. H. and M. D. Schroeder. 1974. The protection of information in computer systems. *Communications of the ACM*, 17(7), 388–402.

Software Assurance Forum for Excellence in Code (SAFECode). February 2008. *Software Assurance: An Overview of Current Industry Best Practices*, Wakefield, MA. http://www.safecode.org/publications/SAFECode_BestPractices0208.pdf.

United States Government Accountability Office (GAO). March 2012. *Report to Congressional Requesters: IT Supply Chain: National Security-Related Agencies Need to Better Address Risks*. Washington, DC: United States Government Accountability Office.

Woody, C., N. R. Mead, and D. Shoemaker. 2012. Foundations for software assurance. *Proceedings of the 45th Hawaii International Conference on System Sciences* (HICSS), Maui, HI. January 4–7, pp. 5368–5374.

Index

B

C

D

E

F

G

I

J

K

N

S

T

W

X

Z